KU-053-400

DICTIONARY OF SCIENCE AND TECHNOLOGY

DICTIONARY OF SCIENCE AND TECHNOLOGY

Revised Edition

Editors

T. C. Collocott M.A.

A. B. Dobson B.Sc. M.Inst.P.

W & R Chambers

W & R Chambers Ltd
11 Thistle Street Edinburgh EH2 1DG

Revised edition 1974
Reprinted 1976

ISBN 0 550 13202 3

Printed in Great Britain by
T & A Constable Ltd
Hopetoun Street Edinburgh

Contents

Preface

Chambers Dictionary of Science and Technology is the natural successor to *Chambers Technical Dictionary*, which for over a quarter of a century from its first publication in 1940 enjoyed wide acceptance as the most useful and comprehensive single-volume work of its kind. It so happened that this particular quarter of a century constituted probably the most explosive period of human history—figuratively as well as literally—revolutionizing man's life, thought and outlook and with that his means of expression. Believing, with Heine, that 'Change goes on and on: only death stands still', the publishers decided during the late 'sixties that the time had come for a complete revision of the earlier work.

The present dictionary is the result. More accurately described today by its longer title, it follows the pattern and traditions of its predecessor. Naturally its vocabulary is much more extensive, for much has happened since 1940; but the aim is essentially the same—to set on record the basic language of communication between those engaged or interested in the numerous branches of scientific activity today.

With this end in view all the existing definitions have been examined by competent authorities. Where necessary, they have been revised, adjusted, replaced, or simply deleted (unless judged to be of historic interest). Many thousands of completely new definitions have been added; most of these are in the newer fields of activity, but many record new developments and ideas in the older subjects. The publishers believe that this present work reflects honestly and adequately the level of vocabulary required by those interested in understanding the scientific and technological developments and problems in our lives today.

Arrangement

Articles are arranged in alphabetical order of their headings. This is not quite the truism that it appears at first sight, as any lexicographer will agree. In a work in which many of the headings consist of two or more words there may be some convenience in grouping the articles under the first component (e.g. *acid*, *gas*, *heat*, etc.); but this leaves some anomalies to be sorted out by the reader (e.g. where to look for *acidosis*, *gaseous discharge*, *heating element*). It was decided, therefore, to treat the whole heading (simple or compound) as one unit for alphabetical arrangement purposes and warn the reader accordingly—which we hereby do. It means that he will find *gas engine* quite considerably separated in space from *gas turbine*, but nevertheless in strict alphabetical order. It means too, that *G-string* and *X-rays* appear, not at the beginning of G and X respectively, but in their strict alphabetical position. It must also be pointed out that, where the same heading has a number of different connotations (whether related or not), the definitions under that heading are arranged in alphabetical order of 'labels': e.g. the various definitions under *scale* appear in the following order of 'labels'—*Acous.*, *Bot.*, *Inst.*, *Met.*, *Photog.*, *Zool.*

Trade Names

Most dictionary makers are beset by the problem of what are (perhaps loosely) referred to as 'trade names', i.e. proprietary names of one kind or another. The position is sometimes complicated by the fact that some such names enjoy different status in different countries; some are misapplied in general usage; and still others belong to proprietors who object to their inclusion in dictionaries (which circumstance accounts for some regrettable omissions from the present work). Where we have knowingly used trade names, we have endeavoured to indicate the fact (e.g. by initial capitals, prefix TN, or some other statement). Where we are thought to have failed, we shall be glad, upon accredited notification, to make suitable amendment in future editions. Further, we must make plain that the use of any trade name in this work must not be taken to imply that the proprietor's rights to that name are in any way unrecognized or impugned by us.

Appendices

The reader's attention is drawn to the very considerable *Appendices.* The text of a dictionary which consists basically of short articles can obviously give no overall picture of, say, the systematic relationships of elements, of the different classes of rocks, and of the innumerable subdivisions of plant and animal life. It is intended that the several tables in the Appendices will go some way towards providing an ordered background for reference purposes.

As this dictionary goes to press, much of the world is in the throes of adapting traditional standards to the new international metric system (Système International d'Unités). The process is far from complete as regards individual countries, and indeed as regards the various industries and activities within countries. For this reason, therefore, we have thought it desirable to place at the beginning of this book a table of Conversion Factors (mainly from traditional to SI units). We believe it will be greatly needed in the immediate years ahead.

A dictionary such as this is obviously the work of many collaborators, all experts in their respective fields. To these, to many others to whom we have applied for advice, and to still others, in all parts of the world, who over the years have, without being asked, given us advice, suggestions and criticisms, we here record our grateful thanks. In particular, and now unfortunately in retrospect, the present editor acknowledges his hard schooling under the original editors of *Chambers Technical Dictionary*, Mr C. F. Tweney and Dr L. E. C. Hughes. The help of the latter in successive revised editions was incalculable, and indeed has been much in evidence in laying the ground work for the present Dictionary.

Obviously, in preparing a work such as this a great many other compilations have been consulted, among them Hughes, Stephens and Brown, *Dictionary of Electronics & Nucleonics*. In particular, indebtedness is acknowledged to the vast corpus of codified information contained in the glossaries and reports of the British Standards Institution (BSI), the International Organization for Standardization (ISO) and the International Union of Pure and Applied Chemistry (IUPAC).

Finally, we thank numerous members of our own editorial staff who have painstakingly marshalled a vast amount of material for publication. We know only too well that we shall be told of sins of omission and commission; for these we apologize in advance, and ask that we may be informed of them with a view to rectification in future editions. Alexander Pope said, 'Index learning turns no student pale, yet holds the eel of science by the tail'. Whatever our shortcomings, we hope at least to have given our readers a more secure grip.

October 1971 **T. C. Collocott**

Preface to Revised Edition

In reprinting the first edition of this Dictionary, the opportunity has been taken to revise a number of definitions, and to include some new terms which have come into use in the three years since the original printing.

Units

Following the universal adoption in education of *Système International* (SI) units of measurement, and anticipating that industry and commerce will adapt themselves completely to the system very soon, all units in the Dictionary have been converted to SI, except where this would be ridiculous for historical reasons (e.g. in referring to 4 ft or 8 ft stops on an organ). No fetish has been made of this, however; where considered necessary, measurements are given in both traditional and SI systems of units. Using the text of the Dictionary in conjunction with the Conversion Tables on pages xiii to xv should enable the reader to cope with almost any metrication problem.

Chemical Nomenclature

In the field of chemistry, a large number of organic compounds has been renamed in a systematic manner, where the name directly indicates the structure of the compound. This involves the disappearance of many of the older so-called trivial names (e.g. *carbon tetrachloride*), and their replacement by more logical, but initially unfamiliar, names (loc. cit. *tetrachloromethane*). Many such compounds are shown under both names in the Dictionary.

It will be obvious that in any one area of knowledge a Dictionary such as this could not possibly cover every technical term in use, not only because of pressure of space but also because many such terms are either ephemeral or localized and highly specialized (often both). The purpose of this volume is to facilitate inter-disciplinary communication not only between the various branches of science and technology but also between non-technical specialists and their technical counterparts. Often the technical terminology which frightens the innocent is merely a shorthand for quite a simple idea, and a few words of explanation can promote quick understanding.

To reduce redundancy to a minimum, much use is made of cross-reference. Within a definition, a reference printed in **bold** characters should be taken as an invitation to refer to the entry under that name for further information. A reference in *italics* indicates that it is defined elsewhere should its own meaning be unknown to the reader.

January 1974 **_Alan Dobson_**

Subjects and Abbreviations

Acoustics (*Acous.*)
Aeronautics (*Aero.*)
Agriculture (*Agric.*)
Anatomy (*Anat.*)
Animal Behaviour (*An. Behav.*)
Architecture (*Arch.*)
Astronomy (*Astron.*)
Automation
Automobiles (*Autos.*)
Bacteriology (*Bacteriol.*)
Biochemistry (*Biochem.*)
Biology (*Biol.*)
Bookbinding (*Bind.*)
Botany (*Bot.*)
Brewing (*Brew.*)
Building (*Build.*)
Cables
Carpentry (*Carp.*)
Chemical Engineering (*Chem. Eng.*)
Chemistry (*Chem.*)
Cinematography (*Cinema.*)
Civil Engineering (*Civ. Eng.*)
Computers (*Comp.*)
Crystallography (*Crystal.*)
Cytology (*Cyt.*)
Ecology (*Ecol.*)
Electrical Engineering (*Elec. Eng.*)
Electricity (*Elec.*)
Electronics
Engineering (*Eng.*)
Forestry (*For.*)
Foundry Practice (*Foundry*)
Fuels
Genetics (*Gen.*)
Geography (*Geog.*)
Geology (*Geol.*)
Geophysics (*Geophys.*)
Glass
Heat, Heating (*Heat.*)
Histology (*Histol.*)
Hydraulic Engineering (*Hyd. Eng.*)
Hydraulics (*Hyd.*)
Instruments (*Instr.*)
Internal-Combustion Engines (*I.C. Engs.*)
Joinery (*Join.*)
Leather
Light
Magnetism (*Mag.*)
Mathematics (*Maths.*)
Mechanics (*Mech.*)
Medicine (*Med.*)
Metallurgy (*Met.*)
Meteorology (*Meteor.*)
Microscopy (*Micros.*)
Mineralogy (*Min.*)
Mineral Processing (*Min. Proc.*)
Mining
Navigation (*Nav.*)
Nuclear Engineering (*Nuc. Eng.*)
Nucleonics (*Nuc.*)
Nutrition (*Nut.*)
Oceanography (*Ocean.*)
Oils
Optics
Painting (*Paint.*)
Papermaking (*Paper*)
Pharmacology (*Pharm.*)
Photography (*Photog.*)
Physics (*Phys.*)
Physiology (*Physiol.*)
Plastics
Plumbing (*Plumb.*)
Powder Metallurgy (*Powder Met.*)
Powder Technology (*Powder Tech.*)
Printing (*Print.*)
Psychiatry (*Psychiat.*)
Psychology (*Psychol.*)
Radar
Radio
Radiology (*Radiol.*)
Railways (*Rail.*)
Sanitary Engineering (*San. Eng.*)
Ships
Space
Spinning
Statistics (*Stats.*)
Surgery (*Surg.*)
Surveying (*Surv.*)
Telecommunications (*Telecomm.*)
Telegraphy (*Teleg.*)
Telephony (*Teleph.*)
Television (*TV*)
Textiles
Tools
Typography (*Typog.*)
Vacuum Technology (*Vac. Tech.*)
Veterinary Science (*Vet.*)
Weaving
Work Study
Zoology (*Zool.*)

Other Abbreviations

The following are the more common abbreviations used in this dictionary. Many others (including contractions, prefixes and symbols) occur in their alphabetical position in the text, especially at the beginning of each letter of the alphabet. For SI and other symbols, see the Tables on pages 1325-1328.

abbrev(s). abbreviation(s)
adj(s). adjective(s)
approx. approximately
asym. asymmetrical
at. no. atomic number
b.p. boiling point
c. century
°*C* degree(s) Celsius (centigrade)
ca. circa
Can. Canada
cf. compare
C.G. centre of gravity
CGS centimetre-gram(me)-second
circum. circumference
CNS central nervous system
colloq. colloquially
conc. concentrated
contr. contraction
CRO cathode-ray oscilloscope
CRT cathode-ray tube
deg. degree(s)
dim. diminutive
E. East
elec. electrical
esp. especially
°*F* degree(s) Fahrenheit
ft foot, feet
g acceleration due to gravity; gram(me)(s)
G constant of gravitation
Gk. Greek
h., hr. hour(s)
in. inch(es)
kg kilogram(me)(s)
km kilometre(s)
K kelvin(s)
L. Latin
lb pound(s)
m metre(s)
min. minute(s)
MKS(A) metre-kilogram(me)-second (-ampere)
m.p. melting point
n. noun
N. North
pfx. prefix
pl(s). plural(s)
q.v. which see; pl. *qqv.*
°*R* degree(s) Rankine
r.a.m. relative atomic mass
rel. d. relative density
r.m.s. root-mean-square
S. South
s., sec. second(s)
SI Système International (d'Unités)
sing. singular
spec. specific
sq. square
s.t.p. standard temperature and pressure
sym. symmetrical
syn. synonym
temp. temperature
TN trade (proprietary) name
TR transmit-receive (tube)
U.K. United Kingdom
U.S.(A.) United States (of America)
W. West
y., yr. year(s)
yd yard(s)

SI Conversion Factors

(Exact values are printed in bold type)

Quantity	Unit		Conversion factor
Length	1 in.	=	25·4 mm
	1 ft	=	0·3048 m
	1 yd	=	0·9144 m
	1 fathom	=	1·8288 m
	1 chain	=	20·1168 m
	1 mile	=	1·609 34 km
	1 International nautical mile	=	1·852 km
	1 UK nautical mile	=	1·853 18 km
Area	1 in.^{2}	=	6·4516 cm^2
	1 ft^2	=	0·092 903 m^2
	1 yd^2	=	0·836 127 m^2
	1 acre	=	4046·86 m^2 = 0·404 686 ha (hectare)
	1 sq. mile	=	2·589 99 km^2 = 258·999 ha
Volume	1 UK minim	=	0·059 193 8 cm^3
	1 UK fluid drachm	=	3·551 63 cm^3
	1 UK fluid ounce	=	28·4131 cm^3
	1 US fluid ounce	=	29·5735 cm^3
	1 US liquid pint	=	473·176 cm^3 = 0·4732 dm^3(= litre)
	1 US dry pint	=	550·610 cm^3 = 0·5506 dm^3
	1 Imperial pint	=	568·261 cm^3 = 0·5683 dm^3
	1 UK gallon	=	1·201 US gallon
		=	4·546 09 dm^3
	1 US gallon	=	0·833 UK gallon
		=	3·785 41 dm^3
	1 UK bu (bushel)	=	0·036 368 7 m^3 = 36·3687 dm^3
	1 US bushel	=	0·035 239 1 m^3 = 35·2391 dm^3
	1 in.^{3}	=	16·3871 cm^3
	1 ft^3	=	0·028 316 8 m^3
	1 yd^3	=	0·764 555 m^3
	1 board foot (timber)	=	0·002 359 74 m^3 = 2·359 74 dm^3
	1 cord (timber)	=	3·624 56 m^3
2nd moment of area	1 in.^{4}	=	41·6231 cm^4
	1 ft^4	=	0·008 630 97 m^4 = 86·3097 dm^4
Moment of inertia	1 lb ft^2	=	0·042 140 1 kg m^2
	1 slug ft^2	=	1·355 82 kg m^2
Mass	1 grain	=	0·064 798 9 g = 64·7989 mg
	1 dram (avoir.)	=	1·771 85 g = 0·001 771 85 kg
	1 drachm (apoth.)	=	3·887 93 g = 0·003 887 93 kg
	1 ounce (troy or apoth.)	=	31·1035 g = 0·031 103 5 kg
	1 oz (avoir.)	=	28·3495 g
	1 lb	=	0·453 592 37 kg
	1 slug	=	14·5939 kg
	1 sh cwt (US hundredweight)	=	45·3592 kg
	1 cwt (UK hundredweight)	=	50·8023 kg
	1 UK ton	=	1016·05 kg
		=	1·016 05 tonne
	1 short ton	=	2000 lb
		=	907·185 kg
		=	0·907 tonne
Mass per unit length	1 lb/yd	=	0·496 055 kg/m
	1 UK ton/mile	=	0·631 342 kg/m
	1 UK ton/1000 yd	=	1·111 16 kg/m
	1 oz/in.	=	1·116 12 kg/m = 11·1612 g/cm
	1 lb/ft	=	1·488 16 kg/m
	1 lb/in	=	17·8580 kg/m
Mass per unit area	1 lb/acre	=	0·112 085 g/m^2 = 1·120 85 $\times 10^{-4}$ kg/m
	1 UK cwt/acre	=	0·012 553 5 kg/m^2
	1 oz/yd^2	=	0·033 905 7 kg/m^2
	1 UK ton/acre	=	0·251 071 kg/m^2
	1 oz/ft^2	=	0·305 152 kg/m^2

SI Conversion Factors (*continued*)

Quantity	Unit		Conversion factor
Mass per unit area *continued*	1 lb/ft²	=	4·882 43 kg/m²
	1 lb/in.²	=	703·070 kg/m²
	1 UK ton/mile²	=	0·392 298 g/m² = 3·922 98 × 10⁻⁴ kg/m²
Density	1 lb/ft³	=	16·0185 kg/m³
	1 lb/UK gal	=	99·7763 kg/m³ = 0·099 78 kg/l
	1 lb/US gal	=	119·826 kg/m³ = 0·1198 kg/l
	1 slug/ft³	=	515·379 kg/m³
	1 ton/yd³	=	1328·94 kg/m³ = 1·328 94 tonne/m³
	1 lb/in.³	=	27·6799 Mg/m³ = 27·6799 g/cm³
Specific volume	1 in.³/lb	=	36·1273 cm³/kg
	1 ft³/lb	=	0·062 428 0 m³/kg = 62·4280 dm³/kg
Velocity	1 in./min	=	0·042 333 cm/s
	1 ft/min	=	**0·005 08** m/s = **0·3048** m/min
	1 ft/s	=	**0·3048** m/s = **1·097 28** km/h
	1 mile/h	=	1·609 34 km/h = **0·447 04** m/s
	1 UK knot	=	1·853 18 km/h = 0·514 773 m/s
	1 International knot	=	**1·852** km/h = 0·514 444 m/s
Acceleration	1 ft/s²	=	**0·3048** m/s²
Mass flow rate	1 lb/h	=	0·125 998 g/s = 1·259 98 × 10⁻⁴ kg/s
	1 UK ton/h	=	0·282 235 kg/s
Force or weight	1 dyne	=	**10⁻⁵** N
	1 pdl (poundal)	=	0·138 255 N
	1 ozf (ounce)	=	0·278 014 N
	1 lbf	=	4·448 22 N
	1 kgf	=	**9·806 65** N
	1 tonf	=	9·964 02 kN
Force or weight per unit length	1 lbf/ft	=	14·5939 N/m
	1 lbf/in.	=	175·127 N/m = 0·175 127 N/mm
	1 tonf/ft	=	32·6903 kN/m
Force (weight) per unit area or pressure or stress	1 pdl/ft²	=	1·488 16 N/m² or Pa (pascal)
	1 lbf/ft²	=	47·8803 N/m²
	1 mm Hg	=	133·322 N/m²
	1 in. H_2O	=	249·089 N/m²
	1 ft H_2O	=	2989·07 N/m² = 0·029 890 7 bar
	1 in. Hg	=	3386·39 N/m² = 0·033 863 9 bar
	1 lbf/in.²	=	6·894 76 kN/m² = 0·068 947 6 bar
	1 bar	=	**10⁵** N/m²
	1 std. atmos.	=	**101·325** kN/m² = **1·013 25** bar
	1 tonf/ft²	=	107·252 kN/m²
	1 tonf/in.²	=	15·4443 MN/m² = 1·544 43 hectobar
Specific weight	1 lbf/ft³	=	157·088 N/m³
	1 lbf/UK gal	=	978·471 N m³
	1 tonf/yd³	=	13·0324 kN/m³
	1 lbf/in.³	=	271·447 kN/m³
Moment, torque or couple	1 ozf in. (ounce-force inch)	=	0·007 061 55 N m
	1 pdl ft	=	0·042 140 1 N m
	1 lbf in	=	0·112 985 N m
	1 lbf ft	=	1·355 82 N m
	1 tonf ft	=	3037·03 N m = 3·037 03 kN m

SI Conversion Factors (*continued*)

Quantity	Unit		Conversion factors
Energy or Heat or Work	1 erg	=	10^{-7} J
	1 hp h (horsepower hour)	=	2·684 52 MJ
	1 thermie = 10^6 cal_{15}	=	4·1855 MJ
	1 therm = 100 000 Btu	=	105 506 MJ
	1 cal_{IT}	=	4·1868 J
	1 Btu	=	1·055 06 kJ
	1 kWh	=	3·6 MJ
Power	1 hp = 550 ft lbf/s	=	0·745 700 kW
	1 metric horsepower (ch, PS)	=	735·499 W
Specific heat Capacity	1 Btu/lb degF 1 Chu/lb degC 1 cal/g degC	=	4·1868 kJ/kg K
Heat flow rate	1 Btu/h	=	0·293 071 W
	1 kcal/h	=	1·163 W
	1 cal/s	=	4·1868 W
Intensity of heat flow rate	1 Btu/ft² h	=	3·154 59 W/m²
Electric stress	1 kV/in.	=	0·039 370 1 kV/mm
Dynamic viscosity	1 lb/ft s	=	14·8816 poise = 1·488 16 kg/m s
Kinematic viscosity	1 ft²/s	=	929·03 stokes = 0·092 903 m²/s
Calorific value or specific enthalpy	1 Btu/ft³	=	0·037 258 9 J/cm³ = 37·2589 kJ/m³
	1 Btu/lb	=	2·326 kJ/kg
	1 cal/g	=	4·1868 J/g
	1 kcal/m³	=	4·1868 kJ/m³
Specific entropy	1 Btu/lb °R	=	4·1868 kJ/kg K
Thermal conductivity	1 cal cm/cm² s degC	=	41·868 W/m K
	1 Btu ft/ft² h degF	=	1·730 73 W/m K
Gas constant	1 ft lbf/lb °R	=	0·005 380 32 kJ/kg K
Plane angle	1 rad (radian)	=	57·2958°
	1 degree	=	0·017 453 3 rad = 1·1111 grade
	1 minute	=	2·908 88 × 10^{-4} rad = 0·0185 grade
	1 second	=	4·848 14 × 10^{-6} rad = 0·0003 grade
Velocity of rotation	1 rev/min	=	0·104 720 rad/s

Based on Ramsay and Taylor: *SI Metrication: Easy to Use Conversion Tables* (Chambers).

The Greek Alphabet

The letters of the Greek alphabet, frequently used in technical terms, are given here for purposes of convenient reference.

Α	α	alpha	=	a
Β	β	bēta	=	b
Γ	γ	gamma	=	g
Δ	δ	delta	=	d
Ε	ε	epsīlon	=	e
Ζ	ζ	zēta	=	z
Η	η	ēta	=	ē
Θ	θ ϑ	thēta	=	th
Ι	ι	iōta	=	i
Κ	κ	kappa	=	k
Λ	λ	lambda	=	l
Μ	μ	mu	=	m
Ν	ν	nu	=	n
Ξ	ξ	xi	=	x
Ο	ο	omīcron	=	o
Π	π	pi	=	p
Ρ	ρ	rho	=	rh, r
Σ	σ ς	sigma	=	s
Τ	τ	tau	=	t
Υ	υ	upsīlon	=	ü
Φ	φ	phi	=	ph
Χ	χ	chi	=	kh
Ψ	ψ	psi	=	ps
Ω	ω	ōmega	=	ō

a-. Prefix signifying *on*. Also shortened form of **ab-, ad-, an-, ap-**.

a. Symbol for: acceleration; relative activity; linear absorption coefficient; amplitude.

a- (*Chem.*). An abbrev. for: (1) asymmetrically substituted; (2) *ana-*, i.e., containing a condensed double aromatic nucleus substituted in the 1.5 positions.

α. See under **alpha**. Symbol for: absorption coefficient; attenuation coefficient; acceleration, angular acceleration; fine structure constant; helium nucleus.

α- (*Chem.*). Symbol for: (1) substitution on the carbon atom of a chain next to the functional group; (2) substitution on a carbon atom next to one common to two condensed aromatic nuclei; (3) substitution on the carbon atom next to the hetero-atom in a hetero-cyclic compound; (4) a stereo-isomer of a sugar.

$[\alpha]_D^t$ (*Chem.*). Symbol for the specific optical rotation of a substance at t°C, measured for the D line of the sodium spectrum.

A. Symbol for: area; ampere; absolute temperature; relative atomic mass (atomic weight); magnetic vector potential; Helmholtz function.

[A] (*Light*). A strong absorption band in deep red of the solar spectrum (wavelength 7621·28 Å) caused by oxygen in the earth's atmosphere. The first of the Fraunhofer lines.

Å (*Phys.*). Abbrev. for *ångström*.

A-amplifier (*Acous.*). One associated with, or immediately following, a high-quality microphone, as in broadcasting studios. *N.B.* Not the same as **class-A amplifier**.

A and B roll printing (*Cinema.*). Method of printing scenes alternately on two rolls, each roll having black spacing equivalent to the omitted scenes, thus facilitating marking up and editing.

A and R display (*Radio*). Expansion of a part of a regular A display.

a.b. (*Build.*). Abbrev. for *as before* in bills of quantities, etc.

ab- (*Elec. Eng.*). Prefix to name of unit, indicating derivation in the CGS system.

abactinal (*Zool.*). See **abambulacral.**

abacus (*Arch.*). The uppermost part of a column capital or pilaster, on which the architrave rests. (*Carp. etc.*) The part of a baluster on which the handrail rests. (*Maths.*) A bead frame. Used as an arithmetical calculating aid.

abambulacral (*Zool.*). Pertaining to that part of the surface of an Echinoderm lacking tube feet.

abampere (*Elec.*). Current which, when flowing round a single-turn coil of 1 cm radius in vacuum, produces a field of 2π oersted at the centre. This is the CGS absolute electromagnetic unit of current and is equal to 10 absolute or SI amperes.

abamurus (*Arch.*). A supporting wall or buttress, built to add strength to another wall.

abandonment (*Mining*). Voluntary surrender of legal rights or title to a mining claim.

abapical (*Zool.*). Pertaining to, or situated at, the lower pole: remote from the apex.

abasia (*Med.*). Absence of correct coordination of muscles in walking.

abatement (*Join.*). Waste of timber in shaping to size.

abatjour (*Arch.*). An opening to admit light, and generally to deflect it downwards; a skylight.

A-battery (*Radio*). U.S. term for low-tension battery.

abatvoix (*Acous.*). A sounding-board or other arrangement over a pulpit or rostrum, to deflect speech downwards and in the direction of those listening.

abaxial (*Bot., Zool.*). Remote from the axis. (*Photog., Optics*). Said of rays of light which do not coincide with the optical axis of a lens system.

AB-battery (*Electronics*). Combination of A (for heating cathodes) and B (anode supply) for portable equipment.

Abbe refractometer (*Chem.*). An instrument for measuring directly the refractive index of liquids.

A.B.C. process (*San. Eng.*). A process of sewage treatment in which alum, blood, clay, and charcoal are used as precipitants.

Abderhalden reaction (*Chem.*). A test for the presence of protective ferments in blood. The prepared albumin and serum are mixed and any change in optical activity noted; or else the mixture is dialysed and the dialysate tested by the ninhydrin reaction. The test can be used to detect pregnancy, malignant disease, etc.

abdominal air sac (*Physiol., Zool.*). Posterior part of the lung in Birds.

abdominal cavity (*Physiol.*). See **peritoneal cavity.**

abdominal gills (*Zool.*). In the aquatic larvae of many Insects, paired segmental leaflike or filamentous expansions of the abdominal cuticle for respiration.

abdominal legs (*Zool.*). See **prolegs.**

abdominal limbs (*Zool.*). Segmented abdominal appendages in certain *Brachiopoda* and most *Crustacea* (*Malacostraca*) which are used for swimming, setting up currents of water or carrying eggs and young. In *Diplopoda*, segmented ambulatory appendages on the abdomen.

abdominal pores (*Zool.*). Apertures leading from the coelom to the exterior in certain Fish and in *Cyclostomata.*

abdominal reflex (*Zool.*). Contraction of the abdominal wall muscles when the skin over the side of the abdomen is stimulated.

abdominal regions (*Anat.*). Nine regions into which the human abdomen is divided by two horizontal and two vertical imaginary planes, i.e., right and left hypochondriac, right and left lumbar, right and left iliac, epigastric, umbilical, and hypogastric.

abdominal ribs (*Zool.*). In certain Reptiles, rods of dermal bone in the ventral body wall, between true ribs and pelvis.

abdominal vein (*Physiol.*). See **epigastric vein.**

abdominohyoideus (*Zool.*). In Amphibians, a muscle continuing the rectus abdominis forward into the throat region.

abducens (*Physiol.*). In Vertebrates, the 6th cranial nerve, purely motor in function, supplying the rectus externus muscle of the eye.

abduction (*Physiol.*). The action of pulling a limb or part away from the median axis.

abductor (*Zool.*). Any muscle that draws a limb or part away from the median axis by contraction; e.g., the abductor pollicis, which moves the thumb outward.

Abegg's rule (*Chem.*). Empirical rule that solubility of salts of alkali metals with strong acids decreases from lithium to caesium, i.e., with increase of rel. at. mass, and those with weak acids follow the opposite order. Sodium chloride is an exception to this rule, being less soluble than potassium chloride.

Abegg's rule of eight (*Chem.*). The sum of the maximum positive and negative valencies of an element is eight, e.g., S in SF_6 and H_2S.

Abel flash-point apparatus (*Chem.*). A petroleum-testing apparatus for determining the flash-point.

Abelian group (*Maths.*). A group in which the group operation is commutative. It is important in the study of rings and vector spaces.

abelite (*Chem.*). Explosive, composed mainly of ammonium nitrate and trinitrotoluene.

Aberdeen granite (*Build.*). A grey or pink granite, widely used as a building-stone for heavy work, or for housing works. See **granite.**

aberrant (*Bot., Zool.*). Showing some unusual difference of structure: having characteristics not strictly in accordance with type.

aberration (*Astron.*). An apparent change of position of a heavenly body, due to the velocity of light having a finite ratio to the relative velocity of the source and the observer. (*Bot.*) Some peculiarity of an individual plant not capable of transmission to offspring, and usually due to some special environmental condition. (*Optics*) In an image-forming system, e.g., an optical or electronic lens, failure to produce a true image, i.e., a point object as a point image, etc. Five geometrical aberrations were recognized by von Seidel, viz., **coma, spherical aberration, astigmatism, curvature of the field, and distortion.**

abhymenial (*Bot.*). Opposite the **hymenium.**

abietic acid (*Chem.*). $C_{20}H_{30}O_{21}$. Acid isolated from rosin or colophonium.

abiogenesis (*Zool.*). See **spontaneous generation.**

abjection (*Bot.*). The forcible projection of spores from the sporophore, or sterigma, by some change in the plant, usually of water content.

abjunction (*Bot.*). The delimitation of a spore from its stalk by means of a septum. *v.* **abjoint.**

ablation (*Geog.*). The wearing and carrying away of the surface of rock or glacier by the action of water. (*Surg.*) Removal of body-tissue by surgery.

ablation shields (*Space*). Protection shields made of laminated glass resin over that portion of a spacecraft or missile which is first exposed to the intense heat (1100°C) generated by the friction caused by re-entry to the atmosphere from space. The shields are vaporized by the heat and carried away.

Abney colour sensitometer (*Photog.*). An apparatus, using a rotary stepped sector-disk behind a number of apertures, for matching the luminosities of different colours.

Abney law (*Light*). If a spectral colour is desaturated by the addition of white light, if its wavelength is less than 570 nm, its hue then moves towards the red end of the spectrum, while if the wavelength is more than 570 nm its hue moves towards the blue.

Abney level (*Surv.*). Hand-held instrument in which angles of steep sights are measured while simultaneously viewing a spirit-level bubble.

Abney mounting (*Light*). A form of mounting for a concave diffraction grating, in which the eyepiece (or photographic plate holder) is fixed at the centre of curvature of the grating and the slit can move around the circumference of the Rowland circle, to bring different orders of spectrum into view.

abnormal glow (*Electronics*). Excess of current over that required to cover completely the cathode with a glow discharge.

abnormal polarization (*Phys.*). Condition in an electromagnetic wave when the alternating magnetic field has a vertical component.

abnormal reflections (*Radio*). Those from the ionosphere for frequencies in excess of *critical frequency.*

abomasitis (*Vet.*). Inflammation of the *abomasum.*

abomasum (*Zool.*). In ruminant Mammals, the fourth or true stomach. Also called **reed, rennet.**

A-bomb (*Nuc.*). See **atomic bomb.**

aboöspore (*Bot.*). An oöspore produced without sexual fusion.

aboral (*Zool.*). Opposite to, leading away from, or distant from, the mouth. See **abambulacral.**

abort (*Space*). To terminate a missile flight by cutting off the flow of fuel or by setting off explosive charges by radio signal.

abortifacient (*Med.*). Anything which causes artificial abortion: a drug which does this.

abortion (*Med.*). Expulsion of the foetus from the uterus during the first 3 months of pregnancy. Abortion may be spontaneous or induced. (*Zool.*) Cessation of development in an organ or foetus.

aboutsledge (*Eng.*). The large hammer used by a blacksmith's mate, turn-about with the smaller hammer of the blacksmith.

abradant (*Eng.*). A substance, usually in powdered form, used for grinding. See **abrasive.**

abrade (*Eng.*). Cut or tear, at two surfaces in contact and relative motion.

Abram's 'law' (*Eng.*). Ratio of water to cement for chemical action to impart strength to concrete is 0·35 : 1.

abranchiate (*Zool.*). Lacking gills.

abrasion (*Geog.*). The wearing away of a part of the earth's surface by the action of ice, water, or wind. (*Med.*) A rubbed-away area of the surface-covering of the body; i.e., of skin or of mucous membrane.

abrasion hardness (*Min. Proc.*). Resistance to abrasive wear, under specified conditions, of metal or mineral.

abrasive (*Chem.*). A substance used for the removal of matter by scratching and grinding (abrasion); e.g., silicon carbide (carborundum).

abreaction (*Psychiat.*). A release of blocked psychic energy attaching to repressed and forgotten memories and phantasies; effected by living through these in feeling or action.

abreuvoir (*Build.*). Mortar joint between two arch-stones, or between stones in a wall.

abrin (*Chem.*). A toxic protein found in seeds of jequirity, or *Abrus precatorius*, an Indian plant, causing agglutination of red blood corpuscles. See also **alvin.**

abrupt (*Bot.*). Truncate. As if cut off transversely.

abruptly pinnate (*Bot.*). Said of a pinnate leaf without a terminal leaflet.

ABS (*Plastics*). A range of copolymers based on cyanoethene/but 1,2 :3,4-diene/phenylethene (acrilonitrile/butadiene/styrene).

abscess (*Med.*). Pus localised in infected tissue and separated from healthy tissue by an abscess wall.

absciss layer (*Bot.*). A layer of parenchymatous cells across the base of a petiole or of a branch or embedded in bark, the dehydration of which causes the leaf, branch, or bark to be separated. Also called **abscission layer, separation layer.**

absciss-phelloid (*Bot.*). The unsuberized cells of an absciss layer situated in bark.

abscissa (*Maths.*). For rectilineal axes of coordinates, the distance of a point from the axis of ordinates measured in a direction parallel to the axis of abscissae, which is usually horizontal. The sign convention is that measurements to the right from the axis of ordinates are positive, measurements to the left negative. *pl.* **abscissae.** Cf. *Cartesian coordinates.*

abscission (*Bot.*). (1) The organized shedding of a part of a plant by means of an absciss layer. (2) The liberation of a fungal spore by the breakdown of a sterile portion of the stalk of the spore.

absence of convection (*Space*). Convection currents cease in zero-g conditions, when air no longer has weight. Breathing and bodily heat exchange processes have then to be aided artificially.

absolute (*Maths.*). In general there are two points at infinity on every line. The assemblage of these points at infinity is a conic (a quadric in three dimensions) called the absolute. Its form determines the metrical properties of the geometrical system being operated. Thus in Euclidean geometry, the absolute is the degenerate conic comprising the line at infinity taken twice, while in non-Euclidean geometry, the absolute is either a real conic (hyperbolic geometry) or an imaginary conic (elliptic geometry).

absolute address (*Comp.*). Code designation of a specific storage register.

absolute alcohol (*Chem.*). Water-free ethanol; rel. d. 0·793 (15·5°C); b.p. 78·4°C; obtained from rectified spirit by adding benzene and refractionating. Very hygroscopic.

absolute ampere (*Elec.*). The standard MKS unit of electric current; replaced the international ampere in 1948. See **ampere.**

absolute block system (*Rail.*). See **block system.**

absolute ceiling (*Aero.*). The height at which the rate of climb of an aircraft, in standard atmosphere, would be zero; the maximum height attainable under standard conditions.

absolute code (*Comp.*). One using absolute addresses and written in machine language.

absolute coefficient (*Maths.*). A coefficient with an absolute value, that is a multiplier which is numerical rather than symbolic.

absolute convergence (*Maths.*). A series Σa_r is absolutely convergent if the series $\Sigma |a_r|$ is convergent.

absolute electrometer (*Elec.*). High-grade attracted-disk electrometer in which an absolute measurement of potential can be made by 'weighing' the attraction between two charged disks against gravity.

absolute humidity (*Meteor.*). The number of grams of water-vapour per 1 m³ of the atmosphere: with vapour pressure e and temperature T (absolute) the absolute humidity is represented by $\delta = 216{\cdot}7\ e/T$ g/m³.

absolute hydraulic gradient (*Hyd.*). An imaginary curve parallel to the *hydraulic gradient* (q.v.) but higher by the amount p_0/ρ, where p_0 = atmospheric pressure intensity, ρ = density of the fluid in the flow system.

absolute instrument (*Phys.*). An instrument which measures a quantity directly in absolute units, without the necessity for previous calibration.

absolute magnitude (*Astron.*). See **magnitudes.**

absolute permeability (*Elec. Eng.*). See **permeability.**

absolute potential (*Chem.*). The theoretical true potential difference between an electrode and a solution of its ions, measured against a hypothetical reference electrode, having an absolute potential of zero, with reference to the same solution.

absolute pressure (*Phys.*). Pressure measured with respect to zero pressure, in units of force per unit of area.

absolute reaction rates (*Chem.*). Theory that if the rate of a chemical reaction is governed by the rate of crossing an energy barrier or of forming an *activated complex* (q.v.), then it can be calculated from statistical thermodynamics. See **Arrhenius theory of dissociation.**

absolute-rest precipitation tanks (*San. Eng.*). Tanks in which a given amount of sewage is dealt with at a time, as opposed to tanks taking a continuous flow. After 2- or 3-hr settlement, the top water is drawn off from above and the precipitated sludge from below.

absolute temperature (*Phys. etc.*). A temperature measured with respect to *absolute zero* which is taken as the zero of a special temperature scale that cannot take negative values. See **kelvin, Rankine.**

absolute transpiration (*Bot.*). The rate of loss of water from a plant, found experimentally.

absolute unit of current (*Elec.*). The current which, flowing in a circular conductor of radius 1 cm, will produce at the centre a magnetic field of strength 2π oersted. The *ampere* is equal to one-tenth of an absolute unit of current.

absolute units (*Elec.*). Those derived directly from the fundamental units of a system and not based on arbitrary numerical definitions. The differences between absolute and international units were small; both are now superseded by the definitions of the S.I.

absolute value (*Maths.*). See **modulus.**

absolute viscosity (*Phys., Eng.*). See **coefficient of viscosity.**

absolute wavemeter (*Electronics*). One in which the frequency of an injected radio-frequency e.m.f. is determined by calculation of the elements of the circuit, or length of a resonant line (cymometer).

absolute weight (*Chem.*). The weight of a body in a vacuum.

absolute zero (*Phys.*). The least possible temperature for all substances. At this temperature the molecules of any substance possess no heat energy. A figure of −273·15°C is generally accepted as the value of absolute zero.

absorbed dose (*Radiol.*). See **dose.**

absorbency tests (*Paper*). These vary for the particular use of the paper. The increase of weight or rise of water, oil, or spirit, in paper are methods usually used.

absorber (*Phys.*). Any material which converts energy of radiation or particles into another form, generally heat. Energy transmitted is not absorbed. Scattered energy is often classed with absorbed energy. See **total absorption coefficient, true absorption coefficient.**

absorber valve (*Teleg.*). That in an absorption modulator which absorbs excess power during troughs of the modulation cycle. In a telegraph transmitter, it is used to stabilize voltages during keying.

absorbing material (*Phys.*). Any medium used for absorbing energy from radiation of any type.

absorbing rod (*Nuc. Eng.*). Alternative name for **control rod.**

absorbing well (*Civ. Eng.*). A shaft sunk through an impermeable stratum to allow water to drain through to a permeable one.

absorptiometer (*Chem.*). An apparatus for determining the solubilities of gases in liquids or the absorption of light.

absorption (*Chem.*). Penetration of a substance

into the body of another. Cf. *adsorption*. (*Med.*) The taking-up of fluids or other substances by the vessels and tissues of the body. (*Nuc., Phys.*) (1) Reduction in intensity of any form of radiated energy resulting from energy conversion in medium. *Broad beam absorption* which is relatively unaffected by scatter, is generally less than *narrow beam attenuation* where much of the energy is lost by scattering processes. See **absorber** and **attenuation.** (2) The incorporation of a bombarding particle into a nucleus with which it interacts. (*Photog.*) Complement of *transmission* (q.v.).

absorption band (*Light*). A dark gap in the continuous spectrum of white light transmitted by a substance which exhibits selective absorption.

absorption capacitor (*Elec. Eng.*). One connected across spark gap to damp the discharge.

absorption coefficient (*Chem.*). The volume of gas, measured at s.t.p., dissolved by unit volume of a liquid under normal pressure (i.e., 1 atmosphere). (*Nuc., Phys.*) (1) At a discontinuity (*surface absorption coefficient*):—(*a*) the fraction of the energy which is absorbed, or (*b*) the reduction of amplitude, for a beam of radiation or other wave system incident on a discontinuity in the medium through which it is propagated, or in the path along which it is transmitted. (2) In a medium (*linear absorption coefficient*):—the natural logarithm of the ratio of incident and emergent energy or amplitude for a beam of radiation passing through unit thickness of a medium. (The *mass absorption coefficient* is defined in the same way but for a thickness of the medium corresponding to unit mass per unit area.) N.B. *True absorption coefficients* exclude scattering losses, *total absorption coefficients* include them. See **absorptivity, atomic-.**

absorption discontinuity (*Phys.*). See **absorption edge.**

absorption dynamometer (*Eng.*). A dynamometer which absorbs and dissipates the power which it measures; e.g., the ordinary rope brake and the Froude hydraulic brake. Cf. *transmission dynamometer*.

absorption edge (*Phys.*). The wavelength at which there is an abrupt discontinuity in the intensity of an absorption spectrum for electromagnetic waves, giving the appearance of a sharp edge in its photograph. This transition is due to one particular energy-dissipating process becoming possible or impossible at the limiting wavelength. In X-ray spectra of the chemical elements the K absorption edge for each element occurs at a wavelength slightly less than that for the K emission spectrum. Also **absorption discontinuity.**

absorption hygrometer (*Meteor.*). An instrument by which the quantity of water vapour in air may be measured. A known volume of air is drawn through tubes containing a drying agent such as phosphorus pentoxide; the increase in weight of the tubes gives the weight of water vapour in the known volume of air.

absorption inductor (*Elec. Eng.*). See **interphase transformer.**

absorption lines (*Phys.*). Dark lines in a continuous spectrum caused by absorption by a gaseous element. The positions (i.e., the wavelengths) of the dark absorption lines are identical with those of the bright lines given by the same element in emission.

absorption meter (*Phys.*). Instrument for measuring light transmitted through transparent sample (liquid or solid) using a detector such as a photocell.

absorption refrigerator (*Eng.*). A plant in which ammonia is continuously evaporated from an aqueous solution under pressure, condensed, allowed to evaporate (so absorbing heat), and then reabsorbed.

absorption spectrum (*Phys.*). The system of absorption bands or lines seen when a selectively absorbing substance is placed between a source of white light and a spectroscope. See **Kirchhoff's law.**

absorption tower (*Met.*). Paulson or scrubbing tower, through which ascending gas meets falling sprays of water or chemical in solution, to transfer constituents.

absorption tubes (*Chem.*). Tubes filled with solid absorbent for the absorption of moisture (e.g. silica gel) and gases (e.g. charcoal).

absorption wavemeter (*Elec. Eng.*). One which depends on a resonance absorption in a tuned circuit, constructed with very stable inductance and capacitance.

absorptivity, absorptive power (*Phys.*). The fraction of the incident radiation which is absorbed by a surface on which it falls.

abstriction (*Bot.*). A general term for the separation of a spore from its stalk; it includes *abjection* and *abscission.*

abundance or **abundance ratio** (*Ecol.*). See **relative-.** (*Nuc.*) For a specified element, the proportion or percentage of one isotope to the total, as occurring in nature.

abutment (*Arch., Civ. Eng.*). See **knapsack-.** (*Eng.*) A point or surface provided to withstand thrust; e.g., end supports of an arch or bridge.

abutment load (*Mining*). In stoping or other deep-level excavation, weight transferred to the adjacent solid rock by unsupported roof.

abutting joint (*Carp.*). A joint whose plane is at right angles to the fibres, the fibres of both jointing pieces being in the same straight line.

abvolt (*Elect.*). Unit of potential difference in the cgs-emu system of units; 10^{-8} volts.

abyssalbenthic zone (*Ecol.*). The lower zone of the sea floor, seaward of the archibenthic zone, and extending from about 2000 metres downwards to 10 000 metres. There is no seasonal variation, the water is relatively still, no light penetrates from above, and temperature is constant at just above 0°C.

abyssal deposits (*Geol.*). The deposits of the deep sea, accumulating in depths of more than 2000 metres of water; they comprise organic oozes, various muds, and the red clay of the deepest regions.

abyssal zone (*Ocean.*). The waters of the ocean beyond the limits of the continental shelf, lying between 2000 and 6000 m below sea level.

Abyssinian pump (*Civ. Eng.*). A pump having a well-tube attached to the suction tube, for use in the *Abyssinian well* (q.v.).

Abyssinian well (*Civ. Eng.*). A tube driven into strata of moderate hardness to obtain a supply of water. The tube is pointed at its lower end, with perforations above the point.

abyssobenthic (*Ocean.*). Relating to that part of the abyssal realm which includes the ocean floor: pertaining to or living on the ocean floor at great depths.

abyssopelagic (*Ocean.*). Relating to that part of the abyssal realm which excludes the ocean floor: floating in the ocean depths.

a.c. (*Elec.*). Abbrev. for *alternating current.*

ac- (*Chem.*). Abbrev. indicating substitution in the alicyclic ring.

Ac (*Chem.*). Symbol for *actinium.*

acacia (*For.*). A member of the *Leguminosae* giving a coarse-textured hardwood, reddish-brown in colour. Used for tool handles, vehicle parts, walking sticks, and turned articles.

acacia gum (*Chem.*). See **gum arabic.**

Acadian (*Geol.*). A series name applied to the Middle Cambrian strata of the Atlantic Province in N. America (Newfoundland, Nova Scotia to eastern Massachusetts).

Acalephae (*Zool.*). See **Scyphozoa.**

acantha (*Bot.*). A prickle or spine. The term occurs in such compounds as *acanthocarpous* (spiny-fruited) and *acanthocladous* (having spiny branches).

acanthin (*Zool.*). A substance which forms the skeleton of some *Radiolaria*; formerly believed to be of an organic nature, but now known to be strontium sulphate.

acanthite (*Min.*). An ore of silver, Ag_2S, crystallizing in the monoclinic system. Cf. *argentite.*

acantho- (*Bot.*). Prefix from Gk. *akantha*, spine, thorn.

acanthocarpous (*Bot.*). Having spiny fruits.

Acanthocephala (*Zool.*). A phylum of elongate worms with rounded body and a protrusible proboscis, furnished with recurved hooks; there is no mouth or alimentary canal; the young stages are parasitic in various Crustaceans, the adults in Fish and aquatic Birds and Mammals. Thorny-headed worms.

acanthosis nigricans (*Med.*). A rare disease characterized by pigmentation and warty growths on the skin, often associated with cancer of the stomach or uterus.

acanthozooid (*Zool.*). In *Cestoda*, the proscolex, or head-portion, of a bladder-worm. Cf. *cystozooid.*

acanthus (*Bot.*). A small genus of Mediterranean plants with white or red flowers. The leaves are the model for an architectural design.

acapnia (*Med.*). Excessive diminution of carbon dioxide in blood.

acariasis (*Med., Vet.*). Contagious skin disease caused by mites (*acari*).

Acarina (*Zool.*). An order of small *Arachnida* with globular, undivided body. The immature stages (hexapod larvae) have 6 legs. A large worldwide group, occupying all types of habitat, and of great economic importance. Many are parasitic. Mites and ticks.

acarodomatium (*Bot.*). A hollow or other protective structure formed by some plants which harbour mites and seem to live in symbiosis with them.

acarophily, acarophytism (*Bot.*). A symbiotic association between plants and mites.

acarus (*Zool.*). A mite; an arachnid of the order *Acarina.* The species called *Sarcoptes scabiei* causes scabies.

acaryallagic (*Bot.*). Of reproduction without nuclear fusion; asexual; vegetative.

acathexis (*Psychiat.*). Lack of emotional attachment or *affect* for a thing which is actually of importance to the subject.

acaudate (*Bot.*). Tailless.

acaulescent (*Bot.*). Having a very short stem.

acauline, acaulose (*Bot.*). Stemless or nearly so.

a.c. balancer (*Elec. Eng.*). An arrangement of transformers or reactors used to equalize the voltages between the wires of a multiple-wire system. Also called a **static balancer.**

a.c. bias (*Electronics*). In magnetic tape or wire recording, the addition of a polarizing alternating current in signal recording to stabilize magnetic saturation.

accelerated ageing test (*Cables*). A stability test using twice normal working voltage. It is claimed this gives quick results that correlate with service records.

accelerated fatigue test (*Eng.*). The application, mechanically, of simulated operating loads to a machine or component to find the limit of safe fatigue life before it is reached in service.

accelerated filtration (*Chem.*). The process of increasing speed of filtration of solid from liquid by applying suction to draw the liquid through the filter, or by applying pressure to force it through.

accelerate-stop distance (*Aero.*). The total distance, under specified conditions, in which an aeroplane can be brought to rest after accelerating to *critical speed* (q.v.) for an engine failure at takeoff.

accelerating chain (*Electronics*). That section of a vacuum system, e.g., CRT or betatron, in which charged particles are accelerated by voltages on accelerating electrodes.

accelerating contactor (*Elec. Eng.*). One of the contactors of an electric-motor control panel which cuts out starting resistance, thereby causing the motor to accelerate.

accelerating electrode (*Electronics*). One in a thermionic valve or CRT maintained at a high positive potential with respect to the electron source. It accelerates electrons in their flight to the anode but does not collect a high proportion of them.

accelerating machine or accelerator (*Electronics*). Machine used to accelerate charged particles to very high energies. See also **betatron, cyclotron, linear accelerator, synchrocyclotron,** and **synchrotron.**

accelerating potential (*Electronics*). That applied to an electrode to accelerate electrons from a cathode.

accelerating pump (*Autos.*). A small cylinder and piston fitted to some types of carburettor, and connected to the throttle so as to provide a momentarily enriched mixture when the engine is accelerated.

acceleration (*Maths.*). Of a point: the rate of change with respect to time of the velocity of the point. In cartesian coordinates the acceleration of the point (x, y) is $(\ddot{x}, \ddot{y})$, where $\ddot{x}$ represents $\frac{d^2x}{dt^2}$ and similarly for $\ddot{y}$. (*Mech.*) The rate of change of velocity, expressed in metres (or feet) per second squared. Certain restricted and special applications of the word occur in Astronomy; e.g., *secular acceleration.*

acceleration due to gravity (*Mech.*). Acceleration with which a body would fall freely under the action of gravity in a vacuum. This varies according to the distance from the earth's centre, but the internationally adopted value is $9{\cdot}806\,65$ m/s² or $32{\cdot}1740$ ft/s². See **Helmert's formula.**

acceleration error (*Aero.*). The error in an airborne magnetic compass due to manoeuvring; caused by the vertical component of the earth's magnetic field when the centre of gravity of the magnetic element is displaced from normal.

acceleration tolerance (*Space*). The maximum of *g* forces an astronaut can withstand before 'blacking out' or otherwise losing control.

accelerator (*Aero.*). A device, similar to a *catapult*, but generally mounted below deck level, for assisting the acceleration of aeroplanes flying off aircraft carriers. Land versions have been tried experimentally. (*Autos., etc.*) A pedal connected to the carburettor throttle valve of a motor vehicle, or to the fuel injection control where oil engines are used. (*Chem.*) (1) A substance which increases the speed of a chemical reaction. See **catalysis.** (2) A substance which increases the efficient action of an enzyme. (3) Any substance effecting acceleration of the vulcanization process of rubber. The principal types are aldehyde derivatives of Schiff's bases,

butyraldehyde-butylidene-aniline,
di-orthotolyl-guanidine,
diphenyl-guanidine,
benzthiazyl disulphide,
tetramethyl-thiuran disulphide,
zinc dimethyl-dithiocarbamate.

(*Civ. Eng., etc.*) Any substance mixed with cement concrete for the purpose of hastening hardening. (*Electronics*) See **accelerating machine.** (*Photog.*) A chemical used to increase the rate of development; e.g., sodium carbonate or borax. (*Plastics*) Substance which increases catalytically the hardening rate of a synthetic resin. (*Zool.*) Any muscle or nerve which increases rate of action.

accelerometer (*Aero.*). An instrument, carried in aircraft for measuring acceleration in a specific direction. Main types are *indicating*, *maximum reading*, *recording* (graphical), and *counting* (digital, totalling all loads above a set value). See **impact-, vertical-gust recorder.** (*Electronics*) Transducer used to give signal proportional to rate of acceleration of body. Used especially in vibration study.

accentuation (*Telecomm.*). A technique for emphasizing particular bands in an audio-amplifier.

acceptance angle (*Electronics*). The solid angle within which all incident light reaches the photocathode of a phototube.

acceptor (*Chem.*). (1) The reactant in an induced reaction whose rate of reaction with a third substance is increased by the presence of the inductor. (2) The atom which contributes no electrons to a co-ordinate bond. (*Electronics*) Impurity atoms introduced in small quantities into a crystalline semiconductor and having a lower valency than the semiconductor from which they attract electrons. In this way 'holes' or positive charge carriers are produced, the acceptor atom becoming negatively charged. The phenomenon is known as *p-type conductivity*. E.g., 100 parts of boron in 10^6 parts of silicon could increase the conductivity of the latter 1 million fold. See also **donor, impurity.**

acceptor circuit (*Telecomm.*). Tuned circuit responding to a signal of one specific frequency.

acceptor level (*Electronics*). See **energy levels.**

access eye (*San. Eng.*). A screwed plug provided in soil, waste, and drain pipes at bends and junctions, so that access is possible to clear a stoppage.

accessibility (*For.*). In logging, the extent to which standing timber is commercially accessible.

accessorius (*Zool.*). A muscle which supplements the action of another muscle: in Vertebrates, the eleventh cranial nerve or spinal accessory.

accessory (*Bot.*). An additional member beyond the normal. (*Zool.*) See **accessorius.**

accessory bodies (*Cyt.*). Minute argyrophil particles, originating from the Golgi body in the cytoplasm of spermatocytes; chromatoid bodies.

accessory bud (*Bot.*). A bud additional to a normal axillary bud.

accessory cell (*Bot.*). A cell associated with the guard cell of a stoma, differing in structure both from it and from the ordinary cells of the epidermis.

accessory character (*Bot., Zool.*). A nonessential character of a species, sometimes used to distinguish one race from another.

accessory chromosome (*Cyt.*). See **sex chromosome.**

accessory gearbox (*Aero.*). A gearbox, driven remotely from an aero-engine, on which aircraft accessories, e.g., hydraulic pump, electrical generator, are mounted.

accessory glands (*Zool.*). Glands of varied structure and function in connexion with genitalia, especially of *Arthropoda*.

accessory hearts (*Zool.*). See **accessory pulsatory organs.**

accessory minerals (*Geol.*). Minerals which occur in small, often minute, amounts in igneous rocks; their presence or absence makes no difference to classification and nomenclature.

accessory multiplication (*Bot.*). Any reproductive process which is not sexual. Sometimes carried out by accessory reproductive structures, e.g., bulbils of some orchids; gemmae of liverworts; sclerotia of some fungi. Also **accessory reproduction.**

accessory nidamental gland (*Zool.*). A paired gland associated with the shell-forming nidamental glands in female nautiloids.

accessory plates (*Min.*). Quartz-wedge, gypsum plate, mica plate. Used with petrological microscope to modify polarized light effects and intensify qualities in translucent minerals which aid in their examination.

accessory pulsatory organs (*Zool.*). In Insects, saclike contractile organs, pulsating independently of the heart, and variously situated on the course of the circulatory system. Also **accessory hearts.**

accessory scapula (*Zool.*). A small bone sometimes found on the outside of the shoulder joint in birds.

accessory shoe (*Photog.*). Grooved fitting on top of a camera which accepts standard types of viewfinders, rangefinders, flash units, and other accessories.

accessory spore (*Bot.*). (1) Aspore of nonsexual origin. (2) A conidium of a type different from that usual in the species.

accessory tarsal (*Zool.*). A bone supporting the spur in *Prototheria.*

access selector (*Teleph.*). A selector used to connect common equipment (e.g., in routine tests of artificial traffic equipment) sequentially to corresponding points in each of a series of identical circuits.

access time (*Comp.*). Maximum time to locate and extract *words* or digits from a *store*, from microseconds in a fast core or film store, to minutes in a long magnetic tape file store.

access to store (*Comp.*). Entry or extraction of data from a storage memory. In a **random access memory** (RAM), each location is independent of all other locations and the locations can be referred to in any order. In a **sequential memory** each location is accessed in turn.

a.c. circuit (*Elec. Eng.*). One which passes a.c., e.g., it may have a capacitor in series, which blocks direct current.

acclimatization (*Chem.*). The change produced in a colloidal sol by the addition of a precipitating (coagulating) agent in small quantities, resulting in less complete precipitation for the addition of a given total amount of precipitant. (*Zool.*) Accustoming a species to a new environment, e.g., a zoo.

accommodation (*Bot.*). The capacity possessed by a plant to adjust itself to new conditions of life, provided the changed conditions come gradually into operation. (*Physiol.*) Ability of the eye to change its effective focal length to see objects distinctly at varying distances. The range of vision for a human eye is from about 250 mm to infinity. Power of accommodation usually diminishes with advancing age. (*Psychol.*) That part of *adaptation* (q.v.) by which development occurs through change in the organism in response to demands of the environment (Piaget).

accommodation coefficient (*Chem.*). Coefficient describing lack of thermal equilibrium between a heated wall and the molecules impinging on it, when the dimensions of the wall are comparable to the mean free path of the molecules (*Knudsen flow*; (q.v.)). Defined as $a=(T_i-T_r)/(T_i-T_w)\leqq 1$—where the temperatures are: T_i of the impinging molecules, T_r of the molecules after reflexion from the wall, and T_w of the wall.

a.c. commutator motor (*Elec. Eng.*). An a.c. motor which embodies a commutator as an essential part of its construction. See **a.c. series motor, compensated induction motor, repulsion motor, Schrage motor.**

accordion (*Arch.*). A folding door or partition working in a manner similar to the bellows of an accordion.

accordion fold (*Bind.*). Method of folding a leaflet or insert so that it opens out and closes in a zig-zag fashion. Also called **concertina fold.**

accoucheur (*Med.*). A physician who practises midwifery.

account book paper (*Paper*). A specially tough paper which will stand erasures with a penknife. Usually azure laid.

accouplement (*Carp.*). A tie or brace of timber.

accrescent (*Bot.*). Enlarged and persistent; usually applied to a calyx which increases in size as the fruit ripens.

accretion (*Zool.*). External addition of new matter: growth by such addition.

accumbent (*Bot.*). Said of any embryo in which the edges of cotyledons lie against the radicle. Resting against another structure.

accumulated temperature (*Meteor.*). The integrated product of the excess of air temperature above 6°C (42°F) and the period in days during which such excess is maintained.

accumulation point (*Maths.*). Of a set of points: one such that every neighbourhood of it includes at least one point of the set. Also called **limit point.**

accumulator (*Comp.*). Name frequently given to register in arithmetic unit of digital computer. (*Elec. Eng.*) Voltaic cell which can be charged and discharged. On charge, when an electric current is passed through it into the positive and out of the negative terminals (according to the conventional direction of flow of current), electrical energy is converted into chemical energy. The process is reversed on discharge, the chemical energy, less losses both in potential and current, being converted into useful electrical energy. Accumulators therefore form a useful portable supply of electric power, but have the disadvantages of being heavy and of being at best 70% efficient. More often known as **battery,** also called **reversible cell, secondary cell, storage battery.** (*Ocean.*) A spring of rubber or steel attached to a trawling warp, to lessen any sudden strain due to the trawl catching.

accumulator box (*Elec. Eng.*). A vessel, usually made of glass, lead-lined wood, or celluloid, for containing the plates and electrolyte of an accumulator.

accumulator grid (*Elec. Eng.*). The lead grid which forms one of the plates of a lead-acid accumulator having pasted plates.

accumulator switchboard (*Elec. Eng.*). A switchboard upon which are mounted all necessary switches and instruments for controlling charging and discharging of a battery of accumulators.

accumulator traction (*Elec. Eng.*). See **battery traction.**

accumulator vehicle (*Elec. Eng.*). An electrically propelled vehicle which derives its energy from an accumulator carried on the vehicle.

AcEm (*Chem.*). The symbol for *actinium emanation* or *actinon.*

acenaphthenequinone (*Chem.*). Chemical, crystallizing in yellow needles, sparingly soluble in water. Forms the basis of scarlet and red vat dyes of the 'Ciba' type.

acentric (*Cyt.*). Having no centromere, applied to chromosomes and chromosome segments.

acentrous (*Zool.*). Having a persistent notochord with no vertebral centra, as in the *Cyclostomata.*

acephalous (*Bot.*). Said of a style which does not terminate in a well-marked stigma. Not terminating in a well-defined head. (*Zool.*) Showing no appreciable degree of cephalization: lacking a head-region, as *Pelecypoda.*

acephalous larva (*Zool.*). The larva of cyclorrhaphan *Diptera,* possessing a vestigial head and no legs.

acerose, acerous (*Bot.*). Needle-shaped.

acervuline (*Zool.*). Having the form of an irregular heap, as the shells of some *Foraminifera.*

acervulus (*Bot.*). A dense cushionlike mass of conidiophores and conidia formed by some fungi. *adj.* **acervulate.**

acervulus cerebri (*Zool.*). See **brain-sand.**

acetabular bone (*Zool.*). In Crocodiles, the separate lower end of the ilium: in some Mammals, as *Galeopithecus,* an additional bone lying between the ilium and pubis and frequently fused with one or the other.

acetabulum (*Zool.*). In *Platyhelminthes, Hirudinea,* and *Cephalopoda,* a circular muscular sucker: in Insects, a thoracic aperture for insertion of a leg: in Vertebrates, a facet or socket of the pelvic girdle with which articulates the pelvic fin or head of the femur: in ruminant Mammals, one of the cotyledons of the placenta.

acetal (*Chem.*). $CH_3CH(OC_2H_5)_2$, b.p. 104°C, 1,1-diethoxy ethane. The term *acetal* is applied to any compound of the type $R{\cdot}CH(OR')_2$, where R and R′ are organic radicals and R may be hydrogen.

acetaldehyde (*Chem.*). Ethanal. CH_3CHO, colourless, ethereal, pungent liquid, b.p. 21°C, m.p. −121°C, rel. d. 0·8, oxidation product of ethanol; intermediate for production of ethanoic acid, important raw material for the synthesis of organic compounds.

acetal resins (*Plastics*). Term applied to stable high polymers of formaldehyde widely used as engineering plastics. See also **Delrin.**

acetals (*Chem.*). The dehydration products of aldehydes, with an excess of alcohols present. Acetals may be termed 1,1-dialkoxyalkanes.

acetamide (*Chem.*). Ethanamide. $CH_3{\cdot}CO{\cdot}NH_2$, needles, soluble in water and alcohol, m.p. 82°C, b.p. 222°C, a primary amide of ethanoic acid.

acetanilide (*Pharm.*). Acetylaminobenzene, *N*-phenyl acetamide (ethanamide), $C_6H_5NH{\cdot}COCH_3$, m.p. 113°–115°C, b.p. 304°–305°C. Formerly used as a mild antiseptic powder. Used pharmaceutically as an antipyretic and analgesic; largely replaced by safer analgesics. Also called **antifebrin.**

acetarsol (*Pharm.*). 3-Acetylamino-4-hydroxy phenyl-arsonic acid, $CH_3{\cdot}CO{\cdot}NH{\cdot}C_6H_3$(OH

$AsO(OH)_2$. White crystals, m.p. 240°–250°C. Intestinal amoebicide; also used in treatment of yaws and certain inflammatory conditions.

acetate film (*Photog.*). One made of cellulose acetate, which is noninflammable.

acetate rayon (*Textiles*). Continuous filament and staple fibre manufactured in various deniers, colours, also staple lengths for blending with other fibres. Fibre forming substance is cellulose acetate produced from cotton linters, and wood pulp treated with acetic anhydride, acetic acid, and sulphuric acid.

acetates (*Chem.*). Ethanoates. Salts of acetic (ethanoic) acid; e.g., sodium acetate. Also the acetylation products of acetic acid.

acetic acid (*Chem.*). Ethanoic acid. $CH_3 \cdot COOH$, synthesized from acetylene (ethyne), also obtained by the destructive distillation of wood; and by the oxidation of ethanol. Acetic acid has a m.p. 16·6°C, b.p. 118°C, rel. d. (20°C) 1·0497.

acetic anhydride (*Chem.*). Ethanoic anhydride. The anhydride of acetic acid, $(CH_3CO)_2O$. Colourless liquid b.p. 137°C. Used industrially for preparation of cellulose acetate and acetylsalicylic acid. Valuable laboratory acetylating agent.

acetic ether (*Chem.*). See **ethyl acetate.**

acetic fermentation (*Chem.*). The fermentation of dilute ethanol solutions by oxidation in presence of bacteria, especially *Bacterium aceti*. Acetic acid is formed.

acetin (*Chem.*). Monoacetin, glyceryl monoacetate, $CH_3COO \cdot C_3H_5(OH)_2$, b.p. 130°C, rel. d. 1·22, a colourless hygroscopic liquid, used as intermediate for explosives, solvent for basic dyestuffs, tanning agent.

acetocarmine fixative (*Micros.*). A combined fixative and stain used for cytological purposes.

acetone (*Chem.*). Propanone. CH_3COCH_3, b.p. 56°C, of ethereal odour, a very important solvent, basis for organic synthesis. Acetone is the simplest type of a saturated ketone. Useful solvent for acetylene.

acetone cyanhydrin (*Chem.*). 2-hydroxy 2-methyl propanonitrile. $(CH_3)_2C(OH)CN$. Addition product of acetone and hydrogen cyanide.

acetone resin (*Chem.*). A synthetic resin formed by the reaction of acetone with another compound, such as phenol or formaldehyde.

acetonuria (*Med.*). See **ketonuria.**

acetophenone (*Chem.*). $C_6H_5COCH_3$, phenyl methyl ketone, m.p. 20°C, b.p. 202°C, rel.d. 1·03, large colourless crystals or liquid, soluble in most organic solvents, insoluble in water; used for organic synthesis and in perfumery.

acetoxyl group (*Chem.*). The group $CH_3 \cdot CO \cdot O$—.

acetylation (*Chem.*). Ethanoylation. Reaction which has the effect of introducing an acetyle radical (CH_3CO) into an organic molecule.

acetylcelluloses (*Chem.*). See **cellulose acetates.**

acetyl chloride (*Chem.*) Ethanoyl chloride. CH_3COCl, m.p. −112°C, b.p. 51°C, rel. d. 1·105. Colourless liquid, of pungent odour, used for synthesis, in particular for introducing the acetyl group into other compounds.

acetylcholine (*Med.*). The substance which brings about the conduction of nerve impulses across the junction of two nerve fibres, and produces effects on the parasympathetic nervous system.

acetyl coenzyme A (*Biochem.*). The S-acetyl derivative of coenzyme A is the precursor for the synthesis of fatty acids and sterols and is the biological acetylating agent in the synthesis of such compounds as acetylcholine. It is derived from the metabolism of fatty acids, carbohydrates and proteins and oxidized through a cyclic process, the *citric acid cycle*, to CO_2 and water while providing electrons to the transport system that reduces oxygen and generates ATP.

acetylene (*Chem.*). Ethyne. HC⋮CH, a colourless, poisonous gas, owing its disagreeable odour to impurities, soluble in ethanol, in acetone (25 times its volume at s.t.p.), and in water. B.p. −84°C, rel. d. 0·91. Prepared by the action of water on calcium carbide and catalytically from naphtha. Used for welding, illuminating, acetic acid synthesis, and for manufacturing derivatives particularly by Reppe synthesis.

acetyl group (*Chem.*). Ethanoyl group. The radical of acetic acid, viz., CH_3CO—.

acetylide (*Chem.*). Ethynide. Carbide formed by bubbling acetylene through a solution of a metallic salt, e.g., cuprous acetylide, Cu_2C_2. Violently explosive compounds.

acetylsalicylic acid (*Chem.*). $C_6H_4(O \cdot COCH_3)COOH$, m.p. approximately 128°C. Used in medical and veterinary practice as an analgesic, antipyretic, and antirheumatic. The acid or its salts are the active components of aspirin.

acetyl value (*Chem.*). Ethanoyl value. A test used chiefly for recognizing and determining oils of the castor-oil group. It indicates the hydrogen in the hydroxyl groups replaceable by the acetyl radical (not in the carboxyl group). Basis for the determination of alcohols, glycerides, hydroxy-acids.

a.c. exciter (*Elec. Eng.*). A term used to denote a commutator machine connected in the rotor circuit of an induction motor to effect power factor improvement.

ACF diagram (*Geol.*). A triangular diagram used to represent the chemical composition of metamorphic rocks. The three corners of the diagram are Al_2O_3, CaO, and FeO + MgO.

a.c. generator (*Elec. Eng.*). An electromagnetic generator for producing alternating e.m.fs. and delivering a.cs. to an outside circuit. See **alternator, induction generator.**

achaenocarp (*Bot.*). See **achene.**

achalasia (*Med.*). Failure to relax.

achalasia of the cardia (*Med.*). Failure to relax on the part of the sphincter round the opening of the oesophagus into the stomach.

achene (*Bot.*). A dry indehiscent, 1-seeded fruit, formed from a single carpel, and with the seed distinct from the fruit wall. Also called **achaenocarp, akene.**

Acheson furnace (*Elec. Eng.*). An electric furnace for production of silicon carbide. The constituents of the compound (coke and sand) are packed round a central core of coke, through which a heavy current is passed.

Acheson graphite (*Eng.*). That made from coke in electric furnace.

Achilles tendon (*Zool.*). In Mammals, the united tendon of the soleus and gastrocnemius muscles; the hamstring.

A.C.H. index (*Nut.*). The arm-girth, depth-of-chest, and width-of-pelvis index used for comparative measurements in studies of nutrition. See **Franzen's index.**

achlamydeous (*Bot.*). Lacking a perianth.

achlorhydria (*Med.*). Absence of hydrochloric acid from gastric juice.

acholuric jaundice (*Med.*). See **spherocytosis.**

achondrite (*Geol.*). A type of stony meteorite which compares closely with some basic igneous rocks such as eucrite.

achondroplasia (*Med.*). Dwarfism characterized by shortness of arms and legs, with a normal body and head. *adj.* **achondroplastic.**

achroglobin (*Zool.*). A colourless respiratory

pigment occurring in some *Mollusca* and some *Urochorda.*
achroite (*Min.*). See **tourmaline.**
achromasie (*Cyt.*). The escape of chromatin from the nucleus. (*Med.*) Lack of the normal skin pigmentation.
achromatic (*Bot.*). Colourless.
achromatic figure (*Cyt.*). The double set of delicate radiating strands, and, when present, the asters, as seen in preparations of dividing nuclei.
achromatic lens (*Light*). A lens designed to minimize chromatic aberration. The simplest form consists of two component lenses, one convergent, the other divergent, made of glasses having different dispersive powers, the ratio of their focal lengths being equal to the ratio of the dispersive powers. Also **antispectroscopic lens.**
achromatic prism (*Optics*). An optical prism with a minimum of dispersion but having a maximum of deviation.
achromatic sensation (*Light*). A sensation of grey.
achromatic spindle (*Cyt.*). In cell-division, a system of apparent fibres joining the poles of the nucleus and diverging towards the equator.
achromatic stimulus (*Light*). One which produces an *achromatic sensation.*
achromatin (*Cyt.*). That part of the nucleus which does not stain with basic dyes, generally comprising the nuclear sap or ground-substance. *adj.* **achromatinic.** Cf. *chromatin.*
Achromycin (*Chem.*). See **tetracycline.**
achroous (*Bot.*). Colourless; hyaline.
achylia gastrica (*Med.*). Complete absence of ferments and hydrochloric acid from the gastric secretion.
acicle (*Bot.*). A stiff bristle, or slender prickle; sometimes with a glandular tip. (*Min.*) A needlelike crystal. *adj.* **acicular.**
aciculate (*Bot.*). Marked on the surface by fine scratches.
aciculum (*Zool.*). A stout internal chaeta in the parapodium of *Polychaeta,* acting as a muscle attachment.
acid (*Chem.*). Generally, a substance which tends to lose a proton. In particular, a substance which dissolves in water with the formation of hydrogen ions, which may be replaced by metals with the formation of salts.
acidaemia (*Med.*). A temporary lowering of the body pH below normal.
acid amides (*Chem.*). A group of compounds derived from an acid by the introduction of the amino group in place of the hydroxyl radical of the carboxyl group.
acid anhydride (*Chem.*). Compound generating a hydroxylic acid on addition of (or derived from the acid by removal of) one or more molecules of water, e.g., sulphur(VI)oxide

$$SO_3+H_2O=H_2SO_4.$$

acid azides (*Chem.*). The acyl derivatives of hydrazoic acid, obtainable from *acid hydrazides* (q.v.) by treatment with nitrous acid. They are very unstable.
acid brittleness (*Met.*). That developed in steel in pickling bath, through evolution of hydrogen.
acid chlorides (*Chem.*). Compounds derived from acids by the replacement of the hydroxyl group by chlorine.
acid clay (*Eng.*). Clay which, when pulped in water, yields hydrogen ions.
acid cure (*Min. Proc.*). In extraction of uranium from its ores, lowering of gangue carbonates by puddling with sulphuric acid before leach treatment.
acid drift (*Mining, Min. Proc.*). Tendency of ores, pulps, products, to become acidic through pick-up of atmospheric oxygen on standing.
acid dyes (*Photog.*). Dyes which have their colour associated with the negative ion or radical.
acid egg (*Chem. Eng.*). A pump for sulphuric acid, of simple and durable construction, with few moving parts. The acid is run into a pressure vessel, usually egg-shaped, from which it can be forcibly expelled by compressed air.
acid esters (*Chem.*). Compounds derived from acids in which part of the replaceable hydrogen has been exchanged for an alkyl radical.
acid fixer (*Photog.*). Fixing solution (hypo) with the addition of an acid (sodium bisulphite or potassium metabisulphite) to prevent staining.
acid glands (*Zool.*). In *Hymenoptera,* a pair of glands opening into the poison sac at the base of the sting. Their secretion causes the painful reaction associated with bee and wasp stings.
acid hydrazides (*Chem.*). Hydrazine derivatives into which an acyl group has been introduced.
acidic (*Chem.*). Said of dyes the colour-base of which is an acid, usually combined with an inorganic base, e.g., acid fuchsin, the sodium salt of a sulphonic acid derived from fuchsin.
acidimetry (*Chem.*). The determination of acids by titration with a standard solution of alkali as in volumetric analysis.
acidity (*Chem.*). (1) Extent to which a solution is acid. See **pH value.** (2) Number of replaceable hydroxyl groups in a molecule of a base. (3) Amount of free acid in vegetable oils, resins, etc.
acid mine water (*Mining.*). Water containing sulphuric acid as a result of the breakdown of the sulphide minerals in rocks. Acid mine water causes corrosion of mining equipment, and may contaminate water supplies into which it drains.
acidolysis (*Chem.*). Acid hydrolysis.
acidophil (*Cyt.*). Said of structures which stain intensely with acid dyes, e.g., acidophil granules in leucocytes.
acidosis (*Med.*). Normal body pH but a decreased amount of buffer base.
acid polishing (*Glass*). A method of polishing cut decorations on glassware by immersing the article in an acid bath for a few minutes, rinsing in water, and brushing out the cut parts.
acid process (*Met.*). A steel-making process, either Bessemer, open-hearth, or electric, in which the furnace is lined with a siliceous refractory, and for which pig iron low in phosphorus is required, as this element is not removed. See **basic process.** (*Paper*) The preparation of pulp by pressure digestion of wood with a liquor containing calcium bisulphite and free sulphur dioxide. Also called the **sulphite process.**
acid radical (*Chem.*). A molecule of an acid minus replaceable hydrogen.
acid refractory (*Met.*). See **silica.**
acid resist foils (*Met.*). Blocking foils for use in etching metal. The foil is stamped on to paper and the excess foil blocked on to the metal rule or other object which is then exposed to an acidic etching fluid such as ferric chloride.
acid rocks (*Geol.*). Those igneous rocks which contain more than 66% of silica, and which, therefore, when crystalline, carry free quartz as an essential constituent.
acid salts (*Chem.*). Salts formed by replacement of part of the replaceable hydrogen of the acid.
acid slag (*Met.*). Furnace slag in which silica and alumina exceed lime and magnesia.
acid sludge (*Chem. Eng.*). Mixture of 'green acid' and 'mahogany soap' salts; reaction product in sulphuric acid treatment of oil refinery heavy fractions, used in froth flotation process for recovering iron minerals.

acid solution (*Chem.*). An aqueous solution containing more hydroxonium ions than hydroxyl ions; one which turns blue litmus red.
acid steel (*Met.*). Steel made by an acid process.
acidum aceticum aromaticum (*Chem.*). Acetic acid flavoured with the essential oils of cloves, lavender, orange, bergamot, sometimes thyme, cinnamon, or camphor.
acid value (*Chem.*). The measure of the free acid content of vegetable oils, resins, etc., indicated by the no. of mg of potassium hydroxide (KOH) required to neutralize 1 g of the substance.
acinaceous (*Bot.*). Full of pips.
acinaciform (*Bot.*). Scimitar-shaped.
Acinetaria (*Zool.*). See **Suctoria.**
aciniform (*Zool.*). Berry-shaped; e.g., in Spiders, the *aciniform* glands producing silk and leading to the median and posterior spinnerets.
acinus (*Bot.*). A drupel, drupelet, i.e., one of a collection of small drupes borne on the receptacle, as in the blackberry, raspberry, etc. (*Cyt.*) Cluster of cells comprising terminal saclike secretory unit of a gland; usually a branched or compound gland. *pl.* **acini.**
acisculis (*Build.*). A small pick with square flat face and a pointed peen.
Ackermann steering (*Autos.*). Arrangement whereby a line extended from the track-arms, when the wheels are set straight ahead, should meet on the chassis centre-line at 2/3 of the wheelbase from the front, allowing the inner stub-axle to move through a greater angle than the outer.
Acker process (*Chem.*). Process for production of sodium hydroxide. Molten sodium chloride is electrolysed, using a molten lead cathode, and the resulting lead-sodium alloy is decomposed by water, yielding pure lead and pure sodium hydroxide. Obsolete.
acknowledgment signal (*Teleph.*). A signal transmitted along a circuit from B to A when triggered by a signal from A to B.
A-class insulation (*Elec. Eng.*). A class of insulating material to which is assigned a temperature of 105°C. See **class-A, -B, -C, etc., insulating materials.**
a.c. magnet (*Elec. Eng.*). Electromagnet excited by a.c., having normally a laminated magnetic circuit. See **shaded pole.**
acme (*Biol.*). The period of maximum vigour, the prime, of an individual, race, or species: the adult period: the ephebic period: the phyloephebic period.
acme screw-thread (*Eng.*). A thread having a profile angle of 29 degrees and a flat crest, sometimes used for lathe lead screw, etc., for easy engagement by a split nut.
acmite (*Min.*). See **aegirine.**
a.c. motor (*Elec. Eng.*). An electric motor which operates from a single or polyphase a.c. supply. See **capacitor motor, induction motor, synchronous motor.**
acne (*Med.*). Inflammation of a sebaceous gland. Pimples in adolescents are commonly due to infection with the acne bacillus.
acne rosacea (*Med.*). A condition in which there is chronic congestion of the superficial vessels of the nose and central part of the face, often associated with dyspepsia in females.
acnode (*Maths.*). See **double point.**
Acoela (*Zool.*). An order of small marine *Turbellaria* in which the gut is not hollow but consists of a syncytium formed by the union of endodermal cells. Acoelomate animals.
acoelomate, acoelomatous (*Zool.*). Without a true coelom.
acoelomate triploblastica (*Zool.*). Animals with three embryonic cell layers but no coelom. They consist of the *Platyhelminthes, Nematoda,* and some minor phyla, i.e., all the helminth phyla.
acoelous (*Zool.*). Lacking a gut-cavity.
acone (*Zool.*). In Insects, said of compound eyes in which the ommatidia contain no cone.
aconitine (*Chem.*). $C_{34}H_{45}O_{11}N$ or $C_{34}H_{47}O_{11}N$, an alkaloid of unknown constitution. It is obtained from *Aconitum napellus.* Highly toxic, affecting heart and respiratory organs.
aconitic acid (*Chem.*). An unsaturated tribasic acid extracted from *Aconitum napellus,* beetroot, and sugar cane. Used as a plasticizer for Buna rubber and plastics, its formula is

$$CH_2(COOH) \cdot C(COOH) = CH(COOH).$$

acontia (*Zool.*). In *Anthozoa,* free threads, loaded with nematocysts, arising from the mesenteries or the mesenteric filaments, and capable of being discharged via the mouth or via special pores.
acorn (*Bot.*). The fruit of the oak.
acorn valve (*Electronics*). A small electronic valve, shaped like an acorn, designed especially for u.h.f., with minimum inter-electrode capacitance and lead inductance.
acotyledonous (*Bot.*). Said of an embryo of a higher plant which has no cotyledons.
acoustextile (*Build.*). A unit of material specially designed for increasing acoustic absorption of walls.
acoustic absorption (*Acous.*). Diminution of energy in a sound-wave during reflection from a surface which is not completely reflecting, mainly because of its lack of hardness.
acoustic absorption factor (or coefficient) (*Acous.*). The measure of the ratio of the acoustic energy absorbed by a surface to that which is incident on the surface. For an open window this can be 1·00: for painted plaster 0·02. The value varies with the frequency of the incident sounds; e.g., for glass fibre 0·04 at 125 Hz, 0·80 at 4000 Hz.
acoustic amplifier (*Acous.*). One amplifying mechanical vibrations.
acoustical compliance (*Acous.*). Reciprocal of acoustical stiffness.
acoustical mass or inertance (*Acous.*). Given by M, where ωM is that part of the acoustical reactance which corresponds to the inductance of an electrical reactance: ω is the pulsatance $= 2\pi \times$ frequency.
acoustical reproduction (*Acous.*). See **electrical reproduction.**
acoustical stiffness (*Acous.*). For an enclosure of volume V, given by $S_A = \rho c^2/V$, where c is velocity of propagation of sound and ρ is density. It is assumed that the dimensions of the enclosure are small compared with the sound wavelength.
acoustic centre (*Acous.*). The effective 'source' point of the spherically divergent wave system observed at distant points in the radiation field of an acoustic transducer.
acoustic concentration (*Acous.*). For a given distance from a radiating source of sound, the ratio of the intensity of the directly radiated sound to the mean spherical intensity. The polar concentration indicates measure of the directivity of a source of sound.
acoustic construction (*Build.*). Building construction which aims at the control of transmission of sound, or of mechanical vibration giving rise to sound, particularly unwanted noises. The parts of the structure are separated

by air-spaces or acoustic absorbing material and decoupled by the interposing of springs.

acoustic dazzle (*Acous.*). Psychological effect associated with high density of radiation of sound from small sources.

acoustic delay (*Telecomm.*). That introduced into sound reproduction of speech or music along a telephone line, by conversion to sound, which is caused to travel a suitable distance along a pipe before reconversion into electric currents. Delay is also obtained through magnetic recording or through wave filters.

acoustic delay line (*Telecomm.*). A device, magnetostrictive or piezoelectric, e.g., a quartz bar or plate of suitable geometry, which reflects an injected sound pulse many times within the body.

acoustic distortion (*Acous.*). Distortion in sound-reproducing systems, due to alteration in the acoustic ratio of sounds arriving at the listener's ears when compared with the original or with a natural ratio.

acoustic feedback (*Telecomm.*). The excitation of the microphone in a public-address system by sound generated by the reproducers of the same system. If this is excessive the system builds up a sustained oscillation. Also, in a radiogram or electric record-reproducer, reception through the pickup, while playing, of vibrations from the loudspeaker, producing a similar effect.

acoustic filter (*Acous.*). One which uses tubes and resonating boxes in shunt and series as reactance elements, providing frequency cutoffs in acoustic wave transmission, as in an electric wave filter.

acoustic grating (*Acous.*). A diffraction grating for production of sound spectra. The spacings are much larger than in optical gratings due to the larger wavelength of sound waves. Both transmission and reflection gratings are used.

acoustic impedance, reactance, and **resistance** (*Acous.*). Impedance is given by complex ratio of sound pressure on surface to sound flux through surface, and has imaginary (reactance) and real (resistance) components, respectively. Unit is the *acoustic ohm*.

acoustic interferometer (*Acous.*). Instrument in which measurements are made by study of interference pattern set up by two sound or ultrasonic waves generated at the same source.

acoustic lens (*Acous.*). A system of slats or disks to spread or converge sound waves.

acoustic ohm (*Acous.*). Unit of acoustic resistance, reactance, and impedance. 10^5 Pa s/m^3.

acousticolateral system (*Zool.*). In Vertebrates, afferent nerve-fibres related to the neuromast organs and to the ear, receptors in aquatic forms of relatively slow vibrations.

acoustic particle size analyser (*Powder Tech.*). An instrument developed by the Armour Research Foundation in which particles are counted and sized by the noise they make when forced at high velocity through a restriction in a glass tube.

acoustic perspective (*Acous.*). The quality of depth and localization inherent in a pair of ears, which is destroyed in a single channel for sound reproduction. It is transferable with 2 microphones and 2 telephone ear-receivers with matched channels, and more adequately realized with 3 microphones and 3 radiating receivers with 3 matched channels.

acoustic plaster (*Build.*). Rough or flocculent plaster which has good acoustic absorbing properties and which can be used for covering walls. It contains metal, such as fine aluminium, which evolves gas on contact with water and so aerates the mass. These tiny holes lower the acoustic impedance and so reduce the reflection of incident sound waves.

acoustic pressure (*Acous.*). See **sound pressure.**

acoustic radiator (*Acous.*). A vibrating area giving off acoustic energy in the form of sound waves, usually of high audio or supersonic frequency; often used in depth-sounding. The radiator may be a steel plate which is hit with a magnetic plunger, or a resonating mass of quartz crystals. See **loudspeaker.**

acoustic ratio (*Acous.*). The ratio between the directly radiated sound intensity from a source, at the ear of a listener (or a microphone), and the intensity of the reverberant sound in the enclosure. The ratio depends on the distance from the source, the polar distribution of the radiated sound power, and the period of reverberation of the enclosure.

acoustic reactance (*Acous.*). See under **acoustic impedance.**

acoustic resistance unit (*Acous.*). A device to control the Q-factor of a loudspeaker enclosure system. It consists of sound-absorbing material placed over the port in a reflex loudspeaker enclosure.

acoustic resonance (*Acous.*). Enhancement of response to an acoustic pressure for a band of frequencies, the response increasing to a maximum and then decreasing as the frequency is increased through the frequency of resonance. It arises from progressive neutralization of the reactive component of the acoustic impedance until, at the frequency of resonance, the acoustic impedance becomes entirely resistive. Prominent in organ and other pipes, Helmholtz resonators, and less so in shallow cavities.

acoustics. The science of mechanical waves including production and propagation properties.

acoustic saturation (*Acous.*). The aural effectiveness of a source of sound amid other sounds; it is low for a violin, but high for a triangle. The relative saturation of instruments indicates the number required in an auditorium of given acoustic properties.

acoustic scattering (*Acous.*). Irregular and multidirectional reflection and diffraction of sound waves produced by multiple reflecting surfaces the dimensions of which are small, cf. a *wavelength*; or by certain discontinuities in the medium through which the wave is propagated.

acoustic shock (*Acous.*). The temporary deafness following a sudden large rush of current in a telephone receiver, e.g., in a telephonist's receiver when connected to a telephone line making contact with a power line.

acoustic spectrometer (*Acous.*). An instrument designed to analyse a complex sound signal into its various wavelength components and measure their frequencies and relative intensities.

acoustic spectrum (*Phys.*). Graph showing frequency distribution of sound energy emitted by source.

acoustic suspension (*Acous.*). Sealed-cabinet system of loudspeakers in which the main restoring force of the diaphragm is provided by the acoustic stiffness of the enclosed air.

acoustic tile (*Build.*). One made of a soft, sound-absorbing substance.

acoustimeter (*Acous.*). The same as **noise meter.**

a.c. pick up (*Telecomm.*). Interfering currents in one channel arising from e.m.fs. induced by currents in other channels, including power mains at power frequencies. Interference also occurs when d.c. is switched.

acquired behaviour (*An. Behav.*). Behaviour which has been experimentally shown to depend on the operation of variables such as the occurrence of reinforcing stimuli encountered in conditioning and learning.

acquired character (*Bot.*). An individual peculiarity in the morphology or physiology of a plant, developed during growth. (*Zool.*) A modification of an organ during the lifetime of an individual and due to use or disuse, e.g. parasite organs; not a result of mutilation.

acquired immunity (*Bacteriol.*). The immunity to subsequent reinfection, acquired by an organism, when as a result of natural or artificial exposure to a pathogen, usually a foreign protein (e.g., bacteria), it develops a specific antibody. Some young animals acquire a temporary immunity from antibodies supplied by the mother via the placenta. See **passive immunity.**

acquired variation (*Biol.*). Any departure from normal structure or behaviour, in response to environmental conditions, which becomes evident as an individual develops.

acquisition (*An. Behav.*). The learning of a particular response by an animal is sometimes referred to as the *acquisition* of this response. Cf. *experimental extinction.*

ACR (*Aero.*). Abbrev. for *approach control radar*.

acrandrous (*Bot.*). Having antheridia at the stem apices, as in some mosses (Bryophyta Musci).

acrasia (*Psychiat.*). Inability to exercise self-restraint.

Acraspeda (*Zool.*). See **Scyphozoa.**

acraspedote (*Zool.*). Of *Coelenterata*, lacking a velum.

acre (*Surv.*). A unit of area, equal to 10 sq. chains (1 chain = 66 ft) or 4840 sq. yd. = 0·4047 hectare. The following are now obsolete:
Cheshire acre, 10,240 sq. yd.
Cunningham acre, 6250 sq. yd.
Irish acre, 7840 sq. yd.
Scottish acre, 6150·4 sq. yd.

acreage meter (*Agric.*). Precision, rotary-type meter for registering the area sown by a drill; essential for accurate rate sowing.

Acree's reaction (*Chem.*). A qualitative test for detecting proteins and tryptophane. It consists in adding to the protein solution an equal volume of a 0·002% methanal solution containing a trace of iron(III)chloride. When concentrated sulphuric acid is introduced below the mixed solutions, a violet ring is formed. Also **Acree-Rosenheim test.**

a.c. resistance (*Electronics*). See **differential anode resistance.**

acridine (*Chem.*). $C_{13}H_9N$, a basic constituent of the crude anthracene fraction of coal-tar. It crystallizes in colourless needles and has a very irritating action upon the epidermis. Chemically it may be considered an analogous compound to anthracene, in which one of the CH groups of the middle ring is replaced by N. Certain amino-acridines have valuable bactericidal powers. Used as an intermediate in the preparation of dyestuffs.

acriflavine (*Pharm.*). 2, 8-Diaminoacridine methochloride:

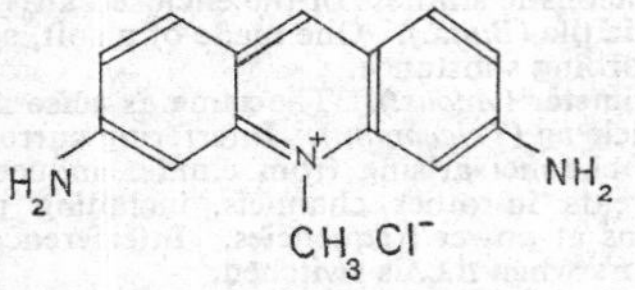

A deep orange, crystalline substance possessing antiseptic (bacteriostatic and bactericidal) properties; used in wound dressings.

Acrilan (*Chem.*). TN for a synthetic polyacrylonitrile fibre obtained by copolymerizing acrylonitrile (85%) with vinyl acetate (15%).

acro-. Prefix from Gk. *akros*, topmost, farthest, terminal.

acroblast (*Zool.*). In the spermatid, a complex composed of idioplasm surrounded by Golgi bodies from which the acrosome is formed.

acrocarpous (*Bot.*). Having fruit at the end of the stem or branch.

acrocentric (*Cyt.*). Having centromere at the end, applied to chromosomes; a rod-shaped chromosome.

acrochordal (*Zool.*). In Birds, an unpaired cartilage of the chondrocranium.

acrocoracoid (*Zool.*). A process of the dorsal end of the coracoid in Birds.

acrocyanosis (*Med.*). A vascular disorder (usually of young women) in which there is persistent blueness of the extremities.

acrodont (*Zool.*). Said of teeth which are fixed by their bases to the summit of the ridge of the jaw.

acrodromous (*Bot.*). Said of venation when the main veins, after running parallel along most of the leaf, unite at the leaf apex.

acrodynia (*Med.*). See **erythroedema.**

acrogenous (*Bot.*). Developing from the apex.

acrogens (*Bot.*). General term to include ferns and mosses (Bryophyta and Pteridophyta).

acrogynous (*Bot.*). Having one or more archegonia at the tip of the axis. Apical growth ceases after production of the archegonia, which sometimes involves the apical cell.

acrolein (*Chem.*). $CH_2 = CH{\cdot}CHO$, acrylaldehyde, propenal, a colourless liquid, b.p. 52·5°C, of pungent odour, obtained by dehydrating glycerine in the presence of a catalyst.

acromegaly (*Med.*). A disease in which occurs enlargement of hands and feet, and thickening of nose, jaw, ears, and brows, due to overactivity of the anterior part of the pituitary gland.

acromion (*Zool.*). In higher Vertebrates, a ventral process of the spine of the scapula. *adj.* **acromial.**

acron (*Zool.*). In Insects, the embryonic, presegmental region of the head.

acroparaesthesia, acroparesthesia (*Med.*). Numbness and tingling of the fingers, tending to persist, in middle-aged women.

acropetal succession (*Bot.*). Development of lateral members in such order that the youngest is nearest the tip of the axis.

acropleurogenous (*Bot.*). Having the spores borne at the tips and along the sides of hyphae.

acropodium (*Zool.*). That part of the pentadactyl limb of land Vertebrates which comprises the digits and includes the phalanges.

acroscopic (*Bot.*). On the side towards the apex.

acrosome (*Zool.*). Structure forming the tip of a mature spermatozoon. *adj.* **acrosomal.**

acrospire (*Brew.*). Young leaf shoot of barley inside the grain, produced during malting.

acrospore (*Bot.*). A spore formed at the tip of a hypha.

acrostichal bristles (*Zool.*). In *Diptera*, a row of bristles along each side of the dorsal median line of the thorax.

acroterium (*Arch.*). A base or mounting, on the apex and/or extremities of a pediment, for the support of an ornamental figure or statuary. *pl.* **acroteria.**

Acrothoracica (*Zool.*). An order of minute *Cirripedia* the females of which possess a flask-shaped mantle; the trunk appendages are reduced in number and the posterior pairs are widely separated from the first pair; they live in hollows excavated in the shells of Molluscs; the males, where known, are dwarfs, without an alimentary canal, living attached to the mantle of the female.

acrotrophic (*Zool.*). In Insects, said of ovarioles in which nutritive cells occur at the apex.

acrylaldehyde (*Chem.*). See **acrolein.**

acrylic acid (*Chem.*). Prop-2-enoic acid. $CH_2=CH{\cdot}COOH$, m.p. 7°C, b.p. 141°C, of similar odour to acetic acid; a very reactive substance, the acid belongs to the series of alkene-monocarboxylic, or oleic, acids.

acrylic ester (*Chem.*). An ester of acrylic acid or of a structural derivative of acrylic acid, e.g., methacrylic acid or its chemical derivatives.

acrylic resins (*Plastics*). Resins formed by the polymerization of the monomeric derivatives, generally esters or amides, of acrylic acid or α-methylacrylic acid. They are transparent, water-white, and thermoplastic; resistant to age, light, weak acids, alkalis, alcohols, alkanes, and fatty oils; but attacked by oxidizing acids, aromatic hydrocarbons, chlorinated hydrocarbons, ketones, and esters. They are mainly used for optical purposes, as lenses and instrument covers. Polymethyl methacrylate is widely known under the TN **Perspex.**

acrylonitrile (*Chem.*). Vinyl cyanide (1-cyanoethene), used as raw material for synthetic acrylic fibres, e.g., Acrilan, Orlon, Courtelle.

a.c. series motor (*Elec. Eng.*). A series motor designed for operation from an a.c. supply; it is characterized by a laminated field structure and usually a compensating winding.

a.c. signalling system (*Teleph.*). Any system of signalling that utilizes a code of signals based on the selective transmission of alternating currents of specified frequency or frequencies.

ACTH (*Biochem., Med.*). Adrenocorticotrophic hormone, or corticotrophin. A protein hormone of the anterior pituitary gland controlling many secretory processes of the adrenal cortex. Used medically for the same conditions as *cortisone* and *corticosterone*.

actin (*Chem.*). Protein combined with myosin to form *actomyosin* (q.v.).

actinal (*Zool.*). In *Echinodermata*, see **ambulacral**: in *Anthozoa*, pertaining to the crown, including the mouth and tentacles; star-shaped.

actinic green (*Glass*). Emerald green glass, normally used for poison bottles.

actinic radiation (*Radiol.*). Ultraviolet waves, which have enhanced biological effect by inducing chemical change; basis of the science of photochemistry.

actinic rays (*Photog.*). Incident electromagnetic waves of wavelengths that can cause a latent image, potentially developable, in a photographic emulsion. They include an extension at each end of the visible spectrum and X-rays.

actinides (*Chem.*). A group name for the series of elements of atomic numbers 89–104 (inclusive), which assumes that they have similar properties to actinium.

actinium (*Chem.*). A radioactive element in the third group of the periodic system. Symbol Ac, at. no. 89, r.a.m. 227; half-life 21·7 years. Produced from natural radioactive decay of the ^{235}U isotope or by neutron bombardment of ^{226}Ra. Gives its name to the actinium (4n+3) series of radioelements.

actino-. Prefix from Gk. *aktis*, ray.

actinobacillosis (*Vet.*). A chronic granulomatous disease of cattle caused by the bacterium *Actinobacillus lignieresii* and characterized by infection of the tongue ('wooden tongue') and occasionally of the stomach, lungs, and lymph glands. In sheep the bacterium causes abscesses involving the head and neck or internal organs.

actinobiology (*Radiol.*). The study of the effects of radiation upon living organisms.

actinoblast (*Zool.*). In *Porifera*, a spicule-forming cell.

actinodermatitis (*Med.*). Inflammation of the skin arising from the action of X-rays.

actinodromous (*Bot.*). Having the main veins radiating from the tip of the petiole.

actinoid (*Zool.*). Star-shaped.

actinolite (*Min.*). A monoclinic amphibole, containing iron, green in colour, and generally showing an elongated or needlelike habit; occurs in schists and altered basic igneous rocks.

actinomere (*Zool.*). A radial segment.

actinomorphic (*Bot., Zool.*). Star-shaped; therefore radially symmetrical, and divisible into two similar parts by more than one plane passing through the centre. See **radial symmetry.**

Actinomycetales (*Bot.*). An order of Bacteria producing a fine mycelium and sometimes arthrospores and conidia. Some members are pathogenic on animals and plants. Some produce antibiotics, e.g., *streptomycin* (q.v.).

actinomycosis (*Med., Vet.*). A chronic granulomatous disease of cattle, swine, and occasionally man, caused by infection by the actinomycete *Actinomyces bovis*. In cattle the lower jaw is commonly affected ('lumpy jaw'); in swine the mammary gland is the common site of infection.

actinomyosin (*Chem.*). See **actomyosin.**

actinon (*Chem.*). Actinium emanation, an isotope of radon produced by the disintegration of actinium; half-life 3·92 sec. Symbol An.

Actinopterygii (*Zool.*). Subclass of *Gnathostomata* in which the basal elements of the paired fins do not project outside the body wall, the fin webs being supported by rays alone. Ganoid scales are diagnostic of the group and the skeleton is fully ossified in most members. The most widespread modern group of fishes, including cod, herring, etc.

actinost (*Zool.*). In bony Fish, one of the bones of a fin-ray.

actinostele (*Bot.*). A pithless stele which has the xylem star-shaped, or approximately so, in cross-section.

actinostome (*Zool.*). In *Asteroidea*, the radially-symmetrical mouth-opening.

actinotherapy (*Med.*). Treatment by means of ultraviolet, infrared, and luminous radiations.

actinotrichia (*Zool.*). In Fish, horny fibres forming the distal part of the fin-rays.

actinotrocha (*Zool.*). In *Phoronidea*, the free-swimming pelagic larval form which possesses characteristic ciliated lobes, and passes into the adult by a remarkable metamorphosis.

actino-uranium (*Chem.*). ^{235}U, radioactive fissile isotope of uranium, of r.a.m. 235; by chain reaction gives neutrons collected by ^{238}U to form plutonium; decays to actinium.

Actinozoa (*Zool.*). See **Anthozoa.**

actinula (*Zool.*). In certain *Hydrozoa* (e.g., in *Tubularia*), an advanced larva, representing a polyp with a short stem.

action (*Acous.*). The mechanism for selecting notes in musical instruments: the hammers, etc., in the piano, and the keyboards and pneumatic controls in the organ. (*Horol.*) The functioning of the escapement of a watch, clock, etc. (*Mech.*) Time integral of kinetic energy (E) of a conservative dynamic system undergoing a change, given by $2\int_{t_1}^{t_2} E\,dt$.

action potential (*Physiol.*). That produced in a nerve by a stimulus. It is a voltage pulse arising from sodium ions entering the axon and changing its potential from −70 mV to +40 mV. With a continuing stimulus the pulses are repeated at up to several hundred times a second, leading in motor nerves to continuous muscular response (tetanus).

activated carbon (*Chem.*). Carbon obtained from vegetable matter by carbonization in the absence of air, preferably in a vacuum. Activated carbon has the property of adsorbing large quantities of gases. Important for gas masks, adsorption of solvent vapours, clarifying of liquids, and in medicine.

activated cathode (*Electronics*). Emitter in thermionic devices comprising a filament of basic tungsten metal, alloyed with thorium, which is brought to the surface by process of activation, such as heating without electric field.

activated charcoal (*Chem.*). *Charcoal* (q.v.) treated with acid, etc., to increase its adsorptive power.

activated complex (*Chem.*). In the Eyring theory of *absolute reaction rates* (q.v.) activated molecules of the reactants collide to form a high energy complex which can lose energy either by decomposing to reform the reactants or by rearranging to form the products.

activated sintering (*Powder Met.*). Sintering of a compact in the presence of a gaseous chemical which will react with part of the metal content, and thus ease and speed the migration necessary for joining the particles into a body. Because the metal salt is unstable, purging of the heated reaction vessel results in reduction to metal. Halides, which also deoxidize the metal particle surfaces, are commonly used for this purpose.

activated sludge (*San. Eng.*). Sludge through which compressed air has been blown, or which has been aerated by mechanical agitation.

activated water (*Chem.*). Water in which ions, atoms, radicals, or molecules are made chemically active by ionizing radiation.

activating agent (*Min. Proc.*). See **activator.**

activation (*Chem.*). (1) The heating process by which the capacity of carbon to adsorb vapours is increased. (2) An increase in the energy of an atom or molecule, rendering it more reactive. (*Nuc. Eng.*) Induction of radioactivity in otherwise nonradioactive atoms, e.g., in a cyclotron or reactor. (*San. Eng.*) The process of sewage purification by intimate admixture with air and activated sludge. (*Zool.*) The process of stimulating an ovum to cleavage, usually performed by a spermatozoon: the liberation of an active enzyme from a nonactive compound, e.g., trypsin from trypsinogen.

activation cross-section (*Nuc.*). Effective cross-sectional area of target nucleus undergoing bombardment by neutrons, etc. Measured in barns.

activation energy (*Chem.*). The excess energy over that of the ground state which an atomic system must acquire to permit a particular process, such as emission or reaction, to occur.

activation level (*Psychol.*). There are two types: *long-term*, the daily cyclical variation of responsiveness with its peak ca. midday and its lowest point during sleep; and *short-term*, in accord with the immediate circumstances.

activator (*Bacteriol.*). Substance used to accelerate bacterial activity. (*Biochem.*) Chemical compound which converts zymogens, proenzymes, or proferments into the true or active enzymes. See also **accelerator (1), kinases.** (*Min. Proc.*) Surface-active chemical used in flotation process to increase the attraction to a specific mineral in an aqueous pulp of collector ions from the ambient liquid and increase its aerophilic quality. Also **activating agent.** (*Phys.*) An impurity, or displaced atom, which augments luminescence in a material, i.e., a sensitizer such as copper in zinc sulphide. (*Zool.*) Any agency bringing about *activation* (q.v.).

active anode (*Ships, etc.*). See **sacrificial anode.**

active area (*Elec. Eng.*). The area of a metal rectifier operating as the rectifying junction.

active array (*Radar*). Aerial array in which the individual elements are separately excited by integrated circuit or transistor amplifiers.

active centres (*Chem.*). Centres of higher catalytic activity formed by peak or loosely bound atoms on the surface of an adsorbent. (*Cryst.*) Areas at solid surfaces which naturally carry excess electrostatic charge and can initiate process changes.

active component (*Elec.*). The accepted term for denoting the component of the vector representing an alternating quantity which is in phase with some reference vector; e.g., the active component of the current; commonly called the active current. See **active current, active voltage, active volt-amperes.**

active current (*Elec.*). That component of a vector representing the a.c. in a circuit which is in phase with the voltage of the circuit. The product of this and the voltage gives power.

active deposit (*Chem.*). The radioactive deposit produced by the disintegration of a radioactive emanation.

active electrode (*Elec. Eng.*). The electrode of an electrical precipitator which is kept at a high potential. Also **discharge electrode.**

active hydrogen (*Chem.*). Atomic form of hydrogen obtained when molecular hydrogen is dissociated by heating, or by electrical discharge at low pressure.

active lattice (*Nuc. Eng.*). Regular pattern of arrangement of fissionable and nonfissionable materials in the core of a lattice reactor.

active lines (*TV*). Those which are effective in establishing a picture.

active mass (*Chem.*). Molecular concentration, generally expressed as moles/dm^3; in the case of gases, active masses are measured by partial pressures.

active materials (*Elec., Electronics*). (1) General term for essential materials required for the functioning of a device, e.g., iron or copper in a relay or machine, electrode materials in a primary or secondary cell, emitting surface material in a valve, electron, or photocell, phosphorescent and fluorescent material forming a phosphor in a CRT, or that on the signal plate of a TV camera. (2) Term applied to all types of radioactive isotopes.

active nitrogen (*Chem.*). Formed when nitrogen is subjected to a silent electrical discharge. Unstable. Can take part in many chemical reactions which do not occur with the ordinary gas. Active nitrogen forms with hydrocarbons hydrocyanic acid. The afterglow of active nitrogen may account for *Aurora Borealis*.

active power (*Elec.*). (1) See **active volt-amperes.** (2) The time average over one cycle of the instantaneous input powers at the points of entry of a polyphase circuit.

active satellite (*Space*). A satellite equipped with instruments and sending out signals. A *passive satellite* is not so equipped.

active site (*Chem.*). That part of an enzyme which can combine with the substrate.

active transducer (*Telecomm.*). Any transducer in

which the applied power controls or modulates locally supplied power, which becomes the transmitted signal, as in a modulator, a radio-transmitter, or a carbon microphone.

active transport (*Chem.*). Processes whereby anions and cations or organic metabolites are accumulated within cells against a concentration gradient by active concentrating mechanisms.

active voltage (*Elec.*). That component of a vector representing the voltage which is in phase with the current in a circuit.

active volt-amperes (*Elec.*). Product of the active voltage and the amperes in a circuit, or of the active current (amperes) and the voltage of the circuit; equal to the power in watts. Also termed **active power.**

activity (*Chem.*). (1) See **optical activity.** (2) The ideal or thermodynamic concentration of a substance the substitution of which for the true concentration permits the application of the law of mass action. (*Elec. Eng.*) The magnitude of the oscillations of a piezoelectric crystal relative to the exciting voltage. (*Radiol.*) Measure of intensity of a radioactive source, expressed as the number of disintegrating atoms per unit time, or a derived quantity such as number of scintillations per second on a screen, or rate of clicks in a Geiger-Müller counter. Often expressed in curies, or röntgens/hour at 1 m from the isolated source. *Specific activity* is curies/gram, or a multiple or submultiple thereof.

activity coefficient (*Chem.*). The ratio of the *activity* (2, q.v.) to the true concentration of a substance.

activity constant (*Chem.*). The *equilibrium constant* (q.v.) written in terms of activities instead of molar concentrations.

actomyosin (*Chem.*). Most important protein of muscle, consisting of two proteins, *actin* and *myosin*, in the form of long threads, which, when stimulated, shorten, causing muscle contraction. Also **actinomyosin.**

actor (*Chem.*). See **donor.**

a.c. transformer (*Elec. Eng.*). A machine which alters the voltage and current of an alternating supply in inverse ratio to one another. It has no moving part and is very efficient.

actuator (*Electronics*). Transducer used to bring electronic equipment into operation.

aculeate (*Bot.*). Bearing prickles, or covered with needlelike outgrowths.

acuminate (*Bot.*). Having a long point bounded by hollow curves; usually descriptive of a leaf-apex. *dim.* **acuminulate.**

acupuncture (*Med.*). Puncture of tissues with needles; used to relieve pain or draw off fluids. It may involve some form of hypnosis.

acutance (*Photog.*). Objective formulation of the sharpness of a photographic image, expressed as:

$$\bar{G}_x/(D_B - D_A)$$

where $$\bar{G}_x^2 = \frac{\Sigma(\Delta D/\Delta x)^2}{N}$$

N = no. of increments between A and B
$D_B - D_A$ = average gradient of density curve
$\Delta D/\Delta x$ = maximum gradient of curve.

acutangular (*Bot.*). Said of a stem which has several sharp edges running longitudinally.

acute (*Bot.*). Bearing a sharp and rather abrupt point: said usually of a leaf-tip. (*Med.*) Said of a disease which rapidly develops to a crisis. Cf. *chronic.*

acute angle (*Maths.*). An angle of less than 90°. Cf. *obtuse angle.*

acute vestibulitis (*Med.*). A condition characterized by slight fever, vertigo, vomiting, and ataxia, resulting in complete deafness; due to an inflammation of the labyrinth and cochlea of the inner ear.

acyclic (*Bot.*). Having the parts of the flower arranged in spirals, not in whorls.

acyclic compounds (*Chem.*). See **aliphatic compounds.**

acylation (*Chem.*). Introduction of an acyl group into a compound, by treatment with a carboxylic acid, its anhydride or its chloride.

acyl group (*Chem.*). Carboxylic acid radical RCO (R being aliphatic), e.g. CH_3CO.

acyloins (*Chem.*). Condensation products of esters brought about by heating the esters in inert solvents with complete exclusion of oxygen and in the presence of finely divided sodium. A typical example is butyroin, $C_4H_9{\cdot}CO{\cdot}CH(OH)C_4H_9$ from butanoic ester.

A.D. (*For.*). Abbrev. for *air-dried timber.*

ad-. Prefix signifying *to, at.*

adamantine clinkers (*Build.*). Bricks similar to *Dutch clinkers* (q.v.) but harder and denser, and having a pale-pinkish colour and smooth surfaces, used for paving purposes.

adamantine compound (*Chem.*). Compound with the same tetrahedral covalent crystal structure as the diamond, e.g. zinc sulphide.

adamantoblast (*Zool.*). See **ameloblast.**

adambulacral (*Zool.*). In *Echinodermata*, adjacent to the ambulacral areas.

adamellite (*Geol.*). A type of granite with approximately equal amounts of alkali-feldspar and plagioclase.

Adam's apple (*Zool.*). In Primates, a ridge on the anterior or ventral surface of the neck, caused by the protuberance of the thyroid cartilage of the larynx.

Adams' catalyst (*Chem.*). A hydrogenation catalyst based on platinum oxide.

Adams sewage lift (*San. Eng.*). An apparatus employed to force sewage from a low-level sewer into a nearby high-level sewer by using the sewage in the latter from a point that will give the air-pressure necessary to secure the lift of sewage.

Adams-Stokes syndrome (*Med.*). Sudden loss of consciousness, with or without convulsions, in heart-block.

adaptation (*An. Behav.*). See **sensory-.** (*Bot.*) Any morphological or physiological characteristic which may be supposed to help in adjusting the organism to the conditions under which it lives. (*Optics*) Of the eye, the sensitivity adjustment effected after considerable exposure to light (*light-adapted*), or darkness (*dark-adapted*). (*Psychol.*) Modification of the organism through interaction with the environment favourable to further development (Piaget). See **accommodation, assimilation, equilibration, functional invariants.** (*Zool.*) The process by which an animal becomes fitted to its environment, external or internal, or to changes in that environment: any structure or habit developed as a result of such a process.

adaptation kit (*Automation*). That which unites sections of a guided weapon, e.g., *safing, arming, fusing,* and *firing* divisions of system.

adapter (*Chem.*). Glass tube, straight or angled, with ground glass sockets at either end, used to join different units of laboratory glassware. (*Elec. Eng.*) Accessory used in electrical installations for connecting a piece of apparatus fitted with one size or type of terminals to a supply point fitted with another size or type. (*Photog.*) (1) An arrangement for using types of photographic material in a camera different from that for which it was designed; e.g., film-pack in a plate camera, or a smaller plate than normal. (2) A device for the interchange of lenses between different types of camera.

adapter transformer (*Elec. Eng.*). A transformer for supplying a single electric lamp; it is arranged so that its primary terminals can fit into an ordinary lampholder while its secondary terminals are brought to a lampholder for a low-voltage lamp.

adaptive branching (*Zool.*). Differences arising within a group of closely related forms whose life habits are in the main similar; e.g., Black and White Rhinoceroses show a different type of lip owing to different feeding habits.

adaptive control (*Automation*). Self-adaptation of controlling system to approach overall optimum, according to a defined requirement.

adaptive polymorphism (*Zool.*). The occurrence at different stages of the life-cycle of an animal of widely differing forms, in response to variations in the conditions of life; e.g., some *Protozoa*.

adaptive radiation (*Zool.*). The development of animals of allied stock along different evolutionary paths in response to differing environments. Also **divergent adaptation.**

adaxial (*Bot.*). The face of a leaf, petal, etc., which is on the side nearest to the axis.

Adcock antenna (*Radio*). Directional receiving antenna consisting of two spaced vertical dipoles or half-dipoles connected by a screened transmission line. It responds only to vertically polarized waves, and is not subject to night error.

addend (*Maths.*). See **addition.**

addendum (*Eng.*). Radial distance between the major and pitch cylinders of an external thread; the radial distance between the minor and pitch cylinders of an internal thread; also the height from the pitch circle to the tip of the tooth on a gear wheel.

adder (*Comp.*). Unit for summation of two quantities, applied to an amplifier in analogue computing or to arithmetic unit of digital computer. In mechanical computing, addition and subtraction can be by rotation of shafts, with differential gears.

Addison's anaemia (*Med.*). See **pernicious anaemia.**

Addison's disease (*Med.*). A disease in which there is progressive destruction of the suprarenal cortex; characterized by extreme weakness, wasting, low blood-pressure, and pigmentation of the skin. Not to be confused with *Addison's anaemia*.

addition (*Maths.*). The process of finding the sum of two quantities, which are called the *addend* and the *augend*. Denoted by the plus sign +. Numbers are added in accordance with the usual rules of arithmetic, but addition of other mathematical entities has to be defined specifically for the entities concerned, e.g., the parallelogram rule for the addition of two vectors.

addition agent (*Elec. Eng.*). A substance added to the electrolyte in an electrodeposition process in order to improve the character of the deposit formed. The agent does not take part in the main electrochemical reaction.

addition reaction (*Chem.*). The reaction $A+B=C$ where reactant molecule A adds on to reactant molecule B to form a single product C.

additive compounds (*Chem.*). Compounds formed by additive reactions, in which a double bond is converted into a single bond by the addition of two more atoms or radicals.

additive constant (*Surv.*). A term used in the computation of distance by tacheometric methods. It is that length (usually constant and small) which must be added to the product of staff intercept and multiplying constant to give the true distance of the object. See also **anallatic lens.**

additive function (*Maths.*). A function $f(x)$ such that $f(x+y)=f(x)+f(y)$. If $f(x+y)<f(x)+f(y)$, the function is said to be *sub-additive*. If $f(x+y)>f(x)+f(y)$, the function is said to be *super-additive*.

additive process (*Photog.*). Colour process in which the desired result is achieved by the combination of 3 images of the subject taken through primary colour filters, and synthesized by white light similarly filtered.

additive property (*Chem.*). One whose value for a given molecule is equal to the sum of the values for the constituent atoms and linkages.

additive reactions (*Chem.*). See **additive compounds.**

address (*Comp.*). Code indicating where data shall be stored in a register, store, or memory. In multi-address codes there are several addresses which can move data between stores.

adduct (*Chem.*). The addition product of a reaction between molecules.

adductor (*Zool.*). A muscle that draws a limb or part inwards, or towards another part; e.g., *adductor mandibulae* in *Amphibia* is a muscle which assists in closing the jaws.

adeciduate (*Zool.*). See **indeciduate.**

adecticous (*Zool.*). Said of insect pupae with nonarticulating mandibles which are often reduced and are not used in escaping from the cocoon.

adelocodonic (*Zool.*). In *Gymnoblastea*, said of medusae which are always attached, and which degenerate after discharging the ripe germ cells.

adelomorphic (*Zool.*). Of indefinite form.

adelphogamy (*Bot.*). A union between two vegetative cells, one the mother cell, the other one of its daughter cells.

adelphous (*Bot.*). Said of an androecium in which the stamens are partly or wholly united by their filaments.

adendritic (*Zool.*). Without dendrites.

adenine (*Chem.*). 6-Aminopurine, a *base*, a purine derivative, in animal and vegetable tissues combined with niacinamide, D-ribose and phosphoric acids, forming coenzyme-active principles:

Pairs with thymine in DNA, uracil in RNA. See **genetic code.**

adenitis (*Med.*). Inflammation of a gland.

adenoblast (*Cyt.*). Embryonic glandular cell.

adenocyte (*Cyt.*). Secretory cell of a gland.

adenohypophysis (*Physiol.*). The glandular lobe of the pituitary gland, derived from buccal ectoderm (Rathke's pouch).

adenoid. Glandlike. (*Anat.*) In some Mammals, a mass of lymphoid tissue occurring on the dorsal wall of the pharynx. Specifically, in *pl.* (*Med.*), infective inflammation of such tissue in the nasopharynx, often causing obstruction.

adenoma (*Med.*). A tumour with a glandlike structure or developed from glandular epithelium.

adenomyoma (*Med.*). See **endometrioma.**

adenopathy (*Med.*). Disease or disorder of glandular tissue. The term is usually used in reference to lymphatic gland dysfunction.

adenophore (*Bot.*). A stalk supporting a gland.

adenophyllous (*Bot.*). Having glandular leaves.

adenose (*Bot.*). Glandular; having glands.

adenosine (*Chem.*). A nucleoside consisting of adenine linked in the C-9 position to C-1 of nitrose by a β-glycosidic linkage.

adenosine triphosphate (ATP) (*Chem.*). An important co-enzyme in various biological reactions, involving the transfer of phosphate bond energy. It is a donor of phosphate in the biological oxidation of carbohydrate and fatty acid. Its conversion to adenosine di- or monophosphate is accompanied by the liberation of energy believed to be used directly in, e.g., muscular contraction. Water intake and respiration in plants are linked by transfer of energy through ATP.

adenotrophic viviparity (*Zool.*). In Insects, a type of viviparity where the larva is retained in the uterus until ready to pupate, seen only in *Diptera Pupipara* and *Glossina* (tsetse fly).

adenyl cyclase (*Chem.*). An enzyme which catalyses the formation of cyclic adenylic acid from ATP.

adenylic acid (*Chem.*). Adenosine phosphate.

adermin (*Chem.*). See **vitamin B complex.**

adesmy (*Bot., Zool.*). Division in an organ usually entire.

adetopneustic (*Zool.*). Of some *Echinodermata*, having dermal gills extending beyond the abambulacral surface.

ADF (*Aero.*). See **automatic direction finder.**

adfrontal (*Zool.*). In *Lepidoptera*, said of a pair of narrow oblique sclerites separating the frons from the median epicranial suture.

adhesion (*Bot.*). The union of members of distinct whorls, as when a calyx is grown up with the ovary wall. *adj.* **adherent, adhering.** (*Build.*) The securing of a bond between plaster and backing, by physical means as opposed to mechanical keys. (*Elec. Eng.*) Mutual forces between two magnetic bodies linked by magnetic flux, or between two charged nonconducting bodies which keeps them in contact. (*Med.*) Abnormal union of two parts which have been inflamed: a band of fibrous tissue which joins such parts. (*Met.*) The intimate sticking together of surfaces under compressive stresses by metallic bonds which form as a function of stress, time, and temperature. The speed of formation is related to slip dislocations, and may occur instantaneously under high shear stresses. See **cold welding.** (*Phys.*) Intermolecular forces which hold matter together, particularly closely contiguous surfaces of neighbouring media, e.g., liquid in contact with a solid. In U.S., **bond strength.**

adhesive cells (*Zool.*). Glandular cells producing a viscous adhesive secretion for attachment use, as on the pedal disk of *Hydra*, the tentacles of *Ctenophora*, and in the epidermis of *Turbellaria*.

adhesive force (*Eng.*). The frictional grip between two contacting surfaces, e.g., between the driving-wheel of a locomotive and the rail; the product of the weight on the wheel and the friction coefficient between wheel and rail.

adiabatic (*Bot.*). Not capable of translocation. (*Phys.*) Without loss or gain of heat.

adiabatic change (*Phys.*). A change in the volume and pressure of the contents of an enclosure without exchange of heat between the enclosure and its surroundings.

adiabatic curve (*Phys.*). The curve obtained by plotting P against V in the adiabatic equation.

adiabatic efficiency (*Eng.*). (1) Of a steam-engine or turbine, the ratio of the work done per pound of steam to the available energy represented by the adiabatic heat drop. (2) Of a compressor, the ratio of that work required to compress a gas adiabatically to the work actually done by the compressor piston or impeller.

adiabatic equation (*Phys.*). PV^{γ}=constant, an equation expressing the law of variation of pressure (P) with volume (V) of a gas during an adiabatic change, γ being the ratio of the specific heat of the gas at constant pressure to that at constant volume. The value of γ is approximately 1·4 for air at S.T.P.

adiabatic expansion (*Heat*). See **adiabatic change.**

adiabatic heat drop (*Eng.*). The heat energy released and theoretically capable of transformation into mechanical work during the adiabatic expansion of unit weight of steam or other vapour.

adiabatic hydrolapse (*Meteor.*). See **hydrolapse.**

adiabatic lapse rate (*Meteor.*). The rate of decrease of temperature which occurs when a parcel of air rises adiabatically through the atmosphere.

adiabatic process (*Phys.*). Process which occurs without interchange of heat with surroundings.

adiactinic (*Phys.*). Said of a substance which does not transmit photochemically-active radiation, e.g., safelights for dark-room lamps.

Adie barometer (*Meteor.*). See **Kew-pattern barometer.**

A-digit hunter (*Teleph.*). In a director system, a hunting selector which, before dialling tone is returned to a calling subscriber, selects and connects a free A-digit selector.

A-digit selector (*Teleph.*). In a director system, a 2-motion selector which returns dialling tone to a calling subscriber, steps vertically to the level indicated by the first digit dialled, then rotates automatically into that level to select and connect the BC-digit selector of a free director.

ad infinitum (*Maths.*). Latin phrase meaning continuing (in a similar manner) indefinitely.

adion (*Chem.*). Ion which can move only on a surface, not away from it.

adipamide (*Chem.*). 1,4-Butanedicarboxamide, $NH_2CO(CH_2)_4CONH_2$. Used in synthetic fibre manufacture.

adipic acid (*Chem.*). Butanedicarboxylic acid, $HOOC{\cdot}(CH_2)_4{\cdot}COOH$. Colourless needles; m.p. 149°C, b.p. 265°C; formed by the oxidation of *cyclo*-hexanone, or by treatment of oleic acid with nitric acid. Used in the preparation of 6.6 *nylon* (q.v.).

adipo-. Prefix from L. *adeps*, fat.

adipocere or mortuary fat (*Med.*). White or yellowish waxy substance formed by the post-mortem conversion of body fats to higher fatty acids.

adipose tissue (*Zool.*). A form of connective tissue consisting of vesicular cells filled with fat and collected into lobules.

adiposis dolorosa (*Med.*). A condition characterized by the development of painful masses of fat under the skin and by extreme weakness. Also **Dercum's disease.**

Adiprene (*Plastics*). TN for a U.S. polyurethane elastomer which combines high abrasion resistance with hardness, resilience, and good load-bearing capacity.

A display (*Radar*). Coordinate display on a CRT in which a level time base represents distance and vertical deflexions of beam indicate echoes.

adit (*Civ. Eng.*). An access tunnel (usually nearly horizontal) leading to a main tunnel, and

frequently used (in lieu of a shaft) in the excavation of the latter. (*Mining*) A level, or nearly level, tunnel driven into a hillside for exploration or drainage.
adjacent channel (*Telecomm.*). One whose frequency is immediately above or below that of the required signal.
adjective dyes (*Chem.*). Dyes which have no direct affinity for the particular textile fibre but can be affixed to it by a *mordant* (q.v.).
adjoint (*Maths.*). Of a square matrix or determinant: the transpose of the matrix or determinant obtained by replacing each element by its cofactor. Some writers use an untransposed adjoint. Also called **adjugate.**
adjugate (*Maths.*) See **adjoint.**
adjustable-pitch airscrew (*Aero.*). See **airscrew.**
adjustable-port proportioning valve (*Heat*). Valve for use with gas or oil burners combined with an air valve, both valves mounted on a common spindle and operated in unison by a single lever linked to a motor actuated by automatic temperature-control apparatus.
adjusting rod (*Horol.*). An instrument for testing the pull of the mainspring. It is a rod having at one end an adjustable clamp for attaching to a fusee or barrel arbor, and provided with sliding weights for balancing the pull exerted by the mainspring. Its use is now confined to the adjustment of chronometers and English full-plate watches.
adjustment (*Bot.*). A response to a stimulus which increases the efficiency of a part or all the plant.
adjustor (*Zool.*). An organ or faculty determining the behaviour of an organism in response to stimuli received: one of the neurones in a reflex arc connecting the receptor neurone with the effector neurone, or with other adjustors.
adjutage or **ajutage** (*Hyd.*). A tube or nozzle through which water is discharged.
adjuvant (*Med.*). A remedy which assists the action of other remedies.
adlacrimal (*Zool.*). The lacrimal bone of Reptiles, so called to indicate that it is not homologous with the lacrimal bone of Mammals.
adminicula (*Zool.*). Spines used for locomotion in certain insect forms.
Admiralty brass (*Met.*). See **Tobin bronze.**
admission (*Mech.*). The point in the working cycles of a steam or I.C. engine at which the inlet valve allows entry of the working fluid into the cylinder.
admittance (*Elec.*). Property which permits the flow of current under the action of a potential difference. The reciprocal of impedance.
admittance gap (*Elec.*). A gap in a resonant cavity to provide for interaction with an electron beam, as in the *klystron.*
adnasal (*Zool.*). In certain primitive Fish, e.g., *Polypterus,* a small bone in front of the nasal.
adnation (*Bot.*). Attachment of one organ to another by its whole length. Sometimes used as equivalent to *adhesion. adj.* **adnate.**
adnexa (*Anat.*). Appendages; usually refers to ovaries and Fallopian tubes. (*Zool.*) Adjacent structures.
adnexed (*Bot.*). Joined to by a narrow margin; as the gills of some agarics.
adobe (clay) (*Build.*). A name for any kind of mud which when mixed with straw can be sun-dried into bricks.
adoral (*Zool.*). Adjacent to the mouth.
adoral wreath (*Zool.*). Row of membranellae encircling the peristome in higher *Ciliata.*
adpressed (*Bot.*). See **appressed.**
ad-radius (*Zool.*). In *Coelenterata,* a radius of the third order: 1 of 8 radii each lying between a per-radius and an inter-radius.
adrectal (*Zool.*). Adjacent to the rectum.
adrenal (*Zool.*). Adjacent to the kidney: pertaining to the adrenal gland.
adrenal gland (*Zool.*). See **suprarenal body.**
adrenaline (*Chem.*). 3,4-$(OH)_2C_6H_3{\cdot}CH(OH){\cdot}CH_2{\cdot}NH{\cdot}CH_3$, a crystalline compound obtained from the suprarenal glands. It has a sympathomimetic action, e.g., increases rate and force of the heart, raises blood pressure, dilates the pupil, as does the sympathetic nervous system. Also **epinephrine.**
adrenergic (*Chem., Physiol.*). Pertaining to or causing stimulation of the sympathetic nervous system; applied to sympathetic nerves which act by releasing an adrenaline-like substance from their nerve endings.
adrenocorticotrophic hormone. See **ACTH.**
ADS. Abbrev. for *air data system.*
adsorbate (*Chem.*). A substance, usually in gaseous or liquid solution, which is to be removed by adsorption. (*Phys.*) Substance adsorbed at a phase boundary.
adsorbent (*Chem.*). The substance, either solid or liquid, on whose surface *adsorption* (q.v.) of another substance takes place.
adsorption (*Chem.*). The taking up of one substance at the surface of another.
adsorption catalysis (*Chem.*). The catalytic influence following adsorption of the reactants, exercised upon many reactions, often upon a specific adsorbent, attributed to the free residual valencies and hence higher reactivity of molecules at an interface (e.g., the ammonia oxidation process).
adsorption chromatography (*Chem.*). See **chromatography.**
adsorption isotherm (*Chem.*). The relation between the amount of a substance adsorbed and its pressure or concentration, at constant temperature.
adsorption potential (*Chem.*). Change of potential in an ion in passing from a gas or solution phase on to the surface of an adsorbent.
adsorption surface area (*Powder Tech.*). The surface area of a particle of powder calculated from data got by a stated adsorption method.
adularescence (*Min.*). A milky or bluish sheen in gemstones, particularly feldspars.
adularia (*Min.*). Variety of triclinic potassium feldspar distinguished by its morphology and restricted occurrence in alpine-type veins.
aduncate (*Bot.*). Hooked.
A-durol (*Photog.*). See **chlorquinol.**
advance (*Rail.*). The length of line beyond a signal which is covered by that signal.
advance ball (*Acous.*). Rounded point riding a surface to be cut by a stylus, to ensure uniformity of depth of cut.
advance metal (*Met.*). Copper-based alloy with 45% nickel.
advance workings (*Mining*). In flattish seams, mining in which the whole face is carried forward, no support pillars being left.
advantage ratio (*Nuc. Eng.*). Ratio between the relative radiation dosage received at any point in a nuclear reactor and that of a reference position for the same time of exposure.
advection (*Meteor.*). The transference of heat by horizontal motion of the air.
adventitia (*Zool.*). Accidental or inessential structures: the superficial layers of the wall of a blood-vessel. *adj.* **adventitious.**
adventitious (*Bot.*). Applied to a plant-part developed out of the usual order or in an unusual position. An *adventitious bud* is any bud

except an *axillary bud*; it gives rise to an *adventitious branch*. An *adventitious root* develops from some part of a plant other than a pre-existing root.

adventive (*Bot.*). Denotes a plant which has not secured a permanent foothold in a given habitat.

adverse (*Bot.*). Facing the main axis.

adynamandrous (*Bot.*). Incapable of self-fertilization.

adze (*Carp.*). A cutting tool with an arched blade at right angles to the handle, used for dressing timber.

adze block (*Carp.*). The part carrying the cutters in a wood-planing machine.

adze-eye hammer (*Tools*). Type of claw hammer with pronounced curve on the claw.

aecial (*Bot.*). A term used in America for *aecidial* (see **aecidium**).

aecidiospore (*Bot.*). A spore of the rust fungi formed in an aecidium.

aecidium (*Bot.*). A spore-producing structure characteristic of many rust fungi. It consists of a cup-shaped peridium of sterile hyphae, containing closely packed chains of aecidiospores. *adj.* **aecidial.**

aedeagus (*Zool.*). In male Insects, a fingerlike evagination of the ventral body-wall, enclosing the terminal section of the ejaculatory duct and forming an intromittent organ.

aegirine (*Min.*). An important member of the pyroxene group of minerals, consisting essentially of silicate of iron and sodium. $NaFeSi_2O_6$; occurs in certain alkali-rich igneous rocks.

aegirine-augite (*Min.*). A common intermediate member of the series between *aegirine* and *augite* (qq.v.).

aegithognathous (*Zool.*). Of Birds, having a type of palate in which the maxillopalatines do not meet the vomer or each other, and in which the vomer is broad and truncate anteriorly; the palatines and pterygoids articulate with the basisphenoid rostrum.

aegophony, egophony (*Med.*). The bleating quality of voice heard through the stethoscope when fluid and air are present in the pleural cavity.

aelosomine (*Zool.*). A green respiratory pigment of some *Oligochaeta* worms.

aenigmatite (*Min.*). A complex silicate of sodium, iron, and titanium; occurs as reddish black triclinic crystals in alkaline igneous rocks. Also known as **cossyrite.**

aeolian deposits (*Geol.*). Rocks or sediments which have accumulated in a nonmarine environment (typically on a desert surface) and consist of essentially windblown sand or dust grains.

aeolian tone (*Acous.*). A musical note set up by vortex action on a stretched string when it is placed in a stream of air.

Aeolight (*Cinema.*). TN for a type of *glowlamp*.

aeolipile(*Phys.*). A contrivance designed to illustrate the reaction of the air upon jets of steam issuing from an otherwise closed vessel, and usually arranged to produce rotary motion of the vessel about a free axis.

aeolotropic (*Phys.*). Having physical properties which vary according to the direction or position in which they are measured. Synonymous with *anisotropic* in this sense.

aerated concrete (*Build.*). Concrete made by adding constituents to the mix which, by chemical reaction, liberate gases which are entrapped in the concrete and thereby reduce its density and increase its insulation value.

aerating root (*Bot.*). A type of root produced by many plants which grow in soft mud. It stands up above the mud and water, is constructed loosely, and acts as a ventilating organ.

aerating tissue (*Bot.*). Loosely constructed tissue, well provided with intercellular spaces, by means of which air can circulate inside a plant.

aeration test burner (*Heat.*). Apparatus by which the combustion characteristics of commercial gases can be correlated and calibrated. Abbrev. **A.T.B.**

aerenchyma (*Bot.*). (1) The same as *aerating tissue*. (2) A tissue of thin-walled, somewhat corky cells, present on the stems of some water-plants at about water level.

aerial (*Radio*). See **antenna.**

aerial cable (*Teleph.*). Cable suspended from a wire between normal telegraph poles, containing pairs of insulated conductors for telephone circuits, the practice being economical until it is worth while to bury a larger cable.

aerial efficiency (*Radio*). Same as **radiation efficiency.**

aerial fog (*Cinema.*). Fog caused by exposure of portions of the film to air in a processing machine.

aerial plant (*Bot.*). See **epiphyte.**

aerial root (*Bot.*). A root, usually adventitious in origin, ordinarily arising from a stem, and commonly serving for climbing; less often it assists in the nutrition of the plant, either by containing chlorophyllous cells capable of photosynthesis or by functioning as a parasitic sucker.

aerial ropeway (*Civ. Eng.*). An apparatus for the overhead transport of materials in carriers running along an overhead cable or cables supported on towers.

aerial surveying (*Surv.*). A process of surveying by photographs taken from the air, the photographs being of two types: (*a*) those giving a vertical or plan view, (*b*) those giving an oblique or bird's-eye view. See **vertical aerial photograph, oblique aerial photograph.**

aerobe (*Biol.*). An organism which can live and grow only in the presence of free oxygen: an organism which uses aerobic respiration.

aerobic respiration (*Biol.*). See **respiration, aerobic.**

aerobiosis (*Biol.*). Existence in the presence of free oxygen.

aerodrome control radar (*Radar*). Surveillance radar showing, on *PPI* presentations, the position of aircraft on the approach, in the circuit and, at large airports, on the runways and taxiways.

aerodrome markers (*Aero.*). Particoloured boards defining areas on an aerodrome, e.g., *boundary markers* which indicate the limits of the landing area, *taxi-channel markers* for taxi tracks, *obstruction markers* for ground hazards, and *runway visual markers*, situated at equal distances, by which visibility is gauged in bad weather.

aerodrome meteorological minima (*Aero.*). The minimum cloud base (vertical) and visibility (horizontal) in which landing or takeoff is permitted at a particular aerodrome. *ICAO* (q.v.) standards: Cat. 1, 200 ft (60 m) height, 2600 ft (800 m) RVR; Cat. 2, 100 ft (30 m) height, 1300 ft (400 m) RVR; Cat. 3, zero height, (*a*) 700 ft (210 m), (*b*) 150 ft (45 m), (*c*) zero RVR. See **runway visual range.**

aerodynamic balance (*Aero.*). (1) A balance, usually but not necessarily in a wind tunnel, designed for measuring aerodynamic forces or moments. (2) Means for balancing air loads on flying control surfaces, so that the pilot need not exert excessive strength, particularly as speed increases. The principle is to use part of the air forces, either directly on a portion of the control surface ahead of the hinge line, or indirectly through a small auxiliary surface with a powerful moment arm, to counterbalance the main airloads. Examples of the first are *horn, inset hinge*, and *frise balances*, and of the second the *balance tab*.

aerodynamic centre (*Aero.*). The point about which the pitching moment coefficient is constant for a range of aerofoil incidence.

aerodynamic damping (*Aero.*). The suppression of oscillations by the inherent stability of an aircraft or of its control surfaces.

aerodynamics (*Aero.*). That part of the mechanics of fluids that deals with the dynamics of gases. Particularly, the study of forces acting upon bodies in motion in air.

aerodyne (*Aero.*). Any form of aircraft deriving lift in flight principally from aerodynamic forces. Commonly called **heavier-than-air aircraft**; e.g., *aeroplane, glider, kite, helicopter*.

aeroelastic divergence (*Aero.*). Aeroelastic instability which occurs when aerodynamic forces, or couples, increase more quickly than the elastic restoring forces or couples in the structure. Generally applied to wing weakness where the incidence at the tips increases under load, so tending to twist the wings off.

aeroelasticity (*Aero.*). The interaction of aerodynamic forces and the elastic reactions of the structure of an aircraft. Phenomena are most prevalent when manoeuvring at very high speed.

aero-engine (*Aero.*). The power unit of an aircraft. Originally a lightweight reciprocating internal-combustion engine, usually Otto cycle, as a general rule either air-cooled radial, in-line, vee, or liquid-cooled vee; gas turbine gradually superseded reciprocating engines from 1945. See **ramjet, turbofan, turbojet, turboprop, turboramjet, turborocket.**

Aerofall mill (*Min. Proc.*). Dry grinding mill with large diameter and short cylindrical length in which ore is mainly or completely ground by large pieces of rock; mill is swept by air-currents which remove finished particles.

Aerofloats (*Min. Proc.*). TN for important range of collector agents used in froth flotation, based on dithiophosphates, cresols, and alcohols.

Aeroflocs (*Min. Proc.*). TN for range of flocculating chemicals used in sedimentation of products from chemical extraction of ores.

aerofoil (*Aero.*). A body shaped so as to produce an aerodynamic reaction (lift) normal to its direction of motion, for a small resistance (drag) in that plane. A wing, plane, aileron, tail plane, rudder, elevator, etc.

aerofoil section (*Aero.*). The cross-sectional shape, or profile, of an aerofoil.

aerography (*Meteor.*). The study of the properties of the atmosphere.

aero-isoclinic wing (*Aero.*). A sweptback wing which has its torsional and flexural stiffness so adjusted that the angle of incidence remains constant as the wing bends under flight loads, instead of decreasing with deflexion towards the tip, which is the normal geometrical effect.

aerolites (*Geol.*). A general name for stony as distinct from iron meteorites.

aerology (*Aero., Meteor.*). The study of the upper air, that part of the atmosphere removed from the effect of surface conditions. The knowledge of these conditions is distributed to aerodromes for the use of pilots proposing to fly at considerable heights.

aeronautical engineering (*Aero.*). That branch of engineering concerned with the design, production, and maintenance of aircraft structures, systems, and power units.

aeronautical fixed services (*Aero.*). A telecommunication service between fixed stations for the transmission of aeronautical information, particularly navigational, safety, and flight planning messages. Abbrev. **AFS.**

aeronautics (*Aero.*). All activities concerned with aerial locomotion.

aeronomy (*Meteor.*). The branch of science dealing with the atmosphere of the Earth and the other planets with reference to their chemical composition, physical properties, relative motion, and reactions to radiation from outer space.

aeroperception (*Zool.*). Ability to perceive the direction of origin of the oxygen supply; e.g., in Tubificid worms.

aerophagy (*Med.*). The swallowing of air, with consequent inflation of the stomach.

aerophare (*Radio*). See **radio beacon.**

aerophyte (*Bot.*). See **epiphyte.**

aeroplane (*Aero.*). Any power-driven heavier-than-air flying machine with fixed wings. Subdivisions: landplane, seaplane (float seaplane and flying-boat), amphibian. See **tractor.**

aeroplane cloth (*Textiles*). Strong, plain-woven fabric made from fine 2-fold cotton, linen, and nylon yarns, formerly used exclusively for wings and fuselages, now used for protective sheets, etc.

aeroplane effect (*Radio*). Error in direction-finding by radio which arises from the tilt of the transmitting aerial on an aircraft, or from any horizontal component in the emitted wave.

aeroplane engine (*Aero.*). See **aero-engine.**

aerosol (*Chem.*). (1) A colloidal system, such as a mist or a fog, in which the dispersion medium is a gas. (2) Pressurized container with built-in spray mechanism used for packaging insecticides, deodorants, paints, etc.

aerosol spectrometer (*Powder Tech.*). Instrument for sampling and classifying airborne particles, using a horizontal *elutriator* (q.v.) in which the particles are deposited on glass slides according to their falling speed.

aerospace (*Space*). The area of possible flight operations for unmanned satellites and for manned spacecraft.

aerostat (*Aero.*). Any form of aircraft deriving support in the air principally from its buoyancy, e.g., a *balloon* or *airship*. (*Zool.*) See **air-sacs.**

aerostation (*Aero.*). The manipulation and flying of such aircraft as are lighter than air, such as balloons and airships.

aerotaxis (*Bot.*). The movement of a whole motile plant, or of a motile gamete, in relation to the concentration of oxygen in the surrounding water. (*Zool.*) Reaction to oxygen, usually manifested by movement towards or away from it. *adj.* **aerotactic.**

aerothermotherapy (*Med.*). Treatment by hot-air currents.

aerotow (*Aero.*). Launching a glider by towing it from an aeroplane to a considerable height before release.

aerotropism (*Bot.*). The curvature by means of one-sided growth of a plant member in relation to oxygen concentration. (*Zool.*) Reaction towards, or sensitivity to, oxygen. *adj.* **aerotropic.**

aeruginose, aeruginous (*Bot.*). Bluish-green, like verdigris.

aesthacytes (*Zool.*). In *Porifera*, cells alleged to be sensory or nervous.

aesthesiometer (*Med.*). An instrument for measuring sensibility to touch.

aesthetascs (*Zool.*). In *Crustacea*, olfactory hairs occurring on most antennules and on many antennae.
aesthetes (*Zool.*). In amphineuran *Mollusca*, sense-organs, some of which are primitive eyes, occurring in canals traversing the shell plates.
aestival (*Bot., Zool.*). Occurring in summer, or characteristic of summer. (*Ecol.*) One of two divisions of summer found useful in temperate regions, meaning early summer. Cf. *serotinal.*
aestival aspect (*Bot.*). The condition of the vegetation of a plant community in summer.
aestivation (*Bot.*). Arrangement of the parts of a flower in the flower-bud: in particular, the arrangement of the calyx and corolla. (*Zool.*) Prolonged summer torpor, as in some Insects. Cf. *hibernation.*
aethalium (*Bot.*). The fruit body of some *Myxomycetes*, consisting of a number of sporangia more or less confluent and incompletely individualized.
aether. See **ether.**
aetiology, etiology (*Med.*). The medical study of the causation of disease.
aetioporphyrin (*Chem.*). 1,3,5,8-Tetramethyl-2,4,6,7-tetraethylporphyrin, obtained by reduction and decarboxylation of haematoporphyrin; also a breakdown product of chlorophyll.
AF (*Telecomm.*). See **audio frequency.**
AFC (*Radio*). Abbrev. for *automatic frequency control.*
AFCS (*Aero.*). Abbrev. for *a*utomatic *f* light *c*ontrol *s*ystem. A category of *automatic pilot* which is essentially for the control of an aircraft while en-route. It is engaged after take-off, can be monitored by speed and altitude data signals, signals from *ILS* and *VOR*, has automatic approach capability and is disengaged before landing. Cf. *autoflare, autoland* and *autothrottle.*
afebrile (*Med.*). Without symptoms of fever.
affectional behaviour (*An. Behav.*). In Monkeys, and possibly other Primates, behaviour involved in social relationships. It is quite distinct from sexual behaviour, and is probably extremely important for the development of social behaviour and organization in the Primates.
afferent (*Zool.*). Carrying towards; as blood vessels carrying blood to an organ or organs, and nerves carrying nervous impulses to the central nervous sytem. Cf. *efferent.*
afferent arc (*Zool.*). The sensory or receptive part of a reflex arc, including the adjustor neurone (s).
affine (*Phys.*). Said of characteristic curves of apparatus when these curves differ only in the scales of one or both coordinates.
affine transformation (*Maths.*). A transformation between two sets of variables in which each variable of either set is expressible as a linear combination of the variables of the other set. Summarized by the matrix equation $x' = Ax + B$. Cf. *linear transformation.*
affinin (*Chem.*). An insecticidal amide isolated from a Mexican plant *Erigeron affinis.*
affinity (*Bot.*). Likeness, especially in relationship. (*Chem.*) Chemical ability to form molecules; measured by the valence properties of atoms. See **valence electrons.**
affixed (*Bot.*). Inserted upon.
affluent (*Geog.*). A tributary; a smaller stream flowing into a larger.
aflagellar (*Zool.*). Lacking flagella.
A-frame (*Eng.*). See **sheers.**
African glanders (*Vet.*). See **epizootic lymphangitis.**
African horse-sickness (*Vet.*). A viral disease of horses and other equidae, occurring in Africa and the Middle East, and transmitted by midges (*Culicoides*). Occurs in an acute, pulmonary form called locally *dunkop* (meaning 'thin head') and due to pulmonary oedema, or in a subacute cardiac form called *dikkop* (meaning 'thick head') which is characterized by oedematous swelling of head, neck, and, sometimes, body.
African mahogany (*For.*). West African hardwood from the genus *Khaya* which seasons without much difficulty. The wood is liable to infestation by powder-post beetles. It works easily and finishes to a good surface.
African swine fever (*Vet.*). An acute, contagious, viral disease of pigs in Africa; characterized by fever, diarrhoea, and multiple haemorrhages. Wart hogs are symptomless carriers of the virus.
African whitewood (*For.*). See **obeche.**
afrormosia (*For.*). A very durable wood from West Africa, used as a substitute for teak.
AFS (*Aero.*). Abbrev. for *aeronautical fixed services.*
aft cg limit (*Aero.*). See **cg limits.**
afterbirth (*Zool.*). See **decidua.**
afterblow (*Met.*). In basic Bessemer process, continuation of blowing of air through molten metal after carbon has been eliminated, to obtain removal of phosphorus.
afterbody (*Aero.*). Rear portion of a flying-boat hull, aft of the main step. (*Space*) Companion object such as a discarded clamp ring, etc., which trails a satellite in orbit.
afterburner (*Aero.*). See **reheat.**
afterburning (*I.C. Engs.*). In an internal-combustion engine, persistence of the combustion process beyond the period proper to the working cycle, i.e., into the expansion period.
afterburst (*Mining*). Delayed further collapse of underground workings after a rockburst.
aftercooler (*Mining*). Chamber in which heat generated during compression of air is removed, allowing cool air to be piped underground.
afterdamp (*Mining*). The noninflammable heavy gas, carbon dioxide, left after an explosion in a coal mine. The chief gaseous product produced by the combustion of coal-gas. See **black damp, firedamp, white damp.**
afterglow (*Electronics*). See **persistence.**
afterheat (*Nuc. Eng.*). That which comes from fusion products in a reactor after it has been shut down.
after-image (*Optics*). Formation of image on retina of eye after removal of visual stimulus, in colour complementary to this stimulus. See **complementary-.**
after-pains (*Med.*). Pains occurring after childbirth, due to contraction of the uterus.
afterpeak (*Ships*). Space abaft the aftermost bulkhead. Lower part frequently used as fresh-water tank, upper part may be used as storeroom.
after-ripening (*Bot.*). The chemical and physical changes which occur inside a seed or other dormant plant-structure, and lead to the development of conditions when growth can be resumed.
aftershaft (*Zool.*). A second shaft, or a tuft of down, arising from the hollow quill of a feather, just proximally to the superior umbilicus.
aft fan (*Aero.*). A free turbine and low pressure ratio compressor mounted at the rear of a turbojet to provide greater air mass flow, thereby improving the propulsive efficiency and specific fuel consumption.
Ag (*Chem.*). The symbol for silver (*argentum*).
agalactia (*Med.*). Failure of the breast to secrete milk. Also **agalacia.**

agalmatolite (*Min.*). See **pagodite.**

agamete (*Zool.*). In *Sporozoa*, a young form which develops directly into an adult, without syngamy. See **merozoite, schizozoite.**

agamic (*Bot.*). Said of reproduction without the cooperation of a male gamete.

agamobium (*Zool.*). In metagenesis, the asexual generation.

agamogenesis (*Bot., Zool.*). Asexual reproduction.

agamogony (*Zool.*). See **schizogony.**

agamohermaphrodite (*Bot.*). Having hermaphrodite and neuter flowers in the same inflorescence.

agamont (*Zool.*). See **schizont.**

agamotropic (*Bot.*). Said of a flower which does not shut after having once opened.

agar-agar (*Chem.*). A dried mucilaginous substance with marked gel-forming properties obtained from certain oriental seaweeds, free from nitrogen, a carbohydrate derivative, used as a substitute for size, as a laxative, as an ingredient of certain foodstuffs, and in the preparation of various media (e.g., 'blood-agar') for the culture of many kinds of bacteria and moulds.

agaric (*Bot.*). A mushroom or toadstool.

Agaricales (*Bot.*). An order of *Basidiomycetes* containing many thousand species, and occurring all over the world when conditions permit plant growth. The basidia develop on a hymenium spread over the surface of gills, pores, or less often a smooth surface, and the hymenium is at first enclosed, but freely exposed at maturity. Mushrooms and toadstools are familiar examples.

Agassiz trawl (*Ocean.*). A small trawl designed to fish whichever way up the trawl falls on the bottom; useful in deep water.

agate (*Min.*). A cryptocrystalline variety of silica, characterized by parallel, and often curved, bands of colour.

agate mortar (*Chem.*). A bowl-shaped vessel in which hard and brittle materials, principally glass, are ground to powder.

AGC (*Radio*). Abbrev. for *automatic gain control.*

age and area theory (*Zool.*). The area occupied by a group of allied species is a measure of their antiquity in evolution.

age distribution (*Ecol.*). The relative frequency in an animal population of individuals of different ages. Generally expressed as a polygon or age pyramid, the number or percentage of individuals in successive age classes being shown by the relative width of horizontal bars.

age equation (*Nuc. Eng.*). See **age theory.**

age hardening (*Met.*). The production of structural change spontaneously after some time; normally it is useful in improving mechanical properties in some respect, particularly hardness.

ageing (*Acous.*). The diminution of effectiveness of a carbon microphone with life, arising from the granules becoming smoother through frictional movement. (*Met.*) Gradual rise in strength due to physical change in metals and alloys, in which there is breakdown from supersaturated solid solution and lattice precipitation over a period of days at atmospheric temperature. (*Phys.*) Change in the properties of a substance with time. A change in the magnetic properties of iron, e.g., increase of hysteresis loss of sheet-steel laminations; also the process whereby the subpermanent magnetism can be removed in the manufacture of permanent magnets. (*Radio*) The preliminary run of a valve, e.g. for 100 hours, before taking it into service.

agenesia, agenesis (*Med.*). Imperfect development (or failure to develop) of any part of the body.

ageotropism (*Bot.*). The condition of not reacting to gravity. *adj.* **ageotropic.**

age theory (*Nuc. Eng.*). In nuclear reactor theory, the slowing-down of neutrons by elastic collisions. The *age equation* relates the spatial distribution of neutrons to their energy. The equation is given by

$$\nabla^2 q - \frac{\partial q}{\partial r} = 0$$

where q is the slowing-down density and r is the spatial coordinate. It was first formulated by Fermi who assumed that the slowing-down process was continuous and so is least applicable to media containing light elements.

agglomerate (*Bot.*). Crowded, or heaped into a cluster. (*Geol.*) An indurated rock built of large angular rock-fragments embedded in an ashy matrix, and resulting from explosive volcanic activity. Occurs typically in volcanic vents. (*Powder Tech.*) Assemblage of particles rigidly joined together, as by partial fusion (sintering) or by growing together.

agglomerating value (*Min. Proc.*). Index of the binding (sintering) qualities of coal which has been subjected to a prescribed heat treatment.

agglomeration (*Zool.*). In *Protozoa*, adherence of forms like Trypanosomes by the aflagellar end of the body (apparently by means of a sticky secretion produced by the kinetonucleus) to form large clumps. Agglomeration always occurs in conditions unfavourable to the parasite, and is possibly due to the formation of agglutinins in the blood of the host.

agglutinate (*Bot.*). Cemented together by sticky material.

agglutination (*Bacteriol.*). The clumping of bacteria (antigen) under the influence of serum antibody developed in response to invasion by the particular antigen. (*Chem.*) The coalescing of small suspended particles to form larger masses which are usually precipitated. (*Zool.*) The formation of clumps by some *Protozoa* and spermatozoa. (*Med.*) Clumping of red blood corpuscles when the plasma of 1 individual is added to the blood of another, or when the blood from 2 individuals of different types is mixed, e.g., bloods of groups A and B (II and III).

agglutinin (*Chem.*). Antibody substance which causes agglutination of blood corpuscles, bacteria, etc. (*Med.*) Specifically a constituent of the blood plasma of 1 individual which causes agglutination by reacting with a specific receptor in the red corpuscles in the blood of another individual.

agglutinogen (*Med.*). The substance in bacteria or in blood cells which stimulates the formation of, and unites with, agglutinin, as in infection and immunization.

aggregate (*Bot.*). Closely packed but not confluent. (*Civ. Eng.*) The sand and broken stone and brick or gravel which together form one of the constituents of concrete, the others being cement and water. See **coarse-, fine-.** (*Geol., Min.*) A mass consisting of rock or mineral fragments. (*Powder Tech.*) Assemblage of particles which are loosely coherent. (*Zool.*) Massed or clustered; e.g., the aggregate silk glands of Spiders, which are connected with the posterior spinnerets, and produce the spiral thread of orb webs and its viscid coating.

aggregate fruit (*Bot.*). (1) The fruit formed by a flower with several free carpels. (2) The fruit formed from several flowers growing close together; e.g., a Pineapple.

aggregate ray (*Bot.*). A group of closely placed, narrow vascular rays.

aggregate recoil (*Nuc.*). Nuclei recoiling after emission of α-particles may carry a small cluster of active atoms away from the surface with them. In strong-activity α-emitters (e.g., polonium, plutonium, etc.) this leads to a build-up of radioactive contamination on surrounding surfaces which may represent a serious health hazard.

aggregate species (*Bot.*). A group of closely related species denoted by a single name.

aggregation (*Ecol.*). A type of animal and plant dispersion which is not random, but in which the organisms tend to occur in groups, or show a contagious distribution. This may be due to environmental or social factors, and such grouping may have advantages. See **Allee's principle.**

aggregen (*Physiol.*). An organized mass of (mammalian) cells growing in agitated culture, resembling an 'aggregate' in having a continuous layer of peripheral cells forming a smooth surface, but differing from it in having the capacity to fragment to form daughter aggregens.

aggressive behaviour (*An. Behav.*). A class of behaviour including both threats and actual attacks on other animals. It sometimes occurs towards inanimate objects.

aggressiveness, aggressivity (*Bot.*). The capacity of a parasite to attack its host.

aggressive resemblance (*Zool.*). See **anticryptic coloration.**

agitator (*Met., Min. Proc.*). Tank, usually cylindrical, near bottom of which is a mixing device such as a propeller or airlift pump. Finely ground mineral slurries (the aqueous component perhaps being a leaching solution) are exposed to appropriate chemicals for purpose of extraction of gold, uranium, or other valuable constituents. Types are *Pachuca* or *Brown, Dorr*, and *Devereux*.

A-glass (*Glass*). A glass containing 10–15% of alkali (calculated as Na_2O) and used for the manufacture of glass fibre.

aglomerular (*Physiol.*). Devoid of glomeruli; as the kidney in certain fishes.

aglossate, aglossal (*Zool.*). Lacking a tongue. *n.* **aglossia**, congenital absence of the tongue.

agmatite (*Geol.*). A mixed rock, in which blocks of metamorphic or sedimentary material are separated by veins of igneous material.

Agnatha (*Zool.*). Fishlike creatures with no true jaw, including *Cyclostomata* and *Osteostraci*.

agnathous, agnathostomatous (*Zool.*). Having a mouth without jaws, as in the Lampreys.

Agnesi (*Maths.*). See **witch of Agnesi.**

agnosia (*Med.*). Loss of the ability to recognize the nature of an object through the senses of the body.

agonic line (*Geog.*). The line joining all places with no magnetic declination, i.e., those where true north and magnetic north coincide on a compass.

agonistic behaviour (*An. Behav.*). A broad class of behaviour patterns including all types of attack, threat, appeasement, and flight.

agoraphobia (*Psychol.*). The fear of being alone in an open space.

agranulocyte (*Cyt.*). A nongranular leucocyte; a lymphoid leucocyte.

agranulocytosis (*Med.*). A pathological state in which there is a marked decrease in the number of granulocytes in the blood.

agraphia (*Med.*). Loss of the power to express thought in writing, as a result of a lesion in the brain.

agressin (*Zool.*). A poisonous substance, produced by disease-causing organisms, which inhibits the defensive reactions of the host.

agrestal (*Bot.*). Growing in cultivated ground, but not itself cultivated; e.g., a weed.

agrimycin (*Biochem.*). A fungicidal antibiotic.

agrostology (*Bot.*). The study of grasses.

AGS (*Aero.*). Abbrev. for *Aircraft General Standard.*

ague (*Med.*). See **malaria.**

agynous (*Bot.*). Said of an abnormal flower in which the gynaeceum has failed to develop.

a.h.m. (*Elec.*). See **ampere-hour meter.**

aiguille (*Tools*). A stone-boring tool.

aileron droop (*Aero.*). The rigging of ailerons so that under static conditions their trailing edges are below the wing trailing-edge line, pressure and suction causing them to rise in flight to the aerodynamically correct position.

ailerons (*Aero.*). Flaps at the trailing edge of the wing, controlled by the pilot, which move differentially to give a rolling motion to the aircraft about its longitudinal axis.

ainhum (*Med.*). A chronic disease of Negro race, terminating in spontaneous amputation of the fingers and toes by the formation of a constricting ring. The fifth toe is the most commonly affected.

air. See **atmosphere (composition of), atmospheric pressure, saturation of the air.**

air bells (*Photog.*). Minute bubbles which have adhered to emulsions during processing, with consequent black or white spots where the emulsion has been protected from the action of chemicals.

air bladder (*Zool.*). In Fish, an air-containing sac developed as a diverticulum of the gut, with which it may retain connexion by the pneumatic duct in later life; usually it has a hydrostatic function, but in some cases it may be respiratory or auditory, or assist in phonation.

air-blast switch (or circuit-breaker) (*Elec. Eng.*). A form of switch (or circuit-breaker) in which the arc is extinguished by a blast of air which removes ions from the gap, thereby eliminating the current path.

air-blowing (*Paint*). The process of bubbling a large stream of air through more or less highly heated vegetable oil. This causes it to polymerize and oxidize, which imparts better solubility, drying, and other paintmaking characteristics.

air brake (*Aero.*). An extendable device, most commonly a hinged flap on wing or fuselage, controlled by the pilot, to increase the drag of an aircraft. Originally a means of slowing bombers to enable them to dive more steeply, it is an essential flight control on clean jet aeroplanes and sailplanes. (*Eng.*) (1) A mechanical brake operated by air-pressure acting on a piston. (2) An absorption dynamometer in which the power is dissipated through the rotation of a fan or airscrew.

air break (*Elec. Eng.*). Term applied to a switch or circuit-breaker which has the contacts in air.

air brick (*Build.*). A perforated cast-iron, concrete, or earthenware brick built into a wall admitting air under the floors or into rooms.

airbus (*Aero.*). A large-capacity short-haul airliner.

air capacitor (*Elec. Eng.*). One in which the dielectric is nearly all air, for tuning electrical circuits with minimum dielectric loss.

air cavity (*Bot.*). (1) A large intercellular space in

a leaf into which stomata open. (2) A cavity in the upper surface of the thallus of some liverworts, opening externally by an *air pore* and containing chains of photosynthetic cells. (3) A large intercellular space in which air is stored in some water plants. Also **air space.**

air cell (*Eng.*). A small auxiliary combustion chamber used in certain types of compression-ignition engines, for promoting turbulence and improving combustion. (*Heat.*) Corrugated asbestos paper cemented in layers to flat asbestos paper to form an insulation sheet.

air classifier (*Min. Proc.*). Appliance in which vertical, horizontal, or cyclonic currents of air sort falling finely ground particles into equal-settling fractions or separate relatively coarse falling material from finer dust which is carried out. Also **air elutriator.**

air cleaner (*I.C. Engs.*). A filter placed at the air intake of an internal-combustion engine to remove dust from the air entering the cylinders.

air compressor (*Eng.*). A machine which draws in air at atmospheric pressure, compresses it, and delivers it at higher pressure. It may be reciprocating, centrifugal, or rotary (vane).

air conduction (*Acous.*). The passing of noise energy along an air path, as contrasted with mechanical conduction of vibrational energy.

air-cooled engine (*I.C. Engs.*). An internal-combustion engine in which the cylinders, finned to increase surface area, are cooled by an airstream. See **cowling.**

air-cooled machine, transformer, etc. (*Elec. Eng.*). A machine, transformer, or other piece of apparatus, in which the heat occasioned by the losses is carried away solely by means of air. The flow of air over the heated surfaces may be due to natural convection or may be produced by a fan.

air cooler (*Heat*). The cold 'accumulator' used in the Linde process of air liquefaction for the preliminary cooling of the air.

air cooling (*Eng.*). The cooling of hot bodies by a stream of cold air; distinct from water cooling.

air-core cable (*Telecomm.*). See **dry-core cable.**

air course (*Mining*). Underground ventilating tunnel or bore.

aircraft engine (*Aero.*). See **aero-engine.**

Aircraft General Standard (*Aero.*). Term referring to small parts or items such as bolts, nuts, rivets, fork joints, etc., which are common to all types of aircraft. Abbrev. **AGS.**

aircraft noise (*Aero.*). Noise from propeller, engine, exhaust, and surface-friction of aircraft, characterized by unstable low frequencies. See **jet noise.**

air crossing (*Mining*). In mine ventilation, ducting which carries one *air course* (q.v.) over another.

air data system (*Aero.*). A centralized unit into which are fed the essential physical measurements for flight, e.g., air speed, *Mach no.*, *pitot* and static pressure, barometric altitude, stagnation air temperature. From this central source data are transmitted to the cockpit dials, to flight and navigational instruments and to computers, etc. Abbrev. **ADS.**

air door (*Mining*). In mine ventilating system, door which admits air or varies its direction.

air dose (*Radiol.*). Radiation dose in röntgens delivered at a point in free air.

airdox (*Mining*). American system for breaking coal in fiery mine by use of injected high-pressure air.

air drag (*Space*). The resistance to the motion of a rocket or artificial satellite caused by the Earth's atmosphere, most serious in the lower atmosphere, causing a reduction in the size of the orbit, so that the satellite spirals in to the surface.

air drain (*Build.*). A cavity in the external walls of a building, designed to prevent damp from getting through to the interior.

air dry (*Min. Proc.*). Said of minerals in which moisture content is in equilibrium with that of atmosphere. (*Paper*) Said of wood pulp the moisture content of which is in equilibrium with the surrounding conditions.

air drying (*Paint.*). The process by which a dry film, able to be handled, is formed from oil or paints under normal atmospheric conditions only. This may occur by solvent evaporation and/or by the oxidation and polymerization of the film constituents on exposure to the oxygen in the air. See also **stoving.**

air ducts (*Eng.*). Pipes or channels through which air is distributed throughout buildings or machinery heating and ventilation.

air ejector (*Eng.*). A type of air-pump used for maintaining a partial vacuum in a vessel through the agency of a high-velocity steam jet which entrains the air and exhausts it against atmospheric pressure.

air elutriator (*Min. Proc.*). See **air classifier.**

air engine (*Eng.*). (1) An engine in which air is used as the working substance, the heat being supplied from an external source. Impracticable except for very small powers. See **hot air engine.** (2) A very small reciprocating engine driven by compressed air.

air-entraining agent (*Eng.*). Resin added to either cement or concrete in order to trap small air bubbles.

air equivalent (*Nuc.*). The thickness of an air column at 15°C and 1 atmosphere pressure which has the same absorption of a beam of radiation as a given thickness of a particular substance.

air escape (*Plumb.*). A contrivance for discharging excess air from a water pipe; it consists of a *ball-cock* (q.v.) which opens the discharging air valve when sufficient air has collected, and closes it in time to prevent loss of water.

air exhauster (*Eng.*). (1) A suction fan. (2) A vacuum pump.

air filter (*I.C. Engs.*). Attachment to the air intake of a carburettor, usually an oil-immersed steel-wool filter, for cleaning air drawn into the engine.

air-float table (*Min. Proc.*). Shaking table in which concentration of heavy fraction in sand-sized feed is promoted by air blown up through the deck, which is porous. Used in desert work.

airflow breakaway (*Aero.*). Term which is generally used for the rapid spread of initial stall across the wing which results in complete loss of lift.

airflow meter (*Aero.*). An instrument, mainly experimental, for measuring the airflow in ducts.

air flue (*Build.*). A flue which is built into a chimney-stack so as to withdraw vitiated air from a room.

airframe (*Aero.*). The complete aircraft structure without the power plant.

air frost (*Meteor.*). A screen temperature below freezing point of water. See **wind frost.**

air-fuel ratio (*Eng.*). The proportion of air to fuel in the working charge of an internal-combustion engine, or in other combustible mixtures, expressed by weight for liquid fuels and by volume for gaseous fuels.

air gap (*Elec.*). Gap with points or knobs, adjusted to breakdown at a specified voltage and hence limit voltages to this value. (*Mag.*) Section of air, usually short, in a magnetic circuit, especially in a motor or generator, a relay, or a choke. The main flux passes through the gap, with leakage outside depending on dimensions and permeability.

air-gap torsion meter (*Eng.*). A device for measuring the twist in a shaft by causing the relative rotation of two sections to alter the air-gap between a pair of electromagnets, the resulting change in the current flowing being indicated by an ammeter.

air gas (*Fuels*). See **producer gas.**

air gate (*Foundry*). Passage from interior of a mould to allow the escape of air and other gases as the metal is poured in. See **riser.**

airglow (*Astron.*). The faint permanent glow of the night sky, due to light-emission from atoms and molecules of sodium, oxygen, and nitrogen, activated by sunlight during the day.

air-hardening steel (*Met.*). Alloy steel which can be hardened by cooling in air instead of in water (which is required for carbon steel); e.g., carbon 0·3%, nickel 4·0%, and chromium 1·5%.

air heater (*Heat*). (1) Direct-fired heaters, in which the products of combustion are combined with the air. (2) Indirect-fired heaters, in which the combustion products are excluded from the air flow. Both can be operated on the recirculation system, by which a proportion of the heated air is returned to and passed through the heating chamber. See **air preheater.**

air hoist (*Mining*). Air winch or other mechanical hoist actuated by compressed air.

airing (*Met.*). Removal of sulphur from molten copper in wirebar furnace, together with slag-forming impurities.

air insulation (*Elec. Eng.*). At normal pressure dry air has a breakdown strength of 30 kV/cm for a uniform field gap of 2 cm. The dielectric strength increases as the gap length is reduced, increases with pressure, and falls for an increase in temperature.

air intake (*Aero.*). Any opening introducing air into an aircraft, but that for the main engine air is usually implied if unqualified. (*Eng.*) Vent in a carburettor through which air is sucked to mix with the petrol vapour from the jet.

air-intake guide vanes (*Aero.*). Radial, toroidal, or volute vanes which guide the air into the compressor of a gas turbine, or the supercharger of a reciprocating engine.

air jig (*Min. Proc.*). In waterless countries, use of pulses of air to stratify crushed ore into heavy and light layers.

air lance (*Min. Proc.*). Length of piping used to work compressed air into settled sand or to free choked sections of process plant, restoring aqueous flow.

air leg (*Mining*). Telescopic cylindrical prop expanded by compressed air, used to support rock drill.

airless injection (*Eng.*). Injection of liquid fuel into the cylinder of an oil engine by a high-pressure fuel pump, so dispensing with the compressed air necessary in the early diesel engines. Also called **solid injection.**

airlift (*Nuc. Eng.*). General use of air or neutral gas to blow material (particularly radioactive), as solid or liquid in processing, to avoid pumps.

airlift pump (*Hyd.*). An apparatus for raising water from a well by the use of compressed air only; the latter is admitted into the lower end of a pipe immersed in the water to be lifted, setting up alternate plugs of water and air which are forced up the pipe by the superior hydrostatic pressure of the water in the open boring. (*Min. Proc.*) An air-operated displacement pump for elevating or circulating pulp in cyanide plants.

airline (*Mag.*) Straight line drawn on the magnetization curve of a motor, or other electrical apparatus, expressing the magnetizing force necessary to maintain the magnetic flux across an air gap in the magnetic circuit. (*Photog.*) Spectral line originated by a spark discharge in air.

air liquefier (*Heat*). A type of gas refrigerating machine based on the 'Stirling' cycle, the cycle of the hot air engine.

air lock (*Civ. Eng.*). Device by which access is obtained to the working chamber (filled with compressed air to prevent entry of water) at base of a hollow caisson. The workman at surface enters and is shut in an air-tight chamber filled with air at atmospheric pressure. Pressure within this air-lock is gradually raised to that used in the working chamber, so that the workman can pass out through another door and communicate with the working chamber. See **caisson.** (*Eng.*) An air pocket or bubble in a pipeline which obstructs the flow of liquid. See **vapour lock.**

air log (*Aero.*). An instrument for registering the distance travelled by an aircraft relative to the air, not to the ground.

air manometer (*Phys.*). A pressure gauge in which the changes in volume of a small quantity of air enclosed by mercury in a glass tube indicate changes in the pressure to which it is subjected.

air mass (*Meteor.*). A part of the atmosphere where the horizontal temperature gradient at all levels within it is very small, perhaps of the order of 1°C/100 km. See **frontal zone.**

air mass flow (*Aero.*). In a gas turbine powerplant, the quantity of air which is ingested by the compressor, normally expressed in pounds or kilograms per second.

air meter (*Eng.*). An apparatus used to measure the rate of flow of air or gas.

air-mileage unit (*Aero.*). An automatic instrument which derives the air distance flown and feeds it into other automatic navigational instruments.

air monitor (*Radiol.*). Radiation (e.g. γ-ray) measuring instrument used for monitoring contamination or dose rate in air.

air plant (*Bot.*). See **epiphyte.**

air pocket (*Aero.*). A colloquialism for a localized region of low air density, a rising or descending air current, also **bump.** Causes an abrupt loss of height as an aircraft passes through it, severity increasing with speed and also with low wing loading. See **vertical gust.**

air pore (*Bot.*). See **stoma, air cavity, lenticel.**

air position (*Aero.*). The geographical position which an aircraft would reach in a given time if flying in still air.

air-position indicator (*Aero.*). An automatic instrument which continually indicates air position, incorporating alterations of course and speed.

air preheater (*Eng.*). System of tubes or passages, heated by flue gas, through which combustion air is passed for preheating before admission to the combustion chamber, thus appreciably raising flame temperatures and returning to the combustion chamber some utilizable heat that would otherwise be lost. See also **recuperative air-heater, regenerative air-heater, Supermiser.**

air pump (*Eng.*). A reciprocating or centrifugal pump used to remove air, and sometimes the condensate, from the condenser of a steam plant. See **air ejector.** Any device used for transferring air from one place to another. A *compressor* (q.v.) is a pump used for increasing the pressure on the high-pressure side; a *vacuum pump* (q.v.) is one in which the object of pumping is to reduce the pressure on the low-pressure side. A *blower* (q.v.) is a pump used for obtaining a rapidly moving air blast.

air receiver (*Eng.*). A vessel into which compressed air is discharged, to be stored until required.

Air Registration Board (*Aero.*). The airworthiness authority of the U.K. until its functions were taken over in 1972 by the *Civil Aviation Authority*. Abbrev. **ARB.**

air route (*Aero.*). In organized flying, a defined route between two aerodromes; usually provided with direction-finding facilities, lighting, emergency-landing grounds, etc. See also **airway.**

air-sac mite (*Vet.*). *Cytoleichus nudus*, a mite (*acarus*) infecting the respiratory passages of gallinaceous birds.

air-sacs (*Zool.*). In Insects, thin-walled distensible dilations of the tracheae, occurring especially in rapid fliers, which increase the oxygen capacity of the respiratory system and otherwise assist the act of flight: in Birds, expansions of the blind ends of certain bronchial tubes, which project into the general body cavity and assist in respiration, as well as lightening the body. Also **aerostats.**

airscrew (*Aero.*). An assembly of radially disposed blades of aerofoil shape which by reason of rotation in air and blade angle of attack produces thrust or power; usually *propeller* for aircraft propulsion and *fan* for accelerating air through a wind tunnel or in some types of turbojet engine. Basic classifications are: (1) a **left-hand airscrew** which rotates counter-clockwise when viewed from the rear; (2) a **right-hand airscrew** which rotates clockwise when it is viewed from the rear; (3) a **pusher airscrew** which is mounted at the rear of an engine; (4) a **tractor airscrew** mounted at the front; (5) a **fixed-pitch airscrew** having wooden or metal blades mounted at fixed angles, although they may not be made in one piece; (6) an **adjustable-pitch airscrew** having blades the angles of which can be altered when the airscrew is stationary; (7) a **variable-pitch airscrew** having blades which are rotatable about their mounting axes while it is revolving, thereby altering the *pitch* to increase the efficiency at different flight speeds and engine rev/min; actuation is usually by hydraulic or electro-hydraulic means but electric motors are used and, on small types, mechanical systems such as centrifugal balance weights may be employed. Abbrev. **v.p. airscrew.** Types of v.p. airscrew: **constant-speed** which maintains the desired and preselected constant rev/min; **controllable-pitch,** with which a desired blade angle may be selected, usually 'fine pitch' (or 'high rev/min') for take-off and 'coarse pitch' (or 'low rev/min') for the rest of the flight, adjustment being made to match the stipulated characteristics of rev/min, *manifold pressure* or *torque*, as appropriate; on some turbo-airscrews a *ground fine pitch* angle may also be selected; a **feathering-airscrew** has an extension of the blade angle to reduce drag when an engine is stopped in flight; a **reverse pitch airscrew** has an additional control which allows the blades to turn to a negative angle as an aid to stopping after landing (also **braking airscrew**); a **swivelling airscrew** has a rotatable axis of rotation so that its thrust direction can be used for control on *airships* or for transition in *VTOL* aircraft. See also **convertiplane, blade angle, feathering pump, helical tip speed, pitch.**

airscrew brake (*Aero.*). A shaft brake to stop, or prevent windmilling of, a turboprop, principally in order to avoid inconvenience on the ground.

airscrew hub (*Aero.*). The detachable metal fitting by which an airscrew is attached to the power-driven shaft.

air seal (*Met.*). Curtain of air maintained in front of kiln or furnace door to aid in retention of heat.

air sextant (*Aero.*). A sextant provided with an artificial horizon, e.g., a bubble or gyroscopic device, and usually fitted with a periscope for extending outside the fuselage.

air shaft (*Civ. Eng.*). An air passage, usually vertical or nearly vertical, which provides for the ventilation of a tunnel or mine.

airship (*Aero.*). Any power-driven *aerostat* (q.v.). Types: *nonrigid airship*, one with the envelope so designed that the internal pressure maintains its correct form without the aid of a built-in structure; small, and used for naval patrol work; *rigid airship*, one having a rigid structure to maintain the designed shape of the hull, and to carry the loads; usually a number of ballonets or gas bags inside the frame; large, used for military purposes in World War I, and having limited commercial use until 1938; *semi-rigid airship*, one having a partial structure, usually a keel only, to distribute the load to, and maintain the designed shape of, the envelope or ballonets; intermediate size.

air shooting (*Mining*). (1) Charging of shot-hole so as to leave pockets of air, thus reducing shatter-effect of blast. (2) In seismic prospecting, explosion in air, above rock formation under examination, as method of propagating seismic wave.

air sinuses (*Zool.*). In Mammals, cavities connected with the nasal chambers and extending into the bones of the skull, especially the maxillae and frontals.

air slaking (*Build.*). The process of exposing quicklime to the air, as a result of which it will gradually absorb moisture and break down into a powder, becoming *hydrated lime*.

air space. (*Aero.*) That part of the atmosphere which lies above a nation and which is therefore under the jurisdiction of that nation. (*Bot.*). See **air cavity.**

air-spaced coil (*Elec. Eng.*). Inductance coil in which the adjacent turns are spaced (instead of being wound close together) to reduce self-capacitance and dielectric loss.

air speed (*Aero.*). Speed measured relative to the air in which the aircraft or missile is moving, as distinct from ground speed. See **equivalent-, indicated-, true-.**

air-speed indicator (*Aero*). Instrument for measuring air speed by the differential between the pressure on a forward-facing intake tube and the ambient pressure taken from a lateral *static vent* (q.v.). See **maximum safe-.**

air split (*Mining*). In mine ventilation, division of main intake air into underground 'districts'.

air standard cycle (*I.C. Engs.*). A standard cycle of reference by which the performance of different internal-combustion engines may be compared, and their relative efficiencies calculated.

air standard efficiency (*I.C. Engs.*). The thermal efficiency of an internal-combustion engine working on the appropriate air standard cycle.

airstrip (*Aero.*). A unidirectional landing area, usually of grass or of a makeshift nature.

air surveying (*Surv.*). See **aerial surveying.**

air-swept mill (*Min. Proc.*). In dry grinding of rock in ball mill, use of modulated current of air to remove sufficiently pulverized material from the charge in the mill.

air table (*Min. Proc.*). See **air-float table.**

air terminal (*Elec.*). Elevated structure acting as a lightning protector, collecting local charge and reducing electric field strength.

air-traffic control (*Aero.*). The organized control, by visual and radio means, of the traffic on air routes and into and out of aerodromes. ATC is divided into general *area control*, including defined *airways*; *control zones*, of specified area and altitude, round busy aerodromes; *approach control* for regulating aircraft landing and departing; and *aerodrome control* for directing aircraft movements on the ground and giving permission for take-off. Air-traffic control operates under two systems, *visual flight rules* (q.v.) and, more severely, *instrument flight rules* (q.v.). Since World War II great advances in radar technology have enabled *air-traffic controllers* to be given very complete 'pictures' of the position of aircraft, not only in flight, but also when manoeuvring on the ground.

air-traffic control centre (*Aero.*). An organization providing (1) air-traffic control in a control area and (2) *flight information* (q.v.) in a region.

air-traffic controller (*Aero.*). One who is licensed to give instructions to aircraft in a control zone.

Air Transport Association (*Aero.*). A U.S. organization noted particularly for its specification setting a standard to which manufacturers of aircraft and associated equipment are required to produce technical manuals for the aircraft operator's use. The specification is accepted by *IATA* as the basis for international standardization. Abbrev. ATA.

air trap (*San. Eng.*). A trap which, by a water-seal, prevents foul air from rising from sinks, wash basins, drains, sewers, etc. Sometimes called **drain trap, stench trap, U-bend.**

air wall (*Nuc. Eng.*). Wall of ionization chamber designed to give same ionization intensity inside chamber as in open space. This means the wall is made of elements with atomic numbers similar to those for air constituents.

airway (*Aero.*). A specified 3-dimensional corridor (the lower as well as the upper boundary being defined) between *control zones* which may only be entered by aircraft in radio contact with *air-traffic control* (q.v.). (*Mining*) Underground passage used mainly for ventilation.

airworthy (*Aero.*). (1) Fit for flight aircraft, aero-engine, instrument, or equipment. (2) Complying with the regulations laid down for ensuring the fitness of an aircraft for flight. (3) Possessing a *certificate of airworthiness*.

Airy disk (*Photog.*). Disk-shaped image of a point source of light, caused by diffraction, increased by stopping down lens. After Sir George Airy (1801–92).

Airy points (*Phys.*). The best points for supporting a bar horizontally so that bending shall be a minimum. The distance apart of the points is equal to $\frac{l}{\sqrt{n^2-1}}$, where l is the length of the bar and n the number of supports.

Airy's differential equation (*Maths.*). One of the form $\frac{d^2y}{dx^2}=xy$.

Airy's integral (*Light*). The factor 1·22, by which the dimensions of the diffraction pattern produced by a slit must be multiplied to obtain the dimensions of the pattern due to a circular aperture.

Airy spirals (*Light*). The spiral interference patterns produced when quartz, cut perpendicularly to the axis, is examined in convergent light circularly polarized.

aisle (*Arch.*). A side division of the *nave* or other part of a church or similar building, generally separated off by pillars: (loosely) any division of a church, or a small building attached: (loosely) a passage between rows of seats.

aisle roof (*Arch.*). See **lean-to roof.**

aitiogenic, aitiogenous (*Bot.*). Said of a reaction by a plant, induced by some external agent, generally a movement of some kind.

aitionastic (*Bot.*). Said of a curvature performed by a plant member in response to a diffuse stimulus.

Aitoff's projection (*Geog.*). A derivative of the *zenithal equal-area projection* (q.v.) which is a reprojection of the projection itself at an angle of 60°, giving an elliptical shape in place of the original circle.

aizoon (*Bot.*). An African genus of plants.

ajacine (*Chem.*). The *N*-acetylanthranilic acid ester of cycactonine, obtained from the seeds of the *Delphinium* species. It crystallizes in colourless needles (m.p. 142°–143°C).

Ajax-Wyatt furnace (*Elec. Eng.*). A form of electric induction furnace having an iron core and a special shape of container, to ensure circulation of the charge; the furnace operates at power frequency.

ajutage (*Hyd.*). See **adjutage.**

akaryote (*Zool.*). A cell lacking a nucleus, or one in which the nucleoplasm is not aggregated to form a nucleus.

akaryote stage (*Bot.*). A stage in the life history of some lower plants, immediately preceding meiosis, when the nuclei are difficult to stain, due to a deficiency of chromatin.

akene (*Bot.*). See **achene.**

åkermanite (*Min.*). The calcium-magnesium end-member of the melilite group of silicates.

akinesia (*Med.*). Partial, complete, or temporary paralysis or muscular weakness. Lack of motor function.

akinete (*Bot.*). A thick-walled cell containing oil and other reserve ood materials, formed by some filamentous algae and fungi serving as a means of vegetative multiplication.

Akroyd engine (*Eng.*). The first compression-ignition (C.I.) oil engine, patented by Akroyd Stuart in 1890. See **compression-ignition engine.**

Akulon (*Plastics*). TN for Dutch nylon-6 polymer used for mouldings and fibres.

Al (*Chem.*). The symbol for *aluminium*.

ala (*Bot.*). (1) One of the side petals of the flower of the pea and its relatives. (2) A membranous outgrowth on a fruit, serving in wind dispersal. (3) A narrow leafy outgrowth down the stem from a decurrent leaf. (*Zool.*) Any flat, wing-like process or projection, especially of bone. *adj.* **alar, alary.**

alabandite (*Min.*). Massive, granular sulphide of manganese occurring in veins in Romania and elsewhere. Also called **manganblende.**

alabaster (*Min.*). A massive form of gypsum, often pleasingly blotched and stained. Because of its softness it is easily carved and polished, and is widely used for ornamental purposes. Chemically it is $CaSO_4{\cdot}2H_2O$. *Oriental alabaster*, onyx marble. A beautifully banded form of stalagmitic calcite occurring in Algeria, Egypt, and elsewhere.

alalia (*Med.*). See **aphonia.**

alanine (*Chem.*). 2-Aminopropanoic acid, $CH_3{\cdot}CH(NH_2){\cdot}COOH$; it occurs combined in proteins, notably in silk fibroin, an amino acid classified biologically as nonessential (for maintenance of growth in rats).

alar membrane (*Zool.*). See **prepatagium.**

alarm flag (*Elec. Eng.*). See **flag indicator.**

alar wall (*Zool.*). Dorsolateral wall of the *prosencephalon* (q.v.).

alary muscles (*Zool.*). In Insects, pairs of striated muscles arising from the terga and spread out fanwise over the surface of the dorsal diaphragm. Also **aliform muscles.**

alaska yarn (*Textiles*). Mixture yarn used to make ladies' dress and coating cloths. Often comprises a blend of combed wool/cotton, wool/rayon, wool/nylon, or wool and other man-made fibres. Full advantage is taken of the different dyeing properties.

ala spuria (*Zool.*). See **bastard wing.**

alassotonic (*Zool.*). Of or against decreasing force (of muscle contraction).

alastrim (*Med.*). Variola minor; a mild form of smallpox differing from it in certain features, mainly nonfatal.

alate (*Bot.*). Winged; applied to stems when decurrent leaves are present. (*Zool.*) Having a broad lip (especially of shells): in *Porifera*, a type of triradiate spicule with unequal angles.

ala temporalis (*Zool.*). In the developing chondrocranium, a small cartilage passing laterally from the side of the trabecula; the alisphenoid bone.

Albada viewfinder (*Photog.*). One with a lightly silvered plano-concave objective which reflects frame marks placed on the eyepiece and at the focus of the mirror.

albedo (*Astron.*). (1) The reflected fraction, in non-specular reflection, of the vertically incident radiation taken as unity. (2) Secondary cosmic ray particles projected upwards. (*Light*) Photometric term for the degree of reflecting power of a matt surface. (*Nuc.*) Ratio of the neutron flow density out of a medium free from sources, to the neutron flow density into it, i.e., reflection factor of a surface for neutrons.

Albers-Schönberg disease (*Med.*). See **osteopetrosis.**

albert (*Paper*). A former standard size of notepaper, 6×4 in.

albertite (*Min.*). A pitch-black solid bitumen of the asphaltite group occurring in veins in oil-bearing strata.

Albertol (*Chem.*). TN for ester gums.

Albert-Precht effect (*Photog.*). Development of a reverse image by first treating the exposed surface with chromic acid, thus destroying the latent image, then exposing to uniform light and redeveloping.

Albian stage (*Geol.*). A division of the Cretaceous System, comprising the rocks between the Aptian stage below and the Cenomanian stage above; approximately equivalent to the English Gault and Upper Greensand.

albinism (*Bot.*). An abnormal condition due to the absence of chlorophyll or other pigments. (*Zool.*) Absence of pigmentation, especially marked in the integument, epidermal outgrowths, and the eyes.

albino (*Bot.*). An abnormal plant of whitish colour due to the more or less complete absence of chlorophyll. (*Zool.*) An individual of any species deficient in pigment in hair, skin, eyes, etc. *adj.* **albinotic.**

albite (*Min.*). The end-member of the plagioclase group of minerals. Ideally a silicate of sodium and aluminium; but commonly contains small quantities of potassium and calcium in addition and crystallizes in the triclinic system. Cf. *monalbite*.

albitite (*Geol.*). A rare type of soda-syenite consisting almost entirely of albite, with a small content of coloured silicates.

albitization (*Geol.*). In igneous rocks, the process by which a soda-lime feldspar (plagioclase) is replaced by albite (soda-feldspar).

Albucid (*Pharm.*). Proprietary name for *sulphacetamide*. See **sulphonamides.**

albuginea (*Cyt.*). Tunica albuginea; a white dense connective tissue surrounding the testis, ovary, corpora cavernosa, spleen, or eye.

albumen (*Bot.*). See **endosperm.** (*Zool.*) White of egg containing a number of soluble proteins, including ovalbumin; the nutritive material which surrounds the yolk in the eggs of higher animals. *adj.* **albuminous.**

albumen process (*Print.*). One in which dichromated albumen is used as a light-sensitive coating when preparing *surface plates* for lithography and line blocks for relief printing.

albuminates (*Chem.*). The alkali compounds of albumins.

albuminoids (*Chem.*). A synonym for *scleroproteins* (q.v.).

albuminous (*Bot.*). Endospermous.

albuminous cell (*Bot.*). A cell rich in contents, associated with the phloem in the leaves and stems of some Gymnosperms, and probably serving to conduct proteins.

albumins (*Chem.*). A group of simple proteins which are soluble in pure water, the solutions being coagulated by heat. The particular albumin found in white of egg is sometimes referred to as egg albumen; other examples are serum albumin (from blood) and lactalbumin (from milk).

albuminuria (*Med.*). Albumin in the urine.

albumoses (*Chem.*). A synonym for *proteoses.*

alburnum (*Bot.*). Sapwood.

Alchlor process (*Chem.*). Used in refining lubricants, by removal of impurities with aluminium chloride.

Alclad (*Met.*). Composite sheets consisting of an alloy of the Duralumin type (to give strength) coated with pure aluminium (to give corrosion resistance).

alcohol (*Chem.*). A general term for compounds formed from hydroxyl groups attached to carbon atoms in place of hydrogen atoms. The general formula is $R{\cdot}OH$, wherein R signifies an aliphatic or an aromatic radical. In particular, *ethanol*. See also **methanol-, absolute-.**

alcohol fuel (*I.C. Engs.*). Volatile liquid fuel consisting wholly, or partly, of alcohol, able to withstand high-compression ratios without detonation.

alcoholic Bouin (*Micros.*). See **Duboscq-Brasil fixative.**

alcoholic fermentation (*Bot.*). The production of alcohol from sugar by enzymes in yeasts. Production is best when the supply of free oxygen is limited (anaerobic respiration). Carbon dioxide and heat are also produced.

alcoholism (*Psychiat.*). Disease produced by addiction to alcohol, manifesting itself in a variety of psychotic disorders, e.g., hallucinosis, delirium tremens.

alcoholometer (*Brew.*). Instrument for measuring the gravity of beer, based on the differential between the b.p. of distilled water and that of liquids containing ethanol.

alcoholometry (*Chem.*). The quantitative determination of alcohol in aqueous solutions.

Alcomax (*Met.*). British equivalent of *Alnico* (q.v.).

alcosol (*Chem.*). A colloidal solution in alcohol.

aldehyde acids (*Chem.*). Products of the partial oxidation of dihydric alcohols, containing both an aldehyde group and a carboxyl group.

aldehyde ammonias (*Chem.*). Crystalline compounds formed by the interaction of aldehydes and ammonia. On distillation with dilute sulphuric acid, the pure aldehydes are regenerated.

aldehyde condensations (*Chem.*). Condensations of aldehydes with aldehydes, ketones, acids, etc., by the elimination of water and the linking of the chains of the reacting compounds.

aldehyde resins (*Plastics*). Highly polymerized resinous condensation products of aldehydes, obtained by treatment of aldehydes with strong caustic soda.

aldehydes (*Chem.*). Alkanals. A group of compounds containing the CO— radical attached to both a hydrogen atom and a hydrocarbon radical, viz. $R{\cdot}CHO$.

Alden power brake (*Eng.*). A form of brake used for measuring the power of an engine; consists of a disk of cast iron revolving between two plates of copper, oil being circulated from the circumference to the centre. Water under pressure circulates outside the copper plates, cooling them, as well as maintaining contact pressure on the moving disk.

alder (*For.*). A tree (*Alnus*) producing a straight-grained fine-textured hardwood noted for its durability under water. It is used for cabinet making, plywood, shoe heels, clogs, bobbins, wooden cogs, and small turned items.

aldimines (*Chem.*). Condensation products of phenols with hydrocyanic acid, formed in the presence of gaseous hydrogen chloride.

Aldoform (*Chem.*). TN for *formaldehyde* (q.v.).

aldohexoses (*Chem.*). Compounds of the formula $OH{\cdot}CH_2{\cdot}(CH{\cdot}OH)_4{\cdot}CHO$, of which numerous stereoisomeric forms are possible. The following aldohexoses are known:

d- and *l*-mannose, *d*- and *l*-glucose,
d- and *l*-gulose, *d*- and *l*-idose,
d- and *l*-galactose, *d*-talose,
d-altrose, *d*-allose.

aldoketenes (*Chem.*). A group comprising ketene, its monoalkyl-substituted derivatives, and carbon suboxide.

aldol (*Chem.*). 2-Hydroxybutanal, a condensation product of acetaldehyde, viz. $H_3C{\cdot}CH(OH){\cdot}CH_2{\cdot}CHO$.

aldol condensation (*Chem.*). The condensation of 2 aldehyde molecules in such a manner that the oxygen of the 1 molecule reacts with the hydrogen of the other molecule, forming a hydroxyl group, with the simultaneous formation of a new link between the 2 carbon atoms. Water is eliminated.

aldoses (*Chem.*). A group of monosaccharides with an aldehydic constitution, e.g. glucose.

aldosterone (*Chem.*, *Med.*). 11β : 21-dihydroxy-3 : 20-dioxo-4-pregnen-18-al, a potent mineralocorticoid (discovered in 1953) isolated from extracts of adrenal gland, used in treatment of Addison's disease.

aldoximes (*Chem.*). Hydroxyimino alkanes. A group of compounds in which the oxygen of the aldehyde group is substituted by the radical $=N{\cdot}OH$, derived from hydroxylamine $H_2N{\cdot}OH$ and an aldehyde by dehydration. The general formula is $R{\cdot}CH=N{\cdot}OH$.

aldrin (*Chem.*). A chloro-derivative of naphthalene (1 : 2 : 3 : 4 : 10 : 10 hexachloro-1 : 4 : 4a : 5 : 8 : 8a-hexahydro-*exo*-1 : 4-*endo*-5 : 8-dimethano-naphthalene), used as a contact insecticide, incorporated in plastics to make cables resistant to termites; used in agriculture, especially against wireworm.

ale (*Brew.*). One of the classes of beers, including pale ale (bitter) with a distinct hop flavour, and mild ale, sometimes sweetened, and not so strongly hopped. Burton and barley wine are strong ales.

alecithal (*Zool.*). Of ova, having little or no yolk.

Alectoromorphae (*Zool.*). A legion of Birds possessing a schizognathous palate; includes the Game-birds, Cranes, Plovers, and Gulls. Obsolete name.

Alençon lace (*Textiles*). A French point lace made by hand with a needle; made in segments and then joined together and stiffened.

alepidote (*Bot.*). Without scales; smooth.

Aleppo galls (*Chem.*). A synonym for nut galls, excrescences on various species of oak, containing gallic and gallitannic acids.

aleuriospore, aleurispore (*Bot.*). A conidium formed laterally on a hypha by some fungi, especially those which cause skin diseases.

aleurone (*Bot.*). Reserve protein material occurring in granules in the *aleurone layer*, a special layer of cells just below the surface of the grains of various cereals, and in the seeds of other plants.

Aleutian disease (*Vet.*). A chronic, fatal disease of mink, with characteristic changes in the liver, kidneys, and other organs; it occurs especially in mink homozygous for the aleutian gene controlling fur colour.

alexandrite (*Min.*). A variety of chrysoberyl, the colour varying, with the conditions of lighting, between emerald green and columbine red.

alexia (*Med.*). Word blindness; loss of the ability to interpret written language, due to a lesion in the brain.

alfalfa (*Agric.*). *Medicago sativa*, better known as *lucerne*. A fodder plant thought to be a native in the Mediterranean region and W. Asia. Long grown by farmers elsewhere and now extensively naturalized on waste ground.

Alfol (*Eng.*). TN for thin corrugated aluminium foil in narrow strips, used for heat insulation, for which it is effective by reason of the numerous small air cells formed when packed.

Alford antenna (*Radio*). One comprising a vertical cylindrical tube with longitudinal slots, often used to transmit VHF or UHF.

Algae (*Bot.*). A large group of simple organisms, mostly aquatic. They contain chlorophyll and/or other photosynthetic pigments, and have simply organized reproductive organs.

algae poisoning (*Vet.*). A form of poisoning affecting farm livestock in the U.S.A. due to the ingestion of toxins in decomposing algae; characterized by nervous symptoms and death.

algal layer, algal zone (*Bot.*). A layer of algal cells lying inside the thallus of a heteromerous lichen. Also called **gonidial layer, gonimic layer.**

algebra (*Maths.*). Originally the abstract investigation of the properties of numbers by means of symbols (x, y, etc.). In this context typical algebraic problems are the solving of equations, the summation of series, permutations and combinations, and matrices. More recently extended to include the study of sets in general with operations. In this context there is the study of mappings, groups, rings, integral domains, fields, and vector spaces. See also **mathematics.**

algebraic function (*Maths.*). One which can be defined by a finite number of algebraic operations, including root extraction, e.g., the quotient of 2 polynomials is an algebraic function, but sin x is not.

algebraic number (*Maths.*). A root of a polynomial equation with rational coefficients.

Algerian onyx (*Min.*). Another name for *oriental alabaster*. See under **alabaster**.

algesimeter (*Med.*). An instrument for measuring sensitivity to pain.

algesis (*Physiol.*). The sense of pain.

alginic acid (*Chem.*). *Norgine* $C_6H_8O_6$, occurs both in the free state and the calcium salt in the larger brown Algae (*Phaeophyceae*). The sodium salt gives a very viscous solution in water even at a concentration of only 2%, and is used in the dyeing, textile, plastics, and explosives industries, in making waterproofing and insulating materials, foodstuffs, adhesives, cosmetics, and in medicine, for its sodium absorption ability by a cation exchange reaction. On hydrolysis, alginic acid gives *d*-mannuronic acid and its molecule consists of units of this acid combined in the same way as the glucose units of cellulose.

algodonite (*Min.*). Arsenide of copper occurring as a white incrustation in the Algodona silver mine, Chile.

Algol (*Astron.*). A star, β, in the constellation of Perseus which is the prototype of the eclipsing binary, where one component passes in front of the other at each revolution, causing an eclipse and a systematic fluctuation of magnitude. (*Chem.*) TN for an anthraquinone vat dyestuff. Algol blue K is the *N*-dimethyl derivative of *indanthrene* (q.v.) blue. (*Comp.*) Universal symbolic machine language, useful for programming scientific calculations, having no reference to the common code of a specific computer. (*Alg*ebraically *o*rientated *l*anguage.)

algology (*Bot.*). The study of algae.

algorithm (*Comp.*). Set of steps to be taken in operations to effect a desired calculation. (*Maths.*) A mathematical rule which when applied repeatedly will produce a result whose degree of accuracy will accord with the number of applications, e.g., if $x_1=1$, the rule $x_{n+1}=\frac{1}{2}\left(\frac{a}{x_n}+x_n\right)$ will produce a sequence of numbers converging to $\sqrt{a}$ and can be used to approximate to $\sqrt{a}$.

alhambra (*Textiles*). Plain or coloured woven quilting featuring large jacquard design, sometimes with a centrepiece (name, etc.), also with borders and fringes, or hemmed. The warp is usually hard-twisted American yarn with cotton condenser or soft spun rayon weft. Two beams are used.

alicyclic (*Chem.*). Abridged term for *aliphatic-cyclic*.

alicyclic hydrogenation (*Chem.*). Hydrogenation in the naphthalene series, wherein hydrogenation takes place only in the substituted benzene ring, e.g., in 2-naphthylamine hydrogenation occurs in the benzene ring containing the amino group.

alidade (*Surv.*). An accessory instrument used in plane-table surveying, consisting of a rule fitted with sights at both ends, which gives the direction of objects from the plane-table station. Also called **sight rule.**

alien (*Bot.*). A plant introduced by man, and maintaining a foothold in its own habitat under natural conditions.

alienicola (*Zool.*). A parthenogenetic, viviparous female aphid developing for the most part on the secondary host.

alienist (*Psychiat.*). One who diagnoses and treats mental illnesses. A psychiatrist, or practitioner of psychological medicine.

alien tones (*Acous.*). Frequencies, harmonic and sum-and-difference products, introduced in sound reproduction because of nonlinearity in some part of the transmission path.

aliform (*Bot.*). Flattened; winglike.

aliform muscles (*Zool.*). See **alary muscles.**

aligned-grid valve. See **beam-power valve.**

alignment (*Civ. Eng.*). (1) A setting in line (usually straight) of, e.g., successive lengths of a railway which is to be constructed. (2) The plan of a road or earthwork. (*Eng.*) The setting in a true line of a number of points, e.g., the centres of the bearings supporting an engine crankshaft. (*Nuc.*) Process of orientating, e.g., spin axes of atoms, during magnetization and similar operations. (*Radio*) Of a superheterodyne receiver, adjustment of preset tuned circuits to give optimum performance.

alignment chart (*Maths.*). See **nomogram.**

alima (*Zool.*). A pelagic larval form of certain stomatopod *Crustacea*, usually distinguished by the possession of a narrow carapace, elongate cephalic region, and the absence of the six posterior thoracic appendages.

alimentary (*Physiol.*). Pertaining to the nutritive functions or organs.

alimentary canal (*Physiol.*). The passage from the mouth to the anus which receives, digests, and assimilates nutrients from foodstuffs: the digestive tract: the gut. Also **alimentary tract.**

alimentary system (*Physiol.*). All the organs connected with digestion, absorption, and nutrition, comprising the digestive tract and associated glands and masticatory mechanisms.

aliphatic acids (*Chem.*). Fatty (alkanoic) acids or acids derived from aliphatic compounds.

aliphatic alcohols (*Chem.*). Alkanes in which one or more hydrogen atoms are replaced by the hydroxyl group.

aliphatic aldehydes (*Chem.*). Alkanals. Compounds of the aliphatic series containing the aldehyde group —CHO.

aliphatic (or acyclic) compounds (*Chem.*). Methane derivatives or fatty compounds; open-chain compounds.

aliquot (*Nuc.*). A small sample of radioactive material assayed to determine the radioactivity of the whole.

aliquot part (*Maths.*). A number or quantity which exactly divides a given number or quantity. (*Min. Proc.*) In sampling for process control, representative fraction, from quantitative analysis of which information as to the assay grade is given.

aliquot scaling, -tuning (*Acous.*). In a piano, the provision of extra wires above the normal wires. These are not struck, but are tuned very slightly above the octave of the struck strings below, so that by sympathetic vibration the musical quality of the note is enhanced.

alisphenoid (*Zool.*). A winglike cartilage bone of the Vertebrate skull, forming part of the lateral wall of the cranial cavity, just in front of the foramen lacerum: one of a pair of dorsal bars of cartilage in the developing Vertebrate skull, lying in front of the basal plate, parallel to the trabeculae: one of the sphenolateral cartilages.

alite (*Chem.*). The ground clinker obtained from sintering mixtures of limestone, clay, and sand in the proportions used in the manufacture of Portland cement.

alitrunk (*Zool.*). In Insects, the thorax fused with the first somite of the abdomen.

alizarin (*Chem.*). $C_{14}H_6O_2(OH)_2$, 1,2-dihydroxyanthraquinone, one of the most important natural and synthetic dyes, red prisms, or needles, m.p. 289°C, soluble in alcohol and ether, very slightly soluble in water, soluble in caustic soda; insoluble stains are formed with the oxides of aluminium, tin, chromium, and iron. Alizarin can be nitrated and forms the basis of a series of other dyestuffs. Has superseded crimson lake as a pigment.

alkalaemia (*Med.*). A condition of the body in which there is a temporary increase in pH above normal.

alkali (*Chem.*). A hydroxide which dissolves in water to form an *alkaline solution* (q.v.), which has pH > 7 and contains hydroxyl ions, OH^-. Alkalis are often spoken of as bases, but the term base has wider significance, denoting all protophil compounds, alkalis included.

alkali disease (*Vet.*). (1) Western duck sickness. A form of botulism causing death of wild duck and other water fowl in America, due to the ingestion of vegetation contaminated with toxin produced by *Clostridium botulinum*, type C. (2) A chronic disease of domestic animals, characterized by emaciation, stiffness, and anaemia, due to an excess of selenium in the diet. An acute form of the disease is called **blind staggers.**

alkali flat (*Geog.*). In an arid region, an alkaline, marshy patch into which desert streams lead. During the dry season it becomes hard mud covered with alkali.

alkali-granite (*Geol.*). An acid, coarse-grained (plutonic) rock carrying free quartz and characterized by a large excess of alkali-feldspar over plagioclase. Cf. *adamellite*, *granodiorite*. In general, the prefix used with a rock name implies a preponderance of soda- or potash-feldspar or feldspathoid over plagioclase, e.g., alkali-dolerite.

alkali metals (*Chem.*). The elements lithium, sodium, potassium, rubidium, caesium and francium, all monovalent metals in the first group of the periodic system.

alkalimetry (*Chem.*). The determination of alkali by titration with a standard solution of acid as in volumetric analysis. See **titration** and **volumetric analysis.**

alkaline earth metals (*Chem.*). The elements calcium, strontium, barium, and radium, all divalent metals in the second group of the periodic system.

alkaline gland (*Zool.*). In *Hymenoptera*, a gland opening into the poison sac at the base of the sting. The function of its secretion is not clear.

alkaline solution (*Chem.*). An aqueous solution containing more hydroxyl ions than hydrogen ions; one which turns red litmus blue.

alkalinity (*Chem.*). The extent to which a solution is alkaline. See **pH value.**

alkaloids (*Chem.*). Natural organic bases found in plants; characterized by their specific physiological action. Alkaloids may be related to various organic bases, the most important ones being pyridine, quinoline, *iso*quinoline, pyrrole, and other more complicated derivatives. Most alkaloids are crystalline solids, others are volatile liquids, and some are gums. They contain nitrogen as part of a ring, and have the general properties of amines.

alkalosis (*Med.*). Condition of the body in which the pH is normal, but there is an increased amount of buffer base.

alkane (*Chem.*). General name of hydrocarbons of the methane series, of general formula C_nH_{2n+2}.

Alkathene (*Chem.*). Trade name for *polyethylene*.

alkene (*Chem.*). General name for unsaturated hydrocarbons of the ethene series, of general formula C_nH_{2n}.

alkyd resins (*Plastics*). Formerly known as glyptal resins, i.e., condensation products derived from glycerol and phthalic anhydride. Now the term also covers diallyl esters and various polyesters used as resin binders in alkyd moulding materials.

alkyl (*Chem.*). A general term for monovalent aliphatic hydrocarbon radicals.

alkylation (*Chem.*). Substitution of an aliphatic hydrocarbon radical for a hydrogen atom in a cyclic organic compound.

alkylene (*Chem.*). A general term for divalent hydrocarbon radicals.

allaesthetic (*Biol.*). Describes an organ, structure or substance which stimulates the sense organs of another organism, and produces a biological effect, e.g., the flower of insect-pollinated plants.

Allan cell (*Elec. Eng.*). A type of electrolytic cell used in America for the production of hydrogen by the electrolysis of an alkaline solution.

allanite or **orthite** (*Min.*). A cerium-bearing epidote occurring as an occasional accessory mineral in igneous rocks.

allantoic (*Zool.*). See **allantois.**

allantoid (*Bot.*). Sausage-shaped.

allantoin (*Physiol.*). Diureide of glyoxylic acid, product occurring in allantoic fluid and excreted in urine of certain Mammals, and, in Insects, by the blowfly larvae.

allantois (*Zool.*). In the embryos of higher Vertebrates, a saclike diverticulum of the posterior part of the alimentary canal, having respiratory, nutritive, or excretory functions. It develops to form one of the embryonic membranes. *adj.* **allantoic.**

Allan valve (*Eng.*). A steam-engine slide-valve, in which a supplementary passage increases the steam supply to the port during admission to reduce wire-drawing. Also **trick valve.**

allargentum (*Min.*). An alloy of silver and antimony, crystallizing in the hexagonal system.

all-burnt (*Aero.*). The moment at which the fuel of a rocket (missile or research vehicle) is completely consumed, after which it coasts under its initial impetus.

Allee's principle (*Ecol.*). Recognition of the fact that in many populations there is a certain degree of clumping due to the aggregation of individuals, and that a particular degree of *aggregation* (q.v.) may be optimal, both undercrowding and overcrowding being detrimental.

alleghanyite (*Min.*). A hydrated manganese silicate, crystallizing in the monoclinic system.

allele (*Gen.*). Abbrev. for *allelomorph*. Any one of the alternative forms of a gene. A single gene may have several alternative forms, called *multiple alleles*. *adj.* **allelic.**

all-electric signalling (*Elec. Eng.*). A railway signalling system in which the signals and points are operated electrically by solenoids or motors, and are also controlled electrically. See **electropneumatic signalling.**

allelocatalysis (*Ecol.*). The favourable conditioning of a habitat by a population due to the elaboration of growth-promoting substances. Mainly studied in *Protozoa*.

allelomimetic behaviour (*An. Behav.*). A type of social behaviour involving imitation of another animal, usually of the same species.

allelomorph (*Gen.*). See **allele.** *adj.* **allelomorphic.**

allemontite (*Min.*). A solid solution of antimony and arsenic occurring in reniform masses at Allemont (France) and elsewhere.

Allen cone (*Min. Proc.*). Conical tank used for continuous sedimentation of liquids at constant level, the solids being removed from the base of the cone and the clear liquid drawn off from the top.

allene (*Chem.*). Propadiene, dimethylene methane, $CH_2=C=CH_2$, obtained by the electrolysis of itaconic acid.

Allen equation (*Min. Proc.*). One applied to sedimentation of finely ground particles intermediate between streamline and turbulent in settling mode.

$$\rho = Kr^n p \mu^{2-n} v^n$$

where ρ is fluid resistance; K, a constant for shape and velocity of fall; r, radius of an equivalent sphere; n, a coefficient of velocity, v; p, density of fluid; and μ, the kinematic viscosity b/p, b being absolute viscosity.

allenes (*Chem.*). Generic term for a series of non-conjugated and di-olefinic hydrocarbons, of which *allene* (q.v.) is the first, and which have the general formula $C_{2n}H_{2n-2}$. They consist mostly of colourless liquids with strong garlic odour.

Allen's law (*Zool.*). An evolutionary generalization stating that feet, ears, and tails of mammals tend to be shorter in colder climates, when closely allied forms are compared.

Allen's loop test (*Elec. Eng.*). A modification of the Varley loop test for localizing a fault in an electric cable; it is particularly suitable for high-resistance faults in short lengths of cable.

allergen (*Med.*). A substance, usually a protein, which, introduced into the body, makes it sensitive to that substance.

allergy (*Med.*). A state in which the cells of the body are supersensitive to substances (*allergens*, q.v.), usually proteins, introduced into it: the reaction of the body to a substance to which it has become sensitive, characterized by oedema, inflammation, and destruction of tissue.

allethrin (*Chem.*). (±)-3-Allyl-2-methyl-4-oxocyclopent-2-enyl(±)-(*cis-*+*trans-*) chrysanthemum monocarboxylate,

```
                                           CH3
                                           |
CH3                                        C
   >C=CH-CH                              /   \\
CH3        \CH·CO·O·CH                  C·CH2·CH=CH2
           /           |                |
  CH3-C                CH2 ------------ CO
      |
      CH3
```

The commercial product is a mixture of 8 optically active isomers; used as an insecticide, toxic to humans.

allette (*Arch.*). (1) A wing of a building. (2) A buttress or pilaster.

all-heart (*For.*). Applied to timber that is heartwood throughout, i.e., free from sapwood.

alliaceous (*Bot.*). Looking, or smelling, like an onion.

alliance (*Bot.*). A subclass, consisting of a number of related families of plants.

alligator (*Min. Proc.*). See **jaw breaker.**

alligatoring (*Paint.*). See **crocodiling.**

alligator wrench (*Plumb.*). A tool with fixed serrated jaws, used for twisting and screwing pipes into position.

all-insulated switch (*Elec. Eng.*). See **shockproof switch.**

all-moving tail (*Aero.*). A one-piece *tailplane*, also controlled by the pilot as is the *elevator*. Also **flying tail, stabilator** and see **T-tail.**

allo-. Prefix from Gk. *allos*, other. (*Chem.*) Used for steroids of the 5 α-series.

alloa (*Textiles*). A heavy, high-quality virgin wool hand knitting yarn. A 3-fold structure produced in one doubling, it is available in self colours and mixtures for glove, etc., manufacture. Originally made in Alloa (Scotland).

allobar (*Nuc.*). A form of element differing in isotopic composition from that occurring naturally.

allobiosis (*Zool.*). The changed reactivity of an organism in a changed internal or external environment.

allocarpy (*Bot.*). Fruiting after cross-fertilization.

allocation (*Comp.*). Process of distributing stored data in digital computer, to keep total access time required for a computation to a minimum.

allochroite (*Min.*). A light brown *andradite*.

allochromatic (*Electronics*). Having photoelectric properties which arise from micro-impurities, or from previous specific irradiation.

allochromy (*Phys.*). Fluorescent reradiation of light of different wavelength from that incident on a surface. See **Stokes's law.**

allochronic speciation (*Bot.*). The development of a species through the passage of time.

allochrous (*Bot.*). Changing colour.

allochthonous (*Ecol.*). In an aquatic community, said of food material reaching the community from outside: more generally, extraneous, exotic, acquired; e.g., *allochthonous* species, characteristics. Cf. *autochthonous*.

allochthonous behaviour (*An. Behav.*). Behaviour not activated by its own drive, sometimes being the result of frustration or conflict of drives. See **displacement activity.** Cf. *autochthonous behaviour*.

allogamy (*Bot.*). See **cross-fertilization.**

alloisomerism (*Chem.*). See **stereoisomerism.**

allomeric (*Chem.*). Having the same crystalline form but a different chemical composition.

allometric coefficient (*Bot.*). The ratio of relative growth rates $d \log Y/d \log X$, where Y is the measurement of part of an organism, and X that of the whole, or of another part of the organism.

allometric growth (*Bot.*). A type of growth in which the size of a part is a constant exponential function of the whole.

allomorphous (*Chem.*). Having the same chemical composition but a different crystalline form.

allopatric speciation (*Bot.*). The development of a species through geographical isolation.

allopelagic (*Zool.*). Of marine organisms, occurring at any depth, and apparently uninfluenced by change of temperature.

allophane (*Min.*). Hydrous aluminium silicate, apparently amorphous, occurring as thin incrustations on the walls of fissures and joints in rocks.

allophores (*Zool.*). Cells bearing a red pigment insoluble in alcohol, occurring in the skin of certain Fishes, Reptiles, and Amphibians.

alloplasm (*Zool.*). That differentiated portion of the cell-protoplasm which does not form independent organelles.

alloplast (*Zool.*). A morphological unit, consisting of more than one kind of tissue.

allopolyploid (*Bot.*). A polyploid possessing unlike sets of chromosomes, these usually in pairs.

Alloprene (*Chem.*). TN for a chlorinated rubber product used in speciality paints where high chemical resistance is required.

all-or-nothing piece (*Horol.*). A piece of the mechanism of a repeating watch which either allows the striking of the hours and quarters or entirely prevents it. Also called **stop slide.**

all-or-nothing response (*Physiol.*). In many irritable protoplasmic systems, response to stimuli is either with full intensity or not at all; e.g., in lower animals, nematocysts; in higher animals, nerve fibres, cardiac and voluntary muscle fibres.

allose (*Chem.*). An aldohexose, an optical stereoisomer of glucose.

allosomal inheritance (*Gen.*). The inheritance of characters carried in an *allosome*.

allosome (*Cyt.*). Any chromosome other than a typical one, e.g., sex chromosome.

allosteric site (*Chem.*). An enzymic site, different from the *active site* (q.v.), which noncompetitively binds molecules other than the substrate, and may influence the activity of the enzyme.

allostoses (*Histol.*). See **membrane bone.**

allothrausmatic (*Geol.*). See **homeothrausmatic.**

Allotriognathi (*Zool.*). An order of *Teleostei* characterized by the possession of a peculiar protractile mouth; includes the oar-fish, the ribbon-fish, and the moon-fish.

allotriomorphic (*Geol.*). A textural term used for igneous rocks describing crystals which show a form related to surrounding previously crystallized minerals rather than to their own internal structure.

allotropous flower (*Bot.*). A flower in which the nectar is accessible to all kinds of insect visitors.

allotropy (*Chem.*). The existence of an element in two or more solid, liquid, or gaseous forms, in one phase of matter, called *allotropes*.

allotter (*Teleph.*). A uniselector used to improve the efficiency of distribution of line finders, by automatically pre-selecting and pre-connecting the first available line finder in the group to which it has access.

allotype (*Zool.*). (1) An additional type-specimen of the opposite sex to the original type-specimen. (2) An animal or plant fossil selected as a species or subspecies, illustrating morphological details not shown in the holotype.

allowable deficiencies (*Aero.*). Aircraft systems or certain items of their equipment, tabulated in the flight or operating manual, which even if unserviceable will not prevent an aircraft from being flown or create a hazard in flight.

allowances (*Aero.*). In airline terminology, fuel reserves are frequently referred to as allowances, and are usually specified as time factors under certain conditions, as distance plus descent, or as a percentage (by weight or volume) of the cruising fuel for a given stage.

allowed band (*Phys.*). Range of energy levels permitted to electrons in a molecule or crystal. These may or may not be occupied.

allowed transition (*Phys.*). Electronic transition between energy levels which is not prohibited by any quantum selection rule.

alloxan (*Chem.*). Mesoxalylurea. A cyclic ureide $(CO)_4(NH)_2$. Oxidation product of uric acid obtained by treatment with dilute nitric acid. Used to destroy pancreatic islet cells to produce diabetes mellitus experimentally.

alloy (*Chem.*). A mixture of a metal and some other element(s), e.g. steel, heated up and cooled to a solid (the alloy). (*Met.*) Metal prepared by adding other metals or nonmetals to a basic metal to secure desirable properties. Solid solution in which maximum difference in radius of component atoms is below 15%. Lower m.p., decreased electrical conductivity, and increased hardness are characteristics.

alloy cast-iron (*Met.*). Cast-iron containing alloying elements. Usually some combination of nickel, chromium, copper, and molybdenum. These elements may be added to increase the strength of ordinary irons, to facilitate heat treatment, or to obtain martensitic, austenitic, or ferritic irons.

alloy junction (*Electronics*). One formed by alloying impurity to an otherwise pure (better than 0·01 ppm impurity) semiconductor crystal.

alloy reaction limit (*Met.*). Concentration in alloy of a specific component, below which corrosion occurs in a given environment.

alloy (or special) steel (*Met.*). A steel to which elements not present in carbon steel have been added, or in which the content of manganese or silicon is increased above that in carbon steels. See **high speed steel, nickel steel, stainless steel**; also **chromium.**

all-pass network (*Telecomm.*). One which introduces a specified phase-shift response without appreciable attenuation for any frequency.

all-sliming process (*Min. Proc.*). Original term for treatment of Rand gold ores by agitation in dilute cyanide, which required grinding sufficiently fine to keep the particles in suspension in the leaching solution. Term in diminishing use, definition by mesh sizes having largely superseded it.

Allström relay (*Elec. Eng.*). See **relay types.**

alluaudite (*Min.*). A phosphate of sodium, iron and manganese, crystallizing in the monoclinic system.

alluring coloration (*Zool.*). Resemblance of an animal to some nonliving object, plant, or other animal, for the purpose of attracting its prey.

alluvial clay (*Geol.*). Sediment of the clay grade which has been transported by rivers from the place of its origin, as distinct from that which has originated *in situ*.

alluvial cone or **fan** (*Geog.*). Sediment usually laid fan-wise by a rapidly flowing stream as it enters an open valley. It is sometimes many miles across and a continuous plain made up of the fan deposits of several neighbouring streams is known as the **Piedmont alluvial plain.**

alluvial deposits (*Geol., etc.*). Earth, sand, gravel, and other material which has been carried in suspension by river or floods, and is deposited at places where for some reason the velocity of flow is insufficient to maintain the materials in suspension. Such sediment of geologically recent age, in the process of deposition by existing rivers, is known as **alluvium.**

alluvial mining (*Mining*). Exploitation of alluvial or placer deposits. Minerals thus extracted include tin, gold, gemstones, rare earths, platinum. Term embraces beach deposits, eluvials, riverine, and offshore workings.

alluvial values (*Mining*). Those shown by panning or assay to be recoverable from an alluvial deposit.

alluvium (*Geol., etc.*). See **alluvial deposits.**

all-watt motor (*Elec. Eng.*). A type of induction motor in which a Scherbius type of phase-advancer is incorporated as an integral part, and which, therefore, operates at almost unity power-factor.

allyl alcohol (*Chem.*). 1-hydroxy prop-2-ene $H_2C{=}CH{\cdot}CH_2OH$, an unsaturated primary alcohol, present in wood spirit, made from glycerine and oxalic acid. M.p. −129°C, b.p. 96°C, rel. d. 0·85, of very pungent odour; an intermediate for organic synthesis.

allyl chloride (*Chem.*). $Cl.CH_2{\cdot}CH{=}CH_2$. Important intermediate in the manufacture of synthetic glycerol. Formed by the high temperature (400–600°C) chlorination of propene.

allyl group (*Chem.*). The unsaturated monovalent aliphatic group $H_2C{=}CH{\cdot}CH_2$—.
allyl resin (*Chem.*). One formed by the polymerization of chemical compounds of the *allyl group* (q.v.).
allyl sulphide (*Chem.*). Oil of garlic; b.p. 139°C; $(CH_2{=}CH{\cdot}CH_2)_2S$. Colourless liquid, found in garlic and largely responsible for its odour. It possesses antiseptic qualities.
allylene (*Chem.*). Propyne, methyl ethyne, $CH_3 . C{\equiv}CH$.
Almagest (*Astron.*). The Arabic form of the title of Claudius Ptolemy's great astronomical treatise, 'The Mathematical Syntaxis', written in Greek about A.D. 140.
Almalec (*Elec. Eng.*). An aluminium alloy having a higher tensile strength than pure aluminium; used for overhead transmission-line conductors.
almandine (*Min.*). Iron-aluminium garnet, occurring in mica-schists and other metamorphic rocks. Commonly forms well-developed crystals, often with 12 or 24 faces.
almandine spinel (*Min.*). See **ruby spinel.**
Almen-Nylander test (*Chem.*). A test for the presence of sugar, consisting in the reduction of a bismuth salt solution to metallic bismuth.
almond oil (*Chem.*). Used for fruit essences, in perfumery and soap making; two grades are known: *bitter almond oil* and *sweet almond oil.*
almucantar (*Astron.*). A small circle of the celestial sphere parallel to the horizontal plane. The term is also applied to an instrument for measuring altitudes and azimuths.
Alnico (*Met.*). U.S. TN for a high-energy permanent magnet material, an alloy of aluminium, nickel, cobalt, iron and copper.
aloin (*Chem.*). Crystalline powder obtained by extraction of aloes; used medically as a purgative or an aperient. Varies in chemical composition according to variety of aloe from which obtained.
aloin test (*Chem.*). A test for the detection of blood in the faeces, consisting in adding an alcoholic aloin solution to an ether extract of faeces and treating the mixture with hydrogen peroxide or ozonized turpentine. The presence of blood will create a cherry-red colour in the ether layer.
alopecia (*Med.*). Baldness.
alopecia areata (*Med.*). A condition in which the hair falls out in patches, leaving smooth, shiny, bald areas.
Aloxite (*Chem.*). TN designating a proprietary fused alumina and associated products.
alpaca (*Textiles*). (1) Fine, silky, hair fibre, 7–14 in. (175-350 mm) in length, from a Peruvian goat in black, white, or brown shades. (2) A type of medium-weight dress or lining cloth woven from a warp of recovered wool and a cotton weft, plain or twill.
Alpax (*Met.*). TN for aluminium-silicon alloy, containing about 13% of silicon. Used mainly for castings. Has good casting properties and corrosion resistance, low relative density (2·66) and satisfactory mechanical properties. Tensile strength 150–190 MPa. Also **Silumin.**
alpha (α) **activation** (*Chem.*). The influence of organic radicals and groups in directing the course of chemical reactions, elg., the carbonyl function in ketones which leads to 2-halogenation predominating.
alpha alloys (*Met.*).
 -beta brass. Copper-zinc alloy containing 38–46% (usually 40%) of zinc. It consists of a mixture of the α-constituent (see **alpha brass**) and the β-constituent (see **beta brass**).
 -brass. A copper-zinc alloy containing up to 38% of zinc. Consists constitutionally of a solid solution of zinc in copper. Commercial alpha brasses of several compositions are made. The most widely used contain 30–37%; others contain 5–20% of zinc. All are used mainly for cold-working.
 -bronze. A copper-tin alloy consisting of the alpha solid solution of tin in copper. Commercial forms contain 4 or 5% of tin. This alloy, which differs from gun metal and phosphor bronze in that it can be worked, is used for coinage, springs, turbine blades, etc.
alphabetic telegraphy (*Teleg.*). Those forms of telegraphy in which literal/numeric text is transferred telegraphically by signals representing individual characters or groups of characters.
Alpha Centauri (*Astron.*). See **Proxima Centauri.**
alpha chamber (*Nuc.*). Ionization chamber for measurements of α-radiation intensity.
alpha counter (*Nuc.*). Tube for counting α-particles, with pulse selector to reject those arising from β- and γ-rays.
alpha counter tube (*Nuc.*). See **alpha chamber.**
alpha cut-off (*Electronics*). Frequency at which the current amplification of a transistor has fallen by more than 3 dB (0·7) of its low-frequency value. See also **cut-off frequency.**
alpha decay (*Nuc.*). Radioactive disintegration resulting in emission of α-particle. Also **alpha disintegration.**
alpha-decay energy, q_A (*Nuc.*). The sum of the kinetic energy of the α-particle (E_A) and that of the recoil of the product atom (E_P), given by

$$E_A=\left(\frac{m_P}{m_P+m_A}\right)q_A \text{ and } E_P=\left(\frac{m_A}{m_P+m_A}\right)q_A,$$

where m_A and m_P are the respective masses of the α-particle and the recoil atom. Also **disintegration energy.**
alpha emitter (*Nuc.*). Natural or artificial radioactive isotope which disintegrates through emission of α-*rays.*
alpha female (*Zool.*). Among Ants (*Formicoidea*), the normal female if it co-exists with the *beta female* (q.v.).
alpha-helix (*Biochem.*). Right-hand helical structure which is the stable configuration, due to internal hydrogen bonding, of many important biological molecules, notably DNA, and some polypeptides.
alpha iron (*Met.*). The polymorphic form of iron, stable below 906°C. Has a body-centred cubic lattice, and is magnetic up to 768°C.
alphanumeric, alphameric (*Comp.*). Said of system which uses character codes which include both alphabetical and numerical.
alphanumeric printers (*Print.*). Printers using characters of the alphabet and the numbers from 0 to 9.
alpha-particle (*Nuc.*). Particle identical in measured properties with high-speed atoms of helium, stripped of their electrons and hence doubly positively charged, ejected from natural or artificial radioisotopes. It is composed, therefore, of two protons and two neutrons. Often written **α-particle.**
alpha pulp (*Paper*). Wood pulp processed so that only a very small percentage of hemicellulose remains. Also called **dissolving pulp.**
alpha-radiation (*Nuc.*). Alpha particles emerging from radioactive atoms.
alpha-rays (*Nuc.*). Streams of α-*particles* (q.v.).
alpha-ray spectrometer (*Nuc. Eng*). Instrument for investigating energy distribution of α-particles emitted by radioactive source.
alpha resin (*Brew.*). See **humulone.**

alpha rhythm (*Physiol.*). The regular electro-encephalographic pattern of ca. 10 Hz obtained from a waking, inactive individual. The pattern is disrupted by such changes as falling asleep, concentrating, etc.

Alphatype (*Typog.*). A *filmsetting system* with electric typewriter as keyboard, recorder unit producing 10-channel magnetic tape, and exposure unit of darkroom design.

alpha wave (*Physiol.*). The principal slow wave (frequency approx. 10 Hz) produced by the human brain and recorded as the electro-encephalogram (EEG).

Alpine (*Geol.*). Pertaining to that period of earth movement in the Tertiary geological period when the Alps and other existing mountain chains came into existence.

Alpine air-swept sieve (*Powder Tech.*). One in which air is blown through a rectangular orifice which sweeps around just below the surface of the sieve. The rest of the sieving system is maintained at a pressure below atmospheric, so that the jet simultaneously cleans the sieve surface and aids the suction of particles through the other parts of the sieve.

alstonite (*Min.*). A double carbonate of calcium and barium found, often with *witherite*, in lead deposits, as on Alston Moor, Northumberland.

altar (*Civ. Eng.*). One of the steps in the stepped face of a dry dock wall; used to hold the ends of the supports which prop the vessel in its upright position.

altar tomb (*Arch.*). A raised tomb or monument usually standing detached or in a position against a wall, and sometimes supporting an effigy. In appearance it resembles a solid altar, but it is never used as one.

altazimuth (*Surv.*). An instrument similar to the *theodolite* (q.v.), but generally larger and capable of more precise work.

alterative (*Med.*). Tending to alter favourably the processes of nutrition: a medicine which does this.

alternant (*Maths.*). A determinant whose, elements are functions of $x_1, x_2, \ldots x_n$, such that interchanging the variables x_i and x_k interchanges the ith row and the kth rows and thus changes the sign of the determinant.

Alternaria solani (*Bot.*). See **early blight.**

alternate (*Bot.*). Said of leaves and branches which are neither opposite nor in whorls, but placed singly on the parent axis.

alternate aerodrome (*Aero.*). One designated in a *flight plan* at which a pilot will land if prevented from alighting at his destination.

alternate angles (*Maths.*). If a transversal cuts two straight lines, the alternate angles lie on either side of it, each angle having one of the lines as an arm, the other arm being in both cases the transversal. If the angles are between the two lines, they are interior alternate angles. If neither of the angles is interior, they are said to be exterior alternate angles.

alternate scanning (*TV*). Method of scanning developed to compensate for the difference in frame frequencies, when televising from cinematograph film.

alternating cleavage (*Zool.*). See **spiral cleavage.**

alternating current (*Elec. Eng.*). Generally abbreviated to **a.c.** Electric current whose flow alternates in direction; the time of flow in one direction is a half-period, and the length of all half-periods is the same. The normal waveform of a.c. is sinusoidal, which allows simple vector or algebraic treatment. Provided by **alternators** or electronic **oscillators.**

alternating function (*Maths.*). A function of two or more variables such that interchanging any two changes the sign but not the absolute value of the function, e.g., $f(x, y, z) = -f(y, x, z)$. Also called **antisymmetric function.**

alternating-gradient focusing (*Electronics*). The principle follows the optical analogy whereby a series of alternate converging and diverging lenses may lead, under suitable conditions, to a net focusing effect since the rays will strike the diverging lenses nearer to the axis. Using magnetic or electrostatic lenses, the idea has been used for the design of electron synchrotrons and ion linear accelerators.

alternating light (*Ships*). A navigation mark identified during darkness by a light showing alternating colours. See **flashing light, occulting light.**

alternating series (*Maths.*). A series whose terms are alternately positive and negative. If the terms of an alternating series decrease monotonically then a necessary and sufficient condition for convergence is that they tend to zero.

alternating stress (*Mech.*). The stress induced in a material by a force which acts alternately in opposite directions.

alternation of generations (*Bot.*). The regular alternation in the life history of two types of plants, differing in their nuclear constitution, and often in their morphology. See **antithetic alternation of generations, homologous alternation of generations.** (*Zool.*) The occurrence, in the typical life cycle of a species, of two or more different forms, produced in a different manner, generally an alternation of a sexually-produced form with an asexually-produced form; as in the Hydrozoan genus *Obelia.*

alternative inheritance (*Gen.*). The kind of inheritance in which one or both of any pair of contrasted characters is/are present in the hybrid, and may be obtained unchanged in offspring from the hybrid.

alternative routing (*Teleph.*). The manual or automatic diversion, to a prearranged secondary route, of traffic which originates at an instant when the primary route is not available.

alternator (*Elec. Eng.*). A type of a.c. generator, driven at a constant speed corresponding to the particular frequency of the electrical supply required from it. Also **synchronous generator.**

alterne (*Bot.*). A sudden change in the nature of the plant covering of a district, related to abrupt change in environmental conditions, e.g., soil.

altimeter (*Aero., Phys., etc.*). An aneroid barometer used for measuring altitude by the decrease in atmospheric pressure with height. The dial of the instrument is graduated to read the altitude directly in feet or metres; the zero being set at ground level. See **encoding, radio-, recording-.**

altitude (*Aero., Surv.*). The height in feet or metres above sea level. For precision, in determining the performance of an aircraft, this must be corrected for the deviation of the meteorological conditions from that of the standardized atmosphere (*International Standard Atmosphere*). See also **cabin-, pressure-.** (*Astron.*) The angular distance of a heavenly body measured on that great circle which passes, perpendicular to the plane of the horizon, through the body and through the zenith. It is measured positively from the horizon to the zenith, from 0° to 90°. (*Maths.*) (1) Of a point above (or below) the horizon: its angular distance above (or below) the horizon. (2) Of a geometrical figure, e.g., a triangle, a cone: its vertical height.

altitude level (*Surv.*). Sensitive spirit level which

ensures that theodolite is truly horizontal with respect to the telescope when vertical angles are measured.

altitudes by barometer (*Phys.*). See **Babinet's formula for altitude.**

altitude switch (*Aero.*). A switching device generally comprising electrical contacts, actuated by an aneroid capsule which in turn is deflected by change in atmospheric pressure. The contacts are adjusted to make or break a warning circuit at the pressure corresponding to a predetermined altitude.

altitude valve (*Aero.*). A manually- or automatically-operated valve fitted to the carburettor of an aero-engine for correcting the mixture-strength as air density falls with altitude.

Altmann's granules (*Cyt.*). See **mitochondria.**

alto-cumulus cloud (*Meteor.*). Rounded masses of cloud in groups or lines, at 3000–7500 m.

altometer (*Surv.*). See **theodolite.**

alto-stratus cloud (*Meteor.*). A dense sheet of cloud of a grey or bluish colour, sometimes forming a compact mass of dull grey colour and fibrous structure; occurs at 3000–7500 m.

altrices (*Zool.*). Birds whose young are hatched in a very immature condition, generally blind, naked, or with down feathers only, unable to leave the nest, fed by the parents; e.g., the Perching Birds, *Passeriformes.*

altrose (*Chem.*). An aldohexose, an optical stereoisomer of glucose.

aludels (*Chem.*). Pear-shaped vessels connected in long rows for the condensation of the mercury vapour liberated from the roasting of cinnabar, i.e., mercury(II)sulphide.

Aludur (*Elec. Eng.*). An aluminium alloy (containing magnesium) which is used for overhead transmission-line conductors on account of its having a higher tensile strength than aluminium.

alula (*Zool.*). (1) In *Diptera*, a free lobe on the posterior margin of the wing near the base. (2) In birds, **bastard wing.**

alumina (*Min.*). The oxide of aluminium, occurring as the mineral corundum. When the compositions of silicate minerals are stated in terms of the component oxides, alumina is found to be important in such groups as the feldspars, feldspathoids, and micas.

aluminate (*Chem.*). Salt of aluminic acid, H_3AlO_3, a tautomeric form of aluminium hydroxide, which acts as a weak acid. Ortho-aluminates have the general formula M_3AlO_3 or $M_3Al(OH)_6$, and meta-aluminates, $MAlO_2$ or $MAl(OH)_4$, where M is a monovalent metal. *Sodium aluminate*, Na_3AlO_3, is used as a coagulant in water purification and softening.

aluminium (*Chem.*). Silver-white metallic element, forming a protective film of oxide. Symbol Al, at. no. 13, r.a.m. 26·9815, m.p. 659·7°C, b.p. 1800°C, rel. d. 2·58. Obtained from bauxite, it has numerous uses and is the basis of light alloys for use in, e.g., structural work; alloyed with silicon for transformer laminations, and iron and cobalt in many types of permanent magnet. Polished aluminium reflects well beyond the visible spectrum in both directions, and does not corrode in sea water. Foil aluminium is much used for capacitors. The metal can be used as a window in X-ray tubes and as sheathing for reactor fuel rods. Aluminium is produced on a large scale where electric power is cheap, e.g., hydroelectric at Kitimat, when bauxite is electrolysed in fused cryolite. ^{28}Al and ^{29}Al are strong γ-ray emitters of very short life. As an electrode in gas-discharge tube, it does not sputter like other metals. In U.S., **aluminum.**

aluminium alloys (*Met.*). Those in which aluminium is the basis (i.e., predominant) metal, e.g., aluminium-copper and aluminium-silicon alloys, Duralumin, *y*-alloy, etc. Also called **light alloys.**

-brass. Brass to which aluminium has been added, to increase its resistance to corrosion. Used for condenser tubes. Contains 1–6% Al, 24–42% Zn, 55–71% Cu.

-bronze. Copper-aluminium alloys which contain 4–11% aluminium, and may also contain up to 5% each of iron and nickel or ½% tin. These alloys have high tensile strength, are capable of being cast or cold-worked, and are resistant to corrosion.

aluminium anode cell (*Elec. Eng.*). One with an aluminium anode immersed in an electrolyte which does not attack aluminium. The cathode may also be of aluminium or some other metal, e.g., lead. Such cells can be used as rectifiers or as high-capacitance capacitors. See **electrolytic capacitor.**

aluminium antimonide (*Electronics*). A semiconducting material used for transistors up to temperature of 500°C.

aluminium arrester (*Elec. Eng.*). A lightning arrester made up of a number of aluminium trays containing electrolyte and arranged to form a number of electrolytic cells in series. An insulating layer is formed on each tray under normal voltage conditions, but this punctures and allows current to pass if the voltage exceeds a certain value. Also called **electrolytic arrester.**

aluminium foil. Very thin aluminium sheet. (*Build.*) Used in conjunction with fibre board for insulation purposes in walls or roofs.

aluminium-steel cable (*Elec. Eng.*). See **steel-cored aluminium.**

aluminized screen (*Electronics*). Cathode-ray tube fluorescent screen the conductivity of which has been improved by application of thin aluminium film. This gives greater contrast, and avoids **ion burn.**

aluminon (*Chem.*). Ammonium aurine-tricarboxylate. Reagent for the colorimetric detection and estimation of aluminium, with which it forms a characteristic red colour.

alumino-silicates (*Chem., Min.*). Compounds of alumina, silica, and bases, with water of hydration in some cases. They include clays, mica, constituents of glass, porcelain, etc.

aluminothermic process (*Chem.*). The reduction of metallic oxides by the use of finely divided aluminium powder. An intimate mixture of the oxide to be reduced and aluminium powder is placed in a refractory crucible; a mixture of aluminium powder and sodium peroxide is placed over this and the mass fired by means of a fuse or magnesium ribbon. The aluminium is almost instantaneously oxidized, at the same time reducing the metallic oxide to metal. This process, also known as the *thermite process*, is used especially for the oxides of metals which are reduced with difficulty (e.g., titanium, molybdenum). On ignition, the mass may reach a temperature of 3500°C. Magnesium incendiary bombs have thermite as the igniting agent.

aluminous cement (*Civ. Eng.*). A cement containing 30–50% of lime, 30–50% of alumina, and not more than 30–50% of silica, iron oxide, etc. The aluminous cements are less susceptible than ordinary Portland cements to low temperature during setting, and to the action of sea-water and acids; and they harden rapidly, due to heat generated in setting.

aluminum (*Met.*). See **aluminium.**

alum leather (*Leather*). Light skins tanned white with alum and salt, or with aluminium salts, flour, and yolk of egg; used for gloves and clothing. This process is known as *tawing*.

alums (*Chem.*). A large number of isomorphous compounds whose general formula is:
$R'R'''(SO_4)_2 \cdot 12H_2O$; or,
$R'_2SO_4 \cdot R_2'''(SO_4)_3 \cdot 24H_2O$,
where R′ represents an atom of a univalent metal or radical—potassium, sodium, ammonium, rubidium, caesium, silver, thallium; and R‴ represents an atom of a tervalent metal—aluminium, iron, chromium, manganese, thallium. See also **pseudoalums.**

alum shale (*Geol.*). Shale containing potassium alum, formerly worked at Whitby in Yorkshire as a source of the latter.

alumstone (*Min.*). See **alunite.**

alunite (*Min.*). A hydrous sulphate of aluminium and potassium, resulting from the alteration of acid igneous rocks by solfataric action; used in the manufacture of alum. Also **alumstone.**

alunogen (*Min.*). Hydrous sulphate of aluminium, occurring as a white incrustation or efflorescence formed in two different ways: either by volcanic action, or by the decomposition of pyrite (iron-sulphide) in carbonaceous or alum shales.

alveola (*Zool.*). See **alveolus.**

alveolar, alveolate (*Bot., Zool.*). Having pits over the surface, and resembling honeycomb.

alveolar gland (*Zool.*). In Insects, compound racemose exocrine gland. It is composed of acini.

alveolar layer (*Zool.*). In ciliate *Protozoa*, a layer of ectoplasm composed of minute regular vacuoles lying immediately beneath the pellicle.

alveolar theory (*Zool.*). The theory that protoplasm is composed of bubbles of a more viscid constituent containing the more fluid constituents.

alveolate (*Zool.*). (1) Said of sessile pedicellariae, in which the jaws are inserted into an *alveolus*, or depression, in the basal plate. (2) See **alveolar.**

alveolitis (*Med.*). Inflammation of the pulmonary alveoli, occurring in paraquat poisoning.

alveolus (*Bot.*). A pit in the surface of a plant member. (*Zool.*) A small pit or depression on the surface of an organ: the cavity of a gland: a small cavity of the lungs: in higher Vertebrates, the tooth socket in the jaw bone: in *Echinodermata*, part of Aristotle's lantern, one of five pairs of grooved ossicles which grasp the teeth: in *Gastropoda*, the glandular end-portion of the tubules of the digestive gland, secreting enzymes.

alveus (*Zool.*). A thin layer of white fibres on the surface of the hippocampus in Eutherian mammals.

alvin (*Chem.*). Mixture of two protein toxins obtained from the seeds of jequirity, *Abrus precatorius*, an Indian plant. Highly toxic, it causes agglutination of red blood corpuscles. (Not to be confused with *abrin*, obtained from the seeds of the same plant.)

alypin hydrochloride (*Chem., Med., Vet.*). See **amydricaine hydrochloride.**

Alzheimer's disease (*Med.*). A rare degenerative brain disease, manifesting itself in premature ageing, with speech disorder.

Am (*Chem.*). A symbol for: (1) the ammonium ion NH_4^+—; (2) the pentyl (amyl) radical C_5H_{11}—; (3) *americium*.

amacrine (*Cyt.*). Having no axon; branched retinal cells in the inner nuclear layer.

amagat (*Phys.*). The unit of density of a gas at 0°C and 1 atmos. pressure; usually 1 amagat = 1 mole/22·4 dm³.

Amagat's law of combining volumes (*Chem.*). The volume of a mixture of gases is equal to the sum of the volumes of the different gases, as existing each by itself at the same temperature and pressure.

amalgam (*Chem.*). The solution of a metal in mercury. (*Min. Proc.*) The pasty amalgam of gold and mercury, about 1/3 gold by weight, obtained from the plates in a mill treating gold ores.

amalgamating table (*Min. Proc.*). Amalgamated plate. Flat sheet of metal to which mercury has adhered to form a thin soft film, used to catch metallic gold as mineral sands are washed gently over it.

amalgamation pan (*Min. Proc.*). Circular cast-iron pan in which finely-crushed gold-bearing ore or concentrate is ground with mercury, the valuable metal thus being amalgamated before separate retrieval.

amalgam barrel (*Min. Proc.*). Small ball mill used to regrind gold-bearing concentrates, and then give them prolonged rubbing contact with mercury.

amalgam retort (or **still**) (*Met.*). Iron vessel in which the mercury is distilled off from gold or silver amalgam obtained in amalgamation.

amaranth (*For.*). See **purpleheart.**

amaurosis (*Med.*). Blindness due to a lesion of the retina, optic nerve, or optic tracts, usually associated with idiocy. See **Tay-Sachs' disease.**

amazia (*Med.*). Nondevelopment of the mammary glands in the female.

amazon (*Textiles*). Dress fabric woven either 2 × 1 warp twill or 5-end warp sateen from a fine worsted (usually merino quality) warp and a woollen weft. On the raising machine, the fibres are laid flat on the surface to impart a lustrous appearance.

amazons (*Zool.*). Among Ants (*Formicoidea*), obligatory slavemakers, entirely dependent on the workers of the subjugated species. See also **dulosis.**

amazonstone, amazonite (*Min.*). A green variety of microcline, sometimes cut and polished as a gemstone.

amber (*Min.*). A fossil resin, found on the shores of the Baltic Sea, containing succinic acid in addition to resin acid and volatile oils. Amber is used, *inter alia*, as a base for photographic varnishes. See **succinite.**

ambergris (*Zool.*). A greyish white fatty substance with a strong but agreeable odour, obtained from the intestines of diseased sperm whales (*Physeter macrocephalus*); sometimes found floating on the surface of the sea. It is used in perfumery as a fixative; on suitable treatment it yields ambreic acid.

amberoid (*Min.*). See **ambroid.**

ambetti (*Glass*). Glass containing minute opaque specks. Also **ambitty.**

ambi-. Latin form of **amphi-** (q.v.).

ambiens (*Zool.*). In certain Birds, a muscle arising from the pectineal process of the pelvis and passing along the inner surface of the thigh to the head of the flexor muscle of the second and third toes; its action causes the toes to grasp the perch.

ambient illumination (*Photog., etc.*). Background uncontrollable light level at a location.

ambient noise (*Acous.*). The noise existing in a room or any other environment, e.g., the ocean. Also called **room noise.**

ambient noise level (*Acous.*). Random uncontrollable and irreducible noise level at a location or in a valve or circuit.

ambient temperature (*Elec. Eng.*). Term used to denote temperature of surrounding air.

ambiguity (*Automation*). The condition of a servo-mechanism in which more than one null point is sought.

ambiguous (*Bot.*). Of doubtful position, or uncertain origin, taxonomically, or anatomically.

ambiophony (*Acous.*). Reproduction of recorded dead studio sound superimposed on itself after very small time delays, thereby giving it atmosphere. This is used in particular to give warmth to orchestral music.

ambiparous bud (*Bot.*). A bud containing young vegetative leaves and young flowers.

ambipolar (*Electronics*). Said of any condition which applies equally to positive and negative ions (including electrons) in a plasma.

ambisexual, ambosexual (*Biol.*). Pertaining to both sexes; activated by both male and female hormones.

ambital (*Zool.*). In *Asteroidea*, pertaining to that part of the skeleton consisting of the inter-ambulacral and antambulacral plates: in *Ophiuroidea*, pertaining to the outer skeleton of the arm.

ambitty (*Glass*). See **ambetti.**

ambitus (*Zool.*). Margin, outer edge: in *Echinoidea*, the outline of the shell viewed from the apical pole.

Amble (*Comp.*). Technique for analogue computing certain differential equations, devised by O. Amble.

amblygonite (*Min.*). Fluophosphate of aluminium and lithium, a rare white or greenish mineral, crystallizing in the triclinic system and found in pegmatites.

amblyopia (*Med.*). Dimness of vision, from the action of noxious agents on the optic nerve or retina, or from hysteria.

ambocepter (*Med.*). A specific antibody or immune body necessary for the action of complement on a toxin or a red corpuscle; a body that is thought to join the complement to the animal or bacterial cell; a lysin. See **side-chain theory.**

ambroid (*Min.*). A synthetic amber formed by heating and compressing pieces of natural amber too small to be of value in themselves. Also called **amberoid, pressed amber.**

ambroin (*Elec.*). A moulded insulating material prepared from copal and silicates.

ambrosia (*Zool.*). Certain Fungi which are cultivated for food by some Beetles (members of the family *Scolytidae*, Ambrosia Beetles): the pollen of flowers collected by social Bees and used in the feeding of the larvae.

ambulacra (*Zool.*). In *Echinodermata*, the radial bands of locomotor tube feet. *adj.* **ambulacral.**

ambulacral grooves (*Zool.*). Radially-arranged grooves containing the tube feet in *Asteroidea.*

ambulacralia (*Zool.*). In *Echinodermata*, the plates of the ambulacral skeleton, through which the locomotor tube feet protrude. See **ambulacral ossicles.**

ambulacral ossicles (*Zool.*). Calcareous plates running down the ambulacral grooves in a double row in *Asteroidea.*

ambulator (*Surv.*). See **perambulator.**

ambulatory (*Arch.*). Covered promenade area, particularly the curved aisle behind the altar in an apse. (*Zool.*) Having the power of walking; used for walking.

ameiosis (*Cyt.*). Nonpairing of the chromosomes in synapsis (meiosis).

amelia (*Med.*). A congenital abnormality in which one or more limbs are completely absent.

amelification (*Zool.*). The formation of enamel.

ameloblast (*Zool.*). A columnar cell forming part of a layer immediately covering the surface of the dentine, and secreting the enamel prisms in the teeth of higher Vertebrates.

amenorrhoea, amenorrhea (*Med.*). Absence or suppression of menstruation.

amensalism (*Biol.*). A category of interaction between two species in which one population is inhibited by the other, but the latter is not affected by the former, e.g., the interaction of *Penicillium* and Bacteria when the former produces the antibiotic penicillin. Cf. *commensalism.*

ament (*Med.*). One suffering from *amentia*; a mentally deficient person.

amentaceous (*Bot.*). Bearing catkins.

amentia (*Med.*). Mental deficiency: failure of the mind to develop normally, whether due to inborn defect, or to injury or disease.

Amentiferae (*Bot.*). The catkin-bearing trees (unacceptable as a taxonomic group).

amentum (*Bot.*). A *catkin adj.* **amentiform.**

American bond (*Build.*). A form of bond in which every 5th or 6th course consists of headers, the other courses being stretchers. Very much used because it can be quickly laid.

American caisson (*Civ. Eng.*). See **stranded caisson.**

American filter (*Met., Min. Proc.*). See **disk filter.**

American red gum (*For.*). *Liquidambar styraciflua*, a tree of silky surface and irregular grain. Not very durable; used mainly for furniture, fittings, and panelling. See **satin walnut.**

American Standard Wire Gauge. See **Brown and Sharpe Wire Gauge.**

American water turbine (*Eng.*). See **mixed-flow water turbine.**

americium (*Chem.*). Transuranic element, symbol Am, at. no. 95, half life 475 years. Of great value as a long life α-particle emitter, free of criticality hazards and γ-radiation, e.g., in (αn) laboratory neutron sources.

ameristic (*Bot.*). Unable to develop completely, owing to poor nourishment.

amerospore (*Bot.*). A dry, single-celled fungus spore.

amesite (*Min.*). A variety of septechlorite rich in magnesium and aluminium.

ametabolic (*Zool.*). Having no obvious metamorphosis.

amethocaine hydrochloride (*Med.*). $C_{15}H_{24}N_2O_2$, HCl = 300·8. A potent local anaesthetic.

amethyst (*Min.*). A purple form of quartz, used as a semiprecious gemstone.

amianthus (*Min.*). A fine silky asbestos.

amicable numbers (*Maths.*). See **friendly numbers.**

amicrons (*Chem.*). Particles, of the order of nanometres, invisible in the ultramicroscope; they act as nuclei for larger submicron particles.

amide plant (*Bot.*). A plant which forms asparagine or glutamine from amino acids.

amides (*Chem.*). Alkanamides. A group of compounds in which the hydroxyl of the carboxyl group of acids has been replaced by the amide group $-NH_2$. They may be regarded as ammonia derivatives in which the hydrogen has been replaced by an acyl group. In primary amides one, secondary amides two, and tertiary amides three, hydrogens have been replaced.

amidine dyestuffs (*Chem.*). Term sometimes used for azo dyes.

amidines (*Chem.*). Compounds derived from amides R·CO·NH$_2$, R·CO·NHR′, and R·CO·

NR'_2, in which the oxygen has been replaced by the divalent imido residue NH or NR. The amidines are crystalline bases forming stable salts, but readily hydrolysed.

amido group (*Chem.*). $CO{\cdot}NH_2$ group in amides, when replacing the hydroxyl in a carboxyl group.

amidol (*Photog.*). Developing agent, 2 : 4-diamino-phenol, fast-acting, but rapidly deteriorating.

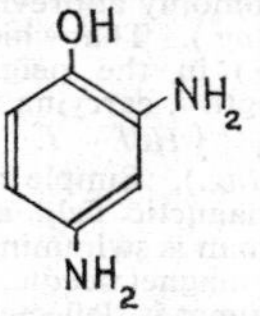

(Structure: benzene ring with OH, NH_2 at position 2, NH_2 at position 4.)

amidopyrine (*Chem.*). Colourless, crystalline compound; m.p. 108°C, with the structure:

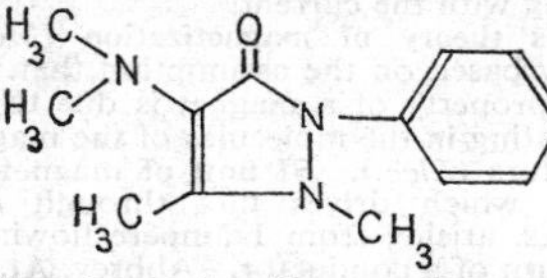

(Structure: pyrazolone ring bearing $(H_3C)_2N$–, C=O, N–phenyl, N–CH_3 and H_3C– groups.)

amines (*Chem.*). Aminoalkanes. Organic derivatives of ammonia NH_3 in which one or more hydrogen atoms are replaced by organic radicals. According to the extent of substitution, amines are classified as primary amines or amino bases, secondary amines or imino bases, tertiary amines or nitrile bases.

aminoacetic acid (*Chem.*). Aminoethanoic acid. NH_2CH_2COOH (glycocoll, glycine), m.p. 230°C, colourless crystals, sweet soluble in water, slightly in alcohol but not in ether.

amino acids (*Chem.*). Aminoalkanoic acids. A group of organic acids in which a hydrogen atom of the hydrocarbon (alkyl) radical is exchanged for the amino group. Some 24 2-amino acids (i.e., having the amino group attached to the carbon next to the carboxyl group) occur in proteins. All except glycine are optically active, and those obtained from animal and plant proteins have the configuration of *l*-2 amino-propionic acid (*l*-alanine). Certain *d*-amino acids have been obtained from micro-organisms. Some amino acids cannot be synthesized in the animal body and therefore must be supplied in food. These are known as *essential amino acids*; they vary somewhat for different species, but for man and most experimental animals are: leucine, *iso*-leucine, valine, threonine, methionine, lysine, histidine, tryptophan, phenylalanine, arginine.

aminoaldehydic resins (*Plastics*). See **urea resins.**

4-aminodiphenyl (*Chem.*). $C_6H_5C_6H_4NH_2$ forming colourless plates, used in the dye and paint industries; known carcinogen.

amino group (*Chem.*). The grouping $-N{<}^{R}_{R'}$, where R and R′ may be hydrogen or organic radicals. If R and R′ are both H, it is a primary amine; if R is H, a secondary amine; if R and R′ are both organic radicals, a tertiary amine.

aminoplastic resin (*Plastics*). One derived from the reaction of urea, thiourea, melamine, or allied compounds (e.g., cyanamide polymers and diaminotriazines) with aldehydes, particularly formaldehyde (methanal).

amitosis, amitotic division (*Cyt.*). Direct division of the nucleus by constriction, without the formation of a spindle and chromosomes; direct nuclear division, occurring in the meganuclei of the *Ciliophora*. Cf. *mitosis, meiosis.*

ammeter (*Elec. Eng.*). An indicating instrument for measuring the current in an electric circuit.

ammines (*Chem.*). Complex inorganic compounds which result from the addition of one or more ammonia molecules to a molecule of a salt or similar compound.

ammodyte (*Bot.*). A plant living in sandy places.

ammonal (*Chem.*). Explosive consisting of a mixture of ammonium nitrate, powdered aluminium and TNT.

ammonia (*Chem.*). NH_3. A colourless, pungent gas, b.p. −33·5°C, extremely soluble in water and very soluble in alcohol. Formed by bacterial decomposition of protein, purines, and urea. Obtained on a large scale from nitrogen obtained from the atmosphere and from the ammoniacal liquor in gas manufacture. Forms salts with most acids, and nitrides with metals. The liquefied gas is used in ice-making plants.

ammonia alum (*Chem.*). See **alums.**

ammonia clock (*Phys.*). An accurate clock which depends for its action on the frequency of vibration of the nitrogen atom in ammonia.

ammonia oxidation process (*Chem.*). An important process for producing nitric acid by the catalytic oxidation of ammonia gas with air on the surface of platinum-rhodium gauze at 900°C.

ammonia plant (*Bot.*). A plant which forms ammonia and organic acids from amino acids.

ammonia-soda process (*Chem.*). See **Solvay's-.**

ammonia synthesis (*Chem.*). See **Haber process.**

ammonite (*Geol.*). An extinct fossil cephalopod found in rocks of Mesozoic age, particularly characteristic of the Lias. Frequently coiled in a plane, or nearly plane, spiral.

ammonium (*Chem.*). The ion NH_4, which behaves in numerous respects like an atom of a monovalent alkali metal.

ammonium chloride (*Chem.*). See **sal ammoniac.**

ammonium dihydrogen phosphate (*Chem.*). Piezoelectric crystal used in microphones and other transducers; it can withstand a temperature higher than can Rochelle salt.

ammonium hydroxide (*Chem.*). NH_4OH. A solution of ammonia in water. Also **ammonia hydrate.**

ammonization (*Bacteriol., Bot.*). The conversion of the complex organic compounds present in decaying material into ammonia, by soil bacteria. First stage in nitrification in the soil.

ammonolysis (*Chem.*). Lyolysis in liquid ammonia solution.

ammono-system (*Chem.*). An ionic system in liquid ammonia.

ammoxidation (*Chem.*). Term used to describe the vapour phase reaction of certain hydrocarbons with ammonia in the presence of oxygen. Used industrially to manufacture nitriles, particularly acrylonitrile.

amnesia (*Med.*). Loss of memory. Common in dissociation states of hysteria. In a concussed patient *retrograde amnesia* is loss of memory of events immediately preceding the concussion.

amnion (*Zool.*). In *Nemertea*, the outer walls of the coalesced amniotic invaginations: in Insects, the inner cell envelope covering and arising from the edge of the germ band: in higher Vertebrates, one of the embryonic membranes, the inner fold of blastoderm covering the embryo, formed of ectoderm internally and somatic mesoderm externally: in *Echinodermata*, the roof of the amniotic invagination.

amnioserosa (*Zool.*). In *Diplura* and *Collembola*

(Insects), the extra-embryonic part of the blastoderm which, in these orders, is not differentiated into an amnion and a serosa.

Amniota (*Zool.*). Those higher Vertebrates which possess an amnion during development, i.e., Reptiles, Birds, and Mammals. *adj.* **amniote.**

amniotic cavity (*Zool.*). In *Amniota*, the space between the embryo and the amnion.

amniotic fluid (*Zool.*). Liquid filling the amniotic cavity.

amniotic folds (*Zool.*). Protrusions round the periphery of the blastoderm which give rise to the amnion and the chorion.

amniotic invagination (*Zool.*). In the pilidium larva of *Nemertea*, four ciliated invaginations of ectoderm on the flattened surface: in *Echinoplutei*, an invagination of ectoderm between the posterodorsal and postoral arms, just above the left hydrocoel.

amniotic isthmus (*Zool.*). See **sero-amniotic connexion.**

amobarbital (*Pharm.*). See **amylobarbitone.**

amoebiasis (*Med.*). Disease caused by a Rhizopod parasite (*Entamoeba histolytica*), producing intestinal lesions and generally dysentery.

Amoebina (*Zool.*). An order of *Sarcodina* the members of which extrude lobose pseudopodia and generally lack a skeleton, or have only a simple shell; their ectoplasm is never vacuolated. Also called **Lobosa.**

amoebocyte (*Zool.*). A metazoan cell having some of the characteristics of an amoeboid cell, especially as regards form and locomotion; in *Porifera*, a wandering cell of varied function; in *Echinodermata*, a wandering coelomic cell of excretory function; a leucocyte.

amoeboid (*Bot., Zool.*). Of a cell, having no fixed form, creeping, and putting out pseudopodia.

amoeboid movement (*Zool.*). Locomotion of an individual cell by means of pseudopodia.

amoebula (*Zool.*). In *Protozoa*, an animal exhibiting amoeboid movement.

amorphous (*Bot.*). Without clearcut shape (especially of lower plants in culture). (*Crystal.*) Noncrystalline.

amorphous sulphur (*Chem.*). Formed when sulphur vapour is cooled quickly. If the product so formed is treated with carbon disulphide and filtered, the amorphous sulphur is left on the filter as a white substance.

amortisseur (*Elec. Eng.*). See **damper.**

amosite (*Min.*). A monoclinic amphibole form of asbestos, the name embodying the initials of the company exploiting this material in the Transvaal, viz. the 'Asbestos Mines of South Africa'.

amp (*Elec.*). Deprecated abbrev. for **ampere.**

amperage (*Elec.*). Current in amperes, more especially the rated current of an electrical apparatus, e.g., fuse or motor.

ampere (*Elec.*). SI unit of current. The force between two parallel conductors each carrying one ampere, one metre apart, is 2×10^{-7} newton/m length, the conductors being infinitely thin, long, and in a vacuum. After Ampère of Paris. See **abampere, international ampere.**

ampere-balance (*Elec.*). See **Kelvin ampere-balance.**

ampere-conductors (*Elec.*). A unit of magnetomotive force occasionally used by machine designers; it is the product of the number of conductors in a slot or on a pole by the current in each. Also called **ampere-wires.**

ampere hour (*Elec.*). Unit of charge, equal to 3600 coulombs, or 1 ampere flowing for 1 hour.

ampere-hour capacity (*Elec.*). Capacity of an accumulator battery measured in ampere-hours, usually specified at a certain definite rate of discharge. Also applicable to primary cells.

ampere-hour efficiency (*Elec.*). Ratio of the ampere hours output during discharge to the ampere hours input during charge, in an accumulator.

ampere-hour meter (*Elec. Eng.*). A meter which records the ampere hours flowing in a circuit. If the voltage is assumed constant the meter can be calibrated in kilowatt hours: this is frequently done when measuring energy in d.c. supplies. Commonly abbreviated to **a.h.m.**

Ampère's law (*Mag.*). That which states the magnetic field (H) in the neighbourhood of a conductor, length l, carrying current I;

$$\oint H dl = I.$$

Ampère's rule (*Mag.*). Simple rule for the direction of the magnetic field associated with a current. If a man is swimming with the current and is facing a magnet needle, the north-seeking pole of the magnet is deflected towards his left hand. Alternatively, the direction of the field is that of an advancing righthand screw when turning with the current.

Ampère's theory of magnetization (*Mag.*). A theory based on the assumption that the magnetic property of a magnet is due to currents circulating in the molecules of the magnet.

ampere-turn (*Elec.*). SI unit of magneto-motive force, which drives flux through magnetic circuits, arising from 1 ampere flowing round one turn of a conductor. Abbrev. **At.**

ampere-turn amplification or gain (*Mag.*). Ratio of the load ampere-turns to the control ampere-turns in a magnetic amplifier.

ampere-turns metre (*Mag.*). A measure of the magnetizing force.

ampere-wires (*Elec.*). See **ampere-conductors.**

Ampex (*TV*). System of magnetic recording of video signals, for storing and editing programmes. High-speed wide magnetic tape is scanned transversely by rotating magnetic heads.

amphetamine (*Pharm.*). Racemic 1-phenyl-2-aminopropane; colourless liquid. The sulphate is used as a drug for its vasomotor, respiratory, stimulant and anorectic effects. TN Drinamyl (dexamphetamine), popularly 'purple hearts'.

amphi- (*Chem.*). Containing a condensed double aromatic nucleus substituted in the 2,6-positions. Gk. prefix, on both sides, around.

amphiaster (*Cyt.*). During cell-division by meiosis or mitosis, the two asters and the spindle connecting them.

amphiastral figure (*Cyt.*). The achromatic figure with asters.

amphibian (*Aero.*). Aeroplane capable of taking off and alighting on land or water; e.g. seaplane or flying boat with retractable landing gear, or landplane with *hydroskis* (q.v.).

amphibious (*Bot., Zool.*). Adapted for both terrestrial and aquatic life.

amphiblastic (*Zool.*). Of ova, showing complete but unequal segmentation.

amphiblastula (*Zool.*). In *Porifera*, a larval form with an anterior flagellate zone, a posterior non-flagellate zone, and a central cavity partially occluded by archaeocytes.

amphiboles (*Min.*). An important group of dark-coloured, rock-forming silicates, including hornblende, the commonest.

amphibolic (*Med.*). Descriptive of a substrate that can supply an organism with energy or precursors in synthetic processes. In established usage it means ambiguous or equivocal. (*Zool.*) Capable of being turned backwards or forwards, as the 4th toe of Owls.

amphibolite (*Geol.*). A crystalline, coarse-grained

rock, containing amphibole as an essential constituent, together with feldspar and frequently garnet; like hornblende-schist, formed by regional metamorphism of basic igneous rocks, but not foliated.

amphicarpic (*Bot.*). Having two kinds of fruits.

amphiclinous progeny, amphiclinous hybrids (*Bot.*). A family, resulting from a cross, in which some of the hybrids resemble one parent and some the other parent.

amphicoelous (*Zool.*). Having both ends concave, as in the centra of the vertebrae of *Selachii* and a few Reptiles.

amphicondylous, amphicondylar (*Zool.*). Having two occipital condyles.

amphicone (*Zool.*). Original undivided outer cusp of molar teeth in extinct mammals.

amphicribal bundle (*Bot.*). A vascular bundle in which a central strand of xylem is surrounded by phloem.

amphidiploid hybrid (*Gen.*). A hybrid which is tetraploid, and contains a diploid chromosome set derived from each of its parents.

amphidisc (*Zool.*). In *Porifera*, a form of spicule consisting of a rod bearing at its distal extremities disklike expansions curved towards the centre and prolonged into toothlike protuberances; occurs especially in the gemmules of fresh water Sponges.

amphigastrium (*Bot.*). One of the members of the ventral row of small leaves present in some liverworts.

amphigenous (*Bot.*). Growing all round; as when a parasitic fungus grows on both sides of a leaf.

amphigony (*Zool.*). Reproduction by fertilization. Cf. *parthenogenesis*.

amphigynous (*Bot.*). Said of fungi having antheridia through which the oogonium grows.

amphikaryon (*Cyt.*). A nucleus with two haploid sets of chromosomes, as after normal fertilization: an amphinucleus. *adj.* **amphikaryotic.**

amphilepsis (*Gen.*). Inheritance such that offspring have characters derived from both parents.

amphimixis (*Bot.*). The fusion of two distinct gametes to form a new individual. (*Zool.*) The mingling of different hereditary tendencies in the same individual, brought about by the union of male and female pronuclei in fertilization.

Amphineura (*Zool.*). A class of bilaterally symmetrical *Mollusca* in which the foot, if present, is broad and flat, the mantle is undivided, and the shell is absent or composed of eight valves. Coat-of-Mail Shells, etc.

amphinucleolus (*Zool.*). A nucleolus comprising both oxyphil and basophil components.

amphinucleus (*Zool.*). A nucleus with a large karyosome; in the theory of binuclearity, representing the kinetic nucleus encapsuled by the trophic nucleus.

amphiont (*Zool.*). See **zygote.**

amphipathic (*Chem.*, *Phys.*). Descriptive of unsymmetrical molecular group, one end being hydrophilic and the other hydrophobic (wetting and nonwetting).

amphiphloic siphonstele (*Bot.*). A stele having a pith and phloem on both sides of xylem.

amphiphloic solenostele (*Bot.*). A tubular stele with a cylinder of xylem coated externally and internally by phloem.

amphiplatyan (*Zool.*). Having both ends flat, as in certain types of vertebral centrum.

amphipneustic (*Zool.*). Possessing both gills and lungs: in dipterous larvae, having the prothoracic and posterior abdominal spiracles only functional.

Amphipoda (*Zool.*). An order of *Peracarida* in which the carapace is absent, the eyes are sessile, and the uropods styliform; the body is compressed. They show great variety of habitat, being found on the shore, in the surface waters of the sea, in fresh water, and in the soil of tropical forests. Some are parasitic. Whale Lice, Sandhoppers, Skeleton Shrimps, etc.

amphipodous (*Zool.*). Having both ambulatory and natatory appendages.

amphiprotic (*Chem.*). Having both protophilic (i.e., basic) and protogenic (i.e., acidic) properties.

amphiprotic solvent (*Chem.*). Solvent capable of showing either protophilic (i.e., acid generating) or protogenic (base generating) properties to different solutes, e.g., water with HCl and NH_3, respectively.

amphirhinal (*Zool.*). Having two external nares.

amphispore (*Bot.*). A thick-walled uredospore produced under dry conditions by some *Uredinales.*

amphistomous (*Zool.*). Having a sucker at each end of the body; as leeches.

amphistyly (*Zool.*). A type of jaw suspension found in the *Heterodontidae* or Bull-headed Sharks, in which the upper jaw fits into a groove in the cranium anteriorly but is still suspended posteriorly by the hyomandibular cartilage.

amphitene (*Cyt.*). The stage of meiosis in which the spireme threads are uniting in pairs.

amphitheatre (*Arch.*). An oval or circular building in which the spectators' seats surround the arena or open space in which the spectacle is presented, the seats rising away from the arena.

amphithecium (*Bot.*). Edge of a lichen amphothecium, similar in structure to the thallus.

amphitoky (*Zool.*). Parthenogenetic reproduction by both males and females. *adj.* **amphitokic.**

amphitrichous (*Bot.*, *Zool.*). Having a flagellum at each end of the cell.

amphitrocha (*Zool.*). In marine *Annelida*, a free-swimming pelagic larval type, having two strings of cilia.

amphitropous (*Bot.*). Said of ovules which may be bent both ways on the stalk.

amphitype (*Photog.*). A slow photographic process, originated by Herschel, in which iron, mercury, or lead compounds are impregnated into paper, which is subsequently bleached and reversed with a hot iron.

amphivasal bundle (*Bot.*). A vascular bundle in which a central strand of phloem is surrounded by xylem.

amphochromatophil (*Cyt.*). See **amphophil.**

amphogenic (*Gen.*). Producing offspring consisting of both males and females.

ampholyte (*Chem.*). An amphoteric electrolyte.

ampholytoid (*Chem.*). An amphoteric colloid.

amphophil (*Cyt.*). Pertaining to cells which stain with either basic or acidic dyes. Also **amphochromatophil.** See also **neutrophil.**

amphoric. Like the sound made by blowing across a narrow-necked vase.

amphoric breathing (*Med.*). Breathing having an amphoric quality; characteristic of an air-containing cavity in the chest.

amphoteric (*Chem.*). Having both acidic and basic properties, e.g., aluminium oxide, zinc oxide, which form salts with acids and with alkalis. (*Nuc.*) Said of a chain molecule charged oppositely at the ends.

amphotericin B (*Pharm.*). A heptaene antifungal antibiotic produced by a strain of *Streptomyces nodosus*. A pale yellow semicrystalline material, molecular weight about 950, optically active; formula is not yet known, tentatively $C_{46}H_{73}NO_{20}$. Used in the treatment of many mycotic infections.

Ampicillin (*Pharm.*). A semisynthetic penicillin with a broad range of activity (against, e.g., bronchitis, pneumonia, gonorrhoea, certain forms of meningitis, enteritis, biliary and urinary tract infections).

amplexicaul (*Bot.*). Said of a sessile leaf with its base clasping the stem horizontally.

amplexiform (*Zool.*). A type of wing-coupling formed in some *Lepidoptera*, whereby the wings are coupled simply by overlapping basally.

amplexus (*Zool.*). See **copulation.**

amplidyne (*Elec. Eng.*). Rotary magnetic amplifier with a high power gain, useful in servo-mechanisms.

amplidyne generator (*Elec. Eng.*). See **cross-field generator.**

amplification factor (*Electronics*). Symbol μ. The number of volts by which the potential of the anode of a thermionic tube must be changed to counteract the effect upon the anode current of a change in grid potential of 1 volt.

amplified A.V.C. (*Radio*). A system of automatic gain control in which the biasing voltage for the variable gain stages is derived from a d.c. amplifier, controlled by the rectified output voltage from the final demodulator.

amplifier (*Elec. Eng., Telecomm.*). Device for controlling power from a source so that more is delivered at the output than is supplied at the input. Source of power may be mechanical, hydraulic, pneumatic, electric, etc. Electric amplifiers may be classified into (1) *valve*, which operates on voltage, (2) *repeater*, especially used for telephone circuits, (3) *transistor*, which operates on current, (4) *magnetic*, which operates on very low-frequency currents, (5) *solid state*, operated by transistor action in a single semiconductor block. (*Photog.*) (1) An additional or supplementary lens for altering the focal-length of a camera lens. (2) Device for increasing the sensitivity of a photoelectric exposure meter.

amplitude (*Maths.*). Of a complex number, its *argument*. Of a periodic curve, the maximum value of its ordinates. (*Phys. etc.*) The maximum value of a periodically varying quantity during a cycle; e.g., the maximum displacement from its position of rest of a vibrating particle, or the maximum value of an alternating current (see **peak value**). See also **double amplitude.**

amplitude discriminator (*Telecomm.*). In a counting circuit, that which passes pulse amplitudes between specified limits, which may include zero and infinity.

amplitude distortion (*Telecomm.*). Distortion of waveform arising from the nonlinear static or dynamic response of a part of a communication system, the output amplitude of the signal at any instant not having a constant proportionality with the corresponding input signal. Can arise in networks, lines, or amplifiers.

amplitude filter (*TV*). One which separates synchronizing signals in a TV signal from the video (picture) signal.

amplitude peak (*Telecomm.*). Maximum positive or negative excursion from zero of any periodic disturbance.

amplosome (*Nut.*). The short, or stocky, type of human figure. Better term than *pyknosome*. See **pyknic type.**

ampoule, ampulla (*Med.*). A small, sealed, glass capsule for holding measured quantities of vaccines, drugs, serums, etc., ready for use.

ampoule tubing (*Glass*). Tubing of special composition suited to the manufacture of ampoules. It must work well in the blowpipe flame, and must resist the action of the materials stored in the ampoule.

ampulla (*Med.*). See **ampoule.** (*Zool.*) Any small membranous vesicle: in Vertebrates, the dilation housing the sensory epithelium at one end of a semicircular canal of the ear: in Mammals, part of a dilated tubule in the mammary gland: in Fish, the terminal vesicle of a neuromast organ: in *Echinodermata*, the internal expansion of a tube-foot, the expansion of the axial sinus below the madreporite: in *Hydrocorallina*, a pit of the skeleton lodging a medusoid: in *Ctenophora*, one of a pair of small sacs forming part of the aboral sense organ. *adj.* **ampullary.**

ampullaceous sensilla (*Zool.*). In Insects, sensory organs consisting of a narrow canal swollen at the extremity into an ampulla, which encloses a hairlike process.

ampulliform (*Bot.*). Flask-shaped.

Amsler vibraphone (*Phys.*). A high-speed, direct stress, fatigue testing machine operating at resonance. The applied load can be static tension or compression with a superimposed alternating load.

Amster cross-sections (*Nuc. Eng.*). Computed mean cross-sections (first compiled by Amster in 1958) for nuclei exposed to the neutron spectrum of hydrogen moderated reactors.

amu (*Nuc.*). Abbrev. for *atomic mass unit*.

amydricaine hydrochloride (*Chem., Pharm.*). The hydrochloride of benzoyl-tetramethyl-diamino-dimethylethyl-carbinol, i.e., $C_{16}H_{27}O_2N_2Cl$. White crystals; m.p. 169°C. Used as a local anaesthetic, particularly in ophthalmic practice. See also **alypin hydrochloride.**

amyelinate (*Zool.*). Of nerve-fibres, nonmedullated, lacking a myelin sheath.

amygdala (*Zool.*). A lobe of the cerebellum: one of the palatal tonsils.

amygdale, amygdule (*Geol.*). An almond-shaped infilling (by secondary minerals such as agate, zeolites, calcite, etc.) of elongated steam cavities in igneous rocks.

amygdalin (*Chem.*). $C_{20}H_{27}O_{11}N$, colourless prisms, m.p. 200°C, a glucoside found in bitter almonds, in peach and cherry kernels.

amygdaloid (*Bot.*). Almondlike.

amylaceous (*Bot.*). Starchy.

amyl acetate (*Chem.*). Pentyl ethanoate. $CH_3 \cdot CO \cdot O \cdot C_5H_{11}$, colourless liquid, of ethereal pear-like odour, b.p. 138°C. *iso*amyl acetate is also known under the name of *pear oil*. It is used for fruit essences and is an important solvent for nitrocellulose. Used as a standard light in photometry when burnt in a special lamp.

amyl alcohol (*Chem.*). Pentyl alcohol. $C_5H_{11}OH$, the fraction of fusel oil that distils about 131°C. There are 8 isomers possible and known, viz. 4 primary, 3 secondary, 1 tertiary amyl alcohol. The important isomers are: *iso*amyl alcohol, *l*-amyl alcohol, tertiary amyl alcohol (amylene hydrate).

amylases (*Chem.*). Enzymes capable of hydrolysing the glycosidic linkages in starch, etc.

amylene (*Chem.*). A synonym for pentene, C_5H_{10}, a higher olefin homologue. Four isomers are known, all low-boiling liquids.

amylene hydrate (*Chem.*). A synonym for *tertiary* amyl alcohol.

amyl group (*Chem.*). The monovalent aliphatic radical C_5H_{11}— .

amyl nitrite (*Chem.*). $C_5H_{11} \cdot O \cdot NO$, the nitrous acid ester of *iso*amyl alcohol, a yellowish liquid, b.p. 98°C, of pleasant odour. Intermediate for the preparation of nitroso- and of diazo-compounds. Used in medicine as a vasodilator, esp. for acute angina.

amylobarbitone (*Pharm.*). 5-Ethyl-5-isopentylbarbituric acid, $C_{11}H_{18}O_3N_2$. Colourless crystals; m.p. 156°C. Much used, especially in its sodium compound, as an intermediate-acting hypnotic and sedative, e.g., for pre-operative and psychotherapeutic purposes. Sometimes called **amobarbital.** The forms **Amytal** and **Sodium Amytal** are TNs.

amylocaine hydrochloride (*Pharm.*). 1-Dimethylaminomethyl-1-methylpropyl benzoate hydrochloride, $C_{14}H_{21}NO_2HCl$. Colourless crystals; m.p. 178°C. Local anaesthetic, formerly used for intraspinal injection.

amyloclastic (*Bot.*). Able to break down starch.

amylo fermentation process (*Chem.*). The use of certain moulds secreting diastase and fermentation enzymes for obtaining ethanol from starchy materials without the use of malt.

amylogenesis (*Bot.*). The building of starch inside the cell of a plant.

amyloid (*Chem.*). A starchlike cellulose compound, produced by treatment of cellulose with concentrated sulphuric acid for a short period. (*Med.*) A waxy substance formed in the body in certain diseases, and composed of a protein and chondroitin sulphuric acid.

amyloid bodies (*Zool.*). In Mammals, starchlike concretions found in the alveoli of the prostate gland of the adult.

amyloid infiltration (*Med.*). The formation of amyloid in the small arteries, capillaries, and tissues of the body, as a result of chronic infection.

amylolytic (*Zool.*). Starch-digesting.

amylopectin (*Chem.*). The gel constituent of starch paste.

amylopsin (*Chem., Zool.*). An enzyme, produced by pancreatic or intestinal glands, which induces complete hydrolysis of starchy matter in the intestinal tract.

amylose (*Bot.*). A general term for starch, inulin, and related carbohydrates. (*Chem.*) The sol constituent of starch paste; polymer of glucose units having α1-4 glucosidic bonds.

amyloses (*Chem., etc.*). Depolymerization products of starch, crystalline substances of the sugar series.

amylostatolith (*Bot.*). A starch grain which acts as a statolith.

amylum (*Chem.*). A synonym for *starch.*

amyotrophic lateral sclerosis (*Med.*). A nervous disease in which atrophy of muscle follows degenerative changes in the motor cells of the spinal cord and brain.

amyotrophy (*Med.*). Wasting or atrophy of muscle.

Amytal (*Pharm.*). See **amylobarbitone.**

an-. Prefix from Gk. *an,* not. See also **ap-.**

An (*Chem.*). The symbol for *actinon.*

ana-. Prefix from Gk. *ana,* up, anew.

anabasine (*Chem.*). $C_{10}H_{14}N_2$; neonicotine. 2-(3-pyridyl) piperidine. An isomer of nicotine:

```
      H     H  H2
   N═C      N─C
  /   \    /    \
HC     C─CH      CH2
 \\   //   \    /
  C──C      C──C
  H  H      H2 H2
```

derived from *Anabasis aphylla,* a chenopodiaceous plant of Central Africa. Colourless liquid, density 1·048, b.p. 28°C, soluble in water. Used as an insecticide.

anabatic (*Meteor.*). A term applied to winds caused by the upward convection of heated air.

anabion (*Biol.*). An organism in which anabolic processes predominate over katabolic processes; e.g., plants.

anabiosis (*Biol., Bot., Zool.*). A temporary state of reduced metabolism in which metabolic activity is absent or undetectable. See also **cryptobiosis.**

anabolism (*Biol.*). The chemical changes proceeding in living organisms with the formation of complex substances from simpler ones, together with the storage of chemical energy.

anabolite (*Biol.*). A substance participating in anabolism.

Anacanthii (*Zool.*). An order of *Teleostei* with flexible jointed finrays; the true caudal fin is absent or reduced, while the pelvic fins are placed far forward on the body; the duct connecting the air bladder and the gullet is lacking. Cod, Whiting, Hake, Haddock, Ling, Pollack, Burbot, Grenadiers.

Anachoropteridales (*Bot.*). An order of the *Primofilices.* The leaves are pinnate with all segments in the same plane. The sporangia are on the flattened margins of the ultimate pinnae. Found in *Carboniferous* rocks.

anacoustic zone (*Space*). One hundred miles (160 km) above the earth's surface, where air molecules are so widely dispersed that sound waves cannot be propagated.

anacromyoidean (*Zool.*). Having the syringeal muscles attached to the dorsal ends of the bronchial cartilages.

anadipsia (*Med.*). Excessive thirst, characteristic of certain diseases, including *manic-depressive psychosis.*

anadromous (*Zool.*). Having the habit of migrating from more dense to less dense water to breed, generally from oceanic to coastal waters, or from salt water to fresh water; as the Salmon.

anaemia, anemia (*Med.*). Diminution of the amount of total circulating haemoglobin in the blood. See **pernicious anaemia** and (*Vet.*) **infectious anaemia.**

anaemia leucopaenia (*Med.*). Disease in which leucocytes are reduced in number.

anaerobe, anaerobiont (*Biol.*). Names applied to an organism for whose life processes a complete or (in some cases) nearly complete absence of oxygen is essential. Facultative anaerobes can utilize free oxygen; obligate anaerobes are poisoned by it. *adjs.* **anaerobic, anaerobiotic.**

anaerobic (*Bot., Zool., etc.*). See **anaerobiosis.**

anaerobic decomposition (*Bot.*). The incomplete breakdown of organic material by bacteria in the absence of free oxygen.

anaerobic respiration (*Biol.*). See **respiration, anaerobic.**

anaerobiosis (*Bot., Zool., etc.*). Existence in the absence of oxygen. *adj.* **anaerobic.**

anaesthesia, anesthesia (*Med.*). Loss of sensibility to touch (loosely, also to pain and temperature): the science and art of administering anaesthetics.

anaesthetic, anesthetic (*Med.*). Insensible to touch (loosely, also to pain and temperature): a drug which produces insensibility to touch, pain, and temperature, with or without loss of consciousness.

anaesthetist (*Med.*). One skilled in the administration of anaesthetic drugs.

anafront (*Meteor.*). A situation at a front, warm or cold, where the warm air is rising relative to the *frontal zone* (q.v.).

anagenesis (*Biol.*). Regeneration of tissues; progressive evolution.

anaglyph (*Photog.*). The use of two images of two colours, red and green, viewed with red and green spectacles (one for each eye) to obtain an

approximation to stereoscopy. Used for both still and motion pictures.
anakinetic (*Biol.*). Leading to the restoration of energy and the formation of reactive, energy-rich substances.
anakinetomeres (*Biol.*). Reactive, energy-rich molecules.
anal (*Zool.*). See **anus.**
anal cerci (*Zool.*). In Insects, appendages of one of the posterior abdominal somites, generally the 11th, retained throughout life.
anal character (*Psychol.*). An adult personality in which there is dominant reflection of the anal phase of pregenital sexual development. Characterized by parsimony, greed, perfectionism, punctuality, extreme neatness, etc.
analcime, analcite (*Min.*). An important hydrated silicate of aluminium and sodium, found in some igneous and sedimentary rocks, but particularly in cavities in lavas.
anal cone (*Zool.*). In larvae of *Entoprocta*, a projection of the ventral surface carrying the proctodaeum.
analeptic (*Med.*). Having restorative or strengthening properties.
anal field (*Zool.*). In Echinoderm and Tornaria larvae, an area surrounded by the posterior loop of the longitudinal ciliated band.
analgesia (*Med.*). Loss of sensibility to pain.
analgesic (*Med.*). A drug which relieves pain.
anal-gland disease (*Vet.*). Inflammation of the perianal glands, and inflammation or impaction of the canine anal sacs.
anallatic lens (*Surv.*). Special lens which, when correctly placed between the object glass and the eyepiece lens of a tacheometric telescope, optically reduces the additive constant for the tacheometer to zero.
anallatic telescope (*Surv.*). One which, when used in tacheometry, has a zero additive constant.
anallatism (*Surv.*). See **centre of-.**
anal lobe (*Zool.*). In Insects, a posterior lobe of the wing. See **anal veins.**
anal loop (*Zool.*). In Echinoderm and Tornaria larvae, the posterior loop of the longitudinal ciliated band.
analogies test (*Psychol.*). A type of mental test of the form: R is to S, as T is to —.
analogous organs (*Bot.*). Organs which are similar in appearance or function, but are not equivalent morphologically.
analog(ue) (*Comp.*). Referring to continuous variation of a magnitude in computation; cf. *digital*. Hence *analogue computer*, one which operates with continuously varying amplitudes, as contrasted with *digital computers*, which use *bits*. (*Meteor.*) A previous *weather map* similar to the current map. The developments following the analogue aid forecasting.
analogue compounds (*Chem.*). Molecules having a similar structure but different elementary composition.
analogue-digital converter (*Elec. Eng.*). Circuit used to convert information in analogue form into digital form (or vice versa), e.g., in a digital voltmeter. The analogue signal may be continuous, or in the form of discrete pulses, the amplitudes of which are to be digitized.
analogy (*Bot., Zool.*). Likeness in function but not in origin, e.g., tendrils, which may be modified leaves, branches, inflorescences; the wings of Birds and of Insects. *adj.* **analogous.** (*Electronics*) Correspondence of pattern or form, e.g., electric wave filter analogy for complex mechanical system, or sheet rubber models for investigation distribution of potentials between electrodes in valves to facilitate conception and calculation from known techniques. U.S. **analagy.**
anal papilla (*Zool.*). In Cephalopod larvae and adult *Crinoidea*, a small papilla at the apex of which the anus is developed.
anal-sadistic (*Psychol.*). Characteristic of destructiveness and aggression, common to the early anal phase of development.
anal suture (*Zool.*). In the posterior wings of some Insects, a line of folding, separating the anal area of the wing from the main area.
anal veins (*Zool.*). In Insects, the 3 posterior elements of the primitive wing venation. When the anal area of the wing is reduced, one or more may be atrophied; when there is an anal lobe, they may be branched.
analyser (*Chem.*). The second Nicol prism in a polarimeter; when rotated 90° relative to polarizer it will not allow polarized light to pass.
analyses (*Chem.*). The expression of the results obtained by *chemical analysis* (q.v.).
analysis (*Maths.*). See **mathematics.**
analysis meter (*Teleph.*). A registering meter used to determine the loading of groups of circuits with calls, particularly for determining the correctness or otherwise of grading.
analysis of variance (*Stats.*). The separation of the *variance* (q.v.), due to one factor (or set of factors), from other factors which affect a set of observations.
analyst (*Chem.*). See **analytical chemist.**
analytical chemist (*Chem.*). A person who carries out the process of analysis by chemical methods. See **chemical analysis.**
analytical geometry (*Maths.*). The application of the methods of algebra to geometry by utilizing the concept of coordinates.
analytical psychology. The school of psychology founded by Jung of Zürich. It deviates from the Freudian school mainly in the method of dream interpretation, the theory of the causation of a neurosis in the present situation instead of the past, by the use of the term *libido* for all instinctual life, not exclusively sexual, by the postulation of a racial or collective unconscious in addition to a personal unconscious, and by an emphasis on spiritual values and a human soul or psyche.
analytical reactions (*Chem.*). The forming of two or more different substances from one substance; e.g., mercury(II)oxide decomposes into mercury and oxygen.
analytical reagent (*Chem.*). See **A.R.**
analytical sampling error (*Powder Tech.*). Errors arising from the fact that only a portion of the particles under examination is measured.
analytic continuation (*Maths.*). A process for extending the range of definition of a function of a complex variable. Starting with a first Taylor series about a point A representing the function, a second Taylor series about a point B within the first circle of convergence can be obtained and providing the line AB does not intersect the first circle of convergence at a singularity, the circle of convergence of the second Taylor series will extend beyond the first circle. This process can be repeated indefinitely and it can be shown that if a particular region can be reached by two different routes, the two Taylor series obtained will be identical. Some functions have their own *natural boundaries* beyond which analytic continuation is impossible. Such a function is

$$g(z) = 1 + \sum_{r=1}^{\infty} z^{2r}$$

which has an infinite number of singularities on every arc of its circle of convergence $|z| = 1$.

analytic function (*Maths.*). A function of a complex variable is analytic in a region if it is single-valued and differentiable at all points of the region. The terms *holomorphic, monogenic* and *regular* are sometimes used. Sometimes a function is said to be analytic in a region if it is single-valued and differentiable at all but a finite number of points (*singularities*) of the region. In this case the term *regular* is frequently contrasted with *analytic* to mean that there are no singularities. When the singularities are all poles the term *meromorphic* is sometimes used. A function that is analytic throughout the whole complex plane, except possibly at a finite number of singularities, is called an *entire* or an *integral* function.

anamnesis (*Med.*). The recollection of past things: the past history of all matters relating to a patient's health.

anamniotic (*Zool.*). Lacking an amnion during development. Also **anamniote.**

anamorpha (*Zool.*). Larvae which do not possess the full number of segments at the time of hatching.

anamorphic lens (*Cinema.*). Lens with cylindrical elements, used in wide-screen cinematography for compressing an elongated image on to standard film stock. A similar lens used in projection restores the image to its original proportions. Also **anamorphote**.

anandrous (*Bot.*). Lacking stamens.

anangian (*Zool.*). Lacking a vascular system.

anaphase (*Cyt.*). The stage in mitotic or meiotic nuclear division when the chromosomes or half-chromosomes move away from the equatorial plate to the poles of the spindle; more rarely, all stages of mitosis leading up to the formation of the chromosomes.

anaphoresis (*Chem.*). The migration of suspended particles towards the anode under the influence of an electric field.

anaphylactic (*Med.*). Being in a state of anaphylaxis: pertaining to anaphylaxis.

anaphylactic shock (*Med.*). The immediate reaction which occurs in the smooth muscle of the body after the injection of a protein to which a person is anaphylactic.

anaphylaxis (*Med.*). Specific supersensitiveness of the smooth or involuntary muscles of the body to a protein previously introduced into it.

anaphysis (*Bot.*). A sterigma-like thread in the apothecium of a lichen.

anaplasia (*Med.*). Loss of the distinctive character of a cell associated with proliferative activity; as in cancer.

anaplasmosis (*Vet.*). Gall-sickness. A disease of cattle caused by infection by protozoa of the genus *Anaplasma,* and characterized by fever, anaemia, and jaundice. The protozoa are found in the red blood corpuscles and are transmitted by ticks, biting flies, and mosquitoes. Infections of sheep and swine also occur.

anaplerotic (*Chem.*). An anaplerotic sequence is any set of enzyme actions that makes intermediates in the central metabolic cycles (e.g., tricarboxylic) from a katabolic product of a substrate needed for the growth of an organism. It thus serves to replenish an intermediate depleted during growth. Also used in pathology of some forms of regeneration.

anapophysis (*Zool.*). In higher Vertebrates, a small process just below the postzygapophysis which strengthens the articulation of the lumbar vertebrae.

anapsid (*Zool.*). Having the skull completely roofed over, i.e., having no dorsal foramina other than the nares, the orbits, and the parietal foramen.

anarthrous (*Zool.*). Without distinct joints.

anasarca (*Med.*). Excessive accumulation of fluid (dropsy) in the skin and subcutaneous tissues.

anaschistic (*Cyt.*). In meiosis, said of tetrads which divide twice longitudinally.

Anaspidacea (*Zool.*). An order of the crustacean *Syncarida.* Few living species, in pools at 1220 m in Tasmania. Mountain shrimps.

anastigmat lens (*Optics.*). A photographic objective designed to be free from astigmatism in at least one extra-axial zone of the image plane.

anastomosis (*Bot.*). Communication by cross-connexions to form a network. (*Med.*) A communication between two blood vessels: an artificial communication, made by operation, between any two parts of the alimentary canal. (*Zool.*) The formation of a meshwork of blood vessels or nerves: the union of blood vessels or nerves arising by the splitting of a common trunk.

anastral (*Cyt.*). Without asters; said of a type of mitosis.

anatase (*Min.*). One of the three naturally occurring forms of crystalline titanium dioxide, of tabular or bipyramidal habit. See also **octahedrite.**

anatomy (*Bot., Zool., etc.*). (1) The study of the form and structure of animals and plants; it includes the study of minute structures, and thus includes *histology.* (2) Dissection of an organized body in order to display its physical structure and economy. The technical term for animal anatomy is *zootomy*; for vegetable anatomy, *phytotomy*; for human anatomy, *anthropotomy.*

anatoxin (*Bacteriol.*). The toxin of diphtheria detoxicated by formalin or by heat.

anatropous (*Bot.*). Said of an ovule in which the body is curved so that the micropyle lies close to the insertion of the funicle.

anaxial (*Zool.*). Asymmetrical.

anchor (*Zool.*). A type of spicule found in the integument of Holothurians.

anchor and collar (*Civ. Eng.*). A type of hinge used to support lock gates, consisting of an *anchor* built into the masonry coping, with a *collar* attached like a clevis to the anchor, the collar forming a socket into which fits the pintle of the heel post of the gate.

anchor bolt (*Build.*). A bolt used to secure frameworks, stanchion bases, etc., to piers or foundations, and having usually a large plate washer built into the latter as anchorage.

anchor clamp (*Elec. Eng.*). A fitting which attaches the conductor of an overhead-transmission line to a strain insulator or support.

anchor ear (*Elec. Eng.*). A fitting attached to the overhead contact wire of a tramway or railway to support the wire, and also to take the longitudinal tension and prevent movement of the wire in a direction parallel to the track.

anchor escapement (*Horol.*). See **recoil escapement.**

anchor-gate (*Civ. Eng.*). A heavy gate, such as a canal lock gate, which is supported at its upper bearing by an anchorage in the masonry such as an *anchor and collar* (q.v.).

anchoring (*Typog.*). A method of fastening plates to mounts when the usual flange is not available; a thin bolt is passed through the mount and secured with a nut. Superseded by sweating to metal mount or by using adhesive film on all kinds of mount.

anchor point (*Chem. Eng.*). A point in a complex pipework system, carrying hot fluids, which is securely fixed so that it cannot move under the influence of thermal expansion, and thereby gives a reference point in design to allow for the expansion elsewhere in the system.
anchor pole (*Elec. Eng.*). See **anchor tower.**
anchor ring (*Elec. Eng.*). Old name for **toroid.** (*Maths.*) A surface shaped like a circular ring. Also called a **torus.**
anchor tower (*Elec. Eng.*). A type of tower placed at intervals along an overhead-transmission line; designed to give longitudinal rigidity. Called an *anchor pole* when the line is supported on wooden poles.
anchylosis (*Med.*). See **ankylosis.**
ancillary shoring (*Build.*). The auxiliary shoring which is necessary in the process of underpinning before any walling is removed, its purpose being to relieve the wall to be underpinned of as much load as possible. It consists of floor, roof, and window strutting, and perhaps raking shoring.
ancillary working (*Teleph.*). A method of operating whereby speed of answering is increased by equipping each line with more than one calling and answering device, so rendering it accessible to more than one operator.
ancipitous (*Bot.*). Two-edged.
ancon (*Arch.*). A console built on each side of a door-opening to carry a cornice. *pl.* **ancones.**
anconeal (*Zool.*). Pertaining to, or situated near, the elbow.
anconeus (*Zool.*). An extensor muscle of the arm attached in the region of the elbow.
Ancylistales (*Bot.*). See **Lagenidiales.**
AND (*Comp., Maths.*). See **logical operations.**
andalusite (*Min.*). One of several crystalline forms of aluminium silicate; a characteristic product of the contact metamorphism of argillaceous rocks. Orthorhombic. See also **chiastolite.**
and element (or gate) (*Comp.*). One producing output signal only if all inputs are energized simultaneously, i.e., output signal is 1 when all input signals are 1.
andesine (*Min.*). A member of the plagioclase group of minerals, with a small excess of sodium over calcium; typical of the intermediate igneous rocks.
andesite (*Geol.*). A fine-grained igneous rock (usually a lava), of intermediate composition, having plagioclase as the dominant feldspar.
AN direction finding (*Electronics*). System in which two signals corresponding to morse A &N (· — and — ·) are received simultaneously. When on course these are of equal intensity and blend into a continuous steady tone.
andiron (*Build.*). A metal support for wood in an open fire. Also called a **firedog.**
andradite (*Min.*). Common calcium-iron garnet. Includes melanite (black garnet), though most andradite is dark brown to yellow or green.
andrase (*Zool.*). An enzyme or hormone tending to produce maleness.
Andreales (*Bot.*). An order of the *Musci*, distinguished by the capsule splitting into four valves at dehiscence. Also called **Andreacales.**
Andreassen apparatus (*Powder Tech.*). An apparatus for measuring cumulative *size distribution* (q.v.) of fine powders ($<50\ \mu$) by the *pipette method* (q.v.). It consists of a graduated glass cylinder fitted with a removable pipette whose tip always rests at the same height in the cylinder.
André-Venner accumulator (*Elec. Eng.*). One in which the anode is silver, the cathode zinc, with potassium hydroxide as electrolyte. It can be made in very small sizes, nearly dry, and for very high rates of charge and discharge.
Andrews elutriator (*Powder Tech.*). A vertical liquid elutriator consisting of 3 conical separating chambers mounted one above the other.
andro-. Prefix from Gk. *aner, andros*, man, male.
androconia (*Zool.*). In certain male *Lepidoptera*, scent scales serving to disseminate the secretion of the odoriferous glands situated at their bases; believed to serve the purpose of sexual attraction.
androcyte (*Bot.*). Cells in an antheridium which will metamorphose to form antherozoids.
androdioecious (*Bot.*). Said of a species in which some of the plants bear staminate flowers, others hermaphrodite flowers.
androecium (*Bot.*). (1) The whole of the stamens in one flower. (2) The group of male organs in mosses. *adj.* **androecial.**
androgen (*Biochem.*). Any substance that stimulates male characteristics; a male hormone.
androgenesis (*Bot., Zool.*). (1) Development from a male cell. (2) Development of an egg after entry of male germ cell without the participation of the egg nucleus.
androgenic (*Physiol.*). Having the effects of the male sex hormone.
androgonidia (*Zool.*). In certain *Protozoa* (as *Volvox*), male gametes occurring after a period of asexual reproduction.
androgynous (*Bot.*). Bearing staminate and pistillate flowers] on distinct parts of the same inflorescence: having the male and female organs on or in the same branch of the thallus.
andrology (*Med.*). That branch of medical science which deals with the functions and diseases peculiar to the male sex.
Andromeda nebula (*Astron.*). The spiral galaxy M31 in Andromeda, visible to the naked eye. Of comparable size with our own Galaxy, it is about 2 million light-years distant.
andromonoecious (*Bot.*). Said of a plant which has staminate and perfect flowers, but no pistillate flowers.
androphore (*Bot.*). An elongation of the receptacle of the flower between the corolla and the stamens. (*Zool.*) In *Siphonophora*, a stalk bearing male gonophores.
androsome (*Cyt.*). A male-limited chromosome.
androsporangium (*Bot.*). A sporangium in which one or more spores (termed *androspores*) able to give rise to male plants are developed.
androsterone (*Biochem.*). A steroid, obtained from the testis and from urine. It has male hormone activity, controlling the secondary male characteristics, development of accessory reproductive organs, growth and distribution of hair, etc.
anecdysis (*Zool.*). A long period during the moulting cycle of *Arthropoda* when there are no signs of either recovery from a moult or of preparations for the next moult.
anechoic room (*Acous.*). One in which internal sound reflections are reduced to an ineffective value by lining with internally pointing pyramids of felt, foam plastic or fibreglass; used for

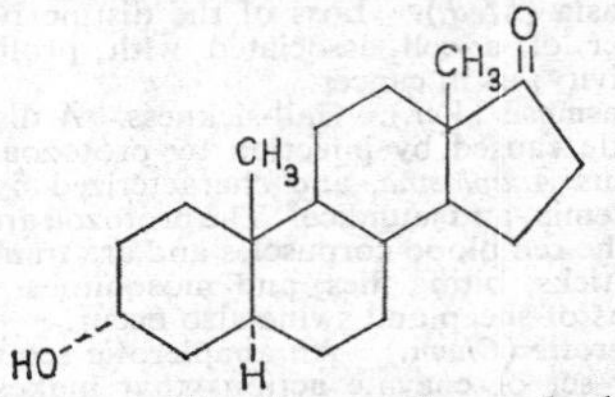

standard measurements of microphones, etc. Also called **dead room.**

anelasticity (*Phys.*). Any deviation from an ideal internal structure of a body which would dampen or attenuate an elastic wave therein.

anelectric (*Elec.*). Term once used for a body which does not become electrified by friction.

anelectrotonus (*Med.*). Diminished activity of a nerve because of electric current, especially near an applied anode.

anemia, anesthesia, etc. (*Med.*). See **anaemia, anaesthesia,** etc.

anemochorous (*Bot.*). Said of plants whose seeds are dispersed by wind, and particularly plants which retain their seeds through the winter, liberating them for wind dispersal in the spring.

anemograph (*Meteor.*). See **anemometer.**

anemometer (*Eng.*). An instrument for measuring the rate of flow of a gas, either by mechanical or electrical methods. (*Meteor.*) An instrument for measuring the speed of the wind. A common type consists of four hemispherical cups carried at the ends of four radial arms pivoted so as to be capable of rotation in a horizontal plane, the speed of rotation being indicated on a dial calibrated to read wind speed directly. An *anemograph* records the speed and sometimes the direction.

anemophily (*Bot.*). Pollination by means of wind. Dispersal of spores by wind. *adj.* **anemophilous.**

anemotaxis (*An. Behav.*). A *taxis* (q.v.) occurring in response to the direction of air currents.

anemotropism (*Biol.*). Active response to the stimulus of an air current.

anencephaly (*Med.*). Developmental defect of the skull and absence of the brain. *adj.* **anencephalic.**

anenteron (*Zool.*). A malformed blastula, having an evaginated archenteron which later disappears entirely; usually produced by exposure to a supranormal temperature.

anenterous, anenteric (*Zool.*). Without a gut.

aner (*Zool.*). A male ant.

anergasia (*Med.*). Mental illness or disorder as a result of an organic brain lesion.

aneroid barometer (*Meteor., Surv.*). One having a vacuum chamber or sylphon bellows of thin corrugated metal, one end diaphragm of which is fixed, the other being connected by a train of levers to a scale pointer which records the movements of the diaphragm under changing atmospheric pressure.

anethole (*Chem.*). *p*-Propenyl anisole; CHMe : CH·C_6H_4·OMe. An ether forming the chief constituent of oil of aniseed, an essential oil.

aneuploidy (*Bot.*). The condition of a nucleus, tissue, individual, or race having a chromosome number which is not an exact multiple of the haploid number.

aneurin (*Chem.*). See **vitamin B complex.**

aneurysm (*Med.*). Pathological dilatation, fusiform or saccular, of an artery.

angel beam (*Arch.*). A horizontal member of a mediaeval roof truss, usually decorated with angels carved on the member.

angelic acid (*Chem.*). $CH_3CH : C(CH_3)COOH$. The cis-form of dimethylacrylic acid, the trans-form being *tiglic acid* (q.v.). M.p. 45°C, b.p. 185°C, used in flavouring extracts.

angels (*Radar*). Stray reflections, e.g., from birds, etc., from the lower atmosphere.

angina pectoris (*Med.*). A condition characterized by the sudden onset of pain in the chest and inner side of the left arm after exertion; due to disease of the coronary arteries. Also called **angina innocens.** See **pseudoangina.**

angio-. Prefix from Gk. *angeion*, denoting a case or vessel.

angioblast (*Zool.*). An embryonic mesodermal cell from which the vessels and early blood cells are derived.

angiocardiography (*Radiol.*). The radiological examination of the heart and great vessels after injection of a *contrast medium* (q.v.).

angiocarpic, angiocarpous (*Bot.*). Applied to the fruit body of a fungus in which the hymenium develops, at any rate at first, within a closed envelope.

angiography (*Radiol.*). The study of the cardiovascular system by means of radio-opaque media.

angiology (*Biol.*). The study or scientific account of the anatomy of blood and lymph vascular systems.

angioma (*Med.*). See **haemangioma.**

angioneurotic oedema (*Med.*). A disorder in which rounded oedematous swellings suddenly appear in the skin or the mucous membranes; Quincke's disease.

Angiospermae (*Bot.*). A major group of flowering plants, probably including over 200,000 species, in which the seeds develop and ripen inside a closed ovary. It includes all the ordinary flowering plants and most other plants of economic importance, except pines and their relatives.

angiostomatous (*Zool.*). Having a narrow or nondistensible mouth, as certain *Mollusca*, certain *Ophidia*.

angiotensin (*Physiol.*). See **hypertensin.**

angle (*Eng.*). Mild steel bar rolled to the cross-section of the letter L, much used for light structural work. Also called **angle bar, angle iron, angle steel.** (*Maths.*) The inclination of one line to another, measured in degrees (of which there are 360 to one complete revolution), or in radians (of which there are 2π to one complete revolution). (*Textiles*) (1) In lace manufacture, the angles of the warp threads with regard to the horizontal perforated steel bars. (2) In spinning, the angle of the yarn from the tip of the spindle to the front roller nip.

angle bar (*Arch.*). The vertical bar between two faces of a polygonal or bow window.

angle bars (*Print.*). On rotary presses, bars at an angle to transfer one or more webs of paper over each other, or the web to the other side of the press, or at right angles to its previous direction. Sometimes called **turner bars.**

angle bead (*Build.*). A small rounded moulding placed at an angle formed by plastered surfaces to protect from damage.

angle bearing (*Eng.*). A shaft-bearing in which the joint between base and cap is not perpendicular to the direction of the load, but is set at an angle.

angle block (*Carp.*). A small wooden block used in woodwork to make joints, especially right-angle joints, more rigid.

angle board (*Carp.*). One used as a gauge by which to plane boards to a required angle between two faces.

angle brace (*Carp.*). (1) Any bar fixed across the inside of an angle in a framework to render the latter more rigid (also called **angle tie, dragon tie**). (2) A special tool for drilling in corners where there is not room to use the cranked handle of the ordinary brace.

angle bracket (*Carp.*). A bracket projecting from the corner of a building beneath the eaves, and not at right-angles to the face of the wall. (*Eng.*) A bracket consisting of two sides set at right angles, often stiffened by a gusset.

angle cleat (*Build.*). A small bracket formed of angle iron, used to support or locate a member in a structural framework.

angle closer (*Build.*). Loose term for *closer* (q.v.) cut at an angle.

angle cutter (*Paper*). A machine for cutting paper at an angle usually for the preparation of envelopes.

angled deck (*Aero.*). The flight deck of an aircraft carrier prolonged diagonally from one side of the ship, so that aeroplanes may fly-off and land-on without interference to or from aircraft parked at the bows. U.S. **canted deck.**

angled draft (*Textiles*). An arrangement of the warp threads in the mails of the healds so that the twill components will produce a 'herring-bone' pattern.

angledozer (*Civ. Eng.*). See **bulldozer.**

angle float (*Build.*). A plasterer's trowel, specially shaped to fit into the angle between adjacent walls of a room.

angle gauge (*Build.*). A tool which is used to set off and test angles in carpenter's, bricklayer's, and mason's work.

angle iron (*Eng.*). See **L-iron.**

angle modulation (*Telecomm.*). Any system in which the transmitted signal varies the phase-angle of an otherwise steady carrier frequency, i.e., phase and frequency modulation.

angle of acceptance (*Light*). The horizontal angle within which light rays should reach a window to ensure adequate penetration.

angle of advance (*Eng.*). (1) The angle in excess of 90° by which the eccentric throw of a steam-engine valve gear is in advance of the crank. (2) The angle between the position of ignition and outer dead centre in a spark-ignition engine; optimizes combustion of the fuel.

angle of approach light (*Aero.*). A light indicating an approach path in a vertical plane to a definite position in the landing area.

angle of arrival (*Radio*). Angle of elevation of a downcoming wave. See **musa.**

angle of attack (*Aero.*). The angle between the *chord line* of an aerofoil and the relative airflow, normally the immediate flight path of the aircraft. Also called **angle of incidence.**

angle-of-attack indicator (*Aero.*). An instrument which senses the true angle of incidence to the relative airflow and presents it to the pilot on a graduated dial or by means of an indicating light.

angle of bank (*Aero.*). See **angle of roll.**

angle of contact (*Eng.*). The angle subtended at the centre of a pulley by that part of the rim in contact with the driving belt. (*Phys.*) The angle made by the surface separating two fluids (one of them generally air) with the wall of the containing vessel, or with any other solid surface cutting the fluid surface. For liquid-air surfaces, the angle of contact is measured in the liquid.

angle of cut-off (*Light*). The largest angle below the horizontal at which a reflector allows the light-source to be visible when viewed from a point outside the reflector.

angle of deflection (*Electronics*). That of the electron beam in a CRT.

angle of departure (*Radio*). Angle of elevation of maximum emission of electromagnetic energy from an antenna.

angle of depression (*Surv.*). The vertical angle measured below the horizontal, from the surveyor's instrument to the point observed; plunge angle.

angle of deviation (*Light*). The angle which the incident ray makes with the emergent ray when light passes through a prism or any other optical device.

angle of dig or drag (*Acous.*). Deviation from normal of angle of cutting stylus in disk sound recording.

angle of dip (*Geol.*). See **dip.**

angle of elevation (*Surv.*). The vertical angle measured above the horizontal, from the surveyor's instrument to the point observed.

angle of flow (*Elec. Eng.*). Angle, or fraction of alternating cycle, during which current flows, e.g., in a thyratron.

angle of friction (*Eng., etc.*). The angle between the normal to the contact surfaces of two bodies, and the direction of the resultant reaction between them, when a force is just tending to cause relative sliding.

angle of heel (*Hyd., Ships*). The angle through which a floating vessel or pontoon tilts owing to eccentric placing of loads, etc.; the angle of inclination of a ship due to 'rolling', or to a 'list'. It is the angle formed between the transverse centre line of the ship when on 'even keel' and when inclined.

angle of incidence (*Aero.*). See **angle of attack.** (*Phys.*) The angle which a ray makes with the normal to a surface on which it is incident.

angle of lag, angle of lead (*Elec. Eng.*). In a.c. circuit theory the phase angle by which the current lags behind, or leads ahead of, the voltage.

angle of lens (*Photog.*). The angular coverage of a lens when exposing a sensitized surface in a camera.

angle of minimum deviation (*Light*). The minimum value of the angle of deviation for a ray of light passing through a prism. By measuring this angle (θ) and also the angle of the prism (α), the refractive index of the prism may be calculated by means of the expression:

$$n=\frac{\sin\frac{1}{2}(\alpha+\theta)}{\sin\frac{1}{2}\alpha}$$

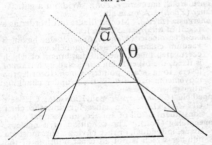

angle of nip (*Chem. Eng.*). The angle between the planes of the crushing surfaces of a crusher and the particle to be crushed. (*Min. Proc.*) The maximum included angle between two approaching faces in a crushing appliance, such as a set of rolls, at which a piece of rock can be seized and entrained.

angle of obliquity (*Eng.*). The deviation of the direction of the force between two gear teeth in contact, from that of their common tangent.

angle of pressure (*Eng.*). The angle between a gear tooth profile and a radial line at its pitch point.

angle of reflection (*Phys.*). The angle which a ray, reflected from a surface, makes with the normal to the surface. The angle of reflection is equal to the *angle of incidence* (q.v.).

angle of refraction (*Phys.*). The angle which is made by a ray refracted at a surface separating two media with the normal to the surface. See **refractive index, Snell's law.**

angle of relief (*Eng.*). The angle between the back face of a cutting tool and the surface of the material being cut.

angle of repose (*Civ. Eng., Powder Tech.*). The greatest angle to the horizontal which is made naturally by the inclined surface of a heap of loose material or embankment.

angle of roll (*Aero.*). The angle through which an aircraft must be turned about its longitudinal axis to bring the lateral axis horizontal. Also horizontal **angle of bank.**

angle of slide (*Mining*). Slope at which heaped rock commences to break away.

angle of stall (*Aero.*). The angle of incidence which corresponds with the maximum lift coefficient.

angle of twist (*Eng.*). The angle through which one section of a shaft is twisted relative to another section when a torque is applied.

angle of view (*Photog.*). The angle subtended at the centre of the lens by the diagonal of the negative.

angle of yaw (*Hyd.*). The angle between the direction of flow of a fluid stream and the direction of pointing of a velocity measuring instrument (such as a Pitot tube) immersed in the flow.

angle plane (*Carp.*). A plane whose cutting-iron shapes an internal angle.

angle plate (*Eng.*). A bracket used to support work on a lathe faceplate or other machine tool.

angle rafter (*Carp.*). The rafter at the hip of a roof. It receives the jack-rafters. Also called **angle ridge, hip rafter.**

angle shaft (*Build.*). An angle bead which is enriched with, e.g., a capital base.

anglesite (*Min.*). Orthorhombic sulphate of lead —a common lead ore; named after the original locality, Anglesey.

angle staff (*Build.*). A strip of wood placed at an angle formed by plastered surfaces to protect from damage. A rounded staff is called an *angle bead* (q.v.).

angle steel (*Eng.*). See **angle.**

angle stone (*Arch.*). A *quoin* (q.v.).

angle support (*Elec. Eng.*). A transmission line tower or pole placed at a point where the line changes its direction. Such a tower or pole differs from a normal tower or pole in that it has to withstand a force tending to overturn it (due to the resultant pull of the conductors).

angle tie (*Carp.*). See **angle brace.**

angola yarn (*Textiles*). A yarn consisting of wool or shoddy and cotton or rayon, coarsely carded and spun on the woollen mule or ringframe. Used for medium and low woollens and for union shirtings.

ångström (*Phys.*). Named after Swedish physicist, A. J. Ångström (1814–74). Unit of wavelength for electromagnetic radiation covering visible light and X-rays. Equal to 10^{-10} m. The international standard of length, the wavelength of the $2p_{10}$–$5d_5$ transition of ^{86}Kr, is 6057·8021 ångström. This unit is also used for inter-atomic spacings. Symbol Å. See **international ångström.** Superseded in SI by nanometre (=10 Å).

angular(e) (*Zool.*). In Vertebrates, membrane bone on the lower margin of the lower jaw, extending up on either side almost to the angle of the jaw.

angular acceleration (*Phys.*). The rate of change of angular velocity; usually expressed in radians per second squared. (*Space*) The acceleration of a spacecraft around an axis, resulting in pitch, roll, or yaw.

angular aperture (*Photog.*). The ratio of the working diameter to the focal-length of a lens, i.e., reciprocal of the *f*-number.

angular contact bearing (*Eng.*). A ball-bearing for radial and thrust loads in which a high shoulder on one side of the outer race takes the thrust.

angular correlations (*Nuc.*). The theory and analysis of the directional distributions of nuclear radiations together with experimental measurements, used to obtain information about the angular momenta and parities of the nuclear states and the transient fields involved.

angular diameter (*Astron.*). The angle which the apparent diameter of a heavenly body subtends at the observer's eye.

angular displacement (*Phys.*). The angle turned through by a body about a given axis, or the angle turned through by a line joining a moving point to a given fixed point.

angular distance of stars (*Astron.*). The apparent distance on the celestial sphere between two stars, measured as an arc of a great circle which passes through them.

angular distribution (*Nuc.*). The distribution relative to the incident beam of scattered particles or the products of nuclear reactions.

angular divergence (*Bot.*). The angle between the lines of insertion of two adjacent leaves.

angular frequency (*Phys.*). Frequency of a steady recurring phenomenon, expressed in radians per second, i.e., frequency in Hz multiplied by 2π. Symbol p or ω; also called **pulsatance, radian frequency.**

angular height (*Radio*). Actual height in wavelengths of an aerial, multiplied by 2π radians or 360°.

angular momentum (*Maths.*). See **momentum.** (*Phys.*) (1) The orbital angular momentum of an electron, measured as above, is *quantized* and can only have values which are exact multiples of the *Dirac unit*. (2) In particle physics, the angular momentum of particles which (appear to) have spin energy is *quantized* to values that are multiples of half the *Dirac unit*. In some cases the physical reality of the spin has not been confirmed.

angular process (*Zool.*). In *Lacertilia*, a backward extension of the articular bone. In Mammals, an inward projection of the mandibular bone at the angle between the horizontal and vertical portions.

angular thread (*Eng.*). See **vee thread.**

angular velocity (*Phys., etc.*). The rate of change of angular displacement, usually expressed in radians per second.

angulosplenial (*Zool.*). In Amphibians, a bone forming most of the inner and lower part of the lower jaw.

Angus-Smith process (*San. Eng.*). An anticorrosion process applied to sanitary ironwork; this is heated to about 316°C immediately after casting, and then plunged into a solution of 4 parts coal-tar or pitch, 3 parts prepared oil, and 1 part paranaphthaline heated to about 149°C. See **Bower-Barff process.**

angustifoliate (*Bot.*). Having narrow leaves.

angustirostrate (*Zool.*). Having a narrow beak, snout, or rostrum.

angustiseptate (*Bot.*). Said of a fruit, especially a silicular, which is laterally compressed, with a septum across the narrowest diameter.

anharmonic (*Electronics*). Said of any oscillation system in which the restoring force is nonlinear with displacement, so that the motion is not simple harmonic.

anharmonic ratio (*Maths.*). See **cross-ratio.**

anhedral (*Aero.*). See **dihedral angle.** (*Geol.*) A term used in petrography to denote a crystal which does not show any crystal faces, i.e., one which is irregular in shape.

anhidrosis (*Med.*). Diminution of the secretion of sweat. (*Vet.*) Nonsweating; drysweating. An affection of horses in humid tropical countries characterized by inability to sweat after exercise; believed to be due to prolonged overstimulation of the sweat glands by adrenaline.

anhydraemia, anhydremia (*Med.*). Loss of water from the blood.

anhydrides (*Chem.*). Substances, including organic compounds and inorganic oxides, which either combine with water to form acids, or which may be obtained from the latter by the elimination of water.

anhydrite (*Min.*). Anhydrous calcium sulphate; alters readily into gypsum. (*Build.*) Naturally occurring anhydrous calcium sulphate from which anhydrite plaster is made by grinding to powder with a suitable accelerator.

anhydrite process (*Chem. Eng.*). A process for the manufacture of sulphuric acid from anhydrite $CaSO_4$. The mineral is roasted with a reducing agent and certain other minerals in large kilns, so that SO_2 gas in relatively low concentration is recovered and after cleaning is passed to a specially designed *contact process* (q.v.). The solid residue is, under normal conditions, readily converted into cement and this forms an economic factor in the process. In Britain the process has a special significance as it provides a large potential of sulphuric acid from an indigenous source of sulphur.

anhydrous (*Chem.*). A term applied to oxides, salts, etc., to emphasize that they do not contain water of crystallization or water of combination.

anhydrous lime (*Build.*). See **lime.**

anilides (*Chem.*). *N*-phenyl amides. A group of compounds in which the hydrogen of the amino group in aniline is substituted by organic acid radicals. The most important compound of this class is *acetanilide.*

aniline (*Chem.*). $C_6H_5NH_2$, phenylamine, a colourless oily liquid, m.p. −8°C, b.p. 189°C, rel. d. 1·024, slightly soluble in water; manufactured by reducing nitrobenzene with iron shavings and hydrochloric acid at 100°C. Basis for the manufacture of dyestuffs, pharmaceutical, plastic (with methanal), and many other products. Also called **aminobenzene.**

aniline black (*Chem.*). An azinedye, produced by the oxidation of aniline on the fabric.

aniline blue stain (*Micros.*). A mixture of basic stains, soluble in alcohol, but not in water. Usually used as aniline blue W.S. (water soluble), which is an acid sulphonated preparation of aniline blue. A constituent of many triple stains used for microanatomical work.

aniline dyes (*Chem.*). A general term for all synthetic dyes having aniline as their base.

aniline foils (*Bind.*). Blocking foils which contain dyestuff; used chiefly for leather.

aniline formaldehyde (methanal) (*Chem.*). Synthetic resin formed by the polycondensation of aniline with formaldehyde.

aniline oil (*Chem.*). A coal-tar fraction consisting chiefly of crude aniline.

aniline point (*Chem.*). The lowest temperature at which an oil is miscible with an equal volume of aniline.

aniline printing (*Print.*). See **flexographic printing.**

aniline salt (*Chem.*). Phenylammonium chloride, $C_6H_5NH_2{\cdot}HCl$, m.p. 198°C, b.p. 245°C, rel. d. 1·22, white crystals, soluble in most organic solvents and water.

anima (*Psychol.*). Term used in Jungian psychology to denote the unconscious feminine component of a male personality.

animal charcoal (*Chem.*). The carbon residue obtained from carbonization of organic matter such as blood, flesh, etc.

animal electricity (*Zool.*). A term used to denote the power possessed by certain animals (e.g., electric eel) of giving powerful electric shocks.

animal field (*Zool.*). In developing blastulae, a region distinguished by the character of the contained yolk granules, and representing the first rudiment of the germ band.

animal pole (*Zool.*). In the developing ovum, the apex of the upper hemisphere, which contains little or no yolk; in the blastula, the corresponding region, wherein the micromeres lie.

animal-sized (*Paper*). Paper which has been hardened by passing the sheet through a bath of gelatine. More costly than engine-sized.

animal starch (*Chem.*). *Glycogen* (q.v.).

animated viewer (*Cinema.*). See **viewer.**

anime (*Bot.*). A hard resin used in paints and varnishes, dug near the roots of some tropical leguminous trees, and extracted from the conifer, *Agathis*, that is tropical/subtropical.

Animikie Series (*Geol.*). An important member of the Pre-Cambrian of the Canadian Shield, extending northwards into the Arctic regions. Perhaps 4250 m in thickness, it includes important iron ores, carbonaceous slate, jaspers, and boulder conglomerates. These succeed the Huronian.

animus (*Psychol.*). A Jungian term denoting the unconscious masculine component of a female personality.

anion (*Phys.*). Negative ion, i.e., atom or molecule which has gained one or more electrons in an electrolyte, and is therefore attracted to an anode, the positive electrode. Anions include all nonmetallic ions, acid radicals, and the hydroxyl ion. In a primary cell, the deposition of anions on an electrode makes it the negative pole. Anions also exist in gaseous discharge.

anionotropy (*Chem.*). Term applied to certain organic chemical reactions in which intermolecular rearrangement (*isomerization*) occurs, e.g., the conversion of linaloöl to geraniol in the presence of mineral acids.

anion respiration (*Bot.*). The absorption and accumulation of free ions by plant roots, against the concentration gradient and the electropotential gradient.

aniridia (*Med.*). See **irideremia.**

anisaldehyde (*Chem.*). 4-methoxybenzaldehyde. Colourless liquid; b.p. 248°C, occurring in aniseed, and used in perfumery. Structure:

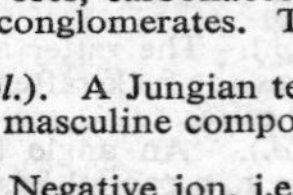

H
C
O
OCH_3

aniseikon (*Electronics*). Arrangement of two photocells, the same external object being focused on both but through separate chequerboard filters, so distorted that only movement is registered by unbalance in a bridge circuit. An electronic movement detector.

anisidines (*Chem.*). Amino-anisoles, methoxyanilines, $CH_3O{\cdot}C_6H_4{\cdot}NH_2$, bases similar to aniline. Intermediates for dyestuffs.

aniso-. Prefix from Gk. *an*, not; *isos*, equal.

anisocercal (*Zool.*). Having the lobes of the tail-fin unequal.

anisochela (*Zool.*). A chela having opposable parts of unequal size.

anisocoria (*Med.*). Inequality in the diameter of the pupils.

anisocotyly (*Bot.*). Inequality in the sizes of the cotyledons in a seedling.

anisodactylous (*Zool.*). Of Birds, having 3 toes turned forward and 1 turned backward when perching, as in the *Passeriformes*.

anisodesmic structure (*Crystal.*). One giving a crystal marked difference between its bond strengths in the intersecting axial planes.

anisogamete (*Biol.*). A gamete differing from the other conjugant in form or size. *adj.* **anisogamous.**

anisogamy (*Biol.*). Fertilization of a gamete by another gamete differing only slightly from it, as in some *Flagellata* and *Chlorophyceae*. A stage in the evolution of oögamy from isogamy. See **heterogamy.**

anisogenomatic (*Bot.*). Said of a chromosome complement made up of unlike sets of chromosomes.

anisole (*Chem.*). Phenyl methyl ether, $C_6H_5 \cdot O \cdot CH_3$, a colourless liquid, b.p. 155°C.

anisomeric (*Chem.*). Not isomeric.

anisomery (*Bot.*). The condition of a flower in which the successive whorls do not all contain the same number of members.

anisophylly (*Bot.*). See **heterophylly.**

anisopleural (*Zool.*). Bilaterally asymmetrical.

anisoploid (*Cyt.*). Having an odd number of chromosome sets in somatic cells.

anisopogonous (*Zool.*). Having the barbs of unequal length on opposite sides of the axis, as in some kinds of feather.

anisospores (*Zool.*). In *Radiolaria*, spores of two kinds found at the same time in the same species, and alleged to be gametes.

anisotonic (*Chem.*). Not isotonic.

anisotropic (*Phys., Min., etc.*). Said of crystalline material for which physical properties depend upon direction relative to crystal axes. These properties normally include elasticity, conductivity, permittivity, permeability, etc. (*Zool.*) Of ova, having a definite polarity, in relation to the primary axis passing from the animal pole to the vegetable pole. *n.* **anisotropy.**

anisotropic coma (*TV*). Distortion in a TV image arising from inclination of the objective, which is the first electric lens adjacent to the electron-emitting cathode.

anisotropic conductivity (*Elec.*). Body which has a different conductivity for different directions of current flow, electric or thermal.

anisotropic dielectric (*Elec.*). One in which electric effects depend on the direction of the applied field, as in many crystals.

anisotropic liquids (*Chem.*). See **liquid crystals.**

anisotropy (*Chem.*). Describes a property of a substance when that property depends on direction as revealed by measurement, e.g., *crystals* and *liquid crystals* (q.v.) in which the refractive index is different in different directions. (*For.*) See **orthotropy.**

ankerite (*Min.*). A carbonate of calcium, magnesium, and iron. Frequently associated with iron ores.

ankylosis, anchylosis (*Med.*). Fixation of a joint by fibrous bands within it, or by pathological union of the bones forming the joint. (*Zool.*) The fusion of two or more skeletal parts, especially bones.

ankylostomiasis (*Med.*). Hookworm disease; miners' anaemia; infection by a parasitic nematode worm in the duodenum (*Ankylostoma duodenale*), which produces a form of pernicious anaemia.

anlage (*Zool.*). See **primordium.**

annabergite (*Min.*). Hydrous nickel arsenate, apple-green monoclinic crystals, rare, usually massive. Associated with other ores of nickel. Also called **nickel bloom.**

annatto (*Chem.*). A natural colouring matter derived from the fruit of *Bixa orellana*. Source of bixin, a carotenoid acid, used to colour margarine.

annealing (*Heat*). Process of maintaining a material at a known temperature to *bleach-out* dislocations, vacancies and other metastable conditions, e.g., glass (see **lehr**), magnetic material.

annealing furnace (*Met.*). Batch-worked or continuous oven or furnace with controllable atmosphere in which metal, alloy, or glass is annealed.

Annelida (*Zool.*). A phylum of metameric *Metazoa*, in which the perivisceral cavity is coelomic, and there is only one somite in front of the mouth; typically there is a definite cuticle and chitinous setae arising from pits of the skin; the central nervous system consists of a pair of preoral ganglia connected by commissures to a postoral ventral ganglionated chain; if a larva occurs it is a trochophore. Ringed Worms.

annihilation (*Nuc.*). Spontaneous conversion of a particle and corresponding antiparticle into radiation, e.g., positron and electron, which yield 2 γ-ray photons each of 0·511 MeV. See also **mass-energy equivalence.**

annihilation radiation (*Nuc.*). The radiation produced by the annihilation of a particle with its corresponding antiparticle.

annihilator (*Maths.*). Of a set, the set of all linear functionals φ, such that $\varphi\alpha = 0$, for every α which is an element of the given set.

annite (*Min.*). The ferrous iron end-member of the biotite series of micas, poorer in alumina than siderophyllite.

annoyance (*Acous.*). The psychological effect arising from excessive noise. There is no absolute measure, but the annoyance caused by specified classes of noise can be correlated.

annual (*Bot.*). A plant which, in the same season that it develops from a seed, flowers, fruits, and dies.

annual equation (*Astron.*). One of the four principal periodic terms in the mathematical expression of the moon's orbital motion. Its period is a year, and it is caused by the varying distance between the earth and the sun.

annual load factor (*Elec. Eng.*). The load factor of a generating station, supply-undertaking, or consumer, taken over a whole year.

annual (or **heliocentric**) **parallax** (*Astron.*). The angle subtended at a star by the radius of the earth's orbit. Measured as a trigonometric parallax for the nearer stars, and for more distant objects as a *spectroscopic parallax* (q.v.). All known stellar parallaxes are less than 25 mm. See **parallactic ellipse, parallax.**

annual ring (*Bot.*). One of the approximately circular bands seen when a branch or trunk is cut across; the band is a section of the cylinder of secondary wood added in one season of growth.

annular (*Bot.*). Having the form of a ring. See also **annulus.**

annular bit (*Carp.*). A bit which cuts an annular (ring-shaped) channel and leaves intact a central cylindrical plug.

annular borer (*Civ. Eng.*). A rock-boring tool which does the work of an *annular bit* (q.v.), and provides a means of obtaining a core showing a section of the strata.

annular cell (*Bot.*). A tracheid of the protoxylem, lignified in discrete rings, thus allowing extension.

annular combustion chamber (*Aero.*). A gas turbine combustion chamber in which the perforated *flame tube* forms a continuous annulus within a cylindrical outer casing.

annular eclipse (*Astron.*). A central eclipse of the sun, in which the moon's disk does not completely cover the sun's disk at the moment of greatest eclipse but leaves a ring of the solar surface visible.

annular gear (*Eng.*). A ring on which gear teeth are cut: e.g., starter ring on automobile flywheel.

annular vault (*Build.*). See **barrel vault.**

annulate (*Bot.*). (1) Shaped like a ring. (2) Having a membranous ring on the stipe.

annulated column (*Arch.*). A column formed of slender shafts clustered together, or sometimes around a central column, and secured by stone or metal bands.

annuli fibrosi (*Histol.*). Rings of connective tissue encircling the openings between the atria and ventricles and between the heart and the blood vessels.

annulus (*Bot.*). (1) A membranous frill present on the stipe of some agarics. (2) A patch or a crest of cells with thickened walls occurring in the wall of the sporangium of ferns, and bringing about dehiscence by setting up a strain as they dry. (3) A zone of cells beneath the operculum of the sporangium of a moss, which break down and assist in the liberation of the operculum. (*Maths.*) A plane surface bounded by two concentric circles, i.e., like a washer. (*Zool.*) Any ring-shaped structure: the fourth digit of a pentadactyl forelimb: in *Arthropoda*, subdivision of a joint forming jointlets; in *Hirudinea*, a transverse ring subdividing a somite externally. *adj.* **annular, annulate.**

annunciator (*Elec. Eng.*). Arrangement of indicators, tripped by relays, for indicating which of a number of circuits has operated a bell, a machine, or one of many units in a plant. Also **indicator.** (*Rail.*) Any device for indicating audibly the passage of a train past a point.

anodal, anodic (*Bot.*). In the upward direction on the genetic spiral.

anode (*Electronics*). In a valve or tube, the electrode held at a positive potential with respect to a cathode, and through which positive current generally enters the vacuum or plasma, through collection of electrons. Also positive electrode of battery or cell. In U.S. **plate.** See **cathode, ultor.**

anode battery (*Electronics*). See **high-tension battery.**

anode bend (*Electronics*). The more or less abrupt curve in the anode-current versus grid-voltage characteristic of a triode, which occurs at small values of anode current. Also called **bottom bend.**

anode-bend detector (*Electronics*). See **plate-bend detector.**

anode-bend rectification (*Electronics*). That dependent on the curvature of the anode-current versus grid-voltage characteristic.

anode breakdown voltage (*Electronics*). That required to trigger a discharge in a cold-cathode glow tube when the starter gap (if any) is not conducting. It is measured with any grids or other electrodes earthed to cathode.

anode brightening (*Met.*). See **electrolytic polishing.**

anode characteristic (*Electronics*). Graph relating anode current and anode voltage for an electron tube.

anode circuit (*Electronics*). Closed circuit formed by the anode-cathode path in the valve, the B-battery, and the coupling impedance or transformer and its load.

anode conductance (*Electronics*). Anode current divided by the anode potential. Frequently, though incorrectly, used for the slope of the anode-current versus anode-voltage characteristic.

anode converter (*Electronics*). A rotary converter or motor generator used for supplying the high voltage required for the anode circuits from a low-voltage source, such as the filament supply battery.

anode current (*Electronics*). Current flowing to the anode of a thermionic tube.

anode-current characteristic (*Electronics*). Curve relating the anode current of a multielectrode tube to the potential of one of the electrodes; e.g., anode current versus grid voltage.

anode-current surface (*Electronics*). Surface geometrically relating the anode current to the potential of two electrodes, usually the anode and grid, of a multielectrode tube. The z (vertical) coordinate represents the anode current, whilst the x and y coordinates represent the grid and anode voltages.

anode dark space (*Electronics*). Dark zone near the anode in a glow-discharge tube.

anode dissipation (*Electronics*). Generally, the energy produced at the anode of a thermionic tube and wasted as heat owing to the bombardment by electrons: specifically, the maximum permissible power which may be dissipated at the anode.

anode drop (or fall) (*Electronics*). Component of the anode-to-cathode potential difference in a gas-filled discharge tube which is independent of the anode current.

anode effect (*Electronics*). In electrolysis, the sudden drop in current due to the formation of a film of gas on the surface of the anode.

anode efficiency (*Electronics*). Ratio of a.c. power in the load to d.c. power supplied to the anode. This may vary between 20% and 80%, depending on conditions of drive; also called **conversion efficiency, plate efficiency.**

anode feed (*Electronics*). Supply of direct current to anode of valve, generally decoupled, so that the supply circuit does not affect the condition of operation of the valve.

anode glow (*Electronics*). Luminous zone on anode side of positive column in a gas-discharge tube.

anode impedance (*Electronics*). Complex a.c. impedance between anode and cathode of thermionic valve, including interelectrode capacitance. Misapplied to differential anode resistance.

anode load impedance (*Electronics*). That between anode and cathode outside a valve, not affected by the feed.

anode load resistance (*Electronics*). External load between anode and steady polarizing voltage.

anode modulation (*Electronics*). Insertion of the modulating signal into the anode circuit of a valve, which is oscillating or is rectifying the carrier. Also called **plate modulation.**

anode mud (*Met.*). See **anode slime.**

anode polishing (*Met.*). Same as **electrolytic polishing.**

anode-ray current (*Electronics*). Current in a

partial vacuum, represented by the movements of positively-charged particles.

anode rays (*Phys.*). See **positive rays.**

anode rectification (*Electronics*). Another name for *anode bend rectification* (q.v.).

anode resistance (*Electronics*). The anode potential divided by the anode current: frequently, though incorrectly, applied to *differential anode resistance.*

anoderm (*Bot.*). Having no skin.

anode saturation (*Electronics*). Limitation of current through the anode of a valve, arising from current, voltage, temperature, or space charge.

anode shield (*Electronics*). Electrode used in high-power gas tubes to shield the anode from damage by ion bombardment.

anode slime (*Met.*). Residual slime left when anode has been electrolytically dissolved. It may contain valuable byproduct metals. Also called **anode mud.**

anode stopper (*Electronics*). Small resistance joined directly to anode contact on thermionic valve and intended to inhibit parasitic oscillation.

anode strap (*Electronics*). Conducting strip connecting segments of magnetron anode.

anode tap (*Electronics*). Tapping point on the inductance coil of a tuned-anode circuit, to which the anode is connected. The position of the tap is adjusted so that the tube operates into the optimum impedance.

anodic etching (*Elec. Eng.*). A method of preparing metals for electrodeposition by making them the anode in a suitable electrolyte and at a suitable current density.

anodic oxidation (*Chem.*). Oxidation, i.e., removal of electrons from a substance by placing it in the anodic region of an electrolytic cell. The substance to be oxidized may be either a part of the electrolyte or the anode itself. See **anodizing.**

anodic protection (*Met.*). System for passivating steel by making it the anode in a protective circuit. Cf. *cathodic protection.*

anodic treatment (*Chem.*). See **anodizing.**

anodized (*Met.*). Said of metal surface protected by chemical or electrolytic action.

anodizing (*Chem.*). A process by which a hard, noncorroding oxide film is deposited on aluminium or light alloys. The aluminium is made the anode in an electrolytic cell containing chromic(VI) or sulphuric acid. Also called **anodic treatment.**

anodontia. (*Zool.*). Absence of teeth.

anoesia (*Psychiat.*). Mental retardation of gross degree. Idiocy.

anoestrus (*Zool.*). In Mammals, a resting stage of the oestrus cycle occurring between successive heat periods.

anolyte (*Electronics*). Ionized liquid in neighbourhood of an anode during electrolysis.

anomalistic month (*Astron.*). The interval (amounting to 27·55455 days) between two successive passages of the moon in its orbit through perigee.

anomalistic year (*Astron.*). The interval (equal to 365·25964 mean solar days) between two successive passages of the sun, in its apparent motion, through perigee.

anomaloscope (*Light*). An instrument for detection and classification of defective colour vision. Two colours are permitted to be mixed, and the result matched with a third.

anomalous dispersion (*Light*). The type of dispersion given by a medium having a strong absorption band, the value of the refractive index being abnormally high on the longer wave side of the band, and abnormally low on the other side. In the spectrum produced by a prism made of such a substance the colours are, therefore, not in their normal order.

anomalous magnetization (*Elec. Eng.*). Irregular distribution of magnetization, e.g., when consequent poles exist as well as main poles on a magnetic circuit.

anomalous scattering (*Nuc.*). See **scattering.**

anomalous secondary thickening (*Bot.*). The production of new vascular tissue by a secondarily formed cambium.

anomalous viscosity (*Phys.*). A term used to describe liquids which show a decrease in viscosity as their rate of flow or velocity gradient increases. Such liquids are also known as **non-Newtonian liquids.**

anomaly. Any departure from the strict characteristics of the type. (*Astron.*) The angle between the radius vector of an orbiting body and the major axis of the orbit, measured from perihelion (periastron, etc.) in the direction of motion.

anomeristic (*Zool.*). Of metameric animals, having an indefinite number of somites.

anomite (*Min.*). A polymorph of biotite, identical with the latter in all characters but optical orientation and structural arrangement.

anomotagmosis (*Zool.*). In metameric animals, the formation of definite regions (*tagmata*) by the differentiation of an indefinite number of somites.

anorectic (*Med.*). Substance that suppresses feeling of hunger, e.g. *amphetamine.*

anorexia (*Med.*). Loss of appetite.

anorexia nervosa (*Med.*). A condition in which loss of appetite from emotional disturbance leads to marked wasting.

anorganic. Used of a tissue (e.g., bone) from which the organic material has been removed.

anorgasmia (*Med.*). Inability to achieve sexual climax in copulation.

anorthic system (*Crystal.*). See **triclinic system.**

anorthite (*Min.*). The calcium end-member of the plagioclase group of minerals; silicate of calcium and aluminium, occurring in some basic igneous rocks, and in hornfels produced by the contact metamorphism of impure calcareous sediments.

anorthoclase (*Min.*). A triclinic sodium-rich high-temperature sodium-potassium feldspar; occurs typically in volcanic rocks, and is also known from a syenite, laurvikite, from S. Norway, which is widely used for facing buildings.

anorthosite (*Geol.*). A coarse-grained rock, consisting almost exclusively of plagioclase, near labradorite in composition. Rare in Britain; but important in certain areas of Pre-Cambrian rocks, e.g., the Canadian Shield.

anosmatic (*Zool.*). Lacking the sense of smell.

anosmia (*Med.*). Loss, partially or completely, of the sense of smell.

Anostraca (*Zool.*). An order of *Branchiopoda* without a carapace; with stalked eyes; with antennae of a fair size but not biramous; and with unjointed caudal rami. Fairy shrimp, brine shrimp.

anoxia, anoxaemia, anoxemia (*Med., Zool.*). Deficiency of oxygen in the blood: any condition of insufficient oxygen supply to the tissues: any condition which retards oxidation processes in the tissues and cells.

anoxybiosis (*Zool.*). Life in absence of oxygen.

Anseriformes (*Zool.*). An order of *Neognathae* having a desmognathous palate and webbed feet. The members of this order are unusual in the possession of an evaginable penis; they are all aquatic forms, living on the animals found living in the mud at the bottom of shallow waters and in marshes; some are powerful fliers. Geese, Ducks, Screamers, Swans.

answer-back code (*Teleg.*). The identifying code combination(s) which a teleprinter automatically transmits in response to an incoming 'who-are-you' signal.

answer-back unit (*Teleg.*). An individually-modified teleprinter subassembly, which transmits identifying code combinations, whenever it is triggered by an incoming 'who-are-you' signal.

answering equipment (*Teleph.*). The equipment provided on telephone switchboards to indicate when calls arise and to provide convenient points of telephonic access to the callers.

answering jack (*Teleph.*). The jack into which an operator inserts the answering cord of a cord circuit, to ascertain the requirements of a calling subscriber.

answer print (*Cinema.*). First print from the edited negative, shown to the producers of the sound film for final approval before release.

answer signal (*Teleph.*). A signal sent backwards along a transmission circuit, indicating that the subscriber has answered the call. It may or may not operate the calling subscriber's meter, or it may be for operator's use only.

anta (*Arch.*). A square pilaster placed at either side of a doorway or the corner of a flank wall. *pl.* **antae.**

anta-cap (*Arch.*). The capital or top of an anta. *pl.* **antae-caps.**

antagonist (*Physiol.*). A muscle which opposes the action of another muscle.

antagonism (*Bot.*). (1) The interactions of plants growing in close association, to the detriment of at least one of them. (2) The interactions of micro-organisms checking the development of parasites in higher plants. (3) The power of one toxic substance to diminish or eliminate the toxic effect of another.

antagonizing screws (*Surv.*). See **clip screws.**

antambulacral (*Zool.*). See **abambulacral.**

antapex (*Astron.*). See **solar-.**

antarticular(e) (*Zool.*). In Vertebrates, a membrane bone of the lower jaw lying inside and below the angulare, with which it usually fuses. Also **dermaticular(e), gonial(e).**

ante-. Prefix from L. *ante*, before.

antebrachium (*Zool.*). The region between the brachium and the carpus in land Vertebrates; the fore-arm.

antecedent (*Maths.*). (1) In a ratio $a : b$, a is the antecedent, b is the consequent. (2) In material implication (if p, then q), p is the antecedent, q is the consequent.

antecedent genome (*Bot.*). The condition of the genome when it plays the principal part in determining inheritance.

antechamber (*Eng.*). A small auxiliary combustion-chamber, used in some compression-ignition engines, in which partial combustion of the fuel is used to force the burning mixture into the cylinder, so promoting more perfect combustion. See **precombustion chamber.**

anteclypeus (*Zool.*). In Insects which have the clypeus divided by a transverse suture, the anterior portion.

antecoxal piece (*Zool.*). In Insects, a sclerite lying between the trochantin and the precoxal bridge, or between the trochantin and the episternum: a sclerite differentiated laterally from that portion of the clypeus which carries the process articulating with the mandible of its side.

antecubital (*Zool.*). In front of the elbow.

antedorsal (*Zool.*). In Fish, situated in front of the dorsal fin.

antefixae (*Arch.*). Ornaments placed at the eaves and cornices of ancient buildings to hide the ends of the roof tiles; sometimes perforated to convey water away from the roof.

antefrons (*Zool.*). In certain Insects, that part of the frons anterior to a line joining the bases of the antennae.

antenna (*Radio*). Exposed conductor for radiation or reception of electromagnetic wave energy. Also **aerial.** (*Zool.*) In *Arthropoda*, one of a pair of anterior appendages, normally many-jointed and of sensory function: in certain Fish of the order *Pediculati*, the elongate first dorsal fin-ray, which bears terminally a skinny flap, used by the fish to attract prey. *pl.* **antennae.** *adjs.* **antennary, antennal.**

antenna changeover switch (*Radio*). Switch used for transferring an antenna from the transmitting to the receiving equipment, and vice versa, protecting the receiver.

antenna downlead (*Radio*). Wire running from the elevated part or conductor of an antenna down to the transmitting or receiving equipment.

antenna earthing switch (*Radio*). One used for disconnecting the antenna from the transmitting or receiving apparatus and connecting it directly to earth, as a protection against lightning.

antenna effect (*Radio*). (1) Error in direction finding arising from nonsymmetry of antenna, interference by downcoming waves when antenna acts as an open aerial. (2) Action of a loop antenna in picking up signals from directions in which it is not normally responsive, due to asymmetrical distribution of capacitance to earth.

antenna feeder (*Radio*). The transmission line or cable by which energy is fed from the transmitter to the antenna.

antenna field (*Radio*). Map showing electromagnetic field strength produced by antenna in the form of contour lines joining points of equal field intensity.

antenna gain (*Radio*). Ratio of maximum energy flux from antenna, to that which would have been received from a single dipole radiating the same power. In U.S. **directivity.**

antenna impedance (*Radio*). Complex ratio of voltage to current at the point where the feeder is connected.

antennal glands (*Zool.*). The principal excretory organs of *Crustacea*. They open at the bases of the appendages from which they take their name. Also called **maxillary glands.**

antenna load (*Elec. Eng.*). Same as **dummy load.**

antenna resistance (*Radio*). Total power supplied to an antenna system divided by the square of a specified current, e.g., in the feeder, or at the earth connection of an open-wire antenna.

antenna-shortening capacitor (*Radio*). That connected in series with an antenna operated at a frequency higher than its first natural frequency, to lower the impedance between the base of the antenna and earth.

antenna system (*Radio*). The whole of the equipment of a radio transmitter or receiver associated with the antenna-to-earth circuit.

antennule (*Zool.*). A small antenna: in some *Arthropoda* (as the *Crustacea*) which possess two pairs of antennae, one of the first pair.

antepetalous, antipetalous (*Bot.*). Inserted opposite to the petals.

anteposition (*Bot.*). Situation opposite, and not alternate to, another plant member.

anterior (*Bot.*). (1) The side of a flower next to the bract, or facing the bract. (2) That end of a motile organism which goes first during locomotion. (*Zool.*) In animals in which cephalization has occurred, nearer the front or cephalad end of the longitudinal axis: in human anatomy, ventral.

antero-. Prefix from *anterior*, former.

antesepalous, antisepalous (*Bot.*). Inserted opposite to the sepals.

anthelion (*Meteor.*). A mock sun appearing at a point in the sky opposite to and at the same altitude as the sun. It is probable that the phenomenon is caused by the reflection of sunlight by ice crystals.

anthelminthic, anthelmintic (*Med.*). Destructive to intestinal worms: a drug used against intestinal worms.

anther (*Bot.*). Fertile part of a stamen, usually containing 4 sporangia, and producing pollen.

antheridiophore, antheridial receptacle (*Bot.*). A special branch bearing one or more antheridia.

antheridium (*Bot.*). The gametangium which produces the male gametes in lower plants. (*Zool.*) In certain flagellate *Protozoa*, a cluster of microgametes. *pl.* **antheridia.** *adj.* **antheridial.**

antherozoid (*Bot.*). A motile male gamete, spermatozoid, or sperm. (*Zool.*) A microgamete occurring in an *antheridium* (q.v.).

anthesis (*Bot.*). The opening of a flower bud: by extension, the duration of life of any one flower, from the opening of the bud to the setting of fruit.

anthocarp (*Bot.*). A fruit consisting of the ripened ovary and its seeds, together with the persistent perianth or other parts of the flower.

anthocaulis (*Zool.*). In certain *Anthozoa*, such as *Fungia*, the pedicle formed by the rest of the trophozooid after the specialisation of the anthocyathus.

Anthocerotales (*Bot.*). An order of the *Bryophyta*. The sporophyte has a meristem between the seta and foot, and has photosynthetic tissue. The chloroplasts are typically large and occur singly in the photosynthetic cells.

anthocyanins (*Bot., Chem.*). The water-soluble colouring matters of many plants and flowers, of glucoside structure.

anthocyathus (*Zool.*). In certain *Anthozoa*, such as *Fungia*, the free discoid adult, formed by the expansion of the upper part of the calycle of the trophozooid.

anthogenesis (*Zool.*). A form of parthenogenesis in which both males and females are produced by asexual forms, as in some *Aphidae*.

anthoid (*Bot.*). Looking like a flower.

antholysis (*Bot.*). Retrograde metamorphosis of a flower.

Anthomedusae (*Zool.*). The medusoid persons of the members of the order *Gymnoblastea* (q.v.), which lack otocysts, but may possess ocelli, and in which the gonads are situated on the manubrium.

anthophilous (*Bot.*). Flower-loving; feeding on flowers.

anthophore (*Bot.*). An elongation of the floral receptacle between the calyx and corolla.

anthophyllite (*Min.*). An orthorhombic amphibole of grey-brown colour, usually massive, and normally occurring in metamorphic rocks; a metasilicate of magnesium and iron.

Anthophyta (*Bot.*). See **Phanerogamae.**

anthotaxy (*Bot.*). The arrangement of floral parts.

anthoxanthin (*Bot.*). Yellow pigment in flowers.

Anthozoa (*Zool.*). A class of *Cnidaria* in which alternation of generations does not occur, the medusoid phase being entirely suppressed; the polyps possess gastral ridges and filaments and a stomodaeal tube, and may be solitary or colonial; the gonads are of endodermal origin. Also **Actinozoa.**

anthracene (*Chem.*). $C_{14}H_{10}$, colourless, blue fluorescent crystals, m.p. 218°C, b.p. 340°C, a valuable raw material for dyestuffs obtained from the fraction of coal-tar boiling above 270°C. Anthracene represents a group of polycyclic compounds with a series of 3 benzene rings condensed together. Carcinogenic. Useful as a scintillator in photoelectric detection of β-particles.

anthracene oil (*Chem.*). A coal-tar fraction boiling above 270°C, consisting of anthracene, phenanthrene, chrysene, carbazole, and other aromatic hydrocarbon oils.

anthraciny (*Bot.*). The breakdown of organic material by fungi, and the further transformation of the results by passage through the alimentary canals of insects and worms, giving a dark coloured soil.

anthracite coals (*Fuels*). These are slow-burning, yield very little ash, moisture, and less than 10% volatiles; generally used in closed stoves.

anthracosis (*Med.*). 'Coal-miner's lung', produced by inhalation of coal dust.

anthraflavine (*Chem.*). An anthraquinone vat dyestuff, which dyes cotton greenish yellow, obtained by heating 2-methylanthraquinone with alcoholic potassium hydroxide at 150°C.

anthranil (*Chem.*). The intramolecular anhydride of anthranilic acid (2-aminobenzoic acid), intermediate in the synthesis of indigo.

anthranilic acid (*Chem.*). $C_6H_4(COOH)\cdot NH_2$, 2-aminobenzoic acid, obtained from phthalimide by the Hofmann reaction, an oxidation product of indigo.

anthranol (*Chem.*). 3-Hydroxy-anthracene.

anthraquinone (*Chem.*). $C_6H_4(CO)_2C_6H_4$, yellow needles or prisms, which sublime easily, m.p. 285°C, b.p. 382°C. Synonym: diphenylene diketone, which name also signifies its closer relation to diketones than to quinones. Obtained by the oxidation of anthracene with sulphuric acid and chromic(VI) acid. Parent substance of an important group of dyes, including alizarin.

anthrax (*Med., Vet.*). An acute infective disease caused by the anthrax bacillus, communicable from animals to man. See also **woolsorter's disease.**

anthraxolite (*Min.*). A member of the asphaltite group.

anthraxylon (*Min.*). One of the constituents of coal, derived from the lignin of the plants forming the seam.

anthrone (*Chem.*). Pale yellow solid, m.p. 155°C. Derivative of anthracene. Tautomeric form of *anthranol* (q.v.).

anthropogenic climax (*Bot.*). A climax of vegetation produced under the influence of human activity.

anthropoid (*Zool.*). Resembling Man: pertaining to, or having the characteristics of, the *Anthropoidea.*

Anthropoidea (*Zool.*). A suborder of Primates with the eyes close together and directed forwards, the nostrils surrounded by naked skin, the upper lip protrusible, and the digits all provided with flat nails. **Marmosets,** monkeys, apes, man.

anthropomorph (*Zool.*). A conventional design of the human figure; resembling a human in form or in attributes.

anthropophyte (*Bot.*). A plant introduced incidentally in the course of cultivation.

anthropotomy. See **anatomy.**

anti-. Prefix from Gk. *anti*, against.

antiae (*Zool.*). In some Birds, feathers at the base of the bill-ridge.

anti-albumoses (*Chem.*). Decomposition products of albuminous matter produced by the action of the enzyme pepsin.

anti-aldoximes (*Chem.*). The stereoisomeric form of aldoximes in which the H and the OH groups are far removed from each other. See **synaldoximes.**

antiar (*Bot.*). The upas, a Javanese tree (*Antiaris toxicaria*). Also, the poison of its latex.

antiauxin (*Biol.*). A compound which regulates or inhibits growth stimulation by auxins.

antibaryon (*Nuc.*). Antiparticle of a baryon. The term baryon is often used generically to include both.

antibiosis (*Biol.*). A state of mutual antagonism. Cf. *symbiosis*.

antibiotics (*Chem.*). Chemical substances developed as metabolic products by many micro-organisms (fungi, bacteria, actino-mycetes, etc.)—and possibly by higher organisms—which inhibit the growth of rival micro-organisms. In particular, these substances, isolated and purified, or synthesized, and used for therapeutic purposes; e.g., *chloramphenicol, penicillin, streptomycin* (qq.v.).

antibody (*Bacteriol.*). A particular form of γ-globulin present in the serum of an animal and developed in response to invasion by an *antigen* (q.v.). It confers immunity against subsequent reinfection by the same antigen.

antibonding orbital (*Nuc.*). Orbital electron of 2 atoms, which increases in energy when the atoms are brought together, and so acts against the closer bonding of a molecule.

antical (*Bot.*). The upper surface of a thallus, stem, or leaf.

anticapacitance switch (*Elec. Eng.*). One designed to have very little capacitance between the terminals when in the open condition.

anticatalyst (*Chem.*). See **catalytic poison.**

anticathode (*Phys., Radiol.*). The anode target of an X-ray tube on which the cathode rays are focused, and from which the X-rays are emitted.

anticer (*Aero.*). Obsolete term. See **anti-icing and de-icing.**

antichlor (*Chem.*). Chemical used to remove excess of chlorine from materials which have been bleached with chlorine compounds; sodium sulphite (sulphate(IV)) is an example. Antichlors are widely used in paper-making.

anticlinal (*Zool.*). In Primates, pertaining to one of the thoracic vertebrae, which has an upright spine towards which the others incline, and which is situated at the centre of motion of the vertebral column.

anticlinal wall (*Bot.*). Line of division of cells perpendicular to the surface of a growing point.

anticline (*Geol.*). A type of fold, comparable with an arch, the strata dipping outwards, away from the fold-axis.

anticlutter (*Radar*). Automatic gain control to diminish display of locally returned echoes.

anticoagulant (*Med.*). Any chemical substance which hinders normal clotting of blood, e.g., heparin, warfarin sodium, phenindione.

anticoherer (*Radio*). See **decoherer.**

anticoincidence circuit (*Comp.*). Unit of a computer with '0' output only when all inputs agree on '1'. (*Electronics*) One which delivers a pulse if one of two pulses is independently applied, but not when both are applied together or non-simultaneously within an assigned time interval.

anticoincidence counter (*Electronics*). System of counters and circuits which record only if an ionizing particle passes through particular counters but not through the others.

anticollision beacon (*Aero.*). A flashing red light which is mounted above and below an aircraft to make it conspicuous when flying in *control zones* or other busy areas.

anticous (*Bot.*). Placed on the anterior side of an organ.

anticryptic coloration (*Zool.*). Resemblance to surroundings to facilitate attack.

anticyclone (*Meteor.*). A distribution of atmospheric pressure in which the pressure increases towards the centre. Winds in such a system circulate in a clockwise direction in the northern hemisphere and in a counterclockwise direction in the southern hemisphere. Anticyclones give rise to fine, calm weather conditions, although in winter fog is likely to develop.

anticyclotron tube (*Electronics*). A type of travelling-wave tube.

antidazzle mirror (*Autos.*). One having a photo-electric control circuit which changes it from a fully reflecting condition to partial reflection from a glass/air interface when actuated by the headlamp beam of a following vehicle.

antidiazo compounds (*Chem.*). The stereoisomeric form of diazo compounds in which the groups attached to the nitrogen atoms are far removed from each other.

antidromic (*Biol.*). Contrary to normal direction, e.g., applied to nerve cells, when the impulse is conducted along the axon towards the cell body.

antidromy (*Bot.*). Left- and right-hand twining in the same species of plant.

antienzymes (*Chem.*). The antibodies of enzymes, which neutralize the action of an enzyme; e.g., the autodigestion of the stomach is prevented by the presence of antiproteases in the walls of the stomach.

antifading antenna (*Radio*). One designed for the optimum ratio of low elevation to high elevation radiation; usually a vertical mast or wire about 0·6 wavelength in height for medium broadcast wavelengths.

antifebrin (*Chem.*). *Acetanilide* (q.v.), used in therapeutics against fever.

antiferromagnetism (*Mag.*). Reversed spin of neighbouring atoms in certain lattices, exhibiting paramagnetism with certain properties of ferromagnetism, usually at low temperatures.

antiflood and tidal valve (*San. Eng.*). A valve consisting of a cast-iron box containing a floating ball, fitted near a drain outlet to prevent back flow.

antifouling composition (*Civ. Eng., etc.*). A substance applied in paint form to ships' bottoms and structures subject to the action of sea water, to discourage marine growths.

antifreeze (*Chem., Eng.*). Solutes which lower the freezing point of water, usually in automobile engine cooling. Most commonly ethylene or propylene glycol, or methanol with a few per cent. of corrosion inhibitor such as phosphates. A 35% solution of ethene glycol or 30% of methanol and water will not freeze at temperatures above $-5°F(-20·6°C)$.

antifriction bearing (*Eng.*). A bearing in which special means (such as the use of narrow wheels or rollers to support the shaft) are adopted to reduce frictional drag.

antifriction metal (*Met.*). See **white metal.**

antigen (*Med.*). Any substance, but usually a protein, foreign to the host, which stimulates the production of antibodies, e.g., bacteria, foreign red blood cells.

antigeny (*Zool.*). Sexual dimorphism; secondary

sexual differences. *adj.* **antigenic.** Also means causing production of antibodies.

antigorite (*Min.*). One of the three minerals which are collectively known as the *serpentines*. It is abundant in the rock type, serpentinite.

anti-g suit (*Aero.*). A close fitting garment covering the legs and abdomen, which is inflated, either automatically or at will by the wearer, so that counter-pressure is applied when blood is centrifuged away from the head and heart during high-speed manœuvres. Colloquially **g-suit.**

anti-g valve (*Aero.*). (1) A spring-loaded mass type of air valve which automatically regulates the inflation of an *anti-g suit* according to the acceleration (g) loads being imposed. (2) A valve incorporated in some aircraft fuel systems to prevent engines being starved of fuel under specific g loads.

antihalation (*Photog.*). The use of backing to reduce halation in plates or films.

antihistamine (*Med.*). A substance or drug which inhibits the actions of histamine by blocking its site of action.

antihormones (*Biol.*). Substances which prevent the effect of hormones.

antihunting (*Telecomm.*). In a d.c. closed loop feedback system, the use of a transformer in series with the load current, so that a rate-of-change signal can be injected into the system to ensure stability and avoid self-oscillation.

anti-icing (*Aero.*). Protection of aircraft against icing by preventing ice formation on e.g. windshield panels, leading edges of wings, tail units and turbine engine air intakes. The most common method is to apply continuous heating either by hot air tapped from an engine or by electrical heating elements. Cf. *de-icing*.

anti-incrustator (*Eng.*). A substance used to prevent the formation of scale on the internal surfaces of steam boilers.

anti-induction network (*Teleg.*). A network connected between circuits to minimize crosstalk.

antiknock substances (*I.C. Engs.*). Substances added to petrol to lessen its tendency to detonate, or 'knock', in an engine; e.g., lead(IV)ethyl.

antiknock value (*I.C. Engs.*). The relative immunity of a volatile liquid fuel from detonation, or 'knocking', in a petrol engine, as compared with some standard fuel. See **knock-rating, octane number.**

antilepton (*Nuc.*). Positron, positive muon or antineutrino. *Lepton* is often used for leptons and antileptons together.

antilift wires (*Aero.*). See **landing wires.**

antilogarithm (*Maths.*). A number whose logarithm is the given number.

antimere (*Zool.*). See **actinomere.**

antimers (*Chem.*). Optical isomers.

antimonial lead (*Chem. Eng.*). A lead-antimony alloy of controlled analysis with much improved chemical strength and retaining good chemical corrosion resistance. For sheet and pipe antimony may be as high as 12%, for machined castings up to 20–25%.

antimonial lead ore (*Min.*). See **bournonite.**

antimoniates(V) (*Chem.*). The antimonic acids give antimoniates with aqueous solutions of potassium hydroxide.

antimonite (*Min.*). See **stibnite.**

antimony (*Chem.*). Metallic element, symbol Sb, at. no. 51, r.a.m. 121·75, m.p. 630°C, rel. d. 6·6. Used in alloys for cable covers, batteries, etc.—also as a donor impurity in germanium. Has several radioactive isotopes which emit very penetrating γ-radiation. These are used in (γn) laboratory neutron sources.

antimony alloys (*Met.*). Antimony is not used as the basis of important alloys, but it is an essential constituent in type metals, bearing metals (which contain 3–20%), in lead for shrapnel (10%), storage battery plates (4–12%), roofing, gutters, and tank linings (6–12%).

antimony black (*Met.*). Finely powdered antimony, gives plaster casts a metallic look.

antimony glance (*Min.*). Obs. name for **stibnite.**

antimony halides (*Chem.*). Antimony(III)fluoride SbF_3, and (V)fluoride SbF_5. (III)chloride $SbCl_3$, and (V)chloride $SbCl_5$. (III)bromide and (III)iodide.

antimony hydrides (*Chem.*). Two hydrides, stibine [(III)hydride] SbH_3, and the solid dihydride Sb_2H_2. See also **stibine.**

antimony oxide (*Paint.*). Chemically antimony (III)oxide (Sb_2O_3), it is the second-best hiding white pigment. Being expensive it is used for special reasons, e.g. to counteract chalking.

antimonyl (*Chem.*). The monovalent radical SbO—. (Sb(III) present).

antimuon (*Nuc.*). See **muon.**

antimutagen (*Gen.*). A compound which inhibits the action of a mutagen.

antineutrino (*Nuc.*). Particle, the emission of which is assumed to accompany radioactive decay brought about by electron capture or positron (i.e., β^+) emission. An apparently different antineutrino is associated in an exactly analogous way with positive muon decay. See **neutrino.**

antineutron (*Nuc.*). Antiparticle with spin and magnetic moment oppositely orientated to those of neutron.

antinode (*Phys.*). At certain positions in a standing wave system of acoustic or electric waves or vibrations, the location of maxima of some wave characteristic, e.g., amplitude, displacement, velocity, current, pressure, voltage. At the *nodes* these would have minimum values.

antinoise microphone (*Acous.*). One in which the diaphragm is exposed to noise on both sides, and on one side to the wanted speech sound, thus improving speech/noise ratio in the output.

antinous release (*Photog.*). A flexible camera release-cable in which the action is transmitted by a steel wire passing through steel beads contained in a fabric tube.

antioxidants (*Chem.*). These are substances which delay the oxidation of paints, plastics, rubbers, etc. Raw vegetable oils contain natural antioxidants which reduce the speed of drying of paints. Deliberately added antioxidants, generally phenol derivatives, delay the skinning of paints in the can at the cost of slightly slower drying. Similar substances added to plastics, rubbers, foods, and drugs delay degradation by oxidation.

antiparallax mirror (*Elec. Eng.*). Mirror positioned on an arc adjacent to the scale of an indicating instrument, so that the parallax error in reading the indication of the pointer is avoided by aligning the eye with the pointer and its image.

antiparallel (*Maths.*). Said of vectors which are parallel, but operate in opposite directions.

antiparticle (*Nuc.*). One which has all the properties of another particle, but reversed in one respect, e.g., the *negatron* (*negative electron*) and the *positron* (*positive electron*). Interaction between particle and antiparticle means simultaneous *annihilation*, with production of great energy, which is radiated.

antipepsin (*Chem., Zool.*). An antienzyme preventing digestion by the proteases of the digestive juices.

antiperistaltic (*Zool.*). Said of waves of contraction passing from anus to mouth, along the alimentary canal; cf. *peristaltic.* *n.* **antiperistalsis.**

antiperthite (*Min.*). An intergrowth of plagioclase and potassium feldspars with plagioclase as the dominant phase. See **perthite.**

antipetalous (*Bot.*). Opposite the petals.

antipleion (*Meteor.*). See under **pleion.**

antipodal cells (*Bot.*). Three cells, consisting each of a nucleus and cytoplasm, but unwalled, lying in the embryo sac at the end remote from the micropyle.

antipodal points (*Maths.*). The points at either end of a diameter of a sphere.

antipodes (*Geog., etc.*). On a sphere, e.g., the earth, points on the surface at either extremity of a diameter.

antipolarizing winding (*Elec. Eng.*). One on a transformer or choke which carries a d.c. to neutralize the magnetizing effect of another d.c., e.g., anode current to a valve.

antipriming pipe (*Eng.*). A pipe placed in the steam space of a boiler, so as to collect the steam while excluding entrained water. See **priming** (1).

antiprothrombin (*Chem.*). An inhibitor against the activation of prothrombin to thrombin; part of the blood clotting process.

antiproton (*Nuc.*). Short-lived particle, half-life 0·05 μsec, identical with proton, but with negative charge; annihilating with normal proton, it yields mesons.

antipyr (*Chem., Photog.*). Same as *formalin* (q.v.).

antipyretic (*Med.*). Counteracting fever: a remedy for fever.

antipyrine (*Pharm.*). Phenazone. 1-Phenyl-2,3-dimethylpyrazolone. Colourless crystals, m.p. 113°C. Used as analgesic and antipyretic.

antiqua (*Typog.*). The German name for *roman* type.

antiquarian (*Paper*). A former size of drawing paper, 53 × 30in. Sometimes called **double atlas.**

antique (*Paper*). Originally applied to machine-made papers made in imitation of handmade printings. The term is now used to describe any good rough-surfaced paper which bulks well. (*Typog.*) A bold type face known as *Antique Roman.* The lines of the letters are almost uniform in thickness.

antiresonance frequency (*Electronics*). Frequency at which the parallel impedance of a tuned circuit rises to a maximum.

antiroll bar (*Autos.*). Torsion bar mounted transversely in the chassis in such a way as to counteract the effect of opposite spring deflections.

antisag bar (*Arch.*). A vertical rod connecting the main tie of a roof truss to the ridge to support it against sagging under its own weight.

antiscreen (*Photog.*). Said of an orthochromatic emulsion with reduced blue-sensitivity.

antisepsis (*Med.*). The inhibition of growth, or the destruction of bacteria in the field of operation by chemical agents: the principle of antiseptic treatment.

antiseptic (*Med.*). Counteracting sepsis or contamination with bacteria: an agent which destroys bacteria or prevents their growth.

anti-set-off spray (*Print.*). To prevent *set-off* the surface of each freshly-printed sheet receives a layer of fine particles, which prevent contact with the succeeding sheet.

anti-set-off tympan cover (*Print.*). A top cover for the second cylinder of any perfecting press, flat-bed or rotary, consisting of a material coated with very small glass beads.

antisinging device (*Teleph.*). See **vodas.**

antisolar glass. Glass which absorbs heat from sunshine and reduces glare, but transmits most of the light.

antispadix (*Zool.*). In *Tetrabranchia*, the smaller of 2 parts into which the right inner tentacular lobe is divided, bearing 4 tentacles.

antispectroscopic lens (*Photog.*). See **achromatic lens.**

antispin parachute (*Aero.*). A small parachute, normally in a canister, which may be fixed to the tail (occasionally to the wing tips) of an aeroplane or glider for release in emergency to lower the nose into a dive and so assist recovery from a spin. It is jettisoned after use. Colloquially **spin chute.**

antispray film (*Elec. Eng.*). An oil film placed on the surface of accumulator cells to prevent the formation of acid spray due to the bursting of gas bubbles during the charging process.

antisquama (*Zool.*). In Insects, when two squamae are present, the one nearest the alula. Also **antitegula, squama.**

antistatic antenna (*Radio*). One in which the receptive portion is placed outside the interfering field as much as possible, and is connected to the receiver by screened leads.

antistatic fluid (*Acous.*). That applied to a direct-recording disk before cutting a record, to obviate swarf adhering to the surface, because of generation of electric charge. Also applied to finished gramophone disks to avoid the attraction of dust, and to sound film to avoid static discharges and consequent scratches on the sound track.

anti-Stokes lines (*Phys.*). Those in scattered or fluorescent light with frequencies greater than that in the incident radiation, because of departure of atoms or molecules from their normal states.

antistyle (*Zool.*). In Insects, a projection at the base of the stylifer.

antisurge valve (*Aero.*). A valve for bleeding off surplus compressor air to suppress the unstable airflow due to *surge* (q.v.).

antisymmetric (*Phys.*). Pattern or waveform in which symmetry is complete except for one particular, e.g., sign of electric charge, direction of current, or of components in waveform.

antisymmetric dyadic (*Maths.*). See **conjugate dyadics.**

antisymmetric function (*Maths.*). See **alternating function.**

antitegula (*Zool.*). See **antisquama.**

antithetic alternation of generations (*Bot.*). (1) The alternation of morphologically (and genetically) unlike generations. (2) The theory that the sporophyte of liverworts and higher cryptogams is of secondary origin, being derived from a germinating zygote during the evolution of terrestrial forms.

antithrombin (*Chem.*). An antienzyme, produced by the liver, preventing the intravascular clotting of the blood.

antitoxins (*Med.*). Substances, produced by the organism, which, by uniting with toxins, prevent their poisonous action.

antitrades (*Meteor.*). Winds, at a height of 900 m or more, which sometimes occur in regions where tradewinds are prevalent, their direction being opposite to that of the tradewinds.

anti-transmit-receive tube (*Electronics*). Gas discharge which blocks a pulse transmitter when pulses are being received. Abbrev. **ATR tube.** See **transmit-receive tube.**

antitrochanter (*Zool.*). In Birds, the iliac articular surface, opposed to the trochanter of the femur.

antitropic (*Bot.*). Left-handed twisting.

antitropous (*Bot.*). Inverted, e.g., of embryos with radicle directed away from hilum (scar). Cf. *orthotropous* (1).

antitrypsin (*Chem., Zool.*). An antienzyme preventing digestion by the proteases of the digestive juices. It is present, e.g., in the blood.

antlia (*Zool.*). In *Lepidoptera* and other Insects, the suctorial proboscis composed of the elongate galeae.

Antonoff's rule (*Chem.*). The interfacial tension between two liquid phases in equilibrium is equal to the difference of the surface tensions of the two phases.

antorbital (*Zool.*). In front of the orbit. In Vertebrates, a small bone in the nasal region.

antorbital capsule (*Zool.*). In developing Vertebrates, a cartilage arising from the outer side of the orbitosphenoid and separating the orbit from the nasal capsule.

antorbital cartilage (*Zool.*). In *Chondrichthyes*, a cartilage in the nasal region of the skull.

antorbital process (*Zool.*). In some Fish and Amphibians, a bony process extending backwards from the anterior margin of the orbit to form part of the lower margin.

antorbital vacuities (*Zool.*). In developing Vertebrates, gaps between certain of the bones in front of the orbit.

ant-proof course (*For.*). See **termite shield.**

antrorse (*Zool.*). Directed or bent forward.

antrum (*Zool.*). A sinus, as the maxillary sinus in Vertebrates: a cavity, as the *antrum of Highmore* (q.v.). *pl.* **antra.**

antrum of Highmore (*Anat.*). An air-containing cavity in the maxilla which communicates with the nasal cavity.

Antrycide (*Pharm., Vet.*). TN. The salts of 1·1′-dimethyl - 4 amino - 6 - (2′ - amino 6′ - methyl - pyrimidyl - 4′ - amino) - quinaldine; the chloride and the methylsulphate salts show low toxicity and highly effective trypanocidal powers in cases of *T. congolense*, *T. vivax*, *T. evansi*, *T. brucei*, *T. simiae*, in cattle and various domestic animals.

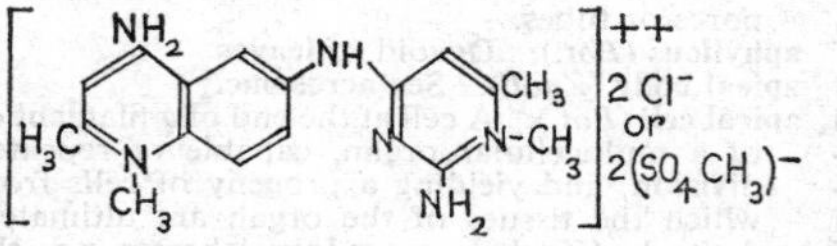

$[\ldots]^{++}$ $2\,Cl^-$ or $2(SO_4CH_3)^-$

antu (*Chem.*). 1-Naphthylthiourea

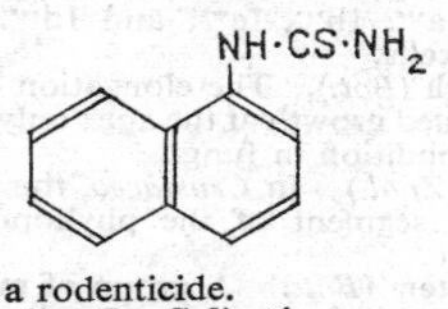

$NH{\cdot}CS{\cdot}NH_2$

used as a rodenticide.

Anura (*Zool.*). See **Salientia.**

anural, anurous (*Zool.*). Without a tail: pertaining to *Anura*.

anuria (*Med.*). Suppression of the secretion of urine.

anus (*Zool.*). The opening of the alimentary canal by which indigestible residues are voided, generally posterior. *adj.* **anal.**

anvil (*Anat.*). One of the three small bones (ossicles) which transmit mechanical vibrations between the outer ear drum and the inner ear. See also **incus.** (*Eng.*) A block of iron, sometimes steel-faced, on which work is supported during forging.

anvil cloud (*Meteor.*). A common feature of a thundercloud, consisting of a wedge-shaped projection of cloud suggesting the point of an anvil.

anvil cutter, anvil chisel (*Eng.*). A chisel with a square shank for insertion in the hardy hole of a smith's anvil, the cutting edge being uppermost.

anxiety (*An. Behav.*). A secondary drive, established by the development of a conditioned avoidance response, the stimulus then acting as negative reinforcement for other responses or eliciting unusual behaviour. (*Psychol.*) A state of mental apprehension and tension experienced by the ego in the face of impending danger. It may be (*a*) normal (or physiological) in relation to an external danger-situation threatening to cause pain to the ego, or (*b*) neurotic in relation to an internal danger-situation of strong instinctive forces threatening pain to the ego.

anxiety hysteria (*Psychol.*). A vague term used to denote a severe state of anxiety, often with hypochondriacal accompaniment and phobias, in which hysterical elements are present. It may also occur when hysteria becomes converted into an anxiety state.

anxiety neurosis (*Psychol.*). Originally classed by Freud as one of the true *neuroses* (q.v.), but later included in the psychoneuroses. The term is used generally to denote a pathological state of anxiety, which is caused by partial failure to repress a strong instinctive force, and which is therefore in close conflict with the conscious aims and ideals of the individual, threatening to undermine these. This state gives rise to feelings of sudden panic, nightmares, etc., with general anxiety symptoms (sweating, starting, palpitation, and excitability).

aorta (*Zool.*). In *Arthropoda*, *Mollusca*, and most Vertebrates, the principal arterial vessel(s) by which the oxygenated blood leaves the heart and passes to the body: in Amphibians, the principal artery by which blood passes to the posterior part of the body, formed by the union of the systemic arteries: in Fish (*ventral aorta*), the vessel by which the blood passes from the heart to the gills, and also (*dorsal aorta*) the vessel by which the blood passes from the gills to the body. *adj.* **aortic.**

aortic arches (*Zool.*). In Vertebrates, a series of pairs of vessels arising from the *ventral aorta*.

aortitis (*Med.*). Inflammation (usually syphilitic) of the aorta.

A.P. (*Surv.*). Abbrev. for *Amsterdamsch Peil* (Amsterdam level), i.e., the datum, or mean level, used as a basis for levels in Holland, Belgium, and N. Germany.

ap-. Another form of *an-* (q.v.).

apandrous (*Bot.*). Having nonfunctional male organs, or lacking male organs altogether.

apatetic coloration (*Zool.*). See **cryptic coloration.**

apatite (*Min.*). Phosphate of calcium, also containing fluoride, chloride, hydroxyl or carbonate ions, according to the variety. It is a major constituent of sedimentary phosphate rocks, of the bones and teeth of vertebrate animals (including Man), and it is usually present as an accessory mineral in igneous rocks.

APC (loop) (*TV*). See **automatic phase control (loop).**

apectometer (*Light*). A device for measuring the numerical aperture of a microscope objective.

aperient (*Med.*). A drug having a laxative or purgative effect.

aperiodic (*Phys.*). Said of any potentially vibrating system, electrical, mechanical, or acoustic, which, because of sufficient damping, does not vibrate when impulsed. Used particularly of the pointers of indicating instruments, which, having no natural period of oscillation, do not oscillate before coming to rest in the final position, and so give their ultimate reading as fast as possible.

aperiodic antenna (*Radio*). One with useful efficiency over a range of radio frequencies, terminated to minimize resonance by reflection, e.g., **Beverage antenna, rhombic antenna.** Also called **non-resonant antenna.**

aperiodic regeneration (*Telecomm.*). That employing direct- or battery-coupled amplifiers in which the degree of regeneration is independent of requency.

aperture (*Arch.*). An opening provided in a wall for a door, a window, an alcove, or for ventilation purposes. (*Optics, Photog.*) The diameter of the circular opening through which light enters an optical system. See **angular-, f-number, numerical-, stop.** (*Radar, Radio*) The effective area over which an aerial extracts power from an incident plane wave. The aperture (A) and gain (G) are related by the equation $G = 4\pi A/\lambda^2$. (*Teleg.*) (1) In a facsimile transmitter, a light-passing opening which effectively controls the size of the scanning spot. (2) In a photo-receiver, the final opening through which light modulated by the received picture signal passes.

apertured disk (*Electronics, TV*). See **Nipkow disk.**

aperture distortion (*TV*). That arising from the scanning spot having finite, instead of infinitely small, dimensions.

aperture error of a sieve (*Powder Tech.*). See **effective sieve aperture size.**

aperture lens (*TV*). Electron lens formed from holes in diaphragms, which are maintained at differing potentials.

aperture number (*Photog.*). Same as ***f*-number.**

aperture plate (*Cinema.*). A plate with a circular hole which determines the aperture of a lens.

Apetalae (*Bot.*). See **Incompletae.**

apetaly (*Bot.*). Absence of petals. *adj.* **apetalous.**

apex. The top or pointed end of anything. (*Anat.*) Said of the root of a tooth, of the top of the upper lobe of a lung, or of the rounded end of the left ventricle of the heart. (*Bot.*) The end of an organ remote from the point of attachment: the tip of a root or stem. (*Met.*) Underflow opening of hydrocyclone through which the coarser equal-settling particles of a mineral pulp flow under influence of mild centrifugal classification. (*Mining*) The 'outcrop' (exposure) or upper edge of a vein reef or lode. (*Photog.*) System of exposure calculation based on *exposure value* (q.v.), subject-brightness and film-speed. *adj.* **apical.**

apex beat (*Anat.*). The point below the left nipple where the heart beat is visible and palpable.

apex inductor (*Teleg.*). In submarine cable working, an inductor fitted at the apex of a duplex network. It not only magnifies received signal reversals, but also mitigates any local transmitter-to-receiver interference produced by cable-to-artificial line unbalance.

apex law (*Mining*). The law entitling the discoverer of an outcrop or exposure of ore to exploit it in depth beyond its lateral boundaries. (Obsolescent.)

apex stone (*Build.*). Triangular stone at the summit of a gable, often decorated with a carved trefoil. Also termed **saddle stone.**

aphagia (*Med.*). Inability to swallow. (*Zool.*) (Of imago of certain insects) inability to feed.

aphakia, aphacia (*Med.*). The condition of the eye when the lens has been removed.

aphaptotropism (*Bot.*). The condition of not reacting to contact stimulus.

aphasia (*Med.*). Loss of, or defect in, the faculty of expressing thought in words, due to a lesion in the brain.

aphelion (*Astron., Space*). The farthest point from the sun on a planet's, comet's, or spacecraft's orbit. *pl.* **aphelia.**

apheliotropic (*Bot., Zool.*). See **aphototropic.**

aphids (*Zool.*). Insects of the family *Aphididae* (order *Hemiptera*, suborder *Homoptera*). Reproduction is either sexual or parthenogenetic, oviparous or viviparous, giving rise to a complex life cycle with polymorphism in different generations. Greenfly.

aphlebia (*Bot.*). A lateral outgrowth from the base of the leaf stalk of some fossil ferns, and of a few living ferns.

aphodus (*Zool.*). In *Porifera*, the opening from a flagellated chamber to an ex-current canal, when it is drawn out into a tube. *adj.* **aphodal.**

aphonia (*Med.*). Loss of voice in hysteria, or in paralysis of the vocal cords, or in laryngitis.

aphotic (*Bot.*). Able to grow with little or no light.

aphotic zone (*Ecol.*). The zone of the sea below about 1500 metres which is essentially dark. Cf. *photic zone*.

aphotometric (*Bot.*). (1) Of a leaf, not reacting to light. (2) Of a motile organism, always directing the same end towards the light.

aphototactic (*Bot.*). Not moving in response to the light intensity.

aphototropic (*Bot.*). Growing away from light. (*Zool.*) Responding actively but negatively to light stimulus. (This is a better term than *apheliotropic*, which means, strictly, growing away from the sun.) *n.* **aphototropism.**

aphtha (*Med.*). A small grey ulcer in the mouth. *pl.* **aphthae.**

aphthous fever (*Vet.*). See **foot-and-mouth disease.**

Aphyllophorales (*Bot.*). An order of the *Basidiomycetes*, which has the basidium lining small pores or tubes.

aphyllous (*Bot.*). Devoid of leaves.

apical body (*Zool.*). See **acrosome.**

apical cell (*Bot.*). A cell at the end of a filament or of a multicellular organ, capable of repeated division, and yielding a progeny of cells from which the tissues of the organ are ultimately derived. (*Zool.*) In some Invertebrates, e.g., the Limpet (*Patella*), during cleavage of the ovum, a quartette of small cells at the apex of the egg, namely, $1a^{\cdots}$, $1b^{\cdots}$, $1c^{\cdots}$, and $1d^{\cdots}$. See also **Verson's cells.**

apical growth (*Bot.*). The elongation of a hypha by continued growth at the apex only; this is the normal condition in fungi.

apical lobe (*Zool.*). In *Crustacea*, the most distal endite or segment of the phyllopod type of appendage.

apical meristem (*Bot.*). A group of meristematic cells at the tip of a stem or root; from it, all the tissues of the mature axis are ultimately formed.

apical nervous system (*Zool.*). In *Echinodermata*, nerves of mesodermal origin, developed from the peritoneum on the aboral side.

apical organ (*Zool.*). In the trochophore larva of *Polychaeta*, a ciliary organ developing from the *apical rosette*. See **apical plate.**

apical placentation (*Bot.*). The condition in which

the ovule or ovules is/are inserted at the top of the ovary.

apical plate (*Zool.*). In various pelagic larval forms, such as trochophores, tornariae, echinoplutei, and larvae of some *Podaxonia* and *Crinoidea*, an aggregation of columnar ectoderm cells at the apical pole, usually bearing cilia.

apical rosette (*Zool.*). The first 4 micromeres to become differentiated in trochophore larvae.

apical sense organ (*Zool.*). In *Ctenophora*, an elaborate sensory structure formed of small otoliths united into a morula, supported on 4 pillars of fused cilia and covered by a roof of fused cilia.

apical string (*Zool.*). In tornariae, the solidified anterior prolongation of a coelomic sac connecting the apical plate with the proboscis-coelom and the oesophagus.

apicolysis (*Surg.*). An operation for compressing or collapsing the apex of the lung; as in the treatment of pulmonary tuberculosis.

apiculate (*Bot.*). Ending in a short, sharp point.

apiculus (*Bot.*). A short, sharp, and usually somewhat hard point, formed by the prolongation of a vein; a short projection on a spore. (*Zool.*) In some *Protozoa*, a minute apical termination.

apiezon oils (*Fuels*). The residue of almost zero vapour pressure left by vacuum distillation of petroleum products.

apileate (*Bot.*). Lacking a pileus.

A.P.I. scale (*Phys.*). American Petroleum Institute scale. Scale of relative density, similar to Baumé scale. Degrees A.P.I. $=(141{\cdot}5/s)-131$ s, where s is the rel. d. of the oil at 15°C against water at 15°C.

apituitarism (*Physiol.*). Absence or deficiency of pituitary gland secretion; also **hypohypophysism.** See also **hypopituitarism.**

Apjohn's formula (*Phys.*). A formula which may be used for determining the pressure of water vapour in the air from readings of the wet and dry bulb hygrometer. The formula is:

$$p_t=p_w-0{\cdot}00075H(t-t_w)[1-0{\cdot}008(t-t_w)],$$

where p_w is the saturated vapour pressure at the temperature (t_w) of the wet bulb, H is the barometric height, and t is the temperature of the dry bulb.

aplacental (*Zool.*). Without a placenta.

Aplacophora (*Zool.*). An order of wormlike *Amphineura* in which the foot is absent or vestigial. The shell is represented by spicules only, and the mantle cavity is perhaps represented by a small cloacal chamber. Gills may be present.

aplanatic (*Optics.*). Said of an optical system which produces an image free from spherical aberration.

aplanatic refraction (*Optics*). Refraction at a surface under conditions in which there is no spherical aberration and in which the sine condition is satisfied.

aplanetism (*Bot.*). Said of fungi having nonmotile spores, when zoopores are typical.

aplanogamete (*Bot., Zool.*). A nonmotile gamete.

aplanospore (*Bot.*). A nonmotile spore.

aplasia (*Med.*). Defective structural development.

aplite (*Geol.*). A fine textured, light coloured microgranite largely made up of quartz and feldspar. Generally occurs in association with granites.

apneusis (*Physiol.*). State of maintained inspiration.

apneustic (*Zool.*). Possessing no organs specialized for respiration; in some aquatic insect larvae, having no functional spiracles, respiration taking place through the general body surface or by means of gills.

apneustic centre (*Zool.*). In the higher Vertebrates, that part of the brain which controls the inflation of the lungs.

apnoea (*Zool.*). In forms with pulmonary respiration, cessation of respiratory movements, due to diminution of carbon dioxide tension in the alveolar air.

apo-. Prefix from Gk. *apo*, away.

apoatropine (*Chem.*). $C_{17}H_{21}NO_2$. An alkaloid of the tropane series, obtained from atropine by dehydration and identical with the naturally occurring atropamine.

apobasidium (*Bot.*). A basidium bearing the spores terminally and symmetrically on the sterigmata.

apocarpous (*Bot.*). Consisting of two or more carpels, all distinct from one another.

apochromatic (*Optics*). Said of a microscope objective in which spherical and chromatic aberrations have been corrected with the greatest possible completeness.

apochromatic lens (*Optics*). A lens so designed that light of 3 wavelengths from an object is brought to the same focus, thus reducing the secondary spectrum.

apocrine (*Med.*). Said of a gland whose product is formed by the breakdown of part of the active cells of the gland.

apocyte, apocytium (*Bot.*). A mass of multinucleate protoplasm not divided up by cell walls.

Apoda (*Zool.*). (1) An order of Amphibians having a cylindrical snakelike body without limbs, reduced eyes, and an anterior sensory tentacle; burrowing forms, living near water and feeding chiefly upon earthworms. Caecilians. (2) An order of *Cirripedia* comprising a minute vermiform animal occurring as a parasite in the mantle-cavity of other Cirripedes; it lacks a mantle and has no trunk appendages; hermaphrodite; found in West Indian waters.

apodal, apodous (*Zool.*). Without feet: without locomotor appendages.

apodeme (*Zool.*). In *Arthropoda*, an ingrowth of the cuticle forming an internal skeleton and serving for the insertion of muscles: in Insects, more particularly, an internal lateral chitinous process of the thorax.

Apodes (*Zool.*). An order of *Teleostei* having a snakelike body without a separate caudal fin; scales are vestigial or absent; there are no pelvic fins. Eels.

apodous larva (*Zool.*). A type of insect larva in which the trunk appendages are completely suppressed; formed in some *Coleoptera*, *Diptera*, *Hymenoptera*, and *Lepidoptera*.

apoenzyme (*Chem.*). The protein component of an enzyme that consists of a protein and a nonprotein portion.

apogamy (*Bot.*). Loss of sexual function without the suppression of the normal products of a sexual act. *adj.* **apogamous.**

apogee (*Astron.*). The point farthest from the earth on the apse line of a central orbit having the earth as a focus. (*Space*) The highest point above the surface of the earth in the orbit of a spacecraft, missile, etc.

apogeny (*Bot.*). Sterility.

apogyny (*Bot.*). Sterility in the female organs.

A-point (*Met.*). Temperature above which steel can be hardened.

Apollonius' circle (*Maths.*). If A and B are two fixed points, then the locus of the point P, which moves so that the ratio $PA : PB$ is a constant, is the *circle of Apollonius.*

Apollonius' theorem (*Maths.*). If the base BC of a triangle ABC is divided by a point P, such that

$\frac{BP}{PC}=\frac{r}{s}$, then

$s.AB^2+r.AC^2=(s+r).AP^2+s.PB^2+r.PC^2$.

apomecometer (*Surv.*). Instrument, based on optical square, for measuring heights and distances.

apomeiosis (*Cyt.*). Sporogenesis without haplosis.

apomixis (*Bot.*). The absence of a sexual fusion: apogamy.

apomorphine (*Pharm.*). $C_{17}H_{17}NO_2$. An alkaloid of the morphine series, obtained from morphine by dehydration. It is not a narcotic, but is an expectorant and emetic.

aponeurosis (*Zool.*). A muscle-tendon forming a broad flat sheet, which may occur not only at the end but also in the middle of a muscle.

apopetalous (*Bot.*). Lacking petals.

apophyge (*Build.*). A concave moulding forming the uppermost part of the base of a column.

apophyllite (*Min.*). A secondary mineral occurring with zeolites in amygdales in basalts and other igneous rocks. Composition: hydrated fluorosilicate of potassium and calcium.

apophysate (*Bot.*). Possessing an apophysis.

apophyseal pits (*Zool.*). In Insects, points of cuticular invagination from which arise a pair of furcal arms which form part of the thoracic endoskeleton.

apophysis (*Bot.*). (1) The swollen distal end of the seta beneath a moss sporangium. (2) An enlargement of the distal end of the scale of a pine cone. (3) A swelling on a fungal hypha beneath a sporangium or other reproductive structure. (*Geol.*) A veinlike offshoot from an igneous intrusion. (*Zool.*) In Vertebrates, a process from a bone, usually for muscle attachment: in Insects, a ventral chitinous ingrowth of the thorax for muscle insertion.

apoplexy (*Med.*). Sudden loss of consciousness, and paralysis as a result of haemorrhage into the brain or of thrombosis of a cerebral artery.

apopyle (*Zool.*) In *Porifera*, the opening by which water escapes from a flagellated chamber.

aporogamy (*Bot.*). The entrance of the pollen tube into the ovule by a path other than through the micropyle.

aposematic coloration (*Zool.*). Warning coloration, such as the gaudy colours of some stinging insects.

A-position (*Teleph.*). In a manual exchange, any individual switchboard section that is equipped for the reception and completion of local subscribers' calls.

aposporogony (*Zool.*). In certain *Sporozoa*, the suppression of sporogony in the life cycle.

apospory (*Bot.*). The elimination of spore formation from the life history, and the formation of the gametophyte from vegetative tissues and not from a spore.

apostilb (*Light*). A unit of surface luminance used in the case of diffusing surfaces, numerically equal to 1/10,000 *lambert* ($1/\pi$ cd m^{-2}).

apostrophe (*Bot.*). The position assumed by chloroplasts in bright light, when they lie against the radial walls of the cells of the palisade layer of the mesophyll.

apothecaries' weight (*Chem.*). Related to avoirdupois and troy systems in being based on the grain. 1 lb apoth. = 5760 gr. = 373·22 g metric.

apothecium (*Bot.*). An open, cup-shaped fructification, or a club-shaped derivative from it, producing ascospores; occurs in *Discomycetes* and *Discolichenes*.

apotropous (*Bot.*). The condition of an anatropous ovule with a ventral raphe.

Appalachian revolution (*Geol.*). A period of intense mountain-building movements in post-Permian, pre-Triassic times during which the deposits in the Appalachian and Cordilleran geosynclines were folded to form the Appalachian and Palaeocordilleran mountains. Equivalent to, but later than, the Armorican and Hercynian movements in Europe.

apparato reticulare (*Cyt.*). See **Golgi apparatus.**

apparent altitude (*Astron., Surv.*). The *altitude* (q.v.) of (usually) a heavenly body as directly observed instrumentally.

apparent cohesion (*Civ. Eng.*). Cohesion of silts and sands due to surface tension in the enclosed films of water, which films tend to pull the silt grains together.

apparent depression of horizon (*Astron., Surv.*). See **dip of the horizon.**

apparent expansion, coefficient of (*Phys.*). See **coefficient of apparent expansion.**

apparent horizon (*Surv.*). See **visible horizon.**

apparent magnitude (*Astron.*). See **magnitudes.**

apparent particle density (*Powder Tech.*). The mass of a particle of powder divided by the volume of the particle, excluding open pores but including closed pores.

apparent place (*Astron.*). The observed position of a star or planet as distinct from the *mean place* (q.v.).

apparent powder density (*Powder Tech.*). The mass of the powder divided by the volume occupied by it under specified conditions of packing.

apparent power (*Elec.*). The volt-amperes, i.e., the product of volts and amperes in an a.c. circuit or system.

apparent resistance (*Elec.*). Obsolete term for impedance.

apparent solar day (*Astron.*). The interval, not constant owing to the earth's elliptic orbit, between two successive transits of the **true sun** over the meridian.

apparent solar time (*Astron.*). The hour angle, at any moment, of the *true*, or *apparent*, *sun* as distinguished from the *mean sun*. Apparent noon, for example, is the instant of the meridian transit of the true sun. Sundials read apparent solar time.

appearing (*Typog.*). Term referring to the depth of the actual printed matter on a page, exclusive of traditional white line at foot.

appeasement behaviour (*An. Behav.*). Behaviour which terminates attack on one animal by another animal of the same species, excluding actual flight.

appendage (*Bot.*). General term for any external outgrowth which does not appear essential to growth or reproduction of the plant. (*Zool.*) A projection of the trunk, as the parapodia and tentacles of *Polychaeta*, sensory tentacle of *Apoda*, fins of Fish, and limbs of land Vertebrates; in *Arthropoda*, almost exclusively one of the paired, metamerically arranged, jointed structures with sensory, masticatory or locomotor function, but also used for the wings of *Insecta*.

appendicectomy, appendectomy (*Surg.*). The surgical removal of the *appendix vermiformis* (q.v.).

appendicitis (*Med.*). Inflammation of the appendix vermiformis.

appendicled (*Bot.*). Bearing small appendages.

appendicular (*Zool.*). Pertaining to, or situated on, an appendage.

appendicularia larva (*Zool.*). See **Ascidian tadpole.**

appendiculate (*Bot.*). (1) Bearing appendages. (2) Retaining fragments of the veil. (3) Having outgrowths at the throat of the corolla.

appendix (*Zool.*). An outgrowth.

appendix interna (*Zool.*). In *Leptostraca, Euphausiacea*, and some *Decapoda*, a process from the base of the inner margin of the pleopods.

appendix vermiformis (*Anat., Zool.*). In some Mammals, the distal rudiment of the caecum of the intestine, which in Man is a narrow, blind tube of gut, from 25–250 mm in length. See **appendicitis**.

appetitive behaviour (*An. Behav.*). Variable behaviour during the introductory phase of an instinctive behaviour pattern, which may be impelled by a drive, and appear purposive. It may consist merely of unoriented locomotion, but can be more variable.

appinite (*Geol.*). A diorite rich in hornblende, originally described from the Appin district of Argyllshire in Scotland. The term **appinite suite** is now applied to an association of mafic and ultramafic plutonic igneous rocks in which hornblende is an abundant constituent.

applanate (*Bot.*). Horizontally expanded.

apple (*Electronics*). Colloquialism for a type of valve used in an audiofrequency amplifier.

apple-and-biscuit microphone (*Acous.*). Omnidirectional microphone in which a moving-coil element is located on the sphere, with an adjacent plate to regulate its response in all directions normal to the axis; also **ball-and-biscuit microphone, thistle microphone.**

Appleby-Frodingham process (*Chem. Eng.*). A fluidized bed process using specially prepared iron oxide for removal of sulphur from coke oven, and coal gas and vapourized oils. Removes $H_2S{\cdot}CS_2$ mercaptans and thiophenes, converting sulphur direct to sulphur dioxide for sulphuric acid.

Applegate diagram (*Electronics*). Presentation of the *bunching* and *debunching* of an electron beam in a velocity-modulation tube, e.g., a klystron.

Appleton layer (*Phys.*). Same as **F-layer.**

applicator (*Elec. Eng.*). Electrodes used in indus trial high-frequency heating or medical diathermy; often specially shaped to fit the sample or body. See also **heating inductor.**

applied (*Bot.*). Lying upon another member by a flat surface.

applied mathematics. Originally the application of mathematics to physical problems, differing from physics and engineering in being concerned more with mathematical rigour and less with practical utility. More recently, also includes numerical analysis, statistics and probability, and applications of mathematics to biology, economics, insurance, etc.

applied power (*Elec.*). That *applied* to an electrical transducer is not equal to the actual power received, because of the reflection arising from nonequality of impedance matching. The *applied power* is the power which would be received if the load matched the source in impedance.

applied pressure (*Elec.*). The potential difference which is applied between the terminals of an electric circuit. Obsolete term.

applied psychology. That part of psychology which puts its knowledge to work in practical situations, e.g., in vocational guidance and assessment, education and industry.

applied stress (*Arch.*). The stress induced in a member under load.

appliqué (*Textiles*). (1) Ornament attached to the face of a fabric for 3-dimensional effect. (2) A lace where the silk, filament rayon, or other man-made fibre body has placed on it a floral motif or similar small design in a separate operation. Formerly handmade, most of it is now manufactured on modern machines.

apposition (*Bot.*). The growth in thickness of the cell wall by the deposition of successive laminae of wall material.

apposition image (*Zool.*). See **mosaic image.**

appressed (*Bot.*). Flattened, and pressed close to, but not united with, another organ.

appressorium (*Bot.*). (1) A flattened outgrowth which attaches a parasite to its host. (2) A modified hypha closely applied to the surface of the host and facilitating the entry of a parasitic fungus into a plant cell.

approach control radar (*Aero.*). A surveillance radar which shows on a CRT display the positions of aircraft in an aerodrome's traffic control area. Abbrev. **ACR.**

approach lights (*Aero.*). Lights indicating the desired approach to a runway, usually of sodium or high-intensity type and laid in a precise pattern of a lead-in line with cross-bars at set distances from the *runway threshold* (q.v.).

approach speed (*Aero.*). The indicated air speed at which an aeroplane approaches for landing.

approximate (*Bot.*). (1) Crowded together but not joined. (2) Of gills of agarics, approaching but not touching the stipe.

approximate integration (*Maths.*). Method of approximating to the area under a curve by considering a small number of its ordinates. See **Simpson's rule, three-eighths rule** and **Weddle's rule.**

appulse (*Astron.*). Name applied in 17th-century astronomy to the conjunction of 2 heavenly bodies; now restricted to cases where angular separation between them is very small.

apron (*Aero.*). A firm surface of concrete or 'tarmac' laid down adjacent to aerodrome buildings to facilitate the movements, loading, and unloading of aeroplanes. (*Carp.*) The wedge securing the bit of a plane. (*Eng.*) In a lathe, that part of the saddle enclosing the gear operated by the lead screw. (*Hyd. Eng.*) (1) The protecting slope on the downstream side of the sluices of a lock gate or dam provided to withstand the force of the falling water. (2) The bags of concrete, blocks of masonry, etc., deposited around the toe of a sea wall to protect its base from scour caused by the returning wave. (*Paper*) A sheet of rubber extending the full width of the paper machine to seal the gap between the breast box and the machine wire. (*Photog.*) Flexible strip used as film support in some types of processing tank. (*Plumb.*) The lead sheeting or other suitable weather resistant material raggled into a wall to divert water into a gutter or other drainage area.

apron cloths (*Textiles*). Cotton or spun rayon cloths, plain weave, colour woven, usually 38 in. wide in blue/white or red/white checks. In some a fancy design or contrasting colours appear on one side to form the apron border.

apron conveyor (*Eng.*). A conveyor for transporting packages or bulk materials, consisting of a series of metal or wood slats (also rubber, cotton, felt, wire, etc.) attached to an endless chain. Also called **slat conveyor.**

apron feeder (*Min. Proc.*). Short endless conveyor belt, sturdily built of articulated plates, used to draw ore at regulated rate from bottom of stockpile or ore bin.

apron lining (*Join.*). A lining of wrought boarding covering the apron piece at a staircase landing.

apron piece (*Join.*). The horizontal timber carrying the upper ends of the carriage pieces or rough-strings of a wooden staircase. Also called a **pitching-piece.**

apron plate (*Mining*). An amalgamated copper or muntz metal plate placed in front of the discharge screens of a stamp battery for crushing gold ores.

aproterodont (*Zool.*). Of Vertebrates, lacking premaxillary teeth.

aprotic (*Chem.*). Term applied to solvents such as ethanonitrile, which have high *relative permittivities* and hence aid the separation of electric charges but do not provide protons.

aprotic solvent (*Chem.*). Solvent which accentuates neither acidic nor basic qualities of the solute. See **Brönsted-Lowry theory, protogenic, protophilic.**

apse (*Arch.*). The semicircular or polygonal recess, either arched or dome-roofed, terminating the choir or chancel of a church. Also called **apsis.**

apse line (*Astron.*). The diameter of an elliptic orbit which passes through both foci and joins the points of greatest and least distance of the revolving body from the centre of attraction. Also called **line of apsides.**

apteria (*Zool.*). In carinate Birds, patches which are naked or covered with down only, lying between the tracts of contour feathers.

apterism (*Zool.*). The condition of winglessness, either primitive or secondary, found in many insects.

apterous (*Zool.*). Without wings.

apterygial (*Zool.*). Without wings: without fins.

Apterygota (*Zool.*). A subclass of Insects of small size and retiring habits. They never exhibit any traces of wings; metamorphosis is slight or absent; the mandibles are similar to those of the *Crustacea*, and a pair of maxillulae occurs.

Aptian Stage (*Geol.*). A division of the Cretaceous System lying between the Neocomian below and the Albian above; approximately equivalent to the English Lower Greensand.

aptyalism (*Med.*). Deficiency or absence of salivary secretion.

APU (*Aero.*). See **auxiliary power unit.**

apyrene (*Zool.*). Said of spermatozoa which develop without maturation divisions and have no nucleus, but possess a number of parallel filaments which are derived from the centrosome of the first spermatocyte; as in certain Gastropods.

apyrexia (*Med.*). Absence of fever.

aq (*Chem.*). A symbol representing a large volume of water.

aqua fortis (*Chem.*). Ancient name for concentrated nitric acid.

aqualung (*Ocean.*). Lightweight apparatus for underwater exploration, comprising face-mask, breathing tube, and compressed air reservoir.

aquamarine (*Min.*). A variety of beryl, of attractive blue-green colour, used as a gemstone.

aquamotrice (*Civ. Eng.*). A dredging implement like the *bag-and-spoon dredger* (q.v.), but having, hinged to the handle in such a way that it can be turned over to release its contents, an iron scoop or bucket instead of the bag.

aqua regia (*Chem.*). A mixture consisting of 1 volume of concentrated nitric acid to 3 volumes of concentrated hydrochloric acid.

aquarium (*Cinema.*). Colloquialism for the booth or soundproof enclosure in which, in sound-film production, mixing is executed.

aquarium reactor (*Nuc. Eng.*). See **swimming-pool reactor.**

aquasol (*Chem.*). Sulphonated castor oil.

aquatint (*Print.*). An intaglio printing process using a copper plate with a ground of resin particles. Half-tone effects are produced by etching and progressive stopping-out.

aqueduct (*Civ. Eng.*). An artificial conduit, generally elevated, used to convey water, generally for long distances, for supply purposes. (*Zool.*) A channel or passage filled with or conveying fluid: in higher Vertebrates, the reduced primitive ventricle of the midbrain.

aqueductus Sylvii (*Zool.*). In Vertebrates, the ventricle of the midbrain; the iter.

aqueductus vestibuli (*Zool.*). In *Craniata*, a narrow tube arising from the auditory sac and opening on the dorsal surface of the head, as in some fishes, or ending blindly. The endolymphatic duct.

aqueous (*Chem.*). Consisting largely of water; dissolved in water.

aqueous humour (*Zool.*). In Vertebrates, the watery fluid filling the space between the lens and the cornea of the eye.

aqueous tissue (*Bot.*). Water-storage tissue, made up of large, thin-walled, hyaline cells.

aquiculture (*Zool.*). Augmentation of aquatic animals of economic importance by direct methods: cultivation of the resources of sea and inland waters as distinct from exploitation.

aquifer (*Geol.*). Rock formation containing water in recoverable quantities.

aquintocubitalism (*Zool.*). See **diastataxy.**

aquo (or aqueous) ion. (*Phys.*). In water solution, ionized molecule which is combined with 1 or more molecules of water.

aquo-system (*Chem.*). An ionic system in water.

ar- (*Chem.*). An abbrev. indicating substitution in the aromatic nucleus.

A.R. (*Chem.*). Abbrev. for *analytical reagent*, indicating a definite standard of purity of a chemical.

Ar (*Chem.*). (1) Symbol for *argon*. (2) A general symbol for an aryl, or aromatic, radical.

arabesque (*Arch.*). An ornamental work used in decorative design for flat surfaces; consists usually of interlocked curves which may be painted, inlaid, or carved in low relief.

arabinose (*Chem.*). L-arabinose, $C_5H_{10}O_5$, is produced by boiling gum-arabic, cherry gum, or beetroot chips with dilute sulphuric acid; prisms soluble in water forming a dextrorotatory solution. Arabinose is a monosaccharide belonging to the pentose group. Used as a culture medium for certain bacteria. See **pentoses.**

arabitol (*Chem.*). A pentahydric alcohol, $OH \cdot CH_2 \cdot (CH \cdot OH)_3 \cdot CH_2 \cdot OH$. This is the alcohol corresponding to *arabinose*, and obtained from it by reduction with sodium tetrahydroborate.

Araceae (*Bot.*). A family of the *Monocotyledons* characterized by the inflorescence being made up of a spadix and spathe, e.g., Arum lily.

arachis oil (*Chem.*). Peanut oil.

arachnactis (*Zool.*). In certain *Anthozoa*, a pelagic larval form.

Arachnida (*Zool.*). A class of mainly terrestrial *Arthropoda* in which the head and thorax are continuous (*prosoma*). The head bears pedipalps and chelicerae but no antennae. There are four pairs of ambulatory legs. Spiders, harvest-men, mites, ticks, scorpions, etc.

arachnidium (*Zool.*). In Spiders, the spinnerets and silk glands.

arachnoid (*Bot., Zool.*). Cobweblike. Formed of entangled hairs or fibres: pertaining to or resembling the *Arachnida*: 1 of the 3 membranes which envelop the brain and spinal cord of Vertebrates, lying between the dura mater and the pia mater.

A.R.A.E.N. (*Teleph.*). A standard telephone transmission reference system that uses comparative articulation tests to assess in decibels (or nepers) the transmission performance of telephone equipment. The term is abbreviated from the French, '*A*ppareil de *R*éférence pour la détermination de l'*A*ffaiblissement *E*quivalent pour la *N*etteté'.

araeostyle (*Arch.*). A colonnade in which the space between the columns is equal to or greater than 4 times the lower diameter of the columns.

araeosystyle (*Arch.*). A colonnade in which the distance between the columns is alternately wide and narrow.

aragonite (*Min.*). The relatively unstable, orthorhombic form of crystalline calcium carbonate, deposited from warm water, but prone to inversion into calcite; also stable at high pressures. See also **flos ferri.**

Arago point (*Light*). The bright spot found along the axis in the shadow of a disk illuminated normally.

Arago's rotations (*Elec. Eng.*). Experiments (conducted by Arago before the discovery of electromagnetic induction by Faraday) in which a rotating copper disk was made to cause rotation of a pivoted magnet.

Araldite (*Plastics*). TN for range of epoxy resins used for adhesives, encapsulation of electrical components, etc., because of its chemical resistance.

Arales (*Bot.*). An order of the *Monocotyledons*, with either 1 to 8 stamens and a few carpels, or cyclic and solitary, with a large spathe. It includes *Araceae* and *Lemnaceae*.

Araneida (*Zool.*). An order of *Arachnida* in which the prosoma is joined to the apparently unsegmented opisthosoma by a waist. Two to four pairs of spinnerets and several kinds of spinning glands occur. The chelicerae are 2-jointed and subchelate. The pedipalps are modified in the male for the transmission of sperm. Spiders. Also **Araneae.**

araneous (*Zool.*). Cobweblike.

Arantius (*Histol.*). See **nodule of-.**

arbor (*Eng.*). Cylindrical or conical shaft on which a cutting tool or part to be machined is mounted. (*For.*) See saw-. (*Horol.*) The axis or shaft upon which a rotatable part is mounted: the shaft upon which a wheel or pinion is mounted. See **barrel-, fusee.** (*Met.*) See **mandrel.**

arbor vitae (*Zool.*). In Mammals, the treelike pattern formed by the arrangement of the white and grey matter, seen in a longitudinal section of the cerebellum.

arbuscle (*Bot.*). (1) A dwarf tree or shrub of treelike habit. (2) A much-branched haustorium formed by some endophytic fungi, and later digested by the cells of the host.

arbutin (*Chem.*). $C_{12}H_{16}O_7$. A glucoside obtained from the bear-berry, has been used as a diuretic.

arc (*Elec. Eng.*). Ionic gaseous discharge maintained between electrodes, characterized by low voltage and high current. See **mercury-arc rectifier.** (*Light*) See **-lamp, enclosed arc lamps.** Also **iron-, mirror-, open-, rotary-, sun-.** (*Maths.*) A portion of a curve.

arc absorber (*Elec. Eng.*). Same as **spark absorber**, but referring to a discharge likely to be destructive if not extinguished.

arcade (*Arch.*). (1) A series of arches, usually in the same plane, supported on columns, e.g., the nave arcades in churches. When filled in with masonry, it becomes a 'blind arcade'. (2) An arched passage, especially one having shops on one or both sides. (*Zool.*) In higher Vertebrates, a bar of bone bounding one of the vacuities or fossae of the skull.

arc-back (*Electronics*). Flow of electrons, opposite to that intended, in a mercury-arc rectifier. Caused by a heated spot on the anode acting as a cathode, leading to possible damage.

arc baffle (*Electronics*). Means of preventing liquid mercury contacting an anode in a mercury-arc rectifier. Also called **splash baffle.**

arc balance (*Horol.*). See **balance arc.**

arc-boutant (*Arch.*). See **flying buttress.**

arc-control device (*Elec. Eng.*). A device fitted to the contacts of a circuit breaker to facilitate the extinction of the arc.

arc converter (*Elec. Eng.*). One which generates a.c. in a circuit when fed with d.c. especially in high-voltage d.c. transmission of power.

arc cos x (*Maths.*). Also written $\cos^{-1} x$. See **inverse trigonometrical functions.**

arc cosh x (*Maths.*). Also written $\cosh^{-1} x$. See **inverse hyperbolic functions.**

arc crater (*Elec. Eng.*). Depression formed in electrodes between which an electric arc has been maintained. In arc welding, the depression which occurs in the weld metal.

arc deflector (*Elec. Eng.*). Magnetic arrangement for controlling the position of the arc in an arc lamp. Also **arc shield.**

arc duration (*Elec. Eng.*). Time during which an arc exists between the contacts of an opening switch or circuit breaker. In a.c. circuits usually measured in cycles, varying between half a cycle and perhaps 20 cycles.

arc furnace (*Elec. Eng.*). An electric furnace in which the heat is produced by an electric arc between carbon electrodes, or between a carbon electrode and the furnace charge. See **direct-, indirect-, Girod furnace, Héroult furnace, Stassano furnace.**

arc generator (*Radio*). A high-frequency generator which depends for its action on the negative-slope resistance of an arc discharge.

arch (*Civ. Eng.*). A form of structure having a curved shape, used to support loads or to resist pressures. (*Zool.*) A curved or arch-shaped skeletal structure supporting, covering, or enclosing an organ or organs, as *haemal arch, neural arch, zygomatic arch*. See **arcualia.**

Archaean (*Geol.*). A term applied to the oldest Pre-Cambrian rocks in a region, normally identified as those which have suffered the greatest degree of metamorphism. Archaean rocks of different regions are not necessarily identical in age.

archaeo-. Prefix from Gk. *archaios*, ancient.

Archaeocyatha (*Geol., Zool.*). A group of extinct marine animals represented by fossils from Lower Palaeozoic rocks, and believed to be intermediate in character between the Porifera and the Coelenterata. The skeleton is calcareous and is conical in shape.

archaeocyte (*Zool.*). In *Porifera*, a granular cell occurring in the blastocoel, later becoming a wandering cell which may develop into a reproductive cell.

Archaeornithes (*Zool.*). A subclass of Birds which comprises only the extinct form *Archaeopteryx* and its allies; characterized by the possession of a long lizardlike tail, teeth in the jaws, and claws at the tips of the fingers. Cf. *Neornithes*.

archaeostomatous (*Zool.*). Having a persistent blastopore, which gives rise to the mouth.

archamphiaster (*Cyt.*). The amphiaster which, during maturation divisions of the germ cells, forms the first or second polar body.

arch brick (*Build.*). A brick having a wedge shape, especially one with a curved face suitable for wells and other circular work.

arch bridge (*Civ. Eng.*). A bridge that depends on the principle of the arch for its stability. See **rigid arch, three-hinged arch, two-hinged arch.**

arch buttress (*Arch.*). See **flying buttress.**

arch dam (*Eng.*). One of which abutments are solid in rock at sides of impounding area.

arche-. Prefix from Gk. *archē*, beginning.

archecentra (*Zool.*). In Vertebrates, centra formed by the enlargement of the bases of the arch-elements which grow around the notochord outside its primary sheath; cf. *chordacentra*. *adj.* **archecentrous, arcicentrous, arcocentrous.**

arched buttress (*Arch.*). See **flying buttress.**

archedictyon (*Zool.*). In Insects of generalized structure, an irregular network of veins found between the principal longitudinal veins of the wing, in which no definite cross veins are present.

Archegoniatae (*Bot.*). In some classifications, one of the main divisions of the plant kingdom, including the *Bryophyta* and *Pteridophyta*. Characterized by the presence of the archegonium as the female organ, and by the regular alternation of gametophyte and sporophyte in the life cycle.

archegoniophore, archegonial receptacle (*Bot.*). A branch of the thallus-bearing archegonia.

archegonium (*Bot.*). A sessile or stalked organ, bounded by a multicellular wall, and flask-shaped in general outline. It consists of a chimney-like neck containing an axial series of neck-canal cells, and a swollen venter below, containing a single egg and a ventral-canal cell. The archegonium is the female organ of *Bryophyta* and *Pteridophyta*, and, in a slightly simplified form, of most *Gymnospermae*.

archencephalon (*Zool.*). In Vertebrates, the primitive forebrain; the cerebrum.

archenteron (*Zool.*). Cavity in the gastrula, enclosed by endoderm. It opens to the exterior at the blastopore.

archesporium (*Bot.*). Spore-producing tissue.

archi-. Prefix from Gk. *archi-*, first, chief.

Archiannelida (*Zool.*). A class of *Annelida*, of small size and marine habit, which usually lack setae and parapodia and have part of the epidermis ciliated; the nervous system retains a close connexion with the epidermis; they resemble the *Polychaeta* in many of their characteristics.

archibenthal, archibenthic (*Ecol.*). Pertaining to, or living on the steep slopes of, the *archibenthic zone*, as the *archibenthal* fauna.

archibenthic zone (*Ecol.*). The zone of the sea floor between the sublittoral and abyssal zones, beginning where the continental shelf begins to slope more rapidly around 200 metres, and ending at about 2000 metres. With this range of depth, pressure varies widely, and though some blue-violet light reaches the upper parts, none penetrates to the lower areas.

archiblast (*Zool.*). The protoplasm of the ovum.

archiblastic (*Zool.*). Exhibiting total and equal segmentation: pertaining to the protoplasm of the egg: pertaining to an *archiblastula* (q.v.).

archiblastula (*Zool.*). A regular spherical blastula, having cells of approximately equal size.

archicarp (*Bot.*). The female branch in *Ascomycetes* or a derivative from it. Also called **ascogonium.**

archicerebrum (*Zool.*). In Invertebrates, a median anterior ganglion forming part of the 'brain' but not associated with any particular somite: the ganglia of the first somite with the ganglia anterior to them: generally, the primitive brain.

Archichlamydeae (*Bot.*). A large subdivision of *Dicotyledons*, including many thousands of species. The flowers are either without a perianth, or have a calyx only, or possess a corolla consisting of a number of distinct petals.

archicoel (*Zool.*). See **blastocoel.**

Archimedean drill (*Eng.*). A drill in which to-and-fro axial movement of a nut on a helix causes an alternating rotary motion of the bit.

Archimedean screw (*Hyd. Eng.*). An ancient water-lifting contrivance—a hollow inclined screw (or a pipe wound in helix fashion around an inclined axis) which had its lower end in water so that, on rotation of the 'screw', water rose to a higher level.

Archimedes' principle (*Phys.*). When a body is wholly or partly immersed in a fluid, it suffers a loss in weight equal to the weight of fluid which it displaces.

Archimedes' spiral (*Maths.*). A spiral with polar equation $r = a\theta$.

Archimycetes (*Bot.*). A group of fungi, with some hundreds of species, which contains the simplest fungi known. The species are mostly aquatic, often of the simplest organization, and commonly multiply by means of zoospores.

archinephric (*Zool.*). In Vertebrates, pertaining to the archinephros (see **pronephros**): in Invertebrates, pertaining to the larval kidney or *archinephridium* (q.v.).

archinephridium (*Zool.*). In Invertebrates, the larval excretory organ, usually a solenocyte.

archinephros (*Zool.*). See **pronephros.**

archipallium (*Zool.*). In Vertebrates, that part of the cerebral hemispheres not included in the olfactory lobes and corpora striata, and comprising the hippocampus and the olfactory tracts and associated olfactory matter: that part of the pallium excluding the neopallium.

archipelago (*Geog.*). A sea area thickly interspersed with islands, originally applied to the part of the Mediterranean which separates Greece from Asia.

archiplasm (*Cyt.*). See **archoplasm.**

archipterygium (*Zool.*). In Vertebrates, a biserial appendage having a jointed central axis and two rows of jointed rays, as the paired fins of *Ceratodus*.

architectural acoustics (*Acous.*). The study of propagation of sound waves in interiors, the results being applied to the design of studios and auditoriums for optimum audition.

architomy (*Zool.*). In *Annelida*, a form of reproduction by fission, in which regeneration takes place after separation. Cf. *paratomy*.

architrave (*Arch.*). The lowest part of an entablature in immediate contact with the abacus on the capital of a column, Also called **epistyle.** (*Build., Join.*) The mouldings surrounding a door or window opening, including the lintel.

architrave block (*Join.*). The block, placed at the foot of the side moulding around a door opening, into which the skirting fits.

architrave cornice (*Arch.*). An *entablature* (q.v.) consisting of only two parts, the architrave or lower part and the cornice or upper projecting part.

architrave jambs (*Build., Join.*). The mouldings at the sides of a door or window opening.

architype (*Zool.*). A primitive type from which others may be derived.

archivolt (*Arch.*). An ornamental moulding carried around the face of an arch.

arch motor (*Elec. Eng.*). An induction machine in which the stator occupies only a fraction of the rotor periphery.

archoplasm, arcoplasm, archiplasm (*Cyt.*). In cell division, the substance of the radiations surrounding the centrosome, consisting in part of hyaloplasm from the cell body, and in part of achromatin from the nucleus: peculiarly modified cytoplasm found in connexion with the Golgi bodies: idioplasm.

archoplasmic apparatus (*Cyt.*). In cell division, the asters, and the spindle-shaped bundle of fibres between them.

arch piece (*Ships*). See **stern frame.**

arch sets (*Mining*). Steel arches used in mine tunnels to support roof and sides and perhaps carry lagging.

arch stone (*Civ. Eng.*). A stone shaped like a wedge, and used as a constituent part on the centre line of an arch. Also called **voussoir.**

arcicentrous (*Zool.*). See **archecentra.**

arciferous (*Zool.*). Having the 2 halves of the pectoral girdle overlapping ventrally, and not firmly united. Cf. *firmisternous.*

arcing contact (*Elec. Eng.*). An auxiliary contact fitted to a switch or circuit breaker, arranged so that it opens after and closes before the main contact, thereby bearing the brunt of any burning due to the arc which occurs when a circuit is interrupted. Designed for easy replacement. Also called **arcing tips.**

arcing-ground suppressor (*Elec. Eng.*). See **arc suppressor.**

arcing ring (*Elec. Eng.*). Circular or oval ring conductor, placed concentrically with a pin insulator or a string of insulators, for deflecting an arc from the insulator surface which could be damaged.

arcing shield (*Elec. Eng.*). See **grading shield.**

arcing tips (*Elec. Eng.*). See **arcing contact.**

arcing voltage (*Elec. Eng.*). That below which a current cannot be maintained between 2 electrodes.

arc lamp (*Light, etc.*). A form of electric lamp which makes use of an electric arc between 2 carbon electrodes as the source of light. It has an extremely high intrinsic brilliance, and is therefore used for searchlights, spotlights, etc. See **carbon-.**

arc-lamp carbon (*Elec. Eng.*). A cylindrical stick of carbon used as the electrode of a carbon arc lamp. The diameter is usually between about ¼ in. and 1 in. (6 mm and 25 mm).

arc-lighting dynamo (*Elec. Eng.*). A special form of dynamo arranged to give a constant current; used for supplying a number of arc lamps in series.

arc modulation (*TV*). In a mechanical scanning system, modulation of the intensity of the light-source by variation of the current in an arc discharge.

arcocentrous (*Zool.*). See **archecentra.**

arc of approach (*Eng.*). The arc on the pitch circle of a gear wheel over which 2 teeth are in contact and approaching the pitch point.

arc of contact (*Eng.*). The arc on the pitch circle of a gear wheel over which 2 teeth are in contact.

arc of recess (*Eng.*). The arc on the pitch circle of a gear wheel over which 2 teeth are in contact while receding from the pitch point.

arcoplasm (*Cyt.*). See **archoplasm.**

arc rectifier (*Elec. Eng.*). A rectifier in which an arc is maintained between 2 electrodes, the cathode being kept at incandescence by the passage of the rectified current. See **atmospheric-arc rectifier, mercury-arc rectifier, Tungar rectifier.**

arc shield (*Elec. Eng.*). See **arc deflector.**

arc sin x (*Maths.*). Also written $\sin^{-1} x$. See **inverse trigonometrical functions.**

arc sinh x (*Maths.*). Also written $\sinh^{-1} x$. See **inverse hyperbolic functions.**

arc spectrum (*Elec., Phys.*). A spectrum originating in the nonionized atoms of an element; usually capable of being excited by the application of a comparatively low stimulus, such as the electric arc. See **spark spectra.**

arc spraying (*Chem.*). Method of fusing refractory ceramic or metal powders by blowing them through an electric arc or plasma.

arc-stream voltage (*Elec. Eng.*). Voltage drop along the arc stream of an electric arc, excluding the voltage drops at the anode and cathode.

arc suppression coil (*Elec. Eng.*). See **Petersen coil.**

arc suppressor (*Elec. Eng.*). A device for automatically earthing the neutral point of an insulated-neutral transmission or distribution line in the event of an arcing ground being set up. Also called **arcing-ground suppressor.**

arc system (*Radio*). A radio communication system using an arc transmitter.

arc tan x (*Maths.*). Also written $\tan^{-1} x$. See **inverse trigonometrical functions.**

arc tanh x (*Maths.*). Also written $\tanh^{-1}x$. See **inverse hyperbolic functions.**

arc therapy (*Radiol.*). That in which the angle of rotation of therapeutic radiation is limited, to avoid certain important regions, e.g., lungs. See **rotational therapy.**

arc-through (*Electronics*). Overflow of electron stream into an intended nonconducting period.

Arctic Circle (*Geog.*). The parallel of latitude 66° 33′ N., bounding the region of the earth surrounding the north terrestrial pole, this parallel being the locus of points where the sun touches the horizon but does not set at the summer solstice.

arcualia (*Zool.*). In developing vertebrae, cartilage elements derived in part from the sclerotomes, in part from the perichordal sheath of the notochord; they give rise by their fusion to the neural and haemal arches, and, in some cases, also to part of the centrum.

arcuate (*Bot., Zool.*). Bent like a bow.

arculus (*Zool.*). A characteristic feature of the wing venation of certain Insects (as zygopterous *Odonata*); it consists of two thickened veins so joined as to form an arc.

arcus senilis (*Med.*). Degeneration of the periphery of the cornea in old people.

arc voltage (*Elec. Eng.*). The total voltage across an electric arc, i.e., the sum of the arc stream voltage, the voltage drop at the anode and the voltage drop at the cathode. The term is frequently used in connexion with arc welding, and with the arc in a switch or circuit breaker.

arc welding (*Elec. Eng.*). Process for joining of metal parts by fusion in which the heat necessary for fusion is produced by an electric arc struck between 2 electrodes or between an electrode and the metal.

ardennite (*Min.*). Silicate and vanadate of manganese and aluminium, crystallizing in the orthorhombic system as yellow prismatic crystals. First discovered in the Ardennes.

ardometer (*Heat*). A type of total radiation pyrometer.

are (*Surv.*). A metric unit of area used for land measurement. 1 are = 100 m² = 119·6 yd². See also **hectare.**

area (*Build.*). The sunken space around the basement of a building, providing access and natural lighting and ventilation. (*Radio*) Space around a transmitter and antenna defined functionally, i.e., grade of service, fading, mush, interference. (*Surv.*) In plane surveying, the superficial content of a ground surface of definite extent, as projected on to a horizontal plane.

areal velocity (*Astron., etc.*). The rate, constant in elliptic motion, at which the radius vector sweeps out unit area.

area-moment method (*Civ. Eng.*). A method of

structural analysis based on the slope and displacement of any part of the structure.

area monitoring (*Radiol.*). The survey and measurement of types of ionizing radiation and dose levels in an area in which radiation hazards are present or suspected.

area opaca (*Zool.*). In developing tetrapods, a whitish peripheral zone of blastoderm, in contact with the yolk.

area pellucida (*Zool.*). In developing tetrapods, a central clear zone of blastoderm, not in direct contact with the yolk.

area rule (*Aero.*). An aerodynamic method of reducing drag at transonic speeds by maintaining a smooth cross-sectional variation throughout the length of an aeroplane. Because of the effect of the wing, this often results in a 'waspwaist' on the fuselage or the addition of bulges to the wing or fuselage. Cf. *nose-to-tail drag curve*.

area vasculosa (*Zool.*). In developing tetrapods, part of the extraembryonic blastoderm, in which the blood vessels develop.

arecaine (*Chem.*). Arecadeine, $C_7H_{11}O_2N{\cdot}H_2O$, an alkaloid of the pyridine group, obtained from the areca or betel-nut palm (*Areca catechu*); colourless 4- or 6-sided plates, m.p. 222°–223°C, soluble in water, insoluble in most organic solvents. It has the following constitution:

COOH
N
CH_3

arenaceous, arenicolous (*Bot.*). Growing best in sandy soil. (*Zool.*) Occurring in sand; composed of sand or similar particles, as the shells of some kinds of *Radiolaria*. See also **arenaceous shell**.

arenaceous rocks (*Geol.*). Sedimentary rocks in which the principal constituents are sand grains, including the various sorts of sands and sandstones.

arenaceous shell (*Zool.*). In some *Foraminifera*, a shell composed of foreign particles.

Arenig Series (*Geol.*). The lowest (oldest) series of rocks in the Ordovician System, taking their name from Arenig mountain in N. Wales, where they were originally described by Adam Sedgwick.

areola (*Bot.*). (1) A small space delimited by lines or cracks on the surface of a lichen. (2) A small pit. (*Zool.*) (1) One of the spaces between the cells and fibres in certain kinds of connective tissue. (2) In the Vertebrate eye, that part of the iris bordering the pupil. (3) In Mammals, the dark-coloured area surrounding a nipple. *pl.* **areolae.**

areolar, areolate (*Bot., Zool.*). (1) Divided into small areas or patches. (2) Pitted. (3) Pertaining to an areola.

areolar tissue (*Zool.*). A type of connective tissue consisting of cells separated by a mucin matrix in which are embedded bundles of white and yellow fibres.

areolation (*Bot.*). The net pattern formed by the boundaries of cells.

areole (*Bot.*). The area occupied by a group of spines or hairs on a cactus.

areopyknometer (*Chem.*). An instrument for the measurement of the density of viscous liquids, e.g., glycerine.

arescent (*Bot.*). Drying.

arfvedsonite (*Min.*). A monoclinic iron-rich alkali-amphibole.

Argand burner (*Light*). A form of gas- or oil-burner in which air is admitted to the inside of a cylindrical wick, ensuring a large area of contact between the flame and the fuel.

Argand diagram (*Elec. Eng.*). The vector diagram for showing the magnitude and phase angle of a vector quantity with reference to some other vector quantity. (*Maths.*) A plane diagram, named after its description by the Swiss J. R. Argand in 1806 but actually first described by the Norwegian Caspar Wessel in 1797, in which every complex number can be represented uniquely by a single point. The diagram represents the complex number $z=x+iy$ by the point (x, y) referred to rectangular cartesian axes Oxy, the axis Ox being called the *real axis* and the axis Oy the *imaginary axis*.

argentaffin (*Chem.*). See **argyrophil.**

argentaffine cells (*Histol.*). Cells located in small numbers in the fundic glands of the stomach, the intestinal mucosa, and occasionally in the mucous lining of the large intestine. Extracts stimulate erythropoiesis in bone marrow cultures.

argentate (*Bot.*). Of silvery appearance.

argentea (*Zool.*). In Fish, a layer of silvery tissue in the eye between the sclerotic and the choroid.

argenteum (*Zool.*). In Fish, a silvery reflecting layer of the dermis containing iridocytes but no chromatophores.

argentic (silver(II)) oxide (*Chem.*). AgO, an oxide of silver.

argentiferous. Containing silver.

argentite (*Min.*). An important ore of silver, having the composition Ag_2S (silver sulphide); crystallizes in the cubic system. Cf. *acanthite.* Also called **silver glance.**

argentophil (*Chem.*). See **argyrophil.**

argentophil tissue (*Zool.*). In Insects, a fibrillar membrane probably derived from degenerating haemocytes. It forms the non-nucleate connective tissue around such viscera as the ovaries, the basement membrane of the hypodermis, and the fibrous noncellular capsule deposited around foreign bodies including parasites.

argentous (silver(I)) oxide (*Chem.*). Formula, Ag_2O. A lower oxide of silver.

argillaceous rocks (*Geol.*). Sedimentary rocks having grain-size of the clay grade. Apart from finely divided detrital matter, they consist essentially of the so-called clay minerals, such as montmorillonite, kaolinite, gibbsite, and diaspore.

argillicolous (*Bot.*). Living on clayey soil.

arginine (*Chem.*). 2-Amino-4-guanidine-valeric acid,

$H_2N{\cdot}C(:NH){\cdot}NH{\cdot}(CH_2)_3{\cdot}CH(NH_2){\cdot}COOH$,

the chief constituent of salmine, a simple protamine. It is formed in the liver and kidney from ornithine, ammonia, and carbon dioxide. It is hydrolysed by the enzyme *arginase* to ornithine and urea, in the formation of which it plays a part. It is also used in the production of creatine. One of the essential amino acids.

argol (*Chem.*). Crude cream of tartar, KH tartrate, which separates in wine vats as a reddish-brown crystalline deposit during the fermentation of grape juice to wine.

argon (*Chem.*). An element which has only the zero oxidation state; it is one of the rare gases. Symbol Ar, at. no. 18, r.a.m. 39·948. A colourless, odourless, monatomic gas; m.p. −189·2°C; b.p. −185·7°C; density

1·7837 g/dm³ at s.t.p. Argon constitutes about 1% by volume of the atmosphere, from which it is obtained by the fractionation of liquid air. It is used in gas-filled electric lamps, radiation counters, fluorescent tubes, etc.

argon laser (*Phys.*). One using singly-ionized argon. It gives strong emission at 488·0, 514·5 and 496·5 nm.

argument (*Maths.*). Of a function $f(x)$: the variable x. Of a complex number $x+iy$: the angle $\tan^{-1}\frac{y}{x}$ which is the angle in the Argand diagram between the line from the origin to the point representing the complex number and the positive direction of the real axis.

Argyll-Robertson pupil (*Med.*). An irregular eccentric pupil which reacts to accommodation, but not light.

argyria (*Med.*). Pigmentation of the skin and the tissues due to taking preparations of silver.

argyrodite (*Min.*). A double sulphide of germanium and silver, the mineral in which the element germanium was first discovered.

argyrophil (*Chem.*). Having an affinity for, easily stained by, silver salts, e.g., reticular fibres. Also **argentophil, argentaffin.**

aril (*Bot.*). An outgrowth on a seed, formed from the stalk or from near the micropyle. It may be spongy or fleshy, or may be a tuft of hairs.

arillate (*Bot.*). Having an aril.

ARINC (*Aero.*). *A*eronautical *R*adio *Inc*orporated, an organisation (U.S.) whose membership includes airlines, aircraft constructors, and *avionics* component manufacturers. It publishes technical papers and agreed standards, and finances research.

arista (*Zool.*). In certain Flies, e.g., *Cyclorrhapha*, a slender bristle borne, usually dorsally, by the terminal joint of the antenna.

aristate (*Bot.*). Bearing a bristlelike outgrowth.

Aristotle's lantern (*Zool.*). In *Echinoidea*, the framework of muscles and ossicles supporting the teeth, and enclosing the lower part of the oesophagus.

arithmetic. The science of numbers, including such processes as addition, subtraction, multiplication, division, and the extraction of roots.

arithmetical continuum (*Maths.*). The aggregate of all real numbers, rational and irrational.

arithmetical progression (*Maths.*). A sequence of numbers, each term being obtained from the preceding one by the addition of the *common difference*, e.g., $a, (a+d), (a+2d) \ldots (a+(n-1)d), \ldots$ where d is the common difference. The sum of the series of the first n terms

$$\sum_{r=0}^{n-1}(a+rd)=\frac{n}{2}\left\{2a+(n-1)d\right\}.$$

arithmetic mean (*Maths.*). Of n numbers a_r: their sum divided by n, i.e., $\frac{1}{n}\sum_{1}^{n} a_r$.

arithmetic unit (*Comp.*). That unit which performs the logical operations in an electronic computer, under the control of the *programme*.

arizonite (*Min.*). An alteration product of ilmenite, probably consisting of a mixture of haematite, ilmenite, anatase, and rutile.

arkose (*Geol.*). A sandstone in which an appreciable proportion of the grains are made of feldspars, as opposed to the commoner varieties in which most of the grains are made of quartz. Arkoses are thought to be derived from the erosion products of granites and gneisses.

arm (*Telecomm.*). See **branch.** (*Zool.*) In *Brachiopoda*, part of the lophophore: in *Echinodermata*, a prolongation of the body in the direction of a radius: in *Cephalopoda*, one of the tentacles surrounding the mouth: in *Pterobranchia*, a hollow, branched, ciliated protrusion of the collar: in bipedal Mammals, one of the upper limbs.

armature (*Bot.*). A group of persistent woody scales on the stems and leaf stalks of *Cyatheaceous* ferns. (*Elec. Eng.*) (1) Moving part which closes a magnetic circuit and which indicates the presence of electric current as the agent of actuation, as in all relays, electric bells, sounders, telephone receivers. (2) Piece of low-reluctance ferromagnetic material (*keeper*) for temporarily bridging the poles of a permanent magnet, to reduce the leakage field and preserve magnetization. (3) The rotating part (*rotor*) of a d.c. motor or generator. (*Teleph.*) See **isthmus-.**

armature bands (*Elec. Eng.*). Steel wire or bands placed round a rotating armature on an electric machine to prevent the conductors being forced out under the action of centrifugal force.

armature bars (*Elec. Eng.*). Rectangular copper bars forming the conductors on the armature in large electric machines having only a few conductors per slot.

armature coil (*Elec. Eng.*). An assembly of conductors ready for placing in the slots of the armature of an electric machine.

armature conductor (*Elec. Eng.*). One of the wires or bars on the armature of an electric machine.

armature core (*Elec. Eng.*). The assembly of laminations forming the magnetic circuit of the armature of an electric machine. The thickness of each lamination is usually of the order of 0·5 mm.

armature core disk (*Elec. Eng.*). A complete circular lamination ready for building up to form the armature core of an electric machine. See **segmental core disk.**

armature ducts (*Elec. Eng.*). Air passages built into the armature core of an electric machine to allow of the flow of cooling air.

armature end connexions (*Elec. Eng.*). The portion of the armature conductors which project beyond the end of the armature core, and which are used for making the connexions among the various conductors. Also called **overhang.**

armature end plate (*Elec. Eng.*). The end plate of a laminated armature core. It is of sufficient mechanical strength to enable the laminations to be clamped together tightly to prevent vibration. Sometimes also called **armature head.**

armature ratio (*Elec. Eng.*). Ratio of distance moved by the spring buffer of an electromagnetic relay, to that moved by the armature.

armature reactance (*Elec. Eng.*). A reactance associated with the armature winding of a machine, caused by armature leakage flux, i.e., flux which does not follow the main magnetic circuit of the machine.

armature reaction (*Elec. Eng., Mag.*). The magnetic field in an electrical machine produced by the armature current.

armature relay (*Elec. Eng.*) A relay operated electromagnetically, thus causing the armature to be magnetically attracted.

armature winding (*Elec. Eng.*). The complete assembly of conductors carried on the armature and connected to the commutator or to the terminals of the machine.

Armco iron (*Met.*). TN for a soft iron made in U.S., with less than 1% impurities. Resistivity 6·2 by volume and 5·4 by mass, compared with copper.

armed (*Bot.*). Protected by prickles or thorns.

armed lodestone (*Elec. Eng.*). One fitted with iron pole-pieces for concentrating flux.

armilla (*Bot.*). See **frill.**

arming press (*Bind.*). A form of blocking press now little used.

Armorican (*Geol.*). In the time sense, the term implies an event dating from the interval between the Carboniferous and the Permian Periods. The Armorican mountain ranges were formed at this time, and the Armorican granites were intruded in a belt of country extending from S.W. England into Brittany and Normandy (Armorica).

armour (*Bot.*). A covering of old leaf bases on the stems of cycads and some pteridophytes.

armour-clad switchgear (*Elec. Eng.*). See **metal-clad switchgear.**

armour clamp (*Cables*). A fitting designed to grip the armouring of a cable where it enters a box. Also called **armour gland, armour grip.**

armouring (*Cables*). The steel wires woven on the outside of submarine cables to protect the more delicate insulation from rocks on the sea bottom and ship's anchors, and to take the strain when the cables are laid or raised. Armouring causes eddy current loss in single core cables, so that certain cables are not armoured. See also **bedding.**

armour plate (*Met.*). Specially heavy alloy steel plate forged in hydraulic presses, hardened on the surface; used for the protection of warships. Approximate composition: C 0·2–0·4%, Cr 1·0–3·5%, Ni 1·5–3·5%, and Mo 0–0·5%.

Armstrong circuit (*Radio*). The original super-regenerative receiving circuit; also the super-sonic heterodyne circuit.

Armstrong oscillator (*Radio*). The original oscillator, in which tuned circuits in the anode and grid circuits of a valve are coupled.

armure (*Textiles*). Originally a silk or worsted cloth with embossed appearance created by a weave that forms a broken figure or wavy effect on the surface either warp- or weft-way. Now made in blended cloths, i.e., worsted/Terylene, worsted/nylon, or mixtures of wool with other fibres.

Arndt-Eistert reaction (*Chem.*). Used for converting a carboxylic acid to a higher homologue. The acid chloride is added to an excess of diazomethane to form a diazoketone. The ketone undergoes catalytic rearrangement to the higher homologue or a derivative.

Arneth count (*Histol.*). A measure of the degree of lobulation of the nucleus of the polymorphonuclear leucocytes of the blood. Changes in the relative proportions of different degrees of lobulation are of clinical interest.

Arnold's test (*Chem.*). A test for diacetic acid and proteins in the urine, consisting in the appearance of a purple or violet colour on treatment with solutions containing 4-aminoacetophenone and sodium nitrite (nitrate(III)), which are prepared in a special way.

Arnott valve (*Build.*). A flap valve fitted near the ceiling in a room to permit the escape of hot vitiated air. The air passes away into an air-flue or chimney, while back-flow is automatically prevented by the closing of the flap.

Arny solutions (*Chem.*). Solutions of stable, compatible coloured, inorganic salts, used as comparison standards in colorimetric analysis.

arolium (*Zool.*). In Insects, a pad borne by the distal joint of the tarsus.

aromatic acids (*Chem.*). Acids in which the carboxyl group is attached to an aryl radical, e.g., benzoic, C_6H_5COOH.

aromatic alcohols (*Chem.*). Derivatives of the aromatic series in which a hydroxyl group has been introduced into the side chain by replacing a hydrogen atom in a CH, CH_2, or CH_3 group, e.g., phenyl methanol, $C_6H_5CH_2OH$.

aromatic aldehydes (*Chem.*). Compounds of the aromatic series containing the group $-CHO$, e.g., benzaldehyde, C_6H_5CHO.

aromatic compounds (*Chem.*). Benzene derivatives.

aromatic hydrogenation (*Chem.*). Hydrogenation in the naphthalene series, of such nature that hydrogenation takes place only in the unsubstituted benzene ring.

aromatic properties (*Chem.*). The characteristic properties of aromatic compounds, viz. reaction with concentrated nitric acid, forming nitro derivatives, reaction with concentrated sulphuric acid, forming sulphonated derivatives. The homologues of benzene differ from alkanes with regard to oxidation by readily forming benzene carboxylic acids. There are many other distinguishing characteristics between aromatic hydrocarbons and alkanes.

aromatic vinegar (*Chem.*). Vinegar obtained by the distillation of copper diethanoate. The aromatic odour is due to the acetone.

aromorphosis (*Zool.*). In *Arthropoda*, an increase in the number of segments in the larva after hatching. See **anamorpha.**

Aron meter (*Elec. Eng.*). A kilowatt-hour meter in which 2 clock movements act on a differential gear connected to the counting train; current in the circuit, being metered, controls the swing of the pendulum; the difference between this and the swing of the other pendulum is a measure of the energy consumption, and is registered through the differential gear.

arousal (*Psychol.*). The diffuse transmission of nerve impulses from the *reticular activating system* to the cortex which results in different levels of consciousness. Some *motivation* theories regard it as the general basis of activity.

arquerite (*Min.*). See **silver amalgam.**

arrastre (*Min. Proc.*). Dragstone mill. A primitive form of grinding mill still used for ores in central America and for cement in Europe. A large boulder is drawn round in a circular stone-lined pit.

array (*Maths.*). A set of values for a particular variate.

arrect (*Bot.*). Rigid and erect.

arrectores pilorum (*Zool.*). In Mammals, unstriated muscles attached to the hair follicles, which cause the hair to stand on end by their contraction.

arrested crushing (*Met., Mining*). Crushing so conducted that the rock falling through the machine is free to drop clear of the zone of comminution when broken smaller than the exit orifice or set.

arrested failure (*Cables*). The taking of a cable off voltage, and examination before failure is complete. This is very instructive in determining the mechanism of breakdown.

arrester (*Elec. Eng.*). See **lightning arrester.**

arrester gear (*Aero.*). (1) A device on aircraft carriers and some military aerodromes, usually consisting of a number of individual transverse cables held by hydraulic shock-absorbers, which stop an aeroplane when its *arrester hook* catches a cable. (2) A barrier net, usually of nylon or webbing attached to heavy drag weights, which stops fast aeroplanes from over-running the end of the runway in an emergency.

arrester hook (*Aero.*). A hook extended from an

aeroplane to engage the cable of an arrester gear, mainly on aircraft carriers.

arrest muscle (*Zool.*). See **catch muscle.**

arrest points (*Met.*). Discontinuities on heating and cooling curves, due to absorption of heat during heating or evolution of heat during cooling, and indicating structural changes occurring in a metal or alloy.

Arrhenius theory of dissociation (*Chem.*). In electrolytic dissociation, the original idea that an acid, base or salt on solution ionizes to some extent, an applied field causing a drift up or down the grade of potential, with associated changes of boiling and freezing points.

arrhenogenic (*Gen.*). Producing mostly or entirely male offspring.

arrhenokaryon (*Cyt.*). A nucleus having a single set of haploid chromosomes, as in dispermy, when the two pairs of centrosomes remain apart without forming a quadripolar figure.

arrhenotoky (*Zool.*). Parthenogenetic production of males.

arrhizal, arrhizous (*Bot.*). Lacking roots.

arrhythmia, arhythmia (*Med.*). Abnormal rhythm of the heart beat.

arris (*Build.*). The (generally) sharp exterior edge formed at the intersection of two surfaces not in the same plane (e.g. the meeting of two sides of a stone block). See also **external angle.**

arris edge (*Glass*). Small bevel, of width not exceeding 1/16 in (1·5 mm), at an angle of approximately 45° to the surface of the glass.

arris fillet (*Build.*). A small strip of wood of triangular cross-section packed beneath the lower courses of slates or tiles on a roof to throw off the water which might otherwise get under the flashing.

arris gutter (*Build.*). A V-shaped gutter, usually made of wood.

arris rail (*Carp.*). A rail, with triangular cross-section, secured to posts for fences in such a manner as to show the arris in front.

arris tile (*Build.*). Purpose-made angular tile used to cover the intersections at hips and ridges in slated and tiled roofs. See also **bonnet tile.**

arris-wise. A term used to describe (*Build.*) the laying of tiles or slates diagonally; (*Carp.*) the sawing of square timber diagonally.

arrival curve (*Teleg.*). The current/time graph of the signal received when a measured voltage step is transmitted into a quiescent line connected between known terminal impedances.

arrow (*Surv.*). Light steel wire pin, bent into ring at one end and perhaps flagged with piece of bright cloth, used to mark measured lengths in chain traversing.

arrowroot (*Bot.*). Starch derived from the roots of plants of the *Maranta* genus; it provides a nutritious and easily digested food.

arsenic (*Chem.*). Symbol As, at. no. 33, r.a.m. 74·9216, oxidation states 3, 5. An element which occurs free and combined in many minerals. An impurity of several commercial metals. Called grey or γ-arsenic to distinguish it from the other allotropic modifications. M.p. 814°C (36 atm.), b.p. 615°C (sublimes), rel. d. 5·73 at 15°C. Used in alloys and in the manufacture of lead shot. It is important as donor impurity in germanium semiconductor devices.

α-. Yellow arsenic, an allotropic modification of arsenic. May be formed by the rapid condensation of arsenic vapour in an inert atmosphere. Rel. d. 2·0 at 18°C.

β-. Black arsenic, an allotropic modification of arsenic. May be formed by the slow condensation of arsenic vapour in an inert atmosphere. Rel. d. 3·70.

white-. The arsenic of commerce, As_2O_3. Arsenious oxide, arsenic(III)oxide. Obtained from the roasting of arsenical ores. It is highly poisonous, and its presence in foods and drinks is subject to severe restriction. Medical uses, once important, have declined, but it is still much used as a herbicide and rodenticide.

arsenic acid (*Chem.*). Formula, H_3AsO_4. Formed by the action of hot dilute nitric acid upon arsenic, or by digesting arsenic(III)oxide with nitric acid. Arsenic acid is also formed when arsenic(V)oxide is dissolved in water.

arsenical copper (*Met.*). Copper containing up to about 0·6% arsenic. This element slightly increases the hardness and strength and raises the recrystallization temperature.

-nickel (*Min.*). See **niccolite.**

-pyrite (*Min.*). See **arsenopyrite.**

arsenic halides (*Chem.*). Arsenic(V)fluoride, AsF_5; arsenic(III)fluoride, AsF_3; arsenic(III)chloride, $AsCl_3$; arsenic(III)bromide, $AsBr_3$; arsenic(III)-iodide, AsI_3.

arsenide (*Chem.*). Arsenic unites with most metals to form *arsenides*; e.g., iron—$FeAs_2$. Arsenides are decomposed by water or dilute acids with the formation of the hydride *arsine* (q.v.).

arsenious acid (*Chem.*). Solution of arsenious oxide. See **white arsenic.**

arsenites (*Chem.*). Arsenates (III). Salts of arsenious acid.

arseniuretted hydrogen (*Chem.*). See **arsine.**

arsenolite (*Min.*). Arsenic oxide, a decomposition product of arsenical ores; occurring commonly as a white incrustation, rarely as octahedral crystals.

arsenopyrite (*Min.*). Sulphide of iron and arsenic, crystallizing in the orthorhombic system; chief ore of arsenic. Also known as **mispickel.**

arsine (*Chem.*). Formula, AsH_3. Arsenic(III)-hydride. Produced by the action of nascent hydrogen upon solutions of the element, or by the action of dilute sulphuric acid upon sodium or zinc arsenide. Very poisonous. Also called **arseniuretted hydrogen.**

arsines (*Chem.*). Organic derivatives of AsH_3 in which one or more hydrogen atoms is replaced by an alkyl radical; other hydrogen atoms may also be replaced by halogen, etc.

artefact. A man-made stone implement. (*Zool.*) Any apparent structure which does not represent part of the actual specimen, but is due to faulty preparation.

arterial drainage (*San. Eng.*). A system of drainage in which the flow from a number of branch drains is led into one main channel.

arterial ring (*Zool.*). In Vertebrates, a blood vessel (formed by the splitting of the basilar artery) which surrounds the hypophysis.

arterial road (*Civ. Eng.*) A specially constructed road between one large town and another.

arterial system (*Zool.*). That part of the vascular system which carries the blood from the heart to the body.

arteriography (*Radiol.*). The radiological examination of arteries following direct injection of a *contrast medium* (q.v.).

arteriole (*Zool.*). A small artery.

arteriosclerosis (*Med.*). Hardening or stiffening of the arteries due to increase of muscular, elastic, or fibrous tissue in the middle coat of the vessel:

arteritis (*Med.*). Inflammation of an artery.

artery (*Zool.*). One of the vessels of the vascular system, that conveys the blood from the heart to the body. *adj.* **arterial.**

Artesian well (*Civ. Eng.*). A well sunk into a permeable stratum which has impervious strata above and below it, and which outcrops at places higher than the place where the well is sunk, so that the hydrostatic pressure of the water in the permeable stratum is alone sufficient to force the water up out of the well. Named from Artois (France).

arthralgia (*Med.*). Pain in a joint.

arthrectomy (*Surg.*). Excision of a joint.

arthritic (*Zool.*). Pertaining to the joints: situated near a joint.

arthritis (*Med.*). Inflammation of a joint.

arthritis deformans (*Med.*). See **osteoarthritis.**

arthrobranchiae (*Zool.*). In *Arthropoda*, gills arising from the arthrodial membranes, at the junction of the limbs with the body.

arthrodesis (*Surg.*). The surgical immobilization of a joint by fusion of the joint surfaces.

arthrodia (*Zool.*). A joint.

arthrodial membranes (*Zool.*). In *Arthropoda*, flexible membranes connecting adjacent body sclerites and adjacent limb joints, and occurring also at the articulation of the appendages.

arthrogenous (*Bot.*). Developed from portions separated off from the parent plant.

arthrography (*Radiol.*). The radiological examination of a joint cavity after direct injection of air or other *contrast media* (q.v.).

arthrophyte (*Bot.*). A member of the *Sphenopsida*, e.g., *Equisetum*, the horse tails.

Arthropoda (*Zool.*). A phylum of metameric animals having jointed appendages (some of which are specialized for mastication) and a well-developed head; there is usually a hard chitinous exoskeleton; the coelom is restricted, the perivisceral cavity being haemocoelic. Centipedes, Millipedes, Insects, Crabs, Lobsters, Shrimps, Spiders, Scorpions, Mites, Ticks, etc.

arthropodin (*Zool.*). The water-soluble fraction of the protein component of insect cuticle.

arthropterous (*Zool.*). Having jointed fin rays.

arthrospore (*Bot.*). A spore formed by segmentation and separation from the parent cell.

arthrostracous (*Zool.*). Having a segmented shell, as *Chiton*.

arthrotomy (*Surg.*). Surgical incision into a joint.

article (*Bot.*). A joint of a stem or fruit, breaking apart at maturity.

articulamentum (*Zool.*). In *Amphineura*, the porcellaneous compact lower layer of the shell. See **tegumentum** (1).

articular(e) (*Zool.*). Pertaining to, or situated at, or near, a joint. In Vertebrates, a small cartilage at the angle of the mandible, derived from the Meckelian, and articulating with the quadrate forming the lower half of the jaw hinge. *pl.* **articularia.**

articular corpuscles (*Zool.*). In Mammals, sensory nerve endings resembling end bulbs, occurring in the neighbourhood of joints.

Articulata (*Zool.*). See **Testicardines.**

articulate, articulated (*Bot.*). Breaking up at maturity, without rupture, into two or more distinct portions.

articulated assembly and rods (*Aero.*). The assembly of master connecting rod and big-end bearing, with articulated connecting rods attached by wrist pins through pairs of lugs on the flanges of the master rod of radial engines.

articulated blade (*Aero.*). A rotorcraft blade which is mounted on one or more hinges to permit flapping and movement about the *drag axis* (q.v.).

articulated connecting rods (*I.C. Engs.*). The auxiliary connecting rods of a radial engine, which work on pins carried by the master rod instead of on the main crank pin. In America called **link rods.** See **articulation** (*Eng.*).

articulation (*Bot.*). A joint at which natural separation may occur in a fruit or stem. (*Eng.*) The connexion of 2 parts in such a way (usually by a pin joint) as to permit of the same relative movement. (*Teleph.*) Percentage of specified speech components (usually *logatoms*) received over a communication system; may be (a) *word*: percentage of words correctly received; (b) *syllable*: percentage number of meaningless syllables correctly recognized; (c) *sound*: percentage number of fundamental speech-sounds (consonant, vowel, initial or final consonant) correctly recognized. (*Zool.*) The movable or immovable connexion between 2 or more bones.

artificial ageing (*Met.*). See **temper-hardening.**

artificial antenna (*Radio*). Combination of resistances, capacitances, and inductances with the same characteristics as an antenna except that it does not radiate energy. It is used in place of the normal antenna for purposes such as repair and checking of a transmitter, or for re-tuning of the transmitter onto a different frequency. Also **dummy antenna, phantom antenna.** See **dumb aerial.**

artificial black (white) (*Teleg.*). A signal corresponding in amplitude or frequency to the normal black (white) signal; usually produced locally in a transmitter or receiver and used for checking, calibrating, or setting up the machine.

artificial character (*Bot.*). A character chosen arbitrarily, without regard to the natural relationships of the plants.

artificial classification (*Bot., Zool.*). A classification based on one or a few arbitrarily chosen characters, and giving no attention to the natural relationships of the organism; the old grouping of plants into trees, shrubs, and herbs was an *artificial classification.*

artificial community (*Bot.*). A plant community kept in existence by artificial means; e.g., a garden habitat or a cloche.

artificial daylight (*Light*). Artificial light having approximately the same spectral distribution curve as daylight, i.e., having a colour temperature of about 4000 K.

artificial disintegration (*Phys.*). The transmutation of nonradioactive substances brought about by the bombardment of the nuclei of their atoms by high-velocity particles, such as α-particles, protons, or neutrons.

artificial ear (*Acous.*). Device for testing earphones which presents an acoustic impedance similar to the human ear and includes facilities for measuring the sound pressure produced at the ear.

artificial earth (*Radio*). See **counterpoise.**

artificial feel (*Aero.*). Simulation of the variation in feed-back of the effects of aerodynamic forces due to air speed on the control surfaces where these are operated by irreversible power controls. See ***q*-feel.**

artificial flags (*Build.*). Paving flags made by mixing cement, ground destructor clinker (or aggregate) and sand with sufficient mortar to form a paste; the mixture is put into moulds, subjected to hydraulic pressure, and finally seasoned in the open air for 6 months before use.

artificial harbour (*Civ. Eng.*). A harbour formed by the construction of breakwaters.

artificial horizon (*Aero.*). See **gyro horizon.** (*Surv.*) An apparatus, for example a shallow trough filled with mercury, used in order to observe altitudes of celestial bodies with a sextant on land, i.e., where there is no visible horizon. The reflection of the object in the artificial horizon is viewed directly and the object itself indirectly by reflection from the index glass of the sextant.

artificial insemination. See **insemination.**

artificial kidney (*Med.*). Machine which is used to replace the function of the body's own organ, when the latter is faulty or has ceased to function. The patient's blood is circulated through sterile semipermeable tubing lying in a suitable solution and is purified by dialysis.

artificial larynx (*Med.*). A reed actuated by the air passing through an opening in front of throat to assist articulation of person who has undergone tracheotomy operation.

artificial line (*Telecomm.*). Repeated network units which have collectively some or all of the transmission properties of a line; also **simulated line.**

artificial planet (*Space*). A space probe which has sufficient velocity to escape from the earth's gravitational field, but not from that of the sun, so that it revolves about that body as a member of the solar system.

artificial pneumothorax (*Med.*). See **pneumothorax** (2).

artificial radioactivity (*Phys.*). Radiation of α-, β-, γ-rays and positrons from isotopes after high-energy bombardment. Discovered by Irene Curie in 1933. Effected generally by neutron bombardment in reactors.

artificial rubber (*Plastics*). See **synthetic rubber.**

artificial satellite (*Space*). A man-made space vehicle whose velocity is sufficient to maintain it in orbit about the earth, moon, etc. Earth satellites are often used as relay stations for telecommunication links; if they amplify signals they are *active*; if they reflect them, *passive*.

artificial stone (*Build.*). A precast imitation of natural stone made in block moulds. The interior of the block is of concrete, the required exterior face of cement mixed with dust or chippings of the natural stone to be imitated.

artificial sunlight (*Light*). Light from special lamps having a large proportion of ultraviolet (health-giving) rays.

artificial traffic (*Teleph.*). Automatically generated calls which are deliberately mixed with subscriber-originated traffic to sample the overall service provided by the switching equipment of an automatic exchange, by recording or holding faults recognized by test equipment.

artificial voice (*Acous.*). Loudspeaker and baffle for simulating speech in testing of microphones.

Artinskian (*Geol.*). A stratigraphical stage in the Lower Permian rocks of Russia and eastern Europe.

artiodactyl (*Zool.*). Possessing an even number of digits.

Artiodactyla (*Zool.*). The one order of the *Paraxonia* (q.v.) containing the 'even-toed' hooved 'Ungulates', i.e. those with a *paraxonic foot* (q.v.). Includes pigs, peccaries, hippopotami, camels, llamas, giraffes, sheep, buffaloes, oxen, deer, gazelles, and antelopes.

art paper (*Paper*). Paper coated on one or both sides with a composition containing china clay. In the manufacture of imitation art paper the clay is added to the pulp.

Artype (*Typog.*). A *transfer lettering system* (q.v.) of adhesive-backed typefaces, screens, etc.

arundinaceous (*Bot.*). Reedlike and thin.

aryl (*Chem.*). A term for aromatic monovalent hydrocarbon radicals; e.g., C_6H_5Cl is an aryl halide.

aryl amines (*Chem.*). Amino derivatives of the aromatic series, e.g., $C_6H_5NH_2$ (aniline).

arylarsinic acids (*Chem.*). A group of acids of the formula $R{\cdot}AsO(OH)_2$, the derivatives of which have great importance on account of their therapeutic value.

arytaenoid (*Zool.*). (1) In Vertebrates, one of a pair of anterior lateral cartilages, forming part of the framework of the larynx. (2) In general, pitcher-shaped.

As (*Chem.*). The symbol for *arsenic.*

as- (*Chem.*). Abbrev. for *asymmetrically substituted.*

A.S. (*For.*). Abbrev. for *air-seasoned timber.*

A.S.A. film speed (*Photog.*). American Standards Association speed rating, based on the $\log_e$ of the point on the characteristic curve of the emulsion at which optimum results are obtained.

asbestine (*Paint.*). An *extender* (q.v.) prepared from fibrous talc; frequently used in *fillers* (q.v. (1)).

asbestos (*Min.*). Two different mineral species are included under this term: (*a*) amphibole, ranging in composition from tremolite to amosite and riebeckite (crocidolite); and (*b*) a form of serpentine. Both types of asbestos occur in veins as fibrous crystals, so extremely thin as to be elastic and capable, in some cases, of being woven into fabric. Withstands high temperatures without change, and hence used in making fireproof curtains, washers, etc.; also used, combined with other materials, in building materials and paints, to impart fire-resistant properties. Asbestos in the form of paper and millboard is used for low-voltage insulation, where high temperatures may occur, and for arc-resisting padding inside air-break circuit-breakers, etc. See **Sindanyo.**

asbestos cement (*Build.*). An inexpensive, but brittle, fire-resisting and weather-proof, non-structural building material, made from Portland cement and asbestos; it is rolled into various forms such as plain sheets, corrugated sheets, roofing slates, rainwater goods, etc.

asbestosis (*Med.*). Disease of the lungs due to inhalation of asbestos particles.

asbestos shingles (*Build.*). A fire-resisting roof-covering, consisting of asbestos cement, made into the form of shingles.

asbolane (or asbolite) (*Min.*). A form of *wad*—soft, earthy manganese dioxide, containing up to about 32% of cobalt oxide.

asbolite (*Min.*). Synonym of **asbolane.**

Ascariales (*Bot.*). An order of the *Myxothallophyta* (the Slime Moulds). The swarm spores are amoeboid, and aflagellate. The germinating spores come together to form a pseudoplasmodium, which is saprophytic, often coprophilous. Spores produced in uncovered masses, which are stalked or sessile.

Ascaridata (*Zool.*). An order of *Phasmidia*, the members of which may be parasitic or free-living; they have normally one dorsal and two subventral lips; males usually have two spicules, but no true bursa.

ascending (*Bot.*). (1) Becoming vertical by means of an upward curve. (2) Said of an ovule which arises obliquely from close to the base of the ovary.

ascending aestivation (*Bot.*). Aestivation in which each petal overlaps the edge of the petal posterior to it.

ascending letters (*Typog.*). Letters the top portions of which rise above the general level of the line; e.g., *b, d, f, h.*

ascensional ventilation (*Mining*). That in which

fresh air is led first to lowest part of mine workings, whence it is directed upward.

Aschheim-Zondek test (*Med.*). A test for pregnancy based on the urinary excretion, during pregnancy, of *chorionic gonadotrophin* (q.v). which acts on the ovaries and causes development of the Graafian follicles. The urine under test is injected into immature female mice which are later killed for examination of the ovaries. In the similar *Friedman's test*, nonpregnant mature rabbits are used. See **pregnancy test.**

Aschoff's nodes (or **Aschoff's bodies**) (*Med.*). Inflammatory nodules found in rheumatic inflammation of the heart.

Ascidiacea (*Zool.*). A class of *Urochorda* in which the adult is tailless and sedentary, with a degenerate nervous system and a dorsal atriopore; the gill clefts are divided by external longitudinal bars. Sea Squirts.

Ascidian tadpole (*Zool.*). In *Urochorda*, a larval form having a tail about 4 times the length of the trunk, containing the notochord and the hollow dorsal nerve cord. Also called **appendicularia larva.**

ascidiozooid (*Zool.*). In *Pyrosomatida*, one of four buds or blastozooids which arise on the stolon of the cyathozooid during development.

ascidium (*Bot.*). A pitcher-shaped leaf or part of a leaf.

asciferous, ascigerous (*Bot.*). Bearing asci (see **ascus**).

asciiform (*Bot.*). Shaped like a hatchet.

ascites (*Med.*). See **hydroperitoneum.**

ascocarp (*Bot.*). The fructification of the *Ascomycetes*, containing asci and ascospores; it may be a cleistocarp, an apothecium or a perithecium.

ascogenous cell (*Bot.*). A cell which gives rise to an ascus.

ascogenous hypha (*Bot.*). A hypha from which one or more asci are formed.

ascogonium (*Bot.*). See **archicarp.**

Ascohymeniales (*Bot.*). A group of *Ascomycetes* producing apothecia or perithecia with true walls.

Ascolichenes (*Bot.*). The main group of lichens, with many hundreds of species, in which the fungal constituent is an Ascomycete.

ascoma (*Bot.*). The fruiting body of *Ascomycetes* having asci.

Ascomycetes (*Bot.*). One of the main groups of fungi, including some 15,000 species. They are characterized by the production of asci and ascospores.

ascon grade (*Zool.*). In *Porifera*, a primitive type of water-vascular system, in which the choanocytes line the whole paragaster.

ascorbic acid (*Chem.*). See **vitamin C.**

ascospore (*Bot.*). A spore formed within an ascus.

ascostome (*Bot.*). The apical pore of an ascus.

Ascothoracica (*Zool.*). An order of parasitic *Crustacea* which, as adults, usually live embedded in the tissue of their hosts; there is a free-swimming larval stage; the mantle contains diverticula of the gut; the trunk appendages are more or less reduced.

ascus (*Bot.*). Enlarged flask-like cell in which spores are formed (commonly in groups of 8). The characteristic spore-forming organ of the *Ascomycetes*; usually elongated and club-shaped, but may be globose or filiform. Asci are often formed in large numbers in apothecia or perithecia. *pl.* **asci.**

asdic (*Acous.*). Underwater acoustic detecting system which transmits a pulse and receives a reflection from underwater objects, particularly submarines, at a distance. Also used by trawlers to detect shoals of fish. Equivalent to U.S. **sonar.** (*A*llied *S*ubmarine *D*etection *I*nvestigation *C*ommittee.)

asemantic (*Biol.*). Used of a molecule that is not produced by an organism and so is not influenced by the presence in the organism of *semantides* (q.v.).

asepalous (*Bot.*). Devoid of sepals.

asepsis (*Med.*). The exclusion of putrefying bacteria from the field of operation, by the use of sterilized dressings and instruments, and the wearing of sterilized gowns and gloves by the surgeon and nurses.

aseptate (*Bot.*). Not divided into segments or cells by septa.

A-service area (*Radio*). Region around a broadcasting station where the electric field-strength is greater than 10 mV/m.

asexual (*Bot., Zool.*). Without sex: lacking, or apparently lacking, functional sexual organs: (with reference to reproduction) parthenogenetic, vegetative, e.g., rhizome, tuber, bulb.

asexual generation (*Bot.*). The spore-bearing plant in mosses, ferns, and their relatives.

asexual reproduction (*Biol.*). Any form of reproduction not depending on a sexual process or on a modified sexual process.

ash (*Chem.*). Nonvolatile inorganic residue remaining after the ignition of an organic compound. (*For.*) A tree (*Fraxinus*) yielding a tough and elastic timber. Typical uses are for ladders, hammer and tool handles, spokes, oars, poles, camp furniture, coffins, and artificial limbs.

ashcroftine (*Min.*). A rare potassium, sodium, and calcium zeolite.

ash curve (*Min. Proc.*). Graph which shows result of sink-and-float laboratory test in form of relationship between specific gravity of crushed small particles and the ash content at that gravity.

Ashgill Series (or **Ashgillian**) (*Geol.*). The highest series of the Ordovician System, comprising only one graptolite zone, and, together with the Caradocian, equivalent to the Bala Series of N. Wales.

ashlar (*Build.*). (1) Masonry work in which the stones are accurately squared and dressed to given dimensions so as to make very good joints over the whole of the touching surfaces. (2) A thin facing of squared stones or thin slabs laid in courses, with close-fitting joints, to cover brick, concrete, or rubble walling.

ashlering (*Build.*). The vertical timbers or quarterings, up to 3 ft (1 m) long, fixed in attics between floor joists and rafters as supports for a partition wall, to cut off the sharp angle under the lower end of the rafter.

asiphonate (*Zool.*). In Insects, having respiratory tubes opening directly to the exterior: lacking siphons.

askania (*Instr.*). Combination of telescope, theodolite and cinecamera for recording track of missiles, together with relevant data, on film; timing marks are recorded by a small discharge lamp.

Askarel (*Chem.*). A dielectric fluid which is both nonflammable and incapable of enduring explosive gas mixtures, even when decomposed by an electric arc, e.g., *Pyrochlor* (q.v.).

asmodular (*Electronics*). *Modular* method of *as*sembly. See **module.**

asparaginase (*Chem.*). An enzyme that causes asparagine to hydrolyze to aspartic acid and ammonia.

asparagine (*Chem.*). A nonessential amino acid; the monoamide of aminosuccinic acid, $NH_2 \cdot CO \cdot CH_2 \cdot CH(NH_2) \cdot COOH$, forming rhombic prisms; found in young leaves, in asparagus, and in other vegetables.

asparagus stone (*Min.*). Apatite of a yellowish-green colour, thus resembling asparagus.

aspartic acid (*Chem.*). Aminosuccinic (amino-butandioic) acid.

$$\begin{array}{c} CH_2{\cdot}COOH \\ | \\ NH_2{-\!\!-}CH{\cdot}COOH \end{array}$$

Prismatic crystals; m.p. 271°C. Has the unusual property of being dextrorotatory in cold water and laevorotatory in hot and alkaline solutions. Obtained by hydrolysis of *asparagine*.

Aspden colour-channel system (*Photog.*). Multiple-flash high-speed photography system using colour filters on flash tubes which are fired in succession.

aspect (*Aero.*). See under **attitude.** (*Bot.*) (1) Degree of exposure to sun, wind, etc., of a plant habitat. (2) Effect of seasonal changes on the appearance of vegetation. (*Rail.*) The indication given by a coloured light signal, as contrasted with that of a semaphore arm signal. A multiple-aspect signal (MAS) conveys more information.

aspection (*Ecol.*). Equivalent to *seasonal succession* (q.v.), if both botanical and zoological events are considered.

aspect ratio (*Aero.*). The ratio of span/mean chord line of an aerofoil (usually in wing); defined as *span*^{2}*/area*. Important for *induced drag* (q.v.) and range/speed characteristics. Normal figure between 6 and 9, lesser values than 6 being *low aspect ratios*, greater than 9 *high aspect ratios*. (*TV*) See **picture ratio.**

aspect society (*Bot.*). A plant community dominated at a given season by a given species or group of species.

asperate, asperous (*Bot.*). Having a rough surface due to short, upstanding stiff hairs.

aspergilliform (*Bot.*). Brushlike, tufted.

aspergillosis (*Vet.*). Pneumomycosis; brooder pneumonia. An infection of the respiratory organs of birds by fungi of the genus *Aspergillus*. In cattle the fungus is a cause of abortion.

asperity (*Phys.*). Actual region of contact between two surfaces, elastically and plastically flattened to take the load (normal force).

aspermia (*Med.*). Complete absence of spermatozoa.

asphalt. The name given to various bituminous substances which may be (*a*) of natural occurrence (see below), (*b*) a residue in petroleum distillation, (*c*) a mixture of asphaltic bitumen and granite chippings, sand, or powdered limestone. Asphalt is used extensively for paving, road-making, damp-proof courses, in the manufacture of roofing felt and paints, and as the raw material for certain moulded plastics. See **bitumen, mastic asphalt.** (*Geol.*) A bituminous deposit formed in oil-bearing strata by the removal, usually through evaporation, of the volatiles. Occurs in the 'tar pools' of California and elsewhere and in the 'pitch lake' in Trinidad, whence enormous quantities are exported.

asphalt blocks (*Civ. Eng.*). Road-surfacing blocks made from a mixture of 8–12% of asphaltic cement and 92–88% of crushed stone. The materials are mixed at a temperature of about 150°C, after which the mixture is placed in moulds and subjected to heavy pressure, the blocks finally being cooled suddenly by plunging them into cold water.

asphaltenes (*Chem.*). Such constituents of asphaltic bitumens as are soluble in carbon disulphide but not in petroleum spirit. See **carbenes, malthenes.**

asphaltite (*Min.*). A group name for the organic compounds albertite, anthraxolite, grahamite, impsonite, libollite, nigrite, and uintaite.

aspherical surface (*Optics*). A lens surface which departs to a greater or less degree from a sphere, e.g., one having a parabolic or elliptical section.

asphyxia (*Med.*). State of suspended animation as a result of deficiency of oxygen in the blood, whether from suffocation or other causes.

Aspidobranchiata (*Zool.*). A suborder of the order *Streptoneura*, having a decentralized nervous system, paired auricles, and multiple central radula teeth; bipectinate ctenidia, which are free distally, usually occur; mainly marine forms, Limpets, Ear Shells, Top Shells, Ormers, and Abalones. Also **Diotocardia.**

Aspidochirotae (*Zool.*). An order of *Holothuroidea*, having shield-shaped buccal tentacles lacking retractor muscles, but possessing ampullae and respiratory trees; bottom-feeding forms, living on comparatively firm ground.

Aspidocotylea (*Zool.*). An order of *Trematoda*, in which the ventral sucker covers most of the ventral surface of the body and is usually subdivided; the genital opening is median, ventral, and anterior; parasitic in the gut and gall bladder of *Chelonia* and Fish, and in various organs of Molluscs.

aspidospermine (*Chem.*). $C_{22}H_{30}O_2N_2$, an alkaloid of unknown constitution, found in the *Aspidosperma* species, crystallizing in needles, m.p. 208°C; almost insoluble in water, soluble in most organic solvents.

aspirated psychrometer (*Meteor.*). A *psychrometer* (q.v.) which uses a forced draught of at least 8 mi/h (12 km/h) over the wet bulb.

aspiration (*Med.*). The removal of fluids or gases from the body by suction.

aspiration pneumonia (*Med.*). See **deglutition.**

aspirator (*Chem.*). A device for drawing a stream of air or oxygen or liquid through an apparatus by suction.

Aspirigera (*Zool.*). See **Holotricha.**

aspirin (*Chem.*). The common name for acetylsalicylic acid.

asplanchnic (*Zool.*). Having no gut.

asplund (*For.*). See under **mechanical pulp.**

asporocystid (*Zool.*). Said of *Sporozoa* in which the zygote divides to form sporozoites, without the formation of a sporocyst.

assay (*Chem.*). The quantitative analysis of a substance to determine the proportion of some valuable or potent constituent, e.g., the active compound in a pharmaceutical or metals in an ore. See **dry assay, wet assay.**

assay balance (*Chem.*). A balance specially made for weighing the small amounts of matter met with in assaying. See also **chemical balance.**

assayer (*Chem.*). A person who carries out the process of assay. See also **dry assay, wet assay.**

assay ton (*Met.*). Used in assaying precious metals. It is equivalent to 29·16 g and 32·67 g for the short and long ton respectively. The number of mg of precious metal in an assay ton of ore indicates the *assay value*, since 1 mg of precious metal per assay ton = 1 troy oz of precious metal per avoirdupois ton of ore.

assay value (*Met.*). Troy ounces of precious metal per avoirdupois ton of ore.

assembler (*Comp.*). Unit which produces a set of machine instructions from a set of symbolic input data. It normally converts specific single-step instructions into machine language and is therefore much more restricted than a *compiler* (q.v.) which can convert single general instructions into complete subroutines for carrying them out.

assigned frequency (*Radio*). That assigned as centre frequency of a class of transmission, with tolerance, by authority.

assigning authority (*Ships*). A body authorized to assign *load lines* (q.v.) to ships. For British ships the Ministry of Transport and, with the minister's approval, Lloyd's Register of Shipping, British Committee of the Bureau Veritas and the British Technical Committee of the American Bureau of Shipping.

assimilate (*Bot.*). Any substance produced in the plant during the processes of food manufacture.

assimilation (*Bot.*). A general term for all the metabolic processes by which nutrient material is built up and utilized by plants. Often used as equivalent to *photosynthesis* (q.v.). (*Geol.*) The incorporation of extraneous material in *igneous magma* (q.v.). (*Psychol.*) (1) Process of changing elements in the milieu so that they may become incorporated into the mental and physical structure of the organism (Piaget). (2) Tendency to perceive, or think of, objects which are similar to a standard, or persons similar to self, as being more similar than is actually the case. See **contrast.** (*Zool.*) (1) Conversion of food material into protoplasm, after it has been ingested, digested, and absorbed. (2) Resemblance of an animal to its surroundings, not only by coloration but also by configuration.

assimilation number (*Bot.*). The amount of carbon dioxide assimilated per hour by a portion of leaf substance containing 1 mg chlorophyll.

assimilation quotient (*Bot.*). The ratio of the volume of carbon dioxide absorbed to that of oxygen set free.

assimilative induction (*Zool.*). In early embryos, induced development of an organ or structure, by the introduction of material from a similar organ or structure in another embryo or from another part of the same embryo.

assistance traffic (*Teleph.*). Calls made to enlist the assistance of an operator on connexions normally completed without such assistance, or to achieve connexion to one of the ancillary services provided by an exchange.

assisted take-off (*Aero.*). Supplementing the full power of the normal engines by auxiliary means, which may or may not be jettisonable. Small turbojet or rocket motor units, powder, or liquid rockets may be used. See **jato, ratog.**

assize (*Build.*). A cylindrical block of stone forming part of a column, or of a layer of stone in a building.

associate Bertrand curves (*Maths.*). See **conjugate Bertrand curves.**

associated emission (*Electronics*). That which brings about equilibrium between incident photons and secondary electrons in ionization.

associated liquid (*Chem., Phys.*). One in which molecules of same kind form complex structure, e.g., water. See **hydrogen bond.**

association (*Bot.*). A plant community usually occupying a wide area, consisting of a definite population of species, having a characteristic appearance and habitat, and stable in its duration. (*Print.*) In rotary printing, the bringing together of separate webs, after printing, to pass through the folder as a complete product. (*Psychol.*) The response to a stimulus idea by a reaction idea in the individual, following the Freudian law of *psychic determinism* (q.v.). (*Zool.*) In certain *Sporozoa*, adherence of individuals without fusion of nuclei: a characteristic set of animals, belonging to a particular habitat.

association fibres (*Zool.*). In the cerebrum of Vertebrates, axons passing from the pyramidal cells of the cortex to the grey matter of other parts of the same hemisphere.

association neurone (*Zool.*). A neurone lying entirely within the central nervous system and connecting a motor neurone with a sensory neurone; it is a constituent of all but the simplest reflexes, which it puts into communication with other regions of the C.N.S.

associative (*Maths.*). An operation is associative if when applied to any two of three things, and then applied to the result and the third thing, the same final result is always produced, e.g., arithmetic addition is associative because when adding three numbers any two can be added first.

associative learning (*An. Behav.*). Learning in which there is the formation of a positive association between a stimulus or situation as opposed to the negative relationship found in *habituation* (q.v.).

associes (*Bot.*). A plant community which may be regarded as in process of development into an association, and is therefore not yet stabilized.

astable circuit (*Radio*). Valve or transistor circuit (with 2 quasi-stable states) which is free-running and self-sustaining in oscillation, e.g., multivibrator. Also **free-running circuit.**

astacene, astacin (*Chem.*). $C_{40}H_{48}O_4$; carotenoid produced by oxidation of *astaxanthin*, a pigment found in the lobster, *Astacus gamm.* and other marine creatures.

A stage (*Plastics*). Stage at which a synthetic resin of the phenol formaldehyde type is fusible and wholly soluble in alcohols and acetone.

astasia, abasia (*Med.*). Inability of an hysterical patient to stand or to walk.

astatic galvanometer (*Mag.*). Moving-magnet galvanometer in which adjustable magnets form an astatic system.

astatic microphone (*Acous.*). Same as **omnidirectional microphone.**

astatic system (*Mag.*). Ideally an arrangement of two or more magnetic needles on a single suspension so that in a uniform magnetic field, such as the earth's field, there is no resultant torque on the suspension.

astatine (*Chem.*). Radioactive element, the heaviest halogen. Symbol At, at. no. 85, mass nos. 202–212, 214–219, half-lives 2×10^{-6} s to 8 hr. Isotopes occur naturally as members of the actinium, uranium, or neptunium series, or may be produced by the α-bombardment of bismuth.

astelic (*Bot.*). Not having a stele.

aster (*Cyt.*). (1) A group of radiating fibrils formed of cytoplasmic granules surrounding the centrosome, seen immediately prior to and during cell division, and usually more prominent in preparations of animal nuclei than in those of plant nuclei. (2) During metaphase, the stellate arrangement of the chromosomes.

Asteraceae (*Bot.*). See **Compositae.**

astereognosis (*Med.*). Loss of ability to recognize, by the sense of touch, the 3-dimensional nature of an object.

asteria (*Min.*). A precious stone which when cut *en cabochon* displays a 6- or 12-rayed star due to *asterism*. Star sapphire and star ruby display this character.

asteriscus (*Zool.*). In *Teleostei*, an otolith lying in the lagena or rudimentary cochlea.

asterism (*Min.*). A light effect due to the presence of minute, almost ultramicroscopic, inclusions arranged in a regular series in some varieties of ruby, sapphire, phlogopite mica, and other minerals. A point source of light viewed through a plate of this form of phlogopite appears as a light star.

asternal (*Zool.*). Of certain ribs, not directly connected with the sternum.

asteroid (*Astron.*). A small planetary body, one of many thousands which revolve in orbits lying mainly between the orbits of Mars and Jupiter. A few exceptional minor planets (e.g., Eros, Icarus) have more eccentric orbits which carry them inside the earth's orbit. See also **astroid.**

Asteroidea (*Zool.*). A class of *Echinodermata*, having a dorsoventrally flattened body of pentagonal or stellate form; the arms merge into the disk; the tube feet possess ampullae and lie in grooves on the lower surface of the arms; the anus and madreporite are aboral, and there is a well-developed skeleton; free-living carnivorous forms. Starfish.

asterospondylous (*Zool.*). Showing partial calcification of cartilaginous vertebral centra, in the form of radiating plates.

aster phase (*Cyt.*). See **metaphase.**

asthenia (*Med.*). Loss of muscular strength.

asthenic type (*Psychol.*). One of Kretschmer's 3 types of individuals, characterized by tall thin men, with hands long in proportion to the trunk. The type shows schizoid character traits.

asthenopia (*Med.*). Weakness of the eye muscles in neurasthenia: eye strain due to errors of refraction.

asthma (*Med.*). A disorder in which there occur attacks of difficult breathing due to spasm of the bronchial muscles.

astichous (*Bot.*). Not arranged in rows.

astigmatism (*Light*). A defect in an optical system on account of which, instead of a point image being formed of a point object, 2 short line images (focal lines) are produced at slightly different distances from the system and at right angles to each other. Astigmatism is always present when light is incident obliquely on a simple lens or spherical mirror. (*Med.*) Unequal curvature of the refracting surfaces of the eye, which prevents the focusing of light rays to a common point on the retina. (*Photog.*) A defect in a lens, causing image blurring in particular directions.

astomatous (*Bot., Zool.*). (1) Lacking stomata. (2) Without a mouth.

Aston dark space (*Electronics*). That in the immediate vicinity of a cathode, in which the emitted electrons have velocities insufficient to ionize the gas.

Aston whole-number rule (*Nuc.*). Empirical observation that relative atomic masses of isotopes are approximately whole numbers. See **mass-spectrograph.**

Astrafoil (*Print.*). A thin, dimensionally stable plastic sheet used for mounting lithographic negatives or positives.

astragal (*Build.*). A specially shaped bar used for connecting together glazing bars or sheets of glass in a window. (*Carp.*) A small convex moulding having a semicircular cross-section, sometimes plain and sometimes carved.

astragal plane (*Carp.*). A plane adapted for cutting astragal mouldings.

astragal tool (*Carp.*). A special tool, with a semicircular cutting edge, used in wood-turning for turning beads and astragals.

astragalus (*Zool.*). In tetrapod Vertebrates, one of the ankle bones, corresponding to the lunar in the wrist.

astrakhan (*Textiles*). (1) Curled woven pile fabric (48 to 70 in. wide) having the appearance of astrakhan lamb fleece. The pile yarn (silk, mohair, lustrous worsted, or one of the bulked synthetics) is curled and heat-set prior to weaving. (2) A weft-knitted fabric in which curled or bulked yarn is incorporated on a tuck-miss arrangement. (3) Warp-knitted structure, in which the yarn is passed over wires to form loops.

astral ray (*Bot.*). One of the fibrils in the cytoplasm which seem to play a part in the delimitation of ascospores.

astringent (*Med.*). Having the power to constrict or contract organic tissues: that which does this.

astrocentre (*Cyt.*). See **centrosome.**

astrocompass (*Aero.*). A nonmagnetic instrument that indicates true north relative to a celestial body.

astrocyte (*Zool.*). A much branched, star-shaped neuroglia cell.

astrodome (*Aero.*). A transparent dome, fitted to some aircraft usually on the top of the fuselage, with calibrated optical characteristics, for astronomical observations.

astroid (*Cyt.*). In cell division, the star-shaped figure formed by the looped chromosomes aggregated round the equator of the nuclear spindle. (*Maths.*) Four-cusped starlike curve with cartesian equation $x^{2/3}+y^{2/3}=a^{2/3}$. Envelope of a straight line whose ends move along the coordinate axes. A hypocycloid in which the radius of the rolling circle is $\frac{1}{4}$ or $\frac{3}{4}$ times the radius of the fixed circle. Also **asteroid.** Cf. *roulette* and *glissette*.

astrolabe (*Astron.*). An instrument used in medieval times for measuring the altitude of heavenly bodies. The modern *prismatic astrolabe* (q.v.) is a precision instrument for the accurate measurement of time and latitude.

astrometry (*Astron.*). The precise measurement of position (of stars, etc.) in astronomy, generally deduced from the coordinates of star images on photographic plates.

astron (*Nuc. Eng.*). Type of chamber used in thermonuclear research.

astronautics (*Space*). The science of space flight.

astronomical clock (*Astron.*). A pendulum clock which reads sidereal time; in its most modern form, electrically driven and controlled.

Astronomical Ephemeris (*Astron.*). The annual ephemeris of sun, moon, and planets, published to high accuracy for the use of astronomers. Formerly (before 1960) called the *Nautical Almanac*, it is identical in content with the *American Ephemeris*.

astronomical telescope (*Astron.*). The general name for a telescope designed for astronomical purposes; the principal forms are the *equatorial* (q.v.) and the *meridian* (*transit*) *circle* (q.v.) telescopes.

astronomical triangle (*Astron.*). Triangle on the celestial sphere formed by a heavenly body S, the zenith Z, and the pole P. The 3 angles are the hour-angle at P, the azimuth at Z, and the parallactic angle at S.

astronomical twilight (*Astron.*). The interval of time during which the sun is between 12° and 18° below the horizon, morning and evening. See **civil twilight, nautical twilight.**

astronomical unit (*Astron.*). The mean distance of the earth from the sun, amounting to 93×10^6 miles or $149{\cdot}6 \times 10^6$ km; used as the principal measure of distance within the solar system. See **solar parallax.** Abbrev. **A.U.**

astronomy. The science of the heavens in all its branches.

astrophyllite (*Min.*). A complex hydrous silicate of potassium, iron, manganese, titanium, and zirconium; occurs in brown laminae in alkaline, igneous rocks.

astrophysics. That branch of astronomy which applies the laws of physics to the study of interstellar matter and the stars, their constitution, evolution, luminosity, etc.

astrosclereide (*Bot.*). A sclereide with radiating branches ending in points.

astrosphere (*Cyt.*). See **attraction sphere.**

astyllen (*Civ. Eng., Mining*). A small dam built across an adit to restrict the flow of water.

asylum switch (*Elec. Eng.*). See **locked-cover switch.**

asymmeter (*Elec. Eng.*). An instrument having 3 movements so arranged that any lack of symmetry when these are connected to a 3-phase system can be observed by a single reading.

asymmetric (*Bot.*). Irregular in form: not divisible into halves about any longitudinal plane.

asymmetrical, dissymmetrical, nonsymmetrical (*Elec. Eng.*). Said of circuits, networks, or transducers when the impedance (image impedance, or iterative impedance) differs in the two directions.

asymmetrical conductivity (*Elec. Eng.*). Phenomenon whereby a substance, or a combination of substances as in a rectifier, conducts electric current differently in opposite directions.

asymmetric atom (*Chem.*). Atom of tri- or higher valency in a chemical compound with three or more different groups, so that the different spatial arrangements give rise to optical isomers. There is no mirror plane of symmetry in the molecule or ion. See **optical isomerism, stereochemistry.**

asymmetric carbon atom (*Chem.*). A carbon atom to which 4 different atoms or groups are attached, resulting in *optical isomerism* (q.v.).

asymmetric conductor (*Elec. Eng.*). Conductor which has a different conductivity for currents flowing in different directions through it, e.g., a diode.

asymmetric flight (*Aero.*). The condition of flying with one or more engines stopped, so that the aeroplane is subject to offset power which requires correction by the flying controls.

asymmetric reflector (*Light*). A reflector in which the beam of light produced is not symmetrical about a central axis.

asymmetric refractor (*Elec. Eng.*). A refractor in which the light is redirected, unsymmetrically, about a central axis.

asymmetric synthesis (*Chem.*). The synthesis of optically active compounds from racemic mixtures. This can be carried out in some cases by chemical methods in which one component is more reactive than the other one. In other cases asymmetric synthesis occurs in the presence of enzymes.

asymmetric system (*Crystal.*). See **triclinic system.**

asymmetric top (*Chem.*). A model of a molecule having no 3- or higher-fold axis of symmetry.

asymmetry. The condition of being *asymmetrical.* (*Zool.*) The condition of the animal body in which no plane can be found which will divide the body into two similar halves; as in Snails.

asymmetry potential (*Elec. Eng.*). The potential difference between the inside and outside surface of a hollow electrode.

asymptote (*Maths.*). Of a given curve: usually a line, but sometimes a curve—frequently called a **curvilinear asymptote**—such that the distance of a point on the given curve from it tends to zero as the point recedes from the origin along the curve. A tangent to the given curve at infinity. If the equation to the curve can be approximated to by

$$y=\sum_0 a_r x^r$$

when x is large, or by

$$x=\sum_0 b_r y^r,$$

when y is large, then $y=a_0+a_1x$ and $x=b_0+b_1y$ are asymptotes to the curve. If $a_0=a_1=b_0=b_1=0$, there are no straight-line asymptotes and the curvilinear asymptotes are $y=a_nx^n$ and $x=b_my^m$, where a_n and b_m are the first nonzero coefficients.

asymptotic breakdown voltage (*Cables*). The voltage which, if applied for a very long time (hundreds of hours), will break down a cable.

asymptotic curve (*Maths.*). One on a surface such that at every point on it, its tangent lies in an asymptotic direction at that point.

asymptotic directions (*Maths.*). See **hyperbolic point on a surface** and **conjugate directions.**

asynapsis (*Gen.*). Absence of pairing of chromosomes in meiosis.

asynchronous (or nonsynchronous) computer (*Comp.*). One in which operations are not all timed by a master clock. The signal to start an operation is provided by the completion of the previous operation.

asynchronous motor (*Elec. Eng.*). See **nonsynchronous motor.**

asynergia, asynergy (*Med.*). Lack of coordinated movement between muscles with opposing actions, due to a lesion in the nervous system.

At (*Chem.*). The symbol for astatine. (*Elec. Eng.*) Abbrev. for **ampere-turn.**

ATA (*Aero.*). Abbrev. for **Air Transport Association.**

atacamite (*Min.*). A hydrated chloride of copper, widely distributed in S. America, Australia, India, etc., in the oxidation zone of copper deposits; occurring also at St. Just (Cornwall).

atactic (*Plastics*). Term used to denote certain linear hydrocarbon polymers in which substituent groups are arranged at random around the main carbon chain. See also **isotactic** and **syndiotactic.**

atactosol (*Chem.*). A colloidal sol not containing *tactoids* (q.v.).

atavism (*Gen.*). The phenomenon in which, by skipping a generation, a particular character in the offspring is unlike the corresponding character in the parents but resembles the corresponding one of grandparents: more generally, occurrence of a characteristic observed in more distant ancestors, but not in the more immediate ancestors. *adj.* **atavistic.**

ataxia, ataxy (*Med.*). Incoordination of muscles, leading to irregular and uncontrolled movements; due to lesions in the nervous system.

ATB (*Ind. Heat.*). Abbrev. for **aeration test burner.**

ATC (*Aero.*). See **air traffic control.**

atelectasis (*Med.*). Collapse of part of a lung, or the whole lung, associated with diminished air entry.

ateleiosis (*Med.*). Dwarfism, with normal bodily proportions; due to disorder of the pituitary.

atelomitic (*Cyt.*). Said of a chromosome having the spindle fibre attached along the side.

athermal solutions (*Chem.*). Solutions formed without production or absorption of heat on mixing the components, e.g., esterifications.

atherosclerosis (*Med.*). Thickening of and rigidity of the intima of the arteries, caused by fatty and calcareous deposits and proliferation of fibrous tissue. Also **atheroma.**

athetosis (*Med.*). Slow, involuntary, spontaneous, repeated, writhing movements of the fingers and of the toes, due to a brain lesion.

athletic type (*Psychol.*). One of Kretschmer's 3 types of individual, characterized by a well-developed skeletal musculature, in which the relation of limbs to trunk is well-proportioned.

athymia (*Med.*). Absence of the thymus gland.

Atkinson cycle (*I.C. Engs.*). A working cycle for internal-combustion engines, in which the expansion ratio exceeds the compression ratio; more efficient than the Otto cycle, but mechanically impracticable.

atlas (*Paper*). A former size of drawing paper, $34 \times 26\frac{1}{2}$ in. (*Zool.*) The first cervical vertebra.

atm (*Phys.*). Abbrev. for **standard atmosphere,** a practical unit of pressure defined as 101·325 kN/m² or 1·013 25 bar.

atmolysis (*Chem., Phys.*). The method of separation of the components of a mixture of two gases, which depends on their different rates of diffusion through a porous partition.

atmophile (*Geol.*). Descriptive of elements which are concentrated in the atmosphere of the Earth, e.g., argon and nitrogen.

Atmos clock (*Horol.*). Clock with self-winding mechanism actuated by the effect of atmospheric changes on a sylphon bellows.

atmosphere microphone (*Acous.*). One used for background noise to a transmission, the adjustable output being added to the main transmission.

atmospheric absorption (*Acous.*). Diminution of intensity of a sound wave in passing through the air, apart from normal inverse square relation, and arising from true absorption. This occurs appreciably for high audiofrequencies only, in addition to apparent absorption by dispersion in aerial strata. (*Astron.*) The absorption of the light of the stars by the earth's atmosphere; it is practically negligible above 45° altitude, but the extinction amounts to about half a magnitude at 20°, 1 magnitude at 10°, and 2 magnitudes at 4° altitude.

atmospheric acoustics (*Acous.*). Concerned with the propagation of sound in the atmosphere, of importance in sound ranging and aircraft noise.

atmospheric arc rectifier (*Elec. Eng.*). A form of high-power high-voltage rectifier consisting of an arc between electrodes in air, the arc being initiated in alternate half-cycles by a pilot spark and extinguished at the end of that half-cycle by rapid cooling. Also called **Marx rectifier.**

atmospheric electricity (*Meteor.*). That causing increasing potential with height, about 100 V/m, in calm conditions, altered considerably by thunder-clouds. See **lightning.**

atmospheric engine (*Eng.*). An early form of steam engine, in which a partial vacuum created by steam condensation allowed atmospheric pressure to drive down the piston.

atmospheric gas-burner system (*Heat.*). A natural-draught burner injector, in which the momentum of a gas stream projected from an orifice into the injector throat inspirates from the atmosphere a part of the air required for combustion.

atmospheric line (*Eng.*). A datum line drawn on an indicator diagram by allowing atmospheric pressure to act on the indicator piston or diaphragm.

atmospheric pressure (*Phys.*). The pressure exerted by the atmosphere at the surface of the earth is due to the weight of the air. Its standard value is $1{\cdot}013\,25 \times 10^5$ N/m², or 14·7 lbf/in². Variations in the atmospheric pressure are measured by means of the barometer. See **barometric pressure, standard atmosphere.**

atmospheric radio wave (*Radio*). One which is propagated by reflections in the atmosphere using either, or both, the ionosphere and the troposphere.

atmospherics (*Radio*). Interfering or disturbing signals of natural origin. Also called **sferics, spherics, strays, X's.** See also **static.**

atmospheric tides (*Meteor.*). The changes of atmospheric pressure arising directly from changes in temperature due to the Earth's rotation. See **diurnal range.**

atmospheric waveguide duct (*Radio*). Under certain atmospheric conditions, a layer of the atmosphere which acts as a true waveguide conductor for radio frequency waves, even down to 20 MHz.

atocia (*Med.*). Sterility in the female.

atokous (*Zool.*). Having no offspring: sterile.

atolls (*Geog.*). Coral reefs, typically found in the Pacific; usually ringlike in shape, enclosing lagoons.

atom (*Chem.*). The smallest particle of an element which can take part in a chemical reaction. See **atomic structure, Dalton's atomic theory.**

atomate (*Bot.*). Having small particles sprinkled over the surface.

atomic absorption coefficient (μ_a) (*Phys.*). For an element, the fractional decrease in intensity per no. of atoms per unit area.

atomic arc welding (*Elec. Eng.*). Same as **atomic hydrogen welding.**

atomic bomb (*Nuc.*). Bomb in which the explosive power, measured in terms of equivalent TNT, is provided by nuclear fissionable material such as uranium 235 or plutonium 239. The bombs dropped on Hiroshima and Nagasaki (1945) were of this type. Also called **A-bomb, atom bomb** and **fission bomb.** See also **hydrogen bomb.**

atomic bond (*Chem.*). Link between adjacent atoms formed by interaction of outer valence electrons from each.

atomic clock (*Phys.*). One which depends for its frequency of operation on the change of spin of the valence electron of caesium. The latest value for this frequency is 9 192 631 770 ± 20 Hz.

atomic diameters (*Chem.*). These vary periodically among the elements, as do their chemical properties. The range of diameters is from 0·6 to $5{\cdot}4 \times 10^{-10}$ m, the rare gases having the smallest diameter in each period and the alkali metals the biggest.

atomic disintegration (*Nuc. Phys.*). Natural decay of radioactive atoms, with radiation, into chemically different atomic products.

atomic distance (*Chem.*). The distance between the atoms (centre to centre) in a solid or liquid.

atomic energy (*Chem.*). Strictly the energy (chemical) obtained from changing the combination of atoms originally in fuels. Misapplied to energy obtained from breakdown of fissile atoms in nuclear reactors. See **nuclear energy.**

atomic frequency (*Phys.*). A natural vibration frequency in an atom used in the atomic clock.

atomic heat (*Chem.*). Product of specific heat capacity and r.a.m. in grams; approx. the same for most solid elements at high temperatures.

atomic hydrogen (*Chem.*). See **active hydrogen.**

atomic hydrogen welding (*Elec. Eng.*). Process in which an electric arc is drawn between tungsten electrodes placed in a jet of hydrogen, which becomes ionized and releases its energy on impact. Also called **atomic arc welding.**

atomicity (*Chem.*). The number of atoms contained in a molecule of an element.

atomic mass unit (*Chem.*). Abbrev. u. Exactly one twelfth the mass of a neutral atom of the most abundant isotope of carbon, ^{12}C.

$$1\ \text{u} = 1{\cdot}660 \times 10^{-27}\ \text{kg}.$$

Before 1960 the u was defined in terms of the mass of the ^{16}O isotope and 1 u was $1{\cdot}6599 \times 10^{-27}$ kg. See **atomic weight.**

atomic number (*Chem.*). The order of an element in the periodic (Mendeleev) chemical classification, and identified with the number of unit positive charges in the nucleus (independent of the associated neutrons). Equal to the number of external electrons in the neutral state of the atom, and determines its chemistry. Symbol Z.

atomic plane (*Phys.*). A solid is crystalline because its atoms are ordered in intersecting planes (*atomic planes*) corresponding to the planes of the crystal. See **Bragg method.**

atomic refraction (*Chem.*). The contribution made by a mole of an element to the molecular refraction of a compound.

atomic scattering (*Nuc.*). That of radiation (usually cathode rays or X-rays) by the individual atoms in the medium through which it passes.

atomic spectrum (*Nuc.*). Emission spectrum arising from electron transitions inside an atom and characteristic of the element concerned.

atomic structure (*Phys.*). All chemical elements have characteristic atoms, and differences in their chemical behaviour arise from differences in the electron configuration of the atom in its normal electrically neutral state. Each atom consists of a heavy nucleus with a positive charge produced by a number of protons equal to its atomic number, and there are normally an equal number of electrons outside the nucleus to balance this charge. (The nucleus also comprises electrically neutral *neutrons*; protons and neutrons are collectively referred to as *nucleons.*) The best picture of the atom is given by Sommerfeld's model modified by the wave mechanical concept of orbitals. Electrons are fermions which must conform to the Pauli exclusion principle, and no two of them in the same atom can have identical energies, while possible energies are all quantized and so differ by discrete amounts. Possible orbitals and their energies are classified by four quantum numbers. The principal quantum number indicates the shell to which the orbital belongs and varies from 1 for the K-shell (that closest to the nucleus) to 7 for the Q-shell (that most remote). In general, the closer an electron to the nucleus the greater the coulomb attraction between them, and therefore the consequent binding energy retaining the electron in the atom. Nuclear binding forces tend to give greatest stability when the neutron number and the proton number are approximately equal. But due to electrostatic repulsion between protons the heavier nuclei are most stable when less than half their nucleons are protons, and elements with more than 83 protons in the nucleus are unstable, eventually undergoing radioactive disintegration. Those with more than 92 protons are not found naturally on earth. They can be synthesized in high-energy laboratories but have limited half-lives. These are the transuranic elements. Most elements exist with several stable isotopes and the chemical atomic weight gives the average weight of a normal mixture of atoms of the various possible isotopes.

atomic transmutation (*Nuc.*). Change of atomic number of atom due to bombardment by high-energy radiation or particles—most easily produced by neutron irradiation. This produces change of chemical nature of element, e.g., gold can be transmuted into mercury—the converse of what ancient alchemists attempted.

atomic volume (*Chem.*). Ratio for an element of the rel. at. mass to the density; this shows a remarkable periodicity with respect to *Z*.

atomic weight (*Chem.*). Relative at. mass. Mass of atoms of an element formerly in *atomic weight units* but now more correctly given on the *unified scale* where 1 u is $1{\cdot}660 \times 10^{-27}$ kg. For natural elements with more than one isotope, it is the average for the mixture of isotopes. Also called **equivalent weight.**

atomic weight unit (*Chem.*). Abbrev. **awu.** Exactly one-sixteenth of the weighted mean of the masses of the neutral atoms of oxygen of isotopic composition found in rain or fresh-water lakes. Superseded. See **atomic mass unit, atomic weight, relative atomic mass.**

atomized powder (*Powder Tech.*). One produced by the dispersion of molten metal or other material by spraying under conditions such that the material breaks down into powder.

atomizer (*Eng.*). A nozzle through which oil fuel is sprayed into the combustion chamber of an oil engine or boiler furnace. Its function is to break up the fuel into a fine mist so as to ensure good dispersion and combustion.

atony (*Med.*). Loss of muscular tone.

Atoxyl (*Pharm.*). TN for sodium salt of *p*-amino-phenyl-arsinic acid (arsanilic acid); it has a therapeutical action against certain veterinary infections.

ATP (*Chem.*). Abbrev. for *adenosine triphosphate.*

atrabiliary (*Med.*). Pertaining to black bile.

atrate, atratous (*Bot.*). Blackened: blackening.

atraton (*Chem.*). 6 - Ethylamino - 4 - isopropyl-amino - 2 - methoxy - 1,3,5 - triazine, used in agriculture as a herbicide.

atrazine (*Chem.*). 2 - Chloro - 6 ethylamino - 4 - isopropylamino-1,3,5-triazine, used in agriculture as a herbicide.

Atremata (*Zool.*). An order of primitive *Ecardines* having a thin horny shell with almost equal valves, and a long peduncle; they live in tubes of sand into which they can withdraw; found in Australian and W. Pacific waters.

atresia (*Med.*). Pathological narrowing of any channel of the body. (*Zool.*) Disappearance by degeneration; as the follicles in the Mammalian ovary. *adjs.* **atresic, atretic.**

atrial (*Zool.*). Pertaining to the *atrium* (q.v.).

atrial siphon (*Zool.*). In *Urochorda*, the papilla at the apex of which is the atriopore.

atricolor (*Bot.*). Inky.

atriopore (*Zool.*). The opening by which the atrial cavity communicates with the exterior.

atrioventricular (*Zool.*). See **auriculoventricular.**

atrium (*Zool.*). In *Platyhelminthes*, a space into which open the ducts from the male and female genital organs: in aquatic *Oligochaeta*, a space lined by the cells of the prostate gland into which the male ducts discharge: in pulmonate *Mollusca*, a cavity into which the vagina and the penis open and which itself opens to the exterior: in ectoproct *Polyzoa*, the cavity of the tentacle sheath: in *Cyphonautes* larvae, the space surrounded by the mantle: in developing entoproct *Polyzoa*, the central invagination, the walls of which later give rise to the nervous system: in certain *Mollusca* larvae, the outer stomodaeum: in metamorphosing *Auriculariae*, the cavity into which the larval mouth opens: in *Protochordata*, the cavity surrounding the respiratory part of the pharynx: in Vertebrates, the anterior part of the nasal tract: in Reptiles and Birds, the cavity connecting the bronchus with the lung chambers: in the developing Vertebrate heart, the division between the sinus venosus and the ventricle, which will later give rise to the auricles.

atrochal (*Zool.*). Of trochophore larvae, possessing an apical tuft of cilia and a general ciliated covering, but lacking a prototroch.

atropamine (*Chem.*). $C_{17}H_{21}NO_2$, a *Solanum*-base alkaloid, identical with *apoatropine* (q.v.).

atrophic rhinitis (*Vet.*). Snuffles. An infectious disease of the pig, characterized by chronic rhinitis and deformity of the snout; the causal agent has not been determined.

atrophy (*Med.*). Wasting of a cell or of an organ of the body. (*Zool.*) Degeneration, i.e., diminution in size, complexity, or function, through disuse.

atropine (*Chem.*). $C_{17}H_{23}O_3N$, an alkaloid of the tropane group, obtained from solanaceous plants in which it occurs in traces, or by treating crude hyoscyamine with dilute alkali, when it undergoes isomerization to atropine. It crystallizes in prisms, m.p. 114–116°C, and sublimes when heated rapidly. It is soluble in alcohol or ether, slightly soluble in water. It causes dilatation of the pupil of the eye. Atropine constitutes the active principle of belladonna (obtained from *Atropa belladonna*, 'deadly nightshade'), which, besides eye dilatation, has other medically valuable properties. It has important effects on the central nervous system, affecting certain bodily secretions and the heart-rate: hence its use before general anaesthesia.

atropous (*Bot.*). See **orthotropous.**

ATR tube (*Electronics*). See **anti-transmit-receive tube.**

attached column (*Arch.*). A column partially built into a wall, instead of standing detached.

attachment constriction (*Bot.*). A constriction in a chromosome to which the spindle fibre is attached.

attachment organ (*Bot.*). (1) A disklike or branched outgrowth from the base of the thallus, attaching an alga to a solid object. (2) A hooked hair or similar structure serving to attach a fruit to an animal, and assisting dispersal.

attachment screw (*Instr.*). A set-screw used to secure two parts of an instrument together to prevent relative movement.

attar of roses (otto de rose) (*Chem.*). Oil distilled from fresh roses for perfumery purposes.

attemperators (*Brew.*). Coils of pipe through which cold water circulates; used to regulate the temperature of beer while fermenting.

attenuate, attenuated (*Bot.*). Tapering off, and so narrowing gradually to a point.

attenuation (*Bot.*). The reduction or loss of virulence of parasitic bacteria, fungi, or viruses. (*Brew.*) Fall in the specific gravity of the wort during fermentation, ideally to 20–25% of the original. (*Nuc., Telecomm.*). General term for reduction in magnitude, amplitude, or intensity of a physical quantity, arising from absorption, scattering, or geometrical dispersion. The latter, arising from diminution by the inverse-square law, is not generally considered as attenuation proper. See also **absorption.**

attenuation coefficient (*Nuc.*). See **total absorption coefficient.**

attenuation compensation (*Telecomm.*). The use of networks to correct for varying attenuation, e.g., in transmission lines. See **pre-emphasis.**

attenuation constant (*Telecomm.*). Defined by

$$\rho = \rho_0 e^{-\alpha x}$$

where α is the constant, which may be complex, x is a distance, and ρ is a physical quantity, e.g., acoustic pressure in a sound wave, vector amplitude in an electromagnetic wave, etc. A more fundamental quantity is

$$\mu = \alpha\lambda,$$

the loss per wavelength *distance of propagation.* See **decibel, neper, propagation constant, transfer constant.**

attenuation distortion (*Telecomm.*). Distortion of a complex waveform resulting from the differing attenuation of each separate frequency component in the signal. This form of distortion is difficult to avoid, e.g., in transmission lines.

attenuator (*Telecomm.*). (1) Arrangement of resistors, capacitors, etc., which introduces known attenuation into a measuring line or circuit. A *variable attenuator* uses switching to vary the attenuation introduced into the circuit. This is often standardized at 600 ohm or 75 ohm impedance levels in ladder attenuators, consisting of a series of T-sections. See **pad** (2). (2) Gas tube used to absorb and hence control a radiofrequency oscillation. (3) Section of waveguide for diminishing intensity of wave transmitted. Geometry may include vane, piston, flap, disk, or a rotary member, in association with absorbing material to avoid reflections.

attitude (*Aero.*). The *attitude* or *aspect* of an aircraft in flight, or on the ground, is defined by the angles made by its axes with the relative airflow, or with the ground, respectively.

attitude angle (*Autos.*). See **slip angle.**

attitude indicator (*Aero.*). A *gyro horizon* which indicates the true attitude of the aircraft in pitch and roll throughout 360° about these axes. See **heading indicator.**

attracted-disk electrometer (*Elec. Eng.*). Fundamental instrument in which potential is measured by the attraction between two oppositely charged disks.

attracted-iron motor (*Elec. Eng.*). A small motor suitable for operating instruments, etc. It consists of a soft iron core which is free to rotate and which is acted upon by magnetizing coils. A laminated iron core is used for motors to operate off an a.c. supply.

attraction cone (*Physiol.*). A structure formed in the cytoplasm of the ovum during fertilization, at the point of contact with the spermatozoon, facilitating the entry of the latter.

attraction particle (*Cyt.*). See **centriole.**

attraction sphere (*Cyt.*). In cell division, the archoplasmic masses, and the striations radiating through them from the centrosomes. Also called **astrosphere, centrosphere.**

attraction spindle (*Cyt.*). In cell division, the terminal portions of the achromatic spindle.

attrition test (*Civ. Eng.*). A test for the determination of the wear-resisting properties of stone, particularly stone for roadmaking. Pieces of the stone are placed in a closed cylinder, which is then rotated for a given time, after which the loss of weight due to wear is found.

Attwood's formula (*Ships*). A formula for determining the moment of statical stability at large angles of heel of a ship. Taking angle of heel θ, and the weight of the ship W, moment

$$= W\left(\frac{v \times hh_1}{V} \pm BG \sin\theta\right) \text{ foot tons,}$$

where v = volume of emerged wedge; hh_1 = distance between C.G.'s of emerged and immersed wedges; V = volume of displacement; B = centre of transverse buoyancy; G = centre of gravity.

at. wt. *Atomic weight*, now **relative atomic mass.**

A-type disk harrow (*Agric.*). A disk harrow comprising two gangs, in tandem, one or both of which may be angled.

A-type pole (*Elec. Eng.*). A type of wooden support for overhead transmission lines consisting of two poles inclined vertically to one another, joined at the top and braced at the centre.

Au (*Chem.*). The symbol for gold (*aurum*).

A.U. (*Astron.*). Abbrev. for **astronomical unit.**

audibility (*Acous.*). Ability to be heard; said of faint sounds in the presence of noise. The extreme range of audibility is 30 to 20 000 Hz in frequency, depending on the applied intensity; and from 2×10^{-5}N/m² (r.m.s.) at 1000 Hz (the zero of the phon scale, selected as the average for good ears), and 120 decibels above this.

audible ringing tone (*Teleph.*). An audible tone fed back to a caller as an indication that ringing current has been remotely extended to solicit attention from the called subscriber (or operator). On PO circuits in U.K. it is heard as a double beat recurring at 2 second intervals. Also called **audible signal.**

AU diode (*Electronics*). See **backward diode.**

audiofrequency (*Acous., Telecomm.*). Frequency which, in an acoustic wave, makes it audible. In general, any wave motion including frequencies in the range of, e.g., 30 Hz to 20 kHz.

audiofrequency amplifier (*Telecomm.*). Amplifier for frequencies within the audible range.

audiofrequency choke (*Elec. Eng.*). Inductor with appreciable reactance at audiofrequencies.

audiofrequency shift modulation (*Telecomm.*). Method of facsimile transmission in which tone values from black to white are represented by a graded system of audiofrequencies.

audiofrequency transformer (*Telecomm.*). Transformer for insertion into a communication channel, so designed that it has a specified, normally uniform, response for signals at all frequencies required for sound reproduction.

audiogenic seizure (*An. Behav.*). Convulsive behaviour induced by sudden loud sounds in some strains of rats and mice, whose inheritance has been widely studied.

audiogram (*Acous.*). Standard graph or chart which indicates the hearing loss (in *bels*) of an individual ear in terms of frequency. See **noise-, sound-level meter.**

audiometer (*Acous.*). Instrument for measurement of acuity of hearing. Specifically to measure the minimum intensities of sounds perceivable by an ear for specified frequencies. See also **gramophone-, noise-.**

audition limits (*Acous.*). Extreme frequencies of sound waves perceivable by the normal ear, and the extent of perception between the maximum tolerable loudness and the minimum perceptible. See **phon, threshold of sound.**

auditory, aural (*Zool.*). Pertaining to the sense of hearing or to the apparatus which subserves that sense: the 8th cranial nerve of Vertebrates, supplying the ear.

auditory canal (*Acous., Physiol.*). Duct connecting the ear drum with the external ear (*pinna*), by which sound waves are transmitted from outer to inner ear.

auditory ossicles (*Zool.*). Three small bones, the *incus*, *malleus*, and *stapes*, bridging the tympanic cavity of the middle ear in mammals.

auditory perspective (*Acous.*). See **stereophony.**

Auerbach's plexus (*Zool.*). In the small intestine of Vertebrates, a close ganglionated plexus of amyelinate nerve fibres belonging to the sympathetic nervous system, and lying between the longitudinal and circular muscle coats; plexus myentericus.

Auer metal (*Met.*). See **mischmetal.**

aufwuchs (*Ecol.*). See **periphyton.**

augend (*Maths.*). See **addition.**

augen-gneiss (*Geol.*). A coarsely crystalline rock of granitic composition, containing lenticular, eye-shaped masses of feldspar or quartz embedded in a finer matrix. A product of regional metamorphism.

auger (*Tools*). A tool for boring holes, especially in wood or in the earth.

Auger effect (*Nuc.*). Nonradiative transition of an atom from an electronic excited state to a lower level, with the emission of an electron. Arises when 2 atomic energy levels belonging to different series lie close enough together for mutual interaction, so that repulsion gives a shift of the energy levels.

Auger yield (*Nuc.*). For a given excited state of an atom of a given element, the probability of de-excitation by Auger process instead of by X-ray emission.

augite (*Min.*). A complex aluminous silicate of calcium, iron, and magnesium, crystallizing in the monoclinic system, and occurring in many igneous rocks, particularly those of basic composition; it is an essential constituent of basalt, dolerite, and gabbro.

augmentation distance (*Nuc.*). Distance between extrapolated boundary and true boundary of a reactor. Also **extrapolation distance.**

augmentation of moon's semi-diameter (*Astron.*). The apparent increase in the angular radius of the moon, due to the observer's being nearer to that body when above the horizon, than is the centre of the earth, the maximum difference in the distance being the earth's radius, 3963 miles (6378 km) approximately.

augmentative-trophic reaction (*Zool.*). In development, the quantitative effect of stimuli on the increase in size of organs or tissues already present.

augmentor (*Zool.*). In Vertebrates, a nerve arising from the sympathetic system and tending, when stimulated, to increase the rate of heart-beat: accelerator: more generally, any nerve which increases the rate-activity of an organ.

Aujesky's disease (*Vet.*). Pseudorabies; mad itch; infectious bulbar paralysis. An encephalomyelitis affecting cattle, sheep, pigs, dogs, cats, and rats, and caused by a virus. Intense itching of the skin is the main nervous symptom in all species except the pig.

aulophyte (*Bot.*). A plant inhabiting a cavity in the body of another plant, but not living as a parasite.

aulostomatous (*Zool.*). Possessing a tubular mouth.

aura (*Med.*). A movement, or sensation, or mental disturbance, which precedes an epileptic convulsion.

aural (*Zool.*). See **auditory.**

aural masking (*Acous.*). See **masking.**

Aurama (*Cinema.*). System of colour and sound control by cues added to stereophonically-recorded music and sound effects on tape.

auramines (*Chem.*). Dyestuffs of the diphenylmethane series.

aureole (*Elec. Eng.*). Luminous glow from outer portion of electric arc. This has different spectral distribution from that from the highly-ionized core. (*Geol.*) Area surrounding an igneous intrusion affected by metamorphic changes. (*Meteor.*) (1) The clear transparent space between the sun or moon and a *halo* (q.v.)

or corona. (2) The bright indefinite ring round the sun in the absence of clouds.

Aureomycin (*Pharm.*). TN for chlortetracycline hydrochloride, an antibiotic metabolic product of *Streptomyces aureofaciens*, with low toxicity, effective against many viruses, rickettsiae, and bacteria.

auric acid, auric oxide (*Chem.*). Gold(III)oxide. Formula, Au_2O_3. An amphoteric oxide of gold.

auricle (*Bot.*). A small ear-shaped lobe at the base of a leaf or other organ. (*Zool.*) A chamber of the heart connecting the afferent blood vessels with the ventricle: the external ear of Vertebrates: any lobed appendage resembling the external ear. Also called **auricula.**

auricled, auriculate. Having auricles.

auricula (*Zool.*). In *Echinoidea*, one of the ossicles of Aristotle's lantern, a radially placed arch arising from the inside of the corona. See also **auricle.**

auricularia (*Zool.*). In *Holothuria*, a pelagic ciliated larva, having the cilia arranged in a single band, produced into a number of short processes.

auricular organs (*Zool.*). In *Rhabdocoela*, areas or grooves on the head with special chemosensory receptors, with short nerve endings projecting just above the surface of the ectoderm. Cilia and rhabdites are absent from these areas.

auriculars (*Zool.*). See **tectrices.**

auriculoventricular (*Zool.*). Pertaining to, or connecting, the auricle and ventricle of the heart; e.g., the *auriculoventricular connexion*, a bundle of muscle fibres which transmits the wave of contraction from the auricle to the ventricle, in higher Vertebrates.

auriferous deposit (*Geol.*). A natural repository of gold, in the general sense, including gold-bearing lodes and sediments such as sands and gravels, or their indurated equivalents, which contain gold in detrital grains or nuggets. See **banket, lode, placers.**

auriferous pyrite (*Min.*). Iron sulphide in the form of pyrite, carrying gold, probably in solid solution.

aurine (*Chem.*). Pararosolic acid $(HO{\cdot}C_6H_4)_2{=}C{=}C_6H_4{=}O$, made from phenol, oxalic acid, and sulphuric acid. It is similar to rosolic acid in properties.

aurines (*Chem.*). A group of dyestuffs derived from aurine.

aurist (*Med.*). A medical man skilled in the treatment of defects of the ear, including the measurement of audition.

aurophore (*Zool.*). In certain *Siphonophora*, a modified medusoid of obscure function, occurring at the side of, and traversed by a canal from, the pneumatophore.

aurora (*Mag.*). Luminous ionization in thin atmosphere near the earth's magnetic poles, excited by particles from sun spiralling in earth's magnetic field.

Aurora Borealis (*Astron.*). The Northern Lights, a phenomenon consisting of luminous arcs, rays, streamers, etc., of green, red, or yellow colour. They are caused by high-speed particles ejected from the sun, and are therefore most frequent at the time of sunspot maximum. They are best seen in the neighbourhood of the North Magnetic Pole. In high southern latitudes the corresponding phenomenon is called **Aurora Australis.**

auroral zone (*Radio*). Zone where radio transmission is limited by aurora.

aurous (*Chem.*). Containing gold(I).

aurum (*Chem.*). See **gold.**

auscultation (*Med.*). The act of listening to the sounds produced in the body.

aussage experiment (*Psychol.*). A test on accuracy in reporting on a picture or on a series of events under cross-examination and in a limited time.

austempering (*Met.*). Heating a steel to transform it to austenite, followed by quenching to a temperature above the martensitic change point, but below the critical range, so that pearlite and bainite are directly formed. It is then cooled to atmospheric temperature.

Austen-Cohen formula (*Radio*). Semi-empirical formula for the field strength at a distance of r kilometres from the transmitting antenna of effective height H metres, and carrying a current of I amperes, the wavelength being λ kilometres. The field strength in microvolts per metre is ca.

$$\frac{377HI}{\lambda r} \exp\left(\frac{-0{\cdot}0015}{\sqrt{\lambda}}\right), \text{ for } \lambda > 1 \text{ km.}$$

austenite (*Met.*). Originally, a solid solution of carbon in γ-iron; now includes all solid solutions based on γ-iron.

austenitic steels (*Met.*). Steels containing sufficient amounts of nickel, nickel and chromium, or manganese to retain austenite at atmospheric temperature; e.g., *austenitic stainless steel* and *Hadfield's manganese steel.*

Austin Moore prosthesis (*Surg.*). A prosthesis which is inserted into the upper end of the femur to reconstruct the hip joint when the original head has been removed.

Australasian region (*Zool.*). One of the primary faunal regions into which the land surface of the globe is divided; includes Australia, New Guinea, Tasmania, New Zealand, and the islands south and east of Wallace's line.

australites (*Min.*). See **tektites.**

Austrian cinnabar (*Chem.*). See **basic lead chromate.**

aut-. Prefix. See **auto-.**

autacoid, autocoid (*Physiol.*). General name for an endocrine secretion; a specific organic substance formed by the cells of one organ, and passed by them into the circulating fluid, to produce effects upon other organs. See **hormones** and **chalones.**

autecology (*Ecol.*). The study of the ecology of any individual species. Cf. *synecology.*

autesthetic (*Psychol.*). Used of an activity (e.g., play or pair-ritual) that rewards the performer as much as other individuals.

authigenic (*Geol.*). Pertaining to minerals which have crystallized in a sediment during or soon after its deposition.

authoritarian personality (*Psychol.*). A constellation of personality characteristics typified by rigidity of thought, strict adherence to principle, even if wrong and in the face of rational argument, and ethnocentrism.

autism (*Psychiat.*). Behavioural abnormality characterized by excessive indulgence in fantasy, and concentration on self. Many sufferers exhibit anti-social behaviour or isolationism which may be erroneously confused with low intelligence, whereas on correct diagnosis suitable treatment is possible.

auto-, aut-. Prefix from Gk. *autos*, self.

autoalarm (*Automation*). See **automatic call device.**

autoallogamy (*Bot.*). The condition of a species in which some individual plants are capable of self-pollination and others of cross-pollination.

Autobasidiomycetes (*Bot.*). A group of about 12 000 species, including mushrooms, toadstools, and many related fungi, characterized by the possession of an *autobasidium* (q.v.).

autobasidium (*Bot.*). A basidium which does not become septate.

autobiotic (*Chem.*). One of a group of substances produced by cells and controlling the behaviour of the producing cells.

autoblast (*Biol.*). A separate, independent cell or micro-organism.

autobrecciation (*Geol.*). The production of a brecciated appearance by processes involved in the consolidation of a rock. Examples of both sedimentary and volcanic autobreccias have been observed.

autocapacitance coupling (*Elec. Eng., Telecomm.*). Coupling of two circuits by a capacitor included in series with a common branch.

autocarp (*Bot.*). A fruit resulting from self-fertilization.

autocatalysis (*Chem.*). The catalysis of a reaction by the product of that reaction, e.g., MnO_4^- and $C_2O_4^{2-}$ by M_n^{2+} ions. (*Zool.*) Reaction or disintegration of a cell or tissue, due to the influence of one of its own products.

autochthonous (*Zool.*). In an aquatic community, said of food material produced within the community: more generally, indigenous, inherited, hereditary (e.g., *autochthonous* species, *autochthonous* characteristics). Cf. *allochthonous*.

autochthonous behaviour (*An. Behav.*). Behaviour activated by its own drive. Cf. *allochthonous behaviour*.

autoclave (*Chem.*). A vessel, constructed of thick-walled steel (usually alloy steel or frequently nickel alloys), for carrying out chemical reactions under pressure and at high temperatures. Pressure gauge, safety valve, and thermometer pocket are provided for control. (*Med.*) An apparatus for sterilization by steam at high pressure.

auto coarse pitch (*Aero.*). The setting of the blades of an airscrew to the minimum drag position if there is a loss of engine power during takeoff.

autocollimator (*Light.*) (1) An instrument for accurately measuring small changes in the inclination of reflecting surfaces. Principally used for engineering metrology measurements. (2) A convex mirror used to produce a parallel beam of light from a reflecting telescope. It is placed at the focus of the main mirror.

autocondensation (*Med.*). Application of high-frequency currents when the patient is one electrode of a capacitor.

autoconduction (*Med.*). Electromagnetic induction of currents when the patient forms the secondary inside a large solenoid.

autoconverter (*Elec. Eng.*). A special form of converter used with certain types of electric-battery vehicle; it is arranged to operate from a constant-voltage battery supply and give an output voltage inversely proportional to the current.

autocorrelation (*Electronics*). Technique for detecting weak signals against strong background level. Signal is subjected to controlled delay, the original delay signals then being fed to the autocorrelation unit which responds strongly only if delay is exact multiple of signal period.

autocyst (*Zool.*). In *Neosporidia*, a thick membrane formed by the parasite, and separating it from the host tissues.

autodiastyly (*Zool.*). In *Craniata*, a type of jaw suspension in which the jaws are attached by ligaments to the neurocranium and the hyoid arch. Only in some extinct Fish.

autodyne (*Radio*). Of an electrical circuit in which the same elements and valves are used both as oscillator and detector. Also called **endodyne, self-heterodyne.**

autodyne oscillator (*Radio*). The valve used in the autodyne receiver which functions as a generator, and for reception (or amplification of the incoming signal).

autodyne receiver (*Radio*). One utilizing the principle of beat reception and including an autodyne oscillator.

autoecious, autoxenous (*Bot.*). A term applied to parasitic fungi which complete the whole of their development upon one species of host plant.

autoemission (*Electronics*). That arising at normal temperatures by a high-voltage gradient, stripping electrons from surface atoms; also called **autoelectric effect, cold emission, field emission.**

autoerotism (*Psychol.*). A condition where sensual pleasure is sought and gratified in one's own person, without the aid of an external love-object; e.g., masturbation. See also **narcissism.**

Autofining (*Chem. Eng.*). Proprietary process for desulphurizing light petroleum fractions by partial *cracking* (q.v.), which produces some hydrogen which catalytically removes the sulphur from the remaining stream.

autoflare (*Aero.*). An aircraft automatic landing system which operates on the *flare-out* (q.v.) stage of the landing, by means of a very critical radio altimeter.

autogamy (*Bot., Zool.*). (1) Self-fertilization. (2) The fusion of sister-cells, or of 2 sister-nuclei. (3) Reproduction, in some *Heliozoa*, by which a nucleus divides into 2 which, after maturation, immediately reunite.

autogenesis, autogeny (*Bot., Zool.*). See **spontaneous generation.**

autogenic movement (*Bot.*). See **autonomic movement.**

autograft (*Biol., Med.*). A mass of tissue, or an organ, moved from one region to another within the same organism.

autoheterodyne (*Radio*). See **autodyne.**

autoheterosis (*Zool.*). Independent modification of a merome, or meromes, without consequent modification of the other meromes in the same somite.

autoicous (*Bot.*). Having staminate and pistillate inflorescences on the same plant.

autoignition (*I.C. Engs.*). The self-ignition or spontaneous combustion of a fuel when introduced into the heated air charge in the cylinder of a compression-ignition engine. See **spontaneous ignition temperature.**

autoimmune diseases (*Med.*). A group of diseases caused by antigen/antibody reactions to the host's own tissues.

autoinductive coupling (*Elec. Eng.*). Coupling of two circuits by an inductance included in series with a common branch.

autoinfection (*Zool.*). Re-infection of a host by its own parasites.

autointoxication (*Med.*). Poisoning of the body by toxins produced within it.

autoiris (*Photog.*). See **automatic diaphragm.**

autoland (*Aero.*). A landing in which the descent, forward speed, *flare-out*, alignment with the runway and touchdown are all automatically controlled. See **autoflare, autothrottle.**

autolithography (*Print.*). The drawing by the artist of his design direct on the stone or plate.

autolysis (*Bot., Zool.*). The breakdown of the cell content of a cell, or of an organ, by the action of enzymes produced in the cells concerned: self-digestion. *adj.* **autolytic.**

automanual exchange (*Teleph.*). A manual switchboard operating in conjunction with one or more automatic exchanges to cater for coin-box and assistance traffic, and to connect manually or semiautomatically all calls requiring extension beyond the area served by the local automatic exchange(s).

automatic alarm (*Automation*). Same as **automatic call device.**

automatic arc lamp (*Light*). An arc lamp in which the feeding of the carbons into the arc and the striking of the arc are done automatically, by electromagnetic or other means.

automatic arc welding (*Elec. Eng.*). Arc welding carried out in a machine which automatically moves the arc along the joint to be welded, feeds the electrode into the arc, and controls the length of the arc.

automatic bias (*Electronics*). Grid bias from a resistor in the cathode circuit of a valve, the grid circuit being returned to the end of the resistor remote from the cathode.

automatic blankets (*Print.*). On letterpress rotary machines, a covering for the impression cylinder with a felt base and a surface of rubber on plastic.

automatic brightness control (*TV*). Circuit used in some television receivers to keep average brightness level of screen constant.

automatic call device (*Automation*). System of relays, responsive to a prearranged set of signals, connected to an unattended receiver, so that an alarm is sounded on operation. Frequently used on ships for detection of distress signals. International alarm signal is twelve 4-second dashes at 1-second intervals followed by SOS call repeated three times. Also **autoalarm, automatic alarm.**

automatic camera (*Photog.*). One in which the aperture or shutter speed is selected automatically by an integral photocell. Such a camera is always ready for instant use by press, or amateurs. See **Photomaton.**

automatic check (*Comp.*). Circuits built into a computer to check correct functioning.

automatic circuit-breaker (*Elec. Eng.*). A circuit-breaker which automatically opens the circuit as soon as certain predetermined conditions (e.g., an overload) occur.

automatic computer (*Comp.*). One operated under the control of a monitor.

automatic contrast control (*TV*). Form of automatic gain control used in video signal channel of television receiver.

automatic control (*Elec., etc.*). (1) Switching system which operates control switches in correct sequence and at correct intervals automatically. (2) Control system incorporating servomechanism or similar device, so that feedback signal from output of system is used to adjust the controls and maintain optimum operating conditions.

automatic cut-out (*Elec. Eng.*). A term frequently applied to a small automatic circuit-breaker suitable for dealing with currents of a few amperes.

automatic data processing (*Comp.*). Complete processing of data by computer, *read out* for immediate use.

automatic diaphragm (*Photog.*). Type of preset iris which selects required aperture when shutter is released and then returns automatically to fully open position ready for focusing. Also called **autoiris.**

automatic direction-finder (*Aero.*). Airborne radio receiver which displays a continuous bearing toward a selected ground radio beacon. Abbrev. **ADF**, also called **radio compass.** (*Automation*) System in which servo motors, controlled by the incoming signal, cause the rotatable loop antenna or goniometer to hunt for the direction of maximum or minimum signal response.

automatic exposure control (*Photog.*). Photoelectric device on a cinecamera which automatically sets the correct lens aperture.

automatic flushing cistern (*San. Eng.*). A tank which discharges its contents, at regular intervals, by a siphonic action started when the water entering the tank rises to a certain level. Used for flushing urinals; also for drains having insufficient fall to ensure self-cleansing.

automatic focusing (*Photog.*). Device whereby the image projected by an enlarger remains in focus whatever its distance from the lens. (*TV*) Electrostatic focusing in a TV tube in which the anode is internally connected via a resistor to the cathode.

automatic frequency control (*Radio*). Electronic or mechanical means for automatically compensating, in a receiver, frequency drifts in transmission carrier or local oscillator. Usually abbrev. to **AFC.**

automatic gain control (*Radio*). Autoadjustment of the gain of the radio amplifier of a radio-receiver, through feedback of the rectified carrier to the input of the amplifier, so that the output of the demodulator is very nearly constant, although the receiver carrier varies over a wide range of amplitudes due to fading, masking of antenna, etc. Abbrev. **AGC.**

automatic gate (door) (*Elec. Eng.*). A gate (or door) on a lift-car, which is automatically opened by the action of the lift when it reaches a landing, and which automatically closes before the car leaves the landing. (*Build.*) A door-system (sliding, hinged, folding) electrically, pneumatically or hydraulically powered for self-opening and closing on impulse triggered by stepping-pad, photoelectric beam or similar device.

automatic gate (door) lock (*Elec. Eng.*). A lock on the gate (or door) of a lift-car or landing, which is arranged so that it can only be released, and the gate or door opened, when the car is in a position of safety at the landing concerned.

automatic generating plant (*Elec. Eng.*). A small generating station, being a petrol (or diesel) engine-driven generator along with a battery; the engine is automatically started up when the battery voltage falls below a certain value and stopped when it is fully charged. The term is also applied to the plant in small unattended hydroelectric generating stations.

automatic grid bias (*Electronics*). Use of potential drop across a cathode resistor for biasing grid or other circuit.

automatic letter facing (*Automation*). System using 1 or 2 phosphorescent ink lines on stamps which are detected photoelectrically.

automatic loom (*Textiles*). Precision-made, high-speed loom equipped with side or centre weft-stop and mechanical or electrical warp-stop motion, and constant tension warp let-off. Automatic weft replenishing arrangements include (*a*) firm changing from circular or horizontal battery and (*b*) shuttle changing mechanism.

automatic meteorological observation station (*Meteor.*). Transistorized and packaged apparatus which transmits weather data for electronic computation.

automatic mixture control (*Aero.*). A device for adjusting the fuel delivery to a reciprocating engine in proportion to air density.

automatic monitor (*Electronics*). Apparatus which compares the quality of transmission at different parts of a system, and raises an alarm if there is appreciable variation.

automatic numbering transmitter (*Teleg.*). An automatic transmitter additionally equipped to transmit a serial number before each message.

automatic observer (*Aero.*). An apparatus for recording, photographically or electronically, the indications of a large number of measuring instruments on experimental or research aircraft.

automatic parachute (*Aero.*). A parachute for personnel which is extracted from its pack by a static line attached to the aircraft.

automatic phase control (*TV*). In reproducing colour TV images, the circuit which interprets the phase of the chrominance signal as a signal to be sent to a matrix.

automatic phase control loop (*TV*). Feedback circuit in which the phase of a local oscillator is controlled by a comparison with that of a reference signal, to obtain a correction voltage for application to the controlled source. Abbrev. **APC loop.**

automatic picture control (*TV*). See **automatic contrast control.**

automatic pilot (*Aero.*). A device for guiding and controlling an aircraft on a given path. It may be set by the pilot or externally by radio control. Colloquially **George.**

automatic programming (*Comp.*). System in which a computer is used for preparing programme instructions in machine form.

automatic quadder (*Print.*). Linotype and Intertype line-casting machines can automatically centre each line or *quad* it to left or right; on the Monotype one depression of the quadder key can result in either 5 or 10 quads.

automatic quiet gain control (*Radio*). Joint use of automatic gain control and muting.

automatic reel change (*Print.*). On rotary machines, equipment to attach a new reel to the old web, without stopping the machine and severing the butt end of the old web. Also called **autopaster, flying paster.**

automatic reperforator switching (*Teleg.*). In a tape relay system, a method of working in which routing characters from the reperforated version of each incoming message are used to initiate and control the selection and connexion of the outgoing circuit on which that message is to be retransmitted.

automatic screw machine (*Eng.*). Fully-automatic single-spindle or multiple-spindle bar stock turret lathe.

automatic shutter (*Cinema.*). In a film projector, the shutter which cuts off the light from the arc if the film should stop, instead of maintaining intermittent motion; without this safeguard, the intense heat from the arc would ignite nitrate film.

automatic signalling (*Rail.*). A system of railway signalling, usually with electric control, in which the signals behind a train are automatically put to 'danger' as soon as the train has passed, and held in that position until the train has attained the next section of line. (*Teleph.*) A method of signalling used on a telephone circuit, whereby the transmission of supervisory signals on the seizure or release of that circuit occurs automatically.

automatic stabilizer (*Aero.*). A form of automatic pilot, operating about one or more axes, adjusted to counteract dynamic instability. Colloquially **autostabilizer.** See also **damper** (*yaw*).

automatic starter (*Elec. Eng.*). A starter for an electric motor which automatically performs the various starting operations (e.g., cutting out steps of starting resistance) in the correct sequence, after being given an initial impulse by means of a push-button or other similar device.

automatic stoker (*Eng.*). See **mechanical stoker.**

automatic substation (*Elec. Eng.*). A substation containing rotating machinery (and, therefore, normally requiring the presence of an attendant), which, as occasion demands, is started and stopped automatically; e.g., by a voltage relay which operates when the voltage falls below or rises above a certain predetermined value.

automatic switching equipment (*Teleph.*). Apparatus for selectively altering the connexions in electrical circuits in response to remotely-generated electrical signals.

automatic synchronizer (*Elec. Eng.*). A device which by means of suitable relays, performs the process of synchronization automatically.

automatic tap-changing equipment (*Elec. Eng.*). A voltage-regulating device which automatically changes the tapping on the winding of a transformer to regulate the voltage in a desired manner.

automatic telegraphy (*Teleg.*). Transmission from equipments which automatically convert data from a storage medium (e.g., punched paper tape) into standard telegraph signals.

automatic telephone exchange (*Teleph.*). An exchange where the switching equipment may be remotely operated by callers to set up connexions between their own telephones and any other telephone on the accessible network.

automatic telephone system (*Teleph.*). A system in which the setting up of a route through the switching equipment to any telephone in the accessible network may be effected, without the service of an operator, by trains of impulses initiated by the caller.

automatic tracking (*Radar*). Servo control of radar system operated by received signal, to keep antenna aligned on target.

automatic train stop (*Elec. Eng.*). A catch, used in conjunction with an automatic signalling system, which engages a trip-cock on the train if the train passes a signal at danger.

automatic transmission (*Eng.*). A power transmission system for road vehicles, in which the approximately optimum engine speed is maintained through mechanical or hydraulic speed changing devices which are automatically selected and operated by reference to the road speed of the vehicle. Additional features meet special requirements for braking, reversing, parking, and unusual driving conditions.

automatic trolley reverser (*Elec. Eng.*). An arrangement of the overhead contact line of a tramway, located at terminal points, which ensures that the trolley collector is reversed when the direction of motion of the car is reversed.

automatic tuning (*Radio*). (1) System of tuning in which any of a number of predetermined transmissions may be selected by means of push-buttons or similar devices. (2) Fine tuning of receiver circuits by electronic means, following rough tuning by hand.

automatic voltage regulator (*Elec. Eng.*). A voltage regulator which automatically holds the voltage of a distribution circuit or an alternator constant within certain limits, or causes it to vary in a predetermined manner. See **automatic tap-changing equipment, moving-coil regulator, Tirill regulator.**

automatic volume compression (*Radio*). Reduction of signal voltage range from sounds which vary widely in volume, e.g., orchestral music. This is necessary before they can be recorded or broadcast but requires corresponding expansion in the reproducing system to compensate.

automatic volume control (*Radio*). Alteration of the contrast (dynamics) of sound during reproduction by any means. By compression (*compounder*) a higher level of average signal is obtained for modulation of a carrier, the expansion (*expander*) performing the reverse function at the receiver. In high-fidelity reproduction, arbitrary expansion can be disturbing because of variation in background noise, if present. Abbrev. AVC.

automatic volume expansion (*Electronics*). Expansion of contrast (dynamics), e.g., keeping the maximum level constant and automatically diminishing the lower levels.

automation. Industrial closed-loop control system in which manual operation of controls is replaced by servo operation.

automatism (*Psychiat.*). An automatic act done without the full cooperation of the personality, which may even be totally unaware of its existence. Commonly seen in hysterical states, such as fugues and somnambulism, but may also be a local condition as in automatic writing.

autometamorphism (*Geol.*). Changes in the mineral composition of an igneous rock consequent upon the action of solutions and vapours which are derived from the same source as the rock itself.

automixis (*Biol.*). The mingling of chromatin from the same source, as in self-fertilization; hence self-fertilization especially in lower plants.

automixte system (*Elec. Eng.*). A system of operation of petrol-electric vehicles in which a battery, connected in parallel with the generator, supplies current during starting and heavy-load periods and is charged by the generator during light-load periods. Also **Pieper system.**

automorphism (*Maths.*). A one-to-one homomorphic mapping from a set on to itself, i.e. an *isomorphism* from a set on to itself. See **homomorphism.**

autonomic, autonomous (*Bot., Zool.*). Independent: self-regulating: spontaneous.

autonomic movement (*Bot.*). A movement of a plant, or of part of a plant, not stimulated by any external condition. Also called **autogenic movement.**

autonomic nervous system (*Zool.*). In Vertebrates, a system of motor nerve fibres supplying the smooth muscles and glands of the body. See **parasympathetic nervous system, sympathetic nervous system.**

autonomics (*Electronics*). Study of self-regulating systems for process control, optimizing performance.

autonomous (*Zool.*). Subject to the same laws, especially of growth and specialization. Cf. *heteronomous*. See also **autonomic.**

autopalatine (*Zool.*). In some Fish, an ossification at the extreme anterior end of the PPQ Bar.

autoparasite (*Bot.*). A parasite which attacks another parasite.

autoparthenogenesis (*Zool.*). Artificially stimulated development of unfertilized eggs.

autopaster (*Print.*). See **automatic reel change.**

autophagic vacuole (*Cyt.*). See **cytosegresome.**

autophagous (*Zool.*). Capable of self-feeding from the moment of birth, as some Birds, which run about and feed as soon as they are hatched.

autophagy (*Zool.*). The eating of a part of the body, usually after its amputation.

autophya (*Zool.*). Elements of the shell or skeleton secreted by the organism itself. See **xenophya.**

autophyte (*Bot.*). A plant which builds up its food substances from simple compounds.

autopilot (*Aero.*). See **automatic pilot.**

autoplasma (*Zool.*). In tissue culture, a medium prepared with plasma from the same animal from which the tissue was taken; cf. *heteroplasma, homoplasma*. *adj.* **autoplastic.**

autoplastic transplantation (*Zool.*). Reinsertion of a transplant or graft from a particular individual in the same individual. Cf. *heteroplastic, homoioplastic, xenoplastic.*

autoplate (*Print.*). A machine which can deliver a curved stereoplate for rotary printing every 15 sec; built to suit the requirements of each particular rotary machine.

autopodium (*Zool.*). In Vertebrates, the hand or foot.

autopolyploid (*Cyt.*). A polyploid having similar sets of chromosomes in its total chromosome content.

autopotamous (*Ecol.*). Originating in rivers and streams.

autopsy. See **necropsy.**

autoradiography (*Chem., Photog.*). In tracer work, the record of a treated specimen on a photographic plate caused by radiations from the radioisotope used. See **tracer elements.** (*Histol., Radiol.*) Technique for the demonstration of substances in tissues and organs, consisting of the introduction of the substances in a radioactive form and the subsequent recording of their distribution on photographic film or emulsion applied to a histological preparation of the tissue. (*Min.*) Photographic examination of radioactive mineral particles made by allowing them to rest on photographic film, perhaps after 'tagging' with radioactive isotopes.

auto-reclose circuit-breaker (*Elec. Eng.*). A circuit-breaker which, after tripping due to a fault, automatically recloses after a time interval which may be adjusted to have any value between a fraction of a second and 1 or 2 min.

autorhythmus (*Zool.*). In metameric animals, repetition of meromes within a somite.

autorotation (*Aero.*). (1) The spin; continuous rotation of a symmetrical body in a uniform air-stream due entirely to aerodynamic moments. (2) Unpowered rotorcraft flight, i.e., a helicopter with engine stopped, in which the symmetrical aerofoil rotates at high incidence parallel with the airflow.

autoset level (*Surv.*). A form of dumpy level for rapid operation, in which the essential features are a quick-levelling head, and an optical device which neutralizes errors of levelling so that the bubbles need not be central while an observation is being made.

autoshaver (*Print.*). A machine which trims the curved stereoplate to the required thickness.

autoskeleton (*Zool.*). A skeleton formed of autophya; usually an endoskeleton.

autosome (*Cyt.*). Any chromosome other than a sex chromosome; a typical chromosome. See **allosome.**

autospasy (*Zool.*). The casting of a limb or part of the body when it is pulled by some outside agent, as when the Slow-worm casts its tail.

autospermatheca (*Zool.*). In *Oligochaeta*, a spermathecal sac, which is a reservoir for the animal's own spermatozoa, which are discharged on its own ova during the formation of the cocoon.

autospore (*Bot.*). A daughter cell formed within

an algal cell, and having all the characters of the parent in miniature before it is set free.

autostabilizer (*Aero.*). See **automatic stabilizer.**

autostoper (*Mining*). Pneumatic rock drill mounted on long cylinder with extendable ram, so as to be held firmly across opening (*stope*) when packed out by compressed air.

autostoses (*Zool.*). See **cartilage bones.**

autostyly (*Zool.*). In *Craniata*, in *Dipnoi* and all tetrapods, a type of jaw suspension in which the hyoid arch is broken up and the hyomandibular attached to the skull. *adj.* **autostylic.**

autosuggestion (*Psychol.*). A condition of self-induced suggestion, brought about when the mind has lessened conscious direct effort and control, and is absorbed only with the suggested idea. In certain cases it may induce a state of light autohypnosis.

autosynchronous motor (*Elec. Eng.*). A term frequently used to denote a *synchronous induction motor*.

autotetraploid (*Cyt.*). A tetraploid with four similar sets of chromosomes in its nuclei.

autothrottle (*Aero.*). A device for controlling the power of an aero-engine to keep the approach path angle and speed constant during an automatic blind landing.

autotilly (*Zool.*). The removal of a limb or part of the body by the animal itself, as in certain Spiders.

autotomy (*Zool.*). Voluntary separation of a part of the body (e.g., limb, tail), as in certain Worms, Arthropods, and Lizards.

autotransductor (*Elec. Eng.*). One in which the same winding is used for power transfer and control.

autotransformer (*Elec. Eng.*). Single winding on a laminated core, the coil being tapped to give desired voltages.

autotransformer starter (*Elec. Eng.*). A starter for squirrel-cage induction motors, in which the voltage applied to the motor at starting is reduced by means of an autotransformer.

autotransplantation (*Zool.*). See **autoplastic transplantation.**

Autotron scanner (*Print.*). An electronically-controlled device for maintaining close register on web-fed printing machines.

autotrophic (*Bot.*). Able to build up food materials from simple substances.

autotrophic bacteria (*Bacteriol.*). Photosynthetic and chemosynthetic bacteria which are able to utilize carbon dioxide in assimilation.

autotropism (*Bot.*). The tendency to grow in a straight line.

autovac (*Autos.*). A vacuum-operated mechanism for raising fuel from a tank situated below the level of the carburettor to a position from which it may be fed to the latter by gravity.

autovalve (*Elec. Eng.*). A form of lightning diverter, consisting of a porous material, in which the discharge is confined to very narrow passages and, therefore, cannot assume the characteristics of an arc. See **lightning arrester.**

autoxenous (*Bot.*). See **autoecious.**

autoxidation (*Chem.*). (1) The slow oxidation of certain substances on exposure to air. (2) Oxidation which is induced by the presence of a second substance, which is itself undergoing oxidation.

autoxidator (*Chem.*). An alkene-oxygen compound acting as a carrier or intermediate agent during oxidation, in particular during *autoxidation* (q.v.).

autozooid (*Zool.*). In *Anthozoa*, order *Alcyonaria*, a secondary polyp which feeds the colony. See **siphonozooid.**

autumnal (*Ecol.*). Referring to the entire autumn season which, unlike spring and summer, is not subdivided into two parts.

autumnal equinox (*Astron.*). The instant at which the sun's apparent longitude is 180°. It then crosses the celestial equator from north to south, the date being about 23rd September. See also **equinox.**

autumn wood (*Bot.*). See **summer wood.**

autunite (*Min.*). Hydrous phosphate of calcium and uranium, resembling torbernite, but yellow.

auxanogram (*Bot.*). A method of determining the nutritional requirements of yeasts by their differential growth in culture.

auxanometer (*Bot.*). An instrument which is used to measure the rate of elongation of a plant member.

auxesis (*Zool.*). Induction of cell division, especially by the influence of a chemical agent.

auxetic (*Zool.*). Inducing or stimulating cell division.

auxiliary air intake (*Aero.*). (1) An air intake for accessories, cooling, cockpit air, etc. (2) Additional intake for turbojet engines when running at full power on the ground, usually spring-loaded so that it will open only at a predetermined suction value.

auxiliary attachment (*Horol.*). A special attachment to a compensation balance, for the purpose of reducing the middle-temperature error.

auxiliary cell (*Bot.*). One of a group of cells which are not part of the sexual organs of red algae, but nevertheless play a part (probably nutritive) in the formation of the fruit.

auxiliary circle (*Maths.*). Of an ellipse, the circle whose diameter is the major axis of the ellipse. Of a hyperbola, the circle whose diameter is the transverse axis.

auxiliary contact (*Elec. Eng.*). See **auxiliary switch.**

auxiliary engine (*Rail.*). Small diesel engine in addition to the prime mover of a diesel locomotive to provide power for battery charging, etc., when the latter is not running.

auxiliary equation (*Maths.*). Of an ordinary linear differential equation: $F(D)x=f(x)$, where D is the differential operator $\frac{d}{dx}$: the algebraic equation $F(m)=0$.

auxiliary grid (*Electronics*). In a pentode, the second grid, maintained at high positive potential.

auxiliary lift-motor (*Elec. Eng.*). A small motor forming part of the driving equipment of an electric lift; used for operating the lift at reduced speeds.

auxiliary plant (*Elec. Eng.*). A term used in generating-station practice to cover the condenser pumps, mechanical stokers, feed-water pumps, and other equipment used with the main boiler, turbine, and generator plant.

auxiliary pole (*Elec. Eng.*). See **compole.**

auxiliary power unit (*Aero.*). An independent airborne engine to provide power for ancillary equipment, electrical services, starting, etc. May be a small reciprocating or turbine engine. Abbrev. **APU.**

auxiliary rotor (*Aero.*). A small rotor mounted at the tail of a helicopter, usually in a perpendicular plane, which counteracts the torque of the main rotor; used to give directional and rotary control to the aircraft.

auxiliary screw arc (*Teleph.*). In certain 2-motion selectors, a metal arc into which contacts may be screwed in each coordinate vertical-rotary position, for sensing by the tips of 2 auxiliary wipers fitted above the head of the shaft; thereby equipping the selector for additional functions (e.g., preventing the return of busy tone from a PBX number until the busy condition has been encountered on every line).

auxiliary spark gap (*Elec. Eng.*). A small spark gap for automobile ignition, placed in series with the main gap of a sparking-plug; it improves the quality and certainty of the spark.

auxiliary store (*Comp.*). See **secondary memory (2)**.

auxiliary switch (*Elec. Eng.*). A small switch operated mechanically from a main switch or circuit-breaker; used for operating such auxiliary devices as alarm bells, indicators, etc.

auxiliary tanks (*Aero.*). See under **fuel tanks.**

auxiliary valency (*Chem.*). That remaining after saturation of the principal valency (oxidation state) of an atom.

auxiliary winding (*Elec. Eng.*). A special winding on a machine or transformer, additional to the main winding.

auxin (*Bot., Chem.*). A general name for a class of substances formed in young actively growing parts of plants, and having the power of affecting their subsequent growth and development. In meristems, one of the best-known auxins, having the general composition $C_{18}H_{32}O_5$, has been isolated from oats or from the germ of maize, and is a plant hormone which causes elongation of individual cells in the growing tips of plants. See **gibberellins, hormones, indoles, kinins, zeatin.**

auxochromes (*Chem.*). A chromophore (or group of atoms) introduced into dyestuffs to give full effectiveness to the colouring properties. The principal auxochromes are Cl, Br, SO_3H, NO_2, NH_2, OH. Auxochromes can also permit, by being present, the formation of salts and the creation of a dyestuff. They have a selective absorption frequency for radiation, and function as colour carriers for a frequency determined mainly by the compound of which they are part.

auxocyte (*Cyt.*). Any cell in which meiosis is started: an androcyte, sporocyte, spermatocyte or oöcyte, during the period of growth.

auxometer (*Optics*). An apparatus for measuring the magnifying power of an optical system.

auxospireme (*Cyt.*). In meiosis, the spireme formed after syndesis.

auxospore (*Bot.*). A resting spore formed by diatoms after a sexual fusion.

auxotonic (*Zool.*). Of muscle contraction, of or against increasing force.

available light photography (*Photog.*). Photography carried out without the use of flash or artificial lighting.

available line (*TV*). Percentage of total length of scanning line on a CRT screen on which information can be displayed.

available power efficiency (*Elec. Eng.*). The ratio of electrical power available at the terminals of an electroacoustic transducer to the acoustical power output of the transducer. The latter should conform with the reciprocity principle so that the efficiency in sound reception is equal to that in transmission.

available power gain (*Telecomm.*). The ratio of the available output power of a transducer to the available signal power at the input.

available power response (*Elec. Eng.*). For an electroacoustic transducer, the ratio of mean square sound pressure at a distance of 1 m, in a defined direction from the 'acoustic centre' of the transducer, to the available electrical power input. The response will be expressed in dB above the reference response of 1 μbar^2/watt of available electrical power.

available water (*Bot.*). The total amount of water in the soil (at any given time) which can be drawn upon by plants.

avalanche (*Nuc.*). Self-augmentation of ionization. See **Townsend avalanche, Zener effect.**

avalanche diode (*Nuc.*). See **Zener diode.**

avascular (*Med.*). Not having blood vessels.

AVC (*Radio*). Abbrev. for **automatic volume control.**

aventurine feldspar (*Min.*). A variety of plagioclase, near albite-oligoclase in composition, characterized by minute disseminated particles of red iron oxide which cause firelike flashes of colour. Also called **sunstone.**

aventurine glass (*Glass*). Glass containing 'spangles' of material, separated out from the main body. It may be produced from glasses containing excessive amounts of copper or chromium compounds, melted under special conditions.

aventurine quartz (*Min.*). A form of quartz spangled, sometimes densely, with minute inclusions of either mica or iron oxide. Used in ornamental jewellery.

average (*Maths.*). General term for **mean, median** and **mode.** (*Ships, etc.*) Loss or damage of marine property, less than total: compensation payment in proportion to amount insured.

average current (*Elec. Eng.*). A term used in connexion with alternating currents to denote the average value of the current taken over half a cycle.

average curvature (*Maths.*). See **curvature (3).**

average deviation (*Maths.*). For a number of like quantities x_j,

$$\frac{1}{n}\sum_{j=1}^{n} | x_j - \bar{x} |,$$

where the mean

$$\bar{x} = \frac{1}{n}\sum_{j=1}^{n} x_j.$$

Also called **mean deviation.**

average haul distance (*Civ. Eng.*). The distance between the centre of gravity of a cutting and that of the embankment formed from material excavated from the cutting.

average life (*Nuc.*). See **mean life.**

average power output (*Telecomm.*). For an amplitude modulation transmitter, the instantaneous output averaged over one cycle of the modulation signal.

average speech power (*Acous.*). The average of the instantaneous acoustic powers of a speaker while speaking.

averse (*Bot.*). Turned back.

aversion (*Bot.*). The tendency shown by elongating fungal hyphae to avoid coming into contact.

aversive (*An. Behav.*). Said of stimuli which, if applied following the occurrence of a particular response, decrease the strength of that response on later occurrences. They generally evoke behaviour associated with fear as well as avoidance.

Aves (*Zool.*). A class of *Craniata* adapted for aerial life. The forelimbs are modified as wings, the sternum and pectoral girdle are modified to serve as origins for the wing-muscles, and the pelvic girdle and hind limbs to support the entire weight of the body on the ground. The body is covered with feathers and there are no teeth. Respiratory and vascular systems are modified for homoiothermy. Birds.

avgas (*Aero.*). See **aviation spirit.**
avian big liver disease (*Vet.*). See **avian leucosis.**
avian diphtheria (*Vet.*). See **fowl pox.**
avian erythroblastosis (*Vet.*). See **avian leucosis.**
avian erythroid leucosis (*Vet.*). See **avian leucosis.**
avian favus (*Vet.*). White comb. A fungal disease of fowls affecting the skin of the head and comb, due to *Trichophyton gallinae*.
avian gout (*Vet.*). A symptom of nephritis in the domestic chicken, in which urates become deposited on the surface of internal organs (*visceral gout*) and in the joints.
avian granuloblastosis (*Vet.*). See **avian leucosis.**
avian leucosis (*Vet.*). The name given to a group of fatal diseases of the chicken, caused by several distinct but related viruses, in which there is an abnormal, neoplastic, proliferation of cells of the lymphocytic series (lymphoid leucosis, visceral lymphomatosis, big liver disease), erythrocytic series (erythroid leucosis, erythroblastosis), or granulocytic series (myeloid leucosis, myeloblastosis, granuloblastosis).
avian leucosis complex (*Vet.*). A term embracing the *avian leucoses* (q.v.) and usually also *fowl paralysis* (q.v.) and *osteopetrosis gallinarum* (q.v.).
avian monocytosis (*Vet.*). Blue comb; pullet disease. An acute or subacute disease of chickens characterized by depression, loss of appetite, and diarrhoea. The main pathological changes are focal necrosis of liver cells, enteritis, nephritis, and an increased number of monocytes in the blood. The cause is unknown.
avian myeloblastosis (*Vet.*). See **avian leucosis.**
avian paratyphoid (*Vet.*). A contagious disease of birds due to infection by *B. aertrycke*.
avian spirochaetosis (*Vet.*). Avian sleeping sickness. An acute and highly fatal septicaemia of domestic, and certain wild birds, due to infection by the spirochaete *Borrelia anserina* (*B. gallinarum*); transmitted by ticks of the genera *Argas* and *Ornithodorus*.
avian typhoid (*Vet.*). A contagious disease of birds due to infection by *Bacterium gallinarum*.
avian visceral lymphomatosis (*Vet.*). See **avian leucosis.**
aviation kerosine (*Aero.*). Finely filtered paraffin for turbine engines, abbrev. **avtur.** See **aviation spirit** and **wide-cut fuel.**
aviation spirit (*Aero.*). A motor fuel with a low initial boiling point and complying with a certain specification, for use in aircraft. Ranges from 73 to 120/130 octane rating; abbrev. **avgas.** See **aviation kerosine, wide-cut fuel.**
avicularium (*Zool.*). In *Ectoprocta*, a modified zooecium, having a movable mandible with powerful muscles acting against a fixed mandible, and capable of snapping movements; its function is mainly defensive.
avionics (*Aero.*). A term (a contraction of *avi*ation and electr*onics*) which refers to the application of the latter to systems and equipment utilized in *aeronautics* and *astronautics*.
avitaminosis (*Med.*). The condition of being deprived of vitamins: any deficiency disease caused by lack of vitamins.
Avogadro number or **constant** (*Chem.*). The number of atoms in 12 g of the pure isotope ^{12}C; i.e. the reciprocal of the *atomic mass unit* (q.v.) in grams. It is also by definition the number of molecules (or atoms, ions, electrons) in a *mole* (q.v.) of any substance and has the value $6{\cdot}022\,52 \times 10^{23}$ mol^{-1}. Symbol N_A or L.
Avogadro's law (*Chem.*). Equal volumes of different gases at the same temperature and pressure contain the same number of molecules.
avoidance conditioning (*An. Behav.*). An experimental procedure in which the making of a response by an animal to a specific stimulus prevents it receiving negative reinforcement (or an aversive stimulus) which it would otherwise receive, the intensity of the response increasing with successive presentations of the stimulus.
Avonian (*Geol.*). A division of Lower Carboniferous time equivalent to the Dinantian, or to the Tournaisian and Viséan. Named after the type-section in the Avon Gorge, Bristol district.
avtag (*Aero.*). See **wide-cut fuel.**
avtur (*Aero.*). See **aviation kerosine.**
avulsion (*Med.*). The tearing away of a part.
AW (*For.*). Abbrev. for *all-widths* (of cut timber).
A-wire (*Teleph.*). In a 2-wire telephone line to a manual exchange, that wire which under normal conditions is nearer to earth potential.
awl (*Carp.*). A small pointed tool for making holes which are to receive nails or screws.
awn (*Bot.*). (1) A long bristle borne on the glumes of some grasses and cereals; e.g., the beards of barley. (2) A long threadlike outgrowth on certain fruits.
awning deck (*Ships*). A superstructure deck, as the name implies. In its simplest form, it is the top deck of a 2-deck ship, and places the ship in a certain category for scantling and freeboard.
awu (*Nuc.*). Abbrev. for **atomic weight unit.** It has now been superseded.
axe (*For.*). See **broad-.** (*Tools*) A pointed hammer used for dressing stone.
axed arch (*Arch.*). An arch built from bricks cut to a wedge shape.
axed work (*Build.*). Hard building stone dressed with an axe to leave a ribbed face.
axerophthol (*Chem.*). See **vitamin A.**
axes *pl.* of *axis* (q.v.) (*Crystal.*). Lines of reference intersecting at the centre of a crystal. Crystal (or morphological) axes, usually three in number, by their relative lengths and attitude, determine the system to which a crystal belongs. (*Maths.*). Of a conic: that pair of conjugate diameters which are mutually perpendicular. For an ellipse they are referred to as *major* and *minor* in accordance with their length. For an hyperbola the one which does not cut the curve is called the *conjugate* axis, and the other the *transverse* axis. Of coordinates: the fixed reference lines used in a system of coordinates.
axial (*Bot.*). Relating to the axis of a plant: of shoot nature.
axial compressor (*Eng.*). A multistage, high-efficiency compressor comprising alternate rows of moving and fixed blades attached to a rotor and its casing respectively.
axial ducts (*Elec. Eng.*). Ducts placed parallel with the shaft in the stator or rotor of an electrical machine to facilitate passage of cooling air.
axial engine (*Aero.*). Turbine engine with an axial-flow compressor.
axial filament (*Zool.*). An internal thread, running down a flagellum and entering the body of the cell, to join the basal granule or blepharoplast: the stiff central thread of a radiate pseudopodium.
axial-flow compressor (*Aero.*). A compressor in which alternate rows of radially-mounted rotating and fixed aerofoil blades pass the air through an annular passage of decreasing area in an axial direction.
axial-flow turbine (*Aero.*). Characteristic aero-engine turbine, usually of 1 to 3 rotating stages, in which the gas flow is substantially axial.
axial girder (*Aero.*). The girder forming the

actual axis of a rigid-airship frame. It connects the hull structure fore and aft with a central fitting of each braced transverse frame.

axial gradient (*Bot.*). The physiological gradient along the axis of a plant, activity being highest at one end and falling off towards the other. (*Zool.*) A gradient of physiological activity, preceding the development of the axiate pattern.

axial organ (*Zool.*). In *Echinodermata*, a structure composed of connective and lacunar tissue, together with genital cells, lying near the axial sinus and stated to be contractile.

axial pitch (*Eng.*). In a screw-thread or helix, the distance from a point on the helix, measured parallel with the axis, to the corresponding point after one complete turn.

axial plane cleavage (*Geol.*). *Cleavage* parallel to the axial plane of a fold.

axial ratio (*Phys.*). Ratio of major to minor axis of polarization ellipse for wave propagated in waveguide, polarized light, etc.

axial response (*Acous.*). The response of a microphone or loudspeaker, measured with the sound-measuring device on the axis of the apparatus being tested.

axial sinus (*Zool.*). In *Echinodermata*, a vertical space, communicating at one end with the exterior or the coelom, via the madreporite, and at the other end with the inner perihaemal ring.

axial skeleton (*Zool.*). The skeleton of the head and trunk: in Vertebrates, the cranium and vertebral column, as opposed to the appendicular skeleton.

axiate pattern (*Zool.*). The morphological differentiation of the parts of an organism, with reference to a given axis.

axil (*Bot.*). The solid angle between a stem and the upper surface of a leaf base growing from it.

axile (*Bot.*). Coinciding with the longitudinal axis.

axile chloroplast (*Bot.*). A chloroplast lying in the axis of the containing cell.

axilemma (*Zool.*). In medullated nerve fibres, the whole of the medullary sheath.

axile placentation (*Bot.*). The condition when the ovules are attached to tissue lying in the axis of the ovary.

axile strand (*Bot.*). A simple condition of a vascular system, with conducting elements in the centre of the plant axis, giving off strands to the leaves; roughly equivalent to a protostele, but also includes the simple strands present in the larger mosses.

axilla (*Anat.*). The arm-pit: the angle between the fore-limb and the body. *adj.* **axillary.**

axillant (*Bot.*). Subtending an angle.

axillary (*Bot.*). Situated in an axil; applied especially to buds, and to shoots developing from them. (*Zool.*) In Insects, one of the articular sclerites of the wing. See also **axilla.**

axillary air sac (*Zool.*). In Birds, one of the paired air sacs, lying in the axillary position. It communicates with the median interclavicular air sac.

axinite (*Min.*). A complex borosilicate of calcium and aluminium, with small quantities of iron and manganese, produced by pneumatolysis and occurring as brown wedge-shaped triclinic crystals.

axiom (*Maths.*). An assumption; usually one of the basic assumptions underlying a particular branch of mathematics. Cf. *postulate.*

axiotron (*Electronics*). Valve in which the electron stream to the anode is controlled by the magnetic field of the heating current.

axis. A line, usually notional, which has a peculiar importance in relation to a particular problem or set of circumstances. Thus the *axis of symmetry* of a figure is a line which divides it symmetrically, and the *axis of rotation* is the line about which a body rotates. Specifically: (*Aero.*) the 3 axes of an aircraft are the straight lines through the centre of gravity about which change of attitude occurs: *longitudinal*, or *drag* axis in the plane of symmetry (roll), *normal*, or *lift*, axis vertically in the plane of symmetry (yaw), and the *lateral*, or *pitch*, axis transversely (pitch). See **wind axes.** (*Bot.*) (1) The line passing through the middle of any organ. (2) The main trunk of a root or shoot. (3) The central stalk in a grass spikelet. (4) Any plant part bearing lateral branches. (*Crystal.*) See **axes.** (*Photog.*) Of a lens, the line of symmetry of the optical system; the line along which there is no refraction. (*Zool.*) In the higher Vertebrates, the second cervical vertebra: the central line of symmetry of an organ or organism. *pl.* **axes.**

axis cylinder (*Zool.*). The excitable core of a medullated nerve fibre. See also **axon.**

axis of symmetry (*Maths.*). See **symmetry.**

axle (*Eng., etc.*). The cross-shaft or beam which carries the wheels of a vehicle; they may be either attached to and driven by it, or freely mounted thereon.

axle-box (*Eng.*). The bearings used for the axles of railway rolling-stock, consisting of an upper half-bearing integral with a box-shaped housing which holds the lubricant.

axle pulley (*Join.*). A pulley set in a sash-frame so that the sash-cord connecting the window and its balancing weight may run over it.

axle weight (*Eng.*). That part of the all-up weight of a vehicle which is borne by the wheels on one particular axle.

Axminster carpet (*Textiles*). A high-class loom-woven carpet with a cut pile surface; named after Axminster (Devon), its place of origin. The four chief Axminster weaves are spool, gripper, gripper/spool, and chenille. The tufts may be all wool, or wool/viscose/nylon blends in cotton/jute foundation.

axon (*Zool.*). The impulse-carrying process of a typical nerve cell or neurocyte, usually giving rise either to a nonmedullated nerve fibre, or to the axis cylinder of a medullated nerve fibre.

axoneme (*Zool.*). In certain *Ciliophora*, the central strand of the stalk: the axial filament of a flagellum.

axon hillock (*Cyt.*). The part of the cell body (*perikaryon*) of a neurone which gives rise to the axon and which is devoid of Nissl granules.

axonometry (*Crystal.*). Measurement of the axes of crystals.

axonost (*Zool.*). In Fish, the basal bone of a fin-ray: interspinal bone.

axopodium (*Zool.*). A pseudopodium, strengthened and stiffened by an axial rod, as in some *Heliozoa* and *Radiolaria.*

axosomatic (*Cyt.*). That type of synapse in which the terminals of an axon make their contact around the cell body of another neurone.

axospermous (*Bot.*). Having axile placentation.

axostyle (*Zool.*). In some *Mastigophora*, a slender flexible skeletal rod of an organic nature.

Aylesford laboratory beater (*Instr.*). A small beater for producing controlled beating conditions for pulp evaluation.

Ayrton-Mather galvanometer (*Elec. Eng.*). A moving-coil galvanometer having no iron within the moving coil.

Ayrton-Mather shunt (*Elec. Eng.*). Design of shunt which can be used with almost any current-sensitive device, to reduce its response by a series of accurately known factors.

Ayrton shunt (*Elec. Eng.*). A relatively large resistance for connecting across a galvanometer; tappings on the resistance are taken at 1/10, 1/100, and 1/1000 of the whole.

azacyanine (*Photog.*). Dye used as a desensitizer.

azaline (*Photog.*). A dye mixture for sensitizing ortho emulsions for red and yellow light.

azan stain (*Micros.*). See **Heidenhain's-**.

azelaic acid (*Chem.*). Heptan 1,7-dicarboxylic acid $COOH{\cdot}(CH_2)_7{\cdot}COOH$, m.p. 106°C. Found in rancid fat. Prepared by the oxidation of oleic acid.

azeotropic distillation (*Chem. Eng.*). A process for separating, by distillation, products not easily separable otherwise. The essential is the introduction of another substance, called the *entrainer*, which then forms an *azeotropic mixture* (q.v.), with one or sometimes both initial substances but using only small quantities, then is distilled off leaving the pure product.

azeotropic mixtures (*Chem.*). Liquid compounds whose boiling point, and hence composition, does not change as vapour is generated and removed on boiling. The boiling po int of the azeotropic mixture may be lower (e.g., water-ethanol) or higher (e.g., water-hydrochloric acid) than those of its components. Also **constant-boiling mixtures.**

azides (*Chem.*). (1) See **acid azides.** (2) Salts of hydrazoic acid. The heavy metal azides are explosive. Used as a coating, which is subsequently reduced to metal, on the filaments of radio receiving valves, to improve electron emission.

azimino compounds (*Chem.*). Heterocyclic compounds containing 3 adjacent nitrogen atoms in 1 ring, very stable. They are prepared by the action of nitrous acid on 1,2-diamines or 1,8-diaminonaphthalenes.

azimuth (*Astron., Maths., Surv.*). The azimuth of a line or celestial body is the angle between the vertical plane containing the line or celestial body and the plane of the meridian, conventionally measured from North through East in astronomical computations, and from South through West in triangulation and precise traverse work. See **azimuth angle.** (*Radio*) See **bearing.**

azimuthal projections (*Geog.*). See **zenithal projections.**

azimuthal quantum number (*Nuc.*). Quantum number associated with eccentricity of elliptic electron orbits, in atom. Now normally replaced by orbital quantum number.

azimuth angle (*Surv.*). Horizontal angle of observed line with reference to true North.

azimuth marker (*Radar*). Line on radar display made to pass through target so that the bearing may be determined.

azimuth stabilized PPI (*Radar*). Form of plan position indicator display which is stabilized by a gyrocompass, so that the top of the screen always corresponds to North.

azines (*Chem.*). Organic bases containing a heterocyclic hexagonal ring of 4 carbon and 2 nitrogen atoms, the nitrogen atoms being in the *para*-position with respect to one another.

azmure (*Textiles*). Originally a silk or worsted cloth with embossed appearance created by a weave that forms a broken figure or wavy effect on the surface either warp- or weft-way. Now made in blended cloths, i.e., worsted/Terylene, worsted/nylon, or mixtures of wool with other fibres.

azobenzene (*Chem.*). $C_6H_5{\cdot}N{=}N{\cdot}C_6H_5$. Orange-red crystals, m.p. 68°C. Prepared by the partial reduction of nitrobenzene.

azo dyes (*Chem.*). Derivatives of azobenzene, obtained as the reaction products of diazonium salts with tertiary amines or phenols (hydroxybenzenes). Usually coloured yellow, red, or brown, they have acidic or basic properties.

azo group (*Chem.*). The group $-N{=}N-$, generally combined with 2 aromatic radicals. The azo group is a chromophore, and a whole class of dyestuffs is characterized by the presence of this group.

azoimide (*Chem.*). Hydrazoic acid, N_3H, a weak unstable acid, prepared by the action of nitrous acid on hydrazine; a colourless liquid, b.p. 37°C. The acid and its derivatives are explosives. See **azides.**

azol (*Photog.*). Developer of the paraminophenol type.

azomethane (*Chem.*). $CH_3{\cdot}N : N{\cdot}CH_3$, b.p. 1·5°C, a yellow liquid, obtained by the oxidation of *sym.* dimethyl(VI)hydrazine with chromic (VI)acid.

azonium bases (*Chem.*). A group of bases including *azines* (q.v.) and *quinoxalines* (q.v.).

azoospermia (*Med.*). Complete absence of spermatozoa in the semen.

azophenine (*Chem.*). 2,5-Dianilo-3-benzaquinone-diphenyldiimine, m.p. 246°C, an intermediate for induline dyes.

azoproteins (*Chem.*). Derivatives obtained by coupling diazotized aromatic amines with serum proteins. On injection into the blood stream they cause the production of antibodies which react with other azoproteins containing the same amine.

azotaemic (azotemic) nephritis (*Med.*). Nephritis in which there is retention of nitrogenous products in the blood.

azote (*Chem.*). The French name for nitrogen.

Azotobacteriaceae (*Bacteriol.*). A family of bacteria included in the order *Eubacteriales*. Contains one genus *Azotobacter*, free-living in soil and water and able to fix free nitrogen in the presence of carbohydrates.

azoturia (*Vet.*). Paralytic equine myoglobinuria. An acute degeneration of muscles in horses of unknown cause, occurring particularly during exercise after a few days of idleness. Characterized by stiffness, lameness and paralysis, hardness of affected muscles, and urine coloured red to dark brown due to myoglobin.

azoxy compounds (*Chem.*). Mostly yellow or red crystalline substances obtained by the action of alcoholic potassium hydroxide upon the nitro compounds, or by the oxidation of azo compounds.

azulenes (*Chem.*). C_8H_{10}. Cyclic nonaromatic hydrocarbons associated with sesquiterpenes. Usually blue in colour, hence the name.

azure quartz (*Min.*). See **sapphire quartz.**

azurite (*Min.*). A basic carbonate of copper, occurring either as deep-blue monoclinic crystals or as kidney-like masses built of closely packed radiating fibres. Also called **chessylite** (from Chessy, in France).

azusa (*Radio*). Radio tracking system for missile guidance.

azygobranchiate (*Zool.*). Having ctenidia or branchiae developed on one side only.

azygomatous (*Zool.*). Lacking a zygomatic arch.

azygos (*Zool.*). An unpaired structure. *adj.* **azygous.**

azygospore (*Bot.*). A structure resembling a zygospore in morphology, but not resulting from a previous sexual union of gametes or of gametangia.

B. Symbol for: boron; susceptance in an a.c. circuit (unit = siemens; measured by the negative of the reactive component of the admittance); magnetic flux density in a magnetic circuit (unit = tesla = $Wb/m^2 = Vs/m^2$).

β- (*Chem.*). A symbol for: (1) substitution on the carbon atom of a chain next but one to the functional group; (2) substitution on a carbon atom next but one to an atom common to two condensed aromatic nuclei; (3) substitution on the carbon atom next but one to the hetero-atom in a heterocyclic compound; (4) a stereo-isomer of a sugar. (*Phys.*) The intermediate refractive index in a biaxial crystal.

β. For β-brass, function, particles, waves, etc., see under **beta.**

β. Symbol for: phase constant; ratio of velocity to velocity of light.

B. and B.B. (*Met.*). Brand-marks signifying *Best* and *Best Best*, placed on wrought iron to indicate the maker's opinion of its quality.

[B] (*Light*). A Fraunhofer line in the red of the solar spectrum, due to absorption by the earth's atmosphere. [B] is actually a close group of lines having a head at wavelength 686·7457nm.

B.A. Abbrev. for *British Association screw-thread* (q.v.).

Ba (*Chem.*). The symbol for *barium.*

Babbitt's metal (*Met.*). A bearing alloy originally patented by Isaac Babbitt, composed of 50 parts tin, 5 antimony, and 1 copper. Modern addition of lead greatly extends range of service. Composition varies widely, with tin 5% to 90%; copper 1·5% to 6%; antimony 7% to 10%; lead 5% to 48½%.

Babcock and Wilcox boiler (*Eng.*). A water-tube boiler consisting in its simplest form of a horizontal drum from which is suspended a pair of headers carrying between them an inclined bank of straight tubes.

Babcock and Wilcox mill (*Eng.*). Dry grinding mill using rotary steel balls.

Babcock apparatus (*Agric.*). Special form of test-tube *centrifuge* (q.v.) for measuring the butterfat content of milk.

Babesia (*Vet., Zool.*). A genus of protozoal parasites which occur in the erythrocytes of mammals.

babesiosis (*Vet.*). Piroplasmosis. Disease caused by infection by protozoa of the genus *Babesia.*

Babinet's compensator (*Light*). A device used, in conjunction with a Nicol prism, for the analysis of elliptically polarized light. It consists of two quartz wedges having their edges parallel and their optic axes at right angles to each other.

Babinet's formula for altitude (*Phys.*).

$$\text{Altitude} = \frac{32(500 + t_1 + t_2)(B_1 - B_2)}{B_1 + B_2}\ \text{metres},$$

t_1 and t_2 being the respective temperatures in degrees Celsius, and B_1 and B_2 the barometric heights at sea level and at the station whose altitude is required.

Babinet's principle (*Phys.*). The radiation field beyond a screen which has apertures, added to that produced by a complementary screen (in which metal replaces the holes, and spaces the metal) is identical with the field which would be produced by the unobstructed beam of radiation, i.e., the two diffraction patterns will also be complementary.

babingtonite (*Min.*). A silicate of iron, calcium, and manganese, crystallizing in the triclinic system. It occurs in greenish to brownish black crystals as a rare mineral in vugs in granitic rocks.

Babinski's reflex (*Med.*). Extension of big toe and fanning of other toes on stimulation of sole of foot; a sign of organic disease of the nervous system. It is normal in infants.

Babo's law (*Phys.*). The vapour pressure of a liquid is lowered when a nonvolatile substance is dissolved in it, by an amount proportional to the concentration of the solution.

baby (*Cinema.*). A small incandescent spotlight, used in sound-film production.

bacca (*Bot.*). A berry formed from an inferior ovary: a berry in general. *adjs.* **baccate,** resembling a berry, pulpy; **bacciferous,** bearing berries; **bacciform,** of berry-like shape.

bacho dust classifier (*Powder Tech.*). A proprietary centrifugal air elutriator in which the particles to be classified are introduced into an air stream flowing radially over a rotating disk. Fine particles follow the stream and are extracted at a jet coaxial with the centre of rotation. Large particles cannot follow the air stream and are collected at the edge of the disk.

B-acid (*Chem.*). 1-amino-8-naphthol-3,5-disulphonic acid. Used as an intermediate in dyestuff manufacture.

Bacillaceae (*Bacteriol.*). A family of bacteria included in the order *Eubacteriales.* Many are able to produce highly resistant endospores; large Gram-positive rods; aerobic or anaerobic; includes many pathological species, e.g., *Bacillus anthracis* (*anthrax*). See **clostridium.**

bacillaemia, bacillemia (*Med.*). Presence of bacilli in the blood.

bacillar (*Bot.*). Shaped like a rod.

Bacillariophyceae (*Bot.*). A class of unicellular algae in the phylum *Chrysophyta.* Characterized by the silicified cell wall, made by two halves; the yellow-brown carotinoid pigments, in addition to chlorophylls *a* and *c*; the storage of oil and leucosin, and the absence of starch. There are hundreds of species, both fresh and salt water, and occasionally inhabiting soil. Also called **diatoms.**

bacillary necrosis (*Vet.*). See **necrobacillosis.**

bacillary white diarrhoea (*Vet.*). See **pullorum disease.**

bacillicide (*Biol.*). A substance that destroys bacilli. *adj.* **bacillicidal.**

bacilluria (*Med.*). Presence of bacilli in the urine.

bacillus (*Bacteriol.*). (1) A rod-shaped member of the *Bacteria.* (2) Genus in the family *Bacillaceae. pl.* **bacilli.**

back. A large vat used in various industries, such as dyeing, soap making, brewing. Also **beck.** (*Bind.*) See **spine.** (*Build.*) (1) The part of a stone or ashlar opposite to its face. (2) The extrados of an arch or vault. (*Carp.*) The upper part of a hand rail, roof rafter, or dome rib. The *back* of a window is the part between the sill of the sash frame and the floor. (*Elec. Eng.*) See **leaving edge.** (*For.*) (1) In felling, the side of a tree opposite to that to which it falls. (2) Of a saw, other than circular, the noncutting edge of the blade.

back ampere-turns (*Elec. Eng.*). That part of the

armature ampere-turns which produces a direct demagnetizing effect on the main poles. Also called **demagnetizing ampere-turns.**

back band (*Build.*). The outside member of a door or window casing.

back bias (*TV*). See **back lighting.**

back boiler (*Build., Plumb.*). A domestic hot-water boiler fitted in the back of an open fire or range; usually made of cast iron or copper.

back cavity (*Bot.*). The widened opening between the lower faces of the two guard cells of a stoma.

back centre (*Eng.*). In a lathe, a pointed spindle (carried by the tailstock) on which the end of the work remote from the chuck is supported.

back cock (*Horol.*). In a clock, the bracket (on the back plate) from which the pendulum is suspended.

back contact (*Teleph.*). Contact in a relay assembly which is isolated when a moving contact separates from it on the operation of the relay.

back coupling (*Telecomm.*). Any form of coupling which permits the transfer of energy from the output circuit of an amplifier to its input circuit. See **feedback, regeneration.**

back cross (*Gen.*). The progeny obtained from mating a first filial generation cross (F_1) to one of the original parental types (P_1).

back cut (*For.*). The cut made on the back side of a tree in felling.

backed cloth (*Textiles*). A woollen or worsted fabric made with two warps and one weft, or with two wefts and one warp, to increase the weight, and obtain a weave or colours different from that on the face. In one example, a cloth may be of fairly heavy type but have a fine face weave, e.g., some suitings.

back edging (*Build.*). A method of cutting a tile or brick by chipping away the biscuit below the glazed face, the front itself being scribed.

back e.m.f. (*Elec. Eng.*). That which arises in an inductance (because of rate of change of current), in an electric motor (because of conversion of energy), or in a primary cell (because of polarization), or in a secondary cell (when being *charged*). Also **counter e.m.f.**

back-e.m.f. cells (*Elec. Eng.*). Cells connected into an electric circuit in such a way that their e.m.f. opposes the flow of current in the circuit.

back emission (*Electronics*). That of electrons from the anode.

back-end man (*Mining*). A man who cleans the cuttings from behind the coal cutting machine, sprags the overhang, and props the roof.

backer (*Build.*). A narrow slate laid on the back of a broad, square headed slate, at the place where a course of slates begins to narrow.

backfall (*Paper*). The ridge in the base of the *hollander* immediately after the roll.

back fill (*Mining*). See **stowing.**

back-fire (*I.C. Engs.*). (1) Premature ignition during the starting of an internal combustion engine, resulting in an explosion before the end of the compression stroke, and consequent reversal of the direction of rotation. (2) An explosion of live gases accumulated in the exhaust system due to incomplete combustion in the cylinder.

back-flap (*Join.*). The part of a shutter which folds up behind; also known as the **back-fold,** or **back-shutter.**

back-flap hinge (*Build.*). A hinge in two leaves, screwed to the face of a door which is too thin to permit of the use of a butt hinge.

back flow (*Hyd., San. Eng.*). Water or sewage flow in a direction contrary to normal.

back focus (*Photog.*). Distance between the rear surface of a lens and the image of an object infinity.

back-fold (*Join.*). See **back-flap.**

back gauge (*Build.*). The distance from the centre of a rivet- or bolt-hole to the back edge of an angle cleat or channel.

back gear (*Eng.*). A speed-reducing gear fitted to the headstock of a belt-driven metal-turning lathe. It consists of a simple layshaft, which may be brought into gear with the coned pulley and mandrel when required.

back geared motor (*Elec. Eng.*). An electric motor in which a geared countershaft is mounted as an integral part of the motor frame to provide speed reduction.

background (*Phys.*). A general problem in physical measurements, which limits the sensitivity of detecting any given phenomenon. Background consists of extraneous signals arising from any cause which might be confused with the required measurements, e.g., in electrical measurements of nuclear phenomena and of radioactivity, it would include counts emanating from amplifier noise, cosmic rays, insulator leakage, etc. Cf. *noise ratio*.

background controls (*TV*). Beam current (brightness) controls for the three electron guns of a colour TV tube.

background (or ground) noise (*Acous.*). Unwanted frequencies entering a wanted frequency band, which cannot be separated from wanted signals. Residual output from microphones, pickups, lines, etc., giving a *signal/noise ratio*.

background radiation (*Radiol.*). That registered by Geiger-Müller counter but coming from sources other than that being observed.

back hearth (*Build.*). That part of a hearth under the grate.

backheating (*Electronics*). Excess heat in cathode due to electron bombardment, as in a magnetron.

backing (*Bind.*). The process by which one half of the sections of a volume are bent over to the right and the other half to the left at the back. The projections formed are *joints*, to which the covers are hinged, by hand or machine. Some machines also perform *rounding* (q.v.). (*Build.*) The coursed masonry built upon and immediately in contact with the extrados of an arch. (*Carp.*) The operation of packing up a joist so that its upper surface shall be in line with those of deeper joists employed under the same floor. (*Meteor.*) The changing of a wind to a counterclockwise direction. Cf. *veering*. (*Photog.*) The coating on the back of film base or glass plates to reduce halation, for which purpose it should attenuate light and have a refractive index the same as the support, to reduce reflection.

backing boards (*Bind.*). Wedge-shaped wooden boards between which an unbound book is held in the lyingpress, while the joints are being formed for attaching the cover.

backing coil (*Elec. Eng.*). See **bucking coil.**

backing-off (*Textiles*). Reversing the rotation of the spindles in a cotton or woollen mule at the completion of twisting on the outward run. This unwinds the length of yarn between cop nose and spindle tip in preparation for the run-in of the carriage to wind the yarn on to the cop.

backing pressure (*Vac. Tech.*). The pressure at the outlet of a vapour diffusion pump.

backing pump (*Vac. Tech.*). Mechanical roughing pump to establish a low pressure, into which a vapour diffusion pump can operate.

backing store (*Comp.*). See **secondary memory (2).**

backings (*Join.*). Wooden battens, secured to

rough walls, for the fixing of wood linings, etc.

backing-up (*Build.*). The use of inferior bricks for the inner face of a wall. (*Print.*) (1) Printing on the second side of a sheet. (2) Backing a printing plate to required height.

back inlet gulley (*San. Eng.*). A trapped gulley in which the inlets discharge under the grating and above the level of the water in the trap, so that splashing and blocking of the grating are avoided.

back iron (*Carp.*). The stiffening plate screwed to the cutting iron of a jack plane. Also called **cap iron, cover iron.**

back joint (*Build.*). The part of the back of a stone step which is dressed to fit into the rebate of the upper step.

back-kick (*Eng.*). Term applied to the violent reversal of an internal combustion engine during starting, due to a *back-fire* (q.v.).

backlash (*Eng.*). The lost motion between two elements of a mechanism, i.e., the amount the first has to move, owing to imperfect connexion, before communicating its motion to the second. (*Radio*) Property of most regenerative and oscillator circuits, by which oscillation is maintained with a smaller positive feedback than is required for inception.

back lighting (*TV*). In camera tubes, illumination of rear surface of mosaic giving greater sensitivity. Also known as **back bias.**

back lining (*Bind.*). See **hollows.** (*Join.*) (1) The part of a cased sash-frame next the wall, opposite the pulley-stile. (2) The piece of framing which forms the back of a recess for boxing shutters.

back lobe (*Telecomm.*). Lobe of polar diagram for antenna, microphone, etc., which points in the reverse direction to that required.

backlocking (*Rail.*). Holding a signal lever partially restored until completion of a predetermined sequence of operations.

back-mixing (*Chem.*). In a chemical reactor, jet engine, or other apparatus through which material flows and thereby undergoes a change in some property, back mixing is said to occur when some of the material is turned back so that it mixes with material which has entered the reactor after it. See **plug flow.**

back observation (*Surv.*). One made with instrument on station just left. Also called **back sight.**

back pitch (*Elec. Eng.*). The winding pitch of the winding of an electrical machine at the end, remote from the commutator.

back-planing (*Print.*). Reducing duplicate plates to the necessary thickness, by shaving metal from the back of stereos and electros, and by grinding, in the case of rubber or plastic plates.

backplate (*TV*). In a TV camera tube, the plate behind, and insulated from, the mosaic which transmits the video signal by capacitance.

backplate lampholder (*Elec. Eng.*). A lampholder fitted with a support which enables it to be screwed on to a flat surface. Also called a **batten-lampholder.**

back porch (*TV*). See **porch.**

back-porch effect (*Telecomm.*). The prolonging of the collector current in a transistor for a brief time after the input signal (particularly if large) has decreased to zero.

back pressure (*Eng.*). The pressure opposing the motion of the piston of an engine on its exhaust stroke: the exhaust pressure of a turbine. Increased by clogged or defective exhaust system. (*Med.*) Proximal pressure produced by an obstruction to fluid flowing through the cardiovascular or urinary systems. (*Plumb.*) Air pressure in pipes exceeding atmospheric pressure.

back-pressure turbine (*Eng.*). A steam turbine from which the whole of the exhaust steam, at a suitable pressure, is taken for heating purposes.

back projection (*Cinema., Photog.*). The use of a translucent background on which synchronized cinematographic pictures or a static scene are projected, so that both this and the action can be photographed together.

back rake (*Eng.*). In a turning tool, the inclination of the top surface or face to a plane parallel to the base of the tool.

back rest (*Textiles*). The steel rod, roller, or oscillating bar at the back of a loom, over which the warp threads pass from beam to healds. Has considerable influence on the cover of cloth being woven. (*Eng.*) See **steady.**

backs (*Mining.*). Hanging wall of mine; the rock above the worker in a stope or other excavation.

back saw (*Carp.*). A saw stiffened by a thickened back; e.g., a tenon saw.

back-sawn (*For.*). Timber converted so that the growth rings meet the face in any part of an angle less than 45°. Also called **bastard-sawn, crown-cut, plain-sawn, slash-sawn.**

back scatter (*Nuc.*). The deflection of radiation or particles by scattering through angles greater than 90° with reference to the original direction of travel.

back-scatter factor (*Nuc.*). The multiplying factor by which dose rate at a point in a beam of radiation is increased when that point is included in incident surface of a body being irradiated.

backsetting (*Build.*). A stone with a boasted or broached face, but with a smooth border all round.

back shift (*Mining*). In colliery, the afternoon shift.

back shore (*Carp.*). One of the outer members of an arrangement of raking shores or props, for supporting temporarily the side of a building. The *back shore* supports the raking shore, or raker, which takes the thrust from the highest part of the building.

back-shunt keying (*Radio*). Keying a radio transmitter between the radiating aerial and a dummy aerial.

back-shutter (*Join.*). See **back-flap.**

back sight (*Surv.*). See **back observation.**

back-stay (*Eng.*). See **steady.**

back-step (*Bind., Print.*). See **black-step marks.**

backstop (*Elec. Eng.*). The structure of a relay which limits the travel of the armature away from the pole-piece or core.

back stopes (*Mining*). *Overhand* (or overhead) *stopes* (q.v.), worked up-slope. Antonym *underhand stopes* (q.v.).

back stops (*Textiles*). In cotton and woollen spinning, buffers which prevent the mule carriage travelling beyond a fixed point on its inward run, as the winding fallers unlock to begin the spinning cycle again.

back stripper (*Mining*). A man who breaks the large lumps of mined coal and fills the tubs at the coal face.

back titration (*Chem.*). In volumetric analysis, the technique of adding a reagent in excess, to exceed the end-point, then determining the end-point by titrating the excess.

back-to-back (*Electronics*). (1) Parallel connection of valves, such that the anode of one is connected to the cathode of the other, for rectification, frequency doubling, etc. (2) A connection used in thyratron (or ignitron) rectifiers to control the alternating current to a load.

back-to-back test (*Elec. Eng.*). The arrangement, originated by Hopkinson, for testing two substantially similar electrical machines on full load, by coupling them mechanically and loading them by regulating the electrical circuit, the total power supply accounting for total losses only; it has been extended to transformers and mechanical gearing.

backwall cell (*Electronics*). Semiconductor photovoltaic cell in which light passes through a polarized grid to an active layer carried on a metal plate.

backward busying (*Teleph.*). Applying busy condition at the incoming end of a trunk or junction (usually during testing or fault-clearance) to indicate at outgoing end that circuit must not be used.

backward diode (*Electronics*). One with characteristic of reverse shape to normal. Also sometimes known as **AU diode.**

backward hold (*Teleph.*). A method of interlocking the links of a switching chain by originating a locking condition in the final link and extending it successively backwards to each of the preceding links.

backward scatter (*Radio*). Scattering backwards of radio signals tangent to lower level of ionosphere. Used to monitor effectiveness of transmission.

backward shift or **backward lead** (*Elec. Eng.*). Movement of the brushes of a commutating machine around the commutator, from the neutral position, and in a direction opposite to that of the rotation of the commutator, so that the brushes short-circuit zero e.m.f. conductors when the load current, through armature reaction, results in a rotation of the neutral axis of the air-gap flux. Shifting the brushes in this way reduces sparking on the commutator.

backward signalling (*Teleph.*). Signalling from the called to the calling end of a circuit.

backward wave (*Electronics*). In a travelling-wave tube, a wave with group-velocity in the opposite direction to the electron stream. See also **forward wave.**

backwash diode (*Electronics*). A diode valve which is connected across the line of a pulse modulator to absorb energy during the voltage pulse reversals. Also called **overswing diode.**

back washer (*Textiles*). (1) A machine consisting of washing-bowls, steam-heated cylinders, and a gill-box, for scouring, drying, and opening out carded slivers in worsted manufacture. (2) A second scouring operation to eliminate further minute impurities from wool slivers or tops.

back-water (*Hyd. Eng.*). Water dammed back in a stream or reservoir by some obstruction.

backwater (*Paper*). Waste water containing fine fibres and other chemicals draining from the paper machine wire.

backwater curve (*Hyd. Eng.*). The longitudinal profile of the water surface in the case of non-uniform flow in an open channel, when the water surface is not parallel to the invert, owing to the depth of water having been increased by the placing of a weir across the flow.

back wave (*Telecomm.*). See **spacing wave.**

backwork (*Mining*). Underground colliery activity behind the working coal face, such as loading, maintenance of travelling ways.

bacteria (*Bacteriol.*). A large group of unicellular or multicellular organisms, lacking chlorophyll, with a simple nucleus, multiplying rapidly by simple fission, some species developing a highly resistant resting ('spore') phase; some species also reproduce sexually, and some are motile. In shape spherical, rodlike, spiral, or filamentous, they occur in air, water, soil, rotting organic material, animals, and plants. Saprophytic forms are more numerous than parasites, but the latter include both animal and plant pathogens. A few species are autotrophic.

bacteria beds (*San. Eng.*). Layers of a filtering medium such as broken stone or clinker, used in the final or oxidizing stage in sewage treatment. See **contact bed, percolating filter.**

bacteriaemia, bacteriemia (*Med.*). Presence of bacteria in the blood.

bacterial leaching (*Min. Proc.*). Use of selected strains of bacteria, including *ferrobacillus ferrooxidans* and *thiobacillus ferrooxidans* to accelerate the acid leach of sulphide minerals. They convert ferrous iron sulphate to ferric, releasing sulphuric acid.

bacteriocide (*Bacteriol.*). A substance capable of destroying bacteria. *adj.* **bacteriocidal.**

bacteriocyte (*Bacteriol., Zool.*). Specialized cell occurring in some Invertebrates, and inhabited by intracellular symbiotic bacteria.

bacterioid, bacteroid (*Bacteriol., Bot.*). Large branching banded rod form of nitrogen-fixing bacteria (*Rhizobium*), which effect the last stages in the penetration of the root and formation of root nodules in *Leguminosae.*

bacteriology. The scientific study of bacteria.

bacteriolysin (*Bacteriol.*). An antibody (or immune body) having a specific action against the particular bacterial species which, by immunization, caused its production. It causes digestion of the bacterial cells with formation of soluble products **(bacteriolysis).**

bacteriophage (*Bacteriol.*). General term for viruses which infect bacteria.

bacteriorrhiza (*Bot.*). A symbiotic relationship between a root and bacteria.

bacteriostatic (*Bacteriol.*). A term used to describe substances able to inhibit bacterial multiplication.

bacteriotoxin (*Biol.*). A toxin destructive to bacteria.

bacteriotropin (*Biol.*). A substance, usually of blood serum, which renders bacteria more subject to the action of antitoxin or more readily phagocytable; e.g., opsonin.

bacteriuria (*Med.*). The presence of bacteria in the urine.

Bacteroidaceae (*Bacteriol.*). A family of pleomorphic bacteria included in the order *Eubacteriales.* Usually Gram-negative rods; anaerobic; inhabit the intestine and mucous membrane of the higher Vertebrates; often associated with gangrenous conditions; includes *Sphaerophorus necrophorus* (*calf diphtheria*).

baculiform (*Bot.*). Of fungal spores, stick-shaped or rod-shaped.

bad colour (*Print.*). An unsatisfactory impression, due to uneven distribution of the ink.

baddeleyite (*Min.*). Monoclinic zirconium dioxide, found concentrated in certain beach sands in Brazil, where it is exploited as a source of zirconium.

badger (*Build.*). An implement used to clear mortar from a drain after it has been laid.

badger plane (*Tools*). Large wooden plane with a rebated offside edge, used for wide rebates.

badger rule (*Phys.*). An empirical relationship between the force constants and the vibrational frequencies of the electronic states of diatomic molecules.

badious, badius (*Bot.*). Chestnut-brown.

Badischer process (*Chem.*). The *contact process* (q.v.) for sulphuric acid, based on a catalyst of platinized asbestos. Catalyst poisons are removed from the feed gases by scrubbing with dilute acid, drying, and filtering. The sulphur (VI)oxide is absorbed in concentrated acid.

Baeyer's tension (or **strain**) **theory** (*Chem.*). This theory assumes that the four valencies of the tetravalent carbon atom are symmetrically distributed in space around the carbon atom. One may predict from this assumption the strain involved in the formation of a ring compound from a chain of carbon atoms, and also estimate the stability of a ring compound from the number of carbon atoms in the ring.

bafertisite (*Min.*). A silicate of barium, iron, and titanium, the mineral being named from the three symbols.

baffle (*Acous.*). Extended surface surrounding a diaphragm of a loudspeaker, so that an acoustic short-circuit is prevented and the diaphragm loaded so as to transmit acoustic energy, especially at low frequencies. (*Electronics*) Internal structure in a tube for controlling discharge or its decay. (*Min. Proc.*) In hydraulic or rake classifier, a plate set across and dipping into the pulp pool; in mechanically agitated flotation cell, one so set as to reduce centrifugal movement in the upper part of the cell.

baffle blankets (*Cinema.*). Blankets temporarily placed about a set, to regulate the distribution of reflected sound during soundfilm production.

baffle chamber (*Met.*). Room or tower so arranged that entering gases impinge on trays, screens or plates over which an absorbing liquid may be flowing or sprayed, to pick up dust or chemical fume before discharge of the washed and cleaned gas.

baffle loudspeaker (*Acous.*). An open-diaphragm loudspeaker, in which the radiation of sound power is enhanced by surrounding it with a large plane baffle, generally of wood.

baffle plate (*Eng., etc.*). A plate used to prevent the movement of a fluid in the direction which it would normally follow, and to direct it into the desired path. (*Telecomm.*) Plate inserted into waveguide to produce change in mode of transmission.

bag and spoon dredger (*Civ. Eng.*). An implement consisting of a leather bag laced to a steel hoop; it is suspended from a chain and guided by a long handle; used to dredge soft material.

bagasse (*Chem.*). Dry refuse (crushed sugar-cane fibres) in sugar making. Used as a fuel in sugar factories, and as a base in paper manufacture.

bagassosis (*Med.*). A lung disease caused by inhalation of bagasse.

Baggy Beds (*Geol.*). A formation occurring in the Upper Devonian Series in North Devon, and comprising green shales and yellowish sandstones, which yield *Lamellibranchiata, Gastropoda*, and plant remains.

bag hides (*Leather*). The upper, or grain, part of split ox- or cow-hides, tanned and dyed.

bag plug (*San. Eng.*). A drain plug consisting of a cylindrical canvas bag which is placed in the drain pipe and inflated.

bag pump (*Hyd. Eng.*). A form of bellows pump, in which the valved disk taking the place of the bucket is connected to the base of the barrel by an elastic bag, distended at intervals by rings.

baguette (*Build.*). A small moulding similar to the *astragal* (q.v.).

baguio (*Meteor.*). A local name for the typhoon arising in the region of the Philippine Islands.

bailer (*Mining*). Length of piping closed at bottom by clack valve, used to remove drilling sands and muds from borehole.

Bailey bridge (*Eng.*) A temporary bridge made by assembling portable prefabricated panels. A 'nose' is projected over rollers across the stream, being followed by the bridge proper, with roadway. Used also over pontoons.

Bailey test for sulphur (*Chem.*). After fusion with sodium carbonate, the aqueous solution gives a blood-red colour with sodium nitroprusside.

bailiff (*Mining*). See **overman (1)**.

Baillarger's lines (*Histol.*). A tract of nerve-fibres in the occipital cortex of the cerebrum.

Baily furnace (*Elec. Eng.*). An electric-resistance furnace in which the resistance material is crushed coke placed between carbon electrodes; used for heating ingots and bars in rolling mills, for annealing, etc.

Baily's beads (*Astron.*). A phenomenon, first observed by Baily, in which, during the last seconds before a solar eclipse becomes total, the advancing dark limb of the moon appears to break up into a series of bright points.

bainite (*Met.*). When austenite is transformed at temperatures intermediate between the pearlite and martensite ranges, bainite forms. It is an acidular mixture of carbide and ferrite, varying in its proportions with the formation temperature.

bait (*Glass*). A vertical iron grille used for drawing molten glass into sheets.

bait crop (*Agric.*). A vegetable attractant planted between rows of crops to attract pests away from main crop.

baize (*Textiles*). Heavy woollen cloth, 68 in. (1·73 m) wide, made from coarse warp and weft, felted and raised, with the nap lying in one direction. Used to cover billiard and other tables, screens, etc.

bajada (*Geog.*). A relatively flat area found in arid regions, formed by the deposition of rock debris as a result of sheet floods.

Bajocian Series (*Geol.*). Broadly. the lower division of the Middle Jurassic rocks, including, in Britain, various sands above the Lias, the Inferior Oölite, and the Dogger, Lower Estuarine Sands, and Millepore Oölite of Yorks.

baked (or **caked**) (*Typog.*). Said of type which, having become stuck together, is difficult to distribute.

baked core (*Foundry*). A dry sand core baked in the oven to render it hard and to fix its shape. See **core sand**.

Bakelite (*Plastics*). TN for synthetic resin (named after L. H. Baekeland), the product of the condensation of cresol or phenol with formaldehyde. First product is *Bakelite A*, a liquid; this is changed by heat into *Bakelite B*, which is a solid that is soft when hot and hard when cold and is thus suitable for moulding. The final product, *Bakelite C*, is obtained by heating *Bakelite B* under pressure. Bakelite is much used for insulating purposes and in the manufacture of plastic products, paints, varnishes, etc. In transparent form, very valuable for polariscopes.

bakelite wood (*Elec. Eng.*). See **Permali**.

bake-out (*Elec. Eng.*). Preliminary heating of the electrodes and container of a mercury-arc rectifier or any other type of tube or valve, to ensure freedom from the later release of gases.

Baker's cyst (*Med.*). A cyst found behind the knee, which may or may not communicate with the synovial joint.

Baker's fixative (*Micros.*). A neutral cytoplasmic fixative, containing formalin, calcium chloride, and calcium carbonate, used particularly when the preservation of cell lipoids is required.

bakerite (*Min.*). A variety of *datolite* (q.v.), occurring in the borax deposits of the Mohave desert, California.

baking soda (*Chem.*). Sodium hydrogen carbonate. Also known as *bicarbonate of soda*.

BAL (*Pharm.*). British anti-lewisite, dithioglycerol—

$$\begin{array}{l} CH_2SH \\ | \\ CH\ SH \\ | \\ CH_2OH \end{array}$$

Oily liquid, b.p. 95°C at 130 N/m^2 pressure, m.p. 77°C. Antidote for poisoning by *lewisite* (q.v.) and other poisons, particularly arsenic and mercury.

balance (*Acous.*). Adjustment of sources of sound in studios so that the final transmission adheres to an artistic standard. (*Biol.*) Equilibrium of the body; governed from the cerebellum, in response to stimuli from the eyes and extremities and especially from the *semicircular canals* (q.v.) of the ears. (*Chem.*) See **chemical-**. (*Elec. Eng.*) A balance in bridge measurements is said to be obtained when the various impedances forming the arms of the bridge have been adjusted, so that no current flows through the detector. See also **current weigher.** (*Horol.*) The vibrating member of a watch, chronometer, or clock (with platform escapement). In conjunction with the balance-spring it forms the time-controlling element.

balance arc (*Horol.*). The portion of the vibration during which the balance is not detached from the escapement.

balance arm (*Horol.*). The portion of the balance connecting the rim to the staff.

balance bar (*Hyd. Eng.*). The heavy beam by which a canal-lock gate may be swung on its pintle, and which partially balances the outer end of the gate.

balance box (*Eng.*). A box, filled with heavy material, used to counterbalance the weight of the jib and load of a crane of the cantilever type.

balance-bridge (*Civ. Eng.*). See **bascule bridge.**

balance cock (*Horol.*). The detachable bracket which carries the upper pivot of the balance staff.

balance-crane (*Eng.*). A crane with two arms, one having counterpoise arrangements to balance the load taken by the other.

balanced amplifier (*Telecomm.*). Same as **push-pull amplifier.**

balanced-armature loudspeaker (*Acous.*). A driving mechanism for loudspeakers, in which the motion is obtained by a pivoting member which is twisted by varying magnetic fluxes.

balanced-armature pick-up (*Acous.*). A pick-up in which the reproducing needle is held by a screw in a magnetic arm, which is pivoted so that its motion diverts magnetic flux from one arm of a magnetic circuit to another, thereby inducing e.m.f.s in coils on these arms.

balanced-beam relay (*Elec. Eng.*). One having two coils arranged to exert their forces on plungers at each end of a beam pivoted about its central point.

balanced circuit (*Elec. Eng.*). For a.c. and d.c., one which is balanced to earth potential, i.e., the two conductors are at equal and opposite potentials with reference to earth at every instant. See **unbalanced circuit.**

balanced draught (*Eng.*). A system of air-supply to a boiler furnace, in which one fan forces air through the grate, while a second, situated in the uptake, exhausts the flue gases. The pressure in the furnace is thus kept atmospheric, i.e., is *balanced.*

balanced hoisting (*Mining*). In mine shaft, arrangement of winding system whereby the weight of the descending cage partly offsets that of the rising one, thus reducing power draft.

balanced line (*Telecomm.*). One in which the impedances to earth of the two conductors are, or are made to be, equal. Also **balanced system.**

balanced load (*Elec. Eng.*). A load connected to a polyphase system, or to a single-phase or d.c. 3-wire system, in such a way that the currents taken from each phase, or from each side of the system, are equal and at equal power factors.

balanced mixer (*Telecomm.*). A waveguide technique to minimize the effect of local oscillator noise.

balanced modulator (*Telecomm.*). Matched pair of valves, operated as modulators, with their anodes connected in push-pull. The carrier voltage is applied to the two grids in phase and the modulating voltage in antiphase, so that the carrier components in the anode currents cancel. Used in suppressed-carrier systems.

balanced network (*Telecomm.*). One arranged for insertion into a *balanced circuit* and therefore symmetrical electrically about the midpoints of its input and output pairs of terminals.

balanced-pair cable (*Telecomm.*). One with two conductors forming a loop circuit, the wires being electrically balanced to each other and earth (shield), e.g. an open-wire antenna feeder. Cf. *coaxial cable.*

balanced pedal (*Acous.*). In an organ console, the foot-operated plate, pivoted so that it stays in any position, for remote control of the shutter of the chambers in which ranks of organ pipes are situated; it also serves for bringing in all the stops in a graded series. See **swell pedal.**

balanced protective system (*Elec. Eng.*). A form of protective system for electric transmission lines and other apparatus, in which the current entering the line or apparatus is balanced against that leaving it; so long as there is no fault on the line or apparatus this balance will be maintained, but an upsetting of the balance owing to a fault causes relays to trip the faulty circuit. Also called **differential protective system** or (coll.) **earth leak relay or earth trip.**

balanced reaction (*Chem.*). See **incomplete reaction.**

balanced sash (*Join.*). See under **sliding sash.**

balanced solution (*Chem.*). A solution of two or more salts, in such proportions that the toxic effects of the individual salts are mutually eliminated; sea water is a *balanced solution.*

balanced step (*Build.*). See **dancing step.**

balanced system (*Telecomm.*). See **balanced line.**

balanced termination (*Telecomm.*). Terminating impedance, the midpoint of which is at earth potential, or connected to earth.

balanced voltage (or current) (*Elec. Eng.*). A term used, in connexion with polyphase circuits, to denote voltages or currents which are equal to all the phases. Also applied to d.c. 3-wire systems.

balance gate (*Hyd. Eng.*). A flood gate which revolves about a central vertical shaft, and which may be made self-opening or self-closing as the current sets in or out of a channel by giving a preponderating area to the inner leaves of the gate.

balance pipe (*Eng.*). A connecting pipe between two points at which pressure is to be equalized.

balance piston (*Eng.*). See **dummy piston.**

balance point (*Civ. Eng.*). Any point where a *mass-haul curve* (q.v.) cuts the datum line, showing that up to this point all excavated material has been used up in embankment.

balancer (*Elec. Eng.*). A device used on polyphase or 3-wire systems to equalize the voltages between the phases or the sides of the system, when unbalanced loads are being delivered.

balancer-booster (*Elec. Eng.*). A balancer set having boosters mounted on the same shaft, to compensate for the voltage drop in feeders.

balancer field rheostat (*Elec. Eng.*). A special form of shunt field rheostat, for controlling the field currents of a balancer set. Cutting the resistance out of one field puts a corresponding amount into the other field.

balance rim (*Horol.*). The circular rim of the balance. It may be monometallic or bimetallic. See **compensation balance.**

balancers (*Zool.*). See **halteres.**

balancer transformer (*Elec. Eng.*). An auto-transformer connected across the outer conductors of an a.c. 3-wire system, the neutral wire being connected to an intermediate tapping.

balance spring (*Horol.*). A very fine metal ribbon, forming a flat spiral, cylindrical, or helical spring round the balance staff. It regulates the movement of the balance.

balance staff (*Horol.*). The staff which carries the balance, and the collet to which the balance spring is attached.

balance tab (*Aero.*). A *tab* whose movement depends upon that of the main control surface. It helps to balance the aerodynamic loads and reduces the *stick forces*. Cf. **servo-, spring-,** and **trimming-.**

balance weights (*Elec. Eng.*). Small weights threaded on radial arms on the movement of an indicating instrument, so adjusted that the pointer gives the same indication whatever the orientation of the instrument. (*Eng.*) A weight used to counterbalance some part of a machine; e.g., the weights applied to a crankshaft to minimize or neutralize the inertia forces due to reciprocating and rotating masses of the engine.

balance wheel (*Horol.*). A *balance* (q.v.) in the form of a wheel, normally combined with a spiral hairspring.

balancing (*Photog.*). Control of the densities of the three printings in a 3-colour subtractive system to get as near as possible to the theoretical requirements. Effected by matching grey tones. (*Surv.*) The process of adjusting a traverse, i.e., applying corrections to the different survey lines and bearings so as to eliminate the closing error. (*Telecomm.*) See **neutralization.**

balancing antenna (*Radio*). Auxiliary reception antenna which responds to interfering but not to the wanted signals. The interfering signals thus picked up are balanced against those picked up by the main antenna, leaving signals more free from interference.

balancing capacitance (*Telecomm.*). That connected between appropriate points in the grid and anode circuits of a valve amplifier to neutralize effects of internal grid-anode capacitance. Also called **neutralizing** or **neutrodyning capacitance.**

balancing flux (*Elec. Eng.*). Flux provided in an electric machine to compensate for the unbalanced magnetic pull.

balancing impedance (*Telecomm.*). See **line balance.**

balancing machine (*Eng.*). A machine for testing the extent to which a revolving part is out of balance, and to determine the weight and position of the masses to be added, or removed, to obtain balance.

balancing network (*Telecomm.*). See **line balance.**

balancing ring (*Elec. Eng.*). See **equalizer ring.**

balancing speed (*Elec. Eng.*). See **free-running speed.**

balanitis (*Med.*). Inflammation of the glans penis.

Balanoglossida (*Zool.*). An order of *Hemichorda.*

balanoposthitis (*Med.*). Inflammation of the glans penis and prepuce.

balanorrhagia (*Med.*). Gonorrhoeal inflammation of the glans penis, with discharge of pus.

balantidiosis (*Vet.*). An enteric infection, especially of swine, by ciliate protozoa of the genus *Balantidium.*

balanus (*Anat.*). The terminal bulbous portion of the penis; the glans penis.

balas ruby (*Min.*). A rose-red variety of the mineral spinel (magnesium aluminate $MgO \cdot Al_2O_3$, crystallizing in the cubic system). See **false ruby.**

balata (*Chem.*). The coagulated latex of the bullet tree of S. America. Similar in properties to gutta-percha, but softer and more ductile. Used extensively in the manufacture of golf balls, and for impregnating cotton duck belting.

Balbach process (*Chem.*). Electrolytic separation of gold and silver from a metal, by making it the anode in a bath of silver nitrate, the silver migrating to the cathode.

balbuties (*Med.*). Stammering, stuttering.

balconet (*Arch.*). A low ornamental railing to a door or window, projecting very little beyond the threshold or sill; mainly used in the Swiss style of architecture.

balcony (*Arch.*). A stage or platform projecting from the wall of a building within or without, supported by pillars or consoles, and surrounded with a balustrade or railing.

baldacchino (*Arch.*). A canopy suspended above altar, tomb, etc., or supported in such position by columns. Also **baldachin, baldaquin.**

bald-eagle (*Zool.*). A common but inaccurate name for the American white-headed eagle.

Baldwin spot (*Agric.*). A virus disease affecting apples.

bale (*Agric.*). A neatly compressed and twine-tied rectangular bundle of hay, straw, cotton, etc.

bale breaker (*Textiles*). A machine that opens up and blends the highly compressed cotton fibres taken from different bales and releases dust, broken leaf, and mote, which is extracted by a fan to filters.

balection moulding (*Join.*). See **bolection moulding.**

baleen (*Zool.*). In certain Whales, horny plates arising from the mucous membrane of the palate and acting as a food strainer.

baler (*Agric.*). A machine for compressing loose bulky material such as hay or cotton and securing it in a form convenient for transport.

Balfour's rule (*Zool.*). The velocity of segmentation in any part of the ovum is, roughly speaking, proportional to the concentration of the protoplasm there; and the size of the segments is inversely proportional to the concentration of the protoplasm.

balk (*Civ. Eng.*). The material between two excavations. Also **baulk.**

Balkan frame (*Med.*). A frame, with pulleys attached, for supporting the leg in the treatment of fractures.

balk back (*Textiles*). A woollen or worsted cloth with a fibrous back and a smart clear surface obtained in raising and shearing processes.

balking (*Elec. Eng.*). See **crawling.**

ball or bloom (*Met., etc.*). (1) A rounded mass of spongy iron, prepared in a puddling furnace. (2) A mass of tempered fireclay, used for forming the crucible in crucible-steel production.

ball or bolus (*Vet.*). A cylindrical-shaped mass of drugs for the curative treatment of horses.

ball-and-biscuit microphone (*Acous.*). See **apple-and-biscuit microphone.**

ball-and-socket head (*Photog.*). Attachment screwed to the top of a camera tripod allowing rotation, tilting, and clamping in any position. Also **tripod head.**

ball-and-socket joint (*Anat.*). Enarthrosis. A joint in which the hemispherical end of one bone is received into the socket of another. (*Eng.*) A joint between two rods, permitting considerable relative angular movement in any plane. A ball formed on the end of one rod is embraced by a spherical cup on the other. Used in light control systems (e.g., in connecting a pair of bell-cranks which operate in planes at right angles) and in the steering mechanism of motor vehicles, in which both ball and cups are of case-hardened metals.

ballast (*Civ. Eng., Rail., etc.*). (1) A layer of broken stone, gravel, or other material deposited above the formation level of road or railway; it serves as foundation for road-metal or permanent-way respectively. (2) Sandy gravel used as a coarse aggregate in making concrete. (*Ships*) Gravel, stone, or other material placed in the hold of a ship to increase her stability when floating without cargo or with insufficient cargo.

ballast car (or wagon) (*Civ. Eng.*). A truck used for the transport and dumping of ballast in road or rail construction.

ballast lamp (*Elec. Eng.*). Normal incandescent lamp used as a ballast resistor, current limiter, alarm, or to stabilize a discharge lamp.

ballast resistance (*Elec. Eng.*). A term used in electric-railway signalling to denote the resistance between the two track rails across the *ballast* on which the track is laid. If allowed to fall too low, it will have the effect of shunting the signal from a train's wheels.

ballast resistor (*Elec. Eng.*). One inserted into a circuit to swamp or compensate changes, e.g., those arising through temperature fluctuations. One similarly used to swamp the negative resistance of an arc or gas discharge. See also **barretter.**

ballast tube (*Elec. Eng.*). Same as **barretter** or **ballast resistor.**

ball-bearing (*Eng.*). A shaft bearing consisting of a number of hardened steel balls which roll between an inner race forced on to the shaft and an outer race carried in a housing. The races carry shallow spherical grooves (*ball-tracks*), and the balls are spaced by a light metal cage. Used to obtain a high load capacity in small compass, to combine the functions of a journal and thrust bearing, or to secure minimum friction. The same principle is applied to thrust bearings.

ball catch (*Join.*). A door-fastening in which a spring-controlled ball, projecting through a smaller hole, engages with a striking plate.

ball centrosome (*Cyt.*). That centrosome of the spermatozoon which gives rise to the long axial fibre of tail.

ball clay (*Geol.*). Fine-textured and highly plastic detrital clay which, on firing, yields pale-coloured or white pottery. Mixed with kaolin, it imparts plasticity to the latter. Also called **potters' clay.**

ball-cock (*Plumb., San. Eng.*). A self-regulating cistern-tap which, through a linkage system, is turned off and on by the rising and falling of a hollow ball floating on the surface of the liquid.

ball-ended magnet (*Elec. Eng.*). A permanent magnet, consisting of a steel wire with a steel ball attached to each end; this gives a close approximation to a unit pole.

ball flower (*Arch.*). An ornament, like a ball enclosed within three or four petals of a flower, set at regular intervals in a hollow moulding.

balling (*Met.*). (1) A process that occurs in the cementite constituent of steels on prolonged annealing at 650°–700°C. (2) The operation of forming balls in a puddling furnace.

balling gun (*Vet.*). An instrument for the administration of *balls* to horses.

Balling scale (*Chem.*). Scale of densities, as marked on hydrometers for measuring density of liquids.

ballistic circuit-breaker (*Elec. Eng.*). A very high-speed circuit-breaker, in which the pressure produced by the fusing of an enclosed wire causes interruption of the circuit.

ballistic fruit (*Bot.*). A fruit which discharges its seeds by some elastic means of propulsion.

ballistic method (*Elec. Eng.*). A method of high-grade testing used in electrical engineering, a ballistic galvanometer being used.

ballistic missile (*Aero., Space*). See **missile.**

ballistic pendulum (*Mech., Phys.*). A heavy block suspended by strings so that its swings are restricted to one plane. If a bullet is fired into the block, the velocity of the bullet may be calculated from a measurement of the angle of swing of the pendulum.

ballistics (*Phys.*). The study of the dynamics of the path taken by an object moving under the influence of a gravitational field.

ballistospore (*Bot.*). A spore attached apically and asymmetrically on a sterigma, and discharged explosively.

ball joint (*Eng., etc.*). Same as **ball-and-socket joint.**

ball lightning (*Meteor.*). A slowly-moving luminous ball, which is occasionally seen at ground level during a thunderstorm. It appears to measure about ½ m (diameter).

ball mill (*Chem. Eng.*). Mill consisting of a horizontal cylindrical vessel in which a given material is ground by rotation with steel or ceramic balls.

ballonnet (*Aero.*). An air compartment in the envelope of an aerostat, used to adjust changes of volume in the filler gas.

balloon (*Aero.*). A general term for aircraft supported by buoyancy and not driven mechanically. (*Arch.*) A spherical ball or globe crowning a pillar, pier, etc.

balloon barrage (*Aero.*). An anti-aircraft device consisting of suitably disposed *balloons* (q.v.).

balloon former (*Print.*). On rotary presses, an additional former mounted above the others, from which folded webs are gathered to make up the sections of multisectioned newspapers or magazines. See **length fold collection.**

balloon framing (*Carp.*). A cheap and rapid method of construction in which all timbers are of light scantling, and are held together entirely by nails and spikes, only the corner posts being tenoned: used in place of braced framing.

ballooning (*Textiles*). Term used to describe the appearance assumed by yarn twisting and running between guide eye and bobbin on ring frames and doublers, and from bobbin to guide on some high-speed winders.

ballotini (*Civ. Eng.*). Small glass balls. Used with road line paints, particularly in the U.S., to increase visibility at night.

ballottement (*Med.*). Method of diagnosing pregnancy by manual displacement of the foetus in the fluid which surrounds it in the uterus.

ball-pane hammer (*Tools*). A fitter's hammer, the head of which has a flat face at one end, and a smaller hemispherical face or pane at the other; used chiefly in riveting.

ball race (*Eng.*). (1) The inner or outer steel ring forming one of the ball-tracks of a ball-bearing. (2) Commonly, the complete *ball-bearing* (q.v.).

ball resolver (*Comp.*). One in which a ball is in contact with three wheels at points, radii of which are mutually perpendicular. Rotation of one wheel rotates the ball and hence the other two wheels at speeds $R \cos \theta$ and $R \sin \theta$, depending on its own speed R and inclination of its axis θ.

ballstone (*Geol.*). The name applied to masses of fine unstratified limestone, occurring chiefly in the Wenlock Limestone of Shropshire, and representing colonies of corals in position of growth.

ball-track (*Eng.*). See **ball-bearing.**

ball valve (*Eng.*). A single nonreturn valve consisting of a ball resting on a cylindrical seating; used in small water and air pumps.

ball warping (*Textiles*). The preparation of cotton warp on a warping mill. The parallel yarns are wound on a frame in the form of untwisted rope and eventually made up, mechanically or by hand, as a ball (*balled warp*).

Balmer series (*Phys.*). A group of lines in the hydrogen spectrum named after the discoverer and given by the formula:

$$\nu = R_H\left(\frac{1}{2^2} - \frac{1}{n^2}\right),$$

where n has various integral values, ν is the wave number and R_H is the hydrogen Rydberg number ($=1{\cdot}096\,775\,8 \times 10^7 m^{-1}$).

balneology (*Med.*). The scientific study of baths and bathing, and of their application to disease.

balneotherapy (*Med.*). Treatment by baths and bathing.

balsamiferous (*Bot.*). Yielding balsam.

balsam of fir (*Chem.*). See **Canada balsam.**

balsam of Peru (*Chem.*). An oleoresin containing esters of benzoic and cinnamic acids, obtained from a South American papilionaceous tree. Used in perfumery and chocolate manufacture.

balsam of Tolu (*Chem.*). An oleoresin containing esters of benzoic and cinnamic acids, obtained from a South American evergreen tree (*Myroxylon toluiferum*). Used in perfumery, chocolate manufacture, and medically as a mild expectorant.

balsa wood (*For.*). The wood of *Ochroma lagopus* ('West Indian corkwood'); it is a highly porous wood, and is valued for its lightness; used for vibration isolation.

Baltic redwood (*For.*). See **Scots pine.**

balun (*Telecomm.*). Abbrev. for *bal*ance to *un*balance transformer, usually a resonating section of transmission line, for coupling balanced to unbalanced lines. Also called **bazooka.**

baluster (*Arch.*). A small pillar supporting the coping of a bridge parapet, or the handrail of a staircase.

balustrade (*Arch.*). A coping or handrail with its supporting balusters.

Bamberger's naphthalene formula (*Chem.*). A centric formula for naphthalene, showing the valencies of the benzene rings pointing towards the centres. This formula does not assume any double bonds as in the Kekulé formula.

bamboo (*Bot.*). A general term for the large woody members of the *Bambuseae*; a tribe of grasses.

B-amplifier (*Telecomm.*). Amplifier following mixers or faders associated with microphone circuits in broadcasting studios, the faders and mixers following the A-amplifiers. *N.B.*—Not the same as **class-B amplifier.**

banak (*For.*). A tree (*Virola*) having a rather featureless wood, pinkish-brown in colour. It grows in central America and is useful for turnery purposes. The wood has a slight tendency to split. It is also used for packing-cases, box shooks, and roofing shingles.

banana (*Bot.*). Member of the monocotyledonous genus Musa; some are edible (the results of *M. acuminata* × *M. balbisiana*). The fruit is a berry.

banana plug (*Elec. Eng.*). A single conductor plug which has a spring metal tip, in the shape of a banana. The corresponding socket or jack is termed a *banana jack*.

band (*Bot.*). A strand of thickened tissue in the thallus of a liverwort; it has strengthening functions. (*Build.*) A flat horizontal member, occasionally ornamented, separating a series of mouldings or dividing a wall surface. (*Telecomm.*) See **frequency band.** (*Teleg.*) Number of unit intervals/second.

band-and-hook hinge (*Carp.*). See **strap hinge.**

band articulation (*Teleph.*). The percentage of speech-bands correctly received over a telephone system, with respect to the total bands transmitted.

band brake (*Eng.*). A flexible band wrapped partially round the periphery of a wheel or drum. One end is anchored, and the braking force is applied to the other.

band chain (*Surv.*). More accurate than ordinary chain. Also called **steel tape.**

band clutch (*Eng.*). A friction clutch in which a fabric-lined steel band is contracted on to the periphery of the driving member by engaging gear. See **friction clutch.**

band (or belt) conveyor (*Eng.*). An endless band passing over, and driven by, horizontal pulleys, thus forming a moving track which is used to convey loose material or small articles.

band cramp (*Tools*). A flexible steel band held in position by clamping screws. Used for shaped work, e.g., chairbacks.

banded column (*Arch.*). A column having *cinctures* at intervals.

banded filter (*Photog.*). A 2- or 3-colour filter for colour photography, with the separate filters side by side.

band-edge energy (*Electronics*). The band of energy between two defined limits in a semiconductor. The lower limit corresponds to the lowest energy required by an electron to remain free, while the upper one is the maximum permissible energy of a freed electron.

banded precipitate (*Chem.*). See **Liesegang phenomenon.**

banded structure (*Geol.*). A structure developed in crystalline rocks, both igneous and metamorphic; due to the alternation of layers differing in texture, composition, or both.

bandeler (*Arch.*). A plain moulding.

bandelet (*Arch.*). A small flat moulding encompassing a column.

band-elimination filter (*Telecomm.*). A filter which highly attenuates currents having frequencies within a specified nominal range and freely passes currents having frequencies outside this range. Also called **band-rejection filter, band-stop filter.**

band expansion factor (*Telecomm.*). Ratio of bandwidth to highest modulation frequency for a frequency-modulation transmission. Equal to twice the deviation ratio.

band-ignitor tube (*Electronics*). A valve of mercury pool type in which the control electrode is a metal band outside the glass envelope. Also termed **capacitron.**

banding plane (*Carp.*). A plane used for cutting out grooves and inlaying strings and bands in straight and circular work.

banding trap (*Agric.*). A band of sacking or straw tied to a tree trunk to induce larvae inimical to health of tree to shelter therein, and finally to be destroyed.

band losses (*Elec. Eng.*). Energy losses due to currents induced in the armature bands of an electric machine.

band merit (*Telecomm.*). Parameter of a valve, the product of the bandwidth and maximum gain (as a ratio) possible; alternatively the mutual conductance divided by 2π times the sum of the grid and anode capacitances.

band-pass filter (*Telecomm.*). One which freely passes currents having frequencies within specified nominal limits, and highly attenuates currents with frequencies outside these limits.

band-pass tuning (*Telecomm.*). Arrangement of two coupled circuits, tuned to same frequency and having substantially uniform response to range of frequencies, instead of marked variation of response with frequency, which characterizes a single-tuned circuit.

band-rejection filter (*Telecomm.*). See **band-elimination filter.**

bands (*Bind.*). The tapes, or cords, placed across the back of a book, to which the sections are attached by sewing. The ends of the bands are subsequently secured to the boards of the cover. See **flexible, raised bands.**

bandsaw (*Eng.*). A narrow endless strip of saw-blading running over and driven by pulleys, as a belt; used for cutting wood or metal to intricate shapes.

band spectrum (*Light*). Molecular optical spectrum consisting of numerous very closely spaced lines which are spread through a limited band of frequencies.

band-spreading (*Telecomm.*). Use of a relatively large fixed capacitor in parallel with a smaller variable capacitor, to reduce the band of frequencies covered by variation of the latter, thereby increasing selectivity.

band-stop filter (*Telecomm.*). See **band-elimination filter.**

B. and S. Wire Gauge. Abbrev. for *Brown and Sharpe Wire Gauge.*

band-switching (*Telecomm.*). System of inductance or capacitance switching, allowing a number of frequency bands to be covered by the same tuning dial.

bandwidth (*Telecomm.*). (1) Section of frequency spectrum within which component frequencies of a signal can pass, ideally with zero attenuation. (2) Arbitrary measure of a radio transmission, the width in Hz within which 99% of the energy is contained. Also used loosely to refer to the maximum frequency accepted by a circuit, assuming the minimum to be d.c.

bang-bang (*Automation*). Control mechanism of servo system with two set points, operation being between these, with no gradation.

Bangiales (*Bot.*). The single order of the *Bangioideae* (q.v.).

Bangioideae (*Bot.*). A subclass of the *Rhodophyceae* (Red algae), characterized by the intercalary growth of the thallus, and the direct formation of carpospores from the zygote.

Bang's bacillus (*Bacteriol.*). *Brucella abortus*; the cause of contagious abortion in animals and of undulant fever in man.

Bang's disease (*Vet.*). See **bovine contagious abortion.**

banister (*Arch., Carp.*). Alternative term for *baluster* (q.v.).

banjo axle (*Autos.*). The commonest form of rear-axle casing, in which the provision of the differential casing in the centre produces a resemblance to a banjo with two necks.

bank (*Civ. Eng.*). See **embankment.** (*Eng., etc.*) A number of similar pieces of equipment grouped in line and connected; e.g., a *bank* of engine cylinders, a *bank* of coke-ovens, a *bank* of transformers. (*Teleph.*) An assemblage of fixed electrical contacts arranged (generally in arc formation) to facilitate the selective connection of a traversing wiper (or wipers). (*Textiles*) A large frame for holding bobbins of yarn when the yarn is being transferred to a warper's beam. See **creel.** (*Typog.*) A bench on which sheets are placed as printed, or on which standing type-matter rests.

Banka drill (*Mining*). One widely used in shallow testing of alluvial deposits. Portable, hand-operated, it consists essentially of an assembly of 5-ft pipes, 4 in. in diameter, which are worked into the ground, the material traversed being recovered by a sand pump or *bailer*.

Banka tin (*Met.*). Tin of high purity (99·75% tin and upwards), produced in Banka in the Indonesian Republic and Malaysia. Also called **Straits tin.**

bank bars (*Mining*). Slabs of timber or reinforced concrete used for lining mine shafts.

banked fire (or boiler) (*Eng.*). A boiler furnace in which the rate of combustion is purposely reduced to a very low rate for a period during which the demand for steam has ceased. See **banking loss, banking-up, dead bank, floating bank.**

banker (*Build.*). A bench upon which bricklayers and stonemasons shape their materials.

banket (*Geol.*). The term originally applied by the Dutch settlers to the gold-bearing conglomerates of the Witwatersrand. It is now used for any metamorphosed conglomerate, containing barren quartz pebbles cemented with siliceous matrix bearing gold and uranium.

banking (*Aero.*). Angular displacement of the wings of an aeroplane about the longitudinal axis, to assist turning. (*Met.*) Process of suspending operation in a smelter, by feeding fuel only into the furnace until as much metal and slag as possible have been removed, after which all air inlets are closed. (*Mining*) The operations involved in removing full trucks, tubs, or wagons and replacing them by empty ones at the top of a shaft.

banking loss (*Eng.*). The fuel used in maintaining a *floating bank* (q.v.), or to maintain a *dead bank* (q.v.), and then raise the steam pressure to normal.

banking pins (*Horol.*). Vertical pins in the bottom plate of a watch which limit the motion of the lever. In the cylinder escapement, the action of the balance is limited by a single banking pin on the balance. In some watches part of the plate itself forms a solid banking.

banking-up (*Eng.*). Reducing the rate of combustion in a boiler furnace by covering the fire with slack or fine coal.

bank multiple (*Teleph.*). The multiple connecting the bank of contacts in selectors or uniselector switches in automatic telephone exchanges. The switches are usually detachable from the contacts and multiples for repair, without disturbing the large number of connexions in the multiple. See **multiple, slipped bank, straight bank.**

bank paper (*Paper*). A thin paper of substance

less than *large post*, 15 lb/500, used for type-writing and correspondence purposes.

bank protector (*Hyd. Eng.*). Any device for minimizing erosion of river banks by water; e.g., groins, pitching, etc.

bank winding (*Elec. Eng.*). Method of coil winding developed to reduce self-capacitance.

bank wires (*Teleph.*). The wires connecting the corresponding contacts in the banks of a shelf of switches. See **slipped bank, straight bank.**

banner (*Bot.*). See **standard.**

bannister harness (*Textiles*). A special harness used for weaving extra wide patterns in fine reeds on a jacquard loom.

bannisterite (*Min.*). A hydrated silicate of manganese, crystallizing in the monoclinic system, and occurring in manganese deposits in North Wales and New Jersey.

Bannockburn tweed (*Textiles*). Originally produced at Bannockburn, this Cheviot tweed is a firmly woven 2/2 twill weave, wool cloth. Single and doubled coloured yarns are employed in 1/1 order warp- and weft-ways.

banquette (*Arch.*). A narrow window-seat in masonry, brickwork, or wood. (*Civ. Eng.*) (1) A raised footway inside a bridge parapet. (2) A ledge on the face of a cutting. See **berm.**

Banti's disease (*Med.*). Anaemia leucopoenia; splenic anaemia; splenomegaly; a chronic disease involving cirrhosis of the liver, anaemia, and bleeding into the alimentary tract. Originally thought to be a separate disease, but now known to be an advanced form of liver cirrhosis.

B.A. ohm (*Elec. Eng.*). A unit of resistance adopted by the British Association in 1865 and now superseded.

bar (*Acous.*). The unit of alternating acoustic (sound or excess) pressure in a freely progressing wave or on a surface; equals 1 r.m.s. dyne/cm^2. (*Civ. Eng., etc.*) A deposit of sand, gravel, or other material in a river, or across the mouth of a river or harbour: a boom of logs, preventing or tending to prevent navigation. (*Eng., etc.*) Material of uniform cross-section, which may be cast, rolled or extruded. (*Horol.*) A narrow detachable plate. (*Meteor., Phys., etc.*) Unit of pressure or stress, 1 bar $=10^5$ N/m^2 or pascals $=750 \cdot 07$ mm of mercury at 0°C and lat. 45°. The *millibar* (1 mbar $=100$ N/m^2 or 10^3 dyn/cm^2) is used for barometric purposes. (N.B. Std. atmos. pressure $=1 \cdot 013\,25$ bar.) The *hectobar* (1 hbar $=10^7$ N/m^2, approx. $0 \cdot 6475$ tonf/in.^{2}) is used for some engineering purposes. (*Rail.*) A pivoted bar, parallel to a running rail, which, being depressed by the wheels of a train, is capable of holding points or giving information about a train's position.

baragnosis (*Med.*). Loss of the ability to judge differences between the weights of objects.

bar-and-yoke (*Elec. Eng.*). Method of magnetic testing in which the sample is in the form of a bar, clamped into a yoke of relatively large cross-section, which forms a low-reluctance return path for the flux.

Bárány's tests (*Med.*). Tests for assessing the functions of the semicircular canals in the inner ear.

barathea (*Textiles*). (1) Botany worsted coating material, usually of twilled hopsack, matt, or broken rib weave. (2) A dress fabric with a spun-silk or man-made fibre warp and a Botany worsted weft.

barb (*Bot.*). A hooked or doubly-hooked hair. (*Zool.*) Any hooked, bristlelike structure: in Birds, one of the lateral processes of the rachis of the feather which form the vane.

Barbados Earth (*Geol.*). A siliceous accumulation of remains of *Radiolaria*, formed originally in deep water and later upraised above sea level.

Barba's law (*Phys.*). Concerned with the plastic deformation of metal test pieces when strained to fracture in a tensile test, it states that test pieces of identical size deform in a similar manner.

barbate (*Bot.*). Bearded; bearing tufts of long hairs.

barbel (*Zool.*). In some Fish, a finger-shaped tactile appendage arising from one of the jaws.

Barber-Greene (*Civ. Eng.*). A proprietary machine for surfacing large road areas and laying the asphalt surface in a continuous operation.

barber's rash (*Med.*). Infection of the beard region of the face with a bacterium or a fungus.

barbicel (*Zool.*). In Birds, one of the minute hooked processes on the lower face of a barbule of a feather.

barbitone (*Pharm.*). Diethyl-malonyl-urea, 5,5-diethylbarbituric acid,

$$\begin{array}{c} \text{CH}_3\text{CH}_2 \\ \text{CH}_3\text{CH}_2 \end{array} > \text{C} < \begin{array}{c} \text{CO-NH} \\ \text{CO-NH} \end{array} > \text{C=O}$$

Prepared by condensing diethymalonic ester with urea in the presence of sodium ethoxide. White crystalline solid; m.p. 191°C. Used as a sedative and hypnotic, usually in the form of the sodium salt, 'soluble barbitone'. Also called **barbital, veronal.**

barbituric acid (*Chem.*). Malonyl urea, $CO(NH \cdot CO)_2CH_2$, crystallizing in large colourless crystals. The hydrogen atoms of the methylene group are reactive and can be replaced by halogen. Basis of important derivatives with therapeutic action.

barbone (*Vet.*). Buffalo disease. A form of *haemorrhagic septicaemia* (q.v.) occurring in buffalo.

barbule (*Bot.*). The inner row of peristome teeth in some mosses. (*Zool.*) In Birds, one of the processes borne on the barbs of a feather, by which the barbs are bound together.

barchan (*Geol.*). See **barkhan.**

Barcoo rot (*Med.*). See **veldt sore.**

bar cramp (*Tools*). A wooden bar with blocks and wedges, for gluing planks edgewise.

Bardac process (*Paper*). The use of melamine formaldehyde resin in the paper stock to improve the retention of fillers, fines, and other beater chemicals.

bare (*Eng., etc.*). A term signifying slightly smaller than the specified dimension. Cf. *full.*

bare carbon (*Light*). An arc lamp carbon not coated with a layer of copper to improve its conductivity.

bare conductor (*Elec. Eng.*). A conductor not continuously covered with insulation, but supported intermittently by insulators, e.g., bus-bars and overhead lines.

bare electrodes (*Elec. Eng.*). Electrodes used in welding that are not coated with a basic slag-forming substance.

bareface tenon (*Carp., Join.*). A tenon which has a shoulder on one face only; used when jointing a rail which is thinner than the stile.

Barfoed's test (*Chem.*). A specific test for monosaccharides, based upon the reduction of copper(II) acetate to red copper(I) oxide.

barge (*Typog.*). A small case with boxes for sorts such as figures and spaces for use when correcting.

barge board (*Build.*). A more or less ornamental board fixed under the gable end of a roof. It hides the ends of the horizontal timbers, and protects from the weather the underside of the *barge-course* (q.v.).

barge-couple (*Carp.*). A beam which is tenoned into another to add strength to a building.

barge-course (*Build.*). (1) That part of the roof of a house which projects slightly over the gable end, and is made up underneath with mortar to keep out rain, etc. (2) A coping course of bricks laid edge-wise and transversely on a wall.

bar generator (*TV*). Source of pulse signals, giving a bar pattern for testing TV cathode-ray tubes.

barite (*Min.*). See **barytes.**

barium (*Chem.*). A heavy metallic element in the second group of the periodic system, an alkaline earth metal. Symbol Ba, at. no. 56, r.a.m. 137·34, m.p. 850°C. Barium salts are used in the coating of thermionic cathodes.

barium carbonate (*Min.*). Witherite.

barium concrete (*Build.*). Concrete containing high proportion of barium compounds used as protective building material against radiation.

barium feldspar (*Min.*). A collective term for barium-bearing feldspars, including celsian and hyalophane.

barium hydroxide (*Chem.*). $Ba(OH)_2$. See **barium oxide.**

barium meal (*Radiol.*). A mixture of barium sulphate used in radiology for outlining the alimentary tract. It is swallowed by the patient, and shows its presence by stopping X-rays.

barium oxide (*Chem.*). BaO. When freshly obtained from the calcined carbonate it is even more reactive with water than calcium oxide and forms barium hydroxide (alkaline). Also called **baryta.**

barium plaster (*Build.*). A cement-sand plaster containing barium salts, used for lining hospital and experimental X-ray rooms to minimise the dangerous reflection of the rays.

barium sulphate (*Chem.*). $BaSO_4$. Formed as a heavy white precipitate when sulphuric acid is added to a solution of a barium salt. One of the most insoluble substances in water known. Although of little pigmentary value, it is much used in paint manufacture and in the preparation of lake pigments. See **barytes, barium meal.**

barium titanate (*Chem.*). $BaTiO_3$; a crystalline ceramic with outstanding dielectric, piezo-electric and ferro-electric properties. Used in capacitors and as a piezo-electric transducer. Has a higher *Curie point* than *Rochelle Salt*.

bark (*Bot.*). Strictly, all tissues external to the cork cambium; popularly, the corky and other material which can be peeled from a woody stem. (*Chem.*) A term which is used in the white-lead industry to indicate the waste tan from tanneries. Used in the manufacture of white lead by the stack process. Now obsolete.

bark allowance (*For.*). The percentage deducted from volume, girth, or diameter measurement over bark to give the underbark measurement.

bark-bound (*Bot.*). An unhealthy condition in a tree when the bark resists splitting and compresses the growing tissues beneath it.

bar keel (*Ships*). See under **keelson.**

Barker index (*Crystal.*). A method of identification of crystalline substances from measurements of interfacial angles.

barkevikite (*Min.*). A member of the amphibole group, resembling basaltic hornblende but having a higher total iron content and a low ferric-ferrous iron ratio. Occurs in alkaline plutonic rocks, as at Barkevik, Norway.

bark gauge (*For.*). An instrument for measuring the thickness of bark.

barkhan (*Geol.*). An isolated crescentic sand-dune. Also **barchan.**

Barkhausen criterion (*Electronics*). The feedback condition which must be satisfied in the design of an oscillator (the product of the complex gain and the feedback factor must be $\leqq -1$).

Barkhausen effect (*Mag.*). The tendency for magnetization to occur in discrete steps rather than by continuous change.

Barkhausen-Kurz oscillator (*Electronics*). Triode valve with anode and grid fed negatively and positively through a Lecher system, the oscillation depending on the retarding field on both sides of the grid.

barking drum (*For.*). A large rotating drum in which logs or billets are tumbled, the bark being removed by friction.

barkometer or **barktrometer** (*Leather*). A hydrometer used for gauging the density of tanning liquors.

bark pocket (*For.*). Bark in a flute or pocket, or associated with a knot, which has been partially or wholly enclosed by the growth of the tree.

bar lathe (*Eng.*). A small lathe of which the bed consists of a single bar of circular, triangular, or rectangular section.

barley wine (*Brew.*). See **ale.**

Barlow lens (*Light*). A negative lens used to decrease the convergence of the beam from the object of a telescope, thus increasing its effective focal length.

Barlow's disease (*Med.*). Infantile scurvy caused by deficiency of vitamin C. See also **scurvy.**

Barlow's wheel (*Elec. Eng.*). A primitive electric motor formed from a pivoted starwheel, intermittently dipping into mercury, torque being obtained from interaction of the radial current and a perpendicular magnetic field supplied by a permanent magnet.

barm (*Brew.*). Yeast from top fermentation, used for flavouring beer.

bar magnet (*Elec. Eng.*). A straight bar-shaped permanent magnet, with a *pole* at each end.

Barmeen machine (*Textiles*). A machine that makes a type of lace resembling hand-made braid. The yarns on king bobbins on carriers are usually plaited together and the carriers are jacquard controlled to give pattern variety.

bar mill (*Met.*). A rolling mill with grooved rolls, for producing round, square, or other forms of bar iron of small section.

bar mining (*Mining*). Alluvial mining of sand-banks, river bars, or submerged deposits.

bar movement (*Horol.*). A watch movement in which the upper pivots are carried in bars.

barn (*Nuc.*). Unit of effective cross-sectional area of nucleus equal to 10^{-28} m^2. So called, because it was pointed out that although one barn is a very small unit of area, to an elementary particle the size of an atom which could capture it was 'as big as a barn door'.

Barnard's star (*Astron.*). A faint star in Ophiuchus, found in 1916 to have the largest proper motion yet measured, amounting to 10 seconds of arc per annum.

barnesite (*Min.*). Hydrated sodium vanadate, crystallizing in the monoclinic system.

Barnett effect (*Mag.*). Magnetization of ferromagnetic material caused by rapid rotation. Also called **Einstein-de Haas effect.**

baroclinic atmosphere (*Meteor.*). An atmosphere which is not *barotropic* (q.v.).

bar of Sanio (*Bot.*). A horizontal rod or band of thickening, consisting of pectic materials or of cellulose, occurring between pits in the walls of tracheides and vessels.

barograph (*Meteor.*). A recording barometer, usually of the aneroid type, in which variations of atmospheric pressure cause movement of a pen which traces a line on a clockwork-driven revolving drum.

barokinesis (*An. Behav.*). A *kinesis* (q.v.) in which activity increases with a change in pressure, as in some planktonic animals.

barometer (*Meteor.*). An instrument used for the measurement of atmospheric pressure. The *mercury barometer* (q.v.) is preferable if the highest accuracy of readings is important, but where compactness has to be considered, the *aneroid barometer* (q.v.) is often used. For *altitudes by barometer*, see **Babinet's formula.**

barometric corrections (*Meteor.*). Necessary corrections to the readings of a mercury barometer for index error, temperature, latitude, and height.

barometric error (*Horol.*). The error in the time of swing of a pendulum due to change of air pressure. Though small, it is avoided in precision clocks by causing the pendulum to swing in an atmosphere of constant (low) pressure.

barometric pressure (*Meteor.*). The pressure of the atmosphere as read by a barometer. Expressed in *millibars* (see **bar**), the height of a column of mercury, or (SI) in kN/m^2.

barometric tendency (*Meteor.*). The rate of change of atmospheric pressure with time. The change of pressure during the previous three hours.

barophil (*Chem.*). Used of organisms that grow and metabolize, as well (or better) at increased pressures as at atmospheric pressure.

barophoresis (*Chem.*). Diffusion of suspended particles at a speed dependent on extraneous forces.

baroque organ (*Acous.*). A type of pipe organ in which low fundamentals are obtained by the subjective difference tones arising from pipes operating at the musical interval of the fifth.

baroreceptor (*Physiol.*). A sensory receptor sensitive to the stimulus of pressure, e.g., receptors located in the walls of arteries and veins responding to changes of intraluminal pressure, such as the carotid and aortic arterial, or right arterial atrial, baroreceptors.

baroscope (*Meteor.*). An instrument giving rough indications of changes in atmospheric pressure.

barostat (*Aero.*). A device which regulates input and output pressure of fuel metering device of a gas turbine to compensate for atmospheric pressure variation with altitude.

barothermograph (*Meteor.*). A self-recording instrument which provides continuous records of atmospheric pressure and temperature; a *meteorograph.*

barotropic atmosphere (*Meteor.*). An atmosphere with zero horizontal temperature gradient at all levels so that the *isopleths* of density and pressure coincide and the *thickness chart* (q.v.) has no pattern.

barrage (*Bot.*). A zone of inhibition between two growing hyphae. (*Hyd. Eng.*) An artificial obstruction placed in a water-course in order to secure increased depth for irrigation, navigation, or some other purpose.

barrage balloon (*Aero.*). A small captive kite balloon, flown at 900–1200 m, the cable of which is intended to destroy low-flying aircraft.

barrage-fixe (*Hyd. Eng.*). A permanent masonry dam.

barrage-mobile (*Hyd. Eng.*). A dam provided with sluices to control the flow of water.

barrage reception (*Radio*). That in which interference of radio signals from any particular direction is minimized by selecting the appropriate directional aerial to give maximum signal to interference ratio.

barred code (*Teleph.*). Any dialled code that automatic exchange apparatus is primed to reject by connecting the caller no further than *N.U. tone.*

barrel (*Brew.*). Cask of 36 gallons capacity (31½ U.S. gallons). (*Chem. Eng.*) U.S. barrel of 42 U.S. gallons (=35 Imperial gallons), frequently employed as a unit of capacity, especially in the oil industry. (*Eng.*) A hollow, usually cylindrical, machine part; often revolving, sometimes with wall apertures. (*Horol.*) The cylindrical container for housing a mainspring. See **fuses, going-, pin-.** (*Telecomm.*) In a system of *transposition* (q.v.), the section in which all possible transpositions are effected in order, before repetition in the next section, to effect balance of circuits, i.e., side impedances to earth.

barrel amalgamation (*Met.*). Recovery of gold from rich concentrates by prolonged gentle tumbling with mercury in a steel barrel.

barrel arbor (*Horol.*). The arbor carrying the barrel; upon its centre portion is coiled the mainspring during winding. In a 'detachable barrel' the arbor and barrel can be withdrawn from between the plates without dismantling the plates.

barrel bolt (*Join.*). Hand-operated door fastening comprising a metal rod sliding in cylindrical guides.

barrel cam (*Eng.*). A cylindrical cam with circumferential or end track.

barrel distortion (*Optics, Photog., TV*). A type of *curvilinear distortion* (q.v.), in which the lines are concave, when the stop on the image side is behind the lens, creating a shape similar to a barrel. See also **pincushion distortion.**

barrel drain (*Build.*). A cylindrical drain.

barrel elevator (*Eng.*). This comprises parallel travelling chains with curved arms projecting. The chains pass over sprocket wheels at the top and bottom of the elevator, and lift barrels from a loading platform to a runway.

barrel hopper (*Eng.*). A machine for unscrambling, orientating, and feeding small components during a manufacturing process, in which a revolving barrel tumbles the components on to a sloping, vibrating feeding-blade.

barrel nipple (*Plumb., etc.*). A *nipple* threaded at both ends and having a plain middle part.

barrel-plating (*Elec. Eng.*). A process of electroplating in which the articles to be plated are placed in a rotating container provided with suitable negative contacts.

barrel-type crankcase (*I.C. Engs.*). A petrol-engine crankcase so constructed that the crankshaft must be removed from one end; in more normal construction the crankcase is split. See **split crankcase.**

barrel vault (*Build.*). A vault of approximately semicircular cross-section, whose length exceeds its diameter. Also called **annular vault, tunnel vault, wagon vault.**

barrel-vault roof (*Civ. Eng.*). A roof formed of reinforced concrete in the shape of an open cylindrical shell, generally with lateral stiffening diaphragms and edge beams. The roof itself is frequently very thin in section.

barrel winding (*Elec. Eng.*). See **drum winding.**

barren (*Bot.*). (1) Lacking pollen. (2) Unable to produce seed; infertile. (*Geol.*) Without fossils.

barren solution (*Met.*). In chemical extraction of metals from their ores, solution left after these have been removed.

barretter (*Elec. Eng.*). Iron-wire resistor mounted in a glass bulb containing hydrogen, and having a temperature variation so arranged that the change of resistance ensures that the current in the circuit in which it is connected remains substantially constant over a wide range of voltage.

barrier (*Cinema.*). Thin black line which separates adjacent frames in the projection print from each other and from the sound-track. (*Elec. Eng.*) (1) In transformers, the solid insulating material which provides the main insulation, apart from the oil. (2) The refractory material intended to localize or direct any arc which may arise on the operation of a circuit-breaker.

barrier layer (*Electronics*). Double electrical layer formed at the surface of substances which have differing work functions, there being a diffusion of electrons up the work function gradient.

barrier-layer cell (*Electronics*). One in which illumination of the barrier layer results in a small e.m.f. because of differing work functions of dissimilar materials.

barrier-layer rectifier (*Electronics*). One using the non-symmetrical properties of a semiconductor *barrier layer* (q.v.).

barrier pillar (*Mining*). A pillar of solid coal left in position to protect a main road from subsidence, or as a division, or to protect workings from flooding.

barrier reef (*Geol.*). A coral-reef developed parallel with the shore-line and enclosing a lagoon between itself and the land. It marks a stage between a fringing-reef and an atoll.

barring gear (*Elec. Eng.*). The arrangement for moving heavy electrical machines or transformers by man power, using crowbars inserted in holes in wheels, thus obtaining a large mechanical advantage.

barring motor (*Elec. Eng.*). A small motor which can be temporarily connected, by a gear or clutch, to a large machine to turn it slowly for adjustment or inspection.

barring-on (*Textiles*). (1) Forcing heavy driving belt on whilst pulley is rotated slowly under power. (2) Transferring knitted fabrics on to the needles of another machine for a further operation.

Barrovian metamorphism, Barrovian zones (*Geol.*). Regional metamorphism of the type first described in the Scottish Highlands by G. Barrow (1893). Zones of increasing metamorphism are characterized by the presence of a series of index minerals: chlorite, biotite, garnet, staurolite, kyanite, sillimanite. This type of metamorphism has since been recognized in many other parts of the world.

barrow pump (*Hyd.*). A combined suction and force pump mounted on a 2-wheeled barrow; much used for agricultural purposes.

bars (*TV*). Artificial appearance of bars or a cross on a TV screen, generated electronically and used for testing TV systems and circuits. See **bar generator.** (*Textiles*) (1) In a lace machine, strips of flexible steel that stretch from one end of the machine to the dropper box at the jacquard end. (2) Yarns connecting the solid portions in needle-point laces.

bars of foot (*Vet.*). Part of the sole of the horse's foot formed by reflexion of the wall on each side of the *frog* (q.v.).

bar suspension (*Elec. Eng.*). A method of mounting the motor on an electrically propelled vehicle. One side of the motor is supported on the driving axle and the other side by a spring-suspended bar lying transversely across the truck. Also called **yoke suspension.**

Bartholin's duct (*Zool.*). An excretory duct of the sublingual gland.

Bartholin's glands (*Zool.*). In some female Mammals, glands (corresponding with Cowper's glands in the male) lying on either side of the upper end of the vagina.

bar timbering (*Mining*). A method of timbering with horizontal bars of wood, which are supported by side legs or walls.

Bartlett force (*Nuc.*). Force acting between two nucleons due to spin exchange. See **short-range forces.**

bartonellosis (*Vet.*). Infection by organisms of the genus *Bartonella*, affecting man, cattle, dogs, and rodents.

Bartonian Series (*Geol.*). The topmost division of the Eocene rocks in Britain; comprising in the type locality the Barton Clay, with its rich molluscan fauna, and the overlying Barton Sands.

bar tracery (*Arch.*). Window-tracery characteristic of Gothic work, resembling more a bar of iron twisted into various forms than stone.

bar-type current transformer (*Elec. Eng.*). A *current transformer* in which the primary consists of a single conductor that passes centrally through the iron core upon which the secondary is wound.

bar warp machines (*Textiles*). Lace machines which make cotton, nylon or Terylene warp net from beams only. Control of pattern is similar in many respects to that on the *Levers machine* (q.v.).

bar winding (*Elec. Eng.*). An armature winding for an electric machine whose conductors are formed of copper bars.

bar-wound armature (*Elec. Eng.*). An armature with large sectioned conductors which are insulated and fixed in position and connected, in contrast with former-wound conductors which are sufficiently thin to be inserted, after shaping in a suitable jig.

barycentric coordinates (*Maths.*). A system of coordinates devised by Möbius and published in 1827, based upon a generalized notion of centre of mass. The barycentric coordinates of a point P in space are the four masses (including negative mass if necessary) that have to be positioned at four reference points in order that their centre of mass shall be at P. For a point in a plane three reference points are sufficient. Such coordinates are homogeneous.

barye (*Phys.*). See **microbar.**

baryon (*Nuc.*). *Nucleon* and *hyperon.* See **anti-baryon.**

baryon number (*Nuc.*). The number of baryons less the corresponding number of anti-baryons taking part in a process. This number appears to be conserved throughout the process.

barysphere (*Geol*). The heavy interior core of the earth (obsolete).

baryta (*Chem.*). See **barium oxide.**

baryta paper (*Paper*). Paper coated on one side with an emulsion of barium sulphate and gelatine. Used in moving pointer recording apparatus and for photographic printing papers.

baryta water (*Chem.*). A suspension of barium hydroxide in distilled water; it is a fairly strong alkali.

barytes or **barite** (*Min.*). Barium sulphate, typically showing tabular orthorhombic crystals. It is a common mineral in association with lead ores, and occurs also as nodules in limestone and locally as a cement of sandstones. Also called **heavy spar.**

barytocalcite (*Min.*). A double carbonate of calcium and barium, $CaCO_3{\cdot}BaCO_3$, crystallizing on the monoclinic system, and occurring typically in lead veins.

basal (*Bot.*). (1) Situated at the base. (2) Said of an ovule which springs from the base of the ovary.

basal anaesthesia (*Med.*). Anaesthesia which acts as a basis for further and deeper anaesthesia.

basalar (*Zool.*). Said of certain small plates (sclerites) located below the articulation of the wings in some Insects.

basal area (*Ecol.*). A measure of the density of trees, being the sum total of the cross-sectional area of the trunks, determined from measurements of the diameter at breast height.

basal body (*Bot., Zool.*). (1) In flagellate, or ciliate, *Protozoa*, zoospores or spermatozoids, a small, deeply-staining granule at the base of the locomotory *organelle* (q.v.). (2) Part of the thallus fixed to the substrate by rhizoids. See also **blepharoplast.**

basal bone (*Zool.*). In *Gymnophiona*, a bone of the skull formed by the fusion of the exoccipitals and the pro-otics with the parasphenoid.

basal cell (*Bot.*). (1) An attaching cell. (2) The lower cell of a *crozier* (q.v.). (3) A uninucleate cell which may be the oögonium in *Uredinales.*

basal conglomerate (*Geol.*). The first stage of sedimentation; lies on plane of *unconformity.*

basal corpuscle (*Bot., Zool.*). See **basal body.**

basal disk (*Zool.*). In corals, the basal area of the ectoderm which secretes the calcareous skeleton.

basale (*Zool.*). In Fish, the proximal segment of a fin-ray. *pl.* **basalia.**

basal ganglia (*Zool.*). In Vertebrates, ganglia connecting the cerebrum with other nerve-centres.

basal granule (*Zool.*). In *Mastigophora*, the cytoplasmic granule to which the axial filament of the flagella is connected. See **basal body.**

basal metabolic rate (*Physiol.*). The minimal quantity of heat produced by an individual at complete physical and mental rest, but not asleep, 12–18 hr after eating, expressed in milliwatts per square metre of body surface.

basal organ (*Zool.*). In *Diptera*, a sense organ in the base of the haltere. See **cupula.**

basal pit (*Bot.*). An undifferentiated zone in the macrosporangial wall of some *Fucales.*

basal placentation (*Bot.*). The condition when the placenta is at the base of the ovary.

basal planes (*Crystal.*). The name applied to the faces representing the terminating *pinacoid* (q.v.) in all the crystal systems exclusive of the cubic system. See also **pedion.**

basal plates (*Zool.*). In the developing Vertebrate skull, a plate of cartilage formed by the fusion of the parachordals and the trabeculae; in *Crinoidea*, certain plates situated at or near the top of the stalk; in *Echinoidea*, certain plates forming part of the apical disk.

basal rim (*Zool.*). In some *Mastigophora*, a row of basal granules underlying an undulating membrane, indicating that the latter is homologous to a row of fused cilia.

basal sclerite (*Zool.*). In *Diptera*, part of the cephalo-pharyngeal skeleton. It forms a trough in which is lodged the pharynx. Also **pharyngeal sclerite.**

basalt (*Geol.*). A fine-grained igneous rock, dark colour, composed essentially of basic plagioclase feldspar and pyroxene, with or without olivine. In the field, the term is generally restricted to lavas, but many minor intrusions of basic composition show identical characters, and therefore cannot be distinguished in the laboratory.

basalt glass (*Geol.*). See **tachylite.**

basaltic hornblende (*Min.*). A variety of hornblende with a high ferric-ferrous iron ratio and a low hydroxyl content, occurring chiefly in volcanic rocks. Also called **oxyhornblende.**

basal walls (*Bot.*). The walls which separate the hypobasal and epibasal halves of a fern embryo when this consists of 8 cells.

basanite (*Geol.*). A basaltic rock containing plagioclase, augite, olivine, and a feldspathoid (nepheline, leucite, or analcite).

bascule bridge (*Civ. Eng.*). A counterpoise bridge which can be rotated in a vertical plane about axes at one or both ends. The roadway over the river rises while the counterpoise section descends into a pit. Also called **balance-bridge.**

base (*Bot.*). That end of a plant member nearest to the point of attachment to another member, usually of different kind. (*Chem.*) Generally, a substance which tends to gain a proton. In particular, a substance which dissolves in water with the formation of hydroxyl ions and reacts with acids to form salts. (*Electronics*) (1) Part of a valve where leads from electrodes are connected to rigid pins which fit into sockets in a holder, so that a valve can be changed quickly. Applicable also to lamps, cathode-ray tubes, capacitors, crystals, etc. (2) Middle region of a transistor into which minority carriers are injected from the external circuit. (*Maths.*) See **cone, cylinder, logarithm, radix.** (*Nav.*) In a navigation chain, the line which joins two of the stations. (*Paint., etc.*) (1) A highly concentrated dispersion of pigments, etc., in a paint medium, which forms a paint when let down with more medium and possibly other liquid ingredients. (2) The inorganic part of a lake or extended dyestuff used as a pigment.

baseband (*Telecomm.*). The frequency band occupied by the signal in modulation.

baseboard (*Build.*). See **skirting board.**

base box (*Met.*). Unit of measurement of tin and terne plate. 112 plates, 20 in. by 14 in., for tin, and up to 195 lb/box. In Britain the commercial basic box is a hundredweight.

base bullion (*Met.*). Ingot base metal containing sufficient silver or gold to repay recovery, e.g., argentiferous lead.

base circle (*Eng.*). The circle used in setting out the profiles of gear-wheel teeth of involute form.

base course (*Build.*). The lowest course of masonry in a building.

Basedow's disease (*Med.*). Exophthalmic goitre. A condition due to overactivity of the thyroid gland; manifested by enlargement of the thyroid gland, protrusion of the eyeballs, rapid pulse, tremors, and nervousness. Also **Graves' disease.**

base exchange (*Chem.*). Chemical method used in soil mechanics to strengthen clays by replacing their H-ions with Na-ions. (*Geol.*) A property by virtue of which certain minerals, notably the zeolites, can exchange atoms of, say, calcium for sodium, when in a suitable environment See **Permutit.**

baseline (*Electronics*). Line on CRO screen from which vertical displacements are measured. (*Nav.*) Great circle line joining two navigational stations, e.g., on *loran* system. (*Surv.*) A survey line the length of which is very accurately measured by precise methods; used as a basis for subsequent triangulation. (*Typog.*) The bottom alignment of type; below it is the *beard* which accommodates the descenders of *f*, *g*, etc.

base load (*Elec. Eng.*). That part of the total load on an electrical power system which is applied, where possible, by the most efficient connected generating stations, the remaining *peak load* (q.v.) being supplied intermittently by the more expensive stations.

basement membrane, membrana propria (*Histol.*). A membrane lying between an epithelium and the underlying connective tissue, and formed usually from the connective tissue by condensation of its intercellular substance.

base metal (*Chem.*). A metal with a relatively negative electrode potential (on IUPAC system). (*Met.*) In electrometallurgy, the metals at the end of the electrochemical series remote from the *noble metals* (q.v.).

baseost (*Zool.*). In Fish, one of the distal elements of a fin-ray.

base-pairs (*Biochem.*). The Watson-Crick model of DNA is an *alpha-helix*, consisting of twin strands of nucleic acid, stabilized by hydrogen bonds between the complementary base-pairs. Thus a purine base on the one strand will pair with its complementary pyrimidine base on the other, and hence the genetic information on each strand is equivalent. Thus, by separation of strands, and further complementation, base-pairing forms the basis of the replication and transcription of genetic information.

base resistance (*Electronics*). Total resistance to base current, including spread.

base transmission factor (*Electronics*). Complex ratio of minority-carrier current arriving at collector to that leaving emitter; also called **base transport factor.**

base unit. The International System of Units (SI) is a coherent system based on seven base units. All derived units are obtained from the base units by multiplication without introducing numerical factors, and approved prefixes are used in the construction of sub-multiples and multiples. There is only one base or derived unit for each physical quantity. The base units are *metre*, *kilogram*, *second*, *ampere*, *kelvin*, *candela*, and *mole*.

base vector (*Maths.*). In a coordinate system, unit vectors taken in the positive direction of the coordinate axes are called base vectors, and any vector may be expressed as a linear combination of base vectors.

basi-. Prefix from Gk. *basis*, base.

Basic (*Comp.*). A simple programming language (*B*eginners' *A*ll-purpose *S*ymbolic *I*nstruction *C*ode).

basichromatin (*Cyt.*). A form of chromatin which stains deeply with basic dyes and contains a fairly high proportion of nucleic acid. Cf. *oxychromatin*.

basicity (*Chem.*). The number of hydrogen ions of an acid which can be neutralized by a base.

basiconic (*Zool.*). In Insects, said of certain sub-conical and immobile sensilla arising from the general surface of the cuticle.

basicranial (*Anat.*). Pertaining to, or situated at, the base of the skull.

basicranial fontanelle (*Zool.*). In Vertebrate embryos an aperture due to the nonunion of the posterior ends of the trabeculae: through it passes the pituitary pouch.

basic cycle (*Comp.*). The time necessary to complete a set of operations for the execution of each instruction.

basic dyes (*Chem.*). Mostly colour bases with hydrochloric acid, or double salts with zinc chloride; chiefly used for printing cotton with a tannin mordant, and also in colour photography.

basic lavas (*Geol.*). The lavas poor in silica (less than 52%). The rocks are typically dark in colour and heavy, and are well represented by the familiar type basalt.

basic lead carbonate (*Chem.*). Approximate composition $2PbCO_3{\cdot}Pb(OH)_2$. See **white lead.**

basic lead chromate (*Chem.*). $PbCrO_4{\cdot}Pb(OH)_2$. Also known as **Austrian cinnabar.** Used as a pigment. Produced when lead chromate is boiled with aqueous ammonia or potassium hydroxide. See **chrome reds.**

basic lead sulphate (*Chem.*). $2PbSO_4{\cdot}PbO$. A fine powder, obtained by roasting galena.

basic loading (*Elec. Eng.*). The limiting mechanical load, per unit length, on an overhead line conductor.

basic network (*Telecomm.*). The 2-terminal network which is a first approximation as a balancing network to the impedance of a line over the working frequency range.

basic number (*Bot.*). The lowest haploid chromosome number in any member of a euploid series formed by the species of a genus.

basic process (*Met.*). A steel-making process, either Bessemer, open-hearth, or electric, in which the furnace is lined with a basic refractory, a slag rich in lime being formed and phosphorus removed.

basic refractory (*Met.*). See **basic lining, basic steel.**

basic rocks (*Geol.*). Igneous rocks with a low silica content. The limits are usually placed at 45% silica, below which rocks are described as ultrabasic, and 52% silica, above which they are described as intermediate. Basic igneous rocks include basalt, the commonest type of lava, and gabbro, its plutonic equivalent.

basic six (*Aero.*). The group of instruments essential for the flight handling of an aeroplane and consisting of the *airspeed indicator*, *vertical speed indicator*, *altimeter*, *heading indicator*, *gyro horizon* and *turn and bank indicator*.

basic slag (*Agric.*, *Met.*). Furnace slag rich in phosphorus (as calcium phosphate) which, with silicate and lime, is produced in steel making, ground and sold for agricultural purposes, e.g., grassland improvement.

basic solvent (*Chem.*). A protophilic solvent, hence one which enhances the acidic (i.e., proton donating) properties of the solute.

basic steel (*Met.*). Steel which has reacted with a basic lining or additive to produce a phosphorus-rich slag and a low-phosphorus steel.

basic T (*Aero.*). A layout of flight instruments standardized for aircraft instrument panels in which four of the essential instruments are arranged in the form of a T. The pitch and roll attitude display is located at the junction of the T flanked by airspeed on the left and attitude on the right. The vertical bar portion of the T is taken up by directional information.

basic time (*Work Study*). The time for carrying out an *element* (q.v.) of work at standard rating, i.e.,

$$\frac{\text{observed time} \times \text{observed rating}}{\text{standard rating}}.$$

basicyte (*Zool.*). A form of ovoidal connective-tissue corpuscle, full of basiphil granules.

basidial (*Bot.*). Relating to a basidium.

basidial layer (*Bot.*). The hymenium in a Basidiomycete.

basidiocarp (*Bot.*). The fruiting body of the *Basidiomycetes*.

basidiogenetic (*Bot.*). Produced upon a basidium.

Basidiomycetes (*Bot.*). A major class of fungi

with about 14 000 species, world-wide in distribution. They have a septate mycelium, and in the higher forms develop complicated sporophores. The principal spores are basidiospores, produced on the outside of basidia, usually in groups of four. The *Basidiomycetes* include the *Ustilaginales*, the *Uredinales*, and many species placed in the *Autobasidiomycetes* (toadstools).

basidiophore (*Bot.*). A fruit body bearing basidia.

basidiospore (*Bot.*). The characteristic spore of the *Basidiomycetes*, formed by meiosis on the outside of the basidium, usually in fours.

basidium (*Bot.*). (1) A row of cells, or more often a single rounded to club-shaped cell, which bears the basidiospores. (2) A sterigma in some moulds (not a good use of the term).

basidorsal (*Zool.*). In developing vertebrae, one of a pair of small cartilages occurring laterodorsally on the notochord.

basifixed (*Bot.*). Said of an anther which is attached by its base to the filament.

basifugal (*Bot.*). Developing in order from the base upwards.

basigamous, basigamic (*Bot.*). Said of an embryo sac in which the synergidae and egg lie towards the base of the cavity, and not at the end nearest to the micropyle.

basihyal (*Zool.*). A broad median plate constituting the basal or median ventral portion of the hyoid arch.

basil (*Leather*). Undyed, tanned sheepskin retanned and finished when intended for bookbinding and fancy-goods.

basilabium (*Zool.*). In Insects, a small sclerite formed by the fusion of the labiostipites.

basilar (*Bot., Zool.*). Situated near, pertaining to, or growing from the base.

basilar cell (*Bot.*). A large, colourless, terminal cell of the filaments of some *Cyanophyceae*.

basilar membrane (*Acous., Zool.*). In Mammals, a flat membrane, part of the partition of the cochlea, containing the collection of auditory nerves in the inner ear which translate mechanical vibrations of differing frequencies into nerve impulses, which are passed to the brain.

basilar plate (*Zool.*). In *Diplopoda*, part of the gnathochilarium formed by the postlabial segment.

basilingual (*Zool.*). In Amphibians and some Reptiles, a broad plate of cartilage forming the main part of the hyoid.

basimandibula (*Zool.*). In some Insects, a small sclerite at the base of the mandible.

basimaxilla (*Zool.*). In some Insects, a small sclerite at the base of the maxilla.

basin (*Geol.*). A large depression in which sediments may be deposited. Alternatively, a gently folded structure in which beds dip inwards from the margin towards the centre.

basinerved (*Bot.*). Having veins proceeding from the base.

basioccipital (*Zool.*). In higher Vertebrates, a cartilage bone, occurring midventrally in the occipital region of the skull.

basion (*Anat.*). The midpoint of the anterior margin of the foramen magnum.

basiophthalamite (*Zool.*). In *Crustacea*, the proximal joint of the eye-stalk.

basiotic (*Zool.*). See **mesotic.**

basipetal (*Bot.*). With each new member of a series developing nearer to the base than the next oldest one.

basipetal sorus (*Bot.*). See **gradate sorus.**

basipharynx (*Zool.*). In Insects, the fused epipharynx and hypopharynx.

basiphil (*Zool.*). Having a marked affinity for basic dyes. Also **basophil(e).**

basiphil cells (*Physiol.*). White blood cells, forming 0–1 % of the granulocytes. Also applied to cells in the anterior lobe of the pituitary gland.

basiphilia (*Med.*). See **basophilia.**

basipodite (*Zool.*). In the appendages of some *Crustacea*, the second or distal joint of the protopodite.

basipodium (*Anat.*). The wrist or ankle.

basiproboscis (*Zool.*). In certain Insects with suctorial mouth-parts, that part of the proboscis consisting of the mentum, submentum, cardines, and stipites.

basipterygium (*Zool.*). In Fish, a bone or cartilage in the pelvic fin, representing the fused basalia.

basipterygoid (*Zool.*). In some Reptiles and Birds, a process of the basisphenoid.

basiscopic (*Bot.*). On the side towards the base.

basis cranii (*Zool.*). In *Craniata*, the floor of the cranium, formed from the basal plate of the embryo.

basisphenoid (*Zool.*). A cranial bone of some Vertebrates, extending forward from the base of the basioccipital to the presphenoid.

basisternum (*Zool.*). One of the three sclerites forming the sternum of each thoracic segment in Insects. The other two are the *presternum* and the *sternellum* (qq.v.).

basis weight (*Paper*). The weight of a ream of paper. Also called **substance.** See **G.S.M.**

basitemporal (*Zool.*). A cranial bone of Birds, formed by the fusion of the otic capsules and the basal plate.

basitonic (*Bot.*). Having the base of the anther against the rostellum, e.g., in orchids.

basiventral (*Zool.*). In developing vertebrae, one of a pair of small cartilages occurring lateroventrally on the notochord.

basivertebral (*Zool.*). Situated on, pertaining to, or emerging from, the posterior surface of a vertebra.

basket cells (*Zool.*). Nerve-cells showing extensive arborization of their axis-cylinder processes.

basket coil (*Elec. Eng.*). One with criss-cross layers, so designed to minimize self-capacitance. Also **duolateral coil.**

basket weave (*Textiles*). Woollen or worsted cloths featuring prominent *matt* or *hopsack weave* (q.v.) to show pronounced basket pattern. Long floats of warp or weft, to give the effect, interlace in plain weave order to impart stability.

basket winding (*Elec. Eng.*). See **chain winding.**

basophil(e) (*Zool.*). See **basiphil.**

basophil cells (*Physiol.*). See **basiphil cells.**

basophilia (*Med.*). An increase of basophil cells in the blood. **Punctate basophilia,** a degeneration of red blood cells, e.g., as in lead-poisoning.

bas relief (*Arch.*). Sculpture or carved work in which the figures project less than their true proportions from the surface on which they are carved.

bass (*Agric.*). Inner bark of lime and some other trees, used for mats, hassocks, and baskets, and for tying plants, flowers, etc. (corruption of bast).

bass boost (*Acous.*). Amplifier circuit adjustment which regulates the attenuation of the lowest frequencies in the audio scale, usually to offset the progressive loss towards low frequencies.

bass compensation (*Acous.*). Differential attenuation introduced into a sound-reproducing system when the loudness of the reproduction is reduced below normal, to compensate for the diminishing sensitivity of the ear towards the lowest frequencies reproduced.

bass frequency (*Acous.*). One towards the lower

limit of frequency in an audiofrequency signal or a channel for such, e.g., below 250 Hz.

basswood (*For.*). A North American tree (*Tilia*) that may grow to over 30 m giving a hardwood with straight-grained fine and uniform texture, creamy white to lightish brown in colour. During seasoning process the wood shrinks considerably but finally stabilizes and is suitable for such purposes as pattern making.

bast (*Agric.*). See **bass.**

bastard. A general term for anything abnormal in shape, size, appearance, etc.

bastard ashlar (*Build.*). (1) Stones, intended for ashlar work, which are merely rough-scabbled to the required size at the quarry. (2) The face-stones of a rubble wall selected, squared, and dressed to resemble ashlar.

bastard cop (*Textiles*). A package of mule-spun cotton yarn of size intermediate between a pin cop and a twist cop.

bastard-cut (*Tools*). Describes file teeth of a medium degree of coarseness.

bastard fount (*Typog.*). See **long-bodied type.**

bastard freestone (*Build.*). A local name for a building-stone quarried from the Inferior Oölite of the Bath district; inferior in quality to the overlying Bath Oölite.

bastard-sawn (*For.*). See **back-sawn.**

bastard size (*Paper*). Paper or board not of a standard size. Bastard sizes are listed and sold by measurement and not by standard names.

bastard thread (*Eng.*). A screw-thread which does not conform to any recognized standard dimensions.

bastard title (*Typog.*). See **half-title.**

bastard tuck pointing (*Build.*). Pointing in which a slight projection is given to the stopping on each joint.

bastard wing (*Zool.*). In Birds, quill feathers, usually three in number, borne on the thumb or first digit of the wing. Also **ala spuria, alula.**

bast cylinder (*Bot.*). See **stereome cylinder.**

bast fibre (*Textiles*). Fibre contained in the inner bark of various plants (jute, etc.). Extracted by *retting* (rotting) and *decorticating* (beating).

Bastian lamp (*Light*). An old form of mercury-vapour lamp in which an electromagnetic tilting device is used to start the discharge.

Bastian meter (*Elec. Eng.*). An electrolytic supply meter, in which the ampere-hours which have passed through a circuit are measured by the amount of liquid electrolysed. Now almost obsolete.

bast island (*Bot.*). See **phloem island.**

bastite (*Min.*). A variety of serpentine, essentially hydrated silicate of magnesium, resulting from the alteration of orthorhombic pyroxenes. It occurs in the serpentine of Baste in the Harz mountains, and in the Cornish and other serpentines. Also known as **schillerspar.**

bat (*Build.*). A portion of a brick, large enough to be used in constructing a wall. See **closer.**

batch (*Glass*). The mixture of raw materials from which glass is produced in the furnace. A proportion of cullet is either added to the mixture, or placed in the furnace previous to the charge. Also called **charge.**

batch box (*Civ. Eng.*). See **gauge box.**

batch distillation (*Chem. Eng.*). See under **continuous distillation.**

batch furnace (*Met.*). A furnace in which the charge is placed and heated to the requisite temperature. The furnace may be maintained at the operating temperature, or heated and cooled with the charge. Distinguished from *continuous furnace*.

batching (*Textiles*). (1) Adding oil or oil/water emulsions to cotton, wool, jute, etc., fibres to minimize dust, static and facilitate drafting and spinning. (2) Folding and laying cloth after inspection. (3) Winding cloth on to roll after finishing or dyeing.

batch mill (*Met., etc.*). Cylindrical grinding mill into which a quantity of material for precise grinding treatment is charged and worked till finished.

batch process (*Eng.*). Any process or manufacture in which operations are completely carried out on specific quantities or a limited number of articles, as contrasted to continuous or mass-production.

bate (*Leather*). A fermenting solution containing enzymes derived from the pancreas, or from synthetically prepared ferments; used for steeping light skins.

batement light (*Arch.*). A window, or one division of a window, having vertical sides, but with the sill not horizontal, as where it follows the rake of a staircase.

Batesian mimicry (*Zool.*). Convergent resemblance between two animals, advantageous in some way to one of them.

bath lubrication (*Eng.*). A method of lubrication in which the part to be lubricated, such as a chain or gear-wheel, dips into an oil-bath.

batho-, bathy-. Prefix from Gk. *bathys*, deep, used esp. with relation to sea-depths.

bathochrome (*Chem.*). A radical which shifts the absorption spectrum of a compound towards the red end of the spectrum.

bathoflare (*Chem.*). A radical which shifts the fluorescence of a compound towards the red end of the spectrum.

batholith (*Geol.*). See **bathylith.**

bathometer or **bathymeter** (*Ocean.*). An instrument used for deep-sea soundings; e.g., the *echo sounder*.

Bathonian Series (*Geol.*). The upper part of the *Lower Oölite* series, often known as the *Great Oölite*. It comprises the Great (or Bath) Oölite below, followed by the Forest Marble, Bradford Clay, and Cornbrash, in ascending stratigraphical order. Also known as **Bradfordian.**

bathophilous (*Zool.*). Adapted to an aquatic life at great depths.

bathotonic (*Chem.*). Tending to diminish surface tension.

B.A. thread (*Eng.*). See **British Association screw-thread.**

Bath stone (*Build.*). A building-stone quarried from the Great Oölite near Bath. Also called **Bath oölite.**

bathyal zone (*Ocean.*). Zone on the continental slope between 200 and 2000 m below sea level; characterized by muddy deposits and occasionally by organic oozes.

bathybic (*Biol.*). Relating to, or existing in, the deep sea, e.g., plankton floating well below the surface.

bathylimnetic (*Zool.*). Living in the depths of lakes and marshes.

bathylith or **batholith** (*Geol.*). A large body of intrusive igneous rock, frequently granite, with no visible floor.

bathymeter (*Ocean*). See **bathometer.**

bathymetric (*Zool.*). Pertaining to the vertical distribution of animals in space.

bathypelagic (*Zool.*). See **abyssopelagic.**

bathyscaphe (*Ocean.*). A free diving vessel for exploring the ocean depths, consisting of a spherical cabin topped by a buoyancy tank filled with petrol and ballast compartments filled with iron shot. Also **bathyscaph.**

bathysmal (*Zool.*). See **abyssal.**

bathysphere (*Ocean.*). A spherical diving-apparatus, made large enough to contain two men and instruments; capable of resisting tremendous pressure, and therefore of descending to great depths; used in oceanography for the investigation of deep-water fauna.

bating (*Leather*). The steeping of light skins in a fermenting solution, prior to tanning, to render them smooth and flexible.

batrachian (*Zool.*). Relating to the *Salientia* (i.e., Frogs and Toads).

batten (*Carp.*). (1) A piece of square-sawn converted timber, 2–4 in. (50–100 mm) thickness and 5–8 in. (125–200 mm) width, used for flooring or as a support for laths. See also **slating and tiling battens.** (2) A bar fastened across a door, or anything composed of parallel boards, to secure them and to add strength and/or reduce warping. (*Light*) A fixed or hanging row of lamps used in stage lighting. (*Textiles*) The reciprocating wooden, steel or alloy frame carrying the shuttle-boxes, race-board, reed and weft-stop motion on a loom. In its backward position the weft in the shuttle is propelled across the loom and the pick is beaten-up into the cloth as the batten returns to front centre.

battenboard (*Join.*). See **coreboard.**

batten door (*Carp.*). A door formed of battens placed side by side and secured by others fastened across them.

battened wall (*Arch.*). See **strapping.**

batten-lampholder (*Elec. Eng.*). See **backplate lampholder.**

batter (*Build.*). Slope (e.g., of the face of a structure) upwards and backwards. (*Typog.*) Broken or damaged type.

batter level (*Surv.*). A form of clinometer for finding the slope of cuttings and embankments.

batter pile (*Civ. Eng.*). A pile which is driven in at an angle to the vertical.

batter post (*Carp.*). One of the inclined side-timbers supporting the roof of a tunnel.

battery (*Elec., Eng., etc.*). General term for a number of objects cooperating together, e.g., a number of accumulator cells, dry cells, capacitors, radars, boilers, etc.

battery booster (*Elec. Eng.*). A motor-generator set used for giving an extra voltage, to enable a battery to be charged from a circuit of a voltage equal to the normal voltage of the battery.

battery-coil ignition (*Autos, etc.*). High-tension supply for sparking-plugs in automobiles, in which the interruption of a primary current from a battery induces a high secondary e.m.f. in another winding on the same magnetic circuit, the high potential being distributed in synchronism with the contact-breaker in the primary circuit and the engine.

battery coupling (*Radio*). Interstage coupling in amplifiers required to transmit very low frequencies or direct currents; the input of one stage is connected to the output of the preceding stage by a battery of the requisite voltage to the grid at its correct operating potential.

battery cut-out (*Elec. Eng.*). An automatic switch for disconnecting a battery during its charge, if the voltage of the charging circuit falls below that of the battery.

battery eliminator (*Radio*). An arrangement for supplying electrical power from supply-mains to a radio receiver designed to operate with batteries.

battery-lamp (*Light*). An electric filament-lamp, usually of 16 volts or under, for use in conjunction with a battery.

battery of tests (*Psychol.*). A set of tests covering various factors which are all relevant to some end purpose, e.g., job selection, and in terms of all of which assessments are made.

battery regulating switch (*Elec. Eng.*). A switch to regulate the number of cells connected in series in a battery.

battery spear (*Elec. Eng.*). A special form of spike used to connect a voltmeter to the plates of the accumulator cells for battery-testing under load. The voltmeter incorporates a low resistance in shunt which simulates a heavy load on the battery, thus testing its work capability. The heavy current passed for this purpose necessitates special heavy duty battery connectors.

battery traction (*Elec. Eng.*). An electric-traction system in which the current is obtained from batteries on the vehicles (*battery vehicles*).

battledore (*Glass*). Implement for shaping the foot of a wine glass.

batyl alcohol (*Chem.*). Colourless, crystalline substance; m.p. 70°C, having the structure
$CH_3{\cdot}(CH_2)_{17}{\cdot}O.CH_2CHOH{\cdot}CH_2OH$.
Found in shark and fish liver oils.

baud (*Teleg.*). Unit of speed of telegraphic code transmission; equal to twice the number of dots evenly sent per second. See **bit.**

Baudot code (*Telecomm.*). One comprising impulses in the time frame of 5 units, devised by J. M. Baudot for mechanical transmission of signals. Used in teleprinters, computers, process controllers. (*Teleg.*) See **five-unit code.**

Baudouin reaction (*Chem.*). A test for certain vegetable oils which give with alcoholic furfural and concentrated HCl, or with $SnCl_2$ and HCl, a characteristic red colour.

baulk (*Civ. Eng.*). See **balk.** (*For.*) A piece of timber square-sawn from the log to a size greater than 6 in. by 6 in. (150 by 150 mm).

Baumann print (*Met.*). Visual method of testing concentration of sulphur on a smooth metal surface. Photographic bromide paper is pressed against it after damping with dilute acid, and silver sulphide stains show, where H_2S has been given off. Also called **sulphur print.**

Baumé hydrometer scale (*Phys.*). The continental Baumé hydrometer has the rational scale proposed by Lunge, in which 0° is the point to which it sinks in water and 10° the point to which it sinks in a 10% solution of sodium chloride, both liquids being at 12·5°C.

Baumé scale of relative density (*Phys.*). For liquids denser than water
$$°Bé = 144{\cdot}3(1-v),$$
less dense than water
$$°Bé = 144{\cdot}3(v-1);$$
where v is the relative specific volume of the liquid at 15°C with respect to water at 15°C.

Baum jig (*Mining*). Pneumatically pulsed *jig* (q.v.) used in coal-washing plants to lift and remove a lighter and low-ash fraction from a denser one containing shale and high-ash material (dirt), which is stratified downward by the effect of pulsed water, and separately withdrawn.

bauxite (*Geol.*). A residual clay, consisting essentially of aluminium hydroxides, formed in tropical regions by the chemical weathering of basic igneous rocks. It is the most important ore of aluminium. See also **laterite.**

bavenite (*Min.*). A hydrated beryllium calcium aluminium silicate, occurring as white fibrous monoclinic crystals in pegmatitic segregations in the granite of Baveno, N. Italy.

bay (*Arch.*). Any division or compartment of an arcade, roof, building, etc.; space from column to column in a building. (*Telecomm.*) (1) Unit of racks designed to accommodate numbers of standard-sized panels, e.g., repeaters or logical units. (2) Unit of horizontally extended antenna, e.g., between masts.

Bayard and Alpert gauge (*Phys.*). One for measuring very low gas pressure by collecting ions on a fine wire inside a helical grid.

Bayer (*Pharm.*). Code prefix for numerous pharmaceutical substances; e.g. Bayer-205 (suramin), Bayer-606 (Salvarsan), Bayer-5360 (metronidazole), etc.

bayerite (*Chem.*). Native aluminium hydroxide, a dimorph of *gibbsite* (q.v.).

Bayer process (*Chem.*). A process for the purification of bauxite, as the first stage in the production of aluminium. Bauxite is digested with a sodium hydroxide solution which dissolves the alumina and precipitates oxides of iron, silicon, titanium, etc. The solution is filtered and the aluminium precipitated as the hydroxide.

Bayes' theorem (*Maths.*). Let $P(c_i)$ be the probability of c_i occurring, where c_i is one of n mutually exclusive events $c_1, c_2 \ldots c_n$, of which one must occur. If E is an observable event, let $P(E|c_i)$ be the probability that E will occur assuming c_i to have occurred, and $P(c_i|E)$ be the probability of c_i occurring assuming E to have occurred.

Then $$P(c_i|E) = \frac{P(c_i) \, . \, P(E|c_i)}{\sum_{j=1}^{n} P(c_j)P(E|c_j)}$$

The theorem is difficult to apply as the $P(c_i)$ are generally not known. [$P(c_i)$ are called the *a priori probabilities* and $P(c_i|E)$ the *a posterior probabilities*.]

bayon (*Geog.*). An ox-bow lake produced when a meander of a river becomes isolated from the main stream. Specifically, those of the Mississippi delta.

bayonet cap (*Elec. Eng.*). (Common abbrev. **B.C.**). A type of cap fitted to an electric lamp, consisting of a cylindrical outer wall fitted with two or three pins for engaging in slots in a lampholder (*bayonet holder*). Within the wall are two contacts connected to the filament, which make contact with two pins in the lampholder. See **centre- contact cap, small bayonet cap.**

bayonet fitting (*Eng.*). A cylindrical fastening in which a plug is pushed and twisted into a socket, against spring pressure. Two or more pins in the plug engage in corresponding L-shaped slots in the socket.

bayonet holder (*Elec. Eng.*). See **bayonet cap.**

bay-stall (*Build.*). See **carol.**

baywood (*For.*). See **Honduras mahogany.**

Bazin's disease (*Med.*). A condition in which indurated ulcers, thought to be tuberculous, appear on the skin of the calves.

bazooka (*Telecomm.*). See **balun.**

B battery (*Elec. Eng.*). U.S. term for **high-tension battery.**

B.C. (*Light*). See **bayonet cap.**

BCD (*Comp.*). See **binary-coded-decimal.**

BCF (*Chem.*). See **bromochloro difluoromethane.**

BCG (*Med.*). Bacillus-Calmette-Guérin, a prophylactic anti-tuberculosis vaccine, developed from a bovine tubercle bacillus by Calmette and Guérin in France in 1909 and first used there in 1921.

B-class insulation (*Elec. Eng.*). A class of insulating material to which is assigned a temperature of 130°C. See **class-A, -B, -C, etc., insulating materials.**

B. converter (*Met.*). See **Bessemer converter.**

Bdelloidea (*Zool.*). An order of *Rotifera*, in which the adults swim freely and possess a telescopic forked tail; they can also move by creeping like a leech, using the dorsal proboscis as a sucker.

B display (*Radar*). Rectangular radar display with target bearing indicated by horizontal coordinate and target distance by the vertical coordinate, the targets appearing as bright spots.

BDV (*Elec. Eng.*). Abbrev. for *breakdown voltage*.

Be (*Chem.*). The symbol for *beryllium*.

beaching gear (*Aero.*). Floatable, detachable, temporary trolleys which enable a seaplane to be run on and off the shore.

beacon (*Aero.*). *Radio beacons* can be of any frequency but are usually VHF, and they can be omni-directional or of directional beam type. *Vertical fan marker* radio beam type beacons are used to identify particular spots in control zones and on approach patterns. A *non-directional beacon* (abbrev. NDB) is a transmitter, the bearing of which can only be determined by an aircraft equipped for direction finding. See **instrument landing system.** (*Hyd. Eng.*) A small tower, generally built on a submerged reef, to warn shipping of minor shoals.

bead (*Join.*). A small convex moulding formed on wood or other material.

bead-and-batten work (*Carp.*). A rough style of work used for partitions, formed of battens with a bead along one of the longitudinal edges.

bead-and-quirk (*Join.*). A bead formed with a narrow groove separating it from the surface which it is decorating. Also called a **quirk-bead.**

bead-butt (*Join.*). Work framed in panels flush with the framing of a door, a bead being used on two sides only of the panel, and being carried up to or butted against the rails.

bead-butt and square-work (*Join.*). Door framing formed with bead-and-butt on one side and square on the other.

bead-flush (*Join.*). Work differing from bead-butt work in that a bead is formed on the framing itself instead of on the panel, and is carried round all the sides.

bead-jointed (*Join.*). Said of that form of jointing in which one of the butting edges has a bead.

bead-plane (*Carp.*). A special plane for cutting beads out of the solid, or for cutting grooves into which separate beading is to be fitted.

bead router (*Carp.*). A stock or body through which passes a beading gouge; for bead-cutting.

bead saw (*Tools*). A light backsaw 4–12 in. (100–300 mm) long with fine teeth, for small work.

bead-tool (*Tools*). A specially shaped cutting-tool used in wood-turning for forming convex mouldings.

bead yarn (*Textiles*). Novelty yarn having bits of pigment or other compositions secured to it at intervals. Beads can also be produced mechanically by over-feeding small portions of hard twisted, coloured yarn and effect thread. See also **knop yarn.**

beak (*Carp.*). The crooked end of a bench hold-fast. (*Zool.*) See **rostrum.**

beaked (*Bot.*). Bearing a beak, i.e., a long, pointed prolongation.

beaking joint (*Carp.*). The joint formed when a number of adjacent heading joints occur in the same straight line.

beak iron, beck iron, bick iron, bickern (*Eng.*). (1) The pointed, or horn-shaped, end of a blacksmith's anvil; used in forging rings, bends, etc. (2) A T-shaped stake, similarly shaped, fitting in the hardy-hole of the anvil.

beam (*Eng.*). (1) A bar which is loaded transversely. (2) Rolled or extruded sections of certain profiles, e.g., I-beam. (*Phys.*) A collimated, or approximately unidirectional, flow of electromagnetic radiation (radio, light, X-rays), or of particles (atoms, electrons, molecules). The angular beam width is defined by the half-intensity points. (*Weaving, etc.*) (1) Wooden roller, steel or alloy cylinder having large flanges at each end. Filled with densely wound thread, it is placed behind the loom as the warp. (2) The beam on which the sheet of threads is wound in beam-warping. (3) One of the hollow metal cylinders carrying the warp threads in lace manufacture.

beam alignment (*TV*). That of the electron beam perpendicular to the mosaic in a TV camera tube.

beam angle (*Electronics*). (1) Angle of response curve of an antenna in which the bulk of energy is transmitted or received. (2) The solid angle subtended by the cone of electrons emerging from the first focus of the beam in an electron gun.

beam antenna (*Electronics*). One with very marked directional properties. Originally a curtain of vertical wires with a similar reflector.

beam balance (*Chem.*). Balance in which the weight of the sample contributes to the balance of moments of a beam about a central fulcrum.

beam bender (*Electronics*). See **ion trap.**

beam compasses (*Instr.*). An instrument for describing large arcs. It consists of a beam of wood or metal carrying two beam heads, adjustable for position along the beam, and serving as the marking points of the compasses. Also called **trammels.**

beam convergence (*TV*). Focusing of the 3 electron beams of a colortron-type triple gun colour TV tube, so that they converge at the shadow-mask aperture.

beam coupling (*Electronics*). That provided between circuits by an electron beam passing between electrodes, as in a beam tetrode.

beam-coupling coefficient (*Elec. Eng.*). The ratio of the a.c. signal current produced to the d.c. beam current in beam coupling.

beam current (*Electronics*). That portion of the gun current in CRT which passes through the aperture in the anode and impinges on the fluorescent screen.

beam-deflection valve (*Electronics*). One in which an electron beam is deflected by side plates and so switched between push-pull anodes.

beam edge (*Light*). The position in a light beam from a searchlight, etc., where the luminous intensity is 10% of that at the centre of the beam.

beam effect (*Acous.*). The differential focusing of high-frequency sound radiation from an open diaphragm (such as that in a loudspeaker), on account of the dimensions of the diaphragm being comparable with or greater than the wavelength of the sound produced.

beam-engine (*Eng.*). A form of construction used in early steam-engines, now obsolete. The vertical steam-cylinder acted at one end of a pivoted beam, the work load being connected to the other.

beam factor (*Light*). Used to define the performance of a searchlight beam. It is equal to the solid angle subtended at source by the region over which the light intensity has dropped to one half of its value at the centre of the beam.

beam-filling (*Build.*). Brick, masonry, or concrete work used to fill in the spaces between the ends of beams or joists carried upon a wall.

beam flux (*Light*). The total light flux in the beam from a projector such as a searchlight.

beam-forming electrode (*Electronics*). Electrode to which a potential is applied to concentrate the electron stream into one or more beams. Used in beam tetrodes and cathode-ray tubes.

beam hole (*Nuc. Eng.*). Hole in shield of reactor, or that around a cyclotron, for extracting a beam of neutrons or γ-rays.

beaming (*Textiles*). Preparation of warp threads for weaving by spreading them, in the desired order for entering, over the cane or warp roller. Also called **turning-on.**

beam jitter (*Electronics*). (1) Random movements of radar beam due to imperfections or wear in mechanical drive of antenna. (2) Random movements of electron beam in CRT due to electronic noise.

beam-power valve (*Electronics*). One in which the control grid and screen are aligned so that a bunching of electrons is equivalent to a suppressor grid without the power loss of the latter. Also called **beam pentode.**

beam relay (*Elec. Eng.*). An electromagnetic relay in which the contacts are mounted on a balanced beam, with energizing coils acting on each end and tending to tilt it one way or the other.

beam rider (or riding) (*Radio*). System in which a guided missile maintains and returns to a course of maximum signal on a radio beam. See **guided missile.**

beam-splitter (*Photog.*). Camera with an arrangement of prisms for taking three negatives of the same object simultaneously. Used, with appropriate filters, for colour separation negatives.

beam suppression (*TV*). Application of a large negative potential to control electrode of a cathode-ray tube, so as to suppress the beam during flyback period between successive scanning lines.

beam switching (*Radar*). **See lobe switching.**

beam system (*Radio*). A point-to-point radio system in which highly directive transmitting and receiving antennae are used.

beam tetrode (*Electronics*). Tetrode having an additional pair of plates, normally connected internally to the cathode, so designed as to concentrate the electron beam between the screen grid and anode, and thus reduce secondary emission effects. See also **beam-power valve.**

beam transadmittance (*Electronics*). See **forward-.**

beam trap (*Electronics*). Bucket-formed electrode mounted in a cathode-ray tube, to catch the electron beam when it is not required to excite fluorescence in the screen.

beam-warping (*Textiles*). Winding warp yarns from cones or cheeses in magazine creels on to a warper's or weaver's beam.

bean (*Agric., Bot.*). (1) Leguminous plant bearing smooth kidney-shaped seed in long pod. (2) The edible seeds themselves, and similar seeds of nonrelated plants, e.g., coffee, cacao.

beard (*Acous.*). A short wooden rod placed across the aperture where an organ pipe is actuated by the blast of air. It modifies the timbre of the sound emitted by the pipe and permits a higher wind-pressure to be used (with consequent increase in sound output), without exciting the octave of the pipe at its fundamental pitch. (*Carp.*) The sharp arris of a square-edged timber. (*Typog.*) The space between the base of a letter and the bottom edge of the type-body.

bearded (*Bot.*). Having an awn; bearing long hairs like a beard.

bearded needle (*Textiles*). See **spring needle.**

Beard-Hunter protective system (*Elec. Eng.*). See **compensated pilot-wire protective system.**

Beard protective system (*Elec. Eng.*). A form of balanced protective system in which the current entering the winding of an alternator is balanced against that leaving it by passing the conductor at the two ends round the core of a single current transformer, in opposite directions, so that there is normally no flux in the transformer core.

bearer cable (*Elec. Eng.*). See **messenger wire.**

bearer ring (*Print.*). See **cylinder bearers.**

bearing (*Build.*). The part of a beam or girder which actually rests on the supports. (*Radio*) Angle of direction in horizontal plane in degrees from true north, e.g., of an arriving radio wave as determined by a direction-finding system. (*Surv.*) The horizontal angle between any survey line and a given reference direction.

bearing classification (*Radio*). Classes considered by operator to be within:

Class A, $\pm 2°$; B, $\pm 5°$; C, $\pm 10°$.

bearing current (*Elec. Eng.*). A stray current, induced by magnetic flux linking the shaft of an electrical machine, that flows between the shaft and bearings and may injure the bearing surfaces.

bearing distance (*Build.*). The unsupported length of a beam between its bearings.

bearing metal (*Met.*). See **white metal.**

bearing metals (*Eng.*). Metals (alloys) used for that part of a bearing which is in contact with the *journal* (q.v.); e.g., bronze or white metal, used on account of their low coefficient of friction when used with a steel shaft.

bearing pile (*Civ. Eng.*). A column which is sunk or driven into the ground to support a vertical load by transmitting it to a firm foundation lower down, or by consolidating the soil so that its bearing power is increased. Generally made of reinforced timber or prestressed concrete.

bearings (*Eng.*). Supports provided to hold a revolving shaft in its correct position.

bearing surface (*Eng.*). That portion of a bearing in direct contact with the journal; the surface of the journal. See **brasses.**

bearing-up stops (*Mining*). Keps or catches used to support a cage at the end of downward travel in mine shaft.

bearing wall (*Civ. Eng.*). The supporting or abutment wall of a bridge or arch.

beat (*Horol.*). The blow given by a tooth of the escape wheel as it strikes the pallets. An escapement is said to be *in beat* when this blow is uniform on both pallets. (*Telecomm.*) Periodic variation in the amplitude of a summation wave containing 2 sinusoidal components of nearly equal frequencies.

beater (*Paper*). A vat containing a heavy cylindrical roll (*beater roll*), fitted with bars, which rotates against a fixed set of bars (*bedplate*). The paper fibres in suspension in water pass between these bars in preparation for sheet making. (*Spinning*) High-speed revolving shaft having two or three arms equipped with blades or steel pins. These beat out the heavy impurities in matted raw cotton in opening and scutching processes.

beater mill (*Min. Proc.*). Hammer mill, disintegrating mill, impactor. In rockbreaking, mill with swinging hammers, disks, or heavy plates, which revolve fast and hit a falling stream of ore with breaking force.

beat frequency (*Radio*). Generally, the difference frequency produced by the intermodulation of two frequencies. Specifically, the supersonic (intermediate) frequency in a supersonic heterodyne receiver.

beat-frequency oscillator (*Radio*). Same as **heterodyne oscillator.** Abbrev. **B.F.O.**

beat-frequency wavemeter (*Radio*). Same as **heterodyne wavemeter.**

beating (*Textiles*). A supply of short threads, placed on a loom, from which warp threads broken during weaving may be replaced. Known as *thrums* in cotton weaving.

beating engine (*Paper*). See **hollander.**

beating oscillator (*Radio*). See **local oscillator.**

beating-up (*Textiles*). Forward movement of the loom sley and reed in which each pick of weft placed in the shed is pushed against the edge of the woven fabric.

beat pins (*Horol.*). The pins projecting from the ends of the gravity arms of the gravity escapement. These pins, one on either side of the pendulum rod, give impulse to the pendulum and enable the pendulum to raise the gravity arms for unlocking.

beat (or beatnote) receiver (*Radio*). See **supersonic heterodyne receiver.**

beat reception (*Radio*). Same as **heterodyne reception.**

Beatrice twill (*Textiles*). A 5-shaft twill weave fabric, usually made 58 in. (1470 mm) wide, from cotton warp and twist-spun alpaca weft, piece-dyed and used for linings, coatings and dress-goods.

beats (*Acous.*). The subjective difference tone when two sound waves of nearly equal frequencies are simultaneously applied to one ear. It appears as a regular increase and decrease of the combined intensity.

beat screws (*Horol.*). Screws which provide for the adjustment of the relative position of the crutch and pendulum, so that the escapement may be brought *in beat.*

Beattie-Bridgeman equation of state (*Phys.*). A semi-empirical equation of state for the compressibility of gases. It is most conveniently stated in the virial form:

$$\frac{pV}{n} = RT + \frac{n\beta}{V} + \frac{n^2\gamma}{V^2} + \frac{n^3\delta}{V^3},$$

where β, γ, and δ are virial coefficients.

Beaufort notation (*Meteor.*). A code of letters used for indicating the state of the weather; for example, *b* stands for *blue sky, o* for *overcast, r* for *rain.*

Beaufort scale (*Meteor.*). A numerical scale of wind force, ranging from 0 for winds less than 1 knot to 17 for winds within the limits 110 to 118 knots. Where V is the mean wind speed in miles/hour, and B is the Beaufort wind force, then $V = 1{\cdot}87\sqrt{B^3} = (1{\cdot}52\ \mathrm{B})^{3/2}$.

beaumontage (*Build., Eng., etc.*). A mixture used as a stopping for holes or other defects in woodwork or metal work.

beaver board (*Build.*). A building-board made of wood-fibre material.

beaver cloth (*Textiles*). Heavy overcoating made from two warps and double-wefted woollen yarns. Specially milled, it has a nap or dressface finish. Cheap imitations are constructed from a cotton warp and low woollen weft with a 4-end weft sateen weave.

Beaver respirator (*Med.*). A small respirator used to assist respiration via a tracheotomy tube.

beavertail antenna (*Radio*). One producing a broad, flat, radar beam.

beavertail beam (*Radio*). (1) A broad, flat, radar beam due to imperfections or wear in mechanical drive of antenna. (2) Similar beam produced by a *beavertail antenna.*

beaverteen (*Textiles*). A heavy cotton fabric with a weft face; it has a coarse weft and is used chiefly for trouserings.

bec carcel (*Light*). A French standard of light intensity, consisting of an Argand colza lamp giving a light equal to 9·6 international candles. Obsolete.

beche (*Civ. Eng.*). A tool used in well-boring when a rod has broken in the bore, the tool being used to grab and remove it.

beck. See **back.**

Beck arc-lamp (*Light*). A special form of high-intensity arc-lamp, in which particularly high-current densities can be used.

Becke line (*Light*). A narrow line of light seen under the microscope at the junction of two minerals (or a mineral and the mount) in contact in a microscope section.

Beck hydrometer (*Phys.*). Hydrometer for measuring the relative density of liquids less dense than water. Graduated in degrees Beck, where °Beck = 200 (1 − rel. d.).

beck iron (*Eng.*). See **beak iron.**

Beckmann apparatus (*Heat*). Apparatus originally designed by Beckmann to be used for measuring the freezing and boiling points of solutions (e.g., in the *cryoscopic method*, q.v.).

Beckmann molecular transformation (*Chem.*). The transformation and rearrangement of ketoxime molecules into acid amides or anilides under the influence of certain reagents, such as acetyl chloride, hydrochloric or sulphuric acid dissolved in glacial acetic acid, benzenesulphonic chloride, etc. An important reaction for determining the configuration of stereo-isomeric *ketoximes* (q.v.).

Beckmann thermometer (*Heat*). A special form of mercury thermometer possessing, because of its larger bulb, great sensitivity but a small range. It is used to measure small changes of temperature with great precision. It contains a mercury reservoir by means of which mercury may be added to or removed from the indicating thread in order that the thermometer may be used at different temperatures.

Becquerel cell. See **photochemical cell.**

Becquerel effect (*Electronics*). Flow of current between two similar metallic electrodes immersed in an electrolyte, which is produced when one of the electrodes is illuminated. (*Photog.*) Strengthening of print-out images on certain types of emulsions, e.g., silver chloride, by exposing to uniform red or yellow light.

becquerelite (*Min.*). Hydrous calcium uranate, found in the uranium deposits of the Congo.

Becquerel rays (*Nuc.*). The penetrating *alpha* (α), *beta* (β) and *gamma* (γ) radiations from uranium, discovered by Becquerel.

bed (*Build.*). The upper or lower surface of a building-stone or ashlar when it is built into a wall; the horizontal surface upon which a course of bricks is laid in mortar. (*Chem.*) A packed, porous mass of solid reagent, adsorbent or catalyst through which a fluid is passed for the purpose of chemical reaction. (*For.*) In felling, the flat, more or less horizontal, surface of the undercut. (*Geol.*) The smallest depositional unit in sedimentary rocks, normally of much greater lateral extent than thickness, bounded above and below by planes of easy parting representing changes or interruptions in the supply of sediment. (*Typog.*) The heavy steel table of a machine or press on which the forme of type or plates is placed for printing. See also **flat bed.**

bedding (*Cables*). See **serving.** (*Geol.*) A term commonly used for *stratification*.

bedding course (*Civ. Eng.*). See **cushion course.**

bedding fault (*Acous.*). A fault in pressing a disk record; caused by material (stock) being forced under the stamper, with consequent irregular pressings.

bedding-in (*Eng.*). The process of accurately fitting a bearing to its shaft by scraping the former until contact occurs uniformly over the surface. (*Foundry*) The preparation of the lower half of a mould in the floor-sand of the foundry instead of in a box.

bedding-stone (*Build.*). The perfectly flat marble slab used by the bricklayer to test the face of a rubbed brick for flatness.

bed dowel (*Build.*). A dowel placed in the centre of a stone bed.

bedeguar (*Bot.*). A soft spongy gall found on roses; also called sweet-brier sponge.

Bedford cord (*Textiles*). A cotton, wool, spun rayon, nylon or blended fabric with ribs or cords running lengthwise in the piece; lighter weights used for summer dresses, shirts, etc., heavier weaves for riding-breeches. The cords may be made more pronounced by adding coarser ends in the warp at intervals.

Bedford limestone (*Build.*). A well-known American building-stone, of the finest quality; named from its shipping point, Bedford (Indiana).

bed joints (*Build.*). The horizontal joints in brickwork or masonry: the radiating joints of an arch.

bed-moulding (*Build.*). Any moulding used to fill up the bare space beneath a projecting cornice.

bedplate (*Eng.*). A cast-iron or fabricated steel base, to which the frame of an engine or other machine is attached. (*Paper*) A plate into which metal bars are inserted; situated beneath the roll in a Hollander beater.

bedrock (*Mining*). Barren formation (seat earth, clay, 'farewell rock') underlying the exploitable part of a mining deposit.

beech (*For.*). A tree (*Fagus*) yielding a hardwood with straight grain and uniform texture. Its colour ranges from whitish to a light reddish-brown. It bends well and is an excellent wood for turnery purposes. Used for pulley-blocks, tool handles, athletic goods, gymnasium equipment and cabinet making.

beef (*Geol.*). Fibrous calcite occurring in veins in sedimentary rocks. Rarely, other minerals with the same structure and occurrence.

beekite (*Min.*). A chemically precipitated form of silica, commonly occurring as an incrustation on pebbles, notably of limestone, in the breccio-conglomerates of E. Devonshire.

beer (*Brew.*). An alcoholic beverage made by the fermentation of a cereal extract, flavoured with a suitable bitter substance. The materials used are water, malt, hops, sugar, and yeast, with certain accessories, such as finings. The proportion of alcohol present in beers ranges from $1\frac{1}{2}$–$8\frac{1}{2}$ g/cm³. See **ale, finings, hops, malt, porter, stout, wort, yeast.** (*Textiles*) A unit used for warp threads. In *linen manufacture* the unit is 40 threads. In *cotton weaving* the number varies, but it is usually a group of 20 dents. In *worsted manufacture* the term indicates a group of 40 threads, or a group of 20 dents, each dent carrying 2 threads. Cf. *porter*.

beerbachite (*Geol.*). A hornfels produced by contact metamorphism of basic igneous rocks, containing olivine.

Beer's law (*Chem.*). The degree of absorption of light varies exponentially with the thickness of the layer of absorbing medium and the molar concentration in the latter.

beer stone (*Build.*). A non-oölitic limestone resembling a hard chalk; used for interior work.

beeswax (*Chem.*). A white or yellowish plastic

substance obtained from honeycomb of the bee, m.p. 63°–65°C. It consists of the myricyl (melissyl) ester of palmitic acid $C_{15}H_{31}\cdot COOC_{30}H_{61}$, free cerotic acid $C_{25}H_{51}COOH$, and other homologues. Used in polishes, modelling, ointments, etc.

beetle (*Textiles*). A machine consisting of a row of wooden hammers, which fall on a roll of damp cloth as it revolves. The operation closes the spaces between the warp and the weft yarns, and imparts a soft glossy finish or a watermark. (*Tools*) A heavy mallet, or wooden hammer, used for driving wedges, consolidating earth, etc. Also called a **mall** or **maul**. (*Zool.*) A member of the insectan order *Coleoptera*.

beetle-head (*Civ. Eng.*). See **monkey.**

beetle-stones (*Geol.*). Coprolitic nodules akin to septaria which, when broken open, give a fancied resemblance to a fossil beetle.

beet sugar (*Chem.*). Sucrose derived from sugar beet. Identical with sucrose derived from sugar cane.

bega- (*Elec. Eng.*). A prefix (obsolete) used to denote 10^9 times. See **giga-**.

Beggiatoales (*Bacteriol.*). Chemosynthetic sulphur-oxidizing bacteria, some of which resemble filamentous algae, and which may in fact be more closely related to them than to the true bacteria. Sulphur granules occur intracellularly.

behaviour (*An. Behav.*). The entire complex of observable, recordable or measurable activities of a living animal, such as muscular contraction causing movement or locomotion, movements of cell organelles (e.g., cilia), changes of colour, glandular secretions, etc., as a result of its relationships with various features of its environment, in other words, anything an animal does.

behavioural trait (*An. Behav.*). Some aspect of an animal's behaviour which can be reliably measured, and is reasonably stable over a period of the life-span, and can thus be studied as a single phenomenon in relation to, e.g., its heredity.

behaviour chain (*An. Behav.*). A sequence of stimuli and responses observed repeatedly in an animal with only minor variations.

behaviourism (*Psychol.*). A school of thought, founded by J. B. Watson, which bases its doctrine solely on objective observation and experiment, denying the validity of all subjective phenomena, such as sensation, emotion, mind, consciousness, will, imagery, etc.

behaviour pattern (*An. Behav.*). A set of responses associated in time and relatively stereotyped in their temporal sequence.

behaviour therapy (*Psychol.*). The application of *learning theory* (q.v.) to the treatment of behaviour disorders, which are thought to be the result of maladaptive learning, e.g., a phobia is a wrongly-conditioned response and not the symptom of any deeper illness. Therapy takes the form of re-learning by associating aversive stimuli with the undesirable response and pleasant stimuli with the desired behaviour.

behenic acid (*Chem.*). Docosoic acid. $CH_3(CH_2)_{20}COOH$, a constituent of ben oil and found in the roots of *Centaurea behen*. Colourless needles, m.p. 80°C, b.p. 304°C. Insoluble in water.

beidellite (*Min.*). A variety of the montmorillonite group (smectites) of clay minerals.

Beilby layer (*Chem., Min. Proc.*). Flow layer produced by polishing a metal or mineral surface, in which the true lattice structure is modified or destroyed by incipient fusion.

Beilstein test (*Chem.*). Negative test for the presence of a halogen in an organic compound. The latter is heated in an oxidizing flame on a copper wire; if no halogen is present there is no green colour to the flame. Volatile N compounds also give a green colour.

bel (*Maths.*). A nondimensional unit used to express the common logarithm of a number, so that the multiplication of the number by a factor can be accomplished by the addition of the logarithm of the factor to the measure in bels. If N is the measure of a in bels, $N = \log_{10}a$, $rN = \log_{10}r + \log_{10}a$, $r^2N = 2\log_{10}r + \log_{10}a$, etc. Useful wherever constant coefficients are applied to variable quantities, e.g. amplifiers or attenuators in electrical circuits. See **decibel, neper.**

belemnite (*Geol.*). An extinct Cephalopod, similar to an Octopus in appearance. The portion commonly found as a fossil is the 'guard', the shape and often the size of a rifle bullet.

belemnoid (*Zool.*). Dart-shaped.

Belemnoidea (*Zool.*). A group of extinct *Cephalopoda* (order *Dibranchiata*, suborder *Decapoda*) known only from fossils found in Mesozoic rocks.

Belfast truss (*Carp.*). A timber bowstring truss having a double bow rafter and a double tie connected by a lattice of cross-members, with a bituminous felt or corrugated iron roof-covering.

Belflix (*Comp.*). Short simple animated films made by a computer process, which involves magnetic tape and a microfilm recorder.

belfry (*Arch.*). A tower, either detached or forming part of a building, containing suspended bells.

Belgian block pavement (*Civ. Eng.*). A *block pavement* (q.v.) formed of stones about $3 \times 6 \times 10$ in. ($75 \times 150 \times 250$ mm).

Belgian truss (*Eng.*). See **French truss.**

Belitzski's reducer (*Photog.*). A simple reducer of contrast, based on ferric oxalate.

bell (*Met.*). See **cone.**

belladonna (*Chem.*). See **atropine.**

bell-and-spigot joint (*Civ. Eng.*). The American equivalent for *spigot-and-socket joint* (q.v.).

bell centre punch (*Eng.*). A centre punch whose point is automatically located centrally on the end of circular work by a sliding hollow bell-shaped guide.

bell chuck (*Eng.*). See **cup chuck.**

bell-crank lever (*Eng.*). A lever consisting of two arms, generally at right angles, with a common fulcrum at their junction.

bell gable (*Arch.*). A gable built above the roof in a church having no belfry, and pierced to accommodate a bell.

bell hanger's gimlet (*Tools*). Large gimlet up to 1 m in length, usually with shell-type shank.

Bellini's ducts (*Zool.*). In the kidney of Vertebrates, ducts formed by the union of the primary collecting tubules and opening into the base of the ureter at the pelvis of the kidney.

Bellini-Tosi antenna (*Radio*). Directional antenna comprising two crossed loops. The direction of maximum reception is controlled by a *radiogoniometer* which varies the relative couplings of the two loops to the receiver.

Bell-jar exhaust (*Vac. Tech.*). A method of electronic valve manufacture in which the electrode assemblies are processed, placed in their envelopes, and the sealing completed entirely within a continuously evacuated enclosure.

bell metal (*Met.*). High tin bronze, containing up to 30% tin and some zinc and lead. Used in casting bells.

bell-metal ore (*Min.*). See **stannite.**

bell-mouthed (*Eng.*). Said of a hole or bore when its diameter gradually increases towards one or both open ends, the bore profile in section being curved. Usually a manufacturing fault.

bellows (*Eng.*). A flexible, corrugated tubular machine element used for pumping, for transmitting motion, as an expansion joint, etc. (*Photog.*) The flexible connexion between parts of a camera or enlarger, necessarily light-tight, to permit delicate adjustments, usually of focusing.

bell-push (*Elec. Eng.*). A switch in the form of a push-button, for operating an electric bell.

Bell's law (*Zool.*). Motor nerve-fibres are without sensory function, while sensory nerve-fibres are equally unable to cause action in any peripheral part.

Bell's palsy (*Med.*). Paralysis of the muscles of the face, due to an affection of the peripheral part of the seventh cranial nerve.

bell transformer (*Elec. Eng.*). A small transformer used for obtaining, from the public power-mains, low-voltage ringing current for house trembler-bells.

bell-type furnace (*Met.*). A portable inverted furnace or heated cover operated in conjunction with a series of bases upon which the work to be heated can be loaded and then left to cool after heat treatment. Used chiefly for bright-annealing of nonferrous metals and bright-hardening of steels.

belly (*Typog.*). That side of a type-letter which bears the nick; placed uppermost in the setting-stick by the compositor.

belly tank (*Aero.*). See **ventral tank.**

belt (*Eng.*). An endless strip of leather, or plastic, fabric, used to transmit rotary motion from one shaft to another by running over pulleys having flat, crowned or grooved rims. (*Masonry*) A projecting course of stones or bricks. Also called **belt course.**

belt conveyor (*Eng.*). See **band conveyor.**

belt dressing (*Eng.*). Substances used to prolong the life and improve the frictional grip of belting.

belt drive (*Eng.*). The transmission of power from one shaft to another by means of an endless *belt* (q.v.).

belted-type cable (*Elec. Eng.*). In this there are three conductors, each insulated with oil-impregnated paper. The triangular interstices are filled with paper packing, and a belt of paper surrounds the whole. A lead sheath surrounds the complete assembly. Obsolete.

belt fastener (*Eng.*). A connecting piece used to join together the ends of a driving belt.

belt fork or **belt striker** (*Eng.*). Two parallel prongs attached at right angles to a sliding rod, used to slide a flat belt from a fast to a loose pulley, and vice versa.

belting (*Eng.*). A general term descriptive of materials from which driving belts are made, e.g., leather, cotton, balata, woven hair, plastics, etc.

belt leakage (*Elec. Eng.*). A leakage flux occurring in slip-ring induction motors when a primary phase group overlaps two secondary phase groups. The term is sometimes used to include the other forms of leakage, e.g., *zig-zag leakage*, occurring in the air-gap.

Belt's corpuscle or **Beltian body** (*Bot.*). A small, orange-yellow, pear-shaped body, consisting of thin-walled cells filled with protein and oil, formed on leaflets of the bull's-horn thorn, and serving as food for ants.

Belt (or Beltian) Series (*Geol.*). A great thickness (perhaps 12 000 m) of younger Pre-Cambrian rocks occurring in the Little Belt Mts., Montana, Idaho, and British Columbia. Argillaceous strata predominate, accompanied by algal limestones. Correlated with the Grand Canyon Series in Colorado and the Uinta Quartzite Series in the Uinta Mts.

belt slip (*Eng.*). The slipping of a driving belt on the face of a pulley, due to insufficient frictional grip to overcome the resistance to motion.

belt striker (*Eng.*). See **belt fork.**

belt transect (*Bot.*). A rectangular strip of ground marked out more or less permanently, so that its vegetation may be mapped and studied.

belvedere (*Arch.*). A room from which to view scenery; it is built on the top of a house, the sides being either open or glazed.

Bence-Jones protein (*Chem.*). A protein with characteristic properties in respect of its solubility, precipitation and coagulation, excreted in the urine of many multiple myeloma patients. It has been identified as the light chain of the myeloma protein in the serum of the same patients.

bench (*Fuels*). The name applied to a complete plant for the manufacture of coal-gas. Also called **retort bench.** (*Telecomm.*) Fixed rails with adjustable and slidable supports for a waveguide system.

benched foundation (*Build., Civ. Eng.*). A foundation which is stepped at the base to safeguard against sliding on sloping sites.

bench hook (*Carp.*). A flat piece of wood having a wooden block at the back edge of the top and a similar block fixed on the underside along the front edge, used to steady the work and prevent injury to the bench top.

benching (*Civ. Eng.*). Concrete sloped up from the concrete-bed foundation on which a pipe-line rests; it slopes up to the sides of the pipe, and gives support along the whole of its length, and to some extent laterally.

benching iron (*Surv.*). A small steel plate sometimes used to provide a solid support for the staff at a change point. It is formed usually of a triangular plate, with the corners turned down so that they may be driven into the ground surface to fix the plate in position, while the staff rests upon a raised central portion.

bench mark (*Surv.*). A fixed point of reference for use in levelling, the reduced level of the point with respect to some assumed datum being known. See **Ordnance Bench Mark.**

bench plane (*Join.*). A plane for use on flat surfaces. See **jack plane, smoothing plane, trying plane.**

bench screw (*Carp.*). The vice fixed at one end of a bench.

bench stop (*Carp.*). A metal stop, adjustable for height, set in the top of a bench, at one end; used to hold work while it is being planed.

bench test (*Eng.*). A complete functional test of a piece of apparatus, when new or after repair, carried out in a workshop or laboratory. It is undertaken to ensure correct and satisfactory operation prior to installation in a situation where repair may be difficult.

bench work (*Carp., Eng.*). Work executed at the bench with hand tools or small machines, as distinct from that done at the machines. (*Eng.*) Small moulds made on a bench in the foundry.

bend (*Eng.*). A curved length of tubing or conduit used to connect the ends of two adjacent straight lengths which are at an angle to one another. (*Leather*) The half of a butt which has been divided longitudinally. (*Telecomm.*) Alteration of direction of a rigid or flexible waveguide. It is *E* or *minor* when electric vector is in plane of arc of bending and *H* or *major* when at right angles to this. Also **corner.**

Ben Day tints (*Print.*). Celluloid sheets with a patterned surface, an inked impression from which is used to lay a *mechanical stipple* (q.v.).

bending iron (or pin) (*Plumb.*). A tool for straightening or expanding lead pipe.

bending moment (*Eng.*). The *bending moment* at any imaginary transverse section of a beam is equal to the algebraic sum of the moments of all the forces to either side of the section.

bending moment diagram (*Eng.*). One representing the variation of bending moment along a beam. It is a graph of bending moment (y-axis) against distance along the beam axis (x-axis).

bending of strata (*Geol.*). See **folding** (*Geol.*).

bending rollers (*Print.*). Rollers at the nose of rotary presses. Also called **forming rollers.**

bending strength (*Eng.*). The ability of a beam, or other structural member, to resist a *bending moment* (q.v.). (*Paper*) Continuous bending, through a double arc under constant tension, of a strip of paper or, more usually, board. Compare with **folding strength.**

bending test (*Eng.*). (1) A test made on a beam to determine its deflection under load. (2) A forge test in which flat bars, etc., are bent through 180° as a test of ductility.

bends (*Med.*). See **caisson disease.**

Benedict's test (*Chem., Med.*). Test for glucose (and for some disaccharides), involving the oxidation of the sugar by an alkaline copper sulphate solution, in the presence of sodium citrate, to give a red copper(I)oxide. Used in testing of urine in treatment of diabetes.

beneficiation (*Min. Proc.*). See **mineral processing or dressing.**

benito (*Nav.*). A continuous-wave navigation system giving bearing and range relative to ground station.

benitoite (*Min.*). A strongly dichroic mineral, varying in tint from sapphire blue to colourless, discovered in San Benito Co., California. Silicate of barium and titanium.

benk (*Mining*). The place underground where coal is being broken from the face of the coal seam. Cf. *stope* in ore mining. See **stall.**

Bennettitales (*Bot.*). An order of extinct *Gymnospermae* characterized by their bisporangiate stroboli arranged in similar ways to the flower of the *Angiospermae* to whose ancestors they may be related. Found in Mesozoic rocks.

Benson boiler (*Eng.*). A high-pressure boiler of the once-through type in which water is pumped through the successive elements of the heating surface, fired by gas, oil, or pulverized coal.

benstonite (*Min.*). Carbonate of barium and calcium, crystallizing in the trigonal system, first discovered in barytes deposits in Arkansas.

bent (*For.*). The posts in a timber trestle which support the transom.

bent chisel (*Tools*). Carving tool for recessing backgrounds, made in three types: right-angled, and right and left corner bent.

bent gouge (*Carp.*). A curved gouge for hollowing out concave work.

benthohyponeuston (*Ecol.*). Sea-bottom organisms that accumulate near the surface at night.

benthon, benthos (*Ecol.*). Collectively, the sedentary animal and plant life living on the sea bottom; cf. *nekton, plankton. adj.* **benthic.**

benthopotamous (*Ecol.*). Living on the bottoms of rivers and streams.

bent knees (*Vet.*). Flexion of the carpus of horses or dogs due to permanent contraction of the flexor tendons or to chronic arthritis.

bentonite (*Geol.*). A valuable clay, similar in its properties to fuller's earth, formed by the decomposition of volcanic glass, under water. Consists largely of montmorillonite. Used as a bond for sand, asbestos, etc.; also in the paper, soap and pharmaceutical industries. Thixotropic properties exploited in oil drilling muds.

bent-tail carrier (*Eng.*). A *lathe carrier* (q.v.) having a bent shank projecting into, and engaged by, a slot in the driving plate or chuck.

benzal chloride (*Chem.*). Dihloromethylbenzene. $C_6H_5{\cdot}CHCl_2$, b.p. 207°C. A chlorination product of toluene, intermediate for the production of benzaldehyde. Also **benzylidene chloride.**

benzaldehyde (*Chem.*). Benzene carbaldehyde. Oil of bitter almonds. $C_6H_5{\cdot}CHO$, m.p. 13°C, b.p. 179°C, rel. d. 1·05, a colourless liquid, with aromatic odour, soluble in alcohol, ether, slightly in water. Flavouring agent.

benzaldoximes (*Chem.*). $C_6H_5{\cdot}CH{=}N{\cdot}OH$, formed from benzaldehyde and hydroxylamine; there are 2 stereoisomeric forms. The *alpha* or *anti*-form, m.p. 35°C, can be transformed by means of acids into the *beta* or *syn*-form, m.p. 125°C.

benzamide (*Chem.*). Benzene carboxamide. $C_6H_5{\cdot}CO{\cdot}NH_2$, the amide of benzoic acid, obtainable from benzoyl chloride and ammonia or ammonium carbonate; lustrous plates, m.p. 130°C.

benzamine (*Chem.*). Also known as **betacaine** and **β-eucaine**, its structure is

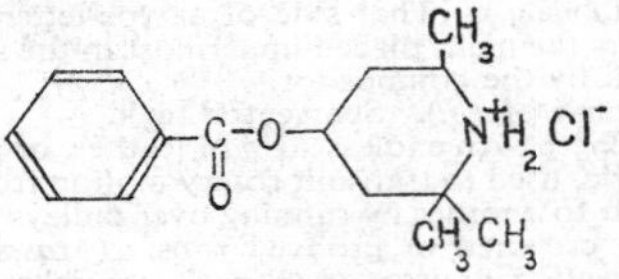

As the hydrochloride or lactate, it was formerly used as a local anaesthetic.

benzanilide (*Chem.*). *N*-phenylbenzamide. $C_6H_5{\cdot}CO{\cdot}NHC_6H_5$, colourless plates, m.p. 158°C; the anilide of benzoic acid, obtained from aniline and benzoic acid or benzoyl chloride.

benzanthrone (*Chem.*). Yellow, crystalline powder, of structure

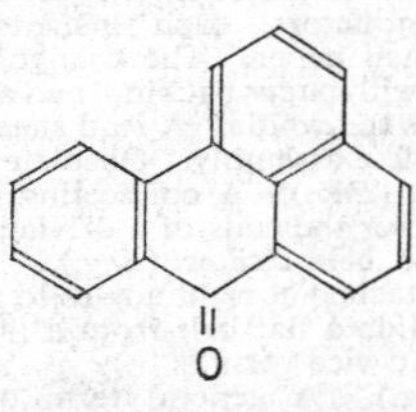

An intermediate widely used in manufacture of vat dyestuffs.

Benzedrine (*Pharm.*). Prop. form of *amphetamine* (q.v.) sulphate, not now marketed.

benzene (*Chem.*). C_6H_6, m.p. 5°C, b.p. 80°C, rel. d. 0·879; a colourless liquid, soluble in alcohol, ether, acetone, insoluble in water. Produced from coal-tar and coke-oven gas; can also be synthesized from open-chain hydrocarbons. Basis for benzene derivatives. A solvent for fats, resins, etc.; very inflammable. Benzene is the simplest member of the aromatic series of hydrocarbons. Carcinogenic. See **benzol.**

benzene carboxylic acids (*Chem.*). Aromatic acids originating from benzene.

benzene formula (*Chem.*). The generally recognized formula for benzene, which takes account of all its characteristics, has been established by Kekulé. It represents a closed chain of 6 carbon atoms, to each of which a hydrogen atom is attached, the carbon atoms being linked alternately by single and by double bonds. It is more stable than this, showing resonance. Resonance energy $\approx$160 kJ/mol.

benzene hexachloride (*Chem.*). Abb. **BHC**. 1, 2, 3, 4, 5, 6-hexachlorocyclohexane. See **Gammexane.**

benzene hydrocarbons (*Chem.*). Homologues of benzene of the general formula C_nH_{2n-6}.

benzene nucleus (*Chem.*). The group of 6 carbon atoms which, with the hydrogen atoms, form the benzene ring. See also **side-chains.**

benzene ring (*Chem.*). See **benzene formula** and **benzene nucleus.**

benzene-sulphonic acids (*Chem.*). Aromatic acids formed from compounds of the benzene series by sulphonation. The acid characteristics are given by the group $—SO_2OH$. Important intermediates for dyestuffs.

benzhydrol (*Chem.*). $(C_6H_5)_2CH{\cdot}OH$. Reduction product of benzophenone prepared by treatment with zinc and aqueous alcoholic alkali.

benzidine (*Chem.*). 4-4′-diamino-biphenyl, $NH_2{\cdot}C_6H_4{\cdot}C_6H_4{\cdot}NH_2$, m.p. 127°C. White to pinkish crystals, soluble in alcohol, ether, insoluble in water. It is an important intermediate for azodyestuffs. It is a known carcinogen.

benzidine transformation (*Chem.*). The transformation of benzene-hydrazo-compounds into benzidine derivatives by strong acids.

benzil (*Chem.*). $C_6H_5{\cdot}CO{\cdot}CO{\cdot}C_6H_2$, m.p. 95°C, large 6-sided prisms, a diketone of the diphenyl group. Synonyms, **bibenzoyl** or **diphenylglyoxal.**

benzine (*Chem.*). Aliphatic petroleum hydrocarbons.

benzocaine (*Chem.*). *Ethyl para-aminobenzoate*:

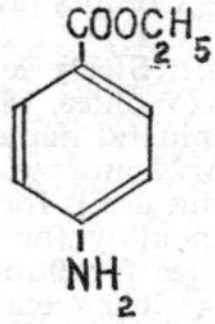

$COOC_2H_5$

NH_2

White crystalline powder, insoluble in water; used as a local anaesthetic and for internal treatment of gastritis.

benzoic acid (*Chem.*). Benzenecarboxylic acid. $C_6H_5{\cdot}COOH$, m.p. 121°C, b.p. 250°C, colourless glistening plates or needles, sublimes readily, volatile in steam. Used as a fruit preservative.

benzoin (*Chem.*). $C_6H_5{\cdot}CHOH{\cdot}CO{\cdot}C_6H_5$, m.p. 137°C, colourless prisms, a condensation product of benzaldehyde. It is both a secondary alcohol and a ketone and can react accordingly. Occurs as a natural resin obtained from a Javanese tree. Chief constituent of Friar's Balsam.

benzol (*Fuels*). Crude *benzene* (q.v.); used as a motor spirit, generally mixed with petrol, and valued for its antiknock properties.

benzol scrubber (*Chem.*). A device for washing gases and absorbing the benzol contained therein by means of a high-boiling mineral oil.

benzonitrile (*Chem.*). Cyanobenzene. $C_6H_5{\cdot}CN$, b.p. 191°C, the *nitrile* (q.v.) of benzoic acid.

benzophenone (*Chem.*). Diphenyl ketone, $C_6H_5{\cdot}CO{\cdot}C_6H_5$, m.p. 49°C, b.p. 307°C, colourless prisms, soluble in alcohol and ether. It is dimorphous, m.p. of the unstable modification 26°C.

1,2-benzopyrene (*Chem.*). A polycyclic hydrocarbon isolated from coal tar as pale yellow crystals, m.p. 177°C. Its structure is

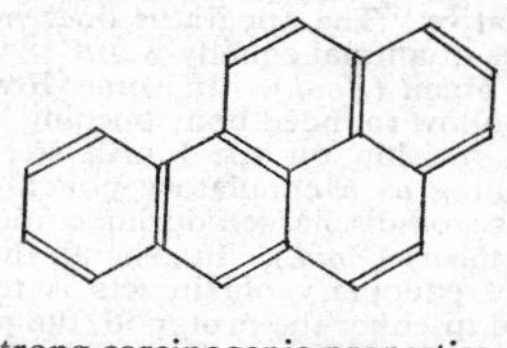

It has strong carcinogenic properties.

benzoquinones (*Chem.*). $C_6H_4O_2$ (2 isomers, see **quinones**).

benzoyl chloride (*Chem.*). Benzene carboxyl chloride. C_6H_5COCl, a colourless liquid, of pungent odour, b.p. 198°C, obtained by the action of PCl_5 on benzoic acid, commercially prepared by chlorinating benzaldehyde.

benzoyl peroxide (*Chem.*). $C_6H_5{\cdot}CO{\cdot}O{\cdot}O{\cdot}CO{\cdot}C_6H_5$. Bleaching agent and catalyst for free radical reactions. M.p. 108°C. Prepared by the action of sodium peroxide on benzoyl chloride.

benzpinacol (*Chem.*). $(C_6H_5)_2C(OH){\cdot}C(OH)(C_6H_5)_2$. Reduction product of benzophenone.

benzyl alcohol (*Chem.*). Phenylmethanol. $C_6H_5{\cdot}CH_2{\cdot}OH$, a colourless liquid, b.p. 204°C, the simplest homologue of the aromatic alcohols.

benzylamine (*Chem.*). Phenylmethylamine. $C_6H_5{\cdot}CH_2{\cdot}NH_2$, colourless liquid, b.p. 183°C, a primary amine of the aromatic series.

benzyl chloride (*Chem.*). (Chloromethyl) benzene. $C_6H_5{\cdot}CH_2Cl$, colourless liquid, b.p. 178°C, obtained by the action of chlorine on boiling toluene. Intermediate for benzyl derivatives.

benzylidene chloride (*Chem.*). (Dichloromethyl) benzene. *Benzal chloride*.

benzylpenicillin (*Pharm.*). Penicillin G. The most important of the *penicillins* (q.v.), effective against a wide range of infections.

benzyne (*Chem.*). C_6H_4. An unstable intermediate formed by the removal of two ortho-hydrogens from benzene.

BEPO (*Nuc. Eng.*). *B*ritish *E*xperimental *P*ile *O*. Principal source of UK radioisotope production until superseded (1969) at Harwell by DIDO.

beraunite (*Min.*). Hydrated phosphate of **iron**, red, crystallizing in the monoclinic system.

berberine (*Chem.*). $C_{20}H_{19}O_5N_1H_2O$, chief alkaloid present in *Hydrastis*. Has been used as an amoebicide and in the treatment of cholera. Also known as **jamaicin, xanthopicrite.**

bergamot oil (*Chem.*). Yellow-green volatile essential oil from the rind of *Citrus bergamia* (*Rutaceae*). Used in perfumery.

Bergius process (*Chem.*). See **hydrogenation of coal.**

Bergmann's law (*Zool.*). In warm-blooded animals, southern forms are smaller than closely-related northern forms.

Bergstrom's method (*Aero.*). A method of assessing the stresses in concrete pavements with particular reference to aerodrome runways and taxi-ing tracks.

beriberi (*Med.*). A disease due to deficiency of vitamin B_1 in the diet; characterized by neuritis and by oedema of the body tissues.

Berkefeld filter (*San. Eng.*). An apparatus for

the domestic filtering of water by passing it through diatomite.

berkelium (*Chem.*). Element, symbol Bk, at. no. 97, synthesized by helium ion bombardment on americium isotope 241.

Berlese funnel (*Ecol.*). A device for extracting mesobiota from soil or litter in which the organisms are forced to move downwards by heat or light until they fall into a vial of preservative. The apparatus does not sample all types of animal equally well.

Berlese's organ (*Zool.*). In some *Hemiptera*, a small hollow rounded body opening by a longitudinal incision on the fourth sternum, and functioning as a copulatory pouch to receive spermatozoa discharged during coition.

Berlese's theory (*Zool.*). In general, the development of endopterygote insects is temporarily arrested in either the protopod, the polypod or the oligopod phase of development, when eclosion from the egg takes place, so that the larva is, so to speak, a protracted, free-living embryo.

Berlin blue (*Paint.*). Prussian blue.

Berlin green (*Paint.*). Ferric ferricyanide, a bright green pigment.

Berlin wool (*Textiles*). Lofty, soft-spun, brightly-dyed hand-knitting yarns, twisted in 4- or 8-ply types.

Berl saddles (*Chem. Eng.*). *Packed column* (q.v.) filling usually consisting of ceramic material in the general form of a saddle, not normally larger than 2 inches (50 mm).

berm (*Civ. Eng.*). A horizontal ledge on the side of an embankment or cutting, to intercept earth rolling down the slopes, or to add strength to the construction. Also called a **bench.**

berm ditch (*Civ. Eng.*). A channel cut along a berm to drain off excess water.

Bernoulli equation (*Maths.*). A differential equation of the form $\frac{dy}{dx}+py=qy^n$, where p and q are functions of x alone.

Bernoulli law (*Phys.*). That the sum of the pressure, potential, and kinetic energies per unit volume is constant at any point in a tube through which liquid is flowing, pressure being smallest at points of the greatest velocity.

Bernoulli's numbers (*Maths.*). The numbers B_1, B_2, B_3, etc., defined by the expansion

$$\frac{x}{1-e^{-x}}=1+\tfrac{1}{2}x+\frac{B_1}{2\,!}x^2-\frac{B_2}{4\,!}x^4+\frac{B_3}{6\,!}x^6-\ldots$$

The first few values are $B_1=\frac{1}{6}$, $B_2=\frac{1}{30}$, $B_3=\frac{1}{42}$, $B_4=\frac{1}{30}$, $B_5=\frac{5}{66}$, after which they continue to increase to infinity.

Bernoulli's polynomials (*Maths.*). The coefficients $\varphi_n(z)$ of $\frac{t^n}{n!}$ in the expansion

$$t\,\frac{e^{zt}-1}{e^t-1}=\sum_{n=1}^{\infty}\varphi_n(z)\frac{t^n}{n!}.$$

Bernoulli's theorems (*Maths.*).

$$(1)\ \sum_{s=1}^{n} s^r=\frac{n^{r+1}}{r+1}+\frac{1}{2}n^r+\frac{r}{2!}\;B_1n^{r-1}-\frac{r(r-1)(r-2)}{4!}\;B_2n^{r-3}+\frac{r(r-1)(r-2)(r-3)(r-4)}{6!}\;B_3n^{r-5}-\ldots$$

there being $\frac{1}{2}(r+3)$ terms if r is odd, and $\frac{1}{2}(r+4)$ if r is even.

$$(2)\ \sum_{s=1}^{\infty}\frac{1}{s^{2m}}=\frac{(2\pi)^{2m}}{2\,.\,(2m)!}B_m,\quad m\geqq 1.$$

where B_1, B_2, B_3, etc., are Bernoulli's numbers.

berry (*Bot.*). A fleshy fruit, without a stone, usually containing many seeds embedded in pulp. It is called *bacca* when formed from an inferior ovary, *uva* when formed from a superior ovary. (*Zool.*) (1) The eggs of lobster, crayfish, and other macruran *Crustacea.* (2) Part of the bill in Swans.

Berry transformer (*Elec. Eng.*). A form of shell-type transformer, in which the iron core is designed in roughly cylindrical shape.

Berthelot's calorimeter (*Heat*). An instrument used by Berthelot for measuring the latent heat of vaporization of liquids. The liquid was boiled in a vessel fitted with an outlet tube, which passed vertically down through the liquid and was connected to a condensing spiral immersed in water in the calorimeter.

Berthon dynamometer (*Light*). A device used to measure the diameter of the exit pencil from a telescope focused at infinity and pointed at the daylight sky.

Bertrand curves (*Maths.*). See **conjugate Bertrand curves.**

bertrandite (*Min.*). A hydrated beryllium silicate, crystallizing in the orthorhombic system, and occurring in pegmatites.

beryl (*Min.*). A silicate of beryllium and aluminium, occurring in pegmatites as hexagonal colourless,green, blue, yellow, or pink crystals. Important ore of beryllium, also used as a gemstone. See **aquamarine, emerald.**

beryllia (*Chem.*). See **beryllium.**

beryllicosis (*Med.*). Chronic beryllium poisoning, the main symptom of which is serious, and usually permanent, lung damage.

beryllides (*Met.*). Compounds of other metals with beryllium.

beryllium (*Chem.*). Steely uncorrodible white metallic element (Wöhler, 1828). Gk. and Lat. *beryl* = the old mineral name. Symbol Be, at. no. 4, r.a.m. 9·0122, m.p. 1281°C, b.p. 2450°C, rel. d. 1·93. Main use is for windows in X-ray tubes and as an alloy for hardening copper. Used as a powder for fluorescent tubes until found poisonous. The metal can be evaporated on to glass, forming a mirror for ultraviolet light. As a slight alloy with nickel, it has the highest coefficient for secondary electron emission, 12·3. Alpha-particles projected into beryllium make it a useful source of neutrons, from which they were discovered by Chadwick in 1932. The oxide (*beryllia*) is a good reflector of neutrons and is also used in thermoluminescent dating; highly toxic. Once called **glucinum.**

beryllium bronze (*Met.*). A copper-base alloy containing 2·25% of beryllium. Develops great hardness (i.e., 300–400 Brinell) after quenching from 800°C followed by heating to 300°C; see **temper-hardening.**

beryllonite (*Min.*). A rare mineral, found at Stoneham, Maine, in decomposed granite, occurring as colourless to yellow crystals. Phosphate of beryllium and sodium.

Berzelius theory of valency (*Chem.*). Chemical affinity is electrical in character, and depends on the mutual attraction of positive (metallic) and negative (non-metallic) elements. A forerunner of the modern *electronic theory of valency* (q.v.).

B.E.S.A. Abbrev. for *British Engineering Stan-*

dards Association, now *British Standards Institution* (q.v.).

Bessel functions (*Maths.*). (1) Of the first kind of order *n*:

$$J_n(x) = \sum_{r=0}^{\infty} \frac{(-1)^r}{r!\Gamma(n+r+1)}\left(\frac{x}{2}\right), \text{ where}$$

Γ indicates the *gamma-function* (q.v.). $J_n(x)$ is a solution of *Bessel's differential equation*.
(2) Of the second kind of order *n*:

$$Y_n(x) = \frac{\cos n\pi J_n(x) - J_{-n}(x)}{\sin n\pi}.$$

There is also Hankel's type of Bessel function of the second kind, namely

$$\mathbf{Y}_n(x) = \frac{\pi e^{n\pi i}}{\cos n\pi} Y_n(x).$$

This is a second independent solution of Bessel's differential equation.
(3) Of the third kind of order *n*:

$$H(x) = J_n(x) + iY_n(x)$$
$$\text{and } H(x) = J_n(x) - iY_n(x).$$
$$\text{where } i = \sqrt{-1}.$$

These are also called *Hankel functions* of the first and second kind respectively.
(4) Modified Bessel functions: (*a*) Of the first kind of order *n*:

$$I_n(x) = \sum_{r=0}^{\infty} \frac{1}{r!(n+r+1)!}\left(\frac{x}{2}\right)^{n+2r}.$$

(*b*) Of the second kind of order *n*:

$$K_n(x) = \frac{\pi}{2}\frac{I_{-n}(x) - I_n(x)}{\sin n\pi}.$$

Bessel functions have application to vibration and heat flow.

Bessel's differential equation (*Maths.*). The equation

$$x^2\frac{d^2y}{dx^2} + x\frac{dy}{dx} + (x^2 - n^2)y = 0.$$

Satisfied by Bessel functions $J_n(x)$ and $\mathbf{Y}_n(x)$.

Bessemer converter (*Met.*). Large barrel-shaped tilting furnace, charged while fairly vertical with molten metal, and 'blown' by air introduced below through tuyères. Discharged by tilting.

Bessemer pig iron (*Met.*). Pig iron which has been dephosphorized in Bessemer converter lined with basic refractory material.

Bessemer process (*Met.*). Removal of impurities from molten metal or matte by blowing air through molten charge in Bessemer converter. Used to remove carbon and phosphorus from steel, sulphur and iron from copper matte.

Besszonoff's reagent (*Chem.*). Complex compound obtained from heating sodium tungstate with phosphoric and phospho-molybdic acids; used for colorimetric detection and estimation of ascorbic acid.

Best and Best Best (*Met.*). See **B.** and **B.B.**

best cokes (*Met.*). A grade of tinplates intermediate between 'coke' plates and common charcoal plates; more heavily tinned than 'coke'.

best selected copper (*Met.*). Metal of a lower purity than high-conductivity copper. Generally contains over 99·75% of copper.

beta (*Electronics*). Complex feed-back factor of electronic circuits.

beta backscattering sedimentometer (*Powder Tech.*). One in which the mass of sediment on the bottom of the sedimentation chamber is measured by the amount of radiation it scatters back from a 1 mCi strontium-90 radioactive source.

beta brass (*Met.*). Copper-zinc alloys, containing 46–49% of zinc, which consist (at room-temperature) of the intermediate constituent (or intermetallic compound) known as β.

betacaine (*Chem.*). See **benzamine.**

beta decay (*Nuc.*). See **beta disintegration.**

beta detector (*Nuc. Eng.*). A radiation detector specially designed to record or monitor beta-radiation.

beta disintegration (*Nuc.*). Radioactive transformation of a nuclide in which the mass number remains unchanged but the atomic number changes by (*a*) +1, with negative β-particle emission or (*b*) −1, with positron emission or electron capture. Also **beta decay.**

beta-disintegration energy (*Nuc.*). For *negatron* emission, it equals the sum of the kinetic energies of the β-particle, the neutrino and the recoil atom. In the case of *positron* emission, there is in addition the energy equivalence of two electron rest masses.

B.E.T. adsorption theory (*Chem.*). *Brunauer, Emmett, and Teller* postulated the building up of multimolecular adsorption layers on the catalyst surface, and extended Langmuir's derivation for single molecular layers to obtain an isothermal equation for multimolecular adsorption.

beta female (*Zool.*). An abnormal queen ant, with excessively long legs and antennae.

betafite (*Min.*). A hydrous, niobate, tantalate, and titanate of uranium: a radioactive mineral described from Betafo in Madagascar.

beta function (*Maths.*). The function $B(p, q)$ defined by $\int_0^1 x^{p-1}(1-x)^{q-1}dx$. Its main property is that $B(p, q) = \frac{\Gamma(p) \, . \, \Gamma(q)}{\Gamma(p+q)}$.

betaine (*Chem.*). $(CH_3)_3N^{+}\cdot CH_2\cdot COO^{-}$ Betaine crystallizes with 1 molecule of H_2O; m.p. of anhydrous betaine 293°C with decomposition. Occurs naturally in plants and animals.

betaines (*Chem.*). A class name for the *zwitterions* exemplified by betaine.

beta-iron (*Met.*). Iron in the temperature range 750°–860° C, in which a change from the magnetic (*alpha*) state to the paramagnetic occurs at about 760° C. With carbon in solution the transition is lowered toward 720° C, and when cooling recalescence is more marked.

beta-particle (*Nuc.*). A *negatron* or a *positron* emitted from a nucleus during *beta disintegration*. The loss of an electron (negatron) from a nucleus has the effect of increasing the atomic number of the atom by one, since the net charge of the nucleus is decreased by −1. Often written *β*-**particle.**

Betaprodine (*Pharm.*). $C_{16}H_{23}NO_2$. Narcotic containing a piperidine ring.

beta-ray gauge (*Paper*). Used for the indication of substance by the measurement of the absorption of β-rays passing through the paper.

beta-ray plaque (*Radiol.*). An extended source of β-radiation (e.g., strontium-90) giving a uniform dose rate over a certain area. It is mounted in a holder, so that it can be applied to the skin of a patient to treat superficial lesions.

beta-rays (*Nuc.*). Streams of β-*particles*.

beta-ray spectrometer (*Nuc. Eng.*). One which determines the spectral distribution of energies of β-particles from radioactive substances or secondary electrons.

beta resin (*Brew.*). See **lupolone.**

beta thickness gauge (*Nuc. Eng.*). Instrument measuring thickness, based on absorption of β-particles from radioactive source.

betatopic (*Nuc.*). Said of atoms differing in atomic number by one unit. One atom can be considered as ejecting an electron (β-particle) to produce the other one.

betatron (*Nuc. Eng.*). Accelerator for high energy beams of electrons and consequently very highly penetrating X-rays. Electrons are accelerated by a rapidly changing magnetic field, orbit being of constant radius. Also **rheotron.** See **cyclotron.**

beta waves (*Physiol.*). Higher frequency waves (15–60 Hz) produced in human brain.

bethanizing (*Met.*). Electrolytic coating of iron or steel with zinc.

Bethe cycle (*Astron.*). See **carbon cycle.** (H. A. Bethe, American astrophysicist.)

Bethell's process (*Build.*). For preserving timber, which is first dried, then subjected to a partial vacuum within a special cylinder, and finally impregnated with creosote under pressure.

Bethenod-Latour alternator (*Radio*). High-frequency alternator in which alternating currents in the stator generate currents of higher frequency in the rotor. By repeating the process, frequencies of the order of 100 kHz were attained in a single machine.

béton (*Civ. Eng.*). A French term, originally applied to lime concrete, but now used for any kind of concrete.

béton armé (*Civ. Eng.*). French equivalent of reinforced concrete.

B.E.T. surface area (*Powder Tech.*). Surface area of a powder calculated from gas adsorption data, by the method devised by *B*runauer, *E*mmett, and *T*eller.

Betts process (*Met.*). An electrolytic process for refining lead after drossing. The electrolyte is a solution of lead silica fluoride and hydrofluorsilicic acid, and both contain some gelatine. Impurities are all more noble than lead and remain on the anode. Gold and silver are recovered from anode sponge.

between-lens shutter (*Photog.*). One located between the components of a camera lens, actuated by a spring-loaded mechanism which opens a series of metal blades pivoted around the periphery of the aperture, the length of exposure being regulated by an air brake or a clockwork escapement which controls the closing action.

between perpendiculars (*Ships*). See **B.P.**

between race (*Bot.*). A race of plants which is intermediate in character between the typical species and one of its well-marked subspecies or varieties.

Betz cells (*Histol.*). Giant pyramidal cells in the motor area of the cerebral cortex; named after the Russian histologist, V. A. Betz.

β-eucaine (*Chem.*). See **benzamine.**

Beutler method (*Photog.*). Use of slow film and fast developer to produce the same results as a fast film and slow fine-grain developer, given the same exposure.

BeV (*Nuc.*). See **GeV.**

Bevatron (*Nuc.*). Accelerator at Berkeley University, California, used to accelerate protons and other atomic particles up to 6 GeV.

bevel (*Carp., Join., etc.*). The sloping surface formed when two surfaces meet at an angle which is not a right angle. See **chamfer.** (*Tools*) A light hardwood stock, slotted at one end to take the blade, which is fastened by a clamping screw passing through the stock and the slot in the blade, enabling the latter to be set at any desired angle to the former. (*Typog.*) The slope on the type from the *face* to the *shoulder*.

bevel gear (*Eng.*). A system of toothed wheels connecting shafts whose axes are at an angle to one another but in the same plane.

bevelled boards (*Bind.*). Boards intended for covers, with the edges at the head, foot, and fore-edge cut at an angle.

bevelled-edge chisel (*Tools*). One with the blade bevelled all the way up to the handle. Used for light work.

bevelled halving (*Carp.*). A halving joint in which the meeting surfaces are not cut parallel to the plane of the timbers but at an angle, so that when they are forced together, the timbers may not be pulled apart by a force in their own plane.

Beverage antenna (*Radio*). See **wave antenna.**

bezel (*Autos.*). (1) A retaining rim, e.g., speedometer outer rim panel light-retaining rim. (2) A small indicator light (e.g., for direction flashers) on instrument panel or dashboard. (*Horol.*). The grooved ring holding the glass of a watch or an instrument dial. (*Tools*) The sloped cutting-edge of a chisel or other cutting tool.

B.F.O. (*Radio*). Abbrev. for *beat frequency oscillator*.

BFPO (*Chem.*). See **dimefox.**

b-group (*Light.*). A close group of Fraunhofer lines in the green of the solar spectrum, due to magnesium.

bhang (*Bot.*). See **cannabis.**

BHC (*Chem.*). See **benzene hexachloride.**

B/H curve (*Elec. Eng.*). See **magnetization curve.**

B/H loop (*Elec. Eng.*). See **hysteresis loop.**

B.H.N. (*Met.*). Abbrev. for *Brinell hardness number*, obtained in the *Brinell hardness test*.

B.H.P. (*Eng.*). Abbrev. for *brake horsepower*.

Bi (*Chem.*). The symbol for *bismuth*.

bialternant (*Maths.*). The quotient of two alternants. A homogeneous symmetric polynomial.

biarticulate (*Bot.*). Having two nodes or joints.

bias (*Comp.*). In a computer, the average of random errors when these are not balanced about zero error. (*Electronics*) Adjustment of a relay so that it operates for currents greater than a given current (against which it is *biased*), or for a current of one polarity.

bias current (*Electronics*). Nonsignal current supplied to electrode of semiconductor device, magnetic amplifier, tape recorder, etc., to control operation at optimal working point.

biased protective system (*Elec. Eng.*). A modification of a balanced protective system, in which the amount of out-of-balance necessary to produce relay operation is increased as the current in the circuit being protected is increased.

biased result (*Surv.*). In observations, sampling, etc., introduction of a systematic error through some malfunction of instrument or weakness in method used, so that error accumulates in series of measurements.

biasing (*Electronics*). Polarization of a recording head in magnetic tape recording, to improve linearity of amplitude response, using d.c., or a.c. much higher than the maximum audio-frequency to be reproduced.

biasing transformer (*Elec. Eng.*). A special form of transformer used in one form of biased protective system.

bias voltage (*Electronics*). Generally, nonsignal or mean potential of any electrode in a thermionic tube, measured with respect to the cathode. Specially applied to that of control grid.

biaxial (*Crystal., Min.*). Said of a crystal having two optical axes. Minerals crystallizing in the orthorhombic, monoclinic and triclinic systems are biaxial. Cf. *uniaxial.*

bib-cock (*Eng.*). A draw-off tap for water-supply, consisting of a plug-cock having a downward curved extension for discharge.

bibenzoyl (*Chem.*). See **benzil.**

Bible paper (*Paper*). Heavily loaded, strong, thin, printing paper.

biblio (*Typog.*). Details of a book's 'history', i.e. original publication date and dates of subsequent reprints and revised editions; usually placed on the verso of the title page.

bib-valve (*Eng., etc.*). A draw-off tap of the kind used for domestic water-supply; closed by screwing down a rubber-washered disk on to a seating in the valve body.

bi-cable (*Civ. Eng.*). The type of *aerial ropeway* (q.v.) in which there are one or more stationary supporting ropes to carry the loads, and an endless hauling rope to move them.

bicalcarate (*Bot.*). Two-spurred.

bicapsular (*Bot.*). Bearing a capsule consisting of two chambers.

bicarbonates (*Chem.*). Hydrogen carbonates. The acid salts of carbonic acid; their aqueous solutions contain the ion $(HCO_3)^-$.

bicarpellary (*Bot.*). Said of an ovary consisting of two carpels.

biceps (*Zool.*). A muscle with two insertions. *adj.* **bicipital.**

Bicheroux process (*Glass*). An intermittent process for making plate glass in which pot glass is cast between rollers on to driven conveyor rollers or a flat moving table.

bidichromate cell (*Elec. Eng.*). A primary cell with a zinc negative and one or more carbon positive electrodes; the electrolyte is dilute sulphuric acid, with potassium bidichromate as a depolariser. See **bottle-battery.**

bidichromate process (*Photog.*). One depending upon the action of light on a colloid treated with ammonium or potassium bidichromate.

biciliate (*Bot.*). See **biflagellate.**

bicipital (*Zool.*). See **biceps.**

bicipital groove (*Zool.*). A groove between the greater and lesser tuberosities of the humerus in Mammals.

bick iron, bickern (*Eng.*). See **beak iron.**

bicollateral (*Bot.*). Said of a vascular bundle with two strands of phloem, one inside and one outside the single strand of xylem.

bicolligate (*Zool.*). Said of Birds, having the feet provided with two stretches of web.

biconcave (*Optics, Photog.*). Said of a lens having both surfaces concave.

biconic (*Bot.*). Having the form of two cones placed base to base.

biconical horn (*Radio*). Two flat cones apex to apex, for radiating uniformly in horizontal directions when driven from a coaxial line.

biconjugate (*Bot.*). Said of a compound leaf when each of the two main ribs bears a pair of leaflets.

biconvex (*Optics*). Said of a lens which is convex on both surfaces.

Bicornes (*Bot.*). The first order of the *Sympetalae.* See **Ericales.**

bicrenate (*Bot.*). Bearing two rounded teeth; not the same as *doubly crenate.* See **crenate.**

bicuspid, bicuspidate (*Bot.*). Having two short hornlike points. (*Zool.*) Having two cusps, as the premolar teeth of some Mammals.

bicuspid valve (*Zool.*). The valve in the left auriculoventricular aperture in the Mammalian heart. Also **mitral valve.**

Bidder's ganglion (*Zool.*). In some *Anura,* a nerve ganglion near the auriculoventricular groove.

Bidder's organ (*Zool.*). In some *Anura,* an aggregation of immature ova, attached to the anterior end of the gonad.

bidentate (*Bot.*). Having two teeth. (*Chem.*) Complexing molecule with two donating ions.

bidirectional microphone (*Acous.*). One, such as the normal open ribbon pressure gradient microphone, which is most sensitive in both directions along one axis.

bidirectional waveform (*Telecomm.*). One which shows reversal of polarity, e.g., a bidirectional pulse generator produces both positive and negative pulses.

bieberite (*Min.*). Hydrated cobalt sulphate, also known as *cobalt vitriol.* Found as stalactites and encrustations in old mines containing other cobalt minerals.

Biebrich scarlet stain (*Micros.*). An acid aniline dye used with haematoxylin as a general stain.

biennial (*Bot.*). A plant that lives for 2 years, flowering in the second season only.

Bier's hyperaemia (*Med.*). Induction of venous congestion of a part by application of an elastic bandage above it.

bifalcial (*Bot.*). Flattened, and having the upper and lower faces of different structure.

bifarious (*Bot.*). Arranged in two rows, one on each side of an axis.

bifid (*Bot., Zool.*). Divided halfway down into two lobes; forked.

bifilar micrometer (*Light*). An instrument attached to the eyepiece end of a telescope to enable the angular separation and orientation of a visual double star to be measured.

bifilar pendulum (*Phys.*). See **bifilar suspension.**

bifilar resistor (*Elec. Eng.*). One formed by winding a resistor with a hairpin-shaped length of resistance wire, thus reducing the total inductance.

bifilar suspension (*Phys.*). The suspension of a body by two parallel vertical wires or threads which give a considerable controlling torque. If the threads are of length l and are distance d apart, the period of torsional vibration of a suspended body of moment of inertia I and mass m is

$$T=4\pi\sqrt{\frac{Il}{mgd^2}}$$

Biflagellatae (*Bot.*). A group of fungi with biflagellate zoospores, e.g., Saprolegniales, Leptomitales, and Perenosporales.

biflagellate (*Bot.*). Bearing two flagella.

bifoliate, bifoliolate (*Bot.*). Said of a compound leaf with two leaflets.

biforous spiracle (*Zool.*). A type of spiracle seen in some *Coleoptera.* There are two slitlike contiguous openings separated by a partition.

bifurcate (*Bot., Zool.*). Twice-forked: forked. *v.* **bifurcate.** *n.* **bifurcation.**

bifurcated rivet (*Eng.*). A rivet with a split shank, used for holding together sheets of light material; it is closed by opening and tapping down the two halves of the shank.

big-bang (*Astron.*). Cosmological hypothesis that the present state of the universe is a continuing expansion from a superdeuce primeval atom, or *ylem.*

bigeminal pulse (*Med.*). A pulse in which the beats occur in pairs, each pair being separated from the other by an interval; due to a disturbed action of the heart.

bigeminate (*Bot.*). In two pairs.

big-end (*I.C. Engs.*). That part of the connecting rod which is attached to the crankshaft.

big-end bolts (*Eng.*). See **connecting-rod bolts.**
bigeneric (*Zool.*). Of hybrids, produced by crossing two distinct genera.
big head disease of horses (*Vet.*). See **osteodystrophia fibrosa.**
big head disease of sheep (*Vet.*). Swelled head. (1) Infection of the head and neck by *Clostridium septicum*. (2) A form of light sensitization occurring in sheep.
big head disease of turkeys (*Vet.*). See **infectious sinusitis of turkeys.**
bigrid valve (*Electronics*). Four-electrode thermionic tube with two control grids, each having approximately the same control on the anode current. Used as a modulating or mixing valve, or as an amplifier operating with low anode voltages.
biguttulate (*Bot.*). Containing two vacuoles or two oil drops.
biharmonic equation (*Maths.*). The equation

$$\frac{\partial^4 z}{\partial x^4}+\frac{2\partial^4 z}{\partial x^2 \partial y^2}+\frac{\partial^4 z}{\partial y^4}=0.$$

The solutions of this equation are called *biharmonic functions.*
bilabiate (*Bot.*). With two lips.
bilabiate dehiscence (*Bot.*). Spontaneous opening by a transverse split across the top.
bilaminar (*Zool.*). See **diploblastic.**
bilateral (*Med.*). Having, or pertaining to, two sides.
bilateral cleavage (*Zool.*). The type of cleavage of the zygote formed in *Chordata.*
bilateral impedance (*Elec. Eng.*). Any electrical or electromechanical device through which power can be transmitted in either direction.
bilateral slit (*Light*). Two metal strips mounted so that their separation can be accurately controlled. Used in a spectrometer or spectrograph.
bilateral symmetry (*Biol.*). The condition when an organism is divisible into similar halves by one plane only.
bilateral tolerance (*Eng.*). A tolerance with dimensional limits above and below the basic size.
bile (*Physiol.*). A viscous liquid produced by the liver. Human bile has an alkaline reaction and possesses a green or golden-yellow colour and a bitter taste. It consists of water, bile salts, mucin and pigments, cholesterol, fats and fatty acids, soaps, lecithin, and inorganic compounds.
bile duct (*Zool.*). The duct formed by the junction of the hepatic duct and the cystic duct, leading into the intestine.
bile pigments (*Physiol.*). The chief are bilirubin (reddish-yellow) and its oxidation product biliverdin (green). Produced by the breakdown of haemoglobin, they consist of an open chain of four substituted pyrrole nuclei joined by two methene (=CH—) bridges and one methylene ($—CH_2—$) bridge.
bile salts (*Physiol.*). Sodium salts of the bile acids, a group of hydroxyl steroid acids, some unsaturated, condensed with taurine or glycine; commonest are the salts of taurocholic and glycocholic acids; secreted in the bile; important in aiding absorption of fats from intestine, due to their surface tension and reducing and emulsification properties.
bilge (*Ships*). (1) The curved part of the shell joining the bottom to the sides. (2) The space inside this, at the sides of the cellular double bottom, into which unwanted water drains.
bilge keel (*Ships*). A projecting fin attached to the shell plating for about half the length of the ship, amidships, at the turn of the bilge, to reduce rolling.
bilge pump (*Ships*). A pump in the engine room which can be connected to suction pipes from any part of the bilge for the removal of bilge water.
Bilgram valve diagram (*Eng.*). See **valve diagram.**
bilharziasis (*Med., Vet.*). Parasitic disease of man and domestic animals caused by *blood flukes* (q.v.). Endemic in Africa and Far East. See **schistosomiasis.**
biliary fever (*Vet.*). A form of piroplasmosis affecting the horse, characterized by fever, jaundice, and haemoglobinuria, and caused by *Nuttallia equi* (*Babesia equi*).
bilicyanin (*Physiol.*). An oxidation product of bilirubin, blue in colour.
bilinear transformation (*Maths.*). A one-to-one transformation between the *w* and *z* planes, given by $w=\dfrac{az+b}{cz+d}$. A special form of linear transformation. Also called *Möbius transformation.*
bilirubin (*Physiol.*). A reddish pigment occurring in bile; believed to be formed as a breakdown product of haemoglobin. See **bile pigments.**
biliverdin (*Physiol.*). A green pigment occurring in bile. It is an oxidation product of *bilirubin* (q.v.). See **bile pigments.**
bill (*Typog.*). A detailed inventory of a fount of type showing the number of each character.
billet (*Carp.*). A piece of timber which has three sides sawn and the fourth left round. (*Met.*) An intermediate product in the rolling of steel. It is between a *bar* and a *bloom* in size. The term is also applied to nonferrous metals; sometimes means ingots of certain shapes.
billet mills (*Met.*). The rolling-mills used in reducing steel ingots to billets. Also called *billet rolls.*
billet split lens (*Light*). One used to produce interference fringes between two displaced parts of the same beam from a slit source.
billiard cloth (*Textiles*). Woollen cloth manufactured from finest quality Merino wool, with a closely cropped dress-face finish to render it perfectly smooth, and damp resistant. A cheaper type is an all-cotton, well-shrunk fabric, 72 in. (183 mm) wide.
billion. In the U.S.A., a thousand million, or 10^9. Elsewhere, a million million, or 10^{12}.
billion-electron-volt (*Nuc.*). See **GeV.**
billitonites (*Geol.*). See **tektites.**
bill of quantities (*Build., Civ. Eng.*). A list of items giving the quantities of material and brief descriptions of work comprised in an engineering or building works contract; it forms the basis for a comparison of tenders.
bilocular (*Bot.*). Consisting of two loculi or chambers.
bilophodont (*Zool.*). Having the two anterior and the two posterior cusps of the grinding teeth joined by ridges.
bimanous (*Zool.*). Having the distal part of the two forelimbs modified as hands, as in some Primates.
bimanual (*Med.*). Performed with both hands, e.g., *bimanual* examination of the female genital organs.
bimastic (*Zool.*). Having two nipples.
bimetal-fuse (*Elec. Eng.*). A fuse element composed of two different metals e.g., a copper wire coated with tin or lead.
bimetallic balance (*Horol.*). A *compensation-balance* (q.v.) in which temperature compensation is obtained by means of a bimetallic strip.

bimetallic plates (*Print.*). Lithographic plates of great durability, in which the ink-accepting and ink-rejecting parts are of two different metals, e.g., copper and chromium; called *polymetallic* when a supporting metal, e.g., steel, is used.

bimetallic strip (*Elec. Eng., Heat.*). Strip of two metals having different temperature coefficients, so arranged that the strip deflects when subjected to a change in temperature; used in thermal switches, thermostats, thermometers, etc.

bimirror (*Light*). A pair of plane mirrors slightly inclined to one another. Used for the production of two coherent images in interference experiments.

bimorph (*Elec. Eng.*). Unit in microphones and vibration detectors in which two piezoelectric plates are cemented together in such a way that application of p.d. causes one to contract and the other to expand, so the combination bends as in a bimetallic strip.

binary (*Astron.*). A double star in which the two components revolve about their common centre of mass under the influence of their gravitational attraction. See **eclipsing-, spectroscopic-, visual-**. (*Chem.*) Consisting of two components, etc. (*Maths.*) See **binary scale**.

binary arithmetic (*Maths.*). Arithmetic using numbers expressed in the binary scale.

binary cell (*Comp.*). An information storage element, which can have one or other of two stable states.

binary-coded-decimal (*Comp.*). A decimal code in which each digit is represented by its binary equivalent. Four binary digits are required to represent each decimal digit.

binary converter (*Elec. Eng.*). An a.c.-to-d.c. converter in which the stator carries both a 3-phase winding and a d.c. exciting winding, while the rotor carries an ordinary commutator winding supplying the d.c. terminals.

binary counter (*Telecomm.*). Flip-flop or toggle circuit which gives one output pulse for two input pulses, thus dividing by two.

binary digit (*Comp.*). See **bit**.

binary fission (*Biol.*). Division of the nucleus into two daughter nuclei, followed by similar division of the cell-body.

binary point (*Comp., Maths.*). The *radix point* in calculations carried out on the *binary scale*.

binary-rate-multiplier (*Comp.*). A circuit with two inputs; one a train of binary pulses, the other a binary control number. The average rate of output pulses is the product of the input rate and the control number.

binary scale (*Comp., Maths.*). Scale of counting with radix 2; used because the figures 1 and 0 can be represented by *on* and *off*, *pulse* or *no-pulse*, in electronic circuits. A specific form of binary code.

binary star (*Astron.*). See **binary**.

binary system and diagram (*Met.*). The alloys formed by two metals constitute a *binary alloy system*, which is represented by the *binary constitutional diagram* for the system.

binary vapour-engine (*Eng.*). The name given to a heat-engine using two separate working fluids, generally mercury vapour and steam, for the high and low temperature portions of the cycle respectively, thus enabling a large temperature range to be used, with improved thermal efficiency.

binate (*Bot.*). Occurring in pairs.

binaural (*Acous.*). Listening with two ears, the result of which is a sense of directivity of the arrival of a sound wave. Said of a stereophonic system with two channels (matched) applying sound to a pair of ears separately, e.g., by earphones. The effect arises from relative phase delay between wavefronts at each ear.

binder (*Acous.*). See **binding agent**. (*Carp.*) A timber or steel beam which supports the bridging joists in a double or framed floor. Also called a **binding joist**. (*Paint.*) The film-forming part of a paint in which, after drying, the pigment and other solid particles are enmeshed. (*Powder Tech.*) Carbon products, organic brake linings, sintered metals, tar macadam, etc., employ binder components in the mix to impart cohesion to the body to be formed. The binder may have cold setting properties, or subsequently be heat-treated to give it permanent properties as part of the body, or to remove it by volatilization. (*Textiles*) (*a*) In silk and rayon, etc., weaving, the tie-binding, long floats of weft. (*b*) Interior warp of pile fabrics binding pile yarns together. (*c*) Attachment to industrial sewing machine for binding edges of cloths.

binding agent or **binder** (*Acous.*). The basic material of coarse-groove records, chiefly shellac, which causes the various materials to adhere together and form, after heating, a solid mass.

binding-beam (*Carp.*). A timber tie serving to bind together portions of a frame.

binding energy (*Electronics*). (1) That required to remove a particle from a system, e.g., electron, when it is the *ionization potential* (q.v.). (2) Energy required to overcome forces of cohesion and disperse a solid into constituent atoms.

binding joist (*Carp.*). See **binder**.

binding wire (*Elec. Eng.*). See **tie wire**.

Binet-Cauchy theorem (*Maths.*). If A is a matrix of order $m \times n$, and B is a matrix of order $n \times p$, then any determinant of order r of the product matrix AB is equal to a sum of terms, each of which is the product of a determinant of order r of A, and a determinant of order r of B (in that order).

Binet's formulae for log Γ(z) (*Maths.*). The first formula is

$$\log \Gamma(z) = (z-\tfrac{1}{2}) \log z - z + \tfrac{1}{2} \log (2\pi) + \int_0^\infty \left(\frac{1}{e^t-1} - \frac{1}{t} + \frac{1}{2}\right) \frac{e^{-tz}}{t}\, dt.$$

The second formula is

$$\log \Gamma(z) = (z-\tfrac{1}{2}) \log z - z + \tfrac{1}{2} \log (2\pi) + 2\int_0^\infty \frac{\arctan \left(\frac{t}{z}\right)}{e^{2\pi t}-1}\, dt.$$

$$(R(z) > 0).$$

Cf. *Stirling's series*.

Binet's test (*Psychol.*). A method of testing a child's intelligence by asking standard questions adapted to the intelligence of normal children at various ages. More fully **Binet-Simon test**. (Named from its originators, Alfred Binet and Théodore Simon.) See also **intelligence quotient**.

Binistor (*Electronics*). N-p-n silicon bistable device for registering or controlled switching.

binocular camera (*Photog.*). See **stereocamera**.

binocular finder (*Photog.*). Viewfinder in which one eye views the field directly and the other views a frame in which the image is focused, the combined effect being that of the frame superimposed on the field.

binoculars (*Optics*). A pair of telescopes for use with both eyes simultaneously. Essential components are an objective, an eyepiece, and some system of prisms to invert and reverse the image.

binode (*Electronics*). 3-electrode thermionic tube having 1 cathode and 2 anodes. Used for full-wave rectification; also **double diode, duo-diode.**

binomial (*Maths.*). An expression containing the sum (or difference) of two terms, e.g., $x+y$.

binomial array (*Radar, Telecomm.*). A linear array in which the current amplitudes are proportional to the coefficients of a binomial expansion. Such an array has no side lobes.

binomial coefficients (*Maths.*). See **binomial theorem.**

binomial distribution (*Stats.*). The distribution

$$y=\binom{n}{x}p^x q^{n-x},$$

where $p+q=1$. It is generated by the binomial $(pt+q)^n$ and it gives the probability of an event, having a probability p of occurring in one trial, occurring x times in n trials.

binomial (or binominal) nomenclature (*Biol.*). The system (introduced by Linnaeus) of denoting an organism by 2 Latin words, the first the name of the genus, the second the specific epithet. The 2 words constitute the name of the species; e.g., *Homo sapiens*; *Bellis perennis*. See **species.**

binomial theorem (*Maths.*). The expansion

$$(1+x)^n=1+\frac{n}{1}x+\frac{n(n-1)}{1\,.\,2}x^2+\frac{n(n-1)(n-2)}{1\,.\,2\,.\,3}x^3+\frac{n(n-1)(n-2)(n-3)}{1\,.\,2\,.\,3\,.\,4}x^4+\dots$$

which is valid either if n is a positive integer (in which case the series terminates) or if $|x|<1$. The coefficients of the powers of x are called *binomial coefficients*, that of x^r being written $\binom{n}{r}$. The series can thus be written

$$(1+x)^n=\sum_{r=0}^{n}\binom{n}{r}x^r.$$

Binomial coefficients are used in the construction of probability models for discrete distributions. See **binomial distribution.**

binormal (*Maths.*). See **moving trihedral.**

binovular twins (*Med.*). Twins resulting from the fertilization of 2 separate ova.

binucleate phase (*Bot.*). Said of *Basidiomycetes* when the cells contain 2 nuclei. See **dicaryon.**

bio-aeration (*San. Eng.*). A system of sewage purification by oxidation; aeration of the crude sewage is effected by passing it through specially designed centrifugal pumps. See also **activated sludge, activation.**

bio-assay (*Pharm.*). Determination of the power of a drug or of a biological product by testing its effect on an animal of standard size.

bioblast (*Bot.*). A chondriosome.

biochemistry. The chemistry of living things; physiological chemistry.

biocides (*Agric.*). Substances that destroy plant life.

bioclimatic rule (*Ecol.*). The rule that in temperate latitudes, periodic phenomena such as flowering, fruiting, and other regular activities of plants or animals, usually come 3–4 days later for each degree of latitude in spring and early summer, and conversely come earlier in late summer and autumn. These phenomena also vary, due to temperature variation, along a given set of meridians.

bioclimatology (*Biol.*). The study of the effects of climate on living organisms.

biocoenosis (*Ecol.*). The association of animals and plants together, especially in relation to any given feeding area; *adj.* **biocoenotic.**

biodegradation (*Bacteriol.*). The breaking-down of substances by bacteria.

bioelectricity (*Biol.*). Electricity of organic origin.

bioengineering (*Med.*). Provision of artificial means (electronic, electrical, etc.) to assist defective body functions, e.g., hearing aids, limbs for thalidomide victims, etc. (*Chem. Eng.*) Engineering methods of achieving biosynthesis of animal and plant products, e.g., for fermentation processes. Also **biological engineering.**

biogenetic law (*Biol.*). See **recapitulation theory.**

biogenous (*Bot.*). Parasitic.

biogeochemical cycle (*Ecol.*). The more or less circular paths followed by the essential chemical elements found in living organisms from the environment to the organism and back to the environment, e.g., the nitrogen and carbon cycles.

biogeographic regions (*Ecol.*). Regions of the world containing recognizably distinct and characteristic endemic fauna or flora.

biological barrier (*Bot.*). Anything due to the activity of organisms which prevents the occupation of an area by plants. Plant communities already present may act as a biological barrier; town-building by man is an extreme example.

biological (R.B.E.) dose (*Radiol.*). See **dose, rem.**

biological form, -race (*Bot.*). A race of a fungus, or bacterium, which is morphologically normal but physiologically different from the normal. The term is used in relation to peculiar host-parasite associations, or to the specific food requirements of a saprophyte.

biological half-life (*Nuc.*). Time interval required for half of a quantity of radioactive material absorbed by a living organism to be eliminated naturally.

biological hole (*Nuc. Eng.*). A cavity within a nuclear reactor in which biological specimens are placed for irradiation experiments.

biological oxygen demand (*Chem.*). An indication of the amount of oxygen required for the aerobic purification of a sample of sewage, measured by its decolorization of potassium permanganate solution. Abbrev. **B.O.D.**

biological race (*Bot.*). See **biological form.** (*Zool.*) A race occurring within a taxonomic species; distinguished from the rest of the species by slight or no morphological differences, but by evident differences of habitat, food-preference, or occupation.

biological shield (*Biol.*). Screen made of material highly absorbent to radiation used to protect human or other life from the effects of such radiation, e.g., X-rays, neutron flux.

biological strain (*Bot.*). A term coextensive with *biological form* (q.v.), but may refer to forms with minor morphological differences.

biological type (*Bot.*). See **life form.**

biological warfare. The use of bacteriological (biological) agents and toxins as weapons. Cf. *chemical warfare.*

bioluminescence (*Biol.*). The production of light by living organisms, as glow-worms, some deep-sea fish, some bacteria, some fungi.

biomass (*Ecol.*). The living mass of an animal or plant population.

biomass, pyramid of (*Ecol.*). A diagrammatic representation of the biomass of a series of organisms, each dependent for food on the one below. It eliminates the 'geometric' factor introduced by the fact that consumer organisms are usually less frequent than producers, and gives a rough picture of the overall effect of ood chain relationships for an ecological group.

biome (*Ecol.*). The largest land community region recognized by ecologists, e.g., tundra, savanna, grassland, desert, temperate and tropical forest.

biometeorology (*Biol.*). The study of the effects of atmospheric conditions on living things.
biometer (*Biol.*). An instrument for measuring the amount of life by assessing the respiration.
biometry (*Biol.*). Statistical methods applied to biological problems.
bion, biont (*Bot.*). An individual plant, independent and capable of separate existence.
bionics. The various phenomena and functions which characterize biological systems with particular reference to electronic systems.
bionomic (*Biol.*). Relating to the environment; ecological.
biophagous, biophilous (*Bot.*). Parasitic.
biophore (*Bot.*). A hypothetical particle of minute size, assumed to be capable of growth and reproduction.
biophysics (*Biol.*). The physics of vital processes; study of biological phenomena in terms of physical principles.
bioplasm (*Biol.*). See **protoplasm.**
biopoiesis (*Biol.*). The creation of living from non-living material as an evolutionary process.
biopsy (*Med.*). Diagnostic examination of tissue (e.g., tumour) removed from the living body.
bios (*Chem.*). A group of substances which act as a growth promoter for yeast. Three components (bios I, IIA, and IIB) are known. Bios I (inositol, possibly a vitamin for mice, protecting against a nutritional alopecia); bios IIA (see **vitamin B complex**); bios IIB (biotin, probably identical with 'co-enzyme R', which stimulates the growth of micro-organisms in the root nodules of certain leguminous plants).
bioseries (*Bot., Zool.*). In evolution, a historical sequence formed by the changes in any one single heritable character.
biosphere (*Ecol.*). That part of the earth (upwards at least to a height of 10 000 m, and downwards to the depths of the ocean, and a few hundred metres below the land surface) and the atmosphere surrounding it, which is able to support life. A term that may be extended theoretically to other planets. Cf. **hydrosphere, lithosphere.**
biosynthesis (*Biol.*). The method of synthesis of complex molecules within the living organism.
biosystematics (*Biol.*). The study of relationships with reference to the laws of classification of organisms; genonomy, taxonomy.
biota (*Biol.*). Fauna and flora of a given region.
biotechnology. See **ergonomics.**
Biot-Fourier equation (*Heat*). The equation representing the nonsteady conduction of heat through a solid. In the one dimensional case,

$$\frac{\partial\varphi}{\partial t}=a\frac{\partial^2\varphi}{\partial x^2},$$

where temperature of a section at right angles to the flow $=\varphi$; distance in flow direction $=x$; time $=t$; and thermal diffusivity ($\frac{k}{\rho s}$ where k is thermal conductivity, ρ is density and s is specific heat capacity) $=a$.
biotic (*Biol.*). Relating to life.
biotic adaptation (*Bot.*). A change in form or in habits, presumed to have arisen as a result of competition with other plants.
biotic barrier (*Ecol.*). Biotic limitations affecting dispersal and/or survival of animals and plants.
biotic climax (*Bot.*). A climax community maintained in a stable condition by some biotic factor, e.g., much grassland, which is prevented by grazing from passing into woodland.
biotic factor (*Bot.*). Any activity of living animals or plants influencing the occurrence of plants.
biotic potential (*Ecol.*). See **intrinsic rate of natural increase.**
biotin (*Chem.*). See **vitamin B complex.**
biotite (*Min.*). A form of black mica widely distributed in igneous rocks (particularly in granites) as lustrous black crystals, with a singularly perfect cleavage. Also occurs in mica-schists and related metamorphic rocks. In composition, it is a complex silicate, chiefly of aluminium, iron and magnesium, together with potassium and hydroxyl.
Biot laws (*Heat*). The rotation produced by optically active media is proportional to the length of path, to the concentration (for solutions) and to the inverse square of the wavelength of the light.
Biot modulus (*Heat*). The heat transfer to a wall by a flowing medium, giving the ratio of heat transfer by convection to that by conduction. Defined as $\alpha\theta/\lambda$ where heat transfer coefficient $=\alpha$, thermal conductivity of medium $=\lambda$, characteristic length of apparatus $=\theta$.
biotope (*Ecol.*). A small habitat in a large community, e.g., a cattle dropping on a grass prairie, whose several short seral stages comprise a microsere.
biotron (*Radio*). Two-valve amplifying circuit, in which high amplification is obtained by aperiodic regeneration.
Biot-Savart's law (*Mag.*). That which states the intensity of the magnetic field produced by a current-carrying conductor. See **Ampère's law.**
biotype (*Biol.*). One individual of a population of genotypically identical organisms; physiological race, i.e., group within a species, morphologically identical to it, but showing differences in physiological or pathogenic characteristics.
bipack (*Photog.*). Two adjacent films with adjacent emulsions, sensitive to different colours and intended to be exposed one through the other.
biparous (*Bot.*). Dichasial (see **dichasium**). (*Zool.*) Having given birth to 2 young.
bipedal (*Zool.*). Using only 2 limbs for walking.
bipenniform (*Zool.*). Having the form of a feather of which the sides of the vane are equal in size.
bi-phase (*Elec. Eng.*). See **two-phase.**
bipinnaria (*Zool.*). A pelagic larva of *Asteroidea*, having a ciliated band separated into pre-oral and post-oral loops, and possessing a large pre-oral lobe.
bipinnate (*Bot.*). Said of a compound pinnate leaf with its main segments pinnately divided.
bipinnatifid (*Bot.*). Said of a pinnatifid leaf when its parts are themselves pinnatifid.
bipolar (*Zool.*). Having two poles: having an axon at each end, as some nerve cells.
bipolar coordinates (*Maths.*). A system of coordinates in which the position of a point in a plane is specified by its two distances r and r' from two fixed reference points.
bipolar electrode (*Elec. Eng.*). An electrode in an electroplating bath not connected to either the anode or cathode; sometimes **secondary electrode.**
bipolar germination (*Bot.*). Germination of a spore by the formation of 2 germ tubes, one from each end.
bipolar transistor (*Electronics*). One in which *p*-type and *n*-type semiconductors are combined. May be *n-p-n* or *p-n-p* junction transistor, or *n*-channel or *p*-channel junction field-effect transistor (*JFET*).
B.I.P.P. (*Med.*). A paste consisting of one part of bismuth subnitrate, two parts of iodoform, and paraffin; used for dressing wounds.
biprism (*Photog.*). A prism with a very obtuse angle, used for beam-splitting.
bipyramid (*Crystal.*). A crystal form consisting of two pyramids on a common base, the one being the mirror-image of the other. Each pyramid is

built of triangular faces, 3, 4, 6, 8, or 12 in number. See **pyramid**.

biquartz (*Light*). Two pieces of quartz, one right-handed and the other left-handed, are cut perpendicular to the axis and are of such a thickness that each rotates yellow light through 90°. The pieces are cemented side by side and together with a Nicol prism form a sensitive analyser for saccharimeters.

biradial symmetry (*Zool.*). The condition in which part of the body shows radial, part bilateral symmetry; as in some *Ctenophora*.

biramous (*Zool.*). Having two branches; forked, as some Crustacean limbs. Cf. *uniramous*.

birch (*For.*). Common British tree (*Betula*), yielding a lightish to very light brown wood which polishes satisfactorily; due to its lack of colour and figuring is useful for imitating superior wood. It also turns well.

birdcage (*Glass*). In the flat-glass industry, the domed end of a tank.

birdie (*Electronics*). Transistorized electronic device for detecting enemy aircraft and directing the fire of local missile batteries on to them. (*B*attery *I*ntegration and *R*adar *Di*splay *E*quipment.)

bird's-eye grain (*Bot.*). The appearance when worked timber shows large numbers of small circular areas dotted about the wood; the rings are due to small, almost dormant, buds, which give rise to thin cylindrical strands of soft tissue lying almost horizontally in the trunk, appearing circular when cut across.

bird's-eye view (*Surv.*). An *oblique aerial photograph* (q.v.) taken for purposes of topographical survey or town-planning work.

bird's-mouth (*Carp.*). A re-entrant angle cut into the end of a timber, so as to allow it to rest over the arris of a cross timber.

birefringence (*Min.*). Literally, the double bending of light by crystalline minerals, causing in extreme cases (e.g., calcite) two images of any object viewed through the mineral. The difference between the greatest and the least refractive index for light passing through a mineral is a measure of its birefringence.

birefringent filter (*Phys.*). One based on the polarization of light which enables a narrow spectral band of $< 0{\cdot}1$ nm to be isolated, i.e., effectively a *monochromator*; for photographing solar flares, etc.

Birkeland and Eyde furnace (*Elec. Eng.*). An electric-arc furnace used for the fixation of nitrogen; the arc, between water-cooled copper electrodes, is drawn out to a circular shape by an arrangement of electromagnets.

Birmingham Gauge, Birmingham Wire Gauge. Systems of designating the diameters of rods and wires by numbers. Obsolescent, being replaced by preferred metric sizes. Abbrev. **B.G., B.W.G.**

birotation (*Chem.*). See **mutarotation**.

birth-mark (*Med.*). See **naevus**.

birth pore (*Zool.*). In some *Platyhelminthes*, the uterine opening when that is distinct from the vaginal opening: in rediae of *Trematoda*, the opening by which daughter rediae escape.

bisaccate (*Bot.*). Having two sepals each with a small pouch at the base.

bisacodyl (*Pharm.*). Di-(4-acetoxyphenyl)-2-pyridylmethane, an effective purgative especially of the large intestine.

bis-azo dyes (*Chem.*). See **disazo dyes**.

bischofite (*Min.*). Hydrated magnesium chloride. A constituent of salt deposits, such as those of Stassfurt in Germany, which decomposes on exposure to the atmosphere.

Bischof process (*Chem.*). A method of making *white lead* more quickly than by stack process.

bise (*Meteor.*). A dry winter wind blowing from the northern sector in the mountains of southern France.

bisect (*Bot.*). A drawing showing the profiles of the shoots and roots of plants growing in their natural positions.

bisector (*Maths.*). The straight line which divides another line, angle or figure into two equal parts.

biseriate (*Bot.*). (1) In two rows. (2) In a double series, as the ascospores in many asci. (3) A vascular ray two cells wide.

biserrate (*Bot.*). Of a leaf margin, having a series of sawlike teeth, which are themselves serrated.

bisexual (*Bot., Zool.*). Possessing both male and female sexual organs. See **hermaphrodite**.

Bismarck brown (*Chem.*). A brown dyestuff obtained by the action of nitrous acid on *m*-phenylenediamine. It contains triamino-azobenzene, $H_2N{\cdot}C_6H_4{\cdot}N{=}N{\cdot}C_6H_3(NH_2)_2$, and a more complex disazo compound, $C_6H_4 = [N{=}N{\cdot}C_6H_3(NH_2)]_2$. (*Paint.*) A mixture of 6 parts of black, 1 of orange and 1 of yellow.

bismite (*Min.*). The monoclinic phase of bismuth trioxide. Cf. *sillénite*, the cubic phase.

bismuth (*Chem.*). Element, symbol Bi, at. no. 83, r.a.m. 208·98, m.p. 268°C. Used as a component of fusible alloys with lead. ^{209}Bi is the heaviest stable nuclide. It is strongly diamagnetic and has a low capture cross-section for neutrons, hence its possible use as a liquid metal coolant for nuclear reactors. A high absorption for γ-rays makes it a useful filter or window for these, while transmitting neutrons.

bismuth glance (*Min.*). See **bismuthinite**.

bismuth(III)hydride (*Chem.*). BiH_3. Volatile, unstable compound. Also called **bismuthine**.

bismuthine (*Chem.*). See **bismuth(III)hydride**.

bismuthinite (*Min.*). Sulphide of bismuth, rarely forming crystals, commonly occurring in shapeless lead-grey masses with a yellowish tarnish. Also called **bismuth glance**.

bismuth ochre (*Min.*). A group name for undetermined oxides and carbonates of bismuth occurring as shapeless masses or earthy deposits.

bismuth spiral (*Elec. Eng.*). Flat coil of bismuth wire used in magnetic flux measurements; the change of flux is measured by observing the change in resistance of the bismuth wire, which increases with increasing fields.

bismuth(III)chloride (*Chem.*). $BiCl_3$. Formed by the direct combination of chlorine and bismuth. Treated with an excess of water, it forms bismuth oxychloride, BiOCl, sometimes used as a pigment under the name of **pearl white**.

bismuth(III)oxide (*Chem.*). Bi_2O_3. Formed when bismuth is heated in air or when the hydroxide, carbonate, or nitrate is calcined. Forms three hydrates which have no acidic properties and do not combine with bases to form salts. It has marked basic properties.

bismutite (*Min.*). Natural bismuth carbonate $(BiO)_2CO_3$, a tetragonal mineral.

bisphenoid (*Crystal., Min.*). A crystal form consisting of 4 faces of triangular shape, 2 meeting at the top and 2 at the base in chisellike edges, at right-angles to one another; hence the name, meaning 'double edged'.

bisphenol A (*Plastics*). 2,2-Bis(4-hydroxyphenyl)-propane. Plastics intermediate for the manufacture of *epoxy resins* (q.v.).

bispheric lens (*Photog.*). One whose elements have a greater radius of curvature at the centre than at the edge.

bisporangiate (*Bot.*). Said of a strobilus which consists of megasporophylls and microsporophylls, with megasporangia and microsporangia.

bispores (*Bot.*). Asexual nonflagellate spores, produced in pairs.

bissagenous (*Zool.*). Pertaining to, or occurring in the byssus gland of Molluscs.

bistable circuit (*Telecomm.*). Valve or transistor circuit which has two stable states which can be decided by input signals; much used in counters and scalers.

bistatic radar (*Radar*). A system in which transmitter and receiver are at different locations, e.g., transmitter on earth, receiver in spacecraft.

bistoury (*Surg.*). A long, narrow surgical knife for cutting abscesses, etc.

bistre (*Paint.*). (1) Beechwood soot, extracted with water and dried to produce an artist's brown pigment. (2) *Vandyke brown* (q.v.) prepared by precipitation.

bisulcate (*Bot.*). Marked by two furrows.

bisulphites (*Chem.*). Hydrogen sulphites, hydrogen sulphates(IV). Acid salts of sulphurous acid. Used as preservatives and as a source of sulphur dioxide. See also **sulphurous acid.**

bisymmetric (*Bot.*). Symmetrical in 2 planes at right angles to one another.

bisynchronous motor (*Elec. Eng.*). A motor similar to an ordinary synchronous motor but capable of being made to run at twice synchronous speed.

bit (*Comp.*). Abbrev. for *bi*nary digi*t*. The smallest unit of information; a unit of storage capacity, i.e., 1 or 0, the normal units for expressing information in a computer corresponding with 'on' and 'off' or 'pulse' and 'no pulse'. (*Mining*) Cutting end of length of drill steel used in boring holes in rock. If detachable, jackbit or ripbit, screwed or taper-fitted on main length or shank. (*Tools*) (1) A boring tool which fits into the socket of a brace, by which it is rotated. (2) The cutting-iron of the plane.

bitch (*Build., Carp., etc.*). A kind of *dog* (q.v.) in which the ends are bent so as to point in opposite directions.

biternate (*Bot.*). Divided into 3 parts, themselves divided ternately.

bit gauge (*Tools*). See **bit stop.**

bit-stock (*Carp.*). See **brace.**

bit stop (*Tools*). An attachment to a bit which limits drilling or boring to a given depth. More correctly **bit gauge.**

bitter almond oil (*Chem.*). *Benzaldehyde* (q.v.); also occurs naturally in *almond oil* (q.v.).

bittern (*Chem.*). The residual liquor remaining from the evaporation of sea water, after the removal of the salt crystals.

Bitter pattern (*Mag.*). Pattern showing boundaries of magnetic domains in ferromagnetic material, formed by powder concentrated along boundaries. Also **powder pattern.**

bitumen (*Chem.*). The nonmineralized substances of coal, lignite, etc., and their distillation residues.

bitumen varnishes (*Chem.*). These contain asphalts, driers, benzine.

bituminous carpeting (*Civ. Eng.*). A road surface of stone chippings bound together with bitumen.

bituminous coals (*Fuels*). Long-flame coals, containing a high proportion of volatile hydrocarbons; classed as *caking* and *non-caking*.

bituminous felt (*Build.*). A manufactured material incorporating asbestos flax or other fibres, and bitumen, generally about ⅛ in. (3 mm) thick. It is produced in rolls, is impervious to water, and is largely used for roof coverings and damp proof courses.

bituminous paint (*Paint.*). Paint in which the base is bitumen instead of the more common oils and resins.

bituminous plastics (*Plastics*). Compositions made from natural bitumens (e.g., Trinidad and gilsonite), petroleum pitches, or certain types of coal-tar pitch, along with a suitable filler; used for accumulator cases, where nonabsorbent acid-resisting material is required, door furniture, and electrical accessories. Cold-moulded plastics are made by adding a drying oil, e.g., linseed oil.

bituminous shale (*Geol., Mining*). Shale rich in hydrocarbons, which are recoverable as gas or oil by distillation.

bi-uniform correspondence or mapping or transformation (*Maths.*). See **one-to-one correspondence.**

biuret (*Chem.*). $NH_2{\cdot}CO{\cdot}NH{\cdot}CO{\cdot}NH_2$, colourless needles, crystallizes with 1 molecule of H_2O; m.p. of the anhydrous compound 190°C. It is formed from urea at 150°–170°C with liberation of NH_3. Has a peptide bond (CONH).

biuret reaction (*Chem.*). The alkaline solution of biuret gives a reddish-violet coloration on the addition of copper(II)sulphate. As biuret is readily formed from urea, this reaction serves to identify the latter.

bivalent (*Chem.*). See **divalent.** (*Cyt.*) One of the pairs of homologous chromosomes present during meiosis.

bivalve (*Zool.*). Having the shell in the form of two plates, as in *Lamellibranchiata*.

bivariant (*Chem.*). Having 2 degrees of freedom.

biventral (*Zool.*). Of muscles, having the ends broad and contractile and joined by a narrow tendon in the middle.

bivium (*Zool.*). In *Echinodermata*, the 2 rays between which the madreporite occurs.

bivoltine (*Zool.*). Having 2 broods in each year.

bixin (*Chem.*). Carotenoid acid, used to colour margarine. Obtained from *annatto* (q.v.).

Bjernim's relation (*Chem.*). A relation, for solutions of strong electrolytes, between the activity coefficient of the solute and the chemical potential of the solvent, viz.:

$$\frac{\partial \ln f}{\partial n_2} = -\frac{j}{n_2} - \frac{\partial j}{\partial n_2}$$

where n_2 = number of moles of solute.

Bk (*Chem.*). The symbol for **berkelium.**

black (*Eng.*). Of parts of castings and forgings not finished by machining; it refers to the dark coating of iron-oxide retained by the surface. (*Typog.*) A blemish on a printed sheet caused by a space or lead rising to the height of type.

black after white (*TV*). Defect produced by ringing in the signal circuit of a TV receiver and leading to white vertical edges being bounded by a dark line. In moderation this improves the apparent horizontal resolution of the receiver.

black ash (*Min. Proc.*). Soda ash; impure sodium carbonate, containing some calcium sulphide and carbon. Important pH regulator in flotation process.

black-band iron-ore (*Met., Min.*). A carbonaceous variety of clay-ironstone, the iron being present as carbonate (chalybite or siderite); occurs in the English Coal Measures.

black bean (*For.*). A tree of the genus *Castanospermum*, having extremely hard wood, liable to warping and collapse, but naturally resistant to the attack of wood-rotting fungi and to termites. Used for internal fittings and high-class furniture.

black body (*Phys.*). A body which completely

absorbs any heat or light radiation falling upon it. A *black body* maintained at a steady temperature is a full radiator at that temperature, since any black body remains in equilibrium with the radiation reaching and leaving it.

black-body radiation (*Phys.*). The quality and quantity of the radiation, depending solely on its temperature, which is emitted by an ideal *black body*. Such radiation is also emitted from the inside of a cavity. See **Stefan-Boltzmann law, Wien's laws.**

black-body temperature (*Phys.*). The temperature at which a *black body* would emit the same radiation as is emitted by a given radiator at a given temperature. The *black-body temperature* of a carbon-arc crater is about 3500°C, whereas its true temperature is about 4000°C.

black box. A generalized colloquial term for a self-contained unit of electronic circuitry; not necessarily black. (*Aero.*) See **flight recorder.**

blackbutt (*For.*). Tree of the genus *Eucalyptus*, giving a hardwood, of interlocked grain and coarse but uniform texture. Colour light brown, with pinkish markings. The wood both looks and feels greasy and yields an abundant supply of oil. Used for heavy-duty flooring, wood blocks, cabinet work, and paneling.

black copper (*Met.*). Impure metal, carrying some iron, lead, and sulphur. Produced from copper ores by blast furnace reduction.

black damp (*Mining*). Mine air which has lost part of its oxygen as result of fire, and is dangerously high in carbon dioxide. See **fire-damp.**

black diamond (*Min.*). A variety of crypto-crystalline massive diamond, but showing no crystal form. Highly prized, for its hardness, as an abrasive. Occurs only in Brazil. Also called **carbonado.**

black disease (*Vet.*). A toxaemic disease of sheep caused by *Clostridium novyi* (*Cl. oedematiens*). Infection is associated with the presence of liver damage caused by immature liver flukes.

black draught (*Med.*). Popular name for a mixture of senna infusion, liquorice extract, magnesium sulphate, etc. Powerful purgative.

black earth (*Geog.*). A type of soil rich in humus, which is developed in semi-arid areas such as the Russian steppes; also known as **chernozem.**

blackening (*Light*). Blackening of the inside of an electric filament bulb, owing to particles being shot off the filament as it disintegrates. (*Paper*) A mottled appearance on the surface when paper of high moisture content is calendered.

blacker than black (*TV*). The range of positive TV or facsimile signal levels in which synchronizing and control signals are transmitted. These cut off the electron beam and so are not visible.

Black feedback (*Telecomm.*). See **bridge feedback.**

black fever (*Med.*). See **kala-azar, Rocky Mountain fever.**

black frost (*Meteor.*). An air frost with no deposit of *hoar frost* (q.v.).

blackhead (*Vet.*). Infectious entero-hepatitis; histomoniasis. An infectious disease affecting turkeys and occasionally domestic and wild fowl, caused by infection of the liver and caecae by the protozoon, *Histomonas meleagridis*.

blackheart (*For.*). An abnormal black or dark brown coloration that may occur in the heartwood of certain timbers.

black hole (*Astron.*). A collapsed star whose density is great enough to cause a gravitational field which allows no radiation to escape.

blacking (*Foundry*). Carbonaceous material applied as a powder or wash to the internal surface of a mould to protect the sand and improve the finish of the casting.

blacking mill (*Foundry*). A small mill in which graphite or other carbonaceous material is ground for the preparation of blacking.

black iron oxide (*Paint.*). The only inorganic black pigment used on any scale, mainly in anticorrosive and antifouling paints. It consists mainly of natural or artificial iron(II, III) (ferro-ferric) oxide Fe_3O_4.

black jack (*Min.*). A popular name for the mineral sphalerite or zinc blende.

black japan (*Paint.*). A semitransparent, quick-drying black varnish, based on asphaltum and drying oil.

black lava glass (*Geol.*). Massive natural glass of volcanic origin, jet black and vitreous.

black lead (*Min.*). A commercial form of *graphite* (q.v.).

blackleg (*Vet.*). Blackquarter; quarter ill. An acute infection of cattle and sheep due to *Clostridium chauvoei*; characterized by fever and usually crepitant swelling of the infected muscles.

black letter (*Typog.*). A style of type including Old English, Ancient Black, Tudor Black, and many others.

black level (*TV*). That percentage of maximum amplitude possible in a positive video signal which corresponds to black in a transmitted picture, a lesser amplitude being concerned with synchronizing. Usually between 30% and 40%.

black liquor (*Paper*). Waste liquid remaining after digestion of rags, straw, and esparto pulps.

black malt (*Brew.*). Malt that has been coloured by the special method of kilning; the colour of stout is due to the proportion of black malt forming the grist.

black mortar (*Build.*). A low strength mortar containing a proportion of ashes. Sometimes used for pointing, where a dark colour is required.

black negative or positive (*TV*). Picture signals in which the voltage corresponding to black is respectively negative or positive in relation to the voltage corresponding to white.

black opal (*Min.*). Includes all opals of dark tint, although the colour is rarely black; the fine Australian blue opal, with flame-coloured flashes, is typical.

blackout (*Med.*). Temporary loss of vision, perhaps with loss of consciousness, due to sudden reduction of blood supply to the brain.

black-out effect (*Electronics*). Temporary loss of sensitivity in vacuum tube after subjection to a strong short pulse.

black peak (*TV*). Maximum excursion of a video signal in the black direction, which may be positive or negative.

black powder (*Mining*). Gun powder used in quarry work. Standard contains 75% potassium nitrate, slow 59% and blasting 40%, the balance being charcoal and sulphur.

blackquarter (*Vet.*). See **blackleg.**

black red heat (*Met.*). Temperature at which hot metal is just seen to glow in subdued daylight—about 540°C.

black sand (*Foundry*). (1) A mixture of sand and powdered coal forming the floor of an iron foundry. (2) Ilmenite, megnebite or other dark heavy minerals found associated with alluvial gold or cassiterite.

black screen (*TV*). Picture tube in which the screen is covered with a light-absorbing neutral filter, or the phosphor is a dark grey colour. This increases contrast by reduction of ambient illumination reflected from the face of the tube.

black-step marks (*Bind., Print.*). Collating marks printed on the fold between the first and last pages of each section, as an alternative or addition to the *signature*. The recommended size is 12-points deep by 5-points wide and the position is stepped down for each successive section, giving an immediate visual check on the *gathering*. Sometimes called **back-step**.

Black's test (*Chem.*). A test for 3-hydroxy-butanoic acid in urine, based upon the oxidation of this acid to diacetic acid, which can be recognized by the iron(III)chloride test.

black tellurium (*Min.*). See **nagyagite.**

black-tongue (*Vet.*). A disease of dogs, due to a deficiency of nicotinic acid in the diet.

blackwater (*Vet.*). See **redwater.**

blackwater fever (*Med.*). Haemoglobinuric (or haematuric) fever. An acute disease prevalent in tropical regions, especially Africa, with feverishness, bilious vomiting, and passages of red or dark-brown urine. Thought to be related to malignant tertian malaria infection (*plasmodium falciparum*).

blackwood (*For.*). Tree of the genus *Acacia*, a native of Australia and Tasmania. It is a hardwood, the grain being either straight, interlocked, or wavy, but the texture is fine and even. Typical uses include furniture-making, interior fittings, high-class joinery, gun stocks, and cooperage.

bladder (*Bot.*). A device which catches small aquatic animals; present in the bladderwort, and regarded as a modified leaf. (*Zool.*) Any membranous sac containing gas or fluid; especially the urinary sac of Mammals.

bladderworm (*Zool.*). See **cysticercus.**

blade (*Bot.*). The flattened part of a leaf, sepal, or petal. The flattened thallus of the larger brown algae. (*Elec. Eng.*) The moving part of a knife-switch which carries the current and makes contact with the fixed jaws.

blade activity factor (*Aero.*). The capacity of an airscrew blade for absorbing power, expressed as a nondimensional function of the surface and expressed by the formula

$$AF=\left[\frac{5}{R}\right]^5 \int_{0{\cdot}2R}^{R} cr^3dr,$$

where R equals diameter, and c equals blade chord at any radius r.

blade angle (*Aero.*). The angle of incidence of an airscrew or rotorcraft blade: it is not constant throughout the radius, because of the higher air speed toward the tip, the incidence being progressively reduced to maintain constant thrust (lift) forces. Also called **blade twist.**

blade loading (*Aero.*). The thrust of a helicopter rotor divided by the total area of the blades.

blade twist (*Aero.*). See **blade angle.**

blaes or **blaze** (*Mining*). A Scottish term for the poorly bituminous sandstone, shale, or fireclay in the Coal Measures.

Blagden's law (*Chem.*). The lowering of the freezing point of a given solvent is proportional to the molar concentration of the solute.

Blaine fineness tester (*Powder Tech.*). *Permeameter* (q.v.) in which the powder bed is connected to the low pressure side of a U-tube manometer. Air is allowed to flow through the powder bed to equalize the pressure in the two arms of the manometer. The time required for a specified movement of the manometer is used to calculate the surface area.

Blake crusher (*Min. Proc.*). See **jaw breaker.**

Blake's hydraulic radius (*Powder Tech.*). When a powder bed of porosity Σ and specific surface S is considered as a single pipe of hydraulic radius M, then $M=\Sigma/S$.

blanc fixe (*Paint., etc.*). An artificial sulphate of barium.

blank (*Acous.*). The lacquer-coated disc ready for placing on a recording machine for making records with a stylus. (*Eng.*) A piece of metal, shaped roughly to the required size, on which finishing processes are carried out.

blanked-off pressure (*Vac. Tech.*). See **ultimate pressure.**

blanket (*Nuc.*). Surround of fertile material to absorb stray neutrons from the core of a reactor and so breed in a small way further fissile material, e.g., by absorption of excess neutrons, thorium becomes uranium. (*Print.*) (1) The rubber-covered clothing of the blanket cylinder on an offset press. (2) The rubber, textile, cork, plastic, or composition covering of the impression cylinder on any rotary press. (*Textiles*) (1) Thick fabric with fibrous surfaces produced by milling and raising. Wool blankets consist entirely of wool; union blankets have a cotton warp, with weft consisting of wool or shoddy, or of wool and cotton scribbled. (2) Cotton blanket made from coarse warp and condenser weft, plain, check, or pastel shades, raised on both faces.

blanket bar (*Print.*). See **blanket wind.**

blanket clamp (*Print.*). On impression or offset cylinders, the means of locating the fixed end of the blanket.

blanket cylinder (*Print.*). The rubber-covered cylinder between the plate cylinder and the impression cylinder on an offset press.

blanket pins (*Print.*). Pins used to attach the blanket to another sheet on the *blanket wind* (q.v.).

blanket strake (*Min. Proc.*). Table or sluice with gentle down-slope, with bottom lining of material on which heavy mineral (e.g., metallic gold) is caught. Originally of rough blanket, now usually corduroy with ribs across flow, or chequered rubber.

blanket wind (*Print.*). The means of securing the loose end of a blanket round the cylinder, usually a rotating **blanket bar.**

blank flange (*Eng.*). A disk, or solid flange, used to blank off the end of a pipe.

blank groove (*Acous.*). Unmodulated groove on disk recording.

blanking (*Electronics*). (1) Blocking or disabling a circuit for a required interval of time. (2) In a CRT, shutting off the beam during flyback.

blanking level (*TV*). Reference level in a video signal, the demarcation between the synchronizing signals and the picture information.

blanking interval (*TV*). Time during which there is no video signal.

blank wall (*Build.*). A wall having no opening in it.

-blast. Suffix from Gk. *blastos*, bud.

blast (*Met.*). Air under pressure, blown into a furnace.

blastema (*Bot.*). The axial part of an embryo, but not the cotyledons. (*Zool.*) Anlage; the protoplasmic part of an egg as distinguished from the yolk.

blasteniospore (*Bot.*). A lichen spore, made of 2 cells, which are joined by a tube.

blast-furnace (*Met.*). Vertical shaft furnace into top of which mineral, fuel, and slag-forming rock is charged. Air, sometimes oxygen-enriched and pre-heated, is blown through from below and products are separately tapped (slag higher and metal lower). Used to smelt iron ore, copper and other minerals.

blast-furnace gas (*Fuels*). A gas of low energy content, a by-product in iron-smelting; used for pre-heating the blast, for steam raising, etc. It may contain up to 30% of carbon monoxide.
blast-furnace Portland cement (*Build., Civ. Eng.*). Cement made by grinding ordinary Portland cement clinker with granulated blast-furnace slag. It has lower setting properties than ordinary Portland cement.
blast action (*Bot.*). A catalytic action exerted by light on a plant, stimulating the division and enlargement of cells.
blasting (*Acous.*). A marked increase in amplitude distortion due to overloading the capacity of some part of a sound-reproducing system; e.g., attempt to exceed 100% depth of modulation in a radio transmitter, or break of continuity in carbon granules in a carbon transmitter. (*Civ. Eng., etc.*) The operation of disintegrating rock, etc., by boring a hole in it, filling with gunpowder or other explosive charge, and firing it.
blasting fuse (*Civ. Eng.*). Compound designed to burn at a regulated speed when closed in a tube, used to ignite detonator or explode blasting charge. Types include 'safety' (slow or instantaneous) and detonating.
blast main (*Eng.*). The main blast air-pipe supplying air to a furnace.
blasto-. Prefix from Gk. *blastos*, bud.
blastochore (*Bot.*). A vascular plant which reproduces vegetatively by offshoots.
blastochyle (*Zool.*). Fluid in the blastocoel.
Blastocladiales (*Bot.*). An order of the *Phycomycetes*, having a true mycelium; reproduces asexually by uniflagellate zoospores, and sexually by anisogamous, uniflagellate zoogametes.
blastocoel (*Zool.*). The cavity formed within a segmenting ovum: cavity within a blastula: primary body cavity: segmentation cavity. Also called **archicoel**.
blastocyst (*Zool.*). In Mammalian development, a structure resulting from the cleavage of the ovum; it consists of an outer hollow sphere and an inner solid mass of cells; germinal vesicle.
blastoderm (*Zool.*). In eggs with much yolk, the disk of cells formed on top of the yolk by cleavage.
blastodermic vesicle (*Zool.*). See **blastula.**
blastodisc (*Zool.*). In a developing ovum, the germinal area.
blastogenesis (*Gen.*). Transmission of inherited characters by means of germ-plasm only. See also **budding.**
blastogenic (*Biol.* Occurring in, arising from, or pertaining to the germ-plasm; pertaining to hereditary characteristics due to the constitution of the germ-plasm.
blastoid (*Geol., Zool.*). An extinct echinoderm, somewhat resembling a crinoid, restricted to rocks of Devonian and Carboniferous age.
Blastoidea (*Zool.*). An extinct class of *Echinodermata*, known by Palaeozoic fossils, with ovoid stalked bodies. The ambulacra are bordered by rows of pinnule-like brachioles, and at the sides of the ambulacral grooves run elongated internal pouches, the hydrospires, which open to the exterior at the oral end by spiracles.
blastokinesis (*Zool.*). Migration of the embryo in heavily-yolked Insect eggs.
blastomere (*Zool.*). One of the cells formed during the early stages of cleavage of the ovum.
blastomycosis (*Med.*). A term applied to a group of diseases due to infection with different species of blastomycetes. They may be generalized, affecting the internal organs, or localized, e.g., **blastomycetic dermatitis.**
blastoparenchymatous (*Bot.*). Said of an alga thallus which consists of filaments united side by side, and not recognizable as separate filaments.
blastophore (*Zool.*). In Birds, the anlage of a plumule; in *Oligochaeta*, central part of spermatocyte mass which remains unchanged during the development of the spermatozoa.
blastopore (*Zool.*). The aperture by which the cavity of the gastrula retains communication with the exterior.
blastosphere (*Zool.*). See **blastula.**
blastospore (*Bot.*). A spore produced by budding.
blastostyle (*Zool.*). In *Hydrozoa*, a zooid bearing gonophores, and having the tentacles and mouth reduced or absent.
blastozoite (*Zool.*). See **blastozooid.**
blastozooid (*Zool.*). In *Urochorda*, a zooid which arises by budding. Cf. *oözooid*. See also **blastostyle.**
blast pipe (*Eng.*). The exhaust steam pipe in the smokebox of a locomotive, which terminates in a nozzle to provide draught by entraining the flue gases in the steam-jet and exhausting them through the chimney. See **jumper-top blast pipe.**
blastula (*Zool.*). A hollow sphere, the wall of which is composed of a single layer of cells, produced as a result of the cleavage of an ovum. Also **blastodermic vesicle, blastosphere.**
blastulation (*Zool.*). A form of cleavage resulting in the production of a blastula.
blast wave (*Phys.*). See **shock wave.**
Blatthaller loudspeaker (*Acous.*). A loudspeaking receiver using a zig-zag ribbon drive behind a flat surface, which generates a sound wave of high intensity.
Blavier's test (*Elec. Eng.*). A method of locating a fault on an electric cable; resistance measurements are taken with the far end of the cable free, and again with it earthed.
B-layer (*Radio*). Weakly reflecting and scattering layer or region 10–30 km above the earth's surface, possibly associated with water vapour or ice in the stratosphere; postulated to explain short-period return signals when these are projected vertically.
blaze (*Surv.*). Temporary survey mark, such as slash on tree trunk, to guide prospector or explorer.
blazing-off (*Eng.*). A rough workshop method of tempering hardened steel by dipping in oil, which on ignition heats the piece to the appropriate tempering temperature.
bleaching (*Paper*). The use of chlorine gas, chlorine dioxide, or bleaching powder (chloride of lime) to bring raw materials to the desired whiteness. (*Photog.*) The removal of reduced silver after development, so that the remaining silver halide, which has not been developed because of its insufficient exposure to light, can be further developed. The resulting image is a positive. (*Textiles*) A series of wet processes, of which there are many variations, for removing small residual impurities, colour and fatty or waxy substances from grey cotton or linen in fibre, yarn, or fabric state. This improves the whiteness and promotes brighter colours after dyeing or printing.
bleaching powder (*Chem.*). Commercial bleaching powder is obtained from calcium hydroxide and chlorine. The commercial value depends on the amount of available chlorine.
bleach-out process (*Photog.*). A system of colour printing involving the decolorizing of dyes by exposing them through transparencies.
bleb (*Med.*). A small vesicle containing clear fluid.
bleed (*Print.*). (1) Illustrations whose edges are

cut away when the page is trimmed after printing are said to *bleed off*. (2) The accidental mutilation of type matter by trimming the paper too closely. (3) The running of printing ink on posters or packages as a result of weather or chemical action.

bleeder resistor (*Elec. Eng.*). Resistor placed across secondary of transformer to regulate its response curve, especially when the transformer is not loaded with a proper terminating resistance. One placed in a power supply or rectifier circuit to control its regulation.

bleeding (*Aero.*). The tapping of air from a gas turbine compressor (1) to prevent *surging* (q.v.) or (2) to feed some other equipment, e.g., cabin pressurization or a de-icing system. (*Bot.*) The exudation of sap from wounds. (*Civ. Eng.*) The oozing to the surface of the grouting medium used in some road surfacings to fill the interstices between the stones; it forms gummy patches in hot weather. (*Eng.*) (1) A method of improving the thermal efficiency of a steam plant by withdrawing a small part of the steam from the higher-pressure stages of a turbine to heat the feed-water. (2) Removing undesirable entrapped air from a hydraulic (e.g., braking) system. (*Paint.*) A defect in a painted or varnished surface, resulting from an undercoat of a different colour, or a surface like tarred road, working through the upper coat, owing to the latter's having partially dissolved the substrate. (*Photog.*) Diffusion of dye from an image. (*Textiles*) In a yarn of fabric composed of two or more colours, or fibres having different dyeing properties, the running of the darker colours, and consequent staining of the lighter colours, during finishing, washing, or solvent cleaning.

bleeding pressure (*Bot.*). See **root pressure.**

blemish (*TV*). A mosaic imperfection which affects the transmission of an image.

blemmatogen, blematogen (*Bot.*). A layer of hyphae, usually with thickened walls, forming the outer covering of the button of an agaric.

blend (*Textiles*). A mixture of different qualities or different kinds of natural or man-made staples, self-coloured or dyed, either in loose stock or sliver form.

blende (or zinc blende) (*Min.*). See **sphalerite.**

blended inheritance (*Gen.*). Inheritance in which the characters of two dissimilar parents appear to be blended in their offspring, e.g., skin colour in mulattoes.

blender (*Textiles*). Person responsible for the accurate mixing of different fibres in blended materials.

blennorrhagia (*Med.*). Discharge of mucus, usually from the genital organs, due to gonorrhoea.

blennorrhoea (*Med.*). See **blennorrhagia. Acute blennorrhoea,** purulent conjunctivitis, due usually to infection with the gonococcus.

blephar-, blepharo-. Prefix from Gk. *blepharon*, eyelid.

blepharism (*Med.*). Spasm of the eyelids.

blepharitis (*Med.*). Chronic inflammation of the eyelids.

blepharochalasis (*Med.*). Laxity of the skin of the eyelid.

blepharoconjunctivitis (*Med.*). Inflammation of the eyelids and of the conjunctiva.

blepharophimosis (*Med.*). Narrowing of the palpebral fissure.

blepharoplast (*Bot., Zool.*). A deeply staining granule having a direct connexion with the base of a locomotor organelle, usually a flagellum. See also **basal body.**

blepharoplegia (*Med.*). Paralysis of the muscles of the eyelid, resulting in drooping of the upper eyelid (**blepharoptosis**).

blepharospasm (*Med.*). Spasm of the orbicular muscle of the eyelid.

BLEU (*Aero.*). The *B*lind *L*anding *E*xperimental *U*nit operated by the Royal Aircraft Establishment which developed a fully automatic blind landing system. See **autothrottle** and **flight director.**

blibe (*Glass*). An elongated bubble larger than *seed* (q.v.).

blimp (*Aero.*). Colloquial for *nonrigid airship*. (*Cinema.*) A temporary cover for apparatus, such as cameras in sound-film studies, to mitigate the effect of noise.

blind apex (*Mining*). The upper edge of a vein or lode, near the surface but overlaid by other formations; a 'suboutcrop'.

blind arcade (*Build.*). See **arcade.**

blind arch (*Arch., Civ. Eng.*). A closed arch which does not penetrate the structure; used for ornamentation, to make one face of a building harmonize with another in which there are actual arched openings.

blind area (*Build.*). A sunken space round the basement of a building, broken up into lengths by small cross-walls, which support the earth-retaining wall but restrict ventilation.

blind blocking (*Bind.*). Applying lettering or design to a book cover without using gold leaf or alternative. See **blocking.**

blind flying, blind landing (*Aero.*). The flying and landing of an aircraft by a pilot who, because of darkness or poor visibility, must rely on the indication of instruments. See **instrument landing system, ground-controlled approach.**

blind flying instruments (*Aero.*). A group of instruments, often on an individual central panel, essential for blind flying. Commonly airspeed indicator, altimeter, vertical speed, turn-and-slip, artificial horizon, directional gyro. See **basic six, basic T.**

blinding (*Bind.*). See **blind tooling.** (*Civ. Eng.*) The process of sprinkling small chippings of stone over a tar dressed road surface.

blind level (*Mining*). (1) One not yet connected with a stope or other underground passages. (2) Drainage level formed by connecting two shafts to create an inverted siphon.

blind lode, blind vein (*Mining*). A lode which does not outcrop to the surface.

blind monitoring (*Radio*). Control of microphone outputs in broadcasting, particularly in outside broadcasts, when the operator is out of sight of the persons originating the transmission.

blind mortise (*Join.*). A mortise which does not pass right through the piece in which it is cut.

blind P (*Typog.*). The *paragraph mark* ¶.

blind page (*Print.*). A page which has no printed folio, but is included in the pagination; usually found in the preliminary matter.

blind rivet (*Eng.*). A type of rivet which can be clinched as well as placed by access to one side only of a structure. Usually based on a tubular or semitubular rivet design, e.g., *Chobert rivet*, *explosive rivet* (qq.v.).

blind spot (*Radio*). Point within normal range of a transmitter at which field-strength is abnormally small. Usually results from interference pattern produced by surrounding objects, or geographical features, e.g., valleys. (*Zool.*) In Vertebrates, an area of the retina where there are no visual cells (due to the exit of the optic nerve), and over which no external image is perceived.

blind staggers (*Vet.*). See **alkali disease.**

blind tooling (*Bind.*). Applying lettering or design to a book cover with hand tools, without using gold. Also called **blinding.**

blind vein (*Mining*). See **blind lode.**

blinking (*Radar*). Modification of a loran transmission, so that a fluctuation in display indicates incorrect operation.

blink microscope (*Astron.*). An instrument in which two photographic plates of the same region are viewed simultaneously, one with each eye, any difference being detected by a device which alternately conceals each plate in rapid succession.

blip (*Radar*). Spot on CRT screen indicating radar reflection.

blister (*Acous.*). A defect in a gramophone record consequent on the release of gases (e.g., water vapour) during pressing. (*Med.*) A thin-walled circumscribed swelling in the skin containing clear or blood-stained serum; caused by irritation. (*Met.*) A raised area on the surface of solid metal produced by the emanation of gas from within the metal while it is hot and plastic. (*Paint.*) See under **blistering.** (*Radar*) See **radome.** (*Vet.*) An irritant drug applied to the skin to cause inflammation and assist in the healing of deep-seated diseased tissues by reflex nervous action.

blister bar (*Met.*). Wrought-iron bars impregnated with carbon by heating in charcoal. Used in making crucible steel.

blister cloth (*Textiles*). A worsted fabric designed to present a raised and irregular surface, as in crepons and crimps. One well-known type is made from two-fold mohair and botany yarns, end and end, warp and weft.

blister copper (*Met.*). An intermediate product in the manufacture of copper. It is produced in a converter, contains 98·5–99·5% of copper, and is subsequently refined to give commercial varieties, e.g., *tough pitch, deoxidized copper*.

blistering (*Paint.*). A paintwork defect in which local areas become detached from their support while still cohering with the rest of the paint. It may be caused by vapour pressure due to entrapped moisture or other reasons, and is accelerated by heat.

blister steel (*Met.*). Wrought-iron bars impregnated with carbon by heating in charcoal. Before 1740 this was the only steel available. Since then, most blister steel has been melted to give crucible steel, most of which is now made, however, from other materials.

bloat (*Vet.*). Dew-blown; hoven. An acute digestive disorder of ruminants in which excessive amounts of gas accumulate in the rumen, associated usually with the ingestion of lush herbage.

Bloch bands (*Phys.*). Sets of discrete but closely adjacent energy levels arising from quantum states when a nondegenerate gas condenses to a solid.

Blochmann's corpuscles (*Zool.*). Minute greenish bodies occurring in many insect eggs, and representing independent organisms capable of cultivation in artificial media.

block (*Comp.*). Group of *words* arranged sequentially on magnetic tape, which form a unit in operations. (*Eng., etc.*) The housing holding the pulley or pulleys over which the rope or chain passes in a lifting tackle. (*Mining*) Rectangular panel of ore defined by drives, raises, and winzes, giving all-round physical access for sampling, testing, and mining purposes. (*Print.*) A term applied to any letterpress printing plate, duplicate or original, brought to type height by mounting.

blockboard (*Join.*). See **coreboard.**

block brake (*Eng.*). A vehicle brake in which a block of cast-iron is forced against the rim of the revolving wheel, either by hand-power, electromagnetic mechanism, or fluid-pressure acting on a piston. See **air brake, electromagnetic brake.**

block caving (*Mining*). Method of mining in which block of ore is undercut, so that it caves in and the fragments gravitate to withdrawal points.

block clutch (*Eng.*). A friction clutch in which friction blocks or shoes are forced inwards into the grooved rim of the driving member, or expanded into contact with the internal surface of a drum. See **friction clutch.**

block code (*Comp., Telecomm.*). Error detecting and correcting code in which several words are used to form a two-dimensional array with parity bits added for each row and column of the array.

block coefficient (*Ships*). The ratio of the underwater volume of a ship to the volume of an enclosing rectangular block.

block diagram (*Eng., etc.*). An illustration in which parts of a machine or a process are represented notionally by blocks or similar symbols.

blocked (*Elec. Eng.*). Descriptive of a grid-controlled mercury-vapour valve when the negative grid voltage inhibits conduction completely.

blocked impedance (*Elec.*). That of the input of a transducer when the output load is infinite, e.g., when the mechanical system, as in a loudspeaker, is prevented from moving.

blocked-out ore (*Mining*). See **block.**

block gauge (*Eng.*). A block of hardened steel having its opposite faces accurately ground flat and parallel and separated by a specified distance, the *gauge distance*. It is used for checking the accuracy of other gauges, etc.

block-holing (*Mining*). The operation of breaking up large boulders or pieces of rock by means of explosive charges, perhaps in small drill holes.

block-in-course (*Build.*). A type of masonry, used for heavy engineering construction, in which the stones are carefully squared and finished to make close joints, and the faces are hammer-dressed.

blocking (*Bind.*). Applying lettering or design to the cover of a book using a press with a heated relief die. Gold leaf is used for special work only; *blocking foil* (q.v.) for *edition binding* (q.v.). The platen may be horizontal and the press hand-operated or partly mechanized, but for long runs a press with a vertical platen and automatic feeding of case and foil(s) is used. (*Carp.*) The operation of securing together two pieces of board by gluing blocks of wood in the interior angle. (*Electronics*) Cutoff of anode current in a valve because of the application of a high negative voltage to the grid; used in *gating* or *blanking*.

blocking action (*Meteor.*). The effect of a well established, extensive high-pressure area in blocking the passage of a *depression* (q.v.).

blocking capacitor (*Elec. Eng.*). One in signal path to prevent d.c. continuity. Also called **buffer capacitor.**

blocking course (*Build.*). A course of stones laid on the top of a cornice.

blocking foil (*Bind.*). A film base with a layer of gold, other metal, or coloured material, used as a substitute for gold leaf. See **blocking.**

blocking oscillator (*Telecomm.*). Valve or transistor tuned oscillator, which has more than sufficient positive feedback for oscillation but in which a condition periodically supervenes to suspend normal oscillation, e.g., by integration of grid current by a capacitor. Discharge of capacitor removes condition and restores initial state. Cycle is continuously repeated, producing sawtooth and pulse waveforms up to very high frequencies.

blocking-out (*Photog.*). The use of Indian ink or other opaque pigment for covering parts of negatives so that they print white.

blocking press (*Bind.*). See **blocking.**

block lava (*Geol.*). See under **ropy lava.**

block pavement (*Civ. Eng.*). A road surfacing formed of blocks of stone or wood or other material quite or nearly rectangular in shape, as distinct from a *sheet pavement* (q.v.).

block plan (*Build.*). A plan of a building site, showing the outlines of existing and proposed buildings.

block plane (*Carp.*). A small plane about 6 in. (150 mm) long which has no cap iron and has the cutting-bevel reversed; used for planing end grain, and fine work.

block prism (*Photog.*). A cube of glass, slit along its diagonal and half-silvered, for splitting the beam in a 3-colour beam-splitting camera.

block rake (*Glass*). Surface imperfection in the form of a chainlike mark on plate glass.

block section (*Rail.*). The length of track in a railway system that is limited by stop signals.

block system (*Rail.*). The system of controlling the movements of trains by signals and by independent communication between block posts, where are situated the instruments indicating the position of trains, condition of the block sections, and controlling levers for signals, points, etc. It is *absolute* if one train alone is permitted within a block section, and *permissive* if trains are allowed to follow into a block section already occupied by a train.

block time (*Aero.*). The time elapsed from the moment an aircraft starts to leave its loading point to the moment when it comes to rest; also **chock-to-chock, buoy-to-buoy** (seaplanes), and **flight time.** It is an important factor in airline organization and scheduling.

block tin (*Plumb.*). Pure tin.

blockwork (*Build., Civ. Eng.*). Walling constructed of pressed or cast blocks with a basic constituent of cement, the other constituents being generally of a nature to improve the insulating qualities of the wall, e.g., climber ash, framed blast furnace slag, or an air entraining agent.

Blondel arc-lamp (*Light*). An enclosed-flame arc-lamp in which a special chamber is provided for the condensation of the fumes.

Blondel oscillograph (*Elec. Eng.*). The earliest form of moving-coil oscillograph, subsequently perfected by Duddell.

Blondel-Rey law (*Light*). Used to assess the apparent point brilliance B of a flashing light.

$$B=B_o[f/(a+f)],$$

where B_o is the point brilliance during the flash, f is the duration of the flash in seconds and a is a constant whose value is about 0·2 secs, when B is near the threshold for white light.

blondin (*Civ. Eng., etc.*). See **cable-way.**

blood (*Physiol.*). A fluid circulating through the tissues of the body, performing the functions of transporting oxygen, nutrients, and hormones, and carrying waste products to the organs of excretion. It plays an important rôle in maintaining a uniform temperature in the body in warm-blooded organisms. Its relative density in Man is about 1·054–1·060, and it has an alkaline reaction. Its chief constituents are: red cells, white cells, platelets, water (77·5–79%), solids including proteins, lipins, nitrogenous and non-nitrogenous extractives, enzymes, hormones and immune bodies, blood sugar, vitamins and inorganic substances (the chlorides of sodium, potassium, magnesium, calcium), organic and inorganic phosphoric acid, organic iron compounds, and gases (oxygen, carbon dioxide, nitrogen). See **blood groups.**

blood albumin (*Med.*). See **serum albumin.**

blood cell (*Physiol.*). See **haematoblast.**

blood-clotting factors (*Med.*). An internationally agreed scale of discernible factors concerned in blood-clotting; indicated by roman numerals, e.g., I, fibrinogen factor; II, prothrombin factor; III, thromboplastin factor; . . VIII, antihaemophilic factor A; IX, antihaemophilic factor B (Christmas factor), etc.

blood corpuscle (*Physiol.*). A cell normally contained in suspension in the blood. See **erythrocyte, leucocyte.**

blood count (*Med.*). The number of red or white corpuscles in the blood.

blood crystals (*Physiol.*). Crystals of haemoglobin, or one of its derivatives, which can be obtained by extracting blood with trichloromethane or ethoxyethane, or by treating it with glacial acetic acid.

blood dust (*Physiol.*). Neutral fats carried by the blood plasma in the form of very fine globules, or fragments of blood corpuscles.

blood flukes (*Zool.*). Trematodes of the genus *Schistosoma*, parasitic on man and domestic animals *via* various species of water snail as intermediate host. They can attack the liver, spleen, intestines, and urinary system, and occasionally the brain. See **bilharziasis.**

blood gills (*Zool.*). In some aquatic Insects, respiratory outgrowths of the body-wall containing blood, but not, as a rule, tracheae.

blood groups (*Physiol.*). A classification of human bloods based on their mutual agglutination reactions. The classification used for transfusion purposes involves four groups, AB(I), A(II), B(III) and O(IV) and depends on the presence or absence of two agglutinins (α and β) in the plasma and two agglutinogens (A and B) in the red cells. The simultaneous presence of α and A, or β and B, causes agglutination. Hence there are four groups as follows:

Group	*Agglutinogens*	*Agglutinins*
AB	A and B	O
A	A	β
B	B	α
O	O	α+β

Other classifications number approximately twenty, including Rh, Mn, Luther, Kell, Duffy, and Lewis groupings. Blood groups are inherited according to Mendelian laws.

blood islands (*Zool.*). In developing Vertebrates, isolated syncytial accumulations of reddish mesoderm cells containing primitive erythroblasts, which give rise respectively to the walls of the blood vessels and to the red corpuscles.

blood plasma (*Physiol.*). See **plasma.**

blood platelet (*Physiol.*). See **thrombocyte.**

blood red heat (*Met.*). Dark red glow from heated metal, in temperature range 550°–630°C.

blood serum (*Physiol.*). See **serum.**

bloodstone (*Min.*). Cryptocrystalline silica, a variety of chalcedony, coloured deep-green, with flecks of red jasper. Also **heliotrope.**

blood transfusion (*Med.*). See **transfusion.**

blood vessel (*Physiol.*). An enclosed space, with well-defined walls, through which blood passes. See **artery, capillary, vein.**

bloom (*Bot.*). A covering of grains, short rods, or crusts of waxy material occurring on the surface of some leaves and fruits. (*Build.*) Efflorescence on a brick wall. (*Chem.*) (1) The colour of the fluorescent light reflected from some oils when illuminated. The colour is usually different from that shown by the oil with transmitted light. (2) Greyish appearance of chocolate due to migration to the surface of certain fatty components. (*Glass*) (1) Deposition on glass of chemical compounds from the vapour phase by a special process (see also **coated lens**). (2) A surface film caused by weathering. (3) A surface film of sulphites and sulphates formed during the annealing process. (*Leather*) A deposit, of greyish colour, formed (by ellagic acid) on the fibres of leather during tanning. (*Met.*) An intermediate product in the rolling of steel. The term is correctly applied when the cross-section is more than 225 cm^2 smaller sizes being called *billets*, but the distinction is not always observed. See also **ball, billet.** (*Mining*) Efflorescence of altered metallic salt at surface of ore exposure, e.g., cobalt bloom. (*Paint.*) See **blooming.**

Bloom gelometer (*Phys.*). A standard apparatus for determining the force required to depress a plunger a specific distance into a jelly, thus indicating the strength of the jelly.

blooming (*Electronics*). (1) Spread of spot on CRT phosphor due to excessive beam current. (2) Coating of dielectric surfaces to reduce reflection of EM waves. (*Optics*) See **coated lens.** (*Paint.*) A paint or varnish defect in which a bloom or cloudy film appears on a newly treated surface; due usually to a damp atmosphere. It is also attributed to an ammonium sulphate deposit on the film. (*Photog.*) Treatment of the glass-air surfaces of a lens with a deposit of magnesium fluoride or other substance, which reduces internal reflection and increases light transmission.

blooming mills (*Met.*). The rolling mills used in reducing steel ingots to blooms. Called *cogging mills* in England, and not always distinguished from *billet* (or *slab*) *mills.*

bloom side (*Leather*). The hair side of a hide.

bloop (*Cinema.*). Dull thud in sound-film reproduction caused by joints in the negative sound-track before printing the positive projection prints. Also **splice bump, split bump.**

blooping patch (*Cinema.*). Black patch painted on the negative sound-track to give a gradual change in the exposure area and so prevent a bloop on projecting the positive print.

blotting-paper (*Paper*). Unsized absorbent rag paper. The cheaper grades are printings which have not been sized.

blow (*Met.*). In Bessemer converter, passage of air through molten charge. (*Mining*) (1) Ejection of part of the explosive charge (unfired) from hole. (2) Sudden rush of gas from coal seam or ore body.

blow-and-blow machines (*Glass*). Machines in which the glass is shaped in two stages, but each time by blowing, as opposed, for example, to pressing or sucking.

blow back (*I.C. Engs.*). The return, at low speeds, of some of the induced mixture through the carburettor of a petrol-engine; due to late closing of the inlet valve during compression, or by worn or sticking valves.

blow-by (*I.C. Engs.*). The gas which leaks past the piston of an internal-combustion engine during the period of maximum pressure, due to seized rings or excessive gap.

blower (*Eng.*). (1) A rotary air-compressor for supplying a relatively large volume of air at low pressure. See **air compressor, supercharger.** (2) A ring-shaped perforated pipe, encircling the top of the *blast pipe* (q.v.) in the smokebox, to which steam is supplied while the engine is standing, the jets providing sufficient draught to keep the fire going. (*Mining*) (1) A fissure or thin seam which discharges a quantity of coal-gas. (2) An auxiliary ventilating appliance, e.g., a fan or venturi tube, for supplying air to subsidiary working places or to dead-ends.

blowfly myiasis (*Vet.*). Strike. Infestation of the skin of sheep, especially in the breech region, by blowfly larvae.

blow-hole (*Geol.*). An aperture near a cliff-top through which air, compressed in a sea-cave by breaking waves, is forcibly expelled.

blowholes (*Build.*). Pock marks existing on the surface of *in situ* or precast concrete when the shuttering is removed. Due to air entrapped during the process of placing the concrete and not released normally due to inadequate vibration. (*Met.*) Gas-filled cavities in solid metals. They are usually formed by the trapping of bubbles of gas evolved during solidification (see **gas evolution**), but may also be caused by steam generated at the mould surface, air entrapped by the incoming metal, or gas given off by inflammable mould dressings.

blowing (*Build.*). A plastering defect in which a conical piece may be blown out of a finished plastered surface owing to moisture getting to an imperfectly slaked particle of quicklime in the work. Also called **pitting.**

blowing current (*Elec. Eng.*). A term used in connexion with fuse links to denote the current (d.c. or r.m.s.) which will cause the link to melt.

blowing engine (*Eng.*). The combined steam- or gas-engine and large reciprocating air-blower for supplying air to a blast-furnace.

blowing-iron (*Glass*). See **blowpipe.**

blowing-out (*Met.*). The operation of stopping down a blast-furnace.

blowing road (*Mining*). In colliery, the main ventilation ingress.

blowing room (*Textiles*). In a cotton-spinning mill, the room containing the bale breakers, openers, and scutchers, in which revolving beaters and exhaust fans remove motes and dust from the compressed cotton.

blowlamp (*Heat*). A portable apparatus for applying intense local heat, used by painters, electricians, and plumbers. In U.S., **blowtorch.**

blow moulding (*Plastics*). Process for 1-piece articles, e.g., bottles, which combines extrusion and blowing with a split mould.

blown (*I.C. Engs.*). (1) Supercharged. See **boost.** (2) Of cylinder head or manifold gaskets, having sprung a leak under pressure.

blown casting (*Foundry.*) Casting spoilt by porosity.

blown flap (*Aero.*). A *flap* (q.v.), the efficiency of which is improved by blowing air or other gas over its surface to prevent or delay the break-up of the airflow.

blown oil (*Chem.*). Linseed oil treated by heating and aeration. Used in the manufacture of linoleum. (*Eng.*) Oil of vegetable origin subjected to partial oxidation by blowing air through it, to increase its viscosity for purposes of lubrication.

blown sand (*Geol.*). Sand which has suffered transportation by wind, the grains in transit developing a perfectly spherical form (*millet-seed sand*); grain-size is dependent upon the wind velocity. See also **sand dunes.**

blowout coil (*Elec. Eng.*). See **magnetic blowout.**

blowout magnet (*Elec. Eng.*). A permanent or an

electromagnet used to extinguish more rapidly the arc (in a switch, etc.) caused by breaking an electric circuit.

blowpipe (*Chem.*). A small laboratory apparatus using a mixture of air under pressure and coal gas in order to give a hot localized flame. It is much used for laboratory glass blowing and glass bending and also in blowpipe analysis. See also **borax bead.** (*Glass*) A metal tube, some 2 m in length, with a bore of 2-4 mm and a thickened nose which is dipped into molten glass and withdrawn from the furnace. The glass is subsequently manipulated on the end of the blowpipe and blown out to shape. Also called **blowing-iron.**

blowtorch (*Heat*). See **blowlamp.**

blub (*Build.*). A swelling on the surface of newly plastered work.

blubber (*Zool.*). In marine Mammals, a thick fatty layer of the dermis.

blubbering (*Leather*). The process of drumming seal and similar skins in warm water, after soaking, to liquefy the fat so that it may be expressed.

blue (*Paint.*). A primary colour. Blue pigments are obtained from vegetable, mineral, and artificial substances.

blue asbestos (*Min.*). A variety of the amphibole riebeckite, hydrous crocidolite, silicate of sodium and iron, occurring in the Asbestos Mountains (Griqualand West, S. Africa) and (rarely) elsewhere.

blue billy (*Met.*). The residue left after burning off the sulphur from iron sulphide ores.

blue bricks (*Build.*). Bricks (made chiefly in Staffordshire and North Wales) which are famous for their strength and durability, and form the best quality engineering bricks.

blue brittleness (*Met.*). Temperature effect due to increase in pearlite in steel between 205° and 315°C. Both brittleness and tensile strength rise.

blue comb (*Vet.*). See **avian monocytosis.**

blue disease (*Med.*). See **Rocky Mountain fever.**

blue glow (*Electronics*). Visible evidence of ionization in thermionic valve, due to gas.

blue-green algae (*Bot.*). See **Myxophyceae.**

blue ground (*Geol., Mining*). Peridotite. Decomposed agglomerate, occurring in volcanic pipes in S. Africa and Brazil; it contains a remarkable assemblage of ultrabasic plutonic rock-fragments (many of large size) and diamonds.

blue gum (*For.*). A strong, brownish-coloured wood from Australia, used for piles, heavy framing, etc.

blueing (*Eng.*). The production of a blue oxide film on polished steel by heating in contact with saltpetre or wood ash; either to form a protective coating, or incidental to annealing. (*Horol.*) Hands, screws, etc., need only have the appearance of blueing, but for springs, where it is necessary to obtain the desired elastic properties, it must be produced by thermal treatment. (*Textiles*) The process of neutralizing a yellowish tint in wool by tinting it with a pale-blue colour, in order to obtain a better white appearance.

blueing salts (*Met.*). Caustic solution of sodium nitrate, used hot to produce a blue oxide film on surface of steel.

blue john (*Min.*). A massive, blue and white, nodular, concentrically banded, variety of the mineral fluorite, occurring near Castleton, in Derbyshire.

blue lead (*Chem.*). A name used in the industry for metallic lead, to distinguish it from other lead products such as white lead, orange lead, red lead, etc.

blue light (*Cinema., TV*). That produced by discharge lamps, which operate with low heat dissipation, for stage or studio: also called **cold light.**

blue-line key (*Print.*). Used when making sets of lithographic plates for colour printing. A key image of each page in position is prepared on Astrafoil using a blue dye, and the appropriate positives or negatives for each page are patched up in register with the key. When printed down to make a plate, the blue does not affect the result.

blue metal (*Met.*). Condensed metallic fume resulting from distillation of zinc from its ore concentrates. Blue tint is due to slight surface-oxidation of the fine particles.

blue nose disease (*Vet.*). A form of photosensitization occurring in the horse, characterized by oedema and purplish discoloration around the nostrils, nervous symptoms, and sometimes sloughing of unpigmented skin.

blue of the sky (*Meteor.*). Sunlight is 'scattered' by molecules of the gases in the atmosphere and by dust particles. Since this scattering is greater for short waves than for long waves, there is a predominance of the shorter waves of visible light (i.e., blue and violet) in the scattered light which we see as the blue of the sky.

blueprint (*Photog.*). Process for reproducing plans and engineering drawings, based on the principle that ferric salts are reduced to ferrous when exposed to light. Now being superseded by *diazo process* (q.v.).

blueprint paper (*Paper*). A paper made from pure rags, coated with a solution of potassium ferricyanide and ammonium ferric citrate bound together with gelatine or gum arabic. Sometimes used by blockmakers to supply a rough blue-on-white or white-on-blue proof before blocks are finished. See **blueprint.**

blue stain (*For.*). A form of sapstain producing a bluish discoloration; caused by the growth of fungi which, however, do not greatly affect the strength of the wood.

bluestone (*Chem.*). See **copper(II)sulphate.**

bluetongue (*Vet.*). Malarial catarrhal fever of sheep. A febrile disease of sheep and cattle, occurring in parts of Africa and the Near East, caused by a virus and transmitted by mosquitoes; characterized by haemorrhagic inflammation of the buccal mucosa and cyanosis and swelling of the tongue.

blue vitriol (*Min.*). Water-soluble oxidation product of copper(II) sulphide ores, corrosive to iron if in underground water. See **chalcanthite.**

blue water gas (*Fuels*). A mixture of approximately equal proportions of carbon monoxide and hydrogen made by passing steam over incandescent coke in special generators; calorific value at 15°C, 11 MJ/m^3 gross, 10 MJ/m^3 net. Usually converted into *carburetted water gas* (q.v.).

Blumlein circuit (*Elec., Radar*). One for producing a square pulse of duration 10^{-8} to 10^{-5} seconds into a matched load by the use of two similar charged transmission lines.

blum reagent (*Chem.*). Acid solutions of manganous chloride or lead(IV)oxide, used in testing for albumin in urine, the albumin producing turbidity.

blunt start (*Eng.*). End of a threaded screw which is rounded or coned to facilitate insertion.

blur (*Acous.*). The introduction of alien frequencies into reproduced sound, so that the sounds are no longer distinct and easily recognizable.

blurb (*Print.*). A short note by the publisher

recommending a book or its author. Usually printed on the dust-jacket or at the beginning of the preliminary matter.

blur factor (*Acous.*). The measure of acoustic blur. It is the root of the ratio of the power of the unwanted tones to the power of the wanted tones in the output of the system.

blur level (*Acous.*). The relative power level, in decibels, of the alien tones to the wanted tones, as a consequence of nonlinear distortion in a sound-reproducing system.

blushing (*Paint.*). A condition in which a cloudy film appears on a newly lacquered surface; due usually to too rapid drying or to a damp atmosphere.

Blyth elutriator (*Powder Tech.*). A liquid elutriator consisting of 6 fractionating chambers, connected in series, whose diameters increase in the ratio n^2. The suspension is syphoned through the equipment.

B.M. (*Surv.*). The common abbrev. for *bench mark*.

B.M.E.P. (*Eng.*). Abbrev. for *brake mean effective pressure*.

B.N.A. (*Anat.*). *Basle Nomina Anatomica*, an international anatomical terminology accepted at Basle in 1895 by the Anatomical Society to standardize terms used in describing parts of the anatomy. Since 1955, Nomina Anatomica.

board (*For.*). Timber cut to a thickness of less than 2 in. (50 mm), and to any width from 4 in. (100 mm) upwards.

board-and-brace work (*Carp.*). Work consisting of boards grooved along both edges, alternating with thinner boards fitting into the grooves.

board drop stamp (*Eng.*). A stamping machine in which the frictional grip of opposed rollers on either side of a vertical board lifts a tup, which falls when the roller pressure is released.

board foot (*Build.*). The unit of measurement in the board-measure system, being a piece of timber of 1 in. thickness by 12 in.^{2}. A *standard* contains 1980 board feet (4·672 m^3).

board hammer (*Eng.*). See **gravity drop hammer.**

boarding (*Leather*). A process for accentuating the natural grain marks on tanned hides. See **graining.**

board measure (*Build., For.*). A method of measuring timber in quantity, the unit being a **board foot.** Cf. *surface measure*.

Board of Trade Panel (*Elec. Eng.*). This is the panel on a traction switchboard which contains the switches and instruments for ascertaining whether the Board of Trade requirements regarding earth-leakage currents, etc., are being fulfilled. Obsolete term.

Board of Trade Unit (*Elec.*). The commercial unit of electrical energy; it is equal to 1 kilowatt-hour.

boards (*Bind.*). A general term for millboards, strawboards, etc., used for book-covers. (*Paper*) Stiff and thick paper ranging from a thickness of 0·2 mm to 5 mm.

boardy feel (*Textiles*). The feel or handle of a fabric that is exceptionally hard, due to excessive twist in warp or weft yarns, over-setting on back of cover during weaving, or faulty finishing techniques.

boart (*Min.*). See **bort.**

boasted ashlar (*Build.*). See **chiselled ashlar.**

boasted joint surface (*Build.*). The surface of a stone which has been worked over with a boasting chisel until it is covered with a series of small parallel grooves, thus forming a key for the mortar at the joint.

boasted work (*Build.*). See **drove work.**

boaster (*Build.*). See **boasting chisel.**

Boas' test (*Chem.*). A test for the detection of free hydrochloric acid in gastric juices, consisting in the formation of a carmine colour in the presence of an alcoholic solution of resorcinol (1,3-dihydroxybenzene) and sucrose.

boasting (*Build.*). The operation of dressing stone with a broad chisel and mallet.

boasting chisel (*Build.*). A steel chisel having a fine broad cutting edge, 2 in. (50 mm) wide; used for preparing a stone surface prior to finish-dressing with a broad tool. Also **boaster.**

boat (*Light*). A structure, extended and suspended, containing a number of lights, the illumination from which is diffused through glass panels.

boat deck (*Ships*). A deck provided on some ships for the purpose of housing lifeboats, though it may be used for additional purposes.

boat scaffold (*Build.*). See **cradle scaffold.**

bobber (*For.*). In floating, a log or other timber, the density of which is so near to that of water that some part of it emerges from time to time.

bobbin (*Elec. Eng.*). A flanged structure intended for the winding of a coil. Also called a **spool.** (*Textiles*) (1) In weaving, a wooden, plastic, alloy or compressed cardboard spool on which yarn is wound. Spools for holding warp yarn have flanges; those for holding weft or knitting yarns are without flanges. (2) In spinning, a parallel spool of similar materials on which roving is wound for successive drafting stages to spinning by mule or ringframe.

bobbin net (*Textiles*). A fine quality of machine-made lace, with a twisted and traversed mesh. Firmly woven in the mesh for durability. Tosca net is a particularly good example.

bobbin winding (*Elec. Eng.*). A term used to denote a transformer winding in which all the turns are arranged on a bobbin, as opposed to a winding in which the turns are in the form of a disk. Generally used for the high-voltage windings of small transformers.

Bobrovska garnet (*Min.*). See **Uralian emerald.**

bob-weight (*Eng.*). A weight used to counterbalance some moving part of a machine. See **balance weights.**

bod (*Foundry*). A ball of clay used to close the tap-hole of a furnace or cupola.

B.O.D. (*San. Eng.*). Abbrev. for *biological oxygen demand*.

Bode plot (or **diagram**) (*Automation*). One in which gain in dB and phase are plotted against frequency, to study margins of control in a servo system.

Bode's law (*Astron.*). A purely empirical numerical relation, discovered by Bode, which expresses with fair approximation the relative mean distances of the planets from the sun, starting with Mercury. It is obtained by adding 4 to each of the series, 0, 3, 6, 12, 24, 48, 96, 192, giving the resulting sequence, 4, 7, 10, 16, 28, 52, 100, 196.

bodkin (*Typog.*). A small steel spike set in a wooden handle. Used to raise individual type letters from the forme when correcting.

Bodoni rules (*Typog.*). See **swelled rules.**

body (*Paint.*). (1) The degree of opacity possessed by a pigment. (2) The apparent viscosity of a paint or varnish. (3) The ability of a paint to give a good, uniform film over an irregular or porous surface. (*Typog.*) (1) The measurement from top to bottom of a type, rule, etc. The unit is the *point*, 72 points amounting to (approx.) 1 in. (2) The solid part of a piece of type below the printing surface or *face*. Also called **shank, stem.** (3) Body of a work, the text of a volume, distinguished from the preliminary matter, such as title and contents, and the end matter, such as appendices and index.

body cavity (*Zool.*). The perivisceral space, or cavity, in which the viscera lie; a vague term, sometimes used incorrectly to mean *coelom* (q.v.).

body cell (*Bot.*). A cell in a pollen grain of *Gymnospermae*, from which the male nuclei are set free. (*Zool.*) Somatic cell. Cf. *germ cells*.

body-centred cubic (*Chem., Crystal.*). A crystal structure whose unit cell is a cube with an atom located at each corner and one at the centre of each cube. See **face-centred cubic**.

body-section radiography (*Radiol.*). See **tomography.**

body stalk (*Zool.*). In some Mammals, a band of mesoderm connecting the chorion with the embryo posteriorly, and representing the commencement of the allantois.

body wall (*Zool.*). The wall of the perivisceral cavity, comprising the skin and muscle layers.

bodywood (*For.*). Cordwood cut from those portions of the boles of trees that are clear of branches.

boehmite (*Min.*). Orthorhombic form of aluminium hydroxide, $Al_2O_3{\cdot}H_2O$. An important constituent of some bauxites.

B.O.E. sill (*Build.*). Abbrev. for *brick-on-edge sill.*

Boettger's test (*Chem.*). A test for the presence of saccharides, based upon the reduction of bismuth oxynitrate to metallic bismuth in alkaline solution.

Boffle (*Acous.*). TN for a box baffle in which the loudspeaking diaphragm is mounted in the centre of one face of the box, the opposite side being open and the interior containing acoustic absorbing material.

Bogen cage (*Psychol.*). A manipulative test, comprising a wooden box, subdivided into compartments with a glass panel and containing a ball which is to be brought out of an opening by means of a stick which may be inserted into the box through various slats.

boghead coal (*Min.*). Coal which is non-banded and translucent, with high yield of tar and oil on distillation. Consists largely of resins, waxes, wind-borne spores, and pollen cases. Originated in deeper, more open parts of the coal swamps than ordinary household coals. Essentially a spore-coal. Also called **parrot coal.** See also **tasmanite, torbanite.**

bogie or bogie truck (*Eng.*). (1) A small truck of short wheel-base running on rails. Commonly used for the conveyance of coal, gold or other ores, concrete, etc. (2) A 4- or 6-wheel undercarriage of short wheelbase, which forms a pivoted support at one or both ends of a long rigid vehicle such as a locomotive or coach.

bogie landing gear (*Aero.*). A main landing gear carrying a pair or pairs of wheels in tandem and pivoted at the end of the shock strut or *oleo*. This arrangement helps to spread the weight of an aeroplane over a larger area and also allows the wheel size to be minimized for easier stowage after retraction.

bogie-type furnace (*Met.*). A furnace provided with a completely portable hearth supported on a bogie or by other means, enabling the work to be loaded and unloaded externally.

bog iron ore (*Min.*). Porous form of *limonite* (q.v.) often mixed with vegetable matter. Found in marshes.

boglame (*Vet.*). See **osteomalacia.**

bog spavin (*Vet.*). Dilatation of the capsule of the tibio-tarsal joint of the horse.

Bohemian garnet (*Min.*). Reddish crystals of the garnet *pyrope*, occurring in large numbers in the Mittelgebirge in Bohemia,

Bohemian gem-stones (*Min.*). These consist of the garnet *pyrope* (q.v.), the false ruby *rose quartz* (q.v.), and the false topaz *citrine* (q.v.), the latter two being gem-cut.

Bohnenbeiger eyepiece (*Light*). An eyepiece used in the measurement of the level error of a meridian circle.

Bohr atom (*Nuc.*). Concept of the atom, with electrons moving in a limited number of circular orbits about the nucleus. These are *stationary states*. Emission or absorption of electromagnetic radiation results only in a *transition* from one orbit (state) to another.

Bohr magneton (*Electronics, Nuc.*). Unit of magnetic moment, for electron, defined by

$$\mu_B = eh/4\pi m_e c,$$

where e = charge,
h = Planck's constant,
m_e = rest mass,
c = velocity of light,

so that $\mu_B = 9{\cdot}27 \times 10^{-24}\ \mathrm{J\,T^{-1}}$.

The **nuclear Bohr magneton** is defined by

$$\mu_N = eh/4\pi Mc = \frac{\mu_B}{1836} = 5{\cdot}05 \times 10^{-27}\ \mathrm{J\,T^{-1}},$$

M being rest mass of the proton.

Bohr radius (*Nuc.*). The radius of any permissible orbit in the Bohr model of the hydrogen atom, i.e.,

$$R_n = \frac{n^2h^2}{4\pi^2 m_e e^2} = 5{\cdot}291 \times 10^{-11}\ n^2\ \mathrm{m},$$

n may have any positive integral value ($n=1$ for the orbit of lowest energy, which is usually what is meant by Bohr radius), h is Planck's constant, m_e is the rest mass of the electron, and e is the electronic charge.

Bohr-Sommerfeld atom (*Nuc.*). Atom obeying modifications of Bohr's laws suggested by Sommerfeld and allowing for possibility of elliptic electron orbits.

Bohr theory (*Nuc.*). A combination of the Rutherford conception of the atom as a central, positively-charged nucleus surrounded by planetary electrons, with the quantum theory, which restricts the permissible orbits in which the electrons can revolve. The jump of an electron to an orbit of smaller radius is accompanied by the emission of monochromatic radiation.

boil (*Acous.*). Extraneous sound accidentally added to the sounds recorded on a blank, before processing. (*Med.*) Infection (with the *Staphylococcus aureus*) of a hair follicle, resulting in a painful, red swelling, which eventually suppurates.

boiled oil (*Paint.*). Linseed-oil raised to a temperature of from 400–600°F (200–300°C) and admixed with driers.

boiler (*Eng.*). A steam-generator consisting of water-drums and tubes which are exposed to the heat of a furnace and arranged so as to promote rapid circulation. (*Vac. Tech.*) Part of a vapour pump in which the working fluid is converted by the supply of heat into vapour to feed the nozzle.

boiler capacity (*Eng.*). The weight of steam, usually expressed in kg or lb/hr, which a boiler can evaporate when steaming at full load output.

boiler compositions (*Eng.*). Chemicals introduced into boiler feed-water to inhibit scale-formation and corrosion, or to prevent priming or foaming. Examples are sodium compounds (such as soda ash), organic matter, and barium compounds.

boiler covering (*Eng.*). See **lagging.**

boiler cradles (*Ships*). See **keelson (2)**.

boiler crown (*Eng.*). The upper rounded plates of a boiler of the shell type.

boiler efficiency (*Eng.*). The ratio of the heat supplied by a boiler in heating and evaporating the feed water to the heat supplied to the boiler in the fuel. It may vary from 60% to 90%.

boiler feed-water (*Eng.*). The water pumped into a boiler for conversion into steam, usually consisting of condensed exhaust steam and 'make-up' fresh water treated to remove air and impurities.

boiler fittings and mountings (*Eng.*). See **feed check-valve, pressure gauge, safety valve, stop valve, water gauge.**

boilermaker's hammer (*Eng.*). One with ball or straight and cross panes; used for caulking, fullering, and scaling boilers.

boiler plate (*Eng.*). Mild steel plate, generally produced by the open-hearth process; used mainly for the shells and drums of steam-boilers.

boiler pressure (*Eng.*). The pressure at which steam is generated in a boiler. It may vary from little over atmospheric pressure for heating purposes, to 1500 lb/in.^{2} (10 000 kN/m^2) and over for high-pressure turbines.

boiler scale (*Eng.*). A hard coating, chiefly calcium sulphate, deposited on the surfaces of plates and tubes in contact with the water in a steam-boiler. If excessive, it leads to overheating of the metal and ultimate failure.

boiler setting (Eng.). The supporting structure on which a boiler rests; usually of brick for land boilers and of steel for marine boilers.

boiler stays (*Eng.*). Screwed rods or tubes provided to support the flat surfaces of a boiler against the bursting effect of internal pressure.

boiler test (*Eng.*). (1) A hydraulic-pressure test applied to check watertightness under pressure greater than the working pressure. (2) An efficiency test carried out to determine evaporative capacity and the magnitude of losses.

boiler trial (*Eng.*). An efficiency test of a steam-boiler, in which the weight of feed-water and of fuel burnt are measured, and various sources of loss assessed.

boiler tubes (*Eng.*). Steel tubes forming part of the heating surface in a boiler. In water-tube boilers the hot gases surround the tubes; in locomotive and some marine boilers (fire-tube boilers), the gases pass through the tube.

boiling (*Heat*). The very rapid conversion of a liquid into vapour by the violent evolution of bubbles; it occurs when the temperature reaches such a value that the saturated vapour pressure of the liquid equals the pressure of the atmosphere.

boiling bed (*Chem.*). In gas fluidized beds, two separate phases are formed at high gas velocities: gas, containing a relatively small proportion of suspended solids, bubbles through a higher density fluidized phase with the result that the system closely resembles in appearance a boiling liquid. The bed in such a condition is referred to as a boiling bed.

boiling hole (*Mining*). A hole dug in the ground for the first stage of making putty from magnesium quicklime, its normal volume being about 30 ft^3 (0·8 m^3).

boiling-off (*Textiles*). The removal of the sericin, or natural gum, from raw or thrown silk by a method of scouring (or *degumming*). This adds to the lustre and softness of the silk but entails loss of weight. Silk lightly scoured is termed *souple* silk.

boiling-point (*Heat*). The temperature at which a liquid boils when exposed to the atmosphere. Since, at the boiling-point, the saturated vapour pressure of a liquid equals the pressure of the atmosphere, the boiling-point varies with pressure; it is usual, therefore, to state its value at the standard pressure of 101·325 kN/m^2. Abbrev. **b.p.**

boiling table (*Elec. Eng.*). A table incorporating in its construction two or more boiling plates.

boiling-water reactor (*Nuc. Eng.*). Nuclear reactor cooled by allowing water to boil.

Bojanus' organ (*Zool.*). In *Lamellibranchiata*, the excretory organ or kidney.

bold face (*Typog.*). A type face heavier than the normal with which it can be used as a companion, e.g., **bold face.**

bole (*Bind.*). A compact clay, a reddish variety of which is used in powdered form (with water and a small quantity of gilding size) as a foundation for gilt edges. (*Bot.*) The trunk of a tree.

bolection (or balection) moulding (*Join.*). A moulding fixed round the edge of a panel and projecting beyond the surface of the framing in which the panel is held.

Boletales (*Bot.*). An order of the *Basidiomycetes*, the members of which are mushroomlike in appearance, but have the hymenium lining pores.

bolide (*Astron.*). A large *meteor* (q.v.), generally one that explodes: a fire-ball.

boll (*Bot.*). The fruit of the cotton plant.

bollard (*Ships, etc.*). On a quay or vessel, a short upright post round which ropes are secured for purposes of mooring. Generally, any similar postused, e.g., to prevent vehicular access.

Bollinger bodies (*Vet.*). Intracytoplasmic inclusion bodies occurring in epithelial cells infected with fowl pox virus; they are made up of aggregates of smaller bodies (*Borrel bodies*), which are the actual masses of the virus collected in the cells.

boll weevil (*Zool.*). A small, greenish coloured beetle (*Anthonomus grandis*), which causes tremendous damage by infesting the ripening bolls of cotton in American plantations.

boll worm (*Zool.*). The larva of a noctuid moth (*Heliothis armigera*) that feeds on cotton bolls.

bolometer (*Elec. Eng.*). Early instrument for measuring radiant energy and for detecting radio-frequency currents, generally by unbalancing a bridge. Modern forms make use of very large change in resistance of metals at very low temperatures, and are very sensitive. Erroneously applied to *barretter* or *thermistor* when used in microwave work.

Bolsover Moor stone (*Build.*). A dolomite of a warm yellowish-brown colour, which is easily worked but is not particularly durable; employed as a building-stone; of Permian (Magnesian Limestone) age.

bolster (*Carp.*). (1) A short piece of timber capping a pillar or post and offering larger bearing to the supported beam. Also called a **corbel-piece.** (2) A synonym for *lagging* (q.v.). (*Civ. Eng.*) The actual support for a truss-bridge, at its abutment. (*Eng.*) (1) A steel block which supports the lower part of the die in a pressing or punching machine. (2) The rocking steel frame by which the bogie of a locomotive supports the load imposed by the weight of the engine. (*Tools*) A form of cold chisel with a broad splayed-out blade, used in cutting stone slabs, etc.

bolt (*Eng., etc.*). A cylindrical, partly screwed bar provided with a head. With a nut, it affords the commonest means of fastening two parts together. (*Join.*) (1) a bar used to fasten a door. (2) The tongue of a lock.

bolted sectional dock (*Civ. Eng.*). A form of *self-locking dock* (q.v.), usually built in three sections of approximately equal length, the two end sections being stepped so as to provide landings for use when carrying out a self-docking operation.

bolting-silk (*Textiles*). A silk cloth of very fine and regular mesh, used, in fishing, in the construction of tow-nets for the smaller members of the surface fauna; and, in photography, for obtaining diffusion effects when stretched in front of a lens.

bolt-making machine (*Eng.*). A machine which forges bolts by forming a head on a round bar.

bolts (*Bind.*). The folded edges at the head, fore-edge, and tail of a sheet before cutting.

Boltzmann equation (*Phys.*). The fundamental particle conservation diffusion equation based on the description of individual collisions, and expressing the fact that the time rate of change of the density of particles in the medium is equal to the rate of production less the rate of leakage and the rate of absorption.

Boltzmann principle (*Phys.*). Statistical distribution of large numbers of small particles when subjected to thermal agitation and acted upon by electric, magnetic or gravitational fields. In statistical equilibrium, number of particles n per unit volume in any region given by

$$n = n_0 \exp(-E/kT),$$

where k = Boltzmann's constant,
T = absolute temperature,
E = potential energy of a particle in given region,
n_0 = number per unit volume when $E = 0$.

Boltzmann's constant (*Phys.*). Given by $k = R/N = 1{\cdot}380\ 5 \times 10^{-23}\ \mathrm{J\ K^{-1}}$, where R = ideal gas constant, N = Avogadro constant.

bolus (*Vet.*). See **ball.**

bombardment (*Nuc.*). Process of directing a beam of neutrons or high-energy charged particles on to a target material in order to produce nuclear reactions.

bomb calorimeter (*Heat*). An apparatus used for determining the calorific values of fuels. The *bomb* consists of a thick-walled steel vessel in which a weighed quantity of the fuel is ignited in an atmosphere of compressed oxygen. The bomb is immersed in a known volume of water, from the rise of temperature of which the calorific value is calculated.

bomb sampler (*Powder Tech.*). Device for obtaining samples of dispersed particles at predetermined depths within a suspension, consisting of a closed cylindrical vessel with an automatic valve which opens when an extension tube hits the bottom of the suspension container. The sampler fills with suspension and then closes when the vessel is lifted.

bombycine (*Bot.*). Silky.

bonanza (*Mining*). Rich body of ore.

Bonawe granite (*Build.*). Fine-grained granite-porphyry quarried at Bonawe near Oban, Scotland. It is durable, used for paving.

bond (*Build.*). The system under which bricks or stones are laid in overlapping courses in a wall in such a way that the vertical joints in any one course are not immediately above the vertical joints of an adjacent course. (*Civ. Eng.*) The adhesion between concrete and its reinforcing steel, due partly to the shrinkage of the concrete in setting and partly to the natural adhesion between the surface particles of steel and concrete. See **mechanical bond.** (*Nuc.*) Link between atoms, considered to be electrical attraction arising from electrons as distributed around the nucleus of atoms thus bonded, controlling properties of such compounds. Represented by a dot (·) or a line (—) between atoms, e.g.,

H—O—H or H·O·H.

bond angle (*Chem.*). That between straight lines joining the nuclei of bonded atoms, e.g., that between the two H—O bonds in water is 105°.

bond distance (length) (*Chem.*). Distance between bonded atoms in a molecule. Specifically used for distances between atoms in a covalent compound. Typical lengths are O—H 0·096nm, C—H 0·107nm, C—C 0·154nm, C=C 0·133nm, C≡C 0·120nm.

bonded fabrics (*Paper*). A material made by fabricating fibres into sheet form with the aid of a binder on a paper machine. Used for polishing cloths, curtains, filter cloths, etc.

bonded wire (*Elec. Eng.*). Enamelled insulated wire also coated with a thin plastic; after forming a coil, it is heated by a current or in an oven or both for the plastic to set and the coil to attain a solid permanent form.

bond energy (*Chem.*). The energy in joules absorbed on the formation of a chemical bond between atoms, and released on its breaking.

bonder (*Build.*). See **bondstone.**

Bonderizing (*Met.*). Proprietary phosphating process for protection against corrosion.

bonding (*Aero.*). The electrical interconnexion of metallic parts of an aircraft normally at earth potential for the safe distribution of electrical charges and currents. Protects against charges due to precipitation, static, and electrostatic induction due to lightning strikes. Reduces interference and provides a low resistance electrical return path for current in earth-return systems. (*Elec. Eng.*) An electrical connexion between adjacent lengths of armouring or lead sheath, or across a joint. See also **cross bonding.**

bonding clip (*Elec. Eng.*). A clip used in wiring systems to make connexion between the earthed metal sheath of different parts of the wiring, in order to ensure continuity of the sheath.

bonding strength (*Paper*). The resistance of a coated or uncoated paper to picking or lifting of the surface fibres while being printed.

bond length (*Civ. Eng.*). The minimum length of reinforcing bar required to be embedded in concrete to ensure that the bond is sufficient for anchorage purposes. (*Chem.*) See **bond distance.**

bond paper (*Paper*). Similar to *bank paper* (q.v.), but weighing 15 lb *large post*/500 sheets and over.

bondstone (*Build.*). A long stone laid as a header through a wall. Also called a **bonder.**

bond strength (*Phys.*). See **adhesion.**

bond-timber (*Build.*). A horizontal large-section timber built into a brick wall and serving as a bond-course.

bone (*Mining*). Coal containing ash (*bone*) in very fine layers along the cleavage planes. Also called **bony coal.** (*Zool.*) A variety of connective tissue in which the matrix is impregnated with salts of lime, chiefly phosphate and carbonate.

bone beds (*Geol.*). Strata, often sands, gravels, or their indurated equivalents, characterized by an extremely high content of fossil remains. Examples are the Ludlow Bone Bed and the Rhaetic Bone Bed, the former consisting largely of the fragmental remains of primitive fossil fishes, and the latter of reptilian bone-fragments. Such abnormal concentration of fossils is due to some sudden change in conditions adversely affecting the fauna, or to the washing together, by wave- or current-action, of scattered bones, shells, etc., into a back-water or sheltered region.

bone black (*Paint.*). A fine black pigment made by calcining bones and containing a high proportion of calcium phosphate and carbonate, and about 10% carbon. The pigment has an intense black colour but poor opacity and tinting strength. Used mainly in polishes and as a decolorizing agent.

bone conduction (*Med.*). The conduction of sound-waves from the bones of the skull to the inner ear, rather than through the ossicles from the outer ear.

bone-dry paper (*Paper*). Paper dried completely to contain no moisture. Also **oven-dry paper.**

bone oil (*Chem.*). See **Dippel's oil.**

bone seeker (*Chem.*). Radioelement similar to calcium, e.g., Sr, Ra, Pu, which can pass into bone where it continues to radiate.

bone-setter (*Med.*). A medically unqualified person who treats disorders of joints by manipulation. See **osteopathy.**

bone tolerance dose (*Radiol.*). The maximum radioactive dose which can safely be given in treatment without bone damage.

Bonetti machine (*Elec. Eng.*). A type of influence machine somewhat similar to the Wimshurst machine but having no sectors and a larger number of brushes. It is not self-starting.

bone turquoise (*Min.*). Fossil bone or tooth, coloured blue with phosphate of iron; widely used as a gemstone. It is not true turquoise, and loses its colour in the course of time. Also called **odontolite.**

boning-in (*Surv.*). The process of locating and driving in pegs so that they are in line and have their tops also in line; carried out by sighting between a near and a far peg previously set in the gradient desired.

boning-rods (*Surv.*). T-shaped rods used, in sets of three, to facilitate the process of boning-in; two of the rods are held on the near and far pegs to establish a line of sight between them in the desired gradient, while the third is used to fix intermediate pegs in line.

Bonne's projection (*Geog.*). A derivative *conical projection* (q.v.) in which the parallels are spaced at true distances along the meridians, which are plotted as curves.

bonnet (*Build.*). A wire-netting cowl covering the top of a ventilating pipe or a chimney. (*Eng., etc.*) A movable protecting cover, e.g., (1) the cap of the valve-box of a pump; (2) the cover plate of a valve chamber; (3) the hood of a forge; (4) the cover over the engine of a motor vehicle. (*Plumb.*) A cover serving as a guide for a valve spindle, and enclosing the valve.

bonnet tile (*Build.*). Special rounded tile used to cover the external angles at hips and ridges on tiled roofs. See **arris tile.**

bony coal (*Mining*). See **bone.**

bookbinding (*Bind.*). Arranging in proper order the sections, etc., of a book and securing them in a cover. The bookbinder distinguishes bound book from cased book, and both from brochure and paperback.

book chase (*Typog.*). A chase with two crossbars, dividing it into quarters; used for bookwork.

book cloth (*Bind.*). The usual covering for edition binding. There are two main kinds: thin, hard-glazed cloth containing starch and other fillings, and matt-surfaced cloth with less starch.

book gill (*Zool.*). See **gill book.**

book lung (*Zool.*). See **lung book.**

book plate (*Bind.*). A label pasted on the inside front cover of a book bearing the owner's name, crest, coat-of-arms or peculiar device.

Boolean algebra (*Comp., Maths.*). An algebra, named after George Boole (1815–64), in which the elements are of two kinds only, and in which the basic operations are the logical AND and OR operations, which are normally symbolized as multiplication and addition respectively. Its main applications are to the design of switching networks and to mathematical logic.

boom (*Acous.*). Enhanced reverberation or resonance in an enclosed space at low frequencies, due to reduced acoustic absorption of the surfaces for low frequencies. (*Cinema., TV*) Mechanical arrangement for swinging the microphone clear of artists and cameras in soundfilm and television studios. (*Eng., Ships, etc.*) Any long beam; more especially (1) the flange of a built-up girder, also **chord;** (2) the main spar of a lifting-tackle; (3) the spar holding the lower part of a fore-and-aft sail; (4) a spar attached to a yard to lengthen it; (5) a barrier of logs to prevent the passage of a vessel; (6) a line of floating timbers used to form a floating harbour; (7) a pole marking a channel.

boomlight (*Photog.*). A studio floodlight attached to a long adjustable arm.

boost (*I.C. Engs.*). The amount by which the induction pressure of a supercharged internal-combustion engine exceeds atmospheric pressure expressed in kN/m^2 or lbf/in^2.

boost control (*Aero.*). A capsule device regulating reciprocating-engine manifold pressure so that supercharged engines are not over-stressed at low altitude.

boost control over-ride (*Aero.*). In a supercharged piston aero-engine fitted with *boost control,* a device (sometimes lightly wire-locked so that its emergency use can be detected), which allows the normal maximum manifold pressure to be exceeded: also **boost control cut-out.**

booster (*Autos.*). See **extractor.**

booster amplifier (*Acous.*). One used specially to compensate loss in mixers and volume-controls, so as to obviate reduction in signal/noise ratio.

booster coil (*Aero.*). A battery-energized induction coil which provides a starting spark for aero-engines.

booster diode (*TV*). See **efficiency diode.**

booster fan (*Eng.*). A fan for increasing the pressure of air or gas; used for restoring the pressure drop in transmission pipes, and for supplying air to furnaces.

booster pump (*Aero.*). A pump which maintains positive pressure between the fuel tank and the engine, thus intensifying the flow. (*Eng.*) A pump which is inserted in a closed-pipe system to increase the pressure of the liquid in some part of the circuit.

booster response (*Automation*). An automatic controller method of operation in which there exists a continuous linear response between rate of change of the controlled variable and the position of the final control element. Also called **rate action** or **time response.**

booster rocket (*Aero.*). See **take-off rocket.**

booster station (*Mining*). In long-distance transport by pipeline of liquids, mineral slurries, or water-carried coal, an intermediate pumping station where lost pressure energy is restored. (*Radio*) One which rebroadcasts a received transmission directly on the same wavelength.

booster transformer (*Elec. Eng.*). A transformer connected in series with a circuit to raise or lower the voltage of that circuit.

boost gauge (*Aero.*). An instrument for measuring the manifold pressure of a supercharged aero-engine in relation to ambient atmosphere or in absolute terms. Also used for racing, and some sports, cars.

boost transformer (*Elec. Eng.*). Same as **buck transformer.**

boot (*Glass*). See **potette.**

booted (*Bot.*). Said of agarics with the lower part of the stipe covered by a volva or veil. (*Zool.*) Having the feet protected by horny scales.

booth (*Cinema.*). See **camera-, monitoring-, projection-.**

Boothroyd Creamer system (*Telecomm.*). A form of time-division multiplex transmission.

bootstrap (*Eng.*). A self-sustaining system in liquid rocket engines by which the main propellants are transferred by a turbo-pump which is driven by hot gases. In turn the gas generator is fed by propellants from the pump.

bootstrap circuit (*Electronics*). Series voltage feedback circuit using a unity gain amplifier to provide an effectively infinite impedance or to improve the linearity of a voltage sweep generator.

bootstrap cold-air unit (*Aero.*). A unit of the compressor-turbine type in which the air charging an aircraft cabin passes through the compressor and, via an *intercooler*, the turbine.

bootstrapping (*Electronics.*). The technique of using bootstrap feedback. See **bootstrap circuit.**

boot tapping (*Ships*). The demarcation line between the two main colours of a ship's paintwork, at or near the waterline.

bora (*Meteor.*). A squally winter wind blowing down upon the northern shores of the Aegean and Adriatic seas.

boracic acid (*Chem.*). See **boric acid.**

boracite (*Min.*). The orthorhombic, pseudocubic form of magnesium borate and chloride, found in beds of gypsum and anhydrite.

borane (*Chem.*). General name for a hydride of boron, or *hydroboron* (q.v.).

borates (*Chem.*). See **boric oxide.**

borax (*Min.*). A mineral deposited by evaporation of the waters of alkaline lakes, notably in California, Nevada, and Tibet. Borax, which is hydrated sodium borate, occurs as a surface efflorescence, or as monoclinic crystals embedded in the lacustrine mud.

borax bead (*Chem.*). Borax, when heated, fuses to a clear glass. Fused borax dissolves some metal oxides giving glasses with a characteristic colour. The use of borax bead in chemical analysis is based on this fact.

borax carmine stain (*Micros.*). A stain made up of a concentrated solution of carmine in borax, and diluted with alcohol. Used for whole mounts of small organisms and for general staining.

Borazon (*Chem.*). A compound (diamond structure) of boron and nitrogen, produced under high pressure at a temperature of ca. 1650°C, characterized by extreme hardness and resistance to oxidation. It is harder than diamond.

bord-and-pillar (*Mining*). A method of mining coal by excavating a series of chambers, rooms, or stalls, leaving pillars of coal in between to support the roof. Also **room-and-pillar.**

Bordeaux B (*Chem.*). An azo-dyestuff derived from 1-naphthylamine coupled with *R-acid.*

bordered pit (*Bot.*). A thin area in the wall between two vessels or tracheids, surrounded by overhanging rims of wall thickening.

border effect (*Photog.*). A faint dark line on the denser side of a boundary between a lightly exposed and a heavily exposed region on a developed emulsion. Also called **edge effect** or **Mackie lines. Cf. fringe effect.**

border parenchyma (*Bot.*). A sheath of one or more layers of parenchymatous cells surrounding a vascular bundle.

border-pile (*Civ. Eng.*). A pile driven to support the sides of a coffer-dam.

border plane (*Join.*). A plane for cutting rebates or grooves along the edge of a piece of timber.

border tie (*Textiles*). A jacquard harness used to weave certain cloths. The arrangement allows one section to weave the centre of the fabric and the other to weave the border or borders.

bordroom-man (*Mining*). A man who removes debris and timbers the roof in old roadways when using the *bord-and-pillar* method.

bore (*Eng.*). The circular hole along the axis of a pipe or machine part; the internal wall of a cylinder; the diameter of such a hole. (*Hyd. Eng.*) A great tide-wave, with crested front, travelling rapidly up a river; it occurs on certain rivers having obstructed channels.

Boreal period (*Bot.*). A postglacial period characterized by extremes of temperature and low relative humidity. In the British Isles, *Hazel scrub* was the dominant vegetation.

borehole (*Civ. Eng.*). A sinking made in the ground by the process of *boring* (q.v.).

borehole survey (*Mining*). Check for deviation from required line of deep borehole. Methods include camera records of compass, plumbline, or gyroscope.

Borg-Warner overdrive (*Autos.*). A method of reducing engine r.p.m. in relation to road speed. The unit is attached at the rear of the gearbox and operates through epicyclic gears.

boric acid (*Chem.*) H_3BO_3. Boric acid is a tribasic acid. On heating it loses water and forms metaboric acid, $H_2B_2O_4$, and on further heating it forms tetraboric acid, or the so-called pyroboric acid, $H_2B_4O_7$. On heating at a still higher temperature it forms anhydrous boron(III)oxide, or boric oxide. It occurs as tabular triclinic crystals deposited in the neighbourhood of fumaroles, and known also in solution in the hot lagoons of Tuscany and elsewhere. Also called **boracic acid, sassolite.**

boric oxide (*Chem.*). Boron(III)oxide. B_2O_3. An 'intermediate oxide' like aluminium oxide, having feeble acidic and basic properties. As a weak acid it forms a series of borates. See **boric acid.**

boride (*Chem.*). Any of a class of substances, some of which are extremely hard and heat-resistant, made by combining boron chemically with a metal.

borine radical (*Chem.*). The radical BH_3, capable of forming compounds through coordination of a lone pair of electrons to the electron deficient boron, e.g., borine carbonyl, BH_3CO.

boring (*Eng.*). The process of machining a cylindrical hole, performed in a lathe, boring machine, or boring mill; for large holes, or when great accuracy is required, it is preferable to *drilling.* (*Mining, etc.*) The drilling of deep holes for the exploitation or exploration of oil fields. The term *drilling* is used similarly in connexion with metalliferous deposits.

boring bar (*Eng.*). A bar clamped to the saddle of a lathe or driven by the spindle of a boring machine, and carrying the boring tool.

boring machine (*Eng.*). A machine on which boring operations are performed, comprising a head, carrying a driving-spindle, and a table to support the work.

boring mill (*Eng.*). A vertical boring machine in which the boring bar is fixed, the work being carried by the rotating table.

boring tool (*Eng.*). The cutting tool used in boring operations. It is held in a boring bar. See **drill rod.**

Borna disease (*Vet.*). An infectious encephalomyelitis of horses, caused by a virus, and characterized by mild fever, a variety of nervous symptoms, and a high death rate. Sheep may occasionally be affected.

Borneo camphor (*Chem.*). *Borneol* (q.v.).

borneol (*Chem.*). $C_{10}H_{17}{\cdot}OH$, m.p. 208°C, b.p. 212°C, crystallizes in hexagonal plates; it has the character of a secondary alcohol, and yields camphor on oxidation; it forms with PCl_5 bornyl chloride, which is identical with pinene hydrochloride. Borneol occurs naturally in a (+), (−) and an inactive form.

bornite (*Min.*). Sulphide of copper and iron, crystallizing in the cubic system; occurs in Cornwall and many other localities. Develops a red and blue tarnish, hence, also called **erubescite, horse-flesh ore, peacock ore, variegated copper ore.**

bornite detector (*Radio*). Early demodulator consisting of a steel point in contact with a bornite crystal, with marked rectifying properties.

bornyl chloride (*Chem.*). $C_{10}H_{17}Cl$, m.p. 148°C, white crystals, identical with pinene hydrochloride, obtained from *borneol* (q.v.) by treatment with PCl_5.

boro-carbon resistors (*Electronics*). Resistors constructed of boron and carbon, so as to be substantially independent of temperature.

boroethane (*Chem.*). See **hydroborons.**

borofluorides (*Chem.*). See **fluoboric acid.**

borolanite (*Geol.*). A basic igneous rock occurring near Loch Borolan, Assynt, in N.W. of Scotland; it consists essentially of feldspar, green mica, garnet, together with conspicuous rounded white aggregates of 'pseudo-leucite', consisting of orthoclase feldspar and altered nepheline.

boron (*Chem.*). Amorphous yellowish-brown element discovered by Davy, 1808, also Gay-Lussac and Thénard. Symbol B, at. no. 5, r.a.m. 10·811, m.p. 2300°C, b.p. 2550°C, rel. d. 2·5. Can be formed into a conducting metal. Most important in reactors, because of great cross-section (absorption) for neutrons; thus, boron steel is used for control rods. The isotope ^{10}B on absorbing neutrons breaks into two charged particles ^{7}Li and ^{4}He which are easily detected, and is therefore most useful for detecting and measuring neutrons.

boron carbide (*Chem.*). B_4C. Boron carbide obtained from B_2O_3 and coke at about 2500°C. Very hard material, and for this reason used as an abrasive in cutting tools where extreme hardness is required. Extremely resistant to chemical reagents at ordinary temperatures.

boron chamber (*Nuc. Eng.*). Counter tube containing boron(III)fluoride, or boron-covered electrodes, for the detection and counting of slow-speed neutrons, which eject α-particles from the isotope ^{10}B.

boron nitride (*Chem.*). BN. When heated in an atmosphere of nitrogen or ammonia, boron forms *boron nitride*. See also **Borazon.**

boron trihalides (*Chem.*). All the four halogens unite with boron to form (III)halides as follows: BF_3, BCl_3, BBr_3, BI_3.

borosilicate glasses (*Glass*). Heat and chemical resisting glasses based on the fusion of silica-borax mixtures to produce complex polymers of sodium borosilicate.

Borrel bodies (*Vet.*). See **Bollinger bodies.**

Borrmann effect (*Phys.*). Light or dark diffraction lines produced when fluorescent X-rays are generated inside a single crystal or metallic plate.

borrow pit (*Civ. Eng.*). Excavation which provides extra material to serve as fill when required.

bort (or **boart**) (*Min.*). A finely crystalline form of diamond in which the crystals are arranged without definite orientation. Possessing the hardness of diamond, bort is exceedingly tough, and is used as the cutting agent in rock drills.

Bosch process (*Chem.*). Production of hydrogen by the catalytic reduction of steam with carbon monoxide at 500°C:

$$CO+H_2O \rightarrow H_2+CO_2.$$

Bose-Einstein distribution law (*Phys.*). A *distribution law* (q.v.) which is applicable to a system of particles with symmetric wave functions, this being the characteristic of most neutral gas molecules. It can be stated as

$$\bar{n}_i = \frac{g_i}{\frac{1}{A} exp\, E_i/kT - 1}$$

where $\bar{n}_i$ = average number of molecules with energy E_i, g_i = degeneracy factor, A = constant.

Bose-Einstein statistics (*Phys.*). Statistical mechanics laws obeyed by a system of particles whose wave function is unchanged when two particles are interchanged.

bosh (*Glass*). A water tank for cooling glass-making tools or quenching glass. (*Met.*) The tapering portion of a blast-furnace, between the largest diameter (at the bottom of the stack) and the smaller diameter (at the top of the hearth).

boson (*Nuc.*). A fundamental particle described by Bose-Einstein statistics and having angular momentum nh, where n is an integer and h is Planck's constant.

boss (*Eng.*). A projection, usually cylindrical, on a machine part in which a shaft or pin is to be supported; e.g., the thickened part at the end of a lever, provided to give a longer bearing to the pin. (*Geol.*) An igneous intrusion of cylindrical form, less than 100 km² in area; otherwise like a bathylith.

bossage (*Build.*). Roughly dressed stones, such as quoins and corbels, which are built in so as to project, and are finish-dressed in position.

bosset (*Zool.*). In Deer, the rudiment of the antlers in the first year.

bossing (*Plumb.*). The operation of shaping malleable metal, particularly sheet-lead, to make it conform to irregularities of the surface it is covering, by tapping with special mallets, known as **bossing mallets.**

bossing stick (*Plumb.*). A wooden tool used to shape sheet-lead into a lining for a tank.

Bostock sedimentation balance (*Powder Tech.*). A *sedimentation technique* (q.v.) in which a torsion balance is used directly to weigh the accumulation of particles. The high sensitivity enables low concentrations to be used, giving increased accuracy.

bostonite (*Geol.*). A fine-grained intrusive igneous rock allied in composition to syenite; essentially feldspathic, and deficient in coloured silicates; type-locality, Boston, Mass.

bostryx (*Bot.*). A cymose inflorescence in which the lateral axes always arise on the same side of the parent axis.

bot (*Vet.*). The larva of flies of the genus *Gastrophilus*; bots parasitize the membrane of the stomach of horses, rarely of other animals.

Botall's duct (*Zool.*). In some Vertebrates, a small blood vessel representing the sixth gill arch dorsal to the origin of the pulmonary arteries, and connecting the systemic with the pulmonary arch; important in the embryonic circulation of *Amniotes* and functional in some *Urodeles*, otherwise vestigial or absent.

Botany (*Textiles*). A term applied to the wool of the Merino sheep, and to tops, yarns, and fabrics made from this. See **quality terms.**

Botany twill (*Textiles*). A dress fabric woven in 2×2, 3×3, and 4×4 twill order from Botany yarns. Woven grey and piece-dyed, it is usually clear finished.

bothridium (*Zool.*). In *Cestoda*, a thin folded flap projecting from the scolex and used as an organ of fixation. Also called **phyllidium.**

bothrium (*Zool.*). In *Cestoda*, a groove-shaped sucker with loose weak musculature.

B.O.T. ohm (*Elec. Eng.*). Board of Trade ohm; a term sometime used to denote the *international ohm.*

botryoidal (*Min.*). Of minerals, formation resembling grapes. (*Zool.*) Shaped like a bunch of grapes; in some *Hirudinea*, said of a tissue surrounding the gut and composed of branched canals, the walls of which are formed of large cells containing black pigment; of unknown function.

botryology. The science of arranging things or concepts in groups and clusters. Gk. *botrys*, a bunch of grapes.

botryomycosis (*Vet.*). A suppurative, granulomatous infection of horses due to *Staphylococcus aureus*. When the spermatic cord becomes infected, following castration, the term *scirrhous cord* is applied.

botryose, botryoid, botrytic (*Bot., Zool.*). Branched; like a bunch of grapes. See **racemose.** See also **botryoidal.**

bottle (*Electronics*). Colloq. for **valve.**

bottle battery (*Elec. Eng.*). A term used to denote a bichromate cell when the electrodes and electrolyte are placed in a glass bottle-shaped container.

bottle glass (*Glass*). Glass used for the manufacture of common bottles, made from a *batch* comprising essentially sand, limestone, and alkali. A typical percentage glass composition may be taken as SiO_2 74·0, Al_2O_3 0·6, CaO 9·0, Na_2O 16·4.

bottle jack (*Eng.*). A screw-jack in which the lower part is shaped like a bottle.

bottle-making machines (*Glass*). These may operate in various ways, the bottle being formed in 2 stages, i.e., the parison and the finished bottle. Wide-mouth ware may be formed by pressing the parison and then blowing, narrow-mouth by blowing and blowing or sucking and blowing. In the last method, the glass is gathered by suction into the parison mould, in the other two it is dropped by hand or more probably by a mechanical-feeding device, hence the terms suction-fed and feeder-fed machines.

bottle-nose drip (*Plumb.*). The shaped edge formed in sheet-lead work at a step on a roof, when jointing the lead across the direction of fall.

bottle-nosed step (*Build.*). A step which has the edge and ends rounded.

bottle-stone (*Min.*). A *tektite* (q.v.) resembling bottle-glass, occurring in Bohemia and Moldavia. Used as a gemstone. Also called **moldavite, water-chrysolite.**

bottom (*Acous.*). A colloquialism for the lower range of audiofrequencies in sounds for recording, which contribute mainly to the loudness and fullness of these sounds.

bottom bars (*Textiles*). In lace manufacture, thin perforated steel strips, having holes punched in them as yarn guides, and which function in the machine, to produce variations in the ground.

bottom bend (*Electronics*). Same as **anode bend.**

bottom board (*Foundry*). A board placed on the underside of a mould during ramming.

bottom dead-centre (*Eng.*). See **outer dead-centre.**

bottom gate (*Foundry*). An *in-gate* (q.v.) leading from the runner into the bottom of a mould.

bottoming (*Civ. Eng.*). The lowest layer of foundation material for a road or other engineering works including structures. (*Rail.*) Ballast in permanent-way.

bottoming tap (*Eng.*). See **plug tap.**

bottom plate (*Horol.*). In a watch, the plate to which the pillars are fixed, generally referred to as the **dial plate.**

bottoms (*Met.*). A term used in connexion with the Orford process for separating nickel and copper as sulphides. When the mixed sulphides are fused with sodium sulphide, the nickel sulphide separates to the bottom. Hence *bottoms* as distinct from *tops*. In reverberatory furnace, the heaviest molten material at bottom of pool.

bottom-samplers (*Ocean.*). Various types of apparatus which, when lowered, are capable of piercing the sea-bottom and retaining a sample of the deposit.

bottomset beds (*Geol.*). See **topset beds.**

bottom shore (*Carp.*). One of the members of an arrangement of raking shores to support temporarily the side of a building; it is the one nearest the wall face.

bottom yeast (*Bot.*). The yeast that collects at the bottom of a vessel in which alcoholic fermentation is proceeding.

botuliform (*Bot.*). Sausage-shaped.

botulism (*Med., Vet.*). Poisoning due to eating flesh, carrion, or decaying vegetation which has been infected with *Bacillus botulinus* (*Clostridium botulinum*).

Bouchard's index (*Nut.*). A French index—weight in kilograms divided by height in decimetres.

Boucherot circuit (*Elec. Eng.*). An arrangement of inductances and capacitances, whereby a constant-current supply is obtained from a constant-voltage circuit.

Boucherot motor (*Elec. Eng.*). A name sometimes given to the double-cage induction motor.

bouchon (*Horol.*). A hollow plug, or bush, inserted in watch or clock plates to form the pivot holes. To repair a worn hole, the hole is enlarged and a bouchon pressed in. In certain cases the jewels are held in bouchons which are a press fit in the plates.

bouchon wire (*Horol.*). Hollow wire, generally of hard brass, which is cut off to the required lengths to form bouchons.

bouclé (*Textiles*). Describes a 3- or 4-fold novelty yarn characterized by a curled or looped effect standing out prominently from the body. This is most noticeable on the face of coating cloths in which the bouclé appears as warp or weft.

boudinage (*Geol.*). Structure found in sedimentary series subjected to folding. It consists of strike-elongated 'sausages' of more rigid rock, enclosed between relatively plastic rocks.

bougie (*Med.*). A tube or a rod for dilating narrowed passages in the body.

bougie-decimale (or **decimal candle**) (*Light*). Obsolete unit. One-twentieth of luminous intensity, viewed normally, of 1 cm^2 of molten platinum at the temperature of solidification. Displaced by the *candela* (q.v.).

Bouguer anomaly (*Geol.*). A gravity anomaly which has been corrected for the station height and for the gravitational effect of the slab of material between the station height and datum. Cf. *free-air anomaly*.

Bouguer law of absorption (*Phys.*). The intensity of a parallel beam of monochromatic radiation entering an absorbing medium is decreased at a constant rate by each infinitesimally thin layer,

$$\frac{-dp}{p}=k\ d\ b,$$

where k is a constant depending on the wavelength, nature of absorber, and concentration, p is the intensity of the beam and db is the absorber thickness.

Bouin's solution (*Micros.*). A microanatomical fixative, suitable for most tissues, and in which material can be preserved for long periods. It consists of a combination of formalin, glacial ethanoic acid and picric acid.

boulder (*Geol.*). The unit of largest size occurring in sediments and sedimentary rocks, the limit between pebble and boulder being placed at 100 mm, though some authorities recognize 'cobbles' between pebbles and boulders. Boulders may consist of any kind of rock, may be subangular or well rounded, may have originated in place or have been transported by running water or ice. Accumulations of boulders are *boulder beds*.

boulder clay (*Geol.*). The characteristic product of glaciation, consisting of stones to the size of boulders, sometimes faceted and striated, embedded in a clay matrix which is essentially rock powder produced by abrasion of the floor over which the glacier moved.

boulder paving (*Build.*). Paving constructed with rounded boulders laid on a gravel foundation.

boulder wall (*Build.*). A wall built of boulders or flints set in mortar.

boule (*Min.*). The pear-shaped or cylindrical drop of synthetic mineral, commonly ruby, sapphire or spinel, produced by the Verneuil flame-fusion process.

bouncing-pin detonation meter (*Eng.*). An apparatus for determining quantitatively the degree of detonation occurring in the cylinder of a petrol-engine; used for fuel testing.

boundary films (*Met.*). Films of one constituent of an alloy surrounding the crystals of another.

boundary layer (*Aero.*). The thin layer of fluid (air) adjacent to the surface in which viscous forces exert a noticeable influence on the motion of the fluid and in which the transition between still air and the body's velocity occurs. (*Chem. Eng.*) When a fluid flows past a solid surface, the layer of fluid next to the solid surface is brought to rest, setting up a viscous motion in adjacent layers. The boundary layer is the total thickness of fluid over which the surface exerts a differential effect. It governs the rate of heat, mass, or momentum between the solid surface and the homogeneous bulk of the fluid. The velocity of the layer differs significantly from that of the main fluid stream, and is therefore of considerable importance in heat transfer problems, as in nuclear reactors.

boundary lights (*Aero.*). Lights defining the boundary of the landing area.

boundary lubrication (*Eng., Phys.*). A state of partial lubrication which may exist between two surfaces in the absence of a fluid oil film, due to the existence of adsorbed mono-molecular layers of lubricant on the surfaces.

boundary markers (*Aero.*). See **aerodrome markers.**

boundary wavelength (*Phys.*). In an X-ray spectrum, the shortest wavelength.

bound book (*Bind.*). Must be hand-sewn on cords or tapes which would be firmly attached to the boards, by lacing-in or using split boards, before applying the covering for which leather would normally be used. Cf. **cased book.** See **full bound, half-bound, quarter-bound, three-quarter-bound.**

bound charge (*Elec. Eng.*). That induced static charge which is 'bound' by the presence of the charge of opposite polarity which induces it. Also, in a dielectric, the charge arising from polarization. Also **surface charge.** See also **free charge.**

bounded function (*Maths.*). A function is said to be bounded above if it has an upper bound, and bounded below if it has a lower bound. If bounded both above and below, the function is said to be bounded. Cf. *bounds of a function.*

bounded set (of numbers) (*Maths.*). A set of numbers lying between two particular numbers.

bounded set (of points) (*Maths.*). A set of points for which the set of distances between pairs of points is a bounded set.

bounds of a function (*Maths.*). M is the upper bound of a function $f(x)$, if $f(x) \leqq M$, and, for at least one value of $x, f(x) > M-\delta$, where δ is any positive number. m is the lower bound of $f(x)$, if $f(x) \geqq m$ and, for at least one value of x, $f(x) < m+\delta$.

bound vector (*Maths.*). A vector with a particular point of application. Also **localized vector.**

bound water (*Bot.*). Water held in organic substances by physical adsorption.

bouquet stage (*Cyt.*). See **pachytene.**

Bourdon gauge (*Eng.*). See **pressure gauge.**

bourette (*Textiles*). (1) Heavy yarn made chiefly from waste silk, but having tufts of wool or other material twisted with it at intervals. (2) Cloth made from silk waste yarns. Plain weave, it has an irregular, uneven surface.

bourgeois (*Typog.*). An old type size, approximately 9-point.

bourne (*Geog.*). An intermittent stream, flowing only during the rainy season.

Bournemouth Beds (*Geol.*). A division of the Bracklesham Beds, of Eocene age, occurring in the Hampshire Basin; it comprises some 135 m of fresh-water beds with plant remains, together with shell-bearing marine beds at the top.

bournonite (*Min.*). Mixed sulphide of general formula $CuPbSbS_3$. Orthorhombic, grey-black streak. Hardness $2\frac{1}{2}$–3; rel. d. 5·7 to 5·9. Commonly occurs as wheel-shaped twins, hence known as **cog-wheel ore,** or **wheel-ore.** Also **antimonial lead ore.**

bouton (*Cyt.*). Specialized terminal (*end bulb*) of the arborization of an axon, forming a synapse with another neurone.

boutons terminaux (*Histol.*). Minute swellings of the terminal branches of the axon of a neurone, which are distributed over the dendrites and cell body of the neurone which they excite, forming a synaptic junction.

bouyoucous hydrometer (*Powder Tech.*). A special hydrometer used in particle size analysis of soil grains. It gives readings direct in concentration of soil colloids per litre of suspension in water.

Bovey Tracey Beds or **Bovey Beds** (*Geol.*). A series of fresh-water clays, sands, and lignites, some 180 m in thickness, occupying a depression near Bovey Tracey in Devonshire; regarded as the site of a lake in Pliocene times, receiving detritus from the surrounding high ground.

bovine acetonaemia (*Vet.*). Bovine ketosis. A metabolic disease of unknown cause, affecting lactating cows, in which ketone bodies accumulate in the tissues and are excreted in the milk, urine, and breath. Characterized by reduction in appetite and milk production, and sometimes nervous symptoms.

bovine brucellosis (*Vet.*). See **bovine contagious abortion.**

bovine contagious abortion (*Vet.*). Bang's disease; bovine brucellosis. A contagious bacterial infection of cattle caused by *Brucella abortus*; the main symptom is abortion due to infection of the placenta.

bovine cutaneous streptothricosis (*Vet.*). A chronic, exudative dermatitis affecting cattle in tropical Africa and elsewhere, caused by a fungus *Dermatophilus congolensis*. A similar disease occurs in goats and horses.

bovine cystic haematuria (*Vet.*). A disease of the cow in which tumourlike growths develop in the mucosa of the bladder and cause haematuria; the cause is not known.

bovine farcy (*Vet.*). A chronic disease of cattle characterized by suppurative nodules beneath the skin, due to infection of the lymphatic vessels and lymph nodes by the bacterium *Nocardia farcinica.*

bovine hyperkeratosis (*Vet.*). X disease. A disease of cattle, characterized by emaciation, loss of hair, and thickening of the skin, due to poisoning by chlorinated naphthalene compounds.

bovine hypomagnesaemia (*Vet.*). Grass disease; grass tetany; grass staggers; lactation tetany; Hereford disease. A metabolic disease of cattle in which the magnesium content of the blood is low; characterized by hyperaesthesia, tetany and death. The disease occurs most commonly in cattle turned out to lush pasture in the spring; the cause is unknown.

bovine infectious petechial fever (*Vet.*). Ondiri disease. A disease of cattle in Kenya characterized by fever, petechial haemorrhages of mucous membranes, diarrhoea, and often death; believed to be caused by a rickettsia-like organism.

bovine ketosis (*Vet.*). See **bovine acetonaemia.**

bovine lipomatosis (*Vet.*). A diffuse growth of lipomata in cattle, usually involving the abdominal mesenteries and viscera.

bovine pasteurellosis (*Vet.*). See **haemorrhagic septicaemia.**

bovine pyelonephritis (*Vet.*). A specific infection of the kidneys of cattle by the bacterium *Corynebacterium renale.*

bow (*Elec. Eng.*). A sliding type of current collector, used on electric vehicles to collect the current from an overhead contact-wire. It consists of a bow-shaped contact strip, mounted on a hinged framework. (*Horol.*) (1) The ring of a pocket-watch case, to which the watch or fob chain is attached. (2) A flexible strip of whalebone or cane, the ends of which are drawn together to give tension to a thread or line which is given a single turn round a pulley of a pair of turns, drill, or mandrel. It is used as a sensitive drive for these tools, and by many it is considered to be the best way to produce very fine accurate pivots.

Bowden-Thomson protective system (*Elec. Eng.*). A form of protective system for feeders, in which special cables, with the cores surrounded by metallic sheaths, are employed; a fault causes current to flow in the sheath and operate a relay to trip the circuit.

Bowditch's rule (*Surv.*). A rule for the adjustment of closed compass traverses, in which it may reasonably be assumed that angles and sides are equally liable to error in measurement. According to this rule, the correction in latitude (or departure) of any line is:

$$\frac{\textit{Length of that line}}{\textit{Perimeter of traverse}} \times \textit{Total error in latitude (or departure)}.$$

Bowdon gauge (*Elec. Eng.*). Form of pressure-sensitive transducer.

bowel oedema disease (*Vet.*). Oedema disease; gut oedema. A disease of pigs characterized by nervous symptoms and oedema in many tissues; believed to be due either to immunological hypersensitivity to *Escherichia coli* or to the production of toxins by this bacterium.

bowenite (*Min.*). A compact, finely granular, massive form of serpentine, formerly thought to be nephrite, and used for the same purposes.

bower anchor (*Ships*). One of the anchors carried at the bow of a ship; used for anchoring and mooring.

Bower-Barff process (*Met.*). Method of blueing steel in which steam heated to 900°C is used to produce an oxidized film. (*San. Eng.*) An anti-corrosion process applied to sanitary ironwork; this, when red-hot, has superheated steam passed over it in a closed space, so that a protective layer of black magnetic oxide is formed on the iron-work. See **Angus-Smith process.**

bowk (*Mining*). A large iron barrel used when sinking a shaft. Also called **kibble.**

bowl (*Typog.*). The enclosed part of letters such as B and R, and the upper portion of g, distinguishing it from the loop or tail. See **counter.**

bowlingite (*Min.*). See **saponite.**

Bowman's capsule (*Zool.*). In the Vertebrate kidney, the dilated commencement of a uriniferous tubule.

Bowman's glands (*Zool.*). In some Vertebrates, serous glands of the mucous membrane of the olfactory organs.

Bowman's membrane (*Zool.*). In Vertebrates, a lamina of homogeneous connective tissue upon which rests the stratified epithelium of the cornea.

bow nut (*Eng.*). See **cap nut.**

bows or **bow compasses** (*Instr.*). See under **spring bows.**

bow-saw (*Tools*). A thin-bladed saw which is kept taut by a bow or special frame.

bow sheaves (*Elec. Eng.*). The sheaves at the bow of a cable-laying ship over which the cable passes when it is being laid in the sea or raised for repair.

Bow's notation (*Eng.*). A method of notation for forces acting at a point, the spaces between the forces being lettered in order, so that any force is described in terms of the letters referring to the two adjacent spaces. By this device the force polygon can be lettered correspondingly. Also called **Henrici's notation.**

bowstring bridge (*Civ. Eng.*). An arched bridge in which the horizontal thrust on the arch is taken by a horizontal tie joining the two ends of the arch.

bowstring suspension (*Elec. Eng.*). A form of suspension for the overhead contact-wire of an electric-tramway system, in which the contact-wire is suspended from a short cross-wire attached to the bracket-arm of the pole.

bow strip (*Elec. Eng.*). See **contact strip.**

box (*For.*). See **boxwood.**

box annealing (*Met.*). See **close annealing.**

box antenna (*Radio*). See **travelling-wave antenna.**

box baffle (*Acous.*). Box, with or without apertures and damping, one side fitted with an open diaphragm loudspeaker unit, generally coil-driven.

boxboard (*Paper*). Thin board used in the manufacture of cartons. Made from mechanical woodpulp and waste.

box calf (*Leather*). Calf skins which have been chrome-tanned, dyed black, and subjected to boarding to produce the box grain.
boxcar (*Comp.*). Converter of digital or sample data to analogue form, output forming a step waveform.
box chronometer (*Ships*). The marine chronometer. The chronometer is normally supported on gimbals, inside a wooden box with a hinged lid.
box cloth (*Textiles*). A woollen fabric woven 2×2 twill, it is milled and finished with a smooth surface like felt; manufactured from fine wools.
box column (*Arch.*). A built-up hollow column of square or rectangular section.
box coupling (*Eng.*). See **muff coupling.**
box culvert (*Civ. Eng.*). A culvert having a rectangular opening.
box dam (*Civ. Eng.*). A coffer-dam built to surround an area in which works are to proceed.
box dock (*Civ. Eng.*). A double-sided floating-dock of channel section.
box drain (*San. Eng.*). A small rectangular section drain, usually built in brickwork or concrete.
boxed frame (*Join.*). A *cased frame* (q.v.).
boxed mullion (*Join.*). A hollow mullion in a sash window-frame, arranged to accommodate the counterweights connected to the vertically moving sashes.
box-frame motor (*Elec. Eng.*). A traction motor in which the frame is cast in one piece instead of being split.
box girder (*Build.*). A cast-iron, or mild steel, girder of hollow rectangular section. See **box plate girder.**
box grain (*Leather*). A small square produced on box calf by treatment while it is damp.
box gutter (*Build.*). A wooden gutter, lined with sheet-lead, zinc, or asphalt, and having upright sides; used along roof valleys or parapets.
box-in (*Typog.*). To surround type with rule, the printed matter appearing in a frame.
boxing (*Civ. Eng.*). A layer of small ballast packed between the sleepers of a railway track. (*Join.*) The part of a window-frame which receives the folded shutter.
boxing shutters (*Join.*). Shutters at the interior side of a window, hung so as to fold back into a recess in the jambs. Also called **folding shutters.**
box junction (*Autos.*). An area at a road junction marked as a square criss-crossed by diagonal lines into which a vehicle may not move unless its exit is clear, or unless it is turning to the right, in which case only one car at a time may rest in the box junction.
box loom (*Textiles*). A loom equipped with several shuttle boxes, circular or drop-type, so that different colours, qualities, or types of weft can be inserted to increase the patterning range.
box nut (*Eng.*). See **cap nut.**
box of tricks (*Spinning*). Speedframe builder motion for regulating the bobbin speed as it builds up in diameter. It also reduces the traverse of the lifter as each layer of slubbing or roving is wound on the bobbin. Also called **differential.**
box plate girder (*Build.*). A built-up steel girder, similar to the plate girder, but having two web plates at a distance apart, so that flanges and webs enclose a rectangular space.
box slip (*Carp.*). A hard boxwood slip secured to the beechwood stock of a tonguing or grooving plane and forming a durable facing at the rubbing surface.
box spanner (*Tools*). One composed of a cylindrical piece of steel shaped at one or both ends to a hexagon in order to fit over the appropriate nut. The spanner is turned by a steel rod (*tommy bar*) inserted through a diametrically drilled hole at the opposite end to the hexagon in use. Used for nuts inaccessible to an ordinary spanner.
box-staple (*Carp.*). The part on a door-post into which the bolt of a lock engages.
box stones (*Geol.*). Hollow concretions. Nodules of sandstone containing molluscan casts found in the Pleistocene deposits of East Anglia.
box tool (*Eng.*). A single-point cutting tool, set radially or tangentially, used in automatic screw machines and in capstan and turret lathes.
box-type brush-holder (*Elec. Eng.*). See **brush-box.**
box-type (or cage-type) negative plate (*Elec. Eng.*). A form of negative plate for an accumulator which is made up by riveting two lead grids together and placing the active material in the spaces between them.
boxwood (*For.*). The pale-yellow, close-grained, hard and tough wood of the box tree, used for drawing scales, tool handles, blocks for woodcuts, etc.; it requires several years of seasoning.
Boyle's law (*Phys.*). The volume of a given mass of gas kept at one uniform temperature varies inversely as the pressure. There are deviations from this law at low and high pressures and according to the nature of the gas.
Boyle temperature (*Heat*). The temperature at which the second virial coefficient of a gas changes sign. Close to this temperature, Boyle's law provides a good approximation to the equation of state of the gas.
Boys' camera. A camera for photographing lightning flashes, gyrating lenses separating the strokes.
b.p. (*Chem.*). Abbrev. for *boiling-point*.
B.P. (*Chem.*). An abbrev. for *British Pharmacopoeia*. (*Ships*) *Between perpendiculars*, i.e., length between *forward perpendicular* (q.v.) and after perpendicular (after side of sternpost).
B-position (*Teleph.*). In a manual exchange, any individual switchboard section that is mainly equipped for the reception and completion of calls originated by subscribers foreign to that exchange.
Br (*Chem.*). The symbol for *bromine*.
braccate (*Zool.*). Of Birds, having feathered legs or feet.
brace (*Carp., etc.*). A tool used to hold a bit and give it rotary motion. The bit is secured axially in a socket at one end, the other end (to which pressure is applied) being in line with it, while the middle part of the brace is cranked out so that the whole may be rotated. Also called a **bit-stock.** (*Eng., etc.*) A tie rod or bar connecting two parts of a structure for stiffening purposes; it is always subjected to a tensile force. (*Typog.*) ⏞ Usually cast to a definite em measurement; sectional braces are built up to the required length; should point towards the lesser number of lines.
braced girder (*Build.*). A girder formed of two flanges connected by a web consisting of a number of bars dividing the girder into triangles or trapeziums and transmitting the vertical loads from one flange to another.
braced-tread (*Autos.*). See **radial-ply.**
brace jaws (*Tools*). The parts of the socket of a brace which clamp upon the shank of the brace bit to secure it while drilling.
Brace polarizer (*Light*). A thin plate of calcite immersed in a liquid of high refractive index. The extraordinary ray is internally reflected by inclining the plate to the incident beam.
Brace spectrophotometer (*Light*). A visual instrument in which a comparison field is produced in

the dispersing prism by means of a second collimator.

brachelytrous (*Zool.*). Said of Insects having short truncate elytra which do not completely cover the abdomen. Also **brachyelytrous.**

brachi-, brachio-. Prefix from L. *brachium*, arm.

brachial (*Zool.*). See **brachium.**

brachial disk (*Zool.*). In certain *Discomedusae*, a horizontal disk formed by the union of the larval arms, and occupying the centre of the subumbrellar surface.

brachial ossicles (*Zool.*). In *Crinoidea*, the ossicles supporting the arms.

brachial skeleton (*Zool.*). In *Testicardines*, the calcareous processes of the dorsal valve which support the lophophore.

brachiate, brachiferous (*Bot., Zool.*). Branched: having widely spreading branches: bearing arms.

brachiocephalic (*Anat.*). Pertaining to the arm and head.

brachiolaria (*Zool.*). A larval stage of some *Asteroidea*, differing from the *bipinnaria* (q.v.) in having three adhesive papillae, and a sucker developed on the pre-oral lobe.

brachioles (*Zool.*). In *Cystoidea*, armlike processes of the oral side of the body which carry the food grooves.

Brachiopoda (*Zool.*). A phylum of solitary nonmetameric *Metazoa*, with a well-developed coelom; sessile marine forms, with a lophophore in the form of a double vertical spiral, and usually with a bivalve shell. Brachiopods range from early geological periods up to the present time; they occur in all seas, often at great depths.

brachium (*Zool.*). The proximal region of the fore-limb in land Vertebrates: a tract of nerve-fibres in the brain: more generally, any armlike structure, as the rays of Starfishes. *adj.* **brachial.**

brachy-. Prefix from Gk. **brachus**, short.

brachyblast (*Bot.*). A short branch of limited growth, bearing leaves and, sometimes, flowers and fruit; a *spur* (q.v.).

brachycephalic (*Anat.*). Short-headed; said of skulls whose breadth is at least four-fifths of the length. Found in some types of mental retardation.

brachycerous (*Zool.*). Having short antennae, as some *Diptera.*

brachycladous (*Bot.*). Having very short branches.

brachydactyly, brachydactylia (*Med.*). Abnormal shortness of fingers or toes.

brachyelytrous (*Zool.*). See **brachelytrous.**

brachyform (*Bot.*). Said of a species of rust fungus in which the aecidium is omitted from the life history, or is replaced by a primary uredosorus.

brachymeiosis (*Cyt.*). A simplified form of meiosis, completed in one division.

brachyodont (*Zool.*). Said of Mammals having low-crowned grinding teeth in which the bases of the infoldings of the enamel are exposed: used also of the teeth. Also **brachydont.** Cf. *hypsodont.*

brachypterism (*Zool.*). In Insects, the condition of having the wings reduced in length. *adj.* **brachypterous.**

brachysclereide (*Bot.*). See **stone cell.**

brachystomatous (*Zool.*). Of Insects, having the proboscis reduced in length.

brachyurous, brachyural (*Zool.*). Said of decapodan *Crustacea*, in which the abdomen is reduced and bent forward underneath the laterally expanded cephalothorax by which it is completely hidden.

bracing (*Civ. Eng.*). The staying or supporting rods or ties which are used in the construction or strengthening of a structure.

bracing wires (*Aero.*). The wires used to brace the wings of biplanes and the earlier monoplanes. See also **drag wires, flying wires,** and **landing wires.**

brack (*For.*). Grades of timber poorer than market grades.

bracken poisoning (*Vet.*). A disease occurring in cattle and horses due to the ingestion of bracken. In cattle, the main symptoms are multiple haemorrhages and high fever, associated with bone marrow damage. In horses, nervous symptoms are shown and the disease is essentially an induced thiamin deficiency.

bracket (*Build.*). A projecting support for a shelf or other part.

bracket arms (*Elec. Eng.*). The transverse projecting arms on the poles, for supporting the overhead contact wire equipment for a tramway or railway system.

bracket baluster (*Build.*). An iron baluster, bent at its foot and fixed into the side of the step, usually when the latter is made of stone or of concrete.

bracketed step (*Join.*). A step supported by a *cut string* (q.v.) which is shaped on its lower edge to form an ornamental bracket.

bracket-fungus (*Bot.*). One of a number of species belonging to the *Basidiomycetes*, occurring on tree stumps and trunks, and projecting in the form of a rounded bracket.

bracketing (*Build.*). The shaped timber supports forming a basis for plasterwork and mouldings of ceilings and parts near ceilings.

bracket scaffold (*Build.*). A scaffold supported on framed brackets carried by **grapplers** (q.v.).

Brackett series (*Phys.*). A group of spectral lines of atomic hydrogen in the infrared given by the formula

$$\nu = R_H\left(\frac{1}{n_1^2} - \frac{1}{n_2^2}\right) \text{ in which}$$

ν = the wave number
$R_H = 1{\cdot}096\,775\,8 \times 10^7$ m^{-1}
$n_1 = 4$
n_2 = various integral values.

Bracklesham Beds (*Geol.*). A series of variable sands and clays of Eocene age occurring above the Lower Bagshot Beds in the Hampshire Basin; of marine origin in the east, but freshwater in the west.

bracon (*Zool.*). A transverse bar on the inner side of the labium in the larvae of most *Coleoptera.*

bract (*Bot.*). The leaf which subtends an inflorescence or a flower; leaves in close association with sporangia or sex organs. (*Zool.*) In *Siphonophora*, a hydrophyllium: in some *Branchiopoda*, a lobe on the side of a trunk outer limb.

bracteal leaf (*Bot.*). A general term for bracts and bracteoles.

bracteate (*Bot.*). Having bracts.

bracteody (*Bot.*). The replacement of other members of the flower by bracts.

bracteole (*Bot.*). A leaf, generally very small, borne on the stalk of a flower. *adj.* **bracteolate.**

bracteomania (*Bot.*). An abnormal condition in which a plant forms an enormous number of bracts, and sometimes fails to form normal flowers.

bracteose (*Bot.*). Having conspicuous bracts.

bract scale (*Bot.*). The small outer scale at the base of the large cone scale in conifers.

brad (*Foundry*). A nail with a small head projecting on one side, or with the head flush with the sides. See **sprig.**

bradawl (*Join.*). A small chisel-edged tool, used to make holes for the insertion of nails and screws.

brad setter (*Join.*). A tool for holding a brad by its head and driving it into position.

bradsot (*Vet.*). See **braxy.**

brady-. Prefix from Gk. *bradys*, slow.

bradyarthria (*Med.*). Abnormally slow delivery of speech.

bradycardia (*Med.*) Slowness of the beating of the heart.

bradykinesia (*Med.*). Abnormal slowness of the movements of the body.

bradyphrenia (*Med.*). Slowness of mental processes.

bradyspore (*Bot.*). A plant from which the seeds are liberated slowly.

Bragg angle (*Phys.*). Angle of incidence of a beam of X-rays which results in wave reflection from the surface of a crystal lattice.

Bragg curve (*Phys.*). Graph giving average number of ions per unit distance along beam of initially monoenergetic α-particles (or other ionizing particles) passing through a gas.

Bragg law (*Phys.*). Conditions for reflecting beam of X-rays with maximum intensity. Equation is:

$$\sin \theta = n\lambda/2d,$$

where θ = angle which the incident and reflected waves make with the crystal planes (Bragg angle), n = integer, λ = wavelength of X-rays, d = spacing of planes or layers of atoms. Also applies for the reflection of de Broglie waves associated with protons, electrons, neutrons.

Bragg method (*Phys.*). A method of investigating crystal structure by means of X-rays, used successfully by Sir Wm. Bragg and latterly by many other workers.

Bragg rule (*Phys.*). An empirical relationship according to which the mass stopping power of an element for α-particles (also applicable to other charged particles) is proportional to (relative atomic mass)$^{-\frac{1}{2}}$.

Bragstad converter (*Elec. Eng.*). See **motor converter.**

braid (*Textiles*). A wide range of narrow, or fancy, fabric woven on smallware looms and used as a trimming for dress material, upholstery or coach and car interiors.

braided stream (*Geod.*). A stream which consists of several channels which separate and join in numerous places. Braided streams occur where the gradient is steep and where seasonal floods are liable to occur. They generally have wide beds filled with loose detritus.

brain (*Zool.*). A term used loosely to describe the principal ganglionic mass of the central nervous system: in Invertebrates, the pre-oral ganglia: in Vertebrates, the expanded and specialized region at the anterior end of the spinal cord, developed from the three primary cerebral vesicles of the embryo.

brain-sand (*Zool.*). In higher Vertebrates, calcareous nodules occurring within the pineal gland, and in the pia mater and its extensions as age advances.

brain stem (*Zool.*). In Vertebrates, regions of the brain conforming to the organization of the spinal cord, as distinct from such suprasegmental structures as the cerebral cortex and the cerebellum.

brain stimulation (*An. Behav.*). An experimental procedure in which rats have micro-electrodes implanted into a certain region of the brain, through which they receive electrical stimulation. This may act either as a positive or a negative reinforcing stimulus, according to the area stimulated.

brain voltage (*Physiol.*). Electric signal waves generated in human brain. Usually classed as alpha, beta and delta waves according to frequency.

brake (*Eng., etc.*). A device for applying resistance to the motion of a body, either (1) to retard it, as with a vehicle brake, or (2) to absorb and measure the power developed by an engine or motor. (*Print.*) The manual or automatic mechanism to control the tension of the reel when a rotary press is running.

brake bands (*Print.*). Strips of leather, fabric, or metal acting on a pulley on the reel shaft of a rotary press.

brake drum (*Eng.*). A steel or cast-iron drum attached to a wheel or shaft so that its motion may be retarded by the application of an external band or internal brake shoes. See **band brake, expanding brake.** (*Print.*) A flat or V-shaped wheel at one end of the reel spindle by which the web tension is controlled.

brake efficiency (*Autos.*). At level road, list percentage reading = $\text{MPH}^2 \div 0{\cdot}3$ of the stopping distance in feet.

brake-fade (*Autos*). Condition caused by overheating due to excessive use, resulting in decreased efficiency and sometimes in complete absence of stopping-power.

brake field (*Electronics*). Same as **retarding field.**

brake horsepower (*Eng.*). The effective or useful horsepower developed by a prime-mover or electric motor, as measured by a brake applied to the driving shaft. 33 000 ft lbf/min = 1 brake horsepower = 745·7 watts. Abbrev. **bhp.**

brake incline (*Mining*). Gravity plane. An incline in which the full trucks descend by gravity and pull up the empty ones.

brake lining (*Eng.*). Strips of asbestos-base friction fabric riveted to the shoes of internal expanding brakes in order to increase the friction between them and the drum and provide a renewable surface. Modern practice is to bond lining to shoe. See **brake shoe.**

brake magnet (*Elec. Eng.*). A permanent magnet or electromagnet which produces a braking effect, either by inducing eddy currents in a moving conductor or by operating a mechanical brake by means of a solenoid.

brake mean effective pressure (*Eng.*). That part of the *indicated mean effective pressure* (q.v.) developed in an engine cylinder which would result in a cylinder output equal to the brake horsepower of the engine; the product of I.M.E.P. and mechanical efficiency. Abbrev. **B.M.E.P.**

brake pads (*Autos.*). The operative members in a disk brake, corresponding to the brake shoes in ordinary drum-type brakes.

brake parachute (*Aero.*). One attached to the tail of some high-performance aircraft and streamed as a brake for landing. Sometimes a ribbon canopy is used for greater strength and on large aeroplanes a cluster of two or three is required to give sufficient area with convenient stowage. Also called **landing parachute.**

brake shoe (*Eng.*). (1) Unit which carries the renewable rubbing surface of a block brake. (2) The segmental member which is pressed against the inner surface of a brake drum.

brake thermal efficiency (*Eng.*). The efficiency of an engine reckoned in terms of the brake horsepower; given by the ratio of the heat equivalent of the brake output to the heat supplied to the engine in the fuel or steam.

brake-wheel arc lamp (*Light*). A form of arc lamp in which the carbons are automatically fed towards the arc, as the current through the arc drops, by release of an electromagnetic brake.

braking airscrew (*Aero.*). See **airscrew.**

braking notches (*Elec. Eng.*). Positions of the handle of a drum-type controller which apply some orm of electric braking.

Bramah's press (*Hyd.*). See **hydrostatic press.**

Bramley Fall stone (*Build.*). A durable dark-brown coarse-grained sandstone quarried from the Millstone Grit of Yorkshire; used for general building, paving, and stone steps, and also for heavy work such as foundations for machinery, piers, and bridges.

brammallite (*Min.*). A variety of illite with sodium as the inter-layer cation.

bran (*Nut.*). The outer protective layers of grain, usually wheat, after removal of the endosperm.

branch (*Comp.*). See **conditional jump.** (*Nuc.*) Alternative modes of radioactive decay. (*Telecomm.*) Electric components comprising a minimum path between junction points of common connection in a network. Also **arm.**

branch abscission (*Bot.*). The shedding of branches by plants, by means of an organized separation layer.

branch circuit (*Elec. Eng.*). A circuit branched off a main circuit.

branch drain (*San. Eng.*). The communicating drain between a gulley, soil pipe, or sanitary fitting and the main drain.

branch exchange (*Teleph.*). See **private — —.**

branch gap (*Bot.*). An area of parenchyma in a stele, below which the *branch trace* arises.

branchia (*Zool.*). In aquatic animals, a respiratory organ consisting of a series of lamellar or filamentous outgrowths; a gill. *adj.* **branchial.**

branchial arch (*Zool.*). In Vertebrates, one of a series of bony or cartilaginous structures lying in the pharyngeal wall posterior to the hyoid arch; it prevents the gill-slits from collapsing.

branchial basket (*Zool.*). (1) In *Cyclostomata* and cartilaginous Fish, the skeletal framework supporting the gills. (2) In the larvae of certain Dragonflies (*Anisoptera*), an elaborate modification of the rectum associated with respiration.

branchial chamber (*Zool.*). In *Urochorda*, cavity of the pharynx.

branchial clefts (*Zool.*). See **gill-slits.**

branchial duct (*Zool.*). In some *Cyclostomata*, the ventral respiratory tube or bronchus.

branchial formula (*Zool.*). A table showing the number and arrangement of the gills.

branchial heart (*Zool.*). In Vertebrates, a heart such as that of *Cyclostomata*, in which all the blood entering the heart is deoxygenated and passes thence directly to the respiratory organs: in *Cephalopoda*, special muscular dilatations which pump blood through the capillaries of the ctenidia.

branchial rays (*Zool.*). Branches of the hyoid and branchial arches which support the gills and gill-septa.

branchicolous (*Zool.*). Parasitic on the gills of Fish.

branchihyal (*Zool.*). One of the skeletal elements composing a branchial arch.

branching (*Nuc.*). The existence of two or more modes by which a radionuclide can undergo radioactive decay, e.g., ^{64}Cu can undergo β^-, β^+ and electron-capture decay.

branching enzyme (*Chem.*). An enzyme concerned with glycogen synthesis demonstrated in liver, muscle and brain, which cleaves fragments of the glycogen chain at α-1→4 linkages and transfers them to the same or another glycogen molecule but in α-1→6 linkage.

branching ratio (*Chem.*). Where a radioactive element can disintegrate in more than one way, the ratio of the quantities of the element undergoing each type of disintegration is called the *branching ratio.*

branchiocardiac vessel (*Zool.*). In *Urochorda*, the vessel given off from the ventral end of the heart, serving the pharynx.

branchiogenital region (*Zool.*). In *Enteropneusta*, the region of the trunk wherein lie the respiratory and genital organs.

branchiomere (*Zool.*). In metameric animals, a somite bearing branchiae.

branchiomerism (*Zool.*). Serial repetition of gill-clefts or gills.

Branchiopoda (*Zool.*). Subclass of *Crustacea*, the members of which are distinguished by the possession of numerous pairs of flattened, leaf-like, lobed swimming feet which also serve as respiratory organs; the mandible is without a palp in the adult; mainly fresh-water forms including the Fairy Shrimps, Brine Shrimps, Tadpole Shrimps, Clam Shrimps, Water Fleas.

branchiostegal (*Zool.*). Pertaining to the gill-covers.

branchiostegal membrane (*Zool.*). In Fish, the lower part of the opercular fold below the operculum.

branchiostegal rays (*Zool.*). Skeletal supporting structures of the branchiostegal membrane.

branchiostege (*Zool.*). See **branchiostegal membrane.**

branchiostegite (*Zool.*). In some *Crustacea*, a lateral extension of the carapace covering the gill-chamber.

branchireme (*Zoo .*). An appendage having both locomotor and respiratory functions, as in *Branchiopoda.*

Branchiura (*Zool.*). Subclass of *Crustacea* in which paired compound eyes occur; the genital openings occur on the fifth trunk somite and the thoracic limbs sometimes possess a flagellum. Fish Lice.

branch jack (*Teleph.*). See **jack.**

branch of a curve (*Maths.*). Any section or portion of a curve which is separated by a discontinuity (sometimes any singularity) from another section, e.g., a hyperbola has two branches.

branch of a function (*Maths.*). See **Riemann surface.**

branch pipe (*Plumb.*). A special pipe having one or more branches.

branch point (*Maths.*). See **Riemann surface.**

branch switch (*Elec. Eng.*). A term used in connexion with electrical installation work to denote a switch of any type for controlling the current in a branch circuit.

branch tendril (*Bot.*). A tendril formed from a modified branch.

branch trace (*Bot.*). The vascular bundle arising from the main stele and extending into a branch.

brandering (*Join.*). The process of nailing small fillets of wood in a counter direction and on the underside of floor joists. Plasterboards or metal lath are fixed to the branders to take plaster. Also called **counter lathing.**

brand fungi (*Bot.*). See **Ustilaginales.**

branding iron (*Civ. Eng.*). See **indenter.**

bran disease (*Vet.*). See **osteodystrophia fibrosa.**

brand spore (*Bot.*). The thick-walled resting spore of the brand fungi; it is black or brown, and forms sooty masses.

brass (*Bind.*). See **binders' brass.** (*Met.*) Primarily, name applied to an alloy of copper and zinc, but other elements such as aluminium, iron, manganese, nickel, tin, and lead are frequently added. There are numerous varieties. (*Mining*) A coal-miner's term for iron pyrites or 'Welsh gold'—a brassy-looking compound of iron and sulphur. Also called **brasses.**

brass bobbins (*Textiles*). Two disks of thin brass, dished and riveted together so as to hold a supply of cotton, nylon or 'Terylene', etc., yarn for curtain, etc., manufacture. The filling of these bobbins collectively from a jack of wood bobbins is called *brass-winding*.

brasses (*Eng.*). Those parts of a bearing which provide a renewable wearing surface; they consist of a sleeve or bored block of brass split diametrally, the two halves being clamped into the bearing block by a cap. (*Mining*) See **brass.** (*Textiles*) (1) In lace-manufacture, brass plates in the form of a parallelogram with rows of holes through which the warp threads pass. (2) In cotton-spinning the upper bearing for mule spindles.

Brassica (*Bot.*). A genus of *Cruciferae* (Brassicaceae), which includes cabbage, radish, wallflower, and many common wildflowers.,

brassidic acid (*Chem.*). Naturally occurring unsaturated C_{22} fatty acid. Transisomer of erucic acid.

brass rule (*Typog.*). Type-high brass strip, preferred to type-metal rule for heavy work.

brattice or brattice cloth (*Mining*). A partition for diverting air, for the purpose of ventilation, into a particular working place or section of a mine.

bratticing, brattishing (*Build.*). See **cresting.**

Braun tube (*Electronics*). Original name for **cathode-ray tube,** after Braun (1850–1918) the inventor.

braunite (*Min.*). A massive, or occasionally well-crystallized, ore of manganese, occurring in India, New South Wales, and elsewhere. Composition $3Mn_2O_3{\cdot}MnSiO_3$.

Bravais lattices (*Crystal.*). The 14 distinct lattices which can be formed by the array of representative points in the study of crystal structure.

braxy (*Vet.*). Bradsot. An acute and fatal toxaemia of sheep due to infection of the abomasum by *Clostridium septicum.*

brayer (*Typog.*). A hand ink-roller.

Brayton cycle (*Eng.*). A constant-pressure cycle of operations used in gas turbines.

brazier (*Build.*). A portable iron container for a lighted fire, used to dry off building work in a room, or as a source of warmth for outside night-watchmen.

Brazilian aquamarines (*Min.*). Fine large blue aquamarines obtained from Minas Novas in the state of Minas Geraes in Brazil.

Brazilian emerald (*Min.*). A pure-green, deeply coloured variety of tourmaline, occurring in Brazil; used as a gemstone. Not a true emerald.

Brazilian kingswood (*For.*). A rare but important timber from the genus *Astronium,* used for high-class furniture, etc.

Brazilian mahogany (*For.*). Hardwood from the genus *Cariniana,* not a true mahogany, though it may be used for the same purposes. It remains very stable after seasoning.

Brazilian pebble (*Min.*). The name applied to Brazilian quartz or rock-crystal, used in the manufacture of spheres for crystal-gazing, lenses, etc.

Brazilian peridot (*Min.*). Green crystals of tourmaline or chrysoberyl from Brazil having the typical colour of peridot (olivine).

Brazilian ruby (*Min.*). Among the many coloured topaz crystals mined in Brazil some are pink (rose topaz), others deep red after heating; these latter are termed Brazilian ruby. The term is also used for a red variety of tourmaline.

Brazilian sapphire (*Min.*). TN for the beautiful clear blue variety of tourmaline mined in Brazil; used as a gemstone, but not a true sapphire.

Brazilian topaz (*Min.*). True topaz varying in colour from pure white to blue and yellow; mined chiefly in the state of Minas Geraes, Brazil.

Brazil resistance (*Elec. Eng.*). See **carbon-dust resistance.**

Brazil wax (*Chem.*). *Carnauba wax* (q.v.).

brazing (*Eng.*). The process of joining two pieces of metal by fusing a layer of brass or spelter between the adjoining surfaces.

brazing solders (*Met.*). Alloys used for brazing. They include copper-zinc (50–55% copper), copper-zinc-silver (16–52% copper, 4–38% zinc, and 10–80% silver), also nickel-silver alloys.

B.R.C. fabric (*Build., Civ. Eng.*). A very open, electrically welded, wire mesh with apertures about 3 by 12 in. (75 × 300 mm), used as a reinforcing medium for concrete roads, floor slabs, etc. (*B*ritish *R*einforced *C*oncrete.)

bread-crust bomb (*Geol.*). A type of *volcanic bomb* (q.v.) having a compact outer crust and a spongy vesicular interior.

breadth coefficient (*Elec. Eng.*). See **distribution factor.**

breadth factor (*Elec. Eng.*). See **distribution factor.**

breadths (*Textiles*). Narrow lace, made in tubular form, the edges being held together by draw-threads which are afterwards withdrawn.

break (*Build.*). (1) Any projection from, or recess into, the surface of a wall. (2) To nail the laths so that the joints are staggered, i.e., not in the same vertical line. See also **break-joint** and **breaking joint** below. (*Elec. Eng.*) The shortest distance between the contacts of a switch, circuit-breaker, or similar apparatus, when the contacts are in the fully open position. (*Mining*) A jointing plane in a coal-seam. (*Min. Proc.*) Optimum range of size to which ore should be ground before concentration in a specific case.

breakaway (*Cinema.*). Any construction in sound-film production which is made so that it can fall to pieces easily.

break-before-make (*Telecomm.*). Classification of switch and relay wipers where existing contacts are opened before new ones close.

breakbone fever (*Med.*). See **dengue.**

breakdown (*Elec. Eng.*). The sudden passage of current through an insulating material at *breakdown voltage* (q.v.). (*Phys.*) Of an emulsion, the reunion of the finely dispersed particles and their separation from the medium with which they form an emulsion. Also called **breaking.**

breakdown crane (*Eng.*). A portable jib crane carried on a railway truck or motor lorry, for rapid transit to the scene of an accident.

breakdown diode (*Electronics*). See **Zener diode.**

breakdown voltage (*Elec. Eng.*). Voltage at which a marked increase in the current through an insulator or semiconductor occurs. Abbrev. **BDV.** See also **disruptive voltage.**

breaker (*Elec. Eng.*). See **circuit-breaker.** (*Paper*) Apparatus, comprising a washing roll fitted with knives, for reducing raw material, other than grasses and wood pulp, to fibres or 'half-stuff'.

break impulse (*Telecomm.*). An impulse formed by interrupting a current in a circuit, e.g., by dialling.

break-in (*Telecomm.*). Attachment of an operator's circuit to a telegraph, telephone line, or radio channel for transmitting on to or taking over control of a circuit already established.

breaking (*Bot.*). The development of striping in flowers due to virus infection. (*Phys.*) See **breakdown.**

breaking capacity (*Elec. Eng.*). The capacity of a switch, circuit-breaker, or other similar device to break an electric circuit under certain specified conditions.

breaking current (*Elec. Eng.*). The maximum current which a switch, circuit-breaker, or other similar device will interrupt without damage to itself.

breaking joint (*Build.*). The principle of laying bricks or building stones in such a manner that joints are not continuous.

breaking length (*Paper*). The length at which a strip of paper would break under its own weight when suspended. Expressed in metres.

breaking of the meres (*Bot.*). The sudden development of large masses of blue-green algae (Cyanophyceae) in small bodies of fresh water.

breaking piece (*Eng.*). An easily replaceable member of a machine subject to sudden overloads; made weaker than the remainder, so that in breaking it protects the machine from extensive damage.

breaking stress (*Eng.*). The stress necessary to break a material, either in tension or compression. See **ultimate tensile stress.**

break iron (*Build.*). Iron dog on which slates are laid for cutting. Also **dressing iron.** (*Carp.*) The iron which is screwed to the cutting iron of a plane, to bend and break the shavings.

break jack (*Teleph.*). See **jack.**

break-joint (*Build.*). See **breaking joint.**

break line (*Typog.*). See **club line.**

break-out (*Foundry*). Failure of mould of furnace wall to retain molten metal which escapes from the containing system.

break point (*Comp.*). Point in a programme where a conditional interruption may be made for a visual check or print-out. (*Work Study*) The instant at which one element in a work cycle ends and another begins.

break rolls (*Eng.*). Grooved chilled-steel rollers, set in pairs, for shearing open wheat before separation of various parts of the grain.

breakthrough (*Min. Proc.*). In industrial ion-exchange recovery of metal (e.g., uranium) from solution, the point at which traces of metal begin to arrive in the last of a series of resin-filled stripping columns.

breakwater (*Civ. Eng.*). A natural or artificial coastal barrier serving to break the force of the waves so as to provide safe harbourage behind; it differs from the bulwark in that it has the sea on both sides of it.

breakwater-glacis (*Civ. Eng.*). An inclined stone paving on piers and breakwaters, designed to take the force of impact of the waves.

breast (*Agric.*). See **mouldboard.** (*Anat.*) An accessory gland of the generative system, rudimentary in the male and secreting milk in the female. Extending from the third to the sixth rib in the front of the chest, it consists of fatty, fibrous, and glandular tissue, the ducts of which end in the nipple. (*Build.*) The wall between a window and the floor. See also **chimney-breast.** (*Carp.*) The underside of a handrail or rafter. (*Mining*) (1) The working coal-face in a colliery. (2) Underground working face. In flat lodes, breast stopes are those from which detached ore will not gravitate without help.

breast beam (*Weaving*). Polished cast-iron plate radiused on one edge. Bolted to the tops of both side frames of the loom, it ensures that the cloth is wound evenly on to the take-up roller.

breast bone (*Zool.*). In higher Vertebrates, the sternum: in the larvae of Gall Midges (*Cecidomyidae*), an elongate sclerite situated mid-ventrally on the thorax.

breast box (*Paper*). The compartment immediately before the machine wire for providing an even distribution of the flow of pulp stock.

breast lining (*Join.*). Panelling between window board and skirting.

breast mouldings (*Carp.*). Mouldings on the part of the wall between a window and the floor.

breast roll (*Paper*). Roll for carrying the wire cloth at the breast box end of the paper machine.

breastsummer (*Build.*). See **bressummer.**

breast wall (*Arch.*). A breast-high parapet or retaining wall.

breastwork (*Arch.*). A parapet on a building.

Breathalyzer (*Chem.*). TN for apparatus designed to measure alcohol content of the blood by a chemical analysis of alveolar air (e.g., by reduction of potassium dichromate in sulphuric acid solution).

breather pipe (*Eng.*). A vent pipe from the crank-case of an internal-combustion engine, to release pressure resulting from blow-by.

breathing (*Zool.*). An activity of many animals, resulting in the rapid movement of the environment (water or air) over a respiratory surface. Now usually referred to the more general concept of *respiration* (q.v.).

breathing apparatus (*Mining*). Mine rescue equipment in which oxygen is fed to a face mask carried by wearer, via a demand valve. See **Weg rescue apparatus.**

breathing root (*Bot.*). A root produced by mangroves and other large plants growing in mud; it projects above the mud and water and air conveys into the roots below.

breccia (*Geol.*). A coarse-grained clastic rock consisting largely of angular fragments of pre-existing rocks. According to its mode of origin, a breccia may be a *fault-breccia*, a *crush-breccia*, an *intrusion breccia*, or a *flow-breccia*.

Bredig's arc process (*Met.*). Process for making colloidal suspensions of metals in a liquid by striking an arc in the liquid between two electrodes of the metal.

breech (*Textiles*). See **britch.**

breech block (*Eng.*). A movable block used for closing and opening an aperture, originally in guns but now also in machines.

breeder reactor (*Nuc. Eng.*). One which produces more fissile material than is consumed in establishing the requisite neutron flux.

breeding cycle (*An. Behav.*). The sequence of behaviour in Birds involved in nest-building and rearing the young, consisting of elaborate types of *appetitive behaviour* (q.v.)

breeding ratio (b_r) (*Nuc.*). The number of fissionable atoms produced per fissionable atom destroyed in a nuclear reactor. ($b_r - 1$) is known as the **breeding gain.**

breeze (*Build.*). A general term for furnace ashes, or for *coke breeze*, *pan breeze*, and *furnace clinker* (qq.v.).

breeze concrete (*Build.*). A concrete made of 3 parts coke breeze, 1 of sand, and 1 of Portland cement. It is cheap and nails can be driven into it, but it has poor fire-resisting qualities.

breeze fixing brick (*Build.*). A brick made from cement and breeze, built into the surface of a wall to take nails.

breezing (*Cinema.*). Said of a projected cinematograph image which is not clear, because of inaccurate focusing in camera, printer, or projector, uneven sprocket holes, or defects in processing.

bregma (*Anat.*). The point of junction of the coronal and sagittal sutures of the skull.

Bréguet spring (*Horol.*). A special form of balance spring, in which the outer coil of the spring is raised above the plane of the spiral, the end of the spring being bent to a special form before it enters the stud.

breithauptite (*Min.*). Nickel antimonide, occurring as bright coppery-red hexagonal crystals, widely distributed in sulphide ore deposits in small amounts.

Breit-Wigner formula (*Nuc.*). An equation which relates the cross-section σ of a particular nuclear reaction to the energy E of the incident particle (A) and the energy E_r of a resonance level, when E is close to E_r. If only one resonance level is involved, then

$$\sigma(A, B) = (2l+1)\frac{\lambda^2}{4\pi} \cdot \frac{W_A W_B}{(E-E_r)^2 + (\frac{1}{2}W)^2}$$

where $\sigma(A, B)$ is the cross-section for the reaction which involves the capture of particle A and the emission of particle B. W is the total width of the energy level, and W_A, W_B are the partial widths respectively of the energy levels for the modes of disintegration in which particle A is re-emitted and B is emitted. λ is the *de Broglie wavelength* of particle A, and l is the orbital angular momentum quantum number of incident particle A.

Bremer arc lamp (*Light*). An early form of inclined-carbon flame arc lamp.

Bremograph (*Light*). An arc projection device for exhibiting moving patterns on a screen, particularly with superposed lettering, in a cinema.

bremsstrahlung (*Nuc.*). Electromagnetic radiation arising from collision or deviation between fast-moving electrons and atoms.

brephnic (*Zool.*). See **neanic.**

bressummer (*Build.*). A beam or lintel spanning a wide opening in a wall with whose surface it is flush. Also called a **breastsummer.**

Bretonian (*Geol.*). The name applied to the Upper Cambrian strata of the Atlantic seaboard of N. America (Cape Breton); cf. *Croixian.*

breunnerite (*Chem., Min.*). Variety of *magnesite* (q.v.) containing some iron; found in Canada, central Europe, and India. Used in manufacture of magnesite bricks.

brevi-. Prefix from L. *brevis*, short.

brevicollate (*Bot.*). Having a short neck.

brevier (*Typog.*). An old type size, approximately 8-point.

brevipennate (*Zool.*). Having short wings.

brevium (*Chem.*). Name sometimes used for uranium(II).

brewer's pound (*Brew.*). Unit of specific gravity used in brewing, expressed as the excess sp. gr. of a liquid over that of water (*brewer's gravity*). Sp. gr. in brewer's pounds = (sp. gr. − 1000) × 0·36.

Brewster angle (*Optics, Radio*). A plane wave polarized in the plane of incidence is totally transmitted when incident on a plane dielectric boundary at the Brewster angle. The Brewster angle (measured from the normal to the boundary) is given by $\tan\theta = \sqrt{\varepsilon_2/\varepsilon_1}$ where ε_1, ε_2 are the permittivities of the two media. Also **polarizing angle.**

brewsterite (*Min.*). A rare strontium-barium zeolite.

Brewster law (*Phys.*). That relating the Brewster angle (θ) to the refractive index (n) of the medium for a particular wavelength, viz.,

$$\tan\theta = n.$$

For sodium light incident on particular glass, $n = 1{\cdot}66$, θ is 51°.

Brewster's bands (*Light*). Interference fringes which are visible when white light is viewed through two parallel and parallel-sided plates, whose thicknesses are in a simple ratio (1 : 1, 2 : 1, 1 : 3, etc.).

Brewster stereoscope (*Light*). A device by which a pair of stereoscopic pictures, the centre separation being greater than the interocular distance, can be viewed.

Brewster windows (*Phys.*). Windows attached in certain designs of gas laser to reduce the reflection losses which would arise from the use of external mirrors. Their operation depends on the setting of the windows at the Brewster angle to the incident light.

Brianchon's theorem (*Maths.*). The lines joining pairs of opposite vertices of a hexagon circumscribed about a conic are concurrent. The dual of *Pascal's theorem* (q.v.).

brick (*Build.*). A shaped and burnt block of special clay, used for building purposes.

brick-and-stud work (*Build.*). See **bricknogging.**

brick-axe (*Build.*). The two-bladed axe used by bricklayers in dressing bricks to special shapes.

brick clay (*Geol.*). An impure clay, containing iron and other ingredients. In industry, the term is applied to any clay, loam, or earth suitable for the manufacture of bricks or coarse pottery. See **brick earths.**

brick-core (*Build.*). Rough brickwork filling between a timber lintel and the soffit of a relieving arch.

brick earths (*Build.*). Earths used for the manufacture of ordinary bricks; they consist generally of clayey silt interstratified with the fluvio-glacial gravels of southern England, frequently exploited in brick manufacture.

bricking (*Build.*). Work on plastered or stuccoed surfaces, in imitation of brickwork.

bricklayer's hammer (*Tools.*). A hammer having both a hammer-head and a sharpened peen; used for dressing bricks to special shapes.

bricklayer's scaffold (*Build.*). A scaffold used in the erection of brick buildings, a characteristic being that one end of the *putlogs* (q.v.) is supported in holes left in the wall.

bricknogging (*Build.*). The type of work used for walls or partitions which are built up of brickwork laid in spaces between timber. Also called **brick-and-stud work.**

brick-on-edge coping (*Build.*). A coping finish to the exposed top of a wall; formed of bricks built on edge in cement in courses 4½ instead of 3 in. high, so that the frogs are concealed and only a few joints are exposed to the weather.

brick-on-edge sill (*Build.*). An external sill to window or door, formed in the manner of the *brick-on-edge coping* (q.v.).

brick-trimmer (*Build.*). See **trimmer arch.**

brick trowel (*Build.*). A flat triangular-shaped tool used by bricklayers for picking up and spreading mortar.

bridge (*Bot.*). See **disjunctor.** (*Elec. Eng.*) Circuit on which two potentials, each divided from a common source, can be made equal, e.g., Wheatstone bridge for d.c., and analogous developments for a.c. Widely used in measurements. (*Horol.*) A raised platform or support, generally with two feet.

bridge board (*Carp.*). See **notch board.**

bridged-T filter (*Telecomm.*). One consisting of

a T-network, with a further arm bridging the two series arms; used for phase compensation.

bridge duplex (*Teleg.*). Duplex system in which the neutrality of the receiving apparatus at each end to the currents sent from that end is secured by a balance of potentials on the Wheatstone bridge principle.

bridge feedback (*Telecomm.*). Original negative feedback amplifier in which a balanced bridge provides a feedback voltage, independent of load impedance. Also called **Black feedback.**

bridge fuse (*Elec. Eng.*). A fuse in which the fusible wire is carried in a holder, supported by spring contacts at its two ends; it is thus easily removable for renewing the fuse wire.

bridge gauge (*Eng.*). A measuring device for detecting the relative movement of two parts of a machine due to wear at bearings, etc.

bridge hanger (*Elec. Eng.*). A form of hanger of small vertical dimensions, for supporting the overhead contact-wire of a traction system under bridges or tunnels.

bridge-megger (*Elec. Eng.*). A portable instrument for measuring large resistances on the Wheatstone-bridge principle. A megger contains a source of e.m.f. and the instrument dial on which the balance is indicated.

bridge network (*Telecomm.*). Same as **lattice network.**

bridge neutralizing (*Elec. Eng.*). Method for overcoming the adverse effects of interelectrode capacitances in thermionic valve amplifiers. Two valves are connected in push-pull with the anodes and grids cross-connected through balancing capacitors, the whole forming a balanced bridge.

bridge oscillator (*Elec. Eng.*). One in which positive feedback and limitation of amplitude is determined by a bridge, which contains a quartz crystal for determining the frequency of oscillation. Devised by Meachan for high stability of operation in crystal clocks, etc.

bridge-pile (*Civ. Eng.*). A pile driven to afford a firm foundation for a bridge-pier.

bridge receiver (*Telecomm.*). One used when two-way working is carried out on one wavelength. By the use of the bridge principle it is sensitive to the distant, but not to the local, transmitter.

bridge rectifier (*Elec. Eng.*). Type of full-wave rectifier employing four rectifiers in the form of a bridge. The alternating supply is connected across one diagonal, and the direct-current output is taken from the other.

bridge-set (*Teleg.*). An arrangement of normal transformers which replaces the hybrid coil, when the associated apparatus of a 4-wire radio-link is joined to the 2-wire exchange and subscribers' lines.

bridge stone (*Build.*). A flat stone spanning a narrow area or gutter.

bridge transformer (*Telecomm.*). The same as **hybrid coil.**

bridge transition (*Elec. Eng.*). A method, employed in connexion with the series-parallel control of traction motors, in which the change from series to parallel is effected without interrupting the main circuit, and without any change in the current flowing in each of the motors.

bridging (*Carp.*). The principle of diminishing lateral distortion of adjacent floor-joists by connecting them together with short cross-pieces. (*Mining, Min. Proc.*) Arching of jammed rock so as to obstruct flow of ore; clogging of filtering septum by tiny particles which are individually small enough to pass, but which form such arches. (*Telecomm.*) (1) Said of wipers in step relays when they do, or do not (**non-bridging**), bridge adjacent contacts when passing from one to the next. (2) Undesirable wear in relay contacts due to a protuberance building up on one of the contacts leading to a false closure of the circuit.

bridging fibrils (*Zool.*). Protoplasmic strands connecting the deep cells of stratified epithelium with one another across the intercellular channels.

bridging floor (*Carp.*). A floor supported by bridging joists, without girders.

bridging host (*Bot.*). A temporary host of a parasitic fungus, by means of which it may pass from one species to another.

bridging joist (*Carp.*). A timber beam immediately supporting the floor-boards in a floor. Also called a **common joist.**

bridle (*Elec. Eng.*). A portion of an overhead contact-wire system. It extends longitudinally between supporting structures and is attached at intervals to the contact-wire, in order to retain the latter in its proper lateral position. (*Paint.*) Whipcord or wire fastened half-way up the bristle of a new brush, when it is too long or flexible.

bridle butts (or **backs**) (*Leather*). Pliable butts, free from defects; used for harness.

bridle joint (*Carp.*). The converse of the *mortise-and-tenon joint* (q.v.). The central part on the first member is cut away to leave two side tongues projecting, and the second member is cut away at the sides to receive these tongues.

brier-tooth saw (*Tools.*). See **gullet saw.**

Briggs logarithms (*Maths.*). See **logarithm.**

bright annealing (*Met.*). The heating and slow cooling of steel or other alloys in a carefully controlled atmosphere, so that oxidation of the surface is reduced to a minimum and the metal surface retains its bright appearance.

bright coal (*Mining*). Coal which includes bands of clean and bright vitrain, clarain, or anthraxylon.

bright emitter (*Electronics*). A thermionic valve with pure tungsten cathode, for emitting electrons at ca. 2600 K. Originally used of any thermionic valve, but now restricted to those of high power.

brightener (*Elec. Eng.*). An addition agent added to an electroplating solution to produce bright deposits.

bright-line viewfinder (*Photog.*). Type of *Albada viewfinder* (q.v.) which reflects a luminous frame outlining the field of view. Also **brilliant-frame finder.**

brightness (*Light*). See **luminance.** As a quantitative term, *brightness* is deprecated.

brightness control (*TV*). Electrical control which alters the brightness and/or contrast (gamma factor) on a cathode-ray tube screen.

brighton (*Textiles*). (1) A cotton fabric of the honeycomb type. (2) Woollen cloth used for vestings, etc. Ornamentation is introduced by bright spots from silk or man-made fibre yarns.

Brighton system (*Elec. Eng.*). A name sometimes given to the maximum-demand method of charging for an electric supply; derived from the name of the town in which it was first used.

bright plating (*Elec. Eng.*). The production of a fairly bright deposit from an electroplating plant. Such surfaces require little finishing.

Bright's disease (*Med.*). A general term for acute and chronic nephritis.

Brightype (*Print.*). A *conversion system* (q.v.) in which type pages or relief blocks are sprayed with a dull black, and are then wiped to produce a bright image on a dark background. This can be photographed to produce a positive that can be used for lithographic or photogravure plate-making.

brilliance (*Acous.*). The presence of considerable numbers of high harmonics in musical tone, or the enhancement of these in sound reproduction. (*TV*) The average illumination over the whole of the picture.

brilliant (*Typog.*). Old type size about 4-point.

brilliant-frame finder (*Photog.*). See **bright-line viewfinder.**

brilliant green (*Chem.*). The sulphate of *tetra-ethyl-diamino-triphenyl-methanol anhydride*; a green dye used as a disinfectant.

brilliantine (*Textiles*). A plain-weave lustre dress material, with a cotton warp and a worsted weft; the lustrous character is produced by the weft. Figured patterns are produced by floating the weft on a plain ground.

brilliant viewfinder (*Photog.*). One comprising a reflector between two small lenses. An inclined mirror gives waist-height viewing; a prism gives eye-level viewing. The image is laterally inverted unless an extra lens called an *erector* is added.

Brillouin formula (*Mag.*). A quantum mechanical analogue in paramagnetism of the Langevin equation in classical theory of magnetism.

Brillouin zone (*Electronics*). Polyhedron in k-space, k being position wave vector of the groups or bands of electron energy-states in the band theory of solids. Often constructed by consideration of crystal lattices and their symmetrics.

brindled bricks (*Build.*). Bricks which, owing to their chemical composition, show a striped surface; when they are otherwise satisfactory they are frequently used in cases where appearance is not an important consideration.

Brinell hardness test (*Met., etc.*). A method of measuring the hardness of a material by measuring the area of the indentation produced by a hard steel ball under standard conditions of loading. Expressed as *Brinell Hardness Number*, which is the quotient of the load on the ball in kgf divided by the area of indentation in mm^2.

brine pump (*Eng.*). The pump used to circulate brine through the evaporator of a refrigerator, the working parts being of corrosion-resisting alloy.

brining (*Agric.*). Treatment of seed corn with brine as protection against *smut disease*. See **Ustilaginales.**

Brin's process (*Chem.*). A now discontinued process for industrial separation of oxygen from air by cyclic operation of oxidizing barium oxide to peroxide in a stream of air followed by higher temperature roasting of the barium peroxide giving free oxygen and barium oxide for next cycle.

briquettes (*Fuels*). Fuel made from finely divided carbonaceous matter in pressure moulds, the shape and size depending on requirements; usually made from low-grade coal or coke breeze, mixed with a binder such as pitch, tar, cement, or asphaltum.

brisket (*Vet.*). The breast or anterior sternal region of an animal.

bristle (*Bot.*). (1) A very stiff, erect hair. (2) A long hollow outgrowth of the cell wall in some algae.

Bristol board (*Paper*). A fine-quality cardboard made by pasting several sheets together, the middle sheets usually being of an inferior grade.

Bristol diamonds (*Min.*). Small lustrous crystals of quartz, i.e., rock crystal, occurring in the Bristol district.

Britannia metal (*Met.*). Alloy series of tin (80–90%) with antimony, copper, lead, or zinc, or a mixture of these.

britch (or **breech**) (*Textiles*). One of the dirtier, shorter grades of wool sheared from the sheep's legs. Often stained, it is a lower quality fibre.

British Association (B.A.) screw-thread (*Eng.*). A system of metric threads, confined to small sizes, used in instrument work, etc. It is designated by numbers from 0 to 25, ranging from 6 mm to 0·25 mm in diameter and from 1 mm to 0·072 mm pitch.

British Columbian pine (*For.*). See **Douglas Fir.**

British Engineering Standards Association. Now *British Standards Institution* (q.v.).

British Standard brass (B.S.B.) thread (*Eng.*). A screw thread of Whitworth profile used for thin-walled tubing; it has 26 threads per inch irrespective of diameter. See **British Standard Whitworth thread.**

British Standard fine (B.S.F.) thread (*Eng.*). A screw-thread of Whitworth profile, but of finer pitch for a given diameter; largely used in automobile work.

British Standard pipe (B.S.P.) thread or British Standard gas thread (*Eng.*). A screw-thread of Whitworth profile, but designated by the bore of the pipe on which it is cut (e.g. ⅜ in. Gas) and not by the full diameter, which is a decimal one, slightly smaller than that of the pipe. See **British Standard Whitworth thread.**

British Standard scale (*Work Study*). Scale in which 100 represents *standard performance*, i.e., production of 1 *standard hour* of work per hour.

British Standards Institution. A national organization for the preparation and issue of standard specifications.

British Standard specification. A specification of efficiency, grade, size, etc., drawn up by the British Standards Institution, referenced so that the material required can be briefly described in a Bill or Schedule of Quantities. The definitions are legally acceptable.

British Standard Whitworth (B.S.W.) thread (*Eng.*). The pre-metric British screw thread, having a profile angle of 55 degrees and a radius at root and crest of 0·1373 × pitch; ⅙ of the thread cut off. The pitch is standardized with respect to the diameter of the bar on which it is cut.

British Standard Wire Gauge (*Eng.*). See **Standard Wire Gauge.**

British thermal unit (*Heat*). The amount of heat required to raise the temperature of 1 lb of water by 1 Fahrenheit degree (usually taken as from 60°–61°F). Abbrev. **Btu.** Equivalent to 252 calories, 778·2 ft lbf, 1055 J. 10^5 Btu = 1 therm.

brittle fracture (*Met.*). Stress failure occurring suddenly in mild-steel vessels, thought to arise from coalescence of multiple dislocations at the boundaries of the component grains.

brittle micas (*Min.*). A group of minerals (the clintonite and margarite group) resembling the true micas in crystallographic characters, but having the cleavage flakes less elastic. Chemically, they are distinguished by containing calcium as an essential constituent.

brittleness (*Met.*). The tendency to fracture without appreciable deformation and under low stress. It is indicated in tensile test by low ultimate tensile stress and very low elongation and reduction in area. The notched-bar test may, however, reveal brittleness in metals that give a high ultimate tensile stress. See **toughness.**

brittle silver ore (*Min.*). A popular name for *stephanite* (q.v.).

Brix (*Chem.*). Scale of densities used in the sugar industry. Hydrometers are marked in 'degrees Brix', representing the density of a corresponding pure sugar solution in units equivalent to the percentage of sugar in the solution, either by volume ('volume Brix') or by mass ('mass Brix').

BRM (*Comp.*). See **binary-rate-multiplier.**

broach (*Arch.*). The sloping timber or masonry pyramid at the projecting corner of the square tower from which springs a *broach spire* (q.v.). (*Eng.*) A metal-cutting tool for machining holes; it consists of a tapered shaft carrying transverse cutting edges, which is driven or pulled through the roughly finished hole. (*Join.*) The locating pin, within a lock, about which the barrel of the key passes.

broache work (*Build.*). The finish given to a building-stone by dressing it with a punch so that broad diagonal grooves are left.

broach-post (*Carp.*). See **king-post.**

broach spire (*Arch.*). An octagonal spire springing from a square tower without a parapet, and having the triangular corners of the tower covered over by short sloping pyramids blending into the spire.

broad (*Cinema.*). A set of Klieg incandescent flood-lights used in studio illumination. Also called **broadside.** (*Tools*) A wood-turning tool, often consisting of a flat disk with sharpened edges fixed at right angles to a stem; used for shaping the insides and bottoms of cylinders.

broad axe (*For.*). (1) In resin-tapping, a wide-bladed axe used to remove bark and make an incision for seating the lip. (2) A broad-edged axe with crooked handle used for rough-dressing timber into sleepers, etc. The sharpening bevel on the blade, as in the case of a chisel, is on one side only.

broadband (*Telecomm.*). Said of a valve or circuit with no resonant elements and therefore capable of operating with similar efficiency over a wide range of frequencies.

broad-base tower (*Elec. Eng.*). A transmission-line tower with each leg separately anchored.

broad beam (*Radiol.*). Said of a gamma- or X-ray beam when scattered radiation makes a significant contribution to the radiation intensity or dose rate at a point in the medium traversed by the beam.

broadcast (*Telecomm.*). (1) Said of signals radiated from antennae for general reception. (2) The paralleling of a number of outgoing channels so that their receivers are simultaneously actuated by a single transmitter. See also **broadcasting.**

broadcast channel (*Telecomm.*). Any specified frequency band used for achieving interference-free and widespread reception from a single transmitting source.

broadcasting (*Telecomm.*). The transmission of a programme of sound, vision, or fascimile for general reception.

broadcasting amplifier or **repeater** (*Telecomm.*). One of superior performance for programmes in a broadcast or programme channel. See **programme repeater.**

broadcast transmitter (*Telecomm.*). Radio transmitter specially designed for broadcasting, the requirements for faithful transmission being higher than those of an ordinary commercial transmitter.

broadcloth (*Textiles*). A woollen cloth, constructed from fine merino yarns, for suitings, woven plain, heavily milled, and finished with a dress face; originally made two yards wide, dyed black or indigo, but now available in pastel shades.

broad gauge (*Rail.*). A railway gauge in excess of the standard 4 ft 8½ in. (1·435 m). In particular, the gauge of 7 ft (2·134 m) laid down by Brunel.

broad irrigation (*San. Eng.*). A process of sewage purification in which the effluent is distributed over a large area of carefully levelled land, and allowed to soak through it and drain away as ordinary subsoil water down the natural water-courses. Cf. *intermittent filtration.*

broad ligament (*Zool.*). In *Lacertilia*, a fold of the peritoneum which attaches the oviduct to the body wall.

broad ray (*Bot.*). A vascular ray, many cells in width, consisting of cells which are rounded in transverse section.

broadsheet (*Print.*). (1) The sheet before it is folded. (2) In rotary printing, the size of newspapers printed *columns around* the cylinder.

broadside (*Cinema.*). See **broad.** (*Typog.*) A large sheet printed on one side, such as a poster.

broadside antenna (*Radio*). Array in which the main direction of the reception or radiation of electromagnetic energy is normal to the line of radiating elements.

broadstone (*Build.*). An *ashlar* (q.v.).

broad tool (*Build.*). A steel chisel having a cutting edge 3½ in. (90 mm) in width, used for finish-dressing stone.

brob (*Carp.*). A pointed spike with a one-sided head, used to support one timber butting against another; a number are driven into the second timber so that their sides give support to the first.

brocade (*Textiles*). Jacquard designed dress or furnishing cloths made from silk, rayon, spun rayon, nylon, Terylene, etc., or any combinations of these. The design is developed by floating the wefts in irregular order.

Broca galvanometer (*Elec. Eng.*). An astatic galvanometer in which the moving element consists of two magnetized needles with consequent poles at their centres.

Broca's area (*Anat.*). The left inferior convolution of the frontal lobe of the brain; the 'speech centre'.

brocatelle (*Textiles*). A silk-and-linen fabric of rich appearance, with a raised satin warp figure; linen weft stiffens the fabric but is not visible. Used as a tapestry in most cases.

brochantite (*Min.*). A basic sulphate of copper occurring in green fibrous masses, or as incrustations; occurs in the oxidation zone of copper deposits.

broché (*Textiles*). In silk manufacture, decorative effects obtained by the use of fancy colours, or additional yarns, which, after floating at the back, are brought to the surface to supply the figuring effect, and give a brocade appearance.

brochonema (*Cyt.*). In cell-division, the spireme thread when it has become arranged in the form of loops.

brochure (*Bind.*). A booklet, pamphlet, or other short work with its pages stitched but not bound.

Brockenspectre (*Meteor*). See **spectre of the Brocken.**

Brock punch (*Surg.*). Instrument used in cardiac surgery for opening stenosed valves.

brockram (*Geol.*). A sedimentary rock occurring in the Permian strata in N.W. England; consists of angular blocks which probably accumulated as scree material.

Brocot suspension (*Horol.*). A form of pendulum suspension in which adjustment to the length of the pendulum can be made from the front of the dial.

Brodie's abscess (*Med.*). A localized abscess in bone due to tuberculous or other infection.

Brodie's reaction (*Chem.*). A method of identifying graphite, which also distinguishes it from amorphous carbon.

Broenner's acid (*Chem.*). See **Brönner's acid.**

brog (*Join.*). An awl.

Broglie wavelength (*Nuc.*). See **de Broglie wavelength.**

broich (*Textiles*). Spindle holding the yarn cop during warping.

broke (*Paper*). Waste paper from the paper making and finishing processes. Usually pulped for re-use.

broken crow twill (*Textiles*). The 4-shaft sateen (or 4-end broken twill) weave.

broken ends (*Textiles*). Warp threads which have broken during weaving owing to defects in the yarn, excessive weighting of the warp beam, or rough places in healds, reed or raceboard.

broken-over (*Bind.*). The term used to indicate that plates or other separate sheets to be inserted in a book have been given a narrow fold on the inner edge, so that they will lie flat and turn easily when fixed.

broken picks (*Textiles*). Defects in weaving due to breaking of the weft carried by the shuttle. Arises from weft trapping in loom, weak weft, slubs, or blocked shuttle-eye.

broken-space saw (*Carp., Join.*). A handsaw having usually 6 teeth to the inch.

broken twills (*Textiles*). Fabrics in which the diagonal line forming the 'twill' is broken, or broken and reversed in direction, at intervals.

broken wind (*Vet.*). A chronic emphysema of the lungs of horses.

brokes (*Textiles*). Short staples of wool that come from the neck and belly parts of a fleece.

broking (*For.*). Hewing between notches made on logs by scoring.

bromal (*Pharm.*). $CBr_3{\cdot}CHO$. B.p. 174°C. Made by treating ethanol with bromine. Can be reduced to $CBr_3{\cdot}CH_2OH$, i.e., tribromoethanol or Bromethol, a rectal anaesthetic.

bromatium (*Zool.*). A fungal enlargement formed by ants; used by them for food.

brom-cresol green (*Chem.*). Indicator used in determination of pH values, suitable for ranges 3·6–5·2.

brom-cresol purple (*Chem.*). Indicator used in determination of pH values within the range 5·2–6·8.

bromelin (*Chem.*). Proteolytic enzyme found in pineapples, similar in properties to *papain* (q.v.).

bromethol (*Pharm.*). See **bromal.**

bromic acid (*Chem.*). $HBrO_3$; with bases it forms *bromates* (v). A powerful oxidizing agent.

bromide paper (*Photog.*). Paper coated with silver bromide emulsion, used chiefly for enlarging but also for contact prints. Made in grades producing different degrees of contrast.

bromides (*Chem.*). Bromides are salts of hydrobromic acid. Silver bromide is extensively used in photography, potassium bromide and, to a less extent sodium, ammonium, and lithium bromides, in medicine.

bromide streaks (*Photog.*). Light streaks in the developed image caused by chemicals produced by developing action.

bromidrosis (*Med.*). Fetid perspiration, especially of the feet.

bromination (*Chem.*). The substitution by bromine in organic compounds.

bromine (*Chem.*). A nonmetallic element in the seventh group of the periodic system, one of the halogens. Symbol Br, at. no. 35, r.a.m. 79·909, oxidation states 1, 3, 5, 7, m.p. −7·3°C, b.p. 58·8°C, rel. d. 3·19. A dark red liquid, giving off a poisonous vapour, Br_2, with an irritating smell. In combination with various metals it is widely but sparingly distributed. The chief commercial source is sea water from which bromine is manufactured by treating the 'bittern' with chlorine. Bromine is used extensively in synthetic organic chemistry, as an anti-knock additive to motor fuel, in medicine, and in halogen-quenched Geiger tubes.

bromochlorodifluoromethane (*Chem.*). BCF. $CHBrClF_2$. B.p. −4°C. Organic substance used as a fire extinguishing fluid, particularly for fires in confined spaces. Low toxicity vapour, 5·7 times as dense as air.

bromoform (*Chem., Min. Proc.*). $CHBr_3$, tribromomethane, m.p. 5°C, b.p. 151°C, rel. d. 2·9; a colourless liquid, of narcotic odour. Much used in laboratory separation of minerals into *floats*, rel. d. less than 2·9, and *sinks*, greater than 2·9.

bromoil process (*Photog.*). Printing process in which a bleached and tanned bromide print is brushed with oil pigment which adheres to the shadow portion and is repelled by the highlights.

bromoil transfer (*Photog.*). Print made by transferring a bromoil print to another sheet of paper by passing the two in contact through a press.

bromothymol blue (*Chem.*). Indicator used in acid-alkali titrations, having a pH range of 6·0 to 7·6, changing from yellow to blue.

bronch-, broncho-. Prefix from Gk. *bronchos*, windpipe.

bronchi (*Zool.*). See **bronchus.**

bronchia (*Zool.*). The branches of the bronchi. *adj.* **bronchial.**

bronchiectasis (*Med.*). Pathological dilatation of the bronchi as a result of weakening of the bronchial wall—usually from infection.

bronchiole (*Zool.*). One of the terminal subdivisions of the bronchia.

bronchiolectasis (*Med.*). Pathological dilatation of the bronchioles.

bronchiolitis (*Med.*). Inflammation of the bronchioles.

bronchitis (*Med.*). Inflammation of the bronchi. (*Vet.*) See **husk.**

bronchography (*Med.*). The radiological examination of the trachea, bronchi, or the bronchial tree after the introduction of a *contrast medium* (q.v.).

bronchophony (*Med.*). Increase of voice sounds heard through the stethoscope when placed over the lungs; indicative of consolidation of lung tissue, as in pneumonia.

bronchoscope (*Med.*). An instrument consisting of a hollow tube and light arranged for inspecting the interior of the bronchi.

bronchus (*Zool.*). One of the two branches into which the trachea divides in higher Vertebrates and which lead to the lungs. *pl.* **bronchi.** *adj.* **bronchial.**

Brönner's acid (*Chem.*). 2,6-Naphthylaminemonosulphonic acid. Intermediate for dyestuffs.

Bronsil Shales (*Geol.*). Grey shales occurring in the Malvern Hills, England; equivalent to the Tremadoc Shales of N. Wales; of Upper Cambrian age.

Bronson resistance (*Electronics*). Resistance between two electrodes in a gas when exposed to a constant source of ionization.

Brönsted-Lowry theory (*Chem.*). Defines as an acid every molecule or ion able to produce a proton and as a base every molecule or ion able to take up a proton. Thus acid ⇄ base + proton. The acid and the corresponding base are called conjugated.

Brönsted's relation (*Chem.*). Expression for the catalytic activity k of acids and bases in terms of their dissociation constants viz.,

$$k_{\text{acid}} = G_a K_a^{\alpha}$$
$$k_{\text{base}} = G_b K_b^{\beta}$$

where G_a or G_b is constant for a series of analogous catalysts of a given reaction in a given solvent and at a given temperature.

brontometer (*Meteor.*). Simultaneously records wind speed, rain and hail fall, frequency of lightning flashes, duration of thunder and barometric pressure during a thunderstorm.

bronze (*Met.*). Primarily an alloy of copper and tin, but the name is now applied to other alloys not containing tin; e.g., aluminium bronze, manganese bronze, and beryllium bronze. For varieties and uses of tin bronze, see **alpha bronze, bell metal, gunmetal, leaded bronze, phospor-bronze.**

bronzed diabetes (*Med.*). See **haemochromatosis.**

bronzing (*Paint.*). Certain blue pigments, particularly of the *Prussian blue* and *Monastral blue* (q.v.) types, exhibit a metallic lustre when ground at fairly high concentrations into paint media. This behaviour is known as bronzing. (*Print.*) Dusting freshly-printed sheets, by hand or machine, with any suitable metallic powder, bronze-coloured or otherwise.

bronzite (*Min.*). A form of orthopyroxene, more iron-rich than enstatite and more magnesian than hypersthene; often has metallic sheen, due to the reflection of light from planes of minute metallic inclusions in the surface layers.

bronzitite (*Geol.*). A rock composed of bronzite with smaller amounts of angite and calcic plagioclase. A common constituent of layered basic igneous intrusions, such as those of the Bushveld, South Africa and Stillwater, Montana.

brood (*Zool.*). A set of offspring produced at the same birth or from the same batch of eggs.

brood bud (*Bot.*). (1) See **soredium.** (2) A bulbil in *Bryophyta.*

brood canals (*Zool.*). In *Strepsiptera*, median ventral ducts through which the already-hatched larvae emerge from the female.

brood cell (*Bot.*). A naked or walled cell, produced asexually, separating from the parent and giving rise to a new plant.

brood gemma (*Bot.*). A multicellular body, formed asexually and separating from the parent, forming a new plant.

brood sac (*Zool.*). The enlarged genital chamber in the females of viviparous cockroaches, in which the eggs are retained until hatching.

brookite (*Min.*). One of the three naturally occurring forms of crystalline titanium dioxide found as flat, red-brown, platy, orthorhombic crystals.

brooming (*Civ. Eng.*). The spreading of the fibres at the head of a timber pile, due to the impact of the monkey.

Broughton countersink (*Tools*). Countersink which can be fixed to a shell bit, thus enabling the screw-hole and countersink to be bored in one operation.

Brouncker's series for $\log_e 2$ (*Maths.*).

$$\log_e 2 = \frac{1}{1\cdot 2} + \frac{1}{3\cdot 4} + \frac{1}{5\cdot 6} + \frac{1}{7\cdot 8} + .$$

brow (*Mining*). The top of the shaft or 'pit'; hence also called **pit-brow.**

brown (*Paint.*). A colour obtained either from natural sources (brown ochre, bitumen) or by mixing black with red, orange, or yellow. (*Textiles*) A term applied by wool-sorters to wool from the haunches and flanks of cross-bred fleeces.

Brown agitator (*Min. Proc.*). See **agitator.**

brown algae (*Bot.*). See **Phaeophyta.**

Brown and Sharpe Wire Gauge. A system of designating, by numbers, the diameter of wires; it ranges from 4/0 (0·46 in.) to 48 (0·00124 in.). Also **American Standard Wire Gauge.**

brown body (*Zool.*). In *Ectoprocta*, a compact mass formed by the degeneration of a polypide: it often comes to lie in the stomach of the new polypide regenerated from the zooecium and is evacuated through the anus.

brown coal (or **lignite**) (*Mining*). Intermediate between peat and true coals, with high moisture content, the calorific value ranging from about 4000 to 8300 Btu/lb (9·5 to 20 MJ/kg).

brown funnels (*Zool.*). In *Cephalochorda*, two tubes lined by pigmented epithelium and projecting into the dorsopharyngeal coelom; of unknown function.

brown haematite (*Min.*). A misnomer, the material bearing this name being *limonite* (q.v.), a hydrous iron oxide, whereas true haematite is anhydrous.

Brownian movement (*Phys.*). Small movements of light suspended bodies such as galvanometer coils, or a colloid in a solution, due to statistical fluctuations in the bombardment by surrounding molecules of the dispersion medium. See **colloidal state.**

brown nose disease (*Vet.*). Copper nose. A form of photosensitization occurring in cattle, characterized by brown discoloration and irritation of the skin of the muzzle and teats.

brown rot (*Bot.*). A disease of plums and other fruit caused by fungi.

Brownstone Series (*Geol.*). A division of the Old Red Sandstone, unfossiliferous and undated, lying above the red marl stage in south-west England.

brown tubes (*Zool.*). In *Sipunculoidea*, the nephridia.

Brucellaceae (*Bacteriol.*). A family of obligate parasites belonging to the order *Eubacteriales.* Gram-negative cocci or rods: aerobic or facultatively anaerobic: many pathogenic species, e.g., *Pasturella pestis* (plague), *Pasturella multocida* (fowl cholera, swine plague, haemorrhagic septicaemia), *Brucella abortus* (undulant fever in man, contagious abortion in cattle, goats and pigs).

brucellosis (*Med., Vet.*). The diseases caused by infection with organisms of the genus *Brucella.* See **undulant fever.**

Bruch's membrane (*Zool.*). In some Vertebrates, a transparent membrane lining the inner surface of the choroid.

brucine (*Chem.*). $C_{23}H_{26}O_4N_2$, $4H_2O$, a strychnine base alkaloid, m.p. of the anhydrous compound 178°C; it contains two methoxyl groups, and is a monoacidic tertiary base. Its physiological action is less than that of *strychnine* (q.v.).

brucite (*Min.*). Hydroxide of magnesium, occurring as fibrous masses in serpentinite and metamorphosed dolomite. See also **periclase.**

brucite-marble (*Geol.*). A product of dedolomitization; a crystalline metamorphic rock formed by the action of intense heat on dolomitic (or magnesian) limestone.

Brückner cycle (*Meteor.*). A recurrence of periods of cold and damp alternating with warm and dry years, the period of a cycle being about 35 years.

bruise (*Med.*). Rupture of blood vessels in a tissue, with extravasation of blood, as a result of a blow which does not lacerate the tissue.

bruit (*Med.*). A sound or murmur heard by auscultation over the lung or heart.

Brunner's glands (*Zool.*). Small racemose glands situated in the submucosa of the duodenum.

brunsvigite (*Min.*). A variety of chlorite, fairly iron-rich.

Brunswick black (*Paint.*). An opaque varnish with a basis of asphaltum, an inferior kind of *black japan* (q.v.).

Brunswick blue (*Paint.*). Prussian blue reduced by precipitation on barium sulphate.

Brunswick green (*Paint.*). A green pigment made from lead chromate and Prussian blue reduced with barytes.

brush (*Comp.*). The wiper which 'reads' punched cards, producing an electric contact through the aperture. (*Elec. Eng.*) A rubbing contact on a commutator, switch or relay. Also **wiper.** (*Mining*) (1) In a coal-mine, a road through the goaf, gob, or worked-out area packed with waste. (2) To clean up fine coal from the floor.

brush arc-lighter (*Elec. Eng.*). An old type of open-coil d.c. generator, used for supplying arc lamps in series.

brush-box (*Elec. Eng.*). That portion of the brush-holder of an electrical machine in which the brush slides or in which it is clamped.

brush coating (*Paper*). The application of coating by a rotary brush directly on to the paper.

brush contact (*Elec. Eng.*). See **laminated contact.**

brush curve (*Elec. Eng.*). The voltage drop between the brush arm and the segment beneath the brush at points along the brush width plotted against brush width as an indication of the correctness of compole flux density in a d.c. machine.

brush(ing) discharge (*Elec. Eng., etc.*). Discharge from a conductor when the p.d. between it and its surroundings exceeds a certain value but is not enough to cause a spark or an arc. It is usually accompanied by hissing high noise. Also *corona* (q.v.). (*Min. Proc.*) Electrical discharge from points along bar charged to between 18 000 and 80 000 volts to create electrostatic field through which mineral particles fall and acquire polarity, in the high-intensity separation process.

brush gear (*Elec. Eng.*). A general term used to denote all the equipment associated with the brushes of a commutating or slip-ring machine.

brush-holder (*Elec. Eng.*). The portion of an electrical machine or other piece of apparatus which holds a brush. See **box-type brush-holder.**

brush-holder arm (*Elec. Eng.*). The rod or arm supporting one or more brush-holders. Also called **brush spindle, brush stud.**

brushing (*Vet.*). Cutting. An injury to the inside of a horse's leg caused by the shoe of the opposite foot.

brushing and steaming (*Textiles*). A machine comprising a steam box, a revolving brush or brushes, and a contrivance for folding; used in finishing woollen and worsted cloths, and woven carpets.

brushing discharge (*Elec. Eng., etc.*). See **brush discharge.**

brush lead (*Elec. Eng.*). See **brush shift.**

brushmarks (*Paint.*). A paintwork defect characterized by visible depressed lines in the direction in which the paint has been brushed on. They are due to insufficient flow or levelling of the liquid paint.

brush-rocker (*Elec. Eng.*). A support for the brushes of an electrical machine which enables them to be moved bodily round the commutator. Also called a **brush-rocker ring.**

brush shift (*Elec. Eng.*). The amount by which the brushes of a commutating machine are moved from the centre of the neutral zone. Also called **brush lead.** See **backward shift, forward shift.**

brush spindle (*Elec. Eng.*). See **brush-holder arm.**

brush spring (*Elec. Eng.*). A spring in a brush-holder which presses the brush against the commutator or slip-ring surface.

brush stud (*Elec. Eng.*). See **brush-holder arm.**

brush yoke (*Elec. Eng.*). A special frame for supporting the brush-rocker or brushes of an electrical machine, when these are not supported from the main frame or the pedestal.

Brussels lace (*Textiles*). (1) Brussels net, a plain net originally made at Brussels. (2) Brussels pillow, a fine pillow lace, the patterns being joined by small loops at the edges. (3) Brussels point, a lace with an open pattern, made partly in open and partly in closed stitch. It has a shaded appearance.

bruxism (*Med.*). Grinding of the teeth during sleep, or without subjective awareness while awake.

bruzze (*Tools*). A vee-shaped edge tool used by wood turners as a parting tool and by wheelwrights for chopping mortises in wheels.

Bryales (*Bot.*). An order of the *Musci* characterized by the capsule having a toothed peristome (a few split open), and being raised on a seta.

Bryophyta (*Bot.*). The liverworts and mosses. One of the main divisions of the plant kingdom, with some thousands of species. The plants are small, rootless, and without organized vascular tissue. They show clear alternation of generations, with a small spore-bearing plant borne on the generation (gametophyte) which bears the archegonia and antheridia.

B.S.B. thread (*Eng.*). Abbrev. for **British Standard brass thread.**

B-service-area (*Radio*). Region surrounding a broadcasting transmitter where the field-strength is between 5 and 10 mV/m.

B.S.F. thread (*Eng.*). Abbrev. for *British Standard fine thread.*

B.S.I. Abbrev. for *British Standards Institution.*

B-side (*Teleg.*). The single-current channels in a quadruplex system.

B.S.P. thread (*Eng.*). Abbrev. for *British Standard pipe thread.*

B stage (*Plastics*). Transition stage through which a thermosetting synthetic resin of the phenol formaldehyde type passes during the curing process, characterized by softening to rubber-like consistency when heated, and insolubility in ethanol or acetone (propanone).

B.S.W. thread (*Eng.*). Abbrev. for *British Standard Whitworth thread.*

BT-cut (*Crystal.*). Special cut of a quartz crystal to obviate temperature effects, such that the angle made with the Z-axis is −40°.

Btu (*Heat*). Abbrev. for **British Thermal Unit.**

B.T.U. (*Elec.*). Board of Trade Unit = 1 kWh.

Bu (*Chem.*). A symbol for the butyl radical, C_4H_9—.

bubble (*Surv.*). The bubble of air and spirit vapour within a *level tube* (q.v.): loosely, the level tube itself.

bubble cap (*Chem. Eng.*). Perforated or slotted cap on the plates of a vertical distillation column, to secure intimate mixing of vapour and condensed liquid.

bubble chamber (*Nuc. Eng.*). Development of *cloud chamber*, in which a vessel is filled with a transparent liquid (for photographing tracks), which is so highly superheated that an ionizing particle passing through starts violent boiling by initiating the development of a string of bubbles along its path.

bubble effect (*Phys.*). When a submarine charge explodes, the very hot, dense mass of gas formed expands violently in bubble-form, sending out a shock wave at speeds much greater than 1500 m/s. After expanding past its equilibrium position the bubble begins to contract, and the water rushing together compresses it to a pressure sufficient to initiate a second expansion and shock wave.

bubble point (*Chem.*). The temperature at which the first bubble appears on heating a mixture of liquids. See **dew point.**

bubbles, pressure in (*Phys.*). See **pressure in bubbles.**

bubble trier, tube (*Surv.*). See **level trier, tube.**

bubo (*Med.*). An inflamed and swollen lymphatic gland, especially in the groin.

bubonic plague (*Med.*). A form of plague in which there is great swelling of the lymphatic glands, especially those in the groin. See **plague.**

bubonocele (*Med.*). A swelling in the groin due to an incomplete hernia.

buccal (*Zool.*). Pertaining to, or situated in or on, the cheek or the mouth. (*Med.*) Administered by mouth. See **bucco-.**

buccal cavity (*Zool.*). The cavity within the mouth opening but prior to the commencement of the pharynx.

buccal funnel (*Zool.*). In *Petromyzontia*, a basin-like depression directed downwards at the anterior end; surrounded by papillae and containing horny teeth on either side of the mouth.

buccal glands (*Zool.*). Glands opening into the buccal cavity in terrestrial *Craniata*; the most important are the salivary glands.

buccal mass (*Zool.*). In Molluscs, the region of the alimentary canal containing the odontophore.

buccal respiration (*Zool.*). See **buccopharyngeal respiration.**

buccal tentacles (*Zool.*). In *Myxinoidea*, cartilage-supported tentacles round the mouth.

buccinator (*Anat.*). A broad, thin muscle at the side of the face, between the upper and lower jaw.

buccinum (*Zool.*). Whelk, genus of *Gastropoda*.

bucco-. Prefix from L. *bucca*, cheek.

buccopharyngeal respiration (*Zool.*). Breathing by means of the moist vascular lining of the mouth cavity or diverticula thereof, as in some Amphibians and certain Fish which have become adapted to existence on land.

bucheron (*For.*). A general term for a forest worker, in Canada.

Buchholz relay (*Elec. Eng.*). A protective relay for use with transformers or other oil-immersed apparatus; it embodies a float which becomes displaced and operates the relay contacts if gas bubbles are generated by a fault within the equipment being protected.

buchite (*Geol.*). A glassy rock which represents the result of partial fusion and recrystallization at very high temperatures of clay and shale material. It often occurs as xenoliths within igneous rocks.

Buchmann-Meyer effect (*Acous.*). The special type of reflection of light from the sound-track on a disk record whereby the lateral velocity of the track can be determined.

Buchmann-Meyer pattern (*Light*). See **optical pattern.**

Buchner funnel (*Chem.*). A stout porcelain funnel having at its base a fixed horizontal perforated plate to act as a support over which a piece of filter paper is placed, thus ensuring a large area of filtration.

buchucamphor (*Pharm.*). Diosphenol:

H_3C— [ring] —$CH\cdot(CH_3)_2$
HO O

Crystalline solid; m.p. 83°C, b.p. 110°C, obtained from oil of buchu leaves (*Barosma* species). Used as urinary antiseptic and mild diuretic.

bucket (*Eng.*). (1) The piston of a reciprocating pump. (2) Any of the cup-shaped vanes attached to the periphery of a *Pelton wheel* (q.v.). (*Hyd. Eng.*) A dredging scoop, usually capable of being opened and shut for convenience in depositing and taking up a load.

bucket conveyor (*Eng.*). A conveyor or elevator consisting of a pair of endless chains running over toothed wheels, and carrying a series of buckets which, on turning over, discharge their contents at the delivery end.

bucket-dredge (*Mining*). System with 2 pontoons between which a chain of buckets digs through alluvium to mineral-bearing sands and delivers these to concentrating appliances on the pontoon decks. Dredge excavates and floats in a pond in which it traverses the deposit.

bucket-ladder dredger (*Civ. Eng.*). A vessel of small draught having a series of buckets moving in a continuous chain reaching down into the material to be dredged, and lifting it for discharge into the vessel itself or into an attendant vessel.

bucket-ladder excavator (*Civ. Eng.*). A mechanical excavator working on the same principle as a *bucket-ladder dredger* (q.v.), but adapted for use on land. Also **dredger excavator.**

bucket valve (*Eng.*). A nonreturn (delivery) valve fitted in the bucket or piston of some types of reciprocating pump.

bucking coil (*Elec. Eng.*). A winding on an electromagnet to oppose the magnetic field of the main winding. Such a device, known as the *hum-bucking coil*, is sometimes used in electromagnetic loudspeakers to smooth out voltage pulsations in the power supply.

bucking kier (*Textiles*). A vessel in which linen cloth is first boiled in lime water and then in alkaline lye, in preparation for full bleaching.

bucking ladder (*For.*). A series of short skids laid parallel at regular intervals; used to support logs while they are being crosscut.

buckle (*Eng.*). (1) To twist or bend out of shape; said usually of plates or of the deformation of a structural member under compressive load. (2) A metal strap. (3) A swelling on the surface of a mould due to steam generated below the surface.

Buckle brush (*Photog.*). A cotton-wool brush made by holding a small wad at the end of a glass tube by a silver wire passing through the tube.

buckle-fold (*Bind.*). A fold made in the paper parallel to its leading edge or to the fold previously made in it; the paper is brought to a sudden stop causing it to buckle into, and be folded by, the folding rollers. Cf. *knife-fold.*

Buckley gauge (*Electronics*). Sensitive pressure gauge, depending on gas ionization.

buckling (*Cinema.*). The irregular motion of film in a camera or projector, causing a jam; due to a break in the sprocket holes or to incorrect threading. (*Elec. Eng.*) A distortion of accumulator plates caused by uneven expansion, usually as a result of heavy discharges or other maltreatment. (*Nuc.*) A term in reactor diffusion theory giving a measure of the curvature of the neutron density distribution. In a homogeneous reactor the buckling factor for the reactor to go critical will depend upon its geometrical configuration.

buck plates (*Met.*). Steel plates used to tie together with tie-rods the brickwork of a furnace.

buckrake (*Agric.*). Attachment (to a tractor) fitted with long tines for mown crops or general transport, or with short tines for the handling of shorter, forage-harvested material.

buckram (*Textiles*). A strong fabric made of jute, linen, cotton, or spun rayon, nylon or Terylene, stiffened by size or latex, or heat set.

buck saw (*Tools*). A large frame-saw having one bar of the frame extended to form a handle.

buckskin cloth (*Textiles*). A fabric, usually woven 8-end sateen weave, from fine quality yarns with a fine raised and milled finish.

buck transformer (*Elec. Eng.*). One with secondary in mains circuit to regulate voltage according to a controlling circuit feeding the primary. Also called **boost transformer.**

buckwheat rash (*Vet.*). See **fagopyrism.**

bud (*Bot.*). The undeveloped stage of a branch. It contains a very short shoot bearing undeveloped leaves, and may contain in addition one or more young flowers.

budding (*Bot.*). (1) The production of daughter cells in the form of rounded outgrowths, characteristic of yeasts and similar fungi. (2) The production of buds in general. (3) A means of artificial propagation, in which a bud taken from one plant is inserted under the bark of another, subsequently developing into a shoot. (*Zool.*) A primitive method of asexual reproduction by growth and specialization and separation by constriction of a part of the parent.

buddle (*Mining*). A shallow annular pit with either a concave or convex bottom for concentrating finely crushed, slimed, base-metal ores. See also **dumb buddle.**

Buddleia (*Bot.*). A genus of plants of the *Loganiaceae*, shrubs and trees with opposite leaves and showy clusters of purple or orange flowers. (Named in honour of Adam Buddle, English botanist, died 1715.)

buddle-work (*Min. Proc.*). Treatment of finely ground tin-bearing sands by gentle sluicing, in which a heavier fraction of the fed pulp is built up (*buddled*) while the lighter fraction flows to discard. This is continued till a satisfactory concentrate is produced.

bud scale (*Bot.*). A simplified leaf or stipule on the outside of a bud, forming part of a covering which protects the contents of the bud.

bud sport or bud variation (*Bot.*). The production of an abnormal branch, inflorescence, or flower, from a bud, as a result of mutation.

Buerger's Disease (*Med.*). See **thrombo-angiitis obliterans.**

buff (*Eng.*). A revolving disk composed of layers of cloth charged with abrasive powder; used for polishing metals.

buffalo disease (*Vet.*). See **barbone.**

buffer (*Comp.*). A memory type of device in a computing system which compensates for any difference in the rates of flow of information in various parts of the system. (*Elec. Eng.*) (1) See **buffer reagent.** (2) An electronic circuit to decouple the output of the buffer from its input, thus avoiding reaction between a driving and a driven circuit. (*Eng.*) A spring-loaded pad attached to the framework of railway rolling-stock to minimize the shock of collision; any resilient pad used for a similar purpose.

buffer action (*Chem.*). The action of certain solutions in opposing a change of composition, especially of hydrogen ion concentration, e.g., salt of weak acid and weak acid.

buffer battery (*Elec. Eng.*). A battery of accumulators arranged in parallel with a d.c. generator to equalize the load on the generator by supplying current at heavy-load periods and taking a charge during light-load periods.

buffer capacitor (*Elec. Eng.*). See **blocking capacitor.**

buffer capacity (*Photog.*). The capacity of an alkali (e.g., sodium metaborate) in a developing solution to maintain a slow rate of decrease in pH value.

buffer circuit (*Acous.*). The resistance-capacitor unit which determines the rate of rise or fall of the envelope of the waveform of emitted sounds which has been generated in electrostatic circuits in electronic organs.

buffer memory (*Comp.*). Same as **temporary memory.**

buffer reagent (*Elec. Eng.*). A substance added to an electrolytic solution which prevents rapid changes in the concentration of a given ion. Also called **buffer.**

buffer resistance (*Elec. Eng.*). See **discharge resistance.**

buffer solution (*Chem.*). A solution of certain salts, usually of a weakly ionized acid or base, whose acidity is not appreciably changed by additions of acid or alkali.

buffer spring (*Eng., Rail.*). The part lending resiliency to a buffer.

buffer stage (*Radio*). An amplifying stage coming between the master oscillator and the modulating stage of a radio transmitter, to prevent the changing load of the modulated output from affecting the frequency of the master drive.

buffer valve (*Electronics*). A valve used in a buffer stage.

buffet boundary (*Aero.*). The maximum *Mach number* (q.v.) at which a subsonic aeroplane may be safely flown without risk of uncontrollability due to *compressibility drag* (q.v.).

buffeting (*Aero.*). An irregular oscillation of any part of an aircraft, caused and maintained by an eddying wake from some other part; commonly, tail buffeting in the downwash of the main planes, which gives warning of the approach of the *stall* (q.v.).

buffing (*Build.*). The grinding down of a surface to remove extrusions or to expose the underlying material.

buff leather (*Leather*). White leather from which the grain surface has been removed; used for army accoutrements.

bug (1) Elusive fault in equipment or programme. (2) Hidden microphone or other device for spying purposes.

bugeye lens (*Cinema.*). Wide-angle (128°) film camera lens used in Todd-AO wide screen process.

bug key (*Teleg.*). A telegraphist's key that permits higher transmission speeds than a normal key. The moving lever, moved horizontally by the hand, makes dashes in one direction when held over. Dots are sent by a spring contact attached to the lever, when the lever is released from sending a dash.

buhl saw (*Tools*). A kind of frame-saw in which the back of the frame is so spaced from the saw itself as to allow the latter to cut well into the work.

buhr (or burr) mill (*Min. Proc.*). One in which material is ground by passage between a fixed and a rubbing surface. Types include old-fashioned flour mill, with a circular grindstone rotating above a fixed lower one, radially grooved to facilitate passage of grist from centre to peripheral discharge. Also, rotating cone in fixed casing, material gravitating through the intervening space. Used for softish material, e.g., grain, food processing, and such minerals as soft limestone.

builders' level (*Build.*). (1) A spirit-level tube set in a long straightedge; for testing and adjusting levels. (2) A simple form of dumpy or tilting level, used on building works or for running the levels of drains.

builders' staging (*Build.*). A robust type of scaffold, formed of square timbers strongly braced together, capable of being used for the handling of heavy materials.

building board (*Build.*). Board manufactured from various materials and supplied with various finishes; used for lining walls and ceilings. See **fibre-board.**

building certificates (*Build.*). Certificates made out by the architect during the progress, or after completion, of the works on a building contract, to enable the contractors to obtain payments on account or in settlement from the employer.

building line (*Build.*). The line beyond which a building may not be erected on any given plot.

building motion (*Textiles*). Precision mechanism in mules or ringframes responsible for winding the yarns into cops or bobbins, having firm base and body with tapering chase (nose).

building paper (*Build.*). Sandwich of fibre and bitumen between two sheets of heavy paper, used in damp-proofing and for insulation between the soil and road surfacing.

build-up (*Radiol.*). Increased radiation intensity in an absorber over what would be expected on a simple exponential absorption model. It results from scattering in the surface layers and increases with increasing width of the radiation beam.

build-up time (*Telecomm.*). See **rise time.**

bulb (*Bot.*). A large underground bud consisting of swollen leaf bases containing much reserve food material, arranged on a short conical stem. (*Elec., etc.*) The gas-tight envelope, usually glass, which encloses the electrodes of a thermionic valve or of an electric discharge lamp, or the filament of an electric filament lamp. (*Zool.*) Any bulb-shaped structure. *adj.* **bulbar.**

bulb bar (*Eng.*). A rolled, or extruded, bar of strip form in which the section is thickened along one edge.

bulb geophyte (*Bot.*). A geophyte which perennates by means of a bulb.

bulbiferous (*Bot.*). Having, on the stem, bulbs or bulbils in place of ordinary buds.

bulbil (*Bot.*). (1) A modified bud consisting of swollen leaves containing food reserves, and able to give rise to a new plant when detached from the parent. (2) Any rounded mass of vegetative cells, detachable from the parent and able to give rise to a new plant. (3) A small sclerotium, usually of rather loose construction. (*Zool.*) (1) A contractile dilatation of an artery. (2) Any small bulblike structure.

bulblet (*Bot.*). (1) A small bulb. (2) See **gemma (1).**

bulblike (*Bot.*). Resembling a bulb, but solid.

bulboeapnine (*Med.*). An alkaloid used in the control of tremors such as those in chorea, etc.

bulbonuclear (*Zool.*). Pertaining to the medulla oblongata and the nuclei of the cranial nerves.

bulbous hair (*Bot.*). A hair having a swollen base.

bulbus (*Bot.*). An enlargement of the base of the stipe of an agaric. (*Zool.*) See below and **bulb.**

bulbus aortae (*Zool.*). In Amphibians, an artery, representing an abbreviated ventral aorta.

bulbus arteriosus (*Zool.*). In many Vertebrates, a strongly muscular region following the conus arteriosus.

bulbus oculi (*Zool.*). The eyeball of Vertebrates.

bulimia (*Med.*). An abnormal increase in the appetite.

bulk (*Paper*). A measure of the (reciprocal) density of paper, being the ratio of thickness to substance. A loose synonym for thickness.

bulk concrete (*Civ. Eng.*). See **mass concrete.**

bulk density (*Powder Tech.*). The value of the apparent powder density when measured under stated freely poured conditions.

bulked yarns (*Textiles*). Yarns composed of man-made filaments subjected to chemical treatment, or false-twisting, heating, tension in a thermal setting zone, final relaxation and crimping to give loftiness.

bulk factor (*Plastics*). Ratio of the density of a sound moulding to the powder density of the moulding material from which it was made.

bulk flotation (*Min. Proc.*). Froth flotation process so applied as to concentrate more than one valuable mineral in one operation.

bulkhead (*Aero.*). In fuselages, a transverse dividing wall providing access between several internal sections, or a strengthened and sealed wall at the front and rear and designed to withstand the differential pressure required for pressurization. In power plant nacelles, a structure serving as a firewall. (*Autos.*) On a public service vehicle, partition at the front between driver and passenger accommodation. (*Civ. Eng.*) A masonry or timber partition to retain earth, as in a tunnel or along a waterfront. (*Ships*) A partition within a ship's hull or superstructure. It may be transverse or longitudinal, watertight, oiltight, gastight, or partially open. It may form part of the ship's subdivision for seaworthiness or otherwise.

bulkhead deck (*Ships*). The uppermost deck up to which watertight transverse bulkheads are carried.

bulkhead-fitting (*Light*). A robust form of electric-light fitting designed for attachment to bulkheads or other situations where space is restricted and where it may be subject to severe treatment.

bulkiness (*Powder Tech.*). A term used to describe the properties of a bed of powder. It is defined as the reciprocal of the apparent density of the powder under the stated conditions.

bulking (*Build., Civ. Eng.*). See **moisture expansion.**

bulk modulus (*Eng.*). Relationship of applied stress and volumetric strain which occurs on the application of a uniform stress to all parts of a body.

bulk sample (*Min. Proc., Powder Tech., etc.*). One composed of several portions taken from different locations within a bulk quantity of the material under test.

bulk supply (*Elec.*). A supply of electricity

purchased by a distribution company from a larger electricity supply company.

bulk temperature (*Heat*). The temperature a gas would reach if it were in thermodynamic equilibrium with the same average energy of the internal degrees of freedom.

bulk test (*Nuc. Eng.*). The large sample usually required for a shield radiation test of a material having a high attenuation.

bulla (*Med.*). A blister or bleb. A circumscribed elevation above the skin, containing clear fluid; larger than a *vesicle*. (*Zool.*) In Vertebrates with a flask-shaped tympanic, the spherical part of that bone which usually forms a protrusion from the surface of the skull.

bullate (*Bot.*). (1) Having a blistered or puckered surface. (2) Bubblelike. (3) Bearing one or more small hemispherical outgrowths.

bull-chain (*For.*). The endless chain used in a log haul-up for conveying logs from the mill pond to the mill.

bulldog calf (*Vet.*). A lethal form of achondroplasia occurring mainly when Dexter cattle are mated together; due to the inheritance of a pair of semi-dominant genes responsible for the achondroplasia trait.

bulldozer (*Civ. Eng.*). A power-operated machine, provided with a blade for spreading and levelling material. Also called **angledozer.**

bullet amplifier (*Telecomm.*). A colloquialism for the amplifier mounted in a cylinder and associated with a condenser microphone hanging therefrom.

bullet catch (*Join.*). See **ball catch.**

bulletwood (*For.*). Timber from a tree of the genus *Mimusops* which also yields *balata* (q.v.), renowned for its strength and durability. It is used for structural work, boat building, furniture and cabinet making, tool handles, wheel spokes and railway sleepers.

Bullhead Bed (*Geol.*). The basal stratum of the Tertiary rocks of S.E. England, composed largely of green-coated flints.

bull-headed rail (*Rail.*). A rail section having the shape roughly of a short dumb-bell in outline, but with unequal heads, the larger being the upper part in use. See **flanged rail.**

bull header (*Build.*). A brick with one corner rounded, laid with the short face exposed, as a quoin or for sills, etc.

bullhead tee (*Plumb.*). A tee having a branch which is longer than the run.

bull-holder (*Vet.*). Forceps for grasping the nasal septum of cattle as a means of restraint.

bulliform cell (*Bot.*). See **motor cell.**

bulling (*Civ. Eng., Mining*). The operation of detaching a piece of loosened rock by exploding blasting charges inserted in the surrounding fissures.

bulling bar (*Civ. Eng., Mining*). An iron bar used to force clay into the crevices in the sides of a bore hole.

bullion (*Met.*). (1) Gold or silver in bulk, i.e., as produced at the refineries, not in the form of coin. (2) The gold-silver alloy produced before the metals are separated. (*Textiles*) (1) A lace constructed of gold or silver threads, other metallic yarns, or plastic threads. (2) A heavy twisted cord fringe, covered with gold, silver, copper or plastic coated aluminium yarns.

bullion content (*Met.*). In parcel of metal or minerals being sold, where the main value is that of the base metal which forms the bulk of the parcel, the contained gold or other precious metal of minor value included in the sale.

bullion point (*Glass*). The centre piece of a sheet of glass made by the old method of spinning a hot glass vessel in a furnace until it opened out under centrifugal action to a circular sheet. The centre piece bears the mark of attachment to the rod used to spin the sheet. The method is obsolete now, but is revived for 'antique' effects.

bull-nose (*Build.*). A purpose-made brick having a rounded corner; for use in positions where sharp arrises might be damaged. (*Join.*) A small metal rebating plane having the mouth for the cutting iron near the front.

bull-nosed step (*Build.*). A step which, in plan, is half-round or quarter-round at the end.

bull-ring (*Elec. Eng.*). A metal ring used in the construction of overhead contact wire systems for electric traction schemes; it forms the junction of three or more straining wires.

bull's-eye arch (*Arch.*). A circular or oval window or opening.

bull's-eye lens (*Photog.*). A small thick lens, used for condensing light from a source.

bull's nose (*Carp.*). A name sometimes given to the salient angle at the intersection of two plane surfaces.

bull stretcher (*Build.*). A brick with one corner rounded, laid, with the long face exposed, as a quoin.

bull wheel (*Mining*). The driving pulley for the camshaft of a stamp battery; one on which bull rope of drilling rig is wound.

bulwark (*Civ. Eng.*). A sea-wall built to withstand the force of the waves; in some cases the reinforcement of the natural *breakwater* (q.v.).

bumblefoot (*Vet.*). A cellulitis of the foot of Birds due to infection by pus-forming organisms.

bump (*Aero.*). See **air pocket.** (*Mining*) See **crump.**

bumping trough (*Mining*). Sheet-metal trough hung on chains from roof of flat stope and swung against a bumping stop so as to convey material shovelled in toward a discharge chute.

bumping-up (*Print.*). Interlaying the half-tone areas of rotary printing plates to provide heavier impression, there being no provision for *make-ready* on the rubber-covered impression cylinder.

bumps (*Cinema.*). Low-frequency extraneous sounds during reproduction, due to irregular motion of sound-track, in recording or reproduction.

Buna (*Plastics*). Synthetic rubber manufactured (at first in Germany) by polymerization of butadiene with sodium (hence the name *Bu+Na*). Buna-N (NBR, GR-N), made from interpolymerization of butadiene with acrylonitrile, has good ageing and oil-resisting properties; Buna-S (SBR, GR-S), made from butadiene and styrene, has good mechanical, electrical, and ageing properties; especially used for tyres.

buncher (*Electronics*). Arrangement which velocity-modulates and thereby introduces bunches in electron space current passed through it. Bunching would be *ideal* if the bunches contained electrons all having the same velocity. Also **input gap, buncher gap.** See **catcher, debunching,** and **rhumbatron.**

bunching (*Electronics*). Collective grouping of electrons along the beam of electrons passing through a rhumbatron.

bunching angle (*Electronics*). Transit delay or phase angle between modulation and extraction of energy in a bunched beam of electrons.

bunch light (*Light*). A group of electric lamps in a portable fitting; used chiefly for stage lighting.

bundle (*Anat.*). Fibres collected into a band in the nervous system or in the heart. (*Bot.*) See **vascular bundle.** (*Textiles*) The commercial unit of yarn or cloth; for *cotton hanks* 10 or 5 lb., for *flax yarn bundles* 60 000 yd, the weight varying according to fineness (i.e., *count*).

bundle conductor (*Elec. Eng.*). Two or more overhead line conductors, suitably spaced to avoid *corona* loss, forming a phase; replaces a single large conductor.

bundle end (*Bot.*). The much simplified termination of a small vascular bundle in the mesophyll of a leaf.

bundle of His (*Physiol.*). See **His's bundle.**

bundle sheath (*Bot.*). A sheath of one or more layers of parenchymatous or of sclerenchymatous cells, surrounding a vascular bundle.

bungalow (*Arch.*). A single-storey house. (*Cinema.*) The same as **blimp.**

Büngner's strands (*Zool.*). Long protoplasmic strands formed in the peripheral part of a cut nerve fibre.

bungum (*Civ. Eng.*). Recent alluvial slit, particularly in the London area. Applied usually to soft clay.

bunion (*Med.*). A bursa formed on the outer side of the big toe where it joins the foot, as a result of deformity of the toe and pressure of tight-fitting shoes or boots.

bunk (*For.*). The cross-member of a log transport vehicle, on which the logs rest.

bunker (*Eng.*). A storage room for coal or oil fuel.

bunker capacity (*Ships*). The capacity of a space in a ship used for carrying fuel (oil, coal, or other combustible material). It is calculated at a fixed rate of stowage per unit volume, according to fuel, and allowances for obstructions are made in percentage.

bunodont (*Zool.*). Said of mammalian teeth in which the cusps remain separate and rounded. Also **bunoid.** Cf. *lophodont*, *selenodont*.

bunolophodont (*Zool.*). Said of teeth showing both *bunodont* and *lophodont* characters.

bunoselenodont (*Zool.*). Said of teeth which show both *bunodont* and *selenodont* characters.

Bunsen burner (*Chem., Light*). A gas burner consisting of a tube with a small gas jet at the lower end, and an adjustable air inlet by means of which the heat of the flame can be controlled; used as a source of heat for laboratory work and formerly, in conjunction with an incandescent mantle, as the usual form of gas burner for illuminating purposes. Invented by Robert Bunsen.

Bunsen cell (*Elec. Eng.*). A double-fluid primary cell yielding 1·9 volt. It consists of a zinc anode dipping into dilute sulphuric acid and a carbon cathode dipping into conc. nitric acid.

Bunsen flame (*Chem.*). The flame produced when a mixture of a hydrocarbon gas and air is ignited in air, as in a Bunsen burner. It consists of an inner cone, in which carbon monoxide is formed, and an outer one, in which it is burnt.

Bunsen photometer (*Light*). See **grease-spot photometer.**

bunt (*Aero.*). A manoeuvre in which an aeroplane performs half an inverted loop, i.e., the pilot is on the outside where he experiences *negative g* (q.v.). (*Bot.*) A parasitic fungus (*Tilletia feotens*), a species of smut, which destroys the grain of wheat by converting the interior portion into a black powder. Mainly confined to Europe.

Bunter Series (*Geol.*). The lowest of the three series into which the rocks of the Triassic System are divided. Well exposed in the English Midlands, it comprises pebble beds with sandstone above and beneath.

buntons (*Mining*). Horizontal timbers in a circular shaft, used to carry the guides for the cage and any pipes. See **dividers.**

buoy (*Hyd. Eng.*). A floating vessel, capable of being illuminated at night, moored in estuaries and ship-canals to mark the position of minor shoals, and to show the navigable channel.

buoyancy (*Aero.*). The vertical thrust on an aircraft due to its immersion, either wholly or partly, in a fluid. Equal to the weight of air displaced by the gas-bags in the case of an airship: equal to the weight of water displaced by the immersed portions of the floats of a seaplane, or the body of a flying-boat. See also **reserve-.** (*Hyd., Phys.*) The loss in weight of a body when immersed in a fluid, due to the resultant upward pressure exerted by the fluid on a body wholly or partly immersed in it. See **Archimedes' principle, correction.**

buoy-to-buoy (*Aero.*). See **block time.**

bupivacaine (*Pharm.*). A powerful local anaesthetic used for regional nerve block anaesthesia, particularly epidural. It is (±)-1-butyl-2-(2,6-xylylcarbamoyl)piperidine hydrochloride.

buran (*Meteor.*). A frequent winter north-easterly wind in Central Asia and Russia.

burden (*Elec. Eng.*). A term used to signify the load on an instrument transformer. It is usually expressed as the normal rated load in volt-amperes, or as the impedance of the circuit fed by the secondary winding. (*Eng.*) See **on-costs.** (*Mining*) (1) Amount of rock to be shattered in blasting between drill-hole and nearest free face. (2) See **overburden.**

Burdizzo pincers (*Vet.*). A castrating instrument which crushes the spermatic cord.

burdo (*Bot.*). A graft hybrid presumed to have arisen by the union of vegetative nuclei derived from the stock and the scion.

Burdorite (*Agric.*). A fungicidal compounded product containing basic copper derivatives.

burette (*Chem.*). A vertical glass tube with a fine tap at the bottom and open at the top, usually holding up to 50 cm^3 of reagent solution; used in *volumetric analysis* (q.v.). The tube is usually graduated in tenths of a cm^3, so that the amount of liquid allowed to run out through the tap can be estimated to one-twentieth of a cm^3.

burgee (*Glass*). Unusable waste sand, containing glass and iron debris, which has been used for grinding plate glass. (*Met.*) Small furnace coal.

Burger's vector (*Crystal.*). Translation vector of crystal lattice, representing a displacement which creates a lattice dislocation.

burglar alarm. Any device whereby a bell is rung unwittingly, or other alarm given (either on the spot or at a distance, e.g., control point, police station) by an intruder; actuated electromechanically or by ultrasonic, photoelectric or laser beam.

burial (*Nuc. Eng.*). Place for the safe deposition, usually in non-corrosive containers, of the highly radioactive products of the operation of nuclear reactors. Also called **graveyard.**

buried antenna (*Telecomm.*). One in which the wires are buried under the ground, the e.m.f. depending on tilt of the ground wavefront.

buried cable (*Telecomm.*). Waterproof cable which contains a number of communication circuits and is buried in the ground. See **aerial cable.**

burkeite (*Min.*). A carbonate and sulphate of sodium, crystallizing in the orthorhombic system.

burlap (*Textiles*). A coarse jute, hemp, or flax textile used as wall-covering, etc.

burl dyeing (*Textiles*). The dying or colouring by separate treatment, of the vegetable matter remaining in some woollen fabrics, to ensure level shade throughout.

burling (*Textiles*). See **mending.**

burling-irons (*Textiles*). Large forceps used to remove burrs and slubs from the cloth face, previous to scouring and finishing.

Burma lancewood (*For.*). A durable wood from the genus *Homalium*, used in India for the making of agricultural implements as well as being a structural timber.

burmite (*Min.*). An amber-like mineral occurring in the upper Hukong Valley, Burma, differing from ordinary amber by containing no succinic acid. A variety of retinite.

burn (*Space*). Controlled expenditure of rocket propellant for course adjustment purposes. (*Electronics*). See **ion burn.**

burnable poison (*Nuc.*). A substance to influence the long-term reactivity variations of a fission reactor. It should have a high neutron capture cross-section and give rise to a capture reaction product of low capture cross-section.

burner firing block (*Heat.*). Unit made from refractory material that fits into a furnace wall at the burner position, having a nozzle-protecting recess at back and a tunnel on the firing side. It is called **quarl** in oil-firing practice.

burner loading (*Heat*). Potential heat that can be liberated efficiently from a burner. Expressed in kilowatts or Btu/hour.

burner turndown factor (*Heat.*). Minimum gas rate at which a burner is capable of stable flame propagation without the flame flashing back to the air-gas mixing point or blowing off from the burner nozzle or head.

Burnett's fluid (*Carp.*). A zinc chloride solution used to preserve timber from wood-boring insects or from dry-rot.

burning (*Met.*). The heating of an alloy to too high a temperature, causing local fusion or excessive penetration of oxide, and rendering the alloy weak and brittle. (*Min. Proc.*) Changing the colour of certain precious stones by exposing them to heat.

burning-in kiln (*Glass*). A kiln in which stain or enamel colour painted on glass-ware or sheet-glass is fired to cause it to adhere more or less permanently; usually of muffle type.

burning-on (*Foundry*). The process of adding a piece to an existing casting by making a mould round the point of juncture and pouring metal into it.

burning voltage (*Electronics*). The minimum voltage between the anode and cathode to maintain the discharge of a cold-cathode electron tube or lamp. This is less than the starting, striking or ignition voltage.

burnishing (*Bind.*). The operation of applying a brilliant finish to gilt or coloured edges by means of a burnishing tool, which is applied under great pressure from the shoulder.

Burnley printer (*Weaving*). See **printer.**

burnout (*Electronics*). Sudden and protracted change in crystal rectifier characteristics as a result of excess voltage.

Burns test (*Brew.*). Flocculation test for yeast in which a small quantity is shaken in an acetate buffer solution of pH 4·6 and the sedimentation measured.

burnt coal. Sooty product of weathering of a coal outcrop.

burnt deposit (*Elec. Eng.*). A loose powdery deposit obtained in electroplating, if the rate of deposition is allowed to be too great.

burnt lime (*Build., Chem.*). See **lime.**

burnt metal (*Eng., Met.*). Metal which has become oxidized by overheating, and so is rendered useless for engineering purposes.

burnt ochre (*Paint.*). Red pigment obtained by calcining yellow ochre, thus dehydrating the iron and removing organic matter.

burnt-out lace (*Textiles*). Embroidery in which a material such as wool, or alginate, forms a temporary ground during production, and is subsequently removed by dissolving in a solution, leaving the design in the unaffected thread intact.

burnt sienna (*Paint.*). Orange-brown pigments produced by carefully calcining sienna to remove water and organic matter.

burnt umber (*Paint.*). Umber which has been calcined at a low temperature to remove water and carbon, resulting in a rich brown pigment.

burnup (*Nuc. Eng.*). In a reactor, consumption (by fission) of fuel rods or other materials.

burr (*Acous.*). The rough edge which the gramophone record has when it is removed from the press. This is removed by spinning the record in a special type of lathe. (*Bot.*) A fruit covered with hooks to aid in dispersal by animals. (*Eng.*) (1) A rough or sharp edge left on metal by a cutting tool. (2) A blank punched from sheet-metal. (3) A small milling cutter used for engraving and dental work. (*For.*) An excrescence in tree growth which, when sliced, produces strong contrasts in the form and colour of markings. (*Textiles*) Seed or other vegetable impurity found in raw wool. It is removed either by chemical action or by passing it through very heavily loaded, accurately ground rollers.

burrs (*Build.*). Lumps of brick, often mis-shapen, which in the burning have fused together, and which are used for rough-walling, artificial rock-work, etc.

bursa (*Zool.*). Any saclike cavity: more particularly, in Vertebrates, a sac of connective tissue containing a viscid, lubricating fluid, and interposed at points of friction between skin and bone and between muscle, ligament, and bone.

bursa copulatrix (*Zool.*). A special genital pouch of various animals acting generally as a female copulatory organ.

bursa Entiana (*Zool.*). In some *Selachii*, a pouch of the gut representing the pyloric caeca.

bursa Fabricii (*Zool.*). In young Birds, a pocket of unknown function developed from the dorsal part of the cloaca.

bursa inguinalis (*Zool.*). The cavity of the scrotal sac in Mammals.

bursa omentalis (*Zool.*). In Mammals, a sac formed by the epiploon or great omentum.

bursa propulsoria (*Zool.*). In some *Oligochaeta*, a muscular sac opening at the male pore, into which the prostate gland discharges.

bursattee (or **bursati**) (*Vet.*). Cutaneous habronemiasis. A disease of the skin of horses caused by nematode larvae of the genus *Habronema*; characterized by granulomatous nodules in the skin.

Bursera (*Bot.*). A tropical American genus of trees of the family Burseraceae, yielding elemi (fragrant resin) and timber. Akin to the rue family. (From Joachim Burser, 17th cent. German botanist.)

bursicon (*Zool.*). In Insects, a hormone produced by the brain and nervous system, but only released from the terminal abdominal nerve ganglion. It brings about the processes involved in *tanning* (q.v.) newly-formed cuticle.

bursiform (*Bot.*). Resembling a bag or pouch.

bursitis (*Med.*). An inflammation of a bursa.

burst (*Nuc.*). Unusually large pulse arising in an

ionization chamber caused by a cosmic-ray shower. (*Radio*) Sudden increase in strength in received radio signals caused by sudden changes in ionosphere.

burst-can detector (*Nuc. Eng.*). An instrument for the early detection of ruptures of the sheaths of fuel elements inside a reactor. Also **leak detector.**

bursting disk (*Chem. Eng.*). A protective device for process vessels in which hazardous operations are performed, consisting of a thin disk of noble or corrosion resisting metal, carefully controlled as to thickness, and designed to burst in event of excess internal pressure, giving a large opening for rapid release of the pressure.

burst pedestal (*TV*). Part of *colour burst* in a colour television signal.

burst signal (*TV*). Component of transmitted signal, which acts as reference for chrominance components. These operate circuits to establish correctly colour elements in the reproduced image.

burst slug (*Nuc. Eng.*). Fuel element with a small leak, emitting fission products. Also **cartridge.**

burst test (*Paper*). Measures the pressure required to rupture the paper by means of a rubber diaphragm being forced against the paper. A widely used test performed usually on a Mullen instrument.

Burton (*Brew.*). See **ale.**

Burton Union system (*Brew.*). Cleansing system in which the wort is run into casks and the yeast carried off into a trough, usually through swan-neck pipes fitted to the bung-holes.

bus (*Comp.*). See **highway.**

bus-bar (*Elec. Eng.*). (1) Sectionalizing switch in power circuits. (2) Supply rail maintained at a constant potential (including zero or earth) in electronic equipment.

bus-coupler switch (*Elec. Eng.*). A switch or circuit-breaker serving to connect two sets of duplicate bus-bars.

bush (*Bot.*). A low woody plant forming a number of branches at ground-level. (*Eng.*) A cylindrical sleeve, usually inserted in a machine part to form a bearing surface for a pin or shaft.

bush adapter (*Photog.*). See **tripod bush.**

bush-hammering (*Civ. Eng.*). The operation of dressing the surface of stone with a special hammer having rows of projecting points on its striking face.

bushing (*Glass*). A small electric melting unit, usually made of platinum, with numerous holes in the base, used for the manufacture of glass fibres. (*Elec. Eng.*) An insulator which enables a live conductor to pass through an earthed wall or tank (e.g., the wall of a switch house or the tank of a transformer). (*Plumb.*) A reducing adapter or screwed piece for connecting together in the same line two pipes of different sizes.

bushing current transformer (*Elec. Eng.*). A current transformer built into a bushing.

bush knife (*For.*). A light chopping or slashing tool in the form of a large knife, originally having the form of a cutlass; used extensively in bush clearing in tropical countries. Also **machete.**

bush metal (*Met.*). Alloy suitable for bearings. Bush bronze is based on copper with some 4% zinc, 4% tin, and 4% lead.

bush sickness (*Vet.*). See **pine.**

bush waxing (*Cables*). Waxing which occurs in conjunction with carbon. There is frequently a waxed area, with streamers at its edges.

bus-line (*Elec. Eng.*). A cable, extending the whole length of an electric train, which connects all the collector shoes of like polarity. Sometimes called a **power line.**

bus-line couplers (*Elec. Eng.*). Plug-and-socket connectors to join the bus-line of one coach of an electric train to that of the next.

Bustamente furnace (*Met.*). Shaft furnace developed for distilling mercury from its ores.

bustamite (*Min.*). A triclinic silicate of manganese, calcium, and iron.

buster (*Mining*). In coal mining, expanding wedge used to break down rock by driving into crack.

bustle pipe (*Met.*). Main air pipe surrounding blast furnace, which delivers low-pressure compressed air to tuyères.

bus-wire coupler (*Elec. Eng.*). A flexible connexion between the coaches of an electric train for maintaining the continuity of any bus-wires which have to run throughout the train-length.

busy (*Teleph.*). A term applied to any line or equipment signifying *engaged.* See **busy tone.**

busy tone (*Teleph.*). An audible signal which is fed back to a caller to indicate that the switching equipment, line, or required subscriber's instrument is already engaged.

butadiene (*Chem.*). Butan 1,2 : 3,4 diene. Divinyl, $CH_2{=}CH{\cdot}CH{=}CH_2$, a di-alkene with conjugate linking. An *isoprene* (q.v.) homologue, an important compound in the manufacture of synthetic rubbers. See **Buna, Butakon, chloroprene, GR-S.**

Butakon (*Plastics*). TN for range of synthetic rubbers based on copolymers of butadiene with a variety of other monomers such as styrene, acrylonitrile, and methyl methacrylate.

butaldehyde (*Chem.*). Butanal. Butyraldehyde, $CH_3{\cdot}CH_2{\cdot}CH_2{\cdot}CHO$. B.p. 76°C. Made by the catalytic dehydrogenation of butan-1-ol.

butane (*Chem.*). C_4H_{10}, an alkane hydrocarbon, b.p. 1°C, rel. d. at 0°C 0·600, contained in natural petroleum, obtained from casing head gases in petroleum distillation. Used commercially in compressed form, and supplied in steel cylinders for domestic and industrial purposes, e.g., Calor gas.

butanoic acids (*Chem.*). See **butyric acids.**

butanol (*Chem.*). See **butyl alcohol.**

butcher's paper (*Paper*). Paper resistant to blood, water and grease.

butenes (*Chem.*). *Butylenes* (q.v.).

Butex (*Nuc. Eng.*). TN for dioxy diethyl ether, used for separating U and Pu from fission products.

buthocrome (*Nuc.*). Particular groups of atoms in organic compounds which have the effect of lowering frequency of the radiation absorbed by these compounds.

butobarbitone (*Pharm.*). A barbiturate hypnotic and sedative; 5-butyl-5-ethylbarbituric acid. TNs **Butomet, Soneryl, etc.**

butt (*Leather*). The back part of a tanned hide: used for boot soles, belting, etc.

butt coupling (*Eng.*). See **muff coupling.**

butte (*Geol.*). A steep-sided flat-topped hill; a small *mesa.*

butted (*Typog.*). Said of (1) rules joined at a corner by bringing the end of one to the inside edge of another, not mitring the joint; (2) two or more slugs joined end to end to form a wide line of type.

butter (*Chem.*). A fat emulsion containing in solution sugar, albumen, salts, whereas fats and casein are present in colloidal dispersion.

butterfly (*Cinema.*). A diffuser used to soften the light from the sun or from lamps; made from silk stretched on a frame.

butterfly circuit (*Elec. Eng.*). One in which both

the inductance and capacitance between opposite re-entrant sections of a punched disk are varied by rotating an insulated coaxial vane of the same shape as the cavity. See **differential capacitor.**

butterfly flower (*Bot.*). (1) A flower pollinated by butterflies. (2) Popular name for *Schizanthus.*

butterfly nut (*Eng.*). See **wing nut.**

butterfly tail (*Aero.*). See **vee-tail.**

butterfly valve (*Eng.*). (1) A disk turning on a diametral axis inside a pipe; used as a throttle valve in petrol and gas engines. (2) A valve consisting of a pair of semicircular plates hinged to a common diametral spindle in a pipe; by hingeing axially, the plates permit flow in one direction only.

buttering (*Build.*). The operation of spreading mortar on the edges of a brick before laying it.

buttering trowel (*Build.*). A flat tool similar to, but smaller than, the *brick trowel*; used for spreading mortar on a brick before placing it in position.

buttermilk (*Chem.*). The aqueous liquid occluded in butter which is removed during the manufacturing process.

butternut (*For.*). A north American timber of the Walnut family, of economic importance for its fruits as well as its wood. It is brownish in colour with normally straight grain but a somewhat coarse texture. It is mostly used for cabinet making and panelling.

butter-rock (*Min.*). See **halotrichite.**

Butterworth filter (*Telecomm.*). Constant-*k* filter designed to give response of maximum flatness through pass band. Cf. *Chebyshev filter*.

butt gauge (*Tools*). One used in fitting the butt hinges on doors, having three markers, one for the thickness of the butt, one for the depth of the door, and one the depth of the jamb.

butt hinge (*Join.*). A hinge formed by 2 leaves, which are secured to the door and door-frame in such a manner that when the door is shut the 2 leaves are folded into contact.

butt joint (*Carp., Join.*). A joint formed between the squared ends of the 2 jointing pieces, which come together but do not overlap. (*Eng.*) A joint between 2 plates whose edges abut, and are covered by a narrow strip or 'strap' riveted or welded to them.

butt-jointed (*Elec. Eng.*). See **joint.**

buttock (*Mining*). Coal from which an undercut has been removed, in readiness for bringing it down.

buttock planes (*Ships*). Longitudinal sectional planes drawn through a ship's form; used for laying-off in the moulding loft, and for calculation of volumes, etc.

button (*Bot.*). A young fruit body of an agaric, before the pileus has spread out and exposed the hymenium. (*Horol.*) The serrated knob by means of which a keyless watch or similar movement is wound.

button-headed screws (*Eng.*). Screws having hemispherical heads, slotted for a screwdriver; known also as **half-round screws.**

buttoning (*Weaving*). Small bunches of short fibres which form on the warp yarns either in front of or behind the reed.

button microphone (*Acous.*). Small microphone which can be fitted in the buttonhole.

buttonwood (*For.*). The N. American equivalent of the sycamore or plane of Europe. It is somewhat difficult to work either by hand or by machine, but it finishes with a good surface and responds to decorative treatment.

buttress (*Civ. Eng.*). A supporting pier built on the exterior of a wall to enable it to resist outward thrust.

buttress root (*Bot.*). A root, often adventitious, which assists in keeping the stem of the plant upright.

buttress screw-thread (*Eng.*). A screw-thread designed to withstand heavy axial thrust in one direction. The back of the thread slopes at 45°, while the front or thrust face is perpendicular to the axis.

butt strap (*Eng.*). The cover plate which overlaps the main plates of a *butt joint* (q.v.).

butt-welded tube (*Met.*). Tube made by drawing mild steel strip through a bell, so that the strip is coiled into a tube, the edges being then pressed together and welded.

butt-welding (*Eng.*). The joining of two plates or surfaces by placing them together, edge to edge, and welding along the seam thus formed. See **welding.**

butyl acetate (ethanoate) (*Chem.*). $CH_3COOC_4H_9$ commercial product has a boiling range 124°–128°C, rel. d. 0·885, a colourless liquid, of fruity odour, soluble in ethanol, ether, acetone, benzene, turpentine, slightly in water. A very important lacquer solvent.

butyl alcohols (*Chem.*). $C_4H_9{\cdot}OH$. There are 4 isomers possible and known, viz.: butan-1-ol $CH_3(CH_2)_3OH$, b.p. 117°C; butan-2-ol $CH_3CH_2CH(OH)CH_3$, b.p. 110°C; 2-methyl propan-1-ol, b.p. 107°C; 2,2-dimethylethanol, m.p. 25°C, b.p. 83°C.

butylenes (*Chem.*). C_4H_8, alkene hydrocarbons, the next higher homologues to propylene. Three isomers are possible and known, normally gaseous, b.p. between −6°C and +3°C.

buty group (*Chem.*). The aliphatic group C_4H_9—, with four isomeric forms: *primary, secondary, iso-* and *tertiary*.

butyl rubber (*Chem.*). Form of synthetic rubber obtained by copolymerizing isobutylene and isoprene. Used chiefly for tubeless tyres and inner tubes, because of very low air permeability.

butyric acids (*Chem.*). Butanoic acids. $C_3H_7{\cdot}COOH$, two isomers, viz., *normal* and *iso*-butyric acid. Only *n*-butyric (butanoic) acid is of importance, m.p. −8°C, b.p. 162°C; it is a thick liquid of rancid odour and occurs in rancid butter.

butyric fermentation (*Chem.*). The fermentation of sugar and starch by fission ferments, e.g., *Bacillus butylicus*, resulting in the formation of butyric acid.

butyrous (*Bot.*). Of a butterlike consistency or colour.

butyryl (*Chem.*). The monovalent acyl radical $C_3H_7{\cdot}CO$—.

Buxbaumiales (*Bot.*). An order of the *Musci*, characterized by the asymmetrical capsule, and a peristome of three to six rows of teeth.

Buxton certification (*Chem., Elec. Eng., Mining*). The certification of the suitability of electrical equipment for use in atmospheres in which fire or explosion hazards are present.

Buys Ballot's law (*Meteor.*). If an observer stands with his back to the wind, the lower atmospheric pressure is on his left in the northern hemisphere, on his right in the southern hemisphere.

buzz (*Aero.*). (1) Severe vibration of a control surface in transonic or supersonic flight caused by separation of the airflow due to compressibility effects. (2) To interfere with an aircraft in flight by flying very close to it.

buzzard (*Zool.*). A large bird of prey (genus *Buteo*) of the Falconiformes.

buzzer (*Telecomm.*). A vibrating reed, actuated by self-interrupted direct current, used to generate a note to indicate the presence of actuating current.

buzzer wavemeter (*Telecomm.*). A wavemeter in which oscillations of known frequency are generated in a resonant circuit connected across the vibrating contacts of a buzzer.

buzz track (*Cinema.*). A sound-track on a normal film on which is registered a constant note; the reproduction from this is used for the alignment of the optical parts of the light-scanning system.

BWD (*Vet.*). See **pullorum disease.**

B.W.G. (*Eng.*). See **Birmingham Wire Gauge.**

B-wire (*Teleph.*). In a 2-wire telephone line to a manual exchange, that wire which under normal conditions is further from earth potential.

bye-channel (*Civ. Eng.*). A channel formed around the side of a reservoir past the end of the dam, to convey flood discharge from the stream above the reservoir into the stream below the dam. Also called **bye-wash, diversion cut, spillway.**

bye-wash (*Civ. Eng.*). See **bye-channel.**

by-pass (*Min. Proc.*). Use of screens to divert undersize round crushing machine or other appliance not set to treat it. (*Plumb.*) Any device for directing flow around a fixture, connexion, or pipe, instead of through it.

by-pass capacitor (*Elec. Eng., Electronics*). A capacitor having a low reactance for frequencies of interest connected in shunt with other components so as to short-circuit them for signal frequency currents.

by-pass ratio (*Aero.*). The ratio of the by-passed airflow to the combustion airflow in a *dual-flow turbojet* (q.v.) having a single air intake.

by-pass road (*Civ. Eng.*). A road specially constructed to serve as a diversion road for traffic which would otherwise pass through a town.

by-pass turbojet (*Aero.*). A turbojet, usually 2-spool, in which part of the compressor delivery is by-passed round the combustion zone and turbine to provide a cool, slow propulsive jet when mixed with the residual efflux from the turbine. See **turbofan.**

by-pass valve (*Elec. Eng.*). A mercury-arc valve normally blocked, connected across the mercury-arc converter of a high-voltage d.c. transmission line to maintain the flow of current in the system whenever the current paths through the main converter valves have to be interrupted. (*Eng.*) A valve by which the flow of fluid in a system may be directed past some part of the system through which it normally flows, e.g., an oil-filter in a lubrication system.

B−Y signal (*TV*). Component of colour TV chrominance signal. Combined with luminance (Y) signal, it gives primary blue component.

bysmalith (*Geol.*). A form of igneous intrusion bounded by a circular fault and having a dome-shaped top; described by Iddings from the Yellowstone Park.

byssaceous, byssoid (*Bot.*). (1) Consisting of a mass of fine threads, and resembling cotton in appearance. (2) Of delicate, filamentous structure.

byssinosis (*Med.*). A lung disease of cotton workers.

byssisede (*Bot.*). Said of a fungus fruiting body growing on a cottonlike mass of mycelium.

byssus (*Zool.*). In *Lamellibranchiata*, a tuft of strong filaments secreted by a gland in a pit (**byssus pit**) in the foot and used for attachment. *adj.* **byssogenous, byssal.**

byte (*Comp.*). A group of *bits* shorter than a *word* and operated on as a unit.

bytownite (*Min.*). A variety of plagioclase feldspar, containing a high proportion of the anorthite molecule; occurs in basic igneous rocks.

Bz (*Chem.*). A symbol for the benzoyl (benzene-carboxyl) radical $C_6H_5{\cdot}CO$.

C

c (*Chem.*). A symbol for *concentration.* (*Phys.*) The symbol used for the velocity of electromagnetic radiation *in vacuo*. Its value, according to the most accurate recent measurements, is $2{\cdot}997\,924\,56 \times 10^8$ m s^{-1}.

c- (*Chem.*). Abbrev. for (1) *cyclo-*, i.e., containing an alicyclic ring. (2) *cis-*, i.e., containing the 2 groups on the same side of the plane of the double bond or ring.

C (*Chem.*). The symbol for *carbon*. (*Elec.*) Coulomb. (*Heat*) When used after a number of degrees thus: 45°C, the symbol indicates a temperature on the Celsius or Centigrade scale.

C (*Chem.*). A symbol for: (1) *concentration*; (2) (with subscript) *molar heat capacity*: C_p, at constant pressure: C_v, at constant volume. (*Elec.*) Capacitance.

C- (*Chem.*). Containing the radical attached to a carbon atom.

[C] (*Light*). One of the Fraunhofer lines in the red of the solar spectrum. Its wavelength is 656·3045 nm; it is due to hydrogen.

Ca (*Chem.*). The symbol for *calcium*.

CAA (*Aero.*). Abbrev. for **Civil Aviation Authority.**

CAB (*Plastics*). Cellulose acetate butyrate.

cabane (*Aero.*). A system of struts in an aircraft arranged in the form of a pyramid or pylon.

cab-horse disease (*Vet.*). Exostosis affecting the inner and posterior surface of the long pastern bone of the foreleg of the horse.

cabin (*Mining*). A fireman's station underground in a coal-mine.

cabin altitude (*Aero.*). The nominal pressure altitude maintained in the cabin of a pressurized aeroplane.

cabin blower (*Aero.*). An engine-driven pump, usually of displacement type, for maintaining an aeroplane cockpit or cabin above atmospheric pressure. Also **cabin supercharger.**

cabin differential pressure (*Aero.*). The pressure by which that in an aircraft cabin is made to exceed that of the surrounding atmosphere to maintain comfortable conditions at high altitude. For an aircraft flying at 9000 m this differential would be about 60 kN/m².

cabinet-file (*Tools*). A single-cut smooth file used by joiners and cabinet makers.

cabinet screwdriver (*Tools*). One with a round shank flattened at the end.

cabinet-work (*Join.*). Fine joinery used in the construction of furniture and fixtures.

cabin hook (*Join.*). A hooked bar and eye, serving as a fastener for doors and casements.

cabin supercharger (*Aero.*). See **cabin blower.**

cable (*Elec. Eng.*). (1) An electrical circuit, comprising one or more conductors surrounded by dielectric and a sheath, suitable for laying on the bed of the ocean for carrying telegraphic signals. (2) The collection of a number of circuits in a lead or other form of waterproof, e.g., jelly-filled (petroleum jelly), sheath, for burying in the ground. (*Eng.*) A general term for rope or chain used for engineering purposes. Specifically, a ship's anchor cable.

cable-angle indicator (*Aero.*). An indicator showing the vertical angle between the longitudinal axis of a glider and its towing cable, also its yaw and roll attitude relative to the towing aeroplane.

cable buoy (*Ships*). A buoy attached to an anchor and serving to mark its position.

cable code (*Teleg.*). A code analogous to the Morse code in its combinations of 'dots' and 'dashes', but differing therefrom by using 'dots' and 'dashes' which are equal in duration, yet opposite in polarity. It is used extensively for telegraph transmission over long submarine cable circuits.

cable drill (*Mining*). Churn drill, percussion drill. Light mechanized rig with cutting tools, *bailer*, etc., suspended from wire rope.

cable ducts (*Elec. Eng.*). Earthenware, steel, plastic or concrete pipes through which cables are drawn, and in which they lie.

cable film (*TV*). Substandard TV film, scanned at low speed for signal transmission up to 4·5 kHz through cables, for reconstruction and video transmission. Cable time is 1000 times viewing time.

cable form (*Elec. Eng.*). The normal scheme of cabling between units of apparatus. The bulk of the cable is made up on a board, using nails at the appropriate corners, each wire of the specified colour identification being stretched over its individual route with adequate *skinner* (q.v.). When the cable is bound with twine and waxed, it is fitted to the apparatus on the racks and the skinners connected, by soldering, to the *tag blocks* (q.v.).

cable grip (*Elec. Eng.*). A flexible cone of wire which is put on the end of a cable. When the cone is pulled, it tightens and bites into the lead sheath of the cable, and can be used to pull the cable into a duct.

cable-laid rope. A rope formed of several strands laid together so that the twist of the rope is in the opposite direction to the twist of the strands.

cable-length. One tenth of a nautical mile. (608 ft, 185 m).

cable railway (*Civ. Eng.*). A means of transport sometimes used for short straight inclines, or other stretches, the motive power coming from a continuous cable, overhead or underground, to which the car may at any time be rigidly connected through a clutch device.

cable release (*Photog.*). Device for releasing a camera shutter in which the trigger is actuated by a length of stiff wire cable in a flexible tube.

cable tramcars (*Civ. Eng.*). Tramcars deriving motive power from an underground cable, in the same manner as the cable railway.

cable wax (*Chem.*). A solid wax formed by the ionic bombardment of the oil in a cable. It is a good insulator, and cables operate very successfully even when much wax is present. It is produced by a condensation process such as $C_6H_{14} + C_5H_{12} \rightarrow C_{10}H_{22} + CH_4$.

cable-way (*Civ. Eng., etc.*). A construction consisting of cables slung over and between two towers, so that a skip suspended from the cables may be raised and lowered and moved to any position along the cables. It is used for transport of spoil and materials. Also **blondin.**

cabling (*Arch.*). A round moulding used to decorate the lower parts of the flutes of columns. (*Teleph.*) The collection of cables required for distributing the power supplies in a telephone exchange. See **trunking.** (*Textiles*)

Twisting together in the opposite direction two or more doubled yarns. The result in most cases is a balanced cord of four, six or more yarns. *Tyre cord* is one example.

cab-tyre sheathing (*Elec. Eng.*). A hard rubber cable-sheath. It may be used in house-wiring without a protecting steel tube.

cachexia (*Med.*). A combination of wasting, weakness, anaemia, and an 'earthy' complexion, characteristic of patients with cancer.

cacodyl (*Chem.*). $As_2(CH_3)_4$, a colourless liquid; b.p. 170°C; of horribly nauseous odour. It combines directly with oxygen, sulphur, chlorine, etc. Cacodyl and cacodyl oxide form the basis for other secondary arsines. Cacodyl derivatives are important as rubber accelerators.

cacosmia (*Med.*). A bad smell.

Cactales, Cactiflorae, Opuntiales (*Bot.*). An order of *Dicotyledons*. The stems are usually fleshy, and the leaves modified as spines. The flowers are regular or zygomorphic. The perianth segments are numerous, with no distinction into sepals and petals, and arranged spirally. The many stamens are epipetallous, and the fruit, which is a berry, is formed from an inferior ovary (except *Pereskia*). Xerophytes of the New World.

cadastral survey (*Surv.*). Land survey, boundary delineation.

cadaverine (*Chem.*). 1,5-diaminopentane. $NH_2{\cdot}(CH_2)_5{\cdot}NH_2$, a colourless syrupy liquid; b.p. 178°C. Formed by the bacterial decomposition of diamino acids. Ring formation occurs by elimination of NH_3 with the formation of the heterocyclic base *piperidine* (q.v.). See also **putrescine.**

cadion (*Chem.*). 4-nitrobenzene-diazoamino-azobenzene:

Used as a reagent for the colorimetric detection and estimation of cadmium (forming a pink colour) and magnesium (blue).

cadmium (*Chem.*). White metallic element, symbol Cd, at. no. 48, r.a.m. 112·40, m.p. 320·9°C, b.p. 767°C, rel. d. 8·648. Cadmium plating is widely used as a corrosion protective to aluminium and its alloys. It is a powerful absorber of neutrons and is used in steel, for the detail control of nuclear reactors, and always for an emergency shutdown. Films of cadmium are photosensitive in the ultraviolet between 250 and 295 nm, with a peak at 260 nm.

cadmium cell (*Electronics*). (1) Vacuum photocell having a cadmium or cadmium-coated cathode, with maximum spectral sensitivity in the ultraviolet range. (2) Old name for **Weston standard cadmium cell.**

cadmium copper (*Met.*). A variety of copper containing 0·7–1·0% of cadmium. Used for trolley, telephone, and telegraph wires because it gives high strength in cold-drawn condition, combined with good conductivity.

cadmium lithopone (*Paint.*). A reduced yellow pigment prepared by co-precipitation and consisting of 38% cadmium sulphide with barium sulphate. Also **cadmopone.**

cadmium red (*Paint.*). A brilliant red pigment consisting of co-precipitated cadmium sulphide, cadmium selenide and barium sulphate. The pigment has low tinting strength and is dirty in tint, but has good resistance to light and mild chemicals. Also **selenium red.**

cadmium red line (*Light*). Line originally proposed as a reproducible standard for length, =643·846 96 nm. See **krypton.**

cadmium-sulphide detector (*Radiol.*). Radiation detector equivalent to a solid-state ionization chamber, but with amplifier effect (due to hole trapping). Its main drawback is a slow response time.

cadmium-sulphide exposure meter (*Photog.*). One using a cadmium-sulphide photoresistor powered by a *Mallory cell* (q.v.). More sensitive than the selenium cell type. Abbrev. CdS meter.

cadmium yellow (*Paint.*). A sulphide of cadmium used as a pale yellow to light orange pigment.

cadmopone (*Paint.*). See **cadmium lithopone.**

cadophore (*Zool.*). In *Doliolida*, a postero-dorsal process to which the zooid buds become attached.

caducibranchiate (*Zool.*). Possessing gills at one period of the life-cycle only, as in some *Caudata*.

caducous (*Bot.*). Soon falling from the plant: lasting for a short time only. (*Zool.*) *Deciduate* (q.v.).

caecostomy, cecostomy (*Surg.*). Formation of an artificial opening into the caecum.

caecum (*Bot.*). An outgrowth from the embryo sac into the endosperm in some plants. (*Zool.*) Any blind diverticulum or pouch, especially one arising from the alimentary canal.

caenogenesis (*Zool.*). A phenomenon whereby features which are adaptations to the needs of the young stages develop early and disappear in the adult stage. *adj.* **caenogenetic.**

Caen stone (*Build.*). A nonoölitic limestone from the Jurassic rocks of Normandy, very suitable for interior carving.

caeoma (*Bot.*). A form of aecidium in which the spores are surrounded by a few sterile hyphae only, and not by a well-formed peridium.

Caesarean (or Cesarean) section (*Surg.*). Delivery of a foetus through the incised abdomen and uterus.

caesious, caesius (*Bot.*). Bearing a bluish-grey waxy covering (bloom).

caesium (*Chem.*). Metallic element, symbol Cs, at. no. 55, r.a.m. 132·905, m.p. 28·6°C, b.p., 713°C, rel. d. 1·88. As a photosensitor it has a peak response at 800 nm. in the infrared, both thermal- and photo-emission being high. Caesium, when alloyed with antimony, gallium, indium, and thorium, is generally photosensitive.

caesium cell (*Elec. Eng.*). One having a cathode consisting of a thin layer of caesium deposited on minute globules of silver; particularly sensitive to infrared radiation, but generally approximating to that of the eye.

caesium clock (*Elec. Eng.*). Frequency-determining apparatus based on caesium-ion resonance of 9 192 631 770 Hz.

caesium-oxygen cell (*Elec. Eng.*). One in which the vacuum is replaced by an atmosphere of oxygen at very low pressure. It is more sensitive to red light than the caesium cell.

caesium unit (*Radiol.*). Source of radioactive caesium (*half-life* 33 yr) mounted in a protective capsule. The operation of a shutter enables a highly collimated beam of γ-radiation to be directed at a patient.

caespitose, cespitose (*Bot.*). Growing from the root in tufts, as many grasses. *dim.* **caespitulose.**

caffeine (*Pharm.*). Theine; 1, 3, 7-trimethyl-2, 6-dihydroxypurine; crystallizes with 1 H_2O in silky needles and loses its water of crystallization completely at 100°C; m.p. of the anhydrous caffeine, 234°C. It has a very marked physiological action on the heart. Can be obtained by extracting tea-dust, or by synthesis from uric acid.

cage (*Civ. Eng.*). The platform on which goods are hoisted up or lowered down a vertical shaft or guides. (*Eng.*) The part of a ball or roller bearing which separates the balls or rollers and keeps them correctly spaced along the periphery of the bearing. (*Mining*) Steel box used in vertical mine shaft to raise and lower men, materials or trucks (trams, tubs). May have two or three decks.

cage antenna (*Radio*). One comprising a number of wires connected in parallel, and arranged in the form of a cage, to reduce the copper losses and increase the effective capacitance.

cage rotor (*Elec. Eng.*). A form of rotor, used for induction motors, having on it a cage winding. Also called **squirrel-cage rotor.**

cage-type negative plate (*Elec. Eng.*). See **box-type negative plate.**

cage winding (*Elec. Eng.*). A type of winding used for rotors of some types of induction motors, and for the starting or damping windings of synchronous machines. It consists of a number of bars of copper or other conducting material, passing along slots in the core and welded to rings at each end. Also called a **squirrel-cage winding.**

Cailletet process (*Heat*). A method for the liquefaction of gases based on the free expansion of a gas from a higher to a lower pressure.

Cailletet's and Mathias' law (*Chem.*). The arithmetical mean of the densities, *d*, of a pure unassociated liquid, *l*, and its saturated vapour, *v*, is a linear function of the temperature, *t*. Mathematically expressed,

$$0{\cdot}5(d_l+d_v)=A+Bt.$$

Cainozoic, Kainozoic (*Geol.*). The word signifies 'recent life', and is applied to the fourth of the great geological eras. It is synonymous with Tertiary plus Quaternary.

cairngorm (*Min.*). Smoky-yellow or brown varieties of quartz, the colouring matter possibly due to some organic compound; named from Cairngorm in the Scottish Grampians, the more attractively coloured varieties being used as semi-precious gemstones. Also **smoky quartz.**

caisson (*Build.*). A deeply recessed sunk panel in a soffit or ceiling. (*Hyd. Eng.*) (1) A watertight air-receiver used to help float a sunken vessel, to which it is attached for this purpose. See also **camel.** (2) A floating structure which may be placed across the entrance to a basin, lock, or dry dock, thereby excluding water from it. (3) A boxlike structure with cutting edges, used in constructing a bridge foundation, and ultimately becoming part of it.

caisson disease (*Med.*). The bends; diver's palsy; diver's paralysis. Pains in the joints and paralysis, occurring in workers in compressed air who are suddenly subjected to atmospheric pressure after compression; it is due to accumulation of bubbles of nitrogen in the nervous system.

caked (*Typog.*). See **baked.**

caking coal (*Mining*). Coal which cakes or forms coke when heated in the absence of air.

Calabrian (*Geol.*). A stage of the Lower Pleistocene, defined by reference to the type section in southern Italy.

calamiferous (*Bot.*). Having a hollow tube-like stem.

calamine (*Min.*). A name formerly used in Britain for *smithsonite* (q.v.) and in America for *hemimorphite* (q.v.).

calamistrum (*Zool.*). In some Spiders, a brushlike series of hairs on the metatarsus of the fourth leg, correlated with the presence of the cribellum.

calamus (*Zool.*). The proximal hollow part of the scapus of a feather; quill. *pl.* **calami.**

calamus scriptorius (*Zool.*). In Mammals, a pointed depression at the posterior end of the median groove in the floor of the fourth ventricle.

calandria (*Chem. Eng.*). That part of evaporating plant in which the tubes carrying the heating fluid are incorporated. (*Nuc.*) Sealed vessel used in core of certain types of reactor.

calathide (*Bot.*). (1) The involucre of a capitulum. (2) The capitulum itself.

calcaneal process (*Zool.*). A long posterior process of the calcaneum in Mammals.

calcaneum (*Zool.*). In some Vertebrates, the fibulare, or large tarsal bone forming the heel; more generally, the heel itself; in Birds, a process of the metatarsus.

calcar (*Zool.*). In Insects, a tibial spine; in Amphibians, the prehallux; in Birds, a spur of the leg, or more occasionally, of the wing; in Bats, a bony or cartilaginous process of the calcaneum supporting the interfemoral part of the patagium.

calcarate (*Bot.*). Bearing one or more spurs.

Calcarea (*Zool.*). A class of *Porifera*, distinguished by the possession of a calcareous skeleton and large choanocytes.

calcareous (*Bot.*). Coated with, or containing, lime (as calcium carbonate). (*Chem.*) Containing compounds of calcium, particularly minerals.

calcareous clay (*Build.*). See **marl.**

calcareous pan (*Bot.*). A hard layer of limy material, more or less impermeable to water, formed below the surface of the soil, and affecting the water supply of plants growing above it.

calcareous ring (*Zool.*). In *Holothuroidea*, a ring of 10 calcareous ossicles surrounding the oesophagus.

calcareous rocks (*Geol.*). The sedimentary rocks containing appreciable quantities of calcium carbonate, or in many cases composed almost entirely of this substance. These include all limestones, magnesian limestones, and dolomites, as well as chalk, and chemically precipitated material such as calc-tufa.

Calceolaria (*Bot.*). A South American genus of *Scrophulariaceae*, with slipper-like flowers.

calc-flinta (*Geol.*). A hard fine-grained rock composed of calcium silicate minerals and produced by the contact metamorphism of an impure limestone.

calcicole, calcicolous (*Bot.*). Flourishing on soils or rocks rich in calcium carbonate.

calcicosis (*Med.*). Lung disease caused by the inhalation of marble dust.

Calciferae (*Bot.*). The group of *Monocotyledons* whose members have a distinct calyx and corolla.

calciferol. See **vitamin D.**

calciferous, calcigerous (*Zool.*). Producing or containing calcium salts.

calcification (*Bot.*). The accumulation of calcium carbonate on or in cell walls. (*Zool.*) The deposition of lime salts, e.g., in diseased or dead tissues such as the walls of arteries.

calcified cartilage (*Zool.*). The cartilages of the skull which, in *Elasmobranchii*, become encrusted by a superficial granular deposit of lime-salts.

calcifuge, calciphobe, calciphobous (*Bot.*). Terms applied to any species of plant which is intolerant of a limy soil.

calcigerous (*Zool.*). See **calciferous.**

calcigerous glands (*Zool.*). In some *Oligochaeta*, a pair of oesophageal glands producing a limy secretion to neutralize the acids in swallowed soil: in some Amphibians, the glands of Swammerdam, calcareous concretions lying on either side of the vertebrae, close to the points of exit of the spinal nerves.

calcimine (*Paint.*). A wash made up of whiting and glue with water, sometimes tinted for use on walls. U.S. term.

calcination (*Chem.*). The result of subjecting a material to prolonged heating at fairly high temperatures. (*Met.*) The operation of heating ores to drive off water and carbon dioxide. Frequently it is not distinguished from *roasting*.

calcine (*Met.*). Ore, carbonate, mineral or concentrate which has been roasted, perhaps in an oxidizing atmosphere, to remove sulphur as SO_2 (*sweet roasting*) or carbon dioxide (*dead roasting*).

calcined powder (*Powder Tech.*). One produced or modified by heating to a high temperature.

calciphile, calciphilous (*Bot.*). Terms applied to any species which occurs more or less exclusively on a limy soil.

calciphobe, calciphobous (*Bot.*). See **calcifuge.**

calcite or **calcspar** (*Min.*). A crystalline form of calcium carbonate, showing trigonal symmetry and a great variety of mineral habits. It is one of the commonest of minerals in association with both igneous and sedimentary rocks.

calcium (*Chem.*). Metallic element, symbol Ca, at. no. 20, r.a.m. 40·08, m.p. 850°C, b.p. 1440°C, rel. d. 1·58. Occurs in nature in the form of several compounds, the form of carbonate predominating. Produced by electrolysis of fused calcium chloride. Used as a reducing metal and as a getter in low-noise valves.

calcium carbide (*Chem.*). Ethynide dicarbide. CaC_2. A compound of calcium and carbon usually prepared by fusing lime and hard coal in an electric furnace. See also **acetylene.**

calcium carbonate (*Chem.*). $CaCO_3$. Very abundant in nature as chalk, limestone and calcite. Almost insoluble in water, unless the water contains dissolved carbon dioxide, when solution results in the form of calcium hydrogen carbonate, causing the temporary hardness of water.

calcium chloride (*Chem.*). $CaCl_2$. Formed by the action of hydrochloric acid on the metal and its common compounds. It absorbs moisture from the atmosphere, and is extensively used for drying gases. Tubes filled with granular calcium chloride are used in industry and laboratories for producing dry air.

calcium fluoride (*Chem.*). CaF_2. In the form of *fluorspar* (q.v.) it is used for the manufacture of hydrofluoric acid. It is also an important constituent of opal glass. It has high thermoluminescent sensitivity.

calcium tungstate screen (*Electronics*). A fluorescent screen used in CRT; it gives a blue and photo-actinic luminescence.

calcospherites (*Zool.*). Calcareous bodies which accumulate in the fat-body of phytophagous *Dipteran* larvae.

calcouranite (*Min.*). See **autunite.**

calcspar (*Min.*). See **calcite.**

calculus (*Maths.*). See **differential-, integral-.** (*Med.*) A concretion of mineral or of organic matter in certain organs of the body, e.g., the kidney, the gallbladder.

calculus of variations (*Maths.*). The determination of one or more functions of one or more independent variables, in order that a definite integral of a known function of these functions and their derivatives shall be a maximum or a minimum.

caldera (*Geol.*). A large volcanic crater produced by the collapse of underground lava reservoirs or by ring fracture, possibly as a surface expression of *cauldron subsidence* (q.v.).

Caledon (*Chem.*). TN for anthraquinone vat dyestuffs.

Caledon blue (*Chem.*). See **indanthrene.**

Caledonian (*Geol.*). Appertaining to the great mountain-building episode of late Silurian-early Devonian time.

Caledonoid direction (*Geol.*). The direction assumed by the Caledonian (Siluro-Devonian) mountainfolds and associated structures in Britain and Scandinavia. Commonly N.E.-S.W., but subject to considerable variations.

calendar month, calendar year. Popularly, a month or year as defined in a calendar, particularly the *Gregorian calendar*. A calendar month differs from the synodic (or lunar) month. The terms are also used for periods equivalent to a month or year, e.g., from July 9 to August 9, or July 9 of one year to July 9 of the next.

calender (*Paper, Textiles, etc.*). A machine, generally consisting of a number of vertical rollers (or bowls), heated or unheated, through which material is passed under pressure, to impart the desired finish (dull, glazed, etc.) or to ensure uniform thickness.

calendered paper (*Paper*). Paper which has been 'finished' in a calender. The varying degrees of gloss are distinguished thus: *low machine* (or *mill*) *finish* (usual printing paper), *high machine* (or *mill*) *finish* (intermediate), and *supercalendered* (highly glazed).

calf (*Bind.*). Superior quality calfskin used for covering books, finished either smooth or rough.

calf diphtheria (*Vet.*). Necrotic or malignant stomatitis; ulceration and necrosis of the mouth and pharynx of calves due to infection by the bacterium *Sphaerophorus necrophorus* (*Fusiformis necrophorus*).

calf scour (*Vet.*). See **white scour.**

calf tetany (*Vet.*). Milk tetany. A form of *bovine hypomagnesaemia* (q.v.), occurring in calves, caused by a deficiency of magnesium.

Calgon (*Chem.*). TN for sodium hexametaphosphate, used in water-softening because of its marked property of forming soluble double salts with calcium. Also used in the treatment of textiles and in laundry work.

caliber. See **calibre.**

calibrated airspeed (*Aero.*). Indicated airspeed corrected for *position error* and instrument error only. Not to be confused with *equivalent airspeed* or *true airspeed* (qq.v.). Abbrev. CAS. Also **rectified airspeed.**

calibration (*Phys., etc.*). The process of determining the absolute values corresponding to the graduations on an arbitrary or inaccurate scale on an instrument.

calibration error (*Nav.*). That in the bearings given by a ship's direction-finder, due to currents in the hull, masts, and rigging. The error is corrected in the initial calibration.

calibre (caliber). The internal diameter or bore of a pipe. (*Horol.*) The arrangement of the various components of a watch or clock.

caliche (*Geol.*). Concretions of calcium carbonate, soda nitre, and other minerals occurring at the surface of the ground in arid regions; due to surface evaporation of subsurface waters.

calico (*Textiles*). A plain grey, or bleached cotton, cloth heavier than muslin. Woven in different qualities and often printed by roller or screen.

calico weave (*Weaving*). See **plain weave.**

California job case (*Typog.*). A *double case* suitable for jobbing type, with a particular layout of boxes.
Californian bees (*Bot.*). See **ginger-beer plant.**
Californian jade (*Min.*). A compact form of green vesuvianite (idocrase) obtained from California, and used as an ornamental stone and in jewellery. Also known as **californite.**
Californian onyx (*Min.*). A term wrongly applied to amber- or brown-tinted aragonite, a soft mineral unsuited for use as a cut stone. Cf. *onyx.*
Californian stamp (*Min. Proc.*). One of five drop-weight crushing pestles which fall some 25 cm and 90 times a minute on ore spread on dies in a mortar box.
californite (*Min.*). See **Californian jade.**
californium (*Chem.*). Manmade element, symbol Cf, at. no. 98, produced in a cyclotron. Its longest lived isotope is ^{251}Cf with 800 years half-life.
caliper (*Horol.*). The size of a watch movement.
calipers. See **callipers.**
calked ends (*Build.*). The ends of built-in iron ties, split and splayed to provide more secure anchorage.
calking (*Civ. Eng.*). See **caulking.**
call (*Teleph.*). An electrically-indicated demand for telephonic connexion, completion and satisfaction.
callainite (*Min.*). A very rare green phosphate of aluminium, resembling turquoise but translucent. A mixture of wavellite and turquoise.
callaite (*Min.*). A little-used name for the mineral turquoise.
Callan cell (*Elec. Eng.*). A modification of the Bunsen type of primary cell, in which the positive electrode is of cast-iron instead of carbon.
Call and Exener bodies (*Histol.*). Groups of cells bordering the fluid-filled central cavity of the ovarian follicles in Mammals.
Callaud cell (*Elec. Eng.*). A form of *Daniell cell* in which the lower part of the container is of copper and the upper part zinc; the zinc sulphate in the top part of the cell floats on the copper sulphate in the lower part, without any porous diaphragm between.
called-party release (*Teleph.*). The operation of automatic telephone switching circuits by which the connecting apparatus is released when the called party replaces his receiver and thereby opens his loop.
Callier coefficient (*Photog.*). Ratio between the specular density and the diffuse density of a photographic emulsion.
Callier effect (*Photog.*). Difference in contrast between the projected image from an enlarger fitted with a condenser and that from a similar one fitted with a diffuser.
call-in (*Comp.*). The transfer control, from the main to a subroutine, during a subsidiary operation in computer programming.
call-indicator (*Elec. Eng.*). An electric indicating device used in conjunction with a system of electric bells, or in a lift, to indicate the point from which a call has been made. (*Teleph.*) The device which indicates to a manual operator a number which has been dialled. Both visual number plates and aural methods are in use.
calling-party release (*Teleph.*). The operation of automatic telephone switching circuits by which the connecting apparatus is released when the calling subscriber replaces his receiver and thereby opens his loop. Also **calling-subscriber release.**
calliper gauge (*Eng.*). A limit gauge having two pairs of jaws, marked 'Go' and 'Not Go' respectively, truly ground to specified distances apart, corresponding to the tolerance allowed on the dimension of the work. See **limit gauge.**
callipers (*Instr.*). An instrument, consisting of a pair of hinged legs, used to measure external and internal dimensions.
calliper splint (*Surg.*). A splint fitted to the broken leg so that the patient may walk without any pressure on the foot, the weight of the body being taken by the hip bone.
call number (*Comp.*). Coding for a subroutine. Also **call word.**
callose (*Bot.*). A carbohydrate, insoluble in cuprammonia, but soluble in the cold in 1% solutions of sodium hydroxide. It occurs in the callus pads which form over sieve plates, and in calcified walls.
callosity (*Bot.*). See **callus** (3). (*Med.*) A thickening of the skin as a result of irritation or friction.
callous, callose (*Bot.*, *Med.*, *etc.*). Hardened, usually thickened, and often like horn in appearance.
call sign (*Telecomm.*). Letter and/or numeral for a transmitting and/or receiving station, or one of its authorized channels. Used for calling or identification.
callus (*Bot.*). (1) A mass of parenchymatous cells formed by plants over or around a wound. (2) A pad of callose formed over a sieve plate either as winter approaches or as the sieve tube ages and degenerates; also called **callus plate.** (3) A deposit of material inside the wall of a cell, around the entering germ tube of a parasitic fungus, which may prevent the parasite establishing itself in the cell: also called **callosity.** (4) The swollen base of the inferior palea of grass flowers. (*Med.*) (1) Callosity. (2) Newly formed bony tissue between the broken ends of a fractured bone. (*Zool.*) In some Insects, a swelling of the mesonotum; in Molluscs, a proliferation of calcareous material within the umbilicus of the shell.
call word (*Comp.*). **See call number.**
Calmuc (*Textiles*). A coarse type of wool, from the Khirghiz district, Central Asia.
calomel (*Chem.*). Mercury(I)chloride, Hg_2Cl_2; found naturally in whitish or greyish masses, associated with cinnabar. Used in medicine as a purgative, and in physical chemistry as a reference electrode (i.e., a half-cell comprising a mercury electrode in a solution of potassium chloride saturated with mercury(I)chloride).
calorescence (*Heat*, *Light*). The absorption of radiation of a certain wavelength by a body, and its re-emission as radiation of shorter wavelength. The effect is familiar in the emission of visible rays by a body which has been heated to redness by focusing infrared heat rays on to it.
Calor gas (*Heat*). See **butane.**
calorie (*Heat*). The unit quantity of heat in the CGS system, replaced by the *joule* in the SI system. The 15°C calorie (cal_{15}) is the quantity of heat required to raise the temperature of 1 g of water by 1 degree C at 15°C; this equals 4·1855 J. By agreement the International Table calorie (cal_{IT}) equals 4·1868 J exactly; the thermochemical calorie equals 4·184 J exactly. There are other designations, e.g., gramme calorie, mean calorie, and large or kilocalorie (=1000 cal; used particularly in nutritional work). See also **pound-calorie.** This proliferation in nomenclature has occasioned the call for the abandonment of the calorie in favour of the joule.

calorific value (*Heat*). The number of heat units obtained by the complete combustion of unit mass of a fuel. The numerical value obtained for the calorific value depends on the units used; e.g., lb-calories/lb, British thermal units (Btu)/lb or MJ/kg for solid and liquid fuels, and Btu/ft^3 or MJ/m^3 for gaseous fuels. In fuels containing hydrogen, which burns to water vapour, there are two heating values, *gross* and *net*. The *gross value* of a fuel is the total heat developed after the products are cooled to the starting point, and the water vapour condensed. The *net value* is the heat produced on combustion of the fuel at any given temperature (for gas, 60°F), with the flue products cooled to the initial temperature, the water vapour remaining uncondensed.

calorifier (*Heat*). An apparatus for heating water in a tank, the source of heat being a separate coil of heated pipes immersed in the water in the tank.

calorigenic (*Heat*). Pertaining to heat production; producing heat and promoting oxygen production.

calorimeter (*Heat*). The vessel containing the liquid used in calorimetry. The name is also applied to the complete apparatus used in measuring thermal quantities.

calorimetry (*Heat*). The measurement of thermal constants, such as specific heat, latent heat, or calorific value. Such measurements usually necessitate the determination of a quantity of heat, by observing the rise of temperature it produces in a known quantity of water or other liquid.

calorizing (*Met.*). A process of rendering the surface of steel or iron resistant to oxidation by spraying the surface with aluminium and heating to a temperature of 800° to 1000°C.

calotte (*Build.*). A small dome in the ceiling of a room, used to increase head room. (*Zool.*) In the larval stage of some *Polyzoa*, a retractile disk bearing motionless sensory cilia.

calotype (*Photog.*). An early wet-plate process, using silver iodide, invented by Talbot.

calvarium (*Anat.*). The dome of the cranium, above the ears, eyes, and occipital protuberance.

Calvert lighting (*Aero.*). An aerodrome approach lighting system, developed at the Royal Aircraft Establishment, consisting of a line of lead-in lights to the runway centreline with a series of cross-bar lights of diminishing width, which indicate the distance of the runway threshold. The cross bars are proportioned so that as the pilot passes over them on his descending path perspective makes each appear the same width.

calvescent, calvous (*Bot.*). Becoming bare.

Calvé's disease (*Med.*). Aseptic necrosis of a vertebral body, usually in children, producing a vertebral collapse.

calving fever (*Vet.*). See **milk fever.**

calvities (*Med.*). Baldness, especially of the anterior and upper part of the head.

calvous (*Bot.*). See **calvescent.**

calx (*Anat.*). Calcaneum; os calcis; the heel. (*Chem.*) Burnt or quick-lime.

calycanthemy (*Bot.*). An abnormal condition in which the calyx becomes coloured, resembling the corolla.

Calyciflorae (*Bot.*). A subclass of *Dicotyledons*, with a corolla usually consisting of distinct petals, and with the stamens perigynous or epigynous.

calyciform cells (*Zool.*). Cells of uncertain function in the enteric epithelium of *Lepidoptera*.

calycine (*Bot.*). Relating to the calyx.

calycle (*Zool.*). A synonym for **calyx** (senses 1-3).

calycule, calyculus (*Bot.*). (1) A group of small leaflike organs placed close beneath the calyx; *adj.* **calyculate.** (2) Sometimes used to denote the *epicalyx* (q.v.). (3) A cupped structure at the base of the sporangium of some *Myxomycetes*.

calymma (*Zool.*). In *Radiolaria*, a layer of jelly containing vacuoles and forming part of the outer layer of the body.

calymmocytes (*Zool.*). In *Urochorda*, follicle cells of the ovary.

calyon (*Build.*). Flint or pebble stone used in wall construction.

calypter (*Zool.*). See **calyptron.**

Calyptoblastea (*Zool.*). An order of *Hydrozoa*, in which the coenosarc is covered by a horny perisarc produced over the nutritive polyps as hydrothecae, and over the reproductive individuals as gonothecae; the medusae are flattened, with gonads on the radial canals, and usually statocysts.

calyptoblastic (*Zool.*). Said of hydroid colonies in which the blastostyles are protected by gonangia. Cf. *gymnoblastic*.

calyptobranchiate (*Zool.*). Having the gills hidden by a gill-cover of some kind.

calyptopsis (*Zool.*). In *Euphausiacea*, a larval form succeeding the metanauplius and resembling the zoaea, from which it differs mainly in having the eyes covered by the carapace.

calyptra (*Bot.*). (1) A membranous covering over the young capsule of a moss or liverwort, derived from part of the archegonium wall. (2) A thickened wall over a terminal cell of a filament in the *Myxophyceae*. (3) The *root cap* (q.v.).

calyptrate (*Zool.*). Of *Diptera*, having the balancers covered by the antitegulae.

calyptrogen (*Bot.*). The group of meristematic cells from which the root cap is formed.

calyptron (*Zool.*). In calyptrate *Diptera*, the enlarged squama which covers the haltere. Also called **calypter.**

calyx (*Bot.*). The outer whorl of the flower, consisting of sepals. It is usually green, and protects the unopened flower-bud. (*Zool.*) (1) A pouch of an oviduct, in which eggs may be stored. (2) In some *Hydrozoa*, the cuplike exoskeletal structure surrounding a hydroid. (3) In *Crinoidea*, the body as distinct from the stalk and the arms. (4) In some Mammals, part of the pelvis of the kidney.

calyx tube (*Bot.*). The hollowed-out receptacle of a perigynous flower, from which the petals and stamens spring. The tube formed by a gamosepalous calyx.

cam (*Eng.*). Sliding mechanical device used to convert rotational to linear motion or vice versa. Widely used in servo systems. (*Weaving*) See **wiper.**

camber (*Aero.*). The curvature of an aerofoil, relative to the chord line. Colloquially, the curved surface of an aerofoil. (*Autos.*) Angle given to a steering swivel so that its axis at road level coincides approximately with the centre line of the wheel. Usually accompanied by converse inclination of wheel itself. (*Civ. Eng.*) An upward curvature to allow for settlement of a structure, to facilitate run-off of water, or for some other purpose. (*Eng.*) A convexity applied for some specific purpose, e.g., to girders to allow for deflection under load, or to road surfaces for drainage. (*Hyd. Eng.*) The recess in the side of the entrance to a basin, lock, or graving dock, accommodating the *sliding caisson* (q.v.). (*Ships*) The convexity of a deck line in a transverse section, normally 2 cm to each metre of breadth. Its purpose is to assist drainage and provide strength. Also called **round of beam.**

camber arch (*Build.*). An arch having a flat horizontal extrados and a cambered intrados, with a rise of about 1 cm per metre of span.

camber-beam (*Carp.*). A beam having an arched upper surface, or one sloping down towards each end, so as to form a support for roof covering on a flat roof.

camber flap (*Aero.*). See **plain flap.**

camber-slip (*Build.*). A strip of wood having a slightly cambered upper surface, upon which the brickwork of a flat arch is laid, so that after settlement the *soffit* (q.v.) shall be straight.

cambiform cell (*Bot.*). A parenchymatous cell occurring in the phloem; it is elongated, and pointed at both ends.

cambium (*Bot.*). A cylinder, strip, or layer of meristematic cells, which divide and give rise to daughter cells from which permanent tissue is ultimately formed.

cambium initial (*Bot.*). One of the permanently meristematic cells of a cambium.

cambogia (*Chem.*). *Gamboge gum* (q.v.).

Cambrian System (*Geol.*). The lowest division of the Palaeozoic rocks. The type area is Wales, where the rocks reach a thickness of 3660 m, but they have been recognized at widely separated localities all over the world. The base of the Cambrian is often taken as the appearance of the earliest fossils, but is generally badly defined.

cambric (*Elec. Eng.*). Varnished cotton-cambric is much used for insulation purposes in an enormous range of electrical equipment, cables, etc. (*Textiles*) A fine linen cloth, used chiefly for handkerchiefs; the name is also applied to a plain weave fine quality cotton cloth.

Cambridge plate (*Print.*). A plastic *duplicate plate* with a hard-wearing face and a resilient backing layer, which was developed at Cambridge University Press.

Cambridge roller (*Agric.*). A roller consisting of loosely mounted ring segments which taper at the periphery to a narrow rim.

came (*Build.*). A bar of lead suitably grooved to hold and connect adjacent panes of glass in a window.

camel (*Hyd. Eng.*). A water-tight air-receiver, used to raise a vessel in the water, to assist its passage in shallow waters. See also **caisson.**

camel hair (*Textiles*). (1) A silky fibre from the haunch and underpart of the camel; used for dress fabrics, warm coverings, etc. (2) Long coarse outer hair which is made into driving belts.

cameo. (1) Carving or modelling in relief. (2) A striated shell or precious stone carved in relief to show different colours in the layers.

camera (*Photog.*). Apparatus for reproducing images on light-sensitive plates or films, or on television.

camera booth (*Cinema.*). In sound-film studios, a portable enclosure for the accommodation of cameras, with optically worked glass windows to avoid distortion of the photographic image; usually provided with wheels.

camera channel (*TV*). In a television studio, the camera, with all its supplies, monitor, control position, and communication to the operator, which forms a unit, with others, for supplying video signal to the control room.

camera lucida (*Optics*). A device for facilitating the drawing of an image seen in a microscope or other optical instrument. In its simplest form, it consists of a thin plate of unsilvered glass, placed above the eyepiece at an angle of 45° with the axis of the instrument, so as to reflect into the eye of the observer an image of the drawing surface, which is seen simultaneously with the microscope image.

camera marker (*Cinema.*). The lamp used, in a cinematograph or sound camera, for fogging the edge of the film and so marking it for future synchronization between the picture track (mute) and the sound track.

camera obscura (*Optics*). A darkened room in which an image of surrounding objects is cast on a screen by a long-focus convex lens.

camera signal characteristic (*TV*). The sensitivity with respect to wavelength, in colour TV, of each of the colour separation channels. Also called **camera spectral characteristic.**

camera timer (*Horol.*). Clock combined with a camera for recording times in races, aerial photography, etc.

camera tube (*TV*). One which converts an image of an external scene into a video signal. Essential component in a TV camera channel. In U.S., **a pick-up tube.**

camouflage (*Geol.*). Describes the relationship between a trace element and a major element whose ionic charge and ionic radius are similar, as a result of which the trace element always occurs in the minerals of the major element but does not form separate minerals of its own. Thus gallium can be considered to be camouflaged by aluminium. (*Radar*) Treatment of objects so that there is ineffective reflection of radar waves.

campaign (*Glass*). The working life of a tank furnace between major repairs which involve cooling down the tank to room temperature.

Campanales (*Bot.*). An order of the *Sympetalae*; the perianth is usually two-whorled, and gamopetalous; ovary is inferior, and the stamens usually epigynous in an irregular flower; the ovary has 2-6 carpels, each having many ovules.

campaniform (*Zool.*). Dome-shaped; as *campaniform sensilla* of certain Insects, which are sensory nerve-endings of unknown function, occurring widely on the body.

campanile (*Arch.*). A bell tower, often detached.

campanula Halleri (*Zool.*). In the eyes of Fishes, a vascular expansion of the falciform process in contact with the lens, the whole structure probably acting as a means of accommodation in the absence of ciliary muscles.

Campanulales (*Bot.*). An order of the sympetalous *Dicotyledons*, including those with one whorl of stamens, the anthers of which have two loculi. The ovary either has one or many carpels, each with one or many ovules, or one loculus and one ovule.

campanulate (*Bot.*). Bell-shaped.

Campbell bridge (*Elec. Eng.*). An electrical network, designed by Campbell, for comparing mutual inductances.

Campbell gauge (*Elec. Eng.*). Electrical bridge, one arm being the filament of a lamp located in the low gas pressure to be measured.

Campbell's formula (*Elec. Eng.*). That which gives the effective attenuation of a coil-loaded transmission line in terms of the constants of the line and the magnitude of the loading.

Campbell-Stokes recorder (*Meteor.*). See **sunshine recorder.**

Campbell (or **Mayo**) **twill** (*Textiles*). A fancy twill, 8-end, used for worsted and woollen coatings.

camp ceiling (*Build.*). A ceiling having two opposite parts sloping in line with the rafters, the middle part being horizontal.

camphane (*Chem.*). $C_{10}H_{18}$, white crystals; m.p.

154°C. It is a saturated terpene hydrocarbon, the parent substance of the camphor group; see **camphor.**

camphene (*Chem.*). $C_{10}H_{16}$, a white solid; m.p. 50°C. An unsaturated terpene hydrocarbon, occurring in various essential oils; it can be prepared from pinene hydrochloride. There are three modifications, viz. the +, −, and the optically inactive form (racemate).

camphor (*Chem.*). Common (or Japan) camphor, $C_{10}H_{16}O$, colourless transparent prisms of characteristic odour; m.p. 175°C, b.p. 204°C, rel. d. 0·985. It can be sublimed readily, is dextrorotatory in alcoholic solution, and is volatile in steam. Natural camphor is obtained from the camphor tree (*Cinnamomum camphora*), widely cultivated in Taiwan, China and Japan. Synthetic camphor (optically inactive) is derived from α-pinene. Camphor is an ingredient of various lotions and liniments; it is used in industry as a plasticizer. See also **borneol, buchucamphor.**

C-amplifier (*Telecomm.*). Line or distribution amplifier following a B-amplifier in broadcasting studios. *N.B.* Not the same as *class-C amplifier* (q.v.).

campodeiform (*Zool.*). Resembling the Thysanuran insect *Campodea*; the term is applied to a primitive type of active insect larva which lacks the lateral abdominal appendages of *Campodea.*

campodeoid larva (*Zool.*). A type of larva without a case found in some *Trichoptera*. Cf. *eruciform.*

cam profile (*Eng.*). The shape of a cam as determined by the form of the flanks and tip; in general, the cam outline.

camp sheathing (*Civ. Eng.*). An earth-retaining wall formed of timber piles placed 2 to 3 m apart and connected by stout timber *walings* (q.v.); often used to support river banks.

camp sheeting (*Build.*). Sheet piling used in foundation work to retain sandy earth.

camptonite (*Geol.*). An igneous rock occurring in minor intrusions, and belonging to the family of the lamprophyres. It consists essentially of plagioclase feldspar and brown hornblende.

campylite (*Min.*). A variety of the pyromorphite-mimetite series crystallizing in barrel-shaped forms.

campylotropous (*Bot.*). Of an ovule, curved in such a way that the chalaza and micropyle do not lie in a straight line.

camshaft (*Eng.*). A shaft on which cams are keyed, or formed integrally; used to operate the valves of internal-combustion engines.

camshaft controller (*Elec. Eng.*). A form of control equipment for electric motors (usually in locomotives), in which the contactors are operated mechanically by cams on a rotating shaft.

cam-type steering-gear (*Autos.*). Steering gear in which the steering column carries a pair of opposed volute cams, which engage with a peg or roller carried by a short arm attached to the drop-arm spindle.

Camus (*Build., Civ. Eng.*). A method of system building developed in France. It consists basically of lateral cross walls with floors spanning them, both the walls and the floors being large precast reinforced concrete panels. To avoid joints, the wall panels are storey height and the floors are frequently of an entire room area, with heating and electrical conduits cast into them. *Cladding* (q.v.) is typically formed of a precast sandwich of concrete, polystyrene, concrete. **See system building.**

can (*Nuc. Eng.*). Cover for reactor fuel rods, often of aluminium or magnox. Also **jacket.**

Canada balsam (*Chem.*). Balsam of fir, or Canada turpentine. A yellowish liquid, of pine-like odour, soluble in ethoxyethane, trichloromethane, benzene; obtained from *Abies balsamica*. Used for lacquers and varnishes, and as an adhesive for lenses, instruments, etc., its refractive index being approximately the same as that of most optical glasses.

Canadian asbestos (*Min.*). See **chrysotile.**

Canadian latch (*Join.*). See **Norfolk latch.**

Canadian Series (*Geol.*). The lowest of the 3 main divisions of the American Ordovician rocks, comprising the Beekmantown and Chazy formations.

Canadian Shield (*Geol.*). The name applied to the vast area of Pre-Cambrian rocks which cover 5 million square kilometres in E. Canada.

Canadian spruce (*For.*). Name for the wood from several trees of the *Picea* genus, the most important commercial timber in Canada. Typical uses include paper, sounding boards for musical instruments, food containers, etc.

Canadian Standard Freeness (**C.S.F.**) (*Paper*). A laboratory control of beating in which diluted pulp is allowed to drain through a sieve. A wet beaten pulp gives a low C.S.F. value.

canal (*Bot.*). An elongated intercellular space, often containing resins, oils, or other similar substances. (*Nuc. Eng.*) Water-filled trench into which the highly-active elements from a reactor core can be discharged. The water acts as a shield against radiation but allows objects to be easily inspected.

canal cell (*Bot.*). One of the short-lived cells present in the central cavity of the neck of an archegonium.

canalette blind (*Build.*). See **Italian blind.**

canalicular apparatus (*Zool.*). See **Golgi apparatus.**

canaliculate (*Bot.*). Marked longitudinally by a channel or groove.

canaliculus (*Zool.*). Any small channel; in the liver, an intercellular bile-channel; in bone, one of the ramified passages uniting the lacunae; in nerve-cells, a fine channel penetrating the cytoplasm of the cell-body. *adj.* **canalicular.**

canalization (*Hyd. Eng.*). The practice, sometimes applied to assist navigation, of dividing the bed of a river into a series of reaches separated by dams or weirs, provided with communicating locks. (*Med.*) The formation of a new channel in a clot blocking the lumen of a blood vessel.

canal rays (*Phys.*). See **positive rays.**

canal system (*Zool.*). See **water-vascular system.**

canard (*Aero.*). See **tail-first aeroplane.**

canaries (*Cinema.*). Extraneous high-frequency noises reproduced from a recording channel.

cancel (*Acous.*). The thumb or toe piston for shutting off groups of stops on a manual or pedals in an organ. See **ventil.** (*Typog.*) A page, or pages, reprinted to replace a defect. A *cancel title* is a new title-page, usually carrying alterations to the imprint.

cancellate (*Bot.*). Latticelike.

cancellous, cancellated (*Zool.*). Having a spongy structure, with obvious interstices.

Cancer (*Geog.*). See **Tropics.**

cancer (*Med.*). See **carcinoma.**

cancerophobia (*Med.*). Morbid fear of contracting cancer.

cancerous (*Med.*). Pertaining to cancer or carcinoma.

can coiler (*Textiles*). A revolving can in which slivers from Noble, Heilman and Nasmith combers are laid in double spiral form to facilitate withdrawal without entanglement at the preceding process, e.g., drawing, etc.

cancrinite (*Min.*). A hydrated silicate of aluminium, sodium and calcium, also containing the carbonate ion. Found in alkaline plutonic igneous rocks. Cf. *vishnevite*.

cancrisocial (*Zool.*). Leading a commensal existence with a crab.

cancrum oris (*Med.*). Noma. A destructive ulceration of the cheek in debilitated children, usually during convalescence from an infectious disease.

candela (*Light*). New candle. Luminous intensity unit in the SI system; symbol **cd**; it is such that the luminance of a black body radiator at the temperature of solidification of platinum (ca. 2042 K) is 60 cd/cm^2. It supersedes the *candle*, *international candle*, etc.

candelilla wax (*Chem.*). A natural wax, of yellow or brown colour; rel. d. 0·983, m.p. 67°–68°C, saponification value 65, iodine value 37. Similar to *carnauba wax* (q.v.).

candidiasis (*Vet.*). See **moniliasis.**

candle (*Light*). Unit of luminous intensity. See **candela.**

candle-fitting (*Light*). An electric-light fitting consisting of an opal glass tube with a lamp at the top imitating an ordinary wax candle.

candle-lamp (*Light*). An electric filament lamp shaped to imitate the flame of a wax candle.

candle power (*Light*). Luminous intensity provided by a candle of specific composition burning at a known rate. Unit, **candela.**

candoluminescence (*Phys.*). Luminescence, beyond normal light radiation depending on temperature alone, evinced by certain slightly impure substances when placed in contact with the flame from the burning of hydrogen.

canescent (*Bot.*). Having a somewhat hoary appearance, due to a covering of short inconspicuous hairs.

cane-sugar (*Chem.*). Sucrose or saccharobiose; $C_{12}H_{22}O_{11}$, a disaccharide carbohydrate; m.p. 160°C; it crystallizes in large monoclinic crystals, is optically active and occurs in beet, sugar-cane, and many other plants. Hydrolyses to glucose and fructose.

caney (*For.*). A term applied to wood (particularly ash) with abnormally narrow growth rings, the end-surface appearing unusually porous.

canicola fever (*Med., Vet.*). The disease of man caused by infection by *Leptospira canicola*, the natural host of which is the dog.

canine (*Zoo .*). Pertaining to, or resembling, a dog; in Mammals, a pointed tooth with a single cusp, adapted for tearing, and occurring between the incisors and the premolars; pertaining to a canine tooth; pertaining to a ridge or groove on the surface of the maxillary.

canine distemper (*Vet.*). Carré's disease. A contagious viral infection of dogs, the main symptoms being fever, oculonasal discharge, respiratory distress, diarrhoea and vomiting, and nervous symptoms. Ferrets, foxes, and mink are also susceptible to infection.

canine hysteria (*Vet.*). A disease of dogs, of unknown cause, characterized by periodic attacks of abnormal mental behaviour varying from somnolence to hysteria.

canine leptospiral jaundice (*Vet.*). See **canine leptospirosis.**

canine leptospirosis (*Vet.*). Infection of dogs by organisms of the genus *Leptospira*. *L. icterohaemorrhagiae* infection (*canine leptospiral jaundice*; *yellows*) causes an acute septicaemia and hepatitis characterized by fever, widespread haemorrhages and jaundice. *L. canicola* causes an acute septicaemia and nephritis, which often develops into subacute or chronic nephritis.

canine typhus (*Vet.*). See **Stuttgart disease.**

canine venereal granulomata (*Vet.*). Infectious venereal sarcoma of dogs. A contagious, viral infection of dogs characterized by tumour-like formations on the genital mucosae. Spread by both venereal and nonvenereal contact.

canker (*Bot.*). A name applied to various diseases of trees, caused by bacteria and fungi; characterized by a sharply defined necrosis of the cortical tissues. (*Vet.*) (1) A chronic inflammation of the keratogenous membrane of the frog and sole of a horse's foot. (2) Chronic eczema of the ear of dogs. (3) Abscess or ulcer in the mouth, eyelids, ear, or cloaca of birds.

cannabis (*Bot.*). The plant *Cannabis sativa* or **Indian hemp.** The dried flowers, exuded resin and leaves are used to produce the drug hashish, marihuana or bhang. Cannabis is widely recognized as a drug of addiction.

canned (*Acous.*). See **canning.**

cannel coal (*Fuels*). A dull variety of coal, breaking with a conchoidal fracture; it is rich in volatile constituents, and burns brightly.

cannibalism (*Vet.*). A vice or depraved appetite of animals and birds due to deficiency of minerals or protein in the food, or to boredom in the case of confined birds and animals. The affected individual injures and eats portions of its own body or those of other individuals.

canning, canned (*Acous.*). Colloquialisms applied respectively to the recording of sound and to the record thereof.

Cannizzaro reaction (*Chem.*). Used for the conversion of aromatic aldehydes to alcohols and acids, by heating with concentrated alkali, e.g., benzaldehyde→benzyl alcohol + a benzoate.

cannon bone (*Zool.*). In the more advanced *Artiodactyla*, the characteristic bone formed by the fusion of the two metapodials in the limb, associated with the reduction of the number of toes to two.

cannon pinion (*Horol.*). A wheel in the motion work; the wheel or pinion with an extended pipe to which the minute hand is attached.

cannon-shot gravel (*Geol.*). A gravel associated with the glacial boulder-clays of East Anglia; it contains large numbers of almost perfectly spherical flint cobbles.

cannula (*Med.*). A tube, usually fitted with a trocar, for insertion into the body for the injection or removal of fluids or gases.

cannular combustion chamber (*Aero.*). A gas turbine combustion system with individual flame tubes inside an annular casing.

canon (*Typog.*). An old type size (about 48-point).

canonical assembly (*Phys.*). Term used in statistical thermodynamics to designate a single assembly of a large number of systems which are such that the number of systems with energies lying between E and $E+dE$ is proportional to $e^{-E\theta}$, where θ is a parameter characteristic of the assembly.

canonical form (*Maths.*). An especially simple or convenient form into which all members of a particular class can be transformed. A standard or normal form, e.g., all cubic equations can be transformed to the form $x^3+ax+b=0$, all quadratic forms can be expressed as a sum of squares, and all square matrices can be reduced to diagonal form.

canopy (*Aero.*). (1) The transparent cover of a cockpit. (2) The fabric (nylon, silk, or cotton) body of a parachute, which provides high air drag. Usually hemispherical, but may be lobed or rectangular in shape. See also **cross para-**

chute, ribbon-parachute. (*Bot.*) The layer of branches, twigs, and leaves formed by woody plants at some distance above the level of the ground. (*Build.*) A roof or balcony projecting from a wall or supported on pillars.

canopy switch (*Elec. Eng.*). A circuit-breaker placed under the canopy of a tramcar, so that the driver can cut off the current in an emergency.

cant (*Bui d.*). A moulding having plane surfaces and angles instead of curves. (*For.*) A log partially square sawn. (*Surv.*) Slope of rail or road curve whereby outer radius is super-elevated, to counteract centrifugal thrust of traffic.

cant bay (*Arch.*). A bay window having three sides, the outer two being splayed from the wall face.

cant board (*Carp.*). A board laid on each side of a valley gutter to support the sheet-lead.

cant brick (*Build.*). A *splay brick* (q.v.).

canted column (*Arch.*). A column having faceted sides instead of curved flutes.

canted deck (*Aero.*). See **angled deck.**

canted wall (*Build.*). A wall built at an angle to the face of another wall.

Canterbury hammer (*Tools*). Type of hammer with thick and shallow-curved claw.

cantharidin (*Pharm.*). A pharmaceutical product ($C_{10}H_{12}O_4$) obtained from the dried elytra of the Spanish Fly (species of *Lytta* and *Mylabris*). A counterirritant, now little used.

cantilever (*Civ. Eng.*). A beam or girder fixed at one extremity and free at the other.

cantilever arm (*Civ. Eng.*). One of the projecting arms of a *cantilever bridge* (q.v.), or other structure.

cantilever bridge (*Civ. Eng.*). A bridge composed of self-supporting projecting arms built inwards from the piers and meeting in the middle of the span, where they are connected together. See also **suspended span.**

cantilever crane (*Civ. Eng.*). A straight steel truss, resting on central supports, used for the transport of excavated materials from the bottom of a cutting to a spoil-bank at one side; for this purpose skips hang from the truss and may be moved along it. Also used in shipbuilding.

cantilever deck (*Civ. Eng.*). A type of *cantilever bridge* (q.v.) in which the loads are carried by the upper chord.

cantilevered steps (*Build.*). See **hanging steps.**

cantilever spring (*Autos.*). A laminated spring, anchored to the frame at its mid-point, and, at its ends, to shackles on the frame and the axle respectively; not a true cantilever.

cantilever-through (*Civ. Eng.*). A type of *cantilever bridge* (q.v.) in which the loads are carried by the lower chord.

canting (*Eng., etc.*). Tilting over from the proper position; as the canting of a piston in its cylinder under the oblique thrust of the connecting-rod.

canting strip (*Build.*). A projecting sloping member of wood or masonry fitted around a building to deflect rain water from the wall. Also called **a water table.**

cantling (*Build.*). The lower of two courses of burnt brick enclosing a clamp for firing bricks.

canton (*Build.*). A pilaster or quoin forming a salient corner, which projects from the wall-face.

canvas (*Textiles*). (1) An openwork cotton fabric, usually with an embroidered design, used for cushion covers, etc. (2) A plain coarse cotton cloth with hard twisted yarns. (3) A coarse linen cloth, frequently made from natural tow yarns; used for interlinings. (4) Sail canvas, made from doubled linen warp threads and strong, coarse, cotton weft.

canyon (*Geol.*). A deep, narrow, steep-sided valley. (*Nuc. Eng.*) Long narrow space with heavy shielding for essential processing of wastes from reactors.

caoutchouc. Raw *rubber* (q.v.).

cap (*Bot.*). (1) See **pileus.** (2) The strand of sclerenchyma often present on the outside of a vascular bundle, and seen as a crescent-shaped mass in a cross-section. (*Build.*) (1) The upper member of a column. (2) A wall coping. (*Carp.*) (1) A door or window lintel. (2) A hand-rail supported on balusters. (*Civ. Eng.*) The head added to a foundation pile. (*For.*) A cone of sheet iron or steel, having a hole in the apex through which a chain passes, which is fitted over the end of a log to prevent its catching on stumps, roots, or other obstacles. (*Meteor.*) (1) The covering of cloud which congregates at the top of a mountain. (2) The transient top of detached clouds above an increasing cumulus. Also **pileus.**

capacitance (self) (*Elec.*). Assuming that all other conductors are sufficiently remote from a conducting body, capacitance of latter is defined as the total electric charge on it divided by its potential. $C=Q/V$. See also **farad, microfarad, micromicrofarad, picofarad;** and **capacity, stray capacitance.**

capacitance bridge (*Elec. Eng.*). An a.c. bridge network for the measurement of capacitance. See **Schering bridge, Wien bridge.**

capacitance coefficients (C_{ij}) (*Elec.*). Charges ($q_1 \ldots q_n$) of system of conductors can be expressed in terms of coefficients of electric induction (C_{ij}) by the following equations:

$$q_1 = C_{1\infty} V_1 + C_{12}(V_1 - V_2) + \ldots + C_{1n}(V_1 - V_n)$$

$$q_2 = C_{21}(V_2 - V_1) + C_{2\infty} V_2 + \ldots + C_{2n}(V_2 - V_n)$$

$$\ldots \quad \ldots \quad \ldots \quad \ldots$$

$$q_n = C_{n1}(V_n - V_1) + C_{n2}(V_n - V_2) + \ldots + C_{n\infty} V_n$$

where $C_{km} = -C_{mk}\ (m \neq k)$,

and $C_{m\infty} = C_{m1} + C_{m2} + \ldots + C_{m(n-1)} + C_{mn}$.

Fundamental relation for partial capacitances of a number of conductors, e.g., electrodes in valves, conductors in cables, variable air-capacitors.

capacitance coupling (*Elec. Eng., Electronics*). Inter-stage coupling through a series capacitance or by a capacitor in a common branch of a circuit.

capacitance current heating (*Elec. Eng.*). See **dielectric heating.**

capacitance grading (*Elec. Eng.*). Grading of the properties of a dielectric, so that the variation of stress from conductor to sheath is reduced. The inner dielectric has the higher permittivity. Ideally, the grading is continuous and the permittivity varies as the reciprocal of the distance from the centre. See **condenser bushing.**

capacitance integrator (*Elec. Eng.*). Resistance-capacitance circuit whose output voltage is approximately equal to the time integral of the input voltage.

capacitance reaction (*Elec. Eng.*). Reaction from the output to the input circuit of an amplifier, through a path which includes a capacitor.

capacitive load (*Elec. Eng.*). Terminating impedance which is markedly capacitive, taking an a.c. leading in phase on the source e.m.f., e.g., electrostatic loudspeaker.

capacitive reactance (*Elec. Eng.*). Negative of the reciprocal of the product of the angular frequency ($\omega = 2\pi \times$ frequency) and capacitance. Measured in ohms when the capacitance is in farads.

capacitor (*Elec. Eng.*). Electric component having *capacitance*; formed by conductors (usually thin and extended) separated by a dielectric, which may be vacuum, paper (waxed or oiled), mica, glass, plastic foil, fused ceramic, or air, etc. Maximum p.d. which can be applied depends on electrical breakdown of dielectric used. Modern construction uses sheets of metal foil and insulating material wound into a compact assembly. Air capacitors, of adjustable parallel vanes, are used for tuning high-frequency oscillators. Formerly **condenser.**

capacitor bushing (*Elec. Eng.*). See **condenser bushing.**

capacitor loudspeaker (*Acous.*). See **electrostatic loudspeaker.**

capacitor microphone (*Acous.*). One with a stretched or slack conducting foil diaphragm, which is polarized at steady potential, or at a high-frequency voltage, both of which are modulated by variations in capacitance due to varying sound pressure.

capacitor modulator (*Acous.*). Capacitor microphone, or similar transducer, which, by variation in capacitance, modulates an oscillation either in amplitude or frequency. See **Vibron.**

capacitor motor (*Elec. Eng.*). A single-phase induction motor arranged to start as a 2-phase motor by connecting a capacitor in series with an auxiliary starting winding. The capacitor may be automatically disconnected when the motor is up to speed (*capacitor-start motor*) or it may be left permanently in circuit for power-factor improvement (*capacitor-start-and-run motor*).

capacitor start (*Elec. Eng.*). Starting unit for electric motor using series capacitance to advance phase of current.

capacitor terminal (*Elec. Eng.*). See **condenser bushing.**

capacitron (*Electronics*). (1) Mercury-arc rectifier which is fired by an externally applied high voltage between an insulated electrode and the cathode. (2) See **band-ignitor tube.**

capacity (*Comp.*). Range of numbers which can be handled by a computer, limited by *storage, register*, or *accumulator*, without *overflow*. (*Elec.*) The output of any electrical apparatus, e.g., that of a motor or a generator in kW. In an accumulator (secondary battery), *capacity* is measured by the ampere-hours of charge it can deliver. Capacity of a switch is the current it can break under specified circuit conditions. Also used for **capacitance.**

cap-and-pin type insulator (*Elec. Eng.*). A special form of the *suspension insulator* (q.v.).

cap cell (*Bot.*). The cell which surmounts the antheridium of a fern, and is thrown off when the antheridium liberates the sperms. (*Zool.*) A cell enclosing the end of the slender distal prolongation of the bipolar neurone in the chordotonal sensilla of insects.

Cape asbestos or Cape blue asbestos (*Min.*). A form of crocidolite, a silicate of sodium and iron, occurring in narrow interbedded veins traversing the Griqua Town series of banded jaspers and ironstones in Griqualand West, S. Africa. Same uses as other asbestos.

Cape diamond (*Min.*). A name used in grading diamonds to designate an off-colour stone of a yellowish tint.

Cape ruby (*Min.*). The fiery red garnet *pyrope* (q.v.), obtained in the diamond-mines in the Kimberley district, mostly from the rocks kimberlite and eclogite. See also **false ruby.**

Cape walnut (*For.*). See **stinkwood.**

capillaceous, capilliform (*Bot.*). Hairlike.

capillariasis (*Vet.*). Inflammation of the alimentary tract of animals or birds due to infection by nematode worms of the genus *Capillaria.*

capillarity (*Phys.*). A phenomenon associated with surface tension, which occurs in fine bore tubes or channels. Examples are the elevation (or depression) of liquids in capillary tubes and the action of blotting paper and wicks. The elevation of liquid in a capillary tube above the general level is given by the formula

$$h = \frac{2T \cos \theta}{\rho g r},$$

where T is the surface tension, θ is the angle of contact of the liquid with the capillary, ρ is the liquid density, g is the acceleration due to gravity, r is the capillary radius.

capillary (*Bot., Zool.*). Of very small diameter; slender, hairlike. (*Zool.*) Any tiny, thin-walled vessel of small diameter, forming part of a network, which aids rapid exchange of substances between the contained fluid and the surrounding tissues; as *bile capillaries, blood capillaries, lymph capillaries.*

capillary activity (*Chem.*). See **surface activity.**

capillary bronchitis (*Med.*). Acute catarrhal bronchitis of the smaller bronchi and bronchioles.

capillary condensation (*Chem.*). Hypothesis that adsorbed vapours can condense under the influence of capillary forces to form liquid inside the pores of the adsorbate.

capillary electrometer (*Chem.*). An instrument in which small electric currents are detected by the movement of a mercury meniscus in a capillary tube.

capillary pressure (*Chem.*). The pressure developed by *capillarity* (q.v.). Mathematically expressed as $p = 2T \cos \theta / r$, where T is the surface tension, θ the angle of contact, and r the radius of the capillary.

capillary pyrite (*Min.*). See **millerite.**

capillary soil water (*Bot.*). Water held between the particles of the soil by capillarity.

capillator (*Chem.*). An apparatus for the colorimetric determination of pH; it involves comparison of solutions in capillary tubes.

capillitium (*Bot.*). (1) A mass of threads. (2) The whole assemblage of threads or tubes, found mingled with the spores in the fructifications of *Myxomycetes* and of *Gasteromycetes*. *pl.* **capillitia.**

cap iron (*Carp.*). See **back iron.**

capital (*Build.*). The upper member of a column, pier, or pilaster.

capitate (*Bot.*). (1) Resembling a pinhead in appearance. (2) Bearing a rounded swelling at the apex. (3) Having flowers grouped in a capitulum or head. (*Zool.*) Having an enlarged tip, as *capitate antennae.*

capitellum (*Zool.*). An enlargement or boss at the end of a bone, for articulation with another bone; more particularly, the smaller of the two articular surfaces on the distal end of the mammalian humerus, for articulation with the radius; the distal knoblike extremity of the haltere in *Diptera.*

capitular facet (*Zool.*). An articular surface for a

rib on either side of each anterior thoracic vertebra, at the junction of the centrum and neural arch, in Reptiles.

capitulate (*Bot.*). Of the nature of a head.

capituliform (*Bot.*). Having the characters of a dense head of flowers.

capitulum (*Bot.*). (1) A racemose inflorescence in which the sessile flowers are crowded on the concave, flat, or convex surface of the enlarged apex of the scape, the whole group being surrounded and covered in bud by an envelope of bracts forming an involucre; the whole group appears to be a single flower, as in the daisy. (2) An apical globular lichen apothecium. (*Zool.*) In pedunculate *Cirripedia*, the whole of the body excluding the stalk: a terminal expansion, as that of some shaft bones, some tentacles, and some hairs.

cap nut (*Eng.*). A nut whose outer end is closed, so protecting the end of the screw, and giving a neat appearance. Also **dome** or **box nut**.

caponizing (*Vet.*). Castration of a cock bird.

capped elbow (*Vet.*). A swelling of the olecranon bursa of animals.

capped hock (*Vet.*). A swelling over the point of the hock of animals.

capping (*Elec. Eng.*). The wooden strip used as a cover for the wood-casing system employed in electrical installation work. (*Mining*) The fixing of a shackle or a swivel to the end of a hoisting rope. Overburden lying above valuable seam or bed of mineral. (*Plumb.*) A copper strip rolled over the ridge and welted to the underlying sheets.

capping-brick (*Build.*). A ***coping brick*** (q.v.).

capping-plane (*Join.*). A plane for giving a slight rounding to the upper surface of a wooden hand-rail.

capreolate (*Bot.*). Having tendrils.

capric acid (*Chem.*). Decanoic acid, $CH_3(CH_2)_8COOH$.

Capricorn (*Geog.*). See **Tropics**.

caprification (*Bot., Zool.*). The fertilization of the flowers of fig-trees by the agency of Fig Insects, a family of Chalcids (*Agaonidae*); the process of hanging caprifigs in the female trees.

caprifier (*Zool.*). A Fig Insect. See under **caprification**.

caprifig (*Bot.*). A race of fig which does not produce edible fruits, but provides food for the wasps which pollinate the figs.

cap rock (*Geol., Mining*). Impervious stratum overlying natural gas or oil deposit.

caproic acid (*Chem.*). Normal hexanoic acid. $CH_3{\cdot}(CH_2)_4{\cdot}COOH$. Oily liquid; solidification point about 2°C; b.p. 205°C. Occurs as glycerides in milk, palm oil, etc.

Caprotti valve-gear (*Eng.*). A locomotive valve-gear using two pairs of vertical double-beat *poppet valves* (q.v.), cam operated, cut-off being adjustable by varying the angular position of the inlet cams.

capryl alcohol (*Chem.*). Octan-2-ol. $CH_3{\cdot}(CH_2)_5{\cdot}CH(OH){\cdot}CH_3$. It is obtained by distilling castor oil with strong alkali. Liquid, colourless, strong smell, b.p. 179°C. Used as a foam reducing agent.

caprylic acid (*Chem.*). Normal octanoic acid, $C_7H_{15}{\cdot}COOH$; b.p. 237°C.

caps (*Mining*). See **girt**.

caps & small caps (*Typog.*). Small capitals with the first letter in capitals. The first word of a chapter is often set in CAPS & SMALL CAPS.

cap screw (*Eng.*). A screw-bolt intended to be used without a nut and to be operated by a spanner. So called because first used to secure the cover of small steam engines.

capsicum (*Pharm.*). Chillies, the dried ripe fruit of *Capsicum minimum*. Contains *capsaicin*:

$$(CH_3)_2CH{-}CH{=}CH{-}(CH_2)_4{\cdot}CO{\cdot}NH{\cdot}CH_2{-}C_6H_3(OCH_3)(OH)$$

Used medicinally as a digestive stimulant and externally as a liniment.

capsomeres (*Biochem.*). Protein molecules which form the shell around the nucleic acid core of a virus.

cap spinning (*Textiles*). A method of spinning used for Botany and fine cross-bred yarns; spun yarn is led on to a bobbin, rotating at high speed, by a cap on the top of the spindle.

capstan (*Eng.*). A vertical drum or spindle on which rope is wound (e.g., for warping a ship alongside a wharf); it is rotated by manpower or by hydraulic or electric motor.

capstan-head screw (*Eng.*). A screw having a cylindrical head provided with radial holes in its circumference. It is tightened by a tommy bar inserted in these holes.

capstan lathe (*Eng.*). A lathe in which the tools required for successive operations are mounted radially in a tool-holder resembling a capstan; by revolving this, each tool in turn may be brought into position in exact location.

capstan nut (*Eng.*). A nut which is tightened in the same way as a *capstan-head screw* (q.v.).

capsule (*Bot.*). (1) The portion of the sporogonium of *Bryophyta* which contains the spores. (2) A fruit formed from a syncarpous gynaeceum, not fleshy when ripe, and splitting at maturity, releasing the seeds. (3) A coating of mucilaginous material outside a bacterial cell. (*Med.*) A soluble case of gelatine in which a dose of medicine may be enclosed. (*Zool.*) Any fibrous or membranous covering of a viscus, e.g., the kidney. The name is applied also to certain areas in the brain which are formed by nerve fibres.

capsulogenous (*Zool.*). Capsule-secreting; capsule-forming.

captacula (*Zool.*). In *Scaphopoda*, sensory and prehensile filaments with expanded adhesive tips, arising from the head.

captan (*Chem.*). *N*-(trichloromethylthio)cyclohex-4-ene-1,2-dicarboxyimide; a fungicide.

caption (*Typog.*). The descriptive wording under an illustration. Often called a **legend**.

captive balloon (*Aero.*). A balloon anchored or towed by a line. Usually the term refers to spherical balloons only. Special shapes for stability, etc., are called *kite balloons*.

captive nut (*Eng.*). A nut (loosely) fastened to an adjacent machine member so as to retain it in position when the corresponding screw element is absent.

captive screw (*Eng.*). A screw (loosely) fastened in position by its head or shank so as to be retained when unscrewed from the matching machine member.

captive tapes (*Print.*). Tapes which operate only at the slow speeds when the web is being threaded on rotary presses.

capture (*Nuc.*). Any process in which an atomic or nuclear system acquires an additional particle. In a nuclear radiative capture process there is an emission of electromagnetic radiation only, e.g., in reactor physics there may be an emission of a γ-ray subsequent to the capture of a neutron by the nucleus. (*Telecomm.*) In frequency and phase modulation, the diminution to zero of a weak signal (noise) by a stronger signal. See **k-capture.**

caput (*Zool.*). An abrupt swelling at the distal end of a structure. *pl.* **capita.** *adj.* **capitate.**

caput medusae (*Med.*). Dilated subcutaneous veins round the umbilicus in cirrhosis of the liver.

car (*Aero.*). In an airship, the part intended for the carrying of the load (crew, passengers, goods, engines, etc.). It may be suspended below, or may be inside, the hull or envelope.

caraate (*Med.*). See **pinta.**

caraboid larva (*Zool.*). A modification of the campodeiform type of insect larva, found in some *Coleoptera.*

caracole (*Build., Carp.*). A helical staircase.

Caradocian (*Geol.*). A stage or series of the Ordovician System in Britain, well represented in the Caradoc region of Salop, Wales, the Lake District, etc.

caramel (*Chem.*). Sugar-dye; the brown dye formed when sucrose is heated above its melting point. Caramel is used for flavouring and colouring in cooking and bakery, and also in confectionery.

carapace (*Zool.*). An exoskeletal shield covering part or all of the dorsal surface of an animal; as the bony dorsal shield of a tortoise, the chitinous dorsal shield of some *Crustacea.*

carat or **karat.** (1) A standard of weight for precious stones. The *metric carat*, standardized in 1932, equals 200 mg (3·086 grain). (2) The standard of fineness for gold. The standard for pure gold is 24 carats; 22 carat gold has 2 parts of alloy; 18 carat gold 6 parts of alloy.

carbamic acid (*Chem.*). $NH_2{\cdot}CO{\cdot}OH$; is not known to occur free, being known only in the form of derivatives, e.g., the ammonium salt, $NH_2\,CO\overset{\ominus}{O}\,\overset{\oplus}{NH_4}$. The esters are known as *urethanes* (q.v.).

carbamide (*Chem.*). *Urea* (q.v.).

carbamyl chloride (*Chem.*). $NH_2{\cdot}CO{\cdot}Cl$; colourless needles of pungent odour; m.p. 50°C, b.p. 61°C; formed by the action of hydrochloric acid (gaseous) upon cyanic acid; it serves for the synthesis of organic acids.

carbanilide (*Chem.*). Diphenylurea, $CO(NHC_6H_5)_2$.

carbanion (*Chem.*). A short-lived, negatively-charged intermediate formed by the removal of a proton from a $\rangle$C—H bond, e.g., in Grignard reactions.

carbazole (*Chem.*). $(C_6H_4)_2NH$; colourless plates: m.p. 238°C; sublimes readily; contained in coal-tar and in crude anthracene oil. It is the imine (intramolecular) of diphenyl and is formed from diphenylamine by passing the vapour through red-hot tubes, or by distilling *o*-aminodiphenyl over lime at about 600°C.

carbenes (*Chem.*). (1) Reactive uncharged intermediates of formula: CXY, where X and Y are organic radicals or halogen atoms. (2) Such constituents of asphaltic material as are soluble in carbon disulphide but not in carbon tetrachloride. See **asphaltenes, malthenes.**

carbethoxy (*Chem.*). The group $—COOC_2H_5$ in organic compounds.

carbides (*Chem.*). Binary compounds of metals with carbon. Carbides of group IV to VI metals (e.g., silicon, iron, tungsten) are exceptionally hard and refractory. In groups I and II, *calcium carbide* (ethynide) (q.v.) is the most useful. See also **cemented carbides, cementite.**

carbinol (*Chem.*). Methanol. The nomenclature of alcohols is often based upon their homologous relation to methanol, e.g., tertiary butyl alcohol, $(CH_3)_3C{\cdot}OH$, is termed *trimethyl carbinol.*

carbitol (*Chem.*). See **diethylene glycol** of which it is the monoethyl ether.

carbocyanine (*Photog.*). Group of dyes used as desensitizers.

carbocyclic (or **isocyclic**) **compounds** (*Chem.*). These are closed-chain or ring compounds in which the closed ring or chain consists entirely of carbon atoms.

carbohydrases (*Chem.*). A group of enzymes which induce digestion of polysaccharides and hydrolyse glycosides and oligo- and polysaccharides generally.

carbohydrates (*Chem.*). A group of compounds represented by the general formula $C_x(H_2O)_y$. Substances found in vegetables and animals; e.g., sugars, starch, cellulose. The carbohydrates also comprise other compounds of a different general formula but closely related to the above substances, e.g., rhamnose, $C_6H_{12}O_5$. Carbohydrates are divided into *monosaccharides, oligosaccharides, polysaccharides* (qq.v.). The carbohydrate element in diet supplies energy, provided by the oxidation of the constituent elements.

carbolic acid (*Chem.*). *Phenol* (q.v.).

carbolic oils (*Chem.*). See **middle oils.**

carboluria (*Med.*). The presence of carbolic acid in the urine, due to carbolic-acid poisoning.

carbomethoxy (*Chem.*). The group $—CH_3O{\cdot}CO$ in organic compounds.

carbon (*Chem.*). Amorphous or crystalline (graphite and diamond) element, symbol C, at. no. 6, r.a.m. 12·011, m.p. above 3500°C, b.p. 4200°C. Its allotropic modifications are diamond and graphite. The assumption that its atom is tetravalent, the bonds being directed towards the apexes of a regular tetrahedron, is the basis of all theoretical organic chemistry (see **carbon compounds**). Near black body, hence good radiator of heat, especially from valve anodes. Widely known as electrode in Leclanché cells, in rods for arcs in cinema projectors and searchlights, and in brushes for electric generators and motors. Alloyed with iron for steels. Colloidal carbon or graphite is used to coat glass in cathode-ray tubes and electrodes in valves, to inhibit photoelectrons and secondary electrons. High-purity carbon, crystallized to graphite in a coke furnace for many days, is used in many types of nuclear reactors, particularly for moderation of neutrons. (*Mining*) *Bort* (q.v.); industrial natural diamond, used in rotary rock-drilling and abrasive metals.

carbonaceous (*Bot.*). Hard, blackened, and appearing as if charred. (*Chem.*) Said of material containing carbon as such or as organic (vegetable or animal) matter.

carbonaceous rocks (*Geol.*). Sedimentary deposits of which the chief constituent is carbon, derived from plant residues. Under this heading are included peat, lignite, or brown coal, and the several varieties of true coal (bituminous coals, anthracite, etc.).

carbonado (*Min.*). See **black diamond.**

carbon anode (*Electronics*). One constructed of carbon, usually in graphite form, to resist the high temperatures encountered under some conditions of operation of valves.

carbon arc (*Elec. Eng.*). An arc between carbon electrodes; usually limited to pure carbon rather than flame carbon electrodes.

carbon-arc lamp (*Light*). A lamp employing as the light source an arc between carbon electrodes.

carbon-arc welding (*Elec. Eng.*). Arc welding carried out by means of an arc between a carbon electrode and the material to be welded.

carbon assimilation (*Bot.*). See **photosynthesis.**

carbonate (*Chem.*). A compound containing the acid radical of carbonic acid (CO_3 group). Bases react with carbonic acid to form carbonates, e.g., $CaCO_3$, calcium carbonate.

carbonate-apatite (*Min.*). A variety of apatite containing appreciable CO_2. See **francolite.**

carbonation (*Chem.*). The process of saturating a liquid with carbon dioxide, or converting a compound to carbonate by means of CO_2.

carbonatite (*Geol.*). An igneous rock composed largely of carbonate minerals (calcite and dolomite). Carbonatites are invariably associated with alkaline igneous rocks such as nepheline-syenite, and are rich in a number of unusual minerals, especially those of the rare earth elements.

carbon bisulphide (*Chem.*). See **carbon disulphide.**

carbon black (*Chem.*). Finely divided carbon produced by burning hydrocarbons (e.g., methane) in conditions where combustion is incomplete. Carbon black is widely used in the rubber, paint, plastics, ink, and other industries. It forms a very fine pigment containing up to 95% of carbon, giving a very intense black; prepared by burning natural gas and allowing the flame to impinge on to a cool surface. Also **channel black.** See also **gas black.**

carbon brush (*Elec. Eng.*). A small block of carbon used in electrical equipment to make contact with a moving surface.

carbon button (*Acous.*). See **carbon microphone.**

carbon compounds (*Chem.*). Compounds containing one or more carbon atoms in the molecule. They comprise all organic compounds and include also compounds, e.g., carbides, carbonates, carbon dioxide, etc., which are usually dealt with in inorganic chemistry. Carbon compounds are the basis of all living matter.

carbon contact (*Elec. Eng.*). In a switch, an auxiliary contact designed to break contact after and make contact before the main contact to prevent burning of the latter; it is of carbon and is made to be easily removable.

carbon core (*Elec. Eng.*). A minute puncture of the first one or two insulating papers, near the conductor. Cores are carbonized, and their diameters lie between 0·05 and 0·2 mm.

carbon cycle (*Astron.*). A nuclear reaction which is believed to be the main source of energy in the larger stars. At the intensely high temperatures in the interior of these stars, hydrogen nuclei combine to form helium nuclei, carbon atoms acting as catalysts. The reactions are:

$$C^{12}+H=N^{13}\rightarrow C^{13}+e^{+}$$
$$C^{13}+H=N^{14}$$
$$N^{14}+H=O^{15}\rightarrow N^{15}+e^{+}$$
$$N^{15}+H=C^{12}+He^{4},$$

the net result being the fusion of 4 protons to give one alpha-particle and two positrons. This is also known as the **Bethe cycle.** (*Biol.*) The biological circulation of carbon from the atmosphere into living organisms and, after their death, back again. See **carbon dioxide, photosynthesis.**

carbon dating (*Nuc.*). Atmospheric carbon dioxide contains a constant proportion of radioactive ^{14}C, formed by cosmic radiation. Living organisms absorb this isotope in the same proportion. After death it decays with a half-life $5{\cdot}57\times10^{3}$ years. The proportion of ^{12}C to the residual ^{14}C indicates the period elapsed since death.

carbon dioxide (*Chem.*). CO_2. A colourless gas with a slight smell of soda water. Density at s.t.p. 1·976 kg/m³, about 1·5 times that of air. Produced by the complete combustion of carbon, by the action of acids on carbonates (e.g., Kipp's apparatus), by the thermal decomposition of carbonates (e.g., lime burning) and during fermentation. It plays an essential part in metabolism, being exhaled by animals and absorbed by plants (see **photosynthesis**). May be liquefied at 20°C at a pressure of 5·7 MN/m², but at atmospheric pressure it sublimes at −78·5°C. Liquid and solid CO_2 are much used as a refrigerant, notably for ice-cream, and in fire extinguishers. CO_2 dissolves in water to form unstable carbonic acids; the pressurized solution produces the effervescent 'sparkle' in carbonated beverages. When solid, it has a convenient temperature for testing electronic components. High-pressure carbon dioxide has found a considerable use as the coolant in carbon-moderated nuclear reactors.

carbon-dioxide (CO_2) laser (*Phys.*). One in which the active gaseous medium is a mixture of carbon dioxide and other gases. It is excited by a glow discharge and operates at 10·6 micrometres wavelength.

carbon disulphide (*Chem.*). CS_2. Sulphur vapour passed over heated charcoal combines with the carbon to form *carbon disulphide.* Used as a solvent for sulphur and rubber. The disagreeable smell associated with commercial carbon disulphide is due to impurities.

carbon-dust resistor (*Elec. Eng.*). A type of resistor made of carbon dust in zig-zag troughs; used as a neutral earthing resistor on account of the fact that its resistance decreases as the current increases.

carbon fibres (*Chem.*). Very fine filaments of the order of 8 micrometres diameter which are used in composite materials, being bound together with epoxy or polyester resins. These fibre-reinforced materials have in general a much higher strength/weight ratio than metals. Boron and glass are alternative fibre materials.

carbon filament (*Light*). The fine conductor, of a carbon-filament lamp, which is heated to incandescence to produce light; the filament is made by heating a thread of cellulose material to convert it to pure carbon.

carbon-filament lamp (*Light*). An incandescent electric lamp consisting of a fine thread of carbon heated to white heat inside an evacuated bulb. Often called simply a **carbon lamp.**

carbon film technique (*Micros.*). One of the methods used in electron microscopy to provide a supporting film for the specimen. The film is prepared by boiling carbon in a vacuum.

carbon gland (*Eng.*). A type of gland used to prevent leakage along the shaft. It consists of carbon rings cut into segments and pressed into contact with the shaft by an encircling helical spring. See **garter spring.**

carbonic acid (*Chem.*). H_2CO_3. A very weak acid formed when carbon dioxide is dissolved in water. It has an acid reaction to litmus.

carbonic acid derivatives (*Chem.*). Carbonic acid forms both normal and acid salts. The esters, chlorides, and amides form two series, viz. normal compounds, in which both hydroxyl groups of the acid are substituted; and acid compounds, in which only one hydroxyl group is substituted. The acid compounds are unstable in the free state, but form stable salts.

carbonic acid esters (*Chem.*). These comprise the methyl, ethyl, and propyl esters of carbonic acid, and also the salts of the acids esters.

carbonic acid gas (*Chem.*). Carbon dioxide effervescing from liquids which have been saturated with carbon dioxide under pressure. The gas escapes when the pressure is withdrawn.

carbonic anhydrase (*Chem.*). Enzyme in blood, which catalyses the decomposition of hydrogen carbonates into carbon dioxide.

carbonic anhydride (*Chem.*). A synonym for *carbon dioxide*. See **carbonic acid.**

Carboniferous System (*Geol.*). One of the chief of the Palaeozoic Systems, comprising, in Britain, the Carboniferous Limestone, the Millstone Grit, and the Coal Measures.

carbonium ion (*Chem.*). A short-lived, positively-charged intermediate formed by the removal of X— from a $\geqslant$C—X bond.

carbonization (*Chem.*). The destructive distillation of organic substances out of contact with air, accompanied by the formation of carbon, in addition to liquid and gaseous products. Coal yields coke, while wood, sugar, etc., yield charcoal. (*Geol.*). The conversion of fossil organic material to a residue of carbon. Plant material is often preserved in this way. (*Met.*) See **cementation.** (*Textiles*) The steeping of burry wool (see **burr**) in a weak solution of sulphuric acid, or its treatment with hydrochloric acid gas (*dry process*). This reduces the burrs to carbon dust and thereby facilitates their removal.

carbonized anode (*Electronics*). Metallic anode coated with carbon, in the form of lampblack, to assist in the radiation of heat and reduce the secondary emission of electrons.

carbonized cloth (*Elec. Eng.*). Cloth, carbonized in a vacuum, which can be used for making variable resistances by arranging for layers of the cloth to be subjected to a varying mechanical pressure.

carbonized filament (*Electronics*). Thoriated tungsten filament coated with tungsten carbide to reduce the loss of thorium from the surface.

carbon lamp (*Light*). See **carbon-filament lamp.**

carbon microphone (*Acous.*). A microphone in which a normally d.c. energizing current is modulated by changes in the resistance of a cavity filled by granulated carbon which is compressed by the movement of the diaphragm. The diameter of the cavity is frequently very much less than that of the diaphragm, and it is then known as a **carbon button.**

carbon monoxide (*Chem.*). CO. Formed when carbon is heated in a limited supply of air, when carbon dioxide is heated with an excess of carbon, when carbon dioxide is passed over some hot metals, or by dehydration of methanoic acid. It is a product of incomplete combustion. Poisonous. Its properties as a reducing agent render it valuable in industrial processes. See also **carbonyl.**

carbon monoxide-haemoglobinaemia (*Med.*). Haemoglobin combines with carbon monoxide instantaneously and is thus deprived of its oxygen-exchanging properties. This leads to poisoning of the body and suffocation.

carbon-nitrogen cycle (*Astron.*). See **carbon cycle.**

carbon paper. Paper coated with waxes containing dyes or carbon black, used for making duplicate copies in typewriting, etc.

carbon pile voltage transformer (*Elec. Eng.*). Variable electrical resistor made from disks or plates of carbon arranged to form a pile.

carbon process (*Photog.*). Printing process using a relief made by the solvent action of warm water on an emulsion of bichromated pigmented gelatine, the image being ultimately transferred to a suitable paper base.

carbon replica technique (*Micros.*). A method used in electron microscopy for making a surface replica of a specimen. It is coated with a structureless carbon film while on its supporting film, and the film and specimen are subsequently removed by dissolving in an appropriate solvent.

carbon resistor (*Elec. Eng.*). Negative temperature coefficient, noninductive resistor formed of powdered carbon with ceramic binding material. Used for low-temperature measurements because of the large increase of resistance as temperature decreases.

carbon rheostat (*Elec. Eng.*). A rheostat consisting of a number of carbon plates through which the current is passed; its resistance can be varied by altering the mechanical pressure on them by means of a screw.

carbon stars (*Astron.*). Certain uncommon giant stars of spectral types *R* and *N*, which show characteristic bands of cyanogen and molecular carbon C_2.

carbon steel (*Met.*). A steel whose properties are determined primarily by the percentage of carbon present. Besides iron and carbon, carbon steels contain manganese (up to 1·6%), silicon (up to 0·6%), sulphur and phosphorus (up to 0·1%), but no chromium, nickel, molybdenum, etc.

carbon suboxide (*Chem.*). Malonic anhydride, C_3O_2, O=C=C=C=O; a colourless liquid or gas; m.p. −107°C, b.p. +7°C; formed by heating malonic acid to 140°–150°C. Example of a *ketene* (q.v.).

carbon tetrachloride (*Chem.*). Tetrachloromethane, CCl_4; a colourless liquid, b.p. 76°C; prepared by the exhaustive chlorination of methane or carbon disulphide. Solvent for fats and oils, widely used in fire extinguishers; it is now little used as a dry-cleaning fluid, because of its toxicity. Not hydrolysed by water.

carbon tetrachloride fuse (*Elec. Eng.*). A fuse immersed in carbon tetrachloride; for use on high-voltage circuits.

carbon tissue (*Photog.*). Paper coated with a mixture of gelatine and a pigment (sometimes carbon powder). Used in the *carbon process* (q.v.). (*Print.*) A photosensitive gelatine layer on a paper backing which is printed down with positives and the photogravure screen, and, after developing, is stripped down on the photogravure cylinder to act as an etchant resist.

carbon value (*Chem.*). The figure obtained empirically as a measure of the tendency of a lubricant to form carbon when in use.

carbonyl (*Chem.*). When carbon monoxide acts as a radical, as it appears to do in many reactions, it is called the *carbonyl* group. Carbon monoxide combines with certain metals to form carbonyls, e.g., $Co(CO)_3$, $Ni(CO)_4$, $Fe(CO)_5$, $Mo(CO)_6$.

carbonyl chloride (*Chem.*). *Phosgene* (q.v.).

carbonyl platinous chlorides (*Chem.*). These are formed when carbon monoxide unites with platinum(II)chloride.

carbonyl powders (*Powder Met.*). Metal powders produced by reacting carbon monoxide with the metal to form the gaseous carbonyl. This is then decomposed by heat to yield powder of high purity.

Carborundum (*Eng., etc.*). TN designating a proprietary range of products, e.g., *silicon carbide.*

Carborundum detector (*Radio*). One consisting of

a point contact between steel and a Carborundum crystal, most sensitive when a small steady voltage is maintained across the contact.

Carborundum wheel (*Eng.*). An abrasive or grinding wheel consisting of Carborundum grains held together and moulded into disk form by a bonding agent. See also **grinding wheel.**

carbostyril (*Chem.*). 2-Hydroxyquinoline, the lactim of 2-amino-cinnamic acid. It crystallizes with 1 H_2O, the m.p. of the anhydrous compound being 201°C.

carboxy-haemoglobinaemia (*Med.*). See **carbon monoxide-haemoglobinaemia.**

carboxylase (*Chem.*). An enzyme which is capable of eliminating CO_2 from 2-ketonic acids, with the formation of aldehydes.

carboxyl group (*Chem.*). The acid group —COOH.

carboxylic acid (*Chem.*). R—$(COOH)_n$. An organic compound having one or more carboxyl radicals. A carboxyl radical is usually designated: $—C{\displaystyle ^{\nearrow O}_{\searrow OH}}$, but shows resonance.

carboxypeptidases (*Chem.*). Proteolytic enzymes which catalyse hydrolysis of peptide bonds adjacent to free carbonyl groups. Pancreatic carboxy peptidase A has little action on C-terminal proline, arginine or lysine residues, whereas pancreatic carboxypeptidase B does liberate C-terminal basic amino acid residues.

carboy (*Glass*). Large narrow-necked container, usually of balloon shape, having a capacity of 20 litres or more.

carbro process (*Photog.*). The printing of an enlarged image by the carbon process, by enlarging the original image on bromide paper and, after sensitizing the latter, bringing it into contact with a tissue, coated with bichromated pigmented gelatine. Also used for three-colour separation prints (see **trichrome carbro).**

carbuncle (*Med.*). Circumscribed staphylococcal infection of the subcutaneous tissues. (*Min.*) A gem variety of garnet cut en cabochon. It has a deep red colour. See **almandine.**

carburation or **carburetion** (*Eng.*). The mixing of air with a volatile fuel to form a combustible mixture for use in an internal-combustion (petrol) engine.

carburetted (or enriched) water gas (*Fuels, etc.*). Blue water gas which has been enriched by passing it through a carburettor into which gas oil is sprayed; calorific value per cubic metre, at 15°C, about 18 MJ. Usually mixed with coal-gas to form *town gas.*

carburettor or **carburetter** or **carburetor** (*Eng.*). A device for mixing air and a volatile fuel in correct proportions, in order to form a combustible mixture. It consists essentially of a jet, or jets, discharging the fuel into the air stream under the pressure difference created by the velocity of the air as it flows through a nozzle-shaped constriction.

carburization or **carbonization** (*Met.*). See **cementation.**

carbylamines (*Chem.*). **See isocyanides.**

carcass (*Build.*). The shell of a house in construction, consisting of walls and roof only, without floors, plastering, or joiner's work.

carcassing timber (*Build.*). Timber for the framing of a building or other structure.

carcass-saw (*Join.*). A saw like the dovetail saw, but of larger size and with fewer teeth per inch.

carcel (*Light*). Unit of luminous flux, formerly used in France.

carcerulus (*Bot.*). A fruit which splits at maturity into several one-seeded portions.

carcinogen (*Chem.*). A substance which induces cancer in a living organism.

carcinogenesis (*Med.*). The production and development of cancer.

carcinoma (*Med.*). A disorderly growth of epithelial cells which invade adjacent tissue and spread via lymphatics and blood vessels to other parts of the body. See also **sarcoma.**

carcinoma en cuirasse (*Med.*). Extensive carcinomatous infiltration of the skin, characterized by hardness and gross thickening.

carcinomatosis, carcinosis (*Med.*). Cancer widely spread throughout the body.

carcinomatous (*Med.*). Of the nature of cancer.

carcinosis (*Med.*). See **carcinomatosis.**

card (*Comp.*). Standard-size card, printed with numbers up to 80, in columns, with holes correspondingly punched, for operation of process controls and computers. (*Electronics*) A board containing a circuit with a plug or connector for attachment to a *card box.* (*Surv.*) The graduated dial or face of a magnetic compass to which the card and needle are firmly connected.

Cardan mount (*Instr.*). Type of gimbal mount used for compasses and gyroscopes.

cardboards (*Paper*). Made from a single layer of pulp (*pulp boards*); several layers (*triplex boards*); or sheets of paper pasted together (*pasteboards*). The term should be applied only to the finer qualities of boards. See also **boxboard, millboards, strawboards.**

card box (*Electronics*). Portion of a complete unit for accommodation of jacked-in replaceable electronic units. Also called **tray.** See **card.**

card chase (*Typog.*). A small chase used for imposing small jobs such as cards.

card clothing (*Textiles*). Strong fillet material comprising a foundation of woven cotton and wool, generally covered with rubber, and filled with masses of wire teeth set closely together; used to cover the cylinder, doffer, flats (or rollers and clearers) on cards.

card deck (*Comp.*). A set of punched cards comprising (part of) a programme.

carded yarns (*Spinning*). Coarse and medium count yarns, made from slivers taken straight from the card through drawframes or flyframes to mules or ringframe. Combing is omitted.

Cardew earthing device (*Elec. Eng.*). An apparatus for earthing a circuit in case of abnormal potential difference; it consists of an earthed metal plate and an aluminium strip, which are electrostatically attracted to each other by the potential difference.

Cardew voltmeter (*Elec. Eng.*). An early form of hot-wire voltmeter, utilizing the expansion of a long platinum wire to operate the needle.

card feed (*Comp.*). The mechanism which moves cards sequentially into read or print units.

cardia, cardiac sphincter (*Anat.*). The physiological sphincter surrounding the opening of the oesophagus into the stomach.

cardiac. Pertaining to the heart.

cardiac aneurysm (*Med.*). A fibrous dilatation of one or other ventricle due to destruction of cardiac muscle.

cardiac arrest (*Med.*). The sudden cessation of the heart's action, now a treatable condition.

cardiac glands (*Histol.*). Simple and branching secreting glands embedded in the stroma of the stomach, near the oesophageal junction.

cardiac massage (*Med., Surg.*). The manual

procedure whereby the heart is restarted after cardiac arrest, externally by pressure on the sternum, or by massaging the heart directly with a hand around it.

cardiac muscle (*Histol.*). The contractile tissue forming the wall of the heart in Vertebrates; the fibres have no well-marked sarcolemma, show only faint transverse striations, frequently branch and anastomose, and have the nuclei in the thickness of the fibre. Cf. *striated muscle*.

cardiac sphincter (*Anat.*). See **cardia**.

cardiac valve (*Zool.*). A valve at the point of junction of the fore- and mid-intestine in many insects. Also **oesophageal valve.**

cardinal (*Zool.*). In Insects, pertaining to the cardo; in *Lamellibranchiata, Brachiopoda*, pertaining to the hinge; more generally, primary, principal, as the *cardinal sinuses* or *veins*, being the principal channels for the return of blood to the heart in the lower Vertebrates.

cardinal number (*Maths.*). A property associated with a set such that two sets X and Y have the same cardinal number if and only if X and Y can be arranged in one-to-one correspondence. Cardinal number of a set differs from the ordinal number in that the former relates to quantity and the latter to order or arrangement. This distinction, however, becomes of real interest only for transfinite numbers, as is suggested by the fact that for finite sets the same symbol is used for both the cardinal and the ordinal number. For a finite set, the cardinal number is just the number of elements in the set. In the content of infinite sets, a set may have the same cardinal number as some of its proper subsets; but this is not the case for finite sets.

cardinal planes (*Light*). In a lens, planes perpendicular to the principal axis, and passing through the cardinal points of the lens.

cardinal points (*Astron., Nav.*). Name given to the four principal points of the horizon—north, south, east, and west—corresponding to *azimuths* (q.v.) 0°, 180°, 90°, and 270°. See **quadrantal points.** (*Light*) In a lens or system of lenses, the two principal foci, the two nodal points, and the two principal points. When the lens is used normally in air, the principal points coincide with the nodal points. For a lens of negligible thickness, the cardinal points coalesce into a single point, at the optical centre of the lens.

carding (*Textiles*). The process of preparing all natural and manmade staple fibres for conversion into sliver.

cardioblast (*Zool.*). A mesodermal cell in an embryo, destined to take part in the formation of the heart.

cardiocentesis (*Surg.*). Puncture of the heart with a needle.

cardiogram (*Med.*). Trace produced by electrocardiograph showing voltage waveform generated during heartbeats.

cardiograph (*Med.*). A recording device to exhibit a waveform determined by heart electromotive forces. The most usual forms are those using either an Einthoven galvanometer or a cathode-ray tube with long afterglow.

cardioid (*Maths.*). A heart-shaped curve with polar equation $r = 2a(1 + \cos\theta)$. An epicycloid in which the rolling circle equals the fixed circle. Cf. *limaçon*.

cardioid reception (*Radio*). Reception with antenna, horizontal response of which is determined by a vertical loop and open vertical wire. Also **heart-shaped reception.**

cardiology (*Med.*). That part of medical science concerned with the function and diseases of the heart. *n.* **cardiologist.**

cardiolysis (*Surg.*). Operative freeing of the heart from the chest wall when it is adherent to it in chronic adhesive pericarditis.

cardiomalacia (*Med.*). Pathological softening of the heart muscle.

cardiorrhexis (*Med.*). Rupture of the heart wall.

cardiospasm (*Med.*). Spasm of the cardia or cardiac sphincter of the stomach.

cardiotachometer (*Med.*). An electronic amplifying instrument for recording and timing the heart rate.

cardiotron (*Med.*). A portable form of electrocardiograph.

cardiovascular (*Med.*). Pertaining to the heart and the blood vessels.

cardiovisceral vessel (*Zool.*). In *Urochorda*, the vessel given off from the dorsal end of the heart, serving the alimentary canal and other viscera.

carditis (*Med.*). Inflammation of the muscle and the coverings of the heart.

card lacing (*Weaving*). Thick twine, sufficiently strong and pliable to permit the laced jacquard cards to assume a folded form easily, when not being acted on in the jacquard mechanism. Also known as **jacquard cord.**

card nippers (*Weaving*). A hand tool used for punching holes in jacquard cards.

cardo (*Zool.*). The hinge of a bivalve shell; in Insects, the proximal segment of the maxilla attached to the head. *pl.* **cardines.**

card puncher (*Weaving*). A machine operated by finger keys which punches jacquard cards using pins operated according to the changes in the drafted pattern.

card reader (*Comp.*). Component part of a computing system which scans punched cards and delivers signals or words corresponding to the information recorded in the holes.

cards (*Weaving*). Punched cardboard strips laced together in a loom jacquard machine to control the cords connecting with the harness mails, which lift or lower the warp yarns to form the desired pattern in the fabric.

caret (*Typog.*). A symbol (∧) used in proof correcting to indicate that something is to be inserted at that point.

Carey-Foster bridge (*Elec. Eng.*). A form of *Wheatstone bridge* of particular use in measuring difference between two nearly equal resistances.

car-floor contact (*Elec. Eng.*). A contact attached to the false floor of an electrically controlled lift; it is usually arranged to prevent operation of the lift by anyone outside the car while a passenger is in the lift.

caridoid facies (*Zool.*). A group of characters which must have been possessed by the ancestors of the *Malacostraca* before the present groups emerged.

caries (*Med.*). (1) Pathological absorption of bone infected by the tubercle bacillus or by syphilis. (2) Decay of teeth due, probably, to a streptococcal infection. *adj.* **carious.**

carina (*Bot.*). The boat-shaped structure formed by the two lower petals in the flower of pea and similar plants; it encloses the stamens and carpel, and plays a part in pollination. (*Zool.*) A median dorsal plate of the exoskeleton of some *Cirripedia*; a ridge of bone resembling the keel of a boat, as that of the sternum of flying Birds.

carinal canal (*Bot.*). A vertical canal running in the region of the metaxylem of living and fossil members of the *Equisetales*. In living members, it is water-filled.

carinate (*Bot., Zool.*). Shaped like a keel; having a projection like a keel. Also **tropeic.**

carious, cariose (*Bot., Med.*). Appearing as if decayed.

carmalum stain (*Micros.*). A red progressive nuclear stain, used as a counterstain with contrasting cytoplasmic stains. It contains potassium alum, carminic acid, salicyclic (2-hydroxybenzenecarboxylic) acid and formalin.

carminative (*Med.*). Relieving gastric flatulence; medicine which does this.

carmine (*Paint.*). A red pigment made from cochineal, if genuine, or from alizarine, if spurious. Due to its poor resistance properties it has now been completely replaced by synthetic pigment dyestuffs.

carnallite (*Min.*). The hydrated chloride of potassium and magnesium, crystallizing in the orthorhombic system; occurring in bedded masses with other saline deposits, as at Stassfurt. Such deposits arise from the desiccation of salt-lakes. It is used as a fertilizer.

Carnarvon arch (*Arch.*). A lintel supported on corbels or shaped into shoulders at its ends.

carnassial (*Zool.*). In terrestrial *Carnivora*, large flesh-cutting teeth derived from the first lower molar and the last upper premolars.

carnauba wax (*Chem.*). Yellow or yellowish-green natural wax; m.p. 84°–86°C, rel. d. 0·995, saponification value 88·3, acid value 2·9, iodine value 13·17; soluble in alkalis, ether, hot alcohol. It is obtained as an exudation from the leaves of the Brazilian wax palm (*Copernica cerifera*). Expensive, but much used for hardness, gloss and waterproofing in polishes, etc. Also called **Brazil wax.**

carnelian (*Min.*). A translucent red variety of chalcedony (silica).

carneous (*Bot.*). Flesh-coloured.

carnitine (*Chem.*). 3-*Hydroxy-4-butyrobetaine.*

$$(CH_3)_3\overset{+}{N}CH_2CH(OH)CH_2CO\bar{O}$$

A base found in muscles.

Carnivora (*Zool.*). An order of carnivorous or omnivorous Mammals, terrestrial or aquatic; usually with three pairs of incisors in each jaw and large prominent canines; the last upper premolar and the first lower molar frequently modified as carnassial teeth; collar-bone reduced or absent; four or five unguiculate digits on each limb. Cats, Lions, Tigers, Panthers, Dogs, Wolves, Jackals, Bears, Raccoons, Skunks, Seals, Sea-Lions, and Walruses.

carnivorous. Flesh-eating.

carnivorous plant (*Bot.*). A plant which traps and digests insects and other small animals.

carnose (*Bot.*). Fleshy in texture.

carnosine (*Chem.*). 3-*Alanyl-histidine*—

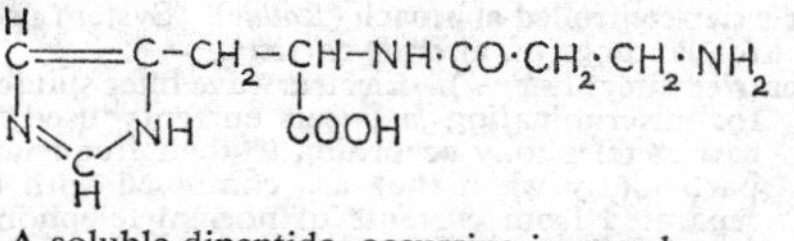

```
H
C═══C–CH₂–CH–NH·CO·CH₂·CH₂·NH₂
|   |      |
N   NH     COOH
 ╲ ╱
  C
  H
```

A soluble dipeptide, occurring in muscles.

Carnot cycle (*Eng.*). An ideal heat engine cycle of maximum thermal efficiency. It consists of isothermal expansion, adiabatic expansion, isothermal compression, and adiabatic compression to the initial state.

carnotite (*Min.*). A hydrated vanadate of uranium and potassium, found (in Colorado) as a yellow impregnation in sandstones. It is an important source of uranium.

Carnot law (*Heat*). No engine is more efficient than a reversible engine working between the same temperatures.

Carnoy fixative (*Micros.*). A nuclear fixative, consisting of absolute alcohol, chloroform, and glacial acetic acid. Cytoplasmic structures are not well preserved.

carol or **caroll** (*Build.*). A seat built into the opening of a bay window. Also called **bay-stall.**

Caro's acid (*Chem.*). *Permonosulphuric acid* (q.v.).

carotenes (*Chem.*). A group of orange-red, crystalline, highly unsaturated hydrocarbons, $C_{40}H_{56}$, to which various animal and vegetable substances owe their yellow colour (e.g., carrots, butter). They are present in the chromatophores of some *Phytomastigina* and in the chloroplasts of plants. They are precursors of vitamin A. All contain an unsubstituted 2-ionine ring, and 2-carotene has two such rings joined by a polyene bridge:

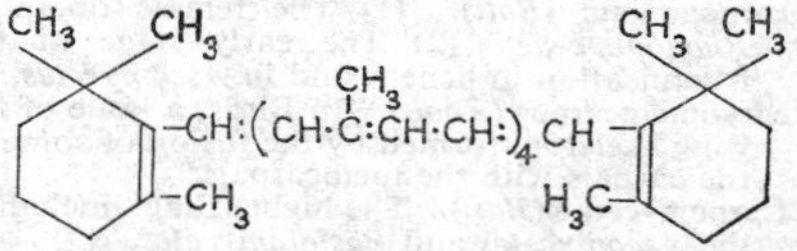

```
 CH₃  CH₃                           CH₃  CH₃
            CH₃
   ╲╱       |                          ╲╱
   ⬡ –CH:(CH·C:CH·CH:)₄CH– ⬡
     –CH₃                 H₃C–
```

carotenoids (*Chem.*). A group of yellow, orange, or red plant pigments whose structure resembles that of the carotenes but does not necessarily include an unsubstituted 2-ionine ring (or even a ring at all). Those with an unsubstituted ring can act as vitamin A precursors (e.g., cryptoxanthin) but those with only a substituted ring (e.g., xanthophyll) or no ring (e.g., lycopene) can not. Carotenoids are also found in animal structures, e.g., feathers.

carotid arteries (*Zool.*). In Vertebrates, the principal arteries carrying blood forward to the head region.

carotid bodies (*Anat., Physiol.*). Two small oval masses situated close to the carotid sinus, richly vascularized and containing chemoreceptor sensory receptors composed of an epitheloid cell and associated endings of sensory fibres of the glossopharyngeal nerve. These chemoreceptors are sensitive to the tension of oxygen and carbon dioxide and to the pH of the arterial blood flowing through them, bringing about important compensatory respiratory and cardiovascular reflexes.

carotid gland (*Zool.*). A network of capillaries situated at the bifurcation of the carotid artery in Vertebrates.

carotid sinus (*Anat., Physiol.*). The dilated portion of the common carotid artery at its division into internal and external branches. The wall of the sinus is richly endowed with sensory endings of the glossopharyngeal nerve which act as baroreceptors, the nerve impulses generated in these sensory receptors producing important compensatory cardiovascular and respiratory reflexes.

carp-, carpo-, -carp, -carpous. Prefix and suffix from Gk. *karpos*, fruit.

carpal, carpale (*Zool.*). One of the bones composing the *carpus* (q.v.) in Vertebrates. *pl.* **carpals, carpalia.**

carpel (*Bot.*). The ovule-bearing structure which, either singly or in association or combination with two or more other carpels, forms the gynaeceum of the flower. A carpel may be compared with a leaf, folded so that its edges come together, and bearing the ovules along the line of junction. A carpel consists of three parts: the *ovary*, a swollen basal portion containing the ovules, the *style*, if present, a filamentous prolongation of the apex of the ovary, and the *stigma*, the specialized tip, on which pollen lodges and germinates.

carpellary scale (*Bot.*). See **bract scale.**

carpelloid (*Bot.*). Said of some other member of a flower which is in part changed into a carpel.

carpentry. The craft of working timber, generally of a building or structural character.

carpet (*Civ. Eng.*). See **wearing course.** (*Radar*) Airborne electronic apparatus for jamming.

carpet strip (*Carp.*). A strip of wood secured to the floor below a door.

carphology, carphologia (*Med.*). Fitful movements of a delirious patient, e.g., plucking at the bedclothes, as in typhoid fever.

carpocerite (*Zool.*). In some *Crustacea*, the fifth antennal joint.

carpogenous (*Bot.*). Of fungi, bacteria, etc., living on fruit.

carpogonium (*Bot.*). (1) The female organ in *Rhodophyceae.* (2) The early stage of the fructification in lichens and in *Ascomycetes.*

carpometacarpus (*Zool.*). In Birds, a bone of the wing skeleton, formed by the fusion of some of the carpals with the metacarpals.

Carpomycetae (*Bot.*). The higher fungi, including the *Ascomycetes* and *Basidiomycetes.*

carpophore (*Bot.*). (1) An elongation of the receptacle of the flower between the insertion of the stamens and that of the carpels. (2) The forked stalk from which the mericarps of the parsley and similar plants are suspended. (3) A general term for the stalk of a fructification, especially in lower plants.

carpopodite (*Zool.*). In some Crustaceans, the third joint of the endopodite of the walking-legs or maxillipeds.

carposporangium (*Bot.*). The sporangium in which one carpospore is formed, characteristic of the *Rhodophyceae.*

carpospore (*Bot.*). A globose, uninucleate, nonmotile spore, formed by *Rhodophyceae* after fertilization and the subsequent development of a fructification containing groups of carposporangia, each yielding one carpospore.

carpostrote (*Bot.*). A plant which spreads by dissemination of the fruit.

carpotropic movement (*Bot.*). A curvature of the fruit-stalk after fertilization, bringing the fruit into a favourable position for ripening the seeds, or liberating them in a place where conditions will favour germination.

carpus (*Zool.*). In land Vertebrates, the basal podial region of the fore-limb; the wrist.

carr (*Bot.*). A community of woody plants growing on drying fenland.

carragheen (*Bot.*). *Irish moss* (q.v.). Carrageenin, a polysaccharide obtained from it, consisting of galactose units, is used as an emulsifier and sizing ingredient, also in certain foodstuffs.

Carrara marble (*Geol.*). A well-known pure white statuary marble, extensively quarried at Carrara; formed by contact metamorphism from ordinary limestone.

Carré's disease (*Vet.*). See **canine distemper.**

carriage (*Carp.*). A timber joist giving intermediate support, between the wall string and the outer string, to the treads of wide wooden staircases. Also called **carriage-piece** or **rough-string.** (*Print.*) The reciprocating assembly on a cylinder printing machine made up of a bed on which the forme lies, bears at each side, and an ink table. (*Textiles*) (1) Part of the mechanism of a lace machine that carries bobbin thread and swings in an arc on the combs containing the warp threads. (2) That section of a mule (cotton or woollen) which contains the spindles and runs out for drafting and twisting, and returns to wind-on the yarn on the cop.

carriage gain (*Spinning*). The distance the mule spindles travel on the outward run in excess of the delivery of roving, e.g.,150:160 cm. This slight stretching of the twisting yarn is the *gain.*

carriage-maker's plane (*Tools*). Special rebate plane for wide rebates, fitted with a back iron to minimize tearing of the grain.

carriage-piece (*Carp.*). See **carriage.**

carriage spring (*Eng.*). A spring used for the suspension of railway rolling-stock and other vehicles. It consists of a number of steel strips of varying length, curved to semi-elliptic form, held together so as to be capable of acting independently, and loaded as a beam. See **laminated spring.**

carriage-type switchgear (*Elec. Eng.*). See **truck-type switchgear.**

carrier (*Bacteriol.*). An organism which is infected with a pathogen, e.g., typhoid bacillus, without showing the clinical symptoms. Frequently a source of infection in a population. Similarly (*Bot.*) a plant showing no symptoms of disease, but containing a concealed virus which can be transmitted to other plants. (*Biol.*) Isotope(s) of an element added to a radioelement for carriage through a chemical or biological process or a precipitation. (*Electronics*) See **carriers.** (*Eng.*) A device for conveying the drive from the face-plate of a lathe to a piece of work which is being turned between centres. It is screwed to the work and driven by a pin projecting from the face-plate. (*Mining*) Container travelling on an aerial ropeway. (*Nuc.*) Nonactive material mixed with, and chemically identical to, a radioactive compound. Carrier is sometimes added to carrier-free material. (*Photog.*) A frame for holding a negative in an enlarger or slides in a projector. (*Telecomm.*) (1) Constant frequency in an amplitude-modulation transmission. It is *quiescent* if present only when modulated; it is *controlled* or *floating* when it varies to a lesser extent with the modulation. (2) Transmission on cable or open-wire circuits in which, after balanced-modulation, the carrier and one sideband are filtered out; at the receiving end *demodulation* is effected after the insertion of a similar carrier. (3) Vehicle for communicating a signal, when the medium cannot convey the signal but can convey the carrier, as in radio transmission, the signal being speech, video, or coded words or figures.

carrier beat (*Telecomm.*). Interference produced by heterodyne note generated between two carriers or between a carrier and a reference oscillator.

carrier-controlled approach (*Radar*). System used for landings on aircraft carriers.

carrier filter (*Teleph.*). Electric wave filter suitable for discriminating between currents used in carrier telephony according to their frequency, particularly when they are combined with or separated from currents of normal telephonic frequency.

carrier metal (*Met.*). Metal, usually a thin film of silver, on which, after surface oxidation molecular layers of caesium are deposited, when making photosensitive surfaces.

carrier mobility (*Elec.*). The mean drift velocity in unit electric field.

carrier noise (*Acous.*). Radiated noise which has

been introduced into the carrier of a transmitter before modulation.

carrier power (*Elec.*). Power radiated by a transmitter in the absence of modulation.

carriers (*Electronics*). In a crystal of semiconductor material thermal agitation will cause a number of electrons to dissociate from their parent atoms; in moving about the crystal they act as *carriers* of negative charge. Other electrons will move from neighbouring atoms to fill the space left behind, thus causing the holes where no electron exists in the lattice to be transferred from one atom to another. As these holes move around they can be considered as *carriers* of positive charge. See **impurity.**

carrier shift (*Teleg.*). (1) Change in carrier amplitude because of non-linearity in modulation. (2) Transmission in which closing of key changes frequency of radiated signal.

carrier suppression (*Telecomm.*). (1) Transmission of a modulated carrier wave with the carrier suppressed, with re-insertion of the carrier at the receiving end. (2) Suppression of the carrier from the radiating system when not required for modulation, especially on board ship to protect the receivers from noise arising from variability of eddy-current paths in the rigging, etc.

carrier telegraphy (*Teleg.*). Use of modulated frequencies, usually in the 5-unit code originated by teleprinters, and transmitted, with others, as a voice-frequency signal in telephone circuits. Also called **voice-frequency telegraphy, wired wireless.**

carrier-to-noise ratio (*Telecomm.*). Ratio of received carrier signal to noise voltage immediately before demodulation or limiting stage.

carrier wave (*Telecomm.*). That component of a modulated wave independent of the modulation; it is normally of considerably higher frequency than that of the modulation which on its own would be unsuitable for transmission.

carrollite (*Min.*). A sulphide of cobalt and copper, with small amounts of iron and nickel, crystallizes in the cubic system.

carry (*Comp.*). In arithmetic and computing, the process of replacing digits whose sum equals the base by a single digit in the column immediately to the left.

carrying capacity (*Elec.*). See **current-carrying capacity.**

carrying current (*Ecol.*). The upper level of a population in a particular environment, beyond which no major increase can occur. Represented by the upper asymptote of the sigmoid curve of population growth. (*Elec. Eng.*) See **instantaneous carrying-current.**

car-shed hanger (*Elec. Eng.*). A hanger of small vertical dimensions used for supporting the overhead contact wire of a traction system in car sheds.

carstone (*Geol.*). Brown sandstone in which the grains are cemented by limonite. Iron-pan.

Carter's coefficient (*Elec. Eng.*). A coefficient used in making calculations on the magnetic circuit of electric machines to allow for the effect of the fringing of the flux in the air gap, due to open or semiclosed slots.

cartesian coordinates (*Maths.*). A system of coordinates in which the position of a point in a plane or in space is specified by its distances from two lines or from three planes respectively. The two lines, and the three lines in which the three planes intersect are called *coordinate axes* and each distance is measured parallel to an axis. If the axes are mutually perpendicular the system is said to be *rectangular*. Devised by Descartes and published in 1637.

Cartesian diver gauge (*Vac. Tech.*). See **Dubrovin gauge.**

Cartesian hydrometer (*Powder Tech.*). An improved form of hydrometer used in the particle size analysis of soils, having a rubber membrane as part of the wall. Adjustment of the pressure above the suspension enables the specific gravity of the hydrometer to be adjusted.

cartesian ovals (*Maths.*). Curves, named after Descartes, defined by the bipolar equation

$$mr \pm nr' = \text{const.}$$

cartilage (*Histol.*). A form of connective tissue in which the cells are embedded in a stiff matrix of chondrin. See **hyaline, fibrocartilage.**

cartilage bones (*Histol.*). Bones formed by the ossification of cartilage; *autostoses.*

cartography (*Surv.*). The preparation and drawing of maps which show, generally, a considerable extent of the earth's surface.

cartouch, cartouche (*Arch.*). (1) An ornamental block supporting the eave of a house. (2) An ornamental scroll to receive an inscription.

cartridge (*Acous.*). (1) Sealed electromechanical transducer unit in gramophone pick-up. (2) A holder for reels of magnetic tape similar to a *cassette* (1, q.v.) but of larger capacity. (*Nuc. Eng.*) See **burst slug, slug.** (*Photog.*) (1) The holder for roll films for daylight loading of cameras. (2) A cardboard tube for the carriage of a small quantity of chemical.

cartridge brass (*Met.*). Copper-zinc alloy containing approximately 30% zinc. Possesses high ductility; capable of being severely cold-worked without becoming brittle. Used for cartridges, tubes, etc.

cartridge fuse (*Elec. Eng.*). A fuse-link enclosed in a cylindrical tube of insulating material, which may or may not be filled with an arc-quenching medium.

cartridge-operated hammer (*Tools*). One using the force of a small explosive cartridge to drive nails and bolts into concrete, brickwork, etc.

cartridge paper (*Paper*). A hard, tough type of paper with a rough surface, used for drawing, wrapping, etc. See also **manilla paper.**

cartridge starter (*Aero.*). A device for starting aero-engines in which a slow-burning cartridge is used to provide gas energy to operate a piston or turbine unit which is geared to the engine shaft.

cartridge system (*Eng.*). See **explosive forming.**

caruncle (*Bot.*). An outgrowth from the neighbourhood of the micropyle of a seed. The seed is said to be *caruncuIate*. (*Med.*) Any small fleshy excrescence; a small growth at the external orifice of the female urethra; (*pl.*) epithelial nodules found at the end of pregnancy on the placenta and the amnion. (*Zool.*) Any fleshy outgrowth; in some *Polychaeta*, a fleshy dorsal sense-organ; in some *Acarina*, a tarsal sucker; in embryo chicks, a horny knob at the tip of the beak.

carunculae hymenales (*Med.*). The small rounded elevations of hymen tissue, the sole remains of this structure after parturition.

caruncula lacrimalis (*Med.*). The small reddish eminence situated at the inner angle of the eye.

carunculate (*Bot.*). Having a caruncle.

carvacrol (*Chem.*). $C_{10}H_{14}O$, 1-methyl-4-isopropyl-2-hydroxy-benzene, an isomer of *thymol* (q.v.); m.p. 0°C, b.p. 236°C; obtained from camphor by heating with iodine; present in *Origanum hirtum.*

carvene (*Chem.*). See **+-limonene.**

carvone (*Chem.*). Δ-6,8-Terpadiene-2-one, an unsaturated ketone of the terpene series, the principal constituent of caraway seed oil; liquid of b.p. 228°C; readily forms *carvacrol*.

cary-, caryo-. See **kary-, karyo-.**

caryallagic reproduction (*Bot.*). Reproduction involving a change in the nucleus, in contrast to vegetative or clonal reproduction.

caryolytes (*Zool.*). In the histolysis which accompanies metamorphosis in some Insects, muscle-fragments containing nuclei.

Caryophanales (*Bacteriol.*). A small order of filamentous bacteria, with visible nuclear bodies, occurring in water containing decomposing organic material, and in the gut.

Caryophyllales (*Bot.*). See **Centrospermae.**

caryophyllatous (*Bot.*). Of flowers, having a claw.

caryophyllenes (*Chem.*). Mixture of isomeric *sesquiterpenes* (q.v.) forming the chief constituents of clove oil.

Caryophyllinae (*Bot.*). An order of the *Polypetalae*. The flowers are regular, made up of two to five (rarely six) sepals and as many petals; there are usually as many, or twice as many stamens. The ovary has one loculus or two to five imperfectly formed, with free central (rarely parietal) placentation.

caryopsis (*Bot.*). A fruit, dry, indehiscent, and containing one seed, of which the testa is united to the fruit wall; e.g., a wheat grain.

CAS (*Aero.*). Abbrev. for **calibrated airspeed.**

cascade (*Civ. Eng., San. Eng.*). The passage down of an outflow in a series of steps. (*Telecomm., etc.*) Number of devices connected in such a way that each operates the next one in turn, as transistors or valves in an amplifier.

cascade control (*Automation*). System in which one or more control systems depend on another central system, which determines the index (or desired) point in operation.

cascade converter (*Elec. Eng.*). See **motor converter.**

cascade generator (*Electronics*). High voltage generator using a series of voltage multiplying stages, especially when designed for X-ray tubes or low energy accelerators.

cascade impactor (*Powder Tech.*). Device for sampling aerosols and dusty air, which automatically fractionates the particles or spray droplets by drawing the gas stream through a series of jet impactors of decreasing nozzle aperture.

cascade motor (*Elec. Eng.*). A single induction motor with a special arrangement of windings, which gives it the properties of two induction motors in cascade connexion.

cascade particle (*Nuc.*). See **Xi particle.**

cascades (*Aero.*). Fixed aerofoil blades which turn the airflow round a bend in a duct, e.g., in wind tunnels or engine intakes.

cascade shower (*Nuc.*). Manifestation of cosmic rays in which high-energy mesons, protons and electrons create high-energy photons, which produce further electrons and positrons, thus increasing the number of particles until the energy is dissipated.

cascading of insulators (*Elec. Eng.*). Flashover of a string of suspension insulators; initiated by the voltage across 1 unit exceeding its safe value and flashing over, thereby imposing additional stress across the other units, and resulting in a complete flashover of the string.

cascara sagrada (*Pharm.*). Dried bark of *Rhamnus Purshiana*, a shrub of N. California, used (as extract) as a cathartic and laxative drug.

cascode amplifier (*Telecomm.*). Thermionic valve circuit in which a grounded-cathode triode followed by a grounded-grid triode provides a low noise amplifier for very high radiofrequencies. Also **Wallman amplifier.**

case (*Build.*). The external facings of a building when these are of better material than the backing. (*Met.*) The surface region of a steel component in which the properties and composition have been modified by special treatment, e.g., carburizing, nitriding. See **case-hardening.** (*Typog.*) A wooden tray divided into many compartments to accommodate individual letters. Originally in pairs, the *upper case* (q.v.) and *lower case* (q.v.), they are now largely replaced by the *double case* (q.v.).

caseation (*Med.*). The process of becoming cheeselike, e.g., in tissue infected with the tubercle bacillus the cells break down into an amorphous cheeselike mass.

case bay (*Carp.*). The space between two binders under a floor.

cased book (*Bind.*). The cover or case is made separately and attached to the book proper as a final operation. The usual method for *publisher's* or *edition binding*.

cased frame (*Join.*). The wooden box-frame containing the sash-weights of a window.

case-hardening (*For.*). In seasoning, a condition in which the surface of timber becomes set in an expanded condition and remains under compression whilst the interior is in tension. (*Met.*) The production of a hard surface layer on steel by heating in a carbonaceous medium to increase the carbon content, then quenching. See **cyanide hardening, pack hardening.** (*Mining*) Cementlike surface on porous rock caused by evaporation.

casein (*Chem.*). The principal albuminous constituent of milk, in which it is present as a calcium salt. Transformed into insoluble paracasein (cheese) by the action of enzymes. Casein is a raw material for thermoplastic materials used for insulators, handles, buttons, artificial fibres and bristles, etc. Also used in adhesives, nerve tonics, and for priming artists' canvases.

caseinogen (*Chem.*). A phosphoprotein found in milk, yielding casein on acidification.

Casella automatic microscope (*Powder Tech.*). A microscope whose stage moves automatically to enable the images of particles to be measured electronically.

casemate (*Join.*). A small hollow moulding subtending about 60° to 90°. Also called a **casement.**

casement (*Arch.*). A window hinged to open about one of its vertical edges. (*Join.*) See **casemate.** (*Textiles*) A plain woven cotton fabric used for curtains. Woven grey, then bleached, mercerized, dyed, or printed by screen or roller methods.

caseous (*Med.*). Cheeselike; having undergone *caseation* (q.v.).

caseous lymphadenitis (*Vet.*). Pseudotuberculosis. A disease of sheep caused by the bacterium *Corynebacterium pseudotuberculosis* (*C. ovis*) and characterized by abscess formation in the lymph glands and lungs.

cashew nut shell oil (*Paint.*). An oil, obtained from the shell of the nut *Anacardium occidentale*, which is the only vegetable oil containing more than minute amounts of phenolic substances. A series of resins with useful paintmaking properties may be derived from the oil by polymerization or by reacting it with methanal.

cashew resin (*Plastics*). Thermosetting resin based on the phenolic fraction of the oil from the shells of cashew nuts. Used for some electrical applications and in friction linings.

cashmere (*Textiles*). Fabric made from wool which forms the under, winter coat of the Cashmere (Kashmir) goat. The term is also used for lightweight fabrics with a silk, cotton, spun rayon, or woollen warp and a fine Botany weft; and for material (shoddy) manufactured from worsted rags.

casing (*Mining*). (1) Piping used to line a drill hole. (2) The steel lining of a circular shaft. (*Paper*) A pre-metrication size of brown paper, 36 × 46 in.

casinghead gasoline (*Fuels*). A very volatile product obtained by condensation of the low-boiling constituents of natural gas derived from oil-wells; blended with gasoline (petrol) to increase starting property.

casket (*Nuc. Eng.*). See **flask.**

casmere (*Textiles*). See **cassimere.**

Casogrande hydrometer (*Powder Tech.*). An improved form of hydrometer used in the particle size analysis of soils. Its main feature is its smooth symmetrical outline as compared with the *bouyoucous hydrometer* (q.v.).

Caspary's band, Casparian strip (*Bot.*). The strip of material, impermeable to water, present in the radial walls of endodermal cells when these are in the primary condition. In transverse section, the strip appears as an elliptical swelling on the wall, and is then known as the *Casparian dot*.

cassava (*Bot.*). Manioc. A tropical plant, the roots of which yield a starchy food.

Cassegrain antenna (*Radio*). One in which radiation from a focus is collimated from one surface, e.g., parabola, and reflected by another surface, e.g., a plane, or in reverse.

Cassegrain telescope (*Astron.*). A form of reflecting telescope in which the rays after reflection at the main mirror fall on a small convex mirror placed inside the prime focus. The rays are thus reflected back through a hole in the main mirror, forming an image beyond it. It is similar in some ways to the *Gregorian telescope* (q.v.).

Cassel brown (*Paint.*). Alternative name for *Vandyke brown*.

Cassel's yellow (*Chem.*). Commercial name for lead oxychloride made by heating lead (II) oxide and ammonium chloride. (*Paint.*) A yellow, artist's pigment consisting of lead oxychloride.

cassette (*Instr.*). (1) Holder for reels of magnetic tape or microfilm, which can be easily clamped and detached from the tape deck of a recorder or reader; avoids handling the tape or film. (2) Lightproof holder for photographic film. (3) Holder for X-ray plate before, during, and after exposure.

cassia oil (*Chem.*). An oil obtained from the bark of *Cinnamomum cassia*, a yellow or brown liquid of cinnamonlike odour; b.p. 240°–260°C, rel. d. 1·045–1·063.

cassideous (*Bot.*). Shaped like a helmet.

cassie (*Paper*). The damaged tops and bottoms of a ream of paper.

cassimere (*Textiles*). A closely woven 2 × 2 twill weave woollen cloth. It is a suiting and trousering material of various weaves and colour effects, light or medium weight. Also known as **casmere.** See **two-and-two twill.**

Cassini's division (*Astron.*). A dark ring concentric with the ring of Saturn and dividing it into two parts; first observed by G. D. Cassini in 1675.

Cassini's ovals (*Maths.*). See **ovals of Cassini.**

Cassini's projection (*Geog.*). A type of *cylindrical projection* (q.v.) in which the cylinder is at right angles to the axis of the globe. Commonly used on large-scale survey maps, including some editions of the Ordnance Survey.

Cassiopeia A (*Astron.*). The strongest source of radio emission in the sky. Identified with the 200-in. telescope as a very faint nebulosity, it is possibly the remnant of a supernova, in which filaments of gas are in violent motion.

cassiopeium (*Chem.*). Rare earth element now known as *lutetium* (q.v.).

cassiterite or **tin-stone** (*Min.*). Oxide of tin, crystallizing in the tetragonal system; it constitutes the most important ore of this metal. It occurs in veins and impregnations associated with granitic rocks; also as 'stream-tin' in alluvial gravels. Tin(IV)oxide.

Cassius (*Chem.*). See **purple of Cassius.**

cast (*Geol.*). A fossil organism, replaced by inorganic matter which has filled the original body cavity after the decay of the soft parts, and which may survive even the destruction of the enclosing skeletal structure. A cast shows the surface features of the organism but nothing of its internal structure. (*Med.*) A mould of cellular or organic matter shed from tubular structures in the body, e.g., the bronchi or the tubules in the kidney. (*Print.*) A general term for duplicate plates of all kinds.

castaneous (*Bot.*). Chestnut brown.

caste (*Zool.*). In some social Insects, one of the types of polymorphic individuals composing the community.

castellated filament (*Light*). A form of electric lamp filament shaped like the battlements of a castle; used in some types of traction lamp.

caster action (*Autos.*). The use of inclined king-pins by which the steerable front wheels of a motor vehicle are given fore-and-aft stability and a self-centring tendency after angular deflection by road shocks, on the principle of the domestic caster. See **trail.**

cast holes (*Foundry*). Holes made in cast objects by the use of cores, in order to reduce the time necessary for machining, and to avoid metal wastage.

casting (*Met.*). (1) The operation of pouring molten metals into sand or metal moulds in which they solidify. (2) A metallic article cast to the shape required, as distinct from one shaped by working. (*Typog.*) (1) Operating the Monotype casting machine, which is controlled by a perforated spool (see **keying**), and produces, from molten metal, single characters assembled into words and lines. (2) Making a stereotype plate from molten metal. (*Vet.*) (1) The process of throwing and securing an animal from the upright into the prone position. (2) The pellet of undigested feathers, fur, or bones disgorged by a raptorial bird.

casting box (*Print.*). A machine for casting stereotype plates, with built-in metal pot or hand-poured. Curved plates require a specific machine. For flat plates the box consists essentially of two heavy iron plates clamped together with interchangeable thickness gauges to enclose the matrix or mould.

casting copper (*Met.*). Metal of a lower purity than *best selected copper* (q.v.). Generally contains about 99·4% of copper.

casting ladle (*Foundry*). A steel ladle, lined with refractory material, in which molten metal is carried from the furnace to the mould in which the casting is to be made.

casting-on (*Foundry*). See **burning-on.**

casting resins (*Plastics*). Term applied to liquid resins which can be introduced into a moulded shape and polymerized *in situ*, with or without heating. Originally the term applied to high methanal ratio phenolic resins but is now applied widely to certain acrylic, polyester, epoxy, and polyurethane resins.

casting wheel (*Met.*). Large wheel on which ingot moulds are arranged peripherally and filled from stream of molten metal issuing from furnace or pouring ladle.

cast-in-situ concrete piles (*Civ. Eng.*). A type of pile formed by driving a steel pipe into the ground and filling it with concrete, using the pipe as a mould. The mould is withdrawn as the concrete is consolidated by heavy tamping. Also known as **moulded-in-place concrete piles.**

cast-iron (*Met.*). An iron-carbon alloy distinguished from steel by its containing substantial amounts of cementite or graphite, which make it unsuitable for working. Carbon content is usually above 2·5%. See **grey iron, mottled iron, white iron.**

castle nut (*Eng.*). A 6-sided nut in the top of which six radial slots are cut. Two of these line up with a hole drilled in the bolt or screw, a split pin being inserted to prevent loosening.

Castner-Kellner process (*Chem.*). The electrolysis of brine to produce sodium hydroxide, chlorine and hydrogen. The cathode is of mercury, in which the sodium dissolves, the amalgam being removed continuously and treated with water to liberate mercury and sodium hydroxide.

Castner's process (*Chem.*). The production of sodium cyanide from sodium. As the temperature is gradually raised, dry ammonia is blown into metallic sodium melted with charcoal, the resulting sodamide reacting quickly with the charcoal to form sodium cyanamide. This is heated to *c* 850°C, and combines further with the carbon to form sodium cyanide.

cast-off (*Typog.*). An estimate of the number of pages or lines that copy will occupy when set up in type.

castor oil (*Chem.*). Oil obtained from the seeds of *Ricinus communis*, a yellow or brown, syrupy, nondrying liquid; m.p. −10°C, rel. d. 0·960–0·970, saponification value 178, iodine value 85, acid value 19·21. Main constituent, ricinoleic acid. Uses: *pure*, in medicine and in hydraulic fluids; *dehydrated*, as a drying oil in paints; *hydrogenated*, as a fat in cosmetics; *sulphonated*, see **Turkey-red oil.**

castration (*Surg., Vet.*). Removal or surgical destruction of the testicles.

castration complex (*Psychol.*). The unconscious mental factor formed around the fear of the loss of the genital organs or their analogue; e.g., the experience of the withdrawal of the mother's breast in nursing. Later privation may become associated with the original experience, resulting in similar emotional reactions to the original fear-situation.

cast steel (*Met.*). Steel as cast, i.e., not shaped by mechanical working. Originally applied to steel made by the crucible process as distinguished from that made by cementation of wrought-iron.

cast stone (*Arch., Build.*). Precast artificially manufactured building components, e.g., blocks, lintels, sills, copes. Basically precast concrete with a facing of fine material and cement intended to look like natural stone.

cast-up (*Typog.*). The overall en content of a job after making necessary additions to the *cast-off* (q.v.) for headlines, folios, etc.

cast welded rail joint (*Rail.*). A joint between the ends of two adjacent rails in position; made by pouring molten metal between them.

casual species (*Bot.*). An alien occurring in a plant community of which it is not a regular inhabitant.

Casuarinales, Casuarinaceae, Casuariniflorae, Unisexales, Verticillatae (*Bot.*). An order of the *Archichlamydeae*, containing the single genus *Casuarina*. Of interest because of its many primitive characteristics. It resembles *Equisetales* in general appearance. The male flowers are in catkins, each flower having a single stamen. The female flowers are made up of 2 fused loculi, 1 of which is sterile. The other has 2 to 4 orthotropous ovules, each of which contains over 20 embryo sacs.

C.A.T. (*Electronics*). A trade abbrev. for *c*ooled-*a*node *t*ransmitting valve.

cata-. See **kata-.**

cata- (*Chem.*). Containing a condensed double aromatic nucleus substituted in the **1,7** positions.

catabions (*Biol.*). See **katabions.**

catabolism (*Biol.*). See **katabolism.**

catabolite (*Biochem.*). See **katabolite.**

catacaustic (*Maths.*). See **caustic curve.**

catacorolla (*Bot.*). A second corolla formed externally to the true one.

catadioptic system (*Optics*). In telescopes, one using a combination of refracting and reflecting surfaces.

catadromous (*Bot., Zool.*). See **katadromous.**

catalan process (*Met.*). Reduction of haematite to wrought iron by smelting with charcoal.

catalase (*Chem.*). An enzyme present in animal and vegetable tissues, characterized by readiness to decompose hydrogen peroxide, while being slowly destroyed itself. It manifests maximum activity at pH 6·5–7.

catalepsy (*Med.*). The condition in which any posture of a limb may be maintained without movement for a period of time longer than normal; occurring in disease of the cerebellum and in hysteria, also in deep hypnotic states and in certain types of schizophrenia. Also used (*Zool.*) to indicate the action known as 'feigning death' which can be induced in some animals by any sudden disturbance. *adj.* **cataleptic.**

catalysis (*Chem.*). The acceleration or retardation of a chemical reaction by a substance which itself undergoes no permanent chemical change, or which may be recovered when the reaction is completed. It lowers the energy of activation.

catalyst (*Chem.*). A substance which catalyses a chemical reaction. See **catalysis.**

catalytic cracking (*Chem.*). A process for breaking down the heavy hydrocarbons of crude petroleum, using silica or aluminium gel as a catalyst. See **cracking.**

catalytic poison (*Chem.*). A substance which inhibits the activity of a catalyst. Also **anticatalyst.**

catamaran (*For.*). A small raft carrying a windlass and grab, used to recover sunken logs.

cataphoresis (*Chem.*). See **electrophoresis.**

cataphorite (*Min.*). See **katophorite.**

cataphyll, cataphyllary leaf (*Bot.*). A simplified form of leaf, a scale leaf on a rhizome, or a bud scale.

cataplasm (*Med.*). A medicated poultice or plaster.

cataplexy (*Med.*). Sudden attack of weakness, following some expression of emotion; the patient falls to the ground, immobile, speechless, but conscious.

catapult (*Aero.*). An accelerating device for launching an aeroplane in a short distance. It may be fixed or rotatable to face into wind. It is usually used on ships which have no landing deck, having been superseded on aircraft carriers by the *accelerator* (q.v.). During World War II, fighters were carried on *CAMS* (catapult *a*rmed *m*erchant *s*hips) for defence against long-range bombers. Land catapults have been tried but have been superseded by *RATOG* (q.v.) and *STOL* (q.v.) aircraft.

catapult mechanism (*Bot.*). A means of seed dispersal, depending on sudden jerks due to the stiffness of the long stalk as the fruit sways in the wind.

cataract (*Med.*). Opacity of the lens of the eye as a result of degenerative changes in it.

catarrh (*Med.*). Inflammation of a mucous membrane, with discharge of mucus, commonly associated with coryza.

catathecium, catothecium (*Bot.*). A flattened, saucer-shaped ascocarp.

catatonic schizophrenia (*Psychiat.*). A type of schizophrenia characterized by states of motor activity and excitement, shown by echolalia and echopraxia, alternating with a state of stupor. Homicidal and suicidal tendencies are strong in the former phase, and are accompanied by delusions and hallucinations; in the stuporous phase all interest is withdrawn from the outside world, and the patient sits mute and idle and often has to be tube-fed. *Flexibilitas cerea* (q.v.) may be present, while a marked feature is negativism—a tendency to do the opposite of what is asked or expected. Also **catatonia, catatony.**

catch bar (*Textiles*). A long bar of steel forming part of a lace machine; it is faced with brass on the part which engages with the carriages.

catch basin (*Civ. Eng.*). See **catch pit.**

catch-bolt (*Join.*). A door-lock having a spring-loaded bolt which is normally in the locking position (i.e., extended), but which is automatically and momentarily retracted in the process of shutting the door.

catcher (*Electronics*). Element in a thermionic valve which abstracts or catches the energy in a bunched electron stream as it passes through it. See **buncher.**

catcher foil (*Nuc. Eng.*). Aluminium sheet used for measuring power levels in nuclear reactor by absorption of fission fragments.

catch feeder (*Hyd. Eng.*). An irrigating ditch.

catching diode (*Electronics*). One used to *clamp* a voltage or current at a preset value. When its anode becomes positive, it prevents the potential rising above any potential applied to its cathode.

catch-line (*Typog.*). A temporary headline inserted on slip proofs, etc.

catchment area (or basin) (*Civ. Eng.*). The area from which water runs off to any given river valley or collecting reservoir.

catch mounts (*Print.*). Individual mounts for bookwork printing to hold duplicate plates, and from which they can be removed and replaced without unlocking the forme.

catch muscle (*Zool.*). A set of smooth muscle-fibres which form part of the adductor muscle in bivalve Molluscs, and are capable of keeping the valves closed by means of a sustained tonus: any set of smooth muscle fibres associated with striated muscle fibres for a similar purpose. Also called **arrest muscle.**

catch net (*Elec. Eng.*). A wire netting placed under high-voltage transmission lines where they pass over public roadways, railways, etc., to prevent danger due to a broken live conductor.

catch pit (*Civ. Eng.*). A small pit constructed at the entrance to a length of sewer or drain pipe to catch and retain matter which would not easily pass through the pipes. Also called a **catch basin.** See also **sump.**

catch plate (*Eng.*). A disk on the spindle nose of a lathe, driving a carrier locked to the work.

catch points (*Rail.*). Trailing points placed on an up-gradient for the purpose of derailing rolling stock accidentally descending the gradient. See **spring points.**

catch props (*Mining*). In a coal mine, props put in advance of the main timbering for safety, i.e., *watch props* or *safety props.*

catch-water drain (*Civ. Eng.*). A drain intercepting water flowing naturally from high land and conducting it in any desired direction, and laid with open joints to take in water in as many places as possible.

catchword (*Typog.*). (1) At the foot of the page, the first word of the next, as in manuscripts and books printed prior to the 19th century. (2) At the head of the page a guide word in such books as dictionaries. (3) A guide word at the head of a galley of type.

catch work (*Hyd. Eng.*). A system of water channels which may be used for flooding land.

cat-cracking. See **catalytic cracking.**

catechol (*Chem.*). 2-hydroxyphenol, $C_6H_4(OH)_2$ (1, 2); colourless crystals; m.p. 104°C, b.p. 240°C. It is 1,2-dihydroxybenzene, a dihydric phenol. Occurs in fresh and fossil vegetable matter and in coal-tar. Important are its derivatives *guaiacol* (q.v.) and *adrenaline* (q.v.). Used in photography as a developing agent with tanning action; used in warm-tone printing, or mixed with 1,4-diaminobenzene to form *Meritol* developer.

OH
OH

Also called **pyrocatechin.**

catecholtannins (*Chem., Leather*). Class of *tannins* (q.v.), containing catechol. They do not deposit bloom on the leather. Sources include hemlock, larch, and mimosa.

catelectrotonus (*Med.*). Enhanced activity of a nerve because of electric current, especially near an applied cathode.

catenarian arch (*Arch.*). An arch having the shape of an inverted catenary, or hanging chain.

catenary (*Maths.*). The curve assumed by a perfectly flexible, uniform, inextensible string when suspended from its ends. Its intrinsic equation is $s=c \tan \psi$ and its Cartesian equation $y=c \cosh x/c$. A string carrying a continuous or discrete load (e.g., as in a suspension bridge) hangs in a parabola if the load is uniformly distributed horizontally.

catenary construction (*Elec. Eng.*). A method of construction used for overhead contact wires of traction systems. A wire is suspended, in the form of a catenary, between two supports, and the contact wire is supported from this by droppers of different lengths, arranged so that the contact wire is horizontal. See **compound catenary construction.**

catenate, catenulate, catenuliform (*Bot.*). Of spores and similar structures, arranged in chains.

catenation (*Cyt.*). The arrangement of chromosomes in chains or in rings.

catenoid (*Maths.*). The surface of revolution generated by the rotation of a catenary about its axis.

Caterpillar (*Eng.*). A device for increasing the tractive effort and mobility of a tractor or other road vehicle. The road wheels are replaced by chain wheels, which carry and drive a pair of endless chains or articulated tracks of large flat steel plates, often provided with removable projecting pieces for increasing the adhesion on soft and irregular ground. (*Zool.*) A type of cruciform larva, found in *Lepidoptera* and *Tenthredinidae*, which typically possesses abdominal locomotor appendages (prolegs).

cat 'flu (*Vet.*). See **feline pneumonitis.**

catgut (*Surg.*). Sterilized strands of sheep's intestine used as ligatures.

catharsis (*Psychol.*). The purging of the effects of a pent-up emotion by bringing them to the surface of consciousness: for example, 'purging' of the mind obtained when a patient freely expresses himself to his physician.

cathartic (*Med.*). Purgative. A drug which promotes evacuation of the bowel.

cathead or **spider** (*Eng.*). A lathe accessory consisting of a turned sleeve having four or more radial screws in each end; used for clamping on to rough work of small diameter and running in the *steady* (q.v.) while centring.

cathelectrotonus (*Physiol.*). Physiological excitability produced in muscle tissue by passage of electric current.

Catherine wheel (*Arch.*). See **rose window.**

catheter (*Med.*). A rigid or flexible tube for admitting or removing gases or liquids through channels of the body, especially for removing urine from the bladder.

cathetometer (*Phys., etc.*). An instrument used for measuring vertical distances not exceeding a few centimetres. It consists of a small telescope held horizontally in a cradle, which can move up and down a vertical pillar carrying a divided scale, the position of the cradle being read by means of a vernier. Images of the two points whose vertical separation is required are brought in succession to the cross-wires of the telescope eyepiece and the separation is obtained from the difference in the vernier readings. Also called a **reading microscope** (or **telescope**).

cathetron (*Elec. Eng.*). A grid-controlled mercury-arc rectifier in which the control grid is placed outside the evacuated tube. Sometimes spelt **kathetron.**

cathexis (*Psychol.*). A charge of mental energy attached to any particular idea or object.

cathode (*Electronics*). (1) Negative terminal of an electrolytic cell, i.e., one at which the positively-charged ions (cations) are discharged. Electrode at which electrons enter the cell. It is the positive terminal of a cell or battery, i.e., one at higher potential, so that the positive current in the external circuit flows from the terminal to the other terminal. See **anode.** (2) Source of electrons, produced by heat or by applying a high electric field (*cold emission*). In a mass spectrometer, secondary emission from a cold cathode is sometimes used as an ion source.

cathode bias (*Electronics*). Grid bias obtained from a resistor in the cathode circuit, which takes the anode current. Negative feedback also results, obviated by a large shunting capacitor.

cathode coating (*Electronics*). Low work-function surface layer applied to cathode. The cathode coating impedance is between the base metal and this layer.

cathode copper (*Met.*). The product of electrolytic refining. Before use the cathodes are melted, oxidized, poled, and cast into wire-bars, cakes, billets, etc.

cathode coupling (*Electronics*). That effected to or from a thermionic valve by an impedance connected between the cathode and negative terminal of the high-tension supply.

cathode efficiency (*Electronics*). Ratio of emission current to energy supplied to cathode. Also **emission efficiency.**

cathode follower (*Electronics*). A valve network in which the output load is in the cathode circuit, the input being applied between the grid and the end of the load remote from the cathode; particularly used for very high frequencies, when the valve becomes an impedance changer.

cathode-follower amplifier (*Electronics*). Amplifier with load impedance connected between cathode and ground. Resulting intrinsic negative feedback limits maximum voltage gain to unity.

cathode glow (*Electronics*). That only near the surface of a cathode, of colour depending on the gas or vapour in the tube. If an arc takes place in a partial vacuum, it may fill the greater part of the discharge tube.

cathode heating time (*Electronics*). That for heating of cathode to give full emission before accelerating voltages are applied. Especially important in thyratron tubes and large valves.

cathode interface impedance (*Electronics*). That between the cathode surface and its support, which may increase with age or through bad coating.

cathode luminous sensitivity (*Electronics*). Ratio of cathode current of photoelectric cell to luminous intensity.

cathode modulation (*Electronics*). Modulation produced by signal applied to cathode of valve through which carrier wave passes.

cathode poisoning (*Electronics*). Reduction of thermionic emission from a cathode as a result of minute traces of adsorbed impurities.

cathode ray (*Electronics*). Stream of negatively charged particles (electrons) emitted normally from the surface of the cathode in a rarefied gas. The velocity of the electrons is proportional to the square root of the potential difference through which they pass, and is equal to 595 km/sec for 1 volt. The beam can be deflected in the direction of an applied electric field, or at right angles to an applied magnetic field. Widely used applications are for oscilloscopes and in TV systems.

cathode-ray camera (*Electronics, Photog.*). Combination of cathode-ray tube and moving-film camera. The photographic film may be internal or external to the vacuum chamber.

cathode-ray direction-finder (*Electronics*). Arrangement by which the direction of arrival of an incoming signal is shown by the inclination of a line on the screen of a cathode-ray tube; this has a circular scale and is actuated by the outputs of receiving antennae, usually Bellini-Tosi loops or a double Adcock.

cathode-ray furnace (*Phys.*). One in which a small specimen is raised to a very high temperature by focusing on it an intense beam of cathode rays.

cathode-ray indicator (*Electronics*). Engine indicator using a cathode-ray tube for recording the diagram. The electron beam is deflected by voltages proportional to cylinder pressure and to time respectively, giving an indicator diagram on a time base; suitable for the highest speeds, being free from all inertia.

cathode-ray oscillograph (*Electronics*). Complete equipment, including high-voltage supply, for registering transient waveforms on a photographic plate within the vacuum of a cathode-ray tube, continuously evacuated.

cathode-ray oscilloscope (*Electronics*). Complete equipment, including a cathode-ray tube, amplifiers, time-base generators, power supply, for observing repeated and transient waveforms of current or voltage, which present a display on a *phosphor* (q.v.).

cathode-ray tube (*Electronics*). An electronic tube in which a well-defined and controllable beam of electrons is produced and directed on to a surface to give a visible or otherwise detectable display or effect. Abbrev. **CRT.**

cathode-ray tuning indicator (*Radio*). See **electric eye.**

cathode-ray voltmeter (*Electronics*). CRT of known deflectional sensitivity, used as a voltmeter. It can indicate crest values of the voltage applied to the deflector plates.

cathode resistor (*Electronics*). One connected between the cathode and the negative supply of a vacuum tube, often to provide bias for the control grid.

cathode spot (*Electronics*). Area on a cathode where electrons are emitted into an arc, the current density being much higher than with simple thermionic emission.

cathodic chalk (*Electronics, Ships*). A coating of magnesium and calcium compounds formed on a steel surface during *cathodic protection* (q.v.) in sea water.

cathodic etching (*Electronics*). Erosion of a cathode by a glow discharge through positive-ion bombardment, in order to show microstructure.

cathodic protection (*Electronics, Ships*). Protection of a metal structure against electrolytic corrosion by making it the cathode (electron receiver) in an electrolytic cell, either by means of an impressed e.m.f. or by coupling it with a more electronegative metal. In ships, corrosion can be prevented by passing sufficient direct current through the sea water to make the metal hull a cathode.

cathodoluminescence (*Phys.*). Excitation of dislodged metal from an anode by a cathode-ray beam, with consequent characteristic radiation, and with possible afterglow.

cathodophone (*Acous.*). Microphone utilizing the silent discharge between a heated oxide-coated filament in air and another electrode. This discharge is modulated directly by the motion of the air particles in a passing sound wave. Also **ionophone.**

cathography, cathodography (*Radiol.*). The practice of X-ray photography.

catholyte (*Elec., etc.*). See **catolyte.**

cation (*Elec., Phys.*). Ion in an electrolyte which carries a positive charge and which migrates towards the cathode under the influence of a potential gradient in electrolysis. It is the deposition of the cation in a primary cell which determines the *positive terminal.*

cationic detergents (*Chem.*). Types of ionic synthetic detergents in which the surface active part of the molecule is the cation, unlike soap and most of the widely used synthetic detergents. Sometimes called *invert soaps*. Typified by the quaternary ammonium salts such as *cetrimide*, benzalkonium chloride, domiphen bromide; cetylpyridinium is similar. All have powerful bactericidal activity, and in addition to skin and utensil cleaning are used in lozenges, creams, mouth-washes, etc.

catkin (*Bot.*). A somewhat specialized inflorescence, consisting of a number of sessile flowers of simple construction, arranged around a common stalk, and usually all staminate or all pistillate. The catkin falls as a unit when pollen has been shed or the seeds have been liberated. Also **amentum.** (*Electronics*) Small receiving valve in which anode also serves as envelope.

catolyte, catholyte (*Electronics*). That portion of the electrolyte of an electrolytic cell which is in the immediate neighbourhood of the cathode.

catoptric element (*Light*). The name given to polished surfaces of metal or glass, or any surface capable of acting as a mirror.

catothecium (*Bot.*). An inverted perithecium in which the asci hang down from the base of the organ.

cat sapphire (*Min.*). Blackish- or greenish-blue oriental sapphire (i.e., true sapphire) of some value as a cut gemstone, but not of characteristic colour.

cat's eye (*Glass*). A crescent-shaped blister. (*Min.*) A variety of fibrous quartz which shows chatoyancy when suitably cut, as an ornamental stone. The term is also applied to crocidolite when infiltrated with silica (see **tiger's eye, hawk's eye**). A more valuable form is *chrysoberyl cat's eye*; see **cymophane.**

cat's whisker (*Radio*). A colloquial term for the fine wire making contact with the crystal in some early forms of crystal detector.

cattierite (*Min.*). Cubic sulphide of cobalt, analogous to the cubic iron sulphide, pyrite.

cattle plague (*Vet.*). See **rinderpest.**

Cauchy-Riemann equations (*Maths.*). The equations

$$\frac{\partial u}{\partial x}=\frac{\partial v}{\partial y} \quad \text{and} \quad \frac{\partial u}{\partial y}=-\frac{\partial v}{\partial x}$$

which must be satisfied for a function of a complex variable $f(z)=u(x, y)+iv(x, y)$, u and v being real, to be differentiable. Functions u and v are said to be *conjugate*.

Cauchy's convergence tests (*Maths.*). (1) If $\sqrt[n]{u_n}\to l$ as $n\to\infty$, then the series of positive terms Σu_n converges if $l<1$ and diverges if $l>1$. (2) If $\log(1/u_n)/\log n\to l$ as $n\to\infty$, then the series of positive terms Σu_n converges if $l>1$ and diverges if $l<1$.

Cauchy's dispersion formula (*Light*).

$$\mu=A+\frac{B}{\lambda^2}+\frac{C}{\lambda^4}+\ldots$$

An empirical expression giving an approximate relation between the refractive index μ of a medium and the wavelength λ of the light; A, B, and C being constants for a given medium.

Cauchy's distribution (*Maths.*). A probability distribution of the form $p(x)=\dfrac{1}{\pi[1+(x-a)^2]}$, where the distribution is symmetric about $x=a$.

Cauchy's inequality (*Maths.*).

$$\left(\sum_1^n a_r b_r\right)^2 \leqq \left(\sum_1^n a_r^2\right)\cdot\left(\sum_1^n b_r^2\right).$$

The equality occurs when $a_r=kb_r$ for all r.

Cauchy's integral formula (*Maths.*). If the function $f(z)$ is analytic within and on a closed contour C, and if a is any point within C, then

$$f(a)=\frac{1}{2\pi i}\oint_C\frac{f(z)dz}{z-a}.$$

Cauchy's theorem (*Maths.*). If a function $f(z)$ is analytic and one-valued inside and on a contour

C, then the integral of *f*(*z*) taken round the contour is equal to zero.

cauda (*Zool.*). The tail, or region behind the anus; any tail-like appendage; the posterior part of an organ, as the *cauda equina*, a bundle of parallel nerves at the posterior end of the spinal cord in Vertebrates. *adj.* **caudal, caudate.**

caudad (*Zool.*). Situated near, facing towards, or passing to, the tail region.

caudal anaesthesia (*Med.*). *Epidural anaesthesia* via the caudal canal (foot of spine).

Caudata (*Zool.*). See **Urodela.**

caudate (*Bot., Zool.*). Bearing a tail-like appendage.

caudate larva (*Zool.*). The vermiform larva, with a caudal outgrowth of varying length, of some symphytan *Hymenoptera.*

caudate nucleus (*Histol.*). A specialized nuclear mass, representing part of the corpus striatum, and situated above the optic thalamus, in the brain of Mammals. Nerve fibres link it with the cerebral cortex and the midbrain.

caudicle (*Bot.*). The stalk composed of mucilaginous threads which attaches the pollen mass of an orchid to the rostellum.

caul (*Zool.*). In the higher Vertebrates, the amnion; more generally, any enclosing membrane.

cauldron subsidence (*Geol.*). The subsidence of a cylindrical mass of the earth's crust, bounded by a circular fault up which lava has commonly risen to fill the cauldron. Good examples have been described from Scotland (Ben Nevis and the Western Isles).

caulescent (*Bot.*). Having a stalk or a stem.

caulicole, caulicolous (*Bot.*). Growing on the stem (especially herbaceous stem) of another plant; usually refers to fungi.

cauline (*Bot.*). (1) Growing from the stem, and not from the base of the plant. (2) Appertaining to the stem. (3) Formed from the internal tissues of the stem.

cauline bundle, cauline vascular bundle (*Bot.*). A vascular bundle formed entirely from the tissues of the stem. A set of such bundles forms the *cauline vascular system.*

caulking (or **calking**) (*Carp.*). See **cogging.** (*Civ. Eng., etc.*) The operation of making a joint or seam tight to withstand pressure; performed by stopping the joint with tow or other filling material and ramming it home. (*Eng.*) The process of closing the spaces between overlapping riveted plates by hammering the exposed edge of the upper plate into intimate contact with the lower plate.

caulking joint (*Join.*). A joint sometimes used to fix a tie-beam to a wall plate; it combines features of the dove-tail and tenon and mortise joints.

caulking (or **calking**) **tool** (*Eng.*). A tool, similar in form to a cold chisel but having a blunt edge, for deforming the metal rather than cutting it.

caulocarpic (*Bot.*). Said of a plant which, after flowering, lives through the winter and flowers again in the next or in a subsequent year.

caulocystidium (*Bot.*). A *cystidium* occurring on the stipe.

caulome (*Bot.*). A general term denoting all organs belonging to the shoot.

causalgia (*Med.*). Intense burning pain in the skin after injury to the nerve supplying it.

causeway (*Civ. Eng.*). A road carried by an embankment or a retaining wall across marshy land or water. A Scottish term for **sett paving.**

caustic (*Med., etc.*). Destructive or corrosive to living tissue; an agent which burns or destroys living tissue.

caustic curve (*Light*). A curve to which rays of light are tangential after reflection or refraction at another curve. (*Maths.*) The caustic, or catacaustic, of a given curve *C*, with a given point *P* as radiant point, is the envelope of the rays from *P* after reflection by *C*. If the rays are retracted, the envelope is called the *diacaustic.*

caustic embrittlement (*Chem.*). The intergranular corrosion of steel in hot alkaline solutions, e.g., in boilers.

caustic lime (*Chem.*). The residue of calcium oxide, obtained from freshly calcined calcium carbonate; it reacts with water, evolving much heat, and producing slaked lime (calcium hydroxide, hydrate of lime, or hydrated lime). Also **quicklime.** See also **lime.**

caustic potash (*Chem.*). Potassium hydroxide. The name *potash* is derived from 'ash' (meaning the ash from wood) and 'pot' from the pots in which the aqueous extract of the ash was formerly evaporated.

caustic soda (*Chem.*). Sodium hydroxide, NaOH, a deliquescent substance, with a soapy feel, whose solution in water is strongly alkaline; it is a common reagent in the laboratory. It is manufactured by treating quicklime with hot sodium carbonate solution; and its main industrial use is in the manufacture of soap. See also **Castner-Kellner process.**

caustic surface (*Light*). A surface to which rays of light are tangential after reflection or refraction at another surface.

Cauvet (*Build., Civ. Eng.*). A method of system building developed in France. It consists basically of large-area hollow reinforced concrete cross-walls and floors. *Cladding* (q.v.) is similar, but insulating material is incorporated and composite precast and *in situ* edge beams support the cladding at each floor. See **system building.**

cave breccias (*Geol.*). Breccias formed of fragments falling from the sides and roofs of caves.

cave earth (*Geol.*). A deposit formed of the finer-grained debris of cave walls and roofs, mixed in some cases with a certain amount of water-borne sediment.

Cavendish experiment (*Elec. Eng.*). Experiment to demonstrate that all charges reside on the surface of a conductor; first performed by Henry Cavendish (1731-1810). (*Phys.*) An experiment carried out by Cavendish to determine the constant of gravitation. A form of torsion-balance was used to measure the very small forces of attraction between lead spheres.

cavern (*Geol.*). A chamber in a rock. Caverns are of varying size, and are due to several causes, the chief being solution of calcareous rocks by underground waters, and marine action.

cavernicolous (*Ecol.*). Cave-living.

cavernosus, cavernous (*Zool.*). Honeycombed; hollow; containing cavities, e.g., *corpora cavernosa.*

cavernous breathing (*Med.*). Low-pitched hollow sound produced in a cavity pathologically formed in the lung.

cavetto (*Arch.*). A hollow moulding, quarter round.

cavil (*Tools*). A small stone axe resembling a *jedding axe* (q.v.).

cavilling (*Mining*). The drawing of lots for working places (usually for 3 months) in the coal-mine.

caving (*Mining*). Controlled collapse of roof in

deep mines to relieve pressure. Undercut stoping.

cavings (*Agric.*). The short straw which is thrown out from a threshing-machine with the chaff.

cavitation (*Eng.*). The formation of a cavity between the downstream surface of a moving body and a liquid normally in contact with it, as, for example, behind the blades of a ship's propeller, or in ultrasonic waves. (*Med.*) The formation of cavities in any structure of the body, especially the lungs.

cavity effect (*Acous.*). The enhancement of response in a microphone due to acoustic resonance in a shallow cavity in front of the diaphragm.

cavity-frequency meter (*Electronics*). One for use in coaxial or waveguide systems. The frequency is related to the wavelength for resonance in the cavity.

cavity magnetron (*Electronics, Radar*). A valve having a circular anode with radial cavities, and a central cathode. An axial magnetic field forces the electrons to interact with the electric fields in the cavities thus producing microwave power.

cavity modes (*Phys.*). Stable electromagnetic or acoustic fields in a cavity which can exhibit resonance. They are *degenerate* if, having similar frequencies, they have fields which differ in pattern from that of main resonance.

cavity radiation (*Phys.*). The radiation emerging from a small hole leading to a constant temperature enclosure. Such radiation is identical with *black-body radiation* (q.v.) at the same temperature, no matter what the nature of the inner surface of the enclosure.

cavity resonance (*Phys.*). The enhancement of air flow for certain frequencies, due to neutralization of the mass (or inertia) reactance with the stiffness reactance of air in a partially enclosed space.

cavity resonator (*Phys.*). Any nearly closed section of waveguide or coaxial line in which a pattern of electric and magnetic fields can be established. Also applies to sound fields.

cavity walls (*Build.*). Hollow walls, normally built of two 4½-in. stretcher-bond walls with a 2-in. gap between, tied together with wall ties. Cavity walls increase the thermal resistance and prevent rain from reaching the inner face.

cavum (*Zool.*). A hollow or cavity: a division of the concha.

cavus (*Med.*). See **pes cavus.**

Caytoniales (*Bot.*). A fossil order of *Gymnospermae*, having the seeds enclosed in an inrolled sporophore, in contrast to a strobilus.

Cb (*Chem.*). The symbol for *columbium.*

C.B. (*Elec. Eng.*). Abbrev. for *central* (or *common*) *battery* (q.v.).

C-battery (*Elec. Eng.*). U.S. term for **grid-bias battery.**

c.b.p. (*Vac. Tech.*). Abbrev. for *critical backing pressure*.

c.c. An abbrev. for *cubic centimetre*, the unit of volume in the CGS metric system. Also cm³.

C.C.I. (*Telecomm.*). Abbrev. for *Comité Consultatif International.*

C.C.I.F. (*Telecomm.*). Abbrev. for *Comité Consultatif International Téléphonique* (*Fernsprech*).

C.C.I.R. (*Telecomm.*). Abbrev. for *Comité Consultatif International des Radiocommunications.*

C.C.I.T. (*Telecomm.*). Abbrev. for *Comité Consultatif International Télégraphique.*

C-class insulation (*Elec. Eng.*). A class of insulating material to which is assigned a temperature of over 180°C. See **class -A, -B, -C, etc., insulating materials.**

cd. Symbol for *candela.*

Cd (*Chem.*). The symbol for *cadmium.*

C.D. (*Elec. Eng.*). Abbrev. for *current density.*

C.D.F. (*Telecomm.*). Abbrev. for *combined distribution frame.*

C display (*Radar*). Display in which bright spot represents the target, with horizontal and vertical displacements representing bearing and elevation, respectively.

CdS meter (*Photog.*). See **cadmium-sulphide exposure meter.**

Ce (*Chem.*). The symbol for *cerium.*

cecidium (*Bot.*). See **gall.**

cedar (*For.*). Genus of tree, *Cedrus*, which is a softwood, with straight grain medium fine and uniform texture. It is used for pattern-making, boat building and cheap furniture. It is of European and Asian origin.

cedar oil (*Micros.*). Oil of cedarwood provides an alternative to xylol, as a clearing agent, in the preparation of specimens for microscopical examination.

cedar-tree laccolith (*Geol.*). A multiple laccolith, i.e., a series of laccoliths, one above the other, forming parts of a single mass of igneous rock.

ceiling (*Ships*). Heavy planking laid on top of the inner double bottom under *hatchways* (q.v.) to protect the tank top.

ceiling fan (*Eng.*). A low-speed electric fan which has a wide sweep and is capable of displacing a large volume of air at low velocity; suspended from the ceiling in rooms where, for comfort, the rate of air change must be increased.

ceiling joist (*Build.*). A joist to which the ceiling is fixed; it is nailed and notched to the binding joists of the floor above.

ceiling plate (*Elec. Eng.*). A metal plate for fixing to a ceiling, from which a pendant electric light fitting may be suspended. Arrangements are embodied in it to enable the flexible cord of the fitting to be connected to the wiring of the installation.

ceiling rose (*Elec. Eng.*). An enclosure of insulating material for attachment to the ceiling; it is equipped with terminals from which a flexible cord carrying an electric light fitting may be suspended. The terminals are also arranged for connexion to the wiring of the installation.

ceiling switch (*Elec. Eng.*). A switch located in the ceiling but operated by a pull on a cord. Also called a **pull switch.**

ceiling voltage (*Elec. Eng.*). A term used to denote the maximum voltage which a machine is capable of giving.

Celastoid (*Plastics*). A proprietary thermoplastic made from cellulose acetate; rel.d. 1·29–1·56, moulding temp. 130–150°C, breakdown value 45 000 V/mm.

Celastrales (*Bot.*). An order of the *Polypetalae*. The flower is regular with a distinct disk, which is adnate to, or lines, the base of the calyx tube. The stamen number is the same as, or fewer than, that of the petals (rarely twice as many). The ovary is usually entire with one to two ovules in each loculus. It is erect with a ventral raphe.

Celdecor process (*Paper*). The preparation of esparto and straw pulp by a mild cooking with caustic soda followed by bleaching with chlorine.

-cele (*Med.*). A suffix derived from the Greek *kēlē*, tumour, hernia.

celestial equator (*Astron.*). The great circle in which the plane of the earth's equator cuts the celestial sphere; the primary circle to which the coordinates right ascension and declination are referred.

celestial mechanics (*Astron.*). The branch of astronomy which deals with the motions of the heavenly bodies under the forces of gravitation. It originated with Newton, and includes the theory of the motions of planets, satellites, comets, etc., within the solar system, and also the orbital motions of binary stars. Also called **gravitational astronomy.**

celestial poles (*Astron.*). The two points in which the earth's axis, produced indefinitely, cuts the celestial sphere.

celestial sphere (*Astron.*). An imaginary sphere, of indeterminate radius, of which the observer is the centre. On the surface all stars, independently of their real distance, are points specified by 2 coordinates, referred to some chosen great circle of the sphere.

celestine (*Min.*). Strontium sulphate, crystallizing in the orthorhombic system; occurs in association with rock-salt and gypsum; also in the sulphur deposits of Sicily, and in nodules in limestone.

Celite (*Chem.*). TN for a form of *diatomite* (q.v.) used as an insulating material or filter-aid.

cell (*Aero.*). U.S. term for aircraft fuel tank. (*Biol.*) One of the specialized units, consisting of nucleus and protoplasm, which compose the bodies of plants and animals; in the former usually surrounded by a nonliving wall: in plants, the term is also used to denote the space occupied by the protoplasm with the wall around it. (*Bot.*) (1) The cavity containing pollen in an anther lobe. (2) One chamber in an ovary. (*Comp.*) Unit of storage. (*Elec. Eng.*) (1) Chemical generator of emf. (2) Small item forming part of experimental assembly, e.g., Kerr cell, dielectric test cell, etc. (*Electronics*) Photoelectric tube. (*Maths.*) Small volume unit on mathematical coordinate system. (*Nuc. Eng.*) (1) Unit of homogeneous reactivity in reactor core. (2) Small storage or work space for 'hot' radioactive preparations. (*Zool.*) One of the spaces into which the wing of an Insect is divided by the veins. *adj.* **cellular.**

cellase (*Chem.*). An enzyme, found in apricot kernels, which hydrolyses *cellobiose* (q.v.).

cell assembly theory (*Psychol.*). That reverberating circuits and assemblies of them form in the C.N.S. through synaptic facilitation, and are the basis of memory, thought, etc.

cell call (*Comp.*). (1) A call number in programming. (2) A set of symbols used in programming to identify a sub-routine.

cell cavity (*Bot.*). See **lumen.**

cell constant (*Chem.*). The conversion factor relating the conductance of a conductivity cell to the conductivity of the liquid in it.

cell division (*Cyt.*). The splitting of a cell into daughter cells. See **amitosis, meiosis, mitosis.**

cell-free system (*Biochem.*). Use of purified combinations of essential cellular machinery to reproduce specific, narrowly-defined functions in molecular biology. An example is the production of a specific polypeptide from its specific m-RNA, using purified ribosomes, enzymes, s-RNA, etc.

cell inclusion (*Cyt.*). Any nonliving material, organic or inorganic, present in the cytoplasm.

cell inspection lamp (*Elec. Eng.*). An electric filament lamp provided with narrow bulb so that it can be used for the examination of an accumulator by insertion between the plates.

cell lineage (*Zool.*). The detailed sequence of events during the cleavage of the ovum, and the history and fate of each of the blastomeres.

cellobiase (*Chem.*). Enzyme which hydrolyses *cellobiose* (q.v.).

cellobiose or **cellose** (*Chem.*). $C_{12}H_{22}O_{11}$, a disaccharide; obtained by incomplete hydrolysis of cellulose. $G-\beta 1 \rightarrow 4-G$ (G = glucose).

cello foils (*Print.*). See **vinyl foils.**

celloidin embedding medium (*Micros.*). A thick solution of celloidin (a pyroxilin derivative) in solvent, e.g. ethanol and ethoxyethane, that later evaporates, used for embedding, when thick sections are to be prepared, or the specimen is brittle or hard. See also **double embedding.**

celloidin-paraffin embedding (*Micros.*). An alternative technique to *double embedding* (q.v.) for very small specimens, which are liable to distortion or disorientation in paraffin wax. The specimen is pre-impregnated with celloidin combined with methyl benzene carboxylate and transferred to paraffin wax via benzene. Also known as **Peterfi's method.**

cellose (*Chem.*). See **cellobiose.**

Cellosolve (*Plastics*). Hydroxy-ether, 2-ethoxyethan 1-ol, $C_2H_5O{\cdot}CH_2{\cdot}CH_2OH$, a colourless liquid used as a solvent in the plastics industry. It is miscible with water, ethanol and ethoxyethane, and boils at 135·3°C.

cell plate (*Bot.*). A delicate membrane formed across the equator of the achromatic spindle as cell division proceeds; it provides the foundation of the wall which separates the daughter cells resulting from the division.

cell sap (*Cyt.*). The fluid constituents of a cell.

cell tester (*Elec. Eng.*). A portable voltmeter for checking voltage of accumulator cells.

cell tissue (*Bot.*). A group of cells formed by division of one or a few original cells, remaining associated and functioning as a whole.

cellular (*Textiles*). The name applied to a fabric featuring an open, or cell-like, structure obtained by the use of a leno or mock-leno weave; used mainly for shirting, blouse and underwear purposes.

cellular concrete (*Build.*). Concrete in the body of which bubbles of air are induced, either by chemical or by mechanical means, in the process of manufacture, thereby producing a concrete of low unit weight.

cellular double bottom (*Ships*). A common construction of the bottom of a ship, where an inner bottom extends throughout the length of the ship between the peak bulkheads and over the width of the ship. The space between the outer and inner bottom plating is divided into cells by transverse floors and fore and aft keelsons, some of which are oil and watertight so that the space may be used for fresh water, water ballast or oil fuel.

cellular glass (*Glass*). Glass foamed to a mass of separate air cells. This is dark in colour with a strong sulphide smell. It is a very moisture-resistant low-temperature insulation material of great value.

cellular horn (*Acous.*). A horn for a high-frequency loudspeaker (tweeter), in which the path from the throat to the outer air is by a number of expanding channels of equal length, so that marked directivity, arising from the short wavelengths in relation to the width of the total opening, is not apparent.

cellular immunity (*Bacteriol.*). An immunity developed at the cellular level, by the response of the white cells of the blood, and the macrophages of the spleen, liver, and lymph glands, to an invading pathogen, e.g., bacteria. These phagocytic cells infest and destroy the pathogen and increase in number to counteract its multiplication.

cellular silica (*Chem.*). Inorganic silicates containing numerous air cells and used for thermal insulation, particularly for cold surfaces, where their low water vapour permeability finds favour.

cellular spore (*Bot.*). A multicellular body which is set free like a spore, and in which each cell is able to germinate separately by a germ tube and to give a new plant.

cellular structure (*Met.*). See **network structure.**

cellular tissue (*Bot.*). A tissue composed of cells.

cellular-type switchboard (*Elec. Eng.*). A switchboard in which each switch with its associated apparatus is contained in a separate cell of fireproof material. Also called a **cubicle-type switchboard.**

cellulase (*Bot.*). See **cytase.**

cellulation (*Zool.*). The reformation of cells in injured tissue.

cellulin (*Bot.*). A refractive substance, probably resembling cellulose in composition, present in the hyphae of some aquatic fungi.

cellulith. A substitute for ebonite, produced by drying wood pulp which has been ground to a homogeneous mass.

cellulitis (*Med.*). A spreading infection of the subcutaneous tissues with pyogenic bacteria. For *pelvic cellulitis* see **parametritis.**

celluloid (*Plastics*). A thermoplastic made from nitrocellulose, camphor, and ethanol; rel. d. 1·35–1·85, moulding temp. 29–49°C, breakdown strength 20 000 to 45 000 V/mm. Highly inflammable. It is elastic and very strong, and can be produced in very thin sheets. Its use has decreased with the introduction of nonflammable thermoplastics.

cellulose (*Chem.*). $(C_6H_{10}O_5)_x$, the most complicated polyose, forming the walls of the cells in all plants. The chief source of cellulose is wood, cotton, and other fibrous materials (e.g., flax, hemp, nettle, etc.). Pure cellulose is obtained by removing all incrustations of lignin resins and other organic and inorganic matter by treatment with alkali, acids, sodium sulphite, etc. Cellulose is soluble in cuprammonium hydroxide (*Schweitzer's reagent* (q.v.)), ammoniacal copper carbonate, a solution of zinc oxide in conc. hydrochloric acid. It is the raw material for the manufacture of paper, rayon, cellulose lacquers, films. Cellulose can undergo many chemical transformations, e.g., strong acids transform it into *amyloid* (q.v.); it can be hydrolysed and oxidized (cellulose hydrates, hydrocelluloses, oxycelluloses) and esterified (*cellulose acetates; cellulose nitrates*; benzyl-cellulose; *cellulose xanthate*. It contains the β1–4 glucosidic linkage.

cellulose acetates (*Chem.*). Acetylcelluloses. These are ethanoic acid esters of cellulose, obtained by the action of glacial ethanoic acid, ethanoic anhydride, and sulphuric acid, upon cellulose. They are considerably less inflammable than cellulose nitrates, and are a raw material for films, windscreens, textile fibres, lacquers, etc.

cellulose esters (*Chem.*). Cellulose derivatives obtained by esterification with nitric acid, acetic acid, etc.

cellulose ether (*Chem.*). The product of partial or complete etherification of the hydroxyl groups in a cellulose molecule. Used in the manufacture of moulding compositions, adhesives and lacquers.

cellulose hydrates (*Chem.*). Cellulose products closely resembling cellulose, extremely hygroscopic, but without reducing properties, obtained by the action of cold concentrated caustic soda on cellulose, e.g., cotton. The latter process, starting with cotton, is used for producing 'mercerized cotton'.

cellulose lacquers (*Chem.*). Lacquers prepared by dissolving nitrocellulose or acetylcellulose in a mixture of suitable solvents, with the admixture of resins and plasticizers and pigments.

cellulose nitrates (*Chem.*). See **nitrocelluloses.**

cellulose propanoate (*Chem.*). An ester of cellulose made by the action of propanoic acid and its anhydride on purified cellulose. It is used as the basis of a thermoplastic moulding material.

cellulose trabecula (*Bot.*). A strand of cellulose crossing the lumen of a cell.

cellulose xanthate (*Chem.*). $[C_6H_8O_3(ONa) \cdot OCS_2Na]n$, an acid salt of cellulose-dithiocarbonic acid, obtained by treating cellulose with concentrated caustic soda, with subsequent dissolution in carbon disulphide. The resulting product is called *viscose*, the intermediate for viscose rayon textile fibres.

cell unit (*Crystal.*). See **unit cell.** (*Elec. Eng.*) A unit which forms the basis of an extended switchboard.

cell wall (*Cyt.*). The membrane or other autogenous structure confining the contents of a cell. See **plasma membrane.**

Celotex (*Build.*). TN for artificial building-board used for acoustic absorption control; made from sugar-cane fibre (bagasse), compressed and baked.

celsian (*Min.*). A barium feldspar, crystallizing in the monoclinic system; barium aluminium silicate. Found in association with some manganese ores.

Celsius scale (*Heat*). The SI name for *Centigrade scale* (q.v.). The original Celsius scale of 1742 was marked zero at the boiling-point of water and 100 at the freezing-point, the scale being inverted by Strömer in 1750. Temperatures on the International Practical Scale of Temperature are expressed in degrees Celsius. See however **Kelvin thermodynamic scale of temperature.**

Celtic twills (*Textiles*). A mat or hopsack weave with a twilled appearance. Also known as **twilled mats.**

Celtic weave (*Textiles*). See **hopsack weave.**

cement (*Build., Civ. Eng., etc.*). A material for uniting other materials or articles. It is generally plastic at the time of application, but hardens when in place. (*Zool.*) In Mammalian teeth, a layer resembling bone covering the dentine beyond the enamel.

cementation (*Bot.*). The union of fungal hyphae by means of a sticky excretion. (*Civ. Eng.*) See **grouting.** (*Met.*) Generally, any process in which the surface of a metal is impregnated by another substance. See **case-hardening.** Also called **carbonization, carburization.**

cement clay (*Geol.*). A clay rock containing a varying quantity of calcium carbonate, hence used for the manufacture of cement.

cement copper (*Met.*). Impure copper, obtained when the metal is precipitated by means of iron from solutions resulting from leaching.

cemented (or **sintered**) **carbides** (*Met.*). Powdered carbides of tungsten, tantalum, or titanium cemented into solid masses by mixing with powdered cobalt or nickel, then compressing and sintering. Used instead of high-speed steel to form cutting tip of cutting-tools, and in parts subjected to heavy wear.

cement fillet (*Build.*). A substitute for metal flashings in the angles between, e.g., a chimney stack and roof, weatherproofing being provided by running in a band of cement mortar.

cement grout (*Civ. Eng.*). A fluid cement mixture for filling crevices.

cement gun (*Civ. Eng.*). An apparatus for spraying fine concrete or cement mortar by pneumatic pressure.

cementing materials in rocks (*Geol.*). The materials which bind any loose sediment into a coherent rock. The commonest cements are ferruginous, calcareous, and siliceous.

cementite (*Met.*). The iron carbide (Fe_3C) constituent of steel and cast-iron (particularly white cast-iron). Very hard and brittle.

cement joggle (*Build.*). A key formed between adjacent stones in parapets, etc., by running cement mortar into a square-section channel cut equally into each of the jointing faces, thereby preventing relative movement.

cement layer (*Zool.*). The outermost layer of the insect cuticle, probably lipoprotein in character.

cement mortar (*Build., Civ. Eng.*). A hydraulic mortar composed of Portland cement (or other siliceous cement) and sand.

cement rock (*Geol.*). Argillaceous limestone containing over 18% of clay.

cement-rubber-latex (*Build.*). A flooring compound composed of cement, aggregate and an elastomer to give flexibility. Also **fleximer.**

cenchri (*Zool.*). Raised bosses on the metanotum which engage with a scaly area on the underside of the forewings to keep them in place when at rest, found in symphytan *Hymenoptera.*

Cenomanian (*Geol.*). The lowest stage of the Upper Cretaceous rocks in western Europe.

censer mechanism (*Bot.*). A means of seed liberation in which the seeds are shaken out of the fruits as the stem of the plant sways in the wind.

censor, censorship (*Psychol.*). A powerful unconscious inhibitive mechanism in the mind, which prevents anything painful to the conscious aims of the individual from emerging into consciousness. It is responsible for the distortion, displacement, and condensation present in dreams. See also **superego.**

censor key (or **switch**) (*TV*). One used in sound or television broadcasting to cut off a channel or change to another channel. Also **cut key.**

cent (*Nuc. Eng.*). Unit of reactivity equal to one-hundredth of a *dollar* (q.v.).

centering (*Civ. Eng.*). See **centring.**

centi-. Prefix meaning *one-hundredth.*

Centigrade heat unit (*Heat*). See **pound-calorie.**

Centigrade scale (*Heat*). The most widely used method of graduating a thermometer. The fundamental interval of temperature between the freezing and boiling points of pure water at normal pressure is divided into 100 equal parts, each of which is a *Centigrade degree*, and the freezing point is made the zero of the scale. To convert a temperature on this scale to the Fahrenheit scale, multiply by 1·8 and add 32; for the Kelvin equivalent add 273·15. See also **Celsius scale.**

centimetre-gramme-second unit. See **CGS unit.**

centipedes (*Zool.*). See **Chilopoda.**

centipoise (*Phys.*). CGS unit of viscosity, being the viscosity of water at 20°C. One cP = 0·01 *poise* (q.v.) = 10^{-3} N s m^2.

centival (*Chem.*). Unit once used in expressing the numeral content of liquids, equal to the gramme equivalent/100 dm^3.

centonate (*Bot.*). Looking like patchwork, because of blotches of diverse colours.

central angle (*Maths.*). An angle bounded by two radii of a circle.

central battery (*Telecomm.*). One large battery provided for power supply for telephone and telegraph circuits, instead of one small supply (e.g., wet batteries on subscribers' premises) for each circuit. Also called **common battery.**

central body (*Bot.*). See **coenocentrum.** (*Cyt.*) See **centrosome.** (*Zool.*) In Insects, the median commissure connecting the protocerebral lobes of the brain.

central capsule (*Zool.*). In *Radiolaria*, a pseudo-chitinous structure of varying shape, enclosing the nucleus and some cytoplasm with oil bubbles.

central cell (*Bot.*). The cell at the base of the archegonium in the *Gymnospermae*, containing the egg and the ventral canal cell. A cell formed in the development of the archegonium of *Bryophyta* and *Pteridophyta*, ultimately forming the egg cell, and canal cells.

central cylinder (*Bot.*). See **stele.**

central dogma (*Biochem.*). Postulate, formulated by Crick in 1958, that genetic information is encoded in the nucleic acids, in the form of a linear sequence of bases, and that these codons ultimately specify the sequence of amino acids in the corresponding proteins. A consequence of this postulate is that genetic information flows in one direction only, i.e., from codon to amino acid.

centrale (*Zool.*). A centrally situated bone of the basipodium. *pl.* **centralia.** See also **navicular bone.**

Centrales (*Bot.*). An order of the diatoms, the valves of which are radially symmetrical with the ornamentation concentric. There is never a raphe or pseudoraphe. Statospores or microspores may be formed, but auxospores are not.

central limit theorem (*Maths.*). The distribution of the sum of n independent variables, of finite variances, tends to normality as n tends to infinity. There are various other forms of this theorem.

central lubricating system (*Eng.*). (1) An oil or grease lubricating system in which the lubricant is distributed by a single pump via pipes to a number of outlets, one actuation of the pump providing a discharge at every outlet. (2) A coolant distribution system in which a battery of machine tools is supplied with cleaned coolant by a single pump/filter unit.

central nervous system (*Zool.*). The main ganglia of the nervous system with their associated nerve cords, consisting usually of a brain or cerebral ganglia and a dorsal or ventral nerve cord which may be double, together with associated ganglia. Abbrev. **C.N.S.**

central processing unit (*Comp.*). The 'heart' of a computer comprising the arithmetic unit, associated special registers and main fast storage, together with circuits for controlling the execution of instructions.

central projection (*Maths.*). One in which a geometrical configuration in one plane P_1 is projected on another plane P_2 by drawing straight lines from a point O (not in P_1 or P_2) through the various points in the configuration, so that these lines intersect P_2.

central quadric (*Maths.*). A quadric surface whose principal axes coincide with the axes of co-ordinates.

central stimulants (*Med.*). Drugs having *sympathomimetic* action, e.g. amphetamine sulphate. These induce various sensations, from euphoria to anorexia, the latter contributing to loss of body weight.

centre (*An. Behav.*). A locus in the nervous system consisting of a number of cell bodies and synapses or a functionally coordinated group of neural structures, whose excitation produces discrete motor or autonomic behaviour patterns, or sensory impressions, and whose destruction is followed by their disappearance or modification. (*Civ. Eng.*) A timber frame built as a temporary support during the construction of an arch or dome. (*Eng.*) A conical support for workpieces, used in lathes and some other machine tools. Live centres, for headstocks, are usually left soft and machined to 60° included angle before use in the lathe. Dead centres, also normally at 60°, are heat-treated or tipped. Antifriction centres are also used. (*Maths.*) Of a conic: the point through which all diameters pass. (*Surv.*) Accurate vertical alignment of centre of rotation of survey instrument over, or under, a fixed point or station.

centre adjustment (*Surv.*). On tripod-mounted survey instrument, device which allows instrument to be moved horizontally till exactly above (or underground, perhaps beneath) survey signal point.

centre arbor (*Horol.*). The arbor in the train of a watch or clock which is planted in the centre of the plates. Usually, this arbor makes one turn per hour.

centre-bit (*Tools*). A wood-boring tool having a projecting central point and two side wings, one of which scribes the boundary of the hole to be cut, while the other removes the material.

centre-contact cap (*Elec. Eng.*). A bayonet cap, fitted to an electric lamp, in which the outer wall forms one of the contacts, the other being a central projection.

centre-contact holder (*Elec. Eng.*). A lamp holder, with a centre spring contact, designed for receiving lamps with centre-contact caps.

centre drill (*Eng.*). A small drill used for drilling the holes in the end of a bar to be mounted between centres in a machine tool; shaped so as to produce a countersunk hole.

centre frequency (*Telecomm.*). (1) Geometric or arithmetic midpoint of the cut-off frequencies of a wave filter. (2) Carrier frequency, when modulated symmetrically.

centre holes (*Comp.*). Line of small holes in centre of punched tape for feeding by star wheel, punched holes for code being in line across tape. Also called **feed holes.**

centre keelson (*Ships*). See under **keelson.**

centre lathe (*Eng.*). A machine tool used for turning, in which the workpiece may be supported between a revolving live centre mounted in the headstock and a stationary dead centre mounted in the tailstock.

centreless grinding (*Eng.*). A method of grinding cylindrical objects. The work is supported on a rest, between a pair of abrasive wheels revolving at different speeds in opposite directions, instead of between centres as in normal practice.

centre margin ring (*Print.*). The bevelled strip round the circumference of the plate cylinder of a letterpress rotary press against which the plates are locked.

centre nailing (*Build.*). A method of nailing slates on a roof; the nail is driven in any one slate just above the line of the head of the slate in the course below. Cf. *head nailing*.

centre notes (*Typog.*). Notes placed between columns, as in bibles, to protect the small type and to economize in space.

centre of action (*Meteor.*). A position occupied, more or less permanently, by an anticyclone or a depression, which largely determines the weather conditions over a wide area. The climate of Europe is dependent on the Siberian anticyclone and the Icelandic depression.

centre of a lens (*Light*). A point on the principal axis of a lens, through which passes any ray whose incident and emergent directions are parallel. Measurements of object and image distance, focal length, etc. are taken from this point.

centre of anallatism (*Surv.*). In a distance-measuring telescope, the point from which the distance to an observed staff is proportional to the staff intercept as seen between the upper and lower stadia lines of the diaphragm.

centre of buoyancy (*Hyd.*). The centroid of the immersed portion of the floating body. In ship construction, a ship being symmetrical about a fore and aft vertical plane, the centre of buoyancy lies in that plane. In vertical position it is calculated from a succession of waterplane areas and its longitudinal position from a succession of transverse section areas.

centre of curvature (*Maths.*). See **curvature.**

centre of flotation (*Hyd.*). The centroid of the waterplane area of a floating body.

centre of gravity, centre of mass (*Maths., Phys.*). The point in an assembly of mass particles which may be regarded as having the entire mass of the assembly concentrated there and the resultant of external forces acting there for considerations not concerning the rotation of the body. It is defined vectorially by

$$\bar{r} = \frac{\sum mr}{\sum m},$$

where m is the mass of a particle at position r, and the summation extends over the whole body. Known also as **centroid.**

centre of inversion (*Maths.*). See **inversion.**

centre of oscillation (*Phys.*). A point in a compound pendulum which, when the pendulum is at rest, is vertically below the point of suspension at a distance equal to the length of the equivalent simple pendulum (that is, the simple pendulum having the same period). If the pendulum is suspended at the centre of oscillation, its period is the same as before. Also called **centre of percussion.**

centre of pressure (*Aero.*). The point at which the resultant of the aerodynamic forces (lift and drag) intersects the chord line of the aerofoil. Its distance behind the leading edge is usually given as a percentage of the chord length. Abbrev. **c.p.** (*Hyd., Phys.*) That point in a surface immersed in a fluid at which the resultant pressure over the immersed area may be taken to act.

centre of symmetry (*Crystal.*). A point within a crystal such that all straight lines that can be drawn through it pass through a pair of similar points, lying on opposite sides of the centre of symmetry and at the same distance from it. Thus, faces and edges of the crystal occur in parallel pairs, on opposite sides of a centre of symmetry. (*Maths.*) See **symmetry.**

centrepin (*Tools*). The centre part of press tooling for ring compacts, which forms the bore.

centre pinion (*Horol.*). The first pinion in a watch or clock train, driven by the great wheel.

centre-point steering (*Autos.*). The relative positioning of the steered wheels and the swivel pins so as to obtain coincidence between the point of intersection of the swivel pin axis with the road and the plane of the wheel.

centre-pole suspension (*Elec. Eng.*). That method of supporting the overhead contact wire of a tramway system in which the supporting poles are placed between the two tracks.

centre punch (*Eng.*). A punch with a conical point, used to mark or 'dot' the centres of holes to be drilled, etc.

centre section (*Aero.*). The central portion of a wing, to which the main planes are attached; in large aeroplanes it is often built into the fuselage and may incorporate the engine nacelles and the main *landing gear*.

centre-slot system (*Elec. Eng.*). A name sometimes given to the conduit system of electric traction, in which the slot of the conduit is placed centrally between the running rails.

centre spread (*Typog.*). A design occupying the area of two pages in the centre opening of a booklet or journal.

centre square (*Eng.*). A device for marking the centres of circular objects and bars. The bar is placed in the angle of the square, which is bisected by a blade that serves as a guide for scribing a diametral line.

centre stitching (*Textiles*). A method of fine warp stitching employed when combining face and backing cloths in double weaves, when these are of different material or colour.

centre wheel (*Horol.*). The wheel mounted on the arbor of the centre pinion. It usually makes one turn per hour, so that any calculations relating to the train or number of vibrations made by the balance are taken from this wheel.

centre-zero instrument (*Elec. Eng.*). An indicating instrument which has the zero at the centre of the scale and can therefore read both positive and negative values of the quantity indicated.

centric diatom (*Bot.*). A diatom in which the valves are built on a radial plan; these diatoms are mostly marine.

centric leaves (*Bot.*). Cylindrical leaves, with the palisade tissue arranged uniformly around the periphery of the cylinder, e.g., Onion.

centric oösphere (*Bot.*). A fungal oösphere in which one or two layers of small globules of oil completely surround the central protoplasm.

centrifugal (*Zool.*). See **efferent**.

centrifugal brake (*Eng.*). An automatic brake used on cranes, etc., in which excessive speed of the rope drum is checked by revolving brake shoes which are forced outwards into contact with a fixed brake drum by centrifugal force.

centrifugal casting (*Foundry, Met.*). The casting of large pipes, cylinder liners, etc., in a rotating mould of sand-lined or water-cooled steel.

centrifugal clutch (*Eng.*). A type of clutch in which the friction surfaces are engaged automatically at a definite speed of the driving member, and thereafter maintained in contact, by the centrifugal force exerted by weighted levers.

centrifugal compressor (*Eng.*). One which passes entering air through a series of low-pressure, high-volume fans via increasingly restricted chambers so that the emerging air attains the required higher pressure; also, small booster fans (superchargers).

centrifugal fan (*Eng.*). A fan with an impeller of paddle-wheel form, in which the air enters axially at the centre and is discharged radially by centrifugal force. Also called **paddle-wheel fan**.

centrifugal-flow compressor (*Aero.*). A compressor in which a vaned rotor inspires air near the axis and throws it toward a peripheral diffuser, the pressure rising mainly through centrifugal forces. The maximum pressure ratio for a single stage is about four; more than one stage is unusual. Centrifugal compressors are universal for aero-engine superchargers and were widely used in the earlier turbojets.

centrifugal force, centripetal force (*Mech.*). A body constrained to move along a curved path reacts against the constraint with a force directed away from the centre of curvature of its path. This force is called the *centrifugal force*. It is equal and opposite to the force directed towards the centre of curvature which is deviating the body from a straight path. This is the *centripetal force*. They are both equal to the product of the mass of the body and its *centripetal acceleration* (q.v.).

centrifugal inflorescence (*Bot.*). See **cyme**.

centrifugal pulp cleaner (*Paper*). A cleaning device for pulp stock which produces a high-speed rotation or rotary vortex motion causing dirt particles to separate out.

centrifugal pump (*Min. Proc.*). Continuously acting pump with rotating impeller, used to accelerate water or suspensions through a fixed casing to peripheral discharge to the desired delivery height.

centrifugal starter (*Elec. Eng.*). A device used with small induction motors; it consists of a centrifugally operated switch on the rotor, which automatically cuts out starting-resistance or performs some other operation as the motor runs up to speed.

centrifugal swing (*An. Behav.*). A force possibly responsible for the behaviour of some animals (ants and rats) in mazes in following a turn in one direction with a turn in the other direction. See also **correcting behaviour**.

centrifugal tension (*Eng.*). The force per unit area of cross-section induced, in consequence of centrifugal force, in the material of a rotating rim, loop, or driving belt.

centrifugal thickening (*Bot.*). The deposition of layers of wall material on the outside of a cell wall, a process possible only when the cell lies free from its neighbours; the sculpturing on walls of pollen grains is formed by centrifugal thickening.

centrifugal xylem (*Bot.*). Xylem in which differentiation proceeds in succession towards the periphery of the stem or root.

centrifuge (*Eng.*). Apparatus rotating at very high speed, designed to separate solids from liquids, or liquids from other liquids dispersed therein. It is essential that there should be a difference in the relative densities of the substances to be separated. Examples of industrial uses: cream from milk, clarifying of lacquers. Also at low speeds, e.g., hydrocyclones, cool centrifuges.

centring or centering (*Civ. Eng.*). The general term applied to centres used in constructional work. (*Eng.*) (1) The marking of the centres of holes to be drilled in a piece of metal. (2) The adjusting of work in a lathe so that its axis coincides with the lathe axis.

centriole (*Cyt.*). A central minute cylindrical rod, sometimes paired within the centrosome. Also called **attraction particle**.

centripetal (*Zool.*). See **afferent**.

centripetal acceleration (*Mech.*). The acceleration, directed towards the centre of curvature of the path, which is possessed by a body moving along a curved path with constant speed. Its value is v^2/R, where v is the speed and R the radius of curvature.

centripetal force (*Mech.*). See under **centrifugal force**.

centripetal inflorescence (*Bot.*). See **raceme**.

centripetal thickening (*Bot.*). The deposition of layers of wall material on the inner side of the wall of a cell—the common process in a developing cell.

centripetal xylem (*Bot.*). Xylem in which differentiation proceeds in succession towards the centre of the axis.

centroacinar (*Cyt.*). Centre of an acinus or alveolus, as in the exocrine pancreas.

centroclinal dip (*Geol.*). A structure in which the rocks dip from all sides towards a central point, giving a basin-like arrangement.

centrodesmose (*Cyt.*). A delicate thread of stainable material connecting the centrosomes at the time of nuclear division.

centrodesmus (*Cyt.*). See **attraction spindle.**

centrodorsal (*Zool.*). In stalkless *Crinoidea*, an ossicle situated in the middle of the aboral side and representing the stump of the stalk.

centroid (*Maths.*). Of a surface: the centre of gravity of the body obtained by coating the surface with a uniform, infinitely thin, layer of matter.

centrolecithal (*Zool.*). Having the yolk in the centre.

centromere (*Cyt.*). The spindle-attachment region of the chromosome, controlling chromosome activity during cell division.

centron (*Zool.*). See **neuron.**

centroplasm (*Bot.*). A mass of plasm of obscure nature lying in the centre of a cell of one of the *Myxophyceae*.

centrosome (*Cyt.*). A minute protoplasmic cell structure near the nucleus and dividing with it during cell division. Centrosomes are of widespread occurrence in the cells of the animal, but they appear to be confined to the Thallophyta among plants. Also called **astrocentre, central body.**

Centrospermae (*Bot.*). An order of the *Archichlamydeae*. The perianth has 3 to 12 segments, which may be divided into sepals and petals. There are as many anthers as perianth segments typically. The gynaeceum is usually superior, containing one campylotropous ovule. Generally the embryo is curved or coiled.

centrosphere (*Cyt.*). See **attraction sphere.**

centrotheca (*Zool.*). See **idiozome.**

centrum (*Bot.*). A group of asci and nutritive cells associated with them, occurring in the perithecia of some *Pyrenomycetes*. (*Zool.*) The basal portion of a vertebra which partially or entirely replaces the notochord, and from which arise the neural and haemal arches, transverse processes, etc.

cepaceous (*Bot.*). Smelling or tasting like onion or garlic.

cephaeline (*Chem.*). $C_{28}H_{38}O_4N_2$, an alkaloid of unknown constitution, occurring in the roots of the Ipecacuanha species; colourless needles, m.p. 115°–116°C, soluble in ethanol or trichloromethane, insoluble in water. It resembles emetine in its physiological action, but it is more toxic.

cephal-, cephalo-. Prefix from Gk. *kephalē*, head.

cephalad (*Zool.*). Situated near, facing towards, or passing to, the head region.

Cephalaspida (*Zool.*). See **Osteostraci.**

cephalic (*Zool.*). Pertaining to, or situated on or in, the head region.

cephalic index (*Anat.*). Ratio of maximum breadth of skull divided by maximum length, multiplied by 100.

cephalin (*Chem.*). A phosphatide found in the brain substance, miscible with water, from which it can be precipitated by propanone. Its constitution is still uncertain. On hydrolysis, it yields glycerine, fatty acids, phosphoric acid, aminoethanol, and a nitrogenous base.

cephalization (*Zool.*). The specialization of the anterior end of a bilaterally symmetrical animal as the site of the mouth, the principal sense-organs, and the principal ganglia of the central nervous system: the formation of a head.

cephalobrachial (*Cyt.*). Said of a chromosome which bears a small, rounded extension at one end.

cephalocele (*Med.*). Protrusion of the membranes of the brain, with or without the substance of the brain, through a hole in the skull.

Cephalochorda (*Zool.*). A subphylum of *Chordata* having a persistent notochord, metameric muscles and gonads, a pharynx having a very large number of gill-slits and surrounded by an atrial cavity, and lacking paired fins, jaws, brain, and skeletal structures of bone or cartilage; there is an asymmetrical larval stage; marine sand-living forms. Lancelets.

cephalodium (*Bot.*). (1) An irregular outgrowth from the thallus of a lichen, usually containing blue-green algae. (2) A group of algal cells different from those characteristic of the lichen, and lying as an intrusion within the thallus.

cephalometry (*Med.*). The radiological measurement of the dimensions of the foetal head in utero.

cephalopharyngeal skeleton (*Zool.*). A characteristic framework of articulated sclerites in the head of cyclorrhaphan *Diptera*.

Cephalopoda (*Zool.*). A class of bilaterally symmetrical marine *Mollusca* in which the anterior part of the foot is modified into arms or tentacles, while the posterior part forms a funnel leading out from the mantle cavity, the mantle is undivided, and the shell is a single internal plate, or an external spiral structure, or absent. Squids, Octopods, and Pearly Nautilus.

cephaloridine (*Pharm.*). Cephalosporin C with the acetoxy group replaced by a pyridinium ion. Antibiotic of wide application and low toxicity.

cephalosporins (*Pharm.*). Group of antibiotics derived from the fungus *Cephalosporium*, in which the side-chains R-CO and R_2 are variable. Cephalosporin C, having one side-chain derived from α-aminoadipic acid and the other (R_2) forming an acetoxy group, is the basis of the clinical antibiotics *cephaloridine* and *cephalothin* (qq.v.).

cephalothin (*Pharm.*). Cephalosporin C having the α-aminoadipic acid side-chain replaced by a thiophene acetyl group. Antibiotic of wide application, especially effective against staphylococci. TN (U.S.) **Keflin.**

cephalothorax (*Zool.*). In some *Crustacea*, a region of the body formed by the fusion of the head and thorax: homologous with the prosoma of the *Arachnida*.

Cepheid parallax (*Astron.*). See **period-luminosity law.**

Cepheid variables (*Astron.*). A class of variable star of short period, whose light curve is of a certain well-defined form, and of which the star δ Cephei is the prototype.

Ceporin (*Pharm.*). TN for **cephaloridine.**

ceraceous (*Bot.*). Resembling crude beeswax in appearance or in colour.

Ceramiales (*Bot.*). An order of the *Rhodophyceae*, characterized by the production of the auxiliary cell subsequent to fertilization.

ceramic capacitor (*Elec. Eng., Electronics*). One using a high permittivity dielectric such as barium titanate to provide a high capitance/unit volume.

ceramic filter (*Chem.*). A deep filter, usually of ceramic, with fine pores in which small particles or bacteria become trapped. Also called **Pasteur filter.**

ceramic insulator (*Elec. Eng.*). An insulator made of a ceramic material, e.g., porcelain; generally used for outdoor installations.

ceramic reactor (*Nuc. Eng.*). One with ceramic

(usually uranium oxide) fuel elements suitable for use at high temperatures.

ceramics (*Chem.*). The art and science of non-organic nonmetallic materials. The term covers the purification of raw materials, the study and production of chemical compounds concerned, their formation into components, the study of structure, constitution, and properties. See **alumina, carbides,** etc.

ceramide (*Chem.*). The N-acylsphingosine portion of cerebrosides.

ceranoid (*Bot.*). Bearing branches shaped like horns.

cerargyrite (*Min.*). A group name for the silver halides of which *chlorargyrite* (q.v.) is the commonest. Also called **hornsilver.**

cerasin (*Bot.*). The insoluble constituent of some gums produced by trees, e.g., Cherry.

cerat- See also **kerat-.**

cerata (*Zool.*). In nudibranch *Gastropoda*, respiratory papillae of the mantle.

ceratobranchial (*Zool.*). An element of a branchial arch lying between the epibranchial and the hypobranchial.

ceratohyal (*Zool.*). An element of the hyoid arch corresponding to the ceratobranchials of the branchial arches.

ceratotrichia (*Zool.*). Unjoined fibrous dermotrichia of horny consistency, occurring in Fish with a cartilaginous skeleton. Cf. *lepidotrichia.*

ceraunograph (*Meteor.*). Electronic instrument for recording thunderstorms.

cercal (*Zool.*). Pertaining to the tail.

cercaria (*Zool.*). The final larval stage of *Trematoda* which develops directly into the adult; usually characterized by the possession of a round or oval body, bearing eye-spots and a sucker, and a propelling tail.

cercariaeum (*Zool.*). A cercaria in which the tail is lacking.

cercoids (*Zool.*). A pair of small pointed structures lying above the cerci at the posterior end of dragonfly nymphs.

cercus (*Zool.*). In some *Arthropoda*, a multi-articulate appendage at the end of the abdomen.

cere (*Zool.*). In Birds, the soft skin covering the base of the upper beak. *adj.* **cerous.**

cerebellar fossa (*Zool.*). See **cerebral fossa.**

cerebellum (*Zool.*). A dorsal thickening of the hind-brain in Vertebrates. *adj.* **cerebellar.**

cerebr-, cerebro-. A prefix from *cerebrum*, brain.

cerebral (*Zool.*). Pertaining to the brain; pertaining to the *cerebrum.*

cerebral eye (*Zool.*). In *Cephalochorda*, an unpaired pigment spot on the anterior wall of the brain, probably not photosensitive.

cerebral flexure (*Zool.*). The bend which develops between the axis of the forebrain and that of the hindbrain, in adult *Craniata.*

cerebral fossa (*Zool.*). In Mammals, a concavity in the cranium corresponding with the cerebrum.

cerebral hemispheres (*Zool.*). See **cerebrum.**

cerebriform (*Bot.*). Shaped like a brain.

cerebroganglion (*Zool.*). The 'brain' or supra-oesophageal ganglia of metameric Invertebrates.

cerebrosides (*Chem.*). A group of substances, found in nervous tissue, which contain no phosphoric acid, but which on hydrolysis yield (among other substances) galactose. The two most important cerebrosides are *phrenosin* and *kerasin.*

cerebrospinal (*Zool.*). Pertaining to the brain and the spinal cord.

cerebrospinal fluid (*Physiol.*). The clear colourless fluid which bathes the surfaces of the brain and spinal cord. It is of alkaline reaction, rel. d. about 1·006–1·008, and contains a small amount of dextrose, albumins, and inorganic salts, as well as traces of globulin.

cerebrum (*Zool.*). A pair of hollow vesicles or hemispheres forming the anterior and largest part of the brain in Vertebrates.

Cereclors (*Chem.*). TN for chlorinated paraffin waxes containing 30–70% chlorine and used as plasticizer extenders in PVC compounds, as paint plasticizers and also in fire retardant finishes.

Cerenkov counter (*Nuc. Eng.*). Radiation counter which operates through the detection of Cerenkov radiation.

Cerenkov radiation (*Nuc.*). Electromagnetic radiation which arises when high-energy particles are ejected into a medium in which the radiation velocity is less than that of the particles. Occurs when the medium has a high refractive index, much greater than unity, as in water.

ceresine wax (*Elec., etc.*). Bleached or refined *ozocerite* (q.v.), obtained by heating the latter with charcoal and concentrated sulphuric acid.

ceriferous (*Bot., Zool.*). Wax-bearing, wax-producing.

cerin (*Chem.*). A triterpenoid, $C_{30}H_{50}O_2$, occurring in bottle cork in the form of needle-shaped crystals.

cerium (*Chem.*). A steel-grey metallic element, one of the 'rare earth' metals. Symbol Ce, at. no. 58, r.a.m. 140·12, rel. d. at 20°C 6·7, m.p. 623°C, electrical resistivity 0·78 microhm metres. When alloyed with iron and several rare elements, it is used as the sparking component in automatic lighters and other ignition devices. It is also a constituent (0·15%) in the aluminium base alloy ceralumin and is photosensitive in the ultraviolet region. It is also used on tracer bullets, for flashlight powders and in gas mantles. It is a getter for noble gases in vacuum apparatus. There are a number of isotopes which are fission products, ^{144}Ce, with half-life of 290 days, being a pure electron emitter of importance.

cernuous (*Bot.*). Drooping, nodding.

ceroma (*Zool.*). A synonym for **cere.**

cerous (*Zool.*). See **cere.**

certation (*Bot.*). Difference in rate of growth of pollen grains which differ in genetic constitution, resulting in competition when pollen of more than one type is placed at the same time on the same stigma.

Certificates of Airworthiness, of Compliance, of Maintenance (*Aero.*). Certifications issued or required by the *CAA* confirming that a civil aircraft is *airworthy* in every respect to fly within the limitations of at least one of six categories (*C of Airworthiness*; **C of A**), that parts of an aircraft have been overhauled, repaired or inspected, etc., to comply with airworthiness requirements (*C of Compliance*), or that an aircraft has been inspected and maintained in accordance with its maintenance schedule (*C of Maintenance*; **C of M**).

Certinal (*Photog.*). A para-amino phenol type of developer.

cerulean blue (*Paint.*). Cobaltous stannate, a light-blue pigment greener in tone than *cobalt blue*, otherwise very similar.

cerumen (*Zool.*). A dark-coloured substance formed by certain Bees (*Melipona, Trigona*), composed of wax mixed with resin or earth: in Mammals, a waxy substance secreted by the ceruminous glands.

ceruminous glands (*Zool.*). Modified sweat glands

occurring in the external auditory meatus of Mammals and producing a waxy secretion.

cerussite (*Min.*). Lead (II) carbonate, crystallizing in the orthorhombic system.

cervical ganglia (*Zool.*). Two pairs of sympathetic ganglia, anterior and posterior, situated in the neck in *Craniata*.

cervicectomy (*Surg.*). Removal of the cervix uteri.

cervicitis (*Med.*). Inflammation of the cervix uteri.

cervicum (*Zool.*). In Insects, the neck or flexible intersegmental region between the head and the pro-thorax: in higher Vertebrates, the neck or narrow flexible region between the head and the trunk. *adj.* **cervical.** Pertaining to the neck or to the cervix uteri.

cervine (*Bot.*). Dark-tawny.

cervix uteri (*Zool.*). Neck of the uterus, situated partly above and partly in the vagina.

Cesarean section (*Surg.*). See **Caesarean section.**

cesium (*Chem.*). See **caesium.**

cesspipe (*San. Eng.*). A discharge pipe for waste water, particularly from a cesspool.

cesspit (*San. Eng.*). See **cesspool.**

cesspool (*Plumb.*). A small square wooden box, lined with lead, which serves as a cistern in a parapet gutter at a point where roof water is discharged into a down pipe. (*San. Eng.*) An underground pit for the reception of sewage from houses not otherwise served in respect of means of sewage disposal. Also called a **cesspit.**

Cestoda (*Zool.*). A class of *Platyhelminthes*, all the members of which are endoparasites; there is a tough cuticle; the alimentary canal is lacking; hooks and suckers for attachment occur at what is considered to be the anterior extremity; eye-spots never occur. Tape Worms.

Cestoidea (*Zool.*). An order of *Tentaculata*, comprising a number of ribbonlike forms which have the body compressed in the stomodaeal plane, and in which tentacles are replaced by tentacular filaments. Venus' Girdles.

Cetacea (*Zool.*). An order of large aquatic carnivorous Mammals; the fore-limbs are fin-like, the hind-limbs lacking; there is a horizontally flattened tail-fin; the skin is thick with little hair, there are two inguinal mammae flanking the vulva, and the neck is very short. Whales, Dolphins, and Porpoises.

cetane (*Chem.*). Normal *hexadecane* (q.v.).

cetane number (*Chem.*). Percentage of cetane in a mixture of cetane and 1-methylnaphthalene that has the same ignition quality as a fuel oil under test; measure of the ignition value of the fuel. Equivalent for diesel fuels of the octane number for petrol.

cetrimide (*Chem.*). One of the quaternary ammonium *cationic detergents*, consisting very largely of alkylmethylammonium bromides. It is a powerful disinfectant for cleansing skin, wounds, etc. TN **Cetavlon.**

cetolith (*Ocean.*). A stony body, sometimes obtained during dredging in deep water, representing the fused otic bones of whales.

cetyl alcohol (*Chem.*). Hexadecan-1-ol. The palmitic ester is the chief constituent of *spermaceti*.

cevadine (*Chem.*). $C_{32}H_{49}O_9N$, an alkaloid isolated from sabadilla seeds. It crystallizes in rhombic prisms with two molecules of ethanol, which it loses at 130°–140°C; m.p. (anhydrous) 205°C.

Ceylon chrysolite (*Min.*). See **chrysolite.**

Ceylon ironwood (*For.*). Hardwood from tree of the genus, *Mesua ferrea*, a native of Ceylon, India, and Malaya. It is naturally durable but is of limited use.

ceylonite (*Min.*). See **pleonaste.**

Ceylon peridot (*Min.*). A yellowish-green variety of the mineral tourmaline, approaching olivine in colour; used as a semiprecious gemstone.

Ceylon ruby (*Min.*). True ruby does occur, rather rarely, in Ceylon, together with much commoner ruby-spinel. Much of the gemstone material sold under this name is spinel.

Ceylon satinwood (*For.*). See **East Indian satinwood.**

Ceylon zircon (*Min.*). True zircon occurs in Ceylon, but frequently this is not differentiated from tourmaline of the same colour. *Ceylonian* (sic) *zircon* is the jewellers' name for fire-red, yellow, yellowish green, and grey zircons.

Cf (*Chem.*). The symbol for *californium*.

C.F.R. engine (*Eng.*). A specially designed petrol engine, standardized by the Co-operative Fuel Research Committee, in which the knock-proneness or detonating tendency of volatile liquid fuels is determined under controlled conditions and specified as an *octane number*. See **detonation, octane number.**

cg limits (*Aero.*). The forward and aft positions within which the resultant weight of an aircraft and its load must lie if balance and control are to be maintained; cf. *loading and cg diagram*.

CGS unit (*Elec. Eng., Phys.*). Abbrev. for *centimetre-gramme-second* unit, a unit in a system based on the centimetre, the gramme, and the second, as *fundamental units* (q.v.). For many purposes superseded by *MKSA* and *SI units* (qq.v.).

chabazite (*Min.*). A white or colourless hydrated silicate of aluminium and calcium, crystallizing in the trigonal system (*rhombohedral habit*) and belonging to the zeolite group.

chad (*Comp.*). Disk of paper removed by punching. Hence *chad tape*, one punched for computer or process control; *chadless tape*, one partially punched, so that there is room for corresponding character printing.

chadless perforation (*Teleg.*). Method of perforating tape in which the holes are only partially perforated, the chads or cuttings remaining attached (hinged) to the tape.

Chaemosiphonales (*Bot.*). An order of the *Myxophyceae*, forming endospores regularly, and tending to form filamentous colonies.

chaeta (*Zool.*). In Invertebrates, a chitinous bristle, embedded in and secreted by an ectodermal pit.

chaeta sac (*Zool.*). The ectodermal pit which secretes a chaeta.

chaetiferous, chaetigerous, chaetophorous (*Zool.*). Bearing bristles.

Chaetognatha (*Zool.*). A phylum of hermaphrodite *Coelomata*, having the body divided into three distinct regions—head, trunk, and tail; the head bears two groups of sickle-shaped setae; small, almost transparent forms, of carnivorous habit, occur in the surface waters of the sea. Arrow Worms.

chaetoplankton (*Bot.*). Small aquatic organisms in which the power of floating is increased by the augmented surface provided by spinous outgrowths from the cells.

Chaetopoda (*Zool.*). A group of *Annelida*, the members of which are distinguished by the possession of conspicuous setae; it includes the *Polychaeta* (q.v.) and the *Oligochaeta* (q.v.).

chaetosema (*Zool.*). A pair of sensory organs replacing the ocelli in some adult *Lepidoptera*.

chaetotaxy (*Zool.*). The arrangement of bristles or chaetae.

chaff (*Agric.*). Separated grain husks; chopped hay or straw. (*Radar*) See **window.**

chaffy (*Bot.*). Covered with tiny, membranous, nongreen scales.

Chagas' disease (*Med.*). See **schizotrypanosomiasis.**

chain. Any series, members of which are related by a specific phenomenon. (*Chem.*) A series of linked atoms, generally in an organic molecule (catenation). Chains may consist of one kind of atoms only (e.g., carbon chains), or of several kinds of atoms (e.g., carbon-nitrogen chains). There are open-chain and closed-chain compounds, the latter also being called ring or cyclic compounds. (*Eng.*) A series of interconnected links forming a flexible cable, used for sustaining a tensile load. (*Surv., etc.*) An instrument used for the measurement of length. It consists of 100 pieces of straight iron or steel wire, looped together end-to-end, and fitted with brass swivel handles at both ends, the overall length being one chain. See **engineer's-, Gunter's-,** also **link.**

chain barrel (*Eng.*). A cylindrical barrel, sometimes grooved, on which surplus chain is wound, as in certain types of crane.

chain block (*Eng.*). A lifting mechanism comprising chains and sheaves in combination, such as a differential chain block, which allows very heavy loads to be hoisted by hand.

chain bond (*Build.*). The bonding together of a stone wall by a built-in chain or iron bar.

chain code (*Comp.*). A sequence of 2^n or fewer binary digits arranged so that the pattern of *n* adjacent digits locates their position uniquely. Normally produced electronically using a shift register with a half-adder providing feedback to the input.

chain conveyor (*Eng.*). Any type of conveyor in which endless chains are used to support slats, apron, pans, or buckets, as distinct from the use of a simple band. See **apron conveyor, bucket conveyor.**

chain coupling (*Eng.*). A shaft coupling allowing slight irregularities in shaft alignment and a small amount of end play, designed for easy disconnection of the shafts, in which the torque is transmitted from a chain sprocket keyed to the driving shaft to a similar sprocket keyed to the driven shaft, via a length of duplex roller chain wrapped round the two sprockets. (*Rail.*) The connecting chain between two coaches, intended to come into use in case of disconnexion of the coupling proper.

chain grate stoker (*Eng.*). A mechanical stoker for boilers, in which the furnace grate consists of an endless chain built up from steel links.

chain harrow (*Agric.*). An implement resembling a flexible mat, formed by a mesh of strong wrought-iron links; flexible tined harrows are similar to chain harrows, but the ends of the links form straight tines. See **harrow.**

chain instincts (*An. Behav.*). The complex chains of activity sometimes found within appetitive behaviour, with an instinctive drive behind all the separate behaviour elements, linking and coordinating them.

chain insulator (*Elec. Eng.*). See **suspension insulator.**

chain lines (*Paper*). The wide continuous lines in the watermark of a laid paper.

chain lockers (*Ships*). The subdivisions within a ship's hull for the housing of the anchor cables. They are usually divided on the centre line to form separate compartments for port and starboard cables.

chain of locks (*Hyd. Eng.*). A system of connected lock chambers, the communicating gates serving in each case as tail gates for the next higher, and as head gates for the next lower, lock-chamber.

chainomatic balance (*Chem.*). Chemical balance with chain in place of smaller weights.

chain pump (*Eng.*). A method of raising water through small lifts by means of disks attached to an endless chain which passes upwards through a tube; a chain alone may be used.

chain reaction (*Chem.*). Chemical or atomic process in which the products of the reaction assist in promoting the process itself, such as ordinary fire or combustion, or atomic fission. Hence chain reactions are characterized by an 'induction period' of comparatively slow reaction rate, followed, as the chain-promoting products accumulate, by a vastly accelerated reaction rate.

chain riveting (*Eng.*). The arrangement of rivets in a plate joint in parallel rows so that rivets in adjacent rows are in line and not staggered.

chain screen (*Glass*). A screen made up of lengths of chain hanging close together from a bar; used for protection against furnace heat whilst bars and other tools are pushed through and manipulated during working on the inside of a furnace for pot-setting or repairs.

chain survey (*Surv., etc.*). A survey in which lengths only are measured (by means of a chain) and no angular measurements are made.

chain timbers (*Build.*). Strengthening timbers used in building brick walls circular in plan.

chain tongs (*Plumb.*). A pipe grip formed by a chain hooked into a bar with a toothed projection.

chain warp (*Weaving*). A warp which, for convenience, has been converted into link formation to facilitate processing through bleaching, dyeing, sizing, etc., operations.

chain wheel (*Eng.*). A toothed disk, solid or with spokes, which meshes with a (roller) chain to transmit motion.

chain winding (*Elec. Eng.*). A winding used on a.c. machines, in which the coils are interlinked with each other like the links of a chain. Also called a **basket winding.**

chair (*Glass*). (1) The 'chair' with long arms on which the glass-maker rolls his blowpipe whilst fashioning the ware. (2) The set of men who work together in the hand process of fabrication. (*Rail.*) The cast-iron support which is spiked to the sleeper and used to secure a rail in position.

chairmaker's saw (*Tools*). See **betty.**

chair rail (*Carp.*). A wooden rail fastened around the wall of a room at such a height as to afford protection against impact and rubbing from the backs of chairs.

chalaza (*Bot.*). The basal portion of the nucellus of an ovule. (*Zool.*) One of two spirally twisted spindlelike cords of dense albumen which connect the yolk to the shell membrane in a bird's egg.

chalazion (*Vet.*). A cystic swelling of the Meibomian gland of the eyelid, seen commonly in dogs.

chalazogamy (*Bot.*). The entry of the pollen tube through the chalaza of the ovule.

chalcanthite (*Min.*). Hydrated copper sulphate, $CuSO_4{\cdot}5H_2O$, crystallizing in the triclinic system; it occurs as an alteration product of copper pyrite and other copper ores. See **blue vitriol.**

chalcedony (*Min.*). A microcrystalline variety of quartz with abundant micropores. It occurs filling cavities in lavas, and in some sedimentary rocks; flint is a variety found in the Chalk.

chalcocite (*Min.*). A greyish-black metallic sulphide of copper, Cu_2S, which crystallizes in the orthorhombic system; it occurs in veins with other copper ores, but is not very abundant. Also called **copper glance, redruthite.**

chalcophile (*Geol.*). Descriptive of elements which have a strong affinity for sulphur and are therefore more abundant in sulphide ore deposits than in other types of rock. Lead is an example.

chalcopyrite or **copper pyrite** (*Min.*). Sulphide of copper and iron, crystallizing in the tetragonal system; the commonest ore of copper, occurring in mineral veins. The crystals are brassy yellow, often showing superficial tarnish or iridescence.

chalcostibite or **wolfsbergite** (*Min.*). Sulphide of copper and antimony, occurring in the orthorhombic system.

chalcotrichite (*Min.*). A red, semitranslucent variety of cuprite, characterized by its capillary habit.

chalet (*Arch.*). A type of country house, distinguished by having a steeply pitched roof, outside balconies, galleries, and staircase; generally built of wood.

chalice (*Zool.*). A flask-shaped gland consisting of a single cell, especially numerous in the epithelia of mucous membranes.

chalicium (*Bot.*). A formation of plants growing on gravel slides.

chalk (*Geol.*). A white, fine-grained and soft limestone, consisting of finely divided calcium carbonate and minute organic remains. In North-West Europe, it forms the upper half of the Cretaceous system. (*Pharm.*) Chalk is employed in suitable mixtures as an antacid; also in the treatment of diarrhoea.

chalk gland (*Bot.*). A secreting organ, present on some leaves, around which a deposit of calcium carbonate accumulates.

chalking (*Build.*). Disintegration of a painted surface by powdering.

chalk line (*Build., Civ. Eng.*). A length of well-chalked string used to mark straight lines on work by holding it taut in position close to the work and plucking it.

Chalk Marl (*Geol.*). The lowest lithological division of the English Chalk, in which the calcareous matter is mixed with up to 30% of muddy sediment.

chalkone, chalcone (*Chem.*). 1,3-diphenylprop-2-en-1-one. Yellow crystals, m.p. 58°C.

chalk overlay (*Print.*). A *mechanical overlay* (q.v.), used for half-tone blocks, made by taking a pull on a special chalk-coated board and dissolving the chalk left exposed with a bleach.

chalones (*Biol., Physiol.*). Internal secretions of inhibitory nature, opposed to hormones.

chalonic (*Zool.*). Inhibitory, depressor; used especially of certain internal secretions (*chalones*).

chalybeate waters (*Geol.*). Natural waters containing iron salts in solution.

chalybite (*Min.*). See **siderite.**

chamaephyte (*Bot.*). A plant which, during the period of perennation, has its buds at or just above the surface of the soil.

chamber (*Chem. Eng.*). A very large box-shaped or cylindrical compartment, lined with lead, in which take place the chemical reactions occurring during the manufacture of sulphuric acid. (*Hyd. Eng.*) The space between the upper and lower gates of a canal lock.

chamber acid (*Chem.*). Sulphuric acid of the concentration and condition in which it leaves the acid chambers. It is of great value in fertilisers.

chamber crystals (*Chem.*). Unwanted crystals of nitrososulphuric acid which condense on the walls of the sulphuric acid chamber under certain conditions of manufacture.

chambered (*Bot.*). Said of an ovary which is not completely divided into separate compartments by the incomplete partitions projecting inwards from the walls.

chambered level tube (*Surv.*). A level tube fitted at one end with an air chamber. By tilting the tube, air may be added to or taken from the bubble, whose length (which tends to be shortened by rise of temperature) may thus be regulated to maintain the sensitivity of the instrument.

chambered organ (*Zool.*). In *Crinoidea*, five radial compartments with nervous walls, situated in the hollow of the centrodorsal ossicle; it forms the centre of the aboral nervous system.

chambering (*Mining*). Enlargement of drill hole by means of small explosive charge, in preparation for use of much heavier one.

chamber process (*Chem. Eng.*). A process for the manufacture of sulphuric acid, in which the reactions between the air, sulphur dioxide, and nitric acid gases necessary to produce the sulphuric acid take place in large lead chambers.

Chambers-Imrie-Sharpe curve (*Radiol.*). Graph of radiation dose against depth of penetration for human body in fall-out field.

chamfer (*Carp., Join., etc.*). The surface produced by bevelling an edge or corner.

chamfer plane (*Carp.*). A plane fitted with adjustable guides to facilitate the cutting of any desired chamfer.

chamois leather (*Leather*). The flesh split of a sheepskin, treated with cod liver oil, squeezed, and then washed and lightly bleached; used for cleaning and polishing, and in clothing. Also called **wash leather.**

chamomile, camomile (*Pharm.*). Derived from the dried flower-heads of *Anthemis nobilis*; formerly a well-known stomachic and tonic, and sometimes used in powdered form in cosmetic preparations and shampoos.

chamosite (*Min.*). A hydrous silicate of iron and aluminium occurring in oölitic and other bedded iron ores.

Champy's fixative (*Micros.*). A fixative consisting of chromic acid, potassium bichromate and osmic oxide, used particularly for *Protozoa*, and similar material.

chance (*For.*). Strictly, any unit of operation in the woods, with many and varied applications, of which the most familiar is *logging* or *cutting chance*. A logging or pulpwood operating unit.

Chance glass (*Photog.*). Deep-blue glass for lenses which are required to transmit ultraviolet rays but to cut out other rays.

chancel (*Arch.*). The eastern part of a church, originally separated from the *nave* by a screen of latticework, so as to prevent general access thereto, though not to interrupt either sight or sound.

chancery (*Typog.*). A style of italic, the original being modelled on the handwriting used in the Papal chancery; the most often seen present-day example is *Bembo italic*.

Chance's Euphon quilt, Chance's glass silk (*Acous.*). See **Euphon quilt.**

chancre (*Med.*). The hard swelling which constitutes the primary lesion in syphilis.

chancroid (*Med.*). Nonsyphilitic ulceration of the genital organs due to venereally-contracted infection.

change-coil instrument (*Elec. Eng.*). A moving-iron electric measuring instrument having two or more replaceable coils to give different ranges.

change face (*Surv.*). To rotate a theodolite telescope about its horizontal axis so as to change from 'face left' to 'face right', or vice versa. See **transit.**

change of state (*Chem.*). A change from solid to liquid, solid to gas, liquid to gas, or vice versa.

change-over (*Cinema.*). The transference of projection from one machine to another at the end of one reel and the start of the next, without an apparent break in the sequence.

change-over contact (*Teleph.*). The group of contacts in a relay assembly, so arranged that, on operation, a moving contact separates from a back contact, is free during transit, and then makes contact with a front contact. See **make-before-break contact.**

change-over switch (*Elec. Eng.*). A switch for changing a circuit from one system of connexions to another.

change point (*Surv.*). A staff station to which two sights are taken, the first a foresight from one set-up of the level, the second a backsight from the next set-up of the level. Sometimes called **a turning point.**

change-pole motor (*Elec. Eng.*). An induction motor with a switch for changing the connexions of the stator winding to give two alternative numbers of poles, so that the motor can run at either of two speeds.

change-speed motor (*Elec. Eng.*). A motor which can be operated at two or more approximately constant speeds, e.g., a change-pole motor.

change wheels (*Eng.*). The gear-wheels through which the lead screw of a screw-cutting lathe is driven from the mandril, the reduction ratio being varied by changing the wheels. See **screw-cutting lathe.**

changing bag (*Photog.*). A light-tight bag to accommodate a camera and sensitive photographic material, so that the former can be loaded or unloaded with the latter in daylight.

channel (*Comp.*). One or more similar highways for transmission of data. (*Eng.*) A standard form of rolled-steel section, consisting of 3 sides at right-angles, in channel form. See **rolled-steel sections.** (*Nuc. Eng.*) Passage through reactor core for coolant, fuel rod or control rod. (*Telecomm.*) (1) Range of frequencies occupied by a modulated transmission. With double sideband amplitude modulation, its width in Hz is twice the highest modulating frequency. A *clear channel* is one occupied by a single transmission, free from interference from other transmissions. (2) General term for a unique transmission path, as for programme, vision, recording, picture, camera, shared, 2-way.

channel black (*Chem.*). See **carbon black.**

channel capacity (*Telecomm.*). The maximum possible information rate for a channel.

channel circuit (*Telecomm.*). Same as **forked circuit.**

channel effect (*Radio*). In a transistor, bypassing the base component by leakage due to surface conduction.

channel gulley (*San. Eng.*). A trapped gulley fitted with a channel, about 45 cm in length, through which sink wastes must pass before reaching the gulley proper. In the process, grease in the wastes congeals on the sides of the channel and may be removed.

channelled (*Bot.*). Elongated and hollowed somewhat like a gutter, as many leaf stalks.

channelling effect (*Nuc. Eng.*). In radiation, the greater transparency of an absorbing material with voids relative to similar homogeneous material. Expressed numerically by the ratio of the attenuation coefficients.

channel pipe (*Build.*). An open drain of half or three-quarter circular section, used in inspection or intercepting chambers.

channel width (*Telecomm.*). (1) See **pulse height selector.** (2) Frequency band allocated to service or transmission.

chantlate (*Build.*). A projecting strip of wood fixed to the rafters at the eaves and supporting the normal roof covering; it serves to carry the drip clear of the wall.

chapel (*Print.*). An association of journeymen printers (and, in certain districts, apprentices), who elect a 'father' to watch their interests.

Chaperon resistor (*Elec. Eng.*). Wirewound resistor of low residual reactance.

chaplash (*For.*). A yellow-brown wood which seasons well and is durable. It is suitable for builders' and joiners' work.

chaplet (*Met.*). Iron support for core placed moulding box in iron foundry.

Chapman equation (*Chem.*). Expression of the viscosity of a gas in terms of the rel. mol. mass, the collision diameter, the mean free path, the average speed and the absolute temperature.

chapters (*Horol.*). The Roman figures used on the dials of watches and clocks to mark the hours.

character (*Biol.*). (1) Any well-marked feature which helps to distinguish one species from another. (2) In genetics, any one readily defined feature which is transmitted from parent to offspring. (*Comp.*) One of a unique set of letters and figures representing data for processing. *Ignore*, *marker* (*sentinel* in U.S.), *key*, *tag*, *sign*, are special characters.

characteristic (*Electronics*). Graph illustrating performance of a particular device, taken under specified conditions, e.g., for a valve, anode current against variable grid voltage, the anode voltage remaining constant.

characteristic curve (*Maths.*). Given a 1-parameter family of surfaces, $\psi(x, y, z, a)=0$, the limiting position of the curve of intersection of two adjacent surfaces of the family is given by

$$\begin{cases} \psi(x, y, z, a)=0 \\ \dfrac{\partial\psi}{\partial a}(x, y, z, a)=0, \text{ for a given value of } a. \end{cases}$$

This curve is a characteristic curve of the family of surfaces. The locus of all the characteristic curves is the envelope of the family. (*Photog.*) Graph plotting the relationship between the density of a developed emulsion and the logarithm of the exposure. Also called **D Log E curve, H and D curve.**

characteristic equation of a matrix (*Maths.*). Only defined for a square matrix. The equation which results when the determinant of the matrix obtained by replacing every diagonal element a_{ii} of the given matrix by $a_{ii}-x$, is equated to zero. Its roots are called either *latent roots* or *eigenvalues*.

characteristic equation of an ordinary differential equation (*Maths.*). See **auxiliary equation.**

characteristic function (*Maths.*). If a random variable x has a frequency function $f(x)$, its characteristic function is

$$\varphi(t)=\int_{-\infty}^{\infty} e^{itx}f(x)\,dx,$$

where t is a real number.

characteristic function of a set (*Maths.*). That which is unity for points of the set, but zero outside the set.

characteristic impedance (*Elec.*). (1) The ratio of voltage to current for a travelling wave on a transmission line. Also **surge impedance.** (2) The impedance level for a filter or network. See **image impedance.**

characteristic of a logarithm (*Maths.*). See **logarithm.**

characteristic points (*Maths.*). (1) Of a 1-parameter family of curves in a plane, $\psi(x, y, a)=0$, the characteristic points or points of contact with the envelope are given by the equations

$$\psi(x, y, a)=0, \qquad \frac{\partial\psi}{\partial a}(x, y, a)=0,$$

for a particular value of the parameter a. The envelope is found by eliminating a from these two equations. (2) Of a 2-parameter family of surfaces, $\psi(x, y, z, a, b)$, the characteristic points, as above, are defined by the equations

$$\psi(x, y, z, a, b)=0,$$

$$\frac{\partial\psi}{\partial a}(x, y, z, a, b)=0,$$

$$\frac{\partial\psi}{\partial b}(x, y, z, a, b)=0,$$

for fixed a and b. The envelope is found by eliminating a and b from these three equations.

characteristic radiation (*Nuc.*). That from an atom and associated with the orbital change of an electron between particular energy levels, the wavelength also depending on the particular atom.

characteristic spectrum (*Nuc.*). Ordered arrangement of radiated (optical or X-ray) wavelengths related to the atomic structure of the material giving rise to them.

characteristic velocity (*Space*). The change of velocity and the sum of changes of velocity (accelerations and decelerations) required to execute a space manoeuvre, e.g., transfer between orbits.

characteristic X-radiation (*Nuc.*). X-radiation consisting of discrete wavelengths which are characteristic of the emitting element. If arising from the absorption of X- or γ-radiation, may be called **fluorescence X-radiation.**

character reading (or recognition) (*Comp.*). Peripheral component in a computing system which recognizes printed or written figures or letters, and displays or provides codings correspondingly. Special alphabets may be required, or magnetic ink.

Charales, Charophyta (*Bot.*). A small group of aquatic plants, showing many algal characters, but of uncertain position. They have main shoots bearing many whorls of lateral branches on which the reproductive organs are developed, these being antheridia liberating motile sperms, and oögonia each with a single egg. The plants occur in fresh and in brackish water, and are often heavily calcified, a circumstance to which they owe the popular name of *stoneworts.*

charcoal (*Chem.*). The residue from the destructive distillation of wood or animal matter with exclusion of air; contains carbon and inorganic matter. (*Pharm.*) *Activated charcoal* has value as an absorbent, especially in cases of alkaloid poisoning, in masks to counter toxic gases and as a palliative for flatulent dyspepsia.

charcoal blacking (*Foundry*). *Blacking* (q.v.) made from powdered charcoal, which is dusted over the surface of a mould to improve the smoothness of the casting.

charcoal iron (*Met., etc.*). Pig-iron made in a blast-furnace using charcoal instead of coke. Sometimes also wrought-iron made from this.

charge (*Chem.*). Electrical energy stored in chemical form in secondary cell. (*Elec.*) Quantity of unbalanced electricity in a body, i.e., excess or deficiency of electrons, giving the body negative or positive electrification, respectively. See **negative and positive.** That of an ion, one or more times that of an electron, of either sign. (*Glass*) (1) Synonym of *batch* (q.v.). (2) The quantity of batch required to fill a *pot* (q.v.). (*Heat*) Load connected to RF heater. (*Met.*) Amounts (ore, fuel, and flux) charged into furnace at one loading, or proportions of these. (*Mining*) Operation of placing explosive, primer, detonator and fuse in drill hole. (*Nuc. Eng.*) Fuel material in nuclear reactor.

charge coupled devices (*Comp.*). Semiconductor logic circuits employing sandwich construction of metal, silicon dioxide, and silicon.

charged (*Elec.*). Said of a capacitor when a working potential difference is applied to its electrodes. Said of a secondary cell or battery (accumulator) when it stores the maximum (rated) energy in chemical form, after passing the necessary ampere-hours of charge through it. Said of a conductor when it is held at an operating potential, e.g., traction conductors or rails, or mains generally.

charged-particle equilibrium (*Nuc.*). See **electronic equilibrium.**

charge exchange (*Phys.*). Exchange of charge between neutral atom and ion in a plasma. After its charge is neutralized, high-energy ions can normally escape from plasma—hence this process reduces plasma temperature.

charge independent (*Nuc.*). Said of nuclear forces between particles, the magnitude and sign of which does not depend on whether the particles are charged. See **nuclear force and short-range forces.**

charge indicator (*Elec. Eng.*). See **potential indicator.**

charge-mass ratio (*Phys.*). Ratio of electric charge to mass of particle, of great importance in physics of all particles and ions.

charge storage (*TV*). Principle whereby image information is stored on a mosaic of photocells or piezoelectric crystals, especially in camera tubes for television.

charging current (*Elec. Eng.*). (1) Current passing through an accumulator during the conversion of electrical energy into stored chemical energy. (2) Impulse of current passing into a capacitor when a steady voltage is suddenly applied, the actual current being limited by the total resistance of the circuit. (3) Alternating current which flows through a capacitor when an alternating voltage is applied.

charging resistor (*Elec. Eng.*). A resistance inserted in series with a switch to limit the rate of rise of current when making the circuit.

charging voltage (*Elec. Eng.*). E.m.f. required to pass the correct charging current through an accumulator, about 2·5 V for each lead-acid cell.

Charles's law (*Heat*). 'The volume of a given mass of any gas, kept at constant pressure, increases by $\frac{1}{273}$ of the volume at 0°C for each degree rise of temperature.' This law, also known as *Gay-Lussac's law,* is equivalent to saying that all gases have the same value for the coefficient of expansion at constant pressure. The law is not exactly followed.

Charlton white (*Paint.*). See **lithopone.**

charnockite (*Geol.*). A coarse-grained, dark granitic rock, consisting of feldspars, blue quartz, and orthopyroxene; it occurs typically in Madras and is named after the founder of Calcutta, Job Charnock.

Charnoid direction (*Geol.*). A NW-SE direction of folding in the rocks of central and eastern England. Exemplified by the Pre-Cambrian rocks of Charnwood Forest.

Charpy test (*Met.*). A notched-bar impact test in which a notched specimen, fixed at both ends, is struck behind the notch by a striker carried on a pendulum. The energy absorbed in fracture is measured by the height to which the pendulum rises.

chartaceous (*Bot.*). Papery in texture.

chart comparison unit (*Radar*). Type of display superimposed upon navigational chart.

Chartered Engineer, abbrev. **C.Eng.** A style or title (U.K.) which may be used by an engineer enrolled on the register of a constituent member Institution of the Council of Engineering Institutions in such class of membership as that Council shall from time to time determine. The member Institutions comprise the Royal Aeronautical Society, the Royal Institution of Naval Architects, the Institute of Marine Engineers, and the Institutions of Chemical Engineers, Civil Engineers, Electrical Engineers, Electronic and Radio Engineers, Gas Engineers, Mechanical Engineers, Mining Engineers, Mining and Metallurgy, Municipal Engineers, Production Engineers, and Structural Engineers. A typical designation would be John Smith, B.Sc., C.Eng., F.I.Mech.E.

Chartered Surveyor (*Build., Civ. Eng.*). A style or title which may be used by a fellow or professional associate of the Chartered Surveyors' Institution.

Chart-Pak (*Typog.*). Rules and border patterns on self-adhesive material.

chase (*Build.*). (1) A trench dug to accommodate a drain-pipe. (2) A groove chiselled in the face of a wall to receive pipes, etc. (*Typog.*) An iron or steel frame into which type is locked by means of wooden or mechanical quoins.

chase-mortising (*Carp., Join.*). A method adopted to frame a timber in between two others already fixed, a sloping chase being cut to the bottom of the mortise so that the cross-piece may be got into position.

chaser (*Eng.*). A cutter of which the edge is serrated to the profile of a screw thread, used for producing or accurately finishing screw threads. Single chasers may be mounted in lathe tool-posts; sets are used in die-heads, collapsible taps, and other screw-cutting tools.

chasmocleistogamous (*Bot.*). Producing both chasmogamous and cleistogamous flowers.

chasmogamous (*Bot.*). Having large conspicuous flowers which open and are pollinated by the wind or by insects.

chasmophyte (*Bot.*). A plant inhabiting rocky places, and rooting in a crevice containing a mixture of inorganic ions and organic compounds.

chassis. Generally, any major part or framework of an assembly, to which other parts are attached. (*Aero.*). The landing gear of an aircraft. (*Autos.*) The main frame of a vehicle (as distinct from the body) to which are attached the engine, steering and transmission gear, wheels, etc. Vehicles in which the steel body, suitably reinforced, itself fulfils this function are sometimes called **chassisless, monocoque** or **unitized.** (*Radio, etc.*) The mounting for circuit components of an electronic or electric circuit.

Chastek paralysis (*Vet.*). Paralytic disease of silver foxes. A disease of foxes, mink, and ferrets characterized by paralysis, convulsions and death, due to thiamin (*vitamin* B_1) deficiency. The deficiency is caused by feeding raw fish, parts of which contain an anti-thiamin factor.

chatoyancy (*Min.*). The characteristic optical effect shown by cat's eye and certain other minerals, due to the reflection of light from minute aligned tubular channels, fibres, or colloidal particles, perhaps 25,000 to the square cm. When cut *en cabochon* such stones exhibit a narrow silvery band of light which changes its position as the gem is turned.

chatter (*Eng.*). Vibration of a cutting tool or of the work in a machine; caused by insufficient rigidity of either, and results in noise and uneven finish. (*Telecomm.*) Rapid closing and opening of contacts on a relay, which reduces their life.

Chatterton's compound (*Chem.*). An adhesive insulating substance consisting largely of gutta-percha; used as a cement or filling, especially in cable jointing.

chaulmoogra (or **hydnocarpus**) **oil** (*Chem.*). Obtained from seeds of certain tropical species of *Flacourtiaceae Warburg*. The two main constituents are optically-active (D) chaulmoogric acid:

$$\begin{array}{l} H_2C-CH-(CH_2)_{12}-COOH \\ H_2C \quad | \\ \quad HC=CH \end{array}$$

and hydnocarpic acid. It has long been used in leprosy treatment. See **dapsone.**

Chebyshev array (*Telecomm.*). A linear array in which the current amplitudes are proportional to the coefficients of a Chebyshev polynomial. Characterized by a uniform side-lobe level.

Chebyshev filter (*Telecomm.*). Third-order constant-*k* filter of which insertion loss is characterized by a Chebyshev polynomial of the sixth order. These show some variation of the residual attenuation in the nominal pass band; but have a more rapid increase of attenuation than equivalent *Butterworth filters* (q.v.) outside the pass band.

Chebyshev inequality (*Maths.*). If $a_r \geqq a_{r+1}$ and $b_r \geqq b_{r+1}$ for all r, then

$$\left(\sum_1^n a_r\right)\cdot\left(\sum_1^n b_r\right) < n\sum_1^n a_r b_r \, .$$

Chebyshev polynomials (*Maths.*). The polynomials $T_n(x)$ in the expansion

$$\frac{1-h^2}{1-2xh+h^2} = \sum_{n=0}^{\infty} T_n(x) \, . \, (2h)^n .$$

Chebyshev response (*Telecomm.*). Minimum fluctuation of response of a network from specification, obtained by a design procedure.

check (*For.*). A separation of the fibres along the grain of the wood, forming a crack or fissure, *not* extending through the piece from one surface to another.

check digit (*Comp.*). Redundant bit in stored word, used in self-checking procedures such as a parity check. If there is more than one check digit they form a **check number.**

checking (*Paint.*). A not very serious fault developed by paintwork to relieve minor stresses in the film, characterized by a network of fine cracks that does not usually go down very deep.

checking motion (*Spinning*). Motion to ease the let-in of the carriage at the end of the inward run. (*Weaving*) Multiple movable boxes at one or both sides of a loom, for holding different colours of weft which may be brought into use as desired.

check-lock (*Join.*). A device for locking in position the bolt of a door lock.

check multiple (*Teleph.*). The multiple for reversed dialling by an operator, to check the number given to her by a subscriber; characterized by the omission of guarding on engagement of line.

check nut (*Eng., etc.*). See **lock nut.**

check programme (*Comp.*). One arranged to reveal faults, or incipient faults, in operation of a computer. Also **test programme.** See **marginal testing.**

check rail (*Rail.*). A third rail laid on a curve alongside the inner rail and spaced a little from it, to safeguard rolling-stock against derailment due to excessive thrust on the outer rail. Also called **guard rail, rail guard, safeguard, safety rail, side rail.**

check receiver (*Radio, TV*). Radio or television receiver for verifying quality or content of a programme. Also called **monitoring receiver.**

check throat (*Join.*). A small groove cut in the face of a short step in the upper surface of a wooden window-sill, just behind the face of the sash. It serves to stop rain from driving up under the sash.

check valve (*Eng., etc.*). A nonreturn valve, closed automatically by fluid pressure; fitted in a pipe to prevent return flow of the fluid pumped through it. See **clack, feed check valve.**

cheddite (*Chem.*). Mixture of castor oil, ammonium chlorate (VII), and 2,4-dinitrotoluene; used as an explosive.

cheek (*Build.*). One of the sides of an opening. (*Carp.*) One of the solid parts on each side of a mortise. (*Hyd. Eng.*) The abutting surface of a mitre-sill. (*Zool.*) In *Trilobita*, the pleural portion of the head; in Mammals, the side of the face below the eye, the fleshy lateral wall of the buccal cavity; in *Diptera*, the genae or parafacials.

cheeks (*Print.*). See **knife cheeks.**

cheese (*Cables*). See **cable wax.** (*Textiles*) (1) Densely wound cylindrical yarn package made on various winding machines for warping, etc., purposes. (2) A small flanged warp beam used in worsted manufacture.

cheese-head screw (*Eng.*). A screw with a cylindrical head, similar in shape to a round cheese, slotted for a screw-driver.

cheilectropion (*Med.*). Turning outwards of the lip.

cheilitis (*Med.*). Inflammation of the lip.

cheilocystidium (*Bot.*). A cystidium formed at the edge of a gill.

cheilognathus (*Med.*). See **harelip.**

cheiropompholyx (*Med.*). A skin disease in which vesicles filled with clear fluid suddenly appear on the hands and feet.

cheiropterygium (*Zool.*). A *pentadactyl limb* (q.v.).

chela (*Zool.*). In *Arthropoda*, any chelate appendage. *adjs.* **cheliferous, cheliform.**

chelate (*Chem.*). Molecular structure in which a metallic or other ion is combined into a ring by residual valencies, i.e., unshared electrons, by an organic compound, known as a *chelating agent*. (*Zool.*) Of *Arthropoda*, having the penultimate joint of an appendage enlarged and modified so that it can be opposed to the distal joint like the blades of a pair of scissors to form a prehensile organ. Cf. *subchelate*.

chelicerae (*Zool.*). In *Arachnida*, the only pair of pre-oral appendages, which are usually chelate.

Chelonethida (*Zool.*). See **Pseudoscorpionidea.**

Chelonia (*Zool.*). An order of Reptiles in which the body is encased in a horny capsule consisting of a dorsal carapace and a ventral plastron, the jaws are provided with horny beaks in place of teeth, and the lower temporal arcade alone is present. Tortoises and Turtles.

chemautotrophy (*Bacteriol.*). The utilization, by an organism, of the energy derived from the oxidation of inorganic compounds, for the assimilation of simple materials, e.g., carbon dioxide and ammonia.

chemical affinity (*Chem.*). See **affinity.**

chemical analysis (*Chem.*). Splitting up of a material into its component parts or constituents by chemical methods to determine the composition of the material.

chemical balance (*Chem.*). An instrument used in chemistry for weighing, to a high degree of accuracy, the small amounts of material dealt with.

chemical bending (*For.*). The bending of timber to a required shape, after softening by impregnation with chemicals.

chemical binding effect (*Nuc.*). Variation of effective cross-section of nucleus for neutron absorption with type of chemical bond.

chemical bond (*Chem.*). The electric forces linking atoms in molecules or single crystals, classified according to the electron configuration of the atoms. Thus, covalent bonds have 2 electrons (each atom providing 1 electron) shared between the atoms, heteropolar bonds produce molecules with resulting dipole moment while homopolar bonds do not; ionic (or electrostatic) bonds are due to electrostatic attraction following ionization, metallic bonds are formed by electrons in lower energy levels leaving the valence electrons free to produce metallic conduction, and coordinate bonds are due to two electrons contributed from the same atom.

chemical closet (*San. Eng.*). A suitably shaped container for use in conjunction with special deodorizing and liquefying chemicals, when running water is not available.

chemical compound (*Chem.*). A substance composed of two or more elements in definite proportions by weight, which are independent of its mode of preparation. Thus the ratio of oxygen to carbon in pure carbon monoxide is the same whether the gas is obtained by the oxidation of carbon, the reduction of carbon dioxide, or by heating methanoic acid, ethandioic acid or potassium hexacyanoferrate (II) with concentrated sulphuric acid.

chemical constitution (*Chem.*). See **constitution.**

chemical elements. See **Table of Chemical Elements in Appendix.**

chemical energy (*Chem.*). The energy liberated in a chemical reaction. See **affinity.**

chemical engineering. Design, construction, and operation of plant and works in which matter undergoes change of state and composition.

chemical equation (*Chem.*). A symbolic representation of the changes occurring in a chemical reaction, based on the requirement that matter is added or removed during the reaction.

chemical fade (*Cinema.*). A fade made by washing out the negative with a chemical (e.g., cyanide), so that, on printing, the effect of a normal fade is obtained.

chemical focus (*Photog.*). The point in an emulsion at which the lens provides a focus; it may differ from the point of visual focus with a lens insufficiently corrected for chromatic aberration.

chemical fog (*Photog.*). General reduction of a proportion of the silver halide in an emulsion, due to chemical decomposition or too vigorous development; it results in reduced contrast and elimination of complete transparency.

chemical hygrometer (*Meteor.*). See **absorption hygrometer.**

chemical kinetics (*Chem.*). The study of the rates of chemical reactions.

chemical lead (*Met.*). Lead of purity exceeding 99·9%; suitable for the lining of vessels used to hold sulphuric acid and other chemicals.

chemically-formed rocks (*Geol.*). Rocks formed by precipitation of materials from solution in water, e.g., calc-tufa, and various saline deposits.

chemical precipitation (*San. Eng.*). The process of assisting settlement of the solid matters in sewage by adding chemicals (such as lime or alumino-ferric) before admitting the sewage to the sedimentation tanks.

chemical pulp (*Paper*). Pulp prepared by a chemical process, as distinguished from mechanical pulp, which is prepared by grinding.

chemical reaction (*Chem.*). A process in which one substance is changed into another.

chemical symbol (*Chem.*). A single capital letter, or a combination of a capital letter and a small one, which is used to represent either an atom or a mole of a chemical element; e.g., the symbol for sodium is Na, for sulphur is S, etc.

chemical toning (*Photog.*). The process of converting the silver image into, or replacing it by, a coloured substance by chemicals other than a dye.

chemical wood-pulp (*Paper*). Wood-pulp, obtained from wood by the sulphite, sulphate, or soda process.

chemiluminescence (*Chem.*). Luminescence arising from chemical processes, e.g., glow-worm.

chemisorption (*Chem.*). Irreversible adsorption in which the adsorbed substance is held on the surface by chemical forces.

chemistry (*Chem.*). The study of the composition of substances and of the changes of composition which they undergo. The main branches of the subject are *inorganic*, *organic*, and *physical chemistry* (qq.v.).

chemokinesis (*An. Behav.*). A *kinesis* (q.v.) occurring in response to olfactory stimuli.

chemonasty (*Bot.*). A change in the position or form of a plant organ in relation to a diffuse chemical stimulus.

chemoreceptor (*Zool.*). A sensory nerve-ending, receiving chemical stimuli.

chemoreflex (*Physiol.*). A reflex initiated by a chemical stimulus.

chemosis (*Med.*). Oedema of the conjunctiva.

chemosmosis (*Chem.*). The occurrence of chemical reactions through semipermeable membranes.

chemosphere (*Astron.*). The lower section of the Earth's atmosphere (below about 80 km) where chemical processes are of importance, as distinct from the *ionosphere*, where electrical effects predominate.

chemosterilants (*Chem.*). Substances, e.g., apholate, which inhibit the breeding processes in Insects, used for pest control.

chemosynthesis (*Bot.*). The formation of organic material by some bacteria by means of energy derived from chemical changes.

chemosynthetic autotroph (*Bacteriol.*). An organism which utilizes the oxidation of inorganic compounds as the energy source, for assimilation of simple materials.

chemotaxis, chemotactism (*An. Behav.*). A *taxis* (q.v.), usually a *tropotaxis*, occurring in response to chemical stimuli, usually detected by the olfactory sense, since the sense of taste only operates on substances in the mouth. (*Bot.*, *Zool.*) Movement of the whole organism in a definite direction in relation to a chemical stimulus.

chemotherapy (*Med.*). Treatment of disease by a chemical compound having a specific bactericidal or bacteriostatic effect against the micro-organism involved but no appreciably harmful effect on body tissues.

chemotropism (*Bot.*, *Zool.*). A growth movement or curvature of part of an organ in a definite direction owing to differential growth, caused by the concentration of some chemical substance.

chemurgy (*Chem.*). (1) Application of chemistry to agriculture. (2) Agriculture pursued for chemical purposes, e.g., potato-growing for making industrial alcohol by fermentation.

chenier (*Geog.*). Long ridges of sand formed along the shore-line of a delta and subsequently incorporated into the succession of deltaic sediments.

chenille (*Textiles*). A self- or multicoloured silk, worsted, or cotton pile fabric specially woven so that it can be accurately cut warp-wise into thin strips. These are twisted into yarns for weaving; *mock chenille* is imitation chenille yarn obtained by twisting three or more yarns on a fancy doubler.

Chenot process (*Met.*). Reduction of iron ore to sponge iron in vertical furnace by heating with coal dust.

cheque paper (*Paper*). Paper chemically treated to show any tampering with documents printed on it. See also **security paper, sensitized paper.**

chequer plate (*Eng.*). Steel plate used for flooring; provided with a raised chequer pattern to give a secure foothold.

cheralite (*Min.*). A radioactive mineral rich in thorium and uranium.

Cherenkov. See **Cerenkov.**

Chernikeef log (*Ships*). A device for measuring distance moved through the water. An impeller is lowered through the bottom of the ship to about 18 in. below the hull. The rotations of the impeller are transmitted electrically to a distance recorder and combined with time to indicate speed.

chernozem (*Geog.*). See **black earth.**

cherry (*For.*). Tree of the genus, *Prunus*, economically not regarded as a timber tree, though the wood has certain minor uses. It carves and turns well, and is used for picture frames, inlaying, tobacco pipes, and walking-sticks.

cherry-picker (*Civil Eng.*). A working platform mounted on a two-limbed hydraulically or electrically operated raising mechanism, on the back of a truck; used for access to elevated inaccessible places, e.g. street lighting fittings. (*For.*) A mobile crane for loading logs within its reach from logging roads directly on to a conveyance without yarding. (*Mining*) Small crane spanning mine tunnel used to transfer empty car to parallel track and thus allow loaded one to be withdrawn.

chert (*Geol.*). A siliceous rock consisting of cryptocrystalline silica, and sometimes including the remains of siliceous organisms such as sponges or radiolaria. It occurs as bedded masses, as well as concretions in limestone.

chessylite (*Min.*). See *azurite*.
chestnut (*For.*). A light- to dark-brown wood resembling oak; much used for fencing, posts, and rails.
chest saw (*Carp.*). A small handsaw without a back, and with 6 to 12 teeth to the inch.
cheval-vapeur (CV) or Pferdestärke (PS) (*Eng.*). The metric unit of horse-power, equivalent to 75 kg m/s, 735·5 W or 0·986 h.p. Also **cheval (ch), force de cheval.**
chevaux de frise (*Build.*). An iron or timber bar with projecting spikes, used as a defence precaution along the tops of walls and elsewhere.
Cheviot (*Textiles*). A quality term for yarns and fabrics manufactured from Cheviot or crossbred wools; also applied to the wool, demi-lustre, from Cheviot sheep.
chevron bones (*Zool.*). In some Vertebrates, V-shaped bones articulating with the ventral surfaces of the caudal vertebrae.
Cheyne-Stokes breathing (*Med.*). Alternating periods of apnoea and dyspnoea in disease of the heart, kidney, or brain, the dyspnoeic phase being gradual in onset and disappearance.
Chezy's formula (*Hyd.*). A formula for computing the velocity of flow in pipes and channels. This is given as proportional to the square root of the product of the hydraulic mean depth and the virtual slope.
chiasma (*Cyt.*). The exchange of material between paired chromosomes during nuclear division. (*Zool.*). A structure in the central nervous system, formed by the crossing over of fibres from the right side to the left side and vice versa.
chiastobasidium (*Bot.*). A club-shaped basidium in which the spindles of the dividing nuclei lie at the same level across the basidium.
chiastolite (*Min.*). A variety of andalusite characterized by cruciform inclusions of carbonaceous matter.
chiastoneury (*Zool.*). A condition, found in some Gastropods, in which the visceral nerve commissures are twisted into a figure of eight.
Chicago rail-bond (*Elec. Eng.*). See **crown rail-bond.**
chickenpox (*Med.*). Varicella. A common mild acute infectious disease in which papules, vesicles, and small pustules appear in successive crops, mainly on the trunk, face, upper arms, and thighs.
chicken wire (*Build.*). Light-gauge wire netting.
Chick-Martin Test (*Chem.*). A standard test of disinfectants, similar to the *Rideal-Walker Test* except that the test is carried out in the presence of a high concentration of organic matter, a suspension of yeast being used.
chicot (*For.*). A dead tree with less than one-half of the stem standing.
chiffon (*Textiles*). A very fine, soft dress material of silk, nylon, etc.; the word is also used in other cloth descriptions to indicate lightness, e.g., chiffon velvet, chiffon taffeta, etc.
chigger or chigoe (*Vet.*). See **harvest mite.**
chilaria (*Zool.*). In *Xiphosura*, a pair of small round plates of doubtful function, situated just behind the mouth, and possibly representing a seventh pair of appendages.
chilblains (*Med.*). See **erythema pernio.**
Child-Langmuir-Schottky equation (*Electronics*). This gives the anode current in a space-charge-limited valve as

$$I=G.V^{3/2}$$

where I=current, G=perveance (a constant for a given material and design), and V=maintained anode voltage.
Chile nitre, Chile saltpetre (*Chem.*). A commercial name for sodium(V)nitrate, $NaNO_3$.
chili (*Meteor.*). A sirocco-type hot and dry southerly wind blowing in Tunisia.
Chilian mill (*Met.*). A type of ore-grinding machine in which crushing is done by heavy rollers running on a circular base plate or die ring.
chill (*Foundry*). An iron mould, or part of a mould, sometimes watercooled; used to accelerate cooling and give great hardness and density to the metal which comes into contact during pouring.
chill crystals (*Met.*). Small crystals formed by the rapid freezing of molten metal when it comes into contact with the surface of a cold metal mould.
chilled iron (*Met.*). Cast-iron cast in moulds constructed wholly or partly of metal, so that the surface of the casting is white and hard while the interior is grey.
Chilognatha (*Zool.*). See **Diplopoda.**
Chilopoda (*Zool.*). Subclass of *Myriapoda* having the trunk composed of numerous somites each bearing one pair of legs; the head bears a pair of uniflagellate antennae, a pair of mandibles, and two pairs of maxillae; the first body somite bears a pair of poison-claws; the genital opening is posterior; active carnivorous forms, some of considerable size and dangerous to Man; some are phosphorescent. Centipedes.
chimaera, chimera (*Bot.*). A plant in which there are at least two kinds of tissue differing in their genetic constitutions.
chime barrel (*Horol.*). The cylinder on whose periphery short vertical pins lift the hammers in a chiming clock, the pins being so arranged as to give the required sequence of notes.
chimney (*Bot.*). An upgrowth of epidermal cells above a stoma, forming a long pore. (*For.*) A vertical opening, sometimes tapering, in a stack of timber, to facilitate air circulation and more uniform drying. (*Mining*) Either an ore shoot or more usually an especially rich steeply inclined part of the lode.
chimney attenuator (*Telecomm.*). Type of coaxial-line attenuator.
chimney bar (*Build.*). An iron bar supporting the arch over a fireplace opening.
chimney bond (*Build.*). A *stretching bond* (q.v.), which is generally used for the internal division walls of domestic chimney-stacks, as well as for the outer walls.
chimney-breast (*Build.*). The part of the chimney between the flue and the room.
chimney jambs (*Build.*). The upright sides of a fireplace opening.
chimney lining (*Build.*). The tile flues, or cement rendering, within a chimney space.
chimney shaft (*Build.*). The part of a chimney projecting above a roof, or a chimney standing isolated like a factory chimney.
chimney stack (*Build.*). The unit containing a number of flues grouped together.
chimonophilous (*Bot.*). Growing chiefly during the winter.
chimopelagic plankton (*Bot.*). Plankton occurring only during winter.
chimyl alcohol (*Chem.*). *Cetyl-glyceryl ether*; testriol. $CH_3{\cdot}(CH_2)_{15}{\cdot}OCH_2{\cdot}CHOH{\cdot}CH_2OH$. Colourless crystalline substance, m.p. 61°C; found in fish liver oils, bull and boar testes.
china clay (*Geol.*). A clay consisting mainly of *kaolinite*, one of the most important raw materials of the ceramic industry. China clay is obtained from kaolinized granite, for example in S.W. England, and is separated from the

other constituents of the granite (quartz and mica) by washing out with high pressure jets of water. Also called **kaolin, porcelain clay.** (*Pharm.*) Kaolin is used as an internal absorbent (e.g. of poisons), for poultices, dusting powers, etc. See also **art paper.**

chinastone (*Geol.*). A kaolinitized granitic rock containing unaltered plagioclase. Also applied to certain limestones of exceptionally fine grain and smooth texture.

China wood oil (*Chem.*). *Tung oil* (q.v.).

chine (*Aero.*). The extreme outside longitudinal member of a flying-boat hull, or of a seaplane float; runs approximately parallel with the keel.

chiné (*Textiles*). A fancy silk material. The pattern is printed on the sheet of warp yarns and, with a self-coloured weft, the design has a distinctive blurred effect.

Chinese binary (*Comp.*). A form of coding for punched cards in which the card is read in columns rather than in rows.

Chinese cotton (*Textiles*). A short, yellowish-brown staple grown in China and Thailand. Improved methods of cultivation provide a staple 1 in. (25 mm) in length.

Chinese oil (*Chem.*). *Cassia oil* (q.v.).

Chinese red (*Paint.*). A basic chromate of lead pigment, also known as **Derby red.**

Chinese white (*Paint.*). Zinc oxide ground in water or oil.

chiniofon (*Chem.*). Mixture of bicarbonate of soda with *7-iodo-8-hydroxyquinoline-5-sulphonic acid*:

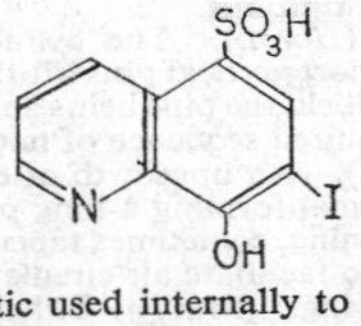

SO$_3$H, N, I, OH

An antiseptic used internally to treat amoebic dysentery.

chinook (*Meteor.*). A föhnlike west wind blowing on the eastern side of the Rocky Mountains.

chip (*Acous.*). See **swarf.** (*Electronics*) A small section of single crystal semiconductor forming the substrate for an integrated circuit.

chip-axe (*Tools*). A small single-handed tool for chipping timber to rough size.

chip board (*Paper*). A board, usually made from waste paper, used in box-making.

chip breaker (*Eng.*). A groove, step, or similar feature in one face of a (lathe) tool or cutter, designed to break long chips, or swarf, into small pieces which will more easily clear the tool and can conveniently be disposed of.

chip log (*Ships*). Quadranted piece of wood, weighted round its edge, attached to the log line. See **nautical log.**

chipping (*Met.*). The removing of surface defects from semifinished metal products by pneumatic chisels. Seams, laps, and rokes are thus eliminated.

chipping chisel (*Tools*). See **cold chisel.**

chirality (*Maths.*). The property of being right-handed or left-handed, usually applied to co-ordinate axes. (*Nuc.*) Relationship between spin vector and momentum vector of certain nuclear particles, especially neutrinos.

Chireix-Mesny antenna (*Radio*). Broadside antenna, with similar reflector, consisting of a square and diagonals.

Chireix transmitter (*Radio*). One with parallel stages which amplify sidebands shifted $\pm 90°$ in phase as class-C, with high efficiency, before combining to form an amplitude-modulated transmission.

chiropody. The care and treatment of minor ailments of the feet.

Chiroptera (*Zool.*). An order of aerial Mammals having the forelimbs specially modified for flight; mainly insectivorous or frugivorous nocturnal forms. Bats.

chiropterophilous (*Bot.*). Pollinated by bats.

chirp radar (*Radar*). A radar system using linear frequency-modulated pulses.

chisel (*Tools*). Steel tool for cutting wood, metal, or stone; it consists of a shank whose end is bevelled to a cutting edge.

chiselled (or **boasted**) **ashlar** (*Build.*). A random-tooled *ashlar* finished with a narrow chisel.

chi-square distribution (χ^2) (*Stats.*). The χ^2 distribution is defined as $\chi^2 = \frac{(n-1)s^2}{\sigma^2}$, where s^2 is the estimated variance $\sum\frac{(x-\bar{x})^2}{n-1}$ ($\bar{x}$ being the mean) of a sample of n values drawn from a normal distribution whose variance is σ^2. It is used as a test of agreement between hypothesis and observation and may be written in the form $\chi^2 = \sum\frac{(O-E)^2}{E}$, where O is the observed frequency, and E the expected frequency. There are χ^2 distributions corresponding to the number of the degrees of freedom, and to every value of χ^2 there is a corresponding value of P, i.e., the probability that χ^2 should be greater than a specified value.

chitin (*Zool.*). A skeletal material found in the majority of groups of Invertebrates but especially in *Arthropoda*. Its empirical formula is $C_{15}H_{26}O_{10}N_2$. On boiling with concentrated hydrochloric acid, it decomposes to glucosamine and acetic acid.

chitinase (*Biochem.*). An enzyme which is capable of hydrolysing *chitin*; secreted by the epidermis of Insects when about to moult, found in digestive juice of snail, in mould fungi, etc.

chitosan (*Biochem.*). A substance produced in van Wisselingh's test for chitin. Chitin is treated with concentrated potassium hydroxide at 160°C for 1·2 ks. The product, *chitosan*, gives a rose-violet colour with 0·2% iodine in 1% sulphuric acid.

Chladni figures (*Acous.*). The visual patterns produced under appropriate conditions by the complicated vibrations of a plate.

chlamydeous (*Bot.*). Having a perianth.

Chlamydobacteriales (*Bacteriol.*). Chemosynthetic bacteria-like organisms, probably closely related to the true bacteria: filamentous: characterized by the deposition of ferric hydroxide in or on their sheaths. Occur in fresh water, particularly moorland bogs, where their oxidation of iron (II) to iron (III) results in the deposition of *bog iron ore.*

chlamydocyst (*Bot.*). A double-walled resting sporangium inside a hypha.

chlamydospore (*Bot.*). A thick-walled fungal spore, capable of a resting period, and often formed inside a hypha. (*Zool.*) A protozoan spore enclosed in a spore-case. Cf. *gymnospore.*

chloanthite, cloanthite (*Min.*). Arsenide of nickel occurring in the cubic system. A valuable nickel ore, often associated with smaltite.

chloasma (*Med.*) A skin disease characterized by yellowish patches.

chloragen (or **chloragogen**) **cells** (*Zool.*). In

Chaetopoda, yellowish flattened cells occurring on the outside of the alimentary canal, and concerned with nitrogenous excretion.

chloral (*Chem.*). Trichloroethanal, $CCl_3{\cdot}CHO$, b.p. 97°C, a viscous liquid, of characteristic odour, obtained by the action of chlorine upon ethanol and subsequent distillation over sulphuric acid.

chloral hydrate (*Chem.*). Trichloroethanal hydrate. $CCl_3{\cdot}CH(OH)_2$, m.p. 57°C, b.p. (with decomposition) 97°C, large colourless crystals, soluble in water, a hypnotic and sedative (especially for children). Obtained from *chloral* and water. One of few compounds having 2 hydroxyl groups attached to the same carbon atom.

chloramines (*Chem.*). Compounds obtained by the action of chlorate (I) solutions on compounds containing the NH or NH_2 groups. Important as disinfectants. Chloramine T, $CH_3{\cdot}C_6H_4{\cdot}SO_2{\cdot}NClNa$, is the active constituent of an ointment employed as an antidote against *vesicant* war gases. It is also used in the chlorination of water and in certain contraceptive foam tablets.

chloramphenicol (*Pharm.*). Antibiotic from some species of *Streptomyces*; also obtained synthetically. Of therapeutic value in treatment of, e.g., typhoid, scrub-typhus, typhus, psittacosis. Unlike penicillin and sulphonamides, it attacks viruses as well as bacteria. An important characteristic is its high degree of penetration into the cerebrospinal fluid. TN **Chloromycetin.**

chloranil (*Chem.*). Tetrachloroquinone, $C_6Cl_4O_2$. A useful oxidizing agent.

chloranthy (*Bot.*). An abnormal condition of a flower, in which all parts have been changed into leafy structures.

chlorapatite (*Min.*). Chlorophosphate of calcium, a member of the apatite group of minerals. See **apatite, fluorapatite.**

chlorargyrite (*Min.*). Silver chloride, crystallizing in the cubic system. It is usually the product of the oxidation of silver ores, and occurs commonly in massive or waxlike forms associated with other silver ores. See also **cerargyrite.**

chlorastrolite (*Min.*). A fibrous green variety of *pumpellyite* (q.v.); it occurs in rounded geodes in basic igneous rocks near Lake Superior. When cut *en cabochon*, it exhibits chatoyancy.

chlorates (V) (*Chem.*). Salts of chloric acid. Powerful oxidizing agents. Explosive when ground or detonated in contact with organic matter.

chlorazide (*Chem.*). N_3Cl; a colourless, highly explosive gas formed by reaction between sodium chlorate (I) and sodium azide.

chlorazine (*Chem.*). 2-Chloro-4,6-bisdiethylamino-1,3,5-triazine, used as a herbicide.

chlorazol black stain (*Micros.*). A progressive nuclear and general stain which normally stains metachromatically. It has a special affinity for cellulose and chitin.

chlordane (*Chem.*). 1,2,4,5,6,7,10,10-Octachloro-4,7,8,9-tetrahydro-4,7-methylene-indane:

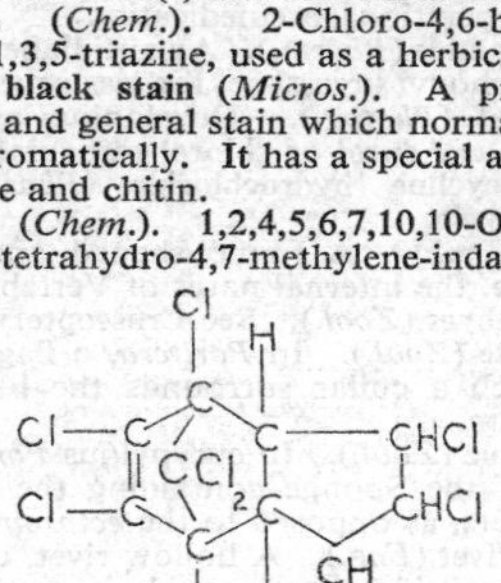

used as an insecticide.

chlordiazepoxide (*Pharm.*). 7-chloro-2-methylamino-5-phenyl-3*H*-1,4-benzodiazepine 4-oxide. A mild tranquillizing drug. TN **Librium.**

chlorella (*Bot.*). Microscopic alga of high protein content grown commercially and used as an additive in confectionery and foodstuffs.

chlorenchyma (*Bot.*). Tissue containing chloroplasts.

Chlorex process (*Chem.*). Removal of waxes and naphthenes from lubricating oils by solvent action of 2,2-*dichloroethoxyethane*, improving their viscosity and ageing properties.

chlorhexidine (*Pharm.*). A disinfectant used as a gluconate solution for skin cleansing, in creams, lozenges, etc.; sometimes with *cetrimide*.

chloric acid (V) (*Chem.*). $HClO_3$. A monobasic acid forming a series of salts, chlorates—where ClO_3 acts as a univalent radical.

chloride of lime (*Chem.*). See **bleaching powder.**

chloride of silver cell (*Elec. Eng.*). A primary cell having electrodes of zinc and silver and a depolarizer of silver chloride.

chlorides (*Chem.*). Salts of hydrochloric acid. Many metals combine directly with chlorine to form chlorides. Very volatile in flame tests.

chloridizing roasting (*Met.*). The roasting of sulphide ores and concentrates, mixed with sodium chloride, to convert the sulphides to chlorides.

chlorimeter, chlorometer (*Chem.*). Apparatus for measuring the available chlorine in bleaching powder, etc.

chlorination (*Chem.*). (1) The substitution or addition of chlorine in organic compounds. (2) The sterilization of water with chlorine, sodium chlorate (I), chloramine, or bleaching powder. (*San. Eng.*) The addition of chlorine to sewage in order to assist the processes of purification in the land treatment method.

chlorine (*Chem.*). Element, symbol Cl, at. no. 17, r.a.m. 35·453, valencies 1−, 3+, 5+, 7+, m.p. −101·6°C, b.p. −34·6°C. The second halogen, chlorine is a greenish yellow gas, with an irritating smell and a destructive effect on the respiratory tract. Produced from the electrolysis of conc. brine, or by oxidation of hydrochloric acid. It is a powerful oxidizing agent, and is widely used, both pure and as *Javel water* (q.v.) or *bleaching powder* (q.v.), for bleaching and disinfecting. Chlorine is used in the organic chemicals industry to produce tetrachloromethane, trichloromethane, PVC, and many other solvents, plastics, disinfectants, and anaesthetics.

chlorine number (*Paper*). Indicates the amount of bleaching agent required by the raw pulp stock.

chlorine oxides (*Chem.*). The (I) *oxide* (Cl_2O) and (IV) *oxide* (ClO_2) are gases, similar to chlorine. The (I) oxide dissolves in water to form hypochlorous acid. The (IV) oxide is used as a bleach. The (VI) *oxide* (Cl_2O_6) and (VII) *oxide* (Cl_2O_7) are liquids. All highly unstable, often explosive.

chlorite (*Min.*). A group of allied minerals which may be regarded as hydrated silicates of aluminium, iron, and magnesium. They crystallize in the monoclinic system, and are typically green in colour. They occur as alteration products of such minerals as biotite and hornblende, and occur in many rock types, but may be important in low grade metamorphic rocks.

chlorite schist (*Geol.*). A schist composed largely of the mineral chlorite, in association with quartz, epidote, etc. Formed from basic igneous rock by dynamothermal metamorphism.

chloritization (*Geol.*). The replacement, by alteration, of ferro-magnesian minerals by chlorite.

chloritoid (*Min.*). A hydrous iron, magnesium, aluminium silicate, crystallizing in the monoclinic or triclinic systems. Characteristic of low and medium grade regionally metamorphosed sedimentary rocks.

chloroacetic (ethanoic) acids (*Chem.*). *Monochloroacetic acid*, $CH_2Cl{\cdot}COOH$, m.p. 62°C, b.p. 186°C; formed by chlorination of acetic acid in the presence of acetic anhydride, sulphur or phosphorus. *Dichloroacetic acid*, $CHCL_2{\cdot}COOH$, b.p. 191°C; formed by heating chloral hydrate with potassium cyanide. *Trichloroacetic acid*, $CCl_3{\cdot}COOH$, m.p. 52°C, b.p. 195°C; formed by oxidizing chloral hydrate with nitric acid.

chloroauric (III) acid (*Chem.*). $HAuCl_4$. A complex acid formed when gold (III) oxide (Au_2O_3) dissolves in hydrochloric acid. Forms a series of complex salts called *chloroaurates* (III).

chlorobromomethane (*Chem.*). $CH_2Cl{\cdot}Br$. Organic liquid used as a fire extinguishing agent.

2-chlorobutadiene (*Chem.*). See **chloroprene.**

Chlorococcales (*Bot.*). An order of the *Chlorophyceae*; the members are unicellular, but may form colonies of uninucleate, or multinucleate, cells, which never divide vegetatively; asexual reproduction by zoospores or autospores, and sexual reproduction by biflagellate gametes.

chlorocruorin (*Zool.*). A green respiratory pigment of certain *Polychaeta*. Conjugated protein containing a prosthetic group similar to, but not identical with, reduced haematin.

chloroform (*Chem.*). Trichloromethane. $CHCl_3$, b.p. 62°C, rel. d. 1·5; a colourless liquid of a peculiar ethereal odour, an anaesthetic, solvent for fats and oils, resins, rubber, and numerous other substances. It is prepared technically from ethanol and calcium chlorate (I).

chlorohydrins (*Chem.*). A group of hydrochloric acid esters of glycols, e.g., *ethylene chlorohydrin* or 1-*hydroxy-2-chloroethane*, $CH_2Cl{\cdot}CH_2OH$.

chlorohydroquinone (*Photog.*). See **chlorquinol.**

chloroma (*Med.*). A greenish tumour in the bones, composed of cells resembling white blood cells; occurs in acute leukaemia.

chlorometer. See **chlorimeter.**

Chloromonadina (*Zool.*). An order of *Phytomastigina*; forms generally passing much of the life-history in the palmella stage; green or colourless; sometimes of amoeboid form; with a gullet but without a transverse groove; having food reserves of oil; with a contractile vacuole.

Chloromycetin (*Pharm.*). See **chloramphenicol.**

chlorophaeite (*Min.*). A mineral closely related to chlorite, dark-green when fresh, but rapidly changing to brown, hence the name (Greek *chloros*, yellowish-green; *phaios*, dun). Described from basic igneous rocks.

chlorophenol red (*Chem.*). *Dichlorophenolsulphonphthalein*, an indicator used in acid-alkali titration, having a pH range of 4·8 to 6·4, over which it changes from yellow to red.

Chlorophyceae (*Bot.*). A large class of algae, inhabiting fresh water, salt water, and less often occurring on soil or objects out of the water. They are green, having a chlorophyll mixture of approximately the same character as that present in higher plants; they form starch as a storage product, and contain cellulose in their walls. Often called **green algae.**

chlorophyll (*Bot.*). The mixture of 2 green and 2 yellow pigments, present in the chloroplasts of all plants which are able to synthesize carbohydrates from carbon dioxide and water, in some fashion enabling the plants to utilize energy derived from light in the synthesis of material. The green pigments, which predominate in the mixture, are *chlorophyll a* ($C_{55}H_{72}O_5N_4Mg$) and *chlorophyll b* ($C_{55}H_{70}O_6N_4Mg$); the yellow pigments are *carotin* ($C_{40}H_{56}$) and *xanthophyll* (q.v.).

chlorophyllase (*Bot.*). An enzyme occurring in association with chlorophyll in plants, and able to decompose it.

chlorophyll corpuscle (*Bot.*). See **chloroplast.**

chlorophyllogen (*Bot.*). Hypothetical forerunner of chlorophyll, formed independently of light, and giving rise to chlorophyll when exposed to light.

chlorophyllous, chlorophyllose (*Bot.*). Containing chlorophyll.

chloroplast (*Bot.*). Plastid containing chlorophyll, with or without other pigments, embedded singly or in considerable numbers in the cytoplasm of a plant cell.

chloroplatinic (IV) acid (*Chem.*). $H_2PtCl_6{\cdot}6H_2O$. Formed when platinum is crystallized from a solution acidified with hydrochloric acid.

chloroprene (*Chem.*). 2-Chlorobutan 1,2 : 3,4-diene. Starting point for the manufacture of *Neoprene*.

chloroquine (*Pharm.*). Widely used synthetic antimalarial drug, 7-chloro-4-(4-diethylamino-1-methylbutylamino) quinoline.

chlorosis (*Bot.*). An unhealthy condition due to deficiency in chlorophyll; shown by yellowing of the plant. (*Med.*) Green sickness. An anaemia in young girls, readily cured by iron; now rare. *Adj.* **chlorotic.**

chlorostatolith (*Bot.*). A chloroplast containing starch which acts as a statolith.

Chlorothene NU (*Chem.*). 1,1,1-*Trichloroethane.*

chlorothiazide (*Pharm.*). 6-Chloro-7-sulphamoyl-2*H*-benzo-1,2,4-thiadiazine 1,1-dioxide, one of the thiazide class of diuretics, used in cases of oedema, hypertension, cirrhosis of the liver.

Chloroxone (*Chem.*). TN for *2,4-dichlorophenoxyacetic* (*ethanoic*) *acid*:

Cl

Cl– ⬡ ·O·CH$_2$·COOH

also known as **2,4-D**, a selective weedkiller.

chlorpromazine (*Pharm.*). A widely used tranquillizing drug, 2-chloro-10-(3-dimethylaminopropyl)phenothiazine hydrochloride, of great value in mental and behavioural disorders, also used as an anti-emetic, analgesic adjuvant, and for preoperational premedication. TN **Largactil.**

chlorpropamide (*Pharm.*). An antidiabetic drug of the sulphonyl urea type, for less severe cases.

chlorquinol (*Photog.*). Developing agent, also known as **Adurol** or **chlorohydroquinone.**

chlortetracycline hydrochloride (*Pharm.*). See **aureomycin.**

choana (*Zool.*). A funnel-shaped aperture; *pl.* choanae, the internal nares of Vertebrates.

Choanichthyes (*Zool.*). See **Crossopterygii.**

choanocyte (*Zool.*). In *Porifera*, a flagellate cell, in which a collar surrounds the base of the flagellum.

choanosome (*Zool.*). In eurypylous *Porifera*, that part of the Sponge containing the flagellated chambers, as opposed to the *ectosome* (q.v.).

Chobert rivet (*Eng.*). A hollow rivet, used where only one side of the work is accessible. It comprises a headed tube having a concentric taper bore of which the smallest diameter is at the shank end. Drawing a wedge-headed mandrel through the bore clinches the rivet by upsetting the shank end on the inaccessible side of the structure. The bore may subsequently be plugged by a parallel sealing pin which also increases the shear strength of the rivet.

chock-to-chock (*Aero.*). See **block time.**

choice (*Textiles*). A term used by woolsorters to denote wool of third quality, taken from the middle in the sides of a fleece.

choice point (*An. Behav.*). The position in a T-maze or other type of maze or any apparatus involved in discrimination training, where an animal can make only one of two or more alternative responses.

choke (*Elec. Eng.*). Colloquialism for inductor. Usually refers to coils of high self-inductance used to limit flow of an a.c. without power dissipation. Any d.c. flowing through the choke will be impeded only by the intrinsic resistance of the winding. (*I.C. Engs.*) (1) The venturi or throat in the air-passage of a *carburettor* (q.v.). (2) A butterfly valve in a carburettor intake which reduces the air supply and so gives a rich mixture for starting purposes. Also called **strangler.** (3) A small piston which, when closed, gives a rich mixture. (*Radar*) Component in waveguide, with two quarter-wavelength sections, one short-circuited.

choke coupling (*Elec. Eng.*). (1) Use of the impedance of a choke for coupling the successive stages of a multistage amplifier. (2) In a waveguide, coupling flanges with quarter-wavelength groove which breaks surface continuity at current node.

choke damp (*Mining*). A term sometimes used for *black damp* (carbon dioxide). More correctly, any mixture of gases which causes choking or suffocation.

choked disk (*Med.*). See **papilloedema.**

choke feed (*Elec. Eng.*). Use of a high inductance path for the d.c. component of the anode current of a valve, the a.c. signal being fed through a capacitor.

choke input filter (*Elec. Eng.*). In a rectifying circuit, a filter which has for its first component a series choke.

choke-jointed (*Elec. Eng.*). See **joint.**

choke modulation (*Elec. Eng.*). System of anode modulation in which the modulating and modulated valves are coupled by means of a high-inductance choke included in the common anode-voltage supply lead.

cholaemia, cholemia (*Med.*). Presence of bile pigments in the blood.

cholagogue (*Med.*). Increasing evacuation of bile; a drug which does this.

cholangitis, cholangeitis (*Med.*). Inflammation of the bile passages.

cholecalciferol (*Chem.*). Vitamin D_3.

cholecystectomy (*Surg.*). Excision of the gall bladder.

cholecystenterostomy (*Surg.*). An artificial opening made between the gall bladder and the upper part of the small intestine.

cholecystitis (*Med.*). Inflammation of the gall bladder.

cholecystography (*Radiol.*). The X-ray investigation of the gall bladder previously filled with a substance opaque to X-rays. A *cholecystogram* is the X-ray photograph so obtained.

cholecystokinin (*Physiol.*). A hormone liberated from the duodenal mucosa, causing contraction of the gall bladder and relaxation of the *sphincter of Oddi.*

cholecystostomy (*Surg.*). The surgical formation of an opening in the wall of the gall bladder.

choledochal (*Zool.*). Containing bile.

choledochitis (*Med.*). Inflammation of the common bile duct.

choledochotomy (*Surg.*). Incision into one of the main bile ducts.

cholelithiasis (*Med.*). The presence of stones in the gall bladder and bile passages.

cholera (*Med.*). An acute bacterial infection in Eastern countries; characterized by severe vomiting and diarrhoea, drying of the tissues and painful cramps; spread by infected food and water.

cholesteatoma (*Med.*). A tumour in the brain, or in the middle ear, composed of cells and crystals of cholesterol.

cholesteraemia, cholesteremia (*Med.*). See **cholesterolaemia.**

cholesterol (*Chem.*). $C_{27}H_{45}OH$, a sterol of the alicyclic series, found in nerve tissues, gall stones, and in other tissues of the body. It is a white crystalline solid, m.p. 148·5°C, soluble in organic solvents and in fats. There are numerous stereoisomers known. Parent compound for many steroids.

cholesterolaemia, cholesterolemia (*Med.*). Excess of cholesterol in the blood.

cholesterosis (*Med.*). Diffuse deposits of cholesterol in the lining membrane of the gall bladder.

choletelin (*Chem.*). A yellow oxidation product of bilirubin.

cholic acid (*Chem.*). $C_{23}H_{39}O_3{\cdot}COOH$, the product of hydrolysis of certain bile acids, is conjugated in the body, forming glycocholic acid with glycine and taurocholic acid with taurine.

choline (*Chem., Physiol.*). Ethylol-trimethyl-ammonia hydrate, $OH{\cdot}CH_2{\cdot}CH_2{\cdot}NMe_3{\cdot}OH$, a strong base, present in the bile, brain, yolk of egg, etc., combined with fatty acids or with glyceryl-phosphoric acid (*lecithin* (q.v.)). It is concerned in regulating the deposition of fat in the liver, and its acetyl ester is intimately connected with the activity of certain nerves. See **neurine.**

cholinergic (*Physiol.*). Activated by acetylcholine, e.g., transmission at preganglionic synapses of the autonomic nervous system and at post-ganglionic endings of the parasympathetic and some sympathetic nerve fibres.

choluria (*Med.*). Bile pigments in the urine.

chomophyte (*Bot.*). A plant growing on rock ledges littered with detritus, or in fissures and crevices where root hold is obtainable.

chondr-, chondrio-, chondro-. Prefix from Gk. *chondros*, cartilage, grain.

chondral (*Histol.*). Pertaining to cartilage.

chondrenchyma (*Zool.*). A form of parenchyma, occurring in Sponges, which closely resembles hyaline cartilage, having the cells embedded in a stiff gelatinous matrix.

chondrification (*Histol.*). Strictly, the formation of chondrin; hence, the development of cartilage.

chondrin (*Histol.*). A firm, elastic, translucent, bluish-white substance of a gelatinous nature, which forms the ground-substance of cartilage.

chondriocont, chondriokont (*Cyt.*). A chondriosome which has the form of a rod or thread.

chondriokinesis (*Cyt.*). Division of chromosomes in mitosis and meiosis.

chondriolysis (*Cyt.*). The dissolution of mitochondria.

chondriome (*Cyt.*). The mitochondria of a cell, collectively.

chondriomite (*Cyt.*). A chondriosome having the form of a chain of granules.

chondriosomal mantle (*Cyt.*). An accumulation

of chondriosomes surrounding a dividing nucleus.
chondriosomes (*Cyt.*). See **mitochondria.**
chondrite (*Min.*). A type of stony meteorite containing *chondrules*—nodule-like aggregates of minerals.
chondroblast (*Histol.*). A cartilage cell which builds up the chondrin matrix.
chondroclast (*Histol.*). A cartilage cell which destroys the chondrin matrix.
chondrocranium (*Zool.*). The primary cranium of *Craniata*, formed by the fusion of the parachordals, auditory capsules and trabeculae.
chondrocyte (*Cyt.*). A cartilage cell.
chondrodite (*Min.*). Hydrated magnesium silicate, crystallizing in the monoclinic system, and occurring in metamorphosed limestones.
chondrodystrophia (*Med.*). See **achondroplasia.**
chondrogenesis (*Histol.*). See **chondrification.**
chondroid (*Bot.*). Said of the medulla of a lichen when it is hard and tough, consisting of thick-walled hyphae in very firm association.
chondroids (*Vet.*). Compact lumps of dried pus commonly found in the exudate of inflamed guttural pouches of the horse.
chondroitin (*Chem.*). Tetrasaccharide consisting of two glucuronic and two chondrosamino groups.
chondroma (*Med.*). A tumour composed of cartilage cells.
chondromucoids (*Biochem.*). A mucoprotein, yielding, on hydrolysis, a protein and chondroitin sulphate; it forms the ground substance of cartilage in combination with collagen.
chondroproteins (*Chem.*). A group of *glucoproteins* (q.v.). They are insoluble in water but soluble in dilute alkalis.
chondrosamine (*Chem.*). $C_6H_{13}O_5N$, or *2-aminogalactose*, is the basis of *chondroitin*, which is the substance of cartilage and similar body tissues.
chondrosarcoma (*Med.*). A malignant tumour composed of sarcoma cells and of cartilage.
chondroskeleton (*Zool.*). The cartilaginous part of the Vertebrate skeleton.
chondrosteosis (*Histol.*). The transformation of cartilage into bone.
chondrule (*Min.*). See under **chondrite.**
Chondrus (*Bot.*). See **Irish moss.**
Chonotricha (*Zool.*). An order of *Ciliata* the members of which are of permanently sessile habit; there is a spiral funnel, ciliated internally, leading to the mouth; other cilia are absent.
chop (*Tools*). Movable jaw of a bench vice.
chopped continuous waves (*Radio*). See **interrupted continuous waves.**
chopped wave (*Elec. Eng.*). A travelling voltage wave which rises to a maximum and then rapidly falls to zero. Such a wave occurs on transmission lines when an ordinary voltage wave has caused an insulator flashover, thereby losing its tail.
chopper (*Elec. Eng.*). (1) Light interrupter used to produce a.c. output from a photocell. (2) Mechanical or semiconductor switch used to modulate a d.c. signal so that it can be amplified by an a.c. amplifier. An output chopper may be operated in synchronism with the input chopper in order to rectify the amplified signal.
chopper amplifier (*Elec. Eng.*). See **chopper-stabilized amplifier, vibrating-reed amplifier.**
chopper disk (*Elec. Eng.*). Rotating toothed or perforated disk which interrupts the light signal falling on a photocell at a desired frequency.
chopper-stabilized amplifier (*Elec. Eng.*). Low drift d.c. amplifier in which a mechanical or semiconductor switch is used to convert the d.c., or low-frequency a.c., signal into an a.c. square wave which can be more readily amplified. A synchronous chopper at the amplifier output may be used as a rectifier. An equivalent input drift voltage of around 0·1 μV/°C can be obtained in this way.
chops (*Horol.*). A clamp. The two flat pieces of metal, usually of brass, between which the end of a pendulum suspension spring is secured.
chord. A straight line drawn between two points on a curve. (*Aero.*) The length of that part of the chord line intercepted by the extremities of the leading and trailing edges of an aerofoil section. (*Eng.*) See **boom.**
chorda (*Zool.*). Any stringlike structure, e.g., the *chordae tendineae*, tendinous cords attaching the valves of the heart; also, the notochord.
chordacentra (*Zool.*). Vertebral centra formed mainly from the notochordal sheaths; cf. *archecentra. adj.* **chordacentrous.**
chordal thickness (*Eng.*). Of a gear-wheel tooth, the length of the chord subtended by the tooth thickness arc.
Chordariales (*Bot.*). An order of the *Heterogeneratae* in which the filamentous sporophyte is not formed into a pseudoparenchymatous thallus.
Chordata (*Zool*). A phylum of the *Metazoa* containing those animals possessing a notochord.
chorda tympani (*Zool.*). The anterior branch of the post-trematic ramus of the seventh cranial nerve (facial) in Mammals.
chorded winding (*Elec. Eng.*). An armature winding in which the span of the coils is less than a pole-pitch.
chordee (*Med.*). Painful distortion of the erect penis as a result of injury or disease, especially gonorrhoea. Cf. *priapism.*
chord line (*Aero.*). A straight line joining the centres of curvature of the leading and trailing edges of any aerofoil section. Historically, it was the common tangent to the curvatures on the lower surfaces, i.e., the line of a straight edge held against the underside of the plane, but with the development of planes with convex lower surfaces, this definition lacks precision.
chord of contact (*Maths.*). Of a conic: the line joining the points of contact of the tangents from an external point.
chordoma (*Med.*). An invasive tumour arising from remains of the notochord in the skull and the spinal column.
chordotomy (*Surg.*). The cutting of the nerve fibres in the spinal cord conveying the sensation of pain; done for the relief of severe pain.
chordotonal organs (*Zool.*). In Insects, sense organs consisting of bundles of scolophores, sensitive to pressure, vibrations, and sound.
chorea (*Med.*). Sydenham's chorea; St Vitus dance. An infection of the brain manifested by uncontrollable, irregular, purposeless movements, closely related to rheumatism.
chorion (*Zool.*). In higher Vertebrates, one of the foetal membranes, being the outer layer of the amniotic fold: in Insects, the hardened egg-shell lying outside the vitelline membrane.
chorion-epithelioma (*Med.*). A malignant tumour of the uterus composed of cells derived from the foetal chorion; it appears during or after pregnancy.
chorionic gonadotrophin (*Biochem.*). A hormone secreted by the placenta with biological effects resembling those of the hypophyseal hormones. It appears in the urine early in pregnancy.
Choriopetalae (*Bot.*). See **Archichlamydeae.**
chorioptic mange (*Vet.*). A mange of horses and

cattle due to mites of the genus *Chorioptes*. In horses, known as **foot mange** or **itchy leg.**

chorioretinitis (*Med.*). Inflammation of the choroid and retina.

choriothete (*Zool.*). A maternal organ which strips the chorion off the egg in the uterus of *Glossina*, the tsetse fly.

choripetalous (*Bot.*). Polypetalous.

chorisis (*Bot.*). Fission into two or more lobes; often applied to abnormalities.

choroid (*Zool.*). The vascular tunic of the Vertebrate eye, lying between the retina and the sclera. *adj.* **choroidal.**

choroiditis (*Med.*). Inflammation of the choroid of the eye.

choroid plexus (*Zool.*). In higher *Craniata*, the thickened, vascularized regions of the *pia mater* in immediate contact with the thin epithelial roofs of the diencephalon and medulla.

chresard (*Bot.*). The total amount of water in the soil which can be drawn upon by plants.

Christmas-tree pattern (*Light*). See **optical pattern.**

Christoffel symbols (*Maths.*). The Christoffel symbols of the first kind are defined as

$$\left[\frac{r}{pq}\right]=\frac{1}{2}\left(\frac{\partial g_{pr}}{\partial x^q}+\frac{\partial g_{qr}}{\partial x^p}-\frac{\partial g_{pq}}{\partial x^r}\right).$$

Also denoted by

$$\left[\frac{pq}{r}\right],\ [pq,r]\quad C\frac{r}{pq},\ \text{and by } \Gamma pqr.$$

The Christoffel symbols of the second kind are defined as

$$\left\{\frac{r}{pq}\right\}=g^{\alpha r}\left[\frac{\alpha}{pq}\right].$$

Also denoted by

$$\left\{\frac{pq}{r}\right\},\ \text{and by } \Gamma\frac{r}{pq}.$$

g_{pq} and g^{pq} are components of the covariant and contravariant fundamental metric tensors.

chrom-, chromo-, chromat-, chromato-. Prefix from Gk. *chrōma, chrōmatos*, colour.

chroma control (*TV*). Control which adjusts colours in colour TV picture. Abbrev. for *chromaticity control.*

chromaffine (*Zool.*). Having an affinity for chromates (VI); as certain cells in Vertebrates which originate in the sympathetic nervous system, and migrate thence to various parts of the body, especially to the ductless glands.

chromaffinoma (*Med.*). See **paraganglioma.**

Chromagraph (*Print.*). Electronic equipment to produce, singly, sets of colour-corrected continuous-tone separations from transparencies. Cf. *Colorgraph.*

chromasie (*Cyt.*). Increase of chromatin.

chromates (VI) (*Chem.*). The salts corresponding to chromium (VI) oxide. The normal chromates, M_2CrO_4, are isomorphous with sulphates and are generally yellow, while the dichromates (VI), $M_2Cr_2O_7$, are isomorphous with disulphates (VI) and are generally orange-red. Chromates are used as pigments and in tanning (see **chrome leather**).

chromatic aberration (*Light, TV*). An enlargement of the focused spot caused (1) in a cathode-ray tube, by the differences in the electron velocity distribution through the beam, and (2) in an optical lens system using white light, by the refractive index of the glass varying with the wavelength of the light.

chromatic adaptation (*Bot.*). A variation in coloration in relation to the amount of light reaching a plant.

chromatic colour (*Light*). A colour which is not a grey.

chromaticity (*Light, TV*). Colour quality of light, as defined by its chromaticity coordinates, or alternatively by its purity (saturation) and dominant wavelength.

chromaticity control (*TV*). See **chroma control.**

chromaticity coordinate (*TV*). Ratio of any one of the tristimulus values of a colour sample to the sum of the three tristimulus values.

chromaticity diagram (*TV*). Plane diagram in which one of three chromaticity coordinates is plotted against another. Generally applied to the CIE (x,y) diagram in rectangular co-ordinates for colour television.

chromaticity flicker (*TV*). Flicker resulting from periodic variations in chromaticity only.

chromatics (*Light*). The science of colours as affected by phenomena determined by their differing wavelengths.

chromatid (*Cyt.*). See under **tetrad.**

chromatin (*Cyt.*). The deeply staining portion of the nucleoplasm.

chromatography (*Chem.*). Method for separating (often complex) mixtures. *Adsorption chromatography* depends on using solid adsorbents which have specific affinities for the adsolved substances. The mixture is introduced on to a column of the adsorbent, e.g., alumina, and the components eluted with a solvent or series of solvents and detected by physical or chemical methods. *Partition chromatography* applies the principle of countercurrent distribution to columns and involves the use of two immiscible solvent systems: one solvent system, the *stationary phase*, is supported on a suitable medium in a column and the mixture introduced in this system at the top of the column; the components are eluted with the other system, the *mobile phase*. See also **gas-liquid chromatography** and **paper chromatography.**

chromatolysis (*Cyt.*). The dissolution of chromatin in injured cells.

chromatophore (*Bacteriol.*). Pigment containing bodies occurring in photosynthetic bacteria, visible only with the electron microscope. (*Cyt.*) (1) A pigment cell; usually a connective-tissue corpuscle containing pigment granules, and capable of changing its shape under the influence of the sympathetic nervous system or of hormones. (2) In *Phytomastigina*, a cup- or plate-shaped mass of protoplasm consisting of green, yellow, or brown pigment for photosynthesis.

chromatoplasm (*Bot.*). The peripheral region of the protoplast in *Myxophyceae*, containing the pigments of the cell.

chromatospherite (*Cyt.*). See **nucleolus.**

chromatron (*TV*). Cathode-ray tube for viewing colour TV images.

chromax (*Met.*). High-temperature iron-based alloys with from 20 to 30% nickel and 15 to 20% chromium.

chrome (*Light*). Term used in Munsell colour system to indicate saturation.

chrome alum (*Chem.*). Potassium chromium sulphate, $K_2SO_4{\cdot}Cr_2(SO_4)_3{\cdot}24H_2O$, purple octahedral crystals obtained by the reduction of potassium dichromate (VI) solution acidified with sulphuric acid. It is used in dyeing, calico printing, and tanning.

chrome brick (*Met.*). Brick incorporating chromite, used as refractory lining in steelmaking furnaces.

chrome gallocyanin stain (*Micros.*). A stain used in electron microscopy to increase the contrast of specimens, e.g., virus particles, by forming local electron-dense deposits.

chrome green (*Paint.*). A series of pigments made from lead chromate coprecipitated, or intimately mixed, with Prussian blue. These are light-resistant pigments of high opacity, widely used for interior and exterior paints.

chrome iron ore (*Min.*). See **chromite.**

chromel (*Met.*). Alloy used in heating elements, based on nickel with less chromium.

chrome leather (*Leather*). Leather that has been tanned with salts of chromium; boot uppers and belting leather are tanned by this method.

chrome oranges (*Paint.*). A series of high opacity orange pigments consisting substantially of basic lead chromate. See also **chrome reds.**

chrome oxide green (*Paint.*). Chromic oxide of up to 99% purity, a high opacity chemical resistant pigment but having a dull appearance which limits its use.

chrome reds (*Paint.*). A series of red pigments resembling the *chrome yellows* (q.v.) chemically, except that they are even more basic. See **basic lead chromate.**

chrome spinel (*Min.*). Another name for the mineral *picotite*, a member of the spinel group. Cf. *chromite*.

chrome yellows (*Paint.*). A series of yellow to orange pigments, with high opacity and tinting strength, but low acid resistance, consisting of *lead chromate* (q.v.) precipitated together with other insoluble lead salts. See **lemon chrome, middle chrome, primrose chrome.**

chromic (VI) acid (*Chem.*). H_2CrO_4. An aqueous solution of chromic anhydride. See **chromium (VI) oxide.**

chromic anhydride (*Chem.*). See **chromium (VI) oxide.**

chromic oxide (*Chem.*). Chromium (III) oxide, Cr_2O_3, an amphoteric oxide corresponding to chromium (III) salts, CrX_3, and to chromites (chromates (III)), $M_2Cr_2O_4$.

chromic salts (*Chem.*). Salts in which chromium is in the (III) oxidation state. They are either green or violet, and are quite stable.

chromidia (*Cyt.*). Fragments of chromatin which lie free in the cytoplasm, and are not massed together to form a nucleus.

chromidium (*Bot.*). An algal cell in the thallus of a lichen. (*Zool.*) The collective name of the numerous basophil cytoplasmic granules present in many *Protozoa*.

chromidrosis (*Med.*). The excretion of sweat containing pigment.

chrominance (*Light, TV*). Colorimetric difference between any colour and a reference colour of equal luminance and specified chromaticity. The reference colour is generally a specified white, e.g., CIE illuminant C for artificial daylight.

chrominance channel (*TV*). Circuit path carrying the chrominance signal in a colour TV system.

chrominance demodulator (*TV*). See **synchronous demodulator.**

chrominance signal (*TV*). In every colour television system, the carrier whose modulation sidebands are added to the monochrome signal in the upper part of the video frequency band to convey colour information.

chrominance subcarrier regenerator (*TV*). In any colour television receiver, the electrical circuit which performs the function of generating a local subcarrier locked in phase with the transmitter burst. See **automatic phase control loop and quadricorrelator.**

chromiole (*Cyt.*). One of the smaller sized, deeply staining granules of which chromatin is composed; a granule within a chromomere.

chromite (*Min.*). A double oxide of chromium and iron (generally also containing magnesium), used as a source of chromium and also as a refractory for resisting high temperatures. It has neutral chemical properties, and is resistant to attack by both acid and basic slags. Chromite occurs as an accessory in some basic and ultrabasic rocks, and crystallizes in the cubic system as lustrous grey-black octahedra; also massive. Also called **chrome iron ore.**

chromium (*Chem.*). Metallic element, symbol Cr, at. no. 24, r.a.m. 51·996, rel. d. at 20°C, 7·138, m.p. 1510°C, electrical resistivity at 20°C, 0·131 microhm metres. Obtained from chromite. Alloyed with nickel in heat-resisting alloys and with iron or iron and nickel in stainless and heat-resisting steels. Also used as a corrosion-resisting plating.

chromium (VI) oxide (*Chem.*). CrO_3. Produced by the action of concentrated sulphuric acid on a concentrated aqueous solution of potassium dichromate (VI). It is deliquescent and a powerful oxidizing agent.

chromium plating (*Met.*). The production of a thin layer of chromium on the surface of another metal by electrodeposition, to protect it against corrosion. Thicker coatings are used to resist wear and abrasion. See **hard plating.**

chromoblast (*Zool.*). An embryonic cell which will develop into a chromatophore.

chromocentre (*Cyt.*). Any accumulation of chromatin in the nuclear reticulum.

chromochondria (*Cyt.*). Mitochondria concerned in pigment formation.

chromocyte (*Cyt.*). Any pigmented cell.

chromogen (*Chem.*). A coloured compound containing a *chromophore*.

chromoisomerism (*Chem.*). The existence of substances having the same chemical composition but different colours.

chromolithography (*Print.*). The preparation of lithographic printing surfaces, whether for single or multicolour printing, by hand methods only, without using a process camera as in *photolithography* (q.v.). For multicolour printing, an outline key drawing is made and transferred to the required number of stones or plates, on each of which the litho artist draws those portions which are required by the colour to be used when printing. See **autolithography, lithography.**

chromomeres (*Cyt.*). The small granules of which chromatin is built up.

chromonema (*Cyt.*). The whole of the threads which make up the nuclear reticulum: one of the threads; a twisted chromatic thread within the chromosome.

chromones (*Chem.*). Compounds having a condensed benzene and pyrone ring.

chromo-optometer (*Light*). An optometer in which the chromatic aberration of the eye is used to determine its refraction.

chromo paper (*Paper*). Paper which is more heavily coated than art paper; used for chromolithography.

chromophil, chromophilic (*Biol.*). Staining heavily in certain microscopical techniques.

chromophobe, chromophobic (*Biol.*). Resisting stains, or staining with difficulty, in certain microscopical techniques.

chromophores (*Chem.*). Characteristic groups which, attached to hydrocarbon radicals in sufficient number, are responsible for the colour

of dyestuffs. Such groups are: C=C, C=O, C=N, N=N, O←N$_2$O=O, and others.
chromophoric electrons (*Electronics*). Valence electrons of elements with excitation levels which lead them to absorb one or more wavelengths of visible light.
chromoplast (*Bot.*). See **chromatophore** (2).
chromoscope (*Photog.*). A type of colorimeter.
chromosomal aberration (*Cyt.*). Any departure from normality in chromosome number or chromosome structure.
chromosomal chimaera (*Bot.*). A chimaera in which not all the nuclei contain the same number of chromosomes.
chromosome (*Cyt.*). One of the deeply staining rodlike bodies, constant in number for any given species, into which the chromatin of the nucleus becomes condensed during meiosis or mitosis. In the human cell there are 22 pairs of chromosomes and 2 sex chromosomes.
chromosome arm (*Cyt.*). One of the two parts of a chromosome to which the spindle fibre is attached along the side.
chromosome complement (*Bot., Zool.*). The set of chromosomes characteristic of the nuclei of any one species of plant or animal.
chromosome cycle (*Cyt.*). The whole of the changes in the chromosomes during the complete life-cycle of an organism or cell.
chromosome map (*Cyt.*). A diagram purporting to show the position of the genes in a chromosome.
chromosome matrix (*Cyt.*). A sheath of weakly staining material surrounding the more stainable substance (*chromonema*) of a chromosome.
chromosome set (*Cyt.*). The whole of the chromosomes present in the nucleus of a gamete, usually consisting of one each of the several kinds that may be present.
chromosphere (*Astron.*). The layer of the sun's atmosphere, just above the reversing layer, which is observed visually during total solar eclipses, and spectroscopically at other times.
chromotherapy (*Med.*). Electrotherapeutic treatment with luminous rays of specific colours.
chromotropy (*Chem.*). See **chromoisomerism.**
chromous salts (*Chem.*). Salts of chromium in the (II) oxidation state; they yield blue solutions with water and are strong reducing agents.
chromyl (*Chem.*). The divalent ion CrO_2^{++}.
chronaxie (*Physiol.*). A time constant in nervous excitation, equal to the smallest time required for excitation of a nerve when the stimulus is an electrical current of twice the threshold intensity required for excitation when the stimulus is indefinitely prolonged.
chronic (*Med.*). Said of a disease which is deep-seated or long-continued. Cf. *acute*.
chronic respiratory disease of fowl (*Vet.*). An infection of the respiratory tract by organisms of the genus *Mycoplasma*. Abbrev. **CRD.**
chronispore (*Bot.*). A resting spore.
chronistor (*Elec. Eng.*). Electrochemical elapsed-time indicator.
chronocyclegraph (*Work Study*). A *cyclegraph* (q.v.) in which the light source is suitably interrupted so that the path appears as a series of pear-shaped spots, the pointed end indicating direction of movement and the spacing indicating speed of movement.
chronograph (*Horol.*). (1) A watch with a centre seconds hand which can be caused to start, stop, and fly back to zero by pressing the button or a push-piece on the side of the case. The chronograph mechanism is independent of the going train, so that the balance is not stopped when the centre seconds hand is stopped. See **stop watch.** (2) Any type of mechanism which gives a record of time intervals, **e.g.,** a *tape chronograph* in which a long paper tape is used, the time intervals being marked on the tape by pens actuated electromagnetically by a chronometer or other suitable time standard.
chronometer (*Horol.*). A precision timekeeper. In U.K. and U.S. the term denotes the very accurate timekeeper kept on board ships for navigational purposes, and fitted with the spring detent escapement. On the Continent the term is also applied to any very accurate clock or watch which may be fitted with the spring detent or lever escapement.
chronometer clock (*Horol.*). A clock fitted with a chronometer escapement.
chronometer escapement (*Horol.*). The spring detent escapement. A highly detached escapement, capable of giving the most exacting performance. Impulse is given to the balance every alternate vibration. Locking is performed by a pallet on a spring detent, and unlocking by a discharging pallet carried on a roller on the balance staff. Also **detent escapement.**
chronopher (*Teleg.*). The arrangement for automatic switching of standard time signals from an observatory to telegraph lines.
chronoscope (*Horol.*). Electronic instrument for precision measurement of very short time intervals.
chronotron (*Electronics*). Device which superimposes pulses on a power line to determine times of phenomena.
Chroococcales (*Bot.*). An order of the *Myxophyceae*, occurring as single cells, or non-filamentous colonies. Reproduction by vegetative division of the cells.
chrysalis (*Zool.*). The pupa of some Insects, especially *Lepidoptera*; the pupa-case.
chrysaniline (*Chem.*). See **phosphine** (2).
chrysarobin (*Chem.*). Yellow powder obtained by extracting araroba with benzene. Its main constituent is *chrysophanic acid*:

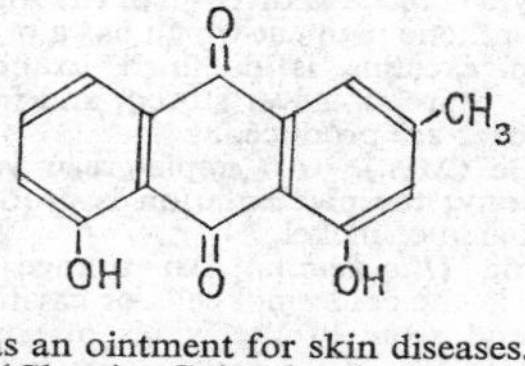

O
CH$_3$
OH O OH

Used as an ointment for skin diseases.
chrysene (*Chem.*). Colourless hydrocarbon found in the highest boiling fractions of coal tar:

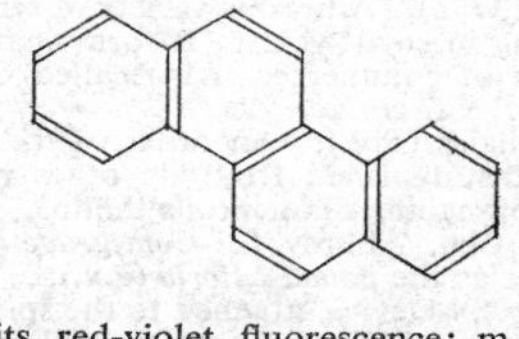

Exhibits red-violet fluorescence; m.p. 254°C, b.p. 448°C.
chrysoberyl (*Min.*). Aluminate of beryllium, crystallizing in the orthorhombic system. The crystals often have a stellate habit and are green to yellow in colour. In the gemstone trade, yellow chrysoberyl is known as *chrysolite*.
chrysoberyl cat's eye (*Min.*). See **cymophane.**
Chrysocapsales (*Bot.*). An order of the *Chrysophyta*, having a palmate thallus of immobile cells in a gelatinous matrix. Cell division

usually takes place throughout the colony. The cells can become motile by direct metamorphosis.

chrysocolla (*Min.*). A hydrated silicate of copper, often containing free silica and other impurities. It occurs in incrustations or thin seams, usually blue and amorphous.

chrysogonidium (*Bot.*). A yellow algal cell in a lichen.

chrysoidine (*Chem.*). 2,4-Diamino-azobenzene hydrochloride, $C_6H_5{\cdot}N{=}N{\cdot}C_6H_3(NH_2)_2{\cdot}HCl$, an orange-red azo-dyestuff for dyeing direct silk and wool.

chrysolite (*Min.*). Mineralogical name for common *olivine* (q.v.), sometimes restricted to the pale yellowish crystals of gemstone quality, or to a member of the olivine series with 70 to 90% of the magnesium orthosilicate component. The term is incorrectly applied to chrysoberyl of golden yellow colour. *Ceylon chrysolite*: TN for the fine golden yellow chrysoberyl obtained in Ceylon and used as a gemstone.

Chrysomonadales (*Bot.*). An order of the *Chrysophyta*, containing members that are motile during the vegetative phases.

Chrysomonadina (*Zool.*). An order of *Phytomastigina*, comprising forms with one or two flagella; yellow, brown, or colourless; often of amoeboid form; without a gullet or a transverse groove; having food-reserves of leucosin and oil, but not of starch; without a contractile vacuole. Considered by botanists to be plants. See **Chrysomonadales.**

Chrysophaeriales (*Bot.*). An order of the *Chrysophyta*. Exist as single cells or nonfilamentous colonies. The protoplast does not become motile directly.

Chrysophyceae (*Bot.*). A class of the *Chrysophyta*, whose members have a small number of yellow-brown chromatophores. The storage products are fats and leucosin, but never starch.

Chrysophyta (*Bot.*). A division of the *Algae*. The photosynthetic pigment which has a preponderance of carotins is confined to chromatophores. Starch is never stored, and frequently statospores are produced.

chrysoprase (*Min.*). An apple-green variety of chalcedony; the pigmentation is probably due to the oxide of nickel.

chrysopterin (*Biochem.*). An orange pigment formed in the epidermal cells or cavities of the scales and setae of the wings of some *Lepidoptera*.

chrysotherapy (*Med.*). Treatment by injections of gold.

chrysotile (*Min.*). A fibrous variety of serpentine, occurring in small veins. It forms part of the asbestos of commerce. Also called **Canadian asbestos.** Cf. *antigorite*.

Chrysotrichales (*Bot.*). An order of the *Chrysophyta*; distinguished from its other members by the branching filamentous thallus.

C.H.U. or **Chu.** Abbrev. for *Centigrade heat unit*, the same as the *pound-calorie* (q.v.).

chuck (*Eng.*). Device attached to the spindle of a machine tool for gripping the revolving work, cutting tool, or drill.

chucking machine (*Eng.*). A machine tool in which the work is held and driven by a chuck, not supported on centres.

chuffs (*Build.*). Bricks which have been rendered useless owing to the presence in them of cracks caused by rain falling on them while they were hot. Also called **shuffs.**

chumship (*Typog.*). The Scottish equivalent of *companionship* (q.v.).

churn drill (*Mining*). See **cable drill.**

churning loss (*Autos.*). In a gear-box, the power wasted in fluid friction through the pumping action of the revolving gears in the oil.

chute (*Hyd.*). An inclined channel for conducting water to a water-wheel. (*Mining*) (1) An inclined trough for the transference of broken coal or ore. (2) An area of rich ore in an inclined vein or lode, generally of much greater vertical than lateral extent. (*San. Eng.*) A special tapered outlet pipe from a deep inspection chamber, employed to make rodding easier.

chute riffler (*Powder Tech.*). Sampling device consisting of a V-shaped trough from which a series of chutes feed two receiving bins. Alternate chutes feed opposite bins. Particles have in theory an equal chance of being retained in either of the bins.

Chvostek's sign (*Med.*). Twitching of the muscles of the face on tapping the facial nerve; a sign of tetany.

chyle (*Zool.*). In Vertebrates, lymph containing the results of the digestive processes, and having a milky appearance due to the presence of emulsified fats and oils.

chylification, chylifaction (*Zool.*). Formation of *chyle* (q.v.).

chylocaulous (*Bot.*). Having a succulent stem.

chylomicrons (*Physiol.*). Minute particles of emulsified fat present in the blood plasma, particularly after digestion of a fatty meal.

chyloperitoneum (*Med.*). The presence of chyle in the peritoneal cavity as a result of obstruction of the abdominal lymphatics.

chylophyllous (*Bot.*). Having succulent leaves.

chylopoiesis (*Zool.*). See **chylification.**

chylothorax (*Med.*). The presence of chyle in the pleural cavity, due to injury to, or pressure on, the thoracic duct.

chyluria (*Med.*). The presence of chyle in the urine.

chyme (*Zool.*). In Vertebrates, the semifluid mass of partially digested food entering the small intestine from the stomach.

chymosin (*Chem.*). An enzyme of the gastric juice which coagulates casein; rennin.

chymotrypsin (*Chem.*). Proteolytic enzymes secreted by the pancreas which are most active in catalysing hydrolysis of peptide bonds in which the carboxyl group of phenylalamine, tyrosine or tryptophan participates.

Chytridiales (*Bot.*). An order of the *Phycomycetes*. The thallus is simple, never forming a true mycelium. Asexual reproduction is by flagellated zoospores, and sexual reproduction by the fusion of flagellated isogametes.

cibarium (*Zool.*). The anterior or dorsal part of the preoral food cavity in Insects with mandibulate mouth parts. It is separated from the posterior or ventral part (the *salivarium*) by the hypopharynx.

cicatricle, cicatricula (*Zool.*). In heavily yolked eggs, the germinating area of blastoderm.

cicatrix (*Bot.*). A scar left on a plant where a member has been shed. (*Med., Zool.*) The scar left after the healing of a wound; one which marks the previous point of attachment of an organ or structure.

cicero (*Typog.*). See **Didot point system.**

Ciconiiformes (*Zool.*). An order of *Neognathae* having a desmognathous palate and usually webbed feet; all are long-legged birds of aquatic habit, living mainly in marshes and nesting usually in colonies. They are powerful flyers and some migrate over long distances. Storks Herons, Ibises, Spoonbills, and Flamingoes.

CIE (*Light*). Abbrev. for *C*ommission *I*nternationale d'*É*clairage. Formed to study problems of illumination.

CIE coordinates (*Light*). Set of colour coordinates specifying proportions of three theoretical additive primary colours required to produce any hue. These theoretical primaries were established by the CIE and form basis of all comparative colour measurement. See **chromaticity diagram.**

C.I. engine (*Eng.*). See **compression-ignition engine.**

cilia (*Zool.*). In Mammals, the eyelashes; in Birds, the barbicels of a feather; in general, small lashlike processes of a cell which beat rhythmically and cause locomotion, or create a current of fluid. *sing.* **cilium.** *adjs.* **ciliated, ciliate.**

ciliary (*Zool.*). In general, pertaining to or resembling cilia; in Vertebrates, used of certain structures in connexion with the eye, as the *ciliary ganglion, ciliary muscles, ciliary process.*

ciliary organ (*Zool.*). A rudiment of the coelomoduct which does not open to the exterior, situated in each segment in some *Polychaeta.*

Ciliata (*Zool.*). A subclass of *Ciliophora*, the members of which possess cilia throughout life; suctorial tentacles are absent.

ciliate, ciliated (*Bot.*). (1) Having a fringe of long hairs on the margin. (2) Having flagella.

ciliated pits (*Zool.*). In some *Platyhelminthes*, pits on the head wherein lie the chemosensory receptors.

ciliograde (*Zool.*). Moving by the agency of cilia.

ciliolate (*Bot.*). Fringed with very short, fine hairs.

Ciliophora (*Zool.*). A class of *Protozoa*, comprising forms which always possess cilia at some stage of the life-cycle, and usually have a meganucleus.

ciliospore (*Zool.*). In *Protozoa*, a ciliated swarm-spore. In *Suctoria*, a bud produced by asexual reproduction.

cilium (*Bot.*). (1) A hairlike appendage to a spore. (2) See **flagellum.** *pl.* **cilia.** (*Zool.*) *Sing.* of *cilia* (q.v.).

Ciment Fondu (*Civ. Eng.*). TN for a type of very rapid-hardening cement made by heating lime and alumina in an electric furnace to incipient fusion, and afterwards grinding to powder.

ciminite (*Geol.*). A volcanic rock found in central Italy; an olivine-bearing trachyte.

cinching (*Cinema.*). Tightening a roll of film by holding the centre and pulling the edge.

cinch marks (*Cinema.*). Marks on film caused by friction or by irregularity in winding.

cinchocaine (*Pharm.*). 2-Butoxy-*N*-(2-diethylaminoethyl) cinchonamide. A long-lasting local anaesthetic. The hydrochloride is used for spinal injection. TN **Nupercaine.**

cinchona bases (*Chem.*). Alkaloids present in cinchona bark, derivatives of quinoline.

cinchonine (*Chem.*). $C_{19}H_{22}ON_2$, an alkaloid of the quinoline group, found in cinchona and cuprea barks; crystallizes in rhombic prisms from alcohol, m.p. 264°C. It behaves as a diacidic base and gives two series of salts.

cinchophen (*Pharm.*). *Quinophan*, or *2-phenylquinoline-4-carboxylic acid*:

COOH

N C_6H_5

An analgesic sometimes used for gout.

Cincinnatian (*Geol.*). Strata in N. America equivalent in age to the Upper Ordovician of N.W. Europe.

cincinnus (*Bot.*). A cymose inflorescence in which the lateral axes arise on alternate sides of the relatively main axis.

cinclides (*Zool.*). In *Anthozoa*, perforations of the body-wall for the extrusion of acontia. *Sing.* **cinclis.**

cincture (*Arch.*). A plain ring or fillet round a column, generally placed at the top and bottom to separate the shaft from capital and base.

cinder coal (*Heat*). Coal altered by heat associated with natural igneous activity. Also **natural coke.**

cinder pig (*Met.*). Pig iron made from a charge containing a considerable proportion of slag from puddling or reheating furnaces.

cinders (*Geol.*). Volcanic *lapilli* composed mainly of dark glass and containing numerous vesicles (air or gas bubbles).

ciné camera (*Cinema.*). Name usually applied to motion-picture cameras using film smaller in width than standard stock, usually 16 and 8 mm; used mainly for amateur and scientific cinematography.

CinemaScope (*Cinema.*). TN for cinematography in which a wide field is compressed in distorted form on to normal-size film by an *anamorphic lens* (q.v.), and rectified by a similar lens in the projector, giving a picture up to 60 ft (18 m) wide.

cinematograph camera (*Cinema.*). See **ciné camera, motion-picture camera.**

cineole (*Chem.*). Eucalyptole, $C_{10}H_{18}O$, b.p. 176°C; optically inactive, considered to be the internal anhydride or ether of terpin. It has no alcoholic or ketonic properties, but forms hydrochlorides, phosphates, arsenates, and additive compounds with hydroxybenzenes.

Cinerama (*Cinema.*). TN for cinematography in which 3 cameras cover adjacent fields of view, giving a total horizontal field of 146°. Using 3 projectors, the combined image is formed on a curved screen 51 ft (15 m) wide.

cinereous (*Bot.*). Grey, like wood ashes.

cingulum (*Bot.*). The attachment between the two valves of the cell wall of a diatom. (*Zool.*) Any girdle-shaped structure. In *Annelida*, the clitellum; in *Rotifera*, the outer post-oral ring of cilia; in Mammals, a tract of fibres connecting the hippocampal and callosal convolutions of the brain; in Mammals, a ridge surrounding the base of the crown of a tooth and serving to protect the gums from the hard parts of food.

cinnabar (*Min.*). Sulphide of mercury, HgS, occurring as red acicular crystals, or massive; the ore of mercury, worked extensively at Almadén, Spain, and elsewhere.

cinnabarine (*Bot.*). Bright, orange-red.

cinnamic acid (*Chem.*). 3-Phenylpropenoic acid. $C_6H_5{\cdot}CH{=}CH{\cdot}COOH$, m.p. 133°C, b.p. 300°C; an unsaturated monobasic aromatic acid, prepared from benzal chloride by heating with sodium acetate. It exhibits a peculiar isomerism, as there are known a *trans*-form and 3 *cis*-forms—Liebermann's *iso*-cinnamic acid, m.p. 57°C; Erlenmeyer's (sen.) *iso*-cinnamic acid, m.p. 38°–46°C; Liebermann's *allo*-cinnamic acid, m.p. 68°C—which, however, are not chemically isomeric, but represent a case of trimorphism.

cinnamic aldehyde (*Chem.*). 3-phenylpropenal $C_6H_5{\cdot}CH{=}CH{\cdot}CHO$, b.p. 246°C; an oil of aromatic odour; the chief constituent of cinnamon oil and of cassia oil.

cinnamon stone (*Min.*). See **hessonite.**

cinquefoil (*Arch.*). A 5-leaved ornament used in panellings, etc.
cipher (*Acous.*). See under **ventil.**
cipher tunnel (*Arch.*). A false chimney built on to a house for symmetrical effect.
cipolin (*Build.*). A white marble with green streaks; used for decorative purposes.
Cipolletti weir (*Hyd.*). A weir in which the notch plate has a trapezoidal opening tapering down to the sill with side slopes of 1 (horizontal) in 4 (vertical).
C.I.P.W. (*Geol.*). A quantitative scheme of rock classification based on the comparison of norms; devised by four American petrologists, *C*ross, *I*ddings, *P*irsson, and *W*ashington.
circadian rhythm (*An. Behav.*). A cyclical variation in the intensity of a metabolic or physiological process, or of some facet of behaviour, of ca. 24 hours. It may be *endogenous*, i.e., originating from within an organism, and persisting when external conditions are held constant. This is thought by some, however, to be a response to similar variation in some geophysical stimulus. Although often not exactly 24 hours in duration, these can be synchronized or entrained by regular 24-hourly changes in, e.g., light intensity or temperature. The stimulus responsible for this is the *zeitgeber* (q.v.). Or it may be *exogenous*, i.e., being the response of an organism to a regular cycle of some external factor. This fails to persist when conditions are held constant.
circinate (*Bot.*). Rolled inwards from the apex towards the base, resembling a crozier in form. (*Med.*) Rounded, circular.
circle (*Maths.*). A plane curve which is the locus of a point which moves so that it is at a constant distance (the *radius*) from a fixed point (the *centre*). The length of the circumference of a circle is $2\pi r$, and its area πr^2, where r is the radius and π is equal to 3·141 593 (to 6 places). The cartesian equation of a circle, with centre (a, b) and radius k, is $(x-a)^2+(y-b)^2=k^2$. (*Textiles*) In lace manufacture, the arc formed by the combs on which the carriages ride.
circle coefficient (*Elec. Eng.*). A term often used to denote the leakage factor of an induction motor.
circle diagram (*Elec. Eng.*). Graphical representation of complex impedances at different points in a transmission system on an orthogonal network. Best known example is a *Smith chart.*
circle of confusion (or **diffusion**) (*Photog.*). The minimum area of a focused bright point of light, the size of which determines the maximum definition possible with a given lens arrangement and stop.
circle of convergence (*Maths.*). Circle within which a complex power series converges. Its radius is called the *radius of convergence.*
circle of curvature (*Maths.*). See **curvature.**
circle of inversion (*Maths.*). See **inversion.**
circle of least confusion (*Photog.*). The image of a point source of light cannot be perfect, and the diameter of this image is a measure of the limit of definition. The maximum permissible diameter for sharp definition is taken to be 0·01 in. (0·25 mm).
circle of position (*Nav.*). A small circle on the earth's surface obtained by any single observation of the altitude of a known star as the locus of the observer's position; the radius of the circle in nautical miles being the zenith distance of the body in minutes of arc, and its centre the substellar point.
circling disease (*Vet.*). See **listeriosis.**
circlip (*Eng.*). A spring washer in the form of an incomplete circle, usually used as a retaining ring (e.g., for ball bearings). It is mounted in a circular groove in a hole or shaft by temporarily distorting it (closing or opening the circular shape).
circuit (*Elec. Eng.*). Series of conductors, forming a partial branched or complete path, which substantially confines the flow of electrons forming a current because of great contrast in conductivity. (*Telecomm.*) The whole or part of the path of transmitted electrical energy in a communication channel.
circuital magnetization (*Mag.*). See **solenoidal magnetization.**
circuit-breaker (*Elec. Eng.*). Device for opening electric circuit under abnormal operating conditions, e.g., excessive current, heat, high ambient radiation level, etc. Also **contact-breaker.** See **oil switch, air-blast switch.**
circuit cheater (*Elec. Eng.*). One which, for test purposes, simulates a component or load. Cf. *dummy load.*
circuit diagram (*Elec. Eng.*). Conventional representation of wiring system of electrical or electronic equipment.
circuit noise (*Electronics*). See **thermal noise.**
circuit parameters (*Elec. Eng.*). Relevant values of physical constants associated with circuit elements.
circulant (*Maths.*). A determinant in which each row is a cyclic permutation by one position of the previous row.
circular cone (*Maths.*). See **cone** (2).
circular error (*Horol.*). The variation in the isochronism of a pendulum due to the path of the bob being that of a circular arc instead of a cycloid. For small amplitudes the error is small, and for precision clocks the amplitude is kept to about two degrees on either side of the line of suspension.
circular form tool (*Eng.*). A ring-shaped profile cutter which is gashed to have a substantially radial surface. The curved surface is shaped, across its width, to correspond to the contour of the part to be produced, the cutting edge being formed by the junction of the radial and the curved surfaces.
circular functions (*Maths.*). The *trigonometrical functions* (q.v.), more particularly when defined with radian argument. Cf. *hyperbolic functions* and *elliptic functions.* All these functions are so named because of their association with the rectification of the similarly named curves.
circular level (*Surv.*). Spirit-level with bubble housed under slightly concave glass.
circular magnetization (*Mag.*). The magnetization of cylindrical magnetic material in such a way that the lines of force are circumferential.
circular measure (*Maths.*). The expression of an angle in *radians*, one radian being the angle subtended at the centre of a circle by an arc of length equal to the radius. There are thus 2π, or approximately 6·283, radians in one complete revolution. 1 radian = 57·2958°; 1 degree = 0·017 453 3 radians.
circular mil (*Elec. Eng.*). U.S. unit for wire sizes, equal to area of wire 1 mil (0·001 in. = 0·025 mm) diameter.
circular mitre (*Carp., Join.*). A mitre formed between a curved and a straight piece.
circular pallets (*Horol.*). *Pallets* (q.v.) equidistant from pallet-staff axis.
circular permutation (*Maths.*). An arrangement of objects in a circle. There are $(n-1)!$ different circular permutations of n objects. See **permutations.**

circular pitch (*Eng.*). The distance between corresponding points on adjacent teeth of a gear-wheel, measured along the *pitch circle*.

circular plane (*Carp.*). A plane adapted (through the use of special irons) for producing curved surfaces, either convex or concave.

circular point on a surface (*Maths.*). A point at which the principal curvatures are equal.

circular points (*Maths.*). Two conjugate imaginary points on the line at infinity in a plane, common to all circles in the plane: denoted by I and J.

circular polarization (*Radio*). State of polarization of an electromagnetic wave when its electric and magnetic fields each contain two equal components, at right angles in space and in phase quadrature.

circular saw (*Eng.*). A steel disk carrying teeth on its periphery, used for sawing wood, metal, or other materials; usually power-driven. See **cold saw, hot saw.**

circular shift (*Comp.*). See **end-around shift.**

circular time base (*Electronics*). Circuit for causing the spot on the screen of a CRT to traverse a circular path at constant angular velocity.

circulating current (*Elec. Eng.*). That which flows round the loop of a complete circuit, as contrasted with *longitudinal current*, which flows along the two sides or *legs* of the same circuit, in parallel.

circulating-current protective system (*Elec. Eng.*). A form of Merz-Price protective system in which the current transformers at the two ends of the circuit to be protected are arranged to circulate a current round the pilots, any difference in the currents from the two transformers passing through a relay.

circulating memory (*Comp.*). One in which the impulses in a sonic delay line store are taken from the output, re-shaped, and re-inserted in the input. The delay line may be a column of mercury, or a fine nickel wire (magnetostriction).

circulating pump (*Eng.*). A pump, usually of centrifugal type, used to circulate cooling water or, more generally, any liquid. See **centrifugal pump.**

circulation (*Aero.*). Commonly used to describe the lift-producing airflow round an aerofoil, but defined as the integral of the component of the fluid velocity along any closed path with respect to the distance round the path. See **super-circulation.** (*Bot.*) A rotatory movement of the protoplasm inside a cell. (*Maths.*) Of a vector, the line integral of the vector along a closed path in the field of vector. (*Physiol.*) The continuous movement of the blood through the heart, arteries, capillaries, and veins.

circulation of electrolyte (*Elec. Eng.*). Movement of the electrolyte in an electroplating bath in order to ensure an even deposit.

circulator (*Radar, Telecomm.*). One of a class of non-reciprocal devices which rely on the tensor form of the permeability in a magnetized ferrite. Commercial devices can operate over an octave band from the UHF to millimetre region with a VSWR $< 1{\cdot}2$, insertion loss $< 0{\cdot}3$ dB and isolation > 20 dB.

circulatory integral (*Maths.*). The line integral of a function over a closed curve.

circulatory system (*Zool.*). A system of organs through which is maintained a constant flow of fluid, which facilitates the transport of materials between the different organs and parts of the body.

circulus (*Zool.*). An annular arrangement of blood vessels, as the *circulus cephalicus* of Fish.

circulus venosus (*Zool.*). In *Gastropoda*, the vessel through which blood from the haemocoele passes to the lung and heart.

circumcentre of a triangle (*Maths.*). The centre of the circumscribed circle. The point at which the perpendicular bisectors of the sides intersect.

circumcircle (*Maths.*). See **circumscribed circle.**

circumcision (*Surg.*). Surgical removal of the prepuce or foreskin in the male or of the labia minora in the female.

circumferential register adjustment (*Print.*). The maintaining on rotary presses of one operation in relation to others in the direction of the web, manual control being superseded by electronic equipment, such as an *Autotron scanner*.

circumferential register screws (*Print.*). Adjustable screws, in place of margin bar, allowing slight movement of the plate around the cylinder on rotary presses.

circumfili (*Zool.*). Conspicuous whorls of hair occurring on the antennae of certain Insects, as Gall Midges (*Cecidomyidae*).

circumfluence (*Zool.*). See **circumvallation.**

circumnutation (*Bot.*). The rotation of the tip of an elongating stem, so that it traces a helical curve in space.

circumpolar stars (*Astron.*). Those stars which, for a given locality on the earth, do not rise and set but revolve about the elevated celestial pole, always above the horizon. To be circumpolar, a star's declination must exceed the co-latitude of the place in question.

circumscissile (*Bot.*). Splitting open by a circular tear.

circumscribed circle (*Maths.*). Of a polygon: the circle which passes through all its vertices. Also called **circumcircle.**

circumvallate papillae (*Histol.*). The largest of the lingual papillae in Man. Many taste buds are found in the stratified epithelium of their walls.

circumvallation (*Zool.*). The action of a phagocytic cell, or of a Sarcodine, in engulfing foreign bodies or food particles by the extrusion of pseudopodia.

circus movements (*An. Behav.*). The continuous turning movements made by some animals if one eye is put out of action by, e.g., paint, and illumination comes from above.

cire perdue (*Met.*). See **investment casting.**

cirque or **corrie** (*Geol.*). A semi-amphitheatre, or 'armchair-shaped' hollow, of large size, excavated in mountain country by, or under the influence of, ice.

cirrals (*Zool.*). In *Crinoidea*, the hollow ossicles supporting the cirri.

cirrate, cirriferous (*Bot., Zool.*). Bearing cirri.

cirrhosis (*Med.*). (1) A disease of the liver in which there is increase of fibrous tissue and destruction of liver cells. (2) Diffuse increase of fibrous tissue in any organ. (Greek *kirros*, orange-tawny.)

Cirripedia (*Zool.*). Subclass of marine *Crustacea*, generally of sessile habit when adult; the young are always free-swimming; the adult possesses an indistinctly segmented body which is partially hidden by a mantle containing calcareous shell plates; there are 6 pairs of biramous thoracic legs; attachment is by the antennules; many species are parasitic. Barnacles.

cirro-cumulus (*Meteor.*). Small rounded masses or white flakes of cloud without shadows, arranged in groups and often in lines. They are composed of ice crystals and occur at heights above 25 000 ft (7500 m). Also called **mackerel sky.**

cirrose (*Bot., Zool.*). Curly, like a waved hair. Consisting of diverging filaments.

cirro-stratus (*Meteor.*). A thin sheet of whitish cloud; sometimes covering the sky completely and giving it a milky appearance (it is then called *cirro-nebula* or *cirrus haze*); at other times presenting more or less distinctly a fibrous structure like a tangled web. This type of cloud, being composed of ice crystals, often produces haloes of radius 22° around the sun and moon.

cirrus (*Bot.*). A *tendril* (q.v.). (*Zool.*) In *Protozoa*, a stout conical vibratile process, formed by the union of cilia; in some *Platyhelminthes*, a copulatory organ formed by the protrusible terminal part of the vas deferens; in *Annelida*, a filamentous tactile and respiratory appendage; in *Cirripedia*, a ramus of a thoracic appendage; in Insects, a hairlike structure on an appendage; in *Crinoidea*, a slender jointed filament arising from the stalk or from the centrodorsal ossicle and used for temporary attachment; in Fish, a barbel. (*Meteor.*) Detached clouds of delicate appearance, fibrous structure, and featherlike form; white in colour; composed of ice crystals; occur at heights between 25 000 and 35 000 ft (7500 and 11 000 m).

cirsoid aneurysm (*Med.*). A mass of newly formed, tortuous, and dilated arteries.

cis- (*Chem.*). A prefix indicating that geometrical isomer in which the 2 radicals are situated on the same side of the plane of a double bond or alicyclic ring.

cissing (*Paint.*). A defect in paint, enamel, or varnish work due to poor adhesion; characterized by the appearance of pinholes, craters, and in serious cases the contraction of the wet paint to form blobs on the surface of the work.

cissoid (*Maths.*). The inverse of a parabola with respect to its vertex. Its cartesian equation is $x(x^2+y^2)=by^2$.

cistron (*Cyt.*). Functional unit of DNA chain (chromosome) controlling protein structure.

citral (*Chem.*). Geranial, $C_{10}H_{16}O$, an alkene-type terpene, formula $CH_2{=}CMe{\cdot}CH_2{\cdot}CH_2{\cdot}CH_2{\cdot}CMe{=}CH{\cdot}CHO$; b.p. 110°–112°C (12 mm); it occurs in the oil of lemons and oranges and in lemon-grass oil. Used as a flavouring and in perfumery.

citrates (*Chem.*). The salts of citric acid.

citrene (*Chem.*). See ***d*-limonene**.

citric acid (*Chem.*). 2-hydroxypropane-1,2,3-tricarboxylic acid. $C_6H_8O_7$, an important hydroxy-tricarboxylic acid, occurs in the free state in the juices of many fruits, especially lemons, but is now prepared commercially largely by fermentation with *Aspergillus*. Much used for flavouring effervescent drinks, and there are also industrial applications.

citric acid cycle (*Biochem.*). This series of reactions for the breakdown of acetyl coenzyme A operates in the mitochondria. It is the major source of the energy required by a living cell. For each revolution of the cycle two moles of CO_2 are produced and electrons provided for the *electron transport chain* which generates ATP and accomplishes the reduction of oxygen. The net process may be considered as

$$CH_3COOH+2O_2 \rightarrow 2CO_2+2H_2O+\text{energy}.$$

oxaloacetic acid → citric acid (+acetyl CoA, $+H_2O$)
citric acid ↓ $-H_2O$ cis-aconitic acid ↓ $+H_2O$ isocitric acid ↓ $-2H$ oxalosuccinic acid
oxalosuccinic acid → $-CO_2$ → α-ketoglutaric acid → $-CO_2$, $+H_2O$, $-2H$ → succinic acid
succinic acid → fumaric acid ↑ $-2H$
fumaric acid → malic acid ↑ $+H_2O$
malic acid → oxaloacetic acid ↑ $-2H$

citriform (*Bot.*). Lemon-shaped.

citrine (*Bot.*). Citron-colour; lemon-coloured. (*Min.*) Not the true *topaz* (q.v.) of mineralogists, but a yellow variety of quartz, which closely resembles it in colour though not in other physical characters; it is of much less value than true topaz. It figures under a variety of geographical names, e.g., *Bohemian topaz*, *Indian topaz*, *Madagascar topaz*, *Madeira topaz*, and *Spanish topaz* (q.v.). Also called **false topaz**. See also **cairngorm**, **Scottish topaz**; and cf. *Brazilian topaz*, the true mineral.

citromycetin (*Chem.*). Yellow crystals, $C_{14}H_{10}O_7$, used as a colouring agent, obtained from *Citromyces* moulds.

citronellal (*Chem.*). Also known as rhodinal; $C_{10}H_{18}O$, an aldehyde forming the main constituent of citronella oil and lemon-grass oil. It is used in perfumery.

citron yellow (*Paint.*). See **zinc yellow**.

citrulline (*Biochem.*). An amino acid, $C_6H_{13}O_3N_3$, occurring as an intermediate product in the formation of urea from ornithine; also found in water-melon.

civery (*Arch.*). One bay of a vaulted ceiling. Also spelt **severy**.

Civil Aviation Authority (*Aero.*). An independent body which controls the technical, economic, and safety regulations of British civil aviation. In 1972 it took over the relevant functions of the Department of Trade and Industry, and those of the Air Registration Board and the Air Transport Licensing Board. It is also responsible for the civil side of the Joint National Air Traffic Control Services. Abbrev. **CAA**.

civil engineering. The design and construction of roads, railways, bridges, aqueducts, canals, docks, ports, harbours, moles, breakwaters, lighthouses, and drainage works.

civil twilight (*Astron.*). The interval of time during which the sun is between the horizon and 6° below the horizon, morning and evening.

Cl (*Chem.*). The symbol for *chlorine*.

clack or **clack valve** (*Eng.*). A check valve admitting water from a feed pump to the boiler of a locomotive. A ball valve is used, the name *clack* being derived from the characteristic sound of the ball striking its seat.

clacking (*Vet.*). Forging; an error of gait in the horse in which the toe of a hind foot strikes the sole or shoe of a fore foot.

cladding (*Build.*). The material used for the external lining of a building. (*Mining*) The material used for the lining of a mine shaft. Traditionally of timber battens or planks, now frequently of precast reinforced concrete slabs. (*Nuc. Eng.*) Thin covering, e.g., of reactor fuel units to contain fission products and of dielectric rods to form light guides.

cladocarpous (*Bot.*). Having the fruit at the end of a lateral shoot.

Cladocera (*Zool.*). A suborder of the order *Diplostraca*. Water Fleas.

Cladocopa (*Zool.*). An order of *Ostracoda*, in which the shell lacks an antennal notch; there are 3 pairs of postoral limbs and the caudal furca have lamellar rami armed with spines.

cladode (*Bot.*). A flattened branch, which looks like, and functions as, a leaf.

cladogenous (*Bot.*). Producing the inflorescences at the ends of branches.

Cladophorales (*Bot.*). An order of the *Chlorophyta*. They form filaments of multinucleate cells, which contain several discoid chloroplasts.
cladophyll (*Bot.*). A cladode.
cladoptosis (*Bot.*). The shedding of branches.
cladosiphonic (*Bot.*). A term applied to a siphonostele which has branch gaps, but not leaf gaps.
Cladoxylales (*Bot.*). An order of fossil ferns; monostelic and, if the leaves are pinnate, the pinnae are at an angle to the plane of the leaf blade. The sporangia are borne in clusters directly on the rachis or on unflattened pinnules.
clagging (*Foundry*). The adhesion of blacking to a trowel or sleeker during smoothing of the surface of a mould.
Clairaut's differential equation (*Maths.*). An equation of the form

$$y=xp+f(p),$$

where $p=\frac{dy}{dx}$. The general solution is

$$y=cx+f(c).$$

clairecolle (*Paint.*). See **clearcole.**
Claisen condensation (*Chem.*). An important synthetic reaction involving condensation between esters, or between esters and ketones, in the presence of sodium ethoxide (ethanolate).
Claisen flask (*Chem.*). A distillation flask used for vacuum distillations; it consists of a glass bulb with a neck for a thermometer, to which another neck with outlet tube is attached.
Claisen reaction (*Chem.*). The elimination of water from an aliphatic and an aromatic aldehyde in the presence of sodium hydroxide, an unsaturated aldehyde being formed, e.g., cinnamic aldehyde is thus formed from ethanol and benzenecarbaldehyde.
clamp (*Build.*). A stack of dried raw bricks built up for burning, together with cinders and coal, over a system of flues roughly formed with burnt bricks. (*Telecomm.*) Circuit in which a waveform is adjusted and maintained at a definite level when recurring after intervals. (*Tools*) See **cramp.**
clamp bar (*Print.*). (1) A bar used to lock or unlock plates to a letterpress rotary. (2) A bar to hold blankets in position on the pins of the impression cylinder of a letterpress rotary or the blanket cylinder of an offset press.
clamp connexion (*Bot.*). A short, backwardly directed hypha, present across the septa of hyphae of some *Basidiomycetes*, formed as the nuclei divide, and making possible the distribution of the daughter nuclei in the two segments formed from the terminal segment of the hypha.
clamping diode (*Electronics*). A *diode* used to clamp a voltage at some point in a circuit.
clamping screw (*Elec. Eng.*). A screw for holding a conductor to a piece of apparatus used in electrical installation work, e.g., a ceiling rose or switch. Also called a **terminal screw.**
clams (*Vet.*). Instruments, usually made of wood, for effecting compression; used commonly for castrating stallions and bulls.
clan (*Bot.*). A small clump formed of individuals of a single species, and developed either as a result of vegetative multiplication or from seed scattered over a small area. (*Geol.*) A suite of igneous rock-types closely related in chemical composition but differing in mode of occurrence, texture, and possibly in mineral contents.
clandestine (*Gen.*). Development of adult characters from ancestral characters; evolution which is not apparent in adult forms.
clap-board (*Carp.*). A form of weather-board which is tongued and rebated and frequently moulded, rather than being feather-edged.
clap-board gauge (*Carp.*). A gauge used in fixing clap-boarding, to ensure that each board is so set as to expose a parallel width.
clapper box (*Eng.*). A slotted tool head carried on the saddle of a planing or a shaping machine. It carries a pivoted block to which the tool is clamped, thus allowing the tool to swing clear of the work on the return stroke of the table, or the head.
clappers (*Cinema.*). A device, usually consisting of two hinged pieces of board, which is closed sharply in front of a camera in sound-film production, thus providing synchronization indication on the picture-track and the sound-track, the noise of the clappers resulting in a characteristic record on the latter.
Clapp oscillator (*Radio*). Low-drift *Colpitts oscillator*.
clap post (*Join.*). The upright post against which the door of a cupboard 'claps' in closing.
clap-sill (*Hyd. Eng.*). See **mitre-sill.**
clarain (*Min.*). A separable constituent of bright coal.
clarendon (*Typog.*). A heavy type-face: **Clarendon.**
Claret-Vuilleumier surface-contact system (*Elec. Eng.*). A system of surface-contact traction in which the supply to the studs is controlled by electromagnetic distributor switches at intervals along the track. Formerly used in Paris.
Clark cell (*Chem.*). A standard cell consisting of a zinc anode and a mercury cathode covered with mercury (I) sulphate paste, both dipping into a saturated solution of zinc sulphate. Used for calibrating sources of e.m.f. as in a potentiometer.
Clark process (*Chem.*). A process for effecting the partial softening of water by the addition of sufficient limewater to convert all the acid carbonates of lime and magnesium into the normal carbonates.
clasmatocyte (*Zool.*). A large actively phagocytic cell occurring in areolar tissue and having a marked tendency to take up vital stains.
Clasp (*Arch.*). A system developed in the U.K. mainly for the construction of schools. It incorporates light steel framework, pin-jointed, and with prefabricated timber decking units which are bolted to the framework so as to form a continuous stressed diaphragm. *Cladding* (q.v.) varies from concrete slabs to curtain walling.
claspers (*Zool.*). In Insects, an outer pair of gonapophyses; in male Selachian Fish, the inner narrow lobe of the pelvic fin, used in copulation; more generally, any organ used by the sexes for clasping one another during copulation.
clasp nail (*Build.*). A square-section cut nail whose head has 2 pointed projections that sink into the wood.
clasp nut (*Eng.*). A nut split diametrally into halves, which may be closed so as to engage with a threaded shaft; used as a clutch between a lathe lead screw and the saddle.
class (*Biol., Bot., Zool.*). In biometry, a group of organisms all falling within the same range, as indicated by the unit of measurement employed; in the animal and plant kingdoms, one of the taxonomic groups into which a phylum (in Botany, more usually a division) is divided, ranking next above an order.
class-A amplifier (*Telecomm.*). One in which the polarizing voltages are adjusted for operation on the linear portion of the characteristic curves of the valve, without grid current. The suffix 1 is added if grid current does not flow during any part of the cycle, and 2 if it does.

(This also applies to class-AB, -B, and -C amplifiers.)
class-AB amplifier (*Telecomm.*). One in which the valve has its grid bias so adjusted that the operation is intermediate between classes A and B, i.e., the anode current is shut off during part of the excitation cycle, but the quiescent anode current is not reduced to a small value.
class-A, -B, -C, etc., insulating materials (*Elec. Eng.*). Classification of insulating materials according to the temperature which they may be expected to withstand.
class-A oils (*Elec.*). See **oils.**
class-B amplifier (*Telecomm.*). One in which the grid bias is adjusted to give the lower cut-off in anode current. Applied, colloquially, also to the combination of 2 such valves in one envelope, the 2 valves being designed to operate with grid current and substantially zero grid bias, and in antiphase.
class-B oils (*Elec.*). See **oils.**
class-C amplifier (*Telecomm.*). One in which the grid bias is greatly in excess of that in class B, and in which the anode output power becomes proportional to the anode voltage for a given grid excitation. In a given valve, the anode current will flow for less than one-half of each cycle of an alternating voltage applied to the grid.
class frequency (*Stats.*). The frequency with which a variate assumes a value in a particular class interval.
classical (*Phys.*). Said of theories based on concepts established before relativity and quantum mechanics, i.e., largely in conformity with Newton's mechanics and Maxwell's electromagnetic theory. U.S. term **non-quantized.**
classical conditioning (*An. Behav.*). An experimental procedure resulting in the acquisition by an animal of the capacity to respond to a given, previously neutral, stimulus (the *conditioned stimulus* or *CS*) with the reflex action or response proper to another stimulus (the *unconditioned stimulus* or *UCS*), when the two stimuli have been presented concurrently for a number of times. The original *unconditioned reflex* (*UCR*) response, when given in response to the CS, is called a *conditioned response* (*CR*).
classical flutter (*Aero.*). See **flutter.**
classical scattering (*Phys.*). See **Thomson scattering.**
classification (*Powder Tech.*). Grading in accordance with particle size, shape, and density, by fluid means.
classification of ships (*Ships*). British passenger ships are classified by the Ministry of Transport according to the nature of the voyages in which they are engaged. *Classification societies* (q.v.) assign ships to a class in the Register Book so long as they are built, equipped, and maintained in accordance with the rules of the society.
classification of the elements (*Chem.*). See **periodic system.**
classification societies (*Ships*). Classification societies exist for the purpose of surveying ships during construction and periodically during service to be able to certify that they are structurally fit for their intended service.
classification test (*Psychol.*). A type of mental test in which a subject is required to mark off words designating objects which belong together.
classifier (*Met.*). A machine for separating the product of ore-crushing plant into 2 portions consisting of particles of different sizes. In general, the finer particles are carried off by a stream of water, while the larger settle. The fine portion is known as the *overflow* or *slime*, the coarse as the *underflow* or *sand.*
class intervals (*Stats.*). The intervals resulting from the division of the range of possible values into suitable groupings.
clastic rocks (*Geol.*). Rocks formed of fragments of pre-existing rocks.
clathrate (*Bot.*). Like lattice-work. (*Chem.*) Form of compound in which one component is combined with another by the enclosure of one kind of molecule by the structure of another; e.g., rare gas in 1,4-dihydroxybenzene. See **molecular sieve.**
Claude process (*Chem.*). A method of liquefying air in stages, the expanding gas being cooled by external work on pistons.
claudication (*Med.*). The action of limping. See **intermittent claudication.**
Claudius' cells (*Zool.*). In Mammals, cuboidal epithelium cells lying on the basilar membrane of the cochlea.
Claus' blue (*Chem.*). A name sometimes applied to the blue solution of rhodium(VI)oxide in alkali hydroxide, considered to be an alkali rhodate.
Clausius-Clapeyron equation (*Chem.*). This shows the influence of pressure on the temperature at which a change of state occurs, and the variation of vapour pressure with temperature.

$$q = T \cdot \Delta v \cdot \frac{dp}{dT},$$

where q is the heat absorbed (latent heat), T is the absolute temperature, Δv is the change of volume.
Clausius-Mosotti equation (*Elec.*). One relating electrical polarizability to permittivity, principally for fluid dielectrics.
Clausius' theorem (*Heat*). For any system undergoing any reversible cycle of changes, in which it finally returns to its initial state, $\oint \frac{\Delta Q}{T} = 0$, where ΔQ represents the infinitesimal quantity of heat absorbed by the system at temperature T kelvins.
Claus kiln (*Chem. Eng.*). The brickwork equipment in which the sulphur recovery stage of the *Claus process* operates.
Claus process (*Chem. Eng.*). A process for recovering elemental sulphur from H_2S, by *desulphurizing* (q.v.). It is a 2-stage process as defined in the 2 chemical equations. The H_2S is divided into 2 streams.

1st step $2H_2S + 3O_2 = 2H_2O + 2SO_2$.
2nd step $2H_2S + SO_2 = 2H_2O + 3S$.

The second step is carried out at a high enough temperature to produce dry sulphur.
claustrophobia (*Med.*). Abnormal fear of being in a confined space.
claustrum (*Zool.*). In *Ostariophysi*, one of the Weberian ossicles, but not forming part of the Weberian chain; in higher Vertebrates, a thin layer of grey matter in the cerebral hemispheres.
clava (*Bot.*). A club-shaped stroma formed by a fungus. (*Zool.*) A gradual swelling at the distal end of a structure, resembling a club.
clavate. *Adj.* from *clava*; shaped like a club, e.g., *clavate* antennae.
clavicle (*Zool.*). In Vertebrates, the collar bone, an anterior membrane bone of the pectoral girdle. *adj.* **clavicular.**
clavicorn (*Zool.*). Having clavate antennae, as some *Coleoptera* and *Lepidoptera.*
clavicularium (*Zool.*). See **epiplastron.**

clavola (*Zool.*). In Insects, the remainder of the antenna, excluding the scape and the pedicel.

clavula (*Zool.*). In *Spatangoida*, a small ciliated spine occurring in a fasciole; in *Porifera*, a rod-shaped spicule bearing a discoidal expansion at one end.

clavulate (*Bot.*). Somewhat club-shaped, but not so distinctly as to be *clavate*.

clavus (*Zool.*). The narrow area of the hardened basal part of the hemelytron of *Hemiptera*, adjacent to the scutellum. Cf. *corium*.

clavus hystericus (*Med.*). Pain in and tenderness of the scalp, with the sensation as if a nail were being driven into the head; it occurs in hysteria.

claw (*Bot.*). The narrow, elongated lower portion of a petal in some plants. (*Carp.*) A small tool with a bent and split end, used for extracting tacks. (*Zool.*) A curved, sharp-pointed process at the distal extremity of a limb; in Vertebrates, a nail which tapers to a sharp point.

claw bolt (*Carp.*). A wrought-iron bolt with a long head flattened in a direction parallel to the length of the bolt, and bent over at right angles near the end.

claw chisel (*Build.*). A chisel, having a 2-in. (50 mm)-long serrated cutting edge, used for rough-dressing building stone.

claw coupling or **claw clutch** (*Eng.*). A shaft coupling in which flanges carried by each shaft engage through teeth cut in their opposing faces, one flange being slidable axially for disengaging the drive.

claw foot (*Med.*). See **pes cavus.**

claw-hammer (*Carp.*). A hammer having a bent and split peen which may be used for extracting a nail by giving leverage under its head.

claw hand (*Med.*). Clawlike position adopted by the hand when the muscles supplied by the ulnar nerve are paralysed.

claw ill (*Vet.*). See **foul in the foot.**

claws (*Cinema.*). The means whereby the motion-picture film is intermittently fed forward through the picture gate, the claws operating on the sprocket holes.

clay (*Geol., etc.*). A fine-textured, sedimentary, or residual, deposit. It consists of hydrated silicates of aluminium mixed with various impurities. Clay for use in the manufacture of pottery and bricks must be fine-grained and sufficiently plastic to be moulded when wet; it must retain its shape when dried, and sinter together, forming a hard coherent mass without losing its original shape, when heated to a sufficiently high temperature. In the mechanical analysis of soil, according to international classification, clay has a grain-size less than 0·002 mm.

Claydon effect (*Photog.*). Partial reversal of an image arising from an intense but brief source, with general exposure to a much weaker light intensity.

C-layer (*Radio*). Reflecting or scattering region between about 35 and 70 km above the earth's surface, postulated to explain return signals sometimes obtained with vertically-radiated waves.

clay gun (*Met.*). Arrangement which shoots a ball of fireclay into a blast furnace's taphole.

claying (*Civ. Eng.*). The operation of lining a blast-hole with clay to prevent the charge from getting damp.

clay ironstone (*Geol.*). Nodular beds of clay and iron minerals, often associated with the Coal Measure rocks.

clay puddle (*Civ. Eng.*). A plastic material produced by mixing clay thoroughly with about one-fifth of its weight of water. It is used in engineering construction to prevent the passage of water, e.g., for cores for earthen reservoir dams.

clay slate (*Geol.*). A hard fissile metamorphic rock, derived from argillaceous material.

claystone (*Geol.*). Clay material cemented by calcite; generally spheroidal and flattened, and sometimes forming large boulders.

clay with flints (*Geol.*). A stiff clay, containing unworn flints, which occurs as a residual deposit in Chalk areas, but which is extensively mixed with other superficial deposits.

cleading (*Hyd. Eng.*). The boarding of a coffer-dam or timber lock-gate.

clean (*Typog.*). Said of a proof sheet containing no corrections.

cleaner (*Foundry*). A small brass tool used by a moulder to improve the finish of the surface of a mould. See **sleekers.**

cleaning eye (*San. Eng.*). See **access eye.**

clean up (*Elec. Eng.*). (1) Improvement in vacuum which occurs in an electric discharge tube or vacuum lamp consequent upon absorption of the residual gases by the glass. (2) The removal of residual gas by a *getter* (q.v.).

clear (*Comp.*). To remove data from stores so that fresh data can be recorded. (*For.*) A term applied to timber that is free from all visible defects. (*Radio*) Message 'in plain', i.e., not *coded* or *scrambled*, so that it is immediately understood. (*Teleph.*) To take down all temporary connexions, in the form of plugs inserted into jacks, on the termination of a call.

clearance (*Eng.*). (1) The distance between two objects, or between a moving and stationary part of a machine. (2) The angular backing-off given to a cutting tool in order that the heel shall clear the work.

clearance rate (*Radiol.*). Rate of elimination of a particular isotope by a given organ.

clearance volume (*Eng.*). In a reciprocating engine or compressor, the volume enclosed by the piston and the adjacent end of the cylinder, when the crank is on the top dead centre. See **compression ratio, cushion steam.**

clear channel (*Telecomm.*). See **channel** (1).

clearcole (*Paint.*). A priming coat composed of size or glue with whiting and sometimes a little alum, used before applying limewashes or distemper, also as a foundation for gold-leaf. Also called **clairecolle, clerecole.**

clearing (*Typog.*). The disposal of letterpress formes after printing, the *furniture* and *quoins* being returned to stock and the type put aside, the machine-set to be remelted, and the hand-set distributed.

clearing agent (*Micros.*). In microscopical technique, a liquid reagent which has the property of rendering objects immersed in it transparent and so capable of being examined by transmitted light.

clearing foot (*Zool.*). In some *Phyllocarida*, a filamentous process of the second maxilla.

clearing hole (*Eng.*). A hole drilled slightly larger than the diameter of the bolt or screw which passes through it.

clearing signal (*Teleph.*). An indication, by means of a lamp, to a distant operator to disengage a circuit.

clear lamp (*Light*). An electric filament lamp in which the bulb is made of clear glass.

clear span (*Build.*). The horizontal distance between the inner extremities of the two bearings at the ends of a beam.

clear water reservoir (*Hyd. Eng.*). See **service reservoir.**

cleat (*Carp.*). A strip of wood fixed to another for strengthening purposes, or as a locating piece to ensure that another piece shall be in its correct position.

cleats (*Mining*). The main cleavage planes or joint planes in a seam of coal.

cleat wiring (*Elec. Eng.*). A system of wiring in which the wires are attached to the wall or other surface by cleats.

cleavage (*Bot., Zool.*). The series of mitotic divisions by which the fertilized ovum is transformed into a multicellular embryo. (*Chem.*) (1) The splitting of a crystal along certain planes parallel to certain actual or possible crystal faces, when subjected to tension. (2) The splitting up of a complex protein molecule into simpler molecules. (*Geol.*) A property of rocks, such as slates, whereby they can be split into thin sheets. Cleavage is produced and oriented by the pressures that have affected rocks during consolidation and earth movements.

cleavage-nucleus (*Zool.*). The nucleus of the fertilized ovum produced by the fusion of the male and female pronuclei: in parthenogenetic forms, the nucleus of the ovum.

cleavelandite (*Min.*). A platy variety of *albite*.

cleaving-saw (*Tools*). See pit-saw.

cleft (*Bot.*). Deeply cut, but with the incisions not reaching to the midrib of the leaf.

cleft chestnut fencing (*Build.*). Fencing formed from cleft poles 2-6 in. (5-15 cm) apart, bound together with twisted wire, the fence being supported by posts about 7 ft (2 m) apart.

cleft palate (*Med.*). A gap in the roof of the mouth as a result of congenital maldevelopment, with or without hare-lip.

cleft timber (*For.*). Timber brought approximately to specified dimensions by cleaving along the grain.

cleidotomy (*Surg.*). The cutting of the clavicles when the shoulders of the foetus prevent delivery in difficult labour.

cleistocarp (*Bot.*). An ascocarp in which the asci and ascospores form and ripen inside a completely closed outer wall, which breaks down, permitting the ultimate escape of the spores.

cleistocarpous (*Bot.*). Said of a capsule which does not open by means of a lid.

cleistogamy (*Bot.*). The production of flowers, often simplified and inconspicuous, which do not open, and in which self-pollination occurs. *adj.* **cleistogamic, cleistogamous.**

cleithrum (*Zool.*). In some Fish, a bone of the pectoral girdle, situated lateral to each clavicle.

Clemmensen reduction (*Chem.*). Reduction of aldehydes and ketones, by heating with hydrochloric acid and zinc amalgam, to the corresponding hydrocarbons, e.g., propanone to propane.

clench (*Eng.*). See **clinch.**

clench nailing (*Carp.*). A method of nailing pieces together in which the end of the nail, after passing right through the last piece, is bent back and driven into this piece, so that it may not be drawn out.

clerecole (*Paint.*). See **clearcole.**

clerk of works (*Build., Civ. Eng.*). The official appointed by the employer to watch over the progress of any given building works, and to see that contractors comply with requirements in connexion with materials and labour.

cleveite (*Min., Chem.*). Variety of *pitchblende* containing uranium oxide and rare earths; often occluding substantial amounts of helium.

Cleveland Iron Ore (*Min.*). An ironstone consisting of iron carbonate, which occurs in the Middle Lias rocks of North Yorkshire near Middlesbrough. The ironstone is oölitic and yields on the average 30% iron.

Cleve's acids (*Chem.*). Mixture of *1-naphthyl-6-sulphonic* and *1-naphthyl-7-sulphonic* acids; used as an intermediate in the preparation of direct black cotton dyestuffs.

clevis (*Eng.*). A hook with a safety pin or latch.

cliché (*Typog.*). International name for printing block; French origin.

click (*Acous.*). Short impulse of sound, with wide frequency spectrum, with no perceptible concentration of energy-giving characterization. (*Horol.*) Pawl or detainer used, in conjunction with a ratchet wheel, to permit rotation in 1 direction only. (*Radio*) Atmospheric disturbances of very short duration arising from lightning.

clicker (*Typog.*). A compositor who receives copy and instructions from the overseer and distributes the work among his companions.

click method (*Telecomm.*). That for determining the resonant frequency of oscillatory circuit, depending upon click produced in telephones of a heterodyne wavemeter when oscillatory circuit coupled thereto is brought into resonance.

click spring (*Horol.*). The spring which holds the click in the teeth of the ratchet wheel.

click stop (*Photog.*). Aperture control which clicks at each reading, enabling it to be set without looking at the figures.

climacteric (*Biol., Physiol.*). A critical period of change in a living organism, e.g., the menopause.

climate (*Meteor.*). The average weather conditions of a place. These comprise the mean daily temperature and the mean daily maximum and minimum temperature, the average humidity, amounts of cloud and sunshine and rainfall, direction and speed of winds, all taken for each month and for the year. Climate depends on latitude, position with respect to oceans and continents, and upon local geographical conditions.

climatic climax (*Bot.*). A stable plant community maintained in any given area by the prevailing climatic conditions.

climatic community (*Bot.*). A plant community determined by the climate of the region in which it exists.

climatic factor (*Bot.*). A condition such as average rainfall, temperature, and so on, which plays a controlling part in determining the features of a plant community and/or the distribution and abundance of animals. Cf. *edaphic factor, biotic factor*.

climatic zones (*Meteor.*). The earth may be divided into zones, approximating to zones of latitude, such that each zone possesses a distinct type of climate. Eight principal zones may be distinguished: a zone of tropical wet climate near the equator; 2 subtropical zones of steppe and desert climate; 2 zones of temperate rain climate; 1 incomplete zone of boreal climate with a great range of temperature in the northern hemisphere; and 2 polar caps of arctic snow climate.

climatology (*Meteor.*). The study of climate and its causes.

climax association (*Bot.*). The plant association which is stable under the influence of any given set of ecological factors.

climax community (*Bot.*). The stable type of vegetation which finally becomes established in any given habitat.

climax dominant (*Bot.*). A species which dominates a climax community.

climax species (*Bot.*). Any species of plant which is a characteristic member of a climax community.

climb cutting (*Eng.*). The method of machining a surface with a rotating multi-toothed cutter in which the workpiece moves in the same direction as the periphery of the cutter at the line of contact. It produces a better surface finish than can be obtained by cutting upwards but has the disadvantage that the workpiece tends to be drawn towards the cutter.

climbing form (*Civ. Eng.*). A type of form sometimes used in the construction of reinforced concrete walls for buildings. The wall is built in horizontal sections, the climbing form being raised, after the pouring of each section, into a position convenient for the pouring of the next higher section.

climograph (*Ecol.*). Chart in which one major climatic factor (e.g., temperature) is plotted against another (e.g., humidity). The use of mean monthly figures gives a 12-sided polygon characteristic of a particular area. Different areas can be compared, and the suitability of particular places for different species determined by its use.

clinandrium (*Bot.*). The part of the column in the flower of an orchid which bears or contains the anther.

clinch (*Eng.*). To set or close a fastener, usually a rivet. Also called **clench.**

clinch nailing (*Carp.*). See **clench nailing.**

C-line (*Phys.*). Fraunhofer line in spectrum of sun at 656·28 nm, arising from ionized hydrogen in its atmosphere.

cline (*Ecol.*). A quantitative gradation in the characteristics of an animal or plant species across different parts of its range associated with changing ecological, geographic, or other factors, e.g., *ecocline, geocline* (qq.v.).

clinic (*Med.*). From the Greek *klinē*, a bed; *klinikos* was a doctor who visited patients in bed, and the word *clinic* has been used in English for both patient and doctor. Today its meanings include: the instruction of medicine in any of its branches at the bedside or in presence of hospital patients (whence *clinical medicine, clinical surgery, clinical psychology*, etc.): an institution, or a department of one, or a group of doctors, for treating patients or for diagnosis (*outpatients clinic, Health clinic*): a private hospital: by extension, any similar instructional and/or remedial meeting.

clink (*Civ. Eng.*). A short pointed steel wedge used for excavation in hard gravel or road surfaces; it is struck with a sledge hammer.

clinker (*Eng.*). Incombustible residue, consisting of fused ash, raked out from coal- or coke-fired furnaces; used for road-making and as aggregate for concrete. See **breeze.**

clinkers (*Build.*). See **klinker brick.**

clinks (*Met.*). Internal cracks formed in steel by differential expansion of surface and interior during heating. The tendency for these to occur increases with the hardness and mass of the metal, and with the rate of heating.

clink-stone (*Geol.*). See **phonolite.**

clinochlore (*Min.*). A variety of chlorite, occurring as tabular monoclinic crystals.

clinograph (*Surv.*). A form of adjustable set-square, the two sides forming the right angle being fixed, while the third side is adjustable; it differs from the adjustable set-square in that no scale is provided to show the angular position of the third side in relation to the other two. Used in surveying deep borehole to check departure from vertical.

clinoid (*Anat.*). Bony processes of the sella turcica that encircle the pituitary gland's stalk.

clinometer (*Surv.*). A hand instrument for the measurement of angles of slope.

clinoptilolite (*Min.*). A silica-rich variety of the zeolite *heulandite* (q.v.).

clinopyroxene (*Min.*). A general term for the monoclinic members of the pyroxene group of silicates.

clino-rhomboidal crystals (*Crystal.*). See **triclinic system.**

clinozoisite (*Min.*). A hydrous calcium aluminium silicate in the epidote group. Differs from zoisite in crystallizing in the monoclinic system.

clintonite (*Min.*). A hydrous aluminium, magnesium, calcium silicate. One of the brittle micas; differs optically from xanthophyllite.

clints (*Geog.*). Ridges on the weathered surfaces of limestones.

clioquinol (*Pharm.*). 5-Chloro-8-hydroxy-7-iodoquinoline; brownish powder, m.p. 180°C. Used as an internal amoebicide, also in creams and ointments (sometimes with hydrocortisone) for certain skin infections. Also called **iodochlorhydroxyquinoline.**

clipper (*For.*). A cutting or shearing machine used to cut veneer to size. (*Mining*) A man who attaches and detaches tubs to and from the rope of an endless rope haulage system.

clipping (*Telecomm.*). Loss of the initial or final speech-sounds in telephone transmission, due to the operation of voice-switching apparatus which is required to prevent the circuit from singing. Obviated by the use of *delay networks* (q.v.), so that the switching has time to operate the transmission circuits before the speech currents pass their contacts.

clipping circuit (*Telecomm.*). That for removing the peaks, tails, or frequency components of pulses in electrical circuits.

clip screws (*Surv.*). The screws by which the two verniers of the vertical circle of a theodolite may be adjusted so as to eliminate index error. Also called **antagonizing screws.**

Clipsham stone (*Build.*). An oölitic limestone, deep-cream or buff in colour, used as a building-stone; quarried from the Inferior Oölite at Clipsham, Oxfordshire.

clitellum (*Zool.*). A special glandular region of the epidermis of *Oligochaeta* which secretes the cocoon and the albuminoid material which nourishes the embryo.

clitoridectomy (*Surg.*). Surgical removal of the clitoris.

clitoridotomy (*Surg.*). Circumcision of the female.

clitoris (*Zool.*). In female Mammals, a small mass of erectile tissue, homologous with the glans penis of the male, situated just anterior to the vaginal aperture.

cloaca (*San. Eng.*). A term for a sewer, arising from the names of those in ancient Rome. (*Zool.*) Generally, a posterior invagination or chamber into which open the anus, the genital ducts, and the urinary ducts; in *Urochorda*, the median dorsal part of the atrium; in *Holothuroidea*, the wide posterior terminal part of the alimentary canal into which the respiratory trees open.

cloacitis (*Vet.*). Inflammation of the cloaca.

cloanthite (*Min.*). See **chloanthite.**

clock (*Comp.*). A fixed frequency source of master timing pulses. (*Elec. Eng.*) See **counter.** (*Horol.*) Timekeeper, other than chronometer or watch.

clock frequency (*Comp.*). That provided by a stable oscillator to regulate the digits (bits) and to keep them synchronous throughout a computer system.

clock gauge (*Eng.*). A length-measuring instrument in which the linear displacement of an anvil is magnified and indicated by the deflection of a needle pointer rotating over a dial. Often used as part of a vertical comparator.

clock meter (*Elec. Eng.*). An energy meter in which the current that passes causes a change in the rate of a clock. See **Aron meter.**

clock stars (*Astron.*). Those stars whose positions are taken as known, in determinations of time, longitude, and latitude by astronomical observations, and whose coordinates are tabulated in astronomical ephemerides.

clock-watch (*Horol.*). A watch which strikes the hours.

clod (*Mining*). A hard earthy clay on the roof of a working place in a coal-seam, often a fireclay.

clomiphene (*Biochem.*). See **fertility drug.**

clone (*Bot.*). The entire stock of plants obtained by budding, grafting, cuttings, or other means of vegetative multiplication from one original seedling: by extension, any stock of plants raised vegetatively from one original parent. In rubber cultivation, a tree which has been developed by grafting a bud from a high yielding rubber tree on to a seedling. (*Zool.*) (1) The descendants of a single individual; a pure line. (2) An asexually produced individual: a form of Sponge spicule.

clonus (*Zool.*). A series of muscular contractions in which the individual contractions are visible.

close annealing (*Met.*). The operation of annealing metal products (e.g., sheets, strip, and rod) in closed containers to avoid oxidation. Also called **pot (or box) annealing.**

close-boarded fencing (*Build.*). Fencing formed of lapped feather-edged boarding posts bound together by two or three rails, the fence being supported by posts about 9 ft (3 m) apart, and having a gravel board along the bottom.

close coupling (*Elec. Eng.*). See **tight coupling.**

closed circuit (*Elec.*). One in which there is zero impedance to the flow of any current, the voltage dropping to zero. (*Radio*) One composed of inductors, capacitors and resistors, all of relatively small dimensions, to load a transmitter instead of an extended antenna and consequent radiation. (*Teleg.*) A circuit which is closed during both marking and spacing signals; if a spacing current is used, continuity is always indicated. (*TV*) System in which transmission (pick-up) and reproduction of TV sound and scene are fixed, the question of broadcast or free reception not arising. Used in industry, medicine, theatre, studio rehearsals, etc., the scanning system being most common.

closed-circuit grinding (*Chem. Eng.*). Size reduction of solids done in several stages, the material after each stage being separated into coarser and finer fractions, the coarser being returned for further size reduction and the finer being passed on to a further stage, so that overgrinding is minimized.

closed-circuit recording (*Radio*). That for subsequent dissemination in broadcasting, and not of a performance being broadcast.

closed-coil winding (*Elec. Eng.*). An armature winding in which the complete winding forms a closed circuit.

closed community (*Bot.*). A plant community which occupies the ground without leaving any spaces bare of vegetation.

closed-core transformer (*Elec. Eng.*). A transformer in which the magnetic circuit is entirely of iron, i.e., with no air gaps.

closed cycle (*Eng.*). Heat engine in which a working substance continuously circulates without replenishment. (*Nuc. Eng.*) Cooling system for reactor in which coolant is continuously circulated through core and heat exchanger.

closed-cycle control system (*Elec. Eng.*). One in which the controller is worked by a change in the quantity being controlled, e.g., an automatic voltage regulator in which a field current is actuated by a deviation of the voltage from a desired value, the *reference voltage* (q.v.).

closed diaphragm (*Acous.*). Diaphragm or cone which is not directly open to the air, but communicates with the latter through a horn, which serves to match the high mechanical impedance of the diaphragm with the low radiation impedance of the outer air.

closed inequality (*Maths.*). One which defines a closed set of points, e.g., $-1 \leqq x \leqq +1$.

closed interval (*Maths.*). An interval which includes its end points. Cf. *open interval.*

closed-jet wind tunnel (*Aero.*). Any wind tunnel in which the working section is enclosed by rigid walls.

closed-loop system (*Telecomm.*). A control system, involving one or more *feedback control loops* (q.v.), which combines functions of controlled signals and of commands, to keep relationships between the two stable. Also **feedback control system.**

closed magnetic circuit (*Elec. Eng.*). The magnetic core of an inductor or transformer without any air gap.

closed pipe (*Acous.*). A *stopped pipe* (q.v.).

closed pore (*Powder Tech.*). A cavity within a particle of powder which does not communicate with the surface of the particle.

closed set (*Maths.*). A set of points which includes all its limit points. Complement of an *open set*

closed shed (*Weaving*). A term indicating that all warp threads in a loom are level or in the same plane.

closed slots (*Elec. Eng.*). In the rotor or stator of an electric machine, slots for receiving the armature winding, which are completely closed at the surface and therefore in the form of a tunnel. Often called **tunnel slots.**

closed stokehold (*Eng.*). A ship's boiler room, closed to allow fans to maintain in it an air pressure slightly above atmospheric, so that forced draught may be provided to the furnaces.

closed traverse (*Surv.*). A traverse in which the final line links up with the first line.

closed vascular bundle (*Bot.*). A vascular bundle which does not include any cambium, and cannot therefore increase in diameter.

close-packed hexagonal structure (*Met.*). An arrangement of atoms in crystals which may be imitated by packing spheres; characteristic of a number of metals. The disposition of the atomic centres in space can be related to a system of hexagonal cells.

close plating (*Elec. Eng.*). The application of a thin sheet of metal by a process of soldering.

closer (*Build.*). Any portion of a brick used, in constructing a wall, to close up the bond next to the end brick of a course. Also **bat, glut.**

close sand (*Foundry*). See under **open sand.**

close string (*Carp.*). See **housed string.**

close timbering (*Build., Civ. Eng.*). Trench or excavation lining in which the boards have no space between them.

closing error (*Surv.*). In a closed traverse, designed to return to starting point, discrepancy revealed on plotting, which may be small enough to allow distribution through the series of measurements.

closing layer (*Bot.*). A sheet of closely packed cells lying across a lenticel, preventing the diffusion of gases and vapours through it.

closing membrane (*Bot.*). A thin wall between the pits in adjacent cells, tracheides, or vessels.

closing-up (*Eng.*). The operation of forming the head on a projecting rivet shank.

clostridium (*Bacteriol.*). An ovoid or spindle-shaped bacterium, specifically one of the anaerobic genus *Clostridium*, which includes several species pathogenic to man and animals: viz., *Cl. botulinum* (botulism), *Cl. chauvei* (blackleg), *Cl. tetani* (tetanus), *Cl. welchii* (gas gangrene).

clot (*Med.*). The semisolid state of blood or of lymph when they coagulate. See also **thrombosis, embolism.**

cloth (*Textiles*). Generic term embracing all fabrics woven, felted, knitted, needled or bonded, using cotton, wool, silk, flax, jute, hemp, hair, or any of the man-made staples or blends of these, to make apparel, furnishing or industrial fabrics.

clothing (*Build.*). The covering (walls, roofing, etc.) applied to the structural framework of a steel-framed building.

clothing leather (*Leather*). Usually sheepskin, tanned, dyed, and treated with a cellulose finish to render it waterproof; used for sports clothing, etc.

clothing wool (*Textiles*). A wool of short fibre and good quality, which can be spun and woven into excellent woollen fabrics, easily milled.

cloth joint (*Bind.*). See **joints.**

clothoid (*Maths.*). See **Cornu's spiral.**

cloud (*Meteor.*). A mass of water droplets remaining more or less at a constant altitude. Cloud is usually formed by condensation brought about by warm moist air which has risen by convection into cooler regions and has been cooled thereby, and by expansion, below its dew point. (*Textiles*) Novelty yarn made by twisting together two threads of different colours. The rate of delivery of each is alternated so that each colour in turn predominates.

cloud and collision warning system (*Aero.*). A primary radar system with forward scanning which gives a cathode-ray display of dangerous clouds and high ground at ranges sufficient to allow course to be altered for their avoidance. A second system is usually necessary to give short-range warning of the presence of other aircraft.

cloudburst (*Meteor.*). An extremely heavy downpour of rain. It arises chiefly from the rapid condensation of rising air, as when a cloud is driven up over a mountain ridge.

cloud chamber (*Nuc. Eng.*). Chamber in which saturated gas or air can be suddenly cooled by expansion, revealing the paths of rays of particles which cause ionization and hence nuclei for condensation.

cloudiness (*Meteor.*). The amount of sky covered by cloud irrespective of type. Estimated in eighths (*oktas*). Overcast, 8; cloudless, 0.

clouding (*Paint.*). See **blooming.**

cloud point (*Chem.*). The temperature at which paraffin wax or other solid substances begin to separate from solution, when a petroleum oil is chilled under definite and prescribed conditions.

cloudy (*Textiles*). (1) Term used to indicate uneven shade in fabrics, due to faulty dyeing or finish. (2) Describes thin places in a web of carded cotton arising from bad machine settings.

cloudy swelling (*Med.*). A mild degenerative change in cells in which the swollen cloudy appearance is due to the presence of small vacuoles containing fluid and indicating damage to the 'sodium pump' of the cell membrane.

cloudy web (*Textiles*). Web of fibres from card doffer, irregular or patchy. Caused by faulty fillet or incorrect machine settings.

clouring (*Build.*). Chiselling or picking small indentations on a wall surface as a key for the finish.

clout nail (*Build.*). A nail having a large, thin, flat head.

clover disease (*Vet.*). A form of photosensitization affecting cattle and horses, believed to be associated with the eating of clover.

clover huller (*Agric.*). A thresher designed for completely threshing and cleaning clover.

cloves oil (*Chem.*). An oil obtained from the flowers of *Eugenia aromatica.* It is a pale yellow, volatile liquid, of strong aromatic odour; b.p. 250°–260°C, rel. d. 1·048–1·070. Used in confectionery, microscopy and dentistry.

club (*Zool.*). In Insects, the distal joints of the antenna when they are enlarged, as in some *Coleoptera* and *Lepidoptera*. *adj.* **clavate, claviform.**

clubbing (*Med.*). Thickening of the ends of the fingers which occurs in chronic disease of the lungs.

club foot (*Med.*). Deformity of the foot. See **talipes.**

club line (*Typog.*). A term used for the last (short) line of a paragraph. In common practice in the upmaking of a book, the occurrence of a club line at the top of a page is to be avoided. Also called **break line.**

club-tooth escapement (*Horol.*). The most widely used escapement for watches. See **escapement.**

clump (*Mining*). (1) A bend in a roadway or passage in a coal-seam. (2) A large fall of roof.

clumps (*Typog.*). Metal spacing material usually 6 pt and 12 pt.

clunch (*Mining*). (1) Sear earth below coal seam of marl or shale. (2) A tough fireclay.

clupanadonic acid (*Chem.*). $C_{22}H_{34}O_2$; a penta-unsaturated fatty acid found in fish oils. Pale yellow liquid with strong fishy odour.

clupeine (*Chem.*). Protein of the *protamine* (q.v.) type found in the sperm of the herring. Hydrolyses to form amino-acid *arginine*.

Clusius column (*Nuc. Eng.*). Device used for isotope separation by method of thermal diffusion.

cluster (*Bot.*). A general term for an inflorescence of small flowers closely crowded together.

cluster cup (*Bot.*). The popular name for an *aecidium*.

clustered column (*Arch.*). A column formed of several shafts bunched together.

cluster variables (*Astron.*). Short-period variable stars first observed in globular clusters; they are typical members of Population II. The periods are less than one day. See **RR Lyrae variables.**

clutch (*Build., Civ. Eng.*). A connecting bar used between adjacent flanges of I-section steel sheet piles to retain them in position. In section, it consists of a web part running between a pair of adjacent piles, with curved 'flanges' which slide over and secure the flanges of the piles. (*Eng.*) A device by which two shafts or rotating members may be connected or disconnected, either while at rest or in relative motion.

clutch arc lamp (*Light*). A type of automatic arc lamp in which the carbons are fed forward by the electromagnetic release of a clutch, as a result of the current falling below a certain predetermined value.
clutch stop (*Autos*.). A small brake arranged to act on the driven member of a clutch when it is fully withdrawn, to facilitate an upward gear change. With modern synchro gear-box this device is used only on heavy gear-boxes.
Clypeastroida (*Zool*.). An order of *Echinoidea* in which the periproct and anus are eccentric, and the dorsal parts of the ambulacra are petaloid; the body is usually flattened; sand-living forms. Cake Urchins.
clypeate (*Bot*.). Shield-shaped.
clypeus (*Bot*.). A stroma formed in the leaf of the host plant by some parasitic *Ascomycetes*. (*Zool*.) A sclerite of the head in Insects, lying immediately anterior to the frons.
Cm (*Chem*.). The symbol for *curium*.
cm³. Abbrev. for *cubic centimetre*.
CMRR (*Electronics*). Abbrev. for **common-mode-rejection ratio**.
CMU (*Chem*.). See **monuron**.
cn (*Maths*.). See **elliptic functions**.
C.N. (*Chem*.). Abbrev. for *coordination number*.
cnemial ridge (*Zool*.). In *Reptilia*, a longitudinal ridge on the anterior (dorsal) edge of the tibia.
cnemidium (*Zool*.). In Birds, the lower part of the leg, bearing usually scales instead of feathers.
cnemis (*Zool*.). The shin or tibia.
C-network (*Telecomm*.). See **network**.
C-neutrons (*Nuc*.). Those with such energies as are conducive to cadmium absorption, e.g., up to ca. 0·3 eV or 5×10^{-20} joules.
Cnidaria (*Zool*.). A subphylum of *Coelenterata*, whose members are distinguished by the possession of stinging-cells or *cnidoblasts* (q.v.), and usually show radial symmetry; locomotion is by muscular action.
cnidoblast (*Zool*.). A thread-cell or stinging-cell, containing a nematocyst; characteristic of the *Cnidaria*. Also **cnida**.
cnidocil (*Zool*.). A short sensory process or trigger-hair developed from the external part of a cnidoblast, stimulation of which causes the discharge of the nematocyst.
cnidosac (*Zool*.). In *Siphonophora*, a battery of cnidoblasts borne on a tentacle, where it gives rise to a coloured swelling.
Cnidosporidia (*Zool*.). An order of *Neosporidia*, in which the spores possess polar capsules.
C.N.S. (*Zool*.). A common abbrev. for *central nervous system*.
Co (*Chem*.). The symbol for *cobalt*.
coacervate (*Bot*.). Massed in a small heap.
coacervation (*Chem*.). The reversible aggregation of particles of an emulsoid into liquid droplets preceding flocculation.
coach bolt (*Eng*.). One having a convex head with a square section beneath it which fits into a corresponding cavity in the material to be bolted and prevents the bolt from turning as the nut is screwed on.
coach screw (*Eng*.). A large wood screw with a square head which is turned by a spanner; used in heavy timber work.
coactions (*Ecol*.). Relationships between organisms in a community and reciprocal effects which these have on their habitat. They may not have survival value for the coacting organisms, and include cooperation, disoperation, and competition.
coadaptation (*Biol*.). Correlated adaptation or change in two mutually dependent organs or organisms.
coagel (*Chem*.). A gel formed by coagulation.
coagulation (*Biol*.). The irreversible setting of protoplasm on exposure to heat or to poisons. (*Chem*.) The precipitation of colloids from solutions, particularly of proteins. (*Med*.) The process of clotting of blood as a result of the interaction of enzymes, calcium, and a protein in the blood, in the presence of tissue injury.
coagulum (*Med*.). A clot.
coak (*Carp*.). A projection on one of the mating surfaces of a scarfed joint, fitting into a corresponding recess in the other surface. Also called **table**.
coal (*Geol*.). A general name for firm brittle *carbonaceous rocks*; derived from vegetable debris, but altered, particularly in respect of volatile constituents, by pressure, temperature, and a variety of chemical processes. The various types are classified on basis of volatile content, calorific value, caking, and coking properties.
coal ball (*Bot*.). A calcareous nodule crowded with petrified plant remains; found in some seams of coal.
coal-cutting machinery (*Mining*). Mechanized systems used in colliery to detach coal from its face, and perhaps gather and load it to transporting device.
coalescent (*Bot*., *Zool*.). Grown together, especially by union of the walls.
coal-gas (*Fuels, etc*.). Gas made by the carbonization of fusible bituminous coal in town gas-works, in coke ovens (high-temperature process), and by low-temperature carbonization. Calorific value of the first ranges from 17·5 to 20·5 MJ/m³ at 15°C; that of low-temperature gas is about 33 MJ/m³. The approximate percentage composition by volume of purified coal-gas is: hydrogen 48, methane 35, carbon monoxide 6, nitrogen 6, olefins 2–4, carbon dioxide 1, with traces of benzene hydrocarbons and acetylene.
Coalite or semicoke (*Fuels*). TN for a smokeless fuel produced by carbonizing coal at a temperature of about 600°C; used for domestic purposes, calorific value per kilogram about 30 megajoules (13 000 Btu/lb).
Coal Measures (*Geol*.). The uppermost division of the *Carboniferous System* (q.v.), consisting of beds of coal interstratified with shales, sandstones, limestones and conglomerates.
Coal Sack (*Astron*.). A large obscuring dust cloud visible to the naked eye as a dark nebula in the Milky Way near the Southern Cross.
coal sizes (*Mining*). Those officially recognized are: large coal, over 6 in.; large cobbles, 6 to 3 in.; cobbles, 4 to 2 in.; trebles, 3 to 2 in.; doubles, 2 to 1 in.; singles, 1 to ½ in.; peas, ½ to ¼ in.; grains, ¼ to ⅛ in.
coal-tar (*Chem*.). The distillation products of the high- or low-temperature carbonization of coal. Coal-tar consists of hydrocarbon oils (benzene, toluene, xylene and higher homologues), phenols (carbolic acid, cresols, xylenols, and higher homologues), and bases, such as pyridine, quinoline, pyrrole, and their derivatives.
co-altitude (*Astron*.). See **zenith distance**.
coal washery (*Min. Proc*.). Cleaning plant where run-of-mine coal is processed to remove shale and pyrite, reduce ash and sort into sizes.
coarctate (*Zool*.). Said of pupae in which the last larval skin is retained as a covering; pressed together; having thorax and abdomen connected by a constricted portion.

coarctation (*Med.*). A narrowing or constriction, especially of the aorta near its origin from the heart.
coarse aggregate (*Civ. Eng.*). Gravel or crushed stone (forming a constituent part of concrete) which when dry will be retained on a sieve having ½ in. diameter holes. See **aggregate.**
coarse-grained (*Geol.*). See **grain-size classification,** and **Table of Igneous Rocks** in Appendix.
coarse scanning (*Radar*). Rapid scan carried out to determine approximate location of any target.
coarse screen (*Photog., Print.*). A term applied to half-tones for use on rough paper. The *screen* may be up to 33 lines to the centimetre.
coarse silt (*Geol.*). See **silt grade.**
coarse stuff (*Build.*). A mixture of lime mortar and hair used as a first coat for plastering internal walls.
coastal bar (*Civ. Eng.*). An accretion of sand or silt formed in estuaries by continual tide movements, and by conflicting or contributary discharges from inflowing rivers; many forms, depending on the hydraulic conditions.
coastal reflection (*Radio*). Reflection of signal by a land mass so that the resultant received signal consists of direct and reflected waves. This effect causes errors in direction-finding.
coastal refraction (*Radio*). Refraction, towards the normal, of waves arriving from sea to land at their incidence with the shoreline, resulting in an appreciable error in radio direction-finding for bearings making a small angle with the shoreline.
coasting (*Elec. Eng.*). A term used in connexion with electric traction to denote running with the supply to the motors cut off and the brakes not applied.
coasting of temperature (*Heat*). Rise above correct predetermined operating temperature in a furnace when the fuel supply has been checked or shut-off; due to excessive thermal storage in the furnace brickwork through prior overheating.
coastline effect (*Radio*). Same as **shore effect.**
coat (*Bot.*). (1) See **integument** (ovule). (2) See **testa** (seed-coat). (*Civ. Eng.*) See **wearing course.**
coated cathode (*Electronics*). One sprayed or dipped with electron-emitting compounds.
coated lens (*Photog., Optics*). Lens in which the air/glass surfaces are *bloomed*, with magnesium fluoride, to reduce reflection and so increase light transmission.
coating (*Elec. Eng.*). The metallic sheets or films forming the plates of a capacitor.
coating machine (*Paper*). A machine which deposits a layer of mineral on paper. Brushes distribute the coating evenly, and the paper is then dried and calendered.
coaxial antenna (*Radio*). Exposed $\lambda/4$ length of coaxial line, with reversed metal cover, acting as a dipole. λ = wavelength.
coaxial cable (*Elec. Eng., Electronics*). One in which a central conductor is surrounded co-axially first by a layer of insulating *dielectric*, then another conductor, usually at earth potential. If more than a few wavelengths long it will present a non-reactive *characteristic impedence* to an imposed a.c., whose value depends on the relative dimensions of the conductors and the *permittivity* of the dielectric.
coaxial circles (*Maths.*). Circles having a common radical axis.
coaxial feeder (*Telecomm.*). Same as **concentric feeder.**
coaxial filter (*Telecomm.*). One in which a section of coaxial line is fitted with re-entrant elements to provide the inductance and capacitance of a filter section.
coaxial line (*Telecomm.*). A *transmission line* coaxial in form, e.g. *coaxial cable.*
coaxial-line resonator (*Telecomm.*). One in which standing waves are established in a coaxial line, short- or open-circuited at the end remote from the drive. Of very high *Q*, these are used for stabilizing oscillators, or for selective coupling between valves. See **coaxial stub.**
coaxial-line tube (*Electronics*). See **Heil oscillator.**
coaxial propellers (*Aero.*). Two propellers mounted on concentric shafts having independent drives and rotating in opposite directions.
coaxial relay (*Telecomm.*). Switching device in which the coaxial circuit on both sides of the contact is maintained at its correct impedance level, thus avoiding wave reflection in the current path.
coaxial stub (*Telecomm.*). Section of coaxial line which is short-circuited at one end and functions as a high impedance at quarter-wave resonance.
cob (*Build.*). An unburnt brick.
cobalt (*Chem.*). Hard, grey metallic element in the eighth group of the periodic system. It is magnetic below 1075°C, and can take a high polish. Symbol Co, at. no. 27, r.a.m. 58·9332, valency 2 or 3, m.p. 1490°C, elec. resist. at 20°C 0·0635 microhm metres. Tensile strength (commercial containing carbon) 450 MN/m². Similar in properties to iron, but harder; used extensively in alloys, and as a source for radiography or industrial irradiation. ^{60}Co is used in cancer treatment.
cobaltammines (*Chem.*). Complex compounds of cobalt salts with ammonia or its organic derivatives, e.g., *hexammino* (III) *compounds*, which contain the group $Co(NH_3)_6^{3+}$.
cobalt black (*Paint.*). Cobaltic oxide, a black pigment of use in heat-resisting paints and in pottery.
cobalt bloom (*Min.*). See **erythrite.**
cobalt blue (*Paint.*). A blue pigment consisting essentially of cobalt aluminate. This is a very stable pigment of low tinting strength; commercial materials vary in cobalt content and in shade. Also **Thénard's blue.**
cobalt bomb (*Nuc.*). Theoretical nuclear weapon loaded with ^{59}Co. The long-life radioactive ^{60}Co, formed on explosion, would make the surrounding area uninhabitable. Also a radioactive source comprising ^{60}Co in lead shield with shutter.
cobalt carbonyl (*Chem.*). $Co_2(CO)_8$, used as a catalyst, esp. in the formation of aldehydes and in diene polymerization.
cobalt glance (*Min.*). See **cobaltite.**
cobaltiferous wad (*Min.*). An impure hydrated oxide of manganese containing up to 30% of cobalt.
cobaltite (*Min.*). Sulphide and arsenide of cobalt (II), crystallizing in the cubic system; usually found massive and compact with smaltite. Also **cobalt glance.**
cobaltomenite (*Min.*). Hydrated cobalt selenite, crystallizing in the monoclinic system.
cobalt (II) oxide (*Chem.*). CoO. Used to produce a deep blue colour in glass, and, in small quantity, to counteract the green tinge in glass caused by the presence of iron.
cobalt steels (*Met.*). Steels containing 5–12% cobalt, 14–20% tungsten, about 4% chromium, 1–2% vanadium, 0·8% carbon and a trace of molybdenum. They have great hardness and cutting power, but are brittle.

cobalt unit (*Radiol.*). Source of radioactive cobalt (half-life 5·3 years) mounted in a protective capsule; operated like the *caesium unit*.

cobalt violet (*Paint.*). Purple and violet pigments prepared by precipitation of cobalt salts with alkaline earth phosphates, and very variable in shade and exact composition.

cobalt vitriol (*Min.*). See **bieberite.**

cobbing (*Min. Proc.*). Removal of selected valuable mineral from lumps of rock by use of light hammer.

cobble (*Geol.*). See **boulder.**

cobbles (*Mining*). See **coal sizes.**

cobblestone (*Civ. Eng.*). A smallish roughly squared stone once used for paving purposes; superseded by setts.

cobbling (*Textiles*). Redyeing a fabric in order to perfect the shade.

Cobb sizing test (*Paper*). A method of measuring the amount of absorption of water in one minute by a sheet of paper under set conditions.

Cobol (*Comp.*). Business machine language, using automatic conversion from normal language to machine code. *C*ommon *b*usiness *o*riented *l*anguage.

cob wall (*Build.*). A wall built of puddled clay and straw, or of a mixture of straw, lime, and earth, laid without shuttering.

cobwebby (*Bot.*). Covered with delicate, long, interlaced hairs.

cocaine (*Pharm.*). A coca-base alkaloid, the methyl ester of benzoyl-L-ecgonine; m.p. 98°C. Used as a local anaesthetic, now mainly for eye, ear, nose and throat surgery. A recognized drug of addiction.

co-carboxylase (*Chem.*). Aneurin diphosphate. One of the three coenzymes controlling the fermentation of sugar solutions by yeast.

coccidioidomycosis (*Vet.*). A chronic infection of cattle, sheep, dogs, cats, and certain rodents, by the fungal organism *Coccidiodes immitis.* Also **coccidiomycosis.**

Coccidiomorpha (*Zool.*). An order of *Telosporidia*, in which the adult trophozoite is an intracellular parasite, and in which schizogony alternates with sporogony.

coccidiosis (*Vet.*). A contagious infection of animals and birds by protozoa of the genera *Eimeria* and *Isospora*, usually affecting the intestinal epithelium and causing enteritis.

coccin (*Photog.*). A red dye for reducing the active value of transparent parts of negatives.

coccineous (*Bot.*). Bright-red.

coccoid (*Bot.*). Unicellular, motionless in the vegetative condition, but liberating motile zoospores or gametes.

coccolith (*Zool.*). In some *Chrysomonadina*, a small calcareous exoskeletal plate.

coccus (*Bacteriol.*). A spherical or near spherical bacterium with a diameter from 0·5–1·25 μm. (*Bot.*) A 1-seeded portion formed by the break-up of a dry fruit.

coccydynia (*Med.*). Severe pain in the coccyx.

coccygectomy (*Surg.*). Excision of the coccyx.

coccygeomesenteric (*Zool.*). In Birds, said of a vein which receives blood from the caudal vein and passes it into the hepatic portal vein; it lies in the mesentery supporting the intestine.

coccygodynia (*Med.*). See **coccydynia.**

coccyx (*Zool.*). A bony structure in *Primates* and Amphibians, formed by the fusion of the caudal vertebrae; urostyle. *pl.* **coccyges.**

cochineal (*Chem.*). The dried bodies of female insects of *Coccus cacti* plus the enclosed larvae. The colouring matter is carminic acid, $C_{17}H_{18}O_{10}$, soluble in water and ethanol.

cochlea (*Acous., Zool.*). In Mammals, the complex spirally coiled part of the inner ear which translates mechanical vibrations into nerve impulses. (*Bot.*) A tightly coiled pod.

cochleariform (*Bot., Zool.*). Rounded and concave, like a spoon.

cochlear potentials (*Acous.*). Electric potentials within the cochlear structures resulting from acoustic stimulation.

cochleate (*Bot., Zool.*). Spirally twisted, like the shell of a snail; cochleariform.

cock (*Horol.*). A carrier or bracket for a pivot. (*Plumb.*) A type of water-valve in which the flow passes through a hole in a plug which is located transversely across the direction of the pipe.

cock-bead (*Join.*). A bead projecting from the surface which it is decorating. Also **cocked bead.**

Cockcroft and Walton accelerator (*Elec. Eng.*). High-voltage machine in which rectifiers charge capacitors in series, the discharge of these driving charged particles through an accelerating tube.

cocked bead (*Join.*). See **cock-bead.**

cocked hat (*Nav.*). Intersection of three bearing lines on chart.

cocket centring (*Civ. Eng.*). Arch centring which leaves some head-room above the springing line.

cocking (*Carp.*). See **cogging.**

cocking rollers (*Print.*). See **slewing rollers.**

cockle-stairs (*Arch.*). Winding stairs.

cockling (*Paper*). Wrinkling of the surface due to (*a*) the moisture content of the paper being out of equilibrium with the atmosphere, or (*b*) the paper being dried too rapidly on the paper machine. (*Textiles*) A defect in the surface of a fabric due to variable counts of yarn, unequal yarn tension in spinning, winding, warping or weaving, or irregular shrinkage during finishing.

cockpit (*Aero.*). The compartment in which the pilot or pilots of an aircraft are seated. It is so called even where it forms a prolongation of the cabin, also *flight deck.*

cockroach (*Typog.*). Typography set entirely in lower-case, without the use of any capitals.

cockscomb or **cockscomb pyrite** (*Min.*). A twinned form of marcasite (sometimes pseudomorphed by pyrite).

cockspur fastener (*Build.*). A bronze or iron fastener for casement windows, used in conjunction with a stay bar and pin.

cock-up (*Typog.*). (1) A large initial that extends above the first line and ranges at the foot. (2) In line-casting machines when the 2-letter matrix is in the cock-up position the companion italic or bold letter is cast.

cocobolo (*For.*). A tree (*Dalbergia*) of tropical American origin giving a hardwood used by the cutlery trade for the making of knife handles, and similar items. It is a dark reddish-brown in colour.

coconut oil (*Chem.*). The oil obtained from the fruit of the coconut palm; a white waxy mass, m.p. 20°–28°C, rel. d. 0·912, saponification value 250–258, iodine value 8·9, acid value 5–50. Chief constituent lauric acid.

cocoon (*Zool.*). In Insects, a special envelope constructed by the larva for protection during the pupal stage; it consists either of silk or of extraneous material bound together with silk. (*Textiles*) Specifically, the envelope spun by the fully grown silkworm round itself as a protective covering when entering the chrysalid state. If intended for silk manufacture, the cocoon is submitted to a heat of 60°-90°C for about 3 hr to destroy the pupa, and the thread is subsequently reeled.

co-current contact (*Chem. Eng.*). The opposite of *countercurrent contact* (q.v.). At the start of the process, fresh *A* is in contact with fresh *B*, and at each succeeding stage progressively spent reagents are in contact, so that the final effective driving force, and hence the overall economy, is lower.

cocuswood (*For.*). A tropical American hardwood of the genus *Brya*, available only in the form of small logs. Typical uses include inlaying, brush backs, turnery, parquet, and musical instruments. Also called **Jamaica ebony, West Indian ebony.**

codan (*Telecomm.*). Circuit for silencing a receiver in the absence of a signal (*c*arrier *o*perated *d*evice *a*nti-*n*oise).

Codd cell (*Chem.*). A single fluid cell with zinc and carbon as electrodes and picric chloride as a depolarizer. The carbon electrode lies horizontally at the bottom of the cell and is covered with an inert powder.

Coddington lens (*Optics*). A magnifying lens cut from a spherical piece of glass, having concentric spherical surfaces of equal radius at the ends, and a V-cut round the centre of its length to act as a stop at the centre of the sphere.

code (*Comp.*). Variations in a unified pattern which represent relevant pieces of information on a one-to-one basis. Thus numbers in a *binary* or *denary* scale of numbers can represent letters or numerals, money, stocks, etc., as data for processing. Such numbers can then be coded into, e.g., a 5-unit code for manipulation.

code delay (*Nav.*). Arbitrary time interval introduced between pulses from master and slave navigational transmitters.

codeine (*Pharm.*). $C_{18}H_{21}O_3N$, an alkaloid of the morphine group, the methyl derivative of *morphine*; a widely used analgesic.

coder (*Teleph.*). Device for translating digit-trains of impulses into marginal currents, for their more rapid transmission and subsequent operation of display panels. (*Telecomm.*) In a pulse-modulation system, the sampler that tests the signal at specified intervals.

codes of construction (*Chem. Eng.*). Written procedures for design, selection of materials, and manufacturing or operating procedures for production of chemical plant items for onerous conditions. Many are internationally acceptable, e.g., *ASME* (*A*merican *S*ociety of *M*echanical *E*ngineers), *TEMA* (*T*ubular *E*xchange *M*anufacturers *A*ssociation), *ABCM* (*A*ssociation of *B*ritish *C*hemical *M*anufacturers).

codes of practice (*Build., Civ. Eng.*). Recommendations (not mandatory) drawn up by a Government-sponsored Committee for design and methods of application thereof.

coding (*Comp.*). Programming of a computer, whether by written instructions on a coding form, by pseudocode, or in machine code. (*Radar*) Alteration of width or gap, or chopping in an IFF system.

cod-liver oil (*Chem.*). Oil obtained from fresh livers of cod fish; a yellow or brown liquid, of characteristic odour, very rich in vitamins, rel. d. 0·992–0·930, saponification value 182–189, iodine value 141–159, acid value 204–207.

codominant (*Bot.*). One of two or more species, which together dominate a plant community.

codon (*Gen.*). Triplet of bases in the DNA of the chromosomes which, via messenger-RNA, determine the genetic code.

coefficient (*Maths.*). In an algebraic expression or equation, a coefficient is the factor multiplying the variable (or quantity) under consideration. It may also be a number or parameter defining a certain characteristic, relationship etc., e.g., coefficient of correlation.

coefficient of absorption (*Phys.*). See **absorption coefficient.**

coefficient of apparent expansion (*Phys.*). The value of the coefficient of expansion of a liquid, which is obtained by means of a dilatometer if the expansion of the dilatometer is neglected. It is equal to the difference between the true coefficient of expansion of the liquid and the coefficient of cubical expansion of the dilatometer.

coefficient of compressibility (*Chem.*). A measure of deviation of a gas from Boyle's law.

coefficient of contraction (*Hyd.*). Of an orifice from which fluid is discharged under pressure, the ratio of the area of the smallest section of the jet to the area of the orifice. See **vena contracta.**

coefficient of discharge (*Hyd.*). Of an orifice from which fluid is discharged under pressure, the ratio of the actual discharge to the theoretical discharge; the product of the coefficient of velocity and contraction.

coefficient of dispersion (*Elec. Eng.*). See **dispersion coefficient.**

coefficient of elasticity (*Mech.*). See **elasticity.**

coefficient of equivalence (*Met.*). A factor used in converting amounts of aluminium, iron, and manganese into equivalent amounts of zinc, in relation to their effect on the constitution of brass.

coefficient of expansion (*Heat*). The fractional expansion (i.e., the expansion of unit length, area, or volume) per degree rise of temperature. Calling the coefficients of linear, superficial, and cubical expansion of a substance α, β, and γ respectively, β is approximately twice, and γ three times, α.

coefficient of fineness of water plane (*Ships*). The ratio of the actual area of the water plane to the area of a rectangle having length and breadth the same as the maximum dimensions of the water plane.

coefficient of friction (*Mech.*). See **friction.**

coefficient of perception (*Light*). A term used in connexion with the effect of glare; equal to the reciprocal of Fechner's constant.

coefficient of performance (*Eng.*). A measure of the efficiency of a refrigerator. It is the ratio of the heat removed from the cold body to the heat equivalent of the work done by the machine.

coefficient of reflection (*Light*). See **reflection factor.**

coefficient of restitution (*Mech.*). The ratio of the relative velocity of two elastic spheres after direct impact to that before impact. If a sphere is dropped from a height on to a fixed horizontal elastic plane, the coefficient of restitution is equal to the square root of the ratio of the height of rebound to the height from which the sphere was dropped. See **impact.**

coefficient of rigidity (*Mech.*). See **modulus of rigidity.**

coefficient of self-induction (*Elec. Eng.*). See **self-inductance.**

coefficient of utilization (*Light*). A term used in lighting calculations to denote the ratio of the useful light to the total output of the installation.

coefficient of variation (*Stats.*). The quotient of the standard deviation and the mean *or*, as a percentage, 100 times this ratio.

coefficient of velocity (Hyd.). Of an orifice discharging a fluid under pressure, the ratio of the actual velocity of discharge to the theoretical velocity.

coefficient of viscosity (*Phys., Eng.*). The value of the tangential force per unit area which is necessary to maintain unit relative velocity between two parallel planes, unit distance apart. Coefficient of viscosity varies inversely with temperature. Unit of measurement in the CGS system is the poise (dyn s cm^{-2}), in SI, N s m^{-2}. Symbol η.

coele-, -coele (*Zool.*). Prefix and suffix derived from Gk. *koilia*, large cavity (of the belly).

Coelenterata (*Zool.*). A phylum of *Metazoa* comprising forms which are of aquatic habit; they show radial or biradial symmetry; possess a single cavity in the body, the enteron, which has a mouth but no anus; and generally have only two germinal layers, the ectoderm and the endoderm, from one of which the germ cells are always developed. Polyps, Corals, Sea-anemones, Jelly-fish, and Hydra.

coeliac (*Zool.*). In Vertebrates, pertaining to the belly or abdomen.

coeliac canals (*Zool.*). In *Crinoidea*, aboral coelomic canals in the arms, derived from the right posterior coelom of the larva.

coeliac disease (*Med.*). A wasting disease of childhood in which failure to absorb fat from the intestines is associated with an excess of this substance in the faeces, due to a sensitivity to wheat gluten.

coeloconic (*Zool.*). Said of papillae which lie in pits, from which the apex does not project; as the *coeloconic sensilla* of Insects.

coelom (*Zool.*). The secondary body cavity of animals, which is from its inception surrounded and separated from the primary body cavity by mesoderm. *adj.* **coelomic, coelomate.**

Coelomata (*Zool.*). A group of *Metazoa*, including all those animals which possess a *coelom* (q.v.) at some stage of their life-history.

coelomere (*Zool.*). In metameric animals, the portion of coelom contained within one somite.

coelomoduct (*Zool.*). A duct of mesodermal origin, opening at one end into the coelom, at the other end to the exterior.

coelomopores (*Zool.*). In living *Tetrabranchiata*, openings by which the pericardium communicates with the exterior.

coelomostome (*Zool.*). In Vertebrates, the ciliated funnel by which the nephrocoel opens into the splanchnocoel.

Coelomycetes (*Bot.*). A large group of *Fungi Imperfecti*, which form their spores within a cavity in the matrix whereupon they grow.

coelostat (*Astron.*). An instrument consisting of a mirror (driven by clock-work) rotating about an axis in its own plane, and pointing to the pole of the heavens. It serves to reflect, continuously, the same region of the sky into the field of view of a fixed telescope.

coelozoic (*Zool.*). Extracellular; living within one of the cavities of the body.

coenecium (*Zool.*). In colonial *Polyzoa*, the common body bearing the zooids.

coenenchyma (*Zool.*). The hard matter filling up the spaces between adjacent polyps of a compound coral.

coenobium (*Bot.*). A group of algal cells, having a definite form and organization, behaving as an individual, and giving rise to daughter coenobia of the same kind.

coenocentrum or central body (*Bot.*). A globular mass of deeply staining material lying round or near the nucleus in the oögonia of some *Oömycetes.*

coenocytia (*Bot., Zool.*). Multinucleate syncytial tissues formed by the division of the nucleus without division of the cell, as striated muscle fibres and the trophoblast of the placenta.

coenocytic (*Bot.*). Containing a number of nuclei, but not divided into separate cells by walls.

coenogametangium (*Bot.*). A gametangium in which a coenogamete is formed.

coenogamete (*Bot.*). A multinucleate gamete.

coenosarc (*Zool.*). In *Hydrozoa*, the tubular common stem uniting the individual polyps of a hydroid colony.

coenosorus (*Bot.*). A sorus probably formed by the fusion of individual sori found in some reticulate ferns.

coenospecies (*Biol.*). Group of species in which direct or indirect hybridization is possible.

coenosteum (*Zool.*). In *Corals* and *Hydrocorallinae*, the common calcareous skeleton of the whole colony.

coenozygote (*Bot.*). The body resulting from the fusion of two coenogametes.

coenuriasis (*Vet.*). Sturdy; gid; turnsick. A nervous disease of sheep caused by the presence in the brain of the cystic stage, *Coenurus cerebralis*, of the tapeworm *Multiceps multiceps.*

coenurus (*Zool.*). A bladderworm, possessing a well-developed bladder with many scolices.

coenzyme (*Chem.*). A substance accompanying, and essential to, the activity of an enzyme. Thus glutathione is a coenzyme of glyoxalase. Coenzymes generally act as acceptors or donors of a grouping of atoms that are removed from, or contributed to, the substrate. Several coenzymes are derived from vitamins. For coenzyme I, coenzyme II see **NAD, NADP.**

coercimeter (*Mag.*). Instrument for measurement of coercive force.

coercive force (*Mag.*). Reverse magnetizing force required to bring magnetization to zero, after ferromagnetic material has been saturated and left with appreciable *residual magnetization.*

coercivity (*Mag.*). Coercive force when the cyclic magnetization reaches saturation.

coesite (*Min.*). A high-pressure variety of silica. Found in rocks subjected to the impact of large meteorites.

C of A (*Aero.*). Abbrev. for Certificate of Airworthiness.

cofactor (*Maths.*). Of an element a_{ij} in the determinant $|A|$ or in the square matrix A: its coefficient A_{ij} in the expansion of $|A|$. It equals $(-1)^{i+j}$ times the minor of a_{ij}. Also called **signed minor.**

coffer (*Arch.*). A sunk panel in a ceiling or soffit. (*Hyd. Eng.*) A canal lock-chamber.

coffer-dam (*Civ. Eng.*). A temporary wall serving to exclude water from any site normally under water, so as to facilitate the laying of foundations or other similar work; usually formed by driving *sheet piling* (q.v.). (*Ships*) A short compartment, 3 to 5 ft in length, separating oil-carrying compartments in an oil tanker from other compartments. Both bulkheads must be oil-tight and the space must be well ventilated.

coffering (*Civ. Eng., Mining*). (1) The operation involved in the construction of dams (see **coffer-dam**) for impounding water. (2) Shaft lining impervious to normal water-pressure.

coffer work (*Build.*). Stone-faced rubble-work.

Coffey-still (*Chem.*). Fractionating apparatus developed (ca. 1830) by Aeneas Coffey of Dublin for obtaining high-grade spirit from alcoholic liquor or fermented wash. Used in grain distillation of whisky and industrial production of rectified spirit.

coffin (*Nuc. Eng.*). See **flask.**

C of M (*Aero.*). Abbrev. for **Certificate of Maintenance.**

cog (*Build.*). See **nib.** (*Carp.*) The solid middle part left between the two notches cut in the lower timber in a cogging joint. (*Eng.*) Wooden tooth, formerly used in gear-wheels.

cogging (*Carp.*). A form of jointing used to connect one beam to another across which it is bearing. Notches as long as the top beam is wide are cut in the top surface of the lower beam opposite one another, so as to leave a solid middle part, and the upper beam has a small transverse groove cut in it, to fit over this solid middle part. Also called **caulking, cocking, corking.** (*Met.*) The operation of rolling or forging an ingot to reduce it to a bloom or billet.

coherence (*Phys.*). Opposite of randomness, especially with reference to radio, light, and acoustic waves.

coherent (*Bot.*). United, but so slightly that the coherent organs can be separated without very much tearing. (*Phys.*) Said of waves controlled in a very narrow frequency band.

coherent oscillator (*Radar*). One which is stabilized by being locked to the transmitter of a radar set for beating with a reflected incoming pulse signal, and used with radar-following circuits. Abbrev. **COHO.**

coherent pulse (*Telecomm.*). Said when individual trains of high-frequency waves are all in the same phase.

coherent sources (*Telecomm.*). Those between which definite phase relationships are maintained, so enabling interference effects to occur.

coherent units. A system of coherent units is one where no constants appear when units are derived from base units.

coherer effect (*Met.*). On an attempted contact between metals, the breakdown of a film by sufficient voltage and the build-up of a conducting bridge of metal by melting.

cohesion (*Bot.*). The union of plant members of the same kind, as when petals are joined in a sympetalous corolla. (*Phys.*) The attraction between the molecules of a liquid which enables drops and thin films to be formed. In gases the molecules are too far apart for cohesion to be appreciable (but see **Joule-Thomson effect**). (*Powder Tech.*) Attraction forces by which particles are held together to form a body, generally imparted by the introduction of a temporary binder, or by compaction. It is measured as green strength.

cohesion mechanism (*Bot.*). Any mechanism in a plant, in particular one concerned with the dehiscence of a sporangium, which depends on the cohesive powers of water (i.e., that a mass of water resists disruption).

cohesive soils (*Civ. Eng.*). Soils possessing inherent strength due to the surface tension of capillary water.

Cohnheim's areas (*Histol.*). Dotted areas, separated by sarcoplasm and formed by cross-sections of groups of fibrils, in a transverse section of striated muscle.

COHO (*Radar*). See **coherent oscillator.**

cohobation (*Chem.*). Successive redistillation.

cohort (*Zool.*). A taxonomic group ranking above a superorder.

Coignet (*Build., Civ. Eng.*). A method of system building developed in France. Similar to *Camus* (q.v.) except that the method of moulding and casting the precast units is more highly mechanized, producing greater precision with a resulting reduction in the necessary erection adjustments. See **system building.**

Coignet pile (*Civ. Eng.*). A type of reinforced concrete pile, similar to the *Hennebique pile*.

coil (*Comp.*). Roll of tape, said to be *blank* when feed holes alone have been punched. (*Elec. Eng.*) Length of insulated conductor wound few or many times around a core which may be *ferromagnetic*. Current passing through the coil will tend to create a magnetic field mainly in the core, thus the coil is an *inductor*.

coil-and-wishbone (*Autos.*). Type of independent front-wheel suspension in which a coil-spring, usually with a telescopic hydraulic damper, is mounted upon a wishbone-shaped frame, the steering-swivels being mounted at the apex and the top of the spring taking the weight of one corner of the chassis.

coil antenna (*Radio*). See **frame antenna.**

coiled-coil filament (*Light*). A spiral filament for an electric lamp which is coiled into a further helix to reduce radiation losses and enable it to be run at a higher temperature.

coiled-coil lamp (*Light*). An electric filament lamp having a coiled-coil filament.

coiler (*Spinning*). In cotton spinning, coiling mechanism on cards and drawframes. The sliver is laid in overlapping coils to make withdrawal easy at the next process.

coil heating (*Build.*). See **panel heating.**

coil ignition (*Autos.*). See **battery-coil ignition.**

coil loading (*Elec. Eng.*). The added inductance, in the form of coils, inserted at intervals along an extended line. Cf. *continuous loading*.

coil-side (*Elec. Eng.*). That part of an armature coil lying in a single slot.

coil-span (*Elec. Eng.*). The distance, measured round the armature periphery, between one side of an armature coil and the other; usually measured in electrical degrees or slots.

coil-span factor (*Elec. Eng.*). A factor introduced into the equation giving the e.m.f. of an electric machine, to allow for the fact that the coils have a fractional pitch and therefore do not embrace the whole flux.

coil-winding machine (*Elec. Eng.*). A machine for automatically, or semi-automatically, winding an electrical conductor coil to a given scheme.

coin analysis (*Comp.*). Breakdown, according to a *programme*, in which the output of a wages-processing is also expressed in specified coins or notes.

coincidence (*Comp.*). Unit with '1' output only when all inputs agree on '1'. (*Nuc. Eng.*) When counting particles, operation of counters apparently simultaneously, or within an assigned time interval.

coincidence counter (*Comp.*). Combination of counters and circuits which record only when all the counters are operated simultaneously or within a specified interval of time.

coincidence gate (*Telecomm.*). Electronic circuit producing an output pulse only when each of two (or more) input circuits receive pulses at the same instant, or within a prescribed time interval.

coincidence tuning (*Radio*). Tuning of all stages to the midband frequency in contradistinction to *staggered tuning* (q.v.).

coin-collector (*Elec. Eng.*). A mechanical device attached to energy meters, in order that the consumer may close a switch after the insertion of a given number of coins.

coining (*Powder Met.*). The impressing or repressing of a component in a die and tool set to impart final shape and accuracy of dimensions. This does not generally imply intentional change to volume and density.

coital exanthema (*Vet.*). Coital vesicular exanthema. A vesicular and papular eruption on the external genital organs of cattle and horses; caused by a virus and transmitted principally through coitus.

coition or coitus (*Zool.*). See **copulation.**

coitus interruptus (*Zool.*). A form of sexual intercourse in which the penis is withdrawn from the vagina before ejaculation occurs.

coke (*Fuels*). The solid residue from the carbonization of coal after the volatile matter of the coal has been distilled off. Coke is used as a fuel, and in metallurgy as a reducing agent for metal oxides.

coke breeze (*Build.*). The smaller grades of coke from coke ovens or gasworks, used in the manufacture of breeze concrete.

coke mill (*Foundry*). A small mill for pulverizing coke, used in the preparation of *blacking* (q.v.).

coke-oven gas (*Fuels*). Gas produced in the manufacture of hard cokes from hard-caking bituminous coals; calorific value per cubic metre, about 19 megajoules at 15°C.

coke ovens (*Fuels*). Large ovens in which hard-caking bituminous coal is subjected to a long process of carbonization at high temperatures. See **metallurgical coke.**

cokes (*Met.*). Originally, tin plates made from wrought-iron produced in a coke furnace. The term is now applied to plates with a thinner tin coating.

coking coals (*Met.*). Those with more than 15% volatile matter and 80% carbon, which can produce a crush-resistant coke.

col (*Meteor.*). The region between two centres of high pressure, or anticyclones.

co-latitude (*Astron.*). The complement of the latitude, terrestrial or celestial. On the celestial sphere it is, therefore, the angular distance between the celestial pole and the observer's zenith, and also the meridian altitude of the celestial equator above the observer's horizon.

Colburn sheet process (*Glass*). A method of making sheet glass by bending the vertically drawn sheet over a driven roller which determines the rate of draw.

Colby's bars (*Surv.*). Compensated bimetallic bars 10 ft in length, arranged to show an unvarying length despite temperature change; used for base-line measurement in the Ordnance Survey.

colchicine (*Chem.*). $C_{22}H_{25}O_6N$, an alkaloid obtained from the root of the autumn crocus, *Colchicum autumnale*; pale yellow needles, m.p. 155°–157°C. Used in treatment of gout. (*Gen.*) Living nuclei dosed with colchicine may subsequently divide abnormally, with increase in the number of chromosomes and the establishment of a temporary condition of polyploidy (see **polyploid**).

cold (*Med.*). Common cold. Coryza. An acute infectious catarrh of the nasal mucous membrane thought to be due to a filter-passing organism.

cold bend (*Eng.*). A test of the ductility of a metal; it consists of bending a bar when cold through a certain specified angle.

cold-blooded (*Zool.*). Having a bodily temperature which is dependent on the environmental temperature; poikilothermal. Cf. *warm-blooded.*

cold cathode (*Electronics*). Electrode from which electron emission results from high-potential gradient at the surface at normal temperatures.

cold-cathode discharge lamp (*Light*). An electric discharge lamp in which the cathode is not heated, the electron emission being produced by a high-voltage gradient at the cathode surface.

cold-cathode rectifier (*Elec. Eng.*). Rectifier for low-frequency a.c., comprising a convex cathode and a concave anode placed close together in low-pressure gas.

cold chisel (*Eng.*). A chisel for chipping or cutting away surplus metal; it is used with a hand hammer. Different forms of cutting edges (e.g., flat, cross-cut, half-round) are used for various purposes. See **cross-cut chisel, flat chisel.**

cold cut (*Paint.*). Said of varnish or lacquer made by dissolving the ingredients in a suitable solvent by mechanical agitation without the use of heat.

cold deck (*For.*). See **deck.**

cold-drawing (*Eng.*). Process of producing bar or wire by drawing through a steel die without heating the material. See **wire-drawing.**

cold emission (*Electronics*). See **autoemission.**

cold front (*Meteor.*). The leading edge of an advancing mass of cold air, usually attended by line-squalls and heavy showers.

cold galvanizing (*Met.*). Coating of iron goods with zinc suspended in an organic liquid. This is evaporated leaving a zinc film on the article.

cold-heading (*Eng.*). The process of forming the heads of bolts or rivets by upsetting the end of the bar without heating the material.

cold insulation mastic (*Build.*). A type of coating prepared from bitumen solution, containing asbestos fibres and other fillers, which is used on cold surfaces where it is necessary to prevent entry of moisture and to provide protection.

cold junction (*Heat*). Junction of thermocouple wires with conductors leading to a thermo-electric pyrometer or other temperature indicator or recorder.

cold light (*Cinema., TV*). See **blue light.**

cold moulding (*Plastics*). A bituminous moulding powder containing a drying oil or varnish and a large percentage of filler is pressed together cold. The moulded article is then baked in an oven to dry, shrink, and harden, after which it may be only slightly thermoplastic.

cold pool (*Meteor.*). See **thickness chart.**

cold pressing (*Acous.*). A defect in gramophone-record pressing, due to the material not having reached a sufficiently high temperature for adequate flow and consequent definition of track. The indication is a lack of lustre on the black surface.

cold riveting (*Eng.*). The process of closing a rivet without previous heating; confined to small rivets.

cold-rolled (*Met.*). Said of metal that has been rolled at a temperature close to atmospheric. The cold-rolling of metal sheets results in a smooth surface-finish.

cold saw (*Eng.*). A metal-cutting circular saw for cold cutting. The teeth may be either integral with the disk or inserted.

cold-set ink (*Print.*). Printing ink formulated to work in a warmed-up inking system and which sets on the cold stock.

cold sett (set or sate) (*Eng.*). A smith's tool similar to a short, stiff cold chisel; used for cutting bars, etc., without heating. It is supported by a metal handle and struck with a sledge-hammer.

cold short (*Met.*). Brittle at atmospheric temperature.

cold shut (*Foundry*). A casting imperfection due to metal entering the mould by different gates or sprues, cooling and failing to unite on meeting.

cold store (*Electronics*). (1) One constructed from

an array of units, each of which depends on persistence of current in a bridge in a hole in a sheet of lead or tin, maintained at a low temperature by liquid helium. (2) Use of cryotron as register. At liquid helium temperature superconductivity is removed in a straight current-carrying wire by axial magnetic field, thus greatly changing its impedance and operating it as a switch. Sn and Pb are the most suitable conductors.

cold tests (*Electronics*). Measurements on electronic circuits, when there is no emission from heated cathodes, to obtain parameters.

cold trough (*Meteor.*). See **thickness chart.**

cold wave (*Meteor.*). The fall of temperature after the passage of a depression.

cold welding (*Eng., Powder Met.*). The forcing together of like or unlike metals at ambient temperature, often in a shearing manner, so that normal oxide surface films are ruptured allowing such intimate metal contact that adhesion takes place.

cold-working (*Met.*). The operation of shaping metals at or near atmospheric temperature by rolling, pressing, drawing, stamping, or spinning. See also **work-hardening.**

Cole-Cole plot (*Elec.*). Graph of real against imaginary part of complex permittivity, theoretically a semicircle.

colectomy (*Surg.*). Excision of the colon.

colemanite (*Min.*). Hydrated calcium borate, crystallizing in the monoclinic system; occurs as nodules in clay found in California and elsewhere.

coleogen (*Bot.*). The layer of meristematic cells from which the endodermis is formed.

Coleoptera (*Zool.*). An order of *Oligoneoptera*, having the fore-wings or elytra thickened and chitinized, meeting in a straight line; the hind-wings, if present, are membranous; the mouth-parts are adapted for biting and the second maxillae are fused; the larvae are either sluggish and grublike, or active and carnivorous. Beetles.

coleopteroid (*Bot.*). Said of a seed or fruit which looks like a beetle.

coleoptile (*Bot.*). The first leaf to appear above ground in a seedling of a grass; it forms a sheath around the younger leaves within it, and contains little or no chlorophyll.

coleorhiza or **coleorrhiza** (*Bot.*). A layer of protective cells on the tip of the radicle of the embryos of some flowering plants.

colibacillaemia or **colibacillemia** (*Med.*). The presence of *B. coli* in the blood.

colibacillosis (*Vet.*). A general term for an infection by the bacterium *Escherichia coli.*

colibacilluria (*Med.*). The presence of *B. coli* in the urine.

colic (*Med., Vet.*). Severe spasmodic pain due to affections of abdominal organs, e.g., kidney, gall-bladder, intestines.

colitis (*Med.*). Inflammation of the colon.

collabent (*Bot.*). Collapsing, falling in, sunken in the middle.

collagen or **collogen** (*Biochem.*). A scleroprotein, occurring in the connective tissue, bone, and cartilage of animals, which is converted into gelatine on boiling with water; ossein. See **scleroproteins.**

collapse (*Med.*). Extreme prostration and depression of vital functions.

collapse of lung (*Med.*). An airless state of the lung, caused by obstruction of a bronchus, or occurring reflexly after abdominal operation.

collapse therapy (*Med.*). The treatment of lung disease by compression of the affected area, e.g., by injecting air between the layers of the pleura.

collapsible tap (*Eng.*). A screw-cutting tap in which the chasers can be withdrawn radially, either for the purpose of producing a tapered thread or, in a more common type, to collapse the tap for quick withdrawal without risk of spoiling the thread produced.

collapsing (*Bot.*). Falling together into a brush-like form.

collar (*Arch.*). A band, either flat or slightly concave, plain or decorated, around a column. (*Bot.*) The junction between the stem and root of a plant, usually situated at soil level. Also called **collet.** (*Eng.*) A ring of rectangular section secured to a shaft to provide axial location with respect to a bearing: a similar ring formed integral with the shaft. (*Mining*) Concrete mat from which shaft linings are suspended at top; entry end of drill hole. (*Zool.*) The rim of a choanocyte; in *Hemichorda*, a collarlike ridge posterior to the proboscis; in *Gastropoda* with a spiral shell, the collarlike fleshy mantle edge protruding beyond the lip of the shell; more generally, any collarlike structure.

collar and clamp (*Civ. Eng.*). A form of lock-gate hinge, otherwise known as *anchor and collar* (q.v.).

collar beam (*Carp.*). The horizontal connecting beam of a collar-beam roof.

collar-beam roof (*Arch.*). A roof composed of 2 rafters tied together by a horizontal beam connecting points about half-way up the rafters.

collar cell (*Zool.*). See **choanocyte.**

collar cloth (*Textiles*). A cotton, spun rayon or nylon fabric, generally formed from a face and a back cloth bound together by a stitching warp; used for collars, cuffs, and double-fronted shirts.

collar-head screw (*Eng.*). A screw in which the head is provided with an integral collar; used where fluid leakage may occur past the threads.

collaring (*Met.*). The term used to indicate that metal passing through a rolling mill follows one of the rolls so as to encircle it.

collars (*Met.*). In rolling mills, the sections of larger diameter separating the grooves in rolls used for the production of rectangular sections.

collate (*Bind.*). To check that the sections of a book are in correct order, after *gathering* (q.v.)

collateral (*Nuc.*). Describing nuclides which decay into one of the main radioactive series, although initially not part of it. (*Zool.*) (1) Running parallel or side by side. (2) Having a common ancestor several generations back.

collateral bud (*Bot.*). A bud inserted by the side of an axillary bud.

collateral bundle (*Bot.*). A vascular bundle with a strand of xylem and a strand of phloem lying externally to it and on the same radius.

collateral siphon (*Zool.*). In *Echiuroidea*, a tube of unknown function which runs parallel to the intestine and opens into it at both ends.

collating machine (*Print.*). A rotary printing press which collects, in correct register and sequence, the components of, e.g., multipart stationery, 'collating' being a misnomer.

collator (*Comp.*). Apparatus for matching or checking cards in separate packs.

collecting cell (*Bot.*). A thin-walled cell of the mesophyll of a leaf, lying below, and in intimate contact with, one or more cells of the palisade layer from which it takes elaborated food material.

collecting cylinder (*Print.*). On rotary presses, the

cylinder for collecting sheets or sections before folding or delivery.

collecting electrode (*Elec. Eng.*). See **passive electrode.**

collecting lens (*Cinema.*). In a multiple condensing lens, the component lens which is nearest the light source.

collecting power (*Light*). The property of a lens to render parallel rays convergent or reduce the divergency of originally divergent rays.

collection (*Print.*). Gathering the sections or pages on a web-fed rotary press.

collective drive (*Elec. Eng.*). A term used in connexion with electric locomotives to denote a drive in which all the driving wheels are coupled and driven by a single motor. Cf. *individual drive.*

collective electron theory (*Mag.*). Assumption that ferromagnetism arises from free electrons; Fermi-Dirac statistics identify the Curie point with the transition from ferro- to paramagnetic states.

collective fruit (*Bot.*). A fruit derived from several flowers, as a mulberry.

collective pitch control (*Aero.*). A helicopter control by which an equal variation is made in the blades of the rotor(s), independently of their azimuthal position, to give climb and descent.

collective unconscious (*Psychol.*). A term used by Jung to denote that part of the unconscious mind which is inherited, and which contains, therefore, the instinct and primitive patterns of thought which are collective rather than personal.

collector (*Electronics*). (1) Any electrode which collects electrons which have already completed and fulfilled their function, e.g., screen grid. (2) Outer section of a transistor which delivers a primary flow of carriers.

collector agent (*Min. Proc.*). Promoter. In froth-flotation, a chemical which is adsorbed by one of the minerals in an ore pulp, causing it to become hydrophobic and removable as a mineralized froth.

collector capacitance (*Electronics*). The capacitance of the depletion layer forming the collector of a transistor.

collector current (*Electronics*). The current which flows at the collector of a transistor on applying a suitable bias.

collector-current runaway (*Electronics*). The continued increase of the collector current arising from an increase of temperature in the collector junction when the current grows. See **thermal runaway.**

collector efficiency (*Electronics*). The ratio of the useful power output to the d.c. power input of a transistor.

collector junction (*Electronics*). One biased in the high-resistance direction, current being controlled by *minority carriers.* The semiconductor junction between the collector and base electrodes of a transistor.

collector rings (*Elec. Eng.*). See **slip rings.**

collector shoe (*Elec. Eng.*). A metal shoe used on the vehicles of an electric traction system to maintain contact with the conductor-rail.

collector strip (*Elec. Eng.*). See **contact strip.**

collect run (*Print.*). A method of increasing the number of pages per copy on web-fed presses by *collection* before final fold and delivery.

College electros (*Print.*). Electrotypes with a plastic backing which eliminates the need for 'slabbing' and results in a pliable plate; developed at the London College of Printing.

Collembola (*Zool.*). An order of *Apterygota* having not more than 6 abdominal somites; the antennae have few joints; anal cerci and abdominal appendages are lacking; there is a ventral adhesive apparatus associated with the first abdominal somite and a saltatorial appendage posteriorly; Malpighian tubules and a tracheal system are absent; usually found under stones and leaves. Spring Tails.

collenchyma (*Bot.*). A mechanical tissue characteristic of petioles and of young stems. It usually lies close to the periphery, and consists of elongated cells with their walls strengthened by longitudinal strips of cellulose thickening material, so that the cell walls are not of uniform thickness. (*Zool.*) A form of parenchyma occurring in Sponges, having scattered stellate cells embedded in a gelatinous matrix.

collencyte (*Zool.*). In Sponges, one of the cells composing the collenchyma.

Colles' fracture (*Surg.*). Fracture of the lower part of the forearm above the wrist.

collet (*Bot.*). See **collar.** (*Eng.*) An externally coned sleeve, slit in two or more planes for part of its length, and arranged to be closed by being drawn into an internally coned rigid sleeve, for the purpose of gripping articles or material. (*Horol.*) A circular flange or collar: the collar held friction-tight on a balance staff to which the inner end of the balance spring is pinned.

collet chuck (*Eng.*). A collet mechanism used for holding work or a tool in a lathe, drilling machine, etc.

colleter (*Bot.*). A plant hair which secretes mucus, often present on the outside of bud scales, and on the petals of some flowers.

colleterial glands (*Zool.*). One or two pairs of accessory reproductive glands, present in most female insects. Their secretion forms the oötheca in *Orthoptera*, a gelatinous investment of the eggs in *Chironomus*, and a cement which fastens the eggs to the substratum in many other insects.

colliculose (*Bot.*). Said of a surface covered with small rounded upgrowths.

colliculus (*Zool.*). A small prominence; as on the surface of the optic lobe of the brain, a rounded process of the arytaenoid cartilage.

colligative properties (*Chem.*). Those properties of solutions which depend only on the concentration of dissolved particles, ions, and molecules, and not on their nature.

collimation. The process of aligning the various parts of an optical system. (The word is falsely derived from the Latin *collineare, -atum,* to bring together in a straight line.) (*Radiol.*) The limiting of a beam of radiation to the required dimensions.

collimation error (*Surv.*). An error produced in levelling or in theodolite work when the line of collimation is out of its correct position; the latter, for the level, is parallel to the bubble line and perpendicular to the vertical axis of rotation of the instrument, and for the theodolite it is also perpendicular to the trunnion axis.

collimation system (*Surv.*). In levelling along a series of instrument set-ups, transfer of *line of collimation* (q.v.) from back to fore sight, followed by adjustment of reduced level shown by change in reading on graduated staff. See **rise and fall system.**

collimator (*Optics*). A device for obtaining a parallel or near parallel beam of radiation. In a spectroscope, the collimator consists of a fine slit at the principal focus of a convex lens. Light from the illuminated slit is rendered parallel by the lens before falling on the prism or grating. Used for testing the focus of a lens at infinity.

collinear array (*Radio*). A directional system consisting of two or more half-wavelength aerial radiators, excited in phase, and placed end to end horizontally. The radiation is a maximum in the perpendicular direction to the line of the array.

collinear transformation of a matrix (*Maths.*). See **congruent transformation.**

collinear vectors (*Maths.*). Two vectors which are parallel to the same line.

collineation (*Maths.*). Analytical transformation having a one-to-one correspondence between points, collinear points being projected into collinear points.

colliquative softening (*Med.*). The absorption of water by diseased tissue, which then disintegrates.

collision (*Nuc.*). With particles, contact is not apparent unless there is *capture*. Collision means such nearness of approach that there is more or less mutual interaction because of the forces associated with the particles, although no impact.

collisional excitation (*Nuc.*). The transfer of energy when an atom is raised to an excited state by collision with another particle.

collision bulkhead (*Ships*). A strong watertight bulkhead, not less than one-twentieth of the vessel's length from the fore end.

collision diameter (*Chem.*). The distance of closest approach of the centres of two colliding molecules.

collision number (*Chem.*). The frequency of collisions per unit concentration of molecules.

colloblast (*Zool.*). See **lasso-cell.**

collodion (*Chem.*, *Pharm.*). A cellulose tetranitrate, soluble in a mixture of ethanol and ethoxyethane (1 : 7); the solution is used for coating materials and, in medicine, for sealing wounds and dressings.

collodion process (*Photog.*). Use of a plate coated with iodized collodion sensitized in silver nitrate solution, exposed while wet and developed with pyrogallol and acetic acid.

collogen (*Biochem.*). See **collagen.**

colloid (*Chem.*). From Gk. *kolla*, glue. Name originally given by Graham to amorphous solids, like gelatine and rubber, which spontaneously disperse in suitable solvents to form lyophilic sols. Contrasted with crystalloids on the one hand and on the other hand with lyophobic sols. The distinctions not being very fine, the term currently denotes any colloidal system.

colloidal electrolyte (*Elec.*). One formed from long-chain hydrocarbon compounds, end radicals of which can ionize, thus providing some properties of electrolytes.

colloidal filament (*Light*). A metal filament for electric-filament lamps which is prepared by the use of colloidal substances.

colloidal fuel (*Eng.*). A mixture of fuel oil and finely pulverized coal, which remains homogeneous in storage; calorific value high; used in oil-fired boilers as substitute for fuel oil alone.

colloidal graphite (*Eng.*). Extremely fine dispersion of ground graphite in oil. Graphite lowers the surface tension of oil without lowering the viscosity; the oil spreads more easily, taking the graphite to rough surfaces where it can build up a smoothness.

colloidal movement (*Chem.*). See **Brownian movement.**

colloidal mud (*Min.*). Thixotropic mixture of finely divided clays with baryte and/or bentonite, used in the drilling of deep oil bores.

colloidal state (*Chem.*). A state of subdivision of matter in which the particle size varies from that of true 'molecular' solutions to that of coarse suspensions, the diameter of the particles lying between 1 nm. and 100 nm. The particles are electrified and can be subjected to cataphoresis, except at the *isoelectric point* (q.v.). They are subject to Brownian movement and have a large amount of surface activity.

colloid equivalent (*Chem.*). The number of atoms per unit charge on the dispersed phase of a colloid.

colloid goitre (*Med.*). Abnormal enlargement of the thyroid gland due to accumulation in it of the viscid iodine-containing colloid.

colloid mill (*Chem. Eng.*). Mill with very fine clearance between the grinding components, operating at high speed, and capable of reducing a given product to a particle size of 0·1 to 1 μm.

colloid rectifier (*Elec. Eng.*). A rectifier with a cathode of colloidal particles in suspension in a liquid which cannot be electrolysed.

collophane (*Min.*). Cryptocrystalline variety of *apatite*.

collophore (*Zool.*). The ventral tube of *Collembola*, held by some to represent an adhesive organ.

collotype (*Print.*). A *planographic process* (q.v.) for the printing of tone subjects in one or more colours without using the *half-tone process* (q.v.). A film of dichromated gelatine is spread on a glass plate, printed down with a continuous tone negative, and developed with the use of glycerine and water. Dark parts of the subject become hardened and ink-accepting, light parts remain watery and ink-rejecting, and areas of intermediate tone accept ink to correspond with their variations in tone. During preparation, the surface develops a fine grain which provides a bite for the ink. Suitable for runs up to one thousand. Used for the finest facsimile reproductions of works of art.

collum (*Zool.*). In *Diplopoda*, the first post-cephalic segment, possessing no appendages or stigmata.

colluviarium (*Civ. Eng.*). An access opening in an aqueduct for maintenance and ventilation.

collyria (*Med.*). Fluids used as isotonic eyewashes.

Collyweston Slate (*Geol.*). A thin stratum of very fissile calcareous sandstone, occurring at the top of the Northampton Sands; used locally for roof-tiling purposes.

coloboma (*Med.*). Congenital defect of development, especially of the lens, the iris, or of the retina.

coloenteritis (*Med.*). Inflammation of the colon and small intestine.

cologarithm (*Maths.*). The logarithm of the reciprocal of a number.

Cologne earth (*Paint.*). Lignite yielding a variety of *Vandyke brown* (q.v.) pigment.

colombier (*Paper*). A standard size of drawing paper, $24 \times 34\frac{1}{2}$ in.

colon (*Zool.*). In Insects, the wide posterior part of the hind-gut; the large intestine of Vertebrates. *adj.* **colonic.**

colonist (*Bot.*). A plant which occurs only on ground which is periodically disturbed by human agency.

colonization (*Bot.*). The occupation of bare ground by seedlings and spores.

colonnade (*Arch.*). A row of columns supporting an entablature.

colony (*Bacteriol.*). The growth of a bacterial inoculum on a solid medium. The resultant macroscopically visible colony is typical of the species inoculated. (*Bot.*) (1) A group of individuals of one species which are invading new ground. (2) A fungal mycelium grown from one spore. (*Zool.*) A collection of individuals living together and in some degree interdependent, as a *colony* of polyps, a *colony* of social Insects: strictly, the members of a colony are in organic connexion with one another.

colopexy or **colopexia** (*Surg.*). The anchoring of part of the colon by sewing it to the abdominal wall.

colophon (*Typog.*). Originally, a device and/or notice by the scribe or printer, placed at the end of a book before title-pages became customary. Replaced today by title page and imprint, although some publishers often append a colophon. In modern practice, a decorative device on the title-page or spine of a book.

colophonic acid (*Chem.*). *Abietic acid* (q.v.), the chief constituent of colophony rosin.

colophony, colophonium (*Chem.*). See **rosin.**

coloptosis (*Med.*). An abnormally low position of the colon in the abdominal cavity.

Coloradoan Stage (*Geol.*). Synonymous with 'Lower Cretaceous' (Cenomanian to Senonian) in the American sense.

Colorado beetle (*Zool.*). A black-and-yellow striped beetle (*Leptinotarsa decemlineata*), which feeds upon potato leaves, causing great destruction.

coloradoite (*Min.*). Mercuric telluride, crystallizing in the cubic system. It usually occurs in the massive state.

Colorado ruby (*Min.*). An incorrect name for the fiery-red garnet (pyrope) crystals obtained from Colorado and certain other parts of U.S.

Colorado topaz (*Min.*). True topaz of a brownish-yellow colour is obtained in Colorado, but quartz similarly coloured is sometimes sold under the same name.

Colorgraph (*Print.*). Electronic equipment to produce, simultaneously, sets of colour-corrected separations, continuous tone or screened, from reflection copy or transparencies. Cf. *Chromagraph.*

colorimeter (*Light*). An instrument used for the precise measurement of the hue, purity, and brightness of a colour.

colorimetric analysis (*Chem.*). Analysis of a solution by comparison of the colour produced by a reagent with that produced in a standard solution.

colorimetric purity (*Photog.*). The ratio of the luminosity of a dominant hue to the total luminosity of a colour. See **saturation.**

colorimetry (*Photog.*). The science of measuring the chromaticity of colours for photographic purposes.

Color-key (*Print.*). A method of producing proofs of colour separations using a film print of each colour mounted in register with the others.

colostomy (*Surg.*). A hole surgically made into the colon for the escape of faeces when the bowel below is obstructed.

colostration (*Med.*). A disease of infants, due to *colostrum.*

colostrum (*Histol.*). A clear fluid accumulated in the mammary glands in the latter part of pregnancy. It precedes the flow of milk at the onset of lactation.

colostrum-corpuscles (*Zool.*). Large cells containing fat-particles, which appear in the secretion of the mammary glands at the commencement of lactation.

colotomy (*Surg.*). An incision into the colon; (loosely) colostomy.

colour (*Light*). Colour depends on the wavelength of the light. In viewing a continuous spectrum, the normal eye perceives a graduation of colour from red at the long-wavelength end, through orange, yellow, green, blue, to violet at the short-wavelength limit. Any of these, or other colours, can be matched by suitably adjusting the stimulations from three primary colours, red, green, and blue-violet, the wavelengths of these (additive) primaries being selected so that a minimum amount of negative primary is required in matching.

colour adjusting (*Print.*). Regulating the flow of ink on printing presses. See **Idotron, Inkatron.**

colour analyser (*Photog.*). A colorimeter for separating the components of a complex colour into their respective wavelengths.

colour and weave (*Textiles*). Small, regular, clear-cut design in dress, suiting and sports cloths. In appearance, usually quite distinct from the weave, e.g., houndstooth is a typical effect in 2 or 3 colours.

colour balance (*Photog.*). See **balancing.** (*TV*) Suitable selection of phosphor efficiencies and electrical and optical properties of a colour TV display for the achromatic reproduction of a grey scale of varying luminance.

colour bar (*Print.*). A standard strip made up of solid and 70% tone to be included by the engraver when pulling *progressive proofs* (q.v.), positioned across the proofing press and printed in full colour, thus guiding the printer to obtain the same result.

colour bleeding (*Photog.*). See **bleeding.**

colour blindness (*Med.*). The lack of one or more of the spectral colour sensations of the eye. The commonest form, Daltonism, consists of an inability to distinguish between red and green. Even persons of normal sight may be colour blind to the indigo of the spectrum. The Edridge-Green lamp, beads, and cards provide the standard means of testing.

colour break-up (*TV*). Transitory separation of colours in an intermittent TV picture arising from motion of the eye of the observer.

colour burst (*TV*). In NTSC/PAL TV system, that part of the composite colour signal consisting of a sine wave at subcarrier frequency present during part of the back porch of the line synchronizing pulse; used to establish a phase reference for demodulating the chrominance signal.

colour cast (*TV*). Predominance of one-colour primary in colour image resulting from lack of colour balance.

colour cell (*TV*). The smallest area in a colour picture tube which includes a complete set of all the primary colours in the repeating pattern.

colour centre (*Crystal.*). Lattice defects in crystal, which interact with electrons so as to produce intense local light absorption. A few such defects give coloured crystals their characteristic hue. (*TV*) Unit region of CRT phosphor, determined by the electron beam and perforated screen in establishing a colour TV image. Also called **U-centre.**

colour coder (*TV*). Apparatus in colour TV to generate chrominance subcarrier and composite colour signal from the camera signals. Also **colourflexer.**

colour contamination (*TV*). Error in colour reproduction caused by incomplete separation of primaries.

colour contrast (*Photog.*). The ratio, or the logarithm, of the intensities of two colours.
colour coordinates (*Light*). Set of numbers representing the location of a hue on a chromaticity diagram.
colour-corrected lens (*Optics., Photog.*). One in which chromatic aberration is eliminated.
colour correction (*Print.*). When making sets of colour printing surfaces, it is necessary to compensate for the inherent faults of the printing inks available. Can be carried out by hand on the separation negatives or positives for any of the printing processes and on the actual plate for relief printing, by *colour masking* (q.v.), and electronically on machines such as the *Colorgraph, Helio-Klischograph, Scan-a-Colour, Scanatron, Springdale Time-Life Scanner, Vario-Klischograph.*
colour decoder (*TV*). Circuit in a TV receiver which extracts, decodes, and separates the 3 constituent colours.
colour difference signal (*TV*). PAL colour TV transmissions comprise a monochrome signal, and 3 colour difference signals. The latter are combined with the monochrome signal in the receiver to produce one of the required primary colour signals.
colour disk (*TV*). Filter disk used in field sequential colour TV systems.
coloured cement (*Build.*). Ordinary Portland cement into which selected pigments are introduced in the grinding process.
colour edging (*TV*). See **fringing.**
colour excess (*Astron.*). The amount by which the colour index of a star exceeds the accepted value for its spectral class; used as a measure of absorption of starlight.
colour fatigue (*Optics*). Changes in the sensation produced by a given colour, when the eye is fatigued by another or by the same colour.
colour filter (*Light*). Film of material selectively absorbing certain wavelengths, and hence changing spectral distribution of transmitted radiation. (*Photog.*) See **filter.**
colourflexer (*TV*). See **colour coder.**
colour gate (*TV*). Circuit in colour TV receiver which allows only primary colour signal, corresponding to excited phosphor, to reach modulation electrode of tube.
colour guides (*Print.*). Term sometimes applied to *progressive proofs* (q.v.).
colour index (*Astron.*). The difference between the photographic and the visual magnitudes of a star, from which may be deduced the effective temperature. (*Geol.*) A number which represents the percentage of dark-coloured heavy silicates in an igneous rock, and is thus a measure of its leucocratic, mesocratic, or melanocratic character. (*Photog.*) The colours of all dyes available commercially, arranged in order. (*Physiol.*) An index of the amount of haemoglobin in the red cells of the blood; obtained by dividing the percentage of red cells of the normal into the percentage of haemoglobin.
colouring pigment (*Paint.*). See **stainer.**
colour killer (*TV*). Circuit rendering the chrominance channel of a colour TV receiver inoperative during the reception of monochrome signals.
colour-light signals (*Civ. Eng., etc.*). A method of signalling adapted to both railway and highway purposes, whereby traffic is controlled by light signals of different colours.
colour-luminosity array (*Astron.*). A variant of the *Hertzsprung-Russell diagram* (q.v.) in which the absolute magnitudes of stars are plotted as a function of their colours.
colour masking (*Print.*). Use of photographic masks in the process camera to compensate for inherent faults in the inks available for colour printing. Separate masks can be prepared, e.g., a low-density positive of the cyan printer is used to modify the magenta printer negative, thus reducing the amount of magenta printing in blue areas; or an all-purpose masking colour film can be used, such as Multi-mask or Trimask. For the best standard of reproduction, a final colour correction by hand may be necessary.
colour mixture curve (*Light*). Representation of the specified three colours which match a given colour.
colour negative (*Photog., TV*). Image in which the correct hues have been replaced by their corresponding complementary colour.
colour phase alternation (*TV*). Sequence of the colour signals in the video signal.
colour photographic sensitivity (*Photog.*). Sensitivity of an emulsion to a specified wavelength.
colour picture signal (*TV*). Monochrome video signal, plus a subcarrier conveying the colour information, which is transmitted with synchronizing signals.
colour plane (*TV*). In a multibeam colour picture tube, the (near) plane containing all the colour centres.
colour primaries (*Photog., Print., TV*). Set of (usually three) colours from which multicolour images are built up.
colour printing (*Print.*). The reproduction of an original subject comprising 2 or more colours. Colour printing is achieved by any of the normal printing processes; each colour is printed separately, in a predetermined order, the superimposed impressions, if accurately registered, building up an image corresponding in colour to the original subject.
colour purity magnet (*TV*). Magnet placed near to neck of colour picture tube to modify path of electron beam and thus improve purity of the displayed colour.
colour pyramid (*Photog.*). See **colour triangle.**
colour reference signal (*TV*). Continuous signal which determines the phase of the burst signal.
colour register (*Print.*). The correct superimposing of two or more colours. On multicolour web-fed machines, manual adjustment has been superseded by electronic equipment, such as the *Autotron scanner.*
colour saturation (*Photog.*). See **saturation.** (*TV*) In colour TV, the complementary and primary hues are termed *saturated hues* or *colours* but in nature very few of these are fully saturated in the same way. Since, therefore, mostly desaturated colours should be displayed on the colour tube, some white colour must be added to the saturated hue.
colour screen (*Photog.*). Either a filter or a mosaic of the primary colours.
colour sensitometer (*Photog.*). See **Abney-.**
colour separation (*Photog.*). The production of 3 separate negatives of the same subject through green, red, and blue filters to record the proportions of the respective colours in the image. The negatives can be used as a basis for colour prints or for half-tone colour blocks. See **three-colour process.** (*Print.*) The printed reproduction of coloured subjects requires the preparation of a set of printing plates. Photographic separations are prepared in the process camera by exposing through colour filters; or on electronic machines such as the *Colorgraph*; or the actual printing surface can be engraved

as the copy is scanned on machines such as the *Vario-Klischograph* and the *Helio-Klischograph*.

colours of thin films (*Light*). When white light is reflected from a thin film, such as a soap bubble or a layer of oil on water, coloured effects are seen which are due to optical interference between light reflected from the upper surface and that from the lower. The colours are not bright unless the film is less than about 10^{-3} mm thick.

colour specification (*Photog.*). The description of a colour in a standard manner, so that it can be duplicated without comparison.

colour standards (*Photog.*). A standard range of colours for reference purposes in making dyes or filters, or for composing colour patterns for colour photography.

colour subcarrier (*TV*). Signal, conveying the colour information as a modulation, added to the monochrome video and synchronizing signals.

colour temperature (*Phys.*). That temperature of a black body which radiates with the same dominant wavelengths as those apparent from a source being described. See **Planck's law.**

colour threshold (*Optics*). The luminance level below which colour differences are indiscernible.

colour transparency (*Photog.*). A colour photograph to be viewed or projected with transmitted light.

colour triangle (*Light, etc.*). That drawn on a chromaticity diagram, representing entire range of chromaticities obtainable from additive mixtures of three prescribed primaries, represented by the corners of the triangle:

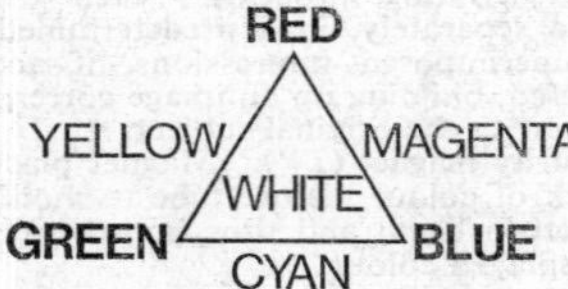

colour vision (*Optics*). No theory is completely satisfactory, but that of Young-Helmholtz is generally followed, i.e., there are 3 colour receptors, for red, green, and blue, perception of colour depending on a combination of receptors stimulated by any particular light.

-colous. Suffix meaning *inhabiting* (L. *colere*, inhabit).

colp-, colpo-. Prefix from Gk. *kolpos*, womb, vagina.

colpitis (*Med.*). Inflammation of the vagina.

Colpitts oscillator (*Electronics*). Tunable self-oscillating circuit in which positive feedback from the output to the control electrode of an amplifier is provided by a tuned circuit whose capacitor is divided to provide bias.

colpocele (*Med.*). A hernia into the vagina.

colpocystitis (*Med.*). Inflammation of the vagina and of the bladder.

colpocystocele (*Med.*). A hernia formed by protrusion of the bladder into the vagina.

colpocystotomy (*Surg.*). Incision of the bladder through the wall of the vagina.

colpoperineoplasty (*Surg.*). Repair of the vagina and perineum by plastic surgery.

colpoperineorrhaphy (*Surg.*). Sewing up of the torn vagina and perineum.

colpoptosis (*Surg.*). Prolapse of the vagina.

colporrhaphy (*Surg.*). Narrowing of the vagina by surgical operation.

colposcope (*Surg.*). An instrument for inspecting the vagina.

colpospasm (*Med.*). Spasm of the vagina.

colulus (*Zool.*). A small pointed appendage just anterior to the spinnerets in some Spiders; of unknown function.

columbarium (*Build.*). A recess left in the face of a wall to support the end of a timber.

columbite (*Min.*). The niobate and tantalate of iron and manganese, crystallizing in the orthorhombic system. It occurs in granitic rocks and pegmatites, and is the chief source of the tantalum used for the metallic filaments of electric lamps.

columbium (*Chem.*). See **niobium.**

columella (*Bot.*). (1) A central column of sterile tissue in the sporangium of a Bryophyte. (2) A dome-shaped wall in the middle of the sporangium of some moulds. (3) A mass of sterile tissue at the base of the fruit bodies of some *Gasteromycetes*. (4) The central part of a root cap, containing statoliths. (*Zool.*) In Mammals, the central pillar of the cochlea; in lower Vertebrates, the auditory ossicle connecting the tympanum with the inner ear; in some lower Tetrapods, the epipterygoid; in spirally coiled gastropod shells, the central pillar; in the skeleton of some Corals, the central pillar. *adj.* **columellar.**

column (*Bot.*). The central portion of the flower of an orchid (probably an outgrowth of the receptacle of the flower), bearing the anther, or anthers, and the stigmas. (*Civ. Eng.*) A vertical shaft supporting an axial load. (*Comp.*) Position in a number corresponding to a power of the *base* or *radix* (usually 10 or 2). Also called **place.** (*Eng.*) A vertical pillar of cast-iron, forged steel, or steel plate in box section, used to support a compressive load. See also **strut.** (*Zool.*) In *Crinoidea*, the stalk; in Vertebrates, a bundle of nerve fibres running longitudinally in the spinal cord; the edge of the nasal septum; more generally, any columnar structure, as the vertebral column.

columnae carneae (*Zool.*). In Mammals, muscular ridges arising from the inner surface of the wall of the right ventricle of the heart.

columnals (*Zool.*). In *Crinoidea*, the ossicles forming the stalk.

column analogy method (*Civ. Eng.*). A method of analysing indeterminate frames of nonuniform section by comparing equilibrium conditions, at any point, with the stresses in an analogous eccentrically loaded column.

columnar crystals (*Met.*). Elongated crystals formed by growth taking place at right angles to the surface of the mould. The extent to which they grow before solidification is completed, by the formation of equi-axed crystals in the interior, is important.

columnar epithelium (*Zool.*). A variety of epithelium consisting of prismatic columnar cells set closely side by side on a basement membrane, generally in a single layer.

columnar structure (*Geol.*). A form of regular jointing, produced by contraction following crystallization and cooling in igneous rocks, especially those of basic composition. The columns are generally roughly perpendicular to the cooling surface.

columnated window stairs (*Arch.*). A staircase whose steps are supported on columns to allow of natural lighting on all sides.

column binary (*Comp.*). See **Chinese binary.**

columns across (*Print.*). Printing newspapers and magazines with the columns imposed across the plate cylinders.

columns around (*Print.*). Printing newspapers and magazines with the columns imposed around the plate cylinders.

column vector (*Maths.*). A matrix consisting of a single column.

colures (*Astron.*). The great circles passing through (*a*) the poles of the celestial equator and ecliptic and through both solstitial points, and (*b*) the poles of the celestial equator and both equinoctial points, these two great circles being the solstitial and equinoctial colures respectively.

colusite (*Min.*). A sulphide of copper, iron, arsenic and tin, crystallizing in the cubic system.

Colymbiformes (*Zool.*). A commonly-used synonym for *Podicipitiformes* (q.v.).

colza oil (*Oils*). Rape oil. Obtained from the seeds of *Brassica campestris*; pale-yellow after refinement by sulphuric acid treatment. Used in some countries as an edible oil and as an illuminant, also, when blown, as a lubricant, and in quenching steel.

COM. Abbrev. for *c*omputer *o*utput *m*icrofilm.

coma (*Bot.*). (1) A tuft of hairs attached to the testa of a seed. (2) A tuft of leaves (the *comal tuft*) at the tip of a moss stem. (*Electronics*) Of a CRT, plumelike distortion of spot arising from misalignment of focusing elements of gun. (*Med.*) A state of complete unconsciousness in which the patient is unable to respond to any external stimulation. In *coma vigil* the patient's eyes are open. (*Optics*) A defect in the image formed by a lens which, when present, appears towards the edges of the field, the images of points being drawn out into small pear-shaped blobs, with their narrow ends directed towards the centre of the field.

comagmatic assemblage (*Geol.*). The name given, by Professor Judd, to a group of igneous rocks which have originated during a definite cycle of igneous activity. They usually possess certain chemical, mineral and textural similarities which suggest common origin and evolution.

comal tuft (*Bot.*). A bunch of leaves at the end of a twig.

Comanchean (*Geol.*). The general name for the Cretaceous strata up to the top of the Albian, as developed in N. America.

comate, comose (*Bot.*). Shaggy, bearing a tuft of hairs at the end.

comate disseminule (*Bot.*). A fruit or seed bearing long silky hairs which aid in dispersal by the wind.

comatose (*Med.*). Being in a state of coma.

comb (*Arch.*). The ridge of a roof. (*Paint.*) A flat flexible, wire- or rubber-toothed instrument used by the painter for graining surfaces. (*Textiles*) To prepare cotton and wool fibres for spinning by separating and straightening them and (for wool) removing any fibres below a specified length. (*Zool.*) In *Ctenophora*, a ctene; the framework of hexagonal wax cells produced by social Bees to shelter the young or for storing food.

combat rating (*Aero.*). See **power rating.**

comb bars (*Textiles*). See **combs.**

comb binding (*Bind.*). A style of loose-leaf binding in which the leaves are held together by a comb, usually of plastic, being passed through slots in the paper and then curved to form a tube.

comb collector (*Elec. Eng.*). A conductor with a row of parallel projecting points, used for collecting the charge in an influence machine.

combed yarns (*Textiles*). Highest quality yarns prepared from carded, combed and drafted fibres that have been mechanically straightened and freed from neps and short fibres. In cotton mills, either of two machines is used, Heilmann or Nasmith. For wool fibres, the machines vary according to quality selected from Noble, Lister, or Holden combs.

combeite (*Min.*). A hydrated silicate of sodium and calcium, crystallizing in the trigonal system.

comber board (*Weaving*). Guide board located between jacquard and loom top. Holes are drilled into it, one for each harness cord. It provides a means for quick adjustment when needed during weaving.

Combescure transformation of a curve (*Maths.*). A one-to-one transformation which maps one space curve on another so that the tangents at corresponding points are parallel.

comb filter (*Telecomm.*). One containing a number of *pass* or *rejection bands* (qq.v.) of frequency.

combination (*Chem.*). Formation of a compound.

combination chuck (*Eng.*). A lathe chuck in which the jaws may be operated all together, as in a universal or *self-centring chuck* (q.v.); or each operated separately for holding work of irregular shape, as in an *independent chuck* (q.v.).

combination colours (*Zool.*). Colour effects produced by pigment in combination with *structural colours* (q.v.), as in certain Butterflies.

combination principle (*Light*). Ritz discovered that the addition or subtraction of the wave-numbers of two spectral lines frequently gave the value of the wave-number of another line in the same spectrum. This rule emerges naturally from the Bohr quantum theory of spectra.

combinations (*Maths.*). The different selections that can be formed from a given number of items, order within each group being immaterial. For n items, all different, taken r at a time, there are $n!/r!.(n-r)!$ combinations. Denoted by ^nC_r or $\binom{n}{r}$. Cf. *permutations*.

combination set (*Eng.*). A fitter's instrument comprising a universal protractor, spirit level, rule and straight-edge, centre head, and square.

combination tank (*Plumb.*). A household hot-water cylinder with a feed tank above it, all within the same outer case.

combination tanned leather (*Leather*). Leather which has been tanned by two or more of the usual processes, e.g., vegetable tanning, mineral tanning, treatment with oils.

combination test (*Psychol.*). See **completion test.**

combination tones (*Acous.*). Additional tones subjectively perceived by the ear when more than one tone is applied, and having frequencies which are sums and differences of the frequencies of the applied tones. Combination tones arise because of amplitude distortion in the ear.

combination turbine (*Eng.*). See **disk-and-drum turbine.**

combine (*Agric.*). Harvesting machine which cuts and threshes grain and ejects the straw on to the land for subsequent collection or burning in.

combine baler (*Agric.*). A machine which gathers hay from the windrows in a field and forms it into bales.

combine drill (*Agric.*). A machine to sow seeds and fertilizer in one operation.

combined carbon (*Met.*). In cast-iron, the carbon present as iron-carbide as distinct from that present as graphite. See **graphitic carbon.**

combined distribution frame (*Telecomm.*). The combination of main and intermediate distribution frames, when the number of circuits is insufficient to warrant the use of separate frames. Abbrev. C.D.F.

combined half-tone and line (*Print.*). A process block on which half-tone and line work are combined and etched.

combined-impulse turbine (*Eng.*). An *impulse turbine* (q.v.) in which the first stage consists of nozzles that direct the steam on to a wheel carrying two rows of moving blades, between which a row of fixed guide blades is interposed.

combined system (*San. Eng.*). A system of sewerage in which only one set of sewers is provided for the removal of both the sewage proper and also rain water. Cf. *separate system.*

combined twills (*Textiles*). Fabrics in which two regular twill weaves are combined, end-and-end, or pick-and-pick, to form a new weave.

combining weight (*Chem.*). See **equivalent weight.**

comb poles (*Elec. Eng.*). Poles for salient-pole electric machines, constructed of laminations and having the alternate laminations made shorter than the others, to improve the flux distributions.

comb-rib or comb (*Zool.*). See **ctene.**

combs (*Textiles*). Strips of brass, cast in leads, forming an arc on which the carriages move in a lace machine. The back and front combs form a well for the warp; *comb bars.* Steel bars extending across a lace machine to support the comb leads.

combustion (*Chem.*). Chemical union of oxygen with gas accompanied by the evolution of light and rapid production of heat (exothermic).

combustion chamber (*Aero.*). The chamber in which combustion occurs: (1) the cylinder of a reciprocating engine; (2) the individual chambers or single annular chamber of a gas turbine; (3) the combustion zone of a ramjet duct; (4) the chamber, with a single venturi outlet, of a rocket. (*Eng.*) (1) In a boiler furnace, the space in which combustion of gaseous products from the fuel takes place. (2) In an internal-combustion engine, the space above the piston (when on its inner dead-centre) in which combustion occurs.

combustion control (*Eng.*). The control, either by an attendant or by automatic devices, of the rate of combustion in a boiler furnace, in order to adjust it to the demand on the boiler.

combustion tube furnace (*Met.*). Laboratory appliance having one or more horizontal refractory tubes heated by gas or electricity; used chiefly for the estimation of carbon content of steels, temperatures 1100° to 1300° C.

come-and-go (*Bind., Print.*). See **fore-and-aft.**

comedo (*Med.*). A blackhead. A collection of cells, sebum, and bacteria, filling the dilated orifices of the sebaceous glands near hair follicles.

comes (*Zool.*). A blood vessel which runs parallel and close to a nerve. *pl.* **comites.**

comet (*Astron.*). A member of the solar system, of small mass, becoming visible as it approaches the sun, partly by reflected sunlight, partly by fluorescence excited by the solar radiation. A bright nucleus is often seen, and sometimes a tail. This points away from the sun, its gases and fine dust being repelled by radiation pressure and the solar wind.

Comité Consultatif International (*Telecomm.*). An international consultative body, comprising representatives from electrical communication administrations, which studies problems of international communication, allocates research work, and makes recommendatoins for improvement in international connexions, charges, routes, etc. The work is divided into 3 departments: *C.C.I.T.*, which deals with telegraphy; *C.C.I.R.*, which deals with radio-communication; and *C.C.I.F.* (*Fernsprech*) which deals with telephony and land-line broadcasting.

comma (*Acous.*). The pitch error, not greater than 80 : 81·1, arising from tuning one note in various ways with natural ratios from a datum note.

commag (*Cinema.*). International code name for a picture film combined with a magnetic sound-track.

command (*Comp.*). (1) Set of signals which bridges programme instruction and intended operation of circuits. (2) U.S. term for *instruction* (q.v.).

command guidance (*Aero.*). The guidance of guided missiles or aircraft by electronic means through signals from an external source controlled by human operator or automatically.

commensalism (*Biol.*). An external, mutually beneficial partnership between 2 organisms (*commensals*). *adj.* **commensal.**

commensurable quantities (*Maths.*). Quantities, each of which is an integral multiple of a common basic quantity or measure, e.g., the numbers 6 and 15 are commensurable but 6 and $\sqrt{3}$ are not.

commentary (*Telecomm.*). The verbal description of an event for instantaneous use, or for subsequent use as material for making up programmes for broadcasting by radio, or in sound-films.

comminator (*Zool.*). In *Echinoidea*, one of the muscles connecting adjacent jaws of the lantern of Aristotle.

comminuted (*Med.*). Reduced to small fragments, e.g., *comminuted* fracture.

comminuted fibres (*Textiles*). Fibres cut or ground to fine proportions for applying to fabric, paper, etc., wholly or partially covered with adhesive; usually sprayed on or deposited electrostatically.

comminuted powder (*Powder Tech.*). Material reduced to a powder by, e.g., attrition, impact, crushing, grinding, abrasion, milling or chemical methods.

comminution (*Mining*). Size-reduction by breaking, crushing, or grinding, e.g., in ore-dressing.

commissural arch (*Bot.*). A loop of vascular tissue joining the ends of the veins of a leaf.

commissure (*Bot.*). (1) A cleft or suture. (2) A surface by which carpels are in union. (3) The line where the antical and postical lobes of the leaf of a liverwort join. (*Build.*) The joint between adjacent courses of stone. (*Zool.*) A joint; a line of junction between two organs or structures; a bundle of nerve-fibres connecting two nerve-centres.

commode step (*Build.*). A step having a riser curved to present a convex surface; used sometimes at and near the foot of a staircase.

common ashlar (*Build.*). A block of stone which is pick- or hammer-dressed.

common-base connexion (*Electronics*). The operation of a transistor in which the signal is fed between base and emitter, the output being between collector and base with the latter earthed. Also **grounded base.**

common battery (*Telecomm.*). The same as **central battery.**

common bond (*Build.*). Like a stretching bond, but with a course of headers every fifth, sixth, or seventh course.

common bricks (*Build.*). A class of brick used in ordinary construction (especially in interior work) for filling in, and to make up the requisite thickness of heavy walls and piers. They usually have plain sides, are not neatly finished, and are much more absorbent and also much weaker than *engineering bricks* (q.v.).

common bundle (*Bot.*). A vascular bundle belonging in part to a stem and in part to a leaf.

common-collector connexion (*Electronics*). The operation of a transistor in which the collector is earthed, input is between collector and base, and output is between emitter and collector. Also **grounded collector, emitter follower.**

common dovetail (*Join.*). An angle-joint between 2 members in which both show end grain.

common-emitter connexion (*Electronics*). The operation of a transistor in which the signal is fed between base and emitter with the latter earthed. The output is between emitter and collector. Also **grounded emitter.**

common-frequency broadcasting (*Radio*). The use of the same carrier frequency by 2 or more broadcast transmitters, sufficiently separated for their useful service areas not to overlap. Also **shared-channel broadcasting.**

common ion effect (*Chem.*). In a solution of two solutes *AB* and *AC*, having the ion *A* in common, an increase in the concentration of *AB* causes a decrease in the ionization of *AC*.

common joist (*Carp.*). See **bridging joist.**

common language (*Comp.*). A machine code which is common to different computers or data-processing devices, e.g., tape-operated print-out units.

common lead (*Met.*). Lead of lower purity than chemical or corroding lead (about 99·85%).

common-mode rejection ratio (*Electronics*). For a differential amplifier, the ratio of the gain for a differential input to that for a common-mode input.

common-mode signal (*Electronics*). A signal applied simultaneously to both inputs of a differential amplifier.

common rafter (*Build., Civ. Eng.*). A subsidiary rafter carried on the purlins and supporting the roof covering. Also **intermediate rafter.**

common-rail injection (*I.C. Engs.*). A fuel-injection system for multi-cylinder C.I. engines; an untimed pump maintains constant pressure in a pipe line (rail), from which branches deliver the oil to the mechanically operated injection valves.

common return (*Telecomm.*). A single conductor which forms the return circuit for 2 or more otherwise separate circuits.

commotio cerebri (*Med.*). Concussion of the brain, in which there is bruising of brain tissue.

communal habitat (*Bot.*). The habitat of any plant community, e.g., a meadow, woodland, ditch.

communications satellite (*Space*). An artificial satellite whose function is to achieve global communication. The passive satellites such as *Echo* are merely reflectors of radio waves; the active satellites have relay systems and may orbit the Earth at low altitudes (e.g., *Telstar*) or at much greater distances. A *synchronous communications satellite* has an orbital period equal to the time of rotation of the Earth on its axis and so remains directly above the same point on the Earth's surface. This facilitates the control of the ground stations operating in conjunction with the satellite. See **satellite.**

community (*Bot.*). Any group of plants growing together under natural conditions, and forming a recognisable unit of vegetation; e.g., woodland, heathland plants. Also (*Zool.*), the animals inhabiting a restricted area, as field or pond: such a community is not necessarily stable.

commutating field (*Elec. Eng.*). The magnetic field under the compoles of a d.c. machine; it induces, in the conductors undergoing commutation, an e.m.f. in a direction to assist in the commutation process.

commutating machine (*Elec. Eng.*). An electrical machine provided with a commutator.

commutating pole (*Elec. Eng.*). See **compole.**

commutating reactor (or reactance) (*Elec. Eng.*). One in the cathode lead of a mercury-arc rectifier, to keep current passing while the arc jumps from one anode to the next anode.

commutation factor (*Electronics*). Product of rate of current decay and rate of voltage rise after a gas discharge, both expressed per microsecond.

commutation switch (*Telecomm.*). That controlling the sequential switching operations required for multichannel pulse communication systems.

commutative (*Maths.*). An operation is commutative if the order in which it is performed on two quantities does not affect the result, e.g., the operation of addition in arithmetic is commutative because a added to b equals b added to a. Cf. *associative, distributive, transitive.*

commutator (*Elec. Eng.*). The part of a motor or generator armature through which electrical connexions are made by rubbing brush contacts. (*Maths.*) An element, c, of a group, such that, if a and b are any other two elements of the group, $bac=ab$.

commutator bar (*Elec. Eng.*). One of the copper bars forming part of a commutator. Also called a **commutator segment.**

commutator bush (*Elec. Eng.*). See **commutator hub.**

commutator face (*Elec. Eng.*). See **commutator surface.**

commutator grinder (*Elec. Eng.*). A portable electric grinding equipment which can be mounted on a commutator machine to grind the commutator surface without removing the armature from the machine.

commutator hub (*Elec. Eng.*). A metal structure used for supporting a commutator. Also called a **commutator bush, commutator shell, commutator sleeve.**

commutator losses (*Elec. Eng.*). Losses occurring at the commutator of an electric machine; they include resistance loss in the segments, in the brushes, and at the contact surface, friction loss due to the brushes sliding on the commutator surface, loss due to sparking, and eddy-current loss in the segment.

commutator lug (*Elec. Eng.*). A projecting piece of metal, connected to or integral with a commutator bar. Also called **commutator riser, commutator tag.**

commutator motor (*Elec. Eng.*). An electric motor which embodies a commutator.

commutator rectifier (*Elec. Eng.*). A device for rectifying alternating currents; it consists of a commutator which reverses the connexions of the circuit at the end of each half-cycle.

commutator ring (*Elec. Eng.*). A ring, usually of cast-iron, made to fit into a dovetail in the commutator segments in order to clamp them firmly in position. The term is also used to denote the insulating rings which have to be placed between the metal of the above ring and the commutator segments.

commutator ripple (*Elec. Eng.*). Small periodic variations in the voltage of a d.c. generator or rotary converter resulting from the fact that there can only be a finite number of commutator segments on the machine.

commutator riser (*Elec. Eng.*). See **commutator lug.**

commutator segment (*Elec. Eng.*). See **commutator bar.**

commutator shell (*Elec. Eng.*). See **commutator hub.**

commutator shrink-ring (*Elec. Eng.*). A steel ring shrunk round the cylindrical surface of commutators on very high speed commutator machines (e.g., d.c. turbogenerators) in order to hold the segments in position against the action of the centrifugal force. The ring must, of course, be insulated from the segments.

commutator sleeve (*Elec. Eng.*). See **commutator hub.**

commutator spider (*Elec. Eng.*). A spider mounted on the shaft of an electrical machine and used for supporting the commutator.

commutator surface (*Elec. Eng.*). The smooth portion of a commutator upon which the current-collecting brushes slide. Also called the **commutator face.**

commutator tag (*Elec. Eng.*). See **commutator lug.**

commutator transformer (*Elec. Eng.*). A device for converting from a low-voltage d.c. to a high-voltage d.c. and vice versa. It consists of a motor-driven commutator which converts from d.c. to a.c., this being then supplied to a transformer and stepped up; another commutator then rectifies the high-voltage a.c. It is used chiefly for obtaining a high-voltage d.c. supply from a small battery.

comopt (*Cinema.*). International code name for a picture film combined with an optical sound-track.

comose (*Bot.*). See **comate.**

compaction (*Build., Civ. Eng.*). The process of consolidating soil or dry concrete by mechanical means, or of wet concrete by vibration.

Compacto (*Typog.*). See **Multifont type tray.**

compander (*Telecomm.*). Device for compressing the volume range of the transmitted signal and re-expanding it at the receiver, thus increasing the signal/noise ratio. (*Com*presser . . . ex*pander*.)

companion cell (*Bot.*). A nucleated cell, associated with a segment of a phloem sieve tube, apparently playing some part in assisting the sieve tube to conduct food material. Companion cells occur only in *Angiospermae*.

companionship (*Typog.*). A number of compositors working together on adjacent portions of the same job. Called in Scotland **chumship.**

comparative psychology (*An. Behav.*). The study of the behaviour of animals, especially albino laboratory rats, pigeons and rhesus monkeys, in a variety of artificial situations. It is especially concerned with the effects of various factors on learning performance and the bearing of this on certain theories about learning. Cf. *ethology*.

comparator (*Phys.*). (1) A form of apparatus used for the accurate comparison of standards of length. It has also been used for measuring the coefficients of expansion of metal bars. (2) A form of colorimeter. (*Telecomm.*) Circuit which compares two sets of impulses, and acts on their matching or otherwise.

comparison lamp (*Light*). A lamp used, when performing photometric tests, for making successive comparisons between the lamp under test and a standard lamp.

comparison prism (*Light*). A small right-angled prism placed in front of a portion of the slit of a spectroscope or spectrograph for the purpose of reflecting light from a second source of light into the collimator, so that two spectra may be viewed simultaneously. See **comparison spectrum.**

comparison spectrum (*Light*). A spectrum formed alongside the spectrum under investigation, for the purpose of measuring the wavelengths of unknown lines. It is desirable that the comparison spectrum should contain many standard lines of known wavelength. The spectrum of the iron arc is often used for this purpose. See **comparison prism.**

comparison surface (*Light*). A surface illuminated by a standard lamp or a comparison lamp; used in photometry.

compass (*Surv.*). Instrument which indicates either (*a*) magnetic bearings by alignment of its needle on the magnetic poles, or (*b*) reference bearing by radio signal, or (*c*) change of orientation in relation to a gyro-maintained line. (*Zool.*) One of five radially disposed, curved ossicles, bifid distally, which overlie the rotulae in the lantern of Aristotle, in *Echinoidea*.

compass brick (*Build.*). A brick which tapers in at least one direction; specially useful for curved work, in arches, parts of furnaces, etc.

compass card (*Nav.*). The pivoted card to which the magnetized needles of a magnetic compass are attached. It may be marked with the points of the compass, with intermediate marks or in degrees, either in quadrants or 000° to 359° clockwise from North.

compass course (*Nav.*). The angle between the *compass meridian* (q.v.) and the direction of the ship's head indicated by the *lubber line* (q.v.).

compass error (*Nav.*). The algebraic sum of the *variation* (q.v.) and the *deviation* (q.v.) for a magnetic compass. The angle between the true meridian and the *compass meridian* (q.v.).

compasses (*Instr.*). An instrument for describing arcs, taking or marking distances, etc.; it consists essentially of two limbs hinged together at one end.

compass meridian (*Nav.*). The great circle in the plane of which the compass needle lies when influenced by the field of the ship in addition to the field of the earth. Cf. *magnetic meridian.*

compass plane (*Tools*). One for use on a curved surface, having a flexible metal sole which can be set concave or convex.

compass roof (*Arch.*). A roof with rafters bent to the shape of an arc.

compass rose (*Nav.*). A graduated circle in the form of a compass card printed on a chart, plan or map to show true or magnetic directions or both.

compass safe distance (*Aero., Nav.*). The minimum distance at which equipment may safely be positioned from a direct-reading magnetic compass, or detector unit of a remote-indicating compass, without exceeding the values of maximum compass deviation change.

compass traverse (*Surv.*). Rapid rough survey method in which the magnetic bearing of each line is measured.

compass window (*Arch.*). A bay or oriel window.

compatibility (*Chem.*). The tolerance of one dissolved substance towards another dissolved substance; e.g., the tolerance of a nitrocellulose solution towards a solution of a resin, mixture of the two solutions not effecting precipitation of the nitrocellulose. (*Comp.*) Requirement that different computers can accept and process data prepared in the same form or by the same peripheral equipment.

compatible (*Bot.*). Capable of self-fertilization.

compatible colour television (*TV*). The technique of transmitting television pictures in colour by a combination of *luminance* and *chrominance* *signals* (qq.v.), compatibility with black-and-white television being preserved as the chrominance elements are virtually disregarded by the monochrome receiver.

compatible equations (*Maths.*). See **consistent equations.**

compensated induction motor (*Elec. Eng.*). An induction motor with a commutator winding on the rotor, in addition to the ordinary primary and secondary windings; this winding is connected to the circuit in such a way that the motor operates at unity or at a leading power factor.

compensated pendulum (*Phys.*). A pendulum made of 2 materials which have different coefficients of expansion and are so chosen that the length of the pendulum remains constant when the temperature varies.

compensated pilot-wire (Beard-Hunter) protective system (*Elec. Eng.*). A modification of the Merz-Price opposed-voltage excess-current protective system, designed to avoid false operation due to capacity currents in the pilot wires. The pilot wires are each enclosed in a conducting sheath, and are so connected that any capacity current flows in the sheath and does not pass through the relays.

compensated repulsion motor (*Elec. Eng.*). A repulsion motor with an additional pair of brushes connected in series with the supply circuit and placed in quadrature with the short-circuited brushes, the object being to obtain an improved power factor.

compensated semiconductor (*Electronics*). Material in which there is a balanced relation between *donors* and *acceptors*, by which their opposing electrical effects are partially cancelled.

compensated series motor (*Elec. Eng.*). The usual type of a.c. series motor, in which a compensating winding is fitted to neutralize the effect of armature reaction and so give a good power-factor. Also called **neutralized series motor.**

compensated shunt box (*Elec. Eng.*). A shunt box for use with a galvanometer, arranged so that on each step a resistance is put in series with the galvanometer, and the total resistance of galvanometer and shunt is not altered.

compensated voltmeter (*Elec. Eng.*). A voltmeter arranged to indicate the voltage at the remote end of a feeder or other circuit, although connected at the sending end. A special winding compensates for the voltage drop in the feeder.

compensated wattmeter (*Elec. Eng.*). A wattmeter in which there is an additional winding, arranged to compensate for the effect of the current flowing in the pressure circuit.

compensating coils (*Elec. Eng.*). Current carrying coils to adjust distribution of magnetic flux.

compensating collar (*Eng.*). A collar fitted on a revolving shaft, for the purpose of compensating for wear at some point that would otherwise cause axial displacement of the shaft.

compensating diaphragm (*Surv.*). A fitment for a tacheometer which, by an adjustment to the stadia interval determined by the vertical angle, enables the horizontal component of a sloping sight to be deduced directly from the staff intercept.

compensating digits (*Teleph.*). Extra trains sent out over certain junctions to operate relays, so that subsequent marginal currents operate in resistances which do not depart markedly from the average.

compensating error (*Surv.*). As opposed to systematic (biased) error in series of observations, one equally likely to be due to over- or under-measurement, and therefore reasonably likely to be compensated by errors of opposite sign.

compensating field (*Elec. Eng.*). A term sometimes used to indicate the field produced by a compensating winding or, occasionally, by a compole.

compensating filter (*Photog.*). A filter used to alter the spectral emission of a light source, or the sensitivity of an emulsion, to a specified response to different wavelengths.

compensating jet (*I.C. Engs.*). An auxiliary petrol jet used in some carburettors to supplement the discharge from the main jet at low rates of air flow, and to keep the mixture strength constant. See **carburettor.**

compensating plate lock-up (*Print.*). A plate lock-up with resilient means to compensate for imperfections in the bevel of plates.

compensating pole (*Elec. Eng.*). See **compole.**

compensating roller (*Print.*). See **jockey roller.**

compensating tongue (*Bot.*). Vascular connexions between the outer and inner siphonosteles in a tricyclic stele.

compensating winding (*Elec. Eng.*). A winding used on d.c. or a.c. commutator machines to neutralize the effect of armature reaction.

compensation (*Acous.*). In a sound-reproducing system, adjustment of an actual frequency response to one specified. (*Med.*) The condition in which, in spite of the presence of heart disease, there is no heart failure; the heart is then said to be *compensated.*

compensation balance (*Horol.*). A balance so constructed as to compensate for the changes of dimensions in the balance and the elastic properties of the balance spring, with changes of temperature. Actually, compensation cannot be complete over a range of temperature. A watch or chronometer adjusted for the extremes of the range will not be correct at the middle of the range. This gives rise to what is known as the 'middle-temperature error'.

compensation level (*Ecol.*). The level in an aquatic habitat where just enough light penetrates for the rate of photosynthesis to equal the rate of respiration.

compensation method (*Elec. Eng.*). See **Poggendorff compensation method.**

compensation pendulum (*Horol.*). A pendulum so constructed that the distance between the centre of oscillation and the point of suspension remains constant with changes of temperature. See **pendulum.**

compensation point (*Bot.*). The light intensity at which, at any given temperature, respiration and photosynthesis just balance in a green plant, so that there is neither liberation nor absorption of carbon dioxide and of oxygen.

compensation ring (*Chem. Eng.*). A piece of metal welded (or riveted) to a pressure vessel shell round an opening to compensate for loss of strength caused by the opening.

compensation sac (*Zool.*). In some *Cheilostomata*, a membranous diverticulum of the ectoderm lying below the calcareous front surface of the zooecium, which fills with water when the tentacles are extruded.

compensation strand (*Bot.*). A strand of vascular tissue passing from an inner to an outer vascular ring in some ferns.

compensation theorem (*Elec.*). That the change in current produced in a network by a small change in any impedance Z carrying a current I is the result of an apparent e.m.f. of $-I.\delta Z$.

compensation water (*Civ. Eng.*). The water which

has to be passed downstream from a reservoir to supply users who, prior to the construction of the dam, took their water directly from the stream.

compensator (*Cinema.*). In a sound-film recording camera, an arrangement for making the speed of the film correspond with a stated number of sprocket holes per second; used because of slight variations in the length of the stated number of sprocket holes. (*Elec. Eng.*) An obsolete name for *autotransformer* (q.v.). (*Photog.*) A device for graduating the light passed through a wide-angle lens, such as an axial rotating star or a plano-convex lens of tinted glass, so that the image of an evenly illuminated surface is also evenly illuminated. (*Print.*) See **jockey roller.**

competition (*Biol.*). The struggle between organisms for the necessities of life (water, light, etc.).

compiler (*Comp.*). (1) See **programmer.** (2) Automatic computer programming unit which transcribes the operator's instructions from a programme-orientated language into the machine language of the specific computer. See also **assembler.**

complanate (*Bot.*). Flattened, compressed.

complement (*Bacteriol.*). A thermolabile substance or mixture of substances present in normal serum, which reacts with any antibody-antigen system, except a toxin-antitoxin system, to lyse the antigen (e.g., bacteria) agglutinated by the antibody. (*Cyt.*) A group of chromosomes derived from one nucleus, and consisting of one, two, or more sets.

complemental air (*Physiol.*). The volume of the deepest inhalation that can be taken in by the lungs at the end of a normal inspiration, which in man is approximately 1500 cm^3.

complemental males (*Zool.*). In some hermaphrodite species, as the Barnacle *Scalpellum*, certain small individuals, lacking ovaries, found living in close proximity to the normal hermaphrodite forms; believed to have the function of effecting cross-fertilization.

complementarity (*Phys.*). The correspondence in quantum mechanics between particles of momentum p and energy E, and wave trains of frequency ν and wavelength λ. This is given by equations $p=h/\lambda$ and $E=h\nu$, where h is Planck's constant and $\nu\lambda=c$ (speed of light).

complementary after-image (*Optics*). The subjective image, in complementary colours, that is experienced after visual fatigue induced by observation of a brightly coloured object.

complementary angles (*Maths.*). Two angles whose sum is 90°. Each is said to be the complement of the other.

complementary colours (*Light*). Pairs of colours which combine to give spectral white.

complementary factor (*Gen.*). A factor in inheritance which, in conjunction with one or more similar factors, leads to the appearance of some character in the offspring.

complementary function (*Maths.*). If $y=u+v$ is a solution of a differential equation, where u is a particular integral and v contains the full number of arbitrary constants, v is called the *complementary function.* The complementary function is the general solution of the auxiliary equation.

complementary society (*Bot.*). A community of two or more species of plants occupying the same soil, but not coming into active competition since they vegetate at different times of the year or develop their roots at different levels in the soil.

complementary symmetry (*Electronics*). Shown by otherwise identical *p-n-p* and *n-p-n* transistors.

complementary tissue (*Bot.*). A loose assemblage of thin-walled, unsuberized cells, fitting loosely together, lying in the cavity of a lenticel, and allowing gases and vapours to diffuse through them.

complementary transistors (*Electronics*). A *n-p-n* and a *p-n-p* transistor pair used to produce a push-pull output using a common signal input.

complement fixation (*Bacteriol.*). A technique used for the detection of antibody or antigen, which can also be modified to determine the quantity of a pathogen, e.g., titre of a virus, present in a preparation. The basis of the complement fixation test is the combination of a normal component of serum (complement) with an antibody-antigen system, thus rendering it unavailable for combination with any other antibody-antigen system, e.g., *Wassermann reaction* for the detection of syphilis.

complement of a set (*Maths.*). The complement of a set A is the set of all members of the universal set which are not members of A, and is sometimes denoted by A'.

complete combustion (*Heat*). Burning fuel without trace of unburnt gases in the products of combustion, usually accompanied by excess air in the flue products. Cf. *perfect combustion.*

complete differential (*Maths.*). The complete differential df of a function $f(x, y, z)$ is

$$\frac{\partial f}{\partial x}\,.\,dx + \frac{\partial f}{\partial y}\,.\,dy + \frac{\partial f}{\partial z}\,.\,dz\,.$$

Similar expressions apply for more or less variables. Also called **total differential.**

complete flower (*Bot.*). A flower which has both calyx and corolla.

complete integral (*Maths.*). See **complete primitive.**

complete operation (*Comp.*). Processing of one complete instruction.

complete primitive (*Maths.*). The solution of a differential equation containing the full number of arbitrary constants. Also called **general** (or **complete**) **solution, general** (or **complete**) **integral.**

complete radiator (*Phys.*). Same as **black body.**

complete reaction (*Chem.*). A reaction which is irreversible and therefore proceeds until one of the reactants has disappeared.

complete set of functions (*Maths.*). See **orthogonal functions.**

complete solution (*Maths.*). See **complete primitive.**

completion test (*Psychol.*). A type of intelligence test in which the candidate has to fill in blanks in sentences, etc.; also called **combination test.**

complex (*Psychol.*). A term introduced by Jung to denote an emotionally toned constellation of mental factors formed by the attachment of instinctive emotions to objects or experiences in the environment, and always containing elements unacceptable to the self. It may be recognized in consciousness, but is usually repressed and unrecognized.

complex hyperbolic functions (*Elec.*). Hyperbolic functions, with complex quantities as variables, which facilitate calculations of electric waves along transmission lines.

complex ion (*Chem.*). Formed by the coordination of a neutral ion molecule with a simple ion, e.g., the cuprammonium ion

$$2NH_3+Cu^{2+}=Cu(NH_3)_2^{2+}.$$

complex number (*Maths.*). A number z of the form $a+ib$, where $i=\sqrt{-1}$ and a and b are real numbers. The number a is called the *real part*, written Rz, and b the *imaginary part*, written Iz. E.g., if $z=3+4i$, $Rz=3$ and $Iz=4$. Cf. *modulus* and *argument*.

complexometric titration (*Chem.*). Titration of a metal ion with a reagent, usually *EDTA* (q.v.), which forms chelate complexes with the metal. The end point is accompanied by a sharp decrease in the concentration of metal ions, and is observed by a suitable indicator.

complexones (*Chem.*). Collective term used to denote chelating reagents (usually organic) used in the analytical determination of metals, e.g., *EDTA* (q.v.).

complexor (*Elec. Eng.*). A representation of a single-frequency, sinusoidally alternating voltage or current, which with similar complexors in a plane can be manipulated as though they were complex numbers on an Argand diagram. Also called **phasor** and **sinor.**

complex Poynting vector (*Elec. Eng.*). See **Poynting vector.**

complex tissue (*Bot.*). A tissue made up of cells or elements of more than one kind.

complex tone (*Acous.*). Strictly, a musical note in which all the separate tones are exact multiples of a fundamental frequency, recognized as the pitch, even when the actual fundamental is absent, as in the lowest octave of a piano. Loosely, a mixed musical chord.

complexus (*Zool.*). A complicated system of organs, e.g., in Primates, a compound muscle of the back.

complex wave (*Phys.*). One with a nonsinusoidal waveform which can be resolved into a fundamental with superimposed harmonics. See **Fourier principle.**

compliance (*Phys.*). A measure of weakness of a spring. Mechanical analogue of capacitance, reciprocal of stiffness.

complicant (*Zool.*). Folding one over another, as the wings of some Insects.

complicate (*Bot., Zool.*). Folded together.

compo (*Build.*). A cement mortar.

compo board (*Build.*). A building-board consisting of narrow strips of wood glued together into one sheet, both sides being faced with a heavy paper.

compole (*Elec. Eng.*). An auxiliary pole, employed on commutator machines, which is placed between the main poles for the purposes of producing an auxiliary flux to assist commutation. Also called **commutating pole, compensating pole, interpole.**

component (*Elec. Eng.*). A term sometimes used to denote one of the component parts into which a vector representing voltage, currents, or volt-amperes may be resolved; the component parts are usually in phase with, or in quadrature with, some reference vector. See **active voltage, active current, active volt-amperes, reactive voltage, reactive component of current, reactive volt-amperes.** (*Mech.*) The resolved part of a force in any particular direction. The component of a force F, along a line making an angle θ with the line of action of F, is $F \cos \theta$. See **resolution of forces.** (*Teleph.*) Any part of a signal (including a space) which is uniform in frequency and amplitude during its allotted duration.

component of a vector (*Maths.*). In general, one of two or more vectors whose sum is the given vector, but unless the contrary is stated the component in a specified direction is the vector in that direction, whose sum with one or more vectors in directions perpendicular to the specified direction is the given vector.

components (*Chem.*). The minimum number of substances required for the establishment of equilibrium, physical or chemical, in a given system. See **phase rule.**

compo pipe (*Gas*). Pipe made from an alloy, as opposed to lead pipe.

compose (*Typog.*). To assemble type matter for printing, either by hand or by type-setting machines.

Compose-a-Forme (*Typog.*). A collection of reproduction proofs of commonly used words, for headings, and both vertical and horizontal rule patterns.

Composertron (*Acous.*). Complex magnetic-tape recording assembly, in which elements of recorded sound can be synthesized and reproduced as a composition.

composing frame (*Typog.*). A wooden or metal structure at which the compositor works. Originally providing accommodation on top for an upper and a lower case with storage below for other cases, there is now a large variety of styles to suit the requirements of the work to be done, with sloping or flat tops, and storage for all kinds of typographic material in addition to cases.

composing machines (*Typog.*). The *Monotype* composes type matter in separate letters, which may be used for hand-setting or correcting; the *Linotype* composes in solid lines, or slugs, which must be reset when correcting. Both machines have a keyboard resembling that of a typewriter. See also **Intertype** and **filmsetting.**

composing rule (*Typog.*). A piece of brass or steel rule with a projecting nose piece, moved up as each line in the stick is completed by the compositor.

composing stick (*Typog.*). A metal or wooden 3-sided boxlike receptacle in which the compositor sets his type letter by letter. The width or measure can be altered as desired.

Compositae (or **Asteraceae**) (*Bot.*). The largest family of flowering plants; except for a few genera, the flowers are in a capitulum; the ovary is inferior with a single basifixed ovule; the anthers are fused to form a tube around the style; the stigma is bifid; the calyx is reduced to a pappus of hairs, and the five petals are fused, forming a tube or a flat strap.

composite balance (*Elec. Eng.*). A modification of the Kelvin balance; the moving coils are of high resistance, enabling the instrument to be used as a wattmeter.

composite beam (*Eng.*). A beam composed of two materials properly bonded together and having different moduli of elasticity, e.g., a reinforced concrete beam.

composite block (*Typog.*). (1) *Combined half-tone and line* (q.v.) block. (2) A block made up from 2 or more originals.

composite cable (*Elec. Eng.*). Cable containing different purpose conductors inside a common sheath.

composite circuit (*Telecomm.*). A telegraph circuit operating over a telephone circuit, without interfering with the latter.

composite conductor (*Elec. Eng.*). One in which strands of different metals are used in parallel.

composite course (*Nav.*). The shortest practical course between two places on the globe, being a series of short rhumb-lines following the great circle course, avoiding land masses.

composite deposits (*Elec. Eng.*). Deposits consisting of two or more layers of different metals, formed by electroplating, the object being protection from corrosion or to obtain a smooth deposit of particular properties.

composite filter (*Telecomm.*). Combination of a number of filter sections, or half-sections, all having the same cut-off frequencies and specified impedance levels.

composite resistor (*Elec. Eng.*). One formed of solid rod of carbon compound.

composite sailing (*Nav.*). Following the shortest route between two points without crossing some particular parallel of latitude. The method involves drawing great circles tangential to the limiting parallel of latitude from both starting point and destination. The middle portion of the route lies along the parallel of latitude.

composite sill (*Geol.*). A sill apparently composed of successive injections of different igneous rocks.

composite truss (*Build.*). A roof truss formed of timber struts and steel or wrought-iron ties (apart from the main tie, which is usually of timber to simplify connexions).

composition (*Chem.*). The nature of the elements present in a substance and the proportions in which they occur.

composition founts (*Typog.*). The smaller sizes of type, up to 14-point, as used for bookwork.

composition nails (*Build.*). Roofing nails made of a cast 60–40 copper-zinc alloy.

composition of atmosphere (*Chem., etc.*). Dry atmospheric air contains the following gases in the proportions (by weight) indicated: nitrogen, 75·5; oxygen, 23·14; argon, 1·3; carbon dioxide, 0·05; krypton, 0·028; xenon, 0·005; neon, 0·000 86; helium, 0·000 056. There are variable trace amounts of other gases incl. hydrogen and ozone. Water content, which varies greatly, is excluded from this analysis.

composition of forces (*Mech.*). The process of finding the resultant of a number of forces, that is a single force which can replace the other forces and produce the same effect. See **parallelogram of forces.**

composition rollers (*Print.*). (1) For letterpress printing, a mixture of glue, glycerine and molasses. (2) For lithographic printing, vegetable oils and rubber, vulcanized.

compositor (*Typog.*). A craftsman whose work consists of setting up type matter by hand, or correcting that set by machine. Skill and judgment in display work are part of his routine.

compound (*Bot.*). Formed of a number of parts which are nearly separate, and usually separate at maturity. (*Chem.*) See **chemical compound.**

compound arch (*Arch.*). An arch having an archivolt receding in steps, so as to give the appearance of a succession of receding arches of varying spans and rises.

compound brush (*Elec. Eng.*). A type of brush used for collecting current from the commutator of an electric machine; the brush has alternate layers of copper and carbon so that the conductivity is greater longitudinally (i.e., in the direction of the main current flow) than laterally.

compound catenary construction (*Elec. Eng.*). A construction used for supporting the overhead contact wire of an electric traction system; the contact wire is supported from an auxiliary catenary which, in turn, is supported from a main catenary, all 3 wires lying in the same plane.

compound curve (*Surv.*). A curve composed of two arcs of different radii, having their centres on the same side of the curve, connecting two straights.

compound dredger (*Civ. Eng.*). A type of dredger combining the suction or suction cutter apparatus with a bucket ladder.

compound engine (*Aero.*). A reciprocating engine having an *exhaust-driven supercharger* (q.v.) in which any surplus turbine power is fed to the airscrew through a fluid, or slip-clutch, drive. (*Eng.*) A steam-engine in which the expansion of the steam from boiler pressure to exhaust pressure is carried out in two stages, necessitating the use of high- and low-pressure cylinders in series.

compound eyes (*Zool.*). Paired eyes consisting of many facets or ommatidia, in most adult *Arthropoda.*

compound fault (*Geol.*). A series of closely spaced parallel or subparallel faults.

compound filled apparatus (*Elec. Eng.*). Electrical apparatus (e.g., bus-bars, potential transformers, switchgear) in which all live parts are enclosed in a metal casing filled with insulating compound.

compound fruit (*Bot.*). A fruit formed from several closely associated flowers.

compound generator (*Elec. Eng.*). See **compound motor.**

compound girder (*Build.*). A rolled-steel joist strengthened by additional plates riveted or welded to the flanges.

compound harness (*Textiles*). An arrangement of cords in a jacquard machine, made to increase the figuring capacity of the loom.

compound horn (*Acous.*). Rectangular horn for electromagnetic waves in which flaring occurs in both planes.

compounding (*Eng.*). The principle, or the use of the principle, of expanding steam in two or more stages, either in reciprocating engines or steam-turbines.

compound-interest law (*Bot.*). The doctrine that the rate of growth of a plant at any time is proportional to the amount of plant material present at the time.

compound leaf (*Bot.*). A leaf, the lamina of which consists of a number of quite distinct leaflets, being divided down to the midrib; the leaflets may fall independently of one another as the leaf dies.

compound lever (*Eng.*). A series of levers for obtaining a large mechanical advantage, the short arm of one being connected to the long arm of the next; used in large weighing and testing machines.

compound magnet (*Elec. Eng.*). A permanent magnet made up of several laminations.

compound microscope (*Light*). See **microscope.**

compound modulation (*Radio*). Use of an already modulated wave as a further modulation envelope. Also called **double modulation.**

compound motor (or generator) (*Elec. Eng.*). One which has both series and shunt field windings.

compound nucleus (*Nuc.*). Highly excited unstable nucleus formed by absorption of bombarding particle.

compound oil (*Oils*). (1) An oil blended from the products of crudes of different characteristics. (2) A blend of mineral and vegetable oils, used to combine certain desirable properties.

compound oösphere (*Bot.*). A multinucleate body occurring in an oögonium, probably composed of a number of female gametes which have not become individualized.

compound pendulum (*Phys.*). Any body capable of rotation about a fixed horizontal axis and in stable equilibrium under the action of gravity. If the centre of gravity is at a distance h from the axis of rotation, and k is the radius of gyration about a horizontal axis through the centre of gravity, the period of small oscillations is:

$$T=2\pi\sqrt{\frac{h+\frac{k^2}{h}}{g}}.$$

See **pendulum.**

compound pillar (*Build.*). A pillar formed of a rolled-steel joist or joists or channels strengthened by additional plates riveted or welded to the flanges.

compound press tool (*Eng.*). A *press tool* (q.v.) which performs two or more operations at the same station at each stroke of the press.

compound pyrenoid (*Bot.*). A pyrenoid made up of two closely associated portions.

compound reflex (*Zool.*). A combination of several reflexes to form a definite coordination, either simultaneous or successive.

compound rest (*Eng.*). Part of a lathe, mounted on the bed to impart movement, in two directions, to the tool.

compound train (*Eng.*). A train of gear-wheels in which intermediate shafts carry both large and small wheels, in order to obtain a large speed ratio in a small space.

compound umbel (*Bot.*). A racemose inflorescence consisting of a number of secondary axes radiating obliquely upwards from the end of a main branch, and bearing in their turn a number of flower stalks similarly arranged. Each flower stalk bears one flower, and, generally, the whole assemblage of flowers forms a flat, circular group.

compressed (*Bot.*). Flattened laterally; pressed together.

compressed air (*Eng.*). Air at higher than atmospheric pressure. It is used (often at about 600 kN/m^2) as a transmitter of energy where the use of electricity or an I.C. engine would be hazardous (e.g. in mining). The exhaust air may be used for cooling or ventilation.

compressed-air capacitor (*Elec. Eng.*). An electric capacitor in which air at several atmospheres' pressure is used as the dielectric, on account of its high dielectric strength at these pressures.

compressed-air disease (*Med.*). See **caisson disease.**

compressed-air inspirator (*Heat*). Injector used with pressure-air burners, by which a stream of compressed air is directed through a venturi throat to inspirate the additional combustion air from surrounding atmosphere, thus reducing the initial pressure.

compressed-air lamp (*Light*). An electric lamp for use in fiery mines; it is supplied from a small compressed-air-driven generator incorporated in the lamp-holder.

compressed-air squeezer (*Eng.*). See **squeezer.**

compressed-air tools (*Eng.*). See **pneumatic tools.**

compressed-air wind tunnel (*Aero.*). See **variable-density wind tunnel.**

compressibility (*Mech., Phys.*). The reciprocal of the bulk modulus. See also **coefficient of compressibility.** (*Powder Tech.*) The property of a powder by which it accepts reduction in volume by pressure. It is measured as the ratio of the volume of loose powder to the volume of the compact, and is related to the pressure applied. This is sometimes referred to as the *compression ratio.*

compressibility drag (*Aero.*). The sharp increase of drag as air speed approaches the speed of sound and flow characteristics change from those of a viscous to those of a compressible fluid, causing the generation of shock waves.

compression (*Acous.*). Increase of density along high pressure wavefront of sound wave. (*Eng.*) In an internal-combustion engine, (*a*) the stroke during which the working agent is compressed, and (*b*) the gas-tightness of the piston during the compression stroke. (*Radio*) Reduction of gain of amplifiers for high signal levels. Normally expressed by

$$10 \log_{10} G_H/G_L,$$

where G_H and G_L are respectively power gains for high and low power levels.

compression cable. See **pressure cable.**

compression flange (*Bot.*). A group of turgescent parenchymatous cells on the convex side of a coiled tendril.

compression-ignition engine (*Eng.*). An internal-combustion engine in which ignition of the liquid fuel injected into the cylinder is performed by the heat of compression of the air charge. See **diesel engine.**

compression moulding (*Plastics*). The material is placed in a heated, hardened, polished steel container and forced down by means of a plunger at a pressure of ca. 20–35 MN/m^2. The disadvantages of the method are the damage done to the mould by the material travelling under a high pressure and the fact that more material than is necessary must be used.

compression plate lock-up (*Print.*). A method of locking plates to the cylinder by movable dogs which press the plate from the angled edge at the side towards the centre of the cylinder. Cf. *tension plate lock-up*.

compression ratio (*Eng.*). In an internal-combustion engine, the ratio of the total volume enclosed in the cylinder at the outer dead-centre to the volume at the end of compression; the ratio of swept volume, plus clearance volume, to clearance volume. See **clearance volume.** (*Powder Tech.*) See under **compressibility.**

compression rib (*Aero.*). See **rib.**

compression spring (*Eng.*). A helical spring with separated coils, or a conical coil spring, with plain, squared or ground ends, made of round, oblong, or square-section wire.

compression test (*Eng.*). A test in which specimens are subjected to an increasing compressive force, usually until they fail by cracking, buckling or disintegration. A stress-strain curve may be plotted to determine mechanical properties, as in a *tensile test* (q.v.). Compression tests are often applied to materials of high compression but low tensile strength, such as concrete.

compression wood (*Bot.*). Wood of dense structure formed at the bases of some tree trunks, and on the undersides of branches.

compressor (*Aero.*). Device for compressing the air supply to a *gas turbine* or a *ramjet*. (*Eng.*) A reciprocating or rotary pump for raising the pressure of a gas. (*Teleph.*) An electronic amplifier designed to reduce the contrast in telephonic speech, for transmission at an average higher level in the presence of interference. (*Zool.*) A muscle which by its contraction serves to compress some organ or structure.

compressor drum (*Aero.*). A cylinder composed of a series of rings or, more usually, disks wherein the blades of an axial compressor are mounted.

Compton absorption (*Nuc.*). That part of the absorption of a beam of X-rays or gamma-rays associated with Compton scattering processes. In general, it is greatest for medium-energy quanta and in absorbers of low atomic weight. At lower energies *photoelectric absorption* is more important, and at high energies *pair production* predominates.

Compton effect (*Nuc.*). Elastic scattering of photons by electrons, i.e., collisions in which both energy and momentum are conserved. If λ_s and λ_i are respectively the wavelengths associated with scattered and incident photons, Compton shift is given by

$$\lambda_s - \lambda_i = \lambda_0(1 - \cos\theta),$$

where θ = angle between directions of the incident and scattered photons, and λ_0 is the Compton wavelength of the electron.

Compton electrometer (*Elec. Eng.*). A sensitive form of quadrant electrometer.

Compton recoil electron (*Nuc.*). An electron which has been set in motion following an interaction with a photon (*Compton effect*).

Compton's rule (*Chem.*). An empirical rule that the melting point of an element in kelvins is equal to half the product of the r.a.m. and the specific latent heat of fusion.

Compton wavelength (*Nuc.*). Wavelength associated with the mass of any particle, such that

$$\lambda = h/mc$$

where λ = wavelength
h = Planck's constant
m = rest mass
c = velocity of light.

compulsion neurosis (*Psychol.*). See **obsessional neurosis.**

computer. Any mechanical, electric, or electronic apparatus or assembly which performs calculations, simple or including integration and differentiation, or simultaneous equations. There are 3 main types: (1) *analogue* (q.v.), which deals with continuous correspondences; (2) *digital* (q.v.), which operates on units; (3) *hybrid* (q.v.) or *quantized*, in which operations are carried out on the quantities processed using both analogue and digital techniques. An *analogue-digital converter* (q.v.) is used to change continuous amplitudes into digital signals and vice versa.

computer typesetting (*Print.*). The use of electronic equipment to process an unjustified input into an output of justified and hyphenated lines. The output can be a new tape to be used on a typesetting or filmsetting machine or can be the final product in some filmsetting systems. Among the simplest and least expensive equipment is the *Linasec* (q.v.) which requires monitoring. More elaborate systems have a memory store to control hyphenation, and operate at speeds up to 10,000 characters per second. Some equipment can accept a random input of items and produce a classified and alphabetically arranged output.

conarium (*Zool.*). In Vertebrates, the pineal gland: in *Coelenterata*, the transparent deep-sea larva of *Velella*.

conation (*Psychol.*). Traditionally one of the three aspects of human psychology, the other two being *cognition* and *affect*. It literally means 'striving' and refers to 'the will', though sometimes more generally to all experienced mental activity Little used at present.

concatenate (*Bot.*). Joined together, forming a chain of similar members.

concatenated connexion (*Elec. Eng.*). See **cascade.**

concatenated motor (*Elec. Eng.*). See **cascade motor.**

Concato's disease (*Med.*). A chronic inflammation of serous membranes attended with effusion of fluid into serous sacs (e.g., peritoneal, pericardial, and pleural cavities).

concave brick (*Build.*). A compass brick.

concave grating (*Light*). A diffraction grating ruled on the surface of a concave spherical mirror, made usually of speculum metal or glass. Such a grating needs no lenses for collimating or focusing the light. Largely on this account it is the most useful means of producing spectra for precise measurement. See **Rowland circle.**

concave lens (*Light*). A *divergent lens* (q.v.).

concave mirror (*Light*). A curved surface—usually a portion of a sphere—the inner surface of which is a polished reflector. Such a mirror is capable of forming real and virtual images, their positions being given by the equation $\frac{2}{R} = \frac{1}{v} + \frac{1}{u}$, R being the radius of curvature of the mirror, u the distance of the object, and v the distance of the image from the mirror. See **convention of signs.**

concave veins (*Zool.*). In Insects, veins which follow the furrows of the wing corrugations.

concealed heating (*Build.*). See **panel heating.**

concentrate (*Bot.*). Arranged around a common centre, or having a common centre. (*Met.*) The products of concentration operations in which a relatively high content of mineral has been obtained and which are ready for treatment by chemical methods. Also called **concentrates.**

concentrated load (*Build.*). A load which is regarded as acting through a point.

concentrated winding (*Elec. Eng.*). A winding in which all the conductors forming one group (e.g., one phase under one pole in an a.c. machine) are placed in one slot instead of being distributed over a number of slots.

concentrating reflector (*Light*). A type of reflector which concentrates a strong beam of light on a particular area; used in industrial work.

concentrating table (*Min. Proc.*). Supported deck, across or along which mineralized sands are washed or moved to produce differentiated products according to the gravitational response of particles of varying size and/or density. Stationary tables include strakes, sluices, buddles; moving tables include shaking tables (e.g., *Wilfley table*), vanners, and rockers.

concentration (*Chem.*). (1) Number of molecules or ions of a substance in a given volume, generally expressed as moles per cubic metre or cubic decimetre. (2) A process in which the *concentration* (1) of a substance is increased, e.g., the evaporation of the solvent from a solution. (*Met.*) Production of *concentrate* (q.v.).

concentration cell (*Elec.*). One with similar electrodes in common electrolyte, the e.m.f. arising from differences in concentration at the electrodes.

concentration plant (*Min. Proc.*). Concentrator mill, reduction works, washing, cleaning plant. Buildings and installations in which ore is processed by physical, chemical and/or electrical methods to retain its valuable constituents and discard as tailings those of no commercial interest.

concentration polarization (*Elec.*). A form of polarization occurring in an electrolytic cell, due to changes in the concentration of the electrolyte surrounding the electrode.

concentric (*Telecomm.*). Term replaced by *coaxial.*

concentric arch (*Arch.*). An arch laid in several courses whose curves have a common centre.

concentric chuck (*Eng.*). See **self-centring chuck**

concentric coils (*Elec. Eng.*). Coils of an armature winding designed to have different spans so that, when placed in the slots, the end connexions shall not have to cross each other.

concentric feeder (*Telecomm.*). Transmission line in which two conductors are concentric tubes. The absence of external fields and freedom from pick-up make it specially suitable for use with directional antenna systems; also called **coaxial feeder.**

concentric plug-and-socket (*Elec. Eng.*). A type of plug-and-socket connexion in which one contact is a central pin and the other is a ring concentric with it.

concentric vascular bundle (*Bot.*). A bundle in which a strand of xylem is completely surrounded by a sheath of phloem, or vice versa.

concentric winding (*Elec. Eng.*). An armature winding, used on a.c. machines, in which groups of concentric coils are used. Also used to denote the type of winding, used on transformers, in which the high-voltage winding is arranged concentrically with the low-voltage winding.

concentric wiring (*Elec. Eng.*). An interior wiring system in which the conductor consists of an insulated central core surrounded by a flexible metal sheath which forms the return lead.

conceptacle (*Bot.*). A flask-shaped cavity in a thallus, opening to the outside by a small pore, and containing reproductive structures (e.g. of Fucus).

concept attainment (*Psychol.*). A mature process of manipulating known or definable concepts in order to find an unknown concept, e.g., 'twenty questions' game. See **concept formation.**

concept formation (*Psychol.*). The process from early life on by which the individual learns to classify the aspects of his world. Much of such learning is verbal and one of the main problems seems to be overgeneralization, e.g., the child has to learn not to call all small four-legged animals 'dog'. See **concept attainment.**

conception (*Physiol.*). The fertilization of an ovum with a spermatozoon.

concertina amplifier (*Telecomm.*). Electronic amplifier which has a load resistor in each polarity of the supply. In this way two outputs of opposite phase are provided, e.g., for driving a push-pull stage.

concertina fold (*Bind.*). See **accordion fold.**

concert pitch (*Acous.*). The recognized pitch, i.e., frequency of the generated sound waves, to which musical instruments are tuned, so that they can play together. The exact value has varied considerably during musical history, but it has recently been internationally standardized so that *A* (above middle-*C*) becomes 440 hertz. Dance-bands are normally pitched a few hertz lower. Allowance must be made for the rise in temperature experienced in concert halls, which alters the pitch in ways peculiar to the different types of instrument.

concha (*Arch.*). The smooth concave surface of a vault. (*Zool.*) In Vertebrates, the cavity of the outer ear; the outer or external ear; a shelf projecting inwards from the wall of the nasal cavity to increase the surface of the nasal epithelium.

conchate or **conchiform** (*Bot.*). Shaped like an oyster shell.

conchiolin (*Zool.*). A horny substance forming the outer layer of the shell in *Mollusca.*

conchitis (*Med.*). Inflammation of a concha.

conchoid (*Maths.*). If, from a fixed point *P*, a line is drawn to meet any curve *C* in the point *Q*, and if *A* and *B* are any two points on the line *PQ* such that $AQ = BQ =$ a constant, then the locus of the points *A* and *B* is a conchoid of *C* with respect to the point *P*. The conchoid of a straight line not passing through *P*, first discussed by Nicomedes, 250 B.C., in connection with the tri-section of an angle, and called the *conchoid of Nicomedes*, has either a node at *P*, a cusp at *P* or no double point at all, depending upon whether the perpendicular length from *P* to the line is greater than, equal to, or less than *AQ* respectively.

Conchostraca (*Zool.*). A suborder of order *Diplostraca*, class *Branchiopoda*, having a carapace, in the form of a bivalve shell, enclosing the head and body; the eyes are sessile and coalescent; the second antennae are biramous and natatory; there are numerous trunk limbs; the caudal furca are clawlike. Clam Shrimps.

concolor or **concolorous** (*Bot., Zool.*). Uniform in colour.

concolorate (*Zool.*). Having both sides the same colour.

concrescence (*Bot.*). Growing together to form a single structure. *adj.* **concrescent.**

concrete (*Bot.*). (1) Grown together to form a solid body. (2) Adhering closely to anything. (*Build., Civ. Eng., etc.*) A mixture of cement, sand, and gravel, with water in varying proportions according to the use which is to be made of it.

concrete blocks (*Civ. Eng., etc.*). Solid or hollow pre-cast blocks of concrete used in the construction of buildings.

concrete mixer (*Civ. Eng., etc.*). An appliance in which the constituents of concrete are mixed mechanically.

concrete operations (*Psychol.*). The thought structures which develop from about 7 years of age until the early teens (Piaget). Concepts are formed and organized into systems so that operations, or internalized responses, can be performed. The operations are still tied to their contents and the two types of reversibility, inverse and reciprocal, are used independently of each other. See **formal operations, operations.**

concretion (*Geol.*). Nodular or irregular concentration of siliceous, calcareous, or other materials, formed by localized deposition from solution in sedimentary rocks. (*Med.*) Collection of organic matter with or without lime salts, or of lime salts, in bodily organs.

concussion (*Med.*). A violent shaking or blow (especially of or to the head), or the condition resulting from it. See **commotio cerebri.**

condensance (*Elec. Eng.*). A term occasionally used to denote the reactance of a capacitor.

condensate (*Eng.*). The liquid obtained as a result of removing from a vapour such portion of the latent heat of evaporation as it may contain.

condensation (*Bot.*). Crowding in a vertical series, due to the absence or suppression of internodes. (*Chem.*) (1) The linking together of two or more molecules, resulting in the formation of long chain compounds. (2) The union of two or more molecules with the elimination of a simpler group, such as H_2O, NH_3, etc. (*Meteor.*) The process of forming a liquid from its vapour. When moist air is cooled below its dew-point, water vapour condenses if there are extended surfaces or nuclei present. These nuclei may be dust particles or ions. Mist, fog, and cloud are formed by nuclear condensation.

condensation coefficient (*Vac. Tech.*). The ratio of the number of molecules condensing on any surface per second to the total number incident on the surface per second.

condensation gutter (*Build.*). A small gutter provided at the curb of lantern lights to carry away condensed water formed on the interior surface of the glazing.

condensation sinking (*Join.*). A groove cut in the bottom rails of skylights to carry away condensed water formed on the interior surface of the glazing.

condensation trails (*Meteor.*). Artificial clouds caused by the passage of an aircraft due either to condensation following the reduction in pressure above the wing surfaces, or to condensation of water vapour contained in the engine exhaust gases. Also **contrails.**

condensed (*Bot.*). Said of an inflorescence in which the flowers are crowded together and nearly or quite sessile. (*Typog.*) See **elongated.**

condensed nucleus (*Chem.*). A ring system in which two rings have one or more (generally two) atoms in common, e.g., naphthalene, phenanthrene, quinoline.

condensed system (*Chem.*). One in which there is no vapour phase. The effect of pressure is then practically negligible, and the *phase rule* may be written $P+F=C+1$.

condenser (*Chem.*). Apparatus used for condensing vapours obtained in distillation. In laboratory practice usually a single tube, either freely exposed to air or contained in a jacket in which water circulates. In industrial operations may be (*a*) surface which usually has large number of small bore tubes with water flowing inside them, all contained in a shell or case into which vapour is introduced, or (*b*) spray in which cold liquid is brought into direct contact with the vapour. (*Elec. Eng.*) A term previously used to denote a *capacitor*, the latter name now being preferred in U.K. and U.S., although the former is still in use on the Continent. (*Eng.*) (1) A chamber into which the exhaust steam from a steam-engine or turbine is delivered, to be condensed by the circulation or the introduction of cooling water; in it a high degree of vacuum is maintained by an air pump. (2) The part of a refrigeration system in which the refrigerant is liquefied by transferring heat to the cooling medium, usually water or air. (*Light*) A large lens or mirror used in an optical projecting system to collect light, radiated from the source, over a large solid angle, and to direct this light on to the object or transparency which is to be focused at a distance by a projection lens. See **mirror arc.** (*Textiles*) A mechanism which divides the web emerging from the waste card into narrow strips. These are rubbed into twistless rovings by passing them between oscillating rubber or leather aprons.

condenser bushing (*Elec. Eng.*). A type of bushing used for terminals of high-voltage apparatus (e.g., transformers and switchgear) in which alternate layers of insulating material and metal foil form the insulation between the conductor and the outer casing; the metal foil serves to improve the voltage distribution. Also **capacitor bushing or terminal.**

condenser circulating pump (*Eng.*). See **circulating pump.**

condenser leathers (*Textiles*). See **rubbing leathers.**

condenser terminal (*Elec. Eng.*). Capacitor terminal. See **condenser bushing.**

condenser tissue (*Paper*). Thin rag paper used as a capacitor dielectric. Also **capacitor tissue.**

condenser tubes (*Eng.*). The tubes through which the cooling water is circulated in a *surface condenser* (q.v.), and on whose outer surfaces the steam is condensed.

condenser yarn (*Spinning*). Yarns spun from clean soft waste material; suitable for cotton blankets, quiltings and towellings.

condensing electroscope (*Elec. Eng.*). A gold-leaf electroscope combined with a small capacitor which can have its upper plate removed to magnify the effect of a charge by decreasing the capacitance of the instrument.

condensing lens (*Photog.*). A simple lens which receives light flux from a source of light at the greatest solid angle possible and concentrates it evenly over the area of a gate, as in a film projector.

con-di nozzle (*Aero., Eng.*). See **convergent-divergent nozzle.**

condition (*Brew.*). Assessment of a beer according to its carbon dioxide content.

conditional jump (*Comp.*). Program instructions involving change of routine if the result of a previous operation satisfies some specific requirement, e.g., when it becomes negative. See **unconditional jump.**

conditionally convergent series (*Maths.*). A series which is convergent but not absolutely convergent, e.g., $1-\frac{1}{2}+\frac{1}{3}-\frac{1}{4}\ldots$ which converges to $\log_e 2$.

conditionally stable (*Telecomm.*). Amplification system which satisfies Nyquist stability criterion until its gain is reduced.

conditionally unstable (*Meteor.*). The atmosphere is said to be conditionally unstable when the *ELR* (q.v.) lies between the *SALR* (q.v.) and the *DALR* (q.v.), since, under these conditions, there is stability if the air is unsaturated and instability if it is saturated.

conditioned reflex (*An. Behav.*). Reflex action by an animal to a previously neutral stimulus as the result of *classical conditioning*.

conditioned response (*An. Behav.*). A response given by an animal after it has undergone any type of conditioning. Abbrev. **CR.**

conditioned stimulus (*An. Behav.*). Also called **conditional stimulus.** See **classical conditioning.**

conditioning (*An. Behav.*). A term which is now somewhat imprecise, having been originally used to describe the process occurring inside an animal by which a *conditioned reflex* (q.v.) was established, but now used with reference to any type of conditioning (e.g. *operant*, *avoidance*, etc.), and also with reference to the experimental procedures resulting in the acquisition of the various types of conditioned response. (*Min. Proc.*) In froth flotation, the treatment of mineral pulp with small additions of chemicals designed to develop specific aerophilic or aerophobic qualities on surfaces of different mineral species as a prelude to their separation. (*Spinning*) The process of adding to yarn, after spinning, the percentage of moisture necessary to bring it up to average conditions; 100 parts of dry cotton will absorb $8\frac{1}{2}$ parts of moisture.

conditions of severity (*Elec. Eng.*). A term used in connexion with the testing of circuit-breakers to denote the conditions (e.g., power factor, rate of rise of restriking voltage, etc.) obtaining in the circuit when the test is carried out.

condor (*Nav.*). Radionavigation system related to *benito* (q.v.).

conductance (*Elec.*). Property of material by which it allows current to flow through when a p.d. is applied; reciprocal of resistance; unit *siemens*, $1/\Omega$, symbol G. (*Vac. Tech.*) Of a system for a given gas (or vapour), the ratio of the throughput of the gas to the partial pressure difference across the system in the steady state. Unit is litre per second. Symbol U.

conductance ratio (*Elec. Eng.*). That between equivalent conductance of a given solution and its value at infinite dilution.

conductimetric analysis (*Chem.*). Volumetric analysis in which the end-point of a titration is determined by measurements of the conductance of the solution.

conducting strand (*Bot.*). See **vascular bundle.**

conduction angle (*Elec. Eng.*). The fraction of a cycle (expressed as an angle) for which a rectifier passes current. A *thyristor* may be used to control the power delivered to e.g. a motor by varying its conduction angle.

conduction band (*Electronics*). In band theory of solids, band which is only partially filled, so that electrons can move freely in it, hence permitting conduction of current.

conduction by defect (*Electronics*). In a doped semiconductor, conduction by *holes* in the valency electron band.

conduction current (*Electronics*). That resulting from flow of electrons or ions in a conducting medium.

conduction electrons (*Electronics*). The electrons situated in the conduction band of a solid, which are free to move under the influence of an electric field.

conduction hole (*Electronics*). In a crystal lattice of semiconductor, conduction obtained by electrons filling holes in sequence, equivalent to a positive current.

conduction of heat (*Heat*). The transfer of heat from one portion of a medium to another, without visible motion of the medium, the heat energy being passed from molecule to molecule. See **thermal conductivity.**

conductivity (*Acous.*). Ratio of air volume current to acoustic potential difference for an acoustic orifice. (*Elec.*) Conductance at a specified temperature between opposite faces of a cube of material, 1 m edge. Reciprocal of resistivity. (*Zool.*) The ability to transmit stimuli—a characteristic property of nervous tissue.

conductivity bridge (*Elec. Eng.*). A form of Wheatstone bridge used for the comparison of low resistances.

conductivity cell (*Elec. Eng.*). Any cell with electrodes for measuring conductivity of liquid or molten metals or salts.

conductivity modulation (*Electronics*). That effected in a semiconductor by varying a charge carrier density.

conductivity test (*Elec. Eng.*). See **fall-of-potential test.**

conductor (*Elec. Eng.*). (1) A material which offers a low resistance to the passage of an electric current. (2) That part of an electric transmission, distribution or wiring system which actually carries the current. (*Heat*) A material used for the transference of heat energy by conduction. All materials conduct heat to a more or less extent, but it can be channelled (like electricity) using materials of different conductivity. (*Plumb.*) A pipe for the conveyance of rain water. Also called **leader.**

conductor load (*Elec. Eng.*). The total mechanical load to which an overhead electric conductor may be subjected, because of its own weight and that of any adhering matter such as snow or ice.

conductor rail (*Elec. Eng.*). In some electric traction systems, a bare rail laid alongside the running rails to conduct the current to or from the train. Also called **contact rail.**

conductor-rail anchor (*Elec. Eng.*). A device for anchoring a conductor rail to prevent the longitudinal movement which might be caused by the sliding of the collector shoes along it.

conductor-rail bond (*Elec. Eng.*). A rail bond used for connecting adjacent lengths of conductor rail at joints to preserve electrical continuity.

conductor-rail insulator (*Elec. Eng.*). An insulator used for supporting a conductor rail and for insulating it from the earth.

conductor-rail ramp (*Elec. Eng.*). A sloping contact surface at the beginning and end of a section of conductor rail; it serves for leading the collector shoe of the train smoothly on to and off the rail.

conductor-rail system (*Elec. Eng.*). A system of electric traction in which current is collected from a *conductor rail* by *collector shoes*.

conduit (*Elec. Eng.*). A trough or pipe containing electric wires or cables, in order to protect them against damage from external causes. (*Hyd. Eng.*) A pipe or channel, usually large, for the conveyance of water.

conduit box (*Elec. Eng.*). A box adapted for connexion to the metal conduit used in electric wiring schemes. The box forms a base to which fittings (e.g., switches or ceiling roses) may be attached, or it may take the place of bends, elbows, or tees, used to facilitate the installation of the wiring.

conduit fittings (*Elec. Eng.*). A term applied to all the auxiliary items, such as boxes, elbows, etc., needed for the conduit system of wiring.

conduit system (*Elec. Eng.*). (1) A system of wiring, used for domestic and other premises, in which the conductors are contained in a steel conduit. (2) A system of current collection used on some electric tramway systems; the conductor rail is laid beneath the roadway, and connexion is made between it and the vehicle by means of a collector shoe passing through a slot in the road surface.

conduplicate (*Bot.*). (1) Folded longitudinally about the midrib, so that the two halves of the upper surface are brought together. (2) Said of a cotyledon which is folded longitudinally about the radicle.

condyle (*Zool.*). A smooth rounded protuberance, at the end of a bone, which fits into a socket on an adjacent bone, as the *condyle* of the lower jaw, the occipital *condyles*. *adj.* **condylar, condyloid.**

condylomata (*Med.*). Inflammatory wartlike papules on the skin round the anus and external genitalia, especially in syphilis. *sing.* **condyloma.**

Condy's crystals (*Chem.*). Common name for *potassium permanganate* (q.v.).

Condy's fluid (*Chem.*). A disinfectant, usually containing calcium permanganate, sodium manganate (VII), and possibly sodium nitrate (V).

cone (*Bot.*). See **strobilus.** (*Maths.*) (1) A (conical) surface generated by a line, one point (the *vertex*) of which is fixed and one point of which describes a fixed plane curve. Any line lying in the surface is called a *generator*. (2) A solid bounded by a conical surface and a plane (the *base*) which cuts the surface. If the fixed curve is a circle the cone is called a *circular cone*, and if also the vertex is perpendicularly over the centre of the circle the cone is called a *right circular cone*. (*Met.*) A device used on top of blast furnaces to enable charge to be put in without permitting gas to escape. Also called **bell.** (*Zool.*) One of the conical percipient structures borne on the outer end of certain visual cells of the Vertebrate retina; principally used for daylight and colour vision. Cf. *rod*.

cone bearing (*Eng.*). A shaft bearing consisting of a conical journal running in a correspondingly tapered bush, so acting as a combined journal and thrust bearing; used for some lathe spindles.

cone capacitor (*Elec. Eng.*). A capacitor consisting of two conducting cones, one inside the other, separated by insulating material. By moving one cone relative to the other a variable capacitance can be obtained.

cone classifier (*Min. Proc.*). Large inverted cone into which ore pulp is fed centrally from above. Coarser material settles to bottom discharge and finer overflows peripherally.

cone clutch (*Eng.*). A friction clutch in which the driving and driven members consist of conical frusta. The externally coned (driven) member may be moved axially in or out of the internally coned member for engaging and disengaging the drive. See **clutch, friction clutch.**

cone diaphragm (*Acous.*). One of paper, plastics or metal foil driven by a circular coil carrying speech currents near its apex; widely used for radiating sound in loudspeaking receivers. Also **cone loudspeaker.**

cone drawing (*Textiles*). A method of drafting wool on the worsted preparation system so that the speed of the bobbin on which it is being wound decreases as the diameter of the bobbin and material enlarge.

cone drive or **cone gear** (*Eng.*). A belt drive between two similar coned or tapered pulleys. A variable speed ratio is obtained by lateral movement (by means of a striker) of the belt along the pulleys.

cone drums (*Spinning*). Mechanism on fly frames controlled by the building motion. It alters the speed of either the flyers or the bobbins as the package builds up.

cone-in-cone structure (*Geol.*). One consisting of a series of cones stacked inside one another, occurring in sedimentary rocks; usually of fibrous calcite, sometimes of other minerals.

cone loudspeaker (*Acous.*). See **cone diaphragm.**

cone of origin (*Zool.*). The conical hyaline area of a nerve-cell body, at the point of origin of an axon.

cone of silence (*Telecomm.*). Cone-shaped space in which signals from a transmitter are virtually undetectable.

cone pulley (*Eng.*). A belt pulley stepped to give two or more diameters; used in conjunction with a similar pulley to obtain different speed ratios. See also **cone drive.**

cone sheets (*Geol.*). Minor intrusions which occur as inwardly inclined sheets of igneous rock, and have the form of segments of concentric cones.

conference system (*Teleph.*). A telephone system used for conference between groups of persons at a distance; high-grade microphones and radiating receivers are used.

conferted (*Bot.*). Near, crowded together.

confervoid (*Bot.*). Consisting of delicate filaments.

configuration (*Chem.*). The spatial arrangement of atoms in a molecule, especially in one containing several asymmetric carbon atoms. (*Psychol.*) See **gestalt.**

configurational stimulus (*An. Behav.*). A stimulus in which the configuration or proportions of the stimulus or its components are important in determining its effectiveness. Sometimes called **a gestalt.**

configuration control (*Nuc. Eng.*). Control of reactivity by alterations to the configuration of the fuel, reflector and moderator assembly.

confinement (*Nuc.*). See **containment.**

conflict (*An. Behav.*). An experimental procedure in which the stimuli for two incompatible responses are presented simultaneously; hence, the state of an animal when two drives associated with incompatible behaviours are equally strong.

confluent (*Bot.*). Said of two or more structures which, as they enlarge, grow together and unite.

confluent filter (*Telecomm.*). An electric wave filter so designed that it has two pass or elimination bands, of which two nominal cut-off frequencies are co-incident.

confocal conics (*Maths.*). Conics with the same foci.

confocal quadrics (*Maths.*). Quadrics with the same foci.

conformable strata (*Geol.*). An unbroken succession of strata. See also **nonsequence, unconformity.**

conformal-conjugate transformation (*Maths.*). A conformal transformation in which a conjugate system on one surface corresponds to a conjugate system on the other surface.

conformal transformation (*Maths.*). If the z-plane is mapped on the w-plane (or vice versa) so that both the magnitude and the sense of the rotation of the angles is the same in either plane, then the transformation is said to be *conformal*. Cf. *isogonal transformation.*

conformational analysis (*Chem.*). A study of the relative spatial positions of atoms within a molecule.

conformation theory (*Chem.*). The theory of the structure of molecules, particularly of the arrangement of atoms in very complex molecules.

confusion (*Photog.*). See **circle of confusion, circle of least confusion.**

congé (*Arch.*). A small circular moulding, either concave or convex, at the junction of a column with its base.

congeneric (*Zool.*). Belonging to the same genus.

congenital (*Zool.*). Dating from birth or from before birth.

congested (*Bot.*). Packed into a tight mass.

congestin (*Zool.*). A toxic substance produced by certain Sea Anemones.

congestion (*Med.*). Pathological accumulation of blood in a part of the body.

congestion call meter (*Teleph.*). A meter which counts the number of calls made over the last choice of outlet in a grading scheme.

congestion traffic-unit meter (*Teleph.*). A meter which registers the traffic flow over the last choice of outlets in a grading scheme.

conglobate (*Bot.*). Heaped together. (*Zool.*) Ball-shaped.

conglobate gland (*Zool.*). An accessory reproductive gland in male cockroaches (*Dictyoptera*). It opens separately from the other accessory glands between the lobes of the phallus.

conglomerate (*Bot.*). Clustered. (*Geol.*) A cemented clastic rock containing rounded fragments, corresponding in their grade sizes to gravel or pebbles.

conglutinate (*Bot.*). United into a mass by a sticky substance.

Congo gum or **Congo copal** (*Chem.*). A natural gum, of yellow or brown colour; rel. d. 1·059–1·080, saponification value 66–175, acid value 35–95. Soluble in benzene, turpentine, trichloromethane, ethoxyethane. Used for varnishes and linoleum.

Congo red (*Chem.*). An important benzidene direct dyestuff (scarlet) produced by coupling diphenyl-bis-diazonium chloride with naphthionic acid; first of a long series of derivatives. Used as a chemical indicator for acid solutions in the pH range 3–5.

congregate (*Bot.*). Collected into a dense group.

congruence (*Maths.*). A statement that two numbers a and b are congruent with respect to a divisor n. Written $a=b$ (mod. n), and meaning that the remainders when a and b are divided by n are equal.

congruent (*Maths.*). Alike in all relevant respects; capable of coincident superposition in Euclidean geometry.

congruent transformation (*Maths.*). If a given matrix A is transformed into B by $B=P'AP$, where P is a nonsingular matrix and P' is its transpose, then B is said to be congruent to A.

conic (*Maths.*). The curve in which a plane cuts a circular (but not necessarily right) cone. The curve is a parabola if the cutting plane is parallel to a generator, otherwise it is an ellipse or a hyperbola according to whether a plane parallel to the cutting plane but through the vertex of the cone wholly contains no generators or two generators respectively. Alternatively, the locus of a point which moves so that its distance from a fixed point (*focus*) is a constant e (*eccentricity*) times its distance from a fixed line (*directrix*). The conic is an ellipse, parabola or hyperbola according to whether e is less than, equal to, or greater than 1. Any second degree equation in two variables represents a conic. By a suitable choice of axes such an equation can be reduced to one of the following forms:

(1) $\dfrac{x^2}{a^2}+\dfrac{y^2}{b^2}=1$, an ellipse

(2) $\dfrac{x^2}{a^2}-\dfrac{y^2}{b^2}=1$, a hyperbola

(3) $y^2=4ax$, a parabola

(4) $ax^2+2hxy+by^2=0$
a line pair if $h^2>ab$
two coincident lines if $h^2=ab$
a single point if $h^2<ab$

Cf. *axes*, *diameter*, *focus* and *vertex*.

conical camber (*Aero.*). An expression applied to the adjustment of the *camber* (q.v.) of a wing across the span to meet the variation of the upflow of the air from the fuselage side to the tip. Used on high-speed aeroplanes, the 'twist' is applied mainly to the leading edge. The name originates from the conic lofting process used in deriving the aerofoil sections. Similar to *wash-out* (q.v.) on a subsonic wing.

conical drum (*Mining*). Winding drum, to which hoisting rope of cage or skip is attached. This shape aids in smooth acceleration from rest to full speed where the rope reaches the flat central part of the compound drum (the full diameter between the cones).

conical horn (*Acous.*). A horn in the form of a cone, the apex being truncated to form the throat.

conical pivot (*Horol.*). (1) A pivot formed as a cone, which runs in a screw with a tapered hole, the angle of taper being greater than that of the conical pivot; used for the balance staff of alarm-clock movements, and for certain watches with pin-pallet escapement. (2) A form of pivot, used in English watches, which has a conical shoulder, the pivot itself being parallel.

conical projections (*Geog.*). In these, the projection is on to an imaginary cone touching the sphere at a given standard parallel. There are equal-area and two standard (cf. *Gall's projection*) versions, as in the cylindrical type. Best for middle latitudes. See **Bonne's projection, Mollweide's projection, polyconic projection, sinusoidal projection.**

conical refraction (*Light*). An effect seen in biaxial crystals where a ray incident in a particular direction is spread on refraction into a hollow cone of rays.

conical scanning (*Radar*). Similar to *lobe switching* (q.v.), but circular. The direction of maximum response generates a conical surface; used in missile guidance.

conical surface (*Maths.*). See **cone.**

conidial (*Bot.*). (1) Referring to, or pertaining to, a conidium. (2) Producing conidia.

conidiole (*Bot.*). A small conidium, usually budded out from another conidium.

conidiophore (*Bot.*). A simple or branched hypha bearing one or more conidia.

conidiosporangium (*Bot.*). A sporangium capable of direct germination, as well as producing zoospores.

conidium (*Bot.*). An asexually formed spore, produced by many species of fungi. It develops at the end of a hypha, usually by a process of abstriction, and it is never enclosed in a sporangium. *pl.* **conidia.**

Coniferae (*Bot.*). The chief class of the *Gymnospermae*, with about 400 species, all woody, and mostly large evergreen trees, forming forests. The main stem usually persists, so that the tree has an elongated conical form. The reproductive organs are borne in cones, which are unisexual. Members of the group, such as pines and spruces, are important timber trees and are now widely used as sources of wood pulp.

coniferin (*Chem.*). $C_{16}H_{22}O_8$, $2H_2O$, a glucoside contained in the cambium sap of the *Coniferae*, yielding glucose and coniferyl alcohol on hydrolysis; serves for the preparation of vanillin.

coniferous (*Bot.*). Cone-bearing; relating to a cone-bearing plant.

conifuge (*Powder Tech.*). A conical type of centrifuge for the collection of aerosol particles.

coniine (*Chem.*). $C_5H_{10}N(C_3H_7)$, dextrorotatory propyl-piperidine; b.p. 167°C; a pyridine alkaloid which is found in *Conium maculatum*. it can be synthesized by reducing allyl-pyridine In alcoholic solution with sodium.

coning and quartering (*Min. Proc., Powder Tech.*) Production of representative sample from a large pile of material such as ore, in which it is first formed into a cone by deposition centrally, and then reduced by removal, one shovelful at a time, into four separate piles drawn alternately from four peripheral opposed points. Two are then discarded and the process (perhaps with intermediate size reduction) is repeated until a manageable hand sample is obtained.

coning angle (*Aero.*). The upward angle adopted by the blades of a helicopter rotor under load in flight.

Conjugales (*Bot.*). An order of the *Chlorophyceae*, characterized by the production of a conjugation tube and nonflagellate gametes during sexual reproduction.

conjugate (*Bot.*). Occurring in pairs.

conjugate acid and base (*Chem.*). These are related, according to the *Brönsted-Lowry theory* (q.v.), by the reversible exchange of a proton; thus Acid$\rightleftarrows$Base$+H^+$.

conjugate algebraic numbers (*Maths.*). Any numbers which are the roots of the same irreducible algebraic equation with rational coefficients.

conjugate angles (*Maths.*). Two angles whose sum is equal to 360°. Each is said to be the *explement* of the other.

conjugate arcs (*Maths.*). Two arcs whose sum is a complete circle.

conjugate axis of hyperbola (*Maths.*). See **axes.**

conjugate Bertrand curves (*Maths.*). Any two curves having the same principal normals. Also called **Bertrand curves** and **associate Bertrand curves.**

conjugate branches (*Elec.*). Any two branches of

an electrical network such that an e.m.f. in one branch produces no current in the conjugate branch.

conjugate complex number (*Maths.*). The conjugate $\bar{z}$ of a complex number $z=x+iy$ is $x-iy$.

conjugate deviation (*Med.*). The sustained deviation of the eyes in one direction as a result of a lesion in the brain.

conjugate diameters (*Maths.*). Two diameters of a conic which are also conjugate lines with respect to the conic are conjugate diameters. Each bisects chords parallel to the other.

conjugate directions (*Maths.*). At a point P on a surface, if Q is a point near P and Q tends to P, then the conjugate directions are the directions in the limiting case of: (*a*) the straight line between P and Q; and (*b*) the line of intersection of the tangent planes at P and Q. If these directions coincide, they are *self-conjugate* or *asymptotic directions*.

conjugate division (*Bot.*). The simultaneous division of a pair of associated nuclei.

conjugate double bonds (*Chem.*). Di-alkene compounds with an arrangement of alternate single and double bonds between the carbon atoms, namely, R·CH=CH·CH=CH·R. Additive reactions take place, inasmuch as atoms or radicals become attached to the two outside carbon atoms of the chain, thus creating a new alkene linkage in the centre.

conjugate dyadics (*Maths.*). Two dyadics such that each can be obtained from the other by reversing the order of the factors in each dyad. If a dyadic is equal to its conjugate it is said to be *symmetric*; if it is equal to the negative of its conjugate it is *antisymmetric*.

conjugate elements of a determinant (*Maths.*). Those which are interchanged when the rows and columns are interchanged, e.g., a_{ij} and a_{ji}.

conjugate elements of a group (*Maths.*). Two elements A and B of a group are conjugate with respect to the group if there exists an element P in the group such that $B=P^{-1}AP$. B is then referred to as the transform of A by P.

conjugate foci (*Light*). Two points such that rays of light diverging from either of them are brought to a focus at the other. For a simple convergent lens an object and its real image are at *conjugate foci*.

conjugate functions (*Maths.*). Two real functions which satisfy the Cauchy-Riemann equations.

conjugate harmonic functions (*Maths.*). Two harmonic functions which satisfy the Cauchy-Riemann equations.

conjugate impedance (*Elec.*). Two impedances are conjugate when their resistance components are equal and their reactances are equal in magnitude but opposite in sign.

conjugate lines (of a conic) (*Maths.*). Two lines are said to be conjugate with respect to a conic if the pole of each lies on the other.

conjugate matrix (*Maths.*). See **transpose of a matrix.**

conjugate planes (*Light*). Planes perpendicular to the axis of an optical system such that any object point in one near the axis is imaged in the other.

conjugate point (*Maths.*). See **double point.**

conjugate points (of a conic) (*Maths.*). Two points are said to be conjugate with respect to a conic if each lies on the polar of the other.

conjugate solutions (*Chem.*). If two liquids A and B are partially miscible, they will produce at equilibrium two conjugate solutions, the one of A in B, the other of B in A.

conjugate subgroup (*Maths.*). Each member of a set of subgroups, obtained by transforming a given subgroup of a group G by every element of the group G, is said to be conjugate to each of the other members. See **conjugate elements of a group.**

conjugate system of curves (*Maths.*). A system consisting of two one-parameter families of curves on a surface, such that where a member of one family intersects a member of the other family, the directions of the tangents to the curves are conjugate directions.

conjugate triangles (*Maths.*). Two triangles are conjugate with respect to a conic if the vertices of one are the poles of the sides of the other. Also called **reciprocal polar triangles.** Cf. *self-polar triangle*.

conjugation (*Biol.*). The temporary or permanent union of two cells or individuals preparatory to the development of new individuals; more particularly, the union of isogametes; in *Protozoa*, a type of syngamy in which the participants remain separate and exchange nuclear material.

conjugation tube (*Bot.*). A tubular outgrowth by means of which the contents of a male gametangium are conveyed into a female gametangium.

conjunction (*Astron.*). Term signifying that two heavenly bodies have the same apparent geocentric longitude. Applied to Venus and Mercury it is subdivided into *inferior conjunction* and *superior conjunction*, according as the planet is between the earth and sun, or the sun between the earth and planet, respectively.

conjunctiva (*Zool.*). In Vertebrates, the modified epidermis of the front of the eye, covering the cornea externally and the inner side of the eyelid.

conjunctive parenchyma (*Bot.*). Parenchymatous tissue occupying the spaces between vascular strands or other specialized tissues.

conjunctivitis (*Med.*). Inflammation of the conjunctiva.

conky (*For.*). Applied to a log or tree bearing fruit bodies of wood-rotting fungi.

connation (*Bot.*). The union of parts of a plant as growth proceeds; applied especially to the union of like parts. *adj.* **connate.**

connected domain (*Maths.*). A **simply-connected domain** is such that any closed curve in it and the region enclosed by the curve lie wholly within the domain. A **multiply-connected domain** has one or more 'holes'. A doubly-connected domain has one 'hole' and so on, an $n$tuply-connected domain having $n-1$ 'holes'.

connected load (*Elec. Eng.*). The sum of the rated inputs of all consumers' apparatus connected to an electric power supply system.

connecting-rod (*Eng.*). In a reciprocating engine or pump, the rod connecting the piston or cross-head to the crank.

connecting-rod bolts (*Eng.*). Bolts securing the outer half of a split big-end bearing of a connecting-rod to the rod itself. Sometimes known as **big-end bolts.**

connecting thread (*Bot.*). A delicate strand of protoplasm passing through a fine perforation in the cell wall, and uniting the protoplasts of contiguous cells.

connective (*Bot.*). (1) Sterile tissue in an anther, lying between the lobes, and usually a prolongation of the filament. (2) See **disjunctor.** (*Zool.*) A bundle of nerve fibres uniting two nerve centres.

connective tissue (*Zool.*). A group of animal tissues fulfilling mechanical functions, developed from the mesoderm and possessing a large quantity of nonliving intercellular matrix, which usually contains fibres; as bone, cartilage, and areolar tissue.

connector bar (*Elec. Eng.*). See **terminal bar.**

connector neurone (*Zool.*). See **association neurone.**

connexion (or connecting-) box (*Elec. Eng.*). A box containing terminals to which are brought a number of conductors of a wiring or distribution system, to facilitate the making of connexions between them.

connivent (*Bot.*). Converging and meeting at the tips.

connotative meaning (*Psychol.*). The meaning of a word or concept apart from what it actually signifies; its affective content, value to an individual, etc. See **denotative meaning.**

consanguinity (*Geol.*). Term applied to rocks having a similarity or community of origin, which is revealed by common peculiarities of mineral and chemical composition and often also of texture.

consciousness (*Psychol.*). A comprehensive state of awareness of the mind of stimuli from the outside world, and of feelings, emotions, and thoughts from within the individual.

consensual (*Zool.*). Said of response to stimuli in which voluntary action and involuntary action are correlated.

consequent (*Maths.*). (1) In a ratio $a : b$, the term b; a is the *antecedent*. (2) In material implication, (if p, then q), q is the consequent, p is the antecedent.

consequent drainage (*Geol.*). A river system directly related to the geological structure of the area in which it occurs.

consequent pole (*Elec. Eng.*). An effective pole not at the end of ferromagnetic material, e.g., at the ends of a diameter of a ring magnet; must occur in pairs. Used in *Broca galvanometer*.

consequent-pole generator (or motor) (*Elec. Eng.*). A generator or motor in which alternate poles do not carry windings, and which are, therefore, consequent poles.

conservancy system (*San. Eng.*). The system of disposing of waste matter from buildings by earth closets and privies, without the use of water to carry it away.

conservation laws (*Chem.*). Usually refers to the classical laws of the separate conservation of mass, of energy and of atomic species, which are sufficiently accurate for most chemical reactions.

conservation of energy law (*Mech.*). That which states that energy cannot be lost or gained in a closed system, but interchanged in form at specific rates of conversion. Relativistically, addition of energy to a mass, when approaching the speed of light, is manifest in an increase of mass, according to Einstein's law, which is completely reversible.

conservation of matter (*Chem.*). See **law of conservation of matter.**

conservation of momentum (*Mech.*). The sum of the momenta in a closed system (that is, one in which no influences act upon it from outside) is constant and is not affected by processes occurring within the system.

conservation of movement of the centre of gravity (*Mech.*). The state of rest or of motion of the centre of gravity of a system can never be altered by the action of internal forces within the system.

conservative field of force (*Maths.*). A field of force such that the work done in moving a particle from A to B is independent of the path followed.

conservative system (*Mech.*). A system such that, in any cycle of operations at the end of which the configuration of the system is the same as it was at the beginning, the work done is zero.

conservator tank (*Elec. Eng.*). A small tank, carried on the top cover of an oil-filled power transformer, to accommodate the change in oil volume with temperature as the load varies, and reduce the oil surface exposed to the air.

conservatory (*Arch.*). A glazed building in which plants may be grown under controlled atmospheric conditions.

Considère pile (*Civ. Eng.*). A type of reinforced-concrete pile having the main bars grouped symmetrically about the centre of the pile and bound together by a close-set spiral of steel wire or small-section steel rod.

consistency (*Paper*). The amount of bone dry pulp in the wet pulp stock, expressed as a percentage.

consistent equations (*Maths.*). Equations having a common solution, e.g., the pair of equations $x+y=3$ and $x-y=1$ are consistent, but the pair $x+y=3$ and $x+y=2$ are not. Sometimes called **compatible equations.**

consociation (*Bot.*). A subdivision of an association dominated by one of the codominants of the association.

consocies (*Bot.*). A developmental stage in the history of a consociation, before conditions have become stabilized.

consol (*Nav.*). Long-range navigational system which can determine a line of position within close limits of accuracy. U.S. **consolan,** German **sonne.**

console. Type of *control panel* (q.v.): central controlling desk in power station, process plant, computer, or reactor, where an operator can supervise operations and give instructions. (*Acous.*) Location of playing controls in an organ, electronic or pipe, including swells, pedals, several manuals, speaking stops and couplers. (*Build., etc.*) (1) A bracket whose shelf is supported by volutes. (2) Any concentration of control mechanism on to a desk-type panel.

consolidation (*Geol.*). Of strata, the drying, compacting, and induration of rocks, as a result of pressures operating after deposition.

consolidation pile (*Civ. Eng.*). A pile driven (with others) into the ground to consolidate the soil and enable it to support heavier loads than would otherwise be possible.

consonance (*Acous.*). The condition where two pure tones blend pleasingly.

consonant articulation (*Acous.*). See **articulation.**

consortism (*Bot.*). See **symbiosis.**

consortium (*Bot.*). (1) The compound thallus of a lichen. (2) The mutual relationships of the fungus and the alga in the lichen thallus.

constancy (*Ecol.*). The percentage of sample plots in a plant community containing a particular species.

constant (*Maths.*). In an algebraic expression or equation, a quantity (or parameter) which remains the same while the variables change. Constants assume different values for different initial and boundary conditions, but are constant for each particular set of conditions. An *absolute* constant, e.g., π, is always the same.

constant-amplitude recording (*Acous.*). Gramophone recording technique whereby the response of recording, i.e., amplitude of the track divided by the root of the applied power, is independent of frequency. Cf. *constant-velocity recording*.

constantan (*Elec. Eng.*). An alloy of about 40% nickel and 60% copper, having a high volume resistivity and almost negligible temperature coefficient; used as the resistance wire in resistance boxes, etc. Also called Eureka.

constant boiling mixture (*Chem.*). An *azeotropic mixture*.

constant-current characteristic (*Elec. Eng., Electronics*). A voltage-current characteristic for which the current is almost independent of the applied voltage, as at the collector of a junction transistor for fixed base-emitter voltage.

constant-current modulation (*Elec. Eng.*). See **choke modulation.**

constant-current motor (*Elec. Eng.*). An electric motor designed to operate at a constant current from a constant-current generator.

constant-current system (*Elec. Eng.*). A system of transmitting electric power in which all the equipment is connected in series and a constant current is passed round the circuit. Variations in power result in a variation of the voltage of the system, constant-current generators being used for the supply. See **series system.**

constant-current transformer (*Elec. Eng.*). A transformer designed to maintain a constant secondary current within a specified working range, for all values of secondary impedance and all values of primary voltage.

constant-frequency oscillator (*Elec. Eng.*). One in which special precautions are taken to ensure that the frequency remains constant under varying conditions of load, supply voltage, temperature, etc.

constant-*k* filter (*Telecomm.*). Simple type formed from a *constant-k network* only.

constant-*k* network (*Telecomm.*). Iterative network for which product of series and shunt impedance is frequency independent.

constant-level chart (*Meteor.*). An upper air chart showing isobars for a particular level. Cf. *constant-pressure chart*.

constant-level tube (*Surv.*). A special form of level tube in which the volume ratio of bubble to liquid is fixed at such a value that decrease in length of the bubble due to expansion of the liquid is exactly counterbalanced by increase in length of the bubble due to diminished surface tension, so that the length of the bubble—and thus the sensitivity of the level tube—remains unaltered by rise in temperature.

constant-luminance transmission (*TV*). Type o colour TV transmission in which transmission primaries consist of one luminance and two chrominance primaries.

constant-mesh gear-box (*Autos.*). A gear-box in which the pairs of wheels providing the various speed ratios are always in mesh, the ratio being determined by the particular wheel which is coupled to the mainshaft by sliding dogs working on splines.

constant of integration (*Maths.*). The arbitrary constant resulting from indefinite integration. It arises because the derivative of a constant is zero, and there is therefore no means of evaluating it without additional information.

constant of inversion (*Maths.*). See **inversion.**

constant-power generator (*Elec. Eng.*). An electric generator which, by variation of the generated voltage, gives a constant power output at varying currents.

constant-pressure chart (*Meteor.*). An upper air chart showing contours of the height above sea level at which a particular pressure occurs. Cf. *constant-level chart*. See **thickness chart.**

constant-pressure cycle (*I.C. Engs.*). See **diesel cycle.**

constant proportions (*Chem.*). See **law of constant (or definite) proportions.**

constant-resistance network (*Telecomm.*). One in which iterative impedance in at least one direction is resistive and independent of the applied signal voltage.

constant-speed airscrew (*Aero.*). See **airscrew.**

constant time-lag (*Elec. Eng.*). See **definite time-lag.**

constant-velocity recording (*Acous.*). In gramophone disk recording, the technique whereby the lateral r.m.s. velocity of the sinuous track is made proportional to the root of the electrical power applied to the recorder, irrespective of the frequency. This criterion is necessary for minimum wear and surface noise on the finished record, but has to be modified to constant-amplitude recording below about 250 Hz as a compromise with playing time.

constant-voltage motor (*Elec. Eng.*). The usual type of electric motor, designed to operate from a constant-voltage system.

constant-voltage system (*Elec. Eng.*). The usual system of transmission of electric power, in which the voltage between the conductors is maintained approximately constant, and all apparatus is connected to the system in parallel across the conductors.

constant-voltage transformer (*Elec. Eng.*). One with or without extra components which gives a constant voltage with varying load current, or with varying input voltage over a specified range.

constant-volume amplifier (*Teleph.*). See **vogad.**

constellation (*Astron.*). A group of stars, not necessarily connected physically, to which have been given a pictorial configuration and a name (generally of Greek mythological origin) which persist in common use although of no scientific significance.

constipate (*Bot.*). Crowded together.

constipation (*Med.*). A condition in which the faeces are abnormally dry and hard: retention of faeces in the bowel: infrequent evacuation.

constituent (*Maths.*). Of a matrix or determinant, one of the numbers in the array which forms the matrix or determinant. Also called an **element.** (*Met.*) Component of alloy or other compound, or of mixture. It may be present as an element or in chemical, physical or intermediate combination.

constituents (*Chem.*). All the substances present in a system.

constitution (*Chem., Met., Phys.*). Structural distribution, linkage and unit cell symmetry of atoms and/or ions composing an alloy, metal or definable regularly coordinated substance. Includes percentage of each constituent and its regularity of occurrence through its material.

constitutional ash (*Min. Proc.*). Ash resulting from combustion of coal, and derived from siliceous matter in the coal-forming plants. 'Fixed ash' as distinct from entrained impurity or 'free ash'.

constitutional changes (*Met.*). Changes in solid alloys which involve the transformation of one constituent to another (as when pearlite is formed from austenite), or a change in the relative proportions of two constituents.

constitutional diagram (*Met.*). Phase diagram which shows effect of temperature and composition on alloy in which two or more metals coexist in solid solution and pass from one crystal lattice to another (e.g., face-central, cubic, body-centred) and also change in general properties.

constitutional formula (*Chem.*). A formula which shows the arrangement of the atoms in a molecule.

constitutional water (*Chem.*). See **water of hydration.**
constraint (*Eng.*). The property which distinguishes a mechanism from a kinematic chain. In a mechanism, the motion of one part is followed by a predetermined motion of the remainder of the mechanism. (*Mech.*) See **principle of least constraint.**
constricted (*Bot., Zool.*). (1) Narrowed suddenly. (2) Narrowed at one or more points along the length, as if compressed by a thread tightened around the member.
constriction (*Bot.*). Any sudden narrowing in a cylindrical member. (*Cyt.*) A narrowed localized region in a chromosome, usually difficult to stain.
constriction resistance (*Elec.*). That across actual area of contact through which current passes from one metal to another, equal to

$$(\rho_1+\rho_2)/4a,$$

where a = radius of circular contact area
ρ_1 and ρ_2 = resistivities of the metals.
constrictor (*Zool.*). A muscle which by its contraction constricts or compresses a structure or organ.
constringence (*Optics*). Inverse of the dispersive power of a medium. Ratio of the mean refractive index diminished by unity to the difference of the refractive indices for red and violet light.
consumers (*Ecol.*). In an ecosystem, the heterotrophic organisms, chiefly animals, which ingest either other organisms or particulate organic matter. Cf. *decomposers, producers.*
consummatory act (*An. Behav.*). A relatively simple fixed action pattern, terminating an instinctive behaviour sequence, e.g., eating.
contabescence (*Bot.*). Abortion in stamens and pollen.
contact (*Elec. Eng.*). (1) That part of either of two conductors which is made to touch the other when it is desired to pass current from one to the other, as in a switch. (2) The juxtaposition of parts, usually of platinum or silver, which, when brought into contact, provide for the passage of a current and, when withdrawn, cause its cessation. For the former, suitable minimum pressure and areas of contact are necessary; for the latter, a sufficiently low recovery voltage, and rapidity of withdrawal of the contacts from each other.
contact action (*Chem.*). See **catalysis.**
contact angle (*Chem.*). The angle between the liquid and the solid at the liquid-solid-gas interface. It is acute for wetting (e.g., water on glass) and obtuse for nonwetting (e.g., water on paraffin wax). (*Min. Proc.*) (θ) That between a bubble of air and the chemically clean, polished and horizontal surface of a specimen of mineral to which it clings, measured between that surface and the side of the bubble. This forms an index to the floatability of the species under prescribed conditions, in which chemicals are added to the water and change in angle observed.
contact bar (*Elec. Eng.*). The conductor of a resistance welding machine, which carries current to the part to be welded and also exerts the necessary mechanical pressure.
contact bed (*San. Eng.*). A tank, filled with material such as broken clinker, used in the final or oxidizing stage in sewage treatment, which consists in charging the filtering medium with the liquid sewage, allowing it to stand for a time, draining it off, and finally keeping the tank empty for a time. See **percolating filter.**
contact bounce (or chatter) (*Telecomm.*). Intermittent opening and closing of relay contacts.
contact breaker (*Elec. Eng.*). Circuit breaker.
contact e.m.f. (*Elec. Eng.*). That which arises at the contact of dissimilar metals at the same temperature, or the same metal at different temperatures.
contact fingers (*Elec. Eng.*). Contacts pressed by springs against the moving contacts of a drum-type controller.
contact flight (*Aero.*). Navigation of an aircraft by ground observation alone.
contact insecticide (*Chem.*). One which kills on contact with insect surface (body, legs, etc.); used against sucking insects (e.g., aphids, mosquitoes) which are not affected by insecticides acting only through the alimentary system; e.g., pyrethrins, rotenone, DDT.
contact ionization (*Electronics*). Loss of electron by an easily ionized atom (e.g., caesium) when it comes into contact with the surface of a metal with affinity for electrons (e.g., tungsten).
contact jaw (*Elec. Eng.*). (1) The clamping device of a resistance welding machine, which secures the parts to be welded and also conducts the current to them. (2) The fixed part of a switch, with which the moving blade makes contact in closing the circuit.
contact lens (*Optics*). A lens, usually of plastic material, worn in contact with the cornea instead of spectacles. Used to aid activity, or for cosmetic purposes.
contact maker (*Elec. Eng.*). Any device used to make an electrical contact, especially a periodical contact, as in *sparking plugs.*
contact metal (*Telecomm.*). That used for contacts on springs of relays, generally silver, platinum, tungsten, etc.
contact metamorphism (*Geol.*). The alteration of rocks caused by their contact with, or proximity to, a body of igneous rock.
contact-modulated amplifier (*Elec. Eng.*). See **vibrating-reed amplifier.**
contact noise (*Elec. Eng.*). Noise voltage arising across a contact, with or without adsorbed gases, arising from differences in work function of contacting metal conductors, one of which may be a semiconductor.
contactor controller (*Elec. Eng.*). A controller in which the various circuits are made and broken by means of contactors.
contactor starter (*Elec. Eng.*). An electric motor starter in which the steps of resistance are cut out, or other operations are performed, by means of contactors.
contactor switching starter (*Elec. Eng.*). A switching starter in which the switching operations are carried out by means of contactors.
contact paper (*Photog.*). Bromide paper so slow in its speed that it can be handled in normal tungsten light for contact printing. Formerly called **gaslight paper.**
contact point (*Elec. Eng.*). The pointed part of the contact bar of a resistance welding machine which actually makes contact with the part to be welded.
contact-point insert (*Elec. Eng.*). A metal tip inserted in the contact point of a resistance welding machine to increase its durability.
contact potential (*Elec. Eng.*). See **contact e.m.f.**
contact-potential barrier (*Electronics*). Potential barrier formed at the junction between regions with different energy gap or carrier concentration.
contact pressure (*Telecomm.*). The pressure between the contacts on relay springs, a minimum pressure being required for certain contact when the circuit is frequently broken.

contact print (*Photog.*). A positive print obtained by exposure through the negative, which is in contact with the sensitive surface.

contact process (*Chem. Eng.*). The most important process for making sulphuric acid. Sulphur dioxide is passed with dry air over a catalyst, usually vanadium (V) oxide, to produce sulphur (VI) oxide which is absorbed in sulphuric acid to which water is added, giving the product. If sulphur dioxide is produced from pyrites, special cleaning is required, usually by electrostatic precipitation, but it can be conveniently made by direct ignition of sulphur in dry air.

contact-radiation therapy (*Radiol.*). Radiation from a very short distance, e.g., 20 mm, with voltages around 50 kV.

contact rail (*Elec. Eng.*). See **conductor rail.**

contact resistance (*Elec. Eng.*). Resistance at surface of contact between two conductors; mainly determined by constriction of current in the materials near the point of contact. Contact area is extended by pressure increasing this resistance.

contact roller (*Elec. Eng.*). A rotating disk used in a seam-welding machine to convey current to the parts to be welded and to exert the necessary mechanical pressure.

contact screen (*Print.*). A vignetted-dot screen on a film base used in the process camera in close contact with the emulsion, as an alternative to the glass half-tone screen; also for same-size screened work by contact.

contact segment (*Elec. Eng.*). Contacts on certain types of motor starter, or other control equipment, which are segment-shaped, so that when in position they form a ring divided by radial gaps.

contact series (*Chem.*). See **electrochemical series.**

contact shoe (*Elec. Eng.*). See **collector shoe.**

contact skate (*Elec. Eng.*). The long conductor placed under an electric vehicle to make contact with the studs used in the surface-contact system of electric traction.

contact spring (*Elec. Eng.*). The flexible metal holder of the contact in a relay. The stiffness of the holder determines the pressure between contacts for a given displacement.

contact strip (*Elec. Eng.*). On a pantograph or bow type of current collector, the renewable metal or carbon strip that actually makes contact with the overhead wire of an electric traction system. Also called a **bow strip,** when used on a bow collector.

contact stud (*Elec. Eng.*). In the surface-contact system of electric traction, the studs in the roadway for making contact with the contact skate on an electric vehicle. The studs are only made alive when the car is actually passing over them.

contact vein (*Mining*). A vein occurring along the line of contact of two different rock formations, one of which may be an igneous intrusion.

contact wire (*Elec. Eng.*). The overhead conductor from which current is collected, by suitable forms of collector gear, for the vehicles of some electric traction systems.

contagion (*Med.*). The communication of disease by direct contact between persons, or between an infected object and a person. *adj.* **contagious.**

contagious agalactia (*Vet.*). A contagious infection of sheep and goats by *Mycoplasma agalactiae*; characterized by fever, mastitis, agalactia, arthritis, and conjunctivitis.

contagious bovine pleuro-pneumonia (*Vet.*). Lung plague. An acute, subacute, or chronic disease of cattle caused by *Mycoplasma mycoides*; characterized by fever, pneumonia, and pleurisy.

contagious catarrh (*Vet.*). See **infectious coryza.**

contagious equine abortion (*Vet.*). A contagious form of abortion in horses due to infection of the placenta by *Salmonella abortivoequina.*

contagious granular vaginitis (*Vet.*). A disease of cattle characterized by small granular eruptions of the mucous membrane of the vagina; believed to be transmitted venereally and caused by a virus.

contagious ophthalmia (*Vet.*). Heather blindness. A contagious disease of sheep, characterized by conjunctivitis and keratitis, caused by a rickettsial organism *Colesiota conjunctivae* (*Rickettsia conjunctivae*).

contagious pustular dermatitis (*Vet.*). Malignant aphtha; orf. A contagious virus disease of sheep and goats characterized by vesicle and pustule formation on the skin and mucous membranes, especially on lips, nose, and feet.

containment (*Nuc.*). Holding together of sufficient density of plasma in strong magnetic fields at a very high temperature, so that fusion nuclear reactions from deuterons and tritons are promoted. Also called **confinement.**

contaminated rocks (*Geol.*). Igneous rocks whose composition has been modified by the incorporation of other rock material.

contamination meter (*Radiol.*). Particular design of Geiger-Müller circuit for indicating for Civil Defence purposes the degree of radioactive contamination in an area, especially for estimating the time for its safe occupation.

contemporaneous erosion (*Geol.*). The removal of sediments immediately after deposition or whilst sediments of identical age are being laid down in another area. See **non-sequence.**

conterminous (*Bot.*). Said of marginal ray cells when they form an uninterrupted row.

context (*Bot.*). The fleshy tissue of the fruit body of *Hymenomycetes,* but not the cortex or hymenium.

contiguous (*Bot.*). In contact but not in organic union.

continental climate (*Meteor.*). A type of climate found in continental areas not subject to maritime influences, and characterized by more pronounced extremes between summer and winter; the winters become colder to a greater degree than the summers become hotter; also relatively small rainfall and low humidities.

continental conditions (*Geol.*). These obtain when portions of the earth's crust are elevated above sea level in large land areas remote from the sea. Characteristics are low rainfall, scarcity of rivers, salt lakes. See also **continental deposition.**

continental crust (*Geol.*). That part of the earth's crust which underlies the continents and continental shelves. It is approximately 30 km thick in most regions but is thicker under mountainous areas. Sedimentary rocks predominate in its uppermost part and metamorphic rocks at depth, but the detailed composition of the lower crust is uncertain. Cf. *oceanic crust.*

continental deposition (*Geol.*). The laying down of rocks under subaerial conditions, or in temporary shallow water areas. See **aeolian rocks, glacial deposits, lacustrine deposits.**

continental drift (*Geol.*). The disruption and drifting apart of land masses and continents.

continental shelf (*Geol.*). The gently sloping offshore zone, extending usually to about 200 metres depth.

Continental system of drawing (*Textiles*). See **porcupine drawing.**

continued fraction (*Maths.*). A terminating or infinite fraction of the form

$$a + \cfrac{b}{c + \cfrac{d}{e + \cfrac{f}{g +}}} \text{ etc.,}$$

usually written $a + \frac{b}{c+} \frac{d}{e+} \frac{f}{g+}$, etc.

The values obtained by ending the fraction at a, at c, at e, etc., are called *convergents*.

continuity (*Cinema.*). The composition and co-ordination of the sequences required in the production of a motion-picture. (*Elec.*) The existence of an uninterrupted path for current in a circuit. (*Radio*) Supervision of succession of items from various sources in making up a broadcast programme.

continuity-bond (*Elec. Eng.*). A rail-bond used to maintain the continuity of the track- or conductor-rail circuit at junctions and crossings.

continuity-fitting (*Elec. Eng.*). A device used in electric wiring installations for ensuring a continuous electric circuit between adjacent lengths of conduit.

continuous (*Bot.*). (1) Without septa. (2) With a smooth surface. (3) Of even uninterrupted outline or contour. (*Maths.*) A function $f(x)$ is continuous at a point x_0 if

$$\lim f(x) = f(x_0) \text{ as } x \to x_0.$$

The function is continuous in an interval (a, b) if it is continuous at every point of (a, b).

continuous beam (*Eng.*). A beam supported at a number of points and continuous over the supports, as distinct from a series of simple independent beams. Also **continuous girder.**

continuous brake (*Eng.*). A brake system used on railway trains, in which operation at one point applies the brakes throughout the train. See **air brake, electropneumatic brake, pneumatic brake, vacuum brake.**

continuous control (*Telecomm.*). System in which controller is supplied with continuously varying actuating signal. Cf. *on-off control.*

continuous creation (*Astron.*). The theory that the universe is in a steady state, showing no overall change, although new systems are continually being formed to replace those carried away by recession. The theory implies that the universe had no beginning and will have no end, evolution being confined to individual stars and galaxies.

continuous current (*Elec. Eng.*). Earlier name for *direct current*, now obsolete in U.K. and U.S. but retained on Continent.

continuous diffusion. Counterflow system of extracting sugar from beet whereby fresh beet slices (cosettes) are extracted by hot dilute sugar solution, and partially extracted slices are finally extracted with fresh hot water.

continuous-disk winding (*Elec. Eng.*). A type of winding used for transformers; the whole winding is made from one continuous length of conductor instead of being split up.

continuous distillation (*Chem. Eng.*). An arrangement by which a fresh distillation charge is continuously fed into the still in the same measure as the still charge is distilled off. The contrary process is known as *batch distillation.*

continuous electrode (*Elec. Eng.*). A type of carbon electrode used in electric furnaces; the electrode is gradually fed forward as the lower part burns away, and the upper part is renewed by adding fresh material. The furnace can thus be worked continuously, without intervals for electrode renewal.

continuous electrophorus (*Elec. Eng.*). The name given to some earlier forms of influence machine which operate on the principle of performing the functions of an *electrophorus* repeatedly.

continuous extraction (*Chem. Eng.*). Extraction of solids or liquids by the same solvent, which circulates through the extracted substance, evaporates, and is condensed again, and continues the same cycle over again; or by exhaustive extraction with solvents in counter-current arrangement.

continuous feeder (*Print.*). An automatic sheet-feeder designed to permit continuous reloading without stopping the machine.

continuous filter (*San. Eng.*). See **percolating filter.**

continuous furnace (*Met.*). A furnace in which the charge enters at one end, moves through continuously, and is discharged at the other.

continuous girder (*Eng.*). See **continuous beam.**

continuous girder bridge (*Civ. Eng.*). A bridge the principal members of which are continuous girders.

continuous impost (*Build.*). An impost which does not project from the general surfaces of the pier and arch.

continuous loading (*Elec. Eng.*). Inductance loading with series inductions at intervals much less than a wavelength.

continuous mill (*Met.*). A rolling-mill consisting of a series of pairs of rolls in which the stock undergoes successive reductions as it passes from one end to the other end of the mill. See also **pull-over mill, reversing mill, three-high mill.**

continuous oscillations (*Radio*). Those generated by alternator, arc, or oscillating circuit which are not broken into individually decaying groups, as opposed to the *damped oscillations* of a spark transmitter, the latter now being obsolete. Also **undamped oscillations.**

continuous printing (*Cinema.*). Contact printing which brings the positive and negative together for printing a cross-line, which progresses along the length of the film as it is drawn continuously by the mechanism.

continuous processing machine (*Photog.*). Apparatus for development of cinematograph film in which the film passes continuously through the chemical and washing baths and drier.

continuous projector (*Cinema.*). A cinematograph projector in which the picture is projected on to the screen by a number of lenses travelling without motion relative to the film, which moves continuously and not intermittently, as is normal.

continuous rating (*Elec. Eng.*). The maximum power dissipation which could be allowed to continue indefinitely. Cf. *intermittent rating.*

continuous sections (*Bind., Print.*). The normal arrangement of the sections for bookwork as distinct from *insetted.* Each section is made up of consecutive pages, its last page being followed by the first page of the next.

continuous spectrum (*Phys.*). One which shows continuous non-discrete changes of intensity with wavelengths or particle energy.

continuous variation (*Bot., Zool.*). Variation as shown by slight individual differences between a number of specimens belonging to the same lineage, so that one may shade into another.

continuous vent (*Plumb.*). Extension of a vertical waste pipe above the point of entry of liquid wastes to a point above all windows, to provide ventilation.

continuous-wave telegraphy. That in which each signal (dot or dash) consists of a train of continuous oscillations.

continuous welded rail (*Rail.*). Rail which has been prewelded in much greater lengths (up to 1200 ft) than the standard (60 ft) before being transported to the site, where successive lengths are welded together to produce unbroken rails many miles long.

continuum (*Ecol.*). The pattern of overlapping populations in a large but definite community with component populations distributed along a gradient of, e.g., altitude. (*Maths.*) See **arithmetical continuum.**

contorted (*Bot.*). Twisted together.

contorted aestivation (*Bot.*). The arrangement of the perianth in the bud, when all the perianth segments overlap by their right-hand edges, or by their left-hand edges.

contortion (*Geol.*). Of strata, the deformation of rocks by directed pressure or torsion.

contortoduplicate (*Bot.*). Twisted and folded.

contour (*Surv.*). The imaginary intersection line between the ground surface and any given level surface: a line connecting points on the ground surface which are at the same height above *datum* (q.v.).

contour acuity (*Light*). The power of the eye to distinguish a displacement between two sections of a line, as in reading a *vernier*.

contour feathers (*Zool.*). See **plumae.**

contour fringes (*Light*). Interference fringes of constant optical thickness.

contour gradient (*Surv.*). A line on the ground surface having a constant inclination to the horizontal.

contourgraph (*Electronics*). Device which uses the three-dimensional aspects of a CRT, viz., vertical signal deflection, intensity modulation and flexibility of time base. Essentially a compact display of non-identical, quasi-periodic data comprising a time-exposure photograph of a number of oscillograph traces. In order to emphasize the similarities or differences between cycles, the data are arranged about an event common to each cycle and collected over extended time periods.

contour interval (*Surv.*). The vertical distance between adjacent contours in any particular case.

contraceptive (*Med.*). Any agent which prevents the fertilization of the ovum with a spermatozoon. See **oral contraceptives.**

contract-demand tariff (*Elec. Eng.*). A form of 2-part tariff in which the fixed charge is made proportional to the maximum kilowatt demand likely to be made.

contractile root (*Bot.*). A somewhat fleshy root which, as it ages, develops transverse corrugations, as it is shortened and drags the plant deeper into the soil.

contractile tissue (*Zool.*). A group of animal tissues which possess the property of contractility; more commonly spoken of as muscle.

contractile vacuole (*Zool.*). In some *Protozoa*, a cavity, filled with fluid, which periodically collapses and expels its contents into the surrounding medium, so ridding the animal of surplus fluid and excreta in solution.

contractility (*Zool.*). The power of becoming reduced in length exhibited by some cells and tissues, as muscle; the power of changing shape.

contraction (*Powder Tech.*). The shrinkage from green dimension after heat treatment to the same dimension when green. It may be expressed as a percentage reduction.

contraction cavities (*Met.*). Porous zones in metal castings caused by contraction of cooling metal, bad mould design, interrupted pouring of molten metal or premature freezing of feeding head.

contraction coefficient (*Elec. Eng.*). A coefficient used in making calculations on the magnetic circuit of an electric machine, to allow for the effect of the fringing of the flux in the air-gap due to open or semiclosed slots. The actual length of the gap is reduced by the coefficient in order to obtain an effective gap length, which is shorter than the actual value. Cf. *extension coefficient*.

contraction in area (*Met.*). See **necking.**

contraction joint (*Civ. Eng.*). An interruption of a structure specifically allowed for contraction in the time following building.

contraction ratio (*Aero.*). The ratio of the maximum cross-sectional area of a wind tunnel to that of the working section.

contract-rate tariff (*Elec. Eng.*). See **two-part tariff.**

contracture (*Med.*). Distortion or shortening of a part, due to spasm or paralysis of muscles, or to the presence of scar tissue. (*Zool.*) Muscular contraction which persists after the stimulus which caused it has ceased.

contradeciduate (*Zool.*). Said of Mammals in which the placenta is largely absorbed by maternal leucocytes at parturition.

contrails (*Meteor.*). See **condensation trails.**

Contran (*Comp.*). *Con*trol *tran*slator. A computer language combining the most desirable features of Fortran IV and Algol 60.

contraries (*Paper*). Anything in the raw pulp stock which is useless for making into paper.

contrast (*Acous.*). The relation, measured in decibels, between the maximum intensity level and the minimum useful intensity level in programme material such as speech or music. (*Psychol.*) The tendency to perceive, or think of, objects which are different from a standard, or persons who are different from oneself, as being more different than is actually the case. See **assimilation.** (*Photog., TV*) The ratio of brightness of two acceptable levels in an image. Range of contrast is the maximum contrast acceptable or possible. See **point gamma.**

contrast amplification (*Acous.*). That in which dynamic contrast in sound reproduction is increased by electronic means, compensating imprecisely for contrast reduction (*control*) which is necessary in most communication systems to avoid intrusion of noise.

contrast control (*TV*). Detailed control of highlights and shadows in a reproduction image.

contrast medium (*Radiol.*). Any suitable substance used in diagnostic radiology in order to give contrast.

contrast photometer (*Light*). A class of photometer in which measurement is made by comparing the illumination produced on two adjacent surfaces by the lamp under test and by a standard lamp.

contrast range (*TV*). See **dynamic range.**

contrast sensitivity (*TV*). Liminal ratio of $\delta B/B$ of a small spot which differs from brightness B of the larger surroundings.

contrate wheel (*Horol.*). (1) A toothed wheel the teeth of which are formed at right-angles to the plane of the wheel; a wheel that transmits motion between two arbors at right angles. (2) The fourth wheel in a watch with the verge escapement.

contra wire (*Elec. Eng.*). Same as *constantan* (q.v.) wire.

control. General term for manual or automatic adjustment (usually by potentiometer, fader, or attenuator) of power level in a transmission within its dynamic range. (*Comp.*) Said to be *sequential* when the instructions are fed to an electronic digital computer in order during an operation. Said to be *dynamic* when the instructions are varied or altered in sequence during, or as a result of, an operation. (*Nuc. Eng.*) Maintenance of power level at desired setting by adjustments to the reactivity of the reactor core through absorbing rods or other means.

control ampere-turns (*Elec. Eng.*). Magnetomotive force applied to a magnetic amplifier.

control-board (*Elec. Eng.*). A switchboard on which are mounted the operating handles, push-buttons, or other devices for operating switch-gear situated remotely from the board. The board usually has mounted on it indicating instruments, key diagrams, and other accessory apparatus.

control characteristic (*Elec. Eng.*). Curve connecting output quantity against control quantity under determined conditions in a magnetic amplifier.

control circuit (*Elec. Eng.*). (1) Separate circuit or line in parallel with a transmission path, for use of operators. (2) All windings and components which determine control current in a magnetic amplifier.

control column (*Aero.*). The lever supporting a hand-wheel or hand-grip by which the ailerons and elevator of an aeroplane are operated. It may be a simple 'joystick', pivoted at the foot and rocking fore-and-aft and laterally; on military aircraft, usually fighters, it is often hinged halfway up for lateral movement; on transports it is usually either 'spectacle' or 'ram's horn' shape.

control current or **voltage** (*Elec. Eng.*). One which, by its magnitude, direction, or relative phase, determines operation of functional apparatus, e.g., that which operates lamps through local relays, traction motors distributed along a train, that in a mercury-arc rectifier, or process control.

control electrode (*Elec. Eng.*). One, e.g., a grid, primary function of which is to control flow of electrons between two other electrodes, without taking appreciable power itself, control being by voltage which regulates electrostatic fields.

control hysteresis (*Elec. Eng.*). Ambiguous control, depending on previous conditions. Jump or snap action arising in electronic or magnetic amplifiers because of excessive positive feedback which occurs under certain conditions of load.

control impedance (*Elec.*). The electrical property of a device which controls power in one direction only, as a gas-filled relay.

control instructions (*Comp.*). The part of a program which governs the manipulation of data and interpretation of instructions, but does not contribute to the actual instructions.

controllable-pitch airscrew (*Aero.*). See **airscrew.**

controlled air space (*Aero.*). Areas and lanes clearly defined in 3 dimensions wherein no aircraft may fly unless it is under radio instructions from *air-traffic control* (q.v.).

controlled carrier (*Telecomm.*). Transmission in which the magnitude of the carrier is controlled by the signal, so that the depth of modulation is nearly independent of the magnitude of the signal.

controlled variable. Quantity or condition which is measured and controlled.

controller (*Elec. Eng.*). An assembly of equipment for controlling the operation of electric apparatus.

control limit-switch (*Elec. Eng.*). A limit-switch connected in the control circuit of the motor whose operation is to be limited.

control-line (*Elec. Eng.*). A train-line used on multiple-unit trains for connecting master controllers or contactor gear on the different coaches.

control magnet (*Elec. Eng.*). A magnet used in electric indicating instruments to provide a force for controlling the movement of the moving system.

control panel. Panel containing full set of indicating devices and remote control units required for operation of industrial plant, reactor, chemical works, etc. Cf. *console*.

control (or **set**) **point** (*Elec. Eng.*). Value of controlled variable, departure from which causes a controller to operate in such a sense as to reduce the *error* and restore an intended steady state.

control ratio (*Elec. Eng.*). In discharge tubes, ratio of change in anode voltage to a compensating change in grid voltage for striking. Generally, the ratio of output parameter to input parameter in any control device.

control register (*Comp.*). That which stores the current instructions which govern the computer for a given cycle.

control relay (*Elec. Eng.*). See **relay types.**

control reversal (*Aero.*). See **reversal of control.**

control rod (*Nuc. Eng.*). Rod moved in and out of reactor core to vary rate of reactivity. May be neutron-absorbing rod, e.g., cadmium, or less often, fuel rod. See **regulating rod, shim rod.**

control system (*Automation*). Servo controller unit operated by error-sensing head through closed loop.

control trailer (*Elec. Eng.*). Same as **driving-trailer.**

control turns (or **windings**) (*Elec. Eng.*). Those wires on the core of a magnetic amplifier or transductor which carry the control current. U.S. term **signal windings.**

control voltage (*Elec. Eng.*). See **control current.**

control word (*Comp.*). One which transmits instructions to a digital computer, e.g., XEQ for execute.

control zone (*Aero.*). An area, precisely defined in plan and altitude, including one or more aerodromes, in which flight rules additional to those in a control area pertain.

conuli (*Zool.*). In certain *Porifera*, conical projections of the surface.

conus (*Zool.*). Any cone-shaped structure or organ.

conus arteriosus (*Zool.*). In some lower Vertebrates, a valvular region of the truncus arteriosus, adjacent to the heart.

conus medullaris (*Zool.*). The conical termination of the spinal cord.

convection current (*Elec.*). Current in which the charges are carried by moving masses appreciably heavier than electrons. (*Heat*) See **convection of heat.**

convection of heat (*Heat*). The trans er of heat from one part of a fluid to another by flow of the fluid from the hotter parts to the colder. Thus if heat is applied to the bottom of a vessel containing a liquid, the hot liquid, being less dense than the cold, will rise, its place being taken by cold liquid moving down. There is thus set up a circulation of liquid (known as a *convection current*), which keeps the temperature more uniform than if the liquid were stagnant.

convective atmosphere (*Meteor.*). Atmosphere characterized by ascending and descending air currents.

convective discharge (*Elec.*). Visible or invisible discharge by charged particles from a high-voltage source.

convective transfer (*Astron.*). The transfer of energy from one part of a star to another by convection.

convector (*Heat*). A type of heater designed to warm the air of a room as it circulates over the heating elements (e.g. electric, gas or steam-heated coils), rather than by direct radiation.

conventional signs (*Civ. Eng., Surv.*). Standard symbols, universally understood, used in the representation on maps and plans of features which would otherwise be difficult or impossible to represent.

convention of signs (*Light*). In lens calculations, all distances must be measured from some origin and, to ensure consistency in the derivation and use of lens formulae, it is necessary to adopt a sign convention. More than one is in use, but the convention usually adopted is that all distances shall be measured from the reflecting or refracting surface being considered, or from the principal planes in the case of a thick lens, and that distances measured in the direction in which the incident light is travelling are given a negative sign, and those which are travelling in the opposite direction are given a positive sign.

convergence (*Maths.*). (1) An infinite sequence $a_1, a_2, \ldots a_n, \ldots$, is said to be convergent to a, if as n tends to infinity, a_n tends to the limit a, i.e. $\lim_{n\to\infty} a_n = a$. (2) An infinite series $a_1 + a_2 + \ldots + a_n + \ldots$, is said to be convergent to the sum A, if the sum of the terms tends to the limit A, as the number of terms tends to infinity, i.e. $\lim_{n\to\infty} \Sigma a_n = A$. (3) An infinite product, $a_1 . a_2 \ldots a_n \ldots$ is said to be convergent to α if as n tends to infinity, the product tends to the limit α, i.e., $\lim_{n\to\infty} \Pi a_n = \alpha$. (4) An infinite integral $\int_a^\infty f(t)dt$, is said to converge to I, if $f(t)$ is continuous for $t \geqq a$, and if $\lim_{x\to\infty} \int_a^x f(t)dt = I$. (*Meteor.*) An accumulation of air over a region caused by lack of uniformity of the winds. Convergence results in upward air-currents, causing cloud and rain. (*TV*) In a colour display tube (tricolour kinescope), meeting or crossing of the three electron beams at a common point. For proper reproduction of colour pictures convergence must be maintained over the whole scanning area. (*Zool.*) See **convergent evolution.**

convergence coils or magnets (*TV*). Used to ensure convergence of the electron beams in a triple gun colour TV tube.

convergence surface (*Electronics*). Ideally, that generated by the point on which the electron beams in a multibeam CRT converge. In practice, this can only be regarded as a point to a first approximation.

convergent (*Maths.*). See **continued fraction.**

convergent-divergent nozzle (*Aero., Eng.*). A venturi type of nozzle in which the cross-section first decreases to a throat and then increases to the exit, such a form being necessary for efficient expansion of steam, as in the steam-turbine, or of other vapours or gases, as in the supersonic aero-engine, ramjet, or rocket. Abbrev. **con-di nozzle.**

convergent evolution (*Zool.*). The tendency of animals living in a simple and uniform environment to develop similar characteristics, whatever their ancestry, owing to the lack of opportunity for variety of habitat or habits.

convergent lens (*Light*). A lens which increases the convergence or decreases the divergence of a beam of light. A simple lens is convergent if it is double-convex, plano-convex, or concavo-convex with the radius of curvature of the concave face greater than that of the convex.

convergent thinking (*Psychol.*). A type of ability which is sampled in standard intelligence tests. Rational, logical thought which works within certain confines, as in solving a mathematical problem. See **creativity, divergent thinking.**

converging (*Bot.*). Having the tips gradually approaching.

converging-beam therapy (*Radiol.*). A form of *cross-fire technique* (q.v.) in which a number of beams of radiation are used simultaneously to treat a particular region through different entry portals.

converging-field therapy (*Radiol.*). Form of *moving field therapy* in which the source of radiation moves in a spiral path.

converse magnetostriction (*Mag.*). The change of magnetic properties, e.g., induction, in a ferromagnetic material when subjected to mechanical stress or pressure.

conversion (*Chem.*). The fraction of some key reagent which has been used up, hence a measure of the rate or completeness of a chemical reaction. (*Comp.*) Component, auxiliary to a computer system, which accepts *decimal* digits and makes a unique conversion to *binary* digits in the output, or vice versa.

conversion coefficient (*Nuc.*). See **internal conversion.**

conversion conductance (*Telecomm.*). Ratio of component of current of intermediate frequency flowing in the output circuit of a conversion detector when working into zero impedance, to the signal frequency input voltage; expressed in mA/V.

conversion detector (*Telecomm.*). See **mixer (2).**

conversion efficiency (*Electronics*). See **anode efficiency.**

conversion electron (*Electronics*). One ejected from inner shell of atom when excited nucleus returns to ground state, and the energy released is given to orbital electron, instead of appearing as quantum.

conversion factor (*Phys.*). Factor by which a quantity, expressed in one set of units, must be multiplied to convert it to another set of units.

conversion gain (*Telecomm.*). Effective amplification of a conversion detector, measured as the ratio of output voltage of intermediate frequency to input voltage of signal frequency. Expressed in dB, the gain is 20 times the logarithm of this ratio.

conversion hysteria (*Psychol.*). The type of hysteria in which the unconscious mental conflict comes to expression, with the partially repressed wish, by physical symptoms. Some manifestations of these are paralysis, blindness, deafness, and pains in various parts of the body. Accompanying this conversion is a relief from anxiety, giving this type of hysteria its characteristic feature of a calm and indifferent mental attitude. Called by Janet *la belle indifférence.*

conversion mixer (*Telecomm.*). Same as **frequency changer.**

conversion ratio (*R*) (*Nuc. Eng.*). Number of

fissionable atoms produced per fissionable atom destroyed in a reactor. Corresponding conversion gain is defined as $R-1$.

conversion systems (*Print.*). Methods of producing plate-making negatives or positives from letterpress pages alternative to the *reproduction proof*. Include **Brightype, Cronapress, Rinco process, Scotchprint.**

conversion transconductance (*Telecomm.*). Ratio of output current of one frequency to corresponding voltage of different frequency in a superheterodyne converter.

conversion transducer (*Telecomm.*). Same as **frequency changer.**

converter (*Comp.*). A device for changing information coded in one form into the same information coded in another, e.g., analogue-digital converter. (*Elec. Eng.*) A machine or device for changing a.c. to d.c. or the converse, particularly transistor oscillator for supplying high-tension from low-tension. See **facsimile modulation, transistor power-pack.** (*Telecomm.*) U.S. for **frequency changer.**

converter reactor (*Nuc. Eng.*). One in which fertile material in reactor core is converted into fissile material different from the fuel material. Cf. *breeder reactor*.

convertible lens (*Photog.*). A compound lens, the units of which can be used separately or in conjunction with other lenses.

converting (*Met.*). See **Bessemer process.**

converting station (*Elec. Eng.*). An electric power system substation containing one or more converters.

convertiplane (*Aero.*). A *VTOL* (q.v.) aircraft which can take-off and land like a helicopter, but cruises like an aeroplane; by swivelling the rotor(s) and/or wing to act as propellers, or by putting the rotor(s) into *autorotation* (q.v.) and using other means for propulsion.

convex lens (*Light*). A *convergent lens* (q.v.).

convex mirror (*Light*). A portion of a sphere of which the outer face is a polished reflecting surface. Such a mirror forms diminished virtual images of all objects in front of it.

convex veins (*Zool.*). In Insects, veins which follow the ridges of the wing corrugations.

conveyor (*Eng.*). A device for the continuous transport of small articles or loose material over a short distance, as between two parts of a workshop, or different levels in a building.

convolute (*Bot.*). Coiled, folded, or rolled, so that one half is covered by the other. (*Zool.*) Having one part twisted over or rolled over another part; twisted; as the cerebral lobes of the brain in higher Vertebrates, gastropod shells in which the outer whorls overlap the inner. *n.* **convolution.**

convolute aestivation (*Bot.*). See **contorted aestivation.**

convolution (*Anat.*). Any elevation of the surface of the brain. See also **convolute.**

convolution integral (*Maths.*). The integral

$$\int_0^x f(x)g(x-t)dt.$$

Also referred to as the *cross-correlation* between $f(x)$ and $g(x)$. When $f(x)=g(x)$ the integral is sometimes referred to as the *autocorrelation* of $f(x)$.

convulsion (*Med.*). Generalized involuntary spasm of the muscles normally under control of the will.

Cooke triplet (*Photog.*). Basic type of 3-element anastigmat lens, comprising two collective components separated by a dispersive component.

cookie cutter tuning (*Electronics*). Tuning of cavity magnetrons, involving changing capacitance of cavities by the introduction of conducting rings between straps. Cf. *crown of thorns tuning*.

coolant (*Eng.*). (1) A mixture of water, soda, oil, and soft-soap, used to cool and lubricate the work and cutting tool in machining operations. See **cutting compound.** (2) A fluid used as the cooling medium in the jackets of liquid-cooled I.C. engines; e.g., water, ethylene glycol (ethan 1,2-diol).

cooled-anode valve (*Electronics*) Large thermionic valve in which special provisions are made for dissipating the heat generated at the anode, effected by circulating water, oil, or air around the anode, or by radiation from its surface.

Coolidge tube (*Electronics*). High-voltage self-rectifying X-ray tube, the first to have a heated cathode as source of electrons.

cooling (*Nuc. Eng.*). The decay of activity of highly radioactive waste before it is processed or disposed of.

cooling coil (*Heat*). An arrangement of pipe or tubing serving to transfer heat from the material or space cooled to the primary refrigerant or brine.

cooling curves (*Met.*). Curves obtained by plotting time against temperature for a metal cooling under constant conditions. The curves show the evolutions of heat which accompany solidification, polymorphic changes in pure metals, and various transformations in alloys.

cooling drag (*Aero.*). That proportion of the total drag due to the flow of cooling air through and round the engine(s).

cooling duct (*Elec. Eng.*). See **duct.**

cooling plate (*Cinema.*). A metal screen to protect mechanisms from the heat from arcs in projectors.

cooling pond (*Eng.*). An open pond in which water, heated through use in an industrial process, or after circulation through a steam-condenser, is, before re-use, allowed to cool through evaporation.

cooling tower (*Eng.*). A tower of wood, concrete, etc., used to cool water after circulation through a condenser. The water is allowed to trickle down over wood slats, thus exposing a large surface to atmospheric cooling.

coombe rock (*Geol.*). Unstratified angular chalk rubble found over the plain tracts of southern England; prob. formed over a frozen subsoil.

cooperation (*Ecol.*). A category of interaction between two species (i.e., a coaction) where each has a beneficial effect on the other, increasing the size or growth rate of the population, but, unlike mutualism, not a necessary relationship. Termed **protocooperation** by some, since its basis is neither conscious nor intelligent, as in man.

Cooper-Hewitt lamp (*Light*). A form of mercury-vapour ultraviolet lamp.

cooperite (*Min.*). Sulphide of platinum crystallizing in the tetragonal system; occurs in minute and irregular grains in basic igneous rocks.

coordinate axes (*Maths.*). See **axes.**

co-ordinate bond (*Chem.*). See **covalent bond.**

coordinated transposition (*Telecomm.*). The reduction of mutual inductive effects in multiline transmission systems (telephony or power) by periodically interchanging positions.

coordinate potentiometer (*Elec. Eng.*). One in which two linear potentiometers carry a.c. currents 90° apart in phase, so that the resultant voltage between tappings can be adjusted in both phase and magnitude.

coordinates (*Maths.*). Numbers which specify the position or orientation of a point or geometric configuration.

coordinating gap (*Elec. Eng.*). A spark-gap, used in power transmission schemes, so arranged that it will break down at a voltage bearing a definite relation to the breakdown voltage of other apparatus in the system, thereby enabling surge voltages to be safely discharged to earth.

coordination number (*Chem.*). The number of atoms or groups (ligands) surrounding the central (nuclear) atom of a complex salt.

cop (*Spinning*). Yarn package built on to mule spindle. Varies in length and diameter according to purpose, e.g., twist, weft, bastard, pincop, etc.

copaiba balsam (*Chem.*). A light-yellow or brown liquid, of peculiar odour, rel. d. 0·940–0·990, saponification value 88–100, acid value 75–100. It is an oleo-resin containing sesquiterpenes. Used in varnishes.

copalite or **copaline** (*Min.*). Synonym for **Highgate resin.**

copals (*Paint.*). A class of natural resins of recent or recent-fossil origin, consisting of resin acids, resenes, and essential oils. They furnish raw material for the varnish and linoleum industries. Important copals are Congo, Kauri, Zanzibar.

Copal square shutter (*Photog.*). Type of focal-plane shutter having metal vanes operating vertically in place of the conventional horizontally-travelling blind.

cope (*Foundry*). The upper half of a mould or moulding box.

Copepoda (*Zool.*). Subclass of *Crustacea*, generally of small size, many of which are parasitic; there is usually no carapace; the abdomen is without appendages but the thorax bears several pairs of biramous oarlike swimming feet; the eggs are usually carried by the female in a pair of sacs formed of a hardened glutinous secretion.

Copernican System (*Astron.*). The heliocentric theory of planetary motion; called after Copernicus, who introduced it in 1543. It eventually superseded the geocentric, or Ptolemaic, System.

coping (*Build.*). (1) A stone or brick covering to the top of a wall exposed to the weather; it is designed to throw off the water, and is preferably wider than the wall, with *drips* (q.v.) cut in its projecting under-surfaces. (2) The operation of splitting stone by driving wedges into it. (*Vet.*) The operation of paring or cutting the beak or claws of a bird, particularly of hawks.

coping brick (*Build.*). Specially shaped brick used for capping the exposed top of a wall; used sometimes with a creasing and sometimes without, in which latter case the brick is wider than the wall and has drips under its lower edges.

coping saw (*Tools*). Small saw with narrow tensioned blade in a D-shaped bow, for cutting curves in wood up to about 15 mm thick (i.e., too thick for *fret-saw* (q.v.)).

coplanar grid valve (*Electronics*). Four-electrode valve containing two control grids intermeshed with, but insulated from, each other, and exercising equal control effects on the anode current. Used as a detector and as a high output amplifier where the anode supply voltage is limited.

coplanar vectors (*Maths.*). Vectors are coplanar if they lie in the same plane.

copolymer (*Chem.*). Polymer formed from the reaction of more than one species of monomer, e.g., poly(styrene-)butan 1,2 : 3,4-diene.

copper (*Met.*). Bright, reddish metallic element, symbol Cu, at. no. 29, r.a.m. 63·54, m.p. 1083°C resistivity at 0°C, 0·016 microhm metre. Native copper crystallizes in the cubic system. It often occurs in thin sheets of plates, filling narrow cracks or fissures. Copper is ductile, with high electrical and thermal conductivity, good resistance to corrosion; it has many uses, notably as an electrical conductor. Basis of brass, bronze, aluminium bronze, and other alloys. Nickel-iron wires with a copper coating are frequently used for *lead-ins* through glass seals, forming vacuum-tight joints. ^{64}Cu is a mixed radiator, of half-life 13 h.

copperas (*Min.*). Iron(II)sulphate, $FeSO_4{\cdot}7H_2O$. See **melanterite.** *White copperas* is goslarite, $ZnSO_4{\cdot}7H_2O$.

copper bit (*Plumb.*). A tool consisting of a pointed piece of copper fastened at the end of an iron rod held in a wooden handle. When the copper part is heated it may be used to melt solder. Also called **soldering iron.**

copper brushes (*Elec. Eng.*). Brushes occasionally used for electric commutator machines where high conductivity is required; they are made of copper strip, wire, or gauze.

copper-chloride process (*Chem. Eng.*). A process for removing mercaptans from petroleum refinery fractions in which the mercaptans are oxidized by air blowing in presence of copper chloride catalyst.

copper circuit (*Elec. Eng.*). One with no d.c. blocking, e.g. capacitor, series component.

copper-clad steel conductor (*Elec. Eng.*). See **steel-cored copper conductor.**

coppered carbons (*Light*). Arc-lamp carbons coated, by electroplating, with copper to improve their conductivity.

copper factor (*Elec. Eng.*). A term used in electric machine design to denote the ratio of the cross-sectional area of the copper in a winding to the total area of the winding, including insulating material and clearance space.

copper glance (*Min.*). A popular name for *chalcocite.*

copper glazing (*Build.*). Glazing formed of a number of individual panes separated by copper strips on the edges of which small flanges of copper have later been formed by deposition to retain the glass. Also **copperlite glazing.**

copper loss (*Elec. Eng.*). The loss occurring in electric machinery or other apparatus due to the current flowing in the windings; it is proportional to the product: $(\text{current})^2 \times \text{resistance}$. (*Radio*) The power dissipated as heat in an antenna, or other oscillatory, circuit due to Joule effect, including that due to eddy currents in the conductors and nearby metallic objects.

copper naphthenate (*Chem.*). Copper derivative of naphthenic acid obtained from petroleum by extraction with alkali. Used widely as a mildew-proofing agent for industrial and military textiles.

copper nickel (*Min.*). See **niccolite.**

copper nose (*Vet.*). See **brown nose disease.**

copper number (*Chem.*). The number of milligrams of copper obtained by the reduction of Fehling's solution by 1 g of a carbohydrate.

copper-oxide rectifier (*Radio*). One depending on the potential barrier between copper and copper (II) oxide or copper sulphide. Small copper (II) oxide rectifiers have suitable characteristics for a.c. indicating meters.

copper pyrite (*Min.*). See **chalcopyrite.**

copper rayon (*Textiles*). See **cuprammonium (tetra(a)ammine copper (II)) rayon.**

copper-sheathed cable (*Elec. Eng.*). See **mineral-insulated cable.**

coppersmith's hammer (*Eng.*). A hammer having a long curved ball-pane head, used in dishing copper plates.

copper (II) sulphate (*Chem.*). $CuSO_4$. Bluestone; a salt, soluble in water, used in copper-plating baths; formed by the action of sulphuric acid on copper; crystallizes as hydrous copper sulphate, $CuSO_4{\cdot}5H_2O$, in deep-blue triclinic crystals. See **blue vitriol, chalcanthite.**

copper (or cupro-) uranite (*Min.*). See **torbernite.**

copper voltmeter (*Elec. Eng.*). An electrolytic cell consisting of two copper plates in a solution of copper sulphate; used to measure current, by weighing the amount of copper deposited on the cathode plate in a given time.

copping (or shaper) rail (*Spinning*). A rail with long and short inclines, whose profile governs the shape and size of the mule cop by means of the poker bar guiding the copping faller in its upward and downward movements.

coprecipitation (*Nuc.*). The precipitation of a radioisotope with a similar substance, which precipitates with the same reagent, and which is added in order to assist the process.

coprodaeum (*Zool.*). That part of the cloaca into which the anus opens, in Birds.

coprolalia (*Psychiat.*). The utterance of filthy words by the insane.

coprolite (*Geol.*). Strictly, the fossilized excreta of animals, but now extended to other phosphatic nodules. Coprolite-rich beds are quarried and used as fertilizers.

coprolith (*Med.*). Hard concretion of faecal matter.

coprophagous (*Zool.*). Dung-eating.

coprophilia (*Psychiat.*). Pleasure or gratification obtained from any dealing with faeces.

coprophilous, coprophilic (*Bot.*). Growing on or in dung.

coprosterol (*Chem.*). $C_{27}H_{47}OH$, a constituent of the faeces, a reduction product of cholesterol.

coprozoic (*Zool.*). Living in dung, as some *Protozoa.*

copula (*Zool.*). A structure which bridges a gap or joins two other structures, as the series of unpaired cartilages which unite successive gill arches in lower Vertebrates.

copularium (*Zool.*). In certain *Gregarinidea*, a cyst formed round two associated gametocytes.

copulation (*Zool.*). In *Protozoa*, a type of syngamy in which the gametes fuse completely; in higher animals, union in sexual intercourse.

copulation path (*Zool.*). The path of the sperm nucleus in the ovum during fertilization.

copulation tube (*Bot.*). See **conjugation tube.**

copulatory spicules (*Zool.*). In male *Nematoda*, chitinous rods which arise from a dorsal pocket of the cloaca and assist the process of copulation.

copy (*Comp.*). The transfer of information from one register to another in a computer without changing the information in the original register. (*Paper*) An obsolete size of writing or drawing paper, 16 × 20 in. Also called **draft.** (*Typog.*) Any matter supplied for typesetting, block-making, etc.

copyholder (*Typog.*). (1) One who reads aloud from the copy as the proof-corrector follows the reading in his proof. (2) A contrivance for holding up sheets of copy on typesetting machines.

copying machine (*Eng.*). A machine for producing numbers of similar objects by an engraving tool or end-cutter, which is guided automatically from a master pattern or template. See also **document copying.**

coquille (*Glass*). Glass in thin curved form used in the manufacture of sun glasses. The radius of curvature is usually 3½ in. (9 cm). Similar glass of 7 in. (18 cm) radius is called **micoquille.**

coquimbite (*Min.*). Hydrated ferric sulphate, crystallizing in the hexagonal system, occurring in some ore deposits and also in volcanic fumaroles such as those of Vesuvius.

coquina (*Geol.*). A limestone made up of coarse shell fragments, usually of molluscs.

Coraciiformes (*Zool.*). An order of *Neognathae*, most of which are short-legged arboreal forms, nesting in holes and having nidicolous young. Mainly tropical and often brightly coloured. Kingfishers, Bee-eaters.

Coraciomorphae (*Zool.*). In a former system of classification, a legion of birds of very ancient origin, including the Cuckoos, Parrots, Kingfishers, Swifts, Swallows, Toucans, Owls, Perching and Singing Birds.

coracoid (*Zool.*). In Vertebrates, a paired posterior ventral bone of the pectoral girdle, or the cartilage which gives rise to it.

coracoid process (*Zool.*). In Mammals, the reduced coracoid bone, which persists only as a small process fused to the ventral end of the scapula.

coral (*Zool.*). The massive calcareous skeleton formed by certain species of *Anthozoa* and some *Hydrozoa*; the colonies of polyps forming this skeleton. *adjs.* **coralline, coralloid, coralliferous, corallaceous, coralliform.**

Corallian (*Geol.*). A division of the Upper Jurassic rocks, comprising sandstones, shelly limestones, and clays, with coral reefs well developed, e.g., at Steeple Ashton.

corallite (*Zool.*). The cuplike portion of a coral surrounding a single polyp.

coralloid (*Bot.*). Having a general appearance like that of a piece of much branched coral.

corallum (*Zool.*). The hard calcareous skeleton of coral polyps.

coral reef (*Geol.*). A calcareous bank formed of the skeletons of corals which live in colonies. The various formations of coral reefs are known as *atolls*, *barrier reefs*, and *fringing reefs*.

coral sand (*Geol.*). A sand made up of calcium carbonate grains derived from eroded coral skeletons, often found in deep water on the seaward side of a coral reef.

coramine (*Chem.*). See **nikethamide.**

corbeille (*Arch.*). Carved work representing a basket, used as a form of decoration.

corbel (*Build.*). Bricks or stones, frequently moulded, projecting from a wall to support a load.

corbelling (*Build.*). Projecting courses of brick or stone forming a ledge used to support a load.

corbel-piece (*Carp.*). See **bolster.**

corbel-table (*Arch.*). A cornice supported by corbels.

corbicula (*Zool.*). The pollen basket of Bees, consisting of the dilated posterior tibia with its fringe of long hairs.

corbie-step gable (*Build.*). A gable having a series of regular steps up each slope. Also called **crow-step gable.**

cor bovinum (*Med.*). Gross enlargement of the heart due to hypertrophy of its muscle.

corbula (*Zool.*). A phylactocarp having the costae well-developed and the hydrothecae suppressed.

cord (*For.*). A non-metric timber measure, 128 ft^3 (8 × 4 × 4 ft, about 3·625 m^3). (*Teleph.*) A flexible cable, usually containing covered tinsel conductors, for use in cord circuits and telephone leads. (*Textiles*) (1) Connexion between

a jacquard hook and the mails controlling the warp threads. (2) A rib effect in a fabric, produced according to plan, in either warp or weft.
Cordaitales (*Bot.*). An extinct order of *Coniferae*; usually large tree forms with simple leaves; the macrosporangia and microsporangia borne in separate strobili; monoecious or dioecious; strobili stalked.
cordate (*Bot.*). Said of a leaf base which has the form of the indented end of a conventional heart.
cord circuit (*Teleph.*). The temporary connexion circuit used by an operator to connect subscribers together, or subscribers to junction lines. In the usual central-battery systems, it provides current for the distant transmitters, supervisory relays and lamps, and means for the operator to speak to subscribers.
cord circuit repeater (*Teleph.*). A cord circuit containing a small-gain repeater, for insertion at trunk exchanges between subscribers' lines or junctions, which together introduce a loss which is greater than tolerable.
cord-de-chêne (*Textiles*). A light-weight worsted dress material with a slight longitudinal cord, made from a Botany warp and a silk weft.
corded way (*Build.*). A sloping path formed with deep sloping steps separated by timber or stone risers.
cordierite or **iolite** (*Min.*). A silicate of aluminium, iron, and magnesium, crystallizing in the orthorhombic system; occurs chiefly in metamorphic rocks.
cordiform (*Bot.*). Said of an ovate leaf with a pointed apex and a cordate base.
cordless board (*Teleph.*). Small exchange boards, usually private, in which all the connexions between lines and subscribers are made with keys and not with flexible cords.
cords (*Bind.*). Lengths of hemp across the back of a book, to which the sections are attached by sewing. See **bands.**
Cordtex (*Civ. Eng.*). A textile detonating fuse containing a core of pentaerythritol tetranitrate, used in the initiation of large explosive charges. Has a high velocity of detonation and will not inspire detonation unless directly connected to the actual explosive. TN.
corduroy (*Textiles*). Strong, hard-wearing cloth having a rounded or flattened cord or rib of weft pile running longitudinally; made entirely from cotton, or cotton warp and a spun rayon pile.
cordwood (*For.*). Tree trunks of medium diameter sawn into uniform lengths.
cordwood saw (*For.*). Transportable circular saw for cutting timber.
core (*Bot.*). (1) See **centrum.** (2) The plant material which forms the inner part of a periclinal chimaera. (*Build.*) The material removed from a mortise. (*Cinema.*) The inner part of the positive arc carbon. This is impregnated with salts, principally cerium fluoride, to increase the brilliance of the crater formed at the termination of the arc. (*Civ. Eng.*) (1) A watertight wall built within a dam or embankment as an absolute barrier to the passage of water. (2) Cylindrical sample of material obtained by driving a hollow-core drill into strata to ascertain the variation of composition. (*Elec. Eng.*) (1) Magnetic material increasing inductance of a coil; it may be wound from tape, moulded from ferro-particles and binder, or of punched laminations; it may be a complete magnetic circuit (divided to contain coil) or simply a rod. (2) The assembly consisting of the conductor and surrounding insulation of a cable or part of a cable, but not including the protective covering. It constitutes the electrical circuit for communication. (*Foundry*) A solid mass of specially prepared sand or loam placed in a mould to provide a hole or cavity in the casting. (*Mining*) Cylindrical rock section cut by rotating drill bit in prospecting, sampling, blasting. (*Nuc.*) In an atom the nucleus and all complete shells of electrons. In the atoms of the alkali metals, the nucleus, together with all but the outermost of the planetary electrons, may be considered to be a core, around which the valency electron revolves in a manner analogous to the revolution of the single electron in the hydrogen atom around the nucleus. In this manner, the simple Bohr theory may be made to give an approximate representation of the alkali spectra. See also **atomic structure.** (*Nuc. Eng.*) That part of a nuclear reactor which contains the fissile material, either dispersed or in cans.
core-balance protective system (*Elec. Eng.*). An excess-current protective system for electric power systems, in which any leakage current to earth in a 3-phase circuit is made to produce a resultant flux in a magnetic circuit surrounding all 3 phases; this flux produces a current in a secondary winding on the magnetic circuit, which operates a relay controlling the appropriate circuit-breakers.
core bar (*Foundry*). (1) An iron bar on which cylindrical loam cores are built up. The bar is supported horizontally and rotated while a loam board is pressed against the core. (2) An iron rod for reinforcing a sand core.
coreboard (*Join.*). Board composed of softwood strips bonded together and sandwiched between two outer layers of veneer whose grain runs across them. Also **battenboard, blockboard, laminboard.**
core box (*Foundry*). A wooden box shaped internally for moulding sand cores in the foundry.
CO_2 recorder (*Eng.*). An instrument which analyses automatically the flue gas leaving a furnace, and records the percentage of carbon dioxide (CO_2) on a chart. See **exhaust-gas analyser.**
cored carbon (*Light*). An arc-lamp carbon which has a core of softer material than that used for the outer part and is designed to have certain special effects on the arc.
cored electrode (*Elec. Eng.*). A metal electrode provided with a core of flux or other material; used in arc welding.
cored hole (*Eng.*). A hole formed in a casting by the use of a core, as distinct from a hole that has been drilled.
cored solder (*Eng.*). Hollow solder wire containing a flux paste, which allows flux and solder to be applied to the work simultaneously.
coreless armature (*Elec. Eng.*). An alternator armature having no iron core; used in certain old types of machine.
coreless induction furnace (*Elec. Eng.*). A high-frequency induction furnace in which there is no iron magnetic circuit other than the charge in the furnace itself.
core losses (*Elec. Eng.*). The losses occurring in electric machinery and equipment owing to hysteresis and eddy-current losses set up in the iron of the magnetic circuit, which are due to an alternating or varying flux.
coremiform (*Bot.*). Forming a tight bundle of elongated elements.
coremium (*Bot.*). (1) A ropelike strand of anastomosing hyphae. (2) A tightly packed

group of erect conidiophores, somewhat resembling a sheaf of corn.
core oven (*Foundry*). A foundry oven used for drying and baking cores before insertion in a mould.
core plate (*Elec. Eng.*). See **lamination.**
core plates (*Foundry*). Disks attached to a *core bar* (q.v.), to reinforce large cores.
core prints (*Foundry*). Projections attached to a pattern to provide recesses in the mould at points where cores are to be supported.
core register (*Foundry*). Corresponding flats or vees formed on cores and core prints, when correct angular location is necessary.
core-sampler (*Ocean.*). A weighted tube for obtaining stratified samples of sea-bed deposits.
core sand (*Foundry*). Moulding sand to which a binding material such as linseed oil has been added to obtain good cohesion and porosity after drying.
core-type induction furnace (*Elec. Eng.*). An induction furnace in which there is an iron core to carry the magnetic flux.
core-type transformer (*Elec. Eng.*). A transformer in which the windings surround the iron core, the former usually being cylindrical in shape.
corf (*Mining*). A basket with an iron loop used in early days for bringing coal to the surface.
coriaceous, corious (*Bot., Zool.*). Firm and tough, like leather in texture.
coring (*Build.*). Process of keeping a core of rags, straw, or shavings in a flue in course of construction, to catch falling mortar or bricks, and of finally passing it right through to clear obstructions. (*Elec. Eng.*) See **carbon core.**
Coriolis effects (*Phys.*). Those arising on the surface of the earth because of rotation, resulting in tangential and centrifugal acceleration, particularly in liquids and gyroscopes. Similarly with particles.
Coriolis force (*Meteor.*). The linear velocity of a point on the surface of a rotating sphere depends on its latitude. If a body is moved from one latitude to another it preserves its initial linear velocity, which is the source of Coriolis force. See **geostrophic force.**
corium (*Zool.*). The broad area of the hardened basal part of the hemelytron of *Hemiptera*, distant from the scutellum; the dermis of Vertebrates.
cork (*Bot.*). Layer of dead cells on the outside of a stem or root, having suberized walls; it is relatively impermeable to air and water. The cells are formed by a special cambium, the phellogen, and the cork protects the living cells inside against desiccation, mechanical injury, and the attacks of parasites.
cork cambium (*Bot.*). See **phellogen.**
cork crust (*Bot.*). A thick layer of corky cells, consisting mainly of large, soft-walled cells, with intermingled narrow strips of flattened cells.
corking (*Carp.*). See **cogging.**
corkscrew rule (*Elec. Eng.*). Rule relating the direction of the magnetic field to the current direction in a conductor. The corkscrew is driven in direction of current and sense of rotation of the handle gives direction of magnetic field.
corkscrew staircase (*Arch.*). A helical staircase built about a solid central newel.
corkscrew weave (*Weaving*). A twill coating, etc., weave on 7, 9, or 11 healds, giving a diagonal warp rib across the fabric at a low angle.
cork wood (*Bot.*). Wood of very low density containing many large, thin-walled parenchymatous cells.
Corliss valve (*Eng.*). A steam-engine admission and exhaust valve in the form of a ported cylinder which is given an oscillating rotary motion over the steam port by an eccentric-driven wrist-plate.
corm (*Bot.*). A rounded, swollen underground stem, with scale leaves, adventitious roots, and buds. (*Zool.*) In certain types of biramous crustacean limbs, an axis formed by the large endopodite with the protopodite, upon which the exopodite stands laterally.
cormidium (*Zool.*). In *Siphonophora*, an assemblage of individuals borne on the coenosarc.
cormophyte (*Bot.*). In former systems of classification, a plant of which the body is differentiated into roots, stems, and leaves.
cormus (*Bot.*). A plant body in which a definite shoot system is developed.
corn (*Bot.*). In Britain, wheat as well as other cereals; in U.S., maize. (*Med.*) Localized overgrowth of the horny layer of the skin due to local irritation, the overgrowth being accentuated at the centre. (*Vet.*) A local inflammation due to bruising or compression of the keratogenous membrane of the posterior portion of the horse's foot; **septic corn,** an abscess localized to the sole of a bird's foot.
Cornbrash (*Geol.*). A rather thin limestone occurring in the Jurassic rocks of Britain. It has an admixture of argillaceous and ferruginous material and yields a good soil for agriculture, as the name implies.
cornea (*Zool.*). In Invertebrates, a transparent area of the cuticle covering the eye, or each facet of the eye; in Vertebrates, the transparent part of the outer coat of the eyeball in front of the eye. *adj.* **corneal.**
corneagen layer (*Zool.*). In the compound eyes of *Arthropoda*, a layer of the hypodermis underlying the cornea, which it produces by secretion.
cornelian (*Min.*). A synonym for *carnelian.*
corneoscute (*Zool.*). One of the horny epidermal scales of Reptiles.
corneous (*Bot., Zool.*). Resembling horn in texture.
corner (*Bind.*). The piece of leather covering each of the outer corners of a half-bound volume. (*Telecomm.*) See **bend.**
corner bead (*Build.*). An *angle staff* (q.v.).
corner chisel (*Carp.*). A special chisel having two straight cutting edges meeting at right-angles; used for cutting the corners of mortises.
corner cramp (*Tools*). See **mitre cramp.**
corner horn (*Acous.*). A horn for coupling a closed or open diaphragm with the outer air, so arranged that its position in a corner of a room utilizes the side-walls as an approximate extension of its flare.
corner reflector (*Radar*). Metal structure of three mutually perpendicular sheets, for returning signals.
corner tool (*Foundry*). A sleeking tool for finishing off the internal corners of a mould.
cornice (*Arch.*). A projecting moulding decorating the top of a building, window, etc.
cornice plane (*Join.*). A plane with shaped sole and iron, used for forming mouldings.
cornicles (*Zool.*). In Aphids, a pair of tubes which secrete a waxy substance as a protection against predaceous enemies.
corniculate, cornute (*Bot., Zool.*). (1) Shaped like a horn. (2) Bearing a horn or hornlike outgrowth.
corniform (*Bot.*). Shaped like a horn.
Cornish boiler (*Eng.*). A horizontal boiler with a cylindrical shell provided with a single longitudinal furnace tube or flue.
Cornish pump (*Mining*). Early form of reciprocating mine pump, in which a string of rods

driven from surface transmitted power to considerable depth.

corn oil (*Chem.*). A pale yellow oil obtained from Indian corn; rel. d. 0·920–0·925, saponification value 188–193, iodine value 111–123, acid value 1·7–20·6. Used as a cooking oil. Also called **maize oil.**

cornstalk disease (*Vet.*). See **haemorrhagic septicaemia.**

cornstone (*Geol.*). An arenaceous or siliceous limestone, particularly characteristic of some of the Palaeozoic Red Sandstone formations.

cornua (*Zool.*). Hornlike processes; as the posterior *cornua* of the hyoid. *adj.* **cornual, cornute.**

cornual cartilage (*Zool.*). In *Cyclostomata*, a small cartilage at the lower end of the styloid process.

Cornu-Hartman formula (*Light*). A formula of empirical type which expresses with some accuracy the relation between the deviation, D, produced by a prism and the wavelength, λ, of the light. The formula is

$$\lambda = \lambda_0 + \frac{C}{D - D_0},$$

λ_0, C, and D_0 being constants for a particular case.

Cornu prism (*Light*). A 60° quartz prism formed of two 30° prisms cemented together, one being of right-handed and the other of left-handed quartz, the optic axes of the two being parallel to the ray passing through the prism at minimum deviation, that is, parallel to the base. This device overcomes a defect due to double refraction.

Cornu's spiral (*Maths.*). A spiral with parametric cartesian equations $x = \int_0^s \cos \tfrac{1}{2}\theta^2 d\theta$ and $y = \int_0^s \sin \tfrac{1}{2}\theta^2 d\theta$. It has the property that its radius of curvature $\rho = \frac{1}{s}$. Also called a **clothoid.**

corolla (*Bot.*). The general name for the whole of the petals of a flower; it is often brightly coloured, and then forms the most conspicuous part of the flower.

Corolliferae (*Bot.*). The group of *Monocotyledons* in which the perianth is merged into one whorl, e.g., lilies.

Corolliflorae (*Bot.*). See **Sympetalae.**

corolline (*Bot.*). Appertaining to the corolla.

corona (*Arch.*). The part of a cornice showing a broad projecting face and throated underneath to throw off the water. (*Bot.*) (1) A trumpetlike outgrowth from the perianth, as in the daffodil. (2) A ring of small leafy upgrowths from the petals, as in campion. (3) A crown of small cells on the oögonium of *Charophyta*. (*Elec.*) Phenomenon of air breakdown when electric stress at the surface of a conductor exceeds a certain value. At higher values, stress results in luminous discharge. See **critical voltage.** (*Meteor.*) A system of coloured rings seen round the sun or moon when viewed through very thin haze. They are caused by diffraction by water droplets. The diameter of the corona is inversely proportional to the size of the droplets. (*Zool.*) In *Echinoidea*, the shell or test; in *Crinoidea*, the disk and arms as opposed to the stalk; in *Rotifera*, the discoidal anterior end of the body; the head or upper surface of a structure or organ. *adj.* **coronal.**

coronagraph (*Astron.*). A type of telescope designed by Lyot in 1930 for observing and photographing the solar corona, prominences, etc., at any time.

coronal (*Zool.*). See **frontal.**

corona radiata (*Zool.*). A layer of cylindrical cells surrounding the developing ovum in Mammals.

coronary (*Zool.*). Crown-shaped; a small conical bone of the lower jaw in Reptiles.

coronary circulation (*Zool.*). In Vertebrates, the system of blood vessels (coronary arteries) which supply the muscle of the heart-wall with blood.

coronary sinus (*Anat.*). A channel opening into the right atrium draining blood from most of the cardiac veins.

coronary thrombosis (*Med.*). Formation of a clot in one of the coronary arteries leading to obstruction of the artery and *infarction* of the area of the heart supplied by it.

coronate (*Bot.*). Having a corona. (*Zool.*) Of shells, having a row of bosses encircling the apex.

corona veneris (*Med.*). A syphilitic rash on the forehead round the margin of the hair.

corona voltmeter (*Elec. Eng.*). Instrument for measuring high voltages by observing the conditions under which a corona discharge takes place on a specially designed wire.

coronene (*Chem.*). A yellow solid hydrocarbon, m.p. 430°C. Chief interest is its structure, being a large planar molecule with seven benzene rings.

coronet (*Zool.*). (1) The junction of the skin of the pastern with the horn of the hoof of a horse. (2) The knob at the base of the antler in deer.

coronium (*Astron.*). A hypothetical element once thought to be responsible for certain bright lines in the spectrum of the solar corona. It is now known that these are due to highly ionized iron, calcium, and nickel.

coronoid (*Zool.*). (1) In some Vertebrates, a membrane bone on the upper side of the lower jaw. (2) More generally, beak-shaped.

corpora allata (*Zool.*). In Insects, endocrine organs behind the brain which secrete *neotenin*, the juvenile hormone. In some species they are paired and laterally placed, but in others they fuse during development to form a single median structure, the *corpus allatum*.

corpora bigemina (*Zool.*). In Vertebrates, the optic lobes of the brain.

corpora cardiaca (*Zool.*). In Insects, paired neurohaemal organs lying behind the brain, and containing the nerve endings of the neurosecretory cells in the brain which produce the brain hormone involved in moulting. This hormone is released into the blood at the *corpora cardiaca*.

corpora cavernosa (*Zool.*). In Mammals, a pair of masses of erectile tissue in the penis.

corpora geniculata (*Zool.*). In the Vertebrate brain, paired protuberances lying below and behind the thalamus.

corporal (*Mining*). The leading man in a batch of men on a contract for mining coal.

corpora lutea (*Histol., Zool.*). See **corpus luteum.**

corpora pedunculata (*Zool.*). In Insects, the *mushroom* or *stalked bodies*, which are the most conspicuous formations in the protocerebral lobes of the brain.

corpora quadrigemina (*Zool.*). The optic lobes of the Mammalian brain, which are transversely divided.

corpus adiposum (*Zool.*). See **fat-body.**

corpus albicans (*Zool.*). See **corpus mamillare.**

corpus callosum (*Zool.*). In the brain of placental Mammals, a commissure connecting the cortical layers of the two lobes of the cerebrum.

corpuscle (*Zool.*). A cell which lies freely in a fluid or solid matrix and is not in continuous contact with other cells.

corpuscular radiation (*Phys.*). A stream of atomic or subatomic particles, which may be charged positively, e.g., α-particles, negatively, e.g., β-particles, or not at all, e.g., neutrons.

corpuscular theory of light (*Phys.*). The view, held by Newton, that the emission of light consisted of the emission of material particles at very high velocity. Although this theory was discredited by observations of interference and diffraction phenomena, which could only be explained on the wave theory, there has been, to some extent, a return to the corpuscular idea in the conception of the photon.

corpus luteum (*Histol, Zool.*). The endocrine structure developed in the ovary from a Graafian follicle after extrusion of the ovum, secreting progesterone; the yellow body.

corpus mamillare (*Zool.*). In the brains of higher Vertebrates, a protuberance on the floor of the hypothalamic region in which the fornix terminates.

corpus spongiosum (*Zool.*). In Mammals, one of the masses of erectile tissue composing the penis.

corpus sterni (*Zool.*). See **gladiolus.**

corpus striatum (*Zool.*). In the Vertebrate brain, the basal ganglionic part of the wall of each cerebral hemisphere.

corpus subthalamicum (*Zool.*). A lens-shaped mass of grey matter, lying in the subthalamus of the Vertebrate brain.

corpus tentorii (*Zool.*). In Insects, the main part of the endoskeleton of the head. See **tentorium.**

corpus trapezoideum (*Zool.*). A short tract of transverse fibres behind the *Vieussens' valve* (q.v.), in the mammalian brain.

corrasion (*Geol.*). This is the work of vertical or lateral cutting performed by a river by virtue of the abrasive power of its load. See **rivers, geological work of.**

correcting behaviour (*An. Behav.*). The tendency for some Invertebrates (including bugs, woodlice and mealworms) to turn in the opposite direction from a prior forced turn. See also **centrifugal swing.**

correction. The correction to a reading of an instrument, or of a measure on a scale, is the magnitude to be added to the perceived magnitude to obtain the true magnitude, i.e., the magnitude which would be observed with an instrument or scale of the highest precision. The use of a *standard* (q.v.) of measurement implies no possibility of correction. Correction is the negative of error. (*Phys.*) In precision weighing, it is necessary to correct for the difference in the buoyancy of the air for the body being weighed and the weights. The correction to be added to the value, w, of the weights (in grams) is:

$$1{\cdot}2w\left(\frac{1}{D}-\frac{1}{\delta}\right)\ \text{milligrams,}$$

where D and δ are the densities of the body and of the weights respectively in g/cm³.

correction of angles (*Surv.*). The process of adjusting the observed angles in any triangle so that their sum shall equal 180°.

correlation (*Biol.*). Mutual relationship. (*Bot.*) The condition of balance existing between the growth and development of various organs of a plant. (*Geol.*) The linking together of strata of the same age occurring in separate outcrops. (*Maths.*) A linear transformation which, in the plane, maps lines into points and points into lines and in space maps points into planes and planes into points. (*Stats.*) The study of the simultaneous variation (*covariation*) of two or more variable quantities. If one variate increases as the other increases and decreases as the other decreases, there is said to be *positive correlation.* If, however, one variate decreases as the other increases and vice versa, there is said to be *negative correlation.*

correlation coefficient (*Stats.*). A measure of the correlation between two variates x and y. If the variates take the values $x_1, x_2 \ldots x_n$, $y_1, y_2 \ldots y_n$, with means $\bar{x}$ and $\bar{y}$ respectively, the coefficient of correlation is given by:

$$r=\frac{\sum_{i=1}^{n}(x_i-x)(y_i-y)}{\sqrt{\sum_{i=1}^{n}(x_i-\bar{x})^2\sum_{i=1}^{n}(y_i-\bar{y})^2}}.$$

If $r=0$, the two variates are completely independent, and if $r=1$, there is complete correlation and one variate may be calculated from the other. Also called **Pearson's coefficient.**

correspondence principle (*Phys.*). The principle that, in the limit of quantum numbers, the predictions of quantum and classical physics always correspond.

corresponding angles (*Maths.*). For a diagram consisting of two straight lines cut by a transversal, corresponding angles are angles on the same side of the transversal, each angle being above its appropriate line or below it. If the two straight lines are parallel then the corresponding angles are equal.

corresponding states (*Phys.*). Substances are said to be in *corresponding states* when their pressures and temperatures are equal fractions of the critical values. A general form of *van der Waals' equation* may then be used which is applicable to all gases.

corridor disease (*Vet.*). A fatal disease of cattle in Africa due to infection by the protozoon *Theileria lawrencei*; transmitted by ticks.

corrie (*Geol.*). See **cirque.**

corrin (*Chem.*). See **vitamin B complex.**

corrosion (*Chem.*). The slow wearing away of solids, especially metals, by chemical attack; in the latter case the mechanism is thought to be electrochemical. (*Geol.*) The modification of crystals formed early in the solidification of an igneous rock by the chemical action of the residual magma.

corrosion-fatigue (*Met.*). Acceleration of weakening of structure exposed to pulsed stress by chemical penetration and attack on metal components.

corrosion voltmeter (*Elec. Eng.*). Instrument which locates and estimates corrosion of materials by measuring e.m.f. arising from electrochemical action between material and corrosive agent.

corrosive sublimate (*Chem.*). *Mercuric chloride.*

corrugate, corrugated (*Bot.*). Having a ridged or wrinkled surface.

corrugated board (*Paper*). Protective wrapping material, its fluted ridges giving resilience.

corrugated iron (*Build.*). Sheet-iron, usually *galvanized*, with corrugations for stiffening.

corrugator (*Zool.*). A muscle which by its contraction produces wrinkling.

corrugmeter (*Civ. Eng.*). See **roughness integrator.**

cortex (*Bot.*). (1) A cylinder of chiefly parenchymatous cells lying between the epidermis and the starch sheath (or the endodermis) in a young stem, and between the piliferous layer and the endodermis in a young root. (2) A similar, but usually less conspicuous, cylinder in older stems and roots. (3) A cellular coating on the outside of the thalli of some algae. (4) The outer layers of the thallus in lichens, and in some fungi. (*Zool.*) The superficial or outer layers of an organ; cf. *medulla*.

cortical (*Bot., Zool.*). (1) Relating to bark. (2) Relating to the cortex. (3) Living on bark.

cortical bundle (*Bot.*). A vascular bundle in the cortex of a stem or root.

corticate (*Bot.*). (1) Having a cortex. (2) Covered with an unbroken sheet of interwoven hyphae.

corticating branch (or filament) (*Bot.*). Found in some of the *Charales*; the branch arises from an initial at the internode growing to cover half the node above and below with the filament. The filaments then constitute the cortex.

cortication (*Bot.*). A covering of cells around the main threads of some algae.

corticolous (*Bot.*). Living on the surface of bark.

corticosterone (*Biochem., Med.*). A hormone of the adrenal cortex, a steroid with the structure:

Other active substances from the cortex include *dehydrocorticosterone* in which the OH attached to the rings is replaced by oxygen, *desoxycorticosterone* in which this OH is replaced by hydrogen (and which can be obtained by partial synthesis and is used therapeutically in Addison's disease, etc.), *hydroxycorticosterone* with an extra hydroxyl group, etc. These substances all prolong the life of adrenalectomized animals, controlling the sodium and water metabolism and being concerned in glycogen formation, etc.

corticotrophic, corticotropic (*Physiol.*). Having a stimulatory influence on the adrenal cortex; adrenocorticotrophic. See **ACTH.**

corticotrophin (*Biochem., Med.*). See **ACTH.**

cortina (*Bot.*). A cobwebby veil hanging from the margin of the pileus of some agarics.

cortinate (*Bot.*). Having a cortina. (*Zool.*) See **craspedote.**

cortisone (*Biochem., Med.*). 17-hydroxy-11-dehydro-corticosterone, a crystalline hormone isolated from the adrenal cortex, with structure:

It is used widely in the treatment of rheumatoid arthritis, and in some inflammatory conditions.

Corti's organ (*Zool.*). In Mammals, the modified epithelium forming the auditory apparatus of the ear, in which nerve fibres terminate.

Corti's rods (*Zool.*). In Mammals, stiff striated rodlike structures of the organ of Corti, forming a double row, with their upper ends in contact and their lower ends resting on the basilar membrane.

Corti's tunnel (*Zool.*). In the organ of Corti, the tubelike space enclosed by the basilar membrane and the rods of Corti.

corundum (*Min.*). Oxide of aluminium, crystallizing in the trigonal system. It is next to diamond in hardness, and hence is used as an abrasive. See also **white sapphire.**

corve (*Mining*). A small tram for carrying minerals underground.

corydaline (*Chem.*). $C_{22}H_{27}O_4N$, an alkaloid of the isoquinoline group, obtained from the root of the *Corydalis* species. It crystallizes in 6-sided prisms: m.p. 135°C.

corymb (*Bot.*). A racemose inflorescence in which the flower stalks become shorter and shorter as they arise closer to the top of the inflorescence axis. As a result, the flowers lie in a flat-topped cluster. *adj.* corymbose.

Corynebacteriaceae (*Bacteriol.*). A family of bacteria belonging to the order *Eubacteriales*. Gram-positive rods; some pleomorphic species, mainly aerobes, occur in dairy products and the soil. Some pathogenic species, e.g., *Erysipelothrix rhusiopathiae* (swine erysipelas), *Corynebacterium diphtheriae* (diphtheria).

coryza (*Med.*). See **cold.** (*Vet.*) See **infectious coryza, malignant catarrhal fever.**

cos, cosine (*Maths.*). See **trigonometrical functions.**

cos φ (*Elec. Eng.*). An expression often used to denote the power factor of a circuit, the power factor being equal to the cosine of the angle (φ) of the phase difference between the current and voltage in the circuit.

cosec, cosecant (*Maths.*). See **trigonometrical functions.**

cosecant antenna (*Radio*). Radiator comprising a surface so shaped that the radiation amplitude pattern is described by a cosecant curve over a wide angle; gives approximately the same intensity for near and far objects. Method also applied to electron beams.

cosh (*Maths.*). See **hyperbolic functions.**

cosine law (*Light*). See **Lambert's cosine law.**

cosine potentiometer (*Elec. Eng.*). Voltage divider in which the output of an applied direct voltage is proportional to the cosine of the angular displacement of a shaft.

Coslettizing (*Met.*). The protection of steel against corrosion by boiling in a solution of phosphoric acid to produce a surface coating of phosphate.

cosmic abundance (*Astron.*). Quantity of an isotope (estimated) in the universe, relative to hydrogen or as a percentage of the whole.

cosmic noise (*Radio*). Interference due to extra-terrestial phenomena, e.g., sun spots.

cosmic rays (*Astron.*). Highly penetrating rays from outer space. The *primary cosmic rays* which enter the earth's upper atmosphere consist mainly of protons with smaller amounts of helium and other heavier nuclei. Cosmic rays of low energy have their origin in the sun, those of high energy in galactic or extragalactic space, possibly as a result of supernovae explosions. Collision with atmospheric particles results in *secondary cosmic rays* and particles of many

kinds, including neutrons, mesons, and hyperons.
cosmic-ray shower (*Astron.*). The simultaneous appearance of a number of downward directed light-ionizing particles as detected by a cloud chamber.
cosmine (*Zool.*). The dentinelike substance forming the outer layer of the cosmoid scales of *Choanichthyes*.
cosmogenic (*Nuc.*). Said of an isotope capable of being produced by the interaction of cosmic radiation with the atmosphere or the surface of the earth.
cosmogony (*Astron.*). The science of the origins of stars, planets, and satellites. It deals with the genesis of the Galaxy and the solar system.
cosmography. A description of the world; the science of the constitution of the universe.
cosmoid scale (*Zool.*). In *Choanichthyes*, the characteristic type of scale consisting of an outer layer of cosmine, coated externally with vitrodentine, a middle bony vascular layer and an inner isopedine layer.
cosmology (*Astron.*). The branch of theoretical astronomy that deals with the known universe as a systematized whole. It comprises our knowledge of the structure, dimensions, and relative connexions of galaxies, star-clusters, nebulae, etc.
cosmotron (*Phys.*). Large proton-synchrotron, using frequency modulation of an electric field; it accelerates the protons to energies greater than 1 GeV.
cossyrite (*Min.*). See **aenigmatite.**
costa (*Bot.*). (1) A general term for a rib or vein. (2) The midrib of the thallus of a liverwort. (3) A rib on the valve of a diatom. (*Zool.*) In Vertebrates, a rib; in *Plumulariidae*, one of a number of protective branches which form a basket-work enclosing the gonangia; in Insects, one of the primary veins of the wing; in *Ctenophora*, one of the meridional rows of ctenes: more generally, any riblike structure. *adj.* **costal, costate.**
costalgia (*Med.*). Pain in the ribs.
costalia (*Zool.*). In *Chelonia*, bony plates of the carapace representing modified ribs; in *Crinoidea*, the primary brachial series of ossicles.
costate (*Bot.*). Ribbed or veined, especially when the ribs are parallel.
costeaning (*Mining*). Prospecting by shallow pits or trenches designed to expose lode outcrop.
costopulmonary muscles (*Zool.*). In Birds, small fanlike muscles arising from the junction of the vertebral and sternal ribs and inserted into the pleura.
cot, cotangent (*Maths.*). See **trigonometrical functions.**
cot bar (*Join.*). A semicircular bar in a sash. Also called **cradle bar.**
cotchell (*For.*). A parcel, or small quantity of timber, sold outside the usual trade channels.
coterminal angles (*Maths.*). Two angles having the same vertex and the same initial line and whose terminal lines are coincident, e.g., 60° and 420°.
coterminous (*Zool.*). Of similar distribution.
coth (*Maths.*). See **hyperbolic functions.**
Cotham Beds (*Geol.*). A series of greenish-yellow marls and limestones occurring in the Rhaetic rocks of the south-west of England.
Cotham Marble (*Geol.*). A member of the Rhaetic rocks of England. It is an impure limestone characterized by arborescent or mosslike markings; a type of landscape marble.
Cotswold Sands (*Geol.*). A local subdivision of the Upper Lias, consisting of yellowish sands occurring beneath the oölites of the Cotswold Hills. See also **Yeovil Sands.**
cotter (*Eng.*). A tapered wedge, usually of rectangular section, passing through a slot in one member and bearing against the end of a second encircling member whose axial position is to be fixed or adjustable.
cotter pin (*Eng.*). A split-pin inserted in a hole in a cotter or other part, to prevent loosening under vibration.
cotter way (*Eng.*). The slot cut in a rod to receive a *cotter* (q.v.).
cotton (*Textiles, etc.*). The downy fibre covering the seeds of the cotton plant (genus *Gossypium*, family *Malvaceae*); it is separated from the seed for textile purposes. The length of fibre ranges from 1-4 cm, 2½ cm being about the average. Besides textile purposes, i.e., yarns and cloths, the fibre has almost limitless uses including surgical, upholstery, bonded fabrics, etc.
Cotton balance (*Elec. Eng.*). An instrument for measuring the intensity of a magnetic field by finding the vertical force on a current-carrying wire placed at right angles to the field.
cotton ball (*Min.*). See **ulexite.**
cotton bleaching (*Textiles*). The processes to which cotton, either in the loose state, as yarn, or fabric, is subjected to render it white. The processes include (1) steeping in a 1% hot solution of caustic soda or sodium carbonate; (2) rinsing; (3) steeping in a cold solution of a hypochlorite; (4) washing; (5) treating with cold dilute solution of a mineral acid, and (6) finally washing until free from acid. There are several bleaching agents, but cotton cloth is always singed before being bleached.
cotton-covered wire (*Elec. Eng.*). An electrical conductor of copper or resistive alloy, insulated with one or more layers of cotton yarn; generally superseded by plastics-insulated wire.
cotton gin (*Textiles*). A machine for separating cotton fibres from the seed. The saw gin and the Macarthy gin are used in America, the latter for long staple cotton. The treated cotton is baled and the seed is pressed for its oil before being made into cattle food and fertilizer.
cotton linters (*Textiles*). See **linters.**
Cotton-Mouton effect (*Elec.*). Effect occurring when a dielectric becomes double-refracting on being placed in a magnetic field H. The retardation δ of the ordinary over the extraordinary ray in traversing a distance l in the dielectric is given by $\delta = C_m \lambda l H^2$, where λ is the light wavelength and C_m is the Cotton-Mouton constant.
cottonseed oil (*Chem.*). Oil from the seeds of *Gossypium herbaceum*, a yellow, brown, or dark-red liquid, m.p. 34°–40°C, rel. d. 0·922–0·930, saponification value 191–196, iodine value 105–114, acid value 0. Used in manufacture of soaps, fats, margarine, etc., and medicinally as a mild laxative.
cotton spinning (*Textiles*). The process whereby cotton fibre, which by preceding processes has been formed into a roving, is simultaneously drafted (drawn out) and twisted into a yarn on a mule or ring frame.
cottonwood (*For.*). A North American species of *poplar*. Its considerable girth provides very wide boards. It is rated as moderately durable, and is rather difficult to impregnate with wood preservatives. It carves and turns well.
cotton wool (*Med.*). Loose cotton which has been bleached and pressed into a sheet; used as an absorbent or as a protective agent. Medicated cotton wool sometimes has a distinguishing colour to indicate its special property.

cotton-wool patches (*Med.*). Areas of white exudate in the retina occurring in nephritis.
cottony (*Bot.*). See **tomentose.**
Cottrell effect (*Phys.*). Gradual loss of elasticity of uranium under conditions of intense radiation.
Cottrell precipitator (*Met.*). System used to remove dust from process gases electrostatically.
cotty wool (*Textiles*). Fleece or skin wool in extremely matted condition; usually due to sheep disease.
cotyledon (*Bot.*). (1) One of the leaves of the embryo in flowering plants. (2) The first leaf developed by a young fern plant.
cotyledonary placentation (*Zool.*). Having the villi in patches, as Ruminants.
cotyliform (*Bot.*). Shaped like a dish or a wheel, and having a distinct upraised rim.
cotyloid (*Zool.*). Cup-shaped; pertaining to the acetabular cavity. In Mammals, a small bone bounding part of the acetabular cavity.
cotype (*Zool.*). An additional type specimen, being a brother or sister of the same brood as the type specimen.
couch (*Paper*). In papermaking, to deposit the web sheet on a felt for pressing and drying.
couching (*Brew.*). The process of spreading the steeped grain (barley) to a depth of 30-35 cm on the cement floor of the malthouse, where germination takes place. See **steeping, flooring.** (*Surg.*) Displacement of the lens in the treatment of cataract.
couch roll (*Paper*). A cylinder covered with felt, used to press out water from the damp web of paper, and to cause the fibres to felt more thoroughly. See also **suction box.**
coudé telescope (*Astron.*). An arrangement by which the image in an equatorial telescope is formed, after an extra reflection, at a point on the polar axis. It is then viewed by a fixed eyepiece looking either down or up the polar axis. This type of mounting is much used for high dispersion spectroscopy with modern large telescopes. Also called **coudé mounting.**
coulisse (*Carp.*). A grooved piece of timber in which another member usually slides. Also called **cullis.**
coulomb (*Elec.*). SI unit of electric charge, realized by 1 ampere flowing through a conductor for 1 second. Symbol Q, abbrev. C.
coulomb energy (*Elec.*). Fraction of binding energy arising from simple electrostatic forces between electrons and ions.
coulomb force (*Elec.*). Electrostatic attraction or repulsion between two charged particles.
coulomb potential (*Nuc.*). One calculated from Coulomb's inverse square law and from known values of electric charge. The term is used particularly in nuclear physics to indicate that component of the potential energy of a particle which varies with position in accordance with an inverse square law, and should be contrasted with, e.g., *Yukawa potential.*
coulomb scattering (*Elec.*). Scattering of particles by action of coulomb force.
Coulomb's law (*Elec.*). Fundamental law which states that the force of attraction or repulsion between two point charges is proportional to the product of their charges and inversely proportional to the square of the distance times the permittivity of the medium between them. If Q_1, Q_2 are point charges at distance d apart, force is:

$$F \propto \frac{Q_1 Q_2}{\varepsilon d^2},$$

where ε = relative permittivity of medium.
Coulomb's law for magnetism (*Mag.*). The force between two isolated point magnetic poles (theoretical abstractions) would be proportional to the product of their strengths and inversely proportional to the square of their distance apart times the permeability of the medium between them. $F \propto \frac{M_1 M_2}{\mu d^2}$, where M_1 and M_2 are the strengths of the two poles, d is their distance apart and μ is the relative permeability of the medium.
coulometer (*Elec. Eng.*). Voltameter or electrolytic cell, designed, e.g., for use in measurement of the quantity of electricity passed.
coulter (*Agric.*). A knife, or a steel disk, which, attached to a plough, makes the vertical cut in the ground to form the furrow slice. See **share.**
Coulter counter (*Powder Tech.*). A proprietary instrument for particle size analysis. A dilute suspension of particles is drawn through an orifice mounted between 2 electrodes and resistance changes caused by the presence of particles are measured and counted electronically.
coumachlor (*Chem.*). 3-(α-acetonyl-4-chlorobenzyl)-4-hydroxy-coumarin, an antiblood-coagulant type of rodenticide.
coumafuryl (*Chem.*). See **fumarin.**
coumalic acid (*Chem.*). Pyrone-5-carboxylic acid, formed by the action of concentrated sulphuric acid on malic acid; m.p. 206°. Pyrone is *coumalin.*
coumaphos (*Chem.*). 3-Chloro-4-methyl-7-coumarinyl diethyl phosphorothionate, used as an insecticide.
coumaric acids (*Chem.*). $HO{\cdot}C_6H_4{\cdot}CH{=}CH{\cdot}COOH$, hydroxy-cinnamic acids.
coumarin (*Chem.*). Odoriferous principle of tonquin beans and woodruff, $C_9H_6O_2$, b.p. 200°C; used for scenting tobacco.
coumarone (*Chem.*). The condensation product of a benzene nucleus with a furan ring. It is a very stable, inert compound, b.p. 169°C; found in coal-tar. Strong acids effect polymerization into para-coumarone and coumarone resins.
coumarone resins (*Chem.*). Condensation and polymerization products obtained from *coumarone* (q.v.); used for varnishes, in printing ink, and as plasticizers for moulding powders. They are neutral and acid- and alkali-resisting.
count (*Nuc. Eng.*). Summation of photons or ionized particles with a counting tube, which passes pulses to counting circuits, including dividers. (*Textiles*) See **count of yarn.**
countable set (*Maths.*). See **denumerable set.**
counter (*Automation*). In industry, a device for counting separate items or events. (*Elec. Eng.*) The part of an integrating electricity meter which indicates the number of revolutions made by the spindle of the meter, this indication being proportional to the amount of energy which has passed through the circuit. Also occasionally called **dial** or **clock.** (*Eng.*) An instrument for recording the number of operations performed by a machine, or the revolutions of a shaft. (*Nuc. Eng.*) (1) Equipment for counting individual events which can be detected in, e.g., a G-M counting tube or a boron chamber for neutrons. (2) Loosely, a complete counting assembly, e.g., counter, scaler and register. (*Ships*) A description applied to a form of ship's stern, implying an overhung portion of deck, abaft the stern post; hence the term 'under the counter'. (*Typog.*) The space between the strokes of a letter; when totally enclosed, called a **bowl.**

counter-arched (*Civ. Eng.*). Said of a revetment having arches turned between counterforts.

counterboring (*Eng.*). The operation of boring the end of a hole to a larger diameter.

counterbracing (*Eng.*). The provision of two diagonal tie-rods in the panels of a frame girder or other structure. Also called **cross-bracing.**

countercurrent contact (*Chem. Eng.*). In processes involving the transfer of heat or mass between two streams *A* and *B*, as in liquid extraction, the arrangement of flow so that at all stages the more spent *A* contacts the less spent *B*, thus ensuring a more even distribution and greater economy than with *co-current contact* (q.v.).

countercurrent distribution (*Chem.*). A repetitive distribution of a mixture of solutes between two immiscible solvents in a series of vessels in which the two solvent phases are in contact. The components are distributed in the vessels according to their partition coefficients.

countercurrent treatment (*Min. Proc.*). Arrangement used in chemical extraction of values from ore, in washing rich liquor away from spent sands, and in many other fields. The stripping liquid enters 'barren' at one end of a typical layout and the rich ore pulp at the other. They pass countercurrent through a series of vessels, the pulp emerging stripped after its final wash with the new 'barren' liquor and the liquor leaving at the far end, now rich with dissolved values and 'pregnant'.

counter efficiency (*Nuc. Eng.*). Ratio of counts recorded by counter to number of incident particles or photons reaching detector. Counts may be lost due to (*a*) absorption in window, (*b*) passage through detector without initiating ionization, (*c*) passage through detector during dead time following previous count.

counter e.m.f. (*Elec. Eng.*). See **back e.m.f.**

counter-flap hinge (*Join.*). A hinge which is arranged, by the provision of separate centres of rotation for each leaf, so that it may fold back to back.

counter-floor (*Carp., Join.*). An inferior floor laid as a base for a better surface (e.g., parquet).

counterflow jet condenser (*Eng.*). A *jet condenser* (q.v.) in which the exhaust steam and air flow upwards to the airpump suction in the opposite direction to that of the descending spray of cooling water.

counterfort (*Civ. Eng.*). A buttress giving lateral support to a retaining wall, to which it is bonded.

counter gauge (*Carp.*). See **mortise gauge.**

counterions (*Chem.*). See **gegenions.**

counter-irritation (*Med.*). Therapeutic irritation for the relief of pain due to inflammation.

counter lathing (*Build.*). See **brandering.**

counter life (*Nuc. Eng.*). The total number of counts a nuclear counter can be expected to make without serious deterioration of efficiency. (Not the *expected operating period.*)

counter-mure (*Build.*). A wall-facing.

counterpoise (*Radio*). A network of conductors placed a short distance above the surface of the ground but insulated from it, and used for the earth connection of an antenna. It will have a large capacity to earth and serves to reduce greatly the earth current losses that would otherwise take place. Also known as a **capacity earth, artificial earth** or **counterpoise antenna.**

counterpoise bridge (*Civ. Eng.*). A bridge, such as a bascule or lift bridge, in which the raising of the platform is assisted by counterpoise weights.

counter range (*Nuc. Eng.*). See **start-up procedure.**

counter recovery time (*Telecomm.*). The minimum time between the start of a recorded pulse and a subsequent one which attains a specified percentage of the amplitude of the first.

countershading (*Zool.*). A type of protective coloration in which animals are darker on their dorsal surface than on their ventral surface, thus ensuring that illumination from above renders them evenly coloured and inconspicuous.

countershaft (*Eng.*). An intermediate shaft interposed between driving and driven shafts in a belt drive, either to obtain a larger speed ratio or where direct connection is impossible.

countersinking (*Eng.*). The provision of a conical enlargement at the end of a hole to receive the head of a screw or rivet. See **counterboring.** (*Build., etc.*) The driving of the head of a screw or nail below the surface so that it may be hidden by a plug.

counter-stern (*Ships*). A type of ship's stern construction. It is virtually an excrescence to the main hull, and is not waterborne.

countersunk head (*Eng.*). A screw or rivet head with a conical base, allowing it to enter the countersunk workpiece so that the top surface of the head is substantially flush with the workpiece surface.

counter-transference (*Psychol.*). State of emotional reaction, partly unconscious, in which there are positive (love) or negative (hate) feelings in the analyst towards the analysand.

counter tube (*Electronics*). Loosely applied to individual scaling tube (e.g., dekatron) in complete counting assembly. (*Nuc. Eng.*) One for detecting ionizing radiation by electric discharge resulting from Townsend avalanche and operating in proportional or Geiger region.

counter-vault (*Civ. Eng.*). An inverted arch.

counterwedging (*Join.*). A method of bringing closely together the butting edges of thin surfaces, such as counter tops, by tightening up parallel wedges between a slot in a batten across the back of one of the surfaces and corresponding slots in two battens across the back of the other surface.

countess (*Build.*). A roofing slate, 20 × 10 in. (508 × 254 mm).

counting glass (*Textiles*). Short-focus, folding magnifying-glass in which the base forms a ¼, ½, or 1 in. square; used when counting the number of threads per given space in fabric, or ascertaining the weave particulars.

counting machine (*Paper*). This counts a pile of paper into reams, inserting a ticket at each ream.

count of yarn (*Textiles*). A number which designates the size of a yarn. Usually a count represents the number of units of length contained in a unit of weight, but in certain classes of yarn the count represents the number of units of weight in a unit of length. The units of length and weight vary in districts. *Woollen*: The *Yorkshire skein* = 256 yd; unit of weight 1 lb. The *American run* = 100 yd; unit of weight 1 oz. *Worsteds*: The *British* and *American hank* = 560 yd; unit of weight 1 lb. *Cotton*: The *cotton hank* = 840 yd; unit of weight 1 lb. The count is the number of hanks of 840 yd which weigh 1 lb. *Linen*: The count is the number of leas of 300 yd which weigh 1 lb. *Jute Yarns*: The count is the number of lb in a spindle of 14,400 yd, the unit of length. Largely superseded by metric units. See **denier system, Tex.**

count ratemeter (*Nuc. Eng.*). One which gives a continuous indication of the rate of count of ionizing radiation, e.g., for radiac survey.

country rock (*Mining*). The valueless rock forming the walls of a reef or lode.

county road (*Civ. Eng.*). Main road maintained by county authorities. Cf. *trunk road*.

coupe (*For.*). A felling area, usually one of an annual succession unless otherwise stated.

couple (*Maths.*). A system of two equal but oppositely directed parallel forces. The perpendicular distance between the two forces is called its *arm* and any line perpendicular to the plane of the two forces its *axis*. The *moment* of a couple is the product of the magnitude of one of its two forces and its arm. A couple can be regarded as a single statical element (analogous to force); it is then uniquely specified by a vector along its axis having a magnitude equal to its moment. Couples so specified combine in accordance with the parallelogram law of addition of vectors.

couple-close roof (*Arch.*). A roof-form derived from the couple roof by connecting the lower ends of the two rafters together with a tie, so as to prevent spreading of the roof under load.

coupled flutter (*Aero.*). See **flutter**.

coupled oscillator (*Electronics*). Circuit in which positive feedback from output circuit to input circuit of an amplifier by mutual inductance is sufficient to initiate or maintain oscillation.

coupled rangefinder (*Photog.*). A rangefinder co-ordinated with the focusing mechanism of a camera lens.

coupled switches (*Elec. Eng.*). See **linked switches.**

coupled wheels (*Eng.*). The wheels of a locomotive which are connected by coupling rods to distribute the driving effort over more than one pair of wheels.

coupler (*Acous.*). A paralleling arrangement in an organ console for playing stops on one manual from the keys on another manual, or from pedals. (*Elec. Eng.*) A short length of tubing for making connexions between adjacent lengths of conduit in an electric wiring system. The term is also used to denote various devices for connecting electric currents. See **bus-coupler switch, bus-wire coupler, plain coupler.**

couple roof (*Build.*). A roof composed of two rafters not braced together.

coupler plug (*Elec. Eng.*). A plug on a jumper cable, such as that used for making connexion between the two coaches of an electric multiple-unit train.

couplers (*Photog.*). Substances which, when added to developers of the *para*-phenylene-diamine or *para*-amino-phenol groups, produce a dye image in addition to the normal silver one. Used in colour photography.

coupler socket (*Elec. Eng.*). A socket for receiving a coupler plug.

couplet (*Gen.*). A pair of alleles.

coupling. Interaction between different systems or different components of a system. (*Eng.*) (1) A device for connecting two lengths of hose, etc. (2) A device for connecting two vehicles. (3) A connexion between two co-axial shafts, conveying a drive from one to the other. (*Gen.*) The tendency for dominant characters to remain in association. (*Plumb.*) A short collar screwed internally at each end to receive the ends of two pipes which are to be joined together.

coupling capacitor (*Telecomm.*). Any capacitor for coupling two circuits, particularly that for coupling antenna to a transmitter or receiver.

coupling coefficient or factor (*Telecomm.*). Ratio of total effective positive (or negative) impedance common to two resonant circuits to geometric mean of total positive (or negative) reactances of two separate circuits.

coupling coil (*Telecomm.*). One whose inductance is a small fraction of the total for circuit of which it forms a part; used for inductive transfer of energy to or from the circuit.

coupling element (*Telecomm.*). The component through which energy is transferred in a coupled system.

coupling factor (*Telecomm.*). See **coupling coefficient.**

coupling loop (*Electronics, etc.*). A loop placed in a waveguide at a position of maximum magnetic field strength in order to extract energy.

coupling probe (*Electronics, etc.*). A probe placed in a waveguide at a position of maximum electric field strength in order to extract energy.

coupling resistance (*Telecomm.*). Common resistance between two circuits for transference of energy from one circuit to the other.

couplings (*Textiles*). The loops which connect the mails to the harness cords and lingoes of a jacquard harness.

coupling spin-orbit (*Electronics*). Interaction between the spin and the orbital angular momentum of an electron.

coupling transformer (*Elec. Eng.*). A transformer used as a coupling element.

courbaril (*For.*). A tall tree (*Hymenaea*) common to the American tropics. It yields *South American copal gum*, and a hardwood timber which is not easy to work. Typical uses are in furniture, cabinet-making, and ship-building.

Courlene (*Chem.*). TN for polythene filaments and yarns, used industrially for protective clothing, filter media, ropes, cords, etc., for making footwear, and in vascular surgery.

Courlose (*Chem.*). TN for sodium carboxy methylcellulose.

course (*Build.*). A horizontal layer of bricks or building-stones running throughout the length and breadth of a wall. See also **cushion course.** (*Ships, etc.*) The angle between some datum line and the direction of the ship's head. (*Surv.*) The known length and bearing of a survey line. (*Weaving*) (1) A series of heald eyes, one on each shaft. (2) A series of mails, one row from back to front in a jacquard harness. (3) One repeat of a pattern.

course and speed error (*Nav.*). The angle that the spin axis of a *gyro compass* (q.v.) makes with the *true meridian* (q.v.) due to the compass being carried over the surface of the earth in the ship.

course correction (*Space*). The firing or burning of a rocket motor, during a ballistic journey, in a controlled direction and for a controlled duration to correct an error in course.

coursed masonry (*Civ. Eng.*). Work consisting of stone blocks laid on their beds in courses; e.g., in some breakwater construction, or on the face of concrete or earth dams.

course line (*Nav.*). Horizontal projection of intended path.

course-line deviation (*Nav.*). The angular difference between track and course.

coursing joint (*Build.*). The mortar joint between adjacent courses of brick or stone.

Courtelle (*Chem.*). TN for synthetic fibres based on polycyanoethene.

courtship behaviour (*An. Behav.*). Behaviour patterns and sequences shown by male and female animals as a preliminary to copulation.

coutil (*Textiles*). A 6-shaft, dobby woven, cotton fabric, usually a 2×1 twill with a 'herring bone' effect. Usually made from warp and weft dyed yarns, or piece-dyed, it is largely used for corsets.

covalency (*Chem.*). The union of 2 atoms by the sharing of a pair of electrons.

covalent bond (*Chem.*). An ordinary chemical

bond between atoms in which each provides 1 electron to a shared pair. When 1 atom provides both electrons it is said to be a *dative* (or *co-ordinate*) *bond*.

cove (*Join.*). A hollow cornice, usually large.

coved ceiling (*Arch.*). A ceiling which is formed at the edges to give a hollow curve from wall to ceiling, instead of a sharp angle of intersection.

covellite (*Min.*). Sulphide of copper crystallizing in the hexagonal system, usually occurring as thin plates. The colour is indigo-blue or darker. Also called **indigo copper**.

cover (*Build.*). In coursed work, the hidden or covered width of a slate or tile. (*Civ. Eng.*) The thickness of concrete between the outer surface of any reinforcement and the nearest surface of the concrete. See **effective depth.** (*Ecol.*) The percentage of the ground surface covered by a plant species.

coverage or covering power (*Photog.*). The area over which a lens can give a sharply focused image.

cover cells (*Bot.*). Cells formed during the formation of the tetrasporangia of some *Rhodophyceae*. They separate to release the tetraspores.

covered electrode (*Elec. Eng.*). A metal electrode covered with a coating of flux; used in arc-welding.

cover flashing (*Plumb.*). A separate flashing fastened into the upright surface and overlapping the flashing in the angle between the surfaces.

covering power (*Paint.*). The area which a given quantity of paint will cover without thinning unduly. Also used for the black-and-white contrasted area a given volume of paint will obliterate. (*Photog.*) See **coverage.**

cover iron (*Carp.*). See **back iron.**

cover paper (*Paper*). A heavy substance printing paper used for pamphlets, booklets.

cover slip (*Micros.*). The thin slip of glass used for covering a specimen that is being observed under a microscope. (*Photog.*) A transparent square of glass bound by the edges to a photographic transparency in the making of a photographic slide; its purpose is to protect the image.

cover stones (*Build.*). Flat stones covering girders, etc., and serving as a foundation for walls above.

covert coating (*Textiles*). Twill weave, all wool or wool/cotton, medium-weight, suiting cloth and, showerproofed, a light overcoating. Made from 2-fold marl or grandrelle warp and single weft.

coverts (*Zool.*). See **tectrices.**

cover unit (*Print.*). A separate printing unit coupled to the main press for producing the cover of a magazine or paper.

coving (*Build.*). (1) The upright splayed side of a fireplace opening. (2) The projection of upper storeys over lower.

co-volume (*Chem.*). The correction term b in van der Waals' equation of state, denoting the effective volume of the gas molecules. It is approximately equal to four times the actual volume of the molecules.

cowl (*Build.*). A cover, frequently louvred and either fixed or revolving, fitted to the top of a chimney to prevent down draught.

Cowles furnace (*Elec. Eng.*). An early form of electrolytic furnace used for the manufacture of aluminium alloys; it consisted of a long trough with carbon electrodes at the ends.

cowl flaps (*Aero.*). See **gills.**

cowling (*Aero.*). The whole or part of the streamlined covering of any aero-engine; in air-cooled engines, designed to assist cooling airflow.

cow-hocked (*Vet.*). Said of horses whose hocks are abnormally close to each other.

Cowper's glands (*Zool.*). In male Mammals, paired glands whose ducts open into the urethra near the base of the penis.

Cowper stoves (*Met.*). See **hot-blast stoves.**

cow-pox (*Med., Vet.*). See **vaccinia.**

coxa (*Zool.*). In Insects, the proximal joint of the leg. *adj.* **coxal.**

coxalgia (*Med.*). Pain in the hip.

coxal glands (*Zool.*). In *Arachnida*, a pair of excretory organs opening on to the surface of the fifth segment or the legs of that segment: formed from the persistent coelomoducts.

coxa valga (*Med.*). A deformity of the hip in which the angle between the neck and the shaft of the femur exceeds 140°.

coxa vara (*Med.*). A deformity of the hip in which the angle between the neck and the shaft of the femur is less than 120°.

coxa vera (*Zool.*). In Insects in which the coxa is divided, the anterior portion.

coxites (*Zool.*). The proximal part of the gonopods in embryonic insects.

coxopodite (*Zool.*). The proximal joint of the protopodite in *Crustacea*.

coxosternum (*Zool.*). The sternum of an abdominal segment when the coxites are fused with it, as occurs in most insects.

cozymase (*Chem.*). See **NAD.**

cP (*Phys.*). Abbrev. for *centipoise*.

c.p. (*Aero.*). See *centre of pressure*.

C.P. (*Chem.*). Abbrev. for *chemically pure*, a grade of chemical reagent for general laboratory use; less pure than *A.R.* (analytic reagent), but purer than *B.P.* (British Pharmacopoeia). (*Nuc.*) Abbrev. for *cosmic-ray produced*. (*Surv.*) Abbrev. for *change point*.

CPBS (*Chem.*). See **fenson.**

C.P.M. (*Eng.*). Abbrev. for *critical path method*. See **critical path planning.**

c.p.s., C.P.S., c/s (*Elec.*). Abbrevs. for *cycles per second*, superseded by *hertz* (*Hz*).

CR (*An. Behav.*). *Conditioned response*.

Cr (*Chem.*). The symbol for *chromium*.

crab (*Eng.*). The travelling lifting-gear of a gantry crane, mounted on a bogie and running on rails.

crab angle (*Nav.*). Angle between track and direction in which vessel or aircraft is heading.

crabbing (*Textiles*). A process applied to worsted or woollen fabrics, usually before scouring, in order to prevent cockling, or pattern distortion in the finishing stages. *Crabbing machine*, a machine, consisting of a trough with a crabbing or boiling roller, tensioning brakes, etc., in which crabbing is carried out. Usually repeated twice, the operation sets the width of the cloth.

Crab nebula (*Astron.*). An expanding nebulosity in Taurus which represents the remains of the supernova of A.D. 1054. It is a powerful source of radio waves and of X-rays. The nebulosity arises from a faint star at the centre, which is a rapid pulsar with a period of 0·033 sec. Both the X-ray and optical radiation show the same pulse, the period of which is slowly increasing.

crabwood (*For.*). Wood from a tree (*Carapa*) of Central and South America. It is inclined to warp and split during seasoning and is not resistant to insect attack, although it is resistant to wood-rotting fungi. Used for making veneers, roofing shingles, and as a structural timber.

crack detector (*Elec. Eng.*). An electromagnetic device for detecting flaws by the gathering of fine magnetic powder along the flaw lines in an iron specimen when magnetized, or by the reflection of ultrasonic waves.

cracked ends (*Textiles*). Threads which break

through the stresses set up in weaving or finishing. Not peculiar to any cloth but frequently happens in worsted pieces.

cracked heels (*Vet.*). See **grease.**

cracking (*Oils*). A thermal process in which petroleum distillates or residues are broken down into products of lower boiling range and of altered chemical constitution, e.g., heavy wax distillates are cracked to obtain motor fuel fractions of improved octane rating for blending purposes. Light straight distillates and naphtha are cracked at high temperatures with steam to give ethylene and propylene gases for chemical manufacture. See also **catalytic cracking.**

crackled (*Glass*). Said of glassware whose surface has been intentionally cracked by water immersion and partially healed by reheating before final shaping.

cradina (*Bot.*). An enzyme able to break down proteins, present in the juice of the stem, leaves and fruit of the fig.

cradle. Generally, any support which partially embraces the object mounted in it, e.g., a gun cradle. (*Build.*) (1) A frame of laths on which scrim is stretched to receive plaster in forming coved or other heavy cornices, etc. (2) See **cradle scaffold.** (*Elec. Eng.*) An earthed metal net placed below a high-voltage overhead transmission line where it crosses a public highway, railway, or telephone circuit; a conductor, if broken, falls on the net and is earthed without doing damage. (*Mining*) (1) The trough-shaped metal support for a mounted pneumatic drill. (2) The means of supporting men and tools during shaft-sinking or repair. (3) See **rocker.** (*Textiles*) Cam blocks which support the catch bar end trucks in a lace machine. (*Vet.*) A frame encircling the neck of a horse; used as a means of restraint.

cradle bar (*Join.*). See **cot bar.**

cradle scaffold (*Build.*). A form of suspended scaffolding consisting of a strong framework fitted with guard rails and boards for the working platform, and slung from two fixed points or from a wire rope secured between two jibs. Also called **boat scaffold.**

cradling (*Build.*). (1) Rough timber work fastened around a beam as a basis for lathing. (2) See **cradle.**

cradling piece (*Carp.*). A short timber fixed at each side of a fireplace hearth, between chimney breast and trimmer, to support the ends of floor boards.

Craftone (*Print.*). TN for *mechanical stipples* (q.v.), on film base, which can be applied to the copy or to the negative when reproducing line work. Similar to **Zip-a-tone.**

crafts (*Textiles*). Plain woven, coarse all-cotton or spun rayon fabric in grey, bleached or piece-dyed form used largely as loin cloths by African peoples; also used for shrouds.

Craf Type (*Typog.*). A paste-up lettering system of self-adhesive typefaces and chemically developed shading media and patterns.

Crag (*Geol.*). A local type of shelly and sandy rocks which have been deposited in relatively shallow water; found in the Pliocene of East Anglia.

crag-and-tail (*Geol.*). A land form consisting partly of solid rock shaped by ice action, with a *tail* of morainic material banked against it on the lee-side.

Craigleith stone (*Build.*). A very hard and durable whitish-grey sandstone which was quarried near Edinburgh, and used for ashlar and building purposes generally.

Cramer's test (*Chem.*). A test for the presence of saccharoses, based upon the reduction of a mercury (II) salt to metallic mercury.

crammed stripe (*Textiles*). A fancy-stripe pattern produced by arranging a larger number of threads in one part of the pattern than in the ground texture.

cramming (*Plumb.*). The operation of temporarily plugging a pipe before making a joint or doing repairs.

cramp (*Build.*). A locking bar of incorrodible metal used to bind together adjacent stones in a course, and having bent ends, one of which is fastened into each stone. Also called a **cramp-iron.** See also **lead plug, slate cramp.** (*Med.*) Painful spasm of muscle. (*Tools*) Adjustable contrivance for holding parts of woodwork together while they are being assembled and glued.

cramp-iron (*Build.*). See **cramp.**

crampon or **crampoon** (*Build.*). An appliance for holding stones or other heavy objects which are to be hoisted by crane. It consists of a pair of bars hinged together like scissors, the points of which are bent inwards for gripping the load, while the handles are connected by short lengths of chain to a common hoist-ring.

cramps (*Med.*). Heat cramps. Painful contractions of muscles in those who work in high temperatures, due to excessive loss of salt in the sweat.

Crampton's muscle (*Zool.*). In Birds, a muscle of the eye which by its contraction decreases the diameter of the eyeball and so aids the eye to focus objects near to it.

crane magnet (*Elec. Eng.*). A lifting magnet.

crane motor (*Elec. Eng.*). A motor specially designed for the operation of a crane or hoist. It should be very robust and have a high starting torque.

crane post (*Eng.*). The vertical member of a jib crane, to the top of which the jib is connected by a tie-rod.

crane rating (*Elec. Eng.*). A term sometimes employed to denote a method of specifying the rating of a motor for intermittent load, such as that of a crane. The maximum power and the load factor are stated.

cranial flexures (*Zool.*). Flexures of the brain in relation to the main axis of the spinal cord, transitory in lower Vertebrates, permanent in higher Vertebrates. See **nuchal flexure, pontal flexure, primary flexure.**

Craniata (*Zool.*). A group of subphyla of *Chordata*, also called *Vertebrata* (q.v.)

cranio-. Prefix from Gk. *kranion*, skull.

cranioclasis (*Surg.*). The instrumental crushing of the foetal skull in obstructed labour.

cranioclast (*Surg.*). An instrument for gripping and crushing the foetal skull in obstructed labour.

craniosacral system (*Zool.*). See **parasympathetic nervous system.**

craniotabes (*Med.*). Thinning of the bone of the skull in rickets or in congenital syphilis.

craniotomy (*Surg.*). Incision of the foetal skull and removal of its contents in obstructed labour.

cranium (*Zool.*). That part of the skull which encloses and protects the brain; the brain-case. *adj.* **cranial.**

crank (*Eng.*). An arm attached to a shaft, carrying at its outer end a pin parallel to the shaft; used either to give reciprocating motion to a member attached to the pin, or in order to transform such motion into rotary motion of the shaft.

crank-brace (*Tools*). A brace having a bent handle by which it may be rotated.
crankcase (*Eng.*). A boxlike casing, usually cast-iron or aluminium, which encloses the crank-shaft and connecting-rods of some types of reciprocating engines, air-compressors, etc.
crank effort (*Eng.*). The effective force acting on the crank pin of an engine in a direction tangential to the circular path of the pin.
crank pin (*Eng.*). The pin which is fitted into the web or arm of a crank, and to which a reciprocating member or connecting-rod is attached.
crankshaft (*Eng.*). The main shaft of an engine or other machine which carries a crank or cranks for the attachment of connecting-rods.
crank throw (*Eng.*). (1) The radial distance from the mainshaft to the pin of a crank, equal to one half the stroke of a reciprocating member attached to the pin. (2) The web or webs and pin of a crank.
crank web (*Eng.*). The arm of a crank, usually of flat rectangular section.
crash (*Textiles*). Plain woven linen cloth of coarse texture, used for towelling, etc. A cheaper, less effective type is made from cotton.
crash line (*Print.*). A mark inscribed on the bed of the machine and sometimes on the machine chase, to indicate that type must be kept outside of this mark to escape damage.
crash recorder (*Aero.*). See **flight recorder.**
craspedote (*Zool.*). Having a velum.
crater (*Geol.*). The orifice of a volcano, usually in the shape of an inverted cone, through which the lavas and gases are emitted during activity.
crateriform (*Bot., etc.*). Hollowed out; like a cup.
crater lamp (*Electronics*). Discharge tube so designed that a concentrated light source arises in a crater in a solid cathode.
craticular stage (*Bot.*). A small immobile form of the pennate diatoms, produced under adverse environmental conditions.
crawl (*Print.*). Continuous running of the press at very low speed. Cf. *inching*.
crawling (*Elec. Eng.*). A phenomenon sometimes observed with induction motors, the motor running up to about only one-seventh of full speed on account of the presence of a pronounced seventh harmonic in the field form. The phenomenon is also observed with other harmonics. Also called **balking.** (*Paint.*) A defect in paint or varnish work, characterized by the appearance of bare patches and paint ridges before drying.
craze or crazing (*Build.*). The minute hair cracks which sometimes appear on the surface of pre-cast concrete work or artificial stone. (*Paint.*) Fissuring of faulty coats of paint or varnish in irregular criss-cross cracks.
crazy chick disease (*Vet.*). See **nutritional encephalomalacia.**
CRD (*Vet.*). Abbrev. for *chronic respiratory disease* of fowl.
creaming of an emulsion (*Phys.*). The concentration without *breaking* (q.v.) of the disperse phase of an *emulsion* (q.v.) through the common rising (or falling, i.e., *downward creaming*) of the droplets under the force of gravity.
cream-laid (*Paper*). White writing-paper made with a laid water-mark.
cream of tartar (*Chem.*). Commercial name for acid potassium tartrate.
cream separator (*Agric.*). A machine, either hand- or power-driven, in which cream is separated from whole milk by centrifugal force; cream, the lighter fraction, passes out of the machine at one outlet, and the heavier skim milk through another outlet.
cream-wove (*Paper*). White writing-paper, in the manufacture of which a wove dandy has been used.
crease-resist (*Textiles*). Finish applied to cotton, linen or rayon, etc., cloths to improve the capacity of the materials to resist and/or recover from creases imparted in wear.
creasing (*Build.*). See **tile creasing.**
creasing resistance (*Paper*). Indicates the extent to which a wrapper retains its position when folded.
creatine (*Chem.*). Methylguanido-ethanoic acid, methylglycocyamine, $HN{=}C(NH_2){\cdot}N(CH_3){\cdot}CH_2{\cdot}COOH$, a weakly basic, crystalline compound, soluble in water and present in muscle. It has been synthesized from cyanamide and methylamino-ethanoic acid. See **arginine.**
creatinine (*Chem.*). Methylglycocyamidine, formed from creatine in acid solution by the elimination of water between the amino and carboxyl groups. The white prisms decompose at 270°C, and are strongly basic. It occurs in muscle and urine, and is the end-product, *in vivo*, of creatine catabolism.
creatinuria (*Med.*). The presence of creatinine in the urine.
creativity (*Psychol.*). The ability to construct original and viable products, ideas, etc., and to go beyond conventional developments. No unitary trait or defining characteristics have been found though *divergent thinking* is thought to be typical. It does not correlate with IQ beyond a base level IQ of between 112–120. See **convergent thinking.**
creatorrhoea, creatorrhea (*Med.*). The abnormal presence of muscle fibres in the faeces.
creel (*Textiles*). The steel or wooden structure which holds the roving bobbins at the spinning-frame, or yarn packages on winding, warping or similar machines.
creep (*Build.*). The gradual alteration in length or size of a structural member or high tensile wire owing to inherent properties of the materials involved. Not to be confused with shrinkage or expansion. (*Chem.*) (1) The rise of a precipitate on the wet walls of a vessel. (2) The formation of crystals on the sides of a vessel above the surface of an evaporating liquid. (*Eng.*) (1) A slow relative movement between 2 parts of a structure. (2) Slow, plastic deformation of metals under stress, particularly at elevated temperatures. (3) The slow relative motion of a belt over the surface of a pulley, due to its continual extension and relaxation as it passes from tight to slack slide. (*Met.*) Continuous deformation of metals under steady load. Exhibited by iron, nickel, copper, and their alloys at elevated temperature, and by zinc, tin, lead, and their alloys at room temperature. Sometimes taken to mean variable deformation. (*Mining*) Gradual rising of the floor in a coal-mine due to pressure. See **crush.**
creeping weasel (*Eng.*). Instrument for measuring the internal details of long, narrow holes, e.g., the fuel channels in reactor cores.
creep limit (*Eng.*). The maximum tensile stress which can be applied to a material at a given temperature without resulting in measurable creep.
creep strength (*Eng.*). The unit stress which will produce plastic deformation in a metal at a specified rate at a specified (elevated) temperature.
creep tests (*Met.*). Methods for measuring the

resistance of metals to creep. Time-extension curves under constant loads are determined.

cremaster (*Zool.*). In the pupae of *Lepidoptera*, an organ of attachment developed from the tenth abdominal somite; in Mammals, a muscle of the spermatic cord; in *Metatheria*, a muscle whose contraction causes the expression of milk from the mammary gland.

cremocarp (*Bot.*). A fruit which splits into two or more 1-seeded portions.

cremorne bolt (*Join.*). See **espagnolette.**

crenate (*Bot.*). Having a margin bearing rounded teeth all more or less of the same size; when the teeth are themselves crenate, the margin is *doubly crenate*.

crenulate (*Bot.*). Having small rounded teeth on the margin.

creosote oil (*Chem.*). A coal-tar fraction, boiling between 240° and 270°C. The crude creosote oil is used as raw material for producing tar acids, etc., or used direct as a germicide, insecticide, or disinfectant in various connexions (e.g., soaps, sheep dips, impregnation of railway sleepers, etc.).

creosoting cylinder (*Civ. Eng.*). A container in which timber may be impregnated with creosote under pressure. See **Bethell's process.**

crêpe (*Textiles*). A dress material of worsted, silk, cotton or man-made fibre. Has a distinctive rough, crinkled appearance because of the special hard-twisted yarns.

crêpe-de-chine (*Textiles*). A plain, jacquard figured or printed fabric with crêpe effect, usually made with botany worsted yarns, warp, and weft; originally made with silk warp and botany worsted weft.

crêpe paper (*Paper*). Formed by doctoring the moist web continuously from a cylinder. Used for decorative work.

crêpe rubber (*Chem.*). Raw, unvulcanized sheet rubber, not chemically treated in any way.

crêpe weaves (*Textiles*). Weaves in which an irregular arrangement of the warp and weft produces a broken surface effect, without twill or rib lines.

crepitation (*Med.*). (1) A crackling sensation felt by the observer on movement of a rheumatic joint. (2) The fine crackling noise made when two ends of a broken bone are rubbed together. (3) The fine crackling sounds heard over the chest in disease of the pleura or of the lungs. Also called **crepitus.** (*Zool.*) The explosive discharge of an acrid fluid by certain Beetles, which use this as a means of self-defence.

crepoline (*Textiles*). A light-weight worsted dress fabric, characterized by a rib-crêpe appearance. It is often made with 64 ends and 46 picks/in.

crepuscular (*Zool.*). Active at twilight or in the hours preceding dawn.

crepuscular rays (*Meteor.*). The radiating and coloured rays from the sun below the horizon, broken up and made apparent by clouds or mountains: also, the apparently diverging rays from the sun passing through irregular spaces between clouds.

crescent (*Horol.*). The circular notch cut in the periphery of the roller of the lever escapement to allow the passing of the guard pin or safety finger. Also known as **passing hollow.**

crescent wing (*Aero.*). A sweptback wing in which the angle of sweep and *thickness ratio* (q.v.) are progressively reduced from root to tips so as to maintain an approximately constant *critical Mach number* (q.v.); a wing of this planform is isoclinic when deflected.

cresol red (*Chem.*). 2-*cresolsulphonphthalein.* Indicator used in acid-alkali titrations, having a pH range of 7·2 to 8·8, over which it changes from yellow to red.

cresol resins (*Plastics*). Resins made from 1 methyl 3-hydroxybenzene and 1 methyl 4-hydroxybenzene and an aldehyde, similar in properties to the phenolics. The 2-compound reacts slowly, and is therefore likely to remain partly unchanged and act as a softener or plasticizer.

cresols (*Chem.*). A technical name for the hydroxytoluenes, $CH_3{\cdot}C_6H_4{\cdot}OH$, monohydric phenols. There are three isomers, viz.:

2-cresol, m.p. 30°C, b.p. 191°C.
3-cresol, m.p. 4°C, b.p. 203°C.
4-cresol, m.p. 36°C, b.p. 202°C.

Only 3- and 4-cresol form nitro-compounds with nitric acid, whereas the 2-cresol is oxidized. Important raw materials for plastics, especially the 3-compound; also used for explosives, as intermediates for dyestuffs, and as antiseptics.

crest (*Build., Civ. Eng., etc.*). The top of a slope or parapet: the ridge of a roof. (*Eng.*) Part of screw thread outline which connects adjacent flanks on the top of the ridge. (*Geol.*) The highest part of an anticlinal fold. A line drawn along the highest points of a particular bed is called the *crest line*, and a plane containing the crest lines of successive folds is called the *crest plane*. (*Zool.*) A ridge or elongate eminence, especially on a bone.

Crestatron (*Electronics*). TN for travelling-wave tube, using the beating-wave principle.

crest factor (*Elec. Eng.*). See **peak factor.**

cresting (*Build.*). Ornamental work along a ridge, cornice, or coping of a building. Also called **bratticing or brattishing.**

crest tile (*Build.*). A purpose-made tile having a V-shape specially suiting it to location astride the ridge-line of a roof.

crest value (*Elec. Eng.*). See **peak value.**

crest voltmeter (*Elec. Eng.*). See **peak voltmeter.**

cresylic acids (*Chem.*). A mixture of the various cresol isomers, also containing higher homologues, e.g., xylenols.

cretaceous (*Bot.*). Chalky.

Cretaceous System (*Geol.*). The rocks which succeed the Jurassic and precede the Tertiary System. The most striking member of this system is the Chalk in Britain.

cretin (*Med.*). One who is affected by **cretinism.** *adjs.* **cretinoid, cretinous.**

cretinism (*Med.*). A congenital condition in which there is failure of mental and physical development, due to absence or insufficiency of the secretion of the thyroid gland.

cretonne (*Textiles*). A plain, or 2×2 woven, cotton or spun rayon material, either all-over printed or made from a printed warp; used for loose covers and curtains.

crevasse (*Geol.*). A fissure, often deep and wide, in a glacier or ice-sheet.

crevasse curve (*Radio*). A curve showing the sharpness of response of a piezoelectric crystal to changes of frequency. It is obtained by taking a resonance curve of a parallel resonant circuit across which the crystal is connected. A sharp 'crevasse' in the curve occurs at the resonant frequency of the crystal.

crib (*Mining*). (1) An interval from work underground for croust, bait, snack, downer, piece, chop, snap, bite, or tiffin. (2) A job. (3) A form of timber support.

cribbing (*Civ. Eng.*). An interior lining for a shaft, formed of framed timbers backed with boards; it is used to support the sides and keep back water.

crib-biting (*Vet.*). A vice acquired by horses, in which some object is grasped with the incisor teeth and air swallowed, which may lead to indigestion.

cribellar glands (*Zool.*). The silk glands which open on the cribellum in certain Spiders.

cribellum (*Zool.*). In certain Spiders, a perforate oval plate, lying just in front of the anterior spinnerets, which produces a broad strip of silk composed of a number of threads.

cribriform (*Zool.*). Perforate, sievelike; as the *cribriform plate*, a perforate cartilaginous element of the developing Vertebrate skull, which later gives rise to the ectethmoid.

cribrose (*Bot.*). Pierced with many holes; resembling a sieve.

cribwork (*Civ. Eng.*). Steel or timber cribs or boxes, sometimes filled with concrete, which are sometimes sunk below water-level to carry the foundations of bridges.

cricoid (*Zool.*). Ring-shaped; as one of the cartilages of the larynx.

Crimp and Bruges formula (*San. Eng.*). A formula giving the rate of discharge through sewers constructed of good brickwork, or of cast-iron pipes in good condition. It states that

$$v = 86\, m^{\frac{2}{3}} i^{\frac{1}{2}},$$

where v = velocity of flow in m/s; m = hydraulic mean depth in m; i = virtual slope.

crimp cloth (*Textiles*). A cotton fabric in which a stripe effect is produced by yarns at different tensions during weaving; or by printing plain cloth with dilute stripes of caustic soda, which causes contraction of the parts printed. Other methods are also used, including weaving a plain cloth from two warp beams, one having greater tension than the other.

crimper (*Civ. Eng.*). See **indenter.**

crimping (*Eng.*). The compressing of a thin metal part on to another one along lines to increase pressure. Common deformation bonds the parts.

crimping pliers (*Mining*). Pliers with serrated jaws used to fasten detonator to fuse when preparing blasting charge.

crimson lake (*Paint.*). A fugitive red pigment obtained when a water extract of cochineal is precipitated with a mixture of tin and aluminium salts.

crinanite (*Geol.*). A basic igneous rock, consisting of intergrown crystals of feldspar, titanaugite, olivine, and analcite. Similar to *teschenite*.

Crinoidea (*Zool.*). A class of *Echinodermata* with branched arms; the oral surface directed upwards; attached for part or the whole of their life by a stalk which springs from the aboral apex; suckerless tube feet; open ambulacral grooves; no madreporite, spines or pedicellariae. Most members extinct.

cripple-timber (*Carp.*). See **jack-timber.**

crisis (*Med.*). (1) A painful paroxysm in tabes dorsalis. (2) The rapid fall of temperature marking the end of a fever.

crispate, crisped (*Bot.*). Having a frizzled appearance.

crispening current (*TV*). One which uses non-linear means to improve definition in a reproduced picture.

criss-cross inheritance (*Gen.*). Occurs when the reciprocal cross of a female with an allelomorphic character X and a male with the corresponding character Y results in all males of the first filial generation having X and all the females Y.

crissum (*Zool.*). In Birds, the region surrounding the cloaca or the feathers situated on that area. *adj.* **crissal.**

crista (*Bacteriol.*). A ridge or membrane running longitudinally along some bacteria. (*Zool.*) A ridge or ridgelike structure; as the projection of the transverse crests of lophodont molars.

crista acustica (*Zool.*). A chordotonal apparatus forming part of the tympanal organ in *Tetigoniidae* and *Gryllidae*. (2) A patch of sensory cells in the ampulla, utricle, and saccule of the vertebrate ear.

cristate (*Bot.*). Bearing a crest.

cristobalite (*Min.*). A high-temperature modification of SiO_2 stable only above 1470°C to its melting-point at 1713°C, though it may exist at lower temperatures. Occurs in volcanic rocks and in slags, etc.

crit (*Nuc.*). Abbrev. for *critical mass.*

crith (*Phys.*). Unit of mass, that of 1 litre of hydrogen at s.t.p. 89·88 milligrams.

crithidial (*Zool.*). Pertaining to, or resembling, the flagellate genus *Crithidia*; said of a stage in the life-cycle of some Trypanosomes.

critical angle (*Phys.*). See **total internal reflection.** (*Radio*) The angle of radiation of a radio wave which will just not be reflected from the ionosphere.

critical backing pressure (*Vac. Tech.*). The value of the backing pressure, at any stated throughput, which, if exceeded, causes an abrupt increase of pressure on the high vacuum side of the pump. In some pumps, the increase does not occur abruptly and this pressure is not precisely determinable. Abbrev. **c.b.p.**

critical coefficient (*Chem.*). Additive property, a measure of the molar volume. It is defined as the ratio of the critical temperature to the critical pressure.

critical corona voltage (*Elec. Eng.*). The voltage at which a corona discharge just begins to take place around an electric conductor.

critical coupling (*Elec. Eng.*). That between two circuits or systems tuned to the same frequency, which gives maximum energy transfer without overcoupling.

critical damping (*Phys.*). Damping in an oscillatory electric circuit or in an oscillating mechanical system (such as the movement of an indicating instrument) just sufficient to prevent free oscillations from arising (ringing).

critical field (*Electronics*). In the case of a *magnetron*, the smallest steady magnetic flux density (at a constant anode voltage) which would prevent an electron (assumed to be emitted at zero velocity from the cathode) from reaching the anode. It can also mean the magnetic field applied to a conductor below which the superconducting transition occurs at a given temperature. Also called **cut-off field.**

critical frequency (*Radio*). Frequency of a radio wave which is just sufficient to penetrate an ionized layer in the upper atmosphere. (*Telecomm.*) Lowest frequency which can be propagated as a periodic wave motion in a waveguide. It will have infinite guide wavelength and phase velocity. The value depends upon the guide dimensions and mode. Also called **cut-off frequency.**

critical humidity (*Chem.*). That at which the equilibrium water vapour pressure of a substance is equal to the partial pressure of the water vapour in the atmosphere, so that it would neither lose nor gain water on exposure.

criticality (*Nuc. Eng.*). State in nuclear reactor when multiplication factor for neutron flux reaches unity and external neutron supply is no longer required to maintain power level.

critical Mach number (*Aero.*). The Mach number at which a subsonic aeroplane becomes uncontrollable due to *compressibility drag* (q.v.); abbrev. M_{crit}

critical mass (or crit) (*Nuc. Eng.*). The minimum mass of fissionable material which can sustain a chain reaction.

critical path planning (*Eng.*). A procedure used in planning a large programme of work. Using a digital computer, it determines the particular sequence of operations which must be followed to complete the overall programme in the minimum time and also determines which events have some 'float' or capacity to reprogramme without affecting the whole.

critical period (*Psychol.*). See **sensitive periods.**

critical point (*Phys.*). The point on the *isothermal* (q.v.) for the critical temperature of a substance at which the pressure and volume have their critical values. At the critical point the densities (and other physical properties) of the liquid and gaseous states are identical. (*Met.*) See **arrest points.**

critical potential (*Phys.*). A measure of the energy (in electron volts) required to ionize a given atom, or raise it to an excited state.

critical pressure (*Phys.*). The pressure at which a gas may just be liquefied at its critical temperature.

critical range (*Met.*). The range of temperature in which the reversible change from austenite (stable at high temperature) to ferrite, pearlite, and cementite (stable at low temperature) occurs. The upper limit varies with carbon content; the lower limit for slow heating and cooling is about 700°C.

critical rate (*Met.*). The rate of cooling required to prevent the formation of pearlite and to secure the formation of martensite in steel. With carbon steel, this means cooling in cold water, but it is reduced by the addition of other elements, hence oil- and air-hardening steels.

critical reaction (*Radio*). The maximum degree of reaction in a regenerative system before self-oscillation commences.

critical size (*Nuc. Eng.*). The minimum size for a nuclear reactor core of given configuration.

critical solution temperature (*Chem.*). The temperature above which two liquids are miscible in all proportions.

critical speed (*Aero.*). The speed during take-off at which it has to be abandoned if the aircraft is to stop in the available space. Cf. *accelerate-stop distance.* Also **decision speed.** Abbrev. V_1. (*Eng.*) The rotational speed of a shaft at which some periodic disturbing force coincides with the fundamental or some higher mode of the natural frequency of torsional or transverse vibration of the shaft and its attached masses.

critical state (*Phys.*). The condition of a gas at its critical point, when it appears to hover between the liquid and gaseous states.

critical temperature (*Elec. Eng.*). The temperature at which magnetic materials lose their magnetic properties; same as **Curie point (temperature).** (*Met.*) The temperature at which some change occurs in a metal or alloy during heating or cooling, i.e., the temperature at which an arrest or critical point is shown on heating or cooling curves. (*Phys.*) The temperature above which a given gas cannot be liquefied. See **liquefaction of gases.**

critical velocity (*Hyd.*). The velocity at which the nature of a given fluid flow in a particular case changes from viscous to eddy flow or vice versa.

critical voltage (*Elec. Eng.*). That which, when applied to a gas-discharge tube, just initiates discharge.

critical volume (*Phys.*). The volume of unit mass of a substance under critical conditions of temperature and pressure.

critical wavelength (*Phys.*). Free space wavelength corresponding to *critical frequency* (q.v.), for waveguide.

crizzling (*Glass*). Fine cracks in the surface of the glass, occasioned by local chilling during manufacture.

CR-law (*Elec. Eng.*). Relates to exponential rise or decay of charge on capacitor in series with a resistor, and, by extension, to signal distortion on long submarine cables.

crochet (*Zool.*). A hook which aids in locomotion, and is associated with the apex of the abdominal legs in Insect larvae.

crocidolite (*Min.*). A silicate of sodium and iron, crystallizing in the monoclinic system and belonging to the amphibole group of rock-forming minerals. A fibrous variety of riebeckite, and an important asbestos mineral. See also **Cape asbestos, tiger's eye.**

crocodile truck (*Eng.*). A high-capacity railway truck consisting of a long open platform carried between two 4-wheeled bogies. The low floor-level facilitates the loading of heavy and bulky freight.

Crocodilia (*Zool.*). An order of Reptiles having upper and lower temporal arcades, a hard palate, an immovable quadrate, loose abdominal ribs, socketed teeth; large powerful amphibious forms. Crocodiles, Alligators, Caimans, Gavials. Also **Loricata.**

crocodiling (*Paint.*). A defect on a painted or varnished surface, characterized by the formation of ridges or cracks, sometimes going back to the surface, in irregular patches. Sometimes known as **alligatoring.**

crocoite, crocoisite (*Min.*). Chromate (VI) of lead (II), crystallizing in the monoclinic system; bright orange-red in colour.

croissant vitellogène (*Zool.*). In the developing oöcyte, a crescentic area surrounding the archoplasm, in which the mitochondria are grouped.

Croixian (*Geol.*). The rocks of Upper Cambrian age in the Pacific Province in N. America, so named from St Croix (Minnesota), where they are typically developed. Of same age as *Bretonian.*

Cronapress (*Print.*). A *conversion system* in which a pressure-sensitive film is placed over a letterpress page in a special frame and bombarded with small steel balls. After simple dye development a negative for plate-making is produced.

Crookes dark space (*Phys.*). A dark region separating the cathode from the luminous 'negative glow' in an electrical discharge in a gas at low pressure. The thickness of the Crookes dark space increases as the pressure is reduced. For air, it is about 1 cm thick at 10 N/m^2 pressure.

Crookes glass (*Glass*). A range of spectacle glasses having low transmission for ultraviolet light, and containing cerium and other rare earths.

crookesite (*Min.*). Selenide of copper and thallium, often with 1%–5% silver. It is massive and compact, and displays metallic lustre.

Crookes radiometer (*Phys.*). A small mica 'paddlewheel' which rotates when placed in daylight in an evacuated glass vessel. Alternate faces of the mica vanes are blackened and the slight rise of temperature of the blackened surfaces caused by the radiation which they absorb warms the air in contact with them and increases the velocity of rebound of the molecules, the sum of whose impulse constitutes the driving pressure.

Crookes tube (*Phys.*). Original gas-discharge tube, illustrating striated positive column, Faraday dark space, negative glow, Crookes dark space, and cathode glow.

crop (*Geol.*). See **outcrop.** (*Met.*) The portion of an ingot cropped off to remove the pipe and other defects. Also called **discard.** (*Zool.*) See **proventriculus** (2).

crop bound (*Vet.*). A term applied to birds suffering from impaction of the crop or ingluvies.

crop loader (*Agric.*). A machine to pick up and deliver cut crops.

cropped (*Bind.*). Said of the edges of a book which have been cut down to an extent that mars the appearance of the pages.

cropper (*Typog.*). A small platen printing machine.

cropping (*Eng.*). An engineering operation closely resembling both blanking and shaving, for cutting strip material to length, sometimes with the additional purpose of producing a radius on the end of the material. (*Met.*) The operation of cutting off the end or ends of an ingot to remove the pipe and other defects. (*Textiles*) *Cutting* (q.v.). (*Vet.*) The operation of amputating a part of the comb or wattles of birds, or of the ears of dogs.

cropping machine (*Textiles*). A machine comprising spiral knives and a ledger plate, used to remove loose ends and knots from linen cloth. On raised wool cloths, it shears the face to a level nap.

cross (*Gen.*). An individual whose parents belong to different breeds or races. (*Plumb.*) A special pipe-fitting having four branches mutually at right angles in one plane.

cross ampere-turns (*Elec. Eng.*). The component of the armature ampere-turns which tends to produce a field at right angles to the main field.

cross arm (*Elec. Eng.*). The horizontal cross-member attached to telegraph poles, or power transmission line poles, for supporting the insulators which carry the conductors.

cross association (*Print.*). Placing in position across the machine the required strips or slit webs before finishing operations.

cross-axle (*Rail.*). A driving axle having cranks mutually at right angles.

cross band (*Textiles*). (1) A breadth of lace, such as flouncings, made across a *Levers machine.* (2) Belt or rope drive crossed instead of running open fashion.

crossband, warp twist, left-hand twine (*Textiles*). Terms used in the woollen industry to indicate yarns with the twist running from right to left; in the worsted trade, usually called **left-hand twist.**

crossbar exchange (*Telecomm.*). An automatic telephone exchange operated electromechanically in which digits are selected on a two-dimensional matrix of contacts.

crossbars (*Typog.*). Book chases are fitted with two crossbars, a long and a short, which quarter the chase and facilitate the lock-up.

crossbar transformer (*Elec. Eng.*). Coupling device between a coaxial cable and a waveguide, the latter having a short transverse rod, to the centre of which the central conductor is connected.

cross-bearing (*Surv.*). Observation on survey point not in the immediate scheme of work, useful for checking purposes.

cross-blast explosion pot (*Elec. Eng.*). An explosion pot in which the pressure generated by the arc in the pot causes a stream of oil to be directed across the arc path at right angles to it.

cross-blast oil circuit-breaker (*Elec. Eng.*). An oil circuit-breaker in which the pressure generated by the arc causes a stream of oil to be forced through ports placed opposite one pair of contacts, thereby cutting across the arc stream.

cross bombardment (*Nuc.*). A method of identification of radioactive nuclides through their production by differing reactions.

cross-bond (*Elec. Eng.*). A rail-bond for connecting together the two rails of a track or the rails of adjacent tracks.

cross-bonding (*Elec. Eng.*). The sheath of cable 1 is connected to that of cable 2 and farther on to cable 3. The total induced e.m.f. vanishes and there are no sheath-circuit eddies.

cross-bracing (*Eng.*). See **counterbracing.**

crossbred (*Textiles*). A term applied to wool obtained from sheep crossed in breed (long-wool lustre class and merino). The wool is coarser than merino, brighter in appearance, and crisp. Used for worsted serges, Cheviot quality woollens, and hosiery.

cross-connexion field (*Teleph.*). **See jumper field.**

cross-coupling (*Elec. Eng.*). Undesired transfer of interfering power from one circuit to another by induction, leakage, etc.

cross cut (*Mining*). In metal mining, a level or tunnel driven through the country rock, generally from a shaft, to intersect a vein or lode.

cross-cut chisel (*Eng.*). A cold chisel having a narrow cutting edge carried by a stiff shank of rectangular section; used for heavy cuts. See **cold chisel.**

cross-cut file (*Eng.*). A file in which the cutting edges are formed by the intersection of two sets of teeth crossing each other.

cross-cut saw (*Carp.*). A saw designed for cutting timber across the grain.

cross direction (*Paper*). See **machine direction.**

cross dyeing (*Textiles*). A method of dyeing a mixture cloth (e.g., wool/rayon, etc.), in which the warp and the weft of blended yarns have different dyeing properties and thus produce two-tone effects in the cloth.

crossed-field multiplier (*Elec. Eng.*). One where a feedback voltage representing the product restores to zero deflection a beam simultaneously deflected by electric and magnetic fields, both representing a magnitude to be multiplied.

crossed lens (*Light*). A simple lens the radii of curvature of which have been chosen to give minimum spherical aberration for parallel incident rays. For a refractive index of 1·5, the radii should be in the ratio 1 : 6, the surface of smaller radius facing the incident light.

crossed Nicols (*Light*). Two Nicol prisms arranged with their principal planes at right angles, in which position the plane-polarized light emerging from one Nicol is extinguished by the other.

crossed shed (*Weaving*). A warp shed made in gauze or leno weaving, one thread being crossed over or under another.

crossette (*Build.*). A projection formed on the flank of a voussoir at the top, giving it a bearing upon the adjacent voussoir on the side towards the springing.

cross fall (*Civ. Eng.*). The difference in vertical height between the highest and lowest points on the cross-section of a road surface. Also, the average rate of fall from one side to the other, or from the crown to a side of a road.

cross-fertilization (*Bot.*). The fertilization of the female gametes of one individual by the male gametes of another individual. Also called **allogamy.**

cross field (*Elec. Eng.*). The component of the flux in an electric machine which is assumed to be produced by the cross ampere-turns.

cross-field generator (*Elec. Eng.*). A type of *rotating amplifier* which uses the armature flux produced by short-circuiting the normal output brushes to develop the machine output across a second pair of brushes halfway between the first. The *amplidyne* (q.v.) embodies the cross-field principle.

crossfire (*Telecomm.*). Interference of signals in one circuit with those in another circuit.

crossfire technique (*Radiol.*). The irradiation of a deep-seated region in the body from several directions so as to reduce damage to surrounding tissues for a given dose to that region.

cross-frogs (*Rail.*). See **crossings.**

cross front (*Photog.*). A sliding front carrying the lens in field and technical cameras; used to avoid the consequence of tilting the axis of a camera away from normality with an object.

cross garnet (*Join.*). A form of *strap hinge* (q.v.).

cross girders (*Eng.*). (1) Short girders acting as ties between two main girders. (2) The members which transmit the weight of the roadway to the main girders of a bridge.

cross-grained float (*Build.*). A float made of a piece of cross-grained wood; used in finishing hard-setting plasterwork.

crosshair (*Surv.*). A spider's thread fixed across the diaphragm of a level or theodolite.

cross-hatch pattern (*TV*). A test pattern of vertical and horizontal lines used on TV picture tube.

crosshead (*Eng.*). A reciprocating block, usually sliding between guides, forming the junction piece between the piston-rod and connecting-rod of an engine. (*Typog.*) A heading or sub-heading centred on the measure.

crossing-over (*Cyt.*). Mutual exchanges between homologous pairs of chromosomes during maturation division.

crossings (*Rail.*). The notches made in rails to allow passage for the wheel-flanges at places where one line crosses another. Also **cross-frogs.**

cross joint (*Build.*). The vertical mortar joint at the sides or back of a brick in position in a wall.

Crosskill roller (*Agric.*). A roller consisting of loosely mounted flat spike rings and rings with projections on both sides. Every second ring may have a very large centre hole.

cross-linking (*Chem.*). The formation of side bonds between different chains in a polymer, thus increasing its rigidity.

cross-magnetizing (*Mag.*). Effect of armature reaction on magnetic field of current generator.

cross modulation (*Radio*). Impression of the envelope of one modulated carrier upon another carrier, due to nonlinearity in the medium transmitting both carriers. See **Luxemburg effect.**

cross neutralization (*Electronics*). A method of neutralization in push-pull amplifiers. Each output is connected by a negative-feedback circuit to the other input.

Crossopterygii (*Zool.*). A subclass of the class *Osteichthyes*, first known as fossils from the Middle Devonian period, and persisting to the present day. They have two dorsal fins, scaly lobed paired fins, the skeleton of the pectoral fin having a basal element attached to the girdle and a branching arrangement at the tip (an arrangement which could easily have led to the evolution of the tetrapod limb), the body is covered with thick rhomboidal cosmoid scales (thinner in later *Dipnoi*), and autostylic jaw suspension. Divided into two orders, the *Rhipidistia* (including the extinct *Osteolepids* and the *Coelocanths*, which still survive) and the *Dipnoi* (Lung-fish) of which there are three types found in Australia, Africa and S. America respectively. Also known as **Choanichthyes.**

cross-over (*Acous.*). In a twin loudspeaker system, the point in the frequency range above which the amplifier output is fed to the treble speaker and below which to the bass speaker. (*Electronics*) In an electron-lens system, location where streams of electrons from the object pass through a very small area, substantially a point, before forming an image. (*Plumb.*) A special pipe-fitting with its middle length cranked out so that one pipe line may be laid across another when the two pipes are to be in the same plane. (*Rail.*) A communicating track between two parallel lines, enabling rolling-stock to be transferred from one line to the other.

cross-over area (*Electronics*). The point at which the electron beam comes to a focus inside the accelerating anode of a cathode-ray tube.

cross-over frequency (*Acous.*). (1) The frequency in a two-channel loudspeaker system at which the high and low frequency units deliver equal acoustic power; alternatively it may be more generally applied to electric dividing networks when equal electric powers are delivered to each of adjacent frequency channels. (2) See **turn-over frequency.**

cross-over network (*Acous.*). Same as **dividing network.**

cross-over unit (*Gen.*). A 1% frequency of interchange between a pair of linked genes.

cross-pane hammer (*Tools*). A hammer with a pane consisting of a blunt chisel-like edge at right angles to the shaft.

cross parachute (*Aero.*). A *brake parachute* (q.v.) in which, for economy, the canopy is composed of rectangular panels forming a cross.

cross-ply (*Autos.*). Term applied to the standard type of motor-tyres with flexible tread and relatively stiff sidewalls. Cf. *radial-ply.*

cross pollination (*Bot.*). The conveyance of pollen from an anther of one flower to the stigma of another, either on the same or on a different plant of the same, or related, species.

cross-product of a vector (*Maths.*). See **vector product.**

cross-ratio (*Maths.*). (1) Of four numbers a, b, c, d, taken in the order a, b, c, d: the ratio $\frac{a-c}{a-d}\Big/\frac{b-c}{b-d}$. Denoted by (ab, cd). (2) Of four points on a straight line A, B, C, D: the cross-ratio is

$$(AB, CD) = \frac{AB}{AD} \cdot \frac{CD}{CB}$$

(the usual convention as to signs being observed). (3) Of a pencil, consisting of four straight lines OA, OB, OC, OD: the cross-ratio is

$$O(AB, CD) = \frac{\sin AOB}{\sin AOD} \cdot \frac{\sin COD}{\sin COB}.$$

Any transversal (i.e., line not through O) cuts the pencil in four points having the same cross-ratio. Some writers associate the above fractions with the order a, c, b, d as opposed to the order a, b, c, d. Also called **anharmonic ratio.**

cross-recording (*Cinema.*). The using of independently generated sounds for mixing with the primary sounds, in recording a complex shot in sound-film production.

cross-section (*Eng., etc.*). The section of a body (e.g., a girder or moulding) at right angles to its length; a drawing showing such a section. (*Maths.*) The section made by a plane which cuts the axis of symmetry or the longest axis at right angles. (*Nuc.*) The measure of the probability of a particular process. The value of the cross-section for any particular process will depend upon the element under bombardment and upon the nature and energy of the bombarding particles. For example, if A is the *capture cross-sectional area* of an element for neutrons, and a volume containing n atoms of the element is exposed to a uniform beam (of sectional area a) of neutrons, then the number of neutrons captured is $\frac{nA}{a} \times N$, where N is total number of neutrons in the beam.

cross-sill (*Rail.*). See **sleeper.**

cross-slide (*Eng.*). That part of a planing machine or lathe on which the toolholder is mounted, and along which it may be moved at right angles to the bed of the machine.

cross-springer (*Build.*). In a groined arch, the rib following the line of a groin.

cross-staff (*Surv.*). An instrument for setting out right angles in the field. It consists of a frame or box having two pairs of vertical slits, giving two lines of sight mutually at right angles.

cross-stone (*Min.*). See **chiastolite, staurolite.**

cross-talk (*Radio*). Interference from an unwanted transmission. (*Teleph.*) The interference caused by energy from one conversation invading another circuit by electrostatic or electromagnetic coupling. See **far-end-, near-end-.**

cross-talk meter (*Teleph.*). An arrangement for measuring the attenuation between circuits which are liable to permit cross-talk.

cross-tie (*Rail.*). See **sleeper.** (*Weaving*) See **London tie.**

cross-ties (*Bot.*). Small veins in the leaf which run in a straight course between larger veins, giving a ladderlike appearance.

cross-tongue (*Join.*). A wooden tongue for a *ploughed-and-tongued joint* (q.v.), cut so that the grain is at right angles to the grooves.

cross-tree (*Ships*). A lateral formation on a ship's mast; its uses are for rigging to top masts hooks, tackle, etc. The term is derived from antique wooden ships.

crotch (*For.*). A small sledge without a tongue, often made from a natural fork of a tree, bridged by a wooden bunk and used as an aid in hauling logs on stony or bare ground. Also **go-devil.**

crotchet (*Zool.*). A hooked or notched chaeta.

crotonaldehyde (*Chem.*). But-2-enal. $CH_3 \cdot CH{=}CH \cdot CHO$, is a liquid, of pungent odour, b.p. 105°C, an unsaturated aldehyde, obtained from ethanol by heating with dilute hydrochloric acid or with a solution of sodium acetate. As an intermediate product, *aldol* is formed.

crotonic acid (*Chem.*). But-2-enoic acid. $CH_3 \cdot CH{=}CH \cdot COOH$, an olefinic monocarboxylic acid. There are two stereoisomers, viz., crotonic acid, m.p. 71°C, b.p. 180°C; and iso- or allo-crotonic acid, m.p. 15°C, b.p. 169°C. The first form is the *cis-*, the latter one the *trans-* form. The crotonic acids are also isomers of methacrylic acid and of vinylacetic acid.

crotonyl (*Chem.*). The group $-CO \cdot CH{=}CH \cdot CH_3$ in organic compounds.

crotyl (*Chem.*). The group $-CH_2CH{=}CHCH_3$ in organic compounds.

croup (*Med.*). Inflammation of the larynx and trachea in children, associated with a peculiar ringing cough; present especially in diphtheria. (*Vet.*) The sacral region of the back of the horse.

croupous inflammation (*Med.*). Inflammatory reaction characterized by an excess of fibrin in the exudate.

crowbar (*Tools*). A round iron bar, pointed at one end and flattened to a wedge shape at the other, used as a lever for moving heavy objects.

Crowe process (*Min. Proc.*). Method of removing oxygen from cyanide solution before recovery of dissolved gold, in which liquor is exposed to vacuum as it flows over trays in a tower.

crown (*Bot.*). A very short rootstock. (*Build., Civ. Eng.*) The highest part of an arch. Also called the **vertex.** (*For.*) The upper branchy part of a tree above the bole. (*Paper*) A pre-metric size of printing paper, 15 × 20 in. (U.S. 15 × 19 in.) (*Zool.*) The part of a polyp bearing the mouth and tentacles; the distal part of a deer's horn; the grinding surface of a tooth; the disc and arms of a Crinoid; crest; head.

crown bar (*Civ. Eng.*). A heavy log, about 30 cm diameter and 5 m in length, fixed along the top of a heading for a tunnel and supporting the poling boards with which the heading is lined. Kept in position by props at each end.

crown cork (*Brew., etc.*). Airtight closure for bottles comprising a metal cap with a cork washer clamped to the mouth.

crown-cut (*For.*). See **back-sawn.**

crowned (*Bot.*). (1) Bearing a terminal outgrowth such as a pappus. (2) Having an appendage on the upper side of the leaf or petal.

crown-gate (*Hyd. Eng.*). A canal-lock headgate.

crown glass (*Glass*). (1) Glass made in disk form by blowing and spinning, having a natural fire-finished surface but varying in thickness with slight convexity, giving a degree of distortion of vision and reflection. (2) Soda-lime-silica glass. See also **optical crown.**

crown leather (*Leather*). A leather which is specially suited for belt laces, washers, etc. The method of manufacture somewhat resembles tawing. Also called **Helvetia leather.**

crown octavo (*Typog.*). A book size, $7\frac{1}{2} \times 4\frac{7}{8}$ in. (metric, 186 × 123 mm).

crown of thorns tuning (*Electronics*). Tuning of cavity magnetrons involving changing inductance of cavities by the introduction of conducting rods along their axes. Cf. *cookie cutter tuning.*

crown-post (*Carp.*). A *king-post* (q.v.).

crown quarto (*Typog.*). A book-size, $9\frac{3}{4} \times 7\frac{3}{8}$in. (metric, 246 × 186 mm).

crown rail-bond (*Elec. Eng.*). A rail-bond consisting of a flexible copper cable with solid terminals which are expanded into holes in the rail by means of drift pins.

crown-tile (*Build.*). An ordinary flat tile. Also called a **plane-tile.**

crown wheel (*Eng.*). The larger wheel of a bevel reduction gear. See **bevel gear.** (*Horol.*) A contrate wheel with pointed teeth forming the escape wheel of a *verge escapement* (q.v.).

crow-step gable (*Build.*). See **corbie-step gable.**

crow twill, swansdown twill (*Weaving*). Terms used to denote the 3-and-1 twill, often employed in woollen and worsted cloths.

croy (*Civ. Eng.*). A protective barrier built out into a stream to prevent erosion of the bank.

crozier (*Bot.*). The young ascus when it is bent in the form of a hook.

crozzle (*Build.*). An excessively hard and misshapen brick which has been partially melted and overheated.

CRT (*Electronics*). Abbrev. for *cathode-ray tube*.

cruciate, cruciform (*Bot.*). Having the form of, or arranged like, a cross.

cruciate basidium (*Bot.*). A septate basidium in which the spindles of the dividing nuclei lie on a level across the basidium.

crucible (*Chem., Met.*). A refractory vessel or pot in which metals are melted. In chemical analysis, smaller crucibles, made of porcelain, nickel, or platinum, are used for igniting precipitates, fusing alkalies, etc.

crucible furnace (*Met.*). A furnace, fired with coal, coke, oil, or gas, in which metal contained in crucibles is melted.

crucible steel (*Met.*). Steel made by melting blister bar or wrought-iron, charcoal, and ferro-alloys in crucibles which hold about 100 lb. This was the first process to produce steel in a molten condition, hence product called *cast steel* (q.v.). Mainly used for the manufacture of tool steels, but now largely replaced by the electric-furnace process.

crucible tongs (*Chem., Met., etc.*). Tongs used for handling crucibles.

crud (*Chem. Eng.*). Undesired impurity or foreign matter arising in a process.

crude fibre (*Bot.*). The residues in the soil derived from the woody parts of plants.

crude oil (*Chem.*). See **petroleum.**

cruise (*For.*). To survey forest land in order to locate timber and estimate its quantity by species assortments, size, products, quality, or other characteristics.

cruiser stern (*Ships*). A type of ship's stern construction. It is integral with the main hull for strength and form, and is partially waterborne. It assists in manœuverability and wave formation, and provides underdeck roominess. See **counter.**

cruising speed (*Autos.*). The optimum travelling speed of a motor-car.

crumena (*Zool.*). A pocket into which the long oval stylets of many *Hemiptera* are withdrawn.

crump (or **bump**) (*Mining*). Rock movement under stress due to underground mining, possibly violent.

crumpled (*Bot.*). See **corrugate.**

crunode (*Maths.*). See **double point.**

cruor (*Zool.*). The coagulated blood of Vertebrates.

crup butt (*Leather*). Leather made from the back part of a horse hide; used for uppers of waterproof boots.

crura (*Zool.*). See **crus.**

crura cerebri (*Zool.*). In Insects, the paraoesophageal connections which unite the brain with the suboesophageal ganglion.

crural (*Zool.*). Pertaining to or resembling a leg. See **crus.**

crural glands (*Zool.*). In *Onychophora*, simple ectodermal glands found on all the legs except the first.

crureus (*Zool.*). A leg muscle of higher Vertebrates.

crus (*Zool.*). The zeugopodium of the hind-limb in Vertebrates; the shank; any organ resembling a leg or shank. *pl.* **crura.** *adj.* **crural.**

crush (*Mining*). The broken condition of pillars of coal in a mine due to pressure of the strata. See **creep.**

crush breccia (*Geol.*). A rock consisting of angular fragments, often recemented, which has resulted from the faulting or folding of pre-existing rocks. See also **crush conglomerate, fault breccia.**

crush conglomerate (*Geol.*). A rock consisting of crushed and rolled fragments, often recemented; it has resulted from the folding or faulting of pre-existing rocks.

crusher (*Min. Proc.*). Machine used in earlier stages of comminution of hard rock. Typically works on dry feed as it falls between advancing and receding breaking plates. Types include jaw, gyratory, stamps, and rolls.

crushing (*Bind.*). See **smashing.** (*TV*) Distortion of contrast in a reproduced picture by diminishing contrast at upper or lower ends of contrast scale (black-to-white). See **gamma.**

crushing strip (*Print.*). A strip sometimes required on fold rollers to increase the nip.

crushing test (*Civ. Eng.*). A test of the suitability of stone to be used for roads or building purposes.

crust (*Civ. Eng.*). See **wearing course.**

crusta (*Zool.*). In the Vertebrate brain, a mass of nerve fibres lying on the ventral aspect of each half of the mesencephalon, lateral and ventral to the tegmentum; any hard coating.

Crustacea (*Zool.*). A class of *Arthropoda*, mostly of aquatic habit and mode of respiration; the second and third somites bear antennae and the fourth a pair of mandibles. Shrimps, Prawns, Barnacles, Crabs, Lobsters, etc.

crustaceous (*Bot.*). (1) Forming a crust on the surface of anything. (2) Thin, and brittle in texture.

crusta petrosa (*Zool.*). A layer of lamellated bone covering the dentine of a tooth beyond the enamel.

crust leather (*Leather*). The name for a light skin which, after tanning, has been shaved on the flesh side and lightly oiled on the grain side, preparatory to finishing.

crust of the earth (*Geol.*). See under **earth.**

crustose (*Bot.*). Forming a more or less interrupted crust.

crut (*Mining*). A short heading or tunnel into the face of a coal-seam.

crutch (*Horol.*). The lever or rod which transmits the impulse from the pallets to the pendulum rod. The end of the crutch may be in the form of a fork to embrace the pendulum rod, or else in the form of a pin which enters a slot in the pendulum rod.

crutching (*Vet.*). The operation of removing the wool from around the tail and quarters of sheep as a preventive of myiasis.

crutchings (*Textiles*). See **daggings.**

cryogenic (*Phys.*). Term applied to low-temperature substances and apparatus.

cryogenic gyro (*Phys.*). One depending on electron spin in atoms at very low temperature.

cryogenics (*Phys.*). The study of materials at very low temperatures.

cryolite or **Greenland spar** (*Min.*). Fluoride of aluminium and sodium, crystallizing in the monoclinic system. It usually occurs as a vein in granite rocks, as in Ivigtut, west Greenland, and is used in the manufacture of aluminium and white porcellanous glass.

cryometer (*Heat*). A thermometer for measuring very low temperatures.

cryophorus (*Heat*). An apparatus to demonstrate the principle of refrigerators.

cryoplankton (*Bot.*). Algae which live on the surface of snow and ice in polar regions and on high mountains.

cryosar (*Electronics*). Low-temperature germanium switch with on-off time of a few nanoseconds.

cryoscope (*Chem.*). Instrument for the determination of freezing- or melting-points.

cryoscopic method (*Chem.*). The determination of the rel. mol. mass of a substance by observing the lowering of the freezing-point of a suitable solvent.

cryostat (*Heat*). Low-temperature thermostat.

cryotherapy (*Med.*). Medical treatment using the application of extreme cold. Also called **crymotherapy.**

cryotron (*Electronics*). Miniature electronic switch, operating in liquid helium, consisting of a short wire wound with a very fine control wire. When a magnetic field is induced via the control wire, the main wire changes from superconductive to resistive. Used as a memory, characterized by exceedingly short access time. See **superconductivity.**

crypt (*Zool.*). A small cavity; a simple tubular gland.

cryptaesthesia (*Psychol.*). A faculty involved in the acquisition of information by means, e.g., telepathy, extrasensory perception, etc., other than the recognized sense organs. See **parapsychology.**

cryptic coloration (*Zool.*). Protective resemblance to some part of the environment or camouflage, from simple *countershading* (q.v.) to more subtle mimicry of, e.g., leaves or twigs. Cf. *aposematic coloration.*

crypto-. Prefix from Gk. *kryptos*, hidden.

cryptobiosis (*Zool.*). The state in which an animal's metabolic activities have come effectively, but reversibly, to a standstill, so that they can, for example, remain viable even if dipped into liquid helium at −270°C. Usually involves more or less complete dehydration. See **anabiosis.**

cryptocrystalline (*Crystal.*). Consisting of very minute crystals.

cryptogam (*Bot.*). In a former system of classification a plant without flowers, and often without distinct stems, leaves, and roots. *adjs.* **cryptogamic, cryptogamous.**

cryptomere (*Gen.*). A character which is inherited but which is not apparent, i.e., 'hidden'.

cryptomerism (*Gen.*). The failure of characters to show in offspring, which nevertheless contain the corresponding hereditary *x* factors.

cryptometer (*Chem.*). An instrument used to determine the obliterating or hiding power, or the opacity, of paints and pigments.

cryptomitosis (*Cyt.*). A form of mitosis occurring in *Protozoa*, in which the chromatin concentrates on the equator of the spindle as a dense mass in which no chromosomes are visible.

Cryptomonadales (*Bot.*). An order of the *Cryptophyceae* containing the motile genera.

Cryptomonadina (*Zool.*). An order of *Phytomastigina*, comprising forms generally with two flagella; green, yellow, brown, or colourless; rarely of amoeboid form; with a gullet or a longitudinal groove, but without a transverse groove; and having food-reserves generally of starch; without a contractile vacuole.

Cryptonemiales (*Bot.*). An order of the *Rhodophyceae* containing members of the tetrasporic *Florideae*, having the auxiliary cell borne on a special filament of the gametophyte.

cryptonephry (*Zool.*). The condition, in many Insects, where the distal ends of the Malpighian tubules are closely applied to the hind gut.

Cryptophyceae (*Bot.*). A class of the *Pyrrophyta*; there are two or more chromatophores with or without pyrenoids; starch is the usual food reserve; motile members are flattened and bear two flagellae of different length, borne terminally or laterally.

cryptophyte (*Bot.*). A plant which forms its resting buds beneath the surface of the soil.

cryptorchid (*Zool.*). Said of testes which remain within the abdomen and do not descend into a scrotal sac. (*Vet.*) An animal so affected.

cryptorchidectomy (*Vet.*). Surgical removal of the testes from a cryptorchid animal.

cryptorhetic (*Zool.*). See **endocrine.**

cryptostoma (*Bot.*). A flask-shaped cavity in the thallus of some large brown seaweeds, containing hairs which secrete mucilage.

cryptozoic (*Zool.*). Living in dark places, as in holes, caves, or under stones and tree-trunks.

crystal. Solid substance showing some marked form of geometrical pattern, to which certain physical properties, angle and distance between planes, refractive index, etc., can be attributed. (*Electronics*) Piezoelectric element, shaped from a crystal in relation to crystallographic axes, e.g., quartz, tourmaline, Rochelle salt, ammonium dihydrogen phosphate, to give *facets* to which electrodes are fixed or deposited for use as transducers or frequency standards. (*Glass*) See **crystal glass.** (*Horol.*) The glass that covers the dial of a watch. (*Min.*) Old name for **quartz.**

crystal anisotropy (*Crystal.*). In general, directional variations of any physical property, e.g., elasticity, thermal conductivity, etc., in crystalline materials. Leads to existence of favoured directions of magnetization, related to lattice structure in some ferromagnetic crystals.

crystal axes (*Crystal.*). The axes of the natural coordinate system formed by the crystal lattice. These are perpendicular to the natural faces for many crystals. See **uniaxial, biaxial.**

crystal boundaries (*Crystal.*). The surfaces of contact between adjacent crystals in a metal. Anything not soluble in the crystals tends to be situated at the crystal boundaries, but in the absence of this, the boundary between two similar crystals is simply the region where the orientation changes.

crystal cell (*Elec. Eng.*). Kerr cell using a quartz or other suitable crystal in place of the more usual nitrobenzine.

crystal counter (*Electronics*). One in which an operating pulse is obtained from a crystal when made conducting by an ionizing particle or wave.

crystal cutter (*Electronics*). A cutter used in gramophone recording, a piezoelectric crystal being the means of initiating the mechanical displacements of the stylus.

crystal detector (*Electronics*). Demodulator incorporating a stable contact, e.g., a cat's whisker, between semiconductors, e.g., germanium, silicon.

crystal diamagnetism (*Met.*). Property of negative susceptibility shown by silver, bismuth, etc.

crystal diode (*Electronics*). See **diode** (2).

crystal dislocation (*Crystal.*). Imperfect alignment between the lattices at the junctions of small blocks of ions ('mosaics') within the crystal. The resulting mobility and opportunity for realignment of the molecules is of importance in crystal growth, plastic flow, sintering, etc.

crystal drive (*Electronics*). System in which oscillations of low power are generated in a *piezoelectric* crystal oscillator, being subsequently amplified to a level requisite for transmission.

crystal electrostriction (*Elec.*). The dimensional changes of a dielectric crystal under an applied electric field. See **electrostriction, magnetostriction.**

crystal face (*Crystal.*). One of the bounding surfaces of a crystal. In the case of small, undistorted crystals, each face is an optically plane surface. A *cleavage face* is the smooth surface resulting from cleavage; in such minerals as mica, the cleavage face may be almost a plane surface, diverging only by the thickness of a molecule.

crystal filter (*Electronics*). Band-pass filter in which piezoelectric crystals provide very sharp frequency-discriminating elements, especially for group modulation of multichannels over coaxial lines.

crystal-gate receiver (*Electronics*). Supersonic heterodyne receiver in which one (or more than one) piezoelectric crystal is included in the intermediate-frequency circuit, to obtain a high degree of selectivity.

crystal glass (*Glass*). A colourless, highly transparent glass of high refractive index, which may be 'lead crystal' or 'lime crystal'. (A somewhat misleading term since it denotes different things in different glassmaking districts.)

crystal goniometer (*Crystal.*). Instrument for measuring angles between crystal faces.

crystal growing (*Electronics*). Technique of forming semiconductors by extracting crystal slowly from molten state. Also called **crystal pulling.**

crystal indices (*Crystal.*). See **Miller indices.**

crystal lattice (*Crystal.*). Three-dimensional repeating array of points used to represent structure of crystal, and classified into fourteen groups by Bravais.

crystalline. Clear, transparent. (*Bot.*) Having a shining appearance. (*Geol.*) Not glassy.

crystalline cone (*Zool.*). The outer refractive body of an ommatidium which is surrounded by the vitrellae.

crystalline form (*Crystal.*). The external geometrical shape of a crystal.

crystalline lens (*Zool.*). The transparent refractive body of the eye in Vertebrates, *Cephalopoda*, etc. It is compressible by muscles and focuses images of objects emitting light on to the retina.

crystalline liquids (*Chem.*). See **liquid crystals.**

crystalline overgrowth (*Crystal.*). The growth of one crystal round another, frequently observed with isomorphous substances. Cf. *cubic system.*

crystalline rocks (*Geol.*). These consist wholly, or chiefly, of mineral crystals. They are usually formed by the solidification of molten rock, by metamorphic action, or by precipitation from solution.

crystalline schists (*Geol.*). A group of rocks which have resulted from heat and pressure. Their structures are controlled by the prevalence of flaky crystals, like mica, and such rocks have a tendency to split in a direction parallel to these flat crystals. See **foliation, schistosity.**

crystalline solid (*Chem.*). A solid in which the atoms or molecules are arranged in a regular manner, the values of certain physical properties depending on the direction in which they are measured. When formed freely, a crystalline mass is bounded by plane surfaces (faces) intersecting at definite angles. See **Bragg method.**

crystalline style (*Zool.*). In *Lamellibranchiata* and some *Gastropoda*, a transparent rod-shaped mass secreted by a diverticulum of the intestine; composed of protein with an adsorbed amylolytic enzyme.

crystallites (*Chem.*). Very small, imperfectly formed crystals. (*Min.*) Minute bodies occurring in glassy igneous rocks, and marking a stage in incipient crystallization.

crystallization (*Chem.*). Slow formation of a crystal from melt or solution.

crystallized (*Paint.*). Said of an enamelled or varnished surface which presents the appearance of galvanized iron.

crystalloblastic texture (*Geol.*). The texture of metamorphic rocks resulting from the growth of crystals in a solid medium.

crystallogram (*Crystal.*). Diffraction pattern from a crystal, whereby its structure can be calculated.

crystallographic axes (*Crystal.*). See **axes.**

crystallographic notation (*Crystal.*). A concise method of writing down the relation of any crystal face to certain axes of reference in the crystal.

crystallographic planes (*Crystal.*). Any set of parallel and equally spaced planes that may be supposed to pass through the centres of atoms in crystals. As every plane must pass through atomic centres, and no centres must be situated between planes, the distance between successive planes in a set depends on their direction in relation to the arrangement of atomic centres.

crystallographic system (*Crystal.*). Any of the major units of crystal classification embracing one or more symmetry classes.

crystallography. Study of internal arrangements (ionic and molecular) and external morphology of crystal species, and their classification into types.

crystalloid (*Bot.*). A crystal of protein, occurring in large numbers in the cells of seeds and other storage organs. (*Chem.*) Obsolete for crystalline substance which dissolves in water and is not a colloid.

crystalloluminescence (*Chem.*). The emission of light during crystallization.

crystal loudspeaker (*Acous.*). One depending on vibrations excited in a *bimorph* piezo crystal which drives, e.g., a small cone diaphragm acting as a *tweeter*.

crystal microphone (*Acous.*). One depending on the generation of an e.m.f. in a bimorph crystal when flexed by the varying pressure in an applied sound wave. Also called **piezoelectric microphone.**

crystal mixer (*Electronics*). Frequency changer in a waveguide system, which is fed from a local oscillator.

crystal nuclei (*Chem.*). The minute crystals whose formation is the beginning of crystallization.

crystal oscillator (*Electronics*). Valve or transistor oscillator in which frequency is held within very close limits by rapid change of mechanical impedance (coupled piezoelectrically) when passing through resonance.

crystal pattern. See **crystallogram.**

crystal pick-up (*Acous.*). Piezoelectric transducer, producing an e.m.f. due to mechanical drive arising from vibration. Widely used for gramophone-disk reproduction and vibration measurements.

crystal pulling (*Electronics*). See **crystal growing.**

crystal receiver (*Electronics*). One using a simple crystal demodulator and hence no external supplies.

crystal rectifier (*Electronics*). One which depends on differential conduction in semiconducting crystals, suitably 'doped', such as Ge or Si.

crystal sac (*Bot.*). A cell almost filled with crystals of calcium oxalate.

crystal set (*Radio*). A simple radio receiver using only a crystal detector.

crystal spectrometer (*Nuc. Eng.*). Instrument employing diffraction of neutron de Broglie waves by crystal lattice to measure the wavelengths of X-rays, γ-rays, and particle waves.

crystal structure. This consists of the whole

assemblage of rows and patterns of atoms, which have a definite arrangement in each crystal. (*Met.*) The arrangement in most pure metals may be imitated by packing spheres, and the same applies to many of the constituents of alloys. See **body-centred cubic, face-centred cubic, close-packed hexagonal structure.**

crystals systems (*Crystal.*). A classification of crystals based on the intercepts made on the crystallographic axes by certain planes.

crystal texture (*Crystal.*). The size and arrangement of the individual crystals in a crystalline mass.

crystal triode (*Radio*). Early name for *transistor*.

crystal violet (*Chem.*). A dyestuff of the rosaniline series, hexamethyl-4-rosaniline.

crystodyne (*Radio*). A type of crystal detector in which a d.c. voltage is maintained across the crystal contact.

c/s (*Elec.*). Cycles per second. See **hertz.**

Cs (*Chem.*). The symbol for *caesium*.

CS (*An. Behav.*). *Conditioned stimulus.*

C-service-area (*Radio*). The region surrounding a broadcasting transmitter where the field strength is between 2·5 and 5 millivolt/metre.

C.S.F. (*Paper*). See **Canadian Standard Freeness.**

CS gas (*Chem.*). Orthochlorobenzylidene malononitrile. A potent tear-gas used for mob dispersal. See **war gas.**

C stage (*Plastics*). Final stage in curing process of phenol formaldehyde resin, characterized by infusibility and insolubility in alcohol or acetone.

ctene (*Zool.*). One of the comb-plates or locomotor organs of *Ctenophora*, consisting of a row of strong cilia of which the bases are fused.

ctenidium (*Zool.*). Generally, any comblike structure; in aquatic Invertebrates a type of gill consisting of a central axis bearing a row of filaments on either side; in Insects, a row of spines resembling a comb.

ctenocyst (*Zool.*). The aboral sense-organ of *Ctenophora*.

ctenoid (*Bot.*). Comblike, pectinate. (*Zool.*) Said of scales which have a comblike free border.

Ctenophora (*Zool.*). A subphylum of *Coelenterata*, the members of which usually do not possess cnidoblasts, and generally show biradial symmetry; they have a system of gastrovascular canals and typically eight meridional rows of swimming plates or ctenes, composed of fused cilia. Sea Acorns, Comb-Bearers.

Cu (*Chem.*). The symbol for *copper*.

cubanite (*Min.*). $CuFe_2S_3$; forms orthorhombic crystals, often with chalcopyrite.

Cuban mahogany (*For.*). Reddish hardwood from *Swietenia*, found in the West Indies and Florida.

cube (*Maths.*). A solid with six square faces. A square parallelepiped. Its volume is the length of a side raised to third power, hence cubed.

cubical antenna (*Radio*). One with radiating elements arranged as sides of adjacent cubes.

cubical epithelium (*Zool.*). A form of columnar epithelium in which the cells are short.

cubic equation (*Maths.*). An algebraic equation of the third degree. The usual standard form is

$$x^3+3ax+b=0,$$

whose solution can be expressed as

$$x=\sqrt[3]{p}-\frac{a}{\sqrt[3]{p}},$$

where $p=-b+\sqrt{b^2+4a^3}$, and where, to obtain all three solutions, all three cube roots of p, including the complex roots, have to be used. Unlike the quadratic, this solution is of little practical value.

cubicle-type switchboard (*Elec. Eng.*). See **cellular-type switchboard.**

cubic system (*Crystal.*). The crystal system which has the highest degree of symmetry; it embraces such forms as the cube and octahedron.

cubing (*Build., etc.*). An approximate method for estimating costs of buildings. The volume of a building is multiplied by a figure known from experience to represent a fair average figure for the cost of unit volume of such building.

cubital (*Zool.*). See **secondary.**

cubital remiges (*Zool.*). The primary quills connected with the ulna in birds.

cubitus (*Zool.*). In Insects, one of the primary veins of the wing. *adj.* **cubital.**

cuboid (*Geom.*). A rectangular parallelepiped. (*Bot.*) Cubical.

Cubomedusae (*Zool.*). An order of subclass *Discomedusae*, class *Scyphozoa*, comprising active marine forms with four perradial tentaculocysts; there is a square manubrium, a broad pseudovelum and other peculiar features; regular alternation of generations occurs.

Cuboni test (*Vet.*). A test for pregnancy in the mare, based on the chemical detection of oestrogens in the urine.

cucullate (*Bot., Zool.*). Hood-shaped.

cucullus (*Zool.*). In *Ricinulei*, a wide, oval plate articulating with the anterior edge of the prosoma: in *Chelonethi*, that part of the carapace anterior to the eyes: any hood-shaped structure. *adj.* **cucullate.**

cucurbitaceous (*Bot.*). Shaped like a gourd.

Cucurbitales (*Bot.*). An order of the *Dicotyledons*. The flowers are sympetalous and regular on a cupped receptacle; the five anthers are free or united; the inferior ovary is usually trilocular, developing into a berry; the seeds are nonendospermic; includes Cucumbers, Pumpkins, Gourds, Melons, etc.

cudbear (*Chem.*). A purplish-red powder, soluble in water, obtained from *Rochelle de Candolle* and other lichens. Used as a dye.

cue mark (*Cinema.*). The mark made by a marker light on the edge of cinematograph film as it passes through the camera, to indicate to the cutter a point of synchronization between strips of film from different cameras and the sound camera.

cuffing (*Med.*). The accumulation of white cells round a blood vessel in certain infections of the nervous system.

cuiller (*Zool.*). In some male Insects, a spoonlike expansion at the end of a clasper.

cuing scale (*Acous.*). A graduated scale attached to gramophone-disk reproducers to enable the needle to be dropped into a selected groove with certainty.

cuirasse respirator (*Med.*). A respirator which is attached, like armour, to the chest wall only, and assists respiration by fluctuations in its own internal pressure.

cull (*For.*). A tree, log or piece of timber of merchantable size but rendered unmerchantable by defects.

cullet (*Glass*). Waste glass used with the 'batch' to improve the rate of melting and to save waste of materials.

cullis (*Carp.*). See **coulisse.**

culm (*Bot.*). The stem of a grass or of a sedge. (*Geol.*) The name given to the rocks of Carboniferous age in the south-west of England, consisting of fine-grained sandstones and shales, with occasional thin bands of crushed coal or 'culm'. (*Mining*) Anthracite dust; more accurately, anthracite which will pass through a screen with $\frac{1}{8}$-in. holes.

culmen (*Zool.*). In Birds, the edge of the upper beak.

culmicole, culmicolous (*Bot.*). Growing on stems of grasses.

culmination (*Astron.*). The highest or lowest altitude attained by a heavenly body as it crosses the meridian. *Upper culmination* indicates its meridian transit above the horizon, *lower culmination* its meridian transit below the horizon, or, in the case of a circumpolar, below the elevated pole.

culms (*Brew.*). The rootlets removed (by screening) from malt after it has been kilned; used as cattle food.

cultellus (*Zool.*). In certain blood-sucking *Diptera*, a sword-shaped organ forming part of the mouth-parts.

cultivator (*Agric.*). A wheeled frame, horse- or tractor-drawn, fitted with two rows of strong curved teeth which enter the ploughed soil and pulverize it. The teeth are generally mounted on springs. See **grubber.**

cultriform (*Bot.*). Shaped like a knife.

culture (*Bot., etc.*). An experimental preparation containing a micro-organism growing on a medium.

culvert (*Civ. Eng.*). Construction for the total enclosure of a drain or watercourse.

Cumacea (*Zool.*). An order of *Peracarida*, in which the carapace is fused dorsally with the first three or four somites; the eyes are usually coalesced and sessile, and the uropods are styliform; small marine forms, most of which burrow in mud.

cumarone (*Chem.*). See **coumarone.**

cumene (*Chem.*). *Iso*propyl (or 2 methylethyl)-benzene, $C_6H_5{\cdot}CH(CH_3)_2$, b.p. 153°C.

cumene hydroperoxide (*Chem.*). Industrial intermediate in the manufacture of phenol and acetone. Prepared by blowing air into cumene in the presence of an alkaline catalyst.

OOH
|
$H_3C—C—CH_3$

cummingtonite (*Min.*). A hydrous magnesium iron silicate, and member of the amphibole group, crystallizing in the monoclinic system. Differs from grunerite in having magnesium in excess of iron.

cumulative distribution (*Chem.*). In an assembly of particles, the fraction having less than a certain value of a common property, e.g., size or energy. Cf. *fractional distribution.*

cumulative dose (*Radiol.*). Integrated radiation dose resulting from repeated exposure.

cumulative errors (*Civ. Eng., Maths., etc.*). See **systematic errors.**

cumulative excitation (*Electronics*). Successive absorption of energy by electrons in collision, leading to ionization.

cumulative frequency function (*Stats.*). See **distribution function.**

cumulative grid detector (*Radio*). See **grid-leak detector.**

cumulative grid rectifier (*Radio*). A thermionic valve rectifier in which the signals are applied to the grid of the valve through a capacitor, shunted by a high-resistance leak. The flow of grid current which occurs each half-cycle results in a progressive depression of the mean grid potential, which in turn decreases the anode current.

cumulatively-compound machine (*Elec. Eng.*). A compound-wound machine in which the series and shunt windings assist each other.

cumulative probability function (*Stats.*). See **distribution function.**

cumulo-nimbus (*Meteor.*). Great masses of cloud rising in the form of mountains or towers or anvils, generally having a veil or screen of fibrous texture at the top and a cloud mass similar to nimbus at the bottom. This is the type of cloud associated with thunderstorms. The top of such a cloud may reach a height of 30,000 ft (9000 m).

cumulo-stratus (*Meteor.*). A cloud combination produced when a cumulus cloud spreads out on top to form a layer of strato-cumulus or alto-cumulus.

cumulus (*Meteor.*). Thick cloud, the upper surface of which is well defined and dome-shaped and exhibits protuberances, while the base is generally horizontal. (*Zool.*) The mass of cells surrounding the developing ovum in Mammals.

cumulus oophorus (*Zool.*). See **zona granulosa.**

cuneate, cuneal, cuneiform (*Bot.*). Shaped like a wedge, and attached to something by its point.

cuneus (*Zool.*). In some *Hemiptera*, a triangular apical portion of the corium.

cunnilingus (*Psychol.*). Oral stimulation of the female genitalia.

Cunningham correction (*Powder Tech.*). Modification of *Stokes's law* applied when particles of an aerosol are small compared to the mean free path of the molecules of the gas through which they are falling.

cup (*Bot.*). An apothecium.

cup chuck (*Eng.*). A lathe chuck in the form of a cup or bell screwed to the mandrel nose. The work is gripped by screws in the walls of the chuck. Also called **bell chuck.**

cupel (*Met.*). A thick-bottomed shallow dish made of bone ash; used in the cupellation of lead beads containing gold and silver, in the assay of these metals.

cupellation (*Met.*). The operation employed in recovering gold and silver from lead. It involves the melting of the lead containing these metals and its oxidation by means of an air-blast.

cup-feed drill (*Agric.*). A drill consisting of a seed-box, mounted on a chassis, from which the seed falls by gravity into a series of chambers; it is conveyed by cups rotating on a spindle to the seed tubes, through which it drops into the shallow trenches formed by the coulters. The rate of feed can be regulated.

cupferron (*Chem.*). Ammonium-nitroso-β-phenylhydrazine. Reagent used in the colorimetric detection and estimation of copper.

cup flow figure (*Plastics*). The time in seconds taken by a mould of standard design to close completely under pressure, when loaded with a charge of phenolic moulding material.

cup head (*Eng.*). A rivet or bolt head shaped like an inverted cup.

cupid's darts (*Min.*). See **flèches d'amour.**

cup joint (*Plumb.*). A joint formed between two lead pipes in the same line by opening out the end of one pipe to receive the tapered end of the other.

cup leather (*Eng.*). A ring o. leather moulded to U-section, used in hydraulic machinery to prevent leakage past plungers, etc.

cupola (*Build.*). A *lantern* (q.v.) constructed on

top of a dome. (*Geol.*) A dome-shaped offshoot rising from the top of a major intrusion.

cupola furnace (*Met.*). A shaft furnace used in melting pig-iron (with or without iron or steel scrap) for iron castings. The lining is firebrick. Metal, coke, and flux (if used) are charged at top, and air is blown in near the bottom.

cupped wire (*Met.*). Wire in which internal cavities have been formed during drawing.

cupping cell (*Bot.*). A swollen hyphal attachment formed by some fungi which parasitize other fungi; in it accumulates nutritive material derived from the host.

cuprammonia (*Chem.*). A solvent for cellulose, prepared by adding ammonium chloride and then excess of caustic soda to a solution of a copper (II) salt, washing and pressing the precipitate, and dissolving it in strong ammonia.

cuprammonium (tetra(a)mmine copper (II)) rayon (*Textiles*). Rayon made from bleached cotton linters by treatment with copper sulphate and ammonia. Also **copper rayon.**

cupric (*Chem.*). Copper (II). Containing divalent copper. Copper (II) salts are blue or green when hydrated, and are stable.

cupriferous pyrite (*Min.*). See **chalcopyrite.**

cuprite (*Min.*). Oxide of copper, crystallizing in the cubic system. It is usually red in colour and often occurs associated with native copper; a common ore.

cupro-nickel (*Met.*). An alloy of copper and nickel; usually contains 15, 20, or 30% of nickel; is very ductile, and has high resistance to corrosion; used for condenser tubes, turbine blades, bullet envelopes.

cupro-uranite (*Min.*). See **torbernite.**

cuprous (*Chem.*). Copper (I) Containing monovalent copper. Copper (I) salts generally form colourless solutions readily oxidized to the copper (II) state.

cuprous (*Chem.*). Copper (I). Containing monovalent copper. Copper (I) salts generally form oxidation of copper.

cup shake (*For.*). A shake between concentric layers. Also called **ring shake.**

cupula (*Zool.*). Any domelike structure, e.g., the apex of the lungs, the apex of the cochlea.

cupule (*Bot.*). (1) A small cup-shaped outgrowth from the thallus of a liverwort, containing gemmae. (2) A cup-shaped envelope more or less surrounding the fruit of some trees, especially obvious in the acorn.

cup wheel (*Eng.*). An abrasive wheel in the form of a cylinder, used mainly for grinding. The cylindrical surfaces may be perpendicular or inclined to the end. Very shallow cup wheels are known as dish wheels or saucers.

curare (*Med.*). South American native poison from the bark of species of *Strychnos* and *Chondodendron*.

curarine (*Med.*). Paralysing toxic alkaloid ($C_{19}H_{26}ON_2$) extracted as *d*-tubocurarine chloride from crude curare; used in anaesthesia as a muscle relaxant.

curb (*Build., Carp.*). A wall-plate carrying a dome at the springings. Also **curb-plate.** (*Civ. Eng.*) A hollow timber or cast-iron cylinder used in sinking and lining a shaft or well, for which purpose it is laid over the site, and then, as earth is excavated from beneath it, the lining is built upon it and sinks with the curb. Also called **cutting curb, drum-curb.** (*Mining*) Framework fixed in rock of mine shaft to act as foundation for brick or timber lining. (*Vet.*) A swelling occurring just below the point of the hock of the horse, usually due to inflammation caused by sprain of the calcaneo-cuboid ligament.

curb pins (*Horol.*). The two vertical pins attached to the index embracing the balance spring, near the point of attachment of the outer coil. By moving the pins nearer to, or farther away from, the point of attachment, a delicate regulation of the time of vibration of the balance is obtained.

curb-plate (*Build., Carp.*). See **curb.**

curb roof (*Build.*). See **mansard roof.**

curdling (*Paint.*). Thickening of varnish in the can, especially the appearance of gelatinous bits.

cure (*Leather, etc.*). A dehydrating agent such as salt, or salts of arsenic, capable of arresting decomposition in skins and hides. See **curing.**

curettage (*Surg.*). The scraping of the walls of cavities (especially of the uterus) with a *curette* (or *curet*), a spoon-shaped instrument.

curie (*Nuc.*). Unit of radioactivity, defined as $3{\cdot}700 \times 10^{10}$ disintegrations/second, roughly equal to the activity of 1 g of radium. Abbrev., Ci.

Curie balance (*Mag.*). A torsion balance for measuring the magnetic properties of nonferromagnetic materials by the force exerted on the specimen in a nonuniform magnetic field.

Curie point (or **temperature**) (*Phys.*). (1) Temperature at which ferromagnetic materials become paramagnetic. Also called **magnetic transition temperature.** (2) Temperature (*upper Curie point*) above which ferroelectric materials lose their polarization. (3) Temperature (*lower Curie point*) below which some ferroelectric materials lose their polarization.

Curies' law (*Phys.*). For paramagnetic substances, the magnetic susceptibility is inversely proportional to the absolute temperature.

Curie-Weiss effect (*Phys.*). Change of ferromagnetism to paramagnetism at the Curie point.

curine (*Chem.*). $(C_{18}H_{19}O_3N)_2$, an alkaloid of the quinoline group, found in curare extract obtained from various *Strychnos* spp. Crystallizes from benzene with 1 molecule of benzene, m.p. 161°C, m.p. (dry) 212°C; forms 4-sided prisms.

curing (*Chem.*). A term applied usually to a fermentation or ageing process of natural products, e.g., rubber, tobacco, etc. (*Civ. Eng.*) A method of reducing the contraction of concrete on setting; the surface is kept covered for a time with damp sacks, or with damp sawdust or sand, or sprayed with water. (*Leather, etc.*) The process of treating rawhides and skins with a *cure* (q.v.), to preserve them during transport or storage. (*Plastics*) The chemical process undergone by a thermosetting plastic by which the hot, liquid resin sets to a solid at the same temperature. Curing generally takes place during the moulding operation, and may require from 45 sec to 30 min for its completion.

curium (*Chem.*). Manmade radioactive element, symbol Cm, at. no. 96, produced from americium. There are several long-lived isotopes (up to $1{\cdot}7 \times 10^7$ years half-life), all α-emitters.

curl (*Elec. Eng.*). A vector of which the flux across a vanishingly small surface is equal to the *circulation* (q.v.) of the given vector round the contour of the surface. Also called **rotation.** (*For.*) A roughly hewn block of timber cut from a crotch and intended for cutting into veneers. (*Maths.*) Of a vector **F**, the vector $\nabla \wedge \mathbf{F}$, where ∇ is the vector operator *del.* Also called **rotation of a vector** or **rot v.** (*Paper*) A paper defect caused by unequal alteration in the dimensions of the top and under sides of the sheet due to changes in the moisture or temperature.

curled toe paralysis (*Vet.*). A disease of chicks

characterized by leg weakness and inward curling of the toes, associated with degenerative changes in the peripheral nerves; caused by a deficiency of riboflavin in the diet.

curl yarn (*Textiles*). A yarn with curls at regular intervals. Made by folding a coarse yarn slackly round a fine one. This 2-fold thread is subsequently folded, in the reverse direction, with another fine thread, to give a balanced yarn with a pronounced effect. Used in dress goods and imitation astrakhan.

curly grain (*For.*). A wavy pattern on the surface of worked timber due to the undulate course taken by the vessels and other elements of the wood.

current. A flow of, e.g., water, air, etc. (*Build.*) The inclination at which a surface is laid, in order that rain water may be carried off. (*Elec.*) Rate of flow of charge in a substance, solid, liquid, or gas. Conventionally, it is opposite to the flow of (negative) electrons, this having been fixed before the nature of the electric current had been determined. Practical unit of current is the *ampere* (q.v.).

current amplification (*Elec. Eng.*). Ratio of output current to input current of valve amplifier, magnetic amplifier, transistor amplifier, or photomultiplier, often expressed in decibels.

current antinode (*Elec. Eng.*). A point of maximum current in a standing-wave system along a transmission line or aerial.

current attenuation (*Elec. Eng.*). Ratio of output to input currents of a transducer, expressed in decibels.

current balance (*Elec. Eng.*). A form of balance in which the force required to prevent the movement of one current-carrying coil in the magnetic field of a second coil carrying the same current is measured by means of a balancing mass. Cf. *magnetic balance*.

current bedding (*Geol.*). The steeply inclined bedding of rocks deposited in shallow water and under strong current action.

current-carrying capacity (*Elec. Eng.*). The current which a cable can carry before the temperature rise exceeds a permissible value (usually 40°C). It depends on the size of the conductor, the thermal resistances of the cable, and surrounding medium.

current circuit (*Elec. Eng.*). The electrical circuit associated with the current coil of a measuring instrument or relay.

current coil (*Elec. Eng.*). A term frequently used in connexion with wattmeters, energy meters, or similar devices, to denote the coil connected in series with the circuit and therefore carrying the main current.

current collector (*Elec. Eng.*). The device used on the vehicles of an electric traction system for making contact with the overhead contact wire or the conductor-rail. See **bow, pantograph, plough, shoe, trolley system.**

current density (*Elec. Eng.*). Current flowing per unit cross-sectional area of conductor or plasma; expressed in amperes (or milliamperes)/square millimetre.

current efficiency (*Elec. Eng.*). The ratio of the mass of substance liberated in an electrochemical process by a given current, to that which should theoretically be liberated according to Faraday's law.

current feed (*Radio*). Delivery of radio power to a current maximum (loop or antinode) in a resonating part of an antenna.

current feedback (*Elec. Eng.*, *Electronics*). In amplifier circuits, a feedback voltage proportional to the load current. It may be applied in series or in shunt with the source of the input signal. See also **negative feedback, positive feedback, voltage feedback.**

current-fender (*Hyd. Eng.*). A protective construction to deflect a current from a bank which might otherwise be undermined by it.

current gain (*Electronics*). In a transistor, ratio of output current to input current. In common emitter configuration it may be as high as 100, whilst in common base not exceed unity.

current generator (*Elec.*). Ideally, a current source of infinite impedance such that the current will be unaltered by any further impedance in its circuit. In practice, a generator whose impedance is much higher than that of its load.

current limiter (*Elec. Eng.*). A component which sets an upper limit to the current which can be passed.

current margin (*Elec. Eng.*). In a relay, difference between steady-state currents corresponding to values used for signalling and for just operating the relay.

current meter (*Hyd.*). An instrument for measuring the velocity of flow of running water; it consists of a wheel which rotates under the impact of the water, the rate of revolution being recorded; this constitutes a measure of the velocity, as the instrument has previously been calibrated.

current node (*Elec. Eng.*). A point of zero electric current in a standing-wave system along a transmission line or aerial.

current regulator (*Elec. Eng.*). Circuit employed to control the current supplied to a unit.

current (or voltage) resonance (*Elec. Eng.*). Condition of a circuit when the magnitude of a current (or voltage) passes through a maximum as the frequency is changed through resonance; obs. **syntony** or **tuning** (Lodge).

current saturation (*Elec. Eng.*). Condition when anode current in triode valve has reached its maximum value.

current sensitivity (*Elec. Eng.*). The magnitude of the deflection of a current-measuring instrument produced by a given change in current. The term is usually applied to galvanometers, and is then generally expressed as the deflection in mm produced by a current of 1 μA.

current transformer (*Elec. Eng.*). (1) One designed to be connected in series with circuit, drawing predetermined current. Sometimes called **series transformer.** (2) Winding enclosing conductor of heavy alternating current; steps down the current in known ratio for measurement.

current weigher (*Elec. Eng.*). See **current balance.**

currying (*Leather*). The process of treating leather with grease mixtures to make it pliable; applied to leather for boot uppers, belting, harness, etc.

cursor (*Instr.*). The adjustable fiducial part of a drawing or other instrument, e.g., the moving limb of beam-compasses, and the transparent slide bearing the reference line and capable of movement along a slide-rule.

cursorial (*Zool.*). Adapted for running.

curtail step (*Build.*). A step which is not only the lowest step in a flight but is also shaped at its outer end to the form of a scroll in plan.

curtain (*Bot.*). See **cortina.** (*Nuc. Eng.*) Neutron-absorbing shield, usually made of cadmium.

curtain antenna (*Radio*). Large number of vertical radiators or reflectors in a plane, as in original beam system.

curtain machine (*Weaving*). A machine for weaving lace curtains. It has three sets of yarn: (1) the main warp (or ground warp); (2) the spool warp, the function of which is similar to that of weft in a woven texture; (3) the brass-bobbin yarn, which acts as a binder between the main and spool warps.

curtain wall (*Build.*). A thin wall whose weight is carried directly by the structural frame of the building, not by the wall below.

curtain walling (*Build.*). Large-area prefabricated framed sections of lightweight material generally also predecorated on the exterior surface.

curtate (*Comp.*). Horizontal division of a standard punched card into *upper* and *lower* areas.

curtate cycloid (*Maths.*). See **roulette.**

curtate trochoid (*Maths.*). See **roulette.**

Curtis winding (*Elec. Eng.*). The winding of low-capacitance and low-inductance resistors in which the wire is periodically reversed.

curvature (*Bot.*). A change in the general direction of an elongated plant member, due to one side growing faster than the other, or to one side containing more water than the other. (*Maths.*) (1) Of a plane curve: the curvature at a point *P* on the curve measures the rate of change (at the point) in the angle ψ, which the tangent makes with a fixed axis, relative to the arc length *s*. It is thus defined as

$$K = \frac{d\psi}{ds} = \left(\frac{d^2y}{dx^2}\right) \Big/ \left\{1 + \left(\frac{dy}{dx}\right)^2\right\}^{3/2}.$$

K is the reciprocal of the radius of curvature, *p*, which is the radius of the circle which touches the curve (on the concave side) at the point in question. This circle is the *circle of curvature*, and its centre is the *centre of curvature* of the curve at the point. The circle of curvature is also called the *osculating circle*. (2) Of a space curve: the curvature at a point on a space curve is the rate of change of direction of the tangent with respect to the arc length, i.e., $K = \frac{1}{p} = \frac{d\theta}{ds}$, where as before *K* is the curvature, *p* the radius of curvature, θ represents the change in direction of the tangent, and *s* is the length. This is called the *first curvature of a space curve*. The second curvature, or *torsion*, of the curve is the corresponding rate of change in the direction of the binormal, i.e., $\lambda = \frac{d\psi}{ds}$, where ψ is the angle through which the binormal turns. $1/\lambda$ is the radius of torsion. Cf. *moving trihedral* and *osculating sphere*. (3) Of a surface: at any point *P* on a surface there is, in general, a single normal line. Planes through this line cut the surface in plane curves called *normal sections*. The curvature at *P* of a normal section is called the *normal curvature at P* in that direction. The maximum and minimum values of the normal curvature at *P* are called the *principal curvatures at P*, and the directions in which they occur, which will be mutually perpendicular, are called the *principal directions at P*. The average of the principal curvatures at *P* is called the *average* or *mean* or *mean normal curvature at P*, and their product, the *total* or *total normal* or *Gaussian curvature at P*. The reciprocals of the normal and of the principal curvatures at *P* are called the *normal* and *principal radii of curvature at P* respectively.

curvature correction (*Civ. Eng.*). A correction used in the calculation of quantities for earthworks following a curved line in plan; the quantities are taken out as if the line were straight, and a curvature correction made to account for the fact that it is not straight.

curvature of field (*Optics*). Said of an optical system in which the surface of maximum definition of the image is not plane but curved.

curvature of spectrum lines (*Light*). In a spectrum produced by a prism the lines are slightly convex towards the red end. Rays from the ends of the slit are inclined at a small angle to the plane at right angles to the refracting edge of the prism, and so suffer a slightly greater deviation than rays from the centre of the slit, appearing bent towards the violet end of the spectrum.

curve (*Civ. Eng.*). A railway, highway, or canal bend. (*Instr.*) An instrument used by the draughtsman for drawing curves other than circular arcs. It consists of a thin flat piece of transparent plastic or other material, having curved edges which are used as guides for the pencil. (*Maths.*) The locus of a point moving with one degree of freedom.

curve ranging (*Surv.*). The operation of setting out on the ground points which lie on the line of a curve of given radius.

curvilinear asymptote (*Maths.*). See **asymptote.**

curvilinear coordinates (*Maths.*). (1) Of a point in space: three systems of surfaces may be defined by the parametric equation $x=x\ (u, v, w)$, $y=y\ (u, v, w)$, $z=z\ (u, v, w)$, and any point in space may be regarded as the intersection of three surfaces, one from each family. The parameters *u, v, w* are then said to be the *curvilinear coordinates* of the point. If the three systems of curves are mutually orthogonal, the coordinates are *orthogonal* curvilinear coordinates. (2) Of a point on a surface: if a surface is given in parametric form $x=x\ (u, v)$, $y=y\ (u, v)$ and $z=z\ (u, v)$, the parameters *u* and *v* are the curvilinear coordinates of a point on the surface.

curvilinear distortion (*Optics, Photog., TV*). Curvature of lines which should be straight, as seen in the outer portions of the image from a stopped-down simple lens. See **barrel distortion, pincushion (or pillow) distortion.**

cusec (*Hyd. Eng.*). A common abbrev. for *cubic feet per second*, a unit of volumetric rate of flow.

Cushing's syndrome (*Med.*). The concurrence of obesity, hairiness, linear atrophy of the skin, loss of sexual function, and curvature of the spine, due to a tumour in the pituitary or adrenal gland.

cushion (*Bot.*). The central portion of the prothallus of a fern; it is several layers of cells in thickness, and bears rhizoids and archegonia. (*Build.*) The capping stone of a pier.

cushion course (*Civ. Eng.*). A layer of sand, or sand and dry Portland cement, or mortar, spread over the foundation of a road to receive a surface of special bricks. Also called **bedding course** or **subcrust** (the preferable term).

cushioncraft (*Ships, etc.*). Name given to certain types of *hovercraft* (q.v.).

cushioning (*Acous.*). The use of resilient material for isolating sensitive devices, such as microphones or recorders, from the influence of external vibrations. (*Eng.*) The compression of a small quantity of steam in a steam-engine cylinder towards the end of the exhaust stroke, to assist in bringing the piston to rest at the dead-centre and minimize inertia forces; or, similarly, of a small quantity of air in a pneumatic cylinder, to minimize inertia forces at the end of the stroke.

cushion steam (*Eng.*). The steam shut in the cylinder of a steam-engine after the closing of the exhaust valve. See **cushioning.**

cusp (*Bot., Zool.*). A sharp-pointed prominence, as on teeth. (*Maths.*) See **double point.** *adj.* **cuspidate.**

cusps (*Astron.*). The horns of the moon or of an inferior planet in the crescent phase.

customs plant (*Min. Proc.*). A crushing or concentrating plant serving a group of mines on a contract basis. It buys ore according to valuable content and complexity of treatment, and relies for profit on sale of products.

cut (*Cinema.*). Sudden ending of a shot, either in the original take or recording, or in the subsequent editing. (*Eng.*) The thickness of the metal shaving removed by a cutting tool. (*For.*) The output of timber (round or sawn) from a forest area for a given period of time; also the actual area felled. (*Hyd. Eng.*) The water-way between the pontoons of a pontoon-bridge. (*Nuc. Eng.*) Proportion of input material to any stage of an isotope separation plant which forms useful product. Also **splitting ratio.** (*Textiles*) (1) In the cotton and wool industries, it denotes length of warp of cloth that has to be woven before removal from the loom; usually plainly marked on the warp sheet. (2) An alternative term for a linen lea, a length of 300 yards. (*Typog.*) A *block* (q.v.).

cut-and-cover (*Civ. Eng.*). A method often adopted in the construction of underground railways at only a moderate depth. A cutting is first excavated to accommodate the railway; it is then covered over to original ground-level by arching supported on side walls.

cut-and-fill (*Civ. Eng.*). A term used to describe any cross-section of highway or railroad earthworks which is partly in cutting and partly in embankment.

cut-and-mitred-string (*Carp.*). A *cut string* (q.v.) which is mitred at the vertical parts of the notches in the upper surface, so that the end grain of the risers may be concealed.

cut-and-mitred valley (*Build.*). A valley formed in a tiled roof by cutting one edge of the tiles on both sides of the valley so that they form a mitre, which is rendered watertight by lead soakers bonded in with the tiles.

cutaneous (*Zool.*). Pertaining to the skin.

cut backs (*Civ. Eng.*). Blends of asphaltic bitumen with various solvents, for use at comparatively low temperatures for road surfacing.

cutch (*For.*). Strictly, a solid extract, containing tannin and catechin, obtained by boiling from the heartwood of *Acacia catechu*. Used for dyeing and in medicine.

cut edges (*Bind.*). Said of the edges of a book when all three are clean cut as distinct from *trimmed edges*, where only the furthest projecting leaves at the tail are trimmed.

cut flush (*Bind.*). Cut after binding, so that the cover does not project.

cuticle (*Bot.*). A deposit of waterproof, waxy material forming the external layer of the outer walls of epidermal cells. (*Zool.*) A nonliving layer secreted by, and overlying, the epidermis.

cuticular diffusion (*Bot.*). The passage of oxygen and of carbon dioxide through the cuticle of a plant.

cuticularization (*Bot.*). The development of a cuticle.

cuticular transpiration (*Bot.*). The loss of water vapour from a plant through the cuticle.

cuticulin (*Zool.*). The innermost layer of the insect epicuticle, consisting of lipoprotein.

cutin (*Bot.*). A mixture of fatty substances, which is deposited on or in the outer layer of cell walls as cuticularization proceeds.

cutinization (*Bot., Zool.*). The formation of cutin; the deposition of cutin in a cell wall to form a cuticle.

cut-in notes (*Typog.*). Notes occupying a rectangular space, set into the text at the outer edge of a paragraph.

cutis (*Zool.*). The dermis or deeper layer of the Vertebrate skin.

cut key (*TV*). See **censor key.**

Cutler feed (*Acous.*). Resonant cavity at the end of waveguides which is used to feed mirror antennae.

Cutleriales (*Bot.*). An order of the *Phaeophyta.* Growth of the thallus is at least in part trichothallic; only unilocular sporangia are produced, and the gametes are heterothallic, producing anisogametes.

cut-off (*Eng.*). The point in an engine cycle, expressed as percentage of stroke, at which the supply of steam, fuel oil, etc., is stopped. (*Print.*) A feature of reel-fed presses; the paper is cut after printing to a size determined by the cylinder periphery; a few models have a selection of cylinder sizes and a consequently variable cut-off. (*Vac. Tech.*) A device in which a liquid surface is used, when required, to separate two parts of a vacuum system. The use of the term cut-off to describe what are really taps or valves is deprecated.

cut-off current (*Electronics*). The residual current flowing in a valve or transistor when the device is biased off in a specified way.

cut-off field (*Electronics*). Same as **critical field.**

cut-off frequency (*Electronics*). The theoretical frequency is that at which the attenuation constant changes from zero to a positive value (or vice versa). See also **alpha cut-off, critical frequency.**

cut-off knife (*Print.*). A plain or serrated blade which severs each copy on a reel-fed rotary. Cf. ***knife***.

cut-off rubbers (*Print.*). The rubber strip set in a cylinder against which the *cut-off knife* presses when cutting the product into copies. Also called **cutting buffer, cutting strip.**

cut-off voltage (*Electronics*). Negative value which must be applied to the grid of a thermionic valve to reduce the anode current substantially to zero; approx. equal to anode voltage divided by amplification factor.

cut-out. Off-switch operated automatically if safe operating conditions are not maintained, e.g., water flow cut-out.

cut-out half-tone (*Print.*). A half-tone from which the background is removed to give prominence to the subject.

cut-over (*Acous.*). The cutting through of the plane surface from one spiral to the next, in cutting originals for disk records. (*Teleph.*) The rapid transfer of large numbers of subscribers' lines from one exchange to another, particularly from a manual to an automatic exchange.

cut pile (*Textiles*). See **pile.**

cut-stone (*Build.*). A stone hewn to shape with a chisel and mallet.

cut string (*Carp.*). A string whose upper surface is shaped to receive the treads and risers of the steps, while the lower surface is parallel to the slope of the stair. Also called an **open string.**

cutter (*Acous.*). The sapphire or diamond point which removes the thread of lacquer in gramophone-disk recording or in a dictaphone. (*Eng.*) Any tool used for severing, often more specifically a milling cutter. See **milling-machine.**

cutter dredge (*Civ. Eng.*). Alluvial dredge which loosens material by means of powered cutting ring, draws it to a pump and delivers it for treatment aboard in adjacent plant.

cutter loader (*Mining*). Coal-cutting machine which both severs the mineral and loads it on to a transporting device such as a face conveyor.

cutters (*Build.*). Bricks which are made soft enough to be cut with a trowel to any shape required, and then rubbed to a smooth face and the correct shape. Also called **rubbers.** (*Mining*) See **backs.**

cutting (*Bot.*). A portion of a plant, usually of a stem or root, which is cut off, induced to form roots, and so serves as a means of vegetative propagation. (*Cinema.*) The last creative operation in selecting the strips of motion-picture required in the final positive, the negatives being then cut to suit before release printing. (*Civ. Eng.*) An open excavation through a hill, for carrying a highway or railroad at a lower level than the surrounding ground. (*Textiles*) In finishing woollen fabrics, the removal of loose or projecting fibres, with revolving blades similar to those on a lawn-mower, in order to secure a clear-finished surface; also known as **shearing.** In some cases (velvets, carpets, etc.), two fabrics woven face to face are cut after weaving to give two cloths. (*Vet.*) See **brushing.**

cutting buffer (*Print.*). See **cut-off rubbers.**

cutting compound (*Eng.*). A mixture of water, oil, and soft soap, etc., used for lubricating and cooling the cutting tool in machining operations. See **coolant.**

cutting curb (*Civ. Eng.*). See **curb.**

cutting cylinder (*Print.*). The cylinder that holds knives to cut the web into separate copies. See **cut-off knife, cut-off rubbers.**

cutting disks (*Print.*). See **slitters.**

cutting edge (*Cinema.*). A colloquialism for the demarcation between the black and transparent parts of the sound-track in variable-area sound recording on film.

cutting gauge (*Join.*). A *marking gauge* (q.v.) fitted with a bevelled cutter in place of a pin. Used for cutting thin wood and for marking across the grain to obviate tearing.

cutting head (*Electronics*). Electrodynamical or electromechanical transducer which deflects the cutting stylus in *direct* gramophone-disk recording. See also **crystal cutter.**

cutting list (*Join.*). A list giving dimensions—sometimes with diagrams of sections—of timber required for any given work.

cutting machines (*Textiles*). (1) Machines for cutting long floating threads in velveteen and corduroy fabrics in order to produce a pile surface. (2) In woollen manufacture, cropping machines for removing fibre from the cloth surface to obtain a clear surface and a clear finish. (3) Machines for cutting, folding and stacking towels, etc.

cutting speed (*Eng.*). The speed of the work relative to the cutting tool in machining operations; usually expressed in feet or metres per minute.

cutting strip (*Print.*). See **cut-off rubbers.**

cutting tools (*Eng.*). Steel tools used for the machining of metals. See **broach, cutter, lathe tools, milling-cutter, planer tools, reamer, screwing die, shaper tools, slotting tools, tap, twist drill.**

cutting-up (*Foundry*). The operation of roughening the surface of a mould at a part where fresh sand is to be added, to assist adhesion.

cuttling (*Textiles*). (1) Operation of folding a fabric to make it convenient to handle; also known as **plaiting.** (2) Permitting wool cloths to relax sufficiently for fibres to recover.

cut-up trade (*Textiles*). Section of the knitting industry dealing with fabric made on a circular knitting-machine. The material is afterwards cut to shape from patterns, pieces being sewn together and pressed to form the final article.

cut-water (*Civ. Eng.*). The angular edge of a bridge-pier, shaped to lessen the resistance it offers to the flow of water.

Cuvierian ducts (*Zool.*). In lower Vertebrates, a pair of large venous trunks entering the heart from the sides.

Cuvierian organs (*Zool.*). In *Holothuroidea*, modified branches of the respiratory trees, covered with a viscous substance which can be extruded by rupture of the body-wall, and which give rise to sticky threads which entangle an enemy.

C-wire (*Teleph.*). The control wire in the Siemens No. 16 system of automatic telephony. Its function is similar to that of the *P-wire* (q.v.), in the standard G.P.O. system.

Cx-band (*Radar*). See **X-band.**

Cy (*Chem.*). A symbol for the cyanogen radical CN—.

cyamelide (*Chem.*). A colourless, insoluble, amorphous polymer of cyanic acid formed rapidly at low temperatures. See also **cyanuric acid.**

cyanamide process (*Chem.*). The fixation of atmospheric nitrogen by heating calcium carbide (ethynide) in a stream of the gas. Calcium cyanamide, $CaCN_2$, is thus formed, and this, on treatment with water, a little sodium hydroxide, and steam under pressure, yields ammonia.

cyanates (*Chem.*). Salts containing the monovalent acid radical CNO′.

cyanhydrins (*Chem.*). A series of compounds formed by the addition of hydrogen cyanide to aldehydes and ketones. Their general formula is $R'\cdot C(OH)(CN)\cdot R''$ and they are useful for the preparation of 2-hydroxy-acids. See **acetone cyanhydrin.** Also **cyanohydrins.**

cyanicide (*Min. Proc.*). Any constituent in ore or chemical product made during treatment of gold-bearing minerals by cyanidation, which attacks or destroys the sodium or calcium cyanide used in the process.

cyanidation vat (*Met.*). A large tank, with a filter bottom, in which sands are treated with sodium cyanide solution to dissolve out gold.

cyanide hardening (*Met.*). Case-hardening in which the carbon content of the surface of the steel is increased by heating in a bath of molten sodium cyanide.

cyanides (*Chem.*). (1) Salts of hydrocyanic acid. (2) See **nitriles.**

cyaniding (*Met.*). The process of treating finely ground gold and silver ores with a weak solution of sodium cyanide, which readily dissolves these metals. The precious metals are obtained by precipitation from solution with zinc.

cyanin (*Chem.*). The colouring matter of the cornflower and the rose. It is an anthocyanin, and on hydrolysis yields cyanidin and two molecules of glucose.

cyanine (*Photog.*). Blue dye, 1,1′-di-n-amyl-4,4′ cyanine:

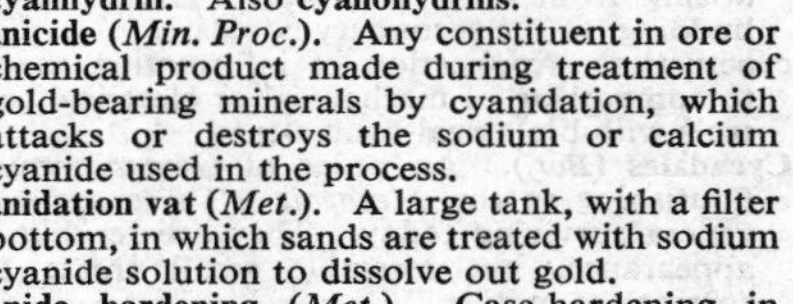

Am–N =CH– N⁺–Am I⁻

Used as sensitizer. Also called **quinoline blue.**

cyanite (*Min.*). See **kyanite.**

cyanizing (*Build.*). See **kyanizing.**

cyanocobalamin (*Biochem.*). Vitamin B_{12}. See **vitamin B complex.**

cyanogen (*Chem.*). A very poisonous, colourless gas with a smell of bitter almonds. It is soluble in 4 volumes of water, ammonium oxalate (ethandioate) being formed on standing. Its formula is C_2N_2, or N≡C—C≡N, and it somewhat resembles the halogens in its chemical behaviour.

cyanohydrins (*Chem.*). See **cyanhydrins.**

Cyanophyceae (*Bot.*). See **Myxophyceae.**

cyanoplast (*Bot.*). A minute pigmented granule in the cytoplasm of cells of the *Myxophyceae*.

cyanosis (*Med.*). Blueness of the skin and the mucous membranes due to insufficient oxygenation of the blood.

cyanotype (*Photog.*). The ferroprussiate process, familiar as blue-printing; it depends on the light reduction of a ferric salt to a ferrous salt, with production of Prussian blue on development.

cyanuric acid (*Chem.*). A tribasic, heterocyclic acid, having the formula $H_3C_3N_3O_3$. The trimer of cyanic acid which is too unstable to exist by itself, an aqueous solution being slowly converted to urea.

cyanuric dyes (*Chem.*). Relatively new class of dyestuffs based on cyanuric chloride ($C_3N_3Cl_3$). Their importance lies in the ability to link chemically with the fabrics or fibres (particularly cellulosic fibres) being dyed. See **reactive dyes.**

cyathiform (*Bot.*). Deeply cup-shaped.

cyathium (*Bot.*). The inflorescence characteristic of spurges. It consists of a number of bracts surrounding a group of staminate flowers each consisting of a single stamen, with a central pistillate flower of a stalked tricarpellary ovary.

cyathozooid (*Zool.*). In *Lucida*, the primary zooid arising from an ovum, which gives rise, by budding, to the secondary zooids.

cybernetics. Application of information theory to comparison of mechanical or electrical controls with biological equivalents.

Cycadales (*Bot.*). An order of *Gymnospermae*, flourishing in the *Mesozoic*. There are nine genera surviving today. They are fernlike in appearance, but resemble small trees; the sperms are motile.

Cycadofilicales (*Bot.*). An extinct order of *Gymnospermae* of the late *Devonian* to early *Triassic*. Superficially like ferns with the macrospores and microspores produced on, or instead of, leaves.

Cycadophyta (*Bot.*). A subclass of the *Gymnospermae*, including the *Bennettitales*, *Cycadales* and *Cycadofilicales*. Generally characterized by an unbranched trunk and large compound leaves. Also **Cycadopsida.**

cyc-arc welding (*Elec. Eng.*). An arc-welding process for attaching studs, etc., to steel plates; an arc is drawn between the stud and the plate, and the two are then pressed together.

cyclamates (*Chem.*). Derivatives of cyclohexylsulphamic acid, having 30 times the sweetening power of sucrose, much used in foods, drinks, and for dietary purposes; banned in some countries because of supposed health hazards.

cyclanes (*Chem.*). A synonym for cycloalkanes, or polymethylenes, hydrocarbons containing saturated carbon rings.

cycle. A series of occurrences in which conditions at the end of the series are the same as they were at the beginning. Usually, but not invariably, a cycle of events is recurrent. (*Acous.*) The complete variation of particle-motion or sound-pressure in a sound-wave of a single frequency or of a continuously repeated stable waveform. (*Comp.*) (1) Sequence of operations which continues until a criterion is reached for stoppage. (2) Interval which must be allowed for minimum (*minor*) or maximum (*major*) length of *words*, or a recurrence in a rotating (cyclic) store. (*Maths.*) For a periodic quantity or function, the set of values that it assumes during a period.

cycle criterion (*Comp.*). Number of times, in programming, that a particular cycle must be performed.

cyclegraph (*Work Study*). A record of a path of movement, usually traced by a continuous source of light on a photograph, preferably stereoscopic.

cycle of erosion (*Geol.*). The definite course of development followed in landscape evolution; it consists of the major stages of youth, maturity, and old age.

cycle reset (*Comp.*). In programming, setting of cycle index back to its original value.

cycle time (*Work Study*). The total time taken to complete the *elements* of a *work cycle*.

cyclic (*Bot.*). Having the parts arranged in whorls not in spirals.

cyclic adenylic acid (*Chem.*). A cyclic nucleotide, adenosine-3′, 5′ phosphate, formed from ATP. It is one factor required for the transition of phosphorylase kinase from the inactive to the active form, which is important in the breakdown of glycogen.

cyclic compounds (*Chem.*). Closed-chain or ring compounds consisting either of carbon atoms only (carbocyclic compounds), or of carbon atoms linked with one or more other atoms (heterocyclic compounds).

cyclic group (*Maths.*). A group in which every element can be expressed as a power of a single element. Cyclic groups are *Abelian*, and those of the same order are *isomorphic.*

cyclic pitch control (*Aero.*). Helicopter rotor control in which the blade angle is varied sinusoidally with the blade azimuth position, thereby giving a tilting effect and horizontal translation in any desired direction.

cyclic quadrilateral (*Maths.*). A four-sided polygon whose vertices lie on a circle.

cyclic test (*Min. Proc., etc.*). See **locked test.**

cycling (*Eng.*). See **hunting.**

cyclitis (*Med.*). Inflammation of the ciliary body of the eye.

cyclized rubber (*Paint.*). A form of *isomerized rubber* (q.v.) in which the chainlike rubber molecules are rearranged into ring structures.

cyclo-. Prefix from Gk. *kyklos*, circle.

cyclo- (*Chem.*). Containing a closed carbon chain or ring.

cycloalkanes (*Chem.*). See **cyclanes.**

cyclobutane (*Chem.*). $(CH_2)_4$. Alicyclic compound; a cycloalkane. B.p. 11°C.

$$\begin{array}{c} CH_2\text{—}CH_2 \\ | \quad\quad | \\ CH_2\text{—}CH_2 \end{array}$$

cyclocoelic (*Zool.*). Having a spirally coiled intestine.

cyclogram (*Electronics*). Graph showing characteristics of negative resistance oscillator and constructed on isocline diagram. It forms a closed loop or any stable oscillator; the loop being a circle when the output is a sine wave.

cyclograph (*Electronics*). An electron optical system in which the electron beam moves in two perpendicular directions.

cyclohexanamine (*Chem.*). See **cyclohexylamine.**

cyclohexane (*Chem.*). C_6H_{12}, m.p. 2°C, b.p. 81°C, rel. d. 0·78, a colourless liquid, of mild ethereal odour.

cyclohexanol (*Chem.*). $C_6H_{11}{\cdot}OH$, m.p. 15°C, b.p. 160°C, rel. d. 0·945, an oily, colourless liquid.

cyclohexanone (*Chem.*). Keto-hexamethylene, b.p. 154°–156°C, rel. d. 0·945, a colourless liquid, of propanone-like odour, solvent for cellulose lacquers.

cyclohexylamine (cyclohexanamine) (*Chem.*). Colourless liquid, b.p. 134°C:

$$\begin{array}{c} CH_2-CH_2 \\ CH_2 \quad\quad CH{\cdot}NH_2 \\ CH_2-CH_2 \end{array}$$

A reduction product of aniline, its derivatives are used in the manufacture of plastics, etc.

cycloid (*Maths.*). An arch-shaped curve with intrinsic equation $s=4a\sin\psi$. Its parametric cartesian equations are $x=a(\theta+\sin\theta)$ and $y=a(1-\cos\theta)$. See **roulette**. (*Psychiat.*) A periodic state of alternating moods, swinging from depression to exhilaration; accompanied by a social extraverted type of personality. It occurs in the *pyknic type* (q.v.) of individual and is a normal disposition; in its extreme form this state may develop into a manic depressive psychosis. (*Zool.*) Evenly curved; said of scales which have an evenly curved free border.

cycloidal teeth (*Eng.*). Gear teeth whose flank profiles consist of cycloidal curves.

cyclometer. A revolution counter calibrated in miles or kilometres, driven by the wheel of a pedal bicycle to record the distance travelled.

cyclomorphosis (*Ecol.*). Cyclic changes in form, or periodic polymorphism, as seen in some *Entomostraca* (*Crustacea*), Malaria Parasites, Aphids, etc. Cyclomorphic species show a life cycle of a population, not just an individual life cycle including metamorphosis.

Cyclomyaria (*Zool.*). See **Doliolidae.**

cyclone (*Meteor.*). (1) Same as **depression.** (2) A *tropical revolving storm* (q.v.) in the Arabian Sea, Bay of Bengal and South Indian Ocean. (*Min. Proc.*) Conical vessel used to classify dry powders or extract dust by centrifugal action. See **hydrocyclone.**

cyclonite (*Chem.*). Hexogen, cyclotrimethylene trinitramine $(CH_2)_3(N{\cdot}NO_2)_3$, a colourless, crystalline solid, m.p. 200–202°C, odourless, tasteless, non-poisonous, soluble in acetone, prepared by oxidative nitration of hexamethylene tetramine. Used, generally with TNT, as an explosive.

cyclo-octadiene (*Chem.*). Dimerization product of butadiene obtained by using *Ziegler catalysts* (q.v.). Used as an intermediate in the preparation of nylon polymers from petrochemical sources.

cycloparaffins (*Chem.*). Same as cyclanes.

cyclopean (*Build.*). A name given to ancient dry-masonry works in which the stones are very large and are irregular in size.

cyclopentane (*Chem.*). See **pentamethylene.**

cyclopentanol (*Chem.*). Cyclic aliphatic alcohol, b.p. 139°C, used as an intermediate in the preparation of perfumery and flavouring esters.

cyclopentanone (*Chem.*). Ketopentamethylene cyclic ketone. Solvent for a wide range of synthetic polymers, particularly PVC.

cyclophon (*Electronics*). Tube which uses the fundamental principle of electron-beam switching.

cycloplegia (*Med.*). Paralysis of the ciliary muscle.

cyclopoid larva (*Zool.*). A larval type of certain *Hymenoptera* which bears a superficial resemblance to the nauplius larva of *Crustacea.*

cyclopropane (*Chem.*). Trimethylene $(CH_2)_3$, a gas given with oxygen to produce anaesthesia.

cyclops larva (*Zool.*). In *Copepoda*, the larval stages succeeding the metanauplius: the first resembles an adult *Cyclops* but lacks appendages behind the third pair of swimming limbs and also the somites of the urosome; in successive stages, the missing somites appear.

Cyclorrhapha (*Zool.*). *Diptera*, e.g., the house fly, whose larvae lack head or legs, and whose pupae are co-arctate, i.e., contained in the puparium.

cyclosis (*Biol.*). The circulation of protoplasm within a cell.

cyclospermous (*Bot.*). Having the embryo coiled round the endosperm.

cyclospondylous (*Zool.*). Showing partial calcification of cartilaginous vertebral centra in the form of concentric rings.

Cyclosporae (*Bot.*). A class of the *Phaeophyta.* There is no alternation of generations. The plant is sporophytic, the spores acting as gametes. Sexual reproduction is oögamous.

cyclosteel (*Met.*). Steel produced by blowing iron-ore powder into a hot gas.

Cyclostomata (*Zool.*). A class of the superclass *Agnatha* (q.v.) of the *Craniata.* Aquatic and gill-breathing, with a round suctorial mouth; buccal cavity contains a muscular tongue bearing horny teeth used to rasp the flesh from the prey; cartilaginous endoskeleton; no fins or limb girdles; slimy skin with no scales. Lampreys and Hagfish.

cyclostrophic wind (*Meteor.*). Wind blowing along a circular track where the *geostrophic force* (q.v.) is small, i.e., in low latitudes. As in *tropical revolving storms* (q.v.).

cyclothymia (*Psychiat.*). Manic-depressive psychosis.

cyclotron (*Nuc.*). Machine in which positively-charged particles are accelerated in a spiral path within *dees* in a vacuum between poles of a magnet, energy being provided by a high-frequency voltage across the dees.

cyclotron frequency (*Nuc.*). In a magnetron, frequency with which an electron orbits in magnetic field in the absence of electric field.

Cycolac (*Plastics*). TN for range of ABS copolymers based on butadiene-styrene and acrylonitrile-styrene copolymers.

cyesis (*Med.*). Pregnancy.

Cygnus A (*Astron., Radio*). A strong source of radio emission in the constellation *Cygnus*; it has been identified as two faint galaxies, possibly in collision.

cylinder (*Eng.*). The tubular chamber in which the piston of an engine or pump reciprocates; the internal diameter is called the *bore*, and the piston-travel the *stroke*. (*Maths.*) (1) A (cylindrical) surface generated by a line which moves parallel to a fixed line so as to cut a fixed plane curve. Any line lying in the surface is called a *generator*. (2) A solid bounded by a cylindrical surface and two parallel planes (the *bases*) which cut the surface. A cylindrical surface is named after its normal sections and a solid cylinder after its bases. The axis of a cylinder (if it has one) is its line of symmetry parallel to its generators or the line joining the midpoints of its bases. A circular cylinder is called a *right circular cylinder* if its bases are normal to its axis, otherwise it is called an *oblique circular cylinder*.

cylinder barrel (*Eng.*). The wall of an engine cylinder, as distinct from the cylinder itself, which term includes the head or covers.

cylinder bearers (*Print.*). At each end of impression and plate cylinders, to provide a datum for packing; may be integral with cylinder or be a removable band (*bearer ring*).

cylinder bit (*Carp.*). A steel drill with helical cutting edge, used for precise boring.
cylinder bore (*Eng.*). See **cylinder.**
cylinder brakes (*Print.*). Mechanism on a fast-running printing press which stops it quickly.
cylinder caisson (*Civ. Eng.*). A caisson formed of hollow cylindrical cast-iron sections arranged one on top of another, so that there is always one above water-level, while the bottom one is a special cutting section. As excavation proceeds within the cylinder, the loaded sections sink, and when they have reached a sufficient depth, the cylinder is filled with concrete.
cylinder collection (*Print.*). On rotary presses, the gathering of the required number of sections or sheets round a cylinder.
cylinder cover (*Eng.*). The end cover of the cylinder of a reciprocating engine or compressor.
cylinder dressing (*Print.*). The layers of board, paper, or other packing under rubbers, automatic blankets, required to produce the necessary height on an impression cylinder.
cylinder-dried (*Paper*). Paper which has been dried by being passed over heated cylinders.
cylinder escapement (*Horol.*). A frictional-rest escapement in which the balance is mounted on a hollow cylinder, and a tooth of the escape wheel gives impulse to the balance by pressing against the lips of the cylinder, the action being that of a wedge. The escape wheel is locked by the tip of the tooth pressing against the outside or inside of the cylinder. The teeth of the escape wheel are mounted on 'stalks' and stand at right angles to the plane of the wheel. To admit the entry of the teeth into the cylinder, about one-half of the cylinder is cut away where the teeth enter. Also known as **horizontal escapement.**
cylinder head (*Eng.*). The closed end of the cylinder of an internal-combustion engine; it may be either integral with the barrel or detachable, and sometimes carries the valves.
cylinder mould machine (*Paper*). A paper machine in which a cylindrical wire mesh rotates in a vat of diluted pulp. The sheet formed is removed at the top of the cylinder by a rubber roll and felt. Also called a **vat machine.**
cylinder oils (*Oils*). Dark or red oils with high viscosity, suitable for use in steam and internal-combustion engine cylinders. See **lubricants, machinery oils.**
cylinder press (*Typog.*). A general term used to distinguish cylinder printing machines from hand presses, platens, and rotary machines. See **stop-cylinder, single-revolution, two-revolution.**
cylinder process (*Glass*). A method of making flat glass in which a gather of molten glass is made into a cylinder which is split lengthwise, reheated and flattened in a flattening kiln.
cylinders (*Textiles*). See **swifts.**
cylinder wrench (*Plumb.*). See **pipe wrench.**
cylindrical. Elongated, and circular in cross-section.
cylindrical coordinates (*Maths.*). Three numbers, r, θ and z, which represent the position of a point in space, the first two numbers r and θ representing, in polar coordinates, the position of the projection of the point on a reference plane, and the third z representing the height of the point above the reference plane. Related to rectangular cartesian by the equations $x=r\cos\theta$, $y=r\sin\theta$ and $z=z$.
cylindrical equal-area projection (*Geog.*). A type of *cylindrical projection* (q.v.) in which the E-W exaggeration is compensated for by N-S reduction.
cylindrical gauge (*Eng.*). A length gauge of cylindrical form whose length and diameter are made to some standard size. See **gauge (3).**
cylindrical grinding (*Eng.*). The operation of accurately finishing cylindrical work by a high-speed abrasive wheel. The work is rotated by the headstock of the machine and the wheel is automatically traversed along it under a copious flow of coolant.
cylindrical lens (*Photog.*). A lens cut in the shape of a segment of a cylinder. Used to obtain a line image, as in sound recording on film or in reproducing from a sound-track.
cylindrical projections (*Geog.*). Best imagined as projected on to a cylinder of paper wrapped around the globe and then flattened into a plane (*developed*). A cylinder concentric with the axis of the globe gives a projection in which all parallels and all meridians are of equal length, making for distortion of scale at points far removed from the equator. Best for representing equatorial areas.
cylindrical record (*Acous.*). The Edison-type of gramophone record, in which the reproducing needle traverses a spiral (helical) record on its surface.
cylindrical rotor (*Elec. Eng.*). A rotor of an electric machine in which the windings are placed in slots around the periphery, so that the surface is cylindrical.
cylindrical wave (*Phys.*). One where equiphase surfaces form coaxial cylinders.
cylindrical winding (*Elec. Eng.*). A type of winding used for core-type transformers; it consists of a single coil of one or more layers wound concentrically with the iron core; it is usually long compared with its diameter.
cylindrite (*Min.*). A complex sulphide of lead, tin, antimony, which has a cylindrical habit.
cyma (*Arch.*). A much-used moulding showing a reverse curve in profile. Also called an **ogee.**
cyma recta (*Arch.*). A cyma which is concave at the top and convex at the bottom.
cyma reversa (or inversa) (*Arch.*). A cyma which is convex at the top and concave at the bottom.
cymbiform (*Bot.*). Shaped like a boat.
cymbium (*Zool.*). In some male Spiders, the cup-shaped tarsus of the pedipalpus, containing the palpal organ.
cyme (*Bot.*). An inflorescence in which the main axis ends in a flower, and in which subsequent flowers are produced at the ends of lateral axes or of successive branches from these. *adj.* **cymose.**
cymene (*Chem.*). $CH_3{\cdot}C_6H_4{\cdot}CH(CH_3)_2$, 1-(1-methylethyl)-4 methylbenzene, b.p. 175°C.
cymography (*Radiol.*). See **kymography.**
cymophane (*Min.*). A variety of the gem-mineral *chrysoberyl* which exhibits chatoyancy; sometimes known as **chrysoberyl cat's eye** or **Oriental cat's eye.**
cymose (*Bot.*). Sympodially branched. See **definite (2).**
cynopodous (*Zool.*). Having nonretractile claws, as dogs.
cyphelia (*Bot.*). A small cup-shaped hollow in the under surface of the thallus of some lichens.
Cyphonautes (*Zool.*). A ciliated pelagic larval form of ectoproct *Polyzoa*, possessing a bivalve shell.
cypress knee (*Bot.*). A vertical upgrowth from the roots of the swamp cypress. It is very loose in structure, and acts as a pneumatophore.
cypris larva (*Zool.*). In *Cirripedia*, the larval stage succeeding the nauplius; it possesses a bivalve shell with an adductor muscle, and a pair of

compound eyes; the antennules bear a disk for attachment to the substratum.

cypsela (*Bot.*). A 1-seeded fruit, formed from a syncarpous, inferior ovary.

cyrtolite (*Min.*). A partially metamict variety of zircon.

cyst (*Zool.*). A nonliving membrane enclosing a cell or cells; any bladderlike structure, as the gall bladder or the urinary bladder of Vertebrates; a sac containing the products of inflammation. *adj.* **cystic, cystoid, cystiform.**

cyst cells (*Zool.*). Cells in the testes of Insects, which enclose, and possibly nourish, masses of developing spermatozoa.

cysteine (*Chem.*). See **cystine.**

cystenchyma (*Zool.*). A form of parenchyma occurring in Sponges, characterized by vesicular vacuolate cells (*cystencytes*) closely packed or embedded in a gelatinous matrix.

cystencyte (*Zool.*). In Sponges, a *collencyte* (q.v.), which has acquired a vesicular structure.

cystic (*Zool.*). Pertaining to the gall bladder; pertaining to the urinary bladder.

cystic adenoma (*Med.*). An adenoma containing numerous cysts.

cystic duct (*Zool.*). The duct from the gall bladder which meets the hepatic duct to form the common bile duct.

cysticercoid (*Zool.*). Larval stage of certain tapeworms in which the bladder is but slightly developed and may possess a tail-like appendage.

cysticercosis (*Med.*). Infection with cysticerci.

cysticercus (*Zool.*). Bladderworm; larval stage in many tapeworms, possessing a fluid-filled sac containing an invaginated scolex.

cystic fibrosis (*Med.*). See **muco-viscidosis.**

cysticolous (*Zool.*). Cyst-inhabiting.

cystidiole (*Bot.*). A sterile structure in the hymenium of *Hymenomycetes.* Wider than paraphyses, and little differentiated.

cystidium (*Bot.*). A swollen, elongated, sterile hypha, occurring among the basidia of the hymenium of some *Hymenomycetes*, usually projecting beyond the surface of the hymenium.

cystine (*Chem.*). A sulphur-containing amino acid, HOOC—CH(NH_2)—CH_2—S—S—CH_2—CH(NH_2)—COOH, present in many proteins, but particularly in the keratins of hair, wool, and skin. It readily undergoes reversible reduction to *cysteine*, HS·CH_2·CH(NH_2)·COOH, a reaction which may be important in relation to protein structure. A nonessential amino acid in respect of growth.

cystitis (*Med.*). Inflammation of the bladder.

cystocarp (*Bot.*). The body which forms after fertilization in the red algae.

cystocele (*Med.*). Hernia of the bladder.

cystocyte (*Zool.*). See **cystencyte.**

cystogenous (*Zool.*). Cyst-forming; cyst-secreting.

cystography (*Radiol.*). The radiological examination of the urinary bladder following the administration intravenously or through the urethra of a *contrast medium* (q.v.).

Cystoidea (*Zool.*). An extinct class of *Echinodermata*, known by Palaeozoic fossils, with sac- or vase-shaped bodies, stalks and food-grooves either on the theca or on special ossicles. The plates of the theca are hexagonal and bear pores either in pairs (diplopores) or in diamond patterns (pore-rhombs).

cystolith (*Bot.*). A stalked mass of calcium carbonate and organic material, present in some plant cells.

cyston (*Zool.*). A dactylozooid specialized for excretion.

cystoscope (*Surg.*). An instrument for inspecting the interior of the bladder.

cystosorus (*Bot.*). A group of sporangia formed after the division of a single protoplast.

cystospore (*Bot.*) An encysted zoospore.

cystostomy (*Surg.*). Formation of an opening in the bladder.

cystotomy (*Surg.*). Incision into the bladder.

cystovarian (*Zool.*). Having the ovaries enclosed in coelomic pouches, and therefore in the form of hollow sacs, e.g., in teleostean Fish.

cystozooid (*Zool.*). In *Cestoda*, the bladder or tail portion of a bladderworm. Cf. *acanthozooid.*

cyt-, cyto-. Prefix from Gk. *kytos*, hollow, cell.

cytarme (*Zool.*). In experimental embryology, flattening of cells against one another.

cytase (*Bot., Zool.*). A general term for an enzyme able to break down the $\beta-1\rightarrow4$ link of cellulose.

cytaster (*Cyt.*). An *aster* (q.v.) which lies distant from the nucleus in the cytoplasm.

cytisine (*Chem.*). $C_{11}H_{14}ON_2$, an alkaloid obtained from the seeds of *Cytisus laburnum.* It crystallizes in large rhombic prisms, m.p. 153°C, b.p. 218°C (250 N/m^2); soluble in water, ethanol, trichloromethane, insoluble in ethoxyethane or benzene. Cystine is extremely poisonous, causing nausea, convulsions, and death by asphyxiation.

cyto-architectonic (*Med.*). Pertaining to the disposition of cells in a tissue.

cytocentrum (*Cyt.*). Centrosome; idiozome.

cytochorismus (*Zool.*). In experimental embryology, partial separation of cells after one division and prior to the next division.

cytochrome (*Chem.*). A group of *haemochromogens* (q.v.) very widely distributed in living cells and of great importance in cell oxidations, acting as intermediate hydrogen carriers.

cytochrome oxidase (*Biochem.*). The terminal member of the cytochrome chain which is the only member capable of reducing oxygen. Its composition is uncertain.

cytode (*Biol.*). A mass of protoplasm without a nucleus.

cytodiaeresis (*Cyt.*). See **mitosis.**

cytogamy (*Cyt.*). *Conjugation* (q.v.).

cytogenesis (*Cyt.*). The formation and development of cells.

cytogenetics (*Gen.*). Genetics in relation to cytology.

cytogenous (*Biol.*). Cell forming; cell producing.

cytolisthesis (*Zool.*). In experimental embryology, sliding or rotary movement of cells over one another.

cytology (*Biol.*). The study of the structure and functions of cells.

cytolosome (*Cyt.*). Enlarged lysosome concerned with cell autolysis. Similar in meaning to *cytosegresome* (q.v.).

cytolymph (*Cyt.*). The fluid part of cytoplasm; cell-sap.

cytolysin (*Cyt.*). A substance causing *cytolysis.*

cytolysis (*Cyt.*). Dissolution of cells.

cytome (*Cyt.*). The whole of the chondriosomes present in a cell.

cytomorphosis (*Cyt.*). The series of changes accompanying the development and specialization of an individual cell and its subsequent decline, or the comparable series of changes in successive generations of cells in the same line of descent.

cytopharynx (*Zool.*). In some *Ciliophora*, an oesophagus-like tube leading from the cytostome into the endoplasm.

cytoplasm (*Cyt.*). The protoplasm of a cell, apart from that of the nucleus. Cf. *nucleoplasm.*

cytoplasmic inheritance (*Gen.*). The transmission of hereditary characters by means of the cytoplasm, instead of by the nucleus.

cytoplasmic stain (*Micros.*). A stain which shows up the cytoplasm and cytoplasmic structures of a cell, as opposed to the nucleus.

cytoproct (*Zool.*). In unicellular or noncellular animals, an opening or weak spot in the ectoplasm and pellicle, by which indigestible residues are discharged; a potential anus. Also called *cytopyge*.

cytosegresome (*Cyt.*). Bodies containing recognizable cytoplasmic organelles, limited by a single or double outer membrane. Also *autophagic vacuole*.

cytosine (*Chem.*). 6-aminopyrimidine-2-one, obtained by the hydrolysis of certain nucleic acids. Pairs with guanine in the *genetic code* (q.v.).

cytosome (*Cyt.*). The whole of the cytoplasm of a cell.

cytostome (*Zool.*). In some *Ciliophora*, the cell-mouth, an aperture in the ectoplasm which opens into the cytopharynx.

cytotaxis (*Cyt.*). Rearrangement of cells as a result of stimulation.

cytotaxonomy (*Bot.*). A scheme of classification based on information gained by a study of cell structure.

cytotoxin (*Biochem., Biol., Cyt.*). A toxin having a specific destructive action on particular cells.

cytotrophoblast (*Zool.*). The inner layer of the trophoblast; layer of Langhans.

cytotropism (*Cyt.*). Reaction or response to the stimulus of mutual attraction between two cells.

cytozoic (*Zool.*). Intracellular; living within a cell.

cytula (*Zool.*). A fertilized ovum.

d Abbrev. for deci-, one tenth; e.g. decimetre = dm.

d- (*Chem.*). Abbrev. for *dextro-rotatory*.

[d] (*Light*). A line in the blue of the solar spectrum, having a wavelength of 437·8720 nm due to iron.

δ- (*Chem.*). Substituted on the fourth carbon atom of a chain.

D Symbol for angle of deviation, electric flux density (displacement), diffusion coefficient.

D (*Chem.*). The symbol for *deuterium*.

[D] (*Light*). A group of 3 Fraunhofer lines in the yellow of the solar spectrum. [D_1] and [D_2], wavelengths 589·6357 and 589·0186 nm, are due to sodium, and [D_3], wavelength 587·5618 nm, to helium.

2,4-D (*Chem.*). See **Chloroxone.**

Δ Symbol for mass excess, difference, e.g., between consecutive tabulated values of a function. (*Chem.*) A prefixed symbol for a double bond beginning on the carbon atom indicated.

dabbing (*Build.*). See **daubing.**

dabrey (*For.*). A small shallow tray in which latex exuding from the balata tree is collected.

dachiardite (*Min.*). A rare hydrous calcium, potassium, sodium, aluminium zeolite.

dacite (*Geol.*). A volcanic rock intermediate in composition between andesite and rhyolite, the volcanic equivalent of a granodiorite.

Dacron (*Chem.*). American equivalent of *Terylene* (q.v.) synthetic fibre.

dacryo-adenitis, dacry-adenitis (*Med.*). Inflammation of the lacrimal gland.

dacryocystitis (*Med.*). Inflammation of the lacrimal sac.

dacryocystorhinostomy (*Surg.*). Formation of a direct opening between the tear sac and the nose.

Dacryomycetales (*Bot.*). An order of the *Basidiomycetes* having two diverging epibasidia on the hypobasidium. The basidia are never covered with sterile tissue.

dacryops (*Med.*). A cystic swelling at the place where the upper lid meets the eyeball, due to blockage of a lacrimal duct.

dactyl (*Zool.*). A digit. *adj.* **dactylar.**

dactyline, dactyloid (*Bot.*). Spreading like outstretched fingers.

dactylitis (*Med.*). Inflammation of a finger or of a toe.

dactylopodite (*Zool.*). In some *Crustacea*, the fifth or distal joint of the endopodite of the walking-legs or maxillipeds.

dactylopore (*Zool.*). In *Hydrocorallinae*, an opening in the common skeleton through which a dactylozooid protrudes.

dactylopterous (*Zool.*). Having the anterior rays of the pectoral fins free and unattached by membrane to the rest of the fin, as in the Gurnards.

dactylozooid (*Zool.*). In *Hydrocorallinae* and *Siphonophora*, a hydroid specialized for catching prey and for defence of the colony; usually with one or more long tentacles richly provided with stinging-cells, and no mouth. Also **palpon.**

dado (*Arch.*). One of the faces of the solid block forming the body of a pedestal. (*Join.*) A border around the lower part of the wall of a room.

dado capping (*Join.*). The name given to the dado rail when the dado occupies as much as two-thirds of the height of the room.

dado plane (*Tools*). Type of grooving plane with two projecting spurs, one on each side at the front, and an adjustable depth stop. Used for making grooves for shelving, etc. The spurs cut across the grain and keep the plane in its correct path. Also **trenching plane.**

dado rail (*Join.*). The moulding capping the dado in a room and separating it from the upper part of the walls. Also called **surbase.**

daedaleous (*Bot.*). A term applied to an irregularly wrinkled, plaited surface.

daft lamb disease (*Vet.*). An hereditary congenital cerebellar atrophy affecting lambs, which are born showing an uncoordinated gait, or unable to stand or walk.

daggings or dags (*Textiles*). Virgin wool formed into clots through adhesion of soil, dung, and other foreign impurities; also **crutchings.**

daguerreotype (*Photog.*). Early process using a silvered copper plate sensitized by fuming with iodine and bromine vapour and developed with mercury vapour.

D.A.H. (*Med.*). Abbrev. for *disordered action of the heart*. See **effort syndrome.**

dailies (*Cinema.*). The same as *rushes* (q.v.).

Dailygraph (*Acous.*). A magnetic recording and reproducing machine, for attachment to telephones.

Dakin's solution (*Chem.*). Dilute solution of sodium hypochlorite and boric acid; used as an antiseptic.

dalapon (*Chem.*). 2,2-Dichloropropanoic acid, used as a weedkiller.

d'Alembert's principle (*Mech.*). On a body in motion, the external forces are in equilibrium with the inertial forces.

d'Alembert's ratio test (*Maths.*). A series of positive terms converges or diverges respectively according to whether the limit of the ratio of a term to its predecessor is less or greater than unity. When the limit is unity the test is inconclusive.

Dalitz pair (*Nuc.*). Electron-positron pair produced by the decay of a free neutral pion (instead of one of the two gamma quanta normally produced).

Dalitz plot (*Nuc.*). A type of graphical diagram used in the study of decay processes leading to particle triplets. Energies of two of the particles are measured for a large number of disintegrations using any suitable track chamber, and are plotted along different axes of the diagram.

DALR (*Meteor.*). Abbrev. for *dry adiabatic lapse rate*.

Dalradian Series (*Geol.*). A very thick and variable succession of sedimentary and volcanic rocks which have suffered regional metamorphism, occurring in the Scottish Highlands approximately between the Great Glen and the Highland Boundary fault. Referred to the Pre-Cambrian System.

Daltonism (*Med.*). See **colour blindness.**

Dalton's atomic theory (*Chem.*). States that matter consists ultimately of indivisible, discrete particles (atoms), and atoms of the same element are identical; chemical action takes place as a result of attraction between these atoms, which combine in simple proportions.

It has since been found that atoms of the chemical elements are not the ultimate particles of matter, and that atoms of different mass can have the same chemical properties (*isotopes* (q.v.)). Nevertheless, this theory of 1808 is fundamental to chemistry. See **atomic structure.**

Dalton's law (*Chem.*). See **law of multiple proportions.**

Dalton's law of partial pressures (*Chem.*). The pressure of a gas in a mixture is equal to the pressure which it would exert if it occupied the same volume alone at the same temperature.

dam (*Civ. Eng.*). An embankment or other construction made across the current of a stream. (*Mining*) (1) A retaining wall or bank for water or tailings. (2) An air-tight barrier to isolate underground workings which are on fire.

damask (*Textiles*). (1) Silk or rayon (viscose or acetate) material, of satin and sateen weaves, in which background and figure have a contrasting effect; used mainly for furnishings. (2) Linen cloth of damask texture, used for tablecloths and towellings; also a cotton cloth of similar nature, used for tablecloths; both fabrics are reversible.

damaskeen. Inlay of metal, or ivory and mother-of-pearl, on metal.

dammar (*Chem.*). Damar; resin Damar; a copal or gum produced by 'dammar' trees: white dammar by *Vateria Indica*; black dammar by *Dammara officinalis*; kauri gum by *Dammara Australis*. Used, *inter alia*, for making special varnishes in photographic work.

dammar pocket (*For.*). See **resin pocket.**

damped balance (*Chem.*). Chemical balance using magnetic or air dash pots to bring it quickly to rest.

damped oscillation (*Phys.*). Oscillation which dies away from an initial maximum asymptotically to zero amplitude, usually with an exponential envelope. E.g. the note from a struck tuning fork. See **damping, critical damping, decay factor,** and cf. **continuous oscillations.**

dampener (*Mech. Eng.*). A device attached to both inlet and outlet sides of large reciprocating machinery to reduce pulsations which, if not reduced, would cause either mechanical damage or noise nuisance or both.

dampening rollers (*Print.*). Fabric-covered rollers on a lithographic printing machine which are kept moist by a water supply, and by means of which the plate is dampened before being rolled by the ink rollers.

damper (*Aero.*). Widely used term applied to devices for the suppression of unfavourable characteristics or behaviour; e.g., *blade damper*, to prevent the hunting of a helicopter rotor; *flame damper*, to prevent visual detection at night of the exhaust of a military aircraft; *shimmy damper*, for the suppression of *shimmy* (q.v.); *yaw damper* suppresses directional oscillations in high-speed aeroplanes, while a *roll damper* does likewise laterally, in both cases the frequency of the disturbances being too high for the pilot to anticipate and correct manually. (*Elec. Eng.*) Energy-absorbing component often used for reducing the transmission of oscillatory energy from a disturbing source. Also called **amortisseur, damper winding, damping grid, damping winding.** (*Eng.*) (1) An adjustable iron plate or shutter fitted across a boiler flue to regulate the draught. (2) A device for damping out torsional vibration in an engine crankshaft, the energy of vibration being dissipated frictionally within the damper. See **vibration dampers.** (3) Device for stiffening the steering of a motor cycle to obviate wheel wobble.

damper weight (*Eng.*). A counterweight used to balance the weight of a damper in a boiler flue.

damper winding (*Elec. Eng.*). See **damper.**

damping (*Aero.*). The capability of an aircraft of suppressing or resisting harmonic excitation and/or flutter. *Internal damping* is intrinsic to the materials, while *structural damping* is the total effect of the built-up structure. See **resonance test.** (*Eng.*) The checking of a (vibratory) motion by friction, etc. The damping capacity of cast iron makes this metal useful for lathe beds and other machine tool castings required to be rigid. (*Paper*) Paper is sometimes subjected to *damping* by contact with a wet cylinder covered with felt to enable it to take a smoother finish. (*Phys.*) Extent of reduction of amplitude of oscillation in an oscillatory system, due to energy dissipation, e.g., friction and viscosity in mechanical system, and resistance in electrical system. With no supply of energy, the oscillation dies away at a rate depending on the *degree of damping* (q.v.). The effect of damping is to increase slightly period of vibrations. It also diminishes sharpness of resonance for frequencies in the neighbourhood of natural frequency of vibrator. See **logarithmic decrement.**

damping down (*Eng.*). The retarding of the rate of combustion in a boiler furnace by covering the fire with fine coal sprinkled with water. (*Met.*) The temporary stopping of a blast furnace by closing all apertures by which air could enter.

damping factor (*Phys.*). See **decay factor.**

damping grid (*Elec. Eng.*). See **damper.**

damping magnet (*Elec. Eng.*). A permanent magnet used to produce damping by inducing eddy currents in a metal disk or other body.

damping-off (*Bot.*). Plant disease caused by parasitic fungi, chiefly *Pythium de barvanum*, resulting in the drooping and subsequent dying of seedlings. Geranium cuttings are also prone.

damping winding (*Elec. Eng.*). See **damper.**

dam plate (*Met.*). Iron vertical plate holding the wall of refractory brick (the *dam stone*) which forms the fore-hearth of a blast furnace.

damp-proof course (*Build.*). A layer of impervious material, as plastic or bituminous sheeting, built into a wall 15 to 25 cm above ground-level, to prevent moisture from the foundations rising in the walls by capillary attraction. Vertical damp courses are also used in suitable cases.

damp-proofing (*Build.*). The process of coating a wall with a special preparation to prevent moisture from getting through.

dan (*Mining*). (1) A tub or barrel for conveying water from a dip place (not reached by a pumping arrangement) for discharge into a pump lodge or a pit sump. (2) A small box or sledge sometimes used in thin seams to convey coals from distant points to the trains or tubs in the roadway.

danalite (*Min.*). An iron-bearing beryllium silicate, also containing sulphur.

danburite (*Min.*). A rare accessory mineral, occurring in pegmatites as yellow orthorhombic crystals. Chemically, danburite is a calcium borosilicate, $CaB_2Si_2O_8$.

dances of bees (*An. Behav.*) Precise behaviour patterns of returning forager honey bees which indicate to other workers the nature and location of any rich source of nectar or pollen. See **round dance** and **waggle dance.**

dancing roller (*Print.*). See **jockey roller.**

dancing step (*Build.*). A step intermediate between a flier and a winder, having its outer end narrower in plan than its inner end. Also called **balanced step.**

dandy roll (*Paper*). A wire-gauze cylinder which comes in contact with paper when in the wet stage. It impresses the ribs in laid paper, also any watermark required.

danger coefficient (*Nuc. Eng.*). That expressing change in reactivity when sample is inserted into reactor.

dangerous semicircle (*Meteor.*). The right-hand half of the storm field in the northern hemisphere, the left-hand half in the southern hemisphere, when looking along the path in the direction a *tropical revolving storm* (q.v.) is travelling. Cf. *navigable semicircle*.

Daniell cell (*Chem.*). Primary cell with zinc and copper electrodes, the zinc rod being inserted in sulphuric acid contained within a porous pot, which is itself immersed in a copper pot containing copper (II) sulphate solution.

Daniell hygrometer (*Phys., etc.*). A form of dew-point *hygrometer* (q.v.), now obsolete, in which dew is formed on the surface of a bulb containing ether which is cooled by evaporation into another bulb, itself cooled by the evaporation of ether on its outer surface.

Danner process (*Glass*). A method of continuously drawing glass tubing or rod horizontally from a rotating mandrel.

Dano composting plant (*San. Eng.*). A method of composting the organic wastes in domestic refuse with sewage sludge by a process of fermentation and grinding, which reduces the materials to a moist granulated condition. The product is a valuable soil conditioner because of its high humus content.

dant (*Mining*). Soft sooty coal found in face and back slips or cleats; fine slack coal.

danty (*Mining*). Broken coal.

Daphnales (*Bot.*). An order of the *Incompletae* (or *Monochlamydeae*). The ovary is usually monocarpellate with one or a few ovules. The sepaloid perianth is complete in one or two whorls. The stamens are perigynous.

daphnite (*Min.*). A variety of chlorite very rich in iron and aluminium.

dapsone (*Pharm*). Di(*p*-aminophenyl) sulphone, $C_{12}H_{12}N_2O_2S$. Widely used for the treatment of leprosy (sometimes along with other compounds, e.g., ditophal, $C_{12}H_{14}O_2S_2$), having largely replaced chaulmoogra oil for that purpose.

daraf (*Elec.*). Unit of elastance, the reciprocal of capacitance in farads. (*Farad* backwards.)

darby (*Tools*). A *derby float* (q.v.).

darcy (*Geol.*). A unit used to express the *permeability coefficient* of a rock, for example in calculating the flow of oil, gas or water. More commonly used is the *millidarcy* (mD), one-thousandth of a darcy (D).

Darcy's law (*Powder Tech.*). A permeability equation which states that the rate of flow of fluid through a porous medium is directly proportional to the pressure gradient causing the flow.

Darimont cell (*Elec. Eng.*). A double-fluid primary cell with zinc and carbon electrodes, an electrolyte of calcium carbonate and sodium chloride, and a depolarizer of ferric chloride.

dark burn fatigue (*Phys.*). Decrease of efficiency of a luminescent material during excitation.

dark current (*Phys.*). Residual current in a photocell when there is no incident illumination. The current depends on temperature. Cf. *light current*.

dark ground illumination (*Micros.*). A method of examining living material, e.g., micro-organisms, tissue culture cells, with a light microscope, fitted with a condenser with a central circular stop, which permits only light scattered or reflected from the specimen to enter the objective. Specimens appear luminous against a dark background.

dark nebulae (*Astron.*). Obscuring clouds of dust and gases, common throughout the Milky Way, and also observed in other galaxies. See **Coal Sack.**

dark red heat (*Met.*). Glow emitted by metal at temperatures between 550° and 630°C.

dark red silver ore (*Min.*). See **pyrargyrite.**

dark resistance (*Electronics*). Resistance of a selenium or other photocell in the dark.

dark-room camera (*Print.*). A built-in process camera controlled completely from inside the dark room, with the copy board and its illumination equipment outside.

dark seed (*Bot.*). A seed which will germinate only if kept in the dark at the time when other conditions (e.g., oxygen, moisture, temperature) make germination possible.

dark slide (*Photog.*). The carrier for plates to be exposed in cameras, loaded in the dark room and uncovered, after attachment to the camera, by withdrawing a slide.

dark space (*Electronics*). See **anode, Aston, Crookes, Faraday dark space.**

dark spot (*TV*). Reproduced spot arising from cloud of electrons released from a mosaic in a TV camera. See **shading.**

dark trace screen (*Electronics*). Screen which yields a dark trace under electron-beam bombardment.

Darley Dale stone (*Build.*). A yellowish-brown sandstone containing mica flakes, quarried at Darley Dale in Derbyshire, used for building.

d'Arsonval galvanometer (*Elec. Eng.*). A moving coil galvanometer in which a light rectangular coil is suspended in the strong field between the poles of a horseshoe magnet. A similar movement is used in commercial ammeters and voltmeters.

d'Arsonvalism (*Med.*). Therapeutic usage of isolated and highly damped oscillatory electric currents.

dart (*Horol.*). See **safety finger.** (*Zool.*) Any dart-like structure, e.g., in certain Snails, a small pointed calcareous rod which is used as an incentive to copulation; in certain *Nematoda*, a pointed weapon used to obtain entrance to the host.

Dartmoor granite (*Build.*). A grey porphyritic granite used in heavy construction. In the geological sense, *Dartmoor granite* comprehends the Armorican granitic complex of Dartmoor, Devon; only a part is porphyritic.

dart sac (*Zool.*). In some *Gastropoda*, a saccular gland, surrounded by muscles, which opens into the vagina and secretes the *dart* (q.v.).

Darwinian theory (*Biol.*). See **natural selection.**

dash pot (*Eng.*). A device for damping-out vibration or for allowing rapid motion in one direction but only much slower motion in the opposite direction. It consists of a piston attached to the part to be controlled (fitted with a nonreturn valve if required) sliding in a cylinder containing liquid to impede motion.

dasypaedes (*Zool.*). Birds which when hatched have a complete covering of down. Cf. *altrices*.

dasyphyllous (*Bot.*). (1) Having crowded leaves. (2) Having thick leaves. (3) Having leaves bearing a thick coat of woolly hairs.

data (*Comp.*). All the *operands* and results of computer operations directed by the detailed *instructions* comprising the *programme*.

data-handling capacity (*Comp.*). The number of bits of information which can be stored at any one time in a computing system and the rate at which these may be fed to the input. (*Telecomm.*) For a telemetering system, the maximum amount of information which can be transmitted and received over a given link.

data-handling system (*Comp.*). Automatic or semi-automatic equipment for collecting, receiving, transmitting, and storing numerical data. It may be handled continuously (as analogue or position signals) or in discrete steps (as digital or binary signals). If necessary the system should be able to carry out mathematical operations on the stored data and to record the results.

data hold (*Comp.*). Equipment for reconstructing a continuous function (analogue) from sampled data by extrapolation.

Dataphone adaptor (*Comp.*). One for operation of a central computer from a remote access point using a telephone line. See **Modem.**

data processing (*Comp.*). Various operations of adding, subtracting, etc., performed by an electronic computer on data supplied, either as punched cards or tape, or on magnetic tape, according to a program.

data reduction (*Comp.*). Rearranging or reducing the quantity of the experimentally obtained data, so permitting the more rapid extraction of conclusions.

data storage (*Comp.*). See **memory capacity.**

Date (*Bot.*). *Phoenix dactylifera*, a member of the *Palmae*. Dioecious, with pinnate leaves. Cultivated in N. Africa, Arabia, etc. The fruit is a berry, but one carpel ripens forming a seed with an endosperm of cellulose. The fruit is eaten, the leaves are used for thatching, etc.

date line (*Geog.*). An imaginary line on the earth's surface for the purpose of fixing the change of date, without ambiguity, for all travellers; it runs approximately along the meridian of longitude 180° from Greenwich, deviating round certain groups of islands for local convenience.

dative bond (*Chem.*). See **covalent bond.**

datolite (*Min.*). A hydrous silicate of boron and calcium occurring as a secondary product in amygdales and veins, usually as distinct prismatic white or colourless monoclinic crystals.

datum (*Aero.*). *Datum level*, or *rigging datum*, is the horizontal plane of reference, in flying attitude, from which all vertical measurements of an aircraft are taken; *cg datum* is the point from which all weight moment arms are measured horizontally when establishing the centre of gravity and loading of an aircraft. (*Eng.*) A point, line or surface to which dimensions are referred on engineering drawings and from which measurements are taken in machining or other engineering operations. (*Surv.*) An assumed surface used as a reference surface for the measurement of reduced levels.

daubing (*Build.*). (1) The operation of dressing a stone surface with a special hammer in order to cover it with small holes. (2) A rough-stone finish given to a wall by throwing a rough coating of plaster upon it. See **rough-cast.**

daughter (*Biol.*). Offspring belonging to the first generation, whether male or female; as *daughter cell, daughter nucleus*.

daughter product (*Nuc.*). Of a given nuclide, any nuclide that originates from it by radioactive decay. Also known as **decay product.**

Davis apparatus. A respiratory apparatus specially designed to permit escape from a pressure-equalizing chamber in a submarine. Oxygen is breathed from a chamber which, embracing the wearer, gives him buoyancy and assists his rise to the surface.

Davis raft (*For.*). A deep-sea raft in which logs are several layers deep and bound together.

Davisson-Germer experiment (*Electronics*). Demonstration of wavelike diffraction patterns from electrons passing through a nickel crystal. Confirmed independently by G. P. Thomson.

Davy lamp (*Mining*). The name of the safety lamp invented by Sir Humphry Davy in 1815.

dawn rocket (*Space*). A rocket booster launched into the orbital motion of the earth, west to east, at dawn, so that the speed of the vehicle is added to earth's orbital velocity (18·5 m/s).

Dawsoniales (*Bot.*). A monogeneric order of the *Eubrya*, with large gametophores, flattened capsule; the peristome is subtended by hairs.

day (*Astron.*). See **apparent solar-, mean solar-, sidereal-.** (*Build.*) The distance between successive mullions in a window.

Day-glo (*Print.*). TN for a range of inks with fluorescent properties much used for posters and other publicity printing.

daylight (*Eng.*). The distance between the bed surface and the bottom of the ram of a press.

daylight factor (*Light*). The ratio of the illumination measured on a horizontal surface inside a building to that which obtains at the same time outside the building, due to an unobstructed hemisphere of sky. Occasionally called **window efficiency ratio.**

daylight lamp (*Light*). A lamp giving light having a spectral distribution curve similar to that of ordinary daylight.

day of clear sky (*Meteor.*). A day on which the average *cloudiness* (q.v.) at the hours of observation is less than 2 oktas.

day of snow (*Meteor.*). A period of 24 hr ending midnight during which snow was observed to fall.

day position (*Bot.*). The posture of the leaves during the day in plants which change the position of their leaves as night comes on.

day sleep (*Bot.*). The folding together of the leaflets of a compound leaf when exposed to bright light, bringing together the surfaces which bear most of the stomata, and doubtless imposing a check on the rate of loss of water from the plant.

dazomet (*Chem.*). Tetrahydro - 3,5 - dimethyl - 2*H*-1,3,5-thiadiazine-2-thione, used as a fungicide and weedkiller. Also **DMTT.**

dB (*Acous.*). Abbrev. for **decibel.**

dBA, dBB, dBC (*Acous.*). Agreed curves of relative response against frequency for weighting the readings of sound level meters to meet practical situations. The dBA curve approximates the human ear.

d.c. (*Elec. Eng.*). Abbrev. for **direct current.** (*Typog.*) Abbrev. for *double column*; *double crown.*

d.c. amplifier (*Elec. Eng.*). One which uses direct coupling between stages to amplify very low frequency changes in voltage or current (zero frequency signals). Chief problem involved is drift stabilization.

d.c. balancer (*Elec. Eng.*). The coupling and connecting of two or more similar direct-current machines, so that the conductors connected to the junction points of the machines are maintained at constant potentials.

d.c. bias (*Elec. Eng.*). In magnetic tape or wire recording, the addition of a polarizing direct current in signal recording to stabilize magnetic saturation.

d.c. bridge (*Elec. Eng.*). A four-arm null bridge energized by a d.c. supply. The prototype is the *Wheatstone bridge* (q.v.), other examples the *metre bridge* and the *Post Office box* (qq.v.).

d.c. component (*TV*). That part of the picture signal which determines the average or datum brightness of the reproduced picture.

d.c. converter (*Elec. Eng.*). A converter which changes direct current from one voltage to another.

d.c. coupling (*Electronics*). See **direct coupling.**

d.c./d.c. converter (*Elec. Eng.*). A d.c. voltage transformer using an inverter and rectifier. Also called **d.c. transformer.**

D.C.F. (*Carp.*). Abbrev. for *deal-cased frame.*

d.c. generator (*Elec. Eng.*). A rotary machine to convert mechanical into direct current power.

d.c. meter (*Elec. Eng.*). One which responds only to d.c. component of a signal, e.g., moving coil instruments.

d.c. picture transmission (*TV*). Television transmission in which the d.c. component of the signal represents the average illumination.

d.c. quadricorrelator (*TV*). Noise-resistant 2-mode type of automatic frequency and phase control circuit.

d.c. resistance (*Elec. Eng.*). The resistance which a circuit offers to the flow of a direct current. Also called **true (or ohmic) resistance.**

d.c. restoration (*TV*). Use of a *catch* or *clamp* diode to restore or hold black *or* white level at the designated pedestal level. Also **reinsertion.**

d.c. restorer (*TV*). Means of restoring a d.c. (or low frequency) component of a signal after transmission in a line which only easily transmits high frequencies.

d.c. testing of cables (*Elec. Eng.*). The application of a d.c. voltage of 5 times the r.m.s. of the working a.c. voltage. Cables which have considerable tracking and are likely to break down in service are broken down by the d.c.; healthy cables are not affected.

d.c. transformer (*Elec. Eng.*). (1) Device to measure large direct currents by means of associated magnetic field. (2) Colloquial name for **d.c./d.c. converter.**

d.c. transmission (*Elec. Eng.*). At very high power levels, the temptation to use higher and higher voltages for transmission along grid lines is met by the difficulty of insulation and flashover. More power can be transmitted for a given maximum voltage if direct current is employed, hence the growing use of d.c. transmission lines operated at very high voltages, of the order of 1 megavolt. (*TV*) Inclusion of d.c. or very low frequency in the transmitted video signal. If omitted, it has to be *restored* in relation to the *pedestal* in the receiver.

D display (*Radar*). *C display* (q.v.) in which targets appear as bright spots, elongated vertically approximately in proportion to the range.

ddt. (*Civ. Eng., etc.*). Abbrev. for *deduct.*

DDT (*Chem.*). Abbrev. for a complex chemical mixture, in which *pp′*-dichlorodiphenyltrichloroethane predominates; a synthetic insecticide remarkable for high toxicity to insects at low rates of application. It functions as *stomach insecticide* and also as *contact insecticide* (qq.v.), having the property (not universally accepted as an advantage) of persistence of activity from residual deposits compared with older natural insecticides. Also **dicophane, chloro-phenothane.**

DDVP (*Chem.*). See **dichlorvos.**

de-accentuator (*Telecomm.*). Circuit arrangement for de-emphasis in FM reception.

deactivation (*Chem.*). The return of an activated atom, molecule, or substance to the normal state. See **activation** (2).

dead (*Acous.*). An enclosure which has a period of reverberation much smaller than usual for its size and audition requirements. Applied to sets in motion-picture production. (*Build.*) Said of materials which have deteriorated.

dead angle (*Eng.*). That period of crank angle of a steam-engine during which the engine will not start when the stop-valve is opened; due to the ports being closed by the slide-valve.

dead and down (*For.*). Standing dead trees and trees on the ground, either dead or living.

dead axle (*Eng.*). An axle which does not rotate with the wheels carried by it. Cf. *live axle.*

dead band (*Telecomm.*). (1) Permitted tolerance in controlled function without operation of automatic control. (2) Range of signal input to a magnetic amplifier which does not effect a change in the output.

dead bank (*Eng.*). A stoker-fired boiler furnace from which the coal feed is shut off, the fire being allowed to burn back as far as possible without going out entirely.

dead-beat compass (*Ships, etc.*). A magnetic compass with a short period of oscillation and heavily damped so that it comes to rest very quickly.

dead-beat escapement (*Horol.*). An escapement in which there is no 'recoil' to the escape wheel. The dead-beat action is obtained by making the locking faces of the pallets arcs of circles, struck from the pallet staff as centre. This escapement is the one used for regulators, and is capable of giving very accurate results.

dead burnt (*Met.*). Descriptive of such carbonates as limestone, dolomite, magnesite, when they have been so kilned that the associated clay is vitrified, part or all of the volatile matter removed and the slaking quality lowered.

dead-centre (*Eng.*). (1) Either of the two points in the crankpin path of an engine at which the crank and connecting-rod are in line and the piston exerts no turning effort on the crank. See **inner (or top) dead-centre, outer (or bottom) dead-centre.** (2) A lathe centre. See **tail-stock and centre.**

dead-centre lathe (*Eng.*). A small lathe (used in instrument-making) in which both centres are fixed, the work being revolved by a small pulley mounted on it.

dead coil (*Elec. Eng.*). A coil in the winding of a machine which does not contribute any e.m.f. to the external circuit, because it is short-circuited or disconnected from the rest of the winding. See also **dummy coil.**

dead earth (*Elec. Eng.*). A connexion between a normally live conductor and earth by means of a path of very low resistance.

dead end (*Plumb.*). The length of pipe between a closed end and the nearest connexion to it, forming a 'dead' pocket in which there is no circulation. (*Radio*) The unused portion of an inductance coil in an oscillatory circuit.

dead-ended feeder (*Elec. Eng.*). See **independent feeder.**

dead-end tower (*Elec. Eng.*). See **terminal tower.**

deadening (*Build.*). (1) The operation of dealing with a surface, so as to give it a dead finish. (2) *Pugging* (q.v.).

dead eye (*Eng., etc.*). (1) A sheaveless block used in setting up rigging. (2) A light type of bearing for supporting a spindle; it may consist merely of a hole in a sheet of metal or other material.

dead finish (*Build.*). A dull or rough finish. (*Paint.*) See **flat finish.**

dead flue (*Build.*). A flue which is bricked in at the bottom.

dead ground (*Mining*). Ground devoid of values: ground not containing veins or lodes of valuable mineral: a barren portion of a coal-seam. Also **deads.**

dead hand (*Med.*). A disease of pneumatic-drill operators, causing cyanosis, anaesthesia of finger-tips, and sometimes bone absorption.

dead heat (*Eng.*). (1) A projecting shank on a casting, formed by the metal which filled the pouring hole or riser. (2) The fixed headstock of a machine tool.

dead knot (*For.*). A knot which is partially or wholly separated from the surrounding wood.

dead letters (*Typog.*). Letters remaining in the type case when setting has to stop because the supply of one *sort* has been exhausted.

dead load (*Eng.*). The permanent loading on a structure, imposing definite fixed strains upon it; it consists of the weight of the structure itself and the fixed loading carried by it. Cf. *live load.*

dead-load safety-valve (*Eng., etc.*). See **dead-weight safety-valve.**

dead lock (*Join.*). A lock the bolt of which is key-operated from one side and handle-operated from the other.

dead-man's handle (*Elec. Eng.*). A form of handle commonly used on the controllers of electric vehicles; designed so that if the driver releases his pressure on the handle, owing to sudden illness or other causes, the current is cut off and the brakes applied. **Dead-man's pedal** is a similar foot-operated safety device, used especially on diesel trains.

dead matter (*Typog.*). Type which has been used and is awaiting distribution.

deadmen (*Civ. Eng.*). The concrete, plate, or other anchorage for land ties.

dead plate (*Eng.*). A stationary bridging surface adjacent to a (belt-)conveyor over which goods or materials are pushed by the line pressure.

dead points (*Eng.*). See **dead-centre (1).**

dead reckoning (*Nav.*). (Derived from *deduced reckoning.*) Estimate of position in navigation made from last known location.

dead rise (*Aero.*). At any cross-section of a flying-boat hull or seaplane float, the vertical distance between the keel and the chine.

dead roasting (*Met.*). Roasting carried out under conditions designed to reduce the sulphur content, or that of the other volatile matter, to the lowest possible value. Distinguished from *partial roasting* and *sulphating roasting* (qq.v.).

dead room (*Acous.*). See **anechoic room.**

deads (*Mining*). (1) Same as *dead ground* (q.v.). (2) Waste (*back fill*) used to support roof.

dead segment (*Elec. Eng.*). A commutator segment which is not connected, either for accidental reasons or for a definite purpose, to the armature winding associated with the commutator.

dead shore (*Carp.*). A vertical timber post used to prop up temporarily any part of a building.

dead-smooth file (*Eng.*). The smoothest grade of file ordinarily used, having 70 to 80 teeth to the inch for files of average length; used for finishing surfaces.

dead sounding (*Build.*). *Pugging* (q.v.).

dead spot (*Radio*). A region where the reception of radio transmissions over a particular frequency range is extremely weak or practically zero.

dead time (*Comp.*). Delay in a controlling system between error signal and response. (*Nuc.*) Time after ionization during which further operation is not possible; reduced by a *quench*, as in G-M counters. When dead time of detector is variable subsequent circuits may incorporate a fixed electronic dead time. Also called **insensitive time.**

dead-time correction (*Nuc. Eng.*). Correction applied to the observed rate in a radiation counter to allow for probable events during the dead time.

dead water (*Eng.*). In a boiler or other plant, water not in proper circulation.

dead weight (*Ships*). The difference, in tons, between a ship's displacement at load draught and light draught. It comprises cargo, bunkers, stores, fresh water, etc.

dead-weight pressure-gauge (*Eng.*). A device in which fluid pressure is measured by its application to the bottom of a vertical piston, the resulting upward force being then balanced by applying weights to the upper end; used for calibrating Bourdon gauges.

dead-weight safety valve (*Eng., etc.*). A safety valve in which the valve itself is loaded by a heavy metal weight; used for small valves and low pressures. See **safety valve.**

dead well (*Civ. Eng.*). An *absorb ng well* (q.v.).

de-aerator (*Eng.*). A vessel in which boiler feed water is heated under reduced pressure in order to remove dissolved air.

deaf aid (*Acous.*). A device used by a deaf person to improve audition of external sounds; either in the form of an acoustic amplifier (collector), or in the form of a microphone-receiver combination, with or without amplifier.

deafening (*Build.*). *Pugging* (q.v.).

deafness (*Med.*). Lack of sensitivity of hearing in one or both ears, with consequent increase in the threshold of minimum audibility, measurement of which is useful in diagnosis.

deal (*For.*). A piece of timber of cross-section roughly 10 in. × 3 in. (250 mm × 75 mm).

dealbate (*Bot.*). Whitened, usually by a coating of hairs.

deal frame (*Carp.*). A reciprocating sawing machine having several parallel saws moving together in a frame; used for cutting deals into thinner boards.

deamination (*Biochem.*). Removal of the amino group $-NH_2$ from an amine, e.g., physiologically, the conversion of ammonium salts into urea (carbamide) in the liver.

Dean and Stark apparatus (*Chem.*). An apparatus for determining the water content of oils. It consists of a distillation flask, a reflux condenser, and a graduated tube attached to the condenser, in which the water is collected and prevented from running back into the flask.

deassimilation (*Bot.*). The utilization of food by the plant.

death (*Biol.*). In a cell or an organism, complete and permanent cessation of the characteristic activities of living matter.

deathnium centre (*Electronics*). Crystal lattice imperfection in semiconductor at which it is believed electron hole pairs are produced or recombine.

death point (*Biol.*). The lethal maximum or minimum limit of any particular factor in the external or internal environment beyond which an organism or cell cannot live, e.g., the temperature above which an organism will die.

death wish (*Psychol.*). Wish, conscious or unconscious, for death for oneself or another.

debacle, débâcle (*Meteor.*). The breaking up of the surface ice of great rivers in spring.

debiteuse (*Glass*). A piece of fireclay with a slit through which molten glass is extruded into sheets.

débridement (*Surg.*). The removal of foreign matter and excision of infected and lacerated tissue from a wound.

de Broglie wavelength (*Nuc.*). That assignable to any particle because of its momentum in motion, i.e., $\lambda = h/mv$, where λ = wavelength, h = Planck's constant, m = relativistic mass, v = velocity.

debug (*Comp., etc.*). To remove all defects from a programme or from operation of a system. (*Slang*) to expose or remove a *bug* (2, q.v.).

debunching (*Electronics*). Tendency for a beam of electrons or a velocity-modulated beam of electrons, to spread because of their mutual repulsion. In gas cathode-ray tube, the maintenance of focus arises from relatively static positive ions in the path of beam counteracting debunching. See **buncher** and **bunching**.

Debye and Scherrer method (*Crystal.*). A method of X-ray crystal analysis applicable to powders of crystalline substances or aggregates of crystals.

Debye-Hückel theory (*Chem.*). A theory of electrolytic conduction which assumes complete ionization and attributes deviations from ideal behaviour to inter-ionic attraction.

Debye length (*Nuc.*). Maximum distance at which coulomb fields of charged particles in a plasma may be expected to interact.

Debye unit (*Elec.*). Unit of *dipole moment*. 1 D.U. $= 10^{-18}$ e.s.u. $= 3{\cdot}34 \times 10^{-30}$ C m.

decade (*Elec.*). Any ratio of 10 : 1. Specifically the interval between frequencies of this ratio.

decade box (*Elec. Eng.*). A resistance (capacitance or inductance) box divided into sections so that each section has 10 switched positions and 10 times the value of the preceding section. The switches can therefore be set to any integral value within the range of the box.

decahydro-naphthalene (*Chem.*). $C_{10}H_{18}$, b.p. 190°C (103 kN/m²), product of complete hydrogenation of naphthalene under pressure and in the presence of a catalyst. Also **decalin**.

decalage (*Aero.*). The angle between the two chords of the upper and lower wings in the plane parallel with that containing the longitudinal and vertical axes of a biplane.

decalcification (*Med.*). The process of absorption of lime salts from bone.

decalcomania paper (*Paper*). A transfer paper for conveying a design on to pottery, etc.

decalescence (*Met.*). The absorption of heat that occurs when iron or steel is heated through the arrest points. See **recalescence**.

decalin (*Chem.*). *Decahydro-naphthalene* (q.v.).

decametric waves (*Radio*). Waveband from 10 to 100 metres.

decandrous (*Bot.*). Having 10 stamens.

decant (*Chem.*). To pour off the supernatant liquor when a suspension has settled.

decaploid (*Cyt.*). Having 10 times the haploid number of chromosomes.

Decapoda (*Zool.*). (1) An order of *Eucarida*, in which the exopodite of the maxilla is large, 3 pairs of thoracic limbs are modified as maxillipeds, and 5 as walking legs. Shrimps, prawns, crabs, lobsters, etc. (2) A suborder of *Dibranchia* having 8 normal arms and 2 longer partially retractile arms; the suckers are pedunculate, there is a well-developed internal shell, and lateral fins are present; actively swimming forms, usually carnivorous. Squids and Cuttlefish.

decapsulation (*Surg.*). Removal of capsule or covering of an organ, especially of the kidney.

decarbonizing (*I.C. Engs.*). The process of removing the solid carbon deposited on the internal surfaces of the combustion chamber and the piston crown of an i.c. engine.

decarburization (*Met.*). Removal of carbon from the surface of steel by heating in an atmosphere in which the concentration of decarburizing gases exceeds a certain value.

decastyle (*Arch.*). A portico having 10 columns.

decatizing (*Textiles*). Process for imparting a permanent finish to worsted and woollen fabrics by forcing steam through them while under tension, to fix or set the structure.

decay constant (*Nuc.*). See **disintegration constant.**

decay factor (*Phys.*). That which expresses rate of decay of oscillations in damped oscillatory system, given by natural logarithm of ratio of two successive amplitude maxima divided by the time interval between them. Calculated from ratio of resistance coefficient to twice mass in a mechanical system, and ratio of resistance to twice inductance in an electrical system. Also called **damping factor.** See **logarithmic decrement.**

decay law (*Phys.*). When the phenomenon is random, and a change depends on a momentary condition, the law of decay is *exponential*, e.g., sound intensity in an enclosure, activity in a radioactive substance, charge into and out of a capacitor, and rise and fall of a current in an inductor (air core).

decay product (*Nuc.*). See **daughter product.**

decay time (*Phys.*). That in which the amplitude of an exponentially decaying quantity reduces to e^{-1} (36·8%) of its original value.

Decca (*Aero., Nav.*). A navigation system of the radio position fixing type using continuous waves. The *fix*, given by the intersection of two hyperbolic position lines, is indicated on meters and can be plotted on a Decca chart or (aircraft) on a flight log which gives a continuous pictorial presentation of position. See **Dectra.**

decelerating electrode (*Electronics*). One which is intended to reduce the velocity of electrons.

deceleration (*Mech.*). Negative acceleration. The rate of diminution of velocity with time. Measured in feet or metres per second squared.

decerebrate (*Zool.*). Lacking a cerebrum.

decerebrate tonus (*Zool.*). A state of reflex tonic contraction of certain skeletal muscles following upon the separation of the cerebral hemispheres from the lower centres.

deci-. Prefix with physical unit, meaning one-tenth.

deci-ampere balance (*Elec. Eng.*). An ampere-balance having a range 0·1 to 10 amperes.

decibel (*Telecomm.*). $\frac{1}{10}$ bel, abbrev. **dB.** Commonly used to compare power levels, as NdB in $N = 10 \log_{10} W_2/W_1$, where W_2 is a power controlled by W_1. Used as a measure of response in all types of electrical communication circuits. See **dBA, etc., sound pressure level.**

decibel meter (*Acous.*). Meter which has a scale calibrated approximately uniformly in logarithmic steps and labelled with decibel units; used for determining the power levels in communication circuits, relative to a datum power level, now 1 mW in 600 ohms.

decidua (*Zool.*). In Mammals, the modified mucous membrane lining the uterus at the point of contact with the placenta, which is torn away at parturition and then ejected; the afterbirth; the maternal part of the placenta.

deciduate (*Zool.*). Said of Mammals in which the maternal part of the placenta comes away at birth. Cf. *indeciduate*.

deciduoma malignum (*Med.*). See **chorionepithelioma.**

deciduous (*Bot.*). Falling off, usually after a lengthy season of growth and functioning, generally before cold or drought sets in. (*Med.*) Falling off; said of the first dentition teeth.

decilog (*Maths.*). Unit of ratio, one-tenth of logarithm to base-10. Same as **decibel**, disused.
decimal candle (*Light*). See **bougie-decimale.**
decimal fraction (*Maths.*). A fraction having a power of ten as denominator. The denominator is not usually written but is indicated by the decimal point, a dot (on the Continent, a comma) placed between the unit figure and the numerator. The number of figures after the decimal point is equal to the power of 10 of the denominator. Thus 42·017 is equal to $42\frac{17}{1000}$. Also **decimal, decimal number.**
decimal system (*Maths.*). A number system whose base is 10 and in which every fraction is expressed as a decimal number. In this system each unit is in theory 10 times the next smaller one (in practice not necessarily so; cf. most decimal coinage systems). The *metric system* and *Système International* (qq.v.) are decimal systems.
decimetre. One-tenth of a metre.
decimetric waves (*Radio*). Waveband having range from 10 cm to 1 metre.
decimolar calomel electrode (*Chem.*). A calomel electrode containing decimolar potassium chloride solution.
decimo octavo (*Print.*). See **eighteenmo.**
decimo sexto (*Print.*). See **sixteenmo.**
decineper (*Elec.*). Unit of voltage and current attenuation in lines and amplifiers, of magnitude one-tenth of the *neper* defined by

$$d = 10 \log_e (x_1/x_2),$$

where d is number of decinepers, x_1 and x_2 are currents (voltages or acoustic pressures). In properly terminated networks, 1 decineper equals 0·8686 dB. Abbrev. **dN.**
decinormal solution (*Chem.*). A standard solution which is one-tenth as concentrated as a normal solution. See **normal solution.**
decision (*Comp.*). Use of processed data in selecting which of two possible operations to perform next.
decision speed (*Aero.*). See **critical speed.**
decit (*Comp.*). Abbrev. for *decimal digit.*
deck (*For.*). Logs collected at a place intermediate between stump and landing. They are called *cold deck* if allowed to accumulate, or *hot deck* if removed immediately. (*Ships*) A platform which forms the top of one horizontal division of a ship and the bottom of that immediately above.
deck beam (*Ships*). A stiffening member of a deck, which may be either transverse or longitudinal. It is supported at extremities by knee connexions to frames or bulkheads or by supporting girders.
deck bridge (*Civ. Eng.*). A bridge in which the track is carried by the upper stringer. Cf. *through bridge.*
deck crane (*Ships*). A crane, either fixed or movable, mounted on deck for use in loading or discharging cargo.
deck forge (*Ships*). A portable forge used by riveters and smiths on board ship.
deck houses (*Ships*). See **top hamper.**
decking (*Build.*). The platform supporting the derrick on a derrick tower gantry.
deckle edge (*Paper*). The feathery edge of hand-made paper due to the 'deckle' or mould frame. In machine-made paper it can be formed by artificial means. Also taken to refer to the untrimmed edge of the machine web.
deckle straps (*Paper*). Continuous rubber straps, at the sides of a paper-making machine, which form a deckle edge on the paper, thus retaining the paper stock in liquid form on the machine wire.
decks (*Print.*). On rotary presses, pairs of horizontal printing couples arranged one above the other.
deck stringer (*Ships*). The main strength portion of a ship's deck, being that portion, on both sides, adjacent and attached to the shell plating. It comprises the stringer strake of plating and the stringer angle section forming such attachment.
declared efficiency (*Elec. Eng.*). The efficiency which the manufacturers of an electric machine or transformer declare it to have, under certain specified conditions.
declinate (*Bot.*). Descending in a curve.
declination (*Astron.*). The angular distance of a heavenly body from the celestial equator measured positively northwards along the hour circle passing through the body. (*Surv., etc.*) See **magnetic declination.**
declination circle (*Astron.*). (1) As applied to the celestial sphere, a synonym for *hour circle* (1) (q.v.). (2) The graduated circle of an equatorial telescope on which declinations are read.
declinimeter (*Elec. Eng.*). An apparatus for determining the direction of a magnetic field with respect to astronomical or survey coordinates.
declining (*Bot.*). Straight, and pointing downwards.
declutch (*Eng.*). The action of separating the driving from the driven member of a clutch to prevent or interrupt the transmission of (rotary) motion.
decoction (*Chem.*). An extract of a substance or substances obtained by boiling.
decoder (*Radio*). A circuit used to derive stereo information from a multiplex signal in a radio tuner.
decoherer (*Radio*). Device for restoring a *coherer* to its sensitive, nonconducting condition after the arrival of a signal; also **anticoherer.** See **tapper.**
decollation (*Zool.*). The dropping off of the upper whorls of a gastropod shell, when the animal has ceased to occupy them.
décollement (*Surg.*). The separation of an organ from tissue to which it adheres.
decolorize (*Chem.*). To remove the coloured material from a liquid by bleaching, precipitation or adsorption.
decolorizers (*Glass*). Materials added to the batch for the express purpose of improving the appearance of the glass by hiding the yellow-green colour due to iron impurities.
decompensation (*Med.*). Failure of a diseased heart which has previously maintained its strength although diseased.
decomposed (*Bot.*). The condition of the cortex of a lichen when composed of gelatinous hyphae. (*Zool.*) Not touching, as the barbs of a feather when they are not in contact.
decomposers (*Ecol.*). In an ecosystem, heterotrophic organisms, chiefly bacteria and fungi, which break down the complex compounds of dead protoplasm, absorbing some of the products of decomposition, but also releasing simple substances usable by producers. Cf. *consumers, producers.*
decomposition (*Chem.*). The more or less permanent breaking down of a molecule into simpler molecules or atoms.
decomposition voltage (*Elec. Eng.*). The minimum voltage which will cause continuous electrolysis in an electrolytic cell.
decompound (*Bot.*). A term applied to a compound leaf having leaflets made up of several distinct parts.

decompounding winding (*Elec. Eng.*). A series winding on a compound-wound d.c. generator, connected in such a way that the m.m.f. produced by it is in opposition to that of the shunt winding.

decompression (*Med., Surg.*). Any procedure for relieving pressure or the effects of pressure.

deconjugation (*Cyt.*). The separation of the paired chromosomes before the end of the prophase of meiosis.

decontamination (*Nuc. Eng.*). Removal or neutralization of radioactive bacteriological or chemical contamination.

decontamination factor (*Nuc. Eng.*). Ratio of initial to final contamination for a given process.

decorticated (*Bot.*). Deprived of bark: devoid of cortex.

decortication of the lung (*Surg.*). Operative removal from the lung of pleura thickened as a result of chronic inflammation.

decoupling (*Electronics*). Reduction of a common impedance between parts of a circuit, e.g., by using a *by-pass capacitor*.

decoupling filter (*Telecomm.*). Simple resistor-capacitor section(s), which decouple feedback circuits and so prevent oscillation or *motor-boating* (relaxation oscillation).

decoupling principle (*Nuc. Eng.*). A means of making an underground explosion produce a smaller signal than normal by letting it off in an air-filled cavity.

decrement (*Elec.*). Ratio of successive amplitudes in a damped harmonic motion.

decremeter (*Elec. Eng.*). An instrument for measuring decrement. It consists essentially of a resonant circuit whose response to a damped impressed e.m.f. is calibrated in terms of the decrement.

decrepitation (*Chem.*). The crackling sound made when crystals are heated, caused by internal stresses and cracking. (*Powder Tech.*) Breakdown in size of the particles of a powder due to internal forces, generally induced by heating.

decticous (*Zool.*). Said of insect pupae which have relatively powerful, sclerotized, articulated mandibles and are exarate; such pupae are found in the more primitive *Endopterygota*.

Dectra (*Aero., Nav.*). A radio position fixing system, based largely on *Decca* principles, designed to cover specific air route segments and trans-oceanic crossings. In addition to fix, location along a track and range information are given: hence, *De*cca *Tr*ack and *Ra*nge.

decubitus (*Med.*). The posture of a patient lying in bed.

decumbent (*Bot.*). Lying flat, except for the tip, which ascends.

decurrent (*Bot.*). (1) Of a leaf, having the base prolonged down the stem as two wings. (2) Of the gills of an agaric, running for some distance down the stipe.

decurved (*Bot.*). Bent downwards.

decussate (*Bot.*). With leaves in pairs, each pair being at right angles to those above and below.

decussate texture (*Geol.*). The arrangement of prismatic or tabular crystals in random orientation in a rock.

decussation (*Zool.*). Crossing over of nerve-tracts with interchange of fibres.

Dedekind cut (*Maths.*). A division of the rational numbers into 2 classes; used to define irrational numbers.

dedendum (*Eng.*). (1) Radial distance from the pitch circle of a gear-wheel to the bottom of the spaces between teeth. See **involute gear teeth, pitch diameter.** (2) Radial distance between the pitch and minor cylinders of an external screw thread. (3) Radial distance between the major and pitch cylinders of an internal thread.

dedifferentiation (*Biol.*). Retrogressive changes in a differentiated tissue, leading to the reduction of all types of cells to a common indifferent form.

dedolomitization (*Geol.*). The recrystallization of a dolomite rock or dolomitic limestone consequent on contact metamorphism; essentially involving the breaking down of the dolomite into its two components, $CaCO_3$ and $MgCO_3$. The former merely recrystallizes into a coarse calcite mosaic; but the latter breaks down further into MgO and CO_2. The magnesium oxide may occur in the rock as periclase, more commonly as brucite, while in the presence of silica, magnesium silicates such as forsterite are formed. See **forsterite-marble.**

deducted spaces (*Ships*). Spaces deducted from the *gross tonnage* (q.v.) to obtain the *net register tonnage* (q.v.). In general, deducted spaces are those spaces required to be used in the working of the ship and accommodating the crew.

dee(s) (*Electronics*). Pair of hollow half-cylinders, i.e., D-shaped, in the vacuum of a cyclotron for accelerating charged particles in a spiral, a high-frequency voltage being applied to them in antiphase.

de-emphasis (*Acous.*). Correction of frequency response distortion originated by *pre-emphasis* (q.v.), by introducing a complementary frequency response; necessary in a form of frequency modulation. Also **post-emphasis.**

de-energize (*Elec. Eng.*). To disconnect a circuit from its source of power.

deep-bar cage winding (*Elec. Eng.*). A form of high-torque rotor winding for induction motors which depends on an increase in effective rotor resistance at starting due to *eddy currents* (q.v.).

deep bead (*Join.*). Piece of timber covering the lower 3 in. or so of movement of the bottom sash in a window, so as to permit ventilation at the meeting rail while keeping bottom of the window to all intents and purposes shut.

deep drawing (*Met.*). The process of cold-working or drawing sheet or strip metal by means of dies into shapes involving considerable plastic distortion of the metal; e.g., automobile mudguards, electrical fittings, etc.

deep etch (*Print.*). A lithographic process in which the image is very slightly etched into the surface of the plate, producing more stability than on a *surface* (or *albumen*) *plate*. A positive is used when printing down.

Deepkill Shales (*Geol.*). The graptolite-shale facies of the Beekmantown Group in the Hudson River valley, equivalent to the Arenig Series of Britain.

deeply-etched (*Print.*). A half-tone block given a supplementary deepening etch to improve its printing and duplicating qualities.

Deep Pictures (*Photog.*). TN for products of a form of stereophotography in which the subject is taken from several contiguous viewpoints, either simultaneously or by traversing camera, through a grid of vertical cylindrical lenticulations, the resultant composite image appearing as a three-dimensional picture when registered with a grid similar to that used in the camera; also used to obtain stereoscopic X-ray images.

deeps (*Surv.*). See **lead-line.**

deep-sea deposits (*Geol.*). Those sediments which accumulate out of reach of ordinary land-derived material; they fall into two categories: (*a*) organic oozes, (*b*) various muds and clays. In the shallower parts of the oceans the oozes are composed of the hard parts of planktonic organisms embedded in a powder arising from their disintegration. In the deeper parts the remains of siliceous organic remains are dominant. See **red clay, diatom ooze, globigerina ooze, pteropod ooze, radiolarian ooze.**

deep-sea lead (*Surv.*). A lead used for attachment to a lead-line measuring beyond 100 fathoms.

deep tank (*Ships*). A large tank extending from the bottom up to the first deck and from side to side. May be used for liquid cargo or for water ballast.

deep therapy (*Radiol.*). X-ray therapy of underlying tissues by *hard radiation* (q.v.) (usually produced at more than 180 kVp) passing through superficial layers.

deep well (*Civ. Eng.*). A shaft sunk through an upper impermeable stratum into a lower permeable one, from which a supply of water may be obtained.

deep-well pump (*Eng.*). A centrifugal pump, generally electrically driven by a submerged motor built integrally with it, placed at the bottom of a deep bore hole for raising water.

deer-fly fever (*Med., Vet.*). See **tularaemia.**

deerite (*Min.*). A black, monoclinic hydrous silicate of iron and manganese.

defaecation, defecation. The ejection of faeces from the body.

defect (*Crystal.*). Lattice imperfection which may be due to the introduction of a minute proportion of a different element into a perfect lattice, e.g., indium into germanium crystal, to form an intrinsic semiconductor for a transistor. A 'point' defect is a 'vacancy' or an 'interstitial atom', while a 'line' defect relates to a dislocation in the lattice. (*For.*) In timber, any feature that lowers its technical quality or commercial value and therefore its grade.

defective equation (*Maths.*). An equation derived from another, but with fewer roots than the original.

defect sintering (*Powder Met.*). Sintering whereby particles are introduced as a fine dispersion in a sintered body, or by chemical action during sintering. Mobility in heat treatment, with or without working, enables the introduced atoms or molecules to migrate through defects giving marked modification of properties.

defect structure (*Crystal.*). Intense localized misalignment or gap in the crystal lattice, due to migration of ions or to slight departures from stoichiometry. The resulting opportunity for mobility is important for semiconductors, catalysis, photography, rectifiers, corrosion, etc. Cf. *dislocation.*

defence mechanism (*Psychol.*). See **ego-defence mechanisms.**

defenders (*Acous.*). See **ear defenders.**

deferent (*Astron.*). See **epicycle.**

defervescence (*Med.*). The fall of temperature during the abatement of a fever: the period when this takes place.

defibration (*For.*). See **pulping.**

defibrillator (*Med.*). Electrical apparatus to arrest ventricular *fibrillation* (q.v.).

deficiency (*Cyt.*). The loss, inactivation or absence of a chromosomal segment or gene. (*Maths.*) The difference between the maximum possible number of double points on a curve and the actual number.

deficiency disease (*Med.*). Any disease resulting from the deprivation of food substances (e.g., vitamins) necessary to good health.

deficient (*Build.*). See **unstable.**

definite (*Bot.*). (1) Ending in a flower and ceasing to elongate. (2) Said of an inflorescence with all its branches of limited growth; the same as *cymose.* (3) Always of the same number in any one species.

definite integral (*Maths.*). If $F(x)$ equals $\int f(x)dx$, then $\int_a^b f(x)dx = F(b) - F(a)$ is the ***definite integral*** over the range a to b.

definite proportions (*Chem.*). See **law of constant (or definite) proportions.**

definite time-lag (*Elec. Eng.*). A time-lag fitted to relays or circuit-breakers to delay their operation; it is quite independent of the magnitude of the current causing that operation. Also called **constant time-lag, fixed time-lag, independent time-lag.**

definition (*Acous.*). The clarity of perception of speech sounds, particularly the transient sounds. Acoustic definition depends largely on the ratio of the direct intensity to the reverberant intensity, when an enclosure is considered. (*Optics.*) See **resolving power.** (*Photog., TV*) Extent to which a television or phototelegraphic system is able to reproduce the detail of the transmitted image; defined as the number of picture elements, or the number of scanning lines into which the picture is divided. For television in Europe (standard) 625 lines (U.K. obsolescent, 405 lines); U.S.A. 525 lines.

definitive (*Zool.*). Final, complete: fully developed: defining or limiting.

deflagrating spoon (*Chem.*). A small spoon-shaped instrument used in chemical laboratories for handling materials which are liable to take fire when exposed to air.

deflagration (*Chem.*). Sudden combustion, generally accompanied by a flame and a crackling sound.

deflagrator (*Electronics*). Primary cell capable of high-peak output currents.

deflecting electrodes (*Electronics*). See **deflector plates.**

deflection (*Eng.*). (1) The amount of bending or twisting of a structure or machine part under load. (2) The movement of the pointer or pen of an indicating or a recording instrument.

deflection angle (*Electronics*). See **angle of deflection.** (*Surv.*) The angle between one survey line and the prolongation of another survey line which meets it. See also **intersection angle.**

deflection defocusing (*Electronics*). Loss of focus of a CRT spot as deflection from the centre of the screen increases.

deflection sensitivity (*Electronics*). (1) Ratio of displacement of spot or angle of an electron beam to the voltage producing it. (2) Ratio of displacement of spot or angle of beam to the magnetic field producing it, or the current in the deflecting coils. Also applies to galvanometer or other instrument for measuring voltage, current or quantity of electricity.

deflection yoke (*Electronics*). Assembly of mounted deflector coil.

deflectometer (*Eng.*). A device for measuring the amount of bending suffered by a beam during a transverse test.

deflector (*Elec. Eng.*). See **arc deflector.** (*Ships, etc.*) An instrument used during the adjustment of magnetic compasses. It measures the strength of the field at the compass.

deflector coil(s) (*Electronics*). Coil(s) so arranged that a current passing through produces a magnetic field which deflects the beam in a cathode-ray tube employing magnetic deflection. Usually applied around the neck of the tube.

deflector plates (*Electronics*). Electrodes so arranged in a cathode-ray tube that the electrostatic field produced by a p.d. deflects the beam. Also **deflecting electrodes.** See **electrostatic deflection.**

deflexed (*Bot.*). Bent outwards and downwards.

deflocculate (*Chem.*). To break up agglomerates and form a stable colloidal dispersion.

defoaming agent (*Chem.*). Substance added to a boiling liquid to prevent or diminish foaming. Usually hydrophobic and of low surface tension, e.g., silicone oils.

defoliant (*Agric.*). Chemical spray or dust which causes vegetation to lose its leaves.

deformation (*Maths.*). See **homotopic mapping.**

deformation potential (*Electronics*). Potential barrier formed in semiconducting materials by lattice deformation.

deformation ratio (*Maths.*). See **magnification.**

deformeter (*Instr.*). An instrument used in the mechanical determination of stresses, to impose a known distortion upon a model of a structure.

degassing (*Electronics*). Removal of last traces of gas from valve envelopes, by pumping or *gettering*, with or without heat, the former with eddy currents, electron bombardment, or simple baking.

degaussing (*Mag.*). Neutralization of the magnetization of a mass of magnetic material, e.g., a ship, by an encircling current. (*TV*) Removal from colour CRT of spurious magnetization which could affect colour purity.

degeneracy (*Elec.*). The condition in a resonant system when two or more modes have the same frequency.

degenerate (*Phys.*). Said of particles or electrons when their energy states are low and equal; also when an atom is stripped of all electrons, as in a star.

degenerate gas (*Electronics*). (1) That which is so concentrated, e.g., electrons in the crystal lattice of a conductor, that the Maxwell-Boltzmann law is inapplicable. (2) Gas at very high temperature in which most of the electrons are stripped from the atoms. (3) An electron gas which is far below its Fermi temperature so that a large fraction of the electrons completely fills the lower energy levels and has to be excited out of these levels in order to take part in any physical processes.

degenerate semiconductor (*Electronics*). One in which the conduction approaches that of a simple metal.

degeneration (*Biol.*). Evolutionary retrogression; the process of returning from a higher or more complex state to a lower or simpler state. (*Bot.*) The loss of morphological or physiological characters by being kept in culture for a long time. (*Telecomm.*) Term used for *negative feedback*.

deglutition (*Zool.*). The act of swallowing.

deglutition pneumonia (*Med.*). Pneumonia induced as a result of aspiration into the lungs of food or drink, or of matter from the nose or throat; also called **aspiration pneumonia.**

degradation (*Nuc.*). Loss of energy of motion solely by collision. Deliberate slowing of neutrons in a reactor is *moderation*. In an isolated system, the *entropy* increases.

degrains (*Leather*). Gloves made from the skins of sheep, gazelles, reindeer, or mocha, dressed on the grain side after the grain has been removed.

degras or moellen (*Chem.*). (1) A semi-oxidized fat expressed from sheep skins after the oiling process and forming a by-product in the manufacture of chamois leather; from it, when purified, *lanolin* (q.v.) is obtained. (2) A similar manufactured substance, commonly known as *sod oil*.

degree (*Heat*). The unit of temperature difference. It is usually defined as a certain fraction of the fundamental interval, which for most thermometers is the difference in temperature between the freezing and boiling points of water. See **Celsius scale, Centigrade scale, Fahrenheit scale, international practical temperature scale, Kelvin thermodynamic scale of temperature.**

degree of a curve (*Surv.*). The angle subtended at the centre of a curve by radii a standard chord length of 100 ft.

degree of damping (*Phys.*). The extent of the damping in an oscillatory system, expressed as a fraction or percentage of that which makes the system critically damped.

degree of dissociation (*Chem.*). The fraction of the total number of molecules which are dissociated.

degree of ionization (*Chem.*). The proportion of the molecules or 'ion-pairs' of a dissolved substance dissociated into charged particles or ions.

degrees of freedom (*Chem.*). (1) The number of variables defining the state of a system (e.g., pressure, temperature) which may be fixed at will. See **phase rule.** (2) Number of independent capacities of a molecule for holding energy, translational, rotational, and vibrational.

degressive (*Bot.*). A change towards simplification or degeneration.

degumming (*Textiles*). See **boiling-off.**

dehiscence (*Bot.*). The spontaneous opening at maturity of a fruit, anther, sporangium, or other reproductive body. (*Zool.*) In general, the act of splitting open: more particularly, in *Porifera*, the splitting of the outer layer of cells of certain types of larva to permit the passage, to the exterior, of the cells primarily occupying the interior; see also **diapedesis.** *adj.* **dehiscent.**

dehorn (*For.*). To re-mark logs for change of ownership.

dehydration (*Chem.*). (1) The removal of H_2O from a molecule by the action of heat, often in the presence of a catalyst, or by the action of a dehydrating agent, e.g., concentrated sulphuric acid. (2) The removal of water from crystals, tars, oils, etc., by heating, distillation, or by chemical action. (*Med.*) Excessive loss of water from the tissues of the body.

dehydrogenase (*Chem.*). An enzyme which catalyses oxidation by the removal of hydrogen.

dehydrogenation (*Chem.*). Removal of hydrogen atoms from molecules. Important industrially, particularly in the petrochemicals field where a wide range of catalysts and operating conditions are used. The production of methanal from methanol is an example of dehydrogenation.

dehydrothiotoluidine (*Chem.*).

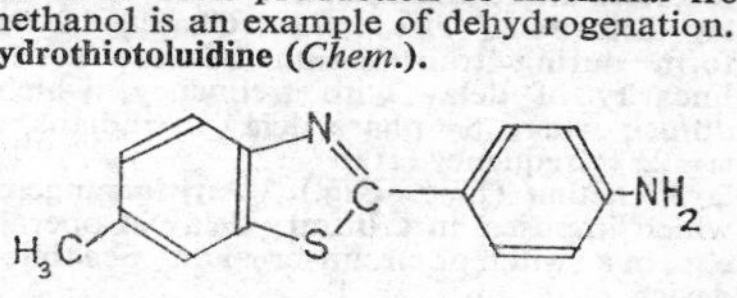

Yellow crystalline substance, m.p. 195°C, used in the preparation of direct cotton azo dyestuffs.

de-icing (*Aero.*). Method of protecting aircraft against icing by removing built-up ice before it assumes dangerous proportions. It may be based on pulsating pneumatic overshoes, chemical applications, or intermittent operation of electrical heating elements. Cf. *anti-icing*.

Deion circuit-breaker (*Elec. Eng.*). TN for a circuit-breaker fitted with an arc-control device in which the arc takes place within a slot in a stack of insulated plates. The plates contain iron inserts or a magnet coil, so that the arc is blown magnetically towards the closed end of the slot, thereby coming into contact with cool oil which has a de-ionizing action and extinguishes the arc.

de-ionization (*Chem.*). Process whereby an ionized gas returns to its normal neutral condition, when sources of ionization have been removed.

de-ionized water (*Chem.*). Water from which ionic impurities have been removed by passing it through cation and anion exchange columns.

Deiter's cells (*Histol.*). In Mammals, supporting cells of the sensory epithelium of the organ of Corti.

deka-ampere-balance (*Elec. Eng.*). An ampere-balance having a current range from 1 to 100 amperes.

Dekatron (*Electronics*). Multi-electrode neon tube in which a set of 10 electrodes function in turn on the application of voltage pulses, thus indicating the number of pulses. Such tubes in cascade constitute a visible counter of impulses in the decimal system.

del (*Maths.*). In cartesian coordinates, the vector operator $\boldsymbol{i}\frac{\partial}{\partial x} + \boldsymbol{j}\frac{\partial}{\partial y} + \boldsymbol{k}\frac{\partial}{\partial z}$. Also called **nabla**. Symbol ∇.

delaine (*Textiles*). Lightweight, plain-weave, all-wool dress fabric manufactured from botany warp and weft. The design is either printed on the warp or on the piece. A cheaper cotton imitation is known as **delainette**.

delamination (*Histol.*). The division of cells in a tissue, leading to the formation of layers.

Delanium (*Chem.*). TN for compacted carbon and graphite materials of high purity developed for use in chemical plant. Resistant to most chemicals except strong oxidizing agents.

De la Rue cell (*Elec. Eng.*). See **chloride of silver cell**.

delay (*Acous.*). That which can be introduced into the transmission of a signal, by recording it magnetically on tape, disk, or wire, and reproducing it at a point traversed later by the record. Used in public-address systems, to give illusion of distance and to coalesce contributions from original source and reproducers. (*Comp.*) Unit which delays signals, e.g., by exactly one *bit*.

delay circuit (*Telecomm.*). One which delays the wavefront of a signal, telephonic or telegraphic, by acoustic tubes, recording and reproduction, or wave filters. Also transmission lines for phasing in feeders between antennae, radio receivers and transmitters.

delay distortion (*Telecomm.*). Change in waveform during transmission because of non-linearity of delay with frequency, which is $d\beta/d\omega$, where β = phase delay in radians, and $\omega = 2\pi \times$ frequency (Hz).

delayed action (*Elec. Eng.*). Any arrangement which imposes an arbitrary delay in operation e.g., in a switch or circuit-breaker. See **time-lag device**.

delayed automatic gain control (*Radio*). That which is operative above a threshold voltage signal in a radio receiver. Provides full amplification in radio receivers for very weak signals, and constant output from detector for signals above the threshold. Also called **quiet automatic volume control**. See also **automatic gain control**.

delayed critical (*Nuc.*). Assembly of fissile material critical only after release of delayed neutrons. Cf. *prompt critical*.

delayed drop (*Aero.*). A live parachute descent in which the parachutist deliberately delays pulling the ripcord.

delayed neutrons (*Nuc.*). Those arising from fission but not released instantaneously. Fission neutrons are all *prompt*; those apparently delayed (up to seconds) arise from breakdown of fission products, not primary fission. Such delay eases control of reactors.

delayed opening (*Aero.*). Delaying the opening of a parachute by an automatic device. In any flight above 40 000 ft (12 500 m), low temperature and pressure require that aircrew must reach lower altitude for survival as rapidly as possible, and it is usual to have a barostatic device to delay opening to a predetermined height, usually 15 000 ft (4500 m).

delayed response (*An. Behav.*). The behaviour of an animal in an experimental situation where it is in some way shown which of two or more alternatives is associated with a reward, this indication then being withdrawn, and a delay enforced before the animal is allowed to choose one of the alternatives.

delay line (*Comp.*). Column of mercury, a quartz plate, or length of nickel wire, in which impressed sonic signals travel at a finite speed and which, by the delay in travelling, can act as a *store*, the signals being constantly recirculated and abstracted (*gated*) as required. (*Telecomm.*) Real or artificial transmission line used to delay a propagated electrical signal.

delay-line store (*Comp.*). One which uses the circulation of impulses along a mercury-tube, quartz plate, or nickel-wire transmission line, with form correction and amplifiers.

delay network (*Telecomm.*). Artificial line of electrical networks, designed to give a specified phase delay in the transmission of currents over a frequency band represented by speech, so that time may be allowed for switches or relays to be operated.

delay period (*Eng.*). The time or crank-angle interval between the passage of the spark and the resulting pressure rise in a petrol or gas engine, or between fuel injection and pressure rise in an oil engine.

delay-time register (*Telecomm.*). An acoustic or electric delay line with its input, output, and circulating circuits.

delay working (*Teleph.*). In trunk operations, when the lines are in such full operation that subscribers have to wait for connexions, their requirements are noted by a trunk-record operator, who routes this to the relevant trunk operator, who in turn calls back the subscriber when a line is free. See **demand working**.

Delbruck scattering (*Phys.*). Elastic coherent scattering of gamma-rays in the coulomb field of a nucleus. The effect is small and so far has not been conclusively detected.

delessite (*Min.*). An oxidized variety of chlorite, relatively rich in iron.

deleted neighbourhood (*Maths.*). Of a point z_0: the set of all points z in the domain $|z - z_0| < \alpha$ (α a constant) excluding the point z_0.

deletion (*Cyt.*). The loss or absence of a portion of a chromosome.

Delhi boil (*Med.*). Tropical sore; Baghdad boil. Oriental sore resulting from infection of the skin with a protozoal parasite.

delignification (*Bot.*). The destruction of lignin in plant material by the action of a fungus.

deliming (*Leather*). The process of removing lime salts from skins and hides, previous to tanning.

delimit (*Comp.*). To fix the boundaries of a sequence of characters or data. *Delimiters* are special characters used to do this.

delinquent (*Psychol.*). An individual, generally a child or adolescent, who shows definite lack of moral and social sense, without evidence of impairment of intellect; in particular, one who commits a social or criminal offence.

deliquescence (*Bot.*). (1) Gelatinization and liquefaction of cell walls, sporangium membranes, etc., at maturity. (2) Of a stem, breaking up into branches. (*Chem.*) The change undergone by certain substances which become damp and finally liquefy when exposed to the air, owing to the very low vapour pressure of their saturated solutions; e.g., calcium chloride.

delirium (*Med.*). A profound disturbance of consciousness occurring in febrile and toxic states; characterized by restlessness, incoherent speech, excitement, delusions, illusions, and hallucinations.

delirium tremens (*Med.*). An acute delirium in chronic alcoholism, characterized by insomnia, restlessness, terrifying hallucinations and illusions, and loss of orientation to time and place.

delivery (*Eng.*). (1) The discharge from a pump or compressor. (2) The withdrawal of a pattern from a mould. (*Print.*) The mechanical arrangement for delivering sheets after printing, there being several designs.

Dellinger fade-out (*Radio*). Complete fade-out (which may last for minutes or hours) and inhibition of short-wave radio-communication because of the formation of a highly absorbing D-layer, lower than the regular E- and F-layers of the ionosphere, on the occasion of a burst of hydrogen particles from an eruption associated with a sun spot.

Delon rectifier (*Elec. Eng.*). A high-voltage rectifier for dealing with small currents; it consists of a system of capacitors in conjunction with a rotating switch.

delph (*Hyd. Eng.*). A drain behind a sea embankment, on the land side.

delphinin (*Chem.*). An anthocyanin responsible for the colour of the flowers of the delphinium. On hydrolysis it decomposes into two molecules of glucose, two of 4-hydroxybenzoic acid, and one of delphinidin.

delphinine (*Chem.*). $C_{33}H_{45}O_9N$. A diterpenoid alkaloid, obtained from the seeds of *Delphinium staphisagria*, closely related in structure to aconitine. Contains one acetate and one benzoyl group and four methoxyl groups. It is intensely toxic, resembling aconitine in its action. Hydrolysis gives **delphonine**, $C_{24}H_{39}O_7N$, which also occurs in nature.

Delrin (*Plastics*). TN for an acetal resin. Highly crystalline stable form of polymerized formaldehyde. Used as an engineering plastic for many applications.

delta (*Geol.*). The more or less triangular area of river-borne sediment deposited at the mouth of rivers heavily charged with detritus. A delta is formed on a low-lying coastline, particularly in seas of low tidal range and in areas where subsidence keeps pace with sediment deposition. The Nile Delta is a good example.

delta connexion (*Elec. Eng.*). The connexion of a 3-phase electrical system such that the corresponding windings of the transformers form a triangle.

deltaic deposits (*Geol.*). The accumulations of sand and clay, with remains of brackish water organisms, drifted plant debris, and animal remains, washed in from the land. Current-bedding is characteristic, with well developed fore-set and top-set beds. The Millstone Grit is a good example of a 'fossil delta'

delta impulse function (*Telecomm.*). Infinitely narrow pulse of great amplitude, such that the product of its height and duration is unity.

delta iron (*Met.*). The polymorphic form of iron stable between 1403°C and the melting point (about 1532°C). The space lattice is the same as that of α-iron and different from that of γ-iron.

delta-matching transformer (*Telecomm.*). A matching network between 2-wire transmission lines and half-wave antennae.

delta network (*Telecomm.*). One with three branches all in series.

delta-particle (*Nuc.*). Very short lived hyperon which decays almost instantaneously through the strong interaction.

delta-ray (*Nuc.*). Any particle ejected by recoil action from passage of ionizing particles, e.g., in a Wilson cloud chamber.

delta-ray spectrometer (*Nuc. Eng.*). See **spectrometer.**

delta voltage (*Elec. Eng.*). Normally synonymous with *mesh voltage* (q.v.), but the term is also used to denote the voltage between alternate terminals or lines of a symmetrical 6-phase system.

delta wave (*Physiol.*). The lower frequency brain waves (1–8 Hz).

delta wing (*Aero.*). A sweptback wing of substantially triangular planform, the trailing edge forming the base. It is longitudinally stable and does not require an auxiliary balancing aerofoil, although tail or nose planes are sometimes fitted to increase pitch control and trim so that landing flaps can be fitted.

delthyrium (*Zool.*). In *Brachiopoda*, the aperture in the ventral valve, or between the beak and the hinge, for the passage of the peduncle.

deltidium (*Zool.*). A small calcareous plate covering the delthyrium in certain species of *Brachiopoda.*

deltoid (*Bot., Zool.*). Having the form of an equilateral triangle: any triangular structure, as the deltoid muscle of the shoulder. (*Maths.*) Recent name for Steiner's three-cusped hypocycloid.

delusion (*Psychol.*). A belief in events for which there is no objective evidence; a false belief peculiar to the believer.

de luxe (*Typog.*). A lavish style of bookwork design, characterized by generous margins.

demagnetization (*Elec. Eng., Mag.*). (1) *Removal* of magnetization of ferromagnetic materials by the use of diminishing saturating alternating magnetizing forces. (2) *Reduction* of magnetic induction by the internal field of a magnet, arising from the distribution of the primary magnetization of the parts of the magnet. (3) *Removal* by heating above the Curie point. (4) *Reduction* by vibration. (*Min. Proc.*) In dense media process using ferro-silicon, passage of the fluid through an a.c. field to deflocculate the agglomerated solid.

demagnetization factor (*Elec. Eng.*). Diminution factor (N) applied to the intensity of magnetization (I) of a ferromagnetic material, to obtain the demagnetizing field (ΔH), i.e., $\Delta H = NI$. N depends primarily on the geometry of the body concerned.

demagnetizing ampere-turns (*Elec. Eng.*). See **back ampere-turns.**

demagnetizing coil (*Electronics*). One used to eliminate residual magnetization from a record or playback head of a tape recorder; powered by mains frequency a.c. Also **degaussing coil.**

demand (*Elec. Eng.*). See **maximum demand.**

demand factor (*Elec. Eng.*). Ratio of the maximum demand on a supply system to the total connected load.

demand indicator (*Elec. Eng.*). See **maximum-demand indicator.**

demand limiter (*Elec. Eng.*). See **current limiter.**

demand meter (*Elec. Eng.*). One reading or recording the loading on an electrical system.

demand working (*Teleph.*). The operation of trunk circuits in such a way that the subscriber's requirements can be met without his waiting an unreasonable time or having to be recalled.

demantoid (*Min.*). Bright-green variety of the garnet andradite, essentially silicate of calcium and iron.

dematioid (*Bot.*). Covered with fine, dark hyphae.

dementia (*Psychol.*). Any form of insanity characterized by the failure or loss of mental powers: the organic deterioration of intelligence, memory, and orientation.

dementia praecox (*Psychiat.*). See **schizophrenia.**

De Meritens alternator (*Elec. Eng.*). An old form of alternator in which the exciting field was produced by a series of rotating permanent magnets.

demersal (*Zool.*). Found in deep water or on the sea bottom; as Fish eggs which sink to the bottom, and 'wet' or midwater and bottom-living Fish as opposed to surface Fish (e.g., Herring) and Shellfish. Cf. *pelagic*.

demeton (*Chem.*). Group of insecticides. The name is commonly applied to a mixture of demeton-O (diethyl 2-(ethylthio) ethyl phosphorothionate) and demeton-S (diethyl *S*-[2-(ethylthio) ethyl] phosphorothiolate).

demi-. Prefix from L. *dimidius*, half.

demifacet (*Zool.*). One of the two half-facets formed when the articular surface for the reception of the capitular head of a rib is divided between the centra of two adjacent vertebrae.

demi-hunter (*Horol.*). See **half-hunter.**

demijohn (*Glass*). Narrow-necked wine or spirit container of more than 2-gal. capacity.

demilune cells (*Zool.*). See **Gianuzzi's crescents.**

demi-lustre (*Textiles*). Term given to a class of wools, in the long fibre class.

demineralization (*Chem. Eng.*). A process for cleaning water in which the anions and cations are removed separately by absorption in synthetic exchange materials, leaving the water free of dissolved salts. The removal cells are regenerated by treatment with alkali and acid.

demisheath (*Zool.*). In some Insects, one of two chitinous sheaths protecting the ovipositor.

demister (*Autos.*). Ducts arranged so that hot dry air is played on the interior of the windscreen to prevent condensation. Heat source may be from electrical heater or heat dissipated from the engine.

demodectic mange (*Vet.*). *Demodectic folliculitis.* Mange of animals caused by mites of the genus *Demodex*, which live in the hair follicles and sebaceous glands.

demodulation (*Radio*). Inverse of *modulation.* Generally effected by passing the modulated carrier, or the high-frequency signal with an added carrier, through a nonlinear system, so that the output currents or voltages contain difference frequencies between the carrier and side frequencies which can be extracted and re-formed into the original modulating signal. Previously known as **detection.** Also previously applied to the reduction of the depth of modulation in a carrier when the latter is partially rectified in a high-frequency amplifier, with or without the presence of a relatively strong alien carrier.

demodulation of an exalted carrier (*Radio*). Same as **homodyne reception.**

demodulator (*Radio*). See **detector, thermionic rectifier.**

demography. The study of population statistics and the estimation of their variation with time.

De Moivre's theorem (*Maths.*). That

$$(\cos\theta + i\sin\theta)^n = \cos n\theta + i\sin n\theta.$$

Expressed in terms of the exponential function, the theorem is $(e^{i\theta})^n = e^{ni\theta}$.

Demospongiae (*Zool.*). A class of *Porifera* usually distinguished by the possession of a skeleton composed of siliceous spicules, or of spongin, or both; triaxial spicules are lacking; the flagellated chambers are small and rounded; the choanocytes are small.

demountable (*Elec.*). Said of X-ray tubes or thermionic valves when they can be taken apart for cleaning and filament replacement, and are continuously pumped during operation.

Dempster positive ray analysis (*Nuc. Eng.*). A technique of separating particles having different charge/mass ratios, using a magnetic field.

demulcent (*Med.*). Soothing; allaying irritation.

demulsification number (*Chem.*). The resistance to emulsification by a lubricant when steam is passed through it; indicated in minutes and half minutes required for the separation of a given volume of oil after emulsification.

demy (*Paper*). A standard size of (*a*) printing paper, $17\frac{1}{2} \times 22\frac{1}{2}$ in.; metric 445×572 mm; (*b*) writing paper, $15\frac{1}{2} \times 20$ in., U.S. 16×21 in.

demy octavo (*Typog.*). A book size, $8\frac{1}{2} \times 5\frac{1}{2}$ in. (metric, 216×140 mm).

demy quarto (*Typog.*). A book-size, $11 \times 8\frac{5}{8}$ in. (metric 279×219 mm).

denaturant (*Nuc.*). Isotope added to fissile material to render it unsuitable for military use.

denaturation (*Chem.*). An alteration in the properties of a protein, e.g., decrease in solubility, loss of crystallizability, or loss of specific activity if the protein is a hormone or enzyme. It can be caused by, e.g. heating, ultraviolet radiation, high concentration of urea.

denatured alcohol (*Chem.*). Alcohol which according to law has been made unfit for human consumption by the admixture of nauseating or poisonous substances, e.g., methyl alcohol, pyridine, benzene, etc.

dendrite (*Crystal.*). A treelike crystal formation. (*Met.*) Metal crystals grow in the first instance by branches developing in certain directions from the nuclei. Secondary branches are later thrown out at periodic intervals by the primary ones and in this way a skeleton crystal, or *dendrite*, is formed. The interstices between the branches are finally filled with solid which in a pure metal is indistinguishable from the skeleton. In many alloys, however, the final structure consists of skeletons of one composition in a matrix of another. (*Zool.*) A branch of a *dendron* (q.v.).

dendritic (*Bot.*). (1) Bearing markings which are treelike or mosslike. (2) Much branched.

dendritic figure (*Zool.*). In experimental embryology, an appearance produced by poisoning an egg under certain conditions; several pseudasters and isolated asters are produced, united by streaks of protoplasm.

dendritic markings (*Geol.*). Treelike markings, usually quite superficial, occurring on joint-faces and other fractures in rocks, frequently consisting of oxide of manganese or of iron. Less frequently the appearance is due to the inclusion of a mineral of dendritic habit in another mineral or rock, e.g., chlorite in silica as in 'moss agate'.

dendritic tentacles (*Zool.*). Much-branched tentacles in some *Holothuroidea*.

dendritic ulcer (*Med.*). A branching ulcer of the cornea, due to herpes of the cornea.

Dendrochirotae (*Zool.*). An order of *Holothuroidea* with dendritic tentacles; retractor muscles but no tentacle ampullae; podia on the trunk; the madreporite internal, and respiratory trees.

dendrograph (*Bot.*). An instrument which is used to measure the periodical swelling and shrinkage of tree trunks.

dendroid (*Bot.*). (1) Tall, with an erect main trunk, as tree-ferns. (2) Freely branched.

dendron (*Zool.*). A nerve-cell process which branches almost from the point at which it leaves the cell-body, as opposed to an axon: one of the terminal twigs into which an afferent axon breaks up at a synapse. Cf. *telodendra.*

denervated (*Med.*). Deprived of nerve supply.

dengue (*Med.*). Breakbone fever; a tropical disease in which the infecting agent is transmitted by mosquito to man; characterized by severe pains in the joints and a rash.

denier system (*Textiles*). The system used in the 'counting' of silk and nylon yarns; designated by the weight in grams of 9000 m of yarn, the unit of length. See **count of yarn, Tex.**

denim (*Textiles*). A strong cotton cloth, employing coloured warp yarns in blue, black, or brown colours, used extensively for overalls, boiler suits, jeans, etc. Usually 3-and-1 twill.

denitrification (*Bacteriol.*). See **nitrate-reducing bacteria.**

denizen (*Bot.*). A species which maintains its footing as a wild plant, though probably introduced by man.

denominational number system (*Maths.*). A system of representing numbers by a combination of digits in which each digit contributes an amount dependent upon both its value and its position. Thus with a radix of *r*, the *n*-digit number $a_{n-1}\,a_{n-2}\ldots a_2a_1a_0$ represents $\sum_{s=0}^{n-1} a_sr^s$, where each digit a_s can have any of *r* values; e.g., in the decimal system $r=10$, and the number 345 represents $3\times10^2+4\times10+5$. Cf. *nondenominational number system.*

denominator (*Maths.*). See **division.**

denotative meaning (*Psychol.*). That which is signified by a word or concept. See **connotative meaning.**

dens (*Zool.*). Any toothlike process, as the distal arms of the furcula in *Collembola*. *pl.* **dentes.**

dense (*Glass*). Of optical glass, having a higher refractive index.

dense-media process (*Min. Proc.*). *Heavy-media* (q.v.) or sink-float process. Dispersion of ferrosilicon or other heavy mineral in water separates lighter (floating) ore from heavier (sinking) ore.

dens epistrophei (*Zool.*). See **odontoid process.**

dense set (*Maths.*). A set of points is dense in itself if every point of the set is a limit point. A set is said to be everywhere dense in an interval, if every subinterval (no matter how small) contains points of the set. A subset X of a set Y is said to be dense in Y if the closure of X is Y. See **closure.**

densification (*Powder Tech.*). All modes of increasing density, including the effect of sintering contraction.

densify (*For.*). To increase the density of wood, e.g., by compression or by impregnating with synthetic resin.

densi-tensimeter (*Chem.*). An apparatus for determining both the pressure and the density of a vapour.

Densithene (*Plastics*). TN for polythene loaded with lead powder in the form of sheet, pipe, etc. Used for radioactive shielding.

densitometer (*Instr.*). Any instrument for measuring the optical transmission or reflecting properties of a material. (*Photog.*) An instrument for measuring the densities of exposed and developed film, particularly in photographic sound-recording.

density (*Photog.*). The logarithm of *opacity.* (*Phys., etc.*) The mass of unit volume of a substance, expressed in such units as kg/m^3, g/cm^3 or lb per cubic foot. See **relative density.**

density bottle (*Phys.*). A thin glass bottle, accurately calibrated, used for the determination of the density of a liquid.

density change method (*Powder Tech.*). Particle-size analysis technique which measures concentration changes within a sedimenting suspension by measuring the pressure exerted by a column of the suspension.

density-dependent factor (*Ecol.*). A factor affecting a population, whose effect on the population varies with population density, either directly or inversely. It includes some mortality factors, such as competition, parasites, predators and disease, and is believed by some ecologists to play a vital part in population control.

density effect (*Phys.*). The variation of mass absorption with density, which occurs for particles of relativistic energies.

density gradient centrifugation (*Chem.*). A technique to establish the density of a polymer or polymer mixture, e.g., DNA. The polymer is centrifuged in an ultracentrifuge in a concentrated solution of CsCl, and collects in bands at those zones of the centrifuge cell where its density and that of the medium are equal.

density-independent factor (*Ecol.*). A factor influencing a population, whose effects on the population are the same regardless of population density. It includes most climatic factors.

density of gases (*Chem., Phys.*). According to the *gas laws* (q.v.), the density of a gas is directly proportional to the pressure and the rel. mol. mass and inversely proportional to the absolute temperature. At standard temperature and pressure the densities of gases range from 0·0899 g/dm^3 for hydrogen to 9·96 g/dm^3 for radon.

dent (*Weaving*). Term denoting one wire in a loom reed; it also refers to the space between two wires, through which warp threads are drawn. The dents per inch is the set of the cloth.

dental formula (*Zool.*). A formula used in describing the dentition of a Mammal to show the number and distribution of the different kinds of teeth in the jaws; thus a Bear has in the upper jaw 3 pairs of incisors, 1 pair of canines, 4 pairs of premolars, and 2 pairs of molars; and in the lower jaw 3 pairs of incisors, 1 pair of canines, 4 pairs of premolars, and 3 pairs of molars. This is expressed by the formula $\frac{3142}{3143}$.

dental surgery. See **dentistry.**

dentale (*Zool.*). See **dentary.**

dentary (*Zool.*). In Vertebrates, a membrane bone of the lower jaw which usually bears teeth. In Mammals, it forms the entire lower jaw.

dentate (*Bot.*). Having a toothed margin; if each tooth bears a subsidiary tooth, the margin is *doubly dentate.*

dentate sclerite (*Zool.*). A cuticular arc which unites the bases of the mandibular sclerites in many *Diptera.*

dentation (*Bot.*). (1) A general name for the toothing of a margin, i.e., when it bears small blunt or pointed outgrowths. (2) The collective name for the ridges of thickened wall material projecting into the lumina of ray tracheides in the wood of pines.

dentelle (*Bind.*). A style of decoration of a toothlike or lacelike character; used in covers. (*Textiles*) A lace edging in the form of a series of small teeth.

denticles (*Zool.*). Any small toothlike structures: the placoid scales of *Selachii.*

denticulate (*Bot.*). Said of a margin bearing small teeth.

denticulated (*Build.*). A term applied to mouldings decorated with dentils.

dentigerous cyst (*Vet.*). A cyst containing teeth; usually a teratomatous cyst on the malar bone of a horse.

dentil (*Build.*). A projecting rectangular block forming one of a row of such blocks under the corona of a cornice.

dentine (*Zool.*). A hard calcareous substance, allied to bone, of which teeth and placoid scales are mainly composed. *adj.* **dentinal.**

dentirostral (*Zool.*). Having a toothed or notched beak.

dentistry or dental surgery. The treatment of diseases and irregularities of the teeth (and often of associated tissues), including conservation, extraction (*exodontia*), and artificial replacement (*dental prosthesis*) of teeth, and rectification of abnormalities in the dentition (*orthodontia*).

dentition (*Zool.*). The kind, arrangement, and number of the teeth; the formation and growth of the teeth; a set of teeth, as the milk dentition.

denudation (*Geol.*). The laying bare (L. *nudus*, naked) of the rocks by chemical and mechanical disintegration and the transportation of the resulting rock debris by wind or running water. Ultimately denudation results in the degradation of the hills to the existing base-level. The process is complementary to sedimentation, the amount of which in any given period is a measure of the denudation. See also **marine denudation.**

denuded quadrat (*Bot.*). A square piece of ground, marked out permanently and cleared of all its vegetation, so that a study may be made of the manner in which the area is reoccupied by plants.

denumerable set (*Maths.*). One that can be put into a one-one correspondence with the positive integers. Also **countable set, enumerable set, numerable set.**

deobstruent (*Med.*). Removing obstruction by opening natural passages of the body; medicine which removes obstruction in this way.

deodar (*For.*). Tree of the genus *Cedrus*, one of the most important Indian softwoods, probably second to teak in importance as native timber. It is not resistant to termite attack although it is resistant to powder-post beetle infestation. It has a distinctive unpleasant smell. In addition to normal joinery purposes it also makes good structural timber.

deodorizing (*Chem. Eng.*). A process extensively used in edible oil and fat refining, in which the oil or fat is held for several hours at high temperatures (varying with product, but 200°C is not uncommon), and low pressure ($\sim 10^4$ N/m^2), during which time steam is blown through to remove the traces of odour-creating substances (usually free fatty acids of high rel. mol. mass).

deoperculate (*Bot.*). (1) Lacking an operculum. (2) Having an operculum which does not come away spontaneously.

deoxidation (*Met.*). The process of reduction or elimination of oxygen from molten metal before casting by adding elements with a high oxygen affinity, which form oxides that tend to rise to the surface.

deoxidized copper (*Met.*). Copper from which the oxygen remaining after poling has been removed by the addition of a deoxidizer, which, remaining in solid solution, lowers the conductivity below that of tough-pitch copper, but the product is more suitable for working operations.

deoxidizer (*Met.*). A substance which will eliminate or modify the effect of the presence of oxygen, particularly in metals. Also **deoxidant.**

deoxyribonucleic acid (*Chem.*). DNA. See **nucleic acids.**

departure time (*Aero.*). The exact time at which an aircraft becomes airborne is an important factor in air traffic control; estimated time of departure (abbrev. ETD).

dependent functions (*Maths.*). A set of functions such that one may be expressed in terms of the others.

dependent variable (*Maths.*). A variable whose values are determined by one or more other (independent) variables.

depersonalization (*Psychiat.*). A condition in which an individual experiences a wide range of feelings of unreality in relation to the self, to the body, or to other people, even extending to the feeling of being dead.

depeter (*Build.*). Plasterwork finished in imitation of tooled stone, small stones being pressed in with a board before the plaster sets. Also called **depreter.**

dephlogisticated air (*Chem.*). The name given by Priestley to oxygen. The term is of historic interest only.

dephosphorization (*Met.*). Elimination, partial or complete, of phosphorus from steel, in basic steel-making processes. Accomplished by forming a slag rich in lime. See **acid process, basic process, Bessemer process, open-hearth process.**

depickling (*Leather*). A process, prior to tanning, for removing from sheepskins the acid and salt-pickle used to preserve them during transport.

depilate (*Med.*). To remove the hair from.

depilation (*Bot.*). The natural loss of a hairy covering from the parts of plants as they mature.

depilatories (*Chem.*). Compounds for removing or destroying hair; usually sulphide preparations.

deplanate (*Bot.*). Flattened, or expanded in a flat surface.

deplasmolysis (*Biol.*). The process of recovery of a cell from a plasmolysed condition by re-acquisition of water.

depleted uranium (*Chem.*). Sample of uranium having less than its content of ^{235}U in nature.

depletion (*Nuc.*). Reduction in the proportion of a specific isotope in a given mixture.

depletion layer (*Electronics*). In semiconductor materials, location where mobile electrons do not neutralize the charge of the donors and acceptors taken together. Has been applied for use as an ultrasonic transducer.

depletion-layer transistor (*Electronics*). One depending on the movement of carriers through a depletion layer.

Deplistor (*Electronics*). Three-terminal component, which can exhibit a negative differential resistance under certain circuit conditions.

depluming itch (*Vet.*). A skin irritation of fowls

due to infestation of the skin around the feather shafts by the mite *Cnemidocoptes gallinae*.

depolarization (*Elec. Eng.*). Reduction of polarization, usually in electrolytes, but sometimes in dielectrics. In the former it may refer to removal of gas collected at plates of cell during charge or discharge.

depolymerization (*Chem.*). The change of a large molecule into simpler ones having the same empirical formula.

deposit (*Elec. Eng.*). (1) The coating of metal deposited electrolytically upon any material. (2) The sediment which is sometimes found at the bottom of a secondary cell owing to gradual disintegration of the electrode material. (*Geol.*) See under **deposition.**

deposit feeders (*Ecol.*). In benthic communities, animals which consume sedimented materials, usually predominating on muddy bottoms. Cf. *filter feeders*.

deposition (*Geol.*). The laying down or placing into position of sheets of sediment (often referred to as *deposits*) or of mineral veins and lodes. Synonymous with *sedimentation* in the former sense.

depreciation factor (*Elec. Eng.*). A term commonly used in the design of floodlighting and similar installations to denote the ratio of the light output when the lighting equipment is clean to that when it is dirty (i.e., after having been in service for some time).

depressant (*Med.*). Lowering functional activity; a medicine which lowers functional activity of the body.

depressed (*Bot.*). (1) Flattened; said especially of the apex of a solid plant member. (2) Somewhat sunken in a concave form.

depressed conductor-rail (*Rail.*). A section of conductor-rail depressed below normal level where contact with the shoes is not desired.

depressing agent (*Min. Proc.*). Wetting agent; one used in froth flotation to render selected fraction of pulp less likely to respond to aerating treatment.

depression (*Meteor.*). A cyclone. That distribution of atmospheric pressure in which the pressure decreases to a minimum at the centre. In the northern hemisphere, the winds circulate in a counter-clockwise direction in such a system; in the southern hemisphere, in a clockwise direction. A depression usually brings stormy unsettled weather. (*Psychol.*) A state of dejection, often combined with feelings of sadness, irritability, or anxiety; usually accompanied by a lowering of psychophysical activity. It can be (1) a normal depression, as a reaction to an unfavourable external event, or (2) pathological, which may be (*a*) endogenous, e.g., melancholia, due to still unknown internal physical changes, or (*b*) reactive, either in relation to external situations or events, or in relation to an internal phantasy-situation. The pathological reaction, in every case, is exaggerated in duration and intensity. (*Zool.*) An unhealthy condition of *Protozoa* which have been prevented from conjugating for many generations; characterized by degeneration of the cell-organs and retardation of division.

depression of freezing point (*Heat*). A solution freezes at a lower temperature than the pure solvent, the amount of the depression of the freezing point being proportional to the concentration of the solution, provided this is not too great. The depression produced by a 1% solution is called the *specific depression*, and is inversely proportional to the molecular weight of the solute. Hence the depression is proportional to the number of moles dissolved in unit weight of the solvent and is independent of the particular solute used.

depression of land (*Geol.*). Depression relative to sea-level may be caused in many ways, including sedimentary consolidation, the superposition of large masses of ice, the migration of magma, or changes of chemical phase at depth. It is generally recognized by the marine transgression produced but is often difficult to distinguish from eustatic changes in sea-level. See **drowned valleys.**

depressomotor (*Zool.*). A depressor nerve.

depressor (*Zool.*). A muscle which by its action lowers a part or organ; a motor nerve which when stimulated checks the activity of the part to which it leads; a reagent which, when introduced into a metabolic system, slows down the rate of metabolism.

depreter (*Build.*). See **depeter.**

depside (*Bot., Chem.*). A product formed from hydroxy-aromatic acids by the condensation of the carboxyl group of one molecule with the phenol group of a second molecule. Depsides are probably concerned with the oxidation of fats and proteins inside plant cells.

depth (*Horol.*). The amount by which the teeth of a wheel intersect the teeth of the mating wheel or pinion. (*Ships*) The depth measured from the top of the keel to the top of some specified deck.

depth dose (*Radiol.*). Ratio or percentage of the surface dose which is received within the body. See **Chambers-Imrie-Sharpe curve.**

depth finder (*Radar*). Radar or ultrasonic depth sounding equipment.

depth gauge (*Tools*). Device clamped to a drill or bit to regulate the depth of the hole bored.

depth localization (*Acous.*). The same as *auditory perspective*. See **stereophony.**

depth of field (*Photog.*). The distance between the nearer and farther planes, in the area photographed, over which the image is in reasonably sharp focus, depending on the type of lens and the stop.

depth of focus (*Photog.*). Distance on the image side of the lens corresponding to the *depth of field* (q.v.).

depth of modulation (*Radio*). A factor indicating extent of modulation of a wave. It is the ratio *difference/sum* of 'peak' and 'trough' values of an amplitude-modulated wave, and of the extreme deviation of carrier-frequency in a frequency-modulated wave. Called **percentage modulation** when expressed as a percentage.

depth of penetration (*Elec. Eng.*). Within a plane conductor the magnitude of an electromagnetic field and the associated current falls off exponentially from the surface. The depth of penetration or *skin depth* (δ metres) is normally considered as the depth for which the field magnitude is $1/e$ of its value at the surface. $\delta = (\pi f \mu \sigma)^{-\frac{1}{2}}$ where f is the frequency, μ is the permeability and σ the conductivity.

depth psychology. The psychology of the unconscious.

depthing tool (*Horol.*). An instrument by means of which two wheels, or a wheel and pinion, can be mounted and their depth adjusted until it is correct, after which the distance apart of their centres can be transferred to the plates for the drilling of the pivot holes.

deputy (*Mining*). (1) The local representative of the owner. (2) In Northumberland and Durham, a man who timbers or props the working places in a coal-mine. (3) An official who holds a certificate under the Coal Mines Act, and is responsible for the working of a district. See **fireman.**

dérailleur (*Eng.*). Variable transmission gear mechanism whereby the driving chain may be ' derailed ' from one sprocket wheel to another of different size, thus changing the driving ratio. Ten or more ratios are possible. Much used in racing bicycles. Cf. *epicyclic gear*.

de-rating (*Elec. Eng.*). Reduction of the maximum performance ratings of equipment when used under unusual conditions, in order to maintain an adequate margin of safety.

derby float (*Tools*). A large trowel consisting of a flat board with two handles on the back.

Derby red (*Paint.*). See **Chinese red.**

Derbyshire neck (*Med.*). Chronic enlargement of the thyroid gland, without signs of overactivity of the gland, due to deprivation of iodine during periods of growth.

Derbyshire spar (*Min.*). A popular name for the mineral *fluorite* or *fluorspar*.

Dercum's disease (*Med.*). See **adiposis dolorosa.**

Deri motor (*Elec. Eng.*). A modification of the ordinary repulsion motor, in which speed control is effected by moving two sets of brushes in opposite directions round the commutator. The angular brush movement to produce a given speed change is twice as great as for the ordinary motor, so that finer control is possible.

derivative equalizer (*TV*). One designed to improve the definition (resolution) in a reproduced picture by adjusting both response and phase relations.

derivative feedback (*Telecomm.*). Feedback signal in control system proportional to time derivative error.

derivative hybrid (*Gen.*). A hybrid obtained by crossing two hybrids, or by crossing a hybrid with one of its parents.

derived fossils (*Geol.*). The remains of organisms entombed in a stratum younger than the fossils themselves. The bed lying above a break in the stratigraphical succession (such as an unconformity) frequently contains fragments of durable rocks (and fossils derived from them) which were laid under contribution during its formation. The Cambridge Greensand is a stratum rich in derived fossils.

derived function (*Maths.*). The derived function f' of the function f is defined by $f'(x)$ = the limit as h tends to zero of

$$\frac{f(x+h)-f(x)}{h}.$$

If $y=f(x)$, other notations for $f'(x)$ are $\frac{dy}{dx}$, y', $D_x y$ and $\frac{d}{dx}f(x)$. The gradient to the curve $y=f(x)$ at the point (a, b) is $f'(a)$. Also called **differential coefficient.** Cf. *partial differential coefficient*.

derived units (*Phys.*). Units derived from the three fundamental units of a system by consideration of the dimensions of the quantity to be measured. See **dimensions.**

Deri winding (*Elec. Eng.*). A form of compensating winding used to neutralize armature reaction; sometimes used on d.c. turbo-generators.

derm (*Zool.*). See **dermis.**

dermal (*Bot.*). Appertaining to the epidermis or other superficial layer of a plant member. (*Zool.*) Pertaining to the skin; more strictly, pertaining to the dermis. Also **dermic.**

dermal appendage (*Bot.*). Any outgrowth from the epidermis.

dermal branchiae (*Zool.*). See **papulae.**

dermarticular(e) (*Zool.*). See **antarticular(e).**

dermatitis (*Med.*). Inflammation of the surface of the skin or epidermis. See also **Dühring's disease, Ritter's disease.**

dermatoblast (*Zool.*). In Insects, one of the cells of the outer layer formed by the neural ridges; it develops into the ventral body wall.

dermatocyst (*Bot.*). A cystidium developing on the cuticle or pellicle.

dermatogen (*Bot.*). A sheet of meristematic cells, one layer thick, covering an apical growing point and giving rise to the epidermis.

dermatographia, dermographia (*Med.*). A sensitive condition of the skin in which pressure with, for example, a pencil point will produce a reddish weal.

dermatology (*Med.*). That branch of medical science which deals with the skin and its diseases.

dermatomyositis (*Med.*). A disease, progressive and usually fatal, which is characterized by acute or subacute inflammation of muscles, dermatitis, oedema over the affected muscles, sweating, and enlargement of the spleen.

dermatophyte (*Bot.*). A parasitic fungus which causes a skin disease in animals.

dermatopsy (*Zool.*). The condition of having a light-sensitive integument.

dermatosclerosis (*Med.*). See **sclerodermia.**

dermatosis (*Med.*). An affection or inflammation of the deeper layers of the skin.

dermatosome (*Bot.*). One of the minute portions into which a cell wall can be dissolved by prolonged treatment with dilute hydrochloric acid, followed by heating for some time at 50°–60°C.

dermentoglossum (*Zool.*). In some Fish, a skeletal plate of the tongue covering the entoglossum and arising from the fusion of the dentinal bases.

dermethmoid (*Zool.*). See **suprethmoid.**

dermis or **derm** (*Zool.*). The inner layers of the integument, lying below the epidermis and consisting of mesodermal connective tissue. *adjs.* **dermal, dermic.**

dermoblast (*Cyt.*). That layer of the mesoblast which develops into the skin.

dermocalyptrogen (*Bot.*). A meristematic layer present in the apex of the root of many *Dicotyledons*; it gives rise to the root cap and to the dermatogen.

dermoccipital (*Zool.*). In lower Vertebrates and embryonic stages of higher Vertebrates, a pair of membrane bones occupying the place of the interparietal.

dermoid (*Med.*). A cyst of congenital origin containing such structures as hair, skin, and teeth; occurs usually in the ovary.

dermomuscular layer (*Zool.*). A sheet of muscular tissue underlying the skin in lower *Metazoa*: it consists of longitudinal and, usually, circular layers.

dermopharyngeal (*Zool.*). A plate of membrane bone supporting the upper or lower pharyngeal teeth in some Fish.

dermoskeleton (*Zool.*). See **exoskeleton.**

dermotrichia (*Zool.*). The horny rays supporting the unpaired fins of Fish.

derotrematous (*Zool.*). Losing the gills but retaining the gill-clefts, as in some *Urodela*.

derrick (*Build., Civ. Eng.*). An arrangement for hoisting materials, distinguished by having a boom stayed from a central post, which in turn is usually stayed in position by guys.

derrick barrel (*Eng.*). The winding drum on which the rope used in derricking or luffing operations in a jib crane is wound or paid out.

derrick crane (*Eng.*). See **derricking jib crane.**

derricking jib crane (*Eng.*). A jib crane in which the inclination of the jib, and hence the radius of action, can be varied by shortening or lengthening the tie-ropes between post and jib.

derris (*Chem.*). An extract of the root of the *Derris* tree, of which rotenone is the chief toxic constituent; effective contact insecticide.

dertrotheca (*Zool.*). In Birds, the horny covering of the maxilla. Also **dertrum.**

derv (*Fuels*). See **diesel oil.**

2,4-DES (*Chem.*). 2-(2,4-Dichlorophenoxy) ethyl hydrogen sulphate, used as a weedkiller. Also **disul, SES, sesin, sesone.**

desalination (*Chem.*). The production of fresh from sea water by one of several processes, including *distillation, electrodialysis*, and reverse *osmosis*.

desaminases (*Chem.*). Enzymes which induce the splitting off of the amino group (NH_2) from amino acids.

desampling (*Automation*). Process of extrapolation in reconstructing a continuous analogue function from samples of a variable.

desaturation (*Optics, Photog.*). The amount of grey in a colour. Cf. **saturation.**

De Sauty's method (*Elec. Eng.*). A Wheatstone bridge method of comparing capacities; the two capacities are placed in two arms of a Wheatstone bridge and resistances are placed in the other two arms, the latter being adjusted until no kick on the galvanometer can be observed when the battery key is depressed.

de-scaling (*Eng.*). The process of (1) removing scale or metallic oxide from metallic surfaces by *pickling* (q.v.); (2) removing scale from the inner surfaces of boiler plates and water tubes.

Descartes' rule of signs (*Maths.*). No algebraic equation $f(x)=0$ can have more positive or negative roots respectively than there are changes of sign from + to − and from − to + in the polynomial $f(x)$ or $f(-x)$.

Descemet's membrane (*Zool.*). In Vertebrates, a homogeneous elastic layer covering the back of the cornea.

descending (*Bot.*). Growing or hanging downwards in a gradual curve. (*Zool.*) Running from the anterior part of the body to the posterior part, or from the cephalic to the caudal region.

descending aestivation (*Bot.*). Aestivation in which each segment overlaps the anterior one.

descending letters (*Typog.*). Letters the lower part of which is below the base line; e.g., *p, q, y*.

descloizite (*Min.*). A hydrous vanadium-lead-zinc ore with the general formula $PbZn(VO_4)OH$. Hardness 3½ (Mohs); sp. gr. 6·1. A variety is $(Zn,Cu)Pb(VO_4)OH$. Important source of vanadium. Occurs in the oxidation zone of lead-zinc deposits.

describer (*Rail.*). The apparatus, either in signal cabins or for public use, which indicates movements, destinations, etc., of trains.

de-seaming (*Met.*). Process of removing surface blemishes, or superficial slag inclusions from ingots or blooms.

desensitization (*Med.*). The method of abolishing the sensitivity of a person to a protein by injecting small amounts of the same protein.

desensitizer (*Photog.*). Substance (e.g., dyes such as phenosafranine or azocyanine) which destroys the sensitivity of an emulsion without affecting the latent image.

desensitizing (or **muting**) (*Electronics*). Suppression of an output of electronic equipment unless there is adequate signal/noise ratio.

desert (*Geog., etc.*). A barren and uninhabited tract of large extent. Deserts are either cold (e.g., the Arctic and Antarctic wastes) or hot (e.g., the Sahara, Kalahari, and Nubian deserts in Africa, those of Chinese Turkestan and Gobi in Asia, and the Great Basin in N. America). Cold deserts are due entirely to the low temperature appropriate to the latitudes in which they occur; hot deserts are due to an excess of evaporation over precipitation, resulting from the physical configuration of the region; in many cases ranges of mountains cut off the moisture-bearing winds.

desert climate (*Geog.*). This type is characterized by aridity and lack of regular precipitation, much radiation and diurnal changes in temperature, with minimum vegetation.

desert rose (*Min.*). A cluster of platy crystals, often including sand grains, formed in arid climates by evaporation. Typically barytes or gypsum. Also called **rock rose.**

desiccants (*Chem.*). Substances of a hygroscopic nature, capable of absorbing moisture and therefore used as drying agents; e.g., anhydrous sodium sulphate, anhydrous calcium chloride, etc.

desiccation (*For.*). The process of seasoning timber by exposing it in an oven to a current of hot air.

desiccator (*Chem.*). Laboratory apparatus for drying substances; it consists of a glass bowl with ground-in lid, containing a drying agent, e.g., concentrated sulphuric acid or anhydrous calcium chloride; a tray for keeping glassware, etc., in position is also provided, and if desired the desiccator can be evacuated.

design (*Textiles*). (1) A point-paper diagram in which the blank and filled-in squares show the interlacing of the threads in a fabric. (2) The pattern woven in or printed on a fabric.

designation marks (*Typog.*). An indication of the title of a book, printed in small letters in the same line with the *signature* on the first page of each section.

desilverization (*Met.*). The process of removing silver (and gold) from lead after softening. See **Parke's process, Pattinson's process.**

desired value (*Automation*). See **index value.**

desk switchboard (*Elec. Eng.*). A form of switchboard panel in which the operating switches, pilot lamps, etc., are mounted on panels inclined to the horizontal like the surface of a desk.

Deslandres equation (*Light*). An empirical expression for the positions of the origins or heads in a band spectrum.

$$\nu=a+bn+cn^2,$$

ν being the wave number of the head, a, b, and c constants and n taking successive integral values.

de-sliming (*Min. Proc.*). Removal of very fine particles from an ore pulp, or classification of it into relatively coarse and fine fractions.

Desmarestiales (*Bot.*). An order of the *Phaeophyta*. The thallus is made up of a filament at each growing apex, having pseudoparenchymatous cortication posteriorly, giving a thallus of a definite form. The gametophyte is microscopic, and oögamous.

desmergate (*Zool.*). A worker ant intermediate in characters between an ordinary worker and a soldier.

Desmids (*Bot.*). Includes two families of the *Conjugales*. The saccoderm Desmids have uninucleate cells, which may be solitary or simple filaments. There are no pores in the cell wall, and regeneration is not by the formation of a half cell. Conjugation is by a conjugation tube in the majority of cases. Except for two genera, the plaeoderm Desmid cell is constricted medianly; half-cells are formed, and the protoplasts escape before conjugation.

desmine (*Min.*). See **stilbite.**

desmochondria (*Zool.*). Cytoplasmic granular projections on the surface of epithelial cells.

Desmodur (*Plastics*). TN for isocyanates used to produce polyurethanes in conjunction with polyester resins.

desmognathous (*Zool.*). In Birds, said of a type of palate in which the vomers are small or wanting and the maxillopalatines meet in the middle line; the palatines and pterygoids articulate with the basisphenoid rostrum.

Desmotiontae (*Bot.*). A class of the ***Pyrrophyta***, distinguished by the division of the cell wall into two valves that are not subdivided. Motile cells have a pair of flattened apical flagella, which move differently from each other.

desmotropism (*Chem.*). A special case of *tautomerism* (q.v.) which consists in the change of position of a double bond, and in which both series of compounds can exist independently; e.g., keto and enol form of acetoacetic ester, malonic ester, phenyl-nitromethane, etc.

desorption (*Chem.*). Reverse process to *adsorption* (q.v.). See **outgassing.**

desoxycorticosterone (*Biochem.*). An adrenocortical hormone, $C_{21}H_{30}O_3$.

desquamation (*Med.*). The shedding of the surface layer of the skin.

destarched (*Bot.*). Said of a plant which has been placed in the dark and which no longer contains starch in its leaves; the treatment stops photosynthesis, and allows translocation to remove the starch from the leaves.

Destriau effect (*Phys.*). A form of electroluminescence arising from localized regions of very intense electric field associated with impurity centres in the phosphor.

destructive distillation (*Chem.*). The distillation of solid substances accompanied by their decomposition. The destructive distillation of coal results in the production of coke, tar products, ammonia, gas, etc.

destructive read-out (*Comp.*). Clearing of a storage location simultaneously with reading its contents.

destructor station (*Elec. Eng.*). An electric generating station in which the fuel used consists chiefly of town or other refuse.

desulphurizing (*Chem. Eng.*). The process of removing sulphur from hydrocarbons by chemical reactions. Many are known by proprietary names listed separately, e.g., *Appleby Frodingham, Claus, Houdry*. See also **autofining, copper-chloride process.**

desuperheater (*Eng.*). A vessel in which superheated steam is brought into contact with a water spray in order to make saturated or less highly superheated steam. See **superheated steam.**

desynapsis (*Cyt.*). Abnormally early breaking up of synapsis in meiosis.

detachable key switch (*Elec. Eng.*). A switch which can be operated only by a special key which is kept under supervisory control.

detached escapement (*Horol.*). An escapement in which there is a minimum of interference with the free vibration of the pendulum or balance.

detaching hook (*Mining*). Safety hook attaching mine hoisting-rope in such a way that the cage becomes detached at the top of the shaft if the safety limit is passed when winding.

detail drawing (*Build.*). A large-scale working drawing (usually of a part only) giving information which does not appear on small-scale drawings of the whole construction.

detailer (*Civ. Eng.*). A draughtsman who draws the details involved in any construction.

detail file (*Comp.*). Data-storage file of which the contents are relatively transient.

detail paper (*Paper*). A translucent tracing paper, usually unoiled.

detector (*Radio*). Circuit component concerned with *demodulation*, i.e., extracting signal which has *modulated* a wave. It is *linear* if there is exact proportionality between output voltage or current and depth of modulation, and *square-law* if output voltage or current is proportional to square of depth of modulation. It is substantially linear for small depths of modulation. See also **demodulation, leakage indicator.** (*Nuc. Eng.*) Device in which presence of radiation induces physical change which is observable.

detector finger (*Print.*). See **web break detector.**

detent (*Eng., Horol.*). A catch which, in being removed, initiates the motion of a machine. In the chronometer escapement of a clock, the detent carries a stone or jewel for locking the escape wheel.

detent escapement (*Horol.*). See **chronometer escapement.**

detergents (*Chem.*). Cleansing agents (solvents, or mixtures thereof, sulphonated oils, abrasives, etc.) for removing dirt, paint, etc. Commonly refers to soapless detergents, containing surfactants which do not precipitate in hard water; sometimes also a protease enzyme to achieve 'biological' cleaning and a whitening agent (see **fluorescent whitening agents**). See also **cationic detergents, nonionic detergents.**

determinant (*Maths.*). A square array of numbers representing the algebraic sum of the products of the numbers, one from each row and column, the sign of each product being determined by the number of interchanges required to restore the row suffices to their proper order, e.g.,

$$\begin{vmatrix} a_1 & b_1 & c_1 \\ a_2 & b_2 & c_2 \\ a_3 & b_3 & c_3 \end{vmatrix} = \begin{matrix} a_1b_2c_3 - a_1b_3c_2 \\ -a_2b_1c_3 + a_2b_3c_1 \\ +a_3b_1c_2 - a_3b_2c_1. \end{matrix}$$

determinate (*Bot.*). (1) With a well-marked edge. (2) Said of an inflorescence which ends in a flower. (*Eng.*) Said of a structure which is a *perfect frame* (q.v.). Cf. *indeterminate.*

determinate cleavage (*Zool.*). A type of cleavage in which the part of the body to be formed by each blastomere is fixed from the first.

de-tinning (*Met.*). Chlorine treatment to remove tin coating from metal scrap.

detonating fuse (*Mining*). Fuse with core of TNT which burns at some 6 km/s and is itself set off by detonator. Varieties include *Primacord, Cordtex*, and *Cordeau Detonnant.*

detonation (*Chem.*). Decomposition of an explosive in which the rate of heat release is great enough for the explosion to be propagated through the explosive as a steep shock front, at velocities above 1 km/s and pressures above 10^9 N/m^2. (*I.C. Engs.*) In a petrol engine, spontaneous combustion of part of the compressed charge after the passage of the spark; the accompanying knock. It is caused by the heating effect of the advancing flame front, which raises the gas remote from the plug to its spontaneous *ignition temperature* (q.v.).

detonation meter (*Eng.*). An instrument for measuring quantitatively the severity and frequency of detonation in a petrol-engine cylinder. See **bouncing-pin detonation meter.**

detorsion (*Zool.*). In *Gastropoda*, partial or complete reversal of torsion, manifested by the untwisting of the visceral nerve loop and the altered position of the ctenidium and anus.

detour behaviour (*An. Behav.*). The behaviour of animals which make a detour round an obstacle without preliminary random trial-and-error. Regarded as evidence of *insight learning* (q.v.).

detoxication (*Biochem.*). The removal of the toxic properties of substances in the body by its conversion into a nontoxic substance, which may be eliminated by excretion, or by the action of antibodies; more generally, the chemical changes which foreign organic compounds undergo in the body.

detrital minerals (*Geol.*). Although, literally, any mineral grains resulting from detrition are detrital, in sedimentary petrology the term is restricted to grains of heavy minerals found in sand and other sediments, and separated therefrom by passing through bromoform or other heavy liquid. See **heavy minerals.**

detrition (*Geol.*). The natural process of rubbing or wearing down strata by blown wind or running water. The product of detrition is *detritus.*

detritus (*Biol.*). Organic material formed from decomposing organisms. (*Geol.*) See **detrition.**

detritus chamber (or **pit**) (*San. Eng.*). A tank through which crude sewage is first passed in order to allow the largest and heaviest of suspended matters to fall to the bottom, from which they can be removed.

detrusor (*Anat., Physiol.*). The muscular coat of the urinary bladder (*detrusor urinae*); sometimes used for the outer of the three muscular coats of the bladder.

Dettol (*Chem.*). TN for nontoxic and nonirritant germicide of which the active principle is chloroxylenol dissolved in a saponified mixture of aromatic oils, e.g., terpineol.

detumescence (*Med.*). The reduction of a swelling.

detuner (*Acous.*). Large concrete structure for receiving the blast from a gas-turbine (jet) engine, so that by baffles and expansion the radiation of noise is diminished. (*Eng.*) See **dynamic damper.**

detuning (*Radio*). Adjustment of a resonant circuit so that its resonant frequency does not coincide with that of the applied e.m.f.

Deuce (*Comp.*). TN for first-generation digital computer. (*D*igital *e*lectronic *u*niversal *c*omputing *e*ngine.

deuteranopic (*Phys.*). Colour blind to green.

deuteration (*Chem.*). The addition of or replacement by deuterium atoms in molecules.

deuterium (*Chem.*). Isotope of element hydrogen having one neutron and one proton in nucleus. Symbol D, when required to be distinguished from natural hydrogen, which is both ^{1}H and ^{2}H. This heavy hydrogen is thus twice as heavy as ^{1}H, but similarly ionized in water. As a chemical, lithium deuteride is the essential element in H-bombs. Used in isotopic 'labelling' experiments (e.g., mechanism of esterification). See **deuteron.**

deuterium reactor (*Nuc. Eng.*). See **heavy-water reactor.**

deutero-. Prefix, in general denoting second in order, derived from. Particularly, (*Chem.*) containing heavy hydrogen (*deuterium*).

deuteroconidium (*Bot.*). A type of spore found in dermatophytes; formed by the division of a hemisphere.

deuterogamy (*Bot.*). Any process which replaces normal fertilization.

deuterognathous (*Zool.*). Having the jaws borne on the second somite of the head.

Deuteromycetes (*Bot.*). See **Fungi Imperfecti.**

deuteron (*Chem.*). Charged particle, D^+, the nucleus of *deuterium*, a stable but lightly-bound combination of one proton and one neutron. Obtained from *heavy water* by electrolysis, mainly used as a bombarding particle accelerated in cyclotrons.

deuteroplasm (*Biol.*). See **metaplasm.**

deuterostoma (*Zool.*). In development, a mouth which arises secondarily, as opposed to a mouth which arises by modification of the blastopore.

deuterotoky (*Zool.*). Parthenogenesis leading to the production of both males and females.

Deuterozoic (*Geol.*). A term, now disused, for the younger Palaeozoic Systems—the Devonian, Carboniferous, and Permian Systems. Cf. *Proterozoic.*

deuterozooid (*Zool.*). A secondary zooid, produced from a primary zooid by budding.

deuthyalosome (*Zool.*). The nucleus of the ovum after the first polar body has been formed.

deutobroque or **deutobroch** (*Cyt.*). In oögenesis, a stage preceding leptotene, during which the chromosomes radiate from the nucleolus, and wind about just under the nuclear membrane.

deutocerebron, deutocerebrum (*Zool.*). In higher *Arthropoda*, as Insects and *Crustacea*, the fused ganglia of the second somite of the head, forming part of the 'brain'.

deutomerite (*Zool.*). In some *Gregarinidea*, the part of the body containing the nucleus.

deutoplasm (*Biol.*). See **metaplasm.**

deutoplasmolysis (*Zool.*). The elimination of the deutoplasm of the Vertebrate ovum, which usually occurs between the formation of the second polar body and the first cleavage.

deutoscolex (*Zool.*). In *Cestoda*, a secondary scolex arising by budding in a bladderworm.

deutovum (*Zool.*). In *Acarina*, a stage in the development, which may come before or after the egg is laid.

Deval attrition test (*Civ. Eng.*). See **attrition test**

developable surface (*Maths.*). That which, without shrinking or stretching, may be rolled out on to a plane.

developed dyes (*Chem.*). Dyes which are developed on the fibre by the interaction of the constituents which produce them. Dyeing with aniline black provides an example of a *developed dye.*

developer (*Photog.*). An aqueous solution used in photography for reducing the exposed silver salts to metallic silver. It generally contains, in addition to the reducing agent, an alkaline *accelerator* (q.v.), a *restrainer* (q.v.), and a preservative.

developer streaks (*Photog.*). Image defect in the form of dark streaks, caused by the flow of fresh developer.

development (*Mining*). Opening of ore body by access shafts, drives, crosscuts, raises and winzes for purpose of proving mineral value (ore reserve) and exploiting it. (*Photog.*) The reduction of exposed silver halides to silver by chemical action, in a sensitive emulsion. Also, the production of a relief by water, after exposing a bichromated gelatine.

Devereux agitator (*Met.*). See **agitator.**

deviascope (*Ships, etc.*). An apparatus consisting of a compass mounted on a turntable. Magnets and pieces of iron are placed on the table to simulate the magnetic character of a steel ship. Used for instruction and examination in compass adjusting.

deviation (*Elec. Eng.*). The difference between the instantaneous quantity and the *index* (or *desired*) *value* (q.v.), e.g., difference between the magnetic north and the setting of a compass needle due to a local magnetic disturbance, or between the actual value of the controlled variable and that to which the controlling mechanism is set in an automatic control system. Also called **error.** (*Radio*) In frequency-modulated transmission, the amount the carrier increases or decreases when modulated.

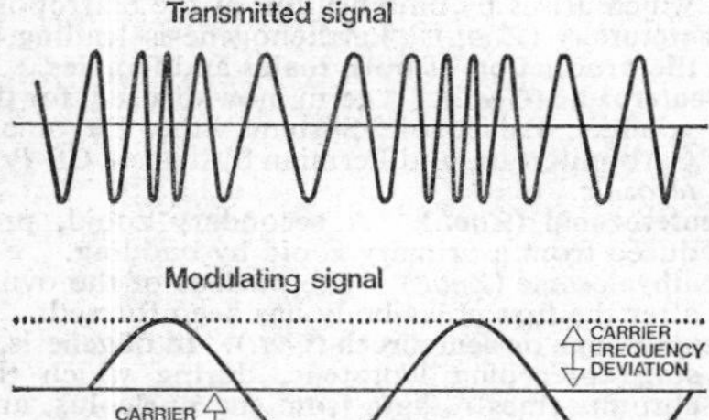

(*Ships, etc.*) The angle between the *magnetic meridian* (q.v.) and the *compass meridian* (q.v.). A *deviation table* is normally prepared showing the deviation for various headings. (*Stats.*) In a set of values, the deviation of x from the average (or mean) $\bar{x}$ is $x-\bar{x}$. The absolute mean deviation is mean of the moduli of the deviations. See **standard deviation, variance.**

deviation absorption (*Radio*). The absorption of radio waves at frequencies near the critical frequency of the ionosphere.

deviation distortion (*Radio*). Consequence of any restriction of bandwidth or linearity of discrimination in the transmission and reception of a frequency-modulated signal.

deviation ratio (*Radio*). Frequency departure of a frequency-modulated carrier from its mean value (that when there is no modulation) divided by this value.

deviation sensitivity (*Nav.*). Ratio of change in course indication to deviation from course line.

devil (*Civ. Eng.*). Travelling fire-box with a horizontal grate for heating or burning road surfaces, and for heating tools used in asphalting. (*Meteor.*) A small whirlwind due to strong convection, which, in the tropics, raises dust or sand in a column. (*Plumb.*) A portable furnace for heating solder and soldering irons.

devil float (*Build.*). A square float having four nails projecting from its working face; used to perform the scoring required in *devilling*.

devilling (*Build.*). The operation of scoring the surface of a plaster coat to provide a key for another coat.

devitrification (*Geol.*). Deferred crystallization which, in the case of glassy igneous rocks, converts obsidian and pitchstone into dull cryptocrystalline rocks (usually termed *felsites*) consisting of minute grains of quartz and feldspar. Such devitrified glasses give evidence of their originally vitreous nature by traces of perlitic and spherulitic textures. (*Glass*) A physical process which causes a change from the glassy state to a minutely crystalline state. This has to be avoided during manufacture. If the change is due to lapse of time, the glass becomes turbid and brittle.

Devonian System (*Geol.*). The rocks formed during the Devonian Period, between the Silurian and the Carboniferous Periods.

dew (*Meteor.*). The deposit of moisture on exposed surfaces which accumulates during calm, clear nights. The surfaces become cooled by radiation to a temperature below the dew-point, thus causing condensation from the moist air in contact with them.

Dewar flask (*Chem.*). A silvered glass flask with double walls, the space between them being evacuated; used for storing, e.g., liquid air.

de-watering (*Civ. Eng.*). The process of pumping water from the interior of a caisson, or of excluding water from an excavation, by the use of pipes sunk round the perimeter, with continuous pumping. (*Min. Proc.*) Partial or complete drainage by thickening, sedimentation, filtration or on screen as process aid, to facilitate shipment or drying of product.

dew-blown (*Vet.*). See **bloat.**

dew claw (*Zool.*). In Dogs, the useless claw on the inner side of the limb (especially the hind limb) which represents the rudimentary first digit.

deweylite (*Min.*). A variety of *antigorite* found in decomposed serpentine-bearing rocks.

dew line (*Radar*). Line of radar missile warning stations along 70th parallel of latitude. (*Distant early-warning line.*)

dew-point (*Meteor., Phys.*). The temperature at which a given sample of moist air will become saturated and deposit dew, if in contact with the ground. Above ground, condensation takes place into water droplets.

dew-point hygrometer (*Meteor.*). A type of hygrometer for determining the dew-point, i.e., the temperature of air when completely saturated. The relative humidity of the air can be ascertained by reference to vapour pressure tables.

dew pond. A shallow pond formed by excavating a suitable area on elevated pastures; the surface temperature of the pond is lowered by insulating it, and water collects at night through condensation.

dexiocardia (*Med.*). See **dextrocardia.**

dexiotropic (*Zool.*). Twisting in a spiral from left to right; spiral cleavage.

dextral. See **dextrorse.**

dextral fault (*Geol.*). A *tear fault* in which the rocks on one side of the fault appear to have moved to the right when viewed from across the fault. The opposite of a **sinistral fault.**

dextran (*Chem.*). A polyglucose formed by micro-organisms in which the units are joined mainly by $\alpha-1\rightarrow6$ links with variable amounts of cross-linking, via $\alpha-1\rightarrow4$, $\alpha-1\rightarrow3$, or $\alpha-1\rightarrow2$. Hydrolysed by dextranases.

dextrin (*Chem.*). Starch gum. A term for a group of intermediate products obtained in the transformation of starch into maltose and *d*-glucose. Dextrins are obtained by boiling starch alone or with dilute acids. They do not reduce *Fehling's solution* (q.v.). Crystalline dextrins have been obtained by the action of *Bacillus macerans*.

dextrinase (*Bot.*). A plant enzyme which hydrolyses dextrin.

dextrocardia, dexiocardia (*Med.*). A developmental anomaly in which the heart lies in the right side of the chest.

dextrorotatory (*Light*). Said of an optically active substance which rotates the plane of polarization in a clockwise direction when viewed in the direction of travel of the light.

dextrorse (*Bot., Zool.*). Turning or twisting in a right-handed direction, or in a spiral from left to right; e.g., gastropod shells. Cf. *sinistrorse*.

dextrose (*Chem.*). See **D-glucose.**

D.H. (*Build.*). Abbrev. for *double-hung*.

dhootie (*Textiles*). Plain weave, lightweight cotton fabric often with coloured borders; worn as loin-cloths by Indian peoples.

diabantite (*Min.*). A variety of chlorite relatively rich in iron and poor in aluminium.

diabase (*Geol.*). In America this term signifies a medium-grained basic igneous rock consisting mainly of plagioclase feldspar and pyroxene, equivalent to the term *dolerite* used in Britain. British petrologists have used the term diabase for dolerites in which the pyroxene has been altered to amphibole; this usage obsolete.

diabetes insipidus (*Med.*). A condition in which there is an abnormal increase in the amount of urine excreted, as a result of disease of, or injury to, the *pars nervosa* of the pituitary gland, reducing the secretion of antidiuretic hormone.

diabetes mellitus (*Med.*). A disorder of metabolism in which excess of sugar appears in the blood and in the urine, associated with loss of weight and excessive thirst.

diacetic acid (*Chem.*). Acetylacetic or acetoacetic acid, $CH_3CO{\cdot}CH_2{\cdot}COOH$.

diacetone alcohol (*Chem.*). $CH_3{\cdot}CO{\cdot}CH_2{\cdot}C(CH_3)_2{\cdot}OH$, 4-hydroxy-2-keto-4-methylpentane, a colourless or light yellow liquid, m.p. −54°C, b.p. 130°–175°C, rel. d. 0·915–0·943. Used as a lacquer solvent.

diacetyl (*Chem.*). Butane 2,3-dione, $CH_3{\cdot}CO{\cdot}CO{\cdot}CH_3$, a yellow-green liquid, b.p. 87°C. It is the simplest diketone, and is obtained by the action of nitrous acid on methyl ethyl ketone. It occurs naturally in butter and is one of the constituents giving butter its characteristic flavour.

diachronism (*Geol.*). The transgression across time planes by a geological formation. A bed of sand, when traced over a wide area, may be found to contain fossils of slightly different ages in different places, as, when deposited during a long-continued marine transgression, the bed becomes younger in the direction in which the sea was advancing. *adj.* **diachronous.**

diacoele (*Zool.*). In *Craniata*, the third ventricle of the brain.

diacranterian, diacranteric (*Zool.*). Having the teeth in a discontinuous row; as some Snakes, in which the anterior and posterior teeth are separated by a gap. Cf. *syncranterian*.

diacritical current (*Elec. Eng.*). The current in a coil to produce a flux equal to half that required for saturation.

diacritical point (*Elec. Eng.*). The point on the magnetizing curve of a sample of iron at which the intensity of magnetization has half its saturation value.

diad (*Chem.*). A divalent atom or radical.

diadelphous (*Bot.*). Said of stamens which are arranged in two bundles.

diadochy (*Geol., Min.*). The replacement of one element by another in a crystal structure.

diadromous (*Bot.*). Said of venation resembling the ribs of a fan.

diagenesis (*Geol.*). Those changes which take place in a sedimentary rock at low temperatures and pressures after its deposition. These include compaction, cementation, recrystallization, etc. Cf. *metamorphism*.

diageotropism (*Bot.*). A *geotropism* (q.v.) which is plagiotropic—i.e., shown by roots or shoots growing horizontally. *adj.* **diageotropic.**

diagnosis (*Bot.*). A formal description of a plant, having special reference to the characters which distinguish it from related species. (*Med.*) The determination of the nature of a disordered state of the body or of the mind; the identification of a diseased state.

diagnostic characters (*Zool.*). Characteristics by which one genus, species, family, or group can be differentiated from another.

diagonal (*Bot.*). Said of any member of a flower situated in a position otherwise than median or lateral. (*Eng.*) A tie or strut joining opposite corners of a rectangular panel in a framed structure. (*Maths.*) (1) Of a polygon: a straight line drawn between 2 nonadjacent vertices. (2) Of a polyhedron: a straight line drawn between 2 vertices which do not lie in the same face. (3) Of a square matrix or determinant matrix: the principal diagonal is that running from the top left-hand corner to the bottom right-hand corner. (*Textiles*) Term applied to heavy cotton, spun rayon or wool fabrics. These have a prominent twill line, usually at a low angle.

diagonal bond (*Build.*). See **raking bond.**

diagonal eyepiece (*Surv.*). One incorporating a right-angle prism for convenience in surveying steep lines with telescope or theodolite.

diagonalizing (*Radio*). Practice of radiating the same programme at different times and on different wavelengths.

diagonal matrix (*Maths.*). A square matrix in which all the elements except those in the leading diagonal are zero.

diagonal pitch (*Eng.*). In zig-zag riveting, the distance between the centres of adjacent rivets.

diagonal winch (*Eng.*). A steam-winch in which, to economize floor space, the engine cylinders are inclined instead of horizontal.

diagram (*Maths., etc.*). (1) An outline figure or scheme of lines, points, and spaces, designed to: (*a*) represent an object or area; (*b*) indicate the relation between parts; (*c*) show the value of quantities or forces. (2) A curve which indicates the sequence of operations in a machine.

diagram factor (*Eng.*). The ratio between the actual mean effective pressure developed in a steam-engine cylinder and the ideal pressure deduced from the hypothetical indicator diagram.

diaheliotropic (*Bot.*). See **diaphototropic,** which is a better term.

diakinesis (*Cyt.*). The last stage of meiotic prophase, in which the nuclear membrane breaks down and the tetrads pass on to the spindle.

dial (*Eng., etc.*). See **counter.** (*Horol.*) Graduated plate immediately behind the hands of a clock or other timekeeper, from which the time is read. (*Mining, Surv.*) A large compass mounted on a tripod, used for surveying or mapping workings in coal-mines. (*Teleph.*) A calling device operated by the rotation of a disk which, on its release, produces the pulses necessary to establish a connexion in an automatic system.

dialdehydes (*Chem.*). Compounds containing two aldehyde groups. The most important one is *glyoxal* (ethan 1,2-dial) (q.v.).

dial feed (*Eng.*). A rotary indexing table used to feed a press remotely with components.

dial foot (*Horol.*). (1) Circular pins, attached to the back of a dial, which enter corresponding holes in the pillar plate to ensure its correct location. (2) In some clocks, the dial feet are pinned to the pillar plate to hold the dial firm.

dial gauge (*Eng.*). A sensitive measuring instrument in which small displacements of a plunger are indicated in 1/100 mm, or similar length units, by a pointer moving over a circular scale.

dialkanones (*Chem.*). See **diketones.**

diallage (*Min.*). An ill-defined, altered monoclinic pyroxene, in composition comparable with augite; occurs typically in basic igneous rocks such as gabbro.

dialling (*Mining, Surv.*). The process of running an underground traverse with a mining *dial* (q.v.).

dialogite (*Min.*). See **rhodochrosite.**

dial plate (*Horol.*). **See bottom plate.**

dial switch (*Elec. Eng.*). A multicontact switch in which the contacts are arranged in the arc of a circle, so that contact can be made by a radial moving arm.
dialycarpic (*Bot.*). See **apocarpous.**
dialyneury (*Zool.*). In some *Gastropoda*, the condition of having the pallial nerves from the pleural ganglion anastomosing with the pallial nerves from the supra-intestinal ganglion or the subintestinal ganglion, or, if they are absent, from the corresponding parts of the visceral nerve. Cf. *zygoneury*.
dialypetalous (*Bot.*). See **polypetalous.**
dialysis (*Chem.*). The separation of a colloid from a substance in true solution by allowing the latter to diffuse through a semipermeable membrane. The process is used in the *artificial kidney*.
dialystelic (*Bot.*). Having several separate steles.
dialyzer (*Chem.*). Dialysis apparatus.
diamagnetism (*Mag.*). Phenomenon in some materials, in which the susceptibility is negative, i.e., the magnetization opposes the magnetizing force, the permeability being less than unity. It arises from precession of spinning charges in a magnetic field.
diameter (*Maths.*). (1) Intercept made by the circumference on a straight line through the centre of a circle. (2) Straight line bisecting a family of parallel chords of a conic. Cf. *conjugate diameters*. (3) Straight line which passes through the centre of parallel sections of a central quadric surface.
diameter of commutation (*Elec. Eng.*). The diametral plane in which the coils of an armature winding that are undergoing commutation should be situated for perfect commutation.
diametral winding (*Elec. Eng.*). An armature winding in which the number of slots is a multiple of the number of poles.
diametrical pitch (*Eng.*). Of a gear-wheel, the number of teeth divided by the pitch diameter.
diametrical tappings (*Elec. Eng.*). Tappings taken on a closed armature, which are diametrically opposite to each other, i.e., displaced from each other by 180 electrical degrees.
diametrical voltage (*Elec. Eng.*). The voltage between opposite lines of a symmetrical 6-phase system, or the voltage between tappings on an armature winding which are diametrically opposite to each other, i.e., displaced from each other by 180 electrical degrees.
diamide (*Chem.*). See **hydrazine hydrate.**
diamine oxidase (*Chem.*). Flavoproteins which catalyse the aerobic oxidation of amines to the corresponding aldehyde and ammonia.
diamines (*Chem.*). Compounds containing two amino groups.
diaminophenol (*Photog.*). See **amidol.**
diamino-pimelic acid (*Bacteriol.*). A cell wall constituent of some bacteria and blue-green algae, not known to occur in any other group.
diamond (*Min.*). One of the crystalline forms of carbon; it crystallizes in the cubic system, rarely in cubes, commonly in forms resembling an octahedron, and less commonly in the tetrahedron. Curved faces are characteristic. It is the hardest mineral (10 in Mohs' scale); hence valuable as an abrasive, for arming rock-boring tools, etc.; its high dispersion makes it valuable as a gemstone. Occurs in the blue ground in the Kimberley District, in river gravels above the Vaal, in shore sands in S.W. Africa, also in Brazil, the Congo, Guyana, and elsewhere. See **black diamond, bort, industrial diamonds, kimberlite.** (*Textiles*) (1) Wool taken from the sides of a Down fleece. (2) A type of fancy worsted twist yarn. (3) Designs in diamond form woven in many cloths. (*Typog.*) An old type size, about 4½-point.
diamond antenna (*Radio*). Same as **rhombic antenna.**
diamond die (*Eng.*). A wire drawing die containing a diamond insert, for reduced wear.
diamond drill (*Mining*). Power-operated drill which drives diamond-crowned bit to required depth in rock at set angle, either recovering solid core sample or producing a sludge.
diamond dust (*Min. Proc.*). The hardest of the substances used for abrasive purposes; used also by lapidaries on rotating wheels.
diamond harrow (*Agric.*). A harrow with a diamond-shaped frame and teeth at the junctions of the frame members.
diamonding (*For.*). The changing of the cross-section of square-sawn timber from square- to diamond-shaped during drying.
diamond mesh (*Build., Civ. Eng.*). A form of *expanded metal* (q.v.) with a diamond-shaped network.
diamond saw (*Tools*). (1) Stone-cutting circular saw used with diamond dust for cutting rock sections. (2) A band or frame saw with diamond-bonded cutting elements, used for cutting stone, concrete (e.g., in ' bump-cutting ' in runways), plastics, etc.
diamond-skin disease (*Vet.*). See **swine erysipelas.**
diamond-type coil (*Elec. Eng.*). A former-wound armature coil in which the overhang is diamond-shaped, i.e., made up of two straight sides meeting at a point.
diamond-work (*Build.*). A wall constructed of lozenge-shaped stones laid in courses.
diandrous (*Bot.*). Having two antheridia or two stamens.
dianysl-trichloroethane (*Chem.*). See **methoxychlor.**
diapause (*Zool.*). In Insects, a state, which may arise at any stage of the life-cycle, in which development is suspended and cannot be resumed, even in the presence of apparently favourable conditions, unless the diapause is first 'broken' by an appropriate environmental change.
diapedesis (*Zool.*). In *Porifera*, the passage to the exterior of cells primarily occupying the interior of certain types of larva: in Vertebrates, the passage of blood leucocytes through the walls of blood vessels into the surrounding tissues.
diaper (*Textiles*). Linen or cotton fabric with a square or diamond pattern, uniform or varied in character, in 5-end sateen weave; used chiefly for table linen.
diaper-work (*Build.*). Paving constructed in a chequered pattern, composed of stones or tiles of different colours.
diaphone (*Acous.*). In an organ, a resonating Helmholtz pipe in which the frequency is determined by a pallet vibrating on a spring, as contrasted with metal-reed and flue pipes.
diaphorase (*Chem.*). A *flavoprotein* (q.v.) which acts as a hydrogen carrier and passes hydrogen to one of the *cytochromes* (q.v.).
diaphoresis (*Med.*). Perspiration.
diaphoretic (*Med.*). Producing perspiration: a medicine which does this.
diaphototropic (*Bot.*). Said of a plant member which grows and comes to a fixed position across the direction of incident light.
diaphragm (*Acous.*). A vibrating membrane as in a loudspeaker, telephone, and similar sound sources, also in receivers, e.g., the human ear-drum. Also **open-diaphragm loudspeaker, pleated-diaphragm loudspeaker.** (*Build.*) A web

across a hollow terracotta block, forming separate compartments. (*Bot.*) (1) A plate of cells with very small intercellular spaces between them, lying here and there across the large intercellular spaces in stems of some aquatic plants. (2) A transverse plate of cells across a stem, generally at a node. (*Elec. Eng.*) A sheet of perforated or porous material placed between the positive and the negative plates of an accumulator cell. (*Optics, Photog.*) Same as **stop** (1). (*Surv.*) A flanged brass ring which is held in place in a surveying telescope tube by means of four screws, and which receives the *reticule* (q.v.). (*Zool.*) Generally, a transverse partition subdividing a cavity. In Mammals, the transverse partition of muscle and connective tissue which separates the thoracic cavity from the abdominal cavity; in *Salientia*, a fan-shaped muscle passing from the ilia to the oesophagus and the base of the lungs; in some *Arachnida*, a transverse septum separating the cavity of the cephalothorax from that of the abdomen; in certain *Polychaeta*, a strongly developed transverse partition dividing the body cavity into two regions.

diaphragm cell (*Chem. Eng.*). An electrolytic cell in which a porous diaphragm is used to separate the electrodes, thus permitting electrolysis of, principally, sodium chloride without recombination in the cell of electrolysis products.

diaphragmless microphone (*Acous.*). The same as *cathodophone* (q.v.).

diaphragm plate (*Build.*). A connecting stiffener between the webs of a box girder.

diaphragm pump (*Eng.*). A pump in which a flexible diaphragm set between two non-return valves replaces a piston or bucket, being clamped round the edge and attached at the centre to a reciprocating rod of short stroke.

diaphysectomy (*Surg.*). Excision of part of the shaft of a long bone.

diaphysis (*Zool.*). The shaft of a long limb bone.

diaphysitis (*Med.*). Inflammation of the diaphysis.

diapophyses (*Zool.*). A pair of dorsal transverse processes of a vertebra, arising from the neural arch.

diapositive (*Photog.*). The description of a positive transparency on glass or film, in contrast to *katapositive*.

diapsid (*Zool.*). Said of skulls in which the supra- and infra-temporal fossae are distinct.

diarch (*Bot.*). Having two xylem strands.

diarrhoea, diarrhea (*Med.*). The frequent evacuation of liquid faeces.

diarthrosis (*Zool.*). A true (as opposed to a fixed) joint between two bones, in which there is great mobility; a cavity, filled with a fluid, generally exists between the two elements.

diaschisis (*Med.*). A disturbance of function of one part of the brain, consequent upon disease of another more remote part which is functionally connected with it.

diaschistic (*Cyt.*). In meiosis, said of tetrads which divide once transversely and once longitudinally.

diaspore (*Bot.*). Any fragment of a fungus used in its dissemination, e.g., a spore or sclerotium. (*Min.*) α-alumina monohydrate, occurring as small platy orthorhombic crystals in clays, notably in some of the bauxites. A dimorph of *boehmite*.

diastase (*Chem.*). An enzyme, or group of enzymes, capable of converting starch into sugar. Produced during the germination of barley in the process of malting.

diastasis (*Surg.*). The separation, without fracture, of an epiphysis from the bone.

diastataxy (*Zool.*). In Birds, absence of the fifth secondary remex or fifth flight feather carried by the ulna. Also **aquintocubitalism.**

diastema (*Zool.*). (1) An equatorial modification of protoplasm preceding cell-division. (2) A gap in a jaw where there are no teeth, as in Mammals lacking the canines.

diaster (*Cyt.*). In cell-division, a stage in which the daughter chromosomes are situated in two groups near the poles of the spindle, ready to form the daughter nuclei.

diastereoisomers (*Chem.*). Stereoisomers which are not mirror images. In order to have diastereoisomers corresponding to a given structural formula, there must be at least two asymmetric centres.

diastole (*Bot., Zool.*). Rhythmical expansion, as of the heart, or of a contractile vacuole. (*Cyt.*) Growth and expansion of the nucleus from the end of one mitosis to the commencement of the next.

diastolic murmur (*Med.*). A murmur heard over the heart during diastole and indicative of valvular disease.

diastrophism (*Geol.*). Relatively rapid and intense change in the configuration of the earth's surface alternating periodically with long spells of time during which the normal processes of denudation and sedimentation take place imperceptibly slowly. See **orogenesis.**

diastyle (*Arch.*). A colonnade in which the space between the columns is equal to 3 times the lower diameter of the columns.

diathermanous (*Phys.*). Relatively transparent to radiant heat.

diathermic coagulation (*Med.*). *Electrodesiccation.*

diathermic surgery (*Surg.*). The use of an electric arc in preference to a knife. This has the advantage of sealing cuts and reducing bleeding.

diathermy (*Med.*). Heat therapy applied to human body by high-frequency currents, generated by specially designed electronic oscillators.

diathesis (*Med.*). The constitutional state of the body which renders it liable to certain diseases.

diatomite or **diatomaceous earth** (*Min.*). A siliceous deposit occurring as a whitish powder consisting essentially of the frustules of diatoms. It is resistant to heat and chemical action, and is used in fireproof cements, insulating materials, as a backing for more refractory materials for furnace walls, and as an absorbent in the manufacture of explosives. Used with asbestos fibres in moulded insulation segments. Also **infusorial earth, kieselguhr, tripolite.**

diatom ooze (*Min.*). A deep-sea deposit consisting essentially of the frustules of diatoms; widely distributed in high latitudes.

diatoms (*Bot.*). See **Bacillariophyceae.**

diatoni (*Build.*). Quoins having two dressed faces projecting from the wall.

diatropism (*Biol.*). The tendency of organs or organisms to orientate themselves with their main axis at right-angles to the line of action of a stimulus.

Diatto surface-contact system (*Elec. Eng.*). A surface-contact electric traction system in which the skate under the vehicle is magnetic and lifts iron plungers which close contacts to make the studs in the road alive.

Diatype (*Typog.*). A *film lettering* machine capable of nongraduated sizes from 4 to 36 point, and suitable for single lines or display.

diaxon (*Zool.*). Having two main axes, as some Sponge spicules.

diaxone (*Zool.*). A bipolar nerve-cell.

diazo (*Print.*). See **dyeline.**

diazo compounds (*Chem.*). Compounds of the

general formula R·N=N·R′, obtained by the action of excess nitrous acid on aromatic amines at temperatures below 10°C. They are important intermediates for dyestuffs.

diazoamino compounds (*Chem.*). Pale yellow crystalline substances obtained by the action of a primary or secondary amine on a diazonium salt. Their general formula is R·N=N·NHR′. They do not form salts and most of them are easily transformed into the isomeric aminoazo compounds.

diazomethane (*Chem.*). CH_2N_2, an aliphatic diazo compound, an odourless, yellow, poisonous gas, very reactive, used for introducing a methyl group into a molecule. It is prepared from nitroso-methyl-urethane by decomposition with alcoholic potassium hydroxide.

diazonium salts (*Chem.*). The acid salts of diazobenzene of the general formula $(RN{\equiv}N^+)Cl^-$, important intermediates for azo-dyestuffs. They are usually prepared only in aqueous solution, by the action of nitrous acid on an aromatic amine at low temperatures in the presence of excess of acid. The —N=N— group can easily be replaced by hydrogen, hydroxyl, halogen, etc., and the diazonium salts can thus be transformed into other benzene derivatives.

diazo process (*Photog.*). A process for obtaining coloured dye images, starting with primuline and sensitizing, then exposing and treating with selected dyes; used chiefly for plan copying.

diazotization (*Chem.*). The process of converting amino into diazo compounds.

dibasic acids (*Chem.*). Acids containing two replaceable hydrogen atoms in the molecule.

dibenzoyl (*Chem.*). See **benzil.**

dibenzoyl-peroxide (*Chem.*). $C_6H_5·CO·O·O·CO·C_6H_5$. Relatively stable organic peroxide used mainly as a catalyst in polymerization and polycondensation reactions.

dibenzyl (*Chem.*). Symmetrical diphenylethane, $C_6H_5·CH_2·CH_2·C_6H_5$. M.p. 52°C.

dibenzyl group (*Chem.*). A synonym for the stilbene group, comprising compounds containing two benzene nuclei linked together by a chain of two or more carbon atoms.

diblastula (*Zool.*). In certain Invertebrates, an embryonic stage consisting of two layers of cells surrounding a central cavity.

diborane (*Chem.*). See **hydroborons.**

Dibranchiata (*Zool.*). An order of *Cephalopoda* with a single pair of ctenidia and kidneys; the internal shell enveloped by the mantle and in varying degrees of reduction; 8 or 10 tentacles; the two halves of the funnel only seen in the embryo; chromatophores and complex eyes.

dibranchiate (*Zool.*). Having 2 gills or ctenidia.

dicaryon, dikaryon (*Cyt.*). A pair of closely associated nuclei which divide at the same time.

dice (*Electronics*). Small regular pieces of semiconductor material for fabrication of devices.

dicentric (*Cyt.*). Having two centromeres.

dicentrine (*Chem.*). $C_{20}H_{21}O_4N$, an alkaloid of the isoquinoline group, obtained from the roots of the *Dicentra* species; it crystallizes in prisms, m.p. 168°C. It affects the heart and the respiratory centres.

dice-pattern (*Textiles*). Term applied to small geometrical designs or check patterns of chessboard type.

dicephalus (*Med.*). A developmental monstrosity in which a foetus is born with 2 heads.

dicephalus tetrabrachius (*Zool.*). In experimental embryology, an abnormal embryo produced by tight constriction in the sagittal plane of the 2-celled stage, when the sagittal plane coincides with the first furrow, and characterized by the possession of 2 pairs of fore-limbs and 2 heads.

dichasial cyme (*Bot.*). See **dichasium.**

dichasium, diachasium (*Bot.*). A cymose inflorescence in which each branch bears two lateral branches, both of about the same strength of development. *adj.* **dichasial.**

dichlamydeous, diplochlamydeous (*Bot.*). Said of a flower having calyx and corolla.

dichlamydeous chimaera (*Bot.*). See **diplochlamydeous chimaera.**

dichlone (*Chem.*). 2,3-Dichloro-1,4-naphthaquinone, used as a fungicide.

dichlorodifluoromethane (*Chem.*). CCl_2F_2, *Freon 12.* Used as a refrigerant, solvent and in fire extinguishers, b.p. −30°C.

dichloroethylenes (*Chem.*). Dichloroethenes, $C_2H_2Cl_2$. Exist in *cis*-form, b.p. 48°C and *trans*-form, b.p. 60°C. Prepared industrially from tetrachloroethane. Used as source material for vinyl chloride, the monomer of PVC.

dichloromethane (*Chem.*). *Methylene chloride*, CH_2Cl_2. Colourless liquid; b.p. 41°C. Used widely as a solvent, e.g., in manufacture of cellulose acetate.

dichlorophen (*Chem.*). Organic compound based on *hexachlorophene* (q.v.), widely used as an antibacterial agent, particularly for water treatment.

dichlorvos (*Chem.*). 2,2-Dichlorovinyl dimethyl phosphate, used as an insecticide. Also DDVP.

dichocephalous (*Zool.*). Said of ribs which have 2 heads, a tuberculum, and a capitulum. Cf. *holocephalous.*

dichogamy (*Bot.*). The condition in which, in a given flower, the stamens and stigmas are not mature at the same time. *adj.* **dichogamous.**

dichophysis (*Bot.*). A dichotomously-branched structure in the hymenium of some fungi.

dichopodium (*Bot.*). A sympodial branch system made up of successive parts of a dichotomizing branch system, of which only one part assists in forming the axis.

dichoptic (*Zool.*). Having the eyes of the two sides distinctly separated.

dichotomy (*Astron.*). The half-illuminated phase of a planet, as the moon at the quarters, and Mercury and Venus at greatest elongation. (*Bot.*) The production of 2 branches of the same size by the apical cell or apical growing point dividing into 2 equal parts, each then growing into a branch. *adj.* **dichotomous.** See **false dichotomy.**

dichroic (*Chem.*). Said of materials, such as solution of chlorophyll, which exhibit one colour by reflected light and another colour by transmitted light.

dichroic fog (*Photog.*). Fog which arises from the formation of an organic compound of silver; so-called because of its reddish coloration by transmitted light, and greenish coloration by reflected light.

dichroic mirror (*Light*). Colour-selective mirror, which reflects a particular band of spectral energy and transmits all others.

dichroism (*Crystal.*). The property possessed by some crystals (e.g., tourmaline) of absorbing the ordinary and extraordinary ray to different extents; this has the effect of giving to the crystal different colours according to the direction of the incident light. (*Min.*) See **pleochroic haloes.**

dichroite (*Min.*). *Cordierite* (q.v.).

dichromates (VI) (*Chem.*). See **chromates (VI).**

dichromatism (*Optics*). Colour blindness in which

power of accurate differentiation is retained for only 2 bands of colour in the spectrum.

dichromophil (*Cyt.*). Having an affinity for, and staining with, both acid and basic dyes.

dichroscope (*Crystal.*). An instrument for testing *dichroism*, using the two complementary rays of polarized light. For the same purpose, a **dichroscopic eyepiece** may be fitted to a polariscope

dichthadiigyne (*Zool.*). A peculiar gynaecoid, found among Driver Ants (*Dorylinae*), which lacks eyes and wings and has exceptionally large gaster and ovaries.

Dicke's radiometer (*Acous.*). An instrument for precise measuring of microwave noise power by reference to a standard source in a waveguide.

dickite (*Min.*). A form of hydrated silicate of aluminium, of the same chemical composition as kaolinite, with which it is grouped, and from which it differs only in the details of atomic structure and in certain physical properties.

Dicksonia (*Bot.*). A tropical and austral genus of (mainly tree-) ferns, named after James Dickson, d. 1822, Scottish botanist.

Dick test (*Bacteriol.*). A test of susceptibility to scarlet fever; the toxins of streptococci obtained from scarlet fever patients are injected into the skin.

dicliny, diclinism (*Bot.*). Having the sexes separate. The term is applied to the condition where the stamens and carpels are produced in separate flowers, either on the same or on a different plant, as well as to that in lower plants where the antheridial and oögonial branches are not obviously closely related in origin. *adj.* **diclinous.**

dicloran (*Chem.*). 2,6-Dichloro-4-nitro=(aminobenzene) used as a fungicide.

dicophane (*Chem.*). See **DDT.**

dicostalia (*Zool.*). The secondary brachial ossicles in *Crinoidea.*

Dicotyledones (*Bot.*). A large group of Angiosperms, containing between 150 000 and 200 000 species. The embryos have 2 cotyledons, the leaves are commonly net-veined, the parts of the flowers are in twos or fives, or multiples of these, and the vascular bundles in the axes usually contain cambium.

Dicranales (*Bot.*). An order of *Mosses*, having an erect gametophore with the lanceolate leaves two or more cells thick. The sporophyte is acrocarpous, and the peristome is simple with sixteen teeth.

dicrotic (*Med.*). Having a double beat or wave; said of the pulse.

dicrotism (*Med.*). In the arterial pulse, the occurrence of a double beat or wave to each beat of the heart.

dicty-, dictyo-. Prefix from Gk. *diktyon*, net.

dictyate stage (*Cyt.*). A resting stage succeeding *diplotene* (q.v.) in oögenesis; the karysome and the chromosomes lose their staining capacity and sharp contours during it, and the nucleus increases in size.

dictyokinesis (*Cyt.*). The division of the Golgi apparatus which accompanies division of the nucleus by karyokinesis, and segregation of dictyosomes to daughter cells.

Dictyosiphoniales (*Bot.*). An order of the *Phaeophyta.* The thallus is much branched with growth by a single apical cell, and the mature parts differentiated into two or three regions. The sporangia are usually unilocular. The gametophytes are isogamous and microscopic.

dictyosome (*Cyt.*). An element of the Golgi apparatus.

dictyosporangium (*Bot.*). A sporangium found in some *Oömycetes*, in which the spores encyst in the sporangium, then emit their contents separately, and leave a network of empty spore walls.

dictyospore (*Bot.*). A multicellular spore divided into segments by both transverse and longitudinal walls.

dictyostele (*Bot.*). A tubular network of vascular tissue, wholly enclosed by an endodermis.

Dictyotales (*Bot.*). An order of the *Phaeophyta.* The erect thallus grows by a single apical cell, or row of marginal cells. Both generations are morphologically identical. The gametophytes are oögamous. The sporophyte produces nonmotile spores.

dictyoxylic (*Bot.*). Having a network of meristeles or of vascular bundles.

dicyclic (*Bot.*). Having the perianth in two whorls. (*Zool.*) Said of the calyx of *Crinoidea* when a row of infrabasals is present. Cf. *monocyclic.*

didactyl (*Zool.*). Having all the toes of the hind feet separate, as in many *Marsupialia.* Cf. *syndactyl.*

didelphic (*Med.*). Pertaining to a double uterus.

Dido (*Nuc. Eng.*). A high flux deuterium moderated nuclear reactor at Harwell. Successor to *Bepo.*

Didot point system (*Typog.*). The continental point system, based on a 12-point *cicero*, the equivalent of the British *pica* but measuring approximately 12·8 British points.

didymospore (*Bot.*). A spore of two cells.

didymous (*Bot.*). Said of a fruit composed of two similar parts slightly attached along one edge.

didynamous (*Bot.*). Having 2 long and 2 short stamens.

die (*Build.*). (1) The body of a pedestal. (2) The enlarged part at either end of a baluster, where it comes into the coping or the plinth. (*Eng.*) (1) A metal block used in stamping operations. It is pressed down on to a blank of sheet-metal, on which the pattern or contour of the die surface is reproduced. (2) The element complementary to the punch in press tool for piercing, blanking, etc. (3) An internally threaded steel block provided with cutting edges, for producing screw threads by hand or machine. (4) A tool made of very hard material, often tungsten, carbide or diamond, with a (bell-mouthed) hole, usually circular, used to reduce the product cross-section by plastic flow, in wire or tube drawing.

die box or **die head** (*Eng.*). The holder into which screw dies are fitted in a screwing machine.

die case (*Typog.*). In the *Monotype system*, the frame in which the matrices of a fount are assembled.

diecasting (*Met.*). Casting of metals or plastics into permanent moulds, made of suitably resistant non-deforming metal. See **gravity-, pressure-.**

diecasting alloys (*Met.*). Alloys suitable for diecasting, which can be relied on for accuracy and resistance to corrosion when cast. Aluminium-base, copper-base, tin-base, zinc-base, and lead-base alloys are those generally used.

diecdysis (*Zool.*). A short period during the moulting cycle of *Arthropoda* when *proecdysis* (q.v.) passes imperceptibly into the succeeding *metecdysis* (q.v.).

die chuck (*Eng.*). A small 2- or 3-jaw *independent chuck* (q.v.).

die fill ratio (*Powder Tech.*). The ratio of uncompacted powder volume to the volume of the green compact.

dieldrin (*Chem.*). A widely-used contact insecticide based on HEOD, a chlorinated naphthalene derivative. Typical use is mothproofing of carpets and other furnishings.

dielectric (*Elec.*). Substance, solid, liquid or gas, which can sustain a steady electric field, and hence an insulator. It can be used for cables, terminals, capacitors, etc.

dielectric absorption (*Elec.*). Phenomenon in which the charging or discharging current of a dielectric does not die away according to the normal exponential law, due to absorbed energy in the medium.

dielectric amplifier (*Elec. Eng.*). One which operates through a capacitor, the capacitance of which varies with applied voltage.

dielectric antenna (*Radio*). One in which required radiation field is principally produced from a nonconducting dielectric.

dielectric breakdown (*Elec. Eng.*). Passage of large current through normally nonconducting medium at sufficiently intense field strengths accompanied by a relative reduction of resistance and, in solids, mechanical damage.

dielectric constant (*Elec. Eng.*). See **permittivity.**

dielectric current (*Elec.*). A changing electric field applied to a dielectric may give rise to a displacement current (always), an absorption current, a conduction current.

dielectric diode (*Elec. Eng.*). A capacitor whose negative plate can emit electrons into, e.g., CdS crystals, so that current flows in one direction.

dielectric dispersion (*Elec. Eng.*). Variation of permittivity with frequency.

dielectric fatigue (*Elec. Eng.*). Breakdown of a dielectric subjected to a repeatedly applied electric stress, insufficient to break down dielectric if applied once or a few times.

dielectric guide (*Elec. Eng.*). Possible transmission path of very-high-frequency electromagnetic energy functionally realized in a dielectric channel, the permittivity of which differs from its surroundings.

dielectric heating (*Elec. Eng.*). Radiofrequency heating in which power is dissipated in a nonconducting medium through dielectric hysteresis. It is proportional to $V^2 fS\varphi/t$, where V is the applied voltage, f the frequency, S the area of the heated specimen and t its thickness. φ is the ' loss factor ' of the material (permittivity × power factor).

dielectric hysteresis (*Elec.*). Phenomenon in which the polarization of a dielectric depends not only on the applied electric field but also on its previous variation. This leads to power loss with alternating electric fields.

dielectric lens (*Elec. Eng.*). By analogy with optics, a lens of dielectric material used for beaming electromagnetic waves from a radar horn.

dielectric loss (*Elec.*). Dissipation of power in a dielectric under alternating electric stress; $W = \omega CV^2\delta$, where W = power loss, V = r.m.s. voltage, C = capacitance, δ = power factor, $\omega = 2\pi \times$ frequency.

dielectric loss angle (*Elec.*). Complement of *dielectric phase angle*.

dielectric phase angle (*Elec.*). That between an applied electric field and the corresponding conduction current vector. The cosine of this angle is the power factor of the dielectric.

dielectric polarization (*Elec.*). Phenomenon explained by formation of doublets (dipoles) of elements of dielectric under electric stress.

dielectric radiation (*Radio*). That measured by dipole moment per unit volume for antenna depending on dielectric material.

dielectric relaxation (*Elec.*). Time delay, arising from dipole moments in a dielectric when an applied electric field varies.

dielectric strain (*Elec.*). See **displacement.**

dielectric strength (*Elec.*). Electric stress (volts per cm or mm) required to break down a dielectric; stress, steady or alternating, is normally maintained for 1 min. when testing.

dielectric viscosity (*Elec.*). Phenomenon in which the polarization lags behind the changes in the applied field, depending on its rate of change.

Diels-Alder reaction. See **diene synthesis.**

die lubricant (*Powder Tech.*). Lubricant applied to reduce friction between the powder and the die walls during compaction; it is usually a solid such as a soap. It may be incorporated in the powder when interparticle friction is also reduced.

diencephalon (*Zool.*). In Vertebrates, the posterior part of the fore-brain connecting the cerebral hemispheres with the midbrain.

diene (*Chem.*). Organic compound containing two double bonds between carbon atoms in its structure.

diene synthesis (*Chem.*). Formation of a cyclic compound by reaction between an aldehyde, acid or ester, and a conjugated diene; also known as the **Diels-Alder reaction.**

diene value (*Chem.*). Measure of the degree of unsaturation in fatty compounds.

die nut (*Eng.*). See **die** (3).

diesel cycle (*I.C. Engs.*). A compression-ignition engine cycle in which air is compressed, heat added at constant volume by the injection and ignition of fuel in the heated charge, expanded (so doing work on the piston), and the products exhausted, the cycle being completed in either two revolutions (4-stroke) or one (2-stroke). See **diesel engine, four-stroke cycle, two-stroke cycle.** (Rudolph *Diesel* (1858-1913), German engineer.)

diesel-electric locomotive (*Rail.*). A locomotive in which the motive power from a diesel engine is used to drive an electric generator which supplies electric motors connected to the driving axles.

diesel engine (*Eng.*). A compression-ignition engine in which the oil fuel is introduced into the heated compressed-air charge by means of a fuel pump which injects measured quantities of fuel to each cylinder in turn. Earlier models sprayed the fuel by means of an air-blast. See **compression-ignition engine.**

diesel generating station (*Elec. Eng.*). A generating station in which the prime-mover is a *diesel engine*.

diesel-hydraulic locomotive (*Rail.*). One whose prime mover is a diesel engine, the power being transmitted through an oil-filled *torque converter*.

diesel knock (*I.C. Engs.*). See **knocking.**

diesel locomotive (*Rail.*). A locomotive powered by a diesel or compression-ignition engine, as distinct from a *steam locomotive* (q.v.). See **diesel-electric, diesel-hydraulic locomotive.**

diesel oil (*Fuels*). *Gas oil* (q.v.) used in compression-ignition engines. When used in vehicle engines it is often known as **derv** (*d*iesel-engine *r*oad *v*ehicle).

die set (*Eng.*). See **suppress.**

die sinking (*Eng.*). The engraving of dies for coining, paper embossing, and similar operations.

die square (*Carp.*). A squared timber intermediate in size between a baulk and a quartering.

die stamping (*Print.*). An *intaglio* method of printing requiring a steel die; used mainly for high-class stationery.

diesterase (*Chem.*). Enzymes which catalyse the hydrolysis of an ester linkage in a phosphodiester. They have been used in the determination of the structure of nucleic acids.

die-stock (*Eng.*). A hand screw-cutting tool, consisting of a holder in which screwing dies can be secured; it is held and rotated by a pair of handles.

diet (*Nut.*). The human intake of food and liquid. According to age, size, and activity, etc., the diet should consist of minimum quantities of proteins, carbohydrates, fats, mineral salts, vitamins, and water, to ensure adequate available energy, repair, and growth of the body tissues, protection from general and specific diseases, potency for reproduction, and sufficient moisture to regulate the body temperature.

dietary standards (*Nut.*). Standards for daily food consumption, usually based upon the number of large calories produced from foods. Another standard specifies that a certain amount of proteins, carbohydrates, and fats should be consumed daily.

Dieterici's equation (*Chem.*). The *van der Waals' equation* (q.v.), modified for the effect of molecules near the boundaries.

Dietert centrifuge (*Powder Tech.*). A proprietary centrifugal air elutriator in which large particles are accelerated towards the periphery, and *fines* (q.v.) are extracted by the air stream leaving the centre of the centrifuge.

diethanolamine (*Chem.*). $CH_2OH{\cdot}CH_2{\cdot}NH{\cdot}CH_2{\cdot}CH_2OH$. One of the *ethanolamines* (q.v.); used industrially as an intermediate in the preparation of emulsifying agents, corrosion inhibitors, etc. Also used as a stripping agent for CO_2 and H_2S in gas streams.

diethyldithiocarbamic acid (*Chem.*). $(C_2H_5)_2{\cdot}N{\cdot}CS{\cdot}SH$. Colourless crystalline solid. Used as a reagent for detecting copper, with which it gives a characteristic brown colour. The zinc salt is used as an accelerator in vulcanization of rubber.

diethylene glycol (*Chem.*). 2,2′-dihydroxyethoxyethane, $(C_2H_4OH)_2O$. Colourless liquid; b.p. 245°C. Used as a solvent, e.g., for cellulose nitrate. Its monoethyl ether is known as **carbitol**, also used widely as a solvent; its derivatives (esters and ethers) are used as plasticizers.

diethyl ether (*Chem.*). Ethoxyethane. Ether, $C_2H_5{\cdot}O{\cdot}C_2H_5$, m.p. −113°C, b.p. 35°C, rel. d. 0·72, a mobile, very volatile liquid of ethereal odour, an anaesthetic and a solvent. It is prepared from ethanol and sulphuric acid.

Dietl's crisis (*Med.*). Severe abdominal pain and vomiting, followed after a time by an increased output of urine, associated with a movable or hydronephrotic kidney.

difference (*Maths.*). See **subtraction.**

difference limen (*Physiol.*). Increase in physiological stimulus which is just detectable in specified percentage of trials. The ratio of this to the original magnitude is the *relative difference limen.*

difference of phase (*Phys., etc.*). See **phase difference.**

difference of potential (*Elec.*). See **magnetic difference of potential, potential difference.**

difference operators (*Maths.*). The three main difference operators used in numerical analysis are defined as follows where u_r is a sequence of numbers, $r=1, 2, 3 \ldots$

$\Delta u_r = u_{r+1} - u_r$ (descending or forward difference)

$\delta u_{r+\frac{1}{2}} = u_{r+1} - u_r$ (central difference)

$\nabla u_{r+1} = u_{r+1} - u_r$ (ascending or backward difference).

Higher order differences are denoted by indices as in the differential calculus. E.g., $\Delta^2 u_r$ is a second order difference.

difference tone (*Acous.*). One of the combination tones produced subjectively when two or more pure tones are applied to the ear; most easily observed when *beats* may be discerned.

differentiable function of a complex variable (*Maths.*). For a function of a complex variable to be differentiable it must satisfy the Cauchy-Riemann equations, and each of the four partial derivatives must be continuous.

differential (*Electronics*). A device or circuit whose operation depends on the difference of two opposing effects. (*Maths.*) An arbitrary increment dx of an independent variable x, or, if $y=f(x)$, the increment dy of y defined by $dy=f'(x)dx$. See **complete differential, derived function.** (*Spinning*) See **box of tricks.**

differential absorption ratio (*Radiol.*). Ratio of concentration of radioisotope in different tissues or organs at a given time after the active material has been ingested or injected.

differential amplifier (*Electronics*). One whose output is proportional to the difference between two inputs. Usually based on the balanced differential pair. Many linear integrated circuits are of this type.

differential analyser (*Comp.*). A computer designed to solve differential equations.

differential anode conductance (*Elec. Eng.*). Reciprocal of *differential anode resistance.*

differential anode resistance (*Elec. Eng.*). The slope of the anode voltage versus anode current curve of a multi-electrode valve, when taken with all other electrodes maintained at constant potentials with reference to the cathode. At high frequencies, resistance values are larger than for d.c. (see **skin effect**) and a separate value may be quoted. Also called **a.c. resistance, slope resistance.**

differential booster (*Elec. Eng.*). A booster in which a series winding on the field is connected in opposition to the shunt winding.

differential calculus (*Maths.*). A branch of mathematics dealing with continuously varying quantities; based on the differential coefficient, or derivative, of one quantity with respect to another of which it is a function.

differential capacitor (*Elec. Eng.*). One with one set of moving plates and two sets of fixed plates so arranged that, as capacitance of the moving plates to one set of fixed plates is increased, that to the other set is decreased.

differential chain block (*Eng.*). A lifting tackle in which a 2-diameter chain wheel carries a continuous chain. Rotation of the chain wheel by a hanging loop shortens a second loop, supporting the load pulley in such a way as to give a large mechanical advantage. Also called **differential pulley block.**

differential coefficient (*Maths.*). See **derived function.**

differential duplex (*Teleg.*). *Duplex* (q.v.) system in which the sent currents divide through two sections of the receiving apparatus at the home end in opposite directions so as to balance their effects, whereas the currents received at the distant end pass mainly through the two sections in the same direction and operate the apparatus.

differential equation (*Maths.*). An equation involving total or partial differential coefficients. Those not involving partial differential coefficients are *ordinary differential equations.*

differential flotation (*Met.*). Production of more than one valuable concentrate by a series of froth flotation treatments of prepared ore pulp.

differential galvanometer (*Teleg.*). A pivoted needle galvanometer operated by two identical coils, so that equal and opposite currents in them neutralize each other in their magnetic effect, and therefore no deflection of the needle results.

differential gear (*Eng.*). A gear permitting relative rotation of two shafts driven by a third. The driving shaft rotates a cage carrying planetary bevel wheels meshing with two bevel wheels on the driven shafts. The latter are independent, but the sum of their rotation rates is constant.

differential grinding (*Min. Proc.*). Comminution so controlled as to develop differences in grindability of constituents of ore.

differential hardening (*Photog.*). Hardening in an emulsion depending on the silver density in the image, or due to the action of light on some special chemical, e.g., bichromate, in the gelatine.

differential ionization chamber (*Nuc. Eng.*). Two-compartment system in which resultant ionization current recorded is the difference between the currents in the two chambers.

differential iron tester (*Elec. Eng.*). An apparatus for iron testing consisting of two magnetic squares, one of the sample to be tested and the other of a standard material. The windings on the squares are connected to a differential wattmeter, so that there will be no deflection when the quality of the two specimens is the same.

differential leakage flux (*Elec. Eng.*). A general term given to the leakage flux occurring in and around the air gap of an induction motor. See **belt leakage, zigzag leakage.**

differentially compound-wound machine (*Elec. Eng.*). A compound-wound d.c. machine in which the magnetomotive forces of the two windings oppose one another.

differentially-wound motor (*Elec. Eng.*). A d.c. motor with series and shunt windings on the field connected so that the series winding opposes the shunt winding, and therefore causes the speed of the motor to rise as load is put on the machine.

differential-mode signal (*Elec. Eng.*). In a balanced 3-terminal system, signal applied between the 2 ungrounded terminals.

differential motion (*Eng.*). A mechanical movement in which the velocity of a driven part is equal to the difference of the velocities of two parts connected to it.

differential permeability (*Elec. Eng.*). Ratio of a small change in magnetic flux density of magnetic material to change in the magnetizing force producing it, i.e., slope of the magnetization loop at the point in question.

differential pressure gauge (*Eng.*). A gauge, commonly of U-tube form, which measures the difference between two fluid pressures applied to it.

differential protective system (*Elec. Eng.*). See **balanced protective system.**

differential psychology. The study of psychological differences between individuals, groups, cultures, etc., or the use of experimental methods in the study of individual differences.

differential pulley block (*Eng.*). See **differential chain block.**

differential relay (*Elec. Eng.*). See **relay types.**

differential resistance (*Elec. Eng.*). Ratio of a small change in the voltage drop across a resistance to the change in current producing the drop, i.e., the slope of the voltage-current characteristic for the material.

differential selsyn (*Elec. Eng.*). One whose stator and rotor are wound to produce rotating magnetic fields, so that a differential angle is introduced into the control system by a change in the rotor position.

differential stain (*Zool.*). A stain which picks out details of structure by giving to them different colours or different shades of the same colour.

differential susceptibility (*Elec. Eng.*). Ratio of a small change in the intensity of magnetization of a magnetic material to the change of magnetizing force producing it, i.e., the slope of the intensity-magnetization (hysteresis) loop.

differential thermal analysis (*Chem.*). The detection and measurement of changes of state and heats of reaction, especially in solids and melts, by simultaneously heating two samples of identical heat capacities and noting the difference in temperature between them, which becomes very marked when one of the two samples passes through a transition temperature with evolution or absorption of heat but the other does not. Abbrev. **DTA.**

differential titration (*Chem.*). Potentiometric titration in which the e.m.f. is noted between addition of small increments of titrant, the end point being where the e.m.f. changes most sharply.

differential winding (*Elec. Eng.*). A winding in a compound motor which is in opposition to the action of another winding.

differentiating circuit (*Elec. Eng.*). Electric circuit comprising resistors, capacitors and/or inductors, whose output waveform represents the rate of change (the differential) of the input waveform. Used generally to detect sudden or rapid changes in an otherwise fairly steady waveform, to provide feedback for servomechanisms or to modify waveforms in pulse circuitry. Also **differentiator.** Cf. *integrating circuit*.

differentiating solvent (*Chem.*). See **levelling solvent, nonaqueous solvents.**

differentiation (*Bot.*). The organization of mature tissues or mature members from generalized rudiments. (*Geol.*) A process by which a homogeneous magma, in solidifying and cooling gives rise to a series of different compositions. (*Maths.*) Operation of finding a differential coefficient. (*Zool.*) (1) The process of removing the excess stain from certain organs in the preparation of specimens for microscopic examination, in order to show up the structure of the whole specimen more clearly. (2) The development of modifications in the structure of tissues or organs owing to the development of division of labour.

differentiator (*Elec. Eng.*). See **differentiating circuit.**

diffluence (*Meteor.*). The spreading apart of isobars.

diffluent (*Bot.*). Readily becoming fluid.

difform, difformed (*Bot.*). Of unusual or irregular form.

diffract (*Bot.*). Said of a surface divided into areolae.

diffraction (*Light*). A small-scale spreading of light beyond the limits of the geometrical shadow, which is observable when the source of light is small. At the edge of the shadow and parallel to it, a few alternately light and dark bands are seen which are called *diffraction fringes*. Diffraction observations established the wave theory of light. (*Phys.*) Property exhibited by electromagnetic and other waves, of bending around edges of obstacles in their path; one factor which accounts for the propagation of radio waves over the curved surface of the earth, and also accounts for the audibility of sound round corners.

diffraction analysis (*Crystal.*). Analysis of the internal structure of crystals by utilizing the diffraction of X-rays caused by the regular atomic or ionic lattice of the crystal.

diffraction angle (*Phys.*). That between the direction of an incident beam of light, sound, or electrons, and any resulting diffracted beam.

diffraction fringes (*Light*). See **diffraction.**

diffraction grating (*Light*). One of the most useful optical devices for producing spectra. In one of its forms, the diffraction grating consists of a flat glass plate, on the surface of which have been ruled, with a diamond, equidistant parallel straight lines, which may be as close as 1000 per millimetre. If a narrow source of light is viewed through such a grating it is seen to be accompanied on each side by one or more spectra, produced by diffraction. See **concave grating**; also **electron diffraction.**

diffraction pattern (*Phys.*). That formed by equal intensity contours as a result of diffraction effects, e.g., in optics or radio transmission.

diffractometer (*Phys.*). An instrument used in the examination of the atomic structure of matter by the diffraction of X-rays, electrons, and neutrons. It is usual to use a monochromatic beam of radiation and to detect the diffracted beams by a suitable counting device.

diffuse (*Bot.*). (1) Said of a prostrate stem which is freely and loosely branched, and spreads widely over the ground. (2) Said of parenchymatous cells scattered throughout the xylem.

diffuse density (*Photog.*). The density of a photographic image as measured with diffuse light instead of with a beam of parallel light. See **specular density.**

diffuse growth (*Bot.*). The growth of the thallus of an alga by the division of any of its cells.

diffuse nebulae (*Astron.*). See **irregular nebulae.**

diffuse nucleus (*Cyt.*). The chromidia sometimes present in non-nucleated cells.

diffuse placentation (*Zool.*). Having the villi developed in all parts of the placenta, except the poles, as in *Lemurs*, most *Ungulates* and *Cetacea.*

diffuse porous (*Bot.*). The condition of xylem when the vessels are scattered uniformly throughout, or when there is little difference between the vessels formed at different parts of the growing season.

diffuse-porous wood (*Bot.*). Wood in which the xylem vessels tend to be evenly distributed, characteristic of evergreens.

diffuser (*Acous.*). Irregular structure, pyramids and cylinder, to break up reflections of sound waves in a studio. (*Aero.*) A means for converting the kinetic energy of a fluid into pressure energy; usually it takes the form of a duct which widens gradually in the direction of flow, also fixed vanes forming expanding passages in a compressor delivery to increase the pressure. (*Cinema., TV*) Translucent material in front of studio lamp to diffuse light and soften shadows. (*Eng.*) A chamber surrounding the impeller of a centrifugal pump or compressor, in which part of the kinetic energy of the fluid is converted to pressure energy by a gradual increase in the cross-sectional area of flow.

diffuse reflection (*Light, etc.*). Same as **nonspecular reflection.**

diffuse-reflection factor (*Light*). The ratio of the luminous flux diffusely reflected from a surface to the total luminous flux incident upon the surface.

diffuse series (*Light*). Series of optical spectrum lines observed in the spectra of alkali metals. Energy levels for which the orbital quantum number is two are designated *d-levels.*

diffuse sound (*Acous.*). That in a (nearly) closed space, the totality of all intensities, phases and directions arising from multiple reflections from the walls, so that the sound at any point does not appear to come from any particular direction; the reverberant component in sound reproduction.

diffuse stimulus (*Bot.*). A stimulus which does not affect the plant from any fixed position.

diffuse tissue (*Bot.*). A tissue consisting of cells which occur in the plant body singly or in small groups intermingled with tissues of distinct type.

diffuse transmittance (*Phys.*). See **transmittance.**

diffusion (*Chem.*). General transport of matter whereby molecules or ions mix through normal thermal agitation. Migration of ions may be directed and accelerated by electric fields, as in *dialysis* (q.v.).

diffusion area (*Nuc.*). See **diffusion length.**

diffusion attachment (*Photog.*). Device for softening the definition of the image, made of glass etched with fine circles.

diffusion barrier (*Phys.*). Porous partition for gaseous separation according to molecular weight and hydrodynamic velocities, especially for separation of isotopes. A fired but unglazed plate.

diffusion capacitance (*Elec.*). The rate of change of injected charge with the applied voltage in a semiconductor diode.

diffusion coefficient (*Chem.*). In the diffusion equation (*Fick's law*), the coefficient of proportionality between molecular flux and concentration gradient. Symbol D. Units, m^2/s, as for thermal diffusivity.

diffusion constant (*Elec.*). The ratio of diffusion current density to the gradient of charge carrier concentration in a semiconductor.

diffusion current (*Chem.*). In electrolysis, the maximum current at which a given bulk concentration of ionic species can be discharged, being limited by the rate of migration of the ions through the diffusion layer.

diffusion flame (*Heat*). Long luminous gas flame holding practically a constant rate of radiation for its designed length of travel, together with uniform precipitation of free carbon, diffusion occurring between adjacent strata of air and gas.

diffusion law (*Chem.*). See **Fick's law, Graham's law.**

diffusion layer (*Chem.*). In electrolysis, the layer of solution adjacent to the electrode, in which the concentration gradient of electrolyte occurs.

diffusion length (*Elec.*). Average distance travelled by carriers in semiconductor between generation and recombination. (*Nuc.*) A length significant to neutron diffusion theory. This length squared (termed the *diffusion area*) is one-sixth of the mean square distance travelled by a thermalized neutron.

diffusion of solids (*Elec.*). In semiconductors, the migration of atoms into pure elements to form surface alloy for providing minority carriers.

diffusion plant (*Nuc. Eng.*). One used for isotope separation by diffusion or thermal diffusion.

diffusion potential (*Elec. Eng.*). The p.d. across the boundary of an electrical double-layer in a liquid.

diffusion pump (*Chem. Eng.*). See **Gaede diffusion pump.**

diffusion theory (*Nuc.*). Simplified neutron migration theory based on *Fick's law* (q.v.). Less accurate than the more detailed *transport theory* (q.v.).

diffusion-transfer reversal (*Photog.*). Process in which an emulsion is developed with a solution containing a silver halide solvent while in contact with another layer which is not light-sensitive and to which unexposed silver halides dissolved by the developer are transferred, forming a positive image thereon by reaction with the chemicals of which it is composed. Abbrev. **DTR.**

diffusivity (*Chem.*). *Diffusion coefficient* (q.v.). (*Phys.*) A quantity which determines the rate of rise of temperature of a point in a bar which is being heated at some other point. It is equal to the thermal conductivity divided by the product of the specific heat capacity and the density, m^2/s.

difluorophosphoric acid (*Chem.*). HPO_2F_2. Formed by partial hydrolysis of phosphoryl fluoride, POF_3, with cold dilute alkali, or preferably by heating phosphoric acid with ammonium fluoride.

DIFO (*Chem.*). See **dimefox.**

digametic (*Zool.*). Having gametes of two different kinds.

digastric (*Zool.*). Of muscles, having fleshy terminal portions joined by a tendinous portion, as the muscles which open the jaws in Mammals.

digenesis (*Zool.*). (1) *Alternation of generations* (q.v.). (2) The condition of having two hosts; said of parasites. *adj.* **digenetic.**

digenetic reproduction. See **sexual reproduction.**

digenite (*Min.*). A cubic sulphide of copper, usually massive and associated with other copper ores. Composition probably near Cu_9S_5.

digester (*Paper*). A receptacle in which fibrous raw materials are heated under pressure with chemicals, in the first stages of papermaking.

digestion (*Min. Proc.*). In chemical extraction of values from ore or concentrate, period during which material is exposed under stated conditions to attacking chemicals. (*Zool.*) The process by which food material ingested by an organism is rendered soluble and assimilable by the action of enzymes. *adj.* **digestive.**

digestive cell (*Bot.*). One of the cells of the cortex of a root, in which the hyphae of an endophytic fungus are killed and digested.

digestive gland (*Bot.*). A glandular hair characteristic of carnivorous plants, producing enzymes by means of which the prey is digested. (*Zool.*) Gland(s) present in many Invertebrates and *Protochordata*, in which intracellular ingestion and absorption take place, as opposed to the alimentary canal proper.

digestive pouch (*Bot.*). A layer of cells on the apex of a lateral root; these cells secrete enzymes which help to bring about the breakdown of the cortical cells of the parent root as the lateral grows through it.

digger plough (*Agric.*). A type of mouldboard plough that leaves the soil over which it has travelled flat and broken, and not in an unbroken furrow slice. See **disk plough, mouldboard plough.**

digit (*Comp.*). Epoch occupied by any character, which may be any letter, figure or sign; one of a limited number in a set for counting, or in which *data* can be *coded.* (*Zool.*) A finger or toe, one of the free distal segments of a pentadactyl limb.

digital (*Zool.*). In Spiders, the distal joint of the pedipalp: pertaining to, or resembling, a digit.

digital computer (*Comp.*). One which performs simple calculations at high speed by impulses, related to digits in a code.

digital differential analyser (*Comp.*). A special electronic computer for solution of differential equations by incremental means.

digital formula (*Zool.*). A figure expressing the number of phalanges in each digit of the metapodium of a pentadactyl limb.

digitaliform (*Bot.*). Shaped like the finger of a glove.

digitalin (*Chem.*). A glucoside present in digitalis.

digitalis (*Pharm.*). Dried leaves of the foxglove, *Digitalis purpurea*, whose steroid glycosides are used in heart disease.

digitalization (*Med.*). Administration of digitalis to a patient with heart disease, in amounts sufficient to produce full therapeutic effect.

digital meter (*Elec. Eng.*). One incorporating an analogue-digital converter and displaying an analogue quantity on digital indicators.

digital remiges (*Zool.*). The wing-quills attached to the phalanges of the second and third digits in Birds.

digitate (*Bot.*). Said of a compound leaf in which the leaflets arise from the top of the petiole and spread like the fingers of a hand.

digit check (*Comp.*). Redundant digit added to *words* in a process to reveal faults.

digitigrade (*Zool.*). Walking on the ventral surfaces of the phalanges only, as terrestrial carnivores.

digitization (*Comp.*). Conversion of an analogue signal to a digital signal, with steps between specified levels. Also called **quantization.**

digitonin (*Chem.*). Saponin glycoside in *Digitalis purpurea*, yielding on hydrolysis the 2 : 3 : 6-trihydroxy-derivative sapogenin, *digitogenin.*

digitoxin (*Chem.*). A vegetable glycoside isolated from the leaves of *Digitalis purpurea.*

digitron (*Electronics*). Glow tube used for numerical read-out. The cathodes are usually shaped to form the digits 0 to 9.

digit sign (*Comp.*). The + or − associated with a *word* read into a computer.

digitule (*Zool.*). Any small fingerlike process.

digonal (*Maths.*). Of symmetry about an axis, such that a half-turn (180°) gives the same figure.

digoneutic (*Zool.*). Producing offspring twice a year.

digonoporous (*Zool.*). Having two genital pores.

digue (*Hyd. Eng.*). An artificial sea-wall, or breakwater, constructed to prevent the encroachment of the water on the land behind.

digynous (*Bot.*). Having two carpels.

dihedral angle (*Aero.*). Angle at which an aerofoil is inclined to the transverse plane of reference. A downward inclination is called *negative dihedral*, sometimes *anhedral.* (*Maths.*) Angle between two planes, as measured in the plane normal to their line of intersection.

diheptal (*Electronics*). Referring to 14 in number, e.g., pins on base of a tube or valve.

dihybrid (*Gen.*). The product of a cross between parents differing in two heritable characters; an organism heterozygous regarding two pairs of alleles.

dihydroxyacetone (*Chem.*). $CH_2OH{\cdot}CO{\cdot}CH_2OH$. The simplest *ketose*; used in sun-tan lotions.

dikaryon (*Cyt.*). See **dicaryon.**

dikaryophase (*Bot.*). That part of the life-history of many *Basidiomycetae* in which the hyphae are made up of segments, each containing two nuclei.

dike (*Hyd. Eng.*). See **dyke.**

diketen (*Chem.*).

$$\begin{array}{l} CH_2{=}C{-}CH_2 \\ \quad\;\;\; | \quad\;\; | \\ \quad\;\; O{-}CO \end{array}$$

Dimer of *ketene* (q.v.). B.p. 127°C. Useful intermediate in preparative organic chemistry.

diketones (*Chem.*). Dialkanones. Compounds containing —CO— groups which, according to their position in the molecule, are termed 1,2-

diketones —CO·CO—, or 1,3-diketones —CO·CH_2·CO—, etc.

dikkop (*Vet.*). See **African horse-sickness.**

dilacerate (*Bot.*). As if torn into strips.

dilambdodont (*Zool.*). Said of teeth in which the paracone and metacone are V-shaped, well separated and placed near the middle of the tooth, as in some *Insectivora.*

dilapidations (*Build.*). A term applied to the damage done to premises during a period of tenancy.

dilatancy (*Chem.*). Property shown by some colloidal systems, of thickening or solidifying under pressure or impact. Cf. *thixotropy.*

dilated (*Bot.*). Expanded and flattened.

dilatometer (*Chem.*). An apparatus for the determination of transition points of solids. It consists of a large bulb joined to a graduated capillary tube, and is filled with an inert liquid. The powdered solid is introduced, and the temperature at which there is a considerable change in volume on slow heating or cooling may be noted; alternatively, the temperature at which the volume shows no tendency to change with time may be found.

dilatometry (*Chem.*). The determination of transition points by the observation of volume changes.

dilator (*Zool.*). A muscle which, by its contraction, opens or widens an orifice. Cf. *sphincter.*

dilsh (*Mining*). A band of inferior coal.

diluent (*Paint.*). A volatile liquid unable to act as a paint solvent on its own, but able to dilute a genuine solvent without disadvantage. It is used to cheapen the formulation.

diluent air (*Build.*). Air admitted or induced into a flue to dilute the noxious effects of combustion.

diluted (*Bot.*). Pale and faint-coloured.

dilution (*Chem.*). (1) Decrease of concentration. (2) The volume of a solution which contains unit quantity of dissolved substance. The reciprocal of *concentration.*

dilution law (*Chem.*). See **Ostwald's dilution law.**

diluvium (*Geol.*). An obsolete term for those accumulations of sand, gravel, etc., which, it was thought, could not be accounted for by normal stream and marine action. In this sense, the deposits resulting from the Deluge of Noah would be *diluvial.*

dimedone (*Chem.*). Dimethyldihydro(1,3-dihydroxybenzene):

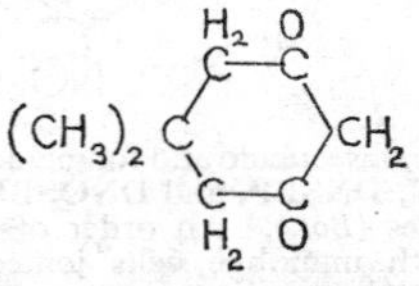

$$\begin{array}{ccc} & H_2\ \ O & \\ & C - C & \\ (CH_3)_2 C & & CH_2 \\ & C - C & \\ & H_2\ \ O & \end{array}$$

Greenish crystalline solid; used as a reagent for detecting the presence of alcohol. The sample tested is distilled with potassium dichromate and sulphuric acid, and the distillate is collected in a solution of *dimedone* at 0°C. If alcohol is present, crystals form with a m.p. of 141·5°C.

dimefox (*Chem.*). Bisdimethylaminofluorophosphine oxide:

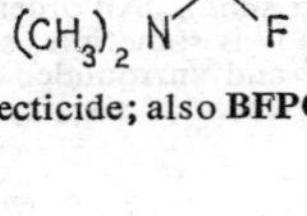

$$\begin{array}{ccc} (CH_3)_2N & & O \\ & P & \\ (CH_3)_2N & & F \end{array}$$

used as an insecticide; also **BFPO** and **DIFO.**

dimegaly (*Zool.*). The condition of having spermatozoa of two different sizes.

dimensional stability (*Paper*). The percentage movement of paper in the machine and cross directions under two different conditions of temperature and relative humidity.

dimension lumber (*For.*). Lumber which is sawn to special (as opposed to stock) sizes.

dimensions (*Phys.*). Measurements of physical quantities need only involve measurements of the fundamental units of the system. The *dimensions* of a quantity are the powers to which these units are involved in measurements of the quantity, e.g., the dimensions of force are mass × length × time^{-2} or [MLT^{-2}.]

dimension stock (*For.*). Timber converted to specified widths, thicknesses and lengths.

dimension stone (*Build.*). A term sometimes used for an *ashlar* (q.v.).

dimer (*Chem.*). Molecular species formed by the union of two like molecules, e.g., carboxylic acids in aprotic solvents:

$$2\,RCOOH \rightleftarrows (RCOOH)_2.$$

dimeric (*Chem.*). Having the same empirical formula but double the molecular formula.

dimerous (*Bot.*). Having 2 members in a whorl.

dimethylformamide (*Chem.*). Solvent of wide industrial application, used in plastics manufacture, in gas separation processes, in the artificial fibre industry.

dimethyl glyoxime (*Chem.*). The compound:

$$\begin{array}{l} CH_3—C{=}NOH \\ \qquad\ \ | \\ CH_3—C{=}NOH. \end{array}$$

Used in analysis as a specific and quantitative precipitant for palladium (weakly acid solution) and nickel (ammoniacal solution).

dimethyl hydrazine (*Chem.*). See **hydrazine.**

dimethyl phthalate (*Chem.*). Dimethyl benzene 1,2-dicarboxylate:

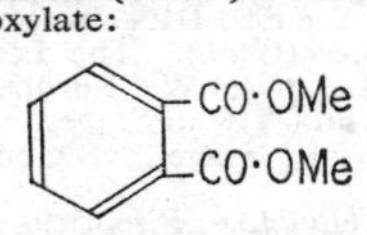

$$C_6H_4(CO{\cdot}OMe)_2$$

High boiling-point ester, b.p. 280°C, widely used as an insect repellent, particularly against midges and mosquitoes.

dimethyl sulphate (*Chem.*). $CH_3O{\cdot}SO_2{\cdot}OCH_3$, b.p. 187–188°C. Widely used as a methylating (i.e., introducing of a methyl (CH_3) group) agent in organic preparations. Manufactured from dimethyl ether and sulphur trioxide.

dimetric system (*Crystal.*). See **tetragonal system.**

dimidiate (*Bot.*). (1) Said of an anther which is lop-sided because one lobe is absent or abortive. (2) Said of a pileus when one side is larger than the other. (3) Said of the perithecium of a lichen when the upper part only is enclosed in a wall. (4) In general, lop-sided.

diminished stile (*Join.*). A door-stile which is narrowed down for a part of its length, as, e.g., in the glazed portion of a sash door. Also called **gunstock stile.**

diminishing courses (*Build.*). See **graduated courses.**

diminishing pipe (*Plumb., San. Eng., etc.*). A tapered pipe length used to connect pipes of different diameters.

dimity (*Textiles*). (1) Strong cotton fabric resembling diaper and damask but with a stripe pattern; used chiefly for mattress coverings. (2) Light plain cotton cloth with cord stripes.

dimmer (*Elec. Eng.*). A light intensity controller using thyristors or a variable resistor. Also called **dimming resistance.** See **liquid dimmer.**

dimmer wheel (*Elec. Eng.*). A handwheel for operating one or more dimmers; commonly used in stage-lighting equipment.

dimming resistance (*Elec. Eng.*). See **dimmer.**

dimonoecious (*Bot.*). Having perfect flowers, as well as staminate, pistillate, and neuter flowers.

dimorphic, dimorphous (*Bot., Zool.*). Existing in 2 forms. (*Chem.*) Capable of crystallizing in 2 different forms.

dimorphic heterostyly (*Bot.*). The condition when flowers of the same species of plant have styles of 2 different lengths.

dimorphism (*Min.*). Crystallization into 2 distinct forms of an element or compound (e.g., carbon as diamond and graphite; FeS_2 as pyrite and graphite). (*Zool.*) The condition of having 2 different forms, as animals which show marked differences between male and female, animals which have 2 different kinds of offspring, and colonial animals in which the members of the colony are of 2 different kinds.

dimple (*Eng.*). (1) A slight conical depression produced by a twist drill after a small initial feed into the work; used as a guide for further drilling. (2) A slight depression in one component of an assembly used to locate a matching component by a corresponding projection in the latter.

Dimyaria (*Zool.*). In some classifications a class of *Nemertinea* in which there are 2 muscle layers in the body wall, an external circular and an internal longitudinal layer, in addition to diagonal muscles.

dimyarian (*Zool.*). Having 2 muscular layers, as certain *Nemertinea.*

DIN. Abbrev. for *Deutsche Industrie Normen*, i.e., the German Industrial Standards. (*Photog.*) Scale of emulsion speeds based upon the exposure required to produce a density of 0·1 above fog. See also **DIN sizes.**

Dinantian Series (*Geol.*). The Lower Carboniferous rocks of N.W. Europe, comprising Tournasian and Visean Stages. Succeeded by the Namurian Stage. Synonymous with **Avonian.**

Dinas bricks (*Build.*). Firebricks made almost entirely of sand with a small amount of lime.

Dinas rock (*Met.*). A natural rock or sand of high silica content, used as an acid refractory.

dinergate (*Zool.*). A worker ant having a very large head and powerful mandibles; a soldier ant.

Dines hygrometer (*Meteor.*). A form of hygrometer in which the dew point is determined by cooling a glass plate by a flow of iced water until dew forms on it, the temperature of the plate being indicated by a thermometer.

dineutron (*Nuc.*). Assumed transient existence of a set of two neutrons in order to explain certain nuclear reactions.

dinex (*Chem.*). 2-Cyclohexyl-4,6-dinitro 1-hydroxybenzene:

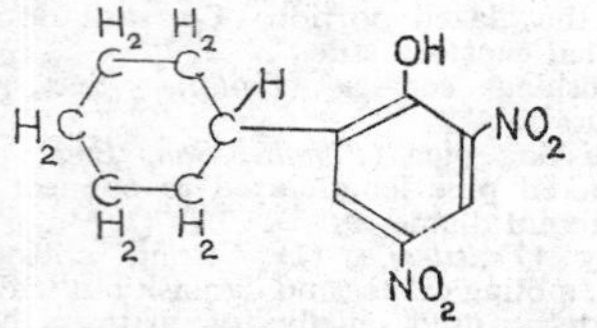

used as an insecticide also known as **DNOHCP**

dinging (*Build.*). Rough plastering for walls, a single coat being put on with a trowel and brush.

dinitrocresol (*Chem.*). See **DNOC.**

dinkey (*For.*). A small logging locomotive.

dinky sheet (*Print.*). A web much narrower than the full width of a rotary press.

dinocap (*Chem.*). 2-(1-Methyl-heptyl)-4,6-dinitrophenyl crotonate; used as a fungicide.

Dinocapsales (*Bot.*). A monotypic order of the *Dinophyceae*, the thallus being palmelloid, and there being a temporary motile gymnodinoid stage.

Dinococcales (*Bot.*). An order of the *Dinophyceae*, whose members are non-motile and unicellular; no vegetative divisions take place, reproduction being by zoospores or aplanospores.

Dinoflagellata (*Zool.*). An order of *Phytomastigina*; forms having two flagella, one directed backwards, the other transverse and lying in a spiral groove, the annulus; yellow, brown, green, or colourless; sometimes of amoeboid form, but usually covered by cellulose plates; usually without a gullet; having food-reserves of starch or oil or both. Known to botanists as **Dinophyceae.**

Dinophyceae (*Bot.*). A class of the *Pyrrophyta*; the cells usually have numerous brown chromatophores, or these may be absent. The motile genera, and the zoospores of nonmotile genera, have a transverse groove around them, with a pair of flagella inserted near it.

dinosam (*Chem.*). 2-(1-Methyl-butyl)-4,6-dinitro 1-hydroxybenzene:

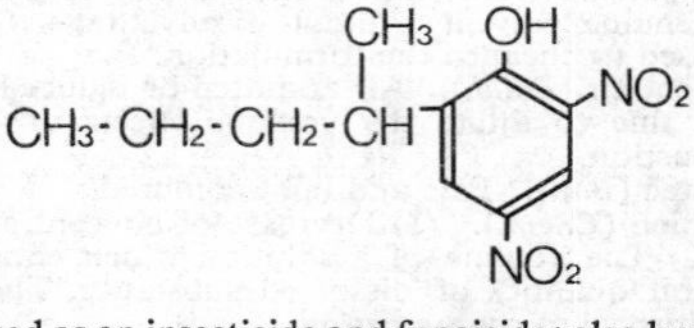

used as an insecticide and fungicide; also known as **DNAP, DNSAP,** and **DNOSAP.**

dinoseb (*Chem.*). 2-(1-Methyl-propyl)-4,6-dinitro 1-hydroxybenzene:

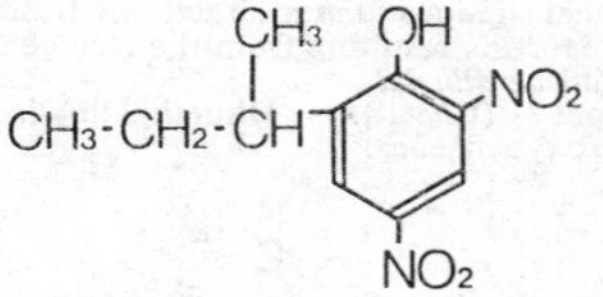

used as an insecticide and fungicide; also known as **DNBP, DNSBP,** and **DNOSBP.**

Dinotrichales (*Bot.*). An order of the *Dinophyceae*, with immobile cells joined to form a branched filament.

DIN sizes (*Paper*). A series of International metric sizes of papers based on a ratio of 1 : 1·414 for the length of sides. Each sheet in the series is half that of the preceding one, e.g., A0 is 841 × 1189 mm, A1 is 594 × 841 mm, up to A6. There is also a B series starting at B0 1000 × 1414 mm. See **DIN.**

dinting (*Mining*). The operation of taking up the floor of a road or level in a coal-mine to increase its height.

diocoel (*Zool.*). The lumen of the diencephalon.

Dioctophymata (*Zool.*). An order of *Nematoda* in which the body is sometimes spiny; the mouth is hexagonal and surrounded by from one to three circles, each of six papillae; the gut is suspended by four well-developed longitudinal muscles inserted in the middle-line of the muscle tracts of the body wall; in the female, there is a single ovary, in the male a caudal bursa without rays and a single copulatory spicule.

diode (*Electronics*). (1) Simplest electron tube, with a heated cathode and anode; used because of unidirectional and hence rectification properties. (2) Semiconductor device with similar properties, evolved from primitive crystal rectifiers for radio reception.

diode characteristic (*Electronics*). A graph showing the anode current as a function of the anode voltage.

diode clipper (*Electronics*). A clipping circuit which employs a diode.

diode isolation (*Electronics*). Isolation of the circuit elements in a micro-electronic circuit by using the very high resistance of a reverse-biased *p-n* junction.

diode-pentode and **diode-triode** (*Electronics*). Thermionic diode in same envelope as a pentode and triode, respectively.

diode voltmeter (*Elec. Eng.*). Voltmeter in which measured voltage is rectified by a valve or semiconductor diode, the rectified voltage then usually being amplified and displayed on a moving-coil meter.

dioecious (*Bot.*). Having the pistillate and staminate flowers on separate plants of the same species, each plant being unisexual. (*Zool.*) Having the sexes separate. *n.* **dioecism.**

dioestrus (*Zool.*). In female Mammals, the growth period following metoestrus.

diolefins (*Chem.*). Dialkenes. Hydrocarbons containing 2 double bonds in their molecules. They exist in 3 forms according to the disposition of the double bonds, i.e., adjacent as in the allenes; separated by 1 single bond in the conjugated compounds such as butadiene; or separated by 2 or more single bonds.

diols (*Chem.*). Dihydric alcohols, chiefly represented by the glycols in which the hydroxyl groups are attached to different carbon atoms.

Dionic (*Elec. Eng.*). TN designating instruments for testing the purity of water by measuring its electrical conductivity.

diophantine equations (*Maths.*). Indeterminate equations for which integral or rational solutions are required.

diopside (*Min.*). A monoclinic pyroxene, ideally consisting of silicate of calcium and magnesium, $CaMgSi_2O_6$, but commonly containing a small content of $Fe_2Si_2O_6$ in addition. Diopside occurs in metamorphosed limestones and dolomites.

dioptase or **emerald copper** (*Min.*). A rare hydrated silicate of copper, crystallizing in the trigonal system and found occasionally, as rich emerald-green crystals, in association with other copper ores. When first found, it was used as a gemstone.

dioptrate (*Zool.*). Having the compound eye transversely divided into an upper and a lower organ, as in *Gyrinidae* (Whirligig Beetles).

dioptre (*Light*). The unit of power of a lens. A convergent lens of 1 metre focal length is said to have a power of +1 dioptre. Generally, the power of a lens is the reciprocal of its focal length in metres, the power of a divergent lens being given a negative sign.

dioptric element (*Optics*). A component in an optical system in which refraction, and not reflection, is used in the process of forming the image.

dioptric mechanism (*Zool.*). A mechanism, consisting of the cornea, aqueous humour, lens and vitreous humour, by which the images of external objects may be focused on the retina of the eye, in *Chordata*.

dioptric system (*Optics*). An optical system, which contains only refracting components.

diordinal crochets (*Zool.*). Hooks of two lengths forming a circle on the apices of the prolegs of some Lepidopteran larvae.

diorite (*Geol.*). A coarse-grained deep-seated (plutonic) igneous rock of intermediate composition, consisting essentially of plagioclase feldspar (typically near andesine in composition) and hornblende, with or without biotite in addition. Differs from granodiorite in the absence of quartz. See also **tonalite.**

Diotocardia (*Zool.*). See **Aspidobranchia.**

dioxan (*Chem.*). 1,4-Diethene:

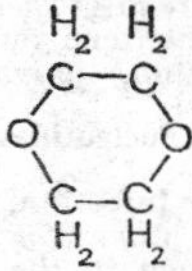

Colourless liquid; m.p. 11°C, b.p. 101°C Used as a solvent for waxes, resins, viscose, etc.

dioxazine dyes (*Chem.*). Range of dyestuffs which are relatively complex sulphonated compounds with an affinity for cellulose. Chiefly derived from chloranil by reaction with amines. The chief dyes in the series are light-fast in the blue range.

dip (*Geol.*). A term implying inclination of strata, measured to the horizontal. The *angle of dip* is measured in the direction of maximum slope of the strata. Indicated on geological maps by a small arrow (the *dip arrow*), with a figure alongside giving the inclination in degrees. (*Hyd. Eng.*) Any departure from the regular slope at which a pipe is laid, when the slope is increased locally. (*Mag.*) The angle measured in a vertical plane between the earth's magnetic field at any point and the horizontal; also called **inclination.** See also **dip of the horizon.**

dip arrow (*Geol.*). See **dip.**

dip circle (*Mag.*). An instrument consisting of a magnetic needle pivoted on a horizontal axis; accurate measurements of magnetic dip can be obtained with it.

dipentene (*Chem.*). Monocyclic terpene isomeric with *d-limonene* (q.v.).

dip fault (*Geol.*). A fracture in the rocks of the earth's crust, which coincides in direction with the dip of the strata and is therefore at right angles to their strike. Cf. *strike fault*.

diphacinone (*Chem.*). 2-Diphenylacetylindane-1, 3-dione, used as a rodenticide. Also **diphenadione.**

diphase (*Elec. Eng.*). A term sometimes used in place of *2-phase*.

diphasic (*Zool.*). Of certain Trypanosomes, having a life-cycle which includes a free active stage. Cf. *monophasic*.

diphenadione (*Chem.*). See **diphacinone.**

diphenyl (*Chem.*). Phenylbenzene, $C_6H_5{\cdot}C_6H_5$; colourless, monoclinic crystals; m.p. 71°C; b.p. 254–255°C; soluble in alcohol and ether. It occurs in coal-tar, and is prepared by heating iodobenzene to 220°C with finely divided copper.

diphenyl ether (*Chem.*). $C_6H_5{\cdot}O{\cdot}C_6H_5$, diphenyl oxide, a liquid of pleasant odour, m.p. 28°C, b.p. 253°C, obtained from phenol by heating with $ZnCl_2$ or $AlCl_3$.

diphenylglyoxal (*Chem.*). See **benzil.**

diphenylguanidine (*Chem.*).

$$C_6H_5\cdot NH\cdot C(=S)\cdot NH\cdot C_6H_5$$

Crystalline solid; m.p. 147°C. Used as an accelerator in vulcanization of rubber.

diphenylmethane (*Chem.*). $(C_6H_5)_2CH_2$, colourless needles, m.p. 26°C. b.p. 262°C, obtained by the action of (chloromethyl)benzene on benzene in the presence of aluminium chloride.

diphenylmethane dyes (*Chem.*). Dyestuffs derived from diphenylmethane. The only important dyestuff of this class is Auramine O, the ketone-imide of diphenylmethane.

diphenylmethane group (*Chem.*). Compounds containing two benzene nuclei attached to a single carbon atom, derivatives of *diphenylmethane* (q.v.).

diphosphopyridine nucleotides (*Chem.*). See **NAD, NADP.**

diphtheria (*Med.*). Infection, usually air-borne, with the bacillus *Corynebacterium diphtheriae*. Bacilli are confined to the throat, producing local necrosis ('pseudo-membrane'), but a powerful *exotoxin* causes damage especially to heart and nerves.

diphtheroid (*Bacteriol.*). A common term for describing many bacilli morphologically similar to the diphtheria bacillus, belonging to the genus *Corynebacterium*, but harmless to man.

diphycercal (*Zool.*). Said of a type of tail-fin (found in Lung-fish, adult Lampreys, the young of all Fish, and many aquatic *Urodela*) in which the vertebral column runs horizontally, the fin being equally developed above and below it

diphygenic (*Zool.*). Having 2 different modes of development.

diphyletic (*Biol.*). Of dual origin: descended from 2 distinct ancestral groups.

diphyodont (*Zool.*). Having 2 sets of teeth—a deciduous or milk dentition and a permanent dentition, as in Mammals.

diplanetary (*Zool.*). Of *Protozoa*, having 2 different kinds of zoospores.

diplanetic (*Bot.*). Swimming for a short time, encysting, and leaving the cyst for a second period of swimming, the second swimming phase differing in morphology from the first.

diplecolobous (*Bot.*). Said of an embryo with an incumbent radicle and with the cotyledons folded twice or more.

diplegia (*Med.*). Bilateral paralysis of like parts of the body.

dipleurula (*Zool.*). A free-swimming ciliated larval stage of eleutherozoan *Echinodermata* which is bilaterally symmetrical and possesses three pairs of coelomic sacs.

diplex (*Teleg.*). See **simplex.**

diplexer (*Telecomm.*). A means of coupling which permits a radar and a communication transmitter to operate with the same aerial.

diplobiont (*Bot.*). A plant which includes in its life-cycle at least 2 kinds of different individuals; if the species is dioecious, there are 3 kinds of individuals. *adj.* **diplobiontic.**

diploblastic (*Zool.*). Having 2 primary germinal layers, namely, ectoderm and endoderm.

diplocaulescent (*Bot.*). Having a main axis with branches upon it.

diplochlamydeous (*Bot.*). See **dichlamydeous.**

diplochlamydeous chimaera (*Bot.*). A pericinal chimaera consisting of an outer skin, two layers of cells thick, representing one constituent, and surrounding a core representing the other.

diplococcus (*Bacteriol.*). A coccus which divides by fission in one plane, the two individuals formed remaining paired.

diplodal (*Zool.*). Of *Porifera*, having a prosodus as well as an aphodus.

diplodesmic (*Bot.*). Having 2 parallel vascular systems.

diplodization, diploidization (*Bot.*). The conversion of a mycelium composed of uninucleate segments into one of binucleate segments.

diploë (*Zool.*). In certain bones of the skull, cancellous tissue between outer and inner layers of compact bone. *adj.* **diploic.**

diplogangliate, diploganglionate (*Zool.*). Having paired ganglia.

diplohaplont (*Bot.*). A plant having a sexual process, and having a morphological alternation between a haploid generation and a diploid generation.

diploid (*Cyt.*). Having the somatic number of chromosomes characteristic of the species. Cf. *haploid.*

diploid apogamety (*Bot.*). See **euapogamy.**

diploid apogamy (*Bot.*). The development of a sporophyte containing diploid nuclei, without any preliminary fusion of gametes, from one or more cells of the gametophyte.

diplokaryotic (*Cyt.*). Having twice the normal diploid number of chromosomes.

diplonema (*Cyt.*). A stage in the meiotic division (*diplotene stage*) at which the chromosomes are clearly double.

diplont (*Bot.*). A plant body containing diploid nuclei.

diplophase (*Biol.*). The period in the life-cycle of any organism when the nuclei are diploid. Cf. *haplophase.*

diplopia (*Med.*). Double vision of objects.

diploplacula (*Zool.*). In the life-history of some Invertebrates, a larval stage resembling a flattened blastula.

Diplopoda (*Zool.*). A subclass of *Myriapoda* having the trunk composed of numerous double somites, each with 2 pairs of legs; the head bears a pair of uniflagellate antennae, a pair of mandibles, and a gnathochilarium representing a pair of partially fused maxillae; the genital opening is in the third segment behind the head; vegetarian animals of retiring habits. Millipedes. Also **Chilognatha.**

diplopores (*Zool.*). Pores which occur in pairs in the plates of the theca in some *Cystoidea.*

diplosis (*Cyt.*). The doubling of the chromosome number in syngamy.

diplosome (*Cyt.*). A paired heterochromosome: a double centrosome lying in the cytoplasm.

diplospondyly (*Zool.*). The condition of having two vertebral centra, or a centrum and an intercentrum, corresponding to a single myotome. *adjs.* **diplospondylic, diplospondylous.**

diplostemonous (*Bot.*). Having twice as many stamens as there are petals, with the stamens in 2 whorls, the members of the outer whorl alternating with the petals.

diplotene (*Cyt.*). The fourth stage of meiotic prophase, intervening between pachytene and diakinesis, in which homologous chromosomes come together and there is condensation into tetrads.

diplozoic (*Zool.*). Bilaterally symmetrical.

Diplura (*Zool.*). An order of *Apterygota*, long, slender, white or yellowish in colour and blind. The *Campodeidae* have two long, segmented, apical abdominal appendages, while the *Japygidae* have two forceps-like structures instead. Living in the soil, they feed on decaying vegetation or on soft, living plant tissues.

dip needle (*Mag., Surv., etc.*). Dipping needle, inclinometer. Magnetic needle on horizontal pivot, which swings in vertical plane. When set in magnetic north-south plane, its inclination shows the *angle of dip*.

Dipnoi or **Dipneusti** (*Zool.*). An order of *Crossopterygii*, in which the air-bladder is adapted to function as a lung, and the dentition consists of large crushing plates. Lung-fish.

dip of the horizon (*Astron., etc.*). The angular difference between the true mathematical and the visible horizon; the difference is due to the curvature of the earth, and its amount depends upon the height of the observer above sea level. Also called **apparent depression of the horizon.**

dipole (*Elec., etc.*). Close combination of equal positive and negative charges of electricity or magnetism. Certain large dielectric molecules also exhibit dipole effects, reflected in the polarization and temperature trends in the dielectric constant of substance.

dipole antenna (*Radio*). One comprising a straight conductor, up to half-wavelength long, attached to a feeder at its centre. Maximum response is for waves polarized along axis of conductor.

dipole molecule (*Elec.*). One which has a permanent moment due to the permanent separation of the effective centres of the positive and negative charges.

dipole moment (*Elec.*). Product of the magnitude of an electric or magnetic charge and the distance between it and its opposite charge; this determines the distant potential, which falls off as the inverse square of the distance.

Dippel's oil (*Chem.*). Bone-oil. A product obtained by the dry distillation of bones from which the fat has not been extracted; contains a mixture of pyridine bases and their homologues.

dipper (or **dipper-bucket**) **dredger** (*Eng.*). A dredger consisting of a single large bucket at the end of a long arm, swung in a vertical plane by gearing. The bucket capacity may be up to 10 m^3.

dipping (*Eng.*). The immersion of pieces of material in a liquid bath for surface treatment such as pickling or galvanizing. (*Paint.*) An industrial process of coating articles by immersion in paint, followed by withdrawal. (*Vet.*) The process of immersing animals in a medicated bath, for the destruction of ectoparasites.

dipping needle (*Mag., Surv., etc.*). See **dip needle.**

dipping refractometer (*Chem.*). A type of refractometer which is dipped into the liquid under examination.

dip road (*Mining*). See **foot rill.**

diprosopus (*Med.*). A developmental monstrosity in which a foetus has 2 faces.

diprosopus triophthalamus (*Zool.*). In experimental embryology, an abnormal embryo produced by constriction of the 2-celled stage in the sagittal plane; characterized by the duplication of the anterior part of the head and the fusion of the median pair of eyes.

diprosthomerous (*Zool.*). Having 2 somites in front of the mouth.

diprotodont (*Zool.*). Having the first pair of upper and lower incisor teeth large and adapted for cutting, the remaining incisor teeth being reduced or absent; pertaining to the *Diprotodontia*. Cf. *polyprotodont*.

dip slope (*Geol.*). A land form developed in regions of gently inclined strata, particularly where hard and soft strata are interbedded. A long gentle sloping surface which coincides with the inclination of the strata below ground.

dip soldering (*Eng.*). The method of soldering previously fluxed components by immersing them in a bath of molten solder. Ideal for bulky assemblies with complicated or multiple joints, and for fast automatic operation.

dipsomania (*Med.*). The condition in which there is a recurring, temporary, and uncontrollable impulse to drink excessively.

dip stick (*Eng.*). A rod inserted in a tank or sump to measure the depth of oil or other liquid.

dipstone trap (*Eng.*). A form of trap in which a stone slab on edge dips below the water surface.

Diptera (*Zool.*). An order of *Oligoneoptera*, having one pair of wings, the hinder pair being represented by a pair of club-shaped balancing organs or halteres; the mouth-parts are suctorial; the larva is legless and sluggish. Flies, Gnats, and Midges.

dipygus (*Med.*). A developmental monstrosity in which a foetus has a double pelvis.

dipyre (*Min.*). A member of the *scapolite* (q.v.) series, containing 20–25% of the meionite molecule.

Dirac's constant (*Phys.*) *Planck's constant* divided by 2π. Usually termed *h-bar*, and written $\hbar$. Unit in which *electron spin* (q.v.) is measured.

Dirac's Theory (*Phys.*). Using the same postulates as the *Schroedinger Equation*, plus the requirement that quantum mechanics conform with the theory of relativity, an electron must have an inherent angular momentum and magnetic moment. (1928.)

direct-acting pump (*Eng.*). A steam-driven reciprocating pump in which the steam and water pistons are carried on opposite ends of a common rod.

direct adaptation (*Bot.*). Any adaptation which does not appear to stand in relation to natural selection.

direct-arc furnace (*Elec. Eng.*). An electric arc furnace in which the arc is drawn between an electrode and the charge in the furnace. See **Heroult furnace.**

direct call (*Teleph.*). In international telephony, a call which involves a single international telephone circuit, i.e., one between two trunk centres and over one frontier.

direct capacitance (*Elec.*). That between two conductors, as if there are no other conductors.

direct circuit (*Teleg.*). A telegraphic circuit going from one station to another without the use of relays at intermediate stations.

direct-conversion reactor (*Elec. Eng.*). One which converts thermal energy directly into electricity by means of thermoelectric elements (usually of silicon-germanium).

direct cooling (*Elec. Eng.*). The cooling of transformer and machine windings by circulating the coolant through hollow conductors.

direct-coupled exciter (*Elec. Eng.*). An exciter for a synchronous or other electric machine, which is mounted on the same shaft as the machine that it is exciting.

direct-coupled generator (*Elec. Eng.*). A generator which is mechanically coupled to the machine which is driving it, i.e., not driven through gearing, a belt, etc.

direct coupling (*Electronics*). Interstage coupling without the use of transformers or series capacitors, so that the d.c. component of the signal is retained; also called **d.c. coupling.**

direct current (*Elec. Eng.*). Current which flows in one direction only, although it may have appreciable pulsations in its magnitude. Abbrev. d.c.

direct-current amplifier, -balancer, etc. See **d.c. amplifier, -balancer, etc.**

direct drive (*Radio*). Transmitting system in which the antenna circuit is directly coupled to the oscillator circuit.
direct effect (*Biochem.*). Effect of primary ionization of vital molecules (such as nucleic acids) in a living cell as a result of exposure to radiation.
direct-fired (*Heat*). Type of furnace in which the fuel is delivered directly into the heating chambers.
direct germination (*Bot.*). The germination of a spore of any kind by means of a hypha or a filament.
direct heating. A system of heating by radiation. Cf. *indirect heating*.
directing stimulus (*An. Behav.*). Stimulus which, though not releasing a component of species-specific behaviour, is important in determining the direction of a response.
direct injection (*Aero.*). The injection of metered fuel (for a spark-ignition engine) into the supercharger eye or the cylinders, which eliminates the freezing and poor high-altitude behaviour of carburettors.
direct-injection pump (*Aero.*). A fuel-metering pump for injecting fuel direct to the individual cylinders.
direction (*Maths.*). The position of one point in space relative to another.
directional antenna (*Radio*). One in which the transmitting and receiving properties are concentrated along certain directions.
directional baffle (*Acous.*). A baffle taking the form of a widely flared horn with the reproducing diaphragm located at its throat.
directional circuit-breaker (*Elec. Eng.*). A circuit-breaker which operates when the current flowing through it is in the direction opposite to normal.
directional coupling (*Elec. Eng.*). That afforded by a device in which power can be transmitted in specified paths and not in others, e.g., line-balancing toroid, valve amplifier, and certain waveguide configurations.
directional derivative (*Maths.*). The rate of change of a function $F(x, y, z)$ with respect to arc length along a given curve (or in a given direction), i.e.,

$$\frac{dF}{ds} = l\frac{\partial F}{\partial x} + m\frac{\partial F}{\partial y} + n\frac{\partial F}{\partial z},$$

where l, m, n are the direction cosines of the tangent to the curve at the point (x, y, z).
directional effect (*Photog.*). Defect in processed film caused by directional movement of used developer in continuous processing apparatus. Characterized by inequality of development between leading and trailing parts of negative, or between neighbouring areas of different contrast.
directional filter (*Telecomm.*). Combination of high- and low-pass filters, which determines route or direction of transmission of signals. Separate filters are not directional, being passive.
directional gain (*Radio*). Ratio (expressed in decibels) of the response (generally along the axis, where it is a maximum) to the mean-spherical (or hemispherical with reflector or baffle) response, of an antenna, loudspeaker or microphone. Also **directivity index.**
directional gyroscope (*Instr.*). A free gyroscope which holds its position in azimuth and therefore indicates any angular deviation from the reference position.
directional homing (*Nav.*). Following of a course so that the objective is always at the same relative bearing.
directional lighting fittings (*Light*). Lighting fittings (often used in street lighting installations) which direct a high proportion of their light output towards a point on the roadway midway between adjacent lamp standards.
directional loudspeaker (*Acous.*). A loudspeaker in which the normal directivity of a moving diaphragm is increased by a horn with a large flare, so that the radiated sound power is mostly directed in a beam.
directional microphone (*Acous.*). One which is directional in response, either inherently (as in the *ribbon microphone*) or with a parabolic reflector of adequate dimensions.
directional radio (*Radio*). System using directional antennae at the transmitting and/or receiving ends, usually to minimize interference.
directional receiver (*Radio*). Receiving system using a directional antenna for discrimination against noise and other transmissions.
directional relay (*Elec. Eng.*). A relay whose operation depends on the direction of the current flowing through it.
directional transmitter (*Radio*). Transmitting system using a directional antenna, to minimize power requirements and to diminish effect of interference.
direction angles (*Maths.*). The 3 angles which a line makes with the positive directions of the coordinate axes.
direction components (*Maths.*). See **direction numbers.**
direction cosines (*Maths.*). The cosines of the 3 direction angles of a line.
direction-finder (*Radio*). A directional receiver in which the directions of maximum and minimum signal response can be varied, in order to determine the direction of arrival of incoming waves.
direction-finding (*Radio*). Principle and practice of determining a bearing by radio means, using a discriminating antenna system and a radio receiver, so that direction of an arriving wave, ostensibly direction or bearing of a distant transmitter, can be determined.
direction numbers (*Maths.*). Of a line, any 3 numbers, not all of which are zero, which are proportional to the direction cosines of the line. Also **direction components, direction ratios.**
direction of a curve (*Maths.*). Direction of tangent to curve at the point.
direction ratios (*Maths.*). See **direction numbers.**
directions image (*Crystal.*). See **interference figure.**
directions test (*Psychol.*). A type of mental test in which the candidate is required to carry out instructions of varying complexity.
direction switch (*Elec. Eng.*). A switch which determines the direction of travel; used on electric lifts or similar equipment.
directive efficiency (*Radio*). Ratio of maximum to average radiation or response of an antenna; the *gain*, in dB, of antenna over a dipole being fed with the same power.
directive force (*Elec. Eng.*). A term used to denote the couple which causes a pivoted magnetic needle to turn into a north and south direction.
directive gain (*Radio*). For a given direction, 4π times the ratio of the radiation intensity in that direction to the total power which is radiated by the aerial.
directive mesenteries (*Zool.*). In *Zoantharia*, two opposite pairs of mesenteries, in which the muscles are on opposite sides.
directive movement (*Bot.*). Movement of orientation.
directives (*Zool.*). See **directive mesenteries.**

directivity (*Acous.*, *Radio*). (1) Measurement, in dB, of the extent to which a directional antenna, loudspeaker, or microphone concentrates its radiation or response in specified directions. (2) See **antenna gain.**

directivity angle (*Radio*). Angle of elevation of direction of maximum radiation or reception of electromagnetic wave by an antenna.

directivity index (*Radio*). Same as **directional gain.**

direct labour (*Build.*, *Civ. Eng.*). A mode whereby labour is employed directly by the client, normally a Local Authority; as opposed to the usual method of working through independent architect, engineer, and surveyor.

direct laying (*Eng.*). Cables are laid in a trench and covered with soil; planks, bricks, tiles, or concrete slabs are put over the cable as protection. Cables used to be armoured, but modern practice is merely to put a serving of bituminized paper or hessian over the lead sheath.

direct lighting (*Light*). A system of lighting in which not less than 90% of the total light emitted is directed downwards, i.e., in the lower hemisphere.

directly-heated cathode (*Electronics*). Metallic (coated) wire heated to a temperature such that electrons are freely emitted. Also called **filament cathode.**

direct metamorphosis (*Zool.*). The incomplete metamorphosis undergone by exopterygote insects, in which a pupal stage is wanting.

direct mixing (*Textiles*). Blending cotton of different types at the bale breaker or hopper feeder in order to obtain homogeneous mixture suitable for spinning yarn of the desired characteristics and strength. See **stack mixing.**

direct nuclear division (*Cyt.*). See **amitosis.**

director (*Teleph.*). The apparatus which obtains a channel, through exchange junctions, to the required exchange. During dialling the trains of impulses are registered, and when the required exchange is found the numerical trains are passed to the required exchange, and operate the selectors, to get the required subscriber. (*Radio*) Free resonant dipole element in front of antenna array which assists the directivity of array in the same direction. See **Yagi antenna.** (*Surg.*) A grooved instrument for guiding a surgical knife.

director circle (*Maths.*). See **orthoptic circle.**

director meter (*Teleph.*). In an automatic switching exchange, a meter which is attached to a director to total the number of times it is taken into operation.

director system (*Teleph.*). An automatic switching system for routing calls between exchanges. It uses a storage mechanism for the numerical impulse trains, while the code impulses are being translated and used to find a route over junctions to the required exchange, which, when found, receives the numerical trains and hence, by step-by-step mechanism, connects to the wanted subscriber.

direct printing (*Print.*). Method in which the print is made directly on the paper as in letterpress or photogravure, as distinct from the usual lithographic method of *offset printing* (q.v.).

direct process (*Met.*). The method originally used for obtaining from ore a form of iron similar to wrought-iron in one operation, i.e., without first making pig-iron.

direct radiation (*Phys.*). See **primary radiation.**

direct ray (*Radio*). That portion of the wave from a transmitter which proceeds directly to the receiver, without reflection from the Heaviside layer. Also **direct wave, ground ray, ground wave.**

direct-reading instrument (*Elec. Eng.*). An instrument in which the scale is calibrated in the actual quantity measured by the instrument, and which therefore does not require the use of a multiplying constant.

direct-recorded disk (*Acous.*). One which, after a groove has been cut into its plastic surface, is immediately ready for *playback* (reproduction).

direct reductive analysis (*Psychol.*). A method of analysis, introduced by Hadfield, which stresses the environmental factor, particularly any specially traumatic or *nuclear* incidents in the life-history of the individual, in the causation of neurosis. It relates the existing emotional attitudes of the patient back to the people for whom they were first felt, thus lessening the phenomenon of transference to the person of the analyst.

directrix (*Maths.*). (1) Of a conic or quadric, the polar line or plane of a focus. See **conic** for alternative definition. (2) A curve of a ruled surface in general through which the generators pass.

direct rope haulage (*Mining*). Engine plane. An ascending truck is partly balanced by a descending one, motive power being applied to the drum round which the haulage rope passes.

direct sound (*Acous.*). The sound intensity arising from the direct radiation from a source to a listener, as contrasted with the reverberant sound which has experienced a large number of reflections between the source and the listener.

direct stress (*Eng.*). The stress produced at a section of a body by a load whose resultant passes through the centre of gravity of the section.

direct stroke (*Elec. Eng.*). When a transmission line or other apparatus is struck by a lightning stroke, it is said to receive a *direct stroke.*

direct-suspension construction (*Elec. Eng.*). A form of construction used for the overhead contact wire on electric traction systems; the contact wire is connected directly to the supports without catenary or messenger wires.

direct-switching starter (*Elec. Eng.*). An electric motor starter arranged to switch the motor directly across the supply, without the insertion of any resistance or the performing of any other current-limiting operation.

direct-trip (*Elec. Eng.*). A term used in connexion with circuit-breakers, starters, or other similar devices, to indicate that the current which flows in the tripping coil is the main current in the circuit, not an auxiliary current obtained from a battery or other source.

direct vernier (*Surv.*). A vernier in which n divisions on the vernier plate correspond in length to $(n-1)$ divisions on the main scale.

direct viewing (*TV*). Reception in which the received image is viewed directly on the screen of a cathode-ray tube, without projection or reflecting devices.

direct-vision prism (*Light*). A compound prism with component prisms of two glasses having different dispersive powers and cemented together so that, in passing through the combination, light suffers dispersion but no deviation.

direct-vision spectroscope (*Light*). A spectroscope employing a *direct-vision prism* (q.v.). Such an instrument is usually in the form of a short straight tube with a slit at one end and an eyepiece at the other; it is used for rough qualitative examination of spectra.

direct-vision viewfinder (*Photog.*). A viewfinder

using a lens and sighting pin or hole with axis parallel to that of the camera lens, or slightly tilted to compensate for parallax.

direct wave (*Radio*). See **direct ray.**

dirigible (*Aero.*). A navigable balloon or airship.

dirt (*Mining*). Broken valueless mineral. Gangue. In U.S. **muck.**

dirt beds (*Geol.*). A general name applied to old fossil soils, particularly to those in the Purbeck Series. The 'Fossil Forest' of the Lulworth District lies immediately upon such a dirt bed.

dis (*Elec. Eng.*). See **discontinuity.**

disaccharides (*Chem.*). A group of carbohydrates considered to be derived from 2 molecules of a monosaccharide (either the same or different) by elimination of 1 molecule of H_2O; e.g., maltose is $G\alpha - 1 \rightarrow 4G$ (G = glucose).

disacryl (*Chem.*). White, semitransparent polymer of *acrolein* (q.v.).

disadvantage factor (*Nuc. Eng.*). Ratio of average neutron flux in reactor lattice to that within actual fuel element.

disappearing-filament pyrometer (*Heat*). An instrument used for estimating the temperature of a furnace by observing a glowing electric-lamp filament against an image of the interior of the furnace formed in a small telescope. The current in the filament is varied until it is no longer visible against the glowing background. From a previous calibration the required temperature is derived from the current value.

disarticulation (*Surg.*). Amputation of a bone through a joint.

disazo dyes (*Chem.*). Dyestuffs containing two azo groups of the type: $C_6H_5{\cdot}N{=}N{\cdot}C_6H_4{\cdot}N{=}N{\cdot}C_6H_4{\cdot}OH$. These dyes are obtained by diazotizing an amino-derivative of azobenzene and then coupling it with a tertiary amine or with a phenol, or by coupling a diamine or dihydric phenol with 2 molecules of a diazonium salt.

disbudding (*Vet.*). The destruction, usually by caustic chemicals or thermo-cautery, of the horn buds of calves, kids, and lambs.

disc. See **disk.**

discal (*Zool.*). Pertaining to or resembling a disk or disklike structure; a wing-cell of various Insects.

discard (*Met.*). See **crop.**

discharge (*Elec.*). (1) The abstraction of energy from a cell by allowing current to flow through a load. (2) Reduction of the p.d. at the terminals (plates) of a capacitor to zero. (3) Flow of electric charge through gas or air due to ionization, e.g., lightning, or at reduced pressure, as in fluorescent tubes. See **field discharge.** (*Hyd.*) The rate of flow of a fluid through a pipe or channel, expressed as either volume or mass per unit time. (*Nuc. Eng.*) Unloading of fuel from a reactor.

discharge bridge (*Elec. Eng.*). Measurement of the ionization or discharge, in dielectrics or cables, depending on the amplification of the high-frequency components of the discharge.

discharge circuit (*Elec. Eng.*). Discharge of a capacitor for safety reasons, e.g., by making a parallel valve or tube conducting by altering its grid bias, or by using a neon lamp.

discharge coefficient (*Hyd.*). See **coefficient of discharge.**

discharge electrode (*Elec. Eng.*). See **active electrode.**

discharge head (*Min. Proc.*). Vertical distance between intake and delivery of pump, *plus* allowance for mechanical friction and other retarding resistances requiring provision of extra power.

discharge lamp (*Elec. Eng.*). One in which luminous output arises from ionization in gaseous discharge. See **discharge (3).**

discharger (*Elec. Eng.*). (1) A device, such as a spark gap, which provides a path whereby a piece of electrical apparatus may be discharged. (2) An apparatus containing an electrically heated wire for firing explosives in blasting.

discharge rate (*Elec. Eng.*). A term used in connexion with the discharge of accumulators. An accumulator has a certain capacity at, e.g., a 1-hr discharge rate, when that capacity can be obtained if the accumulator is completely discharged in 1 hr. If the discharge rate is lower, i.e. the discharge takes more than 1 hr, the capacity obtainable will be higher.

discharge resistance (*Elec. Eng.*). A noninductive resistance placed in parallel with a circuit of high inductance (e.g., the field winding of an electric machine) in order to prevent a high voltage appearing across the circuit when the current in it is switched off. Also called **buffer resistance.**

discharge tube (*Electronics*). Any device in which conduction arises from ionization, initiated by electrons of sufficient energy.

discharge valve (*Eng.*). A valve for controlling the rate of discharge of fluid from a pipe or centrifugal pump.

discharging arch (*Build.*). An arch built in a wall to protect a space beneath from the weight above, and to allow access or discharge.

discharging pallet (*Horol.*). The pallet mounted in the discharging roller of the chronometer escapement, which brings about the unlocking of the escape wheel by removing the locking pallet from one of its teeth.

discharging roller (*Horol.*). The circular disk carrying the discharging pallet mounted on the balance staff in a chronometer escapement.

discharging tongs (*Elec. Eng.*). A pair of metal tongs used for discharging capacitators before they are touched by hand.

Disciflorae (*Bot.*). A group of dicotyledonous families with polypetalous corollas, usually with a disk in the flower bearing the stamens, and with a superior ovary.

discission (*Surg.*). An incision into a part; especially, needling of a cataract.

disclimax (*Ecol.*). A stable community which is not the climatic or edaphic community for a particular place, but is maintained by man or his domestic animals, e.g., a desert produced by overgrazing, where the natural climax would be grassland. The name derives from *dis*turbance *climax*.

discoblastula (*Zool.*). The type of blastula formed by the cleavage of a megalecithal egg consisting of a disklike blastoderm. *adj.* **discoblastic.**

discocarp (*Bot.*). See **apothecium.**

discodactylous (*Zool.*). Having the ends of the digits flattened out to form circular sucking disks.

discoid (*Bot.*). (1) Round and flat. (2) Said of a capitulum which has no ray florets. (3) Said of an open rounded fructification of a lichen. (4) Said of the form of an algal thallus which is closely applied to a substratum, and is one layer of cells in thickness.

discoidal segmentation (*Zool.*). Cleavage of an ovum, leading to the formation of a disklike germinal area, as in the Chick.

Discolichenes (*Bot.*). A group of lichens in which the fungus is a Discomycete.

discolor, discolorous (*Bot.*). Not of the same colour throughout.

Discomedusae (*Zool.*). A subclass of *Scyphozoa*,

comprising active marine forms with eight or more tentaculocysts; regular alternation of generations occurs.

discomposition effect (*Nuc.*). See **Wigner effect.**

Discomycetes (*Bot.*). A group of *Ascomycetes* characterized by the possession of an apothecium as the fructification.

discone antenna (*Radio*). Radiator for VHF transmission, comprising a vertical cone, which acts as a vertical dipole with very constant driving impedance over a wide frequency range.

disconformity (*Geol.*). That type of break in the rock sequence in which there is no angular discordance of dip between the two sets of strata involved. Cf. *unconformity*.

disconnected set (*Maths.*). One which can be divided into two sets having no points in common and neither containing an accumulation point of the other.

disconnecting link (*Elec. Eng.*). See **isolating link.**

disconnection (*Elec. Eng.*). See **discontinuity.**

disconnector (*San. Eng.*). See **interceptor.**

discontinuity (*Elec. Eng.*). A break, whether intentional or accidental, in the conductivity of an electrical circuit; also called **disconnection,** colloquially **dis.** (*Maths.*) A point at which a function is not continuous.

discontinuous distribution (*Biol.*). Isolated distribution of a species, as the Tapir, which is found in the Malay Peninsula and Sumatra and again in Central and South America.

discontinuous variation (*Biol.*). A rare variation; sport; saltation.

disconula (*Zool.*). A larval type of certain *Siphonophora*, having 8 rays, and producing buds on the ventral side of its umbrella.

discoplacenta (*Zool.*). A type of placenta having the villi arranged on a flat disk-shaped area.

discoplankton (*Bot.*). Plankton with cells in the form of thin disks.

discrete (*Bot.*). Remaining separate; said especially of paraphyses of lichens.

discrete radio sources (*Astron.*). Originally known as *radio stars* (q.v.); the sources of cosmic radio waves are, however, not individual stars. The main sources appear to be the centre of the Galaxy, supernovae remnants, spiral galaxies, peculiar galaxies, and the Hydrogen II regions.

discriminant (*Maths.*). (1) Of a polynomial equation $x^n+a_1x^{n-1}+\ldots+a_n=0$: the product of the squares of all the differences of the roots taken in pairs. (2) Of a differential equation: the result of eliminating $p\left(=\dfrac{dy}{dx}\right)$ between the differential equation $F(x, y, p)=0$ and

$$\frac{\partial}{\partial p}F(x, y, p)=0$$

is the *p*-discriminant equation, which represents the *p*-discriminant locus. For a quadratic equation $ax^2+bx+c=0$, the discriminant is b^2-4ac. The roots are equal if $b^2-4ac=0$, real if $b^2-4ac>0$, and imaginary if $b^2-4ac<0$ (a, b, c being real). If the solution of the differential equation is $u(x, y, c)=0$ (c an arbitrary constant), the result of eliminating c between $u(x, y, c)=0$ and $\dfrac{\partial u}{\partial c}(x, y, c)=0$ gives the *c*-discriminant equation.

discriminating circuit-breakers (*Elec. Eng.*). A term sometimes used to denote circuit-breakers which operate only when the current is in a given direction.

discriminating protective system (*Elec. Eng.*). An excess-current protective system which causes to be disconnected only that portion of a power system upon which a fault has actually occurred.

discriminating satellite exchange (*Teleph.*). A small automatic exchange which can decide, without engaging its main exchange, whether or not it can complete a call arising from one of its subscribers.

discriminating selector (*Teleph.*). A selector which discriminates between calls which are to be completed locally and those which are to be completed through other exchanges, working through the absorption of a train of impulses.

discrimination (*An. Behav.*). See **discrimination training.** (*Telecomm.*) The selection of a signal having a particular characteristic, e.g., frequency, amplitude, etc., by the elimination of all the other input signals at the discriminator.

discrimination reversal (*An. Behav.*). An experimental situation in which an animal which has learned to choose one of two stimuli, then has to learn to choose the other, the association of each with positive or negative reinforcement being reversed.

discrimination training (*An. Behav.*). The experimental procedure of reinforcing a response in the presence of one stimulus and not reinforcing it in the presence of other stimuli, resulting in an increase in response intensity in the presence of the particular stimulus, i.e., *discrimination*, if the animal can discriminate between it and the other stimuli.

discriminative stimulus (*An. Behav.*). A stimulus used in operant conditioning, which sets the occasions when a response will be reinforced, the response intensity then becoming greater in the presence of the discriminative stimulus.

discriminator (*Telecomm.*). (1) Circuit which converts a frequency or phase modulation into amplitude modulation for subsequent demodulation. (2) Circuit which rejects pulses below a certain amplitude level, and shapes the remainder to standard amplitude and profile. (*Teleph.*) Device which effects the routing and/or determines the fee units for a call originating at a satellite exchange.

discus (*Bot.*). The hymenium of an apothecium.

discus proligerus (*Zool.*). See **zona granulosa.**

dish (*Radio*). Colloquialism for *parabolic reflector*, solid or mesh, used for directive radiation and reception, especially for radar or radio telescopes.

disharmony (*Zool.*). See **hypertely.**

dished (*Eng.*). Of wheels, especially steering wheels, having the hub inset in a different plane from the rim.

dished-out (*Build.*). Said of the wooden framework or bracketing on which the laths and plastering are fixed in vaults, domes, coved ceilings, and the like.

dished plate (*Eng.*). A plate forged or pressed into a dishlike shape to increase its stiffness when subjected to pressure on the convex side.

dish wheel (*Eng.*). See **cup wheel.**

disincrustant (*Eng.*). See **anti-incrustator.**

disinfectant (*Chem.*). Any preparation that destroys the causes of infection. The most powerful disinfectants are oxidizing agents and chlorinated phenols.

disinfection (*Med.*). The destruction of pathogenic bacteria, usually with an antiseptic chemical or *disinfectant* (q.v.).

disinfestation Generally, the destruction of infesting animals. (*Med.*) The destruction of insects, especially lice.

disintegrating mill (*Eng.*). A mill for reducing lump material to a granular product. It consists of fixed and rotating bars in close proximity, crushing being partly by direct impact and partly by interparticulate attrition. (*Min. Proc.*) One which applies shredding action as a rotating disk, armed with steel pins. See also **beater mill.**

disintegration (*Nuc.*). A process in which a nucleus ejects one or more particles, applied especially, but not only, to spontaneous radioactive decay.

disintegration constant (*Nuc.*). The probability of radioactive decay of a given unstable nucleus per unit time. Statistically, it is the constant λ, expressing the exponential decay $\exp(-\lambda t)$ of activity of a quantity of this isotope with time. It is also the reciprocal of the mean life of an unstable nucleus. Also called **decay constant, transformation constant.**

disintegration energy (*Nuc.*). See **alpha-decay energy.**

disintegration of filament (*Elec. Eng.*). The gradual breaking up of the filament of an electric filament lamp, owing to projection of particles from the filament; the particles adhere to the inner surface of the bulb, causing blackening.

disjunct (*Zool.*). Having deep constrictions between the different tagmata of the body.

disjunction (*Cyt.*). The separation during meiosis (*anaphase*) of the two members of each pair of homologous chromosomes.

disjunctive symbiosis (*Zool.*). A mutually beneficial partnership between 2 animals, in which there is no direct connexion between the partners.

disjunctor (*Bot.*). A portion of wall material forming a link between the successive conidia in a chain, and serving as a weak place where separation may occur. Also called **bridge.**

disk or disc (*Acous.*). See **disk record.** (*Bot.*) (1) An outgrowth from the receptacle of the flower, arising beneath the carpels or stamens, and often secreting honey. (2) The central part of a capitulum. (3) The portion of an apothecium that bears the asci and paraphyses. (*Zool.*) Any flattened circular structure.

disk-and-drum turbine (*Eng.*). A steam-turbine comprising a high-pressure impulse wheel, followed by intermediate and low-pressure reaction blading, mounted on a drum-shaped rotor. Also called **combination turbine** or **impulse-reaction turbine.**

disk anode (*Electronics*). Final anode in an electrostatically focused cathode-ray tube, usually a circular plate containing a central aperture through which the beam passes. See **ultor.**

disk area (*Aero.*). The area of the circle described by the tips of the blades of a rotorcraft; similarly applied to propulsive airscrews.

disk armature (*Elec. Eng.*). One for a motor or generator wound to a large diameter on a short axial length.

disk brakes (*Aero., Autos.*). Type in which two or more pads close by calliper action on to a disk which is connected rigidly to the landing or car wheel-hub; more efficient than drum type, owing to greater heat dissipation.

disk capacitor (*Elec. Eng.*). One in which the variation in capacitance is effected by the relative axial motion of disks.

disk centrifuge (*Powder Tech.*). Apparatus for particle size analysis in which particles are sedimented in a rotating disk.

disk (or plate) clutch (*Eng.*). A friction clutch in which the driving and driven members have flat circular or annular friction surfaces, and consist of either one or a number of disks, running either dry or lubricated. See **single-plate clutch, multiple-disk clutch.**

disk coulter (*Agric.*). A saucer-shaped steel disk; used on a drill to make a shallow trench for the seed.

disk discharger (*Elec. Eng.*). See **rotary spark gap.**

disk filter (*Met., Min. Proc.*). American filter. Continuous heavy-duty vacuum filter in which separating membranes are disks, each revolving slowly through its separate compartment.

disk floret (*Bot.*). One of the regular tubular flowers occupying the central part of a capitulum when this contains 2 kinds of flowers.

disk friction (*Eng.*). The force resisting the rotation of a disk in a fluid. It is of importance in the design of centrifugal machinery as it decreases efficiency and causes a rise in the temperature of the fluid being pumped.

disk harrow (*Agric.*). A harrow consisting of saucer-shaped steel disks mounted on two axles. See **harrow.**

disking (*Aero.*). See **ground fine pitch.**

disk inking (*Print.*). The least efficient arrangement for ink distribution, only suitable for the light *platen machine* (q.v.). Ink is supplied to a slowly rotating disk from which it is picked up by the rollers.

disk loading (*Aero.*). The lift, or upward thrust, of a rotor divided by the disk area.

disk pile (*Civ. Eng.*). A hollow pile having a wide flange at the foot with projecting radial ribs on it; used for piling in sand.

disk plough (or **plow**) (*Agric.*). A plough which cuts a furrow slice by means of a sharp-edged steel disk, of saucerlike shape, set obliquely to the ground surface. See **mouldboard plough.**

disk prism (*Light*). A disk scanner consisting of a glass disk ground in such a way that its periphery forms a series of prisms, so that a ray of light passing through it is deflected through an angle dependent on the position of the disk.

disk record (*Acous.*). The normal type of gramophone record, made of synthetic thermoplastic resin, in which the reproducing needle follows a spiral groove, while the record is rotated at constant speed; devised by Berliner.

disk recording (*Acous.*). The system of recording on circular slabs of lacquer for subsequent disk pressing, as contrasted with recording on cylindrical waxes.

disk ruling (*Bind.*). Method in which the ink is applied to the paper by disks instead of pens, permitting higher speeds on the ruling machine.

disk scanner (*TV*). Rotating disk carrying apertures, lenses, prisms, or other picture-scanning elements used in mechanical scanning systems.

disk-seal tube (*Electronics*). Same as *megatron* (q.v.). The electrodes are annular disks or thick cylinders to reduce lead inductances and inter-electrode coupling.

disk separator (*Eng.*). A machine for separating grain or seeds of different sizes; used in flour-milling. The grain passes between revolving iron disks on a horizontal shaft, their faces carrying numerous small indents which remove the unwanted seed and discharge it at a higher level.

disk valve (*Eng.*). A form of suction and delivery valve used in pumps and compressors; it consists of a light steel or fabric disk resting on a ported flat seating; steel valve disks are usually spring-loaded.

disk wheel (*Eng.*). A wheel in which hub and rim are connected by a solid disk of metal instead of by separate spokes.

disk winding (*Elec. Eng.*). A type of winding, used

for medium and large transformers, in which the turns are made up into a number of annular disks.

dislocation (*Crystal.*). A lattice imperfection in a crystal structure, classified according to type, e.g., edge dislocation, screw dislocation. (*Surg.*) The displacement of one part from another; especially, an abnormal separation of two bones at a joint.

dislocation of meromes (*Zool.*). The passage of a merome from its proper somite to another somite; as the forward movement of the pelvic fins in some Fish.

disomic (*Cyt.*). Relating to two homologous chromosomes or genes.

disoperation (*Ecol.*). A coaction which has deleterious effects on the organisms concerned, either through influence on the habitat, or on contiguous group members, or both. Cf. *cooperation*.

disorientation (*Psychol.*). A mental state in which there is inability to judge the proper relations between events in time and space. Commonly occurs in various physical illnesses, as a result of delirium, etc.

dispatcher (*Comp.*). The section of a digital computer performing the switching operations which determine the sources and destinations for the transfer of words.

dispensable circuit (*Elec. Eng.*). A separate circuit used in a wiring system to which is connected apparatus that can be cut out of circuit at times of heavy load.

dispensary (*Med.*). A place where drugs, etc., are dispensed (i.e., prepared for administration).

dispenser cathode (*Electronics*). One which is not coated, but is continuously supplied with suitable emissive material from a separate electrode element.

dispermy (*Zool.*). Penetration of an ovum by two spermatozoa.

dispersal (*Ecol.*). The active or passive movement of individual plants or animals or their disseminules (seeds, spores, larvae, etc.) into or out of a population or population area. It includes emigration, immigration and migration. Should not be confused with *dispersion* (q.v.).

dispersed phase (*Chem.*). A substance in the colloidal state.

dispersion (*Ecol.*). The internal distribution pattern of the individuals in an animal or plant population, this being random, uniform (more regular than random), or clumped (see **aggregation**). Should not be confused with *dispersal* (q.v.) or with *distribution* (q.v.), which refers to the species as a whole, although the dispersion of a population can be described as following a random, or *Poisson, distribution* (q.v.). (*Light*) Variation of the refractive index of a substance with wavelength (or colour) of the light. It is on account of its dispersion that a prism is able to form a spectrum. For most media, the refractive index increases as the wavelength decreases. See **Cornu-Hartman formula, Cauchy's dispersion formula, anomalous dispersion.** (*Stats.*) General term indicating spread or scatter of a set of values or observations, often about some central value, measured usually by the standard deviation and also by mean deviation or mean difference.

dispersion coefficient (*Elec. Eng.*). A term often used to denote the leakage factor of an induction motor.

dispersion curve (*Light*). A curve obtained by plotting the deviation of light produced by a prism against the wavelength of the light.

dispersion forces (*Chem.*). Weak intermolecular forces, corresponding to the term *a* in *van der Waals' equation*. See **London Forces.**

dispersion hardening (*Met.*). See **precipitation hardening.**

dispersion medium (*Chem.*). A substance in which another is colloidally dispersed.

dispersion parameter (*Powder Tech.*). A term used in specifying a powder. It is defined by the equation:

$$D=\frac{100(D_{75}-D_{25})}{P}$$

where D = dispersion parameter

D_{75} = particle size at the 75 percentile of the cumulative percentage undersize distribution by weight

D_{25} = particle size at the 25 percentile of the cumulative percentage undersize distribution by weight

P = particle-size parameter.

dispersion photometer (*Light*). A photometer for measuring the luminous intensity of strong sources of light. The light from the source is reduced in intensity by means of lenses before being compared with the standard.

dispersive medium (*Phys.*). One in which the phase velocity is a function of frequency.

dispersive power (*Light*). The ratio of the difference in the refractive indices of a medium for the red and violet to the mean refractive index diminished by unity. This may be written

$$\nu=\frac{n_V-n_R}{n-1}.$$

dispersivity quotient *Optics*). The variation of refractive index n with wavelength λ, $dn/d\lambda$.

dispersol colours (*Chem.*). Acetate (ethanoate)-silk dyestuffs, consisting of a paste or suspension of unsulphonated azo-compounds applied directly to the fabric.

dispireme (*Cyt.*). The stage of telophase in which the spireme thread of each daughter nucleus has been formed.

displacement (*Aero.*). The mass of the air displaced by the volume of gas in any lighter-than-air craft. (*Elec.*) Vector representing the electric flux in a medium and given by the product of the absolute permittivity ε and the field strength E. In unrationalized units, $D = 1/4\pi \times \varepsilon E$. Also called **dielectric strain, electric flux density.** (*Eng.*) (1) The volume of fluid displaced by a pump plunger per stroke or per unit time. (2) The swept volume of a working cylinder. (*Hyd. Eng.*) The weight of water displaced by a vessel. It is equal to the total weight of the vessel and contents. See **Archimedes' principle.** (*Psychol.*) A mechanism commonly observed in dreams, whereby a hidden element may be replaced by something more remote; or the accent may be shifted from an important to an unimportant element, the affect also being transferred from one idea to another. See also **distortion.**

displacement activity (*An. Behav.*). The performance of a behaviour pattern out of its normal functional context of behaviour. An example of *allochthonous behaviour* (q.v.).

displacement current (*Elec.*). Integral of the displacement current density through a surface. The time rate of change of the electric flux, D. Current postulated in a dielectric when electric stress or potential gradient is varied. Distinguished from a normal or conduction current in that it is not accompanied by motion of current carriers in the dielectric. Concept introduced by Maxwell for the completion of his electromagnetic equations.

displacement flux (*Elec.*). Integral of the normal component of displacement over any surface in a dielectric. See **displacement.**

displacement law (*Nuc.*). Soddy and Fajans formulation that radiation of an α-particle *reduces* the atomic number by 2 and the mass number by 4, and that radiation of a β-particle *increases* the atomic number by 1, but does not change the mass number. It was later found that emission of a positron *decreases* the atomic number by one, but does not change the mass number. Gamma emission and isomeric transition change neither mass nor atomic number. Displacement laws are summarized as follows:

Type of disintegration	Change in atomic number	Change in mass number
alpha emission	−2	−4
beta electron emission	+1	0
,, positron emission	−1	0
,, electron capture	−1	0
isomeric transition	0	0
gamma emission	0	0

A change in atomic number means displacement in the periodic classification of the chemical elements; a change in mass number determines the radioactive series.

displacement pump (*Mining*). One with pulsing action, produced by steam, compressed air or a plunger, causing non-return valves to prevent return flow of displaced liquid during the retracting phase of the pump cycle.

displacement series (*Chem.*). See **electrochemical series.**

display (*Electronics*). Mode of showing information on a cathode-ray tube screen, especially in radar and navigation. See **scope.**

display behaviour (*An. Behav.*). Movements, postures, and sounds, generally of a conventionalized kind, which have the capacity to initiate specific responses in other animals, especially members of the same species.

display panel (*Teleph.*). The panel, with illuminated numbers, in front of a manual operator who completes connexions arising in an automatic system. The displayed numbers are selected by marginal currents translated from impulse trains by a coder.

display work (*Typog.*). Displayed typesetting (such as title-pages, jobbing work, advertisements), distinguished from solid text composition.

disposable load (*Aero.*). (1) Civil aircraft: crew, fuel, oil, and payload. (2) Military aircraft: fuel, oil and armament.

disposition (*Psychol.*). The mental constitution of an individual, as formed by his reactions to experience and environment.

disruptive discharge (*Elec. Eng.*). That of a capacitor when the discharge arises from breakdown (puncture) of the dielectric by an electric field strength which it cannot withstand.

disruptive strength (*Elec. Eng.*). See **electric strength.**

disruptive voltage (*Elec. Eng.*). That which is just sufficient to puncture the dielectric of a capacitor. A test voltage is normally applied for one minute. See also **breakdown voltage.**

dissected (*Bot.*). Cut deeply into many narrow lobes or leaflets.

dissecting (*Typog.*). The removal of type matter which is to be printed in a second colour, in order to impose it in another chase, position and spacing being carefully regulated.

dissector tube (*Electronics*). One in which the electron pattern of emission from a continuous photoemissive surface is scanned over an aperture which creates the video signal.

disseminate (*Bot.*). Scattered.

disseminated sclerosis (*Med.*). A chronic progressive disease in which patches of thickening appear throughout the central nervous system, resulting in various forms of paralysis; also called **multiple sclerosis.**

disseminated values (*Mining*). Mode of occurrence in which small specks of concentrate are scattered evenly through the gangue mineral.

dissemination (*Bot., Zool.*). The spread or migration of species in plants, usually by means of spores or seeds.

disseminule (*Bot.*). Any part of a plant which serves for the dissemination of the plant.

dissepiment (*Bot.*). (1) A wall dividing the loculi of a syncarpous ovary. (2) See **trama.** (*Zool.*) In *Hydrocorallinae* and certain Corals, an imperfect horizontal calcareous partition stretching between the septa and partially shutting off the lower part of the polyp cup. Cf. *tabula*.

dissimilar terms (*Maths.*). Terms containing different powers of the same variable(s), or containing different variables, e.g., xy, $(xy)^2$; $3x$, $3z$.

dissipation (*Electronics*). Power released from any electrode, depending on electron collection, radiation or conduction of heat received, or ion bombardment, controlled by graphite, fins, or water-cooling. (*Radio*) Loss or diminution of power, leading to flat tuning for steady-state oscillations, and damping in free oscillations. Removes sharpness of cut-off in filters. Dependent on resistance, measured at high frequency, including radiation. It is the dissipation of eddy-current energy in conductors and of displacement polarization energy in dielectrics that makes industrial high-frequency heating possible.

dissipation factor (*Elec. Eng.*). The cotangent of the phase angle (δ) for an inductor or capacitor. For low loss components, the dissipation factor is approximately equal to the power factor,

$\tan\delta \simeq \frac{\sigma}{\omega\varepsilon}$, for a low-loss dielectric, where σ is the electrical conductivity, ε the permittivity of the medium, and ω is $2\pi \times$ frequency.

dissipationless line (*Telecomm.*). A hypothetical transmission line in which there is no energy loss. Also **lossless line.**

dissipation trails (*Meteor.*). Lanes of clear atmosphere formed by the passage of an aircraft through a cloud.

dissipative network (*Telecomm.*). One designed to absorb power, as contrasted with networks which attenuate power by impedance reflection. All networks dissipate to some slight extent, because neither capacitors nor inductors can be made entirely loss-free.

dissociation (*Chem.*). The reversible or temporary breaking-down of a molecule into simpler molecules or atoms. See **Arrhenius theory of-.** (*Psychol.*) A state of temporary loosening of control over consciousness, in which unconscious complexes take control of the personality. In extreme cases, this amounts to a splitting of the personality.

dissociation constant (*Chem.*). The ratio of the product of the active masses of the molecules resulting from dissociation to the active mass of the undissociated molecules, when equilibrium is reached.

dissociation curve (*Zool.*). See **respiratory pigments.**

dissociation of gases (*Heat*). Chemical combustion reaction occurring at the highest temperature of the flame where carbon dioxide and water vapour tend to dissociate into carbon monoxide, hydrogen and oxygen respectively. In high-temperature furnaces the dissociation of gases in this way forms a scaling element which cannot be directly traced in analyses of furnace atmospheres or flue gases.

dissociation of ideas (*Psychol.*). A term used by Janet to denote the splitting off from consciousness of certain ideas with their accompanying emotions; similar to *repression* (q.v.)

dissoconch (*Zool.*). The shell of a veliger larva.

dissogeny (*Zool.*). The condition of having two periods of sexual maturity, one during the larval stage and one during the adult stage.

dissolution (*Chem.*). The taking up of a substance by a liquid, with the formation of a homogeneous solution.

dissolving pulp (*Paper*). See **alpha pulp.**

dissonance (*Acous.*). The playing of two or more musical tones simultaneously to produce an unpleasant effect on a listener. (*Light*) The formation of maxima and minima by the superposition of two sets of interference fringes from light of two different wavelengths. (*Psychol.*) Cognitive dissonance occurs when two mental elements are not in accord with each other. The state is said to have motivating properties inducing the individual to change the situation, his knowledge of it, etc.

dissymmetrical. See **asymmetrical.**

dist. (*Build.*). Abbrev. for (1) *distemper*, (2) distributed.

distal (*Bio .*). Far apart, widely spaced: pertaining to or situated at the outer end: farthest from the point of attachment. Cf. *proximal.*

distalia (*Zool.*). The 5 bones belonging to the third row of carpals or tarsals in the generalized tetrapod limb.

distance (*Maths.*). (1) The length of line joining two points. (2) Angular, between two points: the angle between the two lines from the point of reference to the points in question. (3) The length of the segment between two lines or two planes of the line perpendicular to both lines or both planes. (4) Distance from a line or plane to a point: the length of the perpendicular from the point to the line or the plane.

distance block, distance piece (*Build., Civ. Eng.*). A wooden or other block serving to separate 2 pieces by a desired distance.

distance control (*Elec. Eng.*). See **remote control.**

distance mark (*Electronics*). Mark on the screen of a CRT to denote distance of target.

distance-measuring equipment (*Aero.*). Airborne secondary-radar which indicates distance from a ground transponder beacon. Abbrev. **DME.**

distance meter (*Photog.*). Same as **split-image rangefinder.**

distance piece (*Civ. Eng.*). See **distance block.**

distance protection (*Elec. Eng.*). See **impedance protective system.**

distance receptors (*Zool.*). Exteroceptors which can perceive objects at a distance.

distance relay (*Elec. Eng.*). See **impedance relay.**

distant (*Bot.*). Widely-spaced.

distant-reading compass (*Instr.*). Gyro flux-gate compass in which the indicator is remote from the sensing device.

distant-reading instrument (*Eng.*). A recording or indicating instrument (such as a thermometer or pressure gauge) in which the reading is shown on a scale at some distance from the point of measurement. See **remote control.**

distemper (*Paint.*). A mixture of a dry pigment with size, water and sometimes oil, used as a paint for internal walls and ceilings.

disthene (*Min.*). A less commonly used name for the mineral *kyanite* (q.v.). The name was applied by Haüy on account of the difference in electrical charge developed in different crystals when rubbed, but also because of the differences in hardness, even on the same crystal face, when tested in different directions.

distichalia (*Zool.*). See **dicostalia.**

distichiasis, distichia (*Med.*). A condition in which there are two complete rows of eyelashes in one or both eyelids.

distichous (*Bot.*). Arranged in two opposite vertical rows.

distillation (*Chem.*). A process of evaporation and re-condensation used for separating liquids into various fractions according to their boiling points or boiling ranges. See also **molecular-.**

distillation flask (*Chem.*). A laboratory apparatus, usually made of glass; it consists of a bulb with a neck for the insertion of a thermometer and a side tube attached to the neck, through which the vapours pass into a condenser.

distinct (*Bot.*). (1) Said of a species which has strongly marked characters. (2) Said of plant members which are quite free from one another.

distomiasis or distomatosis (*Vet.*). Infection of the bile ducts by flukes or trematode worms.

distorted wave (*Elec. Eng.*). A term often used in electrical engineering to denote a nonsinusoidal waveform of voltage or current.

distorting network (*Telecomm.*). A network altering the response of a part of a system, and anticipating the correction of response required to restore a signal waveform before actual distortion has occurred, e.g., owing to the inevitable frequency distortion in a line, or to minimize noise interference.

distortion (*Photog.*). Any departure from the criteria for the perfect formation of an image by a lens. (*Psychol.*) A feature of dreamlife whereby, mainly, elements are distorted to make the real meaning of the dream unrecognizable. A mechanism brought about by the censor to exclude painful elements from the dream. (*Telecomm.*) Change of waveform or spectral content of any wave or signal due to any cause.

distortion correction (*Teleg.*). An arrangement for restoring accurate shapes to telegraph signals after their distortion in passing through a circuit.

distortion factor (*Elec. Eng.*). Ratio of the r.m.s. harmonic content to the total r.m.s. value of the distorted sine wave.

distortionless line (*Elec.*). One with constants such that the attenuation (a minimum value) and the delay time are constant in magnitude with variation in frequency. The characteristic impedance is purely resistive. For such a line, $L \cdot G = R \cdot C$, where R=resistance, G=leakance, L=inductance and C=capacitance, all being distributed values per unit length. Also called **distortionless condition.**

distortion of field (*Elec. Eng.*). A term commonly used in connexion with electric machines to denote the change in the distribution of flux in the air gap when the machine is put on load.

distraction display (*An. Behav.*). Behaviour, especially of some female birds, which is sometimes similar to that of an injured individual (hence called **injury-feigning**), generally a response to a predator threatening the eggs or young, and usually effective in diverting its attention.

distrails (*Meteor.*). See **dissipation trails.**

distribute (*Typog.*). To put individual letters and spaces back into their proper compartments in the case after use. Machine-set matter is usually melted down to be used again in the caster.

distributed amplifier (*Telecomm.*). Same as **transmission-line amplifier.**

distributed capacitance (*Elec.*). (1) That distributed along a transmission line, which, with distributed resistance and/or inductance, reduces the velocity of transmission of signals. (2) That between the separate parts of a coil, lowering its inductance; represented by an equivalent lumped capacitor across the terminals, giving the same frequency of resonance.

distributed constants (*Elec.*). Those applicable to real or artificial transmission lines and waveguides, because dimensions are comparable with the wavelength of transmitted energy.

distributed inductance (*Elec.*). Said of a circuit which has an inductance distributed uniformly along it, e.g., a power transmission line, a loaded telephone circuit, or a travelling-wave valve or tube.

distributed winding (*Elec. Eng.*). The winding of an electric machine which is spread uniformly over the stator or rotor surface.

distributing centre (*Elec. Eng.*). In an electric power system, a point at which an incoming supply from a feeder is split up amongst a number of other feeders or distributors.

distributing main (*Elec. Eng.*). See **distributor.**

distributing point (*Elec. Eng.*). See **feeding point.**

distribution (*Ecol.*). The occurrence of a species, considered from a geographical point of view, or with reference to altitude or other factors. Sometimes used as equivalent to *dispersal* (q.v.). Should not be confused with *dispersion* (q.v.), which refers to individuals. (*I.C. Engs.*) The provision of the same quantity and quality of petrol-air mixture to each of the cylinders of a multi-cylinder engine by the carburettor and induction manifold. (*Stats.*) In a population, finite or infinite (more usefully infinite), it is the relative frequency with which a variable quantity or variate assumes particular values. Also called **frequency distribution.** Some distributions are *binomial*, *normal*, and *Poisson* (qq.v.). Cf. *distribution function.*

distribution board (*Elec. Eng.*). An insulating panel carrying terminals and/or fuses, for the distribution of power supplies to repeaters or telegraph circuits.

distribution cable (*Elec. Eng.*). A communication cable extending from a feeder cable into a defined service area.

distribution coefficients (*Phys.*). Chromaticity coordinates for spectral (monochromatic) radiations of equal power, i.e., for the component radiations forming an equal energy spectrum.

distribution factor (*Elec. Eng.*). A factor used in the calculation of the e.m.f. generated in the winding of an a.c. machine, to allow for the fact that the e.m.fs. in each of the individual coils are not in phase with one another. Also called **breadth coefficient, breadth factor.** (*Radiol.*) A *modifying factor* (q.v.) used in calculating biological radiation doses, which allows for the nonuniform distribution of an internally-absorbed radioisotope.

distribution frame (*Teleph.*). A structure with large numbers of terminals, for arranging circuits in specified orders.

distribution function (*Stats.*). If, in a frequency distribution, the variate is discontinuous and takes discrete values x_i, and if the relative frequency of x_i is represented as a function $f(x_i)$ of x_i, the cumulative frequency up to a particular value x_p is given by $F(x_p) = \sum_{i=1}^{p} f(x_i)$. If the variate is continuous, then the cumulative frequency up to x_p is $F(x_p) = \int_{-\infty}^{x_p} f(x)dx$. $f(x)$ is called the *frequency function* (or the *probability density function*). $F(x)$ is the *distribution function* or *cumulative frequency function* or *cumulative probability function* or *probability distribution function.* See **distribution.**

distribution fuse-board (*Elec. Eng.*). A distribution board having fuses in each of the separate circuits.

distribution law (*Chem.*). The total energy in a given assembly of molecules is not distributed equally, but the number of molecules having an energy different from the median decreases as the energy difference increases, according to a statistical law.

distribution pillar (*Elec. Eng.*). A structure in the form of a pillar, containing switches, fuses, etc., for interconnecting the distributing mains of an electric power system.

distribution reservoir (*Hyd. Eng.*). See **service reservoir.**

distribution switchboard (*Elec. Eng.*). A distribution board having a switch in each of the branch circuits.

distributive (*Maths.*). An operation is distributive if the result of applying it to a sum of terms equals the sum of the results of applying it to the terms individually.

distributor (*Elec. Eng.*). The cable or overhead line forming that part of an electric distribution system to which the consumers' circuits are connected. Also called a **distributing main.** (*I.C. Engs.*) A device, driven through the cam-shaft, whereby H.T. current is transmitted in correct sequence to the sparking plugs. (*Teleg.*) An arrangement, generally by a rotating arm with a number of brush contacts, for allocating the line to a number of channels of telegraphic communication so that they each have it for transmitting a signal during a specified fraction of time.

distributor rollers (*Typog.*). In a printing press, the rollers which distribute the ink, as distinct from the inking rollers which supply ink to the forme or plate.

district (*Mining*). An underground section of a coal mine served by its own roads and ventilation ways: a section of a coal mine.

district heating (*Civ. Eng.*). The distribution of energy for heating purposes, usually in the form of steam, from a central boiler house to domestic and commercial premises in the vicinity.

district road (*Civ. Eng.*). One maintained by the district authority. See **county road, trunk road.**

distrix (*Med.*). Splitting of the ends of hairs.

disturbance (*Telecomm.*). Any signal originating from a source other than the wanted transmitter, e.g., atmospherics, unwanted stations, valve noise in the receiver.

disul (*Chem.*). See **2,4-DES.**

disulphide of carbon (*Chem.*). See **carbon disulphide.**

disuse atrophy (*Med.*). Wasting of a part as a result of diminution or cessation of functional activity.

ditch canal (*Hyd. Eng.*). See **level canal.**

ditching (*Aero.*). Emergency alighting of a landplane on water.

diterpenes (*Chem.*). Unsaturated hydrocarbons of the general formula $C_{20}H_{32}$, constituted of two *terpene* (q.v.) molecules.

dithallic (*Bot.*). Said of a mycelium formed by the union of material from two distinct strains.

dithecal (*Bot.*). Said of an anther with two lobes, the usual number.

dither (*Elec. Eng.*). Small continuous signal supplied to servomotor operating hydraulic valve or similar device, and producing a continuous mechanical vibration which prevents sticking.

dithio (*Chem.*). See **sulfotep.**

dithionic acid (*Chem.*). *Hyposulphurous acid* (q.v.).

dithiophosphates (*Min. Proc.*). Chemical basis of important series of collector agents used in froth flotation, marketed as *Aerofloats.*

dithioTEPP (*Chem.*). See **sulfotep.**

dithizone (*Chem.*). Diphenylthiocarbazone:

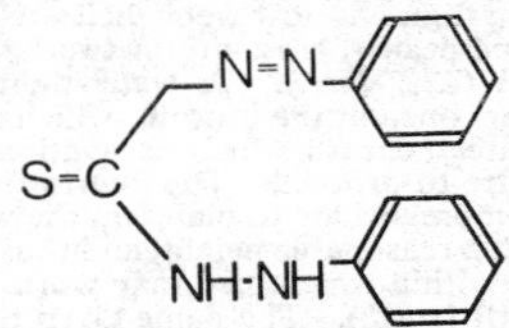

Reagent used in the determination of lead.

Ditisheim balance (*Horol.*). A bimetallic temperature compensation device.

ditrematous (*Zool.*). Of hermaphrodite animals, having the male and female openings separate: of unisexual forms, having the genital opening separate from the anus.

ditrochous (*Zool.*). Having the trochanter divided.

ditróite (*Geol.*). A coarse-grained deep-seated rock, falling in the alkali-syenite subdivision; named from the type locality of Ditró, Transylvania. Consists essentially of alkali-feldspar, together with nepheline (elaeolite), sodalite, and a (usually) small content of soda-amphiboles and/or pyroxenes.

Dittrich's plugs (*Med.*). Small, yellow, foetid plugs of secretion present in the sputum and in the bronchi, in chronic suppurative bronchitis.

Dittus-Boelter equation (*Heat*). An equation for the transfer of heat from tubes to viscous fluids flowing through them.

$$u \propto c\left(\frac{k\Delta s}{d}\right)^{1/3}\left(\frac{vd}{u/s}\right)^{1/12},$$

where u=film transfer factor, k=thermal conductivity, Δ=logarithmic mean temp. difference between tube and liquid, d=thickness of fluid stream, s = relative density of fluid, c = specific heat capacity of fluid, v = mean velocity of fluid in tube, and u/s = kinematic viscosity of fluid.

diuresis (*Med.*). The excretion of urine, especially in excess.

diuretic (*Med.*). Producing diuresis; a drug which does this.

diurnal. During a day. The term is used in astronomy and meteorology to indicate the variations of an element during an average day.

diurnal libration (*Astron.*). The name given to the phenomenon by which, owing to the finite dimensions of both the earth and the moon, an observer can see rather more than half the moon's surface when his observations at different times or from different places on the earth are combined. The effect is one of *parallax* (q.v.), the term *libration* being a misnomer in this case.

diurnal parallax (*Astron.*). See **geocentric parallax.**

diurnal range (*Meteor.*). The extent of the changes which occur during a day in a meteorological element such as atmospheric pressure or temperature.

diurnal rhythm (*An. Behav.*). See **circadian rhythm.**

diurnal variation (*Phys.*). A variation of the earth's magnetic field as observed at a fixed station, which has a period of approximately 24 hours.

diuron (*Chem.*). *N'*-(3,4-dichlorophenyl)-*NN*-dimethylurea, used as a weedkiller.

divalent (*Chem.*). Capable of combining with 2 atoms of hydrogen or their equivalent.

divaricate (*Bot., Zool.*). Spreading widely apart, forked, divergent.

divaricators (*Zool.*). In *Brachiopoda*, muscles passing from the cardinal process to the ventral valve.

dive (*Aero.*). A steep descent, the nose of the aircraft being down. Cf. *spin.*

diver (*Min. Proc.*) Small plummet adjusted to a desired relative density, so that it indicates the density of the fluid in which it is immersed by its up-and-down motion.

dive-recovery flap (*Aero.*). An *air brake* (q.v.) in the form of a flap to reduce the *limiting velocity* of an aeroplane.

divergence (*Aero.*). In aircraft stability, a disturbance that increases without oscillation; *lateral divergence* leads to a spin or an accelerating spiral descent; *longitudinal divergence* causes a nosedive, or a stall. (*Elec.*) The limiting value of the *electric flux* (q.v.), which emanates from a closed surface, divided by the volume of the closed surface when this becomes vanishingly small. (*Meteor.*) The opposite of *convergence* (q.v.). Divergence is usually accompanied by fine dry weather. (*Nuc.*) Initiation of a chain-reaction in a reactor, in which slightly more neutrons are released than are absorbed and lost. The rate and extent of the divergence are normally controlled by neutron absorbing rods, e.g., of cadmium or hafnium.

divergence angle (*Electronics*). Angle of spread of electron beam, arising from mutual repulsion or debunching.

divergence of a vector (*Maths.*). The scalar product $\nabla . V$ of the vector V with the vector operator ∇ (del). Written div V.

divergence speed (*Aero.*). The lowest *equivalent air speed* (q.v.) at which *aeroelastic divergence* (q.v.) can occur.

divergent (*Bot.*). Said of two or more organs which gradually spread so that they are farther apart at their tips than at their bases. (*Nuc.*) Term applied to reactor or critical experiment when multiplication constant exceeds unity.

divergent adaptation (*Zool.*). See **adaptive radiation.**

divergent lens (*Light*). A lens which increases the divergence, or diminishes the convergence, of a beam of light passing through it. Such a lens will be double concave, plano-concave, or convexo-concave, the concave surface having the smaller radius of curvature.

divergent nozzle (*Eng.*). A nozzle whose cross-section increases continuously from entry to exit; used, e.g., in compound impulse turbines.

divergent sequence (*Maths.*). A sequence which does not converge to a finite limit.

divergent series (*Maths.*). A series that does not converge to a finite limit.

divergent strabismus (*Med.*). Squint in which the eyes diverge from each other.
divergent thinking (*Psychol.*). Thinking which is productive, tends to go beyond conventional categories, is original and even profuse with no signs of derangement. It is not taken into account in standard intelligence tests, and may be a mark of *creativity*. Cf. *convergent thinking*.
diversion cut (*Civ. Eng.*). See **bye channel.**
diversity antenna (*Radio*). The antenna system of a diversity receiver.
diversity factor (*Elec. Eng.*). The ratio of the arithmetical sum of the individual maximum demands of a number of consumers connected to an electric supply system, to the simultaneous maximum demand of the group.
diversity ratio (*Light*). The ratio of the maximum illumination on a given plane to the minimum illumination on that plane.
diversity reception (*Radio*). System designed to reduce fading; several antennae, each connected to its own receiver, are spaced several wavelengths apart from one another, the demodulated outputs of the receivers being combined. Alternative systems use antennae orientated for oppositely polarized waves (*polarized diversity*), or independent transmission channels on neighbouring frequencies (*frequency diversity*).
divers' paralysis (*Med.*). See **caisson disease.**
diverter (*Elec. Eng.*). A low resistance connected in parallel with the series winding or the compole winding of a d.c. machine in order to divert some of the current from it, thereby varying the m.m.f. produced by the winding.
diverter relay (*Elec. Eng.*). A relay employed with certain excess-current protective systems; it increases the stability of the protective system by putting resistance in parallel with the tripping relay in the case of a heavy fault.
diverticulitis (*Med.*). Inflammation of diverticula in the colon.
diverticulosis (*Med.*). The presence of diverticula in the colon.
diverticulum (*Anat.*, *Zool.*). (1) Saccular dilatation of a cavity or channel of the body. (2) Lateral outgrowth of the lumen of an organ. (3) Pouchlike protrusion of the mucous membrane of the colon through the weakened muscular wall. (4) A pouchlike side branch on the mycelium of some fungi. *pl.* **diverticula.** See also **Meckel's diverticulum.**
divertor (*Nuc. Eng.*). Trap used in thermonuclear device to divert impurity atoms from entering plasma, and fusion products from striking walls of chamber.
divided (*Bot.*). See **dissected.**
divided bearing (*Eng.*). See **split bearing.**
divided circuit (*Teleg.*). A telegraph circuit extending to more than one receiving station, each receiving the transmitted signals.
divided pitch (*Eng.*). The axial distance between corresponding points on successive threads of a multiple-threaded screw.
divided touch (*Mag.*). The magnetizing of a steel bar by stroking it with the opposite poles of two permanent magnets, these being drawn apart from the centre of the bar to the ends.
divided winding (*Elec. Eng.*). A term proposed for that class of windings (for d.c. machines) usually called multiple or multiplex, in which there are two or more separate windings on the armature, joined in parallel by the brushes.
dividend (*Maths.*). See **division.**
divider (*Elec. Eng.*). Circuit which reduces by an integral factor the number of pulses or alternations per second passing through it by suppressing the unwanted ones.
dividers (*Instr.*). Compasses used only for measuring or transferring distances, and not for describing arcs. (*Mining*) The cross-timbers in a rectangular timbered shaft, other than the end timbers or end plates.
dividing box (*Elec. Eng.*). A box for bringing out separately the cores of a multi-core cable. The insulation of the cable is hermetically sealed and the cores may be brought out either as bare or insulated conductors. See **bifurcating box, trifurcating box.**
dividing engine (*Eng.*). An instrument for marking or engraving accurate subdivisions on scales; it consists of a carriage adjusted by a micrometer screw and holding a marking tool.
dividing fillet (*Elec. Eng.*). See **barrier.**
dividing head (*Eng.*). See **indexing head.**
dividing network (*Acous.*). A frequency-selective network which arranges for the input to be fed into the appropriate loudspeakers, usually two, covering high and low frequencies respectively, Also **loudspeaker, cross-over network.**
diving-bell (*Civ. Eng.*). A water-tight working chamber, open at the bottom, which is lowered into water beneath which excavation or other works are to proceed. The interior is supplied with compressed air to maintain the water-level inside at a reasonable height, and thus leave free a space within which men may work.
divinity calf (*Bind.*). The name given to bindings in dark brown calfskin, with blind tooling; used chiefly for theological works.
division (*Maths.*). (1) For numbers, the operation of ascertaining how many times one number, the *divisor*, is contained in a second, the *dividend*. The result is called the *quotient*, and, if the divisor is not contained an integral number of times in the dividend, any number left over is called the *remainder*. Indicated either by the division sign, ÷, or by a stroke or bar, in which case the expression as a whole is called a *fraction* and the dividend and divisor the *numerator* and *denominator* respectively. Fractions less than 1 are called *common* or *proper* or *vulgar fractions*, and those greater than 1, *improper fractions*. Colloquially, however, a *fraction* is less than 1. (2) For polynomials and other mathematical entities, the inverse operation to multiplication. Where appropriate, nomenclature analogous to that outlined above is used.
division plate (*Eng.*). A plate used for positioning the plunger of an indexing head; provided with several concentric rings of holes accurately dividing the circumference into various equal subdivisions.
division wall (*Build.*). A wall within a building, or serving two houses.
divisor (*Maths.*). See **division.**
divisural line (*Bot.*). The line along which the peristome teeth of a moss split.
dizoic (*Zool.*). Of spores, containing two sporozoites.
dl- (*Chem.*). Containing equimolecular amounts of the dextro-rotatory and the laevo-rotatory forms of a compound; racemic. Now usually written (±).
D-layer (*Radio*). The lowest layer or region of absorbing ionization, up to 90 km above the earth's surface, considered to exist as a consequence of particle radiation of hydrogen bursts from the sun. It brings about complete inhibition of short-wave communication but some improvement in long-wave communication. See also **Dellinger fade-out.**

d-levels (*Light*). See **diffuse series.**
D lines (*Light*). See [D].
D log E curve (*Photog.*). See **characteristic curve.**
D.M. (*San. Eng.*). Abbrev. for *disconnecting manhole*. See **interceptor.**
DMDT (*Chem.*). See **methoxychlor.**
DME (*Aero.*). Abbrev. for *distance-measuring equipment*.
D method or **operator** (*Maths.*). Used in determining the solution of a linear differential equation with constant coefficients. *D* represents *d*/*dx*, and, under certain conditions, it can be manipulated by some of the procedures of ordinary algebra.
DMF. Abbrev. for *dimethylformamide*.
DMTT (*Chem.*). See **dazomet.**
dn (*Maths.*). See **elliptic functions.**
dN (*Telecomm.*). Abbrev. for *decineper*.
DNA (*Chem.*). See **nucleic acids.**
DNAP (*Chem.*). See **dinosam.**
DNBP (*Chem.*). See **dinoseb.**
DNC (*Chem.*). See **DNOC.**
DNOC (*Chem.*). 2-Methyl-4,6-dinitro (1-hydroxybenzene), used as an insecticide and herbicide. Also **DNC, dinitrocresol.**
DNOCHP (*Chem.*). See **dinex.**
DNOSAP (*Chem.*). See **dinosam.**
DNOSBP (*Chem.*). See **dinoseb.**
DNSAP (*Chem.*). See **dinosam.**
DNSBP (*Chem.*). See **dinoseb.**
Doba's network (*Elec. Eng.*). Shaping circuit used in pulse amplifiers where rise times of a few nanoseconds are required.
dobbie or **dobby** (*Weaving*). Mechanism over the top or at the side of a loom. Operated by punched cards or pegged lags, it lifts and lowers the healds to move the warp threads in timed sequence to form the design in the cloth.
Dobie's line (*Zool.*). A dark line which subdivides transversely each of the light areas of a striated muscle fibre; believed to represent a membrane. See **Krause's membrane.**
dock (*Civ. Eng.*). An enclosed basin for the accommodation of shipping. (*For.*) To trim sawn timber or logs to a given length or to remove useless waste ends.
doctor (*Elec. Eng.*). A device used in electroplating for depositing metal on imperfectly plated parts; it consists of an anode of the metal to be deposited, covered with a spongy material saturated with the plating solution. (*Paper*) In a paper-making machine, a device consisting of blades which scrape off pulp, etc., from cylinders. (*Print.*) On a photogravure machine, a blade which wipes superfluous ink from the non-printing area and from the walls between the cells of the cylinder.
doctor test (*Chem.*). A test for sulphur in petroleum using a sodium (II) plumbate solution.
document copying (*Print.*). A variety of methods are in use based on *dye-line*, *dual spectrum*, electrostatic (*xerography*), *photocopying*, and *thermographic* principles, some able to produce lithographic masters for *small offset*.
dodecagon (*Maths.*). A 12-sided polygon.
dodecahedron (*Maths.*). A 12-sided polyhedron.
dodecyl benzene (*Chem.*). Important starting material for anionic detergents derived from petroleum. Based on the tetramer of propene.
dodging (*Photog.*). Manipulation of the light projected through a negative by an enlarger to lighten or darken selected parts of the resultant print.
dodine (*Chem.*). Dodecylguanidine, used as a fungicide.
Dodin rangefinder (*Photog.*). One in which pairs of prisms are located in the front and rear of the focal plane of part of the image, which is displaced laterally when out of focus.
doeskin (*Textiles*). A 5-end warp satin weave, overcoating cloth made from high quality merino yarn. It is shrunk considerably in finishing and a smooth level surface is imparted to it.
doffer (*Textiles*). (1) Young operative who, amongst other duties, removes full cops or other yarn packages from a machine. (2) A fully automatic machine which performs this operation mechanically. (3) Wire-covered cylinder on a cotton or wool card which strips the fibres from the surface of the main cylinder as a fine web; this is immediately converted into a sliver or 'rope' of fibres and deposited in a can.
doffer comb (*Textiles*). High-speed oscillating blade extending across the doffer of a card. It strips off the fine carded web of fibres to be passed through rollers and a trumpet to form a sliver in the can.
dog (*Build., Eng., etc.*). A steel securing-piece used for fastening together 2 timbers, as in the process of shoring, for which purpose it is hooked at each end at right angles to the length, so that the hooked ends may be driven into the timbers. The term is also applied to a great variety of gripping implements, viz.—a clutching attachment for withdrawing well-boring tools; a pawl; an adjustable stop used in machine tools; a spike for securing rails to sleepers; a lathe carrier; a circular clawed object to join members of a roof truss.
dog clutch (*Eng.*). A clutch consisting of opposed flanges, or male and female members, provided with two or more projections and slots, one member being slidable axially for engaging and disengaging the drive. See **clutch.**
dog down (*Ships*). To secure in position by pieces of bent-round iron, driven through holes in a cast-iron slab in such a manner as to be jammed.
dog-eared fold (*Plumb.*). See **dog's ear.**
doggers (*Geol.*). Flattened ovoid concretions, often of very large size, in some cases calcareous, in others ferruginous, occurring in sands or clays. They may be a metre or more in diameter.
dog-hole (*Glass*). See **throat.**
dog-house (*Glass*). A small extension of a glass furnace, into which the *batch* (q.v.) is fed. (*Radio*) Enclosure protecting feed equipment at base of large transmitting antenna.
dog-leg chisel (*Carp.*). See **bent chisel.**
dog-legged stair (*Build.*). A stair having successive flights rising in opposite directions, and arranged without a well-hole.
dog-nail (*Carp.*). A large nail having a head projecting over one side.
dog nose (*Med.*). See **goundou.**
dog's ear (*Plumb.*). The corner of a sheet-lead tray, formed with a folding joint.
dog's tooth (*Build.*). A string course in which bricks are so laid as to have one corner projecting.
dog-tooth spar (*Min.*). A form of calcite in which the scalenohedron is dominant but combined with prism, giving a sharply pointed crystal like a canine tooth.
Doherty amplifier (*Telecomm.*). One in which one section takes low signal amplitudes up to half-maximum, the other section coming in up to full power, the signal being divided and united with 90° phase-shift networks. The result is high efficiency and linearity.
Doherty transmitter (*Telecomm.*). One in which high efficiency of amplification of amplitude-modulated wave is obtained by two valves con-

nected to the load, one directly, the other through a 90° retarding network.

Dohlman's operation (*Surg.*). Diathermy excision of the septum between a pharyngeal pouch and the oesophagus.

dolabriform (*Bot.*). Shaped like a pick-axe.

doldrums (*Meteor.*). Regions of calm in equatorial oceans. Towards the solstices, these regions move about 5° from their mean positions, towards the north in June and towards the south in December.

dolerite (*Geol.*). The general name for basic igneous rocks of medium grain-size, occurring as minor intrusions or in the central parts of thick lava flows; much quarried for road metal. Typical dolerite consists of plagioclase near labradorite in composition, pyroxene, usually augite, and iron ore, usually ilmenite, together with their alteration products.

dolerophanite (*Min.*). A basic copper sulphate, crystallizing in the monoclinic system, occurring as small brown crystals in the fumarolic deposits of Mt Vesuvius.

Dolezalek quadrant electrometer (*Elec. Eng.*). The original quadrant electrometer used for measurement of voltages and charges. It consists of a suspended plate rotating over a metal box divided into four quadrants, two of which are earthed, the other two being charged by the current under measurement. (This is *heterostatic operation. Idiostatic connexion* has the plate joined to one pair of quadrants.)

Dolgelley Gold Belt (*Geol.*). A belt of land lying north of Dolgelley in Merionethshire, N. Wales, traversed by quartz lodes which, where they cut the Middle Cambrian especially, carry gold in rich pockets.

dolichasters (*Zool.*). In some *Neuropteroidea*, modified setae forming fringes on the lateral segmental processes.

dolichocephalic (*Anat.*). Long-headed; said of a skull, the breadth of which is less than four-fifths the length.

dolichocolon (*Anat.*). An excessively long colon.

doliiform, dolioform. Barrel-shaped.

doline (*Geog.*). A swallow-hole, through which surface waters pass into underground streams in limestone areas.

Doliolidae (*Zool.*). An order of *Thaliacea*, having a free tailed larval stage, a free well-formed oözoid, a thin test, 2 rows of stigmata on the posterior wall of the pharynx; the muscular rings of the body wall are complete.

doll (*Eng.*). Small arm or post carrying signalling apparatus, mounted on a gantry or bridge.

dollar (*Nuc. Eng.*). U.S. unit of reactivity corresponding to the prompt critical condition in a reactor. (This is the reactivity contributed by the delayed neutrons.) See also **cent.**

dollar spots (*Vet.*). The skin lesions of dourine.

dolly (*Cinema., TV*). Mobile mounting for a sound-film or television camera and its operator, with or without means of vertical or side motion (crabbing). (*Civ. Eng.*) An object interposed between the monkey of a pile-driver and the head of a pile to prevent damage to the latter by blows from the former. (*Elec. Eng.*) The operating member of a tumbler switch; it consists of a small pivoted lever projecting through the outer cover. (*Eng.*) (1) A heavy hammer-shaped tool for supporting the head of a rivet during the forming of a head on the other end. (2) A shaped block of lead used by panel beaters when hammering out dents. (*Mining*) A counterbalance weight sometimes used in a hoisting shaft.

dolly truck (*Print.*). See **reel bogie.**

dolly tub (or **kieve**) (*Mining*). A large wooden tub used for the final upgrading of valuable minerals separated by water concentration in ore dressing. See **tossing.**

dolomite (*Min.*). The double carbonate of calcium and magnesium, crystallizing in the rhombohedral class of the trigonal system, occurring as cream-coloured crystals or masses with a distinctive pearly lustre, hence the synonym **pearl spar.** An important gangue mineral. (*Met.*) Calcined dolomite is used as a basic refractory for withstanding high temperatures and attack by basic slags in metallurgical furnaces. The name is also used to describe refractories made from magnesian limestone, which does not necessarily contain the mineral dolomite.

dolomite rock (*Geol.*). A limestone consisting entirely of the mineral dolomite and therefore containing $CaCO_3$ and $MgCO_3$ in equal molecular proportions. Cf. *dolomitic limestone* and see **dolostone.**

dolomitic conglomerate (*Geol.*). The local basal bed of the Rhaetic Series or of the Lias where it is banked up against the old limestone ridges of the Mendip Hills and S. Wales. The rock is a limestone-conglomerate.

dolomitic limestone (*Geol.*). A calcareous sedimentary rock containing calcite or aragonite in addition to the mineral dolomite. Cf. *dolomite rock.*

dolomitization (*Geol.*). The process of replacement of calcium carbonate in a limestone or ooze by the double carbonate of calcium and magnesium (dolomite). In *contemporaneous dolomitization* the replacement is effected at the same time as the formation of the sediment; in *subsequent dolomitization* the replacement occurs at any time after the deposition of the sediment, and is effected by solutions working their way along fractures, joints, and bedding planes.

dolostone (*Geol.*). A rock composed entirely of the mineral dolomite. The term is synonymous with *dolomite rock,* but avoids confusion between dolomite rock and the mineral dolomite.

dolphin (*Civ. Eng.*). Permanent moorings for vessels, usually formed by driving a number of timber or steel box piles into the bed of the river or sea and strengthening them against impact or tug by some form of binding or fixing near their heads.

Dolter surface-contact system (*Elec. Eng.*). A form of surface-contact system for electric traction in which the magnetized skate on the car operates switches within the studs, so making these alive when the car passes over them. This system is now obsolete.

domain (*Elec., Mag.*). In ferroelectric or ferromagnetic material, region where there is saturated polarization, depending only on temperature. The transition layer between adjacent domains is the Bloch wall, and the average size of the domain depends on the constituents of the material and its heat treatment. Domains can be seen under the microscope when orientated by strong electric or magnetic fields. (*Maths.*) (1) A set of points. Also called a **region.** (2) Of a function or mapping, the set on which the function or mapping is defined, e.g., for a real function the domain is a subset of the set of real numbers, namely the set of those real numbers for which the function is defined.

domatium (*Bot.*). A cavity or other form of shelter formed by a plant which harbours mites or insects with which it appears to live by commensalism.

dome (*Arch.*). A vault springing from a circular, or nearly circular, base. (*Bot.*) The growing point of the receptacle of a flower. (*Crystal.*) A crystal form consisting of two similar inclined faces meeting in a horizontal edge, thus resembling the roof of a house. The term is frequently incorrectly applied to a four-faced form which is really a prism lying on an edge. (*Eng.*) A domed cylinder attached to a locomotive boiler to act as a steam space and to house the regulator valve. (*Geol.*) A form of igneous intrusion, the roof of which has a domelike shape, that is, the roof-rocks dip gently outwards from a central point. Ideally, the roof of a *bathylith* has this form, so has a *laccolith*; but these two terms can be applied only when the walls and floor can be examined.

dome cell (*Bot.*). A cell cut off apically during the development of the antheridia of some ferns. An apical derivative from it forms the cap of the mature antheridium.

dome nut (*Eng.*). See **cap nut.**

domestic (*Textiles*). (1) Plain or twill weave cotton or spun rayon cloth; it may be grey, coloured, or with a check pattern and made in several widths and weights. (2) A medium quality worsted whipcord.

domical vault (*Arch.*). A vault in which the centre is higher than the sides.

dominance (*Bot.*). (1) The prevalence of one or of a few species in a plant community, determining the physiognomy and influencing the rest of the plant population. (2) See **dominant character.** (*Crystal.*) See under **habit.**

dominance hierarchy (*An. Behav.*). See **social hierarchy.**

dominant (*Bot.*). Said of a species or group of species in a plant community. See **dominance.**

dominant character (*Gen.*). Of a pair of allelomorphic characters, the one which will be manifested if both are present, i.e., a character that appears in the hybrid resulting from the crossbreeding of parents unlike with respect to that character.

dominant mode (*Telecomm.*). (1) The mode with the lowest cut-off frequency in waveguide propagation. (2) In a cavity resonator, the dominant mode is that corresponding to the lowest excitation frequency. This is also true of microwave oscillators (magnetrons, klystrons, etc.) employing resonant cavities.

dominant wavelength (*Light*). The wavelength of monochromatic light which matches a specified colour when combined in suitable proportions with a reference standard light.

Dongola leather (*Leather*). Upper leather produced by a combination of vegetable and mineral tannages, together with alum and oil.

donkey boiler (*Eng.*). A small vertical auxiliary boiler for supplying steam-winches and other deck machinery on board ship when the main boilers are not in use.

donkey pump (*Eng.*). A small steam reciprocating pump independent of the main propelling machinery of a ship; used for general duty.

Donoghue and Bostock centrifuge (*Powder Tech.*). See **stepped bowl centrifuge.**

donor (*Chem.*). That reactant in an induced reaction which reacts rapidly with the inductor. (*Electronics*) Impurity atoms which add to the conductivity of a semiconductor by contributing electrons to a nearly empty conduction band, so making *n*-type conduction possible. See also **acceptor, impurity.**

donor level (*Electronics*). See **energy levels.**

donskoi (*Textiles*). Wool from the Don district in Russia. The fibre is coarse, but, used alone or in blends with others, makes a good pile yarn for carpets.

donut (*Elec. Eng.*). See **doughnut.**

door by-pass switch (*Elec. Eng.*). See **gate by-pass switch.**

door case (*Join.*). The frame into which a door fits to shut an opening.

door check (*Build.*). A device fitted to a door to prevent it from being slammed, and yet to ensure that it closes.

door cheeks (*Join.*). The jambs of a door frame. Also called **door posts.**

door closer (*Elec. Eng.*). See **gate closer.**

door contact (*Elec. Eng.*). A contact attached to a door or gate, so arranged that it closes a circuit when the door is opened or closed, and rings an alarm bell or gives some other signal.

door interlock (*Elec. Eng.*). See **gate interlock.**

door-knob transformer (*Telecomm.*). Device for coupling a coaxial line to a waveguide through a circular hole at the side, the inner conductor terminating on the far side.

door-knob valve or tube (*Electronics*). A thermionic valve in which the electrodes are very small and closely spaced. Designed for V.H.F. transmission.

door operator (*Elec. Eng.*). See **gate operator.**

door posts (*Join.*). See **door cheeks.**

door stop (*Join.*). A device fitted to a door, or to the floor near to a door, to hold it open.

door strip (*Join.*). A strip, often of flexible material, attached to a door to cover the space between the bottom of the door and the threshold. Also **weather strip, draught excluder.**

door switch (*Elec. Eng.*). A switch mounted on a door so that the opening or closing of the door operates the switch. Also called a **gate switch** when used on electric lifts.

dopa (*Biochem.*). Dihydroxyphenylalanine; an amino acid, formed from *tyrosine* by the action of ultraviolet light, and oxidized by the enzyme *dopaoxidase* (*dopase*) to a pigment, which is the precursor of *melanin* (q.v.) in the skin.

dope additive (*Electronics*). Specific impurity added to ultrapure semiconductor to give required electrical properties, e.g., indium to germanium.

doped junction (*Electronics*). One with a semiconductor crystal which has had impurity added during a melt.

doping (*Aero., etc.*). A chemical treatment which is applied to fabric coverings, for the purposes of tautening, strengthening, weatherproofing, etc. (*Electronics*) Addition of known impurities to a semiconductor, to achieve the desired properties in diodes and transistors, i.e., *donor* in *n*-type and *acceptor* in *p*-type.

Doppler broadening (*Phys.*). Frequency spread of radiation in single spectral lines, because of Maxwell distribution of velocities in the molecular radiators. This also broadens the resonance absorption curve for atoms or molecules excited by incident radiation.

Doppler effect (*Phys.*). Change in apparent frequency (or wavelength) because of relative motion of a source of radiation and an observer. Thus, for sound,

$$\text{frequency observed} = \frac{V+W-V_o}{V+W-V_s} \times \left(\text{frequency of source}\right)$$

where V_s, V_o = velocities of source and observer
W = velocity of wind
V = phase velocity of wave.

Effect is observed in the in-line motion of stars, leading to the idea of an expanding universe; in acoustics, e.g., drop in pitch of a horn on a passing train. See **red shift**.

Doppler navigator (*Aero.*). Automated dead reckoning: a device which measures true ground speed by the Doppler frequency shift of radio beam echoes from the ground and computes these with compass readings to give the aircraft's true track and position at any time. Entirely contained in the aircraft, this system cannot be affected by radio interference or hostile jamming.

Doppler radar (*Radar*). Any means of detection by reflection of electromagnetic waves, which depends on measurement of change of frequency of a continuous wave after reflection by a target having relative motion.

doran (*Radar*). Doppler ranging system for tracking missiles. (*Doppler range.*)

doré silver (*Met.*). Silver bullion, i.e., ingots or bars, containing gold.

doria stripes (*Textiles*). Light cotton cloth of plain weave, with crammed stripes or cords lengthways with the piece; doria checks are also made.

doric (*Typog.*). A type face. The same as sans-serif or gothic.

dormancy (*Biol.*). A state of temporarily reduced but detectable metabolic activity, as in seeds.

dormant bolt (*Join.*). A hidden bolt sliding in a mortise in a door; operated by turning a knob or by means of a key.

dormer (*Build.*). A small window projecting from a roof slope.

Dorn effect (*Chem.*). The production of a potential difference when particles suspended in a liquid migrate under the influence of mechanical forces, e.g., gravity; the converse of *electrophoresis* (q.v.).

dorsal (*Anat.*). Said of the back of any part. (*Arch.*) A canopy. Also called **dorse**. (*Bot.*) (1) Said of that side or part of a plant member turned away from the axis. (2) Attached to, or on the back of, a plant member. (*Zool.*) Pertaining to that aspect of a bilaterally symmetrical animal which is normally turned away from the ground.

dorsal fins (*Aero.*). Forward extensions along the top of the fuselage to increase effectiveness of the main fin by smoothing the airflow over it.

dorsalgia (*Med.*). Pain in the back.

dorsalis (*Zool.*). An artery supplying the dorsal surface of an organ.

dorsal light response (*An. Behav.*). The response of keeping the dorsal surface perpendicular to rays of light which normally fall from above, used to maintain some animals' *primary orientation* (q.v.). Cf. *ventral light response*.

dorsal suture (*Bot.*). The midrib of the carpel in cases where dehiscence occurs along it.

dorsal trace (*Bot.*). The median vascular supply to a carpel.

dorse (*Arch.*). See **dorsal**.

Dorset Down (*Textiles*). One of the Down breed of sheep with fine close staple approximately 10 cm long. Used largely for Cheviot quality woollens and for hosiery years.

dorsiferous (*Zool.*). Said of animals which bear their young on their back.

dorsifixed (*Bot.*). Said of an anther when the filament is attached firmly to one point on the dorsal surface; (by some authorities) adnate.

dorsiflexion (*Anat.*). The bending towards the back of a part, e.g., flexion of the toes towards the shin.

dorsigrade (*Zool.*). Walking with the backs of the digits on the ground, as Sloths.

dorsiventral (*Bot.*). Said of a flattened plant member with structural differences between its dorsal and ventral sides.

dorsiventral symmetry (*Bot.*). See **zygomorphy.**

dorsocentral bristles (*Zool.*). A row of setae on either side of the acrostichal bristles, on the inner part of the mesoscutum in *Diptera*.

dorsodynia (*Med.*). Painful fibrositis of the muscles and the ligaments of the back.

dorsulum (*Zool.*). See **mesonotum.**

dorsum (*Zool.*). In *Anthozoa*, the sulcular surface: in *Arthropoda*, the tergum or notum: in Vertebrates, the dorsal surface of the body or back.

Dortmund tank (*San. Eng.*). A deep tank, with conical bottom, to which liquid sewage is supplied by a pipe reaching down nearly to the bottom. The resulting upward flow assists sedimentation of the sludge.

dorylaner (*Zool.*). An abnormally large male ant, with a long cylindrical abdomen and large peculiar jaws, the typical male form in the subfamily *Dorylinae* (Driver Ants).

dose (*Med.*). The prescribed quantity of a medicine or of a remedial agent. (*Radiol.*) A general expression for the quantity of ionizing radiation to which a sample has been exposed. The following terms are preferred: *absorbed dose*, measured in *rad*; *biological* (*rbe*) *dose*, measured in *rem*; *exposure dose*, measured in *röntgens*; *integral dose*: (*a*) for tissue, etc., the total absorbed dose measured in *g rad*; (*b*) for neutron irradiation in reactors, the time integral of the neutron flux through the sample. See also **maximum permissible dose, n-unit.** Also called **dosage.**

dose equivalent (*Radiol.*). A radiation dose corrected for factors which would vary its biological effect (see **modifying factor**) and expressed in *rem*.

dosemeter (*Radiol.*). Instrument for measuring *dose* (q.v.). Commonly takes the form of a pocket electroscope used in radiation survey, hospitals and Civil Defence. A quartz fibre, after being charged, is viewed against a scale and by its movement across the scale on discharge, it gives a measure of the radiation field and dosage experienced. Also **dosimeter.**

dose rate (*Radiol.*). The *absorbed* or *exposure dose* (q.v.) received per unit time.

dose reduction factor (*Radiol.*). A factor giving the reduction in radiation sensitivity for a cell or organism which results from some chemical protective agent.

dot (*Comp., Teleg.*). Unit digital impulse, the basis of telegraphic codes, and electronic computing. See **band, bit.**

dot cycle (*Teleg.*). Minimum period of an *on-off* or *mark-space* (q.v.). See **band.**

dot frequency (*Teleg.*). Number of dots per second in a continuous train of dots. A measure of speed of telegraphic transmission.

dot generator (*TV*). One which provides impulses for a test pattern for adjusting TV images on a cathode-ray tube screen.

Dothideales, Dothidiales (*Bot.*). An order of the *Ascomycetes*; the perithecium is embedded in the mycelium from which it is not separated by a distinct periderm.

dots (*Build.*). (1) Dabs of mortar or wooden blocks laid at intervals on a floor or roof surface to serve as a guide to the level of the screed. (2) Nails partially driven into a wall, so that the projecting lengths may serve as guides to thickness in laying on a coat of plaster, after which they are removed.

dot-sequential (*TV*). Said of colour TV based on the association of individual primary colours in sequence with successive picture elements.

dotting-on (*Build.*). The process of adding together similar items when *taking off* (q.v.).

dotting test (*Psychol.*). A test used in the study of attention and fatigue. The subject has to mark the centres of small circles passing at a given regular speed but irregularly spaced across a slit of adjustable width.

double (*Build.*). A slate size, 13×6 in. (330 ×252 mm).

double-acting engine (*Eng.*). Any reciprocating engine in which the working fluid acts on each side of the piston alternately; most steam-engines and a few internal-combustion engines are so designed.

double-acting pump (*Eng.*). A reciprocating pump in which both sides of the piston act alternately, thus giving two delivery strokes per cycle.

double-action press (*Eng.*). A press fitted with two slides, permitting multiple operations, such as blanking and drawing, to be performed.

double amplitude (*Elec.*). The sum of the maximum values of the positive and negative half-waves of an alternating quantity. Also **peak-to-peak amplitude.**

double bar and yoke method (*Elec. Eng.*). A ballistic method of magnetic testing, in which 2 test specimens are arranged parallel to each other and clamped to yokes at the ends to form a closed magnetic circuit. A correction for the effect of the yokes is made by altering their position on the bars and repeating the test.

double bar gauge (*Join.*). Marking gauge in which the fence slides on 2 independently adjustable bars, each with marking pins.

double-bead (*Join.*). Two side-by-side beads, separated by a quirk.

double-beam cathode-ray tube (*Electronics*). One containing two complete sets of beam-forming and beam-deflecting electrodes operated from the same cathode, thus allowing two separate waveforms to appear on the screen simultaneously (*double-gun tube*). Also can refer to use of a single gun with a beam-switching circuit, i.e., two inputs continuously interchanged, and to single beam tube in which the beam is split, and the two parts are separately deflected, usually in one plane only.

double-beat valve (*Eng.*). A hollow cylindrical valve for controlling high-pressure fluids. The seatings at the two ends exposed to pressure are of only slightly different area, so that the valve is nearly balanced and easily operated.

double-bellied (*Arch.*). A term applied to a baluster which has had both ends turned alike.

double beta decay (*Nuc.*). Energetically possible process involving the emission of two beta-particles simultaneously. Not to be confused with *dual beta decay* (q.v.).

double block (*Teleg.*). Large blocking capacitors used at each end of a submarine cable to improve the definition of the variations in the signal waveform. See **single block.**

double bond (*Chem.*). Covalent bond involving the sharing of 2 pairs of electrons.

double book (*Photog.*). A pair of dark slides hinged together.

double-break (*Elec. Eng.*). Said of switches or circuit-breakers in which the circuit is made or broken at 2 points in each pole or phase.

double bridge (*Elec. Eng.*). Network for measuring low resistances. See **Kelvin bridge.**

double bridging (*Carp.*). Bridging in which 2 pairs of diagonal braces are used to connect adjacent floor joists at points dividing their length into equal parts.

double camera (*Cinema.*). One in which the film registering the outside scene is accompanied by a parallel film for optical or magnetic synchronous recording of the accompanying sound.

double cap (*Paper*). See **double foolscap.**

double-carbon arc lamp (*Elec. Eng.*). See **twin-carbon arc lamp.**

double case (*Typog.*). A case with accommodation for both upper-case and lower-case characters. Wrongly called a *half-case* (q.v.).

double-catenary construction (*Elec. Eng.*). A method of supporting the overhead contact wire of an electric traction system; the contact wire is suspended from two parallel catenary wires so that the three wires are in a triangular formation.

double cloth (*Textiles*). Two distinct cloths woven and bound together simultaneously to obtain greater thickness without affecting the face texture; heavy overcoatings are one type.

double-coated film (*Photog.*). A film base with emulsions on both sides, generally of differing colour sensitivities (except for X-ray photography).

double-coil loudspeaker (*Acous.*). Electrodynamic loudspeaker with two driving coils separated by a compliance, the coils driving one or two open cone diaphragms, thus operating more effectively over a wide range of frequencies.

double-combing (*Textiles*). See **re-combing.**

double-cone loudspeaker (*Acous.*). Large open coil-driven cone loudspeaker with a smaller free-edge cone fixed to the coil former, thus assisting radiation for high audio frequencies.

double contraction (*Foundry*). Total shrinkage allowance necessary to add to the dimensions of a finished casting when making a wood pattern from which a metal pattern is to be cast.

double-cover butt joint (*Eng.*). A butt joint with a cover plate on both sides of the main plates.

double crown (*Paper*). A standard size of printing paper, 20×30 in.

double-current (*Teleg.*). Said of the use of current reversals in changing from mark to space, when transmitting telegraphic signals.

double-current furnace (*Elec. Eng.*). A special form of electric furnace in which direct current is used for an electrolytic process and alternating current for heating, on the principle of the induction furnace.

double-current generator (*Elec. Eng.*). An electric generator which can supply both alternating and direct current.

double decomposition (*Chem.*). A bimolecular reaction in which the atoms are redistributed into different molecules, the number of molecules remaining constant before and after the reaction, i.e., $AB+CD=AD+CB$.

double-delta connexion (*Elec. Eng.*). A method of connecting windings of a six-phase transformer, etc., so that they may be represented diagrammatically by two triangles.

double demy (*Paper*). A standard size of printing paper, $22\frac{1}{2}$×35 in.

double detection receiver (*Radio*). Same as **supersonic heterodyne receiver.**

double diode (*Electronics*). See **binode.**

double-disk winding (*Elec. Eng.*). A form of winding used for transformers, in which two disk coils are wound in such a way that, when placed side by side to form a single coil, the beginning and end of the complete coil are at the outside periphery.

double-dovetail key (*Join.*). A piece of wood used to connect together 2 members lengthwise; shaped and fitted like a *slate cramp* (q.v.).

double driver plate (*Eng.*). A lathe *driver plate* (q.v.) provided with 2 pins for engaging the carrier on the revolving work.

double earth fault (*Elec. Eng.*). A fault on an electric power transmission system, caused by two phases going to earth simultaneously, either at the same point on the system or at different points.

double elephant (*Paper*). A standard size of writing and drawing paper, 27×40 in.

double embedding (*Micros.*). A technique for embedding small objects, liable to distortion or disorientation in paraffin wax, e.g., the specimen may be first orientated and embedded in celloidin and the small celloidin block embedded in hard paraffin wax.

double-ended boiler (*Eng.*). A marine boiler of the shell type provided with a furnace at both ends, each with its independent tubes and uptake.

double-ended bolt (*Eng.*). A bar screwed at each end for the reception of a nut.

double ends (*Textiles*). Two warp threads passing through the same heald eye and reed dent, and weaving as one thread.

double-entry compressor (*Aero.*). A centrifugal compressor with two vaned impellers mounted back-to-back so that air enters from both sides.

double-faced hammer (*Eng.*). A hammer provided with a flat face at each end of the head.

double-faggoted iron (*Met.*). See **faggoted iron.**

double fertilization (*Bot.*). The union, in the embryo sac of an angiosperm, of 1 male nucleus with the egg nucleus, and of the other male nucleus with the 2 polar nuclei, before or after they have themselves fused.

double Flemish bond (*Build.*). A bond in which both exposed faces of a wall are laid in Flemish bond.

double floor (*Carp.*). A floor in which the bridging joists are supported at intervals by binding joists. Cf. *single floor*, *framed floor*.

double flower (*Bot.*). A flower which has more than the normal number of petals.

double flowering (*Bot.*). An abnormal condition exhibited by a plant which flowers in the spring and again in the autumn.

double-flow turbine (*Eng.*). A turbine in which the working fluid enters at the middle of the length of the casing and flows axially towards each end.

double fluid cell (*Elec. Eng.*). A primary cell in which 2 different solutions are used, 1 surrounding the anode and the other surrounding the cathode, the two being separated by a porous vessel or diaphragm. See **Daniell cell.**

double-fluid theory (*Elec.*). See **two-fluid theory.**

double foolscap (*Paper*). A standard size of printing paper, 17×27 in. Also **double cap.**

double frame (*Typog.*). A frame with accommodation for forty cases and a double-sized working top.

double-frequency oscillator (*Radio*). One in which two sets of oscillations, having different frequencies, are simultaneously generated.

double-frill girdle (*For.*). To girdle by making a double series of downward cuts through the bark, cambium and outer wood; usually for the application of poison.

double-glazing (*Build.*). Hermetically sealed glazing where 2 sheets of glass enclose a vacuum. Used where insulation and conservation of heat are important. Term is popularly applied to a **double window** (q.v.).

double gulley (*San. Eng.*). A gulley used in cowshed and stable drainage, for which purpose it is fitted with inlet and outlet in duplicate, so as to permit segregation of the drainage of urine and washing water.

double-gun cathode-ray tube (*TV*). See **double-beam cathode-ray tube.**

double-headed rail (*Rail.*). A rail section similar to the *bull-headed rail* (q.v.) but having equal heads top and bottom, so that a reversal may be effected to increase the life of the rail, when one of the heads has become worn.

double-helical gears (*Eng.*). *Helical gears* (q.v.) in which two sets of teeth having opposite inclinations are cut on the same wheel, thus eliminating axial thrust.

double-hump effect (*Radio*). Property of two coupled resonant circuits, each separately resonant to the same frequency, of showing maximum response to two frequencies disposed about the common resonance frequency.

double-hung window (*Build.*). A window having top and bottom sashes, each balanced by sash cord and weights so as to be capable of vertical movement in its own grooves.

double image (*TV*). See **ghost.**

double-image micrometer (*Micros.*). A microscope attachment for the rapid and precise measurement of small objects. The principle is that 2 images of the object are formed by means which allow the separation of the images to be varied and its magnitude read on a scale. By setting the 2 images edge-to-edge (a very precise setting) and then interchanging them, the difference of the scale-readings is a measure of the diameter of the object.

double-image microscopy (*Micros., Powder Tech.*). An adaptation of the use of the light microscope to produce a double image of the object. The method may be based on the use of birefringent crystals, prisms, an interferometer-like system, or a vibrating mirror, depending on the make of instrument. It can be developed to provide a means of accurate and rapid particle size analysis.

double-image prism (*Photog.*). A prism which divides the rays from a lens in a *beam-splitter*.

double imperial (*Paper*). A brown-paper size 29×45 in., but for writing and printing papers 30×44 in.

double-insulated (*Elec. Eng.*). Descriptive of portable electric appliances (usually domestic) which have two completely separate sets of insulation between the current-carrying parts and any metal accessible to the user. An earth connection is, therefore, unnecessary.

double integral (*Maths.*). A summation of the values of a function $f(x, y)$ over a specified region R of the x, y plane. Written

$$\iint_R f(x, y)\,dx\,dy.$$

double-jointed compasses (*Instr.*). A pair of compasses in which the limbs of the instrument are jointed, so that the parts carrying the marking points may be arranged, when in use, to be always normal, or nearly so, to the plane of the paper.

double junction (*Build.*). A drainage or water-pipe fitting made with a branch on each side.

double large (*Paper*). A standard size of cut card, $4\frac{1}{2} \times 6$ in.

double-large post (*Paper*). A standard size of printing paper, 21×33 in.

double laths (*Build.*). *Wood laths* 1 in. (25 mm) by $\frac{3}{8}$ to $\frac{1}{2}$ in. (10 to 13 mm) thick in section. See **lathing.**

double layer (*TV*). Screen with separate layers of phosphors with differing properties, i.e., colour or persistence.

double-layer winding (*Elec. Eng.*). An armature winding, always used for d.c. machines and frequently for a.c. machines, in which the coils are arranged in 2 layers, one above the other, in the slots.

double limiting (*Radar*). Process of limiting both the positive and the negative amplitudes of a wave.

double lock (*Hyd. Eng.*). A construction consisting of 2 side-by-side lock chambers, across the same canal. They are interconnected through a sluice so that the amount of water lost in lockage is only half that which is lost by a single lock.

double magazine (*Cinema.*). A magazine holding two reels of film for feeding into a camera; one (e.g., a finished positive) is used as a mask for the other (an unexposed negative). Also used for double films in colour photography.

double-margined door (*Join.*). A door hinged as one leaf but having the appearance of a pair of folding doors.

double measure (*Join.*). Joinery work which is moulded on both sides.

double moding (*Electronics*). Irregular switches of frequency by magnetron oscillator, due to changing mode.

double modulation (*Radio*). See **compound modulation.**

double oblique crystals (*Crystal.*). See **triclinic system.**

double partition (*Build.*). A partition having a cavity in which sliding doors may move.

double pica (*Typog.*). An old type size, about 22 points.

double-pipe exchanger (*Chem. Eng.*). Heat exchanger formed from concentric pipes, the inner frequently having fins on the outside to increase the area. The essential flow pattern is that the 2 streams are always parallel.

double-pitch skylight (*Build.*). A skylight having two differently sloped glazed surfaces.

double plains (*Textiles*). Compound fabrics formed from two plain cloths stitched together. Two cloths of different colours will often be made to interchange to form the design.

double plane iron (*Carp.*). A plane iron consisting of two connected parts—the cutting iron and the break iron.

double plating (*Print.*). See **two set.**

double point (*Maths.*). For the curve $f(x, y)=0$: one at which $\frac{\partial f}{\partial x}=\frac{\partial f}{\partial y}=0$ and $\frac{\partial^2 f}{\partial x^2}, \frac{\partial^2 f}{\partial x \partial y}$ and $\frac{\partial^2 f}{\partial y^2}$ are not all zero. A double point at which there are two real and distinct tangents, one real tangent, or no real tangents, is called a *node*, *cusp*, or *isolated* point respectively. A node is sometimes called a *crunode*, a cusp a *spinode*, and an isolated point an *acnode* or *conjugate point*. Nodes and cusps assume a variety of forms but their names have yet to be decided.

double-pole (*Elec. Eng.*). Said of switches, circuit-breakers, etc., which can make or break a circuit on two poles simultaneously.

double post (*Paper*). A paper size, $18\frac{1}{2} \times 29$ in. Sometimes the old size, $19 \times 30\frac{1}{2}$ in., is used.

double precision (*Comp.*). Operation using *words* of double normal length in *bits*.

double printing (*Cinema.*). The process of exposing a positive emulsion in a printing machine successively with more than one negative, resulting in a superposition of two positive images after development.

double-purpose valve (*Radio*). Any valve used for two separate functions simultaneously, e.g., radio- and audio-frequency amplification.

double quirk-head (*Join.*). A bead sunk into a surface so as to leave a quirk on each side.

doubler (*Telecomm.*). See **frequency doubler.**

doubler circuit (*Elec. Eng.*). (1) A form of self-saturating magnetic amplifier having an a.c. output. (2) One driven so that a frequency can be filtered from the output which is double the frequency of the input.

double reaction (*Radio*). Reaction obtained by a combination of electromagnetic (inductive) and electrostatic (capacitive) coupling.

double reception (*Radio*). Simultaneous reception of two signals on different wavelengths by two receivers connected to the same antenna.

double refraction (*Light*). Division of electromagnetic wave in anisotropic media into oppositely polarized components propagated with different velocities. These components are termed the *ordinary ray*, where the wavefronts are spherical so that the normal laws of refraction are obeyed, and the *extraordinary ray*, where the wavefronts are not spherical so that the velocity is a function of the direction of propagation.

double-roller safety action (*Horol.*). The safety action of a lever escapement in which two rollers are used, one carrying the impulse pin, the other being used for the guard finger.

double roof (*Build.*). A roof truss in which the features of more than one type appear.

double roving (*Textiles*). Two rovings attenuated through drafting rollers and spun into one yarn.

double-row ball journal bearing (*Eng.*). A rolling bearing which has two rows of caged balls running in separate tracks. There are *rigid*, *self-aligning*, and *angular contact* double-row ball journal bearings.

double-row radial engine (*Aero.*). A *radial engine* where the cylinders are arranged in two planes, and operate on two crank pins at 180° apart.

double royal (*Paper*). A standard size of printing paper, 25×40 in.; U.S. 24×38 in.

doubles (*Mining*). See **coal sizes.**

double salts (*Chem.*). Compounds having two normal salts crystallizing together in definite molar ratios.

double scalp (*Vet.*). A form of *osteodystrophia* of sheep thought to be caused by phosphorus deficiency; characterized by severe debility, anaemia, and weakening and thinning of the bones, notably the frontal bones.

double series (*Maths.*). A series formed from all the terms of a 2-dimensional array which extends to infinity in both dimensions, i.e.,

$$\sum_{r,\, s=1}^{\infty} a_{rs}\, .$$

double shear (*Eng.*). Arises in a connexion in which the shearing force is opposed by resisting forces in two planes, e.g., in three plates riveted together.

double shrinkage (*Foundry*). See **double contraction.**

double-sideband transmitter (*Radio*). One which transmits both the sidebands produced in the modulation of a radio-frequency carrier wave. Cf. *single-sideband system*.

double silk-covered wire (*Elec. Eng.*). Wire insulated with two layers of silk covering; commonly used for the fine wire coils of delicate instruments, etc. Abbrev. **d.s.c.**

double-six array (*Radio*). A multiple aerial system responsive to a number of channels. It consists of a double array having a folded dipole, each array having in addition one reflector and four directors.

double-small (*Paper*). A standard size of cut card, $3\frac{3}{8} \times 4\frac{3}{4}$ in.; U.S. $3\frac{1}{2} \times 5$ in.

double spread (*Typog.*). Any pair of facing pages designed as one unit.

double squirrel-cage motor (*Elec. Eng.*). A squirrel-cage motor with two cage windings on its rotor, one of high resistance and low reactance, and the other of high reactance and low resistance. The former carries most of the current at starting, and therefore gives a high starting torque, while the latter carries most of the current when running and results in a high efficiency.

double stars (*Astron.*). A pair of stars appearing close together as seen by telescope. They may be at different distances (see **optical double**) or physically connected (see **binary**).

double-stream amplifier (*Electronics*). A type of travelling-wave tube in which the operation depends upon the interaction of two electron beams of differing velocities.

double superheterodyne receiver (*Radio*). Receiver in which amplification takes place at both high and low intermediate frequencies, thus requiring two frequency-changing stages.

double super royal (*Paper*). A standard size of paper, $27\frac{1}{2} \times 41$ in.

double switch call (*Teleph.*). In international telephony, a transit call which includes three international circuits.

doublet (*Electronics*). Nonpolar valence bond between two atoms, formed by one electron from each. (*Light*) A pair of associated lines in a spectrum, such as the two D lines of sodium. The arc spectra of the alkali metals consist entirely of series of doublets. (*Photog.*) A pair of simple lenses designed to be used together, so that optical distortion in one is balanced by reverse distortion in the other.

doublet antenna (*Radio*). One comprising two short straight conductors, essentially less than a quarter-wavelength long, in line and fed at the centre by a transmission line.

double-tariff meter (*Elec. Eng.*). See **two-rate meter.**

double tenons (*Carp., Join.*). Tenons cut at both ends of a member so as to be coaxial.

double-threaded screw (*Eng.*). A screw having two threads, whose distance apart is half the true pitch; also called a **two-start thread.** See **multiple-threaded screw, divided pitch.**

double-throw switch (*Elec. Eng.*). One which enables connexions to be made with either of two sets of contacts.

double thrust-bearing (*Eng.*). A thrust-bearing for taking axial thrust in either direction.

double-tone ink (*Print.*). Ink for half-tone printing, containing a secondary pigment which spreads outward from each dot while the ink is drying to give a richer effect to the illustration.

double touch (*Acous.*). The use of two-step keys on manuals and pedals of an organ, the initial depression playing on one set of stops, the further depression adding further stops. Also for stop keys, the initial depression bringing in the relative stop in addition to all others drawn on the manual, the full depression cancelling all others.

double transfer process (*Photog.*). The use of an additional intermediate support in the *carbon process* (q.v.).

double trenching (*Agric.*). Digging technique whereby the effective depth for root penetration is increased and the upward passage of water facilitated.

double triode (*Electronics*). Thermionic valve with two triode assemblies in the same envelope.

double-trolley system (*Elec. Eng.*). A system of electric traction where, instead of the running rails, a second insulated contact wire is used for the return or negative current. It avoids trouble due to stray earth currents.

double-tuned circuit (*Elec. Eng.*). Circuit containing 2 elements which may be tuned separately.

double-wall coffer-dam (*Civ. Eng.*). A coffer-dam formed of a pair of parallel walls, separated by a mass of clay puddle, carried down to a watertight stratum and to a level below the bottom of the water.

double warp (*Textiles*). Term used to describe cotton of high quality flannelette type such as pyjama fabrics, etc. Made in warp stripe patterns, two ends are often woven as one.

double-webbed girder (*Eng.*). A built-up box girder in which the top and bottom booms are united by two parallel webs.

double-wedge aerofoils (*Aero.*). See **wedge aerofoil.**

double-width press (*Print.*). A newspaper press with a width of four standard pages.

double window (*Build.*). A window arranged with double sashes enclosing air, which acts as a sound and heat insulator. See **double-glazing.**

double-wire system (*Elec. Eng.*). The usual system of electric wiring; it employs separate wires for the go and return conductors, instead of using the earth as a return.

doubling (*Spinning*). See **folded yarns.**

doubling course (*Build.*). A special course of slates laid at the eaves, to ensure that the lowest margin there has two thicknesses of slate throughout its depth. Also **double** (or **eaves**) **course.**

doubling (or **twisting**) **frame** (*Spinning*). A machine (flyer, cap, or ring type) in which yarns are folded or twisted together in a simple manner to give stronger uniform products or to produce a wide variety of fancy effects.

doubling piece (*Build.*). A *tilting fillet* (q.v.).

doubling time (*Nuc. Eng.*). (1) Time required for the neutron flux in a reactor to double. (2) In a breeder reactor, time required for the amount of fissile material to double.

doublures (*Bind.*). The inside of book covers lined with silk or leather and specially decorated.

doubly crenate (*Bot.*). See **crenate.**

doubly-fed series (or **repulsion**) **motor** (*Elec. Eng.*). A single-phase series (or repulsion) motor in which the armature receives its power partly by conduction and partly by induction.

dough-moulding compound (*Plastics*). Alkyd moulding materials in the form of dough or putty.

doughnut (*Elec. Eng.*). (1) Anchor-ring shape used in circular-path accelerator tubes of glass or metal, e.g., *Zeta.* (2) Traditional shape of a pile of annular laminations for magnetic testing, since there is no external field. (3) Traditional shape of loading coils on transmission lines, permitting exact balance in addition to inductance; also **donut, toroid.**

Douglas bag (*Med.*). A specially constructed bag for the collection of air expired from the lung, from the analysis of which the rate of metabolism can be determined.

Douglas fir (*For.*). *Pseudotsuga menziesii*, the most important timber of the North American continent, and one of the best known softwoods in the world. The wood needs to be incised for preservative treatments by pressure processes; it is unaffected by powder-post beetle attack. Also called **British Columbian Pine, Oregon pine, red pine, false hemlock.**

Douglas sea and swell scale (*Meteor.*). A combination of two numbers, each 0–9, the first of which describes the degree of disturbance of the sea and the second the degree of swell.

Douglas's pouch (*Anat.*). In the female pelvis, the pouch of peritoneum between the rectum and the posterior wall of the uterus.

doup (*Textiles*). A half heald used in weaving

gauze and leno fabrics to facilitate the movement of particular warp threads.

dourine (*Vet.*). A contagious infection of breeding horses, characterized by inflammation of the external genital organs and paralysis of the hind limbs; due to *Trypanosoma equiperdum*, which is transmitted through coitus. Also **mal du coit.**

douser (*Cinema.*). The automatic screen which cuts off the light falling on to the film from the projector lamp, when it is not passing intermittently through the gate.

douzième (*Horol.*). A unit of measurement used in the watch trade. See **ligne.**

douzième gauge (*Horol.*). A form of calliper gauge on which the readings are given in douzièmes.

dovap (*Radar*). Doppler system for tracking missiles, in which the missile carrier or transponder returns a double-frequency signal. (*Do*ppler *v*elocity *a*nd *p*osition.)

dovetail (*Carp., Join.*). A joint formed between a flaring tenon, having a width diminishing towards the root, and a corresponding recess or mortise.

dovetail halving (*Carp.*). A form of *halving* (q.v.) in which the mating parts are cut to a dovetail shape.

dovetail hinge (*Join.*). A hinge whose leaves increase in width outwards from the hinge joint.

dovetail key (*Eng.*). A parallel key in which the part sunk in the boss or hub is of dovetail section, the portion on the shaft being of rectangular section. See **key.** (*Join.*) A batten, of dovetail-shaped section, which is driven into a corresponding groove cut across the back of adjacent boards in a panel, and serves to prevent warping.

dovetail mitre (*Join.*). See **secret dovetail.**

dovetail saw (*Join.*). A saw similar to the tenon saw but of smaller size and having usually 12 teeth to the inch.

dowel (*Build.*). A copper, slate, nonferrous metal, or stone pin sunk into opposing holes in the adjacent faces of two stones, when it is required to unite these more strongly than is possible with a mortar joint. (*Eng.*) (1) A pin fixed in one part which, by accurately fitting in a hole in another attached part, locates the two, thus facilitating accurate reassembly. (2) A pin similarly used for locating divided patterns.

dowel bit (*Carp.*). See **spoon bit.**

dowelling jig (*Build.*). A device for directing the bit in drilling holes to receive dowels.

dowel plate (*Join.*). A steel plate having a number of tapered holes in it; used as a pattern for making tapered dowel pins.

dowel screw (*Join.*). See **hand-rail screw.**

Dow-etch plates (*Print.*). Magnesium alloy plates used in *powderless etching* (q.v.).

dowlas (*Textiles*). Originally, a plain linen fabric of coarse texture, used for towelling, etc., but now produced as cotton cloth constructed from very coarse yarns and given a linenlike appearance in finishing.

Dow metals (*Met.*). Magnesium base alloys of the Elektron type, i.e., containing aluminium and manganese and sometimes zinc.

down (*Bot.*). A fine soft coating of hairs on the surface of a plant member.

downcast (*Mining*). A contraction for *downcast shaft*, i.e., the shaft down which fresh air enters a mine. The fresh air may be sucked or forced down the shaft.

downcast (or **downthrow**) **side** (*Geol.*). That side of a fault on which the strata have been displaced downwards in relation to the up-throw or up-cast side.

downcomer (*Build.*). See **downpipe.** (*Chem. Eng.*) A pipe leading the reflux from any tray or plate in a *distillation column* (q.v.) to the tray or plate below it. (*Met.*) See **downtake.**

downdraught (*I.C. Engs.*). Said of a carburettor in which the mixture is drawn downwards in the direction of gravity.

downer (*Print.*). A sudden breaking of the web on a rotary machine, leading to *down time* (q.v.).

down feathers (*Zool.*). See **plumulae.**

downlead (*Radio*). See **antenna downlead.**

down locks (*Aero.*). See **up and/or down locks.**

downpipe (*Build.*). A pipe (usually vertical) for conveying rain water from the gutter to the drain, or to an intermediate gulley. Also called **downcomer, downspout, fall pipe.**

downspout (*Build.*). See **downpipe.**

Down's process (*Met.*). Electrolytic method of producing sodium metal and chlorine from fused salt at 600°C.

Down's syndrome (*Med.*). See **mongolism.**

downstroke press (*Hyd.*). A hydraulic press in which the main ram is situated above the moving table, pressure being applied by a downward movement of the ram.

downtake (*Met.*). The pipe through which the blast-furnace gas is taken down outside the furnace from the top of the furnace to the duct catcher. Also called **downcomer.**

downthrow (*Geol.*). In a *fault*, the vertical displacement of the fractured strata. Indicated on geological maps by a tick attached to the fault line, with (where known) a figure alongside indicating the downthrow in fathoms or metres.

down time (*Eng.*). Time during which a machine, e.g., a computer or printing press (or series of machines), is idle because of adjustment, cleaning, reloading, or other maintenance.

Downtonian Stage (*Geol.*). The lowest stage of the Old Red Sandstone facies of the Devonian System, named from its typical development around Downton Castle, and reaching a thickness of 760 m in the Welsh Borderlands. It commences with the Ludlow Bone Bed, and consists essentially of red marls with thin, chemically formed limestones (cornstones) and sandstones. By some, the highest part of the Silurian and lowest parts of the Devonian are included.

downward modulation (*Radio*). Modulation in which the instantaneous amplitude of the carrier is always smaller than the amplitude of the unmodulated carrier.

downwash (*Aero.*). The angle through which the airflow is deflected by the passage of an aerofoil measured parallel with the plane of symmetry.

Down wools (*Textiles*). Wools of medium fineness produced by the Down breeds of sheep; the staple is crimpy and ranges from 7·5 cm (Shropshire Down) to 15 cm (Oxford Down); Southdown is the best of this class. These staples make excellent worsted yarns.

Dow oscillator (*Electronics*). See **electron-coupled oscillator.**

Dow process (*Met.*). Extraction of magnesium from sea-water by precipitation as hydroxide with lime; also, electrolytic production of magnesium metal from fused chloride.

dowsing. The process of locating underground water by the twitching of a twig (traditionally hazel or witch-hazel) held in the hand. The phenomenon, if it exists, continues to defy rational explanation, and is accordingly frequently scouted.

Dowtherm (*Chem.*). TN for liquid mixture of diphenyl and its oxide, b.p. 258°C, used to transfer heat.

dozer (*Agric.*). An earth-moving blade for attachment to a tractor.

DPN (*Biochem.*). See **NAD.**

dracone (*Ships, etc.*). A flexible sausage-shaped envelope of woven nylon fabric coated with synthetic rubber. Floats by reason of buoyancy of its cargo, oil or fresh water. It is towed.

dracontiasis (*Med.*). The presence of the Guinea worm, *Dracunculus medinensis*, in the subcutaneous and interstitial tissues of the body.

draft. See also under **draught.**

draft (*Paper*). See **copy.** (*Spinning*) Drawing-out or attenuation of textile fibres passing from the feeding end to the delivery end of a card, drawframe, speedframe, or spinning machine. (*Weaving*) The order in which the warp threads are drawn through the eyes of the heald shafts in the loom.

drafting (*Weaving*). Indicating the details of a lace design on squared paper by means of numbers or filled-in squares to show the movement of the different threads.

draft stop (*Build.*). See **fire stop.**

draft tube (*Eng.*). A discharge-pipe from a water-turbine to the tail race. It decreases the pressure at outlet and increases the turbine efficiency.

drag (*Aero.*). Resistance to motion through a fluid. As applied to an aircraft in flight it is the component of the resultant force due to relative airflow measured along the longitudinal, or drag, axis. Colloquially, it is termed **head resistance.** (*Build.*) A steel-toothed tool for dressing stone surfaces or for keying plasterwork. (*Foundry*) The bottom half of a moulding box or flask.

dragaded cullet (*Glass*). See **quenched cullet.**

drag axis (*Aero.*). A line through the centre of gravity of an aircraft parallel with the relative airflow, the positive direction being downwind.

drag-bar (*Eng.*). See **draw-bar.**

drag chains (*Agric.*). Ancillary to multipurpose *drill* (q.v.). Seed covering chains attached individually to each opener unit.

drag classifier (*Min. Proc.*). Endless belt with transverse rakes, which moves upward through an inclined trough so as to drag settled material up and out while slow-settling sands overflow below as pulp arrives for continuous sorting into coarse and fine sands.

drag conveyor or **drag chain conveyor** (*Eng.*). A conveyor in which an endless chain, having wide links carrying projections or wings, is dragged through a trough into which the material to be conveyed is fed.

drag-cup generator (*Elec. Eng.*). A servo unit used to generate a feedback signal proportional to the time derivative of the error.

Dragendorff's reagent (*Chem.*). $KBiI_4$. Used in the estimation of *alkaloids* (q.v.).

dragged work (*Build.*). Stone-dressing done with a drag.

drag harrow (*Agric.*). A heavy harrow with curved teeth.

drag hinge (*Aero.*). The pivot of a rotorcraft's blade which allows limited angular displacement in azimuth.

dragladled cullet (*Glass*). See **quenched cullet.**

dragline excavator (*Civ. Eng.*). A mechanical excavating appliance consisting of a steel scoop bucket which swings on chains from a movable jib; after biting into the material to be excavated, it is dragged towards the machine by means of a wire rope.

drag link (*Autos.*). A link which conveys motion from the drop arm of a steering gear to the steering arm carried by a stub axle, which it connects through ball joints at its ends. Also called **steering rod.** (*Eng.*) A rod by which the link motion of a steam-engine is moved for varying the cut-off. See **link motion.**

dragon-beam or **dragging-beam** (*Build.*). The horizontal timber on which the foot of the hip-rafter rests.

dragon's blood (*Chem.*). A resinous exudation from the fruit of palm trees and the stems of different species of *Dracaena*. It is a red, amorphous substance, m.p. 120°C, soluble in organic solvents. By destructive distillation the resin yields methylbenzene and phenylethene. It is used for colouring varnishes and lacquers; also, in photoengraving, to protect parts of a plate in the etching process.

dragon tie (*Carp.*). An *angle brace* (q.v.).

drag roller (*Cinema.*). A sprocket roller for guiding film through a cinematograph or sound-on-film mechanism, which, by means of an internal friction arrangement, puts tension on the film.

drag struts (*Aero.*). Structural members designed to brace an aerofoil against air loads in its own plane. Also, landing gear struts resisting the rearward component of impact loads.

drag wires (*Aero.*). Streamlined wires, or cables, bracing an aerofoil against drag (rearward) loads. Applicable to biplanes and some early monoplanes.

drain (*Civ. Eng., San. Eng.*). (1) A water-channel to carry away surface or excess waters. (2) A pipe or channel to carry away wastes and liquid sewage. (*Surg.*) Any piece of material, such as a rubber tube, used for directing away the discharges of a wound.

drainage (*Geol.*). The removal of surplus meteoric waters by rivers and streams. The complicated network of rivers is related to the geological structure of a district, being determined by the existing rocks and superficial deposits in the case of youthful drainage systems; but in those that are mature, the courses may have been determined by strata subsequently removed. (*Surg.*) The action of draining discharges from wounds or infected areas.

drainage area (*Civ. Eng.*). See **catchment area.**

drainage coil (*Elec. Eng.*). A coil bridged between the legs of a communication pair, with its electrical midpoint earthed, in order to prevent the accumulation of static charges on the conductors.

drain cock (*Eng.*). A cock placed (*a*) at the lowest point of a vessel or space, for draining off liquid; (*b*) in an engine cylinder, for discharging condensed steam.

drainer (*Paper*). A large vat for draining the water from slush wood or rag pulp through perforated metal or earthenware plates at the base.

drain pipe (*San. Eng.*). See **drain.**

drain plug (*San. Eng.*). A device for closing the outlet of drain pipes. See **bag plug, screw plug.**

drain tiles (*Civ. Eng.*). Hollow clay or concrete tiles laid end to end without joints, to carry off surface or excess water.

drain-trap (*San. Eng.*). See **air-trap.**

drain-well (*Civ. Eng.*). An *absorbing well* (q.v.).

Dralon (*Chem.*). TN for German polyacrylonitrile staple fibre.

Dramamine (*Pharm.*). TN for β-methylaminoethylbenzohydryl ether 8-chlorotheophyllinate, used as a remedy for travel sickness.

Draper effect (*Chem.*). See **photochemical induction.**

draught (*Eng.*). (1) The flow of air through a boiler furnace. (2) A measure of the degree of vacuum inducing air-flow through a boiler furnace. (*Meteor.*) The downward draught of air occurring with the heaviest rain of a thunderstorm and due to the drag of the falling rain or hail. (*Ships*) The depth of water that a ship requires to float freely. More particularly; *draught forward* and *draught aft*, the depths of water required at the forward and after perpendiculars respectively. *Mean draught* is draught at midlength.

draught-bar (*Eng.*). See **draw-bar.**

draught bead (*Join.*). A *deep bead* (q.v.).

draught excluder (*Join.*). See **door strip.**

draught gauge (*Eng.*). A sensitive vacuum gauge for indicating the draught in a boiler furnace or flue.

draughtsman, draftsman (*Eng.*). One who makes engineering drawings (of models, articles to be made, electrical circuits, plans, etc.) from which prints, at one time blue-prints but now usually dye-lines, are made for actual use. See **tracing.** The draughtsman who makes such drawings frequently designs the details, the main design being laid down by an engineer or architect.

dravite (*Min.*). One of the 3 chief compositional varieties of tourmaline; a complex borosilicate of magnesium, aluminium, and sodium; may be referred to as magnesium-tourmaline; sometimes used as a gemstone.

draw (*Foundry*). Internal cavity or spongy area occurring in a casting due to inadequate supplies of molten metal during consolidation. (*Horol.*) The action whereby one part is drawn into another. In a lever escapement, the locking faces of the pallets are formed at an angle relative to the teeth such that a tooth of the escape-wheel, when pressing against the locking face, tends to draw the pallet into the wheel. (*Mining*) (1) To allow ore to run from working places, stopes, through a chute into trucks. (2) To withdraw timber props or sprags from overhanging coal, so that it falls down ready for collection. (3) To collect broken coal in trucks. (*Spinning*) The outward (drafting and twisting) and inward (winding) run of a mule carriage, in cotton spinning. (*Typog.*) When type has not been locked securely in the forme, the rollers *draw* it out during printing.

draw-bar (*Acous.*). The unit which is pulled out by the hand and thereby adjusts the contribution of harmonic tone in a stop of an electronic organ (such as the Hammond). (*Eng.*) The bar by which the tractive effort of a locomotive or tractor is transmitted to the vehicles behind it; it usually terminates in a hook, which engages the coupling link. Also called **drag-bar, draught-bar.** (*Glass*) A bar of fireclay submerged in a glass-making furnace to stabilize the point at which sheet glass is drawn.

draw-bar cradle (*Eng.*). A closed frame or link for connecting the ends of the draw-bars of railway vehicles, so coupling them together.

draw-bar plate (*Eng.*). On a locomotive frame, a heavy transverse plate through which the draw-bar is attached.

draw-bar pull (*Eng.*). The tractive effort exerted, in given circumstances, by a locomotive or tractor.

draw-bit (*Textiles*). In lace manufacture, a hook which connects the steel bar with the jacquard dropper box.

draw-bore (*Carp.*). A hole drilled transversely through a mortise and tenon so that, when a pin is driven in, it will force the shoulders of the tenon down upon the abutment cheeks of the mortise.

draw box (*Textiles*). Two or more pairs of fluted rollers between the doffer and the coiler of a card. The web of fibres from the doffer passes through the draw box funnel and is delivered to the coiler as a sliver or 'rope' of fibres.

draw-bridge (*Civ. Eng.*). A general name for any type of bridge of which the span is capable of being moved bodily to allow of the passage of large vessels.

draw-door weir (*Civ. Eng.*). A weir fitted with doors or gates capable of being raised vertically, to retain or release water when desired.

drawdown (*Eng., Min. Proc.*). Rate of fall of water level in natural system when depletion exceeds rate of replenishment.

drawer-front dovetail (*Join.*). See **lap dovetail.**

drawer-lock chisel (*Tools*). A crank-shaped chisel with two edges, one being parallel to the shank and the other at right angles to it. Used for chopping drawer-lock recesses or other work in restricted positions.

draw-filing (*Eng.*). The operation of finishing a filed surface by drawing the file along the work at right angles to the length of the file.

draw-gate (*Hyd. Eng.*). A name given to the valve controlling a sluice.

draw-in box (or pit) (*Elec. Eng.*). A box or pit to enable cables to be drawn into, or removed from, a conduit or duct.

drawing (*Met.*). The operation of producing wire, or giving rods a good surface and accurate dimensions, by pulling through one or a series of tapered dies. See also **tempering.** (*Spinning*) (1) Running together and attenuation of a number of slivers, usually six, preparatory to making slubbing and roving. (2) The operation in which worsted tops are reduced to a roving.

drawing-down (*Eng.*). The operation of reducing the diameter of a bar, and increasing its length, by forging.

drawing fires (*Eng.*). The operation of raking out fires from boiler furnaces when shutting down.

drawing-in (*Weaving*). Drawing the warp yarns through the eyes of the loom heald in predetermined sequence, conforming with the pattern draft.

drawing of patterns (*Foundry*). The removal of a pattern from a mould; also termed **lifting of patterns.** It is facilitated by the taper or draught of the pattern, and by loosening the pattern by *rapping* (q.v.).

drawing of tubes (*Eng.*). See **tube drawing.**

drawing rollers (*Spinning*). Pairs of steel rollers, each pair running at higher speed to draft the roving passing through. The front pair comprise a fluted bottom roller with the top one covered with leather, rubber, or cork.

drawing temper (*Eng.*). The operation of tempering hardened steel by heating to some specific temperature and quenching to obtain some definite degree of hardness. See **tempering.**

draw-in pit (*Elec. Eng.*). See **draw-in box.**

draw-in system (*Elec. Eng.*). The system whereby the cables are pulled into conduits or ducts of earthenware, concrete, or iron, from one manhole to another. The cables have a serving of hessian but no armouring.

draw knife (*Carp.*). A cutting blade with a handle at each end at right-angles to the blade; used for shaving wood.

draw knob, draw stop (*Acous.*). The knob on the jamb of an organ console which is drawn to bring in the associated stop.

drawn (*Bot.*). Etiolated.

draw-nail (*Tools*). A pointed steel rod driven into a pattern to act as a handle for withdrawing it from the mould. See **draw-screw.**

drawn on (*Bind.*). Said of a book cover which is attached by gluing down the back; if the end-papers are pasted down it is said to be *drawn on solid*.

drawn-wire filament (*Light*). An incandescent lamp filament, made by a wire-drawing process as opposed to a squirting process.

draw-off valve (*Eng., etc.*). A *bib-valve* (q.v.).

draw-out metal-clad switchgear (*Elec. Eng.*). Metal-clad switchgear in which the switch itself can be isolated from the bus-bars, for inspection and maintenance, by moving it away from the bars along suitable guides.

draw plate (*Eng.*). The plate on which dies are supported in *wire-* and *tube-drawing* operations.

draw (or drawing) rollers (*Print.*). A pair of rollers, usually both driven, which control the web tension on rotary presses.

draw-screw (*Tools*). A screwed rod provided with an eye at the end to act as a handle; screwed into a pattern for lifting it from the mould. See **draw-nail.**

draw sheet (*Print.*). (1) The sheet drawn over the completed make-ready on a press before proceeding with the printing. (2) See **shim.**

draw stop (*Acous.*). See **draw knob.**

draw works (*Mining*). Surface gear of rotary drill, which includes drive and pulling arrangements.

dredge (*Civ. Eng.*). Any apparatus used for excavating under water. (*Mining*) Barge or twin pontoons carrying chain of digging buckets, or suction pump, with over gear such as jigs, sluices, trommels, tailing stackers, manœuvring anchor-lines, and power producer. Used to work alluvial deposits of cassiterite, gold, gemstones, etc. (*Ocean.*) A type of bag-net used for investigating the fauna of the sea-bottom where it is too rough to admit of trawling.

dredger excavator (*Eng.*). See **bucket-ladder excavator.**

dredging (*Civ. Eng.*). A form of excavation conducted under water.

dredging well (*Civ. Eng.*). The opening in a dredging vessel through which the bucket ladders work. See **bucket-ladder excavator.**

dreikanter (*Geol.*). A 3-edged stone which has been so shaped by the natural sandblast operating in sandy deserts. The term is often expanded to include wind-shaped pebbles with more than 3 sides: a better general term is *ventifacts*, a term comparable with artefacts.

drenching (*Leather*). A process for removing from light skins (intended for gloves, etc.) any traces of lime present after bating.

drepanium (*Bot.*). A monochasial cyme in which all the branches arise on the same side of the relatively main axis.

drepanocytosis (*Med.*). Sickle-cell anaemia; a severe anaemia, afflicting Negroes, in which red blood-cells of a peculiar sickle shape appear in the blood and interfere with blood circulation, causing great pain. An inherited disease.

Dresinates (*Plastics*). TN for alkali-metal salts of various resins used as emulsifying agents, etc.

dressed timber (*Build.*). Timber which has been planed more or less to size.

dresser (*Eng.*). (1) An iron block used in forging bent work on an anvil. (2) A mallet for flattening sheet-lead. (3) A tool for facing and grooving millstones, or for trueing grinding wheels. (*Plumb.*) A boxwood tool for straightening lead piping and sheet-lead.

dresser sizing (*Textiles*). See **Scotch dressing** (1).

dressing (*Build.*). The operation of smoothing the surface of a stone. (*Leather, etc.*) A process in which fleshed skins are treated with oil to give them a smooth finish. (*Met.*) Fettling of castings, removal of flashes and runners. (*Min. Proc.*) Grinding of worn crushing rolls, to restore cylindrical shape. Rock crushing and screening to required sizes. Preparation of amalgamation plates with liquid mercury for gold recovery. (*Plumb.*) The operation of flattening out sheet-lead with a dresser. (*Surg.*) The application of sterile material, gauze, lint, etc., to a wound or infected part; material so used. (*Textiles*) Stretching lace in a wet condition on a tenter frame to straighten it. (*Typog.*) The operation of fitting furniture around the pages in a forme, preparatory to locking it up. (*Weaving*) (1) Applying a slight coating of starch to warp yarns when making the weaver's beam, to prevent chafing by the healds, reed and shuttle during weaving. (2) Preparing a warp for the loom.

dressing iron (*Build.*). See **break iron.**

dressings (*Build.*). The mouldings, quoins, strings, and like features, in a room or building.

dress linen (*Textiles*). Plain, matt or twill weave linen fabric, generally made from half-bleached yarns and then mercerized and dyed; used for furnishings as well as for dresses. Imitations are made from spun rayon yarns and given a 'linenlike' finish.

Drew's hypothermia apparatus (*Surg.*). A series of machines which take over cardiac function, and at the same time cool the body, in patients undergoing cardiac surgery. The patient's own lungs are used for oxygenation. Cf. *Melrose pump*.

drier (*Met.*). Furnace used to de-water ore products without changing their composition.

driers (*Chem.*). Substances accelerating the drying of vegetable oils, e.g., linseed oil in paints. The most important representatives of this group are the naphthenates, resinates and oleates of lead, manganese, cobalt, calcium, and zinc.

drift (*Aero.*). The motion of an aircraft in a horizontal plane, under the influence of an air current. (*Civ. Eng.*) The direction in which a tunnel is driven. (*Comp.*) Change in output of directly-coupled analogue amplifier, arising from external causes and which cannot be separated from signal after amplification. (*Electronics*) Slow variation in performance, e.g., of thermionic amplifier during warming up or after a long period, or of a magnetic amplifier. (*Eng.*) (1) A tapered steel bar used to draw rivet holes into line. (2) A brass or copper bar used as a punch. (*Geol.*) A general name for the superficial, as distinct from the solid, formations of the earth's crust. It includes typically the Glacial Drift, comprising all the varied deposits of boulder clay, outwash gravel, and sand of Quaternary age. Much of the drift is of fluvio-glacial origin. (*Hyd. Eng.*) The rate of flow of a current of water. (*Mag.*) In a toroid for plasma studies, such as *Zeta*, the marked tendency for the plasma to drift for several reasons, mainly because of nonuniformity of the steady magnetic field across the area of the tube. (*Mining*) (1) A level or tunnel pushed forward underground in a metal mine, for purposes of exploration or exploitation. The inner end of the drift is called a *dead end*. (2) A heading driven obliquely through a coal-seam. (3) A heading in a coal-mine for exploration or ventilation. (4) An inclined haulage road to the surface. (5) Deviation of borehole from planned course.

drift angle (*Aero., Nav.*). The angle between the planned course and the track. Also sometimes used for angle between heading and track.

drift currents (*Meteor.*). Ocean currents produced by prevailing winds.

drifter (*Mining*). A cradle-mounted compressed-air rock drill, used when excavating tunnels (*drifts* or *cross-cuts*).

drifting (*Eng.*). The process of bringing holes into line by hammering a drift through them. (*Mining*) Tunnelling along the strike of a lode, horizontally or at a slight angle.

drifting test (*Eng.*). A workshop test for ductility; a hole is drilled near the edge of a plate and opened by a conical drift until cracking occurs.

drift mobility (*Electronics*). The average drift velocity of the carriers per unit electric field in a homogeneous semiconductor. In general, the mobilities of electrons and holes will be different.

drift-net (*Ocean.*). A form of *gill-net* (q.v.) used for fishing at or near the surface; allowed to drift with the tide; used especially by herring-boats or drifters.

drift plug (*Plumb.*). A wooden plug which is driven through the bore of a lead pipe in order to smooth out a kink.

drift sight (*Aero.*). A navigational instrument for measuring drift angle.

drift space (*Electronics*). (1) Space between the buncher and reflector in a klystron, in which the electrons congregate in longitudinal waves. (2) Space in an electron tube which is free of electric or magnetic fields.

drift transistor (*Radio*). One in which resistivity increases continuously between emitter and collector junctions, improving high-frequency performance.

drift tube (*Electronics*). Section (space) of any tube, e.g., accelerator or klystron, in which electrons pass without significant change in velocity, while the accelerating radiofrequency voltage changes.

drill (*Agric.*). A machine for sowing seed in regularly spaced rows. (*Eng.*) A revolving tool with cutting edges at one end, and having flats or flutes for the release of chips; used for making cylindrical holes in metal. See also **drilling machine.** (*Mining*) (1) Hand drill, *auger* (q.v.). (2) A compressed-air operated rock drill, jack-hammer, *pneumatic drill* (q.v.). (3) Generally, the more elaborate equipment required in power *drilling* (q.v.).

drill bit (*Carp.*). See **grooved bit.** (*Mining, etc.*). The actual cutting or boring tool in a *drill* (q.v.). In rotary drilling it may be of the *drag* variety, with two or more cutting edges (usually hard-tipped against wear), *roller-type* (with rotating hard-toothed rollers), or *diamond-type*, with cutting face (annular if core samples are required) containing suitably embedded *borts* (q.v.).

drill bush (*Eng.*). A hard sleeve inserted in a *drilling jig* (q.v.) to locate and guide a twist drill accurately, in repetition drilling.

drill chuck (*Eng.*). A self-centring chuck usually having three jaws which are contracted on to the drill by the rotation of an internally coned sleeve encasing them.

drill coulter (*Agric.*). A form of coulter used on a drill to make the shallow trench into which the seed drops. See **disk coulter, hoe coulter, Suffolk coulter.**

drilled and strung (*Bind.*). A method of binding in which holes are drilled close to the back and the leaves then secured by thread or cord; superior to and more expensive than either *flat stitching* or *stabbing*.

drillette (*Textiles*). A fine twill cotton cloth of 5-shaft satin weave; printed, it is used as a dress or lining fabric; dyed, as linings.

drill-extractor (*Civ. Eng.*). A tool used to remove from a boring a broken drill or one which has fallen free of the drilling apparatus.

drill feed (*Eng.*). The hand- or power-operated mechanism by which a drill is fed into the work in a drilling machine.

drilling (*Mining*). (1) The operation of tunnelling or stoping. (2) The operation of making short holes for blasting, or deep holes with a diamond drill for prospecting or exploration. Drilling may be percussive (repeated blows on the drilling tool) or rotary (circular grinding; see **drill bit**), or a combination of these. Cf. *boring*.

drilling cramp (*Eng.*). A frame bolted to a large or unwieldy piece of work for the purpose of supporting a portable drilling machine.

drilling jig (*Eng.*). A device used in repetition drilling, which locates and firmly holds the workpiece accurately in relation to a *drill bush* (q.v.) or pattern of drill bushes accurately positioned in the jig.

drilling machine (*Eng.*). A machine tool for drilling holes, consisting generally of a vertical standard, carrying a table for supporting the work and an arm provided with bearings for the drilling spindle. See **pillar drill, radial drill, sensitive drill.**

drilling mud (*Mining*). Slurry of finely ground heavy material (e.g., barite), perhaps rendered thixotropical (e.g., by added bentonite), circulated during deep drilling. It flushes detached rock from bottom of hole and carries it to surface for removal.

drill-rod (*Civ. Eng.*). The long rod reaching down into a boring, and carrying at its lower end the boring-tool proper.

Drinker respirator (*Med.*). A respirator in which the whole body, excluding the head, is placed. It assists respiration by moving the chest. Commonly known as an **iron lung.**

drip (*Build., Civ. Eng.*). A groove in the projecting under-surfaces of a coping brick or stone wider than the wall; designed to prevent water from passing from the coping to the wall. Also called **gorge** or **throat.** (*Plumb.*) A joint made across the direction of fall, between the edges of two lead sheets on a roof surface, the sheets being overlapped over a 2 in. or 3 in. step in the roof.

drip-feed lubricator (*Eng.*). A small reservoir from which lubricating oil is supplied in drops to a bearing, sometimes through a glass tube to render the rate of feed visible. See **sight-feed lubricator.**

drip mould (*Build.*). A projecting moulding arranged to throw off rain water from the face of a wall.

dripping eave (*Build.*). An eave which is not fitted with a gutter and which therefore allows the rain to flow over to a lower roof or to the ground.

drip-proof (*Elec. Eng.*). Said of an electric machine or other electrical equipment which is protected by an enclosure whose openings for ventilation are covered with suitable cowls, or other devices, to prevent the ingress of moisture or dirt falling vertically.

drip-proof burner (*Heat.*). Gas burner designed to prevent the choking of flame ports or nozzles by foreign matter that may drip or fall on to it.

drip sink (*Build.*). A shallow sink at, or just higher than, floor level, to take the drip from a tap.

dripstone (*Build.*). A projecting moulding built in above a doorway, window opening, etc., to deflect rain water.

drip tip (*Bot.*). A marked elongation of the tip of the leaf, said to facilitate the shedding of rain from the surface of the leaf.

dritosan (*Biochem.*). See **chitosan.**

drive (*An. Behav.*). A hypothetical state of an animal, identified by the production of particular changes in behaviour, and experimentally manipulable by deprivation of food, water, etc., alterations of hormone balance, etc. Can also be hypothesized on the basis of observations on the effectiveness of various events as reinforcement. (*Elec. Eng.*) (1) Voltage applied to a vacuum device, especially the grid in a transmitting valve, or base of a transistor. (2) In circuits, the alternating voltage applied to the grid of an amplifying valve. Specially, the master oscillator circuit and its subsequent amplifying stages in a transmitter using independent drive. Cf. *excitation, exciter.* (*Mining*) Tunnel (level) driven along or near a lode, vein or massive ore deposit. (*Radio*) That which controls a master resonator in an oscillator, e.g., *quartz crystal, tuning-fork, resonating line* or *cavity.*

driven elements (*Elec. Eng.*). Those in an antenna which are fed by the transmitter, as compared with reflector, director, or parasitic elements.

driven roller (*Print.*). A roller geared to the press drive and used to take the web from one section of the press to another, maintaining tension in the process.

drive-reduction (*An. Behav.*). The idea that reinforcing stimuli must reduce some drive or need in an animal if learning is to occur.

driver plate (*Eng.*). A disk which is screwed to the mandrel nose of a lathe, and carries a pin which engages with and drives a carrier attached to the work. Also **driving plate.** See **double driver plate.**

driver stage (*Radio*). Bank of amplifiers which drives the final stage in a radio transmitter.

driver unit (*Radio*). Same as **exciter.**

driving (*Mining*). The making of a tunnel or level (a *drive*) in a mineralized lode or vein, as distinct from making one in country rock.

driving axle (*Eng.*). A vehicle axle through which the driving effort is transmitted to the wheels fixed to it. Also called a **live axle.**

driving blade (*Textiles*). In a lace machine, a brass strip on the front edge of a catch bar, or on the rollers of locker bars.

driving chain (*Eng.*). An endless chain consisting of steel links which engage with toothed wheels, so transmitting power from one shaft to another. See **roller chain.**

driving chuck (*Eng.*). A lathe *driver plate* (q.v.) provided with slots by which dogs are attached for gripping the work, instead of a projecting pin.

driving fit (*Eng.*). A degree of fit between two mating pieces such that the inner member, being slightly larger than the outer, must be driven in by a hammer or press.

driving gear (*Eng.*). Any system of shafts, gears, belts, chains, links, etc., by which power is transmitted to another system.

driving plate (*Eng.*). See **driver plate.**

driving point impedance (*Elec. Eng.*). The ratio of the e.m.f. at a particular point in a system to the current at that point.

driving potential (*Electronics*). The positive potential applied to the anode of a photocell to drive the electrons to the anode after they have been released from the cathode by the incident light.

driving side (*Eng.*). The tension side of a driving belt; the side moving from the follower to the driving pulley.

driving signals (*TV*). The original line and frame pulses which synchronize the whole television system.

driving-trailer (*Rail.*). A trailer-coach for use in a multiple-unit electric or diesel train; provided with driver's equipment at one end.

driving wheel (*Eng.*). (1) The first member of a train of gears. (2) The road wheels through which the tractive force is exerted in a locomotive or road vehicle.

drogue (*Aero.*). A sea anchor used on seaplanes and flying-boats; it is a conical canvas sleeve, open at both ends, like a bottomless bucket. Used to check the way of the aircraft.

drogue parachute (*Aero., Space*). A small parachute used to slow down a descending aircraft or spacecraft.

dromaeognathous (*Zool.*). In Birds, said of the type of palate with large basipterygoid processes springing from the basisphenoid, and the vomers large and flat and connected posteriorly with the palatines, which do not articulate with the rostrum.

drone (*Aero.*). Pilotless guided aircraft used as a target or for reconnaissance. (*Zool.*) In social Bees (*Apidae*), a male.

droop snoot (*Aero.*). Cockpit section hinged on to main fuselage to provide downward visibility at low speeds. U.S., see also **leading-edge flap.**

drop (*Comp.*). Digits are said to *drop-in* if they are recorded without a signal, and *drop-out* if not recorded from a signal. (*Elec. Eng.*) A term commonly used to denote *voltage drop* (q.v.). (*Horol.*) The space moved by a tooth of the escape wheel when it is entirely free from contact with the pallets. (*Mining*) The vertical displacement in a downthrow fault: the amount by which the seam is lower on the other side of the fault. (*Typog.*) To unlock a forme and remove the chase after printing. The type matter is then either distributed or tied up and stored.

drop-annunciator (*Elec. Eng.*). A device used in connexion with an electric signalling system, e.g., an electric bell, to indicate the point from which the signal has originated, the indication being given by the dropping of a coloured disk.

drop arch (*Arch.*). An arch similar to, but less pointed than, the *equilateral arch* (q.v.).

drop arm (*Autos.*). A lever attached to a horizontal spindle which receives rotary motion from the steering gear; used to transmit linear motion through the attached steering rod or drag link to the arms carried by the stub axles.

drop black (*Paint.*). A grade of *bone black* (q.v.). The name originated from a fine artists' pigment made on a small scale and dried on wooden palettes, in small drops.

drop-bottom cage (*Mining*). One in which trucks are locked in position as cage leaves its supports in the shaft.

drop box (*Weaving*). A shuttle box constructed to hold 2 to 6 shuttles carrying weft of different colours or sorts, which can be brought into position for picking across the loom according to the pattern changes.

drop-down curve (*Hyd. Eng.*). The longitudinal profile of the water surface in the case of non-uniform flow in an open channel, when the water surface is not parallel to the invert, owing to the depth of water having been diminished by a sudden drop in the invert.

drop elbow (*Plumb.*). A small elbow with ears, by means of which it can be fixed to a support.

drop electrode (*Chem.*). A half-element consisting of mercury dropping in a fine stream through a solution.

drop foot (*Med.*). Dropping of the foot from its normal position, caused by paralysis of the muscles, due to injury or inflammation of the nerves supplying them.

drop forging (*Met.*). The process of shaping metal parts by forging between two dies, one fixed to the hammer and the other to the anvil of a steam or mechanical hammer. The dies are expensive, and the process is used for the mass-production of parts such as connecting-rods, crankshafts, etc.

drop gate (*Foundry*). A pouring-gate or runner leading directly into the top of a mould.

drop hammer (*Eng.*). A gravity-fall hammer or a double acting stamping hammer used to produce drop forgings by stamping hot metal between pairs of matching dies secured to the anvil block and to the top of the drop hammer respectively.

drop-hammer test (*Paper*). A standard weight is released through a certain height on to the paper box to be tested and the compression is measured photoelectrically.

drop-in (*Comp.*). See **drop.**

drop-in winding (*Elec. Eng.*). A term sometimes used for an armature winding which can be dropped into the slots on the armature instead of having to be pushed through from the end.

drop-out (*Comp.*). See **drop.** (*Telecomm.*) Said of a relay when it deoperates, i.e., contacts revert to de-energized condition.

dropped beat (*Med.*). Intermission of a regular pulse wave at the wrist, due to intermission of the heart beat or to an extrasystole.

dropped elbow (*Vet.*). A condition in which there is inability to extend the forelimb, due to paralysis of the radial nerve.

dropped head (*Typog.*). The first page of a chapter, etc., which begins lower down than ordinary pages. As far as possible, the drop should be constant throughout the book.

dropper (*Bot.*). A mature bulb. (*Elec. Eng.*) In catenary constructions for electric traction systems, the fitting used for supporting the contact wire from the catenary wire. (*Textiles*) One of the distance pieces in a lace machine which are lifted into position between the dropper box and the driving blade by the jacquard card.

dropper box (*Textiles*). A movable metal blade on the jacquard in a lace machine; the steel bar is connected to it.

dropping bottle (*Chem.*). A small bottle having usually a ground-glass stopper in which is cut a narrow channel. The bottle neck, in which this stopper fits tightly, has a similar channel, so that the stopper can be rotated until the two channels meet. By this means liquid poured from the bottle and down the channel, drop by drop, can be readily controlled.

dropping mercury electrode (*Chem.*). Used in polarography, a continuously renewed mercury surface being formed at the tip of a glass capillary, the accumulating impurities being swept away with the detaching drops of mercury.

dropping resistor (*Elec. Eng.*). A resistor whose purpose is to reduce a given voltage by the voltage drop across the resistance itself.

drop-point slating (*Build.*). A mode of laying asbestos slates so that one diagonal is horizontal.

drop shaft (*Mining*). A shallow shaft, connecting two coal-seams, through which full trucks are lowered and empties raised by gravity.

drop siding (*Build.*). Weather-boarding (see **weather board**) which is rebated and overlapped.

drop stamping (*Met.*). See **drop forging.**

dropsy (*Med.*). See **oedema.**

drop tank (*Aero.*). A fuel tank designed to be jettisoned in flight. Also **slipper tank.**

drop tee (*Plumb.*). A small tee with ears, by means of which it can be fixed to a support.

drop test (*Elec. Eng.*). (1) A test employed for locating a fault in a commutator winding; the voltage drop between each pair of adjacent segments when a current is passed through the winding is measured. (2) The *fall-of-potential test* (q.v.) for locating a fault in a cable. (*Eng.*) A strength test for steel tyres, consisting in dropping on to a rail from a specified height (dependent on the tyre diameter).

drop tracery (*Arch.*). Tracery which lies partly below the springing of the arch which it decorates.

drop valve (*Eng.*). A conical-seated valve used in some steam-engines; rapid operation by a trip-gear and return spring reduces wire-drawing losses.

drop wrist (*Med.*). Limp flexion of the wrist from paralysis of the extensor muscles, as a result of neuritis or of injury to the nerve supplying them, e.g., in lead poisoning.

drosometer (*Meteor.*). An instrument for measuring the amount of dew deposited.

dross (*Met.*). Metallic oxides that rise to the surface of molten metal in metallurgical processes. (*Mining*) Small coal, inferior or worthless.

drossing (*Met.*). Removal of scums, oxidized films and solidified metals from molten metals.

drought (*Meteor.*). Lack of rain. A *partial drought* is a period of at least 29 days the mean daily rainfall of which does not exceed 0·25 mm. An *absolute drought* is said to exist if, for at least 15 days, the rainfall on each day has been less than 0·25 mm.

drought ring (*For.*). See under **traumatic ring.**

drove (*Hyd. Eng.*). A narrow channel used for irrigation. (*Tools*) A broad-edged chisel for dressing stone.

drove work (*Build.*). Stone dressing done with a boaster, leaving rows of parallel chisel marks on the slant across the face. Also called **boasted work.**

drowned (*Mining, etc.*). Flooded, e.g., *drowned workings*, flooded workings.

drowned valleys (*Geol.*). Literally, river valleys which have become drowned by a rise of sea level relative to the land. This may be due to actual depression of the land, sea level remaining stationary; or to a eustatic rise in sea level, as during the interglacial periods in the Pleistocene, when melting of the ice-caps took place. See also **ria** and **fiords.**

drowning pipe (*Plumb.*). A storage cistern inlet pipe which reaches down below the surface of the water in the tank, the noise of the discharge being thereby lessened.

Drude law (*Optics*). A law relating the specific rotation for polarized light of an optically active material to the wavelength of the incident light.

$$\alpha = \frac{k}{\lambda^2 - \lambda_0^2},$$

where α = specific rotation, k = rotation constant for material, λ_0^2 = dispersion constant for material, λ = wavelength of incident light. The law does not apply near absorption bands.

drug (*Pharm.*). Any substance, natural or synthetic, which has a physiological action on a living body, either when used for the treatment of disease or the alleviation of pain or for purposes of self-indulgence, leading in some cases to progressive addiction.

drum (*Arch.*). (1) A circular wall carrying a cupola. (2) A cylindrical section of a column. (*Eng.*) (1) Any hollow cylindrical barrel; e.g., a metal barrel in which oil is stored. (2) The rotor of a reaction turbine. See **disk-and-drum turbine, winding drum.** (*Join.*) (1) Any timber structure cylindrical in shape. (2) Any cylinder used as a form for bending wood to shape. (*Mining*) Cylinder or cone, or compound of these, on and off which the winding rope is paid when moving cages or skips in a mine shaft. (*Textiles*) (1) The term is applied to various revolving cylinders or pulleys which transmit motion to other parts by surface contact, or belt drive. (2) In lace-making, a cylinder on which threads are wound to definite length, for transfer to brass bobbins. (3) Diagonally split cylinder which both winds and traverses the yarn to make a cheese for warping, etc.

drum armature (*Elec. Eng.*). An armature for an electric machine, having on it a drum winding.

drum breaker starter (*Elec. Eng.*). A drum starter in which a separate circuit-breaker is provided for interrupting the circuit.

drum controller (*Elec. Eng.*). A controller in which the connexions for performing the desired operation are made by means of fixed contact fingers, bearing on metallic contact strips mounted in the form of a rotating cylindrical drum.

drum-curb (*Civ. Eng.*). See **curb.**

drum filter (*Min. Proc.*). Thickened ore pulp is fed to a trough through which a cylindrical hollow drum rotates slowly. Vacuum draws liquid into pipes mounted internally, while solids (filter cake) are arrested on a permeable membrane wrapped round drum circumference, and are removed continuously before re-submergence

drumlin (*Geol.*). An Irish term, meaning a little hill, applied to accumulations of glacial drift moulded by the ice into small hog-backed hills, oval in plan, with the longer axes lying parallel to the direction of ice movement. Drumlins often occur in groups, giving the 'basket of eggs' topography which is seen in many parts of Britain and dates from the last glaciation.

Drumm accumulator (*Elec. Eng.*). A special form of alkaline accumulator capable of high discharge rates; the positive plate contains nickel oxides and the negative plate is of zinc.

drumming (*Leather*). The term covers any of the processes to which raw hides and skins are subjected in a revolving drum to convert them into leather, e.g., soaking, tanning, currying, dyeing.

drum movement (*Horol.*). A clock movement housed in a cylindrical metal case.

drum pump (*Eng.*). See **rotary pump.**

drum scanner (*TV*). A rotating drum carrying a set of apertures, lenses, mirrors, or other picture-scanning elements, used in mechanical-scanning systems.

drum starter (*Elec. Eng.*). A motor starter in which the necessary operations are carried out by fixed contact fingers bearing upon contact strips mounted in the form of a rotating cylindrical drum.

drum washer (*Paper*). A gauze-covered cylinder in the breaking engine which washes the pulp by withdrawing dirty water and replacing it with clean.

drum weir (*Civ. Eng.*). A weir formed by a gate capable of rotation about a horizontal axis in the line of the river bed, by which means the discharge over the weir may be controlled.

drum winding (*Elec. Eng.*). A winding for electrical machines in which the conductors are all placed under the outer surface of the armature core. It is the form of winding almost invariably used. Also called **barrel winding.** (*Textiles*) Method of winding yarn on to a flanged bobbin by means of a driving drum in contact with the bobbin. The yarn is guided by a traversing porcelain eye, or through a groove in the drum.

drunken saw (*Carp.*). A circular saw revolving about an axis which is not absolutely at right angles to the plane of the saw, consequently cutting a wide kerf.

drunken thread (*Eng.*). A screw in which the advance of the helix is irregular in every convolution.

drupaceous (*Bot.*). Resembling a drupe.

drupe (*Bot.*). A succulent fruit formed from a superior ovary, usually one-seeded, with the pericarp clearly differentiated into epicarp, mesocarp, and endocarp.

drupel or **drupelet** (*Bot.*). A small drupe. Drupels usually occur in groups, forming together a larger fruit.

druse (*Bot.*). A globose mass of crystals of calcium oxalate (ethandioate) around a central foundation of organic material; in some plant cells.

drusy (*Mining*). Containing cavities; said of mineralized lodes or veins.

drusy cavities (*Geol.*). See **geodes.**

drusy structure (*Geol.*). See **miarolitic structure.**

dry adiabatic lapse rate (*Meteor.*). The *adiabatic lapse rate* (q.v.) so long as no water vapour condenses. Constant at 10°C/km. Cf. *saturated adiabatic lapse rate.* Abbrev. **DALR.**

dry area (*Build.*). The 2-in. or 3-in. (50 or 75 mm) cavity in the wall below ground-level in basement walls built hollow; the purpose of the cavity is to keep the basement walls dry.

dry assay (*Chem.*). The determination of a given constituent in ores, metallurgical residues, and alloys, by methods which do not involve liquid means of separation. See also **wet assay, scorification, cupellation.**

dry-back boiler (*Eng.*). A shell-type boiler with one or more furnaces passing to a chamber at the back, from which an upper bank of fire-tubes leads to the uptake at the front.

dry battery (*Elec. Eng.*). A battery composed of *dry cells* (q.v.).

dry blowing (*Min. Proc.*). Manual or mechanical winnowing of finely divided sands to separate heavy from light particles, practised in arid regions. (*Textiles*) A process used for setting and lustring fancy woollens and worsteds.

dry bone or **dry bone ore** (*Min.*). See **smithsonite.**

dry box (*Nuc. Eng.*). Sealed box for handling material in low humidity atmosphere. Not synonymous with *glove box* (q.v.).

dry cell (*Elec. Eng.*). A primary cell in which the contents are in the form of a paste. See **Leclanché cell.**

dry compass (*Ships, etc.*). A mariner's magnetic compass which has no liquid in the bowl. Cf. *liquid compass.*

dry construction (*Build.*). In buildings, the use of timber and plasterboard for partitions, lining of walls and ceilings, to eliminate the traditional use of plaster and the consequent necessary drying out period.

dry copper (*Met.*). Copper containing oxygen in excess of that required to give 'tough pitch'. Such metal is liable to be brittle in hot- and cold-working operations.

dry-core cable (*Telecomm.*). A multicore lead-covered core for telephone or telegraph use, the separate conductors to which are covered with a continuous spiral of ribbon-shaped paper.

The paper provides the insulation after being dried with carbon dioxide, which is pumped through the cable and kept under pressure.

dry dock (*Civ. Eng.*). A dock in which ships are repaired. Water is excluded by means of gates or caissons, after the dock has been emptied. Also called a **graving dock.**

dry electrolytic capacitor (*Elec. Eng.*). One in which the negative pole takes the form of a sticky paste, which is sufficiently conducting to maintain a gas and oxide film on the positive aluminium electrode.

dry flashover voltage (*Elec. Eng.*). The breakdown voltage between electrodes in air of a clean dry insulator.

dry flong (*Print.*). Manufactured flong as distinct from the *wet flong* made up by the stereotyper. Used for pressure moulding and dampened slightly before use.

dry flue gas (*Eng.*). The gaseous products of combustion from a boiler furnace, excluding water vapour. See **flue gas.**

dry fruit (*Bot.*). A fruit in which the pericarp does not become fleshy at maturity.

dry hopping (*Brew.*). The addition of a small quantity of raw hops to certain classes of beers, to improve the flavour.

dry ice (*Chem.*). Solid (frozen) carbon dioxide, used in refrigeration (storage) and engineering. At ordinary atmospheric pressures, it evaporates slowly by sublimation.

dry indicator test (*Paper*). A sizing test for paper. Consists of measuring the time of penetration of water shown by the coloration of a dry indicator powder sprinkled on the opposite side of the sheet.

drying cabinet (*Build.*). A heated cabinet for drying or airing clothes. Commonly used in multistorey flats or where alternative drying facilities are not available, and heated by electricity, gas, or forced hot air.

drying cylinder (*Paper*). A hollow cylinder, heated by steam, over which the web of paper is passed to dry it.

drying oils (*Paint.*). Vegetable or animal oils which harden by oxidation when exposed to air.

drying-out (*Elec. Eng.*). (1) The process of heating the windings of electrical equipment to drive all moisture out of the insulation; usually done by passing current through the windings. (2) In electroplating, the process of removing moisture from a metal by passing it through hot water and then through sawdust or a current of hot air.

drying room or **drying loft** (*Paper*). A room in the paper mill in which sheets of hand-made paper are hung to dry before and after sizing.

drying stove (*Foundry*). A large stove or oven in which dry sand moulds and cores are dried. See **core oven, dry sand.**

dry joint (*Elec. Eng.*). A faulty solder joint giving high-resistance contact due to residual oxide film.

dry liner (*Eng.*). See **liner.**

dry moulding (*Foundry*). The preparation of moulds in dry sand, as distinct from the use of greensand or *loam* (q.v.).

dryness fraction (*Eng.*). The proportion, by weight, of dry steam in a mixture of steam and water, i.e., in wet steam.

dry offset (*Print.*). Almost synonymous with *letterset* (q.v.) but chiefly used when relief plates are employed on conventional litho machines, the dampening rollers being unused.

dry pile (*Elec. Eng.*). An early form of primary battery consisting of a pile of disks separated by layers of paper; the disks were alternately of different metals and formed the electrodes, while the electrolyte was the slight moisture contained in the paper.

dry pipe (*Eng.*). A blanked-off and perforated steam-collecting pipe placed in the steam space of a boiler and leading to the stop-valve, for the purpose of excluding water resulting from *priming* (q.v.). See **antipriming pipe.**

dry plate (*Photog.*). The normal glass plate supporting emulsion for photographic purposes; called *dry* to distinguish from obsolete term *wet plate.*

dry-plate rectifier (*Elec. Eng.*). See **metal rectifier.**

dry rectifier (*Elec. Eng.*). Solid state device, in contrast to a thermionic or electrolytic rectifier.

dry rot (*Build.*). A decay of timber due to the attack of certain fungi, esp. *Merulius lacrymans*, the wood becoming light, dry, and friable, and so quite unsuitable for building purposes.

dry sand (*Foundry*). A moulding sand possessing the requisite cohesion and strength when dried. It is moulded in a moist state, then dried in an oven, when a coherent and porous mould results.

Drysdale permeameter (*Elec. Eng.*). An instrument for determining, by a ballistic method, the permeability of a sample of iron; a plug carrying a primary and secondary coil is inserted in an annular hole in the sample of material under test.

Drysdale potentiometer (*Elec. Eng.*). An a.c. potentiometer of the polar type, comprising a phase-shifting transformer and resistive voltage divider. It is calibrated against a standard cell with a direct current and the alternating current is then set to the same value using an electrodynamic indicator.

dry spell (*Meteor.*). A period of 15 consecutive days none of which has 1 mm or more of rain.

dry spinning (*Textiles*). A method of spinning flax which produces a coarse and bulky weft yarn; also used for tow yarns.

dry steam (*Eng.*). Steam free from water, but unsuperheated. Often called **dry saturated steam.**

dry sump (*I.C. Engs.*). An internal-combustion engine lubrication system in which the upper crankcase is kept dry by an oil scavenge pump, which returns the oil to a tank, from which it is delivered to the engine bearings by a pressure pump.

dry-sweating (*Vet.*). See **anhidrosis.**

dry taping (*Textiles*). See **Scotch dressing.**

dry-thread raising (*Textiles*). A process of raising used to produce a rough fibrous type of pile, e.g., moss or blanket type; or for loosening the short fibres on the surface of Saxony and Cheviot woollen fabrics of certain kinds.

dry valley (*Geol.*). A valley produced at some former period by running water, though at present streamless. This may be due to a fall of the water table, to river capture, or to climatic changes.

dry weight (*Aero.*). The weight of an aero-engine, including all essential accessories for its running and the drives for airframe accessories, without oil, fuel, or coolant. (*Space*) The weight of the liquid-propellant booster (minus the fuel) used in spacecraft missiles.

Ds (*Chem.*). An alternative symbol for *dysprosium.*

d.s.c. (*Elec. Eng.*). An abbrev. for *double silk-covered wire.*

D slide-valve (*Eng.*). A simple form of slide-valve, in section like a letter D, sliding on a flat face in which ports are cut. See **slide-valve.**

d state (*Electronics*). That of an orbital electron

when the orbit has angular momentum of two Dirac units.

DTA (*Chem.*). Abbrev. for *differential thermal analysis*.

DTR (*Photog.*). Abbrev. for *diffusion-transfer reversal*.

dual beta decay (*Nuc.*). *Branching* (q.v.) where a radioactive nuclide may decay by either electron or positron emission. Not to be confused with *double beta decay* (q.v.).

dual carriageway (*Civ. Eng.*). A double highway (usually separated and laned) in which the opposing two streams of traffic are confined to one highway each.

dual-channel sound (*TV*). A technique used in television receivers, in which a separate intermediate-frequency stage is used for sound and video signals after the first detector stage.

dual combustion cycle (*Eng.*). An internal-combustion engine cycle sometimes taken as a standard of comparison for the compression-ignition engine, in which combustion occurs in two stages, i.e., partly at constant volume and partly at constant pressure.

dual ignition (*I.C. Engs.*). (1) The provision of two sparking-plugs in each cylinder of an aero engine or other petrol engine, the plugs being supplied by two independent magnetos. (2) A duplicated ignition system comprising battery ignition and magneto ignition with separate sparking-plugs; used on some high-quality automobiles to ensure reliability.

dual ion (*Chem.*). See **zwitterion.**

duality (*Arch., Civ. Eng.*). The repetition of members in the same angular direction in a structure or building.

dualizing (*Maths.*). See **polar reciprocation.**

dual modulation (*Radio*). Simultaneous amplitude and frequency modulation of a carrier.

dual spectrum (*Print.*). A method of *document copying* (q.v.) in which the document, in contact with the copy paper, is heated by infrared and an image of the document is formed on the copy paper, which is then cascaded with a powder which adheres to the surface, is fused, and can be used for *small offset*.

dual track (*Acous.*). Use of two tracks on a magnetic tape, so that recording and subsequent reproduction can proceed along one track and return along the other, thus obviating rewinding.

Duane and Hunt's law (*Phys.*). The maximum photon energy in an X-ray spectrum is equal to the kinetic energy of the electrons producing the X-rays, so that the maximum frequency, as deduced from quantum mechanics, is eV/h, where V = applied voltage, e = electronic charge, and h = Planck's constant.

dubbing (*Acous.*). The combination of two sources of sound into one recording. (*Build.*) The operation of filling in hollows in the surface of a wall with coarse stuff, as a preliminary to plastering. (*Cinema.*) The replacement of the original sound-track of a ciné film, e.g., with one in a different language. (*Leather*) Mixture of tallow and cod liver oil used in hand *stuffing* (q.v.) of leather.

Dubini's disease (*Med.*). Electric chorea. A nervous disease of acute onset, with severe pains and involuntary shocklike movements.

Du Bois balance (*Elec. Eng.*). An instrument used for measuring the permeability of iron or steel rods. The magnetic attraction across an air gap in a magnetic circuit, of which the sample forms a part, is balanced against the gravitational force due to a sliding weight on a beam.

Duboscq-Brasil fixative (*Micros.*). A highly penetrating microanatomical fixative, particularly used for organisms covered by an impervious cuticle. It contains picric acid, glacial acetic acid, formalin and alcohol. Also called **alcoholic Bouin.**

Dubrovin gauge (or **Cartesian diver gauge**) (*Vac. Tech.*). A gauge which makes use of a thin-walled tube sealed at one end and floating vertically in a liquid. The equilibrium position is determined by the forces due to gravity, the gas pressure to be measured, and the weight of liquid displaced by the tube.

Duchemin's formula (*Phys.*). An expression giving the normal wind pressure on an inclined area in terms of that on a vertical area. It states that:

$$N = F\frac{2\sin\alpha}{1+\sin^2\alpha},$$

where F = pressure of wind in N/m^2 of vertical surface; α = angle of the inclined surface with the horizontal; N = normal pressure in N/m^2 of inclined surface.

Duchenne-Erb paralysis (*Med.*). A form of paralysis in which the arm can be neither abducted nor turned outwards nor raised nor flexed at the elbow, as a result of a lesion of the fifth and sixth cervical nerves.

duchess (*Build.*). A slate, 24 × 12 in. (610 × 305 mm) (*Paper*) A notepaper size, $4\frac{5}{8} \times 3\frac{1}{4}$ in.

duck (*Textiles*). A heavy plain bleached cotton cloth, used for tropical suitings; heavier makes are used for sails, tents and conveyor belting, when plied. A similar linen cloth is used for coarse suitings.

duck board (*Build.*). A board which has slats nailed across it at intervals and is used as steps in repair works on roofs, or for walking in excavations.

duck cholera (*Vet.*). An infection of ducks by the bacterium responsible for *fowl cholera* (q.v.).

duckfoot bend (*San. Eng.*). See **rest bend.**

duck-foot quotes (*Typog.*). The chevron-shaped quotation marks used by continental printers.

duck's-bill bit (*Carp.*). See **spoon bit.**

duck virus hepatitis (*Vet.*). An acute and highly fatal virus disease of young ducklings, characterized by liver cell necrosis and sudden death.

duct (*Civ. Eng.*). A lined trench constructed to contain service pipes, cables, etc. (*Elec. Eng.*) An air passage in the core or other parts of an electric machine along which cooling air may pass; also called **ventilating duct** or **cooling duct.** (*Eng.*) (1) A hole, pipe or channel carrying a fluid, e.g., for lubricating, heating or cooling. (2) A large sheet-metal tube or casing through which air is passed for forced-draught, ventilating, or conditioning purposes. (*Print.*) A reservoir holding the ink in a printing machine. The supply is regulated by a number of screws and by a ratchet. (*Zool.*) A tube formed of cells: a tubular aperture in a nonliving substance, through which gases and liquids or other substances (such as spermatozoa, ova, spores) may pass. Also **ductus.**

ducted cooling (*Aero.*). A system in which air is constrained in ducts that convert its kinetic energy into pressure for more efficient cooling of an aero-engine or of its radiator.

ducted fan (*Aero.*). A gas turbine aero-engine in which part of the power developed is harnessed to a multi-bladed propeller or fan mounted inside a duct. Also **turbofan.**

duct height (*Meteor.*). Height above the earth's surface of the lower effective boundary of a tropospheric radio duct.

ductility (*Met.*). Ability of metals and alloys to

retain strength and freedom from cracks when shape is altered. See **work-hardening.**

ductless glands (*Zool.*). Masses of glandular tissue which lack ducts and discharge their products directly into the blood; as the lymph glands and the endocrine glands.

ductule (*Zool.*). A duct with a very narrow lumen: a small duct; the fine terminal portion of a duct.

ductus (*Zool.*). See **duct.**

ductus arteriosus (*Zool.*). See **ductus Botalli.**

ductus Botalli (*Zool.*). In some *Urodela*, a small connecting branch of the dorsal part of the pulmonary artery connecting the latter with the systemic trunk.

ductus caroticus (*Zool.*). In some Vertebrates, a persistent connexion between the systemic and carotid arches.

ductus communis (*Zool.*). The terminal part of the oviduct in *Platyhelminthes*, opening into the genital atrium.

ductus Cuvieri (*Zool.*). See **Cuvierian ducts.**

ductus ejaculatorius (*Zool.*). In many Invertebrates, as the *Platyhelminthes*, a narrow muscular tube forming the lower part of the vas deferens and leading into the copulatory organ.

ductus endolymphaticus (*Zool.*). In lower Vertebrates and the embryos of higher Vertebrates, the tube by which the internal ear communicates with the surrounding medium.

ductus pneumaticus (*Zool.*). In physostomous Fish, a duct which connects the gullet with the air-bladder.

ductus venosus (*Zool.*). In the development of Vertebrates, the left omphalomesenteric vein during the formation of the liver.

duct waveguide (*Phys.*). Layer in the atmosphere which, because of its refractive properties, keeps electromagnetic radiated (or acoustic) energy within its confines. It is *surface* or *ground-based* when the surface of the earth is one confining plane.

duct width or **thickness** (*Meteor.*). Difference in height between the upper and lower boundaries of a tropospheric radio duct.

Duddell arc (*Elec. Eng.*). A d.c. arc which gives rise to an audio-frequency current. If a suitable tuned circuit is connected across the arc, then the arc will sing acoustically. Also called **singing arc.**

Duddell oscillograph (*Elec. Eng.*). An oscillograph operating on the moving-coil principle. The coil consists of two fine wires carrying the current to be observed, and these move in the field of an electromagnet. Mirrors are attached to the wires, so that their movement can be observed by means of a light beam.

Dufaycolor (*Photog.*). Type of colour film employing a three-colour *réseau* (q.v.).

duff (*Mining*). Fine coal too low in calorific value for direct sale.

Duguet's ulcerations (*Med.*). Ulcers in the tonsillar region and the pharynx, occurring in typhoid fever.

Dühring's disease (*Med.*). Dermatitis herpetiformis. A skin disease in which weals or reddish patches, surmounted by vesicles, appear in successive crops on the surface of the body.

Dühring's rule (*Chem. Eng.*). If the temperatures at which two chemically similar liquids have the same vapour pressure be plotted against each other, a straight line results, i.e., $t' = \alpha + \beta t$, where t' and t are the boiling points of the two liquids on the same scale of temperature and at the same pressure, and α and β are constants.

duke (*Paper*). A notepaper size, $7 \times 5\frac{3}{8}$ in.

dukey (*Mining*). (1) A train, or journey of tubs or trains, travelling on an inclined haulage road underground. (2) A carriage or platform on wheels on which tubs or trams are placed in a horizontal position to be lowered on unusually steep self-acting inclines.

dulcin (*Chem.*). Also known as **sucrol.** 4-*ethoxyphenylurea,*

$$C_2H_5O-C_6H_4-NH\cdot CO\cdot NH_2$$

Colourless crystalline substance which is about 200 times as sweet as sugar.

dull (*Med.*). Not resonant to percussion; said of certain regions of the body, especially the chest.

dull-emitter cathode (*Electronics*). One from which electrons are emitted in large quantities at temperatures at which incandescence is barely visible. The emitting surface is the oxide of an alkaline-earth metal.

Dulong and Petit's law (*Chem.*). See **law of Dulong and Petit.**

dulosis (*Zool.*). Among Ants (*Formicoidea*), an extreme form of social parasitism in which the work of the colony of one species is done by captured 'slaves' of another species called *amazons*; slavery, helotism.

dumb aerial (*Radio*). Nonradiating resistive network, similar to an artificial antenna; used for absorbing the output power from a transmitter during the spacing periods in absorber keying.

dumb-bell bone (*Zool.*). See **prevomer.**

dumb buddle (*Mining*). A buddle without revolving arms or sweeps, for concentrating tin ores.

dumb compass (*Ships, etc.*). See **pelorus.**

dumb iron (*Autos.*). Forgings attached to the front of the side-members of the frame, to carry the spring shackles and front cross-member.

dumb pintle (*Eng.*). See **pintle.**

dummy (*Plumb.*). A lump of lead fastened to the end of a cane, to form a mallet which may be used for straightening out lead pipes. (*Print.*) An unprinted volume, generally unbound, made up for the use of publishers in estimating their requirements. Thickness should be measured at the fore-edge and tail. (*Typog.*) See **layout.**

dummy antenna (*Radio*). Same as **artificial antenna.**

dummy coil (*Elec. Eng.*). A coil put on to an armature to preserve mechanical balance and symmetry, but not electrically connected to the rest of the winding.

dummying (*Eng.*). The preliminary rough-shaping of the heated metal before placing between the dies for drop-forging.

dummy load (*Elec. Eng.*). One which matches a feeder or transmitter; so designed to absorb the full load without radiation, particularly for testing. Also called **antenna load.**

dummy piston (*Eng.*). A disk placed on the shaft of a reaction turbine; to one side of it steam pressure is applied to balance the end thrust; sometimes called a **balance piston.**

dummy plates (*Print.*). Blank, small diameter plates used to fill the cylinder of rotary presses for lock-up.

dump (*Comp.*). Holding store for data required for future operations. (*Mining*) The heap of accumulated waste material from a metal mine, or of treated tailings from a mill or ore-dressing plant. Also called **tip.**

dump condenser (*Nuc. Eng.*). Condenser which allows steam from heat exchangers to be bypassed from the turbines in a nuclear-power station, if the electrical load is suddenly taken off.

dumper (*Civ. Eng.*). A wagon used, in the construction of earthworks, for conveying excavated material on site and dumping it where required.

dumpling (*Civ. Eng.*). The soil remaining in the centre of an open excavation which is commenced by sinking a trench around the site; the dumpling is removed later.

dump valve (*Aero.*). An automatic safety valve which drains the fuel manifold of a gas turbine when it stops, or when the fuel pressure fails. Also a valve to release residual cabin pressure after landing, or to release all cabin pressure in an in-flight emergency.

dumpy level (*Surv.*). A type of level in which the essential characteristic is the rigid connexion of the telescope to the vertical spindle.

dune (*Geol.*). An accumulation of sand formed in an area with a prevailing wind in one direction. The principal types are: *barchans*, crescent-shaped dunes which migrate in the direction of the point of the crescent; *seifs*, elongated ridges of sand aligned in the wind direction; *transverse dunes*, at right angles to the wind; and *whaleback dunes*, very large elongated dunes. Fossil sand dunes can be recognized, and they indicate desert conditions during past geological periods.

dune bedding (*Geol.*). That type of current bedding commonly exhibited by sand dunes and interpreted in sandstones as evidence of desert conditions.

dungannonite (*Geol.*). A corundum-bearing diorite containing nepheline, originally described from Dungannon, Ontario.

dungaree (*Textiles*). A cotton cloth, with 4-end twill or satin weave made from coloured warp and coarse weft, generally used for men's overalls.

dunite (*Geol.*). A coarse-grained, deep-seated igneous rock, almost monomineralic, consisting essentially of olivine only, though chromite is an almost constant accessory. In several parts of the world (e.g., Bushveld Complex, S. Africa) it contains native platinum and related metals. Named from Mt Dun, New Zealand.

dunkop (*Vet.*). See **African horse-sickness.**

dunnage (*Ships*). Loose wood laid in the hold to keep the cargo out of the bilge-water, or wedged between parts of the cargo to keep them steady.

Dunning (*Cinema.*). A double-exposure system in motion-picture production; a yellow transparent print of a previously photographed background is used as a synchrononous mask while photographing action illuminated with yellow against a violet plain background.

duode (*Acous.*). Electrodynamic open diaphragm loudspeaker driven by eddy currents in a metal former, the *voice coil* being wound over a rubber compliance; the arrangement gives enhanced width of response with damping of diaphragm resonances.

duodecal (*Electronics*). 12-contact tube base.

duodecimal system (*Maths.*). A number system whose base is 12.

duodecimo (*Print.*). See **twelvemo.**

duodenal ileus (*Med.*). Chronic obstruction of the duodenum from kinking of its wall by anomalously placed blood vessels, associated with visceroptosis.

duodenectomy (*Surg.*). Excision of the duodenum.

duodenitis (*Med.*). Inflammation of the duodenum.

duodenocholecystostomy (*Surg.*). A communication, made by operation, between the duodenum and the gall bladder.

duodenojejunostomy (*Surg.*). A communication surgically made between the duodenum and the jejunum.

duodenum (*Zool.*). In Vertebrates, the region of the small intestine immediately following the pylorus, distinguished usually by the structure of its walls; so called because it is about 12 in. (30 cm) in length in Man. *adj.* **duodenal.**

duodiode (*Electronics*). See **binode.**

duodynatron (*Electronics*). System comprising two resonant circuits, connected to the inner grid and anode of a tetrode, the outer grid being maintained at a higher potential than any of the other electrodes; oscillations of different frequencies are maintained in the two resonant circuits owing to secondary emission from the inner grid and anode.

duolateral coil (*Elec. Eng.*). See **basket coil.**

duophase (*Elec. Eng.*). The use of a choke in the cathode or anode circuit of a valve in an amplifier to obtain a reversed-phase voltage for driving a push-pull output stage.

duotone (*Print.*). Two half-tone plates made from monochrome copy, the key plate at 45° and the tint at 75°, to produce two-tone effect.

duotriode (*Electronics*). Two triode valves in a single glass envelope. Also **double triode.**

duped print (*Cinema.*). A print, usually of reduced photographic quality, from a duplicate negative, which has been made from a good positive to avoid import duties.

Duperrey's lines (*Mag.*). Lines on a magnetic map indicating the direction of the magnetic meridian.

duplet (*Electronics*). Single valence bond between atoms arising from a shared pair of electrons.

duplex (*Cyt.*). Said of a triploid organism which has three homologous chromosomes, two of them carrying a given dominant gene, and the third carrying the corresponding recessive. (*Electronics*) Separate electron systems in a common vacuum, with or without a common cathode, e.g., double diode or duodiode pentode (U.S.). (*Teleg.*) See **simplex.**

duplex balance (*Teleg.*). Telegraph name for *line balance.*

duplex burner (*Aero.*). A gas turbine fuel injector with alternative fuel inlets, but a single outlet nozzle.

duplex carburettor (*I.C. Engs.*). A carburettor in which two barrels or mixing chambers are fed from a single float chamber; used on some aero engines.

duplex chain (*Eng.*). A roller chain construction using 2 sets of rollers and 3 sets of link plates, used where the chain tension exceeds that which can be transmitted by a simple chain. Its use avoids the need for matching 2 simple chains run side by side.

duplex channel (*Telecomm.*). A channel which is used independently in both directions, over the same frequency band.

duplex dialling (*Teleph.*). Dialling over both line wires of a subscriber's loop, with separate batteries and earth return.

duplexer (*Radio*). Circuit using TR switches, which permit the same antenna to be used for transmission and reception, the switch protecting the receiver from the high-powered transmission.

duplex escapement (*Horol.*). An escapement in which the escape wheel has two sets of teeth, one for giving impulse and one for locking. The locking teeth are in the plane of the wheel, and lock by pressing against the outside of a hollow cylinder on the axis of the balance staff. This cylinder has a notch to allow the tooth to escape for unlocking. The impulse teeth are raised above the plane of the escape wheel and give impulse to the balance by striking against a finger fixed to the balance staff; impulse is given every alternate vibration. Although capable of giving very satisfactory results, this escapement is now rarely met with in watches, as it is sensitive and liable to set.

duplex group (*Cyt.*). The diploid outfit of genes and chromosomes.

duplex lathe (*Eng.*). A lathe in which two cutting tools are used, one on each side of the work, either to avoid springing of the latter, or to increase the rate of working. See **multiple tool lathe.**

duplex paper (*Paper*). (1) Paper having 2 differently coloured surfaces. (2) Any paper made up by pasting 2 sheets together.

duplex processes (*Met.*). The combination of two alternative methods in performing one operation; as when steel making is carried out in two stages, first in the open hearth and second in the electric furnace.

duplex pump (*Eng.*). A pump with two working cylinders side by side.

duplex set (*Print.*). A combination of *barring motor* and main drive motor used to provide all speeds from *inching* and *crawl* to full speed.

duplex winding (*Elec. Eng.*). A winding for d.c. machines in which there are two separate and distinct windings on the machine, the two being connected in parallel by the brushes.

duplicate feeder (*Elec. Eng.*). A feeder forming an alternative path to that normally in use.

duplicate plates (*Print.*). Stereotypes, electrotypes, and rubber and plastic plates made from original plates or from type formes.

duplicating papers (*Paper*). Absorbent or semi-absorbent papers used on duplicating machines. They are specially prepared to prevent the ink smearing.

duplication (*Cyt.*). The union of a fragment of a chromosome with a whole chromosome of the same sort.

duplicational polyploid (*Cyt.*). See **autopolyploid.**

duplication check (*Comp.*). Check that the results of two independent performances of the same operation are identical.

duplicator (*Tools*). Machine tool with a mechanical or electronic controlled cutting head, for contour reproduction.

duplicature (*Zool.*). In some *Polyzoa*, a circular fold surrounding the base of the protrusible portion of a polyp.

duplicident (*Zool.*). Having two pairs of incisor teeth in the upper jaw; as Hares and Rabbits.

duplicitas anterior (*Zool.*). In experimental embryology, an abnormal embryo produced by constriction of the 2-celled stage in the sagittal plane; characterized by the duplication of the anterior parts of the head (cerebral hemispheres, epiphysis, hypophysis, and paraphysis, but not the eyes).

Dupuit relation (*Powder Tech.*). A relationship between the apparent linear velocity of a fluid flow through an isotropic powder bed and the actual velocity of flow, which depends on the porosity of the powder bed.

Dupuytren's contraction (*Med.*). Thickening and contraction of the fascia of the palm of the hand, with resulting flexion of the fingers, especially of the ring and little fingers.

durain (*Geol.*). A separable constituent of dull coal; of firm, rather granular structure, sometimes containing many spores.

Duralumin (*Met.*). TN for aluminium base alloy containing copper 3·5–4·5%, magnesium 0·4–0·7%, manganese 0·4–0·7%, and silicon up to 0·7%. Capable of age hardening at room temperature, after quenching from 500°C. Tensile strength, after forging and heat treatment, 390–430 MN/m²; rel. d. 2·7–2·8. Resistivity 5×10^{-8} ohm metres. Widely used in constructional work.

dura mater (*Zool.*). A tough membrane lining the cerebro-spinal cavity in Vertebrates.

duramen (*Bot.*). See **heart wood.**

Duranol dyes (*Chem.*). Acetate (ethanoate) silk (viscose) dyes, derived from amino-anthraquinones.

Duranthrene (*Chem.*). See **indanthrene.**

durchgriff (*Electronics*). *Penetration factor* (q.v.), or inverse of the *amplification factor* (q.v.) of a triode valve.

durchmusterung (*Astron.*). The German word meaning a systematic counting or cataloguing of the stars; often applied in the title of catalogues, e.g., the *Bonner Durchmusterung*.

durene (*Chem.*). 1 : 2 : 4 : 5-tetramethylbenzene, m.p. 79·3°C, b.p. 196°C. See also **mesitylene.**

duricrust (*Geol.*). Very hard soil or sediment near the surface of the ground, due to cementation by material deposited from solution by percolating groundwater.

duriron (*Met.*). Iron alloy high in silicon and resistant to acidic corrosion.

Durosier's murmur (*Med.*). A murmur heard over the main artery in the thigh during diastole of the heart; indicative of disease of the aortic valve.

Dushman equation (*Electronics*). See **Richardson-Dushman equation.**

dusk rocket (*Space*). A rocket booster launched against the orbital motion of the earth, at dusk, so that the speed of the vehicle is subtracted from the earth's orbital velocity (18·5 m/s).

dust (*Powder Tech.*). Particulate material which is or has been airborne and which passes a 200 mesh B.S. rest sieve (76 μm).

dust chamber (*Met.*). Fume chamber in which dust is arrested as dry furnace gases are filtered, cyclozed or baffled.

dust core (*Mag.*). Magnetic circuit embracing or threading a high-frequency coil, made of ferromagnetic particles compressed into an insulating matrix binder, thus obviating losses at high frequency because of eddy currents.

dust counter (*Mining*). Apparatus, usually portable, for collecting dust in mines, for display and check on working conditions underground. (*Meteor.*) An instrument for counting the dust particles in a known volume of air.

dust cover (*Print.*). See **jacket.**

duster (*Agric.*). A machine to apply materials in powder form.

dusters (*Leather*). See **layers.**

dust explosion (*Eng.*). An explosion resulting from the ignition of small concentrations of inflammable dust (e.g., coal dust or flour) in the air.

dust figure (*Elec. Eng.*). See **Lichtenberg figure.**

dust monitor (*Nuc. Eng.*). Instrument which separates airborne dust and tests for radioactive contamination.

dust panel (*Join.*). A thin board fixed horizontally between two drawers.

dust-proof (*Elec. Eng.*). Said of a piece of electrical apparatus which is constructed so as to exclude dust or textile flyings.

Dutch arch (*Arch.*). See **French arch.**

Dutch barn (*Arch.*). A simple open structure, generally of mild steel, the roof support being formed of braced trusses, arch shaped, and carried by slender columns spaced normally at about 4 m intervals. The roof covering may be of corrugated iron or asbestos cement.

Dutch bond (*Build.*). A bond differing from English bond only in the angle detail, the vertical joints of one stretching course being in line with the centre of the stretchers in the next stretching course.

Dutch clinkers (*Build.*). Small, hard, well-burnt yellow bricks, used mostly for paving purposes.

Dutch elm disease (*For.*). See elm.

Dutch gilding (*Paint.*). Transparent yellow lacquer over burnished silver or tinfoil.

Dutch gold (*Met.*). A cheap alternative to *gold-leaf* (q.v.), consisting of copper-leaf, which, by exposure to the fumes from molten zinc, acquires a yellow colour.

dutchman (*Carp.*). A piece of wood driven into a gap left in a joint which has been badly cut.

Dutch pink (*Paint.*). Transparent lake pigments with a deep khaki overtone and a rich green-yellow undertone, prepared from quercitron bark extract and calcium sulphate or carbonate. Used as an artist's pigment and also to modify the shade of chrome green, sienna, umber, etc. Also **English pink, Italian pink.**

Dutch process (*Chem.*). A process of making *white lead* (q.v.) by corroding metallic lead in stacks where fermentation of tan or bark is taking place, in the presence of dilute ethanoic acid.

dutch roll (*Aero.*). Lateral oscillation of an aircraft, in particular an oscillation having a high ratio of rolling to yawing motion. Dutch roll can be countered by yaw dampers.

Dutral (*Plastics*). TN for an ethene-propene synthetic rubber.

duty (*Elec. Eng.*). The cycle of operations which an apparatus is called upon to perform whenever it is used, e.g., with a motor, it is the starting, running for a given period, and stopping; or with a circuit-breaker, it may be closing and opening for a given number of times with given time intervals between; or the prescription of a process timer.

duty factor (*Elec. Eng.*). (1) The ratio of the equivalent current taken by a motor or other apparatus running on a variable load to the full-load current of the motor (continuous rating). (2) Pulse signals giving ratio of pulse duration to space interval.

dwang (*Join.*). A small piece of wood inserted between floor joists to prevent buckling or distortion. (*Tools*) A mason's term for a crowbar.

dwarfism (*Cinema.*). Negative size-distortion in stereoscopic film caused by the camera lens separation being in excess of the normal interocular distance.

dwarf male (*Zool.*). A male animal greatly reduced in size, and usually in complexity of internal structure also, in comparison with the female of the same species; such males may be free-living but are more usually carried by the female, to which they may be attached by a vascular connexion in extreme cases, as some kinds of deep-sea Angler Fish.

dwarf rafter (*Carp.*). A *jack rafter* (q.v.).

dwarf shoot (*Bot.*). See **short shoot.**

dwarf star (*Astron.*). The name given to a low-luminosity star of the *main sequence* (q.v.). See also **White Dwarf.**

dwarf waves (*Electronics*). Those of length less than 1 cm, produced by electronic oscillations in the interelectrode space of a special valve.

dwell (*Eng.*). Of a cam, the angular period during which the cam follower is allowed to remain at a constant lift. (*Plastics*) Term used to describe the pause in the application of pressure in a moulding press, which allows the escape of gas from the moulding material. (*Print.*) The slight pause in the motion of a hand-press or platen when the impression is being made.

Dwight Lloyd machine (*Met.*). A type of continuous *sintering* (q.v.) machine characterized by having air drawn down through the burning bed on a travelling grate into 'wind boxes' and used in roasting pyritic ore so that by segregation and recycling of the gas streams, concentrations of sulphur dioxide sufficiently high for conversion to sulphuric acid are readily obtained.

dy (*Ecol.*). A type of lake-bottom deposit largely composed of plant detritus and having a marked effect on the fauna.

Dy (*Chem.*). The symbol for *dysprosium.*

dyad (*Cyt.*). Half of a tetrad group of chromosomes passing to 1 pole at meiosis. (*Maths.*) A tensor formed from the product of two vectors. For the vectors **A** and **B** it is written **AB** where, unlike scalar and vector multiplication, no multiplication symbol separates the factors.

dyadic (*Maths.*). The sum of two or more dyads. See **conjugate dyadics.**

Dycril plates (*Print.*). See **photopolymer plates.**

dye-coupling process (*Photog.*). One using *couplers.*

dye-destruction (*Photog.*). Colour-printing process in which the tripack layers incorporate dyes of the respective primary colours and the image is differentially bleached in development.

dyeline (*Print.*). A *document-copying* method using *dyeline paper*, printing down with ultraviolet light which neutralizes diazonium salts in nonimage areas, and processing with ammonium which develops the image. Sometimes called *diazo.*

dyeline paper (*Paper*). A paper coated with a diazo compound. The finished product consists of a black or blue image on a white background.

dyestuffs (*Chem.*). Groups of aromatic compounds having the property of dyeing textile fibres, and containing characteristic groups essential to their qualification as dyes. The more important dyestuffs are classified as follows: (*a*) nitroso- and nitro-dyestuffs, (*b*) azo-dyes, (*c*) stilbene, pyrazole, and thiazole dyestuffs, (*d*) di- and triphenylmethane dyes, (*e*) xanthene dyestuffs, (*f*) acridine and quinoline dyestuffs, (*g*) indamine and indophenol dyestuffs, (*h*) azines, oxazines, and thiazines, (*i*) hydroxyketone dyestuffs, (*j*) sulphide dyes, (*k*) vat dyestuffs, indigo, and indanthrenes, (*l*) reactive dyestuffs, (*m*) *acid dyes* (q.v.), used in photography.

dye toning (*Photog.*). The chemical process whereby a dye is made to replace the silver in a normal photographic image, or to adhere to it by mordanting.

dye transfer (*Photog.*). Colour print process (*Kodak*) using three gelatine matrices taken from separation negatives and processed with a tanning developer. These are treated with dye of the appropriate colour and the images transferred in register on to paper.

dying shift (*Mining*). Night (*graveyard*) shift.

Dykanol (*Elec. Eng.*). TN for dielectric material used in paper capacitors.

dyke (*Geol.*). A form of minor intrusion injected into the crust during its subjection to tension, the dyke being thin, with parallel sides, and maintaining a constant direction in some cases for more than 150 km. Some dykes prove less resistant to weathering than the surrounding country rock and therefore form long narrow depressions in the surface of the ground, thus resembling ditches (hence the name); others, on the contrary, stand up like walls. (*Hyd. Eng.*) A wall or embankment of timber, stone, concrete, fascines, or other material, built as training works for river, to confine flow rigidly within definite limits over the length treated.

dyke phase (*Geol.*). That episode in a volcanic cycle characterized by the injection of minor intrusions, especially dykes. The dyke phase usually comes after the major intrusions, and is the last event in the cycle.

dyke swarm (*Geol.*). A series of dykes of the same age, usually trending in a constant direction over a wide area. Occasionally, dykes may radiate outwards from a volcanic centre, as the Tertiary dyke swarm in Rum, Scotland; but usually they are rigidly parallel; e.g., the O.R.S. dyke swarm of S.W. Scotland, of which the trend is north-east to south-west.

Dynamax (*Met.*). TN for high-permeability iron alloy produced as thin strip suitable for use in toroidal cores.

dynamical stability (*Ships*). The work done, usually in ft-tons or megajoules, when a vessel is heeled to a particular angle. See **heel.**

dynamic balance (*Aero.*). The condition wherein centrifugal forces due to rotation of a propeller or *turbine rotor* produce neither couple nor resultant force in the shaft.

dynamic balancing (*Acous.*). The technique of balancing the centrifugal forces in rotating machines so that there is no residual unbalance, and consequent vibration, to give rise to noise.

dynamic-capacity electrometer (*Elec. Eng.*). See **vibrating-reed electrometer.**

dynamic characteristic (*Elec. Eng.*). Generally, any characteristic curve taken under normal working conditions. E.g., applied to the (anode current)/(grid voltage) relationship of a valve, when the effect of the anode load impedance is included.

dynamic convergence (*TV*). The edge corrections necessary after the conditions of static convergence have been applied. It is carried out by passing currents derived from the line and field time bases through the coil windings in the assembly.

dynamic damper or **detuner** (*Eng.*). Supplementary rotating mass driven through springs attached to a crankshaft at a point remote from the node to eliminate a troublesome critical speed.

dynamic drive (*Acous.*). The actuation of a mechanism (e.g., a loudspeaking receiver) by a current in a coil situated in the air-gap of a permanent or coil-excited magnet.

dynamic electricity (*Elec. Eng.*). A term sometimes used to denote electric currents, i.e., electric charges in motion, as opposed to static electricity, in which the charges are normally stationary. Also current electricity.

dynamic focusing (*TV*). The automatic process of varying the voltage on the focusing electrode of a colour TV picture tube, so that the beam spots remain always in focus as they are swept across the flat screen.

dynamic heating (*Aero.*). Heat generated at the surface of a fast-moving body by the bringing to rest of the air molecules in the boundary layer either by direct impact or by viscosity.

dynamic isomerism (*Chem.*). See **tautomerism.**

dynamic loudspeaker (*Acous.*). Open diaphragm, driven by a *voice coil* on a former; intended to be used in a plane (Rice-Kellog) baffle, from the side of a box (box baffle), or at the neck of a large horn (flare).

dynamic metamorphism (*Geol.*). A name, now little used, for that type of metamorphism in which the chief factor is greatly increased pressure. As rise of temperature is, under natural conditions, inseparable from increase of pressure, a better term is **dynamothermal metamorphism.** See **metamorphism.**

dynamic model (*Aero.*). A free-flight aircraft model in which the dimensions and masses are such as to duplicate full-scale behaviour.

dynamic noise suppressor (*Acous.*). One which automatically reduces the effective audio bandwidth, depending on the level of the required signal to that of the noise.

dynamic pressure (*Aero.*). The pressure resulting from the instantaneous arresting of a fluid stream.

dynamic range (*Radio*). Range of intensities in a sample of radio programme, as measured by a programme meter, a peak meter or volume-unit (VU) meter, and expressed in dB. (*TV*) Ratio of maximum to minimum brightness in the original or reproduced image. Also **contrast range.**

dynamic resistance (*Elec. Eng.*). Effective differential resistance over a section of a static characteristic, e.g., arc, valve, tunnel diode. Can be positive or negative, the latter being essential for sustaining self-oscillation.

dynamics. That branch of applied mathematics which studies the way in which force produces motion.

dynamic sensitivity (*Elec. Eng.*). The alternating component of phototube anode current divided by the alternating component of incident radiant flux.

dynamic storage (*Comp.*). One in which signals are circulating and from which they can be extracted, e.g., delay line.

dynamic viscosity (*Phys., Eng.*). See **coefficient of viscosity.**

dynamite (*Chem.*). A mixture of nitroglycerine with kieselguhr, wood pulp, starch flour, etc., making the nitroglycerine safe to handle until detonated. The most common industrial high explosive.

dynamo (*Elec. Eng.*). Electromagnetic machine which converts mechanical energy into a.c. or d.c. electrical supply. See **alternator.**

dynamo-electric amplifier (*Elec. Eng.*). Low- and zero-frequency mechanically rotating armature in a controlled magnetic field; used in servo systems.

dynamometer (*Elec. Eng.*). Instrument for measurement of supply torque. *Electric dynamometers* can be used for measurement of a.c. current, voltage, or power See **Siemens-.** (*Eng.*) A machine for measuring the brake horse-power of a prime-mover or electric motor. See **absorption-, electric-, transmission-, Froude brake, rope brake.**

dynamometer ammeter (*Elec. Eng.*). An ammeter operating on the dynamometer principle, the fixed and moving coils being connected in series and carrying the current to be measured.

dynamometer voltmeter (*Elec. Eng.*). A voltmeter operating on the dynamometer principle, the fixed and moving coils being connected in series, and in series with a high resistance across the voltage to be measured.

dynamometer wattmeter (*Elec. Eng.*). A commonly used type of wattmeter operating on the dynamometer principle, the fixed coil being usually in the current circuit and the moving coil in the voltage circuit.

dynamothermal metamorphism (*Geol.*). See **dynamic metamorphism.**

dynamotor (*Elec. Eng.*). An electric machine having two armature windings, one acting as a generator and the other as a motor, but only a single magnet frame. Also called a **rotary transformer.**

dynatron (*Electronics*). Circuit in which steady-state oscillations are set up in a tuned circuit between screen and anode of a tetrode, the latter exhibiting negative resistance when the anode potential is below the potential of the screen.

dynatron oscillator (*Electronics*). One which uses the normal negative slope resistance of a screened grid valve to sustain oscillations in a shunt-tuned circuit.

dyne (*Mech.*). The unit of force in the CGS system of units. A force of one dyne, acting on a mass of 1 g, imparts to it an acceleration of 1 cm/s^2. Approximately 981 dynes are equal to 1 g weight. 10^5 dynes = 1 newton.

Dynel (*Plastics*). Copolymer synthetic fibre based on chloroethene and cyanoethene. Staple fibre form of *Vinyon N*.

Dynistor (*Electronics*). TN for *p-n-p-n* transistor type component with negative resistance characteristic (*dyn*atron trans*istor*).

dynode (*Electronics*). Intermediate electrode (between cathode and final anode) in photomultiplier or electron multiplier tube. Dynode electrons are those which emit secondary electrons and provide the amplification.

dynode chain (*Electronics*). Resistance potential divider employed to supply increasing potentials to successive dynodes of an electron multiplier.

dynode spots (*TV*). Spurious signals which may be produced in image orthicons.

Dyotron (*Electronics*). TN for simple cavity microwave oscillator.

dys-. Prefix from Gk. *dys-*, =English mis-, un-.

dysadaptation (*Med.*). Marked reduction in rapidity of adaptation of the eye to suddenly reduced illumination, as in vitamin A deficiency.

dysanalyte (*Min.*). The niobian cerian variety of perovskite.

dysarthria (*Med.*). Difficult articulation of speech, due to a lesion in the brain.

dysbasia (*Med.*). Difficulty in walking.

dysbasia angiosclerotica (*Med.*). Pain in the legs on walking, due to thickening of the arteries. See **intermittent claudication.**

dysboulia, dysbulia (*Psychol.*). Psychopathological weakness of the will.

dyschezia (*Med.*). A form of constipation in which the faeces are retained in the rectum, as a result of blunting of a normal reflex, due to faulty habits.

dyscrasia (*Med.*). Any disordered condition of the body, especially of the body fluids.

dyscrasite (*Min.*). Silver ore consisting mainly of a silver antimonide, Ag_3Sb.

dysdiadokokinesia (*Med.*). Inability to perform rapid alternate movements as a result of a lesion in the cerebellum.

dysentery (*Med.*). A term formerly applied to any condition in which inflammation of the colon was associated with the frequent passage of bloody stools. Now confined to *amoebic dysentery*, the result of infection with the *Entamoeba histolytica*; and to *bacillary dysentery*, due to infection with *Bacterium dysenteriae*.

dysgenic (*Zool.*). Causing, or tending towards, racial degeneration.

dysgraphia (*Med.*). Inability to write, as a result of brain damage or other cause.

dyskinesia (*Med.*). A term applied to any one of a number of conditions characterized by involuntary movements which follow a definite pattern, e.g., tics.

dyslalia (*Med.*). Articulation difficulty due to defects in speech organs.

dyslexia (*Med.*). Word blindness; great difficulty in learning to read, write, or spell, of which the cause (not lack of intelligence) has not been established.

dysmelia (*Med.*). Misshapen limbs.

dysmenorrhoea, dysmenorrhea (*Med.*). Painful and difficult menstruation.

dysmetria (*Med.*). Faulty estimation of distance in the performance of muscular movements, due to a lesion in the cerebellum.

dysostosis (*Med.*). Defect in the normal ossification of cartilage.

dyspareunia (*Med.*). Painful or difficult coitus.

dyspepsia (*Med.*). Indigestion: any disturbance of digestion.

dysphagia (*Med.*). Difficulty in swallowing.

dysphasia (*Med.*). Disturbed utterance of speech due to a lesion in the brain.

dysphonia (*Med.*). Difficulty of speaking, due to any affection of the vocal cords.

dysphoria (*Med.*). Unease; absence of feeling of well-being.

dyspituitarism (*Med.*). A condition in which there is disordered function of the pituitary gland.

dysplasia (*Med.*). Abnormality of development.

dyspnea, dyspnoea (*Med.*). Laboured or difficult respiration.

dysprosium (*Chem.*). A metallic element, a member of the rare-earth group. Symbol Dy, at. no. 66, r.a.m. 162·5. It is the most basic of the erbium subgroup (Dy, Ho, Er, Tm) of the yttrium family.

dyssynergia (*Med.*). Incoordination of muscular movements, due to disease of the cerebellum.

dystectic mixture (*Chem.*). A mixture with a maximum melting-point. Cf. *eutectic*.

dystocia, dystokia (*Med.*). Painful or difficult childbirth.

dystrophia adiposogenitalis (*Med.*). Fröhlich's syndrome. A condition characterized by obesity, hairlessness of the body, and underdeveloped genital organs, due to disordered function of the pituitary gland.

dystrophia myotonica (*Med.*). See **myotonia atrophica.**

dystrophic (*Ecol.*). Said of a lake in which the water is rich in organic matter, such as humic acid, but this consists mainly of undecomposed plant fragments, and nutrient salts are sparse.

dystrophy (*Bot.*). The condition in which insects visiting flowers do not enter in the normal fashion, but perforate the perianth and remove the nectar without working the pollination mechanism; the flower is damaged, and the plant obtains no advantage from the visit. (*Med.*) A condition of impaired or imperfect nutrition, as in *muscular dystrophy* (q.v.).

dysuria, dysury (*Med.*). Painful or difficult passage of urine.

Dzierzon's theory (*Zool.*). The theory that a queen Bee can lay at will eggs which will produce either males or females, and that all the eggs will develop into males if unfertilized, into females if fertilized.

e (*Build.*). Symbol for *eccentricity* of a load. (*Maths.*) See ***e***. (*Phys.*) Symbol for *electron*.

e (*Chem.*). With subscript, symbol for single electrode potential; see also E (2). (*Maths.*) A transcendental number, the base of natural (Napierian) logarithms, defined as the limiting value of $\left(1+\frac{1}{m}\right)^m$ as m approaches infinity; $e = 2{\cdot}718\,281\,828\,5\ldots$ to ten places of decimals. Also **e**, *ε*. Eccentricity of a conic. (*Nuc.*) Symbol for elementary charge (of positron).

ε Symbol for *emissivity*; *linear strain*; *permittivity*. (*Chem.*) Symbol for molar extinction coefficient. (*Maths.*) See ***e***.

ε- (*Chem.*). (1) Substituted on the fifth carbon atom. (2) *epi-*, i.e., containing a condensed double aromatic nucleus substituted in the 1.6 positions. (3) *epi-*, i.e., containing an intramolecular bridge.

η A symbol for some specified efficiency. (*Chem.*) A symbol for *electrolytic polarization*, *overvoltage*. (*Phys.*) A symbol for the *coefficient of viscosity*.

E (*Phys.*). A symbol for: (1) *potential difference*, especially electromotive force of voltaic cells; (2) (with subscript) single electrode potential—E_H, on the hydrogen scale; E_O, standard electrode potential. (*Chem. Eng.*) Symbol for *eddy diffusivity*.

[E] (*Light*). One of the Fraunhofer lines in the green of the solar spectrum. Its wavelength is 526·9723 nm, and it is due to iron.

E Symbol for *electromotive force*; *electric field strength*; *energy*; *illumination*; *irradiance*.

Eagle mounting (*Optics*). A compact mounting of a concave diffraction grating based on the principle of the Rowland circle. The mounting suffers from less astigmatism than either the *Rowland* or *Abney mountings*, and is useful for studying higher order spectra.

ear (*Acous.*). The organ of hearing. See below under (*Zool.*). (*Build.*) A *crossette* (q.v.). (*Eng.*) A cast or forged projection integral with, or attached to, an object, for supporting it, or attaching another part to it pivotally; also called **lug**. (*Plumb.*) A projection on a metal pipe, by means of which the pipe may be nailed to a wall. (*Zool.*) Strictly, the sense-organ which receives auditory impressions; in Invertebrates, various sensory structures formerly supposed to be auditory in function, as the statocysts of medusae; in some Birds and Mammals, a prominent tuft of feathers or hair close to the opening of the external auditory meatus; in Mammals, the pinna; more generally, any ear-like structure.

ear defenders (*Acous.*). Plugs of various materials for insertion into, or fitting over, the ear to reduce reception of noise.

ear drum (*Anat.*). The outer termination of the aural mechanism of the ear, consisting of a membrane, in tension, for transferring the acoustic pressures applied from without to the ossicles for transmission to the inner ear.

earing (*Met.*). Excessive elongation along edges and folds of metals being shaped by deep-drawing or rolling.

ear lifters (*Agric.*). Attachment to combines as a grain saver in badly laid crops.

Early Bird (*Telecomm.*). A synchronous *communications satellite* (q.v.) providing regular commercial telecommunications.

early blight (*Bot.*). Fungus disease of potatoes caused by *Alternaria solani*, causing dark spots on the leaves and lesions on the tubers.

early effect (*Electronics*). Variation of junction capacitance and effective base thickness of transistor with the supply potentials.

early-warning radar (*Radar*). A system for the detection of approaching aircraft or missiles at greatest possible distances.

ear microphone (*Acous.*). A contact microphone which is shaped to fit into the ear of the wearer, whose voice is picked up for consequent transmission.

ear muffs (*Acous.*). (1) Pads of rubber or similar material which are placed on head telephone receivers to minimize discomfort during long use. (2) Large pads for reducing the effect of noise on factory workers.

earphone coupler (*Acous.*). A suitably-shaped cavity with an incorporated microphone used for acoustical testing of earphones.

Earth (*Astron.*). The third planet in the solar system, counting outwards from the Sun. It has one satellite, the *Moon* (q.v.). Its mass is 1/332958 that of the Sun's mass. Its mean distance from the Sun is $149{\cdot}6\times10^6$ km. It revolves round the Sun in 1 year, its orbit being slightly elliptical, and it rotates on its own axis from west to east in 1 day. Its axis is inclined to the plane of the ecliptic. (See **obliquity of the ecliptic.**) The origin of the Earth, and of the Sun and the other planets, is attributed to condensation from a rotating mass of gas, or of dust and gas. Older theories of the origin of the planets have proved untenable. *Age of the Earth.* Estimates of this have varied between large limits, but two lines of argument, cosmogonical and geological, now seem to indicate an age of about 4600 million years. *Crust of the Earth.* The outer shell, lying above the *Mohorovičič discontinuity* (q.v.), consists of rocks which have a composition approximating to that of granite; covering this is a thin veneer of sedimentary rocks. See also **lithosphere.** *Density of the Earth.* Average density is ca. 5·6 kg/dm³; but as that of the surface rocks is approx. 2·5 kg/dm³, it is inferred that denser materials lie beneath and include a core of nickel-iron, with a density of ca. 8 kg/dm³. *Form of the Earth.* The true figure is an oblate spheroid, flattened at the poles and having a bulge at the equator; the equatorial and polar radii are respectively 6378 and 6357 km. *Radius of the Earth.* The effective radius for calculations on radio transmission, allowing for a refractive index increasing linearly with height, is 4/3 the actual radius. *Temperature of the Earth.* The temperature of the crust varies seasonally to a depth of ca. 10 m, and diurnally to ca. 0.6 m. Below 10 m, the temperature rises with increasing depth at an average rate of ca. 30°C/km.

earth (*Elec.*). Connexion to main mass of earth by means of a conductor having a very low impedance. (*Radio*) System of plates or wires buried in the ground, to which connexion is made to provide a path to ground for currents flowing in the antenna circuit. U.S. **ground.**

earth (or **ground**) **capacitance** (*Elec.*). The capacitance between an electrical circuit and earth (or conducting body connected to earth).

earth circuit (*Radio*). That part of a radio transmitter or receiver circuit which includes an earth lead or counterpoise.

earth closet (*San. Eng.*). A metal-lined receptacle, usually placed beneath a pierced seat, for receiving human excreta, the latter being deodorized by covering with dry earth, sometimes mixed with chemicals.

earth coil (*Elec. Eng.*). A pivoted coil of large diameter for measuring the strength of the earth's magnetic field; this is done by suddenly changing the position of the coil in this field and observing the throw of a ballistic galvanometer connected to it. Also called an **earth inductor.**

earth colours (*Paint.*). Pigments prepared from natural earths, mostly oxides and silicates, e.g., ochres, umbers, and siennas.

earth continuity conductor (*Elec. Eng.*). A third conductor (with line and neutral) in a mains distribution system, bonded to earth, provided for connection to any metal component not in the electrical circuit; for safety purposes. Abbrev. ECC.

earth currents (*Elec. Eng.*). (1) Currents in the earth which, by electromagnetic induction, cause irregular currents to flow in submarine cables and so interfere with the reception of the transmitted signals. (2) Direct currents in the earth, which are liable to cause corrosion of the lead sheaths of cables; they can be the earth-return currents of power systems.

earth dam (*Civ. Eng.*). One built of local gravels, earth, etc., with impervious clay core which rests on firm ground to avoid undermining.

earth detector (*Elec. Eng.*). See **leakage indicator.**

earthed aerial (*Radio.*). Marconi aerial, in which an elevated wire is earthed at its lower end.

earthed circuit (*Elec. Eng.*). An electric circuit which is intentionally connected to earth at one or more points.

earthed concentric wiring system (*Elec. Eng.*). A 2-wire system for wiring or general distribution, which uses twin-concentric conductors, the outer conductor being earthed.

earthed neutral (*Elec. Eng.*). A neutral point of a polyphase system or piece of electrical apparatus which is connected to earth, either directly or through a low impedance.

earthed pole (*Elec. Eng.*). The pole or line of an earthed circuit which is connected to earth.

earthed switch (*Elec. Eng.*). A switch, used in wiring installations, in which provision is made for earthing all exposed metal parts. Also called a **Home-Office switch.**

earthed system (*Elec. Eng.*). A system of electric supply in which one pole or the neutral point is earthed, either directly or through a low impedance, the former being known as a solidly earthed system.

earth electrode system (*Elec. Eng.*). The totality of conductors, conduits, shields, and screens which are connected to the main earth by low-impedance conductors.

earthenware duct (*Elec. Eng.*). A conduit made of earthenware, for carrying underground cable.

earth fault (*Elec. Eng.*). An accidental connexion between a live part of an electrical system and earth.

earth impedance (*Elec.*). The impedance as normally measured, with all extraneous electromotive forces reduced to zero, between any point in a communicating system or a measuring circuit and earth.

earth inductor (*Elec. Eng.*). See **earth coil.**

earthing autotransformer (*Elec. Eng.*). See **earthing reactor.**

earthing percentage (*Teleg.*). In cable code systems, the percentage of a unit interval during which the cable is earth-connected at the sending end for the purpose of reducing distortion at the distant end.

earthing reactor (*Elec. Eng.*). (1) A reactor connected between the neutral point of an a.c. supply system and earth, in order to limit the earth current which flows on the occurrence of an earth fault. (2) An arrangement of reactors or transformers, so connected to a polyphase system that a neutral point is artificially obtained. Also called **earthing autotransformer, neutral compensator, neutral autotransformer, neutralator.**

earthing resistor (*Elec. Eng.*). A resistance through which the neutral point of a supply system is earthed, in order to limit the current which flows on the occurrence of an earth fault. Also called **earthing resistance.**

earthing switch (*Radio*). One for connecting an antenna to earth when not in use, as a protection against lightning and/or the accumulation of static charge.

earthing tyres (*Aero.*). Tyres for aircraft having an electrically conductive surface in order to discharge static electricity upon landing.

earth lead (*Radio*). Connexion between a radio transmitting or receiving apparatus and its earth.

earth leakage protection (*Elec. Eng.*). Protection system, suitable for domestic use, in which imbalance between live and neutral currents is used to trip a circuit breaker, thus isolating the supply in the event of an earth fault. See **balanced protective system.**

earth movements (*Geol.*). The general term covering change of attitude of strata due to compression and tension in the earth's crust; it includes folding and faulting of different kinds. See **epeirogenic earth movements, eustatic movements, fault.**

earth-pillars (*Geol.*). These occur where sediments consisting of relatively large and preferably flat stones, embedded in a soft, finer-grained matrix, are undergoing erosion, especially in regions of heavy rainfall. As the ground is progressively lowered the flat stones protect the softer material beneath them and are thus left standing on tall, acutely conical pillars.

earth plate (*Elec. Eng.*). A metal plate buried in the earth for the purpose of providing an electrical connexion between an electrical system and the earth.

earth potential (*Elec.*). The electric potential of the earth; usually regarded as zero, so that all other potentials are referred to it. (*Radio*) The potential of any point in a circuit when it does not vary at a radiofrequency, no matter what the steady or low-frequency potential may be. See **earthy.**

earth pressure (*Civ. Eng.*). The pressure exerted on a wall by earth which is retained, i.e., supported laterally by the wall.

earthquake (*Geol.*). A shaking of the earth's crust caused, perhaps in most cases, by displacement along a fault. The place of maximum displacement is the *focus* (cf. *epicentre*). Although the amount of the displacement may be small, a matter of inches only, the destruction wrought at the surface may be very great, due in part to secondary causes, e.g., the severing of gas mains and water mains, as in the great San Francisco earthquake in 1906. Earthquakes are classified as *simple, twin,* or *compound*, according to the nature of the shock, which is recorded by a seismometer. Volcanic phenomena are often accompanied by earthquakes. See **seismology.**

earth resistance (*Elec. Eng.*). The resistance offered by the earth between two points of connexion, and therefore forming a coupling between all circuits making use of the same current path in the earth.

earth-return circuit (*Elec. Eng.*). One which comprises an insulated conductor between two points, the circuit being completed through the earth.

earth satellite (*Space*). Manmade object, with or without radio instruments for transmitting back scientific data, which is placed by multiple rockets in an elliptical orbit about the earth.

earthshine (*Astron.*). The reflected sunlight from the surface of the earth which enables the part of the moon not directly illumined by the sun to be seen faintly against the sky.

earth's magnetic field (*Elec. Eng.*). See **terrestrial magnetism.**

earth system (*Radio*). System of wires, usually buried beneath an antenna system, to increase conductivity of the ground around antenna, making a nearly perfect reflector; also called **ground system.**

earth terminal (*Elec. Eng.*). A terminal provided on the frame of a machine or piece of apparatus to make a connexion to earth.

earth thermometer (*Meteor.*). A thermometer used for measuring the temperature of the earth at depths up to a few metres. *Symons's earth thermometer* (the most commonly used) consists of a mercury thermometer, with its bulb embedded in paraffin wax, suspended in a steel tube.

earth wax (*Chem.*). *Ozocerite* (q.v.).

earthwork (*Civ. Eng.*). A bank or cutting.

earthy (*Elec. Eng.*). Said of (1) circuits when they are connected to earth, either directly (as for direct currents) or through a capacitor (with alternating currents); (2) any point in a communicating system (e.g., the midpoint of a shunting resistance across a balanced line) which is at earth potential, although not actually connected to earth through zero impedance; (3) points of a bridge when reduced to earth potential by a Wagner earth.

earthy cobalt (*Min.*). A variety of wad containing up to about 40% of cobalt oxide. Also called **asbolite.**

E.A.S. (*Aero.*). Abbrev. for *equivalent airspeed.*

easement curve (*Surv.*). See **transition curve.**

easer (*Print.*). A general term for additives mixed with printing ink to produce some particular result, e.g., quicker drying, reduction of stiffness.

easing (*Build., Civ. Eng.*). The shaping of a curve so that there is no abrupt change of curvature in it.

easing centres (*Build., Civ. Eng.*). The process of gradually removing the centring from beneath a newly completed arch, thereby transferring its weight slowly to the arch abutments,

easing motion (*Weaving*). Movable bar over which the crossing ends are passed to facilitate making a crossed shed in weaving gauze or leno fabrics.

easing wedges (*Civ. Eng.*). *Striking wedges* (q.v.).

East Coast fever (*Vet.*). A disease of cattle in Africa due to infection by the protozoan *Theileria parva*; characterized by fever, enlarged lymph glands, respiratory distress, diarrhoea, and loss of condition. Transmitted by the tick *Rhipicephalus appendiculatus.*

East Indian satinwood (*For.*). A hardwood timber with a narrowly interlocked grain, and a fine and uniform texture. From *Chloroxylon Swietenia,* a native of India and Ceylon. It is naturally durable when used in exposed positions, but is regarded as a cabinet wood.

easting (*Surv.*). Departure, or displacement, along an east-west line of a point with reference to a true N–S datum.

easy-care (*Textiles*). Fabrics in which the physical properties have been modified to give better resistance to creasing thus needing minimum ironing.

eau de Javelle (*Chem.*). See **Javel water.**

eave (*Build.*). The lower part of a roof which projects beyond the face of the walls.

eave-board (*Build.*). See **tilting fillet.**

eave-lead (*Build.*). A lead gutter behind a parapet, around the edge of a roof.

eaves course (*Build.*). See **doubling course.**

eaves fascia (*Carp.*). See **fascia** (2).

eaves gutter (*Build.*). A trough fixed beneath an eave to catch and carry away the rain flow from the roof. Also called **shuting.**

eaves plate (*Build.*). A beam carried on piers or posts and supporting the feet of roof rafters in cases where there is no wall beneath.

eaves soffit (*Build.*). The horizontal surface beneath a projecting eave.

ébauche (*Horol.*). A partly finished watch movement, consisting of the dial plate, bridges, and balance cock.

Ebenales (*Bot.*). An order of the *Sympetalae.* Woody plants with simple leaves. The anthers are in 2 to 3 whorls, sometimes one or many. The ovary has axile placentation, made up of several loculi, each containing one or a few ovules.

E-bend (*Telecomm.*). A smooth bend in the axis direction of a waveguide, the axis being maintained in a plane parallel to the polarization direction.

Eberhard effect (*Photog.*). Variation in density between large and small areas of an emulsion which have been given equal exposure and development.

Eber's solution (*Chem.*). A solution of hydrochloric acid in ethoxyethane-ethanol mixture, which produces white fumes in contact with ammonia; used as a test for decayed meat.

ebonite (*Chem.*). A hard insulating material of rubber which has been vulcanized, i.e., the latex molecules have been cross-linked through sulphur atoms.

ebony (*For.*). Heavy hardwood from a tree of the genus *Diospyros,* a native of Nigeria, India, Ceylon, Ghana, Gambia, the East Indies, etc. It is naturally resistant to most diseases and insects. Typical uses are for making brush backs, piano keys, etc.

ebony Sindanyo (*Elec. Eng.*). See **Sindanyo.**

E.B.U. Abbrev. for *European Broadcasting Union* in West Europe.

ebullator (*Heat.*). A heated surface used to impart heat to a fluid in contact. (*Phys.*) A device placed in flooded evaporator tubes to prevent the evaporator from becoming oil-bound.

ebulliometer (*Heat*). A device which enables the true boiling point of a solution to be determined.

ebullioscopy (*Chem.*). The determination of the molecular weight of a substance by observing the elevation of the boiling-point of a suitable solvent.

ebullition (*Heat*). See **boiling.**

eburnation (*Med.*). Ivorylike hardening of bone which occurs at the joint surfaces in osteo-arthritis.

ecad (*Bot.*). A plant form which is assumed to be adapted to the habitat.

ecardinal, ecardinate (*Zool.*). Without a hinge, as some *Brachiopoda*.

Ecardines (*Zool.*). A class of *Brachiopoda* in which there is no internal skeleton to support the lophophore; the shell is without a hinge, and an anus is present.

ecbolic (*Med.*). Aiding expulsion of the foetus.

ECC (*Elec. Eng.*). **Earth continuity conductor.**

eccentric (*Bot.*). (1) Situated to one side. (2) Having fatty drops lying towards one side of a globular structure. Also **excentric.** (*Eng.*) (1) Displaced with reference to a centre; not concentric. (2) A crank in which the pin diameter exceeds the stroke, resulting in a disk eccentric to the shaft; used as a crank, particularly for operating steam-engine valves, pump plungers, etc.

eccentric angle (*Maths.*). Of a point P on an ellipse, the angle θ, where $(a \cos \theta, b \sin \theta)$ are the parametric coordinates of P referred to the axes of the ellipse.

eccentric fitting (*Plumb.*). A fitting in which the centre line is offset.

eccentric groove (*Acous.*). See **locked groove.**

eccentricity (*Build.*). The distance from the centre of application of a load or system of loads to the centroid of the section of the structural member to which it or they are connected. (*Maths.*) See **conic.**

eccentric load (*Build.*). A load which is carried by a structural member at a point other than the centroid of the section.

eccentric oösphere (*Bot.*). A fungal oösphere with one large fatty drop to one side, or several large drops included in the protoplasm to one side, or a crescentic set of small drops lying to one side.

eccentric pole (*Elec. Eng.*). A pole on an electric machine in which the pole face is not concentric with the armature but has a greater air gap at one pole tip than at the other to assist in neutralizing the effect of armature reaction.

eccentric sheave (*Eng.*). The disk of an *eccentric* (q.v.), often formed integral with the shaft.

eccentric station (*Surv.*). One not physically occupied during triangulation, etc., but serving as a fixation point.

eccentric strap (*Eng.*). A narrow split bearing, fitting on to an eccentric sheave and bolted to the end of a valve rod, etc.; corresponds to the 'big end' of a connecting-rod.

eccentric throw-out (*Eng.*). A device for engaging the back gear of a lathe. The back gear shaft runs in eccentric-bored bearings, which are rotated to bring the gears in and out of mesh with those on the mandrel. See **back gear.**

ecchondroma (*Med.*). A tumour composed of cartilage and growing from the surface of bone.

ecchondrosis (*Med.*). An abnormal outgrowth of the joint cartilage, in chronic arthritis.

ecchymoma (*Med.*). A swelling due to extravasation of blood under the skin.

ecchymosis (*Med.*). A large discoloured patch due to extravasation of blood under the skin.

Eccles-Jordan circuit (*Electronics*). Original bistable multivibrator using two triodes or transistors. See **flip-flop.**

Eccrinales (*Bot.*). An order of the *Phycomycetes*, parasitic in the alimentary canal of *Arthropoda*. The simple hyphae are attached to the wall of the alimentary canal by a holdfast.

eccrine (*Med.*). Said of a gland whose product is excreted from its cells.

eccyesis (*Med.*). Ectopic gestation. Fertilization of the ovum and growth of the foetus outside the uterus.

ecdemic (*Zool.*). Foreign; not *indigenous* (q.v.) or *endemic*.

ecdysis (*Zool.*). The act of casting off the outer layers of the integument, as in *Ophidia* and many *Arthropoda*.

ecdyson (*Cyt.*). See **moulting hormone.**

ecesis (*Biol.*). An invasion of an area by a species which is unable to establish itself there, and which dies out after a few generations.

ECG (*Med.*). Abbrev. for *electrocardiogram*. See *electrocardiograph*.

ecgonine (*Chem.*). Tropine-carboxylic acid, a coca-base alkaloid, m.p. 198°C.

echard (*Bot.*). Water present in the soil which cannot be used by plants.

echelon grating (*Light*). A form of interferometer resembling a flight of glass steps, light travelling through the instrument in a direction parallel to the treads of the steps. The number of interfering beams is therefore equal to the number of steps. Owing to the large path difference, $t(\mu - 1)$, where t is the thickness of a step and μ is the *index of refraction*, the order of interference and therefore the resolving power are high, making the instrument suitable for studying the fine structure of spectral lines.

echinate (*Bot.*). Bearing an evenly distributed coating of rather long, stiff, pointed bristles or outgrowths.

echinococcosis (*Vet.*). An infection of sheep, pigs and cattle, and sometimes man, by the intermediate hydatid stage of the tapeworm *Echinococcus granulosus*. The adult worm occurs in dogs and other carnivora.

echinococcus (*Zool.*). A bladderworm possessing a well-developed bladder containing daughter bladders, each with numerous scolices.

Echinodermata (*Zool.*). A phylum of radially symmetrical marine animals, having the body-wall strengthened by calcareous plates; there is a complex coelom; locomotion is usually carried out by the tube-feet, which are distensible finger-like protrusions of a part of the coelom known as the water-vascular system; the larva is bilaterally symmetrical and shows traces of metamerism. Starfish, Sea Urchins, Brittle Stars, Sea Cucumbers, and Sea Lilies.

Echinoidea (*Zool.*). A class of *Echinodermata* having a globular, ovoid, or heart-shaped body which is rarely flattened; there are no arms; the tube-feet possess ampullae and occur on all surfaces, but not in grooves; the anus is aboral or lateral, the madreporite aboral; there is a well-developed skeleton; free-living forms. Sea Urchins. (*Geol.*) Fossil echinoids are found in strata ranging from the Lower Palaeozoic to the present. They are particularly important in the Jurassic (Clypeus Grit, etc.) and Cretaceous, where, in the Chalk, they have proved invaluable indices of horizon, especially the various species of *Micraster* and *Holaster*.

echinopaedium (*Zool.*). See **dipleurula.**

echinopluteus (*Zool.*). In *Echinoidea*, a pelagic ciliated larval form, in which the posterolateral arms, if present, are small and directed outwards or backwards. See also **pluteus.**

echinus (*Arch.*). An ornament in the shape of an egg carved on a moulding, etc.

Echiuroidea (*Zool.*). A class of *Annelida* of sedentary marine habit, in which nearly all trace of metamerism has been lost in the adult, and the setae are much reduced in number; the body is sac-shaped, and feeding is effected by an anterior non-retractile proboscis, bearing a ciliated groove leading to the mouth.

echo (*Acous.*). Received acoustic wave, distinct from a directly received wave, because it has travelled a greater distance due to reflection or refraction. (*Radar*) Return signal in radar, whether from wanted object, or from side or back lobe radiation. (*Radio*) The reception of a signal additional to, and later than, the desired signal; caused by reflection from hills, etc., or travel completely round the earth.

echo box (*Radar*). Adjustable test resonator of high Q for returning a signal to the receiver from the transmitter.

echo chamber (*Acous.*). Same as **reverberation chamber.**

echo checking (*Comp.*). Verification of data transmission by reference back to and matching with original data.

echo flutter (*Radar*). A rapid sequence of reflected radar (or sound) pulses arising from one initial pulse.

echographia (*Med.*). Ability to copy writing, associated with inability to express ideas in writing, due to a lesion in the brain.

echolalia (*Med.*). Senseless repetition of words heard, occurring in disease of the brain or in insanity; often seen in catatonic schizophrenia.

echolocation (*An. Behav.*). Way of locating objects in the dark involving the production of high-frequency sounds, and the detection of their echoes.

echo meter (*Acous.*). An arrangement for recording the short intervals between direct sounds and their echoes from objects. Usually the principle is that of picking up the sounds with a microphone, and applying the output to a recording pen on a rotating drum carrying paper.

echopraxia, echopraxis (*Med.*). Imitation by an insane person of postures or of movements of those near him; commonly present in the catatonic type of schizophrenia.

echo ranging sonar (*Acous.*). Determination of distance and direction of objects, such as submarines, by the reception of the reflection of an ultrasonic pulse under water. See **asdic.**

echo sounding (*Acous.*). Use of echoes of pressure waves sent down to the bottom of the sea and reflected, the delay between sending and receiving times giving a measure of the depth; used also to detect wrecks and shoals of fish.

echo studio (*Acous.*). An enclosure of long reverberation period, used for the artificial introduction of an adjustable degree of reverberation in the main channel of a broadcast programme.

echo suppression (*Nav.*). The desensitizing of navigational equipment for a short period after the reception of a pulse. Pulses travelling in indirect paths are thus rejected. (*Teleph.*) In telephone 2-way circuits, the attenuation of echo currents in one direction which is due to telephone currents in the other direction.

E.C.H.O. viruses (*Med.*). *E*ntero *C*oxsackie *H*uman *O*rphan viruses, a group of Coxsackie viruses, often responsible for enteritis.

eckermannite (*Min.*). A rare monoclinic alkali amphibole; a hydrous sodium lithium aluminium magnesium silicate.

eclampsia (*Med.*). A term now restricted to the acute toxaemia occurring in pregnancy, parturition, or in the puerperium, associated with convulsions and loss of consciousness. *adj.* **eclamptic.**

E-class insulation (*Elec. Eng.*). A class of insulating material to which is assigned a temperature of 120°C. See **class -A, -B, -C, etc., insulating material.**

eclipse (*Astron.*). A name strictly applicable only to cases where a nonluminous body passes into the shadow of another, as the moon does in a lunar eclipse; used in *eclipse of the sun* to mean the interposition of the moon's disk between the observer and the sun.

eclipse seasons (*Astron.*). The two periods, approximately 6 months apart, in which solar and lunar eclipses can occur.

eclipse year (*Astron.*). The interval of time between 2 successive passages of the sun through the same node of the moon's orbit; it amounts to 346·620 03 days.

eclipsing binary (*Astron.*). A binary whose orbital plane lies so nearly in the line of sight that the components pass in front of each other in the course of their mutual revolution.

ecliptic (*Astron.*). The great circle in which the plane containing the centres of the earth and sun cuts the celestial sphere; hence, the apparent path of the sun's annual motion through the fixed stars. See **obliquity of the ecliptic.**

eclogite (*Geol.*). A coarse-grained deep-seated metamorphic rock, consisting essentially of pink garnet, green pyroxene (some of which is often chrome-diopside) and (rarely) kyanite. Good examples occur in the Fichtelgebirge and in the kimberlite pipes in S. Africa.

eclosion (*Zool.*). The act of emergence from an egg or pupa case.

ecoclimate (*Ecol.*). The sum total of meteorological factors within a habitat.

ecocline (*Ecol.*). A *cline* (q.v.) occurring across successive zones of an organism's habitat.

ecological efficiency (*Ecol.*). Ratios between the amount of energy flow at different points along a food chain, e.g., the *primary* or *photosynthetic efficiency* is the percentage of the total energy falling on the earth which is fixed by plants, this being approximately 1%.

ecological equivalents (*Ecol.*). Organisms which occupy the same ecological niche in similar communities of different bio-geographical regions, not necessarily being closely related, e.g., the Cacti of New World deserts, and the similar but totally unrelated Euphorbias of African deserts.

ecological factor (*Bot.*). Anything in the environment which affects the growth, development and distribution of plants, and therefore aids in determining the characters of a plant community.

ecological indicators (*Ecol.*). Organisms whose presence in a particular area indicates the occurrence of a particular set of water and soil conditions, temperature zones, etc. Large species with relatively specific requirements are most useful in this way, and numerical relationships between species, populations, or whole communities are more reliable than a single species.

ecological niche (*Ecol.*). The position or status of an organism within its community or ecosystem. This results from the organism's structural adaptations, physiological responses, and innate or learned behaviour. An organism's niche depends not only on where it lives but also on what it does. An economic term.

ecological pyramids (*Ecol.*). Diagrams in which the producer level forms the base and successive trophic levels the remaining tiers. They include the *pyramids* of *numbers, biomass* and *energy* (qq.v.).

ecological succession (*Ecol.*). See **succession.**

ecological threshold (*Ecol.*). The minimum concentration of an environmental factor for a particular ecological unit or process.

ecology. The scientific study of the interrelations between living organisms and their environment, including both the physical and biotic factors, and emphasizing both interspecific and intraspecific relations; the scientific study of the distribution and abundance of living organisms (i.e., exactly where they occur, and precisely how many there are), and any regular or irregular variations in distribution and abundance, followed by explanation of these phenomena in terms of the physical and biotic factors of the environment.

econometrics (*Stats.*) Statistical analysis of economic data with the aid of electronic computers.

economic ratio (*Civ. Eng.*). In reinforced concrete work, the ratio between steel reinforcement and concrete which allows the full strength of both to be developed at the same time.

economizer (*Eng.*). A bank of tubes, placed across a boiler flue, through which the feed water is pumped, being heated by the otherwise waste heat of the flue gases. See also **Supermiser.** (*Light*) A small hood of refractory material placed over the tip of an arc-lamp carbon. It results in a slower burning-away of the carbon, and also gives an improved light distribution.

Economo's disease (*Med.*). See **Von Economo's disease.**

economy resistance (*Elec. Eng.*). A resistance inserted into the circuit of a contactor coil or other electromagnetic device after its initial operation, in order to reduce the current to a value just sufficient to hold the device closed.

ecophene (*Ecol.*). A type of individual developing as the result of a physiological, as opposed to genetic, response to habitat factors.

ecospecies (*Ecol.*). See **ecophene.**

ecosystem (*Ecol.*). Any area of nature which includes living organisms and nonliving substances interacting to produce an exchange of materials between the living and nonliving parts, e.g., a pond, lake, or forest. It comprises four constituents, abiotic substances, producers, consumers, and decomposers, and is the basic functional unit in ecology.

ecotone (*Bot.*). A boundary between two plant communities of major rank.

ecrustaceous (*Bot.*). Said of a lichen having no well-defined thallus.

E.C.T. (*Psychiat.*). See **electroconvulsive therapy.**

ectadenia (*Zool.*). In male Insects, accessory glands of the genital system, of ectodermal origin. Cf. *mesadenia.*

ectal layer (*Bot.*). A thin membrane at the extreme edge of an excipulum.

ectasia, ectasis (*Med.*). Pathological dilation or distension of any structure of the body. *adj.* **ectatic.**

ectental line (*Zool.*). In a gastrula, the line of union between ectoderm and endoderm at the lip of the blastopore.

ectepicondylar (*Zool.*). In many Reptiles and Mammals, said of a foramen, near the lower end of the humerus, for the median (ulnar) nerve.

ectethmoid (*Zool.*). One of a pair of cartilage bones of the Vertebrate skull, formed by ossification of the ethmoid plate.

ecthoraeum (*Zool.*). The thread of a nematocyst.

ecthyma (*Med.*). Local gangrene and ulceration of the skin as a result of infection, the ulcer being covered by a crust and the skin round it being inflamed.

ectoblast (*Zool.*). See **epiblast.**

ectobronchi (*Zool.*). In Birds, secondary bronchi leading to the air-sacs, being diverticula of the mesobronchium.

Ectocarpales (*Bot.*). An order of the *Phaeophyta.* The alternating generations are morphologically similar. Growth of the filament is trichothallic. The sporophyte produces zoospores, or neutral spores, and the gametophyte isogametes.

ectocarpous (*Zool.*). Said of gonads which originate from ectoderm: having such gonads.

ectochondrosteosis (*Zool.*). Ossification of cartilage, beginning at the inner surface of the perichondrium and gradually invading the cartilage.

ectocoelic (*Zool.*). In *Coelenterata,* situated outside the coelenteron.

ectocrine (*Biol.*). Organic substances, present in minute amounts in the environment of an organism, which exert an effect on its form, development or behaviour.

ectocuneiform (*Zool.*). A bone in the distal row of tarsals in the hind foot of *Tetrapoda,* corresponding to the *magnum* of the fore-limb.

ectocyst (*Zool.*). In *Polyzoa,* the cuticular exoskeleton of a zooecium secreted by the ectoderm: in *Protozoa,* the resistant outer layer of a cyst. Cf. *endocyst.*

ectoderm (*Zool.*). The outer layer of cells forming the wall of a gastrula: the tissues directly derived from this layer.

ectoethmoid (*Zool.*). A paired cartilage bone which forms in the outer walls of the olfactory capsules in some *Craniata.*

ectogenesis (*Bot.*). Variation under the influence of conditions outside the plant. (*Zool.*) Development outside the body.

ectogenous (*Zool.*). Independent; self-supporting.

ectogeny (*Bot.*). The effect of pollen on the tissues of the female organs of the plant.

ectohormones (*Zool.*). A term originally used for *pheromones* (q.v.), but now little used.

ectolecithal (*Zool.*). Said of ova in which the yolk is deposited peripherally.

ectoloph (*Zool.*). The outer border of a Mammalian tooth.

ectomorphy (*Psychol.*). Sheldon's *somatotype* which is characterized by thinness. Said to be associated with more or less introversive personality traits. See **endomorphy, mesomorphy.**

-ectomy. Suffix from Gk. *ektomē,* cutting out, used esp. in surgical terms.

ectoneural system (*Zool.*). A nerve plexus underlying the ectoderm in *Echinodermata.*

ectoparasite (*Bot.*). A parasite feeding on the internal tissues of the host, but having the greater part of the plant body, and the reproduction structures on the surface. (*Zool.*) A parasite living on the surface of the host.

ectopatagium (*Zool.*). In *Cheiroptera,* that part of the patagium borne by the metacarpals and phalanges.

ectophloeodal (*Bot.*). Living on the outside of bark.

ectophloic (*Bot.*). Said of a stem which has no internal phloem.

ectopia, ectopy (*Med.*). Displacement from normal position.

ectopia cordis (*Med.*). Congenital displacement of the heart outside the thoracic cavity.

ectopia vesicae (*Med.*). A congenital abnormality in which the anterior wall of the bladder is absent and the posterior wall opens on to the surface of the abdomen, the lower abdominal wall being also absent.

ectopic gestation (*Med.*). See **eccyesis.**

ectoplasm (*Cyt.*). The layer of a clear non-granular cytoplasm, usually modified at the periphery of a cell, next to the cell wall. Cf. *endoplasm.*

ectoplast (*Bot.*). The outer surface of the cytoplasm of a plant cell, lying against the inner surface of the cell wall.

Ectoprocta (*Zool.*). A phylum of *Metazoa* with the anus outside the lophophore; with a coelomic body cavity and the lophophore retractable into a tentacle sheath and without excretory organs.

ectopterygoid (*Zool.*). In some Fish, a paired ventral bone of the skull, lying between the entopterygoid and the quadrate: in some Reptiles, a bone of the skull extending from the pterygoid to the maxilla.

ectopy (*Med.*). See **ectopia.**

ectosarc (*Cyt.*). See **ectoplasm.**

ectosome (*Zool.*). In eurypylous *Porifera*, the dermal membrane covering the openings of the incurrent sinuses.

ectosporous (*Bot.*). Having exogenous spores.

ectostroma (*Bot.*). The part of the stroma of the *Sphaeria es* growing on the bark, beneath or within the periderm, and consisting of fungus tissue only.

ectotrachea (*Zool.*). A layer of pavement epithelium, outside the chitinous lining of a trachea.

ectotrophic mycorrhiza (*Bot.*). A mycorrhiza in which most of the fungal hyphae lie on the surface of the root of the higher plant.

ectoturbinals (*Zool.*). The longer outer series of ethmoturbinal plates in Mammals. Cf. *entoturbinals.*

ectotympanic (*Zool.*). A ring-shaped bone supporting the tympanum in some primitive Mammals, e.g., *Marsupialia, Insectivores, Edentata.*

ectozoon (*Zool.*). See **ectoparasite.**

ectromelia (*Med.*). (1) Congenital absence of a limb or limbs. (2) An infectious disease of mice, due to a virus.

ectropion, ectropium (*Med.*). Eversion of the eyelid.

eczema (*Med.*). The term generally applied to an itching inflammatory condition of the epidermis occurring as an allergic reaction to certain chemical agents, poisons of some plants, materials used in trades, or as an inherited nervous tendency.

eczematous conjunctivitis (*Med.*). See **phlyctenular conjunctivitis.**

edaphic climax (*Bot.*). A climax community of which the existence is determined by some property of the soil.

edaphic factor (*Bot.*). Any property of the soil, physical or chemical, which influences plants growing on that soil.

edaphology (*Bot.*). Study of the conditions of a plant determined by the soil.

edaphon (*Bot.*). A community of bacteria, algae, fungi, protozoa, worms, crustacea, and insects, living in the soil and influencing its nitrogen content.

eddy. An interruption in the steady flow of a fluid, caused by an obstacle situated in the line of flow; the *vortex* so formed.

eddy-current brake (*Elec. Eng.*). (1) A form of brake for the loading of motors during testing; it consists of a mass of metal rotating in front of permanent magnets so that heavy eddy currents are set up in it. (2) A form of brake, used on tramways, in which the retarding force is produced by the induction of eddy currents in the rail by an electromagnet on the vehicle.

eddy-current heating (*Elec. Eng.*). See **induction heating.**

eddy-current speed indicator (*Elec. Eng.*). A speed indicator consisting of a rotating disk and a spring-controlled magnetic needle; the latter is deflected as a result of eddy currents induced in the disk.

eddy currents (*Elec. Eng.*). Those arising through varying electromotive forces consequent on varying magnetic fields, resulting in diminution of the latter and dissipation of power. They are one of the main causes of heating in motors, transformers, etc. (known also as *iron loss*, cf. *copper loss*). To minimize them the iron cores of such machines are composed of many layers of thin iron sheet (laminations) which are insulated from one another to reduce the currents' ability to flow in the direction in which the field tries to induce them. This in turn slightly increases the reluctance of the magnetic circuit, thereby reducing the overall efficiency of the machine, hence a compromise must be struck. Used for mechanical damping and braking (as in electricity meters) and for induction heating as applied in case-hardening. Also called **Foucault current(s).**

eddy diffusion (*Chem. Eng.*). The migration and interchange of portions of a fluid as a result of their turbulent motion. Cf. *diffusion.*

eddy diffusivity (*Chem. Eng.*). Exactly analogous, for eddy diffusion, to the *diffusivity* (q.v.) for molecular diffusion. Symbol E.

eddy flow (*Phys.*). See **turbulent flow.**

Eddy's theorem (*Eng.*). The bending moment at any point in an arch is equal to the product of the horizontal thrust at the abutment and the vertical distance between the line of action of this thrust and the given point in the arch.

eddy wind (*Meteor.*). See **whirlwind.**

Edeleanu process (*Chem.*). A process for the removal of unsaturated compounds from petroleum distillates by extraction with liquid SO_2.

edelopal (*Min.*). A variety of *opal* with an exceptionally brilliant play of colours.

edema, edematous (*Med.*). See **oedema, oedematous.**

edenite (*Min.*). An end-member compositional variety in the hornblende group of monoclinic amphiboles: a hydrous sodium, magnesium, calcium, and aluminium silicate.

Edentata (*Zool.*). An order of primitive terrestrial Mammals characterized by the incomplete character of the dentition; now subdivided into orders *Xenarthra, Tubulidentata* and *Pholidota*; the testes are abdominal; phytophagous or insectivorous forms. Sloths, ant-eaters, armadillos.

edentulous, edentate (*Zool.*). Without teeth.

edge. See **entering edge, leading edge, leaving edge, trailing edge.**

edge coal (*Mining*). Highly inclined coal-seams.

edged (*Bot.*). Having a margin of a different colour from the rest.

edge effect (*Acous.*). In acoustic absorption measurements, the variations which arise from the size, shape or division of the areas of material being tested—a diffraction effect. (*Ecol.*) The tendency for increased variety and density of organisms at an *ecotone* (q.v.) due to the presence of many of the organisms of both the overlapping communities, as well as some other species characteristic of the ecotone itself. (*Elec. Eng.*) Deviation from parallelism in fields at the edge of parallel plate capacitors, or between poles of permanent or electro-magnets, thus leading to nonuniformity of the field at the edges. (*Photog.*) See **border effect, fringe effect.**

edge filter (*Chem. Eng.*). A type of filter in which a large number of disks are clamped on a perforated hollow shaft joining a cylinder. This is contained in another cylindrical vessel into which liquid is pumped and flows through the narrow spaces between the disks, the solids being trapped on the disk edges. Filtrate leaves via the perforations in the hollow shaft.

edge plane (*Tools*). One with the cutter at the extreme front for working into corners.

edge planing (*Print.*). Trimming and squaring the edges of printing plates, either original or duplicate, by hand- or power-operated tools.

edge rolled (*Bind.*). Said of a pattern on the edges of leather bindings, either blind or decorated, applied by a *roll* (q.v.).

edge runner. A grinding mill used for putty, mortar, pigments, etc.; it consists of cylindrical stones or rollers so mounted as to run on their edges in a circular pan containing the materials to be ground.

edge tones (*Acous.*). Tones produced by the impact of an air jet on a sharply-edged dividing surface, as in an organ pipe mouth.

edge tool (*Eng.*). A hand-worked, mallet-struck or machine-operated cutting tool with one or a regular pattern of cutting edges, i.e., excluding grinding wheels.

edge trimming plane (*Tools*). One with a rebated sole, for squaring up edges.

edge water (*Geol.*). That pressing inward upon the gas or oil in a natural reservoir.

edge winding (*Elec. Eng.*). A form of winding frequently used for the field windings of salient-pole synchronous machines; it consists of copper strip wound on edge around the pole. Such a winding has good heat-dissipating properties.

edgewise instrument (*Elec. Eng.*). A switchboard indicating instrument in which the pointer moves in a plane at right angles to the face of the switchboard. The end of the pointer is bent and moves over a narrow scale.

edingtonite (*Min.*). A rare zeolite; a hydrated barium aluminium silicate.

Edison accumulator (*Elec. Eng.*). *Nickel-iron-alkaline accumulator* (q.v.), or secondary cell.

Edison effect (*Electronics*). The phenomenon of electrical conduction between an incandescent filament and an independent cold electrode contained in the same envelope, when the second electrode is made positive with reference to a part of the filament. Precursor of *Fleming diode* (q.v.). Also called **Richardson effect.**

Edison phonograph (*Acous.*). The original type of gramophone, in which the records were registered on the surface of hollow cylindrical waxes.

Edison screw-cap (*Elec. Eng.*). A lamp cap in which the outer wall forms one of the contacts, and which is in the form of a coarse screw for inserting into a corresponding socket. A central pin forms the other contact. Sizes in descending order include goliath, large, medium, small, miniature and lilliput.

Edison screw-holder (*Elec. Eng.*). A holder for electric lamps with Edison screw-caps.

E display (*Radar*). Display in which target range and elevation are plotted as horizontal and vertical coordinates of the blip.

Ediswan wiring system (*Elec. Eng.*). A wiring system employing flat rubber- or metal-sheathed cables, with special junction boxes and fixing arrangements.

edit (*Comp.*). To prepare data for initial or further processing.

editing (*Cinema.*). In making up a motion picture, the process of determining the shots to be taken, and the cutting of the resulting shots.

edition (*Print.*). A number of copies printed at one time, either as the original issue (*first edition*), or when the text has undergone some change, or the type has been partly or entirely reset, or the format has been altered.

edition binding (*Bind.*). The normal style of binding for hardcover books, highly mechanized, only occasionally sewn on tapes, with case usually cloth-covered and lettered on spine only. Also called **publisher's binding.** Cf. *bound book, library binding.*

editor (*Cinema.*). (1) The person who acts under the director in film production, and who does the cutting and editing of the shots while making up the final sequence in a film. (2) See **viewer.**

edriophthalmic (*Zool.*). Having sessile eyes, as some *Crustacea.*

Edser and Butler's bands (*Light*). Dark bands, having a constant frequency separation, which are seen in the spectrum of white light which has traversed a thin, parallel-sided plate of a transparent material, or a thin parallel-sided film of air between glass plates.

EDT (*Radio*). See **ethene diamine tartrate.**

EDTA (*Chem.*). *E*thene *d*iamine *t*etra-acetic (ethanoic) *a*cid, combines with metals to form chelates. A *complexone*, it is used, e.g., in special soaps, to remove metallic contamination.

```
HO·CO-CH₂\                 /CH₂-CO·OH
           N·CH₂·CH₂·N
HO·CO-CH₂/                 \CH₂-CO·OH
```

Edwards' roaster (*Min. Proc.*). Long horizontal furnace through which sulphide minerals are rabbled counterwise to hot air, to remove part or all of the sulphur by ignition.

EEG (*Med.*). Abbrev. for *electroencephalograph* (or *-gram*).

eel (*Zool.*). Any fish of the Anguillidae, Muraenidae, or other family of the *Apodes*. The name is extended to other fish of similar form, e.g., sand eel.

eel-grass (*Acous.*). A sea plant (*Zostera marina*) whose grasslike leaves are used for sound insulation and correction of acoustical defects.

effective address (*Comp.*). One which has been modified during the process of a computation.

effective antenna height (*Radio*). Height (in metres) which, when multiplied by the field strength (in volts per metre) incident upon the antenna, gives the e.m.f. (in volts) induced therein. It is less than the physical height and differs from the equivalent height in that it is also a function of the direction of arrival of the incident wave.

effective atomic number (*Chem.*). That of a compound or mixture is given by Spier as

$$Z_e = [a_1 Z_1^{2 \cdot 94} + a_2 Z_2^{2 \cdot 94} \ldots]^{1/2 \cdot 94},$$

where a_1, $a_2 \ldots$ are the fractional electron contents of elements of at. no. $Z_1, Z_2, \ldots$.

effective bandwidth (*Telecomm.*). The bandwidth of an ideal (rectangular) band-pass filter which would pass the same proportion of the signal energy as the actual filter.

effective column length (*Build.*). The column length which is used in finding the *ratio of slenderness* (q.v.).

effective depth (*Civ. Eng.*). The depth of a reinforced concrete beam as measured from the surface of the concrete on the compression side to the centre of gravity of the tensile reinforcement. See cover.

effective energy (*Radiol.*). The quantum energy (or wavelength) of a monochromatic beam of X-rays or γ-rays with the same penetrating power as a given heterogeneous beam. Its value depends upon the nature of the absorbing medium. Also known as **effective wavelength.**

effective half-life (*Nuc.*). The time required for the activity of a radioactive nuclide in the body to fall to half its original value as a result of both biological elimination and radioactive decay. Its value is given by

$$\frac{\tau(b\frac{1}{2})\times\tau(r\frac{1}{2})}{\tau(b\frac{1}{2})+\tau(r\frac{1}{2})},$$

where $\tau(b\frac{1}{2})$ and $\tau(r\frac{1}{2})$ are the biological and radioactive half-lives respectively.

effective heating surface (*Eng.*). The total area of a boiler surface in contact with water on one side and with hot gases on the other.

effective mass (*Electronics*). For electrons and/or holes in a semiconductor, effective mass is a parameter which may differ appreciably from the mass of a free electron, and which depends to some extent on the position of the particle in its energy band. This modifies the mobility and hence the resulting current.

effective particle density (*Powder Tech.*). The mass of a particle divided by its volume, including opened and closed pores.

effective porosity (*Powder Tech.*). That portion of the powder porosity which is readily accessible to a fluid moving through a powder compact.

effective radiated power (*Radio*). Actual maximum or unmodulated power delivered to a transmitting aerial (antenna) multiplied by the factor of *gain* in a specified direction in the horizontal plane.

effective radius of the earth (*Astron.*). See **Earth** (*radius of*).

effective range. That part of the scale of an indicating instrument over which a reasonable precision may be expected.

effective resistance (*Elec. Eng.*). The total a.c. resistance covering eddy-current losses, iron losses, dielectric and corona losses, transformed power, as well as conductor loss. For a sinusoidal current it is the component of the voltage in phase with current divided by that current. Measured in ohms.

effective sieve aperture size (*Powder Tech.*). To allow for the size aperture distribution in a real sieve for accurate particle size analysis, sieves have to be calibrated. This is sometimes done by analysing a powder of known distribution. If *A* percentage by weight should be retained on a sieve of nominal aperture size *a*, then *B* percentage is retained on the actual sieve; then from the particle size distribution of the powder *B*, the size at which *B* percentage of the powder is greater than this size, is read off. The effective size of the sieve aperture is then stated to be *B*. The value of *B* will depend upon the size distribution of the powder used to calibrate the sieve. The magnitude of the quantity *a-B* is termed the **aperture error of a sieve.**

effective span (*Build.*). The horizontal distance between the centres of the two bearings at the ends of a beam.

effective SPF (*Nuc. Eng.*). See **simple process factor.**

effective temperature (*Astron.*). The temperature which a given star would have if it were a perfect radiator, or *black body*, with the same distribution of energy among the different wavelengths as the star itself.

effective value. That of a simple parameter which has the same effect as a more complex one, e.g., r.m.s. value of a.c. = d.c. value for many purposes.

effective wavelength (*Light*). See **dominant wavelength.** (*Radiol.*) See **effective energy.**

effector (*Zool.*). A tissue-complex capable of effective response to the stimulus of a nervous impulse, e.g., a muscle or gland.

effector neurone (*Zool.*). A motor neurone.

efferent (*Zool.*). Carrying outwards or away from; as the *efferent branchial vessels* in a Fish, which carry blood away from the gills, and *efferent nerves*, which carry impulses away from the central nervous system. Cf. *afferent*.

effervescence (*Chem.*). The vigorous escape of small gas bubbles from a liquid, especially as a result of chemical action.

efficiency. A non-dimensional measure of the performance of a piece of apparatus, e.g., an engine, obtained from the ratio of the output of a quantity, e.g., power, energy, to its input, often expressed as a percentage. The power efficiency of an ICE is the ratio of the shaft- or brake-horsepower to the rate of intake of fuel, expressed in units of energy content per unit time. It must always be less than 100% which would imply perpetual motion. Not to be confused with **efficacy**, which takes account only of the output of the apparatus, and is not given an exact quantitative definition.

efficiency (or booster) diode (*TV*). Diode used in television receivers to increase the beam deflection voltage of a CRT during the forward part of a cycle of operations.

efficiency index (*Bot.*). The rate at which dry matter accumulates in the plant.

efficiency of impaction (*Powder Tech.*). Ratio of the cross-sectional area of the stream from which particles of a given size are removed, to the total cross-sectional area of the jet stream, both areas measured at the mouth of the jet.

efficiency of screening, numerical index of (*Mining, Powder Tech.*). A quantity used to assess the efficiency of industrial screening procedures, i.e., sieving procedures. It is defined by the equation $E=F(D/B-C/A)$, where

E = the numerical index of efficiency of screening,
F = the percentage of the powder supply passed by the screen,
A = the percentage of difficult oversize in the powder supply,
B = the percentage of difficult undersize in the powder supply,
C = the percentage of the difficult oversize passed by the screen,
D = the percentage of the difficult undersize passed by the screen.

efficiency ratio (*Eng.*). Of a heat engine, the ratio of the actual thermal efficiency to the ideal thermal efficiency corresponding to the cycle on which the engine is operating.

effigurate (*Bot.*). Having a distinct shape.

effiguration (*Bot.*). Outgrowth of the receptacle.

effleurage (*Med.*). The action of lightly stroking in massage.

efflorescence (*Bot.*). Production of flowers; period of flowering. (*Build.*) Formation of a white crystalline deposit on the face of a wall; due to the drying out of salts in the mortar or stone. (*Chem.*) The loss of water from a crystalline hydrate on exposure to air, shown by the formation of a powder on the crystal surface. (*Min.*) A fine-grained crystalline deposit on the surface of a mineral or rock.

effluent (*Nuc. Eng.*). Radioactive waste from atomic plants, etc. (*San. Eng.*) Liquid sewage after having passed through any stage in its purification.

effluent monitor (*Nuc. Eng.*). Instrument for measuring level of radioactivity in fluid effluent.

effluve (*Elec. Eng.*). The corona discharge from an electrostatic machine or high-frequency generator, used to stimulate the human skin.

efflux (*Aero.*). The mixture of combustion products and cooling air which forms the propulsive medium of any jet or rocket engine.

effort syndrome (*Med.*). *D.A.H.* (q.v.). Soldier's heart. A condition in which there is nervousness with symptoms of circulatory inefficiency and exhaustion after exercise, in the absence of heart disease; common in armies.

effuse (*Bot.*). Spread out on a substratum, and often having a vaguely defined edge.

effusiometer (*Chem.*). An apparatus for comparing the relative molecular masses of gases by observing the relative times taken to stream out through a small hole.

effusion (*Med.*). An abnormal outpouring of fluid into the tissues or cavities of the body, as a result of infection or of obstruction to blood-vessels or lymphatics. (*Phys.*) The flow of gases through larger holes than those to which diffusion is strictly applicable; see **Graham's law.** The rate of flow is approximately proportional to the square-root of the pressure difference.

egest (*Zool.*). To throw out, to expel; to defaecate, to excrete. *n.pl.* **egesta.**

egg (*Bot., Zool.*). See **ovum.** (*Vet.*) See **soft-shelled egg.**

egg albumen (*Chem.*). A simple protein from the white of the egg, soluble in water and coagulable by heat.

egg and anchor (*Arch.*). An ornament carved on a moulding, resembling eggs separated by vertical anchors.

egg and dart (*Arch.*). Similar to the above, arrows taking the place of anchors.

egg apparatus (*Bot.*). The egg and the two synergidae in the embryo sac of an angiosperm.

egg-bound (*Vet.*). Said of the oviduct of birds when obstructed by an egg.

egg-box lens (*Radio*). Same as **slatted lens.**

egg-cell (*Zool.*). The ovum, as distinct from any other cells associated with it.

egg-eating (*Vet.*). A vice developed by individual birds, characterized by the eating of their own eggs or of those of other birds.

egg nucleus (*Zool.*). The female pronucleus.

egg-peritonitis (*Vet.*). Septic peritonitis extending from an infected and obstructed oviduct of birds.

egg-shaped sewer (*San. Eng.*). A type of sewer section much used where the flow is a fluctuating one; the section resembles the longitudinal profile of an egg placed with the end of smaller radius at the bottom, and thus increases the velocity of flow at the bottom.

egg-shell finish (*Paper*). A soft dull finish on paper, obtained by omitting the calendering process.

egg-shell gloss (*Paint.*). A subdued gloss in paint finish.

egg sleeker (*Foundry*). A moulder's sleeker with a spoon-shaped end; used for smoothing rounded corners in a mould.

egg tooth (*Zool.*). A sharp projection at the tip of the upper beak of a young bird, by means of which it breaks open the egg shell.

E glass (*Glass*). A glass containing not more than 1% of alkali (calculated as Na_2O) and used for the manufacture of glass fibre.

ego (*Psychol.*). That part of the self formed originally from the instinctual life forces or *id*, which have become modified by contact with reality and the world, and which has tested which part of the *id* can be accepted by the self in conjunction with the demands of reality.

egocentric (*Psychol.*). Piaget's term for the early period of a child's development when he sees the world only from his own point of view, without knowing of the existence of other viewpoints or perspectives, and hence is unaware of his egocentrism.

ego-defence mechanisms (*Psychol.*). The unconscious means by which an individual excludes from consciousness memories, aspects of his personality, etc., which are unacceptable. Such factors remain influential in the unconscious. See **projection, repression.**

ego-ideal (*Psychol.*). The conscious aims and ideals of the individual, formed by his own conscious judgment, education, and culture, and supported by his early moral and ethical standards, based on those of the parents. See **superego.**

egomania (*Psych.*). Excessive self-esteem.

egophony (*Med.*). See **aegophony.**

EGT (*Aero.*). Exhaust *gas temperature.*

Egyptian (*Typog.*). See **slab serif.**

Egyptian cotton (*Textiles*). Cotton of long staple (30 to 40 mm) of excellent quality and therefore used for spinning fine combed counts for shirtings, dress cloths, handkerchiefs, etc., and as doubling wefts for sewing threads, etc.

Egyptian jasper (*Min.*). A variety of jasper occurring in rounded pieces scattered over the surface of the desert, chiefly between Cairo and the Red Sea; used as a broochstone and for other ornamental purposes. Typically shows colour zoning.

E.H.F. (*Radio*). *Extremely high frequencies.*

e.h.p. (*Aero.*). Abbrev. for *total equivalent brake horsepower*. Also **t.e.h.p.**

Ehrlich's haemotoxylin stain (*Micros.*). An acid haemotoxylin stain which is prepared by the addition of glacial acetic acid, glycerol, alum, and distilled water to haemotoxylin dissolved in alcohol. This stain is left to 'ripen', i.e., until it becomes a dark red colour, before use.

Ehrlich's theory (*Biol.*). See **side-chain theory.**

eidetic imagery (*Psychol.*). The ability to reproduce on a dark screen, or when the eyes are closed, a vividly clear picture or visual memory-image of previously seen objects. Commonly present in children up to 14 years, and occasionally persisting into adult life.

eidograph (*Surv.*). An instrument for reducing and enlarging plans.

eidophor system (*TV*). A projection system in which diffraction effects are used to produce the modulation of light for the picture.

Eifelian Stage (*Geol.*). A series of marine Devonian strata, defined in terms of the succession in the Eifel. Small stage-names are now commonly used, *Eifelian* being almost obsolete.

Eiffel wind tunnel (*Aero.*). An open jet, non-return flow wind tunnel.

eigentones (*Phys.*). The natural frequencies of vibration of a system.

eigenvalues (*Phys.*). Possible values for a parameter of an equation for which the solutions will be compatible with the boundary conditions. In quantum mechanics, the energy eigenvalues for the *Schrödinger equation* (q.v.) are possible *energy levels* (q.v.) for the system. (*Maths.*) Of a matrix, see **characteristic equation of a matrix.**

eight bend (*Eng.*). A pipe bend or junction piece for joining 2 pipes inclined at $22\frac{1}{2}°$, i.e., one-eighth of a complete reversal of direction.

eighteenmo (*Print.*). Written 18mo. The 18th of a sheet or a sheet folded to make eighteen leaves or thirty-six pages.

eight-to-pica leads (*Typog.*). Strips of metal, $1\frac{1}{2}$-point in thickness, used to space out lines of type. They are usually called thin leads, two being equivalent to a thick lead, and eight to a pica em.

Eikmeyer coil (*Elec. Eng.*). The name given to the

original type of former-wound armature coil which can be dropped straight into the slots of an electric machine.

Einstein-de Haas effect (*Mag.*). See **Barnett effect.**

Einstein diffusion equation (*Chem.*). As a result of Brownian motion, molecules or colloidal particles migrate an average distance δ in each small time interval τ. Hence the equation for the diffusion of a spherical particle of radius r through a fluid of viscosity η may be written

$$D=\frac{\delta^2}{2\tau}=\frac{RT}{6\pi\eta rN},$$

where D is the diffusion coefficient, N is Avogadro number, R is the gas constant and T the absolute temperature.

Einstein energy (*Phys.*). See **mass-energy equivalence.**

Einstein equation for the specific heat of a solid (*Chem.*). One mole of the solid consists of N molecules, each vibrating with frequency ν in 3 dimensions. Hence, by quantum theory, the molar specific heat may be written:

$$C_\nu=\frac{3R\,x^2\,e^x}{(e^x-1)^2},$$

where $x = h\nu/kT$, R is gas constant, h Planck's constant, k Boltzmann's constant and T the absolute temperaure.

einsteinum (*Chem.*). Artificial element, symbol Es, at. no. 99, produced by bombardment in a cyclotron, but also recognized in H-bomb débris, it having been produced by beta decay of uranium which had captured a large number of neutrons. The longest lived isotope is ^{254}Es, with a half-life of greater than 2 years.

Einstein law of photochemical equivalence (*Chem.*). Each quantum of radiation absorbed in a photochemical process causes the decomposition of one molecule.

Einstein photoelectric equation (*Chem.*). That which gives the energy of an electron, just ejected photoelectrically from a surface by a photon, i.e., $E=h\nu-w$, where E=kinetic energy, h=Planck's constant, ν=frequency of photon, and w=work function.

Einstein shift (*Astron.*). A spectroscopic phenomenon, predicted by the Einstein theory of gravitation and subsequently verified by observation, in which the lines of the solar spectrum are slightly displaced from their normal positions towards the red, owing to the intense gravitational field of the sun.

einstellung (*Psychol.*). See **set.**

Einthoven galvanometer (*Elec. Eng.*). A galvanometer in which the current is carried by a single current-carrying filament in a strong magnetic field, the deflection usually being magnified by a microscope. Also called a **string galvanometer.**

eisenkiesel (*Min.*). A ferruginous quartz, yellow, red, or brown, according to iron content.

eisodal aperture (*Bot.*). The enlargement of the stomatal pore nearest to the surface of the leaf.

ejaculation, ejection (*Bot.*). The forcible expulsion of spores from a sporangium.

ejaculatory duct (*Zool.*). See **ductus ejaculatorius.**

ejection capsule (*Aero.*). A cockpit, cabin, or portion of either, in a high-altitude and/or high-speed military aeroplane which can be fired clear in emergency and which, after being slowed down, descends by parachute.

ejector (*Autos.*). See **extractor.** (*Eng.*) (1) A device for exhausting a fluid by entraining it by a high-velocity steam or air jet, e.g., an *air ejector* (q.v.). (2) A mechanism for removing a part or assembly from a machine at the end of an automatic sequence of operations, e.g., in assembling or machining. (*San. Eng.*) An appliance used for raising sewage from a low-level sewer to a sewer at a higher elevation; worked by compressed air.

ejector seat (*Aero.*). A crew seat for high-speed aeroplanes which can be fired, usually by slow-burning cartridge, clear of the structure in emergency. Automatic releases for the occupant's safety harness and for parachute opening are usually incorporated. Also **ejection seat.**

eka- (*Chem.*). A prefix denoting the element occupying the next lower position in the same group in the periodic system; used in the naming of new elements and unstable radio-elements.

ektrodactylia (*Med.*). Congenital absence of one or more fingers or toes.

elaeoblast (*Zool.*). In certain *Urochorda*, a posterior mass of large mesodermal cells filled with nutritive matter.

elaeodochon (*Zool.*). See **oil gland.**

elaeolite (*Min.*). A massive form of the mineral nepheline, greenish-grey or (when weathered) red in colour, usually shapeless, but in some S. African syenites exhibiting the hexagonal prismatic form of nepheline.

elaeolite-syenite (*Geol.*). See **nepheline-syenite.**

elaidic acid (*Chem.*). $CH_3(CH_2)_7\,CH{=}CH(CH_2)_7{\cdot}COOH$. Mono-unsaturated fatty acid, m.p. 51°C. The trans-form of *oleic acid* (q.v.).

elaioplast (*Bot.*). A plastid which forms oils and fats.

elaiosome (*Bot.*). An outgrowth from the surface of a seed, containing fatty or oily material (often attractive to ants) and serving in seed dispersal.

Elasipoda (*Zool.*). An order of abyssal *Holothuroidea*, characterized by the possession of shield-shaped buccal tentacles, without either ampullae or retractor muscles, and having no respiratory trees or Cuvierian organs.

Elasmobranchii (*Zool.*). A class of *Gnathostomata*, highly developed usually predacious fishes with a cartilaginous skeleton and plate-like gills. Includes sharks, skates, rays, etc.

elastance (*Elec.*). The reciprocal of the capacitance of a capacitor, so termed because of electromechanical analogy with a spring. Unit is the *daraf* (*farad* spelt backwards).

elastica interna (externa) (*Zool.*). In developing Vertebrates, a thin membrane near the notochord, on the inner (outer) side of the fibrous sheath.

elastic bitumen (*Min.*). See **elaterite.**

elastic collision (*Phys.*). Such collision that no change in the total energy or momentum takes place.

elastic constants (*Eng., Phys.*). Quantities, expressed in units of stress (MN/m^2 or tonf/in^2) used to describe the behaviour of a material when subjected to stress in one of three modes: longitudinal, shear or compression. These give rise to *Young's* (elongational), *rigidity* and *bulk moduli* of elasticity respectively. Also includes *Poisson's ratio*. Known alternatively as **elastic moduli, constants or moduli of elasticity.**

elastic fatigue (*Eng., Phys.*). A temporary departure from perfect elasticity shown by some materials, which, after suffering elastic deformation, slowly return to their original form.

elastic fibres (*Zool.*). See **yellow fibres.**

elastic fibrocartilage (*Zool.*). See **yellow fibrocartilage.**

elasticity (*Phys.*). The tendency of a body to return to its original size or shape, after having been stretched, compressed, or deformed. The ratio of the stress called into play in the body by the action of the deforming forces to the strain or change in dimensions or shape is called the *coefficient* (or *modulus*) *of elasticity*. See the following definitions and **Hooke's law**.

elasticity of bulk (*Phys*.). The elasticity for changes in the volume of a body caused by changes in the pressure acting on it. The bulk modulus is the ratio of the change in pressure to the fractional change in volume. See **elasticity**.

elasticity of compression. See **elasticity of bulk**.

elasticity of elongation (*Phys*.). The stress in this case is the stretching force per unit area of cross-section and the strain is the elongation per unit length. The modulus of elasticity of elongation is known as *Young's modulus*. See **elasticity**.

elasticity of flexure (*Phys*.). The elasticity of a bent beam or cantilever which tends to straighten it. For a weightless beam supported at two points distance L apart, the sag produced by a force P applied at the centre is equal to $\frac{PL^3}{4bd^3E}$, where b and d are the breadth and thickness of the beam and E is Young's modulus of elasticity.

elasticity of gases (*Phys*.). If the volume V of a gas is changed by δV when the pressure is changed by δp, the modulus of elasticity is given by $-V\frac{\delta p}{\delta V}$. This may be shown to be numerically equal to the pressure p for isothermal changes, and equal to γp for adiabatic changes, γ being the ratio of the specific heats of the gas.

elasticity of shear (or rigidity) (*Phys*.). The elasticity of a body which has been pulled out of shape by a shearing force. The stress is equal to the tangential shearing force per unit area, and the strain is equal to the angle of shear, that is, the angle turned through by a straight line originally at right angles to the direction of the shearing force. See **elasticity, Poisson's ratio**.

elastic limit (*Phys*.). The limiting value of the deforming force beyond which a body does not return to its original shape or dimensions when the force is removed. (*Met*.) The highest stress that can be applied to a metal without producing a measurable amount of plastic (i.e., permanent) deformation. Usually assumed to coincide with the limit of proportionality.

elastic medium (*Phys*.). A medium which obeys *Hooke's law* (q.v.). No medium is perfectly elastic, but many are sufficiently so to justify the making of calculations which assume perfect elasticity.

elastic moduli (*Eng*.). *Elastic constants*. See **bulk modulus, modulus of rigidity, Young's modulus**.

elastic packing (*Eng*.). Rubber-impregnated canvas material used for packing the glands or stuffing-boxes of water pumps. See **packing**.

elastic scattering (*Nuc*.). See **scattering**.

elastic strain (*Eng*.). The strain or fractional deformation undergone by a material in the elastic state, i.e., a strain which disappears with the removal of the straining force.

elastic tissue (*Zool*.). A form of connective tissue in which elastic fibres predominate.

elastin (*Zool*.). The substance of which elastic fibres are composed; resistant to reagents. See **scleroproteins**.

elastivity (*Elec*.). The reciprocal of the permittivity of a dielectric.

elastomer (*Chem*.). A material, usually synthetic, having elastic properties akin to those of rubber.

elastomeric yarns (*Textiles*). Yarns comprising filament fibres or staples, e.g., polyurethane, treated to give elastic properties varying with the construction and end uses.

elater (*Bot*.). (1) Elongated cell with spiral thickenings on walls, found mixed with the spores of some liverworts. (2) Thread of a capillitium.

elaterite or elastic bitumen (*Min*.). A solid bitumen resembling dark-brown rubber; used in building. Sometimes known as **mineral caoutchouc**; occurs at Castleton, Derbyshire, and in Colorado and Utah.

elaterophore (*Bot*.). The organ which bears elaters.

elaulic (*Hyd*.). A term meaning *oil in pipes*, to correspond with the term *hydraulic* (meaning *water in pipes*).

E-layer (*Phys*.). Most regular of the ionized regions in the ionosphere, which reflects waves from a transmitter back to earth. Its effective maximum density increases from zero before dawn to its greatest at noon, and decreases to zero after sunset, at heights varying between 110 and 120 km. There are at least two such layers. Also called **Heaviside layer** or **Kennelly-Heaviside layer**.

elbaite (*Min*.). One of the three chief compositional varieties of tourmaline; a complex borosilicate of lithium, aluminium and sodium. Most of the gem varieties of tourmaline are elbaites.

elbow (*Build*.). An arch stone whose lower bed is horizontal, while its upper bed is inclined towards the centre of the arch, to correspond with those of the voussoirs. (*Elec. Eng*.) A *sharp-bend* (q.v.). (*Eng*.) A short right-angle pipe connexion, as distinct from a bend which is curved not angular. (*Mining*) A sharp turn in a roadway underground; a sharp bend in a cable or pipe line. (*Plumb*.) A short pipe fitting used to connect 2 pipes at an angle, generally a right angle.

elbow-board (*Carp*.). The window-board beneath a window, in the interior.

elbow linings (*Join*.). The panelling at the sides of a window recess, running from the floor to the level of the window-board. Cf. *jamb linings*.

ELDO (*Space*). Abbrev. for *European Launcher Development Organisation*.

Electra complex (*Psychol*.). Attraction for the father and hostility to the mother; analogous to the Oedipus complex in boys, but applied to girls.

electret (*Elec. Eng*.). Permanently polarized dielectric material, formed by cooling barium titanate from above a Curie point, or waxes (e.g., carnauba) in a strong electric field.

electric. Said of any phenomena which depend essentially on a peculiarity of electric charges.

electrical. Descriptive of means related to, pertaining to, or associated with electricity, but not inherently functional.

electrical absorption (*Elec*.). An effect in a dielectric whereby, after an initially charged capacitor has been once discharged, it is possible after a few minutes to obtain from it another discharge, usually smaller than the first.

electrical analogy (*Acous*.). The correspondence between electrical and acoustical systems, which assists in applying to the latter procedures familiar in the former.

electrical bias (*Telecomm*.). The use of a polarizing winding on a relay core, for adjusting the sensitivity of the relay to signal currents.

electrical chain (*Telecomm*.). Number of circuits (e.g., tuned circuits) coupled together so that energy is transferred from one to the next.

electrical communication. See **telecommunication**.

electrical conductivity (*Elec.*). Ratio of current density to applied electric field. Expressed in siemens per metre (S/m) or ohm^{-1} m^{-1} in SI units. Conductivity of metals at high temperatures varies as T^{-1}, where T is absolute temperature. At very low temperatures, variation is complicated but it increases rapidly (at one stage proportional to T^{-5}), until it is finally limited by material defects of structure.
electrical degrees (*Elec.*). Angle, expressed in degrees, of phase difference of vectors, representing currents or voltages, arising in different parts of a circuit.
electrical dischargers (*Aero.*). See **earthing tyres, static wick dischargers.**
electrical double layer (*Chem.*). The layer of adsorbed ions at the surface of a dispersed phase which gives rise to the *electrokinetic effects* (q.v.).
electrical engineering. That branch of engineering chiefly concerned with the design and construction of all electrical machinery and devices, communications, power transmission, etc.
electrical heating (*Heat*). Heating by electrical means such as current flow through a resistance, induction currents in a conductor, displacement currents in a dielectric, etc.
electrical recording (*Acous.*). Use of amplified currents from microphones for operating electromagnetic or electrodynamic drives for the cutting stylus in disk recording, in contrast to the previous acoustic (mechanical) method of recording, now obsolete. See **tape recording.**
electrical reproduction (*Acous.*). Reproduction of recorded sound by piezo- or electromagnetic devices operated by the tracking needle, as contrasted with *acoustic reproduction*, in which the tracking needle drives the centre of the diaphragm of a sound-box connected to a horn (now largely obsolete).
electrical reset (*Elec. Eng.*). Restoration of a magnetic device, e.g., relay or circuit breaker, by auxiliary coils or relays.
electrical resistivity (*Elec.*). Reciprocal of *electrical conductivity* (q.v.). Units are ohm metres.
electrical resonance (*Elec. Eng.*). Condition arising when a maximum of current or voltage occurs as the frequency of the electrical source is varied; also when the length of a transmission line approximates to multiples of a quarter-wavelength and the current or voltage becomes abnormally large.
electrical thread (*Elec. Eng.*). A form of thread used on screwed steel conduit for electrical installation work.
electric-arc furnace (*Elec. Eng.*). See **arc furnace.**
electric-arc welding (*Elec. Eng.*). See **arc welding.**
electric axis (*Telecomm.*). Direction in a crystal which gives the maximum conductivity to the passage of an electric current. The X-axis of a piezoelectric crystal.
electric balance (*Elec. Eng.*). A name sometimes applied to a type of electrometer, to a current weigher (which establishes the absolute ampere), and to a Wheatstone bridge.
electric bell (*Elec. Eng.*). A bell in which the hammer is operated electrically by means of a solenoid. A single stroke may be given, or, more commonly, a rapid succession of strokes may be maintained by means of a make-and-break contact on the solenoid.
electric braking (*Elec. Eng.*). A method of braking for electrically driven vehicles; the motors are used as generators to return the braking energy to the supply, or to dissipate it as heat in resistances.
electric calamine (*Min.*). See **hemimorphite.**
electric cautery (*Med.*). Burning of parts of the human body for surgical purposes by means of electrically heated instruments. Also called **electrocautery.**
electric chorea (*Med.*). See **Dubini's disease.**
electric circuit (*Elec.*). Series of conductors, forming a partial, branched, or complete path, which substantially confines the flow of electrons forming a current because of great contrast in conductivity.
electric cleaner (*Elec. Eng.*). In electroplating work, a cleansing solution in which the cleansing is accelerated by the passage of an electric current.
electric (or electrostatic) component (*Elec. Eng.*). That component of an electromagnetic wave which produces a force on an electric charge, and along the direction of which currents in a conductor exposed to the field are urged to flow.
electric conduction (*Elec. Eng.*). Transmission of energy by flow of charge along a conductor.
electric connective tissue (*Zool.*). Jellylike connective tissue in which the electroplaxes of certain types of electric organs are embedded.
electric current (*Elec.*). See **current.**
electric dipole (*Elec.*). See **electric doublet.**
electric discharge (*Elec.*). Same as **field discharge.**
electric-discharge lamp (*Light*). A form of electric lamp in which the light is obtained from an electric discharge between two electrodes in an evacuated glass tube. Sometimes called a **gas-discharge lamp.**
electric double layer (*Elec. Eng.*). A positive and a negative layer distribution of electric charge very close together so that effectively the total charge is zero but the 2 layers form an assembly of dipoles, thus giving rise to an electric field.
electric doublet (*Elec.*). System with a definite electric moment, mathematically equivalent to 2 equal charges of opposite sign at a very small distance apart.
electric dynamometer (*Elec. Eng.*). See **dynamometer.** (*Eng.*) An electric generator which is used for measuring power. The stator frame is capable of partial rotation in bearings concentric with those of the armature, and the torque is balanced and measured by hanging weights on an arm projecting from the frame.
electric ear (*Min. Proc.*). Microphone used in grinding circuit to monitor level of noise in ball mill.
electric eye. Photoelectric cell used to monitor plant details, such as (*Min. Proc.*) turbidity of solutions, pile-up at transfer points of ore, levels in ore bins. Interruption of narrow beam of light is registered, and signalled to control point. (*Radio*) Miniature cathode-ray tube used, e.g., in a radio receiver to exhibit a pattern determined by the rectified output voltage obtained from the received carrier, thus assisting in tuning the receiver. Also used for balancing a.c. bridges. Also **cathode-ray tuning indicator, electron-ray indicator tube, magic eye.**
electric field (*Elec.*). Region in which forces are exerted on any electric charge present.
electric field strength (*Elec.*). The strength of an electric field is measured by the force exerted on a unit charge at a given point. Expressed in volts/metre. Symbol E.
electric flux (*Elec.*). Surface integral of the electric field intensity normal to the surface. The electric flux is conceived as emanating from a positive charge and ending on a negative charge without loss. Symbol Ψ.
electric flux density (*Elec.*). See **displacement.**
electric generator (*Elec. Eng.*). See **generator.**

electric harmonic analyser (*Elec. Eng.*). An electrical device for determining the magnitudes of the harmonics in the wave shape of an alternating current or voltage.

electricity. The manifestation of a form of energy associated with static or dynamic electric charges.

electricity meter (*Elec. Eng.*). See **integrating meter.**

electric lamp (*Light*). A lamp in which an electric current is used as the source of energy for radiating light.

electric-light ophthalmia (*Med.*). See **photophthalmia.**

electric locomotive (*Elec. Eng.*). A locomotive in which the driving power is supplied by electric motors, supplied either from a battery (**battery locomotive**) or from a contact wire or rail track *electrification*. Standard supplies are (overhead) 25 kV a.c. single phase, or (3rd rail) 750 V d.c.

electric machine (*Elec. Eng.*). See **electric motor, electromagnetic generator, electrostatic generator.**

electric moment (*Elec. Eng.*). Product of the magnitude of either of 2 equal electric charges and the distance between their centres, with axis direction from the negative to the positive charge. See also **magnetic moment.**

electric motor (*Elec. Eng.*). Any device for converting electrical energy into mechanical torque; occasionally called an **electromotor.**

electric nerve (*Zool.*). A modified motor nerve, serving an electric tissue.

electric organ (*Zool.*). A mass of muscular or epithelial tissue, modified for the production, storage, and discharge of electric energy; occurring mainly in Fish.

electric oscillations (*Radio*). Electric currents which periodically reverse their direction of flow, at a frequency determined by the constants of a resonant circuit. See also **continuous oscillations, electronic oscillations.**

electric polarization (*Elec. Eng.*). The dipole moment per unit volume of a dielectric.

electric potential (*Elec. Eng.*). That measured by the energy of a unit positive charge at a point, expressed relative to *zero potential* (q.v.).

electric resistance welded tube (*Chem. Eng.*). See **e.r.w. tube.**

electric rods (*Zool.*). Small rodlike structures, attached to the electrolemma of an electroplax, through which the electricity is discharged.

electric shielding (*Elec. Eng.*). See **Faraday cage.**

electric spectrum (*Light*). The colour spectrum of an electric arc.

electric storm (*Meteor.*). A meteorological disturbance in which the air becomes highly charged with static electricity. In the presence of clouds this leads to thunderstorms.

electric strength (*Elec. Eng.*). Maximum voltage which can be applied to an insulator or insulating material without sparkover or breakdown taking place. The latter arises when the applied voltage gradient coincides with a breakdown strength at a temperature which is attained through normal heat dissipation.

electric susceptibility (*Elec. Eng.*). The amount by which the relative permittivity of a dielectric exceeds unity—or the ratio of the polarization produced by unit field to the permittivity of free space.

electric tissue (*Zool.*). Modified tissue capable of generating electricity.

electric traction (*Elec. Eng.*). The operation of a railway or road vehicle by means of electric motors, which obtain their power from an overhead contact wire or from batteries mounted on the vehicle.

electric-wave filter (*Telecomm.*). One in which there is a phase retardation for those currents which are passed and an effective time delay for a signal comprising these frequencies, in getting through the filter. Also **frequency-discriminating filter.**

electric wind (*Elec. Eng.*). Stream of air caused by the repulsion of charged particles from a sharply pointed portion of a charged conductor.

electrification (*Elec. Eng.*). (1) Charging a network to a high potential. (2) Conversion of any motive power to electric drive, e.g., trains. (3) Charging a conductor by electric induction from another charged conductor.

electro (*Print.*). Abbrev. for *electrotype.*

electroacoustics (*Acous.*). The branch of technology dealing with the interchange of electric and acoustic energy, e.g., as in a transducer.

electroanalysis (*Chem.*). Electrodeposition of an element or compound to determine its concentration in the electrolyzed solution. See **conductimetric analysis, polarograph, potentiometer, voltameter.**

electroarteriograph (*Med.*). Instrument for recording blood-flow rates.

electroballistics. The study of velocities of projectiles using electronic means.

electrobiology (*Biol.*). The study of electrical phenomena associated with the function of living organisms.

electroblast (*Zool.*). An embryonic cell which will give rise to an electroplax.

electrobrightening (*Met.*). See **electrolytic polishing.**

electrocapillary effect (*Chem.*). The decrease in interfacial tension, usually of mercury, caused by the mutual repulsion of adsorbed ions opposing the attractive force of interfacial tension. See also **capillary electrometer, electrical double layer.**

electrocapillary maximum (*Chem.*). The potential at which a mercury surface in an electrolyte is charge-free and consequently has the maximum interfacial tension; about −0·28 volts.

electrocardiograph (*Med.*). A modification of Einthoven's string galvanometer, used for making *electrocardiograms* (abbrev. *ECG*), i.e., graphic records of the electrical changes during contraction of the muscle of the heart.

electrocardiography (*Med.*). The study of electric currents produced in cardiac muscular activity.

electrocataphoresis (*Phys.*). See **electrophoresis.**

electrocautery (*Med.*). See **electric cautery.**

electrocement (*Civ. Eng.*). Cement made, in an electric furnace, by adding lime to molten slag.

electrochemical constant (*Chem.*). See **faraday.**

electrochemical equivalent (*Elec.*). The mass of a substance deposited at the cathode of a voltameter per coulomb of electricity passing through it.

electrochemical (or **electrode potential**) **series** (*Chem.*). The classification of elements in the order of the electrode potential which is developed when an element is immersed in a solution of molar ionic concentration. Also **contact series.**

electrochemistry (*Chem.*). That branch of technology dealing with the electrical and electronic aspects of chemical laws or processes.

electrochondria (*Zool.*). Granules occurring in the cytoplasm of an electroplax.

electrochronograph (*Elec. Eng.*). The combination of an electrically driven clock and an electromagnetic recorder for recording short time intervals.

electrocoagulation (*Med.*). Coagulation of bodily tissues by high-frequency electric current.

electroconvulsive therapy (*Psychiat.*). A method of treatment of mental disorders which involves the passing of an electric current through the brain. This induces a convulsive reaction and the occasional fracturing of bones, the patient being unconscious at the time and therefore not feeling pain. The method is said to relieve depression and reduce excitation. Abbrev. E.C.T.

electro copper glazing (*Build.*). See **copper glazing.**

electroculture. Stimulation of the growth of plants by electrical means.

electrode (*Elec. Eng.*). (1) Conductor whereby an electric current is led into or out of a liquid (as in an electrolytic cell), or a gas (as in an electric discharge lamp or gas tube), or a vacuum (as in a valve). (*Electronics*) In a semiconductor, emitter or collector of electrons or holes.

electrode admittance (*Elec. Eng.*). Admittance measured between an electrode and earth when all other potentials on electrodes are maintained constant.

electrode boiler (*Elec. Eng.*). A boiler in which heat is produced by the passage of an electric current through the liquid to be heated.

electrode characteristic (*Elec. Eng.*). Graph relating current in electronic device to potential of one electrode, that of all others being maintained constant.

electrode conductance (*Elec. Eng.*). In-phase or real component of an electrode admittance.

electrode current (*Electronics*). The net current flowing in a valve (or tube) from an electrode into the surrounding space.

electrode dark current (*Electronics*). The current which flows in a camera tube or phototube when there is no radiation incident on the photocathode, given certain specified conditions of temperature and shielding from radiation. It limits the sensitivity of the device.

electrode dissipation (*Electronics*). Power released at an electrode, usually an anode, because of electron or ion impact. In large valves the temperature is held down by radiating fins, graphiting to increase radiation, or by water- or oil-cooling.

electrode efficiency (*Elec. Eng.*). The ratio of the quantity of metal deposited in an electrolytic cell to the quantity which should theoretically be deposited according to Faraday's laws.

electrode holder (*Elec. Eng.*). In electric arc-welding, a device used for holding the electrode and leading the current to it.

electrode impedance (*Elec. Eng.*). Impedance (or its components) is the ratio of a small sinusoidal voltage on an electrode to the corresponding sinusoidal current, all other electrodes being maintained at constant potential.

electrodeless discharge (*Electronics*). That arising in a discharge tube because of a sufficiently intense externally applied high-frequency electromagnetic field.

electrodeposition (*Chem.*). Deposition electrolytically of a substance on an electrode, as in electroplating or electroforming.

electrode potential series (*Chem.*). See **electrochemical series.**

electrode resistance (*Elec. Eng.*). Reciprocal of *electrode conductance*.

electrodermal effect (*Physiol.*). Change in skin resistance consequent upon emotional reactions.

electrodesiccation (*Med.*). Destruction of animal tissue by electric sparks, using a moving electrode and a high-frequency generator. Also **diathermic coagulation, fulguration.**

electrodialysis (*Chem.*). Removal of electrolytes from a colloidal solution by an **electric field** between electrodes in pure water outside the two dialysing membranes between which is contained the colloidal solution.

electro-diesel locomotive (*Rail.*). An *electric locomotive* which carries a low power diesel engine and generator for use in sidings, etc., or in emergency, to provide traction current.

electrodisintegration (*Nuc.*). Disintegration of nucleus under electron bombardment.

electrodispersion (*Chem.*). Production of metal in colloidal solution by striking an electric arc between metal electrodes immersed in the solvent.

electrodissolution (*Elec. Eng.*). Dissolving a substance from an electrode by electrolysis.

electrodynamic instrument (*Elec. Eng.*). An electrical measuring instrument which depends for its action on the electromagnetic force between 2 or more current-carrying coils.

electrodynamic loudspeaker (*Acous.*). Open-diaphragm loudspeaker, in which the radiating cone is driven by current in a coil, fixed to the apex of the cone, and moving axially in the radial magnetic field of a pot magnet.

electrodynamic microphone (*Acous.*). See **moving-coil microphone.**

electrodynamics. Science dealing with the interaction, or forces between, currents, or the forces on currents in independent magnetic fields.

electrodynamic wattmeter (*Elec. Eng.*). One for low-frequency measurements. It depends on the torque exerted between currents carried by fixed and movable coils.

electroencephalogram (*Med.*). Trace produced by an *electroencephalograph*. Abbrev. **EEG.**

electroencephalograph (*Med.*). Instrument for study of voltage waves associated with the brain; effectively comprises a sensitive detector (voltage or current), a d.c. amplifier of very good stability and an electronic recording system. Abbrev. **EEG.**

electroendosmosis (*Chem.*). Movement of liquid, under an applied electric field, through a fine tube or membrane. Also **electro-osmosis, electrosmosis.**

electroextraction (*Met.*). The recovery of a metal from a solution of its salts by passing an electric current through the solution.

electrofacing (*Elec. Eng.*). The process of coating, by electrodeposition, a metal surface with a harder metal to render it more durable.

electrofluor (*Elec., Light*). A transparent material which has the property of storing electrical energy and releasing it as visible light.

electrofluorescence (*Light*). See **electroluminescence.**

electroformed sieves (*Powder Tech.*). Fine mesh sieves formed by photograving and electroplating of nickel.

electroforming (*Electronics*). Electrodeposition of copper on stainless steel formers, to obtain components of waveguides with closely dimensioned sections. (*Eng.*) A primary process of forming metals, in which parts are produced by electrolytic deposition of metal on a conductive removable mould or matrix.

electrogen (*Electronics*). A molecule which emits electrons when it is illuminated.

electrogoniometer (*Elec. Eng.*). A name sometimes used to denote a *phase indicator* (q.v.).

electrogram (*Elec. Eng.*). A chart, obtained from an electrograph, of variations in the atmospheric potential gradient.

electrograph (*Elec. Eng.*). A recording electrometer for measuring the potential gradient in the atmosphere.

electrographite (*Powder Tech.*). Carbon can be transformed into graphite by heat. Temperature in excess of 1500°C causes detectable graphite formation. Temperatures in the range 2200°C to 2800°C are commonly used.

electrography (*Elec. Eng.*). Direct electric recording on to teledeltos paper.

electrohydraulic forming (*Eng.*). The discharge of electrical energy across a small gap between electrodes immersed in water, which produces a shock wave. The wave travels through the water, until it hits and forces a metal workpiece into the particular shape of the die which surrounds it, thus, e.g., piercing holes, rock-crushing, etc. Sometimes called *explosive forming*.

electrokinetic effects (*Chem.*). Phenomena due to the interaction of the relative motion with the potential between the two phases in a dispersed system. There are four: *electroendosmosis*, *electrophoresis*, *streaming potential*, and *sedimentation potential*.

electrokinetic potential (*Chem.*). Potential difference between surface of a solid particle immersed in aqueous or conducting liquid and the fully dissociated ionic concentration in the body of the liquid. Concept important in froth flotation, electrophoresis, etc. Also called **zeta potential**.

electrokinetics (*Elec.*). Science of electric charges in motion, without reference to the accompanying magnetic field.

electrokymograph (*Instr.*). An electronic recording instrument used in electrobiology.

electrolemma (*Zool.*). A thin membrane surrounding an electroplax.

electrolier (*Light*). Earlier name for a hanging ornamental *luminaire* with more than one lamp.

electroluminescence (*Light*). Luminescence produced by the application of an electric field to a dielectric phosphor. Also termed **electrofluorescence**. See **Gudden-Pohl effect**.

electrolysis (*Chem.*). Chemical change, generally decomposition, effected by a flow of current through a solution of the chemical, or its molten state, based on ionization. (*Med.*) The removal of hair by applying an electrically charged needle to the follicle.

electrolyte (*Chem.*). Chemical, or its solution in water, which conducts current through ionization.

electrolyte strength (*Chem.*). Extent towards complete ionization in a dilute solution. When concentrated, the ions join in groups, as indicated by lowered mobility.

electrolytic arrester (*Elec. Eng.*). See **aluminium arrester**.

electrolytic capacitor (*Elec. Eng.*). An electrolytic cell in which a very thin layer of nonconducting material has been deposited on one of the electrodes by an electric current. This is known as 'forming' the capacitor, the deposited layer providing the dielectric. Because of its thinness a larger capacitance is achieved in a smaller volume than in the normal construction of a capacitor. In the so-called *dry electrolytic capacitor* the dielectric layer is a gas which however is actually 'formed' from a moist paste within the capacitor.

electrolytic cell (*Chem.*). An assembly of electrodes in an electrolyte, such as a voltameter.

electrolytic copper (*Met.*). Copper refined by electrolysis. This gives metal of high purity (over 99·94% copper), and enables precious metals, such as gold and silver, to be recovered.

electrolytic corrosion (*Met.*). Corrosion produced by contact of two different metals when an electrolyte is present and current flows.

electrolytic depolarization (*Elec. Eng.*). See **depolarization**.

electrolytic dissociation (*Chem.*). The splitting-up (which is reversible) of substances into oppositely-charged ions.

electrolytic instrument (*Elec. Eng.*). An instrument depending for its operation upon electrolytic action, e.g., an electrolytic meter.

electrolytic ion (*Chem.*). Charged current carriers formed by dissociation of ionic compound in polar liquid, such as water. See **electrostatic bonding**.

electrolytic lead (*Met.*). Lead refined by the Betts Process; has purity of about 99·995–99·998% lead.

electrolytic lightning arrester (*Elec. Eng.*). A lightning arrester consisting of a number of electrolytic cells in series; it breaks down, allowing the lightning stroke to discharge to earth, when the voltage across it exceeds about 400 volts/cell.

electrolytic machining (*Eng.*). An electrochemical process based on the same principles as electroplating, except that the workpiece is the anode and the tool is the cathode, resulting in a deplating operation. Sometimes, e.g., in electrolytic grinding, combined with some abrasive action.

electrolytic meter (*Elec. Eng.*). An integrating meter whose operation depends on electrolytic action.

electrolytic polarization (*Elec. Eng.*). Change in the potential of an electrode when a current is passed through it. As current rises, polarization reduces the p.d. between the two electrodes of the system.

electrolytic polishing (*Met.*). By making the metal surface an anode and passing a current under certain conditions, there is a preferential solution so the microscopic irregularities vanish, leaving a smoother surface; also termed **anode brightening, anode polishing, electrobrightening, electropolishing**.

electrolytic rectifier (*Elec. Eng.*). One with electrolyte and electrodes, such that one electrode becomes polarized and opposes the flow of current in one direction.

electrolytic refining (*Met.*). The method of producing pure metals, by making the impure metal the anode in an electrolytic cell and depositing a pure cathode. The impurities either remain undissolved at the anode or pass into solution in the electrolyte.

electrolytic tank (*Elec. Eng.*). A device used to simulate field systems, e.g. the electrostatic field around the electrodes in a CRT. The tank is filled with a poorly conducting fluid in which is immersed a scale model in metal of the desired system. Appropriate voltages are applied between the various parts and the equipotentials are traced out using a suitable probe.

electrolytic wirebar (*Elec. Eng.*). A bar of electrolytically refined copper of suitable dimensions for rolling to form wire.

electrolytic zinc (*Met.*). Zinc produced from its ores by roasting (to convert sulphide to oxide), solution of oxide in sulphuric acid, precipitation of impurities by adding zinc dust, and final electrolytic deposition of zinc on aluminium cathodes. Product has purity over 99·9%.

electromagnet (*Elec. Eng.*). Soft iron core, embraced by a current-carrying coil, which exhibits appreciable magnetic effects only when current passes.

electromagnetic brake (*Elec. Eng.*). A brake in which the braking force is produced by the friction between two surfaces pressed together by the action of a solenoid, or by magnetic attraction, the necessary flux being produced by an electromagnet.

electromagnetic clutch (*Autos., Eng.*). A friction clutch without pressure springs, operated by a solenoid connected to the dynamo. As the current increases the solenoid presses the plates together. A switch in the base of the gear lever breaks the circuit and releases the pressure for gear changes. See also **Smith's coupling.**

electromagnetic component (*Mag.*). The magnetic component of an electromagnetic wave. (*Radio*) Strictly, that component of the combined field surrounding a transmitting antenna which represents the radiated energy.

electromagnetic control (*Elec. Eng.*). A form of remote control for switchgear, etc., in which operation is effected by means of a solenoid.

electromagnetic damping (*Elec. Eng.*). See **magnetic damping.**

electromagnetic deflection (*Electronics*). Deflection of the beam in a cathode-ray tube by a magnetic field produced by a system of coils carrying currents, e.g., for scanning a TV image or for providing a time-base deflection.

electromagnetic field theory (*Phys.*). Theory based on Maxwell's equations and concerned with the interaction between electromagnetic waves and matter.

electromagnetic focusing (*Electronics*). The focusing of a beam of charged particles by magnetic fields associated with current-carrying coils. Used in cathode-ray tube and electron microscope. See **electron lens.**

electromagnetic generator (*Elec. Eng.*). An electric generator which depends for its action on the induction of e.m.f.s in a circuit by a change in the magnetic flux linking with that circuit.

electromagnetic horn (*Radio*). Metal horn designed to radiate a beam of ultra-short-wave energy originated by a dipole within the horn.

electromagnetic induction (*Elec. Eng.*). Transfer of electrical power from one circuit to another by varying the magnetic linkage. See **Faraday's law of induction.**

electromagnetic inertia (*Elec. Eng.*). The energy required to stop or start a current in an inductive circuit. Inductance is analagous to mass in a mechanical system.

electromagnetic instruments (*Elec. Eng.*). Electrical measuring instruments whose action depends on the electromagnetic forces set up between a current-carrying conductor and a magnetic field. See **moving-coil instrument, dynamometer.**

electromagnetic lens (*Electronics*). One using current-carrying coils to focus electron beams.

electromagnetic loudspeaker (*Acous.*). One involving the motion of a magnetic armature.

electromagnetic microphone (*Acous.*). A type of microphone in which the generated electromotive force arises from the motion of a magnetic circuit, so that varying flux generates the electromotive forces in a coil.

electromagnetic mirror (*Elec. Eng.*). A reflecting surface for electromagnetic waves.

electromagnetic pick-up (*Acous.*). One in which the motion of the stylus, in following the recorded track, causes a fluctuation in the magnetic flux carried in any part of a magnetic circuit, with consequent electromotive forces in any coil embracing such magnetic circuit.

electromagnetic pole-piece (*Elec. Eng.*). In a U-shaped core, the pole-pieces are attached at the free end and are often conical in shape to concentrate the magnetic field in the air gap.

electromagnetic prospecting (*Min.*). Method in which distortions produced in electric waves are observed and lead to pinpointing of geological anomalies.

electromagnetic pump (*Elec. Eng.*). Pump designed to conduct fluids, e.g., liquid metals, and to maintain circulation without use of moving parts.

electromagnetic radiation (*Phys.*). The emission and propagation of electromagnetic energy from a source, including long (radio) waves, heat rays, light, X-rays, and (hard) γ-rays.

electromagnetic reaction (*Elec. Eng.*). Reaction between the anode and grid circuits of a valve obtained by electromagnetic coupling; also called **inductive reaction, magnetic reaction.**

electromagnetics. See **electromagnetism.**

electromagnetic separation (*Min. Proc.*). Removal of ferromagnetic objects from town refuse, or 'tramp iron' from bulk materials, as they travel along a conveyor, over a drum, or into a revolving screen, by setting up a magnetic field which diverts the ferromagnetic material from the rest. Concentration of ferromagnetic minerals from gangue. (*Nuc. Eng.*) Isotope separation by *electromagnetic focusing* (q.v.), as in a mass spectrometer.

electromagnetic spectrum (*Phys.*). See **electromagnetic wave.**

electromagnetic switch (*Elec. Eng.*). A switch whose opening and closing is effected by means of electromagnets or solenoids.

electromagnetic theory (*Phys.*). The theory which accounts for the propagation of *electromagnetic waves.*

electromagnetic units. Any system of units based on assigning arbitrary value to permeability of free space. (Unity on CGS electromagnetic system, $4\pi \times 10^{-7}$ H/m on SI system.)

electromagnetic wave (*Phys.*). One formed of time-dependent mutually-perpendicular electric and magnetic fields. The spectrum of electromagnetic waves includes radio waves, heat and light rays, ultraviolet rays, X-rays, and gamma rays.

electromagnetism (*Elec., Mag.*). Science of the properties of, and relations between, magnetism and electric currents. Also called **electromagnetics.**

electromalux (*TV*). A tube using a mosaic, which functions photoelectrically when used as a TV camera.

electromechanical brake (*Elec. Eng.*). A form of electric brake in which the braking force is obtained partly as the result of the attraction of two magnetized surfaces, and partly by mechanical means, as a result of the operation of a solenoid.

electromechanical counter (*Elec. Eng.*). One which records mechanically the number of electric pulses fed to a *solenoid.*

electromechanical recorder (*Elec. Eng.*). One which changes electrical signals into a mechanical motion of a similar form, and cuts or records the shape of the motion in an appropriate medium.

electrometallization (*Met.*). The electrodeposition of a metal on a nonconducting base, either for decorative purposes or to give a protective covering.

electrometallurgy (*Met.*). A term covering the various electrical processes for the industrial working of metals, e.g., electrodeposition, electrorefining, and operations in electric furnaces.

electrometer (*Elec. Eng.*). Fundamental instrument for measuring potential difference, depending on the attraction or repulsion of charges on plates or wires.

electrometer gauge (*Elec. Eng.*). A small attracted-disk electrometer sometimes attached to the needle of a quadrant electrometer to determine whether the needle is sufficiently charged.

electrometer tube or **valve** (*Elec. Eng.*). A valve designed to have a very high insulation resistance to the grid, and to be suitable for use with very high grid input resistance circuits. Currents measured are in the range 10^{-10}–10^{-14} amp.

electrometric titration (*Chem.*). A titration in which the end-point is determined by observing the change of potential of an electrode immersed in the solution titrated.

electromotive force (*Elec. Eng.*). Difference of potential produced by sources of electrical energy which can be used to drive currents through external circuits. Abbrev. **e.m.f.**, symbol *E*.

electromotive intensity (*Elec. Eng.*). Same as **potential gradient.**

electromotive series (*Chem.*). See **electrochemical series.**

electromotor (*Elec. Eng.*). See **electric motor.**

electromyography (*Med.*). The study of electric currents set up in muscle fibres by bodily movement.

electron (*Nuc.*). A fundamental particle with negative electric charge of $1{\cdot}602 \times 10^{-19}$ coulombs and mass $9{\cdot}109 \times 10^{-31}$ kg. Electrons are grouped round the nuclei of atoms in several possible shells, but also exist independently and produce the various electric effects observable in different materials. Due to their small size the wave properties of electrons and their relativistic effects are particularly significant. An equivalent particle but with a positive electric charge is known as the *positron* and is the antiparticle to the electron. The term electron is sometimes used generically to cover both electrons and positrons. Also **negatron.**

electron affinity (*Electronics*). (1) Tendency of certain substances, notably oxidizing agents, to capture an electron. (2) See **work function.**

electron attachment (*Nuc.*). Formation of negative ion by attachment of free electron to neutral atom or molecule.

electron beam (*Electronics*). A stream of electrons moving with the same velocity and direction in neighbouring paths and usually emitted from a single source such as a cathode.

electron beam valve (or tube) (*Electronics*). One in which several electrodes control one or more electron beams.

electron binding energy (*Phys.*). Same as **ionization potential.**

electron camera (*TV*). Generic term for a device which converts an optical image into a corresponding electric current directly by electronic means, without the intervention of mechanical scanning. See **emitron, iconoscope, image dissector.**

electron capture (*Nuc.*). That of a shell electron (K or L) by its own nucleus, decreasing the atomic number of the atom without change of mass.

electron charge/mass ratio (*Electronics*). A fundamental physical constant, the mass being the rest mass of the electron: $e/m = 1{\cdot}759 \times 10^{11}$ C kg^{-1}.

electron cloud (*Chem.*). Electrons which are not bound to any particular atom in a substance, their freedom to diffuse giving rise to the conductivity and lustre of metals and graphite. (*Electronics*) Region in the inter-electrode space of an electron-discharge tube containing large numbers of relatively stationary electrons. See also **space charge, virtual cathode.**

electron concentration (*Chem.*). Ratio of number of valence electrons to number of atoms in a molecule.

electron conduction (*Electronics*). That which arises from the drift of free electrons in metallic conductors when an electric field is applied. See ***n*-type** and ***p*-type semiconductor.**

electron-coupled oscillator (*Electronics*). Valve with four (or more) electrodes, the first three electrodes being used for the production of the oscillations, the output being taken from a subsequent electrode which is electron-coupled to the oscillator circuit proper. Characterized by the very small effect of the load on the frequency of oscillation.

electron coupling (*Electronics*). That between two circuits, due to an electron stream controlled by the one circuit influencing the other circuit. Such coupling tends to be unidirectional, the second circuit having little influence on the first.

electron density (*Nuc.*). The number of electrons per gram of a material. Approx. 3×10^{23} for most light elements. In an ionized gas the equivalent electron density is the product of the ionic density and the ratio of the mass of an electron to that of a gas ion.

electron device (*Electronics*). One which depends on the conduction of electrons through a vacuum, gas or semiconductor.

electron diffraction (*Electronics*). Investigation of crystal structure by the patterns obtained on a screen from electrons diffracted from the surface of crystals or as a result of transmission through thin metal films.

electron discharge (*Electronics*). Current produced by the passage of electrons through otherwise empty space.

electron-discharge tube (*Electronics*). Highly-evacuated tube containing two or more electrodes between which electrons pass.

electron drift (*Electronics*). The actual transfer of electrons in a conductor as distinct from energy transfer arising from encounters between neighbouring electrons.

electronegative (*Chem.*). Said of acid-forming elements, generally nonmetallic, whose atoms have a relatively large affinity for electrons in joining other atoms in forming chemical compounds. (*Elec.*) Carrying a negative charge of electricity. Tending to form negative ions, i.e., having a relatively positive electrode potential.

electron emission (*Electronics*). The liberation of electrons from a surface.

electron gas (*Electronics*). The 'atmosphere' of free electrons in vacuo, in a gas or in a conducting solid. The laws obeyed by an electron gas are governed by Fermi-Dirac statistics, unlike ordinary gases to which Maxwell-Boltzmann statistics apply.

electron gun (*Electronics*). Assembly of electrodes in a cathode-ray tube which produces the electron beam, comprising a cathode from which electrons are emitted, an apertured anode, and one or more focusing diaphragms and cylinders.

electronic. Pertaining essentially to devices depending on electrons, e.g., *electronic computer, electronic control.*

electronic charge (*Nuc.*). The unit in which all nuclear charges are expressed. It is equal to $1{\cdot}602 \times 10^{-19}$ coulombs.

electronic configuration (*Chem.*). Arrangement of

atoms or electrons in various energy levels or orbits in a molecule or crystal.

electronic control (*Elec. Eng.*). General description of a control of machinery, traction, lifts, etc., which includes essential electronic switching and timing of operations, with or without tape triggering.

electronic efficiency (*Electronics*). For a valve, the ratio of the power at the desired frequency delivered by the electron stream to the circuit to the average power supplied to the stream.

electronic engraving (*Print.*). The making of plates direct from the copy without the use of the camera or the etching bath. As the copy is scanned the plates are engraved. Some models produce sets of colour plates, applying the necessary *colour correction* (q.v.) electronically.

electronic equation (*Chem.*). A chemical equation representing the transfer of electrons in a redox reaction, e.g., $2I^- \rightleftharpoons I_2 + 2e$.

electronic equilibrium (*Nuc.*). That between ionization inside and outside the volume of gas in an air-wall ionization chamber when accurate radiation dose measurements are possible. Also **charged-particle equilibrium.**

electronic flash (*Electronics*). Battery or mains device which charges a capacitor, the latter discharging through a tube containing neon or xenon when triggered. Used for photography and stroboscopy.

electronic formula (*Electronics*). One which gives the electronic charges available in radicals and atoms due to the valency electrons in the outer shell. Also called **ionic formula.**

electronic keying (*Teleg.*). Production of telegraphic signals by all-electronic system.

electronic microphone (*Acous.*). One in which the acoustic pressure is applied to an electrode of a valve.

electronic music (*Acous.*). **See electrosonic music.**

electronic oscillations (*Radio*). Those generated by electrons moving between electrodes in an amplifying device, the frequency being determined by the transit time.

electronic pick-up (*Radio*). One in which external vibration affects the grid of a valve and thereby modulates the anode current.

electronic rectifier (*Elec. Eng.*). **See metal rectifier.**

electronic register control (*Print.*). Equipment to maintain lateral and lengthwise register on web-fed printing and paper-converting machines.

electronics. Branch of science that deals with the study and application of electron devices, e.g., electron tubes, transistors, magnetic amplifiers.

electronic sky screen equipment (*Electronics*). See **elsse.**

electronic switch (*Elec. Eng.*). Device for opening or closing electric circuit by electronic means.

electronic theory of valency (*Chem.*). Valency forces arise from the transfer or sharing of the electrons in the outer shells of the atoms in a molecule. The two extremes are complete transfer (*ionic bond*) and close sharing (*covalent bond*), but there are intermediate degrees of bond strength and distance.

electronic tuning (*Electronics*). Changing the operating frequency of a system by changing the characteristics of a coupled electron beam.

electronic voltmeter (*Elec. Eng.*). One which depends on the amplifying and rectifying properties of valves or transistors.

electronic wattmeter (*Elec. Eng.*). One which uses valves or transistors.

electron jet (*Electronics*). Narrow stream of electrons, similar to a beam, but not necessarily focused.

electron lens (*Electronics*). A composite arrangement of magnetic coil and charged electrodes, to focus or divert electron beams in the manner of an optical lens.

electron mass (*Phys.*). A result of relativity theory, that mass can be ascribed to kinetic energy, is that the effective mass (m) of the electron should vary with its velocity according to the experimentally confirmed expression:

$$m = \frac{m_0}{\sqrt{1-\left(\frac{v}{c}\right)^2}},$$

where m_0 is the mass for small velocities, c is the velocity of light, and v that of the electron.

electron micrograph (*Micros.*). A photomicrograph of the image of an object, taken by substituting a photographic plate for the fluorescent viewing screen of an *electron microscope.*

electron microscope (*Micros.*). Tube in which electrons emitted from the cathode are focused, by suitable magnetic and electrostatic fields, to form an enlarged image of the cathode on a fluorescent screen. By passing the electrons through an object, such as a virus, a vastly enlarged image can be obtained on a photographic plate. The instrument has very high resolving power compared with the optical microscope, due to the shorter wavelength associated with electron waves.

electron mirror (*Electronics*). A 'reflecting' electrode in an electron tube, e.g., reflex klystron.

electron multiplier (*Electronics*). Electron tube in which anode is replaced by a series of auxiliary electrodes, maintained at successively increasing positive potentials up to the final anode. Electrons emitted from the cathode impinge on the first of the auxiliary electrodes, from which secondary electrons are ejected and travel to the next electrode, where the process is repeated. With suitable materials for the auxiliary electrodes, the number of secondary electrons emitted at each stage is greater than the number of incident electrons, so that very high overall amplification of the original tube current results. See **dynode.**

electron octet (*Chem., Nuc.*). The (up to) eight valency electrons in an outer shell of an atom or molecule. Characterized by great stability, in so far as the complete shell round an atom makes it chemically inert, and round a molecule (by sharing) makes a stable chemical compound.

electronogen (*Electronics*). Photosensitive molecule which may emit an electron when illuminated.

electron optics (*Optics*). Control of free electrons by curved electric and magnetic fields, leading to focusing and formation of images.

electron pair (*Chem.*). Two valence electrons shared by adjacent nuclei, so forming nonpolar bond.

electron paramagnetic resonance (*Chem.*). The gyromagnetic properties of an unpaired spinning electron, e.g., in a free radical, can be detected as for *nuclear magnetic resonance* (q.v.), but at a frequency of about 10 000 MHz.

electron probe analysis (*Chem.*). A beam of electrons is focused on to a point on the surface of the sample, the elements being detected both qualitatively and quantitatively by their resultant X-ray spectra. An accurate (1%), non-destructive method needing only small quantities (micron size) of sample.

electron radius (*Nuc.*). The classical theoretical value is $2{\cdot}82 \times 10^{-15}$ m, but experimentally the effective value varies greatly with the interaction concerned.

electron-ray indicator tube (*Electronics*). See **electric eye.**

electron relay (*Electronics*). Three (or more) electrode valves operating relays or circuits without taking appreciable input energy.

electron runaway (*Electronics*). Condition in a plasma when the electric fields are sufficiently large for an electron to gain more energy from the field than it loses, on average, in a collision.

electron scanning (*TV*). Scanning or establishing a TV image by an electron beam in a TV camera tube or a CRT (kinescope), normally using a rectangular raster, with horizontal lines.

electron sheath (*Electronics*). Electron space charge around an anode, when the supply of electrons is greater than demanded by the anode circuit. See also **electron cloud.**

electron shell (*Nuc.*). Group of electrons surrounding a nucleus of an atom, and having adjacent energy levels. Radiation is emitted or absorbed by electrons jumping between shells, according to certain rules, the radiation wavelengths being determined by changes of level (energy). See **K- L- M- . . . shells.**

electron sink (*Nuc. Eng.*). In a toroid for plasma studies, the cold electrons which absorb energy from the injected high-speed ions.

electron spin (*Electronics*). See **spin.**

electron-spin resonance (*Electronics*). A branch of microwave spectroscopy in which orbital electrons of a molecular sample resonate with RF fields in a waveguide cavity—the resonance being detected by the resulting energy absorption from the signal.

electron telescope (*Astron.*). Optical instrument, including an electronic image converter, associated with a normal telescope.

electron transport chain (*Biochem.*). The arrangement of mitochondrial components which accepts electrons from citric acid cycle intermediates and brings about the reduction of oxygen. The complete organization is not certain and a simplified version is shown:

$$\left.\begin{array}{l}\text{NADH}\longrightarrow\text{flavoprotein}\searrow\\ \text{succinate}\rightarrow\text{flavoprotein}\nearrow\end{array}\right.\text{cytochrome b}\rightarrow\text{cyt.c}_1\rightarrow\text{cyt.c}\rightarrow\text{cyt.a/a}_3\rightarrow\text{O}_2.$$

The oxidation of NADH produces 3 molecules of ATP per molecule of NADH. The oxidation of succinate produces 2 molecules of ATP per molecule of succinate.

electron trap (*Electronics*). An acceptor impurity in a semiconductor.

electron tube or valve (*Electronics*). In U.S., any electron device in which the electron conduction is in a vacuum or gas inside a gas-tight enclosure.

electron-volt (*Nuc.*). General unit of energy of moving particles, equal to the kinetic energy acquired by an electron losing one volt of potential, equal to $1{\cdot}602 \times 10^{-19}$ J; abbrev. eV.

electron-wave tube (*Electronics*). See **travelling-wave tube.**

electro-optical effect (*Elec. Eng.*). Same as **Kerr effect.** See **Pockel's effect.**

electro-osmosis (*Chem.*). Same as **electro-endosmosis.**

electroparting (*Elec. Eng.*). The electrolytic separation of two or more metals.

electropathology (*Med.*). The study of human diseases using electrical apparatus.

electrophonic effect (*Acous.*). Sensation of hearing arising from the passage of an electric current of suitable magnitude and frequency through the body.

electrophonic music (*Acous.*). Same as **electrosonic music.**

electrophoresis (*Chem.*). Motion of colloidal particles under an electric field in a fluid, positive groups to the cathode and negative groups to the anode. Also **cataphoresis.**

electrophorus (*Elec. Eng.*). Simple electrostatic machine for repeatedly generating charges. A resinous plate, after rubbing, exhibits a positive charge, which displaces a charge through an insulated metal plate placed in partial contact. Earthing the upper surface of this plate leaves a net negative charge on the metal plate when it is removed, a process which can be repeated indefinitely.

electrophysiology (*Biol.*). The science of electrical phenomena associated with living organisms.

electroplane camera (*Cinema.*). An electronic device applied to the lens elements of a motion picture camera to improve the field depth.

electroplating. Deposition of one metal on another by electrolytic action on passing a current through a cell, for decoration or for protection from corrosion, etc. Metal is taken from the anode and deposited on the cathode, through a solution containing the metal as an ion.

electroplating bath (*Elec. Eng.*). Tank in which objects to be electroplated are hung. It is filled with electrolyte at the correct temperature, with anodes of the metal to be deposited on articles which are made cathodes.

electroplating generator (*Elec. Eng.*). A direct-current electric generator, specially designed for electroplating work; it gives a heavy current at a low voltage.

electroplax (*Zool.*). In certain Vertebrates, one of the plates of which the electric organ is built up.

electropneumatic (*Electronics*). Said of control system using both electronic and pneumatic elements.

electropneumatic brake (*Rail.*). A type of brake which can be applied simultaneously throughout the length of the train.

electropneumatic contactor (*Elec. Eng.*). A contactor operated by compressed air but controlled by electrically operated valves.

electropneumatic control (*Elec. Eng.*). A form of remote control in which switches or other apparatus are operated by compressed air controlled by electrically operated valves; commonly used on electric trains.

electropneumatic signalling (*Elec. Eng.*). A signalling system operated by compressed air, the valves which control the latter being operated electrically.

electropolar (*Elec.*). Possessing magnetic poles or positive and negative charges.

electropolishing (*Met.*). See **electrolytic polishing.**

electropositive (*Elec.*). Carrying a positive charge of electricity. Tending to form positive ions, i.e., having a relatively negative electrode potential.

electroradiescence (*Elec.*). Emission of ultraviolet or infrared radiation from dielectric phosphors on the application of an electric field.

electrorefining (*Met.*). See **electrolytic refining.**

electroscope (*Elec. Eng.*). Indicator and measurer of small electric charges, usually 2 gold leaves which diverge because of repulsion of like charges; with 1 gold leaf and a rigid brass plate, indication is more precise.

electroscopic powder (*Elec. Eng.*). A mixture of finely divided materials which can acquire charges by rubbing together, so that, if dusted on to a plate, the different materials adhere to differently charged portions of the plate, forming a figure.

electrosmosis (*Chem.*). See **electroendosmosis.**

electrosol (*Chem.*). Dispersion produced by *Bredig's arc process* (q.v.).

electrosonic music (*Acous.*). Music or other sounds produced by electronic means (e.g., by oscillators, photocells, or generators), then combined electrically and reproduced through loudspeakers; also called **electronic music, electrophonic music.**

electrostatic accelerator (*Elec. Eng.*). One which depends on the electrostatic field due to large d.c. potentials.

electrostatic actuator (*Acous.*). Apparatus used for absolute calibration of microphone through application of known electrostatic force.

electrostatic adhesion (*Elec. Eng.*). That between two substances, or surfaces, due to electrostatic attraction between opposite charges.

electrostatic bonding (*Chem.*). Valence linkage between atoms arising from transfer of one or more electrons from outer shell of one atom to outer shell of another, transfer leading to more near completion of outer shells of both atoms, producing *ions* on dissociation.

electrostatic charge (*Elec. Eng.*). Electric charge at rest on the surface of an insulator or insulated body, and consequently leading to the establishment of an adjacent electrostatic field system.

electrostatic component (*Elec. Eng.*). See **electric component.**

electrostatic coupling (*Electronics*). That between circuit components, one applying a signal to the next through a capacitor.

electrostatic deflection (*Electronics*). Deflection of the beam of a cathode-ray tube by an electrostatic field produced by two plates between which the beam passes on its way to the fluorescent screen.

electrostatic field (*Elec. Eng.*). Electric field associated with stationary electric charges.

electrostatic focusing (*Electronics*). Focusing in high-vacuum cathode-ray tubes by the electrostatic field produced by two or more electrodes maintained at suitable potentials.

electrostatic generator (*Elec. Eng.*). One operating by electrostatic induction, e.g., *Bonetti, Holtz, Lemstrom, Pidgeon, Toepler, Tudsbury, Wimhurst*, and *Wommelsdorf machines* and *Van de Graaff generator* (qq.v.). Also called **frictional machine, induction machine, influence machine, static machine.**

electrostatic induction (*Elec. Eng.*). Movement and manifestation of charges in a conducting body by the proximity of charges in another body. Also the separation of charges in a dielectric by an electric field.

electrostatic instrument (*Elec. Eng.*). An electrical measuring instrument depending for its action on electrostatic forces set up between charged bodies. See **electroscope, electrometer.**

electrostatic Kerr effect (*Elec. Eng.*). Dispersion of the plane of polarization experienced by a beam of plane-polarized light on its passage through a transparent medium subjected to an electrostatic strain. The basis of action of some light-modulation systems.

electrostatic lens (*Electronics*). An arrangement of tubes and diaphragms at different electric potentials.

electrostatic loudspeaker (*Acous.*). One in which the mechanical forces are due to the action of electrostatic fields. Also **capacitor loudspeaker.**

electrostatic machine (*Elec. Eng.*). See **electrostatic generator.**

electrostatic memory (*Electronics*). A device in which the information is stored as electrostatic energy, e.g., storage tube, Williams tube. Also **electrostatic storage.**

electrostatic microphone (*Acous.*). See **capacitor microphone.**

electrostatic oscillograph (*Elec. Eng.*). An oscillograph in which the moving element is actuated by electrostatic attraction or repulsion.

electrostatic photography (*Photog.*). See **xerography.**

electrostatic precipitation (*Elec. Eng.*). Use of an electrostatic field to precipitate solid (or liquid) particles in a gas, e.g., in dust removal.

electrostatic printing (*Print.*). See **xerography.**

electrostatic reaction (*Elec. Eng.*). Capacitance reaction through a separate capacitor. See also **Miller effect.**

electrostatics. Section of science of electricity which deals with the phenomena of electric charges substantially at rest.

electrostatic separator (*Elec. Eng.*). Apparatus in which materials having different permittivities are deflected by different amounts when falling between charged electrodes, and therefore fall into different receptacles.

electrostatic shield (*Elec. Eng.*). Conducting shield surrounding instruments or other apparatus, to prevent their being influenced by external electric fields. (*Radio*) An earthed conducting plate interposed between two circuits to prevent unwanted capacitance coupling between them.

electrostatic storage (*Electronics*). See **electrostatic memory.**

electrostatic units (*Elec. Eng.*). Units for electric and magnetic measurements in which the *permittivity* of a vacuum is taken as unity, with no dimensions on CGS system. Abbrev. **e.s.u.**

electrostatic voltmeter (*Elec. Eng.*). One depending for its action upon the attraction or repulsion between charged bodies. Usual unit of calibration is *kilovolt*. See also **electrometer.**

electrostatic wattmeter (*Elec. Eng.*). One which utilizes electrostatic forces to measure a.c. power at high voltages.

electrostriction (*Elec.*). Change in the dimensions of a dielectric accompanying the application of an electric field.

electrosyntonic switch (*Elec. Eng.*). A switch remotely controlled by means of a high-frequency current superimposed on the main circuit.

electrotaxis (*Zool.*). See **galvanotaxis.**

electrotellurograph (*Elec. Eng.*). An apparatus for the study of earth currents.

electrotherapy, electrotherapeutics (*Med.*). The treatment of diseases involving the use of electricity.

electrothermic (or **electrothermal**) **instrument** (*Elec. Eng.*). One depending on the Joulean heating of a current for its operation. See **bolometer.**

electrothermoluminescence (*Phys.*). Changes in electroluminescent radiation resulting from changes of dielectric temperature. (Some dielectrics show a series of maxima and minima when heated.) The complementary arrangement of observing changes in thermoluminescent radiation when an electric field is applied is termed **thermoelectroluminescence.**

electrotint (*Print.*). A printing block produced by drawing with varnish on a metal plate, and

depositing metal electrically on the parts not covered with the varnish.
electrotomes (*Zool.*). In the electric organ of some *Selachii*, a series of concentric cones into which the mass of electroplaxes is divided by the myotomes of the tail-muscles.
electrotonus (*Physiol.*). The state of a nerve which is being subjected to a steady discharge of electricity.
electrotropism (*Zool.*). See **galvanotaxis.**
electrotype (*Print.*). A hard-wearing printing plate made by depositing a film of copper electrolytically on a mould taken from type or an original plate. The copper shell is backed with a lead alloy. Commonly abbreviated to **electro.**
electrotypograph (*Typog.*). An electrically operated typesetting machine.
electrovalence (*Chem.*). Chemical bond in which an electron is transferred from one atom to another, the resulting ions being held together by electrostatic attraction.
electroviscosity (*Chem.*). Minor change of viscosity when an electric field is applied to certain polar liquids.
electrowinning (*Met.*). See **electroextraction.**
electrum (*Met.*). An alloy of gold and silver (55–88% of gold) used for jewellery and ornaments. Also, nickel-silver (copper 52%, nickel 26%, and zinc 22%); it has the same uses as other nickel-silvers.
electuary (*Med.*). A medicine consisting of the medicinal agent mixed with honey, syrup or jam.
eleidin (*Zool.*). A substance which occurs in the cells of the superficial layer of the rete mucosum of the skin of Vertebrates; it is in the form of granules or droplets, and stains deeply with haematoxylin.
Elektron alloys (*Met.*). TN for magnesium-based light alloys with up to $4\frac{1}{2}$% copper, up to 12% aluminium and perhaps some manganese and zinc.
element (*Chem.*). Simple substance which cannot be resolved into simpler substances by normal chemical means. Because of the existence of *isotopes* (q.v.) of elements, an element cannot be regarded as a substance which has identical atoms, but as one which has atoms of the same atomic number. The more recently discovered are the artificial products, often very short-lived, of nuclear fission in a cyclotron or as in the case of einsteinium (No. 99) and fermium (100) of large-scale military nuclear explosions. (*Comp.*) Unit of a code pattern, e.g., an impulse or no impulse, or a dot or dash, etc. (*Elec. Eng.*) A term often used to denote the resistance wire and former of a resistance type of electric heater. (*Electronics*) Unit of an assembly, especially the detailed parts of electron tubes which affect its operation or performance. (*Maths.*) See **constituent.** (*Meteor.*) See **meteorological-.** (*Optics, Photog.*) Component of a lens, e.g., 6-element lens. (*Work Study*) Distinct part of a specified job selected for convenience of observation, measurement and analysis.
elementary analysis (*Chem.*). *Ultimate analysis* (q.v.), especially referring to an organic compound.
elementary bodies (*Med.*). Particles present in cells of the body in virus infections.
elementary colours (*Light*). See **primary colours.**
elementary particle (*Nuc.*). Particle believed incapable of subdivision. Use of this term is now less general as modern theories cast doubt on which observed particles are really elementary.
element-former (or **-carrier**) (*Elec. Eng.*). A refractory substance upon which the heated wire of a resistance type of electric heater is wound.
elements of an orbit (*Astron.*). The 6 data mathematically necessary to determine completely a planet's orbit and its position in it: (1) longitude of the ascending node, (2) inclination of the orbit, (3) longitude of perihelion, (4) semi-axis major, (5) eccentricity, (6) epoch, or date of planet's passing perihelion. Analogous elements are used in satellite and double star orbits.
elephant (*Paper*). Former paper size: writing, drawing, 23×28 in.; brown paper, 24×34 in.
elephantiasis (*Med.*). Enlargement of the limbs, or of the scrotum, from thickening of the skin and stasis of lymph; due to obstruction of lymphatic channels, especially by filarial worms.
Elephantide pressboard (*Elec. Eng.*). TN for pressboard with a large cotton content, used for insulation of transformers, armature and stator coils. It is specially suitable for use under oil.
elephant trunk (*Civ. Eng.*). A device used for expelling soft material by air pressure from the working chamber of a pneumatic *caisson* (q.v.).
Eleutheroblastea (*Zool.*). See **Hydrida.**
eleutherodactyl (*Zool.*). Having the hind toe free.
eleutheropetalous (*Bot.*). Polypetalous.
Eleutherozoa (*Zool.*). A subphylum of *Echinodermata*, comprising forms of free and active habit. Cf. *Pelmatozoa.*
elevated duct (*Meteor.*). Tropospheric radio duct which has both upper and lower effective boundaries elevated.
elevation (*Arch., Build.*). The view or representation of any given side of a building. (*Eng.*) A view (e.g., side or end elevation) of a component or assembly drawn in projection on a vertical plane. (*Surv.*) *Reduced level* (U.S.).
elevation head (*Hyd.*). The energy possessed per unit weight of a fluid, due to its elevation above some datum. If at a given point the elevation is z feet, the elevation head at this point is said to be z feet. Also called **position head.**
elevation of boiling-point (*Chem.*). The raising of the boiling-point of a liquid by substances in solution.
elevator (*Aero.*). An aerodynamic surface, operated by fore-and-aft movement of the pilot's control column, governing motion in pitch. (*Eng.*) (1) A type of *conveyor* (q.v.) for raising or lowering material which is temporarily carried in buckets or fingers attached to an endless chain or belt. (2) A *lift* (q.v.). (*Zool.*) A muscle which by its contraction raises a part of the body. Cf. *depressor.*
elevator chain (*Eng.*). Chain used to carry a series of buckets or slats and to which the elevator drive is applied. *Cast chain* is used for slow to medium speeds, *precision* (*roller*) *chain* for higher speeds.
elevator dredger (*Civ. Eng.*). A *bucket-ladder dredger* (q.v.).
elevons (*Aero.*). Hinged control surfaces on the wing trailing edge of tailless or delta aeroplanes which are moved in unison to act as elevators and differentially as ailerons.
Elf (*TV*). TN for electroluminescent display screen, having high-speed scanning, good half-tone and brightness characteristics.
eliminant (*Maths.*). See **resultant.**
elimination (*Chem.*). The removal of a simple molecule (e.g., of water, ammonia, etc.) from 2 or more molecules, or from different parts of the same molecule. See **condensation** (2).
elimination filter (*Telecomm.*). See **band-elimination filter.**

Elinvar (*Met.*). A nickel-chromium steel alloy with variable proportions of manganese and tungsten. Used for watch hairsprings because of its constant elasticity at different temperatures.

elittoral zone (*Ocean.*). The portion of the sea bottom which is under 40 metres or more of water.

elixir (*Med.*). A strong extract or tincture.

ell (*Plumb.*). A short L-shaped connecting pipe.

ellagitannins (*Chem., Leather*). Vegetable tannins which deposit bloom (crystalline ellagic acid) on the leather. Sources include oak bark, Valonia and divi-divi.

Elland stone (*Build.*). A fine greyish brown sandstone from the Millstone Grit of Yorkshire.

ellipse (*Maths.*). See **conic.**

ellipsoid (*Histol.*). A specialized connective tissue sheath, acting as a valve, surrounding the blood capillaries of an arteriole in the red pulp of the spleen. (*Maths.*) See **quadric.**

elliptical arch (*Build., Civ. Eng.*). An arch formed to an elliptical curve, or sometimes to a curve which is not a true ellipse but a combination of circular arcs.

elliptical galaxies (*Astron.*). A common type of galaxy of symmetrical form but having no spiral arms; the nearer elliptical galaxies have been resolved into stars, but contain no dust or gas.

elliptical point on a surface (*Maths.*). One at which the curvatures of all normal sections are of the same sign. Cf. *hyperbolic, parabolic,* and *umbilical point on a surface.*

elliptic functions (*Maths.*). In general, $f(z)$ is an elliptic function if it is doubly periodic, i.e., if there exist two numbers w_1 and w_2, $w_1 \neq w_2$, such that $f(z)=f(z+w_1)=f(z+w_2)$. The simplest elliptic functions are the Jacobian *sn, cn* and *dn* functions of u which are defined as follows:

$$\mathrm{sn}\, u = x$$
$$\mathrm{cn}\, u = \sqrt{1-x^2}$$
$$\mathrm{dn}\, u = \sqrt{1-k^2x^2}$$

where u is the elliptic integral of the 1st class.

If x is replaced by sin φ, the following equation can be written from which the basic properties of the Jacobian *sn, cn* and *dn* functions follow:

$$\mathrm{sn}\, u = \sin\varphi$$
$$u = \cos\varphi$$
$$\mathrm{dn}\, u = \sqrt{1-k^2\sin^2\varphi} = \Delta\varphi \text{ (say)}$$
$$\varphi = \mathrm{am}\, u \text{ (say)}.$$

elliptic integral (*Maths.*). All integrals of the type $\int\sqrt{P}\,dx$, where P is a fourth-degree polynomial, can be expressed in one of the following forms called respectively elliptic integrals of the 1st, 2nd and 3rd class:

$$F(k, x) = \int_0^x \frac{dt}{\sqrt{(1-t^2)(1-k^2t^2)}}$$

$$E(k, x) = \int_0^x \sqrt{\frac{1-k^2t^2}{1-t^2}}\,dt$$

$$\Pi(n, k, x) = \int_0^x \frac{dt}{(1+nt^2)\sqrt{(1-t^2)(1-k^2t^2)}}$$

k is called the modulus and may be taken $0<k<1$. Cf. *elliptic functions.*

elliptic polarization (*Light*). This may be regarded as being produced by two mutually perpendicular plane-polarized components which are not in phase. (*Phys.*) That of an electromagnetic wave in which the electric and magnetic fields each contain two unequal components, at right angles in space and in phase quadrature. See **circular polarization.**

elliptic trammel (*Instr.*). An instrument for drawing ellipses, consisting of a straight arm having a pencil point at one end and two adjustable studs (all three of which project at right angles to the arm), and a frame with two grooves crossing one another at right angles. If a stud is placed in each of the grooves, then, as the arm is rotated, the pencil point describes an ellipse.

elm (*For.*). Tree of the genus *Ulmus,* a native of Northern Europe, usually felled when it is between 18 to 30 m in height; the tree may have a life as long as 140 years. It is noted for its durability under water, and is used for wagon making, coffins, agricultural implements, gymnasium equipment, pulley blocks, ship building, etc. Susceptible to **Dutch elm disease** (*Ceratocystis ulmi*), a fungus disease which, spread by bark beetles carrying infected spores, defoliates and kills the tree within weeks.

Elmendorf tear tester (*Paper*). A segment-shaped pendulum which when released tears a standard-size specimen of the paper and records the force required.

Elon (*Photog.*). See **methyl-4-aminophenol.**

elongated (*Typog.*). A narrow form of type, often used in display work. It is commonly known as **condensed.**

elongating stage (*Bot.*). The period of growth when an organ increases in length, before secondary thickening begins.

elongation (*Astron.*). The angular distance between the moon or planets and the sun. The planets Mercury and Venus have maximum elongations of about 28° and 47° respectively. (*Met.*) The percentage extension produced in a tensile test.

elongation ratio (*Powder Tech.*). The ratio of the length of a particle to its breadth.

ELR (*Meteor.*). Abbrev. for *environment lapse rate.*

Elrod (*Typog.*). A machine for producing lead-alloy strip material, spacing, and rule, cut to size or in long lengths.

Elsie (*Electronics*). Letter sorting system in which the address is presented to an operator who operates keys which print phosphorescent dots for photoelectric routing on the envelope. Coding is by first 3 and last 2 letters of the office. (*E*lectronic *l*etter *s*orting and *i*ndicating *e*quipment.)

elsse (*Electronics*). Telemetric system for tracking the flight-path of ground-launched missiles. (*El*ectronic *s*ky *s*creen *e*quipment.)

elution (*Min. Proc.*). Washing of loaded ion-exchange resins to remove captured uranium ions or other seized elements in washing liquor. This liquor is the *eluant* and the enriched solution it becomes is the *eluate.* The resin (or *zeolite*) is regenerated (e.g., by rinsing water softener with brine).

elutriation (*Chem., etc.*). Process for separating into sized fractions finely divided particles in accordance with their rate of gravitation relative to a rising stream of fluid.

elutriator (*Chem., etc.*). An appliance for washing or sizing very fine powders in an upward current of water or air.

elutriator-centrifugal (*Powder Tech.*). Device for fractionating powder particles into sized fractions by means of a suspension undergoing centrifugal motion.

eluvial (eluvium) gravels (*Geol.*). Those gravels formed by the disintegration *in situ* of the rocks which contributed to their formation. Cf. *alluvial deposits*.

elvan (*Geol.*). A term applied by Cornish miners to the dyke rocks associated with the Armorican granites of that county. Elvans are actually quartz-porphyries, micrograntes, and other medium- to fine-textured dyke rocks of granitic composition.

elytra (*Zool.*). In *Coleoptera*, the hardened, chitinized fore-wings which form horny sheaths to protect the hind-wings when the latter are not in use; in certain *Polychaeta*, plate-like modifications of the dorsal cirri, possibly for respiration. *adj.* **elytroid, elytriform.**

e/m (*Phys.*). The ratio of the electric charge to the mass for particles such as electrons and positive rays. For slow-moving electrons, the value of e/m is $1{\cdot}759 \times 10^{11}$ C kg^{-1}. The value decreases with increasing velocity, however, on account of the increase in effective mass. See **mass of the electron.**

em (*Typog.*). The square of the body of any size of type; the 12-point em is the unit of measurement for spacing material and the dimensions of pages. 6 ems of 12-point = 1 inch approximately. See **em quad.**

emanations (*Chem.*). Heavy isotopic inert gases resulting from decay of natural radioactive elements, i.e., *radon* from radium, *thoron* from thorium, *actinon* from actinium; these gases are short-lived and decay to other radioactive elements. They are radioisotopes 222, 220, 219 of element 86 or *radon* (q.v.), sometimes (deprecated) called *emanation*.

Emanueli porosimeter (*Phys.*). A double U-tube arrangement for measuring the porosity of a given sample. One tube has the sample in parallel with it, and the other a known porosity across it.

emarginate (*Bot.*). (1) Having a slight notch at the tip. (2) As if scooped out close to the point of attachment. (3) Lacking a distinct margin. (*Zool.*) Having the tip or margin notched.

emasculation (*Bot.*). The removal of the stamens from an unopened flower. (*Med.*) Removal of testes, or of testes and penis.

emasculator (*Vet.*). An instrument for castrating horses and bulls by crushing the spermatic cord.

embankment (*Civ. Eng.*). A ridge of earth, stones, etc., specially constructed to carry a highway or railroad at a higher level than the surrounding ground; or as a protective bank to prevent water encroachment, etc. Also called a **bank.**

embankment wall (*Civ. Eng.*). A retaining wall from the top of which the supported earth normally rises at a slope.

embattlemented (*Arch.*). A term applied to a building feature (such as a parapet) which is indented along the top like a battlement.

embedded column (*Arch.*). A column which is partly built into the face of a wall.

embedded temperature detector (*Elec. Eng.*). A resistance thermometer or thermocouple built into a machine or other piece of equipment during its construction, in order to be able to ascertain the temperature of a part which is inaccessible under working conditions.

embedding (*Histol.*). The technique of embedding biological specimens in a supporting medium, usually paraffin wax, in preparation for sectioning with a microtome.

embellishment (*Arch.*). Ornamentation applied to building features.

Embioptera (*Zool.*). An order of small, elongate, soft-bodied *Orthopteroidea*, with short anal cerci; the males have two pairs of small wings, the females are wingless; the mouthparts are adapted for biting; the basal joint of the tarsi of the first pair of legs is swollen and contains a silk-gland; gregarious tropical Insects living in silken tunnels.

embolectomy (*Surg.*). Removal of an embolus.

embolic gastrulation (*Zool.*). Gastrulation by invagination.

embolism (*Med.*). The blocking of a blood vessel by a mass carried, to the point of obstruction, from a remote part of the circulation.

embolite (*Min.*). The chief silver ore in some of the Chile mines, occurring as yellow-green incrustations and masses. Chemically, chloride and bromide of silver; a member of the cerargyrite series.

embolium (*Zool.*). In some *Hemiptera*, a narrow strip of the corium, bordering on the costa.

Embolobranchiata (*Zool.*). A group of *Arachnida*, breathing by lung books or tracheae, or both; usually of terrestrial habit. Obsolete name.

embolomerous (*Zool.*). A type of vertebra consisting of a neural arch resting on two notochordal centra, an anterior intercentrum and a posterior pleurocentrum. Found in *Labyrinthodontia*.

embolus (*Med.*). A clot or mass formed in one part of the circulation and impacted in another, to which it is carried by the blood stream. (*Zool.*) The distal portion of the palpal organ of a male Spider, comprising the ejaculatory duct and the conductor.

emboly (*Zool.*). Invagination; the condition of pushing in or growing in. *adj.* **embolic.**

embossed (*Bot.*). Umbonate. (*Build.*) A term applied to any form of ornamentation which is raised from the general surface which it is decorating.

embossed papers (*Paper*). Formed by passing the sheet through rollers, one of which is engraved steel carrying the design, the other of a soft material such as paper or cotton.

embossing (*Print.*). Producing a raised design on paper or board by the use of a die; if the design is unprinted it is called *blind embossing*.

embrasure (*Arch.*). The splayed reveal of a window opening.

embrittlement (*Met.*). Some metals (e.g., steel, wrought iron, zinc or magnesium alloys) show reduced impact toughness while at subnormal temperatures; seasonal cracking of high-zinc brasses is a severe form of embrittlement. Absorption of atomic hydrogen, i.e., in electroplating or pickling, causes hydrogen embrittlement in iron and steel. Embrittlement of boiler plate, attributed to the action of sodium hydroxide, causes cracks along the boiler seams, from rivet to rivet, below the water line.

embrocation (*Med.*). The action of applying or rubbing a medicated liquid into an injured part: the liquid so used.

embroidery (*Textiles*). (1) Lace work consisting of a ground of traverse bobbin net on which an ornamental design has been stitched. (2) Ornamental work done by needle or machine on a cloth, canvas, or other ground.

embryo (*Bot.*). A young plant in a rudimentary state of development, usually contained in a seed or surrounded by protective tissue. (*Zool.*) An immature organism in the early stages of its development, before it emerges from the egg or from the uterus of the mother. *adj.* **embryonic.**

embryocardia (*Med.*). A sign of cardiac weakness, in which the sounds of the heart in an adult resemble those heard in the foetus.

embryogeny (*Biol.*). The processes leading to the formation of the embryo; the study of these processes.

embryology (*Biol.*). The study of the formation and development of embryos.
embryoma (*Med.*). A tumour formed of embryonic or foetal elements.
embryonic fission (*Zool.*). See **polyembryony.**
embryonic tissue (*Bot.*). Meristematic tissue.
embryophore (*Zool.*). In the embryonic stages of certain *Cestoda*, a ciliated envelope which encloses the embryo and is formed from the superficial blastomeres.
embryo sac (*Bot.*). A cavity in the ovule of an angiosperm, formed by the enlargement of the megaspore, and usually containing 8 nuclei, of which the most important is the egg nucleus.
embryotomy (*Med.*). The removal of the viscera or of the head of a foetus, in obstructed labour.
Emdecca (*Build.*). Zinc sheets so treated as to resemble tiling.
emerald (*Min.*). The brilliant green gemstone, a form of beryl; silicate of beryllium and aluminium, crystallizing in hexagonal prismatic forms, occurring chiefly in mica-schists, and rarely in pegmatites.
emerald copper (*Min.*). See **dioptase.**
emerald green (*Chem.*). See **Paris green.**
emerald nickel (*Min.*). See **zaratite.**
emergence (*Bot.*). An outgrowth from a plant, derived from epidermal and cortical tissues, but not containing vascular tissue, and not developing into a stem or leaf. (*Zool.*) An epidermal or subepidermal outgrowth. In Insects, the appearance of the imago from the cocoon, pupa-case, or pupal integument.
emergence of land (*Geol.*). The mere fact of emergence of land from the sea is proved by the occurrence of strata which are obviously indurated marine clays, and sands containing marine shells and fish-remains at many different levels in the stratigraphical column. It is also clearly demonstrated by the occurrence of raised beaches round coasts. It is an open question as to how far these changes are due to isostatic readjustment, to eustatic changes of sea-level, or to other causes.
emergence rhythm (*An. Behav.*). The habit of some insects of emerging from the pupa at a particular time of day or, experimentally, at regular intervals of less than 24 hours.
emergency release-push (*Elec. Eng.*). A switch fitted to an electric lift, to allow the car, in case of emergency, to be moved with the doors open.
emergency stop (*Elec. Eng.*). A switch installed in a lift-car, or other similar piece of equipment, by means of which the power to the operating motor can be cut off. Also called a **safety switch.**
emergency switch (*Elec. Eng.*). A switch placed in a convenient position for cutting off the supply of electricity to a piece of apparatus or to a building, in case of emergency.
emergent ray point (*Radiol.*). See **exit portal.**
emersed (*Bot.*). (1) Protruding upwards. (2) Amphibious.
emersion (*Astron.*). The exit of the moon, or other body, from the shadow which causes its eclipse.
emery (*Min.*). A finely granular intimate admixture of corundum and either magnetite or haematite, occurring naturally in Greece and localities in Asia Minor, etc.; used extensively as an abrasive.
emery paper, emery cloth, emery buff (*Eng.*). Paper, or more often cloth, surfaced with emery powder, held on by an adhesive solution; used for polishing and cleaning metal. See **emery.**
emery wheel (*Eng.*). A *grinding wheel* (q.v.) in which the abrasive grain consists of emery powder, held by a suitable bonding material.
emesis (*Med.*). The act of vomiting.
emetic (*Med.*). Having the power to cause vomiting; a medicament which has this power.
emetine (*Chem.*). $C_{29}H_{40}O_4N_2$ an alkaloid obtained from the roots of Brazilian ipecacuanha. It forms a white amorphous powder, m.p. 74°C, soluble in ethanol, ether or trichloromethane, slightly in water. It is used in medicine as an emetic; its principal use, however, is in the form of emetine bismuthous iodine or emetine hydrochloride, remedies for amoebic dysentery.
e.m.f. (*Elec. Eng.*). Abbrev. for *electromotive force.*
emictory (*Med.*). A drug which excites the excretion of urine.
emigration (*Ecol.*). A category of population dispersal covering one-way movement out of the population area. Cf. *immigration, migration.*
emissary (*Zool.*). Passing out, as certain veins in Vertebrates which pass out through the cranial wall.
emission (*Phys.*). Release of electrons from parent atoms on absorption of energy in excess of normal average. This can arise from (1) *thermal* (thermionic) agitation, as in valves, Coolidge X-ray tubes, cathode-ray tubes; (2) *secondary* emission of electrons, which are ejected by impact of higher energy primary electrons; (3) *photoelectric* release on absorption of quanta above a certain energy level; (4) *field* emission by actual stripping from parent atoms by high electric field.
emission current (*Electronics*). The total electron flow from an emitting source.
emission efficiency (*Electronics*). See **cathode efficiency.**
emission spectrum (*Phys.*). Wavelength distribution of electromagnetic radiation emitted by self-luminous source.
emissive power (*Phys.*). The total energy emitted per unit area of a surface at a specified temperature and under defined surroundings. Also applies to emission of electrons. See **emission.**
emissivity (*Phys.*). The ratio of emissive power of a surface at a given temperature to that of a black body at the same temperature and with the same surroundings. Values range from 1·0 for lampblack down to 0·02 for polished silver. See **Stefan-Boltzmann law.**
emitron (*TV*). Early British TV camera tube similar to *iconoscope* (q.v.).
emitter (*Electronics*). Outer section of a transistor, by which minority carriers enter the base region.
emitter follower (*Electronics*). See **common collector.**
emitter junction (*Electronics*). One biased in the low resistance direction, so as to inject minority carriers into the base region.
emmenagogue (*Med.*). Having the power to stimulate the menstrual flow; such a drug.
emmetrope (*Med.*). One who is emmetropic.
emmetropia (*Med.*). The normal condition of the refractive system of the eye, in which parallel rays of light come to a focus on the retina, the eyes being at rest. *adj.* **emmetropic.**
emotion (*Psychol.*). A complex state involving many bodily changes and the relative overactivity of the sympathetic nervous system. There is a state of mental arousal and feeling, and frequently a tendency towards a definite form of action. If the degree of emotion is extreme there is deterioration of both mental and physical abilities.
empennage (*Aero.*). See **tail unit.**

emperipolesis (*Cyt.*). The behaviour of lymphocytes when, in culture, they become attached to, pass over, or pass into, other cells. Cf. *peripolesis*.
emperor (*Paper*). Before metrication, a standard paper size, 48 × 72 in; U.S., 40 × 60 in..
emphasizer (*Telecomm.*). An audio-frequency circuit which selects and amplifies specific frequencies or frequency bands.
emphysema (*Med.*). (1) The presence of air in the connective tissues. (2) The formation in the lung of bullae or spaces containing air, as a result of distension of alveoli and rupture of weakened alveolar walls.
emphysematous chest (*Med.*). The barrel-shaped, immobile chest which is the result of chronic emphysema and bronchitis.
empire cloth (*Elec. Eng.*). An insulating fabric which is impregnated with linseed oil.
empirical formula. A formula founded on experience or experimental data only, not deduced in form from purely theoretical considerations. (*Chem.*) Formula deduced from the results of analysis which is merely the simplest expression of the ratio of the atoms in a molecule, and may, or may not, show how many atoms of each element the molecule contains: e.g., methanal, CH_2O, ethanoic acid, $C_2H_4O_2$, and lactic acid, $C_3H_6O_3$, have the same percentage composition, and consequently, on analysis, they would all be found to have the same empirical formula.
empiricism. The regular scientific procedure whereby scientific laws are induced by inductive reasoning from relevant observations. Critical phenomena are deduced from such laws for experimental observation, as a check on the assumptions or hypotheses inherent in the theory correlating such laws. Scientific procedure, described by empiricism, is not complete without the experimental checking of deductions from theory.
emplastrum (*Med.*). A medicated plaster for external application.
emplectite (*Min.*). Sulphide of copper and bismuth occurring, at Tannenbaum and elsewhere, as thin, striated grey metallic prisms intimately associated with quartz.
emplectum (*Build.*). An ancient form of masonry, showing a squared stone face, sometimes interrupted by courses of tiles at intervals.
empodium (*Zool.*). In Insects, a bristlelike process of the terminal joint of the tarsus. Cf. *arolium*.
empress (*Build.*). A slate size, 26 × 16 in. (660 × 406 mm).
emprosthotonos (*Med.*). Bending of the body forwards caused by spasm of the abdominal muscles, as in tetanus.
empty band (*Phys.*). See **energy band.**
empty glume (*Bot.*). See **sterile glume.**
empyema (*Med.*). Accumulation of pus in any cavity of the body, especially the pleural.
em quad (*Typog.*). A square quadrat of any size of type. Less than type height, it is used for spacing. Usually called a **mutton,** to distinguish it clearly from an **en** (or **nut**).
em rule (*Typog.*). The dash (—). A thin horizontal line 1 em of the type body in width. Apart from its uses as a mark of punctuation, it is often used to build up rules in tabular work. Also **em score, metal rule, mutton rule.**
emulsified coolants (*Eng.*). These are used as cutting media in (metal) machining. The 3 main types are: (*a*) an emulsion of water, a thick soap solution, and a mineral oil, (*b*) an emulsion of a mineral oil and soft soap or some other alcoholic soap solution, and (*c*) an emulsion of a sulphurized or sulphonated oil neutralized and blended with a soluble oil.
emulsifier (*Chem.*). An apparatus with a rotating, stirring or other device used for making emulsions.
emulsifying agents (*Chem.*). Substances whose presence in small quantities stabilizes an emulsion, e.g., ammonium linoleate, certain benzene-sulphonic acids, etc.
emulsin (*Chem.*). An enzyme which splits lactose into galactose and induces hydrolysis of β-glucosides and β-galactosides.
emulsion (*Build.*). A preparation serving as a retarder. (*Chem.*) A colloidal suspension of one liquid in another, e.g., milk. (*Photog.*) The light-sensitive coating, usually suspensions of silver bromide or chloride in gelatin, on supports, which forms the basis of photography.
emulsion paint (*Paint.*). A water-thinnable paint made from a pigmented emulsion or dispersion of a resin (generally synthetic) in water. The resin may be polyvinyl acetate, polyvinyl chloride, an acrylic resin or the like.
emulsion technique (*Nuc.*). Study of nuclear particles, by means of tracks formed in photographic emulsion.
emulsoid (*Chem.*). See **lyophilic colloid.**
emunctory (*Med.*). Conveying waste matter from the body; any organ or canal which does this.
Emy's roof (*Build.*). A form of roof in which the principal member is built up to a semicircular shape.
en (*Chem.*). 1,2-diamino ethane (*ethene diamine*) $(CH_2{\cdot}NH_2)_2$ in complexes. (*Typog.*) A unit of measurement used in reckoning up composition. It is assumed that the average letter has the width of one *en* and that the average word, including the space following, has the width of six *ens*. See **en quad.**
enabling pulse (*Telecomm.*). One which opens a *gate* (q.v.) which is normally closed.
enamel (*Build.*). *Glaze* (q.v.). (*Elec. Eng.*) A finely ground oil paint containing resin. The oil is thickened by heating linseed oil with china-wood oil for some hours at 300°C. Unrivalled for insulating very fine wires. (*Zool.*) The external calcified layer of a tooth, of epidermal origin and consisting of elongate hexagonal prisms, set vertically on the surface of the underlying dentine; enamel also occurs in certain scales.
enamel back tubing (*Glass*). Glass tubing having a white or coloured ply, interposed between clear plies, extending approximately half-way round the tube.
enamel cell (*Zool.*). See **ameloblast.**
enamel-insulated wire (*Elec. Eng.*). Wire having an insulating covering of *enamel* used for winding small magnet coils, etc.
enamelled brick (*Build.*). A brick having a glazed surface.
enamelled paper (*Paper*). Paper coated on one side with china clay and highly finished. The best quality is used for litho work, cheaper sorts for box covers.
enamelled slate (*Elec. Eng.*). Slate covered with a coating of hard enamel to render it suitable for the construction of switchboards.
enamel organ (*Zool.*). An invagination of the deep layer of the buccal epithelium, which becomes a dental rudiment in *Craniata*.
enamel paint (*Paint.*). A high-grade paint prepared by careful grinding of pigment in an oily medium containing a proportion of resin, and the usual lesser ingredients. It may be anything from glossy to flat.

enanthem, enanthema (*Med.*). An eruption on a mucous membrane.
Enantioblastae (*Bot.*). See **Glumiflorae.**
enantiomerism (*Chem.*). See **optical isomerism.**
enantiomorphism (*Chem.*). Mirror image isomerism. A classical example is that of the crystals of sodium ammonium tartrates which Pasteur showed to exist in mirror image forms. *adj.* **enantiomorphous.**
enantiostylous (*Bot.*). Said of a flower having the style or styles projecting to one side, the stamens often then projecting to the other.
enantiotropy (*Chem.*). The property of a substance of existing in two crystal forms, one being stable below, and the other above, a transition temperature.
enargite (*Min.*). Sulpharsenate of copper, often containing a little antimony. Occurs as greyish black, metallic orthorhombic crystals in several mines in N. and S. America and elsewhere.
enarthrosis (*Zool.*). A ball-and-socket joint.
enation (*Bot.*). (1) A general term for an outgrowth. (2) An outgrowth from the veins of the underside of a leaf, caused by some virus infections.
encallow (*Build.*). The mould which is first removed from the surface of the site where clay for brickmaking is to be obtained, and which is stored in a spoil bank for resurfacing later.
encapsulation. Provision of a tightly fitted envelope of material to protect, e.g., a metal during treatment or use in an environment with which it would otherwise react in a detrimental manner. Similarly, the coating of, e.g., an electronic component in a resin to protect it against the environment. See also **microencapsulation.**
encase (*Build., Join.*). To surround or enclose with linings or other material.
encased knot (*For.*). A *dead knot* (q.v.).
encastré (*Build.*). A term applied to a beam when the end of it is fixed.
encaustic painting (*Paint.*). A process of painting in which hot wax is used as a medium.
encaustic tile (*Build.*). An ornamental coloured tile whose colours are produced by substances added to the clay before firing.
enceph-, encephalo-. Prefix from Gk. *enkephalos*, brain.
encephalalgia (*Med.*). Pain inside the head.
encephalitic (*Med.*). Of the nature of, or affected with, encephalitis.
encephalitis (*Med.*). Inflammation of the brain substance.
encephalitis lethargica (*Med.*). See **Von Economo's disease** and **epidemic encephalitis.**
encephalitis periaxialis, encephalitis periaxialis diffusa (*Med.*). Schilder's disease. A disease characterized by progressive destruction of the nerve fibres composing the central white matter of the brain, causing blindness, deafness, paralysis, and amentia.
encephalocele (*Med.*). Hernial protrusion of brain substance through a defect in the skull.
encephalocoele (*Zool.*). The cerebral ventricle or brain cavity of *Cephalochorda.*
encephalocystocele (*Med.*). Hernia of brain substance which is also distended by cerebrospinal fluid.
encephalogram (*Radiol.*). X-ray plate produced in **encephalography.**
encephalography (*Radiol.*). Radiography of the brain after its cavities and spaces have been filled with air, or dye, previously injected into the space round the spinal cord.
encephaloid cancer (*Med.*). A cancer which is soft and rapidly growing.
encephalomalacia (*Med.*). Pathological softening of the brain.
encephalomyelitis (*Med.*). Diffuse inflammation of the brain and the spinal cord.
encephalon (*Zool.*). The *brain* (q.v.).
encephalopathy (*Med.*). Any general disease of the brain, e.g., *lead encephalopathy* is the brain disorder caused by lead poisoning.
encephalospinal (*Zool.*). See **cerebrospinal.**
enchondroma (*Med.*). A tumour, often multiple, composed of cartilage and occurring in bones.
enchylema (*Cyt.*). A name sometimes given to the liquid part of the cytoplasm in *Protozoa.*
enclitic (*Med.*). Having the planes of the foetal head inclined to those of the maternal pelvis.
enclosed arc lamp (*Light*). A lamp consisting of a carbon arc maintained in a translucent enclosure. The latter is designed to exclude air, so that the arc burns in an atmosphere of the products of combustion.
enclosed-flame arc lamp (*Light*). An enclosed arc lamp employing flame carbons.
enclosed fuse (*Elec. Eng.*). A fuse in which the fuse wire is enclosed in a tube or other covering.
enclosed self-cooled machine (*Elec. Eng.*). An electric machine which is enclosed in such a way as to prevent the circulation of air between the inside and the outside, but in which special provision is made for cooling the enclosed air by some attachment, e.g., an air cooler, forming part of the machine.
enclosed-ventilated (*Elec. Eng.*). Said of electrical apparatus which is protected from ordinary mechanical damage by an enclosure, with openings for ventilation.
encoder (*Comp.*). A network in a computer in which only one input is excited at any one time and each input gives rise to a combination of outputs.
encoding altimeter (*Aero.*). An *altimeter* designed for automatic reporting of altitude to *air-traffic control*. A special encoding disk within the instrument rotates in response to movement of the aneroid capsules, and transmits a signal which is amplified, fed to the aircraft's transponder, and thence automatically to ATC.
encysted (*Med.*). Enclosed in a cyst or a sac.
encystment (*Bot.*). The formation of a walled nonmotile body from a swimming spore. (*Zool.*) The formation by an organism of a protective capsule surrounding itself. Also **encystation.** *v.* **encyst.**
end (*Mining*). Solid rock at end of underground passage. (*Textiles*) (1) A warp thread. (2) In the woollen trade, the name applied to one portion of a standard length of cloth that has been cut into two equal parts.
end-and-end (*Weaving*). An alternate arrangement of warp threads which vary in colour, style, or type, e.g., one thread cotton followed by one thread filament rayon, etc.
endaortitis (*Med.*). Inflammation of the intima of the aorta.
endarch (*Bot.*). Said of a xylem strand having the protoxylem on the edge nearest to the centre of the axis.
end-around shift (*Comp.*). One in which digits drop off one end of a word and return at the other. Also **circular shift, logical shift, ring shift.**
endarteritis (*Med.*). Inflammation of the intima of an artery.
endarteritis obliterans (*Med.*). Obliteration of the lumen of an artery, as a result of inflammatory thickening of the intima.
end-artery (*Anat.*). A terminal artery which does not anastomose with itself or with others.
end bell (*Elec. Eng.*). A strong metal cover placed

over the end-windings of the rotor of a high speed machine, e.g., a turbo-alternator, to prevent their displacement by centrifugal forces.

end bracket (*Elec. Eng.*). An open structure fitted at the end of an electrical machine, for the purpose of carrying a bearing.

end bud (*Zool.*). A sense organ, consisting of an ovoidal group of sensory cells, supplied by a special nerve, in the skin of *Craniata.*

end-bulb (*Zool.*). In Vertebrates, a type of sensory nerve-ending in which the nerve loses its medullary sheath and breaks up into small branches, within a connective-tissue capsule.

end cell (*Elec. Eng.*). See **regulator cell.**

end connexions (*Elec. Eng.*). That part of an armature winding which does not lie in the slots, and which serves to join the ends of the active or slot portions of the conductors.

endecandrous (*Bot.*). Having 11 stamens.

endellionite (*Min.*). Another name for **bournonite.**

endemic (*Med.*). Prevalent in, and confined to, a particular country or district; said of disease. (*Zool.*) See **indigenous.**

enderon (*Zool.*). See **endoderm.**

end-feed magazine (*Eng.*). One of several types of magazine used for the automatic loading of centreless grinding machines. It is suitable for short, conical components.

end-fire aerial array (*Radar, Radio*). See **end-on aerial array.**

end fixing (*Build.*). A term used in referring to the condition of the ends of a column, whether fixed or only partially so.

end gauge (*Eng.*). Gauge consisting of a metal block or cylinder the ends of which are made parallel within very small limits, the distance between such ends defining a specified dimension. See **limit gauge.**

end-hats (*Elec. Eng.*). Metal fitments at the ends of the cathode in certain magnetrons to reduce space charge.

endite (*Zool.*). In *Crustacea*, a lobe on the inner side of a phyllopodium; any process on the inner side of a limb, such as a gnathobase.

end-lap joint (*Carp.*). A *halving* (q.v.) joint formed between the ends of two pieces of timber.

end leakage flux (*Elec. Eng.*). Leakage flux associated with the end connexions of an electric machine.

endless rope haulage (*Mining*). A method of hauling trucks underground by attachment to a long loop of rope, guided by many pulleys along the roads or haulage ways, and actuated by a power-driven drum.

endless saw (*Eng.*). See **bandsaw.**

end links (*Eng.*). The links at either end of a chain; they are made slightly stronger than the remainder to allow for wear when attached to hooks, couplings, etc.

end matter (*Typog.*). The items which follow the main text of a book, i.e., appendices, notes, glossary, bibliography, index.

end measuring instruments (*Eng.*). Measuring instruments (e.g., micrometers, callipers, gauges) which measure length by making contact with the ends of an object.

end mill (*Eng.*). A *milling cutter* (q.v.) having radially disposed teeth on its circular cutting face; used for facing operations.

endo-. Prefix from Gk. *endon*, within.

endoaneurysmorrhaphy (*Surg.*). (1) Obliteration, by suture at either end, of an aneurysm of an artery. (2) Obliteration, by suture, of the aneurysmal sac, with reconstruction of the original arterial channel.

endobiotic (*Bot.*). (1) Growing inside another plant. (2) Formed inside the host cell.

endoblast (*Zool.*). See **hypoblast.**

endocardiac (*Zool.*). Within the heart.

endocarditis (*Med.*). Inflammation of the lining membrane of the heart, especially of that part covering the valves.

endocardium (*Zool.*). In Vertebrates, the layer of endothelium lining the cavities of the heart.

endocarp (*Bot.*). A differentiated innermost layer of a pericarp, usually woody in texture.

endocellular enzyme (*Biol.*). An enzyme which functions within the cell in which it is developed.

endocervicitis (*Med.*). Inflammation of the mucous membrane lining the cervix uteri.

endochondral (*Zool.*). Situated within or taking place within cartilage; as *endochondral ossification*, which begins within the cartilage and works outwards.

endochondrosteosis (*Zool.*). Endochondral ossification.

endochorion (*Zool.*). The inner layer of the chorion in insects.

endocoelar (*Zool.*). See **splanchnopleural.**

endoconidium (*Bot.*). A conidium formed inside a hypha.

endocranium (*Zool.*). In Insects, internal processes of the skeleton of the head, serving for muscle attachment.

endocrine (*Zool.*). Internally secreting; said of certain glands, principally in Vertebrates, which pour their secretion into the blood, and so can affect distant organs or parts of the body. See **hormone.**

endocrinology. The study of the endocrine glands and their secretions.

endocrinopathy (*Med.*). Any disease due to disordered function of the endocrine glands.

endocuticle (*Zool.*). The laminated inner layer of the insect cuticle.

Endocyclica (*Zool.*). An order of *Echinoidea* in which the mouth is central, the ambulacra are not petalloid, and the anus is in the centre of the aboral surface. Regular Urchins.

endocyst (*Zool.*). In *Polyzoa*, the soft part of the body-wall of a zooecium, consisting of ectoderm and mesoderm: in *Protozoa*, the membranous inner layer of a cyst. Cf. *ectocyst*.

endocytosis (*Cyt.*). The uptake by a cell of particles that are too large to diffuse through the wall. Includes both *phagocytosis* and *pinocytosis* (qq.v.).

endoderm (*Zool.*). The inner layer of cells forming the wall of a gastrula and lining the archenteron: the tissues directly derived from this layer.

endodermal pressure (*Bot.*). See **root pressure.**

endoderm disk (*Zool.*). In certain *Malacostraca*, a posterior unpaired thickening on the ventral surface of the blastoderm during early development.

endodermis (*Bot.*). A sheath of cells, one layer thick, at the boundary between the cortex and the stele. The radial walls of the cells develop characteristic bands of thickening (Casparian strip), and may subsequently become heavily thickened over most of their surface. There are no spaces between the cells composing the endodermis. It is thus a firmly built layer, and probably plays a part in controlling the horizontal movement of water in the root or stem.

endoderm lamella (*Zool.*). In certain medusae, a sheet of endoderm lying in the thickness of the mesogloea, in the radial segments bounded by the circular canal, the central enteric cavity, and the radial canals.

endodyne (*Electronics*). See **autodyne.**

endoenzymes (*Chem.*). Enzymes which are endocellular, and do not leave the cell or diffuse through the cell walls. They hydrolyse by stepwise cleavage of the molecule. *Zymase* is a typical endoenzyme.

endoergic process (*Nuc.*). Nuclear process in which energy is consumed. Equivalent to thermodynamic term *endothermic* (q.v.).

***endo-exo* configuration** (*Chem.*). Special case of *cis-trans* isomerism, e.g., borneol shows the *trans-* or *endo*-configuration while *iso*-borneol shows the *cis-* or *exo*-configuration.

end-of-selection signal (*Telecomm.*). One which, normally given by a register, may be transmitted in either direction and can be used to prepare the circuit for speech.

endogamy (*Bot.*). (1) Pollination between two flowers on the same plant. (2) The union of two sister gametes, both female. (*Zool.*) See **inbreeding.**

endogastric (*Zool.*). Said of spirally coiled shells in which the coil is directed on to the posterior face of the animal. Cf. *exogastric.*

endogenous (*Bot.*). Found inside another organ of the plant. (*Zool.*) (In *Sporozoa*) said of forms in which sporulation is effected within the host; (in higher animals) said of metabolism which leads to the building of tissue, or to the replacement of loss by wear and tear. Cf. *exogenous.*

endogenous circadian rhythm (*An. Behav.*). See **circadian rhythm.**

endogenous spore (*Bot.*). A spore formed inside a sporangium.

endogenous thallus (*Bot.*). The thallus of a lichen in which the alga predominates.

endognaths (*Zool.*). In *Crustacea*, the endopodites of the mouth-parts.

endolithic (*Bot.*). Growing within the substance of rocks or stones.

endolymph (*Zool.*). In Vertebrates, the fluid which fills the cavity of the auditory vesicle and its outgrowths (semicircular canals, etc.).

endolymphangial (*Zool.*). Situated within a lymphatic vessel.

endolymphatic (*Zool.*). Pertaining to the membranous labyrinth of the ear in Vertebrates, or to the fluid contained therein.

endometrial (*Anat.*). Pertaining to, or having the character of, endometrium.

endometrioma (*Med.*). Adenomyoma. An endometrial tumour consisting of glandular elements and a cellular connective tissue, occurring in regions where endometrium is normally absent.

endometritis (*Med.*). Inflammation of the endometrium.

endometrium (*Anat.*). The mucous membrane lining the cavity of the uterus.

endomixis (*Zool.*). In *Protozoa*, a process closely resembling conjugation, but taking place in solitary individuals without involving syngamy.

endomorphy (*Psychol.*). Sheldon's *somatotype* which is characterized by general heaviness or stoutness of build. Said to be associated with more or less extraversive personality traits. See **ectomorphy, mesomorphy.**

Endomycetales (*Bot.*). An order of *Ascomycetes*; the thallus is simple; the zygote develops into an ascus directly, e.g., Yeasts. Many are parasitic on man and animal.

endomyocarditis (*Med.*). Inflammation of the heart muscle and of the membrane lining the cavity of the heart.

endomysium (*Zool.*). The intramuscular connective tissue which unites the fibres into bundles.

end-on (*Mining*). Headings or stalls are *end-on* when the coal is being worked away in a direction at right-angles to the natural cleats or main slips.

end-on or end-fire aerial array (*Radio*). A linear array with a progressive phase shift between the elements which is equal to, or greater than, the space phase shift. The principal maximum lies along the axis of the array.

endoneurium (*Zool.*). Delicate connective tissue between the nerve fibres of a funiculus.

endoparasite (*Bot., Zool.*). A parasite living inside the body of its host.

endopeptidases (*Chem.*). Enzymes which catalyse the hydrolysis of the interior peptide bonds of proteins.

endoperidium (*Bot.*). The inner layer of a peridium, when distinguishable.

endophlebitis (*Med.*). Inflammation of the intima of a vein.

endophragmal skeleton (*Zool.*). An internal skeletal framework formed by the elaboration and combination of apodemes and serving for muscle attachment in *Arthropoda.*

endophthalmitis (*Med.*). Inflammation of the internal structures of the eye.

endophyllous (*Bot.*). (1) Said of a parasite living inside a leaf. (2) Said of a plant member developed in the shelter of a sheathing leaf.

endophyte (*Bot.*). A plant living inside another plant, but not necessarily parasitic on it.

endophytic mycorrhiza (*Bot.*). See **endotrophic mycorrhiza.**

endoplasm (*Cyt.*). The granular central portion of the cytoplasm of a cell. Cf. *ectoplasm.*

endoplasmic reticulum (*Cyt.*). System of membranes on which *ribosomes* are arranged, usually best developed in cells engaged in secreting protein, e.g., pancreas or liver cells.

endoplastic reticulum (*Cyt.*). A complex granular matrix in the cell cytoplasm, revealed by the electron microscope.

endopleura (*Bot.*). The inner seed coat of *Cycadales.*

endopleurite (*Zool.*). A lateral apodeme; an infolding of the exoskeleton between the pleurites of the thorax in winged insects.

endopodite (*Zool.*). The inner ramus of a biramous arthropod appendage.

Endoprocta (*Zool.*). See **Entoprocta.**

Endopterygota (*Zool.*). Same as *Oligoneoptera.*

endoradiosondes (*Electronics*). Probes and transducers for insertion into the human body without interfering appreciably with its function, and which transmit data to external receivers.

endorhachis (*Zool.*). In Vertebrates, a layer of connective tissue which lines the canal of the vertebral column and the cavity of the skull.

endorhizoid (*Bot.*). A rhizoid formed at the base of the seta of a Bryophyte, and penetrating the gametophyte.

endosalpingitis (*Med.*). Inflammation of the lining membrane of the Fallopian tube.

endosarc (*Cyt.*). See **endoplasm.**

endoscope (*Instr.*). (1) A tubular instrument for inspecting the cavities of internal organs in medicine. (2) In industry, an instrument used for inspection and photography of remotely situated parts of machinery and equipment.

endoscopic embryology (*Bot.*). The condition when the apex of the developing embryo points towards the base of the archegonium.

endoskeleton (*Zool.*). In *Craniata*, the internal skeleton, formed of cartilage or cartilage bone. In *Arthropoda*, the endophragmal skeleton, i.e., hardened invaginations of the integument forming rigid processes for the attachment of muscles and the support of certain other organs.

endosmosis (*Chem.*). The process of osmosis in an inward direction towards the solution. See **exosmosis, osmosis.**

endosome (*Zool.*). A single central mass in the

nuclei of *Protozoa*; it is usually a nucleolus or an amphinucleolus. See **karyosome.**

endosperm (*Bot.*). A multicellular tissue formed inside a developing seed, serving in the nutrition of the embryo, and sometimes increasing greatly, forming the bulk of the ripe seed. Many endospermous seeds are important sources of food materials, for man and other animals.

endospermous (*Bot.*). Said of a seed having endosperm; albuminous.

Endosporae (*Bot.*). In former classifications, a subclass of the *Myxothallophyta*, including members producing spores inside sporangia.

endospore (*Bot.*). (1) The innermost layer of the wall of a spore. (2) A spore formed inside a mother cell.

endosternite (*Zool.*). In Insects, a sternal apodeme; in *Notostraca*, *Apus* and some decapod *Crustacea*, a mesodermal tendinous plate lying under the anterior part of the alimentary canal; in *Arachnida*, a slightly chitinized cartilaginous plate lying between the nerve ganglia and the alimentary canal.

endosteum (*Zool.*). The periosteum lining the cavities of some kinds of bones.

endostyle (*Zool.*). In some *Protochordata* and in the larvae of *Cyclostomata*, a longitudinal ventral groove of the pharyngeal wall, lined by ciliated and glandular epithelium. *adj.* **endostylar.**

endotergite (*Zool.*). In Insects, a thoracic tergal apodeme, best developed in actively flying insects; phragma.

endothecium (*Bot.*). (1) The inner tissues in a young sporogonium of a moss. (2) The fibrous layer in the wall of an anther.

endotheliochorial placenta (*Zool.*). See **vasochorial placenta.**

endothelioma (*Med.*). A tumour arising from the lining membrane of blood vessels or lymph channels. Also a tumour arising in the pleura, in the peritoneum, or in the meninges.

endothelium (*Zool.*). Pavement epithelium occurring on internal surfaces, such as those of the serous membranes, blood-vessels, lymphatics.

endothermic (*Chem.*). Accompanied by the absorption of heat. (ΔH positive.)

endothiobacteria (*Bacteriol.*). Bacteria which form sulphur inside their cells.

endothion (*Chem.*). *S*-(5 methoxy-4-pyron-2-ylmethyl) dimethyl phosphorothiolate, used as an insecticide.

endothorax (*Zool.*). In *Arthropoda*, the endoskeleton of the thorax, consisting of endotergites, endopleurites and endosternites in Insects.

endotoxin (*Bacteriol.*). Toxin retained within, if not part of, the bacterial protoplasm, and only released on lysis or death of the bacterial cells. Examples of diseases caused by endotoxins are typhoid fever and cholera.

endotrachea (*Zool.*). In Insects, the chitinous innermost layer of a trachea. See also **intima.**

endotracheitis (*Med.*). Inflammation of the mucous membrane of the trachea.

endotrophic (*Zool.*). In Insects, said of the space enclosed within the peritrophic membrane.

endotrophic mycorrhiza (*Bot.*). A mycorrhiza in which the fungus occurs almost entirely inside the body of the associated plant.

endozoic (*Bot.*). (1) Living inside an animal. (2) Said of the method of seed dispersal in which seeds are swallowed by some animal and voided after having been carried for some distance.

end-papers (*Bind.*). Stout papers formed from a folded sheet which is firmly attached to the first and last sections of a volume at the fold. One half of each sheet is securely pasted to the inner side of the front and the back cover, the other half forming a fly-leaf.

end plate (*Electronics*). Disk-shaped electrode, partially or completely closing the end of a concentric cylindrical arrangement of anode and cathode in a thermionic tube, to which a negative potential is applied to prevent the escape of electrons from the end of the tubular anode. (*Zool.*) A form of motor nerve-ending in muscle.

end-plate fins (*Aero.*). Fins mounted at or near the tips of the *tail plane* to increase its efficiency.

end-plate magnetron (*Electronics*). Magnetron fitted with end plates which, in conjunction with a magnetic field directed exactly parallel with the cathode, performs in the same manner as one without end plates, in which the magnetic field is slightly inclined to the axis of the cathode.

end play device (*Elec. Eng.*). A device, used with rotary converters and other commutator machines, which gives a small oscillatory axial movement to the rotor, in order to prevent the wearing of grooves in the commutator.

end-point (*Chem.*). The point in a volumetric titration at which the amount of added reagent is equivalent to the solution titrated.

end product (*Nuc.*). The stable nuclide forming the final member of a radioactive decay series.

end products of weathering (*Geol.*). These vary with the nature of the rocks acted upon; but in general the end products are: quartz sand, which is chemically stable and hard; clay, which is deposited, after transportation by rivers, on the bed of the sea; and salts, which are carried in solution and ultimately added to the amount already present in sea water.

end-quench test (*Eng.*). See **Jominy test.**

endrin (*Chem.*). 1,2,3,4,10,10 - Hexachloro - 6,7 - epoxy - 1,4,4a,5,6,7,8,8a - octahydro - *exo* - 1,4 - *exo* - 5,8 - di-methanonaphthalene, used as an insecticide.

end-sac (*Zool.*). In some *Arthropoda*, the small coelomic vesicle in which a coelomoduct terminates, in the absence of perivisceral coelom.

ends down (*Spinning*). Broken yarns twisting in bunches on the spindle tips. (*Weaving*) Warp threads that have broken during weaving.

end sheet (*Elec. Eng.*). A sheet of insulating material placed between the end section of an accumulator and the lining of the container.

end shield (*Elec. Eng.*). A cover which wholly or partially encloses the end of an electric machine.

end speed (*Aero.*). Naval term for the speed of an aircraft relative to its aircraft carrier at the moment of release from catapult or accelerator.

end spring (*Elec. Eng.*). A small spring of hard lead placed in a lead-acid accumulator, between the end plates and the container to prevent the plates from spreading.

end stone (*Horol.*). A flat jewel which acts as a bearing surface for the end of a pivot, e.g., the balance staff is provided with jewels, with through holes for the parallel portions of the pivots and end stones for the ends of the pivots.

endurance (*Aero.*). The maximum time, or distance, that an aircraft can continue to fly without refuelling, under any agreed conditions.

endurance limit (*Eng.*). In fatigue testing, the alternating stress which just produces fracture after a given number of reversals; often taken as 10 million. (*Met.*) See under **limiting range of stress.** See also **fatigue limit.**

endurance range (*Eng.*). Fatigue testing normally involves stress reversal. This is the maximum total range over which the stress fluctuates, and thus is often approximately double the *endu-*

rance limit. (*Met.*) See under **limiting range of stress.**

end-windings (*Elec. Eng.*). See **armature end connexions.**

end window counter (*Nuc.*). G-M counter designed so that radiation of low penetrating power can enter one end. (This is usually covered with a thin mica sheet.)

endysis (*Zool.*). The formation of new layers of the integument following ecdysis.

enema (*Med.*). Fluid injected into the rectum to promote evacuation of the bowels or to convey nutrition to the body (*nutrient enema*).

energetics (*Chem.*). The abstract study of the energy relations of physical and chemical changes. See **thermodynamics.**

energid (*Biol.*). Any nucleus, together with the cytoplasm which it controls; any living uninucleated protoplasmic unit with or without a cell wall.

energy (*Phys.*). The capacity of a body for doing work. Mechanical energy may be of two kinds: *potential energy*, by virtue of the position of a body; or *kinetic energy*, by virtue of its motion. Both mechanical and electrical energy may be converted into heat, which is itself regarded as another form of energy. An electrical capacitance of magnitude C, holding a charge φ, has an electrical energy $\frac{1}{2}(\varphi^2/C)$, which is resident in the field between the plates. On discharge, an amount of work equal to $\frac{1}{2}(\varphi^2/C)$ will be obtained. *Units of energy*: in the CGS metric system, the *erg* is the work done by a force of 1 dyne moving through 1 cm in the direction of force; the SI unit, the *joule*, is equal to 10^7 ergs and is the work done by a force of 1 newton moving through 1 m in the direction of force. The foot-pound force (ft lbf) of the British system = 1·356 joules. **See kinetic energy, potential energy, mechanical equivalent of heat.**

energy balance (*Chem. Eng.*). A *heat balance* (q.v.). (*Eng.*) A balance-sheet drawn up from an engine test, expressing the various forms of energy produced by the engine (e.g., B.H.P or output power, heat to cooling water, heat carried away in exhaust gases, and heat unaccounted for) as percentages of the heat supplied from the calorific value of the fuel.

energy band (*Phys.*). In a solid, the energy levels of individual atoms interact to form bands of permitted levels with gaps between. Normally there is a valence band with a full complement of electrons and a conduction band which is empty. When these overlap metallic conduction is possible. In semiconductors, there is a small gap and intrinsic conduction occurs only when some electrons acquire the energy necessary to surmount this gap and enter the conduction band. In insulators, the gap is huge and cannot normally be surmounted.

energy barrier (*Chem.*). The minimum amount of free energy which must be attained by a chemical entity in order to undergo a given reaction.

energy component (*Elec. Eng.*). See **active component.**

energy curve (*Photog.*). The spectral distribution of radiated energy in a source of light, e.g., an arc, the ordinate being proportional to the energy contained in a specified small band of wavelengths.

energy density of sound (*Acous.*). In a reverberant sound field, the particle vibration energy per unit volume, averaged over the whole of the enclosure when the distribution has become stable.

energy flow (*Ecol.*). The total assimilation at a particular trophic level, which equals the production of biomass plus respiration.

energy gap (*Phys.*). Range of forbidden energy levels between two permitted bands. See **energy band.**

energy levels (*Electronics*). In semiconductors, a *donor level* is an intermediate level close to the conduction band; being filled at absolute zero, electrons in this level can acquire energies corresponding to the conduction level at other temperatures. An *acceptor level* is an intermediate level close to the normal band, but empty at absolute zero; electrons corresponding to the normal band can acquire energies corresponding to the intermediate level at other temperatures. See **Fermi characteristic-.** Electron energies in atoms are limited to a fixed range of values termed *permitted energy levels*, and represented by horizontal lines drawn against a vertical energy scale. See **eigenvalues, Schrödinger equation.**

energy/mass equivalence (*Nuc.*). Einstein formula, $E = mc^2$, where E = energy, m = mass, and c = velocity of light, which gives the energy liberated when mass is annihilated, as in A and H bombs and the stars. Reversal also occurs.

energy, pyramid of (*Ecol.*). A diagram showing the rate of energy flow, or productivity, at successive trophic levels.

energy unit (*Radiol.*). Energy absorbed from ionizing radiation amounting to $9{\cdot}3 \times 10^{-3}$J/kg of the tissue at the point.

enfleurage (*Chem.*). Process of cold extraction with fat, e.g., of essential oils from flowers. Used in perfumery.

eng (*For.*). A very durable dark reddish-brown wood from Burma, much used for house-building and for parquet flooring. Also called **in.**

engaging speed (*Aero.*). The relative speed of a carrier-borne aircraft to its ship at the moment when the *arrester gear* (q.v.) is engaged.

Engel process (*Plastics*). Patented process for the production of large vessels.

engine (*Eng.*). Generally, a machine in which power is applied to do work: particularly, a machine for converting heat energy into mechanical work; loosely, a locomotive.

engine cylinder (*Eng.*). That part of an engine in which the heat and pressure energy of the working fluid do work on the piston, and so are converted into mechanical energy. See **cylinder, cylinder barrel.**

engineer. One engaged in the science and art of engineering practice. The term is a wide one, but it is properly confined to one qualified to design and supervise the execution of mechanical, electrical, hydraulic, and other devices, public works, etc. See **Chartered Engineer.**

engineering bricks (*Civ. Eng.*). Bricks made of semivitreous materials. They are dense and of high strength and low porosity. Manufactured chiefly in Staffordshire, Lancashire, Sussex, N. Wales, and Yorkshire. Used where severe loading conditions exist or exposure to damp, frost, acid or acid fumes, etc.

engineer's chain (*Surv.*). A 100-ft measuring chain of 100 1-ft links.

engine friction (*Eng.*). The frictional resistance to motion offered by the various working parts of an engine. See **friction horsepower.**

engine pit (*Eng.*). (1) A hole in the floor of a garage to enable a man to work on the underside of a motor vehicle. (2) An engine sump or crank pit; the boxlike lower part of the crankcase. (3) A large pit for giving clearance to the flywheel of any large gas engine or winding engine.

engine plane (*Mining*). In a coalmine, a roadway on which tubs, trucks, or trains are hauled by means of a rope worked by an engine.

engine pod (*Aero.*). A complete turbojet power unit, including cowlings, supported on a pylon, usually under the wings of an aeroplane, an installation method commonly adopted on U.S. types of multi-engined aircraft.

Engine-scope (*Eng.*). TN for an electronic analyser for tracing electrical faults in I.C. engines.

engine-sized paper (*Paper*). Paper hardened by the addition of rosin and alum to the fluid pulp.

engine speed (*Eng.*). In a turbine engine, the revolutions per minute of the main rotor assembly; in a reciprocating engine, those of the crankshaft.

englacial streams (*Geol.*). Streams which follow courses in the ice, after having left the surface of the glacier by plunging into a swallow- or sink-hole. Cf. *subglacial drainage*.

Engler distillation (*Chem.*). The determination of the boiling range of petroleum distillates, carried out in a definite prescribed manner by distilling 100 cm^3 of the substance and taking the temperature after every 5 or 10 cm^3 of distillate has collected. The initial and final boiling points are also measured.

Engler flask (*Chem.*). A 100 cm^3 distillation flask of definite prescribed proportions used for carrying out an *Engler distillation*.

English (*Typog.*). An old type size, approximately 14-point.

English bond (*Build.*). The form of bond in which each course is alternately composed entirely of headers or of stretchers.

English cross bond (*Build.*). A *Dutch bond* (q.v.).

English garden-wall bond (*Build.*). The form of *garden-wall bond* (q.v.) in which a course of headers alternates with 3 courses of stretchers. Also called **Scotch bond.**

English lever (*Horol.*). Type of watch with lever escapement having pointed teeth on the escape wheel.

English pink (*Paint.*). See **Dutch pink.**

English roof truss (*Eng.*). A truss for roofs of large span in which the sloping upper and lower chords are symmetrical about the central vertical, and are connected by vertical and diagonal members.

engorgement (*Med.*). Congestion of a tissue or organ with blood.

engram (*An. Behav.*). A hypothetical neural locus, structure or persistent activity involved in some explanations of learning.

enlargement (*Bot.*). Primary growth in thickness before secondary thickening begins.

enlarging lens (*Photog.*). One corrected specifically for a flat field, for use in an enlarger.

enneandrous (*Bot.*). Having 9 stamens.

enol form (*Chem.*). The carbinol form of a substance exhibiting keto-enol tautomerism, i.e., that form in which the mobile hydrogen atom is attached to the oxygen atom, and therefore has acidic properties. E.g., *ethyl aceto-acetate*.

enophthalmos, enophthalmus (*Med.*). Abnormal retraction of the eye within the orbit.

enostosis (*Med.*). A bony growth formed on the internal surface of a bone.

enprint (*Photog.*). Standard size enlargement produced by D and P firms in lieu of contact print.

en quad (*Typog.*). A type space half an em wide. It is usually called a nut, to distinguish it clearly from an em (or **mutton**).

enriched uranium (*Nuc.*). Uranium in which the proportion of the fissile isotope ^{235}U has been increased above its natural abundance.

enriched water gas (*Fuels, etc.*). See **carburetted water gas.**

enrichment (*Arch., Build.*). Ornamentation applied to building features. (*Mining*) Effect of superficial leaching of lode, whereby part of value is dissolved and redeposited in a lower enriched zone (*secondary enrichment*). (*Nuc. Eng.*) (1) Raising the proportion of fissile nuclei above that for natural uranium in reactor fuel. (2) Raising the proportion of the desired isotope above that present initially in isotope separation.

enrichment factor (*Nuc.*). (1) In the U.K., the abundance ratio of a product divided by that of the raw material. The *enrichment* is the enrichment factor less unity. (2) In the U.S., same as **separation factor.**

enrockment (*Hyd. Eng.*). The layer of stones placed over the face of a dyke or sea-embankment as a protection against the impact of the water.

en rule (*Typog.*). A dash (–) cast on an *en* body. Half the width of an *em* rule, it is often used to divide dates, express range, etc. Also **en score.**

ensiform. Shaped like the blade of a sword.

ensiform process (*Zool.*). See **xiphisternum.**

ensilage. A cattle food formed by the bacteriological breakdown, in portable or fixed silos, of vegetable matter (grasses, clovers, pea haulms, beet tops, etc.), with the admixture of diluted molasses, suitable acids (e.g., formic), etc. Also called **silage.** The process of making such food.

ensor (*Textiles*). A levers net made in imitation of a traverse net.

enstatite (*Min.*). An orthorhombic pyroxene, chemically a silicate of magnesium, $MgSiO_3$; it occurs as a rock-forming mineral, and also as an important end-member component of other, chemically more complex, pyroxenes.

enstyle (*Arch.*). See **eustyle.**

ENT (*Med.*). Abbrev. for Ear, Nose, and Throat as a department of medical practice.

entablature (*Arch.*). The whole of the parts immediately supported upon columns, consisting of an architrave, a frieze, and a cornice. (*Eng.*) An engine framework supported by columns.

entamoebiasis (*Med.*). Infection with *Entamoeba histolytica*.

entasis (*Arch.*). The slight swelling towards the middle of the length of a column to correct for the appearance of concavity in the outline of the column if it had a straight taper.

entasis reverse (or **rule**) (*Arch.*). A templet cut to the profile of an entasis, for the shaping of which it is used as a reference.

entepicondylar (*Zool.*). In some Reptiles and Mammals, said of a foramen near the lower end of the humerus, for the radial nerve and brachial artery.

enteque (*Vet.*). A variety of haemorrhagic septicaemia of cattle, horses, and sheep in South America; osseous metaplasia commonly occurs in the chronic forms of the disease in cattle.

enter-, entero-. Prefix from Gk. *enteron*, intestine.

enteral (*Med.*). Within, or by way of, the intestine. (*Zool.*) Parasympathetic.

enterclose (*Arch.*). A corridor separating two rooms.

enterectomy (*Surg.*). Removal of part of the bowel.

enteric (*Med.*). (1) Pertaining to the intestines. (2) Synonym for *enteric fever* (see **typhoid fever).**

entering edge (*Elec. Eng.*). The edge of the brush of an electrical machine which is first met during revolution by a point on the commutator or slip ring. Also called **leading edge** or **toe of the brush**. Cf. *leaving edge*.

entering pallet (*Horol.*). See **pallet**.

entering tap (*Eng.*). See **taper tap**.

enteritis (*Med.*). Inflammation of the small intestine.

entero-anastomosis (*Surg.*). The operative union of two parts of the intestine; the operation for doing this.

Enterobacteriaceae (*Bacteriol.*). A family of bacteria belonging to the order *Eubacteriales*. Gram-negative rods; aerobes; carbohydrate fermenters; many saprophytic. Includes some gut parasites of animals and some blights and soft rots of plants, e.g., *Salmonella typhi* (typhoid), *Escherichia coli* (some strains cause enteritis in calves and infants).

enterobiasis (*Med.*). Infection by the thread- or pin-worm, *Enterobius vermicularis*.

enterocele (*Med.*). A hernia containing intestine.

enterocentesis (*Surg.*). Operative puncture of the intestine. (*Vet.*) The operation of puncturing the distended intestine of a horse suffering from colic, in order to liberate the gas.

enterocoel (*Zool.*). A coelome formed by the fusion of coelomic sacs which have separated off from the archenteron. In *Chaetognatha*, *Echinodermata*, and *Cephalochorda*.

enterocolitis (*Med.*). Inflammation of the small intestine and of the colon.

enterocolostomy (*Surg.*). A communication, made by operation, between the small intestine and colon.

enterocystocele (*Med.*). Hernia containing intestine and bladder.

entero-enterostomy (*Surg.*). The operative formation of a communication between two separate parts of the small intestine; the operation for doing this.

entero-epiplocele (*Med.*). A hernia containing small intestine and omentum.

enterogastrone (*Physiol.*). A hormone liberated from the duodenal mucosa, by the action of fatty substances in the food, which causes inhibition of gastric motility and secretion.

enterogenous cyanosis (*Med.*). A disorder characterized by chronic cyanosis and by the presence of methaemoglobin or sulphaemoglobin in the blood; due usually to taking drugs, especially aniline derivatives.

enterokinase (*Chem., Zool.*). A kinase of the intestinal juice which activates trypsinogen of the pancreatic juice to trypsin.

enterolith (*Med.*). A concretion of organic matter and lime, bismuth, or magnesium salts, formed in the intestine.

enteron (*Zool.*). The single body cavity of *Coelenterata*; it corresponds to the archenteron of a gastrula: in higher forms, the alimentary canal. See also **archenteron**. *adj.* **enteric**.

Enteropneusta (*Zool.*). See **Hemichorda**.

enteroproctia (*Surg.*). The condition in which an artificial anus is formed from intestine.

enteroptosis (*Med.*). Glénard's disease; visceroptosis. Abnormally low position of the intestines in the abdominal cavity, as a result, partly, of weakness of the abdominal and pelvic muscles.

enterospasm (*Med.*). Spasm of the muscle of the intestine.

enterostomy (*Surg.*). The surgical formation of an opening in the intestine, for the purpose of draining intestinal contents.

enterosympathetic (*Zool.*). Said of that part of the autonomic nervous system which supplies the alimentary tract.

enterotomy (*Surg.*). Incision of the intestinal wall.

enterotoxaemias of sheep (*Vet.*). A group of fatal toxaemic diseases of sheep due to alimentary infection and toxin production by the bacteria *Clostridium perfringens* (*Cl. welchii*), types B, C, or D.

enthalpy (*Heat*). Thermodynamic property of a working substance defined as $H=U+PV$, where U=internal energy, P=pressure and V=volume of system. Useful in studying flow process, replacing total heat.

entire (*Bot.*). Said of the margin of a flattened organ when it is continuous, being neither toothed nor lobed.

entire function (*Maths.*). See **analytic function.**

ento-. Prefix from Gk. *entos*, within.

entobronchia (*Zool.*). In Birds, secondary bronchi leading to the air-sacs, being diverticula of the mesobronchium.

entocoele (*Zool.*). In *Zoantharia*, the portion of coelenteron enclosed by each pair of primary mesenteries. Cf. *exocoele*.

entoconid (*Zool.*). A cusp on the molar teeth of the lower jaw in Mammals.

entocuneiform (*Zool.*). A bone in the distal row of tarsals in the hind foot of *Tetrapods*, corresponding to the *trapezium* of the fore-limb.

entoderm (*Zool.*). See **endoderm.**

entogastric (*Zool.*). Within the stomach or enteron.

entoglossal (*Zool.*). Lying within the substance of the tongue; in some Fish, an anterior extension of the hyoid copula supporting the tongue.

entoglossum (*Zool.*). A structure lying within the tongue; as an anterior extension of the basihyal in some Fish.

entomogenous (*Bot.*). Living on or in insects.

entomology. The branch of zoology which deals with the study of insects.

entomophagous (*Zool.*). Feeding on insects.

entomophily (*Bot.*). Pollination by insects. *adj.* **entomophilous.**

Entomophthorales (*Bot.*). An order of the *Phycomycetes*; includes the genera parasitic, or saprophytic on insects; asexual reproduction is by a conidiosporangium, which is usually discharged explosively, and germinates directly; sexual reproduction by gametangia or aplanospores.

Entomostraca (*Zool.*). An obsolete group of *Crustacea* which comprised the classes *Branchiopoda*, *Copepoda*, *Ostracoda*, *Cirripedia*, and *Branchiura* (qq.v.) and was distinguished by small size, variable number of body somites, situation of the excretory glands in the second maxillary segment, the absence of a gastric mill, and the presence of a simple heart and a persistent nauplius eye.

entomo-urochrome (*Zool.*). A greenish-yellow pigment in the *Malphigian tubules* of some insects, consisting of accumulated riboflavin.

entoplastron (*Zool.*). In *Chelonia*, one of the plates composing the plastron, lying between the hyoplastron and the epiplastron.

Entoprocta (*Zool.*). A phylum of *Metazoa* in which the anus is inside the circlet of tentacles; mainly marine forms. Also **Endoprocta.**

entopterygoid (*Zool.*). In some Fish, a paired ventral bone of the skull, adjoining the palatine.

entosternite (*Zool.*). See **endosternite.**

entoturbinals (*Zool.*). The shorter inner series of ethmoturbinal plates in Mammals. Cf. *ectoturbinals*.

entotympanic (*Zool.*). In some Mammals, a cartilage bone which lies close to and supplements the tympanic bulla.

entovarial (*Zool.*). Within the ovary.

entozoic (*Bot.*). Living inside an animal.

entozoon (*Zool.*). An animal parasite living within the body of the host. *adj.* **entozoic.**

entrain (*Eng.*). To suspend bubbles or particles in a moving fluid.

entrainer (*Chem. Eng.*). See **azeotropic distillation.**

entrainment (*An. Behav.*). The process by which an *endogenous circadian rhythm* becomes a precise 24-hour rhythm. See **circadian rhythm.** (*Chem.*) Transport of small liquid particles in vapour, e.g., when drops of water are carried over in steam.

entrance lock (*Hyd. Eng.*). A lock through which vessels must pass in entering or leaving a dock, on account of difference in level between the water impounded in the dock and that outside.

entrance pupil (*Optics*). The diaphragm, or image of the diaphragm in the object space, which limits the diameter of the cone of rays entering an optical system.

entresol (*Arch.*). A *mezzanine* (q.v.) floor usually between the ground and first floors.

entropy (*Phys.*). In thermodynamic processes this is concerned with the probability of a given distribution of momentum among molecules. Any 'free' physical system will distribute its energy so that the entropy increases but the available energy diminishes. Based on physical conception that if a substance undergoing a reversible change takes in a quantity of heat dQ at absolute temperature T, its entropy, S, is increased by dQ/T. Thus the area under an absolute temperature-entropy graph for a reversible process represents the heat transfer during the process. An adiabatic process is one in which there is no heat transfer, thus its temperature-entropy graph is a vertical straight line, and thus entropy is constant during the process. Any process in which no change of entropy occurs is said to be *isentropic.* (*Telecomm.*) Deprecated term for *information rate.*

entropy of fusion (*Phys.*). A measure of the increased randomness that accompanies the transition from solid to liquid or liquid to gas; equal to the latent heat divided by the absolute temperature. Also called **vaporization.**

entry portal (*Radiol.*). Area through which a beam of radiation enters the body.

Entz booster (*Elec. Eng.*). An automatic battery booster in which the excitation is controlled automatically by a carbon rheostat, according to the load requirements.

enucleation (*Bot., Zool.*). The removal of the nucleus from a cell, e.g., by microdissection. *adj.* and *v.* **enucleate.** (*Surg.*) Removal of any tumour or globular swelling so that it comes out whole.

enumerable set (*Maths.*). See **denumerable set.**

enuresis (*Med.*). Involuntary micturition.

envelope (*Aero.*). The gas-bag of a balloon, or of a nonrigid or semirigid airship. (*Electronics*) The outer containing vessel of a discharge tube. (*Maths.*) Of a family of plane curves: the curve which touches every curve of the family. (*Paper*) A covering (usually of paper folded and gummed) for a letter, etc. Envelope blanks are cut out by dies and folded by hand or by machine. Common sizes are: *commercial,* $5\frac{1}{2} \times 3\frac{1}{4}$ in.; *court,* $4\frac{3}{4} \times 3\frac{3}{4}$ in.; *foolscap,* $8\frac{3}{4} \times 4$ in. (*Telecomm.*) The modulation waveform within which the carrier of an amplitude-modulated signal is contained, i.e., the curve connecting the peaks of successive cycles of the carrier wave.

envelope cell (*Zool.*). In Insects, a cell which, with the cap cell, encloses the distal process of the bipolar neurone in the scolopophores of chordotonal sense organs. In *Cnidosporidia,* a cell which, with its fellow, encloses the sporoblasts in a pansporoblast.

envelope (or **group**) **delay** (*Telecomm.*). The time taken for the envelope of a signal to travel through a transmission system, without reference to the time taken by the individual components.

envelope-delay distortion (*Telecomm.*). Distortion arising when the rate of change of phase shift with frequency for a transmission system is variable over the required frequency range.

envelope velocity (*Telecomm.*). See **group velocity.**

environment. Generally, the physical surroundings of an object or area, e.g., the temperature, humidity, etc., of the enclosure of a piece of machinery. Specifically refers to the natural surroundings of an organized human society, taking account of the effects of that society, reflected back on to its population in both quantifiable and subjective manners. It is recognized that civilization implies consideration for all natural objects, living or not, and their interactions. See **ecology, euthenics, pollution.**

environmental resistance (*Ecol.*). The sum total of environmental limiting factors which prevent a species or population realizing its *biotic potential* (q.v.).

environment lapse rate (*Meteor.*). The rate of change (usually a decrease) of temperature of the atmosphere with height. Abbrev. **ELR.**

enzootic (*Vet.*). Said of a disease prevalent in, and confined to, animals in a certain area; corresponding to *endemic* in man.

enzootic ataxia (*Vet.*). See **swayback.**

enzootic marasmus (*Vet.*). See **pine.**

enzootic ovine abortion (*Vet.*). Kebbing. A contagious form of abortion in sheep, due to infection of the placenta by a virus of the psittacosis-lymphogranuloma group.

enzymes (*Chem.*). Catalysts produced by living cells. They are proteins, often conjugated and consisting of a simple protein (the carrier) joined to a relatively simple substance (the prosthetic group) which may be a metal (e.g., copper), an organo-metallic compound (e.g., haematin) or a purely organic compound (e.g., a nucleotide). They are specific, each enzyme catalysing only one reaction or type of reaction (e.g., esterase effects the hydrolysis only of esters, maltase only of α-glucosides, pepsin only of proteins). A few occur in digestive juices, etc., but most are intracellular. They lower the energy of activation for the process they catalyse.

eobiont (*Biol.*). A general term for a system with some lifelike characteristics evolved from nonliving molecules; a hypothetical stage in biopoiesis.

Eo-Cambrian (*Geol.*). A term applied to rocks whose age is uncertain but lies near the boundary between the *Pre-Cambrian* and *Cambrian.*

Eocene (*Geol.*). The oldest division of Tertiary rocks, now regarded as a series within the Palaeogene system. It includes the London Clay in England.

eolian deposits (*Geol.*). See **aeolian deposits.**

eolith (*Geol.*). Literally 'dawn stone'; a term applied to the oldest-known stone implements used by early man which occur in the Stone Bed at the base of the Crag in E. Anglia and in high-level gravel deposits elsewhere. The workmanship is crude, and some authorities question their human origin, thinking it likely that the chipping has been produced by natural causes.

Eolithic Period (*Geol.*). The time of the primitive men who manufactured and used eoliths: the dawn of the Stone Age. Cf. *Palaeolithic Period, Neolithic Period.*

eosin (*Chem.*). $C_{20}H_6Br_4O_5K_2$, the potassium salt of tetrabromo-fluorescein, a red dye.

eosinophil (*Histol.*). Any cell whose protoplasmic granules readily stain red with the dye eosin, particularly a granulocyte in the blood and the oxyphil cells in the *pars anterior* of the pituitary gland.

eosinophilia (*Med.*). A pathological excess of eosinophils in the blood.

Eötvös balance (*Mining*). Torsion balance sensitive to minute gravitational differences in land masses.

Eötvös equation (*Chem.*). The molecular surface energy of a substance decreases linearly with temperature, becoming zero about 60°C below the critical point.

Eozoic (*Geol.*). A term suggested for the Pre-Cambrian System, but little used. It means the 'dawn of life', and is comparable with *Palaeozoic, Mesozoic,* and *Cainozoic.*

eozoon (*Geol.*). A banded structure found originally in certain Canadian Pre-Cambrian rocks and thought to be of organic origin; now known to be inorganic and a product of dedolomitization, consisting of alternating bands of calcite and serpentine replacing forsterite.

epacme (*Zool.*). The period in the history of an individual or a race when vigour is increasing; the nepionic and neanic periods in the life history of an individual; the phylonepionic and phyloneanic periods in the history of a race. Cf. *paracme.*

epapophysis (*Zool.*). A median process of a vertebral neural arch.

eparterial (*Anat.*). Situated over an artery. *Eparterial bronchus*, the first division of the right bronchus, which passes to the upper lobe of the right lung.

epaulettes (*Zool.*). In *Scyphozoa*, projections of the outer sides of the oral arms; in late *Echinoplutei*, four ciliated projections at the base of the postoral and posterodorsal arms.

epaxial (*Zool.*). Above the axis, especially above the vertebral column, therefore, dorsal; as the upper of two blocks into which the myotomes of fish embryos become divided. Cf. *hypaxial.*

epaxonic (*Zool.*). See **epaxial.**

epeirogenic earth movements (*Geol.*). Continent-building movements, as distinct from mountain-building movements, involving the coastal plain and just-submerged 'continental platform' of the great land areas. Such movements include gentle uplift or depression, with gentle folding and the development of normal tensional faults.

epeirogenic folds (*Geol.*). Folds of continent-building type, as distinct from those of mountain-building type; usually anticlines, synclines, or monoclines of small amplitude, associated with normal faults in the coastal plains of the continental areas of the earth.

epencephalon (*Zool.*). See **cerebellum.**

ependyma (*Zool.*). In Vertebrates, the layer of columnar ciliated epithelium, backed by neuroglia, which lines the central canal of the spinal cord and the ventricles of the brain. *adj.* **ependymal.**

ependymitis (*Med.*). Inflammation of the ependyma.

ependymoma (*Med.*). A tumour within the brain arising in or near the ventricles and containing ependyma-like cells.

epharmonic convergence (*Bot.*). A likeness in external appearance and in structure between plants which are not closely related in their systematic positions.

ephebic (*Zool.*). Pertaining to the period of maximum strength of an individual; adult; mature.

Ephedra (*Bot.*). Sea-grape, a genus of jointed, all but leafless, desert plants of the Gnetaceae. Source of ephedrine.

ephedrine (*Pharm.*). $C_{10}H_{15}ON$, an alkaloid isolated from *Ephedra vulgaris*; it is a colourless crystalline substance, m.p. 40°C, b.p. 255°C, soluble in water, ethanol and ethoxyethane. Isoephedrine has the constitutional formula: $HO{\cdot}CH(C_6H_5){\cdot}CH(CH_3){\cdot}NH(CH_3)$, and is therefore a derivative of an aliphatic amine. Used in medicine as a sympathomimetic agent for many conditions (asthma, allergies), and as an operational adjuvant.

ephemeral fever (*Vet.*). See **three-day sickness.**

ephemeris (*Astron.*). A compilation, published at regular intervals, in which are tabulated the daily positions of the sun, moon, planets and certain stars, with other data necessary for the navigator and observing astronomer. See **Astronomical Ephemeris, Nautical Almanac.**

ephemeris time (*Astron.*). Abbrev. ET. Uniform or Newtonian time, as used in the calculation of future positions of sun and planets. The normal measurement of time by observations of stars includes the irregularities due to the changes in the earth's rate of rotation. The difference ephemeris time minus universal time is adjusted to zero at an epoch in 1900; it amounted to about 40 sec in 1970. The *ephemeris second* is the fundamental unit of time adopted by the International Committee of Weights and Measures, its defined value being 1/31 556 925·974 7 of the tropical year for 1900 January, 0 at 12^h ET.

Ephemeroptera (*Zool.*). An order of *Palaeoptera*, in which the adults have large membranous forewings and reduced hindwings, and the abdomen bears two or three long caudal filaments; the adult life is very short and the mouthparts are reduced and functionless; the immature stages are active aquatic forms. May-flies.

ephippium (*Zool.*). In *Crustacea*, the thickened cuticle of the carapace which separates from the rest at ecdysis, or in some cases (*Cladocera*) is thrown off as an egg-case: in Vertebrates, the pituitary fossa of the sphenoid.

epi- (*Chem.*). (1) Containing a condensed double aromatic nucleus substituted in the 1,6-positions. (2) Containing an intramolecular bridge.

epiandrium (*Zool.*). In *Arachnida*, a plate surrounding the male genital aperture.

epiascidium (*Bot.*). A funnel-shaped abnormal leaf, with the upper surface of the leaf lining the inside of the funnel.

epibasal half (*Bot.*). The anterior portion of an embryo.

epibasidium (*Bot.*). Any structure arising between the hypobasidium and the sterigma.

epibenthos (*Ecol.*). Animals and plants found living below low-tide mark and above the 180 m (100-fathom) line.

epibiosis (*Ecol.*). A relationship in which an organism lives on the surface of another organism without causing it direct harm. Plant epibionts are termed *epiphytes*, animal epibionts, *epizoites.*

epiblast (*Bot.*). A small outgrowth, of obscure nature, from the embryo of a grass. (*Zool.*) The outer germinal layer in the embryo of a metazoan animal, which gives rise to the ectoderm. Cf. *hypoblast.*

epiboly (*Zool.*). Overgrowth; growth of one part or layer so as to cover another. *adj.* **epibolic.**

epibranchial (*Zool.*). An element of a branchial

arch, lying between the pharyngobranchial and the ceratobranchial.

epicalyx (*Bot.*). (1) A group of bracts placed close underneath the calyx of a flower. (2) A series of small sepal-like structures, alternating with the sepals, and growing out from their under sides.

epicanthus (*Anat.*). A semilunar fold of skin above, and sometimes covering, the inner angle of the eye; a normal feature of the Mongolian races.

epicardium (*Zool.*). In Vertebrates, the serous membrane covering the heart: in *Urochorda*, diverticula of the pharynx, which grow out and surround the digestive viscera like a perivisceral cavity. *adj.* **epicardial.**

epicarp (*Bot.*). The superficial layer of the pericarp, especially when it can be stripped off as a skin.

epicentral (*Zool.*). Arising from, or attached to, the vertebral centra; as certain small membrane bones lying in the myosepta in bony Fish.

epicentre. That point on the surface of the earth lying immediately above the focus of an earthquake, or nuclear explosion.

epichil (*Bot.*). The end of an orchid labellum, when distinct from the base.

epichlorhydrin (*Chem.*). B.p. 117°C. A liquid derivative of glycerol formed by reaction with hydrogen chloride to give *dichlorohydrin*, which in turn is treated with concentrated potassium hydroxide solution. Used in the production of *epoxy resins* (q.v.).

epichordal (*Zool.*). Above or upon the notochord; said of the upper lobe of a caudal fin.

epicoele (*Zool.*). In *Craniata*, the cerebellar ventricle or cavity of the cerebellum.

epicoles (*Ecol.*). Animals which grow on the shells or skins of other animals without having any harmful or beneficial effect on the latter.

epicondylar (*Zool.*). Above the condyle; said of a foramen of the humerus. See **ectepicondylar, entepicondylar.**

epicondyle (*Zool.*). The proximal part of the condyle of the humerus or femur.

epicoracoid (*Zool.*). In Vertebrates, a bony or cartilaginous element uniting the ventral ends of the coracoid and the precoracoid.

epicormic (*Bot.*). Said of a branch formed from a dormant bud on the trunk or limb of a tree, which becomes active owing to some mutilation of the tree or other change in the conditions surrounding the tree.

epicotyl (*Bot.*). The part of the axis of a seedling between the cotyledons and the first leaf or whorl of leaves. (*Zool.*) In Birds, the axis of a down-feather.

epicranial suture (*Zool.*). A Y-shaped suture on the head of insects, representing an ecdysial cleavage line along which the head-capsule of the immature insect breaks open at ecdysis.

epicranium (*Zool.*). In Insects, the upper part of the head above and behind the frons.

epicrine (*Cyt.*). Type of gland in which the secretion is discharged without the disintegration of the cells.

epicritic (*Zool.*). Pertaining to sensitivity to slight tactile stimuli.

epictesis (*Bot.*). That property of the living cell which enables it to accumulate soluble salts in a concentration higher than that existing in the solution from which they are diffusing into the cell.

epicuticle (*Zool.*). The thin outermost layer of the insect cuticle.

epicycle (*Astron.*). The term applied in Ptolemaic or geocentric astronomy to a small circle described uniformly by the sun, moon, or planet, the centre of that circle itself describing uniformly a larger circle (the *deferent*), concentric with the earth.

epicyclic gear (*Eng.*). A system of gears, in which one or more wheels travel round the outside or inside of another wheel whose axis is fixed.

epicyclic train (*Eng.*). A system of epicyclic gears, in which at least one wheel axis itself revolves about another fixed axis; used for giving a large reduction ratio in small compass, and for the gearboxes of some motor cars.

epicycloid (*Maths.*). See **roulette.**

epicyte (*Zool.*). The cuticular layer of the ectoplasm in some *Protozoa*.

epideictic (*An. Behav.*). Said of methods for testing population pressure in an animal by means of conventional displays or contests.

epidemes (*Zool.*). In flying Insects, small sclerites associated with the articulation of the wings.

epidemic (*Med.*). An outbreak of an infectious disease spreading widely among people at the same time in any region. Also used as *adj.*

epidemic encephalitis (*Med.*). Sleepy sickness, Von Economo's disease. An acute inflammation of the brain with a filterable virus; characterized by fever and disturbances of sleep, and followed by various persisting forms of nervous disorder (e.g., Parkinsonism), or by changes in character.

epidemic parotitis (*Med.*). See **mumps.**

epidemic tremor (*Vet.*). See **infectious avian encephalomyelitis.**

epidemiology (*Med.*). That branch of medical science concerned with the study of epidemics.

epidermal, epidermatic (*Bot., Zool.*). Relating to the epidermis.

epidermis (*Bot.*). A sheath of closely united cells forming a layer over the surface of the leaves and young stems of a plant. The layer is seldom more than one cell in thickness, and is continuous except where it is perforated by stomata. (*Zool.*) Those layers of the integument which are of ectodermal origin; the epithelium covering the body.

epidermoid cyst (*Med.*). A wen. A cyst lined with epithelium, occurring in the scalp.

epidermolysis bullosa (*Med.*). A congenital defect of the skin, in which the slightest blow produces a blister.

epidiascope (*Light*). A projection lantern which may be used for transparencies or for opaque objects. See **episcope.**

epididymectomy (*Surg.*). Removal of the epididymis.

epididymis (*Zool.*). In the male of *Selachii*, some Amphibians and *Amniota*, the greatly coiled anterior end of the Wolffian duct, which serves as an outlet for the spermatozoa.

epididymitis (*Med.*). Inflammation of the epididymis.

epididymo-orchitis (*Med.*). Inflammation of the epididymis and the testes.

epididymotomy (*Surg.*). Incision into the epididymis.

epidiorite (*Geol.*). A term for altered gabbroic and doleritic rocks in which the original pyroxene has been replaced by fibrous amphibole. Other mineral changes have also taken place, and the rock may be regarded as a first step in the conversion, by dynamothermal metamorphism, of a basic igneous rock into a green schist.

epidosites (*Geol.*). Metamorphic rocks composed of epidote and quartz. See **epidotization.**

epidote (*Min.*). A mineral of low to medium grade regionally-metamorphosed rocks and also a rare accessory, but common secondary, mineral in igneous rocks, covering a wide range of composition. Crystallizes in the monoclinic system, in lustrous green-black to yellowish green crystals. Also called **pistacite**.

epidotization (*Geol.*). A process of alteration, especially of basic igneous rocks in which the feldspar is albitized with the separation of epidote and zoisite representing the anorthite molecule of the original plagioclase.

epidural anaesthesia (*Med.*). Loss of painful sensation in the lower part of the body produced by injecting an anaesthetic into the epidural space surrounding the spinal canal. Cf. *caudal anaesthesia*.

epigamic (*Zool.*). Attractive to the opposite sex; as *epigamic colours*.

epigamous (*Zool.*). Said of the period of acquisition of the heteronereid condition in *Polychaeta*.

epigastric (*Zool.*). Above or in front of the stomach; said of a vein in Birds, which represents the anterior part of the anterior abdominal vein of lower Vertebrates.

epigastrium (*Anat.*). The abdominal region between the umbilicus and sternum. (*Zool.*) In Insects, the sternites of the mesothorax and metathorax.

epigeal, epigaeous (*Bot.*). (1) Germinating with the cotyledons appearing above the surface of the ground. (2) Living on the soil surface.

epigean (*Bot.*). Occurring on the ground.

epigeic (*Bot.*). Said of a plant having stolons on the surface of the soil.

epigenesis (*Biol.*). The theory, now universally accepted, that the development of an embryo consists of the gradual production and organization of parts, as opposed to the theory of preformation, which supposed that the future animal or plant was already present complete, although in miniature, in the germ. **Epigenetics** is the science based on this theory.

epigenetic (*Min.*). See **syngenetic.**

epiglottidectomy (*Surg.*). Excision of the epiglottis.

epiglottis (*Zool.*). In *Ectoprocta*, the epistome; in Insects, the epipharynx; in Mammals, a cartilaginous flap which protects the glottis.

epignathous (*Zool.*). Having the upper jaw longer than the lower jaw, as in Sperm Whales.

epignathus (*Med.*). A foetal monstrosity in which the deformed remnants of one twin project through the mouth of the more developed twin.

epigynous (*Bot.*). Having the calyx, corolla, and stamens inserted on the top of the inferior ovary.

epigynum (*Zool.*). In *Arachnida*, a plate surrounding the female genital aperture.

epihyal (*Zool.*). An element of the hyoid arch, corresponding to the epibranchials of the branchial arches.

Epikote (*Plastics*). TN for range of *epoxy resins* (q.v.), used for castings, encapsulation (potting) and surface coatings.

epilabrum (*Zool.*). In *Myriapoda*, a process at the side of the labrum.

epilation (*Med.*). Removal of hair by the roots.

epilepsy (*Med.*). A general term for a sudden disturbance of cerebral function accompanied by loss of consciousness, with or without convulsion. See **grand mal, petit mal.**

epileptiform (*Med.*). Resembling epilepsy.

epileptogenic (*Med.*). Exciting an attack of epilepsy.

epilimnion (*Ecol.*). The warm upper layer of water in a lake. Cf. *hypolimnion*.

epilithic (*Bot.*). Growing on a rock surface.

epilittoral zone (*Bot.*). A zone on the coast bordering on the ground occupied by plants which cannot withstand exposure to salt.

epiloia (*Med.*). A condition characterized by feeble-mindedness, epileptic fits, sclerosis of the brain, and tumours in the skin and viscera; due to a defect in development. See also **tuberose sclerosis.**

epimeletic behaviour (*An. Behav.*). A type of social behaviour covering the giving of care and attention to the young, generally by the parents, but also by other, often specialized, individuals of the same species (e.g., worker honey bees). See also **etepimeletic behaviour.**

epimenorrhoea (*Med.*). Too frequent occurrence of menstrual periods.

epimere (*Zool.*). In a developing Vertebrate, the dorsal muscle-plate zone of the mesothelial wall.

epimerite (*Zool.*). The region of the body of an Arthropod between the base of an appendage and the pleuron; the pleuron; in some *Amphipoda*, a wide lamellar expansion of the coxopodite of a thoracic leg; in some *Gregarinidea*, part of the body forming an organ of attachment.

epimerization (*Chem.*). A type of asymmetric transformation in organic molecules, e.g., shown by the change of D-gluconic acid into an isomeric mixture of D-gluconic and D-mannonic acids.

epimeron (*Zool.*). The posterior of two sclerites composing each of the thoracic pleura in pterygote insects. Cf. *episternum*. See also **epimerite.**

epimorph (*Min.*). A natural cast of a crystal.

epimorpha (*Zool.*). Larvae which possess the full number of segments at the time of hatching.

epimysium (*Zool.*). The investing connective-tissue coat of a muscle.

epinasty (*Bot.*). (1) The occurrence of stronger growth on the upper than on the under side of a plant member. (2) Excentric thickening of a more or less horizontal branch or root.

epinephrine (*Biochem.*). *Adrenaline* (q.v.).

epinephros (*Zool.*). See **suprarenal body.**

epineural (*Zool.*). In *Echinodermata*, lying above the radial nerve: in Vertebrates, lying above or arising from the neural arch of a vertebra.

epineurium (*Zool.*). The connective tissue which invests a nerve trunk, uniting the different funiculi and joining the nerve to the surrounding and related structures.

epinotum (*Zool.*). See **propodeum.**

epiotic (*Zool.*). A bone of the auditory capsule of the Vertebrate skull, situated above and between the pro-otic and the opisthotic.

epiopticon (*Zool.*). In Insects, the external medullary mass or middle zone of the optic lobe of the brain.

epiparasite (*Zool.*). See **ectoparasite.**

epipetalous (*Bot.*). Arising from the upper surface of a petal.

epipharyngeal bone (*Zool.*). One of the expanded upper elements of the first four gill-arches, in Fish with bony skeletons.

epipharyngeal groove (*Zool.*). In *Cephalochorda*, a groove, similar to the endostyle, extending along the dorsal aspect of the pharynx.

epipharyngeal organ (*Zool.*). In *Hemiptera*, a gustatory organ in the region of the epipharyngeal surface of the labium.

epipharynx (*Zool.*). In Insects, the membranous roof of the mouth which in some forms is produced into a chitinized median fold, and in *Diptera* is associated with the labrum, to form a piercing organ; in *Acarina*, a forward projection of the anterior face of the pharynx. *adj.* **epipharyngeal.**

epiphenomenalism (*Psychol.*). A theory which holds that mental activities have no causal role and are merely the accompaniment of physical activity.

epiphloeodal, epiphloeodic (*Bot.*). Growing on the surface of bark.

epiphora (*Med.*). An overflow of the lacrimal secretion, due to obstruction of the channels which normally drain it.

epiphragm (*Zool.*). In *Gastropoda*, a plate, mostly composed of calcium phosphate, with which the aperture of the shell is sealed during periods of dormancy.

epiphyllous (*Bot.*). Growing upon, or attached to, the upper surface of a leaf; sometimes, growing on any part of a leaf.

epiphysis (*Bot.*). An upgrowth around the hilum of a seed. (*Zool.*) A separate terminal ossification of some bones, which only becomes united with the main bone at the attainment of maturity; the pineal body; in *Echinoidea*, one of the ossicles of Aristotle's lantern. In some *Lepidoptera*, a lamellate spur on the inner surface of the anterior tibiae. *adj.* **epiphysial.**

epiphysis cerebri (*Zool.*). See **pineal organ.**

epiphysitis (*Med.*). Inflammation of the epiphysis of a bone.

epiphyte (*Bot.*). A plant which grows attached to the stems or leaves of another plant, but is not a parasite, that is, it takes no material from the plant to which it is attached. *adj.* **epiphytic.** See **epibiosis.**

epiphytism (*Ecol.*). A form of *commensalism* (q.v.) in which a sessile animal is associated with a larger motile animal, with benefit to the former but not to the latter; e.g., *Temnocephala* on various fresh-water *Crustacea*.

epiplankton (*Ocean.*). Plankton found in depths of less than 180 m (100 fathoms).

epiplasm (*Bot.*). Residual cytoplasm left in the ascus after the ascospores have been delimited; it may play a part in the subsequent nutrition of the developing spores.

epiplastron (*Zool.*). In *Chelonia*, one of the plates composing the plastron, lying anterior to the entoplastron.

epipleura (*Zool.*). In *Coleoptera*, the reflexed sides of the elytra; in Birds, the uncinate process; in bony Fish, upper ribs formed from membrane bone; in *Cephalochorda*, horizontal shelves of membrane arising from the inner sides of the metapleural folds, and forming the floor of the atrial cavity.

epiplocele (*Med.*). A hernia containing omentum (the *epiploon*) only.

epiploic foramen (*Zool.*). See **Winslow's foramen.**

epiploitis (*Med.*). Inflammation of the omentum.

epiploon (*Zool.*). In Mammals, a double fold of serous membrane connecting the colon and the stomach (the great omentum): in Insects, the fat-body. *adj.* **epiploic.**

epipodia (*Zool.*). In some *Gastropoda*, paired lateral lobes of the foot. *adj.* **epipodial.**

epipodite (*Zool.*). In *Crustacea*, a process arising from the protopodite.

epiprecoracoid (*Zool.*). In some Amphibians and Reptiles, one of a pair of small ventral cartilages of the pectoral girdle.

epiproct (*Zool.*). A supra-anal plate over the anus in lower insects, representing the 11th abdominal segment.

epipteric (*Zool.*). Wing-shaped; situated above a wing.

epipterous (*Bot.*). Having a wing at the apex.

epipterygoid (*Zool.*). In some Reptiles, a cartilage bone lying between the pterygoid and the parietal; formed by the ossification of the ascending process of the quadrate.

epipubic (*Zool.*). In front of or above the pubis; pertaining to the epipubis.

epipubis (*Zool.*). See **pelvisternum.**

episclera (*Anat.*). The connective tissue between the conjunctiva and the sclera.

episcleritis (*Med.*). Inflammation of the episclera, sometimes involving the sclera.

episcope (*Light*). A projection lantern which is used for throwing on a screen an enlarged image of a brilliantly illuminated opaque object.

episcotizer (*Optics*). A rotating disk with alternate transparent and opaque sections used with a photocell to reduce the intensity of the light falling on the sensitive surface. See **sector disk.**

episemantic (*Biol.*). The product made as a result of the action of a *semantide* (q.v.).

episematic (*Zool.*). Serving for recognition; as *episematic colours*.

episepalous (*Bot.*). (1) Borne on the sepals. (2) Placed opposite to the sepals.

episiostenosis (*Med.*). Narrowing of the vulvar orifice.

episiotomy (*Surg.*). Cutting the vulvar orifice to facilitate delivery of the foetus.

episomes (*Biol.*). A general term for genetically active particles found in bacteria, but which may occur in other organisms. They can exist in two forms: noninfective and integrated on the chromosome; and infective and not integrated on the chromosome, depending on conditions. They may form a link between fully integrated particles (*genes*) and nonintegrated particles (*viruses*).

epispadias (*Med.*). Congenital defect in the anterior or dorsal wall of the urethra, commoner in the male than in the female.

epispastic (*Med.*). Producing a blister; an agent which does this.

episperm (*Bot.*). The outer part of the seed coat.

epispore (*Bot.*). The outermost layer of a spore wall, often consisting of a deposit forming ridges, spines, or other irregularities of the surface.

epistatic (*Gen.*). Said of a character which is dominant to another to which it is not the allelomorph.

epistaxis (*Med.*). Bleeding from the nose.

episternum (*Zool.*). The anterior of two sclerites composing each of the thoracic pleura in pterygote insects. Cf. *epimeron*. In Reptiles the interclavicle, a bone applied to the anterior end of the sternum.

epistilbite (*Min.*). A white or colourless zeolite; hydrated silicate of calcium and aluminium, crystallizing in the monoclinic system.

epistome (*Zool.*). In *Ectoprocta* and *Phoronidea*, a ridge overhanging the mouth; in decapod *Crustacea*, the sternal region of the body in front of the mouth; in certain *Coleoptera*, the reduced frontoclypeal region; in *Diptera*, the distal border of the face.

epistrophe (*Bot.*). The position of chloroplasts in diffuse light, on the periclinal walls of the palisade cells.

epistropheus (*Zool.*). The axis vertebra.

epistyle (*Arch.*). See **architrave.**

epitaxial transistor (*Electronics*). One made from a semiconductor material by the latter's deposition on a suitable monocrystalline support.

epitaxy (*Crystal.*). Unified crystal growth or deposition of one crystal layer on another.

epithalamus (*Zool.*). In the Vertebrate brain, a dorsal zone of the thalamencephalon.

epithalline (*Bot.*). Growing on a thallus.

epithallus (*Bot.*). The upper layer of fungal hyphae in a lichen thallus.

epitheca (*Bot.*). The older of the two valves forming the wall of a cell of a diatom.

epithecium (*Bot.*). A thin coloured layer over the asci in an apothecium, particularly in lichens, formed from the tips of the paraphyses.

epithelial layer (*Bot.*). A layer of elongated cells set end on to the endosperm in a grain, forming the boundary of the scutellum.

epithelioid (*Anat.*). Resembling epithelium.

epithelioma (*Med.*). A malignant growth derived from epithelium.

epithelioma contagiosa (*Vet.*). See **fowl pox.**

epithelium (*Zool.*). A form of tissue characterized by the arrangement of the cells as an expansion covering a free surface, or as solid masses, by the presence of a basement membrane underlying the lowermost layer of cells, and by the small amount of intercellular matrix: the secretory substance of glands, the tissues lining the alimentary canal and blood-vessels, etc. *adjs.* **epithelial, epitheliomorph.**

epithem (*Bot.*). A group of cells occurring in the mesophyll of a leaf and exuding water.

epithem hydathode (*Bot.*). A hydathode which is directly connected with the vascular system of the leaf.

epithermal (*Nuc.*). Having energy just above the energy of thermal agitation, and comparable with chemical bond energies.

epithermal neutrons (*Nuc.*). See **neutron.**

epithermal reactor (*Nuc.*). See **intermediate reactor.**

epitokous (*Zool.*). Said of the heteronereid stage in *Polychaeta.*

epitrichium (*Zool.*). A superficial layer of the epidermis in Mammals, which consists of greatly swollen cells and is found on parts of the body devoid of hair. *adj.* **epitrichial.**

epitrochlea (*Anat.*). The inner condyle, or bony eminence, on the inner aspect of the lower end of the humerus.

epitrochoid (*Maths.*). See **roulette.**

epitrochoidal engine (*I.C. Engs.*). A rotary type of I.C. engine using a single central rotating member in place of conventional pistons, e.g., the *Wankel engine* (q.v.).

epitrophic (*Bot.*). (1) Having buds on the upper side. (2) Growing more on the upper than on the under side.

epituberculosis (*Med.*). Congestion and inflammation of the area surrounding a tuberculous focus, especially in the lung.

epixylous (*Bot.*). Growing on wood.

epizoic (*Bot.*). (1) Growing on a living animal. See **epibiosis.** (2) Having the seeds or fruits dispersed by animals.

epizoite (*Ecol.*). See **epibiosis.**

epizoon (*Zool.*). An animal which lives on the skin of some other animal; it may be an *ectoparasite* or a *commensal. adj.* **epizoan.**

epizootic (*Vet., Zool.*). Pertaining to an epizoon; appearing as an epidemic among animals; a disease occurring epidemically among animals.

epizootic catarrhal fever (*Vet.*). See **equine influenza.**

epizootic lymphangitis (*Vet.*). A chronic contagious lymphangitis of horses, due to infection by *Cryptococcus farciminosus.*

epoch (*Astron.*). The precise time to which the data of an astronomical problem are referred; thus the elements of an orbit when referred to a specific epoch also implicitly define the obliquity of the ecliptic, the rate of precession and other conditions obtaining only at that instant.

eponychium (*Zool.*). In higher Mammals, the narrow band of cuticle which overlies the base of the nail.

epoöphoron (*Zool.*). A rudimentary structure in the ovary of Vertebrates, homologous with the epididymis of the male.

epoxy (or epoxide) resins (*Chem.*). Polymers derived from epichlorhydrin and bisphenol-A of the general structure

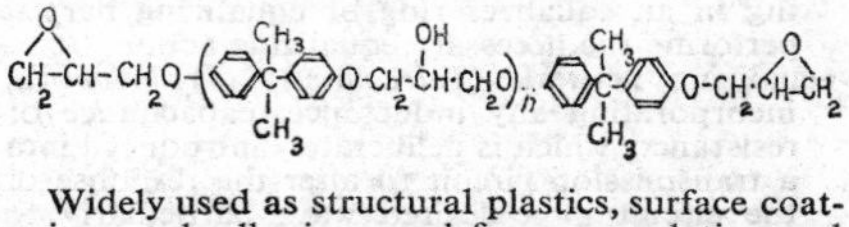

$$\overset{O}{\overbrace{CH_2-CH}}-CH_2O-\left(C_6H_4-\underset{CH_3}{\overset{CH_3}{C}}-C_6H_4-O-CH_2-\underset{}{\overset{OH}{C}}H-CH_2O\right)_n-C_6H_4-\underset{CH_3}{\overset{CH_3}{C}}-C_6H_4-O-CH_2-\overset{O}{\overbrace{CH-CH_2}}$$

Widely used as structural plastics, surface coatings and adhesives, and for encapsulating and embedding electronic components. Characterized by low shrinkage on polymerization, good adhesion, mechanical and electrical strength, and chemical resistance.

E.P.R. (*Chem.*). Abbrev. for *electron paramagnetic resonance.*

epsomite or **Epsom salts** (*Min.*). Hydrated magnesium sulphate, $MgSO_4{\cdot}7H_2O$; occurring in colourless orthorhombic prismatic crystals, botryoidal masses, or incrustations in gypsum mines and limestone caverns; common in solution in mineral waters. The chief commercial source is the salt beds of Stassfurt.

Epstein hysteresis tester (*Elec. Eng.*). An apparatus for measuring the hysteresis and eddy-current losses in a sample of sheet-iron. Strips cut from the sheet are assembled in the form of a square or rectangular magnetic circuit, upon which windings are placed, and the losses are measured by means of a wattmeter. The assembly is sometimes called an **Epstein square.**

epulis (*Med.*). A tumour, innocent or malignant, of the gums, growing from the periosteum of the jaw.

epuré (*Civ. Eng.*). Refined natural asphalt from which volatile matter and water have been sufficiently removed.

Eputmeter (*Electronics*). TN for a counting device for measuring the number of *e*vents *p*er *u*nit *t*ime.

equal-area criterion (*Elec. Eng.*). A term used in connexion with the stability of electric power systems. The stability limit occurs when two areas on the power-angle diagram, governed by the load conditions obtaining, are equal.

equal energy source (*Elec.*). An electromagnetic or acoustic source whose radiated energy is distributed equally over its whole frequency spectrum.

equal falling particles (*Min. Proc.*). Particles possessing equal *terminal velocities* (q.v.); the underflow, oversize product of a classifier.

equalization (*Telecomm.*). Electronically, the reduction of distortion by compensating networks which allow for the particular type of the distortion over the requisite band.

equalization of boundaries (*Surv.*). See **give-and-take lines.**

equalizer (*Telecomm.*). See **equalizing network.**

equalizer (or balancing) ring (*Elec. Eng.*). A conductor on the armature of an electrical machine which serves to connect two points that are normally at the same potential.

equalizer switch (*Elec. Eng.*). A switch for connecting the armature end of the series field winding of a compound-wound generator to an equalizing bar.

equalizing bar (*Elec. Eng.*). A bus-bar connecting the armature ends of the series field windings of a number of compound-wound generators operating in parallel; the series field windings are thus put in parallel, ensuring stable operation.

equalizing bed (*Civ. Eng.*). The bed of fine ballast or concrete laid immediately underneath a pipeline, e.g., in a trench, in cases where the bottom of the excavation is sound but uneven (as in rock, hard chalk, etc.).

equalizing current (*Elec. Eng.*). The current flowing in an equalizer ring or equalizing bar; it performs the necessary equalizing action.

equalizing network (*Telecomm.*). (1) Network, incorporating any inductance, capacitance or resistance, which is deliberately introduced into a transmission circuit to alter the response of the circuit in a desired way; particularly to equalize a response over a frequency range. (2) A similar arrangement incorporated in the coupling between valves in an amplifier. Also called **equalizer.**

equalizing pulse (*TV*). A pulse used in television at twice the line frequency, which is applied immediately before and after the vertical synchronizing pulse. This is done to reduce any effect of line-frequency pulses on the interlace.

equalizing signals (*TV*). Those added to ensure triggering at the exact time in a frame cycle.

equal-signal system (*Aero., Radio*). One in which two signals are emitted for radio-range, an aircraft receiving equal signals only when on the indicated course.

equal- (or equi-) tempered scale (*Acous.*). See **tempered scale.**

equation (*Chem.*). See **chemical equation.** (*Maths.*) (1) A sentence in which the verb is 'is equal to' and which involves a variable (e.g., x) on a set E; e.g., $x^2+6 = 5x$, where x is a variable on the set of integers. Those elements of E which turn the equation into a true statement, when they are put in place of x, constitute the solution set of the equation. For the equation $x^2+6 = 5x$ the solution set is $\{2, 3\}$. If the equation is satisfied by every element of the set E, then the equation is said to be an *identity*; e.g., $\sin^2 x+\cos^2 x = 1$, where x is a variable on the set of real numbers, is an identity, for it is satisfied by every real number. (2) The equation of a curve or surface is an equation satisfied by the coordinates of every point on the curve or surface, but not satisfied by the coordinates of any point not on the curve or surface.

equation division (*Zool.*). Division of chromosomes which results in the production of daughter chromosomes of equal size, form, and potency.

equation of maximum work (*Chem.*). See **Gibbs-Helmholtz equation.**

equation of state (*Chem.*). An equation relating the volume, pressure, and temperature of a given system, e.g., van der Waals' equation.

equation of time (*Astron.*). The difference between the right ascensions of the true and mean sun, and hence the difference between apparent and mean time. In the sense mean time minus apparent time, it has a maximum positive value of nearly 14½ min in February, and a negative maximum of nearly 16½ min in November, and vanishes 4 times a year.

equation solver (*Comp.*). A computer of specific design to solve mathematical equations.

equator. See **celestial-, terrestrial-.**

equatorial (*Astron.*). The name given to an astronomical telescope which is so mounted that it revolves about a polar axis parallel to the earth's axis; when set on a star it will keep that star in the field of view continuously, without adjustment. It has two graduated circles reading *Right Ascension* and *Declination* respectively. (*Cyt.*) Situated or taking place in the equatorial plane; as the *equatorial furrow* which precedes division of an ovum into upper and lower blastomeres, and the *equatorial plate*, which, during mitosis, is the assembly of chromosomes on the spindle in the *equatorial plane*.

equatorial horizontal parallax (*Astron., Surv.*). See **horizontal parallax.**

equatorial radius of the earth (*Astron.*). See under **earth.**

equi-. Prefix from L. *aequus*, equal.

equiangular spiral (*Maths.*). A spiral in which the angle between the tangent and the radius vector is constant. Also called a **logarithmic spiral.** Its polar equation is $\log r = a\theta$.

equidistant locking pallets (*Horol.*). *Pallets* (q.v.) in which locking corners of entering and exit pallets are equidistant from pallet-staff axis.

equilateral arch (*Build.*). An arch in which the two springing points and the crown of the intrados form an equilateral triangle.

equilateral roof (*Build.*). A pitched roof having rafters of a length equal to the span.

equilateral triangle (*Maths.*). A triangle having 3 equal sides.

equilibration (*Eng.*). The production of balance or equilibrium; as in the provision of balance weights for a lift or cage. (*Psychol.*) The process of bringing *accommodation* and *assimilation* into a balanced coordination.

equilibration tissues (*Zool.*). Tissues that record the position of the body, or part of it, in relation to gravity or some other constant factor.

equilibrium (*Chem.*). The state reached in a reversible reaction when the reaction velocities in the two opposing directions are equal, so that the system has no further tendency to change. (*Mech.*) The state of a body which is at rest or is moving with uniform velocity. A body on which forces are acting can be in equilibrium only if the resultant of the forces is zero.

equilibrium constant (*Chem.*). The ratio, at equilibrium, of the product of the active masses of the molecules on one side of the equation representing a reversible reaction to that of the active masses of the molecules on the other side.

equilibrium diagram (*Met.*). See **constitutional diagram.**

equilibrium moisture (*Chem. Eng.*). The percentage water content of a solid material, when the vapour pressure of that water is equal to the partial pressure of the water vapour in the surrounding atmosphere.

equilibrium of floating bodies (*Phys.*). For a body which floats, partly immersed in a fluid, the weight of the body is equal to the weight of fluid which it displaces. Therefore the ratio of the volume of the body to the volume immersed equals the ratio of the density of the fluid to that of the body. See **Archimedes' principle.**

equilibrium ring (*Eng.*). A ring placed in the steam chest of large steam-engines, between the back of the slide-valve and the cover. By connecting the enclosed space to the exhaust, the force on the valve and the corresponding driving effort are much reduced.

equilibrium slide-valve (*Eng.*). A large slide-valve balanced by the use of an *equilibrium ring* (q.v.).

equilibrium still (*Chem. Eng.*). One designed to produce a boiling liquid mixture in complete phase equilibrium with its vapour, for the purposes of physicochemical measurement.

equilibrium time (*Nuc.*). The time required for steady-state working of an isotope separation plant to be established.

equilux spheres (*Optics*). Spherical surfaces which

are concentric with a source of light so that the illumination is constant.

equine contagious catarrh (*Vet.*). See **strangles.**

equine encephalomyelitis (*Vet.*). An acute disease of horses, mules and donkeys, due to a virus which causes an encephalomyelitis; characterized by fever, nervous symptoms, and often death. Transmitted by mosquitoes, ticks and mites. Several strains of virus occur: eastern American, western American, Venezuelan, and Russian. Another form, *Borna disease* (q.v.), is caused by an unrelated virus.

equine infectious anaemia (*Vet.*). Swamp fever. An acute or chronic virus disease of horses characterized by intermittent fever, weakness, emaciation, and anaemia; spread by flies.

equine influenza (*Vet.*). Epizootic catarrhal fever; shipping fever; stable pneumonia; pink eye. The name given to an acute respiratory infection of horses which was formerly considered to be a distinct entity, but which is now believed to have included at least two different diseases. See **equine virus rhinopneumonitis, equine virus arteritis.**

equine thrush (*Vet.*). Inflammation of the frog of the horse's foot, attended with a fetid discharge.

equine virus abortion (*Vet.*). See **equine virus rhinopneumonitis.**

equine virus arteritis (*Vet.*). An acute, respiratory viral disease of horses, characterized by degenerative and inflammatory changes in the small arteries; the main symptoms are fever, respiratory difficulty, oedema of the legs, diarrhoea, and in pregnant mares, abortion.

equine virus rhinopneumonitis (*Vet.*). An acute viral disease of the respiratory tract of horses, characterized by fever and respiratory catarrh. The same virus may also cause abortion in mares, being responsible for the disease known as **equine virus abortion.**

equinoctial (*Bot.*). Said of plants bearing flowers which open and close at definite times.

equinoctial points (*Astron.*). The two points, diametrically opposite each other, in which the celestial equator is cut by the ecliptic; called respectively the *First Point of Aries* and the *First Point of Libra*, from the signs of the Zodiac of which they are the beginning.

equinox (*Astron.*). The instant, occurring twice in each tropical year, at which the sun in its apparent annual motion crosses the celestial equator; so called because the lengths of the day and night are then equal. The term equinox is also used for one of the equinoctial points, more particularly for the *First Point of Aries*. All numerical data are thus referred to the equator (or ecliptic) and equinox of a particular epoch.

equipartition of energy (*Chem.*). The Maxwell-Boltzmann law, which states that the available energy in a closed system eventually distributes itself equally among the *degrees of freedom* (q.v.) present.

equipluve (*Meteor.*). The line on the map joining all places of the same *pluviometric coefficient*.

equipotent (*Zool.*). See **totipotent.**

equipotential cathode (*Electronics*). One in which the electron-emitting surface is carried on a metal or ceramic body, which is heated by an internal wire.

equipotential connection (*Elec. Eng.*). Better term for *equalizer* (*equalizing*) *bar* or *ring* (qq.v.).

equipotential surface (or region) (*Elec. Eng.*). One where there is no difference of potential, and hence no electric field. See **Faraday cage.**

Equisetales (*Bot.*). An order of the *Pteridophyta*. The leaves are scalelike in whorls, alternating with the branches at the nodes. The stems are ribbed, and the sporangia are borne in strobili.

equitant (*Bot.*). A term applied to the condition when a plant member folded inwards about its midrib covers the edges of another similarly folded, that covers a third, and so on; used chiefly of leaves with their bases so folded. (*Zool.*) In spiral shells (as of *Foraminifera* or *Gastropoda*), having the whorls so arranged that each one overlaps the previous one at the sides.

equitonic scale (*Acous.*). The musical scale in which the main notes progress by whole tones, as contrasted with the Pythagorean diatonic, which uses both whole tones and half-tones.

equivalence, photochemical (*Chem.*). See **Einstein law of photochemical equivalence.**

equivalent (*Chem.*). See **equivalent weight.**

equivalent absorption (*Acous.*). Of an object in a room (or the room itself), the equivalent surface area of unity absorption factor which absorbs acoustic energy at the same rate.

equivalent air speed (*Aero.*). *Indicated air speed* (q.v.) corrected for position error (angle of incidence) and air compressibility. Abbrev. **E.A.S.**

equivalent binary digits (*Comp.*). The number required to express a given number of decimal digits or other characters.

equivalent circuit (*Elec. Eng.*). One consisting of resistances, inductances, and capacitances, which behaves, as far as the current and voltage at its terminals are concerned, exactly as some other circuit or piece of apparatus, e.g., a transformer may be represented by an arrangement of resistances and inductances, and an effective self-capacitance.

equivalent conductance (*Chem.*). Electrical conductance of a solution which contains 1 gram-equivalent weight of solute at a specified concentration, measured when placed between two plane parallel electrodes, 1 m apart. *Molar conductance* is more often used.

equivalent diode (*Electronics*). One which would draw the same cathode current as a given valve which has more electrodes.

equivalent electrons (*Electronics*). Those which occupy the same orbit in an atom, hence have the same principal and orbital quantum numbers.

equivalent focal length (*Optics*). The focal length of a thin lens which is equivalent to a thick lens in respect of the size of image it produces.

equivalent free-falling diameter (*Powder Tech.*). The diameter of a sphere which has the same density and the same free-falling velocity in a given fluid as an observed particle of powder.

equivalent height (*Radio*). That of a perfect antenna, erected over a perfectly conducting ground, which, when carrying a uniformly distributed current equal to the maximum current in the actual antenna, radiates the same amount of power.

equivalent hiding power (*Powder Tech.*). Of particles interrupting a light beam, in photo-sedimentation analysis, the weight of a given size, expressed as a fraction or a percentage of the weight of a standard size, which has the same hiding power as unit weight of the standard of the size when in suspension. The standard size is usually the largest particle present.

equivalent lens (*Light*). A simple lens which, substituted for a system of lenses, would give an image of the same size and in the same position.

equivalent network (*Telecomm.*). One identical to another network either in general or at some specified frequency. The same input applied to each would produce outputs identical in both magnitude and phase generated across the same internal impedance.

equivalent points (*Light*). The principal points of a lens that is used with the same medium on both sides. See **cardinal points.**

equivalent proportions (*Chem.*). See **law of equivalent (or reciprocal) proportions.**

equivalent reactance (*Elec. Eng.*). The value which the reactance of an equivalent circuit must have in order that it shall represent the system of magnetic or dielectric linkages present in the actual circuit.

equivalent resistance (*Elec. Eng.*). The value which the resistance of an equivalent circuit must have in order that the loss in it shall represent the total loss occurring in the actual circuit.

equivalent simple pendulum (*Phys.*). See **centre of oscillation.**

equivalent sine wave (*Elec. Eng.*). One which has the same frequency and the same r.m.s. value as a given non-sinusoidal wave.

equivalent surface diameter (*Powder Tech.*). The diameter of a sphere which has the same effective surface as that of an observed particle when determined under stated conditions.

equivalent T-networks (*Telecomm.*). T- or pi-networks equivalent in electrical properties to sections of transmission line, provided these are short in comparison with the wavelength.

equivalent volume diameter (*Powder Tech.*). The diameter of a sphere which has the same effective volume as that of an observed particle when determined under the same conditions.

equivalent weight (*Chem.*). (1) That weight of an element or radical which combines with, displaces, or is in any way equivalent to, a standard unit. (2) See **atomic weight.**

equivalve (*Zool.*). Said of bivalves which have the two halves of the shell of equal size.

Er (*Chem.*). The symbol for *erbium*.

erase (*Comp.*). To remove data in a store or register of a computer.

erase head (*Mag.*). In magnetic tape recording, the head which saturates the tape with high-frequency magnetization, in order to remove any previous recording.

erasion (*Surg.*). Removal of all diseased structures from a joint by cutting and scraping.

erbium (*Chem.*). A metallic element, a member of the rare-earth group. Symbol Er, at. no. 68, r.a.m. 167·26. Found in the same minerals as dysprosium (gadolinite, fergusonite, xenotime), and in euxenite.

erbium laser (*Phys.*). Laser using erbium in YAG (*y*ttrium-*a*luminium-*g*arnet) glass. It has the advantage of operating between 1·53 and 1·64 micrometres, a range in which there is a high attenuation in water. This feature is of particular importance in laser applications to eye investigations, since a great deal of energy absorption will now occur in the cornea and aqueous humour before reaching the delicate retina.

E.R.C. (*Photog.*). Abbrev. for *ever-ready case.*

erect (*Bot.*). Set at right angles to the part from which it grows. (*Zool.*) See **erection.**

erectile tissue (*Physiol.*). Tissue which contains baggy blood-spaces, by the distension of which with blood it can be rendered *turgid*.

erecting prism (*Light*). A right-angled prism used for erecting the image formed by an inverting projection system. The prism is used with its hypotenuse parallel to the beam of light incident on one of the other faces, which is totally reflected at the hypotenuse and emerges from the third face parallel to its original direction.

erecting shop (*Eng.*). That part of an engineering works where finished parts are assembled or fitted together.

erection (*Build.*). The assembly of the parts of a structure into their final positions. (*Zool.*) The *turgid* condition of certain animal tissues when distended with blood; an upright or raised condition of an organ or part. *adj.* **erect.**

erections (*Ships*). See **top hamper.**

erector (*Photog.*). See **brilliant viewfinder.** (*Physiol.*) A muscle which, by its contraction, assists in raising or erecting a part or organ.

eremacausis (*Chem.*). A process of very slow oxidation without the application of heat, e.g., the decay of wood.

eremochaetous (*Zool.*). Having the bristles or chaetae arranged according to no definite plan.

erepsin (*Chem.*). An enzyme of the intestinal juice, acting upon casein, gelatine, and the products of peptic digestion, the end products being polypeptides and amino acids.

erethism (*Med.*). Abnormal irritability of a tissue or organ.

erg (*Phys.*). Unit of work or energy. **See energy.**

ergastic substance (*Cyt.*). A nonprotoplasmic cell inclusion, playing a part in respiration or other metabolic activity of the cell.

ergastoplasm (*Cyt.*). See **archoplasm.**

ergatandromorph (*Zool.*). An ant which has the characters of the worker, combined with those of the male.

ergataner (*Zool.*). An apterous male ant having some of the characteristics of the worker.

ergate (*Zool.*). A sterile female ant or worker.

ergatogyne (*Zool.*). An apterous queen ant.

ergatoid (*Zool.*). Resembling a worker; said of sexually perfect but wingless adults of certain social Insects.

ergometrine (*Chem.*). $C_{19}H_{23}O_2N_3$. An alkaloid, readily soluble in water, and the chief constituent of ergot preparations. Its physiological effects are similar to those of *ergotoxine* (q.v.), but its action is most powerful on uterine muscle. Also called **ergobasine, ergotocin.**

ergon (*Chem.*). The quantum of energy hV of an oscillator, where h is Planck's constant and V is the frequency of the oscillator.

ergonomics (*Work Study*). The study of work in relation to the environment in which it is performed and the personnel who perform it, and the application of scientific knowledge to the improvement of efficiency. U.S. **biotechnology.**

ergonomy (*Physiol.*). The physiological differentiation of functions.

ergosome (*Biol.*). The minimum functional unit of protein synthesis within the cell.

ergosterol (*Chem.*). A *sterol* (q.v.) which occurs in ergot, yeast, and moulds. Traces of it are associated with cholesterol in animal tissues. On irradiation with ultraviolet light, *vitamin* D_2 (q.v.) is produced. Its constitution is:

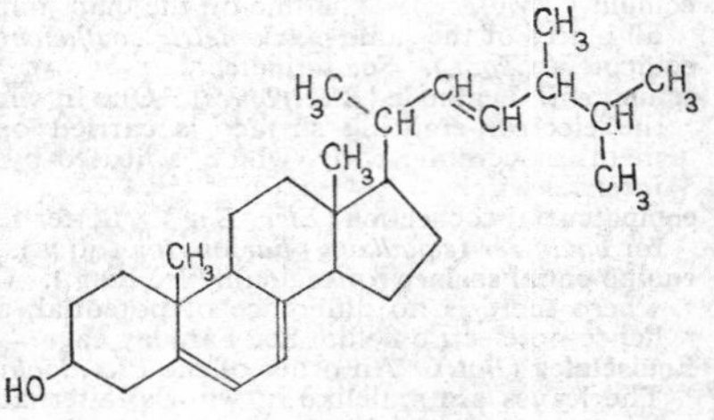

ergotine (*Chem.*). An obsolete term for a crude base obtained from ergot. See **ergotoxine.**

ergotism (*Med.*). A condition characterized by gangrene, or by degenerative changes in the nervous system; due to eating the grains of cereals which are infected by the ergot fungus *Claviceps purpurea.*

ergotocin (*Chem.*). See **ergometrine.**

ergotoxine (*Chem.*) A mixture of the ergot alkaloids ergocornine, ergocristine and ergotcryptine. Produces the characteristic effects of ergot, viz., contraction of the uterus and rise of blood pressure.

Erian Stage (*Geol.*). The higher division of the Middle Devonian in N. America; it includes the Hamilton beds, and marks the maximum extension of the Devonian sea.

erianthin (*Chem.*). $C_{20}H_{20}O_9$. Flavone from *Blumea eriantha.* Yellow prisms, m.p. 154°.

ericaceous (*Bot.*). Heatherlike.

Ericales (*Bot.*). An order of the *Sympetalae*, the leaves simple and flowers regular. The corolla is usually united; the stamens are hypogynous or epigynous, rarely united to the petals. The ovary has two to many loculi, and the ovules have one integument.

ericeticolous (*Bot.*). Growing on a heath.

Erichsen test (*Met.*). A test in which a piece of metal sheet is pressed into a cup by means of a plunger; used to estimate the suitability of sheet for pressing or drawing operations.

erichthus (*Zool.*). A larval stage of *Stomatopoda.*

ericoid (*Bot.*). Having very small tough leaves like those of heather.

ericophyte (*Bot.*). A plant growing on a heath or a moor.

erinite (*Min.*). Basic copper arsenate, $Cu_5(OH)_4(AsO_4)_2$; emerald-green in colour.

Erinoid (*Plastics*). TN used for a wide range of thermoplastic materials including cellulose acetate (ethanoate), polystyrene, PVC and casein.

eriometer (*Phys.*). A device for measuring the size of small particles or fibres, using optical diffraction.

erionite (*Min.*). One of the less common zeolites; a hydrated aluminium silicate of sodium, potassium and calcium.

eriophorous (*Bot.*). Having a thick cottony covering of hairs.

erkensator (*Paper*). A machine which extracts dirt, etc., from the wet pulp by centrifugal force.

erlang (*Teleph.*). International unit of traffic flow in telephone calls.

Erlenmeyer flask (*Chem.*). A conical glass flask with a flat bottom, widely used for titrating, as it is easily cleaned, stood, stoppered, and swirled.

Ernie (*Comp.*). A so-called 'computer' used to select winning numbers in the British Premium Bond lottery. Similar random number generators are used in the *Monte Carlo method* (q.v.). (*E*lectronic *r*andom *n*umber *i*ndicating *e*quipment.)

erogenous zones (*Psychol.*). Sensitive areas of the body, the stimulation of which arouses sexual feelings and responses. They function as substitutes for the genital organs and are associated with stages of development in childhood.

Eros (*Astron.*). The name given to one of the smaller asteroids. In 1931 it came within 16 million miles of the earth, and was used to determine the *solar parallax* (q.v.).

erosion (*Geol.*). The lowering of the land surface by weathering, corrosion, and transportation, under the influence of gravity, wind, and running water. Also, the eating away of the coastline by the action of the sea. Soil erosion may result from factors such as bad agricultural methods, excessive deforestation, overgrazing.

erosion-littoral fauna (*Ecol.*). Animals living a the edges of lakes with open, wave-washed shores.

erratics (*Geol.*). Stones, ranging in size from pebbles to large boulders, which were transported by ice, which, on melting, left them stranded far from their original source. They furnish valuable evidence of the former extent and movements of ice sheets.

error (*Elec. Eng.*). See **deviation.** (*Eng.*) In servo or other control systems, the difference between the actual value of a quantity arising from the process and the adjusted value in the controller.

error-correcting code (*Comp.*). One constructed with redundant elements so that certain types of error can be detected *after* reception and corrected.

error in indication (*Elec. Eng.*). In indicating instruments, the difference between the indication of the instrument and the true value of the quantity being measured. It may be expressed as a percentage of the true value, and a positive value of the error means that the indication of the instrument is greater than the true value.

error of closure (*Surv.*). See **closing error.**

error signal (*Telecomm.*). Feedback signal in control system representing deviation of controlled variable from set value.

ersaeome (*Zool.*). See **eudoxid.**

erubescite (*Min.*). See **bornite.**

erucic acid (*Chem.*). $CH_3[CH_2]_7CH{=}CH[CH_2]_{11}\cdot COOH$. Mono-unsaturated fatty acid in the oleic acid series; m.p. 33°C. Occurs in rapeseed oil and mustard seed oil.

eruciform (*Zool.*). Said of the polypod larvae of *Lepidoptera*, *Trichoptera*, and *Saw-flies*; characterized by well-defined segmentation, abdominal limbs or prolegs, peripneustic tracheal system, and poorly developed antennae and thoracic legs and sluggish habit.

eructation (*Med.*). A belching of gas from the stomach through the mouth.

erumpent (*Bot.*). Developing at first beneath the surface of the substratum, then bursting out through the substratum and spreading somewhat.

eruption (*Med.*). A breaking-out of a rash on the skin or on the mucous membranes; a rash.

eruptive rocks (*Geol.*). A term sometimes used for all igneous rocks; but carefully compare *extrusive rocks.*

e.r.w. tube (*Chem. Eng.*). An *e*lectric *r*esistance *w*elded tube much used in heat exchangers; it is made continuously by forming an accurately rolled strip over a mandrel, and welding the edges by means of electric heating, while held together.

erysipelas (*Med.*). St Anthony's fire. A spreading streptococcal infection of the lymphatics of the skin, especially of the face, neck, forearm and hands, the inflamed area being red and shiny and the edge raised. *adj.* **erysipelatous.**

Erysiphales (*Bot.*). An order of the *Ascomycetes*; superficial parasites on plants; the ascocarp is cleistocarpic and enclosed in a pseudoparenchymatous peridium. Also called **Perisporales.**

erythema (*Med.*). A superficial redness of the skin, due to dilatation of the capillaries.

erythema multiforme (*Med.*). A skin disease in which raised red patches appear and reappear, especially on the upper part of the body, associated with pain and swelling of the joints.

erythema nodosum (*Med.*). A skin disease in which red, painful, oval swellings appear, usually on the shins, associated with fever, joint pains, and sore throat; now believed to be due to a hypersensitive phase in certain infections, such as tuberculosis, sarcoidosis, and streptococci.

erythema pernio (*Med.*). Painful red swellings of the extremities, known as **chilblains.**

erythraemia, erythremia (*Med.*). Polycythaemia vera; Osler's disease; Vaquez's disease. A disease in which persistent increase in the number of red cells in the blood is associated with enlargement of the spleen.

erythrasma (*Med.*). Infection of the horny layer of the skin with the fungus *Microsporon minutissimum*, giving rise to superficial, reddish-yellow patches.

erythrite (*Min.*). Monoclinic arsenate hydrate of cobalt, occurring as pale reddish crystals or incrustations. Also called **cobalt bloom.**

erythritol (*Chem.*). $CH_2OH{\cdot}(CHOH)_2{\cdot}CH_2OH$. A tetrasaccharide carbohydrate found in lichens.

erythroblast (*Zool.*). A nucleated mesodermal embryonic cell, the cytoplasm of which contains haemoglobin, and which will later give rise to an erythrocyte. See also **megaloblast.**

erythroblastosis foetalis (*Med.*). Syn. for **haemolytic disease** of the newborn.

erythrocyte (*Zool.*). One of the red blood corpuscles of Vertebrates; flattened oval, or circular disklike, cells (lacking a nucleus in Mammals), whose purpose is to carry oxygen in combination with the pigment haemoglobin in them, and to remove carbon dioxide.

erythrocytolysis (*Zool.*). See **haemolysis.**

erythrocytosis (*Med.*). Excess in the number of red cells in the blood.

erythrodextrin (*Chem.*). The most complex of all dextrins, obtained by the hydrolysis of starch. It gives a wine-red colour with iodine.

erythroedema, erythredema (*Med.*). Acrodynia; pink disease. A disorder of infants, characterized by swelling and redness of face, fingers, and toes, and by restlessness, weakness, and neuritis. Caused by the administration or application of mercury or mercury salts in teething-powders, ointments, etc.

erythromelalgia (*Med.*). A condition characterized by pain, redness, and swelling of the toes, feet, and hands, due to arterial affection.

erythromycin (*Pharm.*). An antibiotic derived from *Streptomyces erythreus*, useful against some bacterial infections.

erythropenia (*Med.*). Diminution, below normal, of the number of red cells in the blood.

erythrophore (*Zool.*). A chromatophore containing a reddish pigment.

erythrophylls (*Bot.*). See **anthocyanins.**

erythropoiesis (*Med., Physiol.*). The formation of red blood cells.

erythropsia (*Med.*). The state in which objects appear red to the observer, e.g., snow blindness.

erythropsin (*Zool.*). A pigment which occurs in the retinular elements of the eyes of certain night-flying Insects.

erythropterin (*Zool.*). A red heterocyclic compound deposited as a pigment in the epidermal cells or the cavities of the scales and setae of many Insects.

erythrose (*Chem.*). A tetraose of the constitution:

$$\begin{array}{c} \mathrm{CHO} \\ | \\ \mathrm{H{-}C{-}OH} \\ | \\ \mathrm{H{-}C{-}OH} \\ | \\ \mathrm{CH_2OH} \end{array}$$

Es (*Chem.*). Symbol for ***einsteinium.***

Esaki diode (*Electronics*). **See tunnel diode.**

escape (*Bot.*). A cultivated plant growing wild and holding its place more or less successfully against competition.

escape behaviour (*An. Behav.*). Rapid locomotion from a given location following presentation of a specific stimulus which is either in itself noxious, or has been associated with a noxious stimulus. See also **escape conditioning.**

escape conditioning (*An. Behav.*). A procedure in which escape behaviour regularly terminates a negative reinforcement or aversive stimulus, response intensity increasing with successive presentations of the stimulus.

escape factor (*Nuc.*). Fraction of the tracks in a nuclear emulsion which are not complete, due to the escape of the particle from the emulsion.

escape mechanism (*Psychol.*). A mental process by which one evades the unpleasant.

escapement (*Horol.*). A device for converting circular motion into reciprocating motion. Specifically in a watch or clock, it is the mechanism which transmits the power from the weight or spring to maintain the vibration of the pendulum or balance; at the same time it controls the rate at which all the wheels in the train move, the governing factor being the time of the vibration of the pendulum or balance. The ideal, which cannot be attained in practice, is a pendulum or balance vibrating without any external interference. The escapement which provides a minimum of interference is referred to as *highly detached*. Where, for a considerable period, there is contact with the escape wheel, the escapement is referred to as *frictional rest*.

Clock escapements are usually either (*a*) recoil, or (*b*) dead-beat. In the *recoil* type, during part of the action the pendulum pushes the escape wheel back against the power in the train, giving the 'recoil'. Although this is bad from the theoretical standpoint, in practice such an escapement gives very satisfactory results, as it is 'self-correcting', i.e., should the arc of vibration of the pendulum tend to increase, it is checked by the greater amount of recoil. In the *dead-beat* type no recoil is possible as the locking faces of the pallets are arcs struck from the pallet staff centre.

Watch escapements are classified as *frictional-rest* or *detached*. The cylinder is a frictional-rest escapement, and the lever a detached escapement. The chronometer is the most highly detached of all escapements. The majority of modern watches are fitted with the club-tooth lever escapement (straight-line lever) in which the impulse is divided between the pallets and the escape wheel teeth. In the English lever escapement with ratchet tooth escape wheel, the impulse is due entirely to the pallets, whereas in the pin-pallet escapement, an escapement very much used in low-grade watches, the pallets are vertical pins, and impulse is given entirely by the teeth of the escape wheel. (*Eng.*) A mechanism allowing one component at a time to move automatically from one machine to the next in a line. (*Tools*) The cut-away part above the mouth of a plane through which the shavings are voided.

escape (or 'scape) pinion (*Horol.*). The pinion on the arbor of the escape wheel.

escape velocity (*Astron.*). See **parabolic velocity.**

escape (or 'scape) wheel (*Horol.*). The wheel of which the teeth act on the pallets.

escarpment (*Geol.*). A long cliff-like ridge developed by denudation where hard and soft inclined strata are interbedded, the outcrop of each hard rock forming an *escarpment*, such as those of the Chalk (Chiltern Hills, N. and S. Downs) and the Jurassic limestones (Cotswold Hills). Generally an escarpment consists of a short steep rise (the *scarp face*) and a long gentle slope (the *dip-slope*).

eschar (*Med.*). A dry slough produced by burning or by corrosives.

escharotic (*Med.*). A corrosive or caustic chemical agent.

Eschka's reagent (*Chem.*). Mixture of MgO and Na_2CO_3 (2 : 1); used for estimation of the sulphur content of fuels.

escribed circle of a triangle (*Maths.*). One which touches one side of a triangle externally and the extensions of the other two sides.

escutcheon (*Build.*). A perforated plate around an opening, such as a keyhole plate or the plate to which a doorknob is attached.

eserine (*Chem.*). *Physostigmine* (q.v.).

esker (*Geol.*). A long winding gravel ridge, laid down by waters issuing from the front of a retreating glacier, or actually in the bed of a subglacial stream. Known in America as **os**, *pl.* **osar** (from the Swedish *ås, åsar*). See **glacial action**, etc.

Esmarch's bandage (*Med.*). A rubber bandage which, applied to a limb from below upwards, expels blood from the part.

esophagus, etc. (*Zool.*). See **oesophagus, etc.**

esophoria (*Med.*). Latent internal squint, revealed in an apparently normal person by passing a screen before the eye.

espagnolette (*Join.*). Fastening for a French window, having two long bolts which operate in slots at the top and bottom when the door handle is turned. Sometimes called **cremorne bolt.**

esparto (*Paper*). A paper made from a coarse grass of the same name obtained from southern Spain and northern Africa.

esperite (*Min.*). A silicate of lead, calcium and zinc, crystallizing in the monoclinic system.

espundia (*Med.*). An ulcerative infection of the skin, and of the mucous membranes of the nose and the mouth, by the protozoal parasite *Leishmania braziliensis*; occurs in South America.

ESRO (*Space*). Abbrev. for *European Space Research Organization*.

essential minerals (*Geol.*). Components present, by definition, in a rock, the absence of which would automatically change the name and classification of the rock. Cf. *accessory minerals*.

essential oils (*Chem.*). Ethereal oils which are contained in many plants and flowers and give them their characteristic odour. Many of them belong to the terpene group, others are related to benzene derivatives.

essential organs (*Bot.*). Stamens and carpels.

Essex board (*Build.*). A building-board made of layers of compressed wood-fibre material cemented with a fire-resisting cement.

essexite (*Geol.*). A coarse-grained deep-seated igneous rock, essentially an alkali-gabbro, with preponderance of soda. Named from Essex Co., Mass.

Esson coefficient (*Elec. Eng.*). See **specific torque coefficient.**

essonite (*Min.*). Original spelling of *hessonite* (q.v.).

establishment (*Typog.*). A workman who receives a weekly wage is said to be *on the establishment*, in contrast to one who does piecework. Usually abbreviated to *stab*.

establishment charges (*Build., Eng., etc.*). See **on-costs.**

estamenes (*Textiles*). Dress fabrics with a rough fibrous twilled surface, made from crossbred worsted yarns. They are usually of 2-and-2 twill weave.

ester (*Chem.*). Esters are derivatives of acids obtained by the exchange of the replaceable hydrogen for alkyl radicals. Many esters have a fruity smell and are used in artificial fruit essences; also used as solvents.

esterase (*Chem.*). Enzyme which promotes the hydrolysis of esters.

ester gum (*Paint*). One of the class of compounds obtained by esterifying rosin (colophony) or natural gums with a polyhydric alcohol, e.g., glycerol or pentaerythritol. A cheap resin for paints and varnishes.

esterification (*Chem.*). The direct action of an acid on an alcohol, resulting in the formation of esters. An equilibrium is reached between the quantities of acid and alcohol present and the quantities of ester and water formed. Catalysed by hydrogen ions.

ester value (*Chem.*). The number of milligrams of potassium hydroxide required to saponify the fatty acid esters in one gram of a fat, wax, oil, etc.; equal to the saponification value minus the acid value.

ester wax (*Micros.*). A hard translucent material which is used for histological infiltration of tissues. It is based on diethylene glycol distearate and monostearate, with ethyl cellulose, castor oil, and stearic acid, m.p. 48°C. Soluble in hydrocarbons, alcohols, and esters.

esthesiometer (*Med.*). See **aesthesiometer.**

esthiomene (*Med.*). A condition in which there are chronic hypertrophy and destructive ulceration of the external genitals of the female.

Estiatron (*Electronics*). Travelling-wave tube with electrostatic focusing.

estradiol, estrogen, etc. (*Physiol.*). Same as **oestradiol, oestrogen, etc.**

estuarine deposition (*Geol.*). Sedimentation in the environment of an estuary. The deposits differ from those which form in a deltaic environment, chiefly in their relationship to the strata of the adjacent land, and are usually of finer grain and of more uniform composition. Both are characterized by brackish water and by their containing land-derived animal- and plant-remains.

estuarine muds (*Geol.*). So-called *estuarine muds* are, in many cases, silts admixed with sufficient true clay to give them some degree of plasticity; they are characterized by a high content of decomposed organic matter.

estuary (*Geol.*). An inlet of the sea at the mouth of a river; developed especially in areas which have recently been submerged by the sea, the lower end of the valley having been thus drowned. See **fiords** and cf. *delta*.

e.s.u. (*Elec. Eng.*). Abbrev. for *electrostatic unit*.

Et (*Chem.*). A symbol for the ethyl radical C_2H_5—.

ETA (*Aero.*). *E*stimated *T*ime of *A*rrival, as forecast on a *flight plan*, for a civil aircraft or the time of arrival over a target for a military aircraft.

etaerio (*Bot.*). A group of achenes or of drupels.

etalon (*Light*). An interferometer consisting of an air film enclosed between half-silvered plane parallel plates of glass or quartz having a fixed separation. It is used for studying the fine structure of spectral lines. See **Fabry and Pérot interferometer.**

Etard's reaction (*Chem.*). The formation of aromatic aldehydes by oxidizing methylated derivatives and homologues of benzene with chromyl chloride, CrO_2Cl_2.

etchant (*Chem.*). Chemical for removing copper from laminate during production of printed circuits.

etched figures or **etch-figures** (*Crystal.*). Small pits or depressions of geometrical design in the faces of crystals, due to the action of some solvent. The actual form of the figure depends upon the symmetry of the face concerned, and hence they provide invaluable evidence of the true symmetry in distorted crystals.

etching (*Met.*). The process of revealing the structure of metals and alloys by attacking a highly polished surface with a reagent that has a differential effect on different crystals or different constituents. (*Photog.*) The process of (1) dissolving, with an acid, portions of a surface, such as copper or zinc sheet, where it is not protected with a resist; (2) soaking away gelatine differentially to form a relief image.

etching pits (*Met.*). Small cavities formed on the surface of metals during etching.

etching test (*Chem.*). A test used in analytical chemistry for the detection of fluorides. The substance under examination is heated with sulphuric acid in a lead vessel covered with a glass lid. If fluorides are present the glass will be etched owing to the action of hydrogen fluoride produced by the action of the acid on the fluoride.

ETD (*Aero.*). *E*stimated *T*ime of *D*eparture. See **departure time.**

etem (*Chem.*). Hexahydro-2,7-dithio-1,3,6-thiadiazepine, used as a fungicide. Also **UCP/21.**

etepimeletic behaviour (*An. Behav.*). A type of social behaviour shown by young animals to elicit *epimeletic behaviour* (q.v.).

etesian winds (*Meteor.*). In the Mediterranean, winds which blow from the northwest for about 40 days in the summer.

ethanal (*Chem.*). See **acetaldehyde.**

ethanamide (*Chem.*). See **acetamide.**

ethane (*Chem.*). $H_3C{\cdot}CH_3$, a colourless, odourless gas of the alkane series; the critical temperature is +34°C, the critical pressure is 50·2 atm, b.p. −84°C. The second member of the alkane series of hydrocarbons. Chemical properties similar to those of *methane* (q.v.).

ethanoates (*Chem.*). See **acetates.**

ethanoic acid (*Chem.*). See **acetic acid.**

ethanol (*Chem.*). The IUPAC name for **ethyl alcohol.** C_2H_5OH, m.p. −114°C, b.p. 78·4°C; a colourless liquid, of vinous odour, miscible with water and most organic solvents, rel. d. 0·789; formed by the hydrolysis of ethyl chloride or of ethyl hydrogen sulphate; it may be obtained by absorption of ethylene in fuming sulphuric acid at 160°C, followed by hydrolysis with water, by reduction of acetaldehyde, or by direct synthesis from ethylene and water at high temperatures in the presence of a catalyst. It is prepared technically by the alcoholic fermentation of sugar. It forms alcoholates with sodium and potassium.

ethanolamines (*Chem.*). Amino derivatives of ethyl alcohol, *monoethanolamine*, $CH_2OH{\cdot}CH_2{\cdot}NH_2$; *diethanolamine*, $(CH_2{\cdot}CH_2OH)_2NH$; *triethanolamine*, $(CH_2{\cdot}CH_2OH)_3{\cdot}N$; hygroscopic solids with strong ammoniacal smell. Used, in combination with fatty acids, to produce detergents and cosmetic products.

ethanoyl (*Chem.*). See **acetyl.**

ethanoylation (*Chem.*). See **acetylation.**

ethanoyl chloride (*Chem.*). See **acetyl chloride.**

ethene (*Chem.*). The IUPAC name for **ethylene,** $H_2C{=}CH_2$, m.p. −169°C, b.p. −103°C, a gas of the alkene series, contained in illuminating gas and in gases obtained from the cracking of petroleum. Used for synthetic purposes, e.g., polythene and ethylene oxide, and for maturing fruit in storage.

ethenoid resins (*Plastics*). Resins made from compounds containing a double bond between 2 carbon atoms, i.e., the acrylic, vinyl, and styrene groups of plastics.

etheogenesis (*Zool.*). Parthenogenesis of male individuals; development of male gametes without fertilization.

ether (*Phys.*). A hypothetical, nonmaterial entity supposed to fill all space whether 'empty' or occupied by matter. The theory that electromagnetic waves need such a medium for propagation is no longer tenable.

ether (*Chem.*). Alkoxyalkane. (1) Any compound of the type R—O—R′, containing two identical, or different, alkyl groups united to an oxygen atom; they form a homologous seris $C_nH_{2n+2}O$. The term is often erroneously applied to *esters.* (2) Specifically, *diethyl ether* (q.v.).

ethion (*Chem.*). Tetraethyl *SS′*-methylene bis (phosphorothiolothionate); an insecticide.

ethionamide (*Pharm.*). A bright yellow powder (2-ethylisonicotinthionamide), used in the treatment of tuberculosis. TN **Trescatyl.**

Ethiopian region (*Zool.*). One of the primary faunal regions into which the surface of the globe is divided; it includes all of Africa and Arabia south of the tropic of Cancer.

ethmo-. Prefix from Gk. *ethmos*, sieve.

ethmohyostylic (*Zool.*). In some Vertebrates, having the lower jaw suspended from the ethmoid region and the hyoid bar.

ethmoidalia (*Zool.*). A set of cartilage bones (**ethmoids**) forming the anterior part of the brain-case in the Vertebrate skull.

ethmoidectomy (*Surg.*). Surgical removal of the ethmoid cells or of part of the ethmoid bone.

ethmoiditis (*Med.*). Inflammation of the ethmoid cells.

ethmoid plate (*Zool.*). A cartilage element of the developing Vertebrate skull.

ethmoturbinal (*Zool.*). In Mammals, a paired bone or cartilage of the nose, on which are supported the folds of the olfactory mucous membrane.

ethnology (*Gen.*). The branch of science concerned with the study of the races of mankind, in terms of their distribution, activities, relationships and characteristics.

ethogram (*An. Behav.*). A complete inventory of the species-specific behaviour patterns of a particular species.

ethology (*An. Behav.*). The study of animal behaviour, usually with particular reference to the observation of the natural, and largely instinctive, behaviour of animals under natural conditions, but also involving related laboratory experiments. Cf. *comparative psychology.*

ethoxyl group (*Chem.*). The group $-O{\cdot}C_2H_5$.

ethyl (*Chem.*). See **lead tetraethyl.**

ethyl acetate (*Chem.*). Ethanoate. Acetic ether, $CH_3COOC_2H_5$, m.p. −82°C, b.p. 77°C; colourless liquid of fruity odour, used as a lacquer solvent and in medicine.

ethyl aceto-acetate (*Chem.*). $CH_3{\cdot}CO{\cdot}CH_2{\cdot}COO{\cdot}C_2H_5$. One of the best known examples of organic compounds existing in keto and enol forms. Widely used as a chemical intermediate, including the manufacture of pyrazolone dyes and mepacrine. Also **aceto-acetic ester.**

ethyl acrylate (*Chem.*). $CH_2{=}CH{\cdot}COOC_2H_5$. Colourless liquid; b.p. 101°C. Used in the manufacture of plastics.

ethyl alcohol (*Chem.*). See **ethanol.**

ethylamine (*Chem.*). $C_2H_5{\cdot}NH_2$, b.p. 19°C, a

liquid or gas of ammoniacal odour, which dissolves in water, and forms salts; it dissolves $Al(OH)_3$.

ethylene (*Chem.*). See **ethene.**

ethylene diamine tartrate (*Radio*). Chemical in crystal form, exhibiting marked piezoelectric phenomena; used in narrow-band carrier filters. Abbrev. **EDT.**

ethylene diamine tetra-acetic acid (*Chem.*). See **EDTA.**

ethylene dichloride (*Chem.*). 1,2-dichloroethane.

ethylene glycol (*Chem.*). Glycol, $HO{\cdot}CH_2{\cdot}CH_2{\cdot}OH$, b.p. 197·5°C, rel. d. 1·125, a colourless, syrupy, hygroscopic liquid, miscible with water and ethanol. Prepared from ethylene dibromide or ethylene chlorhydrin by hydrolysing with caustic soda. Intermediate for glycol esters, which are solvents and plasticizers for lacquers; used in the textile industry, for printing-inks, foodstuffs, antifreezing mixtures, and for de-icing aeroplane wings.

ethylene oxide (*Chem.*). C_2H_4O, b.p. 13·5°C, a mobile colourless liquid of ethereal odour, obtained by distilling glycol chlorhydrin with concentrated caustic potash. Manufactured directly from ethylene and oxygen in the presence of a catalyst. Very useful organic intermediate in the manufacture of solvents, detergents, etc. Can also be used as a sterilizing medium in the gaseous state.

ethylene-propylene rubber (*Plastics*). A type of synthetic rubber based on readily available petrochemical materials. Developed both in Italy (TN *Dutral*) and U.S. Good all-round mechanical properties but not very resistant to organic solvents.

ethyl group (*Chem.*). The monovalent radical $—C_2H_5$.

ethylidene (*Chem.*). The organic group $CH_3{\cdot}CH{:}$.

ethyl mercaptan (*Chem.*). See **mercaptans.**

ethylpyrophosphate (*Chem.*). See **TEPP.**

ethyne (*Chem.*). See **acetylene.**

ethynide (*Chem.*). See **acetylide.**

etiolation (*Bot.*). The condition of a green plant which has not received sufficient light; the stems are weak, with abnormally long internodes, the leaves are small, yellowish or whitish, and the vascular strands are deficient in xylem. *adj.* **etiolated.**

etiology (*Med.*). See **aetiology.**

E-transformer (*Elec. Eng.*). An electric sensing device in an automatic control system which gives an error voltage in response to linear motion. It consists of coils for detecting small displacements of magnetic armature, which affects balance of currents when off-centre.

Ettinghausen effect (*Mag.*). Effect analogous to the *Hall effect*, concerned with a metal strip carrying an electric current longitudinally and placed in a magnetic field perpendicular to the plane of the strip. Differential temperatures are created between the edges of the strip which are opposite to one another.

etude (*Nuc. Eng.*). Projected improved stellarator, characterized by a variable twist angle on the end loops. Magnetohydrodynamic theory suggests this should be the most stable form of controlled thermonuclear device so far developed.

eu-. Prefix from Gk. *eu*, well, good.

Eu (*Chem.*). The symbol for *europium.*

euapogamy (*Bot.*). The development of the sporophyte from a cell or cells of the gametophyte, not from a zygote resulting from gametic fusion; the sporophyte is diploid.

euapospory (*Bot.*). Complete failure on the part of a plant to form spores.

Euascomycetae (*Bot.*). A subclass of the *Ascomycetes*, in which the asci are formed on ascogenous hyphae growing from a zygote, or from a parthenogenetically formed ascogonium.

Eubacteriales (*Bacteriol.*). One of the two main orders of the true bacteria, distinguished from the *Pseudomonadales* (q.v.) by the peritrichous flagella of the motile members. Spherical or rod-shaped cells, having no photosynthetic pigments; not acid fast; readily stained by aniline dyes.

Eubasidii (*Bot.*). A subclass of the *Basidiomycetes*, in which the basidia develop directly from the diplophase mycelium.

Eubrya (*Bot.*). A subclass of the *Musci*; the leaves have a midrib more than one cell thick; the sporogenous tissue of the capsule does not grow over the columella; there is a complex operculum, and elongation of the sporophyte is by lengthening of the seta.

Eubryales (*Bot.*). An order of the *Eubrya*; the gametophores are perennial with many rows of leaves; the capsule is pendant, having a peristome with well-developed teeth.

eucalyptole (*Chem.*). *Cineole* (q.v.).

Eucarida (*Zool.*). A superorder of *Malacostraca* in which there are 6 abdominal somites and the thoracic limbs have a 2-jointed protopodite; the carapace coalesces dorsally with all the thoracic somites, and the protopodite of the antenna may be 1- or 2-jointed.

eucarpic (*Bot.*). Having both vegetative and reproductive organs, separate and functioning at the same time.

eucephalous (*Zool.*). Having a well-developed head; applied to certain apodous dipteran larvae which possess a head (as Mosquitoes) in contradistinction to the majority of dipteran larvae, which have the head much reduced. Cf. *acephalous, hemicephalous.*

euchroite (*Min.*). Hydrated basic copper arsenic, crystallizing in the orthorhombic system, green in colour, occurring in large crystals at Libethen, Czechoslovakia.

euchromatin (*Cyt.*). The chromatin that constitutes the bulk of the chromosomes and includes the active genes.

euchromosome (*Cyt.*). A typical chromosome, as opposed to a sex chromosome; an autosome.

euclase (*Min.*). A monoclinic mineral, occurring as prismatic, usually colourless, crystals. Chemically it is hydrated silicate of beryllium and aluminium.

eucoiliform (*Zool.*). Said of the protopod larvae of certain parasitic *Hymenoptera*, which possess greatly developed thoracic appendages.

eucolloid (*Chem.*). Colloid of large particle diameter, more than 250 nm.

eucone (*Zool.*). Said of the compound eyes of many insects, in which each ommatidium contains a true crystalline cone.

Eucopepoda (*Zool.*). An order of *Copepoda* in which paired compound eyes are absent, the genital openings occur on the seventh trunk somite, and the thoracic limbs lack a flagellum; includes parasitic forms such as the Sea Lice and Gill Maggots which occur on many food fish, as well as free-living forms.

eucrite (*Geol.*). A coarse-grained, usually ophitic, deep-seated basic igneous rock, containing plagioclase near bytownite in composition, both ortho- and clino-pyroxenes, together with olivine. Eucrite is an important rock type in the Tertiary complexes of Scotland.

eucryptite (*Min.*). A hexagonal lithium aluminium silicate, commonly found as an alteration product of spodumene.

eucyclic (*Bot.*). Said of a flower made up of successive whorls, all with the same number of members.

eudialyte (*Min.*). A pinkish red complex hydrous zirconium silicate of sodium, calcium and iron. It crystallizes in the rhombohedral system and occurs in some alkaline igneous rocks.

eudiometer (*Chem.*). (1) A voltameter-like instrument in which quantity of electricity passing can be found from volume of gas produced by electrolysis. (2) Similar system to determine volume changes in a gas mixture due to combustion.

eudoxid, eudoxome (*Zool.*). In some *Siphonophora*, a free-swimming monogastric stage, without a special nectocalyx. *pl.* **eudoxia.**

eugamic (*Zool.*). Pertaining to the period of maturity.

eugenics (*Gen.*). The branch of science concerned with the conditions tending to improve or impair the qualities of future generations.

eugenol (*Chem.*). $C_6H_3(OH)(OCH_3)(CH_2{\cdot}CH:CH_2)$, a phenol homologue, the chief constituent of oil of cloves and cinnamon leaf oil, b.p. 252°C, rel. d. 1·07; used for manufacturing *vanillin* (q.v.).

Euglenales (*Bot.*). An order of the *Euglenophyta*, in which the motile phase is dominant in the life-cycle.

Euglenoidina (*Zool.*). An order of *Phytomastigina*, comprising forms generally with one flagellum; green or colourless; with a gullet but without a transverse groove; having food-reserves of paramylum, and sometimes also of oil; with a contractile vacuole. Known to botanists as *Euglenales* (q.v.).

euglenoid movement (*Zool.*). A type of movement undergone by many *Protozoa*, which possess a definite body form, by means of contractions of the protoplasm stretching the pellicle. Also called **metaboly.**

Euglenophyta (*Bot.*). A division or phylum of *Algae*. The pigments, if present, are only chlorophyll and carotenoids, which are localized in chloroplast; most members are naked unicellular flagellates; the storage products are paramylum and fats; reproduction is by simple cell division or encystment; sexual reproduction has been established in only one genus.

eugonidium (*Bot.*). A bright-green algal cell forming part of the thallus of a lichen, and belonging to the *Chlorophyceae*.

euhedral crystals (*Geol.*). See **idiomorphic crystals.**

euhymenial (*Bot.*). Having a hymenium in which all the basidia are formed nearly at the same time.

euhyponeuston (*Ecol.*). Organisms living in the upper 50 mm of water for daily periods or for the whole of their life cycle.

eukaryon (*Biol.*). The highly organized nucleus characteristic of higher organisms. Cf. *prokaryon*.

Eulamellibranchiata (*Zool.*). An order of *Lamellibranchiata*, in which the branchial axis of the gill is united to the body throughout its length, and bears parallel, ventrally directed and reflected filaments, which are joined to one another by vascular junctions; similarly, the lamellae of each gill are united by vascular junctions; the byssus gland is variable; ciliary feeders. True Oysters, Cockles, River and Pond Mussels, Clams, Zebra Mussels, and Shipworms.

Eulerian angles (*Maths.*). Three independent angles φ, θ and ψ which specify the direction of a line in space. The line specified can be visualized as follows. Starting with orthogonal cartesian axis *Oxyz*, first rotate the axes about *Oz* through an angle φ. Next rotate the displaced axes about the new position of *Ox* through an angle θ. Finally rotate the axes about the new position of *Oz* through an angle ψ and the resulting position of *Ox* is the line specified by the Eulerian angles φ, θ, ψ.

Euler's constant (*Maths.*). The limit of

$$\sum_1^n \frac{1}{r} - \log_e n, \text{ as } n \to \infty.$$

Denoted by γ. Its value is 0·577 215 66 . . .

Euler's formula (*Civ. Eng.*). A formula giving the collapsing load for a long, thin column of given sizes. It only applies where the load passes down the centroid of the column. It states that

$$P = \frac{\pi^2 EI\ (\text{min.})}{L^2},$$

where P = the collapsing load, E = Young's modulus, I = the least moment of inertia, and L = the length of the pin-jointed column.

Euler's theorem (*Maths.*). $K = A\cos^2\theta + B\sin^2\theta$, where K is the curvature of a normal section of a surface displaced by an angle θ from a principal direction, A is the curvature in that principal direction, and B is the other principal curvature.

eulite (*Min.*). An orthopyroxene containing 70–90% of the iron end-member, and thus 30–10% of the enstatite, or magnesium, end-member.

eulittoral zone (*Ecol.*). The upper part of the littoral zone of the sea bottom, including the intertidal area.

Eumalacostraca (*Zool.*). An obsolete group of *Crustacea*, which comprised the 4 subclasses *Syncarida*, *Peracarida*, *Eucarida*, and *Hoplocarida* (qq.v.). It was distinguished by the possession of 6 abdominal somites, the absence of movable furcal rami and of adductor muscles and the presence of pediform thoracic limbs, with a protopodite of 2 segments.

eumerism (*Zool.*). An aggregation of similar parts.

eumerogenesis (*Zool.*). A form of segmentation in which the parts produced are alike (although dissimilarity may arise later).

eumitosis (*Cyt.*). Typical normal mitosis, in which the chromosomes separate distinctly and clearly divide longitudinally. Cf. *cryptomitosis*, *paramitosis*.

Eumycetes (*Bot.*). The higher fungi, with many thousand species. The group includes the *Ascomycetes* and the *Basidiomycetes*, as well as the *Fungi imperfecti*. The mycelium is usually strongly developed, and consists of branching septate hyphae.

eumydrin (*Surg.*). An atropine-like substance used in the treatment of infantile pyloric stenosis.

eunuch (*Med.*). A male who has no testes.

eunuchoid (*Med.*). A male with testes, but no secondary male characteristics.

eupatheoscope (*Heat*). A device for estimating the heat lost by an individual in an enclosure. Heat loss by radiation, air temperature, and air velocity are taken into account, but humidity is ignored.

eupeptic (*Med.*). Possessing a good digestion.

Euphausiacea (*Zool.*). An order of *Eucarida* in which the exopodite of the maxilla is small; none of the thoracic limbs are modified as maxillipeds; there is a single series of gills, standing upon the coxopodites of the thoracic limbs; and there is no statocyst. In habit, they are marine and pelagic and form part of the diet of whales.

Euphon quilt, Chance's (*Acous.*). An acoustic

absorbing quilt made of matted layers of glass silk threads supported between layers of paper or canvas.

Euphorbiaceae (*Bot.*). A large family of flowering plants, including those of the spurge genus *Euphorbia*; mainly tropical, but also found in temperate zones. Rubber (from *Hevea brasiliensis*), castor oil, and resins are important products of the family.

euphoria, euphory (*Med.*). A feeling of well-being, not necessarily indicative of good health.

euphotic zone (*Ecol.*). See **photic zone.**

euphotometric (*Bot.*). Said of a leaf which occupies a fixed position with its lamina perpendicular to the direction of the strongest diffuse light to which it is exposed.

euploidy (*Cyt.*). Polyploidy in which the chromosome number is an exact multiple of the haploid number; organisms showing this condition are *euploid*.

eupotamous (*Ecol.*). Normally living in rivers and streams. Cf. *tychopotamous*.

eupyrene (*Zool.*). (Of spermatozoa) normal, typical. Cf. *apyrene, oligopyrene*.

Eureka (*Elec. Eng.*). See **constantan.**

eureka (*Radar*). See **rebecca-eureka.**

europium (*Chem.*). Metallic element of the rare-earth group. Symbol Eu, at. no. 63, r.a.m. 151·96, valency 2 and 3. Contained in black monazite, gadolinite, samarskite, xenotime.

Eurotiales (*Bot.*). An order of the *Ascomycetes*, characterized by the small, rounded asci scattered through a peridium.

Eurovision (*TV*). System for relays of television programmes through the EBU between various countries, with or without change of resolution (lines per frame).

euryapsid (*Zool.*). In the skull of Reptiles, the condition where there is one temporal vacuity, this being behind the eye and bordered by the postorbital and squamosal bones. Found in *Protorosaurs* and *Sauropterygii*.

eurybathic (*Ecol.*). Tolerant of a wide range of depth.

euryhaline (*Ecol.*). Normally inhabiting salt water, but adaptable to a wide range of salinity. Cf. *stenohaline*.

euryhydric (*Ecol.*). Tolerating a wide range of soil water content. Cf. *stenohydric*.

euryoecious (*Ecol.*). Tolerating a wide range of habitats. Cf. *stenoecious*.

euryphagic (*Ecol.*). Tolerating a wide range of foods. Cf. *stenophagic*.

Eurypterida (*Geol.*). An order of *Crustaceans* occurring first in the Silurian and Devonian rocks of Britain, and represented by such types as *Eurypterus* and *Stylonurus*, the latter reaching almost 2 m in length.

eurypylous (*Zool.*). In *Porifera*, said of a type of canal system in which the flagellated chambers open directly into the excurrent canals by apopyles.

eurythermal (*Ecol.*). Tolerant of a wide range of temperature.

Euselachii (*Zool.*). A subclass of *Chondrichthyes* with numerous teeth developed in continual succession and with three basal bones, from which a number of pre-axial radials spread out, in the pectoral fins. Sharks and Rays.

eusporangiate fern (*Bot.*). A fern in which the sporangia are developed each from a group of cells. Cf. *leptosporangiate*.

Eustachian tube (*Zool.*). In land Vertebrates, a slender duct connecting the tympanic cavity with the pharynx.

Eustachian valve (*Zool.*). In Mammals, a rudimentary valve separating the openings of the superior venae cavae from that of the inferior vena cava.

eustatic movements (*Geol.*). Changes of sea level, constant over wide areas, due probably to alterations in the volumes of the seas. These may be due to variations in the extent of the polar ice-caps or large-scale crustal movements in ocean basins.

eusternum (*Zool.*). In Insects, a ventral segmental plate of a thoracic segment.

eustomatous (*Zool.*). With a well-defined mouth or opening.

eustyle. A colonnade in which the space between the columns is equal to 2¼ times the lower diameter of the columns.

Eutaw Group (*Geol.*). A subdivision of the Cretaceous System in the southern U.S., lying between the Tuscaloosan, which it overlaps northwards, and the overlying Selma Chalk. The Tombigbee Sand occurs at its top.

eutaxitic (*Geol.*). Descriptive of the streaky banded structure of certain pyroclastic rocks, as contrasted with the smooth layered structure of flow-banded lavas.

eutectic (*Chem.*). Relating to a mixture of two or more substances having a minimum melting point. Such a mixture behaves in some respects like a pure compound.

eutectic change (*Met.*). The transformation from the liquid to the solid state in a eutectic alloy. It involves the simultaneous crystallization of two constituents in a binary system and of three in a ternary system.

eutectic point (*Met., Min.*). The point in the binary or ternary constitutional diagram indicating the composition of the eutectic alloy, or mixture of minerals, and the temperature at which it solidifies.

eutectic structure (*Met.*). The particular arrangement of the constituents in a eutectic alloy which arises from their simultaneous crystallization from the melt. (*Min.*) See **graphic texture.**

eutectic system (*Met.*). A binary or ternary alloy system in which one particular alloy solidifies at a constant temperature which is lower than the beginning of solidification in any other alloy.

eutectic welding (*Eng.*). A metal welding process carried out at relatively low temperature, using the eutectic properties of the metals involved.

eutectoid (*Met.*). Similar to a eutectic except that it involves the simultaneous formation of two or three constituents from another solid constituent instead of from a melt. *Eutectoid point* and *eutectoid structure* have similar meanings to those given for *eutectic*.

eutectoid steel (*Met.*). Steel having the same composition as the eutectoid point in the iron-carbon system (0·87 % C), and which therefore consists entirely of the eutectoid at temperatures below 710°C. See **pearlite.**

eutexia (*Chem., Eng., Met., etc.*). The property of being easily melted, i.e., at a minimum melting point.

euthanasia (*Med.*). Easy or painless death; the action of procuring this.

euthenics (*Biol.*). The study of the scientific control of the environment.

Eutheria (*Zool.*). An infraclass of viviparous Mammals in which the young are born in an advanced stage of development; no marsupial pouch; an allantoic placenta occurs; the scrotal sac is behind the penis, the angle of the lower jaw is not inflexed and the palate is imperforate. Also called **Monodelphia, Placentalia.**

Euthyneura (*Zool.*). A group of hermaphrodite *Gastropoda*, in which the visceral mass and commissure show detorsion, and there are usually two pairs of cephalic tentacles. Sea Slugs, True Snails, Land Slugs, etc.

euthyneural, euthyneurous (*Zool.*). Having a symmetrical nervous system; said especially of certain *Gastropoda* in which the visceral nerve loop is untwisted. *n.* **euthyneury.**

euthyroid (*Med.*). Having a normal level of thyroid activity.

eutrophic (*Ecol.*). Said of a type of lake in which the hypolimnion becomes depleted of oxygen during the summer by the decay of organic matter falling from the epilimnion. A eutrophic lake is usually shallow, with much primary productivity. Cf. *dystrophic*, *oligotrophic*.

eutropic series (*Chem.*). An arrangement of substances in which crystalline form and physical constants show a regular variation.

eutropy (*Chem.*). The regular variation of the crystalline form of a series of compounds with the atomic number of the element.

euxenite (*Min.*). An uncommon mineral containing rare elements; a niobate, tantalate, and titanate of yttrium, erbium, cerium, thorium, and uranium, and valuable on this account. Commonly massive and brownish black in colour; rarely, crystalline.

eV. Abbrev. for *electron-volt*.

E.V. (*Photog.*). Abbrev. for *exposure value*.

evaginate (*Bot., Zool.*). Not having a sheath.

evagination (*Med.*). The turning inside out of an organ. (*Zool.*) Withdrawal from a sheath; the development of an outgrowth; eversion of a hollow ingrowth; an outgrowth; an everted hollow ingrowth. Cf. *invagination*.

evaporation (*Chem. Eng.*). The process of concentrating solutions, either to form crystals or to adjust to a required strength, by removing liquid as vapour. Distinguished from *Phys.* definition (see below), in that it is always done at the boiling point for the pressure condition involved. (*Electronics*) Escape of electrons or other particles having by chance a sufficiently enhanced outward velocity. (*Eng.*) The conversion of water into steam in a boiler. (*Phys.*) The conversion of a liquid into vapour, at temperatures below the boiling point. The rate of evaporation increases with rise of temperature, since it depends on the saturated vapour pressure of the liquid, which rises until it is equal to the atmospheric pressure at the boiling point. Evaporation is used to concentrate a solution.

evaporation coefficient (*Vac. Tech.*). The ratio of the rate of evaporation of molecules from a surface to the rate predicted from vapour pressure calculations.

evaporative capacity (*Eng.*). The mass of steam at a stated temperature and pressure produced in one hour from unit area of heating surface from feed-water at a stated temperature, when steaming at the most economical rate. Measured in lb/ft^{2}hr or kg/m^{2}hr. Also called **evaporation rate.**

evaporative cooling (*Aero., Chem. Eng.*). The process of evaporating part of a liquid by supplying the necessary latent heat from the main bulk of liquid which is thus cooled. Used for some piston aero-engines in the 1930s, some current types of turbine and rocket components, and also for coding purposes in cabin air conditioning systems. Cf *sweat cooling*.

evaporator (*Chem.*). A still designed to evaporate moisture or solvents to obtain the concentrate.

evaporimeter (*Meteor.*). An instrument used for measuring the rate of natural evaporation.

evaporite (*Geol.*). Sedimentary deposit of material previously in aqueous solution and concentrated by the evaporation of the solvent. Normally found as the result of evaporation in lagoons or shallow enclosed seas.

evaporograph (*Phys.*). Device producing direct or photographic images of objects in darkness by focusing infrared radiations from them on to an oil-film, which evaporates in proportion to the amount of radiation, leaving an image.

evapotranspiration (*Ecol.*). The total water loss from a particular area, being the sum of evaporation from the soil and transpiration from vegetation.

evection (*Astron.*). The largest of the four principal periodic inequalities in the mathematical expression of the moon's orbital motion; due to the variable eccentricity of the moon's orbit, with a maximum value of 1° 16′ and a period of 31·81 days.

even-even nuclei (*Nuc.*). Those for which the numbers of protons and neutrons are both even and normally stable.

even function (*Maths.*). f is an even function if $f(-x)=f(x)$. Cf. *odd function*.

evening star (*Astron.*). The name given in popular language to a planet, generally Venus or Mercury, seen in the western sky at or just after sunset. Also used loosely to describe any planet which transits before midnight.

even-odd nuclei (*Nuc.*). Those with an even number of protons and an odd number of neutrons.

even pitch (*Eng.*). In screw-cutting in the lathe, the thread cut is said to be of *even pitch* if its *pitch* is equal to, or an integral multiple of, the pitch of the lead screw.

even small caps (*Typog.*). Small capitals set up without capitals. Sometimes called LEVEL SMALL CAPS.

eventration (*Med.*). Protrusion of the abdominal contents outside the abdomen, e.g., through the diaphragm into the thorax.

even working (*Print.*). When the pages of a book occupy a number of complete *sections* (usually of 16 or 32 pages). If an oddment of 4 or 8 pages is required it is called an **uneven working.**

ever-bearer, ever-bloomer (*Bot.*). A plant which vegetates and flowers for a long period of the growing season, and often bears flowers and fruits at the same time. The flowering of ever-bloomers does not seem to be significantly affected by the duration of the periods of light to which they are exposed, during the ordinary alternation of day and night.

Everest theodolite (*Surv.*). A form of theodolite differing from the transit in that reversal of the line of sight is effected by removing the telescope from its trunnion supports and turning it.

Everite (*Build.*). TN for materials composed principally of asbestos and cement for building products such as corrugated sheets, tubular joists and purlins, rainwater goods.

ever-ready case (*Photog.*). A case from which the camera can operate without being removed. Usually the camera is screwed into the case, whose top and front can hinge down in one piece to expose lens and controls. Abbrev. **E.R.C.**

eversporting race (*Bot.*). A race of plants which does not breed true, and gives mixed progenies.

everted (*Bot.*). Turned outwards abruptly.

evident plasmolysis (*Bot.*). The stage in plasmolysis when the protoplast can be seen to have shrunk away from the cell wall.

Evipan (*Chem.*). See **hexobarbitone sodium.**

evisceration (*Surg.*). Disembowelment; operative removal of thoracic and abdominal contents from the foetus in obstructed labour; operative removal of a structure (e.g., the eye) from its cavity.

evolute (*Biol.*). Having the margins rolled outwards. (*Maths.*) Of a curve: locus of centre of curvature or envelope of normals. The original curve is the *involute* of the evolute.

evolution (*Biol., Gen.*). Changes in the genetic composition of a population during successive generations. The gradual development of more complex organisms from simpler ones. (*Maths.*) Raising a number to a power $1/n$, where n is a positive integer, i.e., finding a root. Cf. *involution* (1).

evolutionary operation (*Min. Proc., etc.*). Introduction of sectional controlled variants into a commercial process, during transfer of laboratory research into better production.

evulsion (*Surg.*). Plucking out by force.

E-wave (*Telecomm.*). See **TM-wave.**

Ewing curve tracer (*Elec. Eng.*). An instrument for throwing a curve representing the hysteresis loop of a sample of iron on to a screen. A mirror is deflected horizontally in proportion to the magnetizing force and vertically in proportion to the flux produced.

Ewing permeability bridge (*Elec. Eng.*). A measuring device in which the flux produced in a sample of iron is balanced against that produced in a standard bar of the same dimensions. The magnetizing force on the bar under test is varied until balance is obtained, and from the value of the force so found the permeability can be estimated.

E.W.T. (*Build., Civ. Eng.*). Abbrev. for *elsewhere taken.*

ex-, e-. Prefixes from L. *ex, e*, out of.

exacerbation (*Med.*). An increase in the severity of a disease, or of its manifestations.

exact equation (*Maths.*). The differential equation $P(x,y)\,dx+Q(x,y)\,dy=0$, where $\frac{\partial P}{\partial y}=\frac{\partial Q}{\partial x}$. The primitive is $u(x, y)=c$, c being a constant, where

$$\frac{\partial u}{\partial x}=P \text{ and } \frac{\partial u}{\partial y}=Q\,.$$

exalbuminous (*Bot.*). Lacking endosperm; exendospermous.

exaltation (*Chem.*). The abnormal increase in the molecular refractivity of a compound produced by the presence of conjugated double bonds ($>C{=}CH{-}CH{=}C<$).

exalted carrier (*Radio*). Addition of a synchronized carrier before demodulation, to improve linearity and to mitigate the effects of fading during transmission.

examiner (*Mining*). See **fireman** (2).

ex. & ct. (*Build., Civ. Eng.*). Abbrev. for *excavate and cart away.*

exanthema, exanthem (*Med.*). An eruption on the surface of the body. *pl.* **exanthemata.** Used specifically for infectious diseases characterized by an exanthem.

exarate (*Zool.*). Said of pupae in which the wings and legs are free, and which are therefore capable of a limited degree of movement.

exarch (*Bot.*). Said of a xylem strand having the protoxylem on the edge remote from the centre of the axis.

exasperate (*Bot.*). Having the surface rough, with hard short points projecting from it.

excalated (*Zool.*). In metameric animals, said of somites which are represented in the embryo, but which atrophy and disappear, leaving no trace in the adult.

excavation (*Civ. Eng.*). The operation of digging material out from the solid mass and depositing it elsewhere. (*Med.*) The process of hollowing out; a part hollowed out.

excavator (*Civ. Eng.*). A power-driven machine for digging away or excavating earth.

exceeding (*Bot.*). Projecting beyond a neighbouring member.

excentric (*Bot.*). (1) See **eccentric.** (2) Said of a pileus in which the stipe is not inserted in the centre.

excess air (*Eng.*). The proportion of air that has to be supplied in excess of that theoretically required for complete combustion of a fuel, because of the imperfect conditions under which combustion takes place in practice.

excess code (*Comp.*). One which increases decimal digits before conversion to binary digits.

excess conduction (*Electronics*). That arising from excess electrons provided by donor impurities.

excess electron (*Electronics*). One added to a semiconductor by, e.g., a donor impurity.

excess feed (*Print.*). When the peripheral speed of a web-driving component is greater than the next ahead too much paper is drawn through; also called *making paper.* Cf. *insufficient feed.*

excess pressure (*Acous.*). Same as **sound pressure.**

excess-3 code (*Comp.*). Binary coding of a decimal number, to which has been added 3. Complements are formed by 1 changed to 0, and 0 changed to 1, in all numerics.

excess voltage (*Elec. Eng.*). See under **over-voltage protective device.**

exchange (*Nuc.*). (1) The interchange of one particle between two others (e.g., a pion between two nucleons), leading to establishment of exchange forces. (2) Possible interchange of state between two indistinguishable particles, which involves no change in the wave function of the system.

exchange force (*Nuc.*). Force arising from exchange of one particle between two others. See **exchange** (1). Both covalent molecular bond (*electron exchange*) and nuclear force between nucleons (*pion exchange*) are examples.

excipient (*Med.*). The inert ingredient in a medicine which takes up and holds together the other ingredients.

exciple, excipulum (*Bot.*). The outer layer of the wall of an apothecium or of a perithecium, especially when it is well developed and distinct.

excipuliform (*Bot.*). Cup-shaped.

excision (*Surg.*). The action of cutting a part out or off; the surgical removal of a part.

excitable tissue (*Zool.*). Tissue which responds to stimulation by activity.

excitant (*Elec. Eng.*). A term occasionally used to denote the electrolyte in a primary cell.

excitation (*Elec. Eng.*). (1) Current in a coil which gives rise to a m.m.f. in a magnetic circuit, especially in a generator or motor. (2) The m.m.f. itself. (3) The magnetizing current of a transformer. (*Nuc.*) Addition of energy to a system, such as an atom or nucleus, raising it above the *ground state.* (*Radio*) Signal which drives any amplifier stage in a transmitter or receiver. (*Zool.*) The setting of a metabolic process into activity or acceleration; cf. *inhibition. adj.* **excitatory.**

excitation anode (*Electronics*). An auxiliary anode used to maintain the cathode spot of a mercury-pool cathode valve.

excitation loss (*Elec. Eng.*). The ohmic loss (I^2R) in the field or exciting windings of an electric machine excited by direct current.

excitation potential (*Electronics*). Potential required to raise orbital electron in atom from one energy level to another. Also called **resonance potential.** See also **ionization potential.**

excitatory cells (*Zool.*). Motor nerve-cells of the autonomic nervous system.

excited atom (*Phys.*). One with more energy than in the normal or ground state. The excess may be associated with the nucleus or an orbital electron.

excited ion (*Phys.*). Ion resulting from the loss of a valence electron, and the transition of another valence electron to a higher energy level.

excited nucleus (*Nuc.*). One with an excess of energy over its normal or ground state.

exciter (*Elec. Eng.*). A small machine for producing the current, usually d.c., necessary for supplying the exciting winding of a larger machine. It is frequently mounted on the shaft of the machine which it is exciting. See **a.c. exciter.** (*Radio*) An original source of high-frequency oscillations in an independent-drive transmitter, comprising the master oscillator and its immediately subsequent amplifying stages; also called **driver unit.**

exciter field rheostat (*Elec. Eng.*). A rheostat in the field of an exciter whereby the voltage of the exciter, and, therefore, the excitation on the main machine, can be controlled.

exciter lamp (*Cinema.*). That which provides the light to be modulated for recording sound photographically on a sound-track, or the light source for modulation by the sound-track in the sound-head of a projector.

exciter set (*Elec. Eng.*). An assembly of one or more exciters with a prime-mover of electric driving motor.

exciting circuit (*Elec. Eng.*). The complete circuit through which flows the current for exciting an electric machine. It comprises the exciter, the windings of the main machine, and possibly a field rheostat and measuring instruments.

exciting coil (*Elec. Eng.*). A coil on a field magnet, or any other electromagnet, which carries the current for producing the magnetic field.

exciting current (*Elec. Eng.*). That drawn by a transformer, magnetic amplifier, or other electric machine under no load conditions.

exciting winding (*Elec. Eng.*). The winding which produces the m.m.f. to set up the flux in an electric machine or other apparatus.

exciton (*Electronics*). Bound hole-electron pair in semiconductor. These have a definite half-life during which they migrate through the crystal. Their eventual recombination energy releases as a photon or, less often, several phonons.

excitor (*Zool.*). Stimulating into activity, as a motor nerve or neurone.

excitron (*Electronics*). Single-anode steel-tank mercury-arc rectifier, the arc being initiated by a jet of mercury from a pool.

exclusion principle. See **Pauli exclusion principle.**

exclusive species (*Bot.*). A species which is confined to a definite plant community.

exconjugant (*Zool.*). An animal which has regained its independence after conjugation.

excoriation (*Med.*). Superficial loss of skin.

excrescence (*Med.*). Any abnormal outgrowth of tissue.

excreta (*Zool.*). Poisonous or waste substances eliminated from a cell, tissue, or organism. *adjs.* **excrete, excreted.** *n.* **excretion.**

excurrent (*Bot.*). Said of a vein which runs out beyond the lamina of the leaf. (*Zool.*) Carrying an outgoing current; said of ducts, and, in certain *Porifera*, of canals leading from the apopyles of the flagellated chambers to the exterior or to the paragaster.

exeiresis, exeresis (*Surg.*). Operative evulsion of a part, especially of a nerve.

exempted spaces (*Ships*). Certain spaces in the ship which are not included in the *gross tonnage* (q.v.), e.g., double bottoms used for water ballast, wheelhouse, galley, W.C.s, etc.

exenteration (*Surg.*). Disembowelment; complete removal of the contents of a cavity.

exeresis (*Surg.*). See **exeiresis.**

Exeter hammer (*Tools*). See **London hammer.**

exflagellation (*Zool.*). The formation of microgametes, in some *Haemosporidia.*

exfoliation. The process of falling away in flakes, layers, or scales, as (*Bot.*) some bark. (*Geol.*) The splitting off of thin folia or sheets of rock from surfaces exposed to the atmosphere, particularly in regions of wide temperature variation. It is one of the processes involved in spheroidal weathering.

exhalant (*Zool.*). Emitting or carrying outwards a gas or fluid; as the *exhalant siphon* in some *Mollusca.*

exhaust (*Eng.*). (1) The working fluid discharged from an engine cylinder after expansion. (2) That period of the cycle occupied by the discharge of the used fluid.

exhaust cone (*Aero.*). In a turbojet or turboprop, the duct immediately behind the turbine and leading to the *jet pipe,* consisting of an inner conical unit behind the *turbine disk* and an outer unit of frustum form connecting the *turbine shroud* to the *jet pipe.*

exhaust-driven supercharger (*Aero.*). A piston-engine supercharger driven by a turbine motivated by the exhaust gases; also **turbo-supercharger.** See **turbo-compound aero-engine.**

exhaust fan (*Eng.*). A fan used in artificial draught systems; placed in the smoke uptake of a boiler to draw air through the furnace and exhaust the flue gases.

exhaust gas (*Eng.*). The gaseous exhaust products of an I.C. engine, containing in general CO_2, CO, O_2, N_2, and water-vapour.

exhaust-gas analyser (*I.C. Engs.*). An instrument which records continuously the mixture strength supplied to a petrol engine by automatic electrical measurement of the thermal conductivity of the exhaust gas. See also **CO_2 recorder, Orsat apparatus.**

exhausting agent (*Chem.*). Substance added to a dye bath to increase the adsorption of the dye by the textile.

exhaustive methylation (*Chem.*). The process of converting bases into their quaternary ammonium salts and subsequent distillation with alkalis, resulting in the formation of simpler unsaturated compounds which can be reduced to known saturated compounds. This method is used for testing the stability of ring compounds and is of particular value for investigating the constitution of alkaloids and other complicated ring systems.

exhaust lap (*Eng.*). Of a slide-valve, the distance moved by the valve from midposition on the port face, before uncovering the steam port to exhaust; sometimes called **inside lap.**

exhaust line (*Eng.*). The lower line of the enclosed area of an *indicator diagram* (q.v.), showing the back pressure on the piston during the exhaust stroke of an engine.

exhaust manifold (*I.C. Engs.*). Branched pipe which channels burnt gases from the combustion chambers to the exhaust system.

exhaust pipe (*Eng.*). The pipe through which the exhaust products of an engine are discharged.

exhaust port (*Eng.*). In an engine cylinder, the

port or opening through which a valve allows egress of the exhaust.

exhaust shaft (*Build.*). A ventilating passage used to convey vitiated air away from rooms.

exhaust stator blades (*Aero.*). An assembly of stator blades, usually in sections to allow for thermal expansion, mounted behind the turbine to remove residual swirl from the gases.

exhaust steam (*Eng.*). See **live steam.**

exhaust stroke (*Eng.*). In a reciprocating engine, the piston stroke during which the *exhaust* (q.v.) is ejected from the cylinder; sometimes called the **scavenging stroke.**

exhaust valve (*Eng.*). The valve controlling the discharge of the exhaust gas in an internal-combustion engine. Risk of overheating presents a design problem, particularly in aircraft engines, which is met by the use of heat-resisting steels, facing with Stellite or a similar hard deposit, or **sodium-cooling.**

exhaust velocity (*Space*). The velocity reached by the burnt gases as they leave the combustion chamber of a rocket.

exhibitionism (*Psychol.*). Generally, extravagant behaviour aimed at drawing attention to self. (*Psychiat.*) A sexual perversion, most frequently seen as a regressive symptom in men of advanced age, in which erotic gratification is obtained by indecent exposure.

exine (*Bot.*). The outermost layer of the wall of a pollen grain or of the spore of a moss.

exinguinal (*Zool.*). In *Arachnida*, the second joint of a walking leg; in land Vertebrates, outside the groin.

exites (*Zool.*). In some *Arthropoda*, lobes on the outer side of a limb; especially, in *Crustacea*, lobes on the outer side of a phyllopodium.

exit pallet (*Horol.*). See **pallet.**

exit portal (*Radiol.*). The area through which a beam of radiation leaves the body. The centre of the exit portal is sometimes called the **emergent ray point.**

exit pupil (*Light*). An imaginary aperture for a telescope or microscope, limiting the emergent beam of light where its cross-sectional area is least. It is usually the image of the objective formed by the eyepiece, and is at the position which should be occupied by the eye of the observer.

Exoascales (*Bot.*). An order of the *Ascomycetes* in which the asci are parallel to each other without a covering periderm. All the members are parasitic.

exobiology (*Biol.*). The study of systems in some ways resembling terrestrial life that may exist elsewhere.

exocardiac (*Zool.*). Outside the heart.

exocarp (*Bot.*). See **epicarp.**

exoccipital (*Zool.*). A paired lateral cartilage bone of the Vertebrate skull, forming the side-wall of the brain-case posteriorly.

exochite (*Bot.*). The firm outer layer of the macrosporangium of the *Fucales*.

exochomophyte (*Bot.*). A plant which forms a dense mat, and occurs on rocky ledges bearing much detritus.

exochorion (*Zool.*). The outer layer of the *chorion* (q.v.) in Insects.

exocoelar (*Zool.*). See **somatopleural.**

exocoele (*Zool.*). In *Zoantharia*, the portion of coelenteron between each pair of primary mesenteries. Cf. *entocoele*.

exocoelom (*Zool.*). The extraembryonic coelom of a developing Bird, Reptile, or Selachian.

exocone (*Zool.*). In Insects, said of compound eyes in which the crystalline cone is replaced by a conical ingrowth from the cornea.

exocrine (*Zool.*). Said of glands the secretion of which is poured into some cavity of the body, or on to the external surface of the body by ducts. Cf. *endocrine*.

exocuticle (*Zool.*). The layer of the insect cuticle, between the epicuticle and endocuticle, which becomes hardened and darkened in most insects.

exocytosis (*Cyt.*). The discharge from a cell of particles too large to diffuse through the wall.

exoderm (*Zool.*). In *Porifera*, the outer or dermal cell layer.

exodermis (*Bot.*). A more or less cuticularized layer ormed from the outer cells of the cortex of a root, and constituting a temporary protective sheath.

exodontia. See under **dentistry.**

exo-electron (*Electronics*). One, emitted from the surface of a metal or semiconductor, which comes from a metastable trap with very low binding energy under conditions such that electrons in their ground state could not be emitted.

exoenzyme (*Biochem.*). Any extracellular enzyme.

exoergic process (*Nuc.*). Nuclear process in which energy is liberated. *Exothermic* is the equivalent thermodynamic term.

exogamete (*Zool.*). A gamete which unites with one from another parent.

exogamy (*Bot., Zool.*). Union between gametes which are not closely related; the mating or conjugation of organisms having different parental stock; outbreeding. Cf. *endogamy*.

exogastric (*Zool.*). Said of spirally coiled shells in which the coil is directed on to the anterior face of the animal. Cf. *endogastric*.

exogenous (*Bot.*). (1) Produced on the outside of another plant member. (2) Developed from superficial tissues. (3) Increasing in thickness by the addition of new layers on the outside. (*Zool.*) In *Sporozoa*, said of forms in which sporulation is effected after the cyst has left the host; in higher animals, said of metabolism which leads to the production of energy for activity. Cf. *endogenous*.

exogenous circadian rhythm (*An. Behav.*). See **circadian rhythm.**

exogenous spore (*Bot.*). A spore formed on the end of a hypha, not inside a sporangium.

exogenous thallus (*Bot.*). The thallus of a lichen in which the fungus predominates.

exogynous (*Bot.*). Having the style projecting beyond the corolla.

exomphalos (*Med.*). A hernia formed by the protrusion of abdominal contents into the umbilicus.

exopeptidases (*Chem.*). Enzymes which catalyse the hydrolysis of peptide bonds adjacent to free amino or carboxyl groups.

exoperidium (*Bot.*). Outer layer of the peridium.

exophoria (*Med.*). Latent external squint revealed in an apparently normal person by passing a screen before the eye.

exophthalmic goitre (*Med.*). See **Basedow's disease.**

exoplasm (*Cyt.*). See **ectoplasm.**

exopodite (*Zool.*). The outer ramus of a crustacean stenopodium.

Exopterygota (*Zool.*). A section of Insects in which wings occur, although sometimes secondarily lost; the change from young form to adult is gradual, the wings developing externally; the young form is usually a generalized nymph.

EX.OR (*Comp., Maths.*). See **logical operations.**

exoscopic embryology (*Bot.*). The condition when the apex of the embryo is turned towards the neck of the archegonium.

exoskeleton (*Zool.*). Hard supporting or protective structures that are external to and secreted by the ectoderm, e.g., in Vertebrates, scales, scutes, nails, and feathers, in Invertebrates, the carapace, sclerites, etc.

exosmosis (*Chem.*). The process of *osmosis* (q.v.) in an outward direction.

exosphere (*Astron.*). The outermost layers of the earth's atmosphere. Studies of *airglow* (q.v.) suggest a temperature of 1500°C in this region, while artificial satellites have shown the density to be greater and the atmosphere more extensive than was formerly supposed.

exospore (*Bot.*). (1) The outer layer of the wall of a spore. (2) A sheath of epiplasm which forms round a young ascospore and plays a part in the formation of the spore wall. (3) A wall formed around some oospores from periplasm.

Exosporae (*Bot.*). A subclass of the *Myxothallophyta* having the spores borne singly on short stalks over surface of sporophore.

exosporous (*Bot.*). Having exogenous spores.

exostosis (*Med.*). A bony tumour growing outwards from a bone.

exothecal (*Zool.*). See **extrathecal.**

exothecium (*Bot.*). The outer layer in the wall of a moss capsule, and of the microsporangium in Gymnosperms.

exothermic (*Chem.*). Accompanied by the evolution of heat.

exotic (*Zool.*). Ecdemic.

exotoxin (*Bacteriol.*). A toxin released by a bacterium into the medium in which it grows. Frequently very toxic, e.g., neurotoxins which destroy cells of the nervous system, haemolytic toxins which lyse red blood cells.

exotype (*Zool.*). A category of individuals dependent on the recognition that a given form is nonheritable.

expanded. Of cellular structure and therefore light in weight. Thus expanded concrete is *lightweight concrete* (q.v.). (*Bot.*) Flattening out and becoming less concave as development proceeds.

expanded metal (*Build., Civ. Eng.*). A metal network formed by suitably stamping or cutting sheet metal and stretching it to form open meshes. It is used as a reinforcing medium in concrete construction, as lathing for plasterwork, and for various other purposes. Cf. *B.R.C. fabric*.

expanded plastics (*Plastics*). Foamed plastic materials, e.g. PVC, polystyrene, polyurethane, polythene, etc., created by the introduction of pockets or cells of inert gas (air, carbon dioxide, nitrogen, etc.) at some stage of manufacture. Used for heat insulation purposes or as core materials in 'sandwich' construction, because of their low densities; also, because of lightness, for packaging, and (as foam rubber) for upholstery cushioning, artificial sponges, etc.

expanded sweep (*Electronics*). A technique for speeding-up the motion of the electron beam in a CRO during a part of the sweep.

expander (*Acous.*). Amplifying apparatus for automatically increasing the contrast in speech modulation, particularly after reception of speech which has had its contrast compressed by a *compressor*. (*Elec. Eng.*) An inert material, such as carbon or barium sulphate, added to the active material in accumulator plates to prevent shrinkage of the mixture.

expanding arbor (*Eng.*). A lathe arbor expandable by blades or keys sliding in taper grooves, which allows work of various bore diameter to be supported and located.

expanding bit (*Carp., etc.*). A boring-bit carrying a cutter on a radial arm, the position of the cutter being adjustable so that holes of different sizes may be cut.

expanding brake (*Eng.*). A brake in which internal shoes are expanded to press against the drum, usually by a cam or toggle mechanism.

expanding cement (*Build., Civ. Eng.*). Cement containing a chemical agent which induces predetermined expansion to minimize the normal shrinkage which occurs in concrete during the setting and drying out process.

expanding mandrel (*Eng.*). See under **mandrel** (1).

expanding metals (*Met.*). Metals or alloys, e.g., 2 parts antimony to 1 part bismuth, which expand in final stage of cooling from liquid; used in type-founding.

expanding plug (*San. Eng.*). A *bag plug* (q.v.).

expanding reamer (*Eng.*). A *reamer* (q.v.) partially slit longitudinally, and capable of slight adjustment in diameter by a coned internal plug.

expanding universe (*Astron.*). The view, based on the evidence of the red-shift, that the whole universe is expanding; supported by relativity theory, in which a static universe would be unstable.

expansion (*Electronics*). A technique by which the effective amplification applied to a signal depends on the magnitude of the signal being larger for the bigger signals. See also **automatic volume expansion.** (*Eng.*) (1) Increase in volume of working fluid in an engine cylinder. (2) Piston stroke during which such expansion occurs. (*Heat*) See **adiabatic-, coefficient of-.**

expansion circuit-breaker (*Elec. Eng.*). A circuit-breaker in which arc extinction occurs as a result of the rapid cooling produced by the expansion of steam or of gases; these result from the arc which arises between the contacts in water or in a small quantity of oil.

expansion curve (or line) (*Eng.*). The line on an *indicator diagram* (q.v.) which shows the pressure of the working fluid during the expansion stroke.

expansion engine (*Eng.*). An engine which utilizes the working fluid expansively.

expansion gear (*Eng.*). That part of a steam-engine valve gear through which the degree of expansion can be varied.

expansion joint (*Civ. Eng., Rail.*). A joint arranged between two parts to allow these parts to expand with temperature rise, without distorting laterally, e.g., the gap left between successive lengths of rail, or the joint made between successive sections of carriageway in road construction. (*Eng.*) A special pipe joint used in long pipelines to allow for expansion, e.g., a horseshoe bend, a corrugated pipe acting as a bellows, a sliding socket joint with a stuffing box.

expansion line (*Eng.*). See **expansion curve.**

expansion of gases (*Phys.*). All gases have very nearly the same coefficient of expansion, namely 0·003 66 per kelvin when kept at constant pressure. See **absolute temperature, gas laws.**

expansion pipe (*Build.*). In a domestic system of heating, a pipe carried up from the hot-water tank to a point above the level of the cold-water tank, where its open end is bent over, so that, if the water boils, it may discharge any water or steam forced out.

expansion ratio (*Aero.*). The ratio between the gas pressure in a rocket combustion chamber, or a jet pipe, and that at the outlet of the propelling nozzle.

expansion rollers (*Eng.*). Rollers on which one end of a large girder or bridge is often carried, to allow of movement resulting from expansion; the other end of the girder, etc., is fixed.

expansion tank (*Build.*). In a hot-water system, the tank connected to, and above, the hot-water cylinder to allow of expansion of the water on heating; often the cold-water feed tank is so used. See **expansion pipe.**

expansion valve (*Eng.*). An auxiliary valve working on the back of the main slide-valve of some steam-engines to provide an independent control of the point of cut-off.

expansive working (*Eng.*). The use of a working fluid expansively in an engine; an essential feature of every efficient working cycle.

expectancy, principle of (*An. Behav.*). Principle assuming that a reinforcement must be such as to confirm an expectancy on the part of an animal, and implying that animal motivation resulting in trial and error learning must involve some reference to an apprehension of time.

expedor phase advancer (*Elec. Eng.*). A phase advancer which injects into the secondary circuit of an induction motor an e.m.f. which is a function of the secondary current. Cf. *susceptor phase advancer.*

experimental embryology (*Zool.*). The experimental study of the physiology and mechanics of development.

experimental extinction (*An. Behav.*). The decrease in the intensity of a conditioned response when it is no longer followed by any reinforcement.

experimental mean pitch (*Aero.*). The distance of travel of an airscrew along its own axis, while making one complete revolution, assuming the condition of its giving no thrust.

expiration (*Zool., etc.*). The expulsion of air or water from the respiratory organs.

expiratory reserve volume (*Physiol.*). See **supplemental air.**

explanate (*Bot.*). Spread out on a surface.

explantation (*Zool.*). In experimental zoology, the culture, in an artificial medium, of a part or organ removed from a living individual, tissue-culture. Cf. *interplantation. n.* **explant.**

explement (*Maths.*). See **conjugate angles.**

expletive (*Build.*). A stone used as a filling for a cavity.

explicit function (*Maths.*). A variable x is an explicit function of y when x is directly expressed in terms of y. Cf. *implicit function.*

exploder (*Mining*). An appliance for firing electrically the explosives used in mining and shaft sinking.

exploding star (*Astron.*). See **nova.**

exploding wire (*Phys.*). The explosive volatilization of a fine wire upon sudden application of a very heavy current. Used in fast fuses and as a source of shock waves.

exploitable girth (*For.*). (1) The minimum girth at breast height at or above which trees are considered suitable for felling or for the purpose specified. (2) The girth down to which all portions of a bole or tree must be exploited as timber or fuel under a permit licence.

exploration (*An. Behav.*). The behaviour of animals when investigating their surroundings. This behaviour appears to be tension-reducing or intrinsically rewarding, and often involves *latent learning* (q.v.). See also **exploratory behaviour.**

exploratory behaviour (*An. Behav.*). Any behaviour which leads to an increase in the rate of change in the stimulation falling on its receptors and is not impelled by any homeostatic or reproductive need. See also **exploration.**

Explorer (*Space*). A series of American artificial satellites used to study the physics of space cosmic rays, temperatures, meteorites, etc.; responsible for the discovery of the *Van Allen radiation belts* (q.v.).

exploring brush (*Elec. Eng.*). A small brush fitted to a d.c. machine for experimental purposes; it can be moved round the commutator in order to investigate the distribution of potential around it.

exploring coil (*Elec. Eng.*). One used to measure magnetic field strengths by finding the change of magnetic linkages on removing the coil from a position in the field to a remote point out of the influence of the field. Also called **search coil.**

explosion (*Chem.*). A rapid increase of pressure in a confined space. Explosions are generally caused by the occurrence of exothermic chemical reactions in which gases are produced in relatively large amounts. For nuclear explosion see **atomic bomb** and **hydrogen bomb.**

explosion engine (*Eng.*). An obsolete term for *internal-combustion engine.*

explosion pot (*Elec. Eng.*). A strong metal container surrounding the contacts of an oil circuit-breaker; the high pressure set up inside the pot when an arc occurs assists in the extinction of the arc.

explosion-proof or flame-proof (*Elec. Eng.*). Said of electrical apparatus so designed that an explosion of inflammable gas inside the enclosure will not ignite inflammable gas outside. Such apparatus is used in mines or other places having an explosive atmosphere.

explosive. A generic term which embraces all materials that can be either exploded or detonated. (*Mining*) There are two main classes—'permitted' and 'non-permitted', i.e., those which are safe for use in coal-mines and those which are not. Ammonium nitrate mixtures are mostly used in coal-mines; nitroglycerine derivatives in metal-mines. ANFO (*a*mmonium *n*itrate and *f*uel *o*il) is now widely used in hard rock mining. Explosives are used as propellants (*low explosives*) and for blasting (*high explosives*), in both civil and military applications.

explosive forming (*Eng.*). One of a range of high-energy rate-forming processes by which parts are formed at a rapid rate by extremely high pressures. Low and high explosives are used in variations of the explosive forming process; with the former, known as the cartridge system, the expanding gas is confined; with the latter, the gas need not be confined and pressure of up to one million atmospheres may be attained. See **electrohydraulic forming.**

explosive rivet (*Eng.*). A type of *blind rivet* (q.v.) which is clinched or set by exploding, electrically, a small charge placed in the hollow end of the shank.

exponent (*Maths.*). See **index.**

exponential baffle (*Acous.*). A baffle approximating to a short section of an exponential horn.

exponential function (*Maths.*). The function e^x. See e and **exponential series.**

exponential growth (*Biol.*). A stage of growth occurring in populations of unicellular micro-organisms when the logarithm of the cell number increases linearly with time.

exponential horn (*Acous.*). One for which the taper or flare follows an exponential law. Ideal acoustically but not fully realizable in practice. See **logarithmic horn.**

exponential reactor (*Nuc.*). One with insufficient fuel to make it diverge, but excited by an external source of neutrons for the determination of its properties.

exponential series (*Maths.*). The series

$$1 + \frac{x}{1!} + \frac{x^2}{2!} + \frac{x^3}{3!} + \ldots + \frac{x^n}{n!} + \ldots$$

which converges to the value e^x for all values of x.

exposed pallets (*Horol.*). Usually a clock movement in which the *pallets* (q.v.) are in front of the dial.

exposure (*Meteor.*). The method by which an instrument is exposed to the elements. The exposure in meteorological stations is standardized so that records from different stations may be comparable. (*Photog.*) (1) The total amount of light allowed to fall upon a sensitive emulsion, i.e., intensity × speed, for a given photographic effect; measured in candela-metre-seconds or lux-seconds. (2) The time during which a camera shutter remains open or an enlarger image is illuminated to give the desired result.

exposure dose (*Radiol.*). See **dose.**

exposure meter (*Photog.*). Instrument, a combination of photocell and current meter, for measuring the light reflected from a subject or falling on a subject (incident light) and expressing it in terms of exposure for a given stop or as an *exposure value* (q.v.). Now often incorporated in the camera.

exposure value (*Photog.*). An index combining both aperture and shutter-speed, expressed as a single figure which remains constant when both are altered, for example to change depth of focus, in such a way that the amount of light reaching the sensitized emulsion is unchanged. Some shutters have stop and speed rings coupled to give a constant E.V. when either is altered. Abbrev. E.V.

expressivity (*Gen.*). The extent to which a gene produces an effect.

expulsion fuse (*Elec. Eng.*). An enclosed fuse-link in which the arc occurring when the link melts is extinguished by the lengthening of the break due to expulsion of part of the fusible material through a vent in the container.

expulsion gap (*Elec. Eng.*). A special form of expulsion fuse connected in series with a gap and placed across insulator strings on an overhead transmission line; a voltage surge breaks down the gap and the resulting arc is quickly broken by the fuse, so that no interruption to the supply need take place.

exscutellate (*Zool.*). Without a scutellum.

exserted. Stretched out; protruded. (*Bot.*) Said of stamens which project beyond the corolla of a flower.

exsiccation (*Geog.*). The draining away of water of precipitation as a result of human activity, e.g., the cutting down (in Africa and elsewhere) of forests, the debris of which retain moisture, the draining of swampy ground, such as the Fens, and the continued ploughing of lands unprotected from winds, which consequently blow the topsoil away and turn the area into a desert.

exstrophy, extrophy (*Med.*). A turning inside out of a hollow organ (especially the bladder).

exsule (*Zool.*). An apterous form in the life-cycle of some *Hemiptera*, produced parthenogenetically and generally reproducing by the same method.

ex. sur. tr. & ct. (*Build., Civ. Eng.*). Abbrev. for *excavate surface trenches and cart away.*

extended delivery (*Print.*). The delivery of a machine lengthened by design to give each printed sheet the maximum time to set before being covered by the next.

extended tube (*Electronics*). See **remote cut-off tube.**

extender (*Paint., Plastics*). (1) A substance added to paint as an adulterant or to give it body, e.g., barytes, china clay, French chalk, gypsum, whiting. (2) In synthetic resin adhesives, a substance (e.g., rye flour) added to reduce the cost of gluing or to adjust viscosity. (3) A noncompatible plasticizer used as an additive to increase the effectiveness of the compatible plasticizer in the manufacture of elastomers.

extenders (*Typog.*). The parts of lower-case type characters which extend above or below the *x-height* (q.v.), i.e., the *ascending* and *descending* parts in, e.g., *b, f, p.*

extension (*Acous.*). The use of one long rank of pipes in an organ for a number of stops of the same class but of different pitches. It is a cheaper system, but volume is lost on thick chords because one pipe can be blown once only. (*Teleph.*) See **subscriber's main station.**

extension coefficient (*Elec. Eng.*). A coefficient applied to the length of the air gap of an electric machine in order to allow for the effect of the teeth when calculating the m.m.f. required for the magnetic circuit. It is greater than unity and therefore increases the effective gap length. Cf. *contraction coefficient.*

extension flap (*Aero.*). A landing flap which moves rearward as it is lowered so as to increase the wing area. See **Fowler flap.**

extension ore (*Mining*). In assessment of reserves, ore which has not been measured and sampled, but is inferred as existing from geological reasoning supported by proved facts regarding adjacent ore.

extension spring (*Eng.*). A helical spring with looped ends which stores energy when in the stretched condition.

extension tubes (*Photog.*). Tubes which increase the distance between the lens and the focal plane of a camera for close-up work. Made in sets to give different degrees of enlargement.

extensometer (*Instr.*). Generally, an instrument for measuring dimensional changes of a material. Specifically, this can be done electrically by measuring the changes of capacitance in a capacitor which is directly influenced by the changes in dimensions of the specimen.

extensor (*Zool.*). A muscle which by its contraction straightens a limb, or a part of the body. Cf. *flexor.*

exterior angle (*Maths.*). The angle between any side produced and the adjacent side (not produced) of a polygon.

external angle (*Build.*). A vertical or horizontal angle forming part of the projecting portion of a wall or feature of a building; also called **salient junction.** See also **arris.**

external characteristic (*Elec. Eng.*). A curve showing the relation between the terminal voltage of an electric generator and the current delivered by it.

external circuit (*Elec. Eng.*). The circuit to which current is supplied from a generator, battery, or other source of electrical energy.

external compensation (*Chem.*). Neutralization of optical activity by the mixture or loose molecular combination of equal quantities of two enantiomorphous molecules.

external conductor (*Elec. Eng.*). The outer earthed conductor of an earthed concentric wiring system.

external digestion (*Zool.*). A method of feeding, adopted by some *Coelenterata, Turbellaria,*

Oligochaeta, Insects, and *Araneida*, in which digestive juices are poured on to food outside the body and imbibed when they have dissolved some or all of the food.

external feedback (*Telecomm.*). Feedback in magnetic amplifiers in which the feedback voltage is derived from the rectified output current.

external firing (*Eng.*). The practice of heating a boiler or pan by a furnace outside the shell; all modern boilers have internal furnaces and flues.

external indicator (*Chem.*). An indicator to which are added drops of the solution in which the main reaction is taking place, away from the bulk of the solution.

external respiration (*Zool.*). See **respiration, external.**

external screw-thread (*Eng.*). A screw-thread cut on the outside of a cylindrical bar. Also called **a male thread.**

external secretion (*Zool.*). A secretion which is discharged to the exterior, or to some cavity of the body communicating with the exterior. Cf. *internal secretion.*

external store (*Comp.*). Supplementary computer store, outside the main unit.

exteroceptor (*Zool.*). A sensory nerve-ending, receiving impressions from outside the body. Cf. *interoceptor.*

extinction (*An. Behav.*). See **experimental-, resistance to-.** (*Psychol.*) The inhibition of a *conditioned reflex* by repeated presentation of the conditioned stimulus without the unconditioned stimulus.

extinction coefficient (*Chem.*). A measure of the absorption of light by a dissolved substance. It is given by the formula $\varepsilon = \frac{1}{cd} \cdot \log \frac{I_o}{I}$, where I_o and I are the intensities of the incident and transmitted light respectively, in a solution d cm thick of molar concentration c.

extinction meter (*Photog.*). See **visual exposure meter.**

extinction voltage (or potential) (*Electronics*). (1) Lowest anode potential which sustains a discharge in a gas at low pressure. (2) In a gas-filled tube, the p.d. across the tube which will extinguish the arc.

extine (*Bot.*). Outer wall-layers of pollen grain.

extra (*Teleg.*). An unwanted marking condition causing incorrect recording in a telegraph receiver.

extra-axillary (*Bot.*). Said of a bud which is formed elsewhere than in an axil.

extra bound (*Bind.*). A book completely bound by hand, expensive, using carefully chosen materials, sewn headbands, special end-papers, gold tooling, etc.

extrabranchials (*Zool.*). A series of slender cartilages, probably modified branchial rays, which support the branchial apertures in *Chondrichthyes.*

extracellular digestion (*Bot.*). The digestion of material by enzymes secreted from the cell and acting outside it.

extracolumella (*Zool.*). The distal element of the auditory ossicles of all land Vertebrates except *Gymnophiona*, *Urodela*, and Mammals.

extract (*Textiles*). Fibrous material made from mixed cotton and wool rags, after the cotton has been extracted by a chemical process (*carbonizing*).

extraction (*Chem.*). A process for dissolving certain constituents of a mixture by means of a liquid with solvent properties for one of the components only. Substances can be extracted from solids, e.g., grease from fabrics with petrol; or from liquids, e.g., extraction of an aqueous solution with exthoxyethane, the efficiency depending on the partition coefficient of the particular substance between the two solvents. (*Comp.*) Isolation of required information *bit* from *word* in computer register, according to machine instruction.

extraction fan (*Eng.*). A fan used to extract foul air, fumes, suspended paint particles, etc., from a working area. See **fan, induced draught.**

extraction instruction (*Comp.*). In a digital computer, an instruction to form a new *word* by a suitable arrangement of selected segments of given *words.*

extraction metallurgy (*Met.*). First stage or stages of ore treatment, in which gangue minerals are discarded and valuable ones separated and prepared for working up into finished metals, rare earths or other saleable products. Characteristically, the methods used do not change the physical structure of these products save by comminution.

extraction thimble (*Chem.*). A porous cylindrical cup containing the solid material to be extracted, usually by placing it under the hot reflux in a still.

extraction turbine (*Eng.*). A steam-turbine from which steam for process work is tapped at a suitable stage in the expansion, the remainder expanding down to condenser pressure.

extractive distillation (*Chem. Eng.*). A technique for improving, or achieving in cases impossible without it, distillation separation processes by the introduction of an additional substance which changes the system equilibrium. It is essentially different from *azeotropic distillation* (q.v.) in that the added substance is not distilled itself but is added as a liquid at some point, usually at or near top in the distillation column, and leaves at the base as liquid.

extractor (*Autos.*). Device fitted to the end of the exhaust pipe to reduce back-pressure. Also **booster, ejector.**

extract ventilator (*Build.*). A cowl-like appliance fitted to the top of a ventilating shaft in a building to induce in it an up-draught.

extradop (*Radar*). Doppler system for tracking missiles.

extrados (*Build., Civ. Eng.*). The back or top surface of an arch. See **intrados.**

extradural (*Anat.*). Situated outside the dura mater.

extra-embryonic (*Zool.*). In embryos developed from eggs containing a great deal of yolk, as those of Birds, pertaining to that part of the germinal area beyond the limits of the embryo.

extra-floral (or extra-nuptial) nectary (*Bot.*). A nectary occurring on or in some part of a plant other than a flower.

extra-galactic nebula (*Astron.*). A *nebula* (q.v.) external to the Galaxy.

extra-heavy (*Plumb.*). Said of pipe which is of greater wall-thickness than the standard pipe.

extra-high voltages (*Elec. Eng.*). A term used in official regulations for voltages above 3·3 kV; but more commonly employed to denote voltages of the order of 100 kV or more.

extra-lateral rights (*Mining*). See **apex law.**

extra material (*Weaving*). Extra warp or weft yarns used in a cloth structure to produce some prominent design effect.

extramatrical (*Bot.*). Said of a fungus which has the greater part of its thallus, and especially the reproductive organs, outside the host cell, or on the surface of the substratum.

extraneous ash (*Min. Proc.*). In raw coal, the so-called 'free dirt' or associated shale and enclosing beds.

extra-nuptial nectary (*Bot.*). See **extra-floral nectary.**

extraordinary ray (*Light.*). See **double refraction.**

extrapolation (*Maths.*). The estimation of the value of a function at a particular point from values of the function on one side only of the point. Cf. *interpolation.*

extrapolation distance (*Nuc.*). See **augmentation distance.**

extras (*Build., Civ. Eng.*). All work the inclusion of which is not expressed or implied in the original contract price. Also called **variations.**

extrasystole (*Med.*). A premature contraction of the heart interrupting the normal rhythm, the origin of the impulse to contraction being abnormally situated.

extrathecal (*Zool.*). In Corals, outside the theca.

extra thirds (*Paper*). A former size of cut card, $1\frac{7}{8} \times 3$ in.

extrauterine (*Anat.*). Situated or happening outside the uterus.

extravasation (*Med., Zool.*). The abnormal escape of fluids, as blood or lymph, from the vessels which contain them. *v.* **extravasate.**

extravascular (*Anat.*). Placed or happening outside a blood vessel.

extraversion (*Med.*). See **exstrophy.** (*Psychol.*) The turning of interest to objects outside the person.

extravert (*Psychol.*). A term given prominence by Jung, though not invented by him, referring to a personality type whose interests are mainly directed outwards to the external environment rather than inwards to the thoughts and feelings of the self. The extravert has been found to be more resistant to conditioning than the *introvert* (q.v.), and it is claimed that there may be a physical basis for the personality differences involving inhibitory and excitatory influences on incoming stimuli.

extreme breadth (*Ships*). The greatest breadth measured to the farthest out part of the structure on each side, including any rubbing strakes or other permanent attachments to the hull or *superstructure* (q.v.) but not to the *top hamper* (q.v.).

extreme dimensions (*Ships*). These dimensions are provided for general information, mainly for docking purposes. They are *length overall, extreme breadth* and *depth* or *summer draught* (qq.v.).

extremely high frequencies (*Radio*). Band between 3×10^4 and 3×10^5 MHz. Abbrev. **E.H.F.** See also **ultra-high frequencies.**

extreme pressure lubricant (*Eng.*). A solid lubricant, such as graphite, or a liquid lubricant with additives which form oxide or sulphide coatings on metal surfaces exposed where, under very heavy loading, the liquid film is interrupted, thus mitigating the effects of dry friction.

extrinsic (*Crystal.*). Said of electrical conduction properties arising from impurities in the crystal. (*Zool.*) Said of appendicular muscles of Vertebrates which run from the trunk to the girdle, or the base of the limb. Cf. *intrinsic.*

extrinsic semiconductor (*Electronics*). See **semiconductor.**

extrophy (*Med.*). See **exstrophy.**

extrorse (*Bot.*). (1) Said of the manner of dehiscence when an anther opens towards the periphery of the flower. (2) Turned so as to face away from the centre of the axis. (*Zool.*) Directed or bent outwards.

extrovate (*Zool.*). The substance which flows out of a punctured cell.

extrovert (*Zool.*). An extrusible proboscis, found in certain aquatic animals. See **lophophore.**

extrusion (*Met., etc.*). Operation of producing rods, tubes, and various solid and hollow sections, by forcing suitable material through a die by means of a ram. Applied to numerous nonferrous metals, alloys, and other substances, notably plastics (for which a screw-drive is frequently used). In addition to rods and tubes, extruded plastics include sheets, film, and wire-coating.

extrusive rocks (*Geol.*). Rocks formed by the consolidation of magma on the surface of the ground, as distinct from *intrusive rocks* which consolidate below ground. Commonly referred to as **lava flows**; normally of fine grain or even glassy.

exudate (*Med.*). The fluid which has escaped (*exudation*) from the blood vessels into the tissues or the cavities of the body as a result of inflammation; it contains protein and many cells, and clots outside the body.

exudation cone (*Zool.*). A cone of clear protoplasm which, after fertilization, protrudes from some types of ovum.

exudation pressure (*Bot.*). See **root pressure.**

exudation theory (*Zool.*). The theory that caste in *Isoptera* depends upon the exudations produced by the nymphs.

exudative diathesis of chicks (*Vet.*). Subcutaneous oedema in chicks associated with excessive capillary permeability due to vitamin E deficiency.

exumbrella (*Zool.*). The upper convex surface of a medusa. *adj.* **exumbrellar.**

exuviae (*Zool.*). The layers of the integument cast off in ecdysis.

exuvial (*Zool.*). Pertaining to, or facilitating, ecdysis.

eye (*Arch.*). (1) The circular opening in the top of a dome. (2) A circular or oval window. (*Eng.*) (1) A loop formed at the end of a steel wire or bolt. See **eye bolt.** (2) The central inlet passage of the impeller of a centrifugal compressor or pump. (*Glass*) The hole in the centre (or elsewhere) of the floor of a pot furnace up which the combustible gases rise as flame to heat the furnace. (*Join.*) The circular centre of a volute scroll. (*Meteor.*) The central calm area of a cyclone or hurricane, which advances as an integral part of the disturbed system. (*Mining*) The mouth or entrance to a pit shaft. (*San. Eng.*) A short branch off a drain-pipe, useful for inspection or clearing purposes. (*Tools*) Of an axe or other tool, the hole or socket in the head for receiving the handle. (*Zool.*) The sense-organ which receives visual impressions.

eye-and-object correction (*Surv.*). A correction applied in precise work to the average angle of elevation read on the vertical circle, in order to compensate for the vertical axis of the theodolite not being truly vertical. The correction is

$$+\frac{\Sigma o - \Sigma e}{4} \cdot \theta,$$

where Σo = object-end reading of the altitude level, Σe = eye-end reading of the altitude level, and θ = angular value of 1 division of altitude level.

eye bolt (*Eng.*). A bolt carrying an eye instead of the normal head; fitted to heavy machines and other parts for lifting purposes.

eyed gneiss (*Geol.*). See **augen-gneiss.**

eye-ground (*Anat.*). The fundus; that part of the cavity of the eyeball which can be seen through the pupil with an ophthalmoscope.

eyepiece (*Phys., etc.*). In an optical instrument,

the lens or lens system to which the observer applies his eye in using the instrument.

eyepiece graticule (*Micros.*). Grid incorporated in the eyepiece for measuring objects under the microscope. Special type used in particle-size analysis consists of a rectangular grid for selecting the particles and a series of graded circles for use in sizing the particles. Also called **micrometer eyepiece, ocular micrometer** (**U.S.**).

eye spot (*Bot., Zool.*). A small mass of light-sensitive pigment found in some lower animals and plants.

eye-stalk (*Zool.*). A paired stalk arising close to the median line on the dorsal surface of the head of many *Crustacea*, bearing an eye.

Eyring formula (*Acous.*). A formula proposed for the period of reverberation of an enclosure, taking into account the time required for waves to travel between successive reflections.

Eytelwein's formula (*Hyd.*). An expression giving the velocity of flow in a pipe as:

$$V = \sqrt{3567\ mi + 0{\cdot}001\ 578} - 0{\cdot}039\ 72,$$

where V = velocity in m/s, m = hydraulic mean depth in metres, and i = virtual slope.

F

f (*Chem.*). A symbol for: (1) activity coefficient, for molar concentration; (2) partition function.

F (*Chem.*). Symbol for *fluorine*. (*Elec. Eng.*) Symbol for *farad*. (*Gen.*) Abbrev. for *filial generation* in work on inheritance; usually distinguished by the addition of a number, thus: F_1, first filial generation; F_2, second filial generation, etc. See also **P.** (*Heat*) Symbol used, following a temperature (e.g. 41°F), to indicate the *Fahrenheit scale* (q.v.).

[F] (*Light*). A Fraunhofer line in the blue of the solar spectrum of wavelength 486·1527 nm. It is the second line in the Balmer hydrogen series, known also as Hβ.

F (*Build.*). Abbrev. for *face* or *flat*. (*Chem.*) Symbol for *free energy*. (*Elec.*) Symbol for *faraday*.

FAA (*Aero.*). Abbrev. for *F*ederal *A*viation *A*dministration, a US Government agency responsible for all aspects of US civil aviation. Cf. *CAA*.

fabella (*Zool.*). In Mammals, one of two small sesamoid bones situated opposite the distal end of the femur on its posterior aspect.

fabric (*Arch.*). Walls, floors, and roof of building.

Fabry and Pérot interferometer (*Light*). An instrument in which circular interference fringes are produced by the passage of monochromatic light through a pair of plane-parallel, half-silvered glass plates, of which one is fixed while the other may be moved by an accurately calibrated screw. By observing the fringes while changing the plate separation, the wavelength of the light may be measured.

façade (*Arch.*). The front elevation of a building.

face (*Bot.*). In general, the upper side of an organ when it has two well-marked sides. (*Build.*) (1) The front of a wall or building. (2) The exposed vertical surface of an arch. (*Comp.*) Readable side of a punched card, which can be passed from a pack in a hopper or magazine, and read face-up or face-down. (*Crystal.*) See **crystal-**. (*Eng.*) The working surface of any part; as the sole of a carpenter's plane, the striking surface of a hammer, the surface of a slide-valve, or the surface of the steam chest on which it slides, the seating surface of a valve, the flank of a gear-tooth, etc. (*Mining*) The exposed surface of coal or other mineral deposit in the working place where mining, winning, or getting is proceeding. (*Typog.*) (1) The actual printing surface of a type as distinct from its shank. (2) The style of any particular type design, there being many in use. See **type face.** (*Zool.*) In Mammals, the portion of the skull anterior to the junction of the presphenoid and the mesethmoid; the fleshy structures overlying this portion of the skull. *adj.* **facial.**

face-airing (*Mining*). The operation of directing a ventilating current along the face of a working place; also called **flushing.**

face-centred cubic (*Chem.*, *Crystal.*). A crystal structure whose unit cell structure has equivalent points occurring at corners as well as at centres of a six-faced shape, such as a cube. Cubical close-packed structures, e.g., copper, silver, have the face-centred cubic lattice.

face chuck or **face plate** (*Eng.*). A large disk which may be screwed to the mandrel of a lathe and is provided with slots and holes for securing work of a flat or irregular shape.

face-hammer (*Tools*). A hammer having a peen which is flat rather than pointed or edged.

face lathe (*Eng.*). A lathe designed for work of large diameter but short length (e.g., large wheels or disks).

face left and face right (*Surv.*). Expressions referring to the pointing of a theodolite telescope when the vertical circle is respectively *left* and *right* of the telescope, as seen from the eyepiece end.

facellite (*Min.*). See **kaliophilite.**

face mark (*Carp.*, *Join.*). A distinguishing mark made on one face of a piece of wood to show that it was used as the basis for truing the other surfaces.

face mix (*Build.*). A mixture of cement and stone dust used for facing concrete blocks in imitation of real stone.

face mould (*Build.*). A templet used as a reference for shaping the face of wood, stone, etc.

face-on (*Mining*). Working of a coal seam in a direction parallel to the natural cleats; cf. *end-on*.

face plate (*Eng.*). (1) See **face chuck.** (2) A *surface plate* (q.v.). (*Electronics*) That part of a cathode-ray tube which carries the phosphor screen, independently of the glass envelope, but mounted therein.

face-plate breaker controller (*Elec. Eng.*). A face-plate controller having a separate contactor for breaking the circuit.

face-plate breaker starter (*Elec. Eng.*). A face-plate starter having a separate interlocked contactor for breaking the circuit.

face-plate controller (*Elec. Eng.*). See **face-plate starter.**

face-plate coupling (*Eng.*). See **flange coupling.**

face-plate starter (*Elec. Eng.*). An electric motor starter in which a contact lever moves over a number of contacts arranged upon a plane surface. Also called a **face-plate controller.**

face right (*Surv.*). See **face left.**

face shovel (*Civ. Eng.*). A mechanical mobile device for cutting into the vertical face of an excavation and depositing soil into vehicles for transporting it elsewhere.

face side (*Carp.*, *Join.*). The side of a piece of wood bearing the face mark.

facet (*Arch.*). A *facette* (q.v.). (*Crystal.*) The flat side of a crystal. (*Zool.*) One of the corneal elements of a compound eye; a small articulatory surface.

facette (*Arch.*). A projecting flat surface between adjacent flutes in a column. Also called a **listel.**

face-wall (*Build.*). The front wall.

facia (*Arch.*). A flat banded projection from the face of a member.

facial (*Zool.*). Pertaining to or situated on the face; the seventh cranial nerve of Vertebrates, supplying the facial muscles and tongue of higher forms, the neuromast organs of the head and snout in lower forms, and the palate in both.

facial carina (*Zool.*). A ridge separating the antennal grooves in some *Diptera*.

facialia (*Zool.*). A pair of ridges which demarcate the lateral margins of the face in *Diptera*.

facial ridges (*Zool.*). See **facialia.**

facial suture (*Zool.*). In *Trilobita*, a suture on the pleural portion of each side of the head.

facies (*Bot.*). The general form and appearance of a plant. (*Geol.*) The sum of the lithological and faunal characters of a sediment is its *facies*.

Lithological facies involves composition, grain-size, texture, colour, as well as such mass characters as current bedding, nature of stratification, ripple-marks, etc. Similarly, metamorphic facies involves the degree of crystallization and the mineral assemblage in a group of metamorphic rocks.

facing (*Civ. Eng., etc.*). An outer covering applied to the exposed face of sea-walls, embankments, brick walls, etc. (*Eng.*) (1) The operation of turning a flat face on a piece of work in the lathe. (2) A raised machined surface to which another part is to be attached. (*Mining*) A front cleat or face slip; situated opposite a back cleat. See **cleats.**

facing bar (*Textiles*). A strip of metal which forms a background to the fabric as it leaves the points in a lace machine. Also called **work bar.**

facing bond (*Build.*). A general term for any bond consisting mainly of stretchers.

facing bricks (*Build.*). A class of brick used for ordinary facing work; of better quality and appearance than common bricks, but not made to withstand heavy loads, as are engineering bricks.

facing gauge (*Eng.*). An instrument for measuring the total head of a stream of fluid; it consists of a small tube which faces upstream in the pipe carrying the fluid and is attached to a manometer. See **Pitot tube.**

facing paviors (*Build.*). A class of hard-burnt bricks used as facing bricks in high-class work.

facing points (*Rail.*). See **points.**

facing sand (*Met.*). Moulding sand with admixed coal dust, used near pattern in foundry flask to give casting smooth surface.

facsimile (*Telecomm.*). The scanning of any still graphic material to convert the image into electrical signals, for subsequent reconversion into a likeness of the original.

facsimile bandwidth (*Telecomm.*). The frequency difference between the highest and lowest components necessary for the adequate transmission of the facsimile signals.

facsimile baseband (*Telecomm.*). The frequency band occupied by the signal before modulation of the carrier frequency to produce the radio or transmission line frequency.

facsimile density (*Phys.*). Measure of the light transmission (or reflection) properties of an area, given by

$$\log_{10}\left(\frac{\text{incident intensity}}{\text{transmitted (or reflected) intensity}}\right).$$

facsimile modulation (*Telecomm.*). The process by which the amplitude, phase or frequency of the wave in facsimile transmission is varied with time in accord with a signal. The device for changing the type of modulation from receiving to transmitting or vice versa is known as a **converter.**

facsimile receiver (*Telecomm.*). One for translating the signals from the communication channel into a facsimile record of the original copy.

facsimile telegraphy (*Teleg.*). Transmission of still pictures over telegraph circuits by photoelectric scanning, modulating a carrier, and consequent reconstruction of the picture by synchronous scanning. A radio link may be included in the transmission circuit. Also **picture telegraphy.**

facsimile transmitter (*Telecomm.*). The means for translating the subject copy into signals suitable for the communication channel. The copy is rotated on a drum and the differences in the brightness of the light reflected from the copy modulate the output of a receiving photocell.

factice (*Chem.*). A substance produced by vulcanizing vegetable oils with sulphur or sulphur chloride; originally used as a rubber substitute but now employed as a compounding ingredient in rubber manufacture, e.g., in the production of rubber-proofed fabrics.

factor (*Psychol.*). A constituent part of a complex whole such as intelligence; a dimension upon which individuals differ from each other. See **factor analysis.**

factor analysis (*Psychol.*). The statistical extraction from data, such as the responses to questionnaires or intelligence tests, of *factors* which account for as much as possible of the *variance* of the scores.

factorial development (*Photog.*). Development procedure in which the time taken for the image to appear is multiplied by a factor, depending on the temperature and the type and dilution of the developer, to obtain the time after which the development is stopped.

factorial *n* (*Maths.*). The product of all whole numbers from a given number (n) down to 1; it is in fact the number of different ways of arranging n objects. Written $n!$ or $\lfloor n$.

$$n! = n(n-1)(n-2)(n-3)\ldots 3.2.1.$$

Cf. *gamma function, Stirling's approximation, subfactorial n.*

factor interaction (*Ecol.*). A subsidiary principle to Liebig's law of the *minimum* (q.v.) taking into account the fact that high concentration of availability of some substance, or the action of some factor other than the minimum one, may modify the rate of utilization of the latter.

factor of merit (*Elec. Eng.*). Of reflecting galvanometers, the deflection, in millimetres, produced on a scale at a distance of 1 m by a current of 1 micro-ampere, the deflection being corrected for coil-resistance and time of swing.

factor of safety (*Build., Eng., etc.*). The ratio, allowed for in design, between the ultimate stress in a member or structure and the safe permissible stress in it. Abbrev. **F.S.**

factor of the habitat (*Bot.*). Anything in the environment which affects, directly or indirectly, the life of a plant.

factor-symbols (*Zool.*). Letters in a genotypic formula which designate separate characters.

factory-fitting (*Elec. Eng.*). An electric-light fitting in which the lamp is housed in a strong protecting glass globe. Also called a **mill-fitting.**

faculae (*Astron.*). The name given to large bright areas of the photosphere of the sun. They can be seen most easily near sunspots and at the edge of the sun's disk; and are at a higher temperature than the average for the sun's surface.

facultative (*Zool.*). Optional; able to live under different conditions. Cf. *obligate.*

facultative anaerobe (*Bot.*). A plant which normally uses free oxygen but can live with little or none of it.

facultative gamete (*Bot.*). A zoospore which can function as a gamete.

facultative parasite (*Bot.*). A saprophyte which may become a parasite under special conditions.

facultative saprophyte (*Bot.*). A parasite which can live as a saprophyte under special conditions. Also **hemiparasite, hemisaprophyte.**

FAD (*Biochem.*). Abbrev. for *flavin-adenine dinucleotide.*

fade, fading (*Radio*). Phenomenon represented by more or less periodic reductions in the received field strength of a distant station, usually as a result of interference between reflected and direct waves from source. Mitigated by using *musa* (q.v.) or by diversity reception.

fade-in (*Cinema.*). The operation of the iris of a cinematograph camera to obtain the gradual appearance of the image in the final projected motion-picture.

fade-out (*Cinema.*). The operation of slowly closing the iris of a cinematograph camera so as to effect the gradual disappearance of the image in the resulting positive print.

fader (*Telecomm.*). Potentiometer device or variable attenuator; used, in a communication channel, for varying a signal level continuously from zero to maximum, or vice versa. It is thus a means of maintaining a constant signal level while one signal is faded out and another is faded in.

fading (*Autos.*). See **brake-fade.** (*Radio*) See **fade.**

fading area (*Radio*). That in which fading is experienced in night reception of radio waves, between the primary and secondary areas surrounding a station transmitting on medium and long waves.

fadometer (*Chem.*). An instrument used to determine the resistance of a dye or pigment to fading.

faeces (*Zool.*). The indigestible residues remaining in the alimentary canal after digestion and absorption of food.

Fagales (*Bot.*). An order of *Dicotyledons*, the nut-bearing trees. They are usually monoecious, rarely dioecious. The staminate flowers are borne in catkins, and the pistillate ones in catkins of two or three, or may be single.

faggot (*Civ. Eng.*). A bundle of brushwood. See **fascine.** (*Met.*) Made by forming a box with four long flat bars of wrought-iron and filling the interior with scrap and short lengths of bar.

faggoted iron (*Met.*). Wrought-iron bar made by heating a faggot to welding heat and rolling down to a solid bar. If the process is repeated *double-faggoted iron* is obtained.

fagopyrism (*Vet.*). Buckwheat rash; a form of photosensitization affecting animals with white or lightly pigmented skin; due to ingestion of a fluorescent substance occurring in buckwheat (*Fagopyrum sagittatum*), with subsequent exposure to strong sunlight.

fagot, fagoted. See **faggot, faggoted.**

fahlerz, fahlore (*Min.*). The grey copper (II) ores *tetrahedrite* and *tennantite* (qq.v.).

Fahrenheit scale (*Heat*). The method of graduating a thermometer in which freezing point of water is marked 32° and boiling point 212°, the fundamental interval being therefore 180°. Fahrenheit is being widely replaced by the Celsius (Centigrade) and Kelvin scales. To convert °F to °C subtract 32 and multiply by 5/9. For the *Rankine* (q.v.) equivalent add 459·67 to °F; this total multiplied by 5/9 gives the *Kelvin* (q.v.) equivalent.

F.A.I. (*San. Eng.*). Abbrev. for *fresh-air inlet.*

faïence (*Build.*). Glazed terra-cotta blocks used as facings for buildings.

faikes or fakes (*Mining*). Shaly sandstone. See **blaes.**

fail safe (*Nuc. Eng., etc.*). Design in which power supply, control or structural failure leads to automatic operation of protective devices.

failure (*Teleg.*). An unwanted spacing condition causing incorrect recording in a receiver.

fair cutting (*Build.*). The operation of cutting brickwork to the finished face of the work. Abbrev. **F.C.**

fair ends (*Build.*). Projecting masonry ends requiring to be dressed to a finished surface.

fairfieldite (*Min.*). A hydrated phosphate of calcium and manganese, crystallizing in the triclinic system as prismatic crystals or fibrous aggregates.

fairing (*Aero.*). A secondary structure added to any part of an aircraft to reduce drag by improving the streamlining. (*Ships*) The process of ensuring that the lines of intersection of all planes with a true ship form are *fair*; the resulting lines are known as *quarter lines* (q.v.).

fairlead (*For.*). In logging, the device containing sheaves or rollers which guides a moving cable so that it can be wound evenly on the drum. Generally, any similar device for reducing friction on rope, etc.

Fair's graticules (*Powder Tech.*). Types of eyepiece graticule marked with rectangles and circles for use in microscope methods of particle size analysis.

fairy ring (*Bot.*). A ring of strongly growing dark-green vegetation, often with a ring of dead plants inside it, the middle being occupied by vegetation more or less normal in appearance. The condition is caused by the spreading, in the soil, of a fungal mycelium which releases compounds of nitrogen at its active edge, and forms a dense mass of hypha just behind that edge. The nitrogenous compounds stimulate the growth of the higher plants, but behind this zone of stimulation the dense mycelium in the soil upsets water movements, so that the plants above it may die of drought. In autumn, the fruit bodies of the fungus may appear at the periphery.

Fajans rule for ionic bonding (*Chem.*). The conditions which favour the formation of ionic (as opposed to covalent) bonds are (*a*) large cation, (*b*) small anion, (*c*) small ionic charge, (*d*) the possession by the cation of an inert gas electronic structure.

Fajans-Soddy law of radioactive displacement (*Chem.*). The atomic number of an element decreases by 2 upon emission of an α-particle, and increases by 1 upon emission of a β-particle.

fakes (*Mining*). See **faikes.**

falciform, falcate (*Bot., Zool.*). Flattened and curved like a sickle; sickle-shaped.

falciform ligament (*Zool.*). In higher Vertebrates, a peritoneal fold attaching the liver to the diaphragm.

falciform process (*Zool.*). In the eyes of many Fish, a muscular and highly vascular structure which enters the retinal cup at the chorioid fissure and extends across to the lens, where it expands into the campanula Halleri; it is believed to be a means of accommodation.

falciform young (*Zool.*). The sickle-shaped sporozoites of some *Eugregarinaria.*

falciphore (*Bot.*). A hooked conidiophore.

Falconbridge process (*Met.*). Method of separating copper from nickel, in which matte is acid-leached to dissolve copper, after which residue is melted and refined electrolytically.

Falconiformes (*Zool.*). An order of *Neognathae* characterized by the possession of a powerful hooked beak, strong talons, and a desmognathous palate; rapacious carnivorous forms; birds of prey. Hawks, Eagles, Falcons, Kites, Buzzards, Vultures, Secretary-birds and Harriers.

falcula (*Zool.*). A sharp curved claw. *adj.* **falculate.**

Falk rail-joint (*Elec. Eng.*). A joint for tramway rails, made by the cast-welding process.

fall (*Civ. Eng., etc.*). The inclination of rivers, streams, ditches, drains, etc., quoted as a fall of so much in a given distance. See **cross fall.** (*Eng.*) A hoisting rope. (*Mining*) (1) The collapse of the roof of a level or tunnel, or of a flat working place or stall; the collapse of the hanging wall of an inclined working place or stope. (2) A mass of stone which has fallen from the roof or sides of an underground roadway, or from the roof of a working place.

fall bar (*Join.*). The part of a latch which pivots on a plate screwed to the inner face of a door, and drops into a hook on the frame.

fallen wool (*Textiles*). (1) Wool procured from sheep that have died. (2) Fibres detached from the fleece before shearing.

fallers (*Mining*). Movable supports for a cage or bond. See **keps.**

falling mould (*Join.*). The development in elevation of the centre line of a handrail.

falling stile (*Join.*). The shutting stile of a gate, especially of a gate so hung that the bottom of the shutting stile falls as the gate closes.

fall-of-potential test (*Elec. Eng.*). A test for locating a fault in an insulated conductor; the voltage drop along a known length of the conductor is compared with the voltage drop between one end of the conductor and the fault. Also called **conductivity test, drop test.**

Fallopian tube (*Zool.*). In Mammals, the anterior portion of the Müllerian duct; the oviduct.

fall-out (*Ecol.*). The natural movement of solid matter through air as dust. (*Nuc.*) Airborne radioactive contamination resulting from distant nuclear explosion, inadequately filtered reactor coolant, etc.

fall pipe (*Build.*). See **downpipe.**

fall-ridder (*Mining*). See **bordroom-man.**

fall table (*Mining*). A hinged shaft-cover.

fall time (*Telecomm.*). The decaying portion of a wave pocket or pulse; usually the time taken for the amplitude to decrease from 90% to 10% of the peak amplitude.

false amethyst, etc. (*Min.*). In naming gemstones those engaged in the trade are guided by the colour of the gem rather than by its composition and physical characters. Thus, to them, all mauve stones are amethyst. Yet many minerals, when quite pure, are without colour; the addition of a minute amount of impurity of the right composition will impart to any of them a mauve tint. Unless the mauve stone is a variety of silica, it is incorrect to call it *amethyst*. For example, some specimens of corundum are mauve; these are known in the trade as *Oriental amethyst*—one type of *false amethyst*.

false amnion (*Zool.*). See **chorion.**

false annual ring (*Bot.*). A second ring of xylem formed in one season, following the defoliation of the tree by the attacks of insects or other accident; oaks are liable to this, as they may be completely stripped of leaves by the oak tortrix.

false axis (*Bot.*). A monochasium which looks like one axis but really consists of a number of successive lateral branches running more or less in a line.

false bands (*Bind.*). Strips of board or leather glued across the spine of hollow-backed books before covering with leather. See **raised bands.**

false bearing (*Build.*). A beam, such as a sill, when not supported under its entire length, is said to have a *false bearing*.

false bedding (*Geol.*). Minor planes of stratification inclined at an angle to the major bedding planes; due to changes in the direction and velocity of currents. Characteristic of sedimentation under deltaic and aeolian conditions.

false berry (*Bot.*). A fleshy fruit, looking like a berry, but with some of the flesh developed from the receptacle of the flower.

false body (*Paint.*). The apparently full-bodied condition of a paint which undergoes a marked reduction in viscosity when agitated and returns to its former condition, either immediately or subsequently when at rest. Cf. *thixotropy*.

false bottom (*Eng.*). (1) A removable bottom placed in a vessel to facilitate cleaning; a casting placed in a grate to raise the fire bars and reduce the size of the fire. (2) Any secondary bottom plate or member used to reduce the volume of a container or to create a secondary container.

false curvature (*Nuc.*). That of particle tracks (e.g., in cloud chambers, bubble chambers, spark chambers or photographic emulsions) which results from undetected interactions and not from an applied magnetic field.

false diamond (*Min.*). Several natural minerals are sometimes completely colourless and, when cut and polished, make brilliant gems. These include zircon, white sapphire, and white topaz. All three, however, are birefringent and can be easily distinguished from true diamond by optical and other tests. See **false amethyst.**

false dichotomy (*Bot.*). Branching in which two lateral branches arise on opposite sides of the main stem and overtop it.

false dissepiment (*Bot.*). A wall which divides the loculus of an ovary into two compartments, but is an ingrowth from the carpel wall and not a wall between one carpel and its neighbour.

false ellipse (*Build.*). An approximate ellipse, composed of circular arcs.

false fruit (*Bot.*). A fruit formed from other parts of the flower in addition to the gynaeceum.

false germination (*Bot.*). An appearance of germination in a dead seed due to swelling of the embryo as it takes up water.

false header (*Build.*). A half-length brick, sometimes used in Flemish bond.

false hemlock (*For.*). See **Douglas fir.**

false hybrid (*Bot.*). A plant developed after cross-fertilization, but possessing characters from one parent only.

false key (*Eng.*). A circular key for attaching a hub to a shaft; it is driven into a hole which is parallel with the shaft axis and has been drilled half in the hub and half in the shaft.

false line lock (*TV*). In a television receiver, a critical setting of the *line-hold* control, which results in a line lock with a part of the picture to the left and a part to the right of a dark and wide vertical bar.

false membrane (*Bot.*). The sterile hyphae limiting the sorus of some smuts.

false pile (*Civ. Eng.*). A length added to the top of a pile which has been driven.

false pregnancy (*Med.*). See **pseudocyesis.** (*Zool.*) See **pseudopregnancy.**

false ribs (*Zool.*). In higher Vertebrates, ribs which do not reach the sternum.

false ruby (*Min.*). Some species of garnet (*Cape ruby*) and some species of spinel (*balas ruby* and *ruby spinel*) possess the colour of ruby, but have neither the chemical composition nor the physical attributes of true ruby. See **false amethyst.**

False Scorpions (*Zool.*). See **Pseudoscorpionidea.**

false septum (*Bot.*). See **spurious dissepiment.**

false tissue (*Bot.*). See **pseudoparenchyma.**

false topaz (*Min.*). A name applied to yellow quartz. See **citrine.**

falsework (*Civ. Eng.*). The temporary work known as *centring* (q.v.); scaffolding, or other temporary supports used in construction.

false-zero test (*Elec. Eng.*). A test, made on a bridge or potentiometer, in which a balance is obtained, not with zero galvanometer reading, but with some definite value caused by a constant extraneous current.

falx (*Zool.*). Any sickle-shaped structure. *adjs.* **falciform, falcate.**

falx cerebri (*Zool.*). A strong fold of the dura mater, lying in the longitudinal fissure between the two cerebral hemispheres.

famatinite (*Min.*). An orthorhombic sulphide of copper and antimony, occurring in the Famatina Mts (Argentina) and in Peru.

family (*Biol.*). A group of similar genera; a group of similar families constitute an order or sub-order. (*Nuc.*) The group of radioactive nuclides which form a decay series.

fan (*Eng.*). (1) A device for delivering or exhausting large volumes of air or gas with only a low pressure increase. It consists either of a rotating paddle-wheel or an airscrew. (2) A small vane to keep the wheel of a wind pump at right angles to the wind. (*Horol.*) A wheel whose velocity is regulated by air resistance; used in some clocks.

fan antenna (*Radio*). Antenna in which a number of vertically inclined wires are arranged in a fanwise formation, the apex being at the bottom.

fan characteristic (*Eng.*). A graph showing the relation between pressure and delivery, used as a basis for fan selection. The characteristic is determined by the shape of the fan blades.

fan cooling (*Autos.*). The use of an engine-driven fan to induce a greater airflow through the radiator at low speeds than would result from the forward motion of the vehicle.

fancy (*Textiles*). One of the wire-covered rollers of a card. It lifts the web of fibres on to the swift, to facilitate removal by the doffer.

fancy yarns (*Textiles*). Yarns made for decorative purposes. The ornamentation of the thread may be due to a variety of reasons, such as (*a*) colour; (*b*) the combination of threads of different types; (*c*) the production of thick and thin places; (*d*) the production of knops, loops, slubs, etc., at suitable intervals. The majority of these fancy yarns are *folded yarns*, two or more threads being combined in some special way in order to produce the desired effect. Special types of ring doubling machines are usually used for their production. The yarns are of many types, e.g., *bead, bouclé, bourette, chenille, cloud, crêpe, curl, diamond, flake, gimp, grandrelle, knickerbocker, knop, loop, marl, melange, mottle, ondé, reany, slub, snarl, spiral, spot*, etc.

fan drift (*Mining*). Ventilating passage along which air is moved by means of a fan.

fang (*Build.*). The part of an iron railing which is embedded in the wall. (*Zool.*) The grooved or perforate poison-tooth of a venomous serpent: one of the cuspidate teeth of carnivorous animals, especially the canine or carnassial.

fang bolt (*Eng.*). A bolt having a nut which carries pointed teeth for gripping the wood through which the bolt passes, so preventing the nut from rotating when the bolt is tightened.

fangs (*Mining*). See **keps.**

fan-guard (*Build.*). A protective parapet formed of boarding secured around the platforms of builders' stagings or gantries, when the platforms are to be used for receiving and distributing materials.

fan marker beacon (*Aero.*). A form of marker beacon radiating a vertical fan-shaped pattern.

fanners (*Mining*). Hand-operated ventilating fans.

fan shaft (*Mining*). Mine shaft or pit at the top of which a ventilating fan is placed.

fan structure (*Geol.*). A complicated arrangement of folds in which the axial planes converge like the ribs of a fan; the normal arrangement of folds in an anticlinorium.

fantail burner (*Eng.*). A pulverized-coal burner which discharges the fuel and primary air vertically downwards into the furnace in a thin flat stream, to meet heated secondary air which is discharged horizontally from the walls.

Fantasound (*Cinema.*). Form of sound recording giving a three-dimensional effect by the use of a multichannel sound-track on a second film synchronized with the picture film.

fan vaulting (*Arch.*). Tracing rising from a capital or a corbel, and diverging like the folds of a fan on the surface of a vault.

farad (*Elec. Eng.*). The practical and absolute SI unit of electrostatic capacitance, defined as that which, when charged by a p.d. of one volt, carries a charge of one coulomb. Equal to 10^{-9} electromagnetic units and 9×10^{11} electrostatic units. Symbol **F.** This unit is in practice too large, and the subdivisions, *microfarad* (μF), *nanofarad* (nF), and *picofarad* (pF), are in more general use.

faraday (*Elec.*). Quantity of electricity associated with one mole of chemical charge, i.e., 96 487 coulombs. Symbol ***F***.

Faraday cage (*Elec. Eng.*). An arrangement of earthed parallel conductors acting as an electrostatic screen; connected so that induced currents cannot circulate. Also **Faraday shield.**

Faraday dark space (*Elec.*). Dark region in a gas-discharge column between the negative glow and the positive column.

Faraday disk (*Elec. Eng.*). Rotating disk in the gap of an electromagnet, so that a low e.m.f. is generated across a radius. Used, e.g., to generate a calculated electromotive force to balance against the drop across a resistance due to a steady current, thus establishing the latter in absolute terms from the calculation of the mutual inductance between the disk and the exciting air-cored coil.

Faraday effect (*Light, Mag.*). Rotation of the plane of polarization of (1) a plane-polarized light beam when propagated through a transparent isotropic medium, (2) a plane-polarized microwave passing through a ferrite along the direction of a magnetic field. If l is the length of path traversed, H is the strength of the magnetic field and θ is the angle of rotation, then $\theta = ClH$, where C is *Verdet's constant*.

Faraday shield (*Elec. Eng.*). See **Faraday cage.**

Faraday's ice-pail experiment (*Elec.*). Classical experiment which consists in lowering a charged body into a metal pail connected to an electroscope, in order to show that charges reside only on the outside surface of conductors.

Faraday's law of induction (*Elec.*). The e.m.f. induced in any circuit is proportional to the rate of change of the number of magnetic lines of force linked with the circuit. Principle used in every practical electrical machine. *Maxwell's field equations* involve a more general mathematical statement of this law.

Faraday's laws of electrolysis (*Elec.*). (1) The amount of chemical change produced by a current is proportional to the quantity of electricity passed. (2) The amounts of different substances liberated or deposited by a given quantity of electricity are proportional to the chemical equivalent weights of those substances.

Faraday tube (*Elec. Eng.*). Tube of force in an electric field, of such magnitude that unit charge gives rise to one tube. Obsolete.

faradic currents (*Med.*). Induced currents

obtained from secondary winding of an induction coil and used for curative purposes.

faradism (*Med.*). The treatment of disease by the use of an interrupted current obtained from an induction coil, the waveform being very peaky.

faradize (*Med.*). To stimulate the muscles or nerves of a living subject with faradic currents.

faradmeter (*Elec. Eng.*). Generic name for direct-reading capacitance meters. Typically, they use the mains voltage in series with an a.c. milli-ammeter.

farctate (*Bot.*). See **stuffed.**

farcy (*Med., Vet.*). See **glanders.**

fardel-bound (*Vet.*). Constipation, especially of cattle and sheep, by the retention of food in the omasum or *psalterium*.

Fareham reds (*Build.*). A form of hand-moulded, sand-faced, red facing bricks.

far-end cross-talk (*Teleph.*). Cross-talk heard by a listener, and caused by a speaker at the distant end of the parallelism.

Farewell Rock (*Geol., Mining*). The highest division of the so-called Millstone Grit of S. Wales, lying immediately beneath the productive Coal Measures. Good refractory material.

farina. Generally, ground corn, meal, starch, etc. (*Textiles*) The name still used in the cotton industry for potato starch; used for sizing warp yarns, to enable them to withstand abrasion during weaving.

farinaceous (*Bot.*). (1) Having a surface covered with particles looking like meal. (2) Of mealy character.

farinose (*Bot.*). Covered with whitish, very short hairs, which are easily detached as whitish dust.

-farious. A suffix meaning *arranged in so many rows.*

Farleigh Down stone (*Build.*). A fine-grained, even-textured, warm cream-coloured Bath oölite; used for building purposes, but weathers poorly.

Farmer's reducer (*Photog.*). A reducing bath for photographic images made by the addition of potassium ferricyanide to hypo.

far point (*Optics*). Object point conjugate to the retina when accommodation is completely relaxed; at infinity in emmetropia, between infinity and the eye in myopia, and behind the eye in hyperopia.

Farrant's medium (*Micros.*). A mixture of glycerine and gum arabic, used particularly for mounting frozen sections, after staining.

Farror's process (*Met.*). Case-hardening by mixture of ammonium chloride, manganese dioxide and potassium ferrocyanide.

fascia (*Arch., Build.*). (1) A wide flat member in an entablature. (2) A board embellishing a gutter around a building. (3) The broad flat surface over a shop front or below a cornice. (*Autos.*) The instrument board of an automobile. (*Zool.*) Any bandlike structure; especially the connective-tissue bands which unite the fasciculi of a muscle. *adj.* **fascial.**

fasciae dentatae (*Zool.*). The specialized lower borders of the hippocampi in Mammals.

fasciation (*Bot.*). An abnormal condition usually shown by marked flattening, and brought about by the union of a number of members, usually all of the same kind, side by side as they develop.

fascicle (*Bot.*). (1) A tuft of leaves crowded on a short stem. (2) A close tuft of branches all arising from about the same place. *adj.* **fascicled.**

fascicular cambium (*Bot.*). The flat strand of cambium between xylem and phloem in a vascular bundle.

fasciculate (*Bot.*). In bunches or bundles consisting of a number of members all of the same kind.

fasciculus (*Zool.*). A small bundle, as of muscle or nerve fibres.

fasciitis, fascitis (*Med.*). Inflammation of fascia.

fascine (*Civ. Eng.*). A bundle of brushwood used to help make a foundation on marshy ground, or to make a wall to protect a shore against erosion by sea or river, or to accumulate sand and silt on the bed of an estuary.

fascine building (*Build.*). A building constructed with logs and boards.

fasciola (*Zool.*). A narrow band of colour; a delicate lamina in the Vertebrate brain.

Fasciola hepatica (*Zool.*). (*Trematoda*; *Malacocotylea.*) The *liver fluke*. The adult lives in the liver of sheep, and the secondary host, entered by the *miracidium* larva, and in which the *redia* and *cercaria* larvae develop, is a water snail, usually of the genus *Limnaea* (*Gastropoda*; *Pulmonata*).

fasciole (*Zool.*). In *Spatangoida*, a tract of ciliated spines (clavulae) which create a current of water.

fascioliasis (*Med., Vet.*). Infection of Man and other animals with the liver fluke *Fasciola hepatica.*

fasciotomy (*Surg.*). Surgical incision of fascia.

fassaite (*Min.*). A monoclinic pyroxene rich in aluminium, calcium, and magnesium, and poor in sodium; found in metamorphosed limestones and dolomites, as in the Fassa valley, Trentino, Italy.

fast (1) (*Mining*). A heading or working place which is driven in the solid coal, in advance of the open places, said to be in the *fast*. (2) A hole in coal which has had insufficient explosive used in it, or which has required undercutting. (3) In shaft sinking, a hard stratum under poorly consolidated ground, on which a wedging crib can be laid. (*Paint.*) Said of colours which are not affected by the conditions of use (i.e., light, heat, chemical action, damp, etc.) to which they are subjected. *n.* **fastness.** (*Photog.*) Contributing to reduction of time of exposure; said of an emulsion or lens.

Fast (*Electronics*). System for testing transistors, incorporating a memory for subsequent readout of parameters. (*F*acility for *a*utomatically *s*orting and *t*esting.)

fast-acting relay (*Teleph.*). A relay designed to act with minimum delay after the application of voltage, usually by increasing the resistance of the circuit in comparison with the inductance, and by minimizing moving masses.

fast coupling (*Eng.*). A coupling which permanently connects two shafts.

fast effect (*Nuc.*). See **fast fission.**

fastener (*Eng.*). An article designed to fasten together two or more other articles, usually in the form of a shaft passing through the articles to be fastened; e.g., nails, screws, rivets, pins.

fast fission (*Nuc.*). ^{238}U has a fission threshold for neutrons of energy about 1 MeV, and the fission cross-section increases rapidly with energy. Fission of this isotope by fast neutrons may cause a substantial increase in the reactivity of a thermal reactor (*fast effect*).

fast fission factor (*Nuc.*). Ratio of total number of neutrons produced by fission in infinite thermal reactor system to the number produced by thermal fission. It represents the fractional increase in neutron population in any generation which results from fast fission, and is usually about 1·028 for natural-uranium graphite-power reactors.

fast head (*Eng.*). The fixed headstock of a lathe.

fastigiate (*Bot.*). Having the branches more or less erect and all more or less parallel, giving an effect like a slender broom.

fastigium (*Build.*). (1) The pediment above a portico. (2) A roof ridge. (*Med.*) The highest point of temperature in a fever.

fast-needle surveying (*Surv.*). See **fixed-needle surveying.**

fastness (*Paint.*). See **fast.**

fast neutrons (*Nuc.*). See **neutron.**

fast pulley (*Eng.*). A pulley fixed to a shaft by a key or set bolt, as distinct from a *loose pulley* (q.v.) which can revolve freely on the shaft.

fast reaction (*Nuc.*). Nuclear reaction involving strong interaction and occurring in a time of the order of 10^{-23} seconds. Due to *strong interaction* forces.

fast reactor (*Nuc. Eng.*). One without a moderator in which chain reaction is maintained almost entirely by fast fission.

fast-reed loom (*Weaving*). A loom with a fixed reed; used for weaving heavy fabrics or those in which weft is very closely picked.

fast sheet (*Build.*). See **stand sheet.**

fast store (*Comp.*). One of small capacity, and hence of quick access, for program or data often required in a calculation, thus increasing overall speed of operation, e.g., short mercury line, or short nickel wire, in which coded information is continuously circulated.

fast-time constant circuits (*Telecomm.*). Those for which the circuit parameters (particularly resistance and capacitance) permit a very rapid response to a step signal. Such circuits should be used exclusively in the case of pulse signals.

fat (*Build.*). Part of a cement mortar mix containing a higher proportion of cement than the rest. This comes to the surface before the mixture has set. (*Chem.*) See **fats.** (*Zool.*) See **adipose tissue.**

fata morgana (*Meteor.*). A complicated mirage caused by the existence of several layers of varying refractive index, resulting in multiple images, possibly elongated. Especially characteristic of the Strait of Messina and Arctic regions.

fat board (*Build.*). A board on which the bricklayer collects the fat during the process of pointing.

fat-body (*Zool.*). In Insects, a mesodermal tissue of fatty appearance, the cells of which contain reserves of fat and other materials and play an important part in the metabolism of the animal; in Amphibians, highly vascular masses of fatty tissue associated with the gonads.

fat coals (*Mining*). Coals which contain plenty of volatile matter (gas-forming constituents).

fat edges (*Paint.*). A defect in paintwork, characterized by the formation of ripples at edges and in angles; due to excess of paint.

fat face (*Typog.*). Heavy types with hairline serifs and main strokes at least half as wide as the height of the letters, *Ultra Bodoni* being a well-known example.

father of the chapel (*Typog.*). A person elected by the associated employees of a printing department to represent them and to watch their interests.

fathom. A unit of measurement. Generally, a nautical measurement of depth = 6 ft. (*For.*) A timber measure = 216 cu. ft = 6 ft by 6 ft by 6 ft. (*Mining*) In general mining, the volume of a 6-ft cube; in gold mining, often a volume 6 ft by 6 ft by the thickness of the reef; in lead mining, sometimes a volume 6 ft by 6 ft by 2 ft. It is the unit of performance of a rock drill—'fathoms per shift'.

fathometer (*Acous.*). An ultrasonic depth-finding device.

fatigue (*Zool.*). The condition of an excitable cell or tissue which, as a result of activity, is less ready to respond to further stimulation until it has had time to recover.

fatigue allowance (*Psychol.*). The allowance made in work study to cover the effect of work on the operator which tends to lower his rate or grade of quality of production.

fatigue limit (*Met.*). The upper limit of the range of stress that a metal can withstand indefinitely. If this limit is exceeded, failure will eventually occur.

fatigue of metals (*Met.*). The phenomenon of the failure of metals under the repeated application of a cycle of stress. Factors involved include amplitude, average severity, rate of cyclic stress and temperature effect. *Notch brittleness* (q.v.) commences at scratch or blemish.

fatigue test (*Eng., Met.*). A test made on a material to determine the range of alternating stress to which it may be subjected without risk of ultimate failure. By subjecting a series of specimens to different ranges of stress, while the mean stress is constant, a stress-number curve is obtained.

fatigue-testing machine (*Eng., Met.*). A machine for subjecting a test piece to rapidly alternating or fluctuating stress, in order to determine its fatigue limit. See **Wöhler test, Haigh fatigue-testing machine.**

fat lime (*Build.*). *Lime* (q.v.) made by burning a pure, or very nearly pure, limestone, such as chalk.

fat liquor (*Leather*). An emulsion consisting of oil, soap, and water, which is added to warm water for drumming skins, before drying.

fat matter (*Typog.*). A composing-room term for easily set portions of the work in hand.

fat-necrosis (*Med.*). The splitting of fat, due to the escape of a fat-splitting enzyme from the pancreas into the abdominal cavity, with death of the fat-containing cells so affected.

fats (*Chem.*). An important group of naturally occurring substances consisting of the glycerides of higher fatty acids, e.g., palmitic acid, stearic acid, oleic acid. Essential component of the human diet, repairing the wastage of human fat, and, by breakdown and oxidation, providing some of the required energy.

fat-splitting (*Chem.*). Term used to describe the hydrolysis of animal and vegetable fats into glycerol and fatty acids. Can be effected in a number of ways but chiefly by using strong alkalis (as in soapmaking) or inorganic acids.

fatty acids (*Chem.*). A term for the whole group of saturated and unsaturated monobasic aliphatic carboxylic acids. The lower members of the series are liquids of pungent odour and corrosive action, soluble in water; the intermediate members are oily liquids of unpleasant smell, slightly soluble in water. The higher members, from C_{10} upwards, are mainly solids, insoluble in water, but soluble in ethanol and in ethoxyethane.

fatty degeneration (*Med.*). Degeneration of the cell substance, accompanied by the appearance in it of droplets of fat, due to the action of poisons or to lack of oxygen.

fatty heart (*Med.*). (1) A heart the muscle of which has undergone fatty degeneration. (2) An increase of fat in parts of the heart where it is normally present, associated with general adiposity.

fauces (*Zool.*). In Vertebrates, the pharynx; in spirally-coiled shells, the aperture.

faucet (*Plumb.*). (1) A small tap or cock. (2) The enlarged or socket end of a pipe at a *spigot-and-socket joint* (q.v.).

faucet ear (*Plumb.*). A projection from the socket of a pipe by means of which the pipe may be nailed to a wall.

faujasite (*Min.*). One of the less common zeolites; a hydrated silicate of sodium, calcium, and aluminium. It exhibits a wider range of molecular absorption than any other zeolite.

faulschlamm (*Ecol.*). A type of lake-bottom deposit composed of organic detritus covered by mineral deposits, which is characterized by the absence of *sediment transporters* (q.v.) and in which anaerobic decomposition takes place.

fault (*Geol., Mining*). A fracture in rocks along which some displacement (the *throw* of the fault) has taken place. The displacement may vary from a few millimetres to thousands of metres. Movement along faults is the common cause of earthquakes.

fault breccia (*Geol.*). A fragmental rock of breccia type resulting from shattering during the development of a fault.

fault current (*Elec. Eng.*). That caused by defects in electrical circuit or device, such as short-circuit in system. The peak value of current is the accepted measure.

fault-finding (*Elec. Eng.*). General description of locating and diagnosing faults, according to a pre-arranged schedule, generally arranged in a chart or table, with or without special instruments. U.S. **trouble-shooting.**

fault rate (*Elec. Eng.*). See **reliability.**

fault resistance (*Elec. Eng.*). A term sometimes used to denote insulation resistance, but more commonly the resistance of an actual fault, e.g., an arc between a conductor and earth.

fauna (*Zool.*). A collective term denoting the animals occurring in a particular region or period. *pl.* **faunas** or **faunae.** *adj.* **faunal.**

faunal region (*Zool.*). An area of the earth's surface characterized by the presence of certain species of animals.

Faure accumulator (*Elec. Eng.*). An accumulator having Faure (pasted) plates.

Faure plate (*Elec. Eng.*). See **pasted plate.**

faveolate, favose (*Bot., Zool.*). Resembling a honeycomb in appearance. *Favous*, said of a surface pitted like a honeycomb. *Favus*, a hexagonal pit or plate.

favus (*Med.*). A contagious skin disease, especially of the scalp, due to infection with the fungus *Achorion Schönleinii.*

fawn foot (*For.*). The swelling at the end of an axe handle to give a better grip.

fax (*Telecomm.*). Abbrev. for *facsimile.*

fayalite or **iron olivine** (*Min.*). A silicate of iron, Fe_2SiO_4, crystallizing in the orthorhombic system, discovered originally at Fayal in the Azores, probably in a slag carried to the island as ballast; but found subsequently in igneous rocks, chiefly of acid composition, including pitchstone, obsidian, quartz-porphyry, rhyolite, and also in ferrogabbro.

faying face (or **surface**) (*Eng.*). That part of a surface of wood or metal specially prepared to fit an adjoining part.

f_c (*Electronics*). Abbrev. for *cut-off frequency.*

F.C. (*Build.*). Abbrev. for *fair cutting.*

F-class insulation (*Elec. Eng.*). A class of insulating material to which is assigned a temperature of 155°C. See **class-A, - B, -C, etc., insulating materials.**

fd. (*Carp.*). Abbrev. for *framed.*

F-diagram (*Chem. Eng.*). The cumulative residence time distribution in a continuous flow system, plotted on dimensionless coordinates. Important in assessing the performance of chemical reactors, kilns, etc.

F display (*Radar*). Type of radar display, used with directional antenna, in which the target appears as a bright spot which is off-centre when the aim is incorrect.

FDS law (*Nuc.*). Fermi-Dirac-Sommerfeld law, which gives the algebraic number of a quantized system of particles which have velocities within a small range.

Fe (*Chem.*). The symbol for *iron.*

fear (*An. Behav.*). Behaviour produced by sudden intense stimuli, including alteration of sphincter control, flight, respiratory changes; also a drive underlying this behaviour.

fearnought (*Textiles*). A machine which opens and mixes woollen or blended materials for carding. Also called **tenterhook willow.**

feather (*Build.*). A steel wedge driven in between a pair of semi-cylindrical plugs placed in a drilled hole in a mass of stone, as a means of splitting the stone. (*Carp.*) A thin tongue along the edge of a board, fitting into a corresponding groove in the edge of another board. See **matched boards.** (*Eng.*) (1) A rectangular key sunk into a shaft to permit a wheel to slide axially, while preventing relative rotation. (2) Iron slips for reducing the friction between a wedge and an object to be split. (*Glass*) A cluster of fine *seed* (q.v.) caused by foreign material entering the glass casting or shaping.

Feather analysis (*Nuc.*). An approximate method of determining the range of β-rays forming part of a combined β-γ spectrum, by comparison of the absorption curve with that for a pure β-emitter.

feather eating (*Vet.*). A vice, acquired by birds, characterized by pecking, plucking, or eating their own plumage or that of other birds. The vice may develop into *cannibalism* (q.v.).

feather-edge brick (*Build.*). A brick similar to a compass brick, used especially for arches.

feather-edged coping (*Build.*). A coping-stone sloping in one direction on its top surface. Also known as **splayed coping.**

feathering (*Paper*). The irregular edge of an ink stroke when the paper is semi-absorbent.

feathering airscrew (*Aero.*). See **airscrew.**

feathering hinge (*Aero.*). A pivot for a rotorcraft blade which allows the angle of incidence to change during rotation.

feathering paddles (*Eng.*). Paddle-wheels so controlled that the floats enter and leave the water at right angles to the surface.

feathering pitch (*Aero.*). The blade angle of an airscrew giving minimum drag when the engine is stopped.

feathering pump (*Aero.*). A pump for supplying the necessary hydraulic pressure to turn the blades of a *feathering airscrew* to and from the feather position.

feather joint (*Join.*). See **ploughed-and-tongued joint.**

feather ore (*Min.*). A plumose or acicular form of the sulphide of lead and antimony, occurring in the Harz Mts and elsewhere. Also called **jamesonite.** The name is also used for *stibnite.*

feather-perforating mite disease (*Vet.*). An infection of the plumage and skin of pigeons due to the feather-perforating mite, *Falculifer rostratus.*

feathers (*Mining*). Compound three-part wedge

used to loosen coal. (*Zool.*) Epidermal outgrowths forming the body covering of Birds; distinguished from scales and hair, to which they are closely allied, by their complex structure, and by the possession of a vascular core which at first projects from the surface.

feather tongue (*Join.*). A wooden tongue for a *ploughed-and-tongued joint* (q.v.), cut so that the grain is diagonal to the grooves.

featherweight paper (*Paper*). A very light antique book paper. It is made from loosely woven esparto, and often three-quarters of its bulk is air space.

febrifuge (*Med.*). Against fever; a remedy which reduces fever.

febrile (*Med.*). Pertaining to, produced by, or affected with fever.

Fechner colours (*Optics*). The visual sensations of colour which are induced by intermittent achromatic stimuli.

Fechner law (*Physiol.*). See **Weber-Fechner law.**

fecundation canal (*Zool.*). In some *Lepidoptera*, a duct connecting the spermatheca with the bursa copulatrix.

fecundity (*Ecol.*). The number of eggs produced by a species or individual. Cf. *fertility*. (*Gen.*) Capacity of a species to undergo multiplication.

feeblemindedness (*Psychiat.*). Term applied generally to individuals whose *IQ* is less than 70.

feebly hydraulic lime (*Build.*). *Lime* (q.v.) made by burning a limestone containing 5–12% clay.

feed (*Eng.*). (1) The rate at which the cutting tool in a machine is advanced. (2) Fluid pumped into a vessel, e.g., feed-water to a boiler. (3) Mechanism for advancing material or components into a machine for processing. See also **hopperfeed.** (*Mining*) Forward motion of drill or coal-cutter. (*Telecomm.*) To offer a programme or signal at some point in a communication network.

feedback (*Acous.*). Reaction of reproduced sound in an enclosure on to the microphone controlling same, resulting in change in response of the system, apparently increased reverberation, and, if excessive, actual oscillation (howling). (*Telecomm.*) Transfer of some output energy of an amplifier to its input, so as to modify its characteristics. *Current* (or *voltage*) *feedback* (q.v.) is when feedback signal depends on current (or voltage) in output respectively. *Bridge feedback* (q.v.) depends on their combination.

feedback admittance (*Telecomm.*). The short-circuit transadmittance from output to input electrode of a thermionic valve.

feedback characteristic (*Electronics*). See **transistor characteristics.**

feedback circuit (*Telecomm.*). (1) Circuit conveying current or voltage feedback from the output to the input of an amplifier. (2) Circuit for injecting a signal from a program source elsewhere.

feedback control loop (*Telecomm.*). A closed transmission path including an active transducer, forward and feedback paths and one or more mixing points. The system is such that a given relation is maintained between the input and output signals of the loop.

feedback control system (*Telecomm.*). U.S. for **closed-loop system.**

feedback factor (*Telecomm.*). In an amplifier using negative feedback, the product of the fraction of the output voltage fed back and the gain of the amplifier before applying feedback.

feedback oscillator (*Telecomm.*). Any oscillator in which the frequency is controlled by the parameters of a positive or negative feedback loop.

feedback path (*Telecomm.*). That from the loop output signal to the feedback signal in a feedback control loop.

feedback ratio (*Telecomm.*). That of the number of turns in the feedback winding to the number in the output winding of, e.g., a simple-series magnetic amplifier using current feedback.

feedback signal (*Telecomm.*). That which is responsive in an automatic controller to the value of the controlled variable.

feedback transducer (*Telecomm.*). One which generates a signal, generally electrical, depending on quantity to be controlled, e.g., for rotation potentiometer, synchro or tacho, giving proportional derivative or integral signals respectively.

feedback windings (*Elec. Eng.*). Those control windings in a saturable reactor to which are made the feedback connexions.

feed-check valve (*Eng.*). Nonreturn valve in the delivery pipe between feed-water pump and boiler.

feed current (*Electronics*). Direct-current component of anode current of a thermionic valve when separated from alternating components.

feeder (*Elec. Eng.*). (1) An overhead or underground cable, of large current-carrying capacity, used in the transmission of electric power; it serves to interconnect generating stations, substations, and feeding points, without intermediate connexions. (2) In electrical circuits, the lines running from the main switchboard to the branch panels in an installation. (*Foundry, etc.*) The runner or riser hole of a mould, containing sufficient molten metal to feed the casting and so compensate for contraction of the solidifying metal. (*Hyd. Eng.*) A natural or artificial channel supplying water to a reservoir or canal. (*Mining*) A mechanical appliance for supplying broken rock or crushed ore, at a predetermined rate, to some form of crusher or concentrator. (*Radio*) Conductor, or system of conductors, connecting the radiating portion of an antenna to the transmitter or receiver. It may be a balanced pair, a quad, or coaxial.

feeder bus-bars (*Elec. Eng.*). In a generating station or main substation, bus-bars to which the outgoing feeders are connected.

feeder ear (*Elec. Eng.*). A type of ear for attaching an overhead contact wire of a tramway system to the supporting wire; it serves also to lead current to the contact wire.

feeder-fed bottle machines (*Glass*). See **bottle-making machines.**

feeder head (*Met.*). See **hot top.**

feeder mains (*Elec. Eng.*). See **feeder (1).**

feeder panel (*Elec. Eng.*). A switchboard panel on which are mounted the switchgear and instruments for controlling one or more feeders.

feeder pillar (*Elec. Eng.*). A pillar containing switches, links, and fuses, for connecting the feeders of an electric power distributing system with the distributors.

feed finger (*Eng.*). A rod used in association with a collet to push or pull the bar of material forward in a capstan lathe or automatic until the bar touches a stop, thus governing the feed length.

feed holes (*Comp.*). Same as **centre holes.**

feeding head (*Foundry, etc.*). Same as **feeder.**

feeding point (*Elec. Eng.*). The junction point between a feeder and a distribution system. Also called a **distributing point.**

feeding rod (*Foundry*). A heated iron rod inserted in the feeder of a mould, and worked with a pumping motion to assist feeding during the cooling of the molten metal.

feeding-up (*Paint.*). A thickening of paints (and sometimes also said of varnishes) in the can, sufficient to make them unsatisfactory for use.
feed mechanism (*Elec. Eng.*). The mechanism which causes the carbons of an arc lamp to move gradually towards the arc at the speed necessary to compensate for the rate at which they burn away, thereby keeping the arc length constant.
feed pipe (*Eng.*). The pipe carrying feed-water from the feed pump to a boiler.
feed reel (*Cinema.*). The reel of film which is being unwound as the film is taken off to pass through the gate in a camera, printer, or projector.
feed rollers (*Print.*). Driven rollers which convey a web into a printing couple. Cf. *idler* or *idling roller*.
feed screw (*Eng.*). A screw used for supplying motion to the feed mechanism of a machine tool.
feedthrough (*Elec. Eng.*). A conductor used to connect patterns on opposite sides of the board of a printed circuit, or an insulated conductor for connection between two sides of a metal earthing screen.
feed-water (*Eng.*). The water, previously treated to remove air and impurities, which is supplied to a boiler for evaporation.
feed-water heater (*Eng.*). An arrangement for heating boiler feed-water by means of steam which has done work in an engine or turbine. It is similar in principle to a steam condenser of either the surface or the jet type.
fee-junction circuit (*Teleph.*). A junction between 2 exchanges which entails more than the charge of a unit fee when used for connecting 2 subscribers.
feel (*Textiles*). Term describing the physical character of a cloth when handled.
feeler (*Weaving*). Mechanical or electrical device, used with automatic weft replenishing motion to determine when the pirn change is necessary.
feeler gauge (*Eng.*). A thin strip of metal of known and accurate thickness, usually one of a set, used to measure the distance between surfaces or temporarily placed between working parts while setting them an accurate distance apart.
feeler switch (*Elec. Eng.*). A switch sometimes forming part of the equipment of an autoreclose circuit-breaker; after the circuit-breaker has opened on a fault, the feeler switch closes a test circuit to determine whether or not the fault has cleared itself, and, if not, it prevents the circuit-breaker from reclosing.
feeling (*Psychol.*). An affective experience, essentially of pleasantness-unpleasantness or of some degree between.
feeling threshold (*Acous.*). Sound level applied to the ear, so loud that it results in a painful perception; about 120–130 on the phon scale.
feet (*Typog.*). The underside of the type, grooved in *foundry type*, but not in *Monotype*. When type is badly locked up, it is said to be *off its feet*.
feet-switch (*Elec. Eng.*). See **tropical switch.**
Fehling's solution (*Chem.*). A solution of cupric sulphate and potassium sodium tartrate (Rochelle salt) in alkali, used as an oxidizing agent. It is an important analytical reagent for aldehydes, glucose, fructose, etc., which reduce it to cuprous oxide.
feint (*Bind.*). The pale, edge to edge, horizontal ruling in account books and notebooks.
feldspar or **felspar** (*Min.*). A most important group of rock-forming silicates of aluminium, together with sodium, potassium, calcium, or (rarely) barium, crystallizing in closely similar forms in the monoclinic and triclinic systems. The chief members are *orthoclase* and *microcline* (*potassium feldspar*); *albite* (*sodium feldspar*); and the *plagioclases* (*sodium-calcium feldspar*). The form *felspar*, though still commonly used, perpetuates a false derivation from the German *fels* (rock); actually it is from the Swedish *feldt* (field).
feldspathic sandstone (*Geol.*). See **arkose.**
feldspathoids (*Min.*). A group of rock-forming minerals chemically related to the feldspars, but undersaturated with regard to silica content, and therefore incapable of free existence in the presence of magmatic silica. The chief members of the group are *haüyne*, *leucite*, *nepheline*, *nosean*, and *sodalite* (qq.v.).
Felici balance (*Elec. Eng.*). An a.c. electrical measuring bridge for determining mutual inductance between windings.
Felici generator (*Elec. Eng.*). A modern form of electrostatic high-voltage generator developed in France and comparable with a Van de Graaff generator.
feline agranulocytosis (*Vet.*). See **feline enteritis.**
feline distemper (*Vet.*). See **feline enteritis.**
feline enteritis (*Vet.*). Feline agranulocytosis; feline distemper; infectious feline panleucopaenia. A highly contagious and often fatal virus disease of cats, characterized by fever, vomiting, and diarrhoea due to enteritis, and a severe fall in the leucocyte content of the blood.
feline influenza (*Vet.*). See **feline pneumonitis.**
feline pneumonitis (*Vet.*). Feline influenza; cat 'flu. A contagious infection of the respiratory tract and lungs of cats, caused by a virus of the *psittacosis-lymphogranuloma* group; characterized by fever, oculo-nasal discharge, coughing and sneezing.
feline viral rhinotracheitis (*Vet.*). An acute viral infection of the upper respiratory tract of the cat.
fell (*Leather*). A skin or hide. (*Weaving*) The edge of the cloth in a loom, where the picks of weft are beaten up by the reed.
felling marks (*Textiles*). Material of a different kind or colour, forming lines across both ends of a length of cloth. Also known as **headings.**
felling subsere (*Bot.*). A developmental series of communities started by the felling of a wood.
fellmonger (*Leather*). A dealer in hides and skins, who prepares them for the tanner by removing the hair or wool.
felloe (*Carp., etc.*). (1) The outer part of the framing for a centre. (2) A segment of the rim of a wooden wheel, about which a tyre is usually shrunk. The term is sometimes applied to the whole rim.
felon or **fellon** (*Med.*). See **paronychia.** (*Vet.*) Suppurative arthritis of cattle; commonly associated with mastitis.
felsite (*Geol.*). An 'omnibus term' for fine-grained igneous rocks of acid composition, occurring as lavas or minor intrusions, and characterized by the felsitic texture—a fine patchy mosaic of quartz and feldspar, resulting from the devitrification of an originally glassy matrix.
felspar (*Min.*). See **feldspar.**
felstone (*Geol.*). An obsolete term for *felsite*.
felt (*Build.*). A fibrous material, treated so as to be rendered watertight, used as underlining for roofs; also as an overlining for roofs when underlaid with asphalt or an asphalt compound. (*Paper*) A woven blanket which carries the web of paper, either into the press rolls to squeeze the moisture from it, or to hold it in contact

with the drying cylinder. (*Textiles*) (1) A densely matted unwoven fabric of wool, wool/rayon or hair that has passed through a felting process of heat, steam, and pressure. (2) Heavily milled woven fabric with fibrous surface. (3) An imitation felt consisting of wool waste and wood pulp, and a bonding agent.

felt chamber (*Zool.*). In dipteran larvae, the spiracular atrium which is lined with fibrous processes to reduce water loss.

felt-grain (*For.*). Grain following the direction from pith to bark.

felting (*Paper*). The binding together of fibres in paper-making. (*Textiles*) The matting together of wool fibres.

female (*Bot.*). A flower having carpels and no stamens. (*Eng.*) See **male and female.** (*Zool.*) An individual the gonads of which produce ova.

female gauge (*Eng.*). See **ring gauge.**

female pronucleus (*Cyt.*). The nucleus remaining in the ovum after maturation.

female thread (*Eng.*). See **internal screw-thread.**

femerell (*Carp.*). A roof lantern having louvres for ventilation.

femic constituents (*Geol.*). Those minerals which are contrasted with the salic constituents in determining the systematic position of a rock in the *C.I.P.W.* scheme of classification. Note that these are the *calculated* components of the 'norm'; the corresponding *actual* minerals in the 'mode' are said to be *mafic*, i.e., rich in magnesium and iron.

femur (*Arch.*). See **meros.** (*Zool.*) The proximal region of the hind limb in land Vertebrates: the bone supporting that region: the third joint of the leg in Insects, *Myriapoda*, and some *Arachnida*. *adj.* **femoral.**

fence (*Eng.*). (1) A guard or stop to limit motion. (2) A guide for material, as in a circular saw or planing machine. (*Tools*) (1) An adjustable grinding edge or plate directing or limiting the movement of one piece with respect to another. (2) An attachment to a plane (cf. *fillister*) which keeps it at a fixed distance from the edge of the work, used especially on rebate and grooving planes.

Fenchel wet expansion test (*Paper*). A strip of paper under tension is immersed in water, and the resultant expansion is indicated by a dial connected to one end of the paper.

fenchenes (*Chem.*). $C_{10}H_{16}$, a series of terpenes derived from *fenchone* (q.v.).

fenchlorphos (*Chem.*). Dimethyl 2,4,5-trichlorophenyl phosphorothionate, used as an insecticide. Also **ronnel.**

fenchone (*Chem.*). A dicyclic ketone:

$$\begin{array}{ccc} CH_2- & CH- & C(CH_3)_2 \\ | & | & | \\ | & CH_2 & | \\ | & | & | \\ CH_2- & C{\cdot}CH_3- & CO \end{array}$$

Optical isomers occur in fennel and lavender oils (dextro-rotatory form), and in thuja oil (laevo-rotatory). Crystalline solid; m.p. 5°C and b.p. 192°C.

fender (*Civ. Eng.*). A timber, bag of old rope, or other object, placed against the edge of a pier, dock, etc., to take the impact from a vessel drawing up alongside. (*Elec. Eng.*) A metal cover attached to the end of the frame of an electric machine in such a way as to prevent accidental contact with live or moving parts. It does not carry a bearing. Also called a **protection cap.**

fender pile (*Civ. Eng.*). An upright timber serving as a *fender* (q.v.) protecting the edge of a dock-wall or wharf.

fender post (*Civ. Eng.*). A protective post on a street refuge to take the impact of a vehicle leaving the road. Also **guard post.**

fender wall (*Build.*). A dwarf brick wall supporting the hearthstone to a ground-floor fireplace.

fenestra (*Arch.*). A window or other opening in the outer walls of a building. *pl.* **fenestrae.** (*Zool.*) An aperture in a bone or cartilage, or an opening between two or more bones; in some *Dictyoptera*, one of the pair of pale spots on the head, thought to represent degenerate ocelli.

fenestra metotica (*Zool.*). In a typical chondrocranium, an opening behind the auditory capsule, through which pass the ninth and tenth cranial nerves and the internal jugular vein.

fenestra ovalis (*Zool.*). In Vertebrates in which the middle-ear is developed, the upper of two openings in the skeletal wall of the ear. Cf. *fenestra rotunda.*

fenestra pro-otica (*Zool.*). In a typical chondrocranium, an opening in front of the auditory capsule, through which pass the fifth, sixth, and seventh cranial nerves.

fenestra rotunda (*Zool.*). In Vertebrates in which the middle-ear is developed, the lower of two openings in the skeletal wall of the ear. Cf. *fenestra ovalis.*

fenestrate, fenestrated (*Bot.*). Perforated and divided into compartments, remotely suggesting a window. (*Zool.*) Perforated or having translucent spots.

fenestrated membrane (*Zool.*). The basement membrane of the compound eye in *Arthropoda*; upon it rest the proximal extremities of the ommatidia.

fenestration (*Build.*). The arrangement of window and other openings in the outer walls of a building. (*Surg.*) A surgical operation to improve hearing which involves a new 'window' being opened to the inner ear.

fenestra tympani (*Zool.*). See **fenestra rotunda.**

fenestra vestibuli (*Zool.*). See **fenestra ovalis.**

fenfluramine hydrochloride (*Pharm.*). *N*-Ethyl-α-methyl-3-trifluoromethylphenethylamine hydrochloride. A sympathomimetic agent, used as an anorectic.

fenoprop (*Chem.*). 2-(2,4,5-trichlorophenoxy) propanoic acid, used as a weedkiller. Also **silvex, 2,4,5-TP.**

fenson (*Chem.*). 4-chlorophenyl benzenesulphonate, used as an insecticide. Also **CPBS, PCPBS.**

fent (*Textiles*). A damaged piece of cloth cut from a length, or a short piece of material; usually sold by weight to wholesalers and market traders.

fentanyl citrate (*Pharm.*). *N*-(1-Phenethylpiperid-4-yl) propionanilide citrate. A powerful analgesic resembling morphine in its action.

fenthion (*Chem.*). Dimethyl 3-methyl-4 methylthiophenyl phosphorothionate, used as an insecticide.

fenuron (*Chem.*). *NN*-dimethyl-N′-phenylurea, used as a weedkiller.

FEP (*Plastics*). *F*luorinated *e*thene *p*ropene. Plastic with many of the properties of *PTFE*, having very good chemical resistance; it is unaffected by moisture and has a wide temperature range of application from −260° to +200°C.

ferbam (*Chem.*). Iron (III) dimethyl-dithiocarbamate: used as a fungicide.

ferberite (*Min.*). A member of the wolframite

group of minerals; theoretically pure tungstate of iron, but usually some of the iron is replaced by manganese.

Feret's diameter (*Powder Tech.*). *Statistical diameter* (q.v.), defined as the mean length of the distance between two tangents on opposite sides of the image of the particle.

fergusite (*Geol.*). An alkaline igneous rock containing large crystals of pseudoleucite in a matrix of aegirine-augite, olivine, apatite, sanidine, and iron oxides.

fergusonite (*Min.*). A rare mineral occurring in pegmatites, it is a niobate and tantalate of yttrium, and may contain small amounts of iron, cerium, calcium, etc.

Fermat's last theorem (*Maths.*). No integral values of x, y and z can be found to satisfy the equation $x^n+y^n=z^n$ if n is an integer greater than 2. This theorem is still unproved.

Fermat's principle of least time (*Phys.*). Principle stating that the path of a ray, e.g., of light, from one point to another (including refractions and reflections) will be that taking the least time.

Fermat's spiral (*Maths.*). See **parabolic spiral.**

ferment (*Chem., etc.*). A substance inducing *fermentation* (q.v.).

fermentation (*Chem.*). A slow decomposition process of organic substances induced by micro-organisms, or by complex nitrogenous organic substances (*enzymes*, q.v.) of vegetable or animal origin; usually accompanied by evolution of heat and gas. Important fermentation processes are the alcoholic fermentation of sugar and starch, and the lactic fermentation.

fermi (*Nuc.*). Femtometre (fm). A very small length unit, i.e., 10^{-15} m. Used in nuclear physics, being of the order of the radius of the proton, viz., 1·2 fm.

Fermi age (*Nuc.*). Slowing-down area for neutron calculated from *Fermi age theory*, which assumes that neutrons, on being slowed down, lose energy continuously and not in finite discrete amounts. This name is deprecated as Fermi age has dimensions of length^2. Also called **neutron age.**

Fermi characteristic energy level (*Electronics*). The highest occupied energy level in a partly filled band at a temperature of absolute zero. All energy levels below this will be occupied. In a semiconductor it will be the energy level for which the Fermi-Dirac distribution function has a value of 0·5.

Fermi constant (*Nuc.*). A universal constant which indicates the coupling between a nucleon and a lepton field. Its value is $1{\cdot}4\times10^{-50}$ J/m^3, and it is important in β-decay theory.

Fermi decay (*Electronics*). Theory of ejection of electrons as β-particles.

Fermi-Dirac gas (*Nuc.*). An assembly of particles which obey Fermi-Dirac statistics.

Fermi-Dirac-Sommerfeld law (*Nuc.*). See **FDS law.**

Fermi-Dirac statistics (*Nuc.*). Statistical mechanics laws obeyed by a system of particles whose wave function changes sign when two particles are interchanged.

Fermi distribution (*Electronics*). The energy distribution of electrons in a metal in which all levels are empty above the Fermi level except for a thin layer ($\sim kT$ in thickness) about the level.

Fermi level (*Electronics*). See **Fermi characteristic energy level.**

fermion (*Nuc.*). Generalized particle treated by Fermi-Dirac statistical theory. Fermions have total angular moments of $(n+\frac{1}{2})\hbar$, where n is 0 or an integer and $\hbar$ (the Dirac constant) is the quantum or unit of angular momentum. Both baryons and leptons are fermions and are subject to the Pauli exclusion principle. Their number always appears to be conserved in any type of interaction. See **baryon, lepton number.**

Fermi plot (*Nuc.*). See **Kurie plot.**

Fermi potential (*Elec.*). The equivalence of the energy of the Fermi level as an electric potential.

Fermi resonance (*Phys.*). Degeneracy in polyatomic molecules when two vibrational levels belonging to different vibrations (or combinations) may have nearly the same energy.

Fermi selection rules (*Nuc.*). A particular set of *nuclear selection rules*, which follow from a definite assumption about the coupling term.

Fermi temperature (*Phys.*). The degeneracy temperature of a Fermi-Dirac gas which is defined by E_F/k, where E_F is the energy of the Fermi level and k is Boltzmann's constant. This temperature is of the order of tens of thousands of kelvins for the free electrons in a metal.

fermium (*Chem.*). Manmade element. Symbol Fm, at. no. 100. Principal isotope ^{257}Fm, half-life 95 days.

ferractor (*Mag.*). Magnetic amplifier with ferrite core.

Ferranti effect (*Elec. Eng.*). The rise in voltage which takes place at the end of a long transmission line when the load is thrown off; it is due to the charging current flowing through the inductance of the line.

Ferranti-Hawkins protective system (*Elec. Eng.*). A discriminative protective system for feeders; core balance transformers are placed at each end, with their secondary windings connected to each other through pilot wires.

Ferranti meter (*Elec. Eng.*). A name often given to the mercury-motor type of supply meter invented by Ferranti.

Ferranti rectifier (*Elec. Eng.*). An old type of rectifier used in conjunction with constant-current transformers for arc lighting; it consists of a synchronously-driven commutator.

Ferraris instruments (*Elec. Eng.*). A name sometimes given to induction-type electrical measuring instruments first developed by Ferraris.

Ferraris motor (*Elec. Eng.*). An early type of 2-phase induction motor.

ferret (*Build.*). Glass smoke container broken in a drain or other pipe to find leaks.

ferri-, ferro-. Prefixes from L. *ferrum*, iron. (*Chem.*) Denoting trivalent and divalent iron respectively.

ferric [iron (III)] chloride (*Chem.*). $FeCl_3$. Brown solid. Deliquescent, soluble in ethanol. Uses: coagulant in sewage and industrial wastes, mordant, photo-engraving, etching of copper, chlorination and condensation catalyst, disinfectant, pharmaceutical, analytic reagent.

ferric [iron (III)] oxide (*Chem.*). Fe_2O_3. The common red oxide of iron. Used in metallurgy, pigments, polishing and theatrical rouge, gas purification, and as a catalyst.

ferric [iron (III)] sulphate (*Chem.*). $Fe_2(S_4O)_3$. A yellowish-white powder which dissolves slowly in water. Uses: pigments, water purification, dyeing, disinfectant, medicine.

ferricyanide, ferrocyanide [hexacyano ferrate (III) and hexacyano ferrate (II)] (*Chem.*). The complex ions $Fe(CN_6{}^{3-}$ and $Fe(CN)_6{}^{3-}$ respectively.

ferrimagnetism (*Mag.*). Antiparallel alignment of the magnetic moments of neighbouring ions, observed in ferrites and similar compounds.

ferrimolybdite (*Min.*). Hydrated molybdate of

iron. Most so-called molybdite is ferrimolybdite. It occurs as a yellowish alteration product of molybdenite.

Ferristor (*Mag.*). TN for a wirewound coil on a ferromagnetic core, encapsulated in epoxy resin.

ferrite (*Met.*). A ceramic iron oxide compound having ferromagnetic properties with a general formula MFe_2O_4, where M is generally a metal such as cobalt, nickel, zinc.

ferrite bead (*Telecomm.*). Small element of ferrite material used particularly for threading on to wire of transmission line to increase the series inductance. See **ferrite-bead memory.**

ferrite-bead memory (*Comp.*). A memory device in which a mixture of ferrite powders is fused into small beads directly to the current-carrying wires of a memory matrix.

ferrite core (*Mag.*). A magnetic core, usually in the form of a small toroid, made of ferrite material such as nickel ferrite, nickel-cobalt ferrite, manganese-magnesium ferrite, yttrium-iron garnet, etc. These materials have high resistance and make eddy-current losses very low at high frequencies.

ferrite-rod antenna (*Radio*). Small reception antenna, using ferrite rod to accept electromagnetic energy, output being from an embracing coil. Also called **loopstick antenna.**

ferritin (*Chem.*). Protein containing iron; found in the liver and spleen, and believed to act as an iron-storing material in the body.

ferro- Prefix. See **ferri-.**

ferroactinolite (*Min.*). An end-member compositional variety in the monoclinic amphiboles; essentially $Ca_2Fe_5Si_8O_{22}(OH, F)_2$, but the name is applied to a member of the actinolite series with more than 80% of this molecule.

Ferrocart (*Radio*). TN for early ferromagnetic material suitable for use at high frequencies on account of its small hysteresis and eddy-current losses, achieved by subdivision of material into fine particles.

ferrochromium (*Met.*). An alloy of iron and chromium (60–72% chromium) used in making additions of chromium to steel and cast-iron.

ferroconcrete (*Civ. Eng.*). *Reinforced concrete* (q.v.).

Ferrocrete (*Build., Civ. Eng.*). TN for a variety of rapid-hardening Portland cement.

ferrodynamometer (*Elec. Eng.*). Any *dynamometer* (q.v.) incorporating ferromagnetic material to enhance the torque.

ferroedenite (*Min.*). An end-member compositional variety in the monoclinic amphiboles; a hydrous sodium, iron, calcium, and aluminium silicate.

ferroelectric materials (*Elec.*). Dielectric materials (usually ceramics) with domain structure, which exhibit spontaneous electric polarization. Analogous to ferromagnetic materials (see **ferromagnetism**). They have relative permittivities of up to 10^5, and show dielectric hysteresis. Rochelle salt was first to be discovered. Others include barium titanite, and potassium dihydrogen phosphate.

ferrogedrite (*Min.*). See **gedrite.**

ferrohastingsite (*Min.*). An end-member compositional variety in the hornblende group of monoclinic amphiboles.

ferromagnetic amplifier (*Elec. Eng.*). A paramagnetic amplifier which depends on the non-linearity in ferroresonance phenomena at high RF power levels.

ferromagnetic resonance (*Elec. Eng.*). A special case of paramagnetic resonance, exhibited by ferromagnetic materials—sometimes termed **ferroresonance.** It involves unpaired spins of electrons which are very close together and so subject to large exchange forces.

ferromagnetism (*Mag.*). Phenomenon in which there is marked increase in magnetization in an independently established magnetic field. Some magnetism is retained in some ferromagnetic substances but can be removed by *demagnetization.* Certain elements (iron, nickel, cobalt) and alloys with other elements (titanium, aluminium) exhibit relative permeabilities much in excess of unity, up to 10^6 (*ferromagnetic materials*); some have marked hysteresis, leading to permanent magnets, storage devices for computers, magnetic amplifiers, etc.

ferromanganese (*Met.*). An alloy of iron and manganese, used in making additions of manganese to steel or cast-iron.

ferrometer (*Elec. Eng.*). An a.c. instrument for measuring the magnetization of a ferromagnet.

ferromolybdenum (*Met.*). An alloy of iron and molybdenum (55–65% molybdenum) used in adding molybdenum to steel and cast-iron.

ferronickel (*Met.*). Alloy of iron and nickel containing more than 30% of nickel. Lower nickel alloys are known as nickel steel. See **Elinvar, Invar, Mumetal, permalloy.**

ferroprussiate paper (*Paper*). A paper specially made for the production of blueprints. It is treated with a sensitizing solution mixed with gum.

ferroresonance (*Elec. Eng.*). See **ferromagnetic resonance.**

ferrosilicon (*Met.*). An alloy of iron and silicon, used in making additions of silicon to steel and cast-iron. When containing 15% silicon, widely used in dense media processes.

ferrospinel (*Met.*). A crystalline material which has the equivalent function to that of the M in a ferrite. See **ferrite.**

ferrotype (*Photog.*). A wet-collodion positive on a plate of darkened metal. Formerly favoured by street-photographers owing to immediate availability of the print. Also **melanotype, tin-type.**

ferrous [iron (II)] oxide (*Chem.*). FeO. Black oxide of iron.

ferrous [iron (II)] sulphate (*Chem.*). $FeSO_4 7H_2O$. See also **copperas.**

Ferroxcube (*Met.*). TN for a ferrite, particularly useful for rod aerials and memory units.

ferroxyl indicator (*Chem.*). A little potassium hexacyano ferrate (III) and phenolphthalein, together with a corroding solution, e.g., of sodium chloride, made into a jelly with agar. It is used to show the presence of anodic and cathodic areas in an apparently uniform piece of iron, by turning blue and pink respectively.

ferruginous clay (*Geol.*). An impure clay rock, with an admixture of iron compounds.

ferruginous deposits (*Geol.*). Sedimentary rocks containing sufficient iron to justify exploitation as iron ore. The iron is present, in different cases, in silicate, carbonate, or oxide form, occurring as the minerals chamosite, thuringite, siderite, haematite, limonite, etc. The ferruginous material may have formed contemporaneously with the accompanying sediment, if any, or may have been introduced later.

ferrule (*Carp.*). The brass ring round the handle of a chisel, or similar tool, at the end where the tang enters. (*Eng.*) (1) A short length of tube. (2) A circular gland nut used for making a joint. (3) A slotted metal tube into the ends of which the conductors of a joint are inserted. The whole is soldered solid. When the conductors are oval, the ferrule is in two parts to allow for the fact that the major axes of the oval sections may not coincide. (*Horol.*) A small grooved pulley around which is wrapped the string of a bow, to give rotary motion. (*Plumb.*) A side opening in a pipe, fitted with a screwed plug giving access for inspection and cleaning.

Ferry (*Elec. Eng.*). A nickel-copper alloy (44% nickel) used for low-temperature resistances; the wire runs at a black heat.

fertile (*Bot.*). Able to produce asexual spores and/or sexual gametes. (*Nuc.*) Capable of conversion into fissile material in a reactor.

fertile flower (*Bot.*). A pistillate flower.

fertilisin (*Zool.*). A substance which is present in the cortex of an ovum and assists in the activation of the latter.

fertility (*Ecol.*). The number or percentage of eggs produced by a species or individual which develop into living young. Cf. *fecundity*.

fertility drug (*Biochem.*). A popular name for any substance, whether synthetic (e.g., clomiphene) or of natural origin (e.g., human chorionic gonadotrophin and FSH), which induces ovulation (sometimes superovulation) in an apparently infertile female. See **gonadotrophins.**

fertilization (*Biol.*). The union of 2 sexually differentiated gametes to form a zygote.

fertilization cone (*Zool.*). A conical projection of protoplasm arising from the surface of an ovum and giving rise to a filamentous process which adheres to a spermatozoon-head and then retracts, dragging the spermatozoon into the cytoplasm of the ovum.

fertilization tube (*Bot.*). See **conjugation tube.**

fertilizer (*Agric.*). Specific chemical for application to the soil to rectify its growth-promoting qualities for various types of plants, and for adjusting its acidity or alkalinity. Typical fertilizers are pure ammonia, sulphates and nitrates of ammonium or sodium, superphosphate of lime, basic slag, lime. See **manure.**

Fery spectrograph (*Optics*). A spectrograph in which the only optical element is a back-reflecting prism with cylindrically curved faces. Considerable astigmatism is experienced with this instrument.

Fessenden detector (*Electronics*). An early form of electrolytic detector, comprising two polarized electrodes immersed in an electrolyte.

Fessenden oscillator (*Acous.*). A low-frequency underwater sound source of the moving-coil type.

festination (*Med.*). Involuntary quick walking with short steps, occurring in certain diseases of the nervous system, e.g., in paralysis agitans.

festoon drier (*Paper*). A drying chamber in which hot air circulates around the web, the paper being carried through in loops about 3 m high on wooden rods.

FET. See **field-effect transistor.**

fetch (*Civ. Eng.*). The distance of the nearest coast in the direction of the strongest and most prevalent winds on a harbour site.

fetishism (*Psychiat.*). A pathological condition or sexual perversion in which sexual attraction and gratification is obtained from various non-genital eroticized areas of the body, or from any object which has become similarly emotionally charged.

fetlock (*Vet.*). The metacarpophalangeal and metatarsophalangeal regions of horses.

fettler (*Textiles*). An operative who clears away the fibrous waste and dirt from card cylinders, doffers, and rollers, etc.

fettling (*Met.*). 'Making good', e.g., ore hearth and walls of an open-hearth, reverberatory or cupola furnace, where erosion or chemical action has damaged the refractory lining.

Feulgen stain (*Micros.*). A specific stain for deoxyribonucleic acid, used particularly for staining the nucleus of higher plants and animals and the chromatinic material in other organisms. The dye used is basic fuchsin, which after reduction reacts with aldehyde groups liberated from deoxyribonucleic acid by hydrochloric acid, giving a violet condensation product.

fever (*Med.*). The complex reaction of the body to infection, associated with a rise in temperature. Less accurately, a rise of the temperature of the body above normal.

f-factor (*Radiol.*). The ratio of absorbed dose (rads) to exposure dose (röntgen) for a given material and X-ray energy.

F.H.P. (*Eng.*). Abbrev. for *friction horsepower.*

Fibonacci numbers, sequence, series (*Maths.*). Sequence of integers, where each is the sum of the two preceding it. 1,1,2,3,5,8,13,21,...

fibre (*Bot.*). (1) A very narrow, elongated, thick-walled cell, tapering to a sharp point at both ends. (2) A very delicate root. (*Met.*) Term for arrangement of the constituents of metals parallel to the direction of working. It is applied to the elongation of the crystals in severely cold-worked metals, to the elongation and stringing out of the inclusions in hot-worked metal, and to preferred orientations. (*Textiles*) Any type of vegetable, animal, regenerated, synthetic, or mineral fibre or filament from which yarns and fabrics are manufactured, by spinning and weaving, or knitting, felting, or bonding.

fibre-board (*Build.*). Building or insulating board made from fibrous material such as wood pulp, waste paper, and other waste vegetable fibre. May be homogeneous, bitumen-bonded or laminated. See **hardboard, insulating board.**

fibre camera (*Instr.*). A type of X-ray camera modified for work with fibrous materials.

Fibreglass. TN used for a range of *glass fibre.*

fibre metallurgy (*Met.*). The metallurgy of the manufacture of the fibres and products made from metallic fibres by sintering.

fibre optics. The technique of using flexible cables composed of bundles of very fine (14 μm diam.), optically insulated glass fibres to inspect areas in, e.g., machinery, the human body, which are difficult to examine by other means. Light is reflected down the inside of the fibres, transmitting an undistorted image from one end to the other. Used also for low-loss wideband transmission of telecommunication signals.

fibre tracheid (*Bot.*). An elongated cell found in wood, which has thicker walls with fewer pits than a tracheid, and thinner walls with more pits than a fibre.

fibril, fibrilla (*Bot.*). (1) A small fibre. (2) A tiny fibrelike branch. (*Zool.*) Any minute threadlike structure, such as the longitudinal contractile elements of a muscle fibre. *adj.* **fibrillar, -ate.**

fibrillation (*Med.*). (1) Twitching of individual muscle fibres, or bundles of fibres, in certain nervous diseases. (2) Incoordinate contraction of individual muscle fibres of the heart, giving rise to an irregular and inefficient action of the heart; especially, *auricular fibrillation.*

fibrillose (*Bot.*). (1) Covered with a loose fibrous coating. (2) Appearing as if made of fibres.

fibrin (*Biochem.*). An insoluble protein, precipitated in the form of a meshwork of fibres when blood coagulates, formed from soluble fibrinogen, present in blood plasma.

fibrinogen (*Biochem.*). A protein, of the keratin-myosin group, contained in the plasma of blood. The coagulation of fibrinogen is responsible for the clotting of blood and the production of *fibrin.*

fibrinolysin (*Biochem.*). An enzyme in the blood which causes breakdown of fibrin in blood clots; a drug having the same effect.

fibrino-purulent (*Med.*). Containing fibrin and pus.

fibro-adenoma (*Med.*). An adenoma in which there is an overgrowth of fibrous tissue.

fibroblasts (*Zool.*). Flattened connective-tissue cells of irregular form, believed to be responsible for the secretion of the white fibres; lamellar cells.

fibrocartilage (*Zool.*). A form of cartilage which has white or yellow fibres embedded in the matrix.

fibrocystic disease (*Med.*). Osteitis fibrosa. A condition in which there may be (*a*) a single cyst in a bone, or (*b*) cysts in many bones; the latter (generalized osteitis fibrosa; Von Recklinghausen's disease of bone) is due to loss of calcium salts from the bone, and is associated with a tumour of the parathyroid glands.

fibroid (*Med.*). Resembling fibrous tissue; a fibromyoma.

fibroid phthisis (*Med.*). Chronic pulmonary tuberculosis, in which there is an overgrowth of fibrous tissue in the lung.

fibroin (*Biochem.*). A tough, elastic protein formed by the denaturation of *fibroinogen* after extrusion.

fibroinogen (*Biochem.*). The substance secreted by the silk glands of *Lepidopteran* and *Trichopteran* larvae; it hardens by denaturation to form *fibroin*.

fibrolipoma (*Med.*). A tumour composed of fibrous and fatty tissue.

fibrolite (*Min.*). A variety of the aluminium silicate *sillimanite* (q.v.) occurring as felted aggregates of exceedingly thin fibrous crystals.

fibroma (*Med.*). A tumour composed of fibrous tissue.

fibromyectomy (*Surg.*). Removal of a fibromyoma.

fibromyoma (*Med.*). Fibroid. A tumour composed of fibrous tissue and unstriped muscle fibres, commonly found in the uterus.

fibromyositis (*Med.*). Inflammation (usually rheumatic) of fibrous tissue in muscle and in the muscle fibres adjacent to it.

fibrose (*Med.*). To form fibrous tissue.

fibrosis (*Med.*). The formation of fibrous tissue as a result of injury or inflammation of a part, or of interference with its blood supply.

fibrositis (*Med.*). Inflammation (especially rheumatic) of fibrous tissue.

fibrotic (*Med.*). Pertaining to fibrosis.

fibrotile (*Build.*). A corrugated tile, 4 ft by 3 ft 10 in. (122 × 117 mm), of asbestos cement.

fibrous concrete (*Build.*). Concrete in which fibrous aggregate, such as asbestos, sawdust, etc., is incorporated either as alternative or additional to the sand and gravel.

fibrous glass (*Heat*). Insulation material consisting of mats of glass fibre bound with phenol-formaldehyde resin.

fibrous layer (*Bot.*). A layer of cells having their walls thickened irregularly by thin bands of material, occurring in the wall of an anther. The walls shrink unevenly as they dry, a strain is set up, and the wall of the anther is torn, releasing the pollen.

fibrous plaster (*Build.*). Prepared plaster slabs formed of canvas stretched across a wooden frame and coated with a thin layer of gypsum plaster.

fibrous tissue (*Zool.*). A form of connective-tissue consisting mainly of bundles of white fibres; any tissue containing a large number of fibres.

fibrovascular bundle (*Bot.*). A vascular bundle accompanied, usually on its outer side, by a strand of sclerenchyma.

fibula (*Build.*). A bent iron bar, used to fasten together adjacent stones. (*Zool.*) In *Trichoptera* and some *Lepidoptera*, a lobe formed from the jugal area of the forewing and engaging with the costal spines of the hindwing. In *Tetrapoda*, the posterior of the two bones in the middle division of the hind-limb.

fibulare (*Zool.*). A bone of the proximal row of the tarsus, in line with the fibula, in *Tetrapoda*.

fiche (*Photo.*). Sheet (usually A6 size) of microfilm containing reduced (20 × to 3000 ×) images of pages of one book, report, etc.

Fick's law of diffusion (*Chem.*). The rate of diffusion in a given direction is proportional to the negative of the concentration gradient, i.e.,

$$\text{molar flux} = -D\frac{\partial C}{\partial x},$$

where D is the diffusion coefficient.

Ficoidales (*Bot.*). An order of the *Polypetalae*; the flowers are usually regular; the ovary has one loculus with parietal placentation, or more than one loculus with axile or basal placentation. The seed may be endospermous with the embryo curved, or nonendospermous with the embryo cyclical or oblique.

fidelity (*Ecol.*). The degree of restriction of a species to a particular situation, a species with high fidelity having a strong preference for a limitation to a particular community. (*Telecomm.*) The exactness of reproducibility of the input signal at the output end of a system transmitting information.

Fidler's gear (*Build., Civ. Eng.*). Hoisting and lowering apparatus adapted to lowering large masonry blocks at any required angle.

fiducial (*Surv., etc.*). Said of a line or point established accurately as a basis of reference.

fiducial points (*Elec. Eng.*). Points on the scale of an indicating instrument located by direct calibration, as contrasted with the intervening points, which are inserted by interpolation or subdivision.

fiducial temperature (*Meteor.*). The temperature at which a sensitive barometer reads correctly, the maker's calibration holding for latitude 45° at the temperature 285 K (12°C) only.

field (*Comp.*). Area obtained by a vertical division of a punched card. (*Maths.*) See **ring**. (*Phys.*) (1) The interaction of bodies or particles is explained in terms of fields, viz., electric, magnetic, gravitational, acoustic, lepton, and baryon fields; see **field theory**. (2) Space in which there are electromagnetic oscillations associated with a radiator. That component which represents the interchange of energy between radiator and space (practically negligible beyond a few wavelengths) is termed *induction field*; *radiation field* represents energy which is lost from the radiator to space. The region where components radiated by antenna elements are parallel is called the *Fraunhofer region*, and where not the *Fresnel region*. The latter will exist between the antenna and the Fraunhofer region, and is usually taken to extend a distance $2D^2/\lambda$, where λ is the wavelength of the radiation and D is the aerial aperture in a given aspect. (*TV*) A set of scanning lines, which, when interlaced with other such sets, construct the complete picture.

field ampere-turns (*Elec. Eng.*). The ampere-turns producing the magnetic field of an electric machine.

field-breaking resistance (*Elec. Eng.*). See **field-discharge resistance**.

field-breaking switch (*Elec. Eng.*). See **field-discharge switch.**

field coil (*Elec. Eng.*). The coil which carries the current for producing the m.m.f. to set up the flux in an electric machine; occasionally called a **field spool.** See also **magnetizing coil.**

field control (*Elec. Eng.*). The adjustment of the field current of a generator or motor to control the voltage or speed respectively.

field copper (*Elec. Eng.*). A term used in the design of electrical machines to denote the total quantity of copper used in the field windings of a machine.

field current (*Elec. Eng.*). The current in the field winding of an electric machine.

field density (*Elec., Mag.*). The number of lines of force passing normally through unit area of an electric or magnetic field.

field discharge (*Elec.*). The passage of electricity through a gas as a result of ionization of the gas; it takes the form of a brush discharge, an arc, or a spark. Also **electric discharge.**

field-discharge resistance (*Elec. Eng.*). A discharge resistance used for connecting across the terminals of the shunt or separately excited field winding of an electric machine, to prevent high induced voltages when interrupting the field circuit. Also called a **field-breaking resistance.**

field-discharge switch (*Elec. Eng.*). A switch for controlling the field circuit of a generator. It is provided with special contacts so that a discharge resistance is connected across the winding at the moment of breaking the field circuit. Also called a **field-breaking switch.**

field-diverter rheostat (*Elec. Eng.*). A rheostat connected in parallel with the series field winding or the compole winding of a d.c. machine to give control of the m.m.f. produced by the winding, independently of the current flowing through the main circuit.

field drains (*Civ. Eng.*). Brick-clay or terracotta pipes laid end-to-end in a trench.

fielded panel (*Join.*). A panel which is moulded, sunk, raised, or divided into smaller panels.

field-effect transistor (*Electronics*). One consisting of a conducting channel formed from a strip of *n*- (or *p*-) type semiconductor with a gate, (typically formed by a collar of *p*- (or *n*-) type material), the voltage applied to which controls the resistivity of the whole strip. The input impedance to the gate is very high and signals applied to this are amplified in a similar way to those applied to the grid of a triode.

field emission (*Electronics*). See **autoemission.**

field-emission microscope (*Phys.*). One in which the positions of the atoms in a surface are made visible by means of the electric field emitted on making the surface the positive electrode in a high-voltage discharge tube containing argon at very low pressure. When an argon atom passes over a charged surface atom, it is stripped of an electron, and thus is drawn toward the negative electrode, where it hits a fluorescent screen in a position corresponding to that of the surface atom.

field enhancement (*Elec.*). Local increase in electrical field strength due to convex curvature of an electrode, or proximity of another electrode.

field form (*Elec. Eng.*). A curve showing the value of the flux density at all points in the air gap of an electric machine.

field-form factor (*Elec. Eng.*). The ratio of average value of the flux density in the air gap of a machine to the maximum value.

field-free (*Electronics*). Said of electron emission when there is no electric field at emitting surface.

field frequency (*TV*). Product of the *frame frequency* and number of *fields* per frame.

field intensity (*Elec.*). Same as **field strength.**

Fieldistor (*Electronics*). TN for transistor which uses an external field for the control of the electron flow.

field lens (*Optics*). Lens placed in or near the plane of an image to ensure that the light to the outer parts of the image is directed into the subsequent lenses of the system, and thus uniform illumination over the field of view is ensured.

field magnet (*Elec. Eng.*). The permanent or electro-magnet which provides m.m.f. for setting up the flux in an electric machine. See **rotating-field magnet.**

field of force (*Elec.*). Principle of *action at a distance*, i.e., mechanical forces experienced by an electric charge, a magnet, or a mass, at a distance from an independent electric charge magnet or mass, because of fields established by these and described by uniform laws.

field of view (*Optics*). The area over which the image is visible in the eyepiece of an optical instrument. It is usually limited by a circular stop in the focal plane of the eye-lens. See **sagittal field, tangential field.**

field oscillator (*Electronics*). Same as **framing oscillator.**

field plates (*Elec. Eng.*). The conductors used in some forms of electrostatic machine for inducing charges on the carriers.

field rheostat (*Elec. Eng.*). A variable resistance (rheostat) connected in series or parallel with the field winding of an electrical machine for the purpose of varying the current in the winding. Also called a **field regulator.**

field rivet (*Eng.*). A rivet which is put in when the work is on the site. Also called **site rivet.**

field-sequential (*TV*). In colour TV, pertaining to the association of individual primary colours with successive fields.

field spider (*Elec. Eng.*). The portion of an electric machine which supports the revolving magnet poles.

field-splitting switch (*Elec. Eng.*). A switch sometimes used for subdividing the field winding of a rotary converter into a number of groups to prevent excessive induced voltages during the starting period.

field spool (*Elec. Eng.*). A spool or bobbin upon which a field coil may be carried. Occasionally used also to denote a *field coil.*

Field's siphon flush tank (*San. Eng.*). A tank used for automatically flushing sewers and drains by a siphoning method.

field strength (*Elec.*). Vector representing the quotient of a force and the *charge* (or *pole*) in an electric (or magnetic) field, with the direction of the force; also called **field intensity.** (*Radio*) Electromagnetic wave, in volts/metre, i.e., that which induces an e.m.f. of one volt in an antenna of one metre effective height. Usually measured by frame or loop antenna.

field suppressor (*Elec. Eng.*). An arrangement for automatically reducing the field current of a generator when a short-circuit or other fault occurs on the machine or its adjacent connexions.

field-synchronizing impulse (*TV*). That transmitted at the end of each field-scanning cycle, to synchronize time-base generator of the receiver with that at the transmitter.

field theory (*Phys.*). As yet unverified attempt to link the properties of all fields (electric, magnetic, gravitational, nuclear) into a unified system.

field tube (*Eng.*). A special form of boiler tube,

consisting of an outer tube which is closed at its lower end and contains a second concentric tube, down which the water passes to return up the annular space between the two.

field winding (*Elec. Eng.*). The winding placed on the field magnets of an electric machine and producing the m.m.f. necessary to set up the exciting flux.

fiery mine (*Mining*). One in which there is a possibility of explosion from gas or coal-dust.

figured twills (*Textiles*). Fabrics having simple figure effects, combined with the diagonal twill lines.

figure-eight wire (*Elec. Eng.*). Wire used for the overhead contact wire of a traction system; it has a cross-section resembling a figure eight, the object being to provide a groove whereby it can be supported from the ears attached to the supporting wires.

figure of loss (*Elec. Eng.*). A term occasionally used in connexion with transformers to denote the energy loss per unit mass of material (iron or copper).

figure of merit (*Elec. Eng.*). General parameter which can be derived specifically for the performance of instruments, e.g., current or power required to deflect the light beam of a reflecting galvanometer 1 mm on a scale at 1 m, or power amplification of a magnetic amplifier.

figure sheet (*Textiles*). The form into which a lace pattern draft is converted, to simplify manufacture in a Levers machine.

filament (*Bot.*). (1) A chain of cells set end to end. (2) The stalk of a stamen. (*Electronics, Light*) A fine wire of high resistance, which is heated to incandescence by the passage of an electric current. In an electric filament-lamp it acts as the source of light, and in thermionic tubes it acts as an emitter of electrons. (*Textiles*) (1) Fibre produced by the silkworm and extracted from the cocoon. (2) Each single strand issuing from the spinneret of a manmade fibre spinning machine, which joins the others to form a thread. (*Zool.*) Any fine threadlike structure; the axis of a down-feather.

filamentary transistor (*Electronics*). One whose length is much greater than its transverse dimensions.

filament cathode (*Electronics*). Same as **directly-heated cathode.**

filament current (*Electronics*). That required to heat a filament in a thermionic valve.

filament efficiency (*Electronics*). Ratio of the current emitted from a filament to that required to heat it.

filament electrometer (*Electronics*). A type of electrometer in which the moving part consists of a fine wire filament, the deflection of which is observed through a microscope.

filament getter (*Electronics*). One which adsorbs gas readily when hot, and is used for this purpose in a high-vacuum assembly.

filament-lamp (*Light*). An electric lamp in which a filament in a glass bulb, evacuated or filled with inert gas, is raised to incandescence by the passage through it of an electric current.

filament limitation (*Electronics*). The limitation of anode current in a thermionic tube by the finite emission from the filament, as distinct from *space-charge limitation* (q.v.); also called **filament saturation.**

filament plate (*Zool.*). A paired sheet of cells connecting the apex of the genital rudiment with the heart rudiment of the same side of the body in female insect embryo.

filament reactivation (*Electronics*). Process of restoring the emission from used thoriated tungsten directly-heated valves, by operating for a short time at a temperature well above normal.

filament saturation (*Electronics*). Same as **filament limitation.**

filament transformer (*Electronics*). One designed specifically to supply the filaments of electron tubes.

fila olfactoria (*Zool.*). The groups of fibres in the sensory system running from the olfactory epithelium of the olfactory sacs to the olfactory bulbs in *Craniata.*

filariasis (*Med., Vet.*). Infestation with nematode worms of the family *Filariidae*, which inhabit the lymphatic channels, often causing elephantiasis.

Filarioidea (*Zool.*). A superfamily of *Nematoda*, comprising a number of parasitic forms in which an intermediate host is always necessary for development; paired lateral lips may occur, and the oesophagus is without a bulb; the eggs hatch in utero or contain embryos when laid.

filar micrometer (*Micros.*). Modified eyepiece graticule with movable scale or movable cross-hair.

file (*Comp.*). General term for a set of data stored sequentially, e.g., along one track of a magnetic memory. (*Eng.*) A metal-cutting tool, blade, or rod of hardened steel on which small teeth have been cut. The file section may be flat, round, triangular, half-round, etc., and either parallel or taper; the degree of fineness of the teeth is specified as rough, middle, bastard, second-cut, smooth, and dead-smooth, corresponding to a range of from about 5 to 50 teeth/cm. See also **float, rasp.**

file card (*Eng.*). A short wire brush used to clean swarf from files.

filet ground (*Textiles*). Any type of square ground-mesh of lace.

filial generation (*Gen.*). The offspring of a cross mating, the second filial generation being the direct offspring of the first filial generation inbred, etc.

Filibranchiata (*Zool.*). An order of *Lame libranchiata* in which the branchial axis of the gill is united to the body throughout its length and bears parallel, ventrally directed, and reflected filaments, which are joined to one another by interlocking cilia; there is a well-developed byssus gland; ciliary feeders. Sea Mussels, Scallops, Pearl Oysters, Thorn Oysters, Hammer Oysters, Wing Shells, etc.

Filicales, Filices (*Bot.*). The ferns. A large order of the class *Felicineae*, with large, often highly compound leaves, bearing the sporangia. They occur chiefly in damp shady places and show a well-marked alternation of generations, a small inconspicuous gametophyte bearing the sexual organs, and a large sporophyte (the ordinary fern plant) producing the spores.

Filicineae (*Bot.*). A class of the *Pteropsida*, in which the sporophyte is divided into stem, leaf, and root. The macrophyllous leaves are arranged spirally on the stem, bearing sporangia on their margins, or underside. If the stem is siphonostelic, there are leaf-gaps.

filicinean (*Bot.*). Relating to ferns.

Filicopsida (*Bot.*). See **Filicineae.**

filiform (*Bot., Zool.*). Threadlike, e.g., *filiform antennae.*

filiform papillae (*Histol.*). The most numerous of the lingual papillae. Each has a cap of cornified epithelium which is particularly well developed in some animals, e.g., cat.

filings coherer (*Radio*). An early detector, comprising 2 electrodes placed end to end in a glas tube and almost touching, the inter-electrode space being partially filled with metallic filings. The inter-electrode resistance remains high until the arrival of a signal, when it falls considerably.

filipendulous (*Bot.*). Having swellings of considerable size along, or at the ends of, thin roots.

Filippi's glands (*Zool.*). In lepidopterous larvae, a pair of accessory glands associated with the silk-glands. Their secretion enables the threads to adhere and also facilitates hardening.

fill (*Build., Civ. Eng.*). Rock or soil dumped to bring a site to the required level. After dumping, the fill has to be progressively consolidated.

filled band (*Chem.*). An energy-level band in which there are no vacancies. Its electrons do not contribute to valence or conduction processes.

filled cloth (*Textiles*). A cloth to which sizing material has been added to increase the weight or improve the appearance.

filled gold (*Met.*). Gold plating, sheet gold being backed by base metal.

filler (*Acous.*). The inert fine-grained material which is added to the binding material of the normal gramophone record to give weight and colour. Examples are carbon black and pigments. (*Civ. Eng.*) A finely divided substance added to bituminous material for road surfacing, in order to reduce it to a suitable consistency. (*Paint.*) (1) A material used to fill in the pores of, or any holes in, wood, plaster, etc., which is to be painted, varnished, or otherwise decorated. (2) An *extender* (q.v.). (*Paper*) Mineral powders added to the paper to improve the quality and in some cases to lower the cost, e.g., China clay, titanium dioxide. (*Plastics*) An inert solid substance added to a synthetic resin either for economy (e.g., wood-flour, clay) or to modify its properties (e.g., mica, aluminium or other metal powder).

filler joist floor (*Build.*). A type of floor in which the principal supporting members are steel joists, the gaps between the joists being filled with concrete.

filler metal (*Eng.*). The metal required to be added at the weld in welding processes in which the fusion temperature is reached. In metal-electrode welding, the electrode, usually fluxed and coated, is melted down to provide the filler metal.

filler rod (*Elec. Eng.*). See **welding rod.**

fillet (*Aero.*). A fairing at the intersection of 2 surfaces intended to improve the airflow. (*Arch.*) (1) A flat and narrow surface separating or strengthening curved mouldings. (2) A *listel.* See **facette.** (*Bind.*) A band or line of gold leaf, or a plain band or line, on a cover. (*Carp.*) A thin strip of wood fixed into the angle between 2 surfaces, to a wall as a shelf support, to a floor as a door stop, etc. (*Eng.*) (1) A narrow strip of metal raised above the general level of a surface. (2) A radius provided, for increased strength, at the intersection of 2 surfaces, particularly in a casting. (*Plastics*) Term used to describe a rounded internal corner in a plastic article, to avoid a possible weakness arising from an abrupt change in cross-section.

filleting. See **cement fillet, fillet.**

fillet weld (*Eng.*). A weld at the junction of 2 parts, e.g., plates, at right angles to each other, in which a fillet of welding metal is laid down in the angle created by the intersection of the surfaces of the parts.

filling (*Mining*). (1) The loading of tubs or trucks with coal, ore, or waste. (2) The filling-up of worked-out areas in a metal-mine. (*Textiles*) (1) Term used in America and Canada for weft yarns. (2) Size added to cloth, either during dyeing and finishing, or subsequently.

filling-in (*Build.*). The operation of building in the middle part of a wall, between the face and the back.

filling pile (*Hyd. Eng.*). A pile serving to retain the sheeting of a coffer-dam.

filling post (*Join.*). A middle post in a timber frame.

fillister (*Carp., Join.*). (1) A groove in the edge of a sash bar, to receive the glass and putty. (2) A rabbeting plane having a movable stop to regulate depth of cut.

fillister-head screw (*Eng.*). A cheese-head screw with a slightly convex upper head surface.

film (*Chem.*). (1) Any thin layer of substance, e.g., that which carries a light-sensitive emulsion for photography, that which carries iron (III) oxide particles in a matrix for sound recording. (2) Thin layer of material deposited, formed or adsorbed on another, down to mono-molecular dimensions, e.g., electroplated films, oxide on aluminium, sputtered depositions on glass or microcomponents. (*Cinema., Photog.*) Transparent cellulose acetate or polyester base coated with light-sensitive emulsion. Cellulose nitrate film, formerly widely used. It is now virtually superseded owing to its inflammability.

film analysis (*Work Study*). Frame-by-frame examination of a film of an operation to determine the state of activity of the subject during each exposure.

film badge (*Radiol.*). Small photographic film used as radiation monitor and dosimeter. Normally worn on lapel, wrist, or finger and sometimes partly covered by cadmium and tin screens so that exposure to neutrons, and to beta and gamma rays, can be estimated separately.

film circuits (*Electronics*). Micro-electronic circuits in which the passive components and their metallic interconnexions are formed directly on an insulating substrate and the active semiconductor devices (usually in wafer form) are subsequently added.

film coefficient (*Chem. Eng.*). The constant of proportionality between the flux (e.g., of mass or heat) and the difference in driving force (e.g., concentration or temperature) across a film. Symbol *h*.

film foils (*Bind.*). Blocking foils with a viscose film substrate. These give better release and finer detail than *paper foils* (q.v.).

film gate (*Cinema.*). The detachable element which guides the film past the exposing aperture and holds it in position during exposure.

film glue. Resin-impregnated paper, or phenolic resin film, used in making resin-bonded plywood.

film recording (*Acous.*). Process of recording sound on a sound-track on the edge of cinematograph film, for synchronous reproduction with the picture.

filmsetting (*Typog.*). The end product is on photographic film or paper ready for platemaking. The 'major' systems are keyboard-operated and have correcting techniques. 'First generation' are *Fotosetter, Fotomatic*, based on *Intertype*, and *Monophoto*, based on *Monotype*. 'Second generation' use new principles and include *Alphatype, ATF Typesetter, Fototronic, Linofilm, Linofilm Quick, Lumitype-Photon, Megatype, P & M Filmsetter, Sapton.* 'Third generation' have enormously greater output and include *Linotron Lumizip.* The 'minor' systems are hand-operated, used for headlines and short text, employ a variety of techniques: using separate photomats, *Hadego*,

Magnoscop, Optiset, the first two having a lens system; using alphabet disks or slides and a lens system, *Additype, Diatype, Faxi-Foto, Letterphot, Luxotyp, Monophoto Photo-Lettering Machine, Photonymograph, Photo-Typositor, Shaken, Starlettograph*; using separate discs or slides for each size, *Alphagraph, Filmatype, Form-o-Type, Foto-Rex, Foto-Riter, Graphatron, Headliner, Protype, Strip Printer, Typro*.

film sizing (*Min. Proc.*). Concentration of finely divided heavy mineral in gently sloped surfaces which may be plane, riffled, or vibrated. See **buddle, Wilfley table.**

film speed (*Cinema.*). The speed at which a cinematograph film passes through a gate; expressed in feet/min or sec, or frames/sec. For standard film, the speed is 90 ft/min, or 24 frames/sec. (*Photog.*) The *speed* (q.v.) of an emulsion.

film store (*Comp.*). One in which magnetic effect is produced in magnetic alloy deposited on a film when currents in grid wires on both sides coincide.

film theory (*Chem. Eng.*). Mass transfer between two well-stirred fluid phases, or between a solid and a well-stirred fluid, takes place by eddy diffusion in the bulk fluid phase, combined with molecular diffusion through a stagnant film at the interface, the latter being the slower process.

Filon (*Plastics*). TN for polyester/glass fibre/nylon sheeting used for constructional purposes, chiefly for roof lights but also for partitions and wall cladding.

filoplasmodium (*Bot.*). A netlike plasmodium, as found in the *Labyrynthulales*.

filoplumes (*Zool.*). Delicate hairlike contour feathers with a long axis and few barbs, devoid of locking apparatus at the distal end.

filopodia (*Zool.*). Fine, threadlike pseudopodia of some *Sarcodina*.

filose (*Zool.*). Said of pseudopodia which are long and slender, and composed entirely of ectoplasm. Cf. *lobose*.

filter (*Chem. Eng.*). Equipment used to separate liquids and suspended solids, either for recovery of solid, classification of liquid, or both simultaneously. The liquid flows through the pores in a cloth, wire mesh, or granular bed, and thus the solid is sieved out. (*Electronics*) (1) Any device, or chain of similar devices, which discriminates between single frequencies or bands of frequency of vibrations of specific phenomena, e.g., *crystal* filter in supersonic heterodyne receivers, *piezoelectric* or *magnetostriction* couplers, plain mechanical *resonators*, acoustic and electric *wave filters*, and *light filters*. (2) Device which discriminates between particles of different energies in the same beam. (*Photog.*) A device, usually consisting of a pigmented glass plate or a sheet of gelatin, interposed across a beam of light for the purpose of altering the relative intensity of the different component wavelengths in the beam.

filterable (or **filtrable**) **virus** (*Bacteriol.*). See **filter-passer.**

filter aid (*Chem. Eng.*). Diatom earth added to solutions, or precoated on to filter before use, to aid separation of difficult solids, usually colloidal, which would otherwise pass through or choke the filter.

filter attenuation (*Telecomm.*). The loss of signal power in its passage through a filter due to absorption, reflection, or radiation. It is usually given in decibels.

filter attenuation band (*Telecomm.*). A frequency band in which there is appreciable attenuation.

filter bed (*San. Eng.*). A *contact bed* (q.v.) or any similar bed used for filtering purposes.

filter cake (*Chem.*). The layer of precipitate which builds up on the cloth of a filter press.

filter chamber (*Zool.*). A chamber, formed of muscle and basement membrane, surrounding the coiled alimentary canal in some *Hemiptera*.

filter circuit (*Elec.*). One, usually composed of *reactors* arranged in resonant circuits, designed to accept certain desired frequencies and to reject all others. Often used to remove noise from a signal, e.g., between a radio transmitter and its antenna. Also **filter network.**

filter cut (*Photog.*). The wavelength at which the relative absorption for light of different wavelengths changes rapidly.

filter factor (*Photog.*). The number of times a given exposure must be increased because of the presence of a filter, which absorbs light and reduces the effective exposure of a lens system.

filter feeders (*Ecol.*). In benthic and planktonic communities, detritus feeders which remove particles from the water. In benthic communities they usually predominate on sandy bottoms. Cf. *deposit feeders*.

filtering basin (*Hyd. Eng.*). A tank through which water passes on its way from the reservoir to the mains, and in which it is subjected to a process of filtration.

filter network (*Telecomm.*). See **filter circuit.**

filter overlap (*Photog.*). The band of wavelengths transmitted by a combination of filters.

filter paper (*Chem.*). Paper, consisting of pure cellulose, which is used for separating solids from liquids by filtration. Filter paper for quantitative purposes is treated with acids to remove all or most inorganic substances, and has a definite ash content.

filter-passer (*Bacteriol., Bot.*). Early term for a micro-organism which can penetrate the pores of a filter which retains microscopically-visible bacteria. Includes certain forms of pleuro-pneumonia and similar organisms, filterable bodies in the *L-forms* of bacteria, and most viruses. See also **virus.**

filter press (*Chem.*). An apparatus used for filtrations; it consists of a set of frames covered with filter cloths into which the mixture which is to be filtered is pumped.

filter-press action (*Geol.*). A differentiation process involving the mechanical separation of the still liquid portion of a magma from the crystal mesh. The effective agent is pressure operating during crystallization.

filter pulp (*Paper*). Rag fibre made up into convenient cakes at the paper mill. It is reduced to fibres again when required.

filter record position (*Teleph.*). The special position at which records of calls on specified lines are made, to ascertain whether they are overloaded or not.

filter transmission band (*Telecomm.*). A frequency band in which there is negligible attenuation. Also **pass band.**

filtrate (*Chem.*). The liquid freed from solid matter after having passed through a filter.

filtration (*Chem., etc.*). The separation of solids from liquids by passing the mixture through a suitable medium, e.g., filter paper, cloth, glass wool, which retains the solid matter on its surface and allows the liquid to pass through. (*Radiol.*) Removal of longer wavelengths in a composite beam of X-rays by the interposition of thin metal, e.g., copper or aluminium.

filum terminale (*Zool.*). In some Vertebrates, a slender non-nervous thread into which the hinder end of the spinal cord is drawn out.

fimbria (*Zool.*). Any fringing or fringe-like structure; the delicate processes fringing the internal opening of the oviduct in Mammals; the ridge of fibres running along the anterior edge of the hippocampus in Mammals; the processes fringing the openings of the siphons in Molluscs.

fimbriate, fimbriated (*Bot.*, *Zool.*). Having a fringed margin.

fimbriocele (*Med.*). Hernia containing the fimbriae of the Fallopian tube.

fimicolous (*Bot.*). Growing on or in dung.

fin (*Aero.*). A fixed vertical surface, usually at the tail, which gives directional stability to a fixed-wing aircraft in motion and to which a rudder is usually attached. In an airship, any fixed stabilizing surface. See **stabilizer.** (*Carp.*) A tongue on the edge of a board. (*Eng.*) (1) One of several thin projecting strips of metal formed integral with an air-cooled engine cylinder or a pump body or gear-box to increase the cooling area. (2) A thin projecting edge on a casting or stamping, formed by metal extruded between the halves of the die; any similar projection. (*Zool.*) In Fish, some *Cephalopoda*, and other aquatic forms, a muscular fold of integument used for locomotion or balancing; supported in the case of Fish by internal skeletal elements.

final approach (*Aero.*). The part of the landing procedure from the time when the aircraft turns into line with the runway until the flare-out is started. Colloquially *finals*.

final limit-switch (*Elec. Eng.*). A limit-switch used on an electric lift; it is arranged to operate in case of failure of the ordinary limit-switch. Also called an **ultimate limit-switch.**

final selector (*Teleph.*). The selector which is operated by two trains of impulses so that the wipers finally rest on contacts which are connected to the required subscriber's line.

finder (*Astron.*, *etc.*). A small auxiliary telescope of low power fixed parallel to the tube of a large telescope for the purpose of finding the required object and setting it in the centre of the field; also used in stellar photography for guiding during an exposure. (*Photog.*) See **view finder.** (*Teleph.*) A uniselector which automatically hunts to find the line of a subscriber, when he lifts his telephone receiver, in order to connect selectors to his circuit. Also **line finder.**

fine (*Textiles*). Term given by wool classers to merino wool of the best quality, from the shoulders.

fine aggregate (*Civ. Eng.*). Sand or the screenings of gravel or crushed stone.

fine boring (*Eng.*). A high-precision final machining process for bores of internal diameters between 5 mm and 750 mm, using a single-point cutter in a very accurate and rugged machine tool in which the cutting stresses, and consequently the clamping distortion, are kept particularly low.

fine etching (*Print.*). The finishing stages in the making of a half-tone block to achieve the required contrast and range of tones.

fine gold (*Met.*). Pure 24-carat gold.

fine-grain developers (*Photog.*). Slow-acting solutions used for miniature negatives or others in which a high degree of enlargement is required. The most widely used include the *para*-phenylene diamine group and MQ with increased sodium sulphite and borax content.

fine-grained (*Geol.*). See **grain-size classification.**

fine machining (*Eng.*). A family of precision finishing processes, including fine boring, milling, grinding, honing, lapping, diamond-turning, and superfinishing, using particularly accurate machine tools with special provision to eliminate vibration and, often, cemented-carbide or diamond cutters to remove small amounts of material in order to attain high accuracy and excellent surface finish.

fineness (*Chem.*). The state of subdivision of a substance. (*Met.*) The purity of a gold or silver alloy; stated as the number of parts per thousand that are gold (or silver).

fineness modulus (*Civ. Eng.*). A numeral indicating the fineness of an aggregate, as determined by ascertaining the percentage residue, by weight or volume, remaining on each of a series of nine sieves with apertures ranging from 1·5 in. (38·1 mm) to 0·0058 in. (0·147 mm), summing, and dividing by 100.

fineness-of-grind gauge (*Powder Tech.*). Device to provide control tests for the presence of large particles of powder in a slurry. It generally comprises a stainless steel block with parallel grooves of gradual decrease of depth from an inch to zero, along which the slurry is distributed by means of a smooth stroke with a rigid blade. When the groove depth is equal to the largest particle present, the smooth surface of the slurry film is marked by score lines.

fineness ratio (*Aero.*). The ratio of the length to the maximum diameter of a streamlined body.

fine papers (*Paper*). Papers of high quality, especially those containing rag and esparto pulps.

finery (*Met.*). Furnace or hearth on which cast iron is converted to malleable, using charcoal as fuel.

fines (*Powder Tech.*). That portion of a powder composed of particles under a specified size.

fine screen (*Photog.*, *Print.*). The term for half-tones suitable for art paper, usually 52 but also 59 lines per centimetre.

fine silt (*Geol.*). See **silt grade.**

fine structure (*Electronics*). Splitting of optical spectrum lines into multiplets due to interaction between spin and orbital angular momenta of electrons in the emitting atoms.

fine stuff (*Build.*). Fine type of plaster, usually composed of lime and plaster of Paris, used for the finishing coat.

fin fan exchanger (*Chem. Eng.*). Heat exchanger in which hot fluid is passed through a bank of tubes with circumferential fins. The battery of tubes is contained in a light-gauge metal case and air is drawn or blown over the tubes by a fan. May be used to cool process liquids, especially in areas short of water, or to heat air especially in air-conditioning applications.

finger (*Comp.*). Probe used to sense presence or absence of a perforation in punched card or tape system.

fingering (*Textiles*). Combed, soft-twisted, worsted yarn of the type generally used for hand-knitting. Dyed in full shades and pastels, also mixed colours.

finger plate (*Join.*). A plate fixed on the side of the meeting stile of a door, near the lock, to prevent damage to the paintwork by finger-marks.

finger stop (*Eng.*). A sliding stop in a press tool, used to locate and position the material to be processed in relation to the tool, usually at the commencement of a production run.

finger-type contact (*Elec. Eng.*). A type of contact which, as usually fitted to drum-type controllers, is in the form of a finger which is pressed against the contact surface by means of a spring.

finial (*Build.*). A term applied to an ornament placed at the summit of a gable, pillar, or spire.

finials (*Zool.*). In *Crinoidea*, the ossicles of the distal rami of the arms.

fining (*Glass*). See **founding.**

fining coat (*Build.*). See **setting coat.**
fining-off (*Build.*). The operation of applying the setting coat.
finings (*Brew.*). A preparation of isinglass and water, a small quantity of which is added to beer to clarify it.
finisher box (*Textiles*). A machine for straightening and levelling the fibres in readiness for drawing and spinning.
finishing (*Bind.*). The lettering and ornamentation of a bound volume by the finisher; the term does not apply to 'cased' work.
finishing coat (*Build.*). See **setting coat.**
finishing cut (*Eng.*). A fine cut taken to finish the surface of a machined workpiece.
finishing house (*Paper*). See **salle.**
finishing stove (*Bind.*). A small gas or electric stove on which the finisher heats the tools required for his work.
finishing tool (*Eng.*). A lathe or planer tool, generally square-ended and cutting on a wide face. Used for taking the final or finishing cut.
fink truss (*Eng.*). See **French truss.**
Finlay process (*Photog.*). A system of colour photography in which a colour screen or a ruled screen is used over a panchromatic plate.
fin rays (*Zool.*). In Fish, the distal skeletal elements which support the fins; in *Cephalochorda*, the rods of connective tissue supporting the dorsal and ventral fins.
Finsen lamp (*Med.*). A form of arc lamp rich in ultraviolet rays; used for medical purposes.
fiords or fjords (*Geol.*). Narrow winding inlets of the sea bounded by mountain slopes; formed by the drowning of steep-sided valleys, which are thought to have been deeply excavated by glacial action; in many cases a rock-bar partially blocks the entrance and impedes navigation.
FIR (*Aero.*). Abbrev. for *Flight Information Region*.
fir (*For.*). Trees of the genus *Abies*, giving a valuable structural softwood; also used for parts of musical instruments and in shipbuilding. The species *A. balsamea*, found in Canada and the Eastern U.S., yields *Canada balsam* (q.v.) and is used for pulp.
fireback boiler (*Build.*). The boiler fitted in a kitchen stove or at the back of any fire.
fireball (*Astron.*). See **bolide.**
fire bank (*Mining*). A slack or rubbish heap or dump, at surface on a colliery, which becomes fired by spontaneous combustion.
fire-bar (*Elec. Eng.*). A heating element fitted to a high-temperature electric radiator.
fire barriers (*Build.*). Fire-resisting doors, enclosed staircases, and similar obstructions to the spread of fire in a building. See **fire stop.**
fire bars (*Eng.*). Cast-iron bars forming a grate on which fuel is burnt, as in domestic fires, boiler furnaces, etc.
fire-box (*Eng.*). That part of a locomotive-type boiler containing the fire; the grate is at the bottom, the walls and top being surrounded by water. See **locomotive boiler.**
firebrick arch (*Eng.*). An arch built at the end of a boiler furnace, either to deflect the burning gases or to assist the combustion of volatile products. Also **flue bridge.**
fire cement (*Build.*). See **refractory cement.**
fireclay (*Geol., Met.*). Clay consisting of minerals containing predominantly SiO_2 and Al_2O_3, low in Fe_2O_3, CaO, MgO, etc. Those clays which soften only at high temperatures are used widely as refractories in metallurgical and other furnaces. Fireclays occur abundantly in the Carboniferous System, as 'seat earths' underneath the coal-seams. See **underclay.**
fire cracks (*Build.*). Fine cracks which appear in a plastered surface, due to unequal contractions between the different coats.
fired (*Electronics*). Said of gas tubes when discharging, particularly TR pulse tubes during the transmission condition.
fire-damp (*Mining*). The combustible gas contained naturally in coal; chiefly a mixture of methane and other hydrocarbons; forms explosive mixtures with air.
fire-damp cap (*Mining*). Blue flame which forms over the flame of a safety-lamp when sufficient fire-damp is present in colliery workings.
firedog (*Build.*). See **andiron.**
fire door (*Build., etc.*). A fire-resisting door of wood, metal, or both, e.g., the door of a boiler furnace.
fire extinguisher. Any of several types of portable apparatus for emergency use against fire. In general, these depend upon the ejection (by rapid chemical reaction or compressed gas, CO_2 or nitrogen) of the fire-inhibiting medium. The latter may be: water or an alkaline solution; a rapidly evolved chemical foam; CO_2 gas and 'snow'; tetrachloromethane, CCl_4; chlorobromomethane, CH_2ClBr, or bromochlorodifluoromethane, CF_2BrCl, both heavy smothering gases; a suitable hydrogen carbonate in dry powder form.
fire foam. Mixture of foaming but non-inflammable substances used to seal off oxygen and to extinguish fire without use of water.
fire load (*Build.*). The heat in MJ/m^2 or Btu/ft^2 of floor area of a building if destroyed by fire (including contents).
fireman (*Mining*). (1) In a metal-mine, a miner whose duty it is to explode the charges of explosive used in headings and working places. (2) In a coal-mine, an official responsible for safety conditions underground. See **deputy.**
fire opal (*Min.*). A variety of opal (cryptocrystalline silica) characterized by a brilliant orange-flame colour. Particularly good specimens, prized as gemstones, are of Mexican origin.
fire plug. A *hydrant* (q.v.) for service in extinguishing fires.
fire point (*Oils, etc.*). The temperature at which sufficient vapour is given off by a heated liquid, under standard test conditions, to maintain combustion. Cf. *flash point*.
fire polishing (*Glass*). The polishing of glassware, decorated with a pressed pattern, by holding it in a *glory-hole* (q.v.).
fireproof aggregates (*Build.*). Materials such as crushed firebricks, fused clinkers, slag, etc., incorporated in concrete to render it fire-resisting.
fire refining (*Met.*). The refining of blister copper by oxidizing the impurities in a reverberatory furnace and removing the excess oxygen by poling. May be used as an alternative to electrolytic refining, and in any case is carried out as a preliminary to this.
fire retardant adhesive (*Eng.*). Heavy duty adhesives intended to fasten lagging around hot surfaces. These are usually alkyd based, but may contain chlorinated paraffins and antimony oxide to render the film noncombustible.
fire ring (*Eng.*). A top piston-ring of a special heat-resisting design, used in some 2-stroke oil engines.
fire sand (*Met.*). Refractory oxide or carbide suitable for lining furnaces.
fire stink (*Mining*). The smell given off underground when a fire is imminent, e.g., in the gob; also, smell of sulphuretted hydrogen from decomposing pyrite.
fire-stone (*Geol.*). A stone or rock capable of

withstanding a considerable amount of heat without injury. The term has been used with reference to certain Cretaceous and Jurassic sandstones employed in the manufacture of glass furnaces.

fire stop (*Build.*). An obstruction across an air passage in a building to prevent flames from spreading further. Also **draft stop.**

fire-trap or magazine valve (*Cinema.*). The pair of rollers through which the film passes in entering and leaving the feed and take-up magazines on a projector; its purpose is to cool the film and prevent the ingress of air to the magazine, should the film ignite in the gate, and so stop the spread of fire to the bulk of the reel in the magazine.

fire-tube boiler (*Eng.*). A boiler in which the hot furnace gases, on their way to the chimney, pass through tubes in the water space, as opposed to a *water-tube boiler* (q.v.). See **marine boiler, locomotive boiler.**

firewall (*Aero.*). See **bulkhead.**

firing (*Electronics*). Establishment of discharge through gas tube. (*Eng.*) (1) The process of adding fuel to a boiler furnace. (2) The ignition of an explosive mixture, as in a petrol or gas engine cylinder. (3) Excessive heating of a bearing. (*Mag.*) Sudden change from unsaturation to saturation of the core in a magnetic amplifier. (*Vet.*) The application of thermo-cautery to the tissues of animals.

firing angle (*Electronics*). Angle of phase of an applied voltage which makes a thyristor or a gas valve conducting, as in a *gating* valve or thyratron.

firing key (*Elec. Eng.*). A key which fires a charge of explosive by completing the electric circuit to a fuse.

firing order (*I.C. Engs.*). The sequence in which the cylinders of a multi-cylinder internal-combustion engine fire, e.g., 1, 3, 4, 2 for a 4-cylinder engine.

firing power (*Electronics*). The minimum RF power required to start a discharge in a switching tube for a specified ignitor current.

firing stroke (*I.C. Engs.*). The power or expansion stroke of an internal-combustion engine.

firing time (*Electronics*). The interval between applying a d.c. voltage to the trigger electrode of a switching transistor or tube and the beginning of the discharge.

firing tools (*Eng.*). Implements (e.g., shovels, rakes, and slicers or slicing bars) used in firing a boiler furnace by hand.

fire top-centre (*I.C. Engs.*). The top dead-centre of an internal-combustion engine, when the piston is about to make its power stroke.

firkin (*Brew.*). Cask holding 9 gal (40 litres).

firmer chisel (*Carp., Join.*). A woodcutting chisel, usually thin in relation to its width ($\frac{1}{16}$ to 2 in. or 3 to 50 mm). Stouter than a paring chisel but less robust than a mortise chisel.

firmer gouge (*Tools*). Standard type of *gouge* (q.v.) with the bevel on the outside (cf. *scribing-gouge*). Used for cutting grooves and recesses.

firmisternous (*Zool.*). Having the 2 halves of the pectoral girdle firmly united, as Frogs. Cf. *arciferous.*

firn (*Geol.*). See **névé.**

firring (*Carp.*). Timber strips of constant width but varying depth, which are nailed to the wood bearers to flat roofs as a basis for roof boarding, to which they give a suitable fall. Also spelt **furring.**

first detector (*Telecomm.*). See **mixer.**

first-order reaction (*Chem.*). One in which the rate of reaction is proportional to the concentration of the reactant, i.e., $\frac{dc}{dt} = -kc$, where c is the concentration of the reagent.

First Point of Aries (*Astron.*). The point in which the ecliptic intersects the celestial equator, crossing it from south to north; the origin from which both right ascension and celestial longitude are measured. See **equinoctial points, equinox.**

First Point of Libra (*Astron.*). See **equinoctial points.**

first runnings (*Chem.*). The first fraction collected from a *fractional distillation* (q.v.) process, usually containing low boiling impurities.

first ventricle (*Zool.*). In Vertebrates, the cavity of the left lobe of the cerebrum.

first weight (*Mining*). The first indications of roof pressure which occur after the removal of coal from a seam.

'fir tree' roots (*Aero.*). A certain type of fixing adopted for turbine blades, the outline form of the roots resembling that of a fir tree.

FIS (*Aero.*). Abbrev. for *Flight Information Service.*

Fischer reagent for water (*Chem.*). The sample is titrated with a methanol solution of iodine, sulphur (IV) oxide, and pyridine, free iodine appearing after the end point, according to the reaction

$$H_2O + I_2 + SO_2 + CH_3OH \rightarrow 2HI + CH_3HSO_4.$$

Fischer-Tropsch process (*Chem.*). Method of obtaining fuel oil from coal, natural gas, etc. Cf. *hydrogenation.* The 'synthesis gas', hydrogen and carbon monoxide in proportional volumes 2 : 1, is passed, at atmospheric or slightly higher pressure and temperature up to 200°C, through contact ovens containing circulating water with an iron or cobalt catalyst. The gases are washed out and the resultant oil contains alkanes and alkenes, from the lower members up to solid waxes; fractionation yields petrols, diesel oils, etc.

fish-bar (*Rail.*). See **fish-plate.**

fish-beam (*Civ. Eng.*). A beam which is *fish-bellied.*

fish-bellied (*Eng.*). Said of (*a*) steel girders with a convex lower edge; (*b*) long straight-edges, which are convex upward. Such a form results in greater resistance to bending.

fishbone antenna (*Radio*). End-fire array of vertical resonators spaced along a transmission line.

fished joint (*Rail.*). A *butt joint* (q.v.) between rails or beams, in which fish-plates or cover-straps are fitted on both sides of the joint and bolted together.

Fisher sub-sieve sizer (*Powder Tech.*). Instrument similar to the *Lea and Nurse permeameter* (q.v.).

fish-eye lens (*Photog.*). Ultra-wide angle lens covering up to 180°. Subject to considerable linear distortion, and so used for freak effects.

fish glue. (1) *Isinglass* (q.v.). (2) Any glue prepared from the skins of fish (esp. sole, plaice), fish-bladders, and offal.

fishing (*Eng.*). Recovering tools dropped from drilling tackle during deep rock drilling operations.

fish-paper (*Elec. Eng.*). A flexible insulating material, usually of varnished cambric, for separating coil windings, etc.

fish-plate (*Rail.*). A steel or wood cover-plate, fitted one to each side of a fished joint between successive lengths of beam or rail. Also called **fish-bar, fish piece, shin, splice piece.**

fish-wire (*Elec. Eng.*). A thin wire drawn into a conduit for electric cables or wires during construction, and subsequently used for drawing in the cables or wires themselves.

Fissidentales (*Bot.*). An order of the *Musci*; the aerial branches have a 2-sided apical cell, so that the leaves develop in 2 rows.

fissile (*Nuc.*). Capable of fission, i.e., breakdown into lighter elements of certain heavy isotopes (^{233}U, ^{235}U, ^{239}Pu), when these capture neutrons of suitable energy. Associated with Einstein release of energy. Also **fissionable.** See **reactor.**

fissilingual (*Zool.*). Having a forked tongue.

fission (*Nuc.*). The spontaneous or induced disintegration of a heavy atom into two or more lighter ones. The process involves a loss of mass which is converted into nuclear energy.

fissionable (*Nuc.*). See **fissile.**

fission bomb (*Nuc.*). Same as **atomic bomb.**

fission chain (*Nuc.*). Atoms formed by uranium or plutonium fission have too high a neutron-proton ratio for stability. This is corrected either by neutron emission (the delayed neutrons) or more usually by the emission of a series of beta-particles, so forming a short radioactive decay chain.

fission chamber (*Nuc. Eng.*). Ionization chamber lined with a thin layer of uranium. This can experience fission by slow neutrons, which are thereby counted by the consequent ionization.

fission fungi (*Bot.*). Bacteria.

fissioning distribution (*Nuc.*). The modification of the neutron energy spectrum which results from allotting to each neutron a weight which is equal to the probability that its next collision will result in fission.

fission neutrons (*Nuc.*). Those released by fission, having a continuous spectrum of energy with a maximum of ca. 10^6 eV.

fission parameter (*Nuc.*). The square root of the atomic number of a fissile element divided by its relative atomic mass.

fission poisons (*Nuc.*). Fission products with abnormally high thermal neutron absorption cross-sections, which reduce the reactivity of nuclear reactors. Principally, xenon and samarium ^{135}Xe, having an absorption cross section of 3·5 million barns for slow neutrons.

fission products (*Nuc.*). Lower mass, often radioactive atoms and particles resulting from fission, e.g., strontium 90, which is the major contributor to radiation in *fall-out* from nuclear explosions.

fission spectrum (*Nuc.*). The energy distribution of neutrons released by fission.

fission yield (*Nuc.*). The percentage of fissions for which one of the products has a specific mass number. Fission yield curves show two peaks of approximately 6% for mass numbers of about 97 and 138. The probability of fission dividing into equal mass products falls to about 0·01%.

fissiped (*Zool.*). Having free digits. Cf. *pinnatiped.*

fissirostral (*Zool.*). Having a cleft beak.

fissure (*Geol.*). A cleft in rock determined in the first instance by a fracture, a joint plane, or fault, subsequently widened by solution or erosion; may be open, or filled in with superficial deposits. See also grike. (*Med.*) (1) Any normal cleft or groove in organs of the body. (2) Linear ulceration of the anus, usually the result of constipation. (*Mining*) Crustal cleft or fissure filled with minerals of pneumatolytic or hydatogenetic provenance.

fissure eruptions (*Geol.*). Throwing-out of lava and (rarely) volcanic 'ashes' from a fissure, which may be many miles in length. Typically there is no explosive violence, but a quiet welling-out of very fluid lava. Recent examples are known from Iceland.

fistula (*Build.*). An ancient name for a water-pipe. (*Med.*) An epithelial lined track connecting two hollow viscera. *adj.* **fistulous.**

fistulous withers (*Vet.*). Abscess and fistula formation in the withers of the horse; *Brucella abortus* infection has been found in some cases, and in others a nematode worm, *Onchocerca cervicalis*, has been found in the affected region.

fit (*Eng.*). The dimensional relationship between mating parts. Limits of tolerances for shafts and holes to result in fits of various qualities, e.g., clearance, transition and interference fits are laid down by British Standard Specifications and others. (*Med.*) A sudden attack of disturbed function of the sensory or of the motor parts of the brain, with or without loss of consciousness. See also **epilepsy.**

fitch (*Paint.*). A small, long-handled, hog's-hair brush, used for fine finishing work. Made with chisel, filbert or round tapered tips.

fitter (*Eng.*). A mechanic who assembles finished parts in an engineering workshop.

fitter's bench (*Eng.*). A heavy wooden bench provided with a vice and a drawer for tools.

fitter's hammer (*Eng.*). A hand hammer having a flat striking face, and either a straight, cross, or ball pane.

Fittig's synthesis (*Chem.*). The synthesis of benzene hydrocarbon homologues by the action of metallic sodium on a mixture of a brominated benzene hydrocarbon and bromo- or iodo-alkane in a solution of dry ether.

fitting (*Elec. Eng.*). A device used for supporting or containing a lamp, together with its holder, and its shade or reflector; part of the equipment of an electric-light installation. (*Eng.*) Hand or bench work involved in the assembly of finished parts by a fitter.

fittings (*Eng.*). (1) Small auxiliary parts of an engine or machine. (2) Boiler accessories, as valves, gauges, etc.

fitting shop (*Eng.*). The department of an engineering workshop where finished parts are assembled. See **erecting shop.**

FitzGerald-Lorentz contraction (*Phys.*). The contraction in dimensions (or time scale) of a body moving through the ether with a velocity approaching that of light, relative to the frame of reference (Lorentz frame) from which measurements are made. Also **Lorentz contraction.**

five-centred arch (*Build.*). An arch having the form of a false ellipse struck from 5 centres.

five-electrode valve. See **pentode valve.**

five-unit code (*Teleg.*). The Baudot code, as used for machine transmission of telegraphic signals in synchronous and start-stop systems.

five-wire system (*Elec. Eng.*). A system used for electrical distribution. It may be a d.c. system in which one wire is at earth potential, two are at a potential of V to earth, and the other two are at a potential of 2V to earth; or it may be a 2-phase 4-wire system, in which the 2 phases are connected and earthed at their midpoints, and a neutral wire is brought out from this point.

fix (*Aero.*). The exact geographical position of an aircraft, as determined by terrestrial or celestial observation or by radio cross-bearing. Cf. *pinpoint.* (*Surv.*) Point accurately established on plan, perhaps by observations for latitude and longitude.

fixation (*An. Behav.*). A response repeatedly shown by an animal, regardless of whether it produces reward or punishment. It is often shown in an *insoluble problem situation* (q.v.).

(*Psychol.*) An emotional arrest of the personality at an earlier stage of development; caused by the maintenance of the instinctual forces in the channels of gratification common to that phase. (*Zool.*) The action of certain muscles which prevent disturbance of the equilibrium or position of the body or limbs; the process of attachment of a free-swimming animal to a substratum, on the commencement of a temporary or permanent sessile existence.

fixation disk (*Zool.*). A disk developed from the preoral lobe in the larvae of *Crinoidea* and *Asteroidea* at the time of metamorphosis for attachment to the substratum.

fixation of nitrogen (*Bot.*). The formation, by soil bacteria, of nitrogenous compounds from elementary nitrogen. (*Chem.*) For commercial purposes (fertilisers, etc.) the *Haber process* (q.v.) is now the most important method of nitrogen; fixation others are the arc process and the heating of calcium carbide in a stream of nitrogen forming calcium cynanimide which on further treatment yields ammonia.

fixation papillae (*Zool.*). Three glandular papillae which develop at the anterior end of the larvae of *Urochorda* for fixation to the substratum prior to metamorphosis.

fixed-action pattern (*An. Behav.*). A highly stereotyped and precise response observable in most members of a species when there have been no experimental manipulations which would have made learning possible.

fixed beam (*Build.*). A beam with fixed ends.

fixed carbon (*Met.*). Residual carbon in coke after removal of hydrocarbons by distillation in inert atmosphere.

fixed-charge collector (*Elec. Eng.*). A device, attached to prepayment meters, which is arranged so that the insertion of coins corresponding to a fixed charge (rental or hire-purchase charge) allows a consumer to close a switch to receive a supply, the switch being opened automatically after a predetermined time.

fixed contact (*Elec. Eng.*). The contact of a switch or fuse which is permanently fixed to the circuit terminal.

fixed eccentric (*Eng.*). An eccentric which is permanently keyed to a shaft, not capable of angular movement, as is a *loose eccentric* (q.v.).

fixed end (*Build.*). The term applied, in theory, to the end of a beam when it is built in or otherwise secured so that the tangent at the end to the curve taken up by the beam when it is deflecting under applied loading remains fixed.

fixed expansion (*Eng.*). A steam-engine in which the cut-off cannot be altered and which thus works with a constant expansion ratio.

fixed handle circuit-breaker. See **fixed-trip**.

fixed lead (*Nav.*). Navigation in which the missile flight path leads the line of sight by a constant angle.

fixed-light position (*Bot.*). The position of a fully developed leaf in respect of the direction of the strongest diffused light that reaches it.

fixed-loop aerial (*Aero.*). A loop aerial, used with a homing receiver, which is fixed in relation to the aircraft's centreline.

fixed-needle surveying (*Surv.*). Traverse with magnetic compass locked, instrument being used like theodolite for measuring azimuth angles, except where there is no nearby iron or steel to affect bearing, when compass needle may be used to give a check reading.

fixed oils (*Chem.*). Vegetable oils which are not changed by heat or distillation, as distinct from *essential oils* (q.v.).

fixed-pitch airscrew (*Aero.*). See **airscrew**.

fixed-point notation (*Comp.*). Representation of a number by a fixed set of digits, so that decimal or radix point has predetermined location.

fixed points (*Heat*). The standard temperatures chosen to define a thermometer scale. Those generally used are the temperature of pure melting ice and that of steam from pure boiling water at one atmosphere pressure. In 1948 the *International Practical Temperature Scale* defined six fixed points, ranging from the 'oxygen point' (−182·97°C) to the gold point (1063°C). See **kelvin, triple point.**

fixed pulley (*Eng.*). A pulley keyed to its shaft. See **fast pulley.**

fixed sash (*Build.*). (1) A *stand sheet* (q.v.). (2) A sash permanently fixed in a solid frame.

fixed star (*Astron.*). The ancient idea that the stars were fixed on the celestial sphere, as opposed to the 'wandering stars' or *planets* (q.v.), whose motion is plainly visible to the naked eye. See **Ptolemaic system.**

fixed time-lag (*Elec. Eng.*). See **definite time-lag.**

fixed-trip (*Elec. Eng.*). A term applied to certain forms of circuit-breaker or motor starter to indicate that the tripping mechanism cannot operate while the breaker or starter is actually being closed. Also called **fixed-handle circuit-breaker.** Cf. *free-trip*.

fixed-tube sheet exchanger (*Chem. Eng.*). See **shell and tube exchanger.**

fixed-type metal-clad switchgear (*Elec. Eng.*). Metal-clad switchgear in which all parts are permanently fixed, no provision being made for easy removal of any part for inspection or maintenance purposes.

fixed word length (*Comp.*). Computing procedure where all word lengths contain the same number of bits regardless of their information content.

fixing (*Photog.*). Process for removing unreduced silver halides after development of an emulsion.

fixing block (*Build.*). A block of material, having the shape of a brick, which can be built into the surface of a wall to provide a substance to which joinery, such as window frames, may be nailed. Fixing blocks are made of porous concrete, of coke breeze, or of a special brick made with a mixture of sawdust which burns away in the kiln to leave a porous brick material. See also **wood brick.**

fixing fillet (or pad) (*Build.*). A *slip* (q.v.).

fixings (*Join.*). Supports, such as grounds and plugs, for securing joinery in position.

fixture (*Build.*). An attachment to a building. (*Eng.*). A device used in the manufacture of (interchangeable) parts to locate and hold the work without guiding the cutting tool.

fjords (*Geol.*). See **fiords.**

flabellate, flabelliform. Shaped like a fan.

flabellum (*Zool.*). Any fan-shaped structure; in *Crustacea*, the distal exite of a phyllopodium; an epipodite; in *Hymenoptera*, a small, spoon-shaped lobe at the apex of the glossa.

flag (*Civ. Eng.*). A flat thin stone, either natural or artificial, used as a paving material or for purposes of providing cover (e.g., for a catch-pit). (*Comp.*) An information bit used to indicate the boundary of a field, or the end of a word. (*Electronics*) Supporter of the *getter*, which is fired by eddy currents after the valve is evacuated. (*Geol.*) Natural flagstones are sedimentary rocks of any composition which can be readily separated, on account of their distinct stratification, into large slabs. They are often fine-grained sandstones interbedded with shaly partings along which they can be split. (*Paper*) A coloured slip of paper inserted in a reel to indicate a join or break in the web so that the printing machine can be stopped. (*TV*) Shield to protect camera lenses from unwanted light.

flag alarm (*Elec. Eng.*). See **flag indicator.**

Flagellata (*Zool.*). See Mastigophora.

flagellate (*Bot., Zool.*). (1) Having flagella. (2) Bearing a long threadlike appendage. (3) A member of the *Mastigophora.*

flagellated chambers (*Zool.*). In *Porifera,* chambers formed by outgrowths of the wall of the vase and lined by choanocytes.

flagellate disseminule (*Bot.*). A zoospore or other motile means of propagation.

flagellispore or **flagellula** (*Zool.*). A zoospore having one or more flagella as locomotor organs.

flagellum (*Bot.*). (1) A delicate filiform branchlet in mosses. (2) A threadlike extension of the protoplast of a cell or of a motile spore. (*Zool.*) (1) A threadlike extension of the protoplasm of a cell, or of a Protozoan, which is capable of carrying out lashing movements. (2) In some *Arachnida,* a group of specialized setae on the proximal segment of a chelicera, believed to be tactile. (3) In other *Arthropoda,* a filiform extension to an appendage, e.g., a crustacean limb, an insect antenna. (4) In *Gastropoda,* one of the male genitalia.

flag indicator (*Elec. Eng.*). In navigational and other instruments, a semaphore-type signal which warns viewer when instrument readings are unreliable. Also called **alarm flag, flag alarm.**

flail joint (*Med.*). A joint in which there is, as a result of disease or of operation, excessive mobility.

flail rotor (*Agric.*). An assembly of high-speed swinging knives or hammers, attached to a horizontal shaft or cylinder.

flake white (*Paint.*). A pigment of pure white lead.

flake yarn (*Textiles*). A fancy yarn, composed of two foundation threads, with portions of short fibred twistless roving twisted in at intervals. Two doublings are necessary. The foundation threads are first twisted together, and the roving which produces the flakes at this stage delivered intermittently. A further twisting the opposite way is necessary to bind the flakes in position.

flakiness ratio (*Powder Tech.*). The ratio of the breadth of a particle to its thickness.

flaking (*Build.*). The breaking away of surface plaster due to nonadhesion with the undercoat or to free lime or impurities in the basic surface. (*Paint.*) A defect in paintwork, the paint film breaking away in greater or lesser areas from the surface it was covering.

flame (*Chem.*). A region in which chemical interaction between gases occurs, accompanied by the evolution of light and heat.

flame-arc (*Elec. Eng.*). An electric arc maintained between carbons containing certain metallic salts, which give a colour to the arc flame.

flame-arc lamp (*Elec. Eng.*). An arc lamp using flame carbons.

flame blow-off factor (*Heat*). Relation between the velocity of combustible mixture and the rate of flame propagation, the latter varying appreciably with gases of different composition. See also **burner firing block, flame retention, piloted head and tunnel burners.**

flame carbons (*Elec. Eng.*). Carbon electrodes containing certain metallic salts, which have the effect of colouring an arc maintained between them. See **flame-cored carbon.**

flame-cell (*Zool.*). See **solenocyte.**

flame-cored carbon (*Elec. Eng.*). A carbon electrode having a central core of a material designed to colour the flame of an arc drawn from it.

flame cutting (*Eng.*). Cutting of ferrous metals by oxidation, using a stream of oxygen from a blow pipe or torch on metal preheated to about 800°C by fuel gas jets in the cutting torch.

flame damper (*Aero.*). See **damper.**

flame failure (*Eng.*). The accidental extinction of a burner flame, e.g., in an oil-fired boiler, which has to be detected in order to stop or otherwise control the supply of further fuel, to prevent explosion or damage.

flame-failure control (*Heat*). Direct-flame thermostat with interconnected relay valve, which provides a constantly burning pilot flame for igniting the main gas burners, and automatically shuts off the gas supply to the main burner in the event of the pilot flame becoming extinguished.

flame-hardening (*Met.*). Hardening of metal surface by heating with oxyacetylene torch, followed by rapid cooling with water or air jet.

flame ionization gauge (*Chem.*). Used in gas chromatography, as a very sensitive detector for the separate fractions in the effluent carrier gas, by burning it in a hydrogen flame, the electrical conductivity of which is measured by inserting two electrodes in the flame with an applied voltage of several hundred volts.

flame lamp (*Elec. Eng.*). A filament lamp having the bulb in the form of a flame.

flame plates (*Eng.*). Those plates of a boiler firebox subjected to the maximum furnace temperature.

flame-proof (*Elec. Eng.*). See **explosion-proof.**

flame retention (*Heat*). Ability to retain a stable flame with gas burners at all rates of gas flow, irrespective of adverse combustion characteristics and conditions. See also **burner firing block, flame blow-off factor, piloted head, tunnel burners.**

flame spectrum (*Phys.*). That imparted to nonluminous flame by substances volatilized in it.

flame temperature (*Heat*). The temperature at the hottest spot of a flame.

flame test (*Chem.*). The detection of the presence of an element in a substance by the coloration imparted to a Bunsen flame.

flame trap (*Aero.*). A device in the induction system of a piston aero-engine which prevents a backfire or blow-back. (*Heat*) Device inserted in pipelines carrying air-gas mixture of a combustible or self-burning nature to arrest the flame in the event of a flash-back (or backfire) occurring at the burner. (*I.C.Engs.*) A gauze or grid of wire, or coiled corrugated sheet, placed in the air intake to a carburettor to prevent the emission of flame from a 'pop-back'.

flame tube (*Aero.*). The perforated inner tubular 'can' of a gas turbine combustion chamber in which the actual burning occurs; cf. *combustion chamber.*

flange (*Eng.*). (1) A projecting rim, as the rim of a wheel which runs on rails. (2) The top or bottom members of a rolled I-beam. (3) A disk-shaped rim formed on the ends of pipes and shafts, for coupling them together; or on an engine cylinder, for attaching the covers. See also **flanged rail.** (*Zool.*) One of the processes of the proximal barbules of feathers, with which the hooklets on the distal barbules engage.

flange coupling (*Eng.*). A shaft coupling consisting of 2 accurately faced flanges keyed to their respective shafts and bolted together.

flanged beam (or girder) (*Eng.*). A rolled-steel joist of I-section.

flanged chuck (*Eng.*). See **face chuck.**

flanged nut (*Eng.*). A nut having a flange or washer formed integral with it. See **collar-head screw.**

flanged pipes (*Eng.*). Pipes provided either with integral or attached flanges for connecting them together by means of bolts.

flanged rail (*Rail., etc.*). A rail section of inverted-T shape, the flange being at the bottom and the end of the stem of the T being at the top—the latter part being enlarged locally to form the head of the rail. Also called a **flat-bottomed rail.**

flanged seam (*Eng.*). A joint made by flanging the ends of furnace tubes and bolting them together between a pair of steel rings.

flange joint (*Eng.*). Any joint between pipes, made by bolting together a pair of flanged ends.

flange protection (*Elec. Eng.*). The rendering of electrical apparatus flame-proof by providing all joints with very wide flanges.

flank (*Build.*). A roof valley. (*Eng.*) (1) That part of a gear-tooth profile which lies inside the pitch-line or circle. (2) The working face of a cam. (3) The straight side connecting the crest and the root of a screw-thread.

flank angle (*Eng.*). An angle between the flanks of a thread and a plane perpendicular to the axis, measured in an axial plane.

flank dispersion (*Elec. Eng.*). See **end leakage flux.**

flanking window (*Build.*). A window located beside an external door.

flanks (*Build., Civ. Eng.*). (1) The parts of the intrados of an arch near to the abutments. Also called **haunches.** (2) The side surfaces of a building stone or ashlar, when it is built into a wall.

flank wall (*Build.*). A side wall.

flannel (*Textiles*). An all-wool material made from fine soft wools, the weave being either plain or twill. The cloth, dyed or undyed, is shrunk and raised.

flannelette (*Textiles*). A cotton fabric of plain or twill weave, raised on both sides and used for pyjamas, nightdresses, sheets, and working shirts.

flanning (*Build.*). The internal splay of a window jamb, or of a fireplace.

flap (*Aero.*). Any surface attached to the wing, usually to the trailing edge, which can be adjusted in flight, either automatically or through controls, to alter the lift as a whole; primarily on fixed-wing aircraft, but occasionally on rotor systems. (*Surg.*) An area of tissue partly separated by the knife from the surface of the body, in connexion with amputation of a limb or for the purpose of grafting skin.

flap angle (*Aero.*). The angle between the chord of the flap, when lowered or extended, and the wing chord.

flap attenuator (*Telecomm.*). One consisting of a strip of absorbing material which is introduced through a non-radiating slot in a waveguide.

flapping angle (*Aero.*). The angle between the tip-path plane of a helicopter rotor and the plane normal to the hub axis.

flapping hinge (*Aero.*). The pivot which permits the blade of a helicopter to rise and fall within limits, i.e., variation of zenithal angle in relation to the rotor head.

flap setting (*Aero.*). The flap angle for a particular condition of flight, e.g., take-off, approach, landing.

flap tile (*Build.*). A purpose-made tile, shaped so as to fit over a hip or valley line, or to catch water.

flap trap (*San. Eng.*). A type of antiflood valve, in which back flow is prevented by a hinged metal flap fitted in an intercepting chamber, to allow of flow in one direction only.

flap-valve (*Eng.*). A nonreturn valve in the form of a hinged disk or flap, sometimes leather- or rubber-faced, used for low pressures.

flare (*Acous.*). The prominent part of the opening of a horn, bell, or trumpet attached to a loud-speaking unit. (*Photog.*) The patches of extraneous light on negatives, caused by reflections from the glass-air surfaces of a lens, flaws in the glass, etc. They are minimized by *blooming* (q.v.). (*TV*) Excess brightness in an image, especially at edges.

flare curve (*Photog.*). Curve showing the relation between the exact luminance scale of a scene and a photographic image of it as affected by internal reflection and other lens defects.

flare factor (*Photog.*). Ratio between the luminance scale of the subject and that of the camera image.

flare header (*Build.*). A brick which has been burnt to a darker colour at one end, so that it may be used with others in facing-work to vary the effect.

flare-out (*Acous.*). Increase in cross-section termination to open-ended waveguide or speaking tube. When the increase is linear in one plane only, a *sectoral horn* (q.v.) is formed. (*Aero.*) Controlled approach path of aircraft immediately prior to landing.

flare stack (*Chem. Eng.*). A stack with an automatic igniter at its top to which vents from petroleum and petrochemical process plants are connected, so that in event of safety devices in the equipment operating to vent vessel contents, these, when combustible, are burned in safe conditions.

flare stars (*Astron.*). A name given to stars whose spectra show evidence of sudden outbursts similar to *solar flares* (q.v.).

flaring A term applied to the end of a pipe, etc., when it is shaped out so as to be of increasing diameter towards the end. See also **flare** (*Acous.*).

flaser structure (*Geol.*). A type of parallel arrangement of mica in thin wavy films in gneisses, typically seen in certain plutonic gneisses. Flaser gabbro is basic plutonic rock in which the original structure is largely destroyed, granular lenses being separated by wavy bands of highly foliated material. See **foliation.**

flash (*Met.*). A thin fin of metal formed at the sides of a forging where some of the metal is forced between the faces of the forging dies. By extension, a similar extrusion in other (e.g., moulded) materials. (*Plastics*) The excess material forced out of a mould during moulding. Sometimes referred to as a *fin*, if the mould is in two halves. (*Telecomm.*) The sudden, emergency, or priority use of a communication channel, such as broadcasting or trunk telephone. Also **flash service.**

flash arc (*Electronics*). See **Rocky Point effect.**

flashback chamber (*Heat*). In certain types of acetylene generators, a compartment filled with water which serves the dual purpose of washing the gas and forming a water seal between the service pipe and the acetylene in the generator.

flashback voltage (*Electronics*). The inverse peak voltage in a gas tube at which ionization occurs.

flash boiler (*Eng.*). A steam boiler consisting of a long coil of steel tube, usually heated by oil burners, in which water is evaporated as it is pumped through by the feed pump. See **steam car.**

flashbulb (*Photog.*). Expendable oxygen-filled electric bulb containing (usually) aluminium foil or filament. Used for flashlight photography. See **flashgun.**

flash-butt welding (*Elec. Eng.*). See **resistance-flash welding.**

flash colours (*Zool.*). Bright colours on the body of an animal which are conspicuous while it is moving, but which are either concealed or merge with the surroundings—and so render it invisible—as soon as it stops.

flash distillation (*Chem.*). The spraying of a liquid mixture into a heated chamber of lower pressure, in order to drive off some of the more volatile constituents. Used in the petroleum industry.

flash drying (*Met.*). Removal of moisture as stream of small particles falls through current of hot gas.

flashed glass (*Glass*). A term sometimes applied to glass coloured by the application of a thin layer of densely coloured glass to a thicker, colourless, base layer.

flasher (*Elec. Eng.*). An oscillator, or thermally operated switch and heater, arranged to switch a lamp on and off repetitively.

flashgun (*Photog.*). A reflector and holder for flashbulbs, operated usually by battery and capacitor, attachable to the camera and synchronized with the shutter release. See **electronic flash, flashbulb.**

flashing (*Build.*). A method of brick burning in which the air supply is periodically stopped in order that the colouring of the bricks shall be irregular. (*Electronics*) Operation in manufacture of thermionic cathode in which it is raised to a very high temperature for a short period. Also used with carbon filaments in carbon atmosphere to obtain a uniform cross-section. (*Hyd. Eng.*) The process of passing a boat across any point in a river where there is a sudden fall; effected by constructing a convergent passage from the high to the low level, shutting it by a sluice gate to allow the water to pond up, and then opening the gate and allowing the boat to be carried through the sluice way by the artificially deepened water. (*Paint.*) Glossy patches or streaks on flat-finished surfaces. (*Plumb.*) A strip of sheet-lead, zinc, or copper which is laid around the junction of two surfaces, e.g., where a chimney projects from a roof, to render the junction watertight. See also **apron.**

flashing board (*Build.*). A board to which flashings (see (*Plumb.*) above) are secured.

flashing compound (*Build.*). Thick nondrying materials used for filling in crevices, e.g., between insulation blocks. Essential properties are impermeability and elasticity.

flashing light (*Ships, etc.*). A navigation mark identified during darkness by a distinctive pattern of flashes of light. See **alternating light, occulting light.**

flashlamp (*Light*). A filament lamp, usually for less than 5 volts, which is intended for use with a dry battery.

flashlight photography (*Photog.*). The use of *flashbulbs* or *electronic flash* (qq.v.), usually synchronized (see **flash-synchronized**) with the camera shutter, when available light is inadequate. Formerly, magnesium ribbon or powder was ignited to produce the flash.

flashover (*Elec. Eng.*). An electric discharge over the surface of an insulator.

flashover test (*Elec. Eng.*). A test applied to electrical apparatus to determine the voltage at which a flashover occurs between any two parts, or between a part and earth. Also called a **sparkover test.**

flashover voltage (*Elec. Eng.*). The highest value of a voltage impulse which just produces flashover.

flash photolysis (*Chem.*). Photolysis induced by light flashes of short duration but high intensity, e.g., from a laser.

flash point (*Phys.*). The temperature at which a liquid, heated in a Cleveland cup (open test) or in a Pensky-Martens apparatus (closed test), gives off sufficient vapour to flash momentarily on the application of a small flame. The *fire point* (q.v.) is ascertained by continuing the test.

flash radiography (*Radiol.*). High-intensity, short-duration X-ray exposure from an X-ray tube fed for example by a Marx high-voltage generator, or a betatron.

flash roasting (*Met.*). The roasting of finely-ground concentrates by introducing them into a large combustion chamber in which the sulphur is burned off as they fall.

flash service (*Telecomm.*). See **flash.**

flash spectrum (*Astron.*). A phenomenon seen at the first instant of totality in a solar eclipse; the dark lines of the Fraunhofer spectrum formed in the chromosphere flash out into bright emission lines as soon as the central light of the sun is cut off.

flash suppressor (*Elec. Eng.*). A device for preventing flashovers on the commutators of d.c. generators; it consists of an automatically operated switch for short-circuiting certain points in the winding, thereby reducing the voltage to zero before a flashover has had time to develop.

flash-synchronized (*Photog.*). Said of shutters which, when released, can close an electrical circuit for firing flash. Two types of setting are generally used: 'M' operating at the moment when the shutter is fully open, and 'X' which delays the shutter opening by about 17 ms to enable certain types of flashbulb to reach peak brightness at exposure.

flash test (*Elec. Eng.*). A test applied to electrical equipment for testing its insulation strength; it consists of the application of a voltage of about twice the working voltage, for a period of not more than about one minute.

flash welding (*Eng.*). An electric welding process similar to butt welding, in which the parts are first brought into very light contact. A high voltage starts a flashing action between the two surfaces, which continues while sufficient forging pressure is applied to the parts to complete the weld.

flask (*Foundry*). A moulding box of wood, cast-iron, or pressed steel, for holding the sand mould in which a casting is made; it may be in several sections. See **cope, drag.** (*Nuc. Eng.*) Lead case for storing or transporting multicurie radioactive sources. Also called **casket, coffin.**

flat (*Cinema.*). (1) The unit panel of which sets are constructed. (2) A *tormentor* (q.v.). (*Elec. Eng.*) A term used to denote a point on the surface of a commutator where the bars are lower than normal, due to wear or displacement. (*Mining*) A cap or cross-piece in a timber for a roof support. See **district.** (*Textiles*) One of a chain of metal bars, covered with fine steel wire points in a cloth and rubber foundation. Arranged to exert a combing action on the fibres over the top of main card cylinder.

flat arch (*Build.*). An arch whose intrados has no curvature and whose voussoirs (laid in parallel courses) are arranged to radiate to a centre. It is used over doorway, fireplace, and window-openings, to relieve the pressure on the beam or lintel below it. Also called a **jack arch** and **straight arch.**

flatback stope (*Mining*). A stope, in *overhand stoping*, worked upwards into a lode more or less parallel to level.

flat band (*Build.*). A square and plain impost stone.

flat bed (*Print.*). A general name distinguishing a cylinder machine with a horizontal flat bed from a *rotary machine* or from a *platen machine*.

flat-bottomed rail (*Rail.*). See **flanged rail.**

flat chisel (*Eng.*). A cold chisel having a relatively broad cutting edge, used in chipping flat surfaces.

flat-compounded (*Elec. Eng.*). Said of a compound-wound generator the series winding of which has been so designed that the voltage remains constant at all loads between no-load and full-load. Also **level-compounded.**

flat finish (*Paint.*). A nonglossy finish, showing no brilliancy of surface. Also called a **dead finish.**

flat foot (*Med.*). See **pes planus.**

flat gouge (*Carp., etc.*). A gouge having a cutting edge shaped to a large radius of curvature. See **middle gouge, quick gouge.**

flat joint (*Build.*). The type of mortar joint made in *flat pointing* (q.v.).

flat-joint jointed (*Build.*). A flat joint which has had a narrow groove struck along the middle of its face by means of a jointer.

flat keel (*Ships*). See under **keelson.**

flat lead (*Plumb.*). Sheet-lead.

flat lighting (*Cinema.*). That resulting from diffuse sources in front of the object, obviating depth or moulding.

flat of keel (*Ships*). The portion of a ship's form actually coinciding with the baseline in a transverse plane.

flat pointing (*Build.*). The method of pointing, used for uncovered internal wall surfaces, in which the stopping is formed into a smooth flat joint in the plane of the wall.

flat random noise (*Acous.*). See **white noise.**

flat-rate tariff (*Elec. Eng.*). A method of charging for electrical energy in which only one single charge is made, e.g., a fixed price per unit consumed.

flat region (*Nuc. Eng.*). Portion of reactor core over which neutron flux (and hence power level) is approximately uniform.

flat roof (*Build.*). A roof surface laid nearly horizontal, i.e., having a fall of only about 1 in 80, provided for drainage.

flats (*Eng.*). (1) Iron or steel bars of rectangular section. (2) The sides of a (hexagonal) unit.

flat sheets (*Mining*). Iron sheets, laid at rail junctions, crossings, and ends underground, on which tubs or trucks can be turned.

flat spot (*I.C. Engs.*). In a carburettor, a point during increase of air flow (resulting from increased throttle opening or speed) at which the air-fuel ratio becomes so weak as to prevent good acceleration.

flat stitching (*Bind.*). The stitching of a book close to the back with wire which passes through from the first page to the last, as distinct from *saddle stitching* (q.v.). Cf. *stabbing, drilled and strung.*

flattener (*Glass*). One who takes a cylindrical piece of glass like a wide tube, cracked longitudinally, and, after heating it to softening in a furnace, flattens it out to form a sheet. An old process only used for making special types of sheet.

flattening material (*Nuc. Eng.*). Neutron absorber or depleted fuel rod used in centre of reactor core to give larger flat region.

flattening of the Earth (*Astron.*). The ratio $f=(a-b)/a$, where a and b are the equatorial and polar radii of the Earth respectively. The accepted value of f is 1/297 but measurements by artificial satellites suggest that 1/298·25 is more correct.

flatter (*Eng.*). (1) A smith's tool resembling a flat-faced hammer, which is placed on forged work and struck by the sledge-hammer. (2) A draw-plate for producing flat wire, such as watch-springs. (*Mining*) A man who uncouples empty tubs or trucks and couples on full tubs, to make up sets at the inbye sidings or putter's flat.

flatting mill (*Met.*). Rolling mill which produces strip metal or sheet.

flatting varnish (*Paint.*). An oil varnish containing resin used as a basis for the final coat of varnish after having been rubbed down with abrasives to give a flatter surface.

flat-top antenna (*Radio*). One in which the uppermost wires are horizontal; also **roof antenna.**

flat-topped wave (*Telecomm.*). A waveform of current, voltage, flux, etc., sometimes met with in alternating-current work, in which the ordinates of the maximum value of the wave are less than those of a corresponding sine wave.

flat tuning (*Radio*). Inability of a tuning system to discriminate sharply between signals having different frequencies.

flat twin cable (*Elec. Eng.*). Cable for wiring work in which two conductors are laid side by side (but not twisted together) and surrounded by a suitable covering or sheath.

flatus (*Med.*). Gas or air accumulated in the stomach or intestines.

flaunching (*Build.*). The slope given to the top surface of a chimney to throw off the rain.

flavanthrone (*Chem.*). A yellow vat dye:

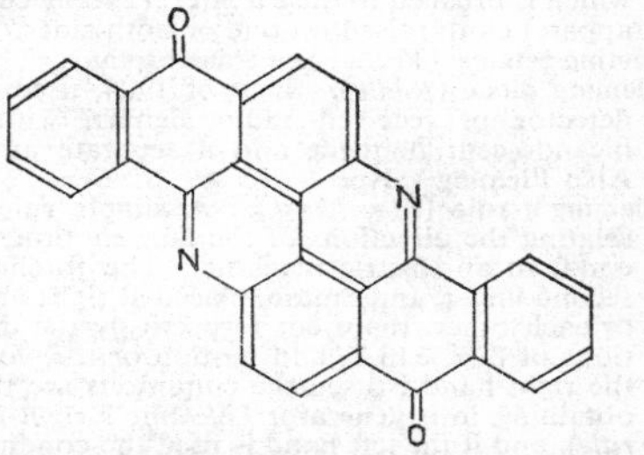

flavescent (*Bot.*). Having yellow-green or yellow spots mingled with the normal surface green.

flavin-adenine dinucleotide (*Biochem.*). A derivative of riboflavin employed as a cofactor in biological oxidations. Abbrev. **FAD.**

flavone (*Chem.*). A yellow plant pigment; the phenyl derivative of chromone, parent substance of a number of natural vegetable dyes.

flavones (*Chem.*). Yellow pigments occurring widely in plants, derivatives of *flavone* (q.v.):

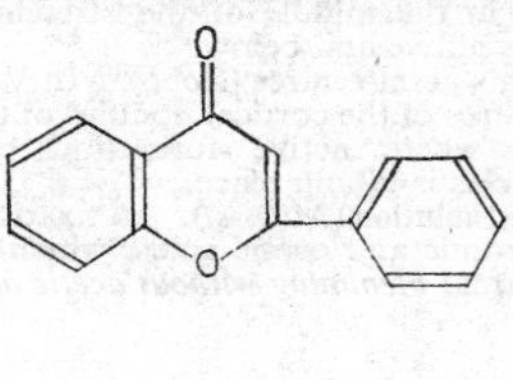

flavoproteins (*Chem.*). Conjugated proteins in which the prosthetic group is riboflavin phosphoric acid or a compound of this with another nucleotide. They function as hydrogen carriers in biological oxidations.

flaw-piece (*For.*). A slab of timber cut from the outer parts of a log.

Flaxedil (*Med.*). See **gallamine.**

F-layer (*Phys.*). Upper ionized layer in the ionosphere resulting from ultraviolet radiation from the sun and capable of reflecting radio waves back to earth at frequencies up to 50 MHz. At a regular height of 300 km during the night, it falls to about 200 km during the day. During some seasons, this remains as the F_1 layer while an extra F_2 layer rises to a maximum of 400 km at noon. Considerable variations are possible during particle bombardment from the sun, the layer rising to great heights or vanishing. It is also known as the **Appleton layer.**

fleaking (*Build.*). Thatching a roof with reeds.

fleam (*Carp.*). Angle of rake between the cutting edge of a sawtooth and the plane of the blade. (*Hyd. Eng.*) An artificial water-course.

fleam-tooth (*Tools.*). A sawtooth having the shape of an isosceles triangle.

flèche (*Arch.*). A slender spire, particularly a timber one, springing from a roof ridge.

flèches d'amour (*Min.*). Acicular, hairlike crystals of rutile, a crystalline form of the (IV) oxide of titanium, TiO_2, embedded in quartz. Used as a semiprecious gemstone. Also called **love arrows** (the literal translation), **cupid's darts,** or **Venus' hair stone.**

flecked (*Textiles*). The term applied to yarn or cloth having a spotted effect. Can be a fault, but often used for a distinctive novelty look.

fleece wool (*Textiles*). Wool obtained from a clip made subsequent to the first clip (which is termed *lamb's wool* or *yearling's wool*).

fleecy fabrics (*Textiles*). (1) Term in the hosiery trade for fabrics having at the back a thick yarn which is brushed to raise a pile. (2) Any cotton apparel cloth raised on one or both sides.

fleeting tetanus (*Vet.*). See **transit tetany.**

Fleming diode (*Radio*). That of 1904, used as a detector of received radio signals, with an incandescent filament and a separate anode. Also **Fleming valve.**

Fleming's rule (*Elec. Eng.*). A simple rule for relating the directions of the flux, motion, and e.m.f. in an electric machine. The forefinger, second finger, and thumb, placed at right angles to each other, represent respectively the directions of flux, e.m.f., and motion or torque. If the right hand is used the conditions are those obtaining in a generator (*Fleming's right-hand rule*), and if the left hand is used the conditions are those obtaining in a motor (*Fleming's left-hand rule*).

Fleming valve (*Electronics*). See **rectifying valve.**

Flemish bond (*Build.*). A bond consisting of alternate headers and stretchers in every course, each header being placed in the middle of the stretchers in the courses above and below.

Flemish garden-wall bond (*Build.*). A bond in which each course consists of three stretchers alternating with a header, each header being placed in the middle of the stretchers in the courses above and below.

Flemming's germ-centres (*Zool.*). In Vertebrates, the centres of the cortical nodules of the lymph-glands where active formation of lymph-corpuscles is taking place.

Flemming' solution (*Micros.*). A fixative containing chromic and osmic acids, which is used in two forms; *Flemming without acetic acid,* which is a cytoplasmic fixative and *Flemming with reduced acetic acid,* which contains glacial acetic acid also, and is used particularly for *Protozoa.*

flesh side (*Eng.*). That side of leather which formed the internal surface of the hide; used next to the pulley in driving belts.

flesh split (*Leather*). The middle split of a hide, or the inner split of a sheepskin.

Fletcher-Munson contours (*Acous.*). Equal-loudness curves for aural perception, measured just outside the ear, extending from 20 to 20 000 Hz, and from the threshold of hearing to the threshold of pain. The basis of the *phon scale* of equivalent loudness level, and of the weighting networks used in sound-level meters.

fletton (*Build.*). A well-known type of brick of a mottled pink and yellow colour and having sharp arrises and a deep frog; made chiefly around Peterborough and Bedford.

f-levels (*Light*). See **fundamental series.**

Flewelling circuit (*Radio*). Original super-regenerative receiving circuit in which quenching oscillations are generated by the same valve as is used for super-regeneration.

Flexibak (*Bind.*). A machine for *unsewn binding.*

flexibilitas cerea (*Med.*). The state in which the limbs will remain in any position in which they are passively placed, e.g., in catalepsy.

flexible (*Bind.*). The term used to indicate that in attaching the sections of a volume together the bands are not let into the back of the sections, the sewing thread passing completely round each band. See **raised bands.**

flexible cable (*Elec. Eng.*). A cable containing one or more cores of such cross-section and fine stranding as to make the whole quite flexible.

flexible cord (*Elec. Eng.*). A flexible cable of small cross-section, consisting usually of a large number of fine wire strands, surrounded by rubber insulation and braiding. Used for connexions to portable domestic apparatus, pendant lamps, etc. See **twin flexible cord.**

flexible coupling (*Eng.*). A coupling used to connect two shafts in which perfectly rigid alignment is impossible; the drive is commonly transmitted from one flange to another through a resilient member, such as a steel spring, or a rubber disk or bushes.

flexible resistor (*Elec. Eng.*). One resembling a flexible cable.

flexible roller bearing (*Eng.*). An antifriction bearing containing hollow cylindrical rollers made by winding strip steel into helical form. The hollow construction permits greater deflection under load. Also called **Hyatt roller bearing.**

flexible support (*Elec. Eng.*). A support for an overhead transmission line, which is designed to be flexible in a direction along the line, but rigid in a direction at right angles to the line.

flexible suspension (*Elec. Eng.*). A method of suspending the contact wire of a traction system so that it has a certain amount of lateral and vertical movement relative to the fixed supports.

flexible wiring (*Elec. Eng.*). The use of flexible cables in wiring an interior installation.

fleximer (*Build.*). See **cement-rubber-latex.**

flexing test (*Paper*). A modified form of tensile strength test for board where the sample is supported horizontally and the load applied at the centre of the strip.

flexion point (*Electronics*). U.S. for upper bend in the characteristic of a diode, depending on temperature limitation of emission, if stable.

flexographic printing (*Print.*). Usually reel-fed rotary, using relief rubber plates and spirit-based aniline inks which dry quickly, leaving no odour; much used for food wrappers.

flexor (*Zool.*). A muscle which by its contraction bends a limb or a part of the body. Cf. *extensor*.

Flexowriter (*Print.*). Keyboard perforator which provides simultaneous page-printing and punched tape. Page-printing is also possible from paper tape, or a new tape from an existing tape.

flexuose, flexuous (*Bot.*). Said of a stem which is zig-zag, usually showing a change of direction at each node.

flexural strength (*Build.*). Cross-breaking or transverse strength. A measure of the resistance to fracture on transverse loading, i.e., the load applied centrally across a simple beam of the material under test.

flexure (*Build.*). The bending of a member, e.g., under load.

flexure crystal (*Crystal.*). Quartz bar crystal, the bending of which is used for stabilizing valve oscillators or exhibiting resonance through ionization of surrounding neon gas.

flex-wing (*Aero.*). A collapsible single surface fabric wing of delta planform investigated first by Ryan (U.S.) for the return of space vehicles as gliders; later applied to army low-performance tactical aeroplanes of collapsible type.

flicker (*Optics*). Visual perception of fluctuation of brightness at frequencies lower than that covered by persistence of vision. Threshold of flicker depends on brightness and angle from optic axis. See also **chromaticity flicker**.

flicker effect (*Electronics*). Irregular emission of electrons from a thermionic cathode due to spontaneous changes in condition of emitting surface; resulting in an electronic noise.

flicker-fusion frequency (*Optics*). The rate of successive intermittent light stimuli just necessary to produce complete fusion, and thereby the same effect as continuous lighting.

flicker photometer (*Light.*). A photometer in which a screen is illuminated alternately and in quick succession by the lamp under test and a standard lamp, thus producing a flickering effect. When the illumination from the two sources of light is equal, the flickering effect disappears.

flicker shutter (*Cinema.*). The rotating shutter that flashes the stationary images on to the screen, as contrasted with the automatic shutter, which cuts the light off the film when intermittent motion ceases.

flick roll (*Aero.*). See **roll**.

flier (*Build.*). A step, rectangular in plan, forming part of a stair. (*Carp.*) See **flying-shore**.

flight (*Eng.*). (1) The helical element in a worm or screw conveyor, usually fabricated from sheet metal, which may be hollow to allow a heating or cooling liquid to pass through it. (2) The helical element in a vibratory bowl feeder over which components pass while being unscrambled and orientated.

flight control (*Space*). Control of vehicle, e.g., spacecraft, missile, or module, so that it attains its target, taking all conditions and corrections into account. Generally done by computer, controlled by signals representing actual and intended path.

flight deck (*Aero.*). A U.S. term now in general use for the *cockpit* (q.v.) of an aircraft which has a separate cabin; usually applied only where the crew is two or more in one cockpit.

flight director (*Aero.*). An aircraft instrument (i.e., blind) flying system in which the dials indicate what the pilot must do to achieve the correct flight path as well as the actual attitude of the aircraft. The dial display is usually an *artificial horizon* with a spot or pointer which must be centred. The equipment can be coupled to a radio- or *gyro-compass* to bring the aeroplane on to a desired heading, preset on the compass, or it can be coupled to the *instrument landing system* to receive these signals. In all cases, the flight director can be made automatic by switching its signals into the autopilot.

flight engineer (*Aero.*). A member of the flying crew of an aircraft responsible for engineering duties, i.e., management of the engines, fuel consumption and power systems, etc.

flight envelope (*Aero.*). A diagram, the boundary of which forms a closed figure and defines the manoeuvering limits of normal acceleration (n) and corresponding equivalent airspeeds (V) specified for the particular aircraft type.

flight fine pitch (*Aero.*). The minimum blade angle, held by a removable stop, which the airscrew of a turboprop engine can reach while in the air, and which provides braking drag for the landing approach.

flight information (*Aero.*). A *flight-information centre* provides a *flight-information service* of weather and navigational information within a specified *flight-information region.*

flight-information region (*Aero.*). An airspace of defined dimensions within which information on air-traffic flow is provided according to the types of airspace therein. Abbrev. **FIR.**

flight-information service (*Aero.*). One giving advice and information to assist in the safe, efficient conduct of flights. Abbrev. **FIS.**

flight level (*Aero.*). *Air traffic control* instructions specify heights at which controlled aircraft must fly and these are given in units of 100 ft for altitudes of 5000 ft and above.

flight Mach number (*Aero.*). Ratio of air speed of an aircraft to speed of sound under identical atmospheric conditions.

flight of ideas (*Psychol.*). A rapid succession of slightly related, or unrelated, ideas as occurs in manic states.

flight path (*Aero.*). The path in space of an aircraft or projectile. (Its *course* is the horizontal projection of this path.)

flight plan (*Aero.*). A legal document filed with *air traffic control* by a pilot before, or during, a flight (by radio), which states his destination, proposed course, altitude, speed, ETA and alternate aerodrome(s) in the event of bad weather, fuel shortage, etc.

flight recorder (*Aero.*). A device which records data on the functioning of an aircraft and its systems on tape or wire. The recorder should be in a crashproof, floatable box which is ejected in case of an accident and is usually fitted with a homing radio beacon and flashlight. Data channels vary from 6 to nearly 100. There are two basic types of flight recorder for: (1) test flights of experimental aircraft. (2) routine flights of airliners so that faults can be traced and maintenance controlled. Popular name, *black box* (frequently a misnomer).

flight time (*Aero.*). See **block time.**

Flinders bar (*Ships*). A bar of soft iron attached to the magnetic compass binnacle to correct for disturbing effects of that part of the ship's field due to vertical soft iron in the fore and aft line.

flint (*Geol.*). Flints are concretions of silica, sometimes tabular, but usually irregular in form, distributed in countless numbers on the bedding planes of the Upper Chalk. Thought to have been formed by the segregation of organic silica derived from siliceous sponges. See also

chert, paramudras. (*Paper*) A coated paper with a hard polished surface produced by a flint burnisher. It is used for box covering.

flint glass (*Glass.*). Originally lead glass; the good quality silica needed to ensure freedom from colour was obtained from crushed flints. Now colourless glass other than flat glass.

flint gravel (*Geol.*). A deposit of gravel in which the component pebbles are dominantly of flint. The Tertiary and fluvioglacial gravels in S.E. England are essentially of this kind.

Flintkote (*Build.*). TN for bitumen-based cold- and hot-type insulation mastics; used as waterproofing agent and as floor covering.

Flintshire process (*Met.*). A process for smelting lead sulphide ores in a reverberatory furnace. Some of the sulphide is first converted to oxide, the temperature is then raised, and the sulphide and oxide combine, producing lead and sulphur-dioxide.

flip-flap (*Cinema.*). Special form of *titler* (q.v.) in which two titles are arranged back to back and changed by turning.

flip-flop (*Telecomm.*). Originally (and still in U.S.), bistable pair of valves or transistors, two stable states being switched by pulses. In U.K., similar circuit with one stable state and one unstable state (monostable) temporarily achieved by pulse. Circuit for binary counting, similar to *multivibrator* without coupling capacitors. See **Eccles-Jordan circuit, toggle.**

flirt (*Horol.*). A device for bringing about the sudden movement of mechanism.

flitch (*For.*). A piece of timber of greater size than 4 × 12 in., intended for reconversion.

flitch beam (*Build.*). A built-up beam formed with an iron plate between 2 timber beams.

flitching (*Mining*). Taking off the sides of an underground roadway or heading.

flit-plug (*Elec. Eng.*). A detachable connecting-box for coupling cables.

float (*Aero.*). (1) The distance travelled by an aircraft between flattening-out and landing. (2) A watertight buoyancy unit which is of combined streamline and hydrodynamic form to reduce air and water resistance; *main floats* are the principal hydrodynamic support of float-planes, while *wing-tip floats*, often retractable, give lateral stability to flying boats. (*Agric.*) An implement consisting of an inclined blade (or blades) used to level a seed bed. (*Build.*) (1) A plasterer's trowel. (2) A polishing block used by marble-workers. Cf. **float stone.** (3) A flat piece of wood with a handle on one side used for *floating* (q.v.). (*Eng.*) A single-cut file (or *float-cut file*), i.e., a file having only one set of parallel teeth, as distinct from a *cross-cut file*. (*For.*) A measure of timber, equalling 18 loads. (*I.C. Engs.*) A small buoyant cylinder of thin brass, steel, or proofed cork, placed in the float chamber of a *carburettor* (q.v.) for actuating a valve controlling the petrol supply from the main tank. (*Mining*) (1) Values so fine that they float on the surface of the water when crushed or washed, e.g., *float gold*. (2) Surfacial deposit of rock or mineral detached from the main dyke or vein, e.g., *float quartz* in the Lake District. (*Plumb.*) The floating, hollow ball in a *ball-cock* (q.v.). (*Weaving*) (1) A thread, either warp or weft, which passes over other threads, to produce the requisite pattern. (2) A defect in a fabric, caused by a thread passing over other threads with which it is designed to interweave. (3) A number of warp ends broken together.

float bowl (*I.C. Engs.*). U.S. term for *float chamber*.

float case (*Hyd. Eng.*). A *caisson* (q.v.).

float chamber (*I.C. Engs.*). In a *carburettor* (q.v.), the petrol reservoir from which the jets are supplied, and in which the fuel level is maintained constant by means of a float-controlled valve, so set to avoid spillage on cambers.

float-cut file (*Eng.*). See **float.**

floated coat (*Build.*). A coat of plaster smoothed with a float.

float glass (*Glass*). A transparent glass, the two surfaces of which are flat, parallel and fire polished so that they give clear, undistorted vision and reflection, manufactured by floating hot glass in ribbon form on a heated liquid of greater density than the glass, e.g., molten tin.

floating (*Build.*). The second of 3 coats applied in plastering, the method of application being with a *float*, to bring the coat level with the *screeds*.

floating address (*Comp.*). A label used to identify a particular word in a program, regardless of its specific location at any given time. Also **symbolic address.**

floating anchor (*Ships*). See **sea anchor.**

floating balance (*Horol.*). Modern type of clock escapement with balance wheel suspended from a cylindrical hairspring.

floating bank (*Eng.*). A stoker-fired boiler to which sufficient coal is fed to keep the boiler under full pressure.

floating battery (*Elec. Eng.*). An electrical supply system in which a storage battery and electrical generator are connected in parallel to share load, so that the former carries the whole load if the generator fails.

floating bay (*Build.*). An area between screeds, which is to be filled in with plaster.

floating bricks (*Build.*). See **Rhenish bricks.**

floating bridge (*Civ. Eng.*). A bridge supported on pontoons instead of on fixed piers.

floating-carrier wave (*Radio*). Transmission in which the carrier is reduced or even suppressed to zero during intervals of no modulation by signals, to economize in power or to avoid interference with reception.

floating compass (*Instr.*). An early type of primitive compass, probably Chinese. The needle was attached to a float of wood or straw and floated in a bowl of water. The directions were marked inside the bottom of the bowl. See **liquid compass.**

floating crane (*Eng.*). A large crane carried on a pontoon; used in fitting-out docks, etc.

floating dam (*Hyd. Eng.*). A *caisson* (q.v.).

floating dry dock (*Civ. Eng.*). A floating structure of iron or steel, with air chambers. It is open at the ends, and can be sunk by admitting water to the air chambers, and raised when a vessel is berthed for repairs, etc., lifting the vessel with it. In some cases the dock is sectional, thus facilitating repair.

floating grid (*Electronics*). One which is unconnected in a valve and so is free to accumulate electrons.

floating gudgeon pin (*Eng.*). One free to revolve in both the connecting-rod and the piston bosses.

floating gyro (*Instr.*). One spinning in enclosure containing fluid, to reduce friction, mainly by buoyancy.

floating harbour (*Hyd. Eng.*). A breakwater formed of booms fastened together and anchored, to afford protection to vessels behind it.

floating-head exchanger (*Chem. Eng.*). See **shell and tube exchanger.**

floating kidney (*Med.*). See **nephroptosis.**

floating point (*Comp.*). A method of representing a large range of numbers by a given number of digits by restricting the number of significant digits in each number. For example, in the decimal system using n digits, every number has to be expressed in the form $10^x \times y$, where x contains r digits, y contains $n-r$ digits and $0{\cdot}1 \leq y < 1$. The number pair x, y is then the floating-point representation. Thus if $n=6$ and $r=2$, the numbers 1 234 567 and 0·000 123 45 would be represented as +07,1235 and −03,1235 respectively (where rounding up has been used for 5).

floating potential (*Electronics*). That appearing on an isolated electrode when all other potentials on electrodes are held constant.

floating respiration (*Bot.*). Respiration using carbohydrates and other reserve materials.

floating ribs (*Zool.*). See **false ribs.**

floating rule (*Build.*). A long straight edge used to form flat surfaces in plaster or cement work.

floating temperature control (*Heat*). The use, in a furnace, of an automatic temperature controller which functions by electrically operated valves.

floating tissue (*Bot.*). A tissue of thin-walled cells, usually containing air. The tissue is very difficult to wet, therefore does not readily become waterlogged; occurs in seeds and fruits which are dispersed by water.

float seaplane (*Aero.*). An aeroplane of the seaplane type, in which the water support consists of floats in place of the main undercarriage, and sometimes at the tail and wing tips. It may be of the *single-* or *twin-float type.*

float stone (*Build.*). A shaped iron block which is rubbed over curved brickwork, such as cylindrical backs, in order to remove marks left on the surface by the rough dressing. (*Min.*) A coarse, porous, friable variety of impure silica, consisting chiefly of the siliceous skeletons of infusoria. On account of its porosity it floats on water until saturated.

float switch (*Elec. Eng.*). A switch operated by a float in a tank or reservoir, and usually controlling the motor of a pump.

floccillation, floccitation (*Med.*). Fitful plucking at the bedclothes by a delirious patient, as in typhoid fever. See **carphology.**

floccose (*Bot.*). Bearing a dense covering of tangled hairs resembling wool, which is easily detached from the plant.

flocculation (*Chem.*). The coalescence of a finely divided precipitate into larger particles. (*Min. Proc.*) Coagulation of ore particles by use of reagents which promote formation of flocs, as a preliminary to settlement and removal of excess water by thickening and/or filtration. (*Nuc.*) Separation of radioactive waste products from water by coagulation.

flocculent (*Chem.*). Existing in the form of cloud-like tufts or flocs.

flocculi (*Astron.*). See **plages.**

flocculus (*Zool.*). In Birds and Mammals, a small lateral outgrowth of the cerebellum. *adj.* **floccular.**

floccus (*Zool.*). In Birds, the downy covering of the young forms of certain species; in Mammals, the tuft of hair at the end of the tail; more generally, a tuft.

flock (*Textiles*). (1) Waste fibres produced in the processes of finishing woollen cloths, used for bedding and upholstery purposes. (2) Short-cut or ground wool, cotton or manmade fibres for spraying on to adhesive coated backings for furniture and upholstery purposes.

flogging (*Carp.*). The operation of rough-dressing a timber to shape, when the material is removed in large pieces.

flogging chisel (*Eng.*). A large heavy cold-chisel used for rough work, now largely superseded by pneumatic chisels.

flong (*Paper, Typog.*). A board made from papier-mâché used for making moulds from which stereo plates are cast. See **dry-, wet-.**

floodable length (*Ships*). The maximum length of that portion of a ship, centred at a given point, which can be flooded without submerging the *margin of safety line* (q.v.).

flood fencing (*Hyd. Eng.*). Fencing which is so anchored as (*a*) to enable it to withstand the force of flood waters, or (*b*) to permit it to hinge over when the water rises sufficiently.

flood fever (*Med.*). See **shimamushi fever.**

flood flanking (*Hyd. Eng.*). The constructing of an embankment by depositing stiff moist clay in separate small loads, so that each shall unite so far as is possible with the others, while the crevices left when the clay has dried out are filled with *sludging* (q.v.).

flooding (*Chem.*). The condition of a fractionating column in which the upward flow of vapour has become great enough to prevent the downward flow of liquid, giving poor performance. (*Med.*) Copious bleeding from the uterus.

floodlighting (*Light*). The lighting of a large area or surface by means of light from projectors situated at some distance from the surface.

floodlight projector (*Light*). The housing and support for a lamp used in a floodlighting scheme; it is designed with a reflector which directs the light from the lamp into a suitable beam.

flookan or **flucan** (*Mining*). A vein of clayey material.

floor (*Foundry*). The bed of sand constituting the floor of a foundry; in it large castings are often made. (*Mining*) The upper surface of the stratum underlying a coal-seam. (*Ships*) Transverse vertical plate connected to the shell plating and to the inner bottom and extending from side to side.

floor contact (*Elec. Eng.*). A switch contact which is attached to the floor of an automatic electric lift and is operated by the passenger stepping into the lift; it is usually arranged to prevent the lift from being operated from any of the landings.

floor cramp (*Join.*). A cramp for closing up the joints of floor-boarding when it is being nailed in position.

floor guide (*Build.*). A groove formed in a floor surface to receive a sliding door or partition and direct its movement.

flooring (*Brew.*). The process of spreading out and turning the germinating grain on the floor of the malthouse to ensure aeration.

flooring saw (*Tools*). One with a curve towards the toe and extra teeth on the back above the toe. Used for cutting floorboards in order to raise them.

floor joist (*Carp.*). A *bridging joist* (q.v.).

floor line (*Join.*). A mark made at the lower end of a door-post, or other finishing, to indicate the level of the floor when the finishing is in position.

floor plan (*Arch.*). A plan drawn for any given floor of a building, normally showing the dimensions of the rooms, etc.

floor sand (*Met.*). Foundry sand in which new and used moulding sand and coal dust are mixed.

floor stop (*Build.*). A *door-stop* (q.v.) projecting from the floor near a door.

floor-switch (*Elec. Eng.*). See **landing switch.**

flop damper (*Build.*). A damper which stays under its own weight in the open or shut position.

flopgate (*Min. Proc.*). Diverting gate which directs moving material into alternative routes. Can be worked by remote control.
flop-over (*Telecomm.*). U.S. for a *flip-flop* (U.K. meaning) or *trigger circuit*.
flora (*Bot.*). (1) The plant population of any area under consideration. (2) A description of the plants of any region.
floral axis (*Bot.*). See **receptacle** (*6a*).
floral diagram (*Bot.*). A conventional plan of the arrangement of the parts of a flower as seen in cross-section.
floral envelope (*Bot.*). The calyx and corolla, or perianth.
floral leaf (*Bot.*). (1) A bract or bracteole. (2) A sepal or petal.
floral mechanism (*Bot.*). The arrangement of the flower parts to ensure either cross- or self-pollination.
Florence flask (*Chem.*). A long-necked rocund flask with a flat bottom.
Florentine (*Textiles*). A twilled cotton cloth used for tropical suitings and army drills. The weave, a 3 up and 1 down twill, is often termed the *Florentine weave*.
Florentine arch (*Arch.*). An arch having a semicircular intrados and a pointed extrados, giving greater strength at the crown.
Florentine blind (*Build.*). An outside roller blind, similar to the Italian blind but having side pieces.
floret (*Bot.*). An individual flower in a crowded inflorescence.
floriated (*Arch.*). Said of an elaborately ornamented building style, particularly with flowers and leaves featured.
Florideae (*Bot.*). The larger of the two subclasses of *Rhodophyceae*, containing many hundreds of species of red seaweeds, many of considerable size, and all characterized by the fact that the cells composing the thalli are united by protoplasmic threads.
floridean starch (*Bot.*). A solid carbohydrate resembling starch, formed by red algae as a product of assimilation. It stains reddish or brown with iodine.
flos ferri (*Min.*). A 'massive' form (as distinct from individual crystals) of the orthorhombic carbonate of calcium, *aragonite*, some of the masses resembling delicate coralline growths; deposited from hot springs.
flotation (*Min. Proc.*). See **froth flotation.**
flotation gear (*Aero.*). A system of air or gas bags, sometimes with hydrovanes, to enable a land aircraft to alight and remain afloat, on water.
flour (*Build., Civ. Eng.*). The fine dust incidentally formed in crushing material to be used as an aggregate.
flourometer (*Civ. Eng.*). An instrument used to determine the proportion of very fine material (*flour*) in a filler for asphalt.
flow (*Eng.*). A pipe by which water leaves a boiler or pressure cistern. Also called **flow pipe.**
flow boxes (*Paper*). See **breast box.**
flowchart (*Comp.*). Chart giving a pictorial representation of the nature and sequence of operations to be carried out according to program in a computer or automatic control system.
flow counter (*Nuc. Eng.*). See **gas-flow counter.**
flow diagram (*Work Study*). A diagram or model which shows the location of specific activities carried out and the routes followed by workers, materials or equipment in their execution.
flower (*Bot.*). A group of closely crowded specialized leaves at the end of a short branch, including one or more of the following kinds of members: sepals, petals, stamens, carpels.
flowering glume (*Bot.*). See **lemma.**
flowers (*Typog.*). Small type ornaments, copied from early designs, used for building up fancy borders, etc.
flowers of sulphur (*Chem.*). A form of sulphur obtained by allowing sulphur vapour to fall on a cold surface.
flow forming (*Met.*). A metal-forming operation used in the aircraft industry and to some extent in the manufacture of domestic holloware, in which thick blanks of aluminium, copper, brass, mild steel, or titanium are made to flow plastically by rolling them under pressure in the direction of roller travel to result in components, often conical, having a wall thickness much less than the original blank thickness.
flow-line production (*Eng.*). A system of mass production, and certain types of batch production, in which machines are arranged in flow lines to enable components to progress from operation to operation in correct sequence.
flow lines (*Met.*). Lines which appear on the surface of iron and steel when stressed to the yield point. They arise from the fact that all parts of a given sample do not yield at the same time; the lines are traces on the surface of the planes along which yielding first occurs.
flowmeter (*Phys.*). Device for measuring, or giving an output signal proportional to, the rate of flow of a fluid in a pipe.
flow-off (*Foundry*). A channel cut from a riser to allow metal to escape when it has reached a pre-determined height.
flow pipe (*Eng.*). A *flow* (q.v.). (*Plumb.*) The pipe conveying hot water from the boiler to the tank in a domestic system of hot-water supply.
flow process (*Glass*). See **gob process.**
flow sheet (*Met., etc.*). A diagram showing the sequence of operations used in a process of production with a given plan, e.g., the extraction and refining of metals.
flow-structure (*Geol.*). A banding, often contorted, resulting from flow movements in a viscous magma, adjacent bands differing in colour and/or degree of crystallization. It is also shown by the alignment of phenocrysts or of minute crystals and crystallites, in the groundmass of lavas and, more rarely, minor intrusions.
flucan (*Mining*). See **flookan.**
fluctuating variation (*Gen.*). Variation as shown by the differences between the individuals of one progeny.
fluctuation (*Bot.*). A change in a plant due to the effect of its environment on it. (*Med.*) The palpable undulation of fluid in any cavity or abnormal swelling of the body.
fluctuation noise (*Acous.*). Noise produced in the output circuit of an amplifier by shot and flicker effects.
flue (*Eng.*). A passage or channel through which the products of combustion of a domestic fire, boiler, etc., are taken to the chimney.
flue baffler (*Heat*). Inverted cowl situated above the outlet of a flue by which the combustion products are exhausted to open atmosphere, enabling the correct degree of draught to be maintained.
flue bridge (*Eng.*). See **firebrick arch.**
flue gas (*Eng.*). The gaseous products of combustion from a boiler furnace, consisting chiefly of CO_2, CO, O_2, N_2 and water vapour, whose analysis is used as a check on the furnace efficiency. See **CO_2 recorder.**
flue-gas temperature (*Heat*). Temperature of flue

gases at the point in the flue where it leaves the furnace.

flue gathering (*Build.*). See **gathering.**

flueing soffit (*Build.*). A flush soffit under a geometrical stair.

flue lining (*Build.*). A fireclay, or fire-resistant concrete pipe, arranged with others within a flue passage to protect the walls.

fluff (*Paper*). Fibres removed from the surface of the paper during printing and converting operations.

fluffing (*Leather*). An operation which produces a velvet finish on the flesh side of leather. (*Print.*) A tendency with soft-sized paper for fibres to be detached from the surface and edges, producing difficulties in printing, particularly if by lithography.

fluid. A substance which flows. It differs from a solid in that it can offer no permanent resistance to change of shape. See **gas, liquid.**

fluid flywheel (*Eng.*). A device for transmitting power through the medium of the change in momentum of a fluid, usually oil. Similar in principle to a Froude brake in which the stator is released and forms the driven member.

fluidics (*Phys., Automation*). The science of liquid flow in tubes, etc., which strongly simulates electron flow in conductors and conducting plasma. The interaction of streams of fluid can thus be used for the control of instruments or industrial processes without the use of moving parts.

fluidity (*Phys.*). The inverse of *viscosity* (q.v.).

fluidization (*Chem. Eng.*). A technique whereby gas or vapour is passed through solids so that the mixture behaves as a liquid, and of special significance when the solid is a catalyst to induce reactions in the fluidizing medium.

fluidized bed (*Chem. Eng.*). If a fluid is passed upward through a bed of solids with a velocity high enough for the particles to separate from one another and become freely supported in the fluid, the bed is said to be fluidized. Then the total fluid frictional force on the particles is equal to the effective weight of the bed. Fluidized beds are extensively used in chemical industry because of the intimate contact between solid and gas, the high rates of heat transfer and the uniform temperatures within the bed, and the high heat transfer coefficients from the bed to the walls of the containing vessel.

fluidized-bed reactor (*Nuc. Eng.*). One in which the active material is carried finely divided in a gas or liquid.

fluid lubrication (*Eng.*). A state of perfect lubrication in which the bearing surfaces are completely separated by a fluid or viscous oil film which is induced and sustained by the relative motion of the surfaces.

fluing (*Build.*). A term applied to window jambs which are splayed. See **splayed jambs.**

fluing arch (*Build., Civ. Eng.*). See **splaying arch.**

fluke (*Zool.*). A semipopular name for worms belonging to the group *Trematoda.*

flukes (*Ships*). The flattened and curving points terminating the arms of an anchor.

flume (*Aero.*). See **wind tunnel.** (*Build., Civ. Eng.*) A metal chute used for the distribution of concrete from a placing plant. (*Hyd. Eng., Mining, etc.*) A flat-bottomed timber trough, or other open channel, generally nowadays formed in concrete, for the conveyance of water, e.g., to ore-washing plant, or as a by-pass.

flume stabilizer (*Ships*). A roll stabilization system using passive fluid tanks fitted athwart ships, and having specifically designed nozzles to cause a phase difference of 90° between the movement of the liquid in the tank and the roll of the ship.

fluoboric acid (*Chem.*). A complex monobasic acid formed by the combination of hydrogen fluoride and boron (III) fluoride. Salts called *borofluorides* or *fluoborates.*

Fluon (*Chem.*). TN for **polytetrafluoroethene.**

fluorapatite (*Min.*). The commonest form of *apatite* (q.v.).

fluorene (*Chem.*). Diphenylenemethane, $(C_6H_4)_2CH_2$; colourless fluorescent plates; m.p. 113°C, b.p. 295°C; contained in coal-tar; produced by leading diphenylmethane through red-hot tubes.

fluorescein (*Chem.*). $C_{20}H_{12}O_5$, (1,3-dihydroxybenzene) phthalein, red crystals which dissolve in alkalis with a red colour and green fluorescence. See also **eosin.**

fluorescence (*Light*). Emission of light, generally visible, from materials *irradiated* (generally from a higher frequency source), or from impact of electrons, as in a phosphor. See **Stokes's law.**

fluorescence microscope (*Micros.*). A modification of the light microscope for the examination of small objects, e.g., bacteria. The organisms are stained with a fluorescent dye and viewed by ultraviolet light.

fluorescent lamp (*Elec. Eng.*). A mercury-vapour electric-discharge lamp having the inside of the bulb or tube coated with fluorescent material so that ultraviolet radiation from the discharge is converted to light of an acceptable colour.

fluorescent screen (*Electronics*). One coated with a layer of luminescent material so that it fluoresces when excited, e.g., by X-rays or cathode rays. See **fluoroscopy, phosphor** (2).

fluorescent whitening agents (*Chem.*). Special dyes widely used to 'whiten' textiles, paper, etc., and sometimes incorporated in detergents. Their effect is based on ability to convert invisible ultraviolet light into visible blue light, giving fabrics greater uniformity of reflectance over the visible part of the spectrum. Also **optical bleaches, optical whites.**

fluorescent yield (*Nuc.*). Probability of a specific excited atom emitting a photon in preference to an Auger electron.

fluoridation (*Chem.*). The addition of inorganic fluorides (usually *sodium fluoride*) to water supplies with the intention of reducing dental decay. The amount added is very small indeed, usually about 1 part/million.

fluorimeter (*Radiol.*). One used for measuring the intensity of fluorescent radiation.

fluorimetry (*Min. Proc.*). Analytical method in which fluorescence induced by ultraviolet light or X-rays is measured. Also **fluoremetry.**

fluorinated ethene propene (*Plastics*). See **FEP.**

fluorination (*Chem.*). The replacement of atoms (usually hydrogen atoms) in molecules by fluorine. Can be carried out catalytically using fluorine in the vapour phase or by using fluorine 'carriers' such as cobaltic fluoride or silver fluoride in solution with increased temperature.

fluorine (*Chem.*). Pale greenish yellow gas, the most electronegative (non-metallic) of the elements and the first of the halogens. Chemically highly corrosive and never found free. Discovered by Scheele in 1771 and isolated by Moissan in 1886. Symbol F, at. no. 9, r.a.m. 18·9984, valency 1, m.p. −223°C, b.p. −187°C, density 1·696 g/dm³ at s.t.p. Used in separating the isotopes of uranium, in the compound uranium (VI) fluoride, which is a gas, and when forced through a porous plug differentiates very slightly between the heavier and lighter isotopes. Combines with carbon to form inert polymers with low coefficient of friction, e.g., Teflon. See **fluorocarbons.**

fluorite or **fluorspar** (*Min.*). Calcium fluoride, CaF_2, crystallizing in the cubic system, commonly in simple cubes. Occasionally colourless, yellow, green, but typically purple; the coloured varieties may fluoresce strongly in ultraviolet light.

fluorocarbons (*Chem.*). Hydrocarbons in which some or all of the hydrogen atoms have been replaced by fluorine. The fluorinated derivatives of methane are widely used as refrigerating agents and propellants for aerosols. See also **Freons** and **polytetrafluoroethene.**

fluorography (*Radiol.*). The photography of fluoroscopic images.

fluorophore (*Chem.*). A group of atoms which give a molecule fluorescent properties.

fluoroscope (*Radiol.*). Measurement system for examining fluorescence optically. Fluorescent screen assembly used in fluoroscopy.

fluoroscopy (*Radiol.*). Examination of objects by observing their X-ray shadow shown on a fluorescent screen; used to ascertain contents of packages without unwrapping, quality of welding, etc. Also called **screening.**

fluorosis (*Med.*). Chronic poisoning with fluorine.

fluorspar (*Min.*). See **fluorite.**

Fluothane (*Chem., Med.*). $CF_3CHClBr$. (1-chloro-1-bromo-2,2,2 trifluoroethane). TN for a nonexplosive anaesthetic, now used extensively, containing 1 atom of fluorine in its molecular structure. More often called **halothane.**

flush (*Bot.*). (1) A period of renewed growth in a woody plant. (2) A limited area watered by a spring or by the run off from rainfall, and distinguished by its luxuriant vegetation. (*Build., Join., etc.*) In the same plane. (*San. Eng.*) To cleanse a space by *flushing* (q.v.); the water used in flushing.

flush bead (*Join.*). A sunk bead, finished so as to be level with the surface which it decorates.

flush boards (*Bind.*). A method of binding in which boards are drawn on and trimmed with the book. The covers are then flush with the page edges.

flush-bolt (*Join.*). A sliding bolt sunk into the side or edge of a door so as to be flush with the surface.

flushing (*Build.*). A crushing of the edges of a stone at a *hollow bed* (q.v.), due to excessive pressure upon them. (*Hyd. Eng., San. Eng.*) The process of cleansing a sewer or other space by suddenly passing through it a quantity of water. (*Mining*) The operation of clearing off accumulation of fire-damp or noxious gases underground by means of air currents. See **face-airing.** (*Vac. Tech.*) A process in the manufacture of lamps or valves in which deleterious gases are removed by successively evacuating the envelope and filling it with a suitable gas.

flushing of ewes (*Vet.*). The practice of increasing the nutritional plane of ewes a few weeks before they are served by the ram, with the object of increasing fertility.

flushing tank (*San. Eng.*). A tank used to accumulate the water for flushing a drain or sewer which is not laid at a self-cleansing gradient. The discharge is often effected automatically by a siphoning device.

flush joint (*Build.*). The type of mortar joint made in *flat pointing* (q.v.).

flush panel (*Join.*). A panel whose surface is in line with the faces of the stiles.

flush-plate (*Elec. Eng.*). See **switch plate.**

flush soffit (*Build.*). The continuous surface formed under any ceiling or stair.

flush-switch (*Elec. Eng.*). A switch which can be mounted flush with the wall. Also called **panel-switch, recessed switch.**

flush valve (*Plumb.*). A valve operating the flushing system for a fixture.

flute (*Build.*). A long vertical groove, usually circular in form, in the surface of a column or other member. (*For.*) A natural longitudinal groove on the bole of a tree.

fluted carbon (*Light*). An arc-lamp carbon with grooved sides, used for heavy-current arcs.

fluting (*Build.*). See **flute.** (*Eng.*) Parallel channels or grooves, longitudinal or helical, cut in a cylindrical object such as a tap or reamer.

fluting plane (*Join.*). A plane for cutting grooves.

flutter (*Acous.*). Rapid fluctuation of frequency or amplitude in radio reception or sound reproduction, especially from disks. See **echo, wow.** (*Aero.*) Sustained oscillation, usually on wing or tail, caused by interaction of aerodynamic forces, elastic reactions and inertia, which rapidly break the structure. *Asymmetrical flutter* occurs where the port and starboard sides of the aircraft simultaneously undergo unequal displacements in opposite directions, as opposed to *symmetrical flutter*, where the displacements and their direction are the same; *classical*, or *coupled flutter* is due solely to the inertial, aerodynamic, or elastic coupling of two or more degrees of freedom. (*Med.*) An abnormality of cardiac rhythm, in which the auricles of the heart contract regularly at a greatly increased frequency (between 180 and 400 beats/min), the ventricles contracting at a slower rate. (*TV*) Variation in brightness of a reproduced picture, arising from additional radio reflection from a moving object, e.g., an aircraft.

flutter echo (*Acous.*). Multiple echo between parallel surfaces.

flutter speed (*Aero.*). The lowest *equivalent air speed* (q.v.) at which flutter can occur.

Fluviales (*Bot.*). See **Naiadales.**

fluviatile (*Bot., Zool.*). Occurring in rivers and streams.

fluviatile deposits (*Geol.*). Sand and gravel deposited in the bed of a river.

fluviomarine (*Zool.*). Able to live in rivers and in the sea, as the Salmon.

fluvioterrestrial (*Zool.*). Found in rivers and on their banks, as the Otter.

flux (*Chem.*). A substance added to a solid to increase its fusibility. In soldering, welding, etc., added to the molten metal to dissolve infusible oxide films which prevent adhesion. (*Maths.*) Through a surface in a vector field: the integral over the surface of the product of the elementary area by the normal component of the vector, i.e., $\int \mathbf{F} \cdot d\mathbf{s}$. (*Met.*) Material added to a furnace charge to combine with constituents not wanted in the final metal produced and which issues as a separate slag. (*Phys.*) The rate of flow of mass, volume, or energy per unit cross-section normal to the direction of flow. See also **electric-, magnetic-.**

flux density (*Nuc.*). The number of photons (or particles) passing through unit area normal to the beam, or the energy of the radiation passing through this area.

flux gate (*Elec. Eng.*). Magnetic reproducing head in which magnetic flux (due to flux leakage from signals recorded on magnetic tape) is modulated by high-frequency saturating magnetic flux in another part of the magnetic circuit.

flux-gate compass (*Instr.*). Device in which the

balance of currents in windings is affected by the earth's magnetic field.

flux guidance (*Elec. Eng.*). Directing the electric or magnetic flux in high-frequency heating by shaped electrodes or magnetic materials, respectively.

flux link or linkage (*Elec. Eng.*). Conservative flux across a surface bounded by a conducting turn. For a coil, *flux linkage* is the integration of the flux with individual turns.

fluxmeter (*Elec. Eng.*). Electrical instrument for measuring total quantity of magnetic flux linked with a circuit; it consists of a search coil placed in the magnetic field under investigation, and a ballistic galvanometer, or an uncontrolled moving-coil element (Grassot) or a semi-conductor probe generating Hall voltage. See also **gaussmeter.**

fly (*Horol.*). An air brake; a fan with two or four blades, used in clocks to maintain uniformity between the blows of the hammers when striking or chiming. (*Print.*) A term indicating a sheet of paper folded once to make 4 pages, with the printing, if any, on the first page only.

fly ash (*Build.*). A fine ash from the pulverised fuel burned in power stations, used in brick-making and as a partial substitute for cement in concrete.

flyback (*Radar, TV*). The return of the beam to its starting point in a radar trace or a line of a TV picture. In the latter the line is blanked out during the process. Also called **retrace.**

flyback action (*Horol.*). In a chronograph or stop-watch, that part of the action which causes the hands to fly back to zero when the button is pressed.

fly-blown (*Vet.*). Affected by myiasis.

fly cutter (*Eng.*). A narrow milling cutter used for cutting slots such as keyways in shafts.

fly disease (*Vet.*). See **nagana.**

flyer (*Build.*). A *flier* (q.v.). (*Carp.*) *Flying shore* (q.v.).

fly frames (*Spinning*). Drafting machines which attenuate roving in preparation for spinning on mules or ring-frames. Also called **speed frames.**

fly hand (*Print.*). A rotary press assistant who removes the printed product from the delivery or the conveyor.

flying-boat (*Aero.*). A seaplane wherein the main body or hull provides water support.

flying bond (*Build.*). See **monk bond.**

flying buttress (*Arch.*). An arched buttress giving support to the foot of another arch. Also called **arc-boutant, arch(ed) buttress.**

flying deck (*Ships*). See **hurricane deck.**

flying levels (*Surv.*). Back-sight and fore-sight readings taken between any 2 points, without reference to bench marks, when only the difference of level of the points is required.

flying paster (*Print.*). See **automatic reel change.**

flying scaffold (*Build.*). A *suspended scaffold* (q.v.).

flying-shore (*Carp.*). A horizontal baulk of timber used to provide temporary support between two opposite walls, usually not more than about 9 m apart. Also **flier, flyer.**

flying speed (*Aero.*). The *maximum flying speed* is the highest attainable speed in level flight, under specified conditions and corrected to standard atmosphere, and the *minimum flying speed* is the lowest speed at which level flight can be maintained.

flying-spot microscope (*Powder Tech.*). Instrument for particle-size analysis, using a television raster placed at the eyepiece of a microscope. By this means a minute spot of light scans the field of view of the microscope slide. The transmitted light is recorded by a photocell and the signals analysed to calculate the size distribution.

flying-spot scanner (*TV*). Device used for converting the light variations over the surface of a picture into a train of electric signals, the original picture being scanned by a small but extremely bright spot of light.

flying tail (*Aero.*). See **all-moving tail.**

flying tuck (*Print.*). A gear-driven rotating folding blade mounted on a cylinder on web-fed presses.

fly leaf (*Bind.*). A blank leaf at the beginning and at the end of a bound volume. It may be part of an end-paper.

fly nut (*Eng.*). See **wing nut.**

flyover (*Civ. Eng.*). An elevated highway providing passage for trunk traffic without interference to local traffic.

fly pinion (*Horol.*). The pinion on the arbor of which a *fly* is mounted.

fly press (*Eng.*). A hand-operated press for punching holes, making driving fits, etc.; it consists of a bed supporting a vertical frame through which a square-threaded screw is fitted. The screw is turned by a cross-piece terminating in one or two heavy steel balls, for giving additional impetus to the descent of the die attached to the bottom end.

fly rail (*Join.*). A flap attached to a table-frame by means of a vertical hinge; it swings out to provide support for a folding leaf.

fly shuttle (*Weaving*). Mechanism, invented by John Kay (Bury, Lancs) in 1733, for propelling the shuttle across the loom. It superseded hand-shuttling.

flywheel (*Eng.*). A heavy wheel attached to a shaft (e.g., an engine crankshaft) either to reduce the speed fluctuation resulting from uneven torque, or to store up kinetic energy to be used in driving a punch, shears, etc., during a short interval.

flywheel effect (*TV*). Maintenance of current in a resonant circuit through electrical inertia of the system during the interval between the exciting impulses. Used in television scanning circuits to control the frame frequency without using synchronizing pulses.

flywheel time base (*TV*). One in which synchronizing pulses control a pulse generator which is connected to the actual CRT. If reception is poor, much better synchronization is achieved because of the *flywheel effect* (q.v.).

flywheel-type alternator (*Elec. Eng.*). An alternator having a heavy spider so that a separate flywheel is not necessary to prevent hunting, when running in parallel with others.

fly wire (*Build.*). A fine woven wire mesh used to cover the joint between adjacent pieces of building-board.

Fm (*Chem.*). Symbol for *fermium.*

FM. Abbrev. for *frequency modulation.*

***f*-number** (*Photog.*). Ratio of the focal length to the true diameter of a lens, controllable in most cameras by adjustment of the iris diaphragm. Also **aperture, stop.**

foam (*Chem.*). See **froth; also fire foam.**

foambacks (*Textiles*). Dress and furnishing fabrics bonded on the back to polyether or polyester foams by adhesive or flame treatment.

foamed plastics (*Plastics*). **See expanded plastics.**

foamed slag (*Build.*). Blast furnace slag aerated while still molten. Used for building blocks and for acoustic and thermal insulation.

foam plug (*Mining*). Mass of foam generated and blown into underground workings to seal off a fire or keep out oxygen, where a fire risk exists.

Foamseal (*Build.*). TN for a flashing compound for use with foamed glass insulation blocks and other rigid foam insulation. Also used as a bedding compound between metal and insulation.

foam separation (*Chem.*). Removal of solutes or ions from a liquid by bubbling air through in the presence of surface active agents which tend to be adsorbed on to the bubbles. Cf. *froth flotation*, for larger particles.

fobbing (*Brew.*). Excessive foaming during bottling process.

focal aperture (*Photog.*). See **f-number.**

focal length of a lens (*Optics*). The distance measured along the principal axis, between the principal focus and the second principal point. In a thin lens both principal points may be taken to coincide with the centre of the lens. See **back focus, equivalent focal length**; also **convention of signs.**

focal plane (*Optics*). The plane, at right angles to the principal axis of a lens or lens system, in which the image of a particular object is formed. The principal focal plane passes through the principal focus, and contains the images of objects at infinity. (*Photog.*) The plane in which light rays from an external object are focused in a camera—the normal location of the sensitive surface of a film or plate, or a ground-glass focusing screen.

focal-plane shutter (*Photog.*). Camera shutter in the form of a blind with a slot, which is pulled rapidly across, and as close as practicable to, the film or plate, speed being varied by adjusting the width of the slot. See also **Copal square shutter.**

focal point (*Phys.*). The focal spot formed on the axis of a lens or curved mirror by a parallel beam of incident radiation. In its general form, this definition includes acoustic lenses, electron lenses, and lenses or mirrors designed for use with radio waves, infrared or ultraviolet radiation.

focal spot (*Phys.*). A spot on to which a beam of light or charged particles converges. See also **X-ray-.**

focometer (*Optics*). An instrument for measuring the focal length of a lens.

focus (*Geol.*). See **earthquake.** (*Optics*) A point to which rays converge after having passed through an optical system, or a point from which such rays appear to diverge. In the first case the focus is said to be *real*; in the second case, *virtual*. The *principal focus* is the focus for a beam of light rays parallel to the principal axis of a lens or spherical mirror. (*Maths.*) Of a conic: a point such that the two lines of every pair of conjugate lines through it are mutually perpendicular. The ellipse and hyperbola each have two real and two imaginary foci, and the parabola has one real focus. See **conic** for alternative definition. Of a quadric: a point not on the quadric, such that the three planes of every set of three mutually conjugate planes through it are mutually perpendicular.

focusing (*Elec.*). The convergence to a point of (*a*) beams of electromagnetic radiation, (*b*) charged particle beams, (*c*) sound or ultrasonic beams. (*Photog.*) The act of adjusting an optical system, as in a camera, by observation of the image on a ground glass screen, through a rangefinder, or by judging the distance.

focusing arc lamp (*Light*). An arc lamp with a feed mechanism arranged in such a way that the position of the arc crater does not alter.

focusing coil (or **electrode**) (*Elec. Eng.*). One used to focus a charged particle beam by magnetic (or electrostatic) field control.

focusing screen (*Photog.*). Screen, usually of ground glass, located in the place of a film or plate, or on the top of a reflex camera, for adjusting the focusing of the lens before exposure. Some reflex cameras have Fresnel screens and/or split-image rangefinders.

focus-skin distance (*Radiol.*). The distance from the focus of an X-ray tube to the surface of incidence on a patient, usually measured along the beam axis. Abbrev. **F.S.D.**

foetal echo (*Med.*). Detection of foetus by ultrasonic techniques; can be used at earlier stage than can radiography.

foetal membranes (*Zool.*). In Reptiles, Birds, and Mammals, outgrowths from the embryo, or the extraembryonic tissue, which surround and protect the foetus and facilitate respiration. See **allantois, amnion, chorion.**

foetus (*Zool.*). A young mammal within the uterus of the mother, from the commencement of organogeny until birth. *adj.* **foetal.**

fog (*Meteor.*). Minute water droplets with radii in the range 1 to 10 μm suspended in the atmosphere and reducing visibility to below 1 km (1100 yd in Britain). (*Photog.*) The general reduction of silver halide, apart from that exposed to the required image; due either to extraneous light (*light fog*), to deterioration of the emulsion or to over-vigorous development (*chemical fog*).

fogbow (*Meteor.*). A bow seen opposite the sun in fog. The bow is similar to the rainbow, but the colours are faint, or even absent, owing to the smallness of the drops, which causes diffraction scattering of the light.

fog fever (*Vet.*). An acute respiratory affection of cattle characterized by oedema of the lungs, respiratory distress and rapid death; some cases are associated with *husk* (q.v.).

fogging (*Paint.*). See **blooming.**

fog signal (*Rail.*). A detonating cap which is placed on a rail before the passage of a train, so that the detonation occurring when a wheel passes over it shall serve as a signal to the driver in bad visibility.

fog-type insulator (*Elec. Eng.*). A type of overhead-line insulator having long leakage distances; specially designed for areas in which fog is prevalent.

föhn wind (*Meteor.*). A warm, dry wind which blows down a mountain side. It is prevalent on the northern slopes of the Alps, and may occur when a cyclonic system passes over mountains.

foil (*Paper*). Wrapping or decorative paper coated with tin, copper, etc., in powder or leaf form.

folacin. See under **vitamin B complex.**

folded dipole (*Radio*). An aerial comprising interconnected parallel dipoles connected together at their outer ends and fed at the centre of one dipole. The dipole separation is a small fraction of the transmitted wavelength.

folded horn (*Acous.*). An acoustic horn which is turned back in itself to reduce necessary space.

folded scale (*Maths.*). On a slide-rule, scale labelled CF or DF calibrated same as principal scales but displaced about half scale length. Used to avoid resetting rule.

folded vernation (*Bot.*). The condition in which the leaf is folded about the midrib, with the two faces brought together.

folded yarns (*Spinning*). Yarns formed from two or more single threads combined for strength or special appearance (*fancy*). Also **doubling.**

folder (*Print.*). The section of a web-fed press where the webs are *associated* (q.v.), folded, cut, delivered, and sometimes stitched.
folder blade (*Print.*). See **folding blade.**
folding (*Geol.*). *Folding* (*bending*) of strata is usually the result of compression that causes the formation of the geological structures known as anticlines, synclines, monoclines, isoclines, etc. The amplitude (i.e., vertical distance from crest to trough) of a fold ranges from a fraction of an inch to thousands of feet.
folding blade (*Print.*). A metal strip which thrusts the web or webs into jaws or rollers to produce a fold. Also **tucker, tucking blade.**
folding cylinder (*Print.*). The cylinder which holds the folder blade or jaw for making a fold on web-fed presses.
folding jaws (*Print.*). The gripping section on a *nip and tuck folder* (q.v.).
folding plates (*Bind.*). (1) Adjustable parts on a folding machine into which the paper travels and, being stopped, receives a *buckle fold* (q.v.). (2) Large sized illustrations which require folding before inclusion in the book.
folding rollers (*Bind.*). Driven rollers on folding machines, the paper being folded as it passes between them. (*Print.*) Similar arrangement on web-fed presses through which the cut web is thrust.
folding shutters (*Join.*). See **boxing shutters.**
folding strength (*Paper*). The number of double folds needed (under definite conditions) to break the paper.
folding wedges (*Build., Civ. Eng., etc.*). *Striking wedges* (q.v.) used for tightening and easing shoring and centring, and in some joint construction in joinery.
fold-out (*Bind.*). A large leaf in a book which must be *folded out* when the book is being used.
foliaceous, foliose (*Bot.*). (1) Flat and leaflike. (2) Bearing leaves.
Folian process (*Zool.*). In Mammals, an anterior process of the malleus which extends into the Glaserian fissure.
foliar gap, trace (*Bot.*). See **leaf gap, leaf trace.**
foliated structure (*Zool., Min.*). A type of crystallite arrangement found in the shells of some molluscs, including oysters. It consists of sheets of overlapping tabular crystals of calcite, arranged rather like roof-tiles.
foliation (*Geol.*). The arrangement of minerals normally possessing a platy habit (such as the micas, chlorites, and talc) in folia or leaves, lying with their principal faces and cleavages in parallel planes; such arrangement is due to development under great pressure during regional metamorphism.
folic acid (*Chem.*). See **vitamin B complex.**
folicole (*Bot.*). Living on leaves, either as a parasite or as an epiphyte.
folie circulaire (*Psychol.*). Manic-depressive psychosis, in which phases of melancholia and mania regularly alternate.
folio (*Typog.*). (1) A sheet of paper folded in half. (2) A book made up of sheets folded once, so having 4 pages to the sheet. (3) The number of a page.
foliobranchiate (*Zool.*). Having leaflike gills.
foliolose (*Bot.*). Made up of minute flattened lobes.
foliose (*Bot.*). Said of a thallus which is flattened and leaflike.
folium of Descartes (*Maths.*). The curve defined by the cartesian equation, $x^3+y^3=3axy$.
follicle (*Bot.*). A fruit formed from a single carpel and containing several seeds; it resembles a pod, but splits open along the ventral suture only. (*Zool.*) Any small saclike structure, such as the pit surrounding a hair-root in Mammals. *adjs.* **follicular, folliculose.**
follicle cells (*Zool.*). In *Cephalopoda*, peritoneal cells which surround the ova and assist in nourishing them.
follicle-stimulating hormone (*Physiol.*). A gonadotrophic mucoprotein hormone secreted by the basophil cells of the pars anterior of the pituitary gland, which stimulates growth of the Graafian follicles of the ovary and spermatogenesis; formerly called **prolan A.** Abbrev. **FSH.**
follicular mange (*Vet.*). See **demodectic mange.**
folliculitis (*Med.*). Inflammation of a follicle, especially of the ovary.
folliculoma (*Med.*). A tumour arising from cells in the Graafian follicle of the ovary.
follower (*Civ. Eng.*). An intermediate length of timber which transmits the blow from the monkey to the pile; used when driving below water-level. (*Horol.*) The driven wheel of a pair of wheels engaging with each other. In clocks and watches the wheels are the drivers and the pinions the followers. In synchronous electric clocks the wheels are the followers. (*Surv.*) A chainman who has charge of the rear end of a chain and is responsible for lining-in the leader at each chain's length.
following dirt (*Mining*). Following stone. A thin bed of loose shale above coal; a parting between the top of a coal-seam and the roof. See **pug.**
following response (*An. Behav.*). See **imprinting.**
follow-rest (*Eng.*). A supporting member attached to the rear of the saddle of a lathe to steady the workpiece or bar material against the cutting pressure.
fomes (*Med.*). Any infected object other than food. *pl.* **fomites.** Fomites such as clothing, bedding, etc., may convey infection from one person to another.
font (*Typog.*). See **fount.**
Fontainebleau Sands (*Geol.*). Marine sands of Oligocene age occurring in France; well known as furnishing the so-called sand-calcites by the local cementation of the sand by calcium carbonate, deposited in such a way as to build up perfect crystals of calcite of rhombohedral form.
fontanelle (*Zool.*). In *Craniata*, a gap or space in the roof of the cranium; in *Isoptera*, a shallow, pale-coloured depression on the surface of the head, surrounding the frontal pore.
food body (*Bot.*). A soft mass of cells, containing oil and other nutrient substances, attached to the outside of the seed coat; it is eaten by ants, which drag the seed along, leave it when they have eaten the food body, and so assist dispersal.
food chain (*Ecol.*). The transfer of food energy from the source in plants through a series of organisms, beginning with a herbivore, which successively depend on each other for food. At each transfer a large proportion of the potential energy is lost as heat. The number of links in the chain is usually only four or five. See **parasite chain, predator chain, saprophyte chain.**
food cycle (*Ecol.*). The total food chains in a given community.
food groove (*Zool.*). A groove along which food is passed to the mouth; median and ventral in *Branchiopoda*; along the edge of each ctenidium in *Lamellibranchiata.*
food pollen (*Bot.*). Pollen formed by some flowers, which attracts insects; it may be incapable of bringing about fertilization and may be formed in special anthers. Insects seeking food pollen help in conveying good pollen to other flowers.

food vacuole (*Zool.*). In the cytoplasm of some *Protozoa*, a space surrounding a food-particle and filled with fluid.

food webs (*Ecol.*). The interlocking patterns formed by a series of interconnected food chains.

foolscap (*Paper*). A superseded size of writing and printing paper $13\frac{1}{2} \times 17$ in. (U.S., 13×16 in.).

foolscap octavo (*Typog.*). A book-size, $6\frac{1}{2} \times 4\frac{1}{8}$ in.

fool's gold (*Min.*). See **pyrite.**

foot (*Acous.*). The pitch of the longest open pipe in a rank which can be operated by the lowest key on a manual is measured by its *footage*; thus an 8-ft stop is in unison with the keys of the manual, while a 4-ft stop sounds an octave higher, etc. For the pedals the unison stops are 16 ft, because that is the length of the open pipe with the lowest unison note. (*Bot.*) A specialized part of the young sporophyte in the *Bryophyta* and *Pteridophyta*, attached to the gametophyte. (*Electronics*) The part of a valve envelope in which the electrodes are mounted. (*Typog.*) The margin at the bottom of a book page. (*Zool.*) A locomotor appendage; in *Crustacea*, any appendage used for swimming or walking; in *Arachnida*, *Myriapoda*, and Insects, the tarsus; in *Echinodermata*, the podia (see **podium**); in *Mollusca*, a median ventral muscular mass, used for fixation or locomotion; in land Vertebrates, the podium of the hind limb, or of all limbs in *Tetrapoda*.

footage (*Acous.*). See **foot.**

footage number (*Cinema.*). Before development, all negative film is exposed in a machine which exposes a sequence number (*footage number*) on the edge of the film, so that every foot of film is identifiable, both as a negative and after printing.

foot-and-mouth disease (*Vet.*). An acute febrile contagious disease of cloven-footed animals, due to infection by a virus; characterized by a vesicular eruption on the mucous membrane and skin, especially in the mouth and in the clefts of the feet. Also **aphthous fever.**

foot block (*Carp.*). An *architrave block* (q.v.).

foot-board (*Rail.*). See **foot-plate.**

foot bolt (*Join.*). A robust form of tower bolt, fixed near the foot of a door in a vertical position.

footbridge (*Civ. Eng.*). A bridge for the use of pedestrians only.

foot-candle (*Light*). Former unit of illumination equivalent to 1 lumen/ft^2 ($=10{\cdot}764$ lux).

foot cell (*Bot.*). (1) A small, thick-walled segment of the hypha from which a conidiophore of a mould (*Aspergillus*) arises. (2) The basal cell of the stalk bearing a uredospore.

footing (*Build., Civ. Eng.*). The lower part of a column or wall, standing immediately upon the foundation; usually enlarged locally in order to distribute the load over a greater area. (*Elec. Eng.*) The foundation which is set in the ground to support a tower of an overhead transmission line. (*Hyd. Eng.*) The lowest and, usually, flattest part of the slope of a sea embankment.

footing resistance (*Elec. Eng.*). The ohmic resistance between a transmission-line tower and the earth.

foot irons (*Build.*). Shaped iron bars which can be built into the joints of a manhole wall, leaving projecting steps for use by workmen descending the manhole. Also called **step irons.**

foot-lambert (*Light*). Unit of luminance of 1 lumen/ft^2. For reflecting surfaces the luminance in foot-lamberts is given by the product of the illumination in foot-candles and the reflectivity of the surface. Obsolete.

foot mange (*Vet.*). See **chorioptic mange.**

foot-plate (*Build.*). A *hammer-beam* (q.v.). (*Rail.*) The platform on which the driver and fireman of a locomotive stand.

foot-pound (force) (*Phys.*). Unit of work in the ft-lb-sec system of units. The work done in raising a mass of one pound through a vertical distance of one foot against gravity $1{\cdot}3558$ J.

foot-rail (*Rail.*). A *flanged rail* (q.v.).

foot rill (*Mining*). Direct road into underground workings, either by adit or downslope; dip road.

foot rot (*Vet.*). A suppurative infection between the horn and the sensitive corium of the hoof of the sheep, caused by *Fusiformis nodosus* and possibly other bacteria; causes lameness.

foot run (*Build., Civ. Eng.*). A term meaning foot of length, as in speaking of a loading or of a price *per foot run.*

foots (*Paint.*). The sludge collecting in oil or varnish containers on standing.

foot screws (*Surv.*). See **plate screws.**

foot-stall (*Build.*). The base of a pillar.

footstep bearing (*Eng.*). A thrust bearing used to support the lower end of a vertical shaft.

foot-stick (*Typog.*). The name given to a *side-stick* when used at the foot of a page.

footstone (*Build.*). The lowest coping-stone over a gable.

foot switch (*Elec. Eng.*). A switch arranged for operation by the foot.

foot-ton (force) (*Phys.*). 2240 *foot-pounds (force)* (q.v.).

foot valve (*Eng.*). (1) The nonreturn or suction valve fitted at the bottom of a pump barrel, or in the valve chest of a pump. (2) A nonreturn valve at the inlet end of a suction pipe.

footwall (*Geol., Mining*). The lower wall of country rock in contact with a vein or lode. The upper wall is the *hanging wall.*

footway (*Mining*). A colliery shaft in which ladders are used for descending and ascending.

forage (*Agric.*). Hay or crops ready for consumption by livestock.

forage crimper (*Agric.*). A machine in which forage crops are subjected to multiple fracturing by means of rollers with corrugated surfaces designed to exert an intermittent crushing action.

forage harvester (*Agric.*). A machine to cut, or pick up, chop, or lacerate fodder and deliver it to a vehicle.

forage mites (*Vet.*). Acari of the family *Acaridae*, which commonly infest the skin of animals and birds.

foramen (*Bot.*). See **micropyle.** (*Zool.*) An opening or perforation, especially in a chitinous, cartilaginous, or bony skeletal structure.

foramen lacerum (*Zool.*). An opening of the Vertebrate skull in the side of the brain-case, which is situated between the alisphenoid and the orbitosphenoid, and through which pass the third, fourth, fifth ophthalmic, and sixth cranial nerves.

foramen magnum (*Zool.*). The main opening at the back of the Vertebrate skull, by which the spinal cord issues from the brain-case.

foramen Panizzae (*Zool.*). In Reptiles, an aperture at the point where the right and left aortic arches cross, placing their cavities in communication.

foramen triosseum (*Zool.*). A large aperture between the three bones of the pectoral girdle in Birds.

Foraminifera (*Zool.*). An order of *Sarcodina*, the members of which have numerous fine anastomosing pseudopodia and a shell which is usually calcareous; the ectoplasm is sometimes vacuolated.

forb (*Bot.*). Any herb other than a grass.

Forbes's zones of depth (*Ocean.*). A series of depth zones having distinct faunae, especially in European seas, i.e., the *littoral zone*, the *laminarian zone*, the *coralline zone*, and the *zone of deep-sea corals*.

forbidden (*Comp.*). Said of a combination of symbols not in an operating code, and which thereby reveals a fault.

forbidden band (*Phys.*). The gap between two bands of allowed energy levels in a crystalline solid; see **energy band, energy gap.**

forbidden lines (*Astron., Phys.*). Spectral lines which cannot be reproduced under laboratory conditions. Such lines correspond to transitions from a metastable state, and occur in extremely rarefied gases, e.g., in the solar corona, and in gaseous nebulae.

forbidden transition (*Phys.*). Transition of electrons between energy states, which, according to Pauli selection rules, have a very low probability in relation to those which have a high probability, and are *allowed*.

force (*Mech.*). That which, when acting on a body which is free to move, produces an acceleration in the motion of the body, measured by rate of change of momentum of body. The unit of force is that which produces unit acceleration in unit mass. See **newton;** also **dyne, poundal.** Extended to denote loosely any operating agency. Electromotive force, magnetomotive force, magnetizing force, etc., are strictly misnomers.

forced-circulation boilers (*Eng.*). Steam boilers in which water and steam are continuously circulated over the heating surface by pumps (as opposed to natural circulation systems) in order to increase the steaming capacity. See **Velox boiler, Löffler boiler.**

forced commutation (*Elec. Eng.*). The usual process of commutation, in which the change of direction of the current in the coil actually undergoing commutation is assisted by flux from a commutating pole.

forced draught (*Elec. Eng.*). Said of electrical apparatus cooled by ventilating air supplied under pressure from some external source. (*Eng.*) An air supply to a furnace driven or induced by fans or steam jets (as opposed to the *natural* draught created by a chimney) in order to obtain a high rate of combustion. See **balanced draught, closed stokehold, induced draught.**

forced-draught furnace (*Eng.*). A furnace, but more particularly a boiler furnace, arranged to work under *forced draught* (q.v.).

force de cheval (*Eng.*). See **cheval.**

forced-flow boilers (*Eng.*). See **forced-circulation boilers.**

force diagram (*Eng.*). A diagram in which the internal forces in a framed structure, assumed pin-jointed, are shown to scale by lines drawn parallel to the members themselves. Also called a **reciprocal** (or **stress**) **diagram.**

forced lubrication (*Eng.*). The lubrication of an engine or machine by oil under pressure. See **force feed, full force feed.**

forced movements (*Physiol.*). See **tropism.**

forced oscillations (*Elec.*). Oscillatory currents whose frequency is determined by factors other than constants of the circuit in which they are flowing, e.g., those flowing in a resonant circuit coupled to a fixed-frequency oscillator. Mechanical forced oscillations occur in a similar way (e.g., in servomechanisms) when controlled by the frequency of a vibrator.

force drying (*Paint.*). A form of *stoving* (q.v.) at low temperatures, generally below about 82°C.

forced vibrations (*Phys.*). Vibrations which result from the application of a periodic force to a body capable of vibrating. The amplitude of forced vibrations becomes very great when resonance occurs, that is, when the frequency of the applied force equals the natural frequency of the vibrator, particularly if the damping is small.

forced vortex (*Hyd.*). The type of vortex formed by storing a liquid in a vessel or by rotating that vessel.

force feed (*Eng.*). Lubrication of an engine by forcing oil to main bearings and through the hollow crankshaft to the big-end bearings.

force on charge (*Phys.*). If a charge Q is moving at velocity V in a magnetic field of intensity B, the normal force is $F=B \,.\, Q \,.\, V \cos\theta$, where θ is the angle between the magnetic field and the direction of motion. If the charge is confined to a conductor, the force on a short length l of the latter is $F=B \,.\, I \,.\, l \cos\theta$, where I is the current.

forceps (*Anat.*). That part of either of the two ends of the *corpus callosum* of the brain which diverges into the adjacent brain tissue on each side. (*Med.*) A pincerlike instrument with two blades, for holding, seizing, or extracting objects. *Obstetrical forceps* have large blades, which, applied to the foetal head, aid delivery. (*Zool.*) In *Dermaptera*, the pincer-shaped cerci; in *Arachnida* and *Crustacea*, the opposable distal joints of the chelae; in *Echinodermata*, the distal opposable jaws of pedicellariae. *adj.* **forcipulate.**

force pump (*Eng.*). A pump which delivers liquid under a pressure greater than its suction pressure. It consists of a barrel fitted with a solid plunger, and a valve chest with suction and delivery valve. (*Plumb.*) An air pump used to clean out gas and other service pipes by blowing air through them.

forcing (*Agric.*). The process of hastening growth artificially, e.g., by frames, glass-houses, soil-heating.

forcipate, forcipulate (*Zool.*). Said of pedicellariae the jaws of which are longer than they are broad.

forcipiform (*Zool.*). Said of pedicellariae in which the jaws cross at their lower ends.

Forcipulata (*Zool.*). An order of *Asteroidea*, in which the dorsal surface is beset with small spines surrounded by numerous forcipulate pedicellariae; the tube-feet terminate in suckers.

fore-and-aft (*Bind., Print.*). The first and last sections of the book are printed and folded in one sheet, the first page being head to head with the last page; the second section is similarly combined with the second last section, and so on. Used for the production of large paper-back editions; there are fewer sheets to handle and the economy of two-on working is maintained throughout all printing and binding operations. Also **come-and-go.**

forebay (*Hyd. Eng.*). A reservoir at the head of a pipeline.

fore-brain (*Zool.*). In Vertebrates, that part of the brain which is derived from the first or anterior brain-vesicle of the embryo, comprising the olfactory lobes, the cerebral hemispheres, and the thalamencephalon; the first or anterior brain-vesicle itself.

forecast (*Meteor.*). A statement of the anticipated weather conditions in a given region, for periods of from 1 hour to 30 days in advance, the longer term being less reliable; made from a study of current synoptic charts.

fore-edge (*Bind., Typog.*). The outside margin of

a book page; the edge opposite to the back; the outer edge of a volume. Cf. *head, tail.*

fore-edge painting (*Bind.*). Pictures painted on the fanned-out edges of a book, unseen when the book is closed, as an additional decoration to an already elaborate binding.

fore-gut (*Zool.*). That part of the alimentary canal of an animal which is derived from the anterior ectodermal invagination or stomodaeum of the embryo.

forehearth (*Met.*). Bay in front of a furnace into which molten products can be run.

fore-kidney (*Zool.*). See **pronephros.**

fore peak (*Ships*). The spaces forward of the *collision bulkhead* (q.v.). Lower part frequently used as freshwater tank and upper part may be used as storeroom.

fore plane (*Join.*). A bench plane intermediate in size between the jack plane and the jointing plane.

forepoling (*Mining*). A method of progressing through loosely consolidated ground by driving poles forward over frames.

fore runner tip (*Bot.*). A leaf tip which becomes active while the rest of the leaf is in the process of developing.

foreset beds (*Geol.*). See **topset beds.**

fore-shore. That area of shore which is uncovered between high water and low water.

fore sight (*Surv.*). The levelling staff reading as taken forward to a station which has not been passed by the instrument. It transfers collimation line from back-sight station to fore-sight station. See also **back observation, intermediate sight.**

forest climax (*Bot.*). A climax community composed of trees.

forest edge (*Ecol.*). An *ecotone* (q.v.) between forest and grass or shrub communities, often maintained or increased by man's activities.

Forest Marble (*Geol.*). A shelly oölitic limestone in the Great Oölite Series of the Cotswold Hills, between the Bradford Clay and Cornbrash.

forfex (*Zool.*). A pair of pincers, as of an earwig.

forficiform (*Zool.*). Said of pedicellariae in which the jaws do not cross.

Forficula (*Zool.*). The common genus of earwigs.

forficulate (*Bot.*). Shaped like scissors.

forge (*Met.*). Open hearth or furnace with forced draught; place where metal is heated and shaped by hammering.

forge pigs (*Met.*). Pig-iron suitable for the manufacture of wrought-iron.

forge scale (*Eng.*). The iron oxide coating which forms on iron and steel during forging.

forge tests (*Eng.*). Rough workshop tests made to check the malleability and ductility of iron and steel.

forging (*Eng.*). The operation of shaping (hot) malleable metals by means of hammers or presses. It includes hand-hammer, steam-hammer, press and drop forging. (*Vet.*) See **clacking.**

forging machines (*Eng.*). Power hammers and presses used for forging and drop forging. See **drop forging.**

fork (*Horol.*). In lever escapements, the end of the lever which receives the impulse pin. (*Mining*) (1) A tool with a long wooden handle and prongs for loading lump coal. (2) A double-pronged clip on a tub or wagon for the haulage rope or chain.

forked circuit (*Telecomm.*). One which divides for either simultaneous or alternative reception from one transmitting system. Also called **channel circuit.**

forked lightning. A popular name given to a lightning stroke; the name derives from the branching of the stroke channel which is commonly observed.

forked tenon (*Join.*). A joint formed by a slot mortise astride a tenon cut across the length of a member.

fork frequency (*Telecomm.*). That generated by a self-sustained tuning-fork for signal carrier in facsimile transmission and reception.

fork-lift truck. A vehicle with power operated prongs which can be raised or lowered at will, for loading, transporting and unloading goods; chiefly used in factories and warehouses. Loads are usually stacked upon stands or pallets with sufficient ground clearance for the prongs to be inserted beneath them.

forkstaff plane (*Join.*). A plane adapted for shaping convex cylindrical work.

fork-tone modulation (*Telecomm.*). A modulation of the picture carrier frequency by the synchronizing (fork) tone. It is used to ensure that the true fork frequency is received for comparison with the local fork.

form (*Civ. Eng.*). See **mould.** (*Crystal.*) A complete assemblage of crystal faces similar in all respects as determined by the symmetry of a particular class of crystal structure. Thus the *cube*, consisting of 6 similar square faces, and the *octahedron*, consisting of 8 faces, each an equilateral triangle, are crystal *forms.* The number of faces in a form ranges from 1 (the pedion) to 48 (the hexakis-octahedron). A natural crystal may consist of one form or many.

formaldehyde (*Chem.*). Methanal. H·CHO, b.p. −21°C, a gas of pungent odour, readily soluble in water, and usually used in aqueous solution. Formaldehyde easily polymerizes to *paraformaldehyde* or metaformaldehyde. It is produced by oxidation of methanol, or by the oxidation of ethene in the presence of a catalyst. It forms with ammonia *hexamethylene-tetramine.* It is a disinfectant and is of great importance in plastics manufacture.

formaldehyde (methanal)-calcium fixative (*Micros.*). See **Baker's fixative.**

formaldehyde (methanal) resins (*Chem.*). Synthetic resins which are condensation products of methanal with hydroxybenzenes, urea, etc. See also **Bakelite, Delrin.**

formalin (*Chem.*). The TN for a commercial 40% aqueous methanal solution.

formal operations (*Psychol.*). The mature thought systems which may be stable by about 15 years of age. The individual can be guided by the form of an argument regardless of content and perform operations on *operations.* Operations have two opposites, inverse and reciprocal. See **concrete operations, operations.**

formanite (*Min.*). Yttrium tantalate, a member of the fergusonite group of minerals, crystallizing in the tetragonal system.

formant (*Acous.*). Speech spectral pattern arising in the vocal cords, which determines the distribution of energy in unvoiced sounds and the reinforcement of harmonics in voiced sounds. It is the formant which leads to recognition of speech sounds by aural perception.

format (*Comp.*). The layout of written data processed by a computer. (*Print.*). The general size, quality of paper, type face and binding.

formates (methanoates) (*Chem.*). The salts of *formic acid.*

formation (*Bot.*). See **association.** (*Geol.*) A noncommittal term for one of the larger stratigraphical divisions, more accurately designated *stage*, *series*, or *system*. (*Paper*) The pattern of the fibres in the paper.

formative region (*Bot.*). The growing point of a stem or root.

formative stage of growth (*Bot.*). The stage in development when a cell is formed from a pre-existing cell.

formative time (*Electronics*). The time interval between the first Townsend discharge in a given gap and the formation there of a self-maintaining glow discharge.

formative-trophic (*Zool.*). In development, pertaining to stimuli which assist indirectly in the production of structures which are qualitatively different, or to the effects of such stimuli. Cf. *augmentative-trophic reaction*.

form drag (*Aero.*). The difference obtained when the *induced drag*, i.e., the fraction of the total drag induced by lift, is subtracted from the *pressure drag* (q.v.). See **drag.**

forme (*Typog.*). Type matter assembled and locked up in a chase ready for printing.

formed plate (*Elec. Eng.*). A type of plate used in lead-acid accumulators; made by electrolytically converting the substance of which the plate is made into active material.

former (*Aero.*). A structural member of a fuselage, nacelle, hull or float to which the skin is attached and having the primary purpose of preserving form or shape. It generally carries structural loads. Cf. **frame.** (*Elec. Eng.*) A tool for giving a coil or winding the correct shape; it sometimes consists of a frame upon which the wire can be wound, the frame afterwards being removed.

forme rollers (*Print.*). Rollers which supply ink to a forme or plate, as distinct from *distributor rollers*.

former-wound coil (*Elec. Eng.*). An armature coil built to the correct shape by means of a *former*, it being then dropped into the slots on the armature.

form factor (*Elec.*). The ratio of the effective value of an alternating quantity to its average value over a half-period.

form genus (*Bot., Gen.*). A group of species which have similar morphological characters, but which are not certainly known to be related in descent. The species are *form species*.

form grinding (*Eng.*). See **profile grinding.**

Formica (*Plastics*). TN used in respect of a series of laminates with a highly resistant, frequently decorative, surface based on melamine resins. Also used for other plastics materials, particularly in extruded forms.

formic (methanoic) acid (*Chem.*). HCOOH, a colourless liquid, of pungent odour, corrosive, m.p. 9°C, b.p. 101°C, formerly prepared by absorption of carbon monoxide in soda-lime at 210°C. Prepared nowadays by passing carbon monoxide and steam at 200°–300°C under pressure over a catalyst.

forming cutter (*Eng.*). See **form tool.**

forming rollers (*Print.*). See **bending rollers.**

formo-acetic-Müller fixative (*Micros.*). A general fixative consisting of a mixture of Müller's fluid, glacial acetic acid and formalin.

formol-dichromate fixative (*Micros.*). A formalin mixture containing potassium dichromate and glacial ethanoic acid, particularly suitable for yolk-laden eggs and embryos. Also called **Smith's fixative.**

formol-silver stain (*Micros.*). A combined fixing and staining method, used particularly for flagellate pellicular systems, trichocysts, cilia and ciliate fibrillar systems. Cultures are fixed in formalin and stained with silver nitrate. Also called **Horváth's stain.**

formol titration (*Chem.*). A method of estimating volumetrically the amount of amino acids present in a solution. It is based upon the fact that amino acids and their derivatives possess both a carboxyl and an amino group which neutralize each other, and that by the addition of methanal the amino group is converted into a methylene derivative without basic properties, by which reaction it becomes possible to titrate subsequently the carboxyl in the usual manner.

formoxy (*Chem.*). The organic radical O·CHO—.

form quality (*Psychol.*). A term from *Gestalt psychology* referring to the quality of a whole but not to its constituent parts.

form species (*Bot.*). See under **form genus.**

form tool or forming cutter (*Eng.*). Any cutting tool which produces a desired contour on the work-piece by being merely fed into the work, the cutting edge having a profile similar to, but not necessarily identical with, the shape produced. See **chaser.**

formula. A fixed rule or set form. (*Chem.*) The representation of the nature and number of the atoms present in a molecule of a compound by means of letters and figures, e.g., H_2SO_4, C_2H_5OH. (*Maths.*) A rule expressed in algebraic symbols. (*Med.*) A prescription.

formwork (*Civ. Eng.*). *Shuttering* (q.v.).

formyl (*Chem.*). The organic radical OCH—.

fornacite, furnacite (*Min.*). A basic chromarsenate of copper and lead, crystallizing in the monoclinic system.

fornicate (*Bot.*). Arched and hoodlike.

fornix (*Zool.*). In the brains of higher Vertebrates, a tract of fibres connecting the posterior part of the cerebrum with the hypothalamus.

forsterite (*Min.*). An end-member of the olivine group of minerals, crystallizing in the orthorhombic system. Chemically, forsterite is a silicate of magnesium, Mg_2SiO_4. It typically occurs in metamorphosed impure dolomitic limestones.

forsterite-marble or ophicalcite (*Geol.*). A characteristic product of the contact metamorphism of magnesian (dolomitic) limestones containing silica of organic or inorganic origin.

Forstner bit (*Carp.*). A patent centre-bit with a small centre-point and a circular rim, for sinking blind holes in timber.

fortified resin (*Paper*). A rosin size for paper, produced by reacting resin with maleic anhydride at a high temperature with a catalyst and then saponifying.

Fortin's barometer (*Meteor.*). A pattern of mercury barometer suited for accurate readings of the pressure of the atmosphere. The zero of the scale is indicated by a pointer inside the mercury cistern, the bottom of which is flexible and may be moved by an adjusting screw until the mercury surface just touches the pointer.

Fortran (*Comp.*). A problem-orientated computer language widely used in scientific work. The machine instructions for a specific computer are prepared from a Fortran program in an automatic compiler forming part of the computer system.

forward eccentric (*Eng.*). On a steam-engine having link motion reverse gear, the eccentric which drives the valve when the engine is going ahead. See **link motion.**

forwarding (*Bind.*). The operations entailed in bookbinding, until a book has been placed in its covers. See **finishing.**

forward lead (*Elec. Eng.*). See forward shift.

forward path (*Telecomm.*). The transmission path from the loop-activating signal to the loop-output signal in a feedback control loop.

forward perpendicular (*Ships*). The forward side of a ship's stem post when this is truly perpendicular to the longitudinal base line; but in cases when the stem post is 'raked', i.e., angled to the baseline, it is the perpendicular intersecting the forward side of the stem post at the summer load water line. Abbrev. **F.P.**

forward scatter (*Radio*). Multiple transmission on centimetric waves, using reflection down and forwards from ionization in troposphere; range about 150 km. (*Radiol.*) Scattering of particles through an angle of less than 90° to the original direction of the beam. Cf. *back scatter*.

forward shift (*Elec. Eng.*). A movement of the brushes of a commutator machine around the commutator, from the neutral position, and in the same direction as that of rotation. Also called **forward lead.**

forward sweep (*Aero.*). See **sweep.**

forward (or beam) transadmittance (*Electronics*). An electron beam tube, which is the complex quotient of the fundamental component of the short-circuit current induced in the second of any two gaps and the fundamental component of the voltage across the first.

forward transfer function (*Telecomm.*). That of the forward path of a feedback control loop.

forward voltage (*Elec. Eng.*). The voltage of the polarity which produces the larger current in an electrical system.

forward wave (*Electronics*). One whose group velocity is in the same direction as the motion of an electron stream in a travelling-wave tube. See **backward wave.**

Fosalsil (*Build.*). A heat-insulating, fire-resisting material made from *moler*, and used as hollow blocks for partitions, and for chimney linings.

fossa (*Zool.*). A ditchlike or pitlike depression, as the *glenoid fossa*.

fossa rhomboidalis (*Zool.*). See **fourth ventricle.**

fossette (*Zool.*). In general, a small pit or depression: in some *Arthropoda*, the socket which receives the base of the antennule.

fossil (*Geol.*). A relic of some former living thing —plant or animal—embedded in, or dug out of, the superficial deposits of past geological periods. Fossils usually occur as the hard parts of organisms, such as bones or shells, and as moulds, casts, and impressions preserved in rocks. Rarely, creatures having no hard parts, such as jelly-fishes, sea-cucumbers, etc., have been recognized in rocks as ancient as the Cambrian; by contrast, complete extinct mammals, including mammoths, occur frozen in the gravels of Pleistocene age in Siberia. See also **palaeontology** and **palaeobotany.**

fossil meal (*Build.*). A diatomaceous earth used in the manufacture of Fosalsil.

fossorial (*Zool.*). Adapted for digging.

Foster-Seeley discriminator (*Elec. Eng.*). One for demodulating FM transmission using a balanced pair of diodes. To these are applied voltages which are sum and difference of limiter signal voltage and half transformer coupled voltage, diode outputs being differenced.

Foster's reactance theorem (*Elec.*). Expression for generalized impedance of a number of tuned circuits, in series or parallel, which indicates that such impedances exhibit in order resonant and anti-resonant frequencies.

Fotomatic (*Typog.*). A filmsetting machine, developed from the *Fotosetter*, using double-letter photomats and operated by tape or keyboard. See **filmsetting, teletypesetting.**

Fotosetter (*Typog.*). A keyboard-operated filmsetting machine based on the Intertype and using single-letter photomats. See **Fotomatic, filmsetting.**

Fototronic (*Typog.*). A filmsetting system making extensive use of electronics.

Föttinger coupling (or **transmitter**) (*Eng.*). A hydraulic coupling, gear, or clutch for transmitting power from, e.g., an engine to a ship's propeller; it consists essentially of an outward-flow water turbine driving an inward-flow turbine, within a common casing.

Föttinger speed transformer (*Eng.*). A hydraulic reduction gear formerly used in marine propulsion, comprising a centrifugal pump and turbine runner in a single unit, giving a speed reduction of 5 : 1.

Foucault currents (*Elec. Eng.*). See **eddy currents.**

Foucault knife-edge test (*Optics*). A method of testing the aberrations of a lens or mirror by placing a knife-edge in the focus of a point source and observing the pattern of light and shade as seen in the lens when an observer places his eye immediately behind the knife-edge.

Foucault's measurement of the velocity of light (*Light*). One of the first successful attempts to obtain an accurate result for this important constant. Foucault, in 1862, made use of a rapidly rotating mirror sending light to a distant fixed concave mirror which reflected it back. Measurement of the displacement of the reflected image gave a value of $2{\cdot}986 \times 10^8$ m/s for the velocity of light *in vacuo*.

Foucault's pendulum (*Astron.*). An instrument devised by Foucault in 1851 to demonstrate the rotation of the earth; it consists of a heavy metal ball suspended by a very long fine wire. The plane of oscillation slowly changes through 15° sin (latitude) per sidereal hour.

foul air flue (*Build.*). A ventilating flue through which vitiated air from a room is drawn.

foul anchor (*Ships*). An anchor which has become entangled in its own cable or with some obstruction on the bottom.

foulard (*Textiles*). A light-weight dress fabric with a printed pattern, made either of silk or man-made filament yarns or of super-quality Sea Island or Egyptian cotton.

foul berth (*Ships*). A situation where anchored ships may collide when swinging round with the change of tidal stream.

foul clay (*Build.*). A brick earth composed of silica and alumina combined with only a small percentage of lime, magnesia, soda, or other salts. Such a clay lacks sufficient fluxing material to fuse its constituents at furnace temperature, and is improved by the addition of sand or loam, lime or ashes. Also called **plastic clay, pure clay, strong clay.**

fouling (*Eng.*). (1) Coming into accidental contact with. (2) Deposition or incrustation of foreign matter on a surface, as of carbon in an engine cylinder, or marine growth on the bottom of a ship or on structures subject to the action of sea-water.

fouling point (*Rail.*). The location before the meeting of two tracks where the loading-gauge outlines come into contact.

foul in the foot (*Vet.*). Infectious pododermatitis; claw ill. Suppuration and necrosis of the interdigital tissues of the foot in cattle due to infection by the bacterium *Sphaerophorus necrophorus* (*Fusiformis necrophorus*); causes lameness.

fouls (*Mining*). The cutting-out of portions of the coal-seam by 'wash outs' or barren ground.

foul solution (*Min. Proc.*). In cyanide process, one contaminated by salts or metals other than gold and silver, and which is not fit to be recirculated in the process.

foundation cylinder (*Eng.*). A large steel or iron cylinder sunk into the ground to provide a solid foundation for bridge piers, etc., in soft ground.

foundation piles (*Civ. Eng.*). Piles driven into the ground to provide an unyielding support for a structure.

foundation ring (*Eng.*). In a locomotive boiler, a rectangular iron ring of rectangular section, to which the lower edges of the inner and outer plates of the fire box are secured.

founded (*Civ. Eng.*). Said of a caisson which has been sunk to a firm level.

foundering (*Geol.*). Subsidence due to two causes: downwarping, resulting in the development of a monoclinal fold; and displacement along faults which hade inwards towards one another, as in rift valleys. *Cauldron subsidence* is special.

founder's type (*Print.*). Founts of type produced by the specialist typefoundries, as distinct from that produced by the printer himself, in hard type metal; used mainly for display lines.

founding (*Glass*). The melting operation in which molten glass is made almost free from undissolved gases. Also **fining, plaining, refining.**

foundry (*Eng.*). A workshop in which metal objects are made by casting in moulds. (*Print.*) That department of a printing establishment where work in connexion with duplicate plates is carried out.

foundry chase (*Print.*). A chase with type-high rims and sometimes with built-in locking-up device, used when taking a mould of type matter.

foundry ladle (*Eng.*). A steel ladle lined with fireclay; used for transporting molten metal from a foundry cupola to the moulds. Small ladles are carried by hand, large ones by a truck or crane. See **hand shank.**

foundry pig-iron (*Eng.*). Bars of cast-iron up to 1 m in length and 1·5 cm in diameter, as bought by an iron foundry.

foundry pit (*Eng.*). A large hole in the floor of a foundry, which serves the purpose of a moulding box for very large or deep castings.

foundry sand (*Met.*). Silica-based sand with either enough natural clay or special additions (oil, molasses, fullers' earth, etc.) to give it some cohesive strength when used in moulding.

foundry stove (*Eng.*). A large oven for drying moulds and cores, heated either externally by hot gases or internally by a fire-basket.

fount or font (*Typog.*). A complete set of type of the same face and size, containing proportionate quantities of the individual characters.

four-bar chain (*Eng.*). A common and simple versatile mechanism comprising four rigid members coupled by lower pairs (turning or sliding) which constrain the rigid members to plane motion.

Fourcault process (*Glass*). Method of making sheet glass by drawing upwards through a *debiteuse* (q.v.).

four-centred arch (*Arch.*). A pointed arch struck from 4 centres.

fourchette (*Anat.*). The poster or junction of the labia minora.

four-colour press (*Print.*). Sheet-fed and web-fed models are available for the main printing processes. Four colours are the minimum for full-colour reproduction but extra units are often added for additional printing or varnishing.

four-colour reproduction (*Print.*). Printing by any of the main printing processes with standard yellow, magenta, cyan, and black inks, not always in that order, from sets of colour plates or cylinders, to produce a reproduction of a coloured subject. See **colour correction, three-colour process.**

fourdrinier (*Paper*). The standard type of paper machine, introduced by the brothers Fourdrinier about 1800.

four-electrode valve (*Electronics*). A valve with 2 grids between cathode and anode, e.g., tetrode.

four-factor formula (*Nuc.*). That giving the multiplication constant of an infinite thermal reactor as the product of fast fission factor, resonance escape probability, thermal utilization factor and neutrons emitted per thermal neutron absorbed for the fuel material.

Fourier analysis (*Elec., Maths.*). The determination of the harmonic components of a complex waveform (i.e., the terms of a Fourier series that represents the waveform) either mathematically or by a wave-analyser device.

Fourier half-range series (*Maths.*). Fourier series with only sine or cosine terms. Valid for x between 0 and π.

Fourier integral (*Elec.*). The expression of a non-repeated and isolated waveform in the form of a summation of adjacent frequency components, from zero to infinity, with a spectral distribution of energy content.

Fourier number (*Phys.*). A dimensionless parameter used for studying heat flow problems. It is the product of *thermal diffusivity* and time divided by the square of the linear dimension.

Fourier principle (*Elec.*). That which shows all repeating waveforms can be resolved into sine wave components consisting of a fundamental and a series of harmonics at multiples of this frequency. It can be extended to prove that nonrepeating waveforms occupy a continuous frequency spectrum.

Fourier series (*Maths.*). The series

$$\tfrac{1}{2}a_0 + \sum_{n=1}^{\infty} (a_n \cos nx + b_n \sin nx)$$

where

$$a_n = \frac{1}{\pi}\int_{-\pi}^{+\pi} f(x) \cos nx \, dx$$

and

$$b_n = \frac{1}{\pi}\int_{-\pi}^{+\pi} f(x) \sin nx \, dx,$$

which, subject only to certain very general restrictions upon $f(x)$, converges to $f(x)$. Fourier series are used in the determination of the component frequencies of vibrating systems. See **Fourier analysis.**

Fourier transform (*Maths., Phys.*). A mathematical relation between the energy in a transient and that in a continuous energy spectrum of adjacent component frequencies. The Fourier transform $F(u)$ of the function $f(x)$ is defined by

$$F(u) = \int_{-\infty}^{+\infty} e^{-iut} f(t) \, dt.$$

Some writers use e^{+iut} instead of e^{-iut}.

four-jaw chuck (*Eng.*). A chuck used in a lathe or in certain other types of machine tools, comprising four jaws disposed at right angles to each other. Usually, the jaws act independently of each other. Used for holding rectangular or irregularly contoured workpieces, or for revolving circular workpieces eccentrically.

four-line mathematics (*Typog.*). A *Monotype system* of machine-setting complex mathematical formulae, the result requiring the minimum adjustment by hand.
four-part vault (*Arch.*). A vault formed at the intersection of 2 barrel vaults.
four-phase system (*Elec. Eng.*). A name sometimes given to a 2-phase system in which the midpoints of the 2 phases are connected to form a neutral point.
four-stroke cycle (*I.C. Engs.*). An engine cycle completed in 4 piston strokes (i.e., in 2 crankshaft revolutions), consisting of suction or induction, compression, expansion or power stroke, and exhaust. See **diesel cycle, Otto cycle.**
four-terminal resistor (*Elec. Eng.*). Laboratory standard fitted with current and potential terminals, arranged so that measurement of the potential drop across the resistor is not affected by contact resistances at the terminals.
fourth pinion (*Horol.*). The pinion on which the fourth wheel is mounted.
fourth rail (*Elec. Eng.*). A conductor-rail on an electric traction system. When there are two running rails and two conductor-rails, the fourth rail generally carries the return current, instead of its being allowed to return along the running rails.
fourth-rail insulator (*Elec. Eng.*). An insulator for supporting a fourth rail in an electric-traction system.
fourth ventricle (*Zool.*). In Vertebrates, the cavity of the hind-brain.
fourth wheel (*Horol.*). The wheel in a watch which drives the escape pinion. If the train is suitable for a seconds hand (i.e., the fourth wheel makes one turn per minute), the hand is carried on an extension of the fourth wheel arbor.
fourth wire (*Elec. Eng.*). A name sometimes given to the neutral wire in a 3-phase, 4-wire distribution system.
four-way tool post (*Eng.*). A supporting and clamping mechanism for cutting tools used in a lathe or certain other machine tools, in which up to 4 tools may be mounted simultaneously but in various positions, the required tool being brought into the operative position by indexing the tool post appropriately.
four-wire circuit (*Telecomm.*). A duplex or quasi-duplex circuit in which the *return* speech currents are separated from the *go* speech currents by segregating them on separate pairs of conductors.
four-wire repeater (*Telecomm.*). A repeater for insertion into a 4-wire telephone circuit, in which the 2 amplifiers, 1 for amplifying in each direction, are kept separate.
four-wire system (*Elec. Eng.*). A system of distribution of electric power requiring 4 wires. In a 3-phase system, the 4 wires are connected to the 3 line terminals of the supply transformer and the neutral point; and in the 2-phase system, the wires are connected to the ends of the 2 transformer windings.
fovea (*Zool.*). A small pit or depression. *adjs.* **foveate, foveolar, foveolate.**
fovea centralis (*Zool.*). A slight depression at the centre of the macula lutea. See **yellow spot.**
foveola (*Zool.*). Same as **fovea.**
fovilla (*Bot.*). The material inside a pollen grain.
fowl cholera (*Vet.*). An acute and usually fatal septicaemia of domestic fowl and other birds, caused by the bacterium *Pasteurella multocida* (*P. aviseptica*). Characterized by dejection, loss of appetite, raised temperature and diarrhoea. Mild and chronic forms of the disease also occur.
fowl coryza (*Vet.*). See **infectious coryza.**
Fowler flap (*Aero.*). A high-lift trailing-edge flap that slides backward as it moves downward, thereby increasing the wing area, also leaving a slot between its leading edge and the wing when fully extended.
Fowler position (*Med.*). The semisitting position in which the patient is placed in bed after an abdominal operation to prevent infective fluids reaching the upper part of the abdominal cavity.
fowl paralysis (*Vet.*). Marek's disease; neurolymphomatosis. A contagious disease of chickens affecting the peripheral nerves and usually characterized by paralysis of the legs or wings. Affected birds may also develop lymphoid tumours. Probably caused by a virus.
fowl pest (*Vet.*). A term usually embracing both *fowl plague* and *Newcastle disease* (qq.v.).
fowl plague (*Vet.*). An acute, contagious virus disease of chickens and other domestic and wild birds; the main symptoms are high temperature, oedema of the head, nasal discharge, and rapid death.
fowl pox (*Vet.*). Avian diphtheria. An acute contagious disease of birds due to infection by a virus; characterized by hyperplastic nodules on the skin or diphtheritic inflammation of the mouth.
fowl typhoid (*Vet.*). A contagious septicaemic disease of chickens and other domesticated birds caused by *Salmonella gallinarum.*
fox encephalitis (*Vet.*). Epizootic fox encephalitis, infectious encephalitis of foxes. An acute infection characterized by severe nervous symptoms, caused by the virus of *infectious canine hepatitis* (q.v.).
foxiness (*For.*). A term applied to a form of decay which affects hardwoods, causing local reddish-brown staining.
fox marks (*Bind.*). Brown stains on the pages of books that have been allowed to get damp.
foxtail saw (*Join.*). A dovetail saw.
foxtail wedging (*Join.*). The tightening-up of a tenon in a blind mortise by inserting small wedges in saw-cuts in the end of the tenon before inserting the latter in the mortise. The operation of driving the tenon into position then forces the wedges into the saw-cuts and spreads the fibres of the tenon, giving a secure hold resisting withdrawal. Also called fox wedging.
foyaite (*Geol.*). A widely distributed variety of nepheline-syenite, described originally from the Foya Hills in Portugal. Typically it contains about equal amounts of nepheline and potassium feldspar, associated with a subordinate amount of coloured mineral such as aegirine.
f.p. (*Heat*). Abbrev. for *freezing-point.*
F.P. (*Ships*). Abbrev. for *forward perpendicular.*
FPS. The system of measuring in *feet, pounds,* and *seconds.*
Fr (*Chem.*). The symbol for *francium.*
fraction (*Maths.*). See **division.**
fractional crystallization (*Chem.*). The separation of substances by the repeated partial crystallization of a solution. (*Geol.*) The formation, at successively lower temperatures, of the component minerals in a magma, coupled with the tendency for the components which crystallize at high temperatures to separate, on account of their high specific gravity, thus giving a concentration in the lower parts of a magma body.
fractional distillation (*Chem.*). Distillation process for the separation of the various com-

ponents of liquid mixtures. An effective separation can only be achieved by the use of *fractionating columns* (q.v.) attached to the still.

fractional distribution (*Chem.*). In an assembly of particles having different values of some common property such as size or energy, the fraction of particles in each range of values is called the fractional distribution in that range. Cf. *cumulative distribution.*

fractional pitch (*Eng.*). Of a screw-thread cut in the lathe, a pitch not an integral multiple or submultiple of the pitch of the lead screw. See **even pitch.**

fractional test meal (*Med.*). See **Rehfuss test.**

fractionating column (*Chem.*). A vertical tube or column attached to a still and usually filled with packing, e.g., Raschig rings, or intersected with bubble plates. An internal reflux takes place, resulting in a gradual separation between the high- and the low-boiling fractions inside the column, whereby the fractions with the lowest boiling-point distil over. The efficiency of the column depends on its length or on the number of bubble plates used, and also on the ability of the packing to promote contact between the vapour and liquid phases.

fractionation (*Chem.*). See **fractional distillation.** (*Radiol.*) A system of treatment commonly used in radiotherapy in which doses are given daily or at longer intervals over a period of 3 to 6 weeks.

fraction of saturation (*Meteor.*). See **relative humidity.**

fracture (*Min.*). The broken surface of a mineral as distinct from its cleavage. The fracture is described, in different cases, as conchoidal (shell-like), platy, or flat, smooth, hackly (like that of cast-iron), or earthy. Thus calcite has a perfect rhombohedral cleavage, but conchoidal fracture. (*Surg.*) Breaking of a bone. Fractures may be *simple* (broken bone only), *compound* (external wound communicating with fracture), *complicated* (additional injury, e.g., to internal organs, blood vessels, etc.), *comminuted* (bone broken in several or many parts), *fissured* (bone cracked, e.g., skull), *impacted* (q.v.), *greenstick* (q.v.).

fragilitas ossium (*Med.*). A condition in which a child is born with abnormally brittle bones, multiple fractures occurring.

fragmental deposits (*Geol.*). These include epiclastic and pyroclastic rocks, i.e., all those which consist of fragments of rocks or minerals covering the whole range of grain size, and resulting from normal disintegration of rocks, or from shattering by volcanic action.

fragmentation (*Bot., Cyt.*). (1) See **amitosis.** (2) The break-up of an algal filament into a number of parts, each capable of growing into a new filament. (3) The separation of a portion from the main body of a chromosome.

fragmentation of the myocardium (*Med.*). Transverse fissuring of the muscle fibres of the heart, occurring after violent death or in a diseased heart.

fraktur (*Typog.*). A *black letter* type peculiar to Germany.

Fram (*Build.*). A patented form of floor incorporating reinforced hollow fireclay blocks, supported by precast reinforced concrete beams.

framboesia, frambesia (*Med.*). *Yaws* (q.v.).

frame (*Aero.*). A transverse structural member of a fuselage, hull, nacelle, or float, which follows the periphery and supports the skin, or the skin-stiffening structure. Cf. **frame**, and see **spar.** (*Build., etc.*) See **framework.** (*Cinema.*) The unit picture in a ciné film, that is locked in position in the picture gate in the camera or projector during photography or projection. (*Typog.*) The compositor's workplace, providing storage for cases of type and provision for mounting them on top. (*TV*) In U.K., obsolete term for the complete scanning of a TV picture. See **field.** In U.S., the total picture, as in sound-film terminology.

frame antenna (*Radio*). One comprising a loop of one or more turns of conductor wound on a frame, its plane being oriented in the direction of the incoming waves, or, in transmission, for the direction of maximum radiation. Also called **coil antenna, loop antenna.**

frame-counter (*Photog.*). Automatic indicator on a camera which shows how many frames of film have been exposed.

framed (*Carp.*). Said of work assembled with mortise and tenon joints.

framed and braced door (*Join.*). A boarded door secured in a frame consisting of two stiles, and top, middle, and bottom rails, with diagonal braces between.

frame development (*Cinema.*). Development of lengths of cinematograph film by winding them round a flat frame, for immersion in the minimum quantity of developer. Also called **rack development.**

framed floor (*Carp.*). A floor in which the bridging joists are supported at intervals by binding joists, which in turn are supported at intervals by girders. Cf. *double floor.*

framed grounds (*Join.*). Grounds used in good work around openings such as door openings, the heads being tenoned into the posts on each side.

frame direction-finding system (*Radio*). Simple type of direction-finder using a loop, preferably screened to obviate *antenna effect*, the polar response of which is a figure of eight; the loop is rotated until the received signal vanishes, when the axis of frame is in line with direction of arrival of wave. Also called **loop direction-finding system.** See **sense.**

framed, ledged, and braced door (*Join.*). A boarded door secured in a frame consisting of two stiles and a top-rail, and braced on one side with middle and bottom rails and diagonal braces.

frame finder (*Photog.*). Viewfinder comprising a wire frame and a peep hole separated by a distance equivalent to the focal length of the lens.

frame frequency (*TV*). The number of complete picture scannings per second. In U.K. 25 Hz, in U.S. 30 Hz, synchronized with mains frequency to minimize the effect of mains interference. Also **picture frequency.**

frame grid (*Electronics*). Said of a rugged high-performance thermionic valve in which the grid is held very rigid and close to the cathode surface by winding it on stiff rods.

frame level (*Tools*). A mason's level.

frame line (*Cinema.*). The thin black line dividing the frames in the positive projection print of a motion picture.

frame noise (*Cinema.*). In sound-film reproduction, noise arising from the film being displaced in the sound-gate, so that the scanning light is interrupted by the frame lines. The resulting modulation contains frequencies which are multiples of the frame frequency, normally 24 per second, and is inseparable from other frequencies recorded on the sound-track.

frames (*Civ. Eng., Join.*). The centering used in concrete construction. Also used of built-up superstructure in any suitable structural material to form the skeleton of a building.

Also used for the surrounding timber or metal members of a door or window.

frame saw (*Tools*). A thin-bladed saw, which is held taut in a special frame. Also called **span saw.**

frame slip (*TV*). Lack of exact synchronization of the vertical scanning and the incoming signal whereby the reproduced picture progresses vertically.

frame-synchronizing impulse (*TV*). An impulse transmitted at the end of each complete frame-scanning operation, to synchronize the framing oscillator at the receiver with that at the transmitter.

frame turner (*Ships*). A tradesman engaged in turning and bevelling ships' frames, when red hot, to the shape of the ship's form.

frame-type switchboard (*Elec. Eng.*). See **skeleton-type switchboard.**

frame weir (*Civ. Eng.*). A type of movable weir consisting of a wooden barrier supported against iron frames placed at intervals across a river, and capable of being lowered on to the bed of the river in flood-time, or of being entirely removed.

framework (*Build., etc.*). The supporting skeleton of a structure.

framing (*Build., etc.*). The operation of assembling into final position the members of a structure. (*Cinema.*) The vertical adjustment of the picture gate in a projector, to get the image on the screen exactly on the desired area without appreciable top or bottom clipping or view of the frame lines. (*TV*) Adjustment of picture-repetition frequency in a TV receiver so as to keep the picture stationary on the screen.

framing chisel (*Carp., Join.*). See **mortise chisel.**

framing oscillator (*Electronics*). That which generates frame-scanning voltage or current. Also **field oscillator.**

framing signal (*TV*). One used in facsimile transmission to adjust the picture to a desired position in the direction of progression.

Francis water turbine (*Eng.*). A reaction turbine in which the water flows radially inwards into guide vanes, and thence into the runner, which it leaves axially.

francium (*Chem.*). The heaviest alkali metal. Symbol Fr, at. no. 87. No stable isotopes exist. ^{223}Fr of half-life 22 min. is most important.

francolite (*Min.*). A variety of apatite containing appreciable CO_2, and more than 1% of fluorine.

Frankfort black (*Paint.*). An alternative term for *drop black*.

franking (*Join.*). The operation of notching a sash-bar to make a mitre joint with a transverse bar.

Franki-pile (*Civ. Eng.*). A proprietary system of reinforced concrete piling.

Franklin antenna (*Radio*). Directive antenna comprising a number of radiating elements uniformly spaced along a line at right angles to the direction of maximum radiation. Each element consists of a vertical wire several half-wavelengths long, the radiation from alternate half-wavelengths being suppressed, to secure maximum radiation along the near horizontal direction. Such a curtain is usually accompanied by a similar, but unexcited, reflector.

franklinite (*Min.*). Iron-zinc spinel, occurring rarely, as at the type-locality, Franklin, New Jersey.

Franzen's index (*Nut.*). Complex index for nutrition studies, involving the relation between the weight of the subject, the muscular development, as exhibited by arm girth, the development of the subcutaneous cellular tissue at the surface of the biceps, the vitality index, and Pryor's index.

Frasch process (*Mining, etc.*). Method of mining elemental sulphur, by drilling into deposit and flushing it out by hot compressed air as a foam of low density. The sulphur is melted by superheated water.

Frasnian (*Geol.*). The lower stage of the Upper Devonian rocks of Europe.

frass (*Zool.*). Faeces; excrement.

frater (*Arch.*). A refectory; sometimes applied in error to a monastic common room or to a chapter house. Also **frater house, fratry.**

Fratercula (*Zool.*). The puffin genus.

Fraunhofer diffraction (*Optics*). The field transmitted through an aperture in an absorbing screen. It is concerned with the field pattern at large distances from the screen, so that waves from the virtual sources at the aperture arrive in phase at a point normal to the screen.

Fraunhofer lines (*Light*). Dark lines in the solar spectrum, produced by selective absorption in the relatively cool gaseous envelope surrounding the incandescent photosphere. See **[A]**, [B], [C], etc.

Fraunhofer region (*Phys.*). See **field.**

frazil ice (*Meteor.*). Ice, in the form of small spikes and plates, formed in rapidly flowing streams, where the formation of large slabs is inhibited.

free (*Bot.*). (1) Not joined laterally to another member of the same kind. (2) Said of gills of agarics which reach the stipe but are not joined to it. (*Elec.*) Said of a transducer when it is not *loaded*, e.g., the input has a *free impedance*. (*Mining*) Descriptive of naturally occurring metallic gold, silver or copper.

free-air anomaly (*Geol.*). A gravity anomaly which had been corrected for the height of the measured station above datum. Cf. *Bouguer anomaly*.

free-air dose (*Radiol.*). A dose of radiation, measured in air, from which secondary radiation (apart from that arising from air, or associated with the source) is excluded.

free association (*Psychol.*). The method used in psychotherapy for making unconscious processes conscious. Associations to ideas are allowed to arise spontaneously in the mind, without conscious direction or selective criticism, when factors previously unknown and unconscious may be revealed; these are often accompanied by an affect of pain or disgust. See **abreaction.**

free atom (*Chem.*). Unattached atom assumed to exist during reactions. See **free radicals.**

free balloon (*Aero.*). Any balloon floating freely in the air, not propelled or guided by any power or mechanism, either within itself or from the ground.

freeboard (*Ships*). An assignment made by law to prevent overloading of a ship; calculated from statutory tables based on the vessel's form. Permanent markings are made on the ship's side to indicate the depth to which a ship may be loaded, and severe penalties are incurred by any overloading.

freeboard deck (*Ships*). Uppermost complete deck having permanent means of closing all openings.

free cell formation (*Bot.*). The formation of daughter cells which do not remain united.

free cementite (*Met.*). Iron carbide in cast-iron or steel other than that associated with ferrite in pearlite.

free central placentation (*Bot.*). The grouping of the ovules on the surface of a placenta which stands up from the base of the ovary and is not united with the walls of the chamber, either at the side or at the top.

free charge (*Elec. Eng.*). An electrostatic charge which is not bound by an equal or greater charge of opposite polarity.

free-cutting brass (*Met.*). α-β Brass containing about 2–3% of lead, to improve the machining properties. Used for engraving and screw machine work.

free-cutting steel (*Met.*). Steel in which the phosphorus is increased to 0·15% and the sulphur to 0·2%, to induce a certain degree of brittleness which facilitates rapid machining.

freedom (*Eng.*). A body free to move in space is said to have 6 degrees of freedom, 3 of translation and 3 of rotation.

free electron theory (*Electronics*). Early theory of metallic conduction based on concept that outer valence electrons, which do not form crystal bonds, are free to migrate through crystal, so forming *electron gas*. Now superseded by **energy band** theory.

free end (*Build.*). The end of a beam which is not fixed or built in.

free energy (*Chem.*). The capacity of a system to perform work, a change in free energy being measured by the maximum work obtainable from a given process.

free fall (*Space*). The motion of an unpropelled body in a gravitational field. In orbit beyond the earth's atmosphere, free fall produces weightlessness (*zero-g*).

free-falling velocity (*Powder Tech.*). The velocity of fall of a particle of powder through a still fluid, at which the effective weight of the particle is balanced by the drag exerted by the fluid on the particle.

free ferrite (*Met.*). Ferrite in steel or cast-iron other than that associated with cementite in pearlite.

free field (*Acous.*). Nonreflecting space required for measurement of the acoustic response of microphones and loudspeakers. Realized down to low frequencies, i.e., order of a few hundred Hz, in an anechoic room.

free-field emission (*Elec.*). That from an emitter when the electric gradient at a surface is zero.

free-flight wind tunnel (*Aero.*). A wind tunnel, usually of up-draught type, wherein the model is not mounted on a support, but flies freely.

free-handle (*Elec. Eng.*). See **free-trip.**

free haul (*Civ. Eng.*). See **overhaul.**

free-hearth electric furnace (*Elec. Eng.*). A direct-arc furnace in which one electrode forms a part of the bottom of the hearth.

free impedance (*Elec.*). That at input terminals of a transducer when its load impedance is zero.

freeing port (*Ships*). An opening in the bulwark to free the deck from large quantities of water which may come on board in heavy weather. Usually with a hinged plate fitted to prevent water coming in.

freelay (*Typog.*). See **Multifont type tray.**

free line signal (*Telecomm.*). A visual signal which is associated with the trunk, toll or junction multiple to indicate circuit to be used next.

freemartin (*Zool.*). In cattle, a sterile female intersex occurring as co-twin with a normal bull-calf.

free milling (*Min. Proc.*). Descriptive term for gold or silver ore, which contains its metal in amalgamable state.

free-needle surveying (*Surv.*). Traverse work done with a compass, the magnetic bearing of each line being read in turn. Cf. *fixed-needle surveying.*

free nuclear division (*Cyt.*). Nuclear division unaccompanied by the formation of cell walls.

free oscillations (*Telecomm.*). (1) Oscillatory currents whose frequency is determined by constants of the circuit in which they are flowing, e.g., those resulting from the discharge of a capacitance through an inductance. (2) Mechanical oscillations governed solely by natural elastic properties of vibrating body.

free path (*Phys.*). See **mean free path.**

free pendulum clock (*Horol.*). Type of *impulse clock* (q.v.) in which a free pendulum swings in a vacuum chamber, receiving impulses from a gravity lever released at regular intervals by an electromagnetically operated slave pendulum.

free-piston engine (*Eng.*). A prime mover of the internal-combustion type, in which a power piston acts directly on a compressor piston working in the opposite direction. May be compounded, as an alternative to a rotary compressor, with a gas turbine.

free pole (*Elec. Eng.*). A magnet pole which is imagined, for theoretical purposes, to exist separately from its corresponding opposite pole.

free radicals (*Chem.*). Usually encountered in organic chemistry, term applies to groups of atoms in particular combinations capable of free existence under special conditions, usually for only very short periods (sometimes only microseconds). The proved existence of free radicals was of great help in explaining the course that certain reactions took in practice. Because they contain unpaired electrons, they are paramagnetic, and this fact has been used in determining the degree of dissociation of compounds into free radicals. The existence of free radicals such as methyl ($CH_3\cdot$) and ethyl ($C_2H_5\cdot$) has been known for many years.

free-running circuit (*Radio*). Same as **astable circuit.**

free-running frequency (*Telecomm.*). That of an oscillator, otherwise uncontrolled, when not locked to a synchronizing signal, especially a time-base generator in a television receiver.

free-running speed (*Elec. Eng.*). The speed which a vehicle or train will attain when propelled by a constant tractive effort, i.e., the speed at which the applied tractive effort exactly equals the forces resisting motion. Also called **balancing speed.**

freesettling (*Min. Proc.*). In classification of finely ground mineral into equal-settling fractions, use of conditions in which particles can fall freely through the environmental fluid, as opposed to the packed conditions of mineral settling.

free space impedance (*Elec.*). For electromagnetic waves the characteristic impedance of a medium is given by the square root of the ratio of permeability to permittivity (or 4π times this in unrationalized systems of units). This gives a free space value of 377·3 ohms.

free-sprung (*Horol.*). A watch is said to be *free-sprung* when no index and curb pins are available for the correction of its rate. The balance and spring are so proportioned and adjusted as to give the best possible performance under all conditions. Chronometers and chronometer watches are always free-sprung.

freestone (*Build.*). A building-stone which can be worked with a chisel without tending to split into definite layers.

free surface (*Ships, etc.*). A free surface exists when any compartment in the ship is partly filled with liquid.

free-trip (*Elec. Eng.*). Said of certain types of circuit-breaker or motor starter in which the tripping mechanism is independent of the closing mechanism, and will therefore allow the switch to trip while the latter is being operated. Also **free-handle.** Cf. *fixed-trip*.

free turbine (*Aero.*). A power take-off turbine mounted behind the main turbine/compressor rotor assembly and either driving a long shaft inside the main rotor to a gearbox at the front of the engine, or a short shaft to a gearbox at the rear of the engine. It can also be a separate unit fed by a remotely produced gas supply.

free vibrations (*Phys.*). The vibrations which occur at the natural frequency of a body when it has been displaced from its position of rest and allowed to vibrate freely without the application of any periodic force.

free vortex (*Hyd.*). The type of vortex which results when a liquid flows through a hole in the bottom of a vessel or other container of liquid.

free-wheel (*Autos.*). A 1-way clutch, usually depending on the wedging action of rollers, placed in the transmission line, so as to transmit torque only when the engine is driving.

free-wheel rectifier (*Electronics*). One placed in shunt with a highly inductive load to stabilize the main rectifier against sudden changes in the load current.

freeze-drying (*Vac. Tech.*). Evaporation to dryness in a vacuum for preservation or storage of a labile solution, which is kept frozen throughout the process. Sometimes known as **lyophilization** or **sublimation from the frozen state.**

freeze-drying fixation (*Micros.*). A method of fixing tissues by instant freezing, e.g., by immersion in isopentane cooled to $-190°C$, followed by dehydration in vacuo or in a stream of dry air.

freeze sinking (*Mining*). Shaft sinking, or penetration of water-logged strata, in which refrigerated brine is circulated to freeze the ground and make establishment of an imperviously lined shaft possible.

freezing (*Heat*). The conversion of a liquid into the solid form. This process occurs at a definite temperature for each substance, this temperature being known as the *freezing-point*. The freezing of a liquid invariably involves the extraction of heat from it, known as *latent heat of fusion*. See **latent heat, depression of freezing-point.**

freezing mixture (*Chem.*). A mixture of two substances, generally of ice and a salt, or solid carbon dioxide with an alcohol, used to produce a temperature below 0°C.

freezing-point (*Heat*). The temperature at which a liquid solidifies, which is the same as that at which the solid melts (the *melting-point*). The freezing-point of water is used as the lower fixed point in graduating a thermometer. Its temperature is defined as 0°C (273·15K). Abbrev. **f.p. See triple point**; also **water** and **depression of freezing-point.** (*Met.*) The temperature at which a metal solidifies. Pure metals, eutectics, and some intermediate constituents freeze at constant temperature; alloys generally solidify over a range.

freezing-point method (*Chem.*). See **cryoscopic method.**

freight car (*Rail.*). The American term for a goods wagon.

fremitus (*Med.*). Palpable vibration, especially of the chest wall, during speech or coughing; variations in intensity are of diagnostic value.

Fremont test (*Met.*). A type of impact test in which a beam specimen notched with a rectangular groove is broken by a falling weight.

Frémy's salt (*Chem.*). Potassium hydrogen fluoride, acid potassium fluoride $K^+(HF_2^-)$.

French arch (*Arch.*). A brick arch, flat at the top and bottom.

French bit (*Carp.*). A boring tool having a flat blade, shaped at the two cutting edges in continuous curves, from the point to and beyond a place of maximum diameter; used in a lathe-head for drilling hard wood.

French chalk (*Min.*). The mineral, talc, ground into a state of fine subdivision, its softness and its perfect cleavage contributing to its special properties when used as a dry lubricant.

French curve (*Instr.*). A drawing instrument used to guide the pen or pencil in drawing curved lines. It consists of a thin flat sheet of transparent plastic or other material cut to curved profiles at the edges.

French door (*Arch.*). *French window* (q.v.).

French drain (*Civ. Eng.*). A drain formed by partly filling a trench at the bottom with loose broken bricks or rubble.

French fliers (*Carp.*). Steps in an *open newel stair* (q.v.) with *quarter-space landings* (q.v.).

French fold (*Print.*). A sheet of paper folded twice to make 8 pages, but left uncut, and printed only on pages 1, 4, 5, and 8.

French gold (*Met.*). Oroide. A copper-based alloy with about 16½% zinc, ½% tin and 0·3% iron.

French joint (*Bind.*). One in which the book cover is given a wide space between board and spine to enable the board to hinge back freely. Also called **French groove.**

frenchman (*Tools*). A joint-trimming tool, used for pointing.

French moult (*Vet.*). A defective development of the first plumage of birds leading to the shedding of the wing and tail primaries; particularly observed in aviary-bred budgerigars.

French pitch (*Phys.*). One for musical instruments, A is 435 Hz at 15°C. See **concert pitch.**

French polish. A solution of shellac dissolved in methylated spirit. Applied to wood surfaces to produce a high polish on them.

French roof (*Build.*). A *mansard roof* (q.v.).

French sewing (*Bind.*). (1) The normal method of present-day machine-sewing, i.e., without tapes. (2) A method of hand-sewing at the edge of the bench without using a frame.

French stuc (*Build.*). Plasterwork finished to present a surface resembling that of stonework.

French system of drawing (*Textiles*). See **porcupine drawing.**

French truss (*Eng.*). A symmetrical roof truss for large spans, composed of a pair of braced isosceles triangles based on the sloping sides of the upper chord, their apices being joined by a horizontal tie. Also called **Belgian truss, Fink truss.**

French window (*Arch.*). A glazed casement, serving as both window and door.

Frenet's formulae (*Maths.*).

$$\begin{aligned} \mathbf{t}' &= \qquad +\frac{1}{\rho}\mathbf{n} \\ \mathbf{n}' &= -\frac{1}{\rho}\mathbf{t} \qquad +\frac{1}{\tau}\mathbf{b} \\ \mathbf{b}' &= \qquad -\frac{1}{\tau}\mathbf{n} \end{aligned}$$

where **t**, **n** and **b** are unit vectors along the tangent, principal normal and binormal respectively at a point on a space curve, ρ and τ are the radii of curvature and torsion respectively, and a dash denotes differentiation with respect to arc length. Also called **Serret-Frenet formulae.**

Frenkel defect (*Chem.*). Disorder in the crystal lattice, due to some of the ions (usually the cations) having entered interstitial positions, leaving a corresponding number of normal lattice sites vacant. Likely to occur if one ion (in practice the cation) is much smaller than the other, e.g., in silver chloride and bromide.

frenotomy, fraenotomy (*Surg.*). Cutting of the frenum of the tongue for tongue-tie.

frenulum (*Zool.*). In some *Lepidoptera*, a bunch of strong bristles arising from the costal border of the hind-wing, which engages the fore-wing and so locks the two wings together during flight; in some *Scyphozoa*, a thickening of the sub-umbrella; more generally, a membranous fold.

frenum (*Zool.*). A membranous or ligamentous structure which checks the movement of a part; in *Cirripedia*, a tegumentary fold at the base of the mantle; in some Insects, a membranous or chitinous fold or ridge extending from the scutellum to the base of the fore-wing; a frenulum. *adj.* **frenate.**

Freons (*Chem.*). TN for compounds consisting of ethane or methane with some or all of the hydrogen substituted by fluorine, or by fluorine and chlorine. Used as refrigerants, in fire-extinguishers, and as aerosol fluids for insecticides, etc., because of their low b.p. and chemical inertness, and for insulating atmospheres in electrical apparatus because of their high breakdown strengths. Principle ones are: Freon 11 trichlorofluoromethane CCl_3F, Freon 12 dichlorodifluoromethane CCl_2F_2, Freon 21 dichlorofluoromethane $CHCl_2F$, Freon 114, dichlorotetrafluoroethane $CClF_2{\cdot}CClF_2$, Freon 142 1-chloro-1 : 1-difluoroethane CH_2CClF_2.

frequency (*Ecol.*). The relative number of any given species in a given place. (*Elec., etc.*) Rate of repetition of a periodic disturbance, measured in hertz (cycles per second). Also called **periodicity.**

frequency allocation (*Telecomm.*). Frequency on which a transmitter has to operate, within specified tolerance. Bands of frequencies for specified services are allocated by international agreement.

frequency band (*Telecomm.*). Interval in the frequency spectrum occupied by a modulated signal. In sinusoidal amplitude modulation, it is twice the maximum modulation frequency, but it is much greater in frequency or pulse modulation.

frequency bridge (*Elec. Eng.*). An a.c. measuring bridge whose balancing condition is a function of the supply frequency, e.g., Robinson bridge.

frequency changer (*Elec. Eng.*). A machine designed to receive power at one frequency and to deliver it at another frequency. See also **frequency transformer.** (*Telecomm.*) A combination of oscillator and modulator circuits used in a superhet receiver to change the incoming signal from its original carrier frequency to a fixed intermediate carrier frequency; also called **conversion mixer, conversion transducer, converter** (U.S.), and **frequency converter.**

frequency converter (*Telecomm.*). Same as **frequency changer.**

frequency demultiplication (*Telecomm.*). See **frequency division.**

frequency departure (*Telecomm.*). Discrepancy between actual and nominal carrier frequencies of a transmitter. Formerly termed **frequency deviation.**

frequency-departure meter (*Telecomm.*). Instrument for measuring frequency departure of transmitter, normally forming part of control console.

frequency deviation (*Telecomm.*). (1) In frequency modulation, maximum departure of the radiated frequency from mean quiescent frequency (carrier). (2) Greatest deviation allowable in operation of frequency modulation. In broadcast systems within the range 88 to 108 MHz, the maximum deviation is ±75 Hz. (3) See **frequency departure.**

frequency-discriminating filter (*Telecomm.*). See **electric-wave filter.**

frequency discriminator (*Telecomm.*). Circuit the output from which is proportional to frequency or phase change in a carrier from condition of no frequency or phase modulation.

frequency distortion (*Acous.*). In sound reproduction, variation in the response to different notes solely because of frequency discrimination in the circuit or channel. Generally plotted as a decibel response on a logarithmic frequency base.

frequency distribution (*Stats.*). See **distribution.**

frequency-distribution curve (*Stats.*). Curve exhibiting a series of observations of values; the base shows the values, the ordinates from which show the frequency of occurrence.

frequency diversity (*Radio*). See **diversity reception.**

frequency division (*Telecomm.*). Dividing a frequency by harmonic locking oscillators or stepping pulse circuits. Specific integral division alone can be obtained; an arbitrary collection of frequencies cannot be divided, except by recording and reproducing at a lower speed; also called **frequency demultiplication.**

frequency doubler (*Telecomm.*). Frequency multiplier in which the output current or voltage has twice the frequency of the input. Achieved by simple push-pull tuned circuits.

frequency doubling (*Telecomm.*). Introduction of marked double-frequency components through lack of polarization in an electromagnetic or electrostatic transducer, in which the operating forces are proportional to the square of the operating currents and voltages respectively.

frequency drift (*Telecomm.*). Change in frequency of oscillation because of internal (ageing, change of characteristic or emission) or external (variation in supply voltages, or ambient temperature) causes. Also **oscillator drift.**

frequency factor (*Bot.*). The percentage occurrence of a species in a plant community. (*Chem.*) The pre-exponential factor in Arrhenius' equation, expressing the frequency of successful collisions between the reactant molecules.

frequency function (*Stats.*). See **distribution function.**

frequency index (*Ecol.*). Percentage of sample plots in which a species occurs.

frequency meter (*Elec. Eng., Electronics*). Circuit for comparing an unknown frequency with a standard frequency derived from a crystal-controlled oscillator or *atomic clock*. See **wavemeter.**

frequency-modulated cyclotron (*Elec. Eng.*). One in which frequency of the voltage applied to the *dees* is varied, so as to keep synchronous orbiting of accelerated particles when their mass increases through relativity effect at the high velocities attained.

frequency modulation (*Elec.*). Variation of frequency of a transmitted wave in accordance with impressed modulation, the amplitude remaining constant. Abbrev. FM.

frequency-modulation receiver (*Electronics, Radio*). One incorporating an FM demodulator. See **frequency discriminator, Foster-Seeley discriminator, ratio detector.**

frequency monitor (*Elec. Eng.*). Nationally or

internationally operated equipment to ascertain whether or not a transmitter is operating within its assigned channel.

frequency multiplier (*Electronics*). Non-linear circuit, usually incorporating power gain, in which the output circuit is tuned to select an harmonic of the input signal. Transistor or varactor diode circuits may be used.

frequency of gyration (*Electronics*). That of electrons about a line indicating direction of magnetic field in ionosphere.

frequency of infinite attenuation (*Telecomm.*). A frequency at which a filter inserted in a communication channel provides a maximum attenuation theoretically infinite with loss-free inductances and capacitances. Such large attenuation is generally provided by an antiresonant series arm, or by an acceptance resonant shunt arm.

frequency of penetration (*Phys.*). That of a wave which just fails to be reflected by ionospheric layer.

frequency overlap (*Telecomm.*). Common parts of frequency bands used, e.g., for the regular video signal and the chrominance signal in colour TV.

frequency pulling (*Telecomm.*). Change in oscillator frequency resulting from variation of load impedance.

frequency relay (*Elec. Eng.*). See **relay types.**

frequency response (*Phys.*). If constant input power is applied to a *transducer* through a range of discrete frequencies, its frequency response over the given range is the *envelope* of the output powers at each of those frequencies. The response may either be measured absolutely in watts against hertz, by implication, e.g., volts or intensity against frequency, or proportionately, e.g., *decibels* below peak output response against frequency. A **flat** or level **response** therefore indicates equal response to all frequencies within the stated range, e.g., for audio equipment an equal response to within, say, 1 dB for the range 20 Hz to 20 kHz.

frequency selectivity (*Radio*). See **selectivity.**

frequency-shift keying (*Teleg.*). In radio telegraphy, altering the carrier by mark-and-space keying.

frequency-shift transmission (*Telecomm.*). A form of modulation used in communication systems in which the carrier is caused to shift between two frequencies denoting respectively *on* and *off* pulse.

frequency stabilization (*Telecomm.*). Prevention of changes produced in frequency of oscillation of a self-oscillating circuit by changes in supply voltage, load impedance, valve parameters, etc. Achieved by resonating crystals, tuned cavities or transmission lines.

frequency standard (*Telecomm.*). Reference oscillator of very high stability. Usually of quartz-ring type, although atomic beam standards provide the final reference.

frequency swing (*Telecomm.*). Extreme difference between maximum and minimum instantaneous frequencies radiated by a transmitter.

frequency tolerance (*Telecomm.*). Extent to which frequency of the carrier of a transmission is permitted to deviate from its allocation.

frequency transformer (*Elec. Eng.*). A static piece of apparatus (e.g., a transformer or mercury-arc converter) which receives power at one frequency and delivers it at another frequency. See **static frequency changer.**

frequency translation (*Telecomm.*). Shifting all frequencies in a transmission by the same amount (not through zero).

frequency tripler (*Electronics*). See **frequency multipli**

fresco (*Paint.*). A method of painting on plastered walls with lime-fast colours while the plaster is still wet.

fresh-water allowance (*Ships*). The difference between the *freeboard* (q.v.) in sea water of 1025, and in fresh water of 1000 oz/ft^3 or g/dm^3.

fresh-water sediments (*Geol.*). These include those of all the main types that accumulate in other environments, and cover the whole range of grain size. Fresh-water conditions were widespread in N.W. Europe in later Tertiary times (Aquitanian-Oligocene), characteristic deposits being the Bovey Tracey pipe-clays and lignites, and the Bembridge fresh-water limestones. These are lacustrine deposits, while fluviatile and fluvioglacial deposits also fall in this main category.

fresnel (*Optics*). A unit of optical frequency, equal to 10^{12} Hz = 1 THz (terahertz).

Fresnel diffraction (*Optics*). The study of the diffracted field at a distance from an aperture in an absorbing screen, which is large compared with the wavelength and the aperture dimensions, but not so large that the phase differences between the waves from the aperture sources can be neglected at points along the normal to the screen.

Fresnel ellipsoid (*Optics*). A method of representing the doubly refracting properties of a crystal, used in crystal optics.

Fresnel lens (*Photog.*). One built up of a number of narrow concentric segments, used where a flat surface is desirable, as in the field lens of viewfinders or the viewing screen of reflex cameras.

Fresnel region (*Phys.*). See **field.**

Fresnel's bi-prism (*Light*). An isosceles prism having an angle of nearly 180°, used for producing interference fringes from the two refracted images of an illuminated slit.

Fresnel's laws (*Light*). These concern the condition of interference of polarized light: (1) two rays of light emanating from the same polarized beam, and polarized in the same plane, interfere in the same way as ordinary light; (2) two rays of light emanating from the same polarized beam and polarized at right angle to each other will interfere only if they are brought into the same plane of polarization; (3) two rays of light polarized at right angles and emanating from ordinary light will not interfere if brought into the same plane of polarization.

Fresnel's mirrors (*Light*). Two plane mirrors inclined at an angle of a little less than 180°, used for producing interference fringes from the two reflected images of an illuminated slit.

Fresnel's reflection formula (*Light*). A formula giving the fraction of the incident light reflected at the surface of a transparent medium. The fraction equals

$$\tfrac{1}{2}\left\{\frac{\sin^2(i-r)}{\sin^2(i+r)}+\frac{\tan^2(i-r)}{\tan^2(i+r)}\right\},$$

where i and r are the angles of incidence and refraction respectively.

Fresnel's rhomb (*Light*). A glass rhomb which is used for obtaining circularly-polarized light from plane-polarized light by total internal reflection. The rhomb is so constructed that two such reflections at an angle of 54° are obtained, each of which introduces a phase difference of one-eighth of a period between the two components obtained from the incident plane-polarized light.

Fresnel zones (*Phys.*). Zones into which a wavefront is divided, according to the phase of the radiation reaching any point from it. The radiation from any one zone will reach the given point half a period out of phase with the adjacent zones.

fret (*Bind., Typog.*). A continuous border design of interlaced bands or fillets, tooled on a book cover or used for typographic decoration. (*Brew.*) Secondary fermentation after casking, due to the presence of sugars left by primary fermentation. Controlled by the use of a porous peg in the cask.

fret-saw (*Carp.*). One with a very thin and narrow blade (usually 5 in. or 125 mm) kept in tension by an elongated metal bow 12–20 in. (300-500 mm) in length. Used for cutting narrow curves in thin wood. There is also a mechanical type.

fretted lead (*Plumb.*). Strip-lead of suitable section for use as a *came* (q.v.).

fretting corrosion (*Met.*). Corrosion due to slight movements of unprotected metal surfaces, left in contact either in a corroding atmosphere or under heavy stress.

fret-work (*Build.*). A mode of glazing in which diamond-shaped panes (quarrels) are connected together by leaden cames to form a window.

Freud's theory of the libido (*Psychol.*). According to this theory, the libido (energy attaching to the sexual instinct) becomes organized at different stages of development along different routes, e.g., oral, anal, phallic, and genital, each phase having its own well-defined characteristics. Mental mechanisms common to these phases are found in psychoneuroses and psychoses.

Freundlich's adsorption isotherm (*Chem.*). The concentration c of adsorbate is related to the equilibrium partial pressure p as $c = ap^b$, where a and b are empirical constants.

Freysinnet (*Civ. Eng.*). A method used for the stretching and anchoring of high tensile wires in a post-tensioned system.

friability test (*Civ. Eng.*). A test for determining the suitability (that is, its resistance to crushing) of any given stone for use in asphalt work. A sample is heated for 15 min in a sand bath at 175°C, and should then not disintegrate on receiving a blow from a hammer.

friable (*Min. Proc.*). Of ore, easily fractured or crumbled during transport or comminution.

friar (*Typog.*). An area of print with too little ink. See **monk**.

friction (*Mech.*). The resistance to motion which is called into play when it is attempted to slide one surface over another with which it is in contact. The frictional force opposing the motion is equal to the moving force up to a value known as the *limiting friction*. Any increase in the moving force will then cause slipping. *Static friction* is the value of the limiting friction just before slipping occurs. *Kinetic friction* is the value of the limiting friction after slipping has occurred. This is slightly less than the static friction. The *coefficient of friction* is the ratio of the limiting friction to the normal reaction between the sliding surfaces. It is constant for a given pair of surfaces. (*Med.*) (1) The sound produced by the rubbing together of two inflamed surfaces, as in pleurisy or pericarditis. (2) Rubbing of a part, as in massage.

frictional damper (*Eng.*). A device consisting of a supplementary mass frictionally driven from a crankshaft at a point remote from a node, which dissipates vibrational energy in heat.

frictional electricity (*Elec.*). Static electricity produced by rubbing bodies or materials together, e.g., an ebonite rod with fur.

frictional machine (*Elec. Eng.*). See **electrostatic generator**.

frictional-rest escapement (*Horol.*). See **escapement**.

friction and windage loss (*Elec. Eng.*). Losses in an electrical machine due to friction of sliding parts (see **friction loss**) and also to air resistance. These losses are frequently considered together in designing and testing electrical machinery.

friction clutch (*Eng.*). A device for connecting or disconnecting two shafts which are in line, in any relative position, through the friction of two surfaces in contact. It consists of a pair of opposed members, between which the drive is transmitted through the friction of their contact surfaces, and which may be separated by a lever system.

friction compensation (*Elec. Eng.*). A small torque, additional to the main torque, provided in a motor-type integrating meter to compensate for the effect of friction of the moving parts.

friction drive (*Eng.*). A drive in which one wheel causes rotation of a second wheel with which it is pressed into contact, through the agency of the friction forces at the contact surfaces.

friction gear (*Eng.*). A gear in which power is transmitted from one shaft to another through the tangential friction set up between a pair of wheels pressed into rolling contact. One of the contacting surfaces is usually fabric-faced. Suitable only for small powers.

friction glazing (*Paper*). A method of glazing in which one or more of the calender cylinders revolves at a speed greater than that of the others. A very high polish is obtained.

friction (horse)power (*Eng.*). That part of the gross or indicated (horse)power developed in an engine cylinder which is absorbed in frictional losses; the difference between the indicated and the brake (horse)power.

friction loss (*Elec. Eng.*). The power absorbed in the bearings, commutator, or slip-ring surfaces, or at any other sliding contacts of an electric machine.

friction pile (*Build., Civ. Eng.*). A pile which supports its load only by the friction on its sides.

friction rollers (*Eng.*). See **antifriction bearing**.

friction welding (*Met.*). Welding in which heat necessary for a weld to form is produced frictionally, e.g., by rotation, and forcing the parts together.

Friedel and Crafts' synthesis (*Chem.*). The synthesis of benzene hydrocarbons or their homologues, and aromatic ketones, by the action of halogenoalkanes or acylhalides on aromatic hydrocarbons in the presence of anhydrous aluminium chloride.

Friedman's test (*Med.*). See **Aschheim-Zondek test**.

Friedreich's ataxia (*Med.*). A hereditary nervous disease in which there are irregular, uncoordinated movements of the voluntary muscles, a slow, staggering, reeling gait, and various deformities—the result of degenerative changes in the nerve tracts of the spinal cord.

friendly numbers (*Maths.*). Pairs of numbers each of which is the sum of the factors of the other, including unity, e.g., 220 and 284. Also called *amicable numbers*. Cf. *perfect number*.

frieze (*Arch., Build.*). (1) The middle part of an entablature, between the architrave and the cornice. (2) The decorated upper part of a wall, below the cornice. (*Textiles*) A heavy woollen material with a rough surface, made from coarse yarns of mixed colours which give a rough tweed effect. Made also from wool and shoddy yarns, loosely set to aid the milling process.

frieze panel (*Join.*). An upper panel in a 6-panel door.

frieze rail (*Join.*). The rail next to the top rail in a 6-panel door.

frig-bob saw (*Tools*). A long handsaw used in Bath stone quarries.

frigidity (*Psychol.*). In women, decrease or absence of the normal sexual response; often dependent on strongly repressed inhibitions.

frill (*Bot.*). A thin sheet of interwoven hyphae forming a horizontal circular flange around the stem of an agaric. Also called **armilla.** (*Textiles*) (1) Bleached linen or cotton fabric with a twill weave, used for tropical suitings, etc. (2) Heavy twilled cotton cloth, 3-end weave, used for sheeting and clothing, when bleached, dyed or printed. There are numerous qualities.

frilled organ (*Zool.*). An attachment organ at the anterior end of tapeworms of the order *Cestodaria*.

frilling (*Photog.*). The crinkling of the emulsion on a plate, resulting in detached folds and wrinkles.

fringe (*Photog.*). A defect in colour photography due to lack of registration of the elementary colours (e.g., owing to parallax). (*Textiles*) Loose threads at the ends of cloth such as a towel, or all round the edges of quilts, table covers, etc.

fringe area (*Radio, TV*). In TV or radio broadcasting, the regions around a transmitter in which good reception is uncertain.

fringe effect (*Photog.*). A faint light line on the low-density side of the boundary between a lightly and a heavily exposed area on a developed emulsion. Also called **edge effect.** See **Mackie lines.**

fringing (*Elec. Eng.*). The spreading of the lines of force of a magnetic field, at the edges of an air gap in a magnetic circuit. (*TV*) Incorrect registration of images in colour TV, so that edges of colour exhibit departure from correct rendering.

fringing coefficient (*Elec. Eng.*). A coefficient used in making magnetic circuit calculations to allow for the effect of fringing of the flux.

fringing reef (*Ocean.*). Platform of coral formation stretching out from the land. See **coral reef.**

frisket (*Print.*). (1) On a hand press, a thin iron frame covered with paper, to hold the sheet to the tympan, prevent it from being soiled, and strip it from the forme after printing. (2) On a platen machine, the adjustable metal fingers which strip the sheet from the forme after printing.

frisking (*Nuc. Eng.*). Searching for radioactive radiation by contamination meter, usually a portable ionization chamber.

fritting (*Met.*). Condition in fire assaying in which the powdered ore, flux and other reagents are in a pasty condition a little below melting-point.

frizing (*Leather*). The process of removing a thin layer from the grain surface, in order to facilitate staining.

Frodingham process (*Chem. Eng.*). See **Appleby-Frodingham process.**

froe or **frow** (*For.*). A blade, about 6 in. (150 mm) long, with a wooden handle at right-angles to it; used to split shakes and staves from bolts.

frog (*Agric.*). Basic component of a plough to which the ploughshare and other working parts are attached. (*Build.*) The depression made in one or both of the larger sides of some bricks in order to form a key for the mortar at the joints. (*Elec. Eng.*) See **trolley-frog.** (*Join.*) Of a plane, the surface against which the blade rests. It determines the *pitch* (q.v.). (*Rail., etc.*) The point of intersection of the inner rails, where a train or tram crosses from one set of rails to another. The frog is in the form of a V. See **turnout.** (*Vet.*) A V-shaped band of horn passing from each heel to the centre of the sole of a horse's foot. (*Weaving*) A metal puffer which stops the loom and prevents the slay beating-up when the shuttle is trapped in the warp shed.

Fröhlich's syndrome (*Med.*). See **dystrophia adiposogenitalis.**

Froin's syndrome (*Med.*). The presence of yellow cerebrospinal fluid, which has a high content of protein but no cells, below the site of obstruction (e.g., by a tumour) of the spinal cord.

frond (*Bot.*). (1) A general term for the leaf of a fern. (2) A flattened expanded thallus of a seaweed. (3) A similar thallus in a liverwort. (4) A general term for a leaflike structure of obscure morphological status. *adjs.* **frondescent, frondose.**

frons (*Zool.*). In Insects, an unpaired sclerite of the front of the head; in higher Vertebrates, the front of the head above the eyes. *adj.* **frontal.**

front (*Carp.*). The sole face of a plane. (*Meteor.*) The line along which a *frontal zone* (q.v.) reaches the earth's surface. See **cold-, warm-.**

frontage line (*Build.*). The *building line* (q.v.).

frontal (*Zool.*). (1) A paired dorsal membrane bone of the Vertebrate skull, lying between the orbits. (2) Pertaining to the frons.

frontal cilia (*Zool.*). In *Lamellibranchiata*, downward-beating cilia, which produce a constant stream of water over the surface of each ctenidium towards its ventral edge.

frontal lobes (*Zool.*). The front part of the cortex traditionally regarded as being the centre of man's higher intelligence. The evidence for this is not conclusive.

frontal organ (*Zool.*). A paired organ consisting of a group of cells, sometimes surrounded by setae, on the front of the head in many *Crustacea*; it is probably sensory.

frontal plane (*Zool.*). The median horizontal longitudinal plane of an animal.

frontal segment (*Zool.*). One of the quasisegments into which the ossified cranium of Mammals is divided, consisting of the presphenoid, the orbito-sphenoids and the frontals.

frontal sinuses (*Zool.*). Air cavities, connected with the nasal chambers, extending into the frontal bones, in Mammals.

frontal zone (*Meteor.*). A narrow sloping zone in the atmosphere marking the boundary between *air masses* (q.v.). The horizontal temperature gradient is high, perhaps of the order of 1°C/10 km. If the leading air mass is the warmer it is a cold frontal zone, if the leading air mass is the colder it is a warm frontal zone.

front cavity (*Bot.*). The opening of a stoma nearest the epidermis.

front clearance (*Eng.*). In a single-point cutting tool, the clearance of the edge below the point; expressed as the angle between this edge as seen in side view on the tool shank and a line through the point and perpendicular to the tool shank base.

front hearth (*Build.*). The part of the hearth extending beyond the chimney breast.

frontispiece (*Print.*). An illustration facing the title-page of a book.

frontoclypeus (*Zool.*). A sclerite of the head in

Insects, formed by the fusion of the frons and the clypeus.

frontogenesis, frontolysis (*Meteor.*). Respectively, the intensification or realization of a front, and its weakening or disappearance.

fronton (*Arch.*). The cornice and pediment, supported on consoles over the entrance to a building.

fronto-parietal (*Zool.*). In *Choanichthyes* and Amphibians, a paired bone, consisting of the fused parietal and frontal, covering the roof of the neurocranium.

front porch (*TV*). See **porch.**

front-to-back ratio (*Radio*). Ratio of effectiveness of a directional antenna or microphone, etc., in the forward and reverse directions.

frontwall cell (*Electronics*). Semiconductor cell in which light passes through a conducting layer to the active layer, which is separated from the base metal by a semiconductor.

frost (*Meteor.*). A *frost* is said to occur when the air temperature falls below the freezing-point of water (0°C or 32°F). See **hoar frost.**

frost day (*Meteor.*). A 24-hour period beginning at 9 a.m. during which the temperature is 0°C or less.

frosted lamp (*Light*). A filament-lamp the bulb of which is etched or sand-blasted to break up any direct rays of light from the filament. See **inside-frosted lamp.**

frostheart (*For.*). An abnormal condition of the heartwood in certain trees, usually accompanied by a deepening of the colour, ascribed to the action of severe frost.

frost hollow (*Meteor.*). A hollow in hilly ground where the air may stagnate and become exceptionally cold due to radiation at night.

frost-point (*Meteor.*). The temperature at which air becomes saturated with respect to ice if cooled at constant pressure.

froth (*Chem., Min. Proc.*). Foam; a gas-liquid continuum in which bubbles of gas are contained in a much smaller volume of liquid, which is expanded to form bubble walls. The system is stabilized by oils, soaps, or emulsifying agents which form a binding network in the bubble walls.

frother (*Met.*). A substance used to promote the formation of a foam in the flotation process.

froth flotation (*Min. Proc.*). Process in dominant use for concentrating values from low-grade ores. After fine grinding, chemicals are added to a pulp (ore and water) to develop differences in surface tension between the various mineral species present. The pulp is then copiously aerated, and the preferred (*aerophilic*) species clings to bubbles and floats as a mineralized froth, which is skimmed off.

frottage (*Psychol.*). A type of sexual perversion in which orgasm is obtained by rubbing against someone.

frotteur (*Psychol.*). One who practises frottage.

Froude brake (*Eng.*). An absorption dynamometer consisting of a rotor inside a casing, itself free to rotate, the space between the two being filled with water. The energy is dissipated in eddy formation and heat, the torque absorbed being measured by the torque necessary to prevent rotation of the casing.

Froude number (*Hyd.*). A dimensionless number used in the study of fluid flow problems with models. It is the ratio of the flow velocity to the square root of a linear dimension of the model times the acceleration due to gravity.

Froude's transition curve (*Surv.*). A transition curve the equation to which is that of a cubic parabola, the offset y from the straight produced being given by $y=\frac{x^3}{6lr}$, where x=distance from tangent point, l=length of transition, r=radius of the circular arc.

frow (*For.*). See **froe.**

frowy (*For.*). Said of timber which is soft and brittle.

frozen (*Cinema.*). Said of arc carbons when they have fused together so that an arc cannot be struck by the mechanism.

frozen bearing (*Eng.*). A seized bearing. See **seizure.**

frozen equilibrium (*Chem.*). The state of a solid at low temperature, which is prevented from attaining the theoretically possible thermodynamic equilibrium, because its molecular motion has become too slow. Cf. *Nernst theory.*

fructicole (*Bot.*). Living on fruits; said of parasitic fungi.

fructification (*Bot.*). (1) A general term for the body which develops after fertilization and contains spores or seeds. (2) Any spore-bearing structure, whether formed after fertilization or by purely vegetative development.

fructosans (*Chem.*). The anhydrides of fructose, e.g., inulin.

***l*-fructose** (*Chem.*). Fruit-sugar or laevulose, $C_6H_{12}O_6$, anhydrous rhombic crystals, m.p. 95°C. A *ketohexose* (q.v.,) is prepared by heating inulin with dilute acids, and is always found together with *d-glucose* (q.v.) in sweet fruit juices.

frue vanner (*Min. Proc.*). Endless rubber belt which is driven gently upslope while finely ground ore is washed gently downslope. Belt is given a side shake to aid distribution, and wash water is so adjusted that heavy material stays on belt, while light gangue is washed down to opposite end of pulley system round which belt circulates.

frugivorous (*Zool.*). Fruit-eating.

fruit (*Bot.*). (1) The same as *fructification.* (2) The structure which develops from the ovary of an angiosperm after fertilization, with or without additional structures formed from other parts of the flower.

fruit body (*Bot.*). A well-defined group of fungal spores and the hyphae which bear and surround them.

frustration (*An. Behav.*). The procedure of preventing an animal from making a response. May elicit displacement activities, hypotheses.

frustule (*Bot.*). The cell of a diatom, consisting of 2 silicified valves fitting one into the other, like a box and its lid, and the living contents.

frustum of a cone (*Maths.*). A cone truncated by a plane parallel to its base.

frutescent (*Bot.*). Shrubby.

fruticose (*Bot.*). (1) Bushy. (2) Said of a lichen thallus which is attached by its base, and stands out from the substratum, branching, and having a bushy appearance.

fry or frying (*Acous.*). (1) Unwanted random noise in the background on a recording, always present but can be minimized. (2) Noise reproduced when excessive current is passed through the carbon granules in a telephone transmitter.

frying arc (*Elec. Eng.*). See **hissing arc.**

F.S. (*Civ. Eng.*). Abbrev. for *factor of safety.*

F.S.D. Abbrev. for *full-scale deflection.* (*Radiol.*) Abbrev. for *focus-skin distance.*

FSH (*Physiol.*). Abbrev. for *follicle-stimulating hormone.*

f state (*Electronics*). That of an orbital electron

when the orbit has angular momentum of three *Dirac units*.
Fucales (*Bot.*). An order of the *Cyclosporae* with which it is coextensive.
fuchsine (*Chem.*). Magenta, the hydrochloride of rosaniline, a basic triphenylmethane dyestuff, dark green crystals, dissolving in water to form a purple-red solution. Used as a disinfectant, especially in certain skin troubles.
fuchsite (*Min.*). A green variety of muscovite (white mica) in which chromium replaces some of the aluminium.
fucivorous (*Zool.*). Seaweed-eating.
fucosterol (*Chem.*). $C_{29}H_{48}O$; the main sterol found in seaweed. Crystalline; m.p. 124°C.
fucoxanthin (*Chem.*). $C_{40}H_{56}O_6$ or $C_{40}H_{60}O_6$; the main carotenoid found in brown algae. Brown crystalline solid; m.p. 168°C.
fudge (*Typog.*). A space reserved in a newspaper for late news. Also known as **stop press.**
fuel (*Nuc. Eng.*). Fissile material inserted in or passed through a reactor; the source of the chain reaction of neutrons, and so of the energy released.
fuel accumulator (*Aero.*). A reservoir which augments the fuel supply when the critical fuel pressure is reached during the starting cycle of a gas turbine.
fuel cell (*Chem.*). A galvanic cell in which the oxidation of a fuel (e.g., methanol) is utilized to produce electricity.
fuel-cooled oil cooler (*Aero.*). A compact oil cooler for high-performance gas turbines in which heat is transferred to fuel passing in the counter bores of the device, instead of to air.
fuel cut-off (*Aero.*). A device which shuts off the fuel supply of a piston aero-engine; also **slow-running cut-out.**
fuel element (*Nuc. Eng.*). The smallest individual unit containing fuel material for a reactor.
fuel grade (*Aero.*). The quality of piston aero-engine fuel as expressed by its *knock rating*.
fuel injection (*I.C. Engs.*). A method of operating a spark-ignition engine by injecting liquid fuel directly into the induction pipe or cylinder during the suction-stroke, thus dispensing with a carburettor; in aero engines it avoids carburettor freezing troubles.
fuel injectors (*I.C. Engs.*). See **injectors.**
fuel jettison (*Aero.*). Apparatus for the rapid emergency discharge of fuel in flight.
fuel manifold (*Aero.*). The main pipe, or gallery, with a series of branch pipes, which distributes fuel to the burners of a gas turbine.
fuel oils (*Chem.*). Oils obtained as residues in the distillation of petroleum; used, either alone or mixed with other oils, for domestic heating and for furnace firing (particularly marine furnaces); also as fuel for I.C. engines.
fuel pump (*I.C. Engs.*). Unit of a diesel engine, or petrol-injection system, which injects measured quantities of fuel to each cylinder in turn on the compression stroke. See **common-rail injection, fuel injection, jerk-pump.**
fuel rating (*Nuc. Eng.*). The ratio of total energy released to initial weight of heavy atoms (U, Th, Pu) for reactor fuel. Usually expressed in megawatts per tonne. U.S. term **specific power**
fuel rod (*Nuc. Eng.*). A body of nuclear fuel in rod form for use in a reactor. Short rods are termed **slugs.**
fuel tanks (*Aero.*). These may be of many forms, for which the names vary. The *main tanks* are normally all those carried permanently and are usually composed either of flexible, self-sealing bags, or cells in wing or fuselage, or are integral with the wing structure; *auxiliary tanks* can be mounted additionally to increase range. See **drop tank.**
fuel trimmer (*Aero.*). A variable-datum device for resetting in flight the automatic fuel regulation, by *barostat*, of a gas turbine to meet changes in ambient temperature.
fugacity (*Chem.*). The tendency of a gas to expand or escape; substituted for pressure in the thermodynamic equations of a real gas. Analogous to **activity.** See **ideal gas.**
fugitive (*Geol.*). Decriptive of the dissolved volatile constituents of magma, which are commonly lost by evaporation when the magma is erupted as lava, and which are partly responsible for metasomatic alteration when the magma is intruded below the surface of the Earth.
fugitometer (*Chem.*). An apparatus for testing the fastness of dyed materials to light.
fugue (*Psychiat.*). 'Flight'. A condition, seen in hysteria and also in organic mental disorder, in which a hypnoidal state develops and the individual attempts to resolve his unconscious conflict by an escape from reality; he may wander about for days without clear knowledge of his actions and behaviour.
fulcrum (*Bot.*). An outgrowth from the wall of the zygospore in some moulds. (*Mech.*) The point of support, or pivot, of a lever. (*Vet.*) An instrument used for obtaining leverage on forceps during the extraction of a horse's teeth. (*Zool.*) In the mouth-parts of *Rotifera*, a segment of the incus; in some *Vorticella*, a small spine; in some *Diptera*, a complex cuticular framework within the rostrum; in some *Palaeopterygii*, a spinelike scale borne in a row on the anterior fin-rays of both median and paired fins.
fulguration (*Med.*). See **electrodesiccation.**
fulgurites or **lightning tubes** (*Min.*). Tubular bodies produced by lightning in loose unconsolidated sand; caused by the vitrification of the sand grains forming silica glass. Although of very narrow cross-section, some specimens have been found to exceed 6 m in length.
fuliginous (*Bot.*). Soot-coloured.
full (*Eng.*). A term signifying slightly larger than the specified dimension. Cf. *bare*.
full-adder (*Comp.*). A functional unit with provision for input of addend, augend and carry bits, and providing sum and carry outputs.
full annealing (*Met.*). Of steel, heating above the critical range, followed by slow cooling, as distinguished from (*a*) annealing below the critical range, and (*b*) normalizing, which involves air-cooling.
full bound (or **whole bound**) (*Bind.*). Said of a volume the sides and back of which are covered with leather or cloth.
full-centre arch (*Arch.*). A semicircular arch or vault.
full compatibility (*TV*). Reception by a colour television receiver of a colour TV signal when the chromaticity information is within the black-and-white video band of frequencies.
Fullerboard (*Elec. Eng.*). An early variety of *pressboard* (q.v.).
Fuller cell (*Elec. Eng.*). A double fluid variety of the bichromate cell.
Fuller faucet (*Plumb.*). A faucet in which a rubber ball is used to close the opening and so to stop the flow.
fullering (*Eng.*). The operation of (*a*) caulking a riveted joint to make it pressure-tight, (*b*) grooving forged work by a *fullering tool* (q.v.).
fullering tool (*Eng.*). A tool for assisting in producing circumferential grooves on circular

work; it consists of a split block internally radiated, which is placed round the work and hammered.

Fullerphone (*Teleg.*). System of telegraphy which uses buzzer signals for keying and listening, and direct currents for actual transmission, thus making the system less easily tappable.

fullers' earth (*Geol.*). A nonplastic clay consisting essentially of the mineral montmorillonite, and similar in this respect to bentonite. Used originally in 'fulling', i.e., absorbing fats from wool, hence the name. The Fullers' Earth of English stratigraphy is a small division of the Jurassic System in the S. Cotswolds, lying immediately above the Forest Marble. A valuable deposit of fullers' earth occurs in the Lower Greensand at Nutfield in Surrey.

full feel (*Textiles*). Cloth term indicating that it is soft and pleasant to handle, usually from being well finished on both sides.

full-force feed (*Eng.*). An engine lubrication system in which oil is forced to main bearings, connecting-rod big-end bearings, and thence, by drilled holes or attached pipes, to the gudgeon pins and cylinder walls.

full gear (*Eng.*). Of a steam-engine valve gear, the position giving maximum valve travel and cut-off for full power.

fulling (*Textiles*). See **milling.**

full-load (*Elec. Eng.*). The normal rated output of an electric machine or transformer. (*Eng.*) The normal maximum load under which an engine or machine is designed to operate.

full moon (*Astron.*). The instant when the geocentric longitudes of the sun and moon differ by 180°; the moon is then opposite the sun, and therefore fully illuminated, appearing as a bright circular disk.

full out (*Typog.*). An instruction to set the matter with no indention.

full-pitch winding (*Elec. Eng.*). An armature winding in which the span of the coils is equal to a pole pitch.

full-plate watch (*Horol.*). A watch in which the top plate is circular and the balance is mounted above the plate.

full radiator (*Phys.*). See **black body.**

full-satellite exchange (*Teleph.*). A small automatic telephone exchange which is entirely dependent for completion of calls on its parent or main exchange.

full shroud (*Eng.*). A gear-wheel in which the shrouding extends up to the tips of the teeth.

full thread (*Eng.*). A screw-thread cut to the theoretically correct depth.

full-wave rectification (*Elec. Eng.*). That in which current flows, during both half-cycles of the alternating voltage, through similar rectifying devices alternately, e.g., in a double diode or bridge rectifier.

fulminates (*Chem.*). $Hg(CNO)_2$. Compounds which explode under slight shock or on heating. Used in detonators, e.g., mercury fulminate.

fulminating gold (*Chem.*). A yellow precipitate formed when a solution of gold (III) chloride is treated with ammonia; explosive.

fumaginous (*Bot.*). Of a smoky colour.

fumaric acid (*Chem.*). (ethene 1,2-dicarboxylic acid) HOOC·CH=CH·OOH, small prisms which do not melt, but sublime at about 200°C, with the formation of maleic anhydride. Fumaric acid being the *trans*-form. Used in ployester resins.

fumarin (*Chem.*). 3 - (α - Acetonylfurfuryl) - 4 - hydroxycoumarin, used as a rodenticide. Also **coumafuryl.**

fumaroles (*Geol.*). Small vents on the flanks of a volcanic cone, or in the crater itself, from which gaseous products emanate.

fume (*Chem., Powder Tech.*). Cloud of airborne particles, generally visible, of low volatility and less than a micrometre in size, arising from condensation of vapours or from chemical reaction.

fume cupboard (*Chem.*). A glass chamber or cupboard where laboratory operations involving obnoxious fumes are carried out under forced ventilation.

fumigants (*Chem.*). Substances which, when volatilized, are capable of destroying vermin, insects, bacteria, moulds, or which act as disinfectants. Examples are hydrogen cyanide and ethylene oxide for vermin and insects, and formaldehyde (disinfectant).

fuming liquids (*Chem.*). Liquids which give off vapours which unite with water to form a mixture or compound with a lower vapour pressure than water.

fuming (or Nordhausen) sulphuric acid (*Chem.*). A solution of sulphur (VI) oxide in concentrated sulphuric acid.

Funariales (*Bot.*). An order of the *Eubrya*; the gametophyte is annual or biennial with a rosette of leaves at the top; the pendant capsule is never cylindrical; the peristome is double or absent.

function (*Biol.*). The normal vital activity of a cell, tissue, or organ. (*Maths.*) A *function*, f, (or *mapping*, f,) from a set X to a set Y, denoted by $f: X \rightarrow Y$, associates with each element x of X a single element y of Y, denoted by $f(x)$.

functional (*Biol.*). Carrying out normal activities; active (as opposed to *passive*).

functional disease (*Bot.*). A plant disease due to some disturbance in the organization or working of the plant, and not to parasitic attack. (*Med.*) Disease characterized by impaired function as opposed to one characterized by impaired structure (*organic disease*).

functional invariants (*Psychol.*). Piaget's term for the means by which an individual develops, development always occurring by the same type of means no matter what the stage of development. See **accommodation, adaptation, assimilation.**

function chamber (*Build., San. Eng.*). A closed chamber, generally of brick and concrete, inserted in a sewer system for accepting the inflow of one or more sewers and allowing for the discharge thereof.

function generator (*Comp.*). Element in an analogue computer capable of generating voltage wave approximately following any desired, single-valued, continuous algebraic function of one variable. (*Telecomm.*) Signal generator with range of alternative non-sinusoidal output waveforms.

function switch (*Comp.*). A network with a number of inputs and outputs connected in a computer so that the output signals give the input information in a different code from that of the input.

fundamental colours. See **primary colours.**

fundamental complex (*Geol.*). A name (really a misnomer) applied to the highly crystalline Pre-Cambrian rocks of N.W. Scotland termed the Lewisian (or Hebridean); it comprises schists and gneisses of several kinds. As some of these were derived from pre-existing rocks, they are not *fundamental.*

fundamental component (*Telecomm.*). The harmonic component of an alternating wave which has the lowest frequency and which usually represents the major portion of the wave.

fundamental crystal (*Electronics*). A crystal which

is designed to vibrate at the lowest order of a given mode.

fundamental dynamical units (*Phys.*). The basic equations of dynamics are designed so as to be correct for any system of fundamental units, the main four of which are tabulated below. They are such that unit force acting on unit mass produces unit acceleration, unit force moved through unit distance does unit work, and unit work in unit time is unit power.

	System				
	ft. lb. sec.	Gravitational	CGS	SI (MKSA)	Dimensions
length	foot (ft.)	foot	centimetre (cm)	metre (m)	L
mass	pound (lb.)	slug	gram (g)	kilogram (kg)	M
time	second (s)	second	second	second	T
velocity	ft./s	ft./s	cm/s	m/s	LT^{-1}
acceleration	ft./s²	ft./s²	cm/s²	m/s²	LT^{-2}
force	poundal (pdl.)	pound force (lbf.)	dyne	newton (N)	MLT^{-2}
work	ft. pdl.	ft. lbf.	erg	joule (J)	ML^2T^{-2}
power	ft. pdl./s	ft. lbf./s	erg/s	watt (W)	ML^2T^{-3}

Notes: (1) There is no name for the unit of power except in the SI system. However, it is customary to specify power in the ft. lb. sec. system and the gravitational system in horsepower (550 ft. lbf./second) and in the CGS system by the watt (10^7 erg/second).

2) The unit of force (*lbf.*) in the Gravitational system is also known as the pound weight (*lbwt.*).

fundamental frequency (*Telecomm.*). In a steady periodic oscillation, a frequency which divides into all components in the waveform. See **harmonic.**

fundamental frequency of antenna (*Telecomm.*). Lowest frequency at which antenna is resonant, when not loaded with terminal inductance.

fundamental interval (*Heat*). The number of degrees between the two fixed points on a thermometer scale.

fundamental metric tensor (*Maths.*). See **metric (1).**

fundamental mode (*Telecomm.*). Mode of oscillation of antenna at its fundamental frequency. An earthed antenna is characterized by a single node of current at extreme end of antenna.

fundamental particle (*Nuc.*). See **elementary particle.**

fundamental series (*Light*). Series of optical spectrum lines observed in the spectra of alkali metals. Energy levels for which the orbital quantum number is three are designated *f-levels*.

fundamental units (*Phys.*). See **fundamental dynamical units.**

fundamental wavelength (*Telecomm.*). (1) Wavelength in free space corresponding to the fundamental frequency of an antenna. (2) The main or operating wavelength in a transmitter radiating harmonics.

fundatrigenia (*Zool.*). Among aphids, an apterous, parthenogenetic, viviparous female, which is the progeny of a fundatrix and lives on the primary host. *pl.* **fundatrigeniae.**

fundatrix (*Zool.*). A stage in the life-cycle of some *Hemiptera*, similar to a *fundatrigenia* but differing in some morphological details.

fundic glands (*Histol.*). Glands occurring in the gastric mucosae, which secrete the gastric juice, a combination of pepsin, hydrochloric acid and mucus.

fundus (*Anat.*). See **eye-ground.** (*Zool.*) The proximal swollen bulb of the palpal organ of a male Spider.

fungal cellulose, fungus cellulose (*Bot.*). A carbohydrate, akin to cellulose, present in the walls of some fungi.

Fungi (*Bot.*). One of the main groups of the *Thallophyta*, distinguished from the algae chiefly by the absence of chlorophyll; they cannot carry on photosynthesis, and live as saprophytes or parasites. The *Fungi* probably include at least 100 000 species showing great diversity in their morphology.

Fungi Imperfecti (*Bot.*). A large assemblage of fungi having a septate mycelium, but, so far as is known, reproducing only by asexual spores. All seem devoid of sexuality or of anything resembling that.

fungicide (*Bot.*, *etc.*). A substance which kills fungi.

fungiform papillae (*Histol.*). Rounded papillae, scattered irregularly on the upper surface of the tongue, covered by a thin stratified epithelium, with some taste buds.

funicle (*Bot.*). The small stalk which unites the ovule to the placenta. (*Zool.*) In some *Hymenoptera*, those joints of the antenna which intervene between the club and the ring-joints, or between the club and the pedicel.

funicular railway (*Civ. Eng.*). A form of *cable railway* (q.v.).

funiculitis (*Med.*). Inflammation of the vas deferens of the testis.

funiculose (*Bot.*). Forming ropes of intertwined hyphae.

funiculus (*Zool.*). In some Invertebrates (as *Ectoprocta*), thickened strands of mesoderm attaching the digestive organs to the body-wall; more generally, any small cord, as a tract of nerve fibres in the central nervous system. *adj.* **funicular.**

funiform (*Bot.*). Ropelike.

funnel (*Zool.*). A modified part of the foot in *Cephalopoda*, protruding from the mantle cavity and acting as an exhalent channel. In *Chætopoda*, a nephrostome.

funnel cell (*Bot.*). A cell in the palisade layer of a leaf, which is widest just beneath the epidermis, and narrows off below.

funnelling (*Meteor.*). The strengthening of a wind blowing along a valley, especially when the valley narrows.

fuor (*Carp.*). A strengthening piece nailed to a decayed rafter.

fur (*Zool.*). In Mammals, the thick undercoat of short, soft, silky hairs.

fural (*Chem.*). See **furfural.**

furan group (*Chem.*). A group of heterocyclic

compounds derived from furan, C_4H_4O, a compound containing a ring of 4 carbon atoms and 1 oxygen atom.

furane resins (*Chem.*). A group of plastics derived from the partial polymerization of furfuryl alcohol, or from condensation of furfuryl alcohol with either furfural or methanal, or of furfural with ketones, and used widely as baked plastic coatings on metal, as adhesives and as resin binders for stoneware.

furca (*Zool.*). Any forked structure; in Insects, a median thoracic apodeme having two arms and a single base; in *Diptera*, a short rod with two arms, which forms the principal skeleton of the oral lobes of the proboscis; in Vertebrates, a divergence of nerve fibres; in *Crustacea*, a pair of divergent processes at the end of the abdomen.

furcula (*Zool.*). In *Collembola*, the leaping apparatus, consisting of a pair of partially fused appendages arising from the fourth abdominal somite; in Birds, the partially fused clavicles; more generally, any forked structure.

furfur (*Med.*). A scale of epidermis; dandruff.

furfuraceous (*Bot.*). Covered with branlike particles; scurfy.

furfural (*Chem.*). Fural or furfuraldehyde, $C_4H_3O{\cdot}CHO$, a colourless liquid, b.p. 162°C, obtained by distilling pentoses with diluted hydrochloric acid. Used as a solvent, particularly for the selective extraction of crude rosin, also as raw material for synthetic resins. Used in petroleum refining for the selective extraction of impurities.

furlong. A distance of 10 Gunter's chains, i.e., 220 yd or one-eighth of a mile.

furnace atmosphere (*Heat*). Three main classes: (*a*) *oxidizing*, produced when air volumes are in excess of fuel requirements; (*b*) *neutral*, when air to fuel ratios are perfectly proportioned; (*c*) *reducing*, due to deficiency of combustion air. See also **protective-**.

furnace brazing (*Met.*). A high-production method of copper-brazing steel, without flux, in a reducing atmosphere, or of brazing steels, copper and copper alloys with brasses or silver-brazing alloys, in continuous or in batch furnaces.

furnace clinker (*Build.*). The final residue from the combustion of coke or coal which has been burnt and reburnt in order to consume the maximum of combustible matter in it. It is useful as an aggregate in the manufacture of concrete.

furnace linings (*Met.*). The interior portions of metallurgical furnaces which are in contact with hot gases and the charge, and must therefore be constructed of materials resistant to heat, abrasion, chemical action, etc. See **refractories.**

furnish (*Paper*). The ingredients from which paper is manufactured.

furniture (*Build.*). A general name for all metal fittings for doors, windows, etc. (*Typog.*) Lengths of wood, plastic or metal, less than type height, used in a forme for making margins, etc. They are made to standard point widths and lengths.

furred (*Plumb.*). A term applied to pipes and boilers in which *furring* (q.v.) has developed.

furring (*Build.*). A wood-strip and plasterwork lining to a wall, which leaves an air space between the plastering and the brickwork. (*Carp.*) See **firring.** (*Plumb.*) The hard lime deposit formed on the inner surface of pipes and boilers in which hard water is heated.

furrowed (*Build.*). A term applied to margin-drafted ashlars having parallel vertical grooves cut in the face.

furrowing (*Bot.*). The formation of a septum by the development of a ring of thickening on the inside of the cell wall, and the gradual closing of this ring by the walls growing in until the cavity of the cell is cut in two.

further outlook (*Meteor.*). A general forecast given for a period additional to that covered by the more detailed forecast.

furuncle (*Med.*). See **boil.**

furunculosis (*Med.*). The condition of having several boils.

fusain (*Min.*). Mineral charcoal, the soiling constituent of coal, occurring chiefly as patches or wedges. It consists of plant remains from which the volatiles have been eliminated.

fuse (*Elec. Eng.*). A device used for protecting electrical apparatus against the effect of excess current; it consists of a piece of fusible metal, which is connected in the circuit to be protected, and which melts and interrupts the circuit when an excess current flows. The term fuse also includes the necessary mounting and cover (if any). (*Mining, etc.*) A thin waterproof canvas length of tube containing gunpowder arranged to burn at a given speed for setting off charges of explosive.

fuseau (*Bot.*). The spindle-shaped macroconidium of dermatophytes.

fuse-board (*Elec. Eng.*). See **distribution-.**

fuse box (*Elec. Eng.*). Term sometimes used for a distribution fuse-board enclosed in a box.

fuse-carrier (*Elec. Eng., etc.*). A carrier for holding a fuse-link; arranged to be easily inserted between fixed contacts, so that a replacement of the fuse-link can be quickly carried out. Also called a **fuse-holder.**

fused junction (*Electronics*). One formed by recrystallization of semiconductor and impurity from a liquid on to pure base crystal.

fused ring (*Chem.*). See **condensed nucleus.**

fused silica (*Glass*). See **vitreous silica.**

fusee (*Horol.*). A spirally grooved pulley of gradually increasing diameter, used to equalize the pull of the mainspring as it runs down by increasing the leverage on a chain or gut-line wound round the fusee and the mainspring barrel.

fuse-element, fuse-link, or **fuse** (*Elec. Eng., etc.*). The essential part of a fusible cut-out.

fuse-holder (*Elec. Eng.*). See **fuse-carrier.**

fuselage (*Aero.*). The name generally applied to the main structural body of a heavier-than-air craft, other than the hull of a flying-boat or amphibian.

fuse-link (*Elec. Eng., etc.*). See **fuse-element.**

fusel oil (*Chem.*). Mainly optically inactive 3-methylbutan-1-ol, $(CH_3)_2{\cdot}CH{\cdot}CH_2{\cdot}CH_2OH$, accompanied by active amyl alcohol, usually occurring in the products of alcoholic fermentation.

fuse rating (*Elec. Eng.*). The maximum current a fuse will carry continuously and/or (less frequently) the minimum current at which it can be relied upon to blow.

fuse-switch (*Elec. Eng.*). A *switch-fuse* (q.v.).

fuse tongs (*Elec. Eng.*). Tongs with insulating handles, used for withdrawing or replacing fuses on high-voltage circuits.

fusible alloys or metals (*Met.*). Alloys of bismuth, lead, and tin (and sometimes cadmium or mercury) which melt in the 248°C-47°C temperature range; used as solders and for safety devices in fire extinguishers and boilers, etc.

fusible cut-out (*Elec. Eng.*). See **fuse.**

fusible plug (*Eng.*). A plug containing a metal of

low melting-point used, for example, in the crown of a boiler fire box to prevent serious overheating of the plates if the water-level falls below them.

fusiform (*Bot.*, *Zool.*). Elongated and tapering towards each end; shaped like a spindle.

fusing factor (*Elec. Eng.*). The minimum current required to blow a fuse, expressed as a ratio to the rated current.

fusing point (*Met.*). See **melting point.**

fusion (*Heat*). The conversion of a solid into the liquid state; the reverse of *freezing*. Fusion of a substance takes place at a definite temperature, the melting-point, and is accompanied by the absorption of latent heat of fusion. (*Nuc.*) The process of forming new elements by the fusion of lighter ones; principally the formation of helium by the fusion of hydrogen and its isotopes. The process involves a loss of mass which appears as *fusion energy* (q.v.).

fusion bomb (*Nuc.*). Same as **hydrogen bomb.**

fusion cones (*Heat.*). See **Seger cones.**

fusion drilling (*Mining*). Method of hard-rock boring with a paraffin-oxygen jet which melts the rock, the slag being decrepitated and flushed out by a water spray.

fusion energy (*Nuc.*). Energy released by nuclear fusion, usually the formation of helium from lighter particles. See **carbon cycle, hydrogen bomb, thermonuclear energy.**

fusion welding (*Elec. Eng.*). A process of welding metals in which the weld is carried out solely by the melting of the metals to be joined, without any mechanical pressure.

fusoid (*Bot.*). Rounded in section, widest in the middle and tapering to each end, and not markedly elongated.

fusospirillosis (*Med.*). Vincent's angina. Infection of the throat with the fusiform bacillus and spirilla described by Vincent.

fusospirochaetosis, fusospirochetosis (*Med.*). Infection with fusiform bacilli and spirochaetes.

fust (*Arch.*). The shaft of a column. (*Build.*) An ancient term for a roof ridge.

fustian (*Textiles*). A term including a number of cotton fabrics differing widely in structure and appearance, but all heavily wefted; they are used for clothing and furnishings. See **corduroy, moleskin, swansdown, velveteen.**

fuzz (*Acous.*). Extraneous noise introduced in recording, made evident on reproduction.

g (*Aero.*). See **load factor.** (*Chem.*) Abbrev. for *gram(me).* (*Psychol.*) Symbol, first employed by Spearman, to represent general intelligence.

g (*Chem.*). Symbol for osmotic coefficient. (*Phys.*) Symbol for *acceleration due to gravity* (q.v.).

γ (*Chem.*). Symbol for: (1) substituted on the carbon atom of a chain next but two to the functional group; (2) substituted on one of the central carbon atoms of an anthracene nucleus; (3) substituted on the carbon atom next but two to the hetero-atom in a heterocyclic compound; (4) a stereoisomer of a sugar. (*Elec.*) Symbol for electrical conductivity. (*Maths.*) See **Euler's constant.** (*Phys.*) Symbol for: (1) ratio of specific heats of a gas; (2) surface tension; (3) propagation coefficient; (4) Gruneisen constant; (5) molar activity coefficient; (6) coefficient of cubic thermal expansion. It is also the greatest refractive index in a biaxial crystal.

Γ (*Maths.*). Symbol for *gamma function* (q.v.).

Γ (*Phys.*). Symbol for *surface concentration excess* (q.v.).

G. Symbol for *giga*, i.e., 10^9. (*San. Eng.*) Abbrev. for *gulley.* (*Nuc.*) See **G-value** (p. 540).

G (*Chem.*). Symbol for: (1) thermodynamic potential: (2) Gibbs function; (3) free energy. (*Elec.*) Symbol for conductance. (*Phys.*) Symbol for (1) the constant of *gravitation* (q.v.); (2) shear modulus, rigidity.

[G] (*Light*). A pair of Fraunhofer lines in the deep blue of the solar spectrum. One, of wavelength 430·8081 nm, is due to iron; the other, of wavelength 430·7907 nm, is due to calcium.

G-11 (*Chem.*). See **hexachlorophene.**

G. 80 (*Build., Civ. Eng.*). A method of system building developed in U.K. Consists of a composition of large precast reinforced concrete wall panels, with precast coffered floor units spanning laterally between them. Composite beams and columns are also incorporated and all components are in multiples of a 4-in. modulus. See **system building.**

Ga (*Chem.*). The symbol for *gallium.*

gab (*Build.*). A pointed tool for working hard stone.

gabardine (*Textiles*). A twill fabric made from worsted warp and cotton weft, the former only appearing on the surface; used for dress and suiting cloths and light showerproof overcoatings. All-cotton gabardine is also used for similar purposes but is not as durable or warm.

gabbart scaffold (*Build.*). Scaffolding in which sawn timbers are used instead of round poles.

gabbro (*Geol.*). The name of a rock clan, and also of a specific igneous rock type. The rock gabbro is a coarse-grained plutonite, consisting essentially of plagioclase, near labradorite in composition, and clinopyroxene, with or without olivine in addition. The gabbro clan includes also norite, eucrite, troctolite, kentallenite, etc.

gabers scaffold (*Build.*). A *gabbart scaffold* (q.v.).

gabion (*Civ. Eng.*). A long wicker or wire basket, containing earth or stones, deposited with others to serve the same purposes as *fascines.*

gable (*Build.*). The triangular part of an external wall at the end of a ridged roof.

gable board (*Build.*). A *barge board* (q.v.).

gable moulding (*Arch.*). A moulding used to decorate a gable.

gable roof (*Arch.*). A ridge roof terminating in a gable-end.

gable shoulder (*Build.*). The projecting masonry or brickwork supporting the foot of a gable.

gable springer (*Build.*). The concrete, brick, or tile corbel supporting the gable shoulder.

gablet (*Arch.*). A small decorated gable over a niche or other opening.

gable tiles (*Build.*). Purpose-made arris tiles to cover the intersection between gable and roof.

gable window (*Arch.*). A window built in the gable of a house.

gablock (*Tools*). See **gavelock.**

gaboon (*For.*). Mahogany-like wood from a tree of the genus *Aucoumea,* found in parts of central and western Africa. Planed surfaces of the wood are silky. Used for veneer and plywood manufacture, furniture and fittings.

Gabriel reaction (*Chem.*). A reaction used in organic syntheses for the preparation of pure primary amines. Based on the use of potassium phthalimide (benzene 1,2-dicarboximide) and halogenoalkanes.

G-acid (*Chem.*). 2-Naphthol-6,8-disulphonic acid; an intermediate for dyestuffs.

gad (*Mining*). A short pointed chisel used for breaking or loosening rock; a steel wedge for breaking coal.

gadding (*Vet.*). The excited behaviour of cattle when irritated by gad-flies.

gad-fly (*Vet.*). A fly of the genus *Hypoderma,* the larva of which parasitizes cattle and is known as a *warble* (q.v.).

gadget (*Glass*). A tool for holding the stem of a piece of ware which is in course of treatment.

gadolinite (*Min.*). A rare accessory mineral occurring in pegmatites as greenish- or brownish-black crystals of composition represented by $Be_2FeY_2Si_2O_{10}$, i.e., silicate of beryllium, iron, and yttrium, often with cerium.

gadolinium (*Chem.*). A rare metallic element; trivalent; a member of the rare earth group. Symbol Gd, at. no. 64, r.a.m. 157·25. Only known in combination; obtained from the same sources as europium.

Gaede diffusion pump (*Chem. Eng.*). Pump using mercury vapour, which entrains molecules of gas from a low pressure established by a backing pump. Oil of low vapour-pressure (apiezon) is a modern alternative.

Gaede molecular pump (*Chem. Eng.*). Rotary pump which ejects molecules of gas by imparting a drift velocity to their random motion.

gag (*Mining*). An obstruction in the clack valve, sump or bucket of a pumping set.

gagger (*Foundry*). See **lifter.**

gahnite (*Min.*). A mineral belonging to the spinel group; occurs as grey octahedral cubic crystals. Also known as **zinc-spinel** (see **spinel**), the composition being zinc aluminate, $ZnAl_2O_4$.

gain (*Carp.*). A notch or mortise cut in a timber to support the end of a beam. (*Mining*) Cross cut in coal seam. (*Telecomm.*) (1) In electric systems, generally provided by insertion of an amplifier into a transmission circuit, or by matching impedances by a loss-free transformer. Measured in decibels or nepers, and defined as the increase in power level in the load, i.e., the ratio of the actual power delivered to that which would be delivered if source were correctly matched, without loss, to the load, in the absence of the amplifier. (2) In a directional antenna, ratio (expressed in decibels) of voltage produced at the receiver terminals by a signal arriving from the direction of maximum sensitivity of the antenna, to that produced by the same signal in an omnidirectional reference antenna (generally a half-wave dipole). In a transmitting antenna, ratio of the field strength produced at a point along the line of maximum radiation by a given power radiated from antenna, to that produced at the same point by the same power from an omnidirectional antenna.

gain amplifier (*Telecomm.*). Thermionic amplifier following the photocell amplifier and preceding the power stages in soundfilm projection equipment.

gain bandwidth (*Telecomm.*). Product of maximum amplifier *gain* (between specified impedances) and the bandwidth possible (between *half-power points*).

gain control (*Telecomm.*). Means for varying the degree of amplification of an amplifier, often a simple potentiometer. See **automatic-**.

gaining (*Carp.*). The operation of cutting notches in timbers.

gait or gaiting (*Spinning*). Restarting and running in a mule or ringframe, etc., after it has been reset and lined-up. (*Weaving*) The operation of preparing a loom for weaving, i.e., placing the warp in position, as well as healds and reed.

gaize (*Build.*). A friable argillaceous sandstone which, under heat treatment, is converted into a pozzolana.

gal (*Geophys.*). See **milligal.**

galactagogue or **galactagog** (*Med.*). Promoting the secretion of milk (Greek, *gala*, gen. *galaktos*); any medicine which does this.

galactans (*Chem.*). The anhydrides of galactose. They comprise many gums, agar, and fruit pectins, and occur in algae, lichens, mosses.

galactic circle (*Astron.*). The great circle of the celestial sphere in which the latter is cut by the galactic plane: hence the primary circle to which the galactic coordinates are referred.

galactic clusters (*Astron.*). Loose clusters of stars within the Galaxy. See **open clusters.**

galactic concentration (*Astron.*). The tendency, first noted by Herschel, of the stars both bright and faint to crowd towards the galactic plane; hence a statistical criterion as to whether or not a given type of celestial object belongs to our stellar system.

galactic coordinates (*Astron.*). Two spherical coordinates referred to the galactic plane; the origin of galactic longitude is at R.A. $17^h42{\cdot}4^m$, dec. $-28°\ 55'$ (1950); galactic latitude is measured positively from the galactic plane towards the north galactic pole.

galactic halo (*Astron.*). An almost spherical aggregation of stars, gas and dust, which is concentric with the Galaxy. It contains stars of Population **II**, and is responsible for much of the background of radio emission from the sky. Similar halos surround other galaxies.

galactic noise (*Radio*). That arriving from outer space, similar to electronic circuit noise, but apparently arising from sources in galaxies.

galactic plane (*Astron.*). The plane passing as nearly as possible through the centre of the belt known as the Milky Way or Galaxy; hence the fundamental plane in sidereal astronomy.

galactic rotation (*Astron.*). The phenomenon of *star-streaming* (q.v.) is caused by the rotation of the Galaxy; at the distance of the sun from the centre (8000 parsecs) the period of rotation is about 200 million years.

galactobolic (*Zool.*). Refers to the action of neurohypophysial peptides which contract mammary myoepithelium and so eject milk.

galactocele (*Med.*). A cystic swelling in the breast, due to retention of milk as the result of a blockage of a milk duct.

galactophoritis (*Med.*). Inflammation of the milk ducts.

galactophorous (*Zool.*). See **lactiferous.**

galactopoiesis (*Med.*). Increase in milk secretion.

galactorrhoea or **galactorrhea** (*Med.*). Excessive secretion of milk by the breast, causing it to overflow through the nipple.

galactosaemia (*Med.*). The presence of galactose in the blood.

galactose (*Chem.*). A hexose of the formula $CH_2OH{\cdot}(CHOH)_4{\cdot}CHO$; thin needles; m.p. 166°C; dextro-rotatory. It is formed together with D-glucose by the hydrolysis of milk-sugar with dilute acids. Stereoisomeric with glucose, which it strongly resembles in properties. Present in certain gums and seaweeds as a polysaccharide *galactan*.

galactosides (*Chem.*). See **cerebrosides.**

galactosis (*Zool.*). See **lactation.**

galactotrophic, galactotropic (*Physiol.*). Stimulating the secretion of milk by the mammary gland.

galatea (*Textiles*). Medium-weight, 2×1 warp twill cotton fabric usually with colour stripe fast to washing; used for shirtings, nurses' uniforms, etc.

galaxite (*Min.*). A rare manganese aluminium spinel, formula $MnAl_2O_4$.

Galaxy (*Astron.*). (1) The name given to the belt of faint stars which encircles the heavens and which is known as the Milky Way. (2) The name is also used for the entire system of dust, gases and stars within which the sun moves; now known to have the typical spiral structure. Still more generally the term signifies any *extra-galactic nebula* (q.v.), each being a vast collection of stars, dust and gas. Probably about a thousand million galaxies are observable; they show a marked tendency to form clusters. (3) Automatic star-plate analyser for measuring brightness and position of high density photographic images of portions of the Galaxy. (*G*eneral *a*utomatic *l*uminosity *a*nd *XY* measuring engine.)

galbulus (*Bot.*). A strobilus with fleshy cone scales, as in juniper.

gale (*Meteor.*). A wind having a velocity of 32 mi/h (51 km/h) or more, at a height of 33 ft (10 m) above the ground. On the Beaufort scale, a gale is a wind of force 8.

galea (*Zool.*). (1) A movable spinneret borne by the chelicerae of certain *Arachnida*. (2) Any helmet-shaped structure; in Insects, the outer distal lobe of the maxilla.

galeate or **galeiform** (*Bot., Zool.*). Shaped like a helmet or a hood.

galena (*Min.*). Lead (II) sulphide, PbS; commonest ore of lead, occurring as grey cubic crystals, often associated with zinc blende, in mineralized veins. Silver (I) sulphide, argentite, that is isomorphous with galena, may be present to the extent of about 1 g/tonne, but may rise to 1%. Galena behaves like a semiconductor and has been used as a crystal rectifier for radio signals. Also **lead glance.** See also **silver lead ore.**

galericulate (*Bot.*). Covered by a caplike lid.

galeriform (*Bot.*). That which is cap-shaped.

galet (*Build.*). See **spall**.

gale warning (*Meteor.*). A notice of the probability of gales issued by the Meteorological Office to certain ports and fishing stations, which then hoist a black cone, apex up if the gale is expected from the north, apex down if from the south.

Galilean binoculars (*Optics*). Those in which the objectives are the usual doublet telescope objectives and the eyepieces are negative lenses.

Galilean finder (*Photog.*). Viewfinder comprising a concave objective viewed through a small convergent lens.

gall (*Bot.*). An abnormal growth formed on a plant following attack by a parasite. (*Glass*) A layer of molten sulphate floating upon the glass in a furnace. Also **salts, salt water**. (*Physiol.*) See **bile**. (*Vet.*) An injury of the skin of animals due to the pressure of harness.

gallamine (*Med.*). A muscle relaxant used extensively in anaesthesia. Also called **Flaxedil**.

gall bladder (*Zool.*) In Vertebrates, a lateral diverticulum of the bile-duct in which the bile is stored.

gallery (*Arch.*). An elevated floor projecting beyond the walls of a building and supported on pillars, brackets, or otherwise, so as to command a view upon the main floor, as at a theatre, etc. Sometimes cantilevered to eliminate obstruction of view by pillars. (*Elec. Eng.*) A device for attaching to a lampholder to provide a support for a glass shade or reflector which is too large or heavy to be supported by the shade-carrier ring. (*Mining*) A tunnel or passage in a mine.

gallery furnace (*Met.*). Type used in distillation of mercury from its ores.

gallet (*Build.*). A splinter of stone.

galleting (*Build.*). See **garreting**.

galley (*Typog.*). A steel tray open at one end, in which type matter is held after setting. Corrections and deletions are more easily made to type in galley form than in page form, and are generally marked on the galley proof or slip, itself commonly referred to as a *galley*.

galley press (*Typog.*). A proofing press which allows for the thickness of the galley.

galley proof (*Typog.*). A proof taken while the type is on a galley and used for checking and pasting-up.

galley rack (*Typog.*). A rack for storing galleys of type.

gall flower (*Bot.*). An imperfectly developed flower of the fig tree, in which the eggs of the gall wasp are laid.

gallic acid (*Chem.*). $C_6H_2(OH)_3COOH$, 3,4,5-trihydroxybenzoic acid; crystallizes with 1 H_2O; thin needles; decomposes at about 200°C into CO_2 and pyrogallol (3,4,5-trihydroxybenzene). Occurs in nut-galls, tea, divi-divi, and other plants; it is obtained from tannin by hydrolysis.

gallicola (*Zool.*). A stage in the life cycle of some *Hemiptera* produced parthenogenetically by fundatrices: it produces a gall on the primary host plant.

Galliformes (*Zool.*). An order of *Neognathae*, possessing a schizognathous palate, a simple rhamphotheca, and 10 carpal remiges; the feather tracts are well defined and the feet well adapted for running. Game birds (which seek their food, berries, seeds, buds, and insects, on the ground), Brush-turkeys, Curassows, Turkeys, Pheasants, Partridge, Grouse, and Quail.

gallium (*Chem.*). A metallic element in the third group of the periodic system. Symbol Ga at. no. 31, r.a.m. 69·72, rel. d. 5·9, oxidation state + 3, m.p. 30·15°C, b.p. 2000°C. Used in fusible alloys and high temperature thermometry. In the form of gallium arsenide it is an an important semiconductor.

gallium arsenide (*Chem.*). Semiconductor in which gallium and arsenic are combined in near *stoichiometric* proportions.

gallon. Liquid measure. One imperial gallon is the volume occupied by 10 lb avoirdupois of water.

1 Imp. gallon = 4·54609 litres = 6/5 U.S. gallon.

1 U.S. gallon = 3·78543 litres = 5/6 Imp. gallon.

galloon (*Arch.*). Decorated work for a band or moulding, to which is applied a row of small round balls. (*Textiles*) Narrow tape woven from fine cotton yarns, or from a silk, rayon or nylon warp and cotton weft; used as bindings on hats or garments.

galloping (*I.C. Engs.*). An American term descriptive of the irregular running of a petrol-engine supplied with an over-rich mixture.

gallotannins (*Chem., Leather*). Tannins of the type found in oak galls, which do not deposit bloom on the leather.

Galloway boiler (*Eng.*). A cylindrical boiler of the Lancashire type, in which the two furnace tubes unite, at a short distance from the grates, into a single arched oval flue, crossed by water tubes.

Galloway tubes (*Eng.*). The inclined water tubes which cross the flue of a Galloway boiler to assist circulation and increase the heating surface.

gallows timber (*Mining*). A timber framework or set for roof support.

gall-sickness (*Vet.*). See **anaplasmosis**.

Gall's projection (*Geog.*). A type of *cylindrical projection* (q.v.), in which the cylinder intersects the globe at the 45th parallels, giving two lines along which scale is true and reducing the areas of marked distortion.

gallstones (*Med.*). Biliary calculi. Pathological concretions in the gall bladder and bile passages. They have not a uniform composition but some constituents may be preponderant, e.g., cholesterol, or calcium carbonate and phosphate.

galon (*Textiles*). A narrow lace resembling insertion but having one or both edges scalloped.

Galton's Laws (*Gen.*). (1) Of ancestral inheritance: 'Any organism of bisexual parentage derives one-half of its inherited qualities from its parents (one-fourth from each parent), one-fourth from its grandparents, one-eighth from its great grandparents, and so on. These successive fractions, whose numerators are one and whose denominators are the successive powers of two, added together equal one, or the total inheritance of the organism, thus: $\frac{1}{2}+\frac{1}{4}+\frac{1}{8}+\ldots=1$.' (2) Of filial regression: 'The offspring of exceptional parents tend to regress toward mediocrity in proportion to the degree of parental exceptionalness' (Lull).

galvanic (*Elec.*). Obsolete for **electric**.

galvanic cell (or **battery**) (*Elec. Eng.*). Obsolete name for a primary cell or a battery of such cells; the name is derived from Galvani.

galvanic current (*Med.*). Steady unidirectional current for therapeutic use.

galvanism (*Elec.*). An obsolete term for the science of electric currents. (*Med.*) Curative treatment by unidirectional currents.

galvanized iron (*Build.*). Iron which has been subjected to *galvanizing* (q.v.); widely used, especially in corrugated form (see **corrugated iron**), for minor roofing purposes, e.g., on wooden buildings, etc. Also for nails, bolts, etc., where moisture may produce corrosion.

galvanizing (*Met.*). The coating of steel or iron with zinc, generally by immersion in a bath of zinc, covered with a flux, at a temperature of 425–500°C. The zinc may alternatively be electrodeposited from cold sulphate solutions. The zinc is capable of protecting the iron from atmospheric corrosion even when the coating is scratched, since the zinc is preferentially attacked by carbonic acid, forming a protective coating of basic zinc carbonates.

galvanocautery (*Surg.*). Cautery by a wire heated by the passage of a galvanic current.

galvanoluminescence (*Elec. Eng.*). Feeble light emitted from the anode in some electrolytic cells.

galvanomagnetic effect (*Elec. Eng.*). See **Hall effect.**

galvanometer (*Elec. Eng.*). Current-measuring device depending on forces on the sides of a current-carrying coil normal to magnetic fields in gaps. In a moving-magnet instrument, the suspended coil is replaced by an astatic magnet system which is magnetically shielded for very sensitive work.

galvanometer constant (*Elec. Eng.*). A number by which the scale reading of a galvanometer must be multiplied in order to give a reading of current in amperes or other suitable units.

galvanometer shunt (*Elec. Eng.*). A shunt connected in parallel with a galvanometer to reduce its sensitivity. See **Ayrton shunt, universal shunt.**

galvanoscope (*Elec. Eng.*). A term sometimes used to denote an instrument for detecting, but not measuring, an electric current.

galvanotaxis (*Zool.*). Tendency of organisms to grow or move into a particular orientation relative to an electric current passing through the surrounding medium. Also **galvanotropism.**

gam-, gamo-. Prefix from Gk. *gamos*, marriage, union.

-gam (*Bot.*). Suffix used for plants having a specified means of reproduction, e.g., *cryptogam.*

gamagarite (*Min.*). A hydrated vanadate of barium and manganese, occurring in the manganese ores of Postmasburg, South Africa.

gamboge gum (*Chem.*). A natural gum, of grey or brown colour; rel. d. 1·03, acid value 68–89, saponification value 115–150: used as a pigment.

gambrel roof (*Build.*). See **mansard roof.**

games, theory of (*Psychol.*). A mathematical formalization of decision and strategic processes involving the probabilities and values of various outcomes of action-choices for the decision-makers.

gametangium (*Biol.*). The organ in which the gametes are formed.

gametes (*Biol.*). Reproductive cells which will unite in pairs to produce zygotes; germ cells. *adj.* **gametal.**

gametic number (*Cyt.*). The number of chromosomes present in the nucleus of a gamete.

gametids (*Zool.*). In *Protozoa*, cells budded off from the body of a sporont and destined to become gametes.

gametocyst (*Zool.*). In *Protozoa*, a cyst in which union of gametes occurs. Also called **gamocyst.**

gametocyte (*Biol.*). A gamete mother-cell. (*Zool.*) In *Protozoa*, a phase developing from a trophozoite and giving rise to gametes.

gametogenesis (*Biol.*). The formation of gametes from gametocytes; gametogeny.

gametogonium (*Biol.*). See **gametocyte.**

gametokinetic (*Biol.*). Stimulating the formation of gametes.

gametonucleus (*Biol.*). The nucleus of a gamete; a nucleus which functions as a gamete.

gametophore (*Bot.*). The upright leafy branch of the Musci, bearing the sex organs.

gametophyte (*Bot.*). The plant which bears the gametangia and gametes. *adj.* **gametophytic.**

gametophytic budding (*Bot.*). The formation of gemmae on a prothallus.

gametropic (*Bot.*). Said of the movements of organs before or after fertilization.

gamma (*Geophys.*). A unit of magnetic field intensity particularly used in magnetic surveys. (*Phys.*) See **microgramme.** (*Photog.*) A development factor (due to Hurter and Driffield) in photography, given by $\gamma = (D_2 - D_1)/\log(E_2/E_1)$, where D_1 and D_2 are the densities of the photographic images, and E_1 and E_2 are the respective exposure times. Usually deduced from linear part of graph relating D with $\log E$. See also **sensitometer.** (*TV*) The relationship between the brightness contrast of two points on the transmitted scene and the brightness contrast of the corresponding points on the receiver screen, given by $\gamma = \dfrac{\log R_2/R_1}{\log S_2/S_1}$. R and S refer to receiver and scene respectively, the subscripts indicating two particular points.

gamma-BHC (*Chem.*). Same as **Gammexane.**

gamma brass, gamma constituent in brass (*Met.*). The γ-constituent in brass is hard and brittle and is stable between 60% and 68% of zinc at room temperature. γ-Brass is an alloy consisting of this constituent.

gamma camera (*Nuc.*). An array of scintillators or semiconductor radiation detectors arranged to form a crude image of a distribution of radioactive material in a body.

gamma control (*Cinema.*). The necessity of adjusting the extent of development (in developing cinematograph films with sound-tracks) within fine limits to minimize amplitude distortion arising from nonlinearity in the γ-curve of the emulsion.

gamma correction (*TV*). Introduction of a nonlinear output-input characteristic to change effective value of γ.

gamma detector (*Radiol.*). A radiation detector specially designed to record or monitor gamma-radiation.

gamma function (*Maths.*). The function $\Gamma(x)$ defined by Euler as $\int_0^\infty e^{-t}t^{x-1}dt$ and by Weierstrass by the equation

$$\frac{1}{\Gamma(x)} = xe^{\gamma x} \prod_1^\infty \left(1 + \frac{x}{n}\right) e^{-x/n}$$

where γ is Euler's constant. Its main properties are

$$\Gamma(1+x) = x\Gamma(x)$$
$$2^{2x-1}\Gamma(x)\Gamma(x+\tfrac{1}{2}) = \sqrt{\pi}\,\Gamma(2x)$$
$$\Gamma(x)\Gamma(1-x) = \frac{\pi}{\sin \pi x}$$
$$\Gamma(\tfrac{1}{2}) = \sqrt{\pi}.$$

gamma globulin (*Biochem., Physiol.*). See **serum globulins.**

gamma infinity (*Photog.*). The maximum γ obtainable with prolonged development.

gamma iron (*Met.*). The polymorphic form of iron stable between 906°C and 1403°C. It has a face-centred cubic lattice and is nonmagnetic. Its range of stability is lowered by carbon, nickel and manganese, and it is the basis of the solid solutions known as ***austenite*** (q.v.).

gamma-radiation (*Nuc.*). Electromagnetic radiation of high quantum energy emitted after nuclear reactions or by radioactive atoms when nucleus is left in excited state after emission of α- or β-particle.

gamma-ray capsule (*Nuc. Eng.*). Usually metal, sealed and of sufficient thickness to reduce β-ray transmission to a safe value.

gamma-ray energy (*Nuc.*). This depends on wavelength and controls depth of penetration. Measured by maximum energy of photo-electrons, or diffraction by certain crystal lattices. See **curie, röntgen.**

gamma-rays (*Nuc.*). See **gamma-radiation.**

gamma-ray source (*Radiol.*). A quantity of matter emitting γ-radiation and in a form convenient for radiology.

gamma-ray spectrometer (*Nuc. Eng.*). Instrument for investigation of energy distribution of γ-ray quanta. Basically a scintillation counter followed by a *pulse height analyser* (q.v.).

Gammexane (*Chem.*). TN applied to the γ isomer of *benzene hexachloride* (*BHC*) (q.v.); a synthetic stomach and contact insecticide of great toxicity to a wide range of pests.

gamobium (*Zool.*). In metagenesis, the sexual generation. Cf. *agamobium.*

gamocyst (*Zool.*). See **gametocyst.**

gamogastrous (*Bot.*). Said of a syncarpous gynaeceum in which the ovaries are fused, the styles and stigmas free.

gamogenesis (*Biol.*). Reproduction by union of sexual elements. *adj.* **gamogenetic.**

gamogenic (*Bot.*). Formed after a sexual act.

gamogony (*Zool.*). In *Sporozoa*, formation of gametes or gametocytes by a gamont; sporogony.

gamont (*Zool.*). In *Sporozoa*, an individual destined to produce gametes; sporont.

Gamopetalae (*Bot.*). See **Sympetalae.**

gamopetaly (*Bot.*). The condition in which the corolla consists of a number of petals united by their edges. *adj.* **gamopetalous.**

gamophyllous (*Bot.*). Having the perianth members united.

gamosepalous (*Bot.*). Having the sepals united by their edges.

gamostely (*Bot.*). The fusion of steles.

gamotropism (*Zool.*). The tendency shown by gametes to attract one another.

Gamow-Teller selection rules (*Nuc.*). See **nuclear selection rules.**

gamut (*Acous.*). The range of frequencies required in a specified type of reproduction, e.g., speech or music.

gang (*Mining*). A train or journey of tubs or trucks.

gang boarding (*Build.*). A board with battens nailed across to form steps; used as a gangway during building operations.

ganged capacitor (*Telecomm.*). Assemblage of two or more variable capacitors mechanically coupled to the same control mechanism.

ganging (*Elec. Eng.*). Mechanical coupling of the movements of variable circuit elements for simultaneous control.

ganging oscillator (*Elec. Eng.*). One giving a constant output, whose frequency can be rapidly varied over a wide range; used for testing accuracy of adjustment of ganged circuits over their tuning range.

ganglia habenulae (*Zool.*). In *Cyclostomata*, a pair of ganglia, the right much larger than the left, on the dorsal border of the lateral wall of the diencephalon, connected with the pineal apparatus.

ganglion (*Med.*). A localized cystic swelling formed in connexion with a tendon sheath, commonly on the back of the hand. (*Zool.*) A plexiform collection of nerve fibre terminations and nerve cells. *pl.* **ganglia.**

ganglionectomy or **gangliectomy** (*Surg.*). Surgical removal of a nerve ganglion, or of a ganglion arising from a tendon sheath.

ganglioneuroma (*Med.*). A tumour composed of ganglion nerve cells, nerve fibres, and fine fibrous tissue, usually arising in connexion with sympathetic nerves, e.g., in the medulla of the suprarenal gland.

ganglion impar (*Zool.*). The unpaired, most posterior of the abdominal sympathetic ganglia.

ganglionitis (*Med.*). Inflammation of a nerve ganglion.

gangliosides (*Chem.*). Class of lipids found mainly in nerve tissue and spleen. They contain a ceramide linked to carbohydrate, several additional moles of carbohydrate, *N*-acetylgalactosamine and *N*-acetylneuraminic acid.

gang milling (*Eng.*). The use of several milling cutters on one spindle to produce a surface with a required profile or to mill the face and sides of the work at one operation.

gang mould (*Build., Civ. Eng.*). A mould in which a number of similar concrete units may be cast simultaneously.

gangosa (*Med.*). A progressive, destructive ulceration of the nose, the palate, and the pharynx, occurring as a late sequel of *yaws* (q.v.).

gang punch (*Comp.*). To punch (the same information) in a number of cards, or (a number of cards) with the same information.

gangrene (*Med.*). Death of a part of the body, associated with putrefaction; due to infection or to cutting off of the blood supply.

gangrenous coryza (*Vet.*). See **malignant catarrhal fever.**

gang rider (*Mining*). Trip rider. See **dukey.**

gang saw (*For.*). An arrangement of parallel saws secured in a frame to operate simultaneously in sawing a log into strips.

gang switches (*Elec. Eng.*). A number of switches mechanically connected together so that they can all be operated simultaneously.

gang tool (*Eng.*). A tool holder having a number of cutters; used in lathes and planes, each tool cutting a little deeper than the one ahead of it.

gangue (*Met.*). The portion of an ore which contains no metal; valueless minerals in a lode.

Ganguillet and Kutter's formula (*Hyd.*). An expression giving the value of the constant of proportionality in *Chezy's formula* (metric) as:

$$\frac{\dfrac{1}{n}+23+\dfrac{0{\cdot}001\ 55}{i}}{0{\cdot}552\left[1+\left(23+\dfrac{0{\cdot}001\ 55}{i}\right)\dfrac{n}{\sqrt{m}}\right]}$$

where n = a coefficient of roughness for the pipe or channel, m = hydraulic mean depth or mean hydraulic radius (metres), i = longitudinal surface slope = fall in surface level per unit length.

gangway (*Build.*). Rough planks laid to provide a footway for the passage of workmen on a site. (*Mining*) Main haulage road, or level underground.

gannister (*Geol.*). A particularly pure and even-grained siliceous grit or loosely cemented quartzite, occurring in the Upper Carboniferous of northern England, and highly prized in the manufacture of silica-bricks.

ganoid (*Zool.*). Formed of, or containing, ganoin. Said of fish scales of rhomboidal form, composed of an outer layer of ganoin and an inner layer of isopedin; (fish) having these scales.

Ganoidei (*Zool.*). A name formerly used to indicate a group of Fish possessing ganoid scales and certain other common features; the group included the Sturgeons and Spoonbills, the Bow-fins, the Gar-pikes, and the Bichirs. See **Ginglymodi.**

ganoin (*Zool.*). A calcareous substance secreted by the dermis and forming the superficial layer of certain fish scales; it was formerly supposed to be homologous with enamel.

gantrees (*Weaving*). Wooden or steel frame supporting the jacquard mechanism and cards over a loom.

gantry (*Build.*). A temporary erection having a working platform used as a base for building operations or for the support of cranes, scaffolding, or materials. (*Space*) The servicing tower beside a rocket on its launching-pad.

Gantt chart (*Eng., etc.*). Graphic construction chart which shows each operation and its connexion and timing as part of the overall scheme.

gap (*Aero.*). The distance from the leading edge of a biplane's upper wing to the point of its projection on to the chord line of a lower wing. (*Comp.*) Digits which separate signals for data or program. (*Elec.*) Space between discharge electrodes. (*Electronics*) Range of energy levels between the lowest of conduction electrons and the highest of valence electrons. (*Mag.*) Air gap in magnetic circuit of recording or erasing head in tape-recorder; allows signal to interact with oxide film. (*Telecomm.*) Region between lobes of polar diagram for which response is inadequate.

gap arrester (*Elec. Eng.*). A lightning arrester consisting essentially of a small air gap connected between the circuit to be protected and earth; the gap breaks down on the occurrence of a lightning surge, and discharges the surge to earth. See **multigap arrester.**

gap bed (*Eng.*). A lathe bed having a gap near the headstock, to permit of turning large flat work of greater radius than the centre height.

gap breakdown (*Elec.*). The cumulative ionization of the gas between electrodes, leading to a breakdown of insulation and a Townsend avalanche.

gap bridge (*Eng.*). A bridge casting of the same cross-section as the bed in a gap-bed lathe, and used to close the gap.

gape (*Min. Proc.*). Aperture below which a crushing machine can receive and work on entering ore. (*Zool.*) The width of the mouth when the jaws are open.

gapes (*Vet.*). See **syngamiasis.**

gap extension coefficient (*Elec. Eng.*). See **extension coefficient.**

gap factor (*Elec. Eng.*). Ratio of energy gain in electron-volts traversing a gap across which an accelerating field acts, to the actual voltage across the gap.

gap lathe (*Eng.*). A lathe with a *gap bed* (q.v.).

gap length (*Mag.*). The distance between adjacent surfaces of the poles in a longitudinal magnetic recording system.

gap press (*Eng.*). A power press with a C-frame, often inclinable, which because of its side openings is particularly suitable for work on long or wide components.

gap waxing (*Cables*). Waxing which occurs at the upper and lower boundaries of the helical space formed by consecutive turns of a paper tape. It is free from carbon.

gap window (*Arch.*). A long and narrow one.

garboard strake (*Ships*). The first strake or line of plating attached to the keel on either side.

garden-wall bonds (*Build.*). Forms of bond used largely for building low boundary walls of single brick thickness, when the load to be carried is that of the wall only and it is desired to show a fair face on both sides of the wall.

garget (*Vet.*). See **mastitis.**

gargle (*Acous.*). A wow which has fluctuation changes ranging between about 30 and 200 Hz.

gargoyle (*Build.*). A grotesquely-shaped spout projecting from the upper part of a building, to carry away the rain water.

gargoylism (*Med.*). Congenital disorder characterized by abnormal development of the bones, cartilage, liver, spleen, the clear part of the eye, the brain and the skin. The basic abnormality is in lipoid metabolism.

garland (*Mining*). A collecting trough for water set in a mine shaft. A frame to heighten sides of coal tub.

garland cells (*Zool.*). The chain of ventral nephrocytes attached at its ends to the salivary glands in Dipteran larvae: the nephrocytes ingest colloidal particles from the blood.

garnet (*Min.*). A group of minerals which crystallize in the cubic system. Some species occur rarely in igneous rocks, but are characteristic of metamorphic rocks, such as garnet-micaschist, garnet gneiss, and eclogite. Some species are of value as gems, rivalling ruby in colour. See also **almandine, andradite, grossular, melanite, pyrope, spessartine.**

garnet hinge (*Join.*). A form of **strap hinge.**

garneting (*Build.*). See **garreting.**

garnetite (*Geol.*). A metamorphic rock composed largely of garnet.

garnet paper (*Join.*). Type of sandpaper having powdered garnet as the abrasive coating.

garnet-peridotite (*Geol.*). A rock composed of olivine, garnet and pyroxenes. Such rocks are relatively uncommon at the Earth's surface but may be the main constituent of the Earth's mantle, giving rise to basaltic magma on melting.

garnierite (*Min.*). A bright-green nickeliferous serpentine, a silicate of magnesium, occurring in ultrabasic rocks as a decomposition product of olivine. In external form, garnierite resembles chalcedony, thus differing from chrysotile.

garnish bolt (*Build.*). A bolt whose head is chamfered or faceted.

garreting (*Build.*). A term applied to the process of inserting small stone splinters in the joints of coarse masonry. Also **galleting, garneting.**

garret window (*Build.*). A skylight of which the glazing lies along the slope of the roof.

Garryales (*Bot.*). An order of the *Dicotyledons.* Woody plants with opposite evergreen leaves. The flowers are in small unisexual panicles. The male flower has a perianth of four separate members, and four anthers; the female flower consists simply of a superior ovary of two to three fused loculi.

garter spring (*Eng.*). An endless band formed by connecting the two ends of a long helical spring; used to exert a uniform radial force on any circular piece round which it is stretched, as in a *carbon gland* (q.v.).

garth (*Arch.*). An enclosed area attached to a building and surrounded usually by a cloister.

Garton lightning arrester (*Elec. Eng.*). A form of gap lightning arrester in which the current flowing when the gap breaks down operates a solenoid which lengthens the gap and extinguishes the arc.

gas (*Fuels*). (1) Fuel for domestic or industrial use, produced from coal or from natural gas fields. See **blast-furnace-, blue-water-, carburetted water-, coal-, coke-oven-, Mond-**

natural-, producer-, semiwater-, sewage-, town-. (2) Abbrev. (used especially in America) for *gasoline* or *petrol* (q.v.). (*Mining*) Explosive mixture of methane and air in 'fiery' colliery. (*Phys.*) (1) A state of matter in which the molecules move freely, thereby causing the matter to expand indefinitely, occupying the total volume of any vessel in which it is contained. (2) The term is sometimes reserved for a gas at a temperature above the critical value. Also defined as a definitely compressible fluid. See **gas laws**, and references at **gases**.

gas amplification (*Electronics*). Increase in sensitivity in a gas-filled photocell, as compared with the corresponding high-vacuum cell, due to ionization of the gas caused by the primary photoelectrons. (*Nuc.*) Increase in sensitivity of a Geiger or proportional counter compared with a corresponding ionization chamber.

gas analysis (*Chem.*). The quantitative analysis of gases by absorption. A measured quantity of gas, 100 cm^3, is brought into intimate contact with the various reagents, and the reduction in volume is measured after each absorption process. Carbon dioxide is absorbed in a concentrated caustic potash solution, oxygen in alkaline pyrogallol solution, carbon monoxide in ammoniacal cuprous chloride solution, unsaturated compounds by absorption in bromine water, etc., hydrogen by combustion with a measured quantity of air over palladium asbestos. Nitrogen is estimated by difference.

gas-and-pressure-air burner (*Heat*). Industrial gas burner designed to operate with low-pressure gas and with air under pressure from fans and compressors.

gas-bag (*Aero.*). Any separate gas-containing unit of a rigid airship.

gas barrel (*Eng.*). Wrought-iron tube, first used for conducting gas from street mains into buildings; now used for a variety of purposes.

gas battery (*Elec. Eng.*). See **Grove gas cell.**

gas black (*Paint.*). A substance produced by carbonizing natural gas; used as an excellent black pigment in paints, plastics, rubbers, etc.

gas-blast circuit-breaker (*Elec. Eng.*). A high-power circuit-breaker in which a blast of gas is directed across the contacts at the instant of separation in order to extinguish the arc. See **air-blast switch.**

gas cable (*Cables*). A paper-insulated power cable in which the paper insulation is not impregnated with compound but which has gas (nitrogen) at a high pressure admitted within the lead sheath to minimize ionization. Also **gas-impregnated cable.**

gas cap (*Mining*). Gas accumulated above oil reservoir or salt dome, encountered when unsuccessfully drilling for oil or natural gas.

gas carbon (*Chem.*). A hard dense deposit of almost pure carbon which slowly collects on the inside of a coal-gas retort.

gas carburizing (*Met.*). The introduction of carbon into the surface layers of mild steel by heating in a reducing atmosphere of gas high in carbon, usually hydrocarbons or hydrocarbons and carbon monoxide.

gas cell (*Chem.*). A galvanic cell in which at least one of the reactants is a gas.

gas checking (*Paint.*). A paintwork defect characterized by the appearance of fine wrinkles. Appears particularly when paints based on undercooked oils dry in the presence of coal-gas combustion products.

gas chromatography (*Chem.*). See **gas-liquid chromatography.**

gas coal (*Fuels*). Coal containing a high percentage of volatile hydrocarbons, gas-yielding constituents; suitable for making town gas.

gas coke (*Fuels*). Solid fuel produced by the carbonization of bituminous coal in a closed chamber known as a retort; the temperature of the charge ranging up to 1100°C; the calorific value is about 28 MJ/kg.

gas colic (*Vet.*). Colic associated with *tympanites.*

gas concrete (*Civ. Eng.*). Lightweight concrete in which bubbles of gas are generated when a metallic additive (powdered aluminium or zinc) reacts chemically with the water and cement in the concrete.

gas conditioning (*Heat*). See **protective furnace atmosphere.**

gas constant (*Phys.*). Its value is $8{\cdot}314\ J\ K^{-1}\ mol^{-1}$. Symbol R.

gas-cooled reactor (*Nuc. Eng.*). One in which cooling medium is gaseous, usually air or carbon dioxide.

gas coulometer (*Elec. Eng.*). One in which the quantity of electricity passing is measured by determining the volume of gas evolved.

gas counter (*Nuc. Eng.*). Geiger counter into which radioactive gases can be introduced.

gas counting (*Nuc. Eng.*). That of radioactive materials in gaseous form. The natural radioactive gases (radon isotopes) and carbon dioxide (^{14}C) are common examples. See also **gas-flow counter.**

gas current (*Electronics*). Positive ions discharged at a negative electrode, e.g., grid of a triode, because of ionization by electrons passing between other electrodes, e.g., cathode to anode. Basis of a vacuum gauge.

gas-cushion cable (*Cables*). A *gas cable* (q.v.) in which the space between the lead sheath and the cable is divided into a large number of separate pockets by a spiral spacer, thus preventing any excessive gas leakage.

gas-discharge lamp (*Light*). See **electric-discharge lamp.**

gas-discharge tube (*Electronics*). Generally, any tube in which an electric discharge takes place through a gas. Specially, a tube comprising a hot or cold cathode, with or without a control electrode (grid) for initiating the discharge, and with gas at an appreciable pressure. See **gas-filled relay, glow tube, grid-glow tube, ignitron, mercury-arc rectifier, phanotron, pool tube, thyratron.**

gas drain (*Mining*). A tunnel or borehole for conducting gas away from old workings.

gas-electric generating set (*Elec. Eng.*). A gas- or petrol-driven electric generating set.

gas electrode (*Elec.*). One which holds gas by adsorption or absorption, so that it becomes effective as an electrode in contact with an electrolyte.

gas engine (*I.C. Engs.*). One (constant volume or Otto cycle) in which gaseous fuel is mixed with air to form a combustible mixture in the cylinder and fired by spark ignition.

gaseous discharge (*Phys.*). Flow of charge arising from ionization of low-pressure gas between electrodes, initiation being by electrons of sufficient energy released from hot or cold cathodes. Various gases give characteristic spectral colours, e.g., mercury, sodium vapour, neon, hydrogen, etc.

gaseous exchange (*Biol.*). See **respiration, external.**

gaseous fuel (*Fuels*). Any combustible gas which can be burned economically in an engine or furnace, such as coal, natural, producer gas, etc.

gases (*Phys.*). See **density of-, expansion of-, kinetic theory of-, liquefaction of-.**

gas evolution (*Met.*). The liberation of gas bubbles

during the solidification of metals. It may be due to the fact that the solubility of a gas is less in the solid and molten metal respectively, as when hydrogen is evolved by aluminium and its alloys, or to the promotion of a gas-forming reaction, as when iron oxide and carbon in molten steel react to form carbon monoxide. See also **blowholes, unsoundness.**

gas exhauster (*Eng.*). A large low-pressure rotary vane pump or centrifugal blower for exhausting gas from the retorts in a gas-works.

gas factor (*Electronics*). Ratio of space current in a gas photoelectric tube to that in a corresponding vacuum one.

gas-filled cable (*Cables*). An impregnated paper-insulated power cable in which gas (nitrogen) at a high pressure is admitted within the lead sheath to minimize ionization.

gas-filled filament lamp (*Light*). One in which the bulb contains an inert gas.

gas-filled photocell (*Electronics*). One in which anode and photocathode are enclosed in atmosphere of gas at low pressure. It is more sensitive than the corresponding high-vacuum cell because of formation of positive ions by collision of the photoelectrons with the gas molecules.

gas-filled relay (*Electronics*). Thermionic tube, usually of the mercury-vapour type, when used as a relay; a thyratron.

gas-flow counter (*Nuc. Eng.*). (1) Counter tube through which gas is passed to measure its radioactivity, e.g., CO_2, which is weakly radioactive because of ^{14}C. (2) One in which a low energy α or β emitter is introduced into the interior of the counter, and to prevent the ingress of air, the counting gas flows through it at a pressure above atmospheric. Also **flow counter.**

gas focusing (*Electronics*). Controlling the beam in a CRT, X-ray tube or other discharge tube by the action of a small amount of residual gas in the envelope, which, on becoming ionized by collision, forms a core of positive ions along the centre of the beam and provides the necessary focusing field. Also called **ionic focusing.**

gas gangrene (*Med.*). Rapidly spreading infection of a wound with gas-forming anaerobic bacteria, causing progressive gangrene of the infected part.

gas generator (*Aero.*). The high-pressure compressor/combustion/turbine section of a gas turbine which supplies a high-energy gas flow for turbines which drive airscrews, fans or compressors. (*Chem.*) Chemical plant for producing gas from coal, e.g., water gas, by alternating combustion of coal and reduction of steam.

gas gland (*Zool.*). A structure in the wall of the air-bladder in certain Fish which is capable of secreting gas into the bladder. See **rete mirabile.**

gas governor (*Gas*). An automatic tap which controls pressure, volume, or temperature of gas supply on consumers' premises.

gash (*Elec. Eng.*). *G*uanidine *a*luminium *s*ulphate *h*exahydrate, a ferroelectric compound with an almost square hysteresis loop suitable for constructing a binary cell.

gasholder (*Gas*). A cylindrical-shaped structure with a domical top floating in a tank of water, in which gas is collected for distribution. When dry gasholders are used, they are sealed with tar or grease.

gas horsepower (*Aero.*). The shaft horsepower recoverable from the efflux of a *gas producer*, assuming expansion through a 100% efficient turbine.

gash veins (*Geol.*). Veins formed in limestone joints that have been somewhat widened by solution prior to the deposition of the vein stuffs which may include blende and galena. The veins run also along the bedding planes. (*Mining*) Mineralized veins of shallow depth, often considerably wide at surface (outcrop).

gas-impregnated cable (*Cables*). See **gas cable.**

gasket (*Eng.*). (1) A layer of packing material firmly held between contact surfaces on two pieces whose joint is to be sealed; in particular, a flat sheet of asbestos compound, sometimes sandwiched between thin copper sheets; used for making gas-tight joints between engine cylinders and heads, etc. (2) Jointing or packing material, such as cotton rope impregnated with graphite grease; used for packing stuffing-boxes on pumps, etc. (*Plumb.*) Hemp or cotton yarn wound round the spigot end of a pipe at a joint, and rammed into the socket of the mating pipe to form a tight joint.

gasket iron (*Plumb.*). A flat tool with blunt end, used for ramming a gasket into position.

gaskin (*Plumb.*). See **gasket.**

gas laws (*Phys.*). Boyle's law, Charles' law, and the pressure law, which are included in the equation $Pv = RT$, where P is the pressure, v the molar volume, T the absolute temperature, and R the gas constant, the value of which is 8·314 J/K mol. A gas which obeys the gas laws perfectly is known as a *perfect* or *ideal gas.* Cf. *van der Waals' equation.*

gaslight paper (*Photog.*). See **contact paper.**

gas lime (*Chem.*). The spent lime from gasworks after it has been used for the absorption of hydrogen sulphide and carbon dioxide in the gas purifiers.

gas-liquid chromatography (*Chem.*). A form of partition or adsorption chromatography in which the mobile phase is a gas and the stationary phase a liquid. Solid and liquid samples are vaporized before introduction on to the column. The use of very sensitive detectors has enabled this form of chromatography to be applied to submicrogram amounts of material. Abbrev. GLC.

gas liquor (*Chem.*). The aqueous solution of ammonia and ammonium salts condensed from coal-gas.

gas mantle or **incandescent mantle** (*Light*). A small dome-shaped structure of knitted or woven ramie or rayon, impregnated with a solution of the nitrates of cerium and thorium, then dried, and the textile fabric burned off.

gas maser (*Electronics*). One in which the interaction takes place between molecules of gas and the microwave signal.

gas mask (*Chem.*). A device for protection against poisonous gases, which are absorbed by activated charcoal or by other reactive substances, e.g., soda-lime. The choice of the absorbing material depends on the nature of the gas to be counteracted.

gas noise (*Acous.*). That arising from ionization in gas-discharge tubes, used for generating 'white' noise test, e.g., acoustic devices.

gas oil (*Fuels*). A petroleum distillate obtained after kerosine and before lubricating-oil, used as C.I.-engine fuel, as an ingredient in petroleum cracking processes, and for carburetting water-gas. See also **diesel oil.**

gasoline (*Fuels*). See **motor spirit.**

gasoline engine (*Eng.*). U.S. term for **petrol engine.**

gasometer (*Gas*). A term widely used, but only outside the industry, for a *gasholder* (q.v.).

Gasparcolor (*Photog.*). A subtractive process for

colour prints by the dye destruction method, using multilayer base.

gas-pipe tongs (*Eng.*). A wrench used for turning pipes when screwing them into, or out of, coupling pieces.

gas pliers (*Eng.*). Stout pliers with narrow jaws, the gripping faces of which are concave and serrated, to provide a secure grip.

gas-pressure cable (*Cables*). See **pressure-cable.**

gas-pressure regulator (*Heat.*). Diaphragm-operated valve or other device actuated by gas pressure and balanced to produce a constant outlet pressure irrespective of fluctuating initial pressures.

gas producer (*Aero.*). A turbo-compressor of which the power output is in the form of the gas energy in the efflux, sometimes mixed with air from an auxiliary compressor. Essentially, the gas producer is mounted remotely from the point of utilization of its energy, e.g., helicopter *rotor-tip jets* (q.v.) or a *free turbine* (q.v.). (*Fuels*) Furnace in which air and steam are passed over incandescent coke to produce water gas.

gas pump (*Eng.*). See **Humphrey gas pump.**

gas purifiers (*Gas*). A section of gasworks plant in which the gas is freed from hydrogen sulphide and certain complex nitrogen compounds of cyanide type by passing it through the layers of hydrated oxide of iron.

gas radiator (*Build.*). A flueless heater resembling a steam or hot-water radiator but equipped with gas burners.

gas ratio (*Electronics*). That of ion current to electron current in a gas discharge.

gas regulator (*Eng.*). (1) Same as **gas-pressure regulator** (q.v.). (2) The throttle valve of a gas engine.

gas scrubber (*Chem.*). An apparatus for the purification of gas from tarry matter and other undesirable impurities, such as ammonia.

gasserectomy (*Surg.*). Excision of the Gasserian ganglion, sometimes performed for the relief of trigeminal neuralgia (tic douloureux).

Gasserian ganglion (*Zool.*). In Vertebrates, a large ganglion of the fifth cranial nerve, near its origin.

gas show (*Geol.*). A surface indication of the escape of natural gas from underground reservoirs; of importance in oilfield exploration.

gassing (*Aero.*). See **inflation.** (*Elec. Eng.*) (1) The evolution of gas which occurs in an accumulator towards the end of its charging period. (2) Evolution of gas from electrodes of voltameter or hard vacuum tube. (*Textiles*) The process of passing yarns through a gas flame or over electrically heated elements to remove outstanding fibres and enhance the appearance.

gassing of copper (*Met.*). A process which denotes the brittleness produced when copper containing oxide is heated in an atmosphere containing hydrogen. The hydrogen diffuses into the metal and combines with oxygen, forming steam which cannot diffuse out. A high steam pressure is built up at the crystal boundaries and the cohesion is diminished.

gas stocks and dies (*Eng.*). Stocks and dies (see **die (2), die-stock**) for cutting screw-threads on gas barrel or piping.

gas tar (*Chem., etc.*). Coal-tar condensed from coal-gas, consisting mainly of hydrocarbons. Distillation of tar provides many substances, e.g., ammoniacal liquor, 'benzole', naphtha, and creosote oils, with a residue of pitch. Dehydrated, it is known as 'road tar', and used as a binder in road-making.

gas temperature (*Aero.*). The temperature of the gas stream resulting from the combustion of fuel and air within a turbine engine. For engine performance monitoring, the temperature may be measured at either of two points signified by the abbrevs. **JPT, TGT. JPT** (jet-pipe temperature) is the measured temperature of the gas stream in the exhaust system, usually at a point behind the turbine where most of the turbulence has been removed; **TGT** (turbine gas temperâture) is the measured temperature of the gas stream between turbine stages. **EGT** (exhaust gas temperature), frequently used, is synonymous with **JPT.**

gaster (*Zool.*). The abdomen proper in *Hymenoptera*, being the region posterior to the first abdominal segment which, in many members, is constricted.

gastero-, gastr-, gastro-. Prefixes from Gk. *gastēr*, gen. *gastros*, stomach.

Gasteromycetes (*Bot.*). A subdivision of the *Autobasidiomycetes*, including about 800 species, in which the hymenium is enclosed in a peridium until after the spores are ripe.

Gasteropoda (*Zool.*). See **Gastropoda.**

gasterospore (*Bot.*). A globular thick-walled asexual spore, characteristic of the fungus *Ganoderma.*

gasterozooid (*Zool.*). In colonial *Hydrozoa*, a hydroid individual of normal structure with a mouth surrounded by tentacles. In *Doliolidae*, food-gathering zooids which develop from the lateral row of buds.

gas thread. See **British Standard pipe thread.**

gastraea (*Zool.*). The original hypothetical ancestor of the *Metazoa*, resembling a gastrula (Haeckel).

gastralgia (*Med.*). Pain in the stomach.

gastral groove (*Zool.*). In most Insects, a groove-like invagination extending over the length of the middle plate in the embryo.

gastralia (*Zool.*). The abdominal ribs of some Vertebrates; sponge spicules occurring in the gastral layer.

gastral layer (*Zool.*). In *Porifera*, the inner cell-layer consisting of choanocytes.

gas transport (*Zool.*). The transport by the blood of oxygen from the site of *external respiration* to cells where it is needed for aerobic respiration (see **respiration, aerobic**), this frequently involving a *respiratory pigment* (q.v.), and the transport away from the respiring cells of any carbon dioxide produced.

gastrectomy (*Surg.*). Removal of the whole, or part, of the stomach.

gastric (*Zool.*). Pertaining to, or in the region of, the stomach.

gastric filaments (*Zool.*). In some *Coelenterata*, endodermal filaments projecting into the enteron and containing cnidoblasts which kill any living prey that enters the stomach.

gastric juice (*Chem.*). Human gastric juice consists principally of water (99·44%), free HCl (0·2%), and small quantities of NaCl, KCl, $CaCl_2$, $Ca_3(PO_4)_2$, $FePO_4$, $Mg_3(PO_4)_2$, and organic matter including enzymes.

gastric mill (*Zool.*). In *Malacostraca*, the proventriculus, which is provided with muscles and ossicles for trituration of food.

gastric shield (*Zool.*). A hard plate in the stomach of *Lamellibranchiata*, against which the crystalline style is worn away.

gastrin (*Biochem., Physiol.*). A hormone liberated from the pyloric mucosa which stimulates the secretion of gastric juice rich in hydrochloric acid. See *histamine.*

gastritis (*Med.*). Inflammation of the mucous membrane of the stomach.

gastrocele (*Med.*). Hernia of the stomach.

gastrocentrous (*Zool.*). Said of vertebrae in which the centrum is composed largely of the pleurocentrum and the intercentrum is reduced or absent; in all *Amniota*.

gastrocnemius (*Zool.*). In land Vertebrates, a muscle of the shank.

gastrocoel (*Zool.*). See **archenteron.**

gastrocolic (*Med.*). Pertaining to, or connected with, the stomach and the colon.

gastrocutaneous pores (*Zool.*). A series of pores which connect the intestine with the external surface in *Enteropneusta*.

gastroduodenal (*Med.*). Pertaining to, or connected with, the stomach and the duodenum.

gastroduodenitis (*Med.*). Inflammation of the mucous membrane of the stomach and the duodenum.

gastroduodenostomy (*Surg.*). A communication between the stomach and the duodenum, made by operation.

gastroenteritis (*Med.*). Inflammation of the mucous membrane of the stomach and the intestines.

gastroenterostomy (*Surg.*). A communication between the stomach and the small intestine, made by operation.

gastrogastrostomy (*Surg.*). A communication, made by operation, between the upper and lower parts of the stomach when these are pathologically separated by a stricture.

gastrojejunal (*Med.*). Pertaining to, or connected with, the stomach and the jejunum.

gastrojejunostomy (*Surg.*). A communication between the stomach and the jejunum, made by operation.

gastrohepatic omentum (*Zool.*). A mesentery connecting the dorsal surface of the liver and the stomach in Reptiles.

gastrolienal (*Med.*). Pertaining to the stomach and the spleen.

gastroliths (*Zool.*). In *Crustacea*, two calcareous masses found in the anterior part of the proventriculus prior to a moult.

gastromyotomy (*Surg.*). Incision of the muscle of the stomach round a gastric ulcer.

gastropexy (*Surg.*). The operation of suturing the stomach to the abdominal wall for the treatment of gastroptosis.

Gastropoda (*Zool.*). A class of *Mollusca* with a distinct head bearing tentacles and eyes, a flattened foot, a visceral lump which undergoes torsion to various degrees and is often coiled, always bilaterally symmetrical to some extent, with the shell usually in one piece. Snails, Slugs, Whelks, etc. Also **Gasteropoda.**

gastropore (*Zool.*). In *Hydrocorallinae*, an opening in the common skeleton through which a gastrozooid protrudes.

gastroptosis (*Med.*). Abnormal downward displacement of the stomach in the abdominal cavity.

gastroscope (*Med.*). An instrument for inspecting the interior of the stomach and by means of which a photograph of the lining of the stomach may be taken.

gastrostaxis (*Med.*). Bleeding or oozing of blood from the stomach, the mucous membrane of which is intact.

gastrostegite (*Zool.*). In *Ophidia*, a ventral scale. Also called **gastrostege.**

gastrostomy (*Surg.*). The operative formation of an opening into the stomach, through which food may be passed when the normal channels are obstructed.

gastrotomy (*Surg.*). Incision of the stomach wall.

Gastrotricha (*Zool.*). A phylum of acoelomate, triploblastic Metazoa; small, vermiform, unsegmented, with ciliary tracts and often bristles and scales on the cuticle; a noncellular hypodermis with adhesive papillae; some members hermaphrodite, others with parthenogenetic females; development direct, cleavage total.

gastrovascular (*Zool.*). Combining digestive and circulatory functions, as the canal system of *Ctenophora* and the coelenteron of *Cnidaria*.

gastrula (*Zool.*). In development, the double-walled stage of the embryo which succeeds the blastula.

gastrulation (*Zool.*). The process of formation of a gastrula from a blastula during development.

gas tube (*Phys.*). Since it is so far impossible to obtain a perfect vacuum, or even one approaching outer space, all tubes and valves are gas tubes; in a so-called gas tube, the pressure of residual gas is sufficiently high to influence the operation. See **gas-filled photocell, gas focusing, thyratron.**

gas turbine (*Eng.*). A simple, high-speed machine used for converting heat energy into mechanical work, in which stationary nozzles discharge jets of expanded gas (usually products of combustion) against the blades of a turbine wheel. Used in stationary power and other plants, locomotives, marine (especially naval) craft, jet aero engines, and experimentally in road vehicles. See **by-pass turbojet, gas producer, shaft turbine, turbofan, turbojet, turboprop.**

gas-turbine plant (*Eng.*). A power plant comprising at least a gas turbine, an air compressor driven by the gas turbine, and a combustion chamber in which liquid (usually) fuel is burnt to form the working medium of the turbine. Intercooling or other thermodynamic modifications may be added to increase the thermal efficiency of the plant.

gas welding (*Eng.*). Any metal welding process in which gases are used in combination to obtain a hot flame. The most commonly used gas welding process employs the oxy-acetylene combination which develops a flame temperature of 3200°C. Some plastics, especially polythene, may be fusion-jointed by a form of low-temperature gas welding without flame.

gas well (*Geol.*). A deep boring, generally in an oil-field, which yields natural gas rather than oil. See **natural gas.**

gate (*Build.*). A hinged or sliding barrier closing an opening to an enclosure. (*Cinema.*) The location of the film in a projector, printer, or camera when it is being acted on or scanned. (*Electronics*) Control electrode for an FET. (*Eng.*) (1) A valve controlling the supply of fluid in a conduit. (2) A frame in which saws are stretched to prevent buckling. (*Foundry*) In a mould, the channel or channels through which the molten metal is led from the runner, down-gate, or pouring-gate to the mould cavity. Also called **geat, git, sprue.** (*Hyd. Eng.*) A movable barrier for stopping or regulating the flow in a channel. (*Nuc. Eng.*) In reactor engineering, movable barrier of shielding material used for closing aperture. (*Plastics*) Term used to denote the small restricted space in an injection mould between the mould cavity and the passage carrying the plastic moulding material. (*Telecomm.*) Electronic circuit which passes impressed signals when permitted by another independent source of similar signals.

gate by-pass switch (*Elec. Eng.*). A switch fitted in an electric lift-car so that the operator can render the gate or door locks inoperative in cases of emergency.

gate-chambers (*Hyd. Eng.*). Recesses in the side

walls of a lock to accommodate the gates when open.

gate-change gear (*Autos.*). A multispeed gearbox in which the control lever is positioned by, e.g., an H-shaped slot or gate, through which it works, the five members of the slot corresponding to three forward speeds, one reverse speed, and neutral.

gate closer (*Elec. Eng.*). A device for automatically closing the gates of a lift-car. Also called **door closer.**

gate current (*Elec. Eng.*). The a.c. or pulsed d.c. which saturates core of a reactor or transductor.

gated-beam tube (*Electronics*). Cathode-ray tube in which beam current can be gated by signals applied to a control electrode.

gate detector (*Electronics*). One whose operation is controlled by an external gating signal.

gated throttle (*Aero.*). A supercharged aero-engine throttle quadrant with restricting stop(s) to prevent the throttle being wrongly used. See **boost control.**

gate-end (*Mining*). Coal-face end of a gateway.

gate-end box (*Elec. Eng.*). A cable box used in mining work for making a joint between one of the main supply cables and a trailing cable for use near the coal face.

gate hook (*Build.*). The part of a gate hinge which is secured to the post and about which pivots the leaf supporting the gate.

gate interlock (*Elec. Eng.*). A combination of a gate or door switch with an automatic gate or door lock.

gate leakage current (*Electronics*). The d.c. gate current which flows in an FET under normal operating conditions.

gate operator (*Elec. Eng.*). A power-operated device for closing the gates of a lift-car.

gate pier (*Build.*). A post similar to a *gate post* but brick, stone or concrete instead of timber.

gate post (*Build.*). A timber post from which a gate is hung, or one against which it shuts. See also **gate pier.**

gate road (*Mining*). See **gateway.**

gate stick (*Foundry*). A stick placed vertically in the cope while it is rammed up; on removal it provides the gate or runner passage into the mould. Also called **runner stick.**

gate switch (*Elec. Eng.*). See **door switch.**

gate valve (*Eng.*). One which can be moved across the line of flow in a pipe line.

gate voltage (*Electronics*). The control voltage for an electronic gate. The voltage applied to the 'gate' electrode of a field-effect transistor.

gateway (*Mining*). A road through the worked-out area (goaf) for haulage in longwall working of coal. Road connecting coal working with main haulage. Also **gate road.**

gate winding (*Elec. Eng.*). That used to obtain gating action in a magnetic amplifier.

gather (*Glass*). A charge of glass picked up by the gatherer.

gatherer (*Glass*). A person who gathers a charge of glass on a blowpipe or gathering iron for the purpose of forming it into ware or feeding a charge to a machine for that purpose.

gathering (*Bind.*). Collecting and arranging in proper sequence the folded sections forming a volume. (*Build.*) The contracting portion of the chimney passage to the flue, situated a short distance above the source of heat. (*Paint.*) Patchy appearance of distempered surfaces owing to irregular absorption.

gathering ground (*Civ. Eng.*). See **catchment area.**

gathering motor (*Mining*). Light electric loco used to move loaded coal trucks from filling points to main haulage system.

gathering pallet (*Horol.*). In the striking mechanism of a clock or repeater watch, a revolving finger which lifts the rack one tooth for each blow struck; a single-toothed wheel.

gating (*Elec.*). Selection of part of a wave on account of time or magnitude. Operation of a circuit when one wave allows another to pass during specific intervals.

Gattermann reactions (*Chem.*). Reactions used in organic syntheses for preparing aromatic aldehydes from hydrocarbons. Based on passing a mixture of carbon monoxide and hydrogen chloride into the hydrocarbon in the presence of cuprous chloride and aluminium chloride. It is possible to carry out the reaction using hydrogen cyanide in place of the carbon monoxide.

Gaucher's disease (*Med.*). A disease which occurs in families and is characterized by anaemia and haemorrhages, associated with enlargement of the spleen, in which peculiar large cells containing lipoid appear.

gauge (*Build.*). (1) The distance between centres (as measured across the courses) of the nails securing the slates on, e.g., a roof. (2) The margin or exposed width of a slate or tile in coursed work. (3) Proportion of gypsum in a lime plaster. (*Eng.*) (1) An object or instrument for the measurement of dimensions, pressure, volume, etc. See **pressure gauge, water gauge.** (2) An accurately dimensioned piece of metal for checking the dimensions of work or less precisely made gauges. See **limit gauge, master gauge, plug gauge, ring gauge.** (3) A tool used for measuring lengths, as a *micrometer gauge* (q.v.). (4) The diameter of wires and rods. See **Birmingham Wire Gauge, Brown & Sharpe Wire Gauge.** (*Join.*) Device for marking lines parallel with an edge. (*Rail.*) Distance between the inside edges of the rails of a permanent way. (*Textiles*) (1) Relates to the fineness of a woven or knitted fabric and generally indicates the number of needles per inch in hosiery. (2) Distance between spindles on spinning, twisting or doubling frames.

gauge box (*Civ. Eng.*). A box which measures a known quantity of material, such as cement, sand, or coarse aggregate or similar substance, for testing or making mixtures. Also called **batch box.**

gauge cocks (*Eng.*). Small test cocks fitted to the side of a vessel to ascertain the liquid level therein, as on many steam boilers.

gauge-concussion (*Rail.*). The lateral outward impact of the wheel flanges against the rails due to centrifugal force.

gauged arch (*Arch.*). An arch built from special bricks cut with a bricklayer's saw and rubbed to exact shape on a stone.

gauged mortar (*Build.*). Mortar made of cement, lime, and sand, to proportions suitable for the bricks, blocks, or other material used.

gauge door (*Mining*). A door underground for controlling the supply of air to part of the mine.

gauged stuff (*Build.*). A stiff plaster used for cornices, mouldings, etc.; made with lime putty to which plaster of Paris is added to hasten setting. Also called **putty and plaster.**

gauge glass (*Eng.*). The glass tube, or pair of flat glass plates, fitted to a water-gauge to provide a visual indication of the water-level in the tank or boiler. See **water gauge.**

gauge number (*Eng.*). An (arbitrary) number denoting the gauge or thickness of sheet metal or the diameter of wire, rod, or twist drills in one of many gauge number systems, e.g., British Standard Wire Gauge (S.W.G.)

Birmingham Wire Gauge (B.W.G.), Brown & Sharpe Wire Gauge in U.S. See also **letter sizes.**

gauge pot (*Build.*). A small receptacle for cement grout, facilitating pouring of small quantities.

gauge pressure (*Eng.*). Of a fluid, the pressure as shown by a pressure gauge, i.e., the amount by which the pressure exceeds the atmospheric pressure, the sum of the two giving the absolute pressure.

gauge rod (*Build.*). A rod used in laying graduated courses of slates.

gauging-board (*Build.*). A platform on which mortar or concrete may be mixed.

gaul (*Build.*). A hollow spot in the *setting* (q.v.).

Gault (*Geol.*). A blue-to-grey clay lying between the Lower Greensand and the Chalk in the Cretaceous System. Minutely zoned by means of the ammonites it contains.

gaults (*Build.*). Very hard, heavy, and durable bricks, white or whitish in colour; made from Gault clay.

Gause's principle (*Ecol.*). The idea that closely related organisms with similar habits, life forms, or ecological requirements will not occur in the same areas, as a result of competition.

gauss (*Elec.*). CGS electromagnetic unit of magnetic flux density; equal to 1 maxwell/cm^2, each unit magnetic pole terminating 4π lines. The SI unit of magnetic flux density is the *tesla* (T) which equals 10^4 gauss.

Gauss' convergence test (*Maths.*). If, for a series of positive terms Σa_n

$$n\left(\frac{a_n}{a_{n+1}}-1\right)=\sigma+0\left(\frac{1}{n^\delta}\right),\quad \delta>0,$$

then Σa_n is convergent if $\sigma>1$ and divergent if $\sigma\leqq 1$. This test is an extension of Raabe's test.

Gauss' differential equation (*Maths.*). The equation

$$x(1-x)\frac{d^2y}{dx^2}+\{c-(a+b+1)x\}\frac{dy}{dx}-aby=0.$$

It is satisfied by the hypergeometric function $F(a; b; c; x)$. Also called **hypergeometric equation.**

Gauss eyepiece (*Light*). A form of eyepiece used in optical instruments, such as spectrometers and refractometers, to facilitate setting the axis of the telescope at right angles to a plane-reflecting surface. Light enters the side of the eyepiece and is reflected down the telescope tube by a piece of unsilvered glass, being then reflected back into the eyepiece by the plane surface.

Gaussian curvature (*Maths.*). See **curvature (3).**

Gaussian distribution (*Stats.*). See **normal distribution.**

Gaussian noise (*Acous.*). That arising with a spectral energy distribution similar to the normal distribution curve of statistics.

Gaussian optics (*Optics*). Refers to simple optical theory which does not consider the aberration of lenses. Practically, it applies only to paraxial rays.

Gaussian points (*Light*). Same as **principal points of a lens.**

Gaussian response (*Phys.*). Response, e.g., of an amplifier, for a transient impulse, which, when differentiated, matches the Gaussian distribution curve.

Gaussian units (*Elec.*). Formerly widely-used system of electric units where quantities associated with electric field are measured in e.s.u. and those associated with magnetic field in e.m.u. This involves introducing a constant c (the free space velocity of electromagnetic waves) into Maxwell's field equations.

Gaussian well (*Nuc. Eng.*). A particular form of potential energy distribution of a nuclear particle in the field of a nucleus or other nuclear particle.

gaussistor (*Elec. Eng.*). Valvelike device of bismuth or other magnetoresistive material, which can amplify or oscillate.

Gauss' laws of electrostatics (and magnetostatics) (*Elec. Eng.*). The surface integral of the normal component of electric displacement (or magnetic flux) over any closed surface in a dielectric is equal to the total electric charge enclosed (or to zero in the magnetic case). Differential forms of these laws comprise two of *Maxwell's field equations*; see also **Poisson's equation.**

gaussmeter (*Elec. Eng.*). Instrument measuring magnetic flux density. This term is most widely used in U.S.

gauze (*Textiles*). A light-weight fabric of open texture in which the crossing ends pass from one side to the other of the standing ends; used as a dress fabric, for mosquito netting, etc. Any type of fine mercerized cotton, spun rayon, or worsted yarn can be used. (See **leno** and (for *imitation gauze*) **mock leno.**)

gauze brushes (*Elec. Eng.*). Brushes made of copper gauze for collecting current from the commutator of an electric machine; they carry a higher current density than carbon brushes, but give less satisfactory commutation and are little used.

gavage (*Vet.*). Forced feeding of birds being fattened for meat production.

gavel (*Tools*). A mallet used for setting stones.

gavelock (*Tools*). An iron crowbar. Also **gablock.**

Gay-Lussac's law (*Chem.*). Of volumes: when gases react, they do so in volumes which bear a simple ratio to one another and to the volumes of the resulting substances in the gaseous state, all volumes being measured at the same temperature and pressure. (*Heat*) *Charles's law* (q.v.).

Gay-Lussac tower (*Chem.*). Name for a tower of a sulphuric acid plant used for the recovery of the nitrogen oxides from the gases which leave the lead chambers.

gay-lussite (*Min.*). A rare grey hydrated carbonate of sodium and calcium, occurring in lacustrine deposits.

gazebo (*Arch.*). A summerhouse resembling a temple in form and commanding a wide view.

G.B. surface-contact system (*Elec. Eng.*). A surface-contact system used for tramway traction; in it, a magnetized contact chain under the car causes the operation of a finger in the stud and makes the stud alive.

GCA (*Aero., Radar*). Abbrev. for *ground-controlled approach.*

GCI (*Aero., Radar*). Abbrev. for *ground-controlled interception.*

G cramp (*Tools*). One in the shape of a G, with a screw passing through one end. The shoe is sometimes swivelled to enable the cramp to be used on tapered surfaces.

Gd (*Chem.*). The symbol for *gadolinium.*

G display (*Radar*). Similar to *F display* (q.v.) but indicating increasing or diminishing range of target by increasing or diminishing lateral extension of the spot.

Ge (*Chem.*). The symbol for *germanium.*

geanticline (*Geol.*). A major geological structure of the largest size, essentially anticlinal, resulting from mountain-building movements operating on the site of a former geosyncline.

gear (*Eng.*). (1) Any system of moving parts transmitting motion, e.g., levers, gear-wheels, etc. (2) A set of tools for performing some particular work. (3) A mechanism built to perform some special purpose, e.g., steering gear, valve gear. (4) The position of the links of a steam-engine valve motion, as astern gear, mid-gear, etc. (5) The actual gear ratio in use, or the gear-wheels involved in transmitting that ratio, in an automobile gearbox, as *first gear*, *third gear*, etc.

gearbox (*Eng.*). Casing containing a *gear train* (q.v.). The term commonly stands also for the casing including its gear train, particularly when applied to gearboxes used with engines or with machine tools.

gear cluster (*Eng.*). A set of gear-wheels integral with, or permanently attached to, a shaft, as on the lay shaft of an automobile gearbox.

gear cutters (*Eng.*). Milling cutters, hobs, etc., having the requisite tooth form for cutting teeth on gear-wheels.

geared lathe (*Eng.*). A lathe provided with a *back gear* (q.v.) or a multi-speed gearbox between the driving motor and the head.

geared locomotive (*Elec. Eng.*). An electric locomotive in which the motors drive the axles through reduction gears.

geared-quill drive (*Elec. Eng.*). A form of quill drive used in electric locomotives; the quill, instead of carrying the motor armature, carries a gear-wheel which is geared to a pinion on the armature shaft.

geared turbo-generator (*Elec. Eng.*). An electric generator driven through a reduction gear from a steam-turbine, the object being to enable both machines to operate at their most economical speeds.

gearing (*Eng.*). Any set of gear-wheels transmitting motion. See gear.

gearing-down (*Eng.*). A reduction in speed between a driving and a driven wheel or unit, e.g., between the engine of an automobile and the road wheels.

gearing-up (*Eng.*). Raising the speed of a driven unit above that of its driver by the use of gears; opposite of *gearing-down*.

gearless locomotive (*Elec. Eng.*). An electric locomotive in which the motor armatures are mounted directly on the driving axle.

gearless motor (*Elec. Eng.*). A traction motor mounted directly on the driving axle of an electric locomotive.

gear lever (*Eng.*). A lever used to move gear-wheels relative to each other to change gear. In a motor car, this lever acts on the gear-wheels indirectly through *selector forks* (q.v.).

gear marks (*Print.*). Slurred streaks or bands across the printed sheet or web caused by uneven rotation of cylinder.

gear pump (*Eng.*). A small pump consisting of a pair of gear-wheels in mesh, enclosed in a casing, the fluid being carried round from the suction to the delivery side in the tooth spaces; used for lubrication systems, etc.

gear-tooth forming (*Eng.*). A family of engineering processes, including casting, plastic moulding, stamping from sheet metal for watch and clock gears, form-cutting, gear shaping, hobbing and other methods of gear-tooth generating.

gear train (*Eng.*). Two or more *gear-wheels* (q.v.), transmitting motion from one shaft to another. With external spur or bevel gears, the velocity ratio is inversely proportional to the number of gear teeth.

gear-wheel (*Eng.*). A toothed wheel used in conjunction with another, or with a rack, to transmit motion.

geat (*Foundry*). See gate.

gecom (*Comp.*). An automatic code to fit all computers to receive instruction in words. (*Ge*neral *com*piler.)

Gedinnian Stage (*Geol.*). The lowest division of the marine Devonian rocks in the Ardennes and Rhineland.

gedrite (*Min.*). An orthorhombic amphibole; more iron-rich than anthophyllite and containing also aluminum. The iron-aluminium end-member has been called *ferrogedrite*. Gedrite occurrences are restricted to metamorphic and metasomatic rocks.

gee (*Nav.*). V.H.F. radio navigation system depending on pulses sent from two or more transmitting stations, difference of timing determining location on a chart with hyperbolic lattice. (*G*round *e*lectronics *e*ngineering.)

gegenions (*Chem.*). The simple ions, of opposite sign to the colloidal ions, produced by the dissociation of a colloidal electrolyte. Also called **counterions.**

gegenschein (*Astron.*). (*Ger.* counter-glow.) A term applied to a faint illumination of the sky sometimes seen in the ecliptic, diametrically opposite the sun, and connected with the zodiacal light.

gehlenite (*Min.*). A tetragonal silicate of calcium and aluminium ($Ca_2Al_2SiO_7$); an end-member of an isomorphous series collectively known as *melilite* (q.v.).

Gehrcke oscilloscope (*Elec. Eng.*). A form of oscilloscope in which the discharge between 2 aluminium electrodes contained in a tube at low pressure is viewed by means of a rotating mirror, which indicates the waveform of the voltage producing the discharge.

Geiger characteristic (*Nuc.*). Plot of recorded count rate against operating potential for Geiger tube interrupting beam of radiation of constant intensity.

Geiger-Müller counter (*Nuc. Eng.*). One for ionizing radiations, with a tube carrying a high-voltage wire in a halogen or organic atmosphere at low pressure, and a valve (or transistor) circuit which quenches the discharge and passes on an impulse for scaling and counting electronically. Also **Geiger counter, G-M counter.** See **Townsend avalanche.**

Geiger-Müller tube (*Nuc. Eng.*). The detector of a G-M counter, i.e., without associated electronic circuits.

Geiger-Nuttall relationship (*Nuc.*). An empirical rule for calculating the half-life of some radioactive elements from the range of the alpha-particle emitted. The relation is now found to be more limited in its application than it was initially thought to be.

Geiger region (*Nuc. Eng.*). That part of the characteristic of a counting tube, where the charge becomes independent of the nature of the ray intercepted. Also called **Geiger plateau,** since in this region the characteristic is almost horizontal.

Geiger threshold (*Nuc. Eng.*). Lowest applied potential for which Geiger tube will operate in Geiger region.

Geissler pump (*Chem.*). A glass vacuum pump which operates from the water supply.

Geissler tube (*Chem.*). Original gas-discharge tube containing various gases, characterized by a central capillary section for concentrating the glow; useful for spectrometer calibrations.

geitonogamy (*Bot.*). Cross-pollination between 2 flowers on the same plant.

gel (*Chem.*). The apparently solid, often jellylike, material formed from a colloidal solution on standing. A gel offers little resistance to liquid diffusion and may contain as little as 0·5% of solid matter. Some gels, e.g., gelatin, may contain as much as 90% water, yet in their properties are more like solids than liquids.

gelatin(e) (*Chem.*). A colourless, odourless, and tasteless glue, prepared from albuminous substances, e.g., bones and hides. Used for foodstuffs, photographic films, glues, etc.

gelatin dynamite (*Mining*). High explosive containing nitroglycerine, sodium nitrate, collodion cotton, and such inert fillers as wood meal and sodium carbonate.

gelatin filter (*Photog.*). See **filter.**

gelatin printing (*Print.*). The original is prepared in a copying ink and transferred to a gelatin surface from which a hundred copies might be expected, hence the term sometimes used **hectographic printing.** A similar principle is the basis of *spirit duplicating*. Little used now.

gelation (*Plastics*). The process whereby plasticized PVC compounds by the application of heat undergo an irreversible change to soft, rubbery thermoplastic materials. The resultant soft solids are referred to as gels.

Gelidiales (*Bot.*). An order of the *Rhodophyceae*. The order produces tetraspores, and the carposporophyte develops directly from the carpogonium.

gelignite (*Chem., Mining*). Explosive used for blasting, composed of a mixture of nitroglycerine (60%), guncotton (5%), woodpulp (10%), and potassium nitrate (25%).

gelometer (*Phys.*). See **Bloom gelometer.**

gemel window (*Arch.*). A 2-bay window.

gem gravels (*Geol.*). Sediments of the gravel grade containing appreciable amounts of gem minerals, and formed by the disintegration and transportation of pre-existing rocks, in which the gem minerals originated. They are really placers of a special type, in which the heavy minerals are not gold or tin, but such minerals as garnets, rubies, sapphires, etc. As most of the gem minerals are heavy and chemically stable, they remain near the point of origin, while the lighter constituents of the parent rocks are washed away, a natural concentration of the valuable components resulting.

geminate (*Bot.*). (1) Paired, twinned. (2) With two branches arising from the same node, on the same side of the stem.

gemini (*Bot.*). Bivalent chromosomes; pairs of paternal and maternal chromosomes at parasyndesis.

gemma (*Bot.*). (1) A small multicellular body, consisting of thin-walled cells, produced by vegetative means, and able to separate from the parent plant and form a new individual. (2) A nonmotile asexual spore in some algae. (3) A thick-walled resting spore formed by some fungi. (*Zool.*) A bud that will give rise to a new individual. *pl.* **gemmae.**

gemmaceous or **gemmiform** (*Bot.*). Like a small bud.

gemma cup (*Bot.*). A cup-shaped or crescent-shaped outgrowth from the thalli of some liverworts, with gemmae in the hollow.

gemmation (*Bot., Zool.*). Budding; gemma-formation.

gemmiferous (*Zool.*). See **gemmiparous.**

gemmiform (*Zool.*). A term applied to pedicellariae having a long stiff stalk, the jaws of which are each provided with a poison gland; found in *Echinoidea*.

gemmiparous (*Zool.*). Producing gemmae.

gemmule (*Zool.*). In fresh water *Porifera*, an aggregation of embryonic cells within a resistant case, which is formed at the onset of hard conditions when the rest of the colony dies down, and which gives rise to a new colony when conditions have once more become favourable.

gen-, geno-. Prefix from Gk. *genos*, race, descent.

-gen, -gene. Suffix meaning *generating, producing*.

gena (*Zool.*). In general, the side of the head; in Insects, a lateral cephalic sclerite extending from the eye to the gular region. *adj.* **genal.**

genal comb (*Zool.*). A paired row of spines on the latero-ventral borders of the head.

genealogy (*Bot., Zool.*). The study of the development of plants and animals from earlier forms.

gene (*Gen.*). See **genes.**

gene mutation (*Gen.*). A heritable variation caused by a spontaneous change in a gene.

general inference (*Meteor.*). A description of the existing pressure distribution over a wide area, together with a general forecast of the weather.

general instruction (*Comp.*). See **macro-instruction.**

general integral (*Maths.*). See **complete primitive.**

generalized root system (*Bot.*). A root system in which the tap root and the lateral roots are all well developed.

general lighting (*Light*). A system of lighting employing fittings which emit the light in approximately equal amounts in an upward and a downward direction.

general paresis (*Med.*). Manifestation of syphilitic infection of long standing, consisting of progressive dementia and generalized paralysis.

general purpose foils (*Bind.*). Blocking foils suitable for marking paper, bookcloths, etc., but not normally used on thermoplastics.

general rainfall (*Meteor.*). The mean depth of rainfall for a given period over a substantial area.

general solution (*Maths.*). See **complete primitive.**

general stain (*Micros.*). A stain which gives the same depth of colour to all structures in the specimen. Cf. *specific stain*.

generating circle (*Eng.*). Any circle in which a point on the circumference is used to trace out a curve when the circle rolls along a straight line or curve.

generating function (*Maths.*). A function which can be regarded as summarizing a sequence of functions which become apparent when the generating function is expanded as a series; e.g. (1) The function $(q+pt)^n$ generates the binomial probability distribution when expanded in powers of t. The coefficient of t^x in the expansion of $(q+pt)^n$ is the probability $\binom{n}{x}p^xq^{n-x}$ of x successes out of n trials. (2) The function $(1-2xh+h^2)^{-1/2}$, when expanded in powers of h, generates the Legendre polynomials $P_n(x)$.

generating line (*Eng.*). A straight or curved line rotated about some axis to generate a surface.

generating set (*Elec. Eng.*). An electric generator, together with the prime mover which drives it.

generating station (*Elec. Eng.*). A building containing the necessary equipment for generating electrical energy.

generation (*Biol.*). Origin; production; the individuals of a species which are separated from a common ancestor by the same number of broods in the direct line of descent.

generation rate (*Electronics*). Rate of production of *electron-hole pairs* in semiconductors.

generation time (*Nuc.*). Average life of fission neutron before absorption by fissile nucleus.

generative apogamy (*Bot.*). See **reduced apogamy.**

generative cell (*Bot.*). A cell in a pollen grain of Gymnosperms which divides to give a stalk cell and a body cell.

generator (*Elec. Eng.*). Electrostatic or electromagnetic device for conversion of mechanical into electrical energy. Also known as **electric generator.** (*Maths.*) See **cone (1), cylinder (1), ruled surface.**

generator bus-bars (*Elec. Eng.*). Bus-bars in a generating station to which all the generators can be connected.

generator-field control (*Elec. Eng.*). See **variable-voltage control.**

generator panel (*Elec. Eng.*). A panel of a switchboard upon which are mounted all the switches, instruments, and other apparatus necessary for controlling a generator.

genes (*Gen.*). In the modern chromosome theory, units which are arranged in linear fashion on the chromosomes, each having a specific effect on the *phenotype* (q.v.).

genesis (*Biol.*). The origin, formation, or development of a group, a species, an individual, an organ, a tissue, or a cell. *adj.* **genetic.**

gene string (*Gen.*). An identifiable component of a chromosome, consisting of a series of genes arranged like a string of beads.

gene substitution (*Gen.*). Replacement of one allele by another, the other genes remaining unchanged.

genetic code (*Gen.*). The informational equivalence between triplet sequences of purine or pyrimidine bases in the nucleic acids, called **codons,** and specific amino-acids in the equivalent polypeptides. For instance, the base-sequence AAA (adenine) on DNA, becomes UUU (uracil) upon transcription to m-RNA, and specifies phenyl alanine. There are codons for all the other amino-acids, and it is thought that there must also be codons for initiating and ending 'messages', i.e., specific polypeptides.

genetic complex (*Gen.*). The sum total of the hereditary factors contained in the nucleus and in the cytoplasm; it is equivalent to the sum of the *genome* (q.v.) and the *plasmon* (q.v.).

genetic equilibrium (*Gen.*). The situation reached when the frequencies of the genes in a population remain constant from one generation to the next.

genetics. Study of heredity and mutation, including effect of radiation on mutation rates.

genetic spiral (*Bot.*). A hypothetical line drawn on a stem passing by the shortest path through the points of insertion of successive leaves.

genetic variation (*Gen.*). Variation due to differences in the gametes.

Geneva movement (*Eng.*). Intermittent movement from a continuous motion by a wheel with slots.

Geneva stop work (*Horol.*). See **stop work.**

genial (*Zool.*). Pertaining to the chin.

genic balance (*Gen.*). The hypothesis that the characters of an organism are each determined by the interactions of a large but unknown number of genes, some affecting development in one direction and some in another, so that the ultimate result is a balance struck between the total effects.

genicular (*Zool.*). Pertaining to, or situated in, the region of the knee.

geniculate (*Bot., Zool.*). Bent rather suddenly, like the leg at the knee; as *geniculate antennae.*

geniculate ganglion (*Zool.*). In *Craniata,* the ganglion from which the seventh cranial nerve arises.

geniohyoglossus (*Zool.*). The muscle which moves the tongue in Vertebrates.

geniohyoid (*Zool.*). In some Vertebrates, a muscle running from the hyoid to the tip of the lower jaw.

genioplasty (*Surg.*). Plastic surgery of the chin.

genital atrium (*Zool.*). In *Platyhelminthes* and some *Mollusca,* a cavity into which open the male and female genital ducts.

genital bursae (*Zool.*). In *Ophiuroidea,* cavities open to the exterior, into which the gonads open.

genital canals (*Zool.*). In *Crinoidea,* the canals in the arms through which the genital cords pass.

genital character (*Psychol.*). The adult personality made up of the traits accumulated during the earlier stages of psychosexual development as modified by social demands.

genital coelom (*Zool.*). A division of the coelom occupying the apex of the visceral hump in *Cephalopoda.*

genital cords (*Zool.*). In *Crinoidea,* cords passing down the arms in the genital canals, from the genital rachis, to the gonads.

genitalia (*Zool.*). The gonads and their ducts and all associated accessory organs.

genital operculum (*Zool.*). In some *Arachnida,* a small plate covering the openings of the genital ducts.

genital plates (*Zool.*). In *Echinoidea,* the inter-radial plates which bear the genital openings at the aboral pole.

genital pleurae (*Zool.*). In *Enteropneusta,* a pair of lateral folds in the branchial region of the trunk, behind which the gonads are situated.

genital rachis (*Zool.*). In *Echinodermata,* a ring of primary genital cells from which the gonads arise.

genitals (*Zool.*). See **genitalia.**

genital sinus (*Zool.*). A paired venous sinus discharging into the posterior cardinal sinuses in *Chondrichthyes.*

genital stage (*Psychol.*). One of two phases of psychosexual development. Some say it occurs between the third and seventh year with the child's growing awareness of his or her genitals, while others say it occurs with the onset of puberty.

genital stolon (*Zool.*). In *Echinodermata,* a collection of primary genital cells in the axial organ, connected with the genital rachis.

genital wings (*Zool.*). The enlarged genital pleurae in some *Enteropneusta.*

genito-urinary (*Med.*). Pertaining to the genital and the urinary organs.

Genklene (*Chem.*). Alternative TN for *Chlorothene NU* (q.v.).

Gennari's band (or fibres or line) (*Anat.*). A layer of nerve fibres in the cerebral cortex.

Genoa cord (*Textiles*). A cotton fabric, corduroy type, with a 3- or 4-end twill back.

Genoa velvet (*Textiles*). A silk fabric with a cut warp pile, formed by means of wires upon a foundation texture similar in appearance to a twill. *Genoa velveteen.* A similarly constructed cloth but made from cotton yarns.

genocentric (*Bot.*). Producing the reproductive structures at the centre of gravity of the thallus.

genom(e) (*Gen.*). The total chromosome content (genetic factors) of the nucleus of a gamete.

genomere (*Gen.*). A hypothetical particle, which, with other similar particles, makes up a gene.

genosome (*Cyt.*). The part of the chromosome on which a gene is located.

genotype (*Gen.*). Genetic or factorial constitution of an individual; a group of individuals all of which possess the same factorial constitution. Cf. *phenotype.*

genotypic (*Gen.*). Determined by the genes, characters arising from hereditary endowment.

genthelvite (*Min.*). A rather rare zinc beryllium silicate also containing sulphur. Found in granite pegmatites and in skarns.

Gentianales (*Bot.*). An order of the *Dicotyledons.* The petals are fused to form a regular hypogenous corolla, bearing the stamens alternating with the corolla lobes. The superior ovary consists of two (rarely one or three) fused carpels. The leaves are usually opposite.

gentian violet (*Chem.*). Mixture of the three dyes, methyl rosaniline, methyl violet, and crystal violet, which is antiseptic and bactericidal. Used as a fungicide and anthelmintic.

gentiobiose (*Chem.*). $C_{12}H_{22}O_{11}$. A disaccharide based on glucose. A reducing sugar which occurs in combination in *amygdalin*. Contains $G \cdot \beta 1 \rightarrow 6 \cdot G$ (G = glucose unit).

Gentopsida (*Bot.*). A group of plants of uncertain position. They bear naked seeds and are probably related to the conifers and cycads.

genu (*Zool.*). A kneelike structure, i.e., a bend in a nerve tract; more particularly, part of the corpus callosum in Mammals.

genuine tissue (*Bot.*). Tissue derived from the division of a mass of related cells, with subsequent differentiation of the daughter cells.

genu recurvatum (*Med.*). The condition in which there is hyperextension of the knee-joint.

genus (*Biol.*). A taxonomic category of closely related forms, which is further subdivided into species. *pl.* **genera.** *adj.* **generic.**

genu valgum (*Med.*). Knock-knee. The angle between the femur and the tibia is so altered that the leg deviates laterally from the midline.

genu varum (*Med.*). Bow leg. The reverse of genu valgum, the altered angle between the femur and the tibia being such that the legs bow outwards at the knee.

genys (*Zool.*). In Vertebrates, the lower jaw.

geobiont (*Ecol.*). True soil organisms which spend all their lives in the soil and have important effects on its formation and structure. Cf. *geocole, geoxene.*

geobiotic (*Zool.*). Terrestrial; living on dry land.

geocarpy (*Bot.*). The ripening of fruits underground, the young fruits being pushed into the soil by a post-fertilization curvature of the stalk.

geocentric (*Astron.*). The term applied to any system or mathematical construction which has as its point of reference the centre of the earth.

geocentric altitude (*Surv.*). The *true altitude* (q.v.) of a heavenly body as corrected for *geocentric parallax* (q.v.).

geocentric latitude. See under **latitude and longitude** (*terrestrial*).

geocentric parallax (*Astron.*). The apparent change of position of a heavenly body due to a shift of the observer by the rotation of the earth; hence only observed in bodies (e.g., the moon and sun) sufficiently close for the earth's radius to subtend a measurable angle when seen from the body. Also **diurnal parallax.** (*Surv.*) The correction which must be applied to the altitude of a heavenly body in the solar system as observed, in order to give the altitude corrected to the earth's centre. Its value is given by

$$p = +P \cdot \cos \alpha,$$

where p = geocentric parallax, P = *horizontal parallax* (q.v.), and α = observed altitude corrected for refraction.

geochemistry (*Chem.*). The study of the chemical composition of the earth's crust.

geocline (*Ecol.*). A *cline* (q.v.) occurring across topographic or spatial features of an organism's range.

geocole (*Ecol.*). Organisms which spend a regular or irregular portion of their lives in the soil, aiding in soil formation, transfer, aeration, and drainage. Cf. *geobiont, geoxene.*

geodes (*Geol.*). Large cavities in rocks, lined with crystals that were free to grow inwards. (*Mining*) Rounded nodules of ironstone with hollow interior.

geodesic (*Maths.*). The shortest path between two points on any surface.

geodesic structures (*Arch.*). Structures consisting of a large number of few identical parts and therefore simple to erect; and whose pressure is loadshed throughout the structure, so that the larger it is, the greater its strength.

geodesy (*Surv.*). Branch of surveying concerned with extensive areas, in which, to obtain accuracy, allowance must be made for the curvature of the earth's surface. Also **geodetic surveying.**

geodetic construction (*Aero.*). A redundant space frame whose members follow diagonal geodesic curves to form a lattice structure, such that compression loads induced in any member are braced by tension loads in crossing members.

geodetic surveying (*Surv.*). See **geodesy.**

geognosy (*Geol.*). An old term for absolute knowledge of the earth, as distinct from geology, which includes various theoretical aspects.

geographical latitude. See under **latitude and longitude** (*terrestrial*).

geographical mile. The length of one minute of latitude, a distance varying with the latitude, and having a mean value of 6076·8 ft. (1852·2 m). (U.S.) One minute of longtitude at the equator, i.e., 6087·1 ft. (1855·3 m).

geographical race (*Zool.*). A collection of individuals within a species, which differ constantly in some slight respects from the normal characters of the species, but not sufficiently to cause them to be classified as a separate species, and which are peculiar to a particular area.

geoid (*Geophys., Surv.*). A gravitational equipotential surface at approximately mean sea level used as the datum level for gravity surveying.

geo-isotherms (*Geol.*). Surfaces of equal temperature in the Earth below ground level.

geological time (*Geol.*). The time extending from the end of the Formative Period of earth history to the beginning of the Historical Period. It is conveniently divided into several Periods, each being the time of formation of one of the Systems into which the Stratigraphical Column is divided. Thus the *Carboniferous Period* is the interval of time during which the rocks including the Carboniferous Limestone, Millstone Grit, and Coal Measures in Britain, and the Mississippian and Pennsylvanian strata in the U.S., were in the process of formation. The complete list of Periods from the oldest to the youngest is: Pre-Cambrian, Cambrian, Silurian, Devonian, Carboniferous, Permian, Triassic, Jurassic, Cretaceous, Tertiary, and Quaternary.

geology. The science which investigates the history of the earth's crust, from the earliest times to the commencement of the Historical Period. It deals with the compositions, arrangement, and origins of the rocks of the earth's crust, and with the processes involved in the evolution of its present structure. Geology is now divided into several branches: *physical geology*, the study of the processes of sedimentation and denudation, the work of the atmosphere, water, ice, rivers, and the sea, the study of rock structures; *petrology*, the study of the nature, composition, textures, and origins of igneous, metamorphic, and sedimentary rocks and of the metallic ores; *mineralogy*, the study of the compositions, physical characters (including crystal form) of the natural minerals; *stratigraphy* or *historical geology*; and *palaeontology* (with *palaeobotany*), which traces the history of life on this planet and the structures and relationships between the several kinds of organisms.

geomagnetic (*Geophys.*). Pertaining to the natural magnetization of the earth.

geometrical attenuation (*Phys.*). Reduction in intensity of radiation on account of the distribution of energy in space, e.g., due to inverse-square law, or progression area along the axis of a horn.

geometrical cross-section (*Nuc.*). Area subtended by a particle or nucleus. This does not usually resemble the interaction cross-section.

geometrical optics (*Optics*). The study of optical problems based on the conception of light rays. See **physical optics.**

geometrical stair (*Build.*). A stair arranged about a well-hole and curved between the successive flights.

geometric capacitance (*Elec.*). That of an isolated conductor *in vacuo*, uninfluenced by dielectric material, and depending only on shape.

geometric distortion (*TV*). Any departure from the original perspective in reproduction (apart from definition). See **keystone distortion.**

geometric mean (*Maths.*). Of *n* positive numbers a_r: the *n*th root of their product,

$$\text{i.e.,}\quad \left(\Pi\, a_r\right)^{\frac{1}{n}}.$$

geometrodynamics. The interpretation of physical phenomena in terms of 4-dimensional space-time geometry.

geometry (*Maths.*). A branch of mathematics concerned generally with the properties of lines, curves and surfaces. Usua ly divided into *pure, algebraic* and *differential geometry* in accordance with the mathematical techniques utilized. (*Nuc.*) In radiation, the experimental limitations introduced by the finite geometry of sources and detectors which make measurements (e.g., of energy) ambiguous.

geometry factor (*Nuc.*). $1/(4\pi)$ of the solid angle subtended by the window or sensitive volume of a radiation detector at the source.

geomorphology (*Geophys.*). The study of the physical features of the earth and of the relation between the physical features and the geological structures beneath.

geonasty (*Bot.*). Curvature towards the ground.

geophagous (*Zool.*). Earth-eating.

geophilic (*Bot.*). Growing in soil.

geophilous (*Bot.*). Having a short stout stem with rather large leaves, borne at soil level. (*Zool.*) Living on, or in, the soil.

geophone (*Geophys.*). A small, portable instrument used in seismic prospecting to record the arrival of shock waves from the artificial source. A number of geophones are normally used simultaneously so that relative times of arrival at measured distances can be observed. See **seismic prospecting.**

geophysical prospecting (*Mining*). Prospecting by measuring difference in the density, electrical resistance, magnetic properties, elastic properties and radioactivity of the earth's crust.

geophysics. Study of physical properties of the earth; it makes use of the data available in geodesy, seismology, meteorology, and oceanography, as well as that relating to atmospheric electricity, terrestrial magnetism, and tidal phenomena. Applied geophysics has, by means of electrical, magnetic, gravitational, seismic, and other methods, achieved many discoveries of geological and economic importance below the earth's surface.

geophyte (*Bot.*). A plant which perennates by means of subterranean buds. *adj.* **geophytic.**

geoplagiotropic (*Bot.*). Growing in a direction at an angle to the ground surface.

George (*Aero.*). Colloquialism for **automatic pilot.**

georgette (*Textiles*). A light-weight silk fabric with a crêpe effect, used as dress material. A similar material is made from fine hard-twisted cotton, nylon staple or spun rayon yarns.

Georgian glass (*Glass*). A reinforced fire-resisting glass.

geostrophic force (*Meteor.*). A virtual force used to account for the change in direction of the wind relative to the surface of the earth, arising from the earth's rotation and the consequent *Coriolis Effects.*

geostrophic wind (*Meteor.*). The theoretical wind arising from the *pressure gradient force* (q.v.) and the *geostrophic force* (q.v.).

geosyncline (*Geol.*). A long, relatively narrow area of marine sedimentation, within which the maximum thickness of sediment is found in the central zone, which also experiences the maximum depression. Thus the floor of the geosyncline is synclinal in form, but ultimately it may be silted up to sea-level, even in the centre. Generally, by the introduction of heat and pressure, geosynclines become the sites of mountain building movements and later mountain ranges. The process is, however, not inevitable.

geotaxis (*Biol.*). The locomotory response of a motile organism or cell to the stimulus of gravity. *adj.* **geotactic.** Cf. *geotropism.*

geotechnical process (*Civ. Eng.*). Means employed to alter the properties of soil for construction purposes. These include compaction and vibration, soil stabilization by the addition of cement, bitumen emulsions, etc., the latter sometimes carried out by injection.

geotectonic (*Geol.*). Relating to the structure of rock masses. **Geotectonics** is the science of structural geology.

geothermal gradient (*Geol.*). The rate at which the temperature of the earth's crust increases with depth.

geotome (*Bot.*). An instrument used for taking soil samples without disturbing the surrounding soil.

geotropism (*Biol.*). The reaction of a plant member or sessile animal to the stimulus of gravity, shown by a growth curvature; cells become more elongated on one side than the other, tending to bring the axis of the affected part into a particular relation to the force of gravity. *adj.* **geotropic.** Cf. *geotaxis.*

geoxene (*Ecol.*). Organisms which occur only accidentally in soil, having little effect on it. Cf. *geobiont, geocole.*

Gephyrea (*Zool.*). A loose term once used by some zoologists to embrace the *Echiuroidea* and the *Sipunculoidea*, and sometimes the *Priapuloidea*. Now obsolete.

gephyrocercal (*Zool.*). Of a secondarily simplified type of tail-fin, resembling the diphycercal type but derived from the homocercal or heterocercal type by reduction.

geranial (*Chem.*). See **citral.**

Geraniales (*Bot.*). An order of the dicotyledons. The perianth in two separate whorls, of separate members, may be absent, but usually in fives. The stamens usually in two whorls of five or the outer whorl absent. The ovary is superior with two to five carpels. There are one to two ovules in each carpel. Trees, shrubs, or herbs.

geraniol (*Chem.*). $C_{10}H_{18}O$. A terpene alcohol forming a constituent of many of the esters used in perfumery.

Gerhardt's test (*Chem.*). A test for detecting diethanoic acid in the urine, based upon the formation of a deep-red colour when iron (III) chloride is added and all precipitated phosphates are filtered off.

Geriach compensator (*Acous.*). The realization of an acoustic pressure by balancing it against the pressure arising in a gold foil in a magnetic field carrying a current which is adjustable in magnitude and phase.

geriatrics (*Med.*). The specialized medical care of the elderly and aged. It is a branch of the science of *gerontology*.

Gerlier's disease (*Med.*). Paralysing vertigo. A disease, endemic in the Canton of Geneva and in the north of Japan, in which a sudden attack of pain in the neck and back is accompanied by dizziness and temporary paralysis.

germ (*Zool.*). The primitive rudiment which will develop into a complete individual, as a fertilized egg or a newly formed bud; a unicellular micro-organism.

germanium (*Chem.*). A metalloid element in the fourth group of the periodic system. Greyish-white in appearance. Symbol Ge, at. no. 32, r.a.m. 72·59, rel. d. 5·47, m.p. about 958°C. Germanium occurs in a few minerals including coal. The chief mineral sources are *argyrodite* (q.v.), also certain zinc and copper concentrates, and to some extent the flue dusts of gasworks. The main use of the metal, which has exceptional properties as a semiconductor, is in the manufacture of solid rectifiers or diodes in microwave detectors, and, in a highly pure state, in transistors.

germanium radiation detector (*Nuc. Eng.*). See **lithium-drift germanium detector, semiconductor radiation detector.**

germanium rectifier (*Elec. Eng., Electronics*). A *p-n* junction diode often with high current rating. It requires a lower forward voltage than a silicon diode, but is has higher reverse leakage current.

German lapis (*Min.*). See **Swiss lapis.**

German measles (*Med.*). See **rubella.**

German nozzle (*Eng.*). See **parabolic nozzle.**

German siding (*Build.*). Weather-boards finished with a hollow curve along the outside of the top edge, and rebated along the inside of the lower edge.

German silver (*Met.*). A series of alloys containing copper, zinc, and nickel within the limits: copper, 25–50%; zinc, 10–35%; and nickel, 5–35%. Also **nickel silver.**

germarium (*Zool.*). The formative area of a testis or ovary, in which the growth and maturation of the germ cells takes place.

germ band (*Zool.*). In Insects, a ventral plate of cells, produced in the egg by cleavage, which later gives rise to the embryo.

germ cells (*Zool.*). In *Metazoa*, special reproductive cells which are liberated by the organism and in which the qualities of the organism are inherent. Gametes; spermatozoa and ova, or the cells which give rise to them.

germ-centre (*Zool.*). See **Flemming's germ-centres.**

germen (*Bot.*). See **ovary** (1). (*Zool.*) The primary mass of undifferentiated cells which will give rise to the germ cells. Cf. *soma*.

germinal aperture or **germinal pore** (*Bot.*). See **germ pore.**

germinal cells (*Zool.*). See **germ cells.**

germinal disk (*Zool.*). The flattened circular region at the top of a megalecithal ovum, in which cleavage takes place.

germinal epithelium (*Zool.*). A layer of columnar epithelium which covers the stroma of the ovary in Vertebrates.

germinal layers (*Zool.*). See **germ layers.**

germinal spot (*Zool.*). The nucleolus of the *germinal vesicle* (q.v.).

germinal vesicle (*Zool.*). The nucleus of the oöcyte, before formation of polar bodies.

germination (*Bot.*). The beginnings of growth in a spore or seed.

germination by repetition (*Bot.*). The production of secondary spores instead of a germ tube.

germiparity (*Zool.*). Reproduction by formation of germs.

germ layers (*Zool.*). The three primary cell-layers in the development of *Metazoa*, i.e., ectoderm, mesoderm, and endoderm.

germ nucleus (*Zool.*). See **pronucleus.**

germ plasm (*Gen.*). In early theories of inheritance, a specific nuclear substance which was supposed to be the bearer of the hereditary characters and to remain unchanged in spite of the differentiation of the body. Also called **gonoplasm, idioplasm.**

germ pore (*Bot.*). (1) A thin, usually rounded area in the wall of a pollen grain through which the pollen tube emerges. (2) A similar spot in a spore wall from which the germ tube develops.

germ sporangium (*Bot.*). A sporangium formed at the end of a germ tube produced by a zygospore.

germ stock (*Zool.*). The stolon of *Urochorda.*

germ theory (*Gen.*). The theory that all living organisms can be produced only from living organisms.

germ track (*Zool.*). The sequence of cell-generations through which the germ cells are connected with the fertilized ovum.

germ tube (*Bot.*). The tubular outgrowth put out by a germinating spore, from which the thallus develops by subsequent branching or on which a germ sporangium is formed.

germ vitellarium (*Zool.*). An organ of some *Turbellaria*, part of which functions as a gonad and part as a vitelline gland. **Germ yolk gland.**

gerontic (*Zool.*). Pertaining to the senescent period in the life history of an individual.

gerontogaeous (*Bot.*). Belonging to, or originating in, the Old World.

gerontology (*Med.*). The scientific study of the processes of aging, and of the problems and diseases of old age. Cf. *geriatrics*.

gersdorffite (*Min.*). Metallic grey sulphide-arsenide of nickel(II), occurring as cubic crystals or in granular or massive forms.

gesso (*Paint.*). A pasty mixture of whiting, prepared with size or glue, applied to a surface as a basis for painting or gilding.

gestalt (*Psychol.*). German, 'form', 'pattern', 'configuration'. An organized whole, e.g., a living organism, a melody, a picture, the solar system, in which each individual part affects every other, the whole being more than a sum of its parts. The *Gestalt School of Psychology* was founded by Wertheimer, Köhler, and Koffka, who conducted experiments, on humans and animals, in the fields of perception, learning, and intelligence. They demonstrated the tendency of the mind to perceive situations as a whole (a 'pattern', *gestalt*), rather than as a number of isolated elements or sensations.

gestation (*Zool.*). In Mammals, the act of retaining and nourishing the young in the uterus; pregnancy.

get (*Mining*). To win or mine.

getter (*Build.*). A familiar name applied to a workman engaged in loosening the earth in an excavation. (*Electronics*) Material (K, Na, Mg,

Ca, Sr, or Ba), used, when evaporated by high-frequency induction currents, for *cleaning* the vacuum of valves, after sealing on the pump line during manufacture.

gettering discharge (*Electronics*). That used to assist the *getter* in the *clean up* of vacuum in valves, through ionization of remaining gas molecules.

GeV (*Nuc.*). Abbrev. for *giga-electron-volt*; unit of particle energy, 10^9 electron volts, $1{\cdot}602 \times 10^{-10}$J. In U.S., sometimes **BeV**.

geyser (*Geol.*). A volcano in miniature, from which hot water and steam are erupted periodically instead of lava and ashes, during the waning phase of volcanic activity. Named from the Great Geyser in Iceland, though the most familiar example is probably 'Old Faithful' in the Yellowstone Park, Wyoming. The eruptive force is the sudden expansion which takes place when locally heated water, raised to a temperature above boiling point, flashes into steam. Until the moment of eruption, this had been prevented by the pressure of the super-incumbent column of water in the pipe of the geyser, which is usually terminated upwards by a sinter crater. Also **gusher.** (*Plumb.*) A water-heating appliance providing supplies of hot water rapidly for domestic purposes, the source of heat being gas or electricity. *Hot-water heater* in U.S.

geyserite (*Min.*). See **sinter.**

G-gas (*Nuc.*). Gaseous mixture (based on helium and isobutane) used in low-energy β-counting (e.g., of tritium) by gas-flow proportional counter.

ghaut (*Build.*). A landing-stage stair on a river-side.

Ghon's focus (*Med.*). The first part of the lung to be infected in pulmonary tuberculosis, constituting the primary lesion of the disease.

ghost (*Cinema.*). Vertical streaks on high-lights in a projected picture, arising from incorrect phasing of the rotary shutter with respect to the moving film. (*Met.*) In steel, a band in which the carbon content is less than that in the adjacent metal and which therefore consists mainly of ferrite. Also **ghost line.** (*TV*) Duplicated image on a television screen, arising from additional reception of a delayed, similar signal which has covered a longer path, e.g., through reflection from a tall building or mast. Also called **double image.**

ghost crystal (*Min.*). A crystal within which may be seen an early stage of growth, outlined by a thin deposit of dust or other mineral deposit.

ghost image (*Light*). The image arising from a mirror when the rays have experienced reflection within the glass between the surface and the silvering.

ghost line (*Met.*). See **ghost.**

giant (*Mining*). Monitor or hydraulic giant. Nozzle mounted on swivel bearings, used to direct stream of high-pressure water in alluvial (placer) mining of soft ground.

giant cells (*Zool.*). Cells of unusual size, as the myeloplaxes of bone-marrow; certain cells of the excitable region of the cerebrum; certain cells sometimes found in lymph-glands; large multinucleate cells of the thymus gland and of the spleen pulp; abnormally large neurocytes in *Annelida*; *osteoclasts* (q.v.)

giant fibres (*Zool.*). In some Invertebrates (e.g., earthworms) and in *Cephalochorda*, certain enlarged motor nerve-fibres of the ventral nerve cord, believed to serve for the occasional violent discharge of nervous energy in case of emergency.

giantism (*Med.*). Uniformly excessive growth of the body, due to overactivity of the anterior lobe of the pituitary gland. Also **gigantism.**

giant powder (*Mining*). Dynamite.

giant source (*Nuc.*). Large source of radioactivity, e.g., 150 000 curies of ^{60}Co, used for industrial sterilization of packed food, or chemical processing (e.g., cross-linking of polymers).

giant star (*Astron.*). A star which is more luminous than the main sequence stars of the same spectral class. Smaller groups of sub-giants and supergiants are recognized.

Gianuzzi's crescents (*Zool.*). Groups of serous or mucous cells lying next to the basement membrane, in the salivary glands of Vertebrates.

giardiasis (*Med.*). Infestation of the intestinal tract with the flagellate protozoon *Giardia lamblia*, sometimes causing severe diarrhoea.

gib (*Eng.*). (1) A metal piece used to transmit the thrust of a wedge or cotter, as in some connecting-rod bearings. (2) A brass bearing surface let into the working face of a steam-engine cross-head. (3) A tapered or parallel strip in bearings for reciprocating slides, used to fit or to clamp the slide in the guide. (*Mining*) A *sprag* (q.v.).

gibberellic acid (*Bot.*). A plant hormone controlling cell elongation (see **auxin**).

gibberellins (*Bot.*). Group of compounds occurring as *auxins* (q.v.) in plants, the first example of which was isolated from the fungus *Gibberella fujikoroi*.

gibbous (*Astron.*). The word applied to the phase of the moon, or of a planet, when it is between either quadrature and opposition, and appears less than a circular disk but greater than a half disk. (*Bot.*) (1) Swollen, especially to one side. (2) Pouched. (3) Convex above and flat below; hump-backed. Also **gibbose.**

Gibbs' adsorption theorem (*Chem.*). Solutes which lower the surface tension of a solvent tend to be concentrated at the surface, and conversely.

Gibbs-Duhem equation (*Chem.*). For binary solutions at constant pressure and temperature, the chemical potentials (μ_1, μ_2) vary with the mole fractions (x_1, x_2) of the two components as follows:

$$\left(\frac{\partial \mu_1}{\partial \ln x_1}\right)_{T,P} = \left(\frac{\partial \mu_2}{\partial \ln x_2}\right)_{T,P}.$$

Gibbs' function (*Chem.*). See **free energy.** Also called **Gibbs' free energy, thermodynamic potential.**

Gibbs-Helmholtz equation (*Chem.*). An equation of thermodynamics,

$$\mathrm{F}-\Delta\mathrm{F}=-\Delta U+T\cdot\frac{\mathrm{d}(-\Delta\mathrm{F})}{\mathrm{d}T},$$

where $-\Delta\mathrm{F}$ = decrease in free energy; $-\Delta U$ = decrease in intrinsic energy; T = the absolute temperature. It is applied to the voltaic cell in the form:

$$yFE=-q+yFT\cdot\frac{\mathrm{d}E}{\mathrm{d}T}$$

where y = number of gram equivalents of chemical change, F = faraday, E = e.m.f. of the cell, and $-q$ = heat of reaction.

gibbsite (*Min.*). Hydroxide of aluminium, $Al(OH)_3$, occurring as minute mica-like crystals, concretional masses, or incrustations. An important constituent of bauxite. Also called **hydrargillite.**

Gibbs-Konowalow rule (*Chem.*). For the phase equilibrium of binary solutions. At constant pressure the equilibrium temperature is a maximum or minimum when the compositions of the two phases are identical, and vice versa, e.g., in eutectics or azeotropes. The corresponding statements hold for pressure at constant temperature.

Gibbs' phase rule (*Chem.*). See **phase rule.**

gib-headed key (*Eng.*). A key for securing a wheel, etc., to a shaft, having a head formed at right angles to its length.

giblet check (*Build.*). An exterior rebate for a door which opens outwards.

Gibraltar fever (*Med.*). See **undulant fever.**

gid (*Vet.*). See **coenuriasis.**

Giemsa stain (*Micros.*). A mixture of methylene blue and eosin, used particularly for staining micro-organisms, blood films, and tissue culture cells.

Gies' biuret reagent (*Chem.*). A reagent for testing the presence of proteins by the biuret reaction; it consists of a solution of 10% KOH and 0·075% copper (II) sulphate.

Giffard's injector (*Eng.*). The original steam *injector* (q.v.).

gig (*Mining*). Winding engine (Scots); 2-deck mine cage; bowk.

giga-. Prefix used to denote 10^9 times, e.g., a gigawatt is 10^9 watts.

giga-electron-volt (*Nuc.*). See **GeV.**

gigantism (*Bot.*). Abnormal increase in size, often associated with polyploidy. (*Med.*) See **giantism.**

gigantoblast. A very large *erythroblast*.

gigantocyte. A very large *erythrocyte*.

Gigartinales (*Bot.*). An order of the *Rhodophyceae*, producing tetraspores; the auxiliary cell is a vegetative cell of the gametophyte.

giggering (*Bind.*). A method of producing lines on the back of a volume by means of a catgut cord.

gig stick (*Build.*). See **radius rod.**

gilbert (*Elec. Eng.*). In the CGS electromagnetic system of units, m.m.f. of an enclosing coil equal to $10/4\pi$ ampere-turn.

gilding (*Paint.*). Operation of finishing painted surfaces with oil gold-size and finally coating with gold-leaf.

gilding metal (*Met.*). Copper-zinc alloy containing zinc up to 10%.

Giles valve (*Elec. Eng.*). A form of lightning arrester in which capacitors are used in conjunction with ordinary spark-gaps.

gill (*Bot.*). One of the vertical plates of tissue that bears the hymenium in an agaric. (*Zool.*) A membranous respiratory outgrowth of aquatic animals, usually in the form of thin lamellae or branched filamentous structures; in *Salpidae*, the dorsal hyperpharyngeal bar which represents the remnant of the branchial chamber.

gill arch (*Zool.*). In Fish, the incomplete jointed skeletal ring supporting a single pair of gill slits; one segment of the branchial basket.

gill bars (*Zool.*). See **gill rods.**

gill basket (*Zool.*). In Fish and Cyclostomes, the skeletal framework which supports the gills and gill slits.

gill book (*Zool.*). The booklike respiratory lamellae of *Xiphosura* borne by the opisthosoma, of which they represent the appendages. Also called **book gill.**

gill cavity (*Bot.*). A ring-shaped hollow in the young fruit body of an agaric, within which the early stages of the organization of the gills are completed.

gill clefts (*Zool.*). See **gill slits.**

gill cover (*Zool.*). See **operculum.**

gillion. 10^9; preferably *giga-* (q.v.) or G.

Gill-Morrell oscillator (*Electronics*). Triode valve with anode and grid fed through a Lecher system, oscillation depending on the *transit time* of the electrons in the valve, but with frequency related to the Lecher line. See **Barkhausen-Kurz oscillator.**

gill net (*Ocean.*). A fixed vertical net, having the head-rope buoyed and the bottom-rope weighted, in the meshes of which fish become entangled by their gill covers.

gill plume (*Zool.*). See **ctenidium.**

gill pouch (*Zool.*). One of the pouchlike gill slits of Cyclostomes and Fish.

gill rakers (*Zool.*). In some Fish, small processes of the branchial arches, which strain the water passing out via the gill slits and prevent the escape of food-particles.

gill rods (*Zool.*). In *Cephalochorda*, skeletal bars which support the pharynx. Also called **gill bars.**

gills (*Aero.*). Controllable flaps which vary the outlet area of an aircooled engine cowling or of a radiator; also **cooling gills, cowl flaps, radiator flaps.** (*Heat*) Ribs which project from heating surfaces and serve to increase the effective radiation area.

gill slits (*Zool.*). In *Chordata*, the openings leading from the pharynx to the exterior, on the walls of which the gills are situated. Also called **branchial clefts, gill clefts.**

gilsonite (*Min.*). See **uintaite.**

Gilson's glands (*Zool.*). Paired segmental excretory glands in the thorax of the larvae of many *Trichoptera*.

gilvous (*Bot.*). Brownish.

gimbal mount (*Instr.*). One giving rotational freedom about two perpendicular axes—as used for gyroscope and nautical compass.

gimbals (*Horol.*). See **gymbals.**

gimlet (*Carp.*). A small hand tool with a wooden cross-handle for boring holes in wood. All types have a rough-pitch screw at the point, but the shank may be of the shell type (see **shell gimlet**), half-twist, or auger type.

gimmick (*Elec. Eng.*). Colloquialism for small capacitor formed by twisting insulated wires.

gimp (*Acous.*). A type of extraneous noise arising in disk recording, becoming apparent on reproduction. (*Textiles*) A fancy yarn consisting of a core of hard twisted thread, the covering being formed by one or more soft spun threads twisted the opposite way.

gimped (*Bot.*). Crenate.

gin (*Eng., Min.*). (1) A hand hoist which consists of a chain or rope barrel supported in bearings and turned by a crank. (2) A portable tripod carrying lifting tackle.

ginger-beer plant (*Bot.*). A symbiotic association of a yeast and a bacterium, which ferments a sugary liquid containing oil of ginger, giving ginger beer. Often known popularly as **Californian bees,** and by similar names.

gingham (*Textiles*). A coloured cotton cloth, generally with a check pattern; used for summer dresses, aprons, blouses, etc.

ginging (*Civ. Eng., Mining*). The process of lining a shaft with bricks or masonry; also the lining itself.

gingival (*Med., Zool.*). In Mammals, pertaining to the gums.

gingivectomy (*Med.*). The cutting back of inflamed or excess gum.

gingivitis (*Med.*). Inflammation of the gums.

Ginglymodi (*Zool.*). An order of *Holostei*, also called **Semionotoidea, Lepisosteiformes.** See **Semionotoidea.**

ginglymus (*Zool.*). An articulation which allows motion to take place in one plane only; a hinge-joint. *adj.* **ginglymoid.**

Ginkgoales (*Bot.*). An order of the Gymnosperms, with one living species, *Ginkgo biloba*. The leaves are borne on spur-shoots. The microspores are produced in lax cones, and the ovules in pairs at the tips of short branches. The plants are monoecious and the microspores are ciliate.

gin-pit (*Mining*). A shallow shaft operated by a gin.

Giorgi system (*Elec.*). System of units proposed in 1904 and later adopted as the MKSA system, the common *practical units* of ohm, volt, ampere, etc., becoming identified with *absolute units* of the same entities. See **SI units.**

GIP (*Paper*). See **imitation parchment.**

girasol (*Min.*). A variety of fire opal of a bright hyacinth-red colour; the finest specimens show a faint bluish opalescence emanating from the centre of the stone.

girder (*Bot.*). An arrangement of the mechanical tissue of a stem or leaf in such a way that effective support is given to the member. (*Eng.*) A beam, usually steel, to bridge an open space. Girders may be rolled sections, built up from plates, or of lattice construction. See also **continuous beam.**

girder bridge (*Civ. Eng.*). A bridge in which the loads are sustained by beams (generally compound) resting across the bridge supports.

girder casing (*Build.*). Material totally enclosing the projecting part of a girder below the general ceiling surface.

girdle (*For.*). A continuous incision which is made all round a bole, cutting through at least bark and cambium, generally with the object of killing the tree. (*Mining*) A thin bed or layer of stone or of coal. (*Zool.*) In Vertebrates, the internal skeleton to which the paired appendages are attached, consisting typically of a U-shaped structure of cartilage or bone with the free ends facing dorsally. In *Amphineura*, the part of the mantle which surrounds the shells.

girdle structure (*Bot.*). A type of leaf structure in which the tubular photosynthetic cells are arranged in radial rows or in a curved pattern converging towards the central strands of vascular tissue.

girdling (*Bot.*). The condition in which a leaf trace arises in the stem on the opposite side from the leaf to which it belongs, and curves widely through the cortex before entering the leaf base.

Girod furnace (*Elec. Eng.*). An old form of d.c. arc furnace in which the arc is maintained between an electrode and the charge.

girt (*Mining*). Timbers set horizontally between *caps* (timbers on hanging walls supported by posts set normal to dip of lode, or vertically when stoping massive deposits). System thus forms a continuous framework of *square sets*.

girth (*For.*). See **exploitable-, mid-, mid-timber-, quarter-, top-.**

girth class (*For.*). One of the intervals into which the range of girth of trees or logs is divided for classification or use; also the trees or logs falling into such an interval.

girt strip (*Carp.*). See **ribbon strip.**

gismondine (*Min.*). A rare zeolite; a hydrated calcium aluminium silicate which occurs in the basaltic lavas of Antrim.

git (*Foundry*). See **gate.**

give-and-take lines (*Surv.*). Straight lines drawn on a plan of any area having irregular boundaries, each line following the trend of a part of the boundary so that any small piece that it cuts off the area is balanced by an equal piece added by it.

Givetian Stage (*Geol.*). A division of the marine Devonian System which includes the massive Middle Devonian limestones, well exposed at the type locality of Givet-sur-Meuse. The limestones of Brixham, Torquay, and Plymouth are of the same age.

gizzard (*Zool.*). See **proventriculus.**

Gl (*Chem.*). The symbol for *glucinum*, an old name for *beryllium*.

glabella (*Zool.*). The middle region of the head in *Trilobita*, separated from the pleural portions by longitudinal grooves.

glabrescent (*Bot.*). (1) Almost but not quite without hairs. (2) Becoming almost hairless as it matures.

glabrous (*Bot.*). Without hairs on the surface; hairless. (*Zool.*) Having a smooth hairless surface.

glacial acetic acid (*Chem.*). Pure concentrated acetic (ethanoic) acid. Owing to its comparatively high m.p. (16·6°C) it solidifies easily, forming icelike crystals.

glacial action (*Geol.*). This comprises: (*a*) the grinding, scouring, plucking, and polishing effected by ice, armed with rock fragments frozen into it; and (*b*) the accumulation of rock debris resulting from these processes. The extent to which melt-waters derived from the ice are responsible for both aspects of glacial action is an open question.

glacial denudation (*Geol.*). Disintegration of rocks consequent upon glacial conditions. The extent to which the enormous amount of erosion in the Pleistocene Period was directly the work of ice is a disputed question, some believing that ice affords a protective covering, and that the erosion is effected by melt-waters, chiefly during the retreat of the ice-sheets.

glacial deposits (*Geol.*). These include spreads of boulder clay, sheets of sand and gravel occurring as outwash fans, outwash deltas, and kames; also deposits of special topographical form, such as drumlins and eskers.

glacial phosphoric acid (*Chem.*). See **metaphosphoric acid.**

glacial sands (*Geol.*). These cover extensive areas in advance of sheets of boulder clay, and together with glacial (largely fluvioglacial) gravels, represent the outwash from great ice-sheets. During the interglacial periods, too, much sand and gravel was deposited during the Pleistocene Period.

glaciation (*Geol.*). The subjection of an area to glacial conditions, with the development of an ice-sheet on its surface. Britain was subjected to glaciation during the Pleistocene Period, the area north of a line joining the Thames to the Severn being covered by ice-sheets which originated in the mountainous areas of Scotland and Wales. Three great ice-sheets, the Cordilleran, the Keewatin, and the Labradoran, covered the northern half of N. America to beyond the Canadian border, with a marked protrusion southwards in the region of the Great Lakes. In this period, four distinct glaciations were experienced, separated by Interglacial Periods. In countries near the Equator, glacial conditions were widespread in the Permo-Carboniferous Period. See **glacial action, glacier.**

glacier (*Geol.*). Usually defined as a river of ice. Three varieties may be recognized: (*a*) the valley glacier, such as the Mer de Glace; (*b*) the corrie glacier; (*c*) the Piedmont glacier, which overflows from a valley that it occupies and spreads out over the plain at its foot. A glacier is constantly fed by the accumulation of snow, which is compressed by pressure into ice that moves slowly downhill, carrying large amounts of detritus on its surface and embedded in the ice. This load of detritus forms the moraines.

glacier lake (*Geol.*). See **lake**.

glacis (*Civ. Eng.*). An inclined bank.

gladiate (*Bot.*). Shaped like a sword blade.

gladiolus (*Zool.*). In some Mammals, a large bone formed by the fusion of the sternebrae.

Gladstone and Dale law (*Phys.*). The law relating the variation in the density and refractive index of a material when it is subjected to pressure or a change of temperature:

$$\frac{n+1}{\rho}=\text{constant},$$

where n = refractive index, ρ = density.

glair (*Bind.*). A preparation made from white of egg and vinegar, used as the adhesive for gold-leaf in gold-finishing and blocking.

glance (*Min.*). Opaque mineral with a resinous or shining lustre.

glancing angle (*Phys.*). The complement of the *angle of incidence* (q.v.).

gland (*Bot.*). A cell or group of cells, inside or on the surface of the plant, or a multicellular outgrowth of special form, secreting some substance, often oily or resinous, sometimes containing digestive enzymes. (*Eng.*) (1) A device for preventing leakage at a point where a rotating or reciprocating shaft emerges from a vessel containing a fluid under pressure. (2) A sleeve or nut used to compress the packing in a *stuffing-box* (q.v.). (*Zool.*) A single epithelial cell, or an aggregation of epithelial cells, specialized for the elaboration of a secretion useful to the organism, or of an excretory product. *adj.* **glandular.**

gland bolts (*Eng.*). Bolts for holding and tightening down a gland.

gland cell (*Zool.*). A unicellular gland, consisting of a single goblet-shaped epithelial cell producing a secretion, usually mucus.

glanders (*Med.*, *Vet.*). A contagious bacterial disease of horses, mules, and asses, due to infection by *Actinobacillus mallei* (*Malleomyces mallei*); inflammatory nodules occur in the respiratory passages and lungs and also in other parts of the body. Infection of the lymphatics under the skin is known as **farcy**. Glanders is communicable to man.

glandular epithelium (*Zool.*). Epithelial tissue specialized for the production of secretions.

glandular fever (*Med.*). See **infectious mononucleosis.**

glandular serrate (*Bot.*). Having a margin consisting of short teeth tipped with glands.

glandular tissue (*Zool.*). See **glandular epithelium.**

glans (*Bot.*). A hard, dry, indehiscent fruit, containing one or a few seeds, derived from an inferior ovary, and more or less surrounded by a cupule; the acorn is a familiar example. (*Zool.*) A glandular structure.

glans penis (*Zool.*). A dilatation of the extremity of the mammalian penis.

glare (*Light*). The visual discomfort experienced by observers in the presence of a visible source of light. The term can also refer to the visual disability produced by the presence of visible sources in the field of view when these sources do not assist the viewing process.

glarimeter (*Paper*). An instrument for measuring the gloss of a paper surface based on light reflectance.

Glaserian fissure (*Zool.*). In Mammals, a fissure of the temporal bone, which receives the Folian process of the malleus.

glass. A hard, amorphous, brittle substance, made by fusing together one or more of the oxides of silicon, boron, or phosphorus, with certain basic oxides (e.g., sodium, magnesium, calcium, potassium), and cooling the product rapidly to prevent crystallization or devitrification. The melting point varies between 800°C and 950°C. The tensile strength of glass resides almost entirely in the outer skin; if this is scratched or corroded, the glass is much more easily broken. See also **natural glass.**

glass blocks (*Arch.*, *Build.*). Hollow blocks made of glass usually with a patterned surface. Used where translucence and decorative effect are required. They also provide insulation against heat and sound. Sometimes called **glass bricks.**

glass-bulb rectifier (*Elec. Eng.*). A mercury-arc rectifier in which the arc takes place within a glass bulb. Cf. *steel-tank rectifier.*

glassed (*Build.*). A term applied to stones such as granite and marble which are highly polished by being held against a revolving disk covered with felt.

Glasser's disease (*Vet.*). A disease of young pigs characterized by fever, swollen joints and lameness; believed to be caused by a mycoplasma-like organism.

glass fibre (*Glass*, *Textiles*). Glass melted and then drawn out by steam through special bushings into fibres of 5–10 micrometres diameter, which may be spun continuously into threads and woven into tapes and cloths by normal process, or may be formed into pads and quiltings, rigid, bitumen-bonded or loose.

glass-fibre paper (*Paper*). A sheet formed by using glass fibres and a binder instead of conventional vegetable fibres.

glassine (*Paper*). A transparent glazed wrapper paper; also used in manifold books. Produced by long beating and high glazing. Also **imitation parchment, transparent parchment.**

glasspaper (*Paper*). Paper coated with glue on which is sprinkled broken glass of a definite grain size; used for rubbing down surfaces. Cf. *sandpaper.*

glass support-rod (*Light*). The glass rod which supports the filament of an electric filament-lamp.

glass tile (*Build.*). A small glass sheet in a roof, bonded in with slates, plain tiles, or pantiles, to admit light within the roof space.

glass wool (*Arch.*, *Build.*, *Chem.*). A felt of fine glass fibres; used as a relatively inert filter, packing, or insulation.

glassy (*For.*). See **wetwood.**

glauberite (*Min.*). Monoclinic sulphate of sodium and calcium, occurring with rock salt, anhydrite, etc., in saline deposits.

Glauber salt (*Min.*). Properly termed *mirabilite* (hydrated sodium sulphate, $Na_2SO_4{\cdot}10H_2O$). A monoclinic mineral formed in salt lakes, deposited by hot springs, or resulting from the action of volcanic gases on sea water. Obtained from Austria and the Great Salt Lake (Utah).

glaucescent (*Bot.*). See **glaucous.**

glaucodot (*Min.*). A tin-white orthorhombic sulph-arsenide of iron and cobalt, occurring with cobaltite in Huasco Province, Chile. Also spelt **glaucodote.**

glaucoma (*Med.*). An eye condition in which, from various causes, the intra-ocular pressure rises, making the eyeball hard and causing partial or total loss of sight.

glauconite (*Min.*). Hydrated silicate of potassium, iron, and aluminium, a green mica mineral occurring almost exclusively in marine sediments, particularly in greensands. It is generally found in rounded fine-grained aggregates of ill-formed platelets. The manner of its formation is somewhat uncertain.

glauconitic sandstone (*Geol.*). See **Greensand.**

glaucophane (*Min.*). A hydrated sodium, magnesium, aluminium silicate, a monoclinic member of the amphibole rock, occurring in regionally metamorphosed rocks subjected to very high pressures during metamorphism.

glaucous (*Bot.*). Covered with a dull greenish grey waxy bloom. *dim.* **glaucescent.**

glaze (*Build.*). A brilliant glass-like surface given to tiles, bricks, etc. (*Paint.*) (1) The colours employed in the operation of graining. (2) A covering of transparent wash on the ground coat of paint.

glazed boards (*Paper*). Similar to cardboards; made from wood pulp, etc., and given a high glaze by rolling.

glazed brick (*Build.*). A brick having a glassy finish to the surface produced by spraying it with special surface preparations before firing.

glazed door (*Build.*). A door fitted with glass panels.

glazed frost (*Meteor.*). A smooth layer of ice which is occasionally formed when rain falls and the temperature of the air and the ground is below freezing-point.

glazed imitation parchment (*Paper*). See **imitation parchment.**

glazed morocco (*Bind.*). Goatskin crushed and polished by rolling.

glazed roller (*Print.*). A hard, smooth-surfaced roller used for rolling up lithographic plates.

glazier (*Build.*). A workman who cuts panes of glass to size and fits them in position.

glazier's putty (*Build.*). A mixture of whiting and linseed oil, sometimes including white lead, forming a plastic substance for sealing panes of glass into frames.

glazing (*Build.*). The operation of fitting panes of glass into sashes. (*Photog.*) The application of a shining surface to photographic prints by drying them in contact with a highly polished surface, e.g., chromium plate. (*Plumb.*) The process of passing a hot iron over the lead of a *wiped joint* (q.v.) to produce a smooth finish.

glazing bead (*Build.*). A bead nailed, instead of putty, to secure a pane.

gleba (*Bot.*). The spore-bearing tissue enclosed within the peridium of the fructification of *Gasteromycetes*, and in truffles.

glebulose (*Bot.*). Bearing rounded humps on the surface of the thallus.

gleet (*Med.*). Chronic discharge from the urethra as a result of gonococcal infection. (*Vet.*) A catarrhal discharge from the nose of the horse usually due to chronic inflammation of the nasal passages and sinuses.

glei soil (*Geol.*). A type of soil formed under the influence of poor drainage.

gleitzeit (*Work Study*). A system of providing employees with working hours which they can arrange for themselves, within specified limits. There are usually two core-times during which all employees are expected to be present; outside these employees come and go as they please during the hours of opening of the place of employment, recording their hours automatically up to the required amount. Also **flexi-time.**

Glénard's disease (*Med.*). See **enteroptosis.**

glenmuirite (*Geol.*). A basic igneous rock. A variety of *teschenite* containing a small amount of alkali-feldspar.

Glenn effect (*Space*). Luminous particles first observed when in orbit by U.S. astronaut John Glenn.

glenoid (*Zool.*). Socket-shaped; any socket-shaped structure; as the cavity of the pectoral girdle which receives the basal element of the skeleton of the fore-limb.

glenoid fossa (*Zool.*). In Mammals, a hollow beneath the zygomatic process.

glia (*Zool.*). Same as **neuroglia.** Gk. *gliā*, meaning glue.

gliadin (*Chem.*). A simple protein, prolamine, obtained from wheat and rye.

glide path (*Aero.*). The approach slope (usually $3\frac{1}{2}°$ or 5°) along which large aircraft are assumed to come in for a landing. The term is imprecise because such aircraft do not glide, but are brought in with a considerable amount of power.

glide path beacon (*Aero.*). A directional radio beacon, associated with an ILS, which provides an aircraft, during approach and landing, with indications of its vertical position relative to the desired approach path.

glide path landing beam (*Aero.*). Radio signal pattern from a radio beacon, which aids the landing of an aircraft during bad visibility.

glider (*Aero.*). A heavier-than-air craft not power-driven within itself, although it may be towed by a power-driven aeroplane. Cf. *sailplane.*

glidesail parachute (*Space*). Flexible wing parachute used to control the descent of spacecraft.

gliding (*Aero.*). (1) Flying a heavier-than-air craft without assistance from its engine, either in a spiral or as an approach glide before flattening-out antecedent to landing. (2) The sport of flying *gliders* (q.v.), which are catapulted into the air, or launched by accelerating with a winch, or towed by car, or towed to height by an aeroplane.

gliding angle (*Aero.*). The angle between the flight path of an aircraft in a glide and the horizontal.

gliding growth (*Bot.*). A process of adjustment in growing tissues in which the ends of elongating cells slide past one another, often becoming interlocked as elongation ceases.

gliding planes (*Crystal.*). In minerals, planes of molecular weakness along which movement can take place without actual fracture. Thus calcite crystals or cleavage masses can be distorted by pressure and pressed into quite thin plates without actual breakage. See **slip planes.**

glimmerite (*Geol.*). A rock composed mainly of mica, usually of the biotite variety.

glint (*Radar*). Apparent random motion of the effective centre of reflection of a target, leading to noise spread.

glioblastoma multiforme (*Med.*). A very malignant tumour of the central nervous system.

glioma (*Med.*). A general term applied to a variety of tumours arising from nervous tissue in the brain and, more rarely, in the spinal cord.

gliomatosis (*Med.*). Diffuse overgrowth of neuroglia in the brain or in the spinal cord.

gliosomes (*Cyt.*). Small cytoplasmic granules found in neuroglial cells.

glissette (*Maths.*). The locus of any point, or envelope of any line, moving with a first curve which slides against two fixed curves. Many curves can be regarded as either glissettes or roulettes, e.g., the astroid.

Glisson's capsule (*Zool.*). In higher Vertebrates, a coat of loose connective tissue enclosing the portal vein, the hepatic artery, and the bile-duct and their branches in the liver.

Glitsch trays (*Chem. Eng.*). Proprietary distillation column trays in which the vapour rises through holes covered with a movable cap retained in a fixed frame to prevent its being displaced from its vertical position but leaving it free over wide range of horizontal positions.

Globar (*Elec. Eng.*). TN for silicon carbide rod which, when heated by a current, acts as radiating black body.

globe photometer (*Light*). See **Ulbricht sphere photometer.**

globigerina ooze (*Geol.*). A deep-sea deposit covering a large part of the ocean floor (one-quarter of the surface of the globe); it consists chiefly of the minute calcareous shells of the foraminifer, *Globigerina*.

globin (*Biochem.*). The protein constituent, a histone, of haemoglobin.

globoid (*Bot.*). A rounded inclusion in an aleurone grain, consisting of a double phosphate of calcium and magnesium, combined with globulins.

globoidal worm gear (*Eng.*). See **Hindley worm gear.**

globular cementite (*Met.*). In steel, cementite occurring in the form of globules instead of in lamellae (as in pearlite) or as envelopes round the crystal boundaries (as in hyper-eutectoid steel). Produced by very slow cooling, or by heating to between 600°C and 700°C.

globular clusters (*Astron.*). Symmetrical clusters in which many thousands of stars are concentrated; they contain short-period variables and Population II stars. More than 100 are known to be distributed about the centre of our Galaxy; others have been found in the nearer spiral galaxies.

globular lightning (*Meteor.*). See **ball lightning.**

globular pearlite (*Met.*). See **granular pearlite.**

globule (*Bot.*). The male fructification of the *Charales*.

globulins (*Chem.*). Simple proteins insoluble in water but soluble in dilute salt solution, from which they can be salted out with magnesium sulphate. Globulin, fibrinogen, fibrin, myosin, legumin, edestin, and conglutin are globulins.

globulites (*Geol.*). Crystallites (i.e., incipient crystals) of minute size and spherical shape occurring in natural glasses such as pitchstones.

globus (*Zool.*). Any globe-shaped structure; as the *globus pallidus* of the Mammalian brain. *adj.* **globate.**

globus hystericus (*Med.*). The sensation as of a lump in the throat experienced in hysteria.

glochidiate (*Bot.*). Bearing bristles with hooked tips.

glochidium (*Bot.*). A hair with a hooked tip, formed on the spore masses of the water fern *Azolla*. (*Zool.*) The modified larval form of *Unionidae* (fresh-water Mussels), characterized by a toothed bivalve shell and a prominent byssus thread; incubated within the gills of the mother, and afterwards parasitic on the gills or fins of Fish. *adj.* **glochidiate.**

gloeocystidium (*Bot.*). A cystidium of horny or of gelatinous consistency.

Gloger's law (*Zool.*). Southern races of warm-blooded animals tend to be dark-coloured, especially black, brown, and dark-red, while northern races tend to be light-coloured and greyish.

glomera carotica (*Anat., Physiol.*). See **carotid bodies.**

glomerulate (*Bot.*). Bearing glomerules.

glomerule (*Bot.*). A small ball-like cluster of spores.

glomerulitis (*Med.*). Inflammation of the glomeruli of the kidney.

glomerulonephritis (*Med.*). Inflammation of the kidney, the glomeruli being mainly affected.

glomerulus (*Bot.*). A cymose inflorescence in the form of a crowded head of small flowers. (*Zool.*) In *Enteropneusta*, an organ containing a blood-plexus, situated in the proboscis, believed to be excretory; a capillary blood-plexus, as in the Vertebrate kidney; a nestlike mass of interlacing nerve-fibrils in the olfactory lobe of the brain. *adj.* **glomerular.**

glomus (*Zool.*). In the pronephros, the glomeruli of the separate somites aggregated to form a single capillary mass.

glomus jugulare tumour (*Surg.*). A very vascular tumour found in the middle ear, locally malignant only.

glonoin (*Chem.*). Spirits of glonoin, a 1% solution of trinitroglycerine in ethanol.

glory (*Meteor.*). A small system of coloured rings surrounding the shadow of the observer's head, cast by the sun on a bank of mist, as in the *spectre of the Brocken*. The *glory* is produced by diffraction caused by the water droplets in the mist.

glory-hole (*Glass, Met.*). (1) A subsidiary furnace, in which articles may be reheated during manufacture. (2) An opening exposing the hot interior of a furnace. (*Mining*) Combination of open pit mining with underground tunnel through which spoil is removed after gravitating down.

gloss-, glosso-. Prefix from Gk. *glōssa*, tongue.

glossa (*Zool.*). In Insects, one of an inner pair of lobes arising from the prementum, or the median structure formed by the fusion of these lobes; in Vertebrates, the tongue; any tongue-like structure. *adj.* **glossate, glossal.**

glossarium (*Zool.*). In certain *Diptera*, the glossa, which is narrow-pointed.

glossectomy (*Surg.*). Removal of the tongue.

glossitis (*Med.*). Inflammation of the tongue.

glossodynia (*Med.*). Pain in the tongue.

glossohyal (*Zool.*). In some Fish, an anterior extension of the basihyal lying within the tongue.

gloss oil (*Paint.*). A cheap, inferior paint medium consisting of a solution of resin in a solvent with little or no oil.

glossophagine (*Zool.*). Securing food by the agency of the tongue.

glossopharyngeal (*Zool.*). Pertaining to the tongue and the pharynx; the ninth cranial nerve of Vertebrates, running to the first gill-cleft in lower forms, to the tongue and the gullet in higher forms.

glossoplegia (*Med.*). Paralysis of the tongue.

glossospasm (*Med.*). Spasm of the muscles of the tongue.

glossotheca (*Zool.*). In some pupal Insects, that part of the integument which covers the proboscis.

gloss paint (*Paint.*). Paint to which varnish is added as an ingredient in the manufacturing process; characterized by a glossy finish.

glottis (*Zool.*). In higher Vertebrates, the opening from the pharynx into the trachea.

glove box (*Nuc. Eng.*). An enclosure in which radioactive or toxic material may be manipulated by the use of gauntlets while isolated from the worker.

glove leather (*Leather*). Usually goat and kid skins, prepared by tanning or chrome-tanning, or by a combination of these processes.

Glover tower (*Chem. Eng.*). A tower of a sulphuric acid plant used to recover the nitrogen oxides from the Gay-Lussac tower, to cool the gases from the burners, to concentrate the acid trickling down the tower, to partly oxidize the gases from the sulphur burners, and to introduce the necessary nitric acid into the chambers by running nitric acid down the tower along with the nitrated acid from the Gay-Lussac tower.

Glover-West retort (*Chem.*). A continuous vertical retort for carbonization; coal enters at the top, coke is withdrawn at the bottom, and gas, tar, and ammonia are distilled off.

glow discharge (*Electronics*). Visible discharge near a cathode when the potential drop is slightly higher than the ionization potential of the gas; consists of luminous spectral bands.

glow-discharge microphone (*Acous.*). One in which the speech signals modulate the current through a glow discharge directly.

glowlamp (*Cinema.*). Helium-filled quartz lamp whose brightness can be varied by modulation from a microphone. Originally used for variable density sound-track on newsreels, etc., owing to its compactness and portability.

glow plug (*Aero.*). In a gas turbine, an electrical igniting plug which can be switched on to ensure automatic re-lighting when the flame is unstable, e.g., under icing conditions.

glow potential (*Electronics*). Potential which initiates sufficient ionization to produce gas discharge between two electrodes, but is below sparking potential.

glow switch (*Electronics*). Tube in which a glow discharge thermally closes a contact, starting fluorescent tubes.

glow tube (*Electronics*). Cold-cathode gas-filled diode, with no space-current control, the colour of glow depending on contained gas.

glucagon (*Biochem.*). A polypeptide hormone secreted by the pancreas which augments blood glucose levels by accelerating glycogen breakdown in the liver.

glucans (*Chem.*). The anhydrides of glucose, e.g., cellulose, starch, dextrin, glycogen, etc.

glucinum (*Chem.*). Old name for **beryllium**.

gluco-. See **glyc-**.

glucocorticoids (*Chem.*). Group of adrenal cortex hormones, which affect carbohydrate metabolism.

D-gluconic acid (*Chem.*). $CH_2OH(CHOH)_4COOH$, an oxidation product (at C—1) of D-glucose.

glucophore (*Chem.*). A group of atoms which causes sweetness of taste.

glucoproteins (*Chem.*). Compounds formed by the conjugation of a protein with a substance containing a carbohydrate group other than a nucleic acid, e.g., mucin.

D-glucosamine (*Chem.*). $CH_2OH(CHOH)_3CHNH_2{\cdot}CHO$, an amino-sugar, it represents a link between the carbohydrates and the proteins. Part of the molecule of *chitin*, and part of the molecule *heparin*.

D-glucose (*Chem.*). Dextrose, grape-sugar; $C_6H_{12}O_6$. It crystallizes from water in 6-sided plates, m.p. 86°C; from methanol in small anhydrous prisms, m.p. 146°C. It is dextrorotatory and is prepared by the hydrolysis of starch and other carbohydrates.

glucosides (*Chem.*). A group of complex organic compounds occurring in nature which are characterized by the formation on hydrolysis of D-glucose. G-O-X (H_2O eliminated, X = a glycone).

glucosuria (*Med.*). See **glycosuria**.

glucuronic acid (*Chem.*). $CHO{\cdot}(CHOH)_4{\cdot}COOH$. A glycuronic acid. It can be prepared by reduction of the lactone of saccharic acid, and occurs in small amounts in the urine. It forms glycosides or esters with phenols or aromatic acids which are removed from the body in this form (6-CH_2OH of glucose→6-COOH).

glue (*Carp., etc.*). A substance used as an adhesive agent between surfaces to be united. Glue is obtained from various sources, e.g., bones, gelatin, starch, resins, etc. Well-dried glue contains about 12–15% of moisture.

glueline (*Elec. Eng.*). High-frequency heating technique for drying glue films in woodwork construction, by applying electric field in line with the film, with specially shaped electrodes. The film should have a high loss factor compared with medium to be 'glued'.

glumaceous (*Bot.*). Thin, brownish, and papery.

Glumales (*Bot.*). A family of the *Glumiflorae*. The grasses. The flowers are reduced, the corolla consists of two (or three) minute lodicules. There are typically three (1–6) stamens, one ovule with two styles and with feathery stigmas. The fruit is a caryopsis. Also **Gramineae.**

glume (*Bot.*). A dry membranous bract associated with the flower of a grass. Several glumes, with the associated flowers, make up the *spikelet* (q.v.).

glumella (*Bot.*). See **palea.**

Glumiflorae (*Bot.*). An order of the Monocotyledons including the members with reduced floral parts, e.g., the Sedges and Grasses.

glut (*Build.*). See **closer.**

glutamic acid (*Chem.*). $HOOC{\cdot}CH_2{\cdot}CH_2{\cdot}CH(NH_2){\cdot}COOH$, α-aminoglutaric acid, a mono-aminodicarboxylic acid, obtained by the hydrolysis of albuminous substances. Nonessential with respect to growth in rats.

glutamine (*Chem.*). The mono-amide of glutamic acid, $(NH_2){\cdot}CO{\cdot}CH_2{\cdot}CH_2{\cdot}CH(NH_2){\cdot}COOH$.

glutaraldehyde (*Chem.*). $CHO{\cdot}(CH_2)_3CHO$, an oil, soluble in water and volatile in steam; used in tanning leather, esp. for clothing, as it imparts resistance to perspiration. Also a disinfectant.

glutathione (*Biochem.*). A sulphur-containing tripeptide, of cysteine, glutamic acid, and glycine, occurring in tissues and taking part in biological oxidation processes by virtue of its property of being alternately oxidized and reduced.

gluteal (*Zool.*). Pertaining to the buttocks.

glutelins (*Chem.*). Simple proteins, insoluble in water and in neutral salt solutions, but soluble in dilute acids and alkalis, e.g., glutenin or oryzenin.

gluten (*Bot.*). (1) A reserve protein found in plants, a product of gliadin and glutenin. (2) A sticky coating on the pilei of some agarics.

glutenin (*Chem.*). A protein of the glutelin group, found in wheat.

gluteus (*Zool.*). In land Vertebrates, a retractor and elevator muscle of the hind-limb.

glycerides (*Chem.*). A term for glycerine esters, the most important of which are the *fats*.

glycerine (*Chem.*). Glycerol, propan-1,2,3-triol, $CH_2OH{\cdot}CHOH{\cdot}CH_2OH$, a syrupy hygroscopic liquid, m.p. 17°C, b.p. 290°C, obtained by the hydrolysis of oils and fats, or by the alcoholic fermentation of glucose in the presence of sodium sulphite solution, which reacts with the aldehydes formed, thus liberating a larger amount of glycerine. It is also prepared synthetically from propylene by chlorination and hydrolysis. Glycerine is a trihydric alcohol, forming alcoholates, esters, and numerous derivatives. Colourless. It is a raw material for *alkyd resins*, *nitroglycerine*, printing inks, foodstuff preparations, etc. Also **glycerin.**

glycerine litharge cement (*Chem.*). A mixture of litharge (lead (II) oxide) and glycerine which rapidly sets to a hard mass.

glycerol (*Chem.*). See **glycerine.**

glycerol-phthalic resins (*Plastics*). See **alkyd resins.**

glycin (*Photog.*). *Para*-hydroxyphenylglycine, a slow-working developing agent used in warm-tone printing:

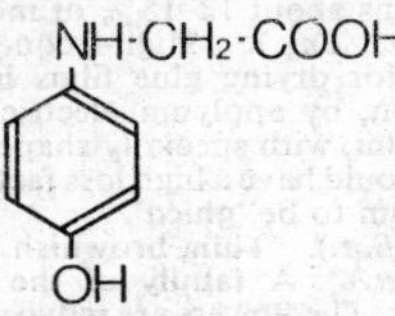

glycine (*Chem.*). Aminoethanoic acid.
glycocoll (*Chem.*). Aminoethanoic acid.
glycogen (*Biochem.*). $(C_6H_{10}O_5)_x$, a polysaccharide also found in plants and in animal cells, e.g., liver and muscle; **animal starch.** Its hydrolysis by acids or by enzymes finally yields glucose. Liver glycogen is a relatively labile store of chemical energy and carbohydrate, whereas muscle glycogen is not readily depleted by fasting. 'Backbone' of α-1→4 glucosidic linkages and α-1→6 branching.
glycogenesis (*Biochem.*). The synthesis of glycogen, essentially in the liver.
glycogen mass (*Bot.*). See **epiplasm.**
glycogenolysis (*Biochem.*). The hydrolysis of glycogen with the formation of glucose.
glycol (*Chem.*). See **ethene glycol.**
glycols (*Chem.*). Dihydric alcohols, of the general formula $C_nH_{2n}(OH)_2$, viscous liquids with a sweet taste or crystalline substances. They give all the alcohol reactions and, having two hydroxyl groups in the molecule, they can also form mixed compounds, e.g., ester-alcohols.
glycolysis (*Biochem.*). The conversion of glucose to lactic acid with the production of 2 moles of ATP. The process is anaerobic and provides a means of obtaining ATP in a relatively anaerobic organ such as muscle.
glyconeogenesis (*Biochem.*). The formation of glycogen from noncarbohydrate substances.
glycoproteins (*Chem.*). See **glucoproteins.**
glycosides (*Chem.*). A group of compounds, derived from the other monosaccharides in the same way as glucosides are derived from glucose, including glucosides as a subclass.
glycosuria, glucosuria (*Med.*). The presence of sugar in the urine.
glycuronic acid (*Chem.*). See **glucuronic acid.**
glyoxal (*Chem.*). Ethan 1,2-dial. CHO·CHO, a dialdehyde, existing in four modifications, viz., polyglyoxal $(CHO{\cdot}CHO)_n$ from which, by heating with P_2O_5, glyoxal CHO·CHO is obtained. This substance is a yellow liquid, m.p. 15°C, b.p. 51°C (776 mm), forming green vapours, but it is not stable, and polymerizes to insoluble paraglyoxal, $(CHO{\cdot}CHO)_x$. There is also known a trimolecular form, $(CHO{\cdot}CHO)_3$.
glyoxalic acid (*Chem.*). Ethan-1-al-2-oic acid $CHO{\cdot}COOH+H_2O$ or $CH(OH)_2{\cdot}COOH$, an aldehyde monobasic acid occurring in unripe fruit; rhombic prisms, soluble in water, volatile in steam, which can be obtained by the oxidation of ethanol with nitric acid, or by the hydrolysis of dichloroacetic or dibromoacetic acid.
glyoxalines (*Chem.*). See **iminazoles.**
glyoxylic acid (*Chem.*). See **glyoxalic acid.**
glyph (*Arch.*). A short upright flute.
glyptal resins (*Plastics*). See **alkyd resins.**
G-M counter (*Nuc. Eng.*). See **Geiger-Müller counter.**
gmelinite (*Min.*). A pseudohexagonal zeolite, white or pink in colour and rhombohedral in form, resembling chabazite. Chemically, it is hydrated silicate of aluminium, sodium and calcium.
Gmelin test (*Chem.*). A test for the presence of bile pigments; based upon the formation of various coloured oxidation products on treatment with concentrated nitric acid.
GMT. Abbrev. for *Greenwich Mean Time.*
gnathic (*Zool.*). Pertaining to the jaws.
gnathites (*Zool.*). Mouth-parts, especially those of Insects.
gnathobase (*Zool.*). In *Arthropoda*, a masticatory process on the inner side of the first joint of an appendage.
Gnathobdellida (*Zool.*). An order of *Hirudinea*, the members of which are all terrestrial or fresh-water forms possessing botryoidal tissue; they have jaws but the proboscis is not protrusible. Red-blooded. Leeches.
gnathochilarium (*Zool.*). In *Diplopoda*, a flat plate forming the lower lip and representing the fused second pair of mouth-parts.
gnathopod (*Zool.*). In *Arthropoda*, any appendage modified to assist in mastication.
gnathopodite (*Zool.*). See **maxilliped.**
gnathos (*Zool.*). A median ventral sclerite forming one of the external genitalia of male *Lepidoptera*, probably representing the sternum of the 10th abdominal segment.
gnathosoma (*Zool.*). In *Acarina*, the segments of the mouth and its appendages.
gnathostegite (*Zool.*). In some *Crustacea*, a covering plate associated with the mouth-parts.
Gnathostomata (*Zool.*). A superclass of *Chordata*, including all Vertebrate animals with upper and lower jaws; comprises a wide range of animals, from Fishes to Tetrapods.
gnathostomatous (*Zool.*). Having the mouth provided with jaws.
gnathotheca (*Zool.*). The horny part of the lower beak of Birds.
gneiss (*Geol.*). A metamorphic rock of coarse grain size, characterized by a mineral banding, in which the light minerals (quartz and feldspar) are separated from the dark ones (mica and/or hornblende). The layers of dark minerals are foliated, while the light bands are granulitic. See also **metamorphism.**
gneissose texture, gneissic texture (*Geol.*). A rock texture in which foliated and granulose (granulitic) bands alternate; typical of rocks which have been recrystallized under directed pressure, during regional metamorphism.
Gnetales (*Bot.*). An order of *Gymnospermae*, characterized by the presence of a perianth, the absence of resin, and the occurrence of vessels in the secondary wood.
gnomon (*Maths.*). The remainder of a parallelogram after a similar parallelogram has been removed from one corner. (*Surv.*) An early instrument for determination of time and latitude, involving the measurement of the shadow of an upright rod as cast by the sun. The pointer of a sundial.
gnomonic projection (*Geog.*). A type of *zenithal projection* (q.v.) having the point of projection at the centre of the globe. Not practicable for a complete hemisphere, but of navigational value because any section of a great-circle course can be plotted on it as a straight line.
gnotobiosis (*Biol.*). Condition of being germ-free or germ-free and then inoculated with known micro-organisms.
go (*Build.*). The *going* (q.v.).
goal box (*An. Behav.*). The part of a maze containing the reinforcement, e.g., food or water.
goal response (*An. Behav.*). The response given by an animal to a reinforcing stimulus.

goat pox (*Vet.*). An epidemic disease of goats due to infection by a virus; characterized by fever and a papulo-vesicular eruption of the skin and mucous membranes.

gob (*Glass*). (1) A measured portion of molten glass as fed to machines making glass articles. (2) A lump of hot glass gathered on a punty or blowing iron. (*Mining*) The space left by the extraction of a coal-seam, into which waste is packed, also loose waste. Also **goaf.**

Gobelin tapestry (*Textiles*). A famous type of tapestry made in Paris since the reign of Louis XIV. Noted for its smooth surface, wonderful blend of colours, and its handsome borders.

gob fire (*Mining*). A fire occurring in a worked-out area, due to ignition of timber or broken coal left in the gob.

gob heading, gob road (*Mining*). A roadway driven through the gob after the filling has settled.

goblet cell (*Zool.*). A goblet- or flask-shaped epithelial gland cell, occurring usually in columnar epithelium.

gobo (*Cinema.*). Sound-absorbing panel used in sound-film production for regulating the reflection of sound-waves on the set; intended to be outside the camera angle.

gob process (*Glass*). One for making hollow ware, in which glass is delivered by an automatic feeder in the form of soft lumps of suitable shape to a forming unit. Also **flow process, gravity process.**

gob stink (*Mining*). A smell indicating spontaneous combustion or a fire in the gob.

godets (*Textiles*). Small glass reels used in rayon, etc., spinning machines.

go-devil (*For.*). See **crotch.** (*Min. Proc.*) Scraper which can be worked through a pipe line to detach scale.

godroon (*Arch.*). An ornamentation taking the form of a cable or bead.

goethite (*Min.*). Orthorhombic hydrated oxide of iron with composition FeO·OH. Dimorphous with lepidocrocite.

Goetz size separator (*Powder Tech.*). Instrument for the size classification of airborne particles, comprising a high-speed rotor with a helical channel along which the particles move in laminar flow and are deposited by centrifugal force on a removable envelope surrounding the rotor.

going (*Build.*). The horizontal interval between consecutive risers in a stair.

going-barrel (*Horol.*). A barrel in which the winding takes place from the arbor. Power is transmitted direct from such a barrel to the train by teeth on the barrel.

going bord (*Mining*). A roadway to the coal face in bord and pillar working.

going fusee (*Horol.*). A fusee with maintaining power.

going light (*Vet.*). A term popularly applied to emaciation of animals.

going part (*Weaving*). The part of a loom known as the *sley* or *batten.* Holding the reed, it is attached by cranks to the top crankshaft and on its forward strokes beats-up the picks of weft to form cloth.

going rod (*Build.*). A rod used for setting out the going of the steps in a flight.

goitre (*Med.*). Morbid enlargement of the thyroid gland. See also **Basedow's disease.**

goitrogenous (*Med.*). Producing, or tending to produce, goitre.

goitrous (*Med.*). Affected with, or pertaining to, goitre.

Golay cell (*Phys.*). Pneumatic cell used as detector of heat radiation, e.g., in infrared spectrometer.

gold (*Chem.*). A heavy, yellow, metallic element in the first group of the periodic system. Symbol Au, at. no. 79, r.a.m. 196·967, rel. d. at 20°C 19·3, m.p. 1062°C, electrical resistivity about 0·02 microhm metres. Most of the metal is retained in gold reserves but some is used in jewellery, dentistry, and for decorating pottery and china. In coinage and jewellery, the gold is alloyed with varying amounts of copper and silver. *White gold* is usually an alloy with nickel, but as used in dentistry this alloy contains platinum or palladium.

gold amalgam (*Min.*). A variety of native gold containing approximately 60% of mercury; discovered in Colombia and occurs also in California.

goldberg wedge (*Optics*). A wedge of tinted glass or gelatin with lampblack in suspension used in photometry to reduce quantitatively light intensity.

gold blocking (*Bind.*). The process of pressing a design upon gold-leaf spread out on the cover of a book, the tools or dies, which are heated, leaving the desired impression. Also carried out by machine, gold-foil being fed from a spool.

gold cushion (*Bind.*). A small board, covered with rough calfskin, which is padded with a soft material. The gold-leaf required for gold blocking is placed on the cushion ready for use.

golden beryl (*Min.*). A clear yellow variety of the mineral beryl, prized as a gemstone. Heliodor is a variety from S.W. Africa. Golden beryl has been used as a name for **chrysoberyl.**

golden number (*Astron.*). A term derived originally from medieval church calendars, and still used to signify the place of a given year in the Metonic cycle of 19 years.

gold-film glass (*Glass*). Glass incorporating a thin gold film which can be electrically heated for demisting and de-icing.

gold grains (*Radiol.*). Small lengths of activated gold wire (half-life 2·70 days) having similar advantages to *radon seeds* (q.v.) and used in a similar manner.

gold-leaf (*Met.*). Pure gold beaten out into extremely thin sheets, so that it may be applied to surfaces which are to be gilded.

gold-leaf electrometer (*Elec. Eng.*). *Gold-leaf electroscope* (q.v.), modified for the measurement of very small currents by observing the rate of movement of the gold leaf through a microscope. The *personal dosimeter* (q.v.) is a development of this instrument. See also **Millikan electrometer.**

gold-leaf electroscope (*Elec. Eng.*). A device for detecting small electric charges which are applied to a piece of thin gold foil usually attached at upper end to a metal electrode. Mutual repulsion between the foil and the similarly charged plate electrode leads to the former being displaced.

gold number (*Chem.*). The weight in milligrams of a lyophilic colloid which is just insufficient to prevent the change from red to blue in 10 cm³ of colloidal gold solution after the addition of 1 cm³ of 10% sodium chloride solution.

gold paints (*Paint.*). Paints made of bronze powders mixed with varnish.

gold rug or gold rubber (*Bind.*). A piece of flannel cloth or soft rubber used for wiping off surplus gold after gilding. It is sold to a refiner after a period of use.

Goldschmidt alternator (*Elec. Eng.*). High-frequency alternator in which the stator and rotor carry a number of windings, each tuned to successively higher frequencies; up to 100 kHz can be thus attained.

Goldschmidt process (*Chem.*). (1) See **aluminothermic process.** (2) Detinning of coated iron by use of chlorine.

gold-size (*Paint.*). Type of *size* (q.v.) used as a basis to secure gold-leaf on to surfaces which are to be gilded, and for other purposes.

Gold slide (*Meteor.*). A slide, named after its designer, which is attached to a mercury barometer to make the corrections for index error, latitude, height, and temperature mechanically.

gold spring (*Horol.*). The delicate spring forming part of the detent escapement. One end is anchored to the detent, and the other end rests against the detent horn and projects just beyond it. On one vibration of the balance, the discharging pallet presses against the end of the gold spring and moves the whole detent, bringing about unlocking of the escape wheel. On the return vibration, the discharging pallet merely lifts the gold spring without causing movement of the detent.

gold toning (*Photog.*). The addition, by chemical means, of a gold film to the surface of the silver image of a photographic print.

gold tooling (*Bind.*). The decorating of book covers by hand, using gold-leaf and heated tools.

Golgi apparatus (*Cyt.*). A cytoplasmic structure which takes the form of scattered particles (*Golgi bodies*) or a continuous network, undergoes changes of form and position in secreting cells, and is actively concerned with cell-metabolism. Also **apparato reticulare.**

Golgi-Mazzoni corpuscles (*Zool.*). In Vertebrates, a type of sensory nerve-ending resembling an end bulb.

Golgi's organs (*Zool.*). In Vertebrates, a type of sensory nerve-ending occurring in tendons near the point of attachment of muscle fibres, and consisting of a terminal arborization with irregular varicosities, in a fibrous capsule.

goliath Edison screw-cap (*Elec. Eng.*). An Edison screw-cap having a screw thread about 1½-in. diameter, with 4 threads/in.; used with large metal-filament lamps.

gomphosis (*Zool.*). A type of articulation in which a conical process fits into a cavity; as the roots of teeth into their sockets.

gon-, gono-. Prefix from Gk. **gonos**, seed, offspring.

gonad (*Zool.*). A mass of tissue arising from the primordial germ cells and within which the spermatozoa or ova are formed; a sex gland; ovary or testis. *adj.* **gonadal.**

gonadotrophic, gonadotropic (*Physiol.*). Pertaining to gonadotrophins.

gonadotrophins (*Biochem.*). Substances having a stimulating effect on the gonads, e.g., as applied to certain hormones of the pituitary gland and placenta. They include the *follicle-stimulating*, *luteinizing*, and *mammotrophic* (see **prolactim**) *hormones*, also *chorionic gonadotrophin* (q.v.), all of which are vitally concerned with sexual characteristics and the reproductive functions. See also **fertility drug, oral contraception.**

gonal (*Zool.*). Forming or giving rise to a gonad: as the *gonal ridge*.

gonangium (*Zool.*). See **gonotheca.**

gonapophyses (*Zool.*). Tubular outgrowths from the medial borders of the coxites of the gonopods in Insects; in females, they are sometimes modified to form an ovipositor or a sting. The term is sometimes used loosely to denote any genital appendages.

Gonell elutriator (*Powder Tech.*). A vertical air elutriator with a long straight fractionating chamber of uniform cross-section.

gones (*Cyt.*). The groups of four nuclei or of four cells which are the immediate results of meiosis.

gone to bed (*Print.*). A newspaper term meaning that the newspaper has been made up and sent to press.

gong (*Horol.*). Rectangular steel strip bent into the form of a spiral which, after blueing, provides a deep note when struck; used in clocks when striking the hours.

goni-. Prefix from Gk. *gōnia*, angle; *gony*, knee. Not to be confused with the prefix *gon-*, *gono-* (q.v.).

gonial(e) (*Zool.*). See **antarticular(e).**

gonidial layer (*Bot.*). See **algal layer.**

gonidiophore (*Bot.*). An old term for **conidiophore.**

gonidium (*Bacteriol.*). A very small flagellate cell produced by some species of bacteria, e.g., *Azotobacter chroococcum*, under certain conditions. (*Bot.*) (1) An algal cell occurring as part of the thallus of a lichen. (2) An old term for **conidium.** (3) A nonmotile spore formed by some *Myxophyceae*. (4) A gemma in some liverworts.

gonimic layer (*Bot.*). See **algal layer.**

gonimium (*Bot.*). A cell of one of the *Myxophyceae* when it occurs as a part of the thallus of a lichen.

gonimoblast (*Bot.*). A short spore-bearing filament formed from the fertilized carpogonium of red algae, and bearing one or more carpospores.

gonioautoicous (*Bot.*). Bearing the antheridium as a budlike outgrowth from a branch bearing archegonia.

goniocyst (*Bot.*). A group of gonidia, found in lichens.

goniometer (*Instr.*). A device for measuring angles or for direction-finding.

gonioscope (*Surg.*). An instrument used in ophthalmology to measure, and to inspect the anterior chamber of the eye (i.e., the region between cornea and iris).

goniosporous (*Bot.*). Producing angled spores.

gonitis (*Vet.*). Inflammation of the stifle joint of animals.

gonnardite (*Min.*). A rare zeolite; a hydrated sodium, calcium, aluminium silicate.

gonoblast (*Zool.*). A reproductive cell.

gonoblastid (*Zool.*). See **blastostyle.**

gonocalyx (*Zool.*). In some Centipedes, a pair of modified appendages belonging to the same somite as the genital opening.

gonocheme (*Zool.*). A sexual medusoid of *Hydrozoa*.

gonochorism (*Zool.*). Sex determination.

gonochoristic (*Zool.*). Having separate sexes.

gonococcus (*Med.*). A Gram-negative diplococcus, the causative agent of *gonorrhoea* (q.v.).

gonocoel (*Zool.*). That portion of the coelom, the walls of which give rise to the gonads: hence, the cavity of the gonads.

gonocyte (*Zool.*). In *Porifera*, a sexual cell which will produce ova or spermatozoa.

gonodendra (*Zool.*). In some of the *Siphonophora*, branched and mouthless zooids bearing gonophores.

gonoduct (*Zool.*). A duct conveying genital products to the exterior; a duct leading from a gonad to the exterior.

gonophore (*Zool.*). A sexual individual in metagenetic *Coelenterata*.

gonoplasm (*Bot.*). In some *Oömycetes*, the protoplasm which passes through the fertilization tube and unites with that of the oösphere. (*Gen.*) See **germ plasm.**

gonopods (*Zool.*). The external organs of reproduction in Insects, associated with the 8th and 9th abdominal segments in females and with the 9th only in males. In *Chilopoda*, a pair of modified appendages borne on the 17th (genital) body segment.
gonopore (*Zool.*). The aperture by which the reproductive elements leave the body.
gonorrhoea, gonorrhea (*Med.*). A contagious infection of the mucous membrane of the genital tract with the gonococcus, contracted usually through sexual intercourse.
gonosome (*Zool.*). In colonial animals, all the individuals concerned with reproduction.
gonosphere (*Bot.*). A zoospore peculiar to the *Chytridiales*.
gonostyle (*Zool.*). In *Siphonophora*, a process of the main stem of the colony bearing gonophores, and sometimes representing a mouthless polyp.
gonostyli (*Zool.*). The third valvulae of stinging *Hymenoptera* forming part of the sting.
go, not-go gauge (*Eng.*). A *limit gauge* (q.v.) of which one section is above and another below the specified dimensional limits of the part to be gauged. If the 'go' section accepts the part and the 'not-go' section interferes with the part, the latter is accurate within the limits specified.
gonotheca (*Zool.*). The vaselike expansion of the perisarc which surrounds a blastostyle and gonophores in *Calyptoblastea*.
gonotokont (*Zool.*). See **auxocyte.**
gonotome (*Zool.*). A somite in an embryo which contains the anlage of a gonad.
gonozooid (*Zool.*). In *Siphonophora*, an individual bearing gonophores. In *Doliolidae*, sexual zooids carried by the phorozooids which develop from the median row of buds.
gonys (*Zool.*). In Birds, the edge of the lower beak, reaching from the angle of the chin to the myxa. *adj.* **gonydial.**
Gooch crucible (*Chem.*). A filter used in laboratories, which consists usually of a small porcelain cup, the bottom of which is perforated with numerous small holes and covered with a thin layer of washed asbestos fibres, which act as a filtering medium. Also **Gooch filter.**
good (*Paper*). Also known as 'perfect' and *insides* (q.v.). A term used to describe the best sheets from the inside of a ream of paper.
good colour (*Print.*). A term which indicates consistent distribution of ink throughout a book. Every page must be of uniform blackness.
goodness (*Electronics*). See **mutual conductance.**
googol (*Maths.*). A name for the number 10^{100}.
gooseberry stone (*Min.*). The garnet *grossular* (q.v.), which was named for the resemblance of this green variety, both in form and colour, to the gooseberry which has the botanical name *Ribes grossularia*.
goose flesh (*Med.*). See **horripilation.**
goose neck (*Eng.*). The passage through which molten metal passes from the pump chamber or cylinder to the dies in some types of pressure diecasting machines.
go-out (*Hyd. Eng.*). A sluice in an embankment impounding tidal waters, which can pass through it when the tide is out.
Gordon's formula (*Civ. Eng.*). An empirical formula giving the collapsing load for a given column. It states that

$$P=\frac{fc \,.\, A}{1+c\left(\frac{l}{d}\right)^2}$$

where P=the collapsing load, fc=safe compressive sizes for very short lengths of the material, A=area of cross-section, l=length of the pin-jointed column, d=the least breadth or diameter of the cross-section, c=a constant for the material and the shape of cross-section.
gore (*Aero.*). One of the sectorlike sections of the canopy of a parachute.
gorge (*Build.*). A *drip* (q.v.). (*Geol.*) A general term for all steep-sided, relatively narrow valleys, e.g., canyons, overflow channels, etc.
goslarite (*Min.*). Hydrated zinc sulphate, a rare mineral precipitated from water seeping through the walls of lead-mines; formed by the decomposition of sphalerite.
Gossage's process (*Chem.*). Manufacture of caustic soda by boiling carbonate of soda solution with lime.
gossan (*Mining.*). The leached, oxidized material found in surface exposures of an ore deposit. Often stained brown by iron oxides and rich in quartz, gossans are an indication of mineral deposits below the surface, although they may not be of any value themselves.
gossypine (*Bot.*). Cottony.
Gothic (*Typog.*). Originally a term applied to black letter, or Old English type, it is now used to include bold sanserif faces.
gothic wing (*Aero.*). A very low *aspect ratio* wing with the double-curvature plan form of a gothic arch of the perpendicular period, used for supersonic aeroplanes to combine low *wave drag* with *separated lift*. See **ogee wing.**
Gothlandian (*Geol.*). See **Silurian.**
go-through machine (*Spinning*). A lace machine of the Levers type but with extended combs, larger bobbins, and greater speed.
Göttingen wind tunnel (*Aero.*). A return-flow wind tunnel in which the working section is open. See **open-jet wind tunnel.**
Gott's method (*Elec. Eng.*). A bridge method of finding the capacitance of a cable. The cable is made to form one arm of the bridge, the others being a standard capacitor and two sections of a slide wire.
gouache (*Paint.*). Opaque colours mixed with water, honey, and gum, applied in impasto style.
gouge (*Bind.*). A hand tool used to form curved lines. (*Carp.*) Similar to a chisel, but having a curved blade and a cutting edge capable of forming a rounded groove. (*Min.*) *Flucan selvage*. (*Surg.*) A hollow chisel for removing and cutting bone.
gouge-bit (*Carp., etc.*). See **shell bit.**
gouge slip (*Carp., etc.*). A shaped piece of oilstone on which the concave side of the cutting edge of a gouge may be rubbed to sharpen it.
goundou (*Med.*). Dog nose; gros nez. Symmetrical bony overgrowth at the sides of the nose, thought to be a late sequel of yaws.
gout (*Med.*). A disorder of metabolism in which there is an excess of uric acid in the blood; this is deposited, as sodium biurate, in the joints, bones, ligaments, and cartilages. In acute gout there is a sudden very painful swelling of the joint, usually of the big toe.
goutte d'eau (*Min.*). Literally 'drop of water'; an old term applied to the whitest of the Brazilian topaz crystals, which, when cut and polished, rival the diamond in brilliancy, but lack its fire.
Goux pail (*San. Eng.*). A pail lined with absorbent material; used in a conservancy closet.
Gouy layer (*Min. Proc.*). Diffuse layer of counterions surrounding charged lattices at surface of particle immersed in liquid.
governor (*Elec. Eng.*). Speed regulator on variable speed motor or rotating machine, e.g., d.c.,

gramophone motor. (*Eng.*) A device for controlling the fuel or steam supply to an engine in accordance with the power demand, so that the speed is kept constant under all conditions of loading.

governor motion (*Spinning*). A mechanism forming part of a mule quadrant. It regulates the speed of the spindles during the inward run of the carriage to compensate for the increasing diameter of the cop bottom. Also **strapping motion.**

GPI (*Med.*). See **paresis.**

G.P.M. (*Hyd.*). An abbrev. for *gallons per minute*.

Graafian follicle (*Zool.*). A vesicle, containing an ovum surrounded by a layer of epithelial tissue, which occurs in the ovary of Mammals.

grab, grab-bucket (*Civ. Eng.*). A steel bucket or cage made of two halves hinged together, so that they dig out and enclose part of the material on which they rest; used in mechanical excavators and dredgers. See **grabbing crane, grab-dredger.**

grabbing crane (*Civ. Eng.*). An excavator consisting of a crane carrying a large grab or bucket in the form of a pair of half-scoops, so hinged as to scoop or dig into the earth as they are lifted.

grab-dredger (*Civ. Eng.*). A dredging appliance consisting of a grab or grab-bucket suspended from the jib-head of a crane, which does the necessary raising and lowering. Also called a **grapple dredger.**

graben (*Geol.*). A geological structure resulting from the subsidence of a strip of country lying between two normal faults, hading towards one another. Cf. *horst faults*.

Graber's organ (*Zool.*). A curious structure of unknown function, but believed to be sensory, found in Tabanid larvae.

Grace (*Teleph.*). Automatic telephone exchange unit used in subscriber trunk-dialling system. (*G*roup *r*outing *a*nd *c*harging *e*quipment.)

gracilis (*Zool.*). A thigh muscle of land Vertebrates.

gradate sorus (*Bot.*). A fern sorus in which the sporangia develop from the apex of the receptacle downwards.

grade (*Civ. Eng.*). The degree of slope, e.g., of a highway or railway. Better term is **gradient.** (*Eng.*) A measure (expressed by a letter of the alphabet) of the hardness of a grinding wheel as a whole, due to the strength of the bond holding the abrasive particles in place.

graded brush (*Elec. Eng.*). A brush for collecting current from the commutator of an electrical machine; made up of layers of different materials, or of material which has different values of lateral and longitudinal resistance.

grade of service (*Teleph.*). The proportion of calls in the busy hour which must fail to be completed through insufficiency of apparatus.

grade pegs (*Surv.*). Pegs driven into the ground as references, to establish gradients in constructional work. Better, **gradient pegs.**

grader (*Civ. Eng.*). A power-operated machine provided with a blade for shaping excavated surfaces to the desired shape or slope.

gradient (*Bot.*). The condition when the intensity of a stimulus acting on a plant gradually increases or decreases towards the position where the plant is. (*Civ. Eng.*) See **grade.** (*Maths.*) Of a plane curve at a given point: the slope of the tangent at the point. Of a scalar function $f(x, y, z)$: the vector $\nabla f(x, y, z)$ where ∇ is the vector operator *nabla.* Written grad. *f.* (*Phys.*) The rate of change of a quantity with distance, **e.g.,** the **temperature** *gradient* in a metal bar is the rate of change of **temperature** along the bar. (*Surv.*) The ratio of the difference in elevation between two given points and the horizontal distance between them, or the distance for unit rise or fall. Also called **incline.**

gradient pegs (or **posts**). *Grade pegs* (q.v.).

gradient post (*Rail.*). A short upright post fixed at the side of a railway at a point of change of gradient. An arm on each side of the post indicates whether the track rises or falls, the figure on it giving the distance to be travelled to rise or fall one of the same unit of distance.

gradient wind (*Meteor.*). The theoretical wind occurring with curved isobars when both *cyclostrophic* (q.v.) and *geostrophic* (q.v.) components are important.

grading (*Build.*). (1) The proportions of the different sizes of stone used in mixing concrete. (2) The selection of these proportions. (*Civ. Eng.*) The operation of preparing a surface to follow a given gradient. (*Teleph.*) (1) The scheme of connecting trunks or outlets so that a group of selectors is given access to individual trunks, while larger groups of selectors share trunks when all the individual trunks are found to be in use. (2) An arrangement of trunks connected to the banks of selectors by the method of grading. See **grade of service.** (*Textiles*) The classification of textile fibres according to their condition, fineness, strength, staple, colour and waste.

grading coefficient (*Elec. Eng.*). A figure denoting the ratio of the lower to the upper limit of current for motor starters.

grading group (*Teleph.*). The group of selectors which are concerned in one grading scheme.

grading instrument (*Surv.*). A general name for any instrument of the *gradiometer* (q.v.) class.

grading shield (*Elec. Eng.*). A circular conductor placed concentric with a string of suspension insulators on an overhead transmission line to equalize the potential across the individual insulator units. Also called an **arcing shield.**

grading, size (*Chem.*). See **size-grading.**

gradiograph (*Build.*). An instrument incorporating a level tube and straightedge; used to measure gradients in laying drainpipes.

gradiometer (*Mag.*). Magnetometer for measurement of gradient of magnetic field. (*Surv.*) An instrument for setting out long uniform gradients; it consists of a level that may be elevated or depressed, by known amounts, by means of a vertical tangent screw.

graduated circle (*Surv.*). A circular plate, marked off in degrees and parts of degrees, used on surveying instruments as a basis for the measurement of horizontal or vertical angles. See **horizontal circle, vertical circle.**

graduated courses (*Build.*). Courses of slates laid so that the gauge diminishes from eaves to ridge.

graduated vessels (*Chem.*). Vessels which are used for measuring liquids and are adapted to measure or deliver definite volumes of liquid.

graffito (*Paint.*). See **sgraffito.**

Graff's 'C' stain (*Micros., Paper*). A microscopical stain for paper fibres, giving a variety of reactions, prepared from aluminium, calcium, and zinc chlorides, together with iodine and potassium iodide.

graft (*Bot.*). A plant consisting of a rooted part (the *stock*) into which another part (the *scion*) has been inserted so as to make organic union. (*Surg.*) A piece of skin, bone, or other tissue, taken from one part of the body and grafted to another.

graft hybrid, graft chimaera (*Bot.*). A plant composed of two sorts of tissue, differing in genetic constitution, and assumed to have arisen as a result of association following grafting.

grafting (*Bot.*). The insertion of a part of one plant into a part of another so that organic union followed by growth ensues. (*Carp.*) The operation of lengthening a timber by jointing another piece on to it.

Graham escapement (*Horol.*). The dead-beat clock escapement.

grahamite (*Min.*). A member of the asphaltite group.

Graham's law (*Chem.*). The velocity of diffusion of a gas is inversely proportional to the square root of its density.

grain (*Bot.*). (1) See **caryopsis.** (2) The pattern on the surface of worked wood due to variations in the size, in the shape and arrangement, and in the composition, of the cells forming the wood. (*Chem.*) Unit of mass. See **apothecaries' weight.** (*Crystal.*) Individual silver bromide crystal in photographic emulsion, which can be reduced to metallic silver on development if at least one photon of light has been absorbed during exposure. Many developers also cause clumping of individual grains, so producing increased granularity in the image. In nuclear research emulsions, individual grain size is *c.* 0·2 μm. (*Geol.*) See **rift and grain.** (*Geol., Met.*) Average size of mineral crystals composing a rock. Direction in which it tends to split. Of a metal, texture of broken surface (smooth-, rough-, etc.). (*Photog.*) The suspension of sensitive particles in an emulsion, the particle size depending on the temperature of deposition. On exposure, each grain becomes completely developable on the absorption of at least one photon of light. Hence the size of the grain and the concentration of silver salt in it largely determines the speed. The size of grain limits the possible magnification of the projected image.

grain agitator (*Agric.*). Rotating shaft with loop projections, which gently agitates the seed above the feed rolls of a grain drill to prevent bridging.

grain growth (*Met.*). Associated with recrystallization, this refers to an increase in the average grain size resulting from some crystals absorbing adjacent ones.

graining (*Leather*). The process of bringing up the natural grain of a tanned skin by rolling it while damp. See **boarding.** (*Paint.*) The operation of brushing, combing, or otherwise marking a painted surface while the paint is still wet, in order to produce an imitation of the grain of wood.

graining boards (*Bind.*). Boards or metal plates with parallel lines in relief, running diagonally. Used to produce a diced effect on covers.

graining comb (*Paint.*). See **comb.**

graining plates (*Print.*). The aluminium or zinc plates for lithography require a grained surface, which will retain moisture and also hold the image areas or 'work'. Plates can be used several times and regrained after each use in a graining machine, where the plate is covered with balls of porcelain, glass, or steel, and, after adding water and an abrasive such as carborundum, is vibrated mechanically to produce the required grain.

grain leather (*Leather*). The grain split of the hide of an ox or cow, grained and oiled; used for uppers of sports shoes of heavy type. See **grain split, graining.**

grain refining (*Met.*). Production of small closely knit grains, resulting in improved mechanical properties. Particularly with aluminium alloys it is achieved by small additions to melts, usually of oxygen, such as boron which cause fine nucleation on casting. For ferrous metallurgy, see also **grain-size control.**

grains (*Brew.*). Insoluble residue remaining in the mash tun after the wort has been run off; used for cattle food and as a manure. (*Mining*) See **coal sizes.**

grain size (*Geol., Met.*). The average size of the grains or crystals in a sample of metal or rock.

grain-size analysis (*Powder Tech.*). An equivalent term for particle-size analysis widely used in America. It is also the literal translation for the German phrase used to specify *particle-size analysis*. Confusion arises because grain-size analysis is used to describe procedures for measuring the size of crystallites in a cast or sintered metal.

grain-size classification (*Geol.*). A scheme of rock classification based upon the average size of certain chosen components: thus each *clan* (q.v.) comprises *coarse-grained*, *medium-grained*, and *fine-grained* members.

grain-size control (*Met.*). Specifically, control of the rate at which the austenite grains grow when steel is heated above the critical range; the control is effected by the addition of aluminium before casting.

grain split (*Leather*). The upper or hair-side section of a split hide or skin. See **flesh split.**

graithe (*Mining*). See **grathe.**

gramicidin (*Chem.*). A polypeptide produced by the soil bacterium, *B. brevis*, with highly bactericidal properties.

graminaceous, gramineous (*Bot.*). Relating to grasses.

Gramineae (*Bot.*). Another name for **Glumales.**

graminicolous (*Bot.*). Living on grasses; esp. of parasitic fungi.

graminivorous (*Zool.*). Grass-eating.

grammatite (*Min.*). Synonym for *tremolite.*

gram(me) (*Phys.*). The unit of *mass* in the CGS system. It was originally intended to be the mass of 1 cm^3 of water at 4°C but was later defined as one-thousandth of the mass of the International Prototype Kilogramme, a cylinder of platinum-iridium kept at Sèvres.

gram(me)-atom (*Chem.*). The quantity of an element whose mass in grams is equal to its relative atomic mass. See **mole.**

gram(me) calorie. See **calorie.**

gram(me)-equivalent (*Chem.*). The quantity of a substance or radical whose mass in grams is equal to its equivalent weight.

gram(me)-ion (*Phys.*). Mass in grams of an ion, numerically equal to that of the molecules or atoms constituting the ion.

gram(me)-molecular volume (*Chem.*). See **molar volume.**

gram(me)-molecule (*Chem.*). The quantity of a substance whose mass in grams is equal to its molecular weight. Now replaced by the *mole* (q.v.)

gram(me)-röntgen (*Phys.*). The real conversion of energy when one röntgen is delivered to 1 g of air (approx. 8·4 microjoules). A convenient multiple is the *mega-gramme-röntgen* (10^6 gramme-röntgens).

gramme winding (*Elec. Eng.*). See **ring winding.**

Gram-negative (*Bacteriol.*). See **Gram-positive.**

gramophone (*Acous.*). An instrument for reproducing sound, using a stylus in contact with a spiral groove on a revolving disk. In U.S.A. the term is proprietary: the general equivalent is *phonograph.*

gramophone audiometer (*Acous.*). A quick method of testing the hearing of a large number of subjects, who are required to write down numbers perceived at diminishing intensities through head-telephones, the sounds being obtained from a gramophone record.

gramophone pick-up (*Radio*). The form of reproducer carried on the tone arm or transcription arm of a gramophone and connected to an audio-frequency amplifier, enabling the record to be reproduced electrically through the loudspeaker.

gramophone record (*Acous.*). See **disk record.**

Gram-positive (*Bacteriol.*). Said of bacteria which stain when treated with a basic dye of the triphenyl-methane group followed by mordanting with iodine and washing in acetone or ethyl-alcohol. Bacteria which do not stain are termed **Gram-negative**. Named after H. C. V. Gram, the Danish bacteriologist, who established the method in 1884.

grandifoliate (*Bot.*). Said of plants in which the leaves are much more conspicuous than the usually shortened stems.

grand mal (*Med.*). General convulsive epileptic seizure, with loss of consciousness. See also **epilepsy, petit mal.**

grand period of growth (*Bot.*). The period in the life of a plant, or of any of its parts, during which growth begins slowly, gradually rises to a maximum, gradually falls off, and comes to an end, external conditions remaining constant.

grandrelle (*Textiles*). Fancy yarns composed of two or more cotton yarns of different colours, twisted the opposite way to the singles; used in shirtings and other striped fabrics.

Grandry's corpuscles (*Zool.*). In Vertebrates, sensory nerve-endings in which the nerve loses its medullary sheath on reaching the capsule, and expands to plates which are inserted between tactile cells enclosed in a connective tissue sheath.

grand swell (*Acous.*). A swell-like balanced pedal for bringing in, as it is depressed, all the stops in an organ in a graded series.

granellae (*Zool.*). In some *Sarcodina*, small oval strongly refracting particles mainly composed of barium sulphate.

granellarium (*Zool.*). In some *Sarcodina*, the system of tubules containing granellae.

granite (*Geol.*). A coarse-grained igneous rock containing megascopic quartz, averaging 25%, much feldspar (orthoclase, microcline, sodic plagioclase), and mica or other coloured minerals. In the wide sense, granite includes alkali-granites, adamellites, and granodiorites, while the *granite clan* includes the medium- and fine-grained equivalents of these rock types.

granite-aplite (*Geol.*). See **aplite.**

granite-porphyry (*Geol.*). Porphyritic microgranite, a rock of granitic composition but with a groundmass of medium grain size in which larger crystals (phenocrysts) are embedded.

granite series (*Geol.*). A series relating the different types of granitic rock with respect to their time and place of formation in an organic belt, starting with early-formed deep-seated autochthonous granites and ending with post-tectonic high-level plutons.

granitic finish (*Build.*). A surface finish, resembling granite, given to cement work by the use of a suitable face mix.

granitic texture (*Geol.*). See **granitoid texture.**

granitization (*Geol.*). A theoretical process by which rocks can be changed into granite *in situ*. It would involve the movement of granitic liquid and vapours and perhaps ionic diffusion, not necessarily associated with very high temperatures.

granitoid texture (*Geol.*). A rock fabric in which the minerals do not possess crystal outlines but occur in shapeless interlocking grains. Such rocks are in the coarse grain-size group. Also called **xenomorphic granular texture.**

granoblastic texture (*Geol.*). An arrangement of mineral grains in a rock of metamorphic origin similar to that of a normal granite but produced by recrystallization in the solid and not by crystallization from a molten condition. The grains show no preferred orientation.

granodiorite (*Geol.*). An igneous rock of coarse grain size, containing abundant quartz and a large excess (more than ⅔ of the total feldspar) of plagioclase over orthoclase, in addition to coloured minerals such as hornblende and biotite. Cf. *diorite*.

granolithic (*Build.*). A rendering of cement and fine granite chippings, used as a covering for concrete floors, on which it is floated in a layer 1 in. to 2 in. (25 mm to 50 mm) in thickness. Used because of its hard-wearing properties.

granophyre (*Geol.*). An igneous rock of medium grain size, in which quartz and feldspar are intergrown as in graphic granite.

Granton Sandstone (*Geol.*). An important sandstone, used as a building-stone.

granular, granulate, granulose (*Bot.*). (1) Having the surface covered by tiny projecting points or very small warts. (2) Composed of, or filled with, minute granules. (*Powder Tech.*) Said of particles having an approximately equidimensional but irregular shape.

granular (or globular) pearlite (*Met.*). Pearlite in which the cementite occurs as globules instead of as lamellae. Produced by very slow cooling through the critical range, or by subsequent heating just below the critical range.

granulating pit (*Met.*). Pit into which molten slag is run, perhaps through a jet of water, and quenched so as to produce granules.

granulation tissue (*Med.*). A new formation of vascular connective tissue which grows to fill up the gap of a wound or ulcer; when healing is completed, a white scar is left.

granule (*Geol.*). A rock or mineral with a grain size between 2 and 4 mm.

granulite (*Geol.*). A granular-textured metamorphic rock, a product of regional metamorphism, similar in composition to one or other of the several kinds of schists or gneisses but lacking the distinctive mineral banding of these rock types.

granulitic texture (*Geol.*). The texture of a granulite, sometimes referred to as *granulose* or *granoblastic*, is an arrangement of shapeless interlocking mineral grains resembling the granitic texture but developed in metamorphic rocks.

granulitization (*Geol.*). The process in regional metamorphism of reducing the components of a solid rock to grains. If the reduction of the size of the particles goes farther, rock flour or mylonite is produced.

granulocytes (*Med.*). A group of blood cells of the *leucocyte* division.

granulocytopenia (*Med.*). An abnormal diminution in the number of granulocytes in the blood.

granuloma (*Med.*). A localized collection of granulation tissue occurring in certain chronic infections, such as tuberculosis and syphilis.

granuloma annulare (*Med.*). A condition in which rings of white cellular nodules appear on the back of the hands, and occasionally elsewhere.

granuloma inguinale (*Med.*). Ulcerating granuloma; granuloma venereum. A chronic disease, occurring in the tropics, in which ulcerating nodules appear on the genital organs, the perineum, and the groins.

granulomatous (*Med.*). Of the nature of, or resembling, granuloma.

granum (*Bot.*). A minute globule of pigment in the colourless stroma of a chromoplast.

grape-sugar (*Chem.*). D-*glucose* (q.v.).

graph (*Maths.*). A drawing depicting the relationship between two or more variables. The relationship may be known mathematically, e.g., the graph of the equation $x^2+y^2=1$ is a circle, or it may be unknown, e.g., the graph of a hospital patient's temperature.

graphecon (*Radar*). A double-ended storage tube used for the integration and storage of radar information and as a translating medium. The recording of the signals is based on the electrical conductivity induced in the target by a high voltage writing beam.

grapher (*Instr.*). See **recorder.**

graphical methods. The name given to those methods in which items, such as forces in structures, are determined by drawing diagrams to scale.

graphical reconstruction (*Zool.*). A method of preparing diagrammatic reconstructions or stereograms of the anatomy of an organ or animal in any desired plane from a given series of serial sections.

graphic formula (*Chem.*). A formula in which every atom is represented by the appropriate symbol, valency bonds being indicated by dashes; e.g., H—O—H, the graphic formula for water.

graphic granite (*Geol.*). Granite of pegmatitic facies, in which quartz and alkali-feldspar are intergrown in such a manner that the quartz simulates runic characters. Also called **runite.**

graphic statics (*Eng.*). A method of finding the stresses in a framed structure, in which the magnitude and direction of the forces are represented by lines drawn to a common scale.

graphic texture (*Geol.*). A rock texture in which one mineral intimately intergrown with another occurs in a form simulating ancient writing, especially runic characters; produced by simultaneous crystallization of two minerals present in eutectic proportions. See **graphic granite.**

graphite (*Chem., Min., etc.*). One of the two naturally occurring forms of crystalline carbon, the other being diamond. Rel. d. 2·23. It occurs as black, soft masses and, rarely, as shiny crystals (of flaky structure and apparently hexagonal) in igneous rocks; in larger quantities in schists particularly in metamorphosed carbonaceous clays and shales, and in marbles; also in contact metamorphosed coals and in meteorites. Its natural occurrence is widespread, notably eastern U.S.A., Austria, Ceylon, Mexico; also Borrowdale (Cumberland). Graphite has very numerous applications in trade and industry (see succeeding articles), now much overshadowing its use in 'lead' pencils. Much graphite is now produced artificially in electric furnaces using petroleum as a starting material. Also called **black lead, plumbago.** See **colloidal graphite.**

graphite brush (*Elec. Eng.*). A brush, made of graphite, for collecting the current from the commutator of an electric machine. It has a higher conductivity and better lubricating properties than an ordinary carbon brush.

graphite paint (*Paint.*). Paint consisting of fine graphite and bitumen solution, with or without oil, and supplied chiefly for the protection of hot wet surfaces such as the insides of industrial boilers.

graphite reactor (*Nuc. Eng.*). One in which fission is produced principally or substantially by slow neutrons moderated in graphite.

graphite resistance (*Elec. Eng.*). A resistance unit consisting of a rod of graphite, which has a high ohmic value; also a variable resistance made up of piles of graphitized disks of cloth under a variable pressure.

graphitic acid (*Chem.*). Graphite which has been treated with nitric acid and potassium (V) chlorate for a prolonged period.

graphitic carbon (*Met.*). In cast-iron, carbon occurring as graphite instead of as cementite.

graphitic lubricants (*Oils*). Graphite used in various forms as a lubricant, especially where film lubrication is not feasible; used in natural flake form, or suspended in oil, grease, distilled water, or other medium in colloidal form.

graphitization (*Chem.*). The transformation from the carbon allotrope to the graphite allotrope brought about by heat. It results in a volume change due to the alteration in atomic lattice layer spacing; see **electrographite.** It is reversible under bombardment; see **Wigner effect.**

graphitized filament (*Elec. Eng.*). See **metallized filament.**

grapnel, grappel (*Eng.*). Any device used to grapple with an object which is obscured to view, such as a submarine object. Grapnels generally take the form of grapnel hooks having several flukes. (*Mining*) An extracting tool used in boring operations.

grapple dredger (*Civ. Eng.*). See **grab-dredger.**

grappler (*Build.*). A wedge-shaped spike with an eye at one end; it is driven into a joint of the brickwork as a support for the hooked end of one of the brackets in a bracket scaffold.

graptolite (*Geol.*). An extinct group of animals, represented by their abundant fossil remains in Palaeozoic rocks, and used for dating Ordovician and Silurian sediments. They are probably related to the Chordata.

grass (*Bot.*). See **Glumales.** (*Radar*) Irregular deflection from the time-base of a radar display, arising from electrical interference or noise. Also **picture noise.**

grass-bleached (*Paper*). A term applied to tissues of a specially white colour. The expression suggests the whiteness obtained in air-bleached linen.

grass disease (*Vet.*). See **bovine hypomagnesaemia.**

grassing (*Textiles*). Method of mild bleaching by spreading linen material on grass and leaving it exposed to the sun and oxygen in the air.

grassland climax (*Bot.*). A climax community consisting of grassland.

Grassot fluxmeter (*Elec. Eng.*). See **fluxmeter.**

grass sickness (*Vet.*). A disease of unknown cause affecting the horse, often occurring soon after the horse is put on to grass, associated with dysfunction of the bowels. Acute cases are characterized by colic and more chronic cases by emaciation. The essential pathological change is believed to be degeneration of the neurones in the ganglia of the autonomous nervous system.

grass staggers (*Vet.*). See **bovine hypomagnesaemia.**

grass table (*Build.*). A *ground table* (q.v.).

grass tetany (*Vet.*). See **bovine hypomagnesaemia.**

grate (*Eng.*). That part of a furnace which supports the fuel. It consists of fire bars or bricks so spaced as to admit the necessary air. See **fire bars.**

grate area (*Eng.*). The area of the grate in a

furnace burning solid fuel; for a boiler furnace, a measure of the evaporative capacity of the boiler.

grathe or graithe (*Mining*). To repair, or put in order, the plant in a coal mine.

graticule (*Surv.*). See **reticule.**

grating (*Phys.*). An arrangement of alternate reflecting and nonreflecting elements, e.g., wire screens or closely spaced lines ruled on a flat (or concave) reflecting surface, which, through diffraction of the incident radiation analyses this into its frequency spectrum. An *optical grating* can contain a thousand lines or more per cm. A *standing-wave* system of high-frequency sound waves with their alternate compressive and rarefied regions can give rise to a diffraction grating in liquids and solids. With a *criss-cross* system of waves, a 3-dimensional grating is obtainable.

grating spectrum (*Optics*). See **prismatic spectrum.**

Grätz rectifier (*Elec. Eng.*). Type of bridge rectifier using 6 rectifying elements for 3 phase supply.

gravel (*Build., Civ. Eng.*). A natural mixture of rounded rock fragments and generally sand used in the manufacture of concrete. (*Geol.*) The name of the aggregate consisting dominantly of pebbles, though usually a considerable amount of sand is intercalated. In the Stratigraphical Column, gravels of different ages and origins occur abundantly, e.g., in S.E. England, where they consist chiefly of well-rounded flint pebbles originally derived from the Chalk. These gravels are mainly of fluviatile and fluvioglacial origin, but marine gravels are also common in the littoral zone. The indurated equivalent of gravel is conglomerate. (*Med., Vet.*) Small calculi in the kidneys, ureters or urinary bladder.

gravel board or gravel plank (*Build.*). A long board standing on its edge at the bottom of a wooden fence, so that the upright boards of the fence do not have to reach down to the ground.

graveolent (*Bot.*). Having a strong rank odour.

Graves's disease (*Med.*). See **Basedow's disease.**

graveyard (*Nuc. Eng.*). See **burial.**

graveyard test (*For.*). A test conducted out of doors on pieces of timber in contact with the ground to determine their durability.

gravid (*Med., Zool.*). Pregnant; carrying eggs or young.

gravimeter (*Geophys.*). Instrument used in gravity surveying to measure relative changes in the value of gravity from place to place. Modern land instruments can detect a change of 0·01 milligals though marine gravimeters have a reduced accuracy.

gravimetric analysis (*Chem.*). The chemical analysis of materials by the separation of the constituents and their estimation by weight.

graving dock (*Civ. Eng.*). See **dry dock.**

gravipause (*Space*). That point or border in space in which the gravitational force of one body is matched by the counter-gravity of another; also **neutral point.**

graviperception (*Bot.*). The perception of gravity by plants.

gravitation (*Phys.*). The name given to that force of nature which manifests itself as a mutual attraction between masses, and whose mathematical expression was first given by Newton, in the law which states: 'Any two particles of matter attract one another with a force directly proportional to the product of their masses and inversely proportional to the square of the distance between them.' This may be expressed by the equation:

$$F=G\frac{m_1m_2}{d^2},$$

where F is the force of gravitational attraction between bodies of mass m_1 and m_2, separated by a distance d. G is the constant of gravitation and has the value $6{\cdot}670\times10^{-11}\text{N m}^2\text{ kg}^{-2}$.

gravitational astronomy (*Astron.*). See **celestial mechanics.**

gravitational differentiation (*Geol.*). The production of igneous rocks of contrasted types by the early separation of crystals such as olivine, pyroxenes, etc., which, sinking on account of their high density, become concentrated in the basal parts of intrusions. The ultramafic rocks such as peridotites and picrites may have been formed in this way.

gravitational field (*Phys.*). That region of space in which appreciable gravitational force exists.

gravitational induction (*Bot.*). The development of a structure from the under side of a plant member.

graviton (*Nuc.*). Hypothetical quantum of gravitational field energy bearing the same relationship to the gravitational field as the photon does to the electromagnetic field. The gravitational interaction between particles is so weak that no discontinuous (quantized) effects have yet been observed experimentally. Hence the existence of the graviton can only be inferred from laws found immutable in other branches of physics.

gravity anomaly (*Geophys.*). Deviation of the corrected value of gravity at a point from the geoid or datum value at that point.

gravity cell (*Elec. Eng.*). A 2-fluid cell in which the electrodes are horizontal and the 2 electrolytes lie in separate layers on account of their difference in relative density.

gravity clock (*Horol.*). One actuated by its own weight, e.g., by moving down a racked pillar.

gravity-controlled instruments (*Elec. Eng.*). Electrical measuring instruments in which the controlling torque is provided solely by the action of gravity.

gravity conveyor (*Eng.*). A conveyor in which the weight of the articles handled is sufficient to effect their transport from a higher to a lower point; as when they are allowed to slide down an inclined runway.

gravity dam (*Civ. Eng.*). A dam which is prevented by its own weight from overturning.

gravity diecasting (*Met.*). A process by means of which castings of various alloys are made in steel or cast-iron moulds, the molten metal being poured by hand. See **diecasting, pressure diecasting.**

gravity drop hammer (*Eng.*). A type of machine hammer used for drop forging, in which the impact pressure is obtained from the kinetic energy of the falling ram and die. Also called a **board hammer.**

gravity escapement (*Horol.*). An escapement in which a constant impulse is given to the pendulum by a weight which is lifted and which always falls through a constant distance. The best-known gravity escapement is the *double 3-legged*, in which pivoted arms are lifted a constant height to give on falling a constant impulse to the pendulum. Used in the clock of Big Ben.

gravity feed tanks (*Aero., I.C. Engs.*). Fuel tanks, situated above the point of delivery to the engine, which feed the engine solely by gravity.

gravity plane (*Mining*). An inclined plane on which the descending full trucks pull up the ascending empty ones.

gravity process (*Glass*). See **gob process.**
gravity prospecting (*Geophys.*). A method of geophysical exploration which uses accurate measurement of the value of gravity to give information about density differences in subsurface rocks.
gravity roller conveyor (*Eng.*). A fabricated framework, usually in longitudinal sections and slightly inclined, carrying freely rotating tubular rollers on which a load may be moved by the force of gravity acting on it.
gravity scale (*Eng.*). A measure of certain characteristics of diesel fuel oil, related to the density of the oil by an arbitrary formula.
gravity separation (*Min. Proc.*). Use of differences between relative densities of roughly sized grains of mineral to promote settlement of denser species while less dense grains are washed away. Machines used include jigs, sluices, shaking tablers, classifies, hydrocyclones, dense-media appliances, Humphreys spirals, vanners, and buddles.
gravity stamp (*Min. Proc.*). One of set of 5 heavy pestles, lifted 90 times a minute by a cam and allowed to fall on ore in mortar-box, for crushing purposes.
gravity tectonics (*Geol.*). Processes of rock deformation and folding which are activated by gravity applied over considerable periods of geological time.
gravity transport (*Geol.*). The movement of material under the influence of gravity. It includes downhill movement of weathering products, and movement of unweathered material in landslides.
gravity water system (*Phys.*). A system in which flow occurs under the natural pressure due to gravity.
gravity waves (*Phys.*). Liquid surface waves controlled by gravity and not surface tension.
gravure (*Print.*). Abbrev. for *photogravure*.
Grawitz's tumour (*Med.*). See **hypernephroma.**
gray. SI unit of absorbed dose. Symbol Gy.
Gray-King Test (*Min. Proc.*). Test of coking quality of coal under prescribed conditions of heating to 600°C.
graywacke (*Geol.*). American spelling of *greywacke*.
grazing angle (*Phys.*). A very small *glancing angle*.
grease (*Oils*). A semisolid form of lubricant, composed of emulsified mineral lubricating oil and soda or lime soap. (*Vet.*) Greasy heels; sore-heels. A mild, localized form of *horse pox* (q.v.), affecting the skin of the flexor surface of joints in the lower part of the leg.
grease cup (*Eng.*). A lubricating device consisting of a cylindrical cup threaded internally and filled with grease. It is screwed down on to a drilled pad which has a nipple screwed into the bearing housing, thus feeding grease into the bearing.
grease gun (*Eng.*). A device for forcing grease into bearings under high pressure. It consists of a cylinder from which the grease is delivered by hand pressure on the piston, intensified by a second plunger which forms the delivery pipe and which is pressed against a nipple screwed into the bearing.
greaseproof paper (*Paper*). (1) A vegetable parchment prepared by chemical treatment. (2) An imitation vegetable parchment, transparency and grease-resisting properties being imparted by prolonged beating of the pulp.
grease-spot photometer (*Light*). A simple means of comparing the intensities of two light sources. A screen of white paper, rendered partially translucent by a spot of grease, is illuminated normally by the two sources, one on each side. The position of the screen is adjusted until the grease spot is indistinguishable from its surround, when the illuminations on the two sides may be assumed to be equal. Also called **Bunsen photometer.** See **photometer.**
grease table (*Min. Proc.*). Sloping table anointed with petroleum jelly, over which diamondiferous concentrate is washed, the diamond adhering strongly while the gangue is worked away.
grease trap (*San. Eng.*). A trapped gulley receiving sink wastes, and specially designed to prevent obstruction of gulley or drain by congealed fatty matter.
greasy heels (*Vet.*). See **grease.**
great (*Acous.*). The principal division of a straight (church or concert) organ. It comprises ranks of pipes, not enclosed in chambers with swell shutters, which are most powerful and brilliant and which are normally played from the *great manual*, usually placed below the *swell manual* and above the *choir* or *accompaniment manual*, if such are present. The name *great* is retained in unit and cinema organs, in which, however, since here all ranks can be drawn on any manual, the word has no real meaning.
great circle (*Maths.*). The intersection of a sphere by a plane passing through its centre. The shortest distance between two points on the surface of a sphere is along the great circle passing through them.
Great Ice Age (*Geol.*). See **Pleistocene Period.**
great primer (*Typog.*). An old type size, approximately 18-point.
great wheel (*Horol.*). The first wheel in the train of a watch or clock; in going barrels it forms part of the barrel. The largest wheel in a watch or clock train.
Greaves-Etchell furnace (*Elec. Eng.*). A form of direct-arc electric furnace used in steel manufacture.
Grecian honeycomb (*Textiles*). A type of honeycomb structure woven from coarse soft cotton yarns; the material is used for towels and quilts.
green (*Civ. Eng.*). A colloquial term for concrete in the hardening stage, after pouring and before setting. (*Powder Tech.*) Describes the unheat-treated condition of preforms held together by the cohesive forces resulting from compaction alone or with the assistance of a temporary binder.
green acids (*Chem.*). Mixtures of sulphonic acids, formed by reaction between petroleum and sulphuric acid, which form detergent soaps with fatty acids.
green algae (*Bot.*). See **Chlorophyceae.**
greenalite (*Min.*). A septechlorite related to the chlorites chemically and to the serpentines structurally. A hydrous, iron (II) silicate of composition $Fe_6Si_4O_{10}(OH)_8$; occurs in the iron ore of the Mesabi Range, Minnesota.
green bricks (*Build.*). Moulded clay shapes which after undergoing a burning process will become bricks.
Greenburg-Smith impinger (*Powder Tech.*). Device for sampling gas streams, consisting of a tall, flat-bottomed flask containing a vertical tube narrowing to a jet at the lower end, which is immersed in dust-free water. The gas stream is sucked through the jet and any dust particles are collected in the water.
green carbonate of copper (*Min.*). See **malachite.**
green cell (*Bot.*). A cell of the alga *Chlorella* living inside certain simple animals.
green chain (*For.*). An endless-chain table over

which all green timber from a sawmill passes for grading and sorting.

green flash (*Astron.*). A phenomenon sometimes seen in clear atmospheres at the instant when the upper rim of the sun finally disappears below the horizon as a bright green blob of light; green is the last apparent colour from the sun, because the more greatly refracted blue is dispersed.

green glands (*Zool.*). The antennal excretory glands of decapod *Crustacea*.

green gun (*TV*). Electron gun used to excite the green phosphor in a 3-gun colour TV tube.

greenheart (*For.*). A very strong yellowish-green timber from South America; largely used for piles and underwater work because of its considerable, resistance to the attack of the teredo.

Greenland spar (*Min.*). See **cryolite.**

greenockite (*Min.*). Crystalline cadmium sulphide, occurring, as small yellow hexagonal crystals exhibiting polar symmetry, in cavities in altered basic lavas as at Bishopton, Renfrewshire, Scotland.

green sand (*Foundry*). Highly siliceous sand, with some alumina, dampened and used with admixed coal or charcoal in moulding.

greensand (*Geol.*). A sand or sandstone with a greenish colour, due to the presence of the mineral glauconite.

green sand casting (*Foundry*). One made in foundry sand not strengthened by kiln-drying before pouring.

Green's theorem (*Maths.*). $\int_s F \cdot ds = \int_v \text{div} \cdot F\, dv$, where F is a vector function and the surface of integration bounds the volume.

greenstick fracture (*Med.*). Fracture of rickety bones, which break like a green stick.

greenstone (*Geol.*). An omnibus term lacking precision and applied indiscriminately to basic and intermediate igneous rocks in which much chlorite has been produced as a result of metamorphism.

green vitriol (*Min.*). *Melanterite* (q.v.).

Greenwich Mean Time (*Astron.*). Abbrev. GMT. *Mean solar time* (q.v.) referred to the zero meridian of longitude, i.e., that through Greenwich. GMT is the basis for scientific and navigational purposes. See **universal time;** also **zulu time.**

gregale (*Meteor.*). A strong wind from the N.W. or N.E. direction in the Mediterranean area, blowing mainly in the cool season.

gregaria phase (*Zool.*). In locusts (*Orthoptera*), a phase characterized by high activity and gregarious tendencies, differing morphologically from the *solitaria phase* (q.v.) with which, under natural conditions, it alternates.

Gregarinidea (*Zool.*). An order of *Telosporidia*, in which the adult trophozoite is an extra-cellular parasite, and in which reproduction is usually by sporogony only.

gregariniform (*Zool.*). Resembling the *Gregarinidea*; said of certain types of swarm-spores which lack organs of locomotion.

gregarinulae (*Zool.*). Zoospores without organs of locomotion.

Gregorian calendar. The name commonly given to the civil calendar now used in most countries of the world. It is the Julian calendar as reformed by decree of Pope Gregory XIII in 1582. The Gregorian reform omitted certain leap years, and brought the length of the year on which the calendar is based nearer to the true astronomical value. See **New Style, Old Style, leap years.**

Gregorian telescope (*Astron.*). A form of reflecting telescope, very similar in principle to the Cassegrainian, in which the large mirror is pierced at the centre and the lighti s reflected back into an eye-piece in this centre by a small concave mirror on the principal axis and a little outside the focus.

gregory formula (*Maths.*). Obtained from the Newton interpolation formula, used for numerical integration. It takes the form

$$\int_a^b f(x)dx = \frac{h}{2}\left(y_0+2y_1+2y_2+ \dots +2y_{n-1}+y_n\right) - \frac{h}{12}\left(\Delta y_{n-1}-\Delta y_0\right) - \frac{h}{24}\left(\Delta^2 y_{n-2}+\Delta^2 y_0\right) - \frac{19h}{720}\left(\Delta^3 y_{n-3}-\Delta^3 y_0\right) - \frac{3h}{160}\left(\Delta^4 y_{n-4}+\Delta^4 y_0\right) - \dots$$

where h is the common interval between successive values of x, and $\Delta^m y_{n-m}$ represent finite differences.

greisen (*Geol.*). A rock composed essentially of mica and quartz, resulting from the alteration of a granite by percolating solutions. Greisens often contain small amounts of fluorite, topaz, tourmaline, cassiterite, and other relatively uncommon minerals, and may be associated with mineral deposits, as in Cornwall. See **pneumatolysis.**

greisenization (*Geol.*). The process by which granite is converted to greisen. Greisenization is a common type of wall rock alteration in areas where granite is traversed by hydrothermal veins. Sometimes **greisening.**

grenz rays (*Radiol.*). X-rays produced by electron beams accelerated through potentials of 25 kV or less. These are generated in many types of electronic equipment; low penetrating power.

grey (*Photog.*). Said of colours which possess only brilliance, i.e., they have no hue and are completely desaturated. (*Textiles*) Term applied to natural yarn or cloth, before bleaching or mercerizing.

grey blibe (*Glass*). An elongated inclusion of sodium sulphate in glass.

grey body (*Phys.*). A body which behaves as a black body in respect of the absorbed fraction of the incident radiation, $E_a = E_i(1-A)$, where E_a is the absorbed fraction, E_i incident energy, and A albedo, and complies with the *Stefan-Boltzmann* equation.

grey copper ore (*Min.*). See **tetrahedrite.**

grey iron (*Met.*). Pig- or cast-iron in which nearly all the carbon not included in pearlite is present as graphitic carbon. See **mottled iron, white iron.**

grey key image (*Photog.*). A neutral image obtained in addition to the 3 images corresponding to the primary colours in colour photography; used to regulate the saturation or assist registration.

grey matter (*Zool.*). The centrally-situated area of the central nervous system, mainly composed of cell bodies. Cf. *white matter*.

grey scale (*Photog.*). A grey *step-wedge* (q.v.).

grey sour (*Textiles*). The name applied to one of the processes in preparing cotton goods for bleaching and dyeing. The material is treated with dilute acid to remove all traces of alkali.

greywacke (*Geol.*). A sandstone containing silt, clay, and rock fragments in addition to quartz grains. It is much more poorly sorted than other types of sandstone, and often occurs on beds which show gradation in grain size from fine at the top to coarse at the bottom.

greywethers (*Geol.*). Grey-coloured rounded blocks of sandstone or quartzite left as residual boulders on the surface of the ground when less resistant material was denuded. From a distance they resemble sheep grazing. See **sarsen.**

grid (*Build.*). The plan layout of the structure of a given building. (*Civ. Eng.*) A timber framework so built that a vessel may be floated in at high water and repairs undertaken as the tide falls. (*Elec. Eng.*) The electrical power distribution system at 132 kV and lower voltages. See **supergrid.** For other *Elec. Eng.* grid articles see below and **accumulator grid, damper, resistance grid.** (*Electronics*) Control electrode having an open structure (e.g., mesh) allowing the passage of electrons; in an electron gun it may be a hole in a plate. (*Eng.*) A grating made up of a number of parallel bars, such as that required to prevent foreign matter from entering a pump intake. (*Surv.*) A network of lines superimposed upon a map and forming squares for referencing, the basis of the network being that each line in it is at a known distance either east or north of a selected origin. In ordnance surveying, triangulation system covering large area.

grid base (*Electronics*). The grid-bias voltage required to produce anode current cut-off for a thermionic valve.

grid bias (*Electronics*). The d.c. negative voltage applied to the control grid of a thermionic valve.

grid-bias battery (*Elec.*). One providing a *pd* for the grid polarization of valves. It usually makes the grid negative with respect to the cathode and so supplies power only if the instantaneous signal swings the grid positive again. U.S. term **C-battery.**

grid-bias modulation (*Radio*). A system of modulation in which the modulating signal is caused to control the grid bias of a valve to the grid of which a high-frequency carrier voltage is applied.

grid-bias resistance (*Radio*). A resistance included between the cathode and the negative terminal of the high-tension supply voltage, to which point the grid circuit is returned. The potential drop produced by the flow of anode current through this resistance furnishes the negative grid bias.

grid capacitor (*Elec. Eng.*). A capacitor between the grid and the remainder of the grid circuit.

grid characteristics (*Electronics*). Graph of grid current against grid potential for a thermionic valve.

grid circuit (*Elec. Eng.*). The circuit connected between the grid and the cathode of a thermionic valve.

grid conductance (*Elec.*). The in-phase component of the grid input admittance, due to grid current, Miller effect, etc.

grid control (*Electronics*). That provided by voltage on grid of a thyratron or mercury-arc rectifier; at a sufficient positive voltage, anode current flows, grid loses control until deionization is effected after loss of anode voltage.

grid-controlled mercury-arc rectifier (*Elec. Eng.*). One in which initiation of arc in each cycle is determined by phase of voltage on a grid for each anode.

grid current (*Electronics*). The current which flows from the grid to the cathode when the grid is made positive, or only slightly negative, with respect to the cathode.

grid-current characteristic (*Electronics*). The curve relating the grid current to the grid potential of a thermionic tube.

grid dip meter (*Elec. Eng.*). Wavemeter in which absorption of energy by tuned circuit at resonance is indicated by decrease of valve grid current.

grid drive (*Elec. Eng.*). Voltage or power required to drive a valve when delivering a specified load.

grid emission (*Electronics*). Thermionic emission of electrons from valve grid, consequent upon serious overheating. Secondary grid emission may also occur as a result of electron bombardment.

grid-glow tube (*Electronics*). Cold-cathode gas-discharge tube in which glow is triggered by a grid.

grid-iron pendulum (*Horol.*). A compensation pendulum with 5 parallel rods of iron and 4 of brass, the total length of each metal being in inverse ratio of its coefficient of linear expansion.

grid leak (*Radio*). High value resistor connected between grid and earth, which biases valve when this carries a 'leakage' current.

grid-leak detector (*Radio*). Demodulator circuit in which modulation wave appears across grid-leak resistor. Preferably termed **cumulative grid detector.**

grid modulation. See **suppressor-grid modulation.**

grid navigation (*Aero., Nav.*). A navigational system in which a grid instead of true North is used for the measurement of angles and for heading reference. The grid is of parallel lines referenced to the 180° meridian which is taken as Grid North. Used principally in polar route flying and, as the system is independent of convergence of meridians, headings remain the same from departure to destination.

grid neutralization (*Elec. Eng.*). The method of neutralization of an amplifier through an inverting network in the grid circuit, which provides the requisite phase shift of 180°. Also called **phase inverter.**

grid pool tube (*Electronics*). A gas-discharge tube with a liquid (or solid) pool-type cathode such as the mercury cathode.

grid ratio (*Radiol.*). In grid therapy, the ratio of the total area of holes to that of the grid.

grid rectification (*Radio*). Rectification of high-frequency currents and voltages by means of the nonlinearity of the grid-current characteristic, more current flowing during the positive half-cycles than during the negative half-cycles.

grid resistance (*Elec. Eng.*). A resistance unit for heavy current work, e.g., starters for railway motors; it is made up of a number of resistance grids placed side by side and mounted in a metal frame.

grid-suppressor resistor (*Elec. Eng.*). A resistor used to suppress parasitic oscillations by connecting it between the control grid and the tuned circuit of an RF amplifier.

grid sweep (*Elec. Eng.*). The range of voltage covered by the grid potential in going from its maximum positive (or minimum negative) to its maximum negative (or minimum positive) value.

grid swing (*Elec. Eng.*). The maximum voltage excursion of the control grid potential from the bias potential when a signal is applied to the grid.

grid therapy (*Radiol.*). A method of treatment in radiotherapy, in which radiation is given through a grid of holes in a suitable absorber, e.g., lead, rubber. By careful positioning, a greater depth dose can be given, and skin breakdown reduced to a minimum.

griffe (*Weaving*). A series of horizontal knives in

a frame, which serves to raise and lower the warp threads in jacquard machines and in some top and side dobbies up to sixteen staves (healds).

Griffin mill (*Min. Proc.*). Pendulum mill in which a hanging roller bears against a stationary bowl as it rotates, crushing passing ore.

Griffith's white (*Paint.*). *Lithopone* (q.v.).

Grignard reagents (*Chem.*). Mixed organometallic compounds prepared by dissolving dry magnesium ribbon or filings in an absolutely dry ethereal solution of an alkyl bromide or iodide. The value of Grignard reagents in the preparation of secondary and tertiary alcohols, and the alkenes, so easily produced therefrom, and in the synthesis of alkyl and aryl derivatives from many halogen compounds, can hardly be overestimated. See **silicones**.

grike (*Geol.*). A fissure in limestone rock caused by the solvent action of rainwater. See **karst**.

grill (*Build.*). A layer of joists in a grillage foundation.

grillage foundation (*Build.*). A type of foundation often used at the base of a column. It consists of one, two, or more tiers of steel beams superimposed on a layer of concrete, adjacent tiers being placed at right angles to each other, while all tiers are encased in concrete.

grille (*Build.*). A plain or ornamental openwork of wood or metal, used as a protecting screen or grating.

Grimmiales (*Bot.*). An order of the *Musci*; the blackish-green gametophore is branched with crowded leaves; the sporophyte is acrocarpous, bearing a symmetrical capsule; the peristome is simple with 16 teeth.

grindability (*Min. Proc.*). Empirical assessment of response of ore to pulverizing forces, applied under specified conditions.

grinder (*Paper*). A large thick circular stone made from grit-stone, used in the manufacture of mechanical wood pulp. (*Radio*) A type of atmospheric disturbance of relatively long duration. Cf. *click*.

grinder's rot (*Med.*). Lung disease caused by inhalation of metallic particles by steel-grinders.

grinding (*Min. Proc.*). Comminution of minerals by dry or, more usually, wet methods, mainly in rod, ball, or pebble mills. (*Textiles*) Sharpening the wire points on card cylinders, doffers, flats, or clearers and strippers, with a revolving emery covered roller.

grinding-in (*Eng.*). The process of obtaining a pressure-tight seal between a conical-faced valve and its seating by grinding the two together with an abrasive mixture such as silicon carbide and oil.

grinding machine (*Eng.*). A machine tool in which flat, cylindrical, or other surfaces are produced by the abrasive action of a high-speed grinding wheel. See **centreless grinding, cylindrical grinding, profile grinding, surface grinding machine, thread grinding.**

grinding medium (*Eng.*). The solid charges (balls, pebbles, rods, etc.) used in suitable mills for grinding certain materials, e.g., cement, pigments, etc., to a fine powder.

grinding slip (*Carp., etc.*). See **gouge slip.**

grinding wheel (*Eng.*). An abrasive wheel for cutting and finishing metal and other materials. It is composed of an abrasive powder, such as silicon carbide or emery, held together by a bond or binding agent, which may be either a vitrified material or a softer material, such as shellac or rubber.

grinning through (*Paint.*). A paintwork defect characterized by the fact that the *hiding power* (q.v.) of the paint is insufficient to obscure completely the surface underneath.

grip (*Build.*). (1) A small channel cut to carry away rain water during construction of foundations. (2) Small channel across the road-side to conduct surface water to a drain. Also **offlet.**

gripper edge (*Print.*). See **grippers.**

gripper-feed mechanism (*Eng.*). A type of strip-feeding mechanism used in presswork, which advances the strip material during its forward stroke, and then returns to its initial position with grippers open, and thus inoperative.

grippers (*Print.*). Attachments which grip the edge (**gripper edge**) of a sheet of paper when it is fed into the printing machine.

grisaille (*Paint.*). A process of painting in shades of grey, to give the appearance of modelling in relief.

griseofulvin (*Chem.*). 7-Chloro-4,6-dimethoxy-coumaran-3-one-2-spiro-1′-(2′-methoxy-6′-methylcyclohex 2′-en-4′-one), used as a fungicide. (*Med.*) An oral antibiotic which is excreted in the cells of the skin, thus destroying any cutaneous fungus infections.

grist (*Brew.*). A mixture of malts sufficient for one brewing.

grist case (*Brew.*). A conical chamber into which grist passes, after grinding, before brewing.

grit (*Eng.*). A measure indicating the sizes of the abrasive particles in a grinding wheel, usually expressed by a figure denoting the number of meshes per linear inch in a sieve through which the particles will pass completely. (*Geol.*) Siliceous sediment, loose or indurated, the component grains being angular. Contrast *sand* and *sandstone*, in which the grains are rounded. (*Powder Tech.*) Hard particles, usually mineral, of natural or industrial origin, retained on a 200 mesh B.S. test sieve (76 μm).

grit blasting (*Eng.*). A process used in preparation for metal spraying, which cleans the surface and gives it the roughness required to retain the sprayed metal particles.

grit cell (*Bot.*). A stone cell occurring in a leaf or in the flesh of a fruit.

grit chamber (*San. Eng.*). A *detritus chamber*.

grivation (*Aero.*). In *grid navigation* the horizontal angle between the grid datum and magnetic meridian at any point. Values are indicated on *isogrivs* as corrections for converting grid headings into magnetic headings.

grizzle (*Mining*). Coal so intermixed with iron pyrites as to make it of little value.

grizzle bricks (*Build.*). Bricks which are underburnt and of bad shape. They are soft inside and unsuitable for good work but are often used for the inside of walls. Also called **place bricks, samel bricks.**

grizzly (*Min. Proc.*). Down-sloped parallel steel bars on which ore is screened as it slides from feed to point where oversize is discharged.

grog (*Build.*). Bricks or waste from a clayworks broken down and added to clay to be used for brick manufacture.

groin (*Arch.*). The line of junction of the 2 constituent arches in a *groined arch* (q.v.).

groined arch (*Arch.*). An arch which is intersected by other arches cutting across it transversely.

groin rib (*Arch.*). A projecting member following the line of a groin.

grommet (*Eng.*). A ring or collar used to line a sharp-edged hole through which a cable or similar material passes.

groove (*Bind.*). The separation between the board and the spine of the book cover, permitting free opening. (*Typog.*) The channel between the feet of a piece of founder's type.

grooved bit (*Carp.*). A wood-boring bit with a

cylindrical shank in which is a helical groove. Also **drill bit.**

grooved wire (*Elec. Eng.*). A special form of wire used for the overhead contact wire of electric traction systems, with grooves into which the supporting ears can be clipped. See **figure-eight wire.**

grooving (*Eng.*). (1) Cracking of the plates of steam boilers at points where stresses are set up by the differential expansion of hot and colder parts. (2) Producing a rectangular, V-shaped or similar groove or channel by milling, shaping, grinding, etc., e.g., to provide a location for a reciprocating slide or as a lubricant reservoir.

grooving plane (*Carp., etc.*). One with a narrow, interchangeable blade and adjustable fence, for cutting grooves.

grooving saw (*Carp., etc.*). A circular saw which may be of the drunken type, used for cutting grooves.

gros nez (*Med.*). See **goundou.**

gross register(ed) tonnage (*Ships*). See **gross tonnage.**

gross tonnage (*Ships*). The cubic capacity (100 $ft^3 = 1$ ton) of all spaces below the freeboard deck and permanently closed-in spaces above that deck except *exempted spaces* (q.v.).

grossular (*Min.*). An end-member of the garnet group, the composition being represented by $3CaO{\cdot}Al_2O_3{\cdot}3SiO_2$; formed in the contact-metamorphism of impure limestone. Commonly contains some iron, and is greenish, brownish, or pinkish. Also **gooseberry stone.**

gross weight (*Aero.*). See **weight.**

gross wing area (*Aero.*). The full area of the wing, including that covered by the fuselage and any nacelles.

grotesque (*Typog.*). The name given to early sanserif types and still used, especially the contraction **grot.**

Grotthus-Draper law (*Chem.*). Only such energy of electromagnetic radiation as is absorbed by tissue is effective in chemical action following ionization in that tissue.

ground (*Mining*). The mineralized deposit and the rocks in which it occurs, e.g., payground, payable reef; barren ground, rock without value. (*Radio*) U.S. for **earth.** (*Textiles*) (1) In lace manufacture, the mesh which forms a foundation for a pattern. (2) The base fabric to which is secured the figuring, etc., threads, e.g., pile or loops in carpet or terry cloths.

ground absorption (*Radio*). The energy loss in radio-wave propagation due to absorption in the ground.

ground air (*San. Eng.*). The air contained in the upper layers of the subsoil; it has a variable composition, including carbon dioxide, ammonia, and other gases resulting from oxidation of organic matters, and may be noxious.

ground auger (*Tools*). An auger specially adapted for boring holes in the ground, for artesian-wells, etc.

ground capacitance (*Elec.*). See **earth capacitance.**

ground clutter (*Radar*). The effect of unwanted ground-return signals on the screen pattern of a radar indicator.

ground-controlled approach (*Aero., Radar*). Aircraft landing system in which information is transmitted by a *ground controller* from a ground radar installation at end of runway to a pilot intending to land. Also called **talk-down.** Abbrev. **GCA.**

ground-controlled interception (*Aero., Radar*). Radar system whereby aircraft are directed on to an interception course by a station on the ground. Abbrev. **GCI.**

ground controller (*Radar*). Man who has control of aircraft over an area, in fog or during landing, by radar. See **ground-controlled approach.**

grounded base (*Electronics*). See **common-base connexion.**

grounded-cathode amplifier (*Electronics*). One with cathode at zero alternating potential, drive on the grid and power taken from the anode; the normal and original use of a triode valve.

grounded circuit (*Elec. Eng.*). A circuit which is deliberately connected to earth at one point or more, for safety or testing.

grounded collector (*Electronics*). See **common-collector connexion.**

grounded emitter (*Electronics*). See **common-emitter connexion.**

grounded-grid amplifier (*Electronics*). One with grid at zero alternating voltage, drive between cathode and earth, output being taken from the anode; there is no anode-grid feedback.

ground engineer (*Aero.*). An individual, selected by the licensing authorities, who has power to certify the safety for flight of an aircraft, or certain specified parts of it. Term now superseded by *licensed aircraft engineer* (q.v.).

ground fine pitch (*Aero.*). A very flat blade angle on a turboprop airscrew, which gives extra braking drag and low airscrew resistance when starting the engine. Colloquially **disking.**

ground frost (*Meteor.*). A temperature of 0°C (32°F) or below, on a horizontal thermometer in contact with the shorn grass tips on a turf surface.

ground joist (*Carp.*). A horizontal timber supported off the ground at a basement or ground-floor level.

ground-level (*Surv.*). The *reduced level* (q.v.) of a survey station with reference to an official bench mark.

ground loop (*Aero.*). An uncontrollable and violent swerve or turn by an aeroplane while taxiing, landing or taking-off.

groundmass (*Geol.*). In igneous rocks which have crystallized in two stages, the groundmass is the finer-grained portion, in which the phenocrysts are embedded. It may consist wholly of minute crystals, wholly of glass, or partly of both.

ground meristem (*Bot.*). Those parts of an apical meristem which give rise to ground tissue.

ground mould (*Civ. Eng.*). A timber piece or frame used as a template to bring earthworks such as embankments to the required form.

ground noise (*Acous.*). See **background noise.**

ground plan (*Build., Civ. Eng.*). A drawing showing a plan view of the foundations for a building or of the layout of rooms, etc., on the ground floor.

ground plate (*Build.*). The bottom horizontal timber to which the frame of a building is secured.

groundplot (*Aero.*). Method of calculating the position of an aircraft by relating groundspeed and time on course to starting position.

ground-position indicator (*Aero.*). An instrument which continuously displays the dead-reckoning position of an aircraft.

ground ray (*Radio*). Same as **direct ray.**

ground reflection (*Radar*). The wave in radar transmission which strikes the target after reflection from the earth.

ground resonance (*Aero.*). A sympathetic response between the dynamic frequency of a rotorcraft's rotor and the natural frequency of the alighting gear which causes rapidly-increasing oscillations.

ground return (*Radar*). The aggregate sum of the radar echoes received after reflection from the earth's surface.

grounds (*Join.*). Strips of wood which are nailed to a wall or partition (fixing plugs being used when necessary) as a basis for the direct attachment of joinery.
ground safety lock (*Aero.*). See **retraction lock.**
ground sill (*Carp.*). A *sleeper* (q.v.).
ground sills (*Hyd. Eng.*). Underwater walls built at intervals across the bed of a channel to prevent excessive scour of the bed or to increase the width of flow.
ground sluicing (*Min. Proc.*). Bulk concentration of heavy minerals *in situ*, by causing a stream of water to flow over unconsolidated alluvial ground with just enough force to flush away the lighter, less valuable sands leaving the heavier ones to be removed for further treatment.
groundspeed (*Aero.*). Speed of aircraft or missile relative to the ground and not to the surrounding medium. Cf. *airspeed.*
ground state (*Nuc.*). State of nuclear system, atoms, etc., when at their lowest energy, i.e., not *excited*. Also **normal state.**
ground system (*Radio*). See **earth system.**
ground table (*Build.*). The course of stones at the foundation of a building. Also **grass table.**
ground tissue (*Bot.*). The general mass of parenchymatous tissue outside and between the vascular strands in a young stem or root.
ground water (*San. Eng.*). Water naturally contained in, and saturating, the subsoil.
ground wave (*Phys.*). See **wave.** (*Radio*) See **direct ray.**
groundwood (*Paper*). Pulp produced by grinding wood. Also termed **mechanical wood.**
group (*Chem.*). (1) A vertical column of the periodic system, containing elements of similar properties. (2) Metallic radicals which are precipitated together during the initial separation in qualitative analysis. (3) A number of atoms which occur together in several compounds. (*Maths.*) A set together with an operation$*$ is said to be a group if it satisfies the four following conditions: (1) The set G is *closed* with respect to $*$; i.e., for all elements x, y in G, $x*y$ is an element of G. (2) The operation $*$ in G is *associative*; i.e., for all elements x, y, z in G, $(x*y)*z = x*(y*z)$. (3) The set G has an *identity element* with respect to $*$; i.e., there is an element in G such that, for all elements x in G, $x*e = x = e*x$. (4) Each element in G has an *inverse* with respect to $*$; i.e., for each element x in G there is an element x' in G such that $x*x' = e = x'*x$. A group $(G;*)$ is said to be an *abelian group* if the operation $*$ in G is commutative; i.e., if for all elements x, y in G, $x*y = y*x$. The real numbers with addition, the non-zero real numbers with multiplication, and the three-dimensional vectors with vector addition are abelian groups.
group automatic operation (*Elec. Eng.*). A method of automatic control sometimes employed with electric lifts; the pressing of a landing pushbutton calls the next available lift.
group delay (*Telecomm.*). See **envelope delay.**
group drive (*Elec. Eng.*). A method of electric motor drive, used in factories, in which a motor drives a group of several machines.
group frequency (*Telecomm.*). The frequency which corresponds to the group velocity of a progressive wave in a waveguide or transmission line.
group mixer or fader (*Telecomm.*). See **mixer.**
group modulation (*Telecomm.*). Use of one carrier for transmitting a group of telephone or telegraph channels, with demodulation on reception and ultimate separation. Side frequencies of the said carrier may represent different groups.
group reaction (*Chem.*). The reaction by which members of a *group* (q.v.) are precipitated.
group selector (*Teleph.*). A selector which is first operated by a train of impulses, so that the wipers are lifted to a desired level of bank contacts, and then hunts by rotating the wipers over the contacts until a free outlet is found.
group theory (*Nuc.*). Approximate method for study of neutron diffusion in reactor core, in which neutrons are divided into a number of groups, all members of a group being assumed to have the same velocity, and are maintained thus for a definite number of collisions, before being transferred into the next group.
group therapy (*Med., Psychiat.*). Therapy in which a small group of patients with similar psychological or physical problems discuss their difficulties under the chairmanship of, e.g., a doctor. They thus learn from the experiences of others and teach by their own.
group (or envelope) velocity (*Phys.*). That of energy propagation for a wave in a dispersive medium. Given by the differential coefficient of the reciprocal of the wavelength with respect to the frequency, i.e., $d\beta/d\omega$, where β is the delay in radians m^{-1}, and ω is 2π frequency.
grouse disease (*Vet.*). A popular term for the specific infection of the intestines of grouse by the nematode worm *Trichostrongylus pergracilis.*
grout (*Civ. Eng.*). See **cement grout.**
grouting (*Build., Civ. Eng.*). The process of injecting cement grout for strengthening purposes. Also called **cementation.** (*Eng.*) Setting a machine foot or base to a required level by filling the space between it and the supporting floor or foundation with cement grout.
groutnick (*Build.*). A groove cut in a masonry joint to give access to grout.
Grove cell (*Elec. Eng.*). A primary cell similar to a Bunsen cell but with a positive electrode of platinum instead of carbon.
Grove gas cell (*Elec. Eng.*). A primary cell with electrodes of platinum immersed in hydrogen and oxygen respectively and an electrolyte of acidulated water.
growing point (*Bot.*). The apical meristem of a growing axis, where active cell divisions occur and the differentiation of tissues begins.
growing zone (*Bot.*). That portion of an organ in which longitudinal elongation proceeds.
growl (*Mining*). Noise heard when coal pillars are about to fall under crushing weight in the mine working (Midlands).
grown-diffusion transistor (*Electronics*). A junction transistor in which the junctions are formed by the diffusion of impurities.
grown junction (*Crystal.*). One formed during growth of crystal when desired impurity is added to molten state.
growth (*Biol.*). An irreversible change in an organism accompanied by the utilization of material, and resulting in increased volume, dry weight or protein content: increase in population or colony size of a culture of microorganisms. (*Nuc.*) (1) Elongation of fuel rods in reactor under irradiation. (Growth is a change of shape and not volume—it must be distinguished from *S-value.*) (2) Build-up of artificial radioactivity in a material under irradiation, or of activity of a daughter product as a result of decay of the parent.
growth curvature (*Bot.*). A curvature in an elongated plant organ, brought about by one side growing faster than the other.
growth form (*Bot.*). See **life form.**
growth hormones (*Biol.*). Secretions of endocrine glands having growth-promoting properties,

e.g., pituitary growth hormone, in plants called *auxins*.

growth ring (*Bot.*). The cylinder of secondary wood added during one season of growth, as seen in cross-section. A growth ring may not be the same as an *annual ring* (q.v.).

groynes (*Hyd. Eng.*). (1) Barrier walls, formed of piling, fascine work, or rubble, built out from river banks at right angles to the flow, in order to reduce the channel and keep the scour of the water within definite bounds. (2) Similar structures built on a sea-shore to check the erosive effects of currents and tides. Also called **jetties.**

grozing iron (*Plumb.*). A tool for smoothing joints made in lead piping.

GR-S (*Plastics*). Abbrev. for *G*overnment *R*ubber-*S*tyrene, the chief synthetic rubber developed in U.S. during World War II. Based on styrene and butadiene. See **Buna.**

grub axe (*Tools*). A tool for digging up roots; it has a broad chisel-shaped point on one side and a flat adze-like blade on the other.

grubber (*Agric.*). A heavy type of cultivator in which the teeth are set rigidly in a frame.

grub saw (*Tools*). A hand-saw for cutting marble, having a steel blade stiffened along the back with wooden strips.

grub screw (*Eng.*). A small headless screw used without a nut to secure a collar or similar part to a shaft.

grummet (*Plumb.*). Hemp and red-lead putty mixed as a jointing material for water-tightness.

grumous (*Bot.*). Having flesh composed of little grains.

grunerite (*Min.*). A monoclinic calcium-free amphibole; a hydrous silicate of iron and magnesium, differing from cummingtonite in having Fe > Mg. Typically found in metamorphosed iron-rich sediments.

gryke (*Geol.*). See **grike.**

Grylloblattodea (*Zool.*). A synonym for *Notoptera* (q.v.).

G.S.M. (*Paper*). A measure of the *substance* of paper expressed in grammes per square metre. An international form which does not depend on the number of sheets in the ream.

G-string (*Elec.*). Colloquialism for single-wire transmission line loaded with dielectric so that surface-wave propagation can be employed. (Named after Dr Groubau.)

g-suit (*Aero.*). See **anti-g suit.**

g-tolerance (*Space*). The tolerance of an object or person to a given value of g-force.

guadalcazarite (*Min.*). A zinc-bearing variety of metacinnabarite.

Guadaloupian (or **Guadloupee**) **Group** (*Geol.*). Strata, referred to the Permian, occurring in Texas and New Mexico; they comprise the Delaware Mount n Series below and the Capitan Limestone above.

guag (*Mining*). The space left after the mineral has been extracted. Also called **gunis.**

guaiac acid test (*Chem.*). One for the presence of blood in body fluids, e.g., in gastric contents, based upon the oxidation of guaiac resin in ethereal solution to guaiaconic acid. The presence of blood is shown by a blue colour in the ether solution.

guaiacol (*Chem.*). $HO{\cdot}C_6H_4{\cdot}OCH_3$, the monomethyl ether of *catechol* (i.e., 1-methoxybenzene 2-hydroxybenzene), found in beechwood tar; a very unstable compound, with strong reducing properties. Used in medicine and in veterinary practice as an expectorant.

guanajuatite (*Min.*). Bismuth selenide, crystallizing in the orthorhombic system.

guanozolo (*Med.*). A synthetic substance, amino-hydroxy-triazolo-pyrimidine, closely resembling **guanine;** used experimentally in controlling cancer by the starving of tumours.

guanidine (*Chem.*). $HN : C(NH_2)_2$, imido-urea or the amidine of amidocarbonic acid; a crystalline compound, easily soluble in water, strongly basic.

guanine (*Biochem.*). A purine base found in all living tissues as a part of the *nucleoproteins.*

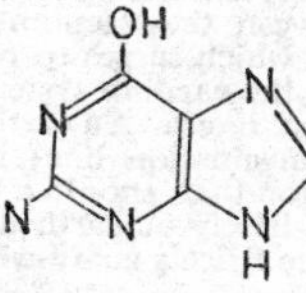

OH N N H₂N N N H

Pairs with cytosine (another purine) in the *genetic code* (q.v.).

guano (*Agric.*). The excrementitious deposit of certain sea-fowl, found on the coasts and islands of Central America, Africa, etc. It consists chiefly of calcium and tricalcium phosphate mixed with ammonium oxalate and ammonium hippurate.

guanophore (*Zool.*). See **xanthophore.**

guard (*Build., etc.*). A protection on a scaffold to prevent persons from falling; on a machine, to prevent injury to the operator or others from gears, cutting tools, etc. (*Civ. Eng.*) See **fender pile.** (*Eng.*) A fence or other safety device fitted to moving parts of machinery to prevent injury to the operator. (*Telecomm.*) Signal which prevents accidental operation by spurious signals, or avoids possible ambiguity. (*Teleph.*) The wire accompanying a speaking pair through an automatic telephone exchange; it is earthed while the speaking pair is being used by subscribers, thus indicating that the speaking pair cannot be engaged by any other circuit. Also **guard wire.** (*Zool.*) In *Belemnoidea*, a thickened part of the shell protecting the phragmocone.

guard band (*TV*). Any additional frequency band on either side of an allocated band (including any *frequency tolerance*), to ensure freedom from interference from other transmissions, e.g., 0·25 MHz between television channels.

guard cell (*Bot.*). One of the 2 specialized epidermal cells which border on the pore of a stoma and together cause it to close or to open.

guard circle (*Acous.*). Inner groove on disk recording which protects stylus from being carried to centre of turntable.

guard cradle (*Elec. Eng.*). A network of wires serving the same purpose as a *guard wire* (q.v.). Also called **guard net.**

guard lock (*Hyd. Eng.*). A lock separating tidal waters from the water in a basin.

guard magnet (*Min. Proc.*). Strong magnet, usually suspended above moving stream of lump ore, to remove steel which has come out with the mineral and which might damage the crushing machine.

guard net (*Elec. Eng.*). See **guard cradle.**

guard pile (*Hyd. Eng.*). A *fender pile* (q.v.).

guard pin (*Horol.*). See **safety finger.**

guard polyp (*Zool.*). See **nematophore.**

guard post (*Civ. Eng.*). A *fender post* (q.v.).

guard rail (*Rail.*). See **check rail.**

guard ring (*Elec. Eng.*). Auxiliary electrode used to avoid distortion of electric (or heat) field pattern in working part of a system as a result of the *edge effect*, or to bypass leakage current through insulator to earth in an ionization chamber.

guard-ring capacitor (*Elec. Eng.*). A *standard* capacitor consisting of circular parallel plates with a concentric ring maintained at the same potential as one of the plates to minimize the edge effect.

guards (*Bind.*). Narrow strips of paper or linen projecting between sections in a book, for the attachment of plates, maps, etc.

guard vacuum (*Vac. Tech.*). An evacuated enclosure between the vacuum system and the atmosphere, which serves to reduce the rate of leakage into the vacuum system.

guard wire (*Elec. Eng.*). An earth wire used on an overhead transmission line; it is arranged in such a position that, should a conductor break, it will immediately be earthed by contact with the wire. See **Price's guard-wire.** (*Teleph.*) See **guard.**

gubernaculum (*Zool.*). In Mammals, the cord supporting the testes, in the scrotal sac: in *Hydrozoa*, an ectodermal strand supporting the gonophore in the gonotheca: in *Mastigophora*, a posterior flagellum used in steering. *adj.* **gubernacular.**

Gucker and Rose aerosol counter (*Powder Tech.*). Instrument for determining the number and size of particles in an aerosol by measuring the number and intensity of pulses of scattered light occurring in a beam through which the stream of particles is passed.

Gudden-Pohl effect (*Light*). A form of electroluminescence which follows metastable excitation of a phosphor by ultraviolet light.

Gudermannian (*Maths.*). A curious function defined by

$$\text{gd } x = \int_0^x \text{sech } \theta \, d\theta.$$

It then follows that if $u = \text{gd } x$, then $\sin u = \tanh x$ and $\cos u = \text{sech } x$, and that

$$x = \int_0^u \sec \theta \, d\theta.$$

gudgeon (*Build.*). A metal pin used for joining adjacent stones. (*Eng.*) A pivot at the end of a beam or axle on which a bell, wheel, etc., works. (*Join.*) The wrought-iron pin which is fastened to a gate-post or door frame, and about which the leaf of a strap-hinge turns.

gudgeon pin (*I.C. Engs.*). The pin connecting the piston with the bearing of the little end of the connecting-rod. Also called **piston pin.** See also **floating-.**

Guerin process (*Eng.*). Used in presswork to cut or form sheet metal by placing it between a die made of a cheap material and a thick rubber pad which adapts itself to the die while under pressure.

guest (*Zool.*). An animal living and/or breeding in the nest of another animal, as a myrmecophile in an ants' nest.

gug (*Mining*). Dip incline; a self-acting inclined roadway in a coal-mine.

guidance. System for missile flight control using electronic means for sensing its own position and that of the target, and adjusting its own velocity, acceleration, orientation, propulsion. Divided into *initial*, *midcourse*, and *terminal*. See **beam riding, guided missile.**

guide (*Civ. Eng.*). A pile driven to indicate a site.

guide bars (*Eng.*). (1) Bars with flat or cylindrical surfaces provided to guide the crosshead of a steam-engine, and so avoid lateral thrust on the piston rod. Also called **motion bars, slide bars.** (2) Any bars used as guides to control one machine element in relation to another.

guide bead (*Join.*). A bead fixed to the inside of a cased frame as a guide for the sliding sash. Also called **inner bead.**

guided missile or weapon (*Aero.*). Tactical unmanned weapon which is guided to its target; propulsion is usually by *rocket* (q.v.), *ramjet* (q.v.), or simplified short-life *turbojet* (q.v.). Guided missiles are broadly divided into classifications, using the initials of the launching and target media (Air, Surface, or Underwater) followed by the letter M for missile: AAM, ASM, AUM, SAM, SSM, UAM, USM. In addition there are the ICBM, intercontinental ballistic missile, IRBM, intermediate range ballistic missile, which, as their names imply, are guided on only part of their journeys, which end in a ballistic descent through the atmosphere. Guidance systems vary greatly: *direct command guidance* is control entirely from the launcher by radio or electric-wire signals; *radar command guidance* is radio signal guidance of the missile from a lock-on radar/computer system on the launcher; a *beam rider* is a missile which follows a radar beam directed from launcher to target; *semi-active homing* is the radar 'illumination' of the target, on the reflections from which the missile 'homes'; *collision-course homing* is similar, but employs an off-set missile aerial to bring it on to a converging course with the target; *fully-active homing* is self-contained, the missile generating its own radar signals and carrying a lock-on or collision-course computer device; *passive homing* is by a sensitive electronic detector on to infrared, heat, sound, static electricity, magnetic, or other wave emissions from the target. Long-range guidance devices include the *celestial*, wherein the missile's automatic astro-navigation equipment is given a preflight programme relating its course to the target and to a fixed star(s) on which its tracking telescope is focused; and the *inertial* system, which depends upon a knowledge of the precise geographical position of the target (many places on the earth's surface are inaccurately charted) so that the course to it can be planned by the variation of inertial forces, which are followed by the internal device which can measure and correct minute variations in gravitational, and other, forces—this is the only known system completely independent of jamming or other external interference.

guided wave (*Elec.*). Electromagnetic or acoustic wave which is constrained within certain boundaries as in a waveguide.

guide field (*Elec. Eng.*). That component of field in a cyclotron or betatron which maintains particles in their intended path.

guide mill (*Met.*). A rolling-mill equipped with guides to ensure that the stock enters the mill at the correct point and angle.

guide piles (*Civ. Eng.*). Stout timber piles used at intervals along both sides of wide excavations where the sides are held firm by sheet piles and strong support is needed for the struts wedging the sides apart.

guide pulley (*Eng.*). A loose pulley used to guide a driving-belt past an obstruction or to divert its direction. Also called **idler pulley.**

guide rail (*Rail.*). A *check rail* (q.v.).

guides (*Mining*). Timbers, ropes, or steel rails at sides of shaft used to steady the cage or skip.

guide track (*Cinema.*). Extra sound-track recorded during the shooting of a film to assist in editing and postsynchronizing, but not used in the final version.

guide-vanes (*Aero.*). A general term for aerofoils which guide the airflow in a duct, also *cascades*. See **impeller-intake-, nozzle-, toroidal-intake-.**

guide wavelength (*Elec.*). Wavelength in a guide operated above the cut-off frequency. $1/\lambda^2_g = 1/\lambda^2_o - 1/\lambda^2_{oc}$, where λ_g is the guide wavelength, λ_o the wavelength in the unbounded medium at the same frequency, and λ_{oc} the wavelength in the unbounded medium at the cut-off frequency for the mode in question.

guiding bed (*Mining*). A thin band of coal or shale which forms a connexion between two parts of a nipped-out seam.

Guignet's green (*Paint.*). A transparent green pigment of hydrated chromium sesquioxide, prepared from sodium dichromate and boric acid, and giving a very clean and light fast colour. Used in distempers and in car finishes. Also called **Guinea green, viridian green.**

Guillemin effect (*Mag.*). The tendency of a bent magnetostrictive rod to straighten in a longitudinal magnetic field.

Guillemin line (*Telecomm.*). A network designed to produce a nearly square pulse with a steep rise and fall.

guillotine (*Paper*). A machine having a heavy steel blade, used for trimming books or cutting stacks of paper. (*Surg.*) An instrument for cutting off tonsils.

Guinea green (*Paint.*). See **Guignet's green.**

guinea-pig paralysis (*Vet.*). A viral infection of cavies characterized by a diffuse meningo-myelo-encephalitis.

gula (*Zool.*). In Vertebrates, the upper part of the throat; in Insects, an unpaired ventral cephalic sclerite lying behind the submentum.

gulamentum (*Zool.*). In most Insects, the ventral sclerite formed by the fusion of the submentum and the gula.

gular (*Zool.*). In some Fish, a bone developed between the rami of the lower jaw: in *Chelonia*, an anterior unpaired element of the plastron.

gular sutures (*Zool.*). In Insects, the forward extensions of the postoccipital sutures which bound the gula laterally.

gulching (*Mining*). The noise which generally precedes a fall or settlement of overlying strata.

Guldberg and Waage's law (*Chem.*). See **law of mass action.**

Guldin's theorems (*Maths.*). See **Pappus' theorems.**

gullet (*Civ. Eng.*). A narrow trench dug the full depth of a proposed cutting (in the case of large cuttings). A track is laid along the bottom of this trench, and wagons carry away the earth as the trench is widened into the full cutting. (*Tools*) A depression cut in the face of a saw in front of each tooth, alternately on each side of the blade. (*Zool.*) The oesophagus; in *Protozoa* the cytopharynx.

gulleting (*Civ. Eng.*). The process of excavating road or railway cuttings in a series of steps worked simultaneously. See also **gullet.**

gullet saw (*Tools*). A saw with gullets cut in front of each tooth. Also called **brier-tooth saw.**

gullet tooth (*Tools*). A saw tooth with a gullet cut away in front of it.

gulley (*San. Eng.*). A fitting installed at the upper end of a drain, to receive the discharge from rain water or waste pipes.

gulley trap (*San. Eng.*). A device installed at a gulley to imprison foul air within the drain pipe. Also called **yard trap**

gulose (*Chem.*). A monosaccharide belonging to the group of aldohexoses.

gum (*Mining*). Small coal, slack or duff. (*For.*) See **American red gum.**

gum arabic (*Chem.*). A fine, yellow or white powder, soluble in water, rel. d. 1·355. It is obtained from certain varieties of acacia, the world's main supply coming from the Sudan and Senegal. Used in pharmacy for making emulsions and pills; also in glues and pastes. Also called **acacia gum, Senegal gum.**

gum-boil (*Med.*). A small abscess on the palate, associated with a carious tooth, or the result of infection following upon local injury.

gum-chloral mountant (*Micros.*). A combined fixative and mountant containing gum arabic, chloral hydrate, and glycerol; used for small specimens, e.g., mites.

gum dichromate process (*Photog.*). The use of gum as a vehicle for pigments and dichromate on printing papers, the exposed image being developed by water.

gum-freezing technique (*Micros.*). A method of section cutting in which the tissue is embedded in gum solution, frozen and cut on a freezing microtome.

gum kino (*For.*). See **kino.**

gum-lac (*Zool.*). An inferior type of lac, containing much wax, produced by some lac insects (*Hemiptera*) in Madagascar.

gumma (*Med.*). A mass of cellular granulation tissue, due to syphilitic infection in the late or tertiary stage. *adj.* **gummatous.**

gummed paper (*Paper*). Paper coated with adhesive (dextrin, gum arabic, etc.). *Self-sealing papers* are coated with a latex adhesive.

gummer (*Mining*). A man who clears the fine coal, gum or dirt from the undercut made by a coal cutting machine.

gummite (*Min.*). Hydrated oxide of uranium and lead. An indefinite generic term for the alteration products of uraninite.

gummosis (*Bot.*). A pathological condition shown by the conversion of cell walls into gum.

gums (*Chem.*). Nonvolatile, colloidal plant products which either dissolve or swell up in contact with water. On hydrolysis, they yield certain complex organic acids in addition to pentoses and hexoses. (*Zool.*) In higher Vertebrates, the thick tissue masses surrounding the bases of the teeth.

gun (*Civ. Eng., Paint., etc.*). A *spray gun* (q.v.). (*Electronics*) Assemblage of electrodes, comprising cathode, anode, focusing and modulating electrodes, from which the electron beam is emitted before being subjected to deflecting fields in a *CRT*.

guncotton (*Chem.*). A *nitrocellulose* (cellulose hexanitrate) (q.v.) with a high nitrogen content. It burns readily and explodes when struck or strongly heated. Used for explosives.

gun current (*Electronics*). Total electronic current flowing to the anode, part of which forms the beam current.

gun drill (*Eng.*). A trepanning drill or a centre-cut drill used for deep-hole drilling.

gunis or **gunnis, gunnice, gunnies** (*Mining*). See **guag.**

gunite (*Chem. Eng.*). A substance similar to that used in civil engineering (see below), but using corrosion-resisting materials rather than sand and cement. (*Civ. Eng.*) A finely graded cement concrete (a mixture of cement and sand), which is sprayed into position under air pressure to produce a dense, impervious adherent layer; used to line or repair existing works, or to build up dense concrete in inaccessible areas of operation.

gunk (*Chem.*). A colloquial term applied to semi-solid material produced in many synthesis processes, which is undersized but for which specific removal measures must be made. It may in certain cases have a small re-use or recovery value.

gunmetal (*Met.*). A copper-tin alloy (i.e., bronze) either Admiralty gunmetal (copper 88, tin 10, zinc 2%), or copper 88, tin 8, and zinc 4%. Lead and nickel are frequently added, and the alloys are used as cast where resistance to corrosion or wear is required, e.g., in bearings, steam-pipe fittings, gears.

Gunn effect (*Electronics*). The production of high-field intensity domains in a semiconductor diode (usually by dipole charges formed across a depletion layer, although other processes, such as charge accumulation, can produce similar effects). These domains can form the basis of negative resistance of microwave semiconductor oscillators.

gunning. The process of applying forcibly refractory, sound insulating, or corrosion insulating, linings, using a gun usually operated by compressed air.

gunstock stile (*Join.*). See **diminished stile.**

Gunter's chain (*Surv.*). A chain having overall length of 66 ft (=20·116 m). One acre equals 10 sq chains; and 1 chain is 1/10th of a furlong.

Gunzberg test (*Chem.*). A test for the presence of free hydrochloric acid in gastric juice, based upon the appearance of a red colour on evaporation with an alcoholic solution of phloroglucinol and vanillin.

Gupta's long arm centrifugal pipette (*Powder Tech.*). Apparatus in which particle-size analysis is carried out in short tubes at the extremities of a long arm centrifuge.

gurley (*Paper*). The number of seconds required for the closed cylinder to fall under its own weight in the *Gurley densimeter* (q.v.).

Gurley densimeter (*Paper.*). An instrument consisting of an open cylinder having one end under water and the other end closed by the paper sample whose porosity is to be determined. The cylinder falls under its own weight and in so doing, pushes air through the paper. If the paper is very dense the air is pushed through slowly and the cylinder takes longer to fall a given distance.

Gurney-Mott hypothesis (*Photog.*). Theory that the latent image is formed in two stages, i.e., a primary electronic stage resulting from photoconductance and a secondary ionic process.

gusher (*Geol.*). See **geyser.**

gusset or gusset plate (*Eng.*). A bracket or stay, cast or built up from plate and angle, used to strengthen a joint between 2 plates which meet at a joint, as the junction of a boiler shell with the front and back plates, or between connecting members of a structure.

gusset piece (*Build.*). A piece of timber covering the triangular end-gap between the roof slope and the horizontal gutter boarding behind a chimney stack.

gustatory calyculus (*Zool.*). See **taste-bud.**

gut (*Zool.*). The alimentary canal.

Gutenberg discontinuity (*Geol.*). The discontinuity separating the mantle of the Earth from the core, at a depth of approximately 2900 km.

gutnik (*Radio*). Small encapsulated transistor radio telemeter for swallowing. Self-contained battery can be charged by rectifying an induction current.

gut oedema (*Vet.*). See **bowel oedema disease.**

gutta (*Bot.*). (1) An oil drop present in a spore or in a fungal hypha. (2) A general term for a vacuole, when small. (*Zool.*) A patch of colour or other marking, resembling a small drop, on the surface of an animal. *pl.* **guttae.** *adj.* **guttulate.**

guttae (*Arch.*). An ornament in the form of a line of truncated cones used to decorate entablatures or hollow mouldings.

gutta-percha (*Chem.*). The coagulated latex of *Isonandra* (or *Palaquium*) *Gutta* and other trees such as *Bassia pallida*, *Mimusops balata*, and *Payena Leerii*, found chiefly in Malaysia and the East Indies. Physical properties of deresinated gutta-percha: rel. d. 0·945, tensile strength about 40 MN/m^2, breakdown strength about 10^5V/cm, m.p. (Wendriner) about 95°C. Having a high resistivity (10 megohm metres) and permittivity of 3·0, and being waterproof, it was especially suitable without lead sheath, for submarine cables before the introduction of plastics.

guttate (*Bot.*). Containing little drops of material.

guttation (*Bot.*). The exudation of drops of fluid from an uninjured part of a plant, commonly from the ends of the main veins of leaves.

gutter (*Build., Civ. Eng.*). A channel along the side of a road, or around the eaves of a building, to collect and carry away rain waters. (*Civ. Eng.*) A trench alongside a canal, for clay puddle. (*Print.*) The spacing material in a forme between the fore-edges of the pages. Loosely applied to the space between two facing pages in a book.

gutter bearer (*Carp.*). A timber about 2×1½ in. (50×38 mm), carrying gutter boarding.

gutter bed (*Plumb.*). A lead sheet fixed behind the eaves gutter and over the tilting fillet to prevent overflow from the gutter from soaking into the wall.

gutter boards (*Build.*). See **snow boards.**

gutter bolt (*Build.*). A securing bolt between the spigot and the socket ends at a joint in a gutter.

gutter-pointed (*Bot.*). Acuminate, and with the point channelled above, forming a spout.

gut-tie (*Vet.*). Strangulation of a loop of intestine which has herniated through a rupture in the peritoneal covering of the right spermatic cord of castrated cattle.

guttural (*Zool.*). Pertaining to the throat.

guttural pouch (*Vet.*). A diverticulum of the Eustachian tube of the horse.

gutturoliths (*Vet.*). See **chondroids.**

Gutzeit test (*Chem.*). A method of determining arsenic, by adding metallic zinc and hydrochloric acid. The evolved gases darken mercury (II) salts.

guy (*Radio*). Thin tension support for antenna mast or similar structure. (*Civ. Eng.*) A rope holding a structure in a desired position.

guy derrick (*Build., Civ. Eng.*). A crane operating from a mast held upright by guy-ropes.

guying (*Civ. Eng.*). The operation of holding or adjusting a structure in position by means of guy-ropes.

guyot (*Geog.*). A flat-topped submarine mountain, occurring in the deep ocean basins, especially the Pacific, and which may represent eroded volcanoes which have sunk below the surface.

G-value (*Nuc.*). A constant in radiation chemistry denoting the number of molecules reacting as a result of the absorption of 100 eV radiation energy.

gyle (*Brew.*). See **wort.**

gymbals, gimbals (*Horol.*). Self-aligning bearings

for supporting a chronometer in its box. Used to ensure that the chronometer is kept level, irrespective of the ship's motion.
gymnetrous (*Zool.*). Of Fish, lacking an anal fin.
gymno-. Prefix from Gk. *gymnos*, naked.
gymnoarian (*Zool.*). Said of gonads which are not enclosed within coelomic sacs. Cf. *cystoarian.*
Gymnoblastea (*Zool.*). An order of *Hydrozoa* in which the polyps are colonial and the skeleton of the colony consists of a perisarc only; the medusae, when set free, are *Anthomedusae*.
gymnoblastic (*Zool.*). Said of hydroid colonies in which the blastostyles are unprotected by gonangia. Cf. *calyptoblastic*.
gymnocarpous (*Bot.*). Having the hymenium exposed from an early stage in its development.
gymnocyte (*Biol.*). A cell without a cell-wall.
Gymnodiniales (*Bot.*). An order of the *Dinophyceae*, including the member having naked protoplasts, or protoplasts covered by a homogenous wall. The vegetative cells are always flagellated.
Gymnolaemata (*Zool.*). A class of *Ectoprocta*, in which the lophophore is circular and there is no epistome; found in estuaries and brackish water, but mainly in the sea.
Gymnophiona (*Zool.*). See **Apoda.**
gymnoplasm (*Biol.*). An amorphous mass of naked protoplasm.
gymnorhinal (*Zool.*). Of Birds, having no feathers on the area surrounding the nostrils.
gymnosomatous (*Zool.*). Having a naked body, as some *Mollusca* which lack both shell and mantle.
Gymnospermae (*Bot.*). One of the two main divisions of seed plants, with about 500 species, mostly conifers. They are distinguished by the production of their ovules and seeds on the surface of a fertile leaf, not enclosed in an ovary.
gymnospermous (*Bot.*). Having the seeds exposed, not contained in an ovary.
gymnospore (*Zool.*). A protozoan spore which is not enclosed by a spore-case. Cf. *chlamydospore*.
gymnostomous (*Bot.*). Lacking a peristome.
gyn-, gyno-, gynaeco-. Prefix from Gk. *gyne*, gen. *gynaikos*, woman.
gyn (*For.*). Any device for loading timber that has a 3-legged structure (tripod) carrying a pulley.
gynaecaner (*Zool.*). A male ant which superficially resembles the female.
gynaeceum (*Bot.*). (1) The group of archegonia in mosses. (2) The carpel or carpels in a flower.
gynaecoid (*Zool.*). An egg-laying worker ant.
gynaecology or **gynecology** (*Med.*). That branch of medical science which deals with the functions and diseases peculiar to women.
gynaecomania (*Psychiat.*). Abnormally increased sexual desire in man. Also **satyriasis, satyromania.**
gynaecomastia or **gynecomastia** (*Med.*). Abnormal enlargement of the male breast.
gynaecophore (*Zool.*). In some *Trematoda*, a canal on the lower side of the male, which accommodates the female. *adj.* **gynaecophoral, gynaecophoric.**
gynandrism (*Zool.*). See **hermaphrodite.**
gynandromorph (*Zool.*). An animal exhibiting male and female characters.
gynandromorphism (*Zool.*). The occurrence of secondary sexual characters of both sexes in the same individual.
gynandrous (*Bot.*). Having the stamens and styles united to form a column, as in the flowers of orchids.
gynase (*Zool.*). A hormone or enzyme which produces femaleness in an organism.
gyne (*Zool.*). A sexually perfect female ant; a queen ant.
gynobasic (*Bot.*). Having the style attached close to the base of the ovary.
gynobasis or **gynophore** (*Bot.*). An elongation of the receptacle of a flower, forming a short stalk to the ovary.
gynodioecious (*Bot.*). Said of a species in which some plants bear perfect flowers and others bear pistillate flowers.
gynoecium (*Bot.*). See **gynaeceum.**
gynogenesis (*Biol.*). A kind of pseudapogamy in which the male gamete enters the egg but nevertheless fails to bring about fertilization, so that all further development is from the egg alone.
gynomonoecious (*Bot.*). Having perfect and pistillate flowers on the same plant.
gynophore (*Bot.*). (1) An elongated receptacle bearing carpels. (2) A multinucleate, developing female structure in some fungi. (3) See **megaspore.**
gynostegium (*Bot.*). A compound organ formed by the union of stamens and style.
gynostemium (*Bot.*). The column, in flowers of orchids.
gyplure (*Zool.*). The name given to a *pheromone* (q.v.) produced by female Gypsy Moths as a sex attractant, and used to bait traps for the control of this pest.
gypsum (*Min.*). Crystalline hydrated sulphate of calcium, $CaSO_4 \cdot 2H_2O$. Occurs massive as alabaster, fibrous as satin spar, and as clear, colourless, monoclinic crystals known as selenite. Used in making plaster of Paris and plaster and plaster-board used in building.
gypsum plate (*Optics*). A thin plate of gypsum used in the determination of the sign of the birefringence of crystals in a polarizing microscope.
gyrate (*Bot.*). See **circinate.**
gyration, radius of (*Maths.*). See **moment of inertia.**
gyrator (*Elec. Eng.*). Electronic component which does not obey reciprocity law. Frequently based on the Faraday effect in ferrites.
gyratory (*Min. Proc.*). A widely used form of rock-breaker, in which an inner cone gyrates in a larger outer hollow cone.
gyro (*Instr.*). See **gyroscope.**
gyrocompass (*Instr.*). A gyroscope, electrically rotated, controlled and damped either by gravity or electrically so that the spin axis settles in the meridian. See **course and speed error.**
gyrodyne (*Aero.*). A form of rotorcraft in which the rotor is power-driven for take-off, climb, hovering and landing, but is in *autorotation* (q.v.) for cruising flight, there usually being small wings further to unload the rotor.
gyro-frequency (*Electronics*). See **frequency of gyration.**
gyro horizon (*Aero.*). An instrument which employs a gyro with a vertical spin axis, so arranged that it displays the attitude of the aircraft about its pitch and bank axes, referenced agginst an artificial horizon. It is normally electrically, but sometimes pneumatically, operated.
gyrolite (*Min.*). A hydrated calcium silicate, formula $Ca_2Si_3O_7(OH)_2 \cdot H_2O$. Often occurs in amygdales with apophyllite.
gyromagnetic compass (*Aero.*). A magnetic compass in which direction is measured by gyroscopic stabilization.
gyromagnetic effect (*Mag.*). See **Barnett effect.**
gyromagnetic ratio (*Nuc. Eng.*). Quantity which

expresses numerical value of gyroscopic effects arising in atomic or nuclear physics, as a result of the spin of charged particles producing interaction with a magnetic field. The classical value is given by the ratio of magnetic moment to angular momentum, but other analogous definitions are used in systems where the classical value does not apply.

gyroplane (*Aero.*). Rotorcraft with unpowered rotor(s) on a vertical axis.

gyroscope (*Instr.*). Spinning body in *gimbal* (or similar) mount, which offers marked resistance to torques tending to alter the alignment of the spin axis, and in which *precession* or *nutation* replace the direct response of static bodies to such applied torques. Used as a compass, as a controlling device in aircraft and torpedoes and, in large sizes, as a ship's stabilizer. Also called **gyro, gyrostat.**

gyroscopic sextant (*Aero.*). See **air sextant.**

gyrose (*Bot.*). Having a folded surface, marked with sinuous lines or ridges.

gyrostat (*Instr.*). See **gyroscope.**

gyrosyn (*Aero., Nav.*). A remote-indicating compass system employing a directional *gyro*scope which is monitored by and *syn*chronized with signals from an element fixed in azimuth and designed to sense its angular displacement from the earth's magnetic meridian. The element is located at some remote point, e.g., wing tips, away from extraneous magnetic influences.

gyrotron (*Instr.*). Rotating vibrating fork used as a gyroscope.

gyrus (*Zool.*). A ridge between two grooves: a convolution of the surface of the cerebrum.

G—Y signal (*TV*). Component of colour TV signal which, when combined with the *Y signal*, produces the green chrominance signal.

H

h Symbol for height. (*Heat.*) Symbol for specific *enthalpy*. (*Phys.*) Symbol for *Planck's constant*. See **Planck's law**.

ℏ or *h-bar* (*Phys.*). Symbol for *Dirac's constant*.

H (*Chem.*). The symbol for *hydrogen*. (*Elec.*) Symbol for *henry*.

[H] (*Light*). One member of the strongest pair (H and K) of Fraunhofer lines in the solar spectrum, almost at the limit of visibility in the extreme violet. Their wavelengths are [H], 396·8625 nm; [K], 393·3825 nm; and the lines are due to ionized calcium.

H (*Elec. Eng.*). Symbol for magnetic field strength. (*Heat*) Symbol for *enthalpy*.

H_α, H_β, H_γ, etc. (*Light*). The lines of the Balmer series in the hydrogen spectrum. Their wavelengths are: H_α, 656·299; H_β, 486·152; H_γ, 434·067; H_δ, 410·194 nm. The series continues into the ultraviolet, where about 20 more lines are observable.

haar (*Meteor.*). A wet sea-fog advancing in summer from the North Sea upon the shores of England and Scotland.

Haas effect (*Acous.*). Phenomenon associated with a long-delayed echo which has been applied to reinforcement systems in auditoria.

habenula (*Zool.*). A strap-like structure; in particular, a nerve-centre of the diencephalon.

Haber process (*Chem.*). Currently the most important process of fixing nitrogen, in which the nitrogen is made to combine with hydrogen under influence of high temperatures (400°–500°C), high pressure (2×10^7 N/m²), and catalyst of finely-divided iron from iron(III) oxide, in large continuous enclaves. Many variants operate at different pressures according to the catalyst. The product ammonia may be dissolved in water or condensed, and unreacted gases recycle.

habit (*Crystal.*). A term used to cover the varying development of the crystal forms possessed by any one mineral. Thus calcite may occur as crystals showing the faces of the hexagonal prism, basal pinacoid, scalenohedron, and rhombohedron. According to the relative development or *dominance* of one or other of these forms, the habit may be prismatic, tabular, scalenohedral, or rhombohedral. (*Zool.*) The established normal behaviour of an animal species.

habitat (*Biol.*). The normal locality or place of abode of an organism, e.g., (*Bot.*) the place inhabited by a plant or by a plant community, together with all influential external factors.

habitat form (*Bot.*). A plant showing features, such as luxuriant growth or dwarfing, which can be related to the place where it is growing.

habitat group (*Bot.*). A set of unrelated plants which inhabit the same kind of situation.

habit formation (*An. Behav.*). A phrase sometimes used for *trial and error learning* (q.v.).

habit spasm (*Med.*). Tic. A repeated, rapidly performed, involuntary, and coordinated movement, occurring in a nervous person.

habituation (*An. Behav.*). The relatively persistent waning of a response as the result of repeated stimulation which is not reinforced.

haboob (*Meteor.*). A line-squall, with dust storms, blowing in the Sudan in the rainy season.

hachure (*Surv.*). Hatching; the use of lines to shade a plan and indicate hills and valleys.

H-acid (*Chem.*). 1,8-Aminonaphthol-3,6-disulphonic acid, an intermediate for dyestuffs.

hack (*Build.*). A long parallel bank, about 6 in. (150 mm) high, of brick, rubbish, and ashes, on which bricks are laid in the course of manufacture, when it is intended to dry them in the open.

hack-barrow (*Build.*). A barrow used to carry green bricks to the hack for drying.

hack-cap (*Build.*). A small timber structure erected to provide cover for a hack.

hacket (*Carp.*). See **hatchet**.

hacking (*Build.*). (1) The operation of piling up green bricks on a hack to dry. (2) The process of making a surface rough, in order to provide a key for plasterwork. (3) A course of stones in a rubble wall, the course being composed partly of single stones of the full height of the course and partly of shallower stones arranged two to the height of the course.

hacking-out knife (*Build.*). A knife used to remove old putty from sash rebates before reglazing.

hackling (*Textiles*). Process of combing scutched flax in the hackling machine, in order to parallelize the long fibres, remove the short ones and impurities.

hackmanite (*Min.*). A fluorescent variety of sodalite, showing on freshly fractured surfaces a pink colour which fades on exposure to light, but which returns if kept in the dark or subjected to X-rays or ultraviolet light.

hack-saw (*Eng.*). (1) A mechanic's hand-saw used for cutting metal. It consists of a steel frame, across which is stretched a narrow saw-blade of hardened steel. (2) A larger saw, similar to the above, but power-driven.

Hackworth valve gear (*Eng.*). A radial gear in which an eccentric opposite the crank operates a link whose other end slides along an inclined guide, the valve rod being pivoted to a point on the link.

hadal zone (*Ocean.*). That part of the ocean lying in the deep-sea trenches below 6000 m. Also **ultra-abyssal zone**.

hade (*Geol., etc.*). The angle of inclination of a fault-plane, measured from the vertical.

Hadego (*Print.*). A hand-operated filmsetting machine analogous to the *Ludlow*; plastic photomats are assembled in a special setting stick and then photographed to the required size.

Hadfield's manganese steel (*Met.*). See **manganese steel**.

hadrocentric vascular bundle (*Bot.*). A concentric vascular bundle in which the xylem is surrounded by phloem.

hadromal, hadromase (*Bot.*). An enzyme present in some fungi which enables them to decompose wood.

hadrome (*Bot.*). The conducting tissues of xylem.

hadron (*Nuc.*). One of a class of elementary particles, including baryons and mesons.

Haeckel's law (*Biol.*). See **recapitulation theory.**

haemad (*Zool.*). Situated on the same side of the vertebral column as the heart.

haemal, haematal, haemic (*Zool.*). Pertaining to the blood or to blood vessels.

haemal arch (*Zool.*). A skeletal structure arising ventrally from a vertebral centrum, which encloses the caudal blood vessels.

haemal canal (*Zool.*). The space enclosed by the centrum and the haemal arch of a vertebra, through which pass the caudal blood vessels.

haemal ridges (*Zool.*). See **haemapophyses.**

haemal spine (*Zool.*). The median ventral vertebral spine formed by the fusion of the haemapophyses, below the haemal canal.

haemal system (*Zool.*). The system of vessels and channels in which the blood circulates.

haemalum stain (*Micros.*). A progressive self-differentiating stain used in conjunction with a counterstain, e.g., eosin, for micro-anatomical work. It contains haematoxylin, sodium iodate, potassium alum, chloral hydrate, and citric acid.

haemangioma (*Med.*). Angioma. A tumour composed of blood vessels irregularly disposed and of varying size.

haemapoiesis (*Zool.*). The process of forming new blood.

haemapophyses (*Zool.*). A pair of plates arising ventrally from the vertebral centrum, and meeting below the haemal canal to form the haemal arch and spine. Also called **haemal ridges.**

haemarthrosis (*Med.*). A joint containing blood which has effused into it.

haematemesis (*Med.*). The vomiting of blood, or of blood-stained contents of the stomach.

haematin (*Chem.*). Protohaematin, oxyhaematin: $C_{34}H_{33}O_5N_4Fe$; a compound of *protoporphyrin* (q.v.) and iron(III), formed by decomposition of haemoglobin. Combines with globin to form *methaemoglobin.* Can be reduced to reduced haematin (*haemochrome, heme*), which contains divalent iron, and unites with globin to give *haemoglobin* (ca. 6% haematin, 94% globulin). It is insoluble in water and in most organic solvents, but dissolves in alkalis and glacial ethanoic acid to form *alkaline* and *acid* haematin respectively.

haematinic (*Med.*). Pertaining to the blood.

haematite (*Min.*). Oxide of iron(III), Fe_2O_3, crystallizing in the trigonal system. It occurs in a number of different forms: kidney iron-ore massive, as found in the iron mines in Lancashire and Cumberland; specular iron-ore in groups of beautiful, lustrous, rhombohedral crystals as, for example, from Elba; bedded ores of sedimentary origin, as in the Pre-Cambrian throughout the world; and as a cement and pigment in sandstones. The Clinton ore is the most important oölitic haematite in the U.S. The Wabana ore in Newfoundland is also haematitic in part, but most of the iron produced in N. America comes from the 'iron ranges' of the Lake Superior district, especially the Mesabi Range, Minn.

haematobium (*Zool.*). An organism living in blood. *adj.* **haematobic.**

haematoblast (*Zool.*). A primitive blood cell, which may develop into an erythrocyte or a leucocyte: a blood platelet.

haematocele (*Med.*). An effusion of blood localized in the form of a cyst in a cavity of the body.

haematochrome (*Bot., Zool.*). A red colouring matter produced by a number of green algae, especially when exposed to drought, and by certain *Phytomastigina.*

haematocolpometra (*Med.*). Accumulation of menstrual blood in the vagina and uterine cavity.

haematocolpos (*Med.*). Accumulation of menstrual blood in the vagina, due to an imperforate hymen.

haematocrit (*Histol.*). A graduated capillary tube of uniform bore in which whole blood is centrifuged, to determine the ratio, by volume, of blood cells to plasma.

haematocryal (*Zool.*). See **cold-blooded.**

haematodocha (*Zool.*). In male Spiders, a distensible blood-cavity which forms part of the palpal organ.

haematogenesis (*Zool.*). See **haemapoiesis.**

haematogenous (*Zool.*). Having origin in the blood.

haematologist (*Med.*). One who specializes in the study of the blood and its diseases.

haematoma (*Med.*). A swelling composed of blood effused into connective tissue.

haematometra (*Med.*). Accumulation of menstrual blood in the uterus, due to blocking of the outlet.

haematomyelia (*Med.*). Haemorrhage into the substance of the spinal cord.

haematophagous (*Zool.*). Feeding on blood.

haematopoiesis (*Med.*). See **haemapoiesis.**

haematoporphyrin (*Chem.*). An iron-free pigment formed by decomposition of haematin, and obtained from haemoglobin by the action of conc. hydrochloric acid. It is 1·3·5·8-tetramethyl-2-4-di-(hydroxyethyl)-6·7-dipropanoic acid-porphin; $C_{34}H_{38}O_6N_4$.

haematorachis (*Med.*). Haemorrhage into the vertebral canal but outside the spinal cord.

haematosalpinx (*Med.*). Collection of blood in a Fallopian tube.

haematosis (*Zool.*). See **haemapoiesis.**

haematothermal (*Zool.*). See **warm-blooded.**

haematoxylin (*Micros.*). A colouring matter extracted from logwood; much used to stain microscopic preparations.

haematozoon (*Zool.*). An animal living parasitically in the blood.

haematuria (*Med.*). Presence of blood in the urine.

haemendothelioma (*Med.*). A tumour composed of cells derived from the lining endothelium of blood vessels.

haemic (*Zool.*). See under **haemal.**

haemin (*Chem.*). The hydrochloride of *haematin* (q.v.), $C_{34}H_{32}N_4O_4FeCl$, brown crystals. Its molecule contains four pyrrole radicals.

haemochromatosis (*Med.*). Bronzed diabetes. A disease in which the iron-containing pigment haemosiderin is deposited in excess in the organs of the body, giving rise to cirrhosis of liver, enlargement of spleen, diabetes, skin pigmentation.

haemochromogen (*Chem.*). (1) A compound of haematin and any nitrogenous (amino) substance. (2) A compound of reduced haematin and denatured globin which polymerizes at pH5 to form haemoglobin, and is formed from haemoglobin by the action of dilute alkali.

haemocoele (*Zool.*). See **primary body cavity.**

haemocoelous viviparity (*Zool.*). A type of reproduction found in *Strepsiptera* and some *Hemiptera*, in which the mature eggs lie free in the haemocoele and embryonic development is completed *in situ.*

haemocyanin (*Zool.*). A blue respiratory pigment, containing copper, in the blood of *Crustacea* and *Mollusca.* It has respiratory functions similar to haemoglobin.

haemocytes (*Zool.*). The corpuscles found floating in haemolymph.

haemocytoblast (*Cyt.*). An embryonic cell from which all blood cells are derived.

haemocytolysis (*Zool.*). See **haemolysis.**

haemocytometer (*Histol.*). An apparatus consisting of a small pipette, a counting slide and a coverslip, used for the estimation of the number of red blood corpuscles or white blood corpuscles per cubic millimetre of blood, or for other small objects in suspension.

haemogenia (*Med.*). See **purpura haemorrhagica** and **thrombocytopenia.**

haemoglobin (*Zool.*). A respiratory pigment occurring in the erythrocytes of all *Craniata*, and in the blood plasma of certain Invertebrates. It belongs to the group of conjugated proteins and has the empirical formula $(C_{759}H_{1208}N_{210}S_2FeO_{204})_4$, and a relative molecular mass of about 68 000. It combines readily with oxygen to form oxyhaemoglobin, but has a still greater affinity for carbon monoxide which destroys it, hence the toxicity of carbon monoxide.

haemoglobinaemia (*Med.*). The abnormal presence of haemoglobin in the blood, as a result of destruction of red blood cells.

haemoglobinometer (*Med.*). An instrument for measuring the percentage of haemoglobin in the blood.

haemoglobinuria (*Med.*). Presence of haemoglobin in the urine, as a result of excessive destruction of red blood cells.

haemolymph (*Zool.*). The watery fluid, containing leucocytes, believed to represent blood, found in the haemocoelic body-cavity of certain Invertebrates.

haemolysins (*Chem.*). Substances produced in, or added to, blood serum, which are capable of causing the dissolution of red blood corpuscles; often immune bodies.

haemolysis (*Zool.*). Disintegration of red blood cells. Also **erythrocytolysis, haemocytolysis.**

haemolytic anaemia (*Med.*). Anaemia due to excessive destruction of red blood cells.

haemolytic disease (*Med., Vet.*). Anaemia and jaundice of the newborn due to destruction of erythrocytes by antibodies, which are produced against the foetal erythrocytes by the mother during pregnancy and conveyed to the offspring via the colostrum or placenta.

haemopericardium (*Med.*). The presence of blood in the pericardial sac.

haemophilia (*Med.*). A hereditary bleeding tendency, with deficiency of a normal blood protein (anti-haemophilic globulin) preventing normal clotting. The classical *sex-linked* disorder. The most common form is (*h.* A), other forms being *h.* B (Christmas disease) and *h.* C.

haemopneumothorax (*Med.*). The presence of blood and air in the pleural cavity.

haemopoiesis (*Zool.*). See **haemapoiesis.**

haemoptysis (*Med.*). The spitting of blood, or of blood-stained sputum.

haemorrhage (*Med.*). Bleeding; escape of blood from a ruptured blood vessel.

haemorrhagic disease (*Vet.*). See **redwater.**

haemorrhagic septicaemia (*Vet.*). Bovine pasteurellosis; shipping fever; shipping pneumonia. A bacterial disease of cattle caused by *Pasteurella multocida* and characterized by high fever, pneumonia and oedematous swelling of the skin. A similar disease occurs in sheep.

haemorrhoid (*Med.*). Pile (usually in plural, *haemorrhoids*). Varicose dilatation of the haemorrhoidal veins at the lower end of the rectum and the anus.

haemosiderosis (*Med.*). Deposition, in the tissues of the body, of the iron-containing pigment *haemosiderin*, after excessive destruction of red blood cells. See also **haemochromatosis.**

haemostasis (*Med.*). The arrest of bleeding.

haemostatic (*Med.*). Arresting or checking bleeding; an agent which does this.

haemotropic (*Zool.*). Affecting blood.

Haerangiomycetes (*Bot.*). A class of *Ascomycetes*. The asci of its members have no definite cell wall, or it dissolves immediately on formation.

Haffkine's vaccine (*Med.*). A prophylactic vaccine for immunization against plague.

hafnium (*Chem.*). A metallic element in the fourth group of the periodic system. Symbol Hf, at. no. 72, r.a.m. 178·49 rel.d. 12·1, m.p. about 2000°C. It occurs in zirconium minerals, where its chemical similarity but relatively high neutron absorption makes it a troublesome impurity in zirconium metal for nuclear engineering. Used to prevent recrystallization of tungsten filaments.

haft (*Bot.*). (1) A leaf stalk with a thin strip of green tissue running along each side forming a wing. (2) The stalk of a spatulate leaf. (3) The claw of a petal. (*Tools*) A tool handle.

hagatalite (*Min.*). A variety of *zircon* containing an appreciable quantity of the rare earth elements.

hagendorfite (*Min.*). A phosphate of sodium, calcium, iron and manganese, crystallizing in the monoclinic system.

häggite (*Min.*). A hydroxide of vanadium, crystallizing in the monoclinic system.

ha-ha (*Build.*). A fence sunk in a ditch below ground-level so as to give an uninterrupted view.

hahnium (*Chem.*). Proposed U.S. name for transuranic element at. no. 105, first claimed to have been isolated in 1970.

haidingerite (*Min.*). Hydrated calcium hydrogen arsenate, crystallizing in the orthorhombic system.

Haigh fatigue-testing machine (*Eng.*). A machine for testing the resistance of materials to fatigue under alternating direct stress; the specimen is loaded by means of a powerful electromagnet and excited by an alternating current.

hail, hailstones (*Meteor.*). Precipitation in the form of hard pellets of ice, which often fall from cumulo-nimbus clouds and accompany thunderstorms. They are formed when raindrops are swept up by strong air-currents into regions where the temperature is below freezing-point. In falling, the hailstone grows by condensation from the warm moist air which it encounters.

hail stage (*Meteor.*). That part of the condensation process taking place at a temperature of 0°C so that water vapour condenses to water liquid which then freezes.

hair (*Anat., Zool.*). A slender, elongate structure, mostly composed of *keratins* (q.v.), arising by proliferation of cells from the Malpighian layer of the epidermis in Mammals; more generally, any threadlike outgrowth of the epidermis. (*Bot.*) See **trichome.**

hair-bulb (*Zool.*). The epidermal cells in immediate contact with the hair-papilla and by whose activity the growth of the hair is affected, in Mammals.

hair cloth (*Textiles*). A material generally composed of coarse hair and cotton yarn; used as a stiffening for coats and upholstery work.

hair compasses (*Instr.*). A drawing instrument consisting of a pair of dividers having needle points, one of which has a fine-screw adjustment for varying the distance between them.

haircords (*Textiles*). Cotton fabrics of light weight, in which fine cords are produced by running two fine threads together at frequent intervals. The warp, for example, often has alternate single and double ends.

hair felt (*Build.*). An insulation material consisting of felted cattle hair.

hair-germ (*Zool.*). The first rudiment of a developing hair, which usually takes the form of a downwardly-projecting outgrowth from the lower mucous layer of the epidermis.

hair hygrometer (*Meteor.*). A form of *hygrometer* which is controlled by the varying length of a human hair with humidity. It is not an absolute instrument, but it can be used at temperatures below freezing-point, and it can be made self-recording.

hairline (*Phys. etc.*). A fine straight line in the optical system of an instrument; used for the positive location of an image or correlation on to a measuring scale.

hair-papilla (*Zool.*). A condensation of dermal tissue which forms beneath a hair-germ in the development of a hair.

hair-pin winding (*Elec. Eng.*). A form of winding used for the armatures of electric machines; it is partly formed and pushed into the slots from one end, the other ends being subsequently welded or otherwise connected together.

hair pit (*Bot.*). See **cryptostoma.**

hair plates (*Zool.*). Groups of articulated sensory hairs occurring near the joints of the appendages in insects and acting as proprioceptors.

hair side (*Leather*). The surface of a skin or hide from which the hair has been removed.

hair space (*Typog.*). The thinnest of the spaces used between words. It is about 1-point wide.

hairspring (*Horol.*). The balance spring.

hake (*Build.*). A *hack* (q.v.) built to dry tiles in the course of their manufacture.

halation (*Electronics*). In a CRT, glow surrounding spot on phosphor arising from internal reflection within the thickness of the glass. Also called **halo.** (*Photog.*) Fogging of an emulsion due to light reflection and dispersion within the emulsion. Reduced by *backing* (q.v.) the glass or film with light-absorbing material having much the same refractive index as the support.

Haldane apparatus (*Chem.*). An apparatus for the analysis of air; used also for the analysis of mine gases.

Haldane's rule (*Gen.*). When two species are crossed it is often found that the offspring of one sex are either inviable or sterile. Haldane's rule states that when one sex is inviable or infertile, it is the heterogametic sex (i.e., the sex with two unlike sex chromosomes, X and Y).

Halden project (*Nuc. Eng.*). The world's first boiling heavy water reactor, erected at Halden, Norway, and intended as part of a joint European research programme.

half-adder (*Comp.*). Circuit having two binary inputs and two (sum and carry) outputs.

half-anchor ear (*Elec. Eng.*). An anchor ear to which only one anchoring wire is attached.

half-bed, half-joint (*Build.*). In pricing the labour charge for stonework, each horizontal surface on a stone is spoken of as a *half-bed*, as it contributes one-half to the cost of preparing each bed joint; similarly, *half-joint* refers to the vertical jointing surfaces.

half-blind dovetail (*Join.*). See **lap dovetail.**

half-bound (*Bind.*). Said of a book having its back, a portion of the sides, and the corners bound in one material (originally leather) and the remainder of the sides in some other material (e.g., cloth or paper).

half-brick wall (*Build.*). A wall built entirely of stretchers and therefore 4½ in. thick.

half-case (*Typog.*). (1) A type case of the usual width, but half the length, for holding display type or special letters. (2) See **double case.**

half-cell (*Elec. Eng.*). See **single-electrode system.**

half-closed slot (*Elec. Eng.*). See **semi-closed slot.**

half-coiled winding (*Elec. Eng.*). A form of single-layer winding in which there is only one group of coils per pole per phase. Also called a **hemitropic winding.**

half-column (*Arch.*). An *embedded column* (q.v.) of which half projects.

half-coupling (*Eng.*). See **flange coupling.**

half-deflection method (*Elec. Eng.*). A method of finding the internal resistance of a cell when the value is known to be high. A second cell, a galvanometer, and a resistance are connected in series with the cell under test, and the value of the resistance required to give a galvanometer deflection of half the value obtained with the cell alone is found.

half-dressed warp (*Weaving*). A warp that has been wound on a beam without separation of the threads by the dents of the reed, etc. (as in a 'dressed' warp).

half-element (*Elec. Eng.*). See **single-electrode system.**

half-frame camera (*Photog.*). One which takes pictures half the normal size for the film used.

half-header (*Build.*). A half-brick used at the corner of a wall to close the course.

half-hour rating (*Elec. Eng.*). A form of rating for electric machinery supplying an intermittent load. It indicates that the machine delivers the specified rating for a period of half an hour without exceeding the specified temperature rises. Cf. *one-hour rating.*

half-hunter (*Horol.*). A form of watch case in which the glass occupies one-half of its hinged cover. Also **demi-hunter.**

half-inferior (*Bot.*). Said of a flower in which the receptacle forms a cup which is adherent to the base of the ovary and partly up its side.

half-landing (*Build.*). A landing extending across the full width of a staircase.

half-lap coupling (*Eng.*). The connexion of two coaxial shafts by cutting away a short length of the ends of each to the diametrical plane, so as to form a *half-lap joint* (q.v.), and either riveting together or enclosing in a keyed-on sleeve or muff.

half-lap joint (*Carp., etc.*). The name of the joint formed by the process of *halving* (q.v.).

half-lattice girder (*Build., Civ. Eng.*). See **Warren girder.**

half-life (*Nuc.*). (1) Time in which half the atoms of a given quantity of radioactive nuclide undergo at least one disintegration. Also **half-value period.** (2) See **biological half-life.** (3) See **effective half-life.**

half-line block (*Print.*). One in which the strength of the lines is reduced to produce a pleasing effect by using in the process camera a coarse cross-line screen or a single-line screen either vertical, horizontal, or diagonal.

half-measure (*Typog.*). Type matter set to half the usual page or column width to accommodate an illustration.

half-normal bend (*Elec. Eng.*). A bend serving to connect two lengths of the conduit used in electrical installation work which are at an angle of 135°.

half-pace (*Build.*). (1) A landing at the end of a flight of steps. (2) A raised floor in a window bay.

half-period zones (*Light*). A conception, due to Huyghens, whereby an optical wavefront is considered to be divided into a number of concentric annular zones, so that, at a given point in front of the wave, the illumination from each zone is half a period out of phase with that from its neighbour. The use of half-period zones facilitates the study of diffraction problems.

half-power (*Acous.*). Condition of a resonant system, electrical, mechanical, acoustical, etc., when amplitude response is reduced to $1/\sqrt{2}$ of maximum, i.e., by 3 dB. The frequency of angular difference between symmetrical half-power points is a measure of *selectivity* of the circuit.

half-principal (*Carp.*). A short rafter which does not reach the ridge of a roof.

half-race (*Bot.*). A race of plants in which only a few of the seedlings show the characters of the race (the rest having the ordinary characters of the species), and in which selection does not lead to the fixing of a pure race.

half-residence time (*Nuc.*). Time in which half the radioactive debris deposited in the stratosphere by a nuclear explosion will be carried down to the troposphere.

half-rip saw (*Join.*). A hand-saw designed for cutting timber along the grain and having slightly smaller teeth than the rip saw.

half-roll (*Aero.*). See **roll.**

half-round chisel (*Eng.*). A *cold chisel* (q.v.) having a small half-round cutting edge; used for chipping semicircular grooves such as oil-ways.

half-round file (*Eng.*). A file whose cross-section has one flat and one convex face.

half-round screws (*Eng.*). See **button-headed screws.**

half-sawn (*Build.*). Said of a stone face as left from the saw.

half-secret dovetail (*Join.*). See **lap dovetail.**

half-section (*Eng.*). On an engineering drawing, a sectional view of a symmetrical workpiece or assembly, terminating at an axis of symmetry. (*Telecomm.*) See **section.**

half-sheet work (*Print.*). Method in which one forme is used to print on each side of the paper which is split after printing, thus producing two copies, each requiring half the sheet of paper. See also **work-and-turn.**

half shroud (*Eng.*). A gear-wheel shroud extending only up to half the tooth height. See **full shroud, shroud.**

half-silvered (*Met.*). Said of a surface of a transparent material when a deposited metallic film, e.g., formed by anode sputtering, reflects a substantial proportion of incident light.

half-socket pipe (*San. Eng.*). A drain-pipe having a socket for the lower half only.

half-speed shaft (*I.C. Engs.*). The camshaft of an internal-combustion engine, which runs at half the speed of the crankshaft.

half-stuff (*Paper*). Raw materials which have been converted into pulp by the breaker.

half-supply voltage principle (*Elec. Eng.*). If the collector-emitter voltage of a power stage transistor is less than half the supply voltage, the circuit will be inherently safe from thermal runaway, as the thermal loop gain will be less than unity.

half-thickness (*Nuc.*). See **half-value thickness.**

half-timber (*For.*). One part of a baulk which has been divided in halves along its length.

half-timbering (*Build.*). An early mode of house-building in which the foundations and principal members were of stout timber, and walls were formed by filling the spaces between members with plaster.

half-title (*Typog.*). The title of a book printed on the leaf preceding the title-page.

half-tone (*Telecomm.*). Description of image in which there is a continuous range of contrast between black and white.

half-tone block (*Print.*). A block for letterpress printing etched in relief on copper for fine screen, or (usually) zinc for coarse screen, after printing down on the metal with a half-tone negative. Also produced by *electronic engraving* (q.v.). See **half-tone process.**

half-tone process (*Photog., Print.*). The reproduction of a subject containing a range of tones by photographing through a cross-line screen which translates these tones into dots of varying size. Used for all the usual printing methods, relief (letterpress), planographic (lithography), intaglio (gravure), and stencil (screen process).

half-twist bit (*Tools*). Bit shaped like a gimlet, used for making screw-holes.

half-uncials (*Typog.*). An early style of lettering in which square forms of letter are mixed with rounded forms; a few type faces are available which copy this style.

half-value layer (*Nuc.*). See **half-value thickness.**

half-value period or half-period (*Nuc.*). See **half-life (1).**

half-value thickness (*Nuc.*). The thickness of a specified substance which must be placed in the path of a beam of radiation in order to reduce the transmitted intensity by one-half. Also **half-thickness, half-value layer.**

half-watt lamp (*Light*). A name sometimes given to gas-filled lamps from the fact that the consumption approaches half a watt per candle-power.

half-wave antenna (*Radio*). An antenna whose overall length is one half-wavelength. The voltage distribution is from a maximum at the top to a minimum in the middle and a maximum at the base.

half-wave plate (*Optics*). A plate of doubly refracting, uniaxial crystal cut parallel to the optic axis, of such thickness that, if light is transmitted normally through it, a phase difference of half a period is introduced between the ordinary and extraordinary waves. A half-wave plate is used in Laurent's polarimeter.

half-wave rectification (*Elec. Eng., Radio*). Rectification in which current flows only during the positive (or negative) half-cycles of the alternating voltage. The commonest form of rectification of radio signals. Also called **single-wave rectification.**

half-wave rectifier (*Elec. Eng., Radio*). One in which there is conduction for part of one-half of the applied alternating cycle of voltage.

half-wave suppressor coil (*Elec. Eng., Radio*). An inductance coil inserted at half-wavelength intervals along an antenna wire; used in some forms of directional antenna to suppress radiation in reverse phase from alternate half-wavelength sections of the wire.

half-wave transmission (*Elec. Eng.*). A method of transmission of electrical energy in which the natural period of oscillation of the transmission line is equal to four times the frequency of the transmitted currency.

half-width (*Phys.*). (1) Half the width of the energy peak in a spectral distribution measured between the two half-amplitude points. It may be expressed either directly or as a percentage of the mean value. (2) Half the width of a response curve between the half-power points.

hali-, halo-. Prefix from Gk. *hals*, salt.

halides (*Chem.*). Fluorides, chlorides, bromides, iodides and astatides.

haliplankton (*Ecol.*). The plankton of the seas.

halisteresis (*Med.*). Softening of bone due to disappearance of lime salts from it.

halite (*Min.*). Common or rock salt. The naturally occurring form of sodium chloride, crystallizing in the cubic system; represented in the Purbeck Series by clay pseudomorphs, and forming deposits of considerable thickness in close association with anhydrite and gypsum in the Permian and Triassic rocks of northern Europe. Deposits of commercial value occur in Cheshire, Lancashire, Co. Antrim, and Somerset, the salt being pumped out as brine from the Keuper Marl, or mined as at Winsford in Cheshire; the Stassfurt deposits in Germany and those of Wieliczka in Poland are famous. In the U.S., valuable salt deposits occur in the Salina beds of Silurian age, worked in Michigan, New York, Ohio, etc.; also in the overlying Mississippian. Salt of Permian age is important in Kansas and Oklahoma. Salt domes occur in Louisiana and Texas.

Hallade recorder (*Eng.*). An instrument for recording vibration of rolling-stock due to track irregularities, etc., in planes parallel, transverse, and perpendicular relative to the track.

Hall coefficient (*Electronics*). This is defined as:

$$R_H = E_y / J_x H_z,$$

where J_x is the current flow in x direction, E_y is electric field developed in transverse direction when a magnetic field H_z is applied in z direction. Also given by:

$$R_H = 1/Nce,$$

where N is the number of free electrons of charge e e.s.u. per unit volume and c is velocity of light, deduced from free electron theory of metals. Values for some metals imply positive charges, explained by band theory of solids. In SI units $R = E_y/BJ_x$ ohm m³/weber $= 1/Ne$ where B is magnetic induction. See **hole.**

Hall effect (*Electronics*). Disturbance of the lines of current flow in a conductor due to the application of a magnetic field, leading to an electric potential gradient transverse to direction of current flow. Also **galvanomagnetic effect.**

Halle-Seydel converter (*Textiles*). See **Pacific converter.**

Haller's organ (*Zool.*). In some *Acarina* (*Ixodidae*), a cavity filled with comb-like teeth and communicating with the exterior by a slit-like opening, situated on the upper side of the tarsus of the first leg; believed to be olfactory.

Hallez's law (*Zool.*). The eggs of Insects are located in the ovarioles in such a position that the cephalic pole of each is directed towards the head of the parent: also, the dorsal and ventral aspects of the egg correspond with those of the parent and of the future embryo.

Hall mobility (*Electronics*). Mobility (mean drift velocity in unit field) of current carriers in a semiconductor as calculated from the product of the Hall coefficient and the conductivity.

halloysite (*Min.*). One of the clay minerals, a hydrated form of kaolinite and member of the kandite group; consists of hydrated aluminium silicate.

Hall process (*Met.*). A process for the extraction of aluminium by the electrolysis of a fused solution of alumina in cryolite at a temperature of approximately 1000°C. The aluminium is molten at this temperature and settles to the bottom of the bath, from which it is drawn off. Contains up to 99·8% aluminium.

hallucination (*Psychol.*). A perception of sensation for which there is no objective reality, such as hearing voices, seeing persons or things, etc., which do not actually exist.

hallucinogen (*Pharm.*). A hallucinatory drug.

hallux (*Zool.*). In land Vertebrates, the first digit of the hind-limb.

hallux flexus (*Med.*). A late stage of hallux rigidus, the big toe being rigidly flexed on the sole.

hallux rigidus (*Med.*). Rigid stiffness of the big toe, due to osteoarthritis of the joint between the toe and the foot.

hallux valgus (*Med.*). A deformity of the big toe, in which it turns towards and comes to lie above the toe next to it; usually associated with bunion.

hallux varus (*Med.*). A rare deformity of the big toe, in which it diverges from the toe next to it.

Hallwachs effect (*Electronics*). Release of electrons from a body when exposed to ultraviolet light. Obsolete for **photoelectric effect.**

halo (*Electronics*). Same as **halation.** (*Meteor.*) A bright ring or system of rings often seen surrounding the sun or moon. The *large halo*, of radius 22°, is due to light refracted at minimum deviation by ice crystals in high cirro-stratus clouds. The *small halo*, termed a corona, a few degrees in radius, is formed by diffraction by water droplets in the atmosphere. (*Mining*) Diffusion into surrounding rock formation of traces of mineral being sought, as shown by geochemical tests.

halobiontic (*Ecol.*). Strictly confined to salt water.

halobiotic (*Zool.*). Living in salt water, especially in the sea.

halochromism (*Chem.*). The formation of coloured salts from colourless organic bases by the addition of acids.

halo effect (*Psychol.*). The tendency to make an overall judgement of an individual (good or bad) on the basis of a few characteristics without giving attention to other relevant aspects.

halogen (*Chem.*). One of the seventh group of elements in the periodic table, for which there is one electron vacancy in the outer energy level, viz., F., Cl, Br, I, At. The main oxidation state is $^-1$.

halogenation (*Chem.*). The introduction of *halogen* atoms into an organic molecule by substitution.

halogen-quench Geiger tube (*Nuc. Eng.*). Low-voltage tube for which halogen gas (normally bromine) absorbs residual electrons after a current pulse, and so quenches the discharge in preparation for a subsequent count.

haloid acids (*Chem.*). A group consisting of hydrogen fluoride, hydrogen chloride, hydrogen bromide, and hydrogen iodide.

halolimnic (*Zool.*). Originally marine but secondarily adapted to fresh water.

halophile (*Ecol.*). A freshwater species capable of surviving in salt water.

halophilic bacteria (*Bacteriol.*). Salt-tolerant bacteria occurring in the surface layers of the sea, where they are important in the nitrogen, carbon, sulphur, and phosphorous cycles. Many are pigmented or phosphorescent and if present in quantity may colour the surface water. Some halophilic bacteria, e.g., halobacterium, are able to live in salted meats.

halophobe (*Bot.*). A plant which will not grow in a soil containing any appreciable amount of salt. Cf. *halophyte*.

halophyte (*Bot.*). A plant which will live in a soil containing an appreciable amount of common salt or of other inorganic salts. Cf. *halophobe*.

halophytic vegetation (*Bot.*). A population consisting of halophytes.

halothane (*Chem.*). See **Fluothane.**

halotrichite (*Min.*). Hydrated sulphate of iron (II) and aluminium, occurring rarely as yellowish fibrous silky colourless crystals in rocks that have been affected by the action of sulphuric acid around fumaroles, and as a weathering product of pyrite and aluminous shales. Also called **butter rock, mountain butter, iron alum.**

halteres (*Zool.*). A pair of capitate threads which take the place of the hind wings in *Diptera*, and are believed to assist the insect to maintain its equilibrium while flying; balancers.

halurgite (*Min.*). A hydrated basic magnesium borate, crystallizing in the monoclinic system.

halvans (*Mining*). Inferior ore; the refuse made during ore mining.

halving (*Carp., etc.*). A method of jointing (e.g., two timbers); it consists in cutting away half the thickness from the face of one, and the remaining half from the back of the other, so that when the two pieces are put together the outer surfaces are flush.

hamate, hamulose (*Bot.*). Said of (1) a narrow leaf hooked at the tip; (2) a trichome similarly bent.

hambergite (*Min.*). Basic beryllium borate, crystallizing in the orthorhombic system.

Hamilton's principle (*Maths.*). The motion of a system from time t_1 to t_2 is such that the line integral $\int_{t_1}^{t_2}(T+W)dt$ is either a maximum or a minimum, where T is the kinetic energy of the system and $W=\sum_i F_i \cdot r_i$, where F_i are the external forces acting at points r_i. This principle can be taken as a basic postulate in place of Newton's laws. For systems to which Newton's laws can be applied it gives the same results, but it is important because it can also be applied to other systems.

hamirostrate (*Zool.*). Having a hooked beak.

hammer-axe (*Tools*). A tool with a double head—axe at one side and hammer at the other side of the handle.

hammer beam (*Build.*). A short cantilever beam projecting into a room or hall from the springing level of the roof, strengthened by a curved strut underneath, and carrying a hammer-beam roof.

hammer-beam roof (*Build.*). A type of timber roof existing in various forms, all affording good headroom beneath. It consists essentially of arched ribs, supported on hammer beams at their feet, and carrying the principal rafters, strengthened sometimes by a collar-beam and/or struts.

hammer blow (*Eng.*). (1) The alternating force between a steam locomotive's driving-wheels and the rails, due to the centrifugal force of the balance weights used to balance the reciprocating masses. (2) The noise due to a pressure wave travelling along a pipe in which the flow of a liquid has been suddenly impeded.

hammer break (*Elec. Eng.*). A name given to the electromagnetic trembler device used on an electric bell or the primary of an induction coil.

hammer-dressed (*Build.*). A term applied to stone surfaces left with a rough finish produced by the hammer.

hammer drill (*Mining*). A compressed-air rock drill in which the piston is not attached to the steel or borer but moves freely.

hammer-drive screw (*Eng.*). A self-tapping screw with a very long pitch, akin to a nail, which can be driven by a hammer or inserted by a press into a plain hole in a relatively soft material.

hammer-headed (*Tools*). A term applied to masons' chisels intended to be struck by a hammer rather than by a mallet.

hammerman (*Eng.*). (1) The operator of a power-driven hammer. (2) A smith's mate.

hammer mill (*Agric.*). A mill in which the foodstuffs for animals are pulverized; they are beaten against the sides of a metal drum by steel hammers. (*Min. Proc.*) See **impact crusher.**

hammer scale (*Eng.*). The scale of iron oxide which forms on work when it is heated for forging.

hammer test (*Met.*). Drop test for impact strength of large metal parts, e.g., rails. Weight is dropped from increasing heights until specified deflection is produced.

hammer toe (*Med.*). A deformity of any toe, especially the second, in which the toe, flexed on itself, is, at its junction with the foot, bent towards the instep.

hammer track (*Nuc.*). Highly characteristic track resembling hammer, formed by decay of ^{8}Li nucleus into two α-particles emitted in opposite directions at right angles to the lithium track.

Hammond organ (*Acous.*). See **organ.**

Hamstead Beds (*Geol.*). The highest division of the Oligocene of the Hampshire Basin, lying above the Bembridge Beds, and consisting of black and green clays containing non-marine shells and plant remains, and a few feet of blue marine clays at the top.

hamula (*Zool.*). The *retinaculum* (q.v.) of *Collembola*.

hamulose (*Bot.*). See **hamate.**

hamulus (*Zool.*). Any small hook-like structure; the hooked end of a barbicel; the row of hooks on the costal margin of the hind wing in *Hymenoptera*: the paired claspers in *Odonata*. *adj.* **hamulate.**

hance (*Arch.*). That part of the intrados, close to the springing of an elliptical or many-centred arch, which forms the arc of smaller radius.

hancockite (*Min.*). A member of the epidote group of minerals containing lead, substituting for calcium.

Hancock jig (*Mining*). One in which ore is jigged up and down with some throw forward in a tank of water, the heavy mineral stratifying down and being separately removed.

hand brace (*Carp.*). See **brace.**

hand-cut overlay (*Print.*). Plies of paper cut out in the highlights and built up in the dark tones and included in the *make-ready* to improve the printing of half-tone blocks. Cf. *mechanical overlay*.

H and D (or Hurter and Driffield) curve (*Photog.*). See **characteristic curve.**

H and D number (*Photog.*). Emulsion speed expressed as $\frac{10}{i}$, where i = *inertia* (q.v.).

hand feed (*Eng.*). The hand operation of the feed mechanism of a machine tool. See **feed.**

hand hole (*Eng.*). A small hole, closed by a removable cover, in the side of a pressure vessel or tank; it provides means of access for the hand to the inside of the vessel.

hand ladle (*Foundry*). A small foundry casting ladle supported by a long handle of steel bar. See also **hand shank.**

hand lead (*Surv.*). Lead plummet used at sea, by attachment to a lead-line measuring within 100 fathoms (183 metres).

handlers (*Leather*). The name for a series of pits in which hides are laid flat after the first stage of tanning.

hand letters (*Bind.*). Letters formed of brass and mounted on a handle, with which the finisher impresses the title on the back or side of a bound volume.

handle-type fuse (*Elec. Eng.*). A fuse in which the carrier containing the fuse-link is provided with a handle to facilitate withdrawal and replacement.

hand level (*Surv.*). Small and light sighting tube above which a spirit level and mirror are arranged, so that the bubble can be seen while sighting on a station.

handmade paper (*Paper*). Paper made by dipping a mould into rag pulp and, by skilful shaking, distributing it into a sheet. The wet sheets are piled up between felts. Handmade paper is used when durability and stylish appearance are required.

hand monitor (*Radiol.*). Radiation monitor designed to measure radioactive contamination on the hands of an operator.

hand picking (*Min. Proc.*). Sorting; removal from passing stream of suitably sized ore of waste wood, detritus or of a special mineral.

hand press (*Print.*). A press operated by hand; now used chiefly for proof-pulling.

hand-rail bolt (*Join.*). A rod which is threaded and fitted with a nut at both ends; used to draw together the mating surfaces of a butt joint, such as that between adjacent lengths of hand-rail. The rod passes through holes in the members, and at one end the nut, a square one, is housed in a square mortise which prevents it from turning. At the other end the nut is circular, with notches cut in its periphery, to afford a means of screwing it into its mortise.

hand-rail plane (*Join.*). A plane having a specially shaped sole and cutting-iron, adapting it to the finishing of the top surface of a hand-rail.

hand-rail punch (*Join.*). A small tool which is inserted in the notches in the periphery of the circular nut of a *hand-rail bolt* (q.v.) and is used to tighten it up.

hand-rail screw (*Join.*). A small rod, taper-threaded at each end, used to connect adjacent lengths of hand-rail as an alternative to the *hand-rail bolt* (q.v.). Also called a **dowel screw.**

hand-regulated arc lamp (*Elec. Eng.*). An arc lamp in which the striking of the arc and the adjustment of the carbons to give the correct arc length are carried out entirely by hand.

hand reset (*Elec. Eng.*). Restoration of a magnetic device, e.g., relay or circuit-breaker, by a manual operation.

hand rest (*Eng.*). A support, shaped like a letter T, on which a turner rests a hand tool, during wood-turning or metal-spinning in a lathe.

hand roller (*Typog.*). A roller used for inking type-matter on the hand press, preparatory to pulling a proof.

hand-rope operation (*Elec. Eng.*). A method of control of electric lifts; a rope is passed through the car and attached to control equipment mounted at the top of the shaft.

Hand-Schüller-Christian disease (*Med.*). A lipoid granuloma of unknown aetiology, affecting mainly the skull. The granulomas contain cholesterol.

handscrew (*Join.*). A wooden frame consisting of two parallel bars connected by two tightening screws; used to hold parts together while a glued joint dries, or to secure work in process of being formed.

handset (*Teleph.*). A rigid combination of microphone and receiver, in a form convenient for holding simultaneously to mouth and ear.

hand shank (*Eng.*). A *foundry ladle* (q.v.) supported at the centre of a long iron bar, formed into a pair of handles at one end for control during pouring; carried by two men.

hand specimen (*Mining, etc.*). Fragment of ore deposit chosen for special study or test as displaying one or more characteristic minerals.

hand tools (*Eng.*). All tools used by fitters when doing hand work at the bench, as hammers, files, scrapers, etc.

hand wheel (*Horol.*). A grooved pulley provided with a cranked handle and mounted on a universal form of vice, used for driving a lathe or other tool by hand.

hand winding (*Elec. Eng.*). The process of winding a machine by the insertion of the coils, turn by turn, into the slots; used in cases where it is not convenient to use former-wound coils.

hangar (*Aero.*). A special construction for the accommodation of aircraft.

hanger (*Build.*). A bracket for the support of a gutter at the eaves. (*Elec. Eng.*) (1) Plates of glass or other material standing on edge in an accumulator cell, and supporting the accumulator-plates by means of their lugs. (2) A fitting used for supporting the overhead contact wire of a traction system from a transverse wire or structure. (*Eng.*) (1) A bracket, usually of cast-iron, bolted to a wall or to the underside of a girder, to hold a bearing for supporting overhead shafting. (2) A bracket, usually of steel strip, used to support a pipe from a roof.

hanger-on (*Mining*). A man who attaches or detaches the tubs or trams on an endless-rope haulage in a coal-mine. A *hitcher* performs a similar task at the shaft bottom. See **clipper.**

hangfire (*Mining*). Unexpected delay or failure of explosive charge to detonate, thus creating a dangerous situation.

hanging (*Met.*). Hang up; in blast furnace, adhesion of partly melted charge to walls, thus upsetting smooth working.

hanging battens (*Elec. Eng.*). A suspended row of lamps used in stage lighting.

hanging bolts (*Mining*). Strong rods, usually of round iron, from which timbers are suspended during sinking of shaft, or which are used to reinforce concrete linings.

hanging buttress (*Build.*). A buttress carried upon a corbel at its base.

hanging drop preparation (*Micros.*). A term used to describe a preparation in which the specimen, in a drop of medium on the undersurface of a coverslip, is suspended over a hollow-ground slide, for microscopic examination. Also describes a type of tissue culture cultivation in which a small fragment of tissue is grown under similar conditions.

hanging figures (*Typog.*). Figures normally supplied with *old style* founts; 1, 2, 0 conform to the *x-height*, 6 and 8 have the height of *ascenders*, and 3, 4, 5, 7 and 9 hang below the *base line*. Also called **old style figures.**

hanging indention (*Typog.*). Layout in which the first line of the paragraph is set *full-out* and the succeeding lines are indented one em or more [as employed in this dictionary].

hanging post (*Build.*). See **hingeing post.**

hangings (*Build.*). A term applied to materials, such as wallpapers, used as wall coverings.

hanging sash (*Join.*). A sash arranged to slide in vertical grooves, and counterweighted so as to be balanced in all positions.

hanging steps (*Build.*). Steps which are built into a wall at one end and are unsupported at the other end. Also known as **cantilevered steps.**

hanging stile (*Join.*). That stile of a door to which the hinges are secured.

hanging valley (*Geol.*). A tributary valley not graded to the main valley. It is a product of large-scale glaciation and may be due to the glacial overdeepening of the main valley or the grading of the tributary to the ice-filled main valley. The two valleys are connected by rapids or waterfalls.

hanging wall (*Mining*). Rock above the miner's head, usually the country rock above the deposit being worked.

Hangman Grits (*Geol.*). A thick series of sandstones and grits, of Old Red Sandstone facies and Devonian age, named from a headland in N. Devon.

hang-over (*Telecomm.*). The delay in restoration of speech-operated switches, as in the *vodas*, to ensure the nonclipping of weak final consonants of words.

hang up (*Met.*). See **hanging**.

hank (*Textiles*). A general term for a reeled length of yarn. In calculating the counts of yarn, a definite length is assigned to the hank for each type of yarn. See **count of yarn, lea.**

Hankel functions of the first and second kind (*Maths.*). See **Bessel functions (of the third kind).**

Hanot's cirrhosis (*Med.*). A cirrhosis of the liver, which is greatly enlarged, associated with jaundice.

Hansa yellow (*Paint.*). Very popular bright yellow azo pigments which have good tinting properties but low opacity, and are used in oil paints and emulsion paints. The pigments are acid and alkali fast and have little tendency to bleed in most media.

Hansen's iron trioxyhaematin stain (*Micros.*). A black nuclear stain; fresh preparations stain cytoplasm very little. It contains iron alum, ammonium sulphate, and haematoxylin. Usually used with a counterstain.

H-antenna (*TV*). Two vertical dipoles, one reflecting, ca. ½ to ⅛ wavelength apart, much used for TV and FM reception in band I.

hapanthous (*Bot.*). Flowering once and then dying.

hapaxanthic (*Bot.*). Single flowering.

haplite (*Geol.*). See **aplite.**

haplo-. Prefix from Gk. *haploos*, single, simple.

haplobiont (*Bot.*). A plant which has only one kind of individual in its life-history. *adj.* **haplobiontic.**

haplocaulescent (*Bot.*). Having a single axis.

haplochlamydeous chimaera (*Bot.*). A periclinal chimaera in which one component is present as a single cell layer, forming the epidermis.

haplodioecious (*Bot.*). Having a haploid thallus, capable of cross-fertilization only. Also **heterothallic.**

haplodiplont (*Bot.*). A sporophyte in which the cells contain the haploid chromosome number.

haplodont (*Zool.*). Having molars with simple crowns.

haplogastrous (*Zool.*). A form of abdomen found in some *Coleoptera*, in which the second abdominal segment exhibits a pleurite and a small lateral plate representing the sternite.

haploid (*Cyt.*). Of the reduced number of chromosomes characteristic of the germ cells of a species, equal to half the number in the somatic cells. Cf. *diploid.*

haploid apogamety, haploid apogamy (*Bot.*). See **reduced apogamy.**

haplometrotic (*Zool.*). Said of colonies of *Hymenoptera* dominated by a single fecundated female or queen and with a clearly differentiated worker caste.

haplomonoecious (*Bot.*). Having a haploid thallus which is self-fertile. Also **homothallic.**

haplont (*Bot.*). A plant which has sexuality, and in which the zygotic nucleus only is diploid; all other nuclei are haploid.

haplophase (*Biol.*). The period in the life-cycle of any organism when the nuclei are haploid. Cf. *diplophase.*

haplosis (*Cyt.*). The halving in the number of the chromosomes at meiosis; reduction and disjunction.

Haplosporidia (*Zool.*). A suborder of *Neosporidia*, most of the members of which infest Fish and aquatic Invertebrates; the spores possess cases but have no polar capsules.

haplostele (*Bot.*). A protostele in which the solid central core of xylem is circular in cross-section.

haplostemonous (*Bot.*). Having a single whorl of stamens.

Haplostichmeae (*Bot.*). A subclass of the *Heterogeneratae*; the microscopic gametophytes are isogamous or oögamous; the sporophytes are trichothallic with free filaments, or may be packed to form a pseudoparenchyma.

haploxylic (*Bot.*). Said of a leaf containing one vascular strand.

hapteron (*Bot.*). A cell or a cellular organ which attaches a plant to a support. The term is used especially in speaking of the attachment organs of the lower plants.

haptotropism (*Bot.*). One-sided growth leading to curvature (*haptotropic curvature*); being the response of an elongated plant organ which has been stimulated by touch or slight pressure.

hard (*Electronics*). Adjective, synonymous with *high-vacuum*, which differentiates thermionic vacuum valves from gas-discharge tubes and solid-state devices, e.g., transistors, diodes, etc., which can perform similar functions. (*Glass*) Having a relatively high softening point. (*Hyd. Eng.*) A layer of gravel or similar materials put down on swampy or sodden ground to provide a way for passage on foot.

hard bast (*Bot.*). Sclerenchyma present in phloem.

hardboard (*Build.*). Fibre-board that has been compressed in drying, giving a material of greater density than *insulating board* (q.v.).

hard bronze (*Met.*). Copper-based alloy used for tough or dense castings; based on 88% copper, plus tin with either some lead or zinc.

hard copy (*Comp.*). A written record of output data to be stored, e.g., on magnetic or punched tape.

hard core (*Build., Civ. Eng.*). Lumps of broken brick, hard natural stone, etc., used to form the basis of a foundation for road or paving or floors to a building.

hard dentine (*Zool.*). A type of dentine consisting of a matrix of animal matter strongly impregnated with calcium carbonate and permeated by delicate parallel tubules containing organic fibrils.

hardenability (*Eng.*). The response of a metal to quenching for the purpose of hardening, measured, most commonly, by the *Jominy (end-quench) test* (q.v.).

hardener (*Photog.*). Chemical (formalin, acrolein, chrome alum, etc.) added to the fixing bath to toughen the emulsion of a film. (*Plastics*) An *accelerator* (q.v.).

hardening (*Met.*). The process of making steel hard by cooling from above the critical range at a rate that prevents the formation of ferrite and pearlite and results in the formation of martensite. May involve cooling in water, oil, or air, according to composition and size of article.

hardening media (*Met.*). Liquids into which steel is plunged in hardening. They include cold water, various oils, and water containing sodium chloride or hydroxide to increase the cooling power.

hardening of oils (*Chem.*). The hydrogenation of oils in the presence of a catalyst, usually finely divided nickel, in which the unsaturated acids are transformed into saturated acids, with the result that the glycerides of the unsaturated acids become hard. This process is of great importance for the foodstuffs industries, e.g., margarine is prepared in this way.

Harder's glands (*Zool.*). In most of the higher Vertebrates, an accumulation of small glands near the inner angle of the eye, closely resembling the lacrimal gland.

hard-facing (*Met.*). (1) The application of a surface layer of hard material to impart, in particular, wear resistance. (2) A surface so formed. The composition is generally of high melting-point metals, carbides, etc., applied by powder, wire, or plasma arc spraying, or by welding.

hard glass (*Chem.*). Borosilicate glass, whose hardness is principally due to boron compounds. Resistant to heat and to chemical action.

hard-gloss paint (*Paint.*). A popular class of paint that dries hard with a high gloss. It always contains some hard resin in the medium.

hard head (*Met.*). Alloy of tin with iron and arsenic left after refining of tin.

hard heading (*Mining*). Sandstone or other hard rock met with in making headings or tunnels in a coal mine.

Hardinge mill (*Min. Proc.*). Widely used grinding mill, made in three sections—a flattish cone at the feed end, a cylindrical drum centrally and a steep cone, the assembly being hung horizontally between trunnions.

hard lead (*Met.*). All antimonial lead; metal in which the high degree of malleability characteristic of pure lead is destroyed by the presence of impurities, of which antimony is the most common.

hardmetal (*Met.*). Sintered tungsten carbide. Used for the working tip of high-speed cutting tools. See **sintering**.

hard metals (*Met.*). Metallic compounds with high melting-points; typified by refractory carbides of the transition metals (4th to 6th groups of the periodic system), notably tungsten, tantalum, titanium, and niobium carbides.

hardness (*Electronics*). Degree of vacuum in an evacuated space, especially of a thermionic valve or X-ray tube. Also penetrating power of X-rays, which is proportional to frequency. (*Met.*) Signifies, in general, resistance to cutting, indentation and/or abrasion. It is actually measured by determining the resistance to indentation, as in *Brinell*, *Rockwell*, *Vickers diamond pyramid*, and *scleroscope hardness tests* (qq.v.). The values of hardness obtained by the different methods are to some extent related to each other, and to the ultimate tensile stress of non-brittle metals. (*Min.*) The resistance which a mineral offers to abrasion. The absolute hardness is measured with the aid of a sclerometer. The comparative hardness is expressed in terms of Mohs' scale, and is determined by testing against ten standard minerals: (1) talc, (2) gypsum, (3) calcite, (4) fluorite, (5) apatite, (6) orthoclase, (7) quartz, (8) topaz, (9) corundum, (10) diamond. Thus a mineral with 'hardness 5' will scratch or abrade fluorite, but will be scratched by orthoclase. Hardness varies on different faces of a crystal, and in some cases (e.g., kyanite) in different directions on any one face.

hard packing (*Typog.*). Hard paper employed to cover the cylinder of a printing press when printing on hard, smooth papers from engravings, etc.; used in order to obtain a sharp impression. Also necessary when printing from plastic or rubber plates.

hard-pad disease (*Vet.*). Paradistemper. A contagious disease of the dog caused by a virus closely related to that causing *canine distemper* (q.v.); characterized especially by hyperkeratosis affecting the footpads and nose and by the frequent occurrence of nervous symptoms.

hard palate (*Zool.*). In Mammals, the anterior part of the roof of the buccal cavity, consisting of the horizontal palatine plates of the maxillary and palatine bones covered with mucous membrane.

hard pan (*Min.*). Lateritic gravels near or at surface of alluvial strata which have been consolidated by weathering and deposition of iron oxide or clay.

hard plaster (*Build.*). Term usually applied to hard-setting forms of gypsum plaster, e.g., *Keene's cement* (q.v.).

hard plating (*Met.*). Chromium plating deposited in appreciable thickness directly on to the base metal, that is, without a preliminary deposit of copper or nickel. The coating is porous, but offers resistance to corrosion and to wear.

hard radiation (*Radiol.*). Qualitatively, the more penetrating types of X-, beta-, and gamma-rays.

hard rock mining (*Mining*). Term used to distinguish between deposits soft enough to be detached by mechanical excavator and those which must first be loosened by blasting.

hard-rock phosphate (*Geol.*). A phosphatic deposit resulting from the leaching of calcium carbonate out of a phosphatic limestone, leaving a phosphate residue. Applied specifically to the phosphate deposits of Florida which have this origin.

hards (*Cinema.*). Colloq. for lights, e.g., arc lamps, which give a hard or harsh type of lighting.

hard soaps (*Chem.*). See **soaps**.

hard solder (*Met.*). One not melting below red heat.

hard steel (*Met.*). One containing more than 0·3% of carbon.

hard stocks (*Build.*). Bricks which are sound but have been overburnt and are not of good shape and colour.

hard twist (*Textiles*). A yarn with more than the standard amount of twist, inserted to secure the desired effects in particular fabrics, e.g., crêpes, voiles, gabardines, etc.

hardware (*Comp.*). General term for all mechanical and electrical component parts of a computer or data-processing system. Cf. *software*.

hard waste (*Spinning*). Waste from single or folded twisted yarns from cop bottoms and waste made during winding, warping, reeling and weaving.

hard water (*Chem.*). Water having magnesium and calcium ions in solution and offering difficulty in making a soap lather. See **permanent hardness, temporary hardness.**

hardwood (*For.*). Dense, close-grained wood from deciduous trees (oak, beech, ash, teak).

Hardy and Schulze 'law'. (*Min. Proc., etc.*). The

efficiency of an ion used as a coagulating agent is roughly proportional to its state of oxidation.

Hardy-Weinberg law (*Gen.*). The gene frequencies in a large population remain constant from generation to generation if mating is at random and there is no selection, migration or mutation. If two alleles A and a are segregating at a locus, and each has a frequency of p and q respectively, then the frequencies of the genotypes AA, Aa, and aa are p^2, $2pq$ and q^2 respectively.

harelip (*Med.*). A congenital cleft in the upper lip, often associated with cleft palate.

Hare's apparatus (*Phys.*). Apparatus for comparing the densities of two liquids. It usually comprises an inverted U-tube, of which each limb is dipped into one of the liquids which rises into it by suction applied to a middle limb. The heights of the liquids above the surface in their reservoir are inversely proportional to their respective densities.

Harker diagram (*Geol.*). A diagram in which chemical analyses of rocks are plotted to show their relationships. The constituents are plotted as ordinates against the silica content as abscissa.

Harlech Dome (*Geol.*). A major geological structure, involving the Cambrian and Ordovician strata of Merionethshire, which dip outwards in all directions from a point near Harlech, the type area of the Harlech Series.

Harlech Series (*Geol.*). A very thick series of alternating massive grits, feldspathic and conglomeratic in part, and shales, the complete succession being (1) Llanbedr Slates, (2) Rhinog Grits, (3) Manganese Shale Group, (4) Barmouth Grits, and (5) Gamlan Shales.

harmattan (*Meteor.*). A dusty, dry north-easterly wind blowing over W. Africa in the dry season.

H-armature (*Elec. Eng.*). See **shuttle armature.**

harmodotron (*Electronics*). V-band or higher frequency (sub-mm) microwave generator, in which resonant cavity similar to rhumbatron is excited to high order mode resonance at a harmonic of the bunching frequency.

harmonic (*Phys., etc.*). Sinusoidal component of repetitive complex waveform with frequency which is an exact multiple of basic repetition frequency (the fundamental). The full set of harmonics forms a Fourier series which completely represents the original complex wave. In acoustics, harmonics are often termed overtones, and these are counted in order of frequency above, but excluding, the lowest of the detectable frequencies in the note; the label of the harmonic is always its frequency divided by the fundamental. The nth overtone is the $(n+1)^{th}$ harmonic.

harmonic absorber (*Phys.*). Arrangement for removing harmonics in current or voltage waveforms, using tuned circuits or a wave filter.

harmonic analysis (*Phys.*). Process of measuring or calculating the relative amplitudes of all the significant harmonic components present in a given complex waveform. The result is frequently presented in the form of a Fourier series, e.g., $A = A_0 \sin \omega t + A_1 \sin (2\omega t + \varphi_1) + A_2 \sin (3\omega t + \varphi_2) + \ldots$ etc., where ω = pulsatance and φ = phase angle.

harmonic antenna (*Radio*). One whose overall length is an integral number (greater than one) of quarter-wavelengths.

harmonic component (*Phys.*). Any term (but the first) in a Fourier series which represents a complex wave.

harmonic conjugate *Maths.*). Two of four collinear points are harmonic conjugates of the other two if the cross-ratio of the four points is harmonic (i.e., has the value minus one). Also used of a pencil of four concurrent lines.

harmonic content (*Phys.*). Nonsinusoidal periodic function with the fundamental extracted.

harmonic dial (*Phys.*). A graphical method of displaying the amplitude and phase of a harmonic component of a periodic variation.

harmonic distortion (*Phys.*). Production of harmonic components from a pure sine wave signal as a result of nonlinearity in the response of a transducer or amplifier.

harmonic drive (*Radio*). See **harmonic excitation.**

harmonic echo (*Acous.*). An echo of a note which is raised in pitch over the original note sounded.

harmonic excitation (*Radio*). (1) Excitation of an antenna on one of its harmonic modes. (2) Excitation of a transmitter from a harmonic of the master oscillator. Also **harmonic drive.**

harmonic filter (*Radio*). One which separates harmonics from fundamental in the feed to an antenna. Also called **harmonic suppressor.**

harmonic function (*Maths.*). A function which satisfies Laplace's equation.

harmonic generator (*Phys.*). Waveform generator with controlled fundamental frequency, producing a very large number of appreciable-amplitude odd and even harmonic components which provide a series of reference frequencies for measurement or calibration. See **multivibrator.**

harmonic interference (*Radio*). That caused by harmonic radiation from a transmitter and outside the specified channel of radio communication.

harmonic means (*Maths.*). (1) Numbers inserted between two numbers such that the resulting sequence is a harmonic progression. (2) The *harmonic mean* of n numbers a_r is the reciprocal of the arithmetic mean of their reciprocals, i.e.,

$$n\Big/\sum_{1}^{n}\frac{1}{a_r}.$$

harmonic progression (*Maths.*). The numbers, $a, b, c, d \ldots$ form a harmonic progression if their respective reciprocals form an arithmetical progression.

harmonic ratio (*Maths.*). If the cross-ratio of four collinear points equals -1, it is harmonic.

harmonic series (*Phys.*). One in which each basic frequency of the series is an integral multiple of a fundamental frequency.

harmonic suppressor (*Radio*). Same as **harmonic filter.**

harmotome (*Min.*). A member of the zeolite group, hydrated silicate of aluminium and barium, crystallizing in the monoclinic system, though the symmetry approaches that of the orthorhombic system. Best known by reason of the distinctive cruciform twin groups that are not uncommon. Occurs at Strontian, Argyll, and elsewhere in mineralized veins.

harness (*Aero.*). (1) The entire system of engine ignition leads, particularly those which are screened to prevent electromagnetic interference with radio equipment. (2) The parallel combination of leads interconnecting the thermocouple probes of a turbine engine exhaust gas temperature indicating system. See **gas temperature.** (*Weaving*) That section of a loom jacquard which specifically operates the warp threads. It consists of strong cords placed in position by a comber board. Also, the healds in a tappet or dobby loom.

harness cord (*Weaving*). Strong varnished

cotton, linen or synthetic fibre cord connecting the figuring hooks with the mails that lift the warp threads in a jacquard loom.

harpactophagous (*Zool.*). Predatory.

harpagones (*Zool.*). The paired claspers of male *Lepidoptera*.

harpes (*Zool.*). Spinelike structures often present on the inner aspect of the claspers of male *Lepidoptera*.

Harris flow (*Electronics*). The flow of electrons in a cylindrical beam subjected to a radial electric field to counteract the beam space-charge divergence.

Harris instability (*Nuc.*). Type of plasma instability due to anisotropic velocity distribution of particles.

Harris process (*Met.*). A method of softening lead. Arsenic, antimony, and tin are oxidized by adding sodium nitrate and lead oxide, and the oxides formed are caused to react with sodium hydroxide and chloride to form arsenates, antimonates, and stannates.

harrow (*Agric.*). Light cultivator without wheels, dragged over soil; used to pulverize and level ground, to break up soil to shallow depth, to kill weeds, cover seeds and mix materials into soil.

harstigite (*Min.*). A basic silicate of beryllium, calcium and manganese, crystallizing in the orthorhombic system, occurring in metamorphosed manganese deposits.

hartite (*Min.*). A naturally occurring hydrocarbon compound, $C_{20}H_{34}$, crystallizing in the triclinic system.

hartley (*Comp.*). Unit of information, equal to 3·32 *bits* and conveyed by one decadal code element.

Hartley oscillator (*Electronics*). Triode valve oscillator circuit consisting essentially of a parallel resonant circuit connected between grid and anode, cathode being connected to the coil by a tapping point, which becomes *earthy*.

Hartley principle (*Telecomm.*). General statement that amount of information which can be transmitted through a channel is the product of frequency bandwidth and time during which it is open, whether time division is used or not. See **information.**

Hartman dispersion formula (*Light*). An empirical expression for the variation of the refractive index n of a material with the wavelength of the light:

$$n=n_0+\frac{c}{(\lambda-\lambda_0)^a},$$

n_0, c, λ_0, and a being constant for a given material. For glass, a may be taken as unity. See also **Cornu-Hartman formula.**

Hartmann oscillator (*Acous.*). A device, consisting basically of a conical nozzle and a cylindrical cavity, used for generating high-intensity sound waves in fluids.

Hartmann test (*Optics*). Test for aberration of a lens, in which a diaphragm containing a number of small apertures is placed in front of the lens and the course of the rays is recorded by photographing the pencils of light in planes on either side of the focus.

Hartnell governor (*Eng.*). An engine governor in which the vertical arms of two or more bell-crank levers support heavy balls, the horizontal arms carrying rollers which abut against the central spring-loaded sleeve operating the engine-governing mechanism.

Hartree equation (*Electronics*). One relating flux density in travelling-wave magnetron to minimum anode potential required for oscillation in any given mode. Graphs representing this relationship for different mode numbers form a **Hartree diagram.**

Hartshill Quartzite (*Geol.*). The basal division of the Cambrian System, occupying the same stratigraphical position as the Wrekin and Malvern Quartzites; extensively quarried for road metal in the neighbourhood of Nuneaton.

Hartwell Clay (*Geol.*). The local equivalent of the Portland Sand occurring in the Aylesbury district.

Harvard classification (*Astron.*). A method of classifying stellar spectra, employed by the compilers of the Draper Catalogue of the Harvard Observatory and now in universal use. See **spectral types.**

Harvard (*Textiles*). A firmly woven cotton shirting with a 2×2 twill weave, and, generally, a coloured fancy stripe; there are many kinds.

Harvard twill (*Textiles*). See **two-and-two twill.**

harvester-thresher (*Agric.*). A machine, mounted on a chassis, which performs the operations of a binder and a thresher.

harvest method (*Ecol.*). The measurement of *primary productivity* (q.v.) by harvesting plants and weighing them. Used with cultivated crops, and in some natural habitats where annual plants predominate.

harvest mite (*Vet.*). The popular name of the parasitic larval stage of the mite *Trombicula autumnalis*, which occurs in the skin of animals and man. The nymphal and adult stages of the mite are free-living.

harvest moon (*Astron.*). The name given in popular language to the full moon occurring nearest to the autumnal equinox, at which time the moon rises on several successive nights at almost the same hour. This retarded rising, due to the small inclination of the moon's path to the horizon, is most noticeable at the time of the full moon, although it occurs for some phase of the moon each month.

Harvey process (*Met.*). Toughening treatment for alloy steels, involving superficial carburization, followed by heating to a high temperature and quenching with water.

harzburgite (*Geol.*). An ultrabasic igneous rock belonging to the peridotite group. It consists almost entirely of olivine and orthopyroxene, usually with a little chromite, magnetite and diopside.

harz jig (*Mining*). Concentrating appliance in which water is pulsed through a submerged fixed screen, across which suitably sized ore moves. Heaviest particles gravitate down and through and lighter ones overflow.

Hashimoto's disease (*Med.*). A diffuse inflammation of the thyroid gland, said to be caused by an antigen/antibody reaction.

hashish (*Bot.*). See **cannabis.**

hasp (*Build.*). A fastening device in which a slotted plate fits over a staple and is secured to it by means of a padlock or peg.

Hassal's corpuscles (*Zool.*). Nests of flattened, concentrically arranged epithelial cells occurring in the medulla of the thymus gland of higher Vertebrates.

hastate (*Bot.*). Having two somewhat out-turned lobes at the base of a leaf; halberd-shaped.

hastingsite (*Min.*). An aluminous variety of the monoclinic amphibole hornblende, intermediate in composition between pargasite and ferrohastingsite.

hatch coaming (*Ships*). The strengthened frame surrounding a *hatchway* (q.v.). Its functions are to replace the strength lost by cutting the beams and deck plating, protect the opening and provide support for the hatch cover.

hatchet (*Carp., etc.*). A small axe used for splitting or rough-dressing timber.

hatchet iron (*Plumb.*). A particular form of *copper bit* (q.v.) having an edged end instead of a point.

hatchet stake (*Eng.*). A smith's tool having a sharp horizontal edge when supported by the anvil; used for bending sheet metal.

hatching enzyme (*Zool.*). An enzyme, secreted by the pleuropodia of some insect larvae, which helps to dissolve part of the egg-shell before hatching.

hatchway (*Ships*). An opening in the deck to allow cargo to be shipped into and removed from the hold.

hat-leather packing (*Eng.*). An L-section leather ring, gripped between discs to form a piston, or similarly attached to the ram of a hydraulic machine to prevent leakage. See **hydraulic packing.**

Hatschek's groove (*Zool.*). In *Cephalochorda*, a ciliated groove running along the roof of the vestibule.

Hatschek's nephridium (*Zool.*). In *Cephalochorda*, an anterior dorsal nephridium opening into the pharynx just behind the mouth.

Hatschek's pit (*Zool.*). In *Cephalochorda*, a small depression arising from the left head cavity and opening into the cavity of the oral hood.

hauerite (*Min.*). A rare brownish-black sulphide of manganese, occurring as small cubic crystals in clay or schist.

haul (*Civ. Eng.*). In the construction of an embankment by depositing material from a cutting, the *haul* is the sum of the products of each load by its haul distance.

haulage level (*Mining*). Underground tramming road in, or parallel to, strike of the ore deposit, usually in footwall. Broken ore gravitates, or is moved, to ore chutes and drawn to trucks in this level.

haul distance (*Civ. Eng.*). The distance, at any particular time, that excavated material has to be carried before deposition in order to form an embankment.

hauling rope (*Civ. Eng.*). See **traction rope.**

haulm (*Bot.*). A stem of a grass.

Haultain infrasizer (*Powder Tech.*). Six-column elutriator using air or other gas instead of liquid to grade fine materials in the minus 150-mesh plus 5-μm range by means of vertical columns of increasing cross-section through which the transporting air is gently blown upward.

haunch (*Join.*). The part forming a stub tenon, left near the root of a haunched tenon. Also called a **hauncheon.**

haunched tenon (*Join.*). A tenon from the width of which a part has been cut away, leaving a haunch near its root.

hauncheon (*Join.*). See **haunch.**

haunches (*Build., Civ. Eng.*). See **flanks.**

haunching (*Civ. Eng.*). A slab of concrete forming the edging to the formation of a road. (*Join.*) A mortise cut to receive the haunch of a haunched tenon.

hausmannite (*Min.*). A blackish-brown crystalline form of manganese (II, III) oxide, occurring with other manganese ores, as in the Lake Superior district. Crystallizes in the tetragonal system but is often found massive.

haustellum (*Zool.*). In *Diptera*, the distal expanded portion of the proboscis. *adj.* **haustellate.**

haustorium (*Bot.*). (1) A lateral hypha produced by the mycelium of a parasitic fungus, which obtains nutriment from the host. (2) A modified root or shoot of a higher plant living as a parasite, serving to attach the parasite to the host, and to obtain nourishment from it. (3) The same as the foot of the embryo of a fern.

haut-pas, haute-pace (*Build.*). See **half-pace.**

haüyne (*Min.*). A feldspathoid, crystallizing in the cubic system, consisting essentially of silicate of aluminium, sodium, and calcium, with sodium sulphate; occurs as small blue crystals, chiefly in phonolites and related rock types.

Haversian canals (*Zool.*). Small channels pervading compact bone and containing blood vessels.

Haversian fringes (*Zool.*). Long finger-shaped projections of the surface of synovial membranes.

Haversian lamellae (*Zool.*). In compact bone, the concentrically arranged lamellae which surround a Haversian canal.

Haversian spaces (*Zool.*). In the development of bone, irregular spaces formed by the internal absorption of the original cartilage bone.

Haversian system (*Zool.*). In compact bone, a Haversian canal with surrounding lamellae.

haversine (*Maths.*). Half of the *versine* (q.v.), i.e., $\frac{1}{2}(1-\cos\theta)$.

hawk (*Build.*). A small square board, with handle underneath, used to carry plaster or mortar.

hawk's eye (*Min.*). A dark-blue form of silicified crocidolite found in Griqualand West; when cut *en cabochon*, it is used as a semi-precious gemstone. Cf. *tiger's eye.*

hawser (or hawse-) pipe (*Ships*). A tubular casting fitted to the bows of a steamer, through which the anchor chain or cable passes.

hay-band (*Foundry*). See **straw-rope.**

Hay bridge (*Elec. Eng.*). An a.c. bridge quite widely used for the measurement of inductance.

hay-elevator (*Agric.*). A mechanically-driven endless band, fitted with prongs, on to which hay is forked for transport to a rick.

hay fever (*Med.*). Allergic rhinitis. Acute congestion and running of the nose, occurring after exposure to an antigen (commonly pollen) to which the individual is sensitive. The tendency to sensitization of this type is inherited.

hay-loader (*Agric.*). A machine towed behind a hay waggon and worked by its land wheels. It picks up the hay, and conveys it to the waggon by means of a trough and reciprocating rake-bars.

hay pole (*Agric.*). A vertical pole with jib, pulleys and ropes to lift hay by means of a hay fork or hay grab.

hay-stacker (*Agric.*). A machine which throws hay on to a rick with an action like a catapult.

haze (*Meteor.*). A suspension of solid particles of dust and smoke, etc., reducing visibility above 1 km.

Hazeltine neutralization (*Elec. Eng.*). A method of anode neutralization used in single-stage power amplifiers.

Hazen and Williams' formula (*Eng.*). An empirical formula relating the flow in pipes and channels to the hydraulic radius and the slope, using a coefficient which varies with the surface roughness.

Hazen's uniformity coefficient (*Powder Tech.*). A measure of the range of particle sizes present in a given powder. It is defined by the equation:

$$H=A/B,$$

where A is the 60th percentile of the cumulative percentage undersize by weight of the powder, and B is the 10th.

Hb (*Chem.*). A symbol for a molecule of haemoglobin minus the iron atom.

h-bar (*Phys.*). See **Dirac's constant.**

H-beam (*Eng.*). A steel beam with a section shaped like the letter H, the cross-piece of which is relatively long. Also called **H-girder, I-beam, rolled steel joist.**

HBN (*Chem.*). A range of potent weedkillers based on 4-hydroxybenzonitrile (4-cyanophenol).

H-class insulation (*Elec. Eng.*). A class of insulating material to which is assigned a temperature of 180°C. See **class -A, -B, -C, etc., insulating materials.**

H display (*Radar*). Modified *B display* (q.v.) to include angle of elevation. The target appears as two adjacent bright spots and the slope of the line joining these is proportional to the sine of the angle of elevation.

He (*Chem.*). The symbol for *helium*.

head (*Arch.*). The capital of a column. (*Bind., Typog.*) The top edge of a volume: the top margin of a page. See **fore-edge, tail.** (*Bot.*) (1) A group of sterigmata and conidia crowded into a dense mass, of rounded outline. (2) A dense inflorescence of small, crowded, usually sessile flowers, surrounded by an involucre. (*Build.*) See **lintel.** (*Elec. Eng.*) Recording and reproducing unit for magnetic tape, containing exciting coils, a laminated core, in ring form with a minute gap. Flux leakage across this gap enters the tape and magnetizes it longitudinally. See **flux gate.** (*Eng.*) (1) Any part having the shape or position of a head, e.g., the *head* of a bolt. (2) Any part or principal part analogous to a head, e.g., the head of a hammer or a lathe. (*Geol.*) A superficial deposit consisting of angular fragments of rock, originating from the breaking up of rock by alternate freezing and thawing of its contained water, followed by downhill movement. Head is found in valleys in periglacial regions, i.e., those formerly near the edge of an ice sheet. In south-eastern England, head is also known as *coombe rock*. (*Hyd. Eng.*) A distance representing the height above a datum which would give unit mass of a fluid in a conduit the potential energy equal to the sum of its actual potential energy, its kinetic energy and its pressure energy. See **Bernoulli Law.** (*Mining*) (1) An advance main roadway driven in solid coal. (2) The top portion of a seam in the coal face. (3) The difference in air pressure producing ventilation. (4) The whole falling unit in a stamp battery, or merely the weight at the end of the stem.

head amplifier (*Cinema.*). That amplifying output of magnetic or photoelectric reproducer head in a cinema projector, when separate from the main or *gain* amplifier. (*Electronics*) Preamplifier which has to be situated geometrically adjacent to signal source. Used especially in conjunction with radiation detectors.

headband (*Bind.*). A decorative band of silk or other material at the head of a book, between the back and the cover.

head-bay (*Hyd. Eng.*). The part of a canal lock immediately above the head-gates.

head-box (*Paper*). A constant level tank situated before the machine wire on the *fourdrinier machine* which controls the amount of paper stock flowing to the machine by means of its stuff tap.

headcap (*Bind.*). The leather at the head and tail of the spine folded over in a curve, the *headband* if present, being left visible.

header (*Build.*). A whole brick which has been laid so that its length is at right angles to the face of the wall. (*Elec. Eng., Electronics*) Base for a relay, or transistor can, with hermetically sealed insulated leads. (*Eng.*) A box or manifold supplying fluid to a number of tubes or passages, or connecting them in parallel.

header joist (*Carp.*). See **trimmer.**

headframe (*Mining*). The steel or timber frame at the top of a shaft, which carries the sheave or pulley for the hoisting rope, and serves such other purposes as, e.g., acting as transfer station for hoisted ore, or as loading station for man and materials.

head-gates (*Hyd. Eng.*). The gates at the high-level end of a lock.

headgear (*Mining*). See **headframe.**

head grit (*Vet.*). See **yellowses.**

heading (*Build.*). A *heading course* (q.v.). (*Civ. Eng.*) A relatively small passage driven in the line of an intended tunnel, the latter being afterwards formed by enlarging the former. (*Mining*) Passage-way through solid coal. (*Nav.*) Horizontal direction in which a vessel or missile points. (This only coincides with the track in the absence of drift and yaw.)

heading bond (*Build.*). The form of bond in which every brick is laid as a header, each $4\frac{1}{2}$ in. face breaking joint above and below; used for footings and corbellings but not for walling.

heading chisel (*Carp., Join.*). See **mortise chisel.**

heading course (*Build.*). An external or visible course of bricks which is made up entirely of headers.

heading indicator (*Aero.*). The development of supersonic fighters, which climb at extremely steep angles, made it essential to have an even more comprehensive instrument than the *attitude indicator* (q.v.). A sphere enables compass heading to be included, thereby giving complete 360° presentation about all three axes.

heading joint (*Build.*). A joint between adjacent stones in the same course. (*Join.*) A joint between the ends of boards abutting against each other.

headings (*Min. Proc.*). 'Head tin'. Concentrate settling nearest to the entry point of a concentrating device such as a sluice or buddle. (*Textiles*) See **felling marks.**

heading tool (*Eng.*). A tool for swaging bolt heads, or for forming bolt heads, etc., in hand forging.

head-kidney (*Zool.*). In some Invertebrates, an anterior larval nephridium: in Vertebrates, the pronephric kidney, usually present in the embryo only.

head-lamp (*Light*). (1) A lamp and reflector for projecting a beam of light ahead of an automobile, locomotive, etc. (2) A lamp for strapping to the forehead.

headline (*Mining*). The line which holds an alluvial dredge up to the front of the paddock as it digs its way forward. (*Typog.*) The line of type placed at the top of a page, giving either the title of the book or the chapter heading.

head motion (*Min. Proc.*). Vibrator, a sturdy device which gives reciprocating movement to shaking tables, used in gravity methods of concentration.

head moulding (*Build.*). A moulding situated above an aperture.

head nailing (*Build.*). The method of nailing slates on a roof in which nails are driven in the slates near their heads or higher edges. Cf. *centre nailing*.

head race (*Hyd. Eng.*). A channel conveying water to a hydraulically operated machine.

head-rail (*Join.*). The horizontal member of a door-case.

head resistance (*Aero.*). See **drag.**

headroom (*Build.*). The uninterrupted height within a building on any floor, or within a staircase, tunnel, doorway, etc.

heads (*Build.*). A term applied to the tiles forming a course around the eaves.

heads-and-feet printing (*Cinema.*). The practice of printing positives with the negative film going backwards as well as forwards, to reduce wear of the negative.

headspace (*Chem.*). Vacant space above the contents of a liquid-filled container, usually left for the purpose of taking up gas.

headstock (*Eng.*). In general, a device for supporting the end or head of a member or part; e.g., (1) the part of a lathe that carries the spindle, (2) the part of a planing machine that supports the cutter or cutters, (3) the supports for the gudgeons of a wheel, (4) the movable head of some measuring machines. (*Textiles*) The part of a machine which contains the main gearing and the drive for a beam.

head tree (*Carp.*). A timber block placed on the top of a post, so as to provide increased bearing-surface.

head-up display (*Aero.*). The projection of instrument information on to the windscreen, in the manner of the *reflector sight* (q.v.), so that the pilot can keep a continual lookout and receive flight data at the same time. Used originally with fighter interception and attack radar, later adapted for blind approach, landing, and general flight data. Abbrev. **HUD.**

head valve (*Eng.*). The delivery valve of a pump, as distinct from the suction or *foot valve* (q.v.).

head wall (*Build.*). A wall built in the same plane as the face of a bridge arch.

headway (*Build.*). *Headroom* (q.v.).

headworks (*Eng.*). In a hydroelectric scheme, a dam forming a reservoir or a low weir across a river or stream, which possesses the necessary intakes and control gear to divert the water into an aqueduct.

heald (*Textiles*). (1) Part of the loom mechanism used to raise and lower the warp in a tappet or dobby loom. Comprises eyes (*mails*), formed of coated twine or wire through which warp ends are drawn. (2) The shaft upon which a large number of healds are mounted.

healding (*Weaving*). See **looming.**

healing (*Build.*). The operation of covering a roof with tiles, lead, etc.

health physics (*Radiol.*). Branch of radiology concerned with health hazards associated with ionizing radiations, and protection measures required to minimize these. Personnel employed for this work are *health physicists* or *radiological safety officers.*

heap leaching (*Min. Proc.*). This, perhaps aided by heap roasting, is the dissolution of copper from oxidized ore by solvation with sulphuric acid. The resulting liquor is run over scrap iron to precipitate out its copper.

heap sampling (*Mining, Min. Proc.*). Reduction of large sample of ore to obtain a small representative fraction for analysis or test. A cone-shaped heap is formed and divided into four equal quarters. Two are discarded, and two, after some crushing, are re-coned. This is repeated until a sufficiently small sample is obtained.

heapstead (*Mining*). The buildings and surface works around a colliery shaft.

hearing (*Acous.*). The subjective appreciation of externally applied sounds.

hearing loss (*Acous.*). At any frequency, the hearing loss of a partially deaf ear is the ratio in decibels of its threshold of hearing to that for a normal ear. (This term is synonymous with *deafness* in scientific work.) 100 times the ratio of the hearing loss to the number of decibels between the normal thresholds of hearing and of feeling is termed the *percentage hearing loss* for the specified frequency.

heart (*Zool.*). A hollow organ, with muscular walls, which by its rhythmic contractions pumps the blood through the vessels and cavities of the circulatory system.

heart-block (*Med.*). The condition in which a lesion of the special tissue that conducts the contraction impulse from the auricle to the ventricle results in a different rate of contraction of these two parts of the heart.

heart bond (*Build.*). A form of bond having no through-stones, headers consisting of a pair of stones meeting in the middle of the wall, the joint between them being covered by another header stone.

heart-burn (*Med.*). A burning sensation in the mid-line of the chest, usually associated with *dyspepsia.*

heart cam (*Horol.*). A cam in the form of a heart, used in stop watches and chronographs to bring the recording hand instantly back to zero, on pressing the button. Also called **heart piece.**

heart failure (*Med.*). Condition in which for some reason (e.g., *atherosclerosis,* q.v.) the heart muscles fail to maintain an adequate flow of blood to all the body tissues.

hearth (*Met.*). Floor of reverberatory, open hearth, cupola or blast furnace, made of refractory material able to support the charge and collect molten products for periodic removal.

hearting (*Build.*). The operation of building the inner part of a wall, between its facings.

heart shake (*For.*). A *shake* starting at the heart of a log.

heart-shaped reception (*Radio*). Same as **cardioid reception.**

heartwater (*Vet.*). A disease of cattle, sheep, and goats, occurring in parts of Africa, caused by *Cowdria ruminantium* (*Rickettsia ruminantium*), and transmitted by ticks of the genus *Amblyomma.* Characterized by fever, nervous symptoms, and often death.

heart wood (*Bot.*). The dense and often dark-coloured wood which lies in the inner part of a trunk or branch, making up the bulk of such a member; *heart wood* furnishes good timber, both for colour and durability.

heat (*Phys.*). Heat is energy in the process of transfer between a system and its surroundings as a result of temperature differences. However, the term is still used also to refer to the energy contained in a sample of matter; also for *temperature*, e.g., forging or welding *heat.* For some of the chief branches in the study of heat see **calorimetry, heat units, internal energy, latent heat, mechanical equivalent of heat, radiant heat, radiation, specific heat capacity, temperature, thermal conductivity, thermometry.** (*Zool.*) The period of sexual desire.

heat balance (*Chem. Eng.*). A statement which relates for a chemical process or factory all sources of heat and all uses. The allowance of heats of reaction, solution and dilution must be included as well as conventional sources for fuel, steam, etc. If done graphically, it is usually called a *Sankey diagram.* Also called **energy balance.** (*Eng.*) See **energy balance.** (*Heat*)

Evaluation of operating efficiency of a furnace or other appliance, the total heat input being apportioned as to heat in the work, heat stored brickwork, or refractory materials, loss by conduction, radiation, unburnt gases in waste products, sensible heat in dry flue gases and latent heat of water vapour, thus determining the quantity and percentage of heat usefully applied and the sources of heat losses.

heat body (*Paint.*). The process of polymerizing and thus thickening varnish oils by heat alone. See **air-blowing, body.**

heat coil (*Elec. Eng.*). A small protective resistance coil, which heats on prolonged excess current in telegraph lines, melts a small quantity of solder, and disconnects the line, so preventing possible damage to apparatus.

heat density (*Heat*). Weight and pressure of live gases in heating chambers of industrial furnaces, upon which the rate of heat transfer depends.

heat detector (*Heat*). An indirect-acting thermostat for operation in conjunction with a gas-flow control valve, and for controlling working temperatures in furnaces and heating appliances up to about 1000°C.

heat drop (*Eng.*). See **adiabatic heat drop.**

heat economizer (*Met.*). Regenerator. See **heat regenerator.**

heater (*Electronics*). Conductor carrying current for heating an equipotential cathode, generally enclosed by the cathode.

heater box (*Bind.*). The heated part of a blocking press. This is usually electrically heated and thermostatically controlled.

heater platen (*Print.*). A grooved platen sliding into the heater box of a blocking press and holding the die.

heat exchange (*Chem. Eng.*). The process of using two streams of fluid for heating or cooling one or the other either for conservation of heat or for the purpose of adjusting process streams to correct processing temperatures.

heat exchanger (*Chem. Eng.*). Process equipment in which the process of *heat exchange* (q.v.) is carried out. See **double pipe exchanger, fin-fan exchanger, fixed-tube sheet exchanger, floating head exchanger, shell and tube exchanger, Votator.** See also **air preheater, calorifier.**

heat filter (*Cinema.*). Shutter device between the condenser and film in a projector in case of accidental stoppage.

heat flush (*Chem.*). Separation of helium-3 and helium-4 by heat flow in fluid helium.

heat flux (*Chem. Eng.*). The total flow of heat in *heat exchange* (q.v.); to be precise this must be qualified, e.g., by expressing a time and if appropriate an area, but these qualifications are varied to suit cases.

heather blindness (*Vet.*). See **contagious ophthalmia.**

heating coefficient (*Heat*). Sometimes, the temperature rise of a piece of electrical apparatus; expressed in degrees per watt of heat dissipated from a surface.

heating curves (*Met.*). Curves obtained by plotting time against temperature for a metal heating under constant conditions. The curves show the absorption of heat which accompanies melting and arrest points marking polymorphic changes in pure metals and various transformations in alloys.

heating depth (*Elec. Eng.*). Thickness of skin of material which is effectively heated by dielectric or eddy-current induction heating, or radiation.

heating element (*Elec. Eng.*). The heating-resistor, together with its former, of an electric heater, electric oven, or other device, in which heat is produced by the passage of an electric current through a resistance.

heating inductor (*Elec. Eng.*). Conductor, usually water-cooled, for inducing eddy currents in a charge, workpiece or load. Also called **applicator, work coil.**

heating limit (*Elec. Eng.*). See **thermal limit.**

heating muff (*Aero.*). A device for providing hot air, consisting of a chamber surrounding an exhaust pipe or jet pipe.

heating resistor (*Elec. Eng.*). The wire or other suitable material used as the source of heat in an electric heater.

heating time (*Electronics*). Time required after switching on, before a valve cathode or other piece of equipment reaches normal operating temperature.

heat-insulating concrete (*Eng.*). High-alumina cement and a lightweight aggregate, e.g. kieselguhr, diatomic earth, or vermiculite, to reduce heat transfer through furnace walls, etc.

heat insulation (*Build.*). The property, possessed in varying degrees by different materials, of impeding the transmission of heat.

heat liberation rate (*Eng.*). A measure of steam boiler performance, expressed in kJ/l h (=26·8 Btu/ft^{3}h).

heat of formation (*Chem.*). The net quantity of heat evolved or absorbed during the formation of one mole of a substance from its component elements in their standard states.

heat of solution (*Chem.*). The quantity of heat evolved or absorbed when one mole of a substance is dissolved in a large volume of a solvent.

heat pump (*Eng.*). A refrigeration machine used primarily for heating. Heat is extracted from a large low-temperature body, e.g., the earth, a well, lake, or stream, by the expanded refrigerant, which is then mechanically compressed and discharged to a heat exchanger in or near the region to be heated.

heat regenerator (*Chem. Eng.*). A matrix of metal which is alternately heated by the waste gases from an industrial process and used to heat up the incoming gases.

heat-resisting steel (*Met.*). A steel with high resistance to oxidation, and moderate strength at high temperatures, i.e., above 500°C. Alloy steels, of a wide variety of compositions, which usually contain large amounts of one or more of the elements chromium, nickel, or tungsten, are used.

heat run (*Elec. Eng.*). A test in which an electric machine or other apparatus is operated at a specified load for a long period to ascertain the temperature which it reaches.

heat-set ink (*Print.*). Printing ink formulated to set by heating the surface of the stock.

heat sink (*Elec. Eng.*). Usually metal plate specially designed (e.g., with matt black fins) to conduct (and radiate) heat from an electrical component, e.g., a transistor.

heat spot (*Zool.*). An area of the skin sensitive to heat owing to the presence of certain nerve-endings beneath the skin.

heat-stroke (*Med.*). Heat hyperpyrexia. The combination of coma, convulsions, a high temperature, and other symptoms, as a result of exposure to excessive heat.

heat transfer (*Heat*). See **conduction of heat, convection of heat, radiation.**

heat transfer coefficient (*Chem. Eng.*). The rate of heat transfer q between two phases may be expressed as

$$q=hA(T_1-T_2),$$

where A is the area of the phase boundary,

$(T_1 - T_2)$ is their difference in temperature, and the heat transfer coefficient *h* depends on the physical properties and relative motions of the two phases. See **Newton's law of cooling.**

heat transfer salt (*Chem. Eng.*). Molten salts used as a heating or quenching medium. Usually mixtures of sodium or potassium nitrate or nitrite, range 200°C to 600°C. Abbrev. H.T.S.

heat treatment (*Met.*). Generally, any heating operation performed on a solid metal, e.g., heating for hot-working, or annealing after cold-working. Particularly, the thermal treatment of steel by normalizing, hardening, tempering, etc.; used also in connexion with precipitation-hardening alloys, such as those of aluminium.

heat units. See **calorie, British thermal unit, joule, pound-calorie;** also **calorific value.**

heave (*Geol., Mining*). (1) The slip along a fault (horizontal components). (2) Rising of the floor of a mine.

heavier-than-air aircraft (*Aero.*). See **aerodyne.**

Heaviside-Campbell bridge (*Elec. Eng.*). An electrical network for comparing mutual and self inductances.

Heaviside layer (*Phys.*). See **E-layer.**

Heaviside-Lorentz units (*Elec.*). Rationalized CGS Gaussian system of units, for which corresponding electric and magnetic laws are always similar in form. See **rationalized units.**

Heaviside unit function (*Telecomm.*). Step change in a magnitude, with an infinite rate of change, required in pulse analysis and transient response of circuits.

heavy-aggregate concrete (*Nuc. Eng.*). That containing very dense aggregate material in place of some or all of the usual gravel, so increasing its gamma-ray absorption coefficient. Also used for appropriate general purposes (e.g., sea walls).

heavy chemicals (*Chem.*). Those basic chemicals which are manufactured in large quantities, e.g. sodium hydroxide, chlorine, nitric acid, sulphuric acid.

heavy ground (*Mining*). Unstable roof rock requiring special care and support.

heavy hydrogen (*Chem.*). Same as **deuterium.**

heavy liquids (*Min. Proc., etc.*). Liquids, organic or solutions of heavy salts, of relative densities adjustable in the range 1·4 to 4·0, used to separate ore constituents into relatively heavy (sink) and light (float) fractions with fair precision. They include tetrachloromethane, saturated calcium chloride solution, bromoform, methyl iodide, and Clerici's solution.

heavy media separation (*Min. Proc.*). Dense media separation. Method of upgrading ore by feeding it into liquid slurry of intermediate density, the heavier fraction sinking to one discharge arrangement and the light ore overflowing. Used to remove shale from coal (*sink-float process*) and to reject waste from ore.

heavy minerals (*Geol.*). In the systematic examination of a sediment, the small quantity of accessory minerals it contains is separated from the predominant quartz by passing the whole through bromoform. Quartz and feldspar are lighter than the latter and therefore float on its surface; but the accessory minerals, such as zircon, rutile, anatase, brookite, kyanite, and iron ores, sink to the bottom of the container and may be separated for detailed examination. The grains heavier than bromoform constitute the *heavy minerals*, and provide valuable evidence of the provenance of the sediment.

heavy particle (*Nuc.*). Fundamental particle heavier than the neutron. See **hyperon.**

heavy spar (*Min.*). See **barytes.**

heavy water (*Chem.*). Deuterium oxide, or water containing a substantial proportion of deuterium atoms (D_2O or HDO).

heavy-water reactor (*Nuc. Eng.*). One using *heavy water* as moderator, achieving sufficiently high neutron flux to manufacture ^{60}Co and other radioisotopes for sources of β- and γ-rays.

heavyweight concrete (*Eng.*). See **heavy-aggregate concrete.**

heazlewoodite (*Min.*). A sulphide of nickel, crystallizing in the trigonal system.

hebephrenia (*Psychiat.*). A common form of *schizophrenia*, characterized by silliness, untidiness, hallucinations, and delusions.

Heberden's nodes (*Med.*). Small bony knobs occurring on the bones of fingers of the old.

hebetate (*Bot.*). Bearing a blunt or soft point.

hebetude (*Med.*). Lethargy and mental dullness, with impairment of the special senses.

heck (*Join.*). A door latch. (*Textiles*) The traverse guide positional between the creel and the beam in a cotton-warping machine.

hectare (*Surv.*). A metric unit of area, equal to 100 *ares* (q.v.). 1 hectare = 2·47 acres; 100 hectares = 1 km^2 = 0·386 mile^2.

hecto-. Metric prefix meaning ×100.

hectobar (*Eng.*). See **bar.**

hectocotylized arm (*Zool.*). In some male *Cephalopoda*, one of the tentacles modified for the purpose of transferring sperm to the female.

hectographic printing (*Print.*). See **gelatin printing.**

hectometric waves (*Radio*). See **medium frequency.**

hectorite (*Min.*). A rare lithium-bearing mineral in the montmorillonite group of clay minerals.

heddle (*Weaving*). The name formerly applied to a heald shaft. Term still used for handlooms.

hedenbergite (*Min.*). An important calcium-iron pyroxene, $CaFeSi_2O_6$, occurring as black crystals, and also as a component molecule in many of the rock-forming clinopyroxenes.

hedleyite (*Min.*). Bismuth telluride, crystallizing in the trigonal system.

Hedley's dial (*Surv.*). A form of compass adapted for taking inclined sights; it consists of a pair of sighting vanes (or a telescope capable of rotation about a horizontal axis), carrying with them a vertical graduated arc moving over a fixed reference mark.

hedonic glands (*Zool.*). In some Reptiles, glands of the skin which secrete a pleasant-smelling substance during the breeding season.

hedrioblast (*Zool.*). See **medusoid.**

heel (*Build.*). The lower end of a timber, as opposed to the head, or upper end. (*Chem. Eng.*) The residual remaining in a vessel after removal of the bulk in batch distillation. (*Elec. Eng.*) See **leaving edge.** (*Eng.*) A projection on a machine component on which the component thrusts or about which it pivots. (*Hyd. Eng.*) The angle of inclination of a floating vessel from the vertical. (*Ships*) An angle of transverse inclination arising from a force external to the ship, e.g., wind. Cf. *list.* (*Tools*) The back end of a plane.

heeling error (*Ships*). *Deviation* (q.v.) at a magnetic compass occurring when a vessel heels. It can be eliminated by a group of vertical permanent magnets.

heel-post (*Hyd. Eng.*). The vertical post at one side of a lock-gate, about which the lock-gate swings. Also called a **quoin-post.**

heel strap (*Build.*). A strap fastening the foot of a principal rafter to the tie-beam in a timber truss. The strap is threaded at the ends to take a cover plate so that the joint is completely encircled.

Heenan dynamatic dynamometer (*Eng.*). An eddy-current dynamometer in which the steel rotor has the shape of a gear-wheel and rotates in a cylindrical stator, with smooth bore, surrounded by d.c. excited field coils.

Heenan hydraulic torque meter (*Eng.*). An arrangement of input and output shafts with vanes forming a concentric annular working compartment through which oil is forced to transmit torque, the differential pressure across the vanes being a measure of the torque.

Hefner candle (*Light*). The standard of luminous intensity formerly used in Germany, being the light produced by a Hefner lamp burning under specified conditions. It was equal to 0·9 international candle (~0·9 candela).

Hefner lamp (*Light*). A flame lamp of specified dimensions burning amyl acetate. See **Hefner candle.**

Hegar's dilators (*Med.*). A series of bougies of varying sizes for dilating the opening of the uterus into the vagina.

Hegman gauge (*Powder Tech.*). A *fineness-of-grind gauge* (q.v.) used widely in the British paint industry. The groove depth from four-thousands of an inch (1/10 mm) to zero is marked off arbitrarily into a linear scale of 8 units. The fineness of a paint pigment is often quoteed by reference to the *Hegman gauge number* at which scoring occurs.

Hehner's test (*Chem.*). A test for the presence of formaldehyde in milk. It is based upon the appearance of a blue or violet ring when the milk is mixed with a dilute ferric chloride solution, and concentrated sulphuric acid is added to form a layer beneath the milk.

Heidenhain's azan stain (*Micros.*). A polychrome microanatomical stain which is a modification of Mallory, but gives more precise differentiation of tissues. Technique includes use of azocarmine, aniline blue W.S. and orange G.

Heidenhain's iron haematoxylin stain (*Micros.*). A microanatomical and cytological stain which can be used as a basis for various polychromatic stains, e.g., Masson's trichrome method; or with a counterstain, e.g., eosin. Iron alum is used as a mordant followed by staining with the haematoxylin and differentiation.

height board (*Carp.*). A gauge for the treads and risers of a timber staircase.

height control (*Radar, TV*). The means of varying the amplitude of the vertical deflection of a *CRO*.

height of a transfer unit (*Chem. Eng.*). A measure of the separating efficiency of packed columns for mass transfer operations. It is the height of packed column used for liquid-liquid extraction in which one theoretical transfer unit takes place. Abbrev. **H.T.U.**

height of instrument (*Surv.*). (1) In levelling or trigonometrical survey work, the vertical distance of the plane of collimation of the level, or the horizontal axis of the theodolite above datum. (2) In tacheometry, the vertical distance of the horizontal axis of the instrument above ground-level. Abbrev. **H.I.**

height power factor (*Aero.*). The ratio of the maximum power, or thrust, of an aero-engine at a given height to that developed at sea-level *standard atmosphere*.

HE igniter (*Aero.*). The spark unit of a *high-energy ignition* (q.v.) system.

heiligtag effect (*Nav.*). Error arising in a bearing because of reception of waves from a transmitter taking different paths. Also called **wave-interference error.**

Heil oscillator (*Electronics*). Early form of velocity-modulated transit time oscillator; forerunner of the **klystron.** Also called **coaxial-line tube, Heil tube.**

Heine-Medin disease (*Med.*). Acute anterior *poliomyelitis* (q.v.).

Heisenberg force (*Nuc.*). Short-range exchange force between nucleons which changes sign if both space or spin coordinates are interchanged. See **short-range forces.**

Heisenberg principle (*Phys.*). See **uncertainty principle.**

Heising modulation (*Electronics*). Constant-current modulation, arising from one valve driven by signal and another valve driven by carrier, having their anodes fed through the same inductor; the modulated carrier is taken from the anode circuit by capacitive or inductive coupling.

Helberger furnace (*Elec. Eng.*). A form of electric crucible furnace.

held water (*Geol.*). That kept above the natural water table through capillary force. U.S. **free water.**

heleoplankton (*Ecol.*). The mixture of true plankton and benthon found in small ponds and pools.

Hele-Shaw apparatus (*Hyd.*). A form of apparatus used to study the motion of a viscous fluid.

heliacal rising, heliacal setting (*Astron.*). The rising or setting of a star or planet, simultaneously with the rising or setting of the sun. It was much observed in ancient times as a basis for a solar calendar for agricultural purposes, but is now obsolete.

helianthine (*Chem.*). $(CH_3)_2N{\cdot}C_6H_4{\cdot}N{=}N{\cdot}C_6H_4{\cdot}SO_3H$, 4 - dimethylamino - azobenzene - 4 - sulphonic acid, a chrysoidine dye. The sodium salt of this acid is methyl orange, used as an indicator.

heli-arc welding (*Eng.*). A welding process in which helium is used to shield the weld area from contamination by atmospheric oxygen and nitrogen.

heliatron (*Electronics*). Voltage-tuned microwave oscillator in which electron beam follows helical path.

helical aerial (*Telecomm., Radar*). A broadband aerial in the form of a helix. Normally used as an end-fire aerial, it can transmit and receive a circularly polarized signal.

helical gears (*Eng.*). Gear-wheels in which the teeth are not parallel with the wheel axis, but *helical* (i.e., parts of a helix described on the wheel face), being therefore set at an angle with the axis. See **double-helical gears.**

helical hinge (*Join.*). A type of hinge used for hanging swing-doors which have to open both ways.

helical spring (*Eng.*). A spring formed by winding wire into a helix along the surface of a cylinder; sometimes erroneously termed a *spiral spring* (q.v.).

helical tip speed (*Aero.*). The resultant velocity of the tip of an airscrew blade, which is a combination of the linear speed of rotation and the flight speed.

Helicidae (*Zool.*). A large family of terrestrial, air-breathing gasteropods.

helicitic (*Geol.*). Descriptive of the S-shaped trails of inclusions found in the minerals of some metamorphic rocks, especially abundant in garnet and staurolite. These inclusion trails are often continuous with the mineral alignments of surrounding crystals and may help to indicate the crystallization history of the rock.

helicoid (*Biol.*). Coiled like a flat spring. **(*Maths.*)**

A surface generated by a screw motion of a curve about a fixed line (axis); i.e., the curve rotates about the axis and moves in a direction of the axis in such a way that the ratio of the rate of rotation and rate of translation is constant. This constant multiplied by 2π is called the *pitch* of the helicoid. See **right helicoid.**

helicoid cyme (*Bot.*). A sympodium in which the branches all develop on the same side of the relatively main axis, but not in the same plane.

helicoid dichotomy (*Bot.*). A branch system with repeated dichotomies, giving each time a weak and a strong branch, the latter always on the same side.

helicopter (*Aero.*). A *rotorcraft* (q.v.) whose main rotor(s) are power driven and rotate about a vertical axis, and which is thus capable of vertical take-off and landing.

helicotrema (*Zool.*). An opening at the apex of the cochlea, by which the scala vestibuli communicates with the scala tympani.

helio-. Prefix from Gk. *hēlios*, sun.

heliocentric parallax (*Astron.*). See **annual parallax.**

heliodor (*Min.*). A beautiful variety of clear yellow beryl occurring in S.-W. Africa.

heliograph (*Surv.*). An instrument similar to the heliostat but fitted with a spring device by which it can be made to flash long or short flashes.

Helio-Klischograph (*Print.*). An electronically operated machine for engraving photogravure cylinders direct from the copy, and incorporating electronic *colour correction* (q.v.).

heliometer (*Astron., etc.*). An instrument for determining the sun's diameter and for measuring the angular distance between two celestial objects in close proximity. It consists of a telescope with its object glass divided along a diameter, the two halves being movable, so that a superposition of the images enables a value of the angular separation to be deduced from a reading of the micrometer.

helion filament (*Elec. Eng.*). An electric-lamp filament consisting of carbon with an outer coating of silicon.

heliophyte (*Bot.*). A plant able to live with full exposure to the sun.

heliosciophyte (*Bot.*). A plant which can live in shade, but does better in the sun.

heliostat (*Astron.*). An instrument designed on the same principle as the coelostat, but with certain modifications that make it more suitable for reflecting the image of the sun than for use on a larger region of the sky; hence used, in conjunction with a fixed instrument, especially for photographic and spectroscopic study of the sun. (*Surv.*) An instrument used to reflect the sun's rays in a continuous beam, so as to serve as a signal enabling a station to be sighted over long distances.

heliotaxis, heliotropism (*Biol.*). Response or reaction of an organism to the stimulus of the sun's rays. *adj.* **heliotactic, heliotropic.**

heliotherapy (*Med.*). The treatment of disease by the exposure of the body to the sun's rays.

heliotrope (*Min.*). See **bloodstone.** (*Surv.*) A form of *heliograph* (q.v.).

heliotropin (*Chem.*). See **piperonal.**

heliotropism (*Biol.*). See **heliotaxis.** More particular response by growth curvature.

Heliozoa (*Zool.*). An order of *Sarcodina*, the members of which generally occur in fresh water and have numerous fine radial axopodia, which do not anastomose; the ectoplasm is usually vacuolated; no shell or central capsule.

Helipot (*Elec. Eng.*). TN for precision multiturn potential divider, resettable to within 0·1%.

helium (*Chem.*). Inert gaseous element, symbol He, at. no. 2, r.a.m. 4·0026, with extremely stable nucleus identical to α-particle. It liquefies at temperatures below 4K, and undergoes a phase change to a form known as *liquid helium II* at 2·2K. The latter form has many unusual properties believed to be due to a substantial proportion of the molecules existing in the lowest possible quantum energy state; see **superfluid.** Liquid helium is the standard coolant for devices working at cryogenic temperatures.

helium diving-bell (*Civ. Eng.*). A *diving-bell* (q.v.) in which the nitrogen in the compressed-air is replaced by helium, thus reducing tendency to the *bends* (q.v.), and permitting effective operation at greater depths than normal. Used in rescuing men from submarines.

helium-neon (He-Ne) laser (*Phys.*). One using a mixture of helium and neon, energized electrically. It can give continuous operation in powers of the order of milliwatts.

helium stars (*Astron.*). Those stars, of spectral type B in the Harvard classification, whose spectrum shows only dark lines, in which those due to the element helium predominate.

helix. A line, thread, wire, or other structure curved into a shape such as it would assume if wound in a single layer round a cylinder; a form like a screw-thread. (*Maths.*) A spiral curve lying on a cone or cylinder and cutting the generators at a constant angle. (*Phys.*) Spiral path as followed by charged particle in magnetic field. (*Telecomm., Radar*) See **helical aerial.**

hell-box (*Typog.*). A container for broken type and, in the type-case, a compartment to hold also wrong founts for correct distribution later.

Heller's test (*Chem.*). A test for the identification of albumins, based upon the fact that many proteins are precipitated when their aqueous solution is floated over conc. mineral acids.

Hellesen cell (*Elec. Eng.*). A dry cell with zinc and carbon electrodes and a depolarizer of manganese dioxide.

Hell system (*Teleg.*). Form of *mosaic telegraphy* (q.v.) in which each character is represented by a fixed number of unit elements, usually 49. The received characters are formed by a rotating spiral, continuously inked, being brought into contact under the control of the unit elements, with a travelling paper tape. Also **Hellschreiber system.**

helm (*Meteor.*). The *helm wind, cloud,* and *bar* constitute a phenomenon on the western slope of Crossfell Range, Cumberland. The cold wind blows down under the cloud and ends under the bar, which is a further parallel and whirling cloud. See **orographic ascent.**

Helmert's formula (*Phys.*). An expression giving the value of g, the acceleration due to terrestrial gravity, for a given latitude and altitude:

$$g = 9{\cdot}806\,16 - 0{\cdot}025\,928 \cos 2\lambda + 0{\cdot}000\,069 \cos^2 2\lambda - 0{\cdot}000\,003\,086H,$$

where λ is the latitude and H is the height in metres above sea-level, g being in m/s^2.

helmet (*Civ. Eng.*). A cast-iron *dolly* (q.v.) used at the head of a reinforced concrete pile, the two being separated by a sand cushion. (*Elec. Eng.*) The iron cap or pole piece sometimes fitted to a lodestone to form an armed lodestone. (*Zool.*) In some species of *Bucerotidae* (Hornbills), a casque-like structure at the base of the upper bill: in Insects, the galea.

Helmholtz coils (*Elec. Eng.*). Pair of identical

compact coaxial coils separated by a distance equal to their radius. These give a uniform magnetic field over a relatively large volume at a position midway between them.

Helmholtz double layer (*Chem., etc.*). Electrical double layer. This assumes an interphase between a relatively insoluble solid and an ambient ionized liquid, in which oppositely charged ions tend to concentrate in layers. See also **electrokinetic potential.**

Helmholtz galvanometer (*Elec. Eng.*). A type of *tangent galvanometer* in which an approximately uniform field is produced by having two coils parallel to each other, a few inches apart.

Helmholtz resonance (*Acous.*). The type of acoustic resonance arising in a flask-shaped cavity (*Helmholtz resonator*) and, by extension, in flat cavities, such as a recessed window or the cavity in front of a microphone diaphragm.

Helmholtz resonator (*Acous.*). An acoustic resonator in which the mass reactance of a short column of air neutralizes at a fairly definite frequency the reactance of the stiffness of the volume contained in an enclosure, which communicates with the open air only through the said column. Used at one time for analysing complex sounds.

Helmholtz's theorem (*Elec. Eng.*). See **Thévenin's theorem.**

Helminthes (*Zool.*). A name formerly used in classification to denote a large group of worm-like Invertebrates now split up into *Platyhelminthes, Nematoda*, and smaller groups.

helminthiasis (*Med.*). Infestation of the body with parasitic worms.

helminthoid (*Bot.*). Shaped like a worm.

helm wind (*Meteor.*). See **helm, orographic ascent.**

Helobiae (*Bot.*). See **Naiadales.**

helophyte (*Bot.*). A bog plant.

helotism (*Zool.*). See **dulosis.**

helve (*Tools*). The handle of an axe, hatchet, or similar chopping tool.

helve hammer (*Eng.*). A power hammer used for plating, etc., in which the tup is secured to one end of a helve or beam, which is supported on a pivot and actuated by a crank or cam mechanism.

Helvellales (*Bot.*). An order of the *Ascomycetes*, in which the ascocarp bears the ascogenous layer freely exposed, and the asci are parallel to each other.

Helvetia leather (*Leather*). See **crown leather.**

helvite (*Min.*). A manganese beryllium silicate also containing sulphur. It crystallizes in the cubic system and occurs in granite pegmatites and skarns.

hem-, hema-, hemo-. Variant (espec. U.S.) of *haem-, haema-, haemo-*.

HEM (*Elec.*). Abbrev. for **hybrid electromagnetic wave.**

hemelytra (*Zool.*). The fore wings of *Heteroptera*, which are thickened at their bases like elytra while the distal portion remains membranous.

hemeralopia (*Med.*). Day-blindness, objects sometimes being better seen in a dull light. The term is wrongly used also to mean night-blindness (see **nyctalopia**).

hemi-. Prefix from Gk. *hemi*, half.

hemianaesthesia, hemianesthesia (*Med.*). Loss of sensibility to touch on one side of the body; usually connotes also loss of sensibility to pain and temperature.

hemianalgesia (*Med.*). Loss of sensibility to pain on one side of the body.

hemiangiocarpic, hemiangiocarpous (*Bot.*). Descriptive of the fruit body of a fungus when the hymenium begins its development inside a closed chamber, but is exposed at maturity.

hemianopia, hemianopsia (*Med.*). Loss of half the field of vision.

hemiascospore (*Bot.*). A spore of a hemiascus.

hemiascus (*Bot.*). An atypical ascus, containing several spores.

hemiataxy, hemiataxia (*Med.*). Loss of co-ordination of the muscles of one side of the body.

hemiatrophy (*Med.*). Wasting of muscles of one side of the body, or of one half of a part of the body.

hemiballism (*Med.*). Involuntary violent twitching and jerking affecting one side of the body. Condition due to a hypothalamic lesion.

Hemibasidii (*Bot.*). A subclass of the *Basidiomycetes* in which the basidia develop from a resting spore, and not from the mycelium.

hemiblastula (*Zool.*). An abnormal blastula produced by the destruction of one of the two primary blastomeres and the cultivation of the other.

hemibranch (*Zool.*). The single row of gill lamellae or filaments, borne by each face of a gill arch in Fish: a gill arch with respiratory lamellae or filaments on one face only.

hemicarp (*Bot.*). See **mericarp.**

hemicephalous (*Zool.*). Said of dipterous larvae which possess a reduced head or jaw capsule which is incomplete posteriorly and can be retracted into the thorax.

Hemichorda (*Zool.*). A subphylum of *Chordata*, lacking any bony or cartilaginous skeletal structures; without tail or atrium; having a reduced notochord in the pre-oral region, and a superficial central nervous system; the three primary coelomic cavities persist in the adult. Also **Hemichordata, Adelochordata.**

hemichorea (*Med.*). Chorea on one side of the body.

hemicolloid (*Chem.*). A particle up to $2{\cdot}5 \times 10^{-8}$ m in length; 20–100 molecules.

hemicrania (*Med.*). See **migraine.**

hemicryptophyte (*Bot.*). A plant which develops its resting buds just above or below the surface of the soil, where they are protected by litter or by a thin layer of soil.

hemicrystalline rocks (*Geol.*). Those rocks of igneous origin which contain some interstitial glass, in addition to crystalline minerals. Cf. *holocrystalline rocks*.

hemicyclic (*Bot.*). Said of a flower which has some of its parts inserted in spirals and some in whorls.

hemiembryo (*Zool.*). An abnormal embryo, of which one half in the sagittal plane is complete and normal while the other half is composed of dead tissue; obtained by the destruction of either of the first two blastomeres and the cultivation of the other.

hemigastrula (*Zool.*). An abnormal embryo representing half of a complete gastrula, together with dead tissue obtained by the destruction of one of the first two blastomeres and the cultivation of the other.

hemignathous (*Zool.*). Having jaws of unequal length.

hemihedral forms (*Crystal.*). Crystal forms, the faces of which are parallel to, and have the same indices as, certain *holohedral* (q.v.) forms; but in consequence of a lower degree of symmetry, half the faces are suppressed. Thus the tetrahedron (four faces) in the cubic system may be regarded as the hemihedral form of the octahedron (eight faces).

hemihydrate plaster. See **plaster of Paris.**

hemikaryotic (*Cyt.*). Having the haploid number of chromosomes. *n.* **hemikaryon.**

hemi-melus (*Med.*). A congenital malformation in which only half the limbs are formed.

Hemimetabola (*Zool.*). See **Exopterygota.**

hemimetabolic (*Zool.*). Having an incomplete or partial metamorphosis.

hemimorphism (*Min.*). The development of polar symmetry in minerals, in consequence of which different forms are exhibited at the ends of bi-terminated crystals. Hemimorphite shows this character in a marked degree.

hemimorphite (*Min.*). An orthorhombic hydrous silicate of zinc; one of the best minerals for demonstrating polar symmetry, the two ends being distinctly dissimilar. U.S. **calamine** or **electric calamine.**

hemimorula (*Zool.*). An abnormal embryo with animal and vegetative cells and a segmentation cavity obtained by the destruction of one of the first two blastomeres and the cultivation of the remaining one.

Hemimyaria (*Zool.*). See **Salpidae.**

hemiparasite (*Bot.*). See **facultative saprophyte.**

hemipenes (*Zool.*). In *Squamata*, the paired eversible copulatory organs.

hemiplegia (*Med.*). Paralysis of one side of the body.

hemipneustic (*Zool.*). Of Insects, having one or more pairs of spiracles closed. Cf. *holopneustic.*

Hemiptera (*Zool.*). An order of *Paraneoptera*, having two pairs of wings of variable character; the mouth-parts are symmetrical and adapted for piercing and sucking, being formed into a beak with fused palpless stylets; in some forms, the females are wingless; many are ecto-parasitic, others feed on plant juices. Bugs, Cicadas, Aphids, Plant Lice, Scale Insects, Leaf Hoppers, White Flies, Black Flies, Green Flies, Cochineal Insects. **See Rhynchota.**

hemipterygoid (*Zool.*). In some Birds, a separated portion of the pterygoid which fuses with the palatine.

hemisaprophyte (*Bot.*). See **facultative saprophyte.**

hemisection (*Surg.*). The cutting through of half of a part, e.g., of the spinal cord.

hemisome (*Zool.*). The lateral symmetrical half of an animal.

Hemisphaeriales (*Bot.*). An order of the *Ascomycetes*, most of whose members are leaf parasites. The fruit-body is a stroma with a hyphal, or pseudoparenchymatous, base and a tougher upper part which bursts to form a pseudo-ostiole.

hemisphere. The half of a sphere, obtained by cutting it by a plane passing through the centre. As applied to the earth, the term usually refers to the *Northern* or the *Southern hemisphere*, the division being by the equatorial plane. (*Zool.*) One of the cerebral hemispheres (see **cerebrum**).

hemispore (*Bot.*). The terminal cell of a fungal hypha, which divides to form deuteroconidia.

hemithyroidectomy (*Surg.*). Removal of one half of the thyroid gland.

hemitropic winding (*Elec. Eng.*). See **half-coiled winding.**

hemitropous (*Bot.*). Of Insects with medium-length tongues, suited to pollinate flowers with completely concealed nectar.

hemorrhage (*Med.*). See **haemorrhage.**

hemorrhoid (*Med.*). See **haemorrhoid.**

hemp (*Bot.*). See **cannabis.** (*Textiles*) General name for a number of different types of fibres (common hemp, sunn, manila, sisal) used in rope, carpet, and sailcloth manufacture.

Hempel burette (*Chem.*). Used for measuring the volume of a gas, e.g., in gas analysis.

hempel quoins (*Typog.*). The simplest type of mechanical quoin, consisting of two separate and identical cast-iron wedges with a rack on the inner side, placed to oppose each other, and operated by turning a ribbed key which fits in the rack of each, causing them to slide and thus increase or decrease their combined width.

hench (*Build.*). The narrow side of a chimney shaft.

hendersonite (*Min.*). Hydrated calcium vanadate, crystallizing in the orthorhombic system.

Henderson process (*Met.*). Roasting of sulphidic copper ores with salt, followed by leach-extraction of the two resulting chlorides.

hendricksite (*Min.*).) A member of the mica group of minerals containing zinc and manganese, first discovered in the skarn deposit of Franklin Furnace, New Jersey.

He-Ne laser (*Phys.*). See **helium-neon laser.**

Henle's layer (*Zool.*). The outermost layer of the inner root sheath of a hair, composed of oblong horny cells with obscure nuclei.

Henle's loop (*Zool.*). The loop formed by a uriniferous tubule when it enters the medulla of the Mammalian kidney and turns round to pass upwards again to the cortex.

Henle's membrane (*Zool.*). A fenestrated membrane which lies immediately outside the endothelium lining a vein or an artery.

Henle's sheath (*Zool.*). In Vertebrates, a prolongation of the perineurium which invests the branches of nerves.

Hennebique pile (*Civ. Eng.*). A type of reinforced concrete pile having the main bars grouped symmetrically about the centre of the pile and bound at intervals by straps or wire lacing.

Henoch's purpura (*Med.*). A disease characterized by purpura, urticaria, swollen joints, and abdominal pain.

Henrici's notation (*Eng.*). See **Bow's notation.**

henrietta (*Textiles*). A dress material made from a silk or acetate rayon warp and fine botany weft; woven with the weft Prunelle twill, the picks per inch greatly exceeding the ends per inch. The original cloth was all silk.

henry (*Elec.*). The SI unit of self and mutual inductance, such that electromotive force of one volt is induced in a circuit by current variation of one ampere per second. Symbol H.

Henry's law (*Chem.*). The amount of a gas absorbed by a given volume of a liquid at a given temperature is directly proportional to the pressure of the gas.

Henry Williams fishplate (*Eng.*). A drop-forged fishplate for insulated rail joints in which the insulating fibre is not subject to mechanical wear, the load on the joint being transmitted by side flanges secured together by bolts.

Hensen's cells (*Zool.*). In Mammals, large pointed cells, forming a single or double row between Deiter's cells and Claudius' cells, in the cochlea.

Hensen's line (*Histol.*). In striated muscle, an indefinite clear line running transversely across the dark area of each fibre.

HEOD (*Chem.*). See **dieldrin.**

heparin (*Chem.*). Complex organic acid containing glucosamine, glucuronic acid, and sulphuric acid. It delays the coagulation of blood, and is used, by intravenous injection, in medicine and surgery, being usually obtained from lung or liver of mammals. Side effects are bleeding, and febrile or allergic reactions.

hepat-, hepato-. Prefix from Gk. *hēpar*, gen. *hēpatos*, the liver.

hepatectomy (*Surg.*). Removal of part of the liver

hepatic (*Bot.*). (1) A liverwort. (2) Dull purplish red. (*Med.*) Pertaining to the liver.
Hepaticae (*Bot.*). A class of the *Bryophyta* in which the plant which bears the sexual organs is either a dichotomosing thallus or a shoot with a dorsiventral arrangement of the leaves. The capsule usually contains elaters mixed with the spores, and seldom contains a columella. Many of the thalloid forms are known as *liverworts*.
hepatic artery (*Zool.*). In *Craniata*, a branch of the coeliac artery which conveys arterial blood to the liver.
hepatic duct (*Zool.*). In *Craniata*, a duct conveying the bile from the liver and discharging into the intestine as the common bile duct.
hepatic portal system (*Zool.*). In *Craniata*, the part of the vascular system which conveys blood to the liver: it consists of the *hepatic portal vein* and the *hepatic artery*.
hepatic portal vein (*Zool.*). In *Craniata*, the vein which conveys blood from the alimentary canal to the liver.
hepatitis (*Med.*). Inflammation of the liver. See **infective hepatitis.**
hepatitis contagiosa canis (*Vet.*). See **infectious canine hepatitis.**
hepatization (*Med.*). Pathological change of tissue so that it becomes liverlike in consistency; as of the lung in pneumonia.
hepatogenous (*Med.*). Having origin in the liver.
hepatolenticular degeneration (*Med.*). Degeneration of the liver and the lenticular nucleus (part of the brain) which occurs in a number of diseases.
hepatolith (*Med.*). A gallstone present in the liver.
hepatoma (*Med.*). A tumour composed of liver tissue.
hepatomegaly, hepatomegalia (*Med.*). Enlargement of the liver.
hepatonephritis (*Med.*). Coincident inflammation of the liver and the kidney.
hepatopancreas (*Zool.*). In many Invertebrates (as *Mollusca, Arthropoda, Brachiopoda*), a glandular diverticulum of the mesenteron, frequently paired, consisting of a mass of branching tubules, and believed to carry out the functions proper to the liver and pancreas of higher Vertebrates.
hepatopexy (*Surg.*). Fixation of the liver by suturing it to the abdominal wall.
hepatoportal system (*Zool.*). See **hepatic portal system.**
hepatoptosis (*Med.*). Displacement of the liver downwards into the abdominal cavity.
hepatorrhexis (*Med.*). Rupture of the liver.
hepatotomy (*Surg.*). Incision of the liver.
hepta- (*Chem.*). Containing seven atoms, groups, etc.
heptachlor (*Chem.*). 1(9) : 4 : 5 : 6 : 7 : 10 : 10-heptachloro-4: 7: 8: 9(1)-tetrahydro-4: 7-methylene indene:

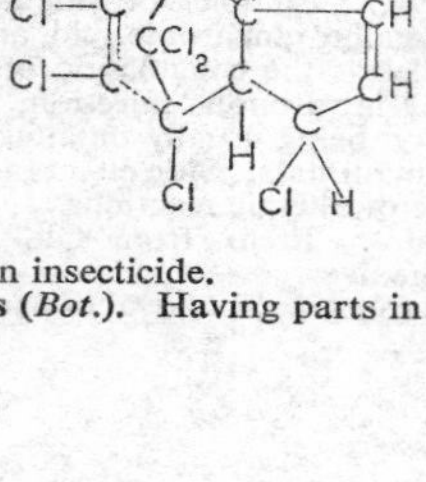

used as an insecticide.
heptamerous (*Bot.*). Having parts in sevens.
heptane (*Chem.*). C_7H_{16}, an alkane hydrocarbon, a colourless liquid, b.p. 98°C, rel. d. 0·68. There are nine isomers. The foregoing properties relate to *normal* heptane, which is a constituent of petrol and resembles hexane in its chemical behaviour.
heptavalent or **septavalent** (*Chem.*). Capable of combining with seven hydrogen atoms or their equivalent.
heptode (*Electronics*). See **pentagrid.**
heptoses (*Chem.*). A subgroup of the monosaccharides containing seven oxygen atoms, of the general formula $HO{\cdot}CH_2{\cdot}(CHOH)_5{\cdot}CHO$.
Heraeus lamp (*Med.*). A special form of mercury-vapour lamp, used for medical purposes on account of its high proportion of ultraviolet rays.
herb (*Bot.*). A flowering plant of small or rather small stature, of which the aerial shoots last only as long as is necessary to develop the flowers and the fruits. *adj.* **herbaceous.**
herbaceous (*Bot.*). (1) Soft and green, containing little woody tissue. (2) See **herb.**
herbaceous perennial (*Bot.*). A herb possessing some more or less modified underground stem system, which lasts for a number of years, and sends up each season one or more flowering shoots which die down after flowering and fruiting is completed.
herbarium (*Bot.*). A collection of dried plants: by extension, the place where such a collection is kept.
herbicide (*Agric.*). A chemical which destroys plant life, esp. a selective weedkiller.
Herbst's corpuscle (*Zool.*). A simple form of *Pacinian corpuscle* (q.v.), with fewer tunics and a core of cubical cells surrounding the nerve expansion.
hercogamy, herkogamy (*Bot.*). The condition of a flower when the stamens and stigmas are so placed that self-pollination is not possible.
hercynite (*Min.*). See under **spinel.**
hereditary (*Biol.*). Inherited; capable of being inherited; passed on or capable of being passed on from one generation to another.
heredity (*Biol.*). The organic relation between successive generations, by which characters persist; germinal constitution.
Hereford disease (*Vet.*). See **bovine hypomagnesaemia.**
Hering furnace (*Elec. Eng.*). A special form of electric furnace for melting metals; in it, circulation of the molten charge is effected by making use of the pinch effect.
hermaphrodite (*Bot.*). See **monoclinous.** (*Zool.*) Having both male and female reproductive organs in one individual. *n.* **hermaphroditism.**
hermaphrodite duct (*Zool.*). In *Gastropoda*, the duct by which the ova and spermatozoa pass from the ovotestis to the albumen gland.
Hermes (*Nuc. Eng.*). Machine (developed by Atomic Energy Research Establishment at Harwell) similar to large mass spectrograph, for electromagnetic separation of isotopes with very similar specific charge. (*H*eavy *e*lement *r*adioactive *m*aterial *e*lectromagnetic *s*eparator.)
hernia (*Med.*). Protrusion of a viscus, or part of a viscus, through an opening or weak spot or defective area in the cavity containing it; especially of an abdominal viscus.
herniotomy (*Surg.*). Cutting operation for hernia.
heroin (*Pharm.*). Diacetylmorphine, an alkaloid prepared by the acetylation of morphine. Heroin is a dangerous drug of addiction, its distribution being limited by law. Its physiological action is narcotic, resembling that of morphine.

Hero's or Heron's formula (*Maths.*). The area of a triangle with sides *a*, *b*, and *c*, is given by $\sqrt{s(s-a)(s-b)(s-c)}$ where $2s=a+b+c$.

Héroult furnace (*Met.*). An electric-arc furnace used in melting steel and other metals. The metal in the hearth acts as conductor for current flowing between electrodes, which enter through the roof. Three electrodes of graphite or amorphous carbon are used, and arcs are formed between each electrode and the bath of metal.

Héroult process (*Met.*). An electrolytic process for the manufacture of aluminium from a solution of bauxite in fused cryolite.

herpes simplex (*Med.*). An eruption of vesicles round the mouth occurring in febrile diseases; due to infection with a virus.

herpes zoster (*Med.*). Shingles. A painful eruption of crops of firm vesicles along the course of a nerve, the posterior root ganglia of which are inflamed.

herring (*Zool.*). Pisces teleostei. A genus of many species of edible fish, both salt and fresh water. Clupea harengus is the common herring.

herring-bone ashlar (*Build.*). Blocks of stone tooled in grooves of herring-bone design.

herring-bone bond (*Build.*). A form of *raking bond* (q.v.) in which the bricks are laid with rake in opposite directions from the centre of the wall, to form a herring-bone pattern. This bond is also used for brick pavings, and has the advantage of making effective bond in the middle.

herring-bone gear (*Eng.*). See **double-helical gears.**

herring-bone structure (*Geol.*). This results from the twinning on the front (ortho-) pinacoid of pyroxenes possessing a schiller structure parallel to the basal pinacoid. The schiller structure appears as a fine striation; the basal pinacoids slope in opposite directions; therefore in appropriate sections, the schiller structure simulates a herring's bones, the twin plane being the 'back-bone'.

herring-bone (or feather) twills (*Weaving*). A fancy twill effect made by stripes in which the diagonal line is run alternately to the right and left. Usually based upon the 2-and-2 twill.

Herschel effect (*Photog.*). Reversal effect obtained when an emulsion is given an initial fogging exposure to diffuse light before camera exposure; used for recording infrared spectra, a direct positive being obtained.

Herschel formula (*Eng.*). A hydraulic formula used to calculate the rate of flow over a drowned nappe.

Herschel fringes (*Optics*). The sharp fringes in a thin film discovered in 1809 by Herschel when observing light emitted near to grazing incidence.

herschelite (*Min.*). One of the zeolite minerals, a variety of chabazite occurring as flat hexagonal prisms showing composite twinning. Herschelite occurs in recent volcanic rocks, notably those of Mt Etna.

Hertwig's rule (*Zool.*). The nucleus always seeks to place itself in the centre of its sphere of activity.

hertz (*Elec.*). SI unit of frequency, indicating number of cycles per second (c/s). Symbol **Hz.**

Hertz antenna (*Radio*). Original half-wave dipole, fed at the centre.

Hertzian dipole (*Elec. Eng.*). Pair of opposite and varying charges, close together, with an electric moment. Also called **Hertzian doublet.**

Hertzian oscillator (*Radio*). Idealized system envisaged by Hertz, comprising two point charges of opposite sign and separated by an infinitesimal distance, whose electric moment varies harmonically with time.

Hertzian radiator (*Elec. Eng.*). Original form of radiator used by Hertz, comprising two parallel flat metal plates, connected by straight conductors to a spark gap placed midway between them. See **dipole.**

Hertzian waves (*Elec.*). Electromagnetic waves from, e.g. 10^4 to 10^{10} Hz, used for communication through space, covering the range from very low to ultra-high frequencies, i.e., from audio reproduction, through radio broadcasting and television to radar.

Hertzsprung-Russell diagram (*Astron.*). The diagram showing the relationship between absolute magnitude and *spectral type* (q.v.), from which is derived the theory of stellar evolution. See **dwarf star, giant star.**

Herzberg stain (*Paper*). A general-purpose microscope stain for identifying paper fibres, consisting of a mixture of zinc chloride, iodine, and potassium iodide.

herzenbergite (*Min.*). A rare decomposition product of tin ores, having the composition, tin sulphide, SnS; described originally from a locality in Bolivia.

hesperidene (*Chem.*). See ***d*-limonene.**

hesperidium (*Bot.*). A fruit like an orange; a fleshy fruit covered by a firm rind, derived from a syncarpous superior gynaeceum, and owing its fleshy material to large numbers of hairs projecting into the loculi and becoming filled with juice as the fruit ripens.

hessian (*Maths.*). The determinant obtained by replacing the *n* functions u_i in the Jacobian by the *n* partial derivatives of a single function *u*. (*Textiles*) A strong plain-weave jute fabric, used for packing material, sacks, in oilcloth and tarpaulin manufacture, and as a furnishing fabric and wall covering. For special purposes, it is laminated with plastics.

hessite (*Min.*). Telluride of silver, a metallic grey pseudo-cubic mineral occurring in silver ores in various parts of the world, notably at Zavodinski in the Altai Mts. in Siberia.

hessonite or **cinnamon stone** (*Min.*). A variety of garnet containing a preponderance of the grossular molecule, and characterized by a pleasing reddish-brown colour.

Hess's law (*Chem.*). The net heat evolved or absorbed in any chemical change depends only on the initial and final states, being independent of the stages by which the final state is reached.

hetaerolite or **heterolite** (*Min.*). A very rare double oxide of zinc and manganese, occurring in ore deposits as black tetragonal and fibrous crystals.

heter-, hetero-. Prefix from Gk. *heteros*, other, different.

heteracanth (*Zool.*). Having dorsal fin spines, which, in the depressed position, turn slightly to one side or the other, alternately. Cf. *homacanth.*

heterandrous (*Bot.*). Having stamens which are not all of the same size.

heteroagglutination (*Physiol., Zool.*). (1) The adhesion of spermatozoa to one another by the action of a substance produced by the ova of another species. (2) The adhesion of erythrocytes to one another when blood of different groups is mixed. Cf. *isoagglutination.*

heteroauxin (*Bot.*). A substance, found in fungi and in urine, which stimulates growth in plants; it appears to be identical with β-indolylacetic acid.

Heterobasidiae (*Bot.*). A subclass of the *Basidio-*

mycetes. The basidia are of various shapes but are either septate, or forked into two prongs, each of which bears a single sterigma and basidiospore.

heterobasidium (*Bot.*). A septate or deeply divided basidium.

heteroblastic (*Zool.*). Showing indirect development.

heterobrachial (*Cyt.*). Said of a chromosome bent into two parts of unequal length.

Heterocapsales (*Bot.*). An order of the *Xanthophyceae*; the palmelloid *Xanthophyceae* whose nonmotile vegetative cells can become motile directly, or the zoospores can divide directly to form new zoospores.

heterocarpous (*Bot.*). Having more than one kind of fruit.

heterocercal (*Zool.*). Said of a type of tail-fin, found in adult Sharks, Rays, Sturgeons, and many other primitive Fish, in which the vertebral column bends abruptly upward and enters the epichordal lobe, which is larger than the hypochordal lobe.

heterochlamydeous (*Bot.*). Having a distinct calyx and corolla.

Heterochloridales (*Bot.*). The order of the *Xanthophyceae* with flagellate vegetative cells.

heterochromatin (*Cyt.*). Chromatin rich in nucleic acid and involved in nuclear and cytoplasmic nucleic acid metabolism.

heterochromia iridis (*Med.*). Difference of colour in the same iris or in the two irides of the same person.

heterochromosome (*Cyt.*). A differentiated chromosome, determining sex; an allosome.

heterochronism (*Zool.*). Departure from the normal time-schedule in development.

heterochrosis (*Zool.*). Abnormal coloration.

Heterococcales (*Bot.*). An order of the *Xanthophyceae*; its members are nonfilamentous and have a cell-wall; the vegetative cells cannot become directly motile.

heterocoelous (*Zool.*). Said of vertebral centra in which the anterior end is convex in vertical section, concave in horizontal section, while the posterior end has these outlines reversed.

Heterocotylea (*Zool.*). An order of *Trematoda* in which the ventral sucker is always posterior, and the genital openings, whether separate or united, are always ventral, and usually anterior; generally ectoparasitic, or in the oral, nasal, or branchial cavities or the urinary bladder of Fish, Amphibians, Reptiles, and Crustacea.

heterocotylized arm (*Zool.*). See **hectocotylized-**.

heterocyclic compounds (*Chem.*). Cyclic or ring compounds containing carbon atoms and other atoms, e.g., O, N, S, as part of the ring.

heterocyst (*Bot.*). An enlarged thick-walled cell occurring in a filament of a member of the *Myxophyceae*; of obscure function.

heterodactylous (*Zool.*). Of Birds, having the first and second toes directed backwards, the third and fourth forwards, as the Trogons.

heterodesmic structure (*Crystal.*). One including two or more types of crystal bonding.

heterodont (*Zool.*). Of teeth, having different forms adapted to different functions.

heterodromous (*Bot.*). The right- and left-hand halves of a flower bud being mirror images of each other.

heterodynamic (*Biol.*). Of unequal potentiality, as *heterodynamic centrosomes*.

heterodynamic hybrid (*Gen.*). A hybrid which resembles in many of its characters either the male or the female parent.

heterodyne (*Radio*). Combination of two sinusoidal RF waves in a nonlinear device with the consequent production of sum and difference frequencies. The latter is the *heterodyne frequency*, and will produce an AF *beat note* when the original two sine waves are sufficiently close in frequency.

heterodyne conversion (*Radio*). Change in the frequency of a modulated carrier wave produced by heterodyning it with a second unmodulated signal. The sum and difference frequencies will carry the original modulation signal and either of these can be isolated for subsequent amplification. The frequency changing stage of a supersonic heterodyne (*superhet*) radio receiver employs this principle, an oscillator being tuned to a fixed amount above the signal frequency so that the difference frequency (*intermediate frequency signal*) remains constant for all incoming signals.

heterodyne-frequency meter (*Radio*). See **heterodyne wavemeter**.

heterodyne interference (*Radio*). That arising from simultaneous reception of two stations the difference between whose carrier frequencies is an audible frequency.

heterodyne oscillator (*Radio*). One which generates a continuously variable frequency by heterodyning two very much higher frequency oscillations, one of which is variable. Also known as **beat-frequency oscillator (B.F.O.)**.

heterodyne reception (*Radio*). Reception of ICW radio telegraphy signals by heterodyning the signal with a locally generated oscillation at a slightly different frequency. The rectified output contains a beat note component which reproduces the incoming signal audibly. Also termed **beat reception**.

heterodyne (or beat-frequency) wavemeter (*Radio*). One in which a continuously variable oscillator is adjusted to give zero beat with an unknown frequency, the value of which then coincides with the calibration of the oscillator.

heterodyne whistle (*Radio*). Variable note heard when tuning an incoming signal with an oscillating receiver and causing interference in reception elsewhere. The constant high-pitched note resulting from two carriers beating.

heteroecious (*Bot.*). Said of a parasitic fungus which forms one or more kinds of spores upon one host, and one or more distinct kinds of spores upon a second host which is not of the same species as the first. (*Zool.*) Said of parasitic forms which have different hosts at different stages in the life-history.

heteroecism (*Bot.*). A condition found in some parasitic fungi, almost all belonging to the *Uredinales*. The parasite lives for a portion of its life-cycle in one host, and a portion in a second host belonging to a distinct species.

heterogametangic (*Bot.*). Having gametangia of more than one kind.

heterogamete (*Biol.*). See **anisogamete**.

heterogamous (*Bot.*). Having pistillate, staminate, perfect, and neuter flowers (or any two or three of these types) present in one inflorescence.

heterogamy (*Biol.*). Metagenesis; alternation of two sexual generations, one being true sexual, the other parthenogenetic; condition of producing, or union of, gametes of different type; anisogamy.

heterogeneous (*Chem.*). Said of a system consisting of more than one phase.

heterogeneous radiation (*Phys.*). Radiation comprising a range of wavelengths or particle energies.

heterogeneous reactor (*Nuc. Eng.*). One in which the fuel and moderator elements in the core form large discontinuous elements arrayed in a lattice structure.

heterogeneous summation (*An. Behav.*). The idea that, if a complex stimulus releasing a response is broken into several parts, the parts in isolation will release as many responses as the whole; cf. *configurational stimulus*.

Heterogeneratae (*Bot.*). A class of the *Phaeophyta* having a microscopic gametophyte and a macroscopic sporophyte.

heterogenesis (*Zool.*). Metagenesis; abiogenesis. *adj.* **heterogenetic.**

heterogeny (*Zool.*). Cyclic reproduction in which several broods of parthenogenetic individuals alternate with one or more broods of sexual forms.

heterogony (*Zool.*). Reproduction by both parthenogenesis and amphigony.

heterograft (*Biol.*). A graft taken from an animal of a different species from the recipient.

heterogynous (*Zool.*). With 2 different types of female.

heteroicous (*Bot.*). Said of *Bryophyta* which have more than 1 kind of arrangement of the antheridia and archegonia on the same plant.

heteroion (*Chem.*). A charged particle produced by the adsorption of a simple ion on a large, complex molecule.

heterokaryon (*Biol.*). Multinucleate cell containing nuclei from 2 different sources.

heterokaryote (*Zool.*). Having 2 different kinds of nuclei.

heterokinesis (*Cyt.*). Qualitative or differential division of chromosomes.

heterokontan (*Bot.*). Said of a motile plant or spore-bearing flagella not all of the same length.

heterolecithal (*Zool.*). With unequally distributed yolk.

heterolite (*Min.*). See **hetaerolite.**

heteromastigote (*Zool.*). Having one or more anterior flagella directed forwards and a posterior flagellum directed backwards.

heteromerism (*Zool.*). In metameric animals, the condition of having unlike somites; cf. *homoeomerism*. *adj.* **heteromeric.**

heteromerous (*Bot.*). Said of a lichen thallus in which a layer of algal cells lies between 2 layers composed of fungal hyphae.

heterometabolic (*Zool.*). Having incomplete metamorphosis.

heterometry (*Chem.*). A process of titration in which precipitation is plotted as an optical density curve.

heteromorphic (*Cyt.*). Said of chromosome pairs which differ in size or form, or both; cf. *homomorphic*. *n.* **heteromorphism.**

heteromorphic alternation (*Bot.*). See **antithetic alternation of generations.**

heteromorphosis (*Zool.*). The regeneration of a part in a different form from the original part; the production of an abnormal structure; cf. *homomorphosis*. *adj.* **heteromorphous.**

heteromorphous (*Bot.*). (1) Existing in more than one form. (2) Having more than one kind of flower on the same plant.

heteromorphous rocks (*Geol.*). Rocks of closely similar chemical composition, but containing different mineral assemblages.

Heteronemertini (*Zool.*). An order of *Anopla* in which the mouth is behind the brain, the proboscis lacks stylets, and the cerebral ganglia and lateral nerves lie between the outer longitudinal and the circular muscles of the body-wall; marine forms. Cf. *Palaeonemertini*.

heteronereis (*Zool.*). In free-swimming *Polychaeta*, a special sexual form with enlarged parapodia.

heteronomous (*Zool.*). Subject to different laws, especially of growth and specialization. Cf. *autonomous*.

heteropelmous (*Zool.*). Having bifid flexor tendons of the digits of the hind foot.

heterophagous (*Zool.*). Of Birds, having young in a very immature condition.

heterophoria (*Med.*). Latent squint revealed by passing a screen before each eye. See also **esophoria** and **exophoria.**

heterophylly (*Bot.*). The occurrence of more than one kind of foliage leaf on the same shoot; also **anisophylly.** *adj.* **heterophyllous, anisophyllous.**

heterophytic (*Bot.*). Of fungi, having the diploid sporophyte unisexual.

heteropic (*Geol.*). Said of 2 formations deposited contemporaneously, but of different facies. Cf. *isopic*.

heteroplasma (*Zool.*). In tissue culture, a medium prepared with plasma from an animal of a different species from that from which the tissue was taken. Cf. *autoplasma, homoplasma*.

heteroplastic (*Zool.*). In experimental zoology, said of a graft which is transplanted to a site different from its point of origin, e.g., epithelial cells of cornea to a skin site.

heteroplasty (*Med.*). The operation of grafting on one person body-tissue removed from another.

heteroploid (*Cyt.*). Possessing an additional chromosome, i.e., above the characteristic number.

heteropolar (*Chem.*). Having an unequal distribution of charge, as in a co-ordinate covalent bond.

heteropolar generator (*Elec. Eng.*). An electromagnetic generator of the usual type, i.e., one in which the conductors pass alternate north and south poles, or in which alternate poles pass the conductors.

heteropolar liquids (*Chem., etc.*). Compounds such as alcohols, amines and organic acids which contain molecules that have localized associated polar groups.

heteropycnosis (*Cyt.*). The tendency of sex chromosomes to undergo precocious condensation in the growth stages, shown by the fact that some portions are densely staining.

Heterosiphonales (*Bot.*). An order of the *Xanthophyceae*, containing the single genus *Botrydium* which is multinucleate and siphonaceous.

heterosis (*Zool.*). Cross-fertilization; in metameric animals, the modification of a merome in form or position, or both, from the type. Cf. *homoeosis*.

heterosporangy (*Bot.*). The formation of more than one kind of sporangium containing more than one kind of spore.

heterospory (*Bot.*). The formation of more than one kind of spore. *adj.* **heterosporous.**

heterostatic operation (*Elec. Eng.*). See **Dolezalek electrometer.**

heterostylous (*Bot.*). Plants on which the lengths of styles vary from flower to flower.

heterosynapsis (*Cyt.*). Pairing of 2 dissimilar chromosomes. Cf. *homosynapsis*.

heterothallic (*Bot.*). Pertaining to *heterothallism*. See **haplodioecious.**

heterothallism (*Bot.*). The existence of physiological differences between the mycelia of many species of fungi, most commonly shown by the inability of the mycelium to manifest sexual activities unless it is brought into contact with another mycelium of distinct character.

heterotopia (*Med.*). Displacement of a group of

cells of an organ from their normal position during the course of development.

Heterotricha (*Zool.*). A synonym for *Spirotricha* (q.v.).

Heterotrichales (*Bot.*). The order of the *Xanthophyceae* in which the cells are joined to form filaments.

heterotrichous (*Zool.*). Having cilia or flagella of two or more different kinds.

heterotrichous thallus (*Bot.*). An algal thallus consisting of a prostrate portion, lying on a substratum, and a series of filamentous branches standing out into the water.

heterotrophic (*Bot.*). Unable to make food from simple beginnings, and therefore dependent for food upon dead or living organisms of another species, and ultimately on the green plant.

heterotropic chromosome (*Cyt.*). See **sex chromosome.**

heterotypic (*Zool.*). Differing from the normal condition. Cf. *homotypic.*

heterotypic division (*Cyt.*). The first (reductional) of two nuclear divisions in meiosis in which the number of chromosomes is halved. The second (equational) is *homotypic division.*

heteroxeny (*Bot.*). See **heteroecism.**

heterozygosis (*Zool.*). See **heterosis.**

heterozygous (*Gen.*). Possessing both the dominant and recessive characters of an allelomorphic pair. Cf. *homozygous. n.* **heterozygote.**

H.E.T.P. (*Chem. Eng.*). *H*eight *E*quivalent to a *T*heoretical *P*late. A measure of the separation efficiency of a distillation column, i.e., the height of packing in a packed distillation column which behaves as a *theoretical plate.*

heulandite (*Min.*). One of the best-known zeolites, often beautifully crystalline, occurring as coffin-shaped monoclinic crystals in cavities in basic igneous rocks. In composition similar to plagioclase, but having a high content of water.

heuristic routine (*Comp.*). Intuitive rather than algorithmic routine for tackling a specific problem with a computer (i.e., one based essentially on trial and error).

heuristics (*Psychol.*). Principles used in making decisions, learning, etc.; e.g., a chess-player cannot work out all the possible effects of all possible moves and therefore uses a number of principles to guide his decision.

Heurtley hot-wire magnifier (*Teleg.*). A device for magnifying weak telegraph currents. The incoming currents deflect hot-wires across a cooling air-blast, the consequent unbalance currents of the Wheatstone bridge becoming the magnified signals.

Heusler alloys (*Met.*). Alloys of manganese, copper, and aluminium (but not containing iron, nickel or cobalt), which are strongly ferromagnetic.

hevea rubber (*Elec. Eng.*). Rubber from *Hevea brasiliensis* tree, affording superior electrical and mechanical properties when used substantially in electrical insulators.

hewettite (*Min.*). A hydroxide of vanadium and calcium, occurring as slender orthorhombic crystals in the oxide zone of the vanadium deposits of Peru, and of the Colorado Plateau.

Hewlett disk insulator (*Elec. Eng.*). A disk-form of suspension-type insulator.

hewn stone (*Build.*). Blocks of hammer-dressed stone.

hex-. Prefix from Gk. *hex,* six. (*Chem.*) Colloquialism for *uranium* (VI) *fluoride,* the compound used in the separation of uranium isotopes by gaseous diffusion.

hexa- (*Chem.*). Containing 6 atoms, groups, etc.

hexacanth (*Zool.*). Having 6 hooks; as a stage in the life-history of some Tapeworms.

hexachlorophane (*Chem.*). 2,2′-methylene bis-(3,4,6 trichloro-hydroxybenzene). White powder. M.p. approx. 160°C. Widely used bactericide in soaps, deodorants and other toilet products. Sometimes referred to as **G-11.**

Hexactinellida (*Zool.*). A class of *Porifera,* usually distinguished by the possession of a siliceous skeleton composed of triaxial spicules, and large thimble-shaped flagellated chambers.

hexadecane (*Chem.*). An alkane hydrocarbon ($C_{16}H_{34}$) found in petroleum, especially that showing normal structure.

hexafluorophosphoric acid (*Chem.*). HPF_6. Produced by the action of strong hydrofluoric acid on difluorophosphoric acid. Also referred to as **phosphorofluoric acid.**

hexagon (*Maths.*). A six-sided polygon.

hexagonal packing (*Crystal., etc.*). System in which many metals crystallize, thus achieving minimum volume. Each lattice point has twelve equidistant neighbours in such a cell construction.

hexagonal system (*Crystal.*). A crystal system in which three equal coplanar axes intersect at an angle of 60°, and a fourth, perpendicular to the others, is of a different length.

hexagon dresser (*Eng.*). A metal disk tool used for dressing grinding wheels.

hexagon voltage (*Elec. Eng.*). The voltage between 2 lines, adjacent as regards phase-sequence, of a 6-phase system.

hexahydrobenzene (*Chem.*). See **cyclohexane.**

hexahydrocresol (*Chem.*). See **methylcyclohexanol.**

hexahydrophenol (*Chem.*). *Cyclohexanol* (q.v.).

hexahydropyridine (*Chem.*). *Piperidine* (q.v.).

hexamerous (*Bot.*). Having parts in sixes.

hexametaphosphates (*Chem.*). Salts of hexa-metaphosphoric acid, $H_6(PO_3)_6$, a polymer of metaphosphoric acid. Cf. *Calgon.*

hexamethylene (*Chem.*). *Cyclohexane* (q.v.).

hexamethylenediamine (*Chem.*). (1,6-diaminohexane) $H_2N(CH_2)_6NH_2$. Important historically as a constituent material of the first nylon (nylon 6.6), which was a condensation polymer of hexamethylenediamine and *adipic acid.*

hexamethylenetetramine (*Chem.*). Hexamine $(CH_2)_6N_4$, a condensation product of methanal with ammonia, a crystalline substance with antiseptic and diuretic properties. Used in the production of *cyclonite* (q.v.), a highly efficient explosive.

hexamitiasis (*Vet.*). A disease of turkeys due to infection by the flagellate protozoon *Hexamita meleagridis,* which causes enteritis.

hexane (*Chem.*). C_6H_{14}. There are five compounds with this formula: normal hexane, a colourless liquid, of ethereal odour, b.p. 69°C, rel. d. 0·66, is an important constituent of petrol and of solvent petroleum ether or ligroin.

hexphase (*Elec. Eng.*). A term sometimes used instead of *6-phase.*

hexaploid (*Gen.*). Possessing 6 sets of chromosomes, 6 times the monoploid number.

hexapod (*Zool.*). Having 6 legs.

hexapterous (*Zool.*). Having 6 winglike processes.

hexarch (*Bot.*). Having 6 strands of protoxylem.

hexastyle (*Arch.*). A portico formed of 6 columns in front.

hexavalent (*Chem.*). Capable of combining with 6 hydrogen atoms or their equivalent.

Hexlet air sampler (*Powder Tech.*). Device for sampling dusty air, using a horizontal elutriator before a dust filter to eliminate the coarser particles.

hexobarbitone sodium (*Chem.*). The monosodium derivative of 5-Δ′-*cyclo*hexenyl-5-methyl-*N*-methylbarbituric acid ($C_{12}H_{15}O_3N_2Na$), used intravenously or intramuscularly as a basic anaesthetic. TN **Evipan.**

hexode (*Electronics*). 6-electrode electron tube comprising an anode, cathode, control electrode, and 3 grid electrodes.

hexoestrol (*Chem.*). Dihydrostilboestrol. A synthetic crystalline derivative of phenol and hexane, $C_{18}H_{22}O_2$, used as an oestrogen.

hexogen (*Chem.*). *Cyclonite* (q.v.)

hexokinases (*Biochem.*). Enzymes which catalyse phosphorylation of *hexoses.*

hexoses (*Chem.*). A subgroup of the monosaccharides containing 6 oxygen atoms, of the general formula $HO{\cdot}CH_2{\cdot}(CHOH)_4{\cdot}CHO$ and $HO{\cdot}CH_2{\cdot}(CHOH)_3{\cdot}CO{\cdot}CH_2OH$. The first formula signifies an *aldohexose* (q.v.), the second one a *ketohexose* (q.v.).

Heyland a.c. generator (*Elec. Eng.*). A self-excited a.c. generator in which the excitation is obtained by a special arrangement of transformers and commutator connected to the armature.

Heyland diagram (*Elec. Eng.*). A particular application of the circle diagram of an a.c. circuit to represent the behaviour of an induction motor.

Hf (*Chem.*). The symbol for *hafnium.*

Hg (*Chem.*). The symbol for *mercury.*

H-girder (*Build., etc.*). See **H-beam.**

HHDN (*Chem.*). 1,2,3,4,10,10-hexachloro-1,4,4a,5,8,8a-hexahydro-*exo*-1,4-*endo*-5,8-dimethanonaphthalene, used as an insecticide.

H-hinge (*Join.*). A hinge which when opened has the shape of the letter H. Also called a **parliament hinge.**

H.I. (*Surv.*). Abbrev. for *height of instrument.*

Hibbert cell (*Elec. Eng.*). A standard cell similar to a Clark cell but having an electrolyte of zinc chloride instead of zinc sulphate.

Hibbert standard (*Elec. Eng.*). A standard of magnetic flux linkage suitable for fluxmeter or galvanometer calibration. It comprises a stabilized magnet producing a radial field in an annular gap, through which a cylinder carrying a multiturn coil can be dropped.

hibernacula (*Zool.*). In freshwater *Ectoprocta*, external buds which are arrested in development and which enable the colony, which has died down at the onset of winter, to regenerate in spring.

hibernating glands (*Zool.*). Vascular reserves of fatty tissue occurring in some hibernating Mammals.

hibernation (*Zool.*). The condition of partial or complete torpor into which some animals relapse during the winter season. *v.* **hibernate.**

hiccup (*Med.*). Sudden spasm of the diaphragm followed immediately by closure of the glottis.

hickie (*Print.*). A blemish on a solid area of print which appears as a spot surrounded by a halo, caused by a small fragment of paper adhering to the plate as a result of *fluffing* or *picking.*

hick joint (*Build.*). A *flat joint* (q.v.) formed in fine mortar when pointing, after the old mortar has been raked out of the joints.

hickory (*For.*). The product of a hardwood tree (*Carya*) common to the eastern U.S. It bends well and may be used for making boats, chairs, tool handles, etc.

Hicks hydrometer (*Elec. Eng.*). A form of hydrometer used for finding the relative density of the electrolyte in an accumulator, to determine the state of its charge; the hydrometer consists of a glass tube containing a number of coloured beads, which float at different relative densities.

hiddenite (*Min.*). See **spodumene.**

hiding place (*Paint.*). Measure of efficiency of paint masking a given test surface, when applied in a specified manner. It is a function of the amount of pigment in the paint, the refractive index of the pigment and the medium, the size of the pigment particles, of their agglomerates and other, lesser properties.

hiding power (*Paint.*). The power of a paint to obscure a black-and-white contrast; generally expressed as the number of square feet per gallon or square metres per litre of paint.

hidrosis (*Zool.*). Formation and excretion of sweat.

Hiduminium R.R. alloys (*Met.*). See **R.R. alloys.**

hiemal aspect (*Bot.*). Appearance and condition of the plants of a community in winter.

hi-fi (*Acous.*). Abbrev. for *high fidelity.*

high (*Typog.*). Type or blocks higher than the rest of the forme are said to be *high.*

high alumina cement (*Eng.*). Made by fusing a mixture of bauxite and chalk or limestone and grinding the resultant clinker. It hardens much more rapidly (24 hr) than Portland cement and has great heat and chemical resistance.

high aspect ratios (*Aero.*). See **aspect ratio.**

high brass (*Met.*). Common brass of 65/35 copper-zinc alloying, as distinct from deep-drawing brass with 66–70% copper.

high by-pass ratio (*Aero.*). Applied to a *turbofan* (q.v.) in which the *air mass flow* (q.v.) ejected directly as propulsive thrust by the fan is 4 or 5 times the quantity passed internally through the *gas generator* (q.v.) section.

high-carbon steel (*Met.*). One such as spring steel, with between ½% and 1½% carbon.

high-conductivity copper (*Met.*). Metal of high purity, having an electrical conductivity not much below that of the international standard, which is a resistance of 0·153 28 ohms for a wire 1 m in length and weighing 1 g.

high-definition (*TV*). A term arbitrarily applied to those systems of television which use 100 or more scanning lines per frame.

high-definition developer (*Photog.*). One which increases the contrast at boundaries between light and dark tones, the light boundary being enhanced by bromide from the heavily exposed part of the image and the dark boundary by comparatively fresh developer from the lightly exposed part.

high-duty cast-iron (*Met.*). Cast-iron with a tensile strength greater than 260 MN/m². Produced by using large proportions of steel scrap, casting in hot moulds, inoculation, superheating, alloying additions, and heat-treatment.

high-energy ignition (*Aero.*). A gas-turbine ignition system using a very high voltage discharge.

high energy rate forming (*Eng.*). Any of a recently developed family of processes, in which metal parts are rapidly compacted, forged, extruded, by the application of extremely high pressures.

higher critical velocity (*Hyd.*). The critical velocity of change from viscous to eddy-flow.

high explosive (*Mining*). One in which the active agent is in chemical combination and is readily detonated by application of shock. Nitrated cotton, nitroglycerine, and ammonium nitrate are widely used, diluted to required explosive strength by inert fillers such as kieselguhr or wood pulp. See also **gelignite, trinitrotoluene.**

high-fidelity (*Acous.*). Said of sound reproduction of exceptionally high quality. Abbrev. **hi-fi.**

high-fidelity amplifier (*Acous.*). One in which the

input signal is reproduced with a very high degree of accuracy.

Highfield booster (*Elec. Eng.*). An automatic battery booster consisting of a generator, a motor and an exciter. Automatic regulation is carried out by balancing the exciter voltage against that of the battery.

high-flux reactor (*Nuc. Eng.*). One designed to operate with a greater neutron flux than normal (at present about 10^{19} neutron/m^{2}s) especially for testing materials. Also **materials testing reactor.**

high frequency (*Acous.*). Any frequency above the audible range, i.e., above 15 kHz, but more especially those frequencies which are used for radio communication.

high-frequency amplification (*Radio*). That at frequencies used for radio transmission. In a receiver, any amplification which takes place before detection, frequency conversion, or demodulation.

high-frequency capacitance microphone (*Acous.*). One which uses audio variation of capacitance to vary the frequency of an oscillator, or response of a tuned circuit.

high-frequency heating (*Heat*). Heating (induction or dielectric) in which frequency of current is above mains frequency; from rotary generators up to, e.g., 3000 Hz and from electronic generators 1–100 MHz. Also **radio heating.**

high-frequency induction furnace (*Met.*). Essentially an air transformer, in which the primary is a water-cooled spiral of copper tubing, and the secondary the metal being melted. Currents at a frequency above about 500 Hz are used to induce eddy currents in the charge, thereby setting up enough heat in it to cause melting. Used in melting steel and other metals. See also **coreless induction furnace.**

high-frequency repeater distribution frame (*Teleph.*). In carrier telephony, a frame providing flexibility primarily in the connecting of apparatus to external cables on which 12-circuit groups are transmitted. It also provides for the interconnection of apparatus used for the assembly of a number of 12-circuit groups before transmission to line.

high-frequency resistance (*Radio*). That of a conductor or circuit as measured at high frequency, greater than that measured with d.c. because of *skin effect*. (Increased losses would mean lower resistance for constant applied p.d.)

high-frequency transformer (*Radio*). One designed to operate at high frequencies, taking into account self-capacitance, usually with band-pass response.

high-frequency treatment (*Med.*). The treatment of diseases by high-frequency currents, usually trains of heavily damped oscillations.

high-frequency welding (*Elec. Eng.*). Welding by RF heating. See **seam welding** (2).

Highgate resin (*Min.*). A popular name for fossil gum-resin occurring in the Tertiary London Clay at Highgate in North London. See **copalite.**

high grading (*Mining*). Selective mining, in which sub-grade ore is abandoned unworked. Also, theft of valuable concentrates or specimens such as nuggets of gold.

high-intensity separation (*Min. Proc.*). Dry concentration of small particles of mineral in accordance with their relative ability to retain ionic charge after passing through an ionizing field.

Highland Boundary Fault (*Geol.*). One of the most important dislocations in the British Isles, extending from the Clyde to Stonehaven and separating the Highlands of Scotland from the Midland Valley. It was initiated in Middle Old Red Sandstone times but movement has taken place along it subsequently.

high-lead bronze (*Met.*). Soft matrix metal used for bearings, of copper/tin/lead alloys in approximate proportions 80, 10 and 10.

high-level firing time (*Radio*). The time required to establish an RF discharge in a switching tube after the application of RF power.

high-level (low-level) modulation (*Radio*). Conditions where modulation of a carrier for transmission takes place at high level for direct coupling to the radiating system, *or* is at lower level than this, with subsequent push-pull or straight amplification. Also known as **high-(low) power modulation.**

high-level RF signal (*Radio*). An RF signal having sufficient power to fire a switching tube.

highlight (*TV*). Very bright area in a screen or image—the converse of shadow.

high-opacity foils (*Bind.*). A type of blocking-foil especially suitable for marking undressed bookcloths and deep-grained materials. These are only made in white and pastel shades and have a considerable weight of pigment to obliterate the surface completely.

high-pass (low-pass) filter (*Telecomm.*). One which freely passes signals of all frequencies above, or below, a reference value known as the cut-off frequency, f_c. (*N.B.*—Beyond f_c attenuation only rises slowly and seldom approaches complete cut-off as implied by this name.)

high-power modulation (*Radio*). See **high-level modulation.**

high-pressure compressor (*Aero.*). In a gas-turbine engine with two or more compressors in series, the last is the high-pressure one. In a dual-flow turbojet this feeds the combustion chamber(s) only. Abbrev. **HP compressor.**

high-pressure cylinder (*Eng.*). The cylinder of a compound or multiple-expansion steam-engine in which the steam is first expanded.

high-pressure gas systems. These utilize coal-gas under high pressure, in conjunction with air at atmospheric pressure, for industrial heating purposes; used where accurate control of furnace temperature is required.

high-pressure hose (*Mining*). 'Armoured hose', reinforced with circumferentially embedded wire, and hence able to withstand moderately high pressure and rough usage.

high-pressure turbine (*Aero.*). The first turbine after the combustion chamber in a gas-turbine engine with two or more turbines in series. Abbrev. **HP turbine.**

high-pressure turbine stage (*Aero.*). The first stage in a *multi-stage turbine* (q.v.). Abbrev. **HP stage.**

high recombination rate contact (*Electronics*). The contact region between a metal and semiconductor (or between semiconductors) in which the densities of charge carriers are maintained effectively independent of the current density.

high-resistance joint (*Elec. Eng.*). See **dry joint.**

high-resistance voltmeter (*Elec. Eng.*). One drawing negligible current and typically having a resistance in excess of 1000 ohms per volt.

high spaces (*Typog.*). The normal height of spacing is about ¾ in. (19 mm) but higher spaces are more convenient when pages are to be stereotyped, and can also be used as a mount for original plates.

high-speed circuit-breaker (*Elec. Eng.*). A

circuit-breaker in which special devices are used to ensure very rapid operation; used particularly on d.c. traction systems.

high-speed steam-engine (*Eng.*). A vertical steam-engine, generally compound, using a piston valve, or valves, whose moving parts are totally enclosed and pressure-lubricated. See **quick-revolution engine.**

high-speed steel (*Met.*). A hard steel used for metal-cutting tools. It retains its hardness at a low red heat, and hence the tools can be used in lathes, etc., operated at high speeds. It usually contains 12 to 18% tungsten, up to 5% chromium, 0·4 to 0·7% carbon, and small amounts of other elements (vanadium, molybdenum, etc.).

high-speed wind tunnel (*Aero.*). A high *subsonic* (q.v.) wind tunnel in which compressibility effects can be studied.

high spot (*Radiol.*). A small volume so situated that the dose therein is significantly above the general dose level in the region treated.

high-stop filter (*Telecomm.*). An electric-wave filter in which currents of frequencies higher than a nominal cut-off frequency are highly attenuated, while those with frequencies below this are passed with minimum attenuation.

high-strength brass (*Met.*). A type of brass based on the 60 copper-40 zinc composition, to which manganese, iron, and aluminium are added to increase the strength. *Manganese bronze* denotes a variety in which manganese is the principal addition, but most varieties now contain all 3 elements.

high-temperature reactor (*Nuc. Eng.*). One designed to attain core temperatures above 660°C.

high-tension (*Elec. Eng.*). See **high-voltage.**

high-tension battery (*Elec. Eng.*). One supplying power for the anode current of valves. U.S. term **B-battery.** Also called **anode battery.**

high-tension ignition (*Elec. Eng.*). An ignition system for internal-combustion engines which employs a spark from a high-tension magneto or an induction coil.

high-tension magneto (*Elec. Eng.*). The usual form of magneto used for producing the high-voltage spark required for the ignition of an internal-combustion engine.

high-tension separation (*Min. Proc.*). Electrostatic separation, in which small particles of dry ore fall through a high-voltage d.c. field, and are deflected from gravitational drop or otherwise separated in accordance with the electric charge they gather and retain.

high-test cast-iron (*Met.*). See **high-duty cast-iron.**

high-vacuum (*Electronics*). A system so completely evacuated that the effect of ionization on its subsequent operation may be neglected. See also **hard.**

high-velocity scanning (*Electronics*). The scanning of a target with electrons of sufficient velocity to produce a secondary emission ratio which is greater than unity.

high-voltage (*Elec. Eng.*). Legally, any voltage above 650 volts. In batteries, etc., often called *high-tension*. See **high-tension battery.**

high-voltage electron microscope (*Micros.*). A modification of the electron microscope, working at a very high voltage (about 1 million volts) and used for the examination of thicker sections (over 250 nm).

high-voltage test (*Elec. Eng.*). The application of a voltage greater than working voltage to a machine, transformer, or other piece of electrical apparatus to test the adequacy of the insulation.

highway (*Comp.*). Defined path for transmission of signals. Also called **trunk,** and in U.S. **bus.** If related, several highways can be a *channel* (q.v.).

high-wing monoplane (*Aero.*). An aeroplane with the wing mounted on or near the top of the fuselage.

Hi-k capacitor (*Elec. Eng.*). One in which the dielectric of barium and strontium titanates has permittivities above 1000.

Hilbert transformer (*Elec. Eng.*). A device for obtaining a phase shift of 90°. It consists of a delay line, fed from a travelling-wave source and terminated by a negligibly small resistor, the p.d. across this forming the 0° output signal. The 90° signal is obtained by integrating the voltage along the line with a weighting function inversely proportional to the distance from the termination.

Hildebrand electrode (*Chem.*). See **hydrogen electrode.**

hile (*Bot.*). See **hilum.**

hill-and-dale recording (*Acous.*). Disk recording in which cutting stylus and reproducing stylus move at right angles to the plane of the disk. Now obsolete. Also called **vertical recording** or **contour recording.** See **stereophonic recording.**

hill-climbing (*Telecomm.*). Continuous or periodic adjustment of self-regulating adaptive control systems to achieve an optimum result.

hill-creep (*Geog.*). The movement of loose material down the slope of a hillside.

hill diarrhoea (*Med.*). A peculiar form of diarrhoea occurring, during the hot season, in Europeans living in high altitudes in India, Ceylon, and elsewhere.

hillebrandite (*Min.*). Dicalcium silicate hydrate. Occurs as white fibrous aggregates in impure thermally metamorphosed limestones and in boiler scale.

hilum (*Bot.*). (1) Lateral depression in which the flagella are inserted in reniform zoospores. (2) Scar left on the testa when the seed separates from its stalk. (3) A small granule in the centre of a starch grain. Also **hile.** (*Zool.*) A small depression in the surface of an organ, which usually marks the point of entry or exit of blood-vessels, lymphatics, or an efferent duct. Also **hilus.**

Hilt's Law (*Geol.*). An expression of the observation that the more deeply buried a coal seam, the higher is the rank of its coal.

hind-brain (*Zool.*). In Vertebrates, that part of the brain which is derived from the third or posterior brain-vesicle of the embryo, comprising the cerebellum and the medulla oblongata: the posterior brain-vesicle itself.

hindered settling (*Min. Proc.*). Hydraulic classification of sand-sized particles in accordance with their ability to gravitate through a column of similar material expanded by a rising current of water. (*Powder Tech.*) Settling of solids in a suspension of a concentration greater than 15% by volume, in which the predominant physical process is the draining of the fluid out of a thick slurry. No particle segregation by size occurs, a clear boundary being formed between the supernatant fluid and the settling suspension.

hind-gut (*Zool.*). That part of the alimentary canal of an animal which is derived from the posterior ectodermal invagination or proctodaeum of the embryo.

Hindley worm gear (*Eng.*). A type of worm gear in which the worm is curved along its face to match the curvature of the worm wheel. Also called **globoidal worm gear.**

hindrance (*Elec.*). Impedances (0 for zero, and 1 for infinite) used in theoretical manipulation of switching.

hinge (*Zool.*). The flexible joint between the 2 valves of the shell in a bivalve Invertebrate, such as a pelecypod Mollusc or a Brachiopod: any similar structure: a joint permitting of movement in one plane only.

hinge-bound door (*Join.*). A door which will not close easily or fully owing to the hinges being too deeply sunk.

hinge fault (*Geol.*). A fault along which the displacement increases from zero at one end to a maximum at the other end.

hingeing post (*Build.*). The post from which a gate is hung. Also called **swinging post.**

hinge ligament (*Zool.*). The tough uncalcified elastic membrane which connects the 2 valves of a bivalve shell.

hinge line (*Zool.*). The line of junction of the two valves in a bivalve shell.

hinge moment (*Aero.*). The moment of the aerodynamic forces about the hinge axis of a control surface, which increases greatly with speed, necessitating *aerodynamic balance* (q.v.).

hinge tooth (*Zool.*). A small sharp projection of the shell near the hinge in bivalves.

Hinkes-Bird bead (*Join.*). See **deep bead.**

hip (*Arch.*). The salient angle formed by the intersection of 2 inclined roof slopes.

Hiperco (*Met.*). TN for a high-permeability, high-saturation, 35% cobalt-iron alloy, with 0·5% chromium, for making magnets, etc.

Hipernik (*Met.*). TN for isotropic alloy of equal parts of iron and nickel, having its grain orientated for making magnets, etc.

Hipersil (*Met.*). TN for silicon-iron alloy, highly grain-orientated.

hip hook (*Build.*). A strap of wrought-iron fixed at the foot of a hip rafter and bent into the form of a scroll, as a support for the hip tiles.

hip iron (*Build.*). See **hip hook.**

hip knob (*Build.*). A finial surmounting the peak of a gable or a hipped roof.

hipped end (*Build.*). The triangular portion of roof covering the sloping end of a hipped roof.

hipped roof (*Build.*). A pitched roof having sloping ends at the gable ends.

hippoboscid (*Zool.*). A parasitic bloodsucking dipteran fly, e.g., a horsefly.

hippocampus (*Zool.*). In the Vertebrate brain, a tract of nervous matter running back from the olfactory lobe to the posterior end of the cerebrum. *adj.* **hippocampal.**

hippocrepiform (*Bot.*). Shaped like a horseshoe.

hippuric acid (*Chem.*). Benzoyl-aminoethanoic acid, $C_6H_5CO{\cdot}NH{\cdot}CH_2{\cdot}COOH$, rhombic crystals, m.p. 187°C, occurring in the urine of many animals, particularly herbivores.

hippus (*Med.*). Rhythmical alternate contraction and dilatation of the pupil of the eye.

hip rafter (*Carp.*). See **angle rafter.**

hip roll (*Carp.*). A timber of circular section with a vee cut out along its length, so as to adapt it for sitting astride the hip of a roof.

hip tile (*Build.*). A form of arris-tile laid across the hip of a roof.

Hirschsprung's disease (*Med.*). A condition occurring in children in which there is great hypertrophy and dilatation of the colon.

hirsute (*Bot., Zool.*). Hairy; having a covering of stiffish hair or hairlike feathers.

hirsuties (*Med.*). Excessive hairiness.

hirudin (*Zool.*). An anticoagulin, present in the salivary secretion of the leech, which prevents blood clotting by inhibiting the action of thrombin on fibrinogen.

Hirudinea (*Zool.*). A class of *Annelida* the members of which are ectoparasitic on a great variety of aquatic and terrestrial animals; they possess anterior and posterior suckers, and most of them lack setae; hermaphrodite animals with median genital openings; the development is direct. Leeches.

hispid (*Bot.*). Well covered with stiff hairs or bristles. (*Zool.*) Hairy; having a covering of stiffish hair or hairlike feathers.

hiss (*Telecomm.*). See **valve noise.**

His's bundle (*Physiol.*). In the Mammalian heart, a bundle of small parallel muscle fibres extending from the wall of the right auricle to the septum between the ventricles: the auriculo-ventricular bundle.

hissing arc (*Elec. Eng.*). An arc, between incorrectly adjusted carbon electrodes, which produces a hissing sound, the current being too great for the size of the arc crater. Sometimes called **frying arc** in U.S.

histamine (*Chem.*). A base, $C_5H_9N_3$, 2-imidazolyl-4 (or 5)-ethylamine. Formed *in vivo* by the decarboxylation of *histidine* (q.v.). It causes the contraction of nearly all smooth muscle, and dilatation of capillaries, which causes a fall of arterial blood pressure and shock after large doses. It stimulates the gastric secretion of hydrochloric acid and this effect is not prevented by antihistamines.

histidine (*Chem.*). 1-Amino-2-imidazole-propanoic acid, a protein derivative belonging to the group of hexone bases. An essential amino acid for most species of animal; precursor of *histamine.*

histioma, histoma (*Med.*). Any tumour derived from fully developed tissue, such as fibrous tissue, cartilage, muscle, blood vessels.

histoblast (*Zool.*). One of the formative cells composing an *imaginal bud* (q.v.).

histochemistry (*Zool.*). The chemistry of living tissues.

histocyte (*Zool.*). A tissue-cell.

histogen (*Bot.*). A more or less well-defined region within a plant where tissues undergo differentiation.

histogenesis (*Zool.*). Formation of new tissues.

histogenous (*Bot.*). Of fungal spores, produced directly from cells or hyphae, i.e., not from conidiophores.

histogram (*Stats.*). Graph in which frequency distribution is shown by means of rectangles.

histohaematin (*Zool.*). A respiratory pigment.

histology (*Zool.*). The study of the minute structure of tissues and organs.

histolysis (*Bot.*). The breakdown, and sometimes liquefaction, of a cell or tissue. (*Zool.*) Dissolution and destruction of tissues, as in the metamorphosis of *Oligoneoptera.*

histoma (*Med.*). See **histioma.**

histones (*Chem.*). A group of simple proteins ranking in complexity between protamines and albumins. They are strongly basic, and often occur combined with nucleic acid, or with haematin. They are soluble in water but insoluble in dilute ammonia. Present in chromosomes, they are believed to act as gene inhibitors.

histoplasmosis (*Vet.*). A disease of animals and man due to infection by the fungal organism *Histoplasma capsulatum.*

historical geology. See under **geology.**

histozoic (*Zool.*). Living in the tissues of the body, amongst the cells.

hit-and-miss ventilator (*Build.*). A ventilating device consisting of a slotted plate over which may be moved another slotted plate, so that the openings for access of air may be restricted as required.

hitch (*Mining*). (1) A fault of minor importance, usually not exceeding the thickness of a seam. (2) Ledge cut in rock face to hold mine timber in place.

hitcher (*Mining*). See **hanger-on.**

hitch feed (*Eng.*). An automatic strip feed for power presses, which advances the strip during the forward stroke of the mechanism and holds it stationary during the idle return stroke.

Hittorf dark space (*Phys.*). See **Crookes dark space.**

Hittorf tube (*Phys.*). Early form of cathode-ray tube.

H.L.B. (*Chem.*). Term used to signify *hydrophilic-lipophilic balance* in emulsifying agents.

H.M.D. (*Hyd.*). An abbrev. for *hydraulic mean depth* (q.v.).

H-network (*Telecomm.*). Symmetrical section of circuit, with one shunt branch and four series branches.

H.N.W. (*Build.*). An abbrev. for *head, nut, and washer.*

Ho (*Chem.*). The symbol for *holmium.*

hoarding or hoard (*Build.*). A close-boarded fence of temporary character erected around a building site.

hoar frost (*Meteor.*). A deposit of ice crystals formed on objects, especially during cold clear nights when the dew-point is below freezing-point. The conditions favouring the formation of hoar-frost are similar to those which produce *dew* (q.v.).

hoary (*Bot.*). Covered with short greyish-white down.

hob (*Eng.*). (1) A hardened master punch, used in die sinking, which is a duplicate of the part to be produced by the die. It is pressed into an unheated die blank to make the die impression. (2) A gear-cutting tool resembling a milling cutter or a worm gear, whose thread is interrupted by grooves so as to form cutting faces. Also **hobbing cutter.**

hobbing machine (*Eng.*). A machine for cutting teeth on gear blanks, for the production of spur, helical, and worm gears by means of a hobbing cutter.

hobbles (*Vet.*). An apparatus applied to the legs of a horse for casting.

Höchstädter cable (*Elec. Eng.*). A high-voltage multi-core cable in which a thin metallized sheath is placed over the insulation of each core, in order to control the distribution of electric stress in the dielectric and ensure that it is purely radial.

Höchstädter paper (*Paper*). See **H-paper.**

hock (*Zool.*). The tarsal joint of a Mammal.

Ho-coefficient (*Vac. Tech.*). The ratio of the speed of the pump, measured at the nozzle, to the conductance of the nozzle clearance area of the pump.

hoctonspheres (*Chem. Eng.*). Storage vessels for low boiling-point liquids stored under pressure at temperatures at which they would at normal pressure vaporize. The vessels are made from segments specially pressed before welding so that the finished vessel is truly spherical. Diameters up to 10 m are not uncommon.

hod (*Build.*). A 3-sided container, supported on a long handle, used for carrying bricks and mortar on the site.

Hodectron (*Electronics*). Mercury-vapour discharge tube, fired by a magnetic pulse.

Hodge's pessary (*Med.*). A ring-shaped pessary for correcting backward displacement of the uterus.

Hodgkin's disease (*Med.*). See **lymphadenoma.**

hodograph (*Mech.*). A curve used to determine the acceleration of a particle moving with known velocity along a curved path. The hodograph is drawn through the ends of vectors drawn from a point to represent the velocity of the particle at successive instants.

hodoscope (*Nuc. Eng.*). Apparatus (e.g., an array of radiation detectors) which is used for tracing paths of charged particles in a magnetic field.

hoe coulter (*Agric.*). A coulter in the form of a small hoe; used on a drill to make a shallow trench for the seed.

Hoesch (*Build., Civ. Eng.*). A method of system building developed in Germany. It is made up of prefabricated light-gauge steel sections and precast reinforced concrete floor beams. Cladding is of lightweight concrete slabs or other suitable material. The system is light and low capital expenditure is required to develop it. See **system building.**

Hoffman binant electrometer (*Elec. Eng.*). Primitive suspended-needle electrometer which, although highly unstable, is the most sensitive d.c. measuring instrument (down to 10^{-19} ampere).

Hofmann degradation (*Chem.*). A process used in organic chemistry in determining the structure of amines. It involves exhaustive methylation, i.e., the use of excess of a methylating agent, such as methyl iodide. Good for alkaloids.

Hofmann's reaction (*Chem.*). A method of preparing primary amines from the amides of acids by the action of bromine and then of caustic soda. The number of carbon atoms in the chain should not be more than six, and the resulting amine has one carbon atom less than the amide from which it has been prepared.

Hofmeister series (*Chem.*). The simple anions and cations arranged in the order of their ability to coagulate solutions of lyophilic colloids.

hog (*For.*). A machine for reducing wood to coarse chips, usually for converting mill waste into fuel. (*Textiles*) Term given to wool from a year-old sheep, i.e., the first clipping. Also **hogget.**

hogback (*Geol.*). Sharp anticlinal rock fold; tilted ridge.

hog-back girder (*Build., Civ. Eng.*). Girder which curves along its top edge to be convex upwards.

hog cholera (*Vet.*). See **swine fever.**

hog-frame (*Build.*). A term applied to some forms of truss which are shaped so as to bulge on the upper side.

hogget (*Textiles*). See **hog.**

hogging (*Build.*). A mixture of gravel and clay, used for paving. Also **hoggin.** (*Ships*) This occurs when the middle of the ship is supported in the crest of the wave while the ends are in troughs or when the ends are more heavily loaded than the middle. If the ship actually bends she is said to be hogged. Cf. *sagging.*

hoghorn (*Radar*). Radar horn in which the waveguide is flared out, so that the electromagnetic wave is reflected from a parabolic surface.

hog-pit (*Paper*). The pit below the couch on the paper machine where waste paper stock is pumped back for re-use.

hogshead (*Brew.*). Cask holding 54 galls (245 litres) of beer.

hohlraum (*Acous.*). Cavity employed as black-body radiator. A *microwave hohlraum* is a similar device covering microwave frequencies and it is employed as wideband noise source.

Hohmann orbit (*Space*). A space trajectory tangential to, or osculating, two planetary orbits at its perihelion and aphelion respectively: it is also the most economical transfer orbit.

hoist (*Mining*). An engine with a drum, used for winding up a load from a shaft or in an underground passage such as a winze.

hoisting (*Eng.*). The process of lifting materials by mechanical means.

hoisting machine (*Eng.*). See **crab, differential chain block, hydraulic lift, jack, jigger, lift, winch.**

hoisting motor (*Elec. Eng.*). See **lift motor.**

hol-. Prefix. See **holo-.**

Holarctic region (*Zool.*). One of the primary faunal regions into which the surface of the globe is divided. It includes North America to the edge of the Mexican plateau, Europe, Asia (except Persia, Afghanistan, India south of the Himalayas, and the Malay peninsula), Africa north of the Sahara, and the Arctic islands.

holard (*Bot.*). The whole of the water contained in the soil.

holaspidean (*Zool.*). Said of Birds which possess a single row of large scales covering the posterior surface of the tarsometatarsus.

holcodont (*Zool.*). Having the teeth inserted in a continuous groove.

hold (*Cinema.*). The retention of an image on the screen longer than is natural; generally obtained by repeated printing of a frame in the negative on the positive. (*Comp.*) The retention of information contained in one storage device after copying it into another. (*Electronics*) The maintenance, in charge-storage tubes, of the equilibrium potential by means of electron bombardment. (*Ships*) A compartment within a ship's hull for the carriage of cargo. Below the lowermost deck it is termed *hold*; above this, *'tween decks*. For identification, the holds are numbered from the fore end of the ship. (*Telecomm.*) Synchronization control, whereby frequency of the local oscillator is adjusted to that of the incoming synchronizing pulses.

holdback (*Mining*). Device used on inclined transport system such as a climbing belt conveyor to prevent the load from running downhill if power fails. (*Nuc. Eng.*) Agent for reducing an effect, e.g., a large quantity of inactive isotope reduces the coprecipitation or absorption of a radioactive isotope of the same element.

Holden permeability bridge (*Elec. Eng.*). A permeability bridge in which the standard bar and the bar under test carry magnetizing coils, and are connected by yokes to form a closed magnetic circuit. The magnetizing currents are varied until there is no magnetic leakage between the yokes.

holder (*Elec. Eng.*). See **lampholder.**

holderbat (*Build.*). A metal collar formed in two half-round parts, capable of being clamped together around a rain-water, soil, or waste pipe, and having a projecting leg on one part for fixing to a wall.

Holder's inequality (*Maths.*). If a_r and b_r are positive, nonproportional, sets, and if $\alpha + \beta = 1$, then

$$(1)\ \sum_{1}^{n} a_r^{\alpha}\, b_r^{\beta} < \left(\sum_{1}^{n} a_r\right)^{\alpha} \left(\sum_{1}^{n} b_r\right)^{\beta},$$

if α and β are positive and

$$(2)\ \sum_{1}^{n} a_r^{\alpha}\, b_r^{\beta} > \left(\sum_{1}^{n} a_r\right)^{\alpha} \left(\sum_{1}^{n} b_r\right)^{\beta},$$

if $\alpha > 1$ or if $\alpha < 0$.

holdfast (*Bot.*). Any organ other than a root which attaches a plant (especially one of the lower plants) to a substratum. (*Join.*) A device for holding down work on a bench, comprising a main pillar which passes through a hole in the bench, and an adjustable clamp with a shoe for placing on the work.

holding altitude (*Aero.*). The height at which a controlled aircraft may be required to remain at a *holding point*.

holding anode (*Electronics*). Auxiliary d.c. anode in a mercury-arc rectifier for maintaining an arc.

holding beam (*Electronics*). Widely-spread beam of electrons used to regenerate charges retained on the dielectric surface of a storage tube or electrostatic memory.

holding pattern (*Aero.*). A specified flight track, e.g., *orbit* (q.v.) or figure-of-eight, which an aircraft may be required to maintain about a holding point.

holding point (*Aero.*). An identifiable point, such as a radio beacon, in the vicinity of which an aircraft under *air-traffic control* may be instructed to remain.

holding-up (*Eng.*). The action of pressing a heavy hammer against the head of a rivet while closing or forming the head on the shank.

hold-on coil (*Elec. Eng.*). An electromagnet which holds the moving arm of a motor starter or other similar device, in the 'on' position; if the current in the coil is reduced or interrupted, the arm returns to the 'off' position under the action of a spring.

hold-up (*Chem. Eng.*). In any process plant, the amount of material which must always be present in the various reactors, etc., to ensure satisfactory operation.

hole (*Civ. Eng.*). (1) A bore-hole. (2) A depression for accommodating a blasting charge. (*Electronics*) Vacancy in a normally filled energy band, *either* as result of electron being elevated by thermal energy to the conduction band, and so producing a hole-electron pair; *or* as a result of one of the crystal lattice sites being occupied by an acceptor impurity atom. Such vacancies are mobile and contribute to electric current in the same manner as positive carriers and mathematically are equivalent to positrons. Also called **negative-ion vacancy.**

hole current (*Electronics*). That part of the current in a semiconductor due to the migration of holes.

hole density (*Electronics*). The density of the holes in a semiconductor in a band which is otherwise full.

hole director (*Mining*). Light handheld frame used in rock drilling to align hole in face being prepared for blasting underground.

hole injection (*Electronics*). Holes can be *emitted* in *n*-type semiconductor by applying a metallic point to its surface.

hole theory of liquids (*Chem.*). Interpretation of the fluidity of liquids by regarding them as disordered crystal lattices with mobile vacancies or holes.

hole through (*Mining*). Meeting of headings in underground tunnelling or winze-arc raise driving.

hole trap (*Electronics*). An impurity in a semiconductor which can release electrons to the conduction or valence bands and so trap a 'hole'.

holiday (*Paint.*). A greater or lesser part of the surface accidentally missed during painting.

holing (*Build.*). The operation of piercing slates to receive nails.

holism (*Psychol.*). The approach to the study of personality which insists on treatment of personality as a whole, taking into account inherited structure, tendencies unique to the individual, cultural background, etc.

hollander or **beating engine** (*Paper*). A trough containing a beating roll with bars set parallel to the axis; it is used for reducing materials prepared in the breaker to the condition requisite for producing a particular class of paper.

hollands (*Textiles*). Linen fabrics of rather coarse texture; used (glazed) principally for window-blinds and linings. Cotton imitations vary widely in quality.

Holley-Mott system (*Chem. Eng.*). A form of liquid-liquid extraction plant in which a series of agitated and non-agitated settling vessels are connected in series on two levels, so that a counter current flow of the two streams is maintained with mixing and settling at each stage.

hollow back (*Bind.*). Type of binding in which the spine of the book is not pasted or glued down, leaving a space between the back of the sections and the leather or cloth when the book is open. Also called **open back**.

hollow bed (*Build.*). A bed joint in which, owing to the surfaces of the stones not being plane, there is contact only at the outer edges.

hollow blocks (or **tiles**) (*Build.*). Hollow burnt-clay or terra-cotta blocks (or tiles) much used as a building material for forming floors, or for partition walls.

hollow-cathode tube (*Electronics*). A gas-discharge tube in which radiation is emitted from a hollow cathode (closed at one end) in the form of a cathode glow.

hollow fusee (*Horol.*). A fusee with its top pivot sunk into the body to reduce the height of the movement.

hollow mandrel lathes (*Eng.*). The term formerly applied to lathes capable of having bar stock fed through the mandrel for repetition work.

hollow newel (*Build.*). The well-hole of a winding stair.

hollow pinion (*Horol.*). A pinion drilled throughout its length.

hollow quoin (*Hyd. Eng.*). A quoin accommodating the heel-post of a lock-gate in a vertical recess.

hollow roll (*Plumb.*). A joint between the edges of two lead sheets on the flat, made by turning up each edge at right angles to the flat surface, bringing the two turned-up parts together, and shaping them over to form a roll.

hollows (*Bind.*). Strips of strong paper, etc., attached to the boards and to the back of a book to strengthen the spine. Also called **back lining.** (*Foundry*) Fillets, or curves of small radius, uniting two surfaces intersecting at an angle; added to a pattern to give strength to the casting and facilitate withdrawal of the pattern from the mould. (*Join.*) Planes having a convex sole and cutting iron for forming hollow surfaces. Cf. *rounds*.

hollow tiles (*Build.*). See **hollow blocks.**

hollow walls (*Build.*). *Cavity walls* (q.v.).

Holmgren's canaliculi (*Zool.*). A system of fine canals which permeate the cytoplasm of the cell-body in some nerve cells.

holmium (*Chem.*). A metallic element, a member of the rare earth group. Symbol Ho, at. no. 67, r.a.m. 164·930. It occurs in euxenite, samarskite, gadolinite, and xenotine.

holmquistite (*Min.*). A rare lithium-bearing calcium-free variety of orthorhombic amphibole.

holo-, hol-. Prefix from Gk. *holos*, whole.

holoaxial (*Crystal.*). A term applied to those classes of crystals characterized by axes of symmetry only; such crystals are not symmetrical about planes of symmetry.

Holobasidae (*Bot.*). A subclass of the *Basidiomycetes*, containing members in which the basidium is not divided into cells.

holobasidium (*Bot.*). See **autobasidium.**

holobenthic (*Zool.*). Passing the whole of the life-cycle in the depths of the sea.

holoblastic (*Zool.*). Said of ova which exhibit total cleavage.

holobranch (*Zool.*). In Fish, a branchial arch carrying two rows of respiratory lamellae or filaments, one on the posterior and one on the anterior face.

Holocaine (*Chem.*). See **phenacaine.**

holocarpic (*Bot.*). Having the whole thallus transformed at maturity into a sporangium or a sorus of sporangia.

Holocene (*Geol.*). The most recent period of geological time, following the Quaternary period, and approximating to the period since the last glaciation. Holocene deposits include recent alluvium, blown sand, peat bogs, and the uppermost layer of marine sediments.

Holocephali (*Zool.*). An order of *Bradyodonti* with holostylic jaw suspension, extra claspers on the head and abdomen and a dentition composed of tooth-plates.

holocephalous (*Zool.*). Said of single-headed ribs.

holocrystalline rocks (*Geol.*). Those igneous rocks in which all the components are crystalline; glass is absent. Cf. *hemicrystalline rocks*.

holoentoblastia (*Zool.*). An abnormal Echinoderm larva, produced by cultivation in strong salt solution, in which the ectoderm is reduced to a small button at the animal pole.

holoenzymes (*Biochem.*). Enzymes which have a protein and a non-protein component.

hologametes (*Biol.*). Gametes as large as the ordinary energids of the species, and not formed by a special act of fission.

hologamy (*Bot.*). A fusion between two mature cells, each of which has been completely changed into a gametangium. (*Zool.*) The condition of having gametes which resemble in size and form the ordinary cells of the species; union of such gametes. Cf. *merogamy*.

hologastrous (*Zool.*). A form of abdomen found in some *Coleoptera*, in which the second abdominal sternite is fully sclerotized and distinct from the third.

hologenic (*Zool.*). Said of induction in which the whole soma is primarily affected.

holognathous (*Zool.*). Having the jaw composed of a single piece.

hologram (*Phys.*). A means of optical 'imaging' without the use of lenses, now a practical reality with the advent of the laser. The laser beam is split into two portions, one part directly illuminating a photographic film (or plate) while the other firstly illuminates the scene. The two portions produce an optical interference pattern on the film, which, when illuminated by a laser beam, will produce two images of the original scene. One of these is virtual, but the other is real and may be viewed without a lens.

holography (*Phys.*). (The technique or process of) making or using holograms.
holohedral (*Crystal.*). Term applied when a crystal is complete, showing all possible faces and angles.
Holomastigina (*Zool.*). An order of *Zoomastigina*, the members of which have many flagella; they are capable of ingesting food by amoeboid action at any point of the body.
holomastigote (*Zool.*). Having numerous flagella scattered evenly over the body.
Holometabola (*Zool.*). See **Oligoneoptera.**
holometabolic (*Zool.*). Showing a complete metamorphosis, as most members of the *Oligoneoptera*. Cf. *hemimetabolic*. *n.* **holometabolism.**
holomorphic function (*Maths.*). See **analytic function.**
holonephros (*Zool.*). A continuous excretory organ of Vertebrates, extending the length of the body cavity; from it, according to some authorities, the parts of the Vertebrate kidney have been derived.
holophytic (*Bot., Zool.*). Able to manufacture food from the simplest beginnings by photosynthesis, and neither parasitic nor saprophytic.
holoplankton (*Ecol.*). Organisms which spend their entire life cycle in the plankton. Cf. *meroplankton.*
holopneustic (*Zool.*). Of Insects, having all the spiracles open. Cf. *hemipneustic.*
holoptic (*Zool.*). Having the eyes of the 2 sides meeting in front.
holorhinal (*Zool.*). Having the posterior margin of the nares rounded.
holoschisis (*Cyt.*). See **amitosis.**
Holostei (*Zool.*). A superorder of *Actinopterygii*, containing ganoid Fishes with well-developed bony skeleton. Gar-pikes, Bow-fins and various extinct genera. Sometimes included with *Teleostei* in the subclass *Neopterygii*.
holostyly (*Zool.*). A type of jaw suspension in which the PPQ bar fuses with the cranium, the hyoid arch playing no part in the suspension; characteristic of *Holocephali*.
Holothuroidea (*Zool.*). A class of *Echinodermata* having a sausage-shaped body without arms; the tube feet possess ampullae and may occur on all surfaces; the anus is aboral, the madreporite internal; the skeleton is reduced to small ossicles embedded in the soft integument; free-living mud-feeders. Sea Cucumbers.
Holotricha (*Zool.*). An order of *Ciliata* the members of which have cilia approximately of equal length all over the body; in some forms, the cilia are restricted to special regions.
holotrichous (*Zool.*). Bearing cilia of uniform length over the whole surface of the body.
holotype (*Zool.*). The original type specimen, from which the description of a new species is established.
holozoic (*Zool.*). Devouring other organisms, as most animals. *n.* **holozoon.**
Holtz machine (*Elec. Eng.*). An early form of influence machine having one fixed and one moving plate; for starting, an initial charge must be given.
homacanth (*Zool.*). Having dorsal fin spines which, in the depressed position, cover one another completely. Cf. *heteracanth.*
Home Office switch (*Elec. Eng.*). See **earthed switch, shockproof switch.**
homeokinesis (*Cyt.*). Mitosis in which the chromosome material is divided equally between the daughter nuclei.
homeomorphic (*Maths.*). Topological spaces T_1 and T_2 are said to be homeomorphic, or topologically equivalent, if there is a mapping f from T_1 on to T_2 which is one-to-one and such that both f and its inverse f^{-1} are continuous. Such a mapping f is said to be a *homeomorphism*. In effect, two geometrical figures are homeomorphic if each can be transformed into the other by a continuous deformation, e.g., the circle, the square and the triangle are homeomorphic, i.e., topologically equivalent.
homeopathy (*Med.*). See **homoeopathy.**
homeostasis (*Zool.*). The tendency for the internal environment of the body to be maintained constant.
homeothrausmatic (*Geol.*). Said of orbicular rocks in which the composition of the cores of the orbs resembles that of the groundmass in which they are embedded. *Allothrausmatic* is the term applied to those rocks in which the two compositions are different. See also **isothrausmatic.**
homeotypic division (*Cyt.*). See **mitosis.**
homer (*Radio*). Any arrangement which provides signals or fields which can be used to guide a vehicle to a specified destination, e.g., a *homing aid* for aircraft, a *leader cable* for ships, or *guidance* system for missiles.
home range (*Ecol.*). A definite area to which individuals, pairs, or family groups of many types of animals restrict their activities.
home recorder (*Acous.*). A simplified type of gramophone-disk recorder, in which a record can be cut on a metal, celluloid or composition surface, and reproduced with simple processing (e.g., rubbing with acetic acid) to harden the surface.
homing (*Nav.*). (1) *Guided missile* (q.v.) guidance system in which a *seeker* is actuated by some influence from target, e.g., infra-red rays from engine exhaust, or radio transmission. (2) Process of directing a vessel straight towards a RDF transmitter. (*Telecomm.*) Said of a stepping relay when its wipers return to a datum set of contacts when relay is released. (*Teleph.*) The operation of a selector in returning to a predetermined normal position following the release of the connection.
homing aid (*Aero.*). Any system designed to guide an aircraft to an aerodrome or aircraft carrier. See **rebecca-eureka.**
homo-. Prefix from Gk. *homos*, same.
homobasidium (*Bot.*). See **autobasidium.**
homobium (*Bot.*). An association between an alga and a fungus, e.g., a lichen.
homoblastic (*Zool.*). Showing direct embryonic development: originating from similar cells.
homocentric (*Phys.*). Term applied when rays are either parallel or pass through one focus.
homocercal (*Zool.*). Said of a type of tail-fin, found in all the adults of the higher Fish, in which the vertebral column bends abruptly upwards and enters the epichordal lobe, which is equal in size to the hypochordal lobe.
homochlamydeous (*Bot.*). Having a perianth consisting of members all of the same kind, not distinguishable into sepals and petals.
homocyclic (*Chem.*). Containing a ring composed entirely of atoms of the same kind.
homocystinuria (*Med.*). A metabolic defect resulting in the excretion of homocystine (the oxidized form of cysteine, an essential amino acid) in the urine.
homodesmic structure (*Crystal.*). Crystal form with only one type of bond (either ionic or covalent).
homodont (*Zool.*). Said of teeth which all have the same characteristics.
homodromous (*Bot.*). Having the leaves inserted on spirals all running in a uniform direction.

homodynamic hybrid (*Gen.*). A hybrid having an equal grouping of characters derived from both its parents, and so differing from both in appearance.

homodyne reception (*Radio*). That using an oscillating valve adjusted to, or locked with, an incoming carrier, to enhance its magnitude and improve demodulation. Also called **demodulation of an exalted carrier.**

homoeomerism (*Zool.*). In metameric animals, the condition of having all the somites alike. Cf. *heteromerism*. *adj.* **homoeomeric.**

homoeopathy or homeopathy (*Med.*). A system of medicine, founded by Dr Samuel Hahnemann (1755–1843), the basic principle of which is *Similia similibus curentur* (=let likes be cured by likes). By experiment, Hahnemann found that (*a*) a disease is characterized by a definite symptom complex, (*b*) it can be effectively treated by the drug which produces in a healthy individual the most similar symptom complex, and (*c*) for a given disease, the proven drug is best administered in extreme dilution (expressed by the term *potency*, e.g., 30th potency).

homoeosis (*Bot.*). A type of variation in which a plant member takes on the characteristics of a member of different nature, as when a petal is changed into a stamen. (*Zool.*) In metameric animals, the assumption by a merome of the characters of the corresponding merome of another somite. Cf. *heterosis.*

homoeroticism (*Psychol.*). Orientation of the libido toward one of the same sex. Occasionally appears as passive *homosexuality*, but usually is well sublimated into normal behaviour.

homogametic (*Zool.*). Having all the gametes alike.

homogamous (*Bot.*). (1) Having all the flowers in an inflorescence alike, all being either perfect, staminate, pistillate, or neuter. (2) Having the anthers and stigmas ripe at the same time.

homogamy (*Zool.*). Inbreeding, usually due to isolation.

homogeneous (*Chem.*). Said of a system consisting of only one phase, i.e., a system in which the chemical composition and physical state of any physically small portion are the same as those of any other portion.

homogeneous coordinates (*Maths.*). A system of coordinates in which any multiple of the coordinates of a point or line represents the same point, e.g., in line-coordinates the line (a, b, c) is the same as the line (ka, kb, kc).

homogeneous function (*Maths.*). An algebraic function such that the sum of the indices occurring in each term is constant, e.g., $x^3+2x^2y+y^3$.

homogeneous ionization chamber (*Nuc. Eng.*). One in which both walls and gas have similar atomic composition, and hence similar energy absorption per unit mass.

homogeneous light (*Phys.*). The same as *monochromatic light.*

homogeneous linear equation (*Maths.*). A differential equation of the type

$$x^n \frac{d^ny}{dx^n} + a_1x^{n-1}\frac{d^{n-1}y}{dx^{n-1}} + \ldots + a_{n-1}x\frac{dy}{dx} + a_ny = f(x),$$

where the a_n are constants.

homogeneous radiation (*Phys.*). That of constant wavelength (monochromatic), or constant particle energy.

homogeneous reactor (*Nuc. Eng.*). One in which the fuel and moderator are finely divided and mixed so as to produce an effectively homogeneous core material. They may both be evenly suspended in a liquid coolant or the moderator itself may be liquid. See, e.g. **slurry reactor.**

homogenesis (*Zool.*). The type of reproduction in which the offspring resemble the parents.

homogenizer (*Phys.*). A device in which coarse and polydisperse emulsions are transformed into nearly monodisperse systems. The liquid is subjected to an energetic shear.

homogeny (*Zool.*). Similarity of individuals or of parts, due to common descent; applied to homograft. *adj.* **homogenous.**

homograft (*Biol.*). A graft taken from an animal of the same species as the recipient.

homoimerous (*Bot.*). Said of a lichen thallus in which the algal and fungal components are mixed and not arranged in layers.

homoioplastic, homoplastic (*Zool.*). In experimental zoology, said of a graft which is transplanted to a site identical with its point of origin, e.g. a skin graft to a skin site. (*Bot.*) See **homoplastic.**

homoiosmotic (*Ecol.*). Of an aquatic animal, able to maintain a steady internal concentration of salts in its body fluids independent of that in its environment. Freshwater animals, and most marine fish, are homoiosmotic. Cf. *poikilosmotic.*

homoiothermal (*Zool.*). See **warm-blooded.**

homologous (*Bot., Zool.*). Of the same essential nature, and of common descent.

homologous alternation of generations (*Bot.*). The doctrine that the sporophyte originated as a modification of the gametophyte, and not as a new phase introduced into the life-cycle.

homologous chromosomes (*Cyt.*). Each pair of chromosomes which associate at synapsis, these being the corresponding members of the two sets derived from the gametes.

homologous series (*Chem.*). A series of organic compounds, each member of which differs from the next by the insertion of a $—CH_2—$ group in the molecule. Such a series may be represented by a general formula and shows a gradual and regular change of properties with increasing molecular weight.

homologous variation (*Bot.*). The occurrence of similar variations in related species.

homolographic projections (*Geog.*). See **cylindrical projections, zenithal equal-area projection.**

homology (*Bot., Zool.*). The state of being homologous. *n.* **homologue.**

homomorphic (*Cyt.*). Said of chromosome pairs which have the same form and size.

homomorphosis (*Zool.*). The regeneration of a part in the same form as the original part. Cf. *heteromorphosis.*

homomorphous (*Bot., Zool.*). Alike in form.

homophyllous (*Bot.*). Having foliage leaves all of the same kind.

homophytic (*Bot.*). Having a bisexual, diploid sporophyte.

homoplasma (*Zool.*). In tissue culture, a medium prepared with plasma from another animal of the same species as that from which the tissue was taken. Cf. *autoplasma, heteroplasma.*

homoplastic (*Bot., Zool.*). (1) Of the same structure and manner of development but not descended from a common source. (2) See **homoioplastic.**

homoplasty (*Zool.*). Similarity between 2 different organs or organisms, due to convergent evolution.

homopolar (*Chem.*). Having an equal distribution of charge, as in a covalent bond.

homopolar generator (*Elec. Eng.*). Low-voltage d.c. generator based on Faraday disk principle which produces ripple-free output without commutation.

homopolar magnet (*Elec. Eng.*). One with concentric pole pieces.

homopolar molecule (*Elec. Eng.*). One without effective electric dipole moment.

homopterous (*Zool.*). Having both pairs of wings similar; pertaining to the *Homoptera.*

homoscedastic (*Stats.*). Applied to distributions having equal variance.

homoseismal line (*Geophys.*). A line drawn through places which are affected by an earthquake at the same time.

homosexuality (*Psychol.*). A very wide term implying attraction to one's own sex; recognized as a phase in normal development, in some individuals it tends to an active eroticism which has distorted the main concept. See **homoeroticism.**

homosporangic (*Bot.*). Having only one kind of sporangium.

homosporous (*Bot.*). Having spores all of the same kind.

homostyled (*Bot.*). Having styles all of the same length.

homosynapsis (*Cyt.*). Pairing of 2 similar chromosomes. Cf. *heterosynapsis.*

homotaxis (*Geol.*). A term introduced by Huxley to indicate that two strata occurring in different areas shared the same faunal characters and were therefore of the same age in the geological sense. A faunal element may originate in, and be dispersed from, a certain locality A. In the course of time it reaches locality B, and, eventually, the more distant C. The sediments accumulating in these three localities which contain the remains of this same organism are *homotaxial* though not contemporaneous.

homothallic (*Bot.*). Pertaining to *homothallism.* See **haplomonoecious.**

homothallism (*Bot.*). The condition in which a fungal mycelium is able to develop functional sexual organs on its own branches, and does not cooperate with another mycelium of the same species but of different physiological nature.

homothermous (*Zool.*). See **warm-blooded.**

homotopic mapping (*Maths.*). Two continuous mappings f and g of a topological space A into a topological space B are said to be homotopic if there is a function $F(x, t)$, representing a continuous mapping of $A \times I$ into B (I being the unit interval), for which $F(x, 0)=f(x)$, $F(x, 1) = g(x)$, for all x in A and for $0 \leqq t \leqq 1$. f and g are also said to be continuously deformed into each other.

homotypic (*Zool.*). Conforming to the normal condition. Cf. *heterotypic.*

homotypic division. See **heterotypic division.**

homozygosis (*Gen.*). The condition of having inherited a given genetical factor from both parents, and therefore of producing gametes of only one kind as regards that factor; genetical stability as regards a given factor. *n.* **homozygote.** *adj.* **homozygous.**

homunculus (*Biol., Med.*). A dwarf of normal proportions; a mannikin or little man created by the imagination; a miniature human form believed by animalculists to exist in the spermatozoon.

Honduras mahogany (*For.*). Reddish-brown hardwood from *Swietenia macrophylla,* found in Central America. Also called **baywood.**

hone, honestone or **whetstone** (*Geol.*). Term applied to fine-textured even-grained indurated sedimentary rocks which may be used as oil-stones for imparting a keen edge to cutting tools. Honestone has been largely replaced now by emery and silicon carbide products.

honey (*Nut.*). The concentrated secretion deposited as a food store by *Apis mellifera*; results from enzyme action upon nectar; rich in invert sugar.

honeycomb (*Aero.*). A gridwork across the duct of a wind tunnel to straighten the airflow; also **straighteners.** (*Textiles*) Fabric with a cell-like appearance resembling honeycomb, used for towels and quilts. Generally woven from coarse soft yarns in compact structures, known respectively as *Brighton honeycomb, Grecian honeycomb,* and *ordinary honeycomb.*

honeycomb bag (*Zool.*). See **reticulum.**

honeycomb base (*Print.*). A metal base for printing plates providing a means of both securing them and moving them into register. The base is drilled with a pattern of holes in which register hooks are placed, there being suitable holes for any position of the plate.

honeycomb coil (*Radio*). One in which wire is wound in a zig-zag formation around a circular former. The adjacent layers are staggered, so that the wires cross each other obliquely to reduce capacitance effects between turns.

honey-combing (*For.*). Separation of the fibres in the interior of timber due to drying stresses.

honeycomb slating (*Build.*). Similar to *drop-point slating* (q.v.); the tiles, however, have their bottom corners removed.

honeycomb structure (*Aero.*). Lightweight, very rigid, material for aircraft skin or floors, usually made from thin light-alloy plates with a bonded foil interlayer of generally honeycomb-like form. For high supersonic speeds, a heat-resistant material is made by brazing together stainless-steel or nickel-alloy skins and honeycomb core.

honeycomb wall (*Build.*). A wall built so as to leave regular spaces, usually entirely with stretchers. See also **sleeper wall.**

honey dew (*Zool.*). A sweet substance secreted by certain *Hemiptera*; emitted through the anus.

honey guide (*Bot.*). A patch or streak of colour on a petal, different from the general colour, possibly serving to direct insects to the nectaries, or as a guide for their landing on the flower.

honey-leaf (*Bot.*). A nectary with a strong development of the lamina of the petal.

honey-stomach (*Zool.*). In some *Hymenoptera,* a dilatation of the oesophagus which serves as a reservoir for the imbibed liquid and regurgitates it when required.

Hong-Kong foot (*Med.*). Ringworm of the foot.

honing (*Eng.*). The process of finishing cylinder bores, etc., to a very high degree of accuracy by the abrasive action of stone or silicon carbide slips held in a head having both a rotatory and axial motion.

honing machine (*Eng.*). A partly hand-operated or wholly automatic machine for *honing.*

hood (*Build.*). (1) A cowl for a chimney. (*Glass.*) See **potette.** (*Zool.*) In *Tetrabranchiata,* a thickened anterior region of the body, in contact with the edge of the shell and forming an operculum when the animal withdraws; in the Cobra, an expansible region just behind the head; the head in any animal if it differs markedly in coloration from the rest of the body; a crestlike eminence on the head. *adj.* **hooded.**

hood jettison (*Aero.*). Mechanism, often operated by explosive bolts or cartridge, for releasing the pilot's canopy in flight. Confined to military aircraft.

hood mould (*Build.*). A projecting moulding above a door or window opening.

hoof (*Zool.*). In Ungulates, a horny proliferation of the epidermis, enclosing the toes.

hook bolt (*Build.*). A galvanized-iron bolt formed out of a rod which is bent at one end into a hook serving as the head, and threaded at the other to take a nut; used for fixing corrugated sheeting.

hook-down (*Typog.*). See **hook-up.**

hooked disseminule (*Bot.*). A fruit, seed, or spore bearing outgrowths in the form of hooks; these become attached to the bodies of animals and assist in dissemination.

hooked joint (*Join.*). A form of joint used between the meeting edges of a door and its case when an airtight or dustproof joint is necessary; the meeting edges on the door have a projection on them fitting into a corresponding recess in the case.

hooked plates (*Bind.*). Single-leaf illustrations sewn in with the *section*, having been printed on paper large enough to allow a narrow fold down the back edge.

hooker cell (*Chem. Eng.*). A commercial type of *diaphragm cell* (q.v.).

Hookeriales (*Bot.*). An order of the Mosses; the gametophores are branched and prostrate. The leaves are asymmetrical, frequently with two midribs and lying in one plane. The sporophyte has a double peristome and is generally pleurocarpous.

Hooker's green (*Paint.*). An artist's pigment consisting of Prussian blue mixed with gamboge gum.

Hooke's joint (*Eng.*). A common form of *universal joint* (q.v.), comprising two forks arranged at right angles and coupled by a cross-piece.

Hooke's law (*Phys.*). 'In an elastic material, strain is proportional to stress.' The value of the stress at which a material ceases to obey Hooke's law is known as the *elastic limit*. See **elasticity.**

hook-out blind (*Build.*). An outside roller blind fitted with metal side arms and bottom rail so that the blind may be supported away from the window.

hook rebate (*Join.*). The S-shaped rebate formed on the meeting edges of a *hooked joint* (q.v.).

hook transistor (*Radio*). A junction transistor which uses an extra *p-n* junction to act as a trap for holes, thus increasing the current amplification.

hook-up (*Min. Proc., etc.*). In pilot plant testing, flexible assembly of machines into continuous flow line before final treatment is arrived at. (*Telecomm.*) A temporary communication channel. (*Typog.*) The end of a line turned over and bracketed in the line above or below (*hook-down*). Often used in setting up poetry.

hookworm disease (*Med.*). See **ankylostomiasis.**

hookworms (*Med., Zool.*). Parasitic strongyloid nematodes with hooklike organs on the mouth for attachment to the host. Humans are attacked chiefly by the genus *Ankylostoma*, which penetrates the bare feet and induces a form of anaemia. See also **ankylostomiasis.**

Hoopes process (*Met.*). A process for the refining of aluminium electrolytically to a purity of 99·99%. Metal made by the Hall process is alloyed with 33% of copper and made the anode in a nonaqueous electrolytic bath composed of alumina and fluorides of barium, sodium, and aluminium. When current is passed between the hearth and carbon electrodes on top of the bath, aluminium dissolves from the anode alloy and pure metal accumulates at the cathodes.

hooping (*Civ. Eng.*). Reinforcing bars for ferro-concrete, bent either to a circular or helical shape.

hoop iron (*Build.*). Thin strip-iron used for securing barrels, and also for various purposes in the building trades, e.g., as reinforcement in brick walls; also for the packaging of bricks or building blocks to enable them to be lifted in quantity. See **hoop-iron bond.**

hoop-iron bond (*Build.*). A bond sometimes formed at the junctions of walls not built at the same time; made by using long strips of hoop iron, usually $2 \times \frac{1}{16}$ in., half built into the old wall and half into the new one.

hoop stress (*Eng.*). The stress in a tube under pressure, acting tangentially to the perimeter of a transverse section.

hoose (*Vet.*). See **husk.**

hop (*Radio*). Distance along earth's surface between successive reflections of a radio wave from an ionized region; also called **skip.**

hop-back (*Brew.*). Vessel containing a filter bed made of hops through which the wort is passed after boiling.

hopeite (*Min.*). Hydrated phosphate of zinc, occurring rarely in zinc mines as orthorhombic grey crystals.

'Hope sapphire' (*Min.*). Synthetic stone having the composition of spinel and a blue colour which turns purple in artificial light. First produced in the attempts to synthesize sapphire.

Hopkins-Cole reaction (*Chem.*). A reaction given by proteins, consisting in the appearance of a reddish-violet ring when a mixture of a protein solution and a solution of glyoxalic acid is brought into contact with concentrated sulphuric acid.

Hopkinson test (*Elec. Eng.*). A method of testing two similar d.c. machines on full load without requiring a large consumption of power from the supply; one machine fed from the supply drives the other as a generator, which returns power to the supply.

Hopkinson-Thring torsion meter (*Eng.*). A torsion meter in which two mirrors are mounted on a flange fixed to one end of the shaft under test, one mirror being fixed and the other so mounted that it tilts as the shaft twists. The separation of light beams reflected by the mirrors is a measure of the shaft twist.

Hoplocarida (*Zool.*). A superorder of *Malacostraca* in which there are 6 abdominal somites, and the anterior thoracic limbs have a 2-jointed protopodite, while in the posterior ones it is 3-jointed; the carapace leaves at least 4 thoracic somites distinct; the ocular and antennular somites of the head are free and movable; the protopodite of the antenna is 2-jointed.

hopper (*Agric.*). (1) A mechanical contrivance for stripping hops from the bines. (2) A vessel from which seed, fertilizer, etc., is spread. (*Build.*) A draught-preventer at the side of a hopper light. (*Mining*) A container or surge bin for broken ore, used to hold small amounts.

hopper barge (*Civ. Eng.*). A vessel divided into compartments fitted with flap doors at the bottom, so that after being loaded by a stationary dredger it can convey the dredged material to a place of deposit.

hopper crystal (*Crystal., Min.*). A crystal which has grown faster along its edges than in the centres of its faces, so that the faces appear to be recessed. This type of skeletal crystallization is often shown by rock-salt.

hopper dredger (*Civ. Eng.*). The type of bucket-ladder or suction dredger which not only dredges material from below but has hopper compartments fitted with flap-doors at the bottom; into these compartments the material is discharged as it is dredged, and from them it is deposited after the vessel has moved to the place of deposit.

hopperfeed (*Eng.*). A machine used for unscrambling identical components from bulk, orienting them, and delivering them in an orderly arrangement to an assembling machine or other mass-production process.

hopper light (*Build.*). A window-sash arranged to open inwards about hinges on its lower edge.

hopper window (*Build.*). A hopper light fitted at the sides with hoppers.

Hoppus foot (*For.*). A unit of volume obtained by using the square of the quarter girth for trees or logs instead of the true sectional area; only applied to timber in the round.

hops (*Brew.*). The amentaceous fruit *strobiles* of the hop (*Humulus lupulus*), a perennial climbing moraceous herb, which is boiled with wort, in order to impart the bitter flavour characteristic of beer; hops also act as a preservative. See **wort.**

hopsack (or matt) weave (*Textiles*). A development of plain weave in which two or more ends and picks interlace in the same order. Twilled varieties are often made. Also known as **Celtic weave.** See also **basket weave.**

hordein (*Chem.*). A prolamine obtained from barley. It is a typical gliadin.

hordeolum (*Med.*). See **sty.**

horizon (*Astron.*). That great circle, of which the zenith and the nadir are the poles, in which the plane tangent to the earth's surface, considered spherical at the point where the observer stands, cuts the celestial sphere. (*Geol.*) That (also called *stratigraphical level*) which has reference to the systematic position of a stratum on the geological time scale. Thus if a certain shale is referred to the horizon of *Didymograptus murchisoni*, it implies that it occurs at a particular level in the Llanvirn Series of the Ordovician System; it will contain the fossil remains of this particular graptolite or, in rocks of different lithology, traces of other creatures living at the same time. (*Optics*) The more or less coloured visual impression experienced subjectively by blind persons. (*Surv.*) A plane perpendicular to the direction of gravity shown by a plumbline at the point of observation.

horizon glass (*Surv.*). See **sextant.**

horizon sensor (*Electronics*). Sensor providing a stable vertical reference level for missiles and depending on the use of a thermistor to detect the thermal discontinuity between earth and space.

horizontal (*Bot.*). Spreading at a right angle to a support.

horizontal antenna (*Radio*). One comprising a system of one or more horizontal conductors, radiating or responsive to horizontally-polarized waves.

horizontal axis (*Surv.*). See **trunnion axis.**

horizontal blanking (*TV*). The elimination of the horizontal trace in a CRT during flyback.

horizontal circle (*Surv.*). The graduated circular plate used for the measurement of horizontal angles by theodolite.

horizontal component (*Elec. Eng.*). Component of earth's magnetic field which acts (i.e., exerts a force on a unit pole placed in it) in a horizontal direction.

horizontal draw-out metal-clad switchgear (*Elec. Eng.*). Metal-clad switchgear in which the switch itself can be isolated by removing it along suitable guides in a horizontal direction from the fixed portion of the panel.

horizontal engine (*Eng.*). Any engine in which the cylinders are horizontal.

horizontal escapement (*Horol.*). See **cylinder escapement.**

horizontal parallax (*Astron., Surv.*). The value of the geocentric parallax for a heavenly body in the solar system when the body is on the observer's horizon. In astronomy, the **equatorial horizontal parallax** of a planet or of the moon is the angle subtended at the centre of that body by the equatorial radius of the earth. See also **solar parallax.**

horizontal polarization (*Elec.*). That of an electromagnetic wave when the alternating electric field is horizontal.

horizontal resolution (*TV*). The number of picture elements resolved along the scanning line in any form of facsimile reproduction.

horizontal sheeting (*Civ. Eng.*). Long horizontal poling boards placed on each side of a trench excavated in bad ground and strutted apart. Short vertical walings are introduced when the trench has been sunk about 1 m.

horizontal stabilizer (*Aero.*). Tail plane. See **stabilizer.**

hormesis (*Bot.*). The stimulus given to an organism by nontoxic concentrations of a toxic substance.

hormic theory (*Psychol.*). Theory which stresses the importance of instinctive impulses and purposive striving.

hormocyst (*Bot.*). A short hormogonium enclosed in a thick stratified sheath.

Hormogonales (*Bot.*). An order of the *Myxophyceae*, in which the cells form definite filaments, and form hormogones.

hormogone (*Bot.*). A short length of filament in the *Myxophyceae*, which breaks free and can grow into a new filament.

hormone (*Biochem.*). An internal secretion produced by the endocrine or ductless glands of the body and exercising a specific stimulatory (Greek *hormōn*, 'urging on', 'stirring up') physiological action on other organs to which it is carried by the blood. Important hormones are thyroxine, adrenaline, corticosterone, insulin, oestradiol, testosterone and progesterone. (*Bot.*) Similar substances (e.g. *auxins* (q.v.)) are formed in the actively growing parts of plants; they diffuse within the plant to other regions of the plant body, and regulate and influence development.

hormospore (*Bot.*). A specialized hormogone, formed terminally in the filament.

horn (*Acous.*). Tube of continuously varying cross-section used in launching or receiving of radiation, e.g., acoustic horn, electromagnetic horn. Horns are best classified by their geometric or other shapes which include: *compound, conical, corner, exponential, folded, logarithmic, pyramidal, re-entrant, sectoral, tractrix* (qq.v.). (*Eng.*) Any projecting part, such as the two jaws of a horn-plate carrying an axle-box. (*Zool.*) Keratin; one of the pointed or branched hard projections borne on the head in many Mammals; any conical or cylindrical projection of the head resembling a horn; in some Birds, a tuft of feathers on the head; in some *Gastropoda*, a tentacle; in some Fish, a spine. *adjs.* **horned, horny.**

horn arrester (*Elec. Eng.*). A lightning arrester consisting of a horn gap which arcs over on the occurrence of a lightning surge, but which rapidly extinguishes the arc on account of the special shape of the electrodes.

horn balance (*Aero.*). An *aerodynamic balance* (q.v.) consisting of an extension forward of the hinge line at the tip of a control surface; it may be *shielded* (i.e., screened by a surface in front) or *unshielded*.

hornbeam (*For.*). Tree of the genus *Carpinus*, yielding a moderately durable wood whose uses are confined to indoor conditions such as inlaying, marquetry, and the making of small novelties. It can also be dyed black to simulate ebony.

hornblende (*Min.*). An important rock-forming mineral of complex composition, essentially silicate of calcium, magnesium, and iron, with smaller amounts of sodium, potassium, hydroxyl, and fluorine; crystallizes in the monoclinic system; occurs as black crystals in many different types of igneous and metamorphic rocks, including hornblende-granite, syenite, diorite, andesite, etc., and hornblende-schist and amphibolite.

hornblende-gneiss (*Geol.*). A coarse-grained metamorphic rock, containing hornblende as the dominant coloured constituent, together with feldspar and quartz, the texture being that typical of the gneisses. Differs from hornblende-schist in grain size and texture only.

hornblende-granite (*Geol.*). A type of granite, usually adamellite or granodiorite, containing hornblende as an essential constituent; with decreasing quartz, grades through tonalite into normal diorite.

hornblende-schist (*Geol.*). A type of green schist, formed from basic igneous rocks by regional metamorphism, and consisting essentially of sodic plagioclase, hornblende, and sphene, frequently with magnetite and epidote. See also **glaucophane.**

horn-break fuse (*Elec. Eng.*). A fuse fitted with arcing horns to assist in the rapid extinction of any arc which may be formed.

horn centre (*Instr.*). A small transparent disk, originally of horn, used by draughtsmen to provide a substance into which the point of the compasses may be placed, in lieu of the paper beneath, at points which are to be much used as centres for describing arcs.

Horner's method (*Maths.*). A method of approximating, with any desired degree of accuracy, to the roots of an algebraic equation. The roots of the equation are successively diminished, each diminished equation giving a further digit of the root sought, so that a further diminution is possible.

Horner's syndrome (*Med.*). The combination of small pupil, sunken eye, and drooping of upper eyelid, due to paralysis of the sympathetic nerve in the region of the neck.

horn feed (*Radar*). The feed to a transmitter in the form of a horn.

hornfels (*Geol.*). A series of rocks which have been partially or completely recrystallized by the heat of igneous masses. See **hornstone.**

horn gap (*Elec. Eng.*). A spark gap of gradually increasing length, such that an arc struck across it gets longer and finally extinguishes itself.

horn gate (*Foundry*). Horn-shaped ingates or sprues, radiating from the bottom of a runner, which supply several small moulds made in the same moulding box. See **ingate.**

horn lead (*Min.*). The translation of the French term *plomb corné*, sometimes applied to the mineral *phosgenite* (q.v.).

horn loudspeaker (*Acous.*). A loudspeaker in which the radiating device is acoustically coupled to the air by means of a horn.

horn press (*Eng.*). A power press used originally for horning or grooving side seams, but suitable for light punching and similar operations when fitted with an (adjustable) table.

horns (*Join.*). The ends of the head in a door or window frame when these project beyond the outer surfaces of the posts.

hornsilver (*Min.*). See **cerargyrite.**

hornstone (*Geol.*). An old name for rocks differing widely in composition and origin, characterized by their flinty, compact appearance.

horn throat and **mouth** (*Radar*). Beginning and end (flare) of metal horn, between which it may be curved, folded, bifurcated.

horological (*Bot.*). Said of a flower which opens and shuts at a definite time of day.

horology. The science of time measurement, or of the construction of timepieces.

horripilation (*Med.*). Erection of the hairs on the skin, giving rise to *goose flesh*.

horse (*Build.*). The wooden backing to a zinc mould used in forming cornices. (*Carp.*) (1) One of the *strings* (q.v.) supporting the treads and risers of a stair. (2) A trestle for supporting a board or timber while it is being sawn. (*Mining*) A mass of barren or country rock occurring in a lode or reef. (*Plumb.*) A wooden finial which is to be covered with lead.

horse-flesh ore (*Min.*). A name applied by Cornish miners to the mineral *bornite* (q.v.) on account of its reddish-brown colour and its red and blue tarnish.

horse latitudes (*Meteor.*). See **trade-winds.**

horse mower (*Agric.*). A cutter bar, 4–5 ft (1·5 m) long, hinged at one end to a 2-wheeled frame, and projecting from the side of the machine. The cutting is done by a scissors-like action, the power being transmitted from the landwheels of the machine.

horse path (*Hyd. Eng.*). A canal towing-path.

horse power (*Eng.*). The mechanical engineering unit of power, equivalent to working at the rate of 33 000 ft lbf/min, or 42·41 B.t.u./min, or 745·70 watts. Abbrev. **h.p.** See also **cheval-vapeur.**

horse pox (*Vet.*). A contagious viral infection of equines. It is characterized by a papulo-vesicular eruption of the skin and mucous membranes.

horseshoe curve (*Surv.*). A curve whose arc subtends an angle of more than 180° at the centre, so that the *intersection point* (q.v.) lies on the same side of the curve as the centre.

horseshoe filament (*Elec. Eng.*). An electric lamp filament in the shape of a single half-turn.

horseshoe magnet (*Elec. Eng.*). Traditional form of an electro- or permanent magnet, as used in many instruments, e.g., meters, magnetrons.

horsing-up (*Build.*). A term applied to the building-up of the mould used in running cornices, etc.

horst faults (*Geol.*). Two parallel normal faults hading outwards and throwing in opposite directions, the resulting structure being termed a *horst*.

Horváth's stain (*Micros.*). See **formol-silver stain.**

hose bit (*Tools*). A type of *shell bit* (q.v.) with a projecting tip which withdraws the waste.

hose-proof (*Elec. Eng.*). Said of a type of enclosure for electrical apparatus so constructed as to exclude water when the apparatus is washed down with a hose.

hosiery (*Textiles*). Term applied formerly to knitted articles intended for footwear, but now applied to all sorts of warp or weft knitted fabrics.

hospital bus-bars (*Elec. Eng.*). A set of bus-bars provided in a power station or substation for temporary or emergency purposes.

hospital switch (*Elec. Eng.*). (1) A switch used on tramway or railway controllers to cut a faulty motor out of circuit. (2) Any switch for changing a circuit over to an emergency supply in case of failure of the main supply.

host (*Biol.*). An organism which, temporarily or permanently, supports another organism (parasite) at its own expense. (*Phys.*) Essential crystal, base material, or matrix of a luminescent material.

host choice (*An. Behav.*). The behaviour of a parasitoid when choosing a specific individual of the host species for oviposition.

host rock (*Mining*). See **country rock.**

hot (*Elec.*). Charged to a dangerously high potential. (*Radiol.*) Colloquial reference to dangerous radioactive substances; hence *hot laboratory*, where extreme precautions against irradiation of humans are taken.

hot-air engine (*Eng.*). One in which the working fluid, air, is alternately heated and cooled by a furnace and regenerator. See **Stirling engine.**

hot-air heater (*Build.*). A heater which supplies warm air through gratings in the floor or openings in the walls.

hot-blast stoves (*Met.*). Large stoves, filled with a brick chequerwork, used for pre-heating the air blown into the blast-furnace. Also called **Cowper stoves.**

hot-bulb ignition (*Eng.*). The method of ignition used in semi-diesel engines. The hot bulb is a mass of metal in the cylinder head which projects into the combustion chamber where it ignites the mixture. The hot bulb has to be heated by a blow torch before starting the engine.

hot cathode (*Electronics*). One in which the electrons are produced thermionically.

hot-cathode discharge lamp (*Light*). A discharge lamp employing a heated cathode to increase its efficiency, improve the starting, and reduce the voltage drop across the tube.

hot-cathode rectifier (*Electronics*). One with emitting cathode heated independently of the rectified current, e.g., a mercury thyratron.

hot deck (*For.*). See **deck.**

hot-die steels (*Met.*). Shock-resistant alloys used in high-temperature forging. In one type, from 3 to 10% chromium is used, and in another, 8 to 10% of tungsten, as main alloying constituents.

hot dip (*Met.*). Galvanizing of iron by immersion in molten zinc or spelter.

hot-drawn (*Met.*). Term describing metal wire, rod or tubing which has been produced by pulling it heated through a constricting orifice.

hot electrons (*Electronics*). Thermal electrons travelling in the solid lattice. As a result of the inability of relaxation processes to carry away the energy received from the electric field, the electrons become heated.

hot-extraction gas analysis (*Vac. Tech.*). Determination of the amount and nature of the gases liberated from a metal in the solid or liquid state when heated in a vacuum.

hot galvanizing (*Eng.*). The process of producing a corrosion-resistant zinc coating on articles by immersion in molten zinc or spelter, as distinct from *electrogalvanizing* (q.v.).

hot-ground pulp (*Paper*). Mechanical wood pulp prepared rapidly from the raw wood, using the minimum of water.

hot insulation mastics (*Build.*). 'Breathing' mastics prepared by emulsifying bitumen in water and filling with asbestos floats, etc. Although the film formed is black and waterproof, it permits the passage of water vapour and allows the lagging around a hot surface to dry out.

hot line (*Teleph.*). A permanent line in a submarine telephone cable system which is leased to governments and military users.

hot press (*Bind.*). See **blocking press.**

hot-pressed (*Paper*). Paper finished by glazing with hot plates.

hot pressing (*Acous.*). A water-wave effect produced on the surface of a gramophone record by exposure to the air before it is sufficiently cool. (*Powder Tech.*) Making a compact by application of heat as well as pressure. The combined effect is general softening and/or the liquefaction of a phase for it to sinter. Far greater amounts of pressure and heat would be required separately.

hot-press stamping (*Bind.*). See **blocking.**

hot-rolled (*Paper*). Paper glazed by means of steam-heated cylinders.

hot saw (*Eng.*). A metal-cutting circular saw used to cut off the ends of heated steel forgings, billets, etc., in a steelwork.

hot-short or red-short (*Met.*). Said of metals that tend to be brittle at temperatures at which hot-working operations are performed.

hot spot (*Electronics*). A small region of an electrode in a valve (or CRO screen) which has a temperature above the average. (*I.C. Engs.*) Part of the wall surface of the induction manifold of a petrol-engine on which the mixture impinges; heated by exhaust gases to assist vaporization and distribution.

hot top (*Met.*). Feeder head, containing a reservoir of molten metal drawn on by a cooling ingot as it solidifies, thus avoiding porosity. See **ingot, ingot mould.**

hot well (*Eng.*). The tank or pipes into which the condensate from a steam-engine or turbine condenser is pumped, and from which it is returned by the feed pump to the boiler.

hot-wire (*Elec. Eng.*). Said of an electrical indicating instrument whose operation depends on the thermal expansion of, or change in resistance of, a wire or strip when it carries a current. See **hot-wire ammeter, etc.**

hot-wire ammeter (*Elec. Eng.*). An ammeter operating on the *hot-wire* (q.v.) principle; of use chiefly for very high frequencies.

hot-wire anemometer (*Meteor.*). An instrument which measures wind velocities by using their cooling effect on a wire carrying an electric current, the resistance of the wire being used as an indication of the velocity.

hot-wire arc lamp (*Elec. Eng.*). A form of clutch arc lamp in which the clutch controlling the movement of the carbons is operated by the expansion of a wire, according to the current passing through it.

hot-wire detector (*Radio*). A fine wire which is heated by the passage of high-frequency currents, producing a change in its d.c. resistance.

hot-wire microphone (*Acous.*). One in which a d.c. signal through a hot wire is modulated by resistance variations consequent upon the cooling effect of an incident sound wave.

hot-wire oscillograph (*Elec. Eng.*). An oscillograph in which a moving mirror is supported by an arrangement of wires carrying the current to be measured, and which is deflected as a result of the thermal expansion of these wires.

hot-wire voltmeter (*Elec. Eng.*). A voltmeter

operating on the *hot-wire* (q.v.) principle. See **Cardew voltmeter.**

hot-wire wattmeter (*Elec. Eng.*). A wattmeter in which the deflection is indicated by means of a mirror mounted on an arrangement of fine wires carrying currents proportional to the voltage, and which is deflected as a result of the thermal expansion of the wires.

hot-working (*Met.*). The process of shaping metals by rolling, extrusion, forging, etc., at elevated temperatures. The hot-working range varies from metal to metal, but it is, in general, a range in which recrystallization proceeds concurrently with the working, so that no strain-hardening occurs.

Houdry process (*Chem.*). Catalytic cracking of petroleum, using activated aluminium hydrosilicate.

hour angle (*Astron.*). The angle, generally measured in hours, minutes, and seconds of time, which the hour circle of a heavenly body makes with the observer's meridian at the celestial pole; it is measured positively westwards from the meridian from 0 to 24 hr.

hour circle (*Astron.*). (1) The great circle passing through the celestial poles and a heavenly body, cutting the celestial equator at 90°. (2) The graduated circle of an equatorial telescope which reads sidereal time and right ascension.

hour-counter (*Elec. Eng.*). See **time-meter.**

hour-glass piston (*I.C. Engs.*). A petrol-engine piston provided with a waisted central portion to reduce the area in contact with the cylinder wall and so reduce friction.

hour-glass stomach (*Med.*). Constriction of the middle part of the stomach, due either to spasm of stomach muscle or to the formation of scar tissue in connexion with a gastric ulcer, the constriction in the latter case being permanent.

hour-meter (*Elec. Eng.*). See **time-meter.**

hour rack (*Horol.*). The toothed quadrant in a striking clock, one tooth of which is picked up by the gathering pallet for each hour struck.

hour wheel (*Horol.*). The wheel in the motion work which carries the hour hand.

house (*Zool.*). In *Protozoa*, a loose-fitting shell with a wide mouth; in *Larvacea*, the test, which is loose-fitting and not attached to the animal; in *Pterobranchiata*, a tube secreted by the proboscis.

house corrections (*Typog.*). The correcting of mistakes made by the compositor when setting by hand or machine, as distinct from author's corrections made by him after checking the proofs.

housed joint (*Carp.*). A fitted joint, such as a tenon in its mortise.

housed string (*Carp.*). A string which has its upper and lower edges parallel to the slope of the stair, and houses, in grooves specially cut in the inner side, the ends of the steps. Also called **a close string.**

house exchange system (*Teleph.*). A *house telephone system* (q.v.) which has means for direct connexion to the public exchange.

housekeeping (*Comp.*). Colloquialism for preparation of memory by programme.

housemaid's knee (*Med.*). Inflammation of the bursa in front of the patella of the knee.

house service meter (*Elec. Eng.*). An integrating meter for measuring the electrical energy consumption of a domestic installation.

house style (*Typog.*). See **style of the house.**

house telephone system (*Teleph.*). A system without a central switchboard, for providing direct connexion between any station and any other station(s) in the system, by means of a multiple cable. The system is not connected to the public exchange.

housing (*Carp., Join.*). A method of jointing two timbers in which the whole of the end of one is fitted into a corresponding blind mortise cut in the other. (*Elec. Eng., etc.*) Containment of apparatus to prevent damage in handling or operation.

hoven (*Vet.*). See **bloat.**

hovercraft (*Ships, etc.*). A craft which can hover over or move across water or land surfaces while being held off the surfaces by a cushion of air. The cushion is produced either by pumping air into a plenum chamber under the craft or by ejecting air downwards and inwards through a peripheral ring of nozzles. Propulsion can be by tilting the craft, or by jet, or air propeller or, over water, by water propeller, or, over land, by low-pressure tyres or tracks.

hovership (*Ships, etc.*). A *hovercraft* (q.v.) intended for use over water only but capable of coming in over a beach to land.

Howard protective system (*Elec. Eng.*). A form of earth-leakage protection sometimes applied to a.c. machines and equipment. It consists of a current transformer connected between the frame of the machine and earth; if a current flows through the transformer a relay is operated, which opens the main circuit-breaker.

Howe factor (*Nuc. Eng.*). One expressing the fractional increase in volume of a reactor fuel as a result of the additional atoms produced during fission.

howieite (*Min.*). A black triclinic hydrous silicate of sodium, manganese and iron.

howl (*Acous.*). A high-pitched audio-tone due to unwanted acoustic (or electrical) feedback. (*Aero.*) See **screeching.**

howler (*Teleph.*). A device which uses acoustic feedback between a telephone transmitter and a telephone receiver to maintain a continuous oscillation and so provides suitable currents for testing telephonic apparatus.

Hoyt's metal (*Met.*). A tin base (91·5%) *white metal* (q.v.), containing also antimony (3·4%), lead (0·25%), copper (4·3%) and nickel (0·55%).

h.p. (*Eng.*). Abbrev. for *horsepower.*

H-paper (*Paper*). Höchstädter paper, a paper which is coated on one side with aluminium foil, the composite sheet being perforated to allow oil to flow freely.

***h*-parameters** (*Electronics*). See **transistor parameters.**

HP compressor (*Aero.*). Abbrev. for *high-pressure compressor.*

H-piece (*Bot.*). A short lateral hypha connecting two longer hyphae, giving an H-like figure, and probably facilitating the flow of nutritive material about the mycelium.

H-plane (*Elec.*). The plane containing the magnetic vector H in electromagnetic waves, and containing the direction of maximum radiation. The electric vector is normal to it.

HP stage (*Aero.*). Abbrev. for *high-pressure turbine stage.*

HP turbine (*Aero.*). Abbrev. for **high-pressure turbine.**

H-radar (*Radar*). Navigation system in which an aircraft interrogates two ground stations for distance.

H2S (*Radar*). An airborne centimetric radar system using a rotational aerial, the radar echoes from the ground being reproduced by intensity modulation on a *plan-position indicator* (q.v.) to 'paint' a contrast map on a cathode-ray tube screen corrected for slant range.

H-section (*Telecomm.*). An electrical network derived from the T-section, in which half of each series arm is placed in the other leg of the circuit, making the section balanced.

H.S.L.-type cable (*Cables*). One in which each core has H-paper wrapping and separate lead sheath.

HTO (*Chem.*). Water in which an appreciable proportion of ordinary hydrogen is replaced by tritium.

H.T.S. (*Chem.*). Abbrev. for *heat transfer salt*.

H.T.U. (*Chem. Eng.*). Abbrev. for *height of a transfer unit*.

H-type cable (*Cables*). A cable having a wrapping of H-paper round each core. There are single-core and 3-core H-type cables. The benefit of H-paper in single-core cables is that no ionization occurs between the core and sheath. In 3-core cables no ionization occurs in the fillings.

H-type pole (*Elec. Eng.*). A type of wooden support for overhead transmission lines, being two vertical poles centrally braced.

hub (*Plumb.*). See **socket.** (*Surv.*) See **change point.**

Hubble's constant (*Astron.*). The constant factor in Hubble's Law, giving the rate of increase of velocity with distance in the expanding universe. The present value of 53 km/s per Mpc leads (assuming a uniform rate of expansion) to a hypothetical period of 18×10^9 years since the expansion began.

Hubble's law (*Astron.*). See **red shift.**

hübnerite or **huebnerite** (*Min.*). Tungstate of manganese, one of the end members of a variable series (the other being *ferberite*, tungstate of iron) commonly known as *wolframite* (q.v.). A product of pneumatolysis; associated with such minerals as scheelite, cassiterite, etc.

huckaback (*Textiles*). A linen or cotton cloth with a spongy structure, used for towels and glass-cloths. Sometimes made from a combination of linen and cotton, or spun rayon.

HUD (*Aero.*). Abbrev. for **head-up display.**

hue (*Optics*). Physiological attribute of colour perception, determined by red, blue, and green wavelengths, which leads to discrimination between colours. Its relation to colour is the same as that of pitch to frequency in hearing.

huebnerite (*Min.*). See **hübnerite.**

hue sensibility (*Optics*). The ability of the eye to distinguish small differences of colour.

Huff separator (*Min. Proc.*). High-tension or electrostatic separator used to concentrate small particles of dry ore.

hull (*Aero.*). The main body (structural, flotation, and cargo-carrying) of a flying-boat or boat amphibian. (*Ships*) A term used in its widest sense to signify the ship itself exclusive of masts, funnels, and *top hamper* (q.v.), but including *superstructure* (q.v.). In a more restricted sense it means the shell of the ship. Also used to distinguish between ship and machinery.

hull fibre (*Textiles*). See **linters.**

hullite (*Min.*). See **chlorophaeite.**

hum (*Acous.*). Objectionable low-frequency components induced from power mains into sound reproduction, caused by inadequate smoothing of rectified power supplies, induction into transformers and chokes, unbalanced capacitances, or leakages from cathode heaters. (*Geog.*) A small isolated hill resulting from the erosion of a limestone terrain.

hum-bucking coil (*Elec., Eng.*). See **bucking coil.**

humectant (*Chem.*). Any hygroscopic substance, e.g., CuO, conc. H_2SO_4; applied to substances added to products to keep them moist, e.g., glycerol in tobacco.

humeral (*Zool.*). In Vertebrates, pertaining to the region of the shoulder; in Insects, pertaining to the anterior basal angle of the wing; in *Chelonia*, one of the horny plates of the plastron.

humerus (*Zool.*). The bone supporting the proximal region of the fore-limb in land Vertebrates. *adj.* **humeral.**

humic acids (*Chem.*). Complex organic acids occurring in the soil and in bituminous substances formed by the decomposition of dead vegetable matter.

humicole, humicolous (*Bot.*). Growing on soil or on humus.

humidifier (*Heat*). An apparatus for maintaining desired humidity conditions in the air supplied to a building.

humidity (*Meteor.*). See **absolute-, relative-, specific-.**

humidity mixing ratio (*Meteor.*). The amount of water vapour in grams which is mixed with one kilogram of dry air.

humification (*Bot.*). The transformation of organic material into humus.

humite (*Min.*). An orthorhombic magnesium orthosilicate, also containing magnesium hydroxide. Found in impure marbles. The humite group also includes chondrodite, clinohumite, and norbergite.

hummer (*Acous.*). A microphone used for balance detection in conductivity measurements.

Hummer screen (*Min. Proc.*). Type used to grade smallish mineral, using a.c. to provide vibration by solenoid action.

hum modulation (*Acous.*). That which results from inadequate smoothing of anode or grid supplies in transmitters or in RF amplifier in a receiver.

hum note (*Acous.*). The pitch of the note of the sound from a bell which persists after the strike note has died away.

humoralism, humorism (*Med.*). The doctrine that diseases arise from some change in the humours or fluids of the body.

humour, humor (*Zool.*). A fluid; as the *aqueous humour* of the eye.

hump (*Bot.*). A tiny outgrowth on the side of a growing point, the rudiment of a future lateral member.

Humphrey gas pump (*Eng.*). A large water-pump in which periodic gas explosions are made to act directly on an oscillating column of water, thereby effecting a pumping cycle; used in waterworks.

Humphreys spiral (*Min. Proc.*). Spiral sluice which combines separation of mineral sands by simple gravitational drag with mild centrifugal action as the pulp cycles downward.

hump speed (*Aero.*). The speed, on the water, at which the water resistance of the floats or boat body of a seaplane or flying-boat is a maximum. After this is past the craft begins to be partially airborne.

humulene (*Chem.*). $C_{15}H_{24}$, a sesquiterpene found in oil of hops.

humulone (*Brew.*). A soft resin contained in hops, which contributes to the flavour of beer and acts as a preservative. Also called **alpha resin.** See **hops.**

humus (*Bot.*). Organic matter present in the soil, and so far decomposed that it has lost all signs of its original structure. It is colloidal, and dark-brown.

humus plant (*Bot.*). A flowering plant, often poorly provided with chlorophyll, which grows in deep humus, with its roots forming a mycorrhiza with a fungus.

Hund rule (*Chem.*). Transition metals which have incomplete inner electron shells have large magnetic moments, as the electron spins in the incomplete shell are self-aligning.

Hungarian cat's-eye (*Min.*). An inferior greenish cat's-eye obtained in the Fichtelgebirge in Bavaria. No such stone occurs in Hungary.

hung sash (*Join.*). See **hanging sash.**

hunter's moon (*Astron.*). The name given in popular language to the full moon which occurs next after the harvest moon, and to which the same phenomena apply in a lesser degree.

hunting (*Aero.*). (1) An uncontrolled oscillation, of approximately constant amplitude, about the flight path of an aircraft. (2) The angular oscillation of a rotorcraft's blade about its drag hinge. (*Eng.*) The tendency of rotating mechanisms which should run at constant speed to pulsate above and below that speed, due to shortcomings in the governor or other speed control systems. Also **cycling, oscillation.** (*I.C. Engs.*) Abnormal time-lag between opening of throttle and increase in engine-speed. (*Teleph.*) The operation of a selector, or other similar device, to establish connexion with an idle circuit of a group.

Huntington mill (*Min. Proc.*). Wet-grinding mill in which 4 cylindrical mullers, hung inside a steel tub, bear outward as they rotate, thus grinding the passing ore.

Huntington's chorea (*Med.*). Hereditary *chorea* (q.v.) occurring in adults; associated with progressive mental deterioration.

hunting tooth (*Eng.*). An extra tooth added to a gear-wheel in order that its teeth shall not be an integral multiple of those in the pinion.

Huppert's test (*Chem.*). A test for the presence of bile based upon precipitation of the bile acids with calcium hydroxide or chloride and ammonium carbonate, with a subsequent colour test of the acidified precipitate in ethanoic acid and in trichloromethane solution.

Huronian System (*Geol.*). A major division of the Pre-Cambrian rocks of the Canadian Shield, typically exposed on the northern shores of Lake Huron and following the Timiskaming Series unconformably. The Huronian comprises two Series, the Bruce below, followed unconformably by the Cobalt.

hurricane (*Meteor.*). (1) A wind of force 12 on the *Beaufort Scale* (q.v.). A mean wind speed of 75 m.p.h. (120 km/h). (2) A *tropical revolving storm* (q.v.) in the North Atlantic, Eastern North Pacific and the Western South Pacific.

hurricane deck (*Ships*). A term, not normally in use, for a superstructure deck. Sometimes termed **flying deck.** It is independent of the ship from the point of view of strength.

hurter (*Build.*). A cast-iron, timber, stone, or concrete block, which is so placed as to protect a quoin from damage from passing vehicles.

Hurter and Driffield (H and D) curve (*Photog.*). See **characteristic curve.**

hush-hush (*Cinema*). Variation in background noise when a noise-reduction system is in use.

hushing or hush (*Mining*). A washing away of the surface soil to lay bare the rock formation for prospecting.

husk (*For.*). The frame supporting the saw arbor and other working parts of a large circular saw. (*Vet.*) Hoose; lungworm disease; parasitic bronchitis. Bronchitis or bronchopneumonia of cattle, sheep, and goats due to infestation of the bronchi by nematode worms of the family *Metastrongylidae.*

hutch (*Mining*). (1) A small **train or wagon.** (2) A basket for coal. (3) A compartment of a jig used for washing ores. The concentrate which passes through a jig screen is called the **hutchwork.**

Hutchinson's teeth (*Med.*). Narrowing and notching of the permanent incisor teeth, occurring in congenital syphilis.

Huxley's layer (*Zool.*). The middle layer of the inner root sheath of a hair, composed of polyhedral nucleated cells containing eleidin.

Huxley's membrana preformativa (*Zool.*). A fine homogeneous membrane supposed to exist between the ameloblasts and the forming enamel, during the development of the teeth in Mammals.

Huygens' eyepiece (*Light*). A combination of two plano-convex lenses placed with their plane sides towards the observer, at a distance apart equal to half the sum of their focal lengths, which are in the ratio of three to one, the shorter focus lens being nearer the observer. Huygens' eyepiece is often used in microscopes, but is not suited for use with cross-wires or an eyepiece scale.

Huygens' principle (*Optics*). The assumption that every element of a wavefront acts as a source of so-called secondary waves. The principle can be applied to the propagation of any wave motion, but it is more frequently used in optics than in acoustics.

H-wave (*Telecomm.*). See **TE-wave.**

hyacinth or jacinth (*Min.*). The reddish-brown variety of transparent zircon, used as a gemstone. The name has also been used for a brownish grossular from Ceylon.

Hyades (*Astron.*). The name of a star cluster, of the 'open' type, situated in the constellation Taurus; visible to the naked eye.

hyal-, hyalo-. Prefix from Gk. *hyalos*, clearstone, glass.

hyaline (*Zool.*). Clear, transparent; without fibres or granules, e.g., *hyaline cartilage.*

hyalite (*Min.*). A colourless transparent variety of *opal* (q.v.), occurring as globular concretions and crusts. Also called **Müller's glass.**

hyaloid (*Zool.*). Clear, transparent; as the *hyaloid membrane* of the eye which envelops the vitreous humour.

hyalophane (*Min.*). One of the rarer feldspars, consisting of the components of orthoclase and celsian (barium-feldspar) in combination, and intermediate in composition between these two minerals. It occurs in colourless monoclinic crystals in manganese deposits in Sweden, South-West Africa and Japan, apparently as a contact mineral.

hyalopilitic texture (*Geol.*). A texture of andesitic volcanic rocks in which the groundmass consists of small microlites of feldspar embedded in glass.

hyaloplasm (*Cyt.*). Clear nongranular protoplasm.

hyalopterous (*Zool.*). Having transparent wings.

hyalosporous (*Bot.*). Having hyaline one-celled spores.

Hyatt roller bearing (*Eng.*). See **flexible roller bearing.**

hybrid (*Gen.*). An organism which is the offspring of a union between 2 different races, species, or genera; heterozygote.

hybrid coil (*Telecomm.*). A coil, comprising four equal windings and an additional winding, used for the separation of incoming and outgoing currents in a two-wire repeater, so that feedback and consequent oscillation is inhibited.

hybrid computer (*Comp.*). One using both analogue and digital elements and techniques.

hybrid electromagnetic wave (*Elec.*). One having longitudinal components of both the electric and magnetic field vectors. Abbrev. HEM.

hybrid integrated circuit (*Electronics*). A complete circuit formed by combining different types of integrated electronic sections.

hybrid junction (*Telecomm.*). A waveguide transducer which is connected to four branches of a circuit, designed to render these branches conjugate in pairs. If in the form of a (*hybrid*) *ring*, it is commonly called a **rat-race.** See **waveguide junction.**

hybrid rocks (*Geol.*). Rocks which originate by interaction between a body of magma and its wall-rock or roof-rock, which may be another igneous rock, or sedimentary, or metamorphic.

hybrid set (*Telecomm.*). Two or more transformers forming a four-pair terminal network. Four impedances may be connected to the pins of these terminals so that the branches containing them may be conjugate in pairs.

hybrid-T (*Radar*). Combination of E- and H-junctions for reflection-free propagation.

hybrid vigour (*Bot.*). The notable increase in strength of growth often exhibited by a hybrid.

hydantoin (*Chem.*). $CH_2{\cdot}NH{\cdot}(CO)_2{\cdot}NH$. Diketotetrahydroglyoxaline. Naturally occurring crystalline substance, m.p. 217–220°C. Its derivatives widely occur as degradation products of proteins.

hydathode (*Bot.*). A water pore, usually at the end of a vein of a leaf, from which liquid water is exuded.

hydatid cyst (*Zool.*). A large sac or vesicle containing a clear watery fluid and encysted immature larval *Cestoda*.

hydatidiform mole (*Med.*). An affection of the chorionic villi (vascular tufts of the foetal part of the placenta) whereby they become greatly enlarged, resembling a bunch of grapes.

hydatogenesis (*Geol.*). Deposition of selected mineral from an aqueous magmatic solution.

hydnocarpus oil (*Pharm.*). See **chaulmoogra oil.**

hydr-, hydro-. Prefix from Gk. *hydōr*, gen. *hydatos*, water.

Hydra (*Zool.*). A solitary nonmetagenetic Coelenterate occurring commonly in fresh water; its name is used in the construction of various terms. See **Hydrozoa.**

hydracrylic acid (*Chem.*). 3-Hydroxypropanoic acid, $CH_2(OH){\cdot}CH_2{\cdot}COOH$. Loses water readily on heating to give *acrylic acid* (q.v.).

hydraemia, hydremia (*Med.*). A watery state of the blood.

hydragogue (*Med.*). Having the property of removing water; a purgative drug which produces watery evacuations.

hydramnios (*Med.*). Excess of fluid in the amniotic sac of the foetus.

hydrant (*Civ. Eng.*). A form of connexion incorporated in a water main to enable a hose to be attached and a continuous supply of water to be obtained for the purpose of extinguishing fires or washing down streets.

hydranth (*Zool.*). In *Hydrozoa*, a nutritive polyp of a hydroid colony.

hydrargillite (*Min.*). See **gibbsite.**

hydrargyrism (*Med.*). The state of being poisoned by mercury and its compounds.

hydrarthrosis (*Med.*). Swelling of the joint, due to the accumulation in it of clear fluid.

hydrated electron (*Chem.*). Very reactive free electron released in aqueous solutions by the action of ionizing radiation.

hydrated ion (*Chem.*). Ion surrounded by molecules of water which it holds in a degree of orientation. Hydronium ion. Solvated H-ion of formula $[H_2O{\rightarrow}H]^+$ or H_3O^+.

hydrate of lime, hydrated lime (*Build.*). See **caustic lime.**

hydrates (*Chem.*). Salts which contain water of crystallization. See also **water of hydration.**

hydration (*Geol.*). The addition of water to anhydrous minerals, the water being of atmospheric or magmatic origin. Thus anhydrite, by hydration, is converted into gypsum, and feldspars into zeolites. (*Paper*) The process of converting raw material into pulp by prolonged beating, thus incorporating water into the fibres. Used for the production of hard or transparent paper. See **rattle.**

hydratuba (*Zool.*). In *Scyphozoa*, the unsegmented polyp stage. Cf. *scyphistoma.*

hydraulic. See **hydraulics.**

hydraulic accumulator (*Aero., Eng.*). A weight-loaded or pneumatic device for storing liquid at constant pressure, to steady the pump load in a system in which the demand is intermittent. In aircraft hydraulic systems an accumulator also provides fluid under pressure for operating components in an emergency, e.g., failure of an engine-driven pump.

hydraulic air compressor (*Mining*). Arrangement in which water falling to the bottom of a shaft entrains air which is released in a tunnel at depth, while the water rises to a lower discharge level.

hydraulic amplifier (*Eng.*). A power amplifier employed in some servomechanisms and control systems, in which power amplification is obtained by the control of the flow of a high-pressure liquid by a valve mechanism.

hydraulic belt (*Eng.*). An endless belt of porous material driven at high speed, with its lower end under water; acts like a *chain pump* (q.v.).

hydraulic blasting (*Mining*). In fiery mines, rock breaking by means of a ram-operated device acting on a hydraulic cartridge.

hydraulic brake (*Eng.*). (1) An absorption dynamometer. See **Froude brake.** (2) A motor-vehicle brake applied by small pistons operated by oil under pressure supplied from a pedal-operated master cylinder.

hydraulic cartridge (*Civ. Eng.*). An apparatus for splitting rock, mass concrete, etc.; it consists of a long cylindrical body which has numerous pistons projecting from one side and moving in a direction at right angles to the body (under hydraulic pressure from within the body), which is placed in a hole drilled to take it.

hydraulic cement (*Build., Civ. Eng.*). A cement which will harden under water.

hydraulic classifier (*Min. Proc.*). Device in which vertically flowing column of water is used to carry up and on light and small mineral particles while heavy and large ones sink.

hydraulic control (*Min. Proc., etc.*). Use of fluid in sensing mechanism to actuate a signalling or a correcting device, in response to pressure changes.

hydraulic coupling (*Eng.*). A traction coupling used for automobiles or for diesel engines and electric motors, in which each half-coupling contains a number of radial vanes and rotation causes a vortex in the hydraulic fluid between them, power being dissipated by the slip.

hydraulic cyclone elutriator (*Powder Tech.*). *Hydrocyclone* fitted with an apex container which serves as a return flow device for recycling the fine particles in suspension to give very efficient fractionation.

hydraulic engineering. That branch of engineering chiefly concerned in the design and production of hydraulic machinery, pumping plants, pipe-lines, etc.

hydraulic fill (*Eng.*). An embankment or other fill in which the materials are deposited in place by a flowing stream of water, gravity and velocity control being used to bring about selected deposition.

hydraulic glue (*Chem.*). A glue which is able partially to resist the action of moisture.

hydraulic gradient (*Hyd.*). In respect of any system of fluid flow, the *hydraulic gradient* is the imaginary curve the ordinate to which at any point is the sum of the position and pressure heads at the point.

hydraulic intensifier (*Eng.*). A device for obtaining a supply of high-pressure liquid from a larger flow at a lower pressure.

hydraulicity (*Build., Civ. Eng.*). The property of a lime, cement, or mortar which enables it to set under water or in situations where access of air is not possible.

hydraulic jack (*Eng.*). A *jack* (q.v.) in which the lifting head is carried on a plunger working in a cylinder, to which oil or water is supplied under pressure from a pump.

hydraulicking (*Mining*). See **hydraulic mining.**

hydraulic leather. A flexible leather prepared by being heavily treated, after tanning, with cod oil and then stoved; while hot, it may be shaped to requirements.

hydraulic lift (*Eng.*). A lift or elevator operated either directly by a long vertical ram, working in a cylinder to which liquid is admitted under pressure, or by a shorter ram through ropes. See **jigger.**

hydraulic main (*Gas*). A large horizontal steel pipe, into which the individual retorts of a gasworks discharge the products of distillation through dip pipes which dip into a layer of liquor maintained at constant level. It acts as a reservoir for tar, and provides a liquid seal between the retorts and the rest of the plant.

hydraulic mean depth (*Hyd.*). The effective average depth of a fluid stream in a channel, used in various empirical formulae, e.g., *Kutter's*. Usually taken as the cross-section area of the stream divided by the perimeter of the channel wetted by the fluid. Also **hydraulic radius.**

hydraulic mining or **hydraulicking** (*Mining*). The operation of breaking down and working a bank of gravel or alluvial deposit by means of jets of water under high pressure.

hydraulic mortar (*Build., Civ. Eng.*). A mortar which will harden under water.

hydraulic motor (*Eng.*). A multicylinder reciprocating machine, generally of radial type, driven by water or oil.

hydraulic packing (*Eng.*). L- or U-section rings providing a self-tightening packing under fluid pressure; used on rams and piston-rods of hydraulic machines. See **hat-leather packing, U-leather.**

hydraulic press (*Eng.*). An upstroke, downstroke, or horizontal press, with one or more rams, working at approximately constant pressure for deep drawing and extruding operations or at progressively increasing pressure for baling, plastics moulding, etc. See **hydrostatic press.**

hydraulic radius (*Hyd.*). Same as **hydraulic mean depth.**

hydraulic ram (*Eng.*). (1) The plunger of a hydraulic press. (2) A device whereby the pressure head produced when a moving column of water is brought to rest is caused to deliver some of the water under pressure.

hydraulic reservoir (*Aero.*). In an aircraft hydraulic system, the header tank which holds the fluid; not to be confused with the *accumulator*, which is a pressure vessel wherein hydraulic energy is stored.

hydraulic riveter (*Eng.*). A machine for closing rivets by hydraulic power.

hydraulics. The science relating to the flow of fluids. *adj.* **hydraulic.**

hydraulic squeezer (*Foundry*). See **squeezer.**

hydraulic stowing (*Mining*). The filling of worked-out portions of a mine with water-borne waste material. The water drains off and is pumped to surface.

hydraulic test (*Eng.*). A test for pressure-tightness and strength applied to pressure vessels, pipe lines, instruments, etc.

hydraulic torque converter (*Eng.*). A variable speed mechanism similar to a hydraulic coupling, in which a decrease in secondary speed is accompanied by an increase in torque.

hydraulic transport (*Mining*). Moving of ore or products by pumping in water through pipe lines.

hydrazides (*Chem.*). The mono-acyl derivatives of hydrazine.

hydrazine (*Chem.*). A fuming, strongly basic liquid, $H_2N{\cdot}NH_2$, b.p. 113°C. A powerful reducing agent, it (also its derivatives, in particular **dimethyl hydrazine**) is used as a high-energy propellant in rockets.

hydrazine hydrate (*Chem.*). $N_2H_4{\cdot}H_2O$. Diacid base. Attacks glass, rubber, and cork. Used in some liquid rocket fuels and as an oxygen scavenger in boiler water treatment. Derivatives are widely used as blowing agents in the production of expanded plastics and rubbers.

hydrazoates (*Chem.*). Salts of hydrazoic acid. See also **azides.**

hydrazo compounds (*Chem.*). Symmetric derivatives of hydrazine, colourless, crystalline, neutral substances, obtained by the reduction of azo compounds.

hydrazoic acid (*Chem.*). N_3H, $HN{-}\overset{-}{N}{\equiv}\overset{+}{N}$. The aqueous solution is a strong monobasic acid and forms azides with many common metals. Used in Schmidt reaction for converting aromatic carboxylic acids into primary amines.

hydrazones (*Chem.*). The condensation products of aldehydes and ketones with hydrazine, water being eliminated from the two molecules.

hydremia (*Med.*). See **hydraemia.**

Hydrida (*Zool.*). An order of *Hydrozoa* in which the medusoid phase is unknown, and the polyps are solitary and develop both male and female gonads from the ectoderm; there is no skeleton. Also **Eleutheroblastea.**

hydrides (*Chem.*). Compounds formed by the union of hydrogen with other elements. Those of the nonmetals are generally liquids or gases, certain of which dissolve in water (oxygen hydride) to form acid (e.g., hydrogen chloride) or alkaline (e.g., ammonia) solutions. The alkali and alkaline earth hydrides are crystalline, saltlike compounds, in which hydrogen behaves as the electronegative element. They contain H^- ions and, when electrolysed, give hydrogen at the anode.

hydriodic acid (*Chem.*). HI. An aqueous solution of hydrogen iodide. Forms salts called *iodides* (q.v.), many with characteristic colours. Easily oxidized.

hydrion (*Chem.*). A synonym for *hydrogen ion.*

hydro-. Prefix. See **hydr-.**

hydroa (*Med.*). A skin disease in which groups of vesicles appear on reddened patches in the skin, associated with intense itching.

hydroborons (*Chem.*). Six hydroborons have been identified—B_2H_6, B_4H_{10}, B_5H_9, B_5H_{11}, B_6H_{10}, and $B_{10}H_{14}$. The simplest hydroboron, B_2H_6, is sometimes referred to as **diborane.**

hydrobromic acid (*Chem.*). HBr. An aqueous solution of hydrogen bromide.

hydrocalumite (*Min.*). A mnemonic name applied by C. E. Tilley to a new mineral occurring in the metamorphic aureole of the dolerite at Scawt Hill, Antrim; it is a hydrated calcium aluminate, and occurs in colourless monoclinic (pseudo-hexagonal) crystals.

hydrocarbons (*Chem.*). A general term for organic compounds which contain only carbon and hydrogen in the molecule. They are divided into saturated and unsaturated hydrocarbons, aliphatic (alkane or fatty) and aromatic (benzene) hydrocarbons. (*Geol.*) See **native-.**

hydrocarpic (*Bot.*). Said of aquatic plants which ripen their fruits under water, after pollination has occurred in the air above the water.

hydrocaulis (*Zool.*). The part of the *coenosarc* (q.v.) which resembles the stem of a plant in a hydroid colony.

hydrocele (*Med.*). A swelling in the scrotum due to an effusion of fluid into the sac (*tunica vaginalis*) which invests the testis.

hydrocelluloses (*Chem.*). Products obtained from cellulose by treatment with cold concentrated acids. They still retain the fibrous structure of cellulose, but are less hygroscopic.

hydrocephalis (*Zool.*). The distal part of a polyp, with the mouth and tentacles.

hydrocephalus (*Med.*). An abnormal accumulation of cerebrospinal fluid in the cavities (ventricles) of the brain, distending them and stretching and thinning the brain tissue over them.

hydrocerussite (*Min.*). A colourless hydrated basic carbonate of lead occurring as an encrustation on native lead, galena, cerussite, and other lead minerals.

hydrochloric acid (*Chem.*). HCl. An aqueous solution of hydrogen chloride gas. Dissolves many metals, forming chlorides and liberating hydrogen. Used extensively in industry for numerous purposes, e.g., for the manufacture of chlorine, pickling, tinning, soldering, etc.

hydrochoric (*Bot.*). Dispersed by water.

hydrocladia (*Zool.*). The hydranth-bearing branches of the main stem of the coenosarc in some *Calyptoblastea.*

hydrocoel (*Zool.*). In *Echinodermata*, the *water-vascular system* (q.v.).

hydrocoles (*Ecol.*). Animals which live in water.

Hydrocorallina (*Zool.*). An order of *Hydrozoa* in which the hydroid phase is predominant and colonial and develops polymorphic forms; the medusae are rudimentary and attached; there is a massive calcareous skeleton.

hydrocortisone (*Chem.*). The most important glucocorticoid produced by the adrenal cortex. Similar in use to cortisone, but in certain cases more effective.

hydrocyanic acid (*Chem.*). An aqueous solution of *hydrogen cyanide* (q.v.). Dilute solution called *prussic acid.* Monobasic. Forms cyanides. Very poisonous.

hydrocyclone (*Min. Proc.*). A small cyclone extractor for removing suspended matter from a flowing liquid by means of the centrifugal forces set up when the liquid is made to flow through a tight conical vortex. Used to separate solids in mineral pulp into coarse and fine fractions. Fluent stream enters tangentially to cylindrical section and coarser sands gravitate down steep-sided conical section to controlled apical discharge. Bulk of pulp, containing finer particles, overflows through a pipe inserted in the central vortex. Classification into fractions is aided by the centrifugal force with which the pulp is delivered to the appliance, which can handle large tonnages. (*Paper*) Pulp-cleaning equipment whereby the paper or pulp stock is caused to revolve in a vortex, so causing dirt particles to separate from the fibres by centrifugal force.

hydrocyst (*Zool.*). See **dactylozooid.**

hydrodynamic governor (*Eng.*). A governor comprising a small centrifugal pump whose pressure head, varying with speed, is caused to act on a piston connected to the regulating valve. See **servomotor.**

hydrodynamic lubrication (*Eng.*). Thick-film lubrication in which the relatively moving surfaces are separated by a substantial distance and the load is supported by the hydrodynamic film pressure.

hydrodynamic power transmission (*Eng.*). A power-transmission system which, in general, employs a hydraulic coupling or a torque converter or a reaction coupling.

hydrodynamic process (*Eng.*). A U.S. patent process for shallow forming and embossing operations, in which high-pressure water presses the blank against a female die, there being no solid punch to conform to the die contour.

hydrodynamics. That branch of dynamics which studies the motion produced in fluids by applied forces.

hydroecium (*Zool.*). In some *Siphonophora*, a cavity at the upper end of the colony, into which the contractile stem with its cormidia can be retracted.

hydroelectric generating set (*Elec. Eng.*). An electric generator driven by a water turbine.

hydroelectric generating station (*Elec. Eng.*). An electric generating station in which the generators are driven by water turbines.

hydroelectric power station (*Elec. Eng.*). A *hydroelectric generating station* (q.v.).

hydroextractor (*Eng.*). See **whizzer.**

hydrofining (*Chem.*). See **hydroforming.**

hydrofluoric acid (*Chem.*). Aqueous solution of hydrogen fluoride. Dissolves many metals, with evolution of hydrogen. Etches glass owing to combination with the silica of the glass to form silicon fluoride, hence it is stored in, e.g., polythene or gutta-percha vessels. See **etching test.**

hydrofluosilicic acid (*Chem.*). H_2SiF_6. Formed when tetrafluorosilane is passed into water. With bases it forms fluosilicates. Easily decomposed into tetrafluorosilane and hydrogen fluoride.

hydrofoil (*Aero.*). An immersed plate-like surface to facilitate the take-off of a seaplane by increasing the hydrodynamic lift. (*Ships*) A fast, light craft fitted with winglike structures (foils) on struts under the hull. These may be extendable and adjustable. Propelled by propeller in water or air, or by jet. Foils may act entirely or partly submerged with hull lifted clear of the water at speed. Steered by water rudder. Cf. *hydroplane.*

hydroforming (*Chem.*). The catalytic refining or reforming of petroleum oils by hydrogen at high temperatures and pressures; also called **hydrofining.** (*Eng.*) A hydraulic forming process in which the shape is produced by forcing the material by means of a punch against a flexible bag partly filled with hydraulic fluid and acting as a die.

hydrofuge (*Zool.*). Water-repelling; said of certain hairs possessed by some aquatic insects and used for retaining a film of air.

hydrogel (*Chem.*). A gel the liquid constituent of which is water.

hydrogen (*Chem.*). The least dense element, forming diatomic molecules H_2, having both nonmetallic and metallic properties. Symbol H, at. no. 1, r.a.m. 1·00797, valency 1. It is a colourless, odourless, diatomic gas, water being formed when it is burnt; m.p. −259·14°C, b.p. −252·7°C, density 0·08988 g/dm³ at s.t.p. It is widely distributed as water, occurs in many minerals, e.g., petroleum, and in living matter. Hydrogen is manufactured chiefly as a by-product of electrolysing caustic soda, water gas and gas cracking. It is used in the Haber process for the fixation of nitrogen, and in the hardening of fats (e.g., in the manufacture of margarine); also for hydrogenation of oils, manufacture of hydrochloric acid, filling small balloons, and as a reducing agent for organic synthesis and metallurgy, oxyhydrogen and atomic hydrogen welding flames. It is of great importance in the moderation (slowing down) of neutrons as hydrogen atoms are the only ones of similar mass to a neutron and are therefore capable of absorbing an appreciable proportion of the neutron energy on collision. Isotopes, with 1 proton and 1 electron are:

Name	Atomic symbol	Relative atomic mass	Atomic number	Neutrons
protium	^{1}H	1·007825	1	0
deuterium	D, ^{2}H	2·01410	2	1
tritium	T, ^{3}H	3·0221	3	2

hydrogen I and II (*Astron.*). The hydrogen of interstellar space, known in two clearly defined states: H I (*neutral hydrogen*) and H II (*ionized hydrogen*). The H I regions, confined mainly to the spiral arms of the galaxy, emit no visible light, and are detected solely by their emission of the 21-cm radio line. The H II regions, found in the gaseous nebulae, emit both visible and radio radiations.

hydrogenation (*Chem.*). Chemical reactions involving addition of hydrogen, present as a gas, to a substance, in the presence of a catalyst. Important processes are: the hydrogenation of coal (but see next article); the hydrogenation of fats and oils; the hydrogenation of naphthalene and other substances.

hydrogenation of coal (*Chem.*). Bergius process: finely powdered coal mixed with heavy oil and hydrogen is heated, with a suitable catalyst, to about 450°C at high pressure ($2-7\times10^{7}$ N/m²), and passed through converters in which the hydrogen combines. The resulting 'middle oil' is further catalytically treated with hydrogen to give diesel oil and petrol, and again to give high-quality aviation spirit. The residual gases yield ammonia, light hydrocarbons, hydrogen and methane, which may be reconverted and passed back to the process. Creosote and tar are also used as starting materials. This process is now obsolete. See also **Fischer-Tropsch process.**

hydrogen bacteria (*Bacteriol.*). Chemosynthetic bacteria which obtain the energy required for carbon dioxide assimilation from the oxidation of hydrogen to water. Organic compounds may also be oxidized by these species under suitable conditions.

hydrogen bomb (*Nuc.*). An *atomic bomb* (q.v.) surrounded by lithium deuteride which, through the atomic bomb temperature, fuses to helium with very great emission of energy because of loss of mass (1–20 megaton TNT equivalent or more, in explosive power, i.e., 1000 times more powerful than Hiroshima atomic bomb). Also called **fusion bomb, H-bomb, thermonuclear bomb.**

hydrogen bond (*Chem.*). Electrostatic intermolecular link between the hydrogen atoms in 'associated' liquids such as water, alcohols, and acids. It leads to the apparent formation of dimeric (e.g., ethanoic acid in benzene) and other molecules, which can cause anomalous results, e.g., in molecular mass determinations.

hydrogen bromide (*Chem.*). HBr. Hydrogen bromide gas can be made by direct combination of the two elements, particularly in the presence of a catalyst. Closely resembles hydrogen chloride.

hydrogen chloride (*Chem.*). HCl. A colourless gas which dissolves in water to form hydrochloric acid. Produced by the action of concentrated sulphuric acid on chlorides, on the industrial scale as a by-product of the chlorination of hydrocarbons, and directly from the elements H_2 and Cl_2, both of which are formed as a by-product of caustic soda manufacture. Uses: to produce hydrochloric acid; chlorination of unsaturated organic compounds, e.g., chloroethene in polymerization, isomerization, and alkylation, as a catalyst.

hydrogen cooling (*Elec. Eng.*). A method of cooling rotating electric machines; the machine is totally enclosed and runs in an atmosphere of hydrogen.

hydrogen cyanide (*Chem.*). HCN. Highly poisonous liquid, m.p. −13·4°C, b.p. 26°C permittivity 95, dissociation constant $1\cdot3\times10^{-9}$ at 18°C. Dissolves in water to form hydrocyanic acid and prussic acid. Faint odour of bitter almonds. Uses: fumigant; medicine; organic synthesis, e.g., acrylic resins; analytical reagent.

hydrogen electrode (*Chem.*). For pH measurement, a platinum-black electrode covered with hydrogen bubbles. Also called **Hildebrand electrode.**

hydrogen embrittlement (*Met.*). Effect produced on metal by sorption of hydrogen during pickling or plating operations.

hydrogen equivalent (*Chem.*). Acidity or alkalinity as defined in terms of reacting H^+ or OH^- ions per molecule.

hydrogen fluoride (*Chem.*). HF. A liquid which fumes strongly in air. Dissolves in water to form hydrofluoric acid. Produced by the action of sulphuric acid on fluorides. Uses: catalyst in organic reactions; preparation of uranium; fluorides and hydrofluoric acid.

hydrogen iodide (*Chem.*). HI. A heavy colourless gas, formed by the direct combination of hydrogen and iodine; fumes strongly in air. Usually made by the decomposition of phosphorus (III) iodide by the action of water. M.p. −50°C, b.p. −35°C. See **hydriodic acid.**

hydrogen ion (*Chem.*). An atom of hydrogen carrying a positive charge, i.e., a proton; in aqueous solution, hydrogen ions are hydrated, H_3O^+.

hydrogen ion concentration (*Chem.*). See **pH value.**

hydrogenous (*Nuc.*). Said of a substance rich in hydrogen and therefore suitable for use as moderator of neutrons.

hydrogen oxide (*Chem.*). Water.

hydrogen peroxide (*Chem.*). H_2O_2. A viscous liquid with strong oxidizing properties. Powerful bleaching agent; and, as its decomposition products are water and oxygen, it is much used as a disinfectant. The strength of an aqueous solution is represented commercially by the number of volumes of oxygen which 100 cm^3 of the solution will give on decomposition. Recent interest in high concentration (ca. 90%) hydrogen peroxide has centred on its use as both a technical and a military oxidizing agent, especially in rocket fuels as an *oxidant* (q.v.). Prepared via the anodic oxidation of hydrogen sulphates to peroxydisulphate followed by hydrolysis and steam distillation.

hydrogen phosphide (*Chem.*). PH_3. See **phosphine** (1).

hydrogen scale (*Chem.*). A system of relative values of electrode potentials, based on that for hydrogen gas, at standard pressure, against hydrogen ions at unit *activity* (2), as zero.

hydrogen sulphide (*Chem.*). H_2S. May be prepared by direct combination of the 2 elements or by the action of dilute hydrochloric or sulphuric acid on iron sulphide. It is readily decomposed. Reacts with bases forming sulphides and with some metals to produce metal sulphides and liberate hydrogen. Poisonous, with a characteristic smell of rotten eggs.

hydrographical surveying (*Surv.*). A branch of surveying dealing with bodies of water at the coastline and in harbours, estuaries, and rivers.

hydrography. The study, determination, and publication of the conditions of seas, rivers, and lakes, viz., surveying and charting of coasts, rivers, estuaries and harbours, and supplying particulars of depth, bottom, tides, currents, etc.

hydrogrossular (*Min.*). A hydrous variety of garnet, close to grossular in composition but with some of the silica replaced by water.

hydrohaematite or **turgite** (*Min.*). $Fe_2O_3 \cdot nH_2O$. Probably a mixture of the two minerals *haematite* and *goethite*, the former being in excess. It is fibrous and red in the mass, with an orange tint when powdered.

hydroid (*Zool.*). An individual of the asexual stage in *Coelenterata* which show alternation of generations.

hydrolapse (*Meteor.*). A fall in the *dew-point* (q.v.) temperature with height.

Hydrolastic (*Autos.*) Said of a proprietary type of suspension in which front and rear suspensions are compensated hydraulically.

hydrolith (*Chem.*). Calcium hydride, $Ca^{2+}(H^-)_2$. Uses: gives off hydrogen with water; analytical reagent; reducing agent; drying agent.

hydrological cycle (*Meteor.*). The evaporation and condensation of water on a world scale.

hydrology. The study of water, including rain, snow, and water on the earth's surface, covering its properties, distribution, utilization, etc.

hydrolysis (*Chem.*). (1) The formation of an acid and a base from a salt by interaction with water; it is caused by the ionic dissociation of water. (2) The decomposition of organic compounds by interaction with water, either in the cold or on heating, alone or in the presence of acids or alkalis; e.g., esters form alcohols and acids; oligo- and polysaccharides on boiling with dilute acids yield monosaccharides.

hydromagnesite (*Min.*). Hydrated magnesium hydroxide and carbonate, occurring as whitish amorphous masses, or rarely as monoclinic crystals in serpentines.

hydromagnetic (*Elec. Eng.*). Pertaining to the behaviour of a plasma in a magnetic field. See also **magnetohydrodynamics**.

hydromagnetic instability (*Nuc.*). See **instability** (*a*).

Hydromedusae (*Zool.*). See **Hydrozoa**.

hydromelia (*Med.*). Dilatation of the central canal of the spinal cord.

hydrometallurgy (*Met., Min. Proc.*). Extraction of metals or their salts from crude or partly concentrated ores by means of aqueous chemical solutions; also electrochemical treatment including electrolysis or ion exchange.

hydrometeor (*Meteor.*). Any weather phenomenon which depends on the moisture content of the atmosphere.

hydrometer (*Phys., etc.*). An instrument by which the relative density of a liquid may be determined by measuring the length of the stem of the hydrometer immersed, when it floats in the liquid with its stem vertical.

hydronephrosis (*Med.*). Distension of the kidney with urine held up as a result of obstruction elsewhere in the urinary tract.

hydrone theory (*Chem.*). The theory that liquid water consists largely of associated molecules, H_4O_2, H_6O_3, etc.

hydronium ion (*Chem.*). See **hydroxonium ion.**

hydropathy (*Med.*). The treatment of disease by water, externally and internally.

hydropericardium (*Med.*). Collection of clear fluid in the pericardial sac.

hydroperitoneum (*Med.*). Ascites. Accumulation of clear fluid in the abdominal cavity.

hydroperoxides (*Chem.*). Intermediate compounds formed during the oxidation of unsaturated organic substances, e.g., fatty oils, such as linseed oil.

hydrophane (*Min.*). A variety of opal which, when dry, is almost opaque, with a pearly lustre, but becomes transparent when soaked with water, as implied in the name.

hydrophilic colloid (*Chem.*). A colloid which readily forms a solution in water.

hydrophilous (*Bot.*). (1) Living in water. (2) Pollinated by water.

hydrophobia (*Med.*). Literally fear of water, a symptom of rabies. Used synonymously with *rabies* (q.v.).

hydrophobic cement (*Build., Civ. Eng.*). Cement into which a special agent is introduced during the grinding process to reduce the deterioration which occurs in other cements when exposed to damp.

hydrophobic colloid (*Chem.*). A colloid which forms a solution in water only with difficulty.

hydrophone (*Acous.*). Electroacoustic transducer used to detect sounds or ultrasonic waves transmitted through water. Also **subaqueous microphone.**

hydrophyllium (*Zool.*). In some *Siphonophora*, a leaflike bract, believed to represent a modified medusoid, which hangs down over the rest of the cormidium.

hydrophyte (*Bot.*). A plant which lives on the surface of, or submerged in, water.

hydrophyton (*Zool.*). In a hydroid colony, the coenosarc together with the hydrorhiza.

hydropic (*Zool.*). Said of the eggs of some parasitic *Hymenoptera*, which are thinly chorionated, small and deficient in yolk when laid but increase in size during development by absorption of food and water through the chorion.

hydroplane (*Ships*). (1) A powered boat which skims the surface of the water; cf. *hydrofoil.* (2) A planing surface which enables a submarine to submerge.

hydroponics (*Bot.*). Technique of growing plants over water instead of in soil, suitable chemicals being dissolved in the water, which feed the plant through the dipping roots.

hydropore (*Zool.*). In the dipleurula larva of

Echinodermata, opening by which the right hydrocoel communicates with the exterior.

hydropote (*Bot.*). A cell, or a group of cells, occurring in a leaf submerged in water, easily permeable to water and dissolved salts.

hydrops folliculi (*Med.*). An ovarian cyst formed by the accumulation of clear fluid in a Graafian follicle.

Hydropteridineae (*Bot.*). A heterosporous subclass of *Filicineae*; the sori or groups of sori are enclosed in an envelope, the whole structure being called a sporocarp. Also **Salviniales, Pilizocarpeae.**

hydropyle (*Zool.*). A modified area of the serosal cuticle of the developing egg of some *Orthoptera* for the uptake or loss of water.

hydroquinone (*Chem., Photog.*). 1,4-dihydroxybenzene, monoclinic plates or hexagonal prisms, m.p. 169°C. It is obtained by the reduction of quinone with sulphurous acid. It is a strong reducing agent and is extensively used as a developer in photography.

Also called **quinol.**

hydrorhizae (*Zool.*). Rootlike processes of the coenosarc by which some hydrozoan colonies are attached to the substratum.

hydrorrhoea (or **hydrorrhea**) **tubae intermittens** (*Med.*). The condition in which fluid from a hydrosalpinx intermittently escapes into the uterus and thence through the vagina.

hydrosalpinx (*Med.*). Accumulation of clear fluid in a Fallopian tube which has become shut off as a result of inflammation.

hydrosere (*Bot.*). A sere beginning in a wet habitat.

hydroskis (*Aero.*). Hydrofoils, usually retractable, fitted to seaplanes without a *planing bottom* (q.v.) as the sole source of hydrodynamic lift. They are also fitted to aeroplane landing gears to make them amphibious (see **pantobase**), in which case a minimum taxiing speed is necessary to keep the aircraft above water.

hydrosol (*Chem.*). A colloidal solution in water.

hydrosome (*Zool.*). A colony of hydranths.

hydrosphere (*Ocean.*). The water on the surface of the earth.

hydrospires (*Zool.*). Elongated internal pouches which run along the sides of the ambulacral grooves and open to the exterior at the oral end by spiracles, in *Blastoidea*.

hydrostatic extrusion (*Met.*). Form of extrusion in which the metal to be shaped is preshaped to fit the die which forms the lower end of a high-pressure container. The container is filled with the pressure-transmitting liquid, and pressure is built up in the liquid by a plunger until the metal is forced through the die.

hydrostatic joint (*Plumb.*). A joint of the spigot-and-socket type, formed in a large water-main by forcing sheet-lead into the socket to ensure security against hydraulic pressure.

hydrostatic press (*Hyd.*). A hydrostatic machine for magnifying force, consisting of 2 connected cylinders (one much larger than the other) fitted with water-tight pistons enclosing water. A small force applied through a linkage to the piston in the smaller cylinder sets up a pressure in the water, and this pressure acting over the much greater area of the other piston gives rise to the magnified force. Also **Bramah's press.** Commonly called **hydraulic press.**

hydrostatic pressing (*Powder Tech.*). The application of liquid pressure directly to a preform which has been sealed, e.g., in a plastic bag. It is characterized by equal pressure in all directions maintaining preform shape to reduced scale.

hydrostatics. Branch of statics which studies forces arising from the presence of fluids.

hydrostatic test (*San. Eng.*). A test to detect leakage in a drain. The latter is plugged at the outlet end and filled with water; any fall of level of the water indicates leakage.

hydrostatic valve (*Eng.*). Apparatus which tends to maintain an underwater body (e.g., a moving torpedo) at the desired depth.

hydrosulphides (*Chem.*). Salts formed by the action of hydrogen sulphide on some of the hydroxides.

hydrosulphuric acid (*Chem.*). An aqueous solution of hydrogen sulphide.

hydrosulphurous acid (*Chem.*). See **hyposulphurous acid (1).**

hydrotaxis (*Biol.*). Response or reaction of an organism to the stimulus of moisture. *adj.* **hydrotactic.**

hydrotheca (*Zool.*). The cuplike expansion of the perisarc which surrounds a hydranth, in *Calyptoblastea.*

hydrotherapy (*Med.*). Treatment of disease by water, either taken internally or more usually by immersion of the body or affected part.

hydrothermal metamorphism (*Geol.*). That kind of change in the mineral composition and texture of a rock which was effected by water under conditions involving high temperatures.

hydrothorax (*Med.*). Clear fluid in the pleural cavity formed by transudation from blood vessels.

hydrotopes (*Chem.*). Co-solvents used in liquid detergent mixtures to prevent any less soluble components, such as CMC (*carboxy methyl cellulose*), from separating out.

hydrotropism (*Biol.*). See **hydrotaxis.** More particular response by growth curvature.

hydrovane (*Aero.*). See **hydrofoil, hydroskis, sponson.**

hydroxides (*Chem.*). Compounds of the basic oxides with water. The term *hydroxide* (a contraction of *hydrated oxide*) is applied to compounds that contain the —OH or hydroxyl group.

hydroxonium ion (*Chem.*). The hydrogen ion, normally present in hydrated form as H_3O^+. Also **hydronium ion.**

hydroxyapatite (*Min.*). The hydroxyl-bearing variety of phosphate apatite.

hydroxyl (*Chem.*). —OH. A monovalent group consisting of a hydrogen atom and an oxygen atom linked together.

hydroxylamine (*Chem.*). Hydroxy-ammonia, NH_2OH, rather explosive, deliquescent colourless crystals which may be obtained by the reduction of nitric oxide, ethyl nitrate, or nitric acid under suitable conditions; m.p. 33°C, b.p. 58°C at 3 kN/m². Its aqueous solution is alkaline, its salts are powerful reducing agents.

hydroxylamines (*Chem.*). Derivatives of hydroxylamine, NH_2OH, in which the hydrogen has been exchanged for alkyl radicals.

hydroxylation theory (*Heat*). Based on the idea that when a hydrocarbon is oxidized there is a natural tendency for its hydrogen atoms to be successively converted into OH groups, thus producing hydroxylated molecules with consequent heat evolution.

hydrozincite or zinc bloom (*Min.*). A monoclinic hydroxide and carbonate of zinc. It is an uncommon ore, occurring with smithsonite as an alteration product of sphalerite in the oxide zone of some lodes.

Hydrozoa (*Zool.*). A class of *Cnidaria*, in which alternation of generations typically occurs; the hydroid phase is usually colonial, and gives rise to the medusoid phase by budding; the polyp is without gastral ridges and filaments; the medusa has a velum and nerve-ring; the gonads are of ectodermal origin. Zoophytes.

Hyeniales (*Bot.*). An extinct order of the *Equisetineae*. The jointed stem bore leaves which were not distinctly verticillate in arrangement, and were narrow and forked. The cylindrical sporangiophores were branched, once or twice recurved, and each branch bore a sporangium. The lax strobili had no sterile appendages.

hyetograph (*Meteor.*). An instrument which collects, measures, and records the fall of rain (Greek *hyetos*, rain).

hyetology (*Meteor.*). The branch of meteorology dealing with rainfall.

hygristor (*Elec. Eng.*). Resistance element sensitive to ambient humidity.

hygro-. Prefix from Gk. *hygros*, wet, moist.

hygrochastic (*Bot.*). Of a fruit, opening by the absorption of water.

hygrodeik (*Meteor.*). A psychrometer in a flame, with indexes for the rapid estimation of the relative humidity.

hygrokinesis (*Biol.*). A *kinesis* (q.v.) occurring in response to humidity.

hygroma (*Vet.*). A fluid-filled swelling, usually associated with a joint, due to distension of a synovial sac.

hygrometer (*Phys., etc.*). Instrument for measuring, or giving output signal proportional to, atmospheric humidity. Electrical hygrometers make use of *hygristors*.

hygrometric movement (*Bot.*). Curvature or other change of form, commonly in dead plant material, caused by the entry of moisture.

hygrometry (*Meteor.*). The measurement of the hygrometric state, or *relative humidity* (q.v.), of the atmosphere.

hygropetrical fauna (*Ecol.*). Animals living in the thin film of water surrounding stones not truly submerged.

hygrophanous (*Bot.*). Darkening in colour following the entry of water into or between the cells; having a soaked appearance.

hygrophile, hygrophilous (*Bot.*). Living where moisture is abundant.

hygrophobe (*Bot.*). Living best in dry situations, where moisture is scanty.

hygrophyte (*Bot.*). A plant adapted to plentiful water-supply.

hygroscopic (*Bot.*). (1) Absorbing water readily, and showing a change of form as a result. (2) Moving as a result of the loss, or of the intake, of water. (*Chem.*) Tending to absorb moisture; in the case of solids, without liquefaction.

hygroscopic (or imbibition) mechanism (*Bot.*). A means of bringing about movement in plant material, depending upon the uneven swelling or shrinking following intake or loss of moisture from cell walls which are not of the same thickness throughout. Dead cell walls most often show this behaviour.

hygrostat (*Chem.*). Apparatus which produces constant humidity.

hygrotaxis (*Biol.*). A *taxis* (q.v.) occurring in response to the direction of a moisture gradient.

hygrotropism (*Biol.*). Reaction or response of an animal to the stimulus of atmospheric water vapour.

hylophagous (*Zool.*). Wood-eating.

hylophyte (*Bot.*). A plant characteristic of damp woods.

hylotomous (*Zool.*). Wood-cutting.

hymen (*Zool.*). In Mammals, a fold of mucous membrane which partly occludes the opening of the vagina in young forms.

hymenial layer (*Bot.*). See **hymenium.**

hymenitis (*Med.*). Inflammation of the hymen.

hymenium (*Bot.*). A layer of asci and paraphyses more or less parallel with one another, in the fructification of an ascomycete; the layer of basidia, paraphyses, and sometimes of cystidia, covering the gills or lining the pores of an autobasidiomycete.

Hymenogastrales (*Bot.*). An order of the *Basidiomycetes*. The fruiting body is closed and subterranean (usually) until mature. The gleba is usually waxy or fleshy and is not digested at maturity.

Hymenomycetes (*Bot.*). In some classifications, a large subclass of the *Basidiomycetes*, with about 12 000 species, in which the hymenium is exposed to the air from an early stage in its development. Coextensive with *Agaricales*.

hymenophore (*Bot.*). Any fungal structure which bears a hymenium.

hymenopodium (*Bot.*). The tissue underlying the hymenium.

Hymenoptera (*Zool.*). An order of *Oligoneoptera* having usually two almost equal pairs of transparent wings, which are frequently connected during flight by a series of hooks on the hind-wing; mandibles always occur but the mouth-parts are often suctorial; the adults are usually of diurnal habit; the larvae show great variation of form and habit. Saw-flies, Gall-flies, Ichneumons, Ants, Bees, and Wasps.

hymenotomy (*Surg.*). Cutting of the hymen.

Hymn-88 (*Mag.*). TN for a magnetic alloy, mostly nickel, with some molybdenum and iron.

hyo-. Prefix from Gk. *hyoeidēs*, U-shaped.

hyoid (*Zool.*). In higher Vertebrates, a skeletal apparatus lying at the base of the tongue, derived from the hyoid arch of the embryo.

hyoid arch (*Zool.*). The second pair of visceral arches in lower Vertebrates and in the embryos of higher Vertebrates, lying between the mandibular arch and the first branchial arch.

hyoid cornu (*Zool.*). The ventral part of a hyoid arch, consisting of epihyal, ceratohyal, and hypohyal elements.

hyoideus (*Zool.*). In Vertebrates, the post-trematic branch of the facial (seventh cranial) nerve, which runs to the mucosa of the mouth and to the muscles of the hyoid region, and, in aquatic forms, to the neuromast organs of the region below and behind the orbit.

hyoid segment (*Zool.*). Third segment of the head of a Vertebrate embryo, lying behind the mandibular somite and in front of the first metotic somite.

hyomandibular (*Zool.*). In *Craniata*, the dorsal element of a hyoid arch.

hyomandibular nerve (*Zool.*). In *Craniata*, the post-trematic branch of the seventh cranial nerve; it divides into anterior and posterior branches.

hyoplastron (*Zool.*). In *Chelonia*, one of the paired plates composing the plastron, lying between the hypoplastron and the epiplastron. Also called **hyosternum.**

hyoscine (*Pharm.*). *Scopolamine* (q.v.).

hyoscyamine (*Chem.*). $C_{17}H_{23}O_3N$, a coca base alkaloid, optically active, stereoisomeric with atropine, forming colourless needles or plates, m.p. 109°C; it can be prepared from *Datura stramonium.*

hyosepticaemia (*Vet.*). See **joint-ill.**

hyostapes (*Zool.*). See **extracolumella.**

hyosternum (*Zool.*). See **hyoplastron.**

hyostyly (*Zool.*). A type of jaw suspension, found in most Fish, in which the upper jaw is attached to the cranium anteriorly by a ligament, posteriorly by the hyomandibular. *adj.* **hyostylic.**

hyp (*Telecomm.*). One-tenth of a *neper*. See **decineper.**

hypabyssal rocks (*Geol.*). Literally, igneous rocks that are not quite abyssal (deep-seated), occurring as minor intrusions. One of the three main divisions (based on mode of occurrence) into which igneous rocks are grouped in some schemes of classification. A more precise classification is based on grain size.

hypaethral (*Arch.*). Said of a building without a roof, or with an opening in its roof.

hypalgesia (*Med.*). Diminished sensitivity to pain.

hypanthium (*Bot.*). The flat or concave receptacle of a perigynous flower.

hypanthodium (*Bot.*). The deeply hollowed receptacle of the fig, which provides the edible material.

hypapophyses (*Zool.*). Paired ventral processes of the vertebrae of many higher *Craniata.*

hyparterial (*Zool.*). Placed beneath an artery.

hypaxial (*Zool.*). Below the axis, especially below the vertebral column, therefore ventral; as the lower of two blocks into which the myotomes of fish embryos become divided.

hyper-. Prefix from Gk. *hyper*, above.

hyperacidity (*Med.*). Excessive acidity, especially of the stomach juices.

hyperacusis (*Med.*). Abnormally increased acuity of hearing.

hyperadrenalism (*Med.*). Abnormally increased activity of the adrenal gland.

hyperaemia, hyperemia (*Med.*). Congestion, or excess of blood, in a part of the body.

hyperaesthesia (*Med.*). Lowered threshold to a given stimulus, e.g., excessive sensitivity to painful or tactile stimulus.

hyperalgesia (*Med.*). Heightened sensitivity to painful stimuli.

hyperaphia (*Med.*). Excessive sensitivity to touch.

hyperapophysis (*Zool.*). A dorsolateral posterior process of a vertebra.

hyperbaric chamber (*Med., Radiol.*). A chamber containing oxygen at high pressures, in which patients are placed to undergo radiotherapy or treatment for certain forms of poisoning.

hyperbarism (*Space*). An agitated bodily condition when the pressure within the body tissues, fluids and cavities is countered by a greater external pressure, such as may happen in a sudden fall from a high altitude.

hyperbilirubinaemia, hyperbilirubinemia (*Med.*). Excess of the bile pigment *bilirubin* in the blood.

hyperbola (*Maths.*). See **conic.**

hyperbolic (*Radio*). Said of any system of navigation which depends on difference in cycles and fractions between locked waves from two or more stations, e.g., *gee, loran.*

hyperbolic functions (*Maths.*). A set of six functions, analogous to the trigonometrical functions sin, cos, tan, etc. The hyperbolic sine and cosine are written *sinh* and *cosh*:

$$\sinh x = \tfrac{1}{2}(e^x - e^{-x}); \cosh x = \tfrac{1}{2}(e^x + e^{-x}).$$

The other four functions, *tanh, cosech, sech* and *cotanh* may be derived from sinh and cosh by the same rules as apply to the trigonometrical forms. In electrical communication they are useful in calculations involving the transmission of currents along wires and in filters.

hyperbolic logarithm (*Maths.*). See **logarithm.**

hyperbolic paraboloid (*Maths.*). A saddle-shaped surface generated by the equation

$$x^2/a - y^2/b = z.$$

hyperbolic point on a surface (*Maths.*). One at which the curvatures of the normal sections are not all of the same sign. The directions which separate normal sections with positive curvature from those with negative curvature are called *asymptotic directions*. Cf. *elliptical, parabolic,* and *umbilical points on a surface.*

hyperbolic spiral (*Maths.*). A spiral with polar equation $r\theta = a^2$.

hyperboloid (*Maths.*). See **quadric.**

hypercalcaemia (*Med.*). Rise in the calcium content of the blood beyond normal limits.

hypercapnia (*Med.*). Excess of carbon dioxide in the lungs or the blood.

hypercharge (*Nuc.*). See **strangeness.**

hyperchimaera (*Bot.*). A chimaera in which the components are intimately mixed.

hyperchlorhydria (*Med.*). Increased secretion of hydrochloric acid by the acid-secreting cells of the stomach.

hypercholesterolaemia (*Med.*). Increase of cholesterol in the blood beyond normal limits.

hypercryalgesia (*Med.*). Abnormally increased sensitivity to cold.

hypercyesis (*Med.*). Superfoetation.

hyperdactyly (*Zool.*). The condition of having more than the normal number (five) of digits, as in *Cetacea.*

hyperdiploidy (*Cyt.*). The condition where the full chromosome complement is present, as well as a portion of one chromosome which has been translocated.

hyperemesis (*Med.*). Excessive vomiting.

hyperemesis gravidarum (*Med.*). Continued vomiting during pregnancy.

hyperemia (*Med.*). See **hyperaemia.**

hyperesthesia (*Med.*). See **hyperaesthesia.**

hyper-eutectoid steel (*Met.*). Steel with more carbon than is contained in pearlite. In carbon steels, one containing more than 0·9% carbon.

hyperfine structure (*Nuc.*). Splitting of spectrum lines into two or more very closely spaced components due to effects such as different isotopic mass (see **isotope structure**); or interaction between the orbital electrons and external fields (see **Stark effect, Zeeman effect**).

hyperfocal distance (*Photog.*). The distance in front of a lens beyond which all objects are substantially in focus, as defined by the focus of a point source of light not exceeding the *circle of confusion* (q.v.).

hypergamesis (*Zool.*). Utilization by a female Insect, during oviposition, of surplus spermatozoa as nutriment.

hypergeometric equation (*Maths.*). See **Gauss' differential equation.**

hypergeometric function (*Maths.*). The function $F(a; b; c; x)$ which is the sum of the *hypergeometric series*

$$1 + \frac{ab}{1c}x + \frac{a(a+1)b(b+1)}{1\,.\,2\,.\,c(c+1)}x^2 + \frac{a(a+1)(a+2)b(b+1)(b+2)}{1\,.\,2\,.\,3\,.\,c(c+1)(c+2)}x^3 + \ldots$$

which converges either if $|x| < 1$ or if $x = 1$ and $a + b < c$.

hypergeometric series (*Maths.*). See **hypergeometric function.**

hyperglycaemia, hyperglycemia (*Med.*). An increase in the sugar content of the blood beyond normal limits. See also **diabetes mellitus.**

hypergol (*Space*). A rocket fuel which ignites when in contact with an oxidizing agent.

hypergonar lens (*Optics*). A cylindrical compression lens used in wide-screen cinematography. See **CinemaScope.**

hyperhidrosis, hyperidrosis (*Med.*). Excessive perspiration.

hypericism (*Vet.*). A form of photosensitization occurring in sheep and cattle following the ingestion of the St John's Wort plant (*Hypericum perforatum*).

hyperinosis (*Med.*). Excess of fibrin in the blood; opp. **hypinosis.**

hyperinsulinism (*Med.*). A condition in which the blood sugar falls below normal limits, due to oversecretion of insulin by the pancreas; usually associated with pancreatic tumours, which provide the excess insulin.

hyperkeratosis (*Med.*). Overgrowth of the horny layer of the skin.

hyperkinesia (*Med.*). Excessive motility of a person, or of muscles.

hypermetamorphic (*Zool.*). Of Insects, passing through two or more sharply distinct larval instars. *n.* **hypermetamorphosis.**

hypermetropia (*Med.*). Long-sightedness. An abnormal condition of the eyes in which parallel rays of light come to a focus behind the retina instead of on it, the eyes being at rest. Also **hyperopia.**

hypermnesia (*Med.*). Exceptional power of memory.

hypernephroma (*Med.*). Grawitz's tumour. A tumour occurring in the kidney, thought by Grawitz to arise from adrenal tissue displaced there, but now known to be a carcinoma of the renal tubular cells.

hyperon (*Nuc.*). Cosmic-ray particle having a mass greater than that of a neutron and less than that of a deuteron.

hyperopia (*Med.*). See **hypermetropia.**

hyperparasitism (*Zool.*). The condition of being parasitic on a parasite. *n.* **hyperparasite.**

hyperphalangy (*Zool.*). The condition of having more than the normal number of phalanges, as in Whales.

hyperpharyngeal (*Zool.*). Above the pharynx as the *hyperpharyngeal band* in *Tunicata*.

hyperpiesia (*Med.*). The condition in which the blood-pressure is persistently raised above normal, in the absence of chronic nephritis.

hyperpiesis (*Med.*). Blood pressure raised above the normal.

hyperpituitarism (*Med.*). Overactivity of the pituitary gland; any condition due to overactivity of the pituitary gland, e.g., acromegaly, gigantism.

hyperplasia (*Med., Zool.*). Excessive multiplication of cells of the body: an overgrowth of tissue due to increase in the number of tissue elements: generally, overgrowth. *adj.* **hyperplastic.**

hyperploid (*Cyt.*). Having a chromosome number slightly exceeding an exact multiple of the haploid number.

hyperpnea, hyperpnoea (*Med.*). Increase in the depth and frequency of respiration; overventilation of the lungs.

hyperpragia (*Psychiat.*). Elation, with *flight of ideas*, as seen in the manic phase of manic-depressive psychosis.

hyperpyrexia (*Med.*). A degree of body temperature greatly above normal (e.g., $> 41°C$ or 105°F). See **heat-stroke.**

hypersensitization (*Photog.*). Treatment of a sensitive emulsion (e.g., by certain dyes or by ammonia) so as to increase its speed.

hypersonic (*Aero., Space*). Velocities of *Mach number* (q.v.) 5 or more. *Hypersonic flow* is the behaviour of a fluid at such speeds, e.g., in a shock tube.

hypersthene (*Min.*). An important rock-forming silicate of magnesium and iron, $(Mg,Fe)SiO_3$, crystallizing in the orthorhombic system; an essential constituent of norite, hypersthene-pyroxenite, hypersthenite, hypersthene-andesite, and charnockite. Strictly, an orthopyroxene containing 50–70% of the enstatite molecule.

hypersthene-gabbro (*Geol.*). A gabbroic igneous rock containing both ortho- and clino-pyroxene, in which the ortho-pyroxene is subordinate to clino-pyroxene.

hypersthenic (*Med.*). Having increased strength or tonicity. **Hypersthenic gastric diathesis.** The constitutional disposition in which the stomach is short and overactive in secretion and movement.

hypersthenite (*Geol.*). A coarse-grained igneous rock, consisting essentially of only one component, hypersthene, together with small quantities of accessory minerals.

hyperstomatic (*Bot.*). Having stomata on the upper surface of the leaf.

hypertelorism (*Med.*). The condition of excessive width between two organs or parts, particularly related to the eyes.

hypertely (*Zool.*). The progressive attainment of disproportionate size, either by a part or by an individual.

hypertensin (*Physiol.*). A polypeptide formed by the action of renin, an enzyme formed in the kidney, upon a plasma protein, hypertensinogen. It causes constriction of arterioles and can produce hypertension. Also called **angiotensin.**

hypertensinogen (*Physiol.*). A blood globulin which, reacting with renin, forms hypertensin.

hypertension (*Med.*). Increase in tension: a blood-pressure higher than normal.

hyperthyroidism (*Med.*). Abnormally high rate of secretion of thyroid hormone by the thyroid gland, resulting in an increase in the basal metabolic rate. See also **Basedow's disease.**

hyperthyrum (*Arch.*). The part of the architrave above a door or window opening.

hypertonic (*Chem.*). Having a higher osmotic pressure than a standard, e.g., that of blood, or of the sap of cells which are being tested for their osmotic properties.

hypertonus (*Med.*). A state of excessive muscular tone; opp. **hypotonus.**

hypertrichiasis, hypertrichosis (*Med.*). Abnormal overgrowth of hair; excessive hairiness.

hypertrophic pyloric stenosis (*Med.*). A disorder in children in which there is hypertrophy of the muscle in the pyloric region of the stomach, leading to obstruction to the passage of food into the small intestine, and vomiting.

hypertrophy (*Bot., Zool.*). An abnormal, usually pathological, enlargement of a plant cell or of a plant member, or of an animal cell or an animal member.

hypervitaminosis (*Med.*). The condition arising when too much of any vitamin (especially vitamin D) has been taken by a person.

hypha (*Bot.*). (1) One of the simple or branched filaments of the thallus (mycelium) of a fungus.

(2) A simple or branched filamentous outgrowth from internal cells in the thallus of a large seaweed.

hyphal body (*Bot.*). A thin-walled multinucleate segment of a hypha, serving for propagation and reproduction in some fungi which live as parasites in insects.

Hyphomicrobiales (*Bacteriol.*). A small order of bacteria characterized by multiplication by budding. In some species, longitudinal fission also occurs.

Hyphomycetes (*Bot.*). See **Moniliales.**

hyphomycetous (*Bot.*). (1) Relating to a Hyphomycete. (2) Mouldlike, cobwebby.

hyphopodium (*Bot.*). A more-or-less lobed outgrowth from a hypha, often serving to attach an epiphytic fungus to a leaf.

hypidiomorphic or **subhedral** (*Geol.*). A term referring to the texture of igneous rocks in which some of the component minerals show crystal faces, the others occurring in irregular grains. Cf. *idiomorphic crystals*.

hypinosis (*Med.*). See **hyperinosis.**

hypnagogic imagery (*Psychol.*). Hallucinatory type of phenomena or fantasy occurring in the drowsy state just before falling asleep.

hypnocyst (*Bot.*). A resting spore formed by some algae.

hypnody (*Zool.*). Resting period in larval forms.

hypnopaedia (*Psychol.*). Absorption of information during sleep, by repetition of sound reproduction from records. It is thought to occur during short periods of wakefulness or semiwakefulness, or in the drowsy state before the onset of sleep.

hypnophrenosis (*Med.*). Any type of disturbance of sleep.

hypnosis (*Psychol.*). A state induced, usually by another person, in which the individual loses much of the conscious control of his own activity and is susceptible to suggestion. There is usually complete relaxation but other physical states can be induced. The subject remains in contact with the hypnotist and can be induced to remember past events, often with the associated affect, which would not otherwise have been possible. It is sometimes used in psychotherapy.

hypnospore (*Bot.*). A thick-walled spore able to live for some time in an inert condition.

hypnotic (*Med.*). Of the nature of, or pertaining to, hypnosis; a medicinal agent which induces sleep.

hypnozygote (*Bot.*). A zygote which remains inert for some time after its formation.

hypo (*Chem., Photog.*). A colloquial abbrev. for *sodium thiosulphate*, the normal fixing solution for silver halide emulsions, the unreduced silver being removed by the hypo.

hypo-. Prefix from Gk. *hypo*, under. (*Chem.*) A prefix which signifies that a compound or group contains fewer radicals or atoms than the normal number.

hypoacidity (*Med.*). A deficiency of acid, especially in the gastric juice.

hypoadrenalism (*Med.*). The condition in which the activity of the adrenal glands is below normal.

hypoaria (*Zool.*). In some Fish, small lobes of the brain lying just below the corpora bigemina.

hypoascidium (*Bot.*). An abnormal cup-shaped outgrowth from a leaf, or a transformation of a leaf; the inner surface corresponds to the lower surface of the leaf.

hypobasal half (*Bot.*). The posterior portion of an embryo.

hypobasidium (*Bot.*). An enlarged cell of a hypha in which a nuclear fusion occurs before the true basidium is formed.

hypobiosis (*Biol.*). Any condition of reduced metabolic activity, whether occurring naturally or induced artificially.

hypoblast (*Zool.*). The innermost germinal layer in the embryo of a metazoan animal, giving rise to the endoderm and sometimes also to the mesoderm. Cf. *epiblast*.

hypobranchial (*Zool.*). The lowermost element of a branchial arch.

hypobranchial space (*Zool.*). The space below the gills in *Decapoda*.

hypocalcaemia, hypocalcemia (*Med.*). A calcium content of the blood below normal limits.

hypocarpogenous (*Bot.*). Flowering and fruiting underground.

hypocaust (*Build.*). A hollow space beneath the floor of a room or bath, serving as a flue for the hot gases from a furnace which, in circulating, give warmth to the room or bath.

hypocentrum (*Zool.*). See **subnotochord.**

hypocercal (*Zool.*). Said of a type of caudal fin, found in *Anaspida* and *Pteraspida*, in which the vertebral column bends downwards and enters the hypochordal lobe which is larger than the epichordal lobe.

hypocerebral ganglion (*Zool.*). A component of the sympathetic nervous system of Insects, lying along the course of the recurrent nerve.

hypochlorhydria (*Med.*). Diminished secretion of hydrochloric acid by the acid-secreting cells of the mucous membrane of the stomach.

hypochlorites (*Chem.*). See **hypochlorous acid.**

hypochlorous acid (*Chem.*). HClO. An aqueous solution of chlorine(I)oxide. Monobasic acid which forms salts called *hypochlorites* (*chlorates* (I)). Weak acid, easily decomposed.

hypochlorous anhydride (*Chem.*). Same as *chlorine* (I) *oxide* (see under **chlorine oxides**).

hypochnoid (*Bot.*). Of a fungus, having a flattened, dry, loose weft of hyphae.

hypochondriasis (*Med.*). Morbid preoccupation with bodily functions and sensations, with the false belief that the latter indicate bodily disease.

hypochord (*Zool.*). See **subnotochord.**

hypochordal (*Zool.*). Below the notochord; said of the lower lobe of a caudal fin.

hypocleidium (*Zool.*). See **interclavicle.**

hypocone (*Zool.*). A fourth cusp arising on the cingulum on the postero-internal side of an upper molar tooth, producing a quadritubercular pattern.

hypoconid (*Zool.*). The innermost of three cusps which arise on the talonid of a lower molar tooth, producing a trituberculo-sectorial pattern.

hypoconulid (*Zool.*). The most posterior of three cusps which arise on the talonid of a lower molar tooth.

hypocotyl (*Bot.*). The part of the axis of a seedling between the insertion of the cotyledons and the radicle.

hypocrateriform (*Bot.*). Having the lower part cylindrical, widening upwards, and with the upper edge expanding more or less horizontally.

Hypocreales (*Bot.*). An order of the *Ascomycetes*; the ascocarp is a perithecium The perithecia are associated with a soft, lightly coloured periderm, which is distinct from the rest of the mycelium.

hypocycloid (*Maths.*). See **roulette.**

hypoderm, hypodermis (*Bot.*). A layer, one or more cells thick, of strongly constructed cells, lying immediately beneath the epidermis and reinforcing it. (*Zool.*) In *Arthropoda* and other Invertebrates with a distinct cuticle, the epithelial cell-layer underlying the cuticle, by which the cuticle is secreted. *adj.* **hypodermal.**

hypodermic (*Med.*). Under the skin; a medical agent injected under the skin.

hypodermoclysis (*Med.*). The injection of fluid (e.g., salt solution) under the skin.

hypo eliminator (*Photog.*). Solution used for removing all trace of fixative after washing prints, thus ensuring permanence. Usually hydrogen peroxide and ammonia.

hypo-eutectoid steel (*Met.*). Steel with less carbon than is contained in pearlite, i.e., the iron-cementite eutectoid. In carbon steels, one containing less than 0·9% carbon.

hypogastrium (*Anat.*). The lower median part of the abdomen; cf. *epigastrium. adj.* **hypogastric.**

hypogeal, hypogaeous (*Bot.*). (1) Living beneath the surface of the ground. (2) Germinating with the cotyledons remaining in the soil.

hypogene (*Geol.*). Said of rocks formed, or agencies at work, under the earth's surface.

hypogenesis (*Zool.*). Direct development without metagenesis.

hypogenous (*Bot.*). Placed on the under side.

hypoglossal (*Zool.*). Underneath the tongue; the 12th cranial nerve of Vertebrates, running to the muscles of the tongue.

hypoglottis (*Zool.*). In Vertebrates, the under part of the tongue; in *Coleoptera*, part of the labium.

hypoglycaemia, hypoglycemia (*Med.*). A reduction in the level of sugar—actually glucose, in the blood.

hypognathous (*Zool.*). Having the under jaw protruding beyond the upper jaw; having the mouth-parts directed downwards.

hypogonadism (*Med.*). The condition in which there is a deficiency of the internal secretion of the gonads.

hypogynous (*Bot.*). (1) Said of a flower in which the calyx, corolla, and androecium, or one or more of these, arise from the receptacle below the gynaeceum. (2) Said of the floral members so placed, provided they are not attached to the calyx or calyx tube. (3) Said of an antheridium which develops in a branch arising from the stalk of the oögonium.

hypohidrosis (*Med.*). Abnormal diminution in the secretion of sweat.

hypohyal (*Zool.*). The lowermost element of a hyoid arch.

hypohypophysism (*Physiol.*). See **apituitarism.**

hypoid bevel gear (*Eng.*). A bevel gear in which the axes of the driving and driven shafts are at right angles but not in the same plane, resulting in some sliding action between the teeth; used in the back-axle drive of some automobiles.

hypoischium (*Zool.*). In some Lizards, an ossification of the posterior part of a ligament, which represents the part of the epipubis between the fenestrae and supports the ventral wall of the cloaca. Also called **os cloacae.**

hypolimnion (*Ecol.*). The cold lower layer of water in a lake. Cf. *epilimnion.*

hypomania (*Med.*). Simple mania. A condition characterized by mental excitement in the absence of mental confusion or of symptoms of insanity.

hypomenorrhea, hypomenorrhoea (*Med.*). The condition in which the interval between 2 menstrual periods is increased to between 35 and 42 days.

hypomeral bones (*Zool.*). In some Fish, slender bones developed in connexion with the hypomeres.

hypomere (*Zool.*). The lateral muscle-plate zone of the mesothelial wall of a developing Vertebrate.

Hypon (*Pharm.*). A proprietary analgesic and antipyretic containing acetylsalicylic acid, phenacetin, codeine, caffeine, etc.

hyponasty (*Bot.*). (1) The more vigorous growth of the under side of a flattened organ, usually causing some change in the position of that organ. (2) Eccentric secondary thickening of the lower side of a stem or root, when this lies in an approximately horizontal posture.

hyponeuston (*Ecol.*). Organisms floating or swimming immediately below the surface of water.

hyponome (*Zool.*). In *Cephalopoda*, the funnel by which water escapes from the mantle cavity.

hyponychium (*Zool.*). In Mammals, the epidermal layer underlying the nail.

hypophare (*Zool.*). In the rhagon type of Sponge colony, the lower basal wall which is without flagellated chambers. Cf. *spongophare.*

hypopharyngeal (*Zool.*). Below the pharynx, as the *hypopharyngeal* groove of *Cephalochorda.*

hypopharyngeal bone (*Zool.*). In some Fish, one of a pair of elements constituting the fifth gill arch.

hypopharyngeal bracon (*Zool.*). In the larvae of some *Coleoptera*, a transverse bar on the inner side of the labium.

hypopharynx (*Zool.*). In Insects, a median tonguelike structure arising from the floor of the mouth.

hypophloeodal (*Bot.*). Growing just within the surface of bark.

hypophosphoric acid (*Chem.*). H_2PO_3, or $H_4P_2O_6$. Obtained by the slow oxidation of phosphorus in moist air. Stable at ordinary temperatures. Hydrolyzed by mineral acids, forming a mixture of phosphoric and phosphorous acids.

hypophosphorous acid (*Chem.*). H_3PO_2. Feeble monobasic acid, which forms a series of salts called *hypophosphites* (phosphates (I)), oxidized to phosphates by oxidizing agents.

hypophyll (*Bot.*). A scale leaf which subtends a cladode.

hypophyllous (*Bot.*). Attached to, or growing from, the under side of a leaf.

hypophyseal cachexia (*Med.*). See **Simmonds' disease.**

hypophysectomy (*Surg.*). Removal of the pituitary gland.

hypophysis (*Bot.*). (1) A cell between the suspensor and the embryo proper, in a flowering plant. (2) A swelling beneath the sporangium. (*Zool.*) A downwardly growing structure; in *Cephalochorda*, the olfactory pit; in Vertebrates, the pituitary body. *adj.* **hypophysial.**

hypopiesis (*Med.*). Very low blood pressure.

hypopituitarism (*Med.*). A general term for any condition caused by diminished activity of the pituitary gland; characterized usually by obesity and imperfect sexual development.

hypoplasia (*Zool.*). Underdevelopment; deficiency. *adj.* **hypoplastic.**

hypoplastron (*Zool.*). In *Chelonia*, one of the paired plates composing the plastron, lying between the xiphiplastron and the hyoplastron.

hypopleuron (*Zool.*). In *Diptera*, a thoracic sclerite lying below the metapleuron and above the coxae of the mesothorax and the metathorax.

hypoploid (*Cyt.*). Having a chromosome number a little less than some exact multiple of the haploid number.

hypopneustic (*Zool.*). Said of the respiratory systems of Insects in which one or more pairs of spiracles have disappeared completely.

hypopteronosis cystica (*Vet.*). An inherited disease of budgerigars and canaries characterized by the formation of dermal cysts containing immature feathers.

hypoptilum (*Zool.*). See **aftershaft.**

hypopus (*Zool.*). In *Tyroglyphidae* (Cheese-mites) and allied forms, a stage in the development which appears when conditions are unfavourable, and which is responsible for the dispersal of the species to more favourable conditions.

hypopygium (*Zool.*). In certain male *Diptera*, an organ formed by the curvature beneath the body of the apical segments of the abdomen.

hypopyon (*Med.*). A collection of pus in the anterior chamber of the eye, between the iris and the cornea.

hyporhachis (*Zool.*). In Birds, a secondary feather shaft arising from the calamus, just proximal to the superior umbilicus.

hypospadias (*Med.*). A congenital deficiency in the floor of the urethra.

hypostasis (*Med.*). Sediment or deposit. Passive hyperaemia in a dependent part owing to sluggishness of the circulation.

hypostatic (*Gen.*). Recessive, when relating to 1 of 2 characters which are not allelomorphs.

hyposthenic (*Med.*). Having diminished strength or tonicity. **Hyposthenic gastric diathesis.** The constitutional disposition in which the stomach is long, sluggishly acting, and secretes little acid.

hyposthenuria (*Med.*). The secretion of a pale urine of unusually low specific gravity.

hypostoma or **hypostome** (*Zool.*). In some *Coelenterata*, the raised oral cone: in Insects, the labrum: in *Crustacea*, the lower lip or fold forming the posterior margin of the mouth: in some *Acarina*, the lower lip formed by the fusion of the pedipalpal coxae.

hypostomatic (*Bot.*). Bearing stomata on the lower surface.

hypostomatous (*Zool.*). Having the mouth placed on the lower side of the head, as Sharks.

hypostroma (*Bot.*). A stroma formed by a parasitic fungus beneath the surface of the epidermis of the host.

hypostyle hall (*Arch.*). A hall having columns to support the roof.

hyposulphuric acid (*Chem.*). Dithionic acid, $H_2S_2O_6$.

hyposulphurous acid (*Chem.*). (1) $H_2S_2O_4$. Unstable. Powerful reducing agent. Salts are hyposulphites. (2) An old term for *thiosulphuric acid*, $H_2S_2O_3$.

hypotarsus (*Zool.*). In Birds, the fibulare.

hypotension (*Med.*). Low blood-pressure.

hypotenusal allowance (*Surv.*). The distance added to each chain length, when chaining along sloping ground, in order to give a length whose horizontal projection shall be exactly one chain. For the 100-link chain the hypotenusal allowance is 100 (sec $\theta - 1$), where θ is the angle of slope of the ground from the horizontal.

hypotenuse (*Maths.*). The side opposite the right angle of a right-angled triangle.

hypothalamus (*Zool.*). In the Vertebrate brain, the ventral zone of the thalamencephalon. In Man the part of the brain which makes up the floor and part of the lateral walls of the third ventricle. The mammillary bodies, tuber cinerium, infundibulum, neurohypophysis and the optic chiasma are also part of the hypothalamus.

hypothallus (*Bot.*). (1) A film of waste material left on the substratum as the plasmodium of a Myxomycete moves about. (2) The first-formed weft of hyphae in the development of the thallus of a lichen, often remaining at the base or edge of the thallus.

hypotheca (*Bot.*). The younger of the 2 valves in the cell-wall of a diatom.

hypothenusal allowance (*Surv.*). See **hypotenusal allowance.**

hypothermia (*Med.*). Subnormal body temperature, especially that induced for purposes of heart and other surgery (see **cryotherapy**).

hypothesis. A statement which although generally believed to be true has not been proved, either definitively or circumstantially. (*An. Behav.*) A term applied to repeated occurrences of a particular response or response pattern when an animal is in a two-choice situation, and, as a result, makes a high proportion of wrong choices.

hypothetical exchange (*Teleph.*). A telephone exchange which, until a new exchange is constructed, is made up from parts of existing exchanges, the subscribers being numbered according to the system of the new exchange.

hypothymia (*Psychiat.*). Low emotional response.

hypothyroidism (*Med.*). The condition accompanying the diminished secretion of the thyroid gland. See also **cretinism** and **myxoedema.**

hypotonic (*Chem.*). Having a lower osmotic pressure than a standard, e.g., that of blood, or of the sap of cells which are being tested for their osmotic properties.

hypotonus (*Med.*). See **hypertonus.**

hypotrachelium (*Arch.*). The junction between the shaft and capital of a column.

Hypotremata (*Zool.*). A suborder of *Euselachii* characterized by the possession of 5 ventral gill clefts, enlarged pectoral fins, a dorso-ventrally flattened body, and the absence of an anal fin. Skates and Rays.

hypotrematic (*Zool.*). In *Cyclostomata*, the lower lateral bar of the branchial basket.

Hypotricha (*Zool.*). A suborder of *Spirotricha*, the members of which are generally of creeping habit; they possess a permanent gullet with undulating membranes, and have a depressed body, with locomotor cilia on the ventral surface only.

Hypotrichiales (*Bot.*). An order of the *Phycomycetes*, closely resembling the *Chytridiales* (q.v.), except that the zoospores have the flagellum anteriorly.

hypotrichous (*Zool.*). Having cilia principally on the lower surface of the body.

hypotrochoid (*Maths.*). See **roulette.**

hypotrophy (*Bot.*). Eccentric thickening of the under side of a horizontal shoot or root.

hypovitaminosis (*Med.*). The condition resulting from deficiency of a vitamin in the diet.

hypoxanthine (*Chem.*). 6-hydroxypurine:

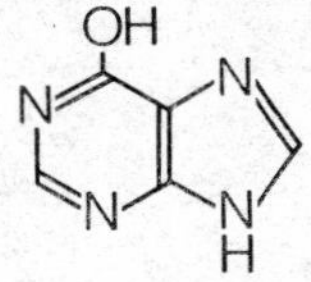

Formed by the breakdown of nucleoproteins by enzymatic action.

hypoxyloid (*Bot.*). Forming a cushion-shaped or crustlike stroma.

hypso-. Prefix from Gk. *hypsos*, height.

hypsochrome (*Chem.*). A radical which shifts the absorption spectrum of a compound towards the violet end of the spectrum.

hypsodont (*Zool.*). Of a Mammalian tooth with a high crown and deep socket. Cf. *brachyodont*.

hypsoflore (*Chem.*). A radical which tends to shift

the fluorescent spectrum of a compound toward shorter wavelengths.

hypsograph (*Telecomm.*). A device for recording the transmission levels on a circuit, either during testing, or continuously during transmission of signals.

hypsometer (*Phys.*). An instrument used for determining the boiling-point of water, either with a view to ascertaining altitude, by calculating the pressure, or for correcting the upper fixed point of the thermometer used.

hypsophyllary leaf (*Bot.*). A bract.

hypural (*Zool.*). Below the tail; one of a set of large plates formed by the fusion of the haemal elements of the caudal vertebrae, and supporting the hypochordal lobe of the tail-fin in some Fish.

Hyracoidea (*Zool.*). An order of small Eutherian Mammals having 4 digits on the fore limb and 3 on the hind limb, pointed incisor teeth with persistent pulps, lophodont grinding teeth, no scrotal sac and 6 mammae; terrestrial African forms. Dassies.

Hy-rib (*Build., Civ. Eng.*). A proprietary form of metal mesh used for ceilings, walls, surroundings of columns, etc., to which plaster is to be applied. Also used as reinforcement for concrete slabs carrying minor loads.

hyster-, hystero-. Prefix from Gk. *hysteria*, womb or *hysteros*, later.

hysteranthous (*Bot.*). Said of leaves which develop after the plant has flowered.

hysterectomy (*Surg.*). Surgical removal of the uterus.

hysteresis (*Phys.*). The retardation or lagging of an effect behind the cause of the effect, e.g., dielectric hysteresis, magnetic hysteresis, etc.

hysteresis coefficient (*Elec. Eng.*). See **Steinmetz coefficient.**

hysteresis error (*Elec. Eng.*). Instruments or control systems may show nonreversibility similar to hysteresis. The maximum difference between the readings or settings obtainable for a given value of the independent variable is the *hysteresis error*.

hysteresis heat (*Elec. Eng.*). That arising from hysteresis loss, in contrast with that from ohmic loss associated with eddy currents.

hysteresis loop (*Elec. Eng.*). A closed figure formed by plotting magnetic induction B in a magnetizable material against magnetizing field H when the latter is taken through a complete cycle of increasing and decreasing values. The area of this loop measures the energy dissipated during a cycle of magnetization. Also called **B/H loop.** Similar effects occur with applied mechanical or electrical stresses, and this is by analogy known as *mechanical* or *electric hysteresis*. Dielectric materials with appreciable hysteresis loss are termed *ferroelectric materials* (q.v.).

hysteresis loss (*Elec. Eng.*). Energy loss in taking unit quantity of material once round a hysteresis loop. It can arise in a dielectric material subjected to a varying electric field or in a magnetic material in a varying magnetic field.

hysteresis motor (*Elec. Eng.*). A synchronous motor which starts by reason of the hysteresis losses induced in its steel secondary by the revolving field of the primary.

hysteresis tester (*Elec. Eng.*). A device, invented by Ewing, for making a direct measurement of magnetic hysteresis in samples of iron or steel.

hysteria (*Psychol.*). A psychoneurosis in which repressed complexes become split off or dissociated from the personality, forming independent units, partially or completely unrecognized by consciousness. It gives rise to hypnoidal states, such as amnesia, fugues, somnambulism, etc., and may also be manifested by various physical symptoms, such as tics, paralysis, blindness, deafness, etc. See **conversion hysteria.** General features of this neurosis are an extreme degree of emotional instability and an intense craving for affection.

Hysteriales (*Bot.*). An order of the *Ascomycetes*; the asci form a definite layer at the base of the ascocarp, which is elongate and opens by a longitudinal slit.

hysteriform, hysteriaeform (*Bot.*). Having the shape of a long narrow ridge, with a longitudinal opening along the top.

hysterocolpectomy (*Surg.*). Removal of the vagina (or part of it) and the uterus.

hysterogenic (*Zool.*). Developing later.

hysteropexy (*Surg.*). The fixation of a displaced uterus surgically.

hysterosoma (*Zool.*). In *Acarina*, the region of the body comprising the metapodosoma and the opisthosoma.

hysterothecium (*Bot.*). An elongated perithecium, remaining closed as it develops, and opening when ripe by a cleft at the top.

hysterotomy (*Surg.*). Incision of the uterus.

hythergraph (*Meteor.*). A graphical representation of the differentiation between various types of climate. Mean monthly temperatures are plotted as ordinates against the mean monthly rainfalls as abscissae, and a closed, twelve-sided polygon, the hythergraph, is obtained. This reveals the type of climate at a glance.

Hz (*Elec.*). Symbol for hertz.

i **(*Chem.*).** A symbol for *van't Hoff's factor*. (*Maths.*) The symbol used by mathematicians to represent the imaginary number whose square equals minus one, i.e., $i^2 = -1$. See *j*.

***i*-** **(*Chem.*).** An abbrev. for: (1) *optically inactive*; (2) *iso*-, i.e., containing a branched hydrocarbon chain.

I (*Chem.*). The symbol for *iodine*.

I **(*Chem.*).** A symbol for *ionic strength*. (*Elec.*) Symbol for *electric current*. (*Eng.*) Symbol for *moment of inertia*. (*Light.*) Symbol for *luminous intensity*.

I.A. (*Phys.*). Abbrev. for *international ångström*.

IAA (*Chem.*). Abbrev. for *indole-3-acetic acid* (see **indoles**).

I.A.S. (*Aero.*). Abbrev. for *indicated air speed*.

IATA (*Aero.*). Abbrev. for *International Air Transport Association*.

iatrochemistry (*Med.*). The study of chemical phenomena in order to obtain results of medical value; practised in the 16th century. Modern equivalents **chemotherapy** or **pharmacology**.

iatrogenic disease (*Med.*). A disease produced by appropriate therapy for another disease, usually occurring as a side-effect of pharmacological agents.

I-beam (*Build., Civ. Eng., etc.*). See **H-beam**.

I.C. (*Telecomm.*). *Information content*.

I.C.A.N. atmosphere (*Meteor.*). *I*nternational *C*ommission for *A*ir *N*avigation atmosphere. Term superseded by *standard atmosphere* (q.v.).

ICAO (*Aero.*). Abbrev. for *International Civil Aviation Organization*.

Icarus (*Astron.*). A minor planet, discovered in 1949, with a small eccentric orbit approaching the sun to within 30 million kilometres, inside the orbit of Mercury. In June 1968 Icarus passed within four million miles of the earth.

ice (*Meteor.*). Ice is formed when water is cooled below its freezing-point. It is a transparent crystalline solid of rel. d. 0·916 and specific heat capacity 0·50. On account of the fact that water attains its maximum density at 4°C, ice is formed on the surface of ponds and lakes during frosts, and thickens downwards.

ice action (*Geol.*). The work and effects of ice on the earth's surface. See **glacial action, glaciation, glacier**.

ice-apron, ice-breaker (*Civ. Eng.*). A construction covering the upstream side of a bridge pier, and serving to break floating ice, or in default of this, to give protection against the thrust of the ice.

iceberg (*Meteor.*). A large mass of ice, floating in the sea, which has broken away from a glacier or ice barrier. Icebergs are carried by ocean currents for great distances, often reaching latitudes of 40° to 50° before having completely melted. Approximately one-tenth of an iceberg shows above the surface.

iceblink (*Meteor.*). A whitish glare in the sky over ice which is too distant to be visible.

ice-breaker (*Civ. Eng.*). (1) An *ice-apron* (q.v.). (2) A projecting pier so arranged in relation to a harbour entrance that floating ice is kept outside. (3) A vessel specially equipped for clearing a passage through ice-bound waters.

ice colours (*Chem.*). Dyestuffs produced on the cotton fibre direct, by the interaction of a second component with a solution of a diazo-salt cooled with ice.

ice contact slope (*Geol.*). The steep slope or back face of a deltaic accumulation originally formed at an ice front and in contact with it.

ice day (*Meteor.*). Same as **frost day**.

ice guard (*Aero.*). A wire-mesh screen fitted to a piston aero-engine intake so that ice will form on it and not inside the intake; a *gapped ice guard* is mounted ahead of the intake so that air can pass round it, while a *gapless ice guard* is inside the intake and an alternative air path comes into use when it ices up.

Iceland agate (*Min.*). A name quite erroneously applied to the natural glass *obsidian* (q.v.).

Iceland spar (*Min.*). A very pure transparent and crystalline form of calcite, first brought from Iceland. It has perfect cleavage, is noted for its double refraction, and hence is used in construction of the Nicol prism.

I.C. engine (ICE). See **internal-combustion engine**.

ichnograph (*Build.*). A view showing the ground plan of a building or part of a building.

ichor (*Geol.*). The name applied by Sederholm to highly penetrating granitic liquids, charged with magmatic vapours (emanations), which he believed to operate in palingenesis. (*Med.*) A thin, watery discharge from a wound or a sore. *adj.* **ichorous**.

ichthy-, ichthyo-. Prefix from Gk. *ichthys*, fish.

ichthyic (*Zool.*). Pertaining to, or resembling, Fish.

Ichthyopterygia (*Zool.*). An extinct subclass of marine Reptiles with a single lateral temporal vacuity, a fishlike body, the head produced into an elongated snout and the limbs in the form of paddles.

ichthyopterygium (*Zool.*). A paddle-like fin, or limb, used for swimming, e.g., pectoral or pelvic fin of Fish.

ichthyosis (*Med.*). Xerodermia. A disease characterized by dryness and scaliness of the skin, due to lack of secretion of the sweat and the sebaceous glands. (*Vet.*) A congenital hardening of the skin of calves.

iconoscope (*TV*). Electron camera comprising a mosaic of photo-emissive material upon which the optical image is focused, and which is scanned by a cathode-ray beam, which restores charge lost through illumination.

Iconotron (*TV*). TN for an *image iconoscope*.

icosahedron (*Maths.*). A twenty-faced *polyhedron* (q.v.). The faces of a *regular icosahedron* are identical equilateral triangles.

icositetrahedron (*Min.*). A solid figure having 24 trapezoidal faces, and belonging to the cubic system. Exemplified by some garnets.

ICSH (*Physiol.*). Abbrev. for *interstitial cell stimulating hormone*. See **luteinizing hormone**.

icteric (*Med.*). Of the nature of, or affected with, jaundice.

icterus (*Med.*). *Jaundice* (q.v.).

ictus (*Med.*). A stroke or sudden attack.

ICW (*Radio*). Abbrev. for *interrupted continuous waves*.

id (*Psychol.*). A term originally introduced by Groddeck and later used by Freud to denote the sum total of the primitive instinctual forces in an individual. It subserves the pleasure-pain principle, in which the activities of the organism are concerned with the immediate increase of pleasurable and reduction of painful stimuli. It is dominated by blind impulsive wishing.

iddingsite (*Min.*). An alteration product of

olivine consisting of goethite, quartz, montmorillonite group clay minerals, and chlorite.

ideal crystal (*Crystal.*). One in which there are no imperfections or alien atoms.

ideal gas (*Chem.*). Gas with molecules of negligible size and exerting no intermolecular attractive forces. Such a gas is a theoretical abstraction which would obey the ideal gas law under all conditions. The behaviour of real gases becomes increasingly close to that of an ideal gas as the pressure on them is reduced. Also called **perfect gas.**

ideal transducer (*Elec. Eng.*). Any transducer which converts without loss all the power supplied to it.

ideal transformer (*Elec. Eng.*). A hypothetical transformer corresponding to one with a coefficient of coupling of unity.

idempotent (*Maths.*). If S is a set and $*$ is an operation in S, then an element e of S is said to be an idempotent with respect to $*$ if $e*e = e$. For the usual addition of real numbers 0 is the only idempotent; for the usual multiplication of real numbers 0 and 1 are the only idempotents. Each group has one and only one idempotent, namely the identity element. There is no restriction on the number of idempotents to be found in a semi-group.

identification dimensions (*Ships*). Same as **registered dimensions.**

identification, friend or foe (*Radar*). Device consisting of automatic transponder carried by vessel in order to modify nature of target blip on radar screen when this has been located by a friendly radar station. If blips are not modified in this way the vessel responsible is assumed hostile. Abbrev. **IFF.**

identity (*Maths.*). (1) See **equation.** (2) If S is a set with an operation $*$, and e is an element in the set S such that, for all elements x in the set S we have $x*e = x = e*x$, then e is said to be an identity element, or neutral element, with respect to the operation $*$ in S. E.g., in the set of real numbers, 0 is an identity element with respect to addition, and 1 is an identity element with respect to multiplication.

identity mapping (*Maths.*). The identity mapping, or identity function, on a set S is the mapping i_s from S on to S defined by $i_s(x) = x$, for all elements x in S.

ideomotor action (*Psychol.*). Action which is a direct result of ideas without any deliberation.

I.D.F. (*Teleph.*). Abbrev. for *intermediate distribution frame.*

idio-. Prefix from Gk. *idios*, peculiar, distinct.

idioblast (*Bot.*). A nonchlorophyllous, thick-walled cell having supporting functions, usually elongated, and occurring among cells containing chlorophyll; more generally, a cell which differs in form, contents and wall-thickening from its neighbours. (*Geol.*) A crystal which developed in metamorphic rocks and is bounded by crystal contours; cf. *idiomorphic crystals. adj.* **idioblastic.** See also **porphyroblastic.**

idiochromatic (*Crystal.*). Said of the properties of a crystal free from impurity. Also **intrinsic.**

idiochromatin (*Cyt.*). A substance within the nucleus which controls the reproduction of the cell.

idiochromidia (*Cyt.*). *Chromidia* (q.v.) which can replace or be reformed into the nucleus; generative chromidia.

idiochromosome (*Cyt.*). A sex chromosome.

idioglossia (*Med.*). The wrong use of consonants by a child, making speech unintelligible.

idiomorphic (euhedral) crystals (*Geol.*). Rock minerals which are bounded by the crystal faces peculiar to the species. Cf. *allotriomorphic* (anhedral) and *hypidiomorphic* (subhedral).

idiomuscular (*Zool.*). Said of a special type of muscular contraction produced by artificial stimulation.

idiopathy (*Med.*). Any morbid condition arising spontaneously, having no known origin. *adj.* **idiopathic.**

idioplasm (*Gen.*). See **germ plasm.**

idiosoma (*Zool.*). In *Acarina*, the prosoma plus the opisthosoma.

idiosphaerotheca (*Zool.*). See **acroblast.**

idiostatic connexion (*Elec. Eng.*). See **Dolezalek electrometer.**

idiot (*Med.*). A person so defective in mind from birth as to be unable to protect himself against ordinary physical dangers; one afflicted with the severest grade of feeble-mindedness with an IQ not above 25.

idiothermous (*Zool.*). See **warm-blooded.**

idiot tape (*Typog.*). A continuous unjustified tape, containing only signals for new paragraphs, which must be processed into a new justified tape before it can control a typesetting or filmsetting machine. **See computer typesetting.**

idioventricular (*Med.*). Pertaining to the ventricle of the heart alone.

idiozome (*Zool.*). The attraction sphere or region of clear protoplasm surrounding the centrosome in spermatogenesis, a separated part of the archoplasm which eventually develops into the head-cap of the spermatozoon.

I display (*Radar*). Display representing the target as a full circle when the antenna is pointed directly at it, the radius being in proportion to the range.

idle component (*Elec. Eng.*). Same as **reactive component.**

idle-current wattmeter (*Elec. Eng.*). A name sometimes given to an electrical instrument for measuring reactive volt-amperes.

idler (*Eng.*). See **idle wheel.** (*Print.*) A free-running roller on a web-fed press. Also **idling roller.**

idler pulley (*Eng.*). See **guide pulley.**

idle wheel (*Eng.*). A wheel interposed in a gear train, either solely to reverse the direction of rotation or also to modify the spacing of centres, without affecting the ratio of the drive. Also **idler.** (*Horol.*) See **intermediate wheel.**

idle wire (*Elec. Eng.*). The part of the armature winding of an electric machine which does not actually cut the lines of force, i.e., that part comprising the end connexions.

idling (*I.C. Engs.*). The slow rate of revolution of an automobile or aero engine, when the throttle pedal or lever is in the closed position.

idling adjustment (*I.C. Engs., etc.*). A setting of the slow-running jet and throttle position of a carburettor, so as to give regular *idling* (q.v.).

idling control valve (*Aero.*). See **minimum burner pressure valve.**

idling roller (*Print.*). **See idler.**

idocrase (*Min.*). See **vesuvianite.**

idose (*Chem.*). A monosaccharide belonging to the group of aldohexoses.

Idotron (*Print.*). Electronic equipment for controlling ink density and viscosity on rotogravure and flexographic presses.

I.F. (*Radio*). *Intermediate frequency.*

iff (*Maths.*). Abbrev. form of 'if and only if'.

IFR (*Aero.*). Abbrev. for ***instrument flight rules.***

Igmerald (*Min.*). TN for synthetic emerald made at Bitterfeld in Germany, the first two letters of the name signifying *I*nteressen*g*emeinschaft.

igneous complex (*Geol.*). A group of rocks,

occurring within a comparatively small area, which differ in type but are related by similar chemical or mineralogical peculiarities. This indicates derivation from a common source.

igneous (or **magmatic**) **cycle** (*Geol.*). The sequence of events usually followed in igneous activity; it consists of an eruptive phase, a plutonic phase, and a phase of minor intrusion.

igneous intrusion (*Geol.*). A mass of igneous rock which crystallized before the magma reached the earth's surface; including *dykes, sills, stocks, bosses* and *bathyliths*.

igneous rocks (*Geol.*). Rock masses generally accepted as being formed by the solidification of magma injected into the earth's crust, or extruded on its surface. In the light of studies of crystal 'mush' movement and ionic diffusion, the magmatic origin of many examples of igneous rocks has been questioned.

ignimbrite (*Geol.*). A pyroclastic rock consisting originally of lava droplets and glass fragments which were so hot at the time of deposition that they were welded together. Also **welded tuffs.**

ignite (*Chem.*). To heat a gaseous mixture to the temperature at which combustion occurs, e.g., by means of an electric spark.

igniter (*Civ. Eng.*). A blasting fuse or other contrivance used to fire an explosive charge.

igniter plug (*Aero.*). An electrical-discharge unit for lighting up gas turbines.

ignition (*Elec. Eng., I.C. Engs.*). The firing of an explosive mixture of gases, vapours, or other substances by means, e.g., of an electric spark.

ignition advance (*I.C. Engs.*). The crank angle before top dead-centre, at which the spark is timed to pass in a petrol or gas engine. See **ignition timing, angle of advance.**

ignition coil (*I.C. Engs.*). An induction coil for converting the low-tension current supplied by the battery into the high-tension current required by the sparking-plugs.

ignition lag (*I.C. Engs.*). Of a combustible mixture in an engine cylinder, the time interval between the passage of the spark and the resulting pressure rise due to combustion.

ignition rating (*Elec. Eng.*). A special rating, in *ampere hours* (q.v.), employed for accumulators used for supplying ignition systems; it is generally twice the continuous rating at a low discharge rate.

ignition rectifier (*Elec. Eng.*). Mercury-arc rectifier in which the cathode spot is initiated by a voltage impulse applied to a special electrode dipping into the mercury pool.

ignition system (*I.C. Engs.*). The arrangement for providing the high-tension voltage required for ignition. See **ignition coil, magneto ignition.**

ignition temperature (*Heat*). In the combustion of gases, the temperature at which the heat loss due to conduction, radiation, etc., is more than counterbalanced by the rate at which heat is developed by the combustion reaction.

ignition timing (*I.C. Engs.*). The crank angle relative to top dead-centre at which the spark passes in a petrol or gas engine. See **angle of advance.**

ignition voltage (*Elec. Eng.*). That required to start discharge in a gas tube.

ignitor (*Electronics*). See **pilot electrode.**

ignitor drop (*Electronics*). The voltage drop between cathode and anode of the ignitor discharge in a switching tube.

ignitron (*Electronics*). Mercury-arc rectifier with *ignitor*, which is an electrode which can dip into the cool mercury pool and draw an arc to start the ionization.

ignore (*Comp.*). A character signifying that no more action is to be taken.

Iguana (*Zool.*). A genus of large thick-tongued arboreal lizards in tropical America. Loosely, others of the same family (**Iguanidae**).

IGV (*Aero.*). Abbrev. for *inlet guide vanes*.

I.H.P. (*Eng.*). Abbrev. for *indicated horsepower*.

ijolite (*Geol.*). A coarse-grained igneous rock, consisting of nepheline, aegirine-augite, with usually melanite garnet as a prominent accessory, occurring in nepheline-syenite complexes in the Kola Peninsula, White Sea, the Transvaal, and elsewhere.

Il (*Chem.*). The symbol for *illinium*.

ileitis (*Med.*). Inflammation of the ileum.

ileocolitis (*Med.*). Inflammation of the ileum and the colon.

ileocolostomy (*Surg.*). The making of a communication between the ileum and the colon by operation.

ileostomy (*Surg.*). An artificial opening in the ileum, made surgically.

ileum (*Zool.*). In Vertebrates, the posterior part of the small intestine.

ileus (*Med.*). Colic due to obstruction in the intestine; obstruction of the intestine. See **paralytic ileus.**

Ilgner system (*Elec. Eng.*). See **Ward-Leonard-Ilgner system.**

iliac region (*Zool.*). The dorsal region of the pelvic girdle in Vertebrates.

iliac veins (*Zool.*). In Fish, the paired veins from the pelvic fins, draining into the lateral veins.

ilio- (*Zool.*). A prefix which refers to that part of the pelvic girdle of a Vertebrate known as the *ilium*; used in the construction of compound terms, e.g., *iliofemoral*, pertaining to the ilium and the femur.

ilium (*Zool.*). A dorsal cartilage bone of the pelvic girdle in Vertebrates. *adj.* **iliac.**

Ilkovic equation (*Chem.*). In polarography, the equation expresses the current to the *dropping-mercury electrode* (q.v.) as,

$$i = AnCD^{1/2}m^{2/3}\tau^{1/6},$$

where i = diffusion current, A = numerical constant, n = ionic charge, D = diffusivity, m = rate of flow of mercury, τ = lifetime of a drop.

illam (*Geol.*). A gem-bearing gravel occurring in Ceylon, and worked extensively for the gem-corundums, spinels, zircons, etc., which it contains; these have been derived from white pegmatite veins in the island.

ill-conditioned (*Surv.*). A term used in *triangulation* to describe triangles of such a shape that the distortion resulting from errors made in measurement and in plotting may be great, the criterion often used being that no angle in a triangle should be less than 30°.

illegitimate pollination (*Bot.*). The transfer of pollen from the anthers to the stigmas of the same flower when the general arrangement of the flower indicates that it is adapted for cross-pollination.

illinium (*Chem.*). Old name for **promethium.**

illite (*Min.*). A monoclinic clay mineral, a hydrous silicate of potassium and aluminium. It is named after the State of Illinois, a source of many clay specimens, and is the dominant clay mineral in shales and mudstones. The illite group of clay minerals also includes glauconite, and other minerals somewhat resembling muscovite.

illuminance (*Light*). See **illumination.**

illuminated diagram (*Elec. Eng.*). A circuit diagram on a switchboard, or a track diagram in a railway signal box, so arranged that lamps behind the diagram illuminate any part of the circuit which is alive or any part of the track upon which a train is standing.

illuminated-dial instrument (*Elec. Eng.*). An electric measuring instrument for switchboard use, having an illuminated scale of translucent glass.

illumination (*Light.*) The quantity of light or luminous flux falling on unit area of a surface. Illumination is inversely proportional to the square of the distance of the surface from the source of light, and proportional to the cosine of the angle made by the normal to the surface with the direction of the light rays. The unit of illumination is the *lux*, which is an illumination of 1 lumen/m^2. Symbol *E*. Also **illuminance.**

illusion (*Psychol.*). A misinterpretation or misperception of a stimulus.

ilmenite (*Min.*). An oxide of iron and titanium, crystallizing in the trigonal system; a widespread accessory mineral in igneous and metamorphic rocks, especially in those of basic composition. A common mineral in detrital sediments, often becoming concentrated in beach sand, e.g., in Florida.

ilmenorutile (*Min.*). A black variety of titanium oxide, containing iron in the form of ferrous titanate, niobate, and tantalate; crystallizes in the tetragonal system.

ILS (*Aero.*). *Instrument landing system.*

ILT (*Vet.*). *Infectious laryngotracheitis.*

ilvaite (*Min.*). (L. *Ilva*, Elba.) Hydrous silicate of iron (II) and iron (III), calcium and manganese (II). It crystallizes in the orthorhombic system.

image (*Optics*). Optical images may be of two kinds, real or virtual. A *real image* is one which is formed by the convergence of rays which have passed through the image-forming device (usually a lens) and can be thrown on a screen, as in the camera and the optical projector. A *virtual image* is one from which rays appear to diverge. It cannot be projected on a screen or on a sensitive emulsion. (*Print.*) The general term to describe the subject on negatives, positives, and plates, at any stage of preparation.

image admittance (*Elec.*). The reciprocal of *image impedance*.

image attenuation constant (*Elec.*). The real part of the image transfer constant of a network.

image charge (*Elec.*). Hypothetical charge used in electrostatic theory as a substitute for a conducting (equipotential) surface. The charge must not modify the field distribution at any point outside this surface.

image converter tube (*Optics*). One in which an optical image applied to a photoemissive surface produces a corresponding image on a luminescent surface.

image curvature (*Optics*). Concerned with the aberrations of the image in an electron microscope. Provided curvature of the field is the only aberration present, a sharp image is formed on a curved surface tangential to the image plane on the axis.

image-dissection camera (*Photog.*). High-speed camera having, in place of a normal lens, an array of parallel light-guides made of flint-glass fibres embedded in an opaque matrix, which transmit light by internal reflection on to successive portions of a moving plate.

image dissector (*TV*). Electron camera in which the optical image is focused on a photoemissive surface, from which electrons are emitted in straight lines in the form of the image. Suitably arranged electric or magnetic fields cause this pattern to sweep across a point anode, which thus effectively scans the image.

image dissector multiplier (*TV*). Combination of image dissector and electron multiplier in one tube or unit.

image force (*Elec.*). That force on an electric (or magnetic) charge between itself and its image induced in a neighbouring body.

image frequency (*Radio*). That which is as much greater (or less) than the local oscillator frequency as the signal frequency is less (or greater) in a supersonic heterodyne receiver. A transmission on this frequency will be accepted, unless rejected by a previous stage tuned to the signal frequency.

image iconoscope (*TV*). Camera tube in which the image is focused on a photoemissive cathode, electrons from which are focused on to a target, which is front-scanned by a high-energy electron beam.

image impedance (*Elec.*). A network is image operated when the input generator impedance equals the network input impedance, and the load impedance equals the output impedance of the network. If Z_{sc} is the short-circuit impedance and Z_{oc} the open-circuit impedance for the network, then the image impedance Z_o is given by $Z_o^2 = Z_{sc} . Z_{oc}$. See **iterative impedance.**

image intensifier (*Radiol.*). A device for enhancing brightness of an image in fluoroscopy, at the same time reducing patient dose. Electrons liberated when an X-ray beam falls on a suitable screen are accelerated and fall on a second screen giving an image of increased brightness but smaller size, which is viewed through a binocular eyepiece.

image interference (*Radio*). That produced by any signal received with a frequency at or near the image frequency.

image orthicon (*TV*). Tube in which the electron image pattern from a continuous emissive surface is focused on a storage photomosaic, which is scanned on the reverse side by a low-energy cathode beam. See **orthicon.**

image phase constant (*Telecomm.*). In a filter section, terminated in both directions, unreal or imaginary part of the (image) transfer constant; phase delay of the section in radians.

image ratio (*Radio*). The ratio of the image frequency signal input at the aerial to the desired signal input, in a heterodyne receiver, for identical outputs.

image reactor (*Nuc. Eng.*). Hypothetical reactor used in theoretical study of neutron flux distribution as a substitute for a neutron reflector. The treatment is analogous to the *image charge* method in electrostatics.

image response (*Radio*). Unwanted response of a superhet receiver to the image frequency.

image-retaining panel (*Radiol.*). An electroluminescent screen which retains a fluorescent image produced by X-rays for periods of up to half an hour, if the energizing voltage on the screen is maintained. This enables the image to be studied or photographed after the X-ray unit has been turned off and the object removed. The record on the screen is wiped off as soon as the applied voltage is interrupted.

image signal (*Radio*). One whose frequency differs from the received signal, in a superheterodyne receiver, by twice the intermediate frequency.

image transfer (*Photog.*). Process in which the original exposed emulsion is used to form a positive image. Used in document copying.

image tube (*Electronics*). One in which an optically focused image on a photo-emissive plate releases electrons which are focused on a phosphor by electric or magnetic means. Used in X-ray intensifiers, infrared telescopes, electron telescopes, and microscopes.

imaginal bud (or disk) (*Zool.*). One of a number of masses of formative cells which are the principal agents in the development of the external organs of the imago, during the metamorphosis of the *Oligoneoptera*.

imaginal rings (*Zool.*). Rings of cells at the inner ends of the stomodaeum and proctodaeum in the larvae of *Coleoptera*; their activity expedites the transition of the alimentary canal from its larval to its adult form by extension.

imaginary axis (*Maths.*). See **Argand diagram.**

imaginary circle (*Maths.*). The name given to a circle defined analytically so as to have an imaginary radius, e.g., the circle defined by $(x-a)^2+(y-b)^2+r^2=0$.

imaginary number (*Maths.*). The product of a real number x and i, where $i^2+1=0$. A complex number in which the real part is zero.

imaginary part (*Maths.*). See **complex number.**

imago (*Zool.*). The form assumed by an Insect after its last ecdysis, when it has become fully mature; final instar. *adj.* **imaginal.**

imbalance (*Med.*). A lack of balance, as between the ocular muscles, or between the activities of the endocrines, or between parts of the involuntary nervous system.

imbecile (*Med.*). A person whose defective mental state (present since birth or an early age) does not amount to idiocy, but who is incapable of managing his own affairs and whose **IQ** is between 25 and 50.

imbibition (*Chem.*). The absorption or adsorption of a liquid by a solid or a gel, accompanied by swelling of the latter. (*Photog.*) The mechanical printing of dye images. A relief or matrix is contacted with an absorbing surface, with a differential transference of dye, forming the positive image.

imbibition matrix (*Photog.*). A relief in gelatine capable of transferring, differentially, a dye to another absorbing surface, and providing a continous print.

imbibition mechanism (*Bot.*). See **hygroscopic mechanism.**

imbowment (*Arch.*). A term sometimes applied to an arch or vault.

imbricate, imbricated (*Bot., Zool.*). Said of leaves, scales, etc., which overlap like the tiles on a roof.

imbricate aestivation (*Bot.*). Aestivation in which the perianth leaves overlap at the edges.

imbricated (*Bot., Zool.*). See **imbricate.** (*Build.*) Said of slates or tiles which are laid so as to overlap.

imbricate structure (*Geol.*). A structure produced in mountain-building by intense pressure, individual blocks of rock being thrust over each other, like a pack of fallen cards.

IMC (*Aero.*). *I*nstrument *m*eteorological *c*onditions, wherein aircraft must conform to *instrument flight rules.*

I.M.E.P. (*Eng.*). Abbrev. for *indicated mean effective pressure.*

Imhoff tank (*San. Eng.*). A form of settling tank to which sewage is passed, the solid matter being exposed to a fermentation process, with the production of methane gas and an inoffensive sludge which can be easily dried.

imidazoles (*Chem.*). Glyoxalines; heterocyclic compounds produced by substitution in a five-membered ring containing two nitrogen atoms on either side of a carbon atom. Benzimidazoles are formed by the condensation of *ortho*-diamines with organic acids, and contain a condensed benzene nucleus:

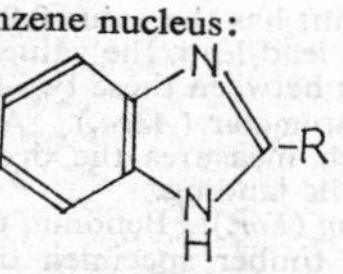

imides (*Chem.*). Organic compounds containing the group —CO·NH·CO—, derived from acid anhydrides.

imino group (*Chem.*). The group $\begin{matrix}R\\R'\end{matrix}\rangle NH$. Imino compounds are secondary amines obtained by the substitution of two hydrogen atoms in ammonia by alkyl radicals.

imitation art paper (*Paper*). See **art paper.**

imitation backed cloths (*Textiles*). Worsted fabrics woven from one warp and one weft, but, owing to the distribution of threads, presenting the appearance of backed cloths; used for coatings and some dress fabrics.

imitation gauze (*Textiles*). Very openly woven fine cloths. See **mock leno.**

imitation parchment (*Paper*). A wood-pulp paper to which strength, transparency, and grease-proof properties have been imparted by prolonged beating of the pulp. When given a high finish, called **GIP** (*g*lazed *i*mitation *p*archment).

imitative behaviour (*An. Behav.*). Behaviour of an animal as a response to a particular non-instinctive action by another animal, consisting of a repetition of this action. Widely found in Primates.

immersed liquid-quenched fuse (*Elec. Eng.*). A fuse in which liquid is used for extinguishing the arc, the fuse-link being totally immersed in the liquid.

immersed pump (*Aero.*). An electrical pump mounted inside a fuel tank.

immersible apparatus (*Elec. Eng.*). Electrical apparatus designed to operate continuously under water.

immersion (*Astron.*). The entry of the moon, or other body, into the shadow which causes its eclipse.

immersion heater (*Elec. Eng.*). An electric heater designed for heating water or other liquids by direct immersion in the liquid.

immigration (*Ecol.*). A category of population dispersal covering one-way movement into the population area. Cf. *emigration, migration.*

immiscibility (*Chem.*). The property of two or more liquids of not mixing and of forming more than one phase when brought together.

immittance (*Elec.*). Combined term covering **impedance** and **admittance.**

immune (*Bot.*). Treatment given to a plant to give it resistance to a parasite, or to increase its powers of resistance. (*Med.*) Protected against any particular infection; one who is in this state, i.e., has **immunity.** *v.* **immunize,** to make immune against infection. *n.* **immunization.**

immune bodies (*Med.*). Antibodies.

immunosuppressive (*Med.*). Applied to drugs which lessen the body's rejection of, e.g., transplanted tissue and organs.

immunotransfusion (*Med.*). The transfusion of blood or plasma containing, in high concentration, the appropriate antibodies for the infection from which the recipient is suffering.

impact (*Mech.*). For the direct impact of two elastic bodies, the ratio of the relative velocity after impact to that before impact is constant and is called the *coefficient of restitution* for the materials of which the bodies are composed.

This constant has the value 0·95 for glass/glass and 0·2 for lead/lead, the values for most other solids lying between these two figures.

impact accelerometer (*Aero.*). An *accelerometer* (q.v.) which measures the deceleration of an aircraft while landing.

impact bending (*For.*). Bending, under an impact load, of a timber specimen or member supported so as to allow it to be stressed; with particular reference to one of the standard impact-bending tests.

impact burner (*Heat*). A nozzle-type burner designed to impact the flame on to a surface of broken refractory material. Principally used for the firing of low-temperature metal-melting furnaces, tinning pots, and galvanizing baths.

impact crusher (*Min. Proc.*). Machine in which soft rock is crushed by swift blows struck by rotating bars or plates. The material may break against other pieces of rock or against casing plates surrounding the rotating hammers.

impacted (*Med.*). Firmly fixed, pressed closely in; said of a tooth which has failed to erupt, or of a fracture in which the broken bones are firmly wedged together.

impacter forging hammer (*Met.*). A horizontal forging machine in which two opposed cylinders propel the dies until they collide on the forging, which is worked equally on both sides.

impact extrusion (*Met.*). A fast, cold-working process for producing tubular components by once hitting with a punch a slug of material placed on the bottom of a die, so that the material squirts up around the punch into the die clearance.

impact ionization (*Electronics*). The loss of orbital electrons by an atom of a crystal lattice which has experienced a high-energy collision.

impaction sampler (*Powder Tech.*). Device for sampling particles of dust and spray droplets from a gaseous stream, which is forced through a jet and directed to a horizontal surface, usually a microscope slide smeared with grease, on which particles are collected by inertial impaction. Also **jet impactor.**

impact parameter (*Nuc.*). The distance at which two particles which collide would have passed if no interaction had occurred between them.

impact strength (*Met.*). Measured resistance of metal specimen to impact loading applied in Izod or Charpy tests.

impact test (*Met.*). Usually means a *notched-bar test* (q.v.), but it may also mean an Izod, Charpy, or Fremont test performed on unnotched specimens, or a test in which the suddenly applied load is in tension instead of bending.

impage (*Join.*). A rail, or horizontal member of a door-frame.

imparipinnate (*Bot.*). Said of a pinnate leaf which has a terminal leaflet.

imparity (*Electronics*). As specially applied to semiconductors and in terms of atoms, an alien substance which adds to or subtracts from the average density of electrons, leading to *n*-conduction by excess electrons, or *p*-conduction by holes.

impedance (*Elec.*). (1) Complex ratio of sinusoidal voltage to current in an electric circuit or component. Its real part is the *resistance* (dissipative or wattful impedance) and its imaginary part *reactance* (nondissipative or wattless), which may be positive or negative according to whether the phase of the current lags or leads on that of the voltage. Resistance, reactance, and impedance are all measured in ohms. Expressed symbolically, the impedance $Z=R+jX$ where R=resistance, X=reactance and $j=\sqrt{-1}$. Sometimes called **apparent resistance.** (2) Component offering electrical impedance. This is preferably termed **impedor.**

impedance bond (*Elec. Eng.*). A special rail-bond of high reactance and low resistance designed to allow the passage of d.c. traction current but not the a.c. used for signalling purposes.

impedance circle (*Elec. Eng.*). Locus of the end of the impedance vector in an Argand (R, X) diagram of a system, e.g., drawn to show the variation of input impedance of an improperly terminated line with frequency.

impedance coupling (*Elec. Eng.*). The coupling of two circuits by means of a tuned circuit or an impedance.

impedance drop (or rise) (*Elec. Eng.*). A drop or rise in the voltage at the terminals of a circuit, caused by current passing through the impedance of the circuit.

impedance factor (*Elec. Eng.*). The ratio of the impedance of a circuit to its resistance.

impedance matching (*Elec. Eng.*). See **matching.**

impedance matching stub (*Elec. Eng.*). See **matching stub.**

impedance matching transformer (*Elec. Eng.*). See **matching transformer.**

impedance protective system (*Elec. Eng.*). A discriminative protective equipment in which discrimination is secured by a measurement of the impedance between the point of installation of the relays (impedance relays) and the point of fault. Also called **distance protection.**

impedance relay (*Elec. Eng.*). A relay, used in discriminative protective gear, whose operation depends on a measurement of the impedance of the circuit beyond the point of installation of the relay; if this falls below a certain value, when a fault occurs, the relay operates. Also called a **distance relay.**

impedance rise (*Elec. Eng.*). See **impedance drop.**

impedance transforming filter (*Elec. Eng.*). A filter network which has differing image impedances, and which can therefore act as a transformer over a band of frequencies.

impedance triangle (*Elec. Eng.*). The right-angled triangle formed by the vectors representing the resistance drop, the reactance drop, and the impedance drop of a circuit carrying an alternating current.

impedance voltage (*Elec. Eng.*). The voltage produced as a result of a current flowing through an impedance.

impedometer (*Elec. Eng.*). Device for measuring impedances in waveguides.

impedor (*Elec.*). Physical realization of an impedance. An inductor, capacitor, or resistor, or any combination of these.

impeller (*Aero.*). The rotating member of a *centrifugal-flow compressor* (q.v.) or *supercharger* (q.v.). (*Eng.*) The rotating member of a centrifugal pump or blower, which imparts kinetic energy to the fluid.

impeller-intake guide-vanes (*Aero.*). The curved extension of the vanes of a centrifugal impeller which extend into the intake eye or throat, and which thereby give the airflow initial rotation.

imperfect (*Build.*). Said of a structural framework which has either more or fewer members than it would require to be determinate.

imperfect dielectric (*Elec.*). One in which there is a loss element resulting in part of the electric energy of the applied field being used to heat the medium.

imperfect flower (*Bot.*). A flower in which either the stamens or the carpels are lacking, or, if present, nonfunctional.

imperfect Fungi (*Bot.*). See **Fungi Imperfecti.**

imperfect hybridization (*Bot.*). An abortive attempt to form zygospores between the hyphae of two distinct species of *Zygomycetes*.

imperfect stage (*Bot.*). The conidium-bearing stage of a fungus.

imperforate (*Med.*). Not perforated; closed abnormally. (*Zool.*) Lacking apertures, especially of shells; said of Gastropod shells which have a solid columella.

imperial (*Arch.*). Of drawing boards, 32 × 23 in. (*Build.*) (1) A slate size, 33 × 24 in. (2) A domed roof shaped to a point at the top. (*Paper*) A pre-metrication paper size, 22 × 30 in.; U.S., 23 × 31 in. (*Textiles*) Heavy cotton fustian, generally drab in colour and occasionally with a raised surface. *Imperial coating*. A 2 × 2 twill weave worsted cloth made from botany yarns and piece-dyed. *Imperial serge*. Has the appearance of coating but employs softer yarns and weave: for dress and suitings.

imperial cap (*Paper*). A pre-metrication size of brown paper, 22 × 29 in.

Imperial Standard Wire Gauge (*Eng.*). See **Standard Wire Gauge.**

impermeable (*Chem., Geol.*). Not permitting the passage of liquids or gases.

impervious (*Build., etc.*). Said of materials which have the property of satisfactorily resisting the passage of water. (*Zool.*) Said of nostrils in which the nasal cavities are separated by a septum.

impetiginous (*Med.*). Resembling, or of the nature of, impetigo.

impetigo (*Med.*). A contagious skin disease, chiefly of the face and hands, due to infection with pus-forming bacteria.

impinger (*Mining*). Small hand-held device used to count the dust in a measured volume of mine air.

implant (*Biol.*). A graft of an organ or tissue to an abnormal position. (*Radiol.*) The radioactive material, in an appropriate container, which is to be imbedded in a tissue for therapeutic use, e.g., needle or seed.

implementation (*Comp.*). The various steps involved in installing and operating a computer data-processing or control system.

implex (*Zool.*). In *Arthropoda*, an inpushing of the integument for muscle attachment.

implicit function (*Maths.*). A variable x is an implicit function of y when x and y are connected by a relation which is not explicit. See **explicit function.**

implosion. Mechanical collapse of a hollow structure, e.g., a cathode-ray tube.

importance function (*Nuc.*). Limiting value, after a long time, of the number of neutrons per generation in the chain reaction initiated by a specified neutron to which this function applies.

imposing stone (*Typog.*). A heavy iron-topped table on which type matter is locked up preparatory to printing. In the early days of printing, level stone-topped tables were used.

imposition (*Typog.*). The process of assembling type pages in their proper order on the stone, arranging appropriate furniture or spacing material, and locking the whole into a chase. The unit is now known as a *forme*, and from it a book section or signature is printed.

impost (*Build.*). The top member of a pier or pillar from which an arch springs.

impregnated carbon (*Light*). An arc-lamp carbon consisting of carbon intimately mixed with some other material in order to produce a flame arc.

impregnating varnish (*Paint.*). The first coat of *insulating varnish* (q.v.) applied to electrical equipment. It fills the voids in the windings.

impregnation (*Powder Tech.*). The partial or complete filling of the pores of a powder product with an organic material, glass, salt, or metal, to make it impervious or impart to it secondary properties. Vacuum, pressure, and capillary forces may be employed. It may include stoving to produce setting. See **infiltration.** (*Zool.*) The passage of spermatozoa from the body of the male into the body of the female.

impression (*Print.*). (1) All copies of a book printed at one time from the same type or plates. (2) The pressure applied to a type forme by the cylinder or platen.

impression cylinder (*Print.*). The cylinder which presses the stock against the printing surface, which may be flat or cylindrical, and either letterpress, gravure or lithographic.

impression eccentric (*Print.*). A means of impression adjustment by rotating the bearing, or the housing in which the cylinder shaft is mounted, out of centre.

imprint (*Print.*). The name of the publisher and/or printer which must appear on certain items, particularly books, periodicals, and election literature.

imprinting (*An. Behav.*). A learning process by which the social preferences or *following response* of many birds become restricted in very early life to a specific class of objects, these usually being large and moving.

improper fraction (*Maths.*). See **division.**

improving (*Met.*). See **softening.**

impsonite (*Min.*). A member of the asphaltite group.

impulse (*Horol.*). The force or blow imparted to the pendulum or balance by the escape wheel through the escapement. (*Maths.*) When two bodies collide, over the period of impact there is a large reaction between them. Such a force can only be measured by its time integral ($\int F\,dt$) which is defined as the impulse of the force, and which equals the change of momentum produced in either body. (*Telecomm.*) Obsolete term for *pulse*.

impulse circuit (*Teleph.*). In an automatic switching exchange, a source of machine-generated impulse trains for operating step-by-step switches, controlled by relays.

impulse circuit-breaker (*Elec. Eng.*). A circuit-breaker, requiring only a small quantity of oil, in which the arc is extinguished by a mechanically produced flow of oil across the contacts.

impulse clock (*Horol.*). One in which a master pendulum clock controls the operation of a number of slave clocks. The vibrations of the pendulum of the master clock are maintained by an electromagnetically reset gravity arm falling on to a pallet on the pendulum rod. When the gravity arm falls for impulse it also closes the circuit to the slave clocks, giving impulse to them. The *free pendulum clock* (q.v.) is also of the impulse type.

impulse excitation (*Electronics*). (1) Exciting the grid of a thermionic tube in which the anode current is allowed to flow for only a very short period during each cycle. (2) Maintenance of oscillatory current in a tuned circuit by pulses synchronous with free oscillations, or at a submultiple frequency.

impulse flashover voltage (*Elec. Eng.*). The value of the impulse voltage which just causes flashover of an insulator or other apparatus.
impulse frequency (*Teleph.*). The number of impulses per second in the impulse trains used in dialling and operating selectors. See **dot frequency.**
impulse function (*Telecomm.*). See **delta impulse function.**
impulse generator (*Elec. Eng.*). Circuit providing single, or a continuous series of, pulses, generally by capacitor discharge and shaping, e.g., by the charging of the capacitors in parallel and the discharging of them in series. Also called **surge generator.**
impulse inertia (*Elec. Eng.*). That property of an insulator by which the voltage required to cause disruptive discharge varies inversely with its time of application.
impulse machine (*Teleph.*). A machine which generates accurately timed impulses for operating selector switches.
impulse period (*Teleph.*). The time between identical phases of a train of impulses: the time between the start of one impulse and the start of the next.
impulse pin (*Horol.*). The vertical pin in the roller of the lever escapement which receives the impulse from the pallets, via the notch in the lever. It also effects the unlocking, on the reverse vibration.
impulse plane (*Horol.*). That part of the pallet upon which a tooth of the escape wheel acts when giving impulse.
impulse ratio (*Elec. Eng.*). The ratio between the breakdown voltage of an insulator or piece of insulating material when subjected to an impulse voltage to the breakdown when subjected to a normal-frequency (50 hertz) voltage. (*Teleph.*) The ratio of the time during an impulse to the total time of impulse plus interval before another impulse.
impulse-reaction turbine (*Eng.*). See **disk-and-drum turbine.**
impulse repeater (*Teleph.*). A relay mechanism for repeating impulses from one circuit into another.
impulse starter (*Aero.*). A mechanism in a magneto which delays the rotor against a spring so that, when released, there is a strong and retarded spark to help starting.
impulse transmission (*Telecomm.*). That involving the use of impulses for signalling, thus reducing the effects of low-frequency interference.
impulse turbine (*Eng.*). A steam turbine in which steam is expanded in nozzles and directed on blades carried by a rotor, in one or more stages, there being no change in pressure as the steam passes the blade-ring.
impulse voltage (*Elec. Eng.*). A transient voltage lasting only for a few microseconds; very frequently used in high-voltage testing of electrical apparatus in order to simulate voltage due to lightning strokes or other similar causes.
impulse wheel (*Eng.*). The wheel of an *impulse turbine* (q.v.). Also used to denote one of the two principal types of turbines, in which the whole available head is transformed into kinetic energy before reaching the wheel. See also **Pelton wheel.**
impulsive current (*Telecomm.*). A current which comprises one or more impulses in one direction round a circuit, as in dialling.
impulsive sound (*Acous.*). Short sharp sound, the energy spectrum of which spreads over a wide frequency range.
impurity (*Electronics*). Small proportion (few $p/10^6$) of foreign matter, e.g., indium, added during a melt of a purer (few $p/10^8$) semiconductor, e.g., silicon, germanium, to obtain required conduction for diodes, transistors. The alien substance in the crystal lattice may add to, or subtract from, the average densities of free electrons and holes in the semiconductor. See **acceptor, carriers, donor.**
impurity levels (*Electronics*). Abnormal energy levels arising from slight impurities, resulting in conduction in semiconductors.
in (*For.*). See **eng.**
in-. Prefix from Latin, meaning either 'in(to)' or 'not'.
In (*Chem.*). The symbol for *indium.*
inactivation (*Chem.*). The destruction of the activity of a catalyst, serum, etc.
inactive component (*Elec. Eng.*). See **reactive component.**
inaequi-hymeniiferous (*Bot.*). Of a mushroom-type fructification with an unequally-developed hymenium.
in-and-in (*Paper*). A method of packing paper too large to travel flat. The ream is divided in half, folded over, and interlocked.
in-and-out movement (*Cinema.*). That part of the intermittent motion in a motion-picture camera which inserts and withdraws the claws which pull the film into position at the gate.
inanition (*Med.*). Exhaustion and wasting of the body from lack of food.
Inarticulata (*Zool.*). See **Ecardines.**
inband (*Build.*). A header stone.
inband rybat (*Build.*). A header stone laid to form the jamb of an opening.
in-between drawings (*Cinema.*). In animated cartoons, drawings which are made to fit between key drawings, in number corresponding to the time intervals demanded by the rhythm, there being 24 frames to be photographed to correspond to the projection time of one second. After the drawings have been made in pencil, over illumination and the previous drawings, they are traced in black ink on transparent acetate, and coloured.
inborn behaviour (*An. Behav.*). See **species-specific behaviour.**
inbreeding (*Zool.*). Breeding within the descendants of a foundation stock of related animals.
inbye (*Mining*). The direction from a haulage way to a working face.
incandescence (*Light*). The emission of light by a substance because of its high temperature, e.g., a glowing electric-lamp filament. In the case of solids and liquids, there is a relation between the colour of the light and the temperature. Cf. *luminescence.*
incandescent lamp (*Light*). A lamp in which light is produced by heating some substance to a white or red heat, e.g., a filament lamp.
incandescent mantle (*Light.*). See **gas mantle.**
incept (*Bot.*). The rudiment of an organ.
incertum (*Build.*). An early form of masonry work in which squared stones were used as a facing, with rubble filling as a backing.
inching (*Eng.*). Very slow, closely controlled step-by-step movement of a usually fast moving machine, e.g. (*Print.*) turning the press by small amounts at a time, using an electrical inching button switch. Cf. *crawl.*
inching starter (*Elec. Eng.*). An electric-motor starter in which provision is made for inching the motor, i.e., running it very slowly for such purposes as the threading of the paper in a printing press.
inch-penny weight (*Mining*). In valuation of gold ore, the width of the lode or reef measured

normal to the enclosing rock, multiplied by the assay value in penny weights per ton.

inch-tool (*Tools*). A steel chisel having a cutting edge 1 in. (25 mm) wide, used by the mason for dressing stone.

inch trim moment (*Ships*). Same as *moment to change trim one inch* (q.v.). Abbrev. **I.T.M.**

incidence, angle of (*Aero.*). See **angle of attack.**

incident beam (*Phys.*). Any wave or particle beam the path of which intercepts a surface of discontinuity.

incipient plasmolysis (*Bot.*). The stage in plasmolysis when the cell wall is fully contracted, but when the protoplast has not yet shrunk away from the wall at any point.

incise (*Arch.*). To cut in; to carve.

incised (*Bot.*). Cut to about the middle.

incised meander (*Geol.*). An intrenched winding or bend of a river, which results from renewed down-cutting at a period of rejuvenation.

incisiform. Shaped like an incisor tooth.

incision (*Surg.*). The act of cutting into something: a cut made by a surgical knife.

incisor process (*Zool.*). In *Malacostraca*, a process of the part of the mandible which projects towards the mouth. Cf. *molar process.*

incisors (*Zool.*). The front teeth of Mammals; they have a single root, are adapted for cutting, and are the only teeth borne by the premaxillae in the upper jaw.

incisura (*Anat.*). A cut or notch. (Various notches in the body are thus designated.)

inclination (*Mag.*). See **dip.**

inclination factor (*Light*). Used in Fresnel's theory of the propagation of light waves, where the disturbance at a point, due to the contributions from the secondary waves, is assumed to depend on the angle θ between the normal to the primary wavefront and the direction to the point. The term containing θ is the *inclination factor.*

incline (*Surv.*). See **gradient.**

inclined-carbon arc lamp (*Elec. Eng.*). An arc lamp in which the carbons are set at an angle to each other, as in some projector lamps and flame-arc lamps.

inclined-catenary construction (*Elec. Eng.*). A catenary construction for the overhead contact wire of an electric traction system; in it, the catenary wire is not placed vertically above the contact wire.

inclined plane (*Mech.*). For a smooth plane inclined at an angle θ to the horizontal, the force parallel to the plane required just to move a mass up the plane is $mg \sin \theta$. The inclined plane may therefore be regarded as a machine having a velocity ratio of cosec θ.

inclined shaft (*Mining*). One which slopes, usually parallel to the dip of the lode and underlying it.

inclined shore (*Carp.*). See **raking shore.**

incline engine (*Mining*). A stationary haulage engine at the top of an inclined haulage road.

inclining experiment (*Hyd. Eng.*). A practical method of determining the metacentric height and the height of the centre of gravity of a floating vessel; accomplished by observing the angle of heel of the vessel resulting from a measured transverse movement of a known weight across the deck.

inclinometer (*Mag., Surv., etc.*). See **dip needle.**

inclusion. A particle or lump of foreign matter embedded in a solid. (*Cyt.*) A body occurring in the cytoplasm of a cell. Inclusions may be of two kinds—*protoplasmic inclusions* (as the nucleus, the Golgi apparatus, etc.) and *deuteroplasmic inclusions* (as granules of secreted material, food particles, etc.). (*Met.*) A particle of alien material retained in a solid metal. Such inclusions are generally oxides, sulphides, or silicates of one or other of the component metals of the alloy, but may also be particles of refractory materials picked up from the furnaces or ladle lining. (*Min.*) A foreign body (gas, liquid, glass, or mineral) enclosed by a mineral. See also **xenolith.**

inclusion bodies (*Med.*). Particulate bodies found in the cells of tissue infected with a virus.

incoherent (*Phys.*). Said of radiation of the same frequency emitted from discrete sources with random phase relationships. All light sources except the *laser* emit incoherent radiation.

incoming feeder (*Elec. Eng.*). A feeder in a substation through which power is received.

incompatibility (*Bot.*). (1) Any difference in the physiological properties of the protoplasts of a host and a parasite which limits or stops the development of the latter. (2) Some difference, usually physiological, which prevents the completion of fertilization.

incompetence (*Med.*). Inability to perform proper function; said especially of diseased valves of the heart which allow the blood to pass in the wrong direction, e.g., *aortic incompetence, mitral incompetence.*

Incompletae or **Apetalae** (*Bot.*). A group of dicotyledons in which the perianth is absent or incompletely developed; the group is not a natural one.

incomplete flower (*Bot.*). A flower in which the calyx and corolla (or one of these) are lacking.

incomplete metamorphosis (*Zool.*). In Insects, a direct metamorphosis, i.e., a more or less gradual change from the immature to the mature state, a pupal stage being absent and the young forms resembling the parents, except in the absence of wings and mature sexual organs and occasionally in the possession of adaptive structures.

incomplete reaction (*Chem.*). A reversible reaction which is allowed to reach equilibrium, a mixture of reactants and reaction products being obtained.

incompressible volume (*Chem.*). See **co-volume.**

Inconels (*Met.*). Nickel-based heat-resistant alloys containing some 13% of chromium, 6% iron, and a little manganese, silicon, or copper. *Inconel X* is much used in gas-turbine blades.

inconsequent drainage (*Geol.*). A river system which is essentially unrelated to the rocks over which it flows.

incontinence (*Med.*). Inability to retain voluntarily natural excretions of the body (e.g., faeces and urine); lack of self-control.

incoordination (*Med.*). Inability to combine muscular movements in the proper performance of an action, the component muscle groups working independently instead of together.

incrassate, incrassated (*Bot.*). Having thickened cell walls.

increaser (*Plumb.*). A coupling piece used to connect a large pipe to a small one. Usually called **reducer.**

incremental hysteresis loss (*Elec. Eng.*). A small pulsation of the magnetic field about a fixed value leading to a small hysteresis loop on the boundary of a full loop.

incremental induction (*Elec. Eng.*). The difference between the maximum and minimum value of a magnetic induction at a point in a polarized material, when subjected to a small cycle of magnetization; cf. *incremental hysteresis loss.*

incremental iron losses (*Elec. Eng.*). A term sometimes used to denote iron losses occurring in an a.c. machine due to frequencies higher than the fundamental, e.g., tooth pulsation losses.
incremental permeability (*Elec. Eng.*). The gradient of the curve relating flux density to magnetizing force (the *B/H curve*). This represents the effective permeability for a small alternating field superimposed on a larger steady field.
incremental resistance (*Elec. Eng.*). The small signal resistance for a component or network, $r = \Delta V/\Delta I$.
incrustation (*Bot.*). A coating of calcium carbonate, or less often of compounds of iron, on or in the walls of some *Algae*. (*Build.*) A term applied to a wall facing which is of different material from that forming the rest of the wall.
incubation (*Med., Zool.*). The period intervening between the infection of a host by bacteria or viruses and the appearance of the first symptoms. (*Zool.*) The process of causing eggs to hatch by the application of heat.
incubous (*Bot.*). Said of the leaf of a liverwort when its upper border (the border towards the apex of the stem) overlaps the lower border of the next leaf above it and on the same side of the stem.
incudate (*Zool.*). Of *Rotifera*, having the mallei reduced and the rami large and hooked.
incudectomy (*Surg.*). Removal of the incus by operation.
incumbent (*Bot.*). Said of a radicle which is bent over and lies on the back of a cotyledon.
incurrent (*Zool.*). Carrying an ingoing current; said of ducts, and, in certain *Porifera*, of canals leading from the exterior to the prosopyles or prosodi of the flagellated chambers.
incurrent ostia (*Zool.*). Lateral inlets into the heart in Insects for the admittance of blood.
incus (*Zool.*). In Mammals, an ear ossicle, derived from the quadrate; in *Rotifera*, one of the masticatory ossicles of the mastax; more generally, any anvil-shaped structure. *pl.* **incudes.**
indamines (*Chem.*). Derivatives of phenylated 1,4-quinone-diimines. Important as intermediates in the production of azine and sulphide dyes.
indanthrene (*Chem.*). $C_{28}H_{14}O_4N_2$, *N*-dihydro-1, 2, 2′, 1′-anthraquinone-azine, an anthraquinone vat dyestuff, a dark-blue powder, practically insoluble in water and organic solvents; it is very stable and can be heated to 470°C without melting or decomposition. Formed by fusion of 2-aminoanthraquinone with caustic potash at 200°–300°C, forming indanthrene A and indanthrene B, of which only the former is valuable. For dyeing purposes, indanthrene is reduced by sodium hydrosulphite to the water-soluble salt of the dihydro derivative, and re-oxidized to indanthrene by exposure to air. TNs **Indanthrone, Caledon blue** and **Duranthrene.**
indeciduate (*Zool.*). Said of Mammals in which the maternal part of the placenta does not come away at birth.
indefinite (*Bot.*). (1) Not fixed in number, but numerous. (2) Not ending in a flower and theoretically capable of continued elongation. (3) Racemose.
indefinite integral (*Maths.*). See **integral.**
indehiscent (*Bot.*). Not opening naturally when ripe.
indene (*Chem.*). An aromatic double-ring liquid hydrocarbon (C_9H_8) occurring in coal-tar. B.p. 182°C. Usually contains coumarone.
indent (*Carp.*). A notch made in a timber. (*Typog.*) To commence a line with a blank space, which in bookwork paragraphs may be 1, 1½, or 2 ems, according to the width of the line.
indentation test (*Build., Civ. Eng.*). A test for a paving, roofing, or roadmaking asphalt, in which a steady load is applied, under constant temperature conditions, to the asphalt surface, through the sector of a wheel resting upon it, the amount of indentation being measured after a fixed time.
indented bar (*Civ. Eng.*). A special type of reinforcing bar used in reinforced concrete work to provide a mechanical band and having for its full length a series of depressions and ridges all round.
indenter (*Civ. Eng.*). A roller having projections from its curved surface, so that, when it is rolled over newly laid asphalt paving, indentations shall be left in the latter surface to render it non-skid. Also called **branding iron, crimper.**
independent axle-drive (*Elec. Eng.*). See **individual axle-drive.**
independent chuck (*Eng.*). A lathe chuck in which each of the jaws is moved independently by a key; used for work of irregular shape, or when very accurate centring is necessary.
independent drive (*Radio*). System in which the frequency of a transmitter is determined by an oscillator whose output is amplified and subsequently delivered to the antenna.
independent equations (*Maths.*). A set of equations none of which can be deduced from a combination of any of the others.
independent feeder (*Elec. Eng.*). A feeder in an electric-power distribution system which is used solely for supply to a substation or a feeding point, and not as an interconnector. Also called **dead-ended feeder, radial feeder.**
independent heterodyne (*Radio*). An oscillator, electrically separate from the detector valve, used for supplying local oscillations used in heterodyne reception.
independently heated cathode (*Electronics*). See **equipotential cathode.**
independent particle model (of a nucleus) (*Nuc.*). Model in which each nucleon is assumed to act quite separately in a common field to which they all contribute.
independent seconds watch (*Horol.*). A watch having an independent train for driving the seconds hand.
independent suspension (*Autos.*). A springing system in which the wheels are not connected by an axle beam, but are mounted separately on the chassis through the medium of springs and guide links, so as to be capable of independent vertical movement.
independent time-lag (*Elec. Eng.*). See **definite time-lag.**
independent trip (*Elec. Eng.*). A tripping device for a circuit-breaker, starter, or similar apparatus, in which the current operating the device is independent of the current flowing in the circuit to which the device is connected.
independent variable (*Maths.*). See **dependent variable.**
indestructibility of matter (*Chem.*). See **law of conservation of matter.**
indeterminacy principle (*Phys.*). See **uncertainty principle.**
indeterminate (*Eng.*). Said of a structure which is *redundant* (q.v.). Cf. *determinate.*
indeterminate equations (*Maths.*). Simultaneous equations which by reason of certain relations between the coefficients, or from insufficient data, have an infinite number of solutions.

index (*Automation*). General term for intermittent motion of tape or mechanism under control. See **Geneva movement.** (*Horol.*) The regulating lever by means of which the rate of a watch may be adjusted. The lever is usually carried on the balance cock, and its short end carries the curb pins, the long end moving over a scale which indicates the amount of movement given to the curb pins. One end of the scale is marked *A* (advance) and the other *R* (retard). Hence, if the watch is losing, the index is moved towards *A*. (*Maths.*) The small number written to the right of and above a number or term to indicate how many times that number or term has to be multiplied by itself, e.g., x in a^x, $(a+b)^x$. Note: $a^0 = 1$, $a^{\frac{1}{2}} = \sqrt{a}$, $a^{x/y} = \sqrt[y]{a^x}$, $a^{-1} = 1/a$. Also called **exponent.** (*Nut.*) A parameter of the human frame used in nutritional and growth studies.

index error (*Surv.*). Difference between the horizontal or vertical angular reading of theodolite and the true line of collimation, with regard to concentric centring of the azimuth and plate circle and accurate engraving of the reading lines.

index fossil (*Geol.*). One which characterizes a particular geological *horizon.*

indexing (*Automation*). Regular step-wise (Geneva) motion (linear or rotary) in automatically controlled mechanisms, timed mechanically, electrically or electronically.

indexing head (*Eng.*). A machine-tool attachment for rotating the work through any required angle, so that faces can be machined, holes drilled, etc., in definite angular relationship.

index mineral (*Geol.*). One whose appearance marks a particular grade of metamorphism in progressive regional metamorphism.

index of refraction (*Phys.*). See **refractive index.**

index value (*Automation*). Preset value of a controlled quantity at which an automatic control is required to aim. Also **desired value.**

Indian cotton (*Textiles*). Cotton of 12 mm to 25 mm staple; chiefly used in India or exported to Japan. It is of low to medium quality.

Indian hemp (*Bot.*). See **cannabis.**

Indian ink. Ink in a solid form made from lamp-black mixed with parchment size or fish glue. Rubbed down in water it produces an intensely black permanent ink, used for line and wash-drawings, etc.

Indian red (*Paint.*). A precipitated red oxide pigment calcined at a high temperature to yield a very deep red which is very durable and has a high opacity.

Indian topaz (*Min.*). See **citrine.**

India paper (*Paper*). A thin, strong, opaque rag paper, made for Bibles and other books where many pages are required in a small compass.

india-rubber (*Chem.*). See **rubber.**

india-rubber cable (or wire) (*Cables*). Cable or wire in which the insulation consists of pure or vulcanized india-rubber.

indican (*Chem.*). (1) A glucoside, $C_{14}H_{17}NO_6$, which hydrolyses to glucose and indoxyl. It forms colourless leaflets, melting at 57°C and soluble in water. It is obtained from the indigo and woad plants and from certain other *Leguminosae*. (2) Indoxyl-sulphuric acid, $C_8H_6N{\cdot}O{\cdot}SO_2{\cdot}OH$, a normal constituent of urine.

indicanaemia, indicanemia (*Med.*). The presence of indican in the blood.

indicanuria (*Med.*). The presence of indican in the urine.

indicated airspeed (*Aero.*). The reading of an air-speed indicator which, as air density falls with altitude, reads low by a factor equal to the square root of the relative air density; abbrev. **I.A.S.**

indicated (horse-)power (*Eng.*). Of a reciprocating engine, the (horse-)power developed by the pressure-volume changes of the working agent within the cylinder; it exceeds the useful or brake (horse-)power at the crankshaft by the power lost in friction and pumping. Abbrev. **I.H.P.**

indicated mean effective pressure (*Eng.*). The average pressure exerted by the working fluid in an engine cylinder throughout the cycle, equal to the mean height of the indicator diagram in kN/m^2 or lbf/in^2. Abbrev. **I.M.E.P.**

indicated ore (*Mining*). Proved limits of deposit, in the light of known geology of mine and economic factors.

indicated thermal efficiency (*Eng.*). The ratio between the indicated power output of an engine and the rate of supply of energy in the steam or fuel.

indicating instrument (*Instr.*). One in which the immediate value only of the measured quantity is visually indicated.

indicator (*Bot.*). A plant which grows under special conditions of climate, or on a particular soil, or in a particular community, and thus, by its presence, indicates the general nature of the habitat. (*Chem.*) (1) A substance whose colour varies with the acidity or alkalinity of the solution in which it is dissolved. (2) Any substance used to indicate the completion of a chemical reaction, generally by a change in colour. (*Elec. Eng.*) Same as **annunciator.** (*Eng.*) An instrument for obtaining a diagram of the pressure-volume or pressure-time changes in an engine or compression cylinder during the working cycle. (*Textiles*) Counting instrument (mechanical) which records the weight or hanks produced on cotton spinning machinery, or the picks woven on looms.

indicator card (*Eng.*). A chart on which the trace of an *indicator* (q.v.) is recorded, producing an *indicator diagram* (q.v.).

indicator diagram (*Eng.*). A graphical representation of the pressure and volume changes undergone by a fluid, while performing a work-cycle in the cylinder of an engine or compressor, the area representing, to scale, the work done during the cycle. See **indicated mean effective pressure, light-spring diagram.**

indicator exponent (*Chem.*). The pH-value at which the change of colour of an *indicator* (1) is most rapid.

indicator gate (*Electronics*). A step signal applied to an indicator tube to control its sensitivity.

indicator range (*Chem.*). The range of pH-values within which an *indicator* (q.v. (1)) changes colour.

indicator tube (*Electronics*). Miniature CRT in which size or shape of target glow varies with input signal.

indicator vein (*Mining*). In prospecting, one associated with the lode or vein being traced, thus guiding the search.

indices of crystal faces (*Crystal.*). See **Miller indices.**

indicial admittance (*Telecomm.*). Transient current response of a circuit to the application of a *step function* of one volt, using Heaviside operational calculus.

indicial response (*Telecomm.*). Output waveform from a system when a step pulse of unit magnitude is applied to the input.

indicolite or indigolite (*Min.*). A blue (either pale or bluish-black) variety of tourmaline.

indifferent (*Zool.*). Said of coloration in animals when it is inherited and is neither useful nor detrimental to the species; as the scarlet coloration of many deep-sea forms.

indifferent species (*Bot.*). A species which occurs in two or more distinct communities.

indigenous (*Zool.*). Native; not imported.

indigestion (*Med.*). A condition, marked by pain and discomfort, in which the normal digestive functions are impeded. The causes are numerous.

indigo (*Chem.*). $C_{16}H_{10}N_2O_2$, a dye occurring in a number of plants, especially in species of *Indigofera*, in the form of a glucoside. It is an indole derivative and its constitution is expressed by the following formula established by Baeyer:

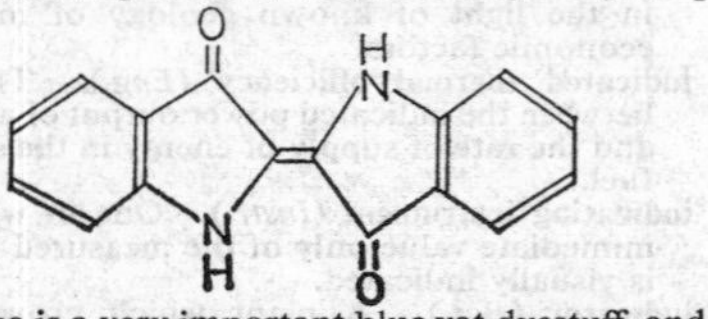

Indigo is a very important blue vat dyestuff, and can be synthesized in various ways: (*a*) from 2-nitrophenyl-ethanoic acid via 2-aminophenyl-ethanoic acid, oxindole, isatin, isatin chloride; (*b*) from 2-nitrobenzaldehyde and propanone via 2-nitrophenyl-lactyl methyl ketone; (*c*) from aniline via phenylaminoethanoic acid and indoxyl; (*d*) from naphthalene via phthalic acid, anthranilic acid, phenylglycine-2-carboxylic acid, indoxylic acid, indoxyl to indigo.

indigo copper (*Min.*). See **covellite**.

indigolite (*Min.*). See **indicolite**.

indirect-arc furnace (*Elec. Eng.*). An electric-arc furnace in which the arc is struck between two electrodes mounted above the charge, the latter being heated chiefly by radiation.

indirect cylinder (*Plumb.*). Hot-water system in which the water from the boiler is used only to heat the supply in the cylinder, which is fed from an independent secondary circuit.

indirect effect (*Nuc.*). Effect of ionizing radiation on a living cell as a result of chemical damage to vital molecules, produced by free radicals released through decomposition of water molecules by the radiation.

indirect-fired furnace (*Met.*). One in which the combustion chamber is separate from the one in which the charge is heated.

indirect fittings (*Light*). Lighting fittings which reflect practically all the light from the lamps contained in them into the upper hemisphere; used, therefore, for *indirect lighting*.

indirect heating (*Heat.*) A system of heating by convection. Cf. *direct heating*.

indirect lighting (*Light*). A system of lighting in which more than 90% of the total light flux from the fittings is emitted in the upper hemisphere.

indirectly-heated cathode (*Electronics*). One with an internal heater, highly insulated from the cathode on a surrounding ceramic cylinder. Also **unipotential cathode**.

indirectly-heated valve (*Electronics*). A valve using an *equipotential cathode* (q.v.).

indirect metamorphosis (*Zool.*). The complex change characterizing the life cycles of *Holometabola*; the young are larvae and the imago is preceded by a pupal instar.

indirect ray or indirect wave (*Radio*). See **ionospheric ray**.

indium (*Chem.*). A silvery metallic element in the third group of the periodic system. Symbol In, at. no. 49, r.a.m. 114·82, m.p. 155°C, b.p. 2100°C, rel. density 7·28 at 13°C, electrical resistivity 9×10^{-8} ohm metres. Found in traces in zinc ores. The metal is soft and marks paper like lead; it forms compounds with carbon compounds. It has a large cross-section for slow neutrons and so is readily activated. Also used in manufacture of transistors and as bonding material for acoustic transducers.

individual (*Bot.*). Strictly, a plant derived from the development of a zygote; loosely, any separate plant. (*Zool.*) A single member of a species; a single zooid of a colony of *Coelenterata* or *Polyzoa*; a single unit or specimen.

individual (or independent) axle-drive (*Elec. Eng.*). A term applied to the arrangement of an electric locomotive in which each driving axle is driven by a separate motor.

individual drive (*Elec. Eng.*). A system used for the electric operation of factories, in which each machine is driven by a separate electric motor. See also **individual axle-drive**.

individual psychology. The system of psychology, founded by Adler of Vienna, which stresses the feeling of inferiority as the main factor in neurosis, and the desire for power as the driving force behind all psychic life, and even sexual life. Adler sees in the goals of the future, rather than in the events in the past, evidence for the causation of neurosis.

individuation (*Psychol.*). Jung's concept of self-realization, comprising not only the emergence of the ego but also the development and coordination of all other aspects of the self. (*Zool.*) The formation of separate functional units which are mutually interdependent; as the formation of the zooids composing a colony.

indole (*Chem.*). Benzpyrrole. C_8H_7N, colourless plates, m.p. 52°C, b.p. (decomposition) 245°C, volatile in steam. Indole forms the basis of the indigo molecule, and its derivatives, e.g., skatole and tryptophane, are of importance in biochemistry as products formed by the decomposition of albuminous matter. It results from the condensation of a benzene nucleus with a pyrrole ring, and is thus:

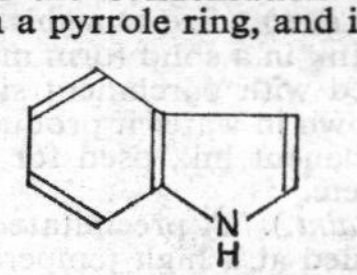

indolent (*Med.*). Causing little or no pain, e.g., *indolent* ulcer.

indoles (*Biochem.*). Derivatives of *indole* (q.v.), occurring as *auxins* (q.v.), including indole-3-acetic (ethanoic) acid (*IAA*), used as a component of rooting powders.

indophenols (*Chem.*). Derivatives of phenylated 4-quinone mono-imines; similar in constitution to the indamines; used for dyeing cotton and wool.

indoxyl (*Chem.*). An indole derivative, an isomer of oxindole, yellow crystals, m.p. 85°C. The formula (keto form) of indoxyl is:

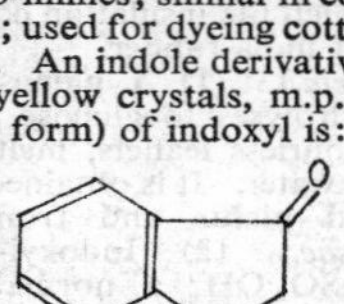

It is a very unstable compound which easily resinifies, and is readily oxidized to indigo by atmospheric oxygen.

induced charge (*Elec.*). That produced on a conductor as a result of a charge on a neighbouring conductor.

induced current (*Elec.*). That which flows in a circuit as a result of induced e.m.f.

induced dipole moment (*Phys.*). Induced moment of an atom or molecule which results from the application of an electric or magnetic field.

induced drag (*Aero.*). The portion of the *drag* (q.v.) of an aircraft attributable to the derivation of lift.

induced draught (*Eng.*). A forced draught system used for boiler furnaces, in which a fan placed in the uptake induces an air-flow through the furnace. See **balanced draught, extraction fan, fan.**

induced e.m.f. (*Elec.*). That which appears in a circuit as a result of changes in the interlinkages of magnetic flux with part of the circuit. The e.m.f. in secondary of a transformer. Discovered by Faraday at Royal Institution in 1831.

induced moving-magnet instrument (*Elec. Eng.*). An instrument whose operation depends on the force exerted by the resultant of the fields produced by a fixed coil carrying a current, and a permanent magnet fixed at an angle thereto, on a movable piece of magnetic material.

induced noise (*Acous.*). That which arises in electrodes because of sufficiently high frequencies in the motions in a space charge in a valve.

induced polarization (*Elec.*). That which is not permanent in a dielectric, but arises from applied fields.

induced radioactivity (*Nuc.*). That induced in non-radioactive elements by neutrons in a reactor, or protons or deuterons in a cyclotron or linear accelerator. X-rays or gamma-rays do not induce radioactivity unless the gamma-ray energy is exceptionally high.

induced reaction (*Chem.*). A chemical reaction which is accelerated by the simultaneous occurrence in the same system of a second, rapid reaction.

inductance (*Elec., Mag.*). (1) That property of an element or circuit which, when carrying a current, is characterized by the formation of a magnetic field and the storage of magnetic energy. (2) The magnitude of such capability. See **inductance coefficient.**

inductance-capacitance filter (*Elec. Eng.*). A circuit arrangement of series inductors and shunt capacitors across the input terminals to eliminate the appropriate ripple frequency from a unidirectional current.

inductance coefficient (*Elec. Eng.*). Property of a component (inductor) in a circuit whereby back e.m.f. arises because of rate of change of current. It is 1 *henry* when 1 volt is generated by rate of change of 1 ampere per second.

inductance coil (*Elec. Eng.*). Coil, with or without an iron circuit, for adding inductance to a circuit. Also called **inductor.**

inductance coupling (*Elec. Eng.*). That between 2 circuits whereby a changing current in one induces a current in the other, proportional to the rate of change of the first current.

inductance factor (*Elec. Eng.*). A term sometimes used to denote the ratio of the reactive current to the total current in an a.c. circuit, i.e., the *sine* of the angle of lag.

induction (*Chem.*). Change in the electronic configuration and hence reactivity of one group in a molecule upon addition of a neighbouring polar group. (*Elec., Mag.*) A term sometimes used to denote the density of an electric or magnetic field. (*Maths.*) See **mathematical induction.** (*Zool.*) The production of a definite condition by the action of an external factor.

induction accelerator (*Nuc. Eng.*). See **betatron.**

induction balance (*Elec. Eng.*). An electrical network, i.e., a bridge, to measure inductance.

induction coil (*Elec. Eng.*). Transformer for producing high-voltage pulses in the secondary winding, obtained from interrupted d.c. in the primary, as for a petrol engine. The original Ruhmkorff induction coil was magnetically open-circuit and self-interrupting, like a buzzer or relay; used for early discharges in gas tubes.

induction compass (*Mag.*). One which indicates the direction of the earth's magnetic field by a rotating coil, in which an e.m.f. is induced.

induction field (*Phys.*). See **field.**

induction flame damper (*Aero.*). See **flame trap.**

induction furnace (*Elec. Eng.*). Application of induction heating in which the metal to be melted forms the secondary of a transformer.

induction generator (*Elec. Eng.*). An electric generator similar in construction and operation to an induction motor; in order to generate, it must be driven above synchronous speed and must be excited from the a.c. supply into which it is delivering power.

induction hardening (*Met.*). Using high-frequency induction to heat the part for hardening. The method is rapid and lends itself to restriction of the hot zone.

induction heating (*Elec. Eng.*). That arising from eddy currents in conducting material, e.g., solder, profiles of gear-wheels, conductor coils around vessels for heating liquids, etc. Generated with a high-frequency source, usually oscillators of high power, operating at 10^6–10^7 Hz. Also **eddy-current heating.**

induction instrument (*Elec. Eng.*). An electrical measuring instrument in which the pointer is moved as the result of the interaction between an alternating flux produced by the quantity to be measured, and currents induced by this flux in a disk.

induction lamp (*Elec. Eng.*). See **neon-.**

induction machine (*Elec. Eng.*). See **electrostatic generator.**

induction (or inlet) manifold (*I.C. Engs.*). In a multi-cylinder petrol-engine, the branched pipe which leads the mixture from the carburettor to the combustion chambers.

induction meter (*Elec. Eng.*). The most common type of a.c. integrating meter; it is a motor meter in which the torque is produced as the result of the interaction between an alternating flux and currents induced in a disk by this flux.

induction motor (*Elec. Eng.*). An a.c. motor in which currents in the primary winding (connected to the supply) set up a flux which causes currents to be induced in the secondary winding (usually the rotor); these currents interact with the flux to produce rotation. Also called **asynchronous motor, nonsynchronous motor.**

induction motor-generator (*Elec. Eng.*). A motor-generator set driven by an induction motor.

induction period (*Chem.*). The interval of time between the initiation of a chemical reaction and its observable occurrence.

induction port, -valve, etc. (*I.C. Engs.*). The port, valve, etc., through which the charge is induced into the cylinder during the suction stroke. Also **inlet port, -valve, etc.**

induction regulator (*Elec. Eng.*). A voltage regulator having a winding connected in series with the supply; voltages are induced in this winding from a primary winding connected across the supply, and regulation of the voltage is carried out by varying the relative position of the two windings.

induction relay (*Elec. Eng.*). A relay, for use in an electrical circuit, in which the contacts are closed as the result of the interaction between an alternating flux and currents induced in a disk by this flux.

induction stroke (*I.C. Engs.*). The suction stroke, charging stroke, or intake stroke, during which the working charge or air is induced into the cylinder of an engine.

inductive (*Elec. Eng.*). Said of an electric circuit or piece of apparatus which possesses self or mutual inductance, which tends to prevent current changes. Always present to some extent, but may often be neglected.

inductive circuit (*Elec. Eng.*). One in which effects arising from inductances are not negligible, the back e.m.f. tending to oppose a change in current, leading to sparking or arcing at contacts which attempt to open the circuit.

inductive drop (*Elec. Eng.*). Voltage drop produced in an a.c. circuit owing to its self or mutual inductance.

inductive load (*Elec. Eng.*). Terminating impedance which is markedly inductive, taking current lagging in phase on the source e.m.f., e.g., electrodynamic loudspeaker or motor. Also **lagging load.**

inductive neutralization (*Elec. Eng.*). An amplifier in which the feedback susceptance of the self capacity of the circuit elements is balanced by the equal and opposite susceptance of an inductor.

inductive pick-off (*Elec. Eng.*). One in which changes in reluctance of a laminated path alter a current or generate an e.m.f. in a winding.

inductive reactance (*Elec. Eng.*). Product of the angular frequency ($=2\pi \times$ frequency) and inductance. Measured in (positive) ohms when the inductance is in henries.

inductive reaction (*Elec. Eng.*). Same as **electromagnetic reaction.**

inductive resistor (*Elec. Eng.*). Wirewound resistor having appreciable inductance at frequencies in use.

inductor (*Chem.*). A substance which accelerates a slow reaction between two or more substances by reacting rapidly with one of the reactants. (*Elec. Eng.*) Any circuit component whose inductance cannot be treated as negligible.

inductor generator (*Elec. Eng.*). An electric generator in which the field and armature windings are fixed relative to each other, the necessary changes of flux to produce the e.m.f. being produced by rotating masses of magnetic material.

inductor loudspeaker (*Acous.*). A cone loudspeaker with an electromagnetic drive, in which the magnetic flux passes across gaps in which the driving-pin attached to the apex of the cone is free to move. On the pin are located magnetic elements slightly displaced from the centre of the gaps, so that fluctuations in magnetic flux pull the pin normally to the direction across the gap.

indulines (*Chem.*). A subgroup of the azine dyestuffs, containing 3 or 4 amino groups, belonging to the class of diphenylamine dyestuffs; it is formed by the interaction of aniline hydrochloride and aminoazobenzene.

indumentum (*Bot.*). The general downy or hairy covering of a plant. (*Zool.*) A covering of hair or feathers.

induplicate aestivation (*Bot.*). A form of valvate aestivation in which the edges of the perianth segments are turned inwards.

induplicate vernation (*Bot.*). The upfolding of the basal section of a leaf pinna, as in some Palms.

indurated. Hardened, made hard. *n.* **induration.**

indusium (*Bot.*). (1) A protective structure associated with the sorus in a fern. It may be a group of hairs or of scales; it may be a cup-shaped structure around the base of the sorus; and it is frequently an umbrella-like scale completely covering the sorus. (2) A netlike structure hanging from the stipe of some fungi. (*Zool.*) In some Insects, a third embryonic envelope lying between the chorion and the amnion in the early stages of development of the egg; a cerebral convolution of the brain in higher Vertebrates; an insect larva case. *adjs.* **indusiate, indusiform.**

industrial diamond (*Min.*). Small diamonds, not of gemstone quality, e.g., *black diamond* and *bort* (qq.v.); used to cut rock in borehole drilling, and in abrasive grinding. Now synthesized on a considerable scale by subjecting carbon to ultra-high pressures and temperature.

industrial foils (*Paint.*). Very thick blocking-foils used for marking rubber belting and hosing and for making silk screens.

industrial frequency (*Elec. Eng.*). A term used to denote the frequency of the alternating current used for ordinary industrial and domestic purposes, usually 50 or 60 hertz. Also **mains** or **power frequency.**

industrialized building (*Build., Civ. Eng.*). See **system building.**

industrial melanism (*Ecol.*). *Melanism* (q.v.) which has developed as a response to blackening of trees, etc., by industrial pollution. This favours melanic forms, especially among moths which rest on trees during the day.

industrial reflector (*Light*). A lighting fitting, of conical shape and whitened or polished internally, suitable for industrial lighting.

industrial television (*TV*). See **closed circuit.**

inelastic collision (*Nuc.*). A collision in which there is change in the total energies of the particles concerned. This results from excitation or de-excitation of one (or both) of the particles involved.

inelastic scattering (*Nuc.*). See **scattering.**

inequality (*Astron.*). The term used to signify any departure from uniformity in orbital motion; it may be (*a*) *periodic*, that is, completing a full cycle within a specific time and then repeating it; or (*b*) *secular*, that is, increasing steadily in magnitude with time. (*Maths.*) A statement as to which is the larger or smaller of two quantities. The statement that a is larger than b is written $a>b$, and the consequential statement that b is smaller than a, $b<a$. The inequality sign, $>$ or $<$, can be coupled with the equality sign, $=$, and is then written $\geqq$ or $\leqq$.

inequipotent (*Zool.*). Possessing different potentialities for development and differentiation.

inequivalve (*Zool.*). Having the 2 valves of the shell unequal.

inert (*Chem.*). Not readily changed by chemical means.

inert anode (*Ships, etc.*). An anode of platinized titanium, used in *cathodic protection* (q.v.). Requires an impressed direct current. Long-lasting. Cf. *virtually inert anode.*

inert cell (*Chem.*). A dry cell containing ingredients which form an electrolyte only when water is added.

inert gases (*Chem.*). Elements helium, neon, argon, krypton, xenon, and radon-222, much used (except the last) in gas-discharge tubes. (Radon-222 has short-lived radioactivity, half-life less than 4 days.) Their outer (valence) electron orbits are complete, thus rendering them inert to all the usual chemical reactions; a property for which argon, the most abundant, finds increasing industrial use. The heavier ones, Rn, Xe, Kr, are known to form a few unstable compounds, e.g., XeF_4. Also called **noble gases, rare gases.**

inertia (*Maths.*). See **moment of inertia.** (*Photog.*) The exposure, in candela-metre-seconds, which is indicated for zero density when the linear portion of the gamma curve for an emulsion is extended. (*Phys.*) The property of a body, proportional to the mass but independent of gravity, which opposes the change in the state of motion of a body. *Centrifugal* and *Coriolis forces* are manifestations of inertia.

inertia governor (*Eng.*). A shaft type of centrifugal governor using an eccentrically pivoted weighted arm, which responds rapidly to speed fluctuations by reason of its inertia, and in such a way as to suppress them.

inertial damping (*Elec.*). That which depends on the acceleration of a system, and not velocity.

inertial guidance (*Nav.*). That based on integration of accelerations of a moving body, e.g., ship, etc., using very accurate gyros, acting with a stable platform. See **guided missile.**

inertial impaction (*Powder Tech.*). Method of collecting small particles of dust and droplets from a fluid stream by allowing them to impinge upon an interposed deflecting surface.

inertia starter (*I.C. Engs.*). A device for turning an aero-engine for starting purposes. A light flywheel is accelerated by hand, through gears, to a high speed, and a slow-speed shaft driven by it is then clutched in to the engine crankshaft.

inertia switch (*Elec. Eng.*). One operated by an abrupt change in its velocity, as for some meters, to avoid overloading.

inert metal (*Nuc.*). Alloy (usually Ti-Zr) for which scattering of neutrons by nuclei is negligible.

I neutrons (*Nuc.*). Those possessing such energy as to undergo resonance absorption by iodine.

infantile paralysis (*Med.*). See **poliomyelitis.**

infantilism (*Med.*). A disturbance of growth, the persistence of infantile characters being associated with general retardation of development.

infarct (*Med.*). That part of an organ which has had its blood supply cut off, the area so deprived undergoing necrosis.

infarction (*Med.*). The formation of an infarct; the infarct itself.

infection (*Med.*). The invasion of body tissue by living micro-organisms, with the consequent production in it of morbid change; a diseased condition caused by such invasion; the infecting micro-organism itself.

infection tube (*Bot.*). The germ-tube which penetrates the host from the germinating spore of a parasitic fungus.

infectious anaemia (*Vet.*). A septicaemia of horses.

infectious avian bronchitis (*Vet.*). An acute, highly contagious respiratory disease of chickens, caused by a virus and associated with inflammation of the respiratory tract, especially the trachea and bronchi; the main symptoms are nasal discharge, gasping, rales, and coughing.

infectious avian encephalomyelitis (*Vet.*). Epidemic tremor. An encephalomyelitis of young chicks, caused by a virus, and characterized by muscular incoordination, muscular tremor, and death.

infectious bovine rhinotracheitis (*Vet.*). An acute virus infection of the upper respiratory tract of cattle.

infectious bulbar paralysis (*Vet.*). See **Aujesky's disease.**

infectious canine hepatitis (*Vet.*). Rubarth's disease; hepatitis contagiosa canis. An acute, contagious and often fatal disease of the dog characterized by fever, diarrhoea, vomiting, and widespread serous or haemorrhagic effusions.

infectious coryza (*Vet.*). Contagious catarrh; fowl coryza; roup. An acute, contagious bacterial infection of the upper respiratory tract of domestic fowl.

infectious hepatitis (*Med.*). See **infective hepatitis.**

infectious icterohaemoglobinuria (*Vet.*). See **redwater.**

infectious jaundice (*Med.*). See **spirochaetosis icterohaemorrhagica.**

infectious keratitis (*Vet.*). See **infectious ophthalmia.**

infectious laryngotracheitis (*Vet.*). A highly contagious, often fatal, virus infection of the respiratory tract of the chicken. Signs include coughing and sneezing. Abbrev. **ILT.**

infectious mononucleosis (*Med.*). Glandular fever. An acute infectious disease characterized by slight fever, enlargement of glands in the neck, and an increase in the white (mononuclear) cells of the blood. The causative virus appears to be identical with that which produces Burkitt's lymphoma in tropical countries.

infectious ophthalmia (*Vet.*). Infectious keratitis; New Forest disease. A contagious form of conjunctivitis and keratitis occurring in cattle; caused by a bacterium *Moraxella bovis* or by a rickettsial organism.

infectious parotitis (*Med.*). See **mumps.**

infectious pig paralysis (*Vet.*). See **Teschen disease.**

infectious pododermatitis (*Vet.*). See **foul in the foot.**

infectious sinusitis of turkeys (*Vet.*). Big head disease of turkeys. A disease of turkeys caused by infection of the infraorbital sinuses of the head by organisms of the genus *Mycoplasma*; characterized by swelling of the face and a discharge from the eyes and nostrils.

infectious synovitis (*Vet.*). A disease of chickens caused by infection by organisms of the genus *Mycoplasma* resulting in exudative synovitis.

infective (infectious) hepatitis (*Med.*). A virus infection of the liver, usually food-borne and often epidemic. *Serum hepatitis* is a related but distinct condition, in which the usual mode of transmission is by injection of infected blood or use of instruments contaminated by such blood.

Inferae (*Bot.*). A series of the *Sympetalae*, having an inferior ovary and the stamens usually the same number as the corolla lobes.

inferior (*Bot.*). (1) Said of the annulus of an agaric when it is placed low down on the stipe. (2) Said of the gynaeceum of a flower when it is enclosed by the receptacle, so that the calyx, corolla, and stamens are developed above it. (3) Said of the calyx, corolla, and stamens of a hypogynous flower. (*Zool.*) Lower; under; situated beneath, e.g., the *inferior* rectus muscle of the eyeball. Cf. *superior*.

inferior conjunction (*Astron.*). See **conjunction.**

inferior figures or letters (*Typog.*). Small figures or letters set below the general level of the line; as in chemical formulae, e.g., C_6H_5.

inferiority complex (*Psychol.*). Generally, a persisting state of feelings of inferiority; specifically (*Adler and the individual psychologists*), feelings of inferiority arising from some organ or limb defect: (*the Freudian school*) feelings of inferiority resulting from deprivation of love in childhood, this being frequently the main factor in the development of a psychoneurosis.

Inferior Oölite (*Geol.*). A term used in Britain for the lower part of the Middle Jurassic.

inferior planets (*Astron.*). See **planet.**

inferior vena cava (*Zool.*). See **postcaval vein.**

inferobranchiate (*Zool.*). Having the gills hidden under the margin of the mantle.

inferred-zero instrument (*Elec. Eng.*). See **suppressed-zero instrument.**

infestation (*Med.*). The condition of being occupied or invaded by parasites, usually parasites other than bacteria.

infibulation (*Med.*). The fastening of the external genital organs with a clasp.

infilling (*Build.*). Material, such as hard core, used for making up levels, e.g., under floors.

infiltration (*Med.*). (1) The accumulation of abnormal substances (or of normal constituents in excess) in cells of the body. (2) The gradual spread of infection in an organ (e.g., tuberculous *infiltration* of the lung). (*Powder Tech.*) Impregnation using capillary forces to soak up the impregnant.

infinite attenuation (*Elec.*). The property of some filters of providing a theoretically infinite attenuation for one or more specified frequencies against which strong discrimination is required.

infinite gamma (*Photog.*). See **gamma infinity.**

infinite line (*Elec.*). A transmission line which is infinitely long, or finite but terminated with its characteristic impedance, and along which there is uniform attenuation and phase delay.

infinite persistence screen (*Electronics*). CRT screen on which displayed traces remain visible (up to weeks) unless removed by electric signal applied to clearing electrode. See **Remscope, Storascope.**

infinite set (*Maths.*). A set that can be put into a one-one correspondence with part of itself, e.g., the positive integers, which can be put into a one-one correspondence with the positive even integers.

infinitesimal (*Maths.*). A vanishingly small part of a quantity, which, although it retains the dimensions or other qualities of the quantity, is thought of as decreasing indefinitely without becoming zero.

infinity (*Maths.*). That which is larger than any quantified concept. For many purposes it may be considered as the reciprocal of zero, and minus infinity equated to plus infinity. Written ∞.

infinity plug (*Elec. Eng.*). A plug in a resistance box which, when withdrawn, breaks the circuit, i.e., introduces an infinite resistance.

inflammation (*Med.*). The reaction of living tissue to injury or to infection, the affected part becoming red, hot, painful, and swollen, due to hyperaemia, exudation of lymph, and escape into the tissue of blood cells.

inflatable aircraft (*Aero.*). A small low-performance aeroplane for military use, in which the aerofoil surfaces (and sometimes the fuselage) are inflated so that it can be compactly packed for transport.

inflation (*Aero.*). The process of filling an airship or balloon with gas. Sometimes called gassing.

inflected arch (*Civ. Eng.*). See **inverted arch.**

inflexion (*Maths.*). See **point of inflexion on a curve.**

inflorescence (*Bot.*). (1) In flowering plants, the part of the shoot which bears flowers. (2) In *Bryophyta*, that part of the plant body differentiated to bear the antheridia and archegonia.

influence line (*Build., Eng.*). An *influence line* for a structure is a curve, the ordinate to which at any point represents the value of some variable (such as the bending moment) at another particular point in the structure, due to the presence of a unit load at the point where the ordinate is taken.

influence machine (*Elec. Eng.*). See **electrostatic generator.**

influenza (*Med.*). An air-borne respiratory virus infection, causing epidemics which are often world-wide, but whose severity varies with the (constantly-changing) virus type. (*Vet.*) See **equine influenza, swine influenza.**

information (*Telecomm.*). This is measured by the number of code elements in use multiplied by the logarithm of the number of values which every code element may take. See **bit, Hartley principle.**

information content (*Telecomm.*). Gross information content is the number of bits or hartleys required to transmit a message via a noiseless system with specified accuracy regardless of redundancy. Net information content is the minimum number which would transmit the essential information. Abbrev. **I.C.**

information rate (*Telecomm.*). The number of *bits* required to represent a message. Same as (deprecated) **entropy.**

information retrieval (*Comp.*). Location of required information previously classified and stored.

information theory (*Telecomm.*). Mathematical analysis of efficiency with which communication channels are employed to find the most efficient system of coding for any channel.

infra-. Prefix from L. *infra*, below.

infrabasal (*Zool.*). In *Crinoidea*, one of a whorl of perradial plates lying below the basals.

infrablack (*TV*). Amplitude in a TV signal beyond the black level of the picture.

infrabranchial (*Zool.*). Below the gills, e.g., part of the mantle chamber in *Lamellibranchiata.*

infraclavicle (*Zool.*). In some Fish, a membrane bone of the pectoral girdle.

infraclavicular (*Zool*). Below the clavicle.

infracostal (*Anat.*). Beneath the ribs.

infradyne (*Radio*). Supersonic heterodyne receiver in which the intermediate frequency is higher than that of the incoming signal.

infra-epimeron (*Zool.*). In Insects, the lower part of the epimeron when it is subdivided.

infra-episternum (*Zool.*). In Insects, the lower part of the epimeron when it is subdivided.

inframarginal (*Zool.*). Below the margin; a marginal structure; in *Chelonia*, one of certain plates of the carapace lying below the marginals; in *Asteroidea*, one series of ossicles situated on the lower margin of each ray.

infraneuston (*Ecol.*). Aquatic animals associated with the under side of the surface film, e.g., some Mosquito pupae. Cf. *supraneuston.*

infraorbital foramen (*Zool.*). In Mammals, a foramen on the outer surface of each maxilla for the passage of the second division of the fifth cranial nerve.

infraorbital glands (*Zool.*). In Mammals, one of the four pairs of salivary glands.

infraproteins (*Chem.*). Protein derivatives obtained by hydrolysis.

infrared detection (*Phys.*). Rays detected and registered photographically with special dyes: photosensitively with a special Cs—O—Ag surface; by photoconduction of lead sulphide and telluride; and, in absolute terms, by bolometer, thermistor, thermocouple or Golay detector.

infrared maser (*Phys.*). One which radiates or

detects signals of mm wavelengths. See also **laser.**

infrared photography (*Photog.*). Use of an emulsion sensitive to infrared light, thus enabling haze-penetration, camouflage and forgery detection, photography in darkness, etc.

infrared radiation (*Phys.*). Electromagnetic waves covering a range from the limit of the visible spectrum (the near region) to the shortest microwaves (the far region). A convenient subdivision is as follows. Near: 0·75 to 2·5 μm; intermediate: 2·5 to 30 μm; far: 30 to 1000 μm. The absorption of the radiation by crystalline materials results from the excitation of lattice vibrations which occur in a relatively narrow frequency band.

infrared spectrometer (*Phys.*). An instrument similar to an optical spectrometer but employing nonvisual detection and designed for use with infrared radiation. The infrared spectrum of a molecule gives information as to the functional groups present in the molecule and is very useful in the identification of unknown compounds.

infrared therapy (*Med.*). Treatment of disease with generators producing electric oscillations of a wavelength between 0·8 and 16 *μ*m.

infrasizer (*Powder Tech.*). See **Haultain infrasizer.**

infrasonic (*Acous.*). Said of frequencies below the usual audible limit, viz. 20 Hz.

infructescence (*Bot.*). The inflorescence after the flowers have fallen and fruits have developed.

infundibuliform (*Bot.*). Tubular below, gradually enlarging upwards, i.e., funnel-shaped.

infundibulum (*Zool.*). A funnel-shaped structure; in Vertebrates, a ventral outgrowth of the brain; a pulmonary vesicle; in *Cephalopoda*, the siphon; in *Ctenophora*, the flattened gastric cavity. *adj.* **infundibular.**

infusible (*Met.*). Refractory; not rendered liquid under specified conditions of pressure, temperature or chemical attack.

infusion (*Chem.*). Solution of the soluble constituents of vegetable matter, obtained by steeping the vegetable matter in liquid, often hot and sometimes under pressure.

infusorial earth (*Min.*). See **tripolite.**

ingate (*Foundry*). The channel, or channels, by which the molten metal is led from the runner hole into the interior of a mould. (*Mining*) The entrance from any point in a shaft to the workings of a colliery.

Ingenhausz's experiment (*Heat*). Rods of different metals are coated with paraffin wax and their ends immersed in hot water. The relative rate at which heat is conducted along the rods is shown by the rate of melting of the wax.

ingestion (*Zool.*). The act of swallowing or engulfing food material (*ingesta*) so that it passes into the body. *v.* **ingest.**

ingluvies (*Zool.*). An oesophageal dilatation of Birds; the crop.

ingluvitis (*Vet.*). Inflammation of the crop, or ingluvies, of Birds.

ingo, ingoing (*Build.*). See **reveal.**

Ingold cutter (*Horol.*). A special cutter used to correct inaccuracies in the teeth of a wheel.

ingot (*Met.*). A metal casting of a shape suitable for subsequent hot working, e.g., rolling.

ingot iron (*Met.*). Iron of comparatively high purity, produced, in the same way as steel, in the open-hearth furnace, but under conditions that keep down the carbon, manganese, and silicon content.

ingot mould (*Met.*). The mould or container in which molten metal is cast and allowed to solidify to form an ingot.

ingot stripper (*Met.*). Mechanism for extracting ingots from ingot moulds.

ingrain carpet (*Textiles*). See **Scotch carpet.**

ingravescent (*Med.*). Gradually increasing in severity.

inguinal (*Zool.*). Pertaining to, or in the region of, the groin, e.g., the *inguinal canal* through which the testes descend in male mammals.

inguinodynia (*Med.*). Pain in the groin.

inhalant (*Zool.*). Pertaining to, or adapted for, the action of drawing in a gas or liquid, e.g., the *inhalant* siphon in some *Mollusca*.

inhalation (*Med.*). The act of breathing in, or taking into the lungs; any medicinal agent breathed into the lungs.

inherent ash (*Min. Proc.*). Noncombustible material intimately bound in the original coal-forming vegetation, as distinct from 'dirt' from extraneous sources.

inherent filtration (*Radiol.*). That introduced by the wall of the X-ray tube as distinct from added primary or secondary filters.

inherent floatability (*Min. Proc.*). Natural tendency of some mineral species to repel water and to become part of the 'float' without preliminary conditioning in the froth-flotation process.

inherent regulation (*Elec. Eng.*). The change in voltage at the output terminals of an electric machine (a.c. or d.c. generator, or a converter) when the load is applied or removed, all other conditions remaining constant. The change in secondary voltage of a transformer when the load is changed between zero and full load.

inherited error (*Comp.*). That in one stage of a multistage calculation which is carried over as an initial condition to a subsequent stage.

inherited memory (*Zool.*). See **instinct.**

inhibited oil (*Elec. Eng.*). Transformer or switch oil which includes an antioxidant to delay the onset of sludge and acid formation.

inhibition (*Zool.*). The stopping or deceleration of a metabolic process; cf. *excitation*. *adj.* **inhibitory.**

inhibitor (*Bot.*). A substance which limits or destroys the catalytic activity of an enzyme. (*Chem.*) Additive which retards or prevents an undesirable reaction, e.g., phosphates, which prevent corrosion by the glycols in *antifreeze* (q.v.) solutions; anti-oxidants in rubber; gastric inhibitors, which reduce gastric secretion of hydrochloric acid and are used in the treatment of ulcers.

inhibitory (*Zool.*). Said of a nerve, the stimulation of which results in the regulation of the activities of a muscle or gland in the direction of decrease.

inhibitory phase (*Chem.*). The protective colloid in a lyophobic sol.

inhour (*Nuc. Eng.*). Unit of reactivity equal to reciprocal of period of nuclear reactor in hours, e.g., a reactivity of two inhours will result in a *period* (q.v.) of half an hour.

Iniomi (*Zool.*). A suborder of *Teleostei* characterized by the possession of black or silvery soft fragile bodies, frequently provided with light organs; the air-bladder is small or absent, and there is usually an adipose fin; abyssal forms. Lantern-fish, Lizard-fish.

inion (*Anat.*). The external bony protuberance on the occiput, at the back of the skull.

initial cell (*Bot.*). A cell which remains meristematic, divides repeatedly, and gives rise to many daughter cells from which, after further divisions, the permanent tissues of the plant are differentiated.

initial conditions (*Comp.*). Values of the parameters of an equation which are used by a computer in calculating a numerical solution.

initial consonant articulation (*Teleph.*). See **articulation.**

initial ionizing event (*Nuc.*). One which starts a pulse in a radiation detector.

initial permeability (*Elec. Eng.*). Limiting value of differential permeability when magnetization tends to zero. Applicable to small alternations of magnetization when there is no magnetic bias.

initial spindle (*Cyt.*). See **netrum.**

initial stability (*Ships*). The moment of the couple tending to return the vessel to her equilibrium position when her *angle of heel* (q.v.) is small. Depends directly on *metacentric height* (q.v.) and is frequently expressed as such.

initial susceptibility (*Mag.*). Limiting value of differential susceptibility when the magnetization tends to zero. Applicable to small alternations of magnetization without magnetic bias.

initiator (*Chem.*). The substance or molecule which starts a chain reaction.

injected (*Bot.*). Having the intercellular spaces filled with water.

injection (*Geol.*). The emplacement of fluid rock matter in crevices, joints, or fissures found in rocks. See **intrusions.** (*I.C. Engs.*) The process of spraying oil-fuel into the cylinder of a compression-ignition engine by means of an injection pump.

injection carburettor (*Aero.*). A pressure carburettor in which the fuel delivery to the jets is maintained by pressure instead of by a float chamber. It is unaffected by negative g in aerobatics or severe atmospheric turbulence.

injection complex (*Geol.*). An assemblage of rocks, partly igneous, partly sedimentary or metamorphic, the former in intricate intrusive relationship to the latter, occurring in zones of intense regional metamorphism. British examples have been described by H. H. Read from the Highlands of Scotland.

injection condenser (*Eng.*). See **jet condenser.**

injection efficiency (*Electronics*). The fraction of the current flowing across the emitter junction in a transistor which is due to the minority carriers.

njection lag (*I.C. Engs.*). In a compression-ignition engine, the time interval between the beginning of the delivery stroke of the fuel injection pump and the beginning of injection into the engine cylinder.

injection moulding (*Plastics*). A method for the fabrication of thermoplastic materials. The viscous resin is squirted, by means of a plunger, out of a heated cylinder into a water-chilled mould, where it is cooled and removed. Method used also with thermosetting moulding powders.

injection valve (*I.C. Engs.*). See **injectors, atomizer.**

njector (*Eng.*). A device by which a stream of fluid, as steam, is expanded to increase its kinetic energy, and caused to entrain a current of a second fluid, as water, so delivering it against a pressure equal to, or greater than, that of the steam; the steam injector is commonly used for feeding boilers. See also **Giffard's injector.**

injector grid (*Radio*). One which injects a modulating voltage into the electron stream of a first detector of a superheterodyne receiver.

injectors (*I.C. Engs.*). Plugs with valved nozzles through which fuel is metered to the combustion chambers in diesel- or fuel-injection engines. Connected with the fuel-pump by a capillary tube. In diesels, the injectors are generally screwed into the cylinder head (corresponding to the sparking-plug in a conventional petrol engine); in petrol-injection engines, they are located beside the inlet valves.

injury-feigning (*An. Behav.*). See **distraction display.**

ink. Writing ink usually consists of a fluid extract of tannin with the addition of solutions of iron salts. The semi-solid ink used in ball-point pens consists of highly-concentrated dyes in a non-volatile solvent. Coloured writing inks are prepared by dissolving suitable dyes in water. Marking inks are made from solutions of silver or copper compounds, aniline being sometimes added. See also **Indian ink, printing ink.** (*Zool.*) See **ink sac.**

Inkatron (*Print.*). Electronic equipment for maintaining ink density on sheet-fed offset litho presses.

inkblot test (*Psychiat.*). See **Rorschach test.**

ink misting (*Print.*). Ink is sometimes reduced to a fine mist, particularly on rotary presses, by the friction between two surfaces rotating in opposite directions, *long inks* being more susceptible than *short inks.*

ink pump (*Print.*). On rotary presses: (1) a special type of pump-operated ink supply system, (2) a circulating pump to agitate liquid inks, particularly for photogravure, and (3) a pump to maintain ink supply to the ink ducts.

ink rail (*Print.*). Part of a pump-operated ink duct which extends across the machine to apply ink to the system.

ink sac (*Zool.*). In some *Cephalopoda,* a large gland, opening into the alimentary canal near the anus, which secretes a dark-brown pigment (sepia).

ink table or ink slab (*Print.*). A flat surface on a printing machine, part of the ink distribution system; also a flat surface used with a hand roller.

inlay (*Join.*). Pieces of metal, ivory, etc., inserted ornamentally (inlaid) in furniture, etc. (*TV*) The electronic mixing of TV images, using masks.

inlet guide vanes (*Aero.*). Radially positioned aerofoils in the annular air intake of axial-flow compressors which direct the airflow on to the first stage at the most efficient angle. The vanes are often rotatable about their mounting axes so that different entry air speeds can be accommodated; they are then called variable inlet guide vanes. Abbrev. **IGV.**

inlet manifold (*I.C. Engs.*). See **induction manifold.**

inlet port (-valve), etc. (*I.C. Engs.*). See **induction port (-valve).**

inlier (*Geol.*). An outcrop of older rocks surrounded by those of younger age.

in-line assembly (*Radio*). Assembly by machine, in which a line of insertion heads inserts components one at a time into a wiring board, on which a circuit has been previously established, by printing or etching.

in-line engine (*I.C. Engs.*). A multicylinder engine, consisting of a bank of cylinders mounted in line along a common crankcase.

innate (*Bot.*). (1) Sunken into the thallus, or originating within the thallus. (2) Said of an anther which has the filament joined to its base only.

innate behaviour (*An. Behav.*). See **species-specific behaviour.**

innate releasing mechanism (*An. Behav.*). A hypothetical neurosensory mechanism invoked to explain the action of sign stimuli, being thought to release the reaction they produce, and to be responsible for the animal's selective susceptibility to a special combination of stimuli. Abbrev. **IRM.**

inner bead (*Join.*). See **guide bead.**

inner bremsstrahlung (*Nuc.*). A process which sometimes occurs whereby, during β-disintegration, a photon is emitted of energy between zero and the maximum energy available in the transition.

inner conductor (*Elec. Eng.*). (1) See **internal conductor.** (2) The neutral conductor of a 3-wire system.

inner dead-centre (*Eng.*). Of a reciprocating engine or pump, the piston position at the beginning of the outstroke, i.e., when the crank-pin is nearest to the cylinder. Also called **top dead-centre.**

inner ear (*Anat.*). Structure encased in bone and filled with fluid, including the cochlea, the semicircular balancing canals and the vestibule.

inner endodermis (*Bot.*). The endodermis internal to the vascular tissues in a solenostele.

inner forme (*Print.*). In *sheet-work printing*, the inner forme includes pages 2 and 3, 6 and 7, 10 and 11, and so on, and is printed on one side of the paper, the outer forme, containing the remaining pages, being printed on the other side.

inner glume (*Bot.*). See **pale.**

inner marker beacon (*Aero.*). A vertically-directed radio beam which marks the aerodrome boundary in a beam-approach landing system, such as the *instrument landing system* (*q.v.*).

innervation (*Zool.*). The distribution of nerves to an organ.

innings (*Civ. Eng.*). Lands reclaimed from the sea.

innocent (*Med.*). Not malignant; not cancerous. See **tumour.**

innominate (*Zool.*). Without a name, e.g., the *innominate artery* of some Mammals, which leads from the aortic arch to give rise to the carotid artery and the subclavian artery, the *innominate vein* of *Cetacea*, *Edentata*, *Carnivora*, and *Primates*, which leads across from the jugular-subclavian trunk of one side to that of the other, the *innominate bone*, which is the lateral half of the pelvic girdle.

innovation (*Bot.*). The portion of a branch added during one season of growth.

inocular (*Zool.*). Having the insertion of the antenna close to the eye.

inoculation (*Bot.*). (1) The conveyance of infection to a host plant by any means of transmission. (2) The entry of the germ tube of a parasitic fungus into the host plant. (3) The placing of spores or portions of growing fungi or bacteria in a culture medium. (*Chem.*) The introduction of a small crystal into a supersaturated solution or supercooled liquid in order to initiate crystallization. (*Med.*) The introduction into an experimental animal, by various routes, of infected material or of pathogenic bacteria: the injection of a vaccine into a person for protection against subsequent infection by the organisms contained in the vaccine. (*Met.*) Modification of crystallizing habit, grain refinement or impartation of alloy qualities to molten metal in furnace or ladle, by addition of small quantities of other metals, deoxidants, etc.

inoculum (*Med.*). The material used in inoculation.

inoperculate (*Bot.*). Lacking a lid to the sporangium.

inordinate (*Bot.*). Not arranged in any definite order.

inorganic chemistry (*Chem.*). The study of the chemical elements and their compounds, other than the compounds of carbon; however, the oxides and sulphides of carbon and the metallic carbides are generally included in inorganic chemistry.

inositol (*Chem.*). $C_6H_6(OH)_6$. Cyclohexanehexol, hexahydroxycyclohexane. There are nine stereoisomers: *myo*-inositol (previously meso-inositol), *scyllo*-inositol, *muco*-inositol, *neo*-inositol, *epi*-inositol, *allo*-inositol, *cis*-inositol, (+)-inositol and (−)-inositol. The last two are enantiomers, the remainder are poticaliy inactive. *Myo*-inositol is essential for the growth of certain yeasts; see also **phytic acid.**

in parallel (*Elec. Eng.*). See **parallel.**

in phase (*Elec. Eng.*). See under **phase.**

in-phase component (*Elec. Eng.*). See **active component.**

in-phase loss (*Elec. Eng.*). See **ohmic loss.**

in-pile test (*Nuc. Eng.*). One in which the effects of irradiation are measured while the specimen is subjected to radiation and neutrons in a reactor.

input (*Comp.*). (1) That part of the computer which is fed with information in coded form on punched cards or punched tape. (2) Vehicle carrying the information. (3) Information itself, which is read in.

input block (*Comp.*). That part of the internal storage reserved for accepting input data.

input capacitance (*Elec.*). The effective capacitance between the input terminals of a network. See **Miller effect.**

input characteristic (*Electronics*). See **transistor characteristics.**

input equipment (*Comp.*). See **input unit.**

input gap (*Electronics*). See **buncher.**

input impedance (*Elec. Eng.*, *Electronics*). The small signal impedance measured between the input terminals of a network.

input signal (*Elec. Eng.*). That connected to the input terminals of any instrument or system (usually electronic).

input transformer (*Elec. Eng.*). One used for isolating a circuit from any d.c. voltage in the applied signal and/or to provide a change in voltage. Also used to match the impedence of an input signal to that of the circuit to maximise power transfer.

input unit or input equipment (*Comp.*). That used to supply coded information to the arithmetic unit of a computer or data processing system.

input voltage (*Elec. Eng.*, *Electronics*). The voltage applied between the input terminals of a network.

inquartation (*Met.*). Quartation. Term associated with dissolution of silver from gold during assay of prill. Before attack with nitric acid, the proportion of silver to gold must be raised by fusion to at least three to one.

inquiline (*Zool.*). A guest animal living in the nest of another animal, or making use of the food provided for itself or its offspring by another animal.

in-sack drier (*Agric.*). A grain drier in which grain in sacks is dried by the insertion of drying elements, e.g., perforated tubes conveying heated air.

insanity (*Psychiat.*). A medico-legal term used for any mental disorder, usually endogenous, which causes an individual to act against the social or legal demands of the society in which he lives. Not identical with *psychosis* (q.v.).

inscribed circle of a triangle (*Maths.*). The circle touching all three sides of a triangle internally. Its centre is the intersection of the angle bisectors.

inscriptio tendinea (*Zool.*). In Amphibians, one of the transverse bands of connective tissue which traverse the extensor dorsi and the rectus abdominis muscles at intervals.

Insecta (*Zool.*). A class of mainly terrestrial *Arthropoda* which breathe by tracheae; they possess uniramous appendages; the head is distinct from the thorax and bears one pair of antennae; there are 3 pairs of similar legs attached to the thorax, which may also bear wings; the body is sharply divided into head, thorax, and abdomen.

insecticides (*Chem.*). Natural (e.g., derris, pyrethrins) or synthetic substances for destroying insects. The widely used synthetic compounds are broadly classified according to chemical composition, viz. chlorinated (e.g., *DDT*), organophosphates (e.g., *malathion*), carbamates, dinitrophenols. See **contact-**, **fumigants**, **stomach-**.

Insectivora (*Zool.*). An order of small, mainly terrestrial Mammals having numerous sharp teeth, tuberculate molars, well-developed collar-bones, and plantigrade unguiculate pentadactyl feet; insectivorous. Shrews, Moles, and Hedgehogs. Sometimes divided into two orders, *Lipotyphla* and *Menotyphla*.

insectorubin (*Zool.*). A reddish-orange redox pigment, probably a pyrrole derivative, which is extractable from some Insects with acidified ethanol.

insectoverdin (*Zool.*). A green chromoprotein pigment of many Insects; of complex structure, its yellow-orange component may have β-carotene, lutein or astaxanthin as its prosthetic group.

inselberg (*Geol.*). A steep-sided eminence arising from a plain tract; often found in the semi-arid regions of tropical countries.

insemination (*Zool.*). The approach and entry of the spermatozoon to the ovum, followed by the fusion of the male and female pronuclei. In animal breeding, *artificial insemination* involves the implanting, by artificial means, of extraneously obtained sperm-containing semen in the female cervix; also, by extension, the transfer of a fertilized ovum (following *superovulation*, q.v.) from the reproductive tract of one female to that of a host mother.

insert (*Bind.*). See **inset**. (*Cinema.*) A simple shot (e.g., close-up of a letter) inserted into a sequence. (*Met.*) In casting, a small metal part which is fitted in the mould or die in such a manner that the metal flows around it to cast it in position. It may provide a hard metal wearing surface, or sintered metal oil-retaining bearing surface, etc. (*Plastics*) A metal core inserted into a plastic article during the (injection) moulding process.

inserted tooth cutter (*Eng.*). A milling cutter in which replaceable teeth, of expensive material, are inserted into slots in the cutter body.

insertion (*Bot.*). (1) The place where one plant member grows out of another, or is attached to another. (2) The manner of attachment. (*Textiles*) Lace used in connexion with a fabric, often at or near the border, for the purpose of ornamentation. It is narrow, with a plain edge to facilitate insertion in the fabric. (*Zool.*) The point or area of attachment of a muscle, mesentery, or other organ.

insertion gain or loss (*Elec. Eng.*). That gain or loss in decibels when a transformer or other impedance matching transducer or network is inserted into a circuit.

insertion head (*Eng.*). An automatic mechanism, which may incorporate cutting, forming or fastening tools, for placing components into a partly completed assembly.

insertion region (*Bot.*). The part of a chromosome where it is attached to a spindle fibre.

insessorial (*Zool.*). Adapted for perching.

inset or insert (*Bind.*). (1) An extra leaf or section loosely inserted in a book. (2) One folded sheet put inside another; an *insetted* (or *inserted*) *book* is one having its sections placed one within the other.

Insetter (*Print.*). Electronically-controlled equipment for registering pre-printed reels into rotary presses to be incorporated in the final product, overprinted if required; with controlled *automatic reel change*; half-width or full-width reels.

inside (*Textiles*). Comprehensive term for the points, combs, guides, jacks, and trucks, which constitute the chief features of a lace machine.

inside callipers (*Eng.*). *Callipers* (q.v.) with the points turned outwards, used for taking inside dimensions.

inside-colour-sprayed lamp (*Light*). An electric filament lamp, the bulb of which is sprayed on the inside with a white or coloured material in order to diffuse the light and give it a coloured effect.

inside crank (*Eng.*). A crank, with two webs, placed between bearings, as distinct from an outside or overhung crank.

inside cylinders (*Eng.*). In a steam locomotive, those cylinders which are fixed inside the frame.

inside-frosted lamp (*Light*). An electric filament lamp, the bulb of which is etched or sand-blasted on the inside to diffuse the light. Also called a **pearl lamp**.

inside lap (*Eng.*). See **exhaust lap**.

insides (*Paper*). The eighteen quires inside a ream of paper, the top and bottom quires being known as *outsides*; composed entirely of 'good' sheets.

inside spider (*Acous.*). A flexible device inserted within a voice coil so as to centre the coil accurately between the pole pieces of a dynamic loudspeaker.

insight (*Psychol.*). The apprehension of the principle of a task, puzzle, etc. *Gestalt psychology* claims that it is a direct awareness of relevant relationships and not arrived at by learning, but this claim is not universally accepted.

insight learning (*An. Behav.*). The sudden production by an animal of a new adaptive response which has not been arrived at by trial and error behaviour.

insistent (*Zool.*). In Birds, said of the hind toe when it is so placed that only the tip touches the ground.

insolation (*Med.*). Sunstroke. It results from exposure of the head and the neck to the direct rays of the sun, and is characterized by high fever, headache, and mental excitement, followed by unconsciousness. (*Meteor.*) The radiation received from the sun. This depends on the position of the earth in its orbit, the thickness and transparency of the atmosphere, the inclination of the intercepting surface to the sun's rays, and the *solar constant* (q.v.).

insolilith (*Geol.*). A pebble, more or less rounded, with a rough surface produced by cracking under exfoliation due to insolation. (Term applied (1934) by F. Raw to Triassic specimens.)

insoluble (*Chem.*). Incapable of being dissolved. Most 'insoluble' salts have a definite, though very limited, solubility.

insoluble electrode (*Elec. Eng.*). Non-ionizing cathode or anode.

insoluble problem situation (*An. Behav.*). An

experimental situation in which things are so arranged that whichever alternative an animal chooses, it receives a reward on half the trials and punishment on half. This generally evokes a *fixation* (q.v.).

inspection chamber (*San. Eng.*). A pit formed at regular points in the length of a drain or sewer to give access for purposes of inspection.

inspection fitting (*Eng.*). A bend, elbow, or tee used in a conduit wiring system, which is fitted with a removable cover to facilitate the drawing-in of the wires and subsequent inspection.

inspection gauges (*Eng.*). Gauges used by the inspection department of a works for testing the accuracy of finished parts.

inspection junction (*San. Eng.*). A special length of drain-pipe having a branch socket into which a short vertical pipe, reaching up to ground-level, can be fitted to provide a means of access for inspection.

inspection plug (*Elec. Eng.*). A plug fitted in the cover of an accumulator, through which the electrolyte can be inserted and its level observed.

inspersed (*Bot.*). Having granules penetrating the substance of the thallus.

inspiration (*Zool.*). The drawing-in of air or water to the respiratory organs.

inspirator (*Heat.*). The injector of a pressure gas burner, combined with a venturi mixing tube, primary combustion air being entrained from surrounding atmosphere by the projection of a gas stream into the injector throat.

inspissation (*Med.*). The thickening of pus as a result of the removal of fluid.

instability (*Aero.*). An aircraft possesses *instability* when any disturbance of its steady motion tends to increase, unless it is overcome by a movement of the controls by the pilot. (*Mech.*) See **equilibrium.** (*Nuc.*) In a plasma, sudden deformation of quasi-static distribution, usually due to either (*a*) weakening of magnetic field as a result of pressure exerted by plasma (hydromagnetic or exchange stability), or (*b*) non-Maxwellian velocity distribution in plasma. (*a*) leads to kinks in discharge track; (*b*) to necking or eventual rupture of track. (*Telecomm.*) Tendency for a circuit to break into unwanted oscillation.

instantaneous automatic gain control (*Radar*). A rapid action system in radar for reducing the clutter.

instantaneous carrying-current (*Elec. Eng.*). Of switches, circuit-breakers, and similar apparatus, the maximum value of current which the apparatus can carry instantaneously.

instantaneous centre (*Eng.*). The imaginary point, not necessarily in the body, about which a body having general motion may be considered to be rotating at a given instant.

instantaneous frequency (*Telecomm.*). That calculated from the instantaneous rate of change of angle of a waveform on a time base. In any oscillation, the rate of change of phase divided by 2π.

instantaneous fuse (*Mining*). Rapid-burning, as distinct from slow fuse. Ignition proceeds at a few kilometres per second, but is slower than that of *detonating fuse.*

instantaneous power (*Elec. Eng.*). For a circuit or component, the product of the instantaneous voltage and the instantaneous current. This may not be zero even for a nondissipative (wattless) system on account of stored energy, although in this case its time integral must be zero.

instantaneous specific heat capacity (*Heat*). Specific heat capacity at any one temperature level; *true s.h.c.* to distinguish from *mean s.h.c.*

instantaneous store (*Comp.*). See **zero-access store.**

instantaneous value. A term used to indicate the value of a varying quantity at a particular instant. More correctly it is the average value of that quantity over an infinitesimally small time interval.

instantaneous velocity of reaction (*Chem.*). Rate of reaction, measured by the change in concentration of some key reagent. For instance, in the first order reaction, $A \rightarrow B$, the ivr at any time t is given by $dC_A/dt = -dC_B/dt = -k.C_A$, where C_A, C_B are the concentrations of reagent and product, respectively, and k is the 'velocity constant' of the reaction. Similarly, for the second order reaction, $A+B \rightarrow C$, the ivr is $dC_A/dt = dC_B/dt = -dC_C/dt = -k.C_A.C_B$.

instar (*Zool.*). The form assumed by an Insect during a particular stadium.

in step (*Elec. Eng.*). See **step.**

instinct (*Psychol.*). An innate or natural impulse underlying certain forms of behaviour, e.g., those pertaining to self-preservation. The role of instinct in human behaviour is by no means clear, but the habit of 'explaining' certain human traits, e.g., 'acquisitiveness', as instincts is not acceptable. Broadly, it is complex co-ordination of reflex actions which results in the achievement of adaptive ends without foresight or experience. See also **species-specific behaviour.**

instinctive centre (*An. Behav.*). A hypothetical neural structure which is presumed to act as a unit when excited, sending out nerve impulses governing the occurrence of a species-specific response.

instron tensile strength tester (*Paper*). A precise instrument for measuring tensile strength, in which the rate of strain is constant and independent of the paper sample.

instruction (*Comp.*). Part of the coded programme which is read in to set the circuits for *processing* subsequently inserted data in a computer. See **address.**

instruction code (*Comp.*). Unique set of codings to regulate the operation of a computer.

instruction number (*Comp.*). One used to label a specific instruction in a programme so that the machine can refer back to it subsequently. Such numbers do not indicate the order or sequence in which instructions must be carried out, or the number of the address at which the instruction will be stored. Also **statement number.**

instrument (*Elec. Eng.*). A term generally employed to denote an indicating instrument but also used to denote other pieces of small electrical apparatus.

instrumental conditioning (*An. Behav.*). See **operant conditioning.**

instrumental sensitivity (*Elec. Eng.*). The ratio of the magnitude of the response to that of the quantity being measured, e.g., divisions per milliamp.

instrument approach (*Aero.*). Aircraft approach made using *instrument landing system* (q.v.).

instrumentation (*Automation*). (1) The measuring and control equipment used in the operation of a plant or process. (2) The study of the application of instruments to industrial control processes.

instrument flight rules (*Aero.*). The regulations governing flying in bad visibility under strict flight control. Abbrev. **IFR.**

instrument landing system (*Aero.*). Method of landing aircraft, using master radar beacons and vertical and lateral guidance without visual reference to ground. Abbrev. **ILS.**

instrument range (*Nuc. Eng.*). The intermediate range of reaction rate in a nuclear reactor, when the neutron flux can be measured by permanently installed control instruments, e.g., ion chambers. See **start-up procedure.**

instrument shunt (*Elec. Eng.*). A resistance of appropriate value connected across the terminals of a measuring instrument to extend its range.

instrument transformer (*Elec. Eng.*). One specially designed to maintain a certain relationship in phase and magnitude between the primary and secondary voltages or currents.

insufficient feed (*Print.*). When too little paper is drawn through the press, there is an increase in tension between units or components. Cf. *excess feed.*

insufflation (*Med.*). (1) The action of blowing gas, air, vapour, or powder into a cavity of the body, e.g., the lungs. (2) The powder, etc., used in insufflation.

insufflator (*Med.*). An instrument used for insufflation.

insula (*Zool.*). See **Reil's island.**

insulance (*Elec. Eng.*). See **insulation resistance.**

insulant. See **insulation (2).**

insulated (*Arch.*). Said of any building or column which stands detached from other buildings.

insulated bolt (*Elec. Eng.*). A bolt having a layer of insulating material around its shank.

insulated clip (*Elec. Eng.*). A clip, incorporating an insulated eye, used for supporting flexible electrical connexions.

insulated eye (*Elec. Eng.*). An eye for supporting flexible connexions; it has an insulating bush to prevent these from making contact with the metal.

insulated hanger (*Elec. Eng.*). A hanger for the contact wire of an electric traction system, which insulates the contact wire from the main supporting system.

insulated hook (*Elec. Eng.*). A hook terminating in an insulated eye through which flexible electric connexions may be passed and supported.

insulated metal roofing (*Build.*). Roofing panels made up of light-gauge aluminium or sheet steel, corrugated for strength, and with a top covering of insulation board and bituminous felt. The ceiling underlining is also frequently prefixed.

insulated neutral (*Elec. Eng.*). A term used to denote: (1) the neutral point of a star-connected generator or transformer when it is not connected to earth directly or through a low impedance; (2) the middle wire of a three-wire distribution system when the wire is an insulated cable.

insulated pliers (*Elec. Eng.*). Pliers having the handles covered with insulating material; used by electricians in order to avoid electric shocks when working on live conductors.

insulated-return system (*Elec. Eng.*). A system of supply for an electric traction system, in which both the outgoing and the return conductors are insulated from earth.

insulated screw-eye (*Elec. Eng.*). A screw terminating in an insulated eye through which flexible electric connexions may be passed.

insulated system (*Elec. Eng.*). A system of electric supply in which each of the conductors is insulated from earth for its normal voltage.

insulated wire (*Elec. Eng.*). A solid conductor insulated throughout its length.

insulating beads (*Elec. Eng.*). Beads of glass or similar material strung over a bare conductor to provide an insulating and heat-resisting covering which is also flexible.

insulating board (*Build.*). Fibre board of low density, about 20 lb/ft^3 or 300 kg/m^3, used for thermal insulation and acoustical control.

insulating compound (*Elec. Eng.*). An insulating material which is liquid at fairly low temperatures so that it can be poured into joint-boxes of cables and other similar pieces of apparatus and then allowed to solidify.

insulating materials, classification (*Elec. Eng.*). See **class-A, -B, -C, etc.**

insulating oils (*Elec. Eng.*). Special types of oil having good insulating properties; used for oil-immersed transformers, circuit-breakers, etc.

insulating tape (*Elec. Eng.*). Tape impregnated with insulating compounds, frequently adhesive; used for covering joints in wires, etc.

insulating varnish (*Paint.*). A type of varnish which has high insulating properties. The lighter forms are often called **insulating lacquer.** See **impregnating varnish.**

insulating water-bottle (*Ocean.*). An instrument used for the accurate determination of the temperature of the sea at moderate depths. Also called **Nansen-Pettersson water-bottle.**

insulation. (1) Any means for confining as far as possible a transmissible phenomenon (e.g., electricity, heat, sound, vibration) to a particular channel or location in order to obviate or minimize loss, damage, or annoyance. (2) Any material (also **insulant**) or means suitable for such a purpose in given conditions, e.g., dry air suitably enclosed, polystyrene and polyurethane foam slab, glass fibre, rubber, porcelain, mica, asbestos, hydrated magnesium carbonate, cork, kapok, crumpled aluminium foil, etc. See also **lagging, shielding.**

insulation resistance or **insulance** (*Elec. Eng.*). The resistance between two conductors, or between a conductor and earth, when they are separated only by insulating material. See **fault resistance.**

insulation test (*Elec. Eng.*). A test made to determine the insulation resistance of a piece of apparatus or of a system of electric conductors.

insulation tester (*Elec. Eng.*). An instrument for measuring *insulation resistance* (q.v.).

insulator (*Elec. Eng.*). Unit of insulating material specifically designed to give insulation and mechanical support, e.g., for telegraph lines, overhead traction and transmission lines, plates and terminals of capacitors.

insulator arcing horn (*Elec. Eng.*). A metal projection placed at the upper and lower ends of a suspension-type or other insulator, in order to deflect an arc away from the insulator surface.

insulator cap (*Elec. Eng.*). A metal cap placed over the top of a suspension insulator, which serves to attach it to the next insulator.

insulator pin (*Elec. Eng.*). The central metal support of a pin insulator, or the metal projection on the under side of a suspension insulator, serving to attach it to the cap of the next unit in the string.

insulator rating number (*Elec.*). The voltage (in kilovolts) used in the 'thirty-seconds rain test'.

insulator strength (*Elec. Eng.*). The maximum mechanical and/or the maximum electrical stress, which can be applied.

insulin (*Biochem., Med.*). A protein hormone which is produced by the islets of Langerhans of the pancreas. Widely used in the treatment of *diabetes mellitus*, insulin injection results promptly in a decline in blood glucose concentration and an increase in formation of products derived from glucose. Insulin shock therapy (cf. *electroconvulsive therapy*), inducing coma or convulsions, may be used in the treatment of psychotic disorders. Insulin was the first protein to have its structure completely established and consists of two peptide chains joined by two disulphide bridges.

intaglio (*Glass*). A form of decoration in which the depth of cut is intermediate between deep cutting and engraving. (*Print.*) A printing process in which the ink-carrying areas of the printing surface are hollows below the surface. The thickness of the layer of ink transferred to the paper varies according to the depth of the hollow, giving very rich effects.

integer (*Maths.*). The set of integers is the set of positive and negative whole numbers with zero; i.e., the set $\{\ldots -4, -3, -2, -1, 0, 1, 2, 3, 4, \ldots\}$.

integral (*Maths.*). The function which, when differentiated, equals the given function. If $\varphi(x)$ is the derivative of $f(x)$, then $f(x)$ is the integral of $\varphi(x)$. More precisely, $f(x)$ is the *indefinite integral* of $\varphi(x)$. Also used generally as the adjective from *integer*.

integral calculus (*Maths.*). The inverse of the *differential calculus* (q.v.), its chief concern being to find the value of a function of a variable when its differential coefficient is known. This process, termed *integration*, is used in the solution of such problems as finding the area enclosed by a given curve, the length of a curve, or the volume enclosed by a given surface.

integral convergence test (*Maths.*). The series of positive terms $\sum_{1}^{n} f(r)$, where $f(r)$ is a monotonic decreasing function, and the integral $\int_{1}^{n} f(x)\,dx$, converge and diverge together.

integral dose (*Radiol.*). See **dose.**

integral function (*Maths.*). See **analytic function.**

integral stiffeners (*Aero.*). The stiffening ridges left when an aircraft skin panel is machined from a solid billet.

integral tripack (*Photog.*). Colour film comprising three superimposed layers of emulsion, each producing a colour-separation image in one of the primary colours. Also **monopack.**

integrand (*Maths.*). A function upon which integration is to be effected.

integrated circuit (*Electronics*). A circuit which is fabricated as an assembly of electronic elements in a single structure, which cannot be subdivided without destroying its intended function.

integrating ammeter, voltmeter, wattmeter, etc. (*Instr.*). Instrument which measures the time integral of the named quantity.

integrating circuit or **network** (*Elec. Eng.*). One comprising resistors, capacitors, and/or inductors, used, e.g., with servomechanisms, the output voltage being proportional to the integral of the input voltage. Cf. *differentiating circuit*.

integrating factor (*Maths.*). A multiplying factor which enables a differential equation to be transformed into an exact equation.

integrating frequency meter (*Elec. Eng.*). A meter which sums the total number of cycles of an a.c. supply in a given time. Also called a **master frequency meter.**

integrating meter (*Elec. Eng.*). An electrical instrument which sums up the value of the quantity measured with respect to time.

integrating motor (*Elec Eng.*). Permanent magnet d.c. motor, angular rotation of which is integration of current in armature.

integrating photometer (*Light*). One which measures the total luminous radiation emitted in all directions.

integration (*Maths.*). See **integral calculus.**

integrator (*Comp.*). Unit which performs the mathematical operation of integration, usually with reference to time. (*Telecomm.*) Any device which integrates a signal over a period of time.

integripallial, integripalliate (*Zool.*). Of Mollusca, having a smooth mantle line, i.e., with siphons small or absent.

integument (*Bot.*). (1) One or more cell layers covering the ovule, leaving only a small pore, the *micropyle*. (2) The seed-coat or testa. (*Zool.*) A covering layer of tissue; especially the skin and its derivatives. *adj.* **integumented.**

intelligence quotient (*Psychol.*). Conventional IQ is the ratio, expressed as a percentage, of an individual's *mental age* to his actual age. As this is an unstable estimate, varying with the subject's age, the *deviation* IQ is now used, treating IQ as a *standard score* with a mean of 100 and a *standard deviation* of 16. Abbrev. **IQ.**

intelligibility (*Teleph.*). The percentage number of simple ideas correctly received over the system.

Intene (*Plastics*). TN for a range of British synthetic rubbers based on *polybutadiene* (q.v.).

intensification (*Photog.*). The increase of density or contrast in a negative or print, obtained by further deposit (e.g., of mercury, silver, or chromium) on the exposed parts. See **subproportional, superproportional.**

intensifier electrode (*Electronics*). One which provides post-deflection acceleration in the electron beam of a cathode-ray tube.

intensifying screen (*Radiol.*). (1) Layer or screen of fluorescent material adjacent to a photographic surface, so that registration by incident X-rays is augmented by local fluorescence. (2) Thin layer of lead which performs a similar function for high-energy X-rays or gamma-rays, as a result of ionization produced by secondary electrons.

intensitometer (*Radiol.*). Instrument for measuring intensities of X-rays during exposures.

intensity level (*Acous.*). Level of power as expressed in decibels above an arbitrary zero power level, e.g., sound intensity level on an audition diagram. See **phon, volume unit.**

intensity modulation (*Radar, TV*). Modulation of the luminosity of the spot on the fluorescent screen of a *kinescope* and/or oscilloscope by variation of the current in the beam.

intensity of field (*Elec. Eng.*). The vector quantity by which an electric or magnetic field at a point is measured. Precisely, it is the same number of units of intensity as the force in newtons on a unit charge (unit electric charge or unit fictitious pole) placed at the point in the electric or magnetic field respectively. Also **field strength.**

intensity of magnetization (*Mag.*). Vector of magnetic moment of an element of a substance divided by the volume of that element.

intensity of pressure (*Phys.*). Former term for **pressure.** Force exerted per unit area by a fluid or a solid in contact with another.

intensity of radiation (*Nuc.*). Energy flux, i.e., of photons or particles, per unit area normal to the direction of propagation.

intensity of sound (*Acous.*). The magnitude of a sound wave, measured in terms of the power transmitted, in watts, through unit area normal to the direction of propagation.

intensity (of wave) (*Elec. Eng.*). The energy carried by any sinusoidal disturbance is proportional to the square of the wave amplitude. See **r.m.s. value.**

intensive reflector (*Elec. Eng.*). A reflector for

incandescent lamps, of such a shape as to produce an intense illumination at the point where it is required.

intention movement (*An. Behav.*). A response which, in a behaviour chain, fairly regularly occurs before another response, is termed an intention movement with respect to the latter.

intention tremor (*Med.*). Tremor of the arms on carrying out a voluntary movement, indicative of disease of the nervous system.

interaction (*Phys.*). (1) Transfer of energy between two particles. (2) Interchange of energy between particles and a wave motion. (3) Between waves, see **interference.**

interaction factor (*Telecomm.*). That allowing for insertion loss due to network connected between source and load. This vanishes where iterative impedance of network equals that of load or source.

interaction gap (*Electronics*). The space in a microwave tube in which the electron beam interacts with the wave system.

interaction loss (*Telecomm.*). The decibel loss in power level due to the *interaction factor*.

interaction space (*Electronics*). That in a vacuum tube where electrons effect transfer of energy between electrodes.

interambulacrum (*Zool.*). In *Echinodermata*, especially *Echinoidea*, the region intervening between two ambulacral areas.

interbase current (*Electronics*). The current flowing between the two base connectors in junction-type tetrode transistors.

interbiotic (*Bot.*). Parasitic on or near one or more living organisms.

interbranchial septa (*Zool.*). The stout fibrous partitions separating the branchial chambers in Fish.

interbreeding (*Gen.*). Experimental hybridization of different species or varieties of animals or plants.

intercalare (*Zool.*). A cartilage or ossification lying between the basiventrals, or between the basidorsals of the vertebral column.

intercalarium (*Zool.*). In *Ostariophysi*, one of the Weberian ossicles.

intercalary (*Bot.*). Lying between other bodies in a row, or placed somewhere along the length of a stem, filament, or hypha. (*Zool.*) See **intercalate.**

intercalary cell (*Bot.*). A small cell between two aecidiospores, which disintegrates as the spores ripen, and breaks down as they are set free.

intercalary meristem (*Bot.*). A meristem located somewhere along the length of a plant member, and by its activity giving *intercalary growth.*

intercalate (*Zool.*). To add, to insert, as an *intercalated somite*. *adj.* **intercalary.**

intercaste (*Zool.*). A rare individual intermediate in form between castes, due to environmental factors, including parasitism, in *Isoptera.*

intercavitary X-ray therapy (*Radiol.*). X-ray therapy in which the appropriate part of suitable X-ray apparatus is placed in a body cavity.

intercellular (*Zool.*). Between cells, as the *intercellular* matrix of connective tissue.

intercellular mycelium (*Bot.*). The mycelium of a parasitic fungus which inhabits the intercellular spaces of the host plant.

intercentra (*Zool.*). See **hypapophyses.**

intercept method (*Powder Tech.*). See **Rosiwal intercept method.**

interceptor or **intercepting trap** (*San. Eng.*). A trap fitted in the length of a house drain, close to its connexion to the sewer, which provides a water seal against foul gases rising up into the drain. Also called a **disconnector.**

interchange (*Cyt.*). The mutual transfer of portions between two chromosomes.

interchondral (*Zool.*). Said of certain ligaments and articulations between the costal cartilages.

interclavicle (*Zool.*). In Vertebrates, a bone lying between the clavicles, forming part of the pectoral girdle.

intercolumniation (*Arch.*). The distance between the columns in a colonnade, in terms of the lower diameter of the columns as a unit.

interconnected star connexion (*Elec. Eng.*). See **zigzag connexion.**

interconnecting (*Teleph.*). The commoning of outlets for the bank multiples of selectors on different shelves, when there is an insufficiency of outlets for full availability. See **grading.**

interconnector or **interconnecting feeder** (*Elec. Eng.*). A feeder which serves to interconnect two substations or generating stations, and along which energy may flow in either direction.

intercooler (*Aero.*). (1) A heat-exchanger on the delivery side of a supercharger which cools the charge heated by compression. (2) A secondary heat exchanger used in a cabin airconditioning system for cooling the charge air from the compressor to the turbine of a *bootstrap cold-air unit*. (*Eng.*) A cooler, generally consisting of water-cooled tubes, interposed between successive cylinders or stages of a multi-stage compressor or blower, to reduce the work of compression.

intercostal (*Zool.*). Between the ribs.

intercrystalline failure (*Met.*). This refers to metal fractures that follow the crystal boundaries instead of passing through the crystals, as in the usual transcrystalline fracture. It is frequently due to combined effect of stress and chemical action, but may be produced by stress alone when the conditions permit a certain amount of recrystallization under working conditions.

interdeferential (*Zool.*). Between the vasa deferentia.

interdentil (*Build.*). The space between successive *dentils* (q.v.).

interdependent functions (*Maths.*). See **dependent functions.**

interdialling (*Teleph.*). The dialling, by subscriber or operator, of the required subscriber's number, preceded by one or more code digits to route the call to a directly-connected automatic exchange.

interdigital cyst (*Vet.*). An abscess occurring between the digits of the paw of the dog due to bacterial infection.

interdigital line (*Electronics*). A combined vacuum tube and folded waveguide such that the phase velocity of the electromagnetic wave relative to the axis of the system is reduced to a value corresponding to the velocity of the electron beam. This principle is used in the linear accelerator and as an alternative to the helix in O-type travelling-wave tubes.

interdorsal (*Zool.*). An intercalary element lying between adjacent basidorsals of the vertebral column.

inter-electrode capacitance (*Electronics*). That of any pair of electrodes in a valve, other electrodes being earthed.

interface (*Min. Proc.*). Sharp contact boundary between two phases, either or both of which may be solid, liquid or gaseous. Differs from interphase lacking a diffuse transition zone.

interfacial film (*Min. Proc.*). Special state developed in emulsification, in which oriented molecules of the emulsifying agent are loosely aggregated in such a way as to surround and enclose droplets of one phase of the emulsified mixture.

interfacial surface tension (*Phys.*). The *surface tension* (q.v.) at the surface separating two non-miscible liquids.

interfascicular cambium (*Bot.*). A strand of cambium between two adjacent vascular bundles. The formation of interfascicular cambium is the first stage in the normal secondary thickening of a stem.

interference (*Aero.*). The aerodynamic influence of one body upon another. Usually, the head resistance, or drag, of two bodies placed close together will be greater than the total of their separate drags, because of interference. (*Phys.*) Interaction between two or more trains of waves of the same frequency emitted from coherent sources. A series of stationary nodes and anti-nodes is established, known as an *interference pattern*. (*Radio*) Any signal (whether generated naturally, e.g., atmospherics, or by radio transmitters or electric machinery) other than that to which it is intended that a radio receiver should respond. (*Telecomm.*) The introduction of electromotive forces, and consequent currents in communication circuits, by electrostatic or electromagnetic induction from external currents generally in power-lines or traction systems, or from disturbances on these.

interference colours (*Light*). See **colours of thin films.**

interference eliminator (*Radio*). Same as **interference filter.**

interference factor (*Teleph.*). See **telephone interference factor.**

interference fading (*Radio*). Fading of signals because of interference among the components of the signals which have taken slightly different paths to the receiver.

interference figure (*Crystal.*). More or less symmetrical pattern of concentric rings or lemniscates, cut by a black cross or hyperbolae, exhibited by a section of anisotropic mineral when viewed in convergent light between crossed Nicol prisms. See also **uniaxial** and **biaxial.** Sometimes called **directions image.**

interference filter (*Optics*). A narrow-band filter in which two sheets of optically worked glass have adjacent surfaces half silvered and spaced at such a distance that only a narrow band of wavelengths can pass through, the rest being partially reflected and thus destroyed by interference. (*Radio*) Means of reducing interference, e.g., a tuned rejector circuit for a single steady transmission, or a band-pass filter to reduce the accepted band of frequencies to the minimum. Also called **interference trap, interference eliminator.**

interference fit (*Mech.*). A negative fit, necessitating force sufficient to cause expansion in one mating part, or contraction in the other mating part, during assembly.

interference fringes (*Light*). Alternate light and dark bands which are seen when two beams of homogeneous light having a constant phase relation overlap and illuminate the same portion of a screen. For methods of producing and utilizing optical interference fringes, see **Fresnel's bi-prism, Lloyd's mirror, Fabry and Pérot interferometer, Newton's rings.** See also **interference** (*Phys.*).

interference inverter (*TV*). A circuit in a TV receiver which reverses the drive to the grid of the CRT when there is a large interference pulse.

interference microscope (*Micros.*). An instrument for the measurement of the optical retardation of a microscopic specimen, e.g., a cell, which by conversion gives its total dry mass.

interference pattern (*Phys.*). See **interference.**

interference trap (*Radio*). Same as **interference filter.**

interfering (*Vet.*). An injury inflicted by a horse's foot on the opposite leg during progression.

interferometer (*Instr.*). Instrument in which an acoustic, optical, or microwave interference pattern of 'fringes' is formed and used to make precision measurements, mainly of wavelength.

interferon (*Chem.*). Antibody produced by cells when infected by viruses.

interfilamentar (*Zool.*). Between the filaments, as the junctions between the filaments of the gills in *Lamellibranchiata*.

interfluve (*Geol.*). A ridge separating two parallel valleys.

interfrontal (*Zool.*). In some Amphibians, a bone lying between the frontal and the nasals; also said of a double row of bristles in front of the ocelli in *Diptera*.

interglacial period (*Geol.*). A period of milder climate between two glacial periods.

intergranular corrosion (*Met.*). Corrosion in a polycrystalline mass of metal, taking place preferentially at the boundaries between the crystal grains. This leads to disintegration of the metallic mass before the bulk of the metal has been attacked by the corrosive agent.

intergranular texture (*Geol.*). A texture characteristic of holocrystalline basalts and doleritic rocks, due to the aggregation of augite grains between feldspar laths arranged in a network.

intergular (*Zool.*). In *Chelonia*, an unpaired shield anterior to the gulars.

interhalogen compound (*Chem.*). Compound of two members of the halogen family, e.g., iodine monochloride ICl, iodine heptafluoride IF_7.

interhyal (*Zool.*). In *Teleostei*, a small bone by which the hyoid cornu is articulated to the cartilaginous interval between the hyomandibular and the symplectic.

interkinesis (*Cyt.*). See **intermitosis.**

interlaced fencing (*Build.*). See **interwoven fencing.**

interlaced scanning (*TV*). Sequential scanning of alternate lines in a television picture, so that definition is improved and flicker is reduced by allowing time for detailed areas of phosphors to discharge.

interlamellar (*Zool.*). Between the lamellae, as the junctions between the lamellae of the gills in *Lamellibranchiata*.

interlay (*Typog.*). Paper inserted between a printing plate and its mount in order to raise the plate to type height. Its thickness may be varied to produce increased or decreased impression where necessary.

interleave (*Comp.*). Alternation of parts from different and unrelated messages on a tape.

interleaved transmission (*TV*). Type of colour TV signal in which luminance and chrominance signals are compressed into the same video band of frequencies.

interlobar (*Med.*). Situated or happening between 2 lobes, especially between 2 lobes of the lung.

interlock (*Cinema.*). Adjustment of furniture or scenery in front of a back projection so that the proper perspective is preserved. (*Elec. Eng.*) Arrangement of controls in which those intended to be operated later are disconnected until the preliminary settings are correct, e.g., it is normally impossible to apply the anode voltage to an X-ray tube before the cooling water is flowing. (*Min. Proc.*) Arrangement of switch-gear by which the controlling source prevents premature loading, starting or continuance during partial malfunction where a series of operations is so interlinked as to require smooth on-line operation. (*Nuc. Eng.*) A mechanical and/or electrical device to prevent hazardous operation of a reactor, e.g., to prevent withdrawal of control rods before coolant flow has been established. (*Textiles*) Smooth, double-faced, knitted underwear fabric made in large quantities from fine two-fold or single cotton or spun yarns.

interlock system (*Cinema.*). The arrangement for the synchronous drive of cameras and sound recorders in the Western Electric system; in it, all such drives are switched on to a special main from a three-phase generator, which is started by a motor and brings all the motors on the line up to speed in synchronism.

intermat (*Textiles*). A term applied to fibres which are liable to become matted easily, e.g., wool.

intermaxilla (*Zool.*). See **premaxillary.**

intermaxillary (*Zool.*). Pertaining to the premaxilla; between the maxillaries.

intermediate (*Chem.*). A general term for any chemical compound which is manufactured from a substance (see **primary**) obtained from raw materials, and which serves as a starting material for the synthesis of another product.

intermediate charged vector boson (*Nuc.*). A hypothetical particle which has been suggested as the agent of weak nuclear interaction forces (as the pion and photon are of strong forces and electromagnetic forces).

intermediate circuit (*Radio*). Closed tuned circuit for coupling an antenna to a transmitter or receiver.

intermediate constituent (*Met.*). A constituent of alloys that is formed when atoms of two metals combine in certain proportions to form crystals with a different structure from that of either of the metals. The proportions of the two kinds of atoms may be indicated by formulae, e.g., CuZn; hence these constituents are also known as *intermetallic compounds.*

intermediate coupling (*Nuc.*). Coupling between spin and orbital angular momenta of valence electrons with characteristics between those of *j.j. coupling* and those of *Russell-Saunders coupling* (qq.v.).

intermediate distribution frame (*Teleph.*). A frame inserted between the main distribution frame and the exchange proper to combine additional circuits with the pairs, before they appear on the switchboard in front of the operators. Abbrev. I.D.F.

intermediate frequency (*Radio*). Output carrier frequency of a frequency changer (first detector) in a supersonic heterodyne receiver, adjusted to coincide with the centre of the frequency pass band of the intermediate amplifier. Abbrev. I.F.

intermediate-frequency amplifier (*Radio*). That, in a supersonic heretodyne receiver, which is between the first (frequency changer) and second (final) demodulator, and which provides the main gain and band pass of the receiver.

intermediate-frequency oscillator (*Radio*). That which, in *heterodyne reception,* is combined with the output of the intermediate amplifier for demodulation in the second (final) detector.

intermediate-frequency response ratio (*Radio*). The ratio in decibels of an input signal at the intermediate frequency to one at the required signal frequency which would produce a corresponding output.

intermediate-frequency strip (*Radio*). The chassis of an intermediate-frequency amplifier. Abbrev. **I.F. strip.**

intermediate-frequency transformer (*Radio*). One specially designed for coupling components (and to provide selectivity) in an intermediate-frequency amplifier, the pass band being determined by antiresonance of the windings and coefficient of coupling.

intermediate host (*Zool.*). In the life-history of a parasite, a secondary host—one which is occupied by the young forms, or by a resting stage between the adult stages in the primary host.

intermediate igneous rocks (*Geol.*). Igneous rocks containing from 55% to 66% silica, and essentially intermediate in composition between the acid (granitic) and basic (gabbroic or basaltic) rocks. See **syenite, syenodiorite, diorite.**

intermediate neutrons (*Nuc.*). See **neutron.**

intermediate oxides (*Chem.*). Also called **amphoteric oxides.** See **amphoteric.**

intermediate pressure compressor (*Aero.*). The section of the axial compressor of a turbofan or by-pass turbojet between the *LP* (q.v.) and *HP* (q.v.) sections. It may be on a shaft of its own with a separate turbine, or it may be mounted on either of the other shafts.

intermediate rafter (*Build., Civ. Eng.*). See **common rafter.**

intermediate reactor (*Nuc. Eng.*). One designed so that the majority of fissions will be produced by the absorption of *intermediate neutrons.* Also **epithermal reactor.**

intermediates (*Paper*). See **nip rolls.**

intermediate sight (*Surv.*). Reading on levelling staff, held at point not to be occupied by level other than back- or fore-sight observation from a levelling point between these last.

intermediate switch (*Elec. Eng.*). A switch for controlling a circuit where more than two positions of control are required; it is connected between the two-way switches which must also be used in such a scheme.

intermediate waves (*Radio*). Those whose wavelengths are in the range 50 to 200 m, bridging the medium and short radio wavelengths.

intermediate (or idle) wheel (*Horol.*). A wheel used to connect two other toothed wheels when their distance apart is such that without the use of an intermediate wheel their diameters would be too large.

intermedin (*Physiol.*). A protein hormone produced by the *pars intermedia* of the pituitary gland, with melanophore expanding properties.

intermedium (*Zool.*). A small bone of the proximal row of the basipodium, lying between the tibiale and fibulare, or between the radiate and ulnare.

intermenstrual (*Med.*). Occurring between two menstrual periods.

intermetallic compounds (*Met.*). See **intermediate constituents.**

intermission (*Med.*). Temporary cessation, as of fever or of the normal pulse.

intermitosis (*Cyt.*). The period between two mitotic divisions of a cell. Also **interphase.**

intermittent claudication (*Med.*). Intermittent lameness. See **dysbasia angiosclerotica.**

intermittent control (*Telecomm.*). Control system in which controlled variable is monitored periodically, an intermittent correcting signal thus being supplied to the controller.

intermittent duty (*Telecomm.*). The conditions of use for a component operated at its intermittent rating.

intermittent earth (*Elec. Eng.*). An accidental earth connexion which is present intermittently.

intermittent fever (*Med.*). See **malaria.**

intermittent filtration (*San. Eng.*). The *land treatment* (q.v.) process of sewage purification, in which the land is drained artificially by ordinary earthenware pipes. Cf. *broad irrigation.*

intermittent jet (*Aero.*). See **pulse jet.**

intermittent loading, intermittently loaded cable (**or circuit**) (*Cables*). A cable, generally for submarine telephony, in which the conductors are loaded continuously for sections of their length only. By staggering the intermittent loading on a number of conductors in a multi-cored cable, economy in weight and cost is effected.

intermittent printing (*Cinema.*). Printing by means of a step-by-step cinematograph printer, which prints a frame at a time.

intermittent rating (*Elec. Eng.*). The specified power handling capacity of a component or instrument under specified conditions of non-continuous usage. See **continuous rating, half-hour rating, one-hour rating.**

intermodillion (*Build.*). The space between successive *modillions* (q.v.).

intermodulation (*Telecomm.*). Undesired modulation of all frequencies with each other in passing through a nonlinear element in a transmission path, giving a blur to a desired signal.

intermodulation distortion (*Acous.*). Amplitude distortion in which the intermodulation products are of greater importance than the harmonic products, as in AF amplifiers for high-quality speech or music.

intermuscular (*Zool.*). Said of the pairs of delicate bones attached by fibrous union to the junctions of the neural arches with the centra, and extending outwards in the septa between the myomeres.

intermyotomic (*Zool.*). Said of vertebrae which arise by the fusion of the cranial elements of one somite with the caudal elements of the preceding somite. Cf. *intrasegmental.*

internal capacitance (*Elec.*). Same as **inter-electrode capacitance.**

internal characteristic (*Elec. Eng.*). A curve showing the relation between the load on an electric generator and the internal e.m.f.

internal-combustion engine (*Eng.*). An engine in which combustion of a fuel takes place within the cylinder, and the products of combustion form the working medium during the power stroke. See **compression-ignition engine, diesel engine, gas engine, petrol engine.**

internal compensation (*Chem.*). Neutralization of optical activity within the molecule by the combination of two enantiomorphous groups, e.g., in neso-tartaric acid.

internal conductor (*Elec. Eng.*). The inner conductor of a concentric cable. Also called **inner conductor.**

internal conversion (*Nuc.*). Nuclear transition where energy released is given to orbital electron which is usually ejected from the atom (conversion electron), instead of appearing in the form of a γ-ray photon. The *conversion coefficient* for a given transition is given by the ratio of conversion electrons to photons.

internal e.m.f. (*Elec. Eng.*). The e.m.f. generated in an electric machine; voltage appearing at the terminals is the internal e.m.f. minus any voltage drop caused by the current in the machine.

internal energy (*Heat*). In a thermodynamic system, the difference between the energy supplied to the system (work done on it) and the energy supplied by the system (work done by it) is conserved by the system as its internal energy. (The First Law of Thermodynamics.) This internal energy takes the form of heat, as measured by its temperature, latent heat, as demonstrated by a change of state, or the repulsive forces between its molecules, seen as expansion.

internal-expanding brake (*Eng.*). See **expanding brake.**

internal factor (*Bot.*). Any factor which depends on the genetic constitution of the plant and which influences its growth and development.

internal flue (*Eng.*). A furnace tube, or fire tube, running through the water space of a boiler.

internal focusing telescope (*Surv.*). One in which focusing is effected by the movement of an internal concave lens fitted between the object glass and the eyepiece, both of which are fixed.

internal friction (*Eng.*). The reaction which opposes flow in liquids and gases and may thus be said to be the phenomenon underlying *viscosity* (q.v.).

internal gear (*Eng.*). A spur gear in which teeth, formed on the inner circumference of an annular wheel, mesh with the external teeth of a smaller pinion. Both wheels revolve in the same direction.

internal grinding (*Eng.*). The grinding of internal cylindrical surfaces by an abrasive wheel, which is either traversed along the revolving work or (in addition) given a planetary motion, the work being fixed.

internal impedance (*Elec.*). The *output impedance* of an amplifier or generator.

internal indicator (*Chem.*). An indicator which is dissolved in the solution in which the main reaction takes place.

internal longitudinal bar (*Zool.*). One of the bars crossing the inner surface of the branchial basket in *Tunicata*, bearing papillae which project into the branchial cavity.

internally fired boiler (*Eng.*). A boiler whose fire box or furnace is inside the boiler and surrounded by water, as in the Lancashire, marine, and locomotive types.

internal memory (*Comp.*). The memory circuits and equipment of a computer which are under its direct control.

internal pair production (*Nuc.*). Production of electron positron pair in the coulomb field of a nucleus. For transitions where the excitation energy released exceeds 1·02 MeV, this process will be competitive with both internal conversion and γ-ray emission. It occurs most readily in nuclei of low atomic number when the excitation energy is several MeV.

internal phloem (*Bot.*). Phloem lying between the xylem and the centre of the stem.

internal resistance (*Elec. Eng.*). That of a valve, battery, photocell, etc., resulting in a drop in terminal voltage when current is taken. See **Thévenin's theorem.** Also **source resistance.**

internal respiration (*Zool.*). See **respiration, internal.**

internal sac (*Zool.*). The characteristic organ by which attachment is effected, prior to metamorphosis in *Ectoprocta.*

internal screw-thread (*Eng.*). A screw-thread cut on the inside of a cylindrical surface, as distinct from an *external screw-thread.* Also called a **female thread.**

internal secretion (*Zool.*). A secretion which is poured into the blood vessels, or into the canal of the spinal cord; a hormone.

internal stress (*Met.*). Residual stress existing between different parts of metal products, as a result of the differential effects of heating, cooling or working operations, or of constitutional changes in the solid metal.

internal voltages (*Electronics*). Those, such as *contact potential* or *work function,* which add an effect to the external voltages applied to the electrodes of a valve.

internasal septum (*Zool.*). See **mesethmoid.**

International Air Transport Association (*Aero*). Association founded in Havana (1945) for the promotion of safe, regular and economic air transport, to foster air commerce and collaboration among operators, and to co-operate with *ICAO* and other international organizations. Abbrev. **IATA.**

international ampere (*Elec.*). The current which, when passed through a solution of silver nitrate in water, will deposit silver at the rate of 0·001 118 000 gramme per second. Replaced by MKSA, followed by SI, **ampere.**

international ångström (*Phys.*). A unit which, although very nearly equal to the ångström unit (10^{-10} m), is defined in a different way. It is such that the red cadmium line at 15°C and 760 mmHg pressure would have a wavelength of 6438·4696 IÅ. Formerly the reference standard for metrology and spectroscopy. Abbrev. **IÅ.**

International Atomic Energy Agency (*Nuc.*). Organization for promoting the peaceful uses of nuclear power. Abbrev. **IAEA.**

international candle or **candle** (*Light*). Former unit of luminous intensity stated in terms of a point source, which was expressed in an equivalence of a standard lamp burning under specified conditions. Replaced by **candela.**

International Civil Aviation Organization (*Aero.*). An intergovernmental coordinating body, which has its headquarters at Montreal, Canada, for the regulation and control of civil aviation and the coordination of *airworthiness* requirements and other safety measures on a world-wide basis. Abbrev. **ICAO.**

international electrical units (*Elec.*). Units (amp, volt, ohm, watt) for expressing magnitudes of electrical quantities, adopted internationally until 1947; replaced first by *MKSA*, later by *SI units*.

international ohm (*Elec.*). Former unit of resistance. It is the resistance offered, at the temperature of melting ice, to an unvarying electric current by a column of mercury 14·4521 g in mass, of uniform cross-sectional area, and 106·300 cm in length.

international paper sizes. See **ISO paper sizes.**

international screw-thread (*Eng.*). A metric system in which the pitch of the thread is related to the diameter, the thread having a rounded root and flat crest.

International Standard Atmosphere (*Aero.*). A standardized atmosphere, adopted internationally, used for comparing aircraft performance: mean sea-level temperature 15°C at 1013·2 millibars, lapse rate 6·5°C/km altitude up to 11 km (ISA tropopause), above which the temperature is assumed constant at −56·5°C in the stratosphere. Abbrev. **ISA.**

international system of units. See **SI units.**

International Telecommunication Union (*Telecomm.*). Organization governing international telephone, telegraph, and radio services.

international practical temperature scale (*Heat*). A practical scale of temperature which is defined to conform as closely as possible with the thermodynamic scale. Various fixed points were defined initially using the gas thermometer, and intermediate temperatures are measured with a stated form of thermometer according to the temperature range involved. The majority of temperature measurements on this scale are now made with platinum resistance thermometers.

international volt (*Elec.*). Former unit of potential difference. It is the potential difference which, when applied to a conductor having a resistance of one international ohm, produces a current of one international ampere.

interneurals (*Zool.*). See **interspinals.**

internode (*Bot.*). The length of a stem between two successive nodes. (*Zool.*) The part of a nerve between two adjacent nodes of Ranvier.

internuncial (*Biol.*). Interconnecting, e.g., neurone of central nervous system interposed between afferent and efferent neurones of a reflex arc.

interoceptor (*Zool.*). A sensory nerve-ending, specialized for the reception of impressions from within the body. Cf. *exteroceptor*.

interopercular (*Zool.*). In Fish, a ventral membrane bone supporting the operculum.

interparietal (*Zool.*). A median dorsal membrane bone of the Vertebrate skull, situated between the parietals and the supraoccipital.

interpenetration twins (*Min.*). Two or more crystals united in a regular fashion, according to a fixed plan (the *twin law*), the individual crystals appearing to have grown through one another. Cf. *juxtaposition twins*.

interphase (*Cyt.*). See **intermitosis.** (*Min. Proc.*) Transition zone between two phases in a system (solid/liquid, liquid/liquid, liquid/gas). In solid/fluid system; zone of shear through which the physical or chemical qualities of the contacting surfaces migrate toward one another.

interphase transformer (*Elec. Eng.*). A centre-tapped, iron-cored reactor, connected between the two star-joints of the supply transformer for a double three-phase mercury-arc rectifier. The centre tap forms the d.c. negative terminal. Also called **interphase reactor, absorption inductor,** and **phase equalizer.**

interplane struts (*Aero.*). In a multiplane structure, those struts, either vertical or inclined, connecting the spars of any pair of planes, one above the other.

interplanetary matter (*Astron.*). See **meteor streams, Poynting-Robertson effect, solar wind, codiacal light.**

interplantation (*Zool.*). In experimental zoology, the culture of a part or organ in a cavity of the body of an older animal; cf. *explantation*.

interpleurite (*Zool.*). In Insects, a pleural intersegmentalium.

interpolation (*Maths.*). The estimation of the value of a function at a particular point from values of the function on either side of the point. Cf. *extrapolation*.

interpolation of contours (*Surv.*). The joining of points derived from spot levelling or in conformity with existing mapped contour lines, assuming that the recorded data disclose all abrupt changes in ground level and configuration.

interpolator (*Comp.*). Apparatus for comparing cards in separate packs and merging them.

interpole (*Elec. Eng.*). See **compole.**

interpole motor (*Elec. Eng.*). An electric motor fitted with *compoles*.

interpreter (*Comp.*). A procedure in digital computers by which a program stored in an arbitrary code is translated into an active language and carries out the operations as they are translated.

inter-radium (*Zool.*). See **inter-radius.**

inter-radius (*Zool.*). In a radially symmetrical animal, a radius which bisects the angle between two adjacent per-radii.

inter-renal body (*Zool.*). In selachian Fish, a ductless gland which lies between the kidneys and corresponds to the cortex of the suprarenal gland of higher Vertebrates.

interrogation (*Telecomm.*). Transmission of a pulse to a *transponder* (pulse repeater).

interrogator responsor (*Telecomm.*). An electronic

combination of an interrogator with the responder which receives and displays the answering pulses.

interrupted (*Bot.*). Said of organs which are not evenly spaced out along an axis.

interrupted continuous waves (*Radio*). Electromagnetic waves radiated from an antenna, driven from an r.f. oscillator the output from which is interrupted periodically at an audible frequency, so that the received signal is directly audible in the telephones after rectification without heterodyning. Abbrev. **I.C.W.** Also called **chopped continuous waves, tonic train.**

interruptedly pinnate (*Bot.*). Said of a pinnate leaf when pairs of small leaflets alternate with pairs of larger leaflets.

interrupter (*Elec. Eng.*). A device which interrupts periodically the flow of a continuous current, such as the mechanical 'make-and-break' of an induction coil.

interruption (*Comp.*). Signal-controlled temporary halt to a specific processing routine used, e.g., in time sharing.

interscapular (*Anat.*). Between the two shoulder blades.

intersection (*Maths.*). The intersection of sets A and B is the set of elements x such that x is in A and x is also in B. (*Surv.*) Plane-table surveying in which the plane table is set up consecutively at each end of a measured base line, rays are drawn on paper at each set-up to show the direction of the point that it is required to fix on the plan, the intersection of these rays giving the position of the point.

intersection angle (*Surv.*). The angle of deflection, as measured at the intersection point between the straights of a railway or highway curve.

intersection point (*Surv.*). The point at which the straights of a railway or highway curve would meet if produced.

intersection theory (*Textiles*). A theory used to determine the number of threads per inch in a woven fabric of compact structure, the number being determined from the diameter of the yarn and the number of interlacings of the threads.

intersegmentalia (*Zool.*). In Insects, small detached sclerites between adjacent segments.

intersegmental membrane (*Zool.*). The flexible infolded portion of the cuticle between adjacent definitive segments to allow the freedom of movement of the body in *Arthropoda.*

intersertal texture (*Geol.*). The texture characterized by the occurrence of interstitial glass between divergent laths of feldspar in basaltic rocks.

inter-setting (*Print.*). The introduction on a rotary press of a pre-printed web in register.

intersex (*Biol.*). An individual which exhibits characters intermediate between those of the male and those of the female of the same species. It has been shown that between these normally accepted limits lies a whole spectrum of sex differentiation dependent upon genetic, hormonal, organic and psychological development, male and female homosexuality and transvestism being merely the more commonly recognized steps in the gradation.

intersheath (*Cables*). Cylindrical electrodes in the interior of a cable dielectric, used for the purpose of keeping the variation of stress at a minimum. The intersheaths must be kept at certain potentials in order to achieve this purpose. See **stress.**

interspace (*For.*). In resin tapping, the space between two faces.

interspecific (*Cyt.*). Said of a cross between two separate species.

interspersed (*Bot.*). Said of marginal ray cells which are scattered among other cells.

interspinals (*Zool.*). In Fish, the basal element or elements of a pterygophore. Also called **interneurals.**

interstage coupling (*Radio*). See **intervalve coupling.**

interstation interference (*Radio*). That from another transmitter on the same or an adjacent wavelength, as distinct from atmospheric interference.

interstation noise suppression (*Radio*). Suppression of radio receiver output when no carrier is received.

interstellar hydrogen (*Astron.*). The commonest element in interstellar space; no absorption lines can be detected, but hydrogen emission lines are visible in the spectra of bright nebulae; neutral hydrogen gives the 21-cm wavelength radio emission, which enables the distance of the hydrogen clouds to be measured.

interstellar matter (*Astron.*). Very low density clouds of dust and gas in interstellar space; detected by absorption lines in stellar spectra and by microwave emission lines. Hydrogen is abundant, but atoms of calcium, sodium and other metals have been identified, and a number of simple radicals of the type OH, CH, CN, NH_3, etc.

intersternite (*Zool.*). In Insects, a sternal intersegmentalium.

interstice (*Chem.*). Space between atoms in a lattice where other atoms can be located, e.g., in close-packed metallic lattices.

interstitial (*Zool.*). Occurring in the interstices between other structures; as the *interstitial* cells of *Coelenterata*, which are small rounded embryonic cells occurring in the interstices between the columnar cells forming the ectoderm and endoderm.

interstitial cell stimulating hormone (*Physiol.*). Abbrev. **ICSH.** See **luteinizing hormone.**

interstitial compounds (*Met.*). Metalloids in which very small atoms of nonmetallic elements (H, B, C, N,) occupy positions in the interstices of metal lattices. In general, the structure of the metal is preserved, though somewhat distorted. They are often (especially those of Group IV and V metals) characterized by exceptionally high melting-points and hardness, by chemical inertness and by metallic lustre and conductivity.

inter-switchboard line (*Teleph.*). See **tie line.**

intertentacular (*Zool.*). Between the tentacles, as the *intertentacular* organ of *Ectoprocta*, consisting of a ciliated tube which lies between, and at the base of, the tentacles leading from the coelom to the exterior, and serves for the escape of the genital products.

intertergite (*Zool.*). In Insects, a tergal intersegmentalium.

interthecial (*Bot.*). Growing between asci.

inter-tie (*Carp.*). In a trussed partition or in half-timbered work, a horizontal timber framed between posts and located between floor levels.

intertrabecula (*Zool.*). In some Birds, a cartilaginous plate lying between the anterior ends of the trabeculae during skull development.

intertrack bond (*Elec. Eng.*). A conductor for connecting electrically the rails of separate tracks on electric railways or tramways to reduce the total resistance of the return path.

intertrigo (*Med.*). Excoriation of the skin in the skin folds, usually due to excessive moisture.

intertrochanteric (*Anat.*). Situated between the two trochanters of the upper part of the femur.

Intertype (*Typog.*). A typesetting machine, using circulating matrices, similar in most respects to the *Linotype* (q.v.). See also **composing machines.**
interval (*Acous.*). Ratio of the frequencies of two sounds; or in some cases its logarithm. See **semitone.** (*Maths.*) See **closed-, open-.**
intervalve coupling (*Radio*). Those components which allow one active device (valve or transistor) to drive another (cascade). Also **interstage coupling.**
intervalve transformer (*Elec. Eng.*). One designed to couple the anode circuit of one valve to the grid circuit of another valve, with a specified frequency response.
interventral (*Zool.*). An intercalary element lying between adjacent basiventrals of the vertebral column.
intervertebral (*Zool.*). Between the vertebrae; said of the fibro-cartilage disks between the vertebral centra in Crocodiles, Birds, and Mammals.
interwoven fencing (*Build.*). Solid wood, or metal, fencing built up of very thin boards interlaced. Also called **interlaced fencing** or **wovenboard.**
interxylary phloem (*Bot.*). A strand of secondary phloem surrounded by secondary xylem.
intestine (*Zool.*). In Vertebrates, that part of the alimentary canal leading from the stomach to the anus; in Invertebrates, that part of the alimentary canal which corresponds to the Vertebrate intestine, or was thought by the early investigators so to correspond. *adj.* **intestinal.**
intima (*Zool.*). The innermost layer of an organ or part, e.g., the innermost layer of the wall of a blood vessel.
intine (*Bot.*). (1) The inner layer of the wall of a pollen grain. (2) The endospore in spores of *Bryophyta.*
intra-alar (*Zool.*). Said of a paired row of bristles on the dorso-lateral aspect of the mesothorax in *Diptera.*
intra-annular tautomerism (*Chem.*). The redistribution of bonds in a ring of carbon or other atoms, as in the benzene nucleus.
intracapsular (*Anat.*). Situated within a capsule, especially within the ligamentous joint capsule enveloping the head and neck of the femur.
intracavitary (*Radiol.*). Applied within cavities of the body; said, e.g., of radium applied in the cavity of the uterus, also of irradiation of part of the body through natural or artificial body cavities.
intracellular (*Biol.*). Within the cell.
intracellular body cavity (*Zool.*). The body cavity in some *Nematoda,* formed of the confluent vacuoles of a small number of large cells.
intracerebral (*Anat.*). Situated in the substance of the brain.
intracervical (*Anat.*). Situated in, or applied to, the canal of the cervix uteri (the lowest part or neck of the uterus).
intracranial (*Anat.*). Situated within the skull.
intradermal (*Anat.*). Situated in, or introduced into, the skin.
intrados (*Build., Civ. Eng.*). The under surface of an arch. Also *soffit* (q.v.). See **extrados.**
intrafusal (*Zool.*). Situated within the spindle, as the fibres in a neurotendinous junction, or the fasciculi in a neuromuscular spindle.
intrahepatic (*Anat.*). Situated or occurring in the substance of the liver.
intralamellar tissue (*Bot.*). See **trama.**
intramammary (*Anat.*). Situated or occurring in the breast.
intramatrical (*Bot.*). Said of a parasitic fungus which lives inside the host cell, or in the matrix.
intramedullary (*Anat.*). In the substance of the medulla oblongata (the brain stem); situated or occurring in the substance of the spinal cord.
intramercurial planet (*Astron.*). A hypothetical planet, to which the name Vulcan was given, believed by some to exist between the Sun and Mercury. Its existence was postulated to explain certain motions of Mercury's perihelion, which have since been accounted for by Einstein's theory of gravitation.
intramolecular respiration (*Bot.*). See **anaerobic respiration.**
intranarial epiglottis (*Zool.*). In *Cetacea,* the epiglottis and arytenoids which are prolonged to form a tube extending into the nasal chambers.
intranuclear forces (*Nuc.*). Those operative between nucleons at close range comprising *short range forces* and *coulomb forces.* According to the hypothesis of charge independence, the former are always the same for two nucleons of corresponding angular momentum and spin regardless of whether these are protons or neutrons. See **isotopic spin.**
intra-ocular (*Anat.*). Situated within the eyeball.
intraperitoneal (*Anat.*). Situated in, or introduced into, the peritoneal cavity.
intrapleural (*Zool.*). Within the thoracic cavity.
intrasegmental (*Zool.*). Said of vertebrae which arise by the fusion of the cranial and caudal elements of the same somite. Cf. *intermyotomic.*
intratarsal (*Zool.*). Within the tarsus; in Reptiles, a joint of the tarsus.
intrathecal (*Anat.*). Within the sheath of membranes investing the spinal cord.
intrathyroid (*Zool.*). Within the thyroid; a cartilage lying between and joining the laminae of the thyroid cartilage, during the early stages of development.
intratracheal (*Anat.*). Within, or introduced into, the trachea.
intratubal (*Anat.*). Situated within a Fallopian tube.
intra-uterine (*Anat.*). Situated within, or developing within, the uterus.
intra-uterine (contraceptive) device (*Med.*). A contraceptive device of relatively high efficacy, inserted in the uterus; consists usually of a spiral or a loop-shaped copper-coated filament. Its action, though not completely understood, may be physiological. Abbrevs. **IUD, IUCD.**
intravenous (*Anat.*). Within, or introduced into, a vein.
intravesical (*Zool.*). Within the bladder.
intra-vitam staining (*Biol., Micros.*). The artificial staining of living cells. Usually by injection of the stain into a circulatory system.
intravitelline (*Zool.*). Within the yolk of an ovum.
intraxylary phloem (*Bot.*). See **internal phloem.**
intrinsic (*Crystal.*). See **idiochromatic.** (*Zool.*) Said of appendicular muscles of Vertebrates which lie within the limb itself, and originate either from the girdle or from the limb-bones. Cf. *extrinsic.*
intrinsic angular momentum (*Nuc.*). The total spin of atom, nucleus, or particle as an idealized point, or arising from orbital motion. When quantized, the former and latter are, respectively,

$$\frac{1}{2}\frac{h}{2\pi} \quad \text{and} \quad \frac{h}{2\pi},$$

where h = Planck's constant.
intrinsic conduction (*Electronics*). That in a semiconductor when electrons are raised from a filled band into the conduction band by thermal energy, so producing hole-electron pairs. It increases rapidly with rising temperature.

intrinsic crystal (*Crystal.*). One whose photoelectric properties do not depend on impurities.

intrinsic energy (*Chem.*). The store of energy possessed by a material system. It is not usually possible to determine its absolute magnitude, but changes in its value can be measured. Changes in the intrinsic energy of a system depend only upon the initial and final conditions, and are therefore independent of the paths of change.

intrinsic equation (*Maths.*). Of a curve: an equation connecting the arc length s of any point on the curve with the gradient ψ at the point. Such an equation is independent of any coordinate system and is an inherent property of the curve.

intrinsic factor (*Biochem.*). One or more mucoproteins produced by the gastric mucosa and essential for the normal absorption of dietary vitamin B_{12}. Defective gastric secretion of *intrinsic factor* results in pernicious anaemia.

intrinsic impedance (*Elec.*). Wave impedance depending on the medium alone, i.e., $Z_0^2=\mu/\varepsilon$, where μ is the absolute permeability and ε the absolute permittivity of the medium.

intrinsic induction (*Mag.*). Synonymous with *intensity of magnetization* (q.v.). (*N.B.*—In unrationalized units, it is 4π times this quantity.)

intrinsic mobility (*Electronics*). The mobility of electrons in an intrinsic semiconductor.

intrinsic rate of natural increase (*Ecol.*). The overall population growth rate under unlimited environmental conditions. The exact value depends on population structure. The maximum value is sometimes termed **biotic potential.**

intrinsic semiconductor (*Electronics*). See **semiconductor.**

introitus (*Anat.*). Entry to a cavity (e.g. vagina).

introjection (*Psychol.*). A mental swallowing. The function of the mind whereby it incorporates objects with their accompanying characteristics from the outside world and adopts them as its own; also an important factor in the development of the *superego* (q.v.).

intromission (*Med.*). The insertion of one part into another, especially of the penis into the vagina.

intromittent (*Zool.*). Adapted for insertion, as the copulatory organs of some male animals.

introrse (*Bot.*). Said of an anther which opens towards the centre of the flower. (*Zool.*) Directed or bent inwards.

introspection (*Psychol.*). The observation by a person of his own mental processes. In fact, it is usually *retrospection* rather than introspection in the strictest sense.

introvert (*Psychol.*). In Jungian theory, a person whose libido is inwardly directed resulting in the shunning of interpersonal relations and absorption in egocentric thoughts. Generally, a person more interested in his thoughts and feelings than in the external world and social activity. See also **extravert.** *n.* **introversion.** (*Zool.*) A structure or part of the body which may be involuted, as the proboscis of a Nemertinean worm, or turned inside-out.

intrusions (*Geol.*). Bodies of igneous rocks of varying size and structure which, in the condition of magma, were intruded into the pre-existing rocks of the earth's crust. Such rocks are *intrusive rocks*. Cf. *extrusive rocks*.

intubation (*Med.*). The introduction of a tube, especially through the larynx into the trachea.

intuitive thought (*Psychol.*). The thought which is typical of a child from about 4–7 years (Piaget). The child is still dominated by perception, overemphasizes various elements of thought and tends to be *egocentric*.

intumescence (*Bot.*). A localized pathological swelling consisting chiefly of parenchyma. (*Chem.*) The swelling of crystals on heating, often with the violent escape of moisture. (*Med.*) The process of swelling; a swelling.

intussusception (*Bot.*). Growth of the cell wall by the interpolation of submicroscopic particles between those already present in the wall. (*Med.*) The pushing down, or invagination, of one part of the intestine into the part below it.

inulin (*Chem., Med.*). $(C_6H_{10}O_5)_x$. A polysaccharide obtained from the tubers of the dahlia and the roots of chicory (*Chicorium intybus*); it is hydrolysed by water into D-fructose. Used in renal function.

inulinase (*Chem.*). An enzyme hydrolysing inulin.

invagination (*Zool.*). Insertion into a sheath; the development of a hollow ingrowth; the pushing-in of one side of the blastula in embolic gastrulation. Cf. *evagination*. *adj.* **invaginate.**

Invar (*Met.*). TN for iron-nickel alloy. Composed of 36% nickel, 63·8% iron and 0·2% carbon. Coefficient of thermal expansion is very small. Used for measuring-tapes, tuning-forks, pendulums, and in instruments.

invariable plane (*Astron.*). A certain plane which remains absolutely unchanged by any mutual action between the planets in the solar system; defined by the condition that the total angular momentum of the system about the normal to this plane is constant. The plane is inclined at 1° 35′ to the ecliptic.

invariant (*Chem.*). Possessing no *degrees of freedom* (q.v. (1)). (*Maths.*) A function of the coefficients of the equation of a curve or surface, unaltered by a *projective transformation* (q.v.).

inventory (*Nuc. Eng.*). Total quantity of material in a reactor, e.g., of neutrons or of fissile material.

inverse (*Bot.*). Said of the condition of an embryo in which the radicle is turned towards a point in the seed at the opposite end to the hilum. (*Maths.*) If S is a set containing an *identity* element e with respect to an operation $*$ in S, then the element x of S is said to have an inverse x' in S, if $x*x' = e = x'*x$. For the usual addition of real numbers, the inverse of x is written $-x$ and is called the negative of x. For the usual multiplication of real numbers, the inverse of the nonzero real number x is written $\frac{1}{x}$ or x^{-1} and is called the reciprocal of x. Note that

$$x+(-x) = 0 = (-x)+x,$$
$$xx^{-1} = 1 = x^{-1}x.$$

inverse current (*Electronics*). See **reverse current.**

inverse function (*Maths.*). If $y=f(x)$ and a function $\varphi(y)$ is found such that $x=\varphi(y)$, then $\varphi(y)$ is the inverse function of $f(x)$, e.g., $\log_e x$ is the inverse function of e^x.

inverse hyperbolic function (*Maths.*). If y is an hyperbolic function of x, then x is the inverse hyperbolic function of y, e.g., if $y=\sinh x$, the inverse hyperbolic function is $x=\text{arc}\sinh y$ or $x=\sinh^{-1} y$, where $\text{arc}\sinh y$ or $\sinh^{-1} y$ is the number whose sinh is x.

inverse networks (*Elec. Eng.*). A pair of two-terminal networks whose impedances are such that their product is independent of frequency.

inverse of a matrix (*Maths.*). For a square matrix, the quotient of the adjoint of the matrix and its determinant provided the determinant is not zero. The inverse of the square matrix A has the same order as A and is denoted by A^{-1}. $AA^{-1} = I = A^{-1}A$, where I is the unit matrix having the same order as A.

inverse of a one-to-one mapping (*Maths.*). If f is a one-to-one mapping from set S on to set T, the inverse of f is the mapping g from T on to S.

inverse power factor (*Elec. Eng.*). A term sometimes used to denote sec φ (=1/cos φ).

inverse segregation (*Met.*). A type of segregation in which the content of impurities, inclusions, and alloying elements in metals tends to decrease from the surface to the centre. See also **segregation, normal segregation.**

inverse-speed motor (*Elec. Eng.*). See **series-characteristic motor.**

inverse square law (*Phys.*). The intensity of a field of radiation is inversely proportional to the square of the distance from the source. Applies to any system with spherical wavefront and negligible energy absorption.

inverse time-lag (*Elec. Eng.*). A time-lag which is approximately proportional to the inverse of the current causing its operation. Also called **inverse time-element, inverse time-limit.**

inverse trigonometrical function (*Maths.*). If y is a trigonometrical ratio of the angle x, then x is the inverse trigonometrical function of y, e.g., if $y = \sin x$, the inverse trigonometrical function is $x = \text{arc} \sin y$ or $\sin^{-1}y$, where arc sin y or $\sin^{-1}y$ is the angle whose *sine* is y.

inverse voltage (*Elec. Eng.*). That generated across a rectifying element during the half cycle when no current flows. Its maximum value is the *peak inverse voltage*.

inversion. (*Bot.*). The turning inside-out during development of a colony in some algae that form coenobia. (*Chem.*) The formation of a laevorotatory solution of fructose and glucose by the hydrolysis of a dextrorotatory solution of sucrose (cane-sugar). (*Cyt.*) The reversal in position of a portion of a chromosome. (*Maths.*) The operation of deriving from a first set of points P a second set of points P' by the rule $OP.\ OP' = a^2$ where O is a fixed point (the *centre of inversion*) and a is a constant (the *radius of inversion*). (*Meteor.*) Inversion of the usual temperature gradient in the atmosphere, the temperature increasing with height. Inversions are of frequent occurrence near the ground on clear nights and in anticyclones, often causing dense smoke fogs over cities. (*Teleph.*) See **inverted speech.**

inversion of relief (*Geol.*). A condition whereby synclinal ridges are separated by anticlinal depressions.

invert (*Civ. Eng., Plumb., etc.*). The lowest part of the inner surface of the cross-section of a non-vertical drain or sewer at any given point, and related to the datum level.

invertase (*Chem.*). A plant enzyme which hydrolyses cane-sugar; it is present in all yeasts except *S. capsularis* and *S. octosporus*, both wine yeasts. Also called **invertin, saccharase, sucrase.**

Invertebrata (*Zool.*). A collective name for all animals other than those in the phylum *Chordata*; that is, all those animals that do not exhibit the characteristics of vertebrates, namely possession of a notochord or vertebral column, ventral heart, eyes, etc. Some animals near the chordate boundary line, e.g., *Hemichorda* and *Urochorda*, are sometimes regarded as invertebrates. The term is not used as a scientific classification.

inverted (*Bot.*). See **anatropous, inverse.**

inverted amplification factor (*Electronics*). The ratio of the differential change in grid voltage, per unit differential change in anode voltage, necessary to maintain a constant grid current in a thermionic triode.

inverted arch (*Civ. Eng.*). An arch having the crown below the line of the springings, e.g., the floor of a tunnel, in order to distribute the pressure on the walls over a greater area. Also called an **inflected arch.**

inverted arc lamp (*Elec. Eng.*). A d.c. arc lamp having the positive carbon above the negative carbon so that the greater part of the light, which comes from the negative crater, shall be directed upwards—as required for indirect lighting.

inverted-brush contact (*Elec. Eng.*). A laminated switch-contact in which the laminations are carried on the fixed, instead of on the moving, contact.

inverted engine (*Aero.*). An in-line engine having its cylinders below the crankshaft. Adopted in certain types of aircraft to improve the forward view of the pilot.

inverted-L antenna (*Radio*). One comprising a vertical up-lead joined to one end of a horizontal conductor.

inverted loop (*Aero.*). An *aerobatic* manoeuvre consisting of a complete revolution in the vertical plane with the upper surface of the aircraft outside, which is started from the inverted position. Also **outside loop.**

inverted machine (*Elec. Eng.*). Any electric machine in which the usual arrangement of the stator and rotor windings is inverted, e.g., an induction motor in which power is supplied to the rotor, the stator winding being short-circuited.

inverted rectifier (*Elec. Eng.*). (1) Circuit arranged to convert d.c. to a.c. (2) Amplifier which inverts polarity. Same as **inverter.**

inverted repulsion motor (*Elec. Eng.*). A repulsion motor in which the supply is taken to the armature through the brushes, the stator winding being short-circuited.

inverted rotary converter (*Elec. Eng.*). A rotary converter which is used to convert from d.c. to a.c.

inverted speech (*Teleph.*). Inversion of the order of speech frequencies before modulation for privacy; effected by modulating with a carrier just above the maximum speech frequency, then discarding this carrier and its upper side-band.

inverted telephoto lens (*Photog.*). One of short focal length and long back focus, comprising a divergent component in front of a convergent one, giving a high-quality wide-angle image.

inverted thermionic valve (*Electronics*). A triode in which the control electrode is on the opposite side of the anode to the cathode, or on the opposite side of the cathode to the anode. The latter is also called a **lunar grid valve.**

inverted triode (*Electronics*). One in which the anode or cathode is in the centre.

inverted-V antenna (*Radio*). Two wires, several quarter-wavelengths long, joined at the top, one lower end being terminated by a resistance, the other end by the transmitter or receiver; the direction of maximum response is horizontal and in the plane of the wires.

inverter (*Elec. Eng.*). A d.c.-a.c. converter using, e.g., a transistor or thyristor oscillator. (*Telecomm.*) Arrangement of modulators and filters for inverting speech or music for privacy. Also **speech inverter.**

invertin (*Chem.*). See **invertase.**

invert soap (*Chem.*). Term sometimes used to denote certain *cationic detergents* (q.v.), e.g., cetyl pyridinium bromide.

invert sugar (*Chem.*). The product obtained by the hydrolysis of cane-sugar with acids; it is a mixture of equal parts of D-fructose and

D-glucose. Most fruits contain invert sugar, and honey averages over 70%.

investing bone (*Histol.*). See **membrane bone.**

investment (*Zool.*). The outer layers of an organ or part, or of an animal.

investment casting (*Met.*). 'Cire perdue', lost wax casting. Pattern is cast in wax, coated with a refractory layer and invested (placed in a sand mould). The wax is then melted and run out, and the mould is heated and filled with molten metal.

invisible glass (*Optics*). Glass which has its surface curved, or has been coated with molecular thickness material, to eliminate surface reflections.

in vitro (*Biol.*). Literally 'in glass'. Used to describe the experimental reproduction of biological processes in isolation from the living organism; cf. *in vivo*.

in vivo (*Biol.*). Used to describe biological processes occurring within the living organism; cf. *in vitro*.

involucel (*Bot.*). The group of bracts sometimes present at the base of a partial umbel.

involucral bract (*Bot.*). One of the leafy members forming an involucre.

involucre (*Bot.*). (1) A short tube around the archegonia and the calyptra in *Bryophyta*. (2) A crowded group of bracts around the base of a capitulum or other dense inflorescence.

involucrum (*Med.*). Sheath of new bone formed round bone which has died as the result of infection of the bone. (*Zool.*) In *Orthoptera*, the metanotum, in some *Hydrozoa*, a cuplike structure into which nematocysts may be retracted.

involuntary muscle (*Zool.*). See **unstriated muscle.**

involute (*Maths.*). See **evolute.** Every curve has many involutes. They are traced out by the different points on a piece of string unwinding from the curve. (*Zool.*) Tightly coiled; said of Gastropod shells.

involute connexion (*Elec. Eng.*). A special form of curved end connexion for the winding of an electric machine.

involute gear teeth (*Eng.*). Wheel teeth whose flank profile consists of an involute curve given by the locus of the end of a string uncoiled from a base circle; the commonest form of tooth.

involution (*Maths.*). (1) Raising a number to a positive integral power. Cf. *evolution*. (2) Let S be a set with $*$ an operation in S. Then an *automorphism* f of S is said to be an involution of S if $f*f = i_S$, where i denotes the identity mapping on S. (*Zool.*) In *Protozoa*, the condition of forms which are structurally deformed owing to an unfavourable environment, but which are capable of recovery if restored to a suitable environment.

involutional psychosis (*Psychiat.*). Mental illness occurring in late middle age. It may be associated with the menopause and may be of depressive or paranoid type.

involution form (*Bacteriol.*). A morphological variant, characteristic of old cultures and usually typical of the species. If viable, will give rise to a normal cell under favourable conditions.

involution period (*Bot.*). The resting period of a spore, seed, or other plant organ which remains inactive for a time.

Io (*Chem.*). The symbol for *ionium*.

iodates (V) (*Chem.*). Salts of iodic acid, HIO_3.

iodazide (*Chem.*). N_3I. An iodine azide.

iodic (V) acid (*Chem.*). HIO_3. Formed by the direct oxidation of iodine with nitric acid. White crystalline solid, which partly melts at 110°C, forming a liquid and a solid. Soluble in water. Forms iodates (V).

iodic anhydride (*Chem.*). See **iodine oxides.**

iodides (*Chem.*). Salts of hydriodic acid. Most metallic iodides liberate free iodine and leave behind the metal or metallic oxide when heated. See **hydriodic acid.**

iodination (*Chem.*). (1) The substitution or addition of iodine atoms in or to organic compounds. (2) The addition of iodine to table salt in districts where thyroid deficiencies are prevalent. See **goitre, Derbyshire neck.**

iodine (*Chem.*). Nonmetallic element in the seventh group of the periodic system, one of the halogens. Symbol I, at. no. 53, r.a.m. 126·9044, valencies 1, 3, 5, 7. It forms blackish scales with a violet lustre and a characteristic smell; m.p. 113·5°C, b.p. 184·35°C, rel. d. 4·95 at 20°C. It is widely but sparingly distributed as iodides, and is a constituent of the thyroid gland. The important commercial sources are crude Chile saltpetre (caliche) and certain seaweeds; it is used in organic synthesis. (*Med.*) It is a powerful disinfectant, and a constituent of *contrast media*. See also **radio iodine.**

iodine monochloride (*Chem.*). ICl. There are two forms, α and β, dependent on the method of cooling. The first occurs when the substance is strongly cooled, red needles, m.p. 27·2°C; the β-form, when slowly cooled, brown-red crystals, m.p. 13·9°C. Prepared by reaction of chlorine with iodine. Used as an iodinating agent in analytical and organic chemistry.

iodine oxides (*Chem.*). Iodine has four oxides with the empirical formulae, I_2O_4, I_4O_9, I_2O_5, IO_4. They differ in marked degree from those of the other halogens. The (V) oxide/I_2O_5 (**iodic anhydride**) has acidic properties; stable, with water yields iodic acid.

iodine pentafluoride (*Chem.*). IF_5. Colourless liquid, formed with incandescence by the direct combination of fluorine and iodine, m.p. −8°C, b.p. 97°C.

iodine trichloride (*Chem.*). ICl_3. Brown rhombic tablets, a powerful disinfectant.

iodine value (*Chem.*). The number of grams of iodine absorbed by 100 grams of a fat or oil. It gives an indication of the amount of unsaturated acids present in fats and oils.

iodism (*Med.*). The condition resulting from overdosage of, or sensitivity to, iodine; characterized by running at the eyes and the nose, salivation, and skin eruptions.

iodochlorhydroxyquinoline (*Pharm.*). See **clioquinol.**

iodo compounds (*Chem.*). Compounds containing the covalent iodine radical —I, e.g., iodobenzene.

iodoform (*Chem.*). Tri-iodomethane. CHI_3, yellow hexagonal plates, of peculiar, saffron odour, m.p. 119°C, volatile in steam, mildly antiseptic. It is prepared by warming alcohol with iodine and alkali; or by an electrolytic method, in which a current is passed through a solution containing KI, Na_2CO_3, and ethanol, the temperature being about 65°C.

iodogorgoic acid (*Chem.*). 3,5-di-iodotyrosine:

I
HO— [ring] —CH₂·CH(NH₂)COOH
I

A constituent of the thyroid hormone.

iodometric (*Chem.*). Measured by iodine.

iodophilic bacteria (*Bacteriol.*). Bacteria which

stain blue with iodine, revealing the presence of starchlike compounds. They occur in large numbers in the rumen where they are associated with cellulose decomposition.

iodopsin (*Optics*). See **visual violet.**

iodoso compound (*Chem.*). Compound containing the iodoso radical —IO.

iodoxy compounds (*Chem.*). Compound containing the radical —IO_2.

iodyl ion (*Chem.*). The monoxide of trivalent electropositive iodine: $(IO)^+$. Forms salts, e.g., iodyl sulphate $(IO)_2SO_4$, which are hydrolysed by water to give iodine, iodic (V) acid, and the acid corresponding to the anion.

iolite (*Min.*). See **cordierite.**

ion (*Phys.*). Strictly, any atom or molecule which has resultant electric charge due to loss or gain of valence electrons. Free electrons are sometimes loosely classified as *negative ions*. Ionic crystals are formed of ionized atoms and in solution exhibit ionic conduction. In gases, ions are normally molecular and cases of double or treble ionization may be encountered. When almost completely ionized, gases form a fourth state of matter known as a *plasma* (q.v.). Since matter is electrically neutral, ions are normally produced in pairs.

ion beam (*Phys.*). A beam of ions moving in the same direction with similar speeds, especially when produced by some form of accelerating machine or mass spectrograph. Also **ionic beam.**

ion burn (*Electronics*). Damage to the phosphor of a magnetic-deflection CRT because of relatively minute deflection of heavy negative ions by the magnetic control which deflects the electron beam. Electric deflection deflects both electrons and heavy negative ions together and there is no burn in an electrostatic CRT.

ion cluster (*Chem.*). Group of molecules loosely bound (by electrostatic forces) to a charged ion in a gas.

ion concentration (*Chem.*). That expressed in moles per unit volume for a particular ion. Also **ionic concentration.** See **pH value.** (*Phys.*) The number of ions of either sign, or of ion pairs, per unit volume. Also **ionization density.**

ion cyclotron frequency (*Phys.*). Ions in a magnetic field follow a circular orbit similar to that in a cyclotron. The orbital frequency is a function of the field and the velocity and specific charge of the ion.

ion exchange (*Chem., Min. Proc., etc.*). Abbrev. **IX.** Use of zeolites, artificial resins or specially treated coal to capture anions or cations from solution. Used in water-softening, separating isotopes, de-salting sea water, and wide range of industrial processes such as chemical extraction of uranium from its ores. Instead of solid particles through which liquid runs, immiscible liquid-liquid systems may be employed to transfer ions from one phase to another.

ion-exchange capacity (*Min. Proc.*). Electrical charge surplus to that uniting the framework of the IX resin or other exchange vehicle.

ion-exchange liquids (*Min. Proc.*). Those immiscible with water (e.g., kerosine) which are rendered active by addition of suitable chemicals, such as tri-lauryl amine. These liquids can be used instead of solid resins in IX, in solvent extraction processes.

ion-exchange resins (*Chem.*). Term applied to a variety of materials, usually organic, which have the capacity of exchanging the ions in solutions passed through them. Different varieties of resin are used dependent on the nature (cationic or anionic) of the ions to be exchanged. Many of the resins in present-day use are based on polystyrene networks cross-linked with divinyl benzene.

ion flotation (*Chem.*). Removal of ions or gels from water by adding a surface active agent which forms complexes. These are floated as a scum by the use of air bubbles. See **froth flotation, Gibbs' adsorption theorem.**

ion getter pump (*Vac. Tech.*). Form of high-vacuum pump capable of achieving pressures lower than 10^{-3} N/m^2.

ionic (*Phys.*). Appertaining to or associated with gaseous or electrolytic ions. *N.B.—Ion* is frequently used interchangeably with *ionic* as an adjective, e.g., in *ion*(*ic*) conduction.

ionic beam (*Phys.*). See **ion beam.**

ionic bombardment (*Electronics*). The impact on the cathode of a gas-filled electron tube of the positive ions created by ionization of the gas. The bombardment may cause electrons to be ejected from the cathode.

ionic bond (*Chem.*). Coulomb force between ion-pairs in molecule or ionic crystal. These bonds usually dissociate in solution.

ionic concentration (*Chem.*). See **ion concentration.**

ionic conduction (*Chem.*). That which arises from the movement of ions in a gas or electrolytic solution. In solids, it refers to electrical conductivity of an ionic crystal, arising from movement of positive and negative ions under an applied electric field. It is a diffusion process and hence very temperature-dependent.

ionic conductivity (*Chem.*). An additive property, symbol l_i; the equivalent *conductivity* (q.v.) of the ion i at infinite dilution. Thus for KCl, $\Lambda_\infty = l_{K^+} + l_{Cl^-}$.

ionic conductor (*Chem.*). One in which conduction is predominantly by ions, rather than by electrons and holes.

ionic crystal (*Crystal.*). Lattice held together by the electric forces between ions, as in a crystalline chemical compound.

ionic current (*Electronics*). Current carried by positively charged ions in a gas at low pressure. Especially applied to the small current which flows from the filament to the grid of a thermionic tube when the grid is made very negative and vacuum is not perfect.

ionic focusing. Same as **gas focusing.**

ionic formula (*Electronics*). Same as **electronic formula.**

ionic-heated cathode (*Electronics*). Hot cathode that is heated primarily by ionic bombardment of the emitting surface.

ionic heating (*Electronics*). That of a cathode when under ionic bombardment during manufacture.

ionic migration (*Chem., etc.*). Transport of ion-bearing particle to an electrode oppositely charged with electricity.

ionic modulation (*Radio*). That of waves propagated through an ionized gas, the degree of ionization being varied in accordance with the modulating signal, which varies absorption of the beam.

ionic potential (*Electronics*). Ratio of an ionic charge to its effective radius, regarded as a capacitor.

ionic product (*Chem.*). The product of the activities (see **activity, (2)**) of the ions into which a pure liquid dissociates.

ionic radius (*Crystal.*). Approximate limiting radius of ions in crystals, ranging, for common metals (including carbon), from a fraction to about one nanometre.

ionic strength (*Chem.*). Half the sum of the terms

obtained by multiplying the *activity* (q.v. (2)) of each ion in a solution by the square of its valency; it is a measure of the intensity of the electrical field existing in a solution.

ionic theory (*Chem.*). The theory that substances whose solutions conduct an electric current undergo electrolytic dissociation on dissolution. This assumption explains both the laws of electrolysis and the abnormal colligative properties, such as osmotic pressure, of electrolyte solutions.

ionic valve (*Electronics*). Cold-cathode rectifier tube comprising two electrodes in the form of a point and a spiral of wire respectively enclosed in a gas at low pressure. Also **Lodge valve.**

ion implantation (*Electronics*). Technique used to produce semiconductor devices. Impurities are introduced by firing high energy ions at the substrate material.

ionium (*Chem.*). The common name of a naturally occurring radioactive nuclide. Symbol Io, at. no. 90. An isotope of thorium.

ionization (*Phys.*). Formation of ions by separating atoms, molecules, or radicals, or by adding or subtracting electrons from atoms by strong electric fields in a gas, or by weakening the electric attractions in a liquid, particularly water.

ionization by collision (*Phys.*). The removal of one or more electrons from an atom by its collision with another particle such as another electron or an *a*-particle. Very prominent in electrical discharge through a rarefied gas.

ionization chamber (*Nuc. Eng.*). Instrument used in study of ionized gases and/or ionizing radiations. It comprises an enclosure with parallel plate or coaxial electrodes, between which ionized gas is formed. For fairly large applied voltages, the current through the chamber is dependent only upon the rate of ion production. For very large voltages, additional ionization by collision enhances the current. The system is then known as a *proportional counter* or *Geiger-Müller counter* (qq.v.).

ionization constant (*Chem.*). The ratio of the product of the activities (see **activity, (2)**) of the ions produced from a given substance to the activity of the undissociated molecules of that substance.

ionization cross-section (*Phys.*). Effective geometrical cross-section offered by an atom or molecule to an ionizing collision.

ionization current (*Phys.*). (1) That passing through ionization chamber. (2) Current passed by an ionization gauge, when used for measuring low gas pressures.

ionization delay (*Nuc. Eng.*). Time between initiation of ionization in an ionization chamber and the resulting current change in the external circuit.

ionization density (*Phys.*). See **ion concentration.**

ionization gauge (*Electronics*). Vacuum gauge formed by small thermionic triode attached to a chamber in which it is desired to measure residual gas pressure. Current is passed from anode to cathode, the grid being made negative; grid current is measured, giving degree of vacuum.

ionization heat (*Heat*). Increase of sensible heat for complete ionization of a mole of a substance.

ionization manometer (*Nuc. Eng.*). Ionization gauge in which a grid-current meter is calibrated to read gas pressure directly.

ionization potential (*Phys.*). Energy in electron-volts (eV) required to detach an electron from a neutral atom, depending on the complexity of the shells; for hydrogen, the value is 13·6 eV. Also **electron binding energy, radiation potential.**

ionization temperature (*Phys., Astron.*). A critical temperature, different for different elements, at which the constituent electrons of an atom will become dissociated from the nucleus; hence a factor in deducing stellar temperatures from observed spectral lines indicating any known stage of ionization.

ionization time (*Electronics*). Delay between the application of ionizing conditions and the onset of ionization, depending on temperature and other factors, e.g., in a mercury-pool rectifier.

ionized (*Chem., Phys.*). (1) Electrolytically dissociated. (2) Converted into an ion by the loss of an electron.

ionized atom (*Phys.*). One with a resultant charge arising from capture or loss of electrons; an *ion* in gas or liquid.

ionized gas detector (*Electronics*). Early form of detector in which discharge through ionized gas is triggered by the arrival of a signal.

ionizing collision (*Phys.*). Interaction in which an ion pair is produced.

ionizing energy (*Phys.*). That required to produce an ion pair in a gas under specified conditions. Measured in eV. For air it is about 32 eV.

ionizing event (*Phys.*). Any interaction which leads to the production of ions.

ionizing particle (*Phys.*). Charged particle which produces considerable ionization on passing through a medium. Neutrons, neutrinos, and photons are not ionizing particles although they may produce some ions.

ionizing radiation (*Phys.*). Any electromagnetic or particulate radiation which produces ion pairs when passing through a medium.

ionizing voltage. See **starting voltage.**

ion migration (*Chem.*). The movement of ions in an electrolyte or semiconductor due to applying a voltage across electrodes.

ion mobility (*Chem.*). *Ion velocity* (q.v.) in unit electric field (1 volt per metre).

ionogenic (*Chem.*). Forming ions, e.g., electrolytes.

ionogram (*Meteor.*). A recording of vertical incidence ionospheric soundings.

ionomer (*Plastics*). Product of ionic bonding action between long-chain molecules, characterized by toughness and high degree of transparency.

ionones (*Chem.*). Terpene compounds related to *irone* (q.v.) in structure, possessing an intense odour of violets; they can be synthesized from citral by condensation with acetone, forming pseudo-ionones, ring formation in which can be effected by boiling with sulphuric acid. There are two isomers, viz.:

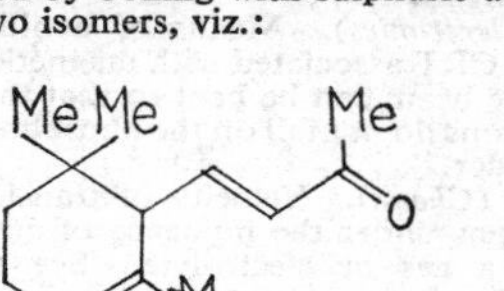

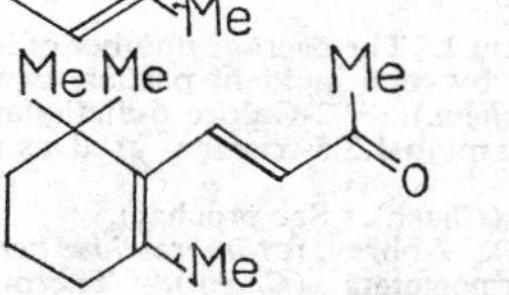

ionophone (*Acous.*). See **cathodophone.**

ionoscope (*TV*). A storage camera tube which

collects the electric charges of the optical image until removed by the scanning electron beam. See also **emitron, orthicon, Vidicon.**

ionosphere (*Meteor.*). That part of the earth's atmosphere (the upper stratosphere) in which an appreciable concentration of ions and free electrons normally exist. This shows daily and seasonal variations. See **E-layer, F-layer, hop.**

ionospheric control points (*Radio*). Points in the ionosphere distant 2000 km and 1000 km from each ground terminal, used respectively to control transmission by way of the F_2- and E-layers.

ionospheric disturbance (*Meteor.*). An abnormal variation of the ion density in part of the ionosphere, commonly produced by solar flares. Has a marked effect on radio communication. See **storm.**

ionospheric forecast (or prediction) (*Meteor.*). The forecasting of ionospheric conditions relevant to communication.

ionospheric path error (*Nav.*). That in bearing arising from irregularity of ionized layer.

ionospheric ray (or wave) (*Radio*). Radiation reflected from an upper ionized region between transmitter and receiver. Also called **indirect, reflected** or **sky ray (or wave).**

ionospheric regions (*Meteor.*). These are: D region, between 50 and 100 km; E region, between 90 and 150 km; F region, over 150 km; all above surface of earth. Internal effective layers are labelled *E, sporadic E,* E_2, *F*, F_1, $F_{1\frac{1}{2}}$, F_2.

ionotropy (*Chem.*). The reversible interconversion of certain organic isomers by migration of part of the molecules as an ion; e.g., hydrogen ion (*prototropy*).

ion pair (*Phys.*). Positive and negative ions produced together by transfer of electron from one atom or molecule to another.

ion rocket (*Space*). A proposed method of rocket propulsion in which charged particles (e.g., lithium or caesium ions) are accelerated electrically, giving a small thrust suitable for planetary journeys.

ion source (*Phys.*). Device for releasing ions as these are required in an accelerating machine. It usually consists of a minute jet of gas or vapour of the required compound which is ionized by bombardment with an electron beam.

ion spot (*Electronics*). The deformation of target or cathode by ion bombardment leading, in camera tubes, to a spurious signal.

iontophoresis (*Chem.*). Migration of ions into body tissue through electric currents. See **electrophoresis.**

ion trap (*Electronics*). Means of avoiding ion burn in a CRT associated with magnetic deflection. The beam can be bent so that the heavy negative ions do not fall on the phosphor. U.S. **beam bender.**

ion velocity (*Chem.*). Velocity of translation of drift of ions under the influence of an electric field, in a gas or electrolyte. See also **ion mobility.**

ion yield (*Phys.*). The average number of ion pairs produced by each incident particle or photon.

ipazine (*Chem.*). 2-Chloro-6-diethylamino-4-isopropylamino-1,3,5-triazine, used as a weed-killer.

IPC, IPPC (*Chem.*). See **propham.**

i.p.h. (*Print.*). Abbrev. for *impressions per hour*.

I.P.T. thermometers (*Chem.*). Thermometers conforming to the standards laid down by the *I*nstitute of *P*etroleum *T*echnologists.

IQ (*Psychol.*). Abbrev. for *intelligence quotient*.

Ir (*Chem.*). The symbol for *iridium*.

I.R. drop (*Elec. Eng.*). The voltage drop due to a current flowing through a resistance.

irid-, irido-. Prefix from Gk. *iris*, gen. *iridos*, rainbow.

Iridaceae (*Bot.*) The iris family, distinguished from the lilies by their inferior ovary and single whorl of stamens. Includes iris, crocus, freesia, gladiolus, etc.

iridalgia (*Med.*). Pain in the iris of the eye.

iridectomy (*Surg.*). Excision of part of the iris of the eye.

iridencleisis (*Med.*). The incarceration of part of the iris in a wound in the cornea.

irideremia (*Med.*). Aniridia. Apparent complete absence of the iris of the eye (a narrow rim always persists).

iridescence (*Phys.*). The production of fine colours on a surface; due to the interference of light reflected from the front and back of a very thin film.

iridescent clouds (*Meteor.*). High clouds of the cirro-cumulus type which show colours, generally delicate pink and green, in irregular patches. It is thought that the effect is caused by the diffraction of sunlight by supercooled water droplets.

iridium (*Met.*). A brittle, steel-grey metallic element of the platinum family. Symbol Ir, at. no. 77, r.a.m. 192·2, rel. d. at 20°C 22·4, m.p. 2375°C, electrical resistivity 6×10^{-8} ohm metres. Alloyed with platinum or osmium to form hard, corrosion-resisting alloys, used for pen-points, watch and compass bearings, crucibles, standards of length. The radioactive isotope ^{192}Ir is a medium-energy gamma-emitter used for industrial radiography.

iridium lamp (*Elec. Eng.*). An early form of electric filament lamp employing iridium wire.

iridochoroiditis (*Med.*). Inflammation of the iris and of the choroid of the eye.

iridocoloboma (*Med.*). Congenital absence of part of the iris, a gap or fissure being present in it.

iridocyclitis (*Med.*). Inflammation of the iris and of the ciliary body of the eye.

iridocyte (*Zool.*). A reflecting cell containing guanin, found in the integument of Fish and of certain Cephalopods, to which it gives an iridescent appearance.

iridodialysis (*Med.*). Separation of the iris from its attachment to the ciliary body of the eye.

iridokeratitis (*Med.*). Inflammation of the iris and of the cornea.

iridoplegia (*Med.*). Paralysis of the sphincter, or circular muscle, of the iris.

iridosmine (*Min.*). An ore of iridium and osmium with Os greater than 35%. Crystallizes in the hexagonal system.

iridotomy (*Surg.*). Surgical cutting of the iris.

iris (*Elec. Eng.*). Apertured diaphragm across a waveguide, for introducing specific impedances. (*Min.*) A form of quartz showing chromatic reflections of light from fractures, often produced artificially by suddenly cooling a heated crystal. Also called **rainbow quartz.** (*Photog.*) Adjustable aperture used in conjunction with camera lens for control of exposure. (*Zool.*) (1) In the Vertebrate eye, that part of the choroid, lying in front of the lens, which takes the form of a circular curtain with a central opening. (2) In Insects, a dense layer of pigment enveloping the margin of the lens and the proximal ends of the visual cells of certain types of ocelli. *adj.* **iridial.**

irisation (*Phys.*). Same as **iridescence.**

iris cells (*Zool.*). Densely-pigmented cells disposed in a circlet surrounding the cells of the crystalline cone and the corneagen layer in each ommatidium of an insect compound eye.

Irish bridge (*Civ. Eng.*). A ford or watersplash treated so as to be permanent.

Irish moss (*Bot.*). Kelp (*Chondrus crispus*) found on the coasts of Ireland and New England. See **carragheen.**

iritis (*Med.*). Inflammation of the iris.

I^2R loss (*Elec. Eng.*). The power loss caused by the flow of a current *I* through a resistance *R*.

IRM (*An. Behav.*). *Innate releasing mechanism.*

iroko (*For.*). A general utility timber from *Chlorophora excelsa*, a tropical African tree, used for structural work, shipbuilding, cabinet work, and furniture.

iron (*Chem.*). A metallic element in the eighth group of the periodic system. It exists in three forms; see **alpha-, gamma-, delta-**. Symbol Fe, at. no. 26, r.a.m. 55·847, rel. d. at 20°C 7·86, m.p. 1525°C, b.p. 2800°C, electrical resistivity $9{\cdot}8 \times 10^{-8}$ ohm metres. As basis metal in steel and cast-iron, it is the most widely used of all metals. See **iron ores.** (*Typog.*) The type-founder's unit when supplying sorts, being a measure of approximately 25 ems.

iron alum (*Min.*). See **halotrichite.**

iron arc (*Optics*). An arc between iron electrodes, used for obtaining light containing standardized lines, for spectrometer and spectrograph calibrations.

iron bacteria (*Bacteriol., Bot., Min.*). Filamentous bacteria which can convert iron (II) oxide to iron (III) hydroxide, deposited on their sheaths. Important in formation of *bog iron ore* (q.v.). One simple bacterium *Ferrobacillus ferrooxidans* is an iron autotroph. See **Chlamydobacteriales.**

ironbark (*For.*). *Eucalyptus leucoxylon*, yielding wood which varies in colour from greyish-brown to reddish-brown. The sapwood is susceptible to powder-post beetle attack, whilst the heartwood is impermeable to treatment with wood preservatives. It is not an easy timber to work.

iron-clad electromagnet (*Elec. Eng.*). One in which the return path for the flux is formed by an iron covering surrounding the winding.

iron-clad switchgear (*Elec. Eng.*). See **metal-clad switchgear.**

iron dust core (*Elec. Eng.*). One used in a high-frequency transformer or inductor to minimize eddy-current losses. It consists of minute magnetic particles bonded in an insulating matrix.

irone (*Chem.*). $C_{14}H_{22}O$, a mixture of isomeric methyl ketones of the terpene group, the odoriferous principle of the iris root and of the violet.

iron-glance (*Min.*). From the German *Eisenglanz*, a name often applied to specular iron-ore (haematite).

ironhat (*Mining*). See **gossan.**

iron loss (*Elec. Eng.*). The power loss due to hysteresis and eddy currents in the iron of magnetic material transformers or electrical machinery.

iron lung (*Med.*). See **Drinker respirator.**

iron meteorites (*Geol.*). One of the two main categories of meteorites, the other being the *stony meteorites.* They are composed of iron and of iron-nickel alloy, with only a small proportion of silicate or sulphide minerals.

iron-nickel accumulator (*Elec. Eng.*). See **nickel-iron-alkaline accumulator.**

iron-olivine (*Min.*). See **fayalite.**

iron ores (*Geol.*). Rocks or deposits containing iron-rich compounds in workable amounts; they may be primary or secondary; they may occur as irregular masses, as lodes or veins, or interbedded with sedimentary strata. See **chamosite, goethite, haematite, limonite, magnetite, siderite** (*Min.* (1)).

iron (II) oxide (*Chem.*). See **ferrous oxide.**

iron (III) oxide (*Chem.*). See **ferric oxide.**

iron pan (*Geol.*). A hard layer often found in sands and gravels; caused by the precipitation of iron salts from percolating waters. It is formed a short distance below the soil surface.

iron pattern (*Foundry*). A pattern made of cast-iron; used when a large number of castings is required from it, and long life is necessary. See **double contraction.**

iron paving (*Civ. Eng.*). A type of road surfacing formed of cast-iron slabs studded on their upper surface to reduce skidding.

iron pentacarbonyl (*Chem.*). $Fe(CO)_5$. Formed at ordinary temperatures when carbon monoxide is passed over finely divided iron. A liquid which readily decomposes.

iron pyrite (*Min.*). See **pyrite.**

iron spinel (*Min.*). See **spinel.**

ironstone (*Geol.*). Iron carbonate, clay, and carbonaceous matter, found in nodules, layers, or beds.

iron tetracarbonyl (*Chem.*). $Fe(CO)_4$. Formed when carbon monoxide is passed over heated finely divided iron. It is gaseous.

ironwood (*For.*). Popular name to describe several very hard and tough timbers, especially that from *Ostrya virginiana* and *Carfinus caroliniana* (Canada and U.S.). Black ironwood (*Krigiodendron*) sinks in water (rel. d. 1·4 approx.). See **Ceylon-.**

irradiance (*Phys.*). See **radiant-flux density.**

irradiation (*Photog.*). Image spread, arising from scatter within the emulsion from the silver halide grains. See **halation.** (*Radiol.*) Exposure of a body to X-rays, gamma-rays or other ionizing radiations.

irrational number (*Maths.*). **A real number which cannot be expressed as the ratio of two integers e.g., $\sqrt{2}$. Cf. *surd*.**

irregular (*Biol.*). (1) Asymmetric, not arranged in an even line or circle. (2) Not divisible into halves by an indefinite number of longitudinal planes.

irregular-coursed (*Build.*). Said of rubble walling built up in courses of different heights.

irregular galaxies (*Astron.*). Small galaxies showing no symmetry; they contain little dust or gas. See **Magellanic Clouds.**

irregular nebulae (*Astron.*). A large number of nebulae of varied shapes, not similar or well defined like the spiral or planetary nebulae. The best-known example is the Great Nebula in Orion. Also called **diffuse nebulae.**

irregular rotation of the earth (*Astron.*). The length of the day is increasing (see **tidal friction**) and is also affected by irregular changes detected by comparison of astronomical time with that kept by atomic clocks.

irregular variables (*Astron.*). See **variable stars.**

irreversible colloid (*Chem.*). See **lyophobic colloid.**

irreversible controls (*Aero.*). A flying control system, almost always hydraulically-operated, wherein there is no feed-back of aerodynamic forces from the control surfaces.

irreversible reaction (*Chem.*). A reaction which takes place in one direction only, and therefore proceeds to completion.

irreversible steering (*Autos.*). A steering gear in which it is impossible for road shocks on the wheels to cause motion at the steering wheel; a condition rarely attained.

irrigation (*Civ. Eng.*). The artificial distribution of water to land, normally by storage, and by diversion of natural waters. (*San. Eng.*) The method of sewage disposal by *land treatment* (q.v.).

irritability (*Biol.*). One of the characteristic properties of living matter, namely, the ability to receive and respond to external stimuli.

irritant (*Biol.*). Any external stimulus which produces an active response in a living organism.

irrorate (*Bot.*). As if covered with dew.

irrotational field (*Elec. Eng.*). A field in which the *circulation* (q.v.) is everywhere zero.

Irwin hot-wire oscillograph (*Elec. Eng.*). See **hot-wire oscillograph.**

ISA (*Aero.*). Abbrev. for *International Standard Atmosphere.*

isallobar (*Meteor.*). The contour line on a weather chart, signifying the location of equal changes in the barometer over a specified period, in contrast with the absolute reduced barometric readings.

isallobaric high and low (*Meteor.*). Centres, respectively, of rising and falling *barometric tendency* (q.v.).

isallobaric wind (*Meteor.*). That component of the wind which is due to the movement of the centre of high pressure towards the *isallobaric high* (q.v.), or the centre of low pressure towards the isallobaric low.

isatin (*Chem.*):

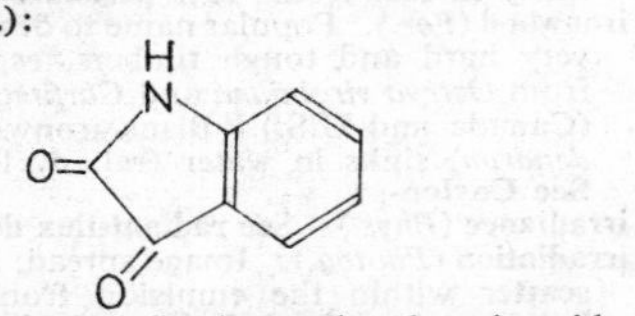

the lactam of 2-amino-benzoylmethanoic acid, obtained by oxidizing indigo or indoxyl with nitric acid; it can be synthesized by numerous methods, and is an important intermediate for the manufacture of indigo. Isatin crystallizes in reddish monoclinic prisms.

Isceons (*Chem.*). TN for range of chlorofluorocarbons similar to the *Freons* (q.v.), used as refrigerants, aerosol propellants and for polyurethane foams.

ischaemia, ischemia (*Med.*). Permanent or temporary deficiency of blood in a part of the body. *adj.* **ischaemic.**

ischiopodite (*Zool.*). The proximal joint of the endopodite of the walking-legs or maxillipeds of certain *Crustacea.*

ischiorectal (*Zool.*). Situated between the rectum and ischium; applied to fossa and muscles.

ischium (*Zool.*). A posterior bone of the pelvic girdle in *Tetrapods. adjs.* **ischial, ischiadic.**

isenthalpic (*Heat*). Of a process carried out at constant enthalpy, or heat function *H*.

isentropic (*Phys.*). See **entropy.**

iserine (*Min.*). Probably a ferruginous rutile, though formerly considered to be a variety of ilmenite. Found at Iserwiese (Bohemia). A doubtful species.

Isherwood system (*Ships*). A method of ship construction of which the dominant feature is longitudinal framing. Named after its originator.

isidium (*Bot.*). A branched outgrowth, recalling coral in form, arising from the thallus of a lichen.

I signal (*TV*). In the NTSC colour TV system, that corresponding to the wideband axis of the chrominance signal.

isinglass (*Chem.*). Fish glue. A white solid amorphous mass, prepared from fish bladders; chief constituent, gelatin. It has strong adhesive properties. Used in various food preparations, as an adhesive, and in the fining of beers, wines, etc.

isinglass gold-size (*Paint.*). Refined isinglass dissolved in hot water, mixed with methylated spirits; used in gilding on glass.

island effect (*Electronics*). The limitation of emission to small areas of the cathode surface of a valve when the grid voltage is less than a minimum value.

island universe (*Astron.*). The name applied to an extra-galactic nebula.

iso- (*Chem.*). A prefix indicating: (1) the presence of a branched carbon chain in the molecule; (2) an isomeric compound.

iso- Prefix from Greek *isos*, equal.

ISO. *I*nternational *S*tandards *O*rganization, which exists to make agreements on international standards.

isoagglutination (*Physiol., Zool.*). (1) The adhesion of spermatozoa to one another by the action of some substance produced by the ova of the same species. (2) The adhesion of erythrocytes to one another within the same blood-group. Cf. *heteroagglutination.*

isoagglutinogen (*Biol.*). See **isohaemagglutinogen.**

isobar (*Chem.*). A curve relating quantities measured at the same pressure. (*Meteor.*) A line drawn on a map through places having the same atmospheric pressure at a given time. (*Nuc.*). In plural, nuclides having the same total of protons and neutrons, with the same *mass number* and approximately the same *atomic mass.*

isobaric spin (*Nuc.*). See **isotopic spin.**

isobarometric (or isobaric) charts (*Meteor.*). Maps on which isobars are drawn. See **isobar.**

isobases (*Geol.*). Lines drawn through places where equal depression of the land mass took place in Glacial times, as a result of the weight of the ice load.

isobilateral (*Bot.*). Divisible into symmetrical halves by two distinct planes.

isoborneol (*Chem.*). A secondary alcohol of the terpene series, m.p. 212°C, very volatile. It is a stereoisomer of *borneol*, $C_{10}H_{17}OH$.

isobrachial (*Bot.*). Said of a chromosome which is bent into two equal arms.

Isobryales (*Bot.*). An order of the *Mosses* having perennial creeping gametophores, with the leaves twisted to appear tobe in two rows. The capsule has a double peristome and is erect (generally), and the sporophyte is generally pleurocarpous.

isobutane (*Chem.*). An isomeric form of *butane* (q.v.) having the structure $(CH_3)_2CH{\cdot}CH_3$, as compared with $CH_3CH_2{\cdot}CH_2CH_3$ for normal butane. Found in natural gas, and produced by cracking petroleum. Used as a refrigerant.

isobutyl alcohol (*Chem.*). 2-methylpropan-1-ol. $(CH_3)_2CH{\cdot}CH_2OH$, b.p. 107°C, partly miscible with water; formed during sugar fermentation.

iso-candle diagram (*Light*). A diagram showing the light distribution due to a particular lighting system by curves drawn on polar coordinates or by more complex diagrams.

isocercal (*Zool.*). Said of a type of secondarily symmetrical tail-fin (in Fish) in which the areas of the fin above and below the vertebral column are equal.

isochela (*Zool.*). A chela having the two opposable joints of approximately equal size.

isochore (*Chem.*). A curve relating quantities measured under conditions in which the volume remains constant.

isochromatic (*Light*). Interference fringe of uniform hue observed with white light source;

especially in photoelastic strain analysis where it joins points of equal phase retardation.

isochrone (*Radio*). Hyperbola (or ellipse) on a chart or map, along which there is a constant difference (or sum) in time of arrival of signals from two stations at ends of a baseline.

isochronism (*Horol.*). For a clock pendulum, isochronism implies that the time of vibration should be the same whatever the amplitude (see **circular error**) and for the balance, that the time of vibration should be the same whatever the arc of vibration. In practice, the balance approaches nearer to true isochronism than does the pendulum. (*Phys.*) Regular periodicity, as the swinging of a pendulum. *adj.* **isochronous.**

isochronous modulation (or **restitution**) (*Teleg.*). Modulation in which the time between any two significant constants is theoretically equal to (or a multiple of) the unit interval.

isoclinal fold (*Geol.*). A fold partially overturned, so that both limbs dip in the same direction. See **folding.**

isocline (*Mag.*). Line on a map, joining points where the angle of dip (or inclination) of the earth's magnetic field is the same. (*Telecomm.*) Contour line on *cyclogram* used in analysis of negative resistance oscillators to determine amplitude and waveform of resulting stable oscillation.

isocoria (*Anat.*). Equality in the size of the pupils of the eye.

isocount contours (*Radiol.*). The curves formed by the intersection of a series of *isocount surfaces* with a specified surface.

isocount surface (*Radiol.*). A surface on which the counting rate is everywhere the same.

isocyanides (*Chem.*). Isonitriles or carbylamines. Isocyano-compounds, R—N≡C. They are colourless liquids, only slightly soluble in water with a feebly alkaline reaction, having a nauseous odour, obtained by the action of trichloromethane and alcoholic potash on primary amines. They are very stable towards alkali, form additive compounds with halogens, HCl, H_2S, etc., and can be hydrolysed into methanoic acid and a primary amine, having one carbon atom less than the original compound.

isocyclic compounds (*Chem.*). *Carbocyclic compounds* (q.v.).

isodactylous (*Zool.*). Having all the digits of a limb the same size.

isodesmic structure (*Crystal., Met.*). Crystal structure with equal lattice bonding in all directions, and no distinct internal groups.

isodiametric, isodiametrical. Of the same length vertically and horizontally.

isodiapheres (*Nuc.*). Nuclides having the same difference between totals of neutrons and protons.

isodimorphous (*Chem.*). Existing in two isomorphous crystalline forms.

isodisperse (*Chem.*). Dispersible in solutions having the same pH value.

isodomon (*Build.*). An ancient form of masonry in which the facing consisted of squared stones laid in courses of equal height, and the filling of coursed stones of smaller size.

isodont (*Zool.*). Having all the teeth similar in size and form.

isodose chart (*Radiol.*). A graphical representation of a number of isodose contours in a given plane, which usually contains the central ray.

isodose curve (or **contour**) (*Radiol.*). The curve obtained at the intersection of a particular *isodose surface* with a given plane.

isodose surface (*Radiol.*). A surface on which the dose received is everywhere the same.

isodrin (*Chem.*). 1,2,3,4,10,10-Hexachloro-1,4,4a,5,8,8a-hexahydro-*exo*-1,4-*exo*-5,8-dimethanonaphthalene, used as an insecticide.

isodulcite (*Chem.*). See **rhamnose.**

isodynamic lines (*Mag.*). Lines on a magnetic map which pass through points having equal strengths of the earth's field.

isoelectric point (*Chem.*). Hydrogen ion concentration in solutions, at which dipolar ions are at a maximum. The point also coincides with minimum viscosity and conductivity. At this pH value, the charge on a colloid is zero, and the ionization of an ampholyte is at a minimum. It has a definite value for each amino acid and protein.

isoelectronic (*Electronics*). Said of similar electron patterns, as in valency electrons of atoms.

Isoetales (*Bot.*). An order of the *Lycopodineae.* The herbaceous sporophyte has secondary thickening in the stem, and large rhizophores. The leaves are microphyllous and legulate. Heterosporous with multiflagellate antherozoids.

isogametangic (*Bot.*). Having gametangia all of the same size and form.

isogamete (*Bot., Zool.*). One of a pair of uniting gametes of similar size and form. *adjs.* **isogamous, isogamic**; *n.* **isogamy.**

Isogeneratae (*Bot.*). A class of the *Phaeophyta,* characterized by the alternation of morphologically similar generations.

isogenetic (*Zool.*). Having a similar origin.

isogenomatic (*Cyt., Gen.*). Said of chromosome complements composed of similar genomes.

isogonal transformation (*Maths.*). A transformation from the *z*-plane to the *w*-plane, in which corresponding curves intersect at the same angle in each plane.

isogonic line (*Mag.*). Line on a map joining points of equal magnetic declination, i.e., corresponding variations from true north.

isogrivs (*Aero., Nav.*). Lines drawn on a *grid navigation system* chart joining points at which convergency of meridians and magnetic variation are equal.

isohaemagglutinogen (*Biol.*). A substance, an agglutinogen (A or B) in erythrocytes, which will produce agglutination of the erythrocytes within the same blood group; also called **isoagglutinogen.**

isohaline (*Ocean.*). A line drawn on a map through points of equal salinity in the sea.

isohel (*Meteor.*). A line drawn on a map through places having equal amounts of sunshine.

isohydric (*Chem.*). Having the same pH value, or concentration of hydrogen ions.

isohyet (*Meteor.*). A line drawn on a map through places having equal amounts of rainfall.

isokeraunic line (*Meteor.*). A line on a chart joining places with equal *thunderstorm* (q.v.) frequency.

isokinetic sample (*Powder Tech.*). Particles from a fluid stream or aerosol entering the mouth of the sampling device at the linear velocity at which the stream was flowing at that point prior to the insertion of the device.

isokontan (*Bot.*). Bearing two (or more) flagella of equal length.

Isolantite (*Elec. Eng.*). TN for an insulation material for high-frequency applications.

isolated essential singularity (*Maths.*). **See pole.**

isolated phase switchgear (*Elec. Eng.*). Switchgear in which all the apparatus associated with each phase is segregated in separate cubicles or on separate floors of the switch-house.

isolated point (*Maths.*). (1) A point of a set in the neighbourhood of which there is no other point. (2) **See double point.**

isolation diode (*Telecomm.*) One used to block signals in one direction but to pass them in the other.
isolation transect (*Bot.*). A belt of land to which grazing animals are admitted under observation, so that the effect of grazing on the vegetation may be studied.
isolation transformer (*Elec. Eng.*). One used to isolate electrical equipment from its power supply.
isolator (*Electronics, Radar*). Non-reciprocal component having a low insertion loss in the forward direction and large loss in the reverse direction. See **gyrator.**
isolecithal (*Zool.*). Said of ova which have yolk distributed evenly through the protoplasm.
isoleucine (*Chem.*). An essential amino acid $CH_3CH_2CH(CH_3)CH(NH_2)COOH$.
isologues (*Chem.*). Compounds with similar molecular structure but containing different atoms of the same valency.
isolux (*Light*). Locus, line, or surface where the light intensity is constant. Also called **isophot.**
isomagnetic lines (*Mag.*). Lines connecting places at which a property of the earth's magnetic field is a constant.
isomastigote (*Zool.*). Having two or four flagella of equal length.
isomerism (*Chem.*). The existence of more than one substance having a given molecular composition and rel. mol. mass but differing in constitution or structure. (See also **optical isomerism.**) The compounds themselves are called *isomers* or *isomerides* (Gr. 'composed of equal parts'). Isobutane and butane have the same formula, C_4H_{10}, but their atoms are placed differently; one type of alkane molecule, $C_{40}H_{82}$, has over 50^{12} possible isomers. Isomerism is frequently met with among organic compounds and complex inorganic salts. (*Nuc.*) The existence of nuclides which have the same *atomic number* and the same *mass number*, but are distinguishable by their energy; that having the lowest (*ground*) state is stable, the others are unstable, and are said to be in *isomeric states.*
isomerized rubber (*Chem.*). Rubber in which the molecules have been rearranged by heating in solution in the presence of suitable catalysts. It is more soluble and more compatible with other varnish constituents than ordinary rubber while maintaining good chemical resistance.
isomerous (*Bot.*). Equal in number to the members of another whorl.
isomers (*Chem.*). See **isomerism.**
isomer separation (*Nuc.*). The chemical separation of the lower energy member of a pair of nuclear isomers.
isometric contraction (*Zool.*). The type of contraction involved when a muscle is held so that it cannot change its length.
isometric system (*Crystal.*). The cubic system.
isomorphic alternation of generations (*Bot.*). See **homologous alternation of generations.**
isomorphic groups (*Maths.*). The groups $(G_1; *)$ and $(G_2; \bigcirc)$ are said to be isomorphic to one another if there exists a one-to-one mapping f from G_1 on to G_2, and that for all elements x, y in G_1 we have $f(x*y) = f(x) \bigcirc f(y)$.
isomorphism (*Biol.*). Apparent likeness between individuals belonging to different species or races. (*Crystal.*) The name given to the phenomenon whereby two or more minerals, which are closely similar in their chemical constitution, crystallize in the same class of the same system of symmetry, and develop very similar forms. *adjs.* **isomorphic, isomorphous.** (*Maths.*) A one-to-one mapping from one algebraic system on to another which shows the systems to have the same abstract structure. (*Psychol.*) In *Gestalt psychology*, a term referring to the (hypothesized) structural correspondence between brain areas activated and conscious content.
isomorphous mixture (*Chem.*). See **mixed crystal.**
isoneph (*Meteor.*). A line drawn on a map through places having equal amounts of *cloudiness.*
isoniazid (*Pharm.*). Pyridine-4-carboxyhydrazide, $C_6H_7N_3O$. Used in the treatment of tuberculosis.
isonitriles (*Chem.*). *Isocyanides* (q.v.).
ISO paper sizes (*Print.*). A, B and C series are defined, always as trimmed sizes, in millimetres, the A series being the one which is most widely used. In each series, the sides have the proportion $1 : \sqrt{2}$ which is therefore maintained for each regular subdivision. A0 measures 1189 mm × 841 mm ($46\frac{3}{4}$ in. × $33\frac{1}{8}$ in.) and has an area of one square metre; A4 measures 297 mm × 210 mm ($11\frac{3}{4}$ in. × $8\frac{1}{4}$ in.) and is a standard size for letterheads, magazines; A6 is 148 mm × 105 mm ($5\frac{7}{8}$ in. × $4\frac{1}{8}$ in.) and is the international postcard size.
isopaque (*Photog.*). The line or contour giving equality of opacity; applied to spectrograms to determine colour-sensitivity of photographic materials.
isopedin (*Zool.*). The thin layer of bone forming the inner layer of the cosmoid scales of *Choanichthyes.*
isoperms (*Met.*). Magnetic alloys containing iron, nickel, and cobalt, used in high-frequency inductors.
isopic (*Geol.*). Said of two formations deposited contemporaneously and of the same facies. Cf. *heteropic.*
isopiestic (*Chem.*). Having equal pressure in the system or conditions described.
isopleth (*Maths.*). See **nomogram.**
isopneustic (*Zool.*). Said of the primitive arrangement of the spiracles, consisting of 2 pairs of thoracic and 8 of abdominal spiracles, all intersegmental, in the embryos of most insects.
Isopoda (*Zool.*). An order of *Peracarida,* in which the carapace is absent, the eyes are sessile or borne on immovable stalks, and the uropods are usually lamellar; the body is depressed and the legs used for walking; they show great variety of form, size, and habit; some are terrestrial, plant-feeders, or ant-guests, others are marine, free-living, and feeding on seaweeds or ectoparasitic on fish. Woodlice, etc.
isopodous (*Zool.*). Having the legs all alike.
isopogonous (*Zool.*). Said of feathers which have the two vanes alike in size and form.
isoprene (*Chem.*). $CH_2{=}C(CH_3){\cdot}CH{=}CH_2$, an alkadiene, a colourless liquid, b.p. 37°C, obtained by the destructive distillation of rubber, by dehydrogenation of 2-methylbutane, from propene, or by several other methods. It is used in the synthesis of rubber. See **butyl rubber.**
isopropyl alcohol (*Chem.*). Propan-2-ol. $(CH_3)_2$ $CHOH$, a colourless liquid, miscible with water, b.p. 81°C.
isopropyl benzene (*Chem.*). See **cumene.**
isopropyl group (*Chem.*). The monovalent radical $(CH_3)_2CH$—.
Isoptera (*Zool.*). An order of social *Polyneoptera* living in large communities which occupy nests excavated in the soil or built up from mud and wood; different polymorphic forms or castes occur in each species; the mouth-parts are adapted for biting and both pairs of wings, if present, are membranous and can be shed by

means of a basal suture; soft-bodied forms with short anal cerci. White Ants, Termites.

isopycnic (*Meteor.*). A line on a chart joining points of equal atmospheric density.

isopyknoscopy (*Chem.*). The determination of the end-point of a volumetric titration from the specific gravity of the titrated solution.

isoquinoline (*Chem.*). Occurs in coal tar. The formula is:

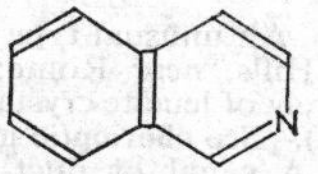

N

It is an isomer of quinoline and a condensation product of a benzene ring with a pyridine ring. It forms colourless crystals, m.p. 23°C, b.p. 240°C.

isoreagent (*Bot.*). A variety or microspecies.

isosceles triangle (*Maths.*). A triangle having two equal sides.

isoseismal line (*Geol.*). A line drawn on a map through places recording the same intensity of earthquake shocks. See **earthquake.**

isospin (*Nuc.*). Contraction of **isotopic spin.**

Isospondyli (*Zool.*). An order of *Teleostei*, characterized by the possession of abdominal pelvic fins, an air-bladder communicating with the gullet by a pneumatic duct, a well-developed mesocoracoid and soft fin-rays. Herring, Salmon, Trout, Sardine, Pilchard, Anchovy, Wide-mouth, Smooth-head, and Lady Fish.

isospores (*Zool.*). In *Radiolaria*, agametes produced in schizogony.

isosporous (*Bot.*). Having asexually produced spores of one kind only.

isostasy (*Geol.*). A condition, which has been presumed to exist in the earth's crust, whereby equal earth masses underlie equal areas down to an assumed level of compensation.

isostemonous (*Bot.*). Having as many stamens as petals.

isostere (*Meteor.*). A line on a chart joining points of equal atmospheric *specific volume* (q.v.).

isosteric (*Chem.*). Consisting of molecules of similar size and shape.

isosthenuria (*Med.*). Condition occurring in chronic renal disease with renal insufficiency in which there is regular secretion of urine of same relative density as protein-free plasma. Rel. d. remains stable (ca. 1·010) irrespective of fluid intake. Inability of kidneys to produce either concentrated or dilute urine.

isotach (*Meteor.*). A line on a chart joining points of equal wind speed.

isotactic (*Plastics*). Term denoting linear-substituted hydrocarbon polymers in which the substituent groups all lie on the same side of the carbon chain. See also **atactic** and **syndiotactic.**

isotaxy (*Chem.*). Polymerization in which the monomers show stereochemical regularity of structure. *adj.* **isotactic.** Cf. *syndyotaxy.*

isoteniscope (*Chem.*). An instrument for the static measurement of vapour pressure by observing the change of level of a liquid in a U-tube.

isotherm (*Meteor.*). A line drawn on a map through places having equal temperatures.

isothermal (*Phys.*). (1) Occurring at constant temperature. (2) A curve relating quantities measured at constant temperature.

isothermal change (*Phys.*). A change in the volume and pressure of a substance which takes place at constant temperature. For gases, Boyle's law applies to isothermal changes.

isothermal efficiency (*Eng.*). Of a compressor, the ratio of the work required to compress a gas isothermally to the work actually done by the compressor.

isothermal layer (*Meteor.*). A layer of the atmosphere with zero *ELR* (q.v.).

isothermal lines (or **curves**) (*Phys.*). Curves obtained by plotting pressure against volume for a gas kept at constant temperature. For a gas sufficiently above its critical temperature for Boyle's law to be obeyed, such curves are rectangular hyperbolas.

isothermal process (*Eng.*). A physical process, particularly one involving the compression and expansion of a gas, which takes place without temperature change.

isothrausmatic (*Geol.*). Said of orbicular igneous rocks in which the compositon of the cores of the orbs is identical with that of the groundmass in which they are embedded. Cf. *homeothrausmatic.*

isotones (*Nuc.*). Nuclei with the same neutron number but different atomic numbers (i.e., those lying in a vertical column of a *Segrè chart*).

isotonic (*Chem.*). Having the same osmotic pressure, e.g., as that of blood, or of the sap of cells which are being tested for their osmotic properties.

isotonic contraction (*Zool.*). The type of contraction involved when a muscle, in shortening, does external work.

isotope (*Nuc.*). One of a set of chemically identical species of atom which have the same *atomic number* but different *mass numbers.* A few elements have only one natural isotope, but all elements have artificially produced radioisotopes. (Gk. *isos*, same; *topos*, place.)

isotope separation (*Nuc.*). Process of altering the relative abundance of isotopes in a mixture. The separation may be virtually complete as in a mass spectrograph, or may give slight enrichment only as in each stage of a diffusion plant. See **Hermes.**

isotope structure (*Nuc.*). Hyperfine structure of spectrum lines resulting from mixture of isotopes in source material. The wavelength difference is termed the **isotope shift.**

isotope therapy (*Radiol.*). Radiotherapy by means of radioisotopes.

isotopic abundance (*Nuc.*). See **abundance ratio.**

isotopic dilution (*Radiol.*). The mixing of a particular nuclide with one or more of its isotopes.

isotopic dilution analysis (*Nuc.*). A method of determining the amount of an element in a specimen by observing the change in isotopic composition produced by the addition of a known amount of radioactive allobar.

isotopic number (*Nuc.*). See **neutron excess.**

isotopic spin (*Nuc.*). Misleadingly named quantum number allocated to nuclear particles to differentiate between members of a multiplet (a group of particles differing only in electric charge) in the same way that different quantum numbers in optical spectroscopy represent the multiplet structure of electron energy levels. It has no connexion with the nuclear spin of the particle. Introduced to distinguish the proton and neutron, both being regarded as the same nucleon with isotopic spin either parallel or anti-parallel with some preferred direction, the proton having isotopic spin $+\frac{1}{2}$ and the neutron $-\frac{1}{2}$. The small mass differences existing between members of multiplets is due to the energy difference associated with their differing charges. This form of nomenclature is in use with all baryons and mesons, the number of members of a multiplet of isotopic spin I being $(2I+1)$. Thus singlets have an isotopic spin 0;

doublets ½ (isotopic quantum numbers +½ and −½); triplets 1 (isotopic spin quantum numbers +1½, +½, −½, and −1½). The justification for this classification is that all members of a multiplet respond identically to the strong nuclear interactions, their charge difference affecting only electromagnetic interactions. Isotopic spin is conserved for all strong interactions and never changes by more than one unit in a weak one. Also **isobaric spin.**

isotopic symbols (*Chem.*). Numerals attached to the symbol for a chemical element, with the following meanings—*upper left*, mass number of atom; *lower left*, nuclear charge of atom; *lower right*, number of atoms in molecule—e.g., $_1^2H_2$, $_{12}^{24}Mg$.

isotron (*Nuc. Eng.*). Device in which pulses from a source of ions are synchronized with a deflecting field. This separates isotopes, because their acceleration varies with mass.

isotropic (*Phys.*). Said of a medium, the physical properties of which, e.g., magnetic susceptibility or elastic constants, do not vary with direction.

isotropic dielectric (*Elec. Eng.*). One in which the electrical properties are independent of the direction of the applied electric field.

isotropic radiator (*Radio*). Same as **omnidirectional radiator.**

isotropic source (*Electronics*). Theoretical source which radiates all its electromagnetic energy equally in all directions.

isotropous (*Zool.*). Of ova, lacking any predetermined axes. *n.* **isotropy.**

isozymes (*Biochem.*). Electrophoretically distinct forms of an enzyme with identical function.

isthmus (*Bot.*). The connecting zone between the semicells of the plaeoderm desmids. (*Zool.*) A neck connecting two expanded portions of an organ; as the constriction connecting the midbrain and the hind-brain of Vertebrates.

isthmus armature (*Teleph.*). Relay armature with a constriction near the end covering the pole of the core. Saturation during operation tends to equalize the operating and release currents.

itacolumite (*Geol.*). A sandstone which, because of the loose relationship between the grains and the cement, is slightly flexible when cut into slices.

itaconic acid (*Chem.*). White crystalline solid, m.p. 163°C, having the formula:

$$H_2C=C\begin{cases}COOH\\CH_2\,COOH\end{cases}$$

Obtained by fermentation of sugar with the *Aspergillus* mould; its esters polymerize to form plastics.

Italian asbestos (*Min.*). A name often given to tremolite asbestos to distinguish it from Canadian or chrysotile asbestos. It is extensively quarried in Piedmont and Lombardy.

Italian blind (*Build.*). An outside roller-blind similar to the hook-out blind but having the side arms attached to the blind and capable of sliding up and down on side rods. Also called a **canalette blind.**

Italian cloth (*Textiles*). (1) Fabric composed of a cotton warp and a fine Botany weft, used for dresses and linings. (2) A grey cotton cloth which has been mercerized, dyed, and schreinered. Weave is 5-end sateen with fewer ends than picks.

Italian pink (*Paint.*). See **Dutch pink.**

Italian roof (*Arch.*). See **hipped roof.**

italic, italics (*Typog.*). A sloping style of type, thus *italic.*

italite (*Geol.*). An unusual type of lava found in the Alban Hills, near Rome; it is made up almost entirely of leucite crystals.

itchy leg (*Vet.*). See **chorioptic mange.**

iter (*Zool.*). A canal or duct, as the reduced ventricle of the mid-brain in higher Vertebrates.

iterated fission expectation (*Nuc.*). Limiting value, after a long time, of the number of fissions per generation in the chain reaction initiated by a specified neutron to which this term applies.

iteration (*Comp.*). One repetition of a group of instructions forming a subroutine which is to be carried out repeatedly in a particular computer program. The repeated application of an algorithm.

iterative impedance (*Elec.*). The *input impedance* of a four-terminal network or transducer when the output is terminated with the same impedance, or when an infinite series of identical such networks are cascaded. See **image impedance.**

I.T.M. (*Ships*). Abbrev. for **inch trim moment.** Same as *moment to change trim one inch* (q.v.).

IUPAC (*Chem.*). Abbrev. for the **International Union of Pure and Applied Chemistry,** a body responsible, among other things, for the standardization of chemical nomenclature.

ivory (*Zool.*). The dentine of teeth, especially the type of dentine composing the tusks of elephants, which in transverse section shows striae proceeding in the arc of a circle and forming, by their decussations, small curvilinear lozenge-shaped spaces.

ivory black (*Chem.*). A name applied to the product obtained by digesting bone black or carbon with hydrochloric acid to remove calcium phosphates. (*Paint.*) A colouring pigment obtained from charred scraps of ivory ground in oil.

ivory boards (*Paper*). A board formed from high-quality papers by starch-pasting two or more together.

ivorywood (*For.*). Rare Australian hardwood from *Siphonadendron*, prized for engraving, turnery, mirror frames, inlaying, etc.

IVP (*Med.*). Abbrev. for *i*ntra*v*enous *p*yelography, i.e. *pyelography* (q.v.) following intravenous injection of suitable fluid.

ivr (*Chem.*). Abbrev. for *instantaneous velocity of reaction.*

I.W. (*Chem.*). An abbrev. for *isotopic weight.*

IX (*Chem., Min. Proc.*). Abbrev. for *ion exchange.*

ixiolite (*Min.*). See **tapiolite.**

Izod test (*Met.*). A notched-bar impact test in which a notched specimen held in a vice is struck on the end by a striker carried on a pendulum; the energy absorbed in fracture is then calculated from the height to which the pendulum rises as it continues its swing.

Izod value (*Met.*). The energy absorbed in fracturing a standard specimen in an Izod pendulum impact-testing machine.

j (*Maths.*). The symbol j is used by electrical engineers in place of the mathematician's i. Its main use is that in circuits carrying sinusoidal current of angular frequency ω, any inductance L and any capacitance C can be replaced by reactances $j\omega L$ and $\frac{1}{j\omega C}$ respectively and Ohm's law can then be used. Also referred to as the 90° operator.

J (*Chem.*). In names of dyestuffs, a symbol for yellow. (*Heat, etc.*) Symbol for *joule*, the unit of energy, work, quantity of heat.

J (*Elec., Mag.*). Symbol for *electric current density*, *magnetic polarization*. (*Eng.*) Symbol for the *polar moment of inertia*.

Jablochkoff candle (*Light*). An early form of arc lamp, in which the carbons were placed side by side and separated by plaster of Paris.

jacaranda (*For.*). See **rosewood.**

jacconette (*Textiles*). See **jaconet.**

J-acid (*Chem.*). 2,5-Aminonaphthol-7-mono-sulphonic acid, an intermediate for dyestuffs, prepared by melting 2-naphthylamine-5,7-disulphonic acid with sodium hydroxide.

jacinth (*Min.*). See **hyacinth.**

jack (*Eng.*). A portable lifting machine for raising heavy weights through a short distance, consisting either of a screw raised by a nut rotated by hand gear and a long lever, or a small hydraulic ram. See **hydraulic-.** (*Teleph.*) Socket whose connexions are short-circuited until a jack plug is inserted. A *break jack* is one which breaks the normal circuit on inserting plug, while *branch jack* is one which does not. (*Textiles*) In lace manufacture, a frame containing horizontal bars that support wires fixed vertically, against which bobbins containing yarn can revolve freely. (*Weaving*) Oscillating lever in a loom dobby; its movement raises or lowers a heald.

jack arch (*Build.*). A *flat arch* (q.v.).

jackbit (*Mining*). Detachable cutting end fitted to shank of miner's rock drill, used to drill short blast-holes. Also called **rip-bit.**

jackblock (*Build., Civ. Eng.*). A method of system building in which the roof and floor slabs are cast on top of each other and hydraulically jacked up to their respective levels, walls being built as and when required.

jack box (*Elec. Eng.*). One containing switches or connexions for changing circuits.

jacket (*Eng.*). An outer casing or cover constructed round a cylinder or pipe, the space being filled with a fluid for either cooling or heating the contents, or with insulating material for keeping the contents at substantially constant temperature, e.g., the water *jackets* of an I.C. engine. (*Nuc. Eng.*) See **can.** (*Print.*) The wrapper, or dust cover, in which a book is enclosed. Bookjackets are usually artistically designed, and executed in colour, as their purpose is to enhance the appeal of the volume as well as to protect it from dust.

jack frame (*Spinning*). The last of three speed-frames used only to prepare rovings for spinning very fine counts.

jackhammer (*Mining*). A hand-held compressed-air hammer drill for rock-drilling.

jacking delivery motion (*Spinning*). Mechanism which provides for the delivery of a small quantity of additional roving by the front rollers, after stoppage at the end of the run of the mule carriage, to compensate for extra twist inserted in hard twisted yarns.

jacking motion (*Spinning*). The mechanism employed to stop the front rollers in a mule just before the carriage completes its outward run; prevents snarling of the twisting yarn.

jacking-up (*Eng.*). The process of lifting an object by means of a jack.

jack-in-the-box (*Spinning*). Term sometimes used for the differential motion that controls the winding speed on a speedframe.

jack off (*Textiles*). In lace manufacture, to wind a length of yarn from a jack of flanged bobbins, for transfer to brass bobbins.

jack plane (*Tools*). A bench plane about 16 in. long, used for bringing the work down to approximate size, prior to finishing with a trying or smoothing plane.

jack-rafter (*Carp.*). A short rafter connecting a hip-rafter and the eaves, or a valley-rafter and the ridge.

jacks (*Textiles*). (1) Levers that, in conjunction with a tappet or dobby motion, raise and lower the heald shafts in a loom. (2) In lace machines, elements (such as interceptor wires, etc.) which control threads.

jack shaft (*Elec. Eng.*). An intermediate shaft used in locomotives having collective drive; the jack shaft is geared to the motor shaft and carries cranks which drive the coupling rods on the driving wheels.

Jacksonian epilepsy (*Med.*). A convulsion of a limited group of muscles spreading gradually from one group to the other, usually without loss of consciousness; the result of a lesion (e.g., tumour) of the brain.

jack-timber (*Carp.*). A timber used in a narrowing situation, such as a rafter in a hip roof, where it has to be shorter than its fellows. Also called a **cripple-timber.**

Jacobian (*Maths.*). Of n functions u_i each of n variables x_j, the determinant whose i, jth element is $\frac{\partial u_i}{\partial x_j}$. Written $\frac{\partial(u_1, u_2, u_3 \ldots u_n)}{\partial(x_1, x_2, x_3 \ldots x_n)}$.

Jacobian elliptic functions (*Maths.*). See **elliptic functions.**

jacobsite (*Min.*). An oxide of manganese and iron (III), often with considerable replacement of manganese by magnesium; crystallizes in the cubic system (usually in the form of distorted octahedra). A spinel.

jacob's ladder (*Eng.*). Vertical belt conveyor with cups or buckets.

Jacobson's cartilage (*Zool.*). In some Vertebrates, a cartilage of the nasal region supporting Jacobson's organs.

Jacobson's commissure (*Zool.*). In Vertebrates, a connexion between the seventh and ninth cranial nerves.

Jacobson's glands (*Zool.*). In some Vertebrates, nasal glands the secretion of which moistens the olfactory epithelium.

Jacobson's organ (*Zool.*). In some Vertebrates, an accessory olfactory organ developed in connexion with the roof of the mouth.

jaconet or jacconette (*Textiles*). (1) Plain, light-bleached cotton cloth, with smooth finish; for Indian markets, etc. (2) Thin material of rubber and linen used for medical dressings when water- or damp-proofing is required.

jacot tool (*Horol.*). A tool used by watchmakers for polishing and burnishing pivots.

jacquard (*Weaving*). Precision-made machine employed over the top of a loom. Enables the most ornamental designs to be woven with practically no limit to the design possibilities. Utilizes punched cards and needles to control lifting and lowering of warp threads. There are several variations; single-lift, single cylinder; double-lift, single-cylinder; double-lift, double-cylinder; cross-border, etc. Named after the French inventor Joseph-Marie Jacquard who introduced the first model in 1801. See **cards**.

jacquard cord (*Weaving*). See **card lacing**.

Jacquet's method (*Met.*). Final polishing of metal surfaces by electrolysis.

jactitation (*Med.*). Restless tossing of a patient severely ill; a twitching or convulsion of a muscle or of a limb.

jacupirangite (*Geol.*). A name applied by Derby in 1891 to a type of nepheline-gabbro consisting of titanaugite, biotite, iron ores, and nepheline, the last being subordinate to the mafic minerals.

jad (*Mining*). Jud. Deep groove cut into bed to detach block of natural stone. To undercut. A *jadder* is a stonecutter and his working tool is a *jadding pick*.

jade (*Min.*). A general term loosely used to include various mineral substances of tough texture and of a green colour. It properly embraces nephrite and jadeite, but green varieties of sillimanite, pectolite, serpentine, vesuvianite, and garnet are sometimes included.

jadeite (*Min.*). A metasilicate of sodium and aluminium which crystallizes in the monoclinic system. It is a member of the pyroxene group and is white to green in colour. The name is from the Spanish *Piedra de jada*, colic stone, from the belief that it could prevent nephritic colic; it is rarer than the nephrite type of jade. The finest material comes from Mogaung, Upper Burma; it also comes from Mexico and Guatemala.

Jäderin wires (*Surv.*). Apparatus once used in accurate *base line* measurement. Largely superseded by use of invar tapes or tellurometry.

jag-bolt (*Eng.*). See **rag-bolt**.

jail fever (*Med.*). See **typhus**.

jalousies (*Build.*). Hanging or sliding wooden sun-shutters giving external protection to a window, and allowing for ventilation through louvres or holes cut in the shutters themselves. Also called **Venetian shutters**.

Jamaica ebony (*For.*). See **cocuswood**.

jamaicin (*Chem.*). See **berberine**.

jamb (*Build.*). The side of an aperture.

jamb linings (*Join.*). The panelling at the sides of a window recess, running from the floor to the level of the window head. Cf. *elbow linings*.

jamb post (*Build.*). An upright member on one side of a doorway opening.

jambs (*Acous.*). The vertical plane surfaces for locating the draw-stops at the sides of the manuals on an organ console.

jamb stone (*Build.*). A stone forming one of the upright sides of an aperture in a wall.

James-Lange theory (*Psychol.*). A theory of emotion which holds that the experience of an emotion is the experience of the relevant bodily changes; e.g., he is angry because he clenches his fists rather than he clenches his fists because he is angry.

jamesonite (*Min.*). See **feather ore**.

Jamin interferometer (*Light*). A form of interferometer in which two interfering beams of light pursue parallel paths a few centimetres apart. The instrument is used to measure the refractive index of a gas, by observing the fringe shift when one of the light beams traverses a tube filled with the gas, while the other traverses a vacuum.

Jamin's chain, jaminian chain (*Bot.*). A series of short threads of water separated by bubbles of air, in the vessels of plants.

jamming (*Radio*). Deliberate interference of a transmission on one carrier by transmission on another approximately equal carrier, with wobble or noise modulation.

Jandus arc lamp (*Light*). A form of enclosed-flame arc lamp, which has an arrangement for circulating the enclosed gases in order to avoid deposits on the globe.

janiceps (*Med.*). A foetal monstrosity in which two heads are fused so that the faces look in opposite directions.

J-antenna (*Radio*). Dipole fed and matched at the end by a quarter-wavelength line.

Janus (*Radio*). Transmitting or receiving antenna which can be switched between opposite directions.

Janus green stain (*Micros.*). A basic aniline dye used as a supravital stain. It has a particular affinity for mitochondria and nerve cells.

japan (*Paint.*). A black, glossy, paint based on asphaltum and drying oils.

Japan camphor (*Chem.*). See **camphor**.

Japanese paper (*Paper*). Japanese hand-made paper prepared from mulberry bark. The surface is similar to that of *Japanese vellum*.

Japanese river fever (*Med.*). See **shimamushi fever**.

Japanese vellum (*Paper*). An expensive hand-made paper, prepared from the inner bark of the mulberry tree. Thicker than *Japanese paper*.

japanners' gold-size (*Paint.*). See **gold-size**.

japanning (*Paint.*). The process of finishing an article with japans, especially the ***stoving*** (q.v.) of japans.

Japan wax (*Chem.*). A natural wax obtained from sumach, m.p. 50°C. It has a high content of palmitin.

jar (*Civ. Eng.*). A special coupling-piece between adjacent boring rods, allowing for relative movement as the rods fall. This permits the drill-bit at the lower end to strike the ground freely, otherwise the weight of the rods, in the case of a deep boring, would be sufficient in itself to break them as they were dropped. (*Elec. Eng.*) Obsolete unit of capacitance, equal to 1/900 μF. See **Leyden jar**.

jargon aphasia (*Med.*). Rapid unintelligible utterance, due to a lesion in the brain.

jargons, jargoons (*Min.*). A name given in the gem trade to the zircons (chiefly of golden-yellow colour) from Ceylon. They resemble diamonds in lustre but are less valuable. See also **hyacinth**.

jarosite (*Min.*). A hydrous sulphate of iron and potassium crystallizing in the trigonal system; a secondary mineral in ferruginous ores.

jarrah (*For.*). A dense wood from Australia. It is of a deep-red colour, and is used for making piles, heavy framing, and wood paving-blocks.

jar-ramming machine (*Foundry*). See **jolt-ramming machine**.

jaspé (*Textiles*). (1) Plain woven cotton or spun rayon cloth with a shaded effect, usually embroidered or printed; used for bedspreads, curtains, etc. The warp ends are in pairs. (2) Fine coloured slub yarn. (3) Hank printed warp yarn.

jaspelline (*Textiles*). A light covert cloth made by a double satin weave from botany warp and fine woollen weft; used for coatings.
jasper (*Min.*). An impure opaque chalcedonic silica, commonly red owing to the presence of iron oxides.
jasp opal (*Min.*). See **opal jasper.**
JATO (*Aero.*). Abbrev. for *jet-assisted take off.*
jatrorrhizine, jateorhizine (*Chem.*). An alkaloid of the isoquinoline group, obtained from the calumba root, *Jatrorrhiza palmata.* It has not been obtained in the free state. Salts, however, are known, e.g., the iodide, $C_{20}H_{20}O_5NI + H_2O$, which crystallizes in reddish-yellow needles, m.p. 208°–210°C.
jaundice (*Med.*). Icterus. Yellow coloration of the skin and other tissues of the body, by excess of bile pigment present in the blood and the lymph. See **spherocytosis, spirochaetosis icterohaemorrhagica.**
jaune brilliant (*Paint.*). An artist's pigment made by mixing cadmium yellow, white lead, and vermilion.
Javel water or **eau de Javelle** (*Chem.*). A mixture of potassium chloride and hypochlorite in solution. Chiefly used for bleaching and disinfecting.
jaw (*Eng.*). (1) One of a pair or group of members between which an object is held, crushed, or cut, as the *jaws* of a vice or chuck. (2) One of several members attached to an object, to locate it by embracing another object.
jaw breaker (*Min. Proc.*). Jaw crusher, Blake crusher, alligator. Heavy-duty rock-breaking machine with fixed vertical, and inclined swing jaw, between which large lumps of ore are crushed.
jaw-foot (*Zool.*). See **maxilliped.**
jaws (*Print.*). See **folding jaws.** (*Zool.*) In gnathostomatous Vertebrates, the skeletal framework of the mouth enclosed by flesh or horny sheaths, assisting in the opening and closing of the mouth, and usually furnished with teeth or horny plates to facilitate seizure of the prey or mastication; in Invertebrates, any similar structures placed at the anterior end of the alimentary tract.
J-carrier system (*Teleph.*). A broadband carrier telephony system providing 12 telephone channels and using frequencies up to 140 kHz.
J display (*Radar*). A modified *A display* (q.v.), with circular time base.
jean (*Textiles*). Strong cotton-twilled material, used for overalls, boot-linings, corsets, etc. Usually of 2-and-1 warp face twill weave.
jeanette or **reversed jeanette** (*Textiles*). A material similar to *jean,* but generally of lighter weight.
jedding axe (*Tools*). An axe having one flat face and one pointed peen.
jejunectomy (*Surg.*). Excision of part of the jejunum.
jejunitis (*Med.*). Inflammation of the jejunum.
jejunocolostomy (*Surg.*). The formation, by operation, of a communication between the jejunum and the colon.
jejunojejunostomy (*Surg.*). The formation, by anastomosis, of a communication between two parts of the jejunum, thus short-circuiting the part in between.
jejunostomy (*Surg.*). The operative formation of an opening into the jejunum.
jejunotomy (*Surg.*). Incision of the jejunum.
jejunum (*Zool.*). In Mammals, that part of the small intestine which intervenes between the duodenum and the ileum.
jelly (*Cinema.*). A gelatine filter which is placed in front of moderately powered lamps to alter the spectral distribution of their light-emission (their colours).
jelutong (*For.*). Lightwood Malayan hardwood from the genus *Dyera.* It is almost white in colour and takes paints better than polishing techniques. The tree also yields a resinous type of rubber.
jemmy, jimmy (*Tools*). A small crowbar.
jerkin head (*Build.*). The end of a pitched roof which is hipped, but not down to the level of the feet of the main rafters, thus leaving a half-gable. Also called **shread head.**
jerk-pump (*I.C. Engs.*). A timed fuel-injection pump in which a cam-driven plunger overruns a spill port, thus causing the abrupt pressure rise necessary to initiate injection through the atomizer.
jerks (*Print.*). Violent intermittent pulls of paper through a web-fed printing press or folder due usually to incorrect or worn drives.
jervine (*Chem.*). $C_{27}H_{39}O_3N$, isolated from white hellebore (*Veratrum grandiflorum*), long prisms, m.p. 238°–242°C. The alkaloid depresses the circulation.
jet. A fluid stream issuing from an orifice or nozzle; a small nozzle, as the *jet* of a carburettor. (*Min.*) A hard coal-black variety of lignite, sometimes exhibiting the structure of coniferous wood.
jet coefficient (*Aero.*). The basic nondimensional thrust-lift relationship of the *jet flap* (q.v.); $C_J = J/\frac{1}{2}\rho \div V^2S$, where J=jet thrust, ρ=air density, V=speed, S=wing area.
jet condenser (*Eng.*). One in which exhaust steam is condensed by jets of cooling water introduced into the steam space.
jet deflexion (*Aero.*). A jet-propulsion system in which the thrust can be directed downwards to assist take-off and landing.
jet drilling (*Mining*). Fusion drilling of rock by means of high-temperature flame which melts the rock, and jet of water which then granulates the product and flushes it out.
jet flap (*Aero.*). A high-lift flight system in which (*a*) the whole efflux of the turbojet engine(s) is ejected downward from a spanwise slot at the wing trailing edge, or (*b*) a large surplus efflux from turboprop engine(s) is so ejected, with the airscrews providing a relative airflow over the wing. The downward ejection of the jet forms a barrier to the passage of air under the wing and induces air to flow downward across the upper surface, so forming a large suction loop and giving very high lift coefficients of the order of 10 and even higher. See also **NGTE rigid rotor.**
jet impactor (*Powder Tech.*). See **impaction sampler.**
jet looms (*Weaving*). High-speed machines in which conventional shuttle is eliminated. Weft is propelled through warp shed by high pressure jet of air or water, cut off at the selvedges and sewn in.
jet mill (*Chem. Eng.*). A mill in which particles are pulverized to micrometre size by the collisions occurring among them when they are swept into a small jet of gas at sonic velocity.
jet noise (*Aero.*). The noise of jet efflux varies as the eighth power of its velocity.
jet pipe shroud (*Aero.*). A covering of heat-insulating material, usually layers of bright foil, round a jet pipe.
jet propulsion (*Aero., Eng.*). Propulsion by reaction from the expulsion of a high-velocity jet of the fluid in which the machine is moving. It has been used for the propulsion of small ships by pumping in water and ejecting it at

increased velocity, but the principal application is to aircraft. See **pulse-jet, ramjet, reaction propulsion, rocket propulsion, turbojet.**

jet pump (*Mining*). Hydraulic elevator, in which a jet of high-pressure water rises in a pipe immersed in a sump containing the water, sands, or gravels which are to be entrained and pumped. Inefficient but cheap and effective where surplus hydraulic power exists.

jet shales (*Geol.*). Shales containing 'jet-rock', found in the Upper Lias of the Whitby district of England.

jet stream (*Meteor.*). A fairly well-defined core of strong wind, perhaps 200–300 miles (320–480 km) wide with wind speeds up to perhaps 200 m.p.h. (320 km/h), occurring in the vicinity of the *tropopause* (q.v.).

jetties (*Hyd. Eng.*). See **groynes.**

jetting-out (*Arch.*). The projection of, **e.g.**, a corbel from the face of a wall.

jet-wave rectifier (*Elec. Eng.*). A form of commutator rectifier in which a jet of mercury impinges on two stationary commutator segments. The requisite deflection of the jet from one segment to the other is effected by electromagnetic means, the jet carrying an alternating current and passing between the poles of a d.c. magnet.

jewel (*Horol.*). Natural ruby or sapphire, or synthetic stone, used for pivot bearings, also for the pallets and impulse pin. Owing to the high polish that can be obtained with such stones, combined with a hardness of surface, they provide wearing surfaces which have a long life and which cause little friction.

jewelled (*Horol.*). Fitted with jewels. In watches (except in the lowest grades), the balance staff is always jewelled with two through holes and two end-stones. The pallet staff is also jewelled, but only the higher grades of watches have end-stones fitted. In watches with the club-tooth escapement, the pallets and impulse pin are invariably jewels. A '15-jewel' watch has the following jewels: balance staff 4, pallet staff 2, escape wheel 2, fourth wheel 2, third 2, pallets 2, impulse pin 1. Watches may have as many as 23 jewels. For platform escapements the holes are generally jewelled. In precision clocks, all the holes may be jewelled, and often the acting faces of the pallets are formed of inset jewels.

jewfish (*Zool.*). A name applied to several species of fish, in particular to the giant Sea Bass Stereolepsis gigas which reaches 200 kg.

JFET (*Electronics*). See **bipolar transistor.**

jib (*Eng.*). The boom of a crane or derrick.

jib barrow (*Eng.*). A wheelbarrow consisting of a platform without sides; used in foundries and workshops.

jib crane (*Eng.*). An inclined arm or jib attached to the foot of a rotatable vertical post and supported by a tie-rod connecting the upper ends of the two. The load rope or chain runs from a winch on the post, and over a pulley at the end of the jib.

jib door (*Join.*). A door which carries and continues the general decoration of the wall.

jig (*Eng.*). A device used in the manufacture of (interchangeable) parts to locate and hold the work and to guide the cutting tool. (*Min. Proc.*) Device for concentrating ore according to relative gravity of its constituent minerals.

jig-back (*Civ. Eng.*). The type of *aerial ropeway* (q.v.) operated with a pair of carriers that travel in reverse directions, and are loaded or brought to rest alternately at the opposite stations, but which do not pass round the terminals. Also called the **to-and-fro aerial ropeway.**

jig borer (*Eng.*). A vertical-spindle machine for accurately boring and locating holes, having a horizontal table which can be precisely positioned by transverse and longitudinal feed motions.

jigger (*Eng.*). A hydraulic lift or elevator in which a short-stroke hydraulic ram operates the lift through a system of ropes and pulleys in order to increase the travel.

jig saw (*Tools*). A mechanical saw with a short narrow reciprocating blade which cuts on the up stroke; used for curved as well as straight cuts.

jig welding (*Elec. Eng.*). See **seam welding.**

Jim Creek transmitter (*Radio*). Very powerful radio transmitter (10^6 watts) of the U.S. Navy at Jim Creek, Washington.

jim-crow (*Eng.*). (1) A rail-bending device, operated by hand or by hydraulic power. (2) A crowbar fitted with a claw. (3) A swivelling tool-head used on a planing machine, cutting during each stroke of the table.

jimmy (*Tools*). See **jemmy.**

jink (*Mining*). A coupling between two mine-tubs or trains in a set or journey.

jink-carrier (*Mining*). A lad employed to carry the loose couplings or jinks from one train of mine-tubs to another.

jitter (*Telecomm.*). Small rapid irregularities in a waveform arising from fluctuations in supply voltages, components, etc.

j.j. coupling (*Nuc.*). Extreme form of coupling between the orbital electrons of atoms. Electrons showing individual spin-orbital coupling also interact with each other. See also **intermediate coupling, Russell-Saunders coupling.**

j.n.d. (*Psychol.*). Abbrev. for ***just noticeable difference.***

job (*Comp.*). Normal term for the compiling (or assembly) and/or processing of a particular programme. *Job control* is used if these are not to be performed sequentially. (*Work Study*) All the work carried out by a worker or a group of workers in the completion of their prescribed duties and grouped together under one title or definition. For special purposes, it may also denote a part of these duties.

jobbing chases (*Typog.*). Plain chases without bars, ranging from card sizes upward to folio.

jobbing founts (*Typog.*). Founts of type used for display purposes.

jobbing machines (*Print.*). The class of machines, usually platens, used for printing commercial or jobbing work.

jobbing work (*Typog.*). Small printed matter such as handbills, billheads, cards, etc.

job case (*Typog.*). A case with accommodation for capitals and small letters, suitable for *jobbing founts.*

jobs (*Textiles*). Defective or damaged cotton fabrics; usually sold by weight to merchants and market traders.

Job's-tears (*Min.*). Rounded grains of chrysolite (olivine) found associated with garnet in certain localities.

jockey pot (*Glass*). A pot of small size that is set on top of another pot, for the purpose of melting special glasses not needed in great quantity. Several such pots may be set in the space of one full-sized pot.

jockey relay (*Teleg.*). A double moving-coil relay in which two independently suspended coils have different degrees of damping, so that when the signal-current is passed through both, antennae fixed to the fast-moving coil operate between antennae fixed to the slow-moving coil. The antennae therefore repeat signals which are the variations in the signal-current, independent of signal-current drift.

jockey roller (*Print.*). On web-fed machines, a roller, usually the first to be traversed by the web, arranged to compensate any uneven tension as the reel unwinds.

jogger (*Print.*). (1) A rapidly-vibrating inclined tray on which sheets of paper are placed to be jogged up to two adjacent edges prior to cutting. (2) An adjustable fitment at the delivery of the printing machine which straightens the sheets as they are delivered.

joggle (*Build.*). (1) A shoulder formed on one structural member to support another member, and to form a weather-proof joint. (2) A piece or pin binding together adjacent stones in a course; similar to a *dowel* (q.v.) or *cramp* (q.v.). (*Carp.*) See **stub tenon.** (*Eng.*) (1) A small projection on a piece of metal fitting into a corresponding recess in another piece, to prevent lateral movement. (2) A lap joint in which one plate is slightly cranked so as to allow the inner edges of the two plates to form a continuous surface. (*Ships*) A sharp distortion in a plate, angle, or other section, made purposely to permit overriding of contacting members. It reduces the amount of steel packing.

joggle beam (*Carp.*). A built beam whose parts are joggled into one another.

joggle joint (*Build.*). A connexion between adjacent ashlars in which *joggles* are used.

joggle-piece, joggle-post (*Carp.*). A *king-post* (q.v.).

joggle-truss (*Carp.*). A roof-truss formed with a king-post.

joggle work (*Build.*). Coursed masonry in which slipping between the stones is prevented by the insertion of *joggles* (q.v.).

johannsenite (*Min.*). A silicate of calcium and manganese, often with some ferrous iron replacing the latter. It is a member of the pyroxene group, crystallizing in the monoclinic system.

Johne's disease (*Vet.*). Paratuberculosis. A chronic disease of cattle, sheep, and goats caused by infection by *Mycobacterium paratuberculosis* (*M. johnei*) which causes a chronic enteritis affecting the small intestine, caecum and colon; the main symptoms are diarrhoea and emaciation.

John Innes composts (*Agric.*). A range of prepared soils prescribed by the John Innes Horticultural Institution, particularly for sowing and potting of seedlings; essentially loam, peat, sand, superphosphate, chalk, etc.

Johnson concentrator (*Min. Proc.*). Machine used to arrest heavy auriferous material flowing in ore pulp. An inclined cylindrical shell rotates slowly, metallic particles being caught in rubber-grooved linings at periphery, lifted and separately discharged.

Johnson-Lark-Horowitz effect (*Electronics*). The resistivity of a metal or degenerate semiconductor, arising from electron scattering by impurity atoms.

Johnson noise. Same as **thermal noise.**

Johnston's organ (*Zool.*). In Insects, a sensory structure situated within the second antennal joint, proprioceptive in function.

join (*Print.*). The adhesion between the ends of two webs to ensure a continuous run of paper.

joiner s chisel (*Join.*). A *paring chisel* (q.v.).

joinery. (1) The craft of working timber to form the finishings of a building, as distinct from *carpentry.* (2) The material worked in this way. *n.* **joiner.**

joint (*Bot.*). A node. Also **joining.** (*Elec. Eng., Radar*) Permanent connexion between two lengths of cable. Waveguide connexion which may be *butt-jointed* (intermetallic contact maintained) or *choke-jointed* (when a half-wavelength short-circuited line is used to provide an effective contact at the guide walls). (*Foundry*) The parting plane in the sand round a rammed mould, to enable the pattern to be withdrawn. It is covered with *parting sand* before the cope or top half is rammed.

joint access (*Teleph.*). Means of allowing access to a circuit from both switchboard and selector levels.

joint chair (*Rail.*). A type of chair used at the joint between successive lengths of rail and providing support for the ends of both lengths.

jointed (*Bot.*). Said of an elongated plant member which is constricted at intervals and ultimately separates into a number of portions by breaking across the constrictions.

joint efficiency (*Eng.*). The ratio, expressed in per cent, of the strength of a section of riveted or welded joint to the strength of an analogous section of solid plate.

jointer (*Build.*). A tool used by bricklayers to form the mortar joint between the courses of bricks in pointing.

jointer plane (*Join.*). See **jointing plane.**

joint fastening (*Rail.*). A fish-plate or other means of fastening together the adjacent ends of successive lengths of rail.

joint hinge (*Join.*). A *strap hinge* (q.v.).

joint-ill (*Vet.*). Navel-ill; pyosepticaemia. A disease of young foals, calves, lambs, and piglets, caused by a variety of bacteria, and characterized by abscess formation in the umbilicus, pyaemia, abscesses in various organs, and arthritis affecting notably the leg joints.

jointing (*Build.*). The operation of making and/or finishing the joints between bricks, stones, timbers, pipes, etc. (*Eng.*) Material used for making a pressure-tight joint between two surfaces; e.g., asbestos sheet, corrugated steel rings, vulcanized rubber, etc. See **gasket.** (*Geol.*). See **joints.**

jointing plane (*Join.*). A bench plane, similar to the jack plane but larger (30 in. or 750 mm long), used for truing the edges of timbers which are to be accurately fitted together. Also called **jointer plane, shooting plane.**

jointing rule (*Build.*). A straightedge about 6 ft (2 m) long, used as a guide when *pencilling.*

jointless flooring (*Build.*). See **magnesite flooring.**

joint-mouse (*Med.*). A hard body, often a piece of cartilage, loose in the joint cavity; found especially in the joints of those suffering from osteoarthritis.

joint runner (*Civ. Eng.*). An incombustible material used as packing in the bottom of the socket of a pipe, at a joint made with lead to prevent the latter from running when molten. See **spigot-and-socket joint.**

joints (*Bind.*). The lateral projections formed on each side of a volume in the process of backing; the cover hinges along it. A linen strip pasted down the fold of the endpaper is called a *cloth joint.* (*Geol.*) Vertical, inclined, or horizontal divisional planes, found in almost all rocks; produced by tension or torsion. Also **jointing.** See **columnar structure, rift and grain.**

joint trunk exchange (*Teleph.*). An exchange restricted primarily to handling calls over toll and trunk circuits at the same positions.

joist (*Build.*). A horizontal beam of timber, reinforced or prestressed concrete, or steel, used with others as a support for a floor and/or ceiling.

Jolly balance (*Chem.*). A spring balance used to measure density by weighing in air and water.

Jolly's apparatus (*Chem.*). Apparatus for the volumetric analysis of air.

jolt- (or jar-) ramming machine (*Foundry*). A moulding machine (see **machine moulding**) in which the box, pattern, and sand are repeatedly lifted by a table operated by air pressure, and allowed to drop by gravity, the resulting jolt or jar packing or ramming the sand in an efficient manner. Also called **jolt-ram machine.**

jolt-squeeze machine (*Foundry*). A moulding machine (see **machine moulding**) used for deep patterns; in it, jolting is used to pack the sand on to the pattern followed by squeezing from the top to complete the ramming.

Joly block photometer (*Optics*). A photometer in which the light sources being compared illuminate two equal blocks of paraffin wax separated by a thin sheet of opaque material.

Joly's steam calorimeter (*Heat*). An instrument used to measure the specific heat capacity of a body, by weighing the amount of steam condensed on it when suspended in a steam chamber.

Jominy (end-quench) test (*Eng.*). A test for determining the relative hardenability of steels, in which one end of a heated cylindrical specimen is quenched, the resulting hardness decreasing towards the unquenched end.

Jona effect (*Cables*). See **stranding effect.**

Joosten process (*Mining*). Use of chemical reaction between solutions of calcium chloride and sodium silicate to consolidate running soils or gravels when tunnelling. A water-resistant gel is formed.

jordanon, Jordan's species (*Bot.*). A true breeding race of a species, but not sufficiently distinct to be given specific rank.

Jordan refiner (*Paper*). A conical plug revolving within a hollow cone, each being fitted with phosphor bronze bars so that pulp passing through is rubbed or beaten in a similar manner to that in a *hollander-engine* (q.v.).

joria (*Textiles*). One of the best types of East Indian wool, produced in the district of Joria.

Joshi effect (*Light*). Change of current in a gas because of light irradiation.

joule (*Phys.*). SI unit of work, energy, and heat, equal to work done when a force of 1 newton advances its point of application 1 m. One joule = 10^7 ergs = 0·2390 calorie = 0·000 948 Btu.

Joule effect (*Elec. Eng.*). Production of heat solely arising from current flow in a conductor. See **Joule's law.** (*Mag.*) Slight increase in the length of an iron core when longitudinally magnetized. See also **magnetostriction.**

Joule magnetostriction (*Mag.*). That for which length increases with increasing longitudinal magnetic field. Also **positive magnetostriction.**

Joule meter (*Elec. Eng.*). An integrating wattmeter whose scale is calibrated in joules.

Joule's equivalent (*Heat*). See **mechanical equivalent of heat.**

Joule's law (*Chem.*). (1) The intrinsic energy of a given mass of gas is a function of temperature alone; it is independent of the pressure and volume of the gas. (2) The molar heat of a solid compound is equal to the sum of the atomic heats of its component elements in the solid state. (*Elec. Eng.*) Heat liberated is $H = I^2Rt$ joules, where I = current in amperes, R = resistance in ohms, t = time in seconds. This is the basis of all electrical heating, wanted or unwanted. With high frequency a.c., R is an *effective resistance* and I may be confined to a thin *skin* of the conductor.

Joule-Thomson (Joule-Kelvin) effect (*Heat*). The slight fall in temperature which occurs when a gas is allowed to expand without doing external work. The effect is due to energy absorbed in overcoming the cohesion of the molecules of the gas. The liquefaction of gases by the Lindé process depends on the Joule-Thomson effect.

journal (*Eng.*). That part of a shaft which is in contact with, and supported by, a bearing.

journey man (or rider) (*Mining*). A dukey-rider, i.e., a man working on a *dukey* (q.v.).

jowl (*Zool.*). In *Diptera*, the lower portion of the gena below the eye.

joystick (*Aero.*). Colloq. for *control column.*

Joy's valve-gear (*Eng.*). A steam-engine valve-gear of the radial type used on some locomotives; in it, motion is taken entirely from a point on the connecting-rod.

jud (*Mining*). See **jad.**

JPT (*Aero.*). See **gas temperature.**

judder (*Telecomm.*). Irregular motion in facsimile transmission or reception, manifested in reproduced picture.

jugal (*Zool.*). A paired membrane bone of the zygoma of the Vertebrate skull, lying between the squamosal and the maxilla.

juggle (*For.*). A log sawn to length, and either split or left in the round.

Juglandales (*Bot.*). An order of the *Monochlamydeae.* Woody plants with the alternate leaves usually primate and without stipules. The male and female flowers are in separate spikes; 3–40 anthers; the unilocular inferior ovary is made of two fused carpels with one basal orthotropous ovule. The fruit is a drupe or nutlike. No endosperm.

jugular (*Zool.*). Pertaining to the throat or neck region, e.g., a *jugular* vein.

jugular nerve (*Zool.*). The posterior branch of the hyomandibular component of the 7th cranial nerve in Vertebrates; it carries the visceromotor component of the hyomandibular branch.

jugulum (*Zool.*). In Insects, the jugum; in Birds, the region of the breast which merges into the neck.

jugum (*Bot.*). A pair of opposite leaves. (*Zool.*) In some *Lepidoptera* and in *Trichoptera*, a finger-shaped process arising from the posterior margin of the fore-wing; it serves to unite the 2 wings during flight. *pl.* **juga.**

julaceous (*Bot.*). Cylindrical and smooth; resembling a catkin.

Julian calendar (*Astron.*). The system of reckoning years and months for civil purposes, based on a tropical year of 365·25 days; instituted by Julius Caesar in 45 B.C. and still the basis of our calendar, although modified and improved by the Gregorian reform.

Julian date (*Astron.*). The number of any given day, in a system of reckoning by successive days, independently of various calendars and chronological epochs; instituted by J. Julius Scaliger in 1582, it has no connexion whatever with the Julian calendar.

Julianiales (*Bot.*). An order of the *Monochlamydeae*; woody plants with alternate, usually pinnate, leaves without stipules; the male and female flowers are separate; the males are numerous in dense panicles, with a perianth and 6–8 anthers; the females naked, unilocular in fours on pendulous spikes; there is one ovule on a broad, hollow funicle; no endosperm.

jumbo (*For.*). A type of light, tongueless, double

sledge with cross-chain coupling, used for short distance hauling. (*Mining*) Carriage on which several drifter rock drills are mounted when advancing a tunnel heading.

jump (*Comp.*). Control of computer so that the next instruction obeyed is not that stored sequentially after the last. (*Phys.*) Deprecated term for transition of electron or nucleus to different energy level.

jump cut (*Cinema.*). A disproportionate break in the action of a film due to careless cutting or interrupted exposure.

jumper (*Build.*). A *through-stone* (q.v.). (*Civ. Eng.*) A pointed steel rod which is repeatedly dropped on the same spot from a suitable height (being turned slightly between blows), and which, by pulverizing the earth, forms a bore-hole. (*Elec. Eng.*) (1) A short section of over-head transmission line conductor serving to form an electrical connexion between two sections of line. (2) Multi-core flexible cable making connexion between the coaches of a multiple-unit electric train. (*Horol.*) A click in the form of a wedge which causes a star wheel to jump forward one space. (*Mining*) The borer, steel, or bit for a compressed-air rock drill. (*Teleph.*) Length of wire used in telephony to re-arrange permanent circuit connexions.

jumper bar (*Mining*). A weighted steel bar with a cutting edge, raised and dropped by hand.

jumper-cable (*Elec. Eng.*). A cable for making electrical connexion between two sections of conductor-rail in an electric traction system.

jumper field (*Teleph.*). The cross-connexion or translation field in a director. Any space devoted to jumpers, i.e., temporary connexions, especially within a distribution frame. Also **cross-connexion field.**

jumper-top blast pipe (*Eng.*). A locomotive *blast pipe* (q.v.) in which the back pressure, and hence the draught, are automatically limited by the lifting of an annular valve, which increases the nozzle area.

jumper wire (*Teleph.*). See **jumper.**

jumping-figure watch (*Horol.*). A watch in which the time is indicated by figures on disks jumping into position in windows in the watch dial. Also **digital watch.**

jumping-up (*Met.*). The operation of thickening the end of a metal rod by heating and hammering it in an endwise direction. Also called **upsetting.**

jump joint (*Met.*). A butt joint made by *jumping-up* (q.v.) the ends of the two pieces before welding them together.

junction (*Electronics*). Area of contact between semiconductor material having different electrical properties, larger than a *point contact*. (*Telecomm.*) Union or division of waveguides in either H- or E-planes, *tee* or *wye*, *tapered* to broaden the frequency response, or *hybrid* to direct flow of wave energy.

junction chamber (*Build.*, *San. Eng.*). A closed chamber, generally of brick and concrete, inserted in a sewer system for accepting the inflow of one or more sewers and allowing for the discharge thereof.

junction circuit (*Teleph.*). One directly connecting two exchanges situated at a distance apart less than that specified for a trunk circuit.

junction coupling (*Elec. Eng.*). In coaxial-line cavity resonators, coupling by direct connexion to the coaxial conductor.

junction diode (*Electronics*). One formed by the junction of *n*- and *p*-type semiconductors, which exhibits rectifying properties as a result of the potential barrier built up across the junction by the diffusion of electrons from the *n*-type material to the *p*-type. Applied voltages, in the sense that they neutralize this potential barrier, produce much larger currents than those that accentuate it.

junction rectifier (*Electronics*). One formed by a *p-n* junction by *holes* being carried into the *n*-type semiconductor. Germanium and silicon power rectifiers have displaced thermionic rectifiers in most applications.

junction transistor (*Electronics*). One in which the base receives minority carriers through *junctions* on each side, which may diminish to *points*. In normal operation, the emitter is forward biased. Signal currents entering the base tend to neutralize reverse bias on the collector junction, thus controlling the relatively large current flowing in collector circuit.

Jungermanneales (*Bot.*). One of the main sub-divisions of the *Hepaticae*. The plant body is either a thallus or a cylindrical branched leafy stem. The thallus is of simple construction, without air pores, and the antheridia and archegonia develop on or (rarely) in it. The capsule usually opens by four valves.

Jungner accumulator (*Elec. Eng.*). A form of nickel-iron accumulator having an alkaline electrolyte.

juniper (*For.*). A conifer, *Juniperus*, yielding an essential oil used mechanically; its fruits are used to flavour gin, and its wood for veneers, pencils, etc.

junk (*For.*). A piece of drawn or hewn timber of large dimensions.

junking (*Mining*). The process of cutting a passage through a pillar of coal.

junk ring (*Eng.*). A metal ring attached to a steam-engine piston for confining soft packing materials; or for similarly holding a cast-iron piston-ring in position.

Jupiter (*Astron.*). The largest planet in the solar system, with a volume 1319 times and mass 318 times those of the earth; its relative density is therefore only 1·33. The oblate shape is easily seen and is due to the rapid rotation in 9 h. 50 min. The belted surface shows variable currents in a very dense atmosphere, in which methane and ammonia have been detected, with hydrogen as the main constituent. The planet revolves at a mean distance of 5·203 astronomical units in a period of 11·86 years; there are 12 satellites, four of which are easily seen in small instruments. Jupiter is the source of irregular outbursts of radio-noise. See **red spot.**

Jurassic System (*Geol.*). The middle division of the Mesozoic era, named after the Jura Mts, where rocks of this age are found.

jury strut (*Aero.*). A strut giving temporary support to a structure. Usually required for folding-wing biplanes, sometimes for naval monoplanes with folding parts.

Justape (*Typog.*). Fully automatic computer to produce a justified tape from *idiot tape* (q.v.) by word-spacing to a predetermined maximum and, if this be insufficient, searching for a suitable word which when letter-spaced will fill the measure; specially suitable for newspaper lines; output up to 6 000 lines per hour. Cf. *Linasec.*

justification (*Typog.*). The correct spacing of words to a given measure of line.

just noticeable difference (*Psychol.*). The smallest difference between two stimuli that enables one to say they are different. Statistically, the difference perceived 50% of the time. Abbrev. **j.n.d.**

Justowriter (*Print.*). A typewriting system, the

keyboard producing a punched tape which when run through the typing unit produces lines of equal length. Used for *near print* (q.v.).

just scale, just temperament (*Acous.*). The same as *natural scale.*

jute (*Textiles*). Strong, brownish, bast fibre from the Asian plants *Corchorus olitorius* and *C. capsularis*, and from the American plant *Abutilon theophrasti* (Indian Mallow). Used in cordage, carpets, canvas, hessian, etc.

jutty (*Mining*). In Derby and Notts, a small tub or truck used for gathering coal in thin seams.

juvenile water (*Geol.*). Water derived from magma, as opposed to meteoric water (derived from rain or snow), or connate water (trapped in sediments at the time of deposition).

juvenile form (*Bot.*). A young plant which has leaves and other features different from those of a mature plant of the same species.

juvenile hormone (*Zool.*). See **neotenin.**

juvenile leaf (*Bot.*). The form of leaf found on a sporeling or seedling, when it differs markedly from the leaf of the adult plant.

juvenile stage (*Bot.*). A special stage in the life-history of some algae, from which the ordinary plant develops as a vegetative outgrowth.

juxta (*Zool.*). In *Lepidoptera*, a sclerotized support at the point where the aedeagus sheath joins the body.

juxtaglomerular (*Physiol.*). Close to the renal glomerulus, e.g., *juxtaglomerular apparatus*, a small group of cells located on the afferent arteriole of the glomerulus of the kidney.

juxtaposition twins (*Min.*). Two (or more) crystals united regularly in accordance with a 'twin law', on a plane (the 'composition plane') which is a possible crystal face of the mineral. Cf. *interpenetration twins.*

K

k (*Chem.*). A symbol for the *velocity constant* of a chemical reaction. (*Chem. Eng.*) Symbol for *mass transfer coefficient*. (*Phys.*) A symbol for: (1) the *Boltzmann constant*; (2) the *radius of gyration*. (*Psychol.*) Symbol representing the group factor of spatial ability.

κ- (*Chem.*). A symbol for: (1) *cata-*, i.e., containing a condensed double aromatic nucleus substituted in the 1.7 positions; (2) substitution on the 10th carbon atom; (3) *electrolytic conductivity*. (*Elec.*, *Mag.*) Symbol for *magnetic susceptibility*. (*Mech.*) Symbol for *compressibility*.

K (*Chem.*). The symbol for *potassium*. (*Heat*) The symbol for *kelvin*.

K (*Chem.*). A symbol for *equilibrium constant*; K_s, solubility product. (*Chem. Eng.*) See **k**. (*Mech.*) Symbol for *bulk modulus*.

[K] (*Light*). A very strong Fraunhofer line in the extreme violet. See **[H]**.

K-acid (*Chem.*). 1,8-Aminonaphthol-4,6-disulphonic acid, an intermediate for dyestuffs.

K aerial (*TV*). A K-shaped TV reception aerial, the 2 short arms forming a dipole and the vertical element acting as a reflector.

kaersutite (*Min.*). A hydrous silicate of calcium, sodium, magnesium, iron, titanium, and aluminium; a member of the amphibole group crystallizing in the monoclinic system. It occurs in somewhat alkaline igneous rocks.

Kahler's disease (*Med.*). Multiple myeloma. A disease in which tumours arising from cells of the bone-marrow appear in various parts of the skeleton.

Kahn's test (*Med.*). A precipitation test for syphilis, diluted antigen being added to inactivated blood serum.

kainite (*Min.*). Hydrated sulphate of magnesium, with potassium chloride, which crystallizes in the monoclinic system. It usually occurs in the upper portions of salt deposits, e.g., at Stassfurt (Germany).

Kainozoic (*Geol.*). See **Cainozoic**.

kaiwekite (*Geol.*). A volcanic rock containing phenocrysts of olivine, titanaugite, barkevikite and anorthoclase, probably a hybrid between basalt and trachyte.

kala-azar (*Med.*). Black fever. A disease due to infection with the protozoon *Leishmania donovani*; characterized by enlargement of the liver and spleen, anaemia, wasting, and fever.

Kalanite (*Cables*). TN for a hard insulating material, not affected by oil, used for spreaders in cable joints.

Kaleoilres tape (*Cables*). TN for an oil-resisting compounded linen tape used in cable joints.

kalinite (*Min.*). Hydrated sulphate of potassium and aluminium, which probably crystallizes in the monoclinic system. It has the same composition as potassium alum.

kaliophilite (*Min.*). Silicate of potassium and aluminium, which crystallizes in the hexagonal system. It has a similar composition to kalsilite and is probably metastable at all temperatures at atmospheric pressure.

kallitron (*Radio*). A periodic combination of two triode valves for obtaining negative resistance.

kalsilite (*Min.*). A potassium-aluminium silicate, $KAlSiO_4$; it is related to nepheline and crystallizes in the hexagonal system.

Kaluza theory (*Phys.*). A unified field theory of gravitation and electromagnetism which is based on a 5-dimensional continuum.

kalymmocytes (*Zool.*). In *Urochorda*, cells of the follicle which pass into the egg after maturation.

kamacite (*Min.*). A variety of nickeliferous iron, found in meteorites; it usually contains about 5·5% nickel. Metallurgically, alpha-iron.

kame (*Geol.*). A mound of gravel and sand which was formed by the deposition of the sediment from a stream as it ran from beneath a glacier. Kames are thus often found on the outwash plain of glaciers.

kames (*Bot.*). See **terfas**.

Kanawha Series (*Geol.*). A group of productive coal measures occurring in the Pennsylvanian of the Appalachian region, and completely developed in Virginia.

kandahar (*Textiles*). An East Indian wool of good quality, used in making Indian carpets, featuring a design consisting of sections of differently coloured squares.

kandite (*Min.*). A collective term for the kaolin minerals or members of the kaolinite group. These include kaolinite, dickite, nacrite, anauxite, halloysite, meta-halloysite, and allophane.

kanthals (*Met.*). Alloys with high electrical resistivity, used as heating elements in furnaces. General composition iron 67%, chromium 25%, aluminium 5%, cobalt 3%.

kaolin (*Geol.*). Clay consisting of the mineral kaolinite; china clay.

kaolinite (*Min.*). A finely crystalline form of hydrated aluminium silicate, $(OH)_4Al_2Si_2O_5$, occurring as minute monoclinic flaky crystals with a perfect basal cleavage, resulting chiefly from the alteration of feldspars under conditions of hydrothermal or pneumatolytic metamorphism, or by weathering. The kaolinite group of clay minerals includes the polymorphs *dickite* and *nacrite*. See **kandite**.

kaolinization (*Geol.*). The process by which the feldspars in a rock such as granite are converted to kaolinite.

kaon (*Nuc.*). See under **meson**.

Kaplan water turbine (*Eng.*). A propeller-type *water turbine* (q.v.) in which the pitch of the blades can be varied in accordance with the load, resulting in high efficiency over a large load-range.

kappa number (*Paper*). A bleachability test for pulps measured by the number of millilitres of 0·1 *N* potassium permanganate solution consumed by 1 g of dry pulp.

Kapp coefficient (*Elec. Eng.*). A coefficient occasionally used in electric-machine design and embodying the form factor, distribution factor, etc., so that, when multiplied by the number of turns in series, the flux and 10^{-8}, it gives the r.m.s. value of the e.m.f.

Kapp line (*Elec. Eng.*). A unit of magnetic flux occasionally used in electric-machine design; equal to 6×10^{-5} Wb = 60 μWb.

Kapp phase advancer, Kapp vibrator (*Elec. Eng.*). A form of phase advancer for use with slip-ring induction motors. It consists of a small armature connected in each phase of the rotor circuit and allowed to oscillate freely in a d.c. field so that it has a leading e.m.f. induced in it.

kapur (*For.*). Wood from *Dryobalanops*, found in Sabah, Sumatra, Sarawak, and the Malay

Peninsula, which also yields Borneo camphor (see **borneol**). When in contact with iron the wood develops bad stains and eventually the metal corrodes.

karat. See **carat.**

Karbate (*Chem. Eng.*). TN for a form of extremely dense carbon having high enough strength to permit its being made into special shapes, e.g., tubes, pumps, and possessing corrosion resistance and good heat conductivity.

karri (*For.*). A dense wood from Australia, similar to *jarrah.* It is of a deep-red colour, and is used for making piles, heavy framing, and wood paving-blocks.

karst (*Geol.*). Any uneven limestone topography, characterized by joints enlarged into criss-cross fissures (*grikes*) and pitted with depressions resulting from the collapse of roofs of underground caverns.

kary-, karyo-. Prefix from Gk. *karyon,* nucleus. Also **cary-, caryo-.**

karyaster (*Cyt.*). A group of chromosomes arranged like the spokes of a wheel.

karyenchyma (*Cyt.*). See **karyolymph.**

karyogamy (*Bot.*). The fusion of the sex nuclei, and interchange of nuclear material, after the fusion of two gametes.

karyokinesis (*Cyt.*). See **mitosis.**

karyolymph (*Cyt.*). A colourless watery fluid, occupying most of the space inside the nuclear membrane; nuclear sap.

karyolysis (*Cyt.*). Dissolution of the nucleus by disintegration of the chromatin; gradual disappearance of the nucleus in a dead cell; liquefaction of nuclear membrane in mitosis.

karyomere (*Cyt.*). A swollen condition sometimes seen in chromosomes towards the end of a nuclear division.

karyomicrosome (*Cyt.*). A nuclear granule.

karyomite (*Cyt.*). See **chromosome.**

karyon (*Biol.*). The cell-nucleus.

karyophans (*Zool.*). In some *Ciliophora,* nuclear granules composing the spironeme and axoneme of the stalk.

karyoplasm (*Cyt.*). See **nucleoplasm.**

karyoplasmatic ratio (*Cyt.*). The ratio between the volume of a nucleus and that of the cytoplasm in the same cell.

karyorrhexis (*Cyt.*). Breaking up of the chromatin of the nucleus into darkly staining granules (in necrosis of the cell).

karyosome (*Cyt.*). A nucleus; a chromosome; an aggregation of chromatin in a resting nucleus; a type of nucleolus well shown by many of the lower plants, which stains with basic dyes and furnishes material for the chromosomes during mitosis. Cf. *plasmosome.*

karyotheca (*Cyt.*). See **nuclear membrane.**

karyotin (*Cyt.*). The substance which makes up the nuclear reticulum. See **chromatin.**

kashgar (*Textiles*). A white silky wool produced in Chinese Turkestan, exported principally to Russia.

Kaspar-Hauser experiments (*An. Behav.*). Experiments in which animals are reared in complete isolation from other animals of their own or other species.

kaspine leather (*Leather*). A white washable leather used for gloves, etc.

kata-. Prefix from Gk. *kata,* down. Also **cata-.**

katabatic (*Meteor.*). Said of a wind which is caused by the downward motion of air due to convection, as when cold air flows down into a valley from high ground.

katabions, catabions (*Biol.*). Organisms in which katabolic processes predominate over anabolic processes, as in animals.

katabolism, catabolism (*Biol.*). Exergonic metabolism, i.e., the breakdown of complex organic molecules by living organisms resulting in the liberation of energy.

katabolite, catabolite (*Biochem.*). A product of katabolism, e.g., urea, carbon dioxide.

katadromous (*Bot.*). Said of the venation in the royal fern, where the first nerves in each leaf segment come off on the basal side of the midrib. (*Zool.*) Of Fish, migrating to water of greater density than that of the normal habitat to spawn, as the Freshwater Eel which migrates from fresh to salt water to spawn. Cf. *anadromous.* Also **catadromous.**

kata-front (*Meteor.*). A situation at a front, warm or cold, where the warm air is sinking relative to the *frontal zone* (q.v.).

katagenesis (*Zool.*). Retrogressive evolution.

katakinetic (*Biol.*). Tending to the discharge of energy. Cf. *anakinetic.*

katakinetomeres (*Biol.*). Unreactive stable protoplasm, atoms or molecules. Cf. *anakinetomeres.*

kataklastic structures (*Geol.*). Structures produced in a rock by the action of severe mechanical stress, during dynamic metamorphism. The constituent minerals generally show deformation and granulation.

katamorphism (*Geol.*). The breaking-down processes of metamorphism, as contrasted with the building-up processes of anamorphism.

kataphase (*Cyt.*). The stages of mitosis from the formation of the chromosomes up to the division of the cell.

kataphorite (*Min.*). See **katophorite.**

kataplexy (*Zool.*). The state of imitation of death, adopted by some animals when alarmed.

katapositive (*Photog.*). The description of a positive on an opaque surface, such as paper, in contrast to a transparency on glass or film.

katathermometer (*Mining*). Instrument used in mine ventilation survey to assess cooling effect of air current. Thermometer bulb is first exposed dry, then when covered with wetted gauze, and time taken for temperature to fall from 100° to 95°F (38° to 35°C) is observed.

Katayama disease (*Med.*). A disease due to invasion of the body by the blood fluke *Schistosoma Japonicum,* characterized by urticaria, painful enlargement of liver and spleen, bronchitis, diarrhoea, loss of appetite, and fever lasting a few days to several weeks.

katepimeron (*Zool.*). In Insects, the lower sclerite of the epimeron when the latter is subdivided.

katepisternum (*Zool.*). In Insects, the lower sclerite of the episternum when the latter is subdivided.

katharometer (*Chem.*). An instrument for the analysis of gases by means of measurements of thermal conductivity.

kathetron (*Electronics*). See **cathetron.**

katophorite, kataphorite, or cataphorite (*Min.*). A hydrous silicate of sodium, calcium, magnesium, iron and aluminium; a member of the amphibole group. It crystallizes in the monoclinic system and occurs in basic alkaline igneous rocks.

katydid (*For.*). See **logging wheels.**

kauri-butanol value (*Chem.*). Term applied to solvents to indicate their dissolving powers for resins. Based on natural kauri gum as the standards.

kauri gum (*Chem.*). A gum found in New Zealand, used for varnishes and linoleum cements. It is the resinous exudation of the kauri pine (*Agathis australis*), a tree whose timber is of value for general joinery and decorative purposes.

Kaye disk centrifuge (*Powder Tech.*). A transparent disk centrifuge with an entry port coaxial with the axis of revolution. The suspension of the power under test is injected into the centrifuge while it is running at the final speed. Concentration changes are measured optically.

kayser (*Phys.*). Unit for *wave number*, the reciprocal of a wavelength (in cm). Obsolete.

Kayser-Fleischer ring (*Med.*). A ring of brownish-yellow pigmentation found in the cornea of patients suffering from *Kinnier-Wilson's disease* (q.v.).

Kazanian (*Geol.*). A stratigraphical stage in the Permian rocks of Russia and eastern Europe.

K-capture (*Nuc.*). Absorption of an electron from the innermost (K) shell of an atom into its nucleus. An alternative to ejection of a *positron* from the nucleus of a radioisotope. Also **K-electron capture**. See also **X-rays**.

kCi (*Nuc.*). Abbrev. for *kilocurie*, i.e., radioactivity equivalent to 1000 curies.

kcps, KCPS or **kc/s** (*Telecomm.*). Abbrev. for **kilocycles per second**. See **kilohertz**.

K display (*Radar*). A form of *A-display* (q.v.) produced with a lobe-switching antenna. Each lobe produces its own peak and the antenna is directly on target when both parts of the resulting double peak have the same height.

keatite (*Min.*). A high-pressure synthetic form of silica not yet found in nature.

kebbing (*Vet.*). See **enzootic ovine abortion.**

Keber's organ (*Zool.*). In some *Lamellibranchiata*, a glandular organ connected with the epithelial lining of the pericardial cavity.

kedge anchor (*Ships*). A small anchor used for steadying and warping purposes.

keelson (*Aero.*). A longitudinal structural member inside the bottom of the hull of a flying-boat. It forms part of the main framework, connecting up the transverse members and bulkheads. (*Ships*) (1) A term descriptive of the longitudinal strength members of a ship, which form the shell-plating stiffeners. A *flat keel* is the lower horizontal member of the ship's backbone; a *centre keelson* is the vertical member thereof; a *bar keel* is similar to the latter, external to the hull; *side keelsons* are vertical members, off the ship's centre line. (2) The wrought-iron saddles or standards which support cylindrical boilers of the Scotch marine type. Sometimes called **boiler cradles.**

Keene's cement (*Build.*). A quick-setting hard plaster, made by soaking plaster of Paris in a solution of alum or borax and cream of tartar.

keep (*Eng.*). See **keeper.**

keep-alive arc (*Elec. Eng.*). Small auxiliary arc in a mercury-arc rectifier to maintain ionization during off-load condition.

keep-alive electrode (*Elec. Eng.*). See **pilot electrode.**

keep down (*Typog.*). A typographic instruction to avoid the use of capital initials so far as possible.

keeper (*Build.*). (1) The part of a Norfolk latch limiting the travel of the fall bar. (2) The socket fitted on a door jamb to house the bolt of the lock in the shut position. (*Elec. Eng.*) See **armature (2).** (*Eng.*) The lower part of the bearing in a railway-truck axle-box, which limits the downward movement of the box due to track irregularities. (*Mag.*) Bar used to close the magnetic circuit of a permanent magnet when not in use, thereby conserving its strength. Also **keep.**

keep standing (*Print.*). An instruction to keep the printing surface intact, particularly to keep the type locked up in chase.

keep up (*Typog.*). (1) A typographic instruction to use capital initials in preference to small letters. (2) A synonym for **keep standing.**

Keesom relationship (*Phys.*). One involving molecular attraction and dipole interaction.

Keewatin Group (*Geol.*). A series of basic pillow lavas associated with sedimentary iron ores (worked in the 'Iron Ranges'); forms part of the Pre-Cambrian succession in the Canadian Shield.

kefir (*Chem.*). A fermentation product of milk in which the lactose has undergone both alcoholic and lactic fermentation simultaneously.

Keflin (*Pharm.*). TN (U.S.) for **cephalothin.**

keilhauite (*Min.*). A variety of sphene containing more than 10% of the rare earths.

Keinbock's disease (*Surg.*). Osteochondritis of the lunate bone in the carpus.

Keith and Flack's node (*Zool.*). A plexiform mass of special connective tissue, situated near the entrance of the superior vena cava, under the endocardium of the right auricle in the Mammalian heart; here the heart contractions originate.

kelat (*Textiles*). A wool from Baluchistan; short and springy, it is used for carpet-making.

K-electron (*Nuc.*). One of the two electrons in the K (innermost) shell of an atom. Its principal quantum number is unity.

K-electron capture (*Nuc.*). See **K-capture.**

Kel-F (*Plastics*). TN (U.S.) for **polytrifluorochloroethene.**

Keller furnace (*Elec. Eng.*). A form of electric furnace for iron smelting, in which part of the heat is produced by the passage of the current through the charge and part by arcs between the electrodes and the charge.

Kelling's test (*Chem.*). A test for the detection of lactic acid in gastric juice, based upon the colouring effect produced by the addition of a few drops of a very dilute neutral iron (III) chloride solution.

kelly (*Mining*). In drilling to depth, top pipe of rotary string of drills, with which is incorporated a flexibly attached swivel through which mud is pumped to bottom of hole.

keloid (*Med.*). A dense new growth of skin occurring in skin that has been injured.

kelp (*Bot.*). (1) A general name for large seaweeds. (2) The calcined ashes of such seaweed, a source of iodine, etc.

kelvin (*Phys.*). The SI unit of thermodynamic temperature. It is 1/273·16 of the temperature of the *triple point* of water above *absolute zero*. The temperature interval of 1 kelvin (K) equals that of 1°C (degree Celsius). See **Kelvin thermodynamic scale of temperature.**

Kelvin ampere-balance (*Elec. Eng.*). Laboratory instrument for measuring current; in it the calculated force between two coils carrying the current to be measured is balanced by the force of gravity on a weight sliding along a beam.

Kelvin bridge (*Elec. Eng.*). An electrical network involving two sets of ratio arms used for accurate measurement of low resistances.

Kelvin compass (*Ships*). A form of ship's compass having a very light card and a number of short parallel needles held by silk cords, as well as other special features. Also called **Thomson compass.**

Kelvin effect (*Elec. Eng.*). Skin effect, whereby varying currents in a conductor tend to concentrate near the surface. Important in radio circuits, and in eddy-current (high-frequency) heating.

Kelvin electrometer (*Elec. Eng.*). Same as *quadrant electrometer* (see **Dolezalek quadrant electrometer**).

Kelvin's law (*Elec. Eng.*). A principle regarding the transmission of electrical energy. It states that the most economical size of conductor to use for a line is that for which the annual cost of the losses is equal to the annual interest and depreciation on that part of the capital cost of the conductor which is proportional to its cross-sectional area.

Kelvin sounder (*Ocean.*). An apparatus which consists of a weight attached to a wire and carries a glass tube of small calibre open at one end and coated inside with a composition sensitive to sea water. This indicates the pressure and therefore the depth of water reached. Cf. *echo sounding*.

Kelvin thermodynamic scale of temperature (*Heat*). Scale of temperature in which measurements are made from absolute zero; the temperature interval corresponds with that of the Celsius (Centigrade) scale, so that the freezing point of water is 273·15 K. Expressed in *kelvins* (symbol K). (Before 1967, expressed in 'degrees Kelvin' (symbol °K).)

Kelvin-Varley slide (*Elec. Eng.*). Constant resistance decade voltage divider of the type used in vernier potentiometers. It consists of a resistor of $2r$ ohms shunting two adjacent units in a series array of eleven resistors, each r ohms. The $2r$ resistor may similarly be divided into eleven equal parts, and shunted if additional decade(s) are required.

kelyphitic rim (*Geol.*). A shell of one mineral enclosing another in an igneous rock, produced by the reaction of the enclosed mineral with the other constituents of the rock. Kelyphitic rims are most common in basic and ultrabasic igneous rocks.

kemp (*Textiles*). A thick fibre, without scales, occurring in badly bred wools. It will not dye but is used for obtaining distinctive surface effects on coating cloths; the kemps show white on a coloured surface.

kenetron (*Electronics*). Large hot-cathode vacuum diode for industrial rectification and X-ray plants. In U.S. a **valve tube**.

Kennack Gneiss (*Geol.*). Banded red and black gneisses which form part of the Lizard Complex of Cornwall; they result from the admixture of acid (granitic) and basic (basaltic) magma, intruded during a period of stress.

Kennametal (*Met.*). TN for hard alloys of tungsten carbide in cobalt matrix.

Kennelly-Heaviside layer (*Phys.*). See **E-layer.**

kentallenite (*Geol.*). A coarse-grained, basic igneous rock, named from the type locality, Kentallen, Argyllshire; it consists essentially of olivine, augite, and biotite, with subordinate quantities of plagioclase and orthoclase in approximately equal amounts.

Kent claw hammer (*Tools*). One with thick, slightly-curved claw and hexagonal head.

kentledge (*Civ. Eng., etc.*). Scrap iron, rails, heavy stones, etc., used as loading on a structure (e.g., upon the top section in sinking a cylinder caisson), or as a counterbalance for a crane.

kentrogon (*Zool.*). A stage in the life-history of certain parasitic *Cirripedia* (e.g., *Sacculina*) which succeeds the Cypris stage and precedes the entry of the parasite into the body of the host.

kenyte (*Geol.*). A fine-grained igneous rock, occurring as lava flows on Mt Kenya, E. Africa, and in the Antarctic; essentially an olivine-bearing phonolite with phenocrysts of anorthoclase.

kephalin (*Chem.*). See **cephalin.**

Kepler's laws of planetary motion (*Astron.*). (1) The planets describe ellipses with the sun at a focus. (2) The line from the sun to any planet describes equal areas in equal times. (3) The squares of the periodic times of the planets are proportional to the cubes of their mean distances from the sun.

keps (*Mining*). Bearing-up stops for supporting a cage or load at the beginning or end of hoisting in a shaft.

keraphyllous (*Zool.*). In ungulate Mammals, said of a layer intervening between the horny part of a hoof and the living tissue.

kerasin (*Chem.*). A *cerebroside* obtained from the brain substance. On hydrolysis it yields a fatty acid (lignoceric acid), galactose and a base (sphingosine).

keratectasia (*Med.*). Local bulging of part of the cornea.

keratectomy (*Surg.*). Excision of part of the cornea.

keratin (*Zool.*). An insoluble scleroprotein formed by the transformation of the eleidin granules in the superficial cells of the stratum granulosum of the Vertebrate integument; it constitutes the horny part of hoofs, nails, and hair, and contains a high percentage of sulphur, part of which is readily split off in the form of H_2S by hydrolysis. Adj. **keratogenous.**

keratitis (*Med.*). Inflammation of the cornea.

keratocele (*Med.*). Protrusion of the innermost layer (Descemet's membrane) of the cornea through a corneal ulcer.

keratoconus (*Med.*). Conical cornea. Cone-shaped deformity of the cornea owing to weakness and thinness of the centre.

keratodermia blennorrhagica (*Med.*). Red patches on the skin which become hard, dry, yellow, and raised above the skin, occurring in gonorrhoea.

keratogenous (*Zool.*). See **keratin.**

keratoglobus (*Med.*). Hemispherical protrusion of the whole cornea, e.g., in glaucoma in infants.

keratohyalin (*Zool.*). In the skin of higher Vertebrates, a hyaline substance present as flakes or droplets in the stratum lucidum.

keratoma (*Med.*). A tumour of the skin in which overgrowth of the horny layer predominates.

keratomalacia (*Med.*). A disease in which the cornea first becomes dry and lustreless and then softens; associated with deficiency of vitamin A in the diet.

keratophyre (*Geol.*). A fine-grained igneous rock of intermediate composition. It is essentially a soda-trachyte, containing albite-oligoclase or anorthoclase in a cryptocrystalline groundmass. The pyroxenes, when present, are often altered to chlorite or epidote.

keratoplasty (*Surg.*). The grafting of a new cornea on to an eye the cornea of which has become opaque.

Keratosa (*Zool.*). An order of *Demospongiae* in which the skeleton is composed of spongin fibres; includes the Bath Sponges.

keratoscleritis (*Med.*). Inflammation of the cornea and of the sclera of the eye.

keratosis (*Med.*). Overgrowth of the horny layer of the skin.

keratotomy (*Surg.*). Incision of the cornea.

keraunograph (*Meteor.*). See **ceraunograph.**

kerf (*Build.*). A heap of mixed clays and ashes left exposed to the action of the weather, sometimes for long periods, in order to produce a mixture suitable for brick-making. (*Carp.*) The cut made by a saw. (*Mining*) In coal winning, undercut made by coal-cutting machine to depth of 1 m or more. Also called **kirve.**

kermes (*Zool.*). A dyestuff prepared from the dried females of *Kermes ilicis*, a Hemipteran insect.

kermesite (*Min.*). Oxysulphide of antimony, which crystallizes in the monoclinic system. It is a secondary mineral occurring as the alteration product of stibnite. Also called **pyrostibite.**

kern (*Typog.*). The portion of some type letters which projects beyond the body and rests on the body of the preceding or following letter, e.g., the tail of an italic *f*.

kernel (*Bot.*). (1) The seed inside the stony endocarp of a drupe. (2) See **centrum.** (*Nuc.*) Mathematical function which defines linear operator. Neutron sources distributed continuously in various configurations may be represented by *kernels* which are used in calculations of flux, and of slowing-down density.

Kernig's sign (*Med.*). A sign of meningitis. When the patient is lying on his back with the thigh bent at right angles to the body, the leg cannot be bent straight at the knee.

kernite (*Min.*). $Na_2B_4O_7 \cdot 4H_2O$, a sodium borate containing 4 molecules of water, found in the Kramer borate district, Kern County, California. Colourless monoclinic crystals, rel. d. 1·908, hardness 2·5.

kerogen (*Geol.*). That part of the organic matter present in a sedimentary rock which cannot be extracted by the use of solvents.

kerosine (*Fuels*). A petroleum fraction with a boiling range of 150°–300°C, rel. d. 0·78–0·82, flash-point not lower than 32°C; used in lamps and heating appliances. Another type of *kerosine* (with higher volatility), known as *vaporizing oil*, is used in some internal-combustion engines; and kerosine-type fuels are of great importance for gas turbine and jet engines.

Kerr cell (*Elec. Eng.*). A light modulator consisting of a liquid cell between crossed polaroids. The light transmission is modulated by an applied electric field. See **Kerr effect.**

Kerr effect (*Elec. Eng.*). Double refraction produced in certain transparent dielectrics by the application of an electric field. Also **electro-optical effect.** (*Light*) Slight rotation of plane of polarization of light when reflected from the polished surface of magnetized material. Also **magneto-optical effect.**

kersantite (*Geol.*). A mica-lamprophyre, named from the type locality Kersanton, near Brest; it consists essentially of biotite and plagioclase feldspar. Cf. *minette*.

kersey (*Textiles*). A milled fabric, woollen or worsted, made from coarse crossbred. Usually 2-and-2 twill weave, there are many varieties.

ketenes (*Chem.*). Compounds of the general formula $R_2C{=}C{=}O$. The ketene series may be considered homologues of carbon monoxide. The first member of the series is ketene, $CH_2{=}CO$, is readily obtainable by passing propanone vapours through a red-hot glass tube filled with broken tile. Propanone is then decomposed into ketene and methane. Ketenes form acids or acid derivatives on adding water, alcohols, ammonia, amines, etc., to one of their double bonds. They are liable to autoxidation; they react with other unsaturated compounds, forming 4-membered rings. They are very unstable and polymerize easily.

keto-enolic tautomerism (*Chem.*). The formation by certain compounds of two series of derivatives, based upon their ketonic or enolic constitution. The enol-form is produced from the keto-form by the migration of a hydrogen atom, which forms a hydroxyl group with the ketone oxygen, accompanied by a change in the position of the double bond. Thus:

$$\underset{\text{keto-form}}{-CH_2-CO-} \rightleftarrows \underset{\text{enol-form}}{-CH{=}C(OH)-}$$

keto form (*Chem.*). That form of a substance exhibiting keto-enol tautomerism which has the properties of a ketone, e.g., ethyl acetoacetate.

ketogenic (*Med.*). Capable of producing ketone bodies, e.g., *ketogenic diet*.

ketohexoses (*Chem.*). $HO{\cdot}CH_2{\cdot}(CHOH)_3{\cdot}CO{\cdot}CH_2OH$, a group of carbohydrates, isomers of *aldohexoses* (q.v.). The formula contains three asymmetric carbon atoms, and there are eight stereoisomers, viz. D- and L-fructose, -sorbose, -tagatose and -allulose. They reduce an alkaline copper solution. Ketohexoses can be oxidized to acids containing fewer carbon atoms in the molecule.

ketonaemia, ketonemia (*Med.*). The presence of ketone bodies in the blood. See **ketosis.**

ketones (*Chem.*). Compounds containing a carbonyl group, —CO—, in the molecule attached to two hydrocarbon radicals. The general formula is

$$R-\underset{\underset{O}{\|}}{C}-R'$$

Ketones are formed by the oxidation of secondary alcohols, by the dry distillation of the calcium or barium salt of an organic acid, by the catalytic condensation of acids or esters, by synthesis with Grignard reagents, and by the action of CO on sodium alkyls. Ketones are very reactive substances, forming additive compounds, e.g., with sodium hydrogen sulphite, etc. Important derivatives are the oximes, obtained by the action of hydroxylamine. Hydrogen reduces ketones to secondary alcohols. The simplest ketone is *acetone* (*propanone*).

ketonuria (*Med.*). The presence of ketone bodies in the urine. See **ketosis.**

ketoses (*Chem.*). A general term for monosaccharides with a ketonic constitution. They always form mixtures of acids on oxidation, containing a smaller number of carbon atoms than the original ketose.

ketosis (*Med.*). The excessive formation in the body (due to incomplete oxidation of fats) of ketone or acetone bodies (aceto-acetic acid and β-hydroxybutyric acid) which are accompanied by ketonaemia and ketonuria; occurs, e.g., in diabetes.

ketoximes (*Chem.*). The reaction products of ketones with hydroxylamine, containing the oximino group $=N{\cdot}OH$ attached to the carbon atom.

kettle (*Met.*). An open-top vessel used in carrying out metallurgical operations on low melting-point metals, e.g., in drossing and desilverizing lead.

Kettleness Beds (*Geol.*). A little-used name for part of the Lias of Yorkshire, consisting chiefly of sandy shales.

kettlestitch (*Bind.*). The stitch which is made at the head and tail of each section of a book to interlock the sections.

Keuper Marl (*Geol.*). A fine-grained siltstone, locally 900 to 1200 m in thickness, forming the higher part of the Triassic System in N.W. Europe. Probably deposited under conditions similar to those applicable to the loess. It consists chiefly of red marls, with dolomite crystals and important salt deposits on different horizons. See also **Tea-green Marl.**

Keuper Series (*Geol.*). The upper series of rocks assigned to the Triassic System in N.W. Europe, lying above the Muschelkalk; it consists of sandstones and marls, deposited in England and Germany under desert conditions, though in the Alps these are represented by marine strata.

keV (*Nuc.*). Abbrev. for kilo-electron-volt; unit of particle energy, 10^3 *electron-volts*.

kevel (*Build.*). A hammer, edged at one end and pointed at the other, used for breaking and rough-hewing stone.

Kew Certificates (*Horol.*). Certificates of performance of watches and chronometers issued by the National Physical Laboratory, Teddington. (This work was originally undertaken at Kew, hence the name.) For an 'absolutely perfect' watch, 100 marks are awarded, made up as follows: 40 for a complete absence of variation of daily rate; 40 for absolute freedom from change of rate with change of position; and 20 for perfect compensation for effects of temperature.

Keweenawan Series (*Geol.*). Conglomerates, arkoses, red sandstones and shales of desert origin, associated with great thicknesses of basic lavas and intrusives. This is the youngest of the Pre-Cambrian formations in the Canadian Shield, and is comparable with the Torridonian of Scotland. The igneous rocks carry copper, silver, nickel, cobalt and platinum.

Kew-pattern barometer (*Meteor.*). The Adie barometer. Specially graduated so that error arising from changes in the free level in the cistern is obviated.

Kew-pattern magnetometer (*Elec. Eng.*). A delicate type of reflecting magnetometer having a photographic recording arrangement for keeping record of changes in the earth's field.

key (*Bind.*). The name applied to a small tool used for securing the bands while a book is being sewn. (*Build.*) In any surface to be plastered, the roughening, lathing, or other preliminary process undertaken in order to give a grip to the coat of plaster and so enable it to adhere more satisfactorily. (*Civ. Eng.*) One of the pair of wedge-shaped hardwood blocks used as *striking wedges* (q.v.). (*Eng.*) A piece inserted between a shaft and a hub to prevent relative rotation. It fits in a key-way, parallel with the shaft axis, in one or both members, the commonest form being the parallel key, of rectangular section. (*For.*) In tree felling, the portion of a tree bole deliberately left uncut until the last moment to lessen the risk of splitting. (*Photog.*) By analogy with music, the psychological fitness of brightnesses in a picture as a whole, or in relation to another. (*Telecomm.*) A hand-operated device for changing circuits. (*Teleg.*) Hand-operated spring-loaded switch used for telegraphy transmissions.

keyboard (*Comp., etc.*). An array of control keys for operating a piece of equipment.

key boss (*Eng.*). A local thickening up of a boss or hub at the point at which a key-way is cut, to compensate for loss of strength due to the cut.

key chuck (*Eng.*). A jaw chuck whose jaws are adjusted by screws turned by a key or spanner. See **self-centring chuck.**

key click (*Telecomm.*). Click produced by *ringing* in key circuit as contact is opened or closed. Normally minimized by use of key-click filter. Also **key chirp.**

key-course (*Civ. Eng.*). The course of stones in an arch corresponding to the keystone.

key drawing (*Cinema.*). In animated cartoon production, *key drawings* indicate situations at special instants, such as at beats in the bar of music, after which the in-between drawings are made to fit with the timing. (*Print.*) In lithography and line-colour block making, an outline drawing which serves as a guide in the making of the separate colour plates.

key drop (*Build.*). A guard plate covering a keyhole and falling into position by its own weight.

keyed pointing (*Build.*). Pointing which is finished with lines or grooves struck on the flat joint. See **flat pointing.**

keyer (*Telecomm.*). A device for changing the output of a transmitter from one frequency (or amplitude) to another according to the intelligence transmitted. Also used for short sound bursts.

keyhole saw (*Tools*). One with stiff, narrow blade 11–16 in. (250–400 mm) long, for internal, curved and small cuts. Also **padsaw.**

key-industry animals (*Zool.*). In an animal community, the small rapidly-reproducing animals at the base of any food-chain, which turn plant protein into animal protein.

keying (*Eng.*). The process of fitting a key to the key-ways in a shaft and boss. (*Telecomm.*) Production of signals by operating of key. Classified by the circuit lead in which keying takes place, i.e., anode, cathode, control grid, screen grid, etc., keying. (*Typog.*) Operating the keyboard of a typesetting or filmsetting machine to produce: lines of type (e.g., *Linotype, Intertype*); immediate exposure of photomats on film (e.g., *Fotosetter*); or a perforated spool of paper from which lines are cast (*Monotype*), or photomats exposed on film (e.g., *Monophoto, Linofilm, Photon*).

keying wave (*Radio*). See **marking wave.**

keyless ringing (*Teleph.*). The normal ringing in a manual exchange, whereby ringing current is applied automatically to a subscriber's line on insertion of a junction plug into a jack in the multiple; it is cut off when the subscriber loops his line and operates a relay.

key light(ing) (*Cinema., TV*). The main illumination on a set.

key plan (*Build., Eng.*). A small-scale plan showing the relative disposition of a number of items in a scheme.

key plate (*Build.*). An *escutcheon* (q.v.).

key print (*Photog.*). The black-and-white print, additional to the primary colour prints, which is used to swamp colour fringes arising from parallax or mis-registration. The key print also controls the saturation.

key-seating (*Eng.*). A key-wave, or the surface on to which a key is bedded.

key-seating machine (*Eng.*). A machine tool for milling key-ways in shafts, etc., by means of an end mill, the work being supported on a table at right angles to the axis of the spindle. Feed is obtained by an automatic traverse of either the tool or the table.

keysender (*Teleph.*). A strip of plunger keys for transmitting marginal currents representing, in a code, dialled impulse trains; the operation of step-by-step switching apparatus is thereby speeded up. See **A-position, B-position.**

keystone (*Civ. Eng.*). The central voussoir at the crown of an arch.

keystone distortion (*TV*). Distortion arising in a camera tube because the length of the horizontal scan line depends upon its vertical displacement, when the electron beam scans the image plate at an acute angle. It results in distortion of the rectangle into a trapezoidal pattern; it can be obviated by special transmitter circuits.

key-way (*Eng.*). A longitudinal slot cut in a shaft or hub to receive a *key* (q.v.).

key-way tool (*Eng.*). A slotting machine tool used for the vertical cutting of key-ways, equal in width to that of the key-way. See **slotting machine.**

keywords (*Comp.*). The most informative or significant words in the title of a document. These are the elements stored in the memory in most information retrieval systems.

k-factor (*Nuc.*). See **specific gamma-ray emission.**

khaki (*Textiles*). A dull, yellowish-brown, piece-dyed fabric known as *drab-mixture*, used for dress and field uniforms in the British Army, in serges, drills, meltons, etc.

khamsin (*Meteor.*). A hot dry wind from the south, which blows over Egypt in front of depressions moving eastward along the Mediterranean.

khorasan (*Textiles*). A long staple Persian wool from the province of Khorasan, suitable for combing; used for high-quality rugs.

kibble (*Mining*). See **bowk.**

kibbler (*Mining*). See **putter.**

kick (*Autos.*). See **back-kick.** (*Build.*) See **frog.**

kick copy (*Print.*). On a web-fed printing press the printed and folded product is counted off in batches, each batch indicated by a displaced copy.

kicker (*Print.*). The mechanism on a web-fed printing press which displaces one copy of the product to indicate the intervals between batches. See **quire spacing.**

Kick's law (*Min. Proc.*). Assumes that the energy required for subdivision of a definite amount of material is the same for the same fractional reduction in average size of the individual particles, i.e., $E = k_K \log_e d_1/d_2$, where E is the energy used in crushing, k_K is a constant, depending on the characteristics of the material and method of operation of the crusher, and d_1 and d_2 are the average linear dimensions before and after crushing.

kicksorters (*Telecomm.*). See **pulse height analyser (1).**

kid (*Hyd. Eng.*). A bundle of brushes serving as a *groyne* (q.v.). (*Leather*) A soft leather made from the skins of young goats (kids) or calves.

Kidderminster carpet (*Textiles*). See **Scotch carpet.**

kidney (*Zool.*). A paired organ for the excretion of nitrogenous waste products in Vertebrates.

kidney machine (*Med.*). See **artificial kidney.**

kidney ore (*Min.*). A form of the mineral haematite, the (III) oxide of iron, which occurs in reniform masses, hence the name (Latin *ren*, kidney).

kidney piece (*Horol.*). A cam, shaped like a kidney, used in perpetual calendar work to give the equation of time.

kidney stone (*Min.*). A name given to *nephrite* (q.v.), which was once supposed to be efficacious in diseases of the kidney (Gk. *nephros*, kidney).

kidney stones (*Med.*). Hard deposits formed in the kidney or bladder. The composition varies, and kidney stones have been found to consist of uric acid and urates, calcium oxalate, calcium and magnesium phosphate, silica and alumina, cystine, xanthine, fibrin, cholesterol, and fatty acids.

kidney worm disease (*Vet.*). Infection of the kidney and other organs of carnivora and occasionally pigs, horses, and cattle, by the nematode worm *Dioctophyma renale*.

kier (*Textiles*). The boiler in which yarn and cloth are treated previous to bleaching or dyeing; it may be of the open or closed type, capable of holding two to three tons of material.

kieselguhr (*Min.*). See **diatomite.**

kieserite (*Min.*). Hydrated sulphate of magnesium which crystallizes in the monoclinic system; found in large amounts in the salt deposits of Germany.

kieve (*Mining*). See **dolly tub.**

kieving (*Min. Proc.*). See **tossing.**

kilderkin (*Brew.*). Cask holding 18 gal (82 litres).

kilhig (*For.*). A short, stout pole used to push a tree in the direction that it should fall.

kill (*Typog.*). In printing, an editorial instruction to delete entirely some item in preparation, derived from the use of the word as an instruction to distribute type.

killed steel (*Met.*). Steel that has been *killed*, i.e., fully deoxidized before casting, by the addition of manganese, silicon, and sometimes aluminium. There is practically no evolution of gas from the reaction between carbon and iron-oxide during solidification. Sound ingots are obtained. See also **rimming steel.**

killer (*Phys.*). See **poison.**

kiln (*Min. Proc., Met.*). Furnace used for: drying ore; driving off carbon dioxide from limestone; roasting sulphide ores or concentrates to remove sulphur as dioxide; reducing iron (II) ores to magnetic state in reducing atmosphere.

kilo-. Prefix denoting 1000; used in the metric system. E.g. 1 kilogram = 1000 grams.

kilocalorie (*Heat*). See **calorie.**

kilocurie source (*Nuc.*). Giant radioactive source —usually in form of ^{60}Co.

kilocycles per second (*Phys.*). See **kilohertz.**

kilo-electron-volt (*Nuc.*). See **keV.**

kilogram(me). Unit of mass in the MKSA(SI) system, being the mass of the *international prototype kilogram*, a cylinder of platinum-iridium alloy kept at Sèvres.

kilogram-calorie (*Heat*). See **calorie.**

kilohertz (*Phys.*). An SI unit of frequency in which there are 1000 complete cycles or oscillations per second. Abbrev. kHz.

kilometric waves (*Radio*). Those with wavelengths between 1000 and 10 000 m.

kiloton (*Nuc.*). Unit of explosive power for nuclear weapons equal to that of 10^3 tons of TNT.

kilovar (*Elec. Eng.*). A unit of reactive volt-amperes equal to 1000 VArs. Abbrev. kVAr.

kilovolt-ampere (*Elec. Eng.*). A commonly used unit for expressing the rating of a.c. electrical machinery and for other purposes; it is equal to 1000 volt-amperes. Abbrev. kVA.

kilowatt (*Elec. Eng.*). A unit of power equal to 1000 watts and approximately equal to 1·34 h.p. Abbrev. kW.

kilowatt-hour (*Elec. Eng.*). The commonly used unit of electrical energy, equal to 1000 watt-hours or 3·6 MJ. Often called simply a *unit*; abbrev. kWh. See **Board of Trade Unit.**

Kimberley horse disease (*Vet.*). Walk-about disease. A disease of horses in Australia due to poisoning caused by eating whitewood (*Atalaya hemiglauca*).

kimberlite (*Geol.*). A type of mica-peridotite, occurring in volcanic pipes in South Africa and elsewhere, and containing xenoliths of many types of ultramafic rocks, and diamonds.

Kimeridge (or Kimmeridge) Clay (*Geol.*). A constant thick bed of black marine clay, stretching across England from the Dorset coast, through the Fens, to the Yorkshire coast near Scarborough; referred to the Upper Jurassic, and lying between the Corallian and the Portlandian Stage.

Kimeridge Coal (*Geol.*). A bed of bituminous shale, occurring in the Kimeridge Clay division of the Upper Jurassic; exposed on the Dorset coast, and capable of yielding oil on destructive distillation.

Kimeridgian (*Geol.*). A stage name of the Upper Jurassic System, applicable to rocks of the same age as the Kimeridge Clay, though not necessarily of the same lithology.

kimzeyite (*Min.*). A zirconium-bearing member of the garnet group of minerals.

kinaesthetic (*Zool.*). Pertaining to the perception of muscular effort.

kinases (*Biochem.*). Substances of biological origin, such as intestinal juice, which convert zymogens, proenzymes, or proferments into the true or active enzymes and thus act as activators.

Kinderscout Grit (*Geol.*). A coarse sandstone, one of the Millstone Grits in the Namurian Stage of the Upper Carboniferous rocks of Derbyshire.

kinematic chain (*Eng.*). A number of links connected to one another to allow motion to take place in combination. It becomes a mechanism when so constructed as to allow constrained relative motion between its links.

kinematics. That branch of applied mathematics which studies the way in which velocities and accelerations of various parts of a moving system are related.

kinematic viscosity (*Phys., Eng.*). The coefficient of viscosity of a fluid divided by its density. Symbol ν. Thus $\nu = \eta/\rho$. Unit in the CGS system is the stokes ($cm^2 s^{-1}$); in SI $m^2 s^{-1}$.

kinesalgia (*Med., Psychol.*). Feeling of pain on movement.

kinescope (*TV*). See **teletube.**

kinesis (*An. Behav.*). *Orientation behaviour* (q.v.) in which an animal reaches its destination without being truly orientated, i.e., its direction is not precisely related to the direction of the source of stimulation. See **klinokinesis, orthokinesis, photokinesis.** Cf. *taxis.*

kinesodic (*Zool.*). Conveying motor impulses.

kinetheodolite (*Space*). An improved form of theodolite used in tracking missiles and artificial satellites; the telescope of the theodolite is replaced by a long-focus ciné camera, in which the film gives a continuous series of photographs of the target, together with recordings of the elevation and azimuth, and the exact time of each exposure. The precision of the readings is of the order of 1/100 of a second of time, and a fraction of a minute of arc.

kinetia (*Zool.*). *Pl.* of *kinety* (q.v.).

kinetic body (*Cyt.*). A tiny granular body lying where a chromosome is attached to the spindle.

kinetic constriction, kinetochore (*Cyt.*). The portion of a chromosome where the attachment is made to a spindle fibre.

kinetic energy (*Mech.*). Energy arising from motion. For a particle of mass m moving with a velocity v it is $\frac{1}{2}mv^2$, and for a body of mass M, moment of inertia I_g, velocity of centre of gravity v_g and angular velocity ω, it is
$$\tfrac{1}{2}Mv_g^2 + \tfrac{1}{2}I_g\omega^2.$$

kinetic friction (*Mech.*). See **friction.**

kinetic head (*Hyd.*). See **velocity head.**

kinetic heating (*Aero.*). See **dynamic heating.**

kinetic pressure (*Aero.*). See **dynamic pressure.**

kinetic theory of gases (*Phys.*). The conception of gas molecules as elastic spheres whose bombardment of the walls of the containing vessel due to their thermal agitation causes the pressure exerted by the gas. The theory gives a simple explanation of the gas laws and has yielded valuable results concerning gaseous viscosity and molecular dimensions.

kinetin (*Biochem.*). Plant growth substance derived from yeast; promotes cell-division:

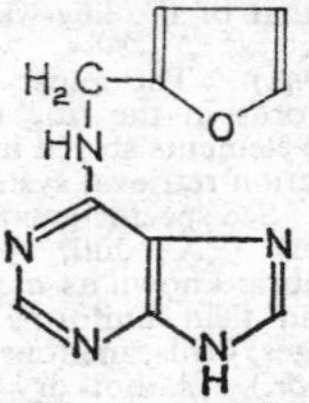

H_2C, O, HN, N, N, N, N, H

kinetoblast (*Zool.*). In certain types of active pelagic larvae, the outer locomotor layer of ciliated cells.

kinetochore (*Cyt.*). See **kinetic constriction.**

kinetodesma (*Zool.*). See **kinety.**

kinetogenesis (*Zool.*). A theory of the mechanism of evolution which postulates that animal structures have been directly or indirectly produced by animal movements.

kinetomeres (*Biol.*). Molecules of protoplasm which may be energy-rich and reactive or energy-poor and stable. See **anakinetomeres, katakinetomeres.**

kinetonucleus (*Zool.*). In blood-living *Mastigophora*, a nucleus-like body situated between the nucleus and the blepharoplast, and believed to represent the parabasal body of other forms. Cf. *trophonucleus.*

kinetosomes (*Zool.*). See **kinety.**

kinety (*Zool.*). A unit of structure in the *Protozoa* comprising the *kinetosomes* (the basal granules of the cilia and flagella) and the *kinetodesma* (a fine strand running to the right of the kinetosomes). In *Flagellata* the line of cell division is parallel to the *kinetia* (symmetrical division), and in *Ciliata* the plane of cleavage cuts across the *kinetia* (percentien division).

king (*Zool.*). In *Isoptera*, a male reproductive which, with the queen, is the original founder of a colony.

king-closer (*Build.*). A three-quarter brick used to maintain the bond of the surface.

kingite (*Min.*). A hydrated basic phosphate of aluminium.

king-piece (*Carp.*). See **king-post.**

king-pile (*Civ. Eng.*). A pile driven down the centre of a wide trench to enable two short struts (butting on opposite sides of the pile) to be used, instead of one long one, for keeping the poling boards on opposite sides of the trench in position.

king-pin (*Autos.*). The pin by which a stub axle is articulated to an axle-beam or steering head; it is inclined to the vertical to provide caster action. Also called **swivel-pin.**

king-post (*Carp.*). A vertical timber tie connecting the ridge and the tie-beam of a roof, shaped at its lower end to afford bearing to two struts supporting the middle points of the rafters. Also called a **broach-post, joggle-piece, joggle-post, king-piece, middle-post.**

king-post truss (*Carp.*). A timber roof-truss in which there is only one vertical member—the central king-post.

king-rod (*Build.*). A vertical steel rod connecting the ridge and tie-beam of a couple-close roof, to prevent sagging of the tie-beam when it is required to support ceiling loads.

Kingston valve (*Eng.*). A sea-valve fitted to a ship's side for the purpose of admitting water to circulating pumps, or flooding or blowing out ballast tanks.

kingswood (*For.*). See **Brazilian kingswood.**

king-tower (*Build.*). In a derrick tower gantry, that one of the three timber towers through which the weight of the derrick itself is transmitted directly to the foundation below.

kinins (*Biochem.*). Group of plant hormones which promote cell division and are used commercially as a preservative for cut flowers. They include *kinetin* and *zeatin* (qq.v.). (*Physiol.*) Specific vasoactive (essentially vasodilator) substances of protein or polypeptide nature, released enzymatically by various tissues, e.g., salivary glands, sweat glands, pancreas. These kinins, e.g., bradykinin, cause hyperaemia and also influence capillary permeability.

kink (*Electronics*). The abrupt reversal of slope as occurs in the anode voltage/anode current characteristic for a tetrode.

kink instability (*Nuc.*). See **instability.**

Kinnier-Wilson's disease (*Med.*). A rare disease in which excessive amounts of copper are deposited in the body.

kino (*For.*). A red resinous exudate, rich in tannins, from various trees, notably *Pterocarpus spp.* Used to a limited extent in medicine and tanning. Also **gum kino.**

kinoplasm (*Cyt.*). Protoplasm which appears to be composed of fibrils and which in cell-division composes the spindle fibres, attraction sphere, and astral rays.

kinsh (*Mining*). A crowbar.

kip (*Eng., etc.*). A unit of mass equivalent to 1000 lb. (*Leather*) The untanned hides of yearling cattle, or of adult cattle of any small breed; used for boot uppers.

Kipp relay (*Telecomm.*). A form of trigger or bistable multivibrator circuit.

Kipp's apparatus (*Chem.*). A generator for hydrogen sulphide, carbon dioxide, or other gases, by reacting a solid with a liquid. By opening or closing the gas outlet, gas pressure automatically fills or empties the solids container with liquid.

Kirchhoff's equation (*Heat*). The rate of change of the heat ΔH of a process with temperature, carried out at constant pressure, is given by:

$$\left(\frac{\partial \Delta H}{\partial T}\right)_p = \Delta\left(\frac{\partial H}{\partial T}\right)_p = \Delta C_p,$$

where ΔC_p is the change in the heat capacity at constant pressure for the same process. Similarly, for a process carried out at constant volume:

$$\left(\frac{\partial \Delta U}{\partial T}\right)_V = \Delta\left(\frac{\partial U}{\partial T}\right)_V = \Delta C_v.$$

Kirchhoff's laws (*Elec. Eng.*). Generalized extensions of *Ohm's law* employed in network analysis. They may be summarized as:

(1) $\Sigma i = 0$ at any junction
(2) $\Sigma E = \Sigma iZ$ round any closed path.

E=e.m.f., i=current, Z=complex impedance. (*Phys.*) A result due to Kirchhoff's researches on radiation by which he established that the ratio of the coefficient of absorption to the coefficient of emission is the same for all substances, and depends only on the temperature and frequency of the rays.

Kirkendall effect (*Met.*). If a piece of a pure metal is placed in contact with a piece of an alloy of that metal and the whole heated, the region of alloy will diffuse into the pure metal.

kirkifier (*Radio*). A form of linear rectifier comprising a triode valve whose grid is maintained at a small positive potential with respect to the filament, the anode being used as the rectifying electrode.

Kirklington (or Kirklinton) Sandstone (*Geol.*). A local sandstone belonging to the Keuper series of rocks, in the Triassic System of Cumberland.

Kirkwood gaps (*Astron.*). Gaps which occur in the belt of asteroids at distances from the sun where the mean motion would be commensurate with that of Jupiter.

kirve (*Mining*). See **kerf.**

kish (*Met.*). Solid graphite which has separated from, and floats on the top of, a molten bath of cast-iron or pig-iron which is high in carbon.

kiss impression (*Print.*). The lightest possible impression on paper so that the type or blocks do not press into the paper, requiring careful *make-ready*; also required when printing rubber or plastic plates to avoid distortion.

kiss of life (*Med.*). The 'mouth-to-mouth' method of resuscitation in which the rescuer places his mouth over the patient's and inflates the latter's lungs by breathing into them.

kitchen midden (*Geol.*). The dump of waste material, largely shells and bones associated with ashes, marking the site of a kitchen in a settlement of early man.

kite (*Aero.*). Any aerodyne anchored or towed by a line, not mechanically or power-driven. Derives its lift from the aerodynamic forces of the relative wind. (*Meteor.*) Box-kites have been used for carrying meteorographs to high altitudes to record the weather conditions there. Altitudes exceeding 6000 m have been reached by using several kites in tandem.

kite winder (*Build.*). A *winder* (q.v.) used at the angle of a change of direction in a stair, and having consequently the shape of a kite in plan.

kitkaite (*Min.*). Nickel selenium telluride, crystallizing in the trigonal system.

kivite (*Geol.*). An alkaline igneous rock, a melanocratic leucite-basanite.

Kjeldahl flask (*Chem.*). A glass flask with a round bottom and a long wide neck. Used in Kjeldahl's method for the estimation of nitrogen.

Kjeldahl's method (*Chem.*). A method for the quantitative estimation of nitrogen in organic compounds, based on the conversion of the organic nitrogen into ammonium sulphate, and subsequent distillation of ammonia after the solution has been made alkaline. The ammonia which distils over can be titrated.

Kjellin furnace (*Elec. Eng.*). A form of electric-induction furnace having an iron core; used for melting ferrous metals.

Klein bottle (*Maths.*). A one-sided surface obtained by pulling the narrow end of a tapering cylinder through the wall of the cylinder and then stretching the narrow end and joining it to the larger end. Cf. *Möbius strip.*

Klein-Gordon equation (*Phys.*). That describing the motion of a spinless charged particle in an electromagnetic field.

Klein-Nishina formula (*Nuc.*). Theoretical expression giving cross-section of electrons for scattering of photons. See **Compton effect.**

kleinpflaster (*Civ. Eng.*). The method of laying road surfacing setts to form segments or arches on plan, as a result of which wheels cross joints obliquely.

Klein-Rydberg construction (*Phys.*). The point by point construction of the potential energy curve of a diatomic molecule from observed rotational and vibrational levels.

klemm's leather. Crown leather.

Klieg eyes (*Cinema.*). The effect of strain on the eyes caused by the brilliance of floodlights in film production.

Klieg light (*Cinema.*). A type of incandescent floodlighting lamp for studio use.

K-lines (*Nuc.*). Characteristic X-ray frequencies from atoms due to excitation of electrons in the K-shell. These give information relating to the atomic numbers of the elements bombarded.

klinker brick (*Build.*). A very hard type of brick much used in Holland and Germany, principally for paving purposes. See **adamantine clinkers, Dutch clinkers, terro-metallic clinkers.**

klinogeotropic (*Biol.*). Said of an organ which takes up a stable position lying at an angle to the vertical.

klinokinesis (*An. Behav.*). A *kinesis* (q.v.) in which an animal moves in a straight line in a favourable environment, but begins to turn when it enters an unfavourable environment, the frequency of the turning being related to the strength of the stimulus, and the result being to lead the animal to a favourable environment. Cf. *orthokinesis.*

klinostat (*Bot.*). An instrument on which a plant may be slowly rotated, and by means of which its reactions to gravity and other stimuli may be investigated.

klinotaxis (*An. Behav.*). A type of *taxis* (q.v.) in which an animal directs itself towards or away from a source of stimulation by comparing the intensity of stimulation on either side by alternative movements of the whole body or by some mobile receptor to the left and right. Cf. *telotaxis, tropotaxis.*

klinotropic (*Biol.*). Placed at a slant to the direction of a given stimulus.

klippen (*Geol.*). A structural term (*plural*) applicable in areas of complex tectonics to thrust masses of strata that map out like outliers; in effect, outliers capping hill-tops and lying on thrust-planes. *sing.* **klippe.**

klirrfaktor (*Telecomm.*). The same as **nonlinear distortion factor.**

Klischograph (*Print.*). An *electronic engraving machine* for same-size plates on metal or plastic, supplied with a choice of screen rulings to suit the user; models available for line work. See also **Vario-Klischograph.**

K-, L-, M- . . . shells (*Nuc.*). Imaginary spheres surrounding the nucleus of an atom, on the surfaces of which groups of electrons may be considered to revolve. Starting with the innermost shell and moving outwards, the shells are called *K, L, M, . . . Q*. The *K*-shell can contain no more than 2 ($=2\times1^2$) electrons, the *L*-shell 8 ($=2\times2^2$), the *M*-shell 18 ($=2\times3^2$). See the articles on the individual shells K-, L-, . . . Q-; also **atomic structure, Bohr theory, Sommerfeld atom.**

K/L ratio (*Nuc.*). The ratio of the number of internal conversion electrons from the K-shell to the number from the L-shell, which are emitted in a particular decay process. See also **L/M ratio.**

klydonogram (*Elec. Eng.*). The record obtained on the photographic plate of a klydonograph.

klydonograph (*Elec. Eng.*). An instrument in which, a photographic plate having been placed between two electrodes, a voltage across the electrodes produces a figure on the plate.

klystron (*Electronics*). General name for class of ultra-high-frequency electron tubes (amplifiers, oscillators, frequency multipliers, cascade amplifiers, etc.), in which electrons in a stream have their velocities varied (velocity modulation) by an ultra-high-frequency field; and subsequently impart energy to it or to other ultra-high-frequency fields, contained in a resonator. Can be a wavelength generator in which a buncher rhumbatron imparts the modulation which excites a catcher rhumbatron cavity. In the Sutton tube, an electron-stream reflector electrode enables one rhumbatron to perform both functions.

klystron frequency multiplier (*Electronics*). A two-cavity klystron whose output cavity is tuned to an integral multiple of the fundamental frequency.

klystron reflection (*Electronics*). See **reflex klystron.**

knapping hammer (*Build.*). A hammer used for breaking and shaping stones and flints. (*Mining*) A special hammer or a machine to break rock and produce a minimum of fine material. Also called **knapper.**

knapsack abutments (*Civ. Eng.*). Abutments provided with an additional integral slab behind them to reduce the overturning pressure.

knaur (*Bot.*). A swollen outgrowth of some size, from the trunk of a tree.

knebelite (*Min.*). An orthosilicate of manganese and ferrous iron, in the fayalite-tephroite series. It crystallizes in the orthorhombic system and occurs as the result of metamorphism and metasomatism.

knee (*Electronics*). The region of maximum curvature on a characteristic curve, e.g., of a valve. (*Eng., Plumb.*) An elbow pipe (see **elbow**). (*Join.*) A sudden rise in a handrail when it is convex upwards. Cf. *ramp.* (*Zool.*) In land Vertebrates, the joint between the femur and the crus.

knee brace (*Build.*). A stiffening member fixed across the inside of an angle in a framework, particularly at the angle between roof and wall in a building frame.

knee-gall (*Vet.*). A distension of the carpal synovial sheath at the back of the carpal joint of the horse. Also called **knee thoroughpin.**

knee-jerk (*Med.*). A normal reflex extension of the leg at the knee joint, obtained by tapping the tendon below the patella.

kneeler (*Build.*). The return of the dripstone at the spring of an arch.

knee rafter (*Carp.*). A rafter having its lower end bent downwards.

knee roof (*Arch.*). A *mansard roof* (q.v.).

knee thoroughpin (*Vet.*). See **knee-gall.**

knib (*Typog.*). The projecting portion of a setting-rule, by which the compositor lifts it out of the composing-stick.

knickerbocker (*Textiles*). A yarn with spots of different colours obtained by flecking the colouring material on to the web during carding.

knife (*Print.*). The blade or blades on an *automatic reel change* (q.v.) that sever the old web after the join is made. Cf. *cut-off knife.*

knife cheeks (*Print.*). Spring-loaded gripping edges to hold the web while it is being severed.

knife edge (*Eng.*). Bearing for a balance beam or similar instrument member, usually in the form of a hardened steel wedge.

knife-fold (*Bind.*). A fold made by pushing the paper into the folding rollers with a blunt knife, a necessary arrangement when the fold is at right angles to the leading edge of the paper, or to the previous fold. Cf. *buckle-fold.*

knife-switch (*Elec. Eng.*). An electric-circuit switch in which the moving element consists of a flat blade which engages with fixed contacts. See **tandem knife-switch.**

knife tool (*Eng.*). A lathe tool having a straight lateral cutting edge, used for turning right up to a shoulder or corner.

Knight shift (*Nuc.*). That of nuclear magnetic resonance frequency due to the effect of the magnetic field of conduction electrons.

knitted plush (*Textiles*). In hosiery manufacture, fabric with a looped (terry) or pile finish made from lustre yarn (rayon or other manmade fibre) on a circular spring needle machine.

knitting frame (or **loom**) (*Textiles*). In hosiery manufacture, flat or circular knitting machine in which the needles are either of the 'bearded' or the 'latch' (self-acting) type.

knob (*Bot.*). A root tuber in some orchids.

knobbing (*Build.*). The operation of breaking projecting pieces off stones in the quarry.

knock (*Eng., etc.*). See **knocking.**

knocked up toe of dogs (*Vet.*). Synovitis and fibrositis affecting the first or second interphalangeal joint of the toe of the dog; caused by sprain and usually met with in racing greyhounds.

knocker-out (*Eng.*). The horns of a planing machine against which the tappets strike to reverse the motion of the table.

knocking, knock (*Eng.*). A periodic noise made by a worn bearing in a reciprocating engine, due to reversal of the load on the shaft or pin. (*I.C. Engs.*) The characteristic metallic noise, often called 'pinking', resulting from *detonation* (q.v.) in a petrol engine, or 'diesel knock', due to excessively rapid pressure rise during combustion in a diesel engine.

knocking-down iron (*Bind.*). A piece of iron, fixed to the lying press, on which joints are beaten out and lacings flattened.

knockings (*Build.*). The stone pieces, smaller than spalls, knocked off in the process of chiselling or hammering. (*Min. Proc.*) See **riddle.**

knock-knee (*Med.*). See **genu valgum.**

knock-on electrons (*Nuc.*). Those which arise from the collision of fast mesons from cosmic rays with the electrons of gaseous atoms in the atmosphere.

knock-out drum (*Chem. Eng.*). A vessel, usually empty, through which gases or vapours are passed when carrying entrained liquid droplets, for the purpose of separating the liquid.

knockrating (*I.C. Engs.*). The measurement of the *antiknock value* (q.v.) of a volatile liquid fuel in terms of the percentage of octane in an octane-heptane mixture of equivalent knock-proneness. See **octane number.**

knop yarn (*Textiles*). Novelty yarn with *knops*, hard spots or lumps, at intervals. Made with one or two foundation threads, and one, two or three effect threads. Temporary stoppage of the delivery of the foundation threads allows the others to form a *knop*. A binder is incorporated by doubling another yarn with it in reverse order, to balance the twists. See also **bead yarn.**

knot (*Bot.*). (1) A node in a grass stem. (2) A hard and often resinous inclusion in timber, formed from the base of a branch which became buried in secondary wood as the trunk thickened. (*Meteor., Ships, Aero.*) Speed of 1 nautical m.p.h. (1·15 m.p.h. or 1·85 km/h) used in navigation and meteorology.

knotter (*Paper*). An appliance for the removal of knots or unbeaten particles from paper pulp.

knotting (*Paint.*). A solution of shellac in spirit used for covering knots in wood, prior to painting, to prevent exudation of resin.

Knott's equations (*Phys.*). The continuity equations governing the energy redistribution of ultrasonic waves incident at an interface between two media.

Knowles cell (*Chem.*). Cell producing hydrogen and oxygen by the electrolysis of an alkaline solution.

knuckle (*Join.*). The parts of a hinge receiving the pin.

knuckle joint (*Eng.*). A hinged joint between 2 rods, in which the ends are formed into an eye and a fork respectively and united by a pin.

knuckle-joint press (*Eng.*). A heavy power press incorporating a knuckle-joint mechanism for actuating the slide. It has a high mechanical advantage near the bottom end of its stroke, which is relatively short, and is therefore used for coining, sizing, and embossing work.

knuckle pins (*I.C. Engs.*). See **wrist pins.**

knuckling (*Vet.*). A state of abnormal flexion of the fetlock joint of the horse, due to a variety of causes, and associated with an inability or lack of desire to extend the joint fully.

Knudsen flow (*Chem. Eng.*). The flow of low pressure gas through a tube whose diameter is much smaller than the mean free path of the molecules of the gas. Under these conditions, the macroscopic concept of viscosity needs to be modified since the resistance to motion is due primarily to molecular collisions with the passage walls. Also **molecular streaming.**

Knudsen gauge (*Vac. Tech.*). A gauge which measures pressure in terms of the net rate of transfer of momentum by molecules between two surfaces maintained at different temperatures and separated by a distance much smaller than the mean free path of the gas molecules.

knurling tools (*Eng.*). Small, hard steel rollers, serrated or cross-hatched on their peripheries, mounted on a pin carried by a holder. They are pressed against circular work in the lathe, to knurl or roughen a surface required to give a grip to the fingers, as the head of a 'milled' screw.

kobellite (*Min.*). A lead, bismuth, antimony sulphide, crystallizing in the orthorhombic system.

Koch resistance (*Chem.*). The resistance of a vacuum photocell or phototube when its active surface is irradiated with light.

Kodak Relief Plate (*Print.*). An easily processed relief plate suitable for either direct or 'dry offset' printing with considerable wearing capacity.

Koepe hoist (*Mining*). Winding drum to which two cages are connected by one rope, to promote partly balanced hoisting.

Kohler's disease (*Med.*). Aseptic necrosis of the navicular bone in the tarsus; a self-limiting disease.

Kohlrausch's law (*Chem.*). The contribution from each ion of an electrolytic solution to the total electrical conductance is independent of the nature of the other ion.

koilonychia (*Med.*). Spoon-shaped depression of the finger-nails.

Kolliker's pit (*Zool.*). In *Cephalochorda*, the olfactory pit situated on the left side of the body at the anterior end.

Kone (*Acous.*). TN of a loudspeaking sound reproducer, consisting of two large paper cones, one supported by the other, and driven by a balanced-armature movement at its apex.

Konel metal (*Met.*). TN of alloy of Ni—Co—Fe—Ti, much used for cathode supports when internally heated.

konimeter (*Powder Tech.*). Type of impaction sampler in which dirty air, sucked through a round orifice, impinges upon a circular glass plate smeared with grease, which collects the particles. In mining, for example, it is used to check dust content of mine atmosphere.

konoscopic observation (*Crystal.*). The investigation of the behaviour of doubly refracting crystal plates under convergent polarized light. Interference pictures obtained are important in explaining crystal optical phenomena.

Kooman's array (*Radio*). A plane array of pinetree aerials, the corresponding dipole elements of the successive pinetrees being substantially collinear.

Koplik's spots (*Med.*). Bluish-white specks on the inner side of the cheek, occurring in measles 2 to 3 days before the rash appears.

Kopp's law (*Chem.*). That the molar heat of a solid compound is approximately equal to the sum of the constituent atomic heats.

körnchenkugeln (*Zool.*). Phagocytic cells distended with tissue débris, especially muscle fragments, occurring in the blood of pupal Insects.

Korndorfer starter (*Elec. Eng.*). A variant of the *autotransformer starter* for 3-phase induction motors which involves no interruption of the supply of power after the starting cycle has been initiated. Used with high-powered and high-voltage machines.

Korsakoff's syndrome, Korsakoff's psychosis (*Med.*). Mental confusion associated with confabulation and disorientation, usually due to chronic alcoholism.

Kossel-Sommerfeld displacement law (*Chem.*). That relating the arc spectrum of an element to spark spectra of elements placed higher in the periodic table.

Kostinsky effect (*Photog.*). Changes in image size and the separation between image areas, arising from changes in concentration of reaction products; diminished by vigorous agitation of the developer.

Kovar (*Met.*). TN for an alloy of Ni—Co—Fe for glass-to-metal seals over working ranges of temperature, when temperature coefficients of expansion coincide.

Kowalevsky's bodies (*Zool.*). In some *Myriapoda*, small glandular masses in which some of the arteries terminate.

Kozeny-Carman equation (*Powder Tech.*). A permeability equation used to calculate the surface area of a powder from the permeability of a packed powder bed to the flow of fluid through it.

Kr (*Chem.*). The symbol for *krypton*.

Kraemer-Sarnow test (*Build., Civ. Eng.*). A test for the determination of the melting-point of a bitumen for use in building or roadmaking. The melting-point is obtained as the temperature at which superincumbent mercury falls through a standard sample of the bitumen, heated under standard conditions.

kraft (*Paper*). Brown paper made from high-class sulphate wood pulp. Used extensively as a dielectric, for gummed tape, wrappings, etc.

Kramer control (*Elec. Eng.*). A form of speed and power factor control for large induction motors in which the slip energy is supplied to a rotary converter, the resulting d.c. power being used to drive a motor either mounted on the main motor shaft or driving an alternator and returning power to the supply.

Kramer's modulus (*Powder Tech.*). Numerical indication of the range of particle sizes present in a given distribution, calculated from the equation:

$$K=\sum_{x_s}^{x_m} x_i n_i \Big/ \sum_{x_s}^{x_l} x_i n_i,$$

where x_s, x_l, x_m are the smallest, largest and median particle sizes, and n_i the number of particles of diameter x_i.

kraton, craton, kratonic region (*Geol.*). A large region of the Earth's crust which acts as a relatively rigid block. Kratons, also known as shields, usually consist of very old igneous and metamorphic rocks, sometimes covered by a thin layer of younger sediments.

kraurosis (*Med.*). Atrophy of the vulva, associated with narrowing of the vaginal orifice.

krausening (*Brew.*). Addition of a small amount of fermenting wort to the conditioning tank in the brewing of lager beer for bottling.

Krause's corpuscles (*Zool.*). In Vertebrates, a type of sensory nerve-ending found in the skin, in which the nerve breaks up into numerous branches, which surround a core of large cells in a connective-tissue capsule.

Krause's membrane (*Zool.*). A transverse line occurring in the middle of each of the clear areas of a striated muscle fibril.

Kremnitz white (*Paint.*). White lead manufactured by the chamber process in Czechoslovakia and at one time used extensively by artists. The pigment was made on a small scale and was a very fine white, but had a low opacity due to an excessive carbonate content.

Kretschmer's types (*Psychol.*). Kretschmer, a German psychiatrist, found a correlation between physique and endogenous psychoses. According to their physique, he classified people into 3 groups: *asthenic*, *athletic*, and *pyknic* (qq.v.), with a fourth group, *dysplastic*, formed from a mixture of the others.

Krilium (*Chem.*). TN for a synthetic soil-conditioning chemical designed to improve soil structure by increasing the water-holding capacity and aeration of soils. It consists of the calcium salt of a copolymer of vinyl acetate and maleic acid: the first product marketed under this name was, however, a sodium polyacrylate.

KR-law (*Elec. Eng.*). See **CR-law**.

krokidolite (*Min.*). See **crocidolite**.

Kroll's process (*Met.*). Reduction to metal of tetrachloride of titanium or zirconium *in vacuo* or by reaction with magnesium in a neutral atmosphere.

Kromayer lamp (*Med.*). Mercury-vapour lamp for generating ultraviolet rays, especially for therapy. It has a quartz mercury-vapour arc and a cold-water cooling device, and is used particularly in treating diseases of the eye, nose, and throat, and in gynaecological conditions. See **ultraviolet therapy**.

Kronecker delta (*Maths.*). A function δ_{ij} of two variables i and j, defined to have the value zero unless $i=j$ when its value is unity.

Krukenberg's tumour (*Med.*). A solid tumour appearing in each ovary and believed to be always secondary to cancer of the stomach.

Kruskal limit (*Nuc.*). Theoretical maximum current for which a plasma discharge can remain stable for any given axial magnetic field.

Kryptol (*Chem.*). A mixture of graphite, silicon carbide, and clay.

krypton (*Chem.*). A zero-valent element, one of the rare gases. Symbol Kr, at. no. 36, r.a.m.

83·80, m.p. −169°C, b.p. −151·7°C, density 3·743 g/dm³ at s.t.p. It is a colourless and odourless monatomic gas, and constitutes about 1-millionth by volume of the atmosphere, from which it is obtained by liquefaction. It is used in certain gas-filled electric lamps.

Krystal crystallizer (*Chem. Eng.*). TN for a form of crystallizing plant which includes a section behaving as elutriator from which very closely controlled uniform size crystals can be produced. Sometimes called an **Oslo crystallizer.**

K-series (*Nuc.*). The shortest wavelengths in the characteristic X-ray spectra of the elements. The *K-series* consist chiefly of two lines, $K\alpha$ and $K\beta$, which may be recognized in the spectra of elements from atomic number 10 to 90, there being a linear relation between the square root of the frequency of the radiation and the atomic number. See **Moseley's law,** also **K-, L-, M- . . . shells.**

K-shell (*Nuc.*). See **K-, L-, M- . . . shells.**

k-space (*Electronics*). Symbol for momentum space or wave-vector space. This is an important concept in semiconductor energy band theory. See **Brillouin zone.**

Kuhm plates (*Chem. Eng.*). TN for plates for distillation columns consisting of concentric channels and caps, and with downcomer pipes so arranged that flow on all plates in a column is in same direction from periphery to centre.

kulaite (*Geol.*). An amphibole-bearing nepheline basalt.

Kultschitsky's cells (*Zool.*). In some Vertebrates, cells containing a large number of strongly basiphil granules, found in the epithelium lining the stomach, intestines, and pancreatic duct; believed to be of pathological origin.

Kümmell's disease (*Med.*). Delayed collapse or crumbling of one or more spinal vertebrae after injury to the spine.

Kummer's convergence test (*Maths.*). If, for a divergent series of positive terms $\Sigma(v_n)^{-1}$ and a series of positive terms Σu_n, the expression

$$v_n \cdot \frac{u_n}{u_{n+1}} - v_{n+1} \rightarrow l,$$ then the series Σu_n converges if $l > 0$, and diverges if $l < 0$.

Kundt constant (*Light*). The ratio of *Verdet's constant* to magnetic susceptibility.

Kundt effect (*Light*). Rotation of plane of polarization of light in certain fluids by a magnetic field.

Kundt's rule (*Light*). The refractive index of a medium on the shorter wavelength side of an absorption band is abnormally low, and on the other side abnormally high. This gives rise to *anomalous dispersion* (q.v.).

Kundt's tube (*Acous.*). Transparent tube in which stationary acoustic waves are established, indicated by lycopodium powder, which congregates in a heap, or even a disk, at the nodes. Used for measuring sound velocities.

kunzite (*Min.*). See **spodumene.**

kupfernickel (*Min.*). See **niccolite.**

Kupffer's cells (*Zool.*). Large isolated phagocytic cells occurring on the walls of the liver sinuses in some Vertebrates.

kurchatovium (*Chem.*). Russian name for transuranic element at. no. 104. See **rutherfordium.**

Kurie plot (*Nuc.*). One used for determining the energy limit of a β-ray spectrum from the intercept of a straight line graph. Prepared by plotting a function of the observed intensity against energy, the intercept on the axis being the energy limit for the spectrum. Also **Fermi plot.**

Kutter's formula (*Hyd.*). See **Ganguillet and Kutter's formula.**

kV (*Elec.*). Abbrev. for *kilovolt.*

kVA (*Elec. Eng.*). Abbrev. for *kilovolt-ampere.*

kVAr (*Elec. Eng.*). Abbrev. for *kilovar.*

kVp (*Radiol.*). Abbrev. for *kilovolts, peak.*

kW (*Elec. Eng.*). Abbrev. for *kilowatt.*

kwashiorkor (*Med.*). A disease of the tropics due to protein and vitamin deficiency in children. There is gross loss of weight, oedema, and pigmentation. Also known as **malignant malnutrition.**

kWh (*Elec. Eng.*). Abbrev. for *kilowatt-hour.*

kyanite or **cyanite** (*Min.*). A silicate of aluminium which crystallizes in the triclinic system. It usually occurs as long-bladed crystals, blue in colour, in metamorphic rocks. See also **disthene.**

kyanizing (*Build.*). The process of impregnating timber with a solution of corrosive sublimate as a preservative. Also called **cyanizing.**

kylite (*Geol.*). An olivine-rich variety of theralite.

kymography (*Radiol.*). A method of recording in a single radiograph the excursions of moving organs in the body, by use of a *kymograph,* which records physiological muscular waves, pulse beats or respirations.

kyphoscoliosis (*Med.*). A deformity of the spine in which dorsal convexity is increased, the spine being also bent laterally.

kyphosis (*Med.*). A deformity of the spine in which the dorsal convexity is increased.

L

l (*Chem.*). An abbrev. for *litre*. (1 dm³ is now used.) (*Chem.*). (1) A symbol for: (*a*) specific latent heat per gramme; (*b*) mean free path of molecules; (*c*) (with subscript) equivalent ionic conductance, 'mobility'. (2) An abbrev. for *laevorotatory*.

L (*Chem.*). Symbol for (1) molar latent heat; (2) Avogadro constant. (*Elec.*) Symbol for self-inductance. (*Mech.*) Symbol for angular momentum. (*Light*) Symbol for luminance.

λ (*Chem.*). Symbol for absolute activity. (*Elec. Eng.*) Symbol for wavelength. (*Heat*) Symbol for (1) linear coefficient of thermal expansion; (2) thermal conductivity. (*Nuc.*) Symbol for radioactive decay constant. (*Phys.*) Symbol for mean free path.

Λ (*Chem.*). A symbol for molar conductance; Λ_0, at infinite dilution.

La (*Chem.*). The symbol for *lanthanum*.

label (*Arch.*). A projecting moulding above a door or window opening.

label-corbel table (*Build.*). A *dripstone* (q.v.) supported by a corbel.

labella (*Zool.*). In *Diptera*, the pair of lobes formed as a distal expansion of the labium.

labelled atom (*Min. Proc.*). One which has been made radioactive, in a compound introduced into a flowline in order to trace progress. (*Nuc.*) That atomic position in a molecular formula which is occupied by an isotopic tracer. See **radioactive tracer.**

labelled molecules (*Nuc.*). Molecules containing an isotopic tracer.

labellum (*Bot.*). The posterior petal of the flower of an orchid; since the flower is usually resupinate, the labellum appears to be anterior. (*Zool.*) A spoon-shaped lobe at the apex of the glossa in Bees; (*pl.*) in certain *Diptera*, a pair of fleshy lobes into which the proboscis (*labium*) is expanded distally. *pl.* **labella.**

labia (*Zool.*). Any structures resembling lips; as the lips of the vulva in *Primates*. See also the *sing.* form **labium.** *adj.* **labial.**

labial cartilage (*Zool.*). In *Craniata*, one of the cartilages frequently formed in association with, and superficial to, the rostrum, olfactory capsules and jaws.

labial glands (*Histol.*). Small auxiliary salivary glands in the mucous membrane of the lips. (*Zool.*) The salivary glands of Insects; they discharge through a median duct which opens on the labium.

labial palp (*Zool.*). In Insects, the palpus of the second maxilla; in some *Lamellibranchiata*, a ciliated grooved muscular flap bordering the mouth.

labial segment (*Zool.*). In Insects, the 6th cephalic segment, bearing the labium.

labial suture (*Zool.*). In Insects, the line of division between the prementum and the postmentum of the labium.

labia majora (*Zool.*). In Mammals, the two prominent folds which form the outer lips of the vulva.

labia minora (*Zool.*). In Mammals, the small inner lips of the vulva; nymphae.

labiate (*Bot.*). Said of a corolla with one or more petals prolonged to form a lip.

labidophorous (*Zool.*). Possessing chelae or other forcipulate organs.

labile (*Chem.*). Unstable, liable to change. (*Electronics*) Said of an oscillator which can be synchronized from a remote source.

lability (*Psychiat.*). The tendency towards quick and spontaneous variations in emotional reactions.

labioglossopharyngeal (*Anat.*). Pertaining to the lips, tongue, and pharynx.

labiostipes (*Zool.*). The prementum of Insects.

labium (*Zool.*). In Insects, the lower lip, formed by the fusion of the second maxillae; in the shells of *Gastropoda*, the inner or columellar lip of the margin of the aperture. See also the *pl.* form **labia.**

laboratory sand-bath (*Heat.*). Gas-heated steel bath containing sand in which crucibles or other vessels are indirectly heated.

Laboulbeniales (*Bot.*). An order of minute *Ascomycetes*, ectoparasitic on insects. The ascogonium has a trichogyne, and fertilization is by spermatia.

labradorescence (*Min.*). A brilliant play of colours shown by some labradorite feldspars. Probably due to a fine intergrowth of two phases.

labradorite (*Min.*). A plagioclase feldspar containing 50 to 70% of the anorthite molecule; occurs in basic igneous rocks; characterized by a beautiful play of colours in some specimens, due to schiller structure. Named from Labrador, whence fine specimens were sent over to Europe by Moravian settlers.

labral nerve (*Zool.*). In Insects, the paired nerve which originates as a branch of the labrofrontal nerve and innervates the labrum.

La Brea Sandstone (*Geol.*). A sandstone which succeeds the oil-bearing sands of Miocene age in Trinidad; it contains the well-known Pitch Lake, 104 acres (42 ha) in extent.

labrofrontal nerve (*Zool.*). In Insects, a paired nerve arising from the tritocerebrum and branching to form the labral nerve and the root of the frontal ganglion.

labrum (*Zool.*). In Insects and *Crustacea*, the platelike upper lip; in the shells of *Gastropoda*, the outer lip or right side of the margin of the aperture. *adj.* **labral.**

labyrinth (*Acous.*). Folded path for loading the rear of loudspeaker diaphragms and microphone ribbons. (*Nuc. Eng.*) See **radiation trap** (2). (*Zool.*) Any convolute tubular structure; especially, the bony tubular cavity of the internal ear in Vertebrates, or the membranous tube lying within it.

labyrinthectomy (*Surg.*). Surgical removal of the labyrinth of the ear.

labyrinthitis (*Med.*). Inflammation of the labyrinth of the ear.

labyrinthodont (*Zool.*). Having the dentine of the teeth folded in a complex manner.

Labyrinthodontia (*Zool.*). An extinct superorder of Amphibians with the dentine of the teeth typically infolded at the base; vertebrae consist of neural arches resting on an intercentrum and, in the early forms, on a pleurocentrum. The order includes the majority of the extinct Amphibians, many of whom reached a large size. Lower Carboniferous—upper Trias.

labyrinth packing (*Eng.*). A type of gland, with radial or axial clearance, used, e.g., in steam engines and turbines.

Labyrinthulales (*Bot.*). An order of the *Myxo-*

mycetes. The amoeboid phase is spindlelike or spherical, sometimes forming a net-plasmodium from which develop stalked or sessile sporangia.

lac (*Zool.*). A resinous substance, an excretion product of certain Coccid insects (*Gascardia*, *Tachardia*, etc.) in certain jungle trees; used in the manufacture of shellac. Chief source, India. See **shellac**, for which the name is frequently loosely used.

laccate (*Bot.*). Having a shining surface, as if varnished.

laccolith (*Geol.*). A form of intrusion with domed top and flat base, the magma having been instrumental in causing the up-arching of the 'roof'. Cf. *phacolith*.

lace (*Textiles*). See **machine-made lace.**

laced valley (*Build.*). A valley formed in a tiled roof by interlacing tile-and-a-half tiles across a valley board.

lace fabrics (*Textiles*). See **openwork.**

Lacertilia (*Zool.*). A suborder of *Squamata* in which the skull has lost one of the temporal vacuities. Limbs are present and are adapted for walking, and the mouth can be opened only to a moderate extent. Lizards, skinks, geckos, monitors, iguanas, chameleons.

lachrymal (*Zool.*). See **lacrimal.**

lacing cord (*Weaving*). A varnished twine used for lacing together the jacquard cards in a lace machine or a loom jacquard.

lacing course (*Build.*). A brickwork bond-course built into rubble or flint walls.

lacing-in (*Bind.*). The operation of attaching the boards to a book by lacing the ends of the bands through holes made in the boards to receive them.

lacinia (*Bot.*). Incision in a leaf, petal, etc.; slender lobe projecting from the margin of a thallus. (*Zool.*) In Insects, the inner distal lobe of the maxilla. *pl.* **laciniae.**

lacinia mobilis (*Zool.*). In *Peracarida*, a movable structure borne behind the incisor process of the mandible.

La Cour converter (*Elec. Eng.*). See **motor converter.**

lacquer (*Paint.*). A solution of film-forming substances in volatile solvents, e.g., a spirit lacquer or varnish consists of shellac or other gums dissolved in methylated spirit. Lacquers may be pigmented or clear. They dry by the evaporation of solvents only, not by oxidation. (This last statement does not apply in U.S.) The most important lacquers are the *cellulose lacquers* (q.v.).

lacrimal (*Zool.*). Pertaining to, or situated near, the tear gland; in some Vertebrates, a paired lateral membrane bone of the orbital region of the skull, in close proximity to the tear gland.

lacrimal duct (*Zool.*). In most of the higher Vertebrates, a duct leading from the inner angle of the eye into the cavity of the nose; it serves to drain off the secretion of the lacrimal gland from the surface of the eye.

lacrimal gland (*Zool.*). In most of the higher Vertebrates, a gland situated at the outer angle of the eye; it secretes a watery substance which washes the surface of the eye and keeps it free from dust.

lactalbumins (*Chem.*). Albumins occurring in milk; like other albumins, they are crystalline substances, soluble in water, are not precipitated by NaCl, but coagulate at 70°–75°C.

lactase (*Chem.*). An enzyme, present in the digestive juices of some animals, which facilitates the conversion of lactose into glucose and galactose. Also found in certain plants.

lactation tetany (*Vet.*). See **bovine hypomagnesaemia.**

lacteals (*Zool.*). In Vertebrates, lymphatics, in the region of the alimentary canal, in which the lymph has a milky appearance, due to minute fat globules in suspension.

lactic (*Zool., etc.*). Pertaining to milk.

lactic acids (*Chem.*). $CH_3{\cdot}CH(OH){\cdot}COOH$ and $CH_2(OH){\cdot}CH_2{\cdot}COOH$, i.e., hydroxy-propanoic acids. There are two isomers, viz., 2-hydroxy-propionic acid, and 3-hydroxy-propionic acid; only the former, known as *lactic acid* or *fermentation lactic acid*, is of importance. **Lactic acid,** a synonym for racemic 2-hydroxy-propanoic acid, $CH_3{\cdot}CH(OH){\cdot}COOH$, hygroscopic crystals, m.p. 18°C, b.p. 83°C at 1 mm, rel. d. 1·248. It is produced commercially by the lactic fermentation of sugars with *Bacillus acidi lactici* or by heating glucose with caustic potash solution of a certain concentration.

lactiferous (*Bot.*). Containing latex. (*Zool.*) Milk-producing, milk-carrying.

Lactobacillaceae (*Bacteriol.*). A family of bacteria belonging to the order *Eubacteriales.* Gram-positive cocci or rods; carbohydrate fermenters, are anaerobic or grow best at low oxygen tensions. Includes several pathogenic organisms, e.g., *Streptococcus pyogenes* (scarlet fever, tonsillitis, erysipelas, puerperal fever), *Diplococcus pneumoniae* (one cause of pneumonia).

lactoflavin (*Chem.*). Old name for *riboflavin.*

lactogenic hormone (*Physiol.*). See **prolactin.**

lactones (*Chem.*). Intramolecular anhydrides of hydroxycarboxylic acids. Most lactones have 5 or 6 membered rings (γ- or δ-lactones) but others are known.

lactose (*Chem.*). $C_{12}H_{22}O_{11}+H_2O$, rhombic prisms which become anhydrous at 140°C, m.p. 201°–202°C with decomposition; when hydrolysed it forms D-galactose and D-glucose. It reduces Fehling's solution and shows *mutarotation* (q.v.). It occurs in milk and is not so sweet as glucose.

lactosuria (*Med.*). The presence of lactose in the urine.

lacuna (*Bot.*). (1) Any depression in a surface of a plant. (2) Any cavity in a plant. (3) A cavity formed in rapidly elongating stems by breakdown of the protoxylem. (4) A large intercellular space. (*Zool.*) A small cavity or space, as one of the cell-containing cavities in bone. *adjs.* **lacunar, lacunose.**

lacunar (*Arch.*). A coffer or panel formed in a ceiling.

lacunar system (*Zool.*). In *Echinodermata*, the vascular system, consisting of a system of strands with intercommunicating spaces.

lacunose (*Bot.*). Having a pitted surface.

lacustrine deposits (*Geol.*). Sediments accumulated in a lake, consisting of marginal deposits of shingle, passing rapidly into clay and limestone in deeper water. May exhibit a well-marked seasonal banding, as in varve clays, while a light colour seems characteristic. The occurrence of fresh-water snails and algae is a certain indication of lacustrine origin of such rocks as the Oligocene Bembridge Limestones.

ladder (*Civ. Eng.*). The continuous line of mud-buckets, carried on an oblique endless chain, in a bucket-ladder excavator or dredger. (*Textiles*) In hosiery manufacture, a defect caused by the breaking of a knitted stitch, which causes the thread to *ladder* in a vertical direction; more liable to occur in fine smooth yarns.

ladder dredger (or excavator) (*Civ. Eng.*). See **bucket-ladder dredger.**

ladder network (*Telecomm.*). One which comprises a number of filter sections, all alike, in series, acting as a transmission line, with attenuation and delay properties.

ladies (*Build.*). Slates 16 × 8 in. (406 × 203 mm).

ladle (*Met.*). An open-topped vessel lined with refractory material; used for conveying molten metal from the furnace to the mould or from one furnace to another.

ladle addition (*Met.*). Addition of alloying metal (e.g., nickel shot) to molten iron in ladle before casting, to give special quality to product.

Laënnec's cirrhosis (*Med.*). Multilobular cirrhosis of the liver, in which degeneration of liver cells is associated with areas of fibrosis enclosing many regenerating lobules of the liver.

laeotropic (*Zool.*). Tending to the left, e.g., *laeotropic* spiral, *laeotropic* division.

laevigate, properly **levigate** (*Bot.*). Having a smooth, polished surface.

laevo-. Prefix from L. *laevus*, left.

laevorotatory (*Light*). Rotating the plane of polarization of a polarized ray of light to the left.

laevulose (*Chem.*). $C_6H_{12}O_6$. Fruit-sugar or D(−)-*fructose*.

laevulosuria (*Med.*). The presence of laevulose in the urine.

lag (*Mining*). To protect a shaft or level from falling rock by lining it with timber (*lagging*). (*Telecomm.*) Time delay between the operation of the transmitting key or relay and the response of the receiving device. (*Weaving*) Small perforated wooden block forming part of the pattern chain for a dobby. This chain is formed by lags connected together, in which wooden or metal pegs have been inserted to lift the healds to form a pattern.

lagena (*Zool.*). In higher Vertebrates, a pocket lined by sensory epithelium and developed from the posterior side of the sacculus, which becomes transformed in Mammals into the scala media or canal of the cochlea.

Lagenidiales (*Bot.*). An order of parasitic *Phycomycetes*. The mycelium is confined to a single host cell. Asexual reproduction is by biflagellate zoospores, and sexual reproduction is oögamous with the fusion of aplanogametes. Also **Ancylistales.**

lageniform (*Bot.*). Having the form of a *Florence flask* (q.v.).

lager beer (*Brew.*). A light beer, low in alcohol, and with a rather high percentage of extract.

lag filter (*Radio*). A radio-frequency filter in close electrical connexion with the contacts of a Morse key, to minimize sparking and transients during the keying of the transmitter.

lagging (*Bot.*). Slow movement towards the poles of the spindle by one or more chromosomes in a dividing nucleus, with the result that these chromosomes do not become incorporated into a daughter nucleus. (*Build., Civ. Eng.*) (1) The wooden boards nailed across the framework of a centre, to form the immediate supporting surface for the arch. (2) Material generally used in floors and roofs to eliminate sound and loss of heat. See **insulation.** (*Eng.*) (1) The process of covering a vessel or pipe with a nonconducting material to prevent either the loss or ingress of heat. (2) The nonconducting material itself, such as a plaster mixture of asbestos and magnesia, moulded sections of magnesia plaster, powdered cork, *Alfol* (q.v.), etc. See **insulation.** (*Mining*) Small round or split poles, slabs, rough timber, or brush placed above the splits and along the sides of pit props. See **lag.**

lagging adhesive (*Plastics*). A very viscous adhesive capable of holding thick cloth in position around lagging, while wet, and of providing mechanical protection when dry. Usually based on well-plasticized PVA, or copolymer emulsions and soluble cellulose derivatives.

lagging-back (*Weaving*). Turning back the pattern chain of a dobby to rectify the fault which would otherwise occur when the loom is stopped on account of a warp or weft breakage.

lagging current (*Elec. Eng.*). An alternating current which reaches its maximum value at a later instant in the cycle than the voltage which is producing it.

lagging load (*Elec. Eng.*). See **inductive load.**

lagging of tides (*Astron.*). The interval between the theoretical moment of maximum tide, which is that of the moon's crossing of the meridian at that place, and the actual moment, which is later by an amount depending on several factors, amongst them the irregular contour of ocean boundaries.

lagging phase (*Elec. Eng.*). A term used in connexion with measuring equipment on 3-phase circuits to denote a phase whose voltage is lagging behind that of one of the other phases by approximately 120°. Also used, particularly in connexion with the 2-wattmeter method of 3-phase power measurement, to denote the phase in which the current at unity power factor lags behind the voltage applied to the meter in which that current is flowing.

Lagomorpha (*Zool.*). An order of *Eutheria* with two pairs of upper, and one of lower, incisor teeth which grow throughout life; there are no canines, and there is a wide diastema between the incisors and the cheek teeth into which the cheeks can be tucked to separate the front part of the mouth from the hind during gnawing. Rabbits and hares.

lagoonal deposition (*Geol.*). The accumulation of sediment in a shallow arm of the sea, which is cut off from the outer ocean by a barrier which prevents free communication. The strata formed under such conditions constitute a *lagoon phase*. Examples in the stratigraphical column occur in the Lower Carboniferous and in the Rhaetic Series in England.

lagophthalmus, lagophthalmos (*Med.*). Imperfect closure of the eyelids when the eyes are shut.

lagopodous (*Zool.*). Having feet covered by hairs or feathers.

Lagrange's dynamical equations (*Maths.*). The equations

$$\frac{d}{dt}\left(\frac{\partial T}{\partial q_i}\right) - \frac{\partial T}{\partial q_i} = Q_i, \quad i=1, 2, 3, \text{etc.},$$

where T is the kinetic energy of a system, q_i the various coordinates which define the position of the system and Q_i the so-called generalized forces, that is, $Q_i\,\delta q_i$ is the work done by the external forces during a small displacement δq_i. These equations are characterized by being independent of any reactions internal to the system.

laid-dry (*Build.*). Said of bricks or blocks which have been laid without mortar.

laid-in moulding (*Join.*). A moulding cut out of a separate strip of wood of the required section, and sunk in a special groove in the surface which it is intended to decorate.

laid-on moulding (*Join.*). A *planted moulding* (q.v.).

laid paper (*Paper*). Writing and printing paper with a ribbed watermark derived from the mould or dandy in which the wires are laid side by side.

Lainer's reducer (*Photog.*). An iodide solution for reducing contrast or density in photographic images.

Laingspan (*Build., Civ. Eng.*). A method of system building developed in U.K., mainly used for schools.

laitance (*Build.*). (1) The milky scum from grout or mortar, squeezed out when tesserae or tiles are pressed into place. (2) The milky scum formed on over-trowelled cement concrete or rendering.

lake (*Geol.*). A body of water lying on the surface of a continent, and unconnected (except indirectly by rivers) with the ocean. Lakes may be *fresh-water lakes*, provided with an outlet to the sea; or *salt lakes*, occurring in the lowest parts of basins of inland drainage, with no connexion with the sea. Lakes act as natural settling tanks, in which silt carried down by rivers is deposited as clay, containing the shells of molluscs, etc. The lakes of former geological periods may thus be recognized by the nature of the sediments deposited in them and the fossils they contain. Lakes occur plentifully in glaciated areas, occupying hollows scooped out by the ice, and depressions lying behind barriers of morainic material.

lakes (*Chem.*). Pigments formed by the interaction of dyestuffs and 'bases' or 'carriers', which are generally metallic salts, oxides or hydroxides. The formation of insoluble lakes in fibres, which are being dyed, is known as *mordanting*, the hydroxides of aluminium, chromium, and iron generally being employed as *mordants*.

lake types (*Ecol.*). See **dystrophic, eutrophic, oligotrophic.**

Lalande cell (*Elec. Eng.*). A primary cell having electrodes of zinc and iron, an electrolyte of caustic soda, and a copper oxide depolarizer.

lalling, lallation (*Med.*). Babbling speech of infants; lack of precision in the articulatory mechanism of the mouth.

lalopathy (*Med.*). Any disorder of speech. Also **logopathy.**

lam (*Weaving*). Wooden bar coupled to a heald. When pulled down, by a tappet depressing its lever, it brings down the heald to form part of the warp shed.

Lamarckism (*Zool.*). A theory of the mechanism of evolution propounded by Lamarck; it postulates that new characters acquired by an organism during its lifetime, as an outcome of use or disuse, may be inherited.

lambda leak (*Phys.*). The leakage of liquid helium II through holes so small that normal liquids cannot pass.

lambda limiting process (*Electronics*). A method of defining a point electron as the limiting case in which a time-like vector tends to zero.

lambda-particle (*Nuc.*). Hyperon with hypercharge 0 and isotopic spin 1.

lambda point (*Phys.*). (1) Transition temperature of helium I to helium II. (2) Temperatures characteristic of second-order phase change, e.g., ferromagnetic Curie point.

lamb dysentery (*Vet.*). An acute and fatal toxaemic disease of newborn lambs caused by an intestinal infection by the bacterium *Clostridium perfringens* (*Cl. welchii*), Type B. The main symptom is diarrhoea due to enteritis.

lambert (*Light*). Unit of luminance or surface brightness of a source or diffuse reflector, emitting 1 lumen/cm^2 $=10^4/\pi$ cd/m^2. The *millilambert* is used for low illuminations. See **lumen.**

Lambert's cosine law (*Light*). The energy emitted from a perfectly diffusing surface in any direction is proportional to the cosine of the angle which that direction makes with the normal.

Lambert's law (*Light*). The illumination of a surface on which the light falls normally from a point source is inversely proportional to the square of the distance of the surface from the source.

Lambert's projection (*Geog.*). See **zenithal equal-area projection.**

lambing sickness (ewe) (*Vet.*). See **milk fever.**

lambskin (*Textiles*). Heavy fustian type of cotton fabric, having more weft than warp, and a well-raised finish.

lamb's wool (*Textiles*). See under **fleece wool.**

Lamé formula (*Eng.*). A formula for calculating the stresses in thick (hydraulic) cylinders under elastic deformation.

lamell-, lamelli-. Prefix from L. *lamella*, thin plate.

lamella (*Bot.*). (1) A plate of cells. (2) The gill of an agaric. (*Zool.*) A structure resembling a thin plate. *adjs.* **lamellar, lamellate.**

lamellar magnetization (*Elec. Eng.*). Magnetization of a sheet or plate distributed in such a way that the whole of the front of the sheet forms one pole and the whole of the back forms the other.

lamellasome (*Bot.*). An intracytoplasmic membranous inclusion consisting of a series of invaginated lamellae enclosed by a common membrane. Such structures appear to be confined to unicellular blue-green *Algae*.

lamellibranch (*Zool.*). Having platelike gills, as members of the group *Lamellibranchiata*.

Lamellibranchiata (*Zool.*). A class of *Mollusca* with a compressed, bilaterally symmetrical body enveloped by the mantle which is divided into two equal lobes, each of which secretes a shell valve; the head is rudimentary; the foot is ventral and wedge-shaped; there are two enlarged ctenidia in the mantle cavity, ciliated and adapted for the collection of food particles. Oysters, cockles, mussels, scallops.

lamelliform (*Zool.*). Having lamellate antennae, as some beetles.

lamellirostral (*Zool.*). Of Birds, having lamelliform ridges on the inner edge of the beak.

lamellose (*Bot.*). Stratified.

Lamiales (*Bot.*). An order of the *Sympetalae*. The hypogynous corolla is usually 2-lipped; the stamens usually 4, may be 2, fixed to the corolla; the ovary of 2-4 loculi, each with a single ovule; the fruit a drupe or nutlets.

lamina (*Bot.*). (1) The flattened blade of a leaf. (2) Any flattened part of a thallus. *pl.* **laminae.** (*Elec. Eng.*) Thin sheet-steel. (*Geol.*) See **lamination.** (*Zool.*) A thin layer; a flat platelike structure. *adjs.* **laminar, laminiform.**

lamina cribrosa (*Zool.*). (1) See **cribriform** *plate.* (2) The area of the sclerotic coat of the eye which is pierced by the bundles of the optic nerve.

lamina fusca (*Zool.*). In the Vertebrate eye, the layer of pigmented connective tissue lining the sclerotic coat.

lamina infra-analis (*Zool.*). In *Odonata*, a paired latero-ventral process on the eleventh abdominal segment, probably representing part of the telson.

lamina perpendicularis (*Zool.*). In Mammals, part of the mesethmoid, which forms the bony partition between the nasal cavities.

lamina reticularis (*Zool.*). A netlike cuticular structure extending over the outer epithelium cells of Corti's organ in the Mammalian ear.

laminar flow (*Phys.*). Airflow in which adjacent

layers do not mix, except at molecular scale. Usually used in connexion with the *boundary layer* (q.v.). See also **streamline flow, viscous flow.**

Laminariales (*Bot.*). An order of the *Phaeophyceae*. The sporophyte is large and differentiated into holdfast, stipe, and blade, with an intercalary meristem usually between the stipe and blade. The kelps.

laminaria tent (*Med.*). A conical plug made of a certain kind of seaweed, for use in dilating the opening of the cervix uteri.

lamina supra-analis (*Zool.*). In *Odonata*, a median dorsal process on the eleventh abdominal segment, probably representing part of the telson.

lamina suprachoroidea (*Zool.*). In the Vertebrate eye, the outermost layer of the choroid coat, consisting of delicate connective tissue containing elastic fibres and pigment cells.

laminate (*Plastics*). See **laminated plastics.**

laminated arch (*Carp.*). An arch formed of successive thicknesses of planking, which are bent into shape and secured together.

laminated bending. The practice of bending several layers of material and at the same time joining them along their surfaces in contact to form a unit.

laminated brush (*Elec. Eng.*). A brush for an electric machine, made up of a number of layers insulated from one another, so that the resistance is greater across the brush than along its length (i.e., in the direction of normal current flow).

laminated-brush switch (*Elec. Eng.*). A switch in which one or both of the contacts are laminated. See **laminated contact.**

laminated bulb (*Bot.*). See **tunicate bulb.**

laminated clay (*Geol.*). A type of clay exhibiting lamination (i.e., very fine stratification); characteristic of accumulation under lacustrine conditions. See also **varve clays.**

laminated conductor (*Elec. Eng.*). A conductor commonly used for armature windings of large machines or for heavy-current bus-bars; it is made up of a number of thin strips, in order to reduce eddy currents in the conductor or to make it more flexible.

laminated contact (*Elec. Eng.*). A switch contact made up of a number of laminations arranged so that each lamination can be pressed into contact with the opposite surface, thereby giving a large area of contact and also a wiping action. Also called **brush contact.**

laminated core (*Elec. Eng.*). A core built up of laminations; this is the usual type of core for an electric machine or transformer.

laminated fabric (*Textiles*). Two or more plies of woven or nonwoven cloths fastened together by adhesives and pressure or a combination of adhesive, stitching, and pressure.

laminated glass. See **safety glass.**

laminated magnet (*Elec. Eng.*). (1) A permanent magnet built up of magnetized strips to obtain a high intensity of magnetization. (2) An electromagnet for a.c. circuits, having a laminated core to reduce eddy currents.

laminated paper (*Paper*). Sheets formed by laminating paper to plastic, paper, or foil by synthetic resins under pressure.

laminated plastics (*Plastics*). Superimposed layers of a synthetic resin-impregnated or coated filler which have been bonded together, usually by means of heat and pressure, to form a single piece. Also **laminate.**

laminated pole (*Elec. Eng.*). A pole for the field windings of an electric machine, having the core built up of laminations to reduce eddy currents caused by flux pulsations in the air gap.

laminated pole-shoe (*Elec. Eng.*). A field magnet pole for an electric machine having the pole-shoe built up of laminations to reduce eddy currents in it caused by flux pulsations in the air gap.

laminated record (*Acous.*). A gramophone record in which the surface material differs from that in the inside, or core, in being finer grained and therefore freer from surface noise. The superior material is carried on a fine sheet of paper, being pressed on to the hot core in the press.

laminated spring (*Eng.*). A flat or curved spring consisting of thin plates or leaves superimposed, acting independently, and forming a beam or cantilever of uniform strength. See **carriage spring.**

laminated yoke (*Elec. Eng.*). A yoke for an electric machine, built up of laminations; used in some forms of a.c. motors.

lamina terminalis (*Zool.*). In *Craniata*, the anterior termination of the spinal cord which lies at the anterior end of the diencephalon.

lamination (*Elec. Eng.*). A sheet-steel stamping shaped so that a number of them can be built up to form the magnetic circuit of an electric machine, transformer, or other piece of apparatus. Also called **core plate, punching, stamping.** (*Geol.*) Stratification on a fine scale, each thin stratum, or *lamina*, being a small fraction of an inch in thickness. Typically exhibited by shales and fine-grained sandstones.

laminboard (*Join.*). See **coreboard.**

laminectomy (*Surg.*). Surgical removal of the posterior arch or arches of one or more spinal vertebrae.

laminiplantar (*Zool.*). Of Birds, having the integument of the metatarsus scaly in front and smooth behind, e.g., in Thrushes.

laminitis (*Vet.*). Inflammation of the sensitive laminae of the hoof of the horse and ox.

laminography (*Radiol.*). See **tomography.**

La Mont boiler (*Eng.*). A once-through or semi-flash boiler in which the feed is supplied to the upper ends of long closely spaced tubes of small diameter, from which the water and steam pass to a separating drum.

Lamont's law (*Elec. Eng.*). That the permeability of steel, at any flux density, is proportional to the difference between that flux density and the saturation value.

lampas (*Vet.*). A swelling of the palatal mucous membrane behind the upper incisor teeth of horses.

lampblack (*Chem.*). The soot (and resulting pigment) obtained when substances rich in carbon (e.g., mineral oil, turpentine, tar, etc.) are burnt in a limited supply of air so as to burn with a smoky flame. The pigment is black with a blue undertone, containing 80–85% carbon and a small percentage of oily material. See also **carbon black, vegetable black.**

lamp cap (*Elec. Eng.*). The cap of brass, plastic, or other material, at the base of a filament or electric-discharge lamp, which contains the terminals, and also serves to support the lamp in the holder; also called **lamp base.** See **bayonet cap, Edison screw-cap.**

Lampén mill (*Paper*). A laboratory beating instrument for pulp evaluation consisting of a spherical rotating chamber containing the pulp and a brass ball which by its falling action beats the pulp.

lampholder (*Elec. Eng.*). A device for supporting an electric lamp; fitted with contacts connected to the source of supply; also called **a holder** or

lamp socket. See **backplate lampholder, bayonet holder, Edison screw-holder.**

lampholder plug (*Elec. Eng.*). A device for connecting a flexible electric cord, instead of a lamp, to an ordinary lampholder. Also called **plug adapter.**

lamphole (*San. Eng.*). A vertical shaft which is sunk in the ground and communicates with the crown of a sewer, thus enabling a lamp to be lowered into it to assist inspection from a nearby manhole.

lamphouse (*Photog.*). That part of an enlarger which contains the source of illumination.

lamping (*Mining, Min. Proc.*). Use of ultraviolet light to detect fluorescent minerals either when prospecting or in checking concentrates during ore treatment.

lamp man (*Mining*). Colliery surface worker in charge of miners' lamps. Works in lamp room or cabin and controls repairs, recharge, issue, lighting, etc., of portable lamps.

lamp oil (*Fuels*). See **kerosine.**

lamp resistance (*Elec. Eng.*). A resistance consisting of one or more electric filament lamps.

lamprophyres (*Geol.*). Igneous rocks usually occurring as dykes intimately related to larger intrusive bodies; characterized by abnormally high contents of coloured silicates, such as biotite, hornblende, and augite, and a correspondingly small amount of feldspar, some being feldspar-free. See also **minette, monchiquite.**

lamp socket (*Elec. Eng.*). See **lampholder.**

lamp working (*Glass*). Making articles, usually from glass tubing or rod, with the aid of an oxy-gas or air-gas flame.

lamziekte (*Vet.*). A form of botulism occurring in cattle in South Africa due to the ingestion, in phosphorus-deficient areas, of bones contaminated with *Clostridium botulinum*, Type D.

lanarkite (*Min.*). A very rare monoclinic sulphate of lead, occurring with anglesite and leadhillite (into which it easily alters), as at Leadhills, Lanarkshire, Scotland.

lanate (*Bot., Zool.*). Covered with long and loosely tangled hairs; pertaining to or having wool; having the appearance of wool or of a woolly coat.

Lancashire boiler (*Eng.*). A cylindrical steam-boiler having two longitudinal furnace tubes containing internal grates at the front. After leaving the tubes the gases pass to the front along a bottom flue, and return to the chimney along side or wing flues.

Lancashire booster (*Elec. Eng.*). A special form of battery booster having four field windings connected respectively across the booster armature, across the main bus-bars, in series with the booster armature, and in series with the main circuit.

lance (*Carp.*). A sharp scribing part of a cutting tool, serving to cut through the grain in advance and on each side of the cutting tool proper.

lancet (*Zool.*). In some *Hymenoptera*, the first valvula, borne on the 8th abdominal sternite, and modified as part of the sting.

lancet arch (*Arch.*). A sharply pointed arch, of greater rise than an equilateral arch of the same span.

lancet window (*Arch.*). A tall narrow window surmounted by a lancet arch.

lancewood (*For.*). Durable straight-grained wood chiefly from *Oxandra lanceolata*, a native of tropical America.

lancinating (*Med.*). Of pain, acute, shooting, piercing, cutting.

land (*Acous.*). The uncut surface between the grooves in the surface of a recording material. (*Eng.*) In a cutting tool, e.g., a twist drill, the surface immediately beyond the cutting edge and before the relief starts.

land and sea breezes (*Meteor.*). Light winds occurring at the coast during fine summer weather. During the day, when the land is hotter than the sea, convection causes a breeze from the sea; at night conditions are reversed, and the sea is warmer than the land, causing a breeze from the land.

Landau theory (*Chem.*). (1) Theory for calculating diamagnetic susceptibility produced by free conduction electrons. (2) That explaining the anomalous properties of *liquid helium II* in terms of a mixture of normal and superfluids; see **helium, superfluid.**

Land camera (*Photog.*). See **Polaroid camera.**

landerite (*Min.*). See **xalostocite.**

Landé splitting factor (*Phys.*). Constant employed in the calculation of the shift of energy levels produced by a magnetic field, and leading to hyperfine structure of spectrum lines (see **Zeeman effect, hyperfine structure**). It is closely related to the gyromagnetic ratio of the electron.

landing (*Mining*). Stage in hoisting shaft at which cages are loaded or discharged.

landing area (*Aero.*). That part of the movement area of an aerodrome intended primarily for take-offs and landings.

landing bar (*Textiles*). Part of the mechanism of a Levers lace machine, on which the tail of the carriage lands at the end of each swing.

landing beacon (*Aero., Radio*). A transmitter used to produce a landing beam.

landing beam (*Aero., Radio*). Field from a transmitter along which an aircraft approaches a landing field during blind landing. See **instrument landing system.**

landing direction indicator (*Aero.*). A device indicating the direction in which landings and take-offs are required to be made, usually a T, toward the cross-bar of which the aeroplane is headed.

landing gear (*Aero.*). That part of an aeroplane which provides for its support and movement on the ground and also for absorbing the shock on landing. It comprises main support assemblies incorporating single or multi-wheel arrangements, and also auxiliary supporting assemblies such as nose-wheels, tail-wheels or skids. See also **bogie-, drag struts, oleo.**

landing ground (*Aero.*). In air transport, any piece of ground that has been prepared for landing of aircraft as required; not necessarily a fully equipped aerodrome. An *emergency landing ground* is any area of land that has been surveyed and indicated to pilots as being suitable for forced or emergency landings. See **air strip.**

landing parachute (*Aero.*). See **brake parachute.**

landing procedure (*Aero.*). The final approach manoeuvres, beginning when the aircraft is in line with the axis of the runway, either for the landing, or upon the reciprocal for the purpose of a procedural turn, until the actual landing is made, or *overshoot* action has to be taken.

landing speed (*Aero.*). The minimum airspeed, with or without engine power, at which an aircraft normally alights.

landing-switch (*Elec. Eng.*). (1) A 2-way lighting switch. (2) A switch operated by a lift car, to effect stoppage at a landing. Also **floor-switch.**

landing wires (*Aero.*). Wires or cables which support the wing structure of a biplane when on the ground, and also the negative loads in flight. Also **anti-lift wires.**

Landolt reaction (*Chem.*). Delay in the formation of the blue colour of iodine with starch when iodic and sulphurous acids react. The delay is reproducible, and is caused by the reaction occurring in three successive steps.

Landolt's fibres (*Zool.*). In some Vertebrates, the free outwardly directed processes of the cone-bipolars of the eye.

Landouzy-Dejernine paralysis (*Med.*). A form of muscular dystrophy in which the muscles of the face, shoulders, and arms are involved.

Landry's paralysis (*Med.*). An acute form of paralysis which starts at the feet and rapidly spreads upwards.

Landsberger apparatus (*Chem.*). An apparatus for the determination of the boiling-point of a solution by using the vapour of the solvent to heat the solution. Prevents errors due to superheating.

landscape (*Typog.*). A term which indicates that the breadth of the page is greater than the depth: e.g., 138×216 mm, the convention being to state the upright edge first. Also applied to a similarly shaped full-page book illustration or map imposed at right-angles to the main text.

landscape lens (*Optics*). Photographic objective of meniscus form, generally with the concave surface towards the object and with the diaphragm on the object side of the lens.

landscape marble (*Geol.*). A type of limestone containing markings resembling miniature trees, etc.; when polished, the surface has the appearance of a sepia drawing.

landside (*Agric.*). A flat plate which is attached to the body of a plough and takes up the horizontal reactions of the ground when working.

landslip (*Geol.*). The sudden sliding of masses of rock, soil, or other superficial deposits from higher to lower levels, on steep slopes. Landslips on a very large scale occur in mountainous districts as a consequence of earthquake shocks, stripping the valley sides bare of all loose material. In other regions landslips occur particularly where permeable rocks, lying on impermeable shales or clays, dip seawards or towards deep valleys. The clays hold up water, becoming lubricated thereby, and the superincumbent strata, fractured by joints, tend to slip downhill, a movement that is facilitated on the coast by marine erosion.

land tie (*Civ. Eng.*). A tie-rod providing horizontal restraint against the lateral pressure exerted by earth retained by a wall.

land treatment (*San. Eng.*). The final or oxidizing stage in sewage treatment, in which the liquid sewage is distributed over an area of land, through which it filters to underdrains. If the land will not permit of easy filtering, the sewage is applied to one plot of land by irrigation, and is then passed on to a second, third, and fourth plot, before final discharge into a stream.

lane (*Nav.*). Space between two hyperbolic lines on a navigation pattern.

Langerhans' cells (*Zool.*). Spindle-shaped cells in the centre of each acinus of the pancreas.

Langerhans' islets (*Zool.*). Irregular masses of hyaline epithelium cells, unfurnished with ducts, occurring in the Vertebrate pancreas; they are responsible for the elaboration of the hormone insulin.

Langevin ion (*Phys.*). Heavy gaseous ion formed by the clustering of molecules.

Langevin theory (*Phys.*). (1) A classical expression for diamagnetic susceptibility produced by orbital electrons of atoms. (The quantum mechanical equivalent of this was derived subsequently by Pauli.) (2) An expression for the resultant effect of atomic magnetic moments which enters into explanations of both *paramagnetism* and *ferromagnetism* (qq.v.).

Langhans' layer (*Zool.*). See **cytotrophoblast.**

langite (*Min.*). A very rare ore of copper occurring in Cornwall, blue to greenish-blue in colour; essentially hydrated copper sulphate, crystallizing in the orthorhombic system.

Lang lay (*Eng.*). A method of making wire ropes in which the wires composing the strands, and the strands themselves, are laid in the same direction of twist.

Langmuir adsorption isotherm (*Chem.*). The fraction θ of the adsorbent surface which is covered by molecules of adsorbed gas is given by

$$\theta = bp/(1 + bp)$$

where p is the gas pressure, and b is a constant.

Langmuir dark space (*Electronics*). Nonglow region surrounding a negative electrode placed in the luminous positive column of a gas discharge.

Langmuir frequency (*Electronics*). The natural frequency of oscillation for electrons in a plasma.

Langmuir law (*Electronics*). Law relating current and voltage in a space-charge limited vacuum tube. $i = GV^{3/2}$, where i = current, V = applied voltage and G = perveance.

Langmuir probe (*Electronics*). Electrode(s) introduced into gas-discharge tube to study potential distribution along the discharge.

Langmuir's theory (*Chem.*). (1) The assumption that the extranuclear electrons in an atom are arranged in shells corresponding to the periods of the periodic system. The chemical properties of the elements are explained by supposing that a complete shell is the most stable structure. (2) The theory that adsorbed atoms and molecules are held to a surface by residual forces of a chemical nature.

Langmuir trough (*Min. Proc., Phys.*). Apparatus in which a rectangular tank is used to measure surface tension of liquid.

language (*Comp.*). In electronic computers a system consisting of a well-defined set of characters (or symbols) with a code for combining these to form words or expressions.

languet (*Zool.*). In *Urochorda*, one of a series of curved processes which are sometimes supposedly formed by subdivision of the dorsal lamina, arising from the hyperpharyngeal band.

languid (*Acous.*). The tongue of an organ or harmonium reed.

laniary (*Zool.*). Adapted for tearing, as a canine tooth.

lanolin (*Chem.*). Fat from wool (*adeps lanae*), a yellowish viscous mass of waxlike consistency, very resistant to acids and alkalis; it emulsifies easily with water and is used for making ointments. It consists of the palmitate, oleate, and stearate of cholesterol.

lansfordite (*Min.*). Hydrated magnesium carbonate, crystallizing in the monoclinic system. Lansfordite occurs in some coal mines but is not stable on exposure to the atmosphere and becomes dehydrated.

lantern (*Arch.*). An erection on the top of a roof, projecting above the general roof level, and usually having glazed sides to admit light, as well as openings for ventilation.

lantern coelom (*Zool.*). In *Echinoidea*, the coelomic space lying within the lantern of Aristotle, and representing an enlarged perihaemal ring.

lantern wheel (or **pinion**) (*Horol.*). A form of pinion consisting of two circular brass disks connected by cylindrical pins, the pins acting as the leaves in an ordinary pinion. Lantern pinions are very satisfactory as followers but should not be used as drivers. Used extensively in cheap clocks, alarm clocks with pin-pallet escapement, and in some turret clocks. The lantern wheel is obsolete in general engineering.

lanthanide contraction (*Chem.*). The peculiar characteristic of the lanthanides that the ionic radius decreases as the atomic number increases, because of the increasing pull of the nuclear charge on the unchanging number of electrons in the two outer shells.

lanthanide series (*Chem.*). Name for the rare earth elements at. nos. 57–71, after lanthanum, the first of the series. Cf. *actinides*.

lanthanum (*Chem.*). A metallic element in the third group of the periodic system, belonging to the rare earths group. Symbol La, at. no. 57, r.a.m. 138·91, m.p. 810°C.

lanthanum glass (*Glass*). Optical glass used for high-quality photographic lenses, etc. Characteristics: high refractive index, low dispersion.

lantharin (*Cyt.*). See **linin.**

lanuginose, lanuginous (*Bot., Zool.*). Bearing a woolly coating; lanate.

lanugo (*Zool.*). In Mammals, prenatal hair.

lap (*Build.*). The length of overlap (from 2½ to 4 in. or 6 to 10 cm) of a slate over the slate next but one below it, in centre-nailed work; or that of a slate over the nail securing the slate next but one below it, in head-nailed work. (*Eng.*) (1) The extent to which one plate overlaps another. See **lap joint.** (2) In steam engines, the amount by which the valve has to move from mid position to open the steam or exhaust port. See **exhaust lap, outside lap.** (3) A piece of soft metal, wood, etc., often in the form of a rotating cylinder or disk, charged with abrasive or polishing powder; used in polishing or finishing metals, gem-cutting, etc. See **lapping, lapping machine.** (*Glass*) (1) A rotating disk or other tool for grinding or polishing glass. (2) A square piece of material, usually rubber, to protect the hands when handling glass. (*Met.*) A surface defect on rolled or forged steel. It is caused by folding a fin on to the surface and squeezing it in; as welding does not occur, a seam appears on the surface. Also, polishing cloth impregnated with diamond dust or other abrasive, used in polishing rock specimens, gemmology, etc. In mining, one coil of rope on the mine hoisting drum. (*Textiles*) (1) Rolled sheet of fibres produced at the delivery end of opening machinery, in preparation for carding, drawing, and spinning. (2) Cloth fent used in gaiting-up new cloth sort in loom. (3) Build-up of fibres on delivery roller after a roving or yarn breaks.

laparotomy (*Surg.*). Cutting into the abdominal cavity. *Exploratory laparotomy*, the operation of cutting into the abdominal cavity so that direct examination of abdominal organs may be made.

lap dissolve (*Cinema., TV*). The same as *dissolve*, the new picture appearing gradually at the same rate as the previous one disappears.

lap dovetail (*Join.*). An angle joint between two members, in which only one shows end grain, a sufficient thickness of wood having been left on this member, in cutting the joint, to cover the end grain of the other member. Also **drawer-front dovetail**, from one of its common uses.

lapel microphone (*Acous.*). Small microphone which does not impede vision of the speaker.

lapidicolous (*Zool.*). Living under stones.

lapiés (*Geog.*). A flat area of bare limestone; a limestone pavement.

lapilli (*Geol.*). Small rounded pieces of lava whirled from a volcanic vent during explosive eruptions; lapilli are thus similar to volcanic bombs but smaller in size, usually about the size of walnuts. *sing.* **lapillus.**

lapillus (*Zool.*). In *Teleostei*, one of the otoliths which lies in the utriculus close to the ampullae of the anterior and horizontal canals of the auditory organ.

lapis lazuli (*Min.*). Original sapphire of ancients, a beautiful blue stone used extensively for ornamental purposes. It consists of the deep-blue feldspathoid *lazurite* (q.v.), usually together with calcite and pyrite.

lap joint (*Carp.*). A joint between two pieces of timber, formed by laying one over the other for a certain length and fastening the two together with metal straps passing around the timbers, or with bolts passing through them. (*Eng.*) A plate joint in which one member overlaps the other, the two being riveted or welded along the seam single, double, or treble.

Laplace linear equation (*Maths.*). A differential equation of the form

$$(a_0+b_0x)\frac{d^ny}{dx^n}+(a_1+b_1x)\frac{d^{n-1}y}{dx^{n-1}}+\ldots+(a_n+b_nx)y=0.$$

Laplace's equation (*Maths.*). The equation $\nabla^2V=0$, where ∇ is the vector operator nabla, i.e., $\frac{\partial^2V}{\partial x^2}+\frac{\partial^2V}{\partial y^2}+\frac{\partial^2V}{\partial z^2}=0$. It is satisfied by electric and gravitational potential functions. Cf. *Poisson's equation.*

Laplace transform (*Maths.*). The Laplace transform $F(p)$ of the function $f(t)$ is defined by

$$F(p)=\int_0^\infty e^{-pt}f(t)dt.$$

In the *two-sided* Laplace transfor the range of integration is from $-\infty$ to $+\infty$.

La Pointe picker (*Min. Proc.*). Small belt conveyor on which ore is so displayed that radioactive pieces are removed as they pass a Geiger-Müller counter.

lappet (*Weaving*). Fabric of a light muslin nature with a zigzag type of surface figuring. This is formed from extra warp threads which give an effect very similar to embroidery. (*Zool.*) Any hanging, lobelike structure, as the ciliated *lappets* of the actinotroch larva of *Phoronidea.*

lapping (*Elec. Eng.*). The final abrasive polishing of a quartz crystal to adjust its operating frequency. Also, smoothing of surface of crystalline semiconductors. (*Eng.*) The finishing of spindles, bored holes, etc., to fine limits, by the use of laps of lead, brass, etc. (*Photog.*) Rubbing one surface against another, generally with an abrasive such as rouge, so that the softer takes up the contour of the harder, e.g., in a lens, or in making optical flats.

lapping machine (*Eng.*). A machine tool for finishing the bores of cylinders, etc., to fine limits by the use of revolving circular laps supplied with an abrasive powder suspended in the coolant.

lapsed intelligence (*Zool.*). See **instinct.**

lapse rate (*Meteor.*). The rate of fall of temperature with increase in height. See **adiabatic-, DALR, ELR, SALR.**

lap winding (*Elec. Eng.*). A form of 2-layer winding for electric machines in which each coil is connected in series with the one adjacent to it. Cf. *wave winding.*

Laramian or Laramie Sands (*Geol.*). A thick formation of sand laid down under continental conditions during the shallowing and retreat of the Cretaceous sea from the central U.S. The formation contains the remains of the last great dinosaurs (the *Ceratops* fauna) in Wyoming and Montana.

Laramide revolution (*Geol.*). A period of earth movement in early Tertiary times during which the interior regions of N. America were folded, producing the Rockies and the Andean and Antillean chains. The Appalachians were uplifted at this time, and the cycle of erosion was initiated which has produced the existing land forms. Volcanic activity occurred from Mexico into Canada.

lardaceous disease (*Med.*). Amyloid or waxy kidney. Amyloid degeneration of the blood vessels of the kidney (and usually elsewhere), associated with the copious excretion of urine of low specific gravity and containing albumin.

lardalite (*Geol.*). See **laurdalite.**

large (*Paper*). A standard size of cut card, $3 \times 4\frac{1}{2}$ in. (76×114 mm).

large-area foils (*Print.*). Blocking foils for display card printing. These are capable of filling in very large areas and yet of giving fine detail for half-tone work.

large bodies (*Bacteriol.*). Morphological variants which result from the subjection of bacterial cultures to a wide range of treatments, e.g., presence of antibiotics, chilling, presence of antiserum. The resultant large and distorted cells revert to normal after brief exposure to the evoking stimulus. See **L-forms.**

large calorie (*Heat*). See **calorie.**

large crown octavo (*Typog.*). A nonstandard book size, larger than crown 8vo and smaller than demy 8vo, varying slightly from publisher to publisher.

large imperial (*Paper*). A standard board size, 22×32 in. (560×810 mm).

large intestine (*Zool.*). See **colon.**

large panel construction (*Build., Civ. Eng.*). See **system building.**

large post (*Paper*). A standard size of printing paper, $16\frac{1}{2} \times 21$ in. (420×530 mm.)

large scale integration (*Electronics*). Production of an integrated circuit with more than 100 gates on a single chip of semiconductor.

larmier (*Build.*). A corona placed over a door or window opening to serve as a dripstone.

Larmor frequency (*Electronics*). The angular frequency of precession for the spin vector of an electron acted on by an external magnetic field.

Larmor precession (*Electronics*). The motion experienced in a small uniform magnetic field by a charged particle (or system of charged particles) when subjected to a central force which is directed towards a common point.

Larmor radius (*Nuc.*). That of the circular or helical path followed by a charged particle in a uniform magnetic field.

larnite (*Min.*). Orthosilicate of calcium, Ca_2SiO_4, discovered in the contact zone of a Tertiary dolerite intrusive into chalk containing flint nodules; formed by reaction between the calcium carbonate of the former and the silica of the latter. Cf. *wollastonite*.

larry (*Tools*). A tool having a curved steel blade fixed to the end of a long handle, to which it is bent normally; used for mixing mortar, or for mixing hair with coarse stuff to form a plaster.

larrying (*Build.*). The process of pouring a mass of mortar upon the wall and working it into the joints; sometimes used in building large masses of brickwork.

Larsen and Nielsen (*Build., Civ. Eng.*). A method of system building developed in Denmark. This is a method in which the components are of room size and put together to form a box frame. The cladding is generally of precast sandwich panels incorporating an insulation layer.

larva (*Zool.*). The young stage of an animal if it differs appreciably in form from the adult; a free-living embryo.

Larvacea (*Zool.*). A class of *Urochorda* in which the adult is a free-living active form possessing a notochord, and closely resembles the larva in its anatomy; there are only 2 gill-slits, which open directly to the exterior; the animal secretes a curious nonadherent 'house' around itself, which is periodically replaced; pelagic forms.

larvikite (*Geol.*). See **laurvikite.**

larviparous (*Zool.*). Giving birth to offspring which have already reached the larva stage.

larvivorous (*Zool.*). Larva-eating.

laryngectomy (*Surg.*). Surgical removal of the larynx.

laryngismus (*Med.*). Spasm of the larynx.

laryngitis (*Med.*). Inflammation of the larynx.

laryngofissure (*Surg.*). Thyrotomy. Surgical exposure of the larynx by dividing the thyroid cartilage (Adam's apple) in the midline.

laryngology (*Med.*). That branch of medical science which treats of abnormal conditions of the larynx and adjacent parts of the upper respiratory tract. *n.* **laryngologist.**

laryngopharyngitis (*Med.*). Inflammation of the larynx and the pharynx.

laryngophone (*Acous.*). See **throat microphone.**

laryngoscope (*Med.*). An instrument used for viewing the larynx.

laryngostenosis (*Med.*). Pathological narrowing of the larynx.

laryngostomy (*Surg.*). The surgical formation of an opening in the larynx.

laryngotomy (*Surg.*). The operation of cutting into the larynx.

laryngotracheal chamber (*Zool.*). In Amphibians, a small chamber into which the lungs open anteriorly, and which communicates with the buccal cavity by the glottis.

larynx (*Acous.*). The fundamental of a speech sound is determined by the tension of the muscles in the larynx, and the spectral distribution of the frequency components in the emitted sound is determined by the acoustic resonance in the mouth and nasal cavities. See **voiced sound, unvoiced sound.** (*Anat.*) The vocal organ in all land Vertebrates except Birds, situated at the anterior end of the trachea. *adj.* **laryngeal.**

laser (*Phys.*). (*L*ight *A*mplification by *S*timulated *E*mission of *R*adiation.) A coherent light source using a crystalline solid (e.g. ruby), liquid or gas-discharge tube, in which atoms are pumped simultaneously into excited states by an incoherent light flash. They return to their ground state with the emission of a light pulse for which the energy flux may be 10 mW/cm^2 (lasting 30 nanosecs) and the beam divergence $< 10^{-2}$ radians. This action is essentially the same as that of a microwave *maser*. The first continuously operating laser was the gas laser produced by Bell Telephone Laboratory (1960), and it utilized a low-energy discharge, in a mixture of neon and helium gases, to produce a sharply defined and intense beam of infrared radiation.

laser gyro (*Instr.*). An integrating-rate gyroscope which combines the properties of the optical oscillator, the laser and general relativity. It differs from conventional gyros in the absence of a spinning mass and so its performance is not affected by accelerations. It has a low power consumption and is not subject to prolonged starting time and hence is an ideal device for space-age aeronautics. Modern versions can measure rates as low as 0·1 degree per hour.

laser threshold (*Phys.*). The minimum pumping power (or energy) required to operate a laser.

lasher-on (*Mining*). A man employed to lash the chains from the tubs to the endless rope, in underground mechanical haulage.

lashing (*Mining*). A South African term for removing broken rock after blasting. Canadian term, **mucking** or **mucking out.**

Lashley jumping stand (*An. Behav.*). A small raised platform from which an animal can jump to one of two doors which generally bear different stimuli. The correct door opens and gives access to food or water, but the wrong door does not open, and the animal falls into a net.

lash-up (*Telecomm.*). The temporary connexion of apparatus, for experimental or emergency use.

Lassaigne's test (*Chem.*). Test for the presence of nitrogen in an organic substance. The sample is heated with metallic sodium in a test tube; the product is placed in water and filtered. Iron(II) sulphate is added and the mixture boiled, cooled, and sodium cyanide and iron (III) chloride added. If nitrogen is present the characteristic *Prussian blue* (q.v.) colour is observed.

lasso-cell or **colloblast** (*Zool.*). In *Ctenophora*, an adhesive cell, with a long anchoring filament, occurring on the tentacles.

lasting (*Textiles*). Strong twill cloth of hard twisted cotton or worsted yarns, or cotton warp and worsted weft; used for boot and bag linings, etc.; sometimes has a glazed finish.

last-subscriber release (*Teleph*). The release of automatic switching plant when the last of both subscribers has replaced his receiver and opened his loop. Also **last-party release.**

latching (*Surv.*). See **dialling.** (*Telecomm.*). Arrangement whereby a circuit is held in position, e.g., in read-out equipment, until previous operating circuits are ready to change this circuit. Also called **locking.**

late-choice call meter (*Teleph.*). A traffic meter so connected as to record the number of calls carried by a late-choice trunk of a grading.

latency (*Comp.*). Delay, in digital computers, between the directing signal and the availability of the required signals from a memory.

latency period (*Psychol.*). The period of *psychosexual development* between the age of 4–5 years and puberty.

latensification (*Photog.*). Increase of effective emulsion speed after exposure.

latent (*Zool.*). In a resting condition or state of arrested development, but capable of becoming active or undergoing further development when conditions become suitable; said also of hidden characteristics which may become evident under the right circumstances.

latent heat (*Heat*). More correctly, **specific latent heat.** The heat which is required to change the state of a substance from solid to liquid, or from liquid to gas, without change of temperature. The numerical value of the specific latent heat is the amount of heat required to change the state of unit mass. Most substances have a latent heat of fusion (melting) and a latent heat of evaporation. In thermodynamics, heat supplied at constant pressure is called *enthalpy*, and thus specific latent heat of evaporation is called *enthalpy of evaporation.*

latent image (*Photog.*). The nondetectable image registered in a sensitive emulsion by ionization of molecules and realized by development.

latent learning (*An. Behav.*). Learning, or the association of previously indifferent stimuli or situations, without patent reward, e.g., the learning of the layout of a maze by rats when they make unrewarded runs through it.

latent magnetization (*Mag.*). The property possessed by certain feebly magnetic metals (e.g., manganese and chromium) of forming strongly magnetic alloys or compounds.

latent neutrons (*Nuc.*). In reactor theory, the delayed neutrons due from (but not yet emitted by) fission products.

latent period (*Radiol.*). That between exposure to radiation and its effect.

latent roots of a matrix (*Maths.*). See **characteristic equation of a matrix.**

latent time (*Bot.*). The period of time between the beginning of stimulation and the first signs of a response.

later (*Build.*). A brick or tile.

lateral. Situated on or at, or pertaining to, a side. (*Bot.*) (1) Arising from the side of the parent axis. (2) Attached to the side of another member. (3) On or near the edge of a thallus or fruit.

lateral axis (*Aero.*). The crosswise axis of an aircraft, particularly that passing through its centre of gravity, parallel to the line joining the wing tips.

lateral canal (*Hyd. Eng.*). A separate navigational canal constructed to follow the lie of a river which does not lend itself to canalization.

lateral cilia (*Zool.*). In *Filibranchiata* cilia on the ctenidial filaments which serve to draw a current of water into the mantle cavity.

lateral deviation (*Nav.*). Error in bearing arising from tilt in the reflecting ionosphere.

lateral instability (*Aero.*). A condition wherein an aircraft suffers increasing oscillation after a rolling disturbance. Also **rolling instability.**

lateral inversion (*TV*). Defect in a reproduced television image, the picture being reversed, the right-hand side appearing on the left, due to a reversal in the connexions from the line-scanning generator.

lateralis (*Zool.*). In Fish, a branch of the tenth cranial nerve, which innervates the lateral line.

lateral line (*Zool.*). In Fish, a line of neuromast organs running along the side of the body. In *Nematoda*, a paired lateral concentration of hypodermis containing the excretory canal.

lateral load (*Eng.*). A force acting on a structure or a structural member in a transverse direction, e.g., wind forces on a bridge or building at right angles to its length, which trusses and girders are not primarily designed to withstand.

lateral mesenteries (*Zool.*). In *Zoantharia*, all mesenteries apart from the *directive mesenteries* (q.v.).

lateral plate (*Zool.*). In *Craniata*, a ventral portion of each mesoderm-band which surrounds the mesenteron in embryo; in Insects, the paired lateral region of the germ-band of the embryo which becomes separated from the opposite member of its pair by the middle plate.

lateral recording (*Acous.*). That in which the cutting stylus removes a thread (*swarf*) from the surface of a blank disk, the modulation being realized as a lateral (radial) deviation as the spiral is transversed. Cf. *hill-and-dale recording, stereophonic recording.*

lateral shift (*Geol.*). The displacement of outcrops in a horizontal sense, as a consequence of faulting. Cf. *throw.*

lateral stability (*Aero.*). Stability of an aircraft's motions out of the plane of symmetry, i.e., side-slipping, rolling and yawing.

lateral traverse (*Eng.*). The longitudinal play given to locomotive trailing axles to permit of taking sharp curves.

lateral vein (*Zool.*). In Fish, a paired vein in the body wall; it receives blood from the paired fins and discharges it into the precaval vein.

lateral velocity (*Aero.*). The rate of *sideslip* of an aircraft, that is, the component of velocity which is resolved in the direction of the lateral axis.

laterigrade (*Zool.*). Moving sideways, as some Crabs.

laterite (*Geol.*). A residual clay formed under tropical climatic conditions by the weathering of igneous rocks, usually of basic composition. Consists chiefly of hydroxides of iron and aluminium, grading through increase of the latter into bauxite.

lateritious (*Bot.*). Brick-red.

laterization (*Geol.*). The process whereby rocks are converted into laterite, the details being imperfectly understood. Essentially, the process involves the abstraction of silica from the silicates. See **laterite**.

laterocranium (*Zool.*). In Insects, the side of the head, comprising the genae and postgenae.

laterofrontal cilia (*Zool.*). In *Lamellibranchiata*, large cilia on the ctenidial filaments which serve to deflect solid particles in the inhalant water current on to the face of the filaments.

laterosphenoid (*Zool.*). The so-called 'alisphenoid' of Fish, Reptiles, and Birds (representing an ossification of the wall of the chondrocranium), as distinct from the alisphenoid of Mammals (developed from the splanchnocranium).

laterosternite (*Zool.*). A paired sclerite sometimes occurring at the side of the eusternum in Insects.

latex (*Bot.*). A milky fluid, present in many higher plants, and in some *Agarics*, usually white, sometimes yellow or reddish; it consists of a mixture of substances, proteins, gums, carbohydrates, etc. (*Chem.*) A milky viscous fluid extruded when rubber trees (e.g., *Hevea Brasiliensis*) are tapped. It is a colloidal system of caoutchouc dispersed in an aqueous medium, rel. d. 0·99, which forms rubber by coagulation. The coagulation of latex can be prevented by the addition of ammonia or formaldehyde. Latex may be vulcanized directly, the product being known as *vultex*. (*Chem. Eng.*) In synthetic rubber manufacture, the process stream in which the polymerized product is produced. (*Paper*) A solution of rubber sometimes used to increase the strength and durability of paper. Natural and synthetic latices are used for impregnation or beater addition to produce imitation leather.

latex duct (*Bot.*). An elongated, branched, aseptate system of anastomosing hyphae present in some of the larger agarics, and containing latex.

latex vessel (*Bot.*). A simple or branched tube, usually anastomosing with other similar tubes, derived by the enlargement and union of a chain of cells, and containing latex. Also **latex cell, latex tube.**

lath (*Build.*). See **lathing.** (*Min.*) A term commonly applied to a lathlike crystal.

lathe (*Eng.*). A machine tool for producing cylindrical work, facing, boring, and screw-cutting. It consists generally of a bed carrying a head-stock and tail-stock, by which the work is driven and supported, and a saddle carrying the slide rest by which the tool is held and traversed.

lathe bed (*Eng.*). That part of a lathe forming the support for the head-stock, tail-stock, and carriage. It consists of a rigid cast box-section girder, supported on legs, its upper face being planed and scraped to provide **true working** surfaces, or 'ways'.

lathe carrier (*Eng.*). A clamp consisting of a shank which is formed into an eye at one end and provided with a set screw. It is attached to work supported between centres and driven by the engagement of the driver-plate pin with the shank or 'tail' of the carrier, which may be straight or bent.

lathe tools (*Eng.*). Turning tools with edges of various shape (round-nosed, side, etc.), and cutting angles, varying with the material worked on, formed by giving clearance to the front of the cutting edge and rake to the top of the tool. See **finishing tool, knife tool, roughing tool, side tool.**

lathe work (*Eng.*). Any work ordinarily performed in the lathe, such as all classes of turning, boring, and screw-cutting.

lathing (*Build.*). Material fixed to surfaces to provide a basis for plaster. Formerly soft wood strips (*laths*) 3–4 ft long were used; nowadays steel meshing or perforated steel sheet is usual.

lathyrism (*Med.*). A disease characterized by stiffness and paralysis of the legs, due to poisoning with certain kinds of chick-pea.

laticiferous (*Bot.*). Conveying latex, e.g., by cells, ducts, tissues.

latiplantar (*Zool.*). Having the posterior tarsal surface rounded.

latirostral (*Zool.*). Having a broad beak.

latiseptate (*Bot.*). Having wide septa.

latitude (*Photog.*) The range of exposure permissible, or range of density usefully obtainable in a photographic emulsion. The range of exposure obtainable with the linear portion of the gamma curve of an emulsion. (*Surv.*) Northing. Distance of a point north (if positive) or south (negative) from point of origin of survey. This, together with departure (easting), locates the point in a rectangular grid orientated on true meridian.

latitude and longitude (*Astron.*). *Celestial.* Spherical coordinates referred to the ecliptic and its poles. **Celestial latitude** is the angular distance of a body north or south of the ecliptic. **Celestial longitude** is the arc of the ecliptic intercepted between the latitude circle and the First Point of Aries, and is measured positively eastwards from 0° to 360°. (*Geog.*) *Terrestrial.* Spherical coordinates referred to the earth's equator and its poles; used to specify a point on the earth's surface. **Terrestrial latitude**: the angular elevation of the celestial pole above a plane tangential to the earth at a given place is known as the *geographical latitude*; the *geocentric latitude* is the angle made with the equatorial plane by the radius of the earth through the given point. The latter is slightly less than the former owing to the oblate form of the earth. **Terrestrial longitude** is the arc of the equator between the meridian through the point and the meridian at Greenwich; generally measured from 0° to 180° east or west of Greenwich.

latitudinal furrow (*Zool.*). In a segmenting ovum, a constriction encircling the ovum above or below, and parallel to, the equatorial furrow.

Latour alternator (*Radio*). See **Bethenod-Latour alternator.**

Latour-Winter-Eichberg motor (*Elec. Eng.*). A

form of compensated repulsion motor sometimes used for traction work.

latterkin (*Plumb.*). A piece of hardwood suitably shaped at one end so that it may be used for clearing the grooves in cames.

lattice (*Bot.*). A weakly developed sieve plate on a lateral wall of a sieve tube, having vaguely defined edges and very minute pores. (*Crystal.*) A regular space arrangement of points as for the sites of atoms in a crystal. (*Maths.*) A partially-ordered set in which any two elements have a least upper bound and a greatest lower bound. (*Nuc. Eng.*) Regular geometrical pattern of discrete bodies of fissionable and nonfissionable material in a nuclear reactor. The arrangement is subcritical or just critical if it is desired to study the properties of the system.

lattice bars (*Eng.*). The diagonal bracing of struts and ties in an *open-frame girder* or *lattice girder* (qq.v.).

lattice bridge (*Eng.*). A bridge of lattice girders.

lattice coil (*Elec. Eng.*). A form of coil used for the armature winding of electric machines, which is arranged so that the end connexions cross over one another in a regular pattern, giving a lattice appearance. (*Radio*) An inductance coil in which the turns are wound so as to cross each other obliquely, to reduce the self-capacitance. See **honeycomb coil**.

lattice diagram (*Elec. Eng.*). A diagram for simplifying the calculation of travelling waves on a transmission line when there are a large number of successive reflections.

lattice energy (*Crystal.*). Energy required to separate the ions of a crystal from each other to an infinite distance.

lattice filter (*Telecomm.*). One or more lattice networks acting as a wave filter.

lattice girder (*Eng.*). A girder formed of upper and lower horizontal members connected by an open web of diagonal crossing members, used in structures such as bridges and large cranes.

lattice network (*Telecomm.*). One formed by two pairs of identical arms on opposite sides of a square, the input terminals being across one diagonal and the output terminals across the other. Also **bridge network, lattice section**.

lattice structure (*Crystal.*). One of three types of crystal structure: (*a*) ionic, with symmetrically arranged ions and good conducting power; (*b*) molecular, covalent, usually volatile and nonconducting; (*c*) layer, with large ions each associated with two small ones, forming laminae weakly held by nonpolar forces.

lattice vibration (*Crystal.*). That of atoms or molecules in a crystal due to thermal energy.

lattice water (*Chem.*). Water of crystallization, which is present in stoichiometric proportions and occupies definite lattice positions, but is in excess of that with which the ions could be bonded. This water apparently fills in holes in the crystal lattice, as with *clathrate* compounds.

lattice winding (*Elec. Eng.*). A winding, made up of lattice coils, for electric machines; always used for d.c. machines and frequently for a.c. machines.

lattice window (*Arch.*). A window in which diamond-shaped panes are supported in a leaden frame consisting of diagonally intersecting cames, the longer axes of the diamonds being vertical.

laudanum (*Med.*). Tincture of opium.

Laue pattern (*Crystal.*). Spot diagram produced on a photographic plate when a heterogeneous X-ray beam is passed through a thin crystal, which acts like an optical grating. Used in the analysis of crystal structure.

laughing gas (*Chem.*). See **nitrous (nitrogen (I) oxide**.

laumontite (*Min.*). A zeolite consisting essentially of hydrated silicate of calcium and aluminium, crystallizing in the monoclinic system; occurs in cavities in igneous rocks and in veins in schists and slates. See **leonhardite**.

launching (*Telecomm.*). Said of the operation of transmitting a signal from a conducting circuit into a waveguide.

launder (*Mining*). Inclined trough for conveying water or crushed ore and water (pulp).

Lauraceae (*Bot.*). Family of leathery-leaved dicotyledons.

laurdalite or lardalite (*Geol.*). A coarse-grained soda-syenite from S. Norway; it resembles laurvikite but contains nepheline (elaeolite) as an essential constituent.

Laurentian Granites (*Geol.*). The oldest granitic intrusives in the Canadian Shield, of post-Keewatin, pre-Timiskaming age; they occur as batholiths elongated N.E. to S.W., and consist of granite, granite-gneiss, and pegmatites worked as a source of potash.

Laurent's expansion (*Maths.*). If the function $f(z)$ is analytic in the annulus formed by two circles C_1 and C_2, of radii r_1 and r_2, respectively, then within this annulus:

$$f(z) = \sum_{1}^{\infty} a_n(z-z_0)^n + \sum_{1}^{\infty} b_n(z-z_0)^{-n},$$

$$\text{where } a_n = \frac{1}{2\pi i}\int_{C_1} \frac{f(z)dz}{(z-z_0)^{n+1}}.$$

$$b_n = \frac{1}{2\pi i}\int_{C_2} (z-z_0)^{n-1} f(z)dz.$$

Laurer-Stieda (or Laurer's) canal (*Zool.*). In *Trematoda*, a canal leading away from the junction of the oviduct and the vitelline duct; it opens either to the exterior of the body dorsally or into the alimentary canal, or ends blindly.

lauic acid (*Chem.*). Dodecanoic acid, $CH_3 \cdot (CH_2)_{10} \cdot COOH$. Crystalline solid; m.p. 44°C. Occurs as glycerides in milk, laurel oil, palm oil, etc.

laurionite (*Min.*). Hydroxide-chloride of lead, exceedingly rare, found in ancient lead slags at Laurion in Greece.

laurite (*Min.*). An iron-black sulphide of ruthenium, with some osmium, occurring as small cubic crystals (octahedra) associated with platinum, in Borneo and Oregon.

Lauritsen electroscope (*Instr.*). Sensitive rugged electroscope using a metallized quartz fibre. See **dosemeter**.

laurvikite or larvikite (*Geol.*). A soda-syenite from S. Norway, very popular as an ornamental stone when cut and polished; widely used for facing buildings, the distinctive feature being a fine blue colour, produced by schiller structure in the anorthoclase feldspars.

lauryl alcohol (*Chem.*). Dodecan-l-ol, $CH_3 \cdot (CH_2)_{10} \cdot CH_2OH$. Insoluble in water, crystalline solid; m.p. 24°C. Used in manufacture of detergents.

lauryl thiocyanate (*Chem.*). $C_{12}H_{25}SCN$. Used as a disinfectant.

lautarite (*Min.*). Monoclinic iodate of calcium, occurring rarely in caliche in Chile.

lava (*Geol.*). The molten rock material that issues from a volcanic vent or fissure and consolidates on the surface of the ground (*subaerial lava*), or on the floor of the sea (*submarine lava*). Chemically, lava varies widely in composition; it may be in the condition of glass, or a holocrystalline rock. See **volcano**; also **basalt, obsidian, pillow structure, pumice.**

lava flows (*Geol.*). See **extrusive rocks.**

lavage (*Med.*). Irrigation or washing-out of a cavity, such as the stomach or the bowel, e.g., gastric *lavage*.

Lavalier cord (*Acous.*). Microphone neck cord used by commentators, television announcers, etc., leaving their hands free.

lavender print (*Cinema.*). The specially dyed print, of low contrast, which is the first print made after editing the negative. It is suitable for making a duplicate negative, should the original negative become defective.

law. A scientific law is a rule or generalization which describes specified natural phenomena within the limits of experimental observation. An apparent exception to a law tests the validity of the law under the specified conditions. A true scientific law admits of no exception. A law is of no scientific value unless it can be related to other laws comprehending relevant phenomena.

law calf (*Bind.*). Calf leather with a rough surface and light in colour; used for account-book bindings, etc.

lawn (*Textiles*). Fine plain cloth, originally made of flax yarns. Some other lightweight fabrics of good quality are now thus named, e.g., sheer lawn, Indian lawn, etc.

lawnmower (*Radar*). Colloquialism for type of radar noise limiter used to reduce *grass* (q.v.) on the display screen.

law of conservation of matter (*Chem.*). Matter is neither created nor destroyed during any physical or chemical change. See **relativity.**

law of constant (or definite) proportions (*Chem.*). Every pure substance always contains the same elements combined in the same proportions by weight.

law of Dulong and Petit (*Chem.*). The atomic heats of solid elements are constant and approximately equal to 26 (when the specific heat capacity is in joules). Certain elements of low atomic mass and high melting-point have, however, much lower atomic heats at ordinary temperatures.

law of effect (*Psychol.*). Important principle of learning that stimulus-response connexions are strengthened by satisfaction and weakened by discomfort or punishment.

laws of electrolysis (*Elec.*). See **Faraday's laws of electrolysis.**

law of equivalent (or reciprocal) proportions (*Chem.*). The proportions in which two elements separately combine with the same weight of a third element are also the proportions in which the first two elements combine together.

law of Guldberg and Waage (*Chem.*). See **law of mass action.**

law of isomorphism (*Chem.*). See **Mitscherlich's law of isomorphism.**

law of least squares (*Maths.*). The law which postulates that the best value to take from a set of observations is that which makes the sum of the squares of the deviations from this value a minimum.

law of mass action (*Chem.*). The velocity of a homogeneous chemical reaction is proportional to the concentrations of the reacting substances.

law of multiple proportions (*Chem.*). When two elements combine to form more than one compound, the amounts of one of them which combine with a fixed amount of the other exhibit a simple multiple relation.

law of octaves (*Chem.*). The relationship observed by Newlands (1863) which arranges the elements in order of atomic weight and in groups of 8 (octaves) with recurring similarity of properties. See **periodic system.**

law of partial pressures (*Chem., Phys.*). See **Dalton's law of partial pressures.**

law of photochemical equivalence (*Chem.*). See **Einstein law of photochemical equivalence.**

law of rational indices (*Crystal.*). A fundamental law of crystallography which states, in the simplest terms, that in any natural crystal the indices may be expressed as small whole numbers.

law of reciprocal proportions (*Chem.*). See **law of equivalent proportions.**

law of volumes (*Chem.*). See **Gay-Lussac's law.**

Lawrence Smith method (*Chem.*). A method used in chemical analysis for the estimation of alkali metals, particularly in the analysis of glass and silicates.

Lawrence tube (*TV*). See **tricolour chromatron.**

lawrencium (*Chem.*). Transuranic element, symbol Lr, at. no. 103. Its only known isotope has a short half-life of 8 seconds.

laws of reflection (*Phys.*). When a ray of light is reflected at a surface, the reflected ray is found to lie in the plane containing the incident ray and the normal to the surface at the point of incidence. The angle of reflection equals the angle of incidence.

lawsonite (*Min.*). An orthorhombic hydrated silicate of aluminium and calcium. It occurs in low-grade regionally metamorphosed schists, particularly in glaucophane schists.

laxator (*Zool.*). A muscle which relaxes or loosens a part of the body without changing the relative position or direction of the axis of the part. Cf. *tensor*.

lay (*Cables*). The axial length of one turn of the helix formed by the core (in a telephone cable) or strand of a conductor (in a power cable). See also **lay ratio.** (*Print.*) See **lay edges.**

lay-away pits (or vats) (*Leather*). See **handlers.**

lay-boys (*Paper*). The automatic sheet catcher on a sheet-cutting machine.

lay-by (*Mining*). Widened section of otherwise single-track underground tramming road, permitting passing of trains of trucks.

Laycock overdrive (*Autos.*). Unit situated at the rear of the gearbox, comprising an epicyclic gear-train used in reverse which drives the propeller shaft at a speed greater than normal direct drive. The input shaft is fitted to the planet carrier and to the driving member of a one-way clutch, the driven half of which also forms the annulus of the epicyclic gear. The sun wheel is free on the input shaft but can be held still by a double-purpose cone-clutch, which slides on splines on the sun wheel bearing, thus giving an overdrive, or locks the epicycled train for direct drive.

lay cords (*Bind.*). Cords with which a book is tied, to prevent the covers from warping while drying.

lay edges (*Print.*). The edges of a sheet of paper which are laid against the guides or lays in a printing or folding machine. The front edge (or *gripper edge*) is laid to the *front lay* and the side of the sheet to the *side lay*.

layer (*Bot.*). A stratum of vegetation, as the shrubs in a wood. (*Build.*) (1) A course in a wall. (2) A bed of mortar between courses. (*Plumb.*) One thickness of roofing felt.

layer board (*Build.*). See **lear board.**

layered map (*Surv.*). See **relief map.**

layering (*Bot.*). (1) See **stratification.** (2) A method of artificial propagation in which stems are pegged down and covered with soil until they root, when they can be detached from the parent plant.

layer lattice (*Crystal.*). Concentration of bonded atoms in parallel planes, with weaker and non-polar bonding between successive planes. This gives marked cleavage, e.g., in graphite, mica.

layer line (*Crystal*). One joining a series of spots on X-ray rotating crystal diffraction photograph. Its position enables the crystal lattice spacing parallel to the axis of rotation to be determined.

layer of Langhans (*Zool.*). See **cytotrophoblast.**

layer-on (*Print.*). A printer's assistant who feeds the sheets to the machine, an occupation largely superseded by mechanical feeders.

layers (*Phys.*). Ionized regions in space which vary vertically and affect radio propagation. See **E-layer, F-layer.** (*Leather*) Pits containing tan liquor, in which heavy hides are laid out with a layer of ground tanning material separating them, after passing through the handlers. Also called **dusters.**

laying (*Build.*). The first coat of plaster in 2-coat work.

laying-in (*For.*). The removal of bark and any rough buttresses round a tree by an axe, preparatory to making the undercut.

laying-off (*Ships*). The process of transferring the design form to full scale, for the purpose of *fairing* and ultimately of fabrication of details.

laying-out (*Eng.*). The marking-out of material, especially plate work, full size, for cutting and drilling.

laying press (*Bind.*). See **lying press.**

laying the bearings (*Acous.*). In tuning fixed-pitch instruments, such as the piano or organ, the technique of tuning the 12 semitones of a central octave. All other notes are then tuned by unison or octaves.

lay light (*Build.*). A window or sash, fixed horizontally in a ceiling, to admit light to a room.

lay marks (*Print.*). The marks to which sheets are laid in printing in order to ensure uniformity of position.

lay of the case (*Typog.*). The position of the characters in the type case.

layout (*Typog.*). (1) The general appearance of a printed page. (2) The art and practice of disposing display (e.g., advertising) matter to the best advantage.

lay panel (*Join.*). A long panel of small height formed in a panelled wall above a doorway, or all round the room immediately below the cornice.

lay ratio (*Cables*). The ratio of the *lay* (q.v.) to the mean diameter of the helix.

Layrub universal joint (*Eng.*). A flexible joint depending on the deflection of a group of rubber bushes. Frequently used where the angle between shafts is not great, say up to 10° on either side of the straight line.

lay shaft (*Eng.*). An auxiliary geared shaft, e.g., a secondary shaft running alongside the main-shaft of an automobile gearbox, to and from which the drive is transferred by gear-wheels of varying ratio.

lazulite (*Min.*). A deep sky-blue, strongly pleochroic mineral, crystallizing in the monoclinic system. In composition essentially a hydrous phosphate of aluminium, magnesium, and iron, with a little calcium. Found in aluminous high-grade metamorphic rocks and in granite pegmatites.

lazurite (*Min.*). An ultramarine-blue mineral occurring in cubic crystals or shapeless masses; it consists of silicate of sodium and aluminium, with some calcium and sulphur, and is considered to be a sulphide-bearing variety of **haüyne.** A constituent of *lapis lazuli* (q.v.).

lazy arm (*Acous.*). Small microphone boom.

lazy coil (*I.C. Engs.*). Coil at the end of a valve-spring having closer spacing than the other coils, acting as a damper against vibration.

lb. Abbrev. for *pound.*

L.B. (*Teleph.*). Abbrev. for *local battery.*

L-band (*Radio*). Frequency band between 390 and 1550 MHz.

lb-cal (*Heat*). See **pound-calorie.**

lb-mol (*Chem.*). Abbrev. for *pound-mol*, the mass of compound equal to its molecular weight in pounds, e.g., 32 lb O_2. Obsolete.

lb. s. t. (*Aero.*). Abbrev. for *static thrust.*

l.c. (*Typog.*). Abbrev. for *lower case.*

L-capture (*Nuc.*). Absorption of an electron from the L-shell into the nucleus, giving rise to X-rays of characteristic wavelength depending on atomic number of the element. See also **K-capture, X-rays.**

LC coupling (*Electronics*). Inductor output load of an amplifier circuit is connected through a capacitor to the input of another circuit.

LCN (*Aero.*). Abbrev. for *load classification number.*

L display (*Radar*). A radar display in which the target appears as two horizontal pulses, left and right from a central vertical time base, varying in amplitude according to accuracy of aim.

lea (*Textiles*). One-seventh of a hank, used for testing counts, strength and elongation; 80 yd in worsted, 120 yd in cotton. In linen, it indicates the size of yarns; or a length of 200 yd.

Lea and Nurse permeameter (*Powder Tech.*). One in which the pressure drop across a powder bed is measured when air is flowing through it under streamlined flow conditions.

leached zone (*Geol., Mining*). Portion of lode above natural water table, from which some ore has been dissolved by penetrating surface water.

leaching (*Bot.*). The removal, by percolating water, of mineral salts from the soil. (*Leather*) The extraction of tannic acid from bark or other tanning material, by passing water through a series of pits (*leaches*) containing the pulverized material. Also **leeching.** (*Met.*) The extraction of a soluble metallic compound from an ore by dissolving in a solvent (frequently sulphuric acid). The metal is subsequently precipitated from the solution.

leaching cesspool (*San. Eng.*). A cesspool which is not watertight.

lead (*Civ. Eng.*). The distance over which earthwork has at any time to be conveyed from a cutting to a place of deposit. (*Elec. Eng.*) (1) A term often used to denote an electric wire or cable. (2) See **backward shift, brush shift, forward shift.** (*Eng.*) Distance parallel to the axis of a screw-thread between consecutive contours on the same helix of the thread; equal to *pitch* in single-start threads. (*Rail.*) The distance from the nose of a crossing to the nose of the switch.

lead (*Chem.*). A metallic element in the fourth group of the periodic system. Symbol Pb, at. no. 82, r.a.m. 207·19, valency 2 or 4, m.p. 327·4°C, b.p. 1750°C, rel. d. at 20°C 11·35. Specific electrical resistivity $20{\cdot}65 \times 10^{-8}$ ohm metres. Occurs chiefly as *galena.* The naturally occurring stable element consists of four isotopes: Pb 204 (1·5%), Pb 206 (23·6%), Pb 207 (22·6%), and Pb 208 (52·3%). Used as the basis for the manufacture of massicot, litharge, white lead, red lead, orange lead, and lead chromes. The metal is bluish-grey, the densest and softest of the common metals. On account of its resistance to corrosion, extensively used for roofing, cable sheathing, and for lining apparatus in the chemical industry. It is used as shielding in X-ray and nuclear work because of its relative cheapness, high density, and nuclear properties. Other principal uses: in storage batteries, sound absorbers, ammunition, foil, and as a constituent of bearing metals, solder, and type-metal. Lead can be hardened by the addition of arsenic or antimony. It occurs very rarely in the native form, and then appears to have been formed by fusion of some simple lead ore accidentally incorporated in lava. (*Plumb.*) The leaden came of a lattice window. (*Surv.*) The leaden sinker secured at one end of a *lead-line* (q.v.). See also **leads**.

lead (of the web) (*Print.*). The reel of paper is torn diagonally to thread it through the press; when the press is started a wrongly threaded path is called a *lost lead* or *wrong lead*.

lead-acid (or lead) accumulator (*Elec. Eng.*). A secondary cell consisting of lead electrodes, the positive one covered with lead dioxide, dipping into sulphuric acid solution. Its e.m.f. is about 2 volts; very widely used.

lead age (*Min.*). The age of a mineral or rock calculated from the ratios of its radiogenic and non-radiogenic lead isotopes.

lead and leave edges (*Print.*). On web-fed presses the first and last edges of the product to be printed.

lead azide (*Chem.*). PbN_6. Explosive like most azides. Sometimes used, instead of mercury fulminate, as a detonator for TNT.

lead burning (*Plumb.*). The process of welding together two pieces of lead, thus forming a joint without the use of solder.

lead (II) carbonate (*Chem.*). $PbCO_3$. Occurs in nature as cerussite. At about 200°C it decomposes into the (II) oxide and carbon dioxide. Readily reduced to metal by CO.

lead (II) chromate (VI) (*Chem.*). $PbCrO_4$. Precipitated when potassium chromate (VI) is added to the solution of a lead salt. Used as pigments, called *chrome yellows* (q.v.). The colour may be varied by varying the conditions under which the precipitation is made.

lead disilicate (*Chem.*). Obtained by fusing lead (II) oxide and silica together. As **lead frit**, it is used as a ready means of incorporating lead oxide in the making of lead glazes.

lead dot (*Build.*). A lead peg or dowel used to fasten sheet-lead to the upper surface of a coping or cornice, for which purpose it is run into a mortise in the stone.

leaded (*Met.*). Term descriptive of copper, bronze, brass, steel, nickel, and phosphor alloys to which from 1% to 4% of lead has been added, mainly to improve machinability.

leaded lights (*Build.*). A window formed of (usually) diamond-shaped panes of glass connected together by leaden cames.

lead equivalent (*Radiol.*). Absorbing power of radiation screen expressed in terms of the thickness of lead which would be equally effective for the same radiation.

leader (*Cinema.*). A blank strip of film, generally white to facilitate threading and for identification purposes, attached or printed at the start of each reel of film, so that the speed of projection may be normal when the projected picture commences. Cf. *trailer*. (*Mining*) A thin mineralized vein parallel to, or so related to, the mineralized vein parallel, or otherwise related, to the main ore-carrying vein, so aiding its discovery. (*Plumb.*) See **conductor**. (*Surv.*) The chainman who has charge of the forward end of a chain. He is directed into line by the follower. (*Typog.*) A group of dots (...) on one em set together to form dotted lines, or spaced at intervals to guide the eye on contents pages, etc.

leader cable (*Nav.*). An under-water cable emitting signals by which vessels with induction-receiving apparatus can find their way into port. A *homer*.

leader-hook (*Build.*). A device, such as a *holderbat* (q.v.), for securing a rainwater pipe to a wall.

lead flare (*Cables*). A flare (or bell) of lead which fits the paper stress cone placed on the core near the cut of the lead sheath.

lead frit (*Chem.*). See **lead disilicate.**

lead glance (*Min.*). See **galena.**

lead glass (*Glass*). Glass containing lead oxide. The amount may vary from 3–4% to 50% or more in special cases. ('English Lead Crystal', used for tableware, contains 33–34%.) Used as transparent radiation shield, esp. for X-rays.

lead grip (*Elec. Eng.*). A bonding device for providing continuity of a lead-sheathed cable.

leadhillite (*Min.*). Hydrous carbonate and sulphate of lead (II), so called from its occurrence with other ores of lead at Leadhills (Scotland).

lead (II) hydroxide (*Chem.*). $Pb(OH)_2$. Dissolves in excess of alkali hydroxides to form plumbites (plumbates (II)).

lead-in (*Acous.*). Unmodulated groove at the start of a recording on a disk, so that the stylus falls into the groove correctly before the start of the modulation. The corresponding final groove after modulation ends is the **lead-out.** (*Radio*) The cable which connects the transmitter or receiver to the elevated part of an aerial.

leading or leading-out (*Typog.*). The process of inserting *leads* between lines of type matter, in order to open them out, thus presenting more white space between the printed lines.

leading current (*Elec. Eng.*). An alternating current whose phase is in advance of that of the applied e.m.f. creating the current. For a pure capacitance in circuit, the phase of the current is $\pi/2$ radians in advance of that of the applied voltage. See **angle of lead, phase angle.**

leading edge (*Aero.*). The edge of a streamline body or aerofoil which is forward in normal motion; structurally, the member that constitutes that part of the body or aerofoil. (*Elec. Eng.*) A term used in connexion with the brushes of electric machines. See **entering edge.** (*Telecomm.*) Rising amplitude portion of a pulse signal. See **rise time, trailing edge.**

leading-edge flap (*Aero.*). A hinged portion of the wing leading edge, usually on fast aeroplanes, which can be lowered to increase the camber and so reduce the stalling speed; U.S. **droop snoot.**

leading-in insulator (or tube) (*Radio*). A tubular form of insulator through which the lead-in enters a building.

leading-in wire (*Elec. Eng.*). The wire conducting the current from the cap contacts of an electric filament or discharge lamp to the filament itself or to the electrodes. (*Radio*) The same as *lead-in*.

leading load (*Elec. Eng.*). See **capacitive load.**

leading note (*Acous.*). The note one semitone below the tonic or key note of the normal musical scale; essential in combining harmonic frequencies.

leading-out wire (*Elec. Eng.*). The flexible insulated wire which is attached to the more delicate insulated wire used for the windings of transformers, etc. It is sufficiently robust for connecting to terminals.

leading phase (*Elec. Eng.*). (1) A term used in connexion with measuring equipment on 3-phase circuits to denote a phase whose voltage is leading upon that of one of the other phases by approximately 120°. (2) Particularly in connexion with the two-wattmeter method of 3-phase power measurement, the phase in which the current at unity power factor leads upon the voltage applied to the meter in which that current is flowing.

leading pole horn (*Elec. Eng.*). The portion of the pole-shoe of an electric machine which is first met by a point on the armature or stator surface as the machine revolves. Hence also **leading pole tip.**

leading ramp (*Elec. Eng.*). The end of a conductor-rail at which the collector-shoe of an electric train first makes contact.

lead joint (*Plumb.*). A joint formed between the spigot and the socket of successive lengths of large water-pipes by pouring molten lead into the annular space between them, or by ramming in lead wool.

lead-line (*Surv.*). A line with which soundings are taken. The depth of water is indicated on the line by 'marks', or by knots in the line, indicating fathoms; fathoms not indicated on the line (e.g., between 7 and 10) are **deeps.**

lead (II) oxide (*Chem.*). PbO. An oxide of lead, varying in colour from pale yellow to brown depending on the method of manufacture. An intermediate product in the manufacture of red lead. See also **litharge** and **massicot.**

lead moulding (*Print.*). The making of an *electrotype* mould in lead; only used for fine screen halftone and colour sets.

lead network (*Telecomm.*). One which provides a signal proportional to rate of change, i.e., time differential or derivative, of error signal.

lead-out (*Acous.*). See **lead-in.**

lead oxychloride (*Chem.*). $PbCl_2 \cdot PbO$. See **Cassel's yellow.**

lead paint (*Paint.*). Ordinary paint, in which the main pigment is white lead.

lead (IV) oxide (*Chem.*). PbO_2. A strong oxidizing agent. Industrial application very limited. Present, in certain conditions, in accumulators or electrical storage batteries as a chocolate-brown powder.

lead plug (*Build.*). A cast-lead connecting-piece binding together adjacent stones in a course; formed by running molten lead into suitably cut channels in the jointing faces.

lead poisoning (*Med.*). Chronic poisoning due to inhalation, ingestion, and skin absorption of lead. Recognized as a hazard, both for young children (through sucking lead-painted articles or toys), and also industrially. Characterized by anaemia, constipation, severe abdominal pain, and perhaps ultimately renal damage.

lead protection (*Radiol.*). Protection provided by metallic lead against ionizing radiation. Joined with other substances to provide further protection, e.g., lead glass, lead rubber.

lead rubber (*Radiol.*). Rubber containing high properties of lead compounds. Used as flexible protective material for, e.g., gloves.

leads (*Eng.*). Lengths of thin lead wire inserted between a very large journal and the bearing cap during assembly to test the clearance. (*Typog.*) Strips of lead placed between lines of type. Usual thicknesses: 1-, 1½-, 2-, and 3-point (*thick*), 6-point (*nonpareil*), 12-point (*pica*) Others are also used, e.g., 4-point.

lead screw (*Acous.*). In disk recording, the screw which feeds the cutting head and its stylus along a radius of the blank disk as it revolves. The rate of feed need not be constant, but can depend on the modulation, or be accelerated for a marker groove. (*Eng.*) A square or acme-thread screw running alongside the bed of a lathe, and driven through *change wheels* (q.v.) from the mandrel, for traversing the slide rest in screw cutting.

lead sulphate (*Chem.*). $PbSO_4$. Formed as a white precipitate when sulphuric acid is added to a solution of a lead salt.

lead (II) sulphide (*Chem.*). Found in nature as *galena*. Formed when lead sulphate is reduced by carbon and when hydrogen sulphide is passed through a solution of a lead salt.

lead telluride (*Heat.*). Thermocouple compound for generating electric power through absorption of heat, e.g., from flame, air, or liquid; this cooling is also the basis of refrigeration units (*Peltier effect*).

lead tetraethyl (*Chem.*). $Pb(C_2H_5)_4$. A colourless liquid, obtained by the action of a zinc or magnesium ethyl halide on lead chloride. Used in motor spirit to increase the *antiknock value*.

lead tree (*Chem.*). Form, in shape of tree, which lead takes after electrodeposition from simple salts.

lead wire (*Cables*). Used in cable joints. The lead (Pb) wire is wound round the lead flare and oil-resisting poultice on to the lead sheath.

lead wool (*Plumb.*). Material used for caulking, packing, and jointing, instead of molten lead.

lead zirconate titanate (*Chem.*). A piezoelectric ceramic with a higher Curie temperature than barium titanate (IV).

leaf (*Bot.*). An outgrowth from the stem of a plant, usually green, and largely concerned with photosynthesis and with transpiration. It consists ordinarily of a leaf base, a petiole or stalk, and a flattened lamina, which in a simple leaf is in one piece; in a compound leaf it is in two or more separate pieces. The lamina is usually conspicuously veined. (*Civ. Eng., etc.*) A term applied to an object which has a large area in relation to its thickness, e.g., one of a pair of lock-gates. (*Textiles*) (1) Impurity in baled cotton. (2) Heald shaft in a loom. (3) One plate of a loom warp shedding tappet (wiper).

leaf cushion (*Bot.*). A swollen leaf base.

leaf divergence (*Bot.*). The angle at the intersection of the planes passing longitudinally through the middles of two successive leaves.

leaf filter (*Chem. Eng., etc.*). A type of filter in which large thin frames with a pipe connexion through the frame are covered in cloth or metal gauze and placed inside a closed cylindrical vessel. Liquid to be filtered is forced under pressure into cylinder and clear filtrate leaves via frame's pipe connexions (Kelly and Sweetland are this type). Similarly ore pulp is filtered by use of vacuum or pressure.

leaf gap (*Bot.*). An interruption of the vascular tissues of a stem, beneath the insertion of a leaf.

leaf incept (*Bot.*). The earliest recognizable rudiment of a leaf.

leafing (*Paint.*). The property of the small metallic flakes used in aluminium, bronze, and similar paints, of floating flat at the surface of the paint, thus increasing the metallic lustre.

leaflet (*Bot.*). One separate portion of the lamina of a compound leaf.

leaf mosaic (*Bot.*). The arrangement of the leaves on a shoot or a plant in such a way that as much leaf surface as possible is exposed to light, and as little as possible is shaded by other leaves.

leaf scar (*Bot.*). The scar left on a stem at the point where a leaf has fallen off; it is commonly covered by a thin sheet of cork.

leaf sheath (*Bot.*). The base of the leaf when it is in the form of a sheath more or less surrounding the stem.

leaf spring (*Eng.*). A machine component comprising one or a group of relatively thin, flexible, or resilient strips, reacting as a spring to forces applied to a main surface.

leaf trace (*Bot.*). The vascular tissue between the stele and the base of the leaf.

leafy raceme (*Bot.*). A raceme in which the bracts differ little, or not at all, from the ordinary foliage leaves of the plant.

leak (*Elec. Eng.*). (1) A path between a conductor and earth (or another conductor) which has a lowered insulation resistance. (2) A high resistance used for the timed discharge of a capacitance, such as the grid leak which is in series with a valve control grid.

leakage coefficient (*Elec. Eng.*). The ratio of the total flux in the magnetic circuit of an electric machine or transformer to the useful flux which actually links with the armature or secondary winding. Also called **leakage factor.**

leakage conductance (*Elec. Eng.*). See **leakance.**

leakage current (*Elec. Eng.*). Small unwanted current flowing through a component such as a capacitor or a reverse-biased diode.

leakage factor (*Elec. Eng.*). See **leakage coefficient.**

leakage flux (*Elec. Eng.*). (1) That which, in any type of electric machine or transformer, does not intercept all the turns of the winding intended to enclose it. (2) That which crosses the air gap of a magnet other than through the intended pole faces. Often called **lost flux.**

leakage indicator (*Elec. Eng.*). An instrument for measuring or detecting a leakage of current from an electric system to earth. Also called an **earth detector.**

leakage protective system (*Elec. Eng.*). A protective system which operates as a result of leakage of current from electrical apparatus to earth.

leakage reactance (*Elec. Eng.*). That, in a transformer, which arises from flux in one winding not entirely linking another winding; measured by the inductance of one winding when the other is short-circuited. Leakage inductance, taken with the effective self-capacitance of a winding, determines the range of frequencies effectively passed by the transformer.

leakance (*Elec. Eng.*). In electrical circuits, the leakage current expressed by the reciprocal of insulation resistance of the circuit. Also called **leakage conductance.**

leak detector (*Nuc. Eng.*). Syn. for **burst-can detector.** (*Phys.*) Device for indicating points of ingress of gas into a high vacuum system.

lean ore (*Mining*). Ore of marginal value; low-grade ore.

lean-to roof (*Arch.*). A roof having only one slope.

leap-frog test (*Comp.*). Program used to test the internal operation of a computer.

leaping weir (*Civ. Eng.*). A special arrangement whereby flood flows may be diverted from a channel into which normal flows would ordinarily pass, the water having to go over a weir set at such a height that flood flows leap beyond the channel to an overflow. Also called a **separating weir.**

leap second (*Astron.*). A periodic adjustment of time signal emissions to maintain synchronism with co-ordinated universal time (UTC). A positive leap second may be inserted, or a negative one omitted, at the end of December or June.

leap years (*Astron.*). Those years in which an extra day, February 29, is added to the civil calendar to allow for the fractional part of a tropical year of 365·2422 days. Since the Gregorian reform of the Julian calendar, the leap years are those whose number is divisible by 4, except centennial years unless these are divisible by 400.

lear board (*Build.*). A board carrying a lead gutter. Also called a **layer board.**

learning (*An. Behav.*). A process which manifests itself by adaptive changes in individual behaviour as the result of experience. See **classical conditioning, habituation, imprinting, operant conditioning.**

learning theory (*Psychol.*). A range of theories which, unlike *depth psychology*, explain behaviour in terms of learning and conditioning. Largely the offspring of *behaviourism*, it takes as its starting point the association of various stimuli and responses. See **behaviour therapy.**

lease rods (*Weaving*). Two rods, across a warp sheet, which separate the yarns, equalize the tension and keep them at the correct height. The yarns are placed end-and-end (1-and-1) or 2-and-2, as required.

least action, principle of (*Phys.*). See **principle of least action.**

least distance of distinct vision (*Optics*). For a normal eye it is assumed that nothing is gained by bringing an object to be inspected nearer than 25 cm, owing to the strain imposed on the ciliary muscles if the eye attempts to focus for a shorter distance.

least energy principle (*Phys.*). That a system is only in stable equilibrium under those conditions for which its potential energy is a minimum.

least time, principle of (*Phys.*). See **Fermat's principle of least time.**

leat (*Mining*). Ditch following contour line used to conduct water to working place.

leather. A universal commodity made by *tanning* (q.v.) and other treatment of the hides or skins of a great variety of creatures (mostly domesticated animals, but including also the whale, seal, shark, crocodile, snake, kangaroo, camel, ostrich, etc.). *Artificial* or *imitation leathers* are commonly based on cellulose nitrate, PVC or polyurethane, suitably treated to simulate leather. See **poromeric.**

leather boards (*Paper*). A board made from leather scrap with usually some added brown mechanical wood pulp.

leathercloth (*Textiles*). See **oilcloth.**

leather hollows (*Eng.*). Strips of leather used by pattern-makers to form the fillets in wood patterning.

leaving edge (*Elec. Eng.*). The edge of the brush of an electric machine which is last met during revolution by a point on the commutator or slip ring. Also called **heel, back, trailing edge.**

Leber's disease (*Med.*). Hereditary optic atrophy; hereditary optic neuritis. A hereditary *sex-linked* disease in which there is gradual loss of sight due to an affection of the optic nerve behind the eyeball.

Lebesgue integral (*Maths.*). An advanced concept based upon the theory of sets whereby functions, not ordinarily integrable, can be said to be integrated, e.g., the function which is one when x is rational and zero elsewhere, is not ordinarily integrable but it does have a Lebesgue integral. Where both an ordinary integral (i.e., that of Riemann) and a Lebesgue integral exist they are equal.

Leblanc connexion (*Elec. Eng.*). Method of connecting transformers for linking a 3-phase to a 2-phase system.

Leblanc phase advancer (*Elec. Eng.*). A d.c. armature with 3 sets of brushes per pole pair on the commutator; the brushes are connected to the 3 slip rings of the induction motor, and the advancer is driven from the motor shaft at an appropriate speed, causing it to take a leading current and improve the motor power factor.

Leblanc process (*Chem.*). A process, formerly of great importance but now obsolete, for the manufacture of sodium carbonate and intermediate products from common salt, coal, limestone, and sulphuric acid.

Lecanorales (*Bot.*). An order of the *Ascomycetes*; the disk lichens. They produce apothecia, and are parasitic on some terrestrial *Chlorophyceae* and *Myxophyceae*.

Le Chatelier-Braun principle (*Chem.*). If any change of conditions is imposed on a system at equilibrium, then the system will alter in such a way as to counteract the imposed change. This principle is of extremely wide application.

lechatelierite (*Min.*). A name sometimes applied to naturally fused vitreous silica, such as that which occurs as fulgurites.

Le Chatelier test (*Build., Civ. Eng.*). A simple method for checking the soundness of cement.

Lecher wires (*Elec. Eng.*). Two insulated parallel stretched wires tunable by means of sliding short-circuiting copper strip. The wires form a microwave electromagnetic transmission line which may be used as a tuned circuit, as an impedance matching device or for the measurement of wavelengths. Also called **parallel wire resonator.**

lechriodont (*Zool.*). Having the teeth borne by the pterygoid and those borne by the vomer forming a transverse row.

lecith-, lecitho-. Prefix from Gk. *lekithos*, yolk of egg.

lecithin (*Chem.*). A phosphatide found in the yolk of the egg and in the brain. It may be described as the choline ester of distearylglycerylphosphoric acid, and has the formula:

$$\begin{array}{l} CH_2{\cdot}OCO{\cdot}C_{17}H_{35} \\ | \\ CH{\cdot}OCO{\cdot}C_{17}H_{35} \\ | \\ CH_2{\cdot}OPO(OH){\cdot}O{\cdot}CH_2{\cdot}CH_2{\cdot}N(CH_3)_3{\cdot}OH. \end{array}$$

Other lecithins contain, besides stearic acid, palmitic or oleic acid. Lecithin is an important constituent of protoplasm.

lecithoblast (*Zool.*). The blastomeres of a segmenting ovum which contain yolk.

lecithocoel (*Zool.*). The segmentation cavity of a holoblastic egg.

lecithoproteins (*Chem.*). Compounds formed from proteins and lecithin.

Leclanché cell (*Elec. Eng.*). A primary cell, good for intermittent use, has a positive electrode of carbon surrounded by a mixture of manganese dioxide and powdered carbon in a porous pot. The pot and the negative zinc electrode stand in a jar containing ammonium chloride solution. The e.m.f. is approximately 1·4 volt. The *dry cell* is a particular form of Leclanché.

lectin (*Bacteriol.*). A naturally occurring substance with specificity similar to that of an antibody but not resulting from immunization.

Lectneriales (*Bot.*). An order of the *Archichlamydeae*. Woody plants with alternate entire leaves, and spikes of dioecious flowers. The male flower is achlamydous with 3–12 stamens. The female flower is haplochlamydous with a single superior ovary with one amphitropous ovule, ultimately forming a drupe. The endosperia is thin; the perianth of small, scaly united leaves.

lectotype (*Biol.*). A substitute for a type specimen, taken from the original material when the type specimen was not designated or has been lost. (*Bot.*) One of the *primary* types taken as typical of a group, but taken later for a group for which the author has given no type originally.

LED (*Electronics*). See **light-emitting diode.**

ledge (*Join.*). One of the battens across the back of a batten door.

ledged-and-braced door (*Join.*). A door similar to a batten door, but framed diagonally with braces across the back, between the battens.

ledged door (*Carp.*). See **batten door.**

ledgement (*Build.*). A horizontal line of mouldings, or a string-course.

ledger (*Build.*). A horizontal pole or member, lashed or otherwise fastened across the standards in a scaffold or in a trench.

ledger board (*Carp.*). See **ribbon strip.**

ledger-line (*Mus.*). A short line added above or below the staff when required. Also **leger-line** (deprecated).

Leduc current (*Telecomm.*). Interrupted direct current in which each pulse is of approximately the same magnitude and in the same sense.

leech (*Zool.*). See **Hirudinea.**

Lee-McCall (*Civ. Eng.*). A system for post-tensioned concrete involving the use of high-tensile steel rods.

leer (*Glass*). See **lehr.**

lee wave (*Meteor.*). See **orographic ascent.**

left-hand airscrew (*Aero.*). See **airscrew.**

left-handed engine (*Aero.*). An aero-engine in which the airscrew shaft rotates counter-clockwise with the engine between the observer and the airscrew.

left-hand rule (*Elec. Eng.*). See **Fleming's rule.**

left-hand thread (*Eng.*). A screw-thread cut in the opposite direction to the normal right-hand. Viewed in elevation, the external thread is inclined upwards from right to left. Used when a normal thread would tend to unscrew.

left-hand tools (*Eng.*). Lathe side-tools with the cutting edge on the right, thus cutting from left to right.

left-hand twine and left-hand twist (*Textiles*). See **crossband.**

leg (*Telecomm.*). One side of a loop circuit, i.e., either the *go* or *return* of an electrical circuit.

legal ohm (*Elec.*). A unit of resistance adopted by the International Congress of Electricians at Paris in 1884 but never given legal sanction. It was equal to the resistance of a column of mercury 106 cm long and 1 mm² in cross-section at 0°C, i.e., equal to 0·997 18 of the later *international ohm.*

legend (*Typog.*). Descriptive matter accompanying an illustration; a caption.

Legendre's differential equation (*Maths.*). The equation

$$(1-x^2)\frac{d^2y}{dx^2}-2x\frac{dy}{dx}+n(n+1)y=0.$$

It is satisfied by the Legendre polynomial $P_n(x)$.

Legendre's polynomials (*Maths.*). The polynomials $P_n(x)$ in the expansion

$$(1-2xh+h^2)^{-\frac{1}{2}}=\sum_{n=0}^{\infty}P_n(x)\,.\,h^n.$$

In particular,

$$P_0(x)=1,\ P_1(x)=x,\ P_2(x)=\tfrac{1}{2}(3x^2-1),$$

and $P_3(x)=\tfrac{1}{2}(5x^3-3x)$. Cf. *Chebyshev polynomials.*

legume (*Bot.*). A fruit formed from a single carpel, splitting along the dorsal and the ventral sutures, and usually containing a row of seeds borne on the inner side of the ventral suture. A pea pod is a familiar example.

lehr (*Glass, etc.*). An enclosed oven or furnace used for annealing or other forms of heat treatment. It is a kind of tunnel down which the glass, hot from the forming process, is sent to cool slowly, so that strain is removed, and cooling takes place without additional strain being introduced. Lehrs may be of the open type (in which the flame comes in contact with the ware) or of the muffle type. Also **lear, leer, lier.**

Leicester wool (*Textiles*). The best quality of the English lustre wools, staple 10 in. (25 cm); used for lining fabrics and sicilians.

Leidig's cell (*Zool.*). A type of connective tissue cell, occurring in *Crustacea*, consisting of a central cytoplasmic mass, with an eccentric nucleus, and a number of radiating branched processes arising from the periphery.

Leighton Buzzard Silver Sand (*Geol.*). Clean white sand used for refractory moulding purposes; it occurs at the top of the Woburn Sands of Lower Cretaceous age, and is named from the locality where it is exploited in Bedfordshire, England.

leiomyoma (*Med.*). A tumour composed of unstriped muscle fibres.

leiosporous (*Bot.*). Having smooth spores.

leishmaniasis, leishmaniosis (*Med.*). A term applied to a group of diseases caused by infection with protozoal parasites of the genus *Leishmania*. See also **kala-azar.**

Leishman's stain (*Micros.*). A particular combination of methylene blue and eosin, prepared as a precipitate, which is then made up in methanol. Used mainly for staining blood films.

leizure (*Textiles*). A term used in the silk trade for *selvedge*.

L electron (*Nuc.*). One of up to 8 in the L-shell of an atom. Their principal quantum number is 2.

LEM, lem (*Space*). Abbrev. for **lunar excursion module.**

Lemberg's stain test (*Geol.*). A black iron (II) sulphide stain, used to distinguish between calcite and dolomite in limestones.

lemma (*Bot.*). A glume subtending a flower of a grass. (*Maths.*) A subsidiary theorem. The proof of a complicated theorem is generally made more clear by proving any subsidiary facts in a series of lemmas.

lemmocytes (*Histol.*). Tubular cells, which form a very thin sheath, the *neurilemma*, round all myelinated and nonmyelinated nerve fibres, except those of the central nervous system.

Lemnaceae (*Bot.*). A suborder of the *Arales*, lacking well-differentiated leaves and stems; aquatic and floating, e.g., duckweed.

lemniscate of Bernoulli (*Maths.*). A figure-eight shaped curve with polar equation $r^2=a^2\cos 2\theta$. Cf. *ovals of Cassini.*

lemniscus (*Zool.*). In *Acanthocephala*, one of a pair of saclike organs situated at the base of the proboscis and containing fat globules.

lemon chrome (*Paint.*). A chrome yellow pigment consisting of lead chromate, lead sulphate, and lead carbonate. A bright lemon colour of high opacity. See **chrome yellows.**

Lemstrom machine (*Elec. Eng.*). An electrostatic generator consisting of concentric cylinders of insulating material revolving in opposite directions and carrying strips of tin foil for collecting the charge.

Lenard rays (*Electronics*). Cathode rays which have passed from a discharge tube through a suitable window.

Lenard tube (*Electronics*). Beam tube in which part of the electron beam can be extracted through a part of the wall of the vacuum chamber.

length (*Comp.*). The number of *bits* in a *word*, or the number of columns or spaces in a *field*.

length between perpendiculars (*Ships*). The length on the summer load waterline from the fore side of the stem to the after side of the stern post or, in a vessel without a stern post, to the centre of the rudder stock.

lengthening bar (*Instr.*). An extension piece which may be fitted to the leg of a pair of compasses to increase its length and so enable an arc of larger radius to be drawn.

lengthening rod (*Civ. Eng.*). A rod with a male thread at one end and a female thread at the other, capable of being used to extend the length of shank carrying a boring piece at its lower end.

length-fast, length-slow (*Min.*). In optically birefringent minerals, terms used to denote whether the fast or the slow vibration direction is aligned parallel or nearly parallel to the length of prismatic crystals.

length fold collection (*Print.*). Bringing together, on rotary presses, folded products from a series of formers sometimes mounted one above another. See **balloon former.**

length of lay (*Cables*). See **lay.**

length overall (*Ships*). The length from the foremost point of the structure, including any figurehead or bowsprit, to the aftermost point of the ship.

Lennard-Jones potential (*Chem.*). Potential due to molecular interaction, also called the *6–12 potential* because the attractive potential varies as the inverse 6th power of the intermolecular distance, and the repulsive potential as the inverse 12th.

leno (*Textiles*). A fabric with an openwork or an embroidered effect, produced by cross-weaving; fabrics of this character that are of regular texture are usually termed *gauze* (q.v.).

leno brocade (*Textiles*). A brocade cotton, or cotton and rayon cloth, produced by a combination of ordinary and cross-weaving; used for dress and furnishing purposes, and for a few specialized industrial cloths.

lens (*Electronics*). Device for focusing radiation. (*Light*) A portion of a homogeneous transparent medium bounded by spherical surfaces. Each of these surfaces may be convex, concave, or plane. If in passing through the lens a beam of light becomes more convergent or less divergent, the lens is said to be *convergent* or *convex*. If the opposite happens, the lens is said to be *divergent* or *concave*. See **chromatic aberration, focal length of a lens, image, lens formula, spherical aberration, thick lens.** (*Zool.*) In *Arthropoda*, the cornea of an ocellus or compound eye; in *Craniata* and *Cephalopoda*, a structure immediately behind the iris which serves to focus light on to the retina. See also **crystalline lens.**

lens antenna (*Electronics*). (1) Symmetrical device, with centre dielectric or a lattice of metal spheres, capable of concentrating a beam of microwaves over a band of frequencies. (2) System of metal slats, adjusted in dimensions to give such phase delays as to concentrate a beam.

lens barrel (*Photog.*). The metal tube in which one or more lenses are mounted.

lens cap (*Photog.*). A temporary light-tight protective covering for the external end of a camera lens; removed during exposure and focusing. It protects the lens when the camera is not in use.

lens cylinder (*Zool.*). A name sometimes given to the cornea and crystalline cone of each ommatidium of the compound eyes of Insects, since these two components function as a lens whose refractive index decreases from the central axis to the periphery.

lens disk (*TV*). A rotating disk having a series of lenses arranged around the periphery, used in some forms of mechanical scanning. The light is projected through the lenses in a direction parallel to the axis of the disk.

lens drum (*TV*). A device similar to a lens disk, except that the lenses are arranged on the surface of a rotating drum, the light being projected radially.

lens formula (*Light*). The equation giving the relation between the image and object distances, v and u, and the focal length f of a lens:

$$\frac{1}{f}=\frac{1}{u}+\frac{1}{v}.$$

See **convention of signs** for the rule for applying the correct signs to f, v, and u.

lens grinding (*Glass*). The process of grinding pieces of flat sheet-glass (or pressed blanks) to the correct form of the lens. Cast-iron 'tools' of the correct curvature, supplied with a slurry of abrasive and water, are used.

lens hood (*Photog.*). A funnel for fixing in front of the lens, in order to exclude stray light from strong sources not in the camera angle.

lens mount (or **mounting**) (*Photog.*). The metal unit in which the lens of a camera is fixed, incorporating the focusing and stop adjustments, the whole being detachable from the rest of the camera.

lentic (*Ecol.*). Associated with standing water; inhabiting ponds, swamps, etc.

lenticel (*Bot.*). A tiny pore in the periderm, which is packed with loose corky cells and allows gaseous diffusion to occur between the interior of the plant and the atmosphere.

lenticle (*Geol.*). A mass of lenslike (lenticular) form. The term may refer to masses of clay in sand, or vice versa, and, in metamorphic rocks, to enclosures of one rock type in another.

lenticonus (*Med.*). Abnormal curvature of the lens of the eye, in which the surface becomes conical instead of spherical.

lenticular (*Bot., Zool.*). Shaped like a double convex lens. (*Min.*) Said of a mineral or rock of this particular shape, embedded in a matrix of a different kind.

lenticular bob (*Horol.*). A pendulum bob whose cross-section corresponds to that of a double convex lens; the usual form of bob for household clocks.

lenticular girder bridge (*Eng.*). A type of girder bridge composed of an arch whose thrust is taken by a suspension system hanging below it, the anchorages of the latter system being provided by the arch thrust.

lentiform nucleus (*Histol.*). A specialized nuclear mass, representing part of the corpus striatum, and lying to one side of the optic thalamus, in the brain of Mammals. Nerve fibres link it with the cerebral cortex and the midbrain. Many of the cells contain yellow pigment.

lentigo (*Med.*). Freckles. *adjs.* **lentiginose, lentiginous.**

Lentz valve gear (*Eng.*). A locomotive valve gear in which the steam is admitted and exhausted through two pairs of *poppet valves* (q.v.), spring-controlled and operated from a camshaft rotating at engine speed.

Lenz's law (*Elec. Eng.*). An induced e.m.f. will tend to cause a current to flow in such a direction as to oppose the cause of the induced e.m.f.

leonhardite (*Min.*). A zeolite; a partially dehydrated variety of *laumontite*.

Leonids (*Astron.*). A swarm of meteors whose orbit round the sun is crossed by the earth's orbit at a point corresponding to about November 17, when a display of more than average numbers is to be expected; the radiant point is in the constellation Leo.

leontiasis ossea (*Med.*). A rare condition characterized by diffuse hypertrophy of the bones of the skull.

leopard wood (*For.*). See **snakewood.**

Lepel discharger (*Radio*). A quenched-spark discharger used in radio systems.

lepido-. Prefix from Gk. *lepis*, gen. *lepidos*, a scale.

Lepidocarpales (*Bot.*). A fossil order of the *Lycopodineae*. The sporangia were heterosporous, with the macrogametophytes being retained within the macrosporangium.

lepidocrocite (*Min.*). An orthorhombic hydrous oxide of iron (III) (FeO·OH) occurring as scaly brownish-red crystals in iron ores. It is often one of the constituents of limonite, together with its dimorph goethite.

Lepidodendrales (*Bot.*). A fossil order of the *Lycopodineae*. The treelike sporophytes were secondarily thickened; the roots formed on large rhizophores; the leaves ligulate and microphyllous; the sporophytes heterosporous and the sporophylls in strobili.

lepidolite (*Min.*). A lithium-bearing mica crystallizing in the monoclinic system as pink to purple scaly aggregates. It is essentially a hydrous silicate of potassium, lithium, and aluminium, often also containing fluorine, and is the most common lithium-bearing mineral. It occurs almost entirely in pegmatites. Also known as **lithia mica.**

lepidomelane (*Min.*). A variety of biotite, rich in iron, which occurs commonly in igneous rocks.

Lepidoptera (*Zool.*). An order of *Oligoneoptera*, having two pairs of large and nearly equal wings densely clothed in scales; the mouth-parts are suctorial, the mandibles being absent and the maxillae forming a tubular proboscis; the larva or caterpillar is active and herbivorous, with biting mouth-parts. Butterflies and Moths.

lepidote (*Bot.*). Said of a surface which bears scalelike hairs. (*Zool.*) Having a coating of minute scales, as the wings of a Butterfly.

lepidotrichia (*Zool.*). Jointed, branched, bony dermotrichia, occurring in Fish with a bony skeleton; believed to have been derived from scales. Cf. *ceratotrichia*.

Lepospondyli (*Zool.*). An extinct subclass of Amphibians, with the neural arches co-ossified with the centra, which are of unknown composition, and frequently with the tabular bones enormously expanded. Lower Carboniferous—lower Permian.

lepospondylous (*Zool.*). Said of vertebral centra in which there is a skeletal ring constricting the notochord in the intervertebral region, with an expansion between each pair of adjacent centra.

lepra reaction (*Med.*). An acute exacerbative episode occurring in leprosy.
leprarioid, leprose (*Bot.*). Having a whitish, mealy, or scurfy surface.
leproma (*Med.*). A nodular lesion of leprosy.
leprosy (*Med.*). Infection with the bacillus *Mycobacterium leprae*; a highly chronic and moderately contagious disease. **Lepromatous** and **neural** forms are recognized, the one with nodular and destructive lesions chiefly of the face and nose, the other, even more chronic, involving chiefly the nerves.
leptite (*Geol.*). An even-grained metamorphic rock composed mainly of quartz and feldspars. Approximately synonymous with *granulite* and **leptynite.**
lepto-. Prefix from Gk. *leptos*, slender.
leptocentric vascular bundle (*Bot.*). A concentric vascular bundle, in which a central strand of phloem is surrounded by xylem.
leptocephalus (*Zool.*). The larval form of the family *Anguillidae* (Eels), a ribbon-shaped transparent animal, with a slender head.
leptocercal, leptocercous (*Zool.*). Having a long slender tail.
leptodactylous (*Zool.*). Having slender digits.
leptodermatous (*Zool.*). Thin-skinned.
leptodermous (*Bot.*). Having a thin wall; said especially of capsules of *Bryophyta* which are soft.
leptokurtosis (*Stats.*). A distribution curve which is sharp-peaked and long-tailed compared with the normal.
leptome (*Bot.*). The conducting elements of the phloem.
Leptomedusae (*Zool.*). See **Calyptoblastea.**
leptomeninges (*Anat.*). The two innermost membranes—the arachnoid and the pia mater—investing the brain and the spinal cord.
leptomeningitis (*Med.*). Inflammation of the leptomeninges.
Leptomitales (*Bot.*). An order of the *Phycomycetes*, having the hyphae constricted at intervals; the sporangia are usually elongated, and usually a single egg is delimited from the periplasm within the oögonium.
lepton (*Phys.*). Electron, negative muon, or neutrino. Positrons, positive muons, and antineutrinos are *antileptons*. Lepton is also used as a generic title for particles of both groups. Leptons do not participate in the strong interactions.
leptonema (*Cyt.*). See **leptotene.**
lepton number (*Phys.*). The number of *leptons* less the number of corresponding antiparticles taking part in a process. The number appears to be conserved in any process.
leptophragmata (*Zool.*). In some Insects exhibiting cryptonephry, small thin areas in the distal parts of the Malpighian tubules, adjoining the haemocoele and giving a strong reaction for chlorides.
leptosome (*Nut.*). The tall or slender type of human figure.
leptospirosis (*Vet.*). A disease of animals or man caused by infection by an organism of the genus *Leptospira*.
leptospirosis icterohaemorrhagica (*Med.*). See **spirochaetosis icterohaemorrhagica.**
leptosporangiate (*Bot.*). Said of ferns in which the sporangium originates from a single cell.
Leptostraca (*Zool.*). A superorder of *Malacostraca* with a large carapace provided with an adductor muscle and not fused with any of the thoracic somites. The eyes are stalked and the thoracic limbs are biramous, normally foliaceous, and all alike, having no oöstegites. There are seven abdominal somites of which the last bears no appendages, and the telson bears caudal rami.
leptotene (*Cyt.*). The first stage of meiotic prophase, in which the chromatin thread acquires definite polarity.
leptotichous (*Bot.*). Of tissue which is thin-walled.
leptus (*Zool.*). A larval form of *Acarina*, having 3 pairs of legs.
leptynite (*Geol.*). See **leptite.**
lesion (*Med.*). Any wound or morbid change anywhere in the body.
Lessing rings (*Chem. Eng.*). *Packed column* (q.v.) filling consisting of open hollow cylinders of diameter equal to length and with an axial rib of full diameter and length.
leste (*Meteor.*). A dry south wind blowing in Madeira and N. Africa in front of a depression.
lethal (*Biol.*). Causing death, e.g., any factor of the environment, normal or abnormal, the presence of which is fatal to an organism, or any hereditary factor fatal to an embryo or tending to retard development, especially of the gametes and zygotes.
lethargic encephalitis (*Med.*). See **epidemic encephalitis** and **Von Economo's disease.**
lethargy (of neutrons) (*Nuc.*). The natural logarithm of the ratio of initial and actual energies of a neutron during the moderation process. The lethargy change per collision is defined similarly in terms of the energy values before and after the collision.
let-off motion (*Textiles*). Mechanism (which may be negative or positive type) behind the loom to regulate the tension and delivery of the sheet of warp threads coming from the beam.
Letraset (*Typog.*). See **transfer lettering systems.**
letterpress (*Print.*). (1) A term applied to printing from relief surfaces (type or blocks), as distinct from lithography, intaglio, etc. See **printing.** (2) The reading matter in a book, apart from illustrations, etc.
letterpress paper (*Paper*). Paper manufactured for the various processes of letterpress printing.
letterset printing (*Print.*). See under **offset printing.**
letter sizes (*Eng.*). A series of drill sizes in which each letter of the alphabet represents a diameter, increasing in irregular increments from A = 0·234 in. (5·94 mm) to Z = 0·413 in. (10·49 mm). Below A, *gauge numbers*, from 1 (0·228 in., 5·77 mm) to 80 (0·0135 in., 0·343 mm), are used in the reverse direction.
letterwood (*For.*). See **snakewood.**
letting down (*Aero.*). The reduction of altitude from cruising height to that required for the approach to landing. (*Eng.*) The process of tempering hardened steel by heating until the desired temperature, as indicated by colour, is reached, and then quenching.
L-leucine (*Chem.*). $(CH_3)_2CH{\cdot}CH_2{\cdot}CH(NH_2)COOH$, 2-amino-4-methyl pentanoic acid, colourless flakes, m.p. 270°C, formed by the decomposition of albuminous substances. An amino acid classified as essential for maintenance of growth in rats.
Leuciscus (*Zool.*). A genus of freshwater cyprinoid fishes, including the roach, minnow, etc.
leucite (*Min.*). A silicate of potassium and aluminium, related in chemical composition to orthoclase, but containing less silica. At ordinary temperatures, it is tetragonal (pseudocubic), but on heating to about 625°C it becomes cubic. Occurs in igneous rocks, particularly potassium-rich, silica-poor lavas of Tertiary and recent age, as, e.g., at Vesuvius.
leucitohedron (*Min.*). See **icositetrahedron.**
leucitophyre (*Geol.*). A fine-grained igneous rock

commonly occurring as a lava, carrying phenocrysts of leucite and other minerals in a matrix essentially trachytic; a well-known example comes from Rieden in the Eifel.

leuco-. Prefix from Gk. *leukos*, white.

leuco-bases or leuco-compounds (*Chem.*). Colourless compounds formed by the reduction of dyes, which when oxidized are converted back into dyes.

leucoblast (*Zool.*). A cell which will give rise to a leucocyte.

leucocrate (*Geol.*). A general name for light-coloured igneous rock rich in felsic minerals and correspondingly poor in mafic constituents. The adjective *leucocratic* is more commonly met with; while the prefix *leuco-* is used with such names as syenite, diorite, etc., to indicate a marked deficiency of coloured silicates. Thus albitite is a soda-leucosyenite consisting almost exclusively of albite.

leucocratic (*Geol.*). A term used to denote a light colour in igneous rocks, due to a high content of felsic minerals, and a correspondingly small amount of dark, heavy silicates.

leucocyte (*Zool.*). A white blood-corpuscle; one of the colourless amoeboid cells occurring in suspension in the blood-plasma of many animals.

leucocythaemia (*Med.*). Obsolete term for *leukaemia*.

leucocytolysis (*Physiol.*). The breaking down of white cells of the blood.

leucocytopenia (*Med.*). Abnormal diminution in the numbers of white cells in the blood.

leucocytosis (*Med.*). An increase in the number of leucocytes in the blood.

leucodermia, leucoderma (*Med.*). Vitiligo; melanodermia. A condition in which white patches, surrounded by a pigmented area, appear in the skin.

leuco-erythroblastic anaemia (*Med.*). A form of anaemia in which immature forms of red and white cell series are found in the peripheral blood.

leucon grade (*Zool.*). In *Porifera*, the third grade of structure, in which the choanocytes are confined to flagellated chambers and exhalant canals occur.

leucopenia (*Med.*). See **leucocytopenia.**

leucophore (*Zool.*). See **iridocyte.**

leucophyre (*Geol.*). An old name, not now used, for dolerites particularly rich in feldspar and therefore very light in colour.

leucoplakia (*Med.*). The stage of a chronically inflamed area at which the surface becomes hard, white, and smooth. On the tongue, it is usually due to syphilitic infection.

leucoplast (*Bot.*). A colourless plastid.

leucopoiesis (*Physiol.*). The production of white cells of the blood.

leucopterin (*Chem.*). White pigment found in the epidermal cells or the scales and setae of some Insects. It is a mixture of heterocyclic pigments related to the purines, with the structure:

```
   HN—CO
   |    |
NH:C    C—NH—CO
   |    ‖
   HN—  C—NH—CO
```

leucorrhoea, leucorrhea (*Med.*). A whitish discharge from the vagina.

leucosapphire (*Min.*). See **white sapphire.**

leucoscope (*Optics*). A device used for testing colour vision. It is based on the dispersion of polarized white light by a solution of optically active sugar.

leucosin (*Bot.*). White, possibly carbohydrate, substance; as lumps, a storage product of Chrysophyceae.

leucosporous (*Bot.*). Having white spores.

leucotomy (*Med.*). Surgical scission of the association fibres between the frontal lobes of the brain and the thalamus to relieve cases of severe schizophrenia and manic-depression. Also called **prefrontal lobotomy.**

leucoxene (*Min.*). An opaque whitish mineral formed as a decomposition product of ilmenite. Normally consists of finely crystalline rutile, but may be composed of finely divided brookite.

leukaemia, (in U.S.) leukemia (*Med.*). Progressive overproduction of any one of the types of white cell of the blood, with suppression of other blood cells and infiltration of organs such as the spleen and liver. Generally regarded as cancer of the bloodcell-forming tissues. Occurs in acute and chronic **myeloid, lymphatic,** and **monocytic** varieties.

leuko-. A variant spelling of **leuco-.**

levan (*Chem.*). A polymer of fructose found in grasses.

levanter (*Meteor.*). An east wind blowing at Gibraltar, the resulting cloud on the Rock being termed the *levant*.

levarterenol (*Chem.*). Laevorotatory *noradrenaline*.

levator (*Zool.*). See **elevator.**

leveche (*Meteor.*). A dry S.W. wind blowing in Spain in front of a depression.

levee (*Hyd. Eng.*). A well-consolidated bank of earth or spoil, having a central core forming an impervious connexion with the natural ground; used as a form of training works to control river flow and prevent flooding of adjacent country. In America the name also signifies a quay, landing-place, or pier.

level (*Acous.*). General term for a magnitude, taken as a ratio to a specified datum magnitude, usually in decibels in a power basis. Thus transmission level is in relation to a *zero level* of 1 milliwatt, and sound levels in relation to 10^{-16} watt/cm^2 (zero phon level). (*Build.*) A *level tube* (q.v.). (*Civ. Eng.*) (1) To reduce a cut or fill surface to an approximately horizontal plane. (2) The state whereby any given surface is at a tangent to the perfect earth's perimeter at any given point. (3) A ditch or channel for drainage, especially in flat country, e.g., the Fens. (*Mining*) An approximately horizontal tunnel in a mine, generally marking a working horizon or level of exploitation, and either in or parallel to the ore body. (*Phys.*) Possible energy value of electron or nuclear particle. (*Surv.*) An instrument for determining the difference in height between two points. See also **level tube.** (*Telecomm.*) Said of a set of contacts in a stepping relay which are operated together. (*Teleph.*) The row of contacts in a selector which is found by the vertical motion of the selector under the action of an impulse train. The wipers then traverse the row of contacts either by impulse, finding, or hunting action.

level canal (*Hyd. Eng.*). A canal which is level throughout. Also called a **ditch canal.**

level-compounded (*Elec. Eng.*). See **flat-compounded.**

level indicator (*Telecomm.*). Voltage indicator on a transmission line, calibrated to indicate decibels in relation to a zero power level; also called **power-level indicator.** See **volume unit.**

levelling (*Surv.*). The operation of finding the difference of elevation between two points.

levelling agent (*Chem.*). Agent added to a dye bath to produce uniform precipitation of the dye on to the fibre.

levelling bulb (*Chem.*). In a gas pipette, the pressure may be equalized to that of the atmosphere by connecting with flexible tubing the liquid in the pipette to that in an open bulb, the level of which is adjusted till the heights of the two liquid interfaces (and hence their hydrostatic pressures) are equal.

levelling solvent (*Chem.*). One which has a high dielectric constant and high polarity, so that most electrolytes appear strong in solution.

levelling staff (*Surv.*). Light extensible system of wooden rods graduated in feet and tenths, used to transfer *line of collimation* (q.v.) of surveyor's level from back sight, on which it is first held vertically, to fore sight, its second vertical station. As the line of collimation is horizontal, change of height is thus measured. See **target rod.**

level-luffing crane (*Eng.*). A jib crane in which, during derricking or luffing, the load is caused to move radially in a horizontal path, with consequent power saving.

level measuring set (*Telecomm.*). Apparatus consisting essentially of a receiving circuit calibrated in decibels or nepers relative to 1 milliwatt in specified impedance(s). It is capable of measuring terminated or through levels. When used to measure through levels the input impedance must be greater than that of the circuit under test. Abbrev. **LMS.**

level multiple (*Teleph.*). That section of a multiple which is concerned with the outlets from a given level in the selector switches.

level response (*Acous.*). See **frequency response.**

level setting (*Telecomm.*). Provision for adjusting the base voltage for an irregular waveform, e.g., in TV scanning circuit voltages and signals. See **clamp.**

level small caps (*Typog.*). See **even small caps.**

level trier (*Surv.*). An apparatus for measuring the angular value of a division on a level tube; it consists of a beam, hinged about a horizontal axis at one end and capable of being moved up or down at the other end by means of a micrometer screw, which records the inclination corresponding to a given number of divisions of movement of the bubble.

level tube or spirit level (*Surv., etc.*). A specially shaped glass tube nearly filled with spirit, so as to leave a 'bubble' of air and spirit vapour, which always rises to the highest part of the tube. The level tube is used to test whether a surface to which it is applied is horizontal. It is an essential feature of many forms of surveying instrument. See **dumpy level, tilting level.**

lever (*Horol.*). The pivoted arm which carries the pallets in the lever escapement. (*Mech.*) One of the simplest machines. It may be considered as a rigid beam pivoted at a point called the *fulcrum*, a load being applied at one point in the beam and an effort, sufficient to balance the load, at another. Three classes of lever may be distinguished: (*a*) fulcrum between effort and load; (*b*) effort between fulcrum and load; (*c*) load between fulcrum and effort. See **machine, mechanical advantage, velocity ratio.**

lever escapement (*Horol.*). The most important type of watch escapement. The impulse from the escape wheel is transmitted to the balance by the equivalent of two levers—a pivoted lever carrying the pallets, and the roller carrying the impulse pin. In the 'English' lever, the escape wheel is planted at right angles to the line joining the pallet staff and balance staff centres. In the 'Swiss' lever or 'straight-line' escapement the escape wheel, pallet staff, and balance staff centres are in a straight line. Strictly speaking, the pin-pallet escapement is a lever escapement, but it is not generally referred to as such.

lever (or **locker**) **jack** (*Textiles*). An accessory part on the under side of the combs of a lace machine, which controls the locker carriage.

lever key (*Teleph.*). A hand-operated key for telephone switchboards; operated by a small lever, which opens and closes one or more spring contacts. May be locking or nonlocking.

lever safety valve (*Eng.*). A safety valve in which the valve is held on its seating by a long lever, loaded by a weight at the other end; a form of *dead-weight safety valve* (q.v.).

Levers machine (*Textiles*). Lace machine originally made by John Levers; the modern machine is used for straight-down fabrics.

lever-type brush-holder (*Elec. Eng.*). A type of brush-holder in which the brush is held at the end of an arm pivoted about the brush spindle.

lever-type starter (*Elec. Eng.*). See **face-plate starter.**

lever-wind (*Photog.*). Device on a camera which transports the film and sets the shutter in one movement.

levigate (*Bot.*). Correct form of *laevigate* (q.v.).

levigation (*Min. Proc.*). Use of sedimentation or elutriation to separate finely ground particles into fractions according to their movement through a separating fluid. Also used in connection with wet grinding.

levitation (*Electronics*). Balance of gravity in a keeper of an electromagnet by current in the energizing coils, control being by interrupted beam of light falling on a photocell. (*Min. Proc.*) In froth flotation, separation of froth carrying selected mineral from remainder of pulp.

levulosuria (*Med.*). See **laevulosuria.**

levyne, levynite (*Min.*). A zeolite; hydrated silicate of calcium and aluminium crystallizing in the trigonal system.

lew (*Build.*). A light covering or roof of straw used to protect bricks on *hacks* (q.v.) during the drying period.

lewis (*Build.*). A truncated steel wedge or dovetail made in 3 pieces, with the larger end downwards and fitting into a similarly shaped hole in the top of a block of masonry; it then provides, by its attached hoist ring, a means of lifting the stone.

lewis bolt (*Civ. Eng.*). A foundation bolt with a tapered and jagged head, which is securely fixed into a hole in the anchoring masonry by having molten lead run round it; also used in concrete foundations.

Lewis formula (*Eng.*). A formula used to calculate the strength of gear teeth.

lewisite (*Chem.*). $ClCH{=}CH{\cdot}AsCl_2$. 2-chloroethenyl dichloro arsine. Dark-coloured oily liquid (colourless when pure) with a strong smell of geraniums. Vesicant having lachrymatory and nose-irritant action. Used as poison gas.

Lewis Jones condenser (*Med.*). Apparatus for testing muscle reactions. See **R.D.**

lewisson (*Build.*). A *lewis* (q.v.).

Lewis's theory (*Chem.*). The assumption that atoms can combine by sharing electrons, thus completing their shells without ionization.

Leyden jar (*Elec. Eng.*). Original form of capacitor, with electrodes of foil on the inner and outer surfaces of a glass jar.

Leydig, interstitial cells of (*Histol.*). Groups of specialized secreting connective tissue cells in the interstitial tissue of the testes.

Leydig's duct (*Zool.*). See **Wolffian duct.**

Leydig's gland (*Zool.*). In male *Elasmobranchii*, the anterior, nonrenal part of the kidney which, together with the spermiduct, constitutes the epididymis.

Leydig's organs (*Zool.*). Minute structures occurring on the antennae of some *Arthropoda*; believed to be olfactory in function.

L-forms (*Bacteriol.*). Morphological variants developed from large bodies by prolonged exposure to various treatments. Consist of colonies of filterable bodies with a cytoplasmic matrix. Frequently revert to the normal form. If blood or serum is supplied in their nutrient medium may breed true for some generations. See **large bodies.**

LGF (*Oils*). Abbrev. for **light gasworks feedstock.** A deprecated term for low flash-point petroleum distillates, which are unsuitable for use in internal combustion engines on account of their properties. Preferably referred to as **light petroleum feedstock (LPF).**

LH (*Physiol.*). Abbrev. for *luteinizing hormone*.

L-head (*I.C. Engs.*). A petrol-engine cylinder head carrying the inlet and exhaust valves in a pocket at one side; resembles an inverted L.

lherzolite (*Geol.*). An ultramafic plutonic rock, a peridotite, consisting essentially of olivine, with both ortho- and clino-pyroxene; named from Lake Lherz in the Pyrenees.

L'Hospital's rule (*Maths.*). A rule for evaluating certain limits. If $f(a)=0$ and $g(a)=0$, and if n is the lowest integer for which $f^n(a)$ and $g^n(a)$ are not both zero, then

$$\lim_{x\to a}\frac{f(x)}{g(x)}=\frac{f^n(a)}{g^n(a)}.$$

Li (*Chem.*). The symbol for *lithium*.

liane (*Bot.*). A woody climber, typical of tropical forests.

Lias (*Geol.*). The lowest division of Jurassic rocks.

Lias Clay (*Geol.*). A thick bed of clay found in the Lower Jurassic rocks of Britain. See **Blue Lias.**

liber (*Bot.*). See **phloem.**

liberation (*Min. Proc.*). First stage in ore treatment, in which comminution is used to detach valuable minerals from gangue.

libethenite (*Min.*). An orthorhombic hydrous phosphate of copper (II), occurring rarely as olive-green crystals in the oxide zone of metalliferous lodes.

libido (*Psychol.*). A term introduced by Freud to denote the energy attached to the sexual impulse, manifesting itself in various forms; subsequently used by some authors (e.g., Jung) to cover vital energy in general, without special reference to the sexual impulse. See **Freud's theory of the libido.**

libollite (*Min.*). A pitchlike member of the asphaltite group.

library (*Comp.*). Collection of *routines* and *subroutines* in computers, which can be reliably used without adjustment.

library binding (*Bind.*). Stronger than the usual *edition binding*; sewn-on tapes which are inserted in *split boards*.

library routine (*Comp.*). A ready-made routine available from a library and which can be used immediately; usually on punched cards or a paper tape.

libration in latitude (*Astron.*). A phenomenon by which, owing to the moon's axis of rotation not being perpendicular to its orbital plane, an observer on the earth sees alternately more of the north and south regions of the lunar surface, and so, in a complete period, more than a hemisphere.

libration in longitude (*Astron.*). A phenomenon by which, owing to the uniform rotation of the moon on its axis combined with its non-uniform orbital motion, an observer on the earth sees more, now on the east and now on the west, of the lunar surface than an exact hemisphere.

librations (*Astron.*). Apparent oscillations of the moon (or other body). The actual physical librations (due to changes in the moon's rate of rotation) are very small; the other librations are librations in latitude or in longitude, and diurnal libration.

libriform fibre (*Bot.*). An elongated thick-walled element of the xylem, formed from a single cell.

Librium (*Pharm.*). TN for chlordiazepoxide.

Libyan glass (*Geol.*). A natural glass consisting of almost pure silica found scattered in lumps across a large area of the Libyan desert. The origin of this glass is uncertain; it is possible that the fragments are *tektites*.

licensed aircraft engineer (*Aero.*). An engineer licensed by the airworthiness authority (in U.K. the *Civil Aviation Authority*) to certify that an aircraft and/or component complies with current regulations.

lichenicole (*Bot.*). Living on lichens; said particularly of parasitic fungi.

Lichens, Lichenes (*Bot.*). A large group of composite plants, consisting of an alga and a fungus in intimate association, and divided into genera and species as if they were independent plants.

Lichtenberg figure (*Elec. Eng.*). A figure appearing on a photographic plate or on a plate coated with fine dust when the plate is placed between electrodes and a high voltage is applied between them.

licker-in (*Textiles*). A toothed roller which slowly feeds the fibre stock to the main cylinder on a carding frame.

lid (*Meteor.*). Temperature inversion in the atmosphere which prevents the mixing of the air above and below the inversion region.

lidar (*Meteor.*). Device for detection and observation of distant cloud patterns by measuring the degree of back scatter in a pulsed laser beam. (*li*ght *d*etection *a*nd *r*anging.)

Lido (*Nuc. Eng.*). Type of swimming-pool reactor at Harwell, used for research into shielding.

lidocaine (*Med.*). See **lignocaine.**

Lie algebra (*Nuc.*). That dealing with groups of quantities subject to relationships which reduce the number that are independent. These are known as **special unitary groups,** and groups SU (2) and SU (3) have proved very valuable in elucidating relationships between fundamental particles.

Lieberkühn's crypts (*Zool.*). Simple tubular glands occurring in the mucous membrane of the small intestine in Vertebrates.

Liebermann-Burchard test (*Chem.*). A colour test for cholesterol, based on a change of colour from red to bluish-green in trichloromethane solution on addition of ethanoic anhydride and concentrated sulphuric acid.

Liebermann-Storch test (*Paper*). A red ring is formed between sulphuric acid and an ethanoic acid extract of paper if rosin is present.

Liebermann test for phenols (*Chem.*). Colour changes observed on treating the phenol with sulphuric acid and sodium nitrite (nitrate (III)), then pouring the solution into excess of aqueous alkali.

Liebig condenser (*Chem.*). The ordinary water

cooled glass condenser used in laboratory distillations.

Liebig's law of the minimum (*Ecol.*). See **minimum, law of the.**

lien (*Zool.*). See **spleen.**

lienal (*Zool.*). Pertaining to the spleen.

lienculus (*Zool.*). An accessory spleen.

lienogastric (*Zool.*). Pertaining to, or leading to, the spleen and the stomach, as the *lienogastric artery* and *vein* in Vertebrates.

lier (*Glass*). See **lehr.**

lierne rib (*Arch.*). A connecting rib between the main ribs in a groined vault.

Liesegang rings (*Chem.*). The stratification, under certain conditions, of precipitates formed in gels by allowing one reactant to diffuse into the other.

life-cycle (*Biol.*). The various stages through which an organism passes, from fertilized ovum to the fertilized ovum of the next generation.

life form (*Bot.*). The form of a plant determined by the position of its resting buds (if any) in respect to the surface of the soil.

life goal (*Psychol.*). In Adlerian psychology, the unconscious drive for superiority which compensates for the feelings of inferiority.

lifetime (*Nuc.*). The mean period between the birth and death of a charge carrier in a semiconductor. See also **half-life, mean life.**

lift. An enclosed platform or car moving in a well to carry persons or goods up and down. In the modern automatic lift the passengers press buttons to register the stops for the floors at which they wish to alight. The doors are automatically opened and closed and the lift stops in turn at each floor registered. For heavy traffic two or more lifts may be used under a collective control pattern, the latter determined either by supervisor or automatically. Also called **elevator.** Cf. ***paternoster.*** (*Aero.*) *Aerodynamic lift* is the component of the aerodynamic forces supporting an aircraft in flight, along the lift axis, due solely to relative airflow, while *total lift* is the component of the resultant forces. (*Hyd. Eng.*) (1) The vertical distance through which a vessel is raised in the process of passing through a lock. (2) A mechanical contrivance for lifting a vessel from one reach of a canal to the next, either by transfer vertically or on an inclined plane. (*Mining*) The plane, parallel to the floor of the quarry, along which the rock is split. (*Textiles*) The amount the warp ends are raised to make the proper top shed for passage of the shuttle. Controlled by (*a*) tappet under loom, (*b*) dobby or spring top motion, (*c*) the griffe in a jacquard machine.

lift axis (*Aero.*). See **axis.**

lift bridge (*Civ. Eng.*). A type of movable bridge which is capable of being lifted bodily through a sufficient vertical distance to allow of the passage of a vessel beneath.

lift coefficient (*Aero.*). A nondimensional factor representing the incremental lift of a body. $C_L = \dfrac{L}{1/2\,\rho\,V^2 S}$, where L = lift, ρ = air density, V = air speed, S = wing area.

lift engine (*Aero.*). An engine used on *VTOL* or *STOL* aircraft, having the primary purpose of providing lifting force.

lifter (*Foundry*). An L- or Z-shaped bar of cast- or wrought-iron, used for supporting the sand in a cope, the upper end being hooked on to a box bar. (*Min. Proc.*) (1) Projecting rib or wave in lining plates of ball or rod mill. (2) Perforated plate in drum washer or heavy-media machine which aids tumbling action or removes floating fraction of ore being treated. (*Mining*) One of the shots fired near the floor when driving a tunnel heading.

lift gate (*Build., Civ. Eng.*). A gate which opens by bodily vertical movement, as distinct from one swinging about an axis at one end.

lifting blocks (*Eng.*). A lifting machine consisting of a continuous rope passing round pulleys mounted in blocks, whereby an effort applied at the free end of the rope lifts a larger weight attached to the lower block. See **mechanical advantage.**

lifting magnet (*Elec. Eng.*). A large electromagnet used, instead of a hook, on cranes or hoists, when lifting iron and steel.

lifting of patterns (*Foundry*). See **drawing of patterns.**

lifting piece (*Horol.*). In the rack-striking work of a striking clock, a cranked lever which carries the warning piece at one end and lifts the rack hook just before the hour.

lifting screw (*Foundry*). An iron rod screwed into a pattern to withdraw it from the mould.

lifting the offsets (*Ships*). The process of measuring the ship's form as 'laid off' on buttocks, waterlines, and sections. These offsets are the permanent record and are used to reproduce the ship's form, initially, for design work and ordering material, and, subsequently, in cases of repair or alterations.

lifting truck (*Eng.*). A truck with three, or four, wheels, drawn by a handle which can be raised and lowered to lift a loaded platform standing on feet. The lift is effected either by leverage or by hydraulic mechanism.

lift-lock (*Hyd. Eng.*). A canal lock serving to lift a vessel from one reach of water to another.

lift motor (*Elec. Eng.*). A motor, sometimes having special characteristics, used for operating an electric lift. Also called **hoisting motor.**

lift-off (*Aero.*). The speed at which a pilot pulls back on the control column to make an aeroplane leave the ground. It is a carefully defined value for large aeroplanes depending upon the weight, runway surface, gradient, altitude and ambient temperature. It is one of the functions established from *WAT curves* (q.v.). Colloquially **unstick.** (*Space*) The point at which the vertical thrust of a space craft exceeds that of local gravity and it begins to rise.

lift-valve (*Eng.*). Any valve consisting of a disk, ball, plate, etc., which lifts or is lifted vertically to allow the passage of a fluid.

ligament (*Zool.*). A bundle of fibrous tissue joining two or more bones or cartilages.

ligamentum denticulatum (*Zool.*). In *Cephalochorda*, the combined atrial/coelomic wall.

ligands (*Chem.*). In a complex ion, the ions surrounding the central (nuclear) ion, e.g., $(CN)^-$ in $Fe(CN)_6{}^{4-}$.

ligasoid (*Phys.*). A colloidal system in which the continuous phase is gaseous and the dispersed phase is liquid.

ligate (*Surg.*). To tie with a ligature. *n.* **ligation.**

ligature (*Med.*). A piece of thread, silk, wire, catgut, or any other material, for tying round blood vessels, etc.: to tie with thread, etc. (*Typog.*) Two or more letters cast together on one body, e.g., fi, ffi, ffl.

light. *Electromagnetic radiation* capable of inducing visual sensation, with wavelengths between about 400 and 800 nm. It is the product of the visibility and the radiant power, the latter being the rate of propagation of radiant energy. See also **illumination, velocity of light.** (*Build.*) (1) A term applied to any glazed opening admitting light to a building. (2) A single division of a window. (*Glass*) Of optical glass, having a lower refractive index.

light-adapted (*Optics*). See **adaptation.**

light alloys (*Met.*). See **aluminium alloys.**

light-band pattern (*Light*). See **optical pattern.**

light-centre length (*Elec. Eng.*). The distance from the geometrical centre of the filament of an electric filament lamp to the contact plate or plates at the end of the lamp cap remote from the bulb. With automobile headlight lamps the measurement is taken from the bulb side instead of from the remote side of the pin.

light-compass reaction (*An. Behav.*). See **menotaxis.**

light current (*Phys.*). A current flowing through a photoelectric device, caused by incident light. See **dark current.**

light-curve (*Astron.*). The line obtained by plotting, on a graph, the apparent change of brightness of a variable star, against the observed times; analysis then divides these stars into *long-* or *short-period variables, Cepheid variables*, etc.

light distribution curve (*Light*). A graph showing the relation between the luminous intensity of a light source and the angle of emission.

light efficiency (*Light*). Measured, e.g., in an electric lamp, by the ratio (total luminous flux/total power input), and expressed in lumens per watt.

light-emitting diode (*Electronics*). Semiconductor diode which radiates in the visible region. Used in *alphanumeric* displays and as a replacement for small filament lamps. Abbrev. **LED.**

lighter-than-air craft (*Aero.*). See **aerostat.**

light face (*Typog.*). A lighter weight of type face, 'medium' being the usual normal in the *type family.*

light filter (*Photog., etc.*). Filter used in photography, etc., to change or control the total (or relative) energy of light distribution from the source. It consists of a homogeneous optical medium (sometimes of a specific thickness as in interference filters) with characteristic light absorption regions.

light flux (*Light*). The measure of the quantity of light passing through an area, e.g., through a lens system. Light flux is measured in lumens. Illumination is light flux per unit area.

light fog (*Photog.*). See **fog.**

lighthouse tube (*Electronics*). Same as **megatron.**

lighting. See **illumination.**

light meter (*Photog.*). Same as **exposure meter.**

light modulation (*TV*). Control of the intensity of light by electrical means, such as a combination of Kerr cell and crossed Nicol prisms. See **positive (or negative-).**

lightness. The degree of illumination of a surface, measured in lumens per square metre.

lightning (*Meteor.*). Luminous discharge of electric charges between clouds, and between cloud and earth (or sea). A path is found by the *leader stroke*, the main discharge following along this ionized path, with possible repetition. See **thunderstorm.**

lightning (or surge) arrester (*Elee. Eng.*). A device for the protection of apparatus from damage by a lightning discharge or other accidental electrical surge. A surge arrester is effectively a gas-filled diode whose discharge changes from a normal glow discharge to an arc discharge under over-voltage conditions. See **autovalve.**

lightning conductor (*Elec. Eng.*). A metal strip connected to earth at its lower end, and its upper end terminated in one or more sharp points where it is attached to the highest part of a building. By electrostatic induction it will tend to neutralize a charged cloud in its neighbourhood and the discharge will pass directly to earth through the conductor. Also called **lightning rod.**

lightning protector (*Elec. Eng.*). See **lightning arrester.**

lightning rod (*Elec. Eng.*). See **lightning conductor.**

lightning tubes (*Min.*). See **fulgurites.**

light oils (*Chem.*). A term for oils with a boiling range of about 100°–210°C, obtained from the distillation of coal-tar.

light-positive (*Elec. Eng.*). Said of a material whose electrical conductivity increases under the action of light. Most photoconductive substances are light-positive, but some are **light-negative.**

light quanta (*Light*). The energy of light appears to be concentrated in packets called *photons.* The energy of each photon $= h\nu$, where ν is the frequency and h is Planck's constant.

light railway (*Rail.*). One of standard or narrow guage, subject to severe restrictions on speed, load, etc., and thereby relieved of many of the operating requirements (signalling, etc.) applied to main line railways.

light ratio (*Astron.*). See **magnitudes.**

light red silver ore (*Min.*). See **proustite.**

light relay (*Telecomm.*). Circuit for operating a relay employing a photoelectric tube. The latter triggers off a switching device, which operates the relay.

light resistance (*Electronics*). The resistance, when exposed to light, of a photocell of the photoconductive type.

light restraint (*Photog.*). The impregnation of a dye in an emulsion, to prevent the deep penetration of light.

light seed (*Bot.*). A seed which requires exposure to light in order that it may germinate.

light-sensitive (*Elec.*). Said of thin surfaces of which the electrical resistance, emission of electrons, or generation of a current, depends on incidence of light.

lightship (*Ships*). A kind of floating lighthouse, consisting of an anchored vessel carrying a powerful light as a warning or guide to shipping.

light ship (*Ships*). A cargo vessel with all necessary stores and fuel on board but no cargo.

light-spring diagram (*Eng.*). An indicator diagram taken by a piston or diaphragm-type engine indicator, using a specially weak control spring or diaphragm in order to reproduce the low-pressure part of the diagram to a large scale.

light table (*Print.*). A ground-glass surface, illuminated from below, there being a variety of sizes and styles, for use in the graphic reproduction processes when working with negatives, positives, and proofs. Also **shiner.**

light time (*Astron.*). The time taken by light to travel from a heavenly body to the observer; the computed position of a planet or comet must be antedated by this amount to give the observed place. See **velocity of light.**

light-trap (*Photog.*). An arrangement for modifying doors or windows of dark-rooms, so that air can pass but light is excluded. (*Zool.*) A device for catching night-flying moths and/or other insects attracted by light.

light-up (*Aero.*). The period during the starting of a *turbojet* or *turboprop* engine when the fuel/air mixture has been ignited.

light valve (*Cinema., etc.*). Device for varying the area for light transmission and exposure on a photographic film sound-track. Early types consisted of two ribbons carrying modulating current which opened or closed because of magnetic field normal to their plane. Other types use the Kerr effect with polarized light.

light water (*Chem.*). Normal water (H_2O) as distinct from *heavy water* (q.v.).

light watt (*Optics*). The photometric radiation equivalent, the emitted radiation being measured in lumens or watts. The ratio of lumens to watts depends on the wavelength and reaches a maximum at $\lambda = 555$ nm. At this wavelength the average value of the light watt is 600 lumens.

lightweight aggregate (*Build., Civ. Eng.*). Aggregate which is used in the manufacture of lightweight concrete. Normally clinker ash or clays or other materials which have been fired to reduce their carbon content and weight.

lightweight concrete (*Build.*). Concrete of low density (20–90 lb/ft^3 or 300–1400 kg/m^3) made by using special aggregates or air entraining processes. Used particularly where lightness and insulation are required rather than load resisting properties.

lightwood (*For.*). Coniferous wood having an abnormally high resin content.

light-year (*Astron.*). An astronomical measure of distance, being the distance travelled by light in space during a year, which is approximately $9{\cdot}46 \times 10^{12}$ km ($5{\cdot}88 \times 10^{12}$ miles).

ligne (*Horol.*). A unit used in the measurement of watch movements. It is equal to 2·256 mm. The *twelfth*, or *douzième* (0·188 mm), is the unit used for the height or thickness of a movement. There are 12 twelfths in a ligne. Ladies' wrist watches vary from 3½ lignes to 8 lignes; gents' wrist watches from 10 to 13, pocket watches from 17 to 19, and deck watches from 20 to 32. Sometimes spelt line. Symbol ‴.

lignicole, lignicolous (*Bot.*). Growing on or in wood, or on trees. (*Zool.*) Living on or in wood, as certain species of Termites.

lignin (*Bot.*). A complicated mixture of substances formed by certain cells of plants and deposited, (**lignification**), in thickened cell walls, particularly in woody tissue.

lignite (*Geol.*). A carbonaceous rock, composed of plant remains, which has not been subjected to as much heat or pressure as ordinary coal and therefore has a lower carbon content and a higher content of volatile constituents. It is commoner than coal in Mesozoic and Tertiary deposits, but not as common in Palaeozoic deposits. It is known as *brown coal* and exploited extensively as a fuel in some countries.

lignivorous (*Zool.*). Wood-eating.

lignocaine, U.S. lidocaine (*Med.*). A powerful comparatively long-lasting local anaesthetic, *N*-diethylaminoacetyl-2,6-xylidine hydrochloride monohydrate.

ligno-celluloses (*Chem.*). Compounds of lignin and cellulose found in wood and other fibrous materials.

ligroin (*Chem.*). A term for a petroleum fraction with a boiling range of from about 90° to 120°C.

ligula (*Zool.*). In Insects, a median structure lying between the labial palps, composed of the paraglossae and the glossae, which may be separate or fused.

ligulate (*Bot.*). (1) Strap-shaped, long, flattened, and narrow. (2) Said of a corolla which has a very short tube and is prolonged above into a flattened group of united petals. (3) Said of a capitulum in which all the flowers have ligulate corollas.

ligule (*Bot.*). A small outgrowth, commonly membranous, from the upper surface of a leaf or leaflike member.

Liliiflorae (*Bot.*). An order of the *Monocotyledons*; herbaceous; the flowers regular with the perianth in two whorls of 3; the anthers usually in two whorls of 3, but may be one whorl, or reduced to a single anther; the ovary is usually superior consisting of 3 loculi; ovules usually anatropous, the endosperm oily, or fleshy.

limaciform (*Zool.*). Sluglike.

limaçon (*Maths.*). A heart-shaped curve with an inner loop at its vertex. An epitrochoid in which the rolling circle equals the fixed circle. Polar equation $r = 2a(k + \cos\theta)$.

limb (*Astron., Surv.*). Term applied to the edge or rim of a heavenly body having a visible disk; used specially of the sun and moon. (*Bot.*) (1) The lamina of a leaf. (2) The widened upper part of a petal. (3) The upper, often spreading, part of a sympetalous corolla. (*Surv.*) Lower horizontal plate of theodolite. (*Zool.*) A jointed appendage, as a leg.

limbate (*Bot.*). Edged in a different colour.

limb darkening (*Astron.*). The apparent darkening of the limb of the sun due to the absorption of light in the deeper layers of the solar atmosphere near the edge of the disk.

limberneck (*Vet.*). *Botulism* (q.v.) of Birds.

limbic (*Anat.*). Marginal, bordering; e.g., lobe of cerebral cortex.

limbous (*Zool.*). Overlapping.

limbric (*Textiles*). Plain weave cotton cloth of medium quality; used for dresses and curtains, etc., after being piece-dyed or printed.

limburgite (*Geol.*). An ultramafic, fine-grained igneous rock occurring in lava flows, similar to the dyke-rock monchiquite, but having interstitial glass between the dominant olivine and augite crystals. Typically, limburgite is feldspar-free; but it does occur in most samples.

limbus (*Zool.*). In Mammals, the thickened, overhanging extremity of the spiral lamina of the cochlea.

lime (*Chem.*). A substance produced by heating limestone to 825°C or more, as a result of which the carbonic acid and moisture are driven off. Lime, which is much used in the building, chemical, metallurgical, agricultural and other industries, may be classified as **high-calcium, magnesian**, or *dolomitic* (q.v.), depending on the composition. Unslaked lime is commonly known as *caustic lime* (q.v.), also **anhydrous lime, burnt lime, quicklime.** (*For.*) A moderately light-weight hardwood, from the genus, *Tilia*, a native of Russia, Germany, and England. Typical uses include carving, inlaying, marquetry work, and cabinet-making.

lime bag (*Foundry*). A bag of powdered lime used for testing the fit of joints. Lime is sprinkled on the parting face, and the cope is lowered and lifted; if the lime adheres to the top face, the joint is good.

lime blast (*Leather*). Dark patches that appear on limed skins while being tanned; caused by carbonate of lime on the skins.

lime blue (*Paint.*). Ultramarine blue reduced on terra alba (calcium sulphate) for use in water paints and distempers.

lime chlorosis (*Bot.*). Yellowing in a plant growing in a soil containing an excess of calcium carbonate; due to chlorophyll deficiency.

lime knot (*Bot.*). A widening in the threads of the capillitium of *Myxomycetes*, containing calcium carbonate.

lime light (*Light*). Intense white light obtained by heating a cylinder of lime in an oxyhydrogen flame.

lime mortar (*Build., Civ. Eng.*). A mortar composed of lime and sand, with the addition sometimes of other material, such as crushed bricks, ground slag, or coke. It is not suitable for use under water; but see **hydraulic lime.**

limen (*Physiol.*). Smallest difference in pitch (frequency) or intensity of a sound or colour which can be perceived by the senses.

lime paste (*Build.*). Slaked lime.

lime powder (*Build.*). The material produced as a result of subjecting quicklime to the process of air-slaking.

lime-silicate rocks (*Geol.*). These result from the contact (high-temperature) metamorphism of limestones containing silica in detrital grains, nodules of flint or chert, or siliceous skeletons, the silica combining with the lime to form such silicates as lime-garnet, anorthite, wollastonite.

lime slurry (*Min. Proc.*). Thick aqueous suspension of finely ground slaked lime, used to control alkalinity of ore pulps in flotation and cyanide process.

lime-soda process (*Chem.*). Standard water-softening process, carried out either hot or cold; uses lime and soda ash to reduce the hardness of the treated water by precipitating the dissolved calcium and magnesium salts as insoluble calcium carbonate and magnesium hydroxide respectively. Often used in conjunction with a *zeolite process* (q.v.). Also used in the preparation of caustic soda, by mixing slaked lime with soda and filtering off the precipitated calcium carbonate. $Ca(OH)_2 + Na_2CO_3 = 2NaOH + CaCo_3$.

limestone (*Geol.*). A sedimentary rock composed mainly of calcium carbonate. Mineralogically, limestone consists of either aragonite or calcite, although the former is abundant only in Tertiary and Recent limestones. Dolomite may also be present, in which case the rock is called *dolomitic limestone*. Limestones may be **organic**, formed from the calcareous skeletal remains of living organisms, **chemically precipitated**, or **detrital**, formed of fragments from pre-existing limestones. The majority are organic, and can be classified according to their texture and the nature of the organisms whose skeletons are incorporated, e.g., oölitic limestone, shelly limestone, algal limestone, crinoidal limestone, etc.

limewash (*Paint.*). A mixture which is prepared by slaking lump lime with about one-third of its weight of water, and then adding sufficient water to make a 'milk'; used as wall covering in cases where a frequent application is necessary. An improved, more water-resistant type embodies tallow.

lime water (*Chem.*). Saturated calcium hydroxide solution.

limicolous (*Zool.*). Living in mud.

liming (*Leather*). The process of soaking hides and skins in milk of lime, which causes them to swell and facilitates removal of the hair. To loosen the wool, sheepskins are painted on the flesh side with a paste consisting of milk of lime and a small amount of alkaline sulphides.

limit (*Maths.*). A sequence of numbers u_r has a *limit L* if, given any positive non-zero number ε, it is possible to find some term u_N such that all subsequent terms differ from L by less than ε, i.e., $|u_r - L| < \varepsilon$ when $r > N$.

limited (*Telecomm.*). Pertaining to amplifiers which have reached a condition of saturation when the output is constant despite increasing input signal. In servo systems, the control motors will be torque- or velocity-limited outside a particular limiting signal.

limited stability (*Telecomm.*). A property of a servomechanism] or a communication system which is characterized by stability only when the input signal falls within a particular range.

limiter (*Telecomm.*). Any transducer in which the output, above a threshold or critical value of the input, does not vary, e.g., a shunt-polarized diode between resistors. Particularly applied to the circuit in a frequency-modulation receiver in which all traces of amplitude modulation in the signal have to be removed before final demodulation.

limiter valve (*Telecomm.*). A valve operated as an amplifier, so biased that the output resulting from a large input voltage is substantially the same as that from a small one.

limit gauge (*Eng.*). A gauge used for verifying that a part has been made to within specified dimensional limits. Limit gauges consist, for example, of a pair of plug gauges on the same bar, one of which should just enter a hole ('go') and the other just not enter ('not go').

limit gauging (*Eng.*). A method of measurement which ensures that pieces intended to fit together shall do so within certain specified limits of clearance, and that similar pieces shall be interchangeable.

limiting conductivity (*Chem.*). The molar conductivity of a substance at infinite dilution, i.e., when completely ionized.

limiting density (*Chem.*). The relative density of a gas at vanishingly low pressures.

limiting factor (*Bot.*). The slowest-acting factor of a group of factors simultaneously affecting a physiological process in a plant. (*Ecol.*) Any environmental condition which approaches or exceeds the limits of tolerance for a species in a particular habitat.

limiting frequency (*Telecomm.*). One for which there is a significant change in response, as contrasted with a cut-off frequency, which (as in a wave filter) may be nominal.

limiting friction (*Mech.*). See **friction.**

limiting gradient (*Civ. Eng.*). *Ruling gradient* (q.v.).

limiting Mach number (*Aero.*). The maximum permissible *flight Mach number* (q.v.) at which any particular aeroplane may be flown, either because of the *buffet boundary* (q.v.) or for structural strength limitations.

limiting range of stress (*Met.*). The greatest range of stress (mean stress zero) that a metal can withstand for an indefinite number of cycles without failure. If exceeded, the metal fractures after a certain number of cycles, which decreases as the range of stress increases. Also called **endurance range**; half this range is the *fatigue limit* or **endurance limit.**

limiting velocity (*Aero.*). The steady speed reached by an aircraft when flown straight, the angle to the horizontal, power output, altitude, and atmospheric conditions all specified. See **max level speed, terminal velocity.**

limit load (*Aero.*). The maximum load anticipated from a particular condition of flight and used as a basis when designing an aircraft structure.

limit of proportionality (*Met.*). The point on a stress-strain curve at which the strain ceases to be proportional to the stress. Its position varies with the sensitivity of the extensometer used in measuring the strain.

limit point (*Maths.*). See **accumulation point.**

limits of audition (*Acous.*). The extreme frequencies of sound waves perceivable by the normal ear, and the extent of perception between the maximum tolerable loudness and the minimum perceptible. See **audibility.**

limits of tolerance (*Ecol.*). See **tolerance, law of.**

limit switch (*Elec. Eng.*). A switch fitted to electric lifts, travelling cranes, etc., to cut off the power supply if the lift-car or moving carriage travels beyond a certain specified limit.

limivorous (*Zool.*). Mud-eating; as certain aquatic Invertebrates which swallow mud to extract from it the nutritious organic matter that it contains.

limnetic zone (*Ecol.*). In ponds and lakes, the open water zone to the depth of effective light penetration, i.e., the *compensation level* (q.v.).

limnobiotic (*Zool.*). Living in fresh water.

limnology (*Geol.*). The study of lakes.

limnophilous (*Zool.*). Living in marshes, especially freshwater marshes.

limnoplankton (*Ecol.*). The plankton of fresh waters, such as ponds, lakes, rivers, and marshes. Cf. *haliplankton.*

+-limonene (*Chem.*). Hesperidene, citrene, carvene. The oil of the orange peel consists almost entirely of this essential oil, b.p. 175°C. −-Limonene is present in the oil of fir cones. The systematic name of these compounds, which are monocyclic terpenes, is 1,8-menthadiene.

limonite (*Min.*). Although originally thought to be a definite hydrated oxide of iron, now known to consist mainly of cryptocrystalline goethite or lepidocrocite along with adsorbed water; some haematite may also be present. It is a common alteration product of most iron-bearing minerals and also the chief constituent of bog iron ore.

limp (*Bind.*). Said of a book having nonrigid sides; described as *limp cloth, limp leather*, according to the covering material.

limpet washer (*Build.*). A form of washer used in fixing corrugated sheeting, for which purpose it is shaped on one side to conform to the curve of a corrugation.

linarite (*Min.*). Hydrous sulphate of lead and copper, found in the oxide zone of metalliferous lodes; a deep-blue mineral resembling azurite, also crystallizing in the monoclinic system.

Linasec (*Typog.*). Special-purpose computer which converts *idiot tape* into 6-level justified tape for use on line-casting machines or *Linofilm Quick*; hyphenation is monitored; speed up to 8000 lines per hour is possible. Cf. *Justape.*

linch pin. A pin placed in a transverse hole on the outside of the axles of a vehicle, to retain a wheel; at the top it has a projection on one side, to prevent it from passing through the hole.

Lincoln index (*Ecol.*). A method of determining the size of an animal population, involving marking a certain number of individuals, releasing them into the population, and later sampling the population again. From the proportion of marked individuals in this sample, the population size is estimated.

Lincolnshire Limestone (*Geol.*). Oölitic limestones, famous as building-stones, which form a prominent escarpment running through Lincolnshire and Northamptonshire; of the same age as the Gloucestershire Inferior Oölite.

Lincoln wool (*Textiles*). The longest of the English lustre wools, with a staple of 12 in. (30 cm); used for sicilians, linings, etc.

Lindeck potentiometer (*Elec. Eng.*). One which differs from most potentiometers in using a fixed resistance variable current to obtain balance.

Lindemann electrometer (*Electronics*). A form of electrometer which uses a metal-coated quartz fibre located between an arrangement of electrodes to which the potentials are applied. The fibre is mounted on and perpendicular to a quartz torsion fibre which provides the controlling couple.

Lindemann glass (*Radiol.*). Low-density glass used for windows of low-voltage X-ray tubes.

Lindé process (*Chem.*). A process for the liquefaction of air and for the manufacture of oxygen and nitrogen from liquid air.

Lindé sieve (*Chem.*). See **molecular sieve.**

line. The 12th part of an inch. (*Acous.*) Said of a microphone or loudspeaker when it is considerably extended in one direction, for directivity in a normal plane. (*Build.*) A cord stretched as a guide to the bricklayer for level and direction of succeeding courses; also for the setting out of foundations. (*Elec. Eng.*) Direction of an electric or magnetic field, as a line of flow or force. See **line of flux (or force).** (*Horol.*) (1) The cord or gut supporting the weight or weights of a weight-driven clock. (2) See **ligne.** (*Maths.*) In a plane: the shortest distance between two points. On a sphere: a portion of a great circle. (*Radiol.*) Single frequency of radiation as in a luminous, X-ray, or neutron spectrum. (*Telecomm.*) Transmission line, coaxial, balanced pair, or earth return, for electric power, signals or modulation currents. (*TV*) Single scan in a facsimile or television picture transmission system. (*Textiles*) Yarn spun from the longer and finer fibres in flax and used in the best cloths.

linea alba (*Anat.*). The tendinous line which extends down the front of the belly, from the lower end of the chest to the pubic bone, and gives attachment to abdominal muscles. (*Zool.*) In Amphibians, a longitudinal band of tendon which unites the right and left abdominal rectus muscles.

lineage (*Gen.*). In evolution, a time-character concept representing a racial complex of lines of descent.

line amplifier (*Telecomm.*). In broadcasting, that which supplies power to the line, either to a control centre or to a transmitter.

line amplitude (*TV*). The amplitude of the voltage generated by the line-scanning generator, or the length of the line on the screen produced thereby.

linea nigra (*Med.*). Pigmented linea alba, occurring in pregnant women.

linear (*Bot.*). Having parallel edges, and at least four to five times as long as broad. (*Phys., Eng.*) Said of any device or motion where the effect is exactly proportional to the cause, as rotation and progression of a screw, current and voltage in a wire resistor (Ohm's law) at constant temperature, output versus input of a modulator (or demodulator). See **nonlinear, ultralinear.**

linear absorption coefficient (*Nuc.*). See **absorption coefficient.**

linear accelerator (*Electronics*). Large device for accelerating electrons or positive ions up to nearly the velocity of light. The particles are accelerated through loaded waveguides by high-frequency pulses or oscillations of the correct phase.

linear amplifier (*Radio*). One for which the output signal level is a constant multiple of the input level.

linear array (*Radar, Telecomm.*). An in-line array of equi-spaced radiators. A long uniform array has a side lobe of −13·5 dB. The side-lobe level can be controlled by tapering the excitation as in the *binomial* and *Chebyshev arrays*, while the direction of the principal maximum can be sheered electronically as in the *phased array.*

linear detector (*Elec. Eng.*). See **detector.**

linear differential equation of order n (*Maths.*). An equation of the form

$$q_0(x)\frac{d^ny}{dx^n}+q_1(x)\frac{d^{n-1}y}{dx^{n-1}}+\ldots+q_{n-1}(x)\frac{dy}{dx}+q_n(x)y=f(x).$$

linear distortion (*Telecomm.*). That which results in the nonlinear response of a system, such as an amplifier, to the envelope of a varying signal, such as speech, without distorting (within acoustic perception) the detailed waveform.

linear energy transfer (*Nuc.*). The linear rate of energy dissipation by particulate or EM radiation while penetrating absorbing media. Abbrev. **LET.**

linearity control (*Electronics*). See **strobe** (2).

linearly dependent (*Maths.*). n quantities u_1, $u_2, \ldots u_n$ are linearly dependent if it is possible to find n numbers $\lambda_1, \lambda_2, \ldots \lambda_n$ (at least one of which is not zero) such that $\lambda_1u_1+\lambda_2u_2+\ldots+\lambda_nu_n=0$. Otherwise the u_n are independent.

linear modulation (*Radio*). Modulation in which the change in the modulated characteristic of the carrier signal is proportional to the value of the modulating signal over the range of the audiofrequency band.

linear momentum (*Maths.*). See **momentum.**

linear motor (*Elec. Eng.*). A particular form of induction motor in which the stator and rotor are linear and parallel instead of cylindrical and coaxial.

linear network (*Telecomm.*). One with electrical elements which are constant in magnitude with varying current.

linear programming (*Comp.*). That which enables a computer to give an optimum result when fed with a number of unrelated variables. (*Eng.*) General term implying that, in a complicated process of operations on an article, the various steps have been mutually planned for maximum economy.

linear rectifier (*Elec. Eng.*). One in which the output current is strictly proportional to the envelope of the applied alternating voltage.

linear resistor (*Elec. Eng.*). One which 'obeys' Ohm's law, i.e., under certain conditions the current is always proportional to voltage. Also called **ohmic resistor.**

linear scan (*Electronics*). The sweeping of a cathode spot across a screen at constant velocity using a deflecting sawtooth waveform. (*Radar*) A radar beam which moves with constant angular velocity. Also a radar scan which is projected and fixed in a straight line in order to increase the intensity of echoes in sector scanning.

linear stopping power (*Nuc.*). See **stopping power.**

linear superpolymer (*Chem.*). Polymer in which the molecules are essentially in the form of long chains with an average molecular weight greater than 10,000.

linear sweep (*Electronics*). The use of a sawtooth waveform to obtain a linear scan.

linear tetrad (*Bot.*). A row of four megaspores, as is usual in flowering plants.

linear time-base oscillator (or generator) (*Electronics*). The electrical apparatus for producing the deflecting voltage (or current) of the linear time base. See **linear scan, sweep circuit.**

linear transformation (*Maths.*). A transformation between two sets of n variables represented by n equations in which the variables of both sets occur linearly. Thus when $n=1$, the equation would be $ax+bxx'+cx'+d=0$. Cf. *bilinear transformation.*

line at infinity (*Maths.*). The line joining the circular points at infinity.

line balance (*Telecomm.*). Matching impedance, equalling the impedance level of the line at all frequencies, for terminating a 2-wire line when it divides through a hybrid coil or bridge set into a 4-wire line. Used at the ends of trunk lines. Also called **balancing impedance, line impedance.**

line bias (*Telecomm.*). The effect of the electrical characteristics of a transmission line on the length of the teletypewriter signals.

line-blanking (*TV*). The reduction of the amplitude of the television video signal to below the black level at the end of each line period. This allows the transmission of the line-synchronizing pulses.

line block (*Print.*). A printing block of a subject consisting of black and white only, without gradations of tone. It is produced by photography and etching into relief on metal, usually zinc. Cf. *line-colour* and *half-tone block.*

line-breaker (*Elec. Eng.*). A contactor on an electric vehicle, arranged for closing or interrupting the main current circuit.

line broadening (*Astron.*). A term used for the increase in width of the lines of a stellar spectrum due to rotation of the star, turbulence in the stellar atmosphere, or the *Stark* or *Zeeman effects.*

line-casting machines (*Typog.*). *Linotype* and *Intertype machines* (qq.v.).

line choking coil (*Elec. Eng.*). An inductor included in an electric power supply circuit in order to protect plant connected to the line from the effect of high-frequency or steep-fronted surges. Also called **screening protector.**

line-colour (*Print.*). A colour-picture produced by superimposed impressions of two or more line blocks printed in different colours. Varied tones are obtainable by the use of stipples.

line coordinates (*Maths.*). A coordinate system in which each set of coordinates specifies a line, usually by reference to cartesian axes. Thus the line coordinates (a, b, c) could specify a line whose ordinary cartesian equation is $ax+by+c=0$. In a line coordinate system a point is specified by a linear equation.

line coupling (*Telecomm.*). Transfer of energy between resonant (tank) circuits in a transmitter, using a short length of line with small inductive coupling at each end. Also called **link coupling.**

line defect (*Crystal.*). See **defect.**

line distortion (*Telecomm.*). That arising in the frequency content or phase distribution in a transmitted signal, as a result of the propagation constant of the line.

line drop (*Telecomm.*). The potential drop between any two points on a transmission line due to resistance, leakage, or reactance.

line equalizer (*Telecomm.*). Device which compensates for attenuation and/or phase delay for transmission of signals along a line over a band of frequencies. Also **lumped loading, phase compensation, phase equalization.**

line finder (*Teleph.*). See **finder.**

line flyback (*Electronics*). (1) The time interval corresponding to the steeper portion of a sawtooth wave. (2) The return time of the image spot from its deflected position to its starting point.

line focus (*Electronics*). CRT in which electron beam meets screen along a line and not at a point. (When unintended this is due to astigmatism.)

line frequency (*TV*). Number of lines scanned per

second in a television image. The international standard is an *interlaced scanning* of 625 lines, 25 times per second; hence the line frequency is 15 625 lines per second. In the U.S. system 525 lines are scanned 30 times per second, hence the line frequency is 15 750 lines per second. The original British system scans 405 lines 25 times per second, hence the line frequency is 10 125 lines per second.

line-frequency generator (*TV*). The generator of the voltage or current which causes the scanning spot to traverse each line of the image.

line hold (*TV*). Variable resistance incorporated in the line time-base generator of a TV receiver. Used to control the time constant of the circuit so that the received picture is maintained in step with the transmitted picture.

line impedance (*Telecomm.*). See **line balance.**

line integral (*Maths.*). Of a function along a path: the limiting sum along the path of the product of the value of the function and an element of length of the path. If the function is a vector function the scalar product of its value with vectorial elements of lengths is taken.

line jump scanning (*TV*). Alternative name for **interlaced scanning.**

linellae (*Zool.*). In some *Sarcodina*, a system of smooth filaments uniting the xenophya.

linen (*Textiles*). Dress, furnishing, etc., fabric woven from yarn prepared from the fibres of the flax plant, ***Linum usitatissimum***, family ***Linaceae***; underpick looms are generally used, and jacquard machines for the very decorative damasks, banners, etc.

line noise (*Teleph.*). Noise in telephone circuits which may arise from cross-talk, babble, or induction from power lines, and which interferes with the normal use of such circuits.

line of action (*Mech.*). The line along which a force acts.

line of apsides (*Astron.*). See **apse line.**

line of centres (*Horol.*). A line passing through two or more centres: the line joining the centres of two or more wheels; in a lever escapement, the line joining the balance and pallet staffs.

line of collimation (*Surv.*). In a surveying telescope, the imaginary line passing through the optical centre of the object glass and the intersection of the cross-hairs in the diaphragm.

line of flux (or force) (*Elec. Eng., Mag.*). A line drawn in a magnetic (or electric) field so that its direction at every point gives the direction of magnetic (or electric) flux (or force) at that point.

line of sight (*Surv.*). Alternative term for **line of collimation.** (*Telecomm.*) Said of a transmission system when there has to be a straight line between transmitting and receiving antennae, as in UHF and radar.

line-of-sight velocity (*Astron.*). The velocity at which a celestial body approaches, or recedes from, the earth. It is measured by Doppler shift of the spectral lines emitted by the body as observed on the earth. Also **radial velocity.**

line oscillator (*Telecomm.*). One which has its frequency stabilized either by a resonant low-loss (high-Q) coaxial line, or by a resistance-capacitance ladder which gives the necessary delay (phase shift) in a feedback loop. Also **phase-shift oscillator.**

line-output transformer (*TV*). The transformer which performs the function of transferring the output of the line time-base generator in a television receiver to the horizontal scanning coils.

line-output valve transistor (*TV*). In a television receiver, the final amplifier in the line time-base generator.

line pad (*Telecomm.*). A resistance-attenuation network which is inserted between the programme amplifier and the transmission line to the broadcasting transmitter. Its purpose is to isolate electrically the amplifier output from the variations of impedance of the line.

line printer (*Print.*). One in which a line of figures is printed mechanically and simultaneously across a roll of paper, the figures being preset while the paper is fed forward one or more lines.

line profile (*Phys.*). A graph showing the fine structure of a spectral line, the intensity of the line, measured with a microphotometer, being plotted as a function of wavelength.

liner (*Eng.*). A separate sleeve placed within an engine cylinder to form a renewable and more durable rubbing surface; in I.C. engines, termed *dry* if in continuous contact with the cylinder wall, and *wet* if supported only at the ends and surrounded by cooling water.

line ranger (*Surv.*). An instrument for locating an intermediate point in line with two distant signals. It consists of two reflecting surfaces so arranged as to bring images of the two signals into coincidence when the instrument is in line with the signals.

line reflection (*Telecomm.*). The reflection of some signal energy at a discontinuity in a transmission line.

liner-off (*Ships*). A tradesman engaged in shipbuilding, whose function it is to 'mark off' by fair lines, using battens to enable plate workers and others to prepare material to fit *in situ.*

line scanning (*TV*). A method of scanning in which the scanning spot repeatedly traverses the field of the image in a series of straight lines.

line screen process (*Photog.*). A colour photographic process in which the screen takes the form of lines ruled on the emulsion.

line-sequential (*TV*). Said of colour TV system in which successive scanning lines generate images in each of the 3 primary colours.

line shafting (*Eng.*). Overhead shafting used in factories to transmit power from an engine or motor to individual machines.

line slip (*TV*). An apparent horizontal movement of part (or all) of a reproduced screen picture due to lack of synchronism between the line frequencies of the signal and the scanning system.

lines of curvature (*Maths.*). Curves on a surface that are tangential to the principal directions at every point of the surface.

line spectrum (*Light*). A spectrum consisting of relatively sharp lines, as distinct from a *band spectrum* (q.v.) or a *continuous spectrum* (q.v.). Line spectra originate in the atoms of incandescent gases or vapours. See **Bohr theory, spectrum.**

line-squall (*Meteor.*). A system of squalls occurring simultaneously along a line, sometimes hundreds of miles long, which advances across the country. It is characterized by an arch or line of low dark cloud and a sudden drop in temperature and rise in pressure. Thunderstorms and heavy rain or hail often accompany these phenomena.

line stabilization (*Telecomm.*). Dependence of an oscillator on a section of transmission line for stabilization of its frequency of oscillation; e.g., a quarter-wave line acts as a rejector circuit of very high Q, thus giving a highly critical change in phase at the resonant frequency.

line standard. A standard of length consisting of a metal bar near whose extremities are engraved fine lines, the standard length being the distance between these lines measured under specified conditions.

line stretcher (*Telecomm.*). A section of an electromagnetic waveguide whose physical length may be varied for adjustment of tuning.

line switch (*Teleph.*). A small uniselector which immediately hunts to seize a free A-digit selector when the subscriber lifts his receiver.

line synchronization (*TV*). Synchronization of the line-scanning generator at the receiver with that at the transmitter so that the scanning spots at the two ends keep in step throughout each line.

line-synchronizing pulses (*Telecomm.*). Negative pulses introduced into modulation signal applied to cathode-ray tube when an image is built up by scanning successive horizontal lines (e.g., TV, radar, etc.). These are filtered out and used to trigger the line time base, thus synchronizing the time base with the transmitted waveform. Since the pulses are at the 'blacker than black' level (tube beam current cut-off), they also eliminate flyback trace on screen.

line transect (*Bot.*). A chart showing the position and names of all the plants occurring on a line drawn across a piece of country.

line-up (*Telecomm.*). Adjustment of a number of circuits in series so that they function in the desired manner when required.

line voltage (*Elec. Eng.*). See **voltage between lines.**

line width (*Telecomm.*). The wavelength spread (or energy spread) of radiation which is normally characterized by a single value. The spread is defined by the separation between the points having half the maximum intensity of the line. In TV, the reciprocal of the number of lines per unit length in the direction of line progression.

lingering period (*Electronics*). The time interval during which an electron remains in its orbit of highest excitation before jumping to the energy level of a lower orbit.

lingo (*Weaving*). One of the metal weights fixed to the foot of the harness cords in a jacquard harness. *pl. lingoes.*

lingua (*Zool.*). Any tonguelike structure; in Insects, the hypopharynx; in *Acarina*, the floor of the mouth.

lingual (*Zool.*). In Arthropods, pertaining to the lingua; in Molluscs, pertaining to the radula; in Vertebrates, pertaining to the tongue.

lingual glands (*Histol.*). Serous and mucous secreting glands situated in the tongue.

lingula (*Zool.*). In some *Hemiptera*, a tongue-shaped organ on the dorsal surface of the last abdominal somite, on which honey-dew accumulates.

Lingula Flags (*Geol.*). Well-stratified rocks, consisting of alternating slaty and sandy layers (the latter known as *ringers*), occurring in the Upper Cambrian of N. Wales and containing large numbers of the lingula-like brachiopod, *Lingulella davisii.* In the Upper Lingula Flags, the Dolgelley Beds, trilobites also are comparatively common.

lingulate (*Bot.*). Tongue-shaped; proportionally shorter and wider than *ligulate*, and somewhat fleshy, with a bluntish apex.

linguodental (*Med.*). Of the tongue and teeth.

linguogingival (*Med.*). Of the tongue and the gums.

linin (*Cyt.*). The more solid, form-conserving part of the nucleus, which holds the chromioles in definite relation to one another.

lining (*Bind.*). (1) The operation of pasting a strip of strong paper down the back of a book after backing. (2) A strip of linen fixed down the middle of a section for strengthening purposes. (*Hyd. Eng.*) A layer of clay puddle covering the sides of a canal, making them watertight.

lining figures (*Typog.*). Said of figures which have the height of the capital letters of a fount, as distinct from *hanging figures*; e.g. 1, 6, 7. Also called **ranging figures.**

lining-papers (*Bind.*). Synonym for **end-papers.**

lining-up (*Eng.*). The operation of arranging the bearings of an engine crankshaft, etc., in perfect alignment.

link (*Eng.*). (1) Any connecting piece in a machine, pivoted at the ends. (2) The curved slotted member of a *link motion* (q.v.). (*Surv.*) The one-hundredth part of a chain. In the Gunter's chain, 1 link = 7·92 in.; in the Engineer's chain, 1 link = 1 ft. (*Telecomm.*) A communication path between two points. (*Teleph.*) A circuit or outlet between one rank of selectors and the next in order of operation, or between such selectors and a manual position. In U.S., a trunk.

linkage (*Chem.*). A chemical bond, particularly a covalent bond in an organic molecule. (*Elec. Eng.*) The product of the total number of lines of magnetic flux and the number of turns in the coil (or circuit) through which they pass. (*Gen.*) The tendency shown by certain genetical characteristics to be inherited together.

linkage group (*Gen.*). A group of hereditary characteristics which remain associated with one another through a number of generations.

linkage map (*Gen.*). A diagram showing the position of the genes in a chromosome or group of chromosomes.

link block (*Eng.*). A sliding block pivoted to the end of the valve rod, and working in the slotted link of a *link motion* (q.v.).

link-by-link signalling (*Teleph.*). A method of signalling in which signals are passed from end to end of a multi-link connection by repetition at intermediate switching points.

link coupling (*Telecomm.*). See **line coupling.**

linked numbering scheme (*Teleph.*). One in which a range of numbers (including the code numbers in a director automatic area) is distributed between the subscribers on several exchanges in a given area.

linked switches (*Elec. Eng.*). Switches mechanically linked, so that they operate together or in a definite sequence. Also called **coupled switches.**

linking (*Automation*). Process whereby articles being machined or manufactured in a transfer line are passed automatically between successive machines, with inspection.

link mechanism (*Eng.*). A system of rigid members joined together with constraints so that motion can be amplified or can be changed in direction.

link motion (*Eng.*). A valve motion, invented by Stephenson, for reversing and controlling the cut-off of a steam-engine. It consists of a pair of eccentrics, set for ahead and reverse rotation, connected to the ends of a slotted link carrying a block attached to the valve rod. Variation of the link position (known as *linking-up*) makes either eccentric effective, and also varies the cut-off.

link resonance (*Telecomm.*). That resulting in a repeated network when the sections are coupled together.

link rods (*I.C. Engs.*). See **articulated connecting rods.**

lin-log receiver (*Radio*). A radio receiver which gives a linear amplitude response for small signals but logarithmic for larger signals.

Linnaean species (*Bot.*). A wide conception of a species, in which many varieties are included.

Linnaean (or Linnean) system (*Biol.*). The system of classification and of *binomial nomenclature* (q.v.) established by the Swedish naturalist Linnaeus.

linocut (*Typog.*). A printing surface cut by hand in linoleum, giving a bold broad effect; not suitable for fine detail or long runs.

Linofilm (*Typog.*). A major system of *filmsetting* consisting of a keyboard which produces a perforated tape; a photographic unit controlled by the tape; a semi-automatic corrector; and a composer which enlarges or reduces line by line for display work.

Linofilm Quick (*Typog.*). A high-speed *filmsetting* machine which operates on 6-level justified tape, mixes 5 point to 18 point typefaces from two, three or four founts; manual change of *matrix grids*; output at least 20 newspaper lines per minute.

linoleic acid (*Chem.*). Unsaturated fatty acid, $CH_3{\cdot}(CH_2{\cdot}CH\ CH)_3{\cdot}(CH_2)_7{\cdot}COOH$. Occurs as glycerides in various vegetable oils, especially linseed oil to which, by forming solid oxidation products on exposure to air, it imparts the 'drying' quality responsible for its utility in paints, etc.

linoleum. A floor-covering material, now largely superseded by *vinyl floor coverings*, made by impregnating a foundation of hessian fabric or waterproofed felt with a mixture of oxidized linseed oil (linoxyn), resins (e.g., kauri gum) and fillers. Abbrev. **lino.**

Linomatic Operating Unit (*Typog.*). Two components, electrical control and mechanical decoder, 'read' 6-level justified tape and control the line-casting machine or Linofilm Quick.

Linomatic Perforator (*Typog.*). Electronic keyboard-operated equipment to produce 6-level justified tape for line-casting or Linofilm Quick machines; 4 counting magazines control up to 8 type faces, $4\frac{3}{4}$ point to 14 point, which can be mixed; character and space widths counted by an electronic unit which can also 'subtract'.

Linomix (*Typog.*). An attachment for *Linofilm* which simplifies the mixing of body sizes in one line, and changes of measure.

Linotron (*Typog.*). A very advanced *filmsetting* system which uses *character generation* by cathode-ray tube, leaves space for *graphics*, composes complete, multi-column, pages at a potential speed of 18 000 characters per minute using magnetic tape input.

Linotype composing machine (*Typog.*). See **composing machines.**

linoxyn (*Chem.*). An elastic substance obtained by the oxidation of linseed oil; used as the basis for making *linoleum* (q.v.).

linseed oil (*Chem.*). An oil obtained from the seeds of flax (*Linum usitatissimum*). It contains solid and liquid glycerides of oleic and other unsaturated acids. Its iodine value is 160-200, which puts it into the class of drying oils. It is easily oxidized and polymerized for forming elastic films. Used for the mixing of paints and varnishes and for the manufacture of linoleum.

linsey (*Textiles*). (1) Dress fabric which is a combination of linen warp and wool weft. (2) Rag-trade term for wool materials containing cotton. (3) Coarse linen fabric.

Linson (*Bind.*). Proprietary name for a range of strong papers used as an alternative to book cloth.

lint (*Med.*). A material made from cotton or linen, with a soft, teased surface; usually rendered antiseptic and used for dressings.

lintel (*Build.*). A beam across the top of an aperture. Also called the **head.**

linters or cotton linters (*Textiles*). Short stiff fibres remaining on cotton seeds after removal of the longer fibres; usually removed before the seeds are crushed (if subsequently, known as **hull fibre**). Linters are used extensively in the manufacture of rayon, guncotton, absorbent cotton, paper, etc.

lintol (*Build.*). Erroneous spelling for **lintel.**

Linville truss (*Eng.*). See **Whipple-Murphy truss.**

Linz-Donawitz process (*Met.*). Method of steel-making in which oxygen is blown into converter-shaped furnace through water-cooled lance.

lionism (*Med.*). A lion-like appearance, often characteristic of lepromatous *leprosy*.

lip (*Horol.*). The edge of the cylinder of the cylinder escapement which receives impulse from the escape wheel.

lipaemia (*Med.*). Excess of fat in the blood.

liparite (*Geol.*). A name suggested in 1860 by Roth to include all the fine-grained granitic rocks occurring as lava flows, such as those found in the Lipari Isles. The synonymous term *rhyolite* is more widely used.

lipase (*Bot.*, *Zool.*). A fat-digesting enzyme.

lip block (*Civ. Eng.*). A block of wood spiked to the end of a strut used in timbering a trench; it overhangs the end of the strut and rests upon the waling, so that the strut is supported and prevented from dropping if the sides of the trench give way.

lipectomy (*Surg.*). Surgical removal of fatty tissue.

lipids, lipoids (*Chem.*). Generic terms for oils, fats, waxes, and related products found in living tissues.

lip microphone (*Acous.*). One constructed in the form of a box shaped to fit the face around the mouth. This reduces extraneous noises as at sporting events, etc.

lipochondria (*Cyt.*). Cytoplasmic lipoid granules occurring in Golgi region.

lipochromes (*Chem.*). Pigments of butter fat.

lipodystrophia progressiva (*Med.*). A rare condition in which there is progressive loss of fat from the subcutaneous tissues of the upper half of the body.

lipogastry (*Zool.*). Temporary disappearance of the gastric cavity or of the paragaster.

lipogenous (*Zool.*). Fat-producing.

lipolytic (*Biochem.*). Having the property of splitting up or dissolving fat.

lipoma (*Med.*). A tumour composed of cells containing fat.

lipomatosis (*Med.*). A term applied to a number of conditions (e.g., Dercum's disease) in which there is an excessive accumulation of fat in the body or part of the body.

lipomerism (*Zool.*). In metameric animals, disappearance of metamerism by overgrowth or rearrangement of somites.

lipoplast (*Bot.*). A fatty globule.

lipoproteins (*Chem.*). Water-soluble proteins conjugated with lipids.

liposome (*Bot.*). A fatty or oily globule in cytoplasm.

lipostomy (*Zool.*). Temporary disappearance of the mouth, or of the osculum.

lipotropic (*Biochem.*, *Physiol.*). Having an effect on fat metabolism by accelerating fat removal or decreasing fat deposition, e.g., of liver fat.

lipoxenous (*Bot.*, *Zool.*). Of parasitic forms, leaving the host before development is completed.

Lippmann process (*Photog.*). Early method of

colour photography using a transparent silver halide emulsion backed with mercury. When viewed at an angle the reflected image appears in natural colours.

lip seal (*Eng.*). An oil-retaining shaft seal, often used adjacent to a bearing, in which an annular rubber sealing element having an aperture slightly smaller than the shaft diameter is deformed into the shape of a sealing lip, the sealing pressure being due partly to the deformation stress and partly to the force supplied by an additional *garter spring* (q.v.).

lip-synchronized shot (*Cinema., TV*). Film or television shot taken so close to a speaker that exact synchronization is essential.

lipuria (*Med.*). The presence of fat in the urine.

liquation (*Met.*). Separation of one constituent from an alloy or ore by partial melting, so that the fused fraction can be run off.

liquefaction of gases (*Phys.*). To liquefy a gas, it must be cooled below its critical temperature and, in some cases, compressed. For the so-called 'permanent' gases, oxygen, nitrogen, hydrogen, and helium, having very low critical temperatures, the problem of liquefaction becomes one of obtaining low temperatures. This is done mainly by allowing the compressed gas to expand through a nozzle, cooling occurring by the Joule-Thomson effect.

liquefaction temperature (*Phys.*). The temperature at which a gas changes state to liquid. Physically the same as the boiling point.

liquefied petroleum gases (*Heat.*). These fall within three main categories, viz., propane, butane, and pentanes. The two sources are natural gas wells and oil refinery separation. Abbrev. **LPG.**

liquid. A state of matter between a solid and a gas, in which the shape of a given mass depends on the containing vessel, the volume being independent. A liquid is practically incompressible.

liquid compass (*Ships, etc.*). A magnetic compass fitted in a bowl containing a suitable liquid. The weight of the card on the pivot is reduced and oscillations of the card are damped out.

liquid counter (*Nuc.*). A counter for measuring the radioactivity of a liquid, usually designed to measure β-, as well as γ-, rays

liquid crystal display (*Electronics*). An *alphanumeric* display based on changes in reflectivity of a liquid crystal cell when an electric field is applied.

liquid crystals (*Chem.*). Certain pure liquids which are turbid and, like crystals, anisotropic over a definite range of temperature above their freezing-points.

liquid dimmer (*Elec. Eng.*). A dimmer making use of liquid resistances and therefore giving a smooth variation in the illumination.

liquid-drop model (*Nuc.*). A model of the atomic nucleus, using the analogy of a liquid-drop, in which the various concepts of surface tension, heat of evaporation, etc., are employed.

liquid-flow counter (*Nuc.*). One for continuous monitoring of radioactivity in flowing liquids.

liquid helium II (*Chem.*). See under **helium.**

liquid limit (*Agric.*). Of a soil, the moisture content corresponding to a specific degree of consistency as indicated by a standardized apparatus.

liquid-liquid extraction (*Chem. Eng.*). Process, both batch and continuous, whereby two non-mixing liquids are brought together to transfer soluble substance from one to the other for useful recovery of these soluble substances.

liquid-metal reactor (*Nuc. Eng.*). (1) Normally used of a reactor designed for liquid-metal (sodium or potassium) cooling. (2) Occasionally used to indicate a liquid-metal fuelled reactor.

liquid oxygen (*Chem.*). See **oxygen.**

liquid paraffin (*Chem.*). A liquid form of *petroleum jelly* (q.v.), colourless and tasteless, used as a mild laxative. An alkane.

liquid phase sintering (*Powder Met.*). Sintering in which a small proportion of the material becomes liquid. It may or may not speed up the sintering process by solution transfer of the phases forming the matrix.

liquid-quenched fuse (*Elec. Eng.*). A fuse in which a liquid is used for quenching the arc. See **semi-immersed-, immersed-.**

liquid resistance (*Elec. Eng.*). A resistance consisting of a liquid of low conductivity, the current being led to and from the liquid by means of suitable electrodes.

liquid rheostat (*Elec. Eng.*). One in which a liquid column is used as the resistive element, the terminals being attached to suitable metal plates, one of which is usually movable, thus providing a continuous variation. Only used where current control need not be too precise.

liquid starter (*Elec. Eng.*). A liquid rheostat arranged to operate as a motor starter.

liquid-state electronics. The biological use of ions in solution to achieve similar results to man-made devices using electrons in solids.

liquidus (*Chem.*). A line in a constitutional diagram indicating the temperatures at which solidification of one phase or constituent begins or melting is completed. See **solidus and solidification range.**

liquor amnii (*Physiol.*). The clear fluid in the amniotic cavity, in which the embryo is suspended.

liquor folliculi (*Histol.*). The fluid lying within the membrana granulosa and surrounding the discus proligerus and its contained ovum, in a Graafian follicle in the ovary of higher Vertebrates.

lirella (*Bot.*). A long narrow apothecium with a ridge in the middle, found in some lichens.

lirelliform (*Bot.*). Like a furrow.

liroconite (*Min.*). A rare secondary ore of copper, sky-blue to green in colour; essentially, hydrated oxide of arsenic (III) combined with hydroxide of copper and aluminium.

L-iron (*Eng.*). A structural member of wrought-iron or rolled steel, having an L-shaped cross-section. Also called **angle iron.**

lisle (*Textiles*). Long-staple, hard-twisted Egyptian cotton yarn, gassed and mercerized to produce a lustrous effect; used especially for stockings.

Lissajous' curves, Lissajous' figures (*Maths., Phys.*). Plane curves formed by the composition of two sinusoidal waveforms in perpendicular directions. The parametric cartesian equations are:

$$x = a \sin (\omega t + \alpha),$$
$$y = b \sin (\Omega t + \beta).$$

The curves embrace a great variety of forms. If the frequencies are commensurable they consist, in general, of a plurality of loops determined by the ratio of the frequencies, and this property is extensively exploited to compare the frequencies of two sinusoidal voltages by applying them to the plates of a cathode-ray tube.

lissencephalous (*Zool.*). Having smooth cerebral hemispheres.

lissoflagellate (*Zool.*). Having the flagellum unprovided with a collar at the base.

list (*Comp.*). Information printed out by a compiler or assembler comprising a list of the actual machine instructions together with data on storage allocations, etc. (*Ships*) An angle of transverse inclination arising from unsymmetrical distribution of internal weight. Cf. *heel*.

listed (*Textiles*). Having defects near the lists or selvedge in a length of cloth.

listel (*Arch.*). See **facette.**

listening key (*Teleph.*). The lever key which the operator throws, to put her head-set on to a cord circuit and speak to a subscriber.

listerellosis (*Vet.*). See **listeriosis.**

listeriosis (*Vet.*). Listerellosis. A bacterial infection of animals and birds by *Listeria monocytogenes* (*Listerella monocytogenes*). The infection occurs in cattle and sheep (**circling disease**) and more rarely in pigs, and causes a febrile meningo-encephalitis. In poultry, and sometimes in young animals, the disease occurs as a septicaemia.

listing (*Carp.*). (1) A narrow edge of a board. (2) The operation of removing the sappy edge of a board.

lists or listing (*Textiles*). See **listed, selvedge.**

literals (*Typog.*). Casual errors of composition, such as one character substituted for another, worn letters, turned letters, etc.

litharge (*Chem., Paint.*). Lead (II) oxide, used in paint-mixing as a drier; used also in the rubber and electrical accumulator industries.

litharge cement (*Chem.*). See **glycerine litharge cement.**

lithia (*Chem.*). Lithium oxide, Li_2O.

lithia emerald (*Min.*). Synonym for *hiddenite*.

lithia mica (*Min.*). See **lepidolite.**

lithiasis (*Med.*). The formation of calculi in the body. The condition in which an excess of uric acid and urates is excreted in the urine—the gouty diathesis.

lithiophilite (*Min.*). Orthorhombic phosphate of lithium and manganese, forming with triphylite a continuously variable series.

lithite (*Zool.*). See **statolith.**

lithium (*Chem.*). An element, symbol Li, r.a.m. 6·939, at. no. 3, m.p. 186°C, b.p. 1360°C, rel. d. 0·585. It is the least dense solid, chemically resembling sodium but less active. It is used in alloys and in the production of tritium; also as a basis for lubricant grease with high resistance to moisture and extremes of temperature; and as an ingredient of high-energy fuels.

lithium aluminium hydride (*Chem.*). Lithium tetrahydroaluminate $LiAlH_4$. Important reducing agent widely used in organic chemistry because of its effectiveness in 'difficult' reductions, e.g., the reduction of carboxylic acid groups (COOH) to primary alcohol groups (CH_2OH).

lithium carbonate (*Med.*). Salts of lithium, used as a tranquillizer, and prophylactically and therapeutically in mental illness.

lithium-drift germanium detector (*Nuc. Eng.*). A semiconductor detector with relatively large sensitive volume which can be used for high resolution gamma-ray spectrometry.

lithium hydride (*Chem.*). LiH. Formed when lithium unites with hydrogen at a red heat, it is a strong reducing agent, used as a hydrogen carrier, e.g., for small balloons at sea.

lithium hydroxide (*Chem.*). LiOH. A strong base, but not so hygroscopic as sodium hydroxide. Used in the form of a coarse, free flowing powder as a carbon dioxide absorber, e.g., in submarines. Also to make lithium stearates, etc., in driers and lubricants.

lithium 12-hydroxy stearate (*Chem.*). A lithium 'soap' widely used in high-performance greases as the main thickening agent. Helps to confer high water-resistance and good low-temperature performance.

litho (*Print.*). A common abbreviation for *lithography*.

lithocyst (*Zool.*). See **statocyst.**

lithodomous (*Zool.*). Living in rocks.

lithogenous (*Zool.*). Rock-building, as certain Corals.

lithographic paper (*Paper*). High machine-finished or supercalendered paper, made so that any stretch occurs the narrow way of the sheet.

lithographic stone (*Geol.*). A compact, porous, fine-grained limestone, often dolomitic, employed in lithography. Pale creamy-yellow in colour, but occasionally grey. Fair samples may be obtained from the Jurassic rocks in Britain, but the finest material comes from Solenhofen and Pappenheim in Bavaria. See also **Solenhofen stone.**

lithographic varnish, lithographic oil or **litho oil** (*Print.*). *Heat-bodied* (q.v.) oil used as binder for lithographic inks.

lithography (*Print.*). Originally the art of printing from stone (see **lithographic stone**), but now applied to printing processes depending on the mutual repulsion of water and greasy ink. Damp rollers pass over the surface, followed by inking rollers. The design, which is greasy, repels the water but retains the ink (also greasy), which is transferred to the paper. In modern practice, a sheet of zinc or aluminium, grained to retain moisture, is commonly used in place of stone, the plate being attached to a cylinder for rotary printing, either sheet-fed or web-fed. Most lithography is offset, the impression from the plate being transferred to a rubber-covered blanket cylinder and thence to the stock which may be paper or board or other flexible sheets including metal. See also **chromolithography, photolithography, offset printing, small offset.**

litholapaxy (*Surg.*). The operation of crushing a stone in the bladder, followed by the washing out of the crushed fragments.

lithology (*Geol.*). The character of a rock expressed in terms of its mineral composition, its structure, the grain-size and arrangement of its component parts; that is, all those visible characters that in the aggregate impart individuality to the rock.

lithophagous (*Zool.*). Stone-eating; said of graminivorous Birds which take small stones into the gizzard, to aid mastication; also of certain Molluscs which tunnel in rock.

lithophile (*Geol.*). Synonymous with *oxyphile*.

lithophilous (*Ecol.*). Using stones as a shelter; said especially of aquatic animals.

lithophyte (*Bot.*). A plant growing on rocks or stones.

lithopone (*Paint.*). A white pigment with fair hiding power, prepared by the coprecipitation of zinc sulphide and barium sulphate, mostly in the ratio of approximately 1 : 2. It is mainly used in distempers and interior paints. Also called **Charlton white, Griffith's white.**

lithosphere (*Geol.*). The outermost, rigid layer of the Earth. Used by some authors to denote the Earth's crust, the term may also be used to describe that thickness of the crust and upper mantle which behaves in a coherent and relatively rigid manner during tectonic movements. See **Earth.**

lithotomous (*Zool.*). Stone-boring, as certain Molluscs.

lithotomy (*Surg.*). Cutting into the bladder or ureter for the removal of a stone or calculus.

lithotrite (*Surg.*). An instrument, with special

blades, adapted for crushing stones in the bladder.

lithotrity (*Surg.*). The operation of crushing stones in the bladder. Cf. **litholapaxy.**

lith process (*Photog.*). Use of a special high-contrast emulsion developed in formaldehyde-hydroquinone to give coarse-grained black and white prints without half-tones, suitable for process-engraving or photolithography.

lithuria (*Med.*). An excess of uric acid and of urates in the urine.

litmus (*Chem.*). A material of organic origin used as an indicator; its colour changes to red for acids, and to blue for alkalis at pH 7. See also **indicator (1).**

lit-par-lit injection (*Geol.*). The injection of fluid or molten material along bedding, cleavage or schistosity planes in a rock. The product is an alternation of apparently igneous and non-igneous material, known as a *migmatite*. Some of the rocks previously considered to have been formed by the process are now thought to be the products of partial melting, the igneous layers representing the fraction of the rock which became liquid.

litre. Unit of volume, equal to 1 cubic decimetre.

litre-atmosphere (*Chem.*). A former unit of work being the pressure energy of 1 litre at 1 atmosphere.

litter (*Bot.*). More or less undecomposed plant residues on the surface of the soil in a wood.

little-end (*I.C. Engs.*). That part of the connecting rod which is attached to the gudgeon pin.

Little's disease (*Med.*). Spastic paralysis of the lower half of the body, due to congenital failure of development of nerve cells in the brain.

littoral conditions (*Geol.*). Conditions which obtain in the littoral zone (i.e., the shore zone); they involve strong current and wave action, causing intense marine abrasion, and the development of much sand and well-rounded pebbles. See also **littoral deposits.**

littoral deposits (*Geol.*). Deposits of the shore zone lying between high- and low-water mark. According to local conditions, these consist of shingle grading into sand, or one of these only. In special areas, the chief or only deposit may be coral sand. Littoral deposits are recognized by their coarse grain, by ripple marking, and by the fauna they contain, which includes in British seas such well-known forms as whelks (*Littorina littorea*), mussels (*Mytilus edulis*), etc.

littoral zone (*Bot.*). The part of the seashore inhabited by plants, below average low-water level. (*Zool.*) Faunal zone bounded by the continental shelf, i.e., down to approximately 200 m.

Littré's glands (*Zool.*). In male Mammals, small mucous glands occurring in the mucous membrane of the urethra.

Littrow mounting (*Light*). The mounting of a plane mirror so that it reflects back the emerging light from a prism spectrometer and the light traverses the prism a second time. By rotating the mirror, the spectrum may be scanned.

Littrow spectrograph (*Light*). A convenient prism spectrograph in which the same lens serves both to collimate the light and to focus the spectrum on the photographic plate, the light being reflected back through the prism and lens by a plane mirror behind the prism. The instrument has several obvious advantages over the type with separate collimator and camera lens.

lituate (*Bot.*). Forked, with outward-turning points.

litz(endraht) wire (*Elec. Eng.*). Multiple-stranded wire, each strand being separately insulated so as to reduce the relative weighting of the *skin effect*, i.e., concentration of high-frequency currents in the surface. Much used in compact low-loss coils in filters and in high-frequency tuning circuits.

live (*Acous.*). Said of an enclosure, or motion-picture set, which is not rendered dead by the presence of sound-absorbing areas, and in which the reverberation is normal or above normal. (*Elec.*) Connected to a voltage source. (*TV*) Direct transmission of sound or television without recording. A *live insert* is that part of an otherwise recorded transmission which is live.

live axle (*Eng.*). A revolving axle to which the road-wheels are rigidly attached, as distinct from a fixed or *dead axle* (q.v.).

live centre (*Eng.*). In machine tools, a centre which rotates with the workpiece.

live load (*Eng.*). A moving load or a variable force on a structure; e.g., that imposed by traffic movement over a bridge, as distinct from a dead weight or load, such as that due to the weight of the bridge.

liver (*Zool.*). In Invertebrates, the digestive gland or *hepatopancreas* (q.v.); in Vertebrates, a large mass of glandular tissue arising as a diverticulum of the gut, which secretes the bile and plays an important part in excretion and other aspects of the general metabolism of the body.

live rail (*Elec. Eng.*). The supply rail of a 3- or 4-rail electric traction system, usually a few hundred volts d.c. above or below earth potential.

liver flukes (*Med., Zool.*). A group of trematode parasites, especially *Fasciola hepatica* (q.v.) and *Clonorchis sinensis*, infecting man via various species of water snail as intermediate hosts, causing damage to the liver and surrounding organs.

live ring (*Eng.*). A large roller-bearing, used for supporting turn-tables and revolving cranes.

livering (*Paint.*). The condition when a paint in bulk becomes jellylike or tough.

live room (*Acous.*). One which has a longer period of reverberation than the optimum for the conditions of performance and listening, with consequent blurring of speech and musical sounds, reduced by introducing additional acoustic damping.

liver opal (*Min.*). A form of opaline silica, in colour resembling liver.

liver rot (*Vet.*). See **distomiasis.**

live steam (*Eng.*). Steam supplied direct from a boiler, as distinct from exhaust steam or steam which has been partly expanded.

lixiviation (*Met.*). See **leaching.**

lizardite (*Min.*). A mineral of the *serpentine* (q.v.) group.

Ljungstrom regenerative air heater (*Eng.*). A heat exchanger largely used in power stations. It consists essentially of a large, vertical, slowly rotating drum filled with a honeycomb arrangement of very thin steel sheet, the flue gases passing upwards through one half of the honeycomb while the air passes downwards through the other.

Llandeilian Series (*Geol.*). A division of the Ordovician System, lying between the Llanvirn Series below and the Caradocian Series above. As described by Murchison from the type locality (Llandeilo, Wales), the series comprises flags and limestones; but, traced westwards into Mid-Wales, these pass into shales containing graptolites. Their equivalents occur also elsewhere in Britain — in Shropshire, S. Scotland, and the Lake District.

Llandovery (or Llandoverian) Series (*Geol.*). The

lowest series in the Silurian System, divisible at the type locality (Llandovery, Wales) into three groups of beds—Lower, Middle, and Upper, defined by the graptolite faunas they contain.

Llanvirn Series (*Geol.*). A division of the Ordovician System, the name being taken from a S. Welsh locality. The series consists of blue-black shales containing a graptolite fauna, the characteristic forms being the 'tuning-fork' graptolites, *Didymograptus bifidus* and *D. murchisoni*.

L-line (*Nuc.*). Characteristic X-ray spectrum line produced when vacancy in L-shell is filled by electron. See also **K-lines, M-line.**

Lloyd's mirror (*Light*). A device for producing interference fringes. A slit, illuminated by monochromatic light, is placed parallel to and just in front of the plane of a plane mirror or piece of unsilvered glass. Interference occurs between direct light from the slit and that reflected from the mirror. See **interference fringes.**

Lloyd's rules (*Ships*). A set of rules laid down by the Classification Society, Lloyd's Register of Shipping, governing the construction of steel ships and their machinery.

lm (*Light*). Abbrev. for **lumen.**

L/M ratio (*Nuc.*). Ratio of the number of internal conversion electrons from the L-shell to the number from the M-shell which are emitted in a particular decay process.

LMS (*Telecomm.*). *Level measuring set*.

L-network, -attenuator or **-filter** (*Telecomm.*). Half an unbalanced T-network. See **L-section.**

load (*Elec. Eng.*). (1) Output of an electric machine or transformer, or a group of such apparatus, e.g., a generating station. See **burden.** (2) Material placed between electrodes for the induction of heat through dielectric loss by means of high-frequency electric fields. (*Mech.*) (1) The weight supported by a structure. (2) Mechanical force applied to a body. (*Telecomm.*) (1) Termination of an amplifier or line which absorbs the transmitted power, which is a maximum when this load *matches* the output impedance. (2) The actual power received by a terminating impedance.

load capacitor (*Elec. Eng.*). That which tunes and maximizes the power to a load in induction or dielectric heating.

load cell (*Eng.*). A load detecting and measuring element utilizing electrical or hydraulic effects which are remotely indicated or recorded.

load characteristic (*Elec. Eng.*). *Instantaneous* voltage-current characteristic at the output of a generator or amplifier under loading conditions. Also called **operating characteristic.**

load classification number (*Aero.*). A number defining the load-carrying capacity of the paved areas of an aerodrome. Abbrev. **LCN.**

load coil (*Elec. Eng.*). The coil in an induction heater used to carry the alternating current which induces the heating current in the specimen or charge.

load curve (*Elec. Eng.*). A curve whose ordinates represent the load on a system or piece of apparatus, and whose abscissae represent time of day, month, or year, so that the curve indicates the value of the load at any time. (*Eng., etc.*) See **influence line.**

load despatcher (*Elec. Eng.*). An engineer who is responsible for the distribution of load over a large interconnected power system.

load displacement (*Ships*). A ship's displacement at load draught, i.e., the draught to the centre of the freeboard disk marking, which is set off to the summer freeboard.

load draught (*Ships*). The draught when loaded to the minimum *freeboard* (q.v.) permitted for the place and season.

loaded antenna (*Radio*). An antenna in which series inductance has been added to increase its natural wavelength.

loaded circuit (*Telecomm.*). A circuit which includes loaded cable and therefore has a reduced and uniform attenuation up to a cut-off frequency, but a lower velocity of transmission.

loaded concrete (*Nuc. Eng.*). Concrete used for shielding nuclear reactors, loaded with elements of high atomic number, e.g., lead or iron shot.

loaded impedance (*Telecomm.*). That of input transducer when output load is connected.

loaded push-pull amplifier (*Telecomm.*). Push-pull stage of valve amplification in which the amplitude distortion arising from grid current is minimized by shunting the grids with resistances low in comparison with the effective nonlinear grid resistance. Also **low-loading amplifier.**

load efficiency (*Telecomm.*). Ratio of the useful power delivered by the output stage to a specified load, and the d.c. input power to the stage.

loader (*Mining*). A mechanical shovel or other device for loading trucks underground.

load-extension curve (*Met.*). A curve, plotted from the results obtained in a tensile test, showing the relations between the applied load and the extension produced.

load factor (*Aero.*). (1) In relation to the structure, it is the ratio of an external load to the weight of an aircraft. Loads may be centrifugal and aerodynamic due to manoeuvring, to gravity, to ground or water reaction; usually expressed as g, e.g., 7g is a load seven times the weight of the aircraft. (2) In aircraft operations, the actual *payload* on a particular flight as a percentage of the maximum permissible payload. (*Elec. Eng.*) The ratio of the average load to peak load over a period. See **plant load factor.** (*Work Study*) The proportion of the overall cycle time required by the worker to carry out the necessary work at standard performance during a machine- (or process-) controlled cycle.

load impedance (*Telecomm.*). Impedance of the device which accepts power from a source, e.g., amplifiers, loudspeakers, magnetizing coils, etc.

loading (*Build., etc.*). See **on-costs.** (*Min. Proc.*) Adsorption by resins of dissolved ionized substances in ion-exchange process. In uranium technology, the loading factor is the mass of U_3O_8 adsorbed by unit volume of the resin. (*Nuc.*) The introduction of fuel into a reactor. (*Paper*) A material, such as clay, added to the pulp in order to produce a smooth surface and solidity. (*Elec. Eng.*) An inductance added to a line for the purpose of improving its transmission characteristics throughout a specified frequency-band. See **coil-, continuous-, intermittent-.** (*Textiles*) Increasing the weight of fabrics by use of starch, size, China clay, or other substances.

loading and cg diagram (*Aero.*). A diagram, usually comprising a side elevation of the aircraft concerned, with a scale and the location of all items of removable equipment, payload and fuel, which is used to adjust the weight of the aircraft so that its resultant lies within the forward and aft *cg limits*.

loading and unloading machine (*Nuc. Eng.*). Structure for introducing and withdrawing fuel elements from a reactor, with safety provision for personnel.

loading capacity (*Min. Proc.*). In ion exchange, saturation limit of resin.
loading coil (*Elec. Eng.*). The coil inserted, in series with a line's conductors, at regular intervals. Also called **Pupin coil.**
loading gauge (*Rail.*). (1) The limiting dimensions governing height, width, etc., of rolling stock to ensure that adequate clearance is obtained for passage under bridges and through tunnels. (2) A shaped bar suspended over a railway track, at the correct height and position, to check compliance of trucks passing underneath with the above limiting dimensions.
load leads (*Elec. Eng.*). The connexions or transmission lines between the power source of an induction (or dielectric) heater and the load coil or applicator.
load-levelling relay (*Elec. Eng.*). A relay used in connexion with certain, temporarily dispensable, forms of apparatus such as storage water-heaters. It automatically switches them off when the demand on the system exceeds a certain value.
load line (*Electronics*). On a set of output characteristic curves for an amplifying device, a line representing the load, straight if entirely resistive, elliptical if reactive.
load lines (*Ships*). A group of lines marked on the outside of both sides of a ship to mark the minimum *freeboard* (q.v.) permitted in different parts of the world and seasons. Sometimes called **plimsoll mark.**
load matching (*Telecomm.*). Adjusting circuit conditions to meet requirements for maximum energy transfer to load.
load-rate prepayment meter (*Elec. Eng.*). A form of prepayment meter in which the charge per unit is changed whenever the load exceeds a certain predetermined value.
loadstone (*Min.*). See **lodestone.**
loam (*Build.*). Earth composed of clay and sand. Also called **mild clay, sandy clay.** (*Foundry*) A clayey sand milled with water to a thin plastic paste, from which moulds are built up on a backing of soft brick; generally swept or strickled to shape without the use of a pattern.
loam board (*Foundry*). See **strickle board.**
loam bricks (*Foundry*). Cakes of loam, or soft building-bricks built up with loam, which form a solid but porous support for the loam forming the wall of the mould. See **loam.**
loamy clay (*Geol.*). An earthy mixture of clay, silt, and sand, with more or less organic matter.
loan (*Paper*). *Animal-sized* rag paper made for bank-notes, bonds, etc.
lobar pneumonia (*Med.*). Inflammation of one or more lobes of the lung, the affected lobes becoming solid; due usually to infection with the pneumococcus. See also **pneumonia.**
Lobata (*Zool.*). A sub-order of the order *Tentaculata*, the members of which are laterally compressed, and possess two lateral lobes in the oral region; the tentacles are nonretractile.
lobe (*Anat.*). (1) A natural division, formed by a fissure, of an organ, e.g., the brain, liver, lungs. See **olfactory lobes, optic lobes.** (2) The soft lower part of the external ear. (*Bot.*) One of the parts into which a flattened plant member is cut when the parts are too large and distinct to be called teeth but not wholly separated from one another. (*I.C. Engs.*) A rounded projection or cam. The term is usually applied to the projections on an ignition contact-breaker, and to the several cams formed on one ring, used in radial aero engines. (*Radio*) Enhanced response of an antenna in the horizontal or vertical plane, as indicated by a lobe or loop in its radiation pattern. The *beam* effect arises from a *major lobe*, generally intended to be along the forward axis. (*Zool.*) A rounded or flap-like projection. *adjs.* **lobate lobed, lobose, lobulate.**
lobeline (*Chem.*). $C_{22}H_{27}O_2N$, a piperidine alkaloid obtained from *Lobelia inflata*. Forms broad needles, m.p. 130°C–131°C. It is monoacidic, and is remarkable for yielding acetophenone when heated with water. Used as a smoking deterrent.
lobe switching (*Radar*). Scanning downcoming waves and switching into circuit networks which accept a chosen direction for optimum signal : noise ratio, as in *musa*. Also **beam switching.**
lobing (*Bot.*). Division of the blade of a leaf, or of a flat thallus, when the separation does not extend much more than half way in, but is deeper than is necessary to cut out teeth.
lobopodia (*Zool.*). Thick, blunt pseudopodia of some *Sarcodina*.
Lobosa (*Zool.*). See **Amoebina.**
lobose (*Zool.*). Said of pseudopodia which are short and blunt and contain both ectoplasm and endoplasm. Cf. *filose*.
lobotomy (*Med.*). See **leucotomy.**
lobule or lobulus (*Zool.*). A small lobe; one of the polyhedral cell masses forming the liver in Vertebrates. *adjs.* **lobular, lobulate.**
local action (*Elec. Eng.*). Deterioration of battery due to currents flowing to and from the same electrode.
local attraction (*Mag.*). See **magnetic anomaly.**
local battery (*Teleph.*). A telephone system in which each subscriber's telephone station is provided with its own battery, wet or dry. Abbrev. **L.B.** Cf. *central battery*.
local carrier (*Radio*). Demodulation with an adequate carrier wave inserted before demodulation.
local exchange (*Teleph.*). The exchange to which a given subscriber has a direct line. Sometimes called **local central office.**
local group of galaxies (*Astron.*). A cluster of galaxies within 10 parsecs, and numbering at least 20; the group includes our own Galaxy, the Andromeda nebulae and the Magellanic clouds.
localization (*Acous.*). See **stereophony.** (*An. Behav.*) An ability of the sense organs of an animal to determine the source of a stimulus in space. Many sense organs can detect its direction, but probably only the eyes and ears or lateral line system, can judge its distance. See also **echolocation.**
localization of respiration (*Zool.*). The specialization of certain parts or areas of the body (as lungs, trachea, etc.) to perform the function of respiration, as opposed to the utilization of the whole surface of the body for this purpose.
localized vector (*Maths.*). See **bound vector.**
localizer beacon (*Aero.*). A directional radio beacon associated with the *ILS*, which provides an aircraft during approach and landing with an indication of its lateral position relative to the runway in use.
local junction circuit (*Teleph.*). A junction between exchanges in the local call area, i.e., where connections are made by local codes as opposed to *S.T.D.*
local Mach number (*Aero.*). The ratio of the velocity of the airflow over a part of a body in flight to the local speed of sound. Usually it is concerned with a part of greater curvature, where the airflow accelerates momentarily, thereby increasing the Mach number above that of the body as a whole, e.g., over the wing.

local oscillations (*Radio*). Those generated within the apparatus which uses them.

local oscillator (*Radio*). That which supplies the frequency for beating in a supersonic heterodyne receiver. It may use electrodes in the first detector or frequency changer, or a separate circuit. Also called **beating oscillator.**

local time (*Astron.*). Applied to any of the 3 systems of time reckoning, sidereal, mean solar, or apparent solar time, it signifies the hour angle of the point of reference in question measured from the local meridian of the observer. The local times of a given instant at 2 places differ by the amount of their difference in longitude expressed in time, the local time at a place east of another being the greater.

local vent (*Plumb.*). A connexion enabling foul air in a room or plumbing fixture to escape to the outer air.

location (*Civ. Eng.*). (1) The exact position of an engineering project, as decided upon in the light of technical and other considerations. (2) The process of determining the above position in the field. (*Comp.*) (1) A position in a store which holds a word or part of a word. (*Eng.*) A geometrical feature, e.g., a projection or a recess, by which one machine component or article may be correctly positioned by engagement in or with a corresponding feature in another, or the correct spatial relationship of machine components or articles.

locator (*Acous.*). See **sound locator.**

locator beacon (*Aero.*). The 'homing' beacon on an airfield used by the pilot until he picks up the localizer signals of the *ILS*.

lochia (*Physiol.*). The normal discharge from the vagina during the first week or two after childbirth.

lock (*Horol.*). The stopping of the escape wheel. Also **locking.** (*Hyd. Eng.*) A communicating channel, having gates at each end, between the higher and lower reaches of a canal. It is used to transfer a vessel from one reach to the other. (*Print.*) A safety device in most control boxes to prevent the press being moved under power; the press can be made free only by cancelling the lock; may also be used during the run to prevent increase of speed. (*Telecomm.*) A relay *locks* when, on operation, it trips a ratchet device which holds it in operation after the operating current ceases; or if, on operation, it *makes* a circuit which maintains in a winding of the relay a current sufficient to keep it operated, after the original operating current has ceased. When the *locking circuit* is opened the relay *unlocks, releases, falls-off,* or *deoperates.*

lockage (*Hyd. Eng.*). Water lost, i.e., transferred from a higher to a lower level, in the operation of passing a vessel through a lock.

lock-and-key theory (*Chem.*). A theory devised by Emil Fischer to explain the specific action of enzymes; based on the assumption that enzyme and substrate must possess a similar geometrical structure in order that a chemical reaction can occur. (*Zool.*) A theory which postulates that the sharply marked differences between the genitalia of different species of Insects act, by preventing interbreeding, as a mechanical means of isolating the species.

lock-bay (*Hyd. Eng.*). The water space enclosed in a *lock-chamber* (q.v.).

lock-chamber (*Hyd. Eng.*). The space between the head-gates and tail-gates of a lock.

locked (*Elec. Eng.*). Said of an oscillator when it is held to a specific frequency by an external source.

locked-coil conductor (*Elec. Eng.*). A form of stranded conductor in which the outer wires are so shaped that they are prevented from having any radial movement.

locked-cover switch (*Elec. Eng.*). A switch which can be operated only after unlocking the cover. Also called **asylum switch, locking switch, secret switch.**

locked (or **eccentric**) **groove** (*Acous.*). Finishing groove on the surface of a gramophone record, motion of the needle in this groove operating *stopping* or *record-changing* mechanisms.

locked oscillator detector (*Telecomm.*). Form of FM detector using a pentode valve which incorporates a self-oscillating tuned circuit connected to the suppressor grid. The FM input signal is applied to a resonant tuned circuit which is connected to the control grid. The tuned circuit of the suppressor grid locks to the frequency of the incoming signal and leads to an AF anode current proportional to the FM input.

locked test (*Min. Proc., etc.*). In preliminary tests on unknown ores, retention of a potentially troublesome fraction of the test product for addition to a new batch of ore, to ascertain whether in continuous work such a build-up would be upsetting. Also called **cyclic test.**

locker jack (*Textiles*). See **lever jack.**

lock-gate (*Hyd. Eng.*). A pair of doors at one end of a lock, serving, in conjunction with a similar pair at the other end, to enclose water within the lock-chamber.

lock-in (*Telecomm.*). Generally, to synchronize one oscillator with another, as in a homodyne or frequency doubler. One oscillator must be free-running and capable of being pulled.

lock-in amplifier (*Telecomm.*). Synchronous amplifier, sensitive to variation of signal of its own frequency.

locking (*Horol.*). See **lock.** (*Telecomm.*) (1) See **latching.** (2) The control of frequency of an oscillating circuit by means of an applied signal of constant frequency.

locking angle or angle of lock (*Horol.*). The angle, measured from the pallet centre, through which the pallets have to move before unlocking can take place.

locking face (*Horol.*). The portion of the pallet upon which the teeth of the escape wheel drop for locking.

locking key (*Teleph.*). A hand-operated telephone key, which, when operated, remains in its operated position until released by the hand.

locking plate (*Horol.*). A circular plate around the periphery of which notches are cut, the distance between the notches regulating the number of hours struck.

locking relay (*Teleph.*). A telephone relay which, when operated, remains in its operated condition when the operating current ceases, either by closing a winding which carries a sustaining current or, more rarely, by mechanical means.

locking switch (*Elec. Eng.*). See **locked-cover switch.**

lockjaw (*Med.*). See **tetanus.**

lock-nut (*Eng., etc.*). (1) An auxiliary nut used in conjunction with another, to prevent it from loosening under vibration. (2) Any special type of nut designed to prevent accidental loosening.

lockover (*Telecomm.*). Circuit which is *bistable.*

lock-paddle (*Hyd. Eng.*). A sluice through which water is passed to fill an empty lock-chamber.

Lockport Limestone and Dolomite (*Geol.*). Marine, highly fossiliferous strata which form the top of the scarp and the lip of the Niagara Falls. Sometimes called **Niagara Limestone.**

lock rail (*Join.*). (1) The door-rail which is level with the lock. (2) The front member of a piano, running along the keyboard, which usually contains the lock.
lockrand (*Build.*). A course of stones laid as bondstones.
lock-saw (*Carp.*). See **compass saw.**
lock seam (*Eng.*). A type of joint, not absolutely tight, produced in the manufacture of sheet metal drums, cans, etc., by simply folding, interlocking and pressing together the longitudinal edge areas of the product.
lock-sill (*Hyd. Eng.*). See **mitre-sill.**
lock stile (*Join.*). That stile of a door in or on which the lock is fastened.
locks, up and/or down (*Aero.*) See **up and/or down locks.**
lock up (*Print.*). To secure type matter firmly in the *chase*, ready for printing.
lock-woven mesh (*Civ. Eng.*). A mechanically woven fabric, made of steel wires crossing at right-angles and secured at the intersections, used in reinforced concrete construction.
locomotive (*Rail.*). A vehicle driven by coal, oil, or electricity, for hauling trucks or carriages on a railway. See **diesel-, diesel-electric-, diesel-hydraulic-, electric-, electro-diesel, steam-.**
locomotive boiler (*Eng.*). The type of boiler used on steam locomotives; it consists of an internal fire-box at one end of the horizontal cylindrical shell, from which the hot gases are led through fire-tubes passing through the water space into the smoke-box at the front of the boiler. See **fire-tube boiler, fire-box.**
locomotor ataxia (or **ataxy**) (*Med.*). See **tabes dorsalis.**
locular, loculatous (*Bot.*). Divided into compartments by septa.
loculicidal (*Bot.*). Said of a fruit which splits open along the midribs of the carpels.
loculus (*Bot.*). (1) One portion of a septate spore. (2) One compartment in a synangium. (3) One compartment in an anther, or in an ovary. (*Zool.*) A small space or cavity; as in *Foraminifera*, a shell chamber.
locus (*Bot.*). The same as the *hilum* (q.v.) of a starch grain. (*Cyt.*) The position of a gene in a chromosome.
locust (*Zool.*). One of several kinds of winged insects of the family *Acridiidae*, akin to grasshoppers, highly destructive to vegetation.
lodar (*Nav.*). Loran RDF system in which responses to ground and sky waves are registered separately, so avoiding night effect.
lode (*Civ. Eng.*). An artificial dyke. (*Mining*) Steeply inclined fissure of nonalluvial mineral enclosed by walls of country rock of different origin. Vein. Stockwork when consisting of several parallel and close veins. Of magmatic or hydatogenetic origin.
lodestone, loadstone (*Min.*). Iron (II, III) oxide. A form of magnetite exhibiting polarity, behaving, when freely suspended, as a magnet. Occurs extensively at Magnet Heights in Sekukuniland (Transvaal) and elsewhere. See **armed lodestone.**
Lodge-Cottrell detarrer (*Chem.*). Plant consisting of a series of pipes with high-tension electrodes for removal of tar particles from gases.
Lodge-Cottrell precipitator (*Chem.*). Plant for carrying out *electrical precipitation* of dust from gases.
Lodge-Muirhead coherer (*Radio*). A coherer consisting of a slowly rotating steel wheel whose periphery dips into a pool of mercury. The arrival of a signal causes a sudden drop in the resistance of the contact.
Lodge valve (*Electronics*). An early form of cold-cathode rectifying valve. See **ionic valve.**
lodicule (*Bot.*). One of two (or rarely three) small scales present below the stamens in the flower of a grass. They become distended with water and assist in the separation of the glumes.
loess (*Geol.*). An aeolian clay, consisting of fine rock-flour (mainly quartz), originating in arid regions and transported by wind. Vast accumulations of loess cover large areas in China, eastern Europe, N. and S. America.
Löffler boiler (*Eng.*). A high-pressure boiler employing forced circulation, by pumping steam through small-diameter tubes. Part of the high-temperature steam is returned to the water drum, to produce the saturated steam supply to the pump.
Lofton-Merritt stain (*Micros., Paper*). A microscopical stain consisting of malachite green and basic fuchsin for distinguishing between unbleached sulphite and sulphate wood pulp fibres.
lofty (*Textiles*). Used to describe loose wool or wool cloths that have bulky but resilient properties.
log (*Geol.*). A record of the rocks encountered by a well or borehole. (*Maths.*) Abbrev. of *logarithm* (q.v.). (*Ships*) See **nautical log.**
logagraphia (*Med.*). Loss of the ability to express ideas in writing.
logarithm (*Maths.*). The logarithm of a number N to a given base b is the power to which the base must be raised to produce the number. It is written as $\log_b N$, or as $\log N$ if the base is implied by the context. There are two systems of logarithms in common use:

(a) *common* or *Briggs's* logarithms, with base 10. If N is any number, it can be written in the form $N=10^n \times M$ (where n is an integer and M is between 1 and 10) so that $\log N=n+\log M$. n is called the *characteristic* of the logarithm and $\log M$, which is obtained from tables, the *mantissa.*

(b) *natural* or *Napierian* or *hyperbolic* logarithms with base e ($=2{\cdot}718\ldots$).

To distinguish them, a logarithm to the base 10 is often written $\log x$; a logarithm to the base e being $\ln x$.

The following conversions are useful:

$$\log_e x = \log_{10} x \times 2{\cdot}30259$$
$$\log_{10} x = \log_e x \times 0{\cdot}43429.$$

logarithmic amplifier (*Acous.*). One with an output which is related logarithmically to the applied signal amplitude, as in decibel meters or recorders.
logarithmic array (*Radar, Telecomm.*). Tapered end-fire array designed to operate over a wide range of frequency.
logarithmic decrement (*Phys.*). Logarithm to base e of the ratio of the amplitude of diminishing successive oscillations.
logarithmic function (*Maths.*). The logarithmic function $\log x$ is defined by the equation

$$\log x = \int_1^x \frac{dt}{t}.$$

It equals $\log_e x$. See **e and exponential function.**
logarithmic horn (*Acous.*). The horn of a loudspeaker which is of *exponential* form. Also a metal horn for microwave work.
logarithmic mean temperature difference (*Chem. Eng.*). Universally used relationship in heat exchange calculations. If temperature difference at one end of exchanger is T_1, and at other t_2, then the mean temperature at which heat

exchange is calculated to occur is the log mean which is

$$\frac{T_1-t_2}{\log_e \dfrac{T_1}{t_2}},$$

and this is less than the arithmetic mean. In complex exchange conditions this may have to be corrected further to allow for special factors and is then usually said to be *weighted*.

logarithmic resistor, capacitor, potentiometer (*Elec. Eng.*). A variable form of resistor, etc. for which the movement of a control is directly or inversely proportional to the fractional change of resistance, etc. Such characteristics are used to counteract opposite characteristics in amplifiers, etc.

logarithmic spiral (*Maths.*). See **equiangular spiral.**

logatom (*Acous.*). Artificial word without meaning, which has a vowel with an initial and/or final consonant, used in articulation testing. See **articulation.**

log-dec. (*Phys.*). Abbrev. for *logarithmic decrement*.

logger (*Comp.*). Colloquialism for recorder or print-out device in control system. (*For.*) One who works in the felling and extraction of timber. Syn. **bucheron** (Can.), **lumberjack** (U.S.). (*Telecomm.*) Arrangement of electronic devices for obtaining an output indication which is proportional to the logarithm of the input amplitude or intensity. Required in modulation and noise meters.

loggia (*Arch.*). A covered gallery or portico built into, or projecting from, the face of a building, and bounded by a colonnade on its open side.

logging wheels (*For.*). A pair of wheels 2–3·5 m (7–12 ft) in diameter for transporting logs, which are slung below the axle. Also **katydid.**

logic (*Comp.*). Basis of operation as designed and effected in a computer, comprising *logical elements*, which perform specified elementary arithmetical functions, e.g., add, subtract, multiply, divide.

logical comparison (*Comp.*). The operation of comparing two numbers in a computer.

logical design (*Comp., etc.*). The basic planning of a data-processing or computer system and/or the synthesizing of a network of logical elements to carry out a particular function.

logical diagram (*Comp.*). Generally a diagram showing the logical elements and interconnections without engineering details.

logical elements (*Comp.*). The smaller building blocks in a data-processing or computer system which can be represented by mathematical operators in symbolic logic.

logical operations (*Comp., Maths.*). The three simplest logical operations that can be performed on binary variable X and Y having the values 0 or 1 are:

X AND Y written $X \,.\, Y$

which is 1 only when both X and Y are 1;

X OR Y written $X+Y$

which is 1 when either X or Y is 1 or when both X and Y are 1; and

NOT X written $\overline{X}$

which is 1 when X is 0 and vice versa. All other logical operations can be expressed in terms of these three. The use of semiconductor elements in computers has led to the popularity of other operations. These include:

X NAND Y, i.e., $\overline{X \,.\, Y}$ which equals $\overline{X}+\overline{Y}$,
X NOR Y, i.e., $\overline{(X+Y)}$ which equals $\overline{X} \,.\, \overline{Y}$,
and X EX. OR Y, i.e., $X \,.\, \overline{Y}+\overline{X} \,.\, Y$,

of which, because of the trend to modular construction, the NAND is the most interesting because (*a*) it can be synthesized from a single transistor and (*b*) all other logical operations can be expressed in terms of it.

logical shift (*Comp.*). See **end-around shift.**

logical symbol (*Comp.*). A graphical representation of a logical element.

logical unit (*Comp.*). Section of a computer which performs a stated operation on data presented in the form of digits, e.g., *add, subtract, gate.*

logopathy (*Med.*). See **lalopathy.**

logotype (*Typog.*). A word, or several letters, cast as one piece of type; also, a block of a trade mark.

log washer (*Min. Proc.*). Trough or tank set at a slope, in which ore is tumbled with water by means of one or more box girders with projecting arms set helically, so that as they rotate coarse lump is cleaned and delivered upslope while mud and sand overflows downslope.

logwood (*For.*). The heartwood of *Haematoxylon campechianum*, having a distinctive sweet taste and a smell resembling that of violets. It yields a black dye, and also *haematoxylin* (q.v.).

loktal base (*Electronics*). A valve base with a centre pin to lock the base securely in an appropriate socket. It has 8 pins which extend directly through the glass envelope of the valve. Also **octal base.**

löllingite, loellingite (*Min.*). Arsenide of iron, $FeAs_2$, occurring as steel-grey crystals, prismatic in habit, belonging to the orthorhombic system.

lomentum (*Bot.*). A fruit, usually elongated, which develops constrictions as it matures, finally breaking across these into 1-seeded portions. *adj.* **lomentose.**

London Basin (*Geol.*). A geographical region of S.E. England, lying between the Chiltern Hills and the North Downs. Structurally the region is a great syncline, pitching towards the east and truncated by the sea.

London Clay (*Geol.*). A blue clay (brown when weathered), up to 180 m in thickness, occurring in the London and Hampshire Basins, of deltaic origin and Eocene in age. Contains septaria, sometimes fossiliferous, and yields the remains of plants (Sheppey), crustaceans, reptiles, fishes, and more prosaic shells on several horizons.

London forces (*Chem.*). Forces arising from the mutual perturbations of the electron clouds of two atoms or molecules, the forces varying as the inverse sixth power of the distance between the molecules. These will be attractive when the molecules are in their ground electronic states.

London hammer (*Tools*). Light type of cross-pane hammer. Also **Exeter hammer.**

London plane (*For.*). *Platanus acerifolia*, yielding a useful general-purpose timber. Also called **sycamore** (chiefly U.S.), but not to be confused with the true *sycamore* (q.v.).

London screwdriver (*Tools*). One with flat blade and tapered end, usually favoured for large sizes.

London-shrunk (*Textiles*). Term in the woollen and worsted trades to indicate that a fabric has been specially treated to prevent shrinkage. Dry fabric is damped by being placed in contact with wet cloth. It is then allowed to dry naturally without tension and/or cold pressed.

London tie (*Weaving*). Method of mounting the harness over a jacquard loom in which the jacquard and the card cylinder lie at right angles to the comber board, in such a way that the pattern cards are on one side of the loom. Also called **cross-tie.**

lone pair (*Chem.*). Pair of valency electrons unshared by another atom. Such lone pairs are responsible for the formation of coordination compounds.

long (*Glass*). Slow-setting.

long-and-short work (*Build.*). A mode of laying quoins and of forming door and window jambs in rubble walling, the stones being alternately laid horizontally and set up on end; the latter stones are usually longer than the former.

long-bodied type (*Typog.*). Type cast on a larger body to avoid leading, e.g., 9-point on 11-point will have the appearance of 9-point leaded 2-point. Also called **bastard fount.**

long column (*Civ. Eng.*). A slender column which fails by bending rather than by crushing. It typically has a length 20 to 30 times its diameter.

long-day plant (*Bot.*). A plant which needs alternating periods of comparatively prolonged illumination and correspondingly reduced darkness for the proper development of flowers and fruit.

long descenders (*Typog.*). The length of the descenders (j, g, etc.) is a feature of type design, long descenders being usually considered to confer elegance. Certain typefaces are supplied with alternative long descenders to improve their appearance for bookwork, among them being Monotype Baskerville, Plantin, Times.

long-distance call (*Teleph.*). In America, any long-distance telephone call. In Europe, an international trunk call.

longeron (*Aero.*). Main longitudinal member of a fuselage or nacelle; usually there are four.

long float (*Build.*). A plasterer's trowel so long as to need two men to handle it.

longi-. Prefix from L. *longus*, long.

longicollous (*Bot.*). Having an elongated beak or neck.

longicorn (*Zool.*). Having elongate antennae, as some Beetles.

long inks, short inks (*Print.*). An empirical method of comparison between inks of differing viscosity, a soft ink being capable of forming a longer thread, when drawn upwards with a knife, than a firm, stiff ink.

longipennate (*Zool.*). Having elongate wings or feathers.

longirostral (*Zool.*). Having a long beak or rostrum.

longitude (*Astron., Geog.*). See **latitude and longitude.** (*Surv.*) The perpendicular distance of the midpoint of a survey line from the reference meridian.

longitudinal (*Aero.*). A girder that runs fore and aft on the outside of a rigid airship frame. Longitudinals connect together the outer rings of the transverse frames.

longitudinal axis (*Aero.*). See **axis.**

longitudinal current (*Elec. Eng.*). See **circulating current.**

longitudinal frame (*Ships*). A stiffening member of a ship's hull disposed longitudinally, as opposed to a transverse frame. It is supported at ends by either bulkheads or web frames, disposed transversely.

longitudinal heating (*Elec. Eng.*). Dielectric heating in which electrodes apply a high-frequency electric field parallel to lamination. See **glueline.**

longitudinal instability (*Aero.*). The tendency of an aircraft's motion in the plane of symmetry to depart from a steady state, i.e., to pitch up or down, to rise or fall, or to vary in horizontal speed.

longitudinal joint (*Carp.*). A joint used to secure two pieces of timber together in the direction of their length.

longitudinal magnetization (*Mag.*). That of magnetic recording medium along an axis parallel to the direction of motion.

longitudinal magnification (*Optics*). The ratio of the length of the image to the length of the object in a lens system when the object is small and lies along the axis of the system.

longitudinal metacentre (*Ships*). The *metacentre* (q.v.) obtained by producing an angle of *trim* (q.v.) about a transverse axis by an external force. Cf. *transverse metacentre*.

longitudinal oscillation (*Aero.*). A periodic variation of speed, height and angle of pitch. See also **phugoid oscillation.**

longitudinal stability (*Aero.*). The steady motion of an aircraft in the plane of symmetry, without variation in forward speed, in vertical velocity, or in pitching motion.

longitudinal valve (*Zool.*). In Amphibians, a large, flaplike valve which traverses the *conus arteriosus* longitudinally and obliquely.

longitudinal wave (*Acous.*). Normal propagating sound wave in which the motions of the relevant particles are in line with the direction of translation of energy.

long letters (*Typog.*). Letters with the long accent added (ā, ē, ī, etc.).

long line (*Ocean.*). A long line having many short subsidiary lines attached to it, each with one or more hooks, the whole being trailed behind a boat, to take certain kinds of midwater fish.

long-line effect (*Telecomm.*). Frequency jumping by an oscillator supplying a load through a long transmission line; due to admittance of line being suitable for oscillation at more than one frequency.

Longmyndian Series (*Geol.*). A thick series of sedimentary rocks occurring in the Longmynd of Shropshire; believed to be Pre-Cambrian in age. The Eastern Longmyndian consists essentially of slaty rocks; the Western, of red feldspathic sandstones (arkoses).

long-oil (*Paint.*). Term applied to varnishes, etc., containing a high proportion, i.e., more than about 60%, of oil.

long-period variables (*Astron.*). See **variable stars.**

long-persistence screen (*Electronics*). CRT screen coated with long afterglow phosphor (up to several seconds).

long plane (*Join.*). A bench plane 27 in. (0·7 m) long, used for planing very straight *stuff*.

longprimer (*Typog.*). An old type-size, approximately 10-point.

long range (*Aero., etc.*). Aircraft, ship or missile capable of covering great distances without refuelling. (*Radio*) Radio or navigational system which is capable of operating over great distances.

long saw (*Tools.*). A *pit-saw* (q.v.).

longshore drift (*Geog.*). Movement of beach sediment along a shoreline on coasts where the predominant direction of wave movement is oblique to the shoreline.

long-shunt compound winding (*Elec. Eng.*). A field winding arrangement for a compound-wound d.c. machine in which the shunt winding is connected across the external terminals, i.e., across the armature and the series winding.

long-sightedness (*Med.*). See **hypermetropia.**

long stick (*Textiles, etc.*). A length of 36½ in.; the extra ½ in. is allowed for folding the cloth.

long superstructure (*Ships*). A superstructure which is sufficiently long to be included in calculations of the ship's main strength, and not simply as an excrescence.

long-tail pair (*Electronics*). Method of obtaining two antiphase outputs from one input signal (e.g., for use in a *push-pull amplifier*). Consists in valve circuitry of two triodes sharing a common cathode resistor, equal bias being applied to both valves. The input signal is applied to one grid, and the antiphase outputs are taken from the anodes.

long-tail tube (*Electronics*). See **remote cut-off tube.**

long tom (*Mining, Min. Proc.*). Portable sluice in which rough concentrates made during treatment of alluvial sands or gravels are sometimes worked up to a better grade.

long ton. A unit of mass, 2240 lb. See **ton.**

longwall coal-cutting machine (*Mining*). A machine which severs coal in mechanized mining.

longwall working (*Mining*). Method of mining bedded deposits, notably coal, in which the whole seam is removed, leaving no pillars. If advancing, work goes outward from shaft and roads must be maintained through excavated areas. If retreating, ground behind workers can be allowed to cave if surface rights are not thus affected.

long waves (*Radio*). Electromagnetic low-frequency waves whose wavelength is more than 1000 metres.

lonk (*Textiles*). Wool that comes from the large type of mountain sheep of the same name, reared on the Lancashire and Yorkshire moors.

look-through (*Paper*). Term for the appearance through the sheet of paper.

loom (*Weaving*). A machine for weaving cloth, in which two sets of threads, *warp* and *weft*, are interlaced. Conventional models employ shuttles to carry the weft through the warp threads. Later types have small carriers or rapiers, or project the weft by jets of air or water. See **jet looms.**

looming (*Meteor.*). The vague enlarged appearance of objects seen through a mist or fog, particularly at sea. (*Optics*) A particular form of mirage in which the images of objects below the horizon appear in a distorted form. (*Weaving*) Drawing the threads of the warp through the eyes of the heald shaft and the reed, in the order arranged for the predetermined pattern; it may also include knotting and twisting. Also called **healding.**

loop (*Aero.*). An aeroplane manoeuvre consisting of a complete revolution about a lateral axis, with the normally upper surface of the machine on the inside of the path of the loop. See also **inverted loop.** (*Cinema.*) The slack which must be left between claws and sprocket wheels in a camera or projector, so that the intermittent motion arising from the one can be taken up before the film takes up the uniform motion of the other. (*Comp.*) Set of instructions used more than once in a program. (*Elec.*) (1) Closed graphical relationship, e.g. hysteresis loop. (2) Same as *antinode* of displacement in standing waves. See **mesh.** (*Telecomm.*) Feedback control system.

loop actuating signal (*Telecomm.*). The combined input and feedback signals in a closed loop system.

loop antenna (*Radio*). See **frame antenna.**

loop-battery pulsing (*Teleph.*). Method of signalling using a loop circuit which is alternately looped and battery connected at the sending end.

loop cable (*Elec. Eng., etc.*). See **twin cable.**

loop coupling (*Elec. Eng.*). Small loop connected between conductor and shield of a coaxial transmission line, for collecting energy from one of the series of resonators in a magnetron or a rhumbatron in a klystron valve.

loop dialling (*Teleph.*). See **loop-disconnect pulsing.**

loop difference signal (*Telecomm.*). Output signal at point in feedback loop produced by input signal applied at the same point.

loop direction-finding system (*Radio*). See **frame direction-finding system.**

loop-disconnect pulsing (*Teleph.*). The normal subscriber dialling, whereby trains of impulses are set up by interrupting the loop current from the exchange. Also **loop dialling.**

looped filament (*Elec. Eng.*). A filament arranged in the form of a large helix of one or more turns; usually employed for carbon filament lamps.

loop feedback signal (*Telecomm.*). The part of the loop output signal fed back to the input to produce the *loop actuating signal* (q.v.).

loop gain (*Telecomm.*). Total gain of amplifier and feedback circuit; if greater than zero, the system is liable to sustain oscillations. See **Nyquist criterion.**

loop galvanometer (*Elec. Eng.*). A sensitive galvanometer in which the moving element is a U-shaped current-carrying loop of aluminium foil, the two sides of the loop being in magnetic fields of opposite direction.

looping-in (*Elec. Eng.*). A term used in wiring work to denote a method of avoiding the use of a tee-joint by carrying the conductor to and from the point to be supplied.

looping roller (*Print.*). A roller on a web-fed press which moves to provide an intermittent flow to the web.

loop input signal (*Telecomm.*). An external signal applied to a feedback control loop in control systems.

loop output signal (*Telecomm.*). The extraction of the controlling signal from a feedback control.

loop pile (*Textiles*). Uncut loops formed on the surface of a firm ground, e.g., moquette. The loops are usually from a second warp beam.

loopstick antenna (*Radio*). Same as **ferrite-rod antenna.**

loop test (*Elec. Eng.*). A method of test used for locating faults in electric cables. The faulty conductor is made to form part of a closed circuit or loop, an adjacent sound conductor usually forming another part. See **Allen's-, Murray loop, Varley-.**

loop tuning error (*Radio*). The error in the bearings given by a direction finder, if the loop is improperly tuned.

loop yarn (*Textiles*). A fancy yarn, with small loops; composed of three threads, one—the *effect*—being fed faster than the others. As the loops form they are twisted in and, at a second doubling, a binder locks them in position.

loose butt hinge (*Join.*). A butt hinge in which one leaf may be lifted from the other, enabling, e.g., a door to be easily removed.

loose centres (*Eng.*). Heads similar to the tail-stock of a lathe; used for supporting work on the table of a planing machine, etc., so that it may be rotated.

loose coupling (*Elec. Eng.*). See **weak coupling.** (*Eng.*) A shaft coupling capable of instant disconnexion, as distinct from a *fast coupling* (q.v.).

loose eccentric (*Eng.*). An eccentric used on small reversing steam-engines. It rides freely on the shaft but is located by either of two stops on the shaft, which position and drive it for ahead- and reverse-running respectively.

loose gland (*Eng.*). A ring used in making an *expansion joint* (q.v.) between hot-water pipes. It slides on the spigot and compresses a rubber ring against the socket, to which it is bolted.

loose-key switch (*Elec. Eng.*). See **detachable-key switch.**

loose-leaf (*Bind.*). A binding system in which separate leaves are held together within a cover by means of a spring, ring, spiral, or other device, by unlocking which a leaf may be removed or inserted at any point.

loose needle survey (*Surv.*). Traverse with miner's dial, in which the magnetic bearing is read at each station.

loose piece (*Eng.*). Part of a foundry pattern which has to be withdrawn from the mould after the main pattern, through the cavity formed by the latter.

loose pulley (*Eng.*). A pulley mounted freely on a shaft; generally used in conjunction with a *fast pulley* (q.v.) to provide means of starting and stopping the shaft by shifting the driving belt from one to the other.

loparite (*Min.*). A titanate of sodium, calcium and the rare earth elements.

loph (*Zool.*). A crest connecting the cusps of a molar tooth.

lophobranchiate (*Zool.*). Having tuftlike, crestlike, or lobelike gills.

lophocercal (*Zool.*). Having the caudal fin ridgelike and without rays.

lophodont (*Zool.*). Of Mammals, having cheek teeth with transverse ridges on the grinding surface.

lophophore (*Zool.*). An extrovert, bearing usually a ring of ciliated tentacles, found surrounding the mouth in certain aquatic animals of sedentary habit.

lophosteon (*Zool.*). A sternal keel, as in *Carinatae*. See **carina.**

lophotrichous (*Biol.*). Said of organisms which have the flagella in one group arising from a single point on the surface of the cell.

lopoliths (*Geol.*). Igneous intrusions having the form of a concavo-convex lens, thickest in centre and tapering away in all directions; formed probably by the sagging downwards of an intrusion of laccolithic form. The great Bushveld intrusion in S. Africa is a good example.

lorac-A (*Nav.*). System with a *master* and two *slave* radio stations. (*Lo*ng *r*ange *ac*curacy.)

lorac-B (*Nav.*). System A with the addition of a reference radio station.

loran (*Nav.*). A tracking system depending on time differences of pulses as received from different but synchronized radio transmitting stations on 2 MHz. The loci of paths with constant time differences are therefore hyperbolic, intersections of two (or more) sets giving a *fix*. (*Lo*ng *ra*nge *n*avigation.) See **gee.**

loran-C (*Nav.*). Loran system on low radio frequencies.

lorandite (*Min.*). Thallium arsenic sulphide, crystallizing in the monoclinic system.

lordoscoliosis (*Med.*). A deformity of the spine in which *lordosis* (q.v.) is associated with lateral curvature of the spine.

lordosis (*Med.*). Deformity of the spine in which there is an increase of the forward convexity of the lower half of the spinal column.

lore (*Zool.*). In Birds, the space between the beak and the eye. *adj.* **loral.**

Lorentz contraction (*Phys.*). See **FitzGerald-Lorentz contraction.**

Lorentz-Lorenz equation (*Chem.*). The equation by which molecular refraction is defined:

$$[R]=\frac{n^2-1}{n^2+2}\cdot\frac{M}{\rho},$$

where $[R]$ is the molecular refraction, n is the refractive index, M is the molecular mass, ρ is the density.

Lorenz apparatus (*Elec. Eng.*). A delicate apparatus for the absolute determination of the value of a resistance. A metal disk is rotated in an accurately known uniform field, and the e.m.f. produced between its centre and its periphery is balanced against the drop of potential caused by the field-producing current in the resistance to be measured.

Lorenzini's ampulla (*Zool.*). In *Selachii*, a large ampulla opening to the exterior by a pore, filled with mucus and innervated by the lateralis nerve; believed to subserve a pressure sense.

lorica (*Zool.*) In *Rotifera* and *Protozoa*, a rigid inflexible case surrounding the body.

Loricata (*Zool.*). See **Crocodilia.**

lorry or lurry (*Mining*). A movable bridge at the top of a sinking shaft.

lorum (*Zool.*). In *Hemiptera*, the mandibular plates; in *Hymenoptera*, the submentum.

Loschmidt number (*Chem.*). (1) Number of molecules in unit volume of an ideal gas at standard temperature and pressure, i.e. $2{\cdot}687\times10^{25}$ m^{-3}. (2) Name sometimes given, esp. on the Continent, to *Avogadro number* (q.v.).

löss. See **loess.**

loss (*Telecomm.*). Opposite of *gain*, the diminution of power, as in a transformer or line. Loss is realized as a standard in a resistive *pad* (*attenuator*), which introduces a known loss into a measuring circuit.

loss angle (*Elec. Eng.*). The difference between 90° and the angle of lead of current over voltage in a capacitor.

Lossev effect (*Electronics*). The radiation due to the recombination of charge carriers injected into a *p-i-n* or *p-n* junction biased in the forward direction.

loss factor (*Telecomm.*). Ratio of average power loss to power loss under peak loading.

lossless line (*Telecomm.*). See **dissipationless line.**

loss of charge method (*Elec. Eng.*). A method of measuring very high resistances. The resistance is placed across the terminals of a charged capacitor and the rate at which the charge leaks away is observed.

loss of vend (*Mining*). Difference in weight between raw (run-of-mine) coal and saleable product leaving washery.

lossy (*Elec. Eng., Telecomm.*). Pertaining to a material or apparatus which dissipates energy, e.g., a dielectric material or a transmission line, with a high attenuation. The attenuation loss will be expressed in *decibels* (dB) while the rate of loss in the dielectric is proportional to its *loss factor*, i.e., to the product of power factor and relative permittivity.

lost flux (*Elec. Eng.*). See **leakage flux.**

lost lead (*Print.*). See **lead (of the web).**

lost wax casting (*Met.*). See **investment casting.**

lotic (*Ecol.*). Associated with running water; inhabiting rivers and streams. Cf. *lentic.*

loud-hailer (*Acous.*). See **megaphone.**

loudness (*Acous.*). The intensity or volume of a sound as perceived subjectively by a human ear.

loudness contour (*Acous.*). Line drawn on the audition diagram of the average ear which indicates the intensities of sounds that appear to the ear to be equally loud. In an objective sound-level meter, any component in the applied sound is referred to the *reftone* by these loudness contours.

loudness level (*Acous.*). That of a specified sound is the intensity of the *reftone*, 1000 Hz, on the *phon* scale, which is adjusted to equal, in apparent loudness, the specified sound. The adjustment of equality is made either subjectively or objectively as in a sound-level meter.

loudspeaker (*Acous.*). An electro-acoustic *transducer* (q.v.) which accepts transmission currents and is particularly designed for radiating (by horn or by conical diaphragm) corresponding sound waves for audition by a number of persons. See **moving-coil-, electrodynamic-, electromagnetic-, piezoelectric-, electrostatic-.**

loudspeaker dividing network (*Acous.*). See **dividing network.**

loudspeaker microphone (*Acous.*). A microphone and dynamic loudspeaker combined which is useful for intercommunication systems.

loudspeaker response (*Acous.*). Response measured under specified conditions over a frequency range, at a specified direction and distance.

loughlinite (*Min.*). A variety of sepiolite (see **meerschaum**) containing sodium.

louping ill (*Vet.*). An encephalomyelitis affecting principally sheep and cattle, transmitted by ticks and caused by a virus, and characterized by acute fever and nervous symptoms.

louvre or louver (*Build., etc.*). A system of covering a space (e.g., window, air vent) with sloping slats of glass, metal, wood, etc., so as to allow ventilation while protecting against rain, wind, etc. The effect is also used architecturally, in indirect lighting, and purely decoratively in furniture design.

love arrows (*Min.*). See **flèches d'amour.**

Lovibond comparator (*Chem.*). Term applied to a range of instruments used in the estimation of concentrations of chemical compounds in solution by comparison of colour intensities with known standards.

Lovibond tintometer (*Chem.*). Instrument for measuring the colours of liquids or solids by transmitted or reflected light and relating the colours to standards. The colour is matched against combinations of glass slides of three basic colours.

Loviton reagent (*Chem.*). Molten ammonium nitrate, used to dissolve copper, zinc, and nickel in metal analysis.

low (*Meteor.*). A region of low pressure, or a *depression* (q.v.). (*Print.*) Type or blocks which are below the level of the forme surface are said to be *low*, and must be adjusted by underlaying or interlaying.

low-angle plane (*Tools*). One with the iron set at 12° level uppermost, for use on end grain.

low aspect ratios (*Aero.*). See **aspect ratio.**

low-carbon steel (*Met.*). Arbitrarily, steel containing from 0·04% to 0·25% of carbon. See **mild steel.**

Lowden drier (*Met.*). Drying floor set above heating furnace, over which ores or concentrates are gently raked, losing moisture between feed-end and discharge.

löweite (*Min.*). A hydrated sulphate of sodium and magnesium, crystallizing in the trigonal system, occurring in salt deposits.

lower bound (*Maths.*). See **bounds of a function.**

lower case (*Typog.*). The small letters as distinct from the capitals, the name deriving from the arrangement (now largely superseded) of the small letters in a lower case with the capitals in an *upper case* (q.v.). Abbreviation: **l.c.**

lower critical velocity (*Hyd.*). The critical velocity of change from eddy to viscous flow.

lower culmination (*Astron.*). See **culmination.**

lower deck (*Ships*). A deck below the weather deck. The term has no legal definition status, and is usually applied to a partial deck which acts simply as a platform and contributes nothing to main longitudinal strength.

Lower Greensand (*Geol.*). Marine strata of Lower Cretaceous age, forming the inner ring of hills in the Weald. They are also present in the Isle of Wight, and have a narrow outcrop extending from Leighton Buzzard to Hunstanton. They consist mainly of sands, which contain bands of sandy limestone (Kentish Rag), chert, and fullers' earth. When unweathered, the beds are greenish in colour, owing to the presence of the mineral glauconite, but in surface exposures they are usually shades of red, buff, and brown, owing to the alteration of the glauconite.

lowering wedges (*Civ. Eng.*). See **striking wedges.**

lower layer (*Zool.*). In developing meroblastic ova, the secondary endoderm, which becomes separated from the under side of the blastoderm.

Lower Limestone Group (*Geol.*). A division of the Lower Carboniferous rocks of northern England; it consists of shales with limestones, delimited by the Dun Limestone at the base and the Oxford Limestone at the top, and is thus equivalent to the lower part of the Upper Bernician.

Lower Limestone Shales (*Geol.*). An obsolete term for the thinly bedded strata occurring at the base of the Carboniferous Limestone in England; not everywhere of the same age.

lower mean hemispherical candle-power (*Light*). See **mean hemispherical candle-power.**

lower-pitch limit (*Acous.*). The minimum frequency for a sinusoidal sound wave which produces a sensation of pitch.

lower transit (*Astron.*). Another name for *lower culmination*.

lowest common multiple (L.C.M.) (*Maths.*). Of a set of numbers, the smallest number which is exactly divisible by all numbers of the set.

low fidelity (*Acous.*). Opposite of *high fidelity*.

low frequency (*Radio*). Vague term widely used to indicate audiofrequencies as distinct from radio frequencies, but more correctly implying radio frequencies between 30 and 300 kHz.

low frequency amplifier (*Teleph.*). An amplifier for audiofrequency signals.

low-frequency compensation (*Telecomm.*). RC network which compensates for signal loss in coupling capacitors.

low hysteresis steel (*Met.*). Steel with from 2·5 to 4·0% silicon, a high permeability and electric resistance, but low loss through hysteresis.

low-level modulation (*Radio*). See **high-level modulation.**

low-loading amplifier (*Telecomm.*). More recent name for a **loaded push-pull amplifier.**

low machine (or mill) finish (*Paper*). See **calendered paper.**

low-melting-point alloys (*Met.*). See **fusible alloys.**

low-pass filter (*Telecomm.*). See **high-pass filter.**

low-power modulation (*Radio*). Same as **low-level modulation.**

low-pressure compressor (*Aero.*). The first compressor in a gas turbine engine with two or more compressors. In a *dual-flow turbojet*, this will feed both the combustion chamber(s) and the cold propulsion-air flow. Abbrev. **LP compressor.**

low-pressure cylinder (*Eng.*). The largest cylinder of a multiple-expansion steam-engine (e.g., the third of a triple-expansion engine), in which the steam is finally expanded.

low-pressure turbine (*Aero.*). The second, or last

turbine in a gas turbine engine with two or more turbines in series. Abbrev. LP turbine.

low-pressure turbine stage (*Aero.*). The last stage in a *multi-stage turbine* (q.v.). Abbrev. LP stage.

low red heat (*Met.*). A temperature between 550° and 700°C.

lowry (*Rail.*). An open form of box-car.

low-stop filter (*Telecomm.*). Same as **high-pass filter.**

low temperature carbonization (*Chem. Eng.*). Processing of bituminous coal at temperatures up to 760°C to produce smokeless fuel, gas, and by-product oil.

low temperature coke (*Fuels*). Type of coke with a much higher percentage of volatile matter than gas and metallurgical cokes (10%) and a lower percentage of fixed carbon, about 83%.

low tension (*Radio*). Term loosely applied to the currents and voltages associated with the filament or heater circuits of a thermionic tube.

low-tension battery (*Radio*). One supplying power for heating the cathodes of valves. U.S. term **A-battery.**

low-tension detonator (*Elec. Eng.*). The usual form of detonator, in which a charge is fired by heating a wire by an electric current.

low-tension ignition (*Elec. Eng.*). Electric ignition of the charge in the cylinder of an I.C. engine by the interruption of a current-carrying circuit inside the cylinder, no special means being used to produce a high-voltage spark as in high-tension ignition.

low-tension magneto (*Elec. Eng.*). A magneto for producing the current impulses necessary in a low-tension ignition system.

low-velocity scanning (*Electronics*). Scanning of target by electron beam under conditions such that the secondary emission ratio is less than one.

low-volt release (*Elec. Eng.*). See **under-voltage release.**

low voltage (*Elec. Eng.*). Legally, any voltage not exceeding 250 volts d.c. or r.m.s.

low warp (*Textiles*). Indicates widely spaced threads in the warp of a woven fabric.

low-water alarm (*Eng.*). An arrangement for indicating that the water-level in a boiler is dangerously low. See **low-water valve.**

low-water valve (*Eng.*). A boiler safety valve which is opened by a float in the water if the level of the latter falls dangerously low; generally fitted to stationary boilers such as the Lancashire.

low-wing monoplane (*Aero.*). A monoplane wherein the main planes are mounted at or near the bottom of the fuselage.

lox (*Chem.*). Abbrev. for liquid *oxygen.*

LP compressor (*Aero.*). Abbrev. for *low-pressure compressor.*

LPG (*Chem.*). *L*iquefied *p*etroleum *g*ases such as propane and butane, used for fuels.

LP stage (*Aero.*). Abbrev. for *low-pressure turbine stage.*

LP turbine (*Aero.*). Abbrev. for *low-pressure turbine.*

L-radiation (*Nuc.*). A particular X-ray series emitted from an element when the electrons of its L-shell are excited.

L-rest (*Eng., etc.*). A lathe rest used in hand turning, shaped like an inverted L.

L-ring (*Elec. Eng.*). A method of tuning a magnetron.

l-s coupling (*Nuc.*). See **Russell-Saunders coupling.**

LSD (*Pharm.*). See **lysergic acid diethylamide.**

L-section (*Telecomm.*). Section or half-section of a wave filter, having shunt and series arms.

L-series (*Nuc.*). Groups of lines in the X-ray spectra of the elements next in order of wavelength to the *K-series* (q.v.), and having their origin in electron transitions to the K-shell.

L-shell (*Nuc.*). See **K-, L-, M- . . . shells.**

LSI (*Electron.*). Abbrev. for **large scale integration.**

L.T. (*Radio*). Abbrev. for *low tension.*

Lu (*Chem.*). The symbol for *lutetium.*

lubber line (*Nav.*). A line on the inside of the compass bowl, or a suitable pointer, to indicate, on the compass card, the direction of the ship's head, i.e., the *compass course* (q.v.).

lubricants. Compounds (solid, plastic or liquid) entrained between two sliding surfaces. 'Wetting' lubricants adhere strongly to one or both such surfaces while leaving the intervening film fairly nonviscous. Solids include graphite, molybdenite, talc. Plastics include fatty acids and soaps, sulphur-treated bitumen, and residues from petroleum distillation. Liquids include oils from animal, vegetable or mineral sources. See **boundary lubrication, fluid lubrication.**

lubricity. Standard of lubricating properties.

Lucas sounder (*Ocean.*). See **sounder.**

Lucas theory (*Chem.*). Theory that substituent groups attract or repel electrons from neighbouring groups and thus alter their reactivity, e.g., in benzene, chlorine attracts electrons and de-activates the benzene ring, hydroxyl and amine lend electrons and activate.

lucerne (*Agric.*). See **alfalfa.**

Luciae (*Zool.*). An order of pelagic *Urochorda*, which reproduce by budding to form hollow cylindrical colonies, open at both ends, with the zooids embedded in the gelatinous wall with their oral apertures open on the outer surface of the cylinder, and their atrial apertures open on the inner surface of the cylinder; there is no larval stage.

luciferase (*Zool.*). An oxidizing enzyme which occurs in the luminous organs of certain animals (e.g., firefly) and acts on luciferin to produce luminosity. Also **photogenin.**

luciferin (*Zool.*). A proteinlike substance which occurs in the luminous organs of certain animals and is oxidized by the action of luciferase. Also **photophelein.**

lucifugous (*Ecol.*). Shunning light.

luciphilous (*Ecol.*). Seeking light.

Lüders lines (*Met.*). See **flow lines.**

Ludlow (*Typog.*). A machine which produces slugs from hand-set matrices, used mainly for headings and display lines for newspapers, etc.

Ludlow Beds (*Geol.*). A rock series in the Silurian System, often referred to as the *Ludlovian series*, lying above the Wenlockian, and divisible in England and Wales into the Lower Ludlow Group, the Aymestry Group, and the Upper Ludlow Group.

Ludlow Bone Bed (*Geol.*). A thin band, usually less than 150 mm in thickness, occurring at the base of the Downtonian stage of the Devonian System, and consisting largely of spines and fragmental hard parts of fossil fishes such as *Onchus*. Occurs in the Ludlow district of S. Shropshire.

Ludwig's angina (*Med.*). Purulent infection in the region of the submaxillary gland below the lower jaw, the infection spreading into the floor of the mouth and towards the chin.

lues (*Med.*). A plague or pestilence. The term is now synonymous with *syphilis* (q.v.). *adj.* **luetic.**

luffer-boarding (*Build.*). Sloping slats arranged as in a *louvre* (q.v.).

luffing-jib crane (*Eng.*). A common form of jib crane, in which the jib is hinged at its lower end to the crane structure, so as to permit of alteration in its radius of action. See **derricking jib crane, level-luffing crane.**

lug (*Elec. Eng.*). On an accumulator plate, a projection to which the electrical connexion is made. See **commutator-, terminal-.** (*Eng.*) See **ear.**

lugarite (*Geol.*). A form of analcite-gabbro, occurring rarely among the basic intrusive rocks of Scotland; named from the Lugar sill in Ayrshire.

lug sill (*Build.*). A sill which is of greater length than the distance between the jambs of the opening, so that its ends have to be built into the wall.

lumachella (*Geol.*). An ornamental marble containing opalized fossil shells.

lumbago (*Med.*). A rheumatic affection of the muscles and ligaments in the lumbar region or lower part of the back.

lumbar (*Zool.*). Pertaining to, or situated near, the lower or posterior part of the back; as the *lumbar* vertebrae.

lumbar puncture (*Med.*). The process of inserting a needle into the lumbar subarachnoid space to obtain a specimen of cerebrospinal fluid.

lumber (*For.*). The term employed in the United States and Canada for sawn wood of all descriptions.

lumberjack (*For.*). U.S. term for logger.

lumbo-sacral plexus (*Zool.*). In *Anura*, a plexus formed by the junction of the 7th, 8th, 9th, and 10th spinal nerves, and which gives off the nerves to the hind limb.

lumen (*Bot.*). The space enclosed by the cell walls; used especially when the contents of the cell have disappeared. (*Light*) Unit of *luminous flux*, being the amount of light emitted in unit solid angle by a small source of output one *candela*. In other words, the lumen is the amount of light which falls on unit area when the surface area is at unit distance from a source of one candela. Abbrev. **lm.** (*Bot., Zool.*) The central cavity of a duct or tubular organ.

lumen-hour (*Light*). Quantity of light emitted by a 1-lumen lamp operating for 1 hour.

lumenmeter (*Light*). An integrating photometer in which the total luminous flux is integrated in a matt white diffusing enclosure and measured through an opening.

lumerg (*Light*). A unit of luminous energy, e.g., 1 erg of radiant energy, which has a luminous efficiency of *l* lumens per watt, is equal to *l* lumergs of luminous energy. Obsolete.

luminaire (*Elec. Eng.*). The British Standard term for any sort of electric-light fitting, whether for a tungsten or a fluorescent lamp.

Luminal (*Pharm.*). See **phenobarbitone.**

luminance (*Light*). Measure of brightness of a surface, e.g., candela per square metre of the surface radiating normally. Symbol *L*.

luminance channel (*TV*). In a colour TV system, any circuit path intended to carry the *luminance signal.*

luminance flicker (*TV*). That of a colour TV picture arising from variations in luminance only. Cf. *chromaticity flicker*.

luminance signal (*TV*). Signal controlling the luminance of a colour TV picture, or of a monochrome picture.

luminescence (*Light*). Emission of light (other than from thermal energy causes) such as *bioluminescence*, *triboluminescence*, *galvanoluminescence*, *photoluminescence*, *cathodoluminescence*, etc. Thermal luminescence in excess of that arising from temperature occurs in certain minerals. Cf. *incandescence*; and see **fluorescence, persistence, phosphor, phosphorescence.**

luminescent centres (*Chem.*). Activator atoms, excited by free electrons in a crystal lattice and giving rise to electroluminescence.

luminophore (*Chem.*). (1) A substance which emits light at room temperature. (2) A group of atoms which can make a compound luminescent.

luminosity (*Astron.*). The measure of the amount of light actually emitted by a star, irrespective of its distance. See **magnitudes.** (*Light*) Apparent brightness of an image. Precisely, density of luminous intensity in a particular direction.

luminosity coefficients (*Light*). Multipliers of the *tristimulus values* of a colour, so that the total represents the luminance of the colour according to its subjective assessment by the eye.

luminosity curve (*Light*). That which gives the relative effectiveness of perception or of sensitivity of vision in terms of wavelength. See **photopic-, scotopic-.**

luminosity factor (*Light*). The ratio of the total luminous flux emitted by a light source at a given wavelength to the total energy emitted.

luminotherapy (*Med.*). See **chromotherapy.**

luminous efficiency (*Light*). The ratio of luminous flux of lamp to total radiated energy flux; usually expressed in lumens per watt for electric lamps. Not to be confused with *overall luminous efficiency*.

luminous flame (*Chem.*). One containing glowing particles of solid carbon due to incomplete combustion.

luminous flux (*Light*). The flux emitted within unit solid angle of one steradian by a point source of uniform intensity of one candela. Unit of measurement is the *lumen*.

luminous flux density (*Light*). The quantity of luminous flux passing through a normal unit area, weighted according to an internationally accepted scale of differential visual sensitivity.

luminous gas flame (*Heat.*). Thermal breakdown of the hydrocarbons into carbon and hydrogen arising from lack of primary combustion air, the carbon being heated to incandescence.

luminous intensity (*Light*). See **candela.**

luminous paints (*Chem.*). Paints which glow in the dark. Based on the salts, e.g., sulphide, silicate, phosphate of, e.g., zinc, cadmium, calcium, with traces of heavy metals, e.g., manganese. Phosphorescent paints glow for longer or shorter periods after exposure to light. Fluorescent paints glow continuously under the action of radioactive additives.

luminous sensitivity (*Light*). In a photoconductive cell, it is the ratio (output current/incident luminous flux), at a constant electrode voltage.

luminous sulphides (*Chem.*). See **luminous paint.**

Lumitype-Photon (*Print.*). A major system of *filmsetting*, making extensive use of electronics, and consisting of a setting console, a control unit, and a high-speed photographic unit.

Lumizip (*Typog.*). An advanced *filmsetting* system, electronic, tape-controlled, using individual flash for each character and optical geometry, capable of thousands of characters per minute.

Lummer-Brodhun photometer (*Light*). A form of contrast photometer in which, by an arrangement of prisms, the surfaces illuminated respectively by the standard source and the source under test are next to one another, thus enabling an easy comparison to be made.

Lummer-Gehrcke interferometer (*Light*). A very accurately worked, plane, parallel-sided glass plate, so arranged that light is internally reflected in a zig-zag path through the plate. The rays which emerge into the air at each reflection are in a condition to produce interference fringes of a very high order. The instrument is used for studying the fine structure of spectral lines. See **interference fringes.**

lump (*Textiles*). (1) Piece of cloth taken straight from the loom. (2) Lightweight fabric which has been made greater than the usual length.

lumped constant (*Elec.*). The electric properties of a component may be taken as *lumped* or *concentrated* when its dimensions are small in comparison with the wavelength of propagation of currents in it. Cf. *distributed constants.* See **primary constants.**

lumped loading (*Telecomm.*). See **line equalizer.**

lumped parameters (*Elec.*). In analysis of electrical circuits, the circuit parameters which, under certain frequency conditions, may be regarded as behaving as localized units of inductance, resistance, capacitance, etc.

lumped voltage (*Electronics*). Fictitious voltage formed by adding to the anode voltage of a multielectrode valve, the sum of the products of the various intermediate electrode voltages and the respective amplification factors associated with these electrodes. The total *space current* is a function of this quantity.

lump lime (*Build.*). The *caustic lime* (q.v.) produced by burning limestone in a kiln.

lumpy jaw (*Vet.*). See **actinomycosis.**

lumpy skin disease (*Vet.*). An acute virus disease of cattle in Africa, characterized by inflammatory nodules in the skin and generalized lymphadenitis.

lumpy withers (*Vet.*). Small swellings on the withers of the horse caused by irritation of the subcutaneous bursae by the collar.

lumpy wool (*Vet.*). See **mycotic dermatitis of sheep.**

lunar (*Zool.*). In tetrapods, the middle member of the three proximal carpals in the fore-limb, corresponding to part of the astragalus in the hind-limb.

lunar bows (*Meteor.*). Bows of a similar nature to *rainbows* (q.v.) but produced by moonlight.

lunar distances (*Astron.*). The name applied to a method, much used in the past at sea, for determining a ship's longitude. It consisted of a comparison between the observed angular distance of the moon from a star, at a known local time, with the tabulated angular distance at a certain Greenwich time.

lunar excursion module (*Space*). Module of a space vehicle used for actually landing on the moon. Abbrevs. **LEM, lem.**

lunar grid valve (*Electronics*). See **inverted thermionic valve.**

lunar month (*Astron.*). See **synodic month.**

lunate, lunulate. Crescent-shaped; shaped like the new moon.

lunation (*Astron.*). See **synodic month.**

lune (*Maths.*). The portion of the surface of a sphere intercepted by two great circles. The remaining portion of a circle when a part is removed by an intersecting circle.

lunette (*Arch.*). (1) A semicircular window or pediment over a doorway. (2) A small arched opening in the curved side of a vault.

lung (*Zool.*). The respiratory organ in air-breathing Vertebrates. The lungs arise as a diverticulum from the ventral side of the pharynx; they consist of two vascular sacs filled with constantly renewed air.

lung book (*Zool.*). An organ of respiration in some *Arachnida* (Scorpions, Spiders), consisting of an air-filled cavity opening on the ventral surface of the body; it contains a large number of thin vascular lamellae, arranged like the leaves of a book.

Lunge nitrometer (*Chem.*). Apparatus devised for the determination of oxides of nitrogen by absorption in a gas burette. May be used for other analytical processes which involve the measurement of a gas.

lung fluke (*Med., Zool.*). A parasitic trematode worm (*Paragonimus westermani*) infecting man usually *via* improperly cooked crayfish or crabs and migrating to the pleural cavity causing pulmonary inflammation.

lung plague (*Vet.*). See **contagious bovine pleuropneumonia.**

lungworm disease (*Vet.*). See **husk.**

luni-solar precession (*Astron.*). See **precession of the equinoxes.**

lunitidal interval. The time interval between the moon's transit and the next high-water at a place.

lunule, lunula (*Zool.*). A crescentic mark. Specifically, the white crescent at the root of the nail. *adj.* **lunular.**

lupinine (*Chem.*). $C_{10}H_{19}ON$, a quinolizidine alkaloid, obtained from the seeds of various species of *Lupinus*; rhombic crystals, m.p. 69°C, b.p. 255°–257°C.

lupinosis (*Vet.*). Poisoning of sheep, goats, cattle, and horses, by plants of the genus *Lupinus.*

lupulone (*Brew.*). A soft resin found in hops which possesses preservative properties and contributes to the flavour of beer. Also called **beta resin.**

lupus erythematosus (*Med.*). A degenerative disease of connective tissue throughout the body, thought to be due to an auto-immune reaction.

lupus vulgaris (*Med.*). Tuberculous infection of skin and mucous membranes, with the formation of nodules which later may ulcerate.

Lurgi process (*Fuels*). A method of obtaining gas from weakly-caking coals that are unsuitable for coke production. The coal is gasified under high pressure, using superheated steam and oxygen, all the carbon being converted into gas. The gas produced is of lower than usual calorific value and is enriched before distribution by the addition of propane, butane or methane.

lurry (*Mining*). See **lorry.**

lustre (*Min.*). This depends upon the quality and amount of light that is reflected from the surface of a mineral. The highest degree of lustre in opaque minerals is *splendent*, the comparable term for transparent minerals being *adamantine* (i.e. the lustre of diamond). *Metallic* and *vitreous* indicate less brilliant lustre, while *pearly, resinous,* and *dull* are self-explanatory terms covering other degrees of lustre.

lustre cloths (*Textiles*). Dress cloths made from cotton warp and weft consisting of mohair, alpaca, or lustre worsted; the cloth is either plain or figured, with lustrous surface.

lustring (*Textiles*). A process of friction-polishing applied to degummed, or nearly degummed, silk yarns, while under tension.

lute (*Build.*). A straightedge for levelling off clay in a brick mould by removing the excess.

luteal (*Zool.*). Pertaining to, or resembling, the *corpus luteum* (q.v.).

lutein (*Chem.*). Leaf xanthophyll. A yellow unsaturated compound occurring in the leaves and petals of various plants; formula, $C_{40}H_{56}O_2$.

Also one of the colouring matters of egg yolk. It belongs to the group of carotenoid pigments.

lutein cells (*Zool.*). Yellowish coloured cells occurring in the corpus luteum of the ovary; they contain fat-soluble substances, developed from either the follicular cells or the theca interna.

luteinizing hormone (*Physiol.*). Abbrev. LH. A gonadotrophic mucoprotein hormone, secreted by the basophil cells of the pars anterior of the pituitary gland, which causes growth of the corpus luteum of the ovary and also stimulates activity of the interstitial cells of the testis; also called **interstitial cell stimulating hormone.**

luteinoma (*Med.*). A tumour occurring in the ovary, composed of cells resembling those of the corpus luteum.

luteotrophic hormone (*Physiol.*). See **prolactin.**

lutetium (*Chem.*). A metallic element, a member of the rare earth group. Symbol Lu, at. no. 71, r.a.m. 174·97. It occurs in bloomstrandite, gadolinite, polycrase, and xenotime. Formerly called **cassiopeium.**

luthern (*Arch.*). A vertical window set in a roof.

lutidines (*Chem.*). Dimethyl-pyridines, which all occur in bone-oil and in coal-tar; general formula, $C_5(CH_3)_2H_3N$.

lux (*Light*). Unit of illuminance or illumination in SI system, 1 lumen per square metre. Abbrev. **lx.**

luxation (*Med.*). Dislocation.

Luxemburg effect (*Radio*). Cross-modulation of radio transmissions during ionospheric propagation, which limits the power radiated. Caused by nonlinearity in motion of the ions.

luxmeter (*Instr.*). An instrument for the measurement of illuminance in *lux*.

luxulianite, luxullianite (*Geol.*). A rare type of granite in which tourmaline, in stellate groups, replaces the normal coloured minerals, the other essential constituents being red orthoclase, partly replaced by tourmaline, and quartz. The rock is named from the original locality in Cornwall.

Lw (*Chem.*). The symbol for **lawrencium.**

lycanthropy (*Psychol.*). An insane delusion of a patient that he or she is a wolf.

lych (or lich) gate (*Arch.*). A roofed gateway entrance to a churchyard.

lychnidiate (*Zool.*). Phosphorescent; light-producing.

Lycoperdales (*Bot.*). An order of the *Basidiomycetes*. The unseptate basidia are enclosed in the sterile tissues of the fruit-body, which may or may not rupture when the spores are ripe. The puff-balls.

lycopersicin, lycopin (*Bot.*). A red pigment, allied to carotin, occurring in tomatoes and red pepper.

Lycopodiales (*Bot.*). An order of the *Lycopsida*. The herbaceous sporophytes have no secondary thickening. The leaves are aligulate and microphyllous. The sporangia are homosporous, and the sporophylls in strobili.

Lycopodineae (*Bot.*). A class of the *Tracheophyta*, The sporophyte is differentiated into root, stem, and leaf. The microphyllous leaves are generally arranged spirally. The sporangia are borne singly near the base on the adaxial surface of the sporophyll. There are no leaf gaps in the vascular cylinder.

Lycopsida (*Bot.*). A subphylum of *Tracheophyta*; the plants have small leaves, and there are no leaf gaps in the stele of the stem.

lyddite (*Chem.*). See **picric acid.**

Lydian stone, lydite, or touchstone (*Min.*). A highly siliceous rock, normally black in colour, although surface alteration may change this to grey. In England lydite occurs as small pebbles in many of the newer sedimentary rocks; for example, in the London Basin. They are uniformly fine-textured and even-grained, and frequently have a high polish. These lydites are rolled fragments of chert, derived from the Lower Carboniferous and Jurassic rocks. In other countries these terms have been applied to silicified argillaceous sedimentary rocks, such as flinty slates. The name *touchstone* has reference to the use of lydite as a streak plate for gold; the colour left on the stone after rubbing the metal across it indicates to the experienced eye the amount of alloy. It does not splinter.

lye (*Chem.*). Strong solution of caustic alkali.

lying panel (*Join.*). A door panel whose width is greater than its height.

lying (or laying) press (*Bind.*). A small portable screw-press in which books are held firmly during various operations.

Lyman series (*Phys.*). One of the hydrogen series occurring in the extreme ultraviolet region of the spectrum. The series may be represented by the formula $\nu = N\left(\frac{1}{1^2} - \frac{1}{n^2}\right)$, $n = 2, 3, 4 \ldots$ (see **Balmer series**), the series limit being at wave number $N = 109\,678$, which corresponds to wavelength 91·26 nm. The leading line, called *Lyman alpha*, has a wavelength of 121·57 nm, and is important in upper atmosphere research, as it is emitted strongly by the sun.

lymph-, lympho-. Prefix from L. *lympha*, water.

lymph (*Zool.*). A colourless circulating fluid occurring in the lymphatic vessels of Vertebrates and closely resembling blood plasma in composition. *adj.* **lymphatic.**

lymphadenitis (*Med.*). Inflammation of lymph glands.

lymphadenoid (*Med.*). Resembling the structure of a lymphatic gland.

lymphadenoma (*Med.*). Hodgkin's disease. A disease characterized by the progressive enlargement of the lymph glands and lymphoid tissue, anaemia, and enlargement of the spleen.

lymphangiectasis (*Med.*). Dilatation and distension of lymphatic vessels, due usually to obstruction.

lymphangiology (*Med.*). The anatomy of the lymphatics.

lymphangioma (*Med.*). A nodular tumour consisting of lymphatic channels.

lymphangitis (*Med.*). Inflammation of a lymphatic vessel or vessels. (*Vet.*). See **epizootic-, ulcerative-.**

lymphatic leukaemia (*Med.*). See **leukaemia.**

lymphatic system (*Zool.*). In Vertebrates, a system of vessels pervading the body, in which the lymph circulates and which communicates with the venous system; lymph glands and lymph hearts are found on its course.

lymph gland (*Zool.*). An aggregation of reticular connective tissue, crowded with lymphocytes, surrounded with a fibrous capsule, and provided with afferent and efferent lymph vessels.

lymph heart (*Zool.*). A contractile portion of a lymph vessel, which assists the circulation of the lymph and forces the lymph back into the veins.

lymphoblast (*Zool.*). A large cell which occurs in a lymph gland and which, by subdivision, gives rise to lymphocytes.

lymphocyte (*Zool.*). A type of leucocyte formed largely in the lymph-glands and the spleen, characterized by hyaline basiphil cytoplasm and slight phagocytic powers.

lymphocythaemia, lymphocythemia (*Med.*). Obsolete term for lymphatic *leukaemia*.

lymphocytopenia (*Med.*). Diminution, below normal, of the number of lymphocytes in the blood.

lymphocytosis (*Med.*). Increase in the number of lymphocytes in the blood, as in certain infections.

lymphogenous (*Zool.*). Lymph-producing.

lymphogranuloma inguinale (*Med.*). Poradenitis venerea. A virus venereal infection, characterized by enlargement of the glands in the groin; common in the tropics.

lymphoid tissue (*Zool.*). The type of reticular connective tissue, containing lymphocytes and lymphoblasts in the interstices, which occurs in lymph glands.

lymphoma (*Med.*). A tumour consisting of lymphoid tissue.

lymphomyelocyte (*Hist.*). A myeloblast.

lymphosarcoma (*Med.*). A sarcoma arising in lymphatic glands or lymphoid tissue.

lymph sinuses (*Zool.*). Part of the *lymphatic system* in some lower Vertebrates, consisting of spaces surrounding the blood vessels and communicating with the coelom by means of small apertures.

lymph-spaces (*Zool.*). In *Cephalochorda*, small cavities separating the fin rays.

lymph vessels (*Zool.*). See **lymphatic system.**

lyocytosis (*Zool.*). Histolysis by the action of enzymes secreted outside the tissue, as in Insect metamorphosis.

lyolysis (*Chem.*). The formation of an acid and a base from a salt by interaction with the solvent; i.e., the chemical reaction which opposes neutralization.

Lyomeri (*Zool.*). A sub-order of the order *Isospondyli*, comprising black, scaleless, abyssal marine forms, having an enormous mouth and a distensible stomach; capable of swallowing prey larger than themselves. Gulper Eels.

lyophilic colloid (*Chem.*). A colloid (*lyophilic*, solvent-loving) which is readily dispersed in a suitable medium, and may be redispersed after coagulation.

lyophilization (*Vac. Tech.*). See **freeze-drying.**

lyophobic colloid (*Chem.*). A colloid (*lyophobic*, solvent-hating) which is dispersed only with difficulty, yielding an unstable solution which cannot be reformed after coagulation.

lyosorption (*Chem.*). The adsorption of a liquid on a solid surface, especially of solvent on suspended particles.

Lyot filter (*Astron.*). See **polarizing monochromator.**

lyotropic series (*Chem.*). Ions, radicals, or salts arranged in order of magnitude of their influence on various colloidal, physiological, and catalytic phenomena, an influence exerted by them as a result of the interaction of ions with the solvent. Cf. *Hofmeister series.*

lyra (*Zool.*). The psalterium in the brain of Mammals; any lyre-shaped structure, as a lyre pattern on a bone.

lyrate (*Bot., Zool.*). Shaped like a lyre. (*Bot.*) Said of a leaf which is pinnately lobed, and has a terminal lobe which is much larger than the lateral lobes.

lyriform organs (*Zool.*). Patches, consisting of well-innervated ridges of chitin, on the legs, palpi, chelicerae, and body of various *Arachnida*; believed to be olfactory in function.

lyse (*Chem., Med.*). To cause to undergo *lysis.*

lysergic acid diethylamide (*Pharm.*). Lysergide. Compound which, when taken in minute quantities, produces hallucinations and thought processes outside the normal range. The results sometimes resemble schizophrenia and the drug is also in trial use in psychotherapy. Abbrev. **LSD.**

Lysholm grid (*Radiol.*). A type of grid interposed between the patient and film in diagnostic radiography in order to minimize the effect of scattered radiation.

Lysholm-Smith torque converter (*Eng.*). A variable-ratio hydraulic gear of the Föttinger type, but in which multistage turbine blading gives high efficiency of transmission over a wide range: used in road and rail vehicles.

lysigenic, lysigenetic, lysigenous (*Bot., Zool.*). Said of a space formed by the breakdown and dissolution of cells. Also **lysogenous.**

lysin (*Zool.*). A substance which will cause dissolution of cells or bacteria.

lysine (*Chem.*). $H_2N{\cdot}CH_2{\cdot}(CH_2)_3{\cdot}CH(NH_2)COOH$, a diamino-caproic (pentanoic) acid, obtained from albuminous substances by decomposition with acids; casein yields 8%. Lysine is essential for animal growth.

lysis (*Chem.*). Decomposition or splitting of cells or molecules, e.g., hydrolysis, lysin. (*Med.*) Destruction of cells or tissues by various pathological processes, e.g., autolysis, necrobiosis.

Lysol (*Chem.*). TN for a solution of cresols in soft soap. It is a well-known disinfectant.

lysosomes (*Biochem.*). Cellular particles intermediate in size between mitochondria and microsomes, containing hydrolytic enzymes.

lysozyme (*Biol.*). A bacteriolytic enzyme present in some plants, animal secretions (e.g., tears), egg-white, etc.

lyssa (*Zool.*). See **lytta.**

lyssophobia (*Med.*). Morbid fear of contracting rabies, leading to the appearance of symptoms of this disease.

lytta (*Zool.*). In *Carnivora*, a rod of cartilage or fibrous tissue embedded in the mass of the tongue. Also **lyssa.**

M

m Abbrev. for *milli-*.

m (*Elec., Mag.*). Symbol for mass of electron; electromagnetic moment. (*Phys., etc.*) Symbol for mass; molality.

m- (*Chem.*). An abbrev. for: (1) *meta-*, i.e., containing a benzene nucleus substituted in the 1.3 positions; (2) *meso-* (1) (q.v.).

μ (*Chem.*). A symbol for: (1) chemical potential; (2) dipole moment. (*Electronics*) Symbol for amplification factor. (*Phys.*) The symbol for: (1) index of refraction; (2) magnetic permeability; (3) the *micron* (Obs.); (4) the prefix *micro-*.

μ- (*Chem.*). A symbol signifying: (1) *meso-* (2) (q.v.); (2) *meso-* (3) (q.v.).

M (*Astron.*). The symbol used for a nebula in Messier's Catalogue of Nebulae. (*Chem.*) A general symbol for a metal or an electropositive radical. Abbrev. for *mega-*, i.e. 10^6.

M (*Chem.*). Symbol for *rel. molecular mass*. (*Elec.*) Symbol for *mutual inductance*. (*Light*) Symbol for *luminous emittance*. (*Phys.*) Symbol for *moment of force*.

macadamized road (*Civ. Eng.*). A road whose surface is formed with broken stones of fairly uniform size rolled into a 6–10 in. (15–25 cm) layer, with gravel to fill the interstices. See **tarmacadam**.

macerals (*Min.*). The elementary microscopic constituents of coal.

maceration (*Zool.*). The process of soaking a specimen in a reagent in order to destroy some parts of it and to isolate other parts.

Machaeridea (*Zool.*). A class of bilateral *Pelmatozoa* known from Cambrian fossils, originally classed as *Cirripedia*, but now placed in the *Echinodermata*, mainly on the affinities of their plate structure with that of other fossil Echinoderms.

Mach angle (*Aero.*). In supersonic flow, the angle between the *shock wave* (q.v.) and the airflow, or line of flight, of the body. The cosecant of this angle is equal to the Mach number.

Mach cone (*Aero.*). In supersonic flow, the conical *shock wave* (q.v.) formed by the nose of a body, whether stationary in a *wind tunnel* (q.v.) or in free flight through the air.

machete (*For.*). See **bush knife**.

machine (*Mech.*). A device for overcoming a resistance at one point by the application of a force at some other point. Typical simple machines are the inclined plane, the lever, the pulley, and the screw. See also **mechanical advantage, velocity ratio**.

machine chase (*Print.*). A chase specially designed for a particular machine and supplied with it.

machine code (*Comp.*). Program instructions coded in the form that is directly obeyed by a specific computer (i.e., without assembly or compiling).

machine direction (*Paper*). The direction of a sheet or web of paper corresponding with the direction of travel of the paper machine. The direction at right angles to this is the *cross direction*.

machine equation (*Comp.*). That which an analogue computer is actually programmed to solve, if not identical with the equation governing the original problem.

machine-finish(ed) (*Paper*). See **MF**.

machine-glazed (*Paper*). See **MG**.

machine instruction (*Comp.*). One which a particular machine is able to recognize and hence carry out.

machine language (*Comp.*). *Words*, etc., acceptable by a computer as *instructions* for processing data.

machine-made lace (*Textiles*). An ornamental fabric such as net, fancy lace, embroidery, and curtain lace, in which the bobbin and warp threads are interwoven, looped, plaited, or twisted together, according to design.

machine moulding (*Eng.*). The process of making moulds and cores by mechanical means, as, for example, by replacing hand-ramming by power-squeezing of the sand, or by the jolting of the box on a vibrating table. See **jolt- (or jar-) ramming machine, jolt-squeeze machine, sandslinger, squeezer**.

machine proof (*Print.*). A proof taken on a printing machine as distinct from a proofing press, and implying a high standard in the result.

machine revise (*Print.*). A proof taken on the printing machine and submitted for a final check before printing.

machine ringing (*Teleph.*). The normal ringing current which is placed on a subscriber's line to attract his attention. The currents are generated by a machine and interrupted by a suitable commutator.

machine riveting (*Eng.*). Clenching rivets by the use of mechanical or compressed-air hammers or hydraulic riveters. See also **hydraulic riveter, pneumatic riveter**.

machine-room (*Print.*). That department of a printing establishment where the actual process of printing (by machine) is carried out.

machinery oils (*Oils*). Pale or red oils suitable for lubricating bearings, shafting, and cylinders (not steam). See also **cylinder oils, lubricants**.

machine switching (*Teleph.*). The same as *automatic telephone system* (q.v.).

machine tools (*Eng.*). Any power-driven, non-portable machine designed primarily for shaping and sizing (metal) parts. See **drilling machine, key-way tool, lathe, milling machine, planing machine, shaping machine**.

machine units (*Comp.*). In electric analogue computing, a machine unit is the voltage used to represent one unit of the simulated variable.

machine variable (*Comp.*). The varying signal in an analogue computer which reproduces the variations of the simulated variable under investigation.

machine wire (*Paper*). See **wire cloth**.

machine word (*Comp.*). The series of *bits* handled in one operation by a digital computer. It may represent a code instruction or a numerical quantity.

machining allowance (*Eng.*). The material provided beyond the finished contours on a casting, forging, or roughly prepared component, which is subsequently removed in machining to size.

machmeter (*Aero.*). A pilot's instrument for measuring *flight Mach number* (q.v.).

Mach number (*Aero., Space*). The ratio of the speed of a body, or of the flow of a fluid, to the speed of sound in the same medium. At *Mach 1*, speed is *sonic*; below *Mach 1*, it is *subsonic*; above *Mach 1*, it is *supersonic*, creating a *Mach* (or *shock*) *wave*. *Hypersonic* conditions in air are reached at Mach numbers between 5 and 10 depending upon altitude and temperature. See **speed of sound.**

machopolyp (*Zool.*). See **nematophore.**

Mach principle. That scientific laws are descriptions of nature, are based on observation and alone can provide deductions which can be tested by experiment and/or observation.

mackerel sky (*Meteor.*). Cirro-cumulus or alto-cumulus cloud arranged in regular patterns suggesting the markings on mackerel.

Mackie lines (*Photog.*). Lines produced in the *border* and *fringe effects* (qq.v.).

mackle (*Typog.*). A defective printed sheet, having a blurred appearance due to incorrect impression. *adj.* **mackled.**

Maclaurin's series (*Maths.*). Modification of *Taylor's series* (q.v.), putting $a=0$.

macle (*Min.*). The French term for a *twin crystal*; in the diamond industry, more commonly used than *twin*, especially for twinned octahedra.

McLeod gauge (*Vac. Tech.*). Vacuum-pressure gauge in which a sample of low-pressure gas is compressed in a known ratio until its pressure can be measured reliably. Used for calibrating direct-reading gauges.

Macleod's equation (*Chem.*). The surface tension of a liquid is given as the difference between the densities of the liquid and its vapour raised to the fourth power, multiplied by a constant.

McNally tube (*Electronics*). A reflex klystron capable of being tuned to a wide range of frequencies.

Maco template (*Join.*). Adjustable template consisting of a clamp which holds a stack of thin brass strips. When the strips are pressed edgewise against a moulding or curved surface, they conform with its shape and can be clamped in position to give an accurate replica.

macr-, macro-. Prefix from Gk. *makros*, large.

macramoeba (*Zool.*). In certain *Sarcodina*, a large amoebula stage probably representing a female gamete. Cf. *micramoeba.*

macraner (*Zool.*). Abnormally large male ant.

macrergate (*Zool.*). Abnormally large worker ant.

macro-axis (*Crystal.*). The long axis in orthorhombic and triclinic crystals.

macrobiota (*Ecol.*). The larger soil organisms, including the roots of plants, the larger insects, earthworms, and other large organisms. Cf. *mesobiota, microbiota.*

macrocephaly, macrocephalia (*Med.*). Abnormal largeness of the head due to excess fluid, resulting in mental retardation. Cf. *hydrocephalus.*

macrochaetae (*Zool.*). In *Diptera*, the differentiated bristles, used in *chaetotaxy* (q.v.).

macrocheilia (*Med.*). Abnormal increase in the size of the lips.

macrocheiria (*Med.*). Abnormally large hands.

macrocode (*Comp.*). Any program code in which a single instruction may initiate a series of individual computer operations. Cf. *microcode.*

macroconjugant (*Zool.*). In certain *Mastigophora*, the larger of a pair of conjugants.

macrocyst (*Bot.*). A very young plasmodium in resting condition; the point of origin of the fertile tissue in *Pyronema* (a soil *Ascomycete*).

macrocyte (*Histol.*). A large uninuclear leucocyte with great powers of mobility and phagocytosis.

macrodactyly, macrodactylia (*Med.*). Congenital hypertrophy of a finger or fingers.

macrogamete (*Zool.*). The larger of a pair of conjugating gametes, generally considered to be the female gamete.

macrogametocyte (*Zool.*). In *Protozoa*, a stage developing from a trophozoite and giving rise to female gametes.

macrogamy (*Zool.*). See **hologamy.**

macrogenitosomia praecox (*Med.*). Excessive growth of the body, associated with precocious sexual development.

macroglia (*Zool.*). A general term for neuroglial cells, including astroglia, oligodendroglia and ependyma.

macroglossia (*Med.*). Abnormal enlargement of the tongue.

macrogols (*Pharm.*). Polyethylene glycols, used as ointment bases, drug solvents, etc.

macrograph (*Photog.*). A photograph or other reproduction of a *macrostructure* (q.v.).

macrogyne (*Zool.*). An abnormally large queen ant.

macro- (or general) instruction (*Comp.*). Basis of operation of an electronic computer, from which a program is derived for a specific operation.

macrolecithal (*Zool.*). See **megalecithal.**

macrolymphocyte (*Zool.*). A large type of lymphocyte occurring in bone-marrow.

macromere (*Zool.*). In a segmenting ovum, one of the large cells which are formed in the lower or vegetable hemisphere.

macromerozoite (*Zool.*). A stage in the life-cycle of certain *Haemosporidia*, arising by schizogony from a macroschizont.

macromolecule (*Chem.*). Term applied to a very large molecule such as that of *haemoglobin* (containing about 10 000 atoms) and sometimes to a polymer of high molecular mass.

macronotal (*Zool.*). Having a large thorax.

macront (*Zool.*). In *Neosporidia*, a stage formed after schizogony which gives rise by fission to macrogametes.

macronucleus (*Zool.*). In *Ciliophora*, the larger of the two nuclei which is composed of vegetative chromatin. Cf. *micronucleus.*

macronutrients (*Ecol.*). Elements and their compounds needed in relatively large amounts by living organisms. Cf. *micronutrients.*

Macropetalichthyida (*Zool.*). An order of the *Placodermi*; heavily armoured fishes known from fossils of the Devonian period. Also called *Petalichthyida.*

macrophage (*Zool.*). A *macrocyte* (q.v.); a large phagocytic cell, fixed or wandering; a histiocyte, a clasmatocyte, pericyte, etc.; a large mononuclear leucocyte.

macrophagous (*Zool.*). Feeding on relatively large particles of food. Cf. *microphagous.*

macrophotography (*Photog.*). The normal process of making enlarged prints from negatives.

macrophylline (*Bot.*). Divided into, or having, large lobes.

macrophyric, macroporphyritic (*Geol.*). A textural term descriptive of medium to fine-grained igneous rocks containing phenocrysts which are more than 2 mm in length. Cf. *microphyric.*

macropsia (*Med.*). The condition in which objects appear to the observer larger than they are, a symptom of hysteria or due to retinal disease.

macroschizont (*Zool.*). A stage in the life-cycle of certain *Haemosporidia.*

macroscopic. Visible to the naked eye.

macroscopic state (*Chem.*). One described in terms of the overall statistical behaviour of the discrete elements from which it is formed. Cf. *microscopic state.*

macrosection (*Met., etc.*). A section of metal, ceramic, etc., mounted, cut, polished, and etched as necessary to exhibit the *macrostructure.*

macrosmatic (*Zool.*). Having a highly developed sense of smell.

macrosome (*Zool.*). A large protoplasmic granule or globule.
macrospheric (*Zool.*). See **megalospheric.**
macrosplanchnic (*Zool.*). Having a large body and short legs; as a Tick.
macrospore (*Bot., Zool.*). See **megaspore.**
macrosporophyll (*Bot.*). See **megasporophyll.**
macrostoma (*Med.*). Abnormal width of the mouth due to a defect in development.
macrostructure (*Met.*). Specifically, the structure of a metal as seen by the naked eye or at low magnification on a ground or polished surface or on one which has been subsequently etched. Generally, the structure of any body visible to the naked eye. Cf. *microstructure*.
macrotous (*Zool.*). Having large ears.
macrotrichia (*Zool.*). Large setae occurring on the wings of certain Insects. Cf. *microtrichia*.
macrurous (*Zool.*). In *Eucarida*, those retaining the caridoid facies, with a long abdomen ending in a tail fan and slender appendages, any of the legs being chelate. Cf. *brachyurous*.
macula, macule (*Bot., Zool.*). A blotch or spot of colour; a small tubercle; a small shallow pit. (*Med.*) A small discoloured spot on the skin, not raised above the surface. *adj.* **macular, maculate, maculiferous, maculiform, maculose.**
macula acustica (*Zool.*). Patches of sensory epithelium in the utriculus, sacculus and lagena of the dogfish and the utricle and saccule of Mammals. In the latter they are organs of static or tonic balance, consisting of supporting cells and hair cells, the hairs being embedded in a gelatinous membrane containing crystals of calcium carbonate, i.e., an otolith.
macula lutea (*Zool.*). See **yellow spot.**
Madagascar aquamarine (*Min.*). A strongly dichroic variety of blue beryl obtained, as gemstone material, from the Malagasy Republic.
Madagascar topaz (*Min.*). See **citrine.**
madder (*Paint.*). A natural colouring matter obtained from the roots of madder plants but now replaced by synthetic alizarin. Madder lakes are used as pigments by inkmakers; they have little opacity but are permanent.
made end-papers (*Bind.*). A term for any of the several varieties of specially assembled end-papers, having cloth joints, doublures, etc., as used in hand binding.
made ground (*Build.*). Ground formed by filling in natural or artificial pits with hardcore or rubbish.
Madeira topaz (*Min.*). A form of *Spanish topaz* (q.v.). See **citrine.**
Madelung's constant (*Chem.*). A constant representing the sum of the mutual potential Coulombic attraction energy of all the ions in a lattice, in the equation for the lattice energy of an ionic crystal.
mad itch (*Vet.*). See **Aujesky's disease.**
Madras muslin (*Textiles*). A woven curtain fabric with opaque figuring on a plain gauze ground; employs extra weft for the figured portions.
madreporic canal (*Zool.*). See **stone canal.**
madreporic vesicle (*Zool.*). In *Echinodermata*, a small cavity which represents the right anterior coelomic sac of the larva.
madreporite (*Zool.*). In *Echinodermata*, a calcareous plate with a grooved surface, perforated by numerous fine canaliculi and situated in an interambulacral position, through which water passes to the axial sinus. *adj.* **madreporic.**
Madura foot (*Med.*). Mycetoma. A disease, endemic in India and occurring elsewhere, in which nodular, ulcerated swellings appear on the foot, due to infection with a fungus.
maelstrom (*Geog.*). A large whirlpool, originally confined to the Moskenstrom, a strong current off the west coast of Norway.
maestro (*Meteor.*). A fine-weather, non-autumn northwest wind in the Adriatic.
Mae West (*Aero.*). Personal lifejacket designed for airmen, inflated by releasing compressed CO_2.
mafic (*Min.*). A mnemonic term for the ferromagnesian and other nonfelsic minerals actually present in an igneous rock.
magamp (*Acous.*). Abbrev. for *magnetic amplifier*.
magazine (*Typog.*). The unit on the Linotype and Intertype slug-casting machines, and on the Fotosetter and Fotomatic filmsetting machines, which houses the matrices, a channel for each sort.
magazine arc-lamp (*Elec. Eng.*). A form of electric arc-lamp having a number of carbons which are automatically brought into operation as the others burn away, so that the lamp can burn for long periods without attention.
magazine projector (*Photog.*). One capable of holding a number of slides, which are fed automatically for projection.
magazine reel stand (*Print.*). A unit on a web-fed press which supports the reel in running position with one or more reels in the ready position.
magazine valve (*Cinema.*). See **fire-trap.**
Magellanic Clouds (*Astron.*). Two irregular galaxies in the southern hemisphere known respectively as the *Larger* and *Lesser Magellanic Cloud*, appearing to the naked eye like detached portions of the Milky Way. They are the two nearest galaxies to our own system.
Magendie's foramen (*Zool.*). In Vertebrates, an aperture in the roof of the fourth ventricle of the brain, through which the cerebrospinal fluid communicates with the fluid in the spaces enclosed by the meningeal membranes.
magenta (*Chem.*). See **fuchsine.**
magenta wax test (*Cables*). The paper tapes are unrolled and the oil extracted with petrol. A magenta water-dye then colours the papers except where wax has been formed. Traces of wax, ordinarily invisible, are shown clearly.
maggot (*Zool.*). An acephalous, apodous, eruciform larva such as that of certain *Diptera*.
maghemite (*Min.*). An iron oxide with the crystal structure of magnetite but the composition of haematite. A cation-deficient spinel, produced by the oxidation of magnetite.
magic eye (*Radio*). See **electric eye.**
magic numbers (*Nuc.*). Certain numbers (e.g., 2, 8, 20, 28, 50, 82, and 126) of protons (atomic number) or neutrons in an atomic nucleus, which result in enhanced stability for a series of isotopes of an element.
magic-T (*Radar*). Combination of an E-plane junction and an H-plane junction, so that it functions as a hybrid, i.e., permitting power to be diverted according to its direction of flow.
magistral (*Met.*). Powdered roasted copper pyrite used in amalgamation of silver ores in Mexican process.
magma (*Geol.*). Molten rock, including dissolved water and other gases. It is formed by melting at depth and rises either to the surface, as lava, or to whatever level it can reach before crystallizing again, in which case it forms an igneous intrusion.
magmatic cycle (*Geol.*). See **igneous cycle.**
magnadur (*Met.*). A ceramic type material, comprising sintered oxides of iron and barium, used for making permanent magnets and as a good electrical insulator, its eddy-current loss in an a.c. field being very small.

magnalium (*Met.*). An aluminium-base alloy. Contains 1·75% copper and 1·75% magnesium.
Magnavolt (*Elec. Eng.*). TN of a range of rotating amplifiers.
magnesia (*Chem.*). See **magnesium oxide.**
magnesia alba (*Chem.*). Commercial basic magnesium carbonate.
magnesia alum (*Min.*). See **pickeringite.**
magnesia cement (*Build.*). See **Sorel's cement.**
magnesia glass (*Glass*). Glass containing usually 3–4% of magnesium oxide. Electric-lamp bulbs have been mainly made from this type of glass since fully automatic methods of production were adopted.
magnesia mixture (*Chem.*). A mixture of magnesium chloride, ammonium chloride, and ammonia solution used in chemical analysis for the estimation of phosphates.
Magnesian Limestone (*Geol.*). The major division of the English Permian System of N.E. England; consists of several hundred feet of dolomitic limestones and dolomites where best developed, in Durham, whence they outcrop continuously southwards to near Nottingham. Also feebly represented locally west of the Pennines.
magnesian spinel (*Min.*). See **spinel.**
magnesia usta (*Chem.*). Commercial name for magnesium carbonate calcined at a low temperature for a long period.
Magnesil (*Met.*). TN for a magnetic (grain-oriented) alloy used for construction of cores of magnetic amplifiers.
magnesioferrite (*Chem.*). $MgO{\cdot}Fe_2O_3$, a *spinel* (q.v.) used as binder in manufacture of magnesite bricks.
magnesite (*Met.*). Carbonate of magnesium, crystallizing in the trigonal system. Magnesite is a basic refractory used in open-hearth and other high-temperature furnaces; it is resistant to attack by basic slag. It is obtained from natural deposits (mostly magnesium carbonate, $MgCO_3$), which is calcined at high temperature to drive off moisture and carbon dioxide, before being used as a refractory. (*Photog.*) Magnesium carbonate is the substance adopted for a standard white surface.
magnesite flooring (*Build.*). A composition of cement, magnesium compounds, sawdust and sand to form a hard continuous screed for floor surfaces. Also called **jointless flooring.**
magnesium (*Chem.*). A light metallic element in the second group of the periodic system. Symbol Mg, at. no. 12, r.a.m. 24·312, m.p. 651°C, rel. d. 1·74, electrical resistivity 42×10^{-8} ohm metres, b.p. 1120°C at 1 atm, specific latent heat of fusion 377 kJ/kg. Found in nature only as compounds. The metal is a brilliant white in colour, and magnesium ribbon burns in air, giving an intense white light, rich in ultraviolet rays. It is used as a deoxidizer for copper, brass, and nickel alloys, and added to several aluminium-base alloys. A basis metal in strong light alloys which are used in aircraft and automobile construction and for reciprocating parts. New alloys with zirconium and thorium are used in aircraft construction.
magnesium carbonate (*Chem.*). $MgCO_3$. See **magnesite.**
magnesium orthodisilicate (*Chem.*). Occurs in nature as *serpentine* (q.v.).
magnesium oxide (*Chem.*). MgO. Obtained by igniting the metal in air. In the form of calcined magnesite and dolomite, it is used as a refractory material. See also **periclase.**
magnesium oxychloride cement (*Build.*). *Sorel's cement* (q.v.), which is used in *magnesite flooring* (q.v.).
magneson (*Chem.*). 4-(4-nitrophenolazo)resorcinol:

$$O_2N-C_6H_4-N{=}N-C_6H_3(OH)-OH$$

used as a reagent for detection and determination of magnesium, with which it forms a characteristic blue colour in alkaline solution.
magnet. A mass of iron or other material which possesses the property of attracting or repelling other masses of iron, and which also exerts a force on a current-carrying conductor placed in its vicinity. See **electromagnet, permanent magnet.**
magnet coil (*Elec. Eng.*). See **magnetizing coil.**
magnet core (*Mag.*). The iron core within the coil of an electromagnet.
magnetic (*Mag.*). Said of all phenomena essentially depending on magnetism. See **diamagnetism, paramagnetism.**
magnetic alloys (*Met.*). Generally, any alloys exhibiting ferromagnetism, e.g., and most importantly, *silicon iron* (q.v.), but also iron-nickel alloys, which may contain small amounts of any of a number of other elements (e.g., copper, chromium, molybdenum, vanadium, etc.), and iron-cobalt alloys. See **permalloy, Supermalloy, Hipernik, Perminvar, Permendur, Mumetal, Radiometal, Rhometal;** also **Alnico, Ferroxcube, ferrite, magnetic ferrites.**
magnetic amplifier (*Acous.*). One in which the saturable properties of magnetic material are utilized to modulate an exciting alternating current, using an applied signal as *bias*; the signal output when rectified becomes a magnification of the input signal.
magnetic annealing (*Met.*). Heat treatment of magnetic alloy in a magnetic field, used to increase its permeability.
magnetic anomaly (*Mag.*). A local deviation of the corrected magnetic component measured in magnetic prospecting, as compared with a larger scale 'smoothed' field or regional gradient.
magnetic armature (*Elec. Eng.*). Ferromagnetic element, the position of which is controlled by external magnetic fields.
magnetic axis (*Mag.*). A line through the effective centres of the poles of a magnet.
magnetic balance (*Elec. Eng.*). A form of fluxmeter in which the force required to prevent the movement of a current-carrying coil in a magnetic field is measured. Cf. *current balance.*
magnetic bearing (*Surv.*). The horizontal angle between a survey line and magnetic north.
magnetic bias (*Acous.*). A steady magnetic field added to the signal field in magnetic recording, to improve linearity of relationship between applied field and magnetic remanence in recording medium.
magnetic blowout (*Elec. Eng.*). A special magnet coil fitted to circuit-breakers or other similar apparatus in order to produce a magnetic field at the point of opening the circuit, so that any arc which is formed is deflected in such a direction as to be lengthened or brought into contact with a cool surface and is rapidly extinguished. Also called **blowout coil.**
magnetic bottle (*Nuc. Eng.*). The containment of a plasma during thermonuclear experiments by applying a specific pattern of magnetic fields.
magnetic braking (*Elec. Eng.*). A method of braking a moving system in which a brake is applied and released by an electromagnet.
magnetic card (*Comp.*). One with a suitable

surface which can be magnetized in selective areas so that data can be stored.

magnetic character reading (*Print.*). Scanning and interpretation of characters printed with magnetic ink on documents, e.g., cheques; special lettering is necessary.

magnetic chuck (*Eng.*). A chuck having a surface in which alternate steel elements, separated by insulating material, are polarized by electromagnets, so as to hold light flat work securely on the table of a grinding machine or other machine tool.

magnetic circuit (*Elec. Eng.*). Complete path, perhaps divided, for magnetic flux, excited by a permanent magnet or electromagnet. The range of reluctance is not so great as resistance in conductors and insulators, so that leakage of the magnetic flux into adjacent nonmagnetic material, especially air, is significant.

magnetic clutch (*Elec. Eng.*). A clutch in which the necessary force to hold the two parts together is provided by an electromagnet.

magnetic compass (*Nav.*). Pivoted magnet used to indicate direction of magnetic north.

magnetic component (*Elec. Eng*). The magnetic field associated with an electromagnetic wave. The magnetic and electric fields are related through the intrinsic impedance of the medium, $E/H = \eta = \sqrt{\mu/\varepsilon}$.

magnetic condenser (*Micros.*). A lens-coil, forming a circular electromagnetic field, and used as the condenser in an electron microscope.

magnetic controller (*Elec. Eng.*). Unit in control system operated by magnetic field applied to magnetic armature.

magnetic core (*Comp.*). Unit for *store* or *memory*, consisting of tiny cores or holes punched in a sheet of ferrite, which are magnetized by wires knitted through them.

magnetic coupling (*Elec. Eng.*). The magnetic flux linkage between one circuit and another.

magnetic course (*Nav.*). The angle between the *magnetic meridian* (q.v.) and the direction of the ship's head.

magnetic creeping (*Elec. Eng.*). A gradual increase in the intensity of magnetization of a piece of magnetic material, after a continued application of the magnetizing force.

magnetic cutter (*Acous.*). A cutter used in recording in which the motions of the recording stylus are operated by magnetic fields.

magnetic damping (*Elec. Eng.*). Damping of motion of a conductor by induced eddy currents in it when moving across a magnetic field, particularly applicable to moving parts of instruments and electricity integrating meters.

magnetic declination (*Surv., etc.*). The angular deviation of a magnetic compass, uninfluenced by local causes, from the true north and south. The declination varies at different points on the earth's surface and at different times of the year. Also called **magnetic deviation.**

magnetic deflection (*Electronics*). That of an electron beam in a cathode-ray tube, caused by a magnetic field established by current in coils where the beam emerges from the electron gun which forms the beam, the deflection being at right angles to the direction of the field.

magnetic delay line (*Comp.*). The use of a magnetostrictive material, e.g., nickel wire, as a delay medium for propagating sound, thus providing a circulating memory or store.

magnetic detector (*Radio*). A generic name for those early forms of detectors of high-frequency currents which depended on the demagnetizing effect of an alternating magnetic field upon a magnetized iron core.

magnetic deviation (*Surv.*). See **magnetic declination.**

magnetic difference of potential (*Elec. Eng.*). A difference in the magnetic conditions at two points which gives rise to a magnetic flux between the points.

magnetic dip. See **dip.**

magnetic discontinuity (*Elec. Eng.*). An air gap, or a layer of non-magnetic material, in a magnetic circuit.

magnetic disk (*Acous.*). Form of recording medium used in some magnetophones.

magnetic-disk memory (*Acous.*). One consisting of a pile of rotating disks of magnetic surface material, so spaced that an arm can swing in for recording or reproduction, with low *access time.*

magnetic displacement (*Elec. Eng.*). An alternative name for **magnetic induction.**

magnetic domain (*Met., etc.*). Aggregated ferromagnetic group of atoms, usually well below one micrometre in diameter, forming part of a system of such groups with no preferred orientation.

magnetic doublet radiator (*Elec. Eng.*). A hypothetical radiator consisting of two equal and opposite varying magnetic poles whose distant field is equivalent to that of a small loop aerial.

magnetic drum (*Comp.*). Cylinder driven very uniformly at high speed, carrying a layer of magnetic material for registering and reproducing impulses in computers, this being effected by magnetic heads just clearing the rotating surface.

magnetic elongation (*Elec. Eng.*). The slight increase in length of a wire of magnetic material when it is magnetized. See **magnetostriction.**

magnetic energy (*Elec. Eng.*). Product of flux density and field strength for points on the demagnetization curve of a permanent magnetic material, measuring the energy established in the magnetic circuit. Normally required to be a maximum for the amount of magnetic material used.

magnetic equator (*Geog.*). That part of the earth's surface where the earth's magnetic field is horizontal. It is a narrow strip around the earth roughly half-way between the magnetic poles of the earth.

magnetic escapement (*Horol.*). One comprising a small horseshoe magnet on the end of a vibrating spring, which actuates the teeth of the escape wheel as they come under the influence of its field.

magnetic ferrites (*Elec. Eng.*). Those having magnetic properties and at the same time possessing good electrical insulating properties by reason of their ceramic structure.

magnetic field (*Elec. Eng.*). Modification of space, so that forces appear on magnetic poles or magnets. Associated with electric currents and the motions of electrons in atoms.

magnetic-field intensity (*Elec. Eng.*). The magnitude of the field strength vector in a medium (i.e., the magnetic strain produced by neighbouring magnetic elements or current-carrying conductors). The SI unit is the ampere per metre and the CGS unit is the oersted. Also called **magnetic-field strength, magnetic intensity, magnetizing force.**

magnetic-field strength (*Elec. Eng.*). See **magnetic-field intensity.**

magnetic-film memory (*Comp.*). See **thin-film memory.**

magnetic flux (*Elec. Eng.*). The surface integral of the product of the permeability of the medium and the magnetic-field intensity normal to the surface. The magnetic flux is conceived, for theoretical purposes, as starting from a positive fictitious north-pole and ending on a fictitious south-pole, without loss. When associated with electric currents, a complete circuit, the *magnetic circuit*, is envisaged, the quantity of magnetic flux being sustained by a magnetomotive force, m.m.f. (coexistent with ampere-turns linked with the said circuit). Permanent magnetism is explained similarly in terms of molecular m.m.fs. (associated with orbiting electrons) acting in the medium. Measured in *maxwells* (CGS) or *webers* (SI).

magnetic flux density. See **magnetic induction.**

magnetic focusing (*Electronics*). That of an electron beam by applied magnetic fields, e.g., in CRT or electron microscope.

magnetic forming (*Eng.*). A fast, accurate production process for swaging, expanding, embossing, blanking, etc., in which a permanent or expendable coil is moved by electromagnetism to act on the workpiece.

magnetic gate (*Elec. Eng.*). A gate circuit used in magnetic amplifiers.

magnetic head (*Acous.*). Recording, reproducing or erasing head in magnetic recorder.

magnetic hysteresis (*Elec. Eng.*). Nondefinitive value of magnetic induction in ferromagnetic medium for given magnetic-field intensity, depending upon magnetic history of sample.

magnetic hysteresis loop. See **hysteresis loop.**

magnetic hysteresis loss (*Elec. Eng.*). The energy expended in taking a piece of magnetic material through a complete cycle of magnetization. The magnitude of the loss per cycle is proportional to the area of the magnetic hysteresis loop.

magnetic induction (*Elec. Eng.*). (1) Induced magnetization in magnetic material, either by saturation, by coil-excitation in a magnetic circuit, or by the primitive method of stroking with another magnet. (2) Magnetic-flux density in a medium. Given by the product of the field intensity and the permeability of the medium. The SI unit is the tesla (T) or weber per square metre (Wb/m^2), the CGS unit is the gauss.

magnetic ink (*Print.*). Ink containing particles of ferromagnetic material, used for printing data so that magnetic character recognition is possible.

magnetic intensity (*Elec. Eng.*). See **magnetic-field intensity.**

magnetic iron-ore (*Min.*). See **magnetite.**

magnetic lag (*Elec. Eng.*). The time required for the magnetic induction to adjust to a change in the applied magnetic field.

magnetic leakage (*Elec. Eng.*). That part of the magnetic flux in a system which is useless for the purpose in hand and may be a nuisance in affecting nearby apparatus.

magnetic lens (*Mag.*). The counterpart of an optical lens—in which a magnet or system of magnets is used to produce a field which acts on a beam of electrons in a similar way to a glass lens on light rays. Comprises current-carrying solenoids of suitable design.

magnetic levitation (*Mag.*). A method of opposing the force of gravity using the mutual repulsion between two like magnetic poles.

magnetic link (*Elec. Eng.*). A small piece of magnet steel placed in the immediate vicinity of a conductor carrying a heavy surge current, e.g., a transmission line tower carrying a lightning stroke current. The magnetization of the link affords a means of estimating the value of the current.

magnetic lock (*Mining*). A locking device, used chiefly on miners' lamps, which can be released only by a magnet.

magnetic loudspeaker (*Acous.*). Loudspeaker basically dependent on magnetic forces to produce the sound waves.

magnetic map (*Mag.*). A map showing the distribution of the earth's magnetic field.

magnetic memory (storage) (*Comp.*). The use of a film of magnetic material on the surface of a drum or flexible film (tape) for registering and recovering information in the form of *bits* (impulses).

magnetic memory plate (*Comp.*). A thin plate of a ferrite material in the form of a rectangular grid. Threaded through the holes are wires which allow pulses of information to be stored by magnetic induction in the material surround of each hole without mutual interference between holes.

magnetic meridian (*Nav.*). The great circle, in the plane of which the compass needle lies when under the influence of the earth's magnetic field only.

magnetic microphone (*Acous.*). A microphone whose action depends upon the induced e.m.f. in a coil (or conductor) when moved relatively to a magnetic field by a sound wave.

magnetic mirror (*Nuc. Eng.*). Device based on the principle that ions moving in a magnetic field tend to be reflected away from higher-than-average magnetic fields. Thus a *magnetic bottle* can be designed for mirror machines.

magnetic modulation (*Radio*). A form of modulation sometimes used for keying continuous-wave transmitters, and formerly for radiotelephony; in it, the inductance of an iron-cored coil is varied by control of a unidirectional polarizing flux.

magnetic modulator (*Elec. Eng.*). One using a magnetic circuit as the modulating element.

magnetic moment (*Mag.*). Vector such that its product with the magnetic induction gives the torque on a magnet in a homogeneous magnetic field. Also called **moment of a magnet.** (*Nuc.*) Dipole moment of atom or nucleus associated with electron orbitals and/or electron and nuclear spin. See **Bohr magneton.**

Magnetic North (*Mag.*). The direction in which the north pole of a pivoted magnet will point. It differs from the Geographical North by an angle called the *magnetic declination* (q.v.).

magnetic objective (*Micros.*). A lens-coil forming a circular electromagnetic field, and used as the objective 'lens' in an electron microscope.

magnetic oxide of iron (*Min.*). See **magnetite.**

magnetic oxides (*Met.*). The iron oxides which are ferromagnetic and which, suitably fabricated from the powder form, provide efficient permanent magnets.

magnetic particle clutch (*Eng.*). A form of hydraulic coupling in which the fluid is a suspension of magnetizable particles, the viscosity of the fluid, and thus the degree of slip of the clutch, being variable by varying the intensity of magnetization.

magnetic particle inspection (*Eng.*). A rapid, nondestructive test for fatigue cracks and other surface and subsurface defects in steel and other magnetic materials, in which the workpiece is magnetized so that local flux-leakage fields are formed at the cracks or other discontinuities and shown up by the concentration at these fields of a magnetic powder applied to the workpiece.

magnetic pendulum (*Elec. Eng.*). Suspended magnet executing torsional oscillations in any horizontal magnetic field. It forms the basis of the horizontal component magnetometer.

magnetic polarization (*Chem.*). The production of optical activity by placing an inactive substance in a magnetic field.

magnetic pole (*Elec. Eng.*). A convenient conception, which cannot exist, deduced from the experimental indication of the direction of the magnetic field arising from a permanent magnet. If the latter is long in comparison with its cross-section and the ends are provided with soft-iron balls, the direction of the magnetic field, as indicated by iron-filings, appears to radiate from the centres of such spheres, called *poles*. Experimentally, such poles appear as magnetic charges, from which are deduced the magnitude of magnetic poles, magnetic-field strength, magnetic flux, magnetic potential, and electrical units.

magnetic potential (*Maths.*). A continuous mathematical function the value of which at any point is equal to the potential energy (relative to infinity) of a theoretical unit north-seeking magnetic pole placed at that point.

magnetic potentiometer (*Elec. Eng.*). A flexible solenoid (wound on a nonmagnetic base) used with a ballistic galvanometer to explore the distribution of magnetic potential in a field, etc.

magnetic printing (*Print.*). Printing ink containing magnetic elements and capable of being read both visually and by computer. (*Telecomm.*) Transfer of recorded signal from one magnetic recording medium or element to another.

magnetic projector (*Micros.*). A lens-coil forming a magnetic lens which magnifies the image in an electron microscope.

magnetic prospecting (*Geophys.*). A method of geophysical exploration in which local variations in the value of the earth's magnetic field are used to draw inferences as to the shape, depth and character of subsurface rocks. Generally only the total and vertical components of the field are used in survey work.

magnetic pumping (*Phys.*). Use of radio-frequency currents in coils over bulges in the tube of a stellarator to modulate the steady axial field and provide heat to the plasma. This process is most efficient when there is resonance between the RF signals and vibrations of the molecules of plasma. The process is then called **resonance heating**.

magnetic pyrite (*Min.*). See **pyrrhotite**.

magnetic quantum numbers (*Phys.*). Those determining the components of orbital and spin angular momentum in the direction of the applied field.

magnetic reaction (*Elec. Eng.*). Same as **electromagnetic reaction**.

magnetic recording (*Acous.*). (1) The magnetic tape, from which the recorded signal may be reproduced. (2) The process of preparing a magnetic recording.

magnetic rigidity (*Nuc. Eng.*). A measure of the momentum of a particle. It is given by the product of the magnetic intensity perpendicular to the path of the particle and the resultant radius of curvature of this path.

magnetic rotation (*Light*). See **Faraday effect.**

magnetic saturation (*Mag.*). The limiting value of the magnetic induction in a medium when its *magnetization* (q.v.) is complete and perfect.

magnetic screen (*Elec. Eng.*). A screen of a high magnetic permeability material such as Mu-metal, used to surround certain electrical and electronic components, in order to protect them from the effect of external magnetic fields.

magnetic separator (*Mag.*). A device for separating, by means of an electromagnet, any magnetic particles in a mixture from the remainder of the mixture, e.g., for separating iron filings from brass filings.

magnetic shell (*Elec. Eng.*). A magnetized body of dimensions identical to a current-carrying coil, which may be considered instead of the latter when considering the forces acting in a system.

magnetic shield (*Elec. Eng.*). Surface of magnetic material which reduces the effect on one side of a magnetic field on the other side. A substantially complete shield is used to protect a.c. indicating instruments from errors arising from external alternating magnetic fields.

magnetic shift register (*Elec. Eng.*). One in which the pattern of settings of a row of magnetic cores is shifted one step along the row by each fresh pulse.

magnetic shunt (*Elec. Eng.*). A piece of magnetic material in parallel with a portion of a magnetic circuit, so arranged as to vary the amount of magnetic flux in that portion of the circuit.

magnetic slot-wedge (*Elec. Eng.*). A slot-wedge of magnetic material which gives the same effect as a closed slot.

Magnetic South (*Mag.*). The direction in which the South pole of a pivoted magnet will point. It differs from the Geographical South by an angle called the *magnetic declination* (q.v.).

magnetic spectrometer (*Nuc. Eng.*). One in which the distribution of energies among a beam of charged particles is investigated by means of magnetic focusing techniques.

magnetic squeezer (*Eng.*). See **squeezer.**

magnetic stability (*Mag.*). A term used to denote the power of permanent magnets to retain their magnetism in spite of the influence of external magnetic fields, vibration, etc.

magnetic storm (*Meteor.*). Magnetic disturbance in the earth, causing spurious currents in submarine cables; probably arises from variation in particle emission from the sun, which affects the ionosphere.

magnetic surface wave (*Radar*). Magnetic wave propagated along the surface of a ferromagnetic garnet substrate. Used in microwave delay lines, filters, etc. Abbrev. MSW.

magnetic susceptibility (*Elec. Eng.*). The amount by which the relative permeability of a medium differs from unity, positive for a paramagnetic medium, but negative for a diamagnetic one. Equals intensity of magnetization ÷ applied field.

magnetic suspension (*Elec. Eng.*). Use of a magnet to assist in the support of, e.g., a vertical shaft in a meter, thereby relieving the jewelled bearings of some of the weight. (*Horol.*) Type of balance-wheel assembly embodying a pair of complementary ring magnets or a magnet above the balance staff to reduce friction.

magnetic tape (*Telecomm., etc.*). Flexible plastic tape, e.g., 0·5 in. wide, on one side of which is a uniform coating of dispersed magnetic material, in which signals are registered for subsequent reproduction. Used for registering television images or sound or computer data.

magnetic tape reader (*Comp.*). One which has a multiple head which transforms the pattern of registered signals into pulse signals.

magnetic track (*Cinema.*). Sound-track on a release cinematograph film (*stripe*). In CinemaScope there are four such tracks, one on each side of each row of sprocket holes.

magnetic transition temperature (*Phys.*). See **Curie point** (1).

magnetic transmission (*Autos.*). A type of clutch

in which ferromagnetic powder takes up the drive between two rotating members when drawn into an annular gap by electromagnets. See also **electromagnetic clutch.**

magnetic tube (*Phys.*). See **trapping region.**

magnetic tuning (*Elec. Eng.*). Control of very-high-frequency oscillator by varying magnetization of a rod of ferrite in the frequency-determining cavity by an externally applied steady field.

magnetic units (*Elec., Mag.*). Units for electric and magnetic measurements in which the *permeability* of a vacuum is taken as unity, with no dimensions.

magnetic variables (*Astron.*). Stars in which strong variable magnetic fields have been detected by the Zeeman effect.

magnetic variations (*Meteor.*). Both diurnal and annual variations of the magnetic elements (dip, declination, etc.) occur, the former having by far the greater range. In the northern hemisphere, the declination moves to the west during the morning and then gradually back, the extreme range being nowhere more than 1°. The dip varies by a few minutes during the day. It is thought that these effects are caused by varying electric currents in the ionized upper atmosphere.

magnetic wire (*Acous.*). A wire of magnetic material used in recording.

magnetism. Science covering magnetic fields and their effect on materials, due to unbalanced spin of electrons in atoms. See **Coulomb's Law for magnetism;** also **paramagnetism, diamagnetism, ferrimagnetism, ferromagnetism, antiferromagnetism.**

magnetite or **magnetic iron-ore** (*Min.*). An oxide of iron, crystallizing in the cubic system. It has the power of being attracted by a magnet, but it has no power to attract particles of iron to itself, except in the form of lodestone.

magnetization (*Mag.*). Orientation from randomness of saturated domains in the body of ferromagnetic material. Denoted by M and related to the field intensity H and the magnetic induction B by $B=\mu_0(H+M)$ teslas, where μ_0 is the permeability of free space.

magnetization curve (*Elec. Eng.*). A curve showing ferromagnetic characteristic of a material. The magnetic induction B is usually plotted as a function of the magnetic field intensity H, since H is proportional to the current in the exciting coil. Also called **B/H curve.** It also denotes the relation between the total m.m.f. and the total flux in a magnetic circuit, such as that of an electric machine. Since m.m.f. is proportional to exciting current and flux is proportional to the e.m.f. generated in a machine, a curve between exciting current and e.m.f. is also called a *magnetization curve.*

magnetize (*Mag.*). (1) To induce magnetization in ferromagnetic material by direct or impulsive current in a coil. (2) To apply alternating voltage to a transformer or choke having ferromagnetic material, generally laminated, in its core.

magnetizing coil (*Elec. Eng.*) A current-carrying coil used to magnetize an electromagnet, such as the field coil of an electric generator or motor. Also called **field coil, magnet coil.**

magnetizing current (*Elec. Eng.*). (1) Current (direct or impulsive) in a coil for the magnetization of ferromagnetic material in its core. (2) The a.c. taken by the primary of a transformer, apart from a load current in the secondary.

magnetizing force. See **magnetic-field intensity.**

magnetizing roast (*Met.*). Reduction of weakly magnetic iron ore or concentrate by heating in a reducing atmosphere to convert it to a more strongly ferromagnetic compound.

magneto (*Elec. Eng.*). A small permanent-magnet electric generator capable of producing periodic high-voltage impulses; used for providing the ignition of internal-combustion engines, firing of explosives, etc.

magneto bell (*Teleph.*). An audible bell actuated by a.c., which causes an armature to vibrate and hit two gongs alternately.

magneto-caloric effect (*Heat*). The reversible heating and cooling of a medium when the magnetization is changed. Also called **thermomagnetic effect.**

magnetochemistry (*Chem.*). The study of the magnetic changes accompanying chemical reactions.

magnetoelectric (*Mag.*). Of certain materials, e.g., chromium oxide, the property of becoming magnetized when placed in an electric field. Conversely, they are electrically-polarized when placed in a magnetic field. May be used for measuring pulse electric or magnetic fields.

magneto generator (*Elec. Eng.*). An electric generator in which the exciting flux is obtained from permanent magnets.

magnetohydrodynamic generator (*Elec. Eng.*). Abbrev. **MHD generator.** See **magnetoplasmadynamic generator.**

magnetohydrodynamics (*Mag., Phys.*). The study of the motion of an electrically conducting fluid in the presence of a magnetic field. The motion of the fluid gives rise to induced electric currents which interact with the magnetic field which in turn modifies the motion. The phenomenon has applications both to magnetic fields in space and to the possibility of generating electricity. If the free electrons in a plasma or high velocity flame are subjected to a strong magnetic field, then the electrons will constitute a current flowing between two electrodes in a flame. Abbrev. **MHD.**

magneto ignition (*Elec. Eng.*). An ignition system for I.C. engines, in which the voltage necessary to produce the spark is generated by a magneto.

magneto-ionic (*Phys.*). Said of components of an electromagnetic wave passing through an ionized region and divided into ordinary and extraordinary waves by magnetic field of the earth.

magnetometer (*Surv., etc.*). Any instrument for measurement either of the absolute value of a magnetic field intensity, or of one component of this, e.g., horizontal component magnetometer for earth's magnetic field. See also **proton precessional-.**

magnetomotive force (*Mag.*). Line integral of the magnetic field intensity round a closed path. Abbrev. **m.m.f.**

magneton (*Electronics, Nuc.*). See **Bohr-.**

magneto-optical effect (*Light*). See **Kerr effect.**

magneto-optic rotation (*Chem.*). See **magnetic polarization.**

magnetophone (*Acous.*). Any recording device involving the magnetization of a medium, e.g., magnetic tape.

magnetoplasmadynamic generator (*Elec. Eng.*). A device which produces electrical energy from an electrically-conducting gas (plasma) flowing through a transverse magnetic field. Abbrev. **MPD generator.** Also called **magnetohydrodynamic generator.**

magnetoresistance (*Elec. Eng.*). The resistivity of a magnetic material in a magnetic field when direction of the current with reference to the field. If parallel to one another, the resistivity increases, but if mutually perpendicular, it decreases.

Magnetoresistor (*Elec. Eng.*). TN for device in which resistance is controlled by a magnetic field.

magnetostatics (*Mag.*). Study of steady-state magnetic fields.

magnetostriction (*Mag.*). Phenomenon of elastic deformation of certain ferromagnetic materials, e.g., nickel, on the application of a magnetizing force. Used in ultrasonic transducers and, in nickel wires, as a *fast store* in electronic computers.

magnetostriction loudspeaker (*Acous.*). Open diaphragm driven at its apex by a nickel rod, in which magnetostriction vibrations are excited by modulation currents in an embracing coil.

magnetostriction microphone (*Acous.*). One depending on the generation of an e.m.f. through magnetostriction reaction.

magnetostriction transducer (*Elec. Eng.*). Any device employing the property of magnetostriction to convert electrical to mechanical oscillations, e.g., by using a rod clamped at centre and passing a.c. through a coil wound around the rod. Also **magnetostrictor.**

magnetostrictive filter (*Elec. Eng.*). A filter network which utilizes magnetostrictive elements, bars, or rods, with their energizing coils.

magnetostrictive oscillation (*Mag.*). One based on the principle of the alteration of dimensions of a bar of magnetic material when the magnetic flux through it is changed. Nickel contracts with an increasing applied magnetic field but iron expands in weak fields and contracts in strong fields. The mechanical oscillatory system, e.g., a bar clamped at its centre, can be coupled magnetically to an amplifier to maintain the oscillation, which may be in the audiofrequency range.

magnetostrictive reaction (*Mag.*). The inverse magnetostrictive effect, i.e., a change in magnetization under applied stresses.

magnetostrictor (*Elec. Eng.*). See **magnetostriction transducer.**

magneto system (*Teleph.*). A local battery system in which the energy required for signalling is obtained from a generator at the subcriber's station.

magnet pole (*Elec. Eng.*). See **pole piece.**

magnetron (*Electronics*). Thermionic valve in which electrons released from a large cathode gyrate in an axial magnetic field before reaching the anode, their energy being collected in a series of slot resonators in the face of the circular anode. The output power, usually impulsive, is taken from a resonator by a small coupling loop.

magnetron arcing (*Electronics*). Arcing between anode and cathode as a result of desorption of gas during operation. Intermittent arcing is not uncommon with high-power pulse magnetrons.

magnetron critical field (*Electronics*). That which would just prevent an electron emitted from the cathode with zero energy from reaching the anode at a given value of anode voltage.

magnetron critical voltage (*Electronics*). The voltage which would just enable an electron emitted from the cathode with zero velocity to reach the anode at a given value of magnetic-flux density.

magnetron effect (*Electronics*). Deflection of electrons emitted from a thermionic filament by the magnetic field produced by the filament-heating current.

magnetron modes (*Electronics*). Different frequencies of oscillation corresponding to different field configurations and selected by strapping the cavities in various ways to control their phase differences.

magnetron oscillator (*Electronics*). One in which radial electrons from a heated cathode are accelerated towards one or more anodes, the electrons being controlled by a near axial magnetic field, and the frequency by resonant cavities in the anodes.

magnet steel (*Met.*). A steel from which permanent magnets are made. It must have a high remanence and coercive force. Steels for this purpose may contain considerable percentages of cobalt (up to 35%) as well as nickel, aluminium, copper, etc. See also **permanent magnet.**

magnettor (*Elec. Eng.*). A second-harmonic type of magnetic modulator which uses a saturable reactor to amplify d.c., or low alternating-frequency, signals.

magnet yoke (*Elec. Eng.*). Sometimes applied to the whole of the magnetic circuit of an electromagnet (or transformer, etc.), but strictly speaking, should refer only to the part which does not carry the windings.

magnification (*Maths.*). In a conformal mapping, given by $w=f(z)$, the magnification varies from point to point but is the same in all directions at a given point. The linear magnification at a point z_0 is given by $|f'(z_0)|$ and the surface magnification by $|f'(z_0)|^2$. (*Optics*) See **magnifying power.**

magnification factor (*Elec.*). See Q.

magnifier (*Radio*). Any thermionic amplifier, especially one used for the amplification of audiofrequencies.

magnifying power (*Optics*). The ratio of the apparent size of the image of an object formed by an optical instrument to that of the object seen by the naked eye. For a microscope, it is necessary to assume that the object would be examined by the naked eye at the least distance of distinct vision, 25 cm. Unless otherwise stated, the *linear* magnification is assumed to be indicated. See also **longitudinal magnification.**

Magnistor (*Elec. Eng.*). TN for type of toroidal reactor, which is used for specific circuit applications, e.g., in gating, switching, etc.

magnitudes (*Astron.*). The scale by which the brightness of stars is measured: (1) *apparent magnitude* is the measure of the brightness on Pogson's logarithmic scale, in which each step of one whole magnitude represents a light ratio of 2·512, and this increases numerically with decreasing brightness. See **telescopic stars.** Apparent magnitudes are measured as *visual* or *photographic*; if the total energy received on all wavelengths is considered, the brightness is given as a *bolometric magnitude.* (2) *absolute magnitude* is the apparent magnitude a given star would have at the standard distance of 10 parsecs.

magnolia metal (*Met.*). A lead-base alloy, containing 78–84% lead; remainder is mainly antimony, but small amounts of iron and tin are present. Used for bearings.

magnon (*Nuc.*). Quantum of spin wave energy in magnetic material.

Magnox (*Met.*). Group of magnesium alloys for canning uranium reactor fuel elements. Best known are Magnox B and Magnox A 12. The latter is Mg with 0·8% Al and 0·01% Be.

magnum (*Zool.*). One of the two inner bones of

the four in the distal row of carpal bones in Mammals.

Magnuminium (*Met.*). A series of magnesium-base alloys, rel. d. 1·8.

Magslip (*Elec. Eng.*). TN for a synchro system for remote control or indication.

Maguel (*Civ. Eng.*). A method used for the stretching and anchoring of high tensile wires in a post-tensioned system.

mahlstick (*Paint.*). A slender stick padded at one end with cloth or leather; used as a support for the hand guiding the brush. Also spelt **maulstick.**

mahogany (*For.*). See **African-, Brazilian-, Cuban-, Honduras-, Uganda-.**

mails (*Textiles*). See **heald.**

main airway (*Mining*). In colliery, one directly connected with point of entry to mine.

main and tail (*Mining*). Single-track underground rope haulage by means of a main rope to draw out the full wagons and a tail rope to draw back the empties.

main anode (*Electronics*). That carrying the load current in mercury-arc rectifiers which have an independent *excitation anode* (q.v.).

main beam (*Build.*). In floor construction, a beam transmitting loads direct to the columns.

main circuit (*Elec. Eng.*). See **current circuit.**

main contacts (*Elec. Eng.*). The contacts of a switch which normally carry the current; cf. *arcing contacts*, which carry the current at the instant when the circuit is being interrupted.

main couple (*Carp.*). The principal truss in a timber roof.

main deck (*Ships*). Usually means the principal deck or the upper of two decks or the second deck from the top when there are more than two decks, excluding the decks covering deck houses or erections. Frequently used to name a deck in a passenger liner.

main distribution frame (*Teleph.*). A frame for rearranging the incoming lines to a telephone exchange into the numerical order required in the exchange. Abbrev. **MDF.**

main exchange (*Teleph.*). One which has other exchanges, such as satellites, dependent on it for extension from them to other exchanges.

main field (*Elec. Eng.*). The chief exciting field in an electric machine, as opposed to an auxiliary field, such as that produced by the compoles.

main float (*Aero.*). The two single, or one central, float(s) which give buoyancy to a seaplane or amphibian.

main gap (*Electronics*). Total gap between the cathode and anode in a gas-discharge tube.

main memory (*Comp.*). The fastest storage device of a computer.

main plane (*Aero.*). One of the principal supporting surfaces, or wing, of an aeroplane or glider, which can be divided into centre, inner, outer, and/or wing-tip sections.

main rope (*Mining*). See **main and tail.**

main rotor (*Aero.*). (1) The principal assembly, or assemblies, of rotating blades which provide lift to a rotorcraft. (2) The assembly of compressor(s) and turbine(s) forming the rotating parts of a gas turbine engine.

main routine (*Comp.*). Coded directions which form the basis of a program in setting an electronic computer to perform a required operation on data, independently supplied.

mains (*Elec. Eng.*). Source of electrical power; normally the electricity supply system.

mains antenna (*Radio*). Electric-supply mains used as a receiving antenna by connexion to the antenna terminal of the receiver through a capacitor, which passes the radiofrequency currents but does not pass those of the supply frequency.

main sequence (*Astron.*). See **spectral types.**

mains frequency (*Elec. Eng.*). Electricity a.c. supply frequency; 50 Hz in U.K., 60 Hz in U.S.A., often 400 Hz in aircraft or ships. Also **power frequency.**

mainspring (*Horol.*). The spring in a watch or spring-driven clock which provides the motive power.

mainspring hook (*Horol.*). The means by which a mainspring is attached to its barrel.

mainspring winder (*Horol.*). A tool for coiling a mainspring prior to its insertion or withdrawal from the barrel.

mains receiver (*Radio*). A receiver which derives its operating power from the public supply mains.

maintained tuning fork (*Acous.*). One associated with an electronic oscillator so that the latter supplies energy continuously to maintain the fork in steady oscillation. The frequency of the oscillation is substantially that of the free fork, and provides a method for establishing frequencies with great accuracy. Also **tuning-fork oscillator.**

maintaining power (*Horol.*). A device which permits of power being transmitted to the train while winding is in progress.

maintaining voltage (*Elec. Eng.*). That which just maintains ionization and discharge.

main tanks (*Aero.*). See under **fuel tanks.**

main tie (*Build.*). The lower tensional members of a roof-truss, connecting the feet of the principal rafters.

main transformer (*Elec. Eng.*). A term sometimes used in connexion with the Scott transformer connexion to indicate the transformer which is connected across 2 of the phases on the 3-phase side.

main wheel (*Horol.*). The first wheel in the train; the great wheel. In a going-barrel, it is integral with the barrel.

maisonette (*Arch., Build.*). A two-storey self-contained dwelling incorporated in a building of three or more storeys.

maize oil (*Chem.*). See **corn oil.**

Majorana force (*Nuc.*). Force between nucleons which changes sign if, in the wave function of the two particles, their space coordinates are interchanged. See **short-range forces.**

major axis (*Maths.*). Of an ellipse: see **axes.**

major cycle (*Comp.*). The access time to a consecutive access memory such as a magnetic disk or tape loop.

majority carrier (*Electronics*). In a semiconductor, the electrons or *holes*, whichever carry most of the measured current. Cf. *minority carrier*.

majority emitter (*Electronics*). Electrode releasing majority carriers into a region of semiconductor.

major lobe (*Telecomm.*). That containing the direction of highest sensitivity or maximum radiation for any form of *polar diagram* (q.v.).

majuscule (*Typog.*). A capital letter as distinct from the small letter or minuscule.

make (*Teleph.*). The operation, partial or complete, of a telephone relay, when current is passed through its windings.

make-before-break contact (*Teleph.*). The group of contacts in a relay assembly so arranged that the one which moves makes contact with a front contact before it separates from a back contact, and so is never free, whether the relay is operated or not.

make-contact (*Teleph.*). A pair of contacts in a relay assembly which are brought together on operation of the relay and so close a circuit.

make even (*Typog.*). To arrange type so that the last word of a portion or 'take' of copy ends a full line.

make impulse (*Telecomm.*). An impulse arising when a circuit is *made*, i.e., closed.

make-ready (*Print.*). See **making-ready.**

make-up (*Typog.*). The arrangement of type-matter and blocks into pages.

making-capacity (*Elec. Eng.*). A term used in connexion with the rating of switchgear to denote the capability of a switch to make a circuit under certain specified conditions.

making-current (*Elec. Eng.*). A term used in connexion with switchgear to denote the maximum peak of current which occurs at the instant of closing the switch.

making paper (*Print.*). See **excess feed.**

making-ready or make-ready (*Print.*). Preparing for printing by underlaying or interlaying blocks and inserting overlays in the packing of the platen or cylinder to produce the required printing pressure.

making-up (*Textiles*). (1) Examination, packing, and ticketing of fabrics before dispatch from the mills. (2) Factory manufacture of dresses and suits, etc., from the cut-out patterns.

Maksutov telescope (*Astron.*). See **meniscus telescope.**

mala (*Zool.*). In some Insects (as many coleopterous larvae), a single lobe borne by the maxilla and possibly homologous with the galea.

malachite (*Min.*). Basic copper(II)carbonate $CuCO_3 \cdot Cu(OH)_2$, crystallizing in the monoclinic system. It is a common ore of copper, and occurs typically in green botryoidal masses in the oxidation zone of copper deposits.

malachite green (*Chem.*). A triphenylmethane dyestuff of the rosaniline group. It has the formula:

$(CH_3)_2N$ — C = $\overset{+}{N}(CH_3)_2$ Cl^-

Obtained by the condensation of benzaldehyde with dimethylaminobenzene in the presence of $ZnCl_2$, HCl, or H_2SO_4, and by the oxidation of the resulting leuco-base with PbO_2.

malacia (*Med.*). Pathological softening of any organ or tissue.

Malacocotylea (*Zool.*). An order of *Trematoda* in which the ventral sucker may be absent, or, if present, is generally anterior in position; the position of the genital pore is variable; the sexual form occurs in connexion with the gut of Vertebrates, the asexual generations in *Mollusca*; all are internal parasites with a complex life history involving more than one host, e.g., *Fasciola hepatica* (q.v.).

malacology (*Zool.*). The study of Molluscs.

malacon (*Min.*). A variety of zircon which is partially *metamict* (q.v.).

malacophily (*Bot.*). Pollination by snails.

malacophyllous (*Bot.*). Xerophytic, and having fleshy leaves containing much water-storing tissue.

malacoplakia (*Med.*). The occurrence of soft, rounded, pale plaques in the wall of the bladder, in chronic inflammation of the bladder.

malacopterous (*Zool.*). Having soft fin-rays.

Malacostraca (*Zool.*). A class of *Crustacea*. On the head are two compound eyes (typically stalked) biramous antennules, antennae with a scalelike exopodite, and three pairs of mouth-parts; the eight-segmented thorax is typically covered by a carapace, on the first three segments are maxillipeds, and on the last five, walking legs; the abdomen has six (rarely seven) segments, the first five bearing biramous swimming appendages (pleopods) and the last a pair of broad uropods which, with the telson, form a tail-fan. No caudal furca present except in *Leptostraca*. Woodlice, Crabs, Prawns, Shrimps, Lobsters, and Crayfish.

malacostracous (*Zool.*). Having a soft shell.

malar (*Zool.*). Pertaining to the *mala* (q.v.); pertaining to, or situated in, the cheek region of Vertebrates; the *jugal* (q.v.).

malaria (*Med.*). Infection with *Plasmodium*, a genus of protozoa that live in red cells of the blood. The recurrent fever occurs in *quartan*, *tertian* or subtertian (malignant) forms according to the species of Plasmodium. Common in many tropical areas; transmitted by anopheline mosquitoes.

malathion (*Chem.*). *S*-[1,2-di(ethoxycarbonyl) ethyl]dimethyl phosphorothiolothionate, used as an insecticide.

malaxation (*Zool.*). A process, adopted by certain Wasps after stinging their prey, repeated compression of the neck by the mandibles.

malchite (*Geol.*). A term applied to rocks which have been described as microdiorite or dioritic lamprophyre.

mal de caderas (*Vet.*). A chronic infectious disease of horses in South America; due to *Trypanosoma equinum*, and characterized by weakness of the hind-quarters.

mal de los pintos (*Med.*). See **pinta.**

mal du coit (*Vet.*). See **dourine.**

male (*Zool.*). An individual of which the gonads produce spermatozoa or some corresponding form of gamete. Cf. *female*.

male and female (*Eng.*). Trade terms applied to inner and outer members respectively of pipe-fittings, threaded pieces, etc. See also **external screw-thread, internal screw-thread.** (*Elec. Eng., Telecomm.*) Applied similarly to a plug and its complementary socket.

male flower (*Bot.*). A flower containing stamens but no pistils.

maleic acid (*Chem.*). *cis*-Butenedioic acid. HOOC·CH : CH·COOH, large prisms or plates, m.p. 130°C, b.p. 160°C, with decomposition into its anhydride and water, readily soluble in water. The *cis*-configuration is ascribed to this acid, whereas its isomer, *fumaric acid* (q.v.), has the *trans*-configuration. Important in polyester resins.

maleic anhydride (*Chem.*). *cis*-Butenedioic anhydride. The anhydride of *maleic acid*. Important in establishing the structure of organic compounds containing conjugated double bonds. Also important industrially, mainly as an intermediate. Manufactured by passing a mixture of benzene vapour and air over heated vanadium pentoxide:

CH—C(=O)
O
CH—C(=O)

maleic hydrazide (*Chem.*). Cyclic hydrazide, m.p. 303°C. Used as a plant growth regulator and sprouting inhibitor:

$$\begin{array}{c} CH = CH \\ OC \qquad\quad CO \\ NH - NH \end{array}$$

male pronucleus (*Cyt.*). Nucleus of the spermatozoon.

male thread (*Eng.*). See **external screw-thread.**

malfunction (*Comp.*). See **mistake.**

malic acid (*Chem.*). $HOOC{\cdot}CH_2{\cdot}CH(OH){\cdot}COOH$, hygroscopic needles, m.p. 100°C., found in unripe fruit; it also occurs in wines. When attacked by certain ferments, butanoic lactic, and propanoic acids are produced. Heat causes the loss of a molecule of water, producing maleic and fumaric acids. It has been synthesized by various methods.

malignant (*Med.*). Tending to go from bad to worse; especially, cancerous (see **tumour**).

malignant aphtha (*Vet.*). See **contagious pustular dermatitis.**

malignant catarrhal fever (*Vet.*). A noncontagious disease of cattle, believed to be caused by a virus; characterized by fever, mucopurulent discharge from nose and eyes, ulceration of gums and inside cheeks, and swollen lymph nodes. Thought to be transmitted by insects. Also called **gangrenous coryza.**

malignant endocarditis (*Med.*). Septic endocarditis, infective endocarditis. A progressive bacterial infection of one or more of the valves of the heart.

malignant oedema (*Vet.*). An acute toxaemia of cattle, sheep, pigs, and horses due to wound infection by *Clostridium septicum* or *Clostridium novyi* (*Cl. oedematiens*).

malignant stomatitis (*Vet.*). See **calf diphtheria.**

malignite (*Geol.*). An alkaline igneous rock, a *melanocratic* variety of nepheline syenite.

mall (*Tools*). See **beetle.**

malleability (*Met.*). Property of metals and alloys which affects their alteration by hammering, rolling, extrusion. Temperature, as it affects crystallization, may alter resistance.

malleable cast-iron (*Met.*). A variety of cast-iron which is cast white, and then annealed at about 850°C to remove carbon (*white-heart process*) or to convert the cementite to rosettes of graphite (*black-heart process*). Distinguished from grey and white cast-iron by exhibiting some elongation and reduction in area in tensile test. See also **pearlitic iron.**

malleable iron (*Met.*). Now usually means malleable cast-iron, but the term is sometimes applied to wrought-iron.

malleable nickel (*Met.*). Nickel obtained by remelting and deoxidizing electrolytic nickel and casting into ingot moulds. Can be rolled into sheet and used in equipment for handling food, for coinage, condensers, and other purposes where resistance to corrosion, particularly by organic acids, is required.

mallein (*Vet.*). A concentrated filtrate of broth cultures of *Actinobacillus mallei* (*Malleomyces mallei*) which have been killed by heat; used as an inoculum for the diagnosis of glanders in horses.

mallenders and sallenders (*Vet.*). Psoriasis affecting the skin at the flexures of the carpus (knee) joint (*mallenders*) and tarsus (hock) joint (*sallenders*) in the horse.

malleolar (*Zool.*). Pertaining to, or situated near, the malleolus; in *Ungulata*, the reduced fibula.

malleolus (*Zool.*). A process of the lower end of the tibia or fibula.

mallet (*Tools*). A wooden hammer, or one made of raw-hide or rubber.

mallet-finger (*Med.*). Permanent flexion of the end-joint of a finger or thumb.

malleus (*Zool.*). In Mammals, one of the ear ossicles; in *Rotifera*, one of the masticatory ossicles of the mastax; more generally, any hammer-shaped structure.

Mallophaga (*Zool.*). An order of the *Psocopteroidea* (*Paraneoptera*); ectoparasites, usually of birds, with reduced eyes, flattened form, tarsal claws and biting mouthparts, e.g., biting lice.

Mallory battery (*Chem.*). TN for type of dry cell with mercury zinc electrodes. These have larger capacity than the conventional dry cell and are less subject to polarization. Their e.m.f. is about 1 volt. Used in cadmium sulphide light meters, radiation monitors and miniature electronic devices.

Mallory's triple stain (*Micros.*). A polychromatic microanatomical stain, containing oxalic acid, aniline blue W.S. and orange G, and used with acid fuchsin. A method used to demonstrate collagen fibres, which stain dark blue.

malm (*Build.*). An artificial imitation of natural marl made by mixing clay and chalk in a wash mill; the product is used as a clay for the manufacture of bricks. Also called **washed clay.**

malm rubber (*Build.*). A soft form of malm brick, capable of being cut or rubbed to special shapes.

malonic acid (*Chem.*). Propanedioic acid. $HOOC{\cdot}CH_2{\cdot}COOH$, soluble in water, alcohol, ether; m.p. 132°C; it decomposes at a slightly higher temperature, giving ethanoic acid; it occurs in beetroot as its calcium salt and can be obtained from malic acid by oxidation with chromic acid. *Malonic ester*, or diethyl malonate, is a liquid of aromatic odour, b.p. 198°C, $CH_2(CO{\cdot}OC_2H_5)_2$. The hydrogen of the methylene group is replaceable by sodium which in turn can be exchanged for an alkyl group. In this way, malonic ester is important for the synthesis of higher dibasic acids. Numerous derivatives of *malonylurea*, $C_4H_4O_3N_2$, barbituric acid, are used as hypnotics and anaesthetics.

Malpighian body (or **corpuscle**) (*Zool.*). In the Vertebrate kidney, the expanded end of a uriniferous tubule surrounding a glomerulus of convoluted capillaries: in the Vertebrate spleen, one of the globular or cylindrical masses of lymphoid tissue which envelops the smaller arteries.

Malpighian cell (*Bot.*). One cell of a layer of closely packed, radially-directed thick-walled cells occurring in the testas of some seeds.

Malpighian layer (*Zool.*). The innermost layer of the epidermis of most Chordates, containing polygonal cells which continually proliferate, dividing by mitosis. Also known as **rete Malpighii, stratum germinativum.**

Malpighian pyramids (*Zool.*). In many Mammals, the conical portions into which the kidney is divided.

Malpighian tubes (*Zool.*). In Insects, *Arachnida*, and *Myriapoda*, tubular glands of excretory function opening into the alimentary canal, near the junction of the mid-gut and hind-gut.

malpresentation (*Med.*). Abnormal posture of the foetus during birth.

malt (*Brew.*). Grain, such as barley, oats, wheat, that has been germinated artificially by means of moisture, and then dried slowly in a kiln.

The rate and the temperature of drying produce malt classed as amber, brown, black, and pale.

Malta fever (*Med.*). See **undulant fever.**

maltase (*Chem.*). An enzyme effecting the hydrolysis of the $\alpha 1 \rightarrow 4$ linkage in maltose and other α-glucosides. Present in many yeasts, in the liver, kidney, pancreas, and other organs, and the digestive juices of many animals.

Maltese cross (*Cinema.*). Device of this shape on a shaft carrying the film sprockets, used for converting continuous rotation to intermittent motion by means of a cam with a striker fin which engages in the slots of the cross, moving it a quarter-turn for every revolution of the cam.

malthenes (*Chem.*). Such constituents of asphaltic bitumen as are soluble in carbon disulphide and petroleum spirit. See **asphaltenes, carbenes.**

malting (*Brew.*). The processes by which barley, a hard vitreous grain, is converted into malt. See **couching, flooring, steeping, withering.**

maltobiose (*Chem.*). See **maltose.**

maltose (*Chem.*). 4-*O*-(α-D-glucopyranosyl)-D-glucopyranose, maltobiose, $C_{12}H_{22}O_{11}$, a disaccharide, a white crystalline mass, dextrorotatory, formed by the action of diastase upon starch during the germination of cereals. It reduces Fehling's solution, and when hydrolysed is converted into D-glucose. In cereal chemistry, the *maltose figure* indicates the natural sugar content and diastatic activity of flour or meal.

malt-sugar (*Chem.*). See **maltose.**

Malus' law (*Optics*). For a plane-polarized beam of light incident on a polarizer,

$$I = I_0 \cos^2 \theta,$$

where I = intensity of transmitted beam; I_0 = intensity of incident beam; θ = angle between plane of vibrations of beam and plane of vibrations which are transmitted by the polarizer.

Malvernian direction (*Geol.*). The name assigned to structures, resulting from earth-movements, having a direction predominantly north-south, after the trend of the structures in the Malvern Hills.

mamilla (*Zool.*). A nipple.

mamillary body (*Zool.*). See **corpus mamillare.**

mamillated (*Geol.*). Describes the structure of an aggregate of crystals displaying a smooth rounded outer surface.

mamma (*Anat., Zool.*). In female Mammals, the milk gland; the *breast* (q.v.). *adj.* **mammary.**

Mammalia (*Zool.*). A class of Vertebrates. The skin is covered by hair (except in aquatic forms) and contains sweat glands and sebaceous glands; they are homoiothermous; the young are born alive (except in the *Prototheria*) and are initially nourished by milk; respiration is by lungs and a diaphragm is present; the circulation is double, and only the left systemic arch is present; dentition heterodont and diphyodont; there is a double occipital condyle; the lower jaw articulates with the squamosal; the long bones and vertebrae have three centres of ossification; there is an external ear and three auditory ossicles in the middle ear; large cerebral hemispheres.

mammato-cumulus (*Meteor.*). Clouds with rounded protuberances on their lower surfaces, something like inverted cumulus. They often occur in thunder-clouds.

mammogenic (*Physiol.*). Promoting growth of the duct and alveolar systems of the mammary gland, e.g., applied to hormones.

mammography (*Radiol.*). The radiological examination of the breast with or without the injection of a *contrast medium* (q.v.).

mamu (*Nuc.*). Abbrev. for *millimass unit.*

Man (*Zool.*). The human race, all living races being included in the genus *Homo*, suborder *Anthropoidea* of the *Primates*. Man's distinguishing features include elaboration of the brain and behaviour, including communication by facial gestures and speech and development of the power of abstract thought; the erect posture; the structure of the limbs (including the opposable thumb) and skull; the dentition with small canines; the long period of postnatal development associated with parental care. The human stock probably diverged from that of the *Anthropoid* apes in early Miocene times.

Manchester yellow (*Chem.*). See **Martius yellow.**

mandelic acid (*Chem.*). Phenyl-glycollic acid; $C_6H_5 \cdot CH(OH) \cdot COOH$, glistening crystals; m.p. 133°C; soluble in water. It occurs naturally in the form of its glycoside, *amygdalin* (q.v.), and can be synthesized by the hydrolysis of benzaldehyde cyanhydrin. Mandelic acid possesses an asymmetric carbon atom, and exists in a (+) and a (−) form and as the **racemic** compound.

mandible (*Zool.*). In Vertebrates, the lower jaw; in *Arthropoda*, a masticatory appendage of the oral somite; in *Polychaeta* and *Cephalopoda*, one of a pair of chitinous jaws lying within the buccal cavity. *adj.* **mandibular, mandibulate.**

mandibular disease (*Vet.*). Shovel beak. A malformation, accompanied by bacterial infection, affecting the beak of young chicks; associated with feeding dry meal.

mandibular glands (*Zool.*). In some Insects, glands opening near the articulation of the mandibles; in some *Lepidopterous larvae*, they function as salivary glands, the true salivary glands secreting silk; also present in the hive bee and other adult *Hymenoptera*.

mandibular groove (*Zool.*). In some *Crustacea*, a groove crossing the cheek immediately behind the mandible, indicating the boundary of the head of the nauplius larva.

mandibular palps (*Zool.*). In *Crustacea*, the distal part of the mandible, which may aid either in locomotion, or in feeding.

mandrel (*Eng.*). (1) An accurately turned shaft on which work already bored is mounted for turning, milling, etc.; if partially split and capable of expansion by a tapered plug, it is called an *expanding mandrel.* (2) The headstock spindle of a lathe. Also **mandril.** (*Glass*) (1) A refractory tube used in the manufacture of glass tubing or rod. (2) A former used in making lamp-blown articles from tubing and in making precision-bore tubing. (*Horol.*) A special faceplate lathe, usually hand driven, for turning or recessing watch plates or similar operations. (*Met.*) Mandril; arbor (U.S.). Accurately turned rod over which metal is forged, drawn or shaped during working so as to create or preserve desired axial cavity.

maneb (*Chem.*). Manganese ethene-1,2-bisdithiocarbamate, used as a fungicide.

maneton (*Aero., I.C. Engs.*). A heavy pinch-bolt used to grip the split end of the separate rear crank web of a radial aero engine on to the crank-pin.

manganates. See **permanganates.**

manganblende (*Min.*). See **alabandite.**

manganepidote (*Min.*). See **piedmontite.**

manganese (*Chem.*). A hard, brittle metallic element, in the seventh group of the periodic system, which exists in 3 polymorphic forms, α, β, and γ, and has a complicated crystal structure. It is brilliant white in colour, with reddish tinge. Symbol Mn, at. no. 25, r.a.m.

54·94, valency 2, 3, 4, 6, 7, rel. d. at 20°C 7·39, m.p. 1220°C, b.p. 1900°C, electrical resistivity $5{\cdot}0 \times 10^{-8}$ ohm metres, hardness in Mohs' scale 6. The principal manganese ores are *pyrolusite*, *hausmannite*, *psilomelane* (qq.v.); pure manganese is obtained electrolytically. Manganese is mainly used in steel manufacture, as a deoxidizing and desulphurizing agent.

manganese alloys (*Met.*). Manganese is not used as the basis of alloys, but is a common constituent in those based on other metals. It is present in all steel and cast-iron, and in larger amount in special varieties of these, e.g., manganese steel, silico-manganese steel; also in many varieties of brass, in aluminium-bronze, and aluminium and nickel base alloys.

manganese bacteria (*Bacteriol.*). Chemosynthetic bacteria which utilize the energy derived from the oxidation of manganous to manganic ions for the assimilation of carbon dioxide.

manganese bronze (*Met.*). Originally an alpha-beta brass containing about 1% of manganese; the term is now applied generally to *high-strength brass* (q.v.), with up to 4% manganese.

manganese dioxide (*Chem.*). Manganese(IV)oxide, MnO_2. Black solid, insoluble in water; occurs in *pyrolusite* (q.v.). Basic and (mainly) acidic. Forms manganites. Oxidizing agent; uses: depolarizer in Leclanché cells, decolorizing oxidant for green ferrous ion in glass, laboratory reagent.

manganese epidote (*Min.*). See **piedmontite.**

manganese garnet (*Min.*). See **spessartine.**

manganese heptoxide (*Chem.*). Manganese(VII) oxide Mn_2O_7. A heavy, dark coloured oil, prepared by adding concentrated sulphuric acid to potassium manganate(VII). Unstable, explosive, strong oxidizing agent. Acidic. Forms manganates(VII).

manganese spar (*Min.*). See **rhodochrosite.**

manganese steel (*Met.*). A term sometimes applied to any steel containing more manganese than is usually present in carbon steel (i.e., 0·3–0·8%), but generally to *austenitic* (*Hadfield's*) *manganese steel*, which contains 11–14%. This steel is resistant to shock and wear. Used for railway crossings, rock crusher parts, etc.

manganin (*Met.*). A copper-base alloy, containing 13–18% of manganese and 1·5–4% of nickel. The electrical resistivity is high (about 38×10^{-8} ohm metres), and its temperature coefficient low; it is therefore suitable for resistors.

manganite (*Min.*). A grey or black hydrous oxide of manganese, crystallizing in the monoclinic system (*pseudo-orthorhombic*). It is a minor ore of manganese.

mangano-manganic oxide (*Chem.*). Manganese(II, III)oxide. Mn_3O_4. Neutral or mixed oxide.

manganophyllite (*Min.*). A phlogopite or biotite mica containing appreciable manganese; it occurs in aggregations of thin scales and has a colour ranging from bronze to copper-red.

manganosite (*Min.*). The protoxide of manganese (MnO) which crystallizes in the cubic system.

manganotantalite (*Min.*). See **tantalite.**

manganous oxide (*Chem.*). Manganese(II)oxide. MnO. Basic. Forms manganese(II) salts. Occurs naturally as *manganosite*.

mange (*Vet.*). Inflammation of the skin of animals due to infection by certain species of acarids or mites. See **chorioptic-, demodectic-, notoedric-, otodectic-, psoroptic-, sarcoptic-.**

mania (*Psychiat.*). The elated phase of manic-depressive psychosis; characterized by euphoric excitement, exaggerated will-power, and flight of ideas of a grandiose nature. See **manic-depressive psychosis.**

manic-depressive psychosis (*Psychiat.*). A type of mental illness characterized by disorders of affect, either of elation or of depression, with intermediate mixed states. The depressed phase (melancholia) may exist by itself, and be repeated after an interval of normal health, or it may alternate with a phase of elation (mania). General characteristics include irritability, suspicion, and, in some cases, a clouding of consciousness, delusions, hallucinations, and disorientations. Also **cyclothymia.**

manifold (*Eng.*). A pipe or chamber with several openings. See **exhaust-, induction-.** (*Paper*) A term describing thin paper used in duplicating; may be waxed.

manifold pressure (*Aero.*). The absolute pressure in the induction manifold of a supercharged reciprocating aero-engine which, indicated by a cockpit gauge, is used together with r.p.m. settings to control engine power output and fuel consumption. See also **supercharger.**

manifold pressure gauge (*Aero.*). See **boost gauge.**

manilla paper (*Paper*). A strong paper made from unbleached wood-pulp, sometimes containing hemp fibre and manilla (about 40% at most). Used for envelopes and folders and impregnated with wax, oil, or varnish, as a dielectric for cables.

manipulation (*An. Behav.*). Behaviour in which an animal moves objects with the digits of its forelimbs. Sometimes thought to give rise to a drive in its own right.

manipulator (*Nuc. Eng.*). Remote handling device used, e.g., with radioactive materials.

man lock (*Civ. Eng.*). An air-lock enabling workmen to pass into, and out of, spaces filled with compressed air.

manna (*Zool.*). See **honey dew.**

mannans (*Chem.*). The anhydrides of mannose.

manned space flight (*Space*). Initial attempts to put man into space led to manned earth satellites, as in the American project *Mercury*, and the Russian *Vostok*. The American project *Apollo* first landed men on the Moon on 21 July 1969.

Mannesmann process (*Met.*). A process for making seamless metal tubing from a solid bar of metal by the action of 2 eccentrically mounted rolls which simultaneously rotate the bar and force it over a mandrel.

Mannheim process (*Chem.*). In manufacture of sulphuric acid, catalysis of SO_2 in two stages, first with iron(III)oxide and then with platinized asbestos.

Manning's formula (*Hyd.*). An empirical expression giving the value of the coefficient C in *Chezy's formula* (metric). It states that

$$C = \frac{m^{\frac{1}{6}}}{n}$$

where m = the hydraulic mean depth in metres, and n = a factor depending on the nature of the surface of the channel. See **Ganguillet and Kutter's formula.**

mannitol (*Chem.*). $HO{\cdot}CH_2{\cdot}(CH{\cdot}OH)_4{\cdot}CH_2{\cdot}OH$; a hexahydric alcohol, fine needles or rhombic prisms, soluble in water and in hot alcohol, m.p. 166°C. It is found in many plants, and is an important diuretic.

mannosans (*Chem.*). See **mannans.**

mannose (*Chem.*). A hexose. D-Mannose can be obtained by the oxidation of *mannitol* (q.v.), and is a stereoisomer of D-glucose.

manocryometer (*Chem.*). A combination manometer and cryoscope for measuring the effect of pressure on melting point.

manocyst (*Bot.*). The receptive papilla in some *Oömycetes*.

manometer (*Phys.*). An instrument used to measure the pressure of a gas or liquid. The simplest form consists of a U-tube containing a liquid (water, oil or mercury), one limb being connected to the enclosure whose pressure is to be measured, while the other limb is either open to the atmosphere or is closed or otherwise (e.g., by float) connected to a registering or recording mechanism. There are, however, numerous more sophisticated variations (e.g. balanced hollow ring, bell chamber, diaphragm, and metallic bellows). See also **pressure gauge, Bourdon gauge.**

manometric flame (*Phys.*). A small gas flame so arranged as to oscillate by the variations of pressure due to sound-waves impinging on a diaphragm forming one wall of a small gas chamber. The oscillations of the flame are detected by viewing it in a rotating mirror.

manoscopy (*Chem.*). The measurement of gas densities.

manostat (*Eng.*). Device for keeping the pressure in an enclosure at a constant level.

Manouvrier's index (*Nut.*). The height of the human frame, divided by the sitting height, less unity. A comparative index, which may indicate a disturbance of nutrition.

manoxylic wood (*Bot.*). Wood of somewhat loose texture, containing a good deal of parenchyma.

mansard roof (*Build.*). A double-sloped pitched roof rising steeply from the eaves, and having a summit of flatter slope on both sides of the ridge. Also called **curb roof, French roof, gambrel roof.**

Mansbridge capacitor (*Elec. Eng.*). See **self-sealing capacitor.**

mantel tree (*Build.*). The lintel of a fireplace.

mantissa (*Maths.*). The positive, fractional part of a logarithm. See **logarithm.**

mantle (*Arch.*). The outer covering of a wall when this is of different material from the inner part. (*Geol.*) That part of the Earth lying between the crust and the core, extending from a depth of approx. 30 km to one of approx. 2900 km. (*Light.*) See **gas mantle.** (*Zool.*) In *Urochorda*, the true body-wall lying below the test and enclosing the atrial cavity; in *Mollusca*, *Brachiopoda*, and *Cirripedia*, a soft fold of integument which encloses the trunk and which is responsible for the secretion of the shell or carapace.

mantle block (*Glass*). (1) Refractory block fitting in the gable wall, over the *dog-house* (q.v.). (2) The upper course of blocks enclosing the Fourcault drawing pit.

mantle cavity (*Zool.*). In *Urochorda*, the atrial cavity: in *Mollusca*, *Brachiopoda*, and *Cirripedia*, the space enclosed between the mantle and the trunk.

mantled gneiss dome (*Geol.*). A dome-like structure consisting of granite at the centre, surrounded by gneiss, and intruded into low-grade regionally metamorphosed rocks. Such structures are found in Pre-Cambrian shield areas.

mantle lobes (*Zool.*). The mantle flaps in *Lamellibranchiata*.

Mantoux test (*Bacteriol.*). See **tuberculin.**

manual hold (*Teleph.*). The holding of the connexion from a subscriber on an automatic exchange by an operator in a manual exchange who is about to complete the call.

manual ringing (*Teleph.*). The application of ringing current to a line by the manual throwing of a key, as performed, for example, by PBX operators.

manuals (*Zool.*). See **primaries.**

manubrium (*Zool.*). Any handle-like structure; the basal part of the furcula in *Collembola*, the anterior sternebra in Mammals; part of the malleus of the ear in Mammals; the pendant oral portion of a medusa; part of the malleus in *Rotifera*.

manufacturing gauge (*Eng.*). A gauge used by machine operators in the production of parts, as distinct from an inspection or master gauge.

manure (*Agric.*). Natural or artificial food material for plants and trees, supplying phosphates, nitrogen, and potash. Natural manure from the farmyard contains these constituents, and is helped out by decayed leaves (compost), animal products, etc. See **fertilizer.**

manure distributor (*Agric.*). A machine for distributing artificial manures. It operates in a manner similar to the *broadcast sower* (q.v.).

manus (*Zool.*). The podium of the fore-limb in land Vertebrates.

manway (*Mining*). Underground ladderway in shaft or through a filled stope.

many-one function switch (*Comp.*). The converse of a *one-many function switch*, in which all of a number of inputs must be excited before the one output is energized. Realization of the logical AND function.

manyplies (*Zool.*). See **psalterium.**

map comparison unit (*Radar*). A device for superimposing a radar screen display on a map or chart of an area, thus facilitating the identification of radar echoes.

maple (*For.*). Various species of the genus *Acer*, found in Europe, Asia, and America, used for cabinet work, flooring, musical instruments, panelling, etc.

maple syrup and **maple sugar** (*Chem.*). Syrup and sugar made from sap tapped from the maple tree. The delicate flavour is probably due to the small protein content.

map measurer (*Surv.*). An instrument used to find the length of a route on a map. It consists of a small wheel which is made to roll over the route, in so doing actuating a needle which records the distance traversed. Also called **opisometer.**

mapping (*Maths., etc.*). In modern mathematical literature generally regarded as equivalent to *function*.

MAR (*Chem.*). An abbrev. for *microanalytical reagent*, a standard of purity which indicates that a reagent is suitable for use in *microanalysis* (q.v.).

maraging (*Met.*). Heat treatment used to harden alloy steels. These include nickel-iron martensites with high toughness and resistance to shock and saline corrosion.

marasmus (*Med.*). Progressive wasting, especially in infants. *adj.* **marasmic.**

marble (*Geol.*). The term strictly applies to a granular crystalline limestone, but in a loose sense it includes any calcareous or other rock of similar hardness that can be polished for decorative purposes.

marble bones (*Med.*). See **osteopetrosis.** (*Vet.*) See **osteopetrosis gallinarum.**

marblewood (*For.*). A straight grained, fine and evenly textured hardwood with figuring of dark black bands on a grey-brown background, from *Diaspyros kurzii*, a native of the Andaman and Nicobar Islands.

marbling (*Bind.*). The operation of decorating book-edges, etc., with a variegated 'marble' effect. Carried out in a trough containing gum (made from carragheen moss), on the surface of which pigments (containing ox-gall) are worked in intricate patterns.

marcasite (*Min.*). (1) White iron pyrite. A disulphide of iron which crystallizes in the orthorhombic system. It resembles iron pyrite, but has a rather lower density, is less stable, and is paler in colour when in a fresh condition. (2) In the gemstone trade, *marcasite* is either pyrite, polished steel (widely used at present in ornamental jewellery in the form of small 'brilliants'), or even white metal.

Marcellus Shale (*Geol.*). A highly bituminous black shale, of Middle Devonian age, occurring at the top of the Ulsterian Division in eastern North America; it is the source rock of the oil stored in the overlying Onondaga Limestone.

marcescent (*Bot.*). Withered but remaining attached to the plant.

marchioness (*Build.*). A slate, 22× 12 in. (558× 305 mm).

Marchi's fluid (*Micros.*). A mixture of Müller's fluid and osmic acid. See **Marchi's method.**

Marchi's method or reaction (*Micros.*). A technique for staining degenerating myelin with Marchi's fluid. The excess unsaturated fatty acids present give a black stain with osmic acid.

Marconi antenna (*Radio*). Original simple vertical wire, fed between base end and earth.

Marconi beam antenna (*Radio*). A directional antenna array comprising a system of tuned vertical radiators and reflectors.

Marconi coherer (*Radio*). An evacuated glass tube containing 2 electrodes in contact with a mixture of iron and nickel filings. The resistance between the electrodes drops suddenly on the application of a high-frequency voltage between them.

Marconi detector (*Radio*). A form of magnetic detector comprising an endless iron or steel wire which is drawn continuously through a coil, connected to the antenna, and situated in the field of a permanent magnet. The arrival of a signal causes a current to be induced in a second coil, wound over the first, which is made audible in telephones that are connected thereto.

marconigram (*Teleg.*). See **radiogram.**

marcus (*Tools*). A large hammer with an iron head.

mare (*Astron.*). See **maria.**

Marechal's test (*Chem.*). A test for the presence of bile acids based on the appearance of a green zone between the bile solution and a dilute solution of iodine.

marekanite (*Geol.*). A rhyolitic perlite broken down into more or less rounded pebbles, named from the type locality, Marekana river, Eastern Siberia.

Marek's disease (*Vet.*). See **fowl paralysis.**

Marezzo marble (*Build.*). An artificial marble made with Keene's cement.

Marform process (*Eng.*). A proprietary production process for forming and deep-drawing irregularly shaped sheet metal parts, in which a confined rubber pad is used on the movable platen of a press and a rigid punch on the fixed platen. The unformed portion of the blank is subjected to sufficient pressure during the progress of the operation to prevent wrinkling.

margaric acid (*Chem.*). Heptadecanoic acid $CH_3(CH_2)_{15}COOH$. A long-chain saturated 'synthetic' fatty acid. It has an odd number of carbon atoms in the molecule in contrast with the even number of almost all the naturally-occurring fatty acids.

margarine (*Chem.*). A butter substitute made from (usually) vegetable fats, suitably treated by heating and cooling, churning with milk, colouring, and adding concentrates of vitamins A and D.

margarite (*Geol.*). An aggregate of minute spherelike crystallites, arranged like beads, found as a texture in glassy igneous rocks. (*Min.*) Hydrated silicate of calcium and aluminium, crystallizing in the monoclinic system, often as lamellae with a pearly lustre; one of the so-called *brittle micas.*

margin (*Build.*). The exposed width of each slate in coursed work. (*Join.*) The flat surface of stiles or rails in panelled framing. (*Telecomm.*) Number of decibels increase in gain necessary to result in oscillation. Applied to feedback amplifiers, telephone repeaters, public address systems, etc. (*Teleg.*) Maximum percentage change in unit of signal before errors arise in decoding.

marginal anchors (*Zool.*). In certain *Scyphozoa*, adhesive organs attached to the edge of the umbrella.

marginal bars (*Join.*). Glazing bars so arranged as to divide the glazed opening into a large central part bordered by narrow panes at the edges.

marginal community (*Bot.*). A plant community bordering on another community of slightly different character.

marginal current (*Teleph.*). The adjusted heavy or light, positive or negative, current which is used for coding impulses in a coder, to expedite the transmission of numericals by key-senders.

marginal ore (*Mining*). Ore which, at current market value of products from its excavation and processing, just repays the cost of its treatment.

marginal ray cell (*Bot.*). A more or less specialized cell occurring with others of the same kind on the edge of a vascular ray.

marginals (*Zool.*). In *Chelonia*, the plates forming the edge of the carapace.

marginal species (*Bot.*). A plant which grows along the edge of woodland.

marginal testing (*Telecomm., etc.*). A form of test where the operation of a piece of equipment is tested with its operating conditions altered to decrease the safety margin against faults, e.g., an amplifier may be required to give a certain minimum gain with reduced supply voltage.

margin draft (*Build.*). A smooth face round a joint in ashlar work.

margin lights (*Build.*). Narrow panes of glass near the edges of a sash.

margin of safety line (*Ships*). A line drawn 3 in. below the upper surface of the bulkhead deck at the side. A passenger ship is subdivided into watertight compartments in such a way that this line is not submerged if two adjacent watertight compartments are flooded.

margin trowel (*Build.*). A box-shaped float for finishing internal angles.

maria (*Astron.*). The Latin designation of the so-called 'seas' on the lunar surface, named before the modern telescope showed their dark areas to be dry planes. Since 1959 spacecraft probes and landings (manned and unmanned) have provided much detailed information but their origin is still problematical. The *sing.* is *Mare* (e.g. *Mare Imbrium*), but there is a marked tendency to delatinize this picturesque terminology (e.g., Sea of Tranquillity, Sea of Fertility, Ocean of Storms, Sea of Moscow, etc.). See **mascons, Moon.**

marialite (*Min.*). Silicate of aluminium and sodium with sodium chloride, crystallizing in the tetragonal system. It is one of the end-members in the isomorphous series of the scapolite group.

marigold window (*Arch.*). See **rose window.**

marigram (*Surv.*). The continuous record of height of tide given by a self-registering tide gauge.

marihuana (*Bot.*). See **cannabis.**

marine boiler (*Eng.*). A cylindrical boiler, of large diameter and short length, provided with two or more furnaces in flue tubes leading to combustion chambers, surrounded by water, at the back. The gases pass through banks of fire tubes to the smoke-box or uptakes at the boiler front. Also called **Scotch boiler.** See also **dry-back boiler, Yarrow boiler.**

marine chronometer (*Horol.*). A specially mounted chronometer for use on board ship in the determination of longitude.

marine compass. See **floating compass, gyro compass.**

marine denudation (*Geol.*). The erosive and sweeping action of the sea. See also **denudation.**

marine deposits (*Geol.*). Rock waste which is laid down under marine conditions.

marine engineering. That branch of mechanical engineering concerned with the design and production of propelling machinery and auxiliary equipment for use in ships.

marine engines (*Eng.*). Engines used for ship propulsion.

marine glue (*Build.*). A form of glue resisting the action of water, and containing rubber, shellac, and oil.

mariner (*Space*). Name for a series of unmanned U.S. space probes designed to undertake preliminary exploration of the planets of the Solar System. The first planets investigated were Venus and Mars, close-up television pictures of the latter being received. Later probes in the series will be aimed at the more remote planets, gaining further information on the nearer ones on the way.

marine screw propeller (*Eng.*). A boss, carrying 2, 3, or 4 blades of helical form, which produces the thrust to drive a ship by giving momentum to the column of water which it displaces in an astern direction.

marine surveying. Hydrographical surveying undertaken in tidal waters.

Mariotte's law (*Phys.*). Same as **Boyle's law.**

mark (*Telecomm.*). Departure (positive or negative) from a neutral or no-signal state (space) in accordance with a code, using equal intervals of time.

mark-and-space (*Teleg.*). The on-and-off of a signal code, the *mark* corresponding to a depression of a telegraph (Morse) key. The actual current may be positive, zero, or negative for mark or space.

mark contact (*Teleg.*). The contact which is made when a telegraph key is depressed and alters the current in a transmitting circuit.

marker (*Aero.*). See **aerodrome-.** (*Cinema.*) See **camera-.** (*Comp.*). See **sentinel.** (*Radar*) Pip on a radar display for calibration of range. (*Teleph.*) Device used in a common control system which tests and selects incoming and outgoing circuits and causes switching apparatus to connect them together.

marker antenna (*Aero., Radio*). One giving a beam of radiation for marking air routes, often vertically for blind or instrument landing.

marker attachment (*Agric.*). Disk-type marker for making accurate joints between bouts.

marker beacon (*Aero., Radio*). A radio beacon in aviation which radiates a signal to define an area above the beacon. See **fan-, inner-, middle-, outer-, Z-, glide-path beacon, localizer beacon, locator beacon, radar beacon, VHF rotating talking-beacon; ramark** and **track guide.**

marker horizon (*Geol.*). A layer of rock in a sedimentary sequence which, because of its distinctive appearance or fossil content, is easily recognized over large areas, facilitating correlation of strata.

marker light (*Rail.*). An indicating light on a signal post, to indicate the position or aspect of the main signal should its light have failed.

marker pulses (*Telecomm.*). (1) Pulses superimposed on a CRT display, e.g., in order to facilitate time measurements. (2) Pulses used to synchronize transmitter and receiver, e.g., in time-division multiplex.

marker system (*Teleph.*). A system of automatic switching in which the setting up of a speech path is controlled by a marker from information supplied by a register.

markfieldite (*Geol.*). A variety of diorite with a porphyritic structure and a granophyric groundmass.

marking gauge (*Carp.*). A tool for marking lines on the work parallel to one edge of it. It consists of a wooden bar having a projecting steel marking pin near one end and a sliding block adjustable for position along the bar.

marking hammer (*For.*). A hammer for marking letters, figures or other devices on trees, logs or converted timber.

marking knife (*Join.*). A small steel tool having a chisel edge at one end and pointed at the other. It is used for setting out fine work.

marking-out (*Build., Surv., etc.*). Setting out boundaries and levels for a proposed piece of work. (*Eng.*) Setting out centre lines and other dimensional marks on material, as a guide for subsequent machining operations. See also **laying-out.**

marking-up (*Typog.*). Adding typographic instructions to the copy or the layout as a guide to setting.

marking (or keying) wave (*Radio*). Wave, slightly different in frequency from the spacing wave, which corresponds to the mark of the signal code.

mark of reference (*Typog.*). A sign which directs the reader to a footnote. The commonest marks of reference are * asterisk, † dagger, ‡ double dagger, § section, ‖ parallel, ¶ paragraph, in order of use. *Superior figures* (q.v.) are now more commonly used.

mark-space multiplier (*Telecomm.*). Circuit which averages square-wave signals, ratio of mark/space in time being determined by one voltage, amplitude by another voltage. Used in analogue computing.

mark/space ratio (*Telecomm.*). The ratio of the time occupied by a mark to that occupied by the space in between marks in a telecommunication channel or recording system using pulses rather than modulated signals.

marl (*Agric.*). A soil consisting of an indefinite mixture of calcareous clay or calcareous loam. (*Build.*) A brick earth which contains a high percentage of carbonate of lime; it is the best clay for making bricks without addition of other substances. Also called **calcareous clay.** (*Geol.*) A general term for a very fine-grained rock, either clay or loam, with a variable admixture of calcium carbonate.

marl yarn (*Textiles*). Yarn of two or more colours

which have been combined in drawing or while the roving is being drafted and spun.

marmatite (*Min.*). A ferruginous variety of *sphalerite* (q.v.); it contains up to 20% of iron.

Marmarosch diamonds (*Min.*). A local name for *rock crystal.*

marmite disease (*Vet.*) An exudative eruption of the skin in piglets, believed to be due to deficiency of the vitamin B complex.

marmoration (*Build.*). A marble casing for a building.

marmoratum or marmoretum (*Build.*). A cement containing pulverized marble and lime.

marmorization (*Geol.*). The recrystallization of a limestone to give a marble, usually the result of heating, as where an igneous rock is intruded into the limestone.

marocain (*Textiles*). A dress fabric with a fine repp, in appearance like a wavy rib, across the width; made from hard twisted yarns of silk, rayon, wool, cotton, or nylon.

marquise (*Arch.*). A projecting canopy over the entrance to a building.

marrow (*Zool.*). The vascular connective tissue which occupies the central cavities of the long bones in most Vertebrates, and also the spaces in certain types of cancellated bone.

marrying (*Cinema.*). The printing of the mute negative and the negative sound-track on the release print. For 'rushes', the mute and sound-track are printed separately.

Mars (*Astron.*). The fourth planet from the sun, revolving in an eccentric orbit at a mean distance of 1·524 A.U. in 687 days. Its diameter is about half and its mass 0·107 that of the earth; its rotation in 24·6 hours and inclination of 24°·8 resemble those of the earth. Space probes have shown that the planet has an extremely rarefied atmosphere and that the surface is pitted with craters which resemble those on the surface of the Moon.

Marsden squares (*Meteor.*). The Marsden system divides the earth's surface into 10° 'squares' by meridians and parallels of latitude 10° apart, with further subdivision to 1°. The system is used by the Meteorological Office for charting meteorological, climatological, and oceanographic data.

Marshall valve gear (*Eng.*). A radial gear of the *Hackworth* (q.v.) type, in which the straight guide is replaced by a curved slot to correct inequalities in steam distribution.

Marsipobranchii (*Zool.*). See **Cyclostomata.**

Marsupiala (*Zool.*). The single order included in the Mammalian group *Metatheria* (q.v.), and having the characteristics of the subclass, e.g., Opossum, Tasmanian wolf, Marsupial mole, Kangaroo, Koala bear.

marsupium (*Zool.*). A pouchlike structure occupied by the immature young of an animal during the later stages of development; as the abdominal pouch of metatherian Mammals. *adj.* **marsupial.**

martempering (*Met.*). See **austempering.**

martensite (*Met.*). A constituent formed in steel when it is cooled at a rate sufficiently rapid to suppress the change from austenite to pearlite. Results from the decomposition of austenite at low temperatures. Consists of a solid solution of carbon in α-iron, and is responsible for the hardness of quenched steel.

Martin's cement (*Build.*). A quick-setting hard plaster, made by soaking plaster of Paris in a solution of potassium carbonate.

Martin's diameter (*Powder Tech.*). A *statistical diameter* (q.v.) defined as the mean length of a line intercepting the profile boundary of the image of the particle and dividing the image into two portions of equal area. The bisecting line is always taken parallel to the direction of traverse, irrespective of the orientation of each particle.

martite (*Min.*). A variety of *haematite* (Fe_2O_3) occurring in dodecahedral or octahedral crystals believed to be pseudomorphous after magnetite, and in part perhaps after iron pyrite.

Martius yellow (*Chem.*). Salts of dinitro-α-naphthol, used as a pigment and for colouring soap; poisonous. Also known as **Manchester yellow.**

Martonite (*Chem.*). A tear gas, a mixture (4 : 1) of monobromopropanone and monochloropropanone.

marver (*Glass*). A flat cast-iron or stone (marble) block upon which glass is rolled during the hand method of working.

Marx-Elmendorf tearing test (*Paper*). See **Elmendorf tear tester.**

Marx rectifier (*Elec. Eng.*). See **atmospheric arc rectifier.**

Marx generator (*Radiol.*). A low to high voltage converter used, e.g. for operating portable X-ray equipment from d.c. supplies. These give high voltage microsecond duration pulses which are usually applied to cold cathode X-ray tubes. Also **surge generator.** See **flash radiography.**

Marzine (*Pharm.*). TN for N-methyl-N-benzhydryl piperazine monochloride, used in prevention of travel sickness.

mascons (*Astron.*). *Mass concentrations.* Regions of high gravity occurring within certain *maria* (q.v.) of the Moon. Their origin is still conjectural.

maser (*Phys., etc.*). (*M*icrowave *a*mplification by *s*timulated *e*mission of *r*adiation.) An extremely stable low-noise amplifier or oscillator, operating by the interaction between radiation (photons) and atomic particles (electrons, atoms, molecules). When an electron in an atom is raised from a lower (ground) to a higher over an intermediate energy state, it has a tendency to drop down to the intermediate state immediately. This tendency can be checked by the application of a low temperature and a powerful magnetic field. In this way energy can be stored in a body acting as a maser (or a laser), and when subsequently it is exposed to photons corresponding to the inhibited transition, they are not absorbed but bounced from atom to atom, causing a rapid release of the stored energy as coherent monochromatic radiation of great intensity as an amplified form of the incident signal.

maser relaxation (*Phys.*). The process by which excited molecules in the higher energy state revert spontaneously or under external stimulation to their equilibrium or ground state. Maser action arises when energy is released in this process to a stimulating microwave field which is thereby reinforced.

mash (*Brew.*). A porridgelike mixture consisting of coarsely ground malt and hot water which have been mixed in a masher.

mashing (*Brew.*). The process of extracting the grist (ground malt) with water, at a temperature of 63–68°C, in a mash tun.

mash seam welding (*Eng.*). An electrical resistance welding process in which the slightly overlapping edges of the workpiece are forged together during welding by broad-faced, flat electrodes.

mash tun (*Brew.*). An insulated metal vessel in which the mash is mechanically mixed and kept at a temperature of about 65°C for a period (about 2 hr) when the sweet wort is run off.

mask (*Cinema., TV*). (1) To obstruct light or sound from an object in a significant manner, as a speaker from a microphone, an actor from an actor or camera. (2) An aperture in a projector which determines the extent of a projected image on a screen. (*Comp.*) A machine word which specifies which part of another word is to be processed, or extracts a selected group of characters from a sequence. (*Photog.*) Opaque material used to define the size and shape of a photographic print or to confine or limit the size of the effective limit. (*Psychol.*) See **persona.** (*Zool.*) A prehensile structure of the nymphs of certain Dragon-flies (*Odonata*), formed by the labium.

masked valve (*I.C. Engs.*). A poppet valve, the head of which is recessed into its seat, so that its outer diameter acts as a piston valve; it allows a lower valve acceleration to be used, and gives a more efficient *valve-opening diagram* (q.v.).

maskelynite (*Min.*). A glass which occurs in colourless isotropic grains in meteorites and has the composition of a plagioclase. It probably represents re-fused feldspar.

masking (*Acous.*). Loss of sensitivity of the ear for specified sounds in the presence of other sounds, one sound *masking* sounds of higher frequency to a marked extent.

masking frame (*Photog.*). Device for holding printing paper in an enlarger, comprising a baseplate to which is hinged a frame forming a border for the print, with movable metal strips for adjusting horizontal and vertical margins.

masking-out (*Typog.*). Covering parts of the forme with a cut-out mask of thin paper so that only the required areas print. Useful when taking proofs in more than one colour from a single forme.

masking paper (*Paper*). A strong kraft paper coated with self-adhesive on one side. Used in the drawing-office trade.

Masonite (*Build.*). TN for a form of very hard building-board used as a lining or surfacing.

masonry cement (*Build.*). A cement incorporating inert material and an air entrainer to increase plasticity in mortar. Specially useful in minimizing possibility of damage in cold weather work.

mason's joint (*Build.*). Pointing finished with a projecting vee.

mason's level (*Build.*). See **plummet level.**

mason's mitre (*Build.*). The name given to an effect similar to the *mitre* (q.v.) but produced (particularly in stonework) by shaping the intersection out of the solid.

mason's putty (*Build.*). A mixture of Portland cement, lime putty, and stone dust, usually in the proportions 2 : 5 : 7, with water. Used for making fine joints, especially in ashlar work.

mason's scaffold (*Build.*). A form of scaffold used in the erection of stone walls, when it is not convenient to leave holes for the support of one end of the putlogs; an inner set of standards and ledgers is used to provide this support.

mason's stop (*Build.*). A *mason's mitre* (q.v.).

mason's trap (*Build.*). A form of trap in which a stone slab on edge dips below the water surface.

mass (*Phys.*). The quantity of matter in a body. Mass may also be considered as the equivalent of inertia, or the resistance offered by a body to change of motion (i.e., acceleration). Masses are compared by weighing them, which amounts to comparing the forces of gravitation acting on them. The mass of nuclear particles cannot be regarded as constant, since the mass of any body is increased if it is moving with a velocity comparable with that of light. For such particles, the usual practice is to quote 'rest masses' and these are frequently given in terms of energy; see **mass-energy equivalence.**

mass absorption coefficient (*Nuc.*). See **absorption coefficient.**

mass action (*Chem.*). See **law of mass action.**

mass balance (*Aero.*). A weight attached ahead of the hinge line of an aircraft control surface to give static balance with no moment about the hinge, and to reduce to zero inertial coupling due to displacement of the control surface. Mass balancing is a precaution against control surface *flutter* (q.v.).

mass concrete (*Civ. Eng.*). Concrete which is placed without reinforcement, and employing large-sized coarse aggregate to reduce the mortar content. Also called **bulk concrete.**

mass control (*Acous.*). Said of mechanical systems, particularly those generating sound waves, when the mass of the system is so large that the compliance and resistance of the system are ineffective in controlling motion.

mass decrement or **mass defect** (*Nuc.*). Both terms may be used for (1) the binding energy of a nucleus expressed in mass units, and (2) the measured mass of an isotope less its mass number. The B.S.I. recommend the use of *mass decrement* for (2) above. In the U.S., it is usually used for (1) and *mass excess* for (2). See also **packing fraction.**

massed trials (*An. Behav.*). An experimental procedure in which animals are given several consecutive trials, there being little or no interval between the end of one trial and the beginning of the next. Cf. *spaced trials.*

mass effect (*Met.*). The tendency for hardened steel to decrease in hardness from the surface to the centre, as a result of the variation in the rate of cooling throughout the section. Becomes less marked as the rate of cooling required for hardening decreases, i.e., as the content of alloying elements increases.

mass-energy equivalence (*Phys.*). Confirmed deduction from relativity theory, such that

$$E = mc^2,$$

where E = energy, m = mass, and c = velocity of light. Also **Einstein energy.**

masseter (*Zool.*). An elevator muscle of the lower jaw in higher Vertebrates. *adj.* **masseteric.**

mass excess (*Nuc.*). See **mass decrement.**

Massey formula (*Nuc.*). One giving the probability of secondary electron emission when an excited atom approaches the surface of a metal.

mass-haul curve (*Civ. Eng.*). A curve used in the design of earthworks involving cuttings and embankments, the abscissae representing chainage along the centre line, and the ordinates the excess of cutting over filling, i.e., the material requiring to be hauled to another position.

massicot (*Min.*). Lead(II)oxide. A rare mineral of secondary origin, associated with galena. (*Paint.*) A yellow pigment, the same chemically as litharge.

mass-luminosity law (*Astron.*). A relationship between the mass and absolute magnitude of stars, the most massive stars being the brightest; applicable to all stars except the white dwarfs.

mass number (*Nuc.*). Total of *protons* and *neutrons* in a nucleus, each being taken as a unit of mass. Also **nuclear, or nucleon, number.**

mass of the electron (*Phys.*). A direct consequence of the special theory of relativity is that the effective mass of the electron varies with its velocity, v, according to:

$$m=\frac{m_0}{\left(1-\frac{v^2}{c^2}\right)^{\frac{1}{2}}},$$

where m_0 = rest mass of electron $9{\cdot}1\times10^{-31}$ kg, c = velocity of light *in vacuo*.

Masson's trichrome stain (*Micros.*). A polychromatic microanatomical stain in which iron haematoxylin, Ponceau-acid fuchsin or xylidene Ponceau and light green are used.

mass provisioning (*An. Behav.*). In certain species of Insects, the provision by the mother (or in the case of social Insects by nurse individuals) of sufficient food to last the offspring through the whole of its larval life. Cf. *nest provisioning*, *progressive provisioning*.

mass radiator (*Electronics*). Radiator which consists of metal particles suspended in a liquid dielectric to provide a source of broad-band electromagnetic radiation up to U.H.F., when a sufficiently high voltage is applied to the dielectric to produce sparking between the particles.

mass ratio (*Aero., Space*). The ratio of the take-off weight of a rocket-propelled vehicle to that when all fuel has been consumed.

mass resistivity (*Elec. Eng.*). The product of the volume resistivity and the density of a given material at a given temperature.

mass screening (*An. Behav.*). An experimental procedure by which a large number of animals, especially Insects, are run simultaneously in a multiple choice apparatus. This gives useful information about the variability of, e.g., the degree of geotaxis in a population.

mass spectrograph (*Phys.*). A vacuum system in which positive rays of various charged atoms are deflected through electric and magnetic fields so as to indicate, in order, the charge-to-mass ratios on a photographic plate, thus measuring the atomic masses of isotopes with precision. System used for the first separation for analysis of the isotopes of uranium. See also **positive rays**.

mass spectrometer (*Phys.*). A *mass spectrograph* in which the charged particles are detected electrically instead of photographically.

mass spectrum (*Phys.*). See **spectrum**.

mass stopping power (*Nuc.*). See **stopping power**.

mass transfer (*Chem. Eng.*). The transport of molecules by convection or diffusion, as in the operations of extraction, distillation, and absorption.

mass transfer coefficient (*Chem. Eng.*). The molecular flux per unit driving force. Symbol k, K.

mass unit (*Phys.*). See **atomic mass unit**.

mast (*Bot.*). The fruit of the beech and related trees. (*Eng.*) A slender vertical structure which is not self-supporting and requires to be held in position by guy-ropes. Cf. *pylon*.

mast antenna (*Radio*). An antenna in which the currents are carried by the metallic structure of the mast itself, instead of in conductors supported thereby.

mastax (*Zool.*). In *Rotifera*, the gizzard or masticatory part of the alimentary canal, between the mouth and the oesophagus.

mast cell (*Zool.*). See **basicyte**.

mastectomy (*Surg.*). Surgical removal of the breast.

master (*Acous.*). In gramophone-record manufacture, the metal (negative) disk obtained by plating the original lacquer surface on which a recording has been cut. (*Eng.*) (1) The term applied to special tools, gauges, etc., used for checking the accuracy of others used in routine work. (2) The chief or key member of a system, as the *master cylinder* of a hydraulic brake mechanism.

master clock (*Comp.*). See **clock**. (*Elec. Eng.*) A clock for use with certain electrical time-keeping equipment. It sends out impulses at predetermined time-intervals for the operation of other clocks or similar equipment which are thereby kept in synchronization.

master connecting-rod (*Aero., I.C. Engs.*). The main member of the master and articulated assembly of a radial aero-engine. It incorporates the crank-pin bearing and the *articulated rods* of the other cylinder oscillate on it by means of wrist pins.

master controller (*Elec. Eng.*). A controller for electrical equipment, the operation of which energizes or de-energizes the contactors which perform the actual switching operations which may be remote from the operator.

master cylinder (*Eng.*). The container holding the fluid which operates the mechanism of hydraulic brakes or clutch.

master factor (*Bot.*). Any powerfully acting ecological factor which plays the main part in determining the occurrence in a given area of a plant community of major rank.

master frequency meter (*Elec. Eng.*). See **integrating frequency meter**.

master gain control (*Acous.*). (1) In a broadcasting studio, the attenuator connected between the programme input and main amplifiers in order to regulate the gain as desired. (2) In a stereo amplifier, the control which adjusts simultaneously the gains of both channels to equalize them.

master gauge (*Eng.*). A standard or reference gauge made to specially fine limits; used for checking the accuracy of inspection gauges.

master haulier (*Mining*). See **pusher-on**.

master oscillator (*Radio*). One, often of low power, which establishes the frequency of transmission of a radio transmitter.

master station (*Radio*). Transmitting station from which one or more 'satellite' stations receive a programme for re-broadcasting.

master switch (*Elec. Eng.*). A switch for controlling the effect of a number of other switches or contactors; e.g., if the master switch is open, none of the other switches is operative.

master tap (*Eng.*). A substandard screw-tap, sometimes used after the plug-tap when great accuracy is required.

master telephone transmission reference system (*Teleph.*). High-quality transmission system provided with means for calibrating its transmitting, attenuating and receiving components in terms of absolute units. Used for comparison of quality between different systems. Located at CCIT laboratories in Geneva.

mastic (*Build.*). A term applied to preparations used for bedding and pointing window frames, bedding wood-block flooring and repairing flat roofs. (*Chem.*) A pale-yellow resin from the bark of *Pistacia lentiscus*, used in the preparation of fine varnishes.

mastic asphalt (*Build.*). A mixture of bitumen with stone chippings or sand, used for roofing, roads, paving, and damp-proof courses.

mastication (*Zool.*). The act of reducing solid food to a fine state of subdivision or to a pulp.

masticator (*Chem.*). An apparatus consisting of 2 revolving and heated cylinders studded with teeth or knives; used for converting rubber into a homogeneous mass.

masticatory (*Zool.*). Pertaining to the trituration of food by the mandibles, teeth, or gnathobases, prior to swallowing.

masticatory stomach (*Zool.*). See **gastric mill.**

mastigonemes (*Zool.*). Small lateral projections found along the length of some flagella.

Mastigophora (*Zool.*). A class of *Protozoa* which possess one or more flagella in the principal phase; may be amoeboid but usually have a pellicle or cuticle; often parasitic but rarely intracellular; have no meganucleus; reproduction mostly by longitudinal binary fission; nutrition may be holophytic, saprophytic, or holozoic. Also **Flagellata.**

mastitis (*Med., Vet.*). Inflammation of the mammary gland.

mastodynia (*Med.*). Pain in the breast.

mastoid (*Bot., Zool.*). Resembling a nipple; as a posterior process of the otic capsule in the Mammalian skull.

mastoidectomy (*Surg.*). Excision of the (infected) air-cells of the mastoid bone.

mastoiditis (*Med.*). Inflammation of the air-cells of the mastoid bone.

masturbation (*Psychol.*). Stimulation, usually by oneself, of the sexual organs by manipulation, etc., so as to produce orgasm.

mat (*Cinema.*). In cinematograph-film printing, a colloquial term for *mask* (q.v.). (*Print.*) The commonly-used contraction for *matrix* (q.v.).

match-boarding or **match-lining** (*Carp.*). *Matched boards* (*q.*v.).

matched boards (*Carp.*). Boards specially cut at the edges to enable close joints to be made, either by tongue and groove or by rebated edge.

matched load (*Telecomm.*). One with impedance equalling the characteristic impedance of a line or waveguide, so that there is no reflection of power, and the received power is the maximum possible. Also **matched termination.** See **termination.**

matching (*Carp.*). See **matched boards.** (*Radar*) Section in a waveguide, for impedance transformation, depending on metal or dielectric therein. (*Telecomm.*) Adjusting a load impedance to match the source impedance with a transformer or network, so that maximum power is received, i.e., so that there is no reflection loss due to *mismatch.* The principle applies to many physical systems, e.g., non-reflecting optical surfaces, use of a horn in loudspeakers to match impedance of vibrator to that of air, matching load on electrical transmission line. Reactance in the load impedance can be neutralized or *tuned-out* by an equal reactance of opposite sign.

matching planes (*Carp.*). See **match planes.**

matchings (*Textiles*). Wool from different fleeces that has been sorted to one quality.

matching stub (*Telecomm.*). Short or open circuited stub line attached to main line to neutralize reactive component of load and so improve matching.

matching transformer (*Telecomm.*). One expressly inserted into a communication circuit to avoid reflection losses because the load impedance differs from the source impedance. In designing for optimum matching, the ratio of the impedances equals the square of the ratio of the turns on the windings.

match planes (*Carp.*). A pair of planes used to cut the tongue and the groove, respectively, on matched boards. Also **matching planes.**

matchwood (*For.*). Billets suitable for manufacturing into matches.

matelassé (*Textiles*). A compound cloth with a brocade or quilted face and, usually, a plain weave back; made from different materials and used for fancy vestings, etc.

materialization (*Phys.*). Reverse of Einstein energy released with annihilation of mass. A common example is by *pair production* (electron-positron) from gamma rays.

material lock (*Civ. Eng.*). An air-lock enabling materials, tools, etc., to be passed into or taken out from spaces filled with compressed air.

materials handling (*Eng.*). The process, which includes mechanical handling, of transporting and positioning raw materials, semifinished and finished products in connexion with industrial operations, by conveyors, cranes, trucks, hopperfeeds, etc.

materials testing reactor (*Nuc. Eng.*). See **high flux reactor.**

maternal deprivation (*Psychol.*). Inadequate mothering of a child either by ignoring it, rejecting, or maintaining a distorted relationship. Such actions are thought to affect the child differently at different ages and may be the cause of the affectionless personality, i.e., one who is incapable of feeling affection, has no conscience, etc.

mathematical check (*Comp.*). A programmed check which depends upon known mathematical relationships between the input data.

mathematical induction (*Maths.*). An argument for verifying identities when the variable is limited to positive integral values. The two steps of the argument are (1) by assuming the correctness of the identity when the variable equals n, to deduce the corresponding identity for when the variable equals $n+1$, (2) to verify that the identity is correct when the variable has a low value such as 1 or 2. Step (1) shows that if the identity is ever correct it is correct for all subsequent values of the variable, and step (2) shows that the identity is correct for some low value of the variable.

mathematical logic. The application of mathematical techniques to logic in an attempt both to deduce new propositions by formal manipulations and to detect any underlying inconsistencies. Its study, by many eminent mathematicians and philosophers, as a means of clarifying the basic concepts of mathematics, has revealed a number of paradoxes, several of which have yet to be resolved. Also called **symbolic logic.** Cf. *Boolean algebra.*

mathematics. May be defined as the study of the logical consequences of sets of axioms or postulates. No completely satisfactory definition however has yet been produced. It comprises two main branches, *analysis* and *algebra,* concerned respectively with continuous and discrete variation. Each of these branches has many subdivisions, e.g., the *calculus* is a branch of analysis and *matrix algebra* of algebra.

mating (*Build., etc.*). Said of surfaces or pieces which come into contact or interlock.

matlockite (*Min.*). Fluorochloride of lead ($PbFCl$), which occurs in tabular tetragonal crystals. Rare.

matrix (*Acous.*). Negative obtained by electrodeposition on the lacquer disk on which a sound recording has been made. Also **solid metal negative.** (*Biol.*) An outer layer of stainable material in a chromosome. (*Bot.*) Any substratum, living or dead, in which a fungus grows. (*Build.*) The lime or cement constituting the cementing material that binds together the aggregate in a mortar or concrete. (*Maths.*) A system of elements of real or complex numbers arranged in a square or rectangular formation, e.g.,

$$\begin{bmatrix} a & b & c \\ d & e & f \\ g & h & j \end{bmatrix}$$

(*Powder Tech.*) The phase or phases which form the continuous skeleton of a powder body, thus forming the cells in which constituents imparting particular qualities may be held. (*Print.*) The mould from which type is cast, produced by an impression from a punch: also, the mould made from a relief surface in stereotyping or electrotyping. Usually contracted to mat. (*TV*) In colour television, an array of coordinates symbolic of a colour coordinate transformation. (*Zool.*) The intercellular ground-substance of connective tissues.

matrix circuit (*TV*). Circuit in which signals can be added without reaction on their respective sources, especially those for controlling colour TV images.

matrixing (*TV*). In colour TV, performing a colour coordinate transformation by computation with electrical or optical methods.

matrix unit (*TV*). Device which performs a colour coordinate transformation by electrical or optical methods.

matroclinous (*Biol.*). Exhibiting the characteristics of the female parent more prominently than those of the male parent. Cf. *patroclinic.*

matromorphic (*Biol.*). Resembling the mother.

matt or **matte**. Smooth but dull; tending to diffuse light; said, e.g., of a surface painted or varnished so as to be dull or flat.

matte (*Met.*). Fusion product consisting of mixed sulphides produced in the smelting of sulphide ores. In the smelting of copper, for example, a slag containing the gangue oxides and a matte consisting of copper and iron sulphides are produced. The copper is subsequently obtained by blowing air through the matte, to oxidize the iron and sulphur.

matter (*Med.*). See **pus.** (*Phys.*) The substances of which the physical universe is composed. Matter is characterized by gravitational properties (on the earth by weight) and by its indestructibility under normal conditions.

Matthiessen hypothesis (*Elec. Eng.*). That the total electrical resistivity of a metal may be equated to the sum of the various resistivities due to the different sources of scattering of free electrons; this also applies to thermal resistivity.

matting (*Cinema.*). Closing the width of the effective variable-density sound-track during recording so that the corresponding area prints black. The reduced area available for the modulation then requires extra modulation in compensation, but the result is an improved speech/noise ratio in sequences where the modulation is low.

mattress (*Civ. Eng.*). (1) Sheet expanded metal for reinforcement of concrete roads. (2) A rectangular mesh of reinforcing rods used normally in reinforced concrete foundations.

matt weave (*Textiles*). See **hopsack weave.**

Matura diamonds (*Min.*). Colourless zircons from Ceylon, which on account of their brilliancy are useful as gemstones. A misleading name.

maturation (*Bot., Zool.*). The final stages in the development of the germ cells.

maturation divisions (*Cyt.*). The divisions by which the germ cells are produced from the primary spermatocyte or oöcyte, during which the number of chromosomes is reduced from the diploid to the haploid number. See **meiosis.**

Mauch Chunk Beds (*Geol.*). A group of red beds of continental origin, partly of Middle, partly of Upper Mississippian age, occurring in Pennsylvania.

Mauchline Lavas (*Geol.*). A series of lava flows and tuffs in Ayrshire, Scotland, which are followed above by red sandstones. The whole is possibly of Permian age.

maul (*Tools*). See **beetle.**

maulstick (*Paint.*). See **mahlstick.**

mauveine (*Chem.*). *Perkin's mauve*, the first synthetic dyestuff,

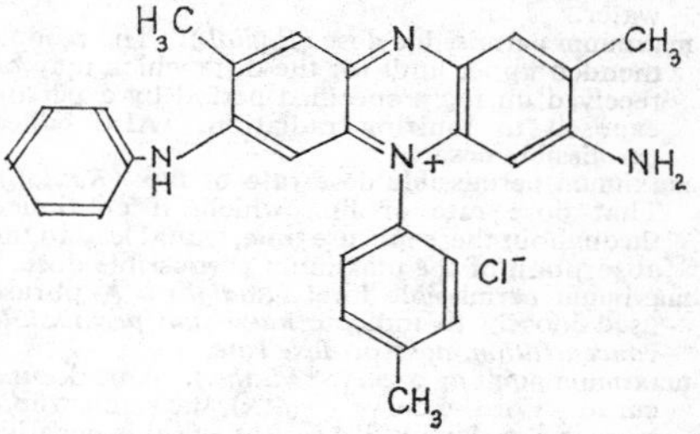

mavar (*Telecomm.*). *M*ixed (or *m*odulating) *a*mplification by *v*ariable *r*eactance. See **parametric amplifier.**

max gross (*Aero.*). See **maximum weight.**

maxilla (*Zool.*). In Vertebrates, the upper jaw: a bone of the upper jaw: in *Arthropoda*, an appendage lying close behind the mouth and modified in connexion with feeding. *pl.* **maxillae.** *adj.* **maxillary, maxilliferous, maxilliform.**

maxillary (*Zool.*). Pertaining to a maxilla: pertaining to the upper jaw: a paired membrane bone of the Vertebrate skull which forms the posterior part of the upper jaw.

maxillary glands (*Zool.*). In some *Crustacea*, excretory glandular organs in the region of the maxillae.

maxilliped (*Zool.*). In *Arthropoda*, especially *Crustacea*, an appendage behind the mouth, adapted to assist in the transference of food to the mouth.

maxilloturbinal (*Zool.*). In Vertebrates, a paired bone or cartilage of the nose which supports the folds of the olfactory mucous membrane.

maxillule (*Zool.*). In *Arthropoda*, one of the first pair of maxillae, if there is more than one pair.

maximally-flat (*Radio*). Said of thermionic amplifiers when designed so that the circuit elements are transformed from wave-filter sections incorporating stray admittances.

maximum and minimum thermometer (*Meteor.*). An instrument for recording the maximum and minimum temperatures of the air between 2 inspections, usually a period of 24 hr. A type widely used is *Six's thermometer* (q.v.).

maximum continuous rating (*Aero.*). See **power rating.**

maximum crest (*Telecomm.*). Same as **amplitude peak.**

maximum demand (*Elec. Eng.*). The maximum load taken by an electrical installation during a given period. It may be expresesd in kW, kVA, or amperes.

maximum-demand indicator (*Elec. Eng.*). An instrument for indicating the maximum demand which has occurred on a circuit within a given period.

maximum-demand tariff (*Elec. Eng.*). A form of charging for electrical energy in which a fixed charge is made, depending on the consumer's maximum demand, together with a charge for each unit (kWh) consumed.

maximum equivalent conductance (*Elec. Eng.*). The value of the equivalent conductance of an electrolytic solution at infinite dilution with its own solvent.

maximum flying speed (*Aero.*). See **flying speed.**

maximum landing weight (*Aero.*). See **weight.**

maximum permissible concentration (*Radiol.*). That for radioactive contaminants in air or water.

maximum permissible dose (*Radiol.*). The recommended upper limit for the dose which may be received during a specified period by a person exposed to ionizing radiation. Also called **permissible dose.**

maximum permissible dose rate or flux (*Radiol.*). That dose rate or flux which, if continued throughout the exposure time, would lead to the absorption of the maximum permissible dose.

maximum permissible level (*Radiol.*). A phrase used loosely to indicate *maximum permissible concentration, dose* or *dose rate*.

maximum point on a curve (*Maths.*). A peak on a curve. For the curve $y=f(x)$, the point where $x=a$ is a maximum if $f(a+h)-f(a)$ is negative for all values of h sufficiently small, i.e., $f'(a)=0$ and the first non-zero higher order derivative at $x=a$ must be of even order and negative.

maximum-reading accelerometer (*Aero.*). See **accelerometer.**

maximum safe air-speed indicator (*Aero.*). A pilot's *air-speed indicator* (q.v.) with an additional pointer showing the *indicated air speed* (q.v.) corresponding to the aeroplane's *limiting Mach number* (q.v.) and also having a mark on the dial for the maximum permissible air speed.

maximum take-off rating (*Aero.*). See **power rating.**

maximum tensile stress (*Met.*). See **ultimate tensile stress.**

maximum traction truck (*Elec. Eng.*). A special form of bogie or truck often used on trams, and arranged so that the greater part of the weight comes on the driving wheels, thereby enabling the maximum tractive effort to be obtained.

maximum usable frequency (*Radio*). That which is effective for long-distance communication, as predicted from diurnal and seasonal ionospheric observation. Varies on an eleven year cycle. See **Wolf's equation.** Abbrev. **MUF.**

maximum value (*Elec. Eng.*). See **peak value.**

maximum weight (*Aero.*). See under **weight.** Also called **max take-off weight,** colloquially **max gross.**

max level speed (*Aero.*). The maximum velocity of a power-driven aircraft at full power without assistance from gravity; the altitude should always be specified.

max take-off weight (*Aero.*). See **maximum weight.**

maxwell (*Elec. Eng.*). The CGS unit of magnetic flux, the MKSA (or SI) unit being the *weber*. One maxwell = 10^{-8} weber.

Maxwell-Boltzmann distribution law (*Phys.*). A law expressing the energy distribution among the molecules of a gas in thermal equilibrium.

Maxwell bridge (*Elec. Eng.*). An early form of a.c. bridge which can be used for the measurement of both inductance and capacitance.

Maxwell experiment (*Photog.*). The demonstration of 3-colour additive synthesis, using 3 black-and-white negatives.

Maxwellian viewing system (*Optics*). In some photometers, spectrophotometers, colorimeters, etc., an arrangement in which the field of view is observed by placing the eye at the focus of a lens, instead of using an eyepiece.

Maxwell primaries (*Photog.*). The colours red, green and blue-violet, used in Maxwell's experiment.

Maxwell relationship (*Phys.*). That between the refractive index (n) and the relative permittivity (ε_r) of a medium. For a non-ferromagnetic medium, $\varepsilon_r=n^2$.

Maxwell's circuital theorems (*Elec.*). Generalized forms of Faraday's law of induction and Ampère's law (modified to incorporate the concept of displacement current). Two of *Maxwell's field equations* are direct developments of the circuital theorems.

Maxwell's circulating current (*Elec. Eng.*). A mesh or cyclic current inserted in closed loops in a complex network for analytical purposes.

Maxwell's demon (*Chem.*). Imaginary creature who, by opening and shutting a tiny door between two volumes of gases, could in principle concentrate slower (i.e. colder) molecules in one and faster (i.e. hotter) molecules in the other, thus reversing the normal tendency toward increased disorder or entropy and breaking the second law of *thermodynamics*.

Maxwell's field equations (*Maths.*). Mathematical formulations of the laws of Gauss, Faraday and Ampère from which the theory of electromagnetic waves can be conveniently derived.

$$\operatorname{div} B=0; \quad \operatorname{curl} H=\frac{\partial D}{\partial t}+j;$$

$$\operatorname{div} D=\rho; \quad \operatorname{curl} E=-\frac{\partial B}{\partial t}.$$

Maxwell's rule (*Elec. Eng.*). A law stating that every part of an electric circuit is acted upon by a force tending to move it in such a direction as to enclose the maximum magnetic flux.

Maxwell's theorem (*Eng.*). See **reciprocal theorem.**

maxwell turns (*Elec. Eng.*). The total flux linkage in a coil expressed in CGS units.

Maycoustic (*Build.*). TN for a pre-cast tile having good acoustic properties.

mayday (*Telecomm.*). Verbal international radiotelephone distress call or signal, corresponding to SOS in telegraphy. Corruption of Fr. *m'aidez*.

Mayer's carmalum stain (*Micros.*). See **carmalum stain.**

Mayo twill (*Textiles*). See **Campbell twill.**

May's graticule (*Powder Tech.*). An eyepiece graticule used in microscopic methods of particle-size analysis, having a rectangular grid for selecting particles, and a series of eight circles for sizing particles. The size of the circle diameters increases by $\sqrt{2}$ progression, and a series of parallel lines is superimposed on the rectangular grid.

maze (*An. Behav.*). An apparatus consisting of a series of pathways in a more or less complicated configuration, beginning with a starting box, possibly including blind alleys, and ending in a goal box which generally contains a reward, this not being visible from the starting box. The simplest mazes are the T- and Y-mazes. (*Nuc. Eng.*) See **radiation trap (2).**

maze-bright (*An. Behav.*). Of rats, able to solve mazes with very few errors. An ability which can be selectively bred over several generations.

maze-dull (*An. Behav.*). Of rats, unable to solve

mazes without making many errors. A lack of ability which can be selectively bred over several generations; cf. *maze-bright*.

M.C., m.c. Abbrev. for *metric carat* (see carat).

McBurney's point (*Med.*). A point situated on a line joining the umbilicus to the bony prominence of the hip-bone at the upper end of the groin, and 38 mm from the latter; a point of maximum tenderness in appendicitis.

McCabe-Thiele diagram (*Chem. Eng.*). A graphical method, based on vapour-liquid equilibrium properties, for establishing the theoretical number of separation stages in a continuous distillation process.

McColl protective system (*Elec. Eng.*). A form of protective system used on electric power networks; it operates on the balanced principle embodying biased beam relays.

McLeod gauge (*Chem.*). Vacuum pressure gauge in which a sample of low-pressure gas is compressed in a known ratio until its pressure can be measured reliably. Used for calibrating direct-reading gauges.

MCPA or **MCP** (*Chem.*). 4-chloro-2-methylphenoxyacetic acid:

$O{\cdot}CH_2{\cdot}COOH$ CH_3 Cl

used as a selective weedkiller. Also called **methoxone.**

MCPB (*Chem.*). 4-(4-chloro-2-methylphenoxy) butanoic acid, used as a weedkiller.

mcps (*Telecomm.*). Abbrev. for *megacycles per second*, replaced in SI units by *megahertz*.

M_{crit} (*Aero.*). See **critical Mach number.**

Mc/s (*Telecomm.*). Abbrev. for *megacycles per second*, replaced in SI units by *megahertz*.

M curve (*Meteor.*). Relationship between the refractive modulus and height above the earth's surface.

Md (*Chem.*). The symbol for *mendelevium*.

m-derived network or filter (*Telecomm.*). Electric wave-filter element which is derived from a normal (constant K) element by transformation, the aim being to obtain more desirable impedance characteristics than is possible in the prototype.

MDF (*Teleph.*). Abbrev. for *main distribution frame*.

M display (*Radar*). A radar *A display* (q.v.) in which a pedestal signal is manipulated by a control calibrated in distance along the baseline until it meets the horizontal of the target break.

Me (*Chem.*). (1) A symbol for the methyl radical $-CH_3$. (2) A general symbol for a metal.

mean (*Maths.*). Generally a number derived from a set of numbers which gives an indication of their order of magnitude. Coupled with their *standard deviation* (q.v.), which gives an indication of the difference between the highest numbers and the lowest numbers of the set, it gives reasonable picture of the set. See **arithmetic-, geometric-, harmonic-.**

mean aerodynamic chord (*Aero.*). See **standard mean chord.**

mean calorie (*Heat*). See **calorie.**

mean chord (*Aero.*). See **standard mean chord.**

mean curvature (*Maths.*). See **curvature** (3).

mean daily motion (*Astron.*). The angle through which a celestial body would move in the course of 1 day if its motion in the orbit were uniform. It is obtained by dividing 360° by the period of revolution.

meander (*Geog.*). A smooth curve or bend in a river or stream. If the main course of the stream cuts across the loop of a particularly sinuous meander, the water may be completely diverted from the meander, leaving a dry section o curved stream bed or an isolated (ox-bow) lake. Such a section is called an **abandoned meander.**

mean deviation (*Stats.*). See **average deviation.**

mean draught (*Ships*). Half of the sum of the forward and after draughts of a vessel; differs slightly from draught at half length.

mean effective pressure (*Eng.*). See **brake-, indicated-.**

mean establishment (*Surv.*). The average value of the lunitidal interval at a place.

mean free path (*Acous.*). Average distance travelled by a sound wave in an enclosure between wall reflections; required for establishing a formula for reverberation calculations. (*Phys.*). The mean distance traversed by a molecule of a gas between successive collisions. The following expressions for the mean free path *L* are derived from the kinetic theory:

$$L=\frac{\eta}{0{\cdot}31\rho U'},\quad L=\frac{2{\cdot}02\eta}{\sqrt{p\rho}},$$

where η is the viscosity, ρ the density, p the pressure, and U the mean molecular velocity. See also **kinetic theory of gases.**

mean free time (*Phys.*). Average time between collisions of electrons with impurity atoms in semiconductors; also of intermolecular collision of gas molecules.

mean hemispherical candle-power (*Light*). The average value of the candle-power in all directions above or below a horizontal plane passing through the source; called the *upper* or *lower* mean hemispherical candle-power according as the candle-power is measured above or below the horizontal plane through the source.

mean horizontal candle-power (*Light*). The average value of the candle-power of a light source in all directions in a horizontal plane through the source.

mean lethal dose (*Radiol*). The single dose of whole body irradiation which will cause death, within a certain period, to 50% of those receiving it. Abbrev. **MLD.**

mean-level AGC (*TV*). AGC system used in receivers, in which AGC potential is obtained from the mean level of the signal at the grid of the synchronizing separator stage.

mean life (*Nuc.*). (1) The average time during which an atom or other system exists in a particular form, e.g., for a thermal neutron it will be the average time interval between the instant at which it becomes thermal and the instant of its disappearance in the reactor by leakage or absorption. Mean life $=1{\cdot}443\times$ *half-life*. Also **average life.** (2) The mean time between the birth and death of a charge carrier in a semiconductor, a particle (e.g., an ion, a pion), etc.

mean noon (*Astron.*). The instant at which the mean sun crosses the meridian at upper culmination at any place; unless otherwise specified, the meridian of Greenwich is generally meant.

mean normal curvature (*Maths.*). See **curvature** (3).

mean place (*Astron.*). The position of a star freed from the effects of precession, nutation, and aberration, and of parallax, proper motion, and orbital motion where appreciable. These corrections can be computed for any future date, and when applied to the mean place give the apparent place.

mean power (*Radio*). Average, over a period, of power supplied by a transmitter to an antenna.

mean residence time (*Nuc.*). Mean period during which radioactive debris from nuclear weapon tests remains in stratosphere.

means, inequalities between (*Maths.*). $A \geqq G \geqq H$, where A is the arithmetic, G the geometric and H the harmonic mean of n positive numbers.

mean sea level (*Surv.*). Ordnance Survey datum level, determined at its Newlyn, Cornwall, station.

mean solar day (*Astron.*). The interval, perfectly constant, between two successive transits of the mean sun across the meridian.

mean solar time (*Astron.*). Time as measured by the hour angle of mean sun. When referred to the meridian of Greenwich it is called Greenwich Mean Time. Before 1925 this began at noon but, by international agreement, is now counted from midnight; it is thus the hour angle of mean sun plus 12 hr, and is identical with *universal time* (q.v.).

mean-spherical candle power (*Light*). The average value of the candle-power of a light source taken in all directions.

mean-spherical response (*Acous.*). That of a microphone or loudspeaker taken over a complete sphere, the radius of which is large in comparison with the size of the apparatus. For a loudspeaker, this response (total response) determines the total output of sound power, and therefore, in conjunction with the acoustic properties of an enclosure, the average reverberation intensity in the enclosure. For a microphone, this response is substantially equal to the response for reverberant sound. See **reverberation response, total response.**

mean-square error (*Maths.*). The square-root of the mean of the squares of the deviations from the mean value when a number of observations are made of a quantity, all known errors having been eliminated, the residual errors being accidental. If n observations provide readings $x_1, x_2 \ldots x_n$, the mean of which is y, the mean-square error is taken as

$$\sqrt{\frac{\Sigma(x-y)^2}{n-1}}.$$

mean stress (*Met.*). The midpoint of a range of stress. When it is zero, the upper and lower limits of the range have the same value but are in tension and compression respectively.

mean sun (*Astron.*). A fictitious point imagined to describe the celestial equator at the same average rate as the true sun's completion of the ecliptic, but uniformly, so that the length of a mean solar day throughout the year is constant.

mean-zonal candle-power (*Light*). The average value of the candle-power of a light source taken in a given zone, the angular limits of the zone being stated.

measles (*Med.*). Morbilli. An acute infectious fever caused by infection with a virus; characterized by catarrh of the respiratory passages, conjunctivitis, *Koplik's spots* (q.v.), and a distinctive rash.

measles of beef (*Vet.*). Infection of beef by the bladderworm stage, *Cysticercus bovis*, of the tapeworm *Taeniarhynchus saginatus* which occurs in man.

measles of pork (*Vet.*). Infection of pork by the bladderworm stage, *Cysticercus cellulosae*, of the tapeworm *Taenia solium* which occurs in man. Hence **measly pork.**

measure (*Typog.*). The width of the type-matter on a page, measured in ems.

measure and a half (*Join.*). Joinery work which is square on one side and moulded on the other.

measured ore (*Mining*). Proved quantity and assay grade of ore deposit as ascertained by competent measurement of exposures and an adequate sampling campaign.

measuring chain (*Build., Surv.*). See **chain.**

measuring frame (*Build.*). A wooden box without top or bottom used as a measure for aggregates in mixing concrete.

measuring instrument (*Elec. Eng.*). A device serving to indicate or record one or more of the electrical conditions in an electrical circuit. Literally, the term also includes integrating meters, but it is not generally used in this connexion.

measuring tape (*Build., Surv.*). See **tape.**

measuring unit (*Instr.*). Device (transducer) which ascertains the magnitude of a quantity to be controlled.

measuring wheel (*Surv.*). See **perambulator.**

meatotomy (*Surg.*). Incision of the urinary meatus to widen it.

meatus (*Zool.*). A duct or channel, as the external auditory *meatus* leading from the external ear to the tympanum.

mecarbam (*Chem.*). S-(*N*-ethoxycarbonyl-*N* methylcarbamoyl-methyl) diethyl phosphorothiolothionate, used as an insecticide.

mechanical advantage (*Mech.*). The ratio of the resistance (or load) to the applied force (or effort) in a *machine* (q.v.).

mechanical analogue (*Eng.*). That which can be drawn between mechanical and electrical systems obeying corresponding equations, e.g., mechanical and electrical resonators.

mechanical bias (*Teleg.*). In a polarized relay, the displacement of the tongue so that unequal marking and spacing currents are required for operation.

mechanical bond (*Civ. Eng.*). A bond used in reinforced concrete construction because the natural bond between the concrete and its reinforcing steel is sometimes inadequate from a strength point of view. Mechanical bond, independent of adhesion, is introduced by the use of plain bars bent into hooks at their ends, or of specially rolled bars with projecting ribs of various forms.

mechanical characteristic (*Elec. Eng.*). See **speed-torque characteristic.**

mechanical depolarization (*Elec. Eng.*). Dissipation, by mechanical means, of the hydrogen bubbles causing polarization of an electrolytic cell.

mechanical deposits (*Geol.*). Those deposits of sediment which owe their accumulation to mechanical or physical processes.

mechanical efficiency (*Eng.*). Of an engine, the ratio of the brake or useful power to the indicated power developed in the cylinders, i.e. the efficiency of the engine regarded as a machine. In other types of machines, the mechanical efficiency similarly accounts for the friction losses.

mechanical engineering. That branch of engineering concerned primarily with the design and production of all mechanical contrivances or machines, including prime movers, vehicles, machine tools, and production machines.

mechanical equivalent of heat (*Phys.*). Originally conceived as a conversion coefficient between mechanical work and heat (4·186 joules = 1 calorie) thereby denying the identity of the concepts. Now recognised simply as the specific heat capacity of water, 4·186 kJ/kg.

mechanical equivalent of light (*Light*). The ratio of the radiant flux, in watts, to the luminous flux, in lumens, at the wavelength for which the *relative visibility factor* (q.v.) is a maximum. Its value is about 0·0015 watts per lumen.

mechanical filter (*Cinema., etc.*). An arrangement of springs and masses interposed in a drive, particularly in sound-cameras, to smooth out variations in the required constant speed.

mechanical impedance (*Mech.*). Ratio of the total force required to move a body, to the velocity resulting, for a specified frequency of motion. It consists of the *real* part, mechanical resistance, which represents the transmission of mechanical power, and the *imaginary* part, which is purely reactive and which represents elastic stored energy.

mechanical line (*Acous.*). Conception and adjustment of the elements in an acoustic system, such as a sound-box or electrical recorder, so that they form the elements of a wave filter in analogy with electric wave filters.

mechanical overlay (*Print.*). An overlay prepared without hand-cutting; there are several varieties in use including chalk, Primaton, 3-M.

mechanical pulp (*Paper*). A type of woodpulp made by grinding wood under a stream of water on a revolving stone in such a way that the fibres of the wood are separated, but remain chemically unchanged; used largely for the production of newsprint. Also called **asplund.**

mechanical rectifier (*Elec. Eng.*). A rectifier in which a rotating or oscillating commutator, operating synchronously with the a.c. supply, is used to rectify alternate half-waves of this supply.

mechanical refrigerator (*Eng.*). A plant comprising a compressor for raising the pressure of the refrigerant, a condenser for removing its latent heat, a regulating valve for lowering its pressure and temperature by throttling, and an evaporator in which it absorbs heat at a low temperature.

mechanical resonance (*Mech.*). Enhanced response to a constant-magnitude disturbing force, as the frequency of this force is increased through a *resonant frequency*, at which the reactance of the inertia balances the reactance of the supporting stiffness of the vibrating system.

mechanical scanning (*TV*). That performed by vibrating or rotating mirrors in simple or primitive systems of TV, universally replaced by electronic methods.

mechanical seals (*Chem. Eng.*). Devices fitted to pump and agitator shafts for onerous duties and for replacing conventional fibrous packing.

mechanical sediments (*Geol.*). Those formed by mechanical processes, e.g. sand and clay, as distinct from those formed by chemical precipitation or organic activity.

mechanical stipple (or tint) (*Print.*). An aid to the making of line blocks and litho plates. The stipple is chosen from a large variety and applied either to the drawing, to the negative made from it or to the block or plate during preparation. When used in line-colour work, a range of tones can be obtained.

mechanical (or automatic) stoker (*Eng.*). A device for stoking or firing a steam boiler by automatic means. It receives fuel continuously by gravity, carries it progressively through the furnace, and deposits or discharges the ash. See **chain grate stoker, overfeed stoker, underfeed stoker.**

mechanical tissues (*Bot.*). Tissues, usually made up of thick-walled cells, which give support to the plant body.

mechanical wood (*Paper*). See **groundwood.**

mechanical woodpulp (*Paper*). See **mechanical pulp.**

mechanics. The study of the action of forces on bodies and of the motions they produce. See **dynamics, kinematics, statics.**

mechanomotive force (*Mech.*). The r.m.s. value of an alternating mechanical force, in newtons, developed in a transducer.

mechanoreceptor (*Physiol.*). A sense organ specialized to respond to mechanical stimuli, such as pressure or deformation.

Meckelian cartilage (*Zool.*). In some Fish, the cartilaginous bar which forms each ramus of the lower jaw.

Meckel's diverticulum (*Med.*). A diverticular outgrowth from the lower end of the small intestine, as a result of the persistence in the adult of the vitelline or yolk-sac duct of the embryo.

meconic acid (*Chem.*). $C_7H_4O_7$, a white crystalline compound present in opium; with iron(III) chloride it gives a dark-red colour, and is a test for opium poisoning.

meconium (*Zool.*). In certain Insects, liquid expelled from the anus immediately after the emergence of the imago; it represents the pupal excreta; also the first faeces of a new-born child or animal.

Mecoptera (*Zool.*). An order of *Mecopteroidea* (*Oligoneoptera*) having an elongated head capsule held vertically with biting mouthparts at its end, two pairs of similar wings with simple venation; larvae terrestrial. Scorpion Flies.

medi-, medio-. Prefix from L. *medius*, middle.

media (*Zool.*). In Insects, one of the primary veins of the wing; in Vertebrates, the middle tissue layer of the wall of a blood vessel.

mediad (*Zool.*). Situated near, or tending towards, the median axis.

median (*Maths.*). That value in a series of observed values which has exactly as many observed values above it as there are below. Cf. *mode*.

mediastinitis (*Med.*). Inflammation of the tissues of the mediastinum.

mediastinotomy (*Surg.*). The surgical exposure of the mediastinal region in the chest. Now used extensively in cardiac surgery.

mediastinum (*Zool.*). In higher Vertebrates, the mesentery-like membrane which separates the pleural cavities of the two sides ventrally; in Mammals, a mass of fibrous tissue representing an internal prolongation of the capsule of the testis.

medical electrolysis (*Med.*). See **galvanism.**

Medina cement (*Build., Civ. Eng.*). A quick setting natural hydraulic cement made by calcining certain nodules found in the Isle of Wight, the bed of the Solent, and Hampshire.

Mediterranean climate (*Geog.*). Climate characterized by warm, dry summers and mild, wet winters, typical of Mediterranean countries and of California, W. and S. Australia, Cape of Good Hope, etc.

Mediterranean fever (*Med.*). See **undulant fever.** (*Vet.*) A disease of cattle and water buffalo in Mediterranean countries due to infection by the protozoon *Theileria annulata*; characterized by fever, oculonasal discharge, anaemia, and diarrhoea, and transmitted by ticks.

medium (*Bot., Zool.*). A nutritive substance,

usually of a pastelike or liquid consistency, on or in which tissues or cultures of microorganisms may be reared; the all-pervading substance in which an animal has its being, as an *aqueous medium*. (*Paint.*) A liquid or a semiliquid vehicle, such as water, oil, spirit, wax, which makes pigment and other components of paint workable. See **vehicle.** (*Paper*) A standard size of printing paper, 18×23 in.

medium Edison screw-cap (*Elec. Eng.*). An *Edison screw-cap* (q.v.) having a diameter of approximately 1 in. and approximately 7 threads per inch. Abbrev. **MES.**

medium frequency (*Radio*). Between 3×10^5 and 3×10^6 Hz (radio frequencies); sometimes known as *hectometric waves.*

medium-grained (*Geol.*). See **grain-size classification.**

medium octavo (*Typog.*). A book-size, $9 \times 5\frac{3}{4}$ in. untrimmed.

medium scale integration (*Electronics*). Production of integrated circuits with around 10 to 100 gates on a single chip of silicon. Abbrev. **MSI.**

medium screen (*Photog., Print.*). The term for half-tones suitable for semi-smooth paper, usually either 100 or 120 lines per inch.

medium voltage (*Elec. Eng.*). Legally, a voltage which is over 250 volts and not greater than 650 volts.

medium waves (*Radio*). Electromagnetic waves of wavelength 200–1000 m.

medulla (*Bot.*). (1) See **pith.** (2) A tangle of loose or moderately loose hyphae in a sclerotium, rhizomorph, or other massive fungal structure. (3) A loose hyphal layer in a thallus of a lichen. (*Zool.*) The central portion of an organ or tissue, as the *medulla* of the Mammalian kidney; bone-marrow. *adj.* **medullary.**

medulla oblongata (*Zool.*). The hind brain in Vertebrates, excluding the cerebellum.

medullary (*Bot.*). Relating to, or belonging to, the pith.

medullary bundle (*Bot.*). A vascular bundle running in the pith.

medullary canal (*Zool.*). The cavity of the central nervous system in Vertebrates; the central marrow cavity of a shaft-bone.

medullary folds (*Zool.*). In a developing Vertebrate, the lateral folds of the medullary plate, by the upgrowth and union of which the tubular central nervous system is formed.

medullary groove (*Zool.*). In a developing Vertebrate, a groove on the surface of the medullary plate which will later become converted into the medullary canal.

medullary plate (*Zool.*). In a developing Vertebrate, the dorsal platelike area of ectoderm which will later give rise to the central nervous system.

medullary ray (*Bot.*). See **vascular ray.**

medullary rays (*Zool.*). Bundles of straight uriniferous tubules passing through the medulla of the Mammalian kidney.

medullary sheath (*Bot.*). The peripheral layers of cells of the pith. The cells are usually small, sometimes thick-walled, and sometimes more or less lignified. (*Zool.*) A layer of peculiar white fatty substance (myelin) which, in Vertebrates, surrounds the axons of the central nervous system and acts as an insulating coat.

medullary stele (*Bot.*). A meristele lying in the central tissues of a fern stem.

medullary velum (*Zool.*). See **Vieussens' valve.**

medullate (*Bot.*). (1) Having pith. (2) See **stuffed.**

medullated nerve fibres (*Zool.*). Axons of the central nervous system which are provided with a *medullary sheath* (q.v.).

medulloblastoma (*Med.*). A malignant and rapidly growing tumour occurring in the cerebellum.

medusa (*Zool.*). In metagenetic *Coelenterata*, a free-swimming sexual individual.

Medusa (*Build.*). TN designating a preparation used to waterproof cement surfaces.

medusoid (*Zool.*). In metagenetic *Coelenterata*, an imperfectly developed sexual individual which remains attached to the parent hydroid colony.

medusoid person (*Zool.*). See **medusa.**

Meehanite (*Met.*). High-silicon cast-iron produced by inoculation with calcium silicide. Resistant to acidic corrosion.

meerschaum (*Min.*). A hydrated silicate of magnesium. It is claylike, and is shown microscopically to be a mixture of a fibrous mineral called parasepiolite and an amorphous mineral β-sepiolite. It is used for making pipes, and was formerly used in Morocco as a soap. Also called **sepiolite.**

meeting post (*Hyd. Eng.*). The vertical post at the outer side of a lock-gate, which is chamfered so as to fit against the corresponding edge of the other gate of a pair when the gates are shut. Also called **mitre post.**

meeting rail (*Join.*). The top rail of the lower sash, or the bottom rail of the upper sash, of a double-hung window.

meeting stile (*Join.*). See **shutting stile.**

mega-. Prefix denoting 1 million, or 10^6, e.g., a frequency of 1 *megahertz* is equal to 10^6 Hz; *megawatt* $= 10^6$ watts, *megavolt* $= 10^6$ volts.

megabit (*Comp.*). Unit of a million *bits*, used in large store calculations.

megachromosomes (*Zool.*). In some *Ciliophora*, the outer set of chromosomes at mitosis, representing the meganucleus.

megacolon (*Med.*). Abnormally large colon.

megacycle (*Telecomm.*). One million cycles.

mega-electron-volt (*Nuc.*). See **MeV.**

megagamete (*Zool.*). See **macrogamete.**

mega-gramme-röntgen (*Phys.*). See **gram(me)-röntgen.**

megahertz (*Telecomm.*). The unit of frequency in which there are one million complete cycles of alternation per second. Used in preference to wavelength when the latter attribute of a wave or oscillation is very short. Abbrev. **MHz.** The unit used to be known as **megacycles per second.**

megakaryocyte (*Zool.*). See **myeloplax.**

megalaesthete (*Zool.*). In *Amphineura*, the larger type of sense-organ occurring in canals traversing the shell, resembling an eye in structure but not proved to be sensitive to light. Cf. *micraesthete.*

megalecithal (*Zool.*). Said of eggs which contain a large quantity of yolk. Cf. *microlecithal.*

megaline (*Elec. Eng.*). A unit often used in connexion with electrical machinery to denote a magnetic flux of 1 million lines or maxwells.

megaloblast (*Zool.*). An embryonic cell which has a large spherical nucleus and of which the cytoplasm contains haemoglobin, which will later give rise to *erythroblasts* (q.v.) by mitotic division within the blood vessels.

megaloblastic anaemia (*Med.*). A form of anaemia in which megaloblasts are found in the bone marrow, and often macrocytes in the peripheral blood. The usual cause is a deficiency in Vitamin B_{12} or in folic acid.

megalocyte (*Med.*). An abnormally large red cell in the blood.

megalocytosis (*Med.*). The presence of many abnormally large red cells in the blood.

megalopore (*Zool.*). One of the large apertures in the shell of an amphineuran Mollusc, containing a megalaesthete.

megalops (*Zool.*). In brachyurous *Decapoda* (Crabs), the last larval stage intervening between the zoaea stages and the adult; characterized by the possession of a broad crablike cephalothorax and a macrurous tail. *adj.* **megalopic.**

Megaloptera (*Zool.*). An order of *Neuropteroidea* (*Oligoneoptera*) with 2 pairs of large broad wings with freely branching longitudinal veins and many cross veins; pterostigma absent or ill-defined; wings held over the back in a rooflike manner at rest; larvae aquatic, carnivorous and with abdominal gills which are jointed and can be moved by intrinsic muscles, e.g., Alder flies (*Sialis*).

megalospheric (*Zool.*). In certain dimorphic species of *Foraminifera*, said of a form in which the initial chamber of the shell is large; cf. *microspheric*. *n.* **megalosphere.**

megamere (*Zool.*). See **macromere.**

meganephridia (*Zool.*). In *Chaetopoda*, the typical nephridia, of which 1 pair occurs in each somite. Cf. *micronephridia.*

meganucleus (*Zool.*). See **macronucleus.**

megaparsec (*Astron.*). Unit used in defining distance of extragalactic objects. 1 Mpc $=10^6$ parsec $=3{\cdot}26\times10^6$ light-years. See **parsec.**

megaphanerophyte (*For.*). A tree over 30 m high.

megaphone (*Acous.*). A horn to direct the voice. Can include microphone, amplifier, and sound reproducer. Then called a *loudhailer*.

megaphyllous (*Bot.*). Having very large leaves.

megascopic. Visible to the naked eye.

megasporangium (*Bot.*). A sporangium which contains megaspores.

megaspore (*Bot.*). A spore which gives rise to a female gametophyte, or its equivalent. (*Zool.*) A large swarm-spore or anisogamete of *Sarcodina*.

megasporophyll (*Bot.*). A leaflike member which bears or subtends one or more megasporangia.

megaton (*Nuc.*). Explosive force equivalent to 1 000 000 tons of TNT. Used as a unit for classifying nuclear weapons.

megatron (*Electronics*). Name for an electron tube having disk-shaped electrodes in dense parallel layers, the edges of the disk electrodes being fused into and projecting through the glass envelope to serve as contacts. Also called **lighthouse tube** or **disk-seal tube.**

megaureter (*Surg.*). Enormous dilatation of a ureter with no demonstrable abnormality.

megavoltage therapy (*Radiol.*). See **supervoltage therapy.**

megawatt days per tonne (*Nuc. Eng.*). Unit for heat output from reactor fuel; a measure of burn-up.

megilp (*Paint.*). A preparation of mastic varnish which has been diluted with linseed oil; used as a medium in the mixing of paints for fine work.

megistotherm (*Bot.*). An organism with a high temperature optimum.

megohmmeter (*Elec. Eng.*). Portable apparatus for indicating values of high resistance, containing a circuit excited by a battery, transistor converter, or d.c. generator. Not to be used for measuring insulation resistance of all types of capacitor.

megrims (*Vet.*). Staggers. A popular term used to describe symptoms such as giddiness, loss of balance and loss of consciousness in animals.

Meibomian glands (*Zool.*). In Mammals, glands on the inner surface of the eyelids, between the tarsi and conjunctiva. Also called **tarsal glands.**

meiomerous (*Bot., Zool.*). Having a small number of parts. *n.* **meiomery.**

meionite (*Min.*). Silicate of aluminium and calcium, together with calcium carbonate, which crystallizes in the tetragonal system. It is an end-member of the isomorphous series forming the scapolite group.

meiosis (*Cyt.*). A process of cell division by which the chromosomes are reduced from the diploid to the haploid number. *adj.* **meiotic.**

meiotaxy (*Bot.*). The failure of a whorl, or whorls, to develop.

meiotic euapogamy (*Bot.*). See **reduced apogamy.**

Meissner circuit (*Radio*). Oscillating valve circuit in which the resonant circuit is inductively coupled to 2 coils included in the anode and grid circuits, respectively.

Meissner effect (*Elec. Eng.*). Apparent expulsion of lines of magnetic induction from a superconductor when cooled below superconducting transition temperature in a magnetic field.

Meissner's corpuscles (*Zool.*). In Vertebrates, a type of sensory nerve-ending found in the skin, in which the nerve breaks up into numerous branches which surround a core of large cells in a connective-tissue capsule. Probably sensitive to touch.

Meissner's plexus (*Zool.*). In Vertebrates, a gangliated plexus of nonmedullated nerve-fibres in the submucous coat of the intestine.

MEK (*Chem.*). Abbrev. for *methyl ethyl ketone* (butan-2-one).

Meker burner (*Chem.*). A Bunsen burner using a wire mesh over its outlet as a stabilizer to support a wide, hot flame.

mekometer (*Instr.*). Device for the accurate measurement of distances by means of a crystal-modulated light beam.

mel (*Acous.*). Unit of pitch in sound, a pitch of 10^3 mels being associated with a simple tone of frequency 10^3 Hz at an intensity of 40 dB above the threshold of the listener.

melaconite (*Min.*). Cupric oxide crystallizing in the monoclinic system. It is a black earthy material found as an oxidation product in copper veins, and represents a massive variety of *tenorite* (q.v.).

melaena (*Med.*). The passage of black, pitchlike faeces due to the admixture of altered blood, the result of haemorrhage in the alimentary tract.

melamine (*Chem.*). A cyclic trimer of cyanamide with the formula $C_3H_6N_6$.

melamine-formaldehyde papers (*Paper*). Papers wet-strengthened with melamine-formaldehyde acid colloid added to the pulp stock. Characterized by high strength when wet, and good permanence.

melamine-formaldehyde methanal resin (*Plastics*). A synthetic resin derived from the reaction of melamine with methanal or its polymers. It is thermosetting and much used for moulding and laminating (see **Formica**). Also **MF resin.**

melan-, melano-. Prefix from Gk. *melas*, gen. *melanos*, black.

melanaemia (*Med.*). The presence in the blood of the pigment melanin.

Melanex (*Plastics*). Thin plastic foil used as light shield over scintillation detectors for α-particles.

melange (*Textiles*). Worsted yarns made from printed slubbing or tops; also known as **vigoureux printing.**

melanin (*Chem.*). A dark-brown or black pigment occurring in hair and skin. Soluble only in alkali, and is formed by the oxidation of tyrosine.

melanism (*Ecol.*). The situation where a proportion of the individuals in an animal population are black or melanic. See **industrial melanism.** (*Zool.*) An abnormal condition caused by overproduction of melanin.

melanite (*Min.*). A dark-brown or black variety of andradite garnet containing appreciable titanium.

melanoblast (*Zool.*). A special connective tissue cell containing melanin.

melanocratic (*Geol.*). A term applied to rocks which are abnormally rich in dark and heavy ferro-magnesium minerals (to the extent of 60% or more). See also **leucocratic, mesocratic.**

melanocyte (*Zool.*). A lymphocyte containing black pigment.

melanodermia or **melanoderma** (*Med.*). See **leucodermia.**

melanoglossia (*Med.*). Black hairy tongue. An overgrowth of the papillae of the tongue, which are stained black as the result either of bacterial action or of chemical action of certain food substances.

melanoma (*Med.*). Strictly, any pigmented tumour; now commonly used as a synonym for *melanotic sarcoma* (q.v.).

melanophore (*Zool.*). A chromatophore containing black pigment.

melanosis (*Med.*). The abnormal deposit of the pigment melanin in the tissues of the body.

melanosporous (*Bot.*). Having black spores.

melanotic sarcoma (*Med.*). Melanoma. Malignant melanoma. A pigmented, malignant tumour arising in the skin or in the choroid of the eye.

melanotype (*Photog.*). See **ferrotype.**

melanterite (*Min.*). Hydrated ferrous sulphate which crystallizes in the monoclinic system. It usually results from the decomposition of iron pyrite or marcasite. Also called **copperas.**

melanuria (*Med.*). The presence in the urine of the pigment melanin.

melaphyre (*Geol.*). An obsolete general term for altered amygdaloidal rocks of basaltic or andesitic types.

Melbourn Rock (*Geol.*). A hard, white, often nodular bed of chalk, found at the base of the Turonian stage of the Chalk in the southern counties of England.

meldometer (*Heat*). A high-temperature thermometer designed by Joly, and used mainly to determine melting points.

M-electron (*Nuc.*). One of 18 in M-shell of atom, having principal quantum number 3.

melena (*Med.*). See **melaena.**

melibiose (*Chem.*). Naturally occurring disaccharide based on glucose and α-galactose. A reducing sugar.

melilite (*Min.*). A complex mineral crystallizing in the tetragonal system and consisting primarily of a mixture of two minerals in isomorphous series—gehlenite (calcium aluminium silicate) and åkermanite (calcium magnesium silicate). Melilite occurs in basic alkaline igneous rocks, in thermally metamorphosed impure carbonate rocks and in slags.

Melinex (*Plastics*). TN for polyethene terephthalate (e.g., Terylene) in film form. Very strong film with extremely good transparency and electrical properties. Similar to *Mylar* (q.v.).

melioidosis (*Vet.*). A glanderslike bacterial disease of wild rodents which also affects horses, sheep, goats, and man, caused by *Pseudomonas pseudomallei* (*Loefflerella whitmori*).

melissic acid (*Chem.*). $CH_3{\cdot}(CH_2)_{29}{\cdot}COOH$, a fatty acid; m.p. 90° C, found in beeswax.

melissyl alcohol (*Chem.*). Also known as myricyl alcohol. $CH_3{\cdot}(CH_2)_{29}{\cdot}CH_2OH$, crystalline solid; m.p. 87° C, found in beeswax, combined with palmitic acid.

melliphagous, mellivorous (*Zool.*). Honey-eating.

mellisugent (*Zool.*). Honey-sucking.

mellitic acid (*Chem.*). Made by oxidizing carbon with strong nitric acid and having the structure:

COOH
HOOC COOH
HOOC COOH
COOH

Condensation with 1,3-ol: hydroxybenzene and aminophenols produces phthalein and rhodamine dyestuffs respectively.

Melloni thermopile (*Heat*). An early device, consisting of blocks of bismuth and antimony, used to measure radiant energy.

mellowing (*Leather*). A lessening of the astringency of a tan liquor by ageing.

meloidogynosis (*Bot.*). A type of damage inflicted in plants by *Nematoda* causing 'root-knot'.

Melrose pump (*Surg.*). A machine which takes over completely all cardio-respiratory function, whilst cooling the body, of patients undergoing cardiac surgery.

meltback transistor (*Electronics*). A junction transistor in which the junction is formed by allowing the molten doped semi-conductor to solidify.

melting point (*Chem., etc.*). The temperature at which a solid begins to liquefy. Pure metals, eutectics, and some intermediate constituents melt at constant temperature. Alloys generally melt over a range. Abbrev. **m.p.** Also called **fusing point.**

melting-point test (*Build., Civ. Eng.*). A test for the determination of the melting point of a bitumen for use in building or roadmaking. Also called **softening-point test.**

melting pot (*Plumb.*). The iron pot in which lead and solder are melted ready for use.

melton (*Textiles*). A strong 2×2 heavily milled woollen fabric used for overcoatings. Made from pure wool or cotton warp and woollen weft, the cloth is raised and cropped to smooth finish.

member (*Bot.*). Any part of a plant considered from the standpoint of morphology. (*Build., Civ. Eng.*) (1) A constituent part of a structural framework. (2) A division of a moulding. (*Zool.*) An organ of the body, especially an appendage.

membrana (*Zool.*). A thin layer or film of tissue; a membrane.

membranaceous, membraniferous, membranous. *Adjs.* from **membrana, membrane.**

membrana granulosa (*Zool.*). In a Graafian follicle, the inner layer which lines the cavity of the follicle.

membrana propria (*Histol.*). See **basement membrane.**

membrana tectoria (*Zool.*). A soft fibrillated membrane overlying Corti's organ.

membrana tympani (*Zool.*). A thin fibrous membrane forming the tympanum or ear-drum.

membrane (*Bot., Zool.*). (1) A thin sheetlike

structure, usually fibrous, connecting other structures or covering or lining a part or organ. (2) The terminal portion of the hemielytrum of some *Hemiptera*.

membrane bone (*Histol.*). Bone formed by the direct ossification of areolar connective tissue, without passing through a cartilaginous stage.

membrane filter (*Chem.*). Filter made of thin layers of cellulose esters, perforated by tiny uniform holes. Also **molecular filter.**

membranella (*Zool.*). In *Ciliophora*, an undulating membrane formed of two or three rows of cilia.

membranous labyrinth (*Zool.*). The soft tubular organ which forms the internal ear of Vertebrates, lining the tubular cavities of the bony labyrinth.

membranula (*Zool.*). A paddle-like organ of some *Ciliata*, formed by several cilia.

memomotion photography (*Photog.*). Time-lapse photography which records activity by a ciné camera adapted to take pictures at longer intervals than normal.

memory (*Comp.*). See **store.**

memory capacity (*Comp.*). The amount of information, usually expressed by the number of words, which can be retained by a *memory*. It may also be expressed in *bits*. The information is produced, unaltered, on command. Also **data storage.** See also **access to store.**

memory location (*Comp.*). A unit storage (holding one computer word) position within the main interval memory.

memory register (*Comp.*). A register in the storage of a computer.

memory tube (*Electronics*). Preferred term **storage tube.**

Memotron (*Electronics*). Cathode-ray tube in which a trace on the screen is reinforced and exhibited continuously.

menaccanite (*Min.*). Variety of detrital *ilmenite* (q.v.) found in sand at Menaccan, Cornwall.

menarche (*Physiol.*). The first menstruation.

Mendeleev's law (*Phys.*). 'Law of octaves'. If elements are listed according to increasing atomic weight, their properties vary but show a general similarity at each period of eight true rises.

Mendeleev's table (*Chem.*). See **periodic system.**

mendelevium (*Chem.*). Manmade element, symbol Md, at no. 101, principal isotope ^{258}Md, half-life 54 days. Named after Mendeleev, Russian chemist, associated with the *periodic table*.

Mendelian character (*Gen.*). A character which is inherited according to the generalization known as *Mendel's law* (q.v.).

Mendelian inheritance, Mendelism (*Gen.*). See **Mendel's laws.**

Mendel's laws (1865) (*Gen.*). Laws dealing with the mechanism of inheritance. 1st law (Law of Segregation) states that the gametes produced by a hybrid or heterozygote contain unchanged either one or the other of any two factors determining alternative unit characters in respect of which its parental gametes differed, i.e., the genes are distributed to each gamete. 2nd law (Law of Recombination) states that the factors determining different unit characters are recombined at random in the gametes of an individual heterozygous in respect of these factors.

mending (*Textiles*). Making good any imperfections in cloth caused during weaving by yarn breakages, etc., or knots; also known as **burling**; see also **picking and perching.** (*Typog.*) A corrected piece which is inserted in a printing plate.

mendipite (*Min.*). Oxychloride of lead(II), $2PbO \cdot PbCl_2$, which crystallizes in the orthorhombic system; found in the Mendip Hills of Somerset.

mendozite (*Min.*). Hydrated sodium aluminium silicate, crystallizing in the monoclinic system.

Ménière's disease (*Med.*). Labyrinthine vertigo. A disorder characterized by attacks of dizziness, buzzing noises in the ears, and progressive deafness; due to chronic disease of the labyrinth of the ear.

menilite (*Min.*). An alternative and more attractive name for *liver opal* (q.v.); it is a grey or brown variety of that mineral.

meninges (*Zool.*). In Vertebrates, envelopes of connective tissue surrounding the brain and spinal cord. *sing.* **meninx** (Gr. 'membrane').

meningioma (*Med.*). A tumour of the meninges of the brain, and, more rarely, of the spinal cord.

meningism, meningismus (*Med.*). The presence of the symptoms of meningitis in conditions in which the meninges are neither diseased nor inflamed.

meningitis (*Med.*). Inflammation of the meninges.

meningocele (*Med.*). The hernial protrusion of the meninges through some defective part of the skull or the spinal column.

meningococcus (*Bacteriol.*). A Gram-negative diplococcus, the causative agent of epidemic cerebrospinal meningitis. Two main types and other less frequent types have been recognized.

meningo-encephalitis (*Med.*). Inflammation of the meninges and of the brain substance.

meningo-encephalocele (*Med.*). A hernia of meninges and brain through some defect in the skull.

meningo-myelitis (*Med.*). Inflammation of the meninges and of the spinal cord.

meningo-myelocele (*Med.*). Hernial protrusion of meninges and spinal cord through a defect in the spinal column.

meningovascular (*Med.*). Pertaining to, or affecting, the meninges and the blood vessels (especially of the nervous system).

meniscus (*Optics*). A lens which is concave on one face and convex on the other. (*Phys.*) The departure from a flat surface where a liquid meets a solid, due to surface tension. The effect can be seen clearly through the wall of a glass tube. The surface of water in air rises up the wall of the tube, while that of mercury is depressed. (*Zool.*) A small interarticular plate of fibrocartilage which prevents violent concussion between 2 bones; as the intervertebral disks of Mammals.

meniscus lens (*Optics, Photog.*). One which is convex on one surface and concave on the other.

meniscus telescope (*Astron.*). A compact instrument, developed by Maksutov in 1941, in which the spherical aberration of a concave spherical mirror is corrected by a meniscus lens. It differs from the Schmidt type in having a correcting plate with 2 spherical surfaces.

menopause (*Med.*). The natural cessation of menstruation in women.

menorrhagia (*Med.*). Excessive loss of blood owing to increased discharge during menstruation.

menostaxis (*Med.*). Excessive loss of blood due to prolongation of the menstrual period.

menotaxis (*An. Behav.*). The orientation of animals in respect of the direction of a source of light, though not necessarily involving movement towards or away from it, and thus differing from other taxes. Also known as the **light-compass reaction.** Also, orientation to complex patterns to maintain a constant visual pattern.

mensa (*Zool.*). The biting surface of a tooth.

menstruation (*Zool.*). The periodical discharge from the uterus. See also **xenomenia.**

mental age (*Psychol.*). An estimation of the degree of general mental development derived from standardized intelligence tests. Not so much used as the deviation IQ. See **intelligence quotient.**

menthol (*Pharm.*). $C_{10}H_{20}O$, a camphor compound of the formula:

$$CH_3{\cdot}CH\langle\begin{matrix}CH_2{\cdot}CH(OH)\\ CH_2 - CH_2\end{matrix}\rangle CH{\cdot}CH(CH_3)_2$$

The (-)isomer is the chief constituent of peppermint oil. M.p. 43° C, b.p. 213° C. It is used as an antiseptic, and local analgesic.

mentomeckelian (*Zool.*). In some Vertebrates, a cartilage bone of the lower jaw formed by the ossification of the tip of the Meckelian cartilage.

mentum (*Zool.*). In higher Vertebrates, the chin; in some *Gastropoda*, a projection between the head and foot; in Insects, the distal sclerite forming the basal portion of the labium, situated between the submentum and the prementum. *adj.* **mental.**

mepacrine (*Pharm.*). A yellow bitter powder, the dihydrochloride of 2-methoxy-6-chloro-9-(1-methyl-4-diethylamino-butyl)-amino acridine. Formerly widely used in the prophylaxis and treatment of malaria; replaced by *chloroquine* (q.v.).

Meprobamate (*Pharm.*). 2-propyl ... 2-methyl-2-propyl-1 : 3-propanediol carbamate:

$$NH_2{-}CO{-}O{-}CH_2{-}\underset{CH_2{\cdot}CH_2{\cdot}CH_3}{\overset{CH_3}{C}}{-}CH_2{-}O{-}CO{-}NH_2,$$

Tranquillizing drug, used in motion sickness and alcoholism.

meralgia paraesthetica (*Med.*). An affection of the nerve supplying the skin of the front and outer part of the thigh; characterized by pain, tingling, and/or numbness.

meranti (*For.*). A Malayan hardwood, from the genus *Shorea*. Three varieties are recognized, red, yellow and white.

Mercalli scale (*Geol.*). A scale of intensity used to measure earthquake shocks. The various observable movements are graded from 1 (very weak) to 12 (catastrophic).

mercaptals (*Chem.*). The condensation products of mercaptans with aldehydes.

mercaptans (*Chem.*). Thio-alcohols. General formula, R·SH. They form salts with sodium, potassium, and mercury, and are formed by warming alkyl halides or sulphates with potassium hydrosulphide in concentrated alcoholic or aqueous solution. Ethyl mercaptan, (ethanethiol) C_2H_5SH; is a liquid of nauseous odour; b.p. 36°C; it is readily oxidized to ethyl disulphide $(C_2H_5)_2S_2$, by exposure to the air. It is an intermediate for *sulphonal* (q.v.) and for rubber accelerators.

mercaptides (*Chem.*). The salts of mercaptans.

mercaptobenzthiazole (*Chem.*). Used as an accelerator for rubber:

[structure: benzene ring fused to a thiazole ring with N and S, C–SH]

mercaptols (*Chem.*). The condensation products of mercaptans with ketones.

mercaptopurine (*Pharm.*). 6-mercaptopurine monohydrate. Used to treat leukaemia in children.

Mercator's projection (*Geog.*). A derivative of the simple *cylindrical projection* (q.v.). The meridian scale is adjusted to coincide with the latitudinal scale at any given point, allowing true representation of shape but not of area, regions towards the poles being exaggerated in size. Long used by navigators, for straight lines drawn on it represent a constant bearing. A transverse form of Mercator is sometimes used in large-scale maps of limited E-W extent.

mercerization (*Textiles*). A process which greatly increases the lustre of cotton yarns and fabrics. It consists of treating the material with concentrated caustic soda lye, which causes swelling and results in the fibres becoming transparent if the material is kept under tension. Increases strength and dye absorption properties.

merchant iron (*Met.*). Bar-iron made by repiling and re-rolling puddled bar. All wrought-iron is treated in this way before being used for manufacture of chains, hooks, etc.

mercurial pendulum (*Horol.*). A compensation pendulum in which mercury is used as the compensating medium.

mercuric bromophenol blue stain (*Micros.*). A stain used in electron microscopy to increase the contrast by forming electron-dense deposits.

mercuric (mercury(II)) chloride (*Chem.*). *Corrosive sublimate.* $HgCl_2$. Prepared by sublimation from a mixture of mercury(II)sulphate and sodium chloride.

mercuric (mercury(II)) iodide (*Chem.*). HgI_2. See also **Nessler's solution.**

mercurous (mercury(I)) chloride (*Chem.*). See **calomel.**

mercury (*Chem.*). A white metallic element which is liquid at atmospheric temperature. Chemical symbol Hg, at. no. 80, r.a.m. 200·59, rel. d. at 20°C 13·596, m.p. −38·9°C, b.p. 356·7°C, electrical resistivity $95{\cdot}8\times10^{-8}$ ohm metres. A solvent for most metals, the products being called *amalgams*. Its chief uses are in the manufacture of drugs and chemicals, fulminate, and vermilion. Used as metal in mercury-vapour lamps, arc rectifiers, power-control switches, and in many scientific and electrical instruments. Also called **quicksilver.** ^{198}Hg is a mercury isotope made from gold in a reactor, for use in a quartz mercury-arc tube, light from which has an exceptionally sharp green line, because of even mass number of a single isotope. This considerably improves comparisons of end-gauges in interferometers.

Mercury (*Astron.*). The planet nearest to the sun, revolving in 88 days at a distance of 0·387 A.U.; diameter 4880 km and mass about 1/20 that of the earth, its relative density (5·5) being almost the same. Previously thought to present the same face to the sun, it is now known to rotate in 58·7 days. Mercury is difficult to observe because of its proximity to the sun, but it appears to have no atmosphere, and its surface may resemble that of the Moon.

mercury arc (*Electronics*). Luminous discharge of wide line spectral content, especially rich in ultraviolet. The gas discharge is widely used to excite coloured radiation in fluorescent powders in both hot- and cold-cathode tubes. Much used for illumination, and in factories, restaurants, and for advertising; known as **mercury discharge lamps.** See also **mercury-vapour tube, mercury vapour lamp.**

mercury-arc converter (*Elec. Eng.*). A *converter* making use of the properties of the *mercury-arc rectifier*.

mercury-arc rectifier (*Elec. Eng.*). One in which rectification arises from the differential migration of electrons and heavy mercury ions in a plasma, formed by evaporation from a *hot spot* on a cathode pool of mercury. This vaporization has to be started by withdrawing an electrode from this pool, thus generating an arc. Used for a.c./d.c. conversion for railways, trams, etc., before the introduction of silicon rectifiers. See **thyratron.**

mercury barometer (*Meteor., Phys.*). An instrument used for measuring the pressure of the atmosphere in terms of the height of a column of mercury which exerts an equal pressure. In its simplest form, it consists of a vertical glass tube about 80 cm long, closed at the top and having its lower open end immersed in mercury in a dish. The tube contains no air, the space above the mercury column being known as a *Torricellian vacuum*.

mercury cell (*Chem.*). (1) Electrolytic cell with mercury cathode. (2) Dry cell employing mercury electrode, e.m.f. *ca.* 1·3 volts; see **Mallory battery, Reuben-Mallory cell.**

mercury delay line (*Telecomm.*). One in which mercury is used as the medium for sound transmission, conversion from and to electrical energy being through suitable transducers at the ends of the mercury column.

mercury discharge lamp (*Electronics*). See under **mercury arc.**

mercury fulminate (*Chem.*). A crystalline solid, $Hg(ONC)_2$, prepared from ethanol, mercury and nitric acid; used as an initiator in detonators and percussion caps.

mercury intrusion method (*Chem.*). Finding the distribution of sizes of capillary pores in a body by forcing in mercury, the radius being found from the pressure, and the percentage from the volume of mercury absorbed at each pressure.

mercury memory (*Telecomm.*). One depending on the velocity of mechanical impulses in mercury, using transducers to connect to the electronic circuits. The impulses are reshaped for continuous circulation, the train thus establishing a store which can be tapped as required.

mercury motor meter (*Elec. Eng.*). A type of motor meter in which the moving part consists of a metal disk rotating in a bath of mercury, Current is led to the disk via the mercury, and interacts with permanent magnets or electromagnets to produce a torque on the disk, thus causing it to rotate.

mercury-pool cathode (*Electronics*). Cold cathode in valve where arc discharge releases electrons from surface of mercury pool, e.g., in *mercury-arc rectifier* (q.v.).

mercury seal (*Chem.*). A device which ensures that the place of entry of a stirrer into a piece of apparatus is gas-tight, while allowing the free rotation of the stirrer.

mercury switch (*Elec. Eng.*). A switch in which the fixed contacts consist of mercury cups into which the moving contacts dip, or in which the mercury is contained in a tube which is made to tilt, thereby causing the mercury to bridge the contacts.

mercury tank (*Telecomm.*). A vessel containing mercury and used to store information by continuous circulation of impulses through the tank. See **mercury memory.**

mercury vapour cycle (*Eng.*). The use of mercury in a closed loop for external combustion turbines. It has certain advantages over water, but is not widely used.

mercury-vapour lamp (*Electronics*). Quartz tube containing a mercury arc, specially designed to provide ultraviolet rays for therapeutic and cosmetic treatment.

mercury-vapour pump (*Chem. Eng.*). See **Gaede diffusion pump.**

mercury-vapour rectifier (*Elec. Eng.*). See **mercury-arc rectifier.**

mercury-vapour tube (*Electronics*). Generally, any device in which an electric discharge takes place through mercury vapour. Specifically, a triode valve with mercury vapour, which is ionized by the passage of electrons, and reduces the space charge and the anode potential necessary to maintain a given current. The grid is effective in controlling the start of the discharge. See **thyratron.**

merging (*Comp.*). Sequencing or interlacing data from separate sources, e.g., punched or magnetic tape, so that they are in required order on new tape.

mericarp (*Bot.*). A 1-seeded portion of a fruit which splits up at maturity.

mericlinal chimaera (*Bot.*). A chimaera in which one component does not completely surround the other; an incomplete *periclinal chimaera* (q.v.).

meridian (*Astron.*). That great circle passing through the poles of the celestial sphere which cuts the observer's horizon in the north and south points, and also passes through his zenith. (*Surv., etc.*) The imaginary plane passing through the earth's polar axis and the point on the earth's surface to which the meridian refers. Called a **meridian of longitude.**

meridian altitude (*Astron.*). The altitude of a heavenly body at the position of upper transit.

meridian circle (*Astron.*). A telescope mounted on a horizontal axis lying due east and west, so that the instrument itself moves in the meridian plane. It is used to determine the times at which stars cross the meridian, and is equipped with a graduated circle for deducing declinations. Also called **transit circle.**

meridian passage (*Astron.*). See **transit (1).**

meridional (*Zool.*). Extending from pole to pole; as a *meridional furrow* in a segmenting egg.

Merilene (*Chem.*). TN for heat-exchange organic fluid based on a mixture of diphenyl and diphenyl oxide.

merino (*Textiles*). Term applied to yarn or knitted goods made from a mixture of merino wool and cotton.

merino wool (*Textiles*). Wool of fine quality from merino sheep; used in saxony quality woollens and botany quality worsted cloths. The name is also used in the low woollen trade for wool fibre recovered from fine worsted clothing rags.

merisis (*Cyt.*). Increase in size due to cell division.

merism (*Bot.*). The development of more than one member of the same kind, usually in such a way that a symmetrical arrangement or pattern is formed.

merispore (*Bot.*). One segment of a multiple spore.

meristele (*Bot.*). A strand of vascular tissue, enclosed in a sheath of endodermis, forming part of a dictyostele.

meristem (*Bot.*). A group of localized dividing, undifferentiated cells, found in regions of active growth, e.g., apical meristem.

meristic (*Zool.*). Segmented; divided up into

parts; pertaining to the number of parts, as *meristic variation* (q.v.). See also **merome.**

meristic variation (*Bot., Zool.*). Variation in the number of organs or parts; as variation in the number of body somites of a metameric animal.

meristogenetic (*Bot.*). Formed from, or by, a meristem.

Meritol (*Photog.*). See ***para*-phenylene diamine.**

Meritol-metol (*Photog.*). Fine-grain developer composed of *para*-phenylenediamine, pyrocatechin, and metol.

Merkel's corpuscles (*Zool.*). See **Grandry's corpuscles.**

merlons (*Arch.*). The projecting parts of a battlement.

mermaid's purse (*Zool.*). A popular name applied to the horny purselike capsule in which the eggs of certain Selachian fish (Sharks, Dogfish, Skates, Rays) are enclosed.

Mermis (*Zool.*). A genus of Nematode worms living parasitically in the bodies of ants.

mermithaner (*Zool.*). A male ant having abnormal characters, as a result of being parasitized by *Mermis.*

mermithergate (*Zool.*). A worker ant having abnormal characters as a result of its being parasitized by *Mermis.*

mermithogyne (*Zool.*). A queen ant having abnormal characters as a result of its being parasitized by *Mermis.*

mero-. Prefix from Gk. *meros*, part.

meroblastic (*Zool.*). Said of a type of ovum in which cleavage is restricted to a part of the ovum, i.e., is incomplete, usually due to the large amount of yolk. Cf. *holoblastic.*

merochrome (*Chem.*). A mixed crystal consisting of 2 differently coloured isomers.

meroerine (*Biol., Histol.*). The type of glandular secretion in which the secretory cells remain intact throughout the process of secretion.

merogamy (*Bot.*). The union of two individualized gametes. (*Zool.*) The condition of having gametes which are smaller than the ordinary cells of the species and are produced by special fission; union of such gametes; microgamy; cf. *hologamy. n.* **merogametes.**

merogenesis (*Zool.*). Segmentation, formation of parts.

merogenic (*Zool.*). Said of induction in which a part only of the soma is affected primarily. Cf. *hologenic.*

merogony (*Zool.*). See **schizogony.**

merohyponeuston (*Ecol.*). Organisms living in the upper 5 cm of water only during the larval stage.

meroistic (*Zool.*). Said of ovaries which produce yolk-forming cells as well as ova. Cf. *panoistic.*

merome (*Zool.*). A body somite or segment of a metameric animal.

meromorphic function (*Maths.*). See **analytic function.**

meron (*Zool.*). In Insects in which the coxa is divided, the posterior portion.

meronephridia (*Zool.*). See **micronephridia.**

meront (*Zool.*). In some *Neosporidia*, a uninucleate phase succeeding the *planont* (q.v.) and multiplying by fission.

meroplankton (*Ecol.*). Organisms which spend only part of their life history in the plankton. Cf. *holoplankton.*

meropodite (*Zool.*). In some *Crustacea*, the second joint of the endopodite of the walking-legs or maxillipeds.

meros (*Arch.*). The surfaces between the channels in a triglyph. Also called **femur.**

merosmia (*Med., Psychiat.*). Temporary deficiency of the sense of smell which may be organic in nature (e.g., sinusitis) or a result of hysteria.

merosomatous (*Zool.*). Said of ascidiozooids which show division into thorax and abdomen.

merosome (*Zool.*). See **merome.**

merosthenic (*Zool.*). Having the hind limbs exceptionally well developed, as Frogs, Kangaroos.

meroxene (*Min.*). One of the classes into which *biotite* (q.v.) was divided by Tschermak. This class includes nearly all ordinary biotite.

Merozoa (*Zool.*). An order of *Cestoda* in which the body is differentiated into a scolex and proglottides, and the genitalia are repeated in each proglottis; there is a complex life-history, the larval form usually occurring in herbivorous animals, the mature form in the gut of various carnivorous, omnivorous, and insectivorous Vertebrates.

merozoite (*Zool.*). In *Protozoa*, a young trophozoite which results from the division of the schizont.

merozoon (*Zool.*). A fragment of a protozoon produced artificially, and containing a portion of the macronucleus.

Merrill-Crowe process (*Met.*). A process for precipitating gold from deoxygenated pregnant cyanide solution by means of zinc dust.

merrythought bone (*Zool.*). The furcula of Birds.

mersalyl (*Pharm.*). A mercurial compound:

$$(HgOH)CH_2\cdot CH\cdot (OCH_3)\cdot CH_2\cdot NH\cdot \overset{O}{\overset{\|}{C}}-C_6H_4-OCH_2\cdot COONa$$

Used medically as a diuretic drug.

Merulius lacrymans (*Build.*). The name of a fungus which commonly occasions *dry rot* (q.v.).

merycism (*Med.*). Rumination. The return, after a meal, of gastric contents to the mouth; they are then chewed and swallowed once more.

Merz-Hunter protective system (*Elec. Eng.*). See **split-conductor protection.**

Merz-Price protective system (*Elec. Eng.*). A form of balanced protective system for electric-power networks, in which the current entering a section of the network is balanced against that leaving it. If a fault occurs on the section this balance is upset, and a relay is caused to operate and trip circuit-breakers to clear the faulty section from the network.

Merz slit (*Optics*). A variable-width bilateral slit for spectrographs. Characterized by the fixed position of the centre of the slit.

mes-, meso-. Prefix from Gk. *mesos*, middle.

mesa (*Electronics*). Type of transistor in which one electrode is made very much smaller than the other, to control bulk resistance. Also, by selective etching, the base and emitter are raised above the region of the collector. (*Geol.*) Flat-topped hill with steep sides. A large *butte.*

mesadenia (*Zool.*). In some male Insects, accessory glands of the genital system, of mesodermal origin. Cf. *ectadenia.*

mesaortitis (*Med.*). Inflammation of the middle coat of the aorta.

mesarch (*Bot.*). Having the protoxylem surrounded by metaxylem.

Mesaxonia (*Zool.*). A superorder of the cohort *Ferungulata* (*Eutheria*; Mammals), containing the one order *Perissodactyla* (q.v.).

mesaxonic foot (*Zool.*). A foot in which the skeletal axis passes down the third digit, as in *Perissodactyla.* Cf. *paraxonic foot.*

mescaline (*Pharm.*). A hallucinogenic drug derived from the Mexican cactus, mescal.
mesectoderm (*Zool.*). Parenchymatous tissue formed from ectoderm cells which have migrated inwards.
mesencephalon (*Zool.*). The mid-brain of Vertebrates.
mesenchyma (*Zool.*). Parenchyma; embryonic mesodermal tissue of spongy appearance. *adj.* **mesenchymatous.**
mesenchyme (*Zool.*). Mesodermal tissue, comprising cells which migrate from ectoderm, or endoderm, or mesothelium into the blastocoele. Cf. *mesothelium.*
mesendoderm (*Zool.*). In ontogeny, endodermal cells which will later give rise to mesoderm.
mesenteric (*Zool.*). Pertaining to the mesenteron; pertaining to a mesentery.
mesenteric caeca (*Zool.*). Digestive diverticula of the mesenteron in many Invertebrates (e.g., in *Arachnida, Crustacea, Echinodermata,* Insects).
mesenteric filament (*Zool.*). In *Anthozoa,* the enlarged free edge of a mesentery.
mesenteron (*Zool.*). See **mid-gut.**
mesentery (*Zool.*). In *Coelenterata,* a vertical fold of the body wall projecting into the enteron. More generally, a fold of tissue supporting part of the viscera. *adjs.* **mesenterial, mesenteric.**
mesethmoid (*Zool.*). A median cartilage bone of the Vertebrate skull, formed by ossification of the ethmoid plate. Also called **internasal septum.**
mesh (*Build., Civ. Eng.*). *Expanded metal* used as a reinforcement for concrete. (*Telecomm.*) A complete electrical path (including capacitors) in the component branches of a complex network. Also called **loop.** (*Textiles*) A completed opening in a lace fabric, formed by the combination of two bobbin threads and two warp threads.
mesh connexion (*Elec. Eng.*). A method of connecting the windings of an a.c. electric machine; the windings are connected in series so that they may be represented diagrammatically by a polygon. The *delta connexion* (q.v.) is a particular example of this method.
mesh network (*Telecomm.*). One formed from a number of impedances in series.
mesh structure (*Geol.*). A term applied to the mode of alteration of olivine to serpentine; this process begins round the margins of the crystal and along the irregular network of cracks which traverse it.
mesh voltage (*Elec. Eng.*). The voltage between any two lines of a symmetrical polyphase system which are consecutive as regards phase sequence. Called *delta voltage* in a three-phase system, and *hexagon voltage* in a six-phase system.
mesiad (*Zool.*). Situated near, or tending towards, the median plane.
mesial, mesian (*Zool.*). In the median vertical or longitudinal plane.
mesic atom (*Nuc.*). Short-lived atom in which negative muon has displaced normal electron.
mesitite (*Met.*). A variety of *magnesite* (q.v.) containing from 30–50% of iron carbonate.
mesitylene (*Chem.*). $C_6H_3(CH_3)_3$, 1,3,5-trimethylbenzene; a colourless liquid, occurring in coal-tar; b.p. 164°C.
mesityl oxide (*Chem.*). $(CH_3)_2C{=}CH{\cdot}CO{\cdot}CH_3$; a colourless liquid of peppermint-like odour; b.p. 122°C. Obtained from propanone by an *aldol condensation* (q.v.) type of reaction induced by ammonia.
meso- (*Chem.*). (1) Optically inactive by intramolecular compensation. Abbrev. ***m-***. (2) Substituted on a carbon atom situated between 2 hetero-atoms in a ring. (3) Substituted on a carbon atom forming part of an intramolecular bridge.
meso-. Prefix. See **mes-.**
mesobenthos (*Zool.*). Fauna and flora of the sea-floor, at depths ranging from 100 to 500 fathoms (200 to 1000 m).
mesobiota (*Ecol.*). The middle-sized soil organisms, including *Nematoda,* small *Oligochaeta,* small Insect Larvae and small adult *Arthropoda.* Cf. *macrobiota, microbiota.*
mesoblast (*Zool.*). The mesodermal or third germinal layer of an embryo, lying between the endoderm and ectoderm. *adj.* **mesoblastic.**
mesoblastic somites (*Zool.*). In developing metameric animals, segmentally arranged blocks of mesoderm, the forerunners of the somites.
mesobronchium (*Zool.*). In Birds, the continuation of the main bronchus through the lung to the abdominal air-sac.
mesocarp (*Bot.*). The middle layer of a pericarp.
mesocerebron (*Zool.*). See **deutocerebron.**
mesocoele (*Zool.*). In Vertebrates, the cavity of the mid-brain; mid-ventricle; Sylvian aqueduct.
mesocolloid (*Chem.*). A particle whose dimensions are 25 to 250 nm containing 100–1000 molecules.
mesocratic (*Geol.*). A term applied to igneous rocks which, in respect of their content of dark silicates, are intermediate between those of leucocratic and melanocratic type, and contain 30–60% of heavy, dark minerals.
mesoderm (*Zool.*). See **mesoblast.**
mesogaster (*Zool.*). In Vertebrates, the portion of the dorsal mesentery which supports the stomach.
mesogloea (*Zool.*). In *Coelenterata,* a structureless layer of gelatinous material intervening between the ectoderm and the endoderm.
mesohepar (*Zool.*). An old name for the *falciform ligament* (q.v.).
mesolecithal (*Zool.*). A type of egg of medium size containing a moderate amount of yolk which is strongly concentrated in one hemisphere, this being the lower one in eggs floating in water. Found in Frogs, *Urodela,* Lungfishes, Lower *Actinopterygii,* and Lampreys. Cf. *oligolecithal, telolecithal.*
mesolite (*Min.*). A zeolite intermediate in composition between *natrolite* and *scolecite.* Crystallizes in the monoclinic system, and occurs in amygdaloidal basalts and similar rocks.
mesomere (*Zool.*). The middle muscle-plate zone of the mesothelial wall of a developing Vertebrate, situated between the epimere and the hypomere.
mesomerism (*Chem.*). (1) See **desmotropism.** (2) See **resonance.**
mesometrium (*Zool.*). The mesentery which supports the uterus and related structures.
mesomitosis (*Cyt.*). Mitosis which occurs within the nuclear membrane, without any cooperation from cytoplasmic elements. Cf. *metamitosis.*
mesomorphous (*Chem.*). Existing in a state of aggregation midway between the true crystalline state and the completely irregular amorphous state. See also **liquid crystals.**
mesomorphy (*Psychol.*). Sheldon's *somatotype* which is characterized by a muscular or athletic build and said to be associated with aggressive and independent personality traits. See **ectomorphy, endomorphy, somatotypes.**
mesomyodian (*Zool.*). Of Birds, having the syringeal muscles attached to the middle of the bronchial hemi-rings.

meson (*Nuc.*). One of a series of unstable particles with masses intermediate between those of electrons and nucleons, and with positive, negative or zero charge. Mesons appear to be related to the nuclear field in a similar way to the relationship between photons and the electromagnetic field (or the hypothetical relationship between gravitons and the gravitational field). They are bosons which, unlike fermions (leptons and baryons), may be created or annihilated freely. They are now classed in 3 groups: η-mesons, π-mesons (*pions*) and κ-mesons (*kaons*). The *muon*, formerly classified as the μ-meson, is now known to be a lepton. Mesons participate with baryons in strong interactions obeying the law of conservation of parity.

mesonephric duct (*Zool.*). See **Wolffian duct.**

mesonephros (*Zool.*). Part of the kidney of Vertebrates, arising later in development than, and posterior to, the *pronephros*, and discharging into the *Wolffian duct* (q.v.); becomes the functional kidney in adult anamniotes. *adj.* **mesonephric.**

meson field (*Nuc.*). That which is considered to be concerned with the interchange of proton and neutron in a nucleus, the mesons transferring energy.

mesonotum (*Zool.*). The notum of the mesothorax in Insects. *adj.* **mesonotal.**

mesophanerophyte (*Bot.*). A tree having a height of 8–30 m.

mesophilic bacteria (*Bacteriol.*). Bacteria which grow best at temperatures of 20–45°C.

mesophragma (*Zool.*). In Insects, an apodeme arising from the postscutellum.

mesophyll (*Bot.*). The parenchymatous tissue between the upper and lower epidermises of a leaf, chiefly concerned with photosynthesis.

mesophyte (*Bot.*). A plant occurring in places where the water supply is neither scanty nor excessive. *adj.* **mesophytic.**

mesophytic environment (*Bot.*). An environment in which the water supply is neither very scanty nor very abundant.

mesoplankton (*Zool.*). Plankton found below a depth of 200 m (100 fathoms).

mesoplastron (*Zool.*). In some *Chelonia*, a bone of the plastron intercalated between the hypoplastron and the hyoplastron.

mesopleuron (*Zool.*). In *Diptera*, a thoracic sclerite lying between the notopleural and sternopleural sutures and in front of the root of the wing.

mesopod (*Bot.*). Said of the fruit body of a fungus which has a central stipe.

mesopodium (*Zool.*). (1) The metacarpus or metatarsus. (2) In *Gastropoda*, that part of the foot between the propodium and the metapodium. *adj.* **mesopodial.**

mesopterygium (*Zool.*). One of the three basal cartilages of the pectoral fin in *Selachii*, lying between the propterygium and the metapterygium.

mesopterygoid (*Zool.*). See **ectopterygoid.**

mesorchium (*Zool.*). In Vertebrates, the mesentery supporting the testis.

mesosaprobe (*Bot.*). A plant living in somewhat foul water.

mesosoma (*Zool.*). In *Arachnida*, the anterior division of the 'abdomen', being the middle tagma of the body, which always bears the genital opening on its first somite. *adj.* **mesosomatic.**

mesosphere (*Astron.*). The region of the earth's atmosphere above the stratosphere, approx. 50 km in width, in which the temperature first rises with height to about +80°C, and then falls to another minimum about −65°C at about 80 km.

mesospore (*Bot.*). (1) A layer in the spore wall, developed inside the first-formed outer layer, and sometimes bearing ridges, pointed outgrowths, or other ornamentation. (2) A teleutospore consisting of one cell.

mesostate (*Zool.*). An intermediate stage in metabolism.

mesosternum (*Zool.*). In Insects, the sternum of the mesothorax; in Vertebrates, the middle part of the sternum, connected with the ribs; the gladiolus.

mesostethium (*Zool.*). See **mesosternum.**

mesotarsal (*Zool.*). In Insects, the tarsus of the second walking leg; in land Vertebrates, the ankle joint or joint between the proximal and distal rows of tarsals.

mesothelium (*Zool.*). Mesodermal tissue comprising cells which form the wall of the cavity known as the coelom. Cf. *mesenchyme.*

mesothorax (*Zool.*). The second of the 3 somites composing the thorax in Insects. *adj.* **mesothoracic.**

mesothorium I and II (*Chem.*). The first and second products of the thorium series. MsThI is a radium isotope, ^{228}Ra, half-life 6·7 y. It decays by beta-emission to MsThII, the actinium isotope ^{228}Ac, half-life 6·13 h.

mesotic (*Zool.*). In Birds, a paired cartilage of the chondrocranium, between the parachordal and the acrochordal. Also **basiotic.**

mesotrochal (*Zool.*). Having an equatorial band of cilia.

mesotympanic (*Zool.*). See **symplectic.**

mesotype (*Min.*). A name given by Haüy to *natrolite*, because its form is intermediate between the forms of stilbite and analcite. S. J. Shand uses the same term as synonymous with *mesocratic* (q.v.).

mesovarium (*Zool.*). In Vertebrates, the mesentery supporting the ovary.

mesoxalic acid (*Chem.*). $CO(COOH)_2+H_2O$ or $C(OH)_2(COOH)_2$; a dibasic ketonic acid, forming deliquescent prisms; it is prepared from dibromomalonic acid by heating with caustic soda solution.

Mesozoic (*Geol.*). The name applied to the era of geological time which includes the periods during which rocks of Triassic, Jurassic and Cretaceous age were deposited.

message (*Comp.*). Collection of *words*, each comprising a number of *bits*, within a pattern.

messenger RNA (*Biochem.*). See **m-RNA.**

messenger wire (*Elec. Eng., etc.*). The strong suspension wire for holding aerial cables, the latter being suspended by leather thongs: a *bearer cable* (q.v.).

messuage (*Build.*). A dwelling-house and its adjacent land and buildings.

mestome (*Bot.*). Conducting tissue, with associated parenchyma, but without mechanical tissue.

meta- (*Chem.*). See ***m-***.

meta-. Prefix from Gk. *meta*, after. (*Chem.*) (1) Derived from an acid anhydride by combination with one molecule of water. (2) A polymer of . . . (3) A derivative of . . . (4) The 1-3 relationship of substituents on a benzene ring, e.g., m-cresol.

meta-aldehyde (*Chem.*). See **metaldehyde.**

metabasipodite (*Zool.*). See **preischiopodite.**

metabiosis (*Biol.*). A condition in which one organism depends upon another for its existence, but in which one lives only after another has died in preparing its environment; a form of commensalism.

metabolic nucleus (*Cyt.*). A nucleus when it is not dividing, and when the chromatin is in the form of a network; the so-called *resting nucleus.*

metabolism (*Biol.*). The sum-total of the chemical and physical changes constantly taking place in living matter. *adj.* **metabolic.**

metabolite (*Zool.*). A substance involved in metabolism, being either synthesized during metabolism or taken in from the environment.

metaboly (*Bot.*). The power possessed by some cells of altering their external form, e.g., as in *Euglenida.* (*Zool.*) See **euglenoid movement.**

metaboric acid (*Chem.*). See **boric acid.**

metacarpal or **metacarpale** (*Zool.*). One of the bones composing the *metacarpus* (q.v.) in Vertebrates.

metacarpus (*Zool.*). In land Vertebrates, the region of the fore-limb between the digits and the carpus.

metacentre (*Phys.*). If a vertical line is drawn through the centre of gravity of a body floating in equilibrium in a liquid, and when the body is slightly displaced from its equilibrium position a second vertical line is drawn through the centre of buoyancy (centre of gravity of the displaced liquid), the two lines meet at a point called the *metacentre.* According to whether this is above or below the centre of gravity of the body, the equilibrium is stable or unstable.

metacentric (*Cyt.*). Condition in which the centromere is at, or near, the middle.

metacentric height (*Phys.*). The distance between the centre of gravity of a floating body and its *metacentre* (q.v.). (*Ships*) The measure of stability of a vessel at small angles of heel, indicative of its behaviour when rolling.

metacercaria (*Zool.*). An encysted cercaria.

metacerebron (*Zool.*). See **tritocerebron.**

metacestode (*Zool.*). The encysted larval stage of a tapeworm (*Cestoda*).

Metachlamydeae (*Bot.*). See **Sympetalae.**

metachromatic (*Micros.*). Showing other than the basic colour constituent after staining. *n.* **metachromasy.**

metachromatic corpuscle or **metachromatic granule** (*Cyt.*). An inclusion in cytoplasm, possibly consisting of *metachromatin,* characterized by its property of metachromasia, i.e., stains in a particular way with basic aniline dyes.

metachromatin (*Biol.*). A complicated substance, stated to be a compound of nucleic acid, occurring in granules in cytoplasm.

metachronal rhythm (*Zool.*). The rhythm shown by beating cilia, or the movements of the ctenes of *Ctenophora,* in which each cilium or ctene beats slightly out of phase with its predecessor, giving the appearance of wave motion.

metachrosis (*Zool.*). The ability, shown by some animals (as the Chameleon) to change colour by expansion or contraction of chromatophores.

metacinnabarite (*Min.*). Mercuric sulphide. In composition similar to *cinnabar* (q.v.), but occurs in black tetrahedral (cubic) crystals (also massive).

metacneme (*Zool.*). In some *Anthozoa,* a secondary mesentery.

metacoele (*Zool.*). In *Craniata,* the cavity of the hind-brain; the fourth ventricle.

metacone (*Zool.*). In Mammals, the postero-external cusp of an upper molar tooth.

metaconid (*Zool.*). In Mammals, the postero-external cusp of a lower molar tooth.

metacromion (*Zool.*). A process terminating the acromion in some *Rodentia.*

metadiscoidal placentation (*Zool.*). Having the villi at first scattered and then restricted to a disk as in Primates.

metadyne (*Elec. Eng.*). See **amplidyne.**

metadyne generator. See **cross-field generator**

metadyne transformer (*Elec. Eng.*). A rotating d.c. machine working on the cross-field principle arranged to give a variable voltage, constant-current output from a constant-voltage input.

metagenesis (*Zool.*). See **alternation of generations.**

metagons (*Zool.*). Gene-initiated cytoplasmic particles found in *Paramecium,* and responsible for the production of mu-particles, found in strains which kill sensitive strains when they mate with them.

meta-isomers (*Chem.*). Isomers differing in the position of a double bond.

metakinesis (*Cyt.*). See **metaphase.**

metal. An element which readily forms positive ions. Metals are characterized by their opacity and high thermal and electrical conductivity. (*Civ. Eng.*) See **road metal.**

metal-arc welding (*Elec. Eng.*). A type of electric welding in which the electrodes are of metal, and melt during the welding process to form filler metal for the weld.

metal-ceramic (*Met.*). See **cermet.**

metal-clad switchgear (*Elec. Eng.*). A type of switchgear in which each part is completely surrounded by an earthed metal casing. Cf. *metal-enclosed switchgear.*

metal-cored carbon (*Elec. Eng.*). An arc-lamp carbon having a core of metal, in order to improve its conductivity.

metaldehyde (*Chem.*). Meta-aldehyde. $(CH_3CHO)_4$, long glistening needles which sublime at 115°C with partial decomposition into ethanal. Acetaldehyde is polymerized to metaldehyde by the action of acids at temperatures below 0°C. It has been used as a convenient portable fuel known as 'meta'-fuel (Tommy cooker).

metal detector (*Eng.*). An instrument, widely used in industrial production, for detecting the presence of stray metal parts embedded, e.g., in food products. It is usually incorporated in a conveyor line and gives visible or audible warning or automatically stops the line.

metal electrode (*Elec. Eng.*). A form of electrode used in *metal-arc welding* (q.v.).

metal-enclosed switchgear (*Elec. Eng.*). A type of switchgear in which the whole equipment is enclosed in an earthed metal casing. Cf. *metal-clad switchgear.*

metal feeder (*Typog.*). A device which maintains a supply of metal to type-casting machines from an ingot fed at the required rate into the metal pot.

metal filament (*Elec. Eng.*). A fine metal conductor heated to incandescence to provide illumination or to act as a source of electrons in a vacuum tube.

metal-filament lamp (*Elec. Eng.*). A filament lamp in which the light is produced by raising a fine wire or filament of metal to white heat.

metal film resistor (*Elec. Eng.*). One formed by coating a high-temperature insulator with a very thin layer of metallic film, e.g., by vacuum deposition.

metal insulator (*Telecomm.*). Waveguide or transmission line an odd number of quarter-wavelengths long, which has a high impedance. Used as a support or anchor without normal insulators at very high frequencies.

metal lathing (*Build.*). *Expanded metal* (q.v.) used to cover surfaces to provide a basis for plaster.

metallic bond (*Chem.*). Theory that the metallic state is due to a structure in which electron-pair bonds resonate between a number of different atoms, so that a de-localized electron cloud is formed, which makes possible the characteristic properties of lustre, electrical and thermal conductivity, toughness, ductility, and density.

metallic circuit (*Teleg.*). Telegraph circuit in which there is a complete copper circuit, with no earth return.

metallic conduction (*Electronics*). That which describes the movement of electrons which are freely moved by an electric field within a body of metal.

metallic-film resistor (*Elec. Eng.*). One formed by coating a high-temperature insulator, such as mica, ceramic, Pyrex glass, or quartz, with a metallic film.

metallic foil (*Print.*). Blocking foil made from powdered metal pigments and giving an imitation gold or silver effect. Used for book spines, greeting cards, pencils, etc.

metallic lens (*Radio*). One with slats or louvres which give varying retardation to a passing electromagnetic wave, so that it is controlled or focused. Can also be used for sound waves.

metallic lustre (*Min.*). A degree of lustre exhibited by certain opaque minerals, comparable with that of polished steel.

metallic packing (*Eng.*). A *packing* (q.v.) consisting of a number of rings of soft metal, or a helix of metallic yarn, encircling the piston rod and pressed into contact therewith by a gland nut.

metalliferous veins (*Geol.*). Cracks and fissures in rocks which are found to contain, among other minerals, the ores of metals. See **lodes.**

metalling (*Civ. Eng.*). See **road metal.**

metallization (*Chem.*). (1) The vacuum deposition of thin metal films on to glass or plastic, for decorative or electrical purposes. (2) The conversion of a substance, e.g., selenium, into a metallic form.

metallized fabrics (*Textiles*). Cloths coated with a very thin insulating layer of metal, either by heat and pressure, or by a vaporization technique.

metallized filament (*Elec. Eng.*). A carbon lamp-filament which has been given special heat treatment in order partially to convert it to graphite, thereby making it almost metallic in its properties. Also called **graphitized filament.**

metallized foils (*Print.*). Blocking foils, usually prepared on polyester or cellulose acetate, consisting of gold or dyed, vacuum-deposited aluminium. These give a very brilliant result comparable to highly burnished metal.

metallized or spray-shielded valve (*Electronics*). One in which the exterior of the envelope is coated with a conducting metallic film which can be connected to cathode or earth, to provide electrostatic shielding of the interior of the valve from external disturbance.

metallizing (*Met.*). Building up worn parts of a machine by spray-coating with molten metal.

metallochrome (*Chem.*). The tinting produced on a metal surface by means of metallic salts.

metallography. The branch of metallurgy which deals with the study of the structure and constitution of solid metals and alloys, and the relation of this to properties on the one hand and manufacture and treatment on the other.

metalloid (*Chem.*). An element having both metallic and non-metallic properties, e.g., arsenic.

metallo-organic compounds (*Chem.*). See **organo-metallic compounds.**

metallurgical balance sheet (*Min. Proc., Met., etc.*). Report in equation form of the products from treatment of a known tonnage of ore of specified assay value, yielding a known weight of concentrate and tailing, to which the head value can be attributed for economic and technical control.

metallurgical coke (*Fuels, Met.*). Coke produced from high-temperature retorting of suitable coal, producing a dense and crush-resistant fuel suitable for use in shaft furnaces. It has a smaller percentage of volatile matter than *gas coke* (0·6% to 0·9–1·6%), and practically the same amount of fixed carbon.

metallurgy. Art and science applied to metals. The term covers extraction from ores, fabrication and refining, alloying, shaping, treating, and the study of structure, constitution, and properties.

metal-oxide semiconductor transistor (*Electronics*). An active semiconductor device in which a conducting channel is induced in the region between two electrodes by applying a voltage to an insulated electrode placed on the surface in this region. It is self-isolating by virtue of its construction, and so can be fabricated in a smaller area than a *bipolar transistor* (q.v.).

metal pattern (*Foundry*). A *pattern* (q.v.) made in cast-iron, brass, or light alloy, in order to ensure durability and permanence of form when a large number of castings are required, as in the case of repetition work on moulding machines.

metal rectifier (*Elec. Eng.*). A form of rectifier making use of the rectifying property of a layer of oxide on a metal disk, e.g., copper oxide on a copper disk. A number of such disks can be connected in series or in parallel to give high-voltage or high-current rectifiers. Generally superseded by semi-conductor rectifying devices. Also called **dry plate rectifier.**

metal rule (*Typog.*). See **rule, em rule.**

metals (*Rail.*). The rails of a railway.

metal spinning (*Eng.*). The shaping of thin sheet-metal disks into cup-shaped forms by the lateral pressure of a steel roller or a stick on the revolving disk, which is gradually pressed into contact with a former on the lathe face-plate.

metal (powder) spraying (*Eng.*). A method of applying protective metal coatings or building up worn parts by spraying molten metal from a gun. The coating metal is supplied as wire or powder, melted by flame, and blown out of the gun as finely divided particles.

metal trim (*Build.*). Architraves and other finishings made out of pressed metal sheeting clipped or screwed in position around door or window openings.

metal valley (*Plumb.*). A gutter, lined with lead, zinc, or copper, between two roof-slopes.

metal V-ring or **metal V-collar** (*Elec. Eng.*). A metal ring used in commutator construction; it has a V-shaped cross-section, so that it will fit into a corresponding recess in the commutator segments.

metamere (*Zool.*). See **merome.**

metameric segmentation (*Zool.*). See **metamerism.**

metamerism (*Chem.*). A form of isomerism occasioned by the attachment of different radicals to the same central polyvalent atom or group, the general chemical behaviour of the compounds being the same. (*Zool.*) Repetition of parts along the long axis of an animal. *adj.* **metameric.**

metamict (*Min.*). A term denoting the glassy amorphous state sometimes produced in several normally anisotropic minerals, probably representing structural breakdown due to radioactivity.

metamitosis (*Cyt.*). Mitosis in which the nuclear membrane disappears and the karyokinetic figure lies free in the cytoplasm. Cf. *mesomitosis.*

metamorphic aureole (*Geol.*). The zone of altered rocks which usually surrounds a large igneous intrusion. See also **metamorphism.**

metamorphic rocks (*Geol.*). Rocks derived from pre-existing rocks by mineralogical, chemical, and structural alterations due to processes operating in the earth's crust. The changes can be sufficient to produce a new rock type.

metamorphism (*Geol.*). A term embracing the changes in rocks which result from heating. Different types of metamorphism result from heating to different temperatures and under different pressures. The two main types are *regional* and *contact metamorphism.*

metamorphosis (*Zool.*). Pronounced change of form and structure taking place within a comparatively short time, as the changes undergone by an animal in passing from the larval to the adult stage. *adj.* **metamorphic.**

metamyelocyte (*Histol.*). An intermediate stage in the development of myelocyte before its transformation into a leucocyte.

metanauplius (*Zool.*). A larval stage in some *Crustacea,* differing from the *nauplius* (q.v.) in having a fourth pair of appendages (representing the maxillae) developed behind the original three pairs.

metanephric duct (*Zool.*). Ureter of amniote Vertebrates.

metanephridia (*Zool.*). Nephridia which open into the coelom, the open end (nephrostome) being ciliated. Cf. *protonephridium.*

metanephromixium (*Zool.*). In *Annelida,* a coelomoduct attached to a metanephridium.

metanephros (*Zool.*). In amniote Vertebrates, part of the kidney arising later in development than, and posterior to, the mesonephros; becomes the functional kidney, with a special *metanephric duct* (q.v.). *adj.* **metanephric.**

metanilic acid (*Chem.*). $C_6H_4(NH_2)(SO_3H)$, *meta*-aminobenzene-sulphonic acid; an intermediate for dyestuffs.

metanotum (*Zool.*). The notum of the metathorax in Insects. *adj.* **metanotal.**

metaphase (*Cyt.*). The stage in mitosis in which the chromosomes aggregated on the equator of the mitotic spindle divide longitudinally and the daughter chromosomes pass outwards towards the poles of the spindle. Also, **aster phase.**

metaphloem (*Bot.*). Completely developed primary phloem, consisting of sieve tubes (with or without companion cells), fibres, and parenchyma.

metaphosphoric acid (*Chem.*). HPO_3. Formed as a viscous solid when phosphorus(v)oxide is left exposed to the air.

metaphragma (*Zool.*). In Insects, an apodeme situated in the metathorax.

metaphysis (*Med.*). The end of the shaft (diaphysis) of a long bone where it joins the epiphysis.

metaplasia (*Zool.*). Tissue transformation, as in the ossification of cartilage.

metaplasis (*Zool.*). The period of maturity in the life-cycle of an individual.

metaplasm (*Biol.*). Any substance within the body of a cell which is not protoplasm, i.e., nonliving material; especially food material, as yolk or fat, within an ovum. *adjs.* **metaplasmic, metaplastic.**

metapleural folds (*Zool.*). In *Cephalochorda,* downgrowths from the dorsolateral body wall which form the side walls of the *atrium* (q.v.).

metapleuron (*Zool.*). In *Diptera,* a thoracic sclerite behind the pteropleuron and above the hypopleuron.

metapneustic (*Zool.*). Of Insects, having only the last pair of abdominal spiracles open, as in larval *Culicidae.*

metapodium (*Zool.*). (1) In Vertebrates, the second podial region; metacarpus or metatarsus; palm or instep. (2) In Insects, that portion of the abdomen posterior to the *petiole* (q.v.). (3) In *Gastropoda,* the posterior part of the foot. *adj.* **metapodial.**

metapodosoma (*Zool.*). In *Acarina,* the segments of the third and fourth pairs of legs.

metapophysis (*Zool.*). In some Mammals, a process of the vertebrae above the prezygapophysis which strengthens the articulation.

metaproteins (*Chem.*). See **infraproteins.**

metapterygium (*Zool.*). In *Selachii,* the innermost or most posterior of the 3 basal cartilages of the pectoral fin.

metapterygoid (*Zool.*). In some Fish, a paired ventral bone of the skull, lying above the quadrate.

metarchon (*Chem.*). An agent which, without being toxic, so changes the behaviour of a pest that its persistence is diminished, e.g., a confusing sex attractant.

metasitism (*Zool.*). Cannibalism.

metasoma (*Zool.*). In *Arachnida,* the posterior part of the abdomen, or hindermost tagma of the body, which is always devoid of appendages. *adj.* **metasomatic.**

metasomatism (*Geol.*). The processes by which one mineral is replaced by another of different chemical composition by the introduction of material from external sources. See also **metamorphism.**

metasome (*Zool.*). A name sometimes given to the mid-body, or the 3 segments behind the cephalothorax, of *Cyclops* (*Copepoda*).

Metaspermae (*Bot.*). See **Angiospermae.**

metastable state (*Chem.*). State which is apparently stable, often because of the slowness with which equilibrium is attained; said, for example, of a supersaturated solution.

metastasic (*Electronics*). Said of electrons which move from one shell to another or are absorbed from a shell into the nucleus.

metastasis (*Med.*). The transfer, by lymphatic channels or blood vessels, of diseased tissue (especially cells of malignant tumours) from one part of the body to another; the diseased area arising from such transfer. (*Zool.*) Transference of a function from one part or organ to another; metabolism.

metastasize (*Med.*). To form metastases.

metastatic (*Zool.*). A term used to describe the life-cycle of a parasite in which metamorphosis into the adult form occurs in the secondary host, which is then eaten by the primary host.

metasternite (*Zool.*). In *Scorpionidea,* a plate between the basal joints of the last pair of walking legs, representing the fused sterna corresponding to those limbs.

metasternum (*Zool.*). In Insects, the sternum of the metathorax; in Vertebrates, the posterior portion of the sternum; xiphisternum.

metastigmate (*Zool.*). Having the tracheal openings situated posteriorly, as certain Mites.

metastoma (*Zool.*). In *Crustacea,* and *Trilobita,* the lower lip.

metasyndesis (*Cyt.*). See **telosynapsis.**

metatarsal or **metatarsale** (*Zool.*). One of the

bones composing the *metatarsus* (q.v.) in Vertebrates.

metatarsalgia (*Med.*). A painful neuralgic condition of the foot, felt in the ball of the foot and often spreading thence up the leg.

metatarsus (*Zool.*). In Insects, the first joint of the tarsus when it is markedly enlarged; in land Vertebrates, the region of the hind-limb between the digits and the tarsus.

Metatheria (*Zool.*). A subclass of viviparous Mammals in which the newly born young are carried in an abdominal pouch which encloses the teats of the mammary glands; an allantoic placenta is usually lacking; the scrotal sac is in front of the penis, the angle of the lower jaw is inflexed, and the palate shows vacuities. Contains only one order, the *Marsupiala.*

metathetely (*Zool.*). In the development of Insects, the condition when development of the imaginal discs and sex organs is inhibited, although the larva continues to grow. Cf. *prothetely.*

metathorax (*Zool.*). The third or most posterior of the 3 somites composing the thorax in Insects. *adj.* **metathoracic.**

metatracheal parenchyma (*Bot.*). Parenchyma occurring in wood scattered throughout the annual ring.

metatroch (*Zool.*). A ciliated band encircling the body of a trochophore posterior to the mouth.

metatropleic (*Biol.*). Utilizing both nitrogenous and carbonaceous organic material for food.

metatympanic (*Zool.*). See **entotympanic.**

metaxenia (*Bot.*). Any effect that may be exerted by pollen on the tissues of the female organs.

metaxylem (*Bot.*). Primary xylem in which the vessels have either reticulate thickening or pitted walls.

Metazoa (*Zool.*). A subkingdom of the animal kingdom, comprising multicellular animals having two or more tissue layers, never possessing choanocytes, usually having a nervous system and enteric cavity, and always showing a high degree of coordination between the different cells composing the body. Cf. *Protozoa, Parazoa.*

metazoaea (*Zool.*). A larval stage of decapod *Crustacea* which differs from the zoaea in showing the rudiments of thoracic limbs behind the maxillipeds.

metecdysis (*Zool.*). The period after a moult in *Arthropoda* when the new cuticle is hardening and the animal is returning to normal in its physiological condition.

metencephalon (*Zool.*). The anterior portion of the hind-brain in Vertebrates, developing into the cerebellum, and, in Mammals, the *pons* (part of the *medulla oblongata*). Cf. *myelencephalon.*

metenteron (*Zool.*). An intermesenteric space in a Coelenterate.

meteor (*Astron.*). A 'shooting star'. A small body which enters the earth's atmosphere from interplanetary space and becomes incandescent by friction, flashing across the sky and generally ceasing to be visible before it falls to the earth. See also **bolide.**

meteor craters (*Astron.*). Circular unnatural craters of which Meteor Crater in Arizona is best known; believed to be caused by the impact of meteorites.

meteoric shower (*Astron.*). A display of meteors in which the number seen per hour greatly exceeds the average. It occurs when the earth crosses the orbit of a meteor swarm and the swarm itself is in the neighbourhood of the **point of section of the two orbits.**

meteorism (*Med., Vet.*). Excessive accumulation of gas in the intestines. See also **tympanites.**

meteorites (*Astron., Min.*). Mineral aggregates of cosmic origin which reach the earth from interplanetary space; cf. *meteor* and *bolide.* See **achondrite, aerolites, chondrite, iron meteorites, pallasite, siderite, stony meteorites.**

meteoritic hypothesis (*Astron.*). A theory that the so-called craters on the moon are due not to volcanic action but to the impact, in a relatively late stage of development, of planetesimal bodies or meteorites on the lunar surface.

meteorograph (*Meteor.*). A collection of meteorological recording instruments, such as the barograph, thermograph, etc., which are attached to kites or small balloons and sent up to record conditions in the upper atmosphere.

meteorological elements (*Meteor.*). Those components (such as temperature, pressure, humidity, wind, rainfall, cloudiness) which determine the state of the weather.

meteorology. The study of the earth's atmosphere in its relation to weather and climate.

meteor streams (*Astron.*). Streams of dust revolving about the sun, whose intersection by the earth causes meteor showers. Some night-time showers have orbits similar to those of known comets; day-time showers, detected by radio-echo methods, have smaller orbits, similar to those of minor planets.

meteotron (*Meteor.*). A system of heat generators to stimulate formation of artificial rain-clouds.

meter. A variant spelling of **metre.** (*Elec. Eng.*). A general term for any electrical measuring instrument, but usually confined to integrating meters. (*Teleph.*) A counting mechanism which indicates the number of times it has been operated by passing a sufficient current through its windings.

meter protection circuit (*Elec. Eng.*). One designed to avoid transient overloads damaging a meter. It may employ a gas-discharge tube which breaks down at a dangerous voltage or a Zener diode or similar device.

meter wire (*Teleph.*). A wire over which the operation of the calling subscriber's meter is controlled. Also **M-wire.**

methadone (*Pharm.*). Methadone hydrochloride. An analgesic with similar but less marked properties, both good and bad, than morphine.

methaemoglobin (*Chem.*). A compound of haemoglobin and oxygen, more stable than oxyhaemoglobin, obtained by the action of oxidizing agents on blood, as in poisoning by nitrites (nitrates(III)) or chlorates(V).

methaemoglobinaemia (*Med.*). The presence of methaemoglobin in the blood, the result usually of the action of drugs derived from aniline.

methaemoglobinuria (*Med.*). The presence of methaemoglobin in the urine.

metham (*Chem.*). *N*-methyldithiocarbamic acid, used as an insecticide, fungicide, and weedkiller.

methanal (*Chem.*). See **formaldehyde.**

methane (*Chem.*). CH_4. The simplest *alkane* (q.v.), a gas, m.p. −186°C, b.p. −164°C; occurs naturally in oil-wells and as marsh gas. *Fire-damp* (q.v.) is a mixture of methane and air; coal-gas contains a large proportion of methane. It can be synthesized from its elements, and prepared by various methods, as by catalytic reduction of CO or CO_2, or by passing CO and H_2O over heated metal oxides, or by the action of water on aluminium carbide (methanide).

methanides (*Chem.*). Carbides, such as aluminium and beryllium carbides, which give methane when decomposed by water.

methanol (*Chem.*). CH_3OH; a colourless liquid,

b.p. 66°C, rel. d. 0·8. It used to be produced by the destructive distillation of wood; is nowadays synthesized from CO and H_2 in the presence of catalysts. It is an important intermediate for numerous chemicals, and is used as a solvent and for denaturing ethanol. Also called **methyl** or **wood alcohol.**

methanoic acid. See **formic acid.**

methene (*Chem.*). See **methylene.**

methine (*Chem.*). The trivalent radical $CH\equiv$.

methine dyes (*Chem.*). Group of dyestuffs important in the development of photography. Consist mainly of dyes containing two quinoline or benzthiazole groups joined by conjugated aliphatic chains.

methionine (*Chem.*). A sulphur-containing amino acid, $CH_3{\cdot}S{-}CH_2{-}CH_2{-}CH(NH_2){-}COOH$, which must, for man as well as other animals, be present in the food proteins. Apparently concerned, *inter alia,* in reactions involving methylation, such as synthesis of adrenalin and creatine. Also used, with serine, for *in vivo* production of cystine.

method of limits (*Psychol.*). A standard procedure in *psychophysics* for determining the *threshold* of a stimulus by taking the average of the intensity at which the stimulus is no longer perceived when it is gradually reduced and that at which it is perceived when it is increased from a sub-threshold level.

method study (*Eng.*). A branch of work study, concerned with determining the best production methods and the corresponding equipment for making an article in the desired quantities at optimum quality and cost.

methoin (*Pharm.*). A chemical substance whose anticonvulsant effects make it effective in the treatment of epilepsy.

methoxone (*Chem.*). See **MCPA.**

methoxychlor (*Chem.*). 1,1,1-Trichloro-2,2-di-(4-methoxyphenyl)ethane:

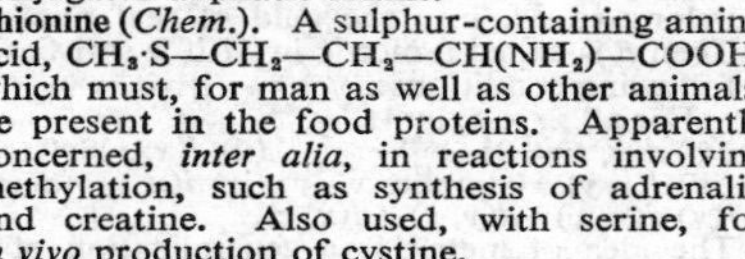

used as an insecticide; also known as **methoxy-DDT, DMDT,** and **dianysl-trichloroethane.**

methoxyl group (*Chem.*). The monovalent radical $-OCH_3$. In certain compounds it can be estimated analytically by *Zeisel's method* (q.v.).

methylal (*Chem.*). $(CH_3O)_2CH_2$, a colourless liquid; b.p. 42°C, widely used as a solvent.

methyl alcohol (*Chem.*). *Methanol.*

methylamines (*Chem.*). *Mono-*, $CH_3{\cdot}NH_2$; *di-* $(CH_3)_2NH$; and *tri-*, $(CH_3)_3N$, may be regarded as ammonia in which one or more of the hydrogen atoms is replaced by the methyl group. Ammoniacal liquids with fishy smell, occurring in herring brine. Used in manufacturing drugs, dyes, etc.

4-methylaminophenol (*Photog.*). Common developing agent, discovered in 1891.

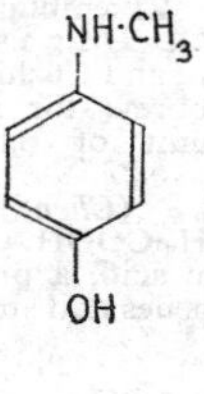

Generally used in the more stable and soluble form of its sulphate, known as metol and also as Elon (TN), a main ingredient of **M.Q.** (or *metol-quinol*) developer.

methylated spirit (*Chem.*). *Methanol*; before sale to general public it is dyed and made unpalatable because of its toxicity.

methylation (*Chem.*). The introduction of methyl (CH_3) groups into organic compounds. Carried out by using methylating agents such as diazomethane or dimethyl sulphate.

methylbenzene (*Chem.*). See **toluene.**

methyl blue-eosin stain (*Micros.*). A polychromatic stain, containing methyl blue and eosin, which distinguishes clearly between basophil and oxyphil components of a specimen.

methyl bromide (*Chem.*). CH_3Br. Bromomethane. B.p. 4°C, rel. d. (0°C) 1·732. Organic liquid widely used as a fire extinguishing medium.

methyl cellulose (*Chem.*). An *ether* prepared from *cellulose* (q.v.), which gives highly viscous solutions in water. It is used in distempers and other water paints, foodstuffs, cosmetics, etc.

methyl chloride (*Chem.*). CH_3Cl. Chloromethane. B.p. −23·7°C, rel. d. (0°C) 0·952. Used as local anaesthetic and refrigerant.

methylcyclohexanol(IV) (*Chem.*). The methyl derivative of *cyclohexanol,* actually a mixture of the ortho-, para- and meta-isomers. Used widely as a solvent for fats, waxes, resins, and in lacquers. Soaps prepared from it are used as detergents. Also **hexahydrocresol, sextol.**

methylene (*Chem.*). The hypothetical compound CH_2. Numerous attempts have been made to obtain it by eliminating, for example, hydrogen and chlorine from methyl chloride, but the resulting CH_2 groups combine together in pairs, yielding ethylene, $H_2C{=}CH_2$.

methylene blue (*Chem.*). A thiazine dyestuff of the formula:

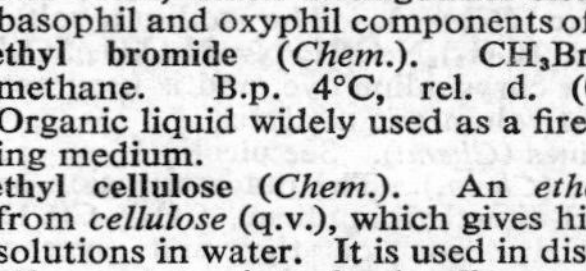

It is prepared by oxidizing dimethyl-*p*-phenylene-diamine and dimethylaniline in the presence of sodium thiosulphate and zinc chloride. It is a very important dye for cotton, upon which it is fixed with the aid of tannin.

methylene dichloride (*Chem.*). Dichloromethane. CH_2Cl_2. Organic solvent widely used in paint strippers. B.p. 41°C. By-product in the manufacture of trichloromethane.

methylene di-iodide (*Chem.*). Di-iodomethane. CH_2I_2. High relative density (3·32) organic liquid. Used in flotation processes for ore separation and in the density determination of minerals.

methylene ditannin (*Chem., Med.*). A reddish powder made by condensing tannic acid with methanal; used in medicine, internally as an antiseptic in treatment of dysentery, and externally for treatment of eczema.

methyl ethyl ketone (*Chem.*). Butan-2-one. $CH_3{\cdot}CO{\cdot}C_2H_5$; a colourless liquid of ethereal odour, b.p. 81°C; prepared by the oxidation of butan-2-ol; an important solvent. Its peroxide is widely used as a catalyst for curing polyester resins. Abbrev. **MEK.**

methyl green-acetic stain (*Micros.*). A fixing and staining solution consisting of methyl green in

1% ethanoic acid, used for staining nuclei in temporary preparations of small organisms, e.g., *Protozoa*.

methyl group (*Chem.*). Monovalent radical CH_3—.

methyl iodide (*Chem.*). CH_3I. Iodomethane. B.p. 42·5°C. Prepared by adding iodine to methanol and red phosphorus. Useful reagent for the preparation of methyl ethers by reaction with primary alcohols in the presence of silver(I) oxide.

methyl methacrylate resins (*Plastics*). See **acrylic resins.**

methylol group (*Chem.*). The group $\cdot CH_2OH$. Found in primary alcohols and glycols.

methyl orange (*Chem.*). The sodium salt of helianthine, $(CH_3)_2N\cdot C_6H_4\cdot N{=}N\cdot C_6H_4\cdot SO_2O^-Na^+$. It is a chrysoidine dye, and is used as an indicator in volumetric analysis.

methyl-pyridines (*Chem.*). See **picolines.**

methyl-rubber (*Chem.*). The polymerization product of 2,3-dimethylbutadiene, $CH_2{=}C(CH_3)\cdot C(CH_3){=}CH_2$, one of the first synthetic rubbers, much inferior to the natural product. It oxidizes easily, and can be vulcanized only by the addition of organic catalysts.

methyl salicylate (*Chem.*).

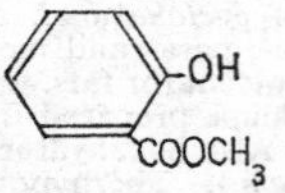

-OH
COOCH$_3$

The main constituent of oil of wintergreen, with a characteristic aromatic odour. Used as an ointment or liniment for treatment of rheumatism.

methyl sulphate (*Chem.*). $(CH_3)_2SO_4$; a colourless syrupy oil, very poisonous, b.p. 188°C; used for introducing the methyl group into phenols, alcohols, and amines.

methyl violet (*Chem.*). A triphenylmethane dyestuff consisting of a mixture of the hydrochlorides of tetra-, penta-, and hexamethylpararosaniline.

Metlex plug (*Build.*). A split hollow plug of soft metal placed in a drilled hole to receive a screw, which opens out the plug against the walls of the hole, thus ensuring a firm fixing.

metochy (*Zool.*). The type of partnership exhibited by a neutral inquiline and the social Insects in the nest of which it makes its home.

metoecious (*Bot., Zool.*). Heteroecious.

metoestrus (*Zool.*). In Mammals, the recuperation period after oestrus.

metol (*Photog.*). See **4-methylaminophenol.**

Metonic cycle (*Astron.*). A period of 19 years, which is very nearly equal to 235 synodic months, this relationship having been introduced in Greece in 433 B.C. by the astronomer Meton; its effect is that after a full cycle the phases of the moon recur on the same days of the year.

metope (*Arch.*). A slab or tablet (generally of marble and ornamented) filling the space between the triglyphs in a Doric frieze.

metotic (*Zool.*). Posterior to the auditory vesicle.

metovum (*Zool.*). An ovum which is surrounded by nutritive material.

metoxenous (*Bol., Zool.*). Heteroecious.

metre. The Système International unit of length. Originally intended to represent 10^{-7} of the distance on the earth's surface between the North Pole and the Equator, and formerly governed by a line standard on a platinum bar, it was redefined in 1960 as 1 650 763·73 vacuum wavelengths of radiation of krypton-86.

metre bridge (*Elec. Eng.*). A Wheatstone bridge in which one metre of resistance wire, usually straight, with a sliding contact is used to form two variable ratio arms in a Wheatstone bridge network.

metre-candle (*Light*). See **lux.**

metre-kilogram(me)-second-ampere (*Elec.*).**MKSA.**

metric (*Maths.*). (1) A differential expression of distance in a generalized vector space, i.e., $ds^2=g_{\alpha\beta}dx^\alpha dx^\beta$. The coefficient $g_{\alpha\beta}$ forms a tensor known as the *fundamental metric tensor*. (2) Let M be any set, and let d be a mapping from the set of ordered pairs (x, y), where x and y belong to M, into the field of real numbers. Then d is called a metric in M if it satisfies the following conditions:

For all x, y, z, in M,
(i) $d(x, x)=0$; (ii) $d(x, y)\geqq 0$;
(iii) $d(x, y)=0$ implies $x=y$; (iv) $d(x, y)=d(y, x)$;
(v) $d(x, y)+d(y, z)\geqq d(x, z)$.

The idea of metric is a generalization of the familiar notion of distance. For any given metric d, the value of $d(x, y)$ is called the distance between x and y. See **metric space.**

metric screw thread (*Eng.*). A standard screw thread in which the diameter and pitch are specified in millimetres.

metric space (*Maths.*). A set M together with a metric d is called a metric space. See **metric.**

metric system. A system of weights and measures based on the principle that each quantity should have one unit whose multiples and sub-multiples are all derived by multiplying or dividing by powers of ten. This simplifies conversion, and eliminates completely the complicated tables of weights and measures found in the traditional British system. Originally introduced in France, it is the basis for the *Système International* (*SI*) now being universally adopted.

metritis (*Med.*). Inflammation of the substance of the uterus.

metrology. The science of measuring.

metromorphic (*Bot.*). Resembling the mother.

metronidazole (*Med.*). Used in the treatment of trichomoniasis of the genito-urinary tract in males and females.

metronomic cell counter (*Instr.*). Instrument which assists the counting of small objects under the microscope (e.g., blood cells) by synchronizing visual counting with an audible rhythmic beat of controllable frequency.

metropathia haemorrhagica (*Med.*). Essential uterine haemorrhage. A condition in which bleeding from the uterus is associated with thickening of the lining of the uterus and with the presence of cysts in the ovaries.

metrorrhagia (*Med.*). Bleeding from the uterus between menstrual periods.

metrosil (*Electronics*). TN for semiconductor devices containing silicon carbide, and having a nonlinear resistance.

metrostaxis (*Med.*). See **metrorrhagia.**

-metry (*Chem., etc.*). A suffix denoting a method of analysis or measurement, e.g., acidimetry, iodimetry, nephelometry.

metuliform (*Bot.*). Resembling a pyramid.

Metyrapone (*Med.*). Drug used in the diagnosis of hypopituarism and Cushing's syndrome.

MeV (*Nuc.*). Abbrev. for *million-* (or *mega-*) *electron-volt*; unit of particle energy, 10^6 electron-volts.

mevalonic acid (*Chem.*). $CH_2OH\cdot CH_2\cdot C(OH)(CH_3)\cdot CH_2\cdot COOH$. 3,5-Dihydroxy-3-methylpentanoic acid, a precursor in the biosynthesis of terpenes and steroids.

Mexican onyx (*Min.*). A translucent, veined, and parti-coloured aragonite, found in Mexico and in the south-western U.S.

mezzanine (*Build.*). An intermediate floor constructed between 2 other floors in a building.

mezzo-rilievo (*Paint.*). Decoration in medium relief.

mezzotint (*Print.*). An intaglio process in which printing is done from a copper plate, grained by rocking a semicircular toothed knife over the surface, the lighter tones being produced by scraping or burnishing away the grain to reduce the ink-holding capacity.

MF (*Paper*). *Machine-finished.* Paper which has been surfaced while on the paper-making machine.

MF resin (*Plastics*). Abbrev. for *melamine-formaldehyde resin.*

μF (*Elec. Eng.*). Abbrev. for *microfarad.*

mg (*Phys.*). Abbrev. for *milligram.*

Mg (*Chem.*). The symbol for *magnesium.*

MG (*Build.*). Abbrev. for *make good.* (*Paper*) *Machine-glazed.* A class of paper which is rough on one side and glazed on the other; used for wrapping, poster work, etc.

MG machine (*Paper*). A machine in which the wet paper is pressed on a polished heated cylinder and dried during one revolution. A high glaze is imparted to one side of the paper, while the other side remains rough. Also known as **Yankee machine.**

Mg point (*Glass*). See **transformation points.**

MHD generator (*Elec. Eng.*). Abbrev. for *magnetohydrodynamic generator.*

mho (*Elec. Eng.*). Name for the reciprocal of the ohm in the CGS system. See **siemens.**

miarolitic structure (*Geol.*). A structure found in an igneous rock, consisting of irregularly shaped cavities into which the constituent minerals may project as perfectly terminated crystals.

mica (*Elec. Eng., Min.*). A group of silicates which crystallize in the monoclinic system; they have similar chemical compositions and highly perfect basal cleavage. Mica is one of the best electrical insulators.

micaceous iron-ore (*Min.*). A variety of specular *haematite* (Fe_2O_3) which is foliated or which simulates mica in the flakiness of its habit.

micaceous sandstone (*Geol.*). A sandstone containing a varying amount of mica flakes.

mica cone (*Elec. Eng.*). See **mica V-ring.**

mica flap valve (*San. Eng.*). A sheet of mica hinged about one edge, so as to permit only unidirectional flow of air through ventilators.

micafolium (*Elec. Eng.*). A composite insulating material consisting of a paper backing covered with mica flakes and varnish. Much used for insulating wire, machine coils, etc.

mica-lamprophyre (*Geol.*). One of the commonest types of *lamprophyre* characterized by an abundant content of mica, originally biotite, but often bleached and altered.

micanite (*Elec. Eng.*). Mica splittings bonded by varnish or shellac into a large sheet; mechanically weak at high temperatures.

mica-schist (*Geol.*). Schist composed essentially of micas and quartz, the foliation being mainly due to the parallel disposition of the mica flakes. See also **schist.**

mica-trap (*Geol.*). An obsolete name for *mica-lamprophyre* (q.v.).

mica V-ring (*Elec. Eng.*). A ring of V-shaped cross-section made of a mica compound and used to insulate a metal V-ring from the bars of the commutator which it supports.

micella (*Bot.*). A hypothetical crystalline structure, an orderly aggregate of chain-like molecules, too small to be seen, which may, with many other similar structures, form the foundation of cell walls, starch grains, etc.

micelle (*Chem.*). A colloidal sized aggregate of molecules, such as those formed by surface active agents.

Michaelis constant (K_m) (*Biochem.*). A relationship between the velocity constants in a kinetic scheme for an enzyme-catalysed reaction. If the same enzyme can attack several substrates K_m values frequently give a useful comparison of its affinity for the different substrates.

Michell bearing (*Eng.*). A thrust or journal bearing in which pivoted pads support the thrust collar or journal in such a way that they tilt slightly under the wedging action of the lubricant induced between the surfaces by their relative motion. The fluid lubrication conditions thus produced result in a very low friction coefficient and power loss in the bearing.

Michelson interferometer (*Light*). One of the earliest interferometers, by means of which much work was done on the fine structure of spectral lines and the evaluation of the standard metre in wavelengths of light. The principle of the instrument is similar to that of the *Fabry and Pérot interferometer* (q.v.).

Michelson-Morley experiment (*Phys.*). An attempt to detect and measure the relative velocity of the earth and the ether by observations of interference fringes with a form of apparatus which can be rotated bodily into different orientations. No such relative motion was detected. The *FitzGerald-Lorentz contraction* was later put forward to explain this result.

micoquille (*Glass*). See **coquille.**

micraesthete (*Zool.*). In *Amphineura*, the smaller type of sense-organ occurring in canals traversing the shell. Cf. *megalaesthete.*

micramoeba (*Zool.*). In certain *Sarcodina*, a small amoebula stage, probably representing a male gamete which fuses with a *macramoeba* (q.v.).

micraner (*Zool.*). An abnormally small male ant.

micrergate (*Zool.*). An abnormally small worker ant.

micro-. Prefix from Gk. *mikros*, small. When used of units it indicates the basic unit $\times 10^{-6}$, e.g., microampere (μA) $= 10^{-6}$ ampere. Symbol μ. (*Geol.*) Applied to names of rocks, it indicates the medium-grained form, e.g., *microdiorite, microsyenite, microtonalite.*

microaerophile (*Bot.*). An organism which does not grow well with ordinary concentrations of oxygen, but only when the oxygen concentration is low. *adj.* **microaerophilic.**

microammeter (*Elec. Eng.*). A most sensitive form of robust current-measuring instrument.

microanalysis (*Chem.*). A special technique of both qualitative and quantitative analysis, by means of which very small amounts of substances may be analysed.

microanalytical reagent (*Chem.*). See **MAR.**

microanatomical fixative (*Micros.*). A fixative used to preserve the microanatomy of specimens, as opposed to the preservation of cytological detail.

microbalance (*Chem.*). Sensitive to one microgram, for use in microanalysis; may be either beam or quartz fibre.

microbar (*Phys.*). Unit of pressure in the CGS system, especially in acoustics, 1 dyne per sq. cm $= \frac{1}{10}$ newton per sq. metre. Sometimes called **barye.**

microbe (*Bacteriol.*). A bacterium which can be seen with the aid of a microscope.

microbiota (*Ecol.*). The smallest soil organisms,

including Soil Algae, Bacteria, Fungi, and *Protozoa*. Cf. *macrobiota*, *mesobiota*.

microblast (*Histol.*). A smaller than normal erythroblast.

microbody (*Cyt.*). Cytoplasmic particles 0·2–1·0 μm in diameter surrounded by a limiting membrane and containing an electron-dense core. Contain oxidases but not the hydrolases found in lysosomes.

microcanonical assembly (*Phys.*). In statistical thermodynamics, designates an assembly which consists of a large number of systems each having the same energy.

microcephaly, microcephalia (*Med.*). Abnormally small size of the head.

microchemistry (*Chem.*). Preparation and analysis of very small samples, usually less than ten milligrams but, in radioactive work, down to a few atoms.

microcircuit isolation (*Elec. Eng.*). The electrical insulation of circuit elements from the electrically conducting silicon wafer. The two main techniques are *oxide isolation* and *diode isolation* (qq.v.).

microcircuits (*Elec. Eng.*). Those with components formed in one unit of semiconductor crystal.

microcline (*Min.*). A silicate of potassium and aluminium which crystallizes in the triclinic system. It resembles orthoclase, but is distinguished by its optical and other physical characters. See also **potassium feldspar.**

Micrococcaceae (*Bacteriol.*). A family of bacteria belonging to the order *Eubacteriales*. Gram-positive cocci; includes free living, parasitic, saprophytic and pathogenic species, e.g., *Staphylococcus aureus* (one cause of food poisoning).

microcode (*Comp.*). Coding in the individual steps obeyed by a specific computer, involving, e.g., the breaking-down of a multiplication routine into successive processes of addition. Cf. *macrocode*.

microconidium (*Bot.*). A small conidium produced by some species of fungi, differing in form as well as in size from the larger conidia characteristic of the species.

microconjugant (*Zool.*). In certain *Mastigophora*, the smaller of a pair of conjugants.

microcosmic salt (*Chem.*). Sodium ammonium hydrogen phosphate, $NaNH_4HPO_4 \cdot 4H_2O$. Present in urine.

microcrystalline texture (*Geol.*). A term applied to a rock or groundmass in which the individual crystals can be seen as such only under the microscope.

Microcyprini (*Zool.*). An order of small *Teleostei*, characterized by the flattened scaly head, protractile mouth, and absence of a distinct lateral line; fresh-water and estuarine forms of the tropics and subtropics, feeding on insects and other small organisms or organic matter in mud; mainly viviparous. Cyprinodonts or Toothed Carps, Killifishes, Four-eyed Fish, Millions Fish.

microcyte (*Histol.*). A term used to describe a small red blood corpuscle, outside the normal range of size variation.

microdensitometer (*Photog.*). Instrument containing a photocell combined with a microscope, used for scanning the emulsion of photographic negatives in order to measure optical density over a small but finite area.

microdiorite (*Geol.*). An intermediate igneous rock of medium grain-size.

microdissection (*Biol.*). A technique for small-scale dissection, e.g., on living cells, using micromanipulators and viewing the object through a microscope.

microelectronics (*Electronics*). Technique of solid circuits in which units of semiconductors are formed into several components.

microencapsulation (*Paper, etc.*). A process whereby a substance in a state of extreme comminution is enclosed in sealing capsules from which the material is released by impact, solution, heat or other means. See **NCR.**

micro-environment (*Ecol.*). The environment of small areas in contrast to large areas, with particular reference to the conditions experienced by individual organisms and their parts (e.g., leaves, etc.).

microfarad (*Elec. Eng.*). A unit of capacitance equal to one-millionth of a farad; more convenient for use than the farad. Abbrev. μF.

microfelsitic texture (*Geol.*). A term applied to the cryptocrystalline texture seen, under the microscope, in the groundmass of quartz-felsites and similar rocks; due to the devitrification of an originally glassy matrix.

microfiche (*Photog.*). A sheet of *microfilm*, usually A6 size with images of 60 or 98 pages reduced 20 to 25 times.

microfilaria (*Zool.*). The early larval stage of certain parasitic *Nematoda*.

microfilm (*Photog.*). A standard non-flam film on which documents, books, pamphlets, etc., are recorded; for viewing, the records are projected on a ground-glass screen by an enlarging machine.

microgamete (*Biol.*). The smaller of a pair of conjugating gametes, generally considered to be the male gamete.

microgametocyte (*Zool.*). In *Protozoa*, a stage developing from a trophozoite and giving rise to male gametes.

microgamy (*Zool.*). In *Protozoa*, syngamy between 2 of the smallest individuals produced by fission or gemmation.

microgap switch (*Elec. Eng.*). A switch, used on low-power, low-voltage circuits, which relies for arc extinction on the lateral spread of the arc stream by mutual repulsion between contacts separated by thousandths of an inch.

microglia (*Zool.*). A small type of neuroglia cell (occurring more frequently in grey matter than in white matter) having an irregular body and freely branching processes which end in terminal spines; can be phagocytic.

microgram(me). Unit of mass equal to one millionth of a gram (10^{-9} kg). Sometimes known as **gamma.**

microgranite (*Geol.*). A medium-grained, microcrystalline, acid igneous rock having the same mineral composition and texture as a granite.

micrographic texture (*Geol.*). A distinctive rock texture in which the simultaneous crystallization of quartz and feldspar has led to the former occurring as apparently isolated fragments, resembling runic hieroglyphs, set in a continuous matrix of the latter mineral.

microgroove records (*Acous.*). Long-playing records with as many as 200–300 grooves/in., and played at rotational speeds of 33⅓ or 45 rev/min.

microgyne (*Zool.*). Abnormally small queen ant.

microgyria (*Med.*). Abnormal smallness of the convolutions of the brain.

micro-incineration (*Biol.*). A technique for examining the distribution of minerals in slide preparations of tissue-sections or cells. The organic material is destroyed by heat and the nature and position of the mineral ash determined by microscopic examination.

microlecithal (*Zool.*). Said of eggs containing very little yolk. Cf. *megalecithal.*

microlite (*Geol.*). A general term for minute crystals of tabular or prismatic habit found in microcrystalline rocks. These give a reaction with polarized light. (*Min.*) A mineral which is essentially a hydrous tantalate of calcium and sodium, but which frequently contains niobium, fluorine, and a variety of bases; a member of the pyrochlore series. It crystallizes in the cubic system, and typically occurs in granite pegmatites.

microlux (*Light.*). A unit for very weak illuminations, equal to one-millionth of a lux.

micro-machine (*Elec. Eng.*). A miniature rotating electrical machine, designed to simulate, in a *micro-system*, the electrical and dynamic characteristics of a full-scale machine.

micro manipulator (*Biol.*). An instrument developed for use in micrurgy, and controlled indirectly by pneumatic, thermal, or mechanical means.

micromazia (*Med.*). Failure of the female breast to develop after puberty.

micromere (*Zool.*). In a segmenting ovum, one of the small cells which are formed in the upper or animal hemisphere.

micromerigraph (*Powder Tech.*). Apparatus for determining the size distribution of small particles by settling them in air. The rates of sedimentation are proportional to the diameters of the particles. See **Sharples.**

micromeritics (*Powder Tech.*). Little used alternative for *powder technology* (q.v.); the science of small particles.

micromerozoite (*Zool.*). A stage in the life-history of certain *Haemosporidia*, arising by schizogony from a microschizont.

micromesh sieves (*Powder Tech.*). See **electroformed sieves.**

micrometeorites (*Astron.*). Meteoric dust too small to be consumed by friction in the atmosphere.

micrometer (*Instr.*). An instrument used mainly for measuring small angular separations visually. It consists of 3 frameworks carrying spider-webs for superposition; 1 of these is fixed, and the 2 others are movable by means of a pair of screws. The micrometer is mounted at the eye end of an equatorial telescope and furnished with an eyepiece; the separation is read from the 2 micrometer heads, and a graduated circle gives the position angle when the object is a double star.

micrometer eyepiece (*Micros.*). See **eyepiece graticule.**

micrometer gauge (*Eng.*). A U-shaped length gauge in which the gap between the measuring faces is adjustable by an accurate screw whose end forms one face. The gap is read off a scale uncovered by a thimble carried by the screw, and by a circular scale which is engraved on the thimble.

micrometer theodolite (*Surv.*). A theodolite equipped with micrometers instead of the usual verniers for reading the horizontal and vertical circles.

micrometre (*Phys., etc.*). One-millionth of a metre. Symbol μm. Also called **micron** in past.

micromicro-. Prefix for 1-millionth-millionth, or 10^{-12}; replaced in SI by *pico-* (p).

micromicrofarad (*Elec. Eng.*). Obs. term for *picofarad.* A unit of capacitance equal to 1-million-millionth of a farad, most convenient for the small capacitances used in radio-frequency circuits.

microminiaturization (*Electronics*). Use of small circuit components as in thin-film and thick-film circuits and integrated circuits.

micromodule (*Elec. Eng.*). Said of circuits of components of minute but specified dimensions which are formed from the same crystal of material, e.g., germanium. This is fashioned into capacitors, resistors, photocells, diodes, transistors, etc.

micromutation (*Gen.*). Mutation at only one gene locus.

micron. Obs. measure of length equal to one millionth of a metre, symbol μ. Replaced in SI by *micrometre*, symbol μm.

micronephridia (*Zool.*). In some *Chaetopoda*, small nephridial tubes, with or without a nephrostome, present in large numbers in a single somite and believed to be derived from *meganephridia* (q.v.).

Micronizer (*Chem. Eng.*). Proprietary name for *jet mill* (q.v.), i.e., apparatus for reducing the particle size of a powder.

micronucleus (*Zool.*). In *Ciliophora*, the smaller of the two nuclei which is composed of reserve generative chromatin. Cf. *macronucleus.*

micronutrients (*Ecol.*). Elements and their compounds which are only required in small quantities by living organisms. Also known as **trace elements.** Cf. *macronutrients.*

micropegmatite (*Geol.*). A term applied to micrographic intergrowths of quartz and feldspar occurring in the groundmass in various igneous rocks. Synonymous with **micrographic texture.**

microperthite (*Min.*). A feldspar which consists of intergrowths of potassium feldspar and albite on a microscopic scale.

microphage (*Zool.*). A small phagocytic cell in blood or lymph, chiefly the polymorphonuclear leucocytes (*neutrophils*). *adj.* **microphagocytic.**

microphagous (*Zool.*). Feeding on small particles of food. Cf. *macrophagous.*

microphanerophyte (*Bot.*). A woody plant from 2–8 m in height.

microphone (*Acous.*). An acousto-electric transducer, essential in all sound-reproducing systems. The *excess pressure* in the sound wave is applied to a mechanical system, such as a ribbon or diaphragm, the motion of which generates an electromotive force, or modulates a current or voltage. Also **mike**; see **transmitter,** see also **carbon-, diaphragmless-, directional-, electromagnetic-, hot-wire-, lapel-, moving-coil-, moving-conductor-, omnidirectional-, Olson-, pressure-, pressure-gradient-, push-pull-, Reiss-, Sykes-, thermal-.**

microphone boom (*Cinema., TV*). A pivoted arm carrying a microphone which can be swung into any position in a studio. See **boom.**

microphone response (*Acous.*). That measured over the operating frequency range, in a particular direction, or averaged over all directions (*reverberant response*). The characteristic response is usually given by the ratio of the open circuit voltage generated by the microphone to the sound pressure (N/m^2) existing in the free progressive wave before introducing the microphone.

microphonic (*Electronics*). Said of a component which responds to aerial vibrations and/or knocks.

microphonic noise (*Acous.*). That in the output of a valve related to mechanical vibration of the electrode system. Also called **microphonicity.**

microphotogram (*Chem.*). A greatly enlarged photograph of a spectrum.

microphotography (*Photog.*). Photography of

normal-sized objects, especially documents, on to abnormally small negatives.

microphylline (*Bot.*). Composed of small scales or lobes.

microphyllous (*Bot.*). Having very small leaves—a condition characterizing many plants living in arid habitats.

microphyric (*Geol.*). A textural term descriptive of medium- to fine-grained igneous rocks containing phenocrysts less than 2 mm in length. Cf. *macrophyric*.

Micropodiformes (*Zool.*). An order of Birds with a short humerus and long distal segments to the wings; Swifts and Hummingbirds.

micropodous (*Zool.*). Having the foot, or feet, small or vestigial.

micropoecilic (or **mikropoikilitic**) **texture** (*Geol.*). A term applied to igneous rocks containing small granular crystals irregularly scattered without common orientation in larger crystals of another mineral as seen under the microscope.

micropore (*Zool.*). One of the small apertures in the shell of an amphineuran Mollusc containing a micraesthete.

microporosity (*Met.*). Minute cavities generally found in heavy sections usually due to lack of efficient feeding or chilling; particularly found in magnesium, when they are highly coloured.

microprism (*Photog.*). Focusing device which breaks the image up into dots when it is not in focus.

microprogram (*Comp.*). One already registered in a fast store to be called on by the main program.

micropsia (*Med.*). The condition in which objects appear to the observer smaller than they actually are; a symptom of hysteria, or it may be due to retinal disease.

micropterous (*Zool.*). Having small or reduced fins; in Insects, having small hind wings, hidden by the fore wings when at rest.

micropyle (*Bot.*). (1) A tiny opening in the integument at the apex of an ovule, through which the pollen tube usually enters. (2) The corresponding opening in the testa of the seed. Also called **foramen.** (*Zool.*) An aperture in the chorion of an Insect egg through which a spermatozoon may gain admittance.

microradiography (*Radiol.*). Exposure of small thin objects to *soft X-rays*, with registration on a fine-grain emulsion and subsequent enlargement up to 100 times. Also used to signify the optical reproduction of an image formed, e.g., by an electron microscope.

microschizont (*Zool.*). A stage in the life-cycle of certain *Haemosporidia*.

microscope (*Optics*). An instrument used for obtaining magnified images of small objects. The *simple microscope* is a convex lens of short focal length, used to form a virtual image of an object placed just inside its principal focus. The *compound microscope* consists of two short-focus convex lenses, the objective and the eyepiece mounted at opposite ends of a tube. For most microscopes, the magnifying power is roughly equal to $450/f_o f_e$, where f_o and f_e are the focal lengths of objective and eyepiece in centimetres. See also **electron, ultraviolet microscope.**

microscope count method (*Powder Tech.*). Technique for measuring the average particle diameter of a powder by microscopic examination of a weighed drop of suspension.

microscopic state (*Chem.*). One in which condition of each individual atom has been fully specified. Cf. *macroscopic state*.

microsection (*Met.*). A section of metal, ceramic, etc., mounted, cut, polished, and etched as necessary to exhibit the microstructure.

microseisms (*Geophys.*). Minute irregular motions of the surface of the earth, of periods of the order 6 seconds and amplitude 1 micrometre, which are continuously recorded on sensitive seismographs; possibly due to fluctuations of pressure on the sea, or of waves on the shore.

microsere (*Ecol.*). The seral sequence in a small habitat within a larger community, e.g., cattle droppings in a grassland community.

microsmatic (*Zool.*). Having a poorly developed sense of smell.

microsome (*Cyt.*). A granular inclusion in the cytoplasm, of very small size, often associated with the endoplasmic reticulum, and intimately connected with certain cytoplasmic enzymes.

microspecies (*Bot.*). A variety of a species.

microspheric (*Zool.*). In certain dimorphic species of *Foraminifera*, said of a form in which the initial chamber of the shell is small. Cf. *megalospheric*. *n.* **microsphere.**

microspherulitic texture (*Geol.*). A texture in which spherulites on a microscopic scale are distributed through the groundmass of an igneous rock.

microsplanchnic (*Zool.*). Having a small body and long legs, as a Harvestman.

microspore (*Bot.*). A spore which gives rise to a male gametophyte, or its equivalent. (*Zool.*) A small swarm-spore or anisogamete of *Sarcodina*.

microsporocyte (*Bot.*). A cell which divides to give microspores, i.e., a microspore mother cell.

microsporophyll (*Bot.*). A leaflike organ, more or less modified, bearing or subtending one or more microsporangia.

microstoma, microstomia (*Med.*). An abnormally small mouth, due either to developmental defect or to contraction of scar tissue.

microstrip (*Telecomm.*). A microwave transmission line consisting basically of a dielectric sheet carrying a conducting strip on one side and an earthed conducting plane on the other. The strip with its image in the plane (see **image charge**) forms a parallel strip transmission line.

microstructure (*Met.*). A term referring to the size, shape, and arrangement with respect to each other (as seen under the microscope) of the constituents present in a metal or alloy, too small to be seen by the naked eye.

microswitch (*Elec. Eng.*). A switch operated by very small movements of a lever, the circuit being made or broken by spring-loaded contacts.

microsyenite (*Geol.*). An intrusive rock of medium grain-size and syenitic composition.

microsyn (*Elec. Eng.*). An a.c. transducer employed in measuring angular rotation and used as an output device in precision gyroscopes and servomechanisms. Its construction is similar to that of an *E-transformer* (q.v.).

micro-system (*Comp.*). An analogue computer comprising model machines, impedances and other apparatus connected, such that the dynamic and electrical parameters, when measured on a per-unit basis, are equal to those of the full-scale system.

microthorax (*Zool.*). The cervicum of Insects, according to certain authorities who regard it as belonging to the prothorax.

microtome (*Bot., Zool.*). An instrument for cutting thin sections of specimens.

microtrichia (*Zool.*). Small bristles occurring on the wings of certain Insects. Cf. *macrotrichia*.

microwave hohlraum (*Acous.*). See **hohlraum.**

microwave resonance (*Elec. Eng.*). One between

microwave signals and atoms or molecules of medium.

microwave resonator (*Telecomm.*). Effective tuned circuit for microwave signal. Usually a cavity resonator but tuned lines are also used.

microwaves (*Radio*). Those electromagnetic wavelengths between 1 mm and 30 cm, i.e., from $0{\cdot}3 \times 10^{12}$ to 10^9 Hz in frequency, thus bridging gap between normal radio waves and heat waves. Also U.H.F. sound waves.

microwave spectrometer (*Phys.*). An instrument designed to separate a complex microwave signal into its various components and to measure the frequency of each; analogous to an *optical spectrometer*. See **spectrometer.**

microwave spectroscopy (*Phys.*). The study of atomic and/or molecular resonances in the microwave spectrum.

microwave spectrum (*Phys.*). The part of the electromagnetic spectrum corresponding to microwave frequencies.

micrurgy (*Biol.*). A technique for the investigation and manipulation of single living cells using micromanipulators.

micturition (*Zool.*). In Mammals, the passing to the exterior of the contents of the urinary bladder.

mid-brain (*Zool.*). In Vertebrates, that part of the brain which is derived from the second or middle brain-vesicle of the embryo; the second or middle brain-vesicle itself dorsally comprises the tectum, including optic lobes (*corpora quadrigemina* in Mammals), and laterally the tegmentum and ventrally the *crura cerebri* (cerebral peduncles).

middle chrome (*Paint.*). A chrome yellow pigment consisting almost entirely of lead chromate. See **chrome yellows.**

middle conductor (*Elec. Eng.*). See **neutral conductor.**

middle gear (*Eng.*). See **mid-gear.**

middle gouge (*Carp., etc.*). A gouge intermediate between the *flat gouge* (q.v.) and the *quick gouge* (q.v.).

middle lamella (*Bot.*). A thin layer of primary wall forming the middle layer of the wall between two sister cells; it often consists largely of pectin, and stains differently from the cellulose wall layers on each side of it.

middle marker beacon (*Aero.*). A marker beacon associated with the *ILS* (q.v.), used to define the second predetermined point during a beam approach.

middle oils (*Chem.*). Carbolic oils, obtained from coal-tar distillation. Their boiling range is from about 210°C to 240°C.

middle-post (*Carp.*). See **king-post.**

middle rail (*Join.*). The rail next above the bottom rail in doors, framing, and panelling.

middle shore (*Carp.*). An inclined shore placed between the bottom and top shores in a set of raking shores.

middle-temperature error (*Horol.*). The error in the time of vibration of a compensation balance due to the fact that compensation for dimensional changes of the balance and the elastic properties of the spring is not complete over a range of temperature. A watch or chronometer regulated to be correct at the extremes of the temperature range over which it is to be used will show an error at temperatures between the extremes.

middle third (*Civ. Eng.*). The middle part of a brickwork or masonry structure, such as an arch or dam. It is equal in width, at any section, to one-third the width of the section, and is centrally disposed. The importance of the middle third is that, providing the line of resultant pressure lies wholly within it, no tensile forces come into play.

middle wire (*Elec. Eng.*). See **neutral conductor.**

middlings (*Min. Proc.*). In ore dressing, an intermediate product left after the removal of clean concentrates and rejected tailings. It consists typically of interlocked particles of desired mineral species and gangue, or of by-product minerals not responding to the treatment used. See also **riddle.**

mid-feather (*Build.*). See **withe.** (*Join.*) (1) See **cross-tongue.** (2) See **parting-slip.** (*Paper*) The partition in the beater which induces circulation of the wet pulp.

mid- (or middle) gear (*Eng.*). The position of a steam-engine link motion or valve gear when the valve motion is a minimum.

mid-girth (*For.*). The girth of a log or bole, measured half-way along its length or height.

mid-gut (*Zool.*). That part of the alimentary canal of an animal which is derived from the archenteron of the embryo.

midland tariff (*Elec. Eng.*). A name sometimes given to that form of tariff for electrical energy in which a fixed charge per year per kVA of maximum demand is made, together with a charge per kWh.

midnight sun (*Astron.*). The term used in popular language for the phenomenon (seen only in the Arctic and Antarctic circles) of the sun's remaining above the horizon all night and being seen at midnight at lower culmination.

mid-point protective system (*Elec. Eng.*). A method of balanced protection used for protecting generators against faults between turns by balancing the voltage of one half of the winding against that of the other.

midrib (*Bot.*). The largest vein of a leaf, running longitudinally through the middle of the lamella.

mid-riff (*Zool.*). See **diaphragm.**

mid space (*Typog.*). A type space cast four to the em.

midsummer growth (*Bot.*). The second period of active growth shown by some trees.

mid-timber girth (*For.*). Girth at half timber height of tree.

midwater zone (*Ocean.*). The depths of the ocean between the surface waters and the abyss.

mid-wing monoplane (*Aero.*). A monoplane wherein the main planes are located approximately midway between the top and bottom of the fuselage.

Miehle (*Print.*). The first successful *two-revolution* (q.v.) printing machine.

Miescher's tubes (*Zool.*). Elongate cylindrical opaque whitish bodies occurring in muscles of domestic animals and representing encysted *Sarcosporidia*. See also **Rainey's corpuscles.**

migmatite (*Geol.*). A rock containing two components, one with the composition of granite and the other with that of a metamorphic rock.

migraine (*Med.*). Hemicrania or paroxysmal headache. A condition in which recurring headaches are often associated with vomiting and disturbances of vision. See also **teichopsia.**

migration (*Chem.*). Movement of ions under influence of electric field and against the viscous resistance of the solvent. Measured by observation in thin tubes. (*Ecol.*) Animal movements which tend to carry individuals away from the normal territory or habitat of the population to which they belong, a habit whose evolution is often correlated with impermanence of the habitat, and which may be obligatory or facultative in different species.

In some cases the movements may cover 1500 kilometres or more, but in other cases they are shorter and there is, in fact, no sharp line between migratory movements and trivial movements inside a normal habitat. The degree to which the direction and distance of migratory movements are under the control of the animals varies, some restricting the term migration to cases where they are entirely controlled, and contrasting migration with passive *dispersal* (q.v.), others restricting the term to cases where there is a regular to and fro movement of a population between two areas at different seasons of the year. See also **immigration, emigration.**

migration area (*Nuc.*). One-sixth of the mean square distance covered by a neutron between creation and capture. Its square root is the *migration length.*

migratory cell (*Zool.*). See **amoebocyte.**

migratory community (*Bot.*). A plant community which occupies a locality for a time and then seems to die out.

mike (*Acous.*). Colloquialism for **microphone.**

mikropoikilitic texture (*Geol.*). See **micropoecilic texture.**

Mikulicz's disease (*Med.*). A chronic inflammation of salivary and lacrimal glands.

mil. A unit of length equal to 10^{-3} in., used in measurement of small thicknesses, i.e., thin sheets. Also (coll.) **thou.**

milarite (*Min.*). A hydrated silicate of aluminium, beryllium, calcium and potassium, crystallizing in the hexagonal system.

mild ale (*Brew.*). Ale of a dark-brown colour, often sweetened, and with less hop flavouring than pale ales; usually sold from the cask.

mild clay (*Build.*). See **loam.**

mildew (*Bot.*). A general term for a number of obligate plant parasites which occur in two fungal orders, the *Peronosporales,* which bear downy mildew, and the *Perisporales,* which bear powdery mildew on the surface of the plant.

mild steel (*Met.*). Hot-rolled steel containing approximately 0·04% of carbon. See **low-carbon steel.**

mile. A unit of length commonly used for distance measurement in the British Commonwealth and the U.S. A *statute mile* = 1760 yd = 1609·34 m. See **geographical-, nautical-.**

mile of standard cable (*Telecomm.*). Old unit of attenuation provided by 1 mile of an arbitrary type of telephone circuit at 800 c/s. Formerly used for estimating possibilities of transmission over circuits. Now displaced by the decibel and neper. Abbrev. MSC.

miliaria (*Med.*). Prickly heat. Inflammation of the sweat glands, accompanied by intense irritation of the skin.

miliary (*Med.*). Like a millet seed; said of lesions which are small, like millet seeds.

miliary tuberculosis (*Med.*). A form of tuberculosis in which small tuberculous lesions are found in various organs of the body, especially in the meninges and in the lungs.

milk fever (*Vet.*). Parturient paresis, parturient fever (cow); lambing sickness (ewe); parturient eclampsia (sow and bitch). A metabolic disease of unknown cause affecting the parturient cow, ewe, goat, sow, and bitch; characterized by hypocalcaemia and muscular weakness and incoordination, tetany, loss of consciousness, and death. The disease also occurs during pregnancy in the ewe and during lactation in the goat.

milk glands (*Zool.*). The mammary glands of a female Mammal: in viviparous Tsetse flies, special uterine glands by which the larva is nourished until it is ready to pupate.

milking generator (*Elec. Eng.*). A low-voltage d.c. generator used for giving one or more cells of a battery a charge independently of the remainder of the cells. Sometimes called by the name milker.

milking machine (*Agric.*). A machine by which a cow is milked. It consists essentially of a vacuum pump, a *pulsator* (q.v.), metal teat cups lined with rubber, and milk pail with an air-tight lid.

milk-leg (*Med.*). See **phlegmasia alba dolens.**

milk-sugar (*Chem.*). Lactobiose or *lactose* (q.v.).

milk teeth (*Zool.*). In diphyodont Mammals, the first or deciduous dentition.

milk tetany (*Vet.*). See **calf tetany.**

Milky Way (*Astron.*). See **Galaxy.**

mill (*Comp.*). The arithmetical unit of a computer. (*Eng.*) Generally, (1) a machine for grinding or crushing, as a *flour mill, paint mill,* etc.; (2) a factory fitted with machinery for manufacturing, as a cotton mill, saw-mill, etc. (*Mining*) In Britain, a crushing and grinding plant. In America, the whole equipment for comminuting and concentrating an ore.

millboards (*Paper*). Boards manufactured from woodpulp, fibre refuse, etc.

mill-dam (*Hyd. Eng.*). A dam built across the current of a stream to raise its level and divert it into a mill-race.

milled (*Eng.*). Having the edge grooved or fluted, as a coin or the head of an adjusting screw. See **knurling tools.**

milled cloth (*Textiles*). Woven or knitted wool or woollen fabric in which the milling action causes the felting properties of wool fibres to interlock. The cloth has a fibrous surface, the threads and structure being indistinguishable. The process involves the application of pressure and friction to the cloth, while it is soapy.

milled lead (*Build.*). Sheet-lead formed from cast slabs by a rolling process.

millefiori glass. Glassware in which a large number of sections of glass rods of various colours form a pattern and are fused together or set in a clear glass matrix.

Miller bridge (*Elec. Eng.*). A particular form used to measure amplification factors of valves.

Miller circuit (*Telecomm.*). Amplifier in which negative feedback from output to input is regulated by a capacitor.

Miller effect (*Electronics*). The change in effective input impedance of an amplifier due to unwanted shunt voltage feedback, which makes the input impedance a function of the voltage gain. In particular the increase in effective input capacitance for a valve or transistor used as a voltage amplifier.

Miller indices (*Crystal.*). Integers which determine the orientation of a crystal plane in relation to three crystallographic axes. The reciprocals of the intercepts of the plane on the axes (in terms of lattice constants) are reduced to the smallest integers in ratio. Also called **crystal indices.**

Miller integrator (*Telecomm.*). Step to saw-tooth converter in which a pentode grid is driven through a high resistance, and the grid to anode capacitance is increased by an external capacitor, the output being taken from the anode fed by a high resistance.

millerite or capillary pyrite (*Min.*). Sulphide of nickel, crystallizing in the trigonal system. It usually occurs in very slender crystals and often in delicately radiating groups.

Miller process (*Met.*). Purification of bullion by removal of base metals as chlorides. Chlorine gas is bubbled through the molten metal.

Miller radiator (*Telecomm.*). Antenna in which the requisite phase differences in the elements are obtained by transmitting the wave through prisms of polythene.

millers' disease (*Vet.*). See **osteodystrophia fibrosa.**

millet-seed sand (*Geol.*). See **blown sand.**

millet-seed sandstone (*Geol.*). A sandstone consisting essentially of small spheroidal grains of quartz; typical of deposits accumulated under desert conditions.

mill-fitting (*Elec. Eng.*). See **factory-fitting.**

mill head (*Min. Proc., Mining*). Ore accepted for processing after removal of waste rock and detritus. Also, assay grade of such ore.

milli-. Prefix from L. *mille*, thousand. When attached to units, it denotes the basic unit $\times \frac{1}{1000}$. Symbol **m.**

milliammeter (*Elec.*). An ammeter calibrated and scaled in milliamperes; used for measuring currents up to about 1 ampere.

millibar (*Meteor.*). See **bar.**

millicurie (*Nuc.*). One-thousandth of a **curie.**

millidarcy. See **darcy.**

milligal (*Geophys.*). The normal unit used in geophysical gravity surveying. It represents one thousandth of a gal, (a gravitational acceleration of 1 cm/s^2), i.e. mGal $= 10^{-5}$ m/s^2.

Millikan electrometer (*Elec. Eng.*). An early form of ionization chamber dosemeter employing a gold-leaf electrometer instead of the modern quartz-fibre type.

millilambert (*Light*). A unit of brightness equal to 0·001 lambert; more convenient magnitude than the lambert.

millilitre (*Chem.*). A unit of volume, one thousandth of a litre, equivalent to 1 cm^3 or 10^{-6} m^3.

millilux (*Light*). A unit of illumination intensity, equal to 1-thousandth of a lux.

millimass unit (*Nuc.*). Equal to 0·001 of atomic mass unit. Abbrev. **mu.**

millimetre. The thousandth part of a metre.

millimetre pitch (*Eng.*). See **metric screw-thread.**

millimicron (*Elec. Eng.*). Obsolete term for nanometre, 10^{-9} metres. Used in the measurement of wavelengths of e.g. light.

milling (*Eng.*). A machine process in which metal is removed by a revolving multiple-tooth cutter, to produce flat or profiled surfaces, grooves, and slots. See also **milling cutter, milling machine.** (*Mining*) Comminution; dressing; removing valueless material and harmful constituents from an ore, in order to render marketing more profitable. (*Textiles*) A preliminary process in the finishing of woollen fabrics, carried out in a milling-machine by the agency of soap, alkali, or acid (depending on the nature of the fabric and the dye), pressure, and friction. The process decreases the surface area of a fabric and imparts a fibrous cover. Also called **fulling.**

milling-cutter (*Eng.*). A hardened steel disk or cylinder on which cutting-teeth are formed by slots or grooves on the periphery and faces, or into which separate teeth are inserted; used in the milling-machine for grooving, slotting, surfacing, etc. See **end mill, milling-machine.**

milling grade and milling width (*Mining, Min. Proc.*). *Milling grade* denotes ore sufficiently rich to repay cost of processing; *milling width* that of the lode which will determine tonnage sent daily from mine to mill.

milling-machine (*Eng.*). A machine tool in which a horizontal arbor or a vertical spindle carries a rotating multi-tooth cutter, the work being supported and fed by an adjustable and power-driven horizontal table. See also **milling-cutter.** (*Textiles*) Rotary milling-machines, which have largely displaced the fulling stocks type, consist of squeezing-rollers, a box channel called the spout, etc., over a large trough, the whole being enclosed.

Millington reverberation formula (*Acous.*). A modified formula for calculating the period of reverberation, taking into account the random disposition of the reflecting and acoustically absorbing surfaces in an enclosure.

million-electron-volt (*Nuc.*). See **MeV.**

millipede (*Zool.*). Any myriapod of the class Chilognatha, vegetarian cylindrical animals with many joints most of which bear two pairs of legs.

Millipore filter (*Chem.*). TN for a type of *membrane filter* (q.v.).

milliradian (*Radar*). Unit for radar or infrared scanning from aircraft, equal to 10^{-3} radian.

mill join (*Print.*). An adhesion between the ends of two similar webs made at paper mill, the joints being sometimes marked for the printer's guidance.

Millman tube (*Electronics*). A slow-wave travelling-wave tube, i.e., one in which the electron beam velocity may be less than that of the electromagnetic wave carrying the signal through the periodic structure which constitutes the interaction region of the tube.

Millon's reaction (*Chem.*). A test for proteins, based on the formation of a pink or dark-red precipitate of coagulated proteins on heating with a solution of mercuric nitrate containing some nitrous acid.

mill race (*Eng.*). Channel or *flume* by which water is led to a mill wheel or waterwheel.

mill rigs (*Textiles*). Creases produced in woollen cloths during milling, as a result of being run twisted through rollers.

mill scale (*Met.*). Layer of oxidized iron on surface of rolled steel.

millstone (*Astron.*). A conventional measure of sensitivity of planetary radar, corresponding to that of the original Millstone Hill radio telescope of the Massachusetts Institute of Technology.

Millstone Grit (*Geol.*). A deltaic facies of the rocks of Upper Carboniferous age, typically developed between the Carboniferous Limestone and the Coal Measures of the North Midlands of England; it consists of alternating grits and marine shales, to a maximum thickness of more than 1500 m.

mill tail (*Eng.*). The channel conveying water away from a mill wheel.

mill wheel (*Eng.*). A water-wheel driving the machinery in a mill.

Milroy's disease (*Med.*). Hereditary oedema. A disorder in which persistent oedema (swelling) of the legs occurs in members of the same family in successive generations.

Mil-Specs (*Aero.*). *Mil*itary *Spec*ification*s* issued in the U.S., which lay down basic requirements to be observed by design teams in the development of aircraft. Abbrev. also **MS.**

milt (*Zool.*). The spleen; in Fish, the testis or spermatozoa; to fertilize the eggs.

mimesis (*An. Behav.*). See **social facilitation.**

mimetic diagram (*Elec. Eng.*). In a control room of a large process plant or an electrical network, the animated diagram which indicates to the controller the state of operations, by coloured lights, recorders, or indicating instruments.

mimetite or **mimetesite** (*Min.*). A chloride-arsenate-phosphate of lead with As > P; cf.

pyromorphite. It crystallizes in the hexagonal system, often in barrel-shaped forms, and is usually found in lead deposits which have undergone a secondary alteration.

mimicry (*Zool.*). The adoption by one species of the colour, habits, sounds, or structure of another species. *adjs.* **mimic, mimetic.**

minaret (*Arch.*). A lofty slender tower rising from a mosque or similar building and surrounded by a gallery near the top.

mind (*Psychol.*). According to Freud, *mind* consists of a relatively small conscious part and a larger unconscious part, each consisting of the processes of thinking, feeling, wishing.

mine (*Mining*). Subterranean excavation made in connexion with exploitation of, or search for, minerals of economic interest. Terms *quarry*, *pit*, and *opencast* are reserved for workings open to daylight.

mine detector (*Instr.*). An electronic device for the detection of buried explosive mines (or buried metal) depending on the change in the electromagnetic coupling between coils in the search head.

mineral (*Min.*). A naturally-occurring substance of more or less definite chemical composition and physical properties. It has a characteristic atomic structure frequently expressed in the crystalline form or other properties.

mineral caoutchouc (*Min.*). See **elaterite.**

mineral dressing (*Min. Proc.*). See **mineral processing.**

mineral flax (*Build.*). A fibrized form of asbestos much used in the manufacture of asbestos-cement sheeting.

mineral-insulated cable (*Elec. Eng.*). One in which the conductor runs in an earthed copper sheath filled with magnesium oxide, which makes it fireproof and able to withstand excess loads. Also **copper-sheathed cable.**

mineralized carbon (*Elec. Eng.*). An arc-lamp carbon impregnated with metallic salts for use in a flame arc-lamp.

mineralogy. The scientific study of minerals.

mineral oils (*Chem.*). Petroleum and other hydrocarbon oils obtained from mineral sources. Cf. *vegetable oils*.

mineral processing or **dressing** (*Min. Proc.*). Crushing, grinding, sizing, classification, separation of ore into waste and value by chemical, electrical, magnetic, gravity, and physico-chemical methods. First-stage extraction metallurgy. Also **ore dressing, beneficiation, préparation mécanique** (*French*).

mineral vein (*Mining*). A fissure or crack in a rock which has been subsequently lined or filled with minerals. See also **lode.**

mineral wool (*Met.*). See **rock wool.**

miners' anaemia (*Med.*). See **ankylostomiasis.**

miner's dip needle (*Mining*). A portable form of *dipping needle* (q.v.) used for indicating the presence of magnetic ores.

miner's lamp (*Mining*). A portable lamp specially designed to be of robust construction and adequate safety for use in mines.

minette (*Geol.*). A lamprophyre composed essentially of biotite and orthoclase, occurring in dykes associated with major granitic intrusions. Originally the term was applied to the Jurassic ironstones of Briey and Lorraine, and it is still so used.

miniature camera (*Photog.*). A term sometimes applied to any small camera, but normally used for the widely used 35-mm size. See also **subminiature camera.**

miniature Edison screw-cap (*Elec. Eng.*). An Edison screw-cap for electric filament lamps, in which the screw-thread has a diameter of about ⅜ in. and about 14 threads per inch.

miniature (or **subminiature**) **valve** (*Electronics*). One in which all the dimensions are reduced to very small values, to keep down the inter-electrode capacitances and the electron transit time. Used in H.F., V.H.F., U.H.F. circuits.

Mini-log (*Electronics*). Unit transistor encapsulated circuit for universal use in process controllers, computers, etc.

minimum access programme (*Comp.*). Programme routine involving minimum loss of time for access to store.

minimum blowing current (*Elec. Eng.*). The minimum current which will cause melting of a fuse link under certain specified conditions.

minimum burner pressure valve (*Aero.*). A device which maintains a safe minimum pressure at the burners of a gas turbine when it is idling.

minimum clearing (*Radio*). Same as **zero clearing.**

minimum deviation (*Light*). See **angle of minimum deviation.**

minimum discernible signal (*Telecomm.*). Smallest input power to any unit which just produces a discernible change in output level.

minimum flying speed (*Aero.*). The minimum speed at which an aeroplane has sufficient lift to support itself in level flight in standard atmosphere. There is a close relationship with the weight, which affects the *wing loading* (q.v.), and the term must be stated with the weight (and altitude if ISA, sea level is not implied) and *T.A.S.* (q.v.) specified.

minimum ionization (*Nuc.*). The smallest possible value of the specific ionization that a charged particle can produce in passing through a given substance. It occurs for particles having velocities $=0{\cdot}95\ c$, where $c=$velocity of light.

minimum, 'law' of the (*Ecol.*). Of all essential materials needed for the growth and reproduction of an organism, that available in amounts most closely approaching the critical minimum needed will tend to be the limiting one determining the organism's distribution and abundance. This was first clearly expressed in 1840 by Liebig. It is sometimes expanded to include factors other than nutrients, but these are better included under *tolerance,'law' of* (q.v.).

minimum pause (*Teleph.*). The interval of lost time which is necessarily introduced into the operation of a dial to ensure that the selectors have time to complete their hunting.

minimum point on a curve (*Maths.*). A trough on a curve. For the curve $y=f(x)$, the point where $x=a$ is a minimum if $f(a+h)-f(a)$ is positive for all values of h sufficiently small, i.e., $f'(a)=0$ and the first non-zero higher order derivative at $x=a$ must be of even order and positive.

minimum sampling frequency (*Telecomm.*). Lowest sampling rate which can provide an accurate reproduction of the signal in a pulse-code-modulation system. Equal to twice the maximum signal frequency.

minimum two-part prepayment meter (*Elec. Eng.*). A 2-part prepayment meter in which the time element is arranged to collect a charge based upon a minimum yearly consumption as well as the usual fixed charge.

minimum wavelength (*Phys.*). Shortest wavelength in an X-ray spectrum, determined by the maximum voltage applied to the X-ray tube. The quantum limit according to the Planck-Einstein quantum equation.

mining dial (*Mining, Surv.*). See **dial.**

mining engineering. That branch of engineering chiefly concerned with the sinking and equipment of mine shafts and workings, and all operations incidental to the winning and preparation of minerals.

mining width (*Mining*). Where the lode is too narrow to permit working access without encroachment on the enclosing walls, width to which excavation must be carried. See **stoping width, overbreak.**

minion (*Typog.*). An old type size, approximately 7-point.

Ministry of Transport loading (*Civ. Eng.*). Statutory loading which is laid down for the structural design of bridges. It takes into account the class of traffic to which the particular bridge may be subjected, and also all possible conditions of loading which that traffic may cause.

minitrack (*Radio*). Phase-comparison angle-tracking radio system, used for tracking satellites.

Minkowski's inequality (*Maths.*).

$$\left(\sum_{1}^{n}|a_r+b_r|^{\alpha}\right)^{\frac{1}{\alpha}} \leqq \left(\sum_{1}^{n}|a_r|^{\alpha}\right)^{\frac{1}{\alpha}} \cdot \left(\sum_{1}^{n}|b_r|^{\alpha}\right)^{\frac{1}{\alpha}},$$

if $\alpha \geqq 1$.

minnesotaite (*Min.*). The iron-bearing equivalent of talc. A major constituent of the siliceous iron ores of the Lake Superior region.

minnikin (*Typog.*). The smallest of the old type sizes, approximately 3-point.

minor (*Maths.*). A sub-determinant, i.e., a determinant contained in another determinant of higher *order* (q.v.). The minor of a particular element of a determinant is the determinant of the elements which remain when the row and column containing the element are deleted.

minor axis (*Maths.*). Of an ellipse: see **axes.**

minor exchange (*Teleph.*). One directly connected to its group centre.

minor intrusions (*Geol.*). Igneous intrusions of relatively small size, compared with *plutonic* (*major*) *intrusions* (q.v.). They comprise dykes, sills, veins, and small laccoliths. The injection of the minor intrusions constitutes the dyke phase of a volcanic cycle.

minority carrier (*Electronics*). In a semiconductor, the electrons or *holes* which carry the lesser degree of measured current. See **majority carrier.**

minor planet (*Astron.*). See **asteroid.**

mint camphor (*Pharm.*). See **menthol.**

minuend (*Maths.*). See **subtraction.**

minus colour (*Photog.*). The complementary colour to a given colour, i.e., the colour which, when added to the given colour, produces white light.

minuscule (*Typog.*). A lower-case, or small, letter.

minus g (*Space*). See **negative *g*.**

minus strain (*Bot.*). One of the two distinct strains of a heterothallic mould; sometimes written (−) strain.

minute. (1) A 60th part of an hour of time. (2) A 60th part of an angular degree. (3) A 60th part of the lower diameter of a column.

minute pinion (or nut) (*Horol.*). The pinion in the motion work that drives the hour wheel.

minute wheel (*Horol.*). The wheel in the motion work driven by the cannon pinion.

minverite (*Geol.*). A basic intrusive rock, in essentials a dolerite, containing a brown, soda-rich hornblende; named from the type-locality, St Minver, Cornwall.

Miocene Period (*Geol.*). The period of geological time which ensued between the Oligocene and Pliocene Periods. In Britain this period was one of erosion and is not represented by any known deposits.

miosis (*Med.*). Contraction of the pupil of the eye.

mipafox (*Chem.*). Fluoro-*bis*-2-propylamino-phosphine oxide:

$(CH_3)_2CH\text{–}NH\text{–}P(=O)(F)\text{–}NH\text{–}CH(CH_3)_2$

Used as an insecticide.

Mipolam (*Plastics*). A proprietary plastic of the polyvinyl chloride type. Noninflammable; of high value for its electrical and acid-resisting qualities.

mirabilite (*Min.*). See **Glauber salt.**

miracidium (*Zool.*). The ciliated first-stage larva of a Trematode.

mirage (*Meteor.*). An effect caused by total reflection of light at the upper surface of shallow layers of hot air in contact with the ground, the appearance being that of pools of water in which are seen inverted images of more distant objects. Other types of mirage are seen in polar regions, where there is a dense, cold layer of air near the ground. See **fata morgana** (*Radio*). Radio signal for which transmission path includes reflection from layer of rarefied air, in analogous manner to the above.

Mira stars (*Astron.*). Long-period variable stars named after Mira Ceti; more than 3000 are known, with periods from 2 months to 2 years, and all are red giant stars.

mirbane, oil of (*Chem.*). See **oil of mirbane.**

mired value (*Photog.*). That of a light source, the colour temperature in kelvins divided by 1 000 000.

mirror (*Phys.*). A highly-polished reflecting surface capable of reflecting light rays without appreciable diffusion. The commonest forms are plane, spherical (convex and concave) and paraboloidal (usually concave). The materials used are glass silvered on the back or front, speculum metal, or stainless steel.

mirror arc (*Cinema.*). A projection arc in which the positive carbon is advanced towards the negative carbon, which is fed along the axis and through the centre of the parabolic metallic mirror; light from a greater radiation angle from the positive crater is thereby collected and focused on to the gate.

mirror drum (or wheel) (*TV*). A rotating drum having a number of mirrors arranged around its circumference; used in some forms of mechanical scanning.

mirror finish (*Eng.*). A very smooth, lustrous surface finish produced, for example, on stainless steels and other metals by electrolytic polishing or lapping, and on electroplated surfaces by mechanical polishing.

mirror galvanometer (*Elec. Eng.*). A galvanometer having a mirror attached to the moving part, so that the deflection can be observed by directing a beam of light on to the mirror and observing the movement of the reflection of this over a suitable scale. Also called **reflecting galvanometer.**

mirror machine (*Nuc. Eng.*). Type of apparatus using magnetic mirror principles for trapping high-energy ions injected into a plasma for fusion studies.

mirror nuclides (*Nuc.*). Those with the same

number of nucleons, but with proton and neutron numbers interchanged.

mirror reflector (*Radio*). Surface or set of metal rods which reflects a wave geometrically, e.g., *dish*.

mirror screw (*TV*). An arrangement of mirrors on a rotating shaft, used for mechanical scanning. The mirrors are set in line along the shaft, each inclined at a small angle to the next.

mirror symmetry (*Phys.*). See **parity.**

mirror wheel (*TV*). See **mirror drum.**

miscarriage (*Med.*). Expulsion of the foetus before the 28th week of pregnancy. Loosely, abortion.

miscella (*Chem.*). Term applied to an oil/solvent solution, particularly in the vegetable oilseed extraction industry.

misch metal (*Met.*). An alloy of cerium, lanthium, neodymium and praseodymium, used as a coating for the cathodes of voltage regulator tubes in order to reduce cathode drop.

miscibility (*Chem.*). The property enabling two or more liquids to dissolve when brought together and thus form one phase.

miscibility gap (*Chem.*). The region of composition and temperature in which two liquids form two layers or phases when brought together.

miser (*Tools*). A large *auger* (q.v.) used for boring holes in the ground in wet situations.

misfire (*Elec. Eng.*). Failure to establish, during an intended conducting period, a discharge or arc in a gas-discharge tube or a mercury-pool rectifier.

misfiring (*I.C. Engs.*). The failure of the compressed charge to fire normally, generally due either to ignition failure or to an over-rich or weak mixture.

mismatch (*Telecomm.*). Load impedance of incorrect value for maximum energy transfer. See **matching.**

mispickel (*Min.*). See **arsenopyrite.**

missile (*Aero., Space*). There are two basic types of missile, in the current sense, *guided* and *ballistic*. The former is controlled from its launch until it hits its target; the latter, always of long-range, surface-to-surface type, is controlled into a precision ballistic path so that its course cannot be deflected by countermeasures. See **guided missile.**

Mississippian System (*Geol.*). Approximately equivalent to the Lower Carboniferous of N.W. Europe; comprises those members of the Carboniferous System which underlie the Coal Measures and are typically exposed in the Mississippi valley.

mist (*Chem.*). A suspension, often colloidal, of a liquid in a gas. (*Meteor.*) A suspension of water droplets (radii less than 1 μm) reducing the visibility to not less than 1 km. See **fog.**

mistake (*Comp.*). Error resulting from a program or operator fault. When one arises from a machine fault, it is a *malfunction*.

mistral (*Meteor.*). A cold, dry, northerly wind of a *katabatic* nature, occurring along the Mediterranean coast of France during fine clear weather.

mitochondria (*Cyt.*). Protoplasmic inclusions of all living cells which take the form of filamentous or rod-like bodies. They are intimately concerned in certain enzyme processes within the cell, e.g., cell respiration, and the corresponding enzymes are either attached to them or make up part of their structure. Also called **Altmann's granules, chondriosomes.**

mitogenetic ray (*Bot.*). A form of radiant energy, of short wavelength, said to be emitted by growing tissues and to stimulate cell division.

mitosis (*Cyt.*). The series of changes through which the nucleus passes during ordinary cell division, and by which each of the daughter cells is provided with a set of chromosomes similar to that possessed by the parent cell.

mitotic index (*Cyt.*). The proportion in any tissue of dividing cells, usually expressed as per thousand cells.

mitral (*Med.*). Pertaining to, or affecting, the mitral valve, or valves, of the heart. (*Zool.*) Mitre-shaped or *mitriform*; as the *mitral valve*, guarding the left auriculo-ventricular aperture of the heart in higher Vertebrates, or the *mitral layer* of the olfactory bulb, composed of mitre-shaped cells.

mitral stenosis (*Med.*). Narrowing of the communication between the left atrium and the left ventricle of the heart, as a result of disease of the mitral valves.

mitral valve (*Zool.*). See **bicuspid valve.**

mitrate (*Bot.*). Descriptive of a rounded, folded fungal fruit body, somewhat bonnet-shaped.

mitre (*Join., etc.*). A joint between two pieces at an angle to one another, each jointing surface being cut at an angle.

mitre block (*Carp., etc.*). A block of wood rebated along one edge and having saw-cuts in the part above the rebate, with the kerfs inclined at 45° to the face of the rebate so as to guide the saw when cutting mouldings for a mitred joint.

mitre board (*Join., etc.*). See **mitre shoot.**

mitre box (*Carp., etc.*). An open-ended box having saw-cuts in the sides at 45° to the length of the box; used like the mitre block but capable of taking deeper mouldings.

mitre cramp (*Tools*). A *cramp* (q.v.) adapted for holding together temporarily the two parts of a mitre joint.

mitre-cut piston-ring (*Eng.*). A piston-ring in which the ends are mitred at the joint, as distinct from stepped or square ends.

mitre dovetail (*Join.*). See **secret dovetail.**

mitred valley (*Build.*). See **cut-and-mitred valley.**

mitre saw (*Join.*). See **tenon saw.**

mitre-saw cut (*Carp., etc.*). A device, such as a mitre block or box, for keeping the saw at the required angle to the work when cutting mouldings for a mitre joint. Also called **mitre-sawing board.**

mitre shoot (*Join., etc.*). A block of wood rebated along one edge as a guide for a jointing plane, and having a pair of wood strips fixed to the top face of the part above the rebate, at 45° to the face of the rebate, so as to hold the mitre face of a moulding at the right angle to the plane while it is being shot. Also called **mitre board.**

mitre-sill (*Hyd. Eng.*). The raised part of the bed of a canal lock against which the lower parts of the gates abut in closing. Also called **clap-sill, lock-sill.**

mitre square (*Carp., etc.*). A tool similar to the bevel, having the blade at 45° to the stock.

mitre wheels (*Eng.*). See **bevel gear.**

mitriform (*Bot.*). Split on two or more sides at the base, in symmetrical manner. (*Zool.*) See **mitral.**

Mitron (*Electronics*). Magnetron which can be tuned over a wide band by voltage control.

Mitscherlich's law of isomorphism (*Chem.*). Salts having similar crystalline forms have similar chemical constitutions.

mixed (*Zool.*). Said of nerve trunks containing motor and sensory fibres.

mixed bud (*Bot.*). A bud containing young foliage leaves and also the rudiments of flowers or of inflorescences.

mixed coupling (*Radio*). Simultaneous inductive and capacitative coupling between two resonant circuits.

mixed crystal (*Crystal.*). A crystal in which certain atoms of one element are replaced by those of another.

mixed-flow (or American) water turbine (*Eng.*). An inward-flow reaction turbine in which the runner vanes are so curved as to be acted on by the water as it enters radially and as it leaves axially. See **Francis water turbine.**

Mixed Forme Base (*Print.*). Precisely-machined units of aluminium alloy, in a selection of point sizes, which are assembled together with type to form a base for mounting an individual plate; in heights to suit *original* and *duplicate* plates.

mixed high technique (*TV*). In colour TV, high-modulation frequencies (representing fine detail in the picture) transmitted as a mixed or achromatic signal in order to conserve band-width.

mixed inflorescence (*Bot.*). An inflorescence in which some of the branching is racemose and some is cymose.

mixed melting point (*Chem.*). Technique used in the identification of chemical compounds, particularly organic, whereby a sample of known identity and melting point is mixed with a purified unknown sample and the melting point determined.

mixed pith (*Bot.*). A pith consisting chiefly of parenchyma, but with isolated tracheides scattered in it.

mixed-pressure turbine (*Eng.*). A steam turbine operated from two or more sources of steam at different pressures, the low-pressure supply, from, for example, the exhaust of other engines, being admitted at the appropriate pressure stage.

mixed service (*Teleph.*). Service provided by a PBX to the main exchange for a number of extension lines only.

mixer (*Build.*, *Civ. Eng.*). See **concrete mixer.** (*Met.*) (1) A large furnace used as a reservoir for molten pig-iron coming from the blast-furnace. The product of several furnaces is thus mixed, and the composition can be kept constant by making suitable additions. Used in connexion with hot-metal steel-making and direct casting of pig-iron. (2) See **agitator.** (*Telecomm.*) (1) Collection of variable attenuators, controlled by knobs, which allows the combination of several transmissions to be independently adjusted from zero to maximum. A *group mixer* or *fader* is one which deals with several groups of trans-missions. Hence *mixer*, the person who operates mixers. (2) Frequency conversion stage in a supersonic heterodyne receiver, in which there is rectification of the incoming signal and a larger signal from a local oscillator. Also **conversion detector, first detector.**

mixer-settlers (*Chem. Eng.*). A countercurrent *liquid-liquid extractor*, consisting of a series of tanks in which the two liquids are alternately dispersed in one another (*mixers*) and separated by gravity (*settlers*).

mixer valve (*Electronics*). A valve in which two currents having different frequencies are combined, generally for the purpose of modula-tion.

mixing (*Nuc.*). In gaseous isotope separation, the process of reducing the concentration gradient for the lighter isotope close to the diffusion barrier. (*Textiles*) Mechanically blending wools or cottons of different types, but of similar staple and colour, to obtain the most suitable material for spinning yarns economic-ally. See **direct-, stack-.** (*TV*) General term for combining transmissions in two or more chan-nels into one channel, usually by *fading*.

mixing efficiency (*Nuc.*). A measure of the effectiveness of the mixing process in isotopes separation.

mixing point (*Telecomm.*). One in a control system where an output is obtained from two independent inputs. If these are added it is a *summing point*, and if multiplied a *multiplication point*.

mixipterygium (*Zool.*). The clasper of *Selachii*. Also **mixopterygium.**

mixochimaera (*Bot.*). A chimaera in fungi, produced experimentally by mixing the contents of two hyphae of different strains of a species.

mixochromosome (*Cyt.*). In syndesis, the new chromosome formed by fusion of a pair of normal chromosomes.

mixonephrium (*Zool.*). In *Annelida*, a coelomo-duct so closely associated with the nephridium that they form an apparently simple funnelled organ. Intermediate forms between this and a *metanephromixium* (q.v.) occur.

mixopterygium (*Zool.*). See **mixipterygium.**

mixotrophic (*Zool.*). Combining two or more fundamental methods of nutrition, as certain *Mastigophora* which combine holophytic with saprophytic nutrition, or as a partial parasite.

mixture (*Acous.*). The fixed combination of a number of ranks of pipes, containing octaves and mutations, on one stop, which is therefore specially loud and brilliant. (*I.C. Engs.*) The combined inflammable gas and air constituting the explosive charge. (*Textiles*) (1) A mixture of different qualities of material, the combined material being termed the *blend*. (2) Mixture of colours or staple lengths. (3) Mixture of yarn. (*Typog.*) An extra charge for composition if three or more type faces are used.

mixture control (*Aero.*). An auxiliary control fitted to a carburettor to allow of variation of mixture strength with altitude. May be manu-ally operated or automatic.

mizzonite (*Min.*). One of the series of minerals forming the scapolite group, consisting of a mixture of the meionite and marialite mole-cules. Mizzonite includes those minerals with 50–80% of the meionite end-member molecule. Found in metamorphosed limestones and in some altered basic igneous rocks.

MKSA (*Elec.*). *M*etre-*k*ilogram(me)-*s*econd-*a*mpere system of units, adopted by the Inter-national Electrotechnical Commission, in place of all other systems of units. See also **SI units.**

ml. Abbrev. for *millilitre*.

MLD (*Radiol.*). Abbrev. for *mean lethal dose*.

M-line (*Nuc.*). Characteristic X-ray spectrum line produced when vacancy in M-shell is filled by electron. See also **K-lines, L-line.**

mm. Abbrev. for *millimetre*.

μμ (*Elec. Eng.*). Abbrev. *for micromicro*. Obsolete.

MMA (*Chem.*). Monomethyl aniline, N-methyl-aminobenzene, $C_6H_5{\cdot}NH{\cdot}CH_3$. Used occa-sionally as an *anti-knock substance*, as an alternative to lead tetraethyl.

m.m.f. (*Mag.*). Abbrev. for *magnetomotive force*.

μμF (*Elec. Eng.*). Abbrev. for *micromicrofarad*.

MMM overlay (*Print.*). A special material on which an impression of the entire letterpress forme is taken and which, after heat treatment, becomes a *mechanical overlay* (q.v.).

Mn (*Chem.*). The symbol for *manganese*.

Mne (*Aero.*). Abbrev. for the maximum per-missible indicated *Mach number*: a safety limitation, the suffix means 'never exceed' because of strength or handling considerations. The symbol is used mainly in operational instructions, generally for flight levels above 7600 m.

mnemic principle (*Zool.*). The principle which explains heredity, development, and evolution on the basis of the inherited memory of past generations.

mnemonics (*Psychol.*). Systems for aiding the memory, usually by consciously coding fairly specific types of information.

mnemotaxis (*An. Behav.*). See **pharotaxis.**

Mno (*Aero.*). Abbrev. for *normal operating Mach numbers*, usually of a jet airliner, the term being used mainly in flight operation instructions for flight levels above 7600 m.

M/N ratio (*Nuc.*). (1) The ratio of conversion fractions for the M- and N-shells; analogous to the K/L ratio. (2) In radiation chemistry, the ion yield.

mo. (*Build.*). An abbrev. for *moulded.*

Mo (*Chem.*). (1) The symbol for *molybdenum.* (2) A symbol for *morphine*, $C_{17}H_{19}O_3N$.

mobile phase (*Chem.*). See **chromatography.**

mobility (*Electronics*). Description of drift of ions (including electrons and holes in semiconductors) under applied electric fields, additional to thermal agitation. Measured in cm^2/s V, and related to conductivity and Hall effect.

Möbius strip (*Maths.*). The one-sided surface formed by joining together the two ends of a long rectangular strip, one end being twisted through 180° before the join is made.

Möbius transformation (*Maths.*). See **bilinear transformation.**

moccasins (*For.*). Special wide runners for sledges used on snow roads, consisting of a wide steel or wooden shoe installed between the wooden runner and its metal shoe.

Mocha stone (*Min.*). See **moss agate.**

mock chenille (*Textiles*). See **chenille.**

mock grandrelle (*Textiles*). A 2-colour thread produced by spinning a single yarn in which two rovings of different colours are incorporated.

mock leno (*Textiles*). A fabric in which openwork effect is produced by a grouping of threads, which, however, do not cross, as they do in leno and gauze fabrics. Also called **imitation gauze.**

mock moons (*Meteor.*). Lunar images similar to *mock suns* (q.v.). Also called **paraselenae.**

mock suns (*Meteor.*). Images of the sun, not usually very well defined, seen towards sunset at the same altitude as the sun and 22° from it on each side. They are portions of the 22° ice *halo* (q.v.) formed by ice crystals which, for some reasons, are arranged with their axes vertical. Also called **parhelia.**

modal value (*Maths.*). See **mode.**

mode (*Geol.*). The actual mineral composition of a rock expressed quantitatively in percentages. Cf. *norm.* (*Maths.*) The value of a series of observed values that is the most frequently observed, as exhibited by a frequency-distribution curve. Also **modal value.** Cf. *median.* (*Phys., etc.*) (1) One of several electromagnetic wave frequencies which a given oscillator may generate, or to which a given resonator may respond, e.g., magnetron modes, tuned line modes. In a waveguide, the mode gives the number of half-period field variations parallel to the transverse axes of the guide. Similarly, for a cavity resonator, the half-period variations parallel to all three axes must be specified. In all cases, different modes will be characterized by different field configurations. (2) Similarly, one of several frequencies of mechanical vibration which a body may execute or with which it may respond to a forcing signal. (3) A well-defined distribution of the radiation amplitude in a cavity which results in the corresponding distribution pattern in the laser output beam. In a multimodal system the beam will tend to diverge.

mode jump (*Telecomm.*). Switch of an oscillator from one mode to another; also **mode shift.**

modem (*Telecomm.*). A unit which can be used to transmit and receive binary data as a frequency modulated tone over the public telephone network.

mode number(s) (*Electronics*). These indicate the mode in which devices capable of operating with more than one field configuration are actually being used, e.g., in a cavity resonator, the mode numbers indicate the number of half wavelengths in the field pattern parallel to the three axes; in a magnetron, the mode number gives the number of cycles through which the phase shifts in one circuit of the anode; and in a klystron, it gives the number of cycles of the field which occur while an electron is in the field-free drift space.

moderating ratio (*Nuc.*). That of the slowing-down power of a moderator to the macroscopic adsorption cross-section.

moderation (*Nuc.*). See **degradation.**

moderator (*Nuc.*). Material, such as heavy water, graphite, or beryllium, used to slow down neutrons in a nuclear reactor. See **lethargy (of neutrons)** and **slowing-down power.**

moderator control (*Nuc.*). Control of a reactor by varying the position or quantity of the moderator.

modern face (*Typog.*). A style of type with contrasting thick and thin strokes, serifs at right angles, curves thickened, etc. See **type.**

mode separation (*Telecomm.*). The frequency difference between operation of a microwave tube in adjacent modes. See **mode jump.**

mode shift (*Telecomm.*). See **mode jump.**

modification (*Bot.*). A change in a plant brought about by environmental conditions and lasting only as long as the operative conditions last.

modified refractive index (*Meteor.*). Sum of the refractive index of the atmosphere at a given height and the ratio of the height to the radius of the earth.

modifier (*Comp.*). A code element used to alter the address of an operand. (*Min. Proc.*) Modifying agent used in froth flotation to increase *either* the wettability *or* the water-repelling quality of one or more of the minerals being treated. (*Gen.*) A gene which influences the operation of another.

modifying factor (*Radiol.*). General term applied to any factor used for calculating a biological radiation dose (or *dose equivalent*) in rem from the corresponding physical dose in rad. See **distribution factor, quality factor.** A *special modifying factor* is one which is used only under particular specified circumstances, e.g., for irradiation of a particular organ by particular types of radiation (there is a SMF of 3 for irradiation of the eye by high LET radiation) or under particular conditions (such as pregnancy, for example).

modiolus (*Zool.*). The conical central pillar of the cochlea.

modular (*Electronics*). Form of construction in which units, often with differing functions, are built into identical size *modules*, which are therefore quickly interchangeable.

modular ratio (*Civ. Eng.*). The ratio between Young's modulus for steel and that for the concrete in any given case of reinforced concrete.

modulated amplifier (*Telecomm.*). An amplifier stage during which modulation of the signal is carried out. Also **modulated stage.**

modulated amplifier valve (*Radio*). The valve in an anode-modulation system, to the grid of which the high-frequency carrier voltage is applied, and whose mean anode potential (over a high-frequency cycle) is varied, in accordance with the impressed modulation, through the coupling to the modulator valve.

modulated carrier or wave (*Telecomm.*). A frequency which can be transmitted or received through space or a transmission circuit, with a superposed information signal, which, by itself, could not be effectively transmitted and received.

modulated continuous wave (*Telecomm.*). Transmission in which a carrier is modulated by a tone or interrupted by keying. Abbrev. MCW.

modulated stage (*Telecomm.*). See **modulated amplifier.**

modulating electrode (*Electronics*). That of a thermionic valve to which a voltage is applied to control the size of the beam current.

modulation (*Acous.*). Changing from one key to another in music. The continual change from one fundamental frequency to another in speech. (*Radio*) Process of impressing a signal on to a radio-frequency carrier at *high-power* if the modulator is directly connected to a load, e.g., antenna, or at *low-power* if there is high-power amplification, e.g., class-A or class-B, between modulator and load. (*Telecomm.*) Signals for recording on tape, which can be *dipole* or *phase*, depending on number reversals of magnetization for each signal. (*Teleg.*) The variation in time of one or more given characteristics of an electromagnetic wave or of a direct current brought about directly (in facsimile telegraphy) or by a code (in alphabetic telegraphy) according to the content of the document to be transmitted.

modulation capability (*Telecomm.*). The maximum percentage modulation which can be used without exceeding a specified distortion level.

modulation condition (*Telecomm.*). The condition of voltages and currents in an amplifier for a modulated signal when the carrier is steadily modulated to a stated degree, e.g., 100%.

modulation depth (*Radio*). Factor indicating extent of modulation of a wave. *Difference ÷ sum* of peak and trough values of an amplitude-modulated wave, and of extreme deviations of carrier frequency in a frequency-modulated wave. Often expressed as a percentage.

modulation distortion (*Telecomm.*). When a carrier is modulated, any departure from invariability of carrier amplitude and/or addition of side frequencies proportional in amplitude to the corresponding frequencies in the signal, with phases balanced with respect to the initial phase of the carrier.

modulation frequency (*Radio*). One impressed upon a carrier wave in a modulator.

modulation hum (*Radio*). Low frequencies which enter a carrier-frequency amplifier and appear as interference after demodulation.

modulation index (*Telecomm.*). For a single modulation frequency, the ratio of the frequency deviation of the carrier to this frequency.

modulation meter (*Telecomm.*). That placed in shunt with a communication channel, giving indication that interprets, in a stated way, instant-to-instant power level in varying modulation currents. See **VU meter.**

modulation pattern (*Telecomm.*). That on a CRO when the modulated wave is connected to the Y deflection system and the modulation signal to the X deflection plates. The result is a trapezoidal pattern which enables the modulation depth to be measured.

modulation suppression (*Telecomm.*). Reduction of modulation in wanted signal in presence of an unwanted signal. Originally but erroneously termed *demodulation.*

modulation transformer (*Telecomm.*). One which applies the modulating signal to the carrier-wave amplifier in a transmitter.

modulator (*Acous.*). Circuit in an electronic organ which changes the pitch of the notes. (*Cinema.*) Any device, *light-valve, galvanometer, vibrator,* which regulates the light falling on a photographic sound-track. (*Radio*) Any circuit unit which modulates a radio carrier, at *high level* directly for transmission, or at *low level* for amplification of the modulated carrier before transmission.

modulator valve (*Radio*). The valve in an anode-modulation system to the grid of which the modulating signal is applied, and which impresses the variations in its anode potential upon the anode circuit of the modulated amplifier valve.

module (*Arch.*). The radius of the lower end of the shaft of a column. (*Build.*) Unit of size used in the standardized planning of buildings and design of components. (*Elec. Eng.*) Geometrical framework in which circuits can be realized, largely from units of specified dimensions. (*Eng.*) Of a gear-wheel, the pitch diameter divided by the number of teeth. The reciprocal of *diametrical pitch* (q.v.). (*Space*) A separate, and separable, compartment of a space vehicle.

modulo check (*Comp.*). One which anticipates remainder when a processed number is divided by a number which is also carried with data through the process. (*Maths.*) The base of counting for a particular arithmetic. E.g. the modulo of everyday decimal numbers is 10, that of binary numbers is 2.

modulus. Constant for units conversion between systems. (*Eng.*) Various moduli determine the deflection of a material under stress. Each is the ratio of stress to strain, which is a constant for a given material up to the elastic limit stress. See **elasticity** and **bulk modulus.** (*Maths.*) The absolute value of a number. The modulus of the complex number $z = x + iy$ is $|z| = +\sqrt{x^2+y^2}$. The modulus of a real number is its positive value. Sometimes called **absolute value norm.** See also **amplitude, complex number.**

modulus of elasticity (*Phys.*). For a substance, the ratio of stress to strain within the elastic range, i.e., where Hooke's law is obeyed. Measured in units of stress, e.g. GN/m^2. See also **elasticity.**

modulus of rigidity (*Eng.*). See **elasticity of shear.**

modulus of rupture (*Eng.*). A measure of the ultimate strength or the breaking load per unit area of a specimen, as determined from a torsion or a bending test.

Moebius process (*Met.*). An electrolytic process for parting gold-silver bullion. The electrolyte is silver nitrate. Bullion forms the anode. Silver passes into solution and is deposited on the cathode. Gold remains on the anode.

moellen (*Chem.*). See **degras.**

moellon (*Build.*). A rubble filling between the facing walls of a structure, sometimes laid in mortar.

Moerner's test (*Chem.*). A test for the presence of tyrosine, based on the appearance of a green colour when the solution is heated with a mixture of formalin and sulphuric acid.

mofette (*Geol.*). A volcanic opening through which emanations of carbon dioxide, nitrogen, and oxygen pass. It marks the last phase of volcanic activity.

Mogadon (*Pharm.*). See **nitrazepam.**

mohair (*Textiles*). Formerly, the long fine hair from the Angora goat, *Capra Mircus*; now, hair obtained for commercial purposes from crossbred animals. *Mohair lustre*: Plain or figured dress fabric made from fine cotton warp and mohair weft.

Mohawkian Series (*Geol.*). The middle of the main divisions of the Ordovician System in N. America, comprising the Utica, Lorraine, Richmond, and Gamache formations.

mohlaine (*Textiles*). See **permo.**

Mohorovičić discontinuity (*Geophys.*). A seismic discontinuity, named after its discoverer, which lies at a depth of between 10 and 40 km beneath the earth's surface and which marks the lower boundary of the *crust*. See **seismology.** Usually called **Moho.**

Mohr balance (*Chem.*). Balance used to determine density by weighing a solid when suspended in air and in a liquid.

Mohr-Coulomb theory (*Min. Proc.*). The resistance of rock to crushing is due to internal friction *plus* cohesion of bonding materials.

Mohr's litre (*Chem.*). The volume occupied by 1 kg of distilled water when weighed in air against brass weights; it is approximately equal to 1·002 true litres.

Mohr's salt (*Chem.*). Ammonium iron(II) sulphate. $(NH_4)_2SO_4{\cdot}FeSO_4{\cdot}6H_2O$.

Mohs' scale of hardness (*Min.*). A scale introduced by Mohs to measure the hardness of minerals. See **hardness.**

moil (*Glass*). (1) Glass left on a punty or blowing iron after the gather has been cut off or after a piece of ware has been blown and severed. (2) Glass originally in contact with the blowing mechanism or head, which becomes *cullet* (q.v.) after the desired article is severed from it.

Moine Schists (or Series) (*Geol.*). Flaggy granulites and pelitic schists of sedimentary origin, occurring in northern Scotland; of Pre-Cambrian age.

moiré effect (*Photog.*). A 'watered-silk' *pattern* (q.v.) arising from interference between 2 line-screens; a defect for which occasional uses are found.

moisture expansion (*Build., Civ. Eng.*). Increase in the volume of a material from absorption of moisture. Also called **bulking.**

mol (*Chem.*). See **mole.**

molality (*Chem.*). The concentration of a solution expressed as the number of moles of dissolved substance per kilogram of solvent.

molal specific heat capacity (*Phys., etc.*). The *specific heat capacity* of 1 mole of an element or compound. Also called **volumetric heat** (for gases).

molar conductivity (*Elec.*). The electrical conductivity of an electrolyte having 1 mole of solute in 1 dm^3 of solution. Measured in S cm^2/mol.

molar heat. See **molecular heat.** Cf. *C*.

molarity (*Chem.*). The concentration of a solution expressed as the number of moles of dissolved substance per dm^3 of solution.

molar process (*Zool.*). In *Malacostraca*, one of two processes of the part of the mandible which projects towards the mouth.

molars (*Zool.*). The posterior grinding or cheek teeth of Mammals which are not represented in the milk dentition.

molar surface energy (*Chem.*). The surface energy of a sphere containing 1 mole of liquid; equal to $\gamma V^{2/3}$, where γ is the surface tension and V the molar volume. It is zero near the critical point, and its temperature coefficient is often a colligative property. See **Eötvös equation.**

molar volume (*Chem.*). The volume occupied by one mole of a substance under specified conditions. That of an ideal gas at s.t.p. is $2{\cdot}2414 \times 10^{-2}$ m^3 mol^{-1}.

molasse (*Geol.*). Sediments produced by erosion of mountain ranges following the final stages of an *orogeny*. Sandstones and other detrital rocks are the dominant products of this type of sedimentation.

molasses (*Chem.*). Residual sugar syrups from which no crystalline sugar can be obtained by simple means. An important raw material for the manufacture of ethyl and other alcohols.

mold, molding, etc. A variant spelling of *mould, moulding*, etc.

moldavite (*Min.*). A type of *tektite* (q.v.). See also **bottle-stone.**

mole (*Chem.*). The amount of substance that contains as many entities (atom, molecules, ions, electrons, photons, etc.) as there are atoms in 12g of ^{12}C. It replaces in SI the older terms *gram-atom, gram-molecule*, etc., and for any chemical compound will correspond to a mass equal to the relative molecular mass in grams. Abbrev. **mol.** See **Avogadro number.** (*Civ. Eng.*) A *breakwater* (q.v.) or a masonry pier. (*Med.*) (1) *Naevus* (q.v.). (2) A haemorrhagic mass formed in the Fallopian tube as a result of bleeding into the sac enclosing the embryo.

molectronics (*Electronics*). See **mole-electronics.**

molecular (*Chem.*). (1) Pertaining to a molecule or molecules. (2) Pertaining to 1 mole.

molecular association (*Chem.*). The relatively loose binding together of the molecules of a liquid or vapour in groups of two or more.

molecular beam (*Chem.*). Directed stream of un-ionized molecules issuing from a source and depending only on their thermal energy.

molecular biology (*Biol.*). The study of biological macromolecules.

molecular compound (*Chem.*). A compound formed by the combination of two or more molecules which are capable of independent existence.

molecular conductivity (*Chem.*). The conductivity of a volume of electrolyte containing 1 mole of dissolved substance.

molecular depression of freezing-point (*Chem.*). The drop in the freezing-point of a liquid which would be produced by the dissolution of 1 mole of a substance in 100 g of the solvent, if the same laws held as in dilute solutions.

molecular distillation (*Vac. Tech.*). The distillation, in a vacuum, of a labile material, in which the temperature and time of heating are minimized by providing a condensation surface at such a distance from the evaporating surface that molecules can reach the condenser without intermolecular collision.

molecular drag pump (*Vac. Tech.*). A mechanical vacuum pump in which a flow of gas towards a backing pump is produced by causing the gas molecules to collide with, and receive momentum from, a moving surface separated from a fixed surface by a distance which is small in comparison with the mean free path of the molecules.

molecular electronics (*Electronics*). See **mole-electronics**.

molecular elevation of boiling-point (*Chem.*). The rise in the boiling-point of a liquid which would be produced by the dissolution of 1 mole of a substance in 100 g of the solvent, if the same laws held as in dilute solutions.

molecular filter (*Chem.*). Alternative name for **membrane filter**.

molecular formula (*Chem.*). The formula showing the number of atoms in a molecule of the compound, e.g., benzene C_6H_6, methanal $H{\cdot}CHO$, aluminium chloride $AlCl_3$. In organic chemistry, elements are usually given in the following order, C, H, O, N, S, halogen, etc. Functional groups are written separately, e.g., Prontosil $(NH_2)_2C_6H_3{\cdot}N{=}N{\cdot}C_6H_4{\cdot}SO_2NH_2$, not $C_{12}H_{13}O_2N_5S$.

molecular genetics (*Gen.*). Interpretation of inheritance and mutation in terms of changes in the linear sequence of bases which specify the codons of the nucleic acid (usually DNA) strands in the genes.

molecular heat (*Chem.*). The product of the specific heat capacity of a substance and its relative molecular mass.

molecular layer (*Zool.*). The outer layer of the cerebral cortex in higher Vertebrates; the outer layer of the cerebellar cortex.

molecular orbital (*Chem.*). Theory that some of the electron orbitals in a molecule enclose the molecule as a whole, rather than particular atoms.

molecular refraction (*Chem.*). See **Lorentz-Lorenz equation**.

molecular rotation (*Chem.*). One-hundredth of the product of the specific rotation and the relative molecular mass of an optically-active compound.

molecular sieve (*Chem.*). *Clathrate* compound, usually a synthetic zeolite, used to absorb or separate molecules. The molecules are trapped in 'cages' formed by the shape of the crystal lattice, the sizes of which can be selected to suit a given molecule.

molecular solution (*Chem.*). A true solution, in which molecules of the dissolved substance are separated from one another by molecules of solvent.

molecular stopping power (*Nuc.*). The energy loss per molecule per unit area normal to the motion of the particle in travelling unit distance. It is *approximately* equal to the sum of the atomic stopping powers of the constituent atoms.

molecular streaming (*Chem. Eng.*). See **Knudsen flow**.

molecular structure (*Chem.*). The way in which atoms are linked together in a molecule.

molecular volume (*Chem.*). The volume occupied by 1 mole of a substance in gaseous form at standard temperature and pressure (approx. $2{\cdot}2414\times10^{-2}$ m^3).

molecular weight (*Chem.*). Better **relative molecular mass**. (1) The mass of a molecule of a substance referred to that of an atom of ^{12}C taken as 12·000. (2) The sum of the relative atomic masses of the constituent atoms of a molecule.

molecule (*Chem.*). Smallest part of an element or compound which exhibits all the chemical properties of that specific compound or element.

mole drainer (*Agric.*). Pointed cylinder on the lower edge of a blade, which is drawn longitudinally through soil to form a drainage channel.

mole-electronics (*Electronics*). Technique of growing solid-state crystals so as to form transistors, diodes, resistors in one mass, for microminiaturization. Also **molectronics, molecular electronics.**

mole fraction (*Chem.*). Fraction, of the total number of molecules in a phase, represented by a given component.

moler (*Build.*). A diatomaceous earth used in the manufacture of *Fosalsil.*

moleskin (*Textiles*). A heavy fustian type of fabric, with smooth face and twill back; used for working clothes.

Molisch test (*Chem.*). A test for certain proteins, based on the appearance of a violet ring between two layers of a protein solution containing some α-naphthol and concentrated sulphuric acid. The reaction is due to the presence of a carbohydrate group in the protein molecule, and the same reaction is also given by carbohydrates in general.

molleton or **mollitan** (*Textiles*). A heavy type of woollen or cotton fabric, plain or figured, raised on both sides; used for dressing-gowns.

Mollier diagram (*Chem. Eng.*). A diagram, for any substance relating total heat and entropy, in which heat and work quantities are represented by line segments, not areas. Used to calculate the efficiency of steam engines or refrigeration cycles.

mollities ossium (*Med.*). See **osteomalacia**.

Moll's glands (*Histol.*). Modified sweat glands, found in the eyelid, lying between the hair follicles of the lashes.

Mollusca (*Zool.*). Unsegmented coelomate Invertebrates with a head (usually well developed), a ventral muscular foot and a dorsal visceral hump; the skin over the visceral hump (the *mantle*) often secretes a largely calcareous shell and encloses a mantle cavity into which open the anus and kidneys and in which are the *ctenidia* (q.v.), originally used for gaseous exchange. There is usually a *radula* (q.v.); some have haemocyanin as a respiratory pigment; well-developed blood and nervous systems; often with larva of the *trochophore* type. Chitons, Slugs, Snails, Mussels, Whelks, Limpets, Squids, Cuttlefish, and Octopods.

molluscum contagiosum (*Med.*). A contagious condition in which small, white, waxy nodules appear on the skin; believed to be due to infection with a virus.

molluscum fibrosum (*Med.*). Von Recklinghausen's disease; multiple neurofibromata. A disease characterized by the appearance on the skin (and elsewhere in the body) of soft nodules composed of fibrous tissue and believed to develop from nerve fibres.

Mollweide's projection (*Geog.*). An elliptical equal-area projection for the whole globe capable of 'interruption' as in the *sinusoidal projection* (q.v.), but less liable to distortion of shape, if not of scale. All parallels of latitude and the central meridian are straight lines parallel to each other, but other meridians are curved, the curvature increasing towards the marginal meridians.

Molpadida (*Zool.*). An order of *Holothuroidea* having small simple buccal tube-feet with ampullae, and with or without retractor muscles; there are no tube-feet on the trunk; respiratory trees occur; the madreporite is internal. Burrowing forms.

molten metal dyeing (*Textiles*). Fabric impregnated with a solution of dyestuff is passed through a bath of molten metal which develops the shade and helps to dry the cloth.

molybdates(VI) (*Chem.*). Salts of molybdic acid.

molybdenite (*Min.*). Disulphide of molybdenum, crystallizing in the hexagonal system. It is the most common ore of molybdenum, and occurs in lustrous lead-grey crystals in small amounts in granites and associated rocks.

molybdenosis (*Vet.*). Teart. A disease of cattle and sheep caused by an excess of molybdenum in the diet; characterized by chronic diarrhoea and emaciation. Occurs where high levels of molybdenum are present in the soil.

molybdenum (*Chem.*). A metallic element in the sixth group of the periodic system. Symbol Mo, at. no. 42, r.a.m. 95·94, rel. d. at 20°C 10·2, hardness 147 (Brinell), m.p. 2625°C, b.p. 3200°C, resistivity *c.* 5×10^{-8} ohm metres. Its physical properties are similar to those of iron, its chemical properties to those of a nonmetal. Used in the form of wire for filament supports, hooks, etc., in electric lamps and radio valves, for electrodes of mercury-vapour lamps, and for winding electric resistance furnaces. Is added to a number of types of alloy steels, certain types of Permalloy, and Stellite. It seals well to Pyrex, spot-welds to iron and steel.

molybdenum blues (*Chem.*). A variety of reducing agents (including *molybdenum* (III)) convert molybdates into colloidal molybdenum blues, the compositions of which approach $Mo_8O_{23} \cdot xH_2O$.

molybdenum(VI)oxide (*Chem.*). MoO_3. Behaves as an acid anhydride, forming molybdic acid. The essential starting-point in the manufacture of molybdenum metal, and most other compounds of molybdenum.

molybdic acid (*Chem*). H_2MoO_4. See also **molybdenum(VI)oxide.**

molybdite (*Min.*). Strictly, molybdenum oxide. Much so-called molybdite is ferrimolybdite, a hydrous ferric molybdate which crystallizes in the orthorhombic system. It is commonly impure and occurs in small amounts as an oxidation product of molybdenite. Also called **molybdic ochre.**

moment (*Aero.*). See **hinge-, pitching-, rolling-, yawing-.** (*Elec. Eng.*) See **electric-.** (*Maths.*) (1) Of a couple, see **couple.** (2) Of a force or vector about a point, the product of the force or vector and the perpendicular distance of the point from its line of action. In vector notation, $\mathbf{r} \wedge \mathbf{F}$, where $\mathbf{r}$ is the position vector of the point, and $\mathbf{F}$ is the force or vector. (3) Of a force or vector about a line, the product of the component of the force or vector parallel to the line and its perpendicular distance from the line.

moment distribution (*Civ. Eng.*). A method of analysing the forces in a continuous structure by adjusting between the imposed loads, spans, and sectional properties of the supporting members concerned.

moment of a magnet (*Elec. Eng.*). See **magnetic moment.**

moment of inertia (*Maths.*). Of a body about an axis: the sum Σmr^2 taken over all particles of the body where m is the mass of a particle and r its perpendicular distance from the specified axis. When expressed in the form Mk^2, where M is the total mass of the body ($M = \Sigma m$), k is called the *radius of gyration* about the specified axis. Also used erroneously for *second moment of area.*

moment to change trim one inch (*Ships*). The moment, taken about the *centre of flotation* (q.v.) which will change the *trim* (q.v.) by 1 in. Expressed in foot-tons.

momentum (*Maths.*). A dynamical quantity, conserved within a closed system. A body of mass M and whose centre of gravity G has a velocity v has a *linear momentum* of Mv. It has an *angular momentum* about a point O defined as the moment of the linear momentum about O. About G this reduces to $I\omega$ where I is the moment of inertia about G and ω the angular velocity of the body.

mon-, mono-. Prefix from Gr. *monos*, alone, single. (*Chem.*) Containing one atom, group, etc., e.g., monobasic.

monacid (*Chem.*). Containing one hydroxyl group, replaceable by an acid radical, with the formation of a salt.

Mona Complex (*Geol.*). The rock groups of Pre-Cambrian age exposed in Anglesey and parts of peninsular Caernarvonshire.

monad (*Zool.*). A *flagellispore* (q.v.); a primitive organism. *adj.* **monadiform.**

monadelphous (*Bot.*). Having all the stamens in the flower joined together by their filaments.

monalbite (*Min.*). A high-temperature albite feldspar with monoclinic symmetry.

monandrous (*Bot.*). (1) Having 1 antheridium. (2) Having 1 stamen.

monangial (*Bot.*). Said of a sorus consisting of a single sporangium.

monarch (*Bot.*). Having a single strand of protoxylem in the stele.

Monastral blue (*Chem.*). Widely used TN for *phthalocyanine blue* pigments. Also **Monastral fast blue B.S.** See **phthalocyanines.**

Monastral green (*Chem.*). Widely used TN for *phthalocyanine green* pigments. See **phthalocyanines.**

monaural (*Acous.*). Pertaining to use of one ear instead of two (cf. *binaural*). In sound-recording, the (inaccurate) opposite of *stereophonic*. See **monophonic.**

monaxon (*Zool.*). Having only 1 axis; said of Sponge spicules.

Monaxonida (*Zool.*). An order of *Demospongiae* in which the skeleton is composed of monaxial spicules only, in some cases cemented together by spongin.

monazite (*Min.*). Monoclinic phosphate of the rare earth metals, $CePO_4$, containing cerium as the principal metallic constituent, and also some thorium. Monazite is exploited from beach sands in Travancore, India, where it is relatively abundant, and is one of the principal sources of rare earths and thorium.

monchiquite (*Geol.*). A dark-coloured lamprophyric igneous rock, microcrystalline or porphyritic, and containing abundant mafic minerals, with little or no feldspar, in an isotropic base consisting of analcite. Some varieties contain olivine, nepheline, or leucite.

Mönckeberg's sclerosis (or **degeneration**) (*Med.*). Degeneration of the middle coat of medium-sized arteries in old people, characterized by the deposit of lime salts in it.

Monday morning disease (*Vet.*). See **sporadic lymphangitis.**

Mond gas (*Chem.*). The gas produced by passing air and a large excess of steam over coal-slack at about 650°C. See also **semiwater gas.**

Mond process (*Met.*). A process used by Mond Nickel Co. in extracting nickel from a matte consisting of copper-nickel sulphides. The matte is roasted to obtain oxides, the copper is leached out with H_2SO_4, the nickel oxide is reduced to nickel with hydrogen, then the nickel is caused to combine with CO to form a carbonyl, which is decomposed by heating.

Monel metal (*Met.*). A nickel-base alloy containing nickel 68%, copper 29%, and iron, manganese, silicon, and carbon 3%. Has high strength (about 500 MN/m²), good elongation (about 45%), and high resistance to corrosion. Used for condenser tubes, propellers, pump fittings, turbine blades, and for chemical and food-handling plant.

mongolism (*Med.*). A form of congenital mental defect due to a chromosomal abnormality, trisomy 21, the face being slightly mongoloid. Also called **Down's syndrome.**

mongrel (*Bot., Zool.*). The offspring of a cross between varieties or races of a species.

monial (*Join.*). Same as **mullion.**

Moniliales (*Bot.*). A group of fungi belonging to the *Fungi Imperfecti*, all bearing conidia, either on mycelia or on conidiophores, and including many parasites and saprophytes of economic importance, e.g., *Candida albicans* (moniliasis) *Aspergillus niger* (used in the preparation of citric acid). These non-sporing yeastlike organisms are classified by some authorities with the asco-sporogenous yeasts.

moniliasis (*Med., Vet.*). Infection with any of the species of the fungus *Monilia* (*Candida*). See **thrush.** (*Vet.*) Candidiasis. Infection of the mucosa of the mouth and other parts of the digestive tract of birds and animals by yeastlike organisms of the genus *Monilia.*

monilicorn (*Zool.*). Having moniliform antennae.

moniliform (*Zool.*). Resembling a necklace or chain of beads, as *moniliform antennae.*

monimostyly (*Zool.*). In Vertebrates, the condition of having the quadrate immovably united to the squamosal; cf. *streptostyly.*

monitor (*Automation*). Person or apparatus whose purpose is to ascertain that equipment, materials, or output, fulfil prescribed conditions or operate within prescribed limits. **(*Comp.*)** Unit in large computers which prepares the machine instructions from the source program, using built-in compiler(s) for one or more program languages, and feeds these into the processing and output units in sequence once compiling is completed. It also controls time-sharing procedures. **(*Mining, Min. Proc.*)** In hydraulic mining, high-pressure jet of water used to break down loosely consolidated ground in open-cast work. In ore treatment, a device which checks part of the process and sounds a signal, makes a record, or initiates compensating adjustment if the detail monitored requires it. **(*Radiol.*)** Ionization chamber or other radiation detector arranged to give a continuous indication of intensity of radiation, as in radiation laboratories, radiation protection from fallout contamination, industrial operations, or X-ray exposure. **(*Telecomm.*)** An arrangement for reproducing and checking any transmission without interfering with the regular transmission.

monitoring (*Radiol.*). Periodic or continuous determination of the amount of specified substances, e.g., toxic materials or radioactive contamination, present in a region or a person, or of a flow rate, pressure, etc., in a (continuous) process, as safety measures. **(*Telecomm.*)** Tapping a communication circuit to ascertain that transmission is as required, without interfering with it. This may be with headphones, amplifier and loudspeaker or a TV receiver.

monitoring booth (*Cinema.*). A portable enclosure on wheels for the use of the recordist on the floor of a sound-film studio. It is arranged so that he sees the action and hears on a monitoring receiver what is being recorded.

monitoring loudspeaker (*Acous.*). One of high quality and exactly matching others, for verifying quality of programme transmission before radiation or recording. Located in a monitoring room adjacent to film or broadcasting studio.

monitoring receiver (*Radio, TV*). See **check receiver.**

monitoring station (*Radio*). National service for verifying the frequencies of emission of radio transmitters allocated within prescribed bands. See **wavemeter.**

monitor position (*Teleph.*). The special position at any exchange at which operators deal with queries and complaints.

monitor tube (*TV*). A cathode-ray tube at a TV transmitting station to check the quality of the transmitted picture.

monk (*Typog.*). An area of printing with too much ink. See **friar.**

monk bond (*Build.*). A modification of the *Flemish bond* (q.v.), each course consisting of two stretchers and one header alternately. Also called **flying bond.**

monkey (*Civ. Eng.*). The falling weight of a pile-driver. Also called **beetle-head, tup.**

monkey-chatter (*Telecomm.*). Interference between a wanted carrier and the nearer sideband (modulation) of an unwanted transmission in radio reception.

monkey tail (*Join.*). A vertical scroll at one end of a hand-rail.

monkey-tail bolt (*Join.*). A long-handled bolt for the top of a door, capable of operation from the floor.

mono (*Acous.*). Abbrev. for **monophonic.**

mono-. Pfx. See **mon-.**

monobasic (*Chem.*). Containing 1 hydrogen atom replaceable by a metal with the formation of a salt.

monobath (*Photog.*). Solution which develops and fixes an emulsion in one stage.

monobloc (*I.C. Engs.*). The integral casting of all the cylinders of an engine, i.e., in the same cylinder block.

monocable (*Civ. Eng.*). The type of *aerial ropeway* (q.v.) in which a single endless rope is used both to support and to move the loads.

monocardian (*Zool.*). Having a completely undivided heart.

monocarpellary (*Bot.*). Having, or consisting of, a single carpel.

monocarpic (*Bot.*). (1) Forming a single fruit and then dying. (2) Dying after one flowering season.

monocerous (*Zool.*). Having a single horn.

monochasium (*Bot.*). A cymose inflorescence in which each successive branch bears one branch in its turn.

Monochlamydeae (*Bot.*). See **Incompletae.**

monochlamydeous (*Bot.*). Having a perianth of one whorl of members.

monochlamydeous chimaera (*Bot.*). See **haplochlamydeous chimaera.**

monochord (*Acous.*). A laboratory apparatus for demonstrating the properties of single stretched wires and the sounds generated thereby.

monochromatic (*Phys.*). By extension from *monochromatic light*, any form of oscillation or radiation characterized by a unique or very narrow band of frequency.

monochromatic filter (*Photog.*). A filter which transmits light of a single wavelength, or, in practice, a very narrow band of wavelengths.

monochromatic light (*Phys.*). Light containing radiation of a single wavelength only. For many purposes, the light from a sodium flame (i.e., a Bunsen flame coloured yellow by the introduction of common salt) is sufficiently nearly monochromatic. The wavelengths it gives are 589·0 and 589·6 nm. See also **cadmium red line.**

monochromatic radiation (*Phys.*). Electromagnetic radiation (originally visible) of one single frequency component. By extension, a beam of particulate radiation comprising particles all of the same type and energy. *Homogeneous* or *monoenergic* is preferable in this sense.

monochromator (*Phys.*). Device for converting heterogeneous beam of radiation (electromagnetic or particulate) to homogeneous beam by absorption or refraction of unwanted components.

monochrome (*Photog.*). A photographic print in one colour of varying brightness. (*TV, etc.*) Signal or picture of one colour or hue.

monochrome receiver (*TV*). TV receiver which reproduces a black-and-white transmission, or a colour transmission in black-and-white. See **compatible colour television.**

monochrome signal (*TV*). That part of a TV signal controlling luminance only; cf. *chrominance signal.* In compatible systems, this signal provides the black-and-white picture.

monochrome transmission (*TV*). Transmission of a signal controlling the luminance values in the picture, but not the chromaticity.

monocle (*Photog.*). An uncorrected simple lens, similar to a spectacle lens, for soft focus work.

monocline (*Geol.*). A fold with a single limb which produces a sudden steepening of the dip; the rocks, however, soon approximate to horizontality on either side of this flexure.

monoclinic system (*Crystal.*). The style of crystal architecture in which the 3 crystal axes are of unequal lengths, having 1 of their intersections oblique and the other 2 at right angles. Also called **oblique system.**

monoclinous (*Bot.*). Having stamens and pistils in the same flower.

monocoque (*Aero.*). A fuselage or nacelle in which all structural loads are carried by the skin. In a *semimonocoque*, loads are shared between skin and framework, which provides local reinforcement for openings, mountings, etc. See **stressed-skin construction.** (*Autos.*) See **chassis.**

Monocotyledones (*Bot.*). One of the two main groups included in the Angiospermae, with many thousands of species. The plants are mostly herbs with parallel venation in the leaves, flowers with parts in threes, and stems without a clearly defined pith, and containing scattered vascular bundles without a cambium (with few exceptions), e.g., cereals. The embryo has one cotyledon.

monocotyledonous (*Bot.*). (1) Belonging to the *Monocotyledones.* (2) Said of an embryo or seedling having a single cotyledon.

monocule (*Zool.*). An animal possessing a single eye, as the Water Flea *Daphnia. adj.* **monocular.**

monoculture (*Agric.*). The growing of one kind of crop only, or the large area over which it is grown.

monocyclic (*Bot.*). (1) Said of an annual plant. (2) Having each kind of member forming one whorl. (*Zool.*) Said of the calyx of *Crinoidea* when the row of infrabasals is lacking.

monocyte (*Zool.*). A type of large uninuclear nongranular leucocyte; possesses phagocytic properties.

monodactylous (*Zool.*). Having only a single digit.

Monodelphia (*Zool.*). See **Eutheria.**

monodesmic (*Bot.*). Said of a petiole which contains one vascular strand.

monodisperse system (*Chem.*). A colloidal dispersion having particles all of effectively the same size.

monodont (*Zool.*). Having a single persistent tooth, as the male Narwhal.

monoecious (*Bot.*). Having separate staminate and pistillate flowers on the same individual plant. (*Zool.*) See **hermaphrodite.**

monoenergic (*Nuc.*). See **monochromatic radiation.**

monoestrous (*Zool.*). Exhibiting only one oestrous cycle during the breeding season. Cf. *polyoestrous.*

monofilament (*Plastics*). A single filament of indefinite length. Used for ropes, surgical sutures, etc., and, in finer gauge, for textiles.

monoformer (*Telecomm.*). A *photoformer* (q.v.) in which the shadow mask is built into the CRT; used where it is only necessary to generate one specific function.

monogastric (*Zool.*). Of *Siphonophora*, having a single gastric cavity.

monogenetic (*Chem.*). Producing only one colour on fabrics. (*Zool.*) Multiplying by asexual reproduction; showing a direct life-history; of parasites, having a single host.

monogenetic gravel (*Geol.*). A loose detrital sediment in which the predominant size of the particles is 2–10 mm; it consists of one type of constituent.

monogenic function (*Maths.*). See **analytic function.**

monogerm (*Bot.*). Used in agriculture to describe a seed which produces only one seedling, e.g., monogerm varieties of sugar beet.

monogonoporous (*Zool.*). In hermaphrodite forms, having a single genital opening through which both male and female genital products are discharged.

monogony (*Zool.*). Asexual reproduction.

monohybrid (*Gen.*). The result of a cross between 2 individuals in which the inheritance of a single pair of characters (from different alleles of one given gene) is being investigated.

monohydric alcohols (*Chem.*). Alcohols containing one hydroxyl group only.

monoicous (*Bot.*). Said of mosses which have the antheridia and archegonia borne on the same plant, but in separate groups.

monokaryon (*Cyt.*). A nucleus with only one centriole.

monolayer (*Chem.*). See **monomolecular layer.**

monolayer culture (*Histol.*). A tissue culture technique whereby thin sheets of cells are grown, on glass, in a nutrient medium.

monolith (*Build.*). A single detached column or block of stone. (*Civ. Eng.*) A hollow foundation piece of concrete, brickwork, or masonry with a number of open wells passing through it. It is sunk in a manner similar to the *cylinder caisson* (q.v.), the wells being finally filled with concrete to form a solid foundation.

monolithic (*Build., Civ. Eng.*). The term applied to a structure made of a continuous mass of material.

monolithic integrated circuit (*Electronics*). Electronic circuit formed by diffusion or ion-implantation on a single crystal of semiconductor, usually silicon.

monomer (*Chem.*). A substance consisting of single molecules, in contrast to *associated molecules* and *polymers.* (*Plastics*) See **polymer.**

monomerosomatous (*Zool.*). Having all the somites of the body fused to form a single tagma, as in Mites.

monomethylparamino phenol (*Photog.*). See **4-methylaminophenol.**

monomineralic rocks (*Geol.*). Rocks consisting

essentially of one mineral, e.g., dunite and anorthosite.

monomolecular layer, monolayer (*Chem.*). A film of a substance 1 molecule in thickness.

monomolecular reaction (*Chem.*). A reaction in which only one species is involved in forming the activated complex of the reaction.

monomorphic (*Zool.*). Showing little change of form during the life-history; showing only a single type of bodily structure. Cf. *polymorphic*.

monomorphous (*Crystal.*). Existing in only one crystalline form.

mononychous (*Zool.*). Possessing a single claw.

monopack (*Photog.*). See **integral tripack.**

Monopetalae (*Bot.*). See **Sympetalae.**

monopetaly (*Bot.*). See **gamopetaly.**

monophagous (*Bot.*). Said of a fungal parasite having a thallus which attacks one host cell only. (*Zool.*) Feeding on one kind of food only; as *Sporozoa*, living always in the same cell, or phytophagous Insects, with only one food-plant. Also **monotrophic.**

monophasic (*Zool.*). Having an abbreviated life-cycle, without a free active stage; said of certain Trypanosomes. Cf. *diphasic*.

monophonic (*Acous.*). Single-channel sound reproduction, recreating the acoustic source, as compared with *stereophonic*, which uses two or more identical channels for auditory perspective. See **monaural.**

Monophoto (*Print.*). A major *filmsetting* system developed from the Monotype hot-metal system, the caster being replaced by a photographic unit.

monophyletic (*Gen.*). Descended from a single parent form.

monophyodont (*Zool.*). Having only a single set of teeth, the permanent dentition. Cf. *polyphyodont*.

Monoplacophora (*Zool.*). An order of *Amphineura*, which are almost bilaterally symmetrical with internal metamerism; a single shell; well-developed coelomic cavity.

monoplane (*Aero.*). A heavier-than-air aircraft, either an aeroplane or glider, having one main supporting surface.

monoplanetic (*Bot.*). Having one period of locomotion.

monoplegia (*Med.*). Paralysis of an arm or of a leg.

monoploid (*Cyt.*). True haploid.

monopodium (*Bot.*). A branch system in which each or any branch continues to increase in length by apical extension and bears similar lateral branches in acropetal succession.

monopropellant (*Space*). A liquid rocket propellant in which the fuel and oxidizing agent are mixed prior to combustion. Cf. *multi-propellant*.

monoprosthomerous (*Zool.*). Having one somite in front of the mouth.

monopulse (*Radar*). Radar using spaced antennae to give phase differences on received signal. This arrangement can give precise directional information.

monorail (*Civ. Eng.*). A railway system in which carriages are suspended from, and run along, a single continuous elevated rail. (*Photog.*) Said of technical cameras having the focusing movement of the bellows accommodated on a single rail, obviating the need for a baseboard.

monorhinal (*Zool.*). Having only a single nostril.

monosaccharides (*Chem.*). The simplest group of carbohydrates, classified into tetroses, pentoses, hexoses, etc., according to the number of carbon atoms, and into aldoses and ketoses depending on whether they contain a potential aldehyde or ketone group.

monoscope (*TV*). An electron-beam tube used to produce a picture signal from a fixed object; used for test purposes. Also **monotron.**

monosepalous (*Bot.*). See **gamosepalous.**

monosodium glutamate (*Chem.*). Sodium hydrogen glutamate. $HOOC{\cdot}CH(NH_2)CH_2CH_2COONa$, a derivative of glutaric acid, used as a condiment or flavouring.

monosome (*Cyt.*). The unpaired accessory or X-chromosome. See **sex chromosome.**

monospermy (*Zool.*). Fertilization of an ovum by a single spermatozoon.

monosporous (*Bot.*). (1) Containing 1 spore. (2) Derived from 1 spore.

monostable (*Telecomm.*). Of a circuit or system, fully stable in one state only but *metastable* in another to which it can be driven for a fixed period by an input pulse.

monostichous (*Bot.*). Forming 1 row.

monosy (*Bot.*, *Zool.*). Separation of parts normally fused.

monosymmetrical (*Bot.*). See **zygomorphic.**

monosymmetric system (*Crystal.*). See **monoclinic system.**

monothalamous (*Zool.*). In *Protozoa*, said of a shell or test which always has a single chamber.

monotocous (*Zool.*). Producing a single offspring at a birth.

monotonic (*Maths.*). The property of either never increasing or never decreasing. Applied to a function or sequence. (*Telecomm.*) Said of transient response when it increases continuously with time.

Monotremata (*Zool.*). The single order included in the Mammalian group *Prototheria* (q.v.) and having the characteristics of the subclass, e.g., Spiny Ant-eater (*Tachyglossus*), Duck-billed Platypus (*Ornithorhynchus*).

monotrichous (*Bot.*). Bearing a single flagellum.

monotrochous (*Zool.*). Having the trochanter undivided.

monotron (*TV*). See **monoscope.**

monotrophic (*Zool.*). See **monophagous.**

monotropic (*Chem.*). Existing in only one stable crystalline form, the other forms being unstable under all conditions.

Monotype (*Typog.*). See **composing machines.**

Monotype set system (*Typog.*). To make the best interpretation of a type design, the em quad for each fount is allotted a set, there being ½-point intervals between each. There are a few 'normal set' founts in which the width of the em quad is the same as the depth of the body, but in the majority of founts the set is narrower, e.g., 10 pt Times, 9¾ set. See also **Monotype unit system.**

Monotype unit system (*Typog.*). Each character of a fount is allotted an appropriate number of units out of the 18 which represent the em quad of the 'set'. See also **Monotype set system.** An arrangement of this kind is essential for all tape-operated type-setting and filmsetting systems.

monotypic (*Bot.*, *Zool.*). Said of a species or genus which is exemplified by only a single type.

monovalent (*Chem.*). Capable of combining with 1 atom of hydrogen or its equivalent.

monoxenous (*Bot.*). Said of a parasitic fungus which is restricted to one species of host plant.

Monozoa (*Zool.*). An order of *Cestoda* in which the body is not differentiated into a scolex and proglottides, and the genitalia are not repeated; most of the adults are parasitic in the gut of various Fish, but one form occurs in an Annelid.

monozygotic (*Biol.*). Developing from one fertilized ovum.

Monro's foramen (*Zool.*). A narrow canal connecting the first or second ventricle with the third ventricle in the brain of Vertebrates.

monsoon (*Meteor.*). Originally winds prevailing in the Indian Ocean, which blow S.W. from April to October and N.E. from October to April; now generally winds which blow in opposite directions at different seasons of the year. Similar in origin to land and sea breezes, but on a much larger scale, both in space and time. Particularly well developed over southern and eastern Asia, where the wet summer monsoon from the S.W. is the outstanding feature of the climate.

monsoon climate (*Geog.*). Type of climate in which there is a seasonal reversal in the direction of the prevailing winds. It is characteristic of S. and E. Asia, particularly India.

monsoon current (*Geog.*). Ocean current of N. Indian Ocean flowing east in summer and west in winter. It flows strongly between Sumatra and the concave African coast north of Madagascar setting up a subsidiary circulation in the Bay of Bengal.

mons pubis (*Anat.*). A convex formation of subcutaneous fatty tissue over the pubic symphysis.

monster (*Biol.*). An abnormal form of a species.

montage (*Cinema.*). A composition of shots or strips of images so that the whole idea conveyed to the viewer is something more than the sum of the material in the shots.

montage photograph (*Photog.*). A composite photograph, for decoration or publicity, made from the juxtaposition of cut-up photographs arranged in a pattern.

montant (*Join.*). A vertical member in panelling or framing.

montan wax (*Chem.*). A bituminous wax extracted under high temperature and pressure from lignite. It is dark brown in colour, and can be bleached. M.p. 80°–90°C; soluble in benzene, trichloromethane, tetrachloromethane.

montasite (*Min.*). An asbestiform variety of the amphibole grunerite. It differs from *amosite* (q.v.) in having less harsh and more silky fibres.

montebrasite (*Min.*). A variety of *amblygonite* (q.v.) in which the amount of hydroxyl exceeds that of fluorine.

Monte Carlo method (*Stats.*). Statistical procedure when mathematical operations are performed on random numbers.

month (*Astron.*). See **anomalistic-, sidereal-, synodic-, tropical-.**

monticellite (*Min.*). A silicate of calcium and magnesium which crystallizes in the orthorhombic system. It occurs in metamorphosed dolomitic limestones and, more rarely, in some ultrabasic igneous rocks.

montmorillonite (*Min.*). A hydrated silicate of aluminium, one of the important clay minerals and the chief constituent of bentonite and fuller's earth. The montmorillonite group of clay minerals are also collectively termed **smectites.**

monuron (*Chem.*). *N′*-(4-chlorophenyl)-*NN*-dimethylurea, used as a weedkiller. Also **CMU.**

monzonite (*Geol.*). A coarse-grained igneous rock of intermediate composition, characterized by approximately equal amounts of orthoclase and plagioclase (near andesine in composition) together with coloured silicates in variety. Named from Monzoni in the Tyrol. Often referred to as **syenodiorite.**

mood (*An. Behav.*). A state of tension ready to activate an animal. See also **drive.**

Moon (*Astron.*). (1) The satellite which revolves about the Earth in a variable orbit at a mean distance of 384 400 km in a period of 1 month. Its diameter is 3476 km, and it rotates in the same period, thus presenting the same face to the earth (but see *librations*). Its density is 3/5, gravity 1/6 and mass 1/81·3 that of the earth. The surface is heavily marked with *maria*, mountain ranges, valleys, clefts and innumerable craters from a few cm to over 200 km in diameter. Surface exploration confirms the absence of water or atmosphere, and there is no measurable magnetic field. Rock specimens collected give evidence of formation by both volcanic forces and meteoric impact, with ages ranging from 3300 to 4500 million years. The internal structure of the Moon appears to be quite different from that of the earth, and it is now regarded as unlikely that the Moon once formed part of the earth. Photographs from spacecraft have shown that there are no large maria on the far side of the moon, which has now been mapped in some detail.

moon blindness (*Med., Vet.*). See **nyctalopia, periodic ophthalmia.**

moonstone (*Min.*). A variety of alkali feldspar, or sometimes plagioclase, which possesses a bluish pearly opalescence attributed to lamellar micro- or crypto-perthitic intergrowth. It is used as a gemstone.

Moore lamp (*Elec. Eng.*). A type of electric discharge lamp, usually in the form of a long tube, using a gas other than mercury or sodium vapour.

mooring anchor (*Ships*). See **bower anchor.**

mooring tower (*Aero.*). A permanent tower for the mooring of airships. Provided with facilities for the transference of passengers and freight and arrangements for replenishing ballast, gas, and fuel.

mopboard (*Build.*). See **skirting board.**

mop-stick hand-rail (*Join.*). A timber hand-rail having a circular section flattened on the under side.

mor (*For.*). A type of humus layer, practically unmixed with the mineral soil, usually showing a well-defined line of demarcation from it.

moraine (*Geol.*). A deposit of *boulder clay* laid down by moving ice. Moraines are found in all areas which have been glaciated, and are of several types. Terminal moraines are irregular ridges of material marking the farthest extent of the ice and representing debris pushed along in front of the ice. Lateral moraines are found along the sides of present-day and former glaciers. Medial moraines form by the combination of two lateral moraines when two glaciers join. Ground moraine is an irregular sheet of boulder clay laid down beneath an ice sheet.

morbid (*Med.*). Diseased: pertaining to, or of the nature of, disease.

morbidity (*Med.*). The state of being diseased: the sick-rate in a community.

morbilli (*Med.*). See **measles.**

morbus maculosus (*Med.*). See **purpura haemorrhagica.**

mordanting (*Photog.*). Adding to a silver image, or replacing a silver image by, a substance having the requisite affinity for a specified dye, which silver in itself has not.

mordants (*Chem.*). Substances, chiefly the weakly basic hydroxides of aluminium, chromium, and iron, which combine with and fix a dyestuff on the fibre in cases where the fibre cannot be dyed direct. The product formed by the action of a dye on a mordant is called a *lake* (q.v.). (*Paint.*) Preparations applied to surfaces to assist paint or gold-leaf to adhere thereto.

mordenite (*Min.*). A hydrated sodium, potassium, calcium, aluminium silicate. A zeolite crystallizing in the orthorhombic system and occurring in amygdales in igneous rocks and as a hydration product of volcanic glass.

moreen (*Textiles*). A plain cotton cloth with fine cord running lengthwise and moiré finish; used for linings.

more hug (*Print.*). An increase in the wrap of the paper round the cylinder(s) of web-fed presses.

Morgagni's columns (*Histol.*). Longitudinal folds in the mucous coat lining the rectum.

Morgagni's ventricle (*Zool.*). In the higher Mammals, a paired pocket of the larynx, anterior to the vocal cords, and acting in the *Anthropoidea* as a resonator.

morganite (*Min.*). A pink or rose-coloured variety of beryl (named after J. Pierpont Morgan), obtained chiefly from California and Madagascar; used as a gemstone.

Morgan's canon (*An. Behav.*). The rule that in no case should any aspect of an animal's behaviour be interpreted as the outcome of the existence of a higher psychical faculty if it can be interpreted as the outcome of one standing lower on the psychological scale.

morgen (*Surv.*). S. African area measure; 1 morgen is 2·12 acres (0·85 ha).

moriform (*Bot.*). Mulberry-like.

morion (*Min.*). A variety of smoky-quartz which is almost black in colour.

morning glory column (*Build.*). A slim tapering hollow concrete shaft topped by a shallow flaring circular cap which supports the floor above.

morning star (*Astron.*). Popularly, a planet, generally Venus or Mercury, seen in the eastern sky at or about sunrise; also, loosely, any planet which transits after midnight.

morocco (*Leather*). Tanned goatskins, finished by glazing or polishing; used in upholstery and high-class bookbinding, etc.

moron (*Psychiat.*). A mentally retarded person with an IQ between 50 and 70.

morph-, morpho-, -morph. Prefix and suffix from Gk. *morphē*, form.

morphallaxis (*Zool.*). Change of form during regeneration of parts, as the development of an antenna in certain *Crustacea* to replace an eye; gradual growth or development.

morphine (*Pharm.*). $C_{17}H_{19}NO_3$, the principal alkaloid present in opium. Characterized by containing a phenanthrene nucleus in addition to a nitrogen ring. M.p. 247°–254°C. Extensively used as a hypnotic to obtain relief from pain. Its formula is:

morpho-. See **morph-**.

morphoea, morphea (*Med.*). Localized scleroderma. A disease in which thickened pinkish patches appear on the skin of the trunk.

morphogenesis (*Zool.*). The origin and development of a part, organ, or organism. *adj.* **morphogenetic.**

morpholine (*Chem.*). A 6-membered heterocyclic compound of the formula:

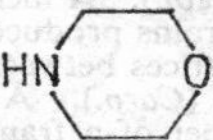

It is a liquid, b.p. 128°C, with strong basic properties. Miscible with water and used as a solvent in organic reactions.

morphology (*Bot., Zool.*). (1) The study of the structure and form of organisms, as opposed to the study of their functions. (2) By extension, the nature of a member. (*Geol.*) The study of the shapes and contours of objects, especially of the surface of the earth. *adj.* **morphological.**

morphopoiesis (*Zool.*). The formation of a structure, e.g., an organelle or virus, from a limited number of sub-units. The distinction from *morphogenesis* (q.v.) depends on the limitation.

morphosis (*Zool.*). The development of structural characteristics; tissue formation. *adj.* **morphotic.**

morphothion (*Chem.*). Dimethyl *S*-(morpholinocarbonylmethyl) phosphorothiolothionate, used as an insecticide.

morphotropy (*Chem.*). The change in crystalline form produced by replacing certain atoms or radicals in a crystal by others.

Morquio-Brailsford disease (*Med.*). An hereditary disease characterized by dwarfism, kyphosis, and skeletal defects in the hip joint.

Morrison Series (*Geol.*). Part of the Comanchean, consisting of flood-plain, fluvatile deposits occurring in the Colorado region.

Morse code (*Teleg.*). System used in signalling or telegraphy, which consists of various combinations of dots and dashes.

Morse equation (*Nuc.*). An equation which relates the potential energy of a diatomic molecule to the internuclear distance.

Morse key (*Teleg.*). A hand-operated device that opens and closes contacts which modulate currents with coded telegraph signals. See also **bug key.**

Morse taper (*Eng.*). A system of matching tapered shanks and sockets used for holding tools in lathes, drills, etc. There are six sizes, with tapers approx. 3°.

mortality (*Ecol.*). The death of individuals in a population. Ecological or realized mortality, i.e., the loss of individuals under particular conditions, varies with these conditions, but a theoretical minimum is constant for a population, representing the loss under ideal conditions when all individuals die of old age. Cf. *natality*.

mortar (*Build., Civ. Eng.*). A pasty substance formed normally by the mixing of cement, sand, and water or cement, lime sand, and water in varying proportions. Used normally for the binding of brickwork or masonry. Hardens on setting and forms the bond between the bricks or stones. (*Chem.*) A bowl, made of porcelain, glass, or agate, in which solids are ground up with a pestle.

mortar board (*Build.*). See **hawk.**

mortar fruit (*Bot.*). A structure consisting of a persistent calyx from which the true fruits are thrown out by the wind or by shaking caused by animals.

mortar mill (*Build.*). An appliance in which the ingredients of a mortar mix are crushed mechanically by two rollers running on the ends of a horizontal bar rotating about a central vertical axis, the rollers running around a shallow pan containing the ingredients. Also called **pan-mill mixer.**

mortar structure (*Geol.*). A mechanical structure in which small grains produced by granulation occupy the interstices between larger grains.

mortise or **mortice** (*Carp.*). A rectangular hole cut in one member of a framework to receive a corresponding projection on the mating member.

mortise-and-tenon joint (*Carp.*). A framing joint between a *mortise* (q.v.) and a *tenon* (q.v.).

mortise bolt (*Join.*). A bolt which is housed in a mortise in a door so as to be flush with its edge.

mortise chisel (*Carp., Join.*). A more robust type of chisel than the *firmer*, for use in cutting mortises and therefore designed to withstand blows from a mallet. Also called **framing chisel, heading chisel.**

mortise gauge (*Carp.*). A tool similar to the *marking gauge* (q.v.) but having an additional marking pin, which is adjustable for position along the bar and allows parallel lines to be set out in marking tenons and mortises. Also called **counter gauge.**

mortise joint (*Carp.*). See **mortise-and-tenon joint.**

mortise lock (*Join.*). A lock sunk into a mortise in the edge of a door.

Morton wave current (*Med.*). Interrupted current from the positive electrode of an electrostatic machine, being limited for therapeutic use by a shunt spark-gap.

mortuary fat (*Med.*). See **adipocere.**

morula (*Zool.*). A solid spherical mass of cells resulting from the cleavage of an ovum.

Morvan's disease (*Med.*). A form of *syringomyelia* (q.v.) in which there is complete loss of sensibility in, and wasting of muscles of, hands and feet, accompanied by cyanosis of these parts and, in the early stages, severe pain.

mosaic (*Bot.*). (1) The arrangement of leaves in a pattern in such a way that each leaf is very little covered by its immediate neighbours. (2) A virus-caused disease, indicated chiefly by chlorosis or mottling of the leaves. (*Build.*) Inlaid work on plaster or stone, formed with small cubes (tesserae) or irregular-shaped fragments of marble, glazed pottery, or glass. (*Chem.*) Structure of crystalline solids due to dislocations, consisting of an irregular matrix or mosaic of otherwise perfect crystallites. (*Nuc.*) The reconstruction of the track of a nuclear particle through a stack of photographic emulsions. (*TV*) Photoelectric surface made up of a large number of infinitesimal granules of photoemissive material deposited on an insulating support. Used as emitting electrode in some forms of electron camera, e.g., iconoscope. Mosaics of piezoelectric crystal elements are employed in ultrasonic cameras. (*Zool.*) Animal chimaera.

mosaic development (*Zool.*). Development when the eventual fate of cells of the developing embryo is determined, e.g., on formation of the blastomeres in the spirally segmented eggs of some Invertebrates.

mosaic egg (*Zool.*). Egg which shows mosaic development, i.e., where areas of future functional development are determined in the early stages of cleavage.

mosaic gold (*Chem.*). Complex tin(IV)sulphide obtained by heating dry tin amalgam, ammonium chloride, and sulphur in a retort. Sometimes used as a pigment.

mosaic image (*Zool.*). The type of image formed in a compound eye by apposition of the separate images formed by the various facets. Also **apposition image.** Cf. *superposition image.*

mosaicism (*Med.*). A condition descriptive of individuals who possess cell populations with mixed numbers of chromosomes, e.g., some cells have 45, others 47, making a mosaic of 45/47.

mosaic screen (*Photog.*). The particoloured screen of fine pattern used for separating the component colours in colour photography.

mosaic structure (*Met.*). Discontinuous structure of a compound or metal into minute domains, each bounded by its discontinuity lattice at the interface with other domains of like composition.

mosaic telegraphy (*Teleg.*). System of alphabetic telegraphy in which the characters are formed as mosaics made up from units transmitted as individual signal elements.

mosaic vision (*Zool.*). The mode of vision of a compound eye when the pigment is extended. A mosaic or apposition image is formed which is composed of as many points of light as there are visual elements.

Moscicki capacitor (*Elec. Eng.*). A capacitor on the principle of the Leyden jar; sometimes used on transmission lines to act as a protective device against effects of high-frequency surges.

Moseley's law (*Nuc.*). The frequencies of the characteristic X-rays of the elements show a strict linear relationship with the square of the atomic number. This result of Moseley's researches stressed the importance of atomic number rather than of atomic weight. See **K-series.**

mosquito (*Zool.*). The family *Culicidae* of the *Diptera*; have piercing probosces and suck blood, often transmitting diseases (e.g., malaria, yellow fever, elephantiasis) while doing so; larvae and pupae aquatic, e.g., Anopheles, Culex, Aedes.

moss agate (*Min.*). A variegated cryptocrystalline silica containing visible impurities, as manganese(IV)oxide, in mosslike or dendritic form. Also called **Mocha stone.**

Mössbauer effect (*Nuc.*). Modification of gamma-ray absorption spectrum due to change in frequency of radiation produced by motion of source (analogous to *Doppler effect* in sound). The effect is also shown when the source is stationary and the absorber moves.

mossite (*Min.*). See **tapiolite.**

most economical range (*Aero.*). The range obtainable when the aircraft is flown at the height, air speed, and engine conditions which give the lowest fuel consumption for the aircraft weight and the wind conditions prevailing.

moth (*Zool.*). The Heterocera, popular but unscientific division of Lepidoptera, inexactly distinguished from butterflies.

mother (*Acous.*). Metal positive which is made from the master in gramophone-record manufacture. See **matrix.**

mother cell (*Bot., Zool.*). A cell which divides to give daughter cells; the term is applied particularly to cells which divide to give spores, pollen grains, gametes, and blood corpuscles.

mother liquor (*Chem.*). The solution remaining after a salt has been crystallized out.

mother of emerald (*Min.*). A variety of *prase*, a leek-green quartz owing its colour to included fibres of actinolite; thought at one time to be the mother-rock of emerald.

mother rod (*Eng.*). See **master connecting-rod.**

mother set (*Typog.*). A set of printing plates

(e.g., of a standard reference work) kept solely for the purpose of electro- or stereotyping, as required, further sets therefrom. Not used for printing.

motional impedance (*Telecomm.*). In an electro-mechanical transducer, e.g., a telephone receiver or relay, that part of the input electrical impedance due to the motion of the mechanism; the difference between the input electrical impedance when the mechanical system is allowed to oscillate and the same impedance when the mechanical system is stopped from moving, or blocked.

motion bars (*Eng.*). See **guide bars.**

motion block (*Eng.*). In some steam-engine valve-gears, a block attached to the valve rod, held by or constrained to move in a circular path by a curved slotted link. See **Joy's valve-gear, Walschaert's valve-gear.**

motion-picture camera (*Cinema.*). A camera, with intermittent motion, in which frames are exposed successively, for subsequent projection after processing. Normal operating-speed is 24 frames per second, for amateur films 16 frames per second is acceptable.

motion study (*Work Study*). A technical study of the essential movements of a workman in performing a given piece of work.

motion work (*Horol.*). The auxiliary train of wheels, normally under the dial, which gives the correct relative motion to the hour and minute hands.

motivation (*An. Behav.*). The broadest class of nonstimulus variables controlling behaviour, including drive. (*Psychol.*) That which induces action. Much modern theory rejects reference, in human behaviour, to instinct and many regard the problem as having two parts; generalized drive and the direction of action. The identification of the former with *arousal* is also subject to controversy. See **reticular activating system.**

motoneuron(e) (*Biol.*). A motor neuron(e).

motor (*Bot., Zool.*). Pertaining to movement; as nerves which convey movement-initiating impulses to the muscles from the central nervous system. (*Eng.*) A machine used to transform power into mechanical form from some other form.

motor areas (*Zool.*). Nerve-centres of the brain concerned with the initiation and correlation of movement.

motor-boating (*Radio*). Very-low-frequency relaxation oscillation in an amplifier, arising from inadequate decoupling of common sources of current supply.

motor-bogie (*Elec. Eng.*). A bogie or truck on a railway locomotive or motor-coach which carries one or more electric motors.

motor cell (*Bot.*). One of a number of cells which together can expand or contract and so cause movement in a plant member.

motor converter (*Elec. Eng.*). A form of converter in which an induction motor, to which the a.c. supply is connected, is combined with a synchronous converter which is connected to the d.c. circuit, the armature winding of the induction motor being connected directly to that of the synchronous converter.

motor end plates (*Zool.*). The special end-organ in which a motor nerve terminates in a striated muscle.

motor generator (*Elec. Eng.*). A converter consisting of a motor connected to a supply of one voltage, frequency, or number of phases, and a generator providing output power to a system of different voltage, frequency, or number of phases, the motor and generator being mechanically connected.

motor-generator locomotive (*Elec. Eng.*). A type of electric locomotive on which is mounted a motor generator for converting current of one type supplied by the contact wire, e.g. 1-phase a.c., to current of another type, e.g. d.c., for supply to the traction motors.

motorium (*Zool.*). See **motor areas.**

motorized fuel valve (*Heat*). See **adjustable-port proportioning valve.**

motor meter (*Elec. Eng.*). An integrating meter embodying a motor whose speed is proportional to the power flowing in the circuit to which it is connected, so that the number of revolutions made by the spindle is proportional to the energy consumed by the circuit. See also **mercury motor meter, induction meter.**

motor oculi (*Zool.*). See **oculomotor.**

motor-operated switch (or circuit-breaker) (*Elec. Eng.*). A large switch or circuit-breaker which is closed by means of an electric motor. Cf. *solenoid-operated switch.*

motor spirit (*Chem., Eng.*). Normal British term for a light hydrocarbon liquid fuel for spark-ignition engines. Other terms for such fuel are petrol, gasoline, motor fuel. Modern motor spirits are blends of several products of petroleum, such as straight-run distillate, thermal and catalytically-cracked gasoline alkylates, isomerates, benzene (benzole) and, rarely, alcohol. It boils in the range of 30–200°C. An important measurement point is the engine test or octane rating of the fuel.

motor starter (*Elec. Eng.*). A device for operating the necessary circuits for starting and accelerating to full speed an electric motor, but not for controlling its speed when running. See **controller.**

motor system (*Bot.*). A general name for the tissues and structures concerned in the movements of plant members.

motorway (*Civ. Eng.*). A *dual carriageway* (q.v.), with hard shoulders; subject to various legal regulations and vehicle restrictions.

mottle (*Textiles*). Two yarns of different colours but similar counts folded together, the opposite way.

mottled clay (*Geol.*). Variegated clay rock, found, for example, in the Keuper Marls. The mottling results from oxidation of iron compounds in localized patches.

mottled iron (*Met.*). Pig-iron in which most of the carbon is combined with iron in the form of cementite (Fe_3C) but in which there is also a small amount of graphite. The fractured pig has a white crystalline fracture with clusters of dark spots, indicating the presence of graphite.

mottled sandstone (*Geol.*). Variegated sandstones found in the Bunter Series of the Triassic System in the Midlands and North of England.

mottramite (*Min.*). Descloizite in which the zinc is almost entirely replaced by copper.

mould (*Bot.*). A popular name for any of numerous small *fungi* (q.v.) appearing on bread, jam, cheese, etc., as a fluffy or woolly growth. See also **dry rot, penicillin.** (*Build.*) Zinc sheet or thin board cut to a given profile; used in running cornices, etc. (*Civ. Eng.*) A temporary construction to support setting concrete in position. Also **form.** (*Paper*) A frame covered with woven or laid wire, on which hand-made paper is made. (*Print.*) An impression of the type or blocks made in flong or plastic, and used for making stereotypes, electrotypes, or rubber or plastic duplicates.

mouldboard (*Agric.*). In the general purposes

plough, a long, gently curving implement which turns over the furrow slice cut by the coulter and the share. In the digger plough, the mouldboard is short, with an abrupt curve. Also called **breast.**

mouldboard plough (*Agric.*). An implement which by means of a coulter and a share detaches a slice of soil vertically from the undersoil. This is turned over by the *mouldboard* (q.v.), leaving an unbroken furrow slice, or, in the case of the digger plough, a flat broken surface. See also **disk plough, tractor plough.**

moulded breadth (*Ships*). The breadth over the frames of a ship, i.e., heel of frame to heel of frame. It is the breadth termed *B* by Lloyd's Register, and is the line faired in the moulding loft. Measured at the widest part of the hull.

moulded depth (*Ships*). The depth of a ship from the top of the keel to the top of the beam at side, measured at midlength; referred to as Lloyd's *D*.

moulded dimensions (*Ships*). The dimensions used in the calculation and tabulation of the scantlings (sizes) of the various members of the structure. They are *length between perpendiculars, moulded breadth,* and *moulded depth* (qq.v.).

moulded-in-place concrete piles (*Civ. Eng.*). See **cast-in-situ concrete piles.**

moulded-intake belt course (*Build.*). An *intake belt course* (q.v.) shaped along the projecting corner to a more or less ornamental profile.

moulding. See **foundry, machine-, moulding sands, pipe-, plate-.** (*Build.*) A more or less ornamental band projecting from the surface of a wall or other surface. (*Plastics*) The moulding powder is weighed carefully into small containers or preformed in pill machines to pellets. The material is then placed in a steel die, heated to about 160°C, and subjected to pressure of 20–35 MN/m². Wood moulds are used to make articles from heated thermoplastic sheets. See also **blow-, cold-, compression-, injection-, rotational-, transfer-.**

moulding box (*Foundry*). See **flask.**

moulding cutter (*Join.*). A specially shaped cutting tool which, when revolving about its own axis, is capable of cutting a desired moulding profile.

moulding machines (*Foundry*). See **machine moulding.**

moulding plane (*Tools*). Plane with a shaped sole and blade for cutting mouldings. There are two types—*English* (held at an angle) and *Continental* (held upright).

moulding powder (*Plastics*). The finely ground mixture of binder, accelerator, colouring matter, filler, and lubricant which is converted under pressure into the final moulding.

mouldings (*Join.*). Strips of wood cut to a given cross-sectional profile and applied to surfaces required to be decorated.

moulding sands (*Foundry*). Siliceous sands (containing clay or aluminium silicate as a binding agent) possessing naturally or by blending the qualities of fineness, plasticity, adhesiveness, strength, permeability, and refractiveness. See **dry sand, green sand, loam.**

mould machine (*Paper*). See **cylinder-.**

mould oil (*Build., Civ. Eng.*). A substance applied to shuttering to prevent adherence of the concrete.

Mouldrite (*Plastics*). TN for a thermosetting plastic of the urea-formaldehyde type.

Moullin voltmeter (*Radio*). A thermionic voltmeter employing the anode bend principle, in which the anode is connected to one end of the filament, the potential drop across which provides the anode voltage supply.

moult (*Zool.*). See **ecdysis.**

moulting hormone (*Cyt.*). In Insects, a substance produced by the cells of the thoracic gland in response to stimulation by another hormone produced by the neurosecretory cells in the brain, which is released at the corpora cardiaca. It initiates growth of the epidermal cells and the formation of a new cuticle in preparation for moulting. Known as **ecdyson.**

mound (*Civ. Eng.*). An undisturbed hillock left on an excavated site as an indication of the depth of the excavation.

mound breakwater (*Civ. Eng.*). A breakwater formed by depositing rubble to form a mound which eventually rises above water.

mounding (*Med.*). See **myoidema.**

mount (*Print.*). The base on which a printing plate is mounted to bring it to type height. There is a large variety: wood is now considered suitable only for small blocks, there being a preference for all-metal, a mixture of wood and metal, plastic, or the patent precision mounts such as **honeycomb base, Mixed Forme Base, Register Sets.** (*Telecomm.*) That part of a switching tube or cavity which enables it to be connected to a waveguide.

mountain climate (*Geog.*). Climate influenced by altitude and relief. Rarefaction of the atmosphere with altitude reduces the absorption of the sun's rays and reduces the variation between day and night temperatures. Relief causes variations in mean temperatures owing to shading from the sun, causing air currents to rise and cool, producing rainfall on the windward slopes and often warm dry winds on the leeward slopes.

mountain cork (*Min.*). A variety of asbestos which consists of thick interlaced fibres. It is light and will float, and is of a white or grey colour.

mountain effect (*Radio*). Error or uncertainty produced in radio direction finding by reflection of signals from mountains.

mountain leather (*Min.*). A variety of asbestos which consists of thin flexible sheets made of interlaced fibres.

mountain wood (*Min.*). A compact fibrous variety of asbestos looking like dry wood.

mountants (*Photog.*). Special adhesives for fixing prints on mounts, free from chemicals which might attack the silver or other image during the course of time.

mouse roller (*Print.*). A small extra roller used to obtain better distribution of the ink on a machine.

mousseline-de-laine (*Textiles*). An all-wool muslin-type fabric made from yarn spun on the worsted system. The weave can be plain or 2×1 twill. Printed by screen or roller methods. Cheaper type has cotton warp.

mouth (*Acous.*). See **flare.** (*Carp., etc.*) The slot in the body of a plane into which the cutting iron fits.

mouth-parts (*Zool.*). In *Arthropoda*, the appendages associated with the mouth.

movable types (*Typog.*). Single types, as distinguished from Linotype slugs or from blocks.

movable weir (*Civ. Eng.*). A temporary weir capable of being removed from the river channel in times of flood.

move (*Glass*). A fixed number of articles to be made for a given rate of pay by a *chair* (q.v.).

movement (*Horol.*). The mechanism of a clock or watch, not including the case or dial.

movement area (*Aero.*). That part of an aerodrome reserved for the take-off, landing, and movement of aircraft.

Movieflood lamp (*Cinema.*). Lamp similar to *Photoflood* (q.v.), used in motion picture photography.

Movietone (*Cinema.*). The original method of recording sound on film, in which the variable light from a *glowlamp* (q.v.) illuminates a constant slit sliding on the surface of the film, with a resulting variable-density track.

Movil (*Plastics*). Synthetic fibre based on polyvinyl chloride (PVC). Its relatively low softening point (about 100°C) restricts its use as a textile fibre, but it is widely used on the Continent.

moving-coil galvanometer (*Elec. Eng.*). A galvanometer depending for its action on the movement of a current-carrying coil in a magnetic field. Cf. *moving-magnet galvanometer*.

moving-coil instrument (*Elec. Eng.*). An electrical measuring instrument depending for its action upon the force on a current-carrying coil in the field of a permanent magnet.

moving-coil loudspeaker (*Acous.*). The inverse of *moving-coil microphone* except that the coil is attached to a radiating cone.

moving-coil microphone (*Acous.*). One in which an e.m.f. is induced in a coil attached to a diaphragm, the coil being in the radial field of a small pot magnet. See **Sykes microphone.** Also called **dynamic microphone, electrodynamic microphone.**

moving-coil regulator (*Elec. Eng.*). A type of voltage regulator, for use on a.c. circuits, in which a short-circuited coil is made to move up and down the iron core of a specially arranged autotransformer.

moving-coil transformer (*Elec. Eng.*). A type of transformer, occasionally used in constant-current systems, in which one coil is made to move relatively to the other for regulating purposes.

moving-coil voltmeter (*Elec. Eng.*). One constructed like a galvanometer and used for d.c. measurements.

moving-conductor microphone (*Acous.*). One in which the conductor is stretched between the wedge poles of a permanent magnet; also **ribbon microphone.**

moving-field therapy (*Radiol.*). A form of cross-fire technique in which there is relative movement between the beam of radiation and the patient, so that the entry portal of the beam is constantly changing. See **converging-field therapy, pendulum therapy, rotational therapy.**

moving-flap liquid meter (*Eng.*). An impact flow-meter consisting of a weighted gate pivoted on a horizontal axis and surrounded by a mechanical shield, so that for any flow rate the gate assumes a definite position which may be transmitted to a recorder.

moving form (*Civ. Eng.*). See **climbing form.**

moving-iron (*Elec. Eng.*). Descriptive of one-time drive in microphones and loudspeakers, which depends on the motion of magnetic material which is part of a magnetic circuit.

moving-iron voltmeter (*Elec. Eng.*). One used for a.c. measurements, depending on attraction of a soft iron vane into the magnetic field due to current.

moving load (*Civ. Eng.*). A variable loading on a structure, consisting of the pedestrians or vehicles passing over it. Also **live load.**

moving-magnet galvanometer (*Elec. Eng.*). A galvanometer depending for its action on the movement of a small permanent magnet in the magnetic field produced by the current to be measured. Cf. *moving-coil galvanometer*.

moving period (*Cinema.*). In describing an intermittent mechanism, the period (expressed in degrees or as a fraction of the whole cycle of operation) during which the film is actually in motion.

moving-target indicator (*Radar*). A device for restricting the display of information to moving targets. Abbrev. MTI.

moving trihedral (*Maths.*). Three mutually orthogonal unit vectors **t, n** and **b** along the tangent, principal normal and binormal respectively at a point P on a space curve. The three planes perpendicular to the three vectors are called respectively the *normal, tangent* and *osculating* planes. The triad is determined by two limits, i.e., (1) the limiting position of a chord PQ as Q approaches P, which determines **t**, and (2) the limiting position of a plane PQR as Q and R approach P, which determines the osculating plane. Cf. *curvature*. $\mathbf{t}=\mathbf{n}\wedge\mathbf{b}$, $\mathbf{n}=\mathbf{b}\wedge\mathbf{t}$.

mower (*Agric.*). See **horse mower, tractor mower.**

Mowlem (*Build., Civ. Eng.*). A proprietary form of low-level system building.

m.p. (*Chem.*). Abbrev. for *melting point*.

Mpc (*Astron.*). Abbrev. for *megaparsec*.

m.p.h. Abbrev. for *miles per hour*.

MPD (*Elec. Eng.*). Abbrev. for *magneto-plasmadynamic*. (*Radiol.*) Abbrev. for *maximum permissible dose*.

M.Q. (*Photog.*). An abbrev. for *metol-quinol* or *metol-hydroquinone* developers. See **4-methyl-aminophenol.**

m-RNA (*Biochem.*). A short-lived, transient form of RNA, called **messenger RNA,** which serves to carry genetic information from the DNA of the genes to the ribosomes where the requisite proteins are made. See **transcription.**

MS (*Aero.*). See **Mil-Specs.**

MSA particle-size analyser (*Powder Tech.*). A proprietary centrifugal particle-size analyser based on the Whitby column technique. Suspension is floated on to the tops of clear fluid in a centrifuge tube which has a capillary tube base. Amounts sedimented after various times of centrifuging are calculated from the height of the slurry in this capillary.

MSC (*Build., Civ. Eng.*). A method of system building developed in U.K. It combines structural steelwork with *in situ* concrete floors which have a galvanized sheet metal deck both as shuttering and reinforcement. Cladding can be composed of prefabricated timber framing with coloured asbestos cement, fixed externally, and with polystyrene, aluminium foil and plasterboard internally. See **system building.** (*Telecomm.*) Abbrev. for *mile of standard cable*.

M-shell (*Nuc.*). See **K-, L-, M- . . . shells.**

MsTh (*Chem.*). The symbol for *mesothorium*.

MSW (*Radar*). Abbrev. for *magnetic surface wave*.

M-synchronization (*Photog.*). See **flash-synchronized.**

M-test of Weierstrass for uniform convergence (*Maths.*). Both the series $\Sigma u_n(x)$ and the product $\Pi(1+u_n(x))$ are uniformly convergent in the interval (a, b) if it is possible to find a convergent series of positive constants ΣM_n, such that $|u_n(x)| \leqq M_n$, for all values of x in the interval (a, b).

MTI (*Radar*). Abbrev. for *moving-target indicator*.

mucedineous, mucedinous (*Bot.*). Mouldlike; white and cottony.

mucic acid (*Chem.*). HOOC·(CHOH)$_4$·COOH, an acid obtained by the oxidation of galactose.

mucid (*Bot.*). Musty or mouldy.

mucigen (*Zool.*). A substance occurring as granules or globules in chalice cells and later extruded as mucin.

mucilages (*Chem.*). Complex organic compounds related to the polysaccharides, of vegetable origin, and having gluelike properties.

mucilaginous (*Bot., Zool.*). Pertaining to, containing, or resembling mucilage or mucin.

mucinogen (*Zool.*). A substance producing or being the precursor of mucin, e.g., granules of mucous gland cells.

mucins (*Chem.*). A group of glycoproteins occurring in mucus and saliva and widely distributed in nature. They are acid in character and are not coagulable by heat.

mu-circuit (*Telecomm.*). The part of a feedback amplifier in which the vectorial sum of the input signal and that of the feedback portion of the output is amplified.

muck (*Mining*). See **dirt.**

mucking or mucking out (*Mining*). See **lashing.**

mucocele (*Med.*). A localized accumulation of mucous secretion, in, e.g., a hollow organ, the outlet of which is blocked.

mucoids (*Chem.*). A group of glycoproteins.

mucomembranous colic (*Med.*). A condition in which constipation is associated with abdominal pain and the passing in the stools of membranes or casts of mucus.

muconic acid (*Chem.*). (but 1,2:3,4 dien 1,4-dicarboxylic acid.) HOOC·CH=CH·CH=CH·COOH. Often obtained by cleavage of naturally occurring aromatic amino acids. A white crystalline solid; m.p. about 260°C, soluble in ethanol and in ethanoic acid.

mucopolysaccharides (*Chem.*). Heteropolysaccharides each containing a hexosamine in its characteristic repeating disaccharide unit.

mucoproteins (*Chem.*). Conjugated proteins containing a carbohydrate group, which may include hexoses, hexosamines, or glucuronic acids.

mucopurulent (*Med.*). Consisting of mucus and pus.

Mucorales (*Bot.*). An order of fungi included in the *Phycomycetes* and characterized by a mycelium of rapidly growing sparsely branched hyphae. Commonly known as moulds. Includes many saprophytic and a few weakly parasitic species, e.g., *Mucor*.

mucosa (*Zool.*). See **mucous membrane.**

mucosal disease (*Vet.*). An acute or chronic, often fatal, virus disease of cattle manifested by inflammation and ulceration throughout the alimentary tract, and characterized by fever, nasal discharge and diarrhoea.

mucosanguineous (*Med.*). Consisting of mucus and blood.

mucous glands (*Zool.*). Glands secreting or producing mucus.

mucous membrane (*Zool.*). A tissue layer found lining various tubular cavities of the body (as the gut, uterus, trachea, etc.). It is composed of a layer of epithelium containing numerous unicellular mucous glands and an underlying layer of areolar and lymphoid tissue, separated by a basement membrane. Also **mucosa.**

muco-viscidosis (*Med.*). An inherited, recessive type disease of young children in which all mucus glands in the body produce a thick viscid secretion obstructing the ducts. Also called **cystic fibrosis.**

mucro (*Bot.*). A short sharp point formed by a continuation of the midrib. (*Zool.*) Any sharp spiny structure, as the clawlike process of each dens of the furcula in *Collembola*.

mucronate (*Bot.*). Said of a leaf tipped by a short sharp point of similar texture to the leaf.

mucus (*Zool.*). The viscous slimy fluid secreted by the mucous glands. *adjs.* **mucous, mucoid, muciform.**

mud (*Geol.*). A fine-grained unconsolidated rock, of the clay grade, often with a high percentage of water present. It may consist of several minerals.

mud acid (*Mining*). Inorganic acids and chemicals used to promote flow in oil wells by acid treatment.

mud drum (*Eng.*). A vessel placed at the lowest part of a steam-boiler: a similar plant to intercept and retain insoluble matter or sludge, as the lowest drum of a *water-tube boiler* (q.v.).

mud fever (*Vet.*). See **grease.**

mud hole (*Eng.*). A *hand hole* (q.v.) in a mud drum, or in the bottom of a boiler, for the removal of scale and sludge.

mudline (*Min. Proc.*). Separating line between clear overflow water and settled slurry in dewatering or thickening plant.

mud pump (*Mining*). One used in drilling of deep holes (e.g., oil wells) to force thixotropic mud to bottom of hole and flush out rock chips.

mudstone (*Geol.*). An argillaceous sedimentary rock characterized by the absence of obvious stratification. Cf. *shale*.

mud volcano (*Geol.*). A conical hill formed by the accumulation of fine mud which is emitted, together with various gases, from an orifice in the ground.

MUF (*Radio*). Abbrev. for *maximum usable frequency*.

mu-factor (*Electronics*). *Voltage amplification factor* of grid-to-anode (or another grid) of a valve, current and all other voltages being constant.

muff (or box) coupling (*Eng.*). A shaft coupling consisting of a sleeve, split longitudinally, which embraces and is keyed to the shafts, the halves being bolted together. Also **butt coupling.**

muffle furnace (*Met.*). A furnace in which heat is applied to the outside of a refractory chamber containing the charge.

muffler (*I.C. Engs.*). U.S. for **silencer.**

mugearite (*Geol.*). A dark, finely crystalline, basic igneous rock which contains oligoclase, orthoclase, and usually olivine in greater amount than augite. Occurs typically at Mugeary in Skye.

mule (*Spinning*). Machine, intermittent in action, which first drafts and twists the yarn on the outward run, then on the inward run winds it in a spindle. It can spin very fine counts but there are also woollen mules for heavy yarns and condenser mules for waste yarns. (Invented by Samuel Crompton, Bolton, ca. 1770.)

mull (*For.*). A type of humus layer with a loose crumb structure, mingled with the mineral soil and associated with relatively low acidity, high nutrient supplies, and rich soil fauna. (*Textiles*) A lightweight plain open cotton fabric, generally pure sized and bleached. Used for clothing and in bookbinding.

Mullen burst test (*Paper*). The *burst test* (q.v.) when performed on a Mullen instrument.

muller (*Min. Proc.*). Pestle, dragstone, iron wearing plate or shoe used to crush and/or abrade rock entrained between it and a base-plate over which it is moved.

Müllerian duct (*Zool.*). A duct which arises close beside the oviduct, or which, by the actual longitudinal division of the archinephric duct, in many female Vertebrates becomes the oviduct.

Müllerian mimicry (*Zool.*). Resemblance in colour between two animals, both benefiting by the resemblance. Cf. *Batesian mimicry*.
Müller's fibres (*Zool.*). Long neuroglial cells occurring in the retina of the Vertebrate eye.
Müller's fluid (*Micros.*). A slow acting fixative, which also acts as a mordant. It contains potassium bichromate and sodium sulphate.
Müller's glass (*Min.*). See **hyalite.**
Müller's larva (*Zool.*). The planktonic larva of some Polyclad Turbellarians, which has eight ciliated lappets by which it swims for a few days before settling down to crawl on the sea bottom.
Müller's law (*Zool.*). Each nerve of special sense gives rise to its own peculiar sensation, however excited.
Müller's muscle (*Zool.*). The circular ciliary muscle of the Vertebrate eye.
Müller's organ (*Zool.*). In *Acridiidae* (*Orthoptera*), a swelling formed by numerous scolophores applied to the inner surface of the tympanum.
mullion (*Join.*). A vertical member of a window frame separating adjacent panes. Also called **monial, munnion.**
mullion structure (*Geol.*). A structure found in severely folded sedimentary and metamorphic rocks in which the harder beds form elongated fluted columns. Mullions are parallel to fold axes.
mullite (*Min.*). A silicate of aluminium, rather similar to sillimanite but with formula close to $3Al_2O_3.2SiO_2$. It occurs in contact-altered argillaceous rocks.
multiaddress (*Comp.*). (1) Technique whereby *words* are directed to temporary locations before being placed in the main *store*; used to reduce the size of the main store. (2) See **multiple-address.**
multiarticulate (*Zool.*). Many-jointed.
multi-break switch (or **circuit-breaker**) (*Elec. Eng.*). A switch or circuit-breaker in which the circuit is broken at two or more points in series on each pole or phase.
multicavity magnetron (*Electronics*). A magnetron in which the anode comprises several cavities, e.g., the *rising-sun magnetron*, in which the resonators are of two different frequencies and are arranged alternately to obtain mode separation.
multicellular (*Biol.*). Consisting of a number of cells.
multicellular voltmeter (*Elec. Eng.*). A form of electrostatic voltmeter in which a number of moving vanes mounted upon the spindle are drawn into the spaces between a corresponding number of fixed vanes.
multichannel (*Telecomm.*). Any system which divides the frequency spectrum of a signal into a number of bands which are separately transmitted, with subsequent recombination.
multichip integrated circuit (*Electronics*). An electronic circuit comprising two or more semiconductor wafers which contain single elements (or simple circuits). These are interconnected to produce a more complex circuit and are encapsulated within a single pack.
multicipital (*Zool.*). Many-headed.
multicuspidate (*Zool.*). Teeth with many cusps.
multi-electrode valve (*Electronics*). One comprising two or more complete electrode systems having independent electron streams in the same envelope. There is sometimes a common electrode, e.g., double-diode.
multi-exchange system (*Teleph.*). A group of local exchanges in an exchange area.
multifee metering (*Teleph.*). Successive operations of a subscriber's meter to correspond with the multi-unit fee chargeable for the call.
multifilament lamp (*Elec. Eng.*). An electric filament-lamp having more than one filament in the same bulb, so that failure of one filament will not cause the lamp to be extinguished.
multifoliolate (*Bot.*). Having many leaflets.
Multifont type tray (*Typog.*). The first British adaptation of the American *Rob Roy* System, using 108 small interchangeable plastic boxes, one for each sort, and arranging them on trays which can be stored in a galley rack, thus taking up much less room than the ordinary type case, either when in use or in store. Several colours of box are used, to distinguish one fount, or range of sorts, from another. *Compacto* is a similar system; *Freelay* is a development of it, using three sizes of box to house an appropriate quantity of each character, and suitable for the larger sizes of type.
multifrequency generator (*Teleg.*). The multifrequency inductor generator which is used for the multichannel voice-frequency telegraph system operated by teleprinter and transmitting over normal telephone lines.
multigap discharger (*Elec. Eng.*). A discharger for use in a spark-type radio transmitter, making use of a number of gaps in series.
multigravida (*Med.*). A woman who has been pregnant more than once.
multigroup theory (*Nuc.*). Theoretical reactor model in which presence of several energy groups among the neutrons is allowed for. Cf. *two-group theory*.
multilayer winding (*Elec. Eng.*). A type of cylindrical winding, used chiefly for transformers, in which several layers of wire are wound one over the other with layers of insulation between.
multilocular, multiloculate (*Bot.*). Having a number of compartments.
Multi-mask (*Print.*). An all-purpose film used for *colour masking* (q.v.).
multinomial distribution (*Stats.*). See **binomial distribution.**
multinucleate (*Bot., Zool.*). With many nuclei.
multiparasitism (*Zool.*). The situation which results when more than one species of parasitoid attack the same host simultaneously. Some, or all, the immature parasitoids die. Cf. *superparasitism, hyperparasitism*.
multiparous (*Zool.*). Bearing many offspring at a birth.
multiple (*Maths.*). If an integer a is a divisor of an integer b, then b is said to be a multiple of a. Formally, if a, b and x are integers such that $ax = b$, then b is a multiple of a and a is a divisor of b. (*Teleph.*) The face of the telephone switchboard in a telephone exchange, containing the outgoing jacks to subscribers' lines, each of which appears within the reach of every operator. The jacks are multipled round the positions by connecting the relevant jacks, which appear in order, in parallel. See **check-.**
multiple-address (*Comp.*). Having several addresses, for the better handling of data.
multiple-circuit winding (*Elec. Eng.*). See **lap winding.**
multiple decay or **multiple disintegration** (*Nuc.*). Branching in a radioactive decay series.
multiple-disk clutch (*Eng.*). A friction clutch similar in principle to the *single-plate clutch* (q.v.), but in which a smaller diameter is obtained by using a number of disks, alternately splined to the driving and driven members, loaded by springs, and usually run in oil. See **friction clutch.**

multiple duct (*Elec. Eng.*). A cable duct having a number of tunnels for the reception of several cables.

multiple echo (*Acous., Radio, etc.*). Perception of a number of distinct repetitions of a signal, because of reflections with differential delays of separate waves following various paths between the source and observer.

multiple effect evaporation (*Chem. Eng.*). A system of evaporation in which the hot vapour produced in one vessel instead of being condensed by cooling water is condensed in the calandria of another evaporator to produce further evaporation. To achieve this, the pressure in second evaporator must be artificially lowered. Can, according to conditions, work as double, triple, or quadruple effect.

multiple-expansion engine (*Eng.*). An engine in which the expansion of the steam or other working fluid is divided into two or more stages, which are performed successively in cylinders of increasing size. See **compound engine, triple-expansion engine, quadruple-expansion engine.**

multiple factors (*Gen.*). Genes which produce joint or cumulative effects.

multiple feeder (*Elec. Eng.*). A feeder consisting of a number of cables connected in parallel; used where a single cable to carry the load would be prohibitively large.

multiple fission (*Zool.*). A method of multiplication found in *Protozoa*, in which the nucleus divides repeatedly without corresponding division of the cytoplasm, which subsequently divides into an equal number of parts leaving usually a residuum of cytoplasm. Cf. *binary fission.*

multiple fruit (*Bot.*). A fruit formed from the flowers of an inflorescence, and not from one flower.

multiple-hearth furnace (*Met.*). A type of roasting furnace consisting of a number of hearths (six or more). The charge enters on the top hearth and passes downwards from hearth to hearth, being rabbled by rotating arms from centre to circumference and circumference to centre of alternate hearths.

multiple hop transmission (*Radio*). Transmission which uses multiple reflection of the sky wave by the ground and ionosphere.

multiple intrusions (*Geol.*). Minor intrusions formed by several successive injections of approximately the same magma.

multiple modulation (*Radio*). The use of a modulated wave for modulating a further independent carrier of much higher frequency.

multiple neuritis (*Med.*). See **polyneuritis.**

multiple-operator welding-unit (*Elec. Eng.*). An electric-arc welding generator or transformer designed to supply current to two or more welding arcs operating in parallel.

multiple personality (*Psychol.*). Rare condition in which two or more differentiated personalities exist in the same person. Each of the personalities takes over for varying lengths of time and may or may not be aware of the other(s). Also called *split personality* which is sometimes taken as referring to schizophrenia, although the lack of integration of the latter makes it an unsuitable name.

multiple point (*Maths.*). A point on the curve $f(x, y)$ at which the first non-zero derivative of f is of order n, is a multiple point of order n. Cf. *double point.*

multiple proportions (*Chem.*). See **law of multiple proportions.**

multiple-retort underfeed stoker (*Eng.*). A number of underfed inclined retorts arranged side by side with tuyères between, resulting in a fuel bed the full width of the furnace walls. See **single-retort underfeed stoker.**

multiple sclerosis (*Med.*). See **disseminated sclerosis.**

multiple-spark system (*Radio*). A form of quenched spark system in which the spark discharge occurs across a series of gaps formed between a number of closely spaced metallic plates.

multiple-spindle drilling machine (*Eng.*). A drilling machine having two or more vertical spindles for simultaneous operation.

multiple-spot scanning (*TV*). Facsimile transmission in which image is divided into two or more sections which are scanned simultaneously.

multiple star (*Astron.*). A system in which three or more stars are so united by gravitational forces as to revolve about a common centre of gravity.

multiple-switch starter (or controller) (*Elec. Eng.*). A starter (or controller) for an electric machine in which the steps of resistance are cut out, or other operations performed, by hand-operated switches.

multiplet (*Bot.*). One of several individuals derived by the segmentation of an ovum.

multiple-threaded screw (*Eng.*). A screw of coarse pitch in which two or more threads are used to reduce the size of thread and maintain adequate core strength. Also called **multi-start thread.** See also **divided pitch, multistart worm.**

multiple T-maze (*An. Behav.*). A series of T-shaped runways with the starting box at the base of the first T, the base of the second T at the end of one arm of the first T, and so on until reaching a goal box at the end of one arm of the final T.

multiple-tool lathe (*Eng.*). A heavy lathe having two large tool posts, one on either side of the work, each carrying several tools operating simultaneously on different parts of the work.

multiplets (*Nuc.*). (1) Optical spectrum lines showing fine structure with several components (i.e., triplets or still more complex structures). (2) See under **isotopic spin.**

multiple-tuned antenna (*Radio*). A transmitting antenna system comprising an extensive horizontal 'roof' with a number of spaced vertical leads, each connected to earth through appropriate tuning circuits, and all tuned to the same frequency, the connexion to the transmitter being made through one of them.

multiple-twin cable (*Telecomm.*). A lead-covered cable in which there are numbers of cores, each comprising two pairs twisted together.

multiple unit or **valve** (*Electronics*). Envelope containing a number of valves, with or without common cathode.

multiple-unit control (*Elec. Eng.*). The method of control by which a number of motors operating in parallel can be controlled from any one of a number of points; used on multiple-unit trains.

multiple-unit steerable antenna (*Radio*). Antenna which can be rotated by varying the phases of the individual units. Abbrev. **musa.**

multiple-unit train (*Elec. Eng.*). An electric or diesel train consisting of a number of motor-coaches and trailers, all controlled from one driving position at the front or rear of the train.

multiple valve (*Electronics*). See **multiple unit.**

multiplex (*Telecomm.*). Use of one channel for several messages by *time-division multiplex* or *frequency division.*

multiplex printing (*Teleg.*). A system which provides for printing the messages received over a multiplex system.

multiplex transmission (*Radio*). Transmission in which two or more signals modulate the carrier wave.

multiplex winding (*Elec. Eng.*). A 2-layer armature winding sometimes used on d.c. machines. It has more than two parallel paths per pole-pair between the positive and negative terminals.

multiplicand (*Maths.*). See **multiplication.**

multiplication (*Bot.*). Increase by vegetative means. (*Electronics*) Secondary emission of electrons due to impact of primary electrons on a surface. It varies from zero with graphite to 1·3 with Ni and up to 12·3 with Ni-Be alloy. (*Maths.*) The process of finding the *product* of two quantities which are called the *multiplicand* and the *multiplier*. Denoted by the multiplication sign ×, or by a dot, or by the mere juxtaposition of two quantities. Some branches of mathematics use special signs, e.g., the vector product of **a** and **b** is indicated by $\mathbf{a} \wedge \mathbf{b}$. Numbers are multiplied in accordance with the usual rules of arithmetic, but multiplication of other mathematical entities has to be defined specifically for the entities concerned. (*Nuc.*) Ratio of neutron flux in subcritical reactor to that supplied by the neutron source alone. Equal to $1/(1-k)$ where k is the *multiplication constant* (q.v.).

multiplication constant (*Nuc.*). The ratio of the average number of neutrons produced by fission in one neutron lifetime to the total number of neutrons absorbed or leaking out in the same interval. Also called **reproduction constant** and denoted by k.

multiplication constant, infinite ($k\infty$) (*Nuc.*). The value of k for an infinitely large reactor (i.e., ignoring leakage).

multiplication point (*Telecomm.*). See **mixing point.**

multiplier (*Comp.*). Unit in analogue computer used to obtain the product of two varying signals. (*Electronics*) See **electron multiplier.** (*Maths.*) See **multiplication.**

multi-ply (*Carp.*). Plywood formed of more than 3 layers of wood.

multiply-connected domain (*Maths.*). See **connected domain.**

multiplying camera (*Photog.*). A camera for taking a number of small exposures on one negative, using a deflecting mirror or a number of lenses which can be traversed.

multiplying constant (*Surv.*). A factor in the computation of distance by tacheometric methods (see **tacheometer**). It is that constant value for the particular instrument by which the staff intercept must be multiplied in order to give the distance of the staff from the focus of the object glass. If the distance from the centre of the instrument is required, it is necessary to add (see **additive constant**) the distance between the focus of the object glass and the centre of the instrument.

multipolar (*Zool.*). Said of nerve cells having many axons. Cf. *bipolar*, *unipolar*.

multipole moments (*Phys.*). These are magnetic and electric, and are measures of the charge, current and magnet (via intrinsic spin) distributions in a given state. These *static* multipole moments determine the interaction of the system with weak external fields. There are also *transition* multipole moments which determine radiative transitions between two states.

multiposition (*Telecomm.*). Said of a system in which output or final control can take on 3 or more preset values.

multiprogramming (*Comp.*). Technique of handling several programs simultaneously by interleaving their execution through time-sharing.

multipropellant (*Space*). Rocket propellant made up of two or more constituent liquids which are kept separate prior to combustion. Cf. *monopropellant*.

multirow radial engine (*Aero.*). A radial *aero-engine* (q.v.) with two or more rows of cylinders.

multiseriate (*Bot.*). (1) Said of ascospores arranged in several rows in the ascus. (2) Said of a vascular ray several to many cells wide.

multispeed supercharger (*Aero.*). A gear-driven *supercharger* (q.v.) in which a clutch system allows engagement of different ratios to suit changes in altitude.

multispindle automatic machine (*Eng.*). A fully-automatic, high-speed, special-purpose lathe primarily for bar work, in which the work sequence is divided so that a portion of it is performed at each of the several spindle stations, one part being completed each time the tools are withdrawn and the spindles indexed.

multistage (*Electronics*). Said of a tube in which electrons are progressively accelerated by anode rings at increasing potentials, also of an amplifier with transistors or valves in series, or cascade. (*Space*) Said of a space-vehicle having successive rocket-firing stages, each capable of being jettisoned after use.

multistage compressor (*Aero.*). A gas turbine compressor with more than one stage; each row of blades in an *axial-flow compressor* (q.v.) is a stage, each *impeller* (q.v.) is a stage in a *centrifugal-flow compressor* (q.v.). In practice, all axial compressors are multistage, while almost all centrifugal compressors are single stage. Occasionally an initial axial stage is combined with a centrifugal delivery stage.

multistage supercharger (*Aero.*). A supercharger with more than one impeller in series.

multistage turbine (*Aero.*). A turbine with two or more disks joined and driving one shaft.

multistart thread (*Eng.*). See **multiple-threaded screw.**

multistart worm (*Eng.*). A worm in which two or more helical threads are used to obtain a larger pitch and hence a higher velocity ratio of the drive.

multitone (*Acous.*). Generator, thermionic or mechanical, which produces a spectrum of currents, i.e., with a large number of components, equally spaced in frequency.

Multitron (*Electronics*). Power amplifier for very-high-frequency pulse operation.

Multituberculata (*Zool.*). An order of *Mesozoic* Mammals (Jurassic to Eocene); herbivorous, and sometimes quite large, with rodentlike incisors and a diastema.

multituberculate (*Zool.*). Said of tuberculate teeth with many cusps: having many small projections.

multi-turn current transformer (*Elec. Eng.*). A current transformer in which there are several turns on the primary winding. Cf. *bar-type current transformer*.

multivibrator (*Telecomm.*). Original Eccles-Jordan free-running circuit, with two stages of resistance-capacitance valve (and now transistor) amplification, back-coupled. Period of oscillatory waveform is ca. sum of time constants of two grid circuits. Easily locked to tuning-fork or sinusoidal drive, producing many harmonics. Normal symmetrical waveform can be markedly altered by varying a grid bias. Without coupling capacitors, circuit becomes bistable and with asymmetrical back coupling may also be monostable.

multivincular (*Zool.*). In Bivalves, said of the hinge of the shell when it is provided with several ligaments.

multivoltine (*Zool.*). Having more than one brood in a year. Cf. *univoltine*.

multiwire antenna (*Radio*). An antenna consisting of a number of horizontal wires in parallel.

multungulate (*Zool.*). Having the hoof divided into three or more parts.

mu-meson (*Nuc.*). An elementary particle once thought to be a *meson* but now known to be a *lepton*. See **muon.**

Mumetal (*Met.*). TN of high-permeability, low-saturation magnetic alloy of about 80% nickel, requiring special heat-treatment to achieve special low-loss properties. Useful in nonpolarized transformer cores, a.c. instruments, small relays. Also for shielding devices from external field as in CROs.

Mummery's plexus (*Zool.*). A network of fine nerve-fibrils lying between the odontoblasts and the dentine in a tooth.

mumps (*Med.*). Epidemic or infectious parotitis. An acute infectious disease characterized by a painful swelling of the parotid gland; caused by a virus.

mundic (*Min.*). See **pyrite.**

mungo (*Textiles*). A low grade of waste recovered from old wool rags, tailor's cuttings, old felt, etc., cleaned and spun again as woollen yarns.

Munsell scale (*Optics*). A scale of chromaticity values giving approximately equal magnitude changes in visual hue.

munting or **muntin** (*Join.*). The vertical framing piece separating the panels of a door.

Muntz metal (*Met.*). Alpha-beta brass, 60% copper and 40% zinc. Stronger than alpha brass and used for castings and hot-worked (rolled, stamped, or extruded) products.

muon (*Nuc.*). Subatomic particle with rest mass equivalent to 106 MeV (the heaviest known *lepton*). Has unit negative charge (*antimuons* have positive charge) and half-life of about 2 μs. Decays into electron, neutrino, and antineutrino (*antimuon* into positron, neutrino, and antineutrino). Participates only in weak interactions without conservation of parity.

muramic acid (*Chem.*). Occurs in bacterial cell walls and consists of N-acetylglucosamine joined by an ethereal linkage to D(−)-lactic acid.

murexide (*Chem.*). The dye ammonium purpurate, used as an indicator for calcium in *complexometric titration* using EDTA.

murexide test (*Chem.*). A test for uric acid in which the substance is heated with nitric acid. The residue of alloxantin thus produced turns purple-red on adding ammonia, owing to the formation of murexide, the ammonium salt of purpuric acid, $C_8H_5N_5O_6$.

Murex process (*Min. Proc.*). Magnetic separation of desired mineral from pulp, in which the ore is mixed with oil and magnetite and then pulped in water. Magnetite clings selectively to the desired mineral.

Murgatroyd belt (*Glass*). A circumferential band in the side wall of a container, extending from the bottom upwards for about 20 mm. In this band, maximum stresses due to thermal shock occur.

muriatic acid (*Chem.*). See **hydrochloric acid.**

muricate (*Bot.*). Having a surface roughened by short, sharp points.

muriform (*Bot.*). Said of a spore made up of a mass of cells formed by divisions in three intersecting planes.

murmur (*Med.*). An irregular sound which follows, accompanies, or replaces the normal heart sounds and often indicates disease of the valves of the heart; similar sound heard over blood-vessels. The sound heard over the lungs during respiration.

Murphree efficiency (*Chem. Eng.*). In the performance of a single plate in a *fractionating column* (q.v.), the ratio of the actual enrichment of the vapour leaving the plate, to the enrichment which would have occurred if there had been sufficient time for the vapour to reach equilibrium with the liquid leaving the plate.

Murray loop (*Telecomm.*). A bridge test circuit used for detecting earth-leakage points in telephone and other similar signal lines.

musa (*Radio*). Abbrev. for *multiple-unit steerable antenna.*

Muschelkalk (*Geol.*). A marine shelly limestone occurring between the rocks of Bunter and Keuper age in the Trias of France and Germany.

Musci (*Bot.*). One of the two main groups of the *Bryophyta*, with some thousands of species. The plants are small, attached to the substratum by rhizoids and not by roots. The stem bears a number of small leaves, both members being of very simple internal structure, containing no woody material. The sexual organs are antheridia and archegonia, borne on the leafy moss plant, and the zygote gives rise to a small spore-bearing plant, a stalked capsule known as the sporogonium, which lives parasitically on the leafy moss plant.

muscle (*Zool.*). A tissue whose cells can contract; a definitive mass of such tissue. See **cardiac-, striated-, unstriated-, voluntary-.**

muscle-spindle (*Histol.*). A sensory receptor in skeletal muscle, consisting of sensory nerve endings and intrafusal muscle fibres within a spindle-shaped connective tissue sheath arranged in parallel with the extrafusal muscle fibres and responding to stretch. The afferent nerve impulses so initiated by the muscle spindle produce spinal reflexes resulting in increased muscle tension of extrafusal fibres (myotatic or stretch reflexes) which are the basis of posture.

muscovite (*Geol., Min.*). The common or white mica; a hydrous silicate of aluminium and potassium, crystallizing in the monoclinic system. It occurs in many geological environments but deposits of economic importance are found in granitic pegmatites. It can be used as an insulator (not above 600°C, when its water of composition is driven off; see *phlogopite*), as a lubricant, or for non-inflammable windows.

muscovite-granite (*Geol.*). A granite which contains a fairly large proportion of muscovite.

Muscovy glass (*Min.*). Formerly a popular name for muscovite.

muscular dystrophy (*Med.*). A hereditary disease marked by hypertrophy or wasting of pelvic and limb muscles. See **myotonia atrophica, pseudohypertrophic paralysis.**

musculature (*Zool.*). The disposition and arrangement of the muscles in the body of an animal.

musculocutaneous (*Zool.*). Pertaining to the muscles and the skin.

mush (*Radio*). Fading and distortion in reception in an area through the interaction of waves from two or more synchronized radio transmitters.

mushroom bodies (*Zool.*). Paired nerve centres

of the protocerebrum in Insects, regarded as the principal association areas. Also known as **corpora pedunculata.**

mushroom construction (*Civ. Eng.*). Reinforced concrete construction composed only of columns and floor slabs, the columns having a spread at their head to counteract shear forces.

mushroom follower (*Eng.*). A cam follower in the form of a mushroom, i.e., with a domed surface, as distinct from a roller-type follower.

mushroom gland (*Zool.*). In male *Orthoptera*, the large compact mushroom-shaped mass of the accessory genital glands.

mushroom loudspeaker (*Acous.*). One in which a vertical horn is fitted with directing baffles, so that the radiated sound power is directed uniformly over a limited area in its neighbourhood.

mushroom valve (*I.C. Engs.*). See **poppet valve.**

mush winding (*Elec. Eng.*). A type of winding used for a.c. machines; the conductors are placed one by one into partially closed ready-lined slots, the end connexions being subsequently insulated separately.

mushy (*Textiles*). Said of wool that is lacking in grease and uneven in staple, the result of dry climate and light soil.

musical echo (*Acous.*). One in which the interval between the reception of successive echoes is so small that the impulses received appear to have the quality of a musical tone.

muskeg (*Bot.*). Mossy growth found in high latitudes, usually as deep and treacherous swamp, bog or marsh. Unstable peat formation.

musk glands (*Zool.*). In some Vertebrates, glands, associated with the genitalia, the secretion of which has an odour of musk.

muslin (*Textiles*). Lightweight, plain cotton cloths of open texture and soft finish, bleached, dyed, or printed. There are also plain white butter, cheese, or meat, etc., muslins for wrapping purposes.

mustard gas (*Chem.*). $(CH_2Cl{\cdot}CH_2)_2S$, dichlorodiethyl sulphide; a poison gas manufactured from ethene and S_2Cl_2. See also **nitrogen mustards.**

mustard oils (*Chem.*). *Iso*-thiocyanates.

mustine (*Chem.*). See **nitrogen mustards.**

mutagen (*Gen.*). A substance that produces mutations.

mutamerism (*Chem.*). The formation of an equilibrium mixture of two isomers from a freshly prepared solution of one of them.

mutant (*Gen.*). See **mutation.**

mutarotation (*Chem.*). The change with time of the optical activity of a freshly prepared solution of an active substance. The sugars are the best-known class to exhibit the phenomenon.

mutation (*Gen.*). A change in the characteristics of an organism produced by an alteration of the hereditary material.

mutation rate (*Gen.*). The frequency of gene mutations in a given species.

mutations (*Acous.*). Stops on an organ manual or pedal which sound notes with intervals other than multiples of an octave with respect to the nominal pitches of the keys.

mute (*Cinema.*). A rush print of the exposed film in a motion-picture camera, i.e., one printed without the sound-track; it is independently printed on another positive and reproduced synchronously in another projection machine when the rushes are viewed.

mute shot (*Cinema.*). A shot taken in a motion-picture studio without sound being recorded.

muticate (*Bot.*). Without a point.

muticous (*Bot.*). Muticate. (*Zool.*) Lacking defensive structures.

muting (*Electronics*). See **desensitizing.**

muting circuit or switch (*Electronics*). Arrangement for attenuating amplifier output unless paralysed by a useful incoming carrier; used with automatic tuning devices.

muton (*Gen.*). The smallest element of a gene capable of giving rise to a new form by mutation.

mutton (*Typog.*). See **em quad.**

mutton rule (*Typog.*). See **em rule.**

mutual capacitance (*Elec.*). Capacitance calculated from the displacement current flowing between two conducting bodies when all adjacent conducting bodies are earthed.

mutual conductance (*Electronics*). Transconductance specifically applied to a thermionic valve. Differential change in a space or anode current divided by differential change of grid potential which causes it. Colloquially termed *slope, slope conductance*, or *goodness* of a valve, measuring the effectiveness of the valve as an amplifier in normal circuits. Expressed in milliamperes per volt, and denoted by g_m.

mutual coupling (*Elec. Eng.*). See **transformer coupling.**

mutual impedance (*Elec. Eng.*). See **transfer impedance.**

mutual inductance (*Elec. Eng.*). Generation of e.m.f. in one system of conductors by a variation of current in another system linked to the first by magnetic flux. The unit is the henry, when the rate of change of current of one ampere per second induces an e.m.f. of one volt.

mutual inductor (*Elec. Eng.*). A component consisting of two coils designed to have a definite mutual inductance (fixed or variable) between them.

mutualism (*Biol.*). Any association between 2 organisms which is beneficial to both and injurious to neither. See **symbiosis.**

muzzle (*Zool.*). See **rhinarium.**

MWE, MWe (*Nuc. Eng.*). Abbrev. for *megawatt electric*; unit for electric power generated by a nuclear reactor.

M-wire (*Teleph.*). See **meter wire.**

myalgia (*Med.*). The sensation of pain in muscle.

myarian (*Zool.*). Based on musculature, as a system of classification: pertaining to the musculature.

myasthenia (*Med.*). Muscular weakness.

myasthenia gravis (*Med.*). A malady of adult life in which there is a variable paralysis of muscles on exertion, with slow recovery during rest, and a permanent paralysis of muscles which succeeds the variable paralysis.

Mycalex (*Elec.*). TN for mica bonded with glass. It is hard, and can be drilled, sawn and polished; has a low power factor at high frequencies, and is a very good insulating material at all frequencies.

mycetocytes (*Zool.*). Cells containing symbiotic micro-organisms occurring in the *pseudovitellus* (q.v.).

mycetoma (*Med.*). See **Madura foot.**

mycetome (*Bot., Zool.*). Pseudovitellus: special organ in some species of Insects, Ticks and Mites, inhabited by intracellular symbionts, e.g., Bacteria, Fungi, Rickettsias.

mycetophagous (*Zool.*). Fungus-eating.

Mycetozoa (*Bot.*). See **Myxomycetes.** (*Zool.*) An order of semiterrestrial *Sarcodina*, the members of which possess numerous blunt pseudopodia but have no shell, skeleton, or central capsule; they form cellulose-coated cysts and spores.

myco-, myc-, myceto-. Prefix from Gk. *mykēs*, gen. *mykētos*, fungus.

mycocecidium (*Bot.*). A gall caused by a fungus.

mycocriny (*Bot.*). The decomposition of plant material by fungi.

mycoderma (*Bot.*). A name sometimes applied to the saccharomyces or *yeasts* (q.v.).

mycology (*Bot.*). The study of fungi.

mycophthorous (*Bot.*). Said of a fungus which is parasitic on another fungus.

Mycoplasmatales (*Bacteriol.*). Organisms of this group are associated with pleuropneumonia in cattle and contagious agalactia in sheep: very variable in form, consisting of cocci, rods, etc., and a cytoplasmic matrix: very similar to the L-forms of true bacteria: nonpathogenic morphologically similar organisms have also been found in animals and man.

mycoplasmosis (*Vet.*). Infection of animals and birds by organisms of the genus *Mycoplasma*, e.g., chronic respiratory disease of fowl, contagious agalactia of sheep and goats, contagious pleuropneumonia of cattle.

mycorrhiza (*Bot.*). A symbiotic association between a fungus and a higher plant, most often consisting of an intimate relation between the roots of the higher plant and the mycelium of the fungus.

mycosis (*Vet.*). A disease of animals caused by a fungus.

mycosis fungoides (*Med.*). A chronic, and usually fatal, disease in which funguslike tumours appear in the skin.

Mycostatin (*Pharm.*). TN for **nystatin.**

mycotic dermatitis of sheep (*Vet.*). Lumpy wool. A dermatitis of sheep due to infection of the skin by the fungus *Dermatophilus dermatonomus*.

mycotrophic plant (*Bot.*). A plant which lives in symbiosis with a fungus, i.e., has mycorrhizas.

mydriasis (*Med.*). Extreme dilatation of the pupil of the eye.

mydriatic (*Med.*). Producing dilatation of the pupil of the eye: any drug which does this.

mydriatic alkaloids (*Med.*). Alkaloids which cause dilatation of the pupil of the eye, e.g., atropine.

myel-, myelo-. Prefix from Gk. *myelos*, marrow.

myelencephalon (*Zool.*). The part of the hind-brain in Vertebrates forming part of the *medulla oblongata*. Cf. *metencephalon*.

myelin (*Zool.*). A white fatty substance which forms the medullary sheath of nerve fibres.

myelination (*Zool.*). Formation of a myelin sheath.

myelin sheath (*Zool.*). See **medullary sheath.**

myelitis (*Med.*). (1) Inflammation of the spinal cord. (2) Inflammation of the bone marrow, but the term *osteomyelitis* is generally used for this.

myelo-. See **myel-.**

myelocele (*Med.*). A condition in which the spinal cord protrudes on to the surface of the body; due to a defect of the spinal vertebrae.

myelocoel (*Zool.*). The central canal of the spinal cord.

myelocyte (*Zool.*). A bone-marrow cell; a large amoeboid cell found in the marrow of the long bones of some higher Vertebrates, and believed to give rise, by division, to leucocytes.

myelography (*Radiol.*). The radiological examination of the space between the theca and the spinal cord, following the injection of air or other contrast media.

myeloid tissue (*Histol.*). Haemopoietic tissue which is responsible for the development of granular leucocytes and erythrocytes.

myeloma (*Med.*). A tumour composed of bone-marrow cells, appearing usually in the marrow of several bones. See also **Kahler's disease.**

myelomalacia (*Med.*). Pathological softening of the spinal cord.

myelomatosis (*Med.*). The occurrence of myelomata in several bones. See **myeloma.**

myelomeningocele (*Med.*). Protrusion of the spinal cord and spinal membranes; due to a defect in the spinal column.

myelopathy (*Med.*). Any disease affecting spinal cord or myeloid tissues.

myeloplast (*Zool.*). A leucocyte of bone marrow.

myeloplax (*Zool.*). A giant cell of bone marrow and other blood-forming organs, sometimes multinucleate, and usually having a central group of centrioles; believed to give rise to the blood-platelets. Also called **megakaryocyte.**

myenteric (*Zool.*). Pertaining to the muscles of the gut; as a sympathetic nerve plexus controlling their movements.

myiasis (*Vet.*). Parasitism of the tissues of animals by the larvae of flies of the suborder *Cyclorrhapha*.

myiophilous plants (*Bot.*). Plants with inconspicuous and often ill-smelling flowers, which are pollinated by flies.

Mylar (*Plastics*). TN for U.S. polyethylene terephthalate polyester film. Very strong with extremely good transparency and electrical properties. See also **Melinex.**

Mylius' test (*Chem.*). A colour test for bile acids, based on the appearance of a red colour between the layers of bile acids mixed with a trace of furfural and concentrated sulphuric acid.

mylohyoid (*Zool.*). Pertaining to, or situated near, the mandible and the hyoid: a muscle of this region.

mylonite (*Geol.*). A hard compact rock with a streaky or banded structure which is produced along thrust-planes in mountain-building movements.

mylonitization (*Geol.*). The process by which rocks are granulated and pulverized and formed into mylonite.

myo-, my-. Prefix from Gk. *mys*, gen. *myos*, muscle.

myoblast (*Zool.*). An embryonic muscle cell which will develop into a muscle fibre.

myocardial (*Med.*). Pertaining to, or affecting, the myocardium.

myocarditis (*Med.*). Inflammation of the muscle of the heart.

myocardium (*Zool.*). In Vertebrates, the muscular wall of the heart.

myochondria (*Zool.*). Granules occurring in irregular masses in sarcoplasm.

myoclonia congenita (*Vet.*). Trembling. Spasmodic muscular twitching in newborn pigs and lambs; the cause is unknown.

myoclonus (*Med.*). (1) Paramyoclonus multiplex. A condition in which there occur sudden shocklike contractions of muscles, often associated with epilepsy and progressive mental deterioration. (2) A sudden shocklike contraction of a muscle (see **clonus**).

myocoel (*Zool.*). The coelomic space within a myotome.

myocomma (*Zool.*). A partition of connective tissue between 2 adjacent myomeres. Also **myoseptum.**

myocyte (*Zool.*). A muscle cell: a deep contractile layer of the ectoplasm of certain *Protozoa*.

myodome (*Zool.*). In some *Neopterygii*, a chamber in which the muscles of the eye are situated.

myo-edema (*Med.*). See **myoidema.**

myoepithelial (*Zool.*). A term used to describe the epithelial cells of *Coelenterata* which are provided with taillike contractile outgrowths at the base.

myofibrillae (*Zool.*). The contractile fibrils of a muscle.

myogen (*Chem.*). An albumin, soluble in water, found in muscle.

myogenic (*Zool.*). Said of contraction arising spontaneously in a muscle, independent of nervous stimuli. cf. *neurogenic.*

myoglobin, myohaemoglobin (*Biochem., Histol.*). A respiratory pigment found in the red fibres of muscle; a form of haemoglobin, differing from haemoglobin in its absorption spectrum, content of globin and rate of combination with gases. Its oxygen dissociation curve is a rectangular hyperbola.

myohypertrophy (*Med.*). Increase in the size of muscle fibres.

myoidema or myo-edema (*Med.*). Mounding. A localized swelling of wasting muscle obtained when the muscle is lightly struck.

myolemma (*Zool.*). See **sarcolemma.**

myology (*Zool.*). The study of muscles.

myoma (*Med.*). A tumour composed of unstriped (leiomyoma) or striped (rhabdomyoma) muscle fibres.

myomalacia cordis (*Med.*). Pathological softening of the muscle of the heart.

myomectomy (*Surg.*). Surgical removal of a myoma, especially of a fibromyoma of the uterus.

myomere (*Zool.*). In metameric animals, the voluntary muscles of a single somite.

myometrium (*Histol.*). The muscular coat of the uterus.

myoneme (*Zool.*). In ***Protozoa,*** a contractile fibril of the ectoplasm.

myoneural (*Zool.*). Pertaining to muscle and nerve, as the junction of a muscle and a nerve.

myopathy (*Med.*). Any one of a number of conditions in which there is progressive wasting of skeletal muscles from no known cause.

myophore (*Zool.*). A structure to which muscle attachments are led; as an apodeme.

myophrisk (*Zool.*). In some *Radiolaria*, contractile structures surrounding the spines at their junction with the *calymma* (q.v.), and used to regulate the diameter of the body.

myopia (*Med.*). Short-sightedness. A condition of the eye in which, with the eye at rest, parallel rays of light come to a focus in front of the retina. *adj.* **myopic.**

myoplasm (*Cyt.*). The contractile part of a muscle cell.

myosarcoma (*Med.*). A malignant tumour composed of muscle cells and sarcoma cells.

myoseptum (*Zool.*). See **myocomma.**

myosin (*Biochem.*). Chief protein of muscle, which combines with actin to form *actomyosin* (q.v.).

myosinogen (*Chem.*). One of the chief proteins contained in the living muscle.

myosis (*Med.*). See **miosis.**

myositis (*Med.*). Inflammation of striped muscle.

myositis ossificans progressiva (*Med.*). A condition in which there is progressive ossification of the muscles of the body.

myotasis (*Physiol.*). Muscular tension. *adj.* **myotatic.**

myotome (*Zool.*). A muscle merome; one of the metameric series of muscle masses in a developing segmented animal.

myotonia atrophica (*Med.*). Dystrophia myotonica. A disease characterized by wasting of certain groups of muscles, difficulty in relaxing muscles after muscular effort, and general debility.

myotonia congenita (*Med.*). Thomsen's disease. A rare and congenital malady characterized by extreme slowness in relaxation of muscles after voluntary effort.

myriametric waves (*Radio*). Waves of frequency less than 30 kHz or of length greater than 10 km.

Myriapoda (*Zool.*). A class of terrestrial *Arthropoda* which breathe by tracheae; they possess uniramous appendages; the head is distinct from the thorax and bears one pair of antennae; the legs are numerous, all alike, and extend the whole length of the body. *Chilopoda* (Centipedes) and *Diplopoda* (Millipedes).

myricyl alcohol (*Chem.*). See **melissyl alcohol.**

myringitis (*Med.*). Inflammation of the drum of the ear.

myringoscope (*Med.*). An instrument for viewing the drum of the ear.

myringotomy (*Surg.*). Incision of the drum of the ear.

myriospored or myriosporous (*Bot.*). Having many spores.

myristic acid (*Chem.*) tetradecanoic acid, $CH_3{\cdot}(CH_2)_{12}{\cdot}COOH$. Crystalline solid; m.p. 58°C. Found (as glycerides) in milk and various vegetable oils.

myrmecochory (*Bot.*). The distribution of seeds by ants (Gk. *myrmex*, ant). *adj.* **myrmecochorous.**

myrmecophagous (*Zool.*). Feeding on ants.

myrmecophile (*Zool.*). See **guest.**

myrmecophilous (*Bot.*). Pollinated by ants.

myrmecophily (*Bot.*). A symbiotic association between plants and ants.

myrmekite (*Min.*). A wartlike intergrowth of plagioclase and vermicular quartz, usually associated with potassium feldspar.

myrtenol (*Chem.*). A monoalcohol derived from α-pinene:

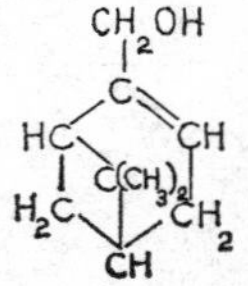

CH_2OH
C
HC CH
$C(CH_3)_2$
H_2C CH_2
CH

the dextrorotatory isomer occurs in myrtle oil.

Mysidacea (*Zool.*). An order of *Peracarida* retaining more or less completely the primitive *caridoid facies* (q.v.). The carapace covers most or all of the thoracic somites; eyes (when present) stalked; scale of the antenna well developed; exopodites on most or all of the thoracic limbs; well formed tail fan; mostly marine and pelagic, but a few 'relicts' or immigrants in fresh water, e.g., *Mysis*.

mysophobia (*Med.*). Morbid fear of being contaminated.

Mysophyceae (*Zool.*). See **Lecanorales.**

mytiliform (*Bot.*). Resembling a sea shell in form.

myxa (*Zool.*). In Birds, the fused extremities of the rami of the lower beak.

myxamoeba (*Bot.*). The product of germination of a spore of a Myxomycete. It is a uninucleate, naked protoplast, which creeps like an amoeba, and may develop a flagellum and swim as a zoospore.

myxo-. Prefix from Gk. *myxa*, mucus, slime.

Myxobacteriales (*Bacteriol.*). A group of bacteria characterized by the absence of a rigid cell wall, gliding movements and the production of slime. Some genera produce characteristic fruiting bodies.

myxoedema (*Med.*). A condition due to deficiency of thyroid secretion; characterized by loss of hair, increased thickness and dryness of the skin, increase in weight, slowing of mental processes, and diminution of metabolism.

myxoma (*Med.*). A term applied to a tumour composed of a clear jellylike substance and star-shaped cells.

myxomatosis (*Vet.*). A highly contagious and fatal virus disease of rabbits characterized by tumourlike proliferation of myxomatous tissue especially beneath the skin of the head and body.

Myxomycetes (*Bot.*). A group of very simple organisms mostly living on or in rotten wood or in soil and having some plant and some animal characters. The thallus is a naked creeping mass of protoplasm, containing many nuclei and known as a *plasmodium.* Sporangia are developed and liberate spores which germinate to yield myxamoebae. Also called **Mycetozoa, Myxothallophyta, slime moulds.**

Myxophyceae (*Bot.*). Blue-green algae. A class of algae, unicellular or filamentous, without a well-defined nucleus or chromatophore, and usually bluish green in colour. They appear to have no sexuality, have no zoospores, and propagate chiefly by vegetative means. There are many fresh water and marine species, including many waterblooms. Also called **Cyanophyceae.**

Myxospongiae, Myxospongida (*Zool.*). An order of *Demospongiae* with no skeleton.

Myxothallophyta (*Bot.*). See **Myxomycetes.**

myxoviruses (*Bot.*). One of eight families of viruses which attack Vertebrates. Includes the influenza virus and related species, which are subject to great variability, e.g., Asian strain of the influenza virus. Affinity for mucins on surfaces of red cells.

Myzostomida (*Zool.*). A class of *Annelida*, the members of which are parasitic in or on *Crinoidea*; they have an oval depressed body provided with 5 pairs of ventral parapodia, each with a hooked seta, and 10 pairs of marginal cirri; the epidermis is ciliated, and there are ventral suckers for attachment.

n Symbol for *nano-*. (*Nuc.*) Symbol for *neutrino*.
n (*Chem.*). Symbol for amount of substance. (*Light*) Symbol for index of refraction.
n- (*Chem.*). An abbrev. for *normal*, i.e., containing an unbranched carbon chain in the molecule.
ν (*Phys.*). Symbol for: (1) *frequency*; (2) *Poisson's ratio*; (3) *kinematic viscosity*.
N (*Chem.*). Symbol for *nitrogen*. (*Eng., Phys.*) Symbol for *newton*. (*Eng.*) A symbol often used for *modulus of rigidity*.
N (*Chem.*). Symbol for: (1) *Avogadro number*; (2) *number of molecules*. (*Elec.*) Symbol for *number of turns*. (*Nuc.*) Symbol for *neutron number*.
N- (*Chem.*). Symbol indicating substitution on the nitrogen atom.
Na (*Chem.*). Symbol for *sodium*.
N_A (*Phys.*). Symbol for *Avogadro number*.
NA (*Eng.*). An abbrev. for *neutral axis*. (*Optics*) Abbrev. for *numerical aperture*.
nab (*Join.*). The keeper part of a door-lock.
nabam (*Chem.*). Disodium ethylene-1,2-bisdithiocarbamate, used as a fungicide.
nabla (*Maths.*). See **del**.
nacelle (*Aero.*). A small streamlined body on an aircraft, distinct from the fuselage, housing engine(s), special equipment, or crew.
nacre or **nacreous layer** (*Zool.*). The iridescent calcareous substance, mostly calcium carbonate, composing the inner layer of a Molluscan shell, which is formed by the cells of the whole of the mantle. Mother of pearl.
nacreous (*Min.*). A term applied to the lustre of certain minerals, usually on crystal faces parallel to a good cleavage, the lustre resembling that of pearls.
nacreous clouds (*Meteor.*). Clouds composed of ice crystals in 'mother of pearl' formations, found at a height of 25 to 30 km.
nacrite (*Min.*). A species of clay mineral, identical in composition with kaolinite, from which it differs in certain optical characters and in atomic structure.
NAD (*Biochem.*). Nicotinamide adenine dinucleotide. A coenzyme employed as acceptor of electrons from a substrate. It is readily dissociable from the dehydrogenase. Reoxidation requires the participation of another group of enzymes. Sometimes abbreviated **DPN** from the trivial name diphosphopyridine nucleotide and formerly called *coenzyme I*.
NADH (*Biochem.*). See **electron transport chain**.
nadir (*Astron.*). That pole of the observer's horizon which is vertically below his feet; hence, the point on the celestial sphere diametrically opposite the *zenith* (q.v.).
NADP (*Biochem.*). Nicotinamide adenine dinucleotide phosphate. A coenzyme which functions similarly to *NAD* (q.v.). Sometimes abbreviated **TPN** from the trivial name triphosphopyridine nucleotide.
naevus (*Med.*). Birthmark; mole. (1) A pigmented tumour in the skin. (2) A patch or swelling in the skin composed of small dilated blood vessels.
nagana (*Vet.*). Fly disease; tsetse fly disease. A name embracing a group of diseases of animals in Africa, due to infection by trypanosomes.
nagatelite (*Min.*). A phosphorus-bearing *allanite*.
Nagelfluh (*Geol.*). A group of massive conglomerates of Miocene age which form the Rigi and Rossberg in Switzerland.
nagyagite (*Min.*). A sulpho-telluride of lead and gold (sometimes containing antimony) crystallizing in thin tabular or massive habit; black tellurium.
naiad (*Zool.*). An aquatic nymph of a hemimetabolic insect.
Naiadales (*Bot.*). An order of the *Monocotyledons*. Aquatic or semi-aquatic species. The flowers are cyclic or hemicyclic. The perianth is absent or in one or two whorls; hypogynous or epigynous. The number of stamens varies from many to one, and the gynaeceum consists of many to one carpels which are free or united. Also **Fluviales**.
nail (*Zool.*). In higher Mammals, a horny plate of epidermal origin taking the place of a claw at the end of a digit.
nail punch (*Join.*). A small steel rod tapering at one end almost to a point; used to transmit the blow from a hammer to drive a nail in so that its head is beneath the surface. Also **nail set**.
NaK (*Met.*). Acronym for sodium (Na) and potassium (K) alloy used as coolant for liquid-metal reactor. It is molten at room temperature and below.
naked (*Bot.*). (1) Lacking a perianth. (2) Without any appendages. (3) Not enclosed in a pericarp. (4) Bractless.
naked-light mine (*Mining, etc.*). Nonfiery mine, where safety lamps are not required.
Namurian (*Geol.*). A stratigraphical stage at the base of the Upper Carboniferous in Europe.
NAND (*Comp., Maths.*). See **logical operations**.
nand element (or **gate**) (*Comp.*). One producing an output signal except when all inputs are energized simultaneously, i.e., output signal is 0 when all input signals are 1, otherwise it is 1.
nanism (*Med., etc.*). The condition of being a dwarf (Latin *nanus*); dwarfism.
nankeen (*Textiles*). A drab 3-end twill cotton cloth used for pocket linings, corsets, etc. *Nankeen twill* is a 2-and-1 twill weave.
nano-. Prefix for 10^{-9}, i.e., equivalent to millimicro or one thousand millionth. Symbol **n**.
nanophanerophyte (*Bot.*). A plant from 25 cm to 2 m in height, with its resting buds above the soil level.
nanoplankton (*Zool.*). Plankton of microscopic size.
nanosoma (or **nanosomia**) **pituitaria** (*Med.*). Dwarfism due to hypofunction of the pituitary gland.
Nansen-Pettersson water-bottle (*Ocean.*). See **insulating water-bottle**.
nap (*Textiles*). A woolly surface on fabrics produced by the finishing process of raising. *Napping* is raising by means of a revolving cylinder covered with stiff wire brushes, or rollers covered by teazles.
napalm (*Chem.*). A gel of inflammable hydrocarbon oils with soaps, useful in warfare because of its cheapness, ease of application and ability to stick to the target while burning.
naphtha (*Chem.*). A mixture of light hydrocarbons which may be of coal tar, petroleum or shale oil origin. *Coal tar naphtha* (boiling range approx. 80–170°C) is characterized by the predominant aromatic nature of its hydrocarbons. Generally *petroleum naphtha* is a cut

between gasoline and kerosine with a boiling range 120–180°C, but much wider naphtha cuts may be taken for special purposes, e.g., feedstock for high temperature cracking for chemical manufacture. The hydrocarbons in petroleum naphthas are predominantly aliphatic.

naphthalene (*Chem.*). $C_{10}H_8$; consists of 2 condensed benzene rings:

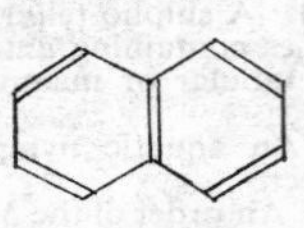

glistening plates, insoluble in water, slightly soluble in cold ethanol and ligroin, readily soluble in hot ethanol and ethoxyethane; m.p. 80°C, b.p. 218°C; sublimes easily and is volatile in steam. It occurs in the coal-tar fraction boiling between 180°C and 200°C. It forms an additive compound with picric acid (trinitrophenol). Naphthalene is more reactive than benzene, and substitution occurs in the first instance in the *alpha* position. It is an important raw material for numerous derivatives, many of which play a rôle in the manufacture of dyestuffs.

naphthalene derivatives (*Chem.*). Substitution products of naphthalene. The monosubstituted products form 2 series of isomers according to the position of the substituents in the benzene rings. The disubstituted derivatives can form 10 isomers with identical substituents or even more with different substituents. The positions taken by substituents in the naphthalene molecule are marked as follows:

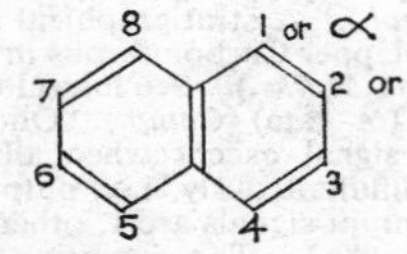

8 1 or α
7 2 or β
6 3
5 4

naphthaquinones (*Chem.*). $C_{10}H_6O_2$. Three isomers are known, viz.: α, or 1,4-*naphthaquinone*,

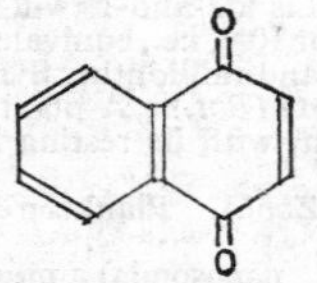

O
O

yellow rhombic plates, m.p. 125°C β, or 1,2-*naphthaquinone*,

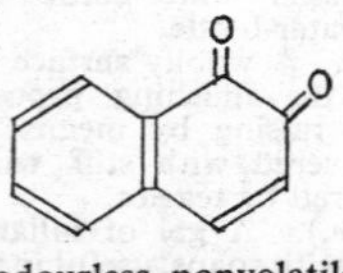

O
O

red needles, odourless, nonvolatile, decomposes at 120°C; *amphi*-, or 2,6-*naphthaquinone*,

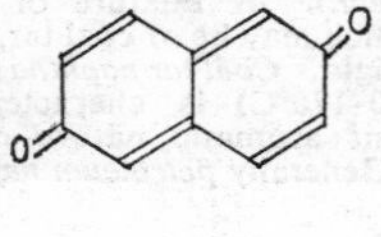

O
O

crystallizes in small red prisms, and is not stable in the presence of water, acids, alkalis and alcohols.

naphthenates (*Chem.*). Metal salts of the naphthenic acids which occur naturally in crude petroleum. Mainly used as wood and textile fungicides, e.g., copper naphthenate, and also as paint driers.

naphthenes (*Chem.*). Cycloalkanes; polymethylene hydrocarbons. Many occur in petroleum.

naphthionic acid (*Chem.*). 1,4-Naphthylamine-monosulphonic acid, $H_2N{\cdot}C_{10}H_6{\cdot}SO_3H$, obtained by the sulphonation of 1-naphthylamine. Intermediate for azo-dyes.

naphthoic acids (*Chem.*). $C_{10}H_7{\cdot}COOH$, naphthalene carboxylic acids. There are two isomers, of which the 1-naphthoic acid crystallizes in fine needles, m.p. 160°C. On distillation with lime, they are decomposed into naphthalene and CO_2.

naphthols (*Chem.*). $C_{10}H_7OH$. There is an 1-naphthol, m.p. 95°C, b.p. 282°C, and a 2-naphthol, m.p. 122°C, b.p. 288°C. Both are present in coal-tar and can be prepared from the respective naphthalene-sulphonic acids or by diazotizing the naphthylamines. 1-Naphthol is prepared on a large scale by heating α-naphthylamine under pressure with sulphuric acid. They have a phenolic character. 2-Naphthol is an antiseptic and can be used as a test for primary amines.

naphthylamines (*Chem.*). $C_{10}H_7{\cdot}NH_2$. 1-Naphthylamine forms colourless prisms or needles, m.p. 50°C, b.p. 300°C, soluble in ethanol. It is of unpleasant odour, sublimes readily, and turns brown on exposure to the air. 2-Naphthylamine, odourless, m.p. 112°C, b.p. 294°C, forms colourless plates. Originally used as an anti-oxidant in rubber manufacture, but found to be a cause of bladder cancer.

Napierian logarithm (*Maths.*). See **logarithm.**

Napier's analogies (*Maths.*). Formulae for solving spherical triangles.

Napier's compasses (*Instr.*). A form of compasses having a needle-point and a pencil-holder pivoted at the end of one limb, and a needle-point and a pen pivoted at the end of the other, both limbs being jointed for safe carrying.

napiform (*Bot.*). Shaped like a turnip.

Naples yellow (*Paint.*). Originally lead antimonate, an artists' pigment, now usually substituted by a mixture of zinc oxide and tinters.

napoleonite (*Geol.*). A diorite containing spheroidal structures, about 1 in (2·5 cm) in diameter, which consist of alternating shells essentially of hornblende and feldspars.

nappe (*Geol.*). A major structure of mountain chains such as the Alps, consisting essentially of a great recumbent fold with both limbs lying approximately horizontally. It is assumed to be produced by a combination of compressional earth movements and sliding under gravity, resulting in translation of the folded strata over considerable horizontal distances. (*Maths.*) A sheet; one of the two sheets on either side of the vertex forming a cone.

napping and friezing machine (*Textiles*). A machine used in the woollen trade to produce small beads (frieze effect) or ripples (wave effect) on the surface of a fabric.

nap roller (*Print.*). A hand roller with a rough surface used for lithographic transfer work.

naptalam (*Chem.*). *N*-1-naphthylphthalamic acid, used as a weedkiller. Also **panala, NP, NPA.**

narcissism (*Psychol.*). See **autoerotism.**

narcolepsy (*Med.*). A condition characterized by sudden attacks of an uncontrollable desire to sleep and/or by cataplectic attacks. See **cataplexy.**

narcosis (*Med.*). A state of unconsciousness produced by a drug; the production of a narcotic state.

narcotic (*Med.*). Tending to induce sleep or unconsciousness; a drug which does this.

narcotine (*Chem.*). $C_{22}H_{23}O_7N$, an alkaloid of the *iso*quinoline series; occurs in opium; forms colourless needles; m.p. 176°C.

narcotize (*Med.*). To subject to the influence of a narcotic.

nares (*Zool.*). Nostrils; nasal openings; as the internal or posterior *nares* to the pharynx, the external or anterior *nares* to the exterior. *adj.* **narial, nariform.**

naricorn (*Zool.*). A scale of the nasal region; the terminal part of the nostril when that is horny.

narrow-base tower (*Elec. Eng.*). A tower for overhead transmission lines having a base sufficiently small to be supported on a single foundation. Cf. *broad-base tower.*

narrow gauge (*Rail.*). A railway gauge less than the standard 4 ft 8½ in. (1·435 m).

NASA. See **National Aeronautics and Space Administration.**

nasal (*Zool.*). Pertaining to the nose; a paired dorsal membrane bone covering the olfactory region of the Vertebrate skull.

nasal sinusitis (*Med.*). See **sinusitis.**

nascent (*Chem.*). Just formed by a chemical reaction, and therefore very reactive. Nascent gases are probably in an atomic state.

nasil (*An. Behav.*). A social signal between members of the same species made during nonsexual encounter and tending to inhibit conflict.

Nasmyth pile-driver (*Civ. Eng.*). A form of pile-driver in which the monkey is raised by steam pressure acting on an attached piston, facilitating rapid operation.

Nasmyth's membrane (*Zool.*). A thin horny membrane covering the enamel of unworn teeth.

nasolacrimal canal (*Zool.*). A passage through the skull of Mammals, passing from the orbit to the nasal cavity, and through which the tear duct passes.

nasopalatine duct (*Zool.*). In some Reptiles and Mammals, a duct piercing the secondary palate and connecting the vomeronasal organs with the mouth.

nasopharyngeal duct (*Zool.*). The posterior part of the original vault of the Vertebrate mouth which, in Mammals, due to the development of a secondary palate, carries air from the nasal cavity to the pharynx.

nasopharyngitis (*Med.*). Inflammation of the nasopharynx.

nasopharynx (*Zool.*). In Vertebrates, that part of the pharynx continuous with the internal nares.

nasosinusitis (*Med.*). Inflammation of the air-containing bony cavities in communication with the nose.

nasoturbinal (*Zool.*). In Vertebrates, a paired bone or cartilage of the nose which supports the folds of the olfactory mucous membrane.

nastic movement (*Bot.*). Plant movement in response to a stimulus, which is independent of its direction; usually a growth movement, but may involve change in turgidity, e.g., wilting of Mimosa pudica on contact.

nasus (*Zool.*). See **clypeus.**

natal (*Med., Zool.*). (1) Pertaining to birth. (2) Pertaining to the buttocks.

natality (*Ecol.*). The inherent ability of a population to increase. *Maximum* or *absolute natality* is the theoretical maximum production of new individuals under ideal conditions. *Ecological* or *realized natality* is the population increase occurring under specific environmental conditions. Cf. *mortality.*

natamycin. See **pimaricin**

natatory or **natatorial** (*Zool.*). **Adapted for swimming.**

nates (*Med.*). The buttocks.

National Aeronautics and Space Administration. U.S. body established by federal act in 1958 to supervise, control, coordinate civil developments in space exploration. Abbrev. NASA.

National Bureau of Standards. U.S. federal department set up in 1901 to promulgate standards of weights and measures and generally investigate and establish data in all branches of physical and industrial sciences.

National Physical Laboratory. National authority for establishing basic units of mass, length, time, resistance, frequency, radioactivity, etc. Founded by the Royal Society in 1900; now government-controlled and engaged in a very wide range of research. Abbrev. **N.P.L.**

native (*Min., Mining*). Said of naturally occurring metal; e.g., *native* gold, *native* copper.

native hydrocarbons (*Geol.*). A series of compounds of hydrogen and carbon formed by the decomposition of plant and animal remains, including the several types of coal, mineral oil, petroleum, paraffin, the fossil resins, and the solid bitumens occurring in rocks. Many which have been allotted specific names are actually mixtures. By the loss of the more volatile constituents as natural gas, the liquid hydrocarbons are gradually converted into the solid bitumens, such as *ozocerite.* See also **asphalt, bitumen, coal, mineral oils.**

natrojarosite (*Min.*). Hydrous sulphate of sodium and iron crystallizing in the trigonal system.

natrolite (*Min.*). Hydrated silicate of sodium and aluminium crystallizing in the orthorhombic system. A sodium-zeolite. It usually occurs in slender or acicular crystals, and is found in cavities in basaltic rocks and as an alteration product of nepheline or plagioclase.

natural abundance (*Nuc.*). See **abundance.**

natural background (*Nuc.*). Radiation due to natural radioactive minerals and gases, cosmic rays, also often contamination and fallout.

natural cement (*Build., Civ. Eng.*). A cement similar to a hydraulic lime, made from a natural earth, with but little preparation.

natural classification (*Bot.*). One based on the presumed relationships of plants in descent.

natural coke (*Heat*). See **cinder coal.**

natural draught (*Eng.*). The draught or air flow through a furnace induced by a chimney and dependent on its height and the temperature difference between the ascending gases and the atmosphere.

natural evaporation (*Meteor.*). The evaporation that takes place at the surface of ponds, rivers, etc., which are exposed to the weather; it depends on the relative humidity, and also, to some extent, on the strength of the wind.

natural frequency (*Telecomm.*). That of free oscillations in a system. See **natural period.**

natural frequency of antenna (*Radio*). The lowest frequency at which an antenna system is resonant when directly earthed, without the addition of loading inductance.

natural gas (*Geol.*). Any gas found in the earth's crust, including gases generated during volcanic activity (see **pneumatolysis, solfatara**). The term, however, is particularly applied to natural hydrocarbon gases which are associated with the production of petroleum. These gases are principally methane and ethane, sometimes with propane, butane, nitrogen, CO_2 and sulphur compounds (notably H_2S). The gas is found both above the petroleum and dissolved in it, but many very large gas fields are known which produce little or no petroleum. Natural gas has largely replaced *town gas* (q.v.) in many countries where it occurs abundantly (U.S.A., Canada, Algeria, W. Europe); since 1967 important finds in the North Sea have provided supplies for the British gas-grid system.

natural glass (*Geol.*). Magma of any composition is liable to occur in the glassy condition if cooled sufficiently rapidly. Acid (i.e., granitic) glass is commoner than basic (i.e., basaltic) glass; the former is represented among igneous rocks by pumice, obsidian, and pitchstone; the latter by tachylite. Natural quartz glass occurs in masses lying on the surface of certain sandy deserts (e.g., the Libyan Desert); while both clay rocks and sandstones are locally fused by basic intrusions. See also **buchite, tektites.**

naturalized (*Bot.*). Introduced from another region, reproducing freely by seed and maintaining its position in competition with wild plants. (*Zool.*) Said of introduced species which compete successfully with the native fauna.

natural load (*Elec. Eng.*). A resistive load impedance, numerically equal to the characteristic impedance of the transmission line which it terminates, or the power which the line would transmit if it were so loaded.

natural magnet. See **lodestone.**

natural modes (*Telecomm.*). See **normal modes.**

natural number (*Maths.*). Term used to describe any element of the set {1,2,3, . . .} or the set {0,1,2,3, . . .}. The inclusion of zero is a matter of definition.

natural period (*Telecomm.*). Time of one cycle of oscillation arising from free oscillation, depending on inertia and elastance of a system. Reciprocal of natural frequency.

natural radioactivity (*Nuc.*). That which arises from isotopes which are found in nature and which are groupable in families.

natural resonance (*Telecomm.*). Response of system to signal with period equal to its own natural period.

natural scale (*Acous.*). The musical scale in which the frequencies of the notes within the octave are proportional to 24, 27, 30, 32, 36, 40, 45, and 48, and which can be realized in continuously variable pitch instruments, such as the human voice and stringed instruments, but not in keyboard instruments, which use the *tempered scale* (q.v). Also **just scale, just temperament.** (*Surv.*) A term applied to a section drawn with equal vertical and horizontal scales.

natural selection (*Biol.*). A theory of the mechanism of evolution which postulates the survival of the best-adapted forms, with the inheritance of those distinctive characteristics wherein their fitness lies, and which arise as small uncontrolled variations; it was first propounded by Charles Darwin, and is often referred to as *Darwinism* or the *Darwinian Theory*.

natural slope (*Civ. Eng.*). The maximum angle at which soil in cutting or bank will stand without slipping.

natural uranium (*Chem.*). That with its natural isotopic abundance.

natural-uranium reactor (*Nuc. Eng.*). One in which natural, i.e., unenriched, uranium is the chief fissionable material.

natural wavelength of antenna (*Radio*). The free-space wavelength corresponding to the natural frequency of an antenna.

naupliiform larva (*Zool.*). Same as **cyclopoid larva.**

nauplius (*Zool.*). The typical first larval form of *Crustacea*; egg-shaped, unsegmented, and having three pairs of appendages and a median eye; found in some members of every class of *Crustacea*, but often passed over, becoming an entirely embryonic stage.

nauplius eye (*Zool.*). A median unpaired eye similar to that found in a nauplius larva.

Nauta mixer (*Chem. Eng.*). A proprietary mixer comprising a stationary cone, with point vertically downwards, within which a screw arm rotates on its own axis parallel to the conical surface. The lower end of the screw is fixed in a universal joint at its lower end, and to an arm at its upper end, by means of which the whole screw, while rotating, is moved round the interior surface of the cone. Much used for mixing small quantities of one constituent with large quantities of another and mostly used on dry solids.

Nautical Almanac (*Astron., etc.*). An astronomical ephemeris published annually, some years in advance, by the Lords Commissioners of the Admiralty, for navigators and astronomers. First published in 1767, it is now (since 1960) called the *Astronomical Ephemeris* (q.v.). An abridged version, for the use of seamen, is now given the original title of *The Nautical Almanac*.

nautical log (*Ships*). A device for estimating the speed of a vessel. In the old-fashioned log, a line divided into equal spaces (knots) runs freely off a reel and is attached to a *chip log*, which is stationary in the water as the vessel travels. Time is measured by a log-glass. The modern *patent* (or *taffrail*) *log* mechanically indicates the rate of travel by means of a submerged fly or rotator, whose revolutions are conveyed to a register on the rail of the vessel by a braided hemp line secured to the rotator.

nautical mile. One-sixtieth of a degree of latitude, a distance varying with latitude. The *UK nautical mile* is 6080 ft (1853·18 m), differing slightly from the *international nautical mile*, 1852 m.

nautical twilight (*Astron.*). The interval of time during which the sun is between 6° and 12° below the horizon, morning and evening. See **astronomical twilight, civil twilight.**

Nautiloidea (*Zool.*). The only living suborder of *Tetrabranchiata* (q.v.), having a wide central siphuncle, a membranous protoconch, a shell with simple sutures, and interlocular septa which are concave on the side nearest the aperture. Abundant from early Cambrian to late Cretaceous, but now represented by one genus, *Nautilus*, which lives in tropical seas. All chambers except the terminal living chamber contain gas which buoys up the heavy shell.

naval brass (*Met.*). See **Tobin bronze.**

navarho (*Nav.*). Low-frequency, long-range radio-navigation system for aircraft.

nave (*Arch.*). The middle or main body of a basilica, rising above the *aisles*: the main part of a church, generally west of the crossing, including or excluding its aisles.

navel (*Zool.*). In Mammals, the point of attachment of the umbilical cord to the body of the foetus.

navel-ill (*Vet.*). See **joint-ill.**

navicular bone (*Zool.*). In Mammals, one of the tarsal bones, also known as the **centrale** or **scaphoid.**

navicular disease (*Vet.*). Chronic osteitis of the navicular bone in the foot of the horse.

navigable semicircle (*Meteor.*). The left-hand half of the storm field in the northern hemisphere, the right-hand half in the southern hemisphere, when looking along the path in the direction a *tropical revolving storm* (q.v.) is travelling. Cf. *dangerous semicircle*.

navigation. The art of directing a vessel or vehicle by terrestrial or stellar observation, or by radio and radar signals. (*Hyd. Eng.*) A name frequently given to a canalized river the flow of which is more or less under artificial control.

navigational planets (*Nav.*). Those which can be used in celestial navigation (Venus, Mars, Jupiter, and Saturn).

navigational system (*Radio*). Any system of obtaining bearings and/or ranges for navigational purposes by radio techniques.

navigation flame float (*Aero.*). A pyrotechnic device, dropped from an aircraft, which burns with a flame while floating on the water. Used to determine the drift of the aircraft at night.

navigation lights (*Aero.*). Aircraft navigation lights consist of red, green and white lamps located in the port wing tip, starboard wing tip and tail respectively.

navigation smoke float (*Aero.*). A pyrotechnic device, dropped from an aircraft, which emits smoke while floating on the water. Used for ascertaining the direction of the wind or the drift of the aircraft.

Nb (*Chem.*). The symbol for *niobium*.

NCR (*Paper*). '*N*o *c*arbon *r*equired.' TN for stationery, especially for office machine use, in which simultaneous duplicate copies are obtained without the use of carbon paper. Microcapsules containing dyes on the verso of the 'top' copy are fractured by the impact of the writing medium (pen, typewriter, etc.), the image being then transferred to the under copy or copies, the receiving surface of which has a coating of attapulgite with a starch or latex binder. See **microencapsulation.**

Nd (*Chem.*). The symbol for *neodymium*.

NDB (*Aero., Radio*). Abbrev. for non-directional beacon. See **beacon.**

N display (*Radar*). A radar *K display* (q.v.) in which the target produces two breaks on the horizontal time base. Direction is proportional to the relative amplitude of the breaks, and range is indicated by a calibrated control which moves a pedestal signal to coincide with the breaks.

Ne (*Chem.*). The symbol for *neon*.

neanic (*Zool.*). Said of the adolescent period in the life-history of an individual.

neap tides (*Astron.*). High tides occurring at the moon's first or third quarter, when the sun's tidal influence is working against the moon's, so that the height of the tide is below the maximum in the approximate ratio 3 : 8.

Nearctic region (*Zool.*). One of the subrealms into which the Holarctic region is divided; it includes N. America and Greenland.

near-end cross-talk (*Telecomm.*). That occurring in a signal channel and propagated in the reverse direction to the signal in the interfering channel. (*Teleph.*) Cross-talk between 2 parallel circuits when both the listener and the speaker originating the inducing currents are at the same end of the parallelism. See **far-end cross-talk.**

near point (*Optics*). The nearest position to the eye at which an object can be seen distinctly. The object point conjugate to the retina when accommodation is exerted to its fullest extent.

near print (*Print.*). A name coined to describe printing of poor typographic standards, such as is produced by justifying typewriters.

nearside (*Autos.*). The side of a vehicle nearer the verge when observing the rule of the road.

neat (*Textiles*). The name applied by wool-sorters to wool taken from the sides of a lustre fleece of average quality.

neat cement (*Build., Civ. Eng.*). A cement mortar mixture made up with water only, without addition of sand.

neat size (*Build.*). The net or exact size after preparation.

neat work (*Build.*). The brickwork above the footings.

Nebitype (*Typog.*). A machine similar to the *Ludlow* which produces slugs from handset matrices.

nebula (*Astron.*). Any faint luminous patch seen among the stars; the true nebulae are gaseous and lie mainly in the plane of the Milky Way. The extragalactic nebulae are now known to be galaxies. (*Med.*) (1) A slight opacity in the cornea of the eye. (2) An oily preparation for use in an atomizer or nebulizer (e.g., a nasal spray).

nebular hypothesis (*Astron.*). One of the earliest scientific theories of the origin of the solar system, stated by Laplace. It supposed a flattened mass of gas extending beyond Neptune's orbit to have cooled and shrunk, throwing off in the process successive rings which in time coalesced to form the several planets.

nebule (*Arch.*). An ornamental moulding characterized by a wavy lower edge.

nebulium (*Astron.*). A hypothetical element at one time thought to be responsible for certain emission lines in nebular spectra. These are now known to be due to ionized oxygen and nitrogen.

neck (*Arch.*). The narrow moulding separating the capital of a column from the shaft. (*Bot.*) (1) The upper tubular part of an archegonium, and of a perithecium. (2) The lower part of the capsule of a moss, just above the junction with the seta. (*Eng.*) See **necking.** (*Geol.*) A plug of volcanic rock representing a former feeder channel of an extinct volcano.

neck canal cell or neck cell (*Bot.*). One of the cells in the central canal in the neck of an archegonium.

necking (*Arch.*). See neck. (*Eng.*) In a material tensile test just prior to fracture, the cross-sectional area is reduced over a short length; this is the neck. *Necking* quantitatively indicates disinclination to strain-harden under cold work.

neck-mould (*Arch.*). The neck round the top of the shaft of a column.

necro-. Prefix from Gk. *nekros*, a dead body.

necrobacillosis (*Vet.*). Bacillary necrosis. Infection of animals by *Sphaerophorus necrophorus* (*Fusiformis necrophorus*).

necrobiosis (*Med.*). The gradual death, through stages of degeneration and disintegration, of a cell in the living body.

necrogenic abortion (*Bot.*). The speedy death of the tissues of a plant close under the point of attack of a parasite, checking the spread of the latter.

necrogenous (*Biol.*). Living or developing in the bodies of dead animals.

necron (*Bot.*). Dead plant material not rotted into humus.

necrophagous (*Zool.*). Feeding on the bodies of dead animals.

necrophorous (*Zool.*). Carrying away the bodies of dead animals; as certain Beetles, which usually afterwards bury the bodies.

necropsy (*Med.*). Autopsy. A postmortem examination of the body.

necrosis (*Biol.*). Death of a cell (or of groups of cells) while still part of the living body. *adj.* **necrotic.** *v.* **necrose.**

necrotic enteritis of swine (*Vet.*). A chronic inflammation of the large, and sometimes small, intestine of pigs. Infection by *Salmonella cholerae suis* may be incriminated in some outbreaks, but in others the aetiology is obscure.

necrotic stomatitis (*Vet.*). See **calf diphtheria.**

necrotoxin (*Bacteriol.*). A bacterial toxin, easily separable from the bacterial cells producing it and having a necrotic effect on the tissue cells of the host, e.g., the necrotoxin of *Clostridium perfringens* (gas gangrene).

nectar (*Bot.*). A sugary fluid exuded by plants, usually from some part of the flower, occasionally from somewhere else on the plant; it attracts insects, which assist in pollination.

nectarivorous (*Zool.*). Nectar-eating.

nectary (*Bot.*). A glandular organ or surface from which nectar is secreted.

necto-. Prefix from Gk. *nēktos*, swimming.

nectocalyx (*Zool.*). In *Siphonophora*, a medusoid modified as a swimming organ and lacking a manubrium.

nectocyst (*Zool.*). The cavity of a nectocalyx.

necton (*Bot., Zool.*). See **nekton.**

nectonic benthos (*Bot.*). Small organisms floating at the bottom of the water.

nectophore (*Zool.*). A *nectocalyx* (q.v.); that portion of the common stem of a siphonophoran colony bearing the nectocalyces. Also **nectozooid.**

nectopod (*Zool.*). An appendage adapted for swimming.

nectosome (*Zool.*). The upper part of a siphonophoran colony to which the nectocalyces and pneumatophores are attached.

nectozooid (*Zool.*). See **nectophore.**

Nectridia (*Zool.*). An order of Amphibians known from late Palaeozoic fossils.

need (*An. Behav.*). A condition which, if continued, will lead to the death of an individual or the disappearance of the species. A state of physiological disequilibrium and homeostatic imbalance.

Needham's sac (*Zool.*). In certain Cephalopods, a spacious sac, being an expansion of the lower part of the vas deferens, in which spermatophores are stored.

needle (*Acous.*). See **stylus.** (*Bot.*) A long, narrow, stiffly constructed leaf, tending to hold water; characteristic of pine trees and related plants. (*Build.*) A timber or steel beam used in the process of underpinning. It is laid horizontally at right angles to the wall (through which it passes) and is supported on both sides by dead shores, so as to take the load of the upper part of the walls. (*Civ. Eng.*) The timbers used in a *needle weir* (q.v.). (*Elec. Eng.*) The moving magnet of a compass or galvanometer of the moving-magnet type. Sometimes also the moving element of an electrostatic voltmeter. (*Print.*) See **needles.**

needle beam (*Civ. Eng.*). A transverse floor-beam supported across the chords of a bridge.

needle chatter (*Acous.*). That arising from vibrations of the gramophone needle being transferred to the tone arm and radiated as noise.

needle lubricator (*Eng.*). A crude form of lubricator consisting of an inverted stoppered flask attached to a bearing and containing a wire loosely fitting a hole in the stopper and touching the shaft.

needle machine (*Textiles*). (1) A type of single-needle embroidery machine. (2) A modern development for making needle felt, or needled unwoven fabrics, by a stitching technique.

needle paper (*Paper*). An acid-free, black paper for wrapping needles, pins, etc.

needle pick-up (*Acous.*). A pick-up in which the sole moving part is the magnetic needle, which by its motion diverts magnetic flux and induces electromotive forces in coils on the magnetic circuit.

needle roller bearing (*Eng.*). A *roller bearing* (q.v.) without cage, in which long rollers of small diameter are used, located endwise by a lip on the inner or outer race.

needles (*Print.*). Removable points on web-fed presses, either fixed or retractable by cam action, used to control and convey the cut leading edge of the web or copy until it is severed, transferred, folded, or stitched, as required; can be mounted on the folding, transfer, cutting, or collection cylinders.

needle scaffold (*Build.*). A scaffold which is supported on cantilever or needle beams jutting out from an intermediate height in the building, thereby avoiding the necessity for erection from ground level.

needle scratch (*Acous.*). Noise emanating from gramophone, specifically due to irregularities in the contact surface of the groove. Also called **surface noise.**

needle stone (*Min.*). A popular term for clear quartz containing acicular inclusions, usually of rutile, but in some specimens, of actinolite. Also called **rutilated quartz.** The name has also been used for various acicular zeolites.

needle talk (*Acous.*). Direct sound output from transducer of gramophone pick-up.

needle traverse (*Surv.*). A traverse in which the angles between successive lines, or the directions of the lines, are measured by means of a magnetic compass.

needle valve (*Eng.*). A slender pointed rod working in a hole or circular seating; operated by automatic means, as in a carburettor float chamber, or by a screw, for the control of fluid.

needle wear (*Acous.*). Wear of the rounded point of a gramophone reproducing needle caused by sliding along the track on the record. In principle, the needle has to fit itself into the track during the first few (blank) revolutions of the disk and not wear shoulders during the remainder of the playing.

needle weir (*Civ. Eng.*). A form of frame weir in which the wooden barrier consists of upright square-section timbers placed side-by-side against the iron frames.

needling (*Build.*). The process of underpinning in which needles are used in the support of the upper part of the building. (*Surg.*) *Discission* (q.v.). Cutting with a needle the lens of the eye in the treatment of cataract.

Neelium (*Met.*). See **thermoelectric materials.**

Néel temperature (*Phys.*). The temperature at which the susceptibility of an antiferromagnetic material has a maximum value.

negater (*Comp.*). Type of inverter which interchanges '1' and '0'.

negative (*Elec.*). Designation of electric charge, introduced by Franklin, now known to be exhibited by the electron, which, in moving, forms the normal electric current. (*Photog.*) The black-and-white reversed image obtained by developing a sensitized and exposed photographic emulsion; so called because the greater the brilliance of the light arriving from the object, the blacker is the image, i.e., the greater is the reduction and retention of silver after fixing. See also **negative colour film.** (*Weaving*)

Relates to various loom mechanisms. See also **positive**.

negative after-image (*Optics*). The image of complementary colour arising after visual fatigue from viewing a coloured object and then a white surface.

negative bias (*Electronics*). Static potential, negative with reference to earth, applied to electrode of valve or transistor, usually to prevent collection of free electrons.

negative booster (*Elec. Eng.*). A series-wound booster used in connexion with an earthed-return power supply system, e.g., for a tramway. It is connected between two points on the earthed return path to reduce the potential between them and minimize the possibility of electrolysis due to leakage currents.

negative carbon (*Elec. Eng.*). The carbon of a d.c. arc lamp which is connected to the negative terminal of the supply. It is usually of smaller diameter than the positive carbon as it burns away more slowly.

negative catalysis (*Chem.*). The retardation of a chemical reaction by a substance which itself undergoes no permanent chemical change.

negative colour film (*Photog.*). Film which reproduces the negative image in complementary colours to facilitate the making of positive prints as opposed to transparencies.

negative conductance (*Radio, etc.*). (1) A property which is similar to *negative resistance* (q.v.), but in which an increase in voltage produces a decrease in current. A device possessing such a property can maintain oscillations in a resonant circuit with which it is connected in parallel. A typical example is the dynatron. (2) In semiconductors, the use of 'hot' electrons in some form of a two-terminal negative conductance, forming the basis of both avalanche transit-time devices and those devices relating to the *Gunn effect* (q.v.). In the former, e.g., using silicon, the negative conductance arises from a phase shift (greater than 90° and preferably near to 180°) between current and voltage. In the Gunn devices it arises within the Ga As crystal in a strong electric field, the local current density decreasing whenever the local electric field exceeds a certain threshold.

negative coupling (*Elec. Eng.*). See **positive coupling**.

negative crystal (*Electronics*). Birefringent material for which the velocity of the extraordinary ray is greater than that of the ordinary ray.

negative electricity (*Elec.*). Phenomenon in a body when it gives rise to effects associated with excess of electrons. See **positive electricity**.

negative electrode (*Electronics*). The *anode* of a primary cell (the electrode by which conventional current returns to the cell), but the *cathode* of a valve or voltameter (connected to the negative side of the power supply).

negative electron (*Electronics*). See **negatron**.

negative feedback (*Telecomm.*). Reverse of *positive feedback*, reducing gain of an amplifier by feeding part of output signal back to input out of phase with incoming signal. Gives more uniform performance, greater stability, and reduced distortion. Also called **degeneration, reverse coupling**.

negative feeder (*Elec. Eng.*). In the power supply to an electric traction system, the feeder connecting the track rails or negative conductor-rail to the negative bus-bars at the substation. Also called **return feeder**.

negative g (*Aero., Space*). (1) In a manœuvring aircraft, any force acting opposite to the normal force of gravity. (2) The force exerted on the human body in a gravitational field or during acceleration so that the force of inertia acts in a foot-to-head direction, causing considerable blood pressure on the brain. Also called **minus g**. *Negative g tolerance*, in practice, is the degree of tolerance 3*g* for 10–15 sec.

negative glow (*Electronics*). In a medium-pressure Crookes tube, the glow between the cathode and Faraday dark space.

negative group (*Chem.*). (1) An acid radical. (2) A group of atoms whose introduction into an organic molecule tends to give it an acidic character.

negative image (*TV*). A reproduced TV image in which the light parts of the original scene appear dark, and vice versa.

negative impedance amplifier (*Teleph.*). A telephone amplifier that comprises one or more negative impedances. Each negative impedance is usually produced by the combination of a negative impedance converter and a passive network.

negative impedance converter (*Teleph.*). Active network for which a positive impedance connected across one pair of terminals produces a negative impedance across the other pair.

negative ion (*Electronics*). Radical, molecule or atom which has become negatively charged through the gain of one or more electrons. See **ion**.

negative-ion vacancy (*Electronics*). Same as **hole**.

negative light modulation (*TV*). See under **positive light modulation**.

negative mineral (*Light*). A doubly refracting mineral in which the ordinary refractive index is greater than the extraordinary. Calcite is a negative mineral, for which the values of ω and ε are 1·66 and 1·48 respectively. See **optic sign**.

negative mutual inductance (*Elec. Eng.*). See under **positive coupling**.

negative phase sequence (*Elec. Eng.*). A 3-phase system in which the voltages or currents in the three phases reach their maximum values in the nonstandard order, i.e., in the order red, blue, yellow. See **phase sequence**.

negative phase-sequence component (*Elec. Eng.*). The symmetrical component of an unbalanced 3-phase system of voltages or currents in which the phase-sequence is in the opposite order to standard, i.e., it is in the order red, blue, yellow.

negative phase-sequence relay (*Elec. Eng.*). A relay which operates when any negative phase-sequence components of current or voltage appear in the circuit to which it is connected.

negative plate (*Elec. Eng.*). The plate of an accumulator or primary cell which is normally at the lower potential and to which the current from the circuit during discharge is said to return.

negative proton (*Nuc.*). See **antiproton**.

negative reaction (*Biol.*). A tactism or tropism in which the organism moves, or the member grows, from a region where the stimulus is stronger to one where it is weaker.

negative reinforcing stimulus (*An. Behav.*). A *reinforcing stimulus* (q.v.), usually aversive, which increases the intensity of escape or avoidance behaviour. Also called **punishment**.

negative resistance (*Electronics*). Effective property of most forms of discharge, in that an increase in voltage accompanies decrease of current, as in a carbon arc. This negative resistance can overcome the positive resistance of a tuned circuit and set up continuous oscillations at the resonant frequency, e.g., the **magnetron and the dynatron oscillators**.

negative scanning (*TV*). Scanning a photographic negative, with reversal in the circuits, so that the reproduced image is the normal positive. This saves time in transmitting televised newsfilm.

negative stagger (*Aero.*). See **stagger.**

negative staining (*Micros.*). A technique for the examination of organisms or structures which do not easily stain by conventional methods. A dye, e.g., nigrosin, is used to stain the background and the colourless specimen is shown in outline. Modifications of this technique are used in electron microscopy.

negative taxis (*Biol.*). See **taxis.**

negative transconductance (*Electronics*). Property of certain valves whereby an increase in positive potential on one electrode accompanies a decrease in current to another electrode.

negative transference (*Psychiat.*). *Transference* (q.v.) in which the emotions displaced on to the analyst are those of hostility and hate.

negative tropism (*Biol.*). The phenomenon of movement away from a stimulus.

negative video signal (*TV*). A video signal in which increasing amplitude corresponds to decreasing light-value in the transmitted picture. Black is taken as 100%, white about 30%, of the maximum amplitude in the signal. Now used as standard. Signals below 30% are known as whiter-than-white.

negative well (*Civ. Eng.*). A shaft sunk through an impermeable stratum to allow water to drain through to a permeable one.

negativism (*Psychiat.*). General refusal to cooperate and tendency to do the opposite of what other people wish. This may be passive, resembling extreme withdrawal, or active, as in resistance, impulsiveness or even homicidal or suicidal tendencies.

negatron (*Electronics*). (1) A four-electrode thermionic tube for obtaining negative resistance, comprising an anode and grid on one side of a cathode, and an anode on the other. (2) Term sometimes used to distinguish the *negat*ive elec*tron* from the positron.

negri bodies (*Vet.*). Specific inclusion bodies found in the cytoplasm of nerve cells in the brain and spinal cord in animals affected with rabies.

Neher-Pickering circuit (*Electronics*). A widely-used thermionic valve-quenching circuit for Geiger-Müller counters.

neighbourhood (*Maths.*). The neighbourhood of a point z_0 is the set of all points z such that $|z-z_0|<\alpha$, where α is a given positive number.

Neisseriaceae (*Bacteriol.*). A family of bacteria belonging to the order *Eubacteriales*. Gram-negative, characteristically occurring as paired spheres, parasitic in mammals, e.g., *Neisseria gonorrhoeae* (gonorrhoea).

nekton (*Bot., Zool.*). Actively swimming aquatic organisms, as opposed to the passively drifting organisms or *plankton* (q.v.). Also **necton.**

N-electron (*Nuc.*). One from the *N-shell*, which, when complete, totals 32 electrons.

nemathecium (*Bot.*). A cushionlike projection formed on the thallus of a seaweed and bearing reproductive organs.

Nemathelminthes (*Zool.*). A phylum which formerly comprised the groups *Nematoda, Acanthocephala*, and *Nematomorpha*; now obsolete.

nematic (*Phys.*). Said of a mesomorphous substance whose molecules or atoms are oriented in parallel lines. Cf. *smectic.*

nematoblast (*Zool.*). A cell which will develop a *nematocyst* (q.v.).

nematocalyx (*Zool.*). See **nematophore.**

nematocyst (*Zool.*). An independent effector found in most *Coelenterata* (and a few *Protozoa*), consisting of a chitinous sac filled with fluid and produced at one end into a long narrow pointed hollow thread, which normally lies inverted and coiled up within the sac, but can be everted when the *cnidocil* (q.v.) is stimulated.

Nematoda (*Zool.*). A phylum of unsegmented worms with an elongate rounded body pointed at both ends; marked by lateral lines and covered by a heavy cuticle composed of protein and secreted by a thin sheet of noncellular hypodermis; have a mouth and alimentary canal; have only longitudinal muscles, which are divided into four quadrants; nervous system consists of a circum-pharyngeal ring and a number of longitudinal cords; perivisceral cavity formed by the confluence of the vacuoles of a few giant cells; cilia absent; sexes separate; development direct, the larvae resembling the adults; many species are of economic importance; mostly free-living but some are parasitic. Round Worms, Thread Worms, Eel Worms.

Nematomorpha (*Zool.*). A small phylum of nonmetameric worms with an elongate rounded body pointed at both ends, without lateral lines; in the adults the mouth and alimentary canal are occluded and functionless; there is a dorsal 'brain' and a single ventral nerve cord; the young are parasitic in various insects, the adults being free-living in fresh water.

nematoparenchymatous thallus (*Bot.*). An algal thallus composed of united threads, which are, however, still recognizable as individuals.

nematophore (*Zool.*). A small, secondary type of polyp of certain *Hydrozoa* which lacks a mouth but is capable of engulfing organisms by means of pseudopodia. Also called **nematocalyx.**

nematozooid (*Zool.*). See **dactylozooid.**

Nembutal (*Pharm.*). TN for sodium ethylmethylbutyl barbiturate. White crystalline powder, used as a sedative and hypnotic drug.

Nemertea (*Zool.*). A phylum of apparently nonmetameric worms with an elongate flattened body, a ciliated ectoderm, and a dorsal eversible proboscis not connected with the alimentary canal; no perivisceral cavity; alimentary canal with mouth and anus; there is a vascular system; the sexes are separate; some forms have a pilidium larva; most are marine, but some are fresh water or terrestrial. Ribbon Worms Also **Nemertinea.**

neoarsphenamine (*Pharm.*). See **Neosalvarsan.**

neoblasts (*Zool.*). In many of the lower animals (*Annelida, Ascidia*, etc.), large amoeboid cells widely distributed through the body which play an important part in the phenomena of regeneration.

neocerebellum (*Anat.*). Phylogenetically, the more recently developed part of the posterior lobe of the cerebellum, receiving predominantly partine fibres via the pontocerebellar tract.

Neocomian (*Geol.*). The name given to the lowest stage of the Cretaceous System which precedes the Aptian stage. See also **Wealden Series.**

neo-Darwinism (*Zool.*). A revival of Darwin's theory of evolution by natural selection, using additional facts discovered since his time.

neodymium (*Chem.*). A metallic element, a member of the rare earth group. Symbol Nd, at. no. 60, r.a.m. 144·24, rel. d. 6·956, m.p. 840°C. The metal is found in cerite, monazite, and orthite. Neodymium glass is used for solid-state lasers and light amplifiers

neogallicola (*Zool.*). A form of *Viteus vitifolii*, the Vine Phylloxera, developing from eggs laid by gallicolae which form leaf galls, and whose progeny may either also be gall formers, or pass to the roots, becoming radicolae.

neogamous (*Zool.*). In *Protozoa*, showing precocious association of gametocytes.

Neognathae (*Zool.*). A super-order of the *Neornithes* in which the sternum has a keel, the coracoid and scapula are not ankylosed, the quadrate articulates with the skull by 2 facets, the vomer is small and the palatines converge behind it, being movably attached by a ball and socket joint, and the barbs of the feathers have hooklets.

neohexane (*Chem.*). 2,2-dimethylbutane. B.p. 50°C. Antiknock fuel prepared by the addition of ethene to 2-methylpropane under pressure at increased temperatures.

neo-Lamarckism (*Zool.*). A revival of the evolution theories of Lamarck, which postulated that differences between species arose by the inheritance of the effects of use and disuse.

Neolite (*Plastics*). U.S. TN for hard-wearing synthetic rubber compounds used for shoe soles and heels, based on butadiene-styrene compounds.

Neolithic Period (*Geol.*). The later portion of the Stone Age, characterized by well-finished, polished stone implements, made by men of the same species as ourselves. Cf. *Palaeolithic Period*.

neologism (*Psychiat.*). Verbal construction such as occurs in schizophrenia, manic-depressive psychosis and some aphasias, in which the patient uses coined words, which may have meaning for him but not for others, or else gives inappropriate meanings to ordinary words.

neomorphosis (*Zool.*). A type of regeneration in which the part reformed does not resemble any existing part of the body. *n.* **neomorph.**

neomycin (*Pharm.*). An antibiotic derived from *Streptomyces fradiae*; its sulphate is especially effective against external staphylococcal infections (skin, eyes, etc.); also used internally.

neon (*Chem.*). Light, gaseous, inert element, recovered from atmosphere. Symbol Ne, at. no. 10, r.a.m. 20·179, m.p. −248·67°C, b.p. −245·9°C. Historically important in that J. J. Thomson, through his parabolas for charge/mass of particles, found two isotopes in neon, the first nonradioactive isotopes to be recognized. Used in many types of lamp, particularly to start up sodium vapour discharge lamps. Pure neon was the first gas to be used for high-voltage display lighting, being bright orange in colour. Much used in cold-cathode tubes, reference tubes, Dekatrons.

neon induction lamp (*Light*). A lamp consisting of a small tube containing neon at low pressure; luminescence is produced by the action of high-frequency currents in a few turns surrounding the tube.

neon time base (*Elec. Eng.*). Relaxation oscillator produced by charging capacitor through high resistance and allowing periodic discharge through neon glow tube.

neon voltage regulator (*Elec. Eng.*). One using the property of a neon lamp so that when the applied voltage exceeds the ionization potential the potential difference between the electrodes will remain constant.

neonychium (*Zool.*). A pad of soft tissue enclosing a claw of the foetus during the development of many Mammals, to eliminate the risk of ripping the foetal membranes.

neopallium (*Zool.*). In Mammals, that part of the cerebrum occupied with impressions from senses other than the sense of smell.

neoplasm (*Med.*). A new formation of tissue in the body; a *tumour* (q.v.). *adj.* **neoplastic.**

Neoprene (*Chem.*). Polychloroprene; the first commercial synthetic rubber (U.S. 1931). Chloroprene (3 chlorobut 1,2:3,4-diene), the monomer, $CH_2{=}CCl{\cdot}CH{=}CH_2$, is derived from acetylene and hydrochloric acid.

Neopterygii (*Zool.*). A subclass of the *Actinopterygii* comprising the *Holostei* and *Teleostei*, and having fins without basal supports outside the body wall.

Neornithes (*Zool.*). A subclass of Birds with a greatly shortened tail ending in a pygostyle, around which the retrices, when present, are arranged in a semicircle. Except in a few extinct forms (*Odontognathae* (q.v.)) there are no teeth; the metacarpals are fused with the distal carpals to form a carpometacarpus; with one exception not more than 2 digits of the hand bear claws, and these are usually absent; essentially bipedal. Cf. *Archaeornithes*.

neorophags (*Zool.*). Phagocytic cells which engulf and destroy senile nerve cells.

Neosalvarsan (*Pharm.*). No. 914. The Na salt of 3,3′ - diamino - 4,4′ - dihydroxyarsenobenzene - *N*-methylene-sulphinic acid. It has a therapeutic action against syphilis. See **Salvarsan.**

Neosporidia (*Zool.*). A subclass of *Sporozoa* in which the adult is a syncytium and forms spores continuously; the spore cases are complex and usually contain only a single sporozoite.

neossoptiles (*Zool.*). The down-feathers found on a newly hatched Bird.

neoteinic (*Zool.*). In a state of arrested development, as certain castes of *Isoptera*.

neotenin (*Zool.*). In Insects, the *juvenile hormone* produced by the corpora allata, which suppresses the development of adult characteristics at each moult except the last, when the corpora allata become inactive and metamorphosis occurs.

neoteny (*Zool.*). The retention of larval characters beyond the usual period, as in some Amphibians which may still have the appearance of Tadpoles when they have reached sexual maturity.

Neotropical region (*Zool.*). One of the primary faunal regions into which the surface of the globe is divided. It comprises South America, the West Indian islands, and Central America south of the Mexican plateau.

neotype (*Biol.*). A substitute specimen for the type specimen when all the original material has been lost or destroyed.

neovitalism (*Zool.*). The theory which postulates that a complete causal explanation of vital phenomena cannot be reached without invoking some extra-material concept.

Neozoic (*Geol.*). The name (= 'new life') sometimes given to the Tertiary and Post-Tertiary rocks.

neper (*Telecomm.*). Unit of attenuation, adopted by CCI as of equal status with the decibel. If current I_1 is attenuated to I_2 so that

$$I_2/I_1 = e^{-N},$$

then N is attenuation (the βl of the Continent). In circuits matched in impedance,

$$1 \text{ neper} = 8{\cdot}686 \text{ dB}.$$

After John Napier (Lat. *Nepero*), Scottish scientist, inventor of natural logarithms.

nepheline, nephelite (*Min.*). Silicate of sodium and aluminium, $NaAlSiO_4$, but generally with some potassium partially replacing sodium, which crystallizes in the hexagonal system. It is frequently present in igneous rocks with a high sodium content and a low percentage of silica, i.e., the undersaturated rocks. See **elaeolite.**

nepheline-basalt (*Geol.*). A general name for any

basic lava carrying nepheline as an essential constituent. See **nephelinite,** also **basanite, tephrite.** By some petrologists the term is restricted to those nepheline-basalts which carry olivine but no feldspar. A more accurate name for the latter is *olivine-nephelinite.*

nepheline-syenite (*Geol.*). A coarse-grained igneous rock of intermediate composition, undersaturated with regard to silica, and consisting essentially of nepheline, a varying content of alkali-feldspar, with soda amphiboles and/or soda-pyroxenes. Common hornblende, augite, or mica are present in some varieties. See, for example, **foyaite, laurdalite.**

nephelinite (*Geol.*). A fine-grained igneous rock normally occurring as lava flows, and resembling basalt in general appearance; it consists essentially of nepheline and pyroxene, but not of olivine or feldspar. The addition of the former gives *olivine-nephelinite,* and of the latter, *nepheline-tephrite.*

nephelometric analysis (*Chem.*). A method of quantitative analysis in which the concentration or particle size of suspended matter in a liquid is determined by measurement of light absorption. Also called **photoextinction method, turbedimetric analysis.**

nephograph (*Photog.*). An instrument comprising electrically controlled cameras for photographing clouds, etc., in order that their position in the sky may subsequently be determined.

nephoscope (*Meteor.*). An instrument for observing the direction of movement of a cloud and its angular velocity about the point on the earth's surface vertically beneath it. If the cloud height is also known its linear speed may be calculated.

nephr-, nephro-. Pfx. from Gk. *nephros*, kidney.

nephrectomy (*Surg.*). Removal of a kidney.

nephric (*Anat., Zool.*). Pertaining to the kidney.

nephridiopore (*Zool.*). The external opening of a nephridium or nephromixium.

nephridium (*Zool.*). In Invertebrates and lower *Chordata*, a segmental excretory organ consisting of an intercellular duct of ectodermal origin leading from the coelom to the exterior; more generally, an excretory tubule. *adj.* **nephridial.** Cf. *coelomoduct*.

nephrite (*Min.*). One of the minerals grouped under the name of *jade*; consists of compact and fine-grained tremolite or actinolite. It has been widely used for ornaments in the Americas and the East.

nephritis (*Med.*). Inflammation of the substance of the kidney. *adj.* **nephritic.**

nephrocoel (*Zool.*). The coelomic cavity of a nephrotome.

nephrocystitis (*Med.*). Inflammation both of the kidney and of the bladder.

nephrocyte (*Zool.*). A cell which has the property of storing up substances of an excretory nature, as in *Porifera* and *Insecta*.

nephrodinic (*Zool.*). Using the same duct for the discharge of both excretory and genital products.

nephrogenic tissue (*Zool.*). In the embryonic development of Vertebrates, a relatively small intermediate region of the mesoderm lateral and ventral to the somites, from which derive the kidney tubules, their ducts, and the deeper tissues of the gonads. May be segmented, forming nephrotomes, or a continuous band of tissue.

nephrogonoduct (*Zool.*). Especially in Invertebrates, a common duct for genital and excretory products.

nephroid (*Bot.*). See **reniform.**

nephrolithiasis (*Med.*). The presence of stones in the kidney.

nephrolithotomy (*Surg.*). Removal of stones from the kidney through an incision in the kidney.

nephromixium (*Zool.*). See **mixonephrium.**

nephropathy (*Med.*). Any disease of the kidneys.

nephropexy (*Surg.*). The fixation, by operative measures, of a kidney which is abnormally movable. Cf. *nephrorrhaphy.*

nephropore (*Zool.*). See **nephridiopore.**

nephroptosis (*Med.*). Movable kidney; floating kidney. An abnormally mobile kidney, associated with general displacement downwards of other abdominal organs.

nephropyelitis (*Med.*). Inflammation both of the substance of the kidney and of its pelvis.

nephrorrhaphy (*Surg.*). The fixation, by suture, of a displaced kidney. Cf. *nephropexy.*

nephros (*Zool.*). A kidney. *adj.* **nephric.**

nephrosclerosis (*Med.*). Hardening and contraction of the kidney as a result of arteriosclerosis and of general increase of fibrous tissue in the substance of the kidney.

nephrosis (*Med.*). A degeneration of the tubules of the kidney, associated with oedema of the tissues and albuminuria.

nephrostome (*Zool.*). The ciliated funnel by which some types of nephridia and nephromixia open into the coelom.

nephrostomy (*Surg.*). Formation of an opening into the pelvis of the kidney for the drainage of urine.

nephrotome (*Zool.*). That region of the coelomic epithelium in each somite of a metameric animal that gives rise to the excretory tubes.

nephrotomy (*Surg.*). The making of an incision into the kidney.

nephrotoxin (*Med.*). A poison or toxin which specifically affects the cells of the kidney.

nepionic (*Zool.*). Said of the embryonic period in the life-history of an individual.

neps (*Textiles*). (1) Term applied in the cotton industry to small entanglements of fibres that cannot be unravelled; generally formed during the ginning process. (2) Small lumps formed in cotton or wool fibres during carding.

Neptune (*Astron.*). The most distant of the giant planets, discovered in 1846 as a result of independent predictions by Adams and Leverrier based on perturbations of Uranus. A recent measure of the diameter gives 48 450 km; its mass is 17·2 times that of the earth, and it revolves at a distance of 30·06 A.U. from the sun in a period of 164·8 years. The planet rotates in 15·8 hours and the atmosphere appears to be similar to that of Uranus; there are two satellites.

neptunium (*Chem.*). Element, at. no. 93, symbol Np; named after planet Neptune; produced artificially by nuclear reaction between uranium and neutrons. Has principal isotopes 237 and 239.

neptunium series (*Chem.*). Series formed by the decay of artificial radioelements, the first member being *plutonium-241* and the last *bismuth-209*, of which *neptunium-237* is the longest lived (half-life $2{\cdot}2 \times 10^6$ years).

neritic zone (*Geol., Ocean.*). That portion of the sea floor lying between low-water mark and the edge of the continental shelf, at a depth of about 180 metres. Sediments deposited here are of *neritic facies*, showing rapid alternations of the clay and sand grades; ripple marks, etc., indicate accumulation in shallow water.

Nernst bridge (*Elec. Eng.*). An a.c. bridge for capacitance measurements at high frequencies.

Nernst effect (*Electronics*). Analogous to *Hall*

effect (q.v.), a difference of potential arising between opposite edges of a metal strip, due to heat flow along the strip when a transverse magnetic field is applied.

Nernst heat theorem (*Chem.*). As the absolute temperature of a homogeneous system approaches zero, so does the specific heat and the temperature coefficient of the free energy.

Nernst lamp (*Light*). One depending on the electric heating of a rod of zirconia in air, giving infrared rays for spectroscopy. The rod must be heated separately to start lamp, as the material is insulating at room temperature.

Nernst's distribution law (*Chem.*). When a single solute distributes itself between two immiscible solutes, then for each molecular species (i.e., dissociated, single, or associated molecules) at a given temperature, there exists a constant ratio of distribution between the two solvents, i.e., $(C_1/C_2)_i = K_i$ for each molecular species, *i*.

Nernst theory (*Chem.*). An explanation of the development of electrode potentials, based on the supposition that an equilibrium is established between the tendency of the electrode material to pass into solution and that of the ions to be deposited on the electrode.

nervation, nervature (*Bot.*). See **venation.** (*Zool.*) See **nervuration.**

nerve (*Anat., Zool.*). One of the branches of the central nervous system passing to an organ or part of the body; a nervure. (*Arch.*) See **nerves.** (*Bot.*) A general name for the midrib and the larger veins of a leaf. *adj.* **nervous, neural.**

nerve block (*Med.*). Production of insensibility of a part by injecting an anaesthetic into the nerve or nerves supplying it.

nerve canal (*Zool.*). An aperture in the root of a tooth through which the nerve may pass to the pulp.

nerve cell (*Zool.*). See **neuron.**

nerve centre (*Zool.*). An aggregation of nerve cells assoc. with a particular sense or function.

nerve ending (*Zool.*). The free distal end of a nerve or nerve fibre, generally with accessory parts forming a complex end-organ.

nerve fibre (*Zool.*). An axon.

nerve gas (*Chem.*). One which, by inhibiting the enzyme cholinesterase, rapidly and fatally acts on the nervous system; derivative of fluorophosphoric acid. Several have been made for military use, and some for insecticidal purposes.

nerve impulse (*Zool.*). The disturbance passing along a nerve fibre when it has been stimulated, resulting in a temporary breakdown of the membrane surrounding the nerve fibre which allows sodium ions to move in and potassium ions to move out. The former move in before the latter move out, creating the momentary action potential.

nerve net (*Zool.*). The primitive type of nervous system found in *Coelenterata*, consisting of numerous multipolar neurons which form a net underlying and connecting the various cells of the body wall.

nerve plexus (*Zool.*). A network of interlacing nerve fibres.

nerve root (*Zool.*). The origin of a nerve in the central nervous system.

nerves (*Arch.*). The projecting ribs on a vault surface. Also called **nervures.**

nerve trunk (*Zool.*). A bundle of nerve fibres united within a connective-tissue coat.

nervicolous (*Bot.*). Said of a parasitic fungus which grows on the veins of a leaf.

nervi nervorum (*Zool.*). Sensory nerve fibres received by a nerve trunk and usually terminating within the epineurium as end-bulbs.

nervous system (*Zool.*). The whole system of nerves, ganglia, and nerve endings of the body of an animal, considered collectively.

nervule (*Zool.*). A small branch of a nervure.

nervuration, nervation, nervature (*Zool.*). The arrangement of the nervures in an Insect wing.

nervure (*Zool.*). One of the chitinous struts which support and strengthen the wings of an Insect.

nervures (*Arch.*). See **nerves.**

nervus lateralis (*Zool.*). See **lateralis.**

NESA panel (*Aero.*). A laminated glass windshield panel containing a thin film of transparent conductor (usually stannic oxide). When electric current is passed through it, its resistance causes heating of the panel. Used for *anti-icing* and for improving the impact strength of the panel. Abbrev. of *n*on-*e*lectrostatic *s*hield formulation '*A*'.

nesistor (*Electronics*). Transistor depending on a bipolar field effect.

nesosilicate (*Min.*). A silicate mineral whose atomic structure contains isolated groups of silicon-oxygen tetrahedra.

Nessler's solution (*Chem.*). Used in the analysis of water for determination of free and combined ammonia. A solution of mercury(II)iodide in potassium iodide, made alkaline with sodium or potassium hydroxide. With a trace of ammonia it gives a yellow colour, but with larger amounts, a brown precipitate.

nest (*Glass*). A cushion upon which glass is placed to be cut with a diamond. (*Zool.*) In most Birds, a structure ranging in complexity from a hole in the ground to an elaborate built or woven structure of sticks, moss, leaves, or feathers, in which the eggs are laid and incubated, and in which the young of altricial species spend the early part of their lives. In Insects, the habitations of some social insects, e.g., termites, ants, wasps, and bees, built of earth, sand, or wood pulp.

nest epiphyte (*Bot.*). An epiphyte which develops a tangle of stems, roots and leaves among which humus collects, and is utilized by the plant.

nestocalyx (*Zool.*). The bell-shaped swimming-organ of a hydrozoan, such as a medusa or jellyfish.

nest provisioning (*An. Behav.*). The behaviour, especially of some solitary Wasps, of returning regularly to a particular nest or series of nests to bring food for the developing larvae. Cf. *mass provisioning, progressive provisioning.*

net airscrew efficiency (*Aero.*). See **propeller efficiency.**

net efficiency (*Aero.*). The net thrust power divided by the torque power.

net knot (*Cyt.*). A small accumulation of chromatin, particularly at the intersections of the nuclear reticulum.

net register tonnage (*Ships*). Theoretically, the earning space of the ship, and the figure on which payment of harbour dues, etc., is based. It is the *gross tonnage* less *deducted spaces.*

netrum (*Cyt.*). A minute spindle which arises within the centrosome during the division of the centriole.

net tonnage (*Ships*). See **net register tonnage.**

net transport (*Nuc. Eng.*). The difference between the actual *transport* (q.v.) in an isotope separation plant and that which would be obtained by the same plant with raw material of natural isotopic abundance.

net wing area (*Aero.*). The *gross wing area* (q.v.) minus that part covered by the fuselage and any nacelles.

network (*Telecomm., etc.*). (1) Any arrangement

of electric components, *active* or *passive*, the former implying an internal source of e.m.f. or current. (2) Circuit element described by the following letters, e.g., *L*, having 1 shunt and 1 series element; *C*, 1 shunt followed by 1 series element in each leg; *T*, 2 series and 1 shunt element from junction; *H*, 4 series and 1 shunt element; π, 1 shunt, 1 series, and 1 shunt element; *O*, 1 shunt, 1 series in each leg and 1 shunt element; *lattice* (*bridge*), element between each input terminal to both output terminals; *bridged-T*, T-network, with series arms bypassed by fourth element; *twin-T*, 2 T-networks overlaying each other. All networks have 4 terminals (*quadripole*) in 2-wire line, are *balanced* or *unbalanced* to earth potential, and elements or *branches* may be complex.

network analysis (*Elec. Eng.*). Process of calculating theoretically transfer constant of a network.

network calculator (*Elec. Eng.*). Combination of resistors, inductors, capacitors, and generators used to simulate electrical characteristics of a power generation system, so that the effects of varying different operating conditions can be studied in computers. Also **network analyser.**

network polymer (*Chem.*). One in which *cross-linking* (q.v.) occurs.

network structure (*Met.*). The type of structure formed in alloys when one constituent exists in the form of a continuous network round the boundaries of the grains of the other. Even if the grains included in the cells are themselves duplex, they are regarded as individual grains.

network synthesis (*Elec. Eng.*). Process of formulating a network with specific electrical requirements.

network transfer function (*Maths.*). Mathematical expression giving the ratio of the output of a network to its input. The natural logarithm of its value at any one frequency is the *transfer constant*, and gives the attenuation and phase shift for the signal propagated through the network.

Neumann function (*Maths.*). A *Bessel function* (q.v.) of the second kind.

Neumann lamellae (*Met.*). Straight, narrow bands parallel to the crystallographic planes in the crystals of metals that have been subjected to deformation by sudden impact. They are actually narrow twin bands, and are most frequently observed in iron.

Neumann principle (*Crystal.*). Physical properties of a crystal are never of lower symmetry than the symmetry of the external form of the crystal. Consequently tensor properties of a cubic crystal, such as elasticity or conductivity, must have cubic symmetry, and the behaviour of the crystal will be isotropic.

neur-, neuro-. Prefix from Gk. *neuron*, nerve.

neurad (*Zool.*). Situated on the same side of the vertebral column as the spinal cord; hence, dorsal.

neural (*Zool.*). See **nerve.**

neural arch (*Zool.*). The skeletal structure arising dorsally from a vertebral centrum, formed by the neurapophyses and enclosing the spinal cord.

neural canal (*Zool.*). The space enclosed by the centrum and the neural arch of a vertebra, through which passes the spinal cord.

neural crest (*Zool.*). In a Vertebrate embryo, a band of cells lying parallel and close to the nerve cord which will later give rise to the ganglia of the dorsal roots of the spinal nerves.

Neural folds (*Zool.*). See **medullary folds.**

neuralgia (*Med.*). Paroxysmal intermittent pain along the course of a nerve. *adj.* **neuralgic.** See also **tic douloureux.**

neural gland (*Zool.*). In *Urochorda*, a structure on the ventral side of the nerve ganglion, sometimes correlated with the hypophysis of the *Craniata*, and having a duct opening into the cavity of the pharynx.

neural groove (*Zool.*). See **medullary groove.**

neural lamella (*Zool.*). In Insects, a layer surrounding the ganglia and nerves, secreted by a syncytial membrane (*epineurium*) around the nerves, and probably forming a barrier between the potassium ions of the haemolymph and the neurons.

neural plate (*Zool.*). In *Chelonia*, a median row of bony plates forming part of the carapace; medullary plate; neurapophysis.

neural spine (*Zool.*). The median dorsal vertebral spine, formed by the fusion of the neurapophyses above the neural canal.

neural tube (*Zool.*). A tube formed dorsally in the embryonic development of Vertebrates by the joining of the 2 upturned neural folds formed by the edges of the ectodermal neural plate, giving rise to the brain and spinal nerve cord.

neuraminic acids (*Biochem.*). A nine carbon, 3-deoxy-, 5-amino-sugar, formula $CH_2OH.(CHOH)_3.CHNH_2.CH(OH).CH_2.CO.COOH$.

neurapophyses (*Zool.*). A pair of plates arising dorsally from the vertebral centrum, and meeting above the spinal cord to form the neural arch and spine. *sing.* **neurapophysis.**

neurasthenia (*Med.*). Nervous debility, causing extreme tiredness and lassitude, accompanied by physical symptoms such as headache, backache, indigestion; possibly caused by mental strain or overwork.

neurectomy (*Surg.*). Excision of part of a nerve.

neurenteric canal (*Zool.*). A temporary passage connecting the cavity of the nerve cord and the archenteron in the early embryonic stages of some *Chordata*.

neurilemma (*Anat.*, *Zool.*). See **neurolemma.**

neurine (*Chem.*). $CH_2{=}CH{\cdot}N(CH_3)_3OH$, trimethylvinylammonium hydroxide, obtainable from brain substance and from putrid meat; related to *choline* (q.v.), into which it can be transformed. It is a ptomaine base.

neuritis (*Med.*). Inflammation of a nerve. For *multiple neuritis*, see **polyneuritis**; for *optic neuritis*, see **papillitis.**

neuroblastoma (*Med.*). A malignant tumour composed of primitive nerve cells, arising in the adrenal gland or in connexion with sympathetic nerve cells.

neuroblasts (*Zool.*). Cells of ectodermal origin which give rise to neurons.

neurocirrus (*Zool.*). In *Polychaeta*, a cirrus borne by the neuropodium.

neurocoel (*Zool.*). The spinal canal.

neurocranium (*Zool.*). The brain-case and sense capsules of a Vertebrate skull. Cf. *viscerocranium*.

neurocrine (*Biol.*). Neurosecretory; secretory property of nervous tissue.

neurocyte (*Zool.*). See **neuron.**

neuroectoderm (*Zool.*). In the embryonic development of Vertebrates, an area on the dorsal and posterior surface of the gastrula giving rise to the neural plate.

neuroelectricity (*Med.*). The electricity generated in the nervous system of an animal or human being.

neuro-epithelium (*Zool.*). A layer of superficial cells specialized for the reception of stimuli.

neurofibrils (*Zool.*). Fine fibrils which run in

parallel bundles in the cell-processes and form an intricate plexus in the cell-body.

neurofibroma (*Med.*). A tumour composed of fibrous tissue derived from the connective-tissue sheath of a nerve. See **molluscum fibrosum.**

neurofibromatosis (*Med.*). Von Recklinghausen's disease; multiple neurofibromata. See **molluscum fibrosum.**

neurofibrositis (*Med.*). Interstitial neuritis. Inflammation of the connective tissue which surrounds and binds together large nerve trunks (e.g., the sciatic nerve).

neurogenesis (*Zool.*). The development and formation of nerves.

neurogenic (*Zool.*). Activity of a muscle or gland which is dependent on continued nervous stimuli. Cf. *myogenic*.

neuroglia (*Zool.*). The supporting tissue of the brain and spinal cord of Vertebrates, composed of much-branched fibrous cells which occur among the nerve cells and fibres. Also **glia.**

neurohaemal organs (*Zool.*). Organs which serve as a gateway for the escape of the products of neurosecretory cells from the neurons into the circulating blood, e.g., the *corpora cardiaca* of Insects.

neurohumour (*Zool.*). A chemical substance given off by the tips of nerve fibres, and possibly connected with the transmission of nerve impulses across synapses. In the *peripheral* nervous system and the *parasympathetic* nervous system of Vertebrates it is acetylcholine; noradrenaline and, to a lesser degree, adrenaline, occur in the *sympathetic* nervous system.

neurohypophysis (*Zool.*). See **pars nervosa.**

neurokeratin (*Chem.*). A protein forming part of the nerve and brain substance.

neurolemma or **neurilemma** (*Zool.*). A thin homogeneous sheath investing the medullary sheath of a medullated nerve fibre; sheath of Schwann.

neurology. The study of the nervous system.

neurolymphomatosis (*Vet.*). See **fowl paralysis.**

neuromasts (*Zool.*). Small sense organs consisting of a cluster of bottle-shaped sense cells, each bearing a sense hair, these hairs being embedded in a jellylike cupula. Neuromasts occur either on the surface (superficial neuromasts) as in Cyclostomes, many bony Fish, and all aquatic Amphibians, or in canals of the lateral line system, as in most Fish. Displacement of the cupula deforms the sensory hairs and causes changes in the basic discharge frequency. Modified neuromasts form ampullary organs which appear to be electroreceptors, but ordinary neuromasts detect and locate moving animals at short range on the basis of disturbances of the water.

neuromere (*Zool.*). In metameric animals, the portion of the central nervous system contained within one somite.

neuromuscular (*Zool.*). Pertaining to nerve and muscle, as a myoneural junction.

neuron, neurone (*Zool.*). A nerve cell and its processes, regarded as an anatomically independent element. Also **neurocyte.**

neuronemes (*Zool.*). In some *Ciliophora*, filaments, believed to be of a nervous nature, running parallel with and external to the myonemes.

neuronephroblast (*Zool.*). In a segmenting ovum, a cell which will later contribute to the nerve cord and nephridia.

neuronitis (*Med.*). Inflammation (or degeneration) of neurons.

neuronophagia (*Med.*). The destruction of diseased nerve cells by leucocytes and by microglial cells.

neuropathy (*Med.*). Functional or structural disorders of the nervous system.

neuropil (*Zool.*). In Vertebrates, a maze of interlacing nerve fibres diffusely distributed through the brain.

neuropile (*Zool.*). In Insects, the central medullary substance of nerve ganglia, formed by the fine twigs of the axons of the cortical neurons, held together by a variable amount of neuroglia.

neuroplasm (*Cyt.*). The protoplasm of a nerve cell.

neuropodium (*Zool.*). In *Polychaeta*, the ventral lobe of a parapodium.

neuropore (*Zool.*). The anterior opening by which the cavity of the central nervous system communicates with the exterior.

Neuropteroidea (*Zool.*). A superorder of the *Oligoneoptera*, comprising the orders *Megaloptera*, *Raphidioptera*, and *Planipennia*, formerly suborders in the order *Neuroptera*.

neurosecretory cell (*Histol.*). A special type of neurone in which the axon terminates against the wall of a blood vessel or sinus and communicates with it by the passage of a hormone.

neurosis (*Med.*). Any one of a group of diseases thought to be due to disordered function of the involuntary nervous system, shown by instability of the circulatory system. (*Psychiat.*) A diagnosis given to a person who persistently exhibits nonadaptive behaviour and attitudes, or anxiety states. Generally regarded as a conflict state in which there is an unconscious attempt to achieve a compromise with the unconscious conflict—an attempt to come to terms with reality rather than to escape from it as in *psychosis* (q.v.). See **anxiety-, obsessional-, hysteria**; also **psychoneurosis.**

neurosurgery (*Surg.*). That part of surgical science which deals with the nervous system.

neurosynapse (*Zool.*). See **synapse.**

neurosyphilis (*Med.*). Syphilitic infection of the nervous system.

neurotic paradox (*Psychiat.*). The persistence of behaviour which is against one's own interest, as seen by others, in spite of the recognition of this. It is interpreted as being an attempt to punish oneself.

neurotropic (*Med.*). Having a special affinity for nerve cells, e.g., *neurotropic* virus.

neurotropism (*Zool.*). The tendency to mutual attraction shown by masses of nervous tissue under suitable conditions.

neurula (*Zool.*). In the embryonic development of Vertebrates, the stage after the gastrula in which the processes of organ formation begin, with the formation of the neural tube, mesodermal somites, notochord and archenteron.

neuston (*Ecol.*). Aquatic animals associated with the surface film.

neuter (*Bot.*). (1) Apparently sexless, said of certain strains of fungi which normally show sexuality. (2) Said of a flower in which the androecium and gynaeceum are nonfunctional or absent. (*Zool.*) Without sex; lacking functional sexual organs; neither male nor female; a sexless animal. Also **neutral.**

neutral (*Elec.*). (1) Exhibiting no resultant charge or voltage. (2) Return conductor of a balanced power supply, nearly at earth potential, if without a local earth connexion. See **neutral point, neutral conductor.** (*Glass*) Of high chemical durability. (*Photog.*) Possessing no colour or hue; grey. (*Phys.*) The normal state of atoms and the universe. (*Zool.*) See **neuter.**

neutralator (*Elec. Eng.*). See **earthing reactor.**

neutral autotransformer (*Elec. Eng.*). See **earthing reactor.**

neutral axis (*Elec. Eng.*). A term used to denote the diametral plane in which the brushes of a commutator machine should be situated to give perfect commutation. (*Eng.*) In a beam subjected to bending, the line of zero stress—a transverse section of the longitudinal plane, or neutral surface, which passes through the centre of area of the section.

neutral compensator (*Elec. Eng.*). See **earthing reactor.**

neutral (or middle) conductor (*Elec. Eng.*). The middle wire of a d.c. *three-wire system* or a distribution system, or the wire of a polyphase distribution system which is connected to the neutral point of the supply transformer or alternator. Sometimes called the **neutral (or middle) wire** or the **neutral.** Cf. *outer.*

neutral element (*Maths.*). See **identity** (2).

neutral equilibrium (*Mech.*). The state of equilibrium of a body when a slight displacement does not alter its potential energy.

neutral (density) filter (*Photog.*). One which attenuates all colours uniformly so that the relative spectral distribution of the energy of the transmitted light has been unaltered.

neutral flame (*Met.*). In welding, flame produced by a mixture at the torch of acetylene and oxygen in equal volumes.

neutral flux (*Met.*). One used to modify fusibility of furnace slags, but which exerts neither basic nor acidic influence, e.g., calcium fluoride.

neutral inversion (*Elec. Eng.*). A condition in which the phase to neutral voltages of a 3-phase, star-connected system are unbalanced. Commonly due to *ferromagnetic* resonance.

neutralism (*Ecol.*). A type of population interaction in which neither population is affected by the other.

neutralization (*Chem.*). The interaction of an acid and a base with the formation of a salt. In the case of strong acids and bases, the essential reaction is the combination of hydrogen ions with hydroxyl ions to form water molecules. (*Telecomm.*) Counteracting a tendency to oscillation through feedback via anode-grid capacitance. Reversed feedback is provided by a balancing capacitance to which is applied voltage equal and in antiphase to that on the anode. Also called **balancing.**

neutralized series motor (*Elec. Eng.*). See **compensated series motor.**

neutralizing (*Paint.*). Chemical treatment of cement surfaces, before painting, to neutralize lime; rarely used now, since good alkali-resistant paints, esp. *emulsion paints* exist.

neutralizing capacitance (*Telecomm.*). See **balancing capacitance.**

neutralizing indicator (*Elec. Eng.*). One indicating the degree of neutralization present in an amplifier.

neutralizing voltage (*Telecomm.*). That fed back in the process of neutralization.

neutral point (*Elec. Eng.*). The point at which the windings of a polyphase star-connected system of windings are connected together. Also, the midpoint of the neutral zone of a d.c. machine. Also called **neutral.** (*Mag.*) A point in the field of a magnet where the earth's magnetic field (usually the horizontal component) is exactly neutralized. (*Meteor.*) Any point at which the axis of a wedge of high pressure intersects the axis of a trough of low pressure. Sometimes called a **saddle point.** (*Space*) See **gravipause.**

neutral relay (*Telecomm.*). In U.S. **nonpolarized relay.**

neutral solution (*Chem.*). An aqueous solution which is neither acidic nor alkaline. It therefore contains equal quantities of hydrogen and hydroxyl ions and has a pH value of 7.

neutral state (*Mag.*). Said of ferromagnetic material when completely demagnetized. Also **virgin state.**

neutral surface (*Eng.*). See **neutral axis.**

neutral-tongue relay (*Telecomm.*). Relay for radio telegraphy in which the tongue is maintained in a central position by springs, being moved to either of the contacts by a current in the appropriate direction.

neutral wedge filter (*Photog.*). A neutral grey filter, originally a wedge of grey glass, which introduces a continuously variable attenuation of light, depending on the density or thickness introduced into the beam, without altering the relation between hues in the transmitted light.

neutral wire (*Elec. Eng.*). See **neutral conductor.**

neutral zone (*Elec. Eng.*). That part of the commutator of a d.c. machine in which, when the machine is running normally, the voltage between adjacent commutator bars is approximately zero.

neutrino (*Nuc.*). Subatomic particle emitted simultaneously with emission of positron from atomic nuclei. Its existence was first predicted by Pauli in 1933 to avoid β-decay infringing the laws of conservation of energy and angular momentum. It has very weak interactions with matter and was not observed experimentally until 1956. An apparently different neutrino with similar properties is also associated with the decay of a positive muon. Both neutrinos would appear to have approximately zero mass, no charge and spin of $\frac{1}{2}\left(\frac{h}{2\pi}\right)$. All reactors emit numbers of neutrinos. Often used generically to indicate both neutrinos and antineutrinos. Symbol **n.** See also **antineutrino.**

neutrodyning capacitance (*Telecomm.*). See **balancing capacitance.**

neutron (*Nuc.*). Uncharged subatomic particle, mass approximately equal to that of a proton, which enters into the structure of atomic nuclei. Interacts with matter primarily by collisions. Neutrons with the lowest energy distribution are termed *thermal neutrons,* those with energies between 10^{-2} and 10^2 eV are known as *epithermal neutrons,* between 10^2 and 10^5 eV as *intermediate neutrons* and the most energetic as *fast neutrons.* In a nuclear reactor, *slow* (*thermal*) *neutrons* have energy of 0·05 eV; *fast neutrons* of about 1 MeV on fission. Spin quantum number of neutron $=\frac{1}{2}$, rest mass $=$ 1·008 665 atomic mass unit, charge is zero and magnetic moment $=$ $-$1·9125 nuclear Bohr magnetons. Although stable in nuclei, isolated neutrons decay by beta emission into protons, with a half-life of 11·6 minutes.

neutron absorption cross-section (*Nuc.*). The cross-section for a nuclear reaction initiated by neutrons. This is expressed in *barns.* For many materials this rises to a large value at particular neutron energies due to resonance effects, e.g., a thin sheet of cadmium forms an almost impenetrable barrier to thermal neutrons.

neutron age (*Nuc.*). See **Fermi age.**

neutron balance (*Nuc. Eng.*). For a constant power level in a reactor, there must be a balance between the rate of production of both prompt and delayed neutrons, and their rate of loss due to both absorption and leakage. To assist design calculations two constants are widely used. The *neutrons per absorption* (symbol η) is the average number of neutrons emitted (including delayed neutrons) per neutron absorbed by the fissile material; the *neutrons per fission* (symbol γ) is the average number emitted (including delayed neutrons) per atom undergoing fission. Thus $\gamma = \eta$ multiplied by the probability of fission following absorption.

neutron current (*Nuc.*). The net rate of flow of neutrons per unit area through a surface perpendicular to the direction in which they are migrating. Neutron current is a *vector*, whereas *neutron flux* is a *scalar*, based on the concept of random movement through a system in equilibrium assuming negligible migration.

neutron detection (*Nuc. Eng.*). Observation of charged particle recoils following collisions of neutrons with protons, etc.; or of charged particles produced by interaction of neutrons with atomic nuclei, e.g., in a boron-trifluoride *counter*.

neutron energy (*Nuc.*). (1) The binding energy of a neutron in a nucleus, usually several MeV. (2) The energy of a free neutron which in a reactor will be classed in several groups: high-energy neutrons—energy > 10 MeV, fast neutrons—energy 10 MeV–20 keV, intermediate neutrons—energy 20 keV–100 eV, epithermal neutrons—energy 100 eV–0·025 eV, thermal or slow neutrons—energy approx. 0·025 eV. See **multigroup theory.**

neutron excess (*Nuc.*). The difference between the neutron number and the proton number for a nuclide. Also called **isotopic number.**

neutron flux (*Nuc.*). Number of neutrons passing through 1 cm^2 in any direction in 1 sec. Power reactors have above 10^{13}, while zero energy reactors have of the order of 10^8. See **neutron current.**

neutron gun (*Nuc. Eng.*). Block of moderating material with a channel used for producing a beam of fast neutrons.

neutron hardening (*Nuc.*). Increasing the average energy of a beam of neutrons by passing them through a medium which shows preferential absorption of slow neutrons.

neutron leakage (*Nuc. Eng.*). Escape of neutrons in a reactor from the core containing the fissile material; reduced by using a reflector.

neutron logging (*Nuc.*). The use of neutrons to detect the presence of water or oil in subsoil.

neutron number (*Nuc.*). The number of neutrons in a nucleus. Equal to the difference between the rel. atomic mass (the total number of nucleons) and the atomic number (the number of protons).

neutron radiography (*Radiol.*). Radiography by beam of neutrons from nuclear reactor which then produces an image on a photoelectric image intensifier following neutron absorption. It has advantages over X-ray radiography where the mass absorption coefficients for neutrons are very different for different parts of the specimen, although the atomic numbers (and hence X-ray absorption) are very similar, e.g. specimens containing hydrogenous material.

neutron shield (*Nuc. Eng.*). Radiation shield erected to protect personnel from neutron irradiation. In contradistinction to gamma-ray shields, it must be constructed of very light hydrogenous materials which will quickly moderate the neutrons.

neutron source (*Nuc. Eng.*). One giving a high neutron flux, e.g., for chemical activation analysis. Apart from reactors, these are chemical sources such as a radium-beryllium mixture emitting neutrons as a result of the (αn) reaction, and accelerator sources in which deuterium nuclei are usually accelerated to strike a tritium-impregnated titanium target, thus releasing neutrons by the (DT) reaction. The former are continuously active but the latter have the advantage of becoming inert as soon as the accelerating voltage is switched off.

neutron spectrometer (*Nuc. Eng.*). Instrument for investigation of energy spectrum of neutrons. See **crystal spectrometer, time-of-flight spectrometer, helium three, lithium six, proton recoil counter.**

neutron star (*Astron.*). A small body of very high density (ca. 10^{12} kg/dm^3) resulting from a supernova explosion in which a massive star collapses under its own gravitational forces, the electrons and protons combining to form neutrons.

neutron therapy (*Med.*). Use of neutrons for medical treatment.

neutron velocity selector (*Nuc. Eng.*). In a simple type, the detector is shielded from neutrons of energy less than 0·3 eV by a cadmium shield. (Cadmium is an almost perfect absorber of thermal neutrons.) More advanced systems employ crystal diffraction or time-of-flight methods. See **neutron spectrometer.**

neutropenia (*Med.*). Abnormal diminution in the number of neutrophil leucocytes in the blood.

neutrophil (*Physiol.*). (1) Stainable by neutral dyes. (2) A white blood cell, a polymorphonuclear leucocyte, whose granular protoplasm is stainable with neutral dyes. See also **amphophil.**

neutrosonic receiver (*Radio*). Supersonic heterodyne receiver using neutralized triode amplifiers for intermediate frequency.

Nevadian (Nevadan) orogeny (*Geol.*). An *orogeny* of Jurassic age affecting western U.S.A.

nevadite (*Geol.*). An acid lava (rhyolite) containing an abnormally large quantity of phenocrysts, with correspondingly little groundmass.

névé (*Geol.*). A more or less compacted snow-ice occurring above the snowline; it consists of small rounded crystalline grains formed from snow crystals. Also called **firn.**

Nevile and Winther's acid (*Chem.*). 1-Naphthol-4-sulphonic acid, an intermediate for dyestuffs.

Newall system (*Eng.*). A commonly used system of limits and fits, using a hole basis.

Newcastle disease (*Vet.*). Pseudo fowl plague; pneumoencephalitis; Ranikhet disease. An acute, highly contagious viral disease of chickens and other domestic and wild birds; symptoms include loss of appetite, diarrhoea, and respiratory and nervous symptoms.

newel (*Join.*). An upright post fixed at the foot of a stair or at a point of change of direction and used as a support for a balustrade.

newel cap, drop (*Join.*). Ornamental finish planted on the upper (lower) end of a newel post.

newel joints (*Join.*). The joints connecting the newel and the handrail or string.

newer drift (*Geol.*). Glacial deposits formed during the latest (Würm) stage of the Quaternary ice age in the British Isles.

New Forest disease (*Vet.*). See **infectious ophthalmia.**

Newlyn datum (*Surv.*). See **mean sea level.**

new moon (*Astron.*). The instant when sun and moon have the same celestial longitude; the illuminated hemisphere of the moon is then invisible.

New Red Sandstone (*Geol.*). A name frequently applied to the combined Permian and Triassic Systems, and particularly applicable in N.W. England, where the palaeontological evidence is insufficient to allow of their separation. The term reflects the general resemblance between the rocks comprising these two systems and the Old Red Sandstone of Devonian age.

newsprint (*Paper*). Cheap-quality printing paper, usually supplied in reels, for newspaper purposes; composed largely of mechanical or ground woodpulp with a relatively small admixture of strong sulphite woodpulp.

new star (*Astron.*). See **nova.**

New Style (*Astron., etc.*). A name given to the system of date-reckoning established by the *Gregorian calendar*. Abbrev. NS.

newton (*Elec. Eng.*). Symbol N. The unit of force in the SI system, being the force required to impart, to a mass of 1 kilogram, an acceleration of 1 metre per second per second. 1 newton = 0·2248 pounds force.

Newtonian telescope (*Astron.*). A form of reflecting telescope due to Newton, in which the object is viewed through an eyepiece in the side of the tube, the light reflected from the main mirror being deflected into it by a small plane-mirror inclined at 45° to the axis of the telescope and situated just inside the principal focus.

Newton's disk (*Optics*). Motor-driven disk with sectors of primary colours, which appears white on fast rotation and with white illumination, demonstrating the synthesis of colour vision.

Newton's law of cooling (*Heat*). The rate of cooling of a hot body which is losing heat both by radiation and by natural convection is proportional to the difference in temperature between it and its surroundings. The law does not hold for large temperature excesses.

Newton's laws of motion (*Phys.*). Briefly paraphrased, these are: (1) Every body either remains at rest or moves with constant speed in a straight line unless it is acted upon by a force. (2) A force applied to a body accelerates the body by an amount proportional to the force. (3) Every action is opposed by an equal and opposite reaction. The laws were first stated by Newton in his *Principia*, 1687. Classical mechanics consists merely in the application of these laws.

Newton's rings (*Light*). Circular concentric interference fringes seen surrounding the point of contact of a convex lens and a plane surface. Interference occurs in the air film enclosed between the two surfaces. If r_n is the radius of the *n*th ring, R the radius of curvature of the lens surface, and λ the wavelength, $r_n = \sqrt{nR\lambda}$.

New Zealand greenstone (*Min.*). Nephritic 'jade' of gemstone quality, from New Zealand.

NGTE rigid rotor (*Aero.*). A helicopter self-propelling rotor system evolved at the National Gas Turbine Establishment which uses the principle of the *jet flap* (q.v.) to obtain very high lift coefficients. Gas turbine exhaust efflux is ejected downward from a narrow spanwise slit in a tubular, oval-section, rotor blade, both to cause it to revolve and to induce a large downwash airflow. The tubular rotor is rigid because the balance of lift between the advancing and retreating halves of the blade, normally provided by the rotor blade hinges, is maintained by automatic variation of the gas flow through the slits. *Cyclic* and *collective pitch control* (qq.v) are replaced by variation of the gas supply to the slits.

Ni (*Chem.*). The symbol for *nickel.*

niacin (*Chem.*). See **vitamin B complex.**

Niagara Series or **Niagaran** (*Geol.*). Strata of Silurian age typically exposed in the Niagara Gorge section; includes the Medina Group below, and the Clinton, Rochester, and Lockport beds.

nib (*Build.*). A small projection, sometimes continuous, formed on the underside at the top of a tile, enabling the tile to be hung on battens. Also called a cog. (*Tools*) The point of a crowbar.

nibbling (*Eng.*). A sheet-metal cutting process akin to sawing, in which a rapidly reciprocating cutter slots the sheet along any desired path.

nibs (*Paint.*). Specks of solid matter in varnish.

Nicaloi (*Met.*). TN for a high-permeability magnetic alloy containing approximately equal amounts of nickel and iron.

niccolite (*Min.*). Arsenide of nickel, crystallizing in the hexagonal system. It usually contains a little iron, cobalt, and sulphur and is one of the chief ores of metallic nickel. Also called **copper nickel, kupfernickel.**

niche (*Ecol.*). See **ecological niche.**

niched column (*Arch.*). A column set back in a wall with a clear space between it and the wall.

Nicholson hydrometer (*Phys.*). A hydrometer of the constant-displacement type, used for determining the density of a solid.

Nichrome (*Met.*). TN for nickel-chromium alloy largely used for resistance heating elements because of its high resistivity ($\sim 110 \times 10^{-8}$ ohm metres) and its ability to withstand high temperatures.

nick (*Typog.*). The groove(s) in the shank of a type letter.

nickel (*Chem.*). Silver-white metallic element. Symbol Ni, at. no. 28, r.a.m. 58·71, m.p. 1450°C, b.p. 3000°C, electrical resistivity at 20°C, 10·9 $\times 10^{-8}$ ohm metres, rel. d. 8·9. Used for structural parts of valves. It is magnetostrictive, showing a decrease in length in an applied magnetic field and, in the form of wire, is much used in computers for small stores, the data circulating and extracted when required. Used pure for electroplating, coinage, and in chemical and food-handling plants. See also **nickel alloys.**

nickel alloys (*Met.*). Nickel is the main constituent in Monel metal, Permalloy, and nickel-chromium alloys. It is also used in cupro-nickel, nickel-silver, various types of steel and cast-iron, brass, bronze, and light alloys.

nickel antimony glance (*Min.*). See **ullmanite.**

nickel arsenic glance (*Min.*). See **gersdorffite.**

nickelates (*Chem.*). Salts of nickel oxide acting as an acid radical, with alkaline or alkaline earth oxides, e.g., potassium nickelate, K_2NiO_4.

nickel bloom (*Min.*). See **annabergite.**

nickel-cadmium accumulator (*Elec.*). Battery using nickel and cadmium compounds in potassium hydroxide electrolyte. Characterized by extremely low self-discharge rate; therefore used in emergency lighting systems, electronic flash apparatus, as a power supply for transistor TV receivers, etc.

nickel carbonyl (*Chem.*). $Ni(CO)_4$. A volatile compound of nickel, formed by passing carbon monoxide over the heated metal. The compound is decomposed into nickel and carbon monoxide by further heating. Used on a large scale in industry for the production of nickel from its ores by the Mond process.

nickel-chromium steel (*Met.*). Steel containing nickel and chromium as alloying elements. 1·5% to 4% nickel and 0·5% to 2% chromium are added to produce an alloy of high-tensile strength, hardness, and toughness. Used for highly stressed automobile and aero-engine parts, for armour plate, etc.

nickel electro (*Print.*). An electro in which a thin layer of nickel is first deposited, the shell being completed by depositing copper. Superior to *nickel-faced electro.*

nickel-faced electro (*Print.*). An electro which has

been given a facing of nickel to improve its quality. Cf. *nickel electro.*

nickeline (*Min.*). An old term for niccolite.

nickel-iron-alkaline accumulator or **cell** (*Elec. Eng.*). An accumulator in which the positive plate consists of nickel hydroxide enclosed in perforated steel tubes, and the negative plate consists of iron or cadmium also enclosed in perforated steel tubes. The electrolyte is potassium hydrate, the e.m.f. 1·2 volts per cell. It is lighter than the equivalent lead accumulator. Also **Edison accumulator, Ni-Fe accumulator.**

nickel silver (*Met.*). See **German silver.**

nickel steel (*Met.*). Steel containing nickel as an alloying element. Varying amounts, between 0·5% and 6·0%, are added to increase the strength in the normalized condition, to enable hardening to be performed in oil or air instead of water, or to increase the core strength of carburized parts.

nicker (*Carp.*). The side wing of a *centre-bit* (q.v.), scribing the boundary of the hole to be cut.

Niclad (*Met.*). Composite sheets made by rolling together sheets of nickel and mild steel, to obtain the corrosion resistance of nickel with the strength of steel.

Niclausse boiler (*Eng.*). A French marine boiler which consists of a horizontal water and steam drum from which vertical double headers are suspended, carrying banks of *field tubes* (q.v.) slightly inclined downwards.

Nicol prism (*Optics*). A device for obtaining plane-polarized light. It consists of a crystal of Iceland spar which has been cut and cemented together in such a way that the ordinary ray is totally reflected out at the side of the crystal, while the extraordinary plane-polarized ray is freely transmitted. Largely superseded by Polaroid.

nicopyrite (*Min.*). See **pentlandite.**

nicotine (*Chem.*). An alkaloid of the pyridine series, $C_{10}H_{14}N_2$, having the constitutional formula:

[Structural formula: pyridine ring linked to N-methylpyrrolidine ring; labels N, N, CH$_3$]

It occurs in tobacco leaves, and is extremely poisonous. It is a colourless oil, of nauseous odour; b.p. 246°C at 97 kN/m².

nicotinic acid (*Chem.*). See **vitamin B complex.**

nicrosilal (*Met.*). Cast-iron alloy of austenitic structure, containing nickel (18%) and silicon (about 5%); used particularly in construction of parts for high-temperature operations.

nictitating (*Zool.*). Winking; said of the third eyelid of the Vertebrates, which by its movements keeps clean the surface of the eye.

nidamental (*Zool.*). Said of glands which secrete material for the formation of an egg-covering.

nidation (*Zool.*). In the oestrous cycle of Mammals, the process of renewal of the lining of the uterus between the menstrual periods.

nidged ashlar (*Build.*). See **nigged ashlar.**

nidicolous (*Zool.*). Said of Birds which remain in the parental nest for some time after hatching. Cf. *nidifugous.*

nidification or **nidulation** (*Zool.*). The process of building or making a nest.

nidifugous (*Zool.*). Said of Birds which leave the parental nest soon after hatching. Cf. *nidicolous.*

nidus (*Zool.*). A nest; a small hollow resembling a nest; a nucleus.

niello (*Met.*). A method of decorating metal. Sunk designs are filled with an alloy of silver, lead, and copper, with sulphur and borax as fluxes, and fired.

Niemann-Pick disease (*Med.*). An inborn error of metabolism in which large amounts of lecithin are deposited in the reticulo-endothelial tissue.

Ni-Fe accumulator (*Elec. Eng.*). TN of a **nickel-iron-alkaline accumulator.**

niger morocco (*Bind.*). A goatskin tanned and dyed in Nigeria; limited range of colours, flexible, used for good quality binding.

nigged ashlar (*Build.*). A block of stone dressed with a pointed hammer.

nigger (*Cinema.*). Any adjustable panel which is used to cut off stray light which might enter the camera lens and cause fog.

night blindness (*Med.*). See **nyctalopia.**

night bolt (*Join.*). See **night latch.**

night error or **effect.** See **polarization error.**

night latch (*Join.*). A lock whose bolt is key-operated from the outside and knob-operated from the inside, but which is fitted with a device for preventing operation from either side.

nightshade (*Bot.*). Deadly nightshade (*Atropa belladonna*) is a plant containing various isomeric and closely related alkaloids, including *atropine* (q.v.).

nigrescent (*Bot.*). Becoming blackish.

nigrite (*Min.*). A pitchlike member of the asphaltite group.

nigropunctate (*Bot.*). Marked with black dots.

nigrosines (*Chem.*). Diphenylamine dyestuffs, used as black pigments, prepared by heating nitrobenzene or nitrophenol, aniline, and phenylammonium chloride with iron filings.

Ni-hard (*Met.*). Cast-iron to which a ladle addition of about 4½% nickel has been made, to render the alloy martensitic and abrasion-resistant.

nikethamide (*Chem.*). N-diethylnicotinamide, also known more generally as **coramine** (TN):

[Structural formula: pyridine ring with C(=O)–N(C$_2$H$_5$)$_2$ group; labels O, C, N(C$_2$H$_5$)$_2$, N]

Colourless oily liquid which freezes at about 22°C to crystalline solid. Used medically as a respiratory stimulant in treatment of shock, heart failure, etc.

nile (*Nuc. Eng.*). 1 nile corresponds to a *reactivity* of 0·01. In indicating reactivity changes, it is more usual to use the smaller unit, the **millinile,** equal to a change in reactivity of 10^{-5}.

nimbus (*Meteor.*). A dense layer of dark shapeless cloud with ragged edges, from which steady rain or snow usually falls.

nimonic (*Met.*). Alloy used in such high-temperature work as that called for in gas turbine blades. About 80% nickel and 20% chromium, with some titanium and aluminium.

nine-point circle (of a triangle) (*Maths.*). The circle which passes through the midpoints of the three sides, the points of intersection of each side with the perpendicular to that side from the opposite vertex, and, if *C* is the point where the three perpendiculars intersect, the three points midway between *C* and the respective vertices.

ninhydrin (*Chem.*). Tri*keto*-hydrindene-hydrate:

[structural formula: 2,2-dihydroxyindane-1,3-dione, with two C=O groups and two OH groups]

Used as a reagent to detect proteins or amino acids, with which it forms a characteristic blue colour on heating.

niobite (*Min.*). See **columbite.**

niobium (*Chem.*). Rare metallic element, symbol Nb, at. no. 41, r.a.m. 92·9064, m.p. 2500°C, used in high-temperature engineering products (e.g., gas turbines and nuclear reactors) owing to the strength of its alloys at temperatures of 1200°C+. Combined with tin (Nb_3Sn) it has a high degree of superconductivity. Formerly called **columbium** (Cb).

nip (*Chem. Eng.*). See **angle of nip.** (*Eng.*) The manufacturing practice of giving different curvatures to the individual leaves of a laminated spring before assembling, so as to attain the most favourable distribution of working stresses. (*Glass*) (1) The gap between rollers in a sheet glass rolling machine. (2) Bottle of small size for mineral water.

nip and tuck folder (*Print.*). A type of folder on a rotary press in which a fold at right angles to the web is formed by a blade thrusting the web between folding jaws.

Nipkow disk (*TV*). Rotating disk having a series of apertures arranged in the form of a spiral around the circumference; used for mechanical scanning and in the original Baird system.

nippers (*Build.*). See **stone tongs.**

nipping (*Bind.*). See **smashing.**

nipple (*Eng., Plumb., etc.*). (1) A short length of externally threaded pipe for connecting two lengths of internally threaded pipe. (2) A small drilled bush, sometimes containing a nonreturn valve, screwed into a bearing for the supply of lubricant by a *grease-gun* (q.v.). (3) A nut for securing (bicycle) wheel spokes. (*Zool.*) The mamma or protuberant part of the mammary gland in female Mammals, bearing the openings of the milk-forming glands.

nip rolls (*Paper*). A pair of steel rolls situated before the last section of drying cylinders of the paper machine to give a high finish to the paper whilst still moist. Also called **intermediates.**

Ni-resist (*Met.*). A cast-iron consisting of graphite in a matrix of austenite. Contains carbon 3%, nickel 14%, copper 6%, chromium 2%, and silicon 1·5%. Has a high resistance to growth, oxidation and corrosion.

nisin (*Pharm.*). A mixture of several polypeptides, crystalline. Produced in the growth of *Streptococcus lactis.* Food preservative.

Nissl bodies (*Zool.*). Angular bodies of granular appearance occurring in the cell-body of a neuron; they stain deeply with basic dyes.

Nissl degeneration (*Zool.*). Degenerative changes associated with chromatolysis, occurring in the cell-body of a neuron the axis-cylinder of which has been severed.

nit (*Comp.*). Unit of information equal to 1·44 *bits* (q.v.). (*Light*) Unit of luminance, 1 candela/m². Symbol **nt.** See **candela.**

niter. See **nitre.**

nitidous (*Bot.*). Lustrous.

nitometer (*Light*). An instrument for measuring the brightness of small light sources.

Nitralloy (*Met.*). Steel specially developed for nitriding (which is less effective with ordinary steels). Contains carbon 0·2–0·3%, aluminium 0·9–1·5%, chromium 0·9–1·5%, and molybdenum 0·15–0·25%.

nitramines (*Chem.*). Amines in which an amino-hydrogen has been replaced by the nitro group. They have the general formula $R{\cdot}NH{\cdot}NO_2$.

nitrate bacteria (*Bot.*). Soil-inhabiting bacteria which convert nitrites into nitrates.

nitrate-reducing bacteria (*Bacteriol.*). Facultative aerobes able to reduce nitrates to nitrites, nitrous oxide, or nitrogen under anaerobic conditions, e.g., *Micrococcus denitrificans.* This process is termed *denitrification.* A few bacteria use such reduction processes as hydrogen-acceptor reactions and hence as a source of energy; in this case the end product is ammonia.

nitrates (*Chem.*). Salts or esters of nitric(v)acid. Metal nitrates are soluble in water; decompose when heated. The nitrates of polyhydric alcohols and the alkyl radicals explode with violence. Uses: explosives, fertilizers, chemical intermediates.

nitration (*Chem.*). The introduction of nitro-(NO_2) groups into organic compounds. Usually carried out by using mixed concentrated nitric and sulphuric acids. The final stage of nitrification in the soil.

nitrazepam (*Pharm.*). 2,3-Dihydro-7-nitro-5-phenyl-1H-1,4-benzodiazepin-2-one; in yellow powder widely used non-barbiturate hypnotic; TN: Mogadon.

nitre (*Chem.*). *Potassium nitrate* (v), crystallizing in the orthorhombic system. Also called **saltpetre.** See also **Chile nitre, soda nitre.**

nitric(V)acid (*Chem.*). NHO_3, a fuming unstable liquid, b.p. 83°C, m.p. −41·59°C, rel. d. 1·5, miscible with water. Old name **aqua fortis.** Prepared in small quantities by the action of conc. sulphuric acid on sodium nitrates and on large scale by the oxidation of nitrogen or ammonia. An important intermediate for fertilizers, explosives, organic synthesis, metal extraction, and sulphuric acid manufacture. See **nitrates, chamber process.**

nitric anhydride (*Chem.*). Nitrogen(v)oxide. N_2O_5. Dissolves in water to give nitric acid.

nitric [nitrogen(II)] oxide (*Chem.*). NO. Colourless gas. In contact with air it forms reddish-brown fumes of nitrogen dioxide.

nitrides (*Chem.*). Compounds of metals with nitrogen. Usually prepared by passing nitrogen or gaseous ammonia over the heated metal. Those of Group I and II metals are ionic compounds which react with water to release ammonia. Those of Group III to V exist as interstitial compounds, having great hardness and refractoriness, e.g., boron, titanium, iron.

nitriding, nitrogen case-hardening (*Met.*). A process for producing hard surface on special types of steel by heating in gaseous ammonia. Components are finish-machined, hardened and tempered, and heated for about 60 hr at 520°C. Case is about 0·5 mm. deep and surface hardness is 1100 V.P.N. See **Nitralloy.**

nitrification (*Bacteriol., Bot.*). The conversion of nitrogenous matter into nitrates by bacteria, especially in the soil. See **ammonization, nitration, nitrozation.** (*Chem.*) The treatment of a material with nitric acid.

nitrifying bacteria (*Bacteriol., Bot.*). See **Nitrobacteriaceae.**

nitriles (*Chem.*). Alkyl cyanides of the general formula $R{\cdot}C{\equiv}N$. When hydrolysed they yield carboxylic acids or the corresponding ammonium salts; reduced, they yield **amines.**

nitrite bacteria (*Bacteriol., Bot.*). Soil-inhabiting

bacteria which form nitrites (nitrates(III)) from ammonia.

nitrites [nitrates(III)] (*Chem.*). Salts or esters of nitrous acid, O=NOH.

nitroanilines (*Chem.*). The 4- and 3-isomers are used as important intermediates in the preparation of azo dyestuffs.

Nitrobacteriaceae (*Bacteriol., Bot.*). Family of bacteria belonging to the order *Pseudomonadales*. Important in nitrification processes in the soil and fresh water. Autotrophic bacteria which derive energy from oxidation processes: *Nitrosomonas* from the oxidation of ammonia to nitrites: *Nitrobacter* from the oxidation of nitrites to nitrates. Also **nitrifying bacteria.**

nitrobenzene (*Chem.*). $C_6H_5{\cdot}NO_2$, a yellow liquid with an odour of bitter almonds, m.p. 5°C, b.p. 211°C, rel. d. 1·2. It is obtained by the action of a mixture of concentrated sulphuric acid and nitric acid on benzene. When reduced it yields aniline.

nitrocelluloses (*Chem.*). Cellulose nitrates. They are the nitric acid esters of cellulose formed by action of a mixture of nitric and sulphuric acids on cellulose. The cellulose can be nitrated to a varying extent ranging from 2 to 6 nitrate groups in the molecule. Nitrocelluloses with a low nitrogen content, up to the tetranitrate, are not explosive, and are used in lacquer and artificial silk and imitation leather book cloth manufacture. They dissolve in ether-alcohol mixtures, and in so-called lacquer solvents. A nitrocellulose with a high nitrogen content is gun-cotton, an explosive (see **pyroxylins**).

nitro derivatives (*Chem.*). Aliphatic or aromatic compounds containing the group $—NO_2$. The aliphatic nitro derivatives are colourless liquids which are not readily hydrolysed, but have acidic properties, e.g., the primary and secondary aliphatic nitro derivatives can form metallic compounds. Aromatic nitro derivatives are easily formed by the action of nitric acid on aromatic compounds. The nitro groups substitute in the nucleus and only exceptionally in the side chain.

nitrogen (*Chem.*). Gaseous element, colourless and odourless. Symbol N, at. no. 7, r.a.m. 14·0067, m.p. −209·86°C, b.p. −195°C. It was the determination by Rayleigh and Ramsay of its rel. at. mass that led to the discovery of the inert gases in the atmosphere, argon, etc. Approx. 80% of the normal atmosphere is nitrogen, which is also widely spread in minerals, the sea, and in all living matter. Used as a neutral filler in filament lamps, in sealed relays, and in Van de Graaff generators; and in high-voltage cables as an insulant. Liquid is a coolant. See **nitric acid, ammonia, nitrates.**

nitrogen balance (*Biochem., Physiol.*). The state of equilibrium of the body in terms of intake and output of nitrogen; positive nitrogen balance indicates intake exceeds output, negative nitrogen balance denotes output exceeds intake.

nitrogen bases (*Chem.*). See **amines.**

nitrogen case-hardening (*Met.*). See **nitriding.**

nitrogen chlorides (*Chem.*). Three nitrogen chlorides, NH_2Cl, $NHCl_2$, and NCl_3 produced by the chlorination of ammonium ions. Unstable and explosive.

nitrogen cycle (*Bacteriol.*). The sum total of the transformations undergone by nitrogen and nitrogenous compounds in nature in relation to living organisms.

nitrogen [dioxide](IV) oxide (*Chem.*). NO_2. Oxidizing agent; formed when 1 volume of oxygen is mixed with 2 volumes of nitric oxide. Below 17°C the formula is N_2O_4 and the molecule is colourless. Decomposed by water, forming a mixture of nitric and nitrous acids.

nitrogen fixation (*Bot., Chem.*). See **fixation of nitrogen.**

nitrogen mustards (*Chem.*). Derivatives of *mustard gas* (q.v.). They are important reagents in *molecular genetics*, being powerful chemical *mutagens*. Nitrogen mustard itself has the formula $CH_3N(C_2H_5Cl)_2$, and acts by eliminating chlorine to form cross-links between units of *guanine* in *nucleic acids*, thus causing errors in the *genetic code*. The hydrochloride (*mustine*) is a potent cytotoxin; used in certain cases of Hodgkin's disease, leukaemia, etc. See **radiomimetic, carcinogen.**

nitrogen narcosis (*Med.*). 'Rapture of the deep.' Intoxicating and anaesthetic effect of too much nitrogen in the brain, experienced by divers at considerable depths.

nitrogen pentoxide (*Chem.*). See **nitric anhydride.**

nitrogen [tetroxide](IV) oxide (*Chem.*). N_2O_4. See **nitrogen dioxide.**

nitroglycerine (*Chem.*). $C_3H_5(ONO_2)_3$, a colourless oil; m.p. 11°–12°C; insoluble in water; prepared by treating glycerine with a cold mixture of concentrated nitric and sulphuric acids. It solidifies on cooling, and exists in 2 physical crystalline modifications. In thin layers it burns without explosion, but explodes with tremendous force when heated quickly or struck. See also **dynamite, explosive.** Used medically in solution and tablets for angina.

nitrol compound (*Chem.*). Organic compound having a *nitroso* group —NO and a *nitro* group $—NO_2$ on the same carbon atom.

nitrolic (*Chem.*). Having a *nitro* group $—NO_2$ and an *oxime* group =NOH on the same carbon atom.

Nitrolime (*Chem.*). TN for an artificial fertilizer consisting of calcium cyanamide.

nitro-metals (*Chem.*). Metals which unite with nitrogen peroxide to give metallic nitroxyls.

nitromethane (*Chem.*). $CH_3{\cdot}NO_2$, a liquid; b.p. 99°–101°C; prepared from chloroacetic acid and sodium nitrite.

nitron (*Chem.*). 1,4-Diphenyl-3,5-*end*anilodihydrotriazole. It forms an insoluble salt with nitric acid, and serves as a reagent for the quantitative estimation of nitric acid.

nitrophilous (*Bot.*). Said of plants which occur characteristically in places where there are good supplies of nitrogenous compounds.

nitroplatinate (*Chem.*). The anion $Pt(NO_2)_4^{2-}$. Similar complex nitrates are formed by other members of the platinum family.

nitroprussides (*Chem.*). Formed by the action of nitric acid on either hexacyano ferrates (II) or (III). Also called **nitrosoferricyanides.**

nitroso compounds (*Chem.*). Compounds containing the monovalent radical —NO.

nitroso-dyes (*Chem.*). Dyestuffs resulting from reaction between phenols and nitrous acid.

nitrosoferricyanides (*Chem.*). See **nitroprussides.**

Nitrosomonas (*Bacteriol., Bot.*). See **Nitrobacteriaceae.**

nitrososulphuric acid (*Chem.*). See **chamber crystals.**

nitrotoluenes (*Chem.*). $CH_3{\cdot}C_6H_4{\cdot}NO_2$. On nitration of toluene a mixture of 2- and 4-nitrotoluene is obtained with very little 3-nitrotoluene. 2-Nitrotoluene, a liquid, has a b.p. 218°C; 4-nitrotoluene crystallizes in large crystals; m.p. 54°C, b.p. 230°C.

nitrous [nitric(III)] acid (*Chem.*). The pale-blue unstable solution obtained by precipitating barium nitrite with dilute sulphuric acid is supposed to contain nitrous acid, HNO_2.

nitrous oxide (*Chem.*). Laughing gas, N_2O. A colourless gas with a sweetish odour and taste, soluble in water, alcohol, ether and benzene. Nitrous oxide is endothermic, and supports combustion better than air. The gas is manufactured by the decomposition of ammonium nitrate by heat, and purified by passage through solutions of ferrous sulphate, caustic potash and milk of lime. It is used for producing anaesthesia of short duration.

nitroxyl (*Chem.*). The radical $—NO_2$ when attached to a halogen atom or a metal. Compounds containing the group are *nitroxyls*.

nitrozation (*Bacteriol., Bot.*). The conversion of ammonia into nitrites by the action of soil bacteria (*Nitrosomonas*), being the second stage in the nitrification in the soil.

Nivarox (*Horol.*). Alloy of iron and nickel with a small addition of beryllium, nonmagnetic, rustless, and of controllable elasticity, used for hairsprings.

nm. Abbrev. for **nanometre** = 10 Ångstroms.

NMR (*Chem., Nuc.*). Abbrev. for *nuclear magnetic resonance*.

***n-n* junction** (*Electronics*). One between crystals of *n*-type semiconductors, having different electrical properties.

No (*Chem.*). The symbol for *nobelium*.

No. 606 (*Pharm.*). Salvarsan. **No. 914.** Neo-salvarsan.

nobelium (*Chem.*). Manmade element. At. no. 102; symbol No. Principal isotope is 254.

noble gases (*Chem.*). See **inert gases.**

noble metals (*Met.*). Metals, such as gold, silver, platinum, etc., which have a relatively positive electrode potential, and which do not enter readily into chemical combination with non-metals. They have high resistance to corrosive attack by acids and corrosive agents, and resist atmospheric oxidation. Cf. *base metal*.

no bottom, no top (*Acous.*). Said of sounds which, when reproduced, are attenuated in the low or/and high audiofrequency range.

nocardiasis (*Med.*). Infection (usually of the lungs) with any one of a number of spore-forming fungi of the genus *Nocardia*.

nociceptive (*Zool.*). Sensitive to pain.

noctilucent (*Zool.*). Phosphorescent: light-producing.

noctilucent clouds (*Meteor.*). Clouds of dust visible only against a night sky drifting at a height of 80 km and at a speed of 640 km/h.

Noctovision (*TV*). TN for a system of television in which the light-sensitive elements respond to infrared light, and which can therefore be operated in apparent darkness.

nodal gearing (*Eng.*). The location of gear-wheels, e.g., between a turbine and propeller shaft, at a nodal point of the shaft system with respect to torsional vibration.

nodalizer (*Telecomm.*). One which adjusts to minimum disturbing effects in an electrical circuit, such as hum in an amplifier, or interfering current in a bridge.

nodal point or **node** (*Radio*). Point in a high-frequency circuit where current is a maximum and voltage a minimum, or vice-versa.

nodal points of a lens (*Light*). Two points on the principal axis of a lens or lens system such that an incident ray of light directed towards one of them emerges from the lens as if from the other, in a direction parallel to that of the incident ray. For a lens having the same medium on its two sides, the nodal points coincide with the principal points.

node (*Acous.*). The location where the interference between two or more progressive sound waves results in a standing wave, with either the sound pressure or particle velocity zero or a minimum. (*Bot.*) The place where a leaf is attached to a stem; it is sometimes swollen. (*Elec. Eng.*) Point of minimum disturbance in a system of waves in tubes, plates or rods. The amplitude cannot become zero, otherwise no power could be transmitted beyond the point. (*Maths.*) See **double point.** (*Radio*) See **nodal point.** (*Telecomm.*) In electrical networks, a terminal common to two or more branches of a network or to a terminal of any branch.

node of Keith and Flack (*Zool.*). See **Keith and Flack's node.**

node of Ranvier (*Zool.*). See **Ranvier's node.**

nodes (*Astron.*). The two points, diametrically opposite each other, in which the orbit of a heavenly body cuts some great circle, the ecliptic in the case of the moon and planets. The *ascending* and *descending* nodes are distinguished by the body's passage to north or south of the reference plane in question.

node voltage (*Telecomm.*). That of some point in an electrical network with reference to a node.

nodon rectifier (*Electronics*). An electrolytic device in which ammonium phosphate is the electrolyte, aluminium is the cathode and lead is the anode.

nodose (*Bot.*). Bearing knotlike swellings.

nodular (*Bot.*). Bearing local thickenings; said especially of an elongated plant member. Also **nodulose.** (*Powder Tech.*) Of particles, having a rounded, irregular shape.

nodular structures (*Geol.*). Spheroidal, ovoid, or irregular bodies often encountered in both igneous and sedimentary rocks, and formed by segregation about centres. See, for example, **clay ironstone, doggers, flint, septaria.**

nodule (*Bot.*). (1) Any small rounded structure on a plant. (2) A swelling on a root inhabited by symbiotic bacteria.

nodule of Arantius (*Histol.*). A fibrous tissue thickening in the middle of the free edge of the semilunar valves of the heart.

nodulizing (*Met., Min. Proc.*). Aggregation of finely divided material such as mineral concentrates, by aid of binder and perhaps kilning, into nodules sufficiently strong and heavy to facilitate subsequent use, such as charging into blast furnaces.

nodus (*Zool.*). In Insects, a joint or swelling formed by the union of two nervuers.

no-fines concrete (*Build., Civ. Eng.*). Concrete in which the fine aggregate has been omitted. Therefore open textured, with a comparatively low strength, but other advantages. It can be cast *in situ* using simple, even open-mesh shuttering, sets rapidly, has little capillary attraction and low moisture movement, and can be plastered and rendered readily owing to its texture. Normally used for low building or as in-filling to framed high buildings. Has however been used as load-bearing up to 12 storeys and research into mix proportions will probably increase the possible load-bearing capacities.

nog (*Build.*). A block of wood built into a wall to provide a substance to which joinery, such as skirtings, may be nailed.

nogging (*Build.*). The filling in with bricks of the spaces between timbers in walls and partitions.

nogging-piece (*Build.*). A timber in a brick-nogging wall or partition.

noil (*Textiles*). (1) Short or broken fibres from waste silk after opening and combing. (2) Short fibre removed from wool during combing; used to enhance the quality in woollen blends.

noise (*Acous.*). (1) Socially unwanted sounds.

(2) Interference in a communication channel. (3) Clicks arising from radiation due to sudden changes of current. See also **frying, hum, loudness level, phon.**

noise audiogram (*Acous.*). That taken in the presence of a specified masking noise.

noise audiometer (*Acous.*). An audiometer which measures the threshold of hearing of a deaf person's ear, the other ear, or both ears, being subjected to a standardized noise in addition to the test sound.

noise background (*Acous.*). In reproduction or recording, the total noise in the system, with that due to the signal absent.

noise bandwidth (*Acous.*). That of an ideal power gain-frequency response curve (i.e., one with a completely flat response over the bandwidth concerned) with the same gain as the maximum of, and subtending the same area as, the actual response curve of an amplifier. It is given by the integral:

$$\int_0^\infty \frac{P_f}{P_{max}}\,df,$$

where P_f is the power gain as a function of frequency f and P_{max} is the maximum power gain.

noise current (*Telecomm.*). That part of a signal current conveying noise power.

noise diode (*Elec. Eng.*). One operating as noise generator, under temperature-limited conditions.

noise factor or figure (*Acous.*). Ratio of noise in a linear amplification system to thermal noise over the same frequency band and at the same temperature.

noise field (*Acous.*). The electromagnetic radiation field due to all sources which can interfere with reception of the required signal.

noise generator (*Elec. Eng.*). Device for producing a controlled noise signal for test purposes. Also **noise source.** See **noise diode.**

noise intensity (*Acous.*). That of the noise field in a specific frequency band.

noiseless recording (*Acous.*). The practice of making the sound track of a positive sound-film as dense as possible, consistent with the accommodation of the modulation, in order to keep the photographic noise level as far below the recorded level as possible.

noise level (*Acous.*). The *loudness level* (q.v.) of a noise signal.

noise limiter (*Acous.*). Device for removing the high peaks in a transmission, thus reducing contribution of clicks to the noise level and eliminating acoustic shocks. Effected by biased diodes or other clipping circuits.

noise meter (*Acous.*). See **objective noise meter.**

noise power (*Electronics*). That dissipated in a system by all noise signals present. See **Nyquist noise theorem.**

noise ratio (*Acous.*). Noise power in a terminated filter or circuit divided into the signal power; expressed in decibels. Also **signal/noise ratio, speech/noise ratio.**

noise reduction (*Acous.*). Procedure in photographic recording whereby the average density in the negative sound-track is kept as low as possible, and hence in the positive print as high as possible, both consistent with linearity, so that dirt, random electrostatics and scratches produce less noise.

noise resistance (*Acous.*). One for which the thermal noise would equal the actual noise signal present, usually in a specific frequency band.

noise source (*Acous.*). The object or system from which the noise originates. (*Elec. Eng.*) See **noise generator.**

noise suppressor (*Aero.*). A turbojet propelling nozzle fitted with fluted members which induct air to slow and break up the jet efflux, thereby reducing the noise level. (*Radio*) Circuit which suppresses noise between usable channels when these are passed through during tuning. An *automatic gain control* which cuts out weak signals and levels out loud signals.

noise temperature (*Electronics*). Temperature at which the thermal noise power of a passive system per unit bandwidth is equal to the noise at the actual terminals. The standard reference temperature for noise measurements is 290 K. At this temperature the available noise power per unit bandwidth (see **Nyquist noise theorem**) is 4×10^{-21} watts.

noise transmission impairment (*Acous.*). The transmission loss in dB which would impair intelligibility of a telephone system to the same extent as the existing noise signal. Abbrev. **NTI.**

noise voltage (*Acous.*). A noise signal measured in r.m.s. volts.

noisy blacks (*Acous., TV*). The non-uniformity of the black areas of a picture due to the level changes arising from noise.

Noland's stain (*Micros.*). A combined fixative and staining method used for cilia and flagella in temporary preparations. It contains phenol, formalin, glycerol, and gentian violet.

no-load characteristic (*Elec. Eng.*). See **open-circuit characteristic.**

no-load current (*Elec. Eng.*). The current taken by a transformer when it is energized but is giving no output, or by a motor when it is running but taking no mechanical load.

no-load loss (*Elec. Eng.*). The losses occurring in a motor or transformer when it is operating but giving no output. Also **open-circuit loss.**

noma (*Med.*). See **cancrum oris.**

nomadism (*Ecol.*). The habit of some animals of roaming irregularly without regularly returning to a particular place. Cf. *migration.*

Nomag (*Met.*). TN for nonmagnetic, high-resistance cast-iron alloy containing nickel (10–12%) and manganese (5%).

nomeristic (*Zool.*). Of metameric animals, having a definite number of somites.

nominal section (*Telecomm.*). Network which is equivalent to section of transmission line, based on the assumption of lumped constants.

nomogram (*Maths.*). Chart or diagram of scaled lines or curves for facilitating calculations. Those comprising three scales in which a line joining values on two determines a value on the third are frequently called *alignment charts.* Also called **nomograph.**

nomotagmosis (*Zool.*). In metameric animals, the formation of definite regions (tagmata) of the body by the differentiation of a definite number of somites.

non-, nona- (*Chem.*). Containing 9 atoms, groups, etc.

nonagon (*Maths.*). A nine-sided polygon.

nonane (*Chem.*). C_9H_{20}, a paraffin hydrocarbon; m.p. −51°C, b.p. 150°C, rel. d. 0·72.

nonaqueous solvents (*Chem.*). May be classed broadly into (*a*) waterlike or levelling solvents, which are highly polar and form strong electrolytic solutions with most ionizable solutes, (*b*) differentiating solvents, which bring out differences in the strength of electrolytes. Examples of (*a*) are $NH_3(NH_4^+—NH_2^-)$, $SO_2(SO^{++}—SO_3^{--})$, $N_2O_4(NO^+—NO_3^-)$. Example of (*b*) are weak amines or acids, ethers, halogenated hydrocarbons.

nonaqueous titration (*Chem.*). Certain substances such as weak acids or bases, or compounds sparingly soluble in water, are better titrated in *nonaqueous solvents*. The strengths of weak acids are enhanced in *protophilic* solvents, and of weak bases in *protogenic* solvents, making a sharp *end-point* possible. The use of *amphiprotic* or *aprotic* solvents, or mixtures, makes available a wide choice of *levelling* or *differentiating* properties.

nonassociation cable (*Elec. Eng.*). Cable which is not manufactured or designed in accordance with the standards of the Cable Makers' Association.

nonbearing wall (*Build.*). A wall carrying no load apart from its own weight.

nonbridging (*Telecomm.*). See **bridging (1)**.

noncaducous (*Zool.*). See **indeciduate**.

nonconductor (*Electronics*). Under normal conditions, an electrical insulator in which there are very few free electrons.

nonconjunction (*Cyt.*). The complete failure of syndesis; failure of chromosome pairing.

nondegenerate gas (*Electronics*). That which is insufficiently concentrated, e.g., hot-cathode electrons or ordinary gases, so that the *Maxwell-Boltzmann distribution law* (q.v.) is applicable.

nondenominational number system (*Maths.*). Any system which is not denominational. The best-known is the Roman number system but a large variety have been devised for use in digital computers.

nondirectional microphone (*Acous.*). Same as **omnidirectional microphone.**

nondisjunction (*Cyt.*). Failure of 2 chromosomes to disjoin in meiosis.

nondissipative network (*Telecomm.*). One designed as if the inductances and capacitances are free from dissipation, and as if constructed with components of minimum loss.

nonerasable storage (*Comp.*). Data storage in a form (e.g., punched card or tape) which cannot be lost during the normal course of computation.

nonessential organs (*Bot.*). Sepals and petals.

Nonex (*Glass*). TN for brand of glass effective in transmitting ultraviolet light.

nonferrous alloy (*Met.*). Any alloy based mainly on metals other than iron, i.e., usually on copper, aluminium, lead, zinc, tin, nickel, or magnesium.

nonflam film (*Cinema.*). See **safety film.**

Nonidet (*Chem.*). TN for *nonionic detergent*, based on the condensation products of polyglycols with octyl or nonyl hydroxybenzenes.

noninductive capacitor (*Telecomm.*). One specially designed for impulse work to have its inherent inductance reduced to a minimum.

noninductive circuit (*Elec. Eng.*). One in which effects arising from associated inductances are negligible.

noninductive load (*Elec. Eng.*). See **nonreactive load.**

noninductive resistor (*Elec. Eng.*). Resistor having a negligible inductance, e.g., comprising a ceramic rod, or a special design of winding of resistance wire.

nonionic detergents (*Chem.*). Series of detergents in which the molecules do not ionize in aqueous solution, unlike soap and the sulphonated alkylates. Typical examples are the detergents based on condensation products of long-chain glycols and octyl or nonyl hydroxybenzenes.

nonisolated essential singularity (*Maths.*). If a function $f(x)$ has an infinite set of singularities at the points $a_0, a_1, \ldots$, which tend to a limit point α, α is a nonisolated essential singularity.

nonleakage probability (*Nuc.*). For neutron in reactor, the ratio of the actual multiplication constant to the infinite multiplication constant.

nonlinear (*Elec. Eng.*). The relationship between instantaneous voltage applied to a component or network and the resulting current is dependent on amplitude.

nonlinear distortion (*Telecomm.*). That in which alien tones are introduced by nonlinear response of part of a communicating system.

nonlinear distortion factor (*Telecomm.*). Square root of ratio of the powers associated with alien tones to the powers associated with wanted tones in the output of a nonlinear distorting device.

nonlinearity (*Elec. Eng., Telecomm.*). Lack of proportionality between output and input currents and voltages of a network, amplifier, or transmission line, resulting in distortion of a passing signal. Marked nonlinearity is a property of diodes, triodes, transistors, etc.

nonlinear network (*Elec. Eng.*). A network in which the electrical elements are not all linear with varying current, as rectifying semiconductor diodes.

nonlinear resistance (*Elec. Eng.*). Nonproportionality between potential difference and current in an electric component.

nonlinear resistor (*Elec. Eng.*). One which does not 'obey' Ohm's law, in that there is departure from proportionality between voltage and current. In semiconductor crystals, ratio between forward and backward resistance may be 1 : 1000. See **diode, thyrite, transistor.**

nonlocking relay or key (*Telecomm.*). Key or relay which returns to its unoperated condition when the hand is removed, or the current ceases, usually by the action of a spring, which is extended on operation. The key is said to *operate* and *restore*, and the relay to *make* and *release*, or *operate* and *de-operate* or *fall off*.

nonmagnetic steel (*Elec. Eng.*). One containing ca. 12% Mn, exhibiting no magnetic properties. Stainless steel is almost nonmagnetic.

nonmagnetic watch (*Horol.*). A watch so constructed that its performance is not affected by magnetic fields. Usually, the balance, balance spring, roller, and fork are made of a nonmagnetic alloy.

nonmedullated (*Zool.*). See **amyelinate.**

nonmetal (*Chem.*). An element which readily forms negative ions, often in combination with other nonmetals. Nonmetals are generally poor conductors of electricity.

nonmetallic inclusions (*Met.*). See **inclusions.**

non-Newtonian liquids (*Phys.*). See **anomalous viscosity.**

nonohmic (*Elec.*). See **ohmic.**

nonoperable instruction (*Comp.*). One whose only effect is to advance the instruction index counter. Often written as 'continue'.

nonoses (*Chem.*). A group of monosaccharides, containing nine oxygen atoms in the molecule, e.g., $HO{\cdot}CH_2{\cdot}(CHOH)_7{\cdot}CHO$.

nonpareil (*Typog.*). An old type size approximately *6-point*, for which it is still synonymous.

nonpolarized relay (*Telecomm.*). One in which there is no magnetic polarization. Operation depends on the square of the current in the windings, and is therefore independent of direction, as in telephone and a.c. relays.

nonquantized (*Phys.*). See **classical.**

nonreactive load (*Elec. Eng.*). A load in which the current is in phase with the voltage across its terminals. Also called **noninductive load.**

nonreactive power (*Elec. Eng.*). The root-mean-square value of the active power and of the distortion power in an electric system.

nonrelativistic (*Phys.*). Said of any procedure in which effects arising from relativity theory are absent or can be disregarded, e.g., properties of particles moving with low velocity, e.g., 1/20th that of light propagation.

nonresonant antenna (*Radio*). See **aperiodic antenna.**

nonreturn-flow wind tunnel (*Aero.*). A straight-through wind tunnel in which the air flow is not recirculated.

nonsensibility (*Bot.*). The ability of a plant to support the development of a parasite without showing marked signs of disease.

nonsequence (*Geol.*). A break in the continuity of the stratigraphical column, less important and less obvious than an unconformity, and deduced generally on palaeontological evidence.

nonsingular matrix (*Maths.*). A square matrix the determinant of which is not equal to zero.

nonspeaking stop (*Acous.*). See **speaking stop.**

nonspectral colour (*Light*). That which is outside the range which contributes to white light, but which affects photocells.

nonspecular reflection (*Light, etc.*). Wave reflection of light or sound from rough surfaces, resulting in scattering of wave components, depending on relation between wavelength and dimensions of irregularities. Also called **diffuse reflection.**

nonstoichiometric compounds (*Chem.*). Some solid compounds do not possess the exact compositions which are predicted from Daltonic or electronic considerations alone (e.g. iron(II)sulphide is $FeS_{1\cdot1}$), a phenomenon which is associated with the so-called *defect structure of crystal lattices.* Often show semi-conductivity, fluorescence, and centres of colour.

nonstorage tube (*TV*). Camera tube in which output signal depends on incident light intensity, and not on its integration over a defined period of time between scannings. See **iconoscope.**

nonsweating (*Vet.*). See **anhidrosis.**

nonsymmetrical (*Elec. Eng.*). See **asymmetrical.**

nonsynchronous computer (*Comp.*). See **asynchronous computer.**

nonsynchronous motor (*Elec. Eng.*). An a.c. motor which does not run at synchronous speed, e.g., an induction motor or an a.c. commutator motor. Also called **asynchronous motor.**

nontension joint (*Elec. Eng.*). A joint in an overhead transmission line conductor which is designed to carry full-load current but not to withstand the full mechanical tension of the conductor.

nontronite (*Min.*). A clay mineral in the montmorillonite group (smectites), containing appreciable ferric iron replacing aluminium.

non-uniform flow (*Hyd.*). Flow in a channel when the water surface is not parallel to the invert.

nonvalent (*Chem.*). See **zero-valent.**

nonviable (*Bot., Zool.*). Incapable of surviving.

nonvolatile memory (*Comp.*). One in computer which holds data even if power has been disconnected. A magnetic memory is nonvolatile, an ultrasonic one is volatile (it does not comply with the above criterion).

nonwoven fabrics (*Textiles*). Cloth formed from a random arrangement of natural or synthetic fibres by adhesives, heat and pressure, or needling techniques.

noon (*Astron.*). The instant of the sun's upper culmination at any place. See also **mean-, sidereal-.**

No-Pack (*Print.*). A stereotype flong with a thick backing layer which eliminates the need to pack the back before casting.

NOR (*Comp., Maths.*). See **logical operations.**

noradrenaline (*Chem.*). A sympathomimetic substance obtained from the adrenal glands, having similar properties to *adrenaline* (q.v.) and differing in structure only by not having the amino group substituted with a methyl group.

norbergite (*Min.*). Magnesium orthosilicate with magnesium fluoride or hydroxide. Crystallizes in the orthorhombic system and occurs in metamorphosed dolomitic limestones.

Nordhausen sulphuric acid (*Chem.*). See **fuming sulphuric acid.**

nordmarkite (*Geol.*). A type of quartz-bearing soda-syenite, described originally from Nordmarken in Norway; consists essentially of microperthites, aegirine, soda amphibole, and accessory quartz.

nor element (or **gate**) (*Comp.*). An *or element* followed by a negater since the output is 1 when neither of the inputs is 1.

norepinephrine (*Chem.*). See **noradrenaline.**

Norfolk (or **Canadian**) **latch** (*Join.*). A latch in which the fall bar is actuated within the limits of the keeper by a lifting lever passing through a slot in the door and operated by the pressure of the thumb at one end.

norgine (*Chem.*). See **alginic acid.**

norite (*Geol.*). A coarse-grained igneous rock of basic composition consisting essentially of plagioclase (near labradorite in composition) and orthopyroxene. Other coloured minerals are usually present in varying amount, notably clinopyroxene, which, however, must not exceed half of the total pyroxene content.

norm (*Geol.*). The composition of an igneous rock expressed in terms of standard mineral molecules, calculated from the chemical analysis as stated in terms of percentages of oxides. Cf. *mode*. (*Maths., Stats.*) (1) The average. The value of a quantity or of a state which is statistically most frequent. (2) See **modulus.**

normal (*Maths., etc.*). The *normal* to a line or surface is a line drawn perpendicular to it.

normal (*Chem.*). Containing an unbranched chain of carbon atoms; e.g., *normal* propyl alcohol is $CH_3{\cdot}CH_2{\cdot}CH_2{\cdot}OH$, whereas the isomeric *iso*propyl alcohol, $(CH_3)_2{=}CH{\cdot}OH$, has a branched chain. This system of naming is now slowly disappearing in favour of *IUPAC.* Abbrev. *n-*.

normal axis (*Aero.*). See **axis.**

normal bend (*Elec. Eng.*). A section of conduit bent to a moderately large radius; used in an electrical installation for connecting two other pieces of conduit which are at right angles to each other.

normal calomel electrode (*Chem.*). A calomel electrode containing normal potassium chloride solution.

normal curvature (*Maths.*). See **curvature (3).**

normal distribution (*Stats.*). The distribution

$$y=\frac{1}{\sqrt{2\pi}}\,e^{-x^2/2}.$$

normal electrode potential (*Chem.*). See **standard electrode potential.**

normal fault (*Geol.*). A fracture in rocks along which relative displacement has taken place under tensional conditions, the *fault* (q.v.) hading to the downthrow side. Cf. *reversed fault.*

normal flight (*Aero.*). All flying other than aerobatics including straight and level, climbing, gliding, turns and *sideslips* (q.v.) for the loss of height or to counteract drift; a licensing category for certifying whether *airworthy* (q.v.).

normal form (*Maths.*). See **canonical form.**

normal functions (*Maths.*). See **orthogonal functions.**

normal induction (*Elec. Eng.*). That represented by the curve of normal magnetization.

normality (*Chem.*). The concentration of a solution expressed in gram-equivalents of active material per dm^3 of solution.

normalizing (*Met.*). A heat treatment applied to steel. Involves heating above the critical range, followed by cooling in air. Is performed to refine the crystal structure and eliminate internal stress.

normal magnetization (*Elec. Eng.*). Locus of the tips of the magnetic hysteresis loops obtained by varying the limits of the range of alternating magnetization.

normal modes (*Telecomm.*). In a linear system, the least number of independent component oscillations which may be regarded as constituting the free, or natural, oscillations of the system. Also called **natural modes.**

normal mole volume (*Chem.*). See **molar volume.**

normal order of crystallization (*Geol.*). A term which perpetuates a misconception. There is no generally applicable order of crystallization, but for any particular rock, the study of thin sections enables the observer to deduce the order in which the minerals *finished* crystallizing. The order in many rocks is: accessories, mafic minerals in the order olivine, orthopyroxenes, clinopyroxenes, amphiboles, micas, overlapping with plagioclase, orthoclase, quartz; but there are many exceptions.

normal polarization (*Radio*). That of an electromagnetic wave radiated from a vertical antenna, when measured on the ground not too far from the transmitter. Electric component is vertical, and magnetic component horizontal.

normal pressure (*Chem.*). Standard pressure, $101{\cdot}325\ kN/m^2 = 760$ torr, to which experimental data on gases are referred.

normal radius of curvature (*Maths.*). See **curvature.**

normal salts (*Chem.*). Salts formed by the replacement by metals of all the replaceable hydrogen of the acid.

normal section of a surface (*Maths.*). A section of the surface made by a plane which contains a normal to the surface.

normal segregation (*Met.*). A type of segregation in which the content of impurities and inclusions tends to increase from the surface to the centre of cast metals. Of special importance in steel, in which phosphorus, sulphur, and oxide inclusions segregate in this way. See also **inverse segregation.**

normal solution (*Chem.*). A solution made by dissolving the gram-equivalent mass of a substance in sufficient distilled water to make a dm^3 of solution.

normal state (*Nuc.*). See **ground state.**

normal temperature and pressure (*Chem.*). Earlier term for **s.t.p.**

normative composition (*Geol.*). The chemical composition of a rock expressed in terms of the standard minerals of the *norm.*

normoblast (*Zool.*). A stage in the development of an erythrocyte from an erythroblast when the nucleus has become reduced in size and the cytoplasm contains much haemoglobin.

normocyte (*Cyt.*). A fully developed red blood corpuscle (erythrocyte) normal in size, shape and colour.

norte (*Meteor.*). See **norther.**

norther or **norte** (*Meteor.*). A dry cold wind blowing from the north over the Gulf of Mexico, Valparaiso, or Table Bay.

Northern Lights (*Astron.*). See **Aurora Borealis.**

northing (*Surv.*). A north latitude. Northerly displacement of a point with reference to observer's station.

north light roof (*Arch.*). A pitched roof with unequal slopes, of which the steeper is glazed and arranged in such a way as to receive light from the north.

north pole (*Mag.*). See **pole.**

Northrup furnace (*Elec. Eng.*). See **coreless induction furnace.**

North Sea gas (*Geol.*). See **natural gas.**

Norton's theorem (*Elec. Eng.*). That the source behind two accessible terminals can be regarded as a constant-current generator, the current being the short-circuit current arising in an infinite impedance source, in shunt with an admittance which is that measured between the terminals with no source current. See **Thévenin's theorem.**

Norwegian quartz (*Build.*). A white translucent quartz found in Norway (a similar type occurs in the Isle of Man), now frequently used for the facing of *cladding* (q.v.) slabs.

Norwich (or straight) tie (*Weaving*). The tying up of the harness in a jacquard loom so that the machine and card cylinder are parallel to the comberboard. See **London tie.**

nosean or **noselite** (*Min.*). Silicate of sodium and aluminium with sodium sulphate, crystallizing in the cubic system. Occurs in extrusive igneous rocks which are rich in alkalis and deficient in silica, e.g., phonolite.

nose bit (*Tools*). A type of *shell bit* (q.v.) with a projecting tip for extracting the waste.

nose dive (*Aero.*). See **dive.**

nose heaviness (*Aero.*). The state in which the combination of the forces acting upon an aircraft in flight is such that it tends to pitch downwards by the nose.

nose key (*Join.*). One of the small wedges used in the operation of foxtail wedging.

nose ribs (*Aero.*). Small intermediate ribs, usually from the front spar to the leading edge only, of planes and control surfaces. They maintain the correct wing contour under the exceptionally heavy air load at that part of the aerofoil.

nose suspension (*Elec. Eng.*). A method of mounting a traction motor, by supporting one side of it on the axle and the other side on the framework of the truck.

nose-to-tail drag curve (*Aero.*). A plot of the cross-sectional area throughout the length of a high-speed aeroplane used, in connection with *area rule*, to eliminate abrupt changes, or 'bumps', that have a bad effect on drag, particularly at *transonic* speeds. The ideal plot resembles the longitudinal half-section of an egg.

nose-wheel landing gear (*Aero.*). See **tricycle landing gear.**

nosing (*Build.*). The exposed edge of the tread of a step, often rounded and projecting beyond the riser. (*For.*) One edge of a board worked into the form of a semicircle. (*Join.*) A bead on the edge of a board, making it half-round.

nosing motion (*Textiles*). Part of a mule building motion. Fixed on the quadrant, it is operated by hand adjustment or automatic means. It gradually increases the spindle speed during winding to compensate for the taper of the spindle.

nosology (*Med.*). Systematic classification of diseases: the branch of medical science which deals with this.

nosophobia (*Med.*). Morbid fear of contracting disease.

nostoc (*Zool.*). A genus of blue-green algae, beaded filaments forming gelatinous colonies on damp earth, etc.

nostomania (*Psychiat.*). Extreme homesickness.

nostrils (*Anat., Zool.*). The external nares.

not (*Paper*). The unglazed surface of hand-made or mould-made drawing-papers. See **hot-pressed and rough.**

NOT (*Comp., Maths.*). See **logical operations.**

notal processes (*Zool.*). In Insects, two processes (anterior and posterior) on each side of the notum of each thoracic segment bearing wings, with which the wings articulate through the articular sclerites.

not and element (*Comp.*). An *and element* followed by a negater.

notation. An agreed set of characters, figures, symbols, and abbreviations used to convey information or data.

notaulices (*Zool.*). In some *Hymenoptera*, a pair of sutures near the midline of the mesonotum.

notch (*Elec. Eng.*). A term often used to denote any of the various positions of a controller.

notch board (*Carp.*). A notched board carrying the treads and risers of a staircase.

notch brittleness (*Met.*). Susceptibility to fracture as disclosed by Izod or Charpy test, due to weakening effect of a surface continuity.

notched-bar test (*Met.*). A test in which a notched metal specimen is given a sudden blow by a striker carried by a pendulum or a falling weight, and the energy absorbed in breaking the specimen is measured. Also called **impact test.** See also **Izod test, Charpy test, Fremont test.**

notching (*Civ. Eng.*). The method of excavating cuttings for roads or railways in a series of steps worked at the same time. (*Eng.*) A press work process similar to punching or piercing, in which material is removed from the edge of a strip or sheet, often as a means of location or registering.

notching (or linking) up (*Eng.*). Movement of the gear-lever of a locomotive or steam-engine towards the centre of a notched quadrant, to decrease the valve travel and shorten the cut-off.

notch plate (*Civ. Eng.*). A vertical barrier, with a notch, either V-shaped or *rectangular* (q.v.), cut in its upper edge, placed across the current of a stream. The flow over the notch is gauged by measurement of the head of water above it.

notch sensitivity (*Met.*). The extent to which the endurance of metals, as determined on smooth and polished specimens, is reduced by surface discontinuities, such as tool marks, notches, and changes in section, which are common features of actual components. It tends to increase with the hardness and endurance limit.

notch toughness (*Met.*). The energy in joules or ft lbf required to break standard specimens under the standard conditions realized in the Izod or the Charpy test. It may also mean the opposite of *notch brittleness* (q.v.).

note (*Acous.*). An identifiable musical tone, whether pure or complex. See **hum-, strike-, wolf-.**

note tuning (*Radio*). Obtaining additional freedom from interference in radio-telegraphic receivers for ICW signals, by tuning post-detector circuits for audio-signals.

Notgrove Freestone (*Geol.*). A limestone of Middle Jurassic age occurring in the Cotswold Hills in Gloucestershire; has been much used for building purposes.

Nothosauria (*Zool.*). A suborder of *Sauropterygia*. Small, primitive, amphibious. Known as fossils from the Middle and Upper Trias.

no-throw (*Textiles*). The name for silk thread to which the twist imparted has been just sufficient to bind together the filaments composing it.

notochord (*Zool.*). In *Chordata*, skeletal rod formed of turgid vacuolated cells and originating from the endoderm of the middorsal line of the archenteron. *adj.* **notochordal.**

notocirrus (*Zool.*). In *Polychaeta*, a cirrus borne by the notopodium.

notoedric mange (*Vet.*). Mange on the face and ears of cats and rabbits, due to *Notoedres cati*.

notonectal (*Zool.*). Swimming in an inverted position with the ventral surface uppermost.

no top (*Acous.*). See **no bottom.**

notopleural bristles (*Zool.*). In *Diptera*, a pair of thoracic bristles found on each side between the humeral callus and the wing base.

notopleural suture (*Zool.*). In *Diptera*, a thoracic suture running from the humeral callus to the wing base, separating the mesonotum from the pleuron.

notopodium (*Zool.*). In *Polychaeta*, the dorsal lobe of a parapodium.

Notoptera (*Zool.*). A small order of the *Orthopteroidea*; wingless; biting mouthparts; multi-jointed cerci; ovipositor present; intermediate between crickets and cockroaches.

Notostraca (*Zool.*). An order of mainly fresh-water *Branchiopoda* in which there is a broad shieldlike carapace and a pair of sessile compound eyes set close together; antennules and antennae much reduced; there are numerous pairs of trunk limbs; the caudal furca are long and jointed. Tadpole Shrimps.

Notoungulata (*Zool.*). A varied order of extinct herbivorous Mammals, abundant in S. America throughout the Tertiary era.

Notting quoin (*Typog.*). The original screw-and-wedge-pattern quoin, all steel, in which the two sides expand by a wedge drawn upward as a screw is turned by the quoin key.

notum (*Zool.*). The tergum of Insects.

nova (*Astron.*). A star which makes a sudden appearance in the sky, generally decreasing rapidly in brightness. Although called a *new* or *temporary star*, it must have existed previously, probably as a faint star whose surface erupts explosively. See also **supernovae.**

novaculite (*Geol.*). A fine-grained or crypto-crystalline rock composed of quartz or other forms of silica; a form of chert. Used as a whetstone.

novakite (*Min.*). Copper arsenide, crystallizing in the tetragonal system.

Novocain (*Chem.*). TN for $H_2N{\cdot}C_6H_4{\cdot}CO{\cdot}O{\cdot}CH_2{\cdot}CH_2{\cdot}N(C_2H_5)_2$, HCl, the hydrochloride of diethylaminoethyl *p*-aminobenzoate. Colourless needles; m.p. 156°C; a widely used local anaesthetic. Also known as **procaine.**

Novoid (*Build.*). TN for a powder having a colloidal silica basis; used for making cement surfaces water-, oil-, and acid-proof.

novolaks (*Chem.*). Soluble, fusible, resinous products obtained by the condensation of hydroxy-benzenes with methanal.

no-voltage release (*Elec. Eng.*). A relay or similar device which causes the circuit to a motor or other equipment to be opened automatically if the supply voltage falls.

nox (*Light*). When a perfect diffuser has an equivalent luminance of 1 *skot*, the illumination is said to be 1 *nox*. Since *skot* is the equivalent luminance of 10^{-3} *apostilb* and 1 *apostilb* is $1/\pi$ *candela*/m², 1 *nox* $=10^{-3}$ *lux*.

noy (*Acous.*). Unit of perceived noisiness by which equal-noisiness contours, e.g. for 10 noys, 20 noys, etc., replace equal-loudness contours.

nozzle (*Aero.*). See **propelling nozzle.** (*Electronics*) End of a waveguide, which may be contracted in area. (*Eng.*) (1) In impulse turbines, specially shaped passages for expanding the steam, thus creating kinetic energy of flow with minimum loss. (2) In oil engines, orifices, open or controlled by the injection valve, through which the fuel is sprayed into the cylinder. See also **convergent-divergent nozzle.** (3) Generally, a convergent or convergent-divergent tube attached to the outlet of a pipe or a pressure chamber to direct efficiently the pressure of a fluid into velocity.

nozzle guide vanes (*Aero.*). In a gas turbine, a ring of radially positioned aerofoils which accelerate the gases from the combustion chamber and direct them on to the first rotating turbine stage.

Np (*Chem.*). Symbol for *neptunium.*

NP, NPA (*Chem.*). See **naptalam.**

N-part commutator (*Elec. Eng.*). A commutator having N-bars.

N.P.D. (*Astron.*). See **polar distance.**

n-p-i-n **transistor** (*Electronics*). Similar to *n-p-n* transistor with a layer of intrinsic semiconductor (germanium or silicon of high purity) between the base and the collector to extend the high-frequency range.

n-p **junction** (*Electronics*). A semiconductor junction with electron and hole conductivities on respective sides of the junction.

N.P.L. type wind tunnel (*Aero.*). The *closed-jet, return-flow* type is often called the original N.P.L. type, and the *closed-jet, nonreturn-flow* the standard N.P.L. type, as they were first used by the *National Physical Laboratory* (q.v.).

n-p-n **transistor** (*Electronics*). A junction transistor with a thin slice of *p*-type forming the base between two pieces of *n*-type semiconductor which are the collector and emitter, the conduction being electronic.

NRME (*Build.*). Abbrev. for *notched, returned, and mitred ends.*

NS (*Astron., etc.*). Abbrev. for *New Style.*

N-shell (*Nuc.*). The fourth (after K-, L- and M-) shell of electrons in an atom, containing $32(=2\times4^2)$ electrons when complete. Transitions among electrons in this shell are associated with the optical spectrum of the atom.

nt (*Light*). Symbol for *nit.*

N-terminal pair network (*Telecomm.*). One having N-terminal pairs in which one terminal of each pair may coincide with a node.

NTI (*Acous.*). Abbrev. for *noise transmission impairment.*

NTP (*Chem.*). *Normal temperature and pressure.* Previous term for **s.t.p.**, i.e., 0°C and 101·325 kN/m².

N-truss (*Eng.*). See **Whipple-Murphy truss.**

NTS. Abbrev. for *not to scale*; on plans, etc.

NTSC (*TV*). U.S. *N*ational *T*elevision *S*ystem *C*ommittee.

***n*-type semiconductor** (*Electronics*). One in which the electron conduction (negative) exceeds the hole conduction (absence of electrons), the *donor* impurity predominating.

Nubian Sandstone (*Geol.*). The basal member of the Cretaceous System in Egypt, deposited in the southern part of the Tethys. By its disintegration it has formed the Libyan desert.

nucellar budding (or embryony) (*Bot.*). The formation of an embryo from cells of the nucellus and not from the fertilized egg.

nucellus (*Bot.*). A mass of thin-walled cells occupying the middle of an ovule, protected by the integument, and containing the embryo sac.

nuchal (*Zool.*). Pertaining to, or situated on, the back of the neck.

nuchal bone (*Zool.*). In Sturgeons and their allies, a posterior unpaired membrane bone of the skull situated just behind the dermoccipital.

nuchal cartilage (*Zool.*). In some *Cephalopoda*, a thin plate of cartilage at the back of the neck.

nuchal crest (*Zool.*). A transverse bony ridge forming across the posterior margin of the roof of the vertebrate skull for attachment of muscles and ligaments supporting the head.

nuchal flexure (*Zool.*). In developing Vertebrates, the flexure of the brain occurring in the hinder part of the medulla oblongata, which bends in the same direction as the primary flexure.

nuchal organ (*Zool.*). In *Polychaeta*, one of a pair of ciliated pits on the surface of the prostomium, believed to be olfactory in function: in branchiopod *Crustacea*, a group of sensory cells containing refractive bodies occurring on the upper side of the head, the function of which is unknown.

nuchal plate (*Zool.*). In *Chelonia*, the most anterior median plate of the carapace.

nucivorous (*Zool.*). Nut-eating.

nuclear battery (*Nuc.*). One in which the electric current is produced from the energy of radioactive decay, either directly by collecting beta particles or indirectly, e.g., by using the heat liberated to operate a thermojunction. In general, nuclear batteries have very low outputs (often only microwatts) but long and trouble-free operating lives.

nuclear Bohr magneton (*Nuc.*). See **Bohr magneton.**

nuclear breeder (*Nuc. Eng.*). A nuclear reactor in which in each generation there is more fissionable material produced than is used up in fission.

nuclear budding (*Cyt.*). Production of 2 daughter nuclei of unequal size by constriction of the parent nucleus.

nuclear charge (*Nuc.*). Positive charge arising in the atomic nucleus because of protons, equal in number to the atomic number.

nuclear chemistry (*Chem.*). The study of reactions in which new elements are produced.

nuclear conversion ratio (*Nuc.*). That of the fissile atoms produced to the fissile atoms consumed, in a breeder reactor.

nuclear disintegration (*Nuc.*). Fission, radioactive decay, internal conversion or isomeric transition. Also *nuclear reaction* (q.v.).

nuclear emission (*Nuc.*). Emission of gamma ray or particle from nucleus of atom as distinct from emission associated with orbital phenomena.

nuclear emulsion (*Nuc.*). Thick photographic coating in which the tracks of various fundamental particles are revealed by development as black traces.

nuclear energy (*Nuc.*). In principle, the binding energy of a system of particles forming an atomic nucleus. More usually, the energy released during nuclear reactions involving regrouping of such particles (e.g., fission or fusion processes). The term *atomic energy* is deprecated as it implies rearrangement of atoms rather than of nuclear particles.

nuclear field (*Nuc.*). Postulated short-range field within a nucleus, which holds protons and neutrons together, possibly in shells.

nuclear fission (*Nuc.*). Spontaneous or induced splitting of atomic nucleus.

nuclear force (*Nuc.*). That which keeps neutrons and protons together in a nucleus, differing in nature from electric and magnetic forces, gravitational force being negligible. The force is of short range, is practically independent of charge and is associated with *mesons.* Yukawa has postulated that the force arises in the interchange of mesons between two nucleons. See **meson** and **short-range forces.**

nuclear fragmentation (*Cyt.*). The formation of two or more portions from a nucleus by direct break-up, and not by mitosis.

nuclear fusion (*Nuc.*). Creation of new nucleus by merging two lighter ones.

nuclear isomer (*Nuc.*). A nuclide existing in an excited metastable state. It has a finite half-life after which it returns to the ground state with the emission of a γ-quantum or by internal conversion. Metastable isomers are indicated by adding *m* to the mass number.

nuclear magnetic resonance (*Chem., Nuc.*). The nuclei of certain isotopes, e.g., ^{1}H, ^{19}F, ^{31}P, behave like small bar magnets and will line up when placed in a strong d.c. magnetic field. The direction of alignment of the nuclei in a sample may be altered by RF irradiation in a surrounding coil whose axis is at right angles to the magnetic field. The frequency at which energy is absorbed and the direction of alignment changes is known as the resonant frequency and varies with the type of nucleus; e.g., in a magnetic field of 14 000 gauss ^{1}H resonates at 60 MHz, ^{19}F at 56 MHz, and ^{31}P at 24·3 MHz. Nuclei with even mass and atomic number, e.g., ^{12}C, ^{16}O, have no magnetic properties and do not exhibit this behaviour. Abbrev. **NMR.**

nuclear magnetic resonance spectroscopy (*Chem.*). The resonant frequencies for identical nuclei in a constant external field vary slightly with the chemical environment of the nucleus (about 1-10 p.p.m. for ^{1}H). By keeping the external magnetic field constant and varying the RF radiation over a small range, 1000 Hz, for a solution of a given molecule, an absorption spectrum for the compound under examination is obtained from which information may be deduced on the structure of the molecule.

nuclear magneton (*Electronics*). See **Bohr magneton.**

nuclear membrane (*Cyt.*). The delicate bounding membrane of the nucleus.

nuclear model (*Nuc. Eng.*). One giving some explanation of certain nuclear properties, e.g., independent particle model, liquid drop model, one-particle model, optical model, shell model, unified model.

nuclear number (*Nuc.*). Same as **mass number.**

nuclear pacemaker (*Med.*). A *pacemaker* (q.v.) powered by a nuclear battery, whose long life (at least 10 years) obviates the need for frequent surgery to replace the conventional type of battery.

nuclear paramagnetic resonance (*Chem., Nuc.*). See **nuclear magnetic resonance.**

nuclear photoeffect (*Nuc.*). See **photodisintegration.**

nuclear plate (*Cyt.*). The aggregation of chromosomes in the equatorial plane during mitosis or meiosis.

nuclear potential (*Nuc.*). The potential energy of a nuclear particle in the field of a nucleus as determined by the short-range forces acting. Plotted as a function of position it will normally represent some sort of *potential well* (q.v.).

nuclear power (*Nuc.*). That obtained by the release of nuclear energy.

nuclear propulsion (*Space*). A suggested method of rocket propulsion in which a nuclear reactor would supply the heat necessary to vaporize a working fluid.

nuclear radius (*Nuc.*). The somewhat indefinite radius of a nucleus within which the density of *nucleons* (protons and neutrons) is experimentally found to be nearly constant. The radius in metres is $1{\cdot}2\times10^{-15}$ times the cube root of the nuclear number (atomic mass). It is not a precise determinable quantity.

nuclear reaction (*Nuc.*). The interaction of photon or particle with nuclear structure.

nuclear reactor (*Nuc. Eng.*). See **reactor.**

nuclear reticulum (*Cyt.*). A meshwork of delicate threads of chromatin seen in stained preparations of metabolic nuclei.

nuclear sap (*Cyt.*). See **karyolymph.**

nuclear selection rules (*Phys.*). Those specifying the transitions of electrons or nucleons between different energy levels which may take place (*allowed transitions*). The rules may be derived theoretically through wave mechanics but are not obeyed rigorously—so-called *forbidden transitions* merely being highly improbable.

nuclear spindle (*Cyt.*). The fusiform structure, composed of fine fibrils, which appears in the cytoplasm of a cell surrounding the nucleus during mitosis and meiosis.

nuclear stain (*Micros.*). A stain which will pick out the nuclei in a tissue or organ.

nucleases (*Chem.*). Enzymes inducing hydrolysis of nucleic acid.

nucleating agent (*Meteor.*). Substance used for seeding clouds to control rainfall and fog formation. See **rainmaking.**

nuclei (*Met.*). Centres or 'seeds' from which crystals begin to grow during solidification. In general, they are minute crystal fragments formed spontaneously in the melt, but frequently nonmetallic inclusions act as nuclei. See also **crystal nuclei.**

nucleic acids (*Biochem.*). The nonprotein constituents of nucleoproteins. They are complex organic acids of high molecular weight consisting of chains of alternate units of phosphate and a pentose sugar which has a purine and pyrimidine base attached to it. In ribonucleic acid (RNA) the sugar is ribose, in deoxyribonucleic acid (DNA) it is 2-deoxyribose. In DNA the bases are thymine, cytosine, guanine, and adenine. In RNA uracil is substituted for thymine. Nucleic acids play a central role in protein synthesis and in the transmission of hereditary characteristics, since they embody the *genetic code* (q.v.). See also **alpha-helix, cell-free system, central dogma, codon, molecular genetics, m-RNA, ribosomes, s-RNA, transcription, transformation, Watson-Crick model.**

nucleochylema (*Cyt.*). Nuclear sap. See **karyolymph.**

nucleogenesis (*Nuc.*). Theoretical process(es) by which nuclei could be created from possible fundamental dense plasma. See **ylem.**

nucleohyaloplasm (*Zool.*). Nuclear sap. See **karyolymph.**

nucleolate (*Bot.*). Said of a spore which contains one or more conspicuous oil-drops.

nucleolini (*Cyt.*). Special particles within the nucleolus which do not disappear during mitosis.

nucleolo-centrosome (*Cyt.*). A prominent deeply-staining body found in the nucleus in some lower plants, which may act as a centrosome during mitosis.

nucleolus (*Cyt.*). A homogenous spherical body occurring within a cell nucleus. *adj.* **nucleolar.** Also called **plasmosome, karyosome.**

nucleome (*Biol.*). The whole of the nuclear substance in a protoplast.

nucleonics (*Phys.*). The science and technology of nuclear studies.

nucleon number (*Nuc.*). See **mass number.**

nucleons (*Nuc.*). Proton and neutrons in a nucleus of an atom.

nucleophilic reagents (*Chem.*). Term applied to anions which are attracted to positive centres, e.g., the hydroxyl ion (OH^-) in replacing Cl^- from a C—Cl link in which the C atom can be considered as being the positive centre.

nucleoplasm (*Cyt.*). The dense protoplasm composing the nucleus of a cell. Cf. *cytoplasm*.

nucleoplasmic ratio (*Cyt.*). The ratio between the volume of the nucleus and of the cytoplasm in any given cell.

nucleoproteins (*Biochem.*). A group of compounds containing a protein molecule combined with nucleic acid; the mode of linkage probably varies in different cases. They are important constituents of cell nuclei (hence the name) but also occur in the cytoplasm. Viruses seem to consist almost entirely of nucleoprotein.

nucleor (*Nuc.*). The hypothetical core of a nucleon. It is suggested that it is the same for protons and neutrons.

nucleoside (*Chem.*). A purine or pyrimidine base linked to ribose or deoxyribose.

nucleotide (*Biochem.*). A compound of a base (derived from purine, pyrimidine or pyridine), a pentose (ribose or deoxyribose) and phosphoric acid. The term is applied to certain coenzymes (e.g., *NAD*, q.v.), to compounds produced by partial hydrolysis of nucleic acids and also to the nucleic acids themselves, which may be considered as polynucleotides.

nucleus (*Astron.*). The term applied to the denser core of a tenuous body such as a comet, and to the starry condensations seen in many nebulae. (*Biol.*) The chief organ of the cell. It is a usually spheroidal mass in the cytoplasm, bounded by a nuclear membrane and containing, bathed in nuclear sap, a complicated system of proteins disposed in the form of a network and/or of rounded nucleoli. This system of proteins appears to control the activities of the cell and to determine the transmission of inheritable characters when the nucleus divides. (*Nuc.*) This is composed of protons (positively charged) and neutrons (no charge), and constitutes practically all the mass of the atom. Its charge equals the atomic number; its diameter is from 10^{-15} to 10^{-14} m. With protons equal to the atomic number and neutrons to make up the atomic mass number, the positive charge in the protons is balanced by the same number of electrons orbiting in shells at a distance. (*Zool.*) (1) Any nut-shaped structure. (2) A nerve centre in the brain. (3) In some *Hemimyaria*, a relatively small dark mass at the hinder end of the body, representing the oesophagus, stomach, and intestine. (4) A collection of nerve cells on the course of a nerve or tract of nerve fibres.

nucleus pulposus (*Zool.*). A gelatinous mass in the centre of an invertebral disk, representing the remnant of the notochord in the adults of higher Vertebrates.

nuclide (*Nuc.*). Isotope of an element as distinguished by the constitution of the nucleus of its atoms. Including natural and man-made radioisotopes, there are about 1400 nuclides among the known elements. See **mirror-**, **Wigner-**.

Nuda (*Zool.*). A class of *Ctenophora* the members of which do not possess tentacles. Includes only one genus.

nudation (*Bot.*). The formation, by natural or artificial means, of an area bare of plants.

nudibranchiate (*Zool.*). Having the gills exposed, not within a branchial chamber.

nudicaudate (*Zool.*). Having the tail uncovered by fur or hair, as Rats.

null electrode (*Chem.*). Electrode having zero potential with respect to the electrolyte.

null indicator (*Elec. Eng.*). Any device, such as head telephones, a cathode-ray oscillograph, or a sensitive galvanometer, for determining zero current, or voltage, in a specified part of an electric circuit, e.g., in a radio goniometer.

nullipara (*Med.*). A woman who has never given birth to a child. *adj.* **nulliparous.**

nulliplex (*Gen.*). Polyploid, but lacking the dominant of any given pair of allelomorphs.

null method (*Elec. Eng.*). Same as **zero method.**

number (*Maths.*). An attribute of objects or labels obtained according to a law or rule of counting.

number sense (*An. Behav.*). The ability of some animals to abstract the quality of 'numerical identity' from groups of the same number of different objects.

number sizes (*Eng.*). A series of drill sizes in which gauge numbers represent diameters, the higher numbers representing the smaller diameters, in irregular increments.

numbers, pyramid of (*Ecol.*). The relative decrease in numbers at each stage in a food chain, characteristic of animal communities.

numerable set (*Maths.*). See **denumerable set.**

numerator (*Maths.*). See **division.**

numerical analysis (*Maths.*). The derivation of a particular numerical solution or set of solutions arising from the input of a single set of data into a mathematical model, as opposed to a general algebraic analytical solution.

numerical aperture (*Optics*). Product of the refractive index of the object space and the sine of the semiaperture of the cone of rays entering the entrance pupil of the objective lens from the object point. The resolving power is proportional to the numerical aperture. Abbrev. NA.

numerical control (*Eng.*). The operation of machine tools from numerical data which have been recorded and stored on magnetic or punched tape or on punched cards.

numerical forecasting (*Meteor.*). A method of weather forecasting based upon a large number of observations throughout the *troposphere* (q.v.) and the calculation, electronically, from these, of the conditions which should follow in accordance with the known laws of physics.

numerical selector (*Teleph.*). A group selector which connects the caller to the called subscriber within the latter's own exchange.

nummulation (*Zool.*). Formation of rouleaux by erythrocytes.

nummulites (*Geol.*). A group of extinct *Foraminifera* which were important rock-forming organisms in the early Tertiary period. A nummulitic limestone is one which is composed mainly of their skeletal remains.

nunatak (*Geol.*). An isolated mountain peak which projects through an ice sheet.

n-unit (*Nuc.*). That quantity of fast neutrons which will produce in a Victoreen 100r dosimeter the same reading as 1 röntgen of X-rays.

nun's veiling (*Textiles*). A lightweight dress fabric, of plain weave, made from single botany worsted yarns of very good quality.

nuptial chamber (*Zool.*). The small burrow formed by a pair of sexual Termites initiating a new colony.

nuptial flight (*Zool.*). The flight of a virgin Queen Bee, during which she is followed by a number of males, copulation and fertilization taking place in mid-air.

nurse (*Zool.*). A budding form in *Cyclomyaria*.

nurse cells (*Zool.*). Cells surrounding, or attached to, an ovum, probably to perform a nutritive function.

Nusselt number (*Heat*). The significant non-dimensional parameter in convective heat loss problems, defined by $Qd/k\Delta\theta$, where Q is rate of heat lost per unit area from a solid body, $\Delta\theta$ is temperature difference between the body and its surroundings, k is the thermal conductivity of the surrounding fluid and d is the significant linear dimension of the solid.

Nut (*Bot.*). A hard, dry, indehiscent fruit formed from a syncarpous gynaeceum, and usually containing one seed. The term is used loosely for any fairly-large to large hard, dry, one-seeded fruit. (*Eng.*) A metal collar, screwed internally, to fit a bolt; usually hexagonal in shape, but sometimes square or round. See **castle-, lock-**. (*Typog.*) See **en quad.**

nutant (*Bot.*). Hanging with the apex downwards; nodding.

nutating feed (*Radar*). That to a radar transmitter which produces an oscillation of the beam without change in the plane of polarization. The resulting radiation field is a *nutation field*.

nutation. The periodic variation of the inclination of the axis of a spinning top (or gyroscope) to the vertical. (*Astron.*) An oscillation of the earth's pole about the mean position. It has a period of about 19 years, and is superimposed on the precessional movement. (*Bot.*) The rotation in space of the apex of an axis which is growing in length.

nutation field (*Radar*). See **nutating feed.**

nut galls (*Chem.*). See **Aleppo galls.**

nutlet (*Bot.*). A 1-seeded portion of a fruit which fragments as it matures.

N.U. tone (*Teleph.*). Number unobtainable tone.

nutrient (*Med.*). Conveying, serving as, or providing nourishment. Nourishing food.

nutrient solution (*Bot.*). An artificially prepared solution containing some or all of the mineral substances used by a plant in its nutrition.

nutrition (*Zool.*). The process of feeding and the subsequent digestion and assimilation of food-material. *Adj.* **nutritive.**

nutritional encephalomalacia (*Vet.*). Crazy chick disease. A nervous disease of young chicks characterized by loss of balance, inability to stand, and other nervous symptoms, associated with oedema, haemorrhages and degenerative changes in the brain; caused by vitamin E deficiency.

nutritional roup (*Vet.*). An affection of the upper respiratory tract of the chicken, caused by vitamin A deficiency, and characterized by a discharge from the eyes and nostrils.

nut runner (*Eng.*). A power tool or head fitted with a socket bit and adapted to drive and tighten a nut, often at controlled torque.

Nutsch filter (*Chem. Eng.*). A large open-topped, false-bottomed vessel with a perforated plate for false bottom. This is covered with cloth and the vessel filled. Filtrate passes cloth and perforated bottom and is removed via pipe in the lower space. Can be assisted by applying vacuum to lower space.

nux vomica (*Bot.*). A seed, yielding *Strychnine*, produced by the East Indian tree *Strychnos nux-vomica*; family loganiaceae.

Nuvistor (*Electronics*). TN for a subminiature electron tube.

NVM (*Chem.*). An abbrev. for *nonvolatile matter.*

NW (*Build.*). An abbrev. for *narrow widths.*

nyctalopia (*Med.*). Night blindness; moon blindness. Abnormal difficulty in seeing objects in the dark; due often to deficiency of vitamin A in the diet.

nyctanthous (*Bot.*). Said of flowers which open at night.

nyctinastic movements (*Bot.*). Movements of plants associated with the alternation of day and night, due to changes in temperature and illumination.

nyctipelagic (*Zool.*). Found in the surface waters of the sea at night only.

nyctitropic (*Bot.*). A term referring to the position assumed by leaves, etc., at night.

Nylander solution (*Chem.*). An alkaline solution of bismuth subnitrate and Rochelle salt, giving a black colour on boiling with glucose; used to detect glucose in urine.

Nylatron (*Plastics*). TN for a range of filled nylon compounds reinforced for engineering purposes.

nylon (*Plastics*). Generic name for any long-chain synthetic polymeric amide which has recurring amide groups as an integral part of the main polymer chain, and which is capable of being formed into a filament in which the structural elements are oriented in the direction of the axis.

nylon plates (*Print.*). See **photopolymer plates.**

nymph (*Zool.*). In *Acarina*, the immature stage intervening between the period of acquisition of four pairs of legs and the attainment of full maturity: in *Insecta*, a young stage of *Exopterygota* intervening between the egg and the adult, and differing from the latter only in the rudimentary condition of the wings and genitalia.

nymphae (*Zool.*). See **labia minora.**

Nyquist criterion (*Telecomm.*). If for a quadripole the complex transfer ratio is plotted for an infinite range of frequencies, and the point $-1+j0$ is enclosed, the system is unstable; if excluded, the system is stable. Such a plot is a **Nyquist diagram.**

Nyquist limit (or rate) (*Telecomm.*). Maximum rate of transmitting pulse signals through a system. If B is the effective bandwidth in Hz, then $2B$ is the maximum number of code elements (*bands*) per sec which can be received with certainty. $1/2B$ is known as the **Nyquist interval.**

Nyquist noise theorem (*Electronics*). One by which the thermal noise power P in a resistor at any frequency can be calculated from Boltzmann and Planck's constants k and h and the temperature. At normal temperatures it reduces to $P = kTdf$, where T is the temperature in K and df is the frequency interval.

nystagmoid (*Med.*). Resembling nystagmus.

nystagmus (*Med.*). An abnormal and involuntary movement backwards-and-forwards of the eyeballs.

nystatin (*Pharm.*). An antifungal antibiotic, produced by a strain of *Streptomyces noursei*. An amphoteric crystalline polyene, empirical formula $C_{46}H_{77}NO_{19}$. Used in treatment of moniliasis and candidiasis and as a fungicide in tissue culture preparations. TN **Nystan, (U.S.) Mycostatin.**

O

o, O, ~ (*Maths.*). These symbols are defined as follows:

$$f(x)=o\{\varphi(x)\} \quad \text{if} \quad \frac{f(x)}{\varphi(x)}\to 0$$

$$f(x)=O\{\varphi(x)\} \quad \text{if} \quad |f(x)|<K\varphi(x)$$

$$f(x)\sim\varphi(x) \quad \text{if} \quad \frac{f(x)}{\varphi(x)}\to 1,$$

where the limiting value of x is either stated explicitly or implied. E.g., if $f(x)=x^2+\frac{1}{x}$, then $f(x)=o\left(\frac{1}{x}\right)$ if x is small, and $f(x)=O(x^2)$ if x is large.

o- (*Chem.*). An abbrev. for *ortho-*, i.e., containing a benzene nucleus substituted in the 1.2 positions.

ω (*Chem.*). A symbol for specific magnetic rotation. (*Light*) Symbol for dispersive power. (*Maths.*) Symbol for solid angle. (*Phys., etc.*) Symbol for: (1) angular frequency; (2) angular velocity; (3) pulsatance.

ω- (*Chem.*). A symbol indicating: (1) substitution in the side chain of a benzene derivative; (2) substitution on the last carbon atom of a chain, farthest from a functional group.

O (*Chem.*). Symbol for *oxygen*.

***O*-** (*Chem.*). A symbol indicating that the radical is attached to the oxygen atom.

Ω (*Elec., Mag.*). Symbol for *ohm*. (*Maths.*) Symbol for solid angle. (*Phys., etc.*) Symbol for angular velocity.

—Ω (*Chem.*). A symbol for the ultimate disintegration product of a radioactive series.

oak (*For.*). A strong, tough, and heavy hardwood, very durable in exposed positions. Commonly used in constructional work for timber bridges, heavy framing and piles, as well as for joinery.

oakum. Tarred, untwisted rope or hemp used for caulking joints.

oast (*Brew.*). A hop-drying kiln.

oatmeal cloths (*Textiles*). Fabrics having a resemblance to oatmeal and a rough appearance; made with a crêpe type of weave from crossbred worsted yarns, or in cotton from a fine warp and coarse weft.

ob- (*Bot.*). Prefix meaning *reversed, turned about*. Thus, *obclavate* is reversed *clavate*, i.e., attached by the broad and not the narrow end.

Obach cell (*Elec. Eng.*). A dry primary cell having ingredients similar to the Leclanché cell, i.e., a carbon positive electrode surrounded by a depolarizer of manganese dioxide paste, and having the electrolyte in the form of a paste of sal-ammoniac, plaster of Paris, and flour.

Obach process (*Chem. Eng.*). A process for deresinating raw gutta-percha; the solvent is lythene, a low-boiling fraction of petroleum.

obcompressed (*Bot.*). Flattened from front to back.

obconic, obconical (*Bot.*). Cone-shaped but attached by the point.

obdiplostemonous (*Bot.*). Having two whorls of stamens, the members of the outer whorl placed opposite to the petals and not alternating with them.

obeche (*For.*). *Triplochiton Scleroxylon*, a tree yielding lightweight wood used for food-containers, etc. Also called **African whitewood, wawa.**

obelisk (*Arch.*). A slender stone shaft, generally monolithic, square in section, and tapering towards the top, which is surmounted by a small pyramid.

obex (*Anat.*). A triangular area of grey matter and thickened ependyma in the *medulla oblongata*, in the roof of the fourth ventricle over the region of the *calamus scriptorius*.

object computer (*Comp.*). One on which an object deck (or object program) is processed, if this is different from one used to compile the object deck from the source deck. Cf. *source computer*.

object deck (*Comp.*). A stack of punched cards forming a program in machine language. Usually prepared from an equivalent source deck by the compiler for the machine.

objective (lens) (*Light*). Usually the lens of an optical system nearest the object.

objective noise meter (*Acous.*). Sound-level meter in which noise level to be measured operates a microphone, amplifier, and detector, the last-named indicating noise level on the phon scale. The apparatus is previously calibrated with known intensities of the *reference tone*, 1 kHz, suitable weighting networks and an integrating circuit being incorporated in the amplifier to simulate relevant properties of the ear in appreciating noise.

objective prism (*Astron.*). One of the forms of prism spectroscope in which the prism is placed in front of instead of behind the objective, the light from the star passing through the prism and being drawn out into a spectrum without the use of a slit; hence the alternative term for this instrument, **slitless spectroscope.**

object language (*Comp.*). That of the object deck. The basic *machine language* of the computer.

oblate. Globose, but noticeably wider than long.

oblate ellipsoid (*Maths.*). An ellipsoid obtained by rotating an ellipse about its minor axis, i.e., like a squashed sphere. Cf. *prolate ellipsoid*.

obligate (*Bot., Zool.*). Compelled; without the power of choice of mode of life; as an *obligate* parasite, which is incapable of leading a free-living existence. Cf. *facultative*.

obligate saprophyte (*Biol.*). An organism which lives on dead organic material and cannot attack a living host.

oblique aerial photograph (*Surv.*). A photograph taken from the air, for purposes of aerial survey work, with the optical axis of the camera inclined from the vertical, generally at some predetermined angle.

oblique arch (*Civ. Eng.*). An arch whose axis is not normal to the face.

oblique axes (*Maths.*). Coordinate axes which are not mutually perpendicular.

oblique circular cylinder (*Maths.*). See **cylinder.**

oblique incidence transmission (*Telecomm.*). Short-wave transmission using reflection from ionosphere.

oblique septum (*Zool.*). In Birds, paired sheets of fibrous tissue covering the free ventral walls of the thoracic air sacs and dividing the coelom into two compartments.

oblique system (*Crystal.*). See **monoclinic system.**

obliquity of the ecliptic (*Astron.*). The angle at which the earth's orbital plane, or the ecliptic, is inclined to the earth's equatorial plane, or the celestial equator; it amounted in 1973 to 23° 26′ 34″ and is slowly decreasing.

obliquus (*Zool.*). An asymmetrical or obliquely placed muscle.

obliteration (*Bot.*). The crushing and closing up of tubular elements within a plant by the pressure set up by new elements as they develop.

obliterative shading (*Zool.*). A type of coloration shown by many animals that are usually subject to light striking them from above; the upper parts of the body are dark-coloured (to counteract the greater intensity of light on those parts), and shade gradually into the light-coloured lower parts.

Oblivon (*Pharm.*). TN for **methylpentynol,** used medically for treatment of insomnia and relief of emotional stress and mental tension.

oblongata (*Zool.*). See **medulla oblongata.**

oblongum (*Zool.*). In some *Coleoptera,* a cell formed in the veins of the hind-wing by two transverse veins joining the first and second branches of the media.

O.B.M. (*Surv.*). Common abbrev. for *Ordnance Bench Mark.*

obnubilation (*Med.*). A state of mental disturbance just before loss of consciousness.

oboe (*Radar*). Distance-measuring system using two ground stations and aircraft transponder.

obovate (*Bot.*). Having the general shape of the longitudinal section of an egg, not exceeding twice as long as broad, and with the greatest width slightly above the middle; hence, attached by the narrower end.

obovoid (*Bot.*). Solid, egg-shaped, and attached by the narrower end.

obscuration (*Phys.*). Fraction of incident radiation which is removed in passing through a body or a medium.

obscure (*Bot.*). Said of venation which is very little developed, so that hardly more than the midrib can be seen.

obscured glass (*Glass*). Glass which is so treated as to render it translucent but not transparent.

obsession (*Psychiat.*). The morbid persistence of an idea in the mind, against the wish of the obsessed person.

obsessional neurosis (*Psychiat.*). A functional nervous disorder characterized by a need to perform routine patterns of behaviour, in an attempt to solve the strong underlying unconscious conflicts—e.g., repeated handwashing, repeated confirmation of locked doors, ceremonial routines performed periodically each day. A high degree of doubt is present about everything in the minds of those individuals. This neurosis arises from an emotional arrest at a pregenital anal phase of development, and shows the corresponding character traits. See **anal character, anal-sadistic.**

obsessive-compulsive reaction (*Psychiat.*). A syndrome involving a maladaptive preoccupation with one idea or desire. The individual may be concerned over the fact that he is wasting time in this way but is unable to change his ways.

obsidian (*Geol.*). A volcanic glass of granitic composition, generally black with vitreous lustre and conchoidal fracture; occurs at Mt Hecla in Iceland, in the Lipari Isles, and in Yellowstone Park, U.S. A green silica glass found in ploughed fields in Moravia is cut as a gemstone and sold under the name *obsidian.* True obsidian is used as a gemstone and is often termed **Iceland agate.**

obstetrician (*Med.*). A medically qualified person who practises obstetrics.

obstetrics (*Med.*). That branch of medical science which deals with the problems and management of pregnancy and labour.

obstipation (*Med.*). Severe and intractable constipation.

obstruction lights (*Aero.*). Lights fixed to all structures near airports which constitute a hazard to aircraft in flight.

obstruction markers (*Aero.*). See **aerodrome markers.**

obstruent (*Med.*). Obstructing; that which obstructs; an astringent drug.

obtect (*Zool.*). Said of pupae which have the wings and legs glued to the body and which are therefore incapable of movement.

obturator (*Zool.*). Any structure which closes off a cavity, e.g., all the structures which close the large oval foramen formed by the ischio-pubic fenestra; the foramen itself.

obturator notch (*Zool.*). In Birds, the space between the pubis and ischium.

obtuse angle (*Maths.*). An angle greater than 90° and less than 180°.

obtusilingual (*Zool.*). Having a short blunt tongue.

obvolvent (*Zool.*). Folded downwards and inwards, as the wings in some Insects.

occasional species (*Bot.*). A species which is found from time to time in a given plant community but is not a regular member of that community.

occidental topaz (*Min.*). A rather misleading TN for certain yellow-coloured varieties of quartz used as semiprecious gemstones. See also **Spanish topaz.**

occipital (*Zool.*). A bone of the Vertebrate skull, one of the *occipitalia* (q.v.); pertaining to the occiput.

occipital condyle (*Zool.*). In Insects, a projection from the posterior margin of the head which articulates with one of the lateral cervical sclerites. In *Craniata,* one or two projections from the skull which articulate with the first vertebra.

occipitalia (*Zool.*). A set of cartilage bones forming the posterior part of the brain-case in the Vertebrate skull.

occiput (*Zool.*). In Insects, the part of the epicranium between the vertex and the neck: the occipital region of the Vertebrate skull forming the back of the head. *adj.* **occipital.**

occlusion (*Bot.*). The blocking of a stoma by the ingrowth of parenchymatous cells into the substomatal cavity. (*Chem.*) The retention of a gas or a liquid in a solid mass or on the surface of solid particles; especially the retention of gases by solid metals. (*Meteor.*) The coming together of the *warm* and *cold fronts* (qq.v.) in a depression so that the warm air is no longer in contact with the earth's surface. If the *warm frontal zone* is cut off from contact with the surface it is a *cold occlusion.* If the *cold frontal zone* is cut off from contact with the surface it is a *warm occlusion.* (*Zool.*) Closure of a duct or aperture, e.g., the upper and lower teeth of a Vertebrate.

occlusor (*Zool.*). A muscle which by its contraction closes an operculum or other movable lidlike structure.

occultation (*Astron.*). The hiding of one celestial body by another interposed between it and the observer, as the hiding of the stars and planets by the moon, or the satellites of a planet by the planet itself.

occulting light (*Ships*). A navigation mark identified during darkness by a light extinguished for short periods in a distinctive pattern. See **alternating light, flashing light.**

ocean (*Geog.*). A major expanse of salt water lying between the continents and outside the area of the continental shelf.

ocean depths (*Geog.*). The greatest depth (over 10 400 m) discovered in the various oceans lies in the Pacific; depths of over 8500 m and 7000 m have been recorded in the Atlantic and the Indian Ocean respectively.

oceanic crust (*Geol.*). That part of the earth's crust which is normally characteristic of oceans. In descending vertical section, it consists of approximately 5 km of water, 1 km of sediments, and 5 km of basaltic rocks.

oceanite (*Geol.*). A type of basaltic igneous rock occurring typically in the oceanic islands as lava flows; characterized by a higher percentage of coloured silicates (olivine and pyroxene), and a lower percentage of alkalis, than in normal basalt.

ocean temperatures (*Geog.*). The mean surface temperature of the Pacific and Atlantic Oceans is 17°C; that of the Indian Ocean 18°C. Maximum temperatures are respectively 32°C, 30°C, and 35°C.

ocellar bristles (*Zool.*). In *Diptera*, a pair of cephalic bristles in the triangle formed by the ocelli.

ocellar lobes (*Zool.*). In the brain of Insects, four small lobes between the mushroom bodies, from each of which an ocellar nerve originates.

ocellar triangle (*Zool.*). On the head of *Diptera*, the triangular region bearing the ocelli, which is often bounded by grooves or depressions.

ocellate (*Bot.*). Marked by a round patch different in colour from the ground.

ocellus (*Bot.*). (1) An enlarged discoloured cell in a leaf. (2) A swelling on the sporangiophore in some fungi; it may be able to perceive light. (*Zool.*) A simple eye or eye-spot in Invertebrates: an eye-shaped spot or blotch of colour. *adj.* **ocellate.**

ocher (*Paint.*). See **ochre.**

ochlophobia (*Psychiat.*). Morbid fear of crowds.

ochre or ocher (*Paint.*). A natural earth pigment consisting of iron oxides and silicates which mixes with clay and stains it yellow. The dried and ground clay is mixed with a suitable vehicle to form paint.

ochrea, ocrea (*Bot.*). A cup-shaped structure around a stem, formed from united stipules or united leaf bases.

ochroleucous (*Bot.*). Yellowish-white.

ochronosis (*Med.*). A rare condition in which a dark-brown or black pigment is deposited in the cartilages and other tissues of the body; associated often with alkaptonuria (in which homogentisic acid is present in the urine), and sometimes following the prolonged use of dressings of carbolic acid.

ochrophore (*Zool.*). See **xanthophore.**

ochrosporous (*Bot.*). Having yellow or yellow-brown spores.

octa- (*Chem.*). Containing 8 atoms, groups, etc.

octahedral system (*Crystal.*). See **cubic system.**

octahedrite (*Min.*). A form of *anatase* (q.v.), crystallizing in tetragonal bipyramids (*not* in octahedra—thus the name is a misnomer). It is usually of secondary origin, derived from other titanium-bearing minerals.

octahedron (*Crystal.*). A form of the cubic system which is bounded by 8 similar faces, each being an equilateral triangle with plane angles of 60°. *pl.* **octahedra.**

octal base (*Electronics*). International version of **loktal base.**

octamerous (*Bot.*). Having parts in eights.

octane (*Chem.*). C_8H_{18} alkane hydrocarbon. There are 18 compounds of this formula. The normal octane, a colourless liquid, b.p. 126°C, rel. d. 0·702 at 20°, is found in petroleum. It has been synthesized.

octane number (*I.C. Engs.*). The percentage, by volume, of *iso*-octane (2,2,4-trimethylpentane) in a mixture of *iso*-octane and *normal* heptane which has the same knocking characteristics as the motor fuel under test; it serves as an indication of the knock-rating of a motor fuel. See also **lead tetraethyl.**

octant division (*Bot.*). The division of an embryonic cell by walls at right angles, giving 8 cells.

octastyle (*Arch.*). A building having a colonnade of 8 columns in front.

octavalent (*Chem.*). Capable of combining with 8 atoms of hydrogen or their equivalent.

octave (*Acous.*). The interval between any 2 frequencies having the ratio 2 : 1. (*Chem.*) See **law of octaves.**

octave analyser (*Acous.*). A filter in which the upper cut-off frequency is twice the lower.

octave filter (*Acous.*). Bank of filters for analysing the spectral energy content of complex sounds and noises, using adjacent octave bands of frequency over the whole audio range. One-half and one-third octave filters are also used.

octavo (*Print.*). The eighth of a sheet, or a sheet folded 3 times to make 8 leaves or 16 pages and written 8vo.

octet (*Chem.*). An extremely stable group of 8 electrons, formed in most cases of combination between atoms.

octode (*Electronics*). Valve containing cathode, anode, and 6 intermediate electrodes, used as a frequency changer in superheterodyne receivers. The first 3 electrodes, starting from the cathode, form the local oscillator system; the remaining 5 constitute a pentode system whose emission varies at the frequency of the local oscillator.

octodecimo (*Print.*). See **eighteenmo.**

octopod (*Zool.*). Having 8 feet, arms, or tentacles.

octoses (*Chem.*). A group of monosaccharides, containing 8 oxygen atoms in the molecule, e.g., $HO{\cdot}CH_2{\cdot}(CHOH)_6{\cdot}CHO$.

octosporous (*Bot.*). Containing 8 spores.

octyl alcohol (*Chem.*). See **capryl alcohol.**

ocular (*Zool.*). Pertaining to the eye; capable of being perceived by the eyes.

ocular micrometer (*Micros.*). U.S. for **eyepiece graticule.**

ocular plate (*Zool.*). In *Echinoidea*, a small plate occurring at the aboral end of each radius or ambulacral area, bearing the opening of the terminal tentacle.

ocular sclerites (*Zool.*). Annular sclerites surrounding the compound eye of each side in Insects.

oculate (*Zool.*). Possessing eyes; having markings which resemble eyes.

oculist (*Med.*). One skilled in the knowledge and treatment of refractive diseases of the eye.

oculomotor (*Zool.*). Pertaining to, or causing movements of, the eye: the third cranial nerve of Vertebrates, running to some of the muscles of the eyeball.

oculus (*Arch.*). A round window.

odd-even check (*Telecomm.*). Verification of correctness of operation by using only even numbers of impulses, so that presence of an odd number reveals a fault.

odd-even nuclei (*Nuc.*). Nuclei containing an odd number of protons and an even number of neutrons.

odd function (*Maths.*). f is an odd function if $f(-x) = -f(x)$. Cf. *even function.*
oddity discrimination (*An. Behav.*). An experimental situation where an animal is given positive reinforcement for choosing the 'odd' stimulus out of three.
odd legs (*Eng.*). Callipers with one straight leg and one leg curved like a leg of outside callipers, used principally for locating centres and for layout work.
odd-odd nuclei (*Nuc.*). Nuclei with an odd number of both protons and neutrons. Very few are stable.
odd sorts (*Typog.*). A general term for any unusual characters required for technical or foreign language setting.
Oden sedimentation method (*Powder Tech.*). The first procedure devised for using a *sedimentation balance* (q.v.).
odograph (*Nav.*). Instrument plotting ship's track direct on chart. (*Surv.*) A *pedometer*.
odometer (*Ocean.*). A recording sheave used with line and weight sounders and other machines to determine how much warp or wire has been paid out. (*Surv.*) See **perambulator.**
Odonata (*Zool.*). The only order of the *Odonatopteroidea*. Predacious insects with biting mouthparts; 2 similar pairs of wings with reticulate venation, having a pterostigma, nodus, and arculus; prominent eyes; small antennae; young stages aquatic and carnivorous, having a modified labium known as the mask. Dragonflies, Damsel Flies.
Odonatopteroidea (*Zool.*). A superorder of the *Palaeoptera*, which contains only one order *Odonata*.
odontalgia (*Med.*). Toothache.
odontic (*Anat.*). Pertaining to the teeth.
odontoblast (*Zool.*). A dentine-forming cell, one of the columnar cells lining the pulp-cavity of a tooth.
odontoclast (*Zool.*). A dentine-destroying cell, one of the large multinucleate cells which absorb the roots of the milk-teeth in Mammals.
odontogeny (*Zool.*). The origin and development of teeth.
Odontognathae (*Zool.*). A superorder of the Neornithes containing toothed marine diving birds known from Cretaceous fossils.
odontograph (*Eng.*). An approximate but practical guide for setting out the profiles of involute gear teeth, in which a pair of circular arcs is substituted for the true involute curve.
odontoid (*Bot., Zool.*). Toothlike.
odontoid process (*Zool.*). A process of the anterior face of the centrum of the axis vertebra which forms a pivot on which the atlas vertebra can turn.
odontolite (*Min.*). See **bone turquoise.**
odontology (*Med.*). Study of the physiology, anatomy, pathology, etc., of the teeth.
odontoma (*Med.*). Any of a variety of tumours that arise in connexion with the teeth.
odontophore (*Zool.*). In Molluscs, the radula and radula sac, with muscles and cartilages: sometimes confined to the prominence on the floor of the buccal cavity which carries the radula.
odontostomatous (*Zool.*). Having jaws which bear teeth.
odorimetry (*Chem.*). Measurement of the intensity and permanency of odours. Also called **olfactometry.**
odoriphore (*Chem.*). A group of atoms which confer an odour on a compound.
oedema (*Bot.*). A large mass of unhealthy parenchyma. Also **edema.** (*Med.*) Dropsy. Pathological accumulation of fluid in the tissue spaces and serous sacs of the body; sometimes the term is restricted to such accumulation in tissue spaces only. See also **Milroy's disease.**
oedema disease (*Vet.*). See **bowel oedema disease.**
oedematous (*Med.*). Affected by oedema.
Oedipus complex (*Psychol.*). Freudian name for a complex, present in all boys at an early age (often persisting into adult life), characterized by an unconscious rivalry for the mother's love, resulting in hostility to the father. Named from a circumstance in the legend of the Greek hero Oedipus. Cf. *Electra complex*.
O-electron (*Nuc.*). One occupying place in O-shell of atom.
oenanthal (*Chem.*). Oenanthic aldehyde or heptan-1-al, $C_6H_{13}{\cdot}CHO$. B.p. 155°C. Obtained from castor oil by distillation under reduced pressure.
oenocytes (*Zool.*). In Insects, cells which arise from metameric groups of ectodermal cells in the abdomen, sometimes being associated with the spiracles. They often undergo a cycle of activity during each instar, and may produce the lipoproteins from which the cuticulin of the epicuticle is formed.
oenocytoids (*Zool.*). Rounded cells with homogenous acidophil protoplasm which do not exhibit phagocytosis, occurring in the haemolymph of Insects.
oersted (*Elec.*). Unit of field strength in the CGS electromagnetic units system, such that 2π oersted is field produced at the centre of a circular conductor, 1 cm in radius, carrying 1 abampere (10 amperes). In the MKSA (or SI) system 1 oersted $= \frac{1000}{4\pi}$ ampere/metre.
oesophageal bulbs (*Zool.*). In *Nematoda*, the 2 swollen portions of the oesophagus. The posterior one (also called the *pharynx*) shows rhythmical pumping movements.
oesophageal pouches (*Zool.*). In Earthworms, diverticula of the oesophagus, the cells of which secrete calcium carbonate.
oesophageal sclerite (*Zool.*). In *Mallophaga*, a sclerite forming the posterior part of the hypopharynx.
oesophageal valve (*Zool.*). In Insects, a structure at the junction of the fore and mid-intestines, formed by the wall of the former being prolonged into the cavity of the stomach and then becoming reflected upon itself and joining the stomach wall. Probably reduces regurgitation.
oesophagectasis, oesophagectasia (*Med.*). Pathological dilatation of the oesophagus.
oesophagectomy (*Surg.*). Removal of the oesophagus or part of it.
oesophageocutaneous duct (*Zool.*). See **pharyngocutaneous duct.**
oesophagismus (*Med.*). See **achalasia of the cardia.**
oesophagitis (*Med.*). Inflammation of the oesophagus.
oesophagoscope (*Med.*). An instrument for viewing the interior of the oesophagus.
oesophagospasm (*Med.*). See **achalasia of the cardia.**
oesophagostenosis (*Med.*). Pathological constriction of any part of the oesophagus.
oesophagostomiasis (*Med.*). Infestation of the intestine with nematode worms of the genus *Oesophagostomum*; occurs in the tropics.
oesophagostomy (*Surg.*). Formation of an artificial opening into the oesophagus.
oesophagotomy (*Surg.*). Incision into the oesophagus.
oesophagus, esophagus (*Zool.*). In Vertebrates, the section of the alimentary canal leading from the pharynx to the stomach; usually lacking a

serous coat and digestive glands; the corresponding portion of the alimentary canal in Invertebrates. *adj.* oesophageal.

oestradiol (*Physiol.*). The principal hormone secreted by the ovarian follicle (follicular hormone) and responsible for the development of the sexual characteristics of the female. Oestradiol ($C_{18}H_{24}O_2$) is a sterol having the following structural formula:

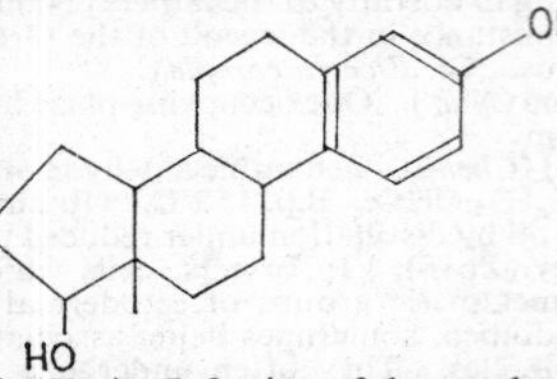

oestriasis (*Vet.*). Infection of the nasal cavity and sinuses of sheep and goats by larvae of the sheep nasal fly *Oestrus ovis*. A form of *myiasis* (q.v.).

oestriol (*Physiol.*). A female sex hormone (related to the sterols) found in the urine of pregnancy, and consisting of one phenolic hydroxyl radical and two secondary alcohol groups ($C_{18}H_{24}O_3$). With potassium hydrogen sulphate it yields *oestrone* (q.v.). Structural formula:

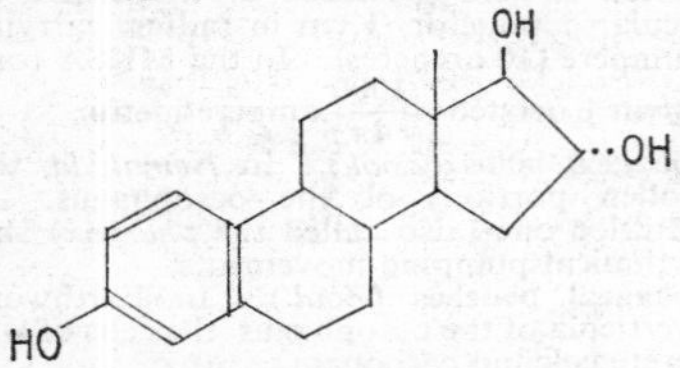

oestrogen (*Physiol.*). The generic term for a group of female sex hormones, which induce oestrus. Oestradiol (ethinyl-oestradiol) and mestranol are the common oestrogen components of some oral contraceptives. Cf. *androgen, progestogen.*

oestrone (*Physiol.*). The urinary excretion product of oestradiol. Oestrone ($C_{18}H_{22}O_2$) has the following structural formula:

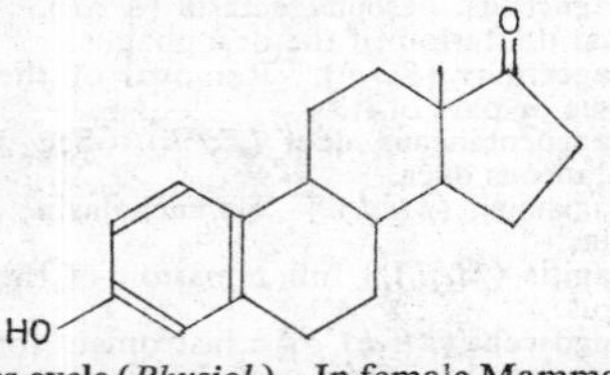

oestrous cycle (*Physiol.*). In female Mammals, the succession of changes in the genitalia commencing with one oestrous period and finishing with the next.

oestrus, oestrum (*Zool.*). In female Mammals, the period of sexual excitement and acceptance of the male occurring between pro-oestrum and metoestrum; more generally, the period of sexual excitement. *adj.* oestral.

official, officinal (*Bot.*). Used in medicine.

off-its-feet (*Typog.*). Describes type which is lying at an angle and does not give a level print.

off-lap (*Geol.*). The dispositional arrangement of a series of conformable strata laid down in the waters of a shrinking sea, or on the margins of a rising landmass, so that the successive strata cover smaller areas than their predecessors. Cf. *overlap.*

offlet (*Build., Civ. Eng.*). See **grip.**

off-line (*Comp.*). Serial usage of units of a computer with data delay (magnetic or punched tape) for subsequent operations under independent control.

off-peak load (*Elec. Eng.*). Load on a generating station or power supply system taken at times other than the time of the system peak load, e.g., during the night.

offprints (*Print.*). Separately printed copies of articles that have appeared in periodicals.

offset (*Automation*). Sustained deviation of the control point from the *index* (or *desired*) *value.* (*Bot.*) See **stolon.** (*Build.*) A ledge formed at a place where part of a wall is set back from the face. (*Surv.*) The horizontal distance measured to a point from a main survey line, in a direction at right angles to a known point on the latter.

offset blanket (*Print.*). The rubber-covered cylinder on which the image is taken from the plate cylinder and from which it is offset to the stock being printed, being the distinguishing feature of an offset press, usually litho, but occasionally letterpress. See **offset printing, printing.**

offset disk harrow (*Agric.*). A disk harrow with two gangs in tandem, capable of being offset to either side of the centre line of a tractor.

offset printing (*Print.*). Process in which the ink from a plate is received on a rubber-covered blanket cylinder from which it is transferred to the paper or other material, including sheet metal, the rubber surface making good contact even with rough surfaces. The standard method of lithographic printing, it is now being used to a lesser extent for relief printing, then called **letterset printing.**

offset rod (*Surv.*). A wooden pole painted in bands of different colours of suitable length so that the pole may be used for the measurement of short distances.

offset scale (*Surv.*). Short graduated scale used in plotting detail from field notes of *offsets* (q.v.).

offset tone-arm (*Acous.*). In a gramophone record reproducer, one which is bent or canted to minimize the effect of the arc which the reproducing stylus normally traverses, the path of the original cutting stylus being radial.

off-shore dock (*Civ. Eng.*). A form of *self-docking dock* (q.v.) built in 2 equal-length sections of L-shaped end elevation, the side wall being connected to the shore by girders hinged at each end. Self-docking is effected by carrying one section in the lap of the other.

offside (*Autos.*). The side of a vehicle further from the verge when observing the statutory rule of the road.

OFHC (*Met.*). Abbrev. for *oxygen-free high-conductivity copper.*

OG (*Arch.*). Abbrev. for *ogee.*

ogee (*Arch.*). See **cyma.**

ogee (or ogival) arch (*Arch.*). A pointed arch of which each side consists of a reverse curve.

ogee wing (*Aero.*). A wing of ogee plan form and very low aspect ratio which combines low *wave drag* in supersonic flight with high lift at high incidence through separation vortices at low speed.

ogival arch (*Arch.*). See **ogee arch.**

ogive (*Stats.*). A smooth curve drawn through the plot of the cumulative frequency of a distribution. It allows interpolations and, with suitable precautions, extrapolations to be made more accurately than linearly.

ohm (*Elec.*). SI unit of electrical resistance, such that 1 ampere through it produces a potential difference across it of 1 volt. Symbol Ω.

ohm-cm (*Elec.*). The CGS unit of *resistivity*.

ohmic, nonohmic (*Elec. Eng.*). A resistor is said to be *ohmic* or *nonohmic*, according to whether or not its resistance is described by Ohm's law. See **linear resistor**.

ohmic contact (*Elec. Eng.*). One in which part of the p.d. is proportional to current across contact. With metals, this arises from diverging current from contact point, which, if clean, has no resistance at all.

ohmic loss (*Elec. Eng.*). Power dissipation in a circuit arising from resistance, apart from loss from back e.m.f. in electrolytes or motor, or conversion by transformer. Also **in-phase loss** or **wattful loss**.

ohmic resistance (*Elec. Eng.*). See **d.c. resistance.**

ohmic resistor (*Elec. Eng.*). See **linear resistor.**

ohm-metre (*Elec.*). SI unit of *resistivity*.

Ohm's law (*Elec.*). Law, applicable strictly only to electrical components carrying direct current, also for practical purposes to those of negligible reactance carrying alternating current, which states that, in metallic conductors, at a constant temperature and zero magnetic field, the current I flowing through the component is proportional to the potential difference V between its ends, the constant of proportionality being the *conductance* $1/R$ of the component. Hence $I = V/R$ or $V = IR$. Extended by analogy to any physical situation where a pressure difference causes a flow through an impedance, e.g. heat through walls, liquid through pipes.

Ohm's law of hearing (*Acous.*). A simple harmonic motion of the air is appreciated as a simple tone by the human ear. All other motions of the air are analysed into their harmonic components which the ear appreciates as such separately.

-oid. Suffix after Gk., *-oides*, from *eidos*, form.

oidium (*Bot.*). One of a number of similar spores formed in a chain by the development of transverse septa in a hypha and the subsequent separation of the spores across the septa.

oikoplast (*Zool.*). In *Larvacea*, one of the glandular ectoderm cells by which the test or capsule is secreted.

oil. See **oils.**

oil absorption (*Chem.*). A term usually applied to pigments. The amount of linseed oil a pigment will absorb to reach a given consistency as determined by certain standards.

oil-blast circuit-breaker (*Elec. Eng.*). A form of oil circuit-breaker in which pressure set up by the gases produced as a result of the arc causes a blast of oil across the contact space, which ensures rapid extinction of the arc.

oil body (*Bot.*). A rounded mass of oily material found in the cell contents of many *Hepaticae*.

oil-break (*Elec. Eng.*). A term applied to switches, circuit-breakers, fuses, etc., to indicate that the circuit is opened under oil.

oilcloth (*Textiles*). A waterproof material obtained by coating a cotton fabric with oxidized linseed or other oils, or synthetic resins, such as polyvinyl chloride. Embossed with a leather pattern it is called **leathercloth.**

oil-cooled (*Eng., etc.*). Said of apparatus which is immersed in oil to facilitate cooling.

oil cooler (*Aero.*). See **fuel-cooled-**. (*I.C. Engs.*) A small air-cooled radiator used in aircraft and racing cars for cooling the lubricant after its return from the engine and before delivery to the oil tank.

oil-dilution system (*Aero.*). In a reciprocating aero-engine, a device for diluting the lubricant with fuel as the engine is stopped so that there is less resistance when starting in cold weather.

oil drop (*Bot.*). Any small droplet of oily substance included in the cytoplasm.

oiled cotton (*Textiles*). See **oilcloth.**

oiled paper (*Paper*). Paper impregnated with nondrying oil, e.g., oiled wrapping, or with linseed oil, e.g., oiled manilla.

oiled-silk tape (*Cables*). Silk tape varnished with oil (linseed); used in cable joints for keeping spreaders in position.

oil engine (*I.C. Engs.*). See **compression-ignition engine.**

oil-filled cable (*Cables*). Cable with a central duct, formed by an open spiral of steel tape, through which oil is fed to it. Gaseous voids and the consequent ionization are thus eliminated. Used up to the highest voltages. Development is due to Emanueli.

oilgas (*Chem.*). A gas of high energy value, obtained by the destructive distillation of high-boiling mineral oils. It consists chiefly of methane, ethene, ethyne, benzene, and higher homologues.

oil gland (*Zool.*). The preen gland or uropygial gland of Birds, a cutaneous gland forming an oil secretion used in preening the feathers.

oil hardening (*Eng.*). The hardening of cutting tools, etc., of high carbon content by heating and quenching in oil, resulting in a cooling less sudden than is effected by water, and reduced risk of cracking.

oil-hardening steel (*Met.*). Alloy steel which can be hardened by cooling in oil instead of in water. A typical example is carbon 0·3%, nickel 3·0%, and chromium 0·75%.

oil-immersed (*Elec. Eng.*). Said of electrical apparatus which is immersed in oil. See **oil-break, oil-cooled, oil-insulated.**

oil-immersion objective (*Micros.*). The use of a thin film of oil, with the same refractive index as glass, between the objective of a microscope and the specimen. This extends the resolving power and gives the maximum useful magnification obtainable with a light microscope.

oil-impregnated paper (*Cables*). That used for low and high voltage cables; the oil has resin in it to increase viscosity at working temperatures.

oiling (*Textiles*). Spraying wool, woollen blends, cotton, jute, and some manmade fibres with oil or oil/water emulsion during opening operations. It facilitates drawing, drafting, and spinning, and reduces static and waste.

oiling ring (*Eng.*). A simple device used to feed oil to a journal bearing. It consists of a light metal ring, larger in diameter than the shaft, and riding loosely thereon, located at the mid-point of the brasses, the upper brass being slotted to receive it. The ring dips into an oil reservoir in the base of the housing, and as it rotates, feeds oil to the brasses.

oil-insulated (*Elec. Eng.*). Said of electrical apparatus which is immersed in oil to facilitate its insulation.

oil length (*Paint.*). Ration of oil to resin in an alkyd or varnish medium.

oil-less circuit-breaker (*Elec. Eng.*). A circuit-breaker which does not use oil either as the quenching medium or for insulation purposes.

oil-modified resin (*Paint.*). One which contains, chemically combined therein, an oil selected from the class comprising polyhydric alcohol esters of saturated or unsaturated long chain fatty acids.

oil of bitter almonds (*Chem.*). See **bitter almond oil.**

oil of cloves (*Chem.*). See **cloves oil.**

oil of mirbane (*Chem.*). A commercial name for *nitrobenzene* (q.v.).

oil of turpentine (*Chem.*). See **turpentine.**

oil of vitriol (*Chem.*). An old name for **sulphuric acid.**

Oilostatic cable (*Cables*). **TN** for a paper-insulated power cable operated under hydrostatic pressure by means of oil contained in an outer steel pipe, in order to minimize ionization. Cf. *pressure cable.*

oil paints (*Paint.*). Pigments with or without extenders ground in oil as a vehicle or medium. There are auxiliary ingredients such as solvents, driers, etc. Paints based on *alkyd resins* (q.v.) contain from 35% to 80% linseed, soya, or other oil.

oil-poor circuit-breaker (*Elec. Eng.*). A circuit-breaker which employs a very much smaller quantity of oil (about 10%) than an ordinary oil circuit-breaker; e.g., certain types of impulse and expansion breakers.

oil pump (*I.C. Engs.*). A small auxiliary pump, driven from an engine crankshaft, which forces oil from the sump or oil tank to the bearings; often of the gear type. See **gear pump.**

oil quench (*Met.*). Immersion, for purpose of tempering, of hot metal in oil.

oil-quenched fuse (*Elec. Eng.*). A liquid-quenched fuse having oil as the quenching medium.

oil-reactive resin (*Paint.*). A synthetic resin which is assumed to react chemically with a drying oil during the process of solution.

oil red O (*Chem.*). An acid monoazo dye; a fat stain.

oil ring (*I.C. Engs.*). See **scraper ring.**

oils (*Chem.*). A group of neutral liquids comprising 3 main classes: (*a*) *fixed*(*fatty*) *oils*, from animal, vegetable and marine sources, consisting chiefly of glycerides and esters of fatty acids; (*b*) *mineral oils*, derived from petroleum, coal, shale, etc., consisting of hydrocarbons; (*c*) *essential oils*, volatile products, mainly hydrocarbons, with characteristic odours, derived from certain plants. (*Elec. Eng*.) (Mineral oils, as used in transformers, cables, switchgear.) Class A oils have a maximum sludge of 0·1% and may be used in transformers above 80°C and in oil-switches above 70°C. Class B oils have a maximum of 0·8% and may be used in transformers up to 75°C.

oil-shale (*Geol.*). One of the argillaceous rocks of sedimentary origin containing diffused hydrocarbons in a state suitable for distillation into paraffin and other mineral oils by the application of heat. See **shale oils.**

oilsilk (*Textiles*). Silk fabric impregnated with oxidized linseed or other oil.

oil sink (*Horol.*). The spherical recess around a pivot hole in a watch or clock plate. Its purpose is to act as a reservoir for the oil.

oil-soluble resin (*Paint.*). One which will react, if necessary by heating, with oils selected from predominantly polyhydric alcohol esters of unsaturated long chain fatty acids.

oilstone. A smooth stone used to impart a fine keen edge to a cutting tool, for which purpose it is first moistened with oil.

oilstone slip (*Tools*). See **gouge slip.**

oil sump (or pan) (*I.C. Engs.*). The lower part of the crankcase of an automobile engine, which usually acts as an oil reservoir.

oil switch (or circuit-breaker) (*Elec. Eng.*). The usual type of switch or circuit-breaker used on high and medium power a.c. circuits; the contacts are immersed in oil for insulating and arc-rupturing purposes.

oil tankers (*Ships*). Ships constructed specially for the carriage of oil in bulk. They are subdivided internally into a number of tanks by longitudinal and transverse bulkheads and have a pumping installation for discharging.

oil varnishes (*Paint.*). Varnishes containing a drying oil, resin, and driers. The oils used are linseed, China wood, soya bean, poppy-seed, cotton-seed, and castor-oil.

okaite (*Geol.*). An alkaline igneous rock containing melilite, haüyne and biotite.

okta (*Aero., Meteor.*). One-eighth of the sky area used in specifying cloud cover for aerodrome weather condition reports.

Old English (*Typog.*). A black-letter type in which early books were printed.

old face (*Typog.*). The earliest form of roman type, as: Bembo.

Oldham coupling (*Eng.*). A coupling permitting misalignment of the shafts connected. It consists of a pair of flanges whose opposed faces carry diametrical slots, and between which a floating disk is supported through corresponding diametral tongues which are arranged at right angles.

oldhamite (*Min.*). Sulphide of calcium, usually found as cubic crystals in meteorites.

Old Red Sandstone (*Geol.*). The continental facies of the Devonian System in the British Isles, comprising perhaps 12 000 m of red, brown, or chocolate sandstones, red and green marls, cornstones, breccias, flags, and conglomerates, yielding on certain horizons the remains of archaic fishes, eurypterids, plants, and rare shelly fossils. See **Brownstone Series.**

old style (*Typog.*). A type-face imitating the old style of roman letter used until the end of the seventeenth century, as: Old Style No. 2.

Old Style (*Astron., etc.*). A name given to the system of date-reckoning superseded by the adoption of the *Gregorian calendar.*

old style figures (*Typog.*). See **hanging figures.**

olecranon (*Zool.*). In land Vertebrates, a process at the upper end of the ulna which forms the point of the elbow.

olefin (*Textiles*). Synthetic fibre. The fibre-forming substance is a long-chain polymer comprising over 80% by weight of propylene or ethylene, etc.

olefins (alkenes) (*Chem.*). Hydrocarbons of the ethene series, having the general formula C_nH_{2n}. They contain a double bond and are very reactive substances.

oleic acid (*Chem.*). $C_{18}H_{34}O_2$, a colourless liquid, m.p. 14°C. It is an unsaturated acid of the formula $CH_3{\cdot}(CH_2)_7{\cdot}CH{=}CH{\cdot}(CH_2)_7{\cdot}COOH$, with *cis* configuration about the double bond, and occurs as the glycerine ester in fatty oils. Oleic acid oxidizes readily on exposure to the air, turns yellow and becomes rancid.

oleic acid series (*Chem.*). A group of unsaturated aliphatic acids having the general formula $C_nH_{2n-1}{\cdot}COOH$. Its most important members are acrylic acid, the crotonic acids, angelic acid, tiglic acid, oleic acid and erucic acid.

olein (*Chem.*). A glycerine ester of oleic acid.

oleo (*Aero.*). Main structural member of the support assemblies of an aeroplane's *landing gear*. Of telescopic construction and containing oil so that on landing the oil is passed under pressure through chambers at a controlled rate thereby absorbing the shock. Also referred to as **oleo leg, shock strut and shock absorber.**

oleocyst (*Zool.*). In some *Siphonophora*, a diverticulum of a nectocalyx containing a drop of oil.

oleo-resin (*Paint.*). Oily resinous sap of plants, partially deprived of volatile constituents.

oleoresinous (*Paint.*). Paints and varnishes the binder of which contains both oils and resins.

oleosome (*Bot.*). A large fatty inclusion in the cytoplasm of a cell.

oleothorax (*Med.*). Injection of oil into the pleural cavity (between the lung and chest-wall) to compress the diseased lung.

oleum (*Chem.*). A commercial name for *fuming sulphuric* (*disulphuric*) *acid*, $H_2S_2O_7$.

olfactometer (*An. Behav.*). An apparatus in which an animal is presented with two currents of air bearing different scents, and demonstrates its ability to discriminate between them and also any preference for one by moving towards the source of the appropriate air stream.

olfactometry (*Chem.*). See **odorimetry.**

olfactory (*Zool.*). Pertaining to the sense of smell; the first cranial nerve of Vertebrates, running to the olfactory organ.

olfactory lobes (*Zool.*). Part of the forebrain in Vertebrates, which is concerned with the sense of smell, and from which the olfactory nerves originate.

olfactory pit (*Zool.*). In Invertebrates, a small depression lined by sensory cells and subserving the sense of smell; in developing Vertebrates, a depression which will later give rise to one of the external nares.

olig-, oligo-. Prefix from Gk. *oligos*, few, small.

oligaemia, oligemia (*Med.*). Diminution in the volume of the blood.

oligiste-iron (*Min.*). Specular iron-ore. See **haematite.**

Oligocene (*Geol.*). The period of geological time which followed the Eocene period in the Tertiary era. Rocks of the Oligocene System are restricted in Britain to the Hampshire Basin and Isle of Wight.

Oligochaeta (*Zool.*). A class of *Annelida* with relatively few chaetae, not situated on parapodia; have a definite prostomium which usually has no appendages; always hermaphrodite; have only 1 or 2 pairs of male and female gonads in fixed segments of the anterior part of the body; reproduction involves copulation and cross-fertilization; the eggs are laid in a cocoon and develop directly; terrestrial and aquatic. Earthworms.

oligoclase (*Min.*). One of the plagioclase feldspars, consisting of the Albite (Ab) and Anorthite (An) molecules combined in the proportions of Ab_9, An_1 to Ab_7, An_3. It is found especially in the more acid igneous and metamorphic rocks.

oligocythaemia, oligocythemia (*Med.*). Diminution in the number of red cells in the blood.

oligodendroglia (*Zool.*). A type of small neuroglia cell occurring more frequently in white matter than in grey matter, containing no fibres and having no feet; more generally, a neuroglia cell.

oligodendroglioma (*Med.*). A cerebral tumour composed of oligodendroglia.

oligogenic (*Gen.*). Applied to heritable characteristics controlled by a few genes.

oligogenics (*Gen., Med.*). Practice and study of birth control.

oligolecithal (*Zool.*). A type of egg in which there is little yolk, what there is being somewhat more concentrated in one hemisphere. Found in *Amphioxus* (a genus of *Cephalochorda*) and Mammals. Cf. *mesolecithal, telolecithal.*

oligolectic (*Zool.*). Of Bees, showing restricted choice of flowers when in search of nectar.

oligomenorrhoea, oligomenorrhea (*Med.*). Scantiness of the discharge which occurs during menstruation.

oligomerous (*Bot.*). Consisting of only few parts.

Oligoneoptera (*Zool.*). A section of Insects, including forms which fold their wings when resting, are holometabolic and endopterygote, and usually have few Malpighian tubes. It comprises the superorders *Coleopteroidea, Neuropteroidea, Mecopteroidea, Siphonapteroidea,* and *Hymenopteroidea.* Also **Holometabola, Endopterygota.**

oligopneustic (*Zool.*). Among Insect larvae, those having either only prothoracic and posterior abdominal spiracles (*amphipneustic*) or only the former (*propneustic*) or only the latter (*metapneustic*). Generally an adaptation to life in a liquid or semiliquid medium. Cf. *apneustic, holopneustic, peripneustic.*

oligopod (*Zool.*). (1) Having few legs or feet. (2) Said of a phase in the development of larval Insects in which the thoracic limbs are large while the evanescent abdominal appendages of the polypod phase have disappeared. Cf. *polypod, protopod.*

oligopyrene (*Zool.*). Said of spermatozoa in which the number of chromosomes is subnormal. Cf. *apyrene, eupyrene.*

oligosaccharides (*Chem.*). Carbohydrates containing a small number (2 to 10) of monosaccharide units linked together, with elimination of water.

oligospermia (*Med.*). Diminution, below the average, of the quantity of semen voided in an ejaculation.

oligosporous (*Bot.*). Containing only few spores.

oligotokous (*Zool.*). Bearing few offspring. Cf. *polytokous.*

oligotrophic (*Ecol.*). Said of a type of lake having steep and rocky shores and scanty littoral vegetation, and in which the hypolimnion does not become depleted of oxygen in the summer. The hypolimnion is generally larger than the epilimnion, and primary productivity is low. Cf. *dystrophic, eutrophic.*

oligotrophophyte (*Bot.*). A plant growing in a soil poor in soluble mineral salts.

oligozoospermia (*Med.*). Diminution of the number of spermatozoa in the semen.

oliguria (*Med.*). Abnormally diminished secretion of urine.

oliphagous (*Zool.*). Feeding on few different kinds of food; as phytophagous Insects which are limited to a few related food-plants. Cf. *polyphagous.*

olivary nucleus (*Zool.*). A wavy band of grey matter within the medulla oblongata, corresponding to the position of the *olive.*

olive (*For.*). *Olea europaea*, found in the Mediterranean regions of Europe, important for its fruit which produces edible oil when compressed, and is itself edible after processing with dilute NaOH to neutralize the bitter glucoside content. (*Zool.*) In the medulla oblongata of higher Vertebrates, a prominence on the lateral aspect of each pyramid, corresponding to a wavy band of grey matter (the *olivary nucleus*) in the substance of the brain.

olivenite (*Min.*). A hydrous arsenate of copper which crystallizes in the orthorhombic system. It is a rare green mineral of secondary origin found in copper deposits.

olive oil (*Chem.*). A pale-yellow or greenish oil obtained from the fruit of *Olea europaea*; rel. d. 0·91–0·92, acid value 1·9–5, saponification value 185–196, iodine value 77–88.

oliver filter (*Min. Proc.*). Drum filter used in large-scale dewatering or filtration of mineral pulps, usually after *thickening.* Pulp is drawn by vacuum to filtering membrane as drum rotates slowly through a trough, and filtrate is drawn off while moist filter cake is scraped from down-running side of drum.

olivine (*Min.*). Orthosilicate of iron and magnesium, crystallizing in the orthorhombic system, which occurs widely in the basic and ultramafic igneous rocks, including olivine-gabbro, olivine-dolerite, olivine-basalt, peridotites, etc. See **chrysolite.** The clear-green variety is used as a gemstone under the name *peridot*. For *iron olivine* see **fayalite,** and for *magnesium olivine*, see **forsterite.**

olivine-nodules (*Geol.*). Xenolithic masses of olivine found in some basic igneous rocks, such as the Carboniferous basalts in Derbyshire.

Ollier's disease (*Med.*). Multiple cartilaginous tumours in the small bones of the carpus and tarsus.

Olson microphone (*Acous.*). The original *ribbon microphone*, using a battery-excited magnet.

omasitis (*Vet.*). Inflammation of the omasum.

omasum (*Zool.*). See **psalterium.**

ombré (*Textiles*). Woven stripes in which warp yarns, of the same colour but different tones, produce a shaded effect in the cloth.

ombrometer (*Meteor.*). A *rain-gauge* (q.v.).

ombrophile (*Bot.*). A plant which thrives in a place where rain is abundant.

ombrophyte (*Bot.*). A plant inhabiting rainy places.

omega-particle (*Nuc.*). The heaviest (1672 MeV) hyperon, detected in 1964. Its existence provides strong experimental evidence for the classification system developed from *Lie algebra* (q.v.) (group theory).

omegatron (*Instr.*). Device for detailed examination of the residual gas in a vacuum tube.

omental bursa (*Zool.*). In some Mammals, a pouch formed ventrally and dorsally to the stomach by the mesentery supporting the stomach.

omentopexy (*Surg.*). The stitching of the omentum to the abdominal wall in the treatment of cirrhosis of the liver.

omentum (*Zool.*). In Vertebrates, a portion of the serosa connecting two or more folds of the alimentary canal. *adj.* **omental.**

ommateum (*Zool.*). A compound eye.

ommatidium (*Zool.*). One of the visual elements composing the compound eyes of *Arthropoda*.

ommatochromes (*Biol.*). Eye-pigments.

ommatoids (*Zool.*). In some *Arachnida*, 2 or 4 white patches on the last somite of the opisthosoma; their nature and function are unknown.

ommatophore (*Zool.*). An eye-stalk.

omnibus-bar (*Elec. Eng.*). The original term from which the commonly used expression *bus-bar* (q.v.) is derived.

omnidirectional antenna (*Radio*). One receiving or transmitting equally in all directions in horizontal plane, with little effect outside.

omnidirectional microphone (*Acous.*). One whose response is essentially independent of the direction of sound incidence. Also called **astatic microphone and nondirectional microphone.**

omnidirectional radiator (*Radio*). One which radiates energy uniformly in all directions. Also **isotropic radiator, spherical radiator.**

omnidirectional radio beacon (*Aero.*). A VHF radio beacon radiating through 360 degrees upon which an aircraft can obtain a bearing. Used for navigation by *VOR* with *DME* (qq.v.). Abbrev. **ORB.**

omnidirectional ranging (*Nav.*). A system which gives the bearing of an aircraft from a transmitter independently of the direction in which the aircraft is heading.

omnigraph (*Instr.*). A copying instrument used to reproduce drawings on the same or some other scale.

omnivorous (*Bot.*). Said of a parasitic fungus which attacks several or many species of host plant. (*Zool.*) Including both animal and vegetable tissue in the diet.

omoideum (*Zool.*). In Birds, the pterygoid.

omosternum (*Zool.*). An anterior element of the sternum in Amphibians.

omphacite (*Min.*). An aluminous sodium-bearing pyroxene, near diopside in composition, occurring in eclogites as pale-green mineral grains; in a thin section, colourless to pale green.

omphalectomy (*Surg.*). Removal of the umbilicus.

omphalic (*Zool.*). Pertaining to the umbilicus

omphalitis (*Med.*). Inflammation of the umbilicus

omphaloid (*Zool.*). Navel-shaped.

omphalophlebitis (*Vet.*). Inflammation of the navel or umbilical cord. See **joint ill.**

once-through boiler (*Eng.*). See **Benson boiler, flash boiler.**

onchocerciasis (*Med.*). Infestation of the skin and subcutaneous tissues with the nematode worm of the genus *Onchocerca volvulus* in West Africa. It causes skin tumours and eye diseases, including blindness.

onchosphere, oncosphere (*Zool.*). In *Cestoda*, a larval form characterized by the possession of 3 pairs of hooks.

oncogenous (*Med.*). Inducing, or tending to induce, the formation of tumours.

oncology (*Med.*). That part of medical science dealing with new growths (tumours) of body tissue.

oncometer (*Med.*). An apparatus for measuring variations in the size of bodily organs.

oncosphere (*Zool.*). See **onchosphere.**

on-costs (*Build., Eng., etc.*). All items of expenditure that cannot be allocated to a definite job, i.e., all expenses other than the prime cost. Also called **overhead expenses, establishment charges, loading, burden.**

Ondiri disease (*Vet.*). See **bovine infectious petechial fever.**

ondograph (*Elec. Eng.*). A recording device for tracing out a.c. waveforms by a step-by-step method. A capacitor is connected to the voltage under investigation at successive points in the cycle, and is made to discharge through a ballistic galvanometer, whose deflection will be proportional to the voltage at that point.

ondoscope (*Electronics*). Glow tube operated by strong microwave radiation fields, and used, e.g., in tuning transmitters.

ondule (*Weaving*). A wave effect running the length of a fabric; developed by alternately spacing out and closing up certain warp threads by employing a special reed in the loom.

one-at-once winding (*Textiles*). The operation of filling up a brass bobbin with a special kind of yarn.

one-electron bond (*Chem.*). Bond formed by resonance of a single electron between 2 atoms or radicals of similar energy e.g., the H_2^+ ion, which has a bond energy of $\Delta H = -247$kJ/mol for the following resonance structure:

$$H.H^+ \rightleftharpoons H^+.H.$$

one-group theory (*Nuc.*). Greatly simplified reactor model in which all neutrons are regarded as having the same energy; cf. *multigroup theory*.

one-hour rating (*Elec. Eng.*). A form of rating commonly used for electrical machinery supplying an intermittent load, e.g., traction motors for suburban service. It indicates that the machine will deliver its specified rating for a period of 1 h without exceeding the specified temperature rises.

oneiric, oniric (*Psychol.*). Pertaining to dreams.

one-many function switch (*Comp.*). One in which one input only is excited at a time and each input gives a combination of outputs. Cf. *many-one function switch*.

one-particle model of a nucleus (*Nuc.*). A form of *shell model* (q.v.) in which nuclear spin and magnetic moment are regarded as associated with one resident nucleon.

one-phase (*Elec. Eng.*). See **single-phase.**

one-pipe system (*Plumb.*). A plumbing system in which both soil and waste are carried by a common pipe, the fittings being protected with specially deep seals.

one-shot circuit (*Telecomm.*). See **single circuit.**

one-shot multivibrator (*Telecomm.*). Same as **flip-flop** as used in U.K.

one-sided inheritance (*Gen.*). Inheritance through a chromosome which occurs in one sex only, e.g., a Y-chromosome.

one-to-one correspondence or transformation (*Maths.*). Two sets S and T are said to be in one-to-one correspondence if there exists a mapping f from S to T with the properties: (1) distinct elements of S have distinct images in T; (2) each element of T is the image of some element in S. Also called **one-to-one mapping, bi-uniform correspondence or mapping or transformation.** Two finite sets can be arranged in one-to-one correspondence if, and only if, the two sets have the same number of elements. If two infinite sets can be arranged in one-to-one correspondence, then the two sets are said to have the same cardinal number.

one-to-one transformer (*Elec. Eng.*). A transformer having the same number of turns on the primary as on the secondary; used in circuits when it is desired to insulate one part of a circuit from another.

O-network (*Telecomm.*). One consisting of 4 impedances connected in O-formation, with 2 adjacent junction points connected to the input terminals and the other 2 to the output. The balanced form of the π-network is an O-network.

one-way switch (*Elec. Eng.*). A switch providing only one path for the current. Cf. *double-throw switch* (2-way switch).

onion skin paper (*Paper*). A highly glazed translucent paper having an undulating surface caused by the special beating and drying of the paper.

onium acids (*Chem.*). Salts of the ammonium type BH^+X^- behave as acids at elevated temperatures. For instance, fused ammonium chloride behaves like hydrochloric acid in soldering fluxes; fused ammonium nitrate dissolves metal carbonates and oxidizes metals in the same way as nitric acid; their mixture behaves like high temperature *aqua regia*.

on-lap (*Geol.*). An *unconformity* above which successively younger beds are of increasing extent. Largely synonymous with *overlap*, and the reverse of *off-lap*.

on-line (*Comp.*). Serial usage of units of a computer, without delay or storage between units.

on-line operation (*Comp.*). Procedure such that results for each stage are available immediately.

on-line processing (*Comp.*). Automatic data ordering from data as and when received.

on-off control (*Telecomm.*). A simple control system which is either on or off, with no intermediate positions. See **continuous control.**

on-off keying (*Telecomm.*). That in which the output from a source is alternately transmitted and suppressed to form signals.

onomatomania (*Psychiat.*). An obsessional state of mind in which forgetfulness of certain words is associated with an irresistible impulse to repeat other words, often obscene.

Onsager equation (*Chem.*). An equation based on the Debye-Hückel theory, relating the *equivalent conductance* Λ_c at concentration c to that at zero concentration, of the form

$$\Lambda_c = \Lambda_0 - (A + B\Lambda_0)\sqrt{c},$$

where A and B are constants depending only on the solvent and the temperature.

ontoanalysis (*Psychiat.*). School of psychiatry that joins existentialism and psychoanalysis.

ontogeny, ontogenesis (*Biol.*). The history of the development of an individual. Cf. *phylogeny*. *adj.* **ontogenetic.**

on-top altitude clearance (*Aero.*). *Air-traffic control* clearance for *visual flight rules* (q.v.) flying above cloud, haze, smoke, or fog.

onych-, onycho-. Prefix from Gk. *onyx*, gen. *onychos*, a nail or claw.

onychia (*Med.*). Inflammation of the nail-bed.

onychium (*Zool.*). In Insects, a pulvillus: in some Spiders, an extension of the tarsus between the paired claws.

onychocryptosis (*Med.*). Ingrowing toe-nail.

onychogenic (*Zool.*). Nail-forming, nail-producing; as a substance similar to eleidin occurring in the superficial cells of the nail-bed.

onychogryphosis, onychogryposis (*Med.*). Thickening, twisting, and overgrowth of the nails (usually of the toes) as a result of chronic infection and irritation.

onychomycosis (*Med.*). A disease of the nails due to a fungus.

Onychophora (*Zool.*). A class of *Arthropoda* having trachea; a soft thin cuticle; body wall consisting of layers of circular and longitudinal muscles; head not marked off from the body, and consisting of 3 segments, 1 preoral, bearing preantennae and 2 postoral bearing jaws and oral papillae respectively; a pair of simple vesicular eyes; all body segments similar, each bearing a pair of parapodialike limbs which end in claws and containing a pair of excretory tubules; spiracles scattered irregularly over the body; cilia in the genital region; development direct, e.g., the genus *Peripatus*.

onyx (*Min.*). A cryptocrystalline variety of silica with layers of different colour, typically whitish layers alternating with brown or black bands. Used in cameos, the figure being carved in relief in the white band with the dark band as background.

onyx marble (*Min.*). Oriental alabaster (see under **alabaster**).

oö-. Prefix from Gk. *oon*, egg.

oöblastema (*Zool.*). A fertilized egg.

oöcyst (*Zool.*). In certain *Protozoa*, the cyst formed around two conjugating gametes; in *Sporozoa*, the passive phase into which an oökinete changes in the host.

oöcyte (*Zool.*). An ovum prior to the formation of the first polar body; a female gametocyte.

oöcytin (*Biol.*). Substance occurring in spermatozoa and having both fertilizing and agglutinating action on ova of the same species.

oœcium (*Zool.*). A brood pouch.

oögamy (*Biol.*). (1) The union of gametes of dissimilar size, usually of a relatively large non-motile egg and a small active sperm. (2) In *Protozoa*, anisogamy in which the female gamete is a hologamete.

oögenesis (*Biol.*). The origin and development of ova.

oögonial branch (*Bot.*). A hypha on or in which an oögonium develops.
oögonium (*Bot.*). The female sexual organ in algae and fungi. It has a noncellular wall, and contains one or more oöspheres, which are converted into oöspores at fertilization and set free before they germinate. (*Zool.*) An egg-mother-cell or oöcyte.
oökinete (*Zool.*). In certain *Protozoa*, an active vermiform stage developed from the zygote.
oölemma (*Zool.*). See **vitelline membrane.**
oölite (*Geol.*). A sedimentary rock composed of oöliths. In most cases oöliths are composed of calcium carbonate, in which case the rock is an oölitic limestone, but they can also be made of chamosite or limonite, in which case the rock is an oölitic ironstone. Written by itself, the word can be assumed to refer to an oölitic limestone.
oölith (*Geol.*). A more or less spherical concretion of calcium carbonate, chamosite, or dolomite, not exceeding 2 mm in diameter, usually showing a concentric-layered and/or a radiating fibrous structure.
oölitic (*Geol.*). A textural term applicable to sedimentary rocks, of several different kinds, which consist largely of oöliths. Cf. *pisolitic*.
oölogy. The study of ova.
oölysis (*Bot.*). The conversion of carpels and ovules into leafy structures.
Oömycetes (*Bot.*). Group of *Phycomycetes*, with usually well-developed mycelium, on which oögonia are formed. Thick walled resting spores (oöspores) are characteristic. Sometimes classified as a subclass of *Biflagellatae*.
oöphorectomy (*Surg.*). Removal of an ovary.
oöphoritis (*Med.*). Inflammation of an ovary.
oöphorosalpingectomy (*Surg.*). Removal of an ovary and a Fallopian tube.
oöphorostomy (*Surg.*). The formation of an opening into an ovarian cyst.
oöphorotomy (*Surg.*). Incision of an ovary.
oöphyte (*Bot.*). The gametophyte, in *Bryophyta* and *Pteridophyta*.
oöplasm (*Bot.*). The central cytoplasm, producing the oösphere, in the oögonium of some *Oömycetes*.
oösperm (*Zool.*). See **oöblastema.**
oösphere (*Biol.*). The unfertilized female gamete or its homologue.
oöspore (*Bot.*). A thick-walled spore, which normally germinates only after a period of inactivity, formed after the fertilization of the oösphere in lower plants. (*Zool.*) A fertilized ovum; in *Protozoa*, an encysted zygote.
oöstegite (*Zool.*). In some higher *Crustacea*, a brood pouch formed by plates upon the thoracic limbs.
oöstegopod (*Zool.*). In some higher *Crustacea*, one of the thoracic limbs bearing the brood pouch.
oötheca (*Zool.*). An egg-case, as in the Cockroach.
oötocoid (*Zool.*). Bringing forth the young in an immature condition and allowing them to complete their early development in a marsupium.
oötocous (*Zool.*). Oviparous.
oötype (*Zool.*). A section of the oviduct in which the egg-shell is secreted; in *Platyhelminthes*, a chamber, situated at the junction of the oviduct and vitelline ducts, into which the shell-glands open.
ooze (*Geol.*). A fine-grained, soft, deep-sea deposit, composed of shells and fragments of foraminifera, diatoms, and other organisms.
oözooid (*Zool.*). A *zooid* (q.v.) which arises directly from an ovum, as opposed to a *blastozooid*, which arises by budding from another individual.
opacity (*Light, Photog.*). Reciprocal of *transmission ratio*. (*Paint.*) The *hiding power* (q.v.) of a paint or other pigmented compound.
opal (*Min.*). A cryptocrystalline or colloidal variety of silica with a varying amount of water. The transparent coloured varieties, exhibiting opalescence, are highly prized as gemstones.
opal agate (*Min.*). A variety of opal, of different shades of colour and agatelike in structure.
opalescence (*Chem.*). The milky, iridescent appearance of a solution or mineral, due to the reflection of light from very fine, suspended particles. (*Min.*) The play of colour exhibited by precious opal, due to interference at the surfaces of minutely thin films, the thicknesses of the latter being of the same order of magnitude as the wavelength of light.
opal glass (*Glass*). Glass which is opalescent or white; made by the addition of fluorides (e.g., fluorspar, cryolite) to the glass mixture.
Opalina (*Zool.*). An order of *Zoomastigina*, containing forms previously placed in the *Ciliata*. Many flagella; never amoeboid; several similar nuclei; cleavage during cell division parallel to the *kinetia*.
Opalite (*Build.*). A composition, obtainable in many colours, which is applied to walls to produce the surface appearance of glazed bricks.
opal jasper (*Min.*). Opal containing some yellow iron(IV)oxide and other impurities; it has the colour of yellow jasper with the lustre of common opal. Also called **jasp opal.**
opal lamp (*Glass*). An electric filament lamp having the bulb made of opalescent glassware, so that the filament itself is not directly visible.
opaque. Totally absorbent of rays of a specified wavelength, e.g., wood is opaque to visible light but slightly transparent to infrared rays, and completely transparent to X-rays and waves for radio communication.
opaque enema (*Radiol.*). One of barium sulphate, used in radiography to outline the colon.
open account (*Comp.*). One of a large number of accounts registered on magnetic tape, all of which are re-recorded with *up-dating* at regular intervals, e.g., daily.
open aerial (*Radio*). An aerial which is open-ended and able to support a standing wave of current.
open aestivation (*Bot.*). Aestivation in which the perianth leaves neither overlap nor meet by their edges.
open arc (*Light*). A carbon arc to which there is free access of the external atmosphere, any enclosure being for light diffusion or to give protection from draughts.
open back (*Bind.*). See **hollow back.**
openband (or **right-hand**) **twine** (*Textiles*). The term used in the woollen trade to denote the direction (S) of twist in a yarn produced by open-spindle bands; synonymous with the term *right-hand twist* used in the worsted industry.
opencast (*Mining*). Quarry. Open cut. Mineral deposit worked from surface and open to daylight.
open channel (*Hyd.*). A channel for the conveyance of liquid, the free surface of which is always within the channel and at atmospheric pressure. Open channels are frequently covered over in practice.
open circuit (*Elec. Eng.*). A circuit providing infinite impedance. See **short-circuit.**
open-circuit characteristic (*Elec. Eng.*). A term commonly used to denote the curve obtained by

plotting the e.m.f. generated by an electric generator on open circuit against the field current. Also called **no-load characteristic.**

open-circuit grinding (*Chem. Eng., Min. Proc.*). Size reduction of solids in which the material to be crushed is passed only once through the equipment, so that all the grinding has to be done in a single step. Generally less efficient than *closed-circuit grinding* (q.v.).

open-circuit impedance (*Elec.*). Input or driving impedance of line or network when the far end is *free*, *open-circuited*, not *grounded* or *loaded*.

open-circuit loss (*Elec. Eng.*). See **no-load loss.**

open-circuit transition (*Elec. Eng.*). A method used, in the series-parallel control of traction motors, for changing the connexions of the motors from series to parallel; the circuit is broken while the reconnexion is being made. Cf. *bridge transition.*

open-circuit voltage (*Elec. Eng.*). The voltage appearing across the terminals of an electric generator or transformer when it is delivering no load. The term also refers to the voltage appearing across the electrodes of an arc-welding plant when no current is flowing.

open clusters (*Astron.*). Galactic clusters of stars of a loose type containing at most a few hundred stars; the stars of a cluster have a common motion through space, and are associated with dust and gas clouds. See **Hyades, Pleiades, Praesepe, Ursa Major cluster.**

open community (*Bot.*). A plant community which does not occupy the ground completely, so that bare spaces are visible.

open-diaphragm loudspeaker (*Acous.*). Most common domestic type, in which a paper, plastic or doped-fabric cone diaphragm is driven by a circular coil at its apex. Mounted in a baffle or box, with or without ports or labyrinth.

open-field test (*An. Behav.*). An experimental procedure in which an animal is released into an open area with no obstacles, and features of its behaviour, such as defecation, urination, and locomotion are observed, these sometimes being related to emotionality.

open floor (*Build.*). A floor which is not covered by a ceiling, the joists being therefore on view.

open-frame girder (*Eng.*). A girder consisting of upper and lower booms connected at intervals by (usually) vertical members, and not braced by any diagonal members.

open fuse (*Elec. Eng.*). A fuse in which the mounting is such that the fuse link is fully exposed, except for any external containing-case.

open-hearth furnace (*Met.*). A furnace, of reverberatory type, used in steel-making. The charge is contained on a shallow hearth, and the furnace fired with gas or oil.

open-hearth process (*Met.*). Siemens-Martin process (1866). Cold pig-iron and steel scrap are charged to open hearth and melted by preheated gases with flame temperatures up to 1750°C produced by regeneration as they enter through brick labyrinth, where heat is maintained by periodic reversal between these and exit gases. After melting (up to 1600°C) most of impurities are in covering slag. Remainder are removed by addition of iron ore, scale, or limestone before tapping into ladles, where ferrosilicon, ferromanganese, or aluminium may be added as deoxidants. See also **acid process, basic process, steel-making.**

open inequality (*Maths.*). One which defines an open set of points, e.g., $-1 < x < +1$.

opening (*Textiles*). Preliminary beating and opening-up of compressed cotton after leaving the bale-breaking machine. Large, high-speed spiked cylinders beat out sand, dust, broken leaf, etc., into dust chambers, leaving the cotton in a fluffy state for lap formation. (*Typog.*) The appearance of a pair of facing pages, given careful consideration in the best typography.

open interval (*Maths.*). An interval, such as $a < x < b$, the points of which form an open set. Cf. *closed interval.*

open-jet wind tunnel (*Aero.*). A *wind tunnel* (q.v.) in which the working section is not enclosed by a duct.

open mortise (*Join.*). See **slot mortise.**

open newel stair (*Build.*). A stair having successive flights rising in opposite directions, and arranged about a rectangular well hole.

open pipe (*Acous.*). A pipe which is partially or completely open at the upper end, so that the wavelength of the fundamental resonance is approximately double the length of the air column.

open pore (*Powder Tech.*). A cavity within a particle of powder which communicates with the surface of the particle.

open roof (*Build.*). A roof which is not covered in by a ceiling, the trusses being exposed.

open sand (*Foundry*). (1) A sand of good porosity or permeability, as distinct from a *close* sand. (2) The process of casting in an open mould when the finish of the top surface is immaterial.

open set (*Maths.*). A set of points such that a neighbourhood of each point lies within the set.

open shed (*Weaving*). A passage between upper and lower lines of warp threads; the individual threads remain stationary until required to move up or down. Only those threads move which are required to change position. Employed mainly on high-speed tappet and dobby looms, and, to a lesser extent, on jacquards.

open slating (*Build.*). Slating in which gaps of 1–4 in. (25-100 mm) are left between adjacent slates in any course.

open slot (*Elec. Eng.*). A type of slot, used in the armatures of electric machines, in which the opening is the same width as the rest of the slot. Cf. *semiclosed slot.*

open string (*Carp.*). See **cut string.**

open traverse (*Surv.*). One which does not close on the point of origin, and is therefore not directly checked for accuracy or closing error.

open vascular bundle (*Bot.*). A bundle including cambium.

open well (*Arch.*). A stair enclosing a vertical opening between the outer sides of the flights.

open window unit (*Acous.*). Same as *sabin.*

open-wire feeder (*Telecomm.*). One supported from insulators on poles, forming a pole route between antenna and transmitter or receiver.

openwork or lace fabrics (*Textiles*). Openwork pattern produced in flat and circular knitted fabrics by a transference of certain stitches from a needle or needles to adjacent needles; this process lightens the weight of the fabric. Openwork patterns may also be arranged by a disposal of stitches.

operand (*Comp., Maths.*). Quantity on which an operation is to be performed.

operant conditioning (*An. Behav.*). An experimental procedure in which an animal is given reinforcement after it spontaneously makes a particular response, and the intensity of the response then increases. Also known as **instrumental conditioning** on the view that the response is instrumental in producing reinforcement for the animal.

operant level (*An. Behav.*). The rate of occurrence

of an operant response before it has been experimentally reinforced.

operant response (*An. Behav.*). A response which is identified by its consequences.

operating characteristic (*Elec. Eng.*). See **load characteristic.**

operating duty (*Elec. Eng.*). A term applied to a switch or circuit-breaker to denote the series of making and breaking operations used in specifying its performance.

operating factor (*Elec. Eng.*). The ratio which the time during which an intermittently run motor is actually running bears to the length of the duty cycle.

operation (*Psychol.*). In Piaget's terms a mental operation is an internalized action which is reversible. The two types of reversibility are inversion or negation, and reciprocity. See **concrete operations, formal operations.**

operational amplifier (*Electronics*). High gain, high stability d.c. amplifier used with external feedback path. These were originally used to perform one specific mathematical operation in analogue computing but are now more widely used as the basic building blocks of electronic control systems.

operational sphericity (*Powder Tech.*). A *shape factor* (q.v.) defined as the cube root of the ratio of the volume of a particle of powder to the volume of the circumscribing sphere.

operation time (*Comp.*). That required by computer to complete one specific operation. (*Electronics*) That which elapses between application of electrode voltages to valve and output current reaching specified value.

operator (*Maths.*). Mathematical symbol representing a specific operation to be carried out on a particular operand, e.g., the differential operator $D(\equiv d/dx)$.

opercular apparatus (*Zool.*). In Fish, the operculum, together with the branchiostegal membrane and rays.

opercular cell (*Bot.*). A lid cell by means of which some antheridia open.

operculare (*Zool.*). In Fish, a dorsal membrane bone of the operculum.

opercular fold (*Zool.*). In the embryo of most Birds, a fold growing back from the hyoid arch to cover the second and third branchial clefts.

operculate (*Bot., Zool.*). (1) Possessing a lid. (2) Opening by means of a lid.

operculum (*Bot.*). A cover or lid which opens to allow the escape of spores from a sporangium or other container. (*Zool.*) In the eggs of some Insects, a differentiated area of the chorion which lifts up when the larva emerges from the egg. In the higher Fish, a fold which articulates with the hyoid arch in front of the first gill slit, and extends backwards, covering the branchial clefts. A similar structure occurs in the larvae of Amphibians. In some tubicolous *Polychaeta*, an enlarged branch of a tentacle closing the mouth of the tube when the animal is retracted; in *Xiphosura*, the united anterior pair of abdominal appendages bearing the genital apertures; in Spiders, a small plate partially covering the opening of a lung book; in some *Cirripedia*, plates of the carapace which can be closed over the retracted thorax; in *Pterobranchiata*, a ventral and lateral projection of the collar forming a lower lip; in some *Gastropoda*, a plate of chitinoid material, strengthened by calcareous deposits, which fits across the opening of the shell.

operon (*Gen.*). A group of genes lying together in a chromosome and regulating the production of an enzyme.

ophicalcite (*Geol.*). See **forsterite-marble.**

ophicephalous (*Zool.*). Snake-headed; in *Echinoidea*, said of small pedicellariae having a flexible stalk and short, broad jaws with toothed edges, which lack poison-glands but have a special articulating device.

ophimottling (*Geol.*). A textural character of certain basic gabbroic rocks in which large pyroxenes enclosing small plagioclase crystals ophitically are embedded in a groundmass essentially feldspathic.

ophiopluteus (*Zool.*). In *Ophiuroidea*, a pelagic ciliated larval form, in which the posterolateral arms are large and are directed forwards. See also **pluteus.**

ophitic texture (*Geol.*). A texture characteristic of dolerites in which relatively large pyroxene crystals completely enclose smaller, lath-shaped plagioclases. See also **poikilitic texture.**

Ophiuroidea (*Zool.*). A class of *Echinodermata* with a dorsoventrally flattened star-shaped body; arms sharply differentiated from the disk and not containing caecae of the alimentary canal; tube feet lack ampullae and suckers and lie on the lower surface, although not in grooves; no anus; madreporite aboral; well-developed skeleton; no pedicellariae; free-living. Brittle-stars.

ophthalmectomy (*Surg.*). Excision of an eye.

ophthalmia (*Med.*). Inflammation of various parts of the eye, especially of the conjunctiva.

ophthalmic (*Zool.*). Pertaining to or situated near the eye, as the *ophthalmic nerve*, which passes along the back of the orbit in lower Vertebrates.

ophthalmodynamometer (*Med.*). An instrument used to measure the intraocular arterial pressure by means of external pressure on the eye.

ophthalmology (*Med.*). The study of the eye and its diseases. *n.* **ophthalmologist.**

ophthalmoplegia (*Med.*). Paralysis of one or more muscles of the eye.

ophthalmoscope (*Med.*). An instrument for inspecting the interior of the eye by light reflected from a mirror.

Opiliones (*Zool.*). See **Phalangida.**

opisthial aperture (*Bot.*). The opening between the base of the stomatal pore and the substomatal cavity.

Opisthobranchiata (*Zool.*). An order of *Gastropoda*, descended from *Prosobranchiata* (q.v.) which had undergone torsion, but which themselves show detorsion; mantle cavity (where present) tending to be posterior, the shell to become smaller, internal, or absent, and the single ctenidium to disappear, either being replaced by accessory respiratory organs, or the whole external surface becoming a respiratory organ. Sea hares, sea slugs, sea butterflies.

opisthocoelous (*Zool.*). Concave posteriorly and convex anteriorly; said of vertebral centra.

opisthodetic (*Zool.*). In Bivalves, posterior to the beak.

opisthoglossal (*Zool.*). Having the tongue attached anteriorly, free posteriorly, as in Frogs.

opisthognathous (*Zool.*). The condition, found in some Insects, where the head is directed backwards so that the specialized mouthparts arise between the anterior legs. Cf. *orthognathous, prognathous.*

opisthogoneate (*Zool.*). Having the genital opening at the hind end of the body. Cf. *progoneate.*

opisthomere (*Zool.*). A postoral somite.

Opisthomi (*Zool.*). A suborder of the order *Isospondyli* of the *Teleostei*; of small size and eel-like form, with pelvic fins absent, and spiny dorsal fin; freshwater carnivorous forms of Africa and Southern Asia. Spiny Eels.

opisthonephros (*Zool.*). In Vertebrates, a type of excretory organ corresponding to the mesonephros together with the metanephros. *adj.* **opisthonephric.**

opisthosoma (*Zool.*). In *Acarina*, the segments of the body posterior to the last pair of legs; in *Arachnida*, the abdomen, comprising the mesosoma and metasoma.

opisthotic (*Zool.*). A posterior bone of the auditory capsule in the Vertebrate skull.

opisthotonos (*Med.*). Extreme arching backwards of the spine and the neck as a result of spasm of the muscles in these regions, e.g., in tetanus.

opium (*Pharm.*). The latex from unripe capsules of Papaver somniferum, dried and powdered. Contains about 10% morphine, hence its analgesic and narcotic uses.

opotherapy (*Med.*). (Literally, treatment by juices.) Treatment by administration of extracts of animal organs, especially of ductless glands. See also **organotherapy.**

Oppenheimer-Phillips (O.P.) process (*Nuc.*). A form of *stripping* (q.v.) in which a deuteron surrenders its neutron to a nucleus without entering it.

OPP film (*Plastics*). Term used for oriented polypropene film which has been biaxially stretched to improve its physical properties. Widely used for packaging because of its high gloss and clarity, high impact strength and low moisture permeability.

opponens (*Zool.*). A muscle which, by its contraction, assists the opposition of the digits.

opposed-cylinder engine (*I.C. Engs.*). An engine with cylinders, or banks of cylinders, on opposite sides of the crankcase in the same plane, their connecting-rods working on a common crankshaft placed between and below them.

opposed-voltage protective system (*Elec. Eng.*). A form of Merz-Price protective system in which the secondary voltages of current transformers, situated at each end of the circuit to be protected, are balanced against each other, so that there is normally no current on the pilots connecting them.

opposite (*Bot.*). (1) Said of leaves inserted in pairs at each node, with one on each side of the stem. (2) Said of a stamen which stands opposite to the middle of a petal.

opposition (*Astron.*). The instant when the geocentric longitude of the moon or of a planet differs from that of the sun by 180°.

O.P. process (*Nuc.*). See **Oppenheimer-Phillips process.**

opsin (*Biochem.*). The protein portion of *rhodopsin* (q.v.).

opsonin (*Biol.*). See **bacteriotropin.**

optic (*Zool.*). Pertaining to the sense of sight; the second cranial nerve of Vertebrates.

optical activity (*Chem., Phys.*). A property possessed by many substances whereby plane-polarized light, in passing through them, suffers a rotation of its plane of polarization, the angle of rotation being proportional to the thickness of substance traversed by the light. In the case of molten or dissolved substances it is due to the possession of an asymmetric molecular structure, e.g. no mirror plane of symmetry in the molecule. See **specific rotation.**

optical ammeter (*Elec. Eng.*). An instrument used for measuring the r.m.s. values of currents with unusual waveforms. The current is passed through a lamp filament and the light output is compared photometrically with that obtained using a d.c. or sine wave supply.

optical bench (*Optics*). A rigid bed along which optical components, mounted in suitable holders, may be moved. It is a device used in experimental work on linear optical systems.

optical black (*Optics*). A body when it absorbs all radiation falling on its surface. No substance is completely black in this context.

optical bleaches (*Chem.*). See **fluorescent whitening agents.**

optical centre of a lens (*Light*). That point on the principal axis of a lens or lens system through which passes a ray whose incident and emergent directions are parallel.

optical character recognition (*Comp.*). See **character reading.**

optical constants (*Optics*). The refractive index (n) and the absorption coefficient (k) of an absorbing medium. Together these determine the complex refractive index ($n-ik$) of the medium.

optical crown (*Glass*). Glass of low dispersion made for optical purposes. See under **soda-lime-silica glass.**

optical distance or **path** (*Optics*). The distance travelled by light (d) multiplied by the refractive index of the medium (n). Length of equivalent path in air (strictly vacuum).

optical double (*Astron.*). A pair of stars which appear in close proximity owing to the perspective in which they are seen, but which have no physical connexion.

optical flat (*Eng.*). A flat glass disk having very accurately polished surfaces, used for testing by interferometry the flatness of gauge anvils and other plane surfaces. (*Photog.*) A surface, generally of glass, which has been lapped by rubbing on an optically flat surface so that deviations from a true plane surface are small in comparison with the wavelength of light.

optical flint (*Glass*). Glass of high dispersion made for optical purposes.

optical glass (*Glass*). Glass made expressly for its optical qualities. The composition varies widely both as to constituents and amounts, and the requirements as regards freedom from streaks and bubbles are very exacting.

optical indicator (*I.C. Engs.*). An engine indicator in which a ray of light is deflected successively by mirrors in directions at right angles, proportionately first to cylinder pressure, then to piston displacement, being finally focused on a ground-glass screen or photographic plate, on which it traces the *indicator diagram* (q.v.).

optical isomerism (*Chem.*). The existence of isomeric compounds which differ in their optical activity.

optical lever (*Optics*). A device for measuring the small relative displacement of two objects by means of angular displacement of a light beam.

optical maser (*Phys.*). One in which the stimulating frequency is visible or infrared radiation; see **laser.**

optical-mechanical system (*TV*). Any system of TV using mechanical scanning.

optical model of a nucleus (*Nuc.*). One explaining interactions by treating nucleus as sphere of constant refractive index and absorption coefficient.

optical pattern (*Light*). Light thrown back from the side of the groove of a track cut on a disk, or reproduced in a finished gramophone record, the width being proportional to the lateral recorded velocity of the track. Also **Buchmann-Meyer pattern, Christmas-tree pattern, light-band pattern.**

optical printing (*Cinema.*). The process of printing positive frames by optical arrangements which can reduce, enlarge, reshape, fade, or wipe the image on the negative. Used extensively for trailers and general editing.

optical pyrometer (*Heat*). An instrument which measures the temperatures of furnaces by estimating the colour of the radiation, or by matching it with that of a glowing filament.

optical range (*Radio*). That which a radio transmitter or beacon would have if it were radiating visible light.

optical recording (*Cinema.*). Technique of recording sound on a track on photographic film, so that variations in average density and in average area along the track modulate a light beam. After conversion by a photocell into a current, this is amplified and acoustically reproduced synchronously with the cinematograph film.

optical reproducer (*Cinema.*). That part of a cinematograph projector (*head*) concerned with reproducing optically recorded soundtracks, by focusing a slit of light on to the film, receiving the fluctuations in a photocell and so providing voltages for amplification.

optical rotation (*Optics*). The rotation of the plane of polarization of a beam of light when passing through certain materials.

optical rotatory dispersion (*Chem.*). The change of optical rotation with wavelength. Rotatory dispersion curves may be used to study the configuration of molecules. Abbrev. ORD.

optical scratch (*Cinema.*). A scratch on a soundtrack caused by a particle of dirt which has entered the recording mechanism.

optical spectrometer (*Optics*). A spectroscope fitted with a graduated circle for measuring the frequencies of the different colour components emitted from a light source.

optical spectrum (*Phys.*). The visible radiation emitted from a source separated into its component frequencies.

optical square (*Surv.*). A hand instrument for setting out right angles in the field. It works on the principle of the *sextant* (q.v.), the 2 reflecting surfaces being arranged in this case to yield lines of sight at a fixed angle of 90° apart.

optical track (*Cinema.*). Photographic soundtrack on film, composed of *variable area* or *variable density* (qq.v.) optical recording of sound signals.

optical transmission (system) (*Telecomm.*). Same as **line of sight.**

optical whites (*Chem.*). See **fluorescent whitening agents.**

optic atrophy (*Med.*). The condition of the optic disk (where the nerve fibres of the retina pass through the eyeball) resulting from degeneration of the optic nerve.

optic axial angle (*Min.*). The angle between the two optic axes in biaxial minerals, usually denoted as 2V (when measured in the mineral) or 2E (in air).

optic axis (*Crystal.*). Direction(s) in a doubly refracting crystal for which both the ordinary and the extraordinary rays are propagated with the same velocity. Only one exists in uniaxial crystals, two in biaxial. (*Light*) The line which passes through the centre of curvature of a lens' surface, so that the rays are neither reflected nor refracted. Also called **principal axis.**

optic lobes (*Zool.*). In Vertebrates, part of the mid-brain, which is concerned with the sense of sight, and from which the optic nerves originate.

optic neuritis (*Med.*). See **papillitis.**

opticon (*Zool.*). In Insects, the internal medullary mass or inner zone of the optic lobe of the brain.

optics. The study of light. *Physical optics* deals with the nature of light and its wave properties; *geometrical optics* ignores the wave nature of light and treats problems of reflection and refraction from the ray aspect.

optic sign (*Min.*). Anisotropic minerals are either optically positive or negative, indicated by + or − in technical descriptions. See **negative mineral, positive mineral.**

optimal damping (*Instr.*). Adjustment of damping just short of *critical*, which allows a little overshoot. This attains the ultimate reading of an indicating instrument most rapidly while indicating it is moving freely.

Optimat (*Automation*). TN for transistor system of process control and programming, whereby adjustments are made to detailed controls in a plant so as to approach its best or set performance.

optimum programming (*Comp.*). Coding program so that *latency* (delays in getting data from register) is minimized. See **minimum access program.**

optocoele (*Zool.*). The cavity of one of the optic lobes of the Vertebrate brain.

optometer (*Optics*). Instrument designed to measure the refractive state of the human eye.

optomotor response (*An. Behav.*). The response of moving the head or eyes if they are held stationary and a striped screen is moved past them. Also called **optokinetic response.** See also **menotaxis.**

optophone (*Electronics*). A photoelectric device for training the blind by converting printed words into sounds.

Opuntiales (*Bot.*). See **Cactales.**

OR (*Comp., Maths.*). See **logical operations.**

oral (*Zool.*). Pertaining to the mouth.

oral cone (*Zool.*). In polyps of *Hydrozoa* (*Coelenterata*), a conical area surrounded by tentacles, and bearing the mouth at its apex. Cf. *oral disk.*

oral contraception (*Med.*). The use of synthetic hormones (oestrogen and progestogen steroids in varying proportions), taken orally in pill form, to prevent conception by reacting on the natural *luteinizing* and *follicle-stimulating hormones* (qq.v.) and so inhibiting ovulation and/or fertilization. *Colloq.* **the pill.**

oral disk (*Zool.*). In polyps of *Actinozoa* (*Coelenterata*), a circular flattened area surrounded by tentacles with the mouth situated in the centre. Cf. *oral cone.*

oral siphon (*Zool.*). In *Urochorda*, the projection of the body which terminates in the mouth.

oral stage (*Psychol.*). Preoccupation during early infancy with the experience of feeding, sucking, and biting. The associated emotions may be incorporated into the personality.

oral valves (*Zool.*). In *Crinoidea*, five triangular flaps which separate the ambulacral grooves.

orange G stain (*Micros.*). An acid aniline dye used as a counter stain, e.g., with haematoxylin.

orange lead (*Chem.*). Lead(II, II, IV)oxide. Pb_3O_4. Obtained by heating white lead (basic lead(II) carbonate) in air at approximately 450°C. Commercial varieties contain up to approximately 35% PbO_2.

orange peel (*Paint.*). A defect of *lacquers* (q.v.) (but not of paints) which have been applied by spray gun, characterized by more or less pronounced small depressions in the film to produce an appearance resembling an orange skin.

ora serrata (*Zool.*). The edge of the retina.

ORB (*Aero.*). Abbrev. for *omnidirectional radio beacon.*

orbicular (*Bot.*). Flat, with a circular or almost circular outline.

orbiculares (*Zool.*). Muscles which surround an aperture; as the muscles which close the lips and eyelids in Mammals.

orbicular structure (*Geol.*). A structure exhibited by those igneous rocks which contain spherical orbs up to several centimetres in diameter, each showing a development of alternating concentric shells of different minerals, so deposited by rhythmic crystallization.

orbit (*Aero.*). An aircraft circling a given point is said to orbit that point and *air-traffic control* (q.v.) instructions incorporate the term. See also **holding pattern.** (*Astron.*) The path of a heavenly body (and, by extension, an artificial satellite, spacecraft, etc.) moving about another under gravitational attraction. In unperturbed motion the orbit is a conic. See **elements of an orbit.** (*Zool.*) A space lodging an eye; in Vertebrates, the depression in the skull containing the eye; in *Arthropoda*, the hollow which receives the eye or the base of the eye-stalk; in Birds, the skin surrounding the eye.

orbital (*Phys.*). Motion of an electron in a field of force as described by a *wave function*, which expresses the probability of finding the electron in a region, there being one orbital for each quantized orbit.

orbital engine (*I.C. Engs.*). Axial 2-stroke engine having curved pistons in a circular cylinder-block which rotates around a fixed shaft.

orbital quantum number (*Electronics*). The second for an orbital electron, indicating the angular momentum associated with the orbit. 0 for *s states*; 1 for *p states*; 2 for *d states*; and 3 for *f states*. See also **principal quantum number.**

orbitosphenoid (*Zool.*). A paired cartilage bone of the Vertebrate skull, forming the side wall of the brain-case in the region of the presphenoid.

Orbitron (*Radiol.*). TN for large apparatus containing giant source for the radiotherapy treatment of cancer (surface). The shield, with the requisite window, is carried round a large hollow wheel, the patient being on a stretcher which can be inserted in the wheel. Rotation minimizes the dose elsewhere than in required region.

orbit shift coils (*Nuc. Eng.*). These placed on the magnet pole faces of a betatron or synchrotron in the region of the stable orbit so that by passing a current pulse the particles may be momentarily displaced to strike a target placed outside of the stable orbit.

Orcadian Series (*Geol.*). A name sometimes applied to the Middle Old Red Sandstone of Scotland, including as the chief member the Caithness Flagstone Group, overlying barren red sandstones without fossils. The name *Orcadian* is also applied to the basin in which the Old Red Sandstone of Caithness, Moray and the Orkney and Shetland Islands was deposited.

orcein stain (*Micros.*). A dye obtained from lichen which is particularly absorbed by elastic fibres. It is used in solution with alcohol and hydrochloric acid.

orchic, orchitic (*Zool.*). Pertaining to the testis.

orchidalgia (*Med.*). Pain in a testis.

orchi(d)ectomy (*Surg.*). Removal of a testis.

orchidopexy (*Surg.*). The operation of stitching an undescended testis to the scrotum.

orchiepididymitis (*Med.*). Inflammation of the testis and the epididymis.

orchitis (*Med.*). Inflammation of a testis.

orcinol (*Chem.*). $CH_3 \cdot C_6H_3 \cdot (OH)_2$ (1,3,5), 3-dihydroxy (methyl benzene), a dihydroxybenzene, colourless crystals, m.p. 107°C, b.p. 288°C.

orculiform (*Bot.*). Said of a 2-celled spore having a thick septum pierced by a connecting-tube.

ORD (*Chem.*). Abbrev. for *optical rotatory dispersion.*

order (*Bot., Zool.*). A group of closely related plants or animals forming a subdivision of a *class*, and itself further subdivided into *families.* (*Chem.*) If $A + B \rightarrow$ products, and the rate of reaction is $k(A)^x(B)^y$, the order is x with respect to A and y with respect to B. (*Comp.*) An instruction in a program. (*Maths.*) Of a curve or surface: the number of points, real, coincident, or imaginary, in which the curve or surface is intersected by a straight line. Of a differential equation: the order of the highest order derivative that occurs. Of a matrix: (1) a square matrix of n rows and n columns is of order n; (2) a rectangular matrix of n rows and m columns is of order $n \times m$, or n by m.

order-disorder transformation (*Met.*). In solid solution (e.g., a brass alloy) the ordered state can occur below the temperature at which lattice sites are occupied by single atomic species, and the disordered state can ensue above a critical temperature, with two or more elements in one primitive lattice.

ordered state (*Met.*) In solid solution of alloy, repeated and regular atomic arrangement.

order number (*Eng.*). Of a torque impulse or a vibration, as the torsional oscillation of an engine crankshaft, the number of impulses or vibrations during one revolution of the shaft.

order of reaction (*Chem.*). A classification of chemical reactions based on the index of the power to which concentration terms are raised in the expression for the instantaneous velocity of the reaction, i.e., on the apparent number of molecules which interact.

order-wire circuit (*Teleph.*). A circuit between operators in different exchanges serving for exchange of information in setting up subscribers' calls.

ordinal number (*Maths.*). Number derived from the notion of counting, and possessing the fundamental property of order or position in an aggregate.

ordinary differential equation (*Maths.*). See **differential equation.**

ordinary ray (*Light*). See **double refraction.**

ordinate (*Maths.*). See **cartesian coordinates.**

Ordnance Bench Mark (*Surv.*). A bench mark officially established with reference to the Ordnance datum. Abbrev. **O.B.M.**

Ordnance datum (*Surv.*). Arbitrary zero height, which is assumed to be the mean sea level at Newlyn, Cornwall. All official bench marks in Britain are referred to this for height above sea level.

Ordnance Survey (*Surv.*). Originally a triangulation of Great Britain made in 1791 by the military branch of the Board of Ordnance. Now a Civil Service survey which is concerned with land, buildings, roads, etc. Published as maps on which the National Grid is superimposed, giving the whole country a unified reference system. Abbrev. **OS.**

Ordovician (*Geol.*). The middle period in the lower Palaeozoic era, which followed the Cambrian period. It was named by Prof. Chas. Lapworth after the Ordovices, an old British tribe of the Welsh Border, in which area rocks of this age are found. See **Arenig Series Ashgill Series, Bala Series, Caradocian, Llandeilian Series, Llanvirn Series, Ordovicic.**

Ordovicic (*Geol.*). An obsolete synonym for *Ordovician.*

ore (*Min.*). A term applied to any metalliferous mineral from which the metal may be profitably extracted. It is extended to nonmetals and also to minerals which are potentially valuable.

ore bin (*Min. Proc., Mining*). Storage system (usually of steel or concrete) which receives ore intermittently from the mine. A fine ore bin holds material crushed to centimetric size and keeps from 1 to 3 days' milling supply.

ore body (*Mining*). Deposit, seam, bed, lode, reef, placer, lenticle, mass, stockwork, according to geological genesis.

ore dressing (*Min. Proc.*). See **mineral processing.**

Oregon pine (*For.*). See **Douglas fir.**

or element (or **gate**) (*Comp.*). One for which output is energized if one or more of the inputs are energized.

ore reserves (*Mining*). Ore whose grade and tonnage has been established by drilling, etc., with reasonable assurance.

orf (*Vet.*). See **contagious pustular dermatitis.**

Orford process (*Met.*). A process once used by the International Nickel Co. for separating the copper and nickel in the matte obtained by bessemerizing. The matte, which consists of copper-nickel sulphides, is fused with sodium sulphide, and a separation into 2 layers, the top rich in nickel and the bottom rich in copper, is obtained. Now superseded by a method based on froth-flotation.

organ (*Acous.*). A musical instrument, comprising ranks of pipes which radiate sound when blown by compressed air, the operation of the pipes being controlled by manuals or keyboards and by a set of pedals. Hence, any musical instrument producing synthetically (e.g. electronically) tones similar to those from pipes and operated from keyboards, e.g., the *Electrone* and *Hammond*. (*Bot., Zool.*) A part of the body of an animal or a plant adapted and specialized for the performance of a particular function.

organdie (*Textiles*). Plain woven, stiff finished, light dress fabric of cotton, spun rayon or nylon. Often has crimped or crammed stripes warpways.

organelle (*Cyt.*). A specialized protoplasmic part of a cell having a particular function, as the Golgi apparatus or a mitochondrion.

organic chemistry (*Chem.*). The study of the compounds of carbon. Owing to the ability of carbon atoms to combine together in long chains (catenate), these compounds are far more numerous than those of other elements. They are the basis of living matter.

organic disease. See under **functional disease.**

organic phosphor (*Radiol.*). Organic chemical used as solid or liquid scintillator in radiation detection.

organism (*Biol.*). A living animal or plant: includes also microorganisms not readily classifiable as either, but having the properties of living matter, e.g., *Rickettsiae.*

organized (*Biol.*). Showing the characteristics of an organism; having the tissues and organs formed into a unified whole.

organizer (*Zool.*). An organization centre in development; a part of the body from which pass out organizing influences to other regions of the embryo, e.g., the anterior lip of the blastopore.

organogeny, organogenesis (*Biol.*). The study of the formation and development of organs.

organography (*Bot.*). A descriptive study of the external form of plants, with relation to function.

organoleptic (*Chem., Psychol.*). Denotes any method of systematically testing or assessing the effects of a substance on the human senses, particularly taste or smell.

organo-magnesium compounds (*Chem.*). *Grignard reagents* (q.v.), compounds of the type R.Mg.I.

organo-metallic compounds (*Chem.*). Wide range of compounds in which carbon atoms are linked directly with metal atoms, including the alkali metals, e.g., sodium phenyl, $C_6H_5{\cdot}Na$; lead, e.g., lead tetraethyl, $Pb(C_2H_5)_4$; zinc, e.g., zinc dimethyl, $Zn(CH_3)_2$. The compounds are useful reagents in preparative organic chemistry. See **Grignard reagents.**

organonomy (*Biol.*). The laws relating to living organisms.

organonymy (*Biol.*). The nomenclature of organs.

organosilicone (*Chem.*). Synthetic resin characterized by long thermal life and resistance to thermal ageing. See also **silicones.**

organosol (*Chem., Plastics*). A coating composition based on a dispersed PVC resin mixed with a plasticizer and a diluent; a colloidal solution in any organic liquid.

organotherapy (*Med.*). Treatment of disease by administration of animal organs or extracts of them, especially of ductless gland extracts. See also **opotherapy.**

organo-tin compounds (*Chem.*). Used as pesticides, which decompose, after use, into totally harmless inorganic substances. See **tributyl tin compound.**

organ pitch (*Acous.*). See **pitch of organ pipe stops.**

organzine (*Textiles*). Silk yarn intended as warp; two or more threads are run together with a slight twist imparted to withstand the stresses and abrasion of weaving.

orgasm (*Zool.*). Culmination of sexual excitement. *adj.* **orgastic.**

Orgatron (*Acous.*). An electronic musical instrument using the pneumatic action of a reed organ. The electrical current for the operation of radiating loudspeakers is obtained by electrostatic pick-ups operated by the motion of the languids of air-operated reeds, using adequate amplifiers.

oriel (*Arch.*). A projecting window suppor ed upon corbels or brackets.

Oriental alabaster (*Min.*). See under **alabaster**

Oriental almandine (*Min.*). A name sometimes used for *corundum*, of gemstone quality, which is deep-red in colour, resembling true almandine (a garnet) in this, but no other, respect.

Oriental amethyst, etc. (*Min.*). See **false amethyst,** etc.

Oriental cat's eye (*Min.*). See **cymophane.**

Oriental emerald (*Min.*). A name sometimes used for *corundum*, of gemstone quality, resembling true emerald in colour.

oriental region (*Zool.*). One of the primary faunal regions into which the land surface of the globe is divided. It includes the southern coast of Asia east of the Persian Gulf, the Indian subcontinent south of the Himalayas, southern China and Malaysia, and the islands of the Malay Archipelago north and west of Wallace's line.

Oriental ruby (*Min.*). See **ruby.**

Oriental topaz (*Min.*). A variety of *corundum*, resembling topaz in colour.

orientation (*Biol.*). The position, or change of position, of a part or organ with relation to the whole; change of position of an organism under stimulus. (*Chem.*) (1) The determination of the position of substituent atoms and groups in an organic molecule, especially in a benzene nucleus. (2) The ordering of molecules, particles, or crystals so that they point in a definite direction. (*Met.*) The position of important sets of planes in a crystal in relation to any fixed system of planes. See **pure metal crystal.** (*Surv.*) Fixing of line or plan in azimuth, with respect to true north.

orientation behaviour (*An. Behav.*). Behaviour of an animal which (*a*) guides it into its *primary* *orientation* (q.v.) and/or (*b*) guides it into its *secondary orientation* (q.v.). See **kinesis, taxis.**

orienting curvature (*Physiol.*). See **tropism.**

orifice (*Hyd.*). A small opening intended for the passage of a fluid.

orifice gauge (*Eng.*). A flow gauge consisting of a thin orifice plate clamped between pipe flanges, with pressure take-offs drilled into the adjacent pipes, or of a thick orifice plate similarly clamped but containing its own pressure take-offs.

origin (*Maths.*). A fixed point with respect to which points, lines, etc., are located; the point common to the axes of coordinates. (*Zool.*) That end of a skeletal muscle which is attached to a portion of the skeleton which remains, or is held, rigid when the muscle contracts, the other end, which is attached to a part of the skeleton which moves as a result of the contraction, being known as the *insertion* (q.v.).

original gravity (*Brew.*). Specific gravity of the wort from which beer is brewed.

origin distortion (*Electronics*). Distortion of waveform displayed by gas-focused cathode-ray tube employing electrostatic deflection, due to nonlinear relation between angular deflection and deflecting voltage at low values of the latter. This results in flattening a waveform where it crosses zero line.

origin of hills (*Geol.*). In the general lowering of ground-level by denudation, rivers work quickly downward in grading themselves to the existing base level. Between the river valleys, higher ground is left upstanding as hills. Again, some areas may be afforded protection by a capping of rock which is resistant to denudation; such areas will form hills as the surrounding ground is more rapidly lowered. Some hills are directly of volcanic origin, formed by the eruption of lava and ash. Even when the volcanic rocks of an area have been completely covered up by the sediments of a later geological period, in the process of re-excavation the igneous rocks tend to form hills on account of their durability, as in the Lake District and North Wales. See also **escarpment.**

origin of mountains (*Geol.*). Most mountains, other than those of volcanic origin, have originated by the dissection of uplifted areas of the earth's crust.

O-ring (*Eng.*). A toroidal ring, usually of circular cross-section, made of neoprene or similar materials, used, e.g., as an oil or air seal.

orleans (*Textiles*). (1) Plain woven fabrics constructed from fine cotton warp and botany worsted weft. (2) A good type of soft American cotton shipped from New Orleans.

Orlon (*Chem.*). TN for synthetic fibre based almost wholly on cyano ethene with a small amount of a different monomer to serve as a dye receptor. Widely used in knitted fabrics, as imitation fur and in carpets.

orlop deck (*Ships*). The lowermost deck in a ship of several decks. It is simply a platform, and contributes nothing to main longitudinal strength; usually of small extent.

ormolu (*Met.*). An alloy of copper, zinc, and sometimes tin, used (especially in the 18th century) for furniture mountings, decorated clocks, etc.

ornis (*Zool.*). A Bird fauna. *adj.* **ornithic.**

ornithine (*Chem.*). 2-6-Diaminovaleric acid. It is concerned in urea formation in the animal body (see **arginine**), and a derivative, *ornithuric acid*, is found in the excrement of Birds.

Ornithischia (*Zool.*). An order of herbivorous *Archosauria*, including the 'birdlike' dinosaurs (cf. *Saurischia*), with a pelvis in which the pubis was parallel to an extended ischium, and often with a predentary bone in the mandible which may have supported a heavy beak. Jurassic and Cretaceous forms, often heavily armoured, including Iguanodon and Triceratops.

ornithocopros (*Zool.*). The excrement of Birds.

ornithology (*Zool.*). The study of Birds.

ornithophily (*Bot.*). Pollination by Birds.

ornithopter (*Aero.*). Any flying machine that derives its principal support in flight from the air reactions caused by flapping motions of the wings, this motion having been imparted to the wings from the source of power being carried.

ornithosis (*Vet.*). See **psittacosis.**

oro-. (1) Prefix from L. *os*, gen. *oris*, mouth. (2) Prefix from Gk. *oros*, mountain.

oroanal (*Zool.*). Connecting, pertaining to, or serving as, mouth and anus.

orogenesis (*Geol.*). The processes which take place during an *orogeny*.

orogenic belt (*Geol.*). A region of the earth's crust, usually elongated, which has been subjected to an *orogeny*. Recently formed orogenic belts correspond to mountain ranges, but older belts have often been eroded flat.

orogeny (*Geol.*). A period during which rocks are severely folded, metamorphosed and uplifted, as contrasted with the longer periods of geological time during which these processes are not taking place very actively. Igneous activity is greater during such periods than at other times. Different periods of orogeny are given specific names, e.g. **Caledonian, Alpine orogeny.**

orographic ascent (*Meteor.*). The upward displacement of air blowing over a mountain. The subsequent downward movement is known as a *helm wind*. The repetition of this wave downwind is a *lee wave*, and the reversed wind which may occur under a lee wave is the *rotor wind*.

orographic rain (*Meteor.*). Rain caused by moisture-laden winds impinging on the rising slopes of hills and mountains. Precipitation is caused by the cooling of the moist air consequent upon its being forced upwards.

oroide (*Met.*). See **French gold.**

oronasal (*Zool.*). Pertaining to or connecting the mouth and the nose.

orpiment (*Min.*). Trisulphide of arsenic, which crystallizes in the monoclinic system; commonly associated with realgar; golden-yellow in colour and used as a pigment.

orrery (*Astron.*). A mechanical model of the solar system showing the relative motions of the planets by means of clockwork; much in vogue in the 18th century. (Named after Charles Boyle, Earl of Orrery.)

Orsat apparatus (*Chem.*). A portable apparatus used in the analysis of flue, furnace, and exhaust gases. A sample of the gas is successively scrubbed by solutions which selectively absorb the CO_2, CO, and oxygen.

orthicon (*TV*). Camera tube in which the external image is focused on to a mosaic, which is scanned from behind by a low-power electron beam; output signal is taken from reaction of this. See **image orthicon.**

orthite (*Min.*). See **allanite.**

ortho (*Photog.*). See **orthochromatic.**

ortho-. Prefix from Gk. *orthos*, straight. (*Chem.*) (1) Derived from an acid anhydride by combination with the largest possible number of water molecules, e.g., *orthophosphoric acid* (q.v.). (2) Consisting of diatomic molecules with parallel nuclear spins and an odd rotational quantum number, e.g., *orthohydrogen* (q.v.).

ortho- (*Chem.*). Containing a benzene nucleus substituted in the 1.2 positions.

ortho film stock (*Cinema.*). Fine-grained positive film stock for printing in black-and-white, i.e., when the registration of the relative luminosities of colours is not in question.

orthoaluminic acid (*Chem.*). A term used to emphasize the acidic nature of aluminium hydroxide.

orthocaine (*Pharm.*). Methyl 3-amino-4-hydroxy benzene carboxylate:

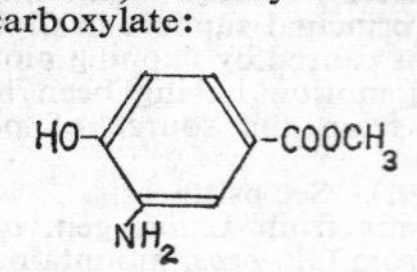

HO– –$COOCH_3$

NH_2

White crystalline solid; m.p. 142°C; used as a local anaesthetic.

orthocentre (*Maths.*). Of a triangle: the point of intersection of the perpendiculars from the vertices to their opposite sides.

orthochromatic (*Photog.*). Said of emulsions which register visual luminosities correctly, apart from colour. Actually such emulsions are sensitive to all colours except red. Abbrev. **ortho.**

orthocladous (*Bot.*). Having long, straight branches.

orthoclase (*Min.*). Silicate of potassium and aluminium, $KAlSi_3O_8$, crystallizing in the monoclinic system; a feldspar, occurring as an essential constituent in granitic and syenitic rocks, and as an accessory in many other rock types. See also **microcline, sanidine.**

orthodiagraph (*Radiol.*). An X-ray apparatus for recording exactly the size and form of organs and structures inside the body.

orthodox sleep (*Psychol.*). This comprises three of the four stages of sleep, the part excluded being *rapid eye movement* or *paradoxical sleep*. It was at one time thought that orthodox sleep was the deeper type but it is in fact easier to awake a person from this sleep than from paradoxical sleep.

orthoferrosilite (*Min.*). The ferrous iron end-member of the orthopyroxene group of silicates.

orthogeotropism (*Bot.*). Growth of a stem vertically upwards, or of a root vertically downwards, in relation to gravity.

orthognathous (*Zool.*). With the long axis of the head at right angles to that of the body, and the mouth directed downwards. Cf. *prognathous*.

orthogneiss (*Geol.*). Term applied to gneissose rocks which have been derived from rocks of igneous origin. Cf. *paragneiss*.

orthogonal (*Bot.*). The manner of arrangement of four members of a flower when two are median and two lateral.

orthogonal cutting (*Eng.*). A cutting process used to analyse the forces occurring in machining, in which a straight-edged cutting tool moves relatively to the workpiece in a direction perpendicular to its cutting edge.

orthogonal functions (*Maths.*). A set of functions $f_r(x)$ such that, over the range considered, $\int f_r(x)f_s(x)dx=0$, except when $r=s$. If the integral equals unity when $r=s$ the functions are also said to be *normal*. If also the equation $\int F(x) \, . \, f_r(x)dx=0$ for all r implies that $F(x)$ is identically zero, then the set is said to be *complete*.

orthogonal matrix (*Maths.*). One equal to the inverse of its transpose. Any two rows, or any two columns, will be orthogonal vectors.

orthogonal vectors (*Maths.*). Two vectors whose scalar product is zero.

orthograph (*Arch.*). A view showing an elevation of a building or of part of a building.

orthographic projection (*Eng.*). A method of representing solid objects on 3 mutually perpendicular plane surfaces, using parallel rays or projectors perpendicular to the surfaces; commonly used as the basis of engineering drawing. (*Geog.*) A type of *zenithal projection* (q.v.), in which the view point is at infinity.

orthohydrogen (*Chem.*). Hydrogen molecule in which the two nuclear spins are parallel, forming a triplet state.

orthojector circuit-breaker (*Elec. Eng.*). A form of circuit-breaker, requiring only a small quantity of oil, in which the arc is extinguished by a flow of oil across the contacts, which flow is produced partly by the high pressure set up by the arc itself.

orthokinesis (*An. Behav.*). A *kinesis* (q.v.) in which the average speed of an animal's movement, or the frequency of its activity, depends on the intensity of a stimulus. While the stimulus acts, the animal generally moves, and when it ceases to act, the animal comes to rest, and tends to aggregate. Cf. *klinokinesis*.

orthokinetic (*Chem.*). Migrating in the same direction.

orthomorphic projection (*Geog.*). A type of map projection in which the shapes truly picture the shapes on earth.

orthopaedics, orthopedics (*Med.*). That branch of surgery which deals with deformities arising from injury or disease of bones or of joints. *adj.* **orthopaedic.**

ortho-**phenylene diamine** (*Photog.*). Developing agent:

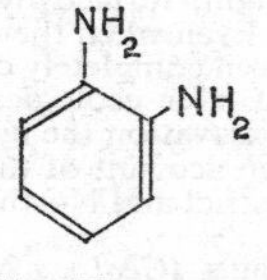

NH_2

NH_2

orthophoria (*Med.*). The state of normal adjustment and balance of the muscles of the eye.

orthophosphoric acid (*Chem.*). H_3PO_4. Formed when phosphorus pentoxide is dissolved in water and the solution is boiled. The highest hydrated stable form of phosphoric acid.

orthophyre (*Geol.*). A little-used term for a form of microsyenite, consisting essentially of orthoclase. Better used, if at all, as a textural term.

orthophyric (*Geol.*). A textural term applied to medium- and fine-grained syenitic rocks consisting of closely packed orthoclase crystals of stouter build than in the typical trachytic texture. The term actually implies the presence of porphyritic orthoclase crystals.

orthopnoea, orthopnea (*Med.*). Dyspnoea so severe that the patient is unable to lie down; a symptom of heart failure. *adj.* **orthopnoeic.**

orthopsychiatry (*Psychiat.*). The branch of psychiatry concerned with the prevention and correction of incipient mental illness.

Orthoptera (*Zool.*). An order of the *Orthopteroidea*. Large insects with an orthognathous head; biting mouthparts; posterior legs with enlarged femora for jumping; forewings toughened (tegmina) and overlapping when folded; unjointed cerci; well developed ovipositor; possess a variety of stridulatory organs. Grasshoppers, Locusts, Crickets.

Orthopteroidea (*Zool.*). A superorder of the

Polyneoptera, containing the orders *Plecoptera*, *Notoptera*, *Cheleutoptera*, *Orthoptera*, and *Embioptera*.
orthopterous (*Zool.*). Having the posterior pair of wings straight folded.
orthoptic circle (*Maths.*). Of a conic: the locus of a point from which tangents to the conic are perpendicular to each other. Except for the parabola, this locus is a circle concentric with the conic. For the parabola, the locus is its directrix. Also called **director circle.**
orthoptic treatment (*Med.*). The nonoperative treatment of squint by specially devised stereoscopic exercises.
orthopyroxene (*Min.*). A group of pyroxene minerals crystallizing in the orthorhombic system, e.g., enstatite, hypersthene.
orthoradial (*Zool.*). Said of a type of segmentation in which the lines of cleavage are symmetrically arranged with relation to the main axis of the ovum.
orthorhombic system (*Crystal.*). The style of crystal architecture which is characterized by three crystal axes, at right angles to each other and all of different lengths. It includes such minerals as olivine, topaz, and barytes.
orthoselection (*Biol.*). Modification resulting from the elimination of all other lines of variation through the selective struggle.
orthosilicic acid (*Chem.*). $Si(OH)_4$.
orthostatic (*Med.*). Associated with or caused by the erect posture, e.g., *orthostatic* albuminuria.
orthostichous (*Zool.*). Arranged in straight rows; as the fin skeleton in Fish when the peripheral elements are parallel. Cf. *rachiostichous, rhipidostichous.*
orthostichy, orthostichies (*Bot.*). A vertical rank of leaves on a stem.
orthostyle (*Arch.*). A colonnade formed of columns arranged in a straight line.
orthotropous (*Bot.*). (1) Said of an ovule which is straight, i.e., with the micropyle in a straight line with the funicle. Also called **atropous.** (2) Said of organs which show a sharp positive or negative tropism in respect to a given stimulus.
orthotropy (*Phys.*). The phenomenon whereby the elastic properties of a material vary in different planes, e.g., timber.
Ortmann's coastal regions (*Ocean.*). A series of faunistic regions into which the coastal waters of the world have been divided.
oryzenin (*Chem.*). A protein of the *glutelin* (q.v.) group found in rice.
os (*Geol.*). See **esker.** (*Zool.*) (1) An opening, as the *os uteri.* (L. *os*, gen. *oris*, mouth.) *pl.* **ora.** (2) A bone, as the *os coccygis.* (L. *os*, gen. *ossis*, bone.) *pl.* **ossa.**
Os (*Chem.*). The symbol for *osmium.*
OS (*Astron., etc.*). Abbrev. for *Old Style.* (*Build.*) Abbrev. for *one side.* (*Surv.*) Abbrev. for *Ordnance Survey.*
OS & W (*Build.*). Abbrev. for *oak, sunk and weathered.*
osazones (*Chem.*). The diphenylhydrazones of monosaccharides, obtained by the action of 2 molecules of phenylhydrazine on one molecule of the monosaccharide. They are sparingly soluble in water, can be purified by recrystallization, and serve to identify the respective monosaccharides.
oscillating capacitor. See **vibrating capacitor.**
oscillating neutral (*Elec. Eng.*). A phenomenon occurring in 3-phase, unearthed, star-connected systems, due to third harmonic voltages, which results in a distorted phase voltage waveform.
oscillating sequence, oscillating series (*Maths.*). **See divergent sequence, divergent series.**
oscillation (*Aero.*). See **longitudinal-, phugoid-.** (*Eng.*) See **hunting.** (*Phys.*) See **centre of-.** (*Radio*) Sustained and very stable periodic alteration of current in a tuned circuit (inductance and capacitance, or quarter-wave stub). Maintained by the supply of synchronous pulses of energy from a transistor or valve, to compensate energy lost through dissipation and output.
oscillation constant (*Radio*). Square-root of product of inductance (henry) and capacitance (farad) of a resonant circuit.
oscillation frequency (*Radio*). That determined by the balance between the inertia reactance and the elastic reactance of a system, e.g., open or short-circuited transmission line, cavity, resonant circuit, quartz crystal.

If C = capacitance
L = self-inductance of circuit,

then frequency $f = 1/(2\pi\sqrt{LC})$. In a mechanical oscillating system $f = 2\pi\sqrt{M/S}$, where M = mass, and S = restoring force per unit displacement.
oscillator (*Radio*). A source of alternating current of any frequency, which is sustained in a circuit by a valve or transistor using positive feedback principle. There are 2 types: (*a*) *stable-type*, in which frequency is determined by a line or a tuned (LC) circuit, waveform being substantially sinusoidal, and (*b*) *relaxation-type*, in which frequency is determined by resistors and capacitors, waveform having considerable content of harmonics. Also applies to mechanical systems, velocities being equivalent to currents. See **multivibrator.**
oscillator crystal (*Radio*). A piezoelectric crystal used in an oscillator to control the frequency of oscillation.
oscillator drift (*Telecomm.*). See **frequency drift.**
oscillatory discharge (*Radio*). That of capacitor through inductor when the resistance of circuit is sufficiently low and current persists after the capacitor has completely discharged, so that it charges again in the reverse direction. This process is repeated until all initial energy is dissipated in resistance, including radiation.
oscillatory scanning (*TV*). That in which the scanning spot moves repeatedly to and fro across the image, so that successive lines are scanned in opposite directions.
oscillatory zoning (*Min.*). The compositional variation within a crystal which consists of alternating layers rich in the two end-members of an isomorphous solid-solution series.
oscillatron (*Electronics*). Normal CRT for displaying or registering waveforms by deflecting a beam of electrons.
oscillogram (*Electronics*). Record of a waveform obtained from any oscillograph. Usually photograph of CRT display.
oscillograph (*Electronics*). In U.S. and U.K., an oscilloscope equipment, with addition of a photographic recording system to register waveforms displayed. In France, equipment for direct recording of a trace by the beam on a photographic plate in vacuum.
oscilloscope (*Electronics*). (1) Equipment incorporating a cathode-ray tube, time-base generators, triggers, etc., for the delineation of a wide range of waveforms by electron beam. (2) Mechanical or optical equipment with a corresponding function, e.g., Duddell oscilloscope.
os cloacae (*Zool.*). See **hypoischium.**
os coccygis (*Zool.*). See **urostyle.**
os cordis (*Zool.*). In *Ungulata*, a visceral bone at the base of the heart.
osculant, osculate (*Zool.*). Intermediate in characteristics between 2 groups, genera, or species.

osculating circle (*Maths.*). See **curvature.**
osculating orbit (*Astron.*). The name given to the instantaneous ellipse whose elements represent the actual position and velocity of a comet or planet at a given instant (*epoch of osculation*).
osculating plane (*Maths.*). See **moving trihedral.**
osculating sphere (*Maths.*). At a point P on a space curve, a sphere that has four-point contact with the curve at P, i.e., the limiting sphere through four neighbouring points P, Q, R and S, on the curve, as Q, R and S tend to P. Its centre and radius are called respectively the *centre* and *radius* of *spherical curvature*.
osculation (*Maths.*). See **point of osculation.**
osculum (*Zool.*). In *Porifera*, an exhalant aperture by which water escapes from the canal system. *adjs.* **oscular, osculiferous.**
Osgood-Schlatter disease (*Surg.*). Osteochondritis of the tibial tubercle.
O-shell (*Nuc.*). The collections of electrons characterized by the principal quantum number 5, starting with the element rubidium which has one electron in its O-shell.
Osler's disease, Osler-Vaquez disease (*Med.*). See **erythraemia.**
Osler's nodes (*Med.*). Painful papules on the digits in cases of bacterial endocarditis.
Oslo crystallizer (*Chem. Eng.*). See **Krystal crystallizer.**
osmeterium (*Zool.*). In the larvae of certain *Papilionidae* (*Lepidoptera*), a bifurcate sac exhaling a disagreeable odour which can be protruded through a slitlike aperture in the first thoracic segment.
osmic acid (*Chem.*). An erroneous name for *osmium*(VIII)*oxide*, OsO_4, yellow crystals which give off an ill-smelling, poisonous vapour. Its aqueous solution is used as a histological stain for fat, as a catalyst in organic reactions, and as a fixative in electron microscopy.
osmiophilic (*Chem.*). Having an affinity for, staining readily with osmic acid, e.g., certain components and organelles of cells.
osmiridium (*Met.*). A very hard, white, naturally occurring alloy of osmium (17–48%) and iridium (49%) containing smaller amounts of platinum, ruthenium, and rhodium. Used for the tips of pen-nibs.
osmium (*Chem.*). A metallic element, a member of the platinum group. Symbol Os, at. no. 76, r.a.m. 190·2, m.p. 2700°C. Osmium is the densest element, rel. d. at 20°C 22·48. Like platinum it is a powerful catalyst for gas reactions, and is soluble in aqua regia, but unlike platinum when heated in air it gives an oxide, volatile OsO_4. Alloyed with iridium it forms an extremely hard material. See **osmiridium.**
osmium tetroxide (*Chem.*). Osmium(VIII)oxide. See **osmic acid.**
osmometer (*Chem.*). An apparatus for the measurement of osmotic pressures.
osmophore (*Chem.*). See **odoriphore.**
osmoreceptors (*Physiol.*). Cells specialized to react to osmotic changes in their environment, e.g., cells which react to osmotic changes in the blood or tissue fluid and which are involved in the regulation of secretion of antidiuretic hormone by the neurohypophysis.
osmoregulation (*Zool.*). The process by which animals regulate the amount of water in their bodies, and the concentration of various solutes and ions in their body fluids.
osmosis (*Chem.*). Diffusion of a solvent through a semipermeable membrane into a more concentrated solution, tending to equalize the concentrations on both sides of the membrane.
osmotic coefficient (*Chem.*). The quotient of the van't Hoff factor and the number of ions produced by the dissociation of 1 molecule of the electrolyte.
osmotic pressure (*Chem.*). The pressure exerted by a dissolved substance in virtue of the motion of its molecules. It may be measured by the pressure which must be applied to a solution in order just to prevent osmosis into the solution.
osone (*Chem.*). The oxidation product of an osazone, obtained by the elimination of phenylhydrazine and oxidation by the action of HCl. Osones contain the group $O{=}\overset{|}{C}{-}CH{=}O$. When reduced, the aldehyde group only is converted into a hydroxyl group, and by this method it is possible to convert an *aldose* (q.v.) into a *ketose* (q.v.), e.g., glucose into fructose.
osophone (*Acous.*). Headphone for deaf persons employing bone conduction of sound.
os penis (*Zool.*). A bone developed in the middle line of the penis in some Mammals, as Bats, Whales, some Rodents, Carnivores, and Primates.
osphradium (*Zool.*). A sense-organ of certain aquatic *Mollusca*, consisting usually of a patch of columnar ciliated epithelium, richly innervated from the visceral commissure; formerly believed to be of olfactory function, but now suspected of being concerned in the assessment of suspended silt in the water entering the mantle chamber. *adj.* **osphradial.**
ossa (*Zool.*). See **os** (2).
osseous (*Zool.*). Bony; resembling bone.
ossicle (*Zool.*). A small bone; in *Echinodermata*, one of the skeletal plates; in *Crustacea*, one of the calcified toothed plates of the gastric mill.
ossification (*Zool.*). The formation of bone; transformation of cartilage or mesenchymatous tissue into bone. *v.* **ossify.**
Ostariophysi (*Zool.*). A large order of the *Teleostei*, characterized by a peculiar chain of bones, the Weberian ossicles, connecting the air bladder with the ear region. One of the largest orders of fish, including most freshwater fish such as Carp, Roach, Goldfish, Catfishes.
oste-, osteo-. Prefix from Gk. *osteon*, bone.
Osteichthyes (*Zool.*). A class of *Gnathostomata* including the vast majority of fish. The skeleton is bony, and the body is covered with scales. First found in the Devonian period. Comprise the two subclasses *Actinopterygii* (ray-finned fish) and *Crossopterygii* (Coelocanth and lungfish).
osteitis (*Med.*). Inflammation of a bone.
osteitis deformans (*Med.*). See **Paget's disease.**
osteitis fibrosa (*Med.*). See **fibrocystic disease.**
osteoarthritis, ostearthritis (*Med.*). Arthritis deformans. A form of chronic arthritis in which the cartilages of the joint and the bone adjacent to them are gradually worn away.
osteoarthropathy (*Med.*). Strictly, any disease affecting both bones and joints. Specifically, symmetrical enlargement of the bones of the hands and the feet with thickening of the fingers and toes, associated especially with chronic diseases of the lungs or of the heart.
osteoblast (*Zool.*). A bone-forming cell.
osteochondritis (*Med.*). Inflammation of both bone and cartilage. See also **Perthe's disease.**
osteochondritis dessicans (*Med.*). A rare disease characterized by avascular necrosis of bone and cartilage which eventually resolves spontaneously.
osteochondroma (*Med.*). A tumour composed of bony and of cartilaginous elements.
osteoclasis (*Med.*). The absorption and destruction of bone tissue by osteoclasts. (*Surg.*) The

fracture of a bone for the correction of deformity.

osteoclast (*Zool.*). A bone-destroying cell, especially one which breaks down any preceding matrix, chondrified or calcified, during bone-formation.

osteocranium (*Zool.*). The bony brain-case which replaces the chondrocranium in higher Vertebrates.

osteocyte (*Histol.*). A bone cell derived from an osteoblast.

osteodentine (*Zool.*). A form of dentine structurally resembling bone.

osteodermis (*Zool.*). An ossified or partially ossified dermis; membrane bones formed by ossification of the dermis. *adj.* **osteodermal.**

osteodystrophia fibrosa (*Vet.*). Osteofibrosis. A disease of the skeletal system in animals in which excessive amounts of calcium and phosphorus are withdrawn from the bones and replaced by fibrous tissue. Caused by an excess of phosphorus or a deficiency of calcium in the diet in the horse (*bran disease, millers' disease, big head of horses*), calcium deficiency in the pig (snuffles), and as a complication of chronic nephritis in the dog (rubber jaw).

osteofibrosis (*Vet.*). See **osteodystrophia fibrosa.**

osteogenesis (*Zool.*). See **ossification.**

osteogenesis imperfecta (*Med.*). See **fragilitas ossium.**

osteoid (*Med.*). Resembling bone.

osteology (*Zool.*). The study of bones.

osteoma (*Med.*). A tumour composed of bone.

osteomalacia (*Med.*). Mollities ossium. A condition in which softening of the bones, as a result of absorption of calcium salts from them, occurs, especially in pregnant women; thought to be causally related to deficiency in vitamin D in the diet. (*Vet.*) Boglame; stiff sickness. A disease of the skeletal system affecting cattle in which excessive amounts of calcium and phosphorus are withdrawn from the bones; caused by a deficiency of phosphorus in the diet.

osteomyelitis (*Med.*). Inflammation of the bone-marrow and of the bone.

osteopathy (*Med.*). A method of healing, based on the fact that abnormalities in the human framework (bones, muscles, ligaments, etc.) ultimately cause disease by interfering with the blood and nerve supply to the body, thereby allowing other factors in ill-health to exert their influence unduly. These abnormalities are often the direct single cause of much suffering and they can be removed by skilled manual adjustment.

osteopathyrosis (*Med.*). See **fragilitas ossium.**

osteopetrosis (*Med.*). Albers-Schönberg disease; congenital osteosclerotic anaemia; marble bones. A rare condition in which the bones become solid as a result of obliteration of the bone-marrow by bone, associated with enlargement of the liver and of the spleen, and with anaemia.

osteopetrosis gallinarum (*Vet.*). Marble bone disease; thick leg disease. A chronic virus infection of chickens in which there is an excessive stimulation of bone formation, leading to thickening and deformation of the bones.

osteophagia (*Vet.*). A depraved appetite for bones and dead animals, exhibited by herbivorous animals suffering from a deficiency of phosphorus and calcium salts in the diet.

osteophyte (*Med.*). A bony excrescence or outgrowth from the margin of osteoarthritic joints or from diseased bone.

osteoporosis (*Zool.*). Decrease in bone density and mass, often occurring in old age.

osteosarcoma (*Med.*). A malignant tumour derived from osteoblasts, composed of bone and sarcoma cells.

osteosclereide (*Bot.*). A thick-walled idioblast which is shaped something like a thighbone.

osteosclerosis (*Med.*). Abnormal thickening of bone. See also **osteopetrosis.**

osteoscute (*Zool.*). In Vertebrates, a flat dermal ossification.

Osteostraci (*Zool.*). A class of *Agnatha* with an expanded flattened head protected by a bony carapace, a body protected by bony scales and bearing a pair of anterior fins, and a heterocercal tail. Known from fossils from the Silurian to Devonian periods. Also known as **Cephalaspida.**

osteotomy (*Surg.*). Cutting of a bone.

ostiolate (*Bot.*). Having an opening.

ostiole (*Bot.*). (1) The opening by means of which spores, etc., escape from a conceptacle or a perithecium. (2) A general term for a pore.

ostium (*Zool.*). A mouthlike aperture; in *Porifera*, an inhalant opening on the surface; in *Arthropoda*, an aperture in the wall of the heart by which blood enters the heart from the pericardial cavity; in Mammals, the internal aperture of a Fallopian tube. *adj.* **ostiate.**

Ostracoda (*Zool.*). A subclass of the *Crustacea*, with or without compound eyes; having a bivalve shell with adductor muscle; cephalic appendages well developed and complex; not more than 2 pairs of trunk limbs, which are not phyllopodia; often parthenogenetic, the males of some species being unknown, e.g., *Cypris.*

Ostracodermi (*Zool.*). In some systems of classification, a group of *Agnatha* comprising the *Osteostraci*, the *Anaspida* and the *Heterostraci*, all of which have heavily armoured bodies.

Ostwald colour atlas (*Photog.*). A system of colour relations arranged according to hue, luminosity, and saturation.

Ostwald ripening (*Photog.*). See **ripening.**

Ostwald's dilution law (*Chem.*). The application of the law of mass action to the ionization of a weak electrolyte, yielding the expression $\frac{\alpha^2}{(1-\alpha)V}=K$, where α is the degree of ionization, V the *dilution* (2, q.v.), and K the ionization constant, for the case in which two ions are formed.

Ostwald's theory of indicators (*Chem.*). The assumption that all *indicators* (1, q.v.) are either weak acids or weak bases, in which the colour of the ionized form differs markedly from that of the undissociated form.

osumilite (*Min.*). A silicate closely related in structure and chemistry to *cordierite* (q.v.). It crystallizes, however, in the hexagonal system and appears to be restricted to volcanic rocks and ejectamenta.

ot-, oto-. Prefix from Gk. *ous*, gen. *otos*, ear.

otalgia (*Med.*). Earache.

otic (*Physiol.*). Pertaining to the ear or to the auditory capsule; one of the cartilage bones of the auditory capsule.

otitis (*Med.*). Inflammation of the ear. *Otitis externa*, a term for various inflammatory conditions of the external ear. *Otitis media*, inflammation of the middle ear.

otoconia (*Zool.*). Numerous small concretions which, in some *Mollusca* (e.g., most *Euthyneura*), replace the single large concretion or otolith in the otocyst.

otocyst (*Zool.*). In many aquatic Invertebrates, a sac lined by sensory hairlets, filled with fluid, and containing a calcareous concretion (*otolith*) which subserves the equilibristic sense, and was formerly believed also to serve the sense of hearing; in Vertebrates, part of the internal ear which is similarly constructed (as the utriculus).

otodectic mange (*Vet.*). Mange affecting the external ear canal, particularly of dogs and cats, caused by *Otodectes cyanotis*.

otolith (*Zool.*). The calcareous concretion which occurs in an otocyst.

otology (*Med.*). That part of surgical science dealing with the organ of hearing and its diseases. *n.* **otologist.**

otorhinolaryngology (*Surg.*). That part of surgical science which deals with diseases of the ear, nose, and throat.

otorrhoea, otorrhea (*Med.*). A discharge, especially of pus, from the ear.

otosclerosis (*Med.*). The formation of spongy bone in the capsule of the labyrinth of the ear, associated with progressive deafness.

otoscope (*Med.*). An instrument for inspecting the external canal of the ear and the eardrum.

otter, otter boards. Oblong boards, bound with iron, attached to the sides of a trawl net eccentrically to the towing warps; they keep the mouth of the net open. As used in minesweeping, the otter is a heavy steel frame with horizontal vanes.

Otto cycle (*I.C. Engs.*). The working cycle of a 4-stroke engine—suction, compression, explosion at constant volume, expansion, and exhaust, occupying 2 revolutions of the crankshaft.

ottoman rib (*Textiles*). Warp rib dress fabrics made from botany quality yarns, generally hard twisted. In *ottoman* and *soleil*, the ribs are of equal size, but in *ottoman cord* they are of 2 different sizes, usually arranged alternately.

ottrelite (*Min.*). A manganese-bearing chloritoid mineral, a product of the metamorphism of certain argillaceous sedimentary rocks, named from Ottrez in the Ardennes (Belgium).

ottrelite-slate (*Geol.*). A metamorphic argillaceous rock characterized by abundant crystals of ottrelite.

Oudin resonator (*Elec. Eng.*). Coil, adjustable in number of turns to maximize coupling, for applying *effluve* (q.v.) to a patient.

out-and-in bond (*Build.*). The mode of laying ashlar quoins, so that they will be headers and stretchers alternately.

out-and-out (*Typog.*). Spacing out pages to their exact finished size with no allowance for trim between them.

outband (*Build.*). A jamb stone laid as a stretcher and recessed to take a frame.

outburst bank (*Hyd. Eng.*). The middle part of the slope of a sea embankment, above the footing and below the swash-bank.

outcrop (*Geol.*). An occurrence of a rock at the surface of the ground.

outer (*Elec. Eng.*). Either of the two conductors of a 3-wire distribution system which are respectively at a voltage above and below earth. Cf. the *neutral* (or *middle*) *conductor*, which is at approximately earth potential.

outer conductor (*Elec. Eng.*). See **external conductor.**

outer dead-centre (*Eng.*). The position of the crank of a reciprocating engine or pump when the piston is at the end of its outstroke, i.e., when the piston is nearest the crankshaft. Also **bottom dead-centre.**

outer forme (*Typog.*). The forme of type from which the outside of a sheet is printed, viz., pages 1; 4, 5; 8, 9, etc.

outer marker beacon (*Aero.*). A marker beacon, associated with the *I.L.S.* or with the *standard beam approach system*, which defines the first predetermined point during a beam approach.

outer section (*Aero.*). See **main plane.**

outer space (*Astron.*). The space beyond the solar system; also, more commonly but less accurately, the space beyond the influence of the earth's atmosphere.

outer string (*Join.*). The *string* (q.v.) farthest from the wall.

outer tympan (*Typog.*). The larger tympan of a hand press, into which the smaller one fits.

outfall (*San. Eng.*). The discharge point of a sewer.

outfall sewer (*San. Eng.*). The main sewer carrying away sewage material from a town to the place where it is to be purified, discharged or otherwise dealt with.

outgassing (*Chem.*). Removal of occluded, absorbed or dissolved gas from a solid or liquid. For metals and alloys, done by heating *in vacuo*. (*Electronics*) Removal of maximum amount of residual gas in a valve envelope by baking the whole valve before sealing.

out-gate (*Foundry*). See **riser.**

outgoing feeder (*Elec. Eng.*). A feeder along which power is supplied from a substation or generating station.

outgoing jack (*Teleph.*). A jack which is directly connected to a junction to another exchange, or with a trunk line.

outlier (*Geol.*). A remnant of a younger rock which is surrounded by older strata. It normally forms a hill, often capped by a durable rock, and occurs in front of an escarpment. Cf. *inlier*.

outline letters (*Typog.*). Display types in which the outline only of the letters is shown; sometimes issued as part of a type family, but may also be a specifically designed type style on its own.

out of balance (*Eng.*). Said of a rotating machine element which is imperfectly balanced, or of a mechanism or machine which contains such an element.

out of phase (*Elec. Eng., Telecomm.*). See under **phase.**

out of wind (*Carp., etc.*). A term applied to a flat surface; a surface which is not twisted. (*For.*) A plane surface free from warp or twist.

output (*Comp.*). (1) Data either in coded or printed form after processing by an electronic computer. (2) Vehicle, punched tape or printed page, by which processed data leaves a computer. (*Telecomm.*) Audio, electric or mechanical signal delivered by instrument or system to a load.

output capacitance (*Electronics*). The anode-cathode impedance of a thermionic valve or the capacitive component of the output impedance of a transducer.

output characteristic (*Electronics*). See **transistor characteristics.**

output coefficient (*Elec. Eng.*). See **specific torque coefficient.**

output gap (*Electronics*). An interaction gap through which an output signal can be withdrawn from an electron beam.

output impedance (*Elec. Eng.*). That presented by the device to the load and which determines *regulation* (voltage drop) of source when current is taken. In a linear source, the backward impedance when the e.m.f. is reduced to zero. See **Thévenin's theorem**; also called **source impedance.**

output meter (*Elec. Eng.*). That which measures output voltage of an oscillator, amplifier, etc.

Calibrated in *volts*, or *power level* in dB in relation to *zero power level* (1 milliwatt) when circuit is properly terminated. See **volume unit.**

output noise See **thermal noise.**

output regulation (*Elec. Eng.*). Of a power supply, the variation of voltage with load current.

output transformer (*Elec. Eng.*). One which couples last stage in an amplifier to the load, e.g., a loudspeaker or line.

output valve (*Electronics*). One designed for delivering power to a load, e.g., line or loudspeakers, voltage gain not being relevant. Final stage of any multivalve amplifier. Also called **power valve.**

output variation (*TV*). That of the peak amplitude of a TV video signal due to noise, etc., and usually taken over a period of one frame.

output winding (*Elec. Eng.*). That from which power is withdrawn in a transformer, transductor, or magnetic amplifier.

outrigger (*Build.*). A projecting beam carrying a suspended scaffold.

outset (*Bind.*). A section placed on the outside of the main section, and sometimes called a *wrap round.*

outside crank (*Eng.*). An overhung, or single-web crank attached to a crankshaft outside the main bearings.

outside cylinders (*Eng.*). The steam cylinders carried outside the frame of a locomotive, working on to crank pins in the driving wheels.

outside gouge (*Carp., etc.*). A firmer gouge having the bevel ground upon the convex side of the cutting edge.

outside lap (*Eng.*). The amount by which the slide-valve of a steam-engine overlaps the edge of the steam ports when in mid-position. Also called **steam lap.**

outside lining (*Join.*). The external member of a cased frame.

outside loop (*Aero.*). See **inverted loop.**

outside reams (*Paper*). Reams of paper made up of defective sheets (*outsides*).

outsides (*Paper*). See under **insides.**

outside sorts (*Typog.*). Accented letters, technical and mathematical characters, etc., extraneous to the usual alphabets, and accommodated separately from the normal type case, die case, or magazine. Also **side sorts.**

out-to-out (*Join., etc.*). A term applied to an overall measurement across a piece of framing.

outturn sheet (*Paper*). A representative sample sheet of a particular batch of paper.

outwash fan (*Geol.*). A sheet of gravel and sand, lying beyond the margins of a sheet of boulder clay, deposited by melt-waters from an ice-sheet or glacier.

ova (*Bot., Zool.*). *Pl.* of *ovum.*

ovalbumin (*Chem.*). See **egg albumen.**

Ovalizing balance (*Horol.*). Bimetallic balance which tends to become elliptical with use in temperature, this characteristic being exploited to provide temperature compensation.

oval pistons (*I.C. Engs.*). (1) Pistons, originally round, worn oval through friction at the thrust faces. (2) Pistons purposely turned slightly oval, to compensate for the unequal diametral expansion.

ovals of Cassini (*Maths.*). Curves defined by the bipolar equation $rr'=k^2$. Each consists of either two ovals, one surrounding each reference point, or a single oval surrounding both reference points, according as $k<c$ or $k>c$ respectively, where c is the distance between the two reference points. If $k=c$, it reduces to the *lemniscate of Bernoulli* (q.v.).

oval window (*Zool.*). See **fenestra ovalis.**

ovarian (*Anat.*). Pertaining to or connected with the ovary.

ovariole (*Zool.*). In Insects, one of the egg-tubes of which the ovary is composed.

ovariotestis (*Zool.*). See **ovotestis.**

ovaritis (*Med.*). Inflammation of an ovary.

ovary (*Anat., Zool.*). A female gonad; a reproductive gland producing ova. *adj.* **ovarian.** (*Bot.*) (1) The basal enlarged part of a carpel or of a syncarpous gynaeceum, containing the ovules. (2) Loosely used as meaning the *pistil.*

oven-dry paper (*Paper*). See **bone-dry paper.**

oven-type furnace (*Heat.*). Industrial heat-treatment furnace fired under the hearth, the live gases flowing directly into the heating chamber through live-gas flues disposed along each side of the hearth. Also **semimuffle-type furnace.**

overall efficiency (*Elec. Eng.*). Ratio of useful output to total input power.

overall luminous efficiency (*Light*). Ratio of luminous flux of lamp to total energy input. Not to be confused with **luminous efficiency.**

overall merit (*Telecomm.*). System of rating a communication channel (especially radio broadcasting) on scale 0–5, derived from signal strength, fading, interference, modulation depth, distortion.

overblowing (*Met.*). Continuation of forcing oxygen through molten steel in *Bessemer process* after the carbon has been removed, resulting in oxidizing of iron.

overbreak (*Mining*). Removal of more rock than is called for in tunnelling or surface work, as called for by the 'neat lines' or cross-section on the plan.

overburden (*Build.*). The encallow, or overlying stratum of soil. Generally applied to brickfields or to potential area of gravel for quarrying. (*Mining*) Earth or rock overlying the valuable deposit. In smelting, a furnace is *overburdened,* when the ratio of ore to flux or fuel is too high.

overcast (*Meteor.*). See **cloudiness.**

overcast day (*Meteor.*). A day on which the average *cloudiness* (q.v.) at the hours of observation is more than 6 oktas.

overcasting (*Bind.*). The method of sewing used to make separate leaves into sections for binding. Also called **whipping, whipstitching.**

overcheck (*Textiles*). In a check cloth, it is the more prominent, larger check effect easily distinguished from the ground check by its stronger colours. There are limitless variations.

overcloak (*Plumb.*). When the overlapping edge in a roll extends over to the flat surface beyond the roll, it is called an *overcloak.*

overcoil (*Horol.*). The last coil of a balance spring, which is raised above the plane of the spring and then bent to form a terminal curve.

overcompounded generator (*Elec. Eng.*). A compound-wound d.c. generator in which the series winding is so designed that the voltage rises as the load increases.

overcompounded motor (*Elec. Eng.*). A compound-wound d.c. motor in which the series winding is so designed that the speed rises with an increase in load.

overcompression (*Nuc. Eng.*). A technique used in the operation of expansion cloud chambers to reduce the waiting time between expansions.

overcurrent relay (*Elec. Eng.*). A relay which operates as soon as the current exceeds a certain predetermined value. Also called **overload relay.**

overcurrent release (*Elec. Eng.*). A device for tripping an electric circuit when the current in it exceeds a certain predetermined value. Also called **overload release.**

overcutting (*Acous.*). Too great amplitude of radial motion in cutting the original track on the lacquer blank or direct recording disk, so that one track cuts into the adjacent track.

overdoor (*Build.*). (1) An ornamental doorhead. (2) A pediment.

overdrive (*Autos.*). A device for reducing gear ratio; to lower fuel consumption, reducing wear and fatigue.

over-exposure (*Photog.*). Excess of exposure of any sensitive surface, above that required for the proper gradation of light and shade. Over-exposure is indicated by lack of detail and prevalence of fog after development or by excessive contrast.

overfeed stoker (*Eng.*). A *mechanical stoker* (q.v.) consisting of a hopper from which the fuel is continuously fed on to the bars of an inclined stepped grate, mechanically oscillated or rocked to cause the burning fuel to descend towards an ash table.

overflow (*Comp.*). (1) Coded data which cannot be handled immediately and which is relegated temporarily to a *store* until it can be dealt with. (2) Numbers out of the capacity of a computer which stop operations and raise an alarm.

overflow activity (*An. Behav.*). See **specific action potential.**

overfold (*Geol.*). A fold with both limbs dipping in the same direction, but one more steeply inclined than the other. Cf. *isoclinal fold.* (*Print.*) A lip or overhang formed by the leading edge of the section or copy when the fold is out of centre. Over-adjustment of the folding mechanism will result in *underfold*.

overgassing (*Heat*). Condition occurring in a gas-heated furnace or appliance when the burners are calibrated for, and operated at, a higher gas rate than that actually required.

overgraining (*Paint.*). A coat of graining colour applied over grained work so as to produce shades across the work.

overgrowth (*Crystal.*). See **crystalline overgrowth.**

overhand stopes (*Mining*). *Stopes* in which severed ore from an inclined seam or lode gravitates downward to tramming level.

overhang (*Aero.*). (1) In multiplanes, the distance by which the tip of one of the planes projects beyond the tip of another. (2) In a wing structure, the distance from the outermost supporting point to the extremity of the wing tip. (*Elec. Eng.*) See **armature end connexions.**

overhaul (*Civ. Eng.*). The excess of the actual *haul* (q.v.) above the *free haul*, i.e., the haul which is conveyed without extra cost.

overhaul period (*Aero.*). See **time between overhauls.**

overhead camshaft (*I.C. Engs.*). A *camshaft* (q.v.) running across the top of the cylinder-heads of an engine, usually driven by a bevel-shaft from the crankshaft, the cams operating on rockers or directly on the valve-stems.

overhead-contact system (*Elec. Eng.*). The method of supplying current to the vehicles of an electric traction system whereby the current is collected from a contact wire suspended above the track by means of current-collectors mounted on the roof of the vehicle. The term may also refer to the actual contact wire and its supporting structure.

overhead crossing (*Elec. Eng.*). A device used on the overhead-contact wire system of an electric railway or tramway, to allow the crossing of 2 contact wires and permit the passage of a current-collector along either wire.

overhead expenses (*Build., etc.*). See **on-costs.**

overhead projector (*Light*). See under **projector.**

overhead transmission line (*Elec. Eng.*). A transmission line in which the conductors are supported on poles, towers, or other similar structures at a considerable height above the earth.

overhead travelling-crane (*Eng.*). A workshop crane consisting of a girder along which a wheeled crab can be traversed. The girder is mounted on wheels running on rails fixed along the length of the shop, near the roof. Also called **shop traveller.**

overhead valves (*I.C. Engs.*). In a vertical petrol- or oil-engine, inlet and exhaust valves working in the surface of the head opposite the piston, either in a vertical or inclined position.

overheated (*Met.*). Said of metal which has been heated in preparation for hot-working, or during a heat-treating operation, to a temperature at which rapid grain growth occurs and large grains are produced. The structure and properties can be restored by treatment, and in this respect it differs from burning (see **burnt metal**).

overlag (*TV*). Electronic mixing of foreground and background images.

overlap (*Elec. Eng.*). The period during which successive anodes in a multi-anode rectifier conduct simultaneously due to anode circuit inductance. (*Geol.*) The relationship between conformable strata laid down during an extension of the basin of sedimentation (e.g., on the margins of a slowly sinking landmass), so that each successive stratum extends beyond the boundaries of the one lying immediately beneath. Cf. *off-lap* and *overstep*.

overlap changeover (*Acous.*). Disk recordings made in such a way that there is about 30 sec of overlap, during which they can be aligned for reproduction and a smooth changeover can be effected.

overlapping covers (*Bind.*). The term is used only for pamphlet binding, where the cover may overlap as distinct from being *cut flush* (q.v.).

overlap span (*Elec. Eng.*). See **section gap.**

overlap test (*Elec. Eng.*). A test used for locating a fault in a cable; the resistance between the cable and earth is measured, first with the far end of the cable earthed, and then with it free.

overlay (*Typog.*). To adjust the impression surface of a machine in order to increase the pressure on dark tones and decrease it on light tones.

overload (*Elec. Eng.*). One exceeding the level at which operation can continue satisfactorily for an indefinite period. Overloading may lead to distortion or to overheating and risk of damage depending on the type of circuit or device. In many cases temporary overloads are permissible. See **overload capacity.**

overload capacity (*Elec. Eng.*). Excess capacity of a generator over that of its *rating*, generally for a specified time.

overload protective system (*Elec. Eng.*). A system of protecting an electric power network by means of overcurrent relays. To provide discrimination, the relays have time lags, graded so that the relays more remote from the supply point have shorter lags.

overload relay (*Elec. Eng.*). See **overcurrent relay.**

overload release (*Elec. Eng.*). See **overcurrent release.**

overman (*Mining*). (1) An underground manager of one or more ventilating districts in a coal-mine. (2) An umpire appointed to an arbitration board in a mine dispute.

overmodulation (*Radio*). Attempted modulation to depth exceeding 100%, i.e., to such a degree that amplitude falls to zero for an appreciable fraction of the modulating cycle, with marked distortion.

overpick (*Weaving*). Term describing the picking mechanism. Moving in an arc over the shuttle box, it propels the shuttle by a leather picking band, tugger and picker.

overplugging (*Teleph.*). The action of plugging into a multiple jack, ignoring the busy condition.

overpoled copper (*Met.*). See **poling.**

overproof. See **proof.**

overreach (*Vet.*). An error of gait in the horse, in which the toe of the hindfoot strikes the heel of the forefoot.

override (*Automation*). Said of any electronic controlling equipment, such as autopilot ('George') or throttle control when, under ordinary or fault conditions, its operation can be superseded by manual control.

overrigid (*Eng.*). See **redundant.**

overrun (*For.*). The excess of the amount of timber actually sawn from logs over the estimated volume or log scale, usually expressed in per cent of the log-scale volume. (*Typog.*) To carry words from the end of one line of type to the beginning of the next, and so on until the matter fits. Insertions or deletions frequently necessitate overrunning.

overs (*Print.*). Extra sheets allowed to a job to provide for ordinary *spoilage* (q.v.).

oversailing courses (*Build.*). Brick or stone courses projecting from a wall for the sake of appearance only, as distinct from corbels, which are normally load-carrying.

overscanning (*TV, etc.*). Deflection of electron beam beyond the phosphor in a television reproducer (kinescope) or a cathode-ray oscilloscope.

overshoot (*Automation*). The extent to which a servo system carries the controlled variable past its final equilibrium position. (*Telecomm.*) For a step change in signal amplitude *undershoot* and *overshoot* are the maximum transient signal excursions outside the range from the initial to the final mean amplitude levels.

overshooting (*Cinema.*). The application of too high transmission level to the film recorder, so that the modulation exceeds the available maximum amplitude in variable-area recording and the latitude of the emulsion in variable-density recording.

overshot duct (*Print.*). An ink duct using a blade mounted above the roller which rotates in a trough of ink.

overshot ink fountain (*Print.*). See **overshot duct.**

overshot wheel (*Eng.*). A water-wheel in which the discharge flume or head-race is at the top, the water flowing tangentially into the bucket near the top of the wheel.

oversite concrete (*Build.*). A concrete layer covering a building site within the external walls, serving to keep out ground air and moisture, and also providing a foundation for the floor.

overspeed protection (*Elec. Eng.*). Protection, usually by means of a centrifugally-operated device, against excessive speed of an electric machine; used on inverted rotary converters and also on d.c. motors in certain cases.

overspun wire (*Acous.*). A wire from a musical instrument, e.g., for a low note in a piano, round which a loading wire is tightly spun to lower its fundamental pitch.

oversteer (*Autos.*). Tendency of a vehicle to exaggerate the degree of turn applied to the steering wheel. Cf. *understeer*.

overstep (*Geol.*). The structural relationship between an unconformable stratum and the outcrops of the underlying rocks, across which the former transgresses. Cf. *overlap*.

overstrain (*Eng.*). The result of stressing an elastic material beyond its *yield point* (q.v.); a new and higher yield point results, but the elastic limit is reduced.

overswing diode (*Electronics*). See **backwash diode.**

overthrust fault (*Geol.*). A fault of low hade along which one slice or block of rock has been pushed bodily over another, during intense compressional earth-movements. The horizontal displacement along the *thrust plane* (q.v.) may amount to several kilometres.

overtone (*Acous.*). In a complex tone, any of the components above the fundamental frequency.

overtone crystal (*Electronics*). A piezoelectric crystal operating at a higher frequency than the fundamental for any given mode of vibration.

overvoltage (*Electronics*). The amount by which the operating voltage on a Geiger counter exceeds the threshold value.

overvoltage protective device (*Elec. Eng.*). A device giving protection to electrical apparatus against the possibility of damage caused by an excess voltage, i.e., a voltage above normal.

overvoltage release (*Elec. Eng.*). A device arranged to trip an electrical circuit when the voltage in it exceeds a certain predetermined value.

ovicell (*Zool.*). In *Ectoprocta*, a modified zooecium which acts as a brood-pouch.

oviduct (*Zool.*). The tube which leads from the ovary to the exterior and by which the ova are discharged. *adj.* **oviducal.**

oviferous, ovigerous (*Zool.*). Used to carry eggs, as the *ovigerous* legs of *Pycnogonida*.

oviparous (*Zool.*). Egg-laying; cf. *viviparous*.

oviposition (*Zool.*). The act of depositing eggs.

ovipositor (*Zool.*). In some Fish (as the Bitterling), a flexible tube formed by the extension of the edges of the genital aperture in the female; in female Insects, the egg-laying organ formed by the three pairs of gonapophyses.

ovisac (*Zool.*). A brood-pouch; an egg receptacle.

ovism (*Gen.*). Theory that the germ of all future generations is contained within the ovum.

Ovitron (*Electronics*). Liquid valve in which a tantalum film is broken by d.c. control on a platinum electrode.

ovolo (*Arch.*). A quarter-round convex moulding.

ovomucoid (*Biochem.*). A mucoid substance contained in egg albumen.

ovotestis (*Zool.*). A genital gland which produces both ova and spermatozoa, as the gonad of the Snail.

ovoviviparous (*Zool.*). Producing eggs which hatch out within the uterus of the mother.

ovulation (*Zool.*). The formation of ova; in Mammals, the process of escape of the ovum from the ovary.

ovule (*Bot.*). (1) The nucellus containing the embryo sac and enclosed by one or two integuments, which, after fertilization and subsequent development, becomes the seed. (2) A young seed in course of development.

ovuliferous scale (*Bot.*). One of the scales of a fertile cone in *Coniferae*; it bears the ovules, and later, the seeds, and the sum total of all the ovuliferous scales present makes up the greater part of the mature cone.

ovum (*Bot., Zool.*). A nonmotile, female gamete.

Owen bridge (*Elec. Eng.*). An a.c. bridge of the four arm (or Wheatstone) type, used for the measurement of inductance.

Owen's jet sampler (*Powder Tech.*). An impaction sampler which exploits the condensation of moisture from a gas stream expanded through a jet.

Owen's organ (*Zool.*). In female *Nautiloidea*, an oval structure with numerous closely set ridges lying between the groups of tentacles.

owl (*Powder Tech.*). Apparatus for determining the size of particles in monodisperse aerosols. The state of polarization of the light scattered by the particles is used to measure sizes of 0·1–0·4 micrometres. For larger particles the colour distribution of the scattered light is examined.

own ends (*Bind.*). See **self end-papers.**

oxalates (ethandioates) (*Chem.*). Salts and esters of oxalic acid.

oxalic acid (*Chem.*). $HOOC{\cdot}COOH,2H_2O$, a dibasic acid which crystallizes with 2 molecules of water in monoclinic prisms, m.p. 101°C, m.p. (anhydrous) 190°C; it sublimes readily, occurs in many plants, is obtainable by the oxidation of many organic substances and is prepared technically by the fusion of sawdust with potassium manganate(VII) at about 220°C. NaOH or KOH oxdizes it to CO_2, which reaction is used in volumetric analysis.

oxaluria (*Med.*). The presence of crystals of oxalates in the urine.

oxalyl (*Chem.*). The bivalent acid radical

$$O{=}\overset{|}{C}{\cdot}\overset{|}{C}{=}O.$$

oxamic acid (*Chem.*). $H_2N{\cdot}CO{\cdot}COOH$, a crystalline powder, m.p. 210°C (with decomposition).

oxamide (ethandiamide) (*Chem.*). $H_2N{\cdot}CO{\cdot}CO{\cdot}NH_2$, the normal amide of oxalic acid, a crystalline powder which sublimes when heated.

oxazoles (*Chem.*). Derivatives of oxazole:

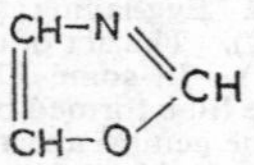

CH–N
‖ ⟩CH
CH–O

oxeote (*Zool.*). Rod-shaped. *n.* **oxea.**

ox-eye (*Arch.*). An oval-shaped dormer window.

Oxford hollow (*Bind.*). A *hollow back* (q.v.), in which the cover is kept separated from the back of the sections by a flattened tube which is glued to each.

Oxford shirting (*Textiles*). A plain-weave, warp striped shirting in which two ends weave as one.

oxidant (*Aero.*). The oxygen-bearing component in a bipropellant rocket, usually liquid oxygen, high-test hydrogen peroxide, or nitric acid.

oxidase (*Bot., Zool.*). One of a group of enzymes occurring in plant and animal cells and promoting oxidation.

oxidates (*Geol.*). Those sedimentary rocks and weathering products whose composition and geochemical behaviour is mainly determined by the oxidation process. This category includes mainly the minerals and rocks containing iron and manganese.

oxidation (*Chem.*). The addition of oxygen to a compound. More generally, any reaction involving the loss of electrons from an atom.

oxidation number (*Chem.*). The number of electrons which must be added to a cation or removed from an anion to produce a normal atom, e.g., +2 for Mg^{++}, −1 for Br^-.

oxidation-reduction indicators (*Chem.*). Substances which exist in oxidized and reduced forms having different colours, used to give approximate values of oxidation-reduction potentials.

oxidation-reduction potential (*Chem.*). See **standard oxidation-reduction potential.**

oxidative phosphorylation (*Biochem.*). See **electron transport chain.**

oxide-coated cathode (*Electronics*). One coated with oxides of the alkali and alkaline-earth metals, to produce thermionic emission at relatively low temperatures.

oxide-coated filament (*Electronics*). Incandescent filament, coated with oxides of alkali metals, which acts as an oxide-coated cathode.

oxide-film arrester (*Elec. Eng.*). A lightning arrester using lead peroxide (a good conductor) which changes to red lead (a good insulator) when heated by the passing of a current.

oxide isolation (*Electrons*). The isolation of the circuit elements in a microelectronic circuit by forming a layer of silicon oxide around each element.

oxides (*Chem.*). Compounds of oxygen with another element. Oxides are formed by the combination of oxygen with most other elements, particularly at elevated temperatures, with the exception of the noble gases and some of the noble metals.

oxidizer (*Chem.*). A substance that combines with another to produce heat and, in the case of a rocket, a gas.

oxidizing agent (*Chem.*). A substance which is capable of bringing about the chemical change known as oxidation.

oxidizing flame (*Chem.*). The outer cone of a nonluminous gas flame, which contains an excess of air over fuel.

oxidizing roast (*Met.*). Heating of sulphide ores or concentrates to burn off part or all of the sulphur as dioxide.

oximes (*Chem.*). Compounds obtained by the action of hydroxylamine on aldehydes or ketones, containing the bivalent oximino group $=N{\cdot}OH$ attached to the carbon. The oximes of aldehydes are termed *aldoximes* and those obtained from ketones *ketoximes*. Collectively, hydroximino-alkanes.

oximide (ethandi-imide) (*Chem.*). $\begin{matrix}CO\\CO\end{matrix}\Big\rangle NH$, colourless prisms. It is prepared by the action of P_2O_5 upon oxamic acid.

oxindole (*Chem.*). $C_6H_4\Big\langle\begin{matrix}NH\\CH\end{matrix}\Big\rangle CO$, the lactam of 2-aminophenyl-ethanoic acid, colourless needles, m.p. 120°C. Oxindole is amphoteric, dissolving both in alkalis and in acids. The imino-hydrogen is exchangeable for other groups, e.g., ethyl, acetyl, etc.

oxine (*Chem.*). 8-Hydroxy-quinoline:

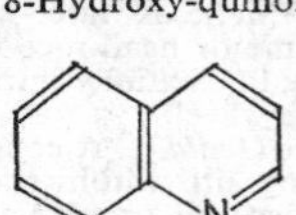

N
OH

Used widely as a reagent in analysis of metals. When the H in the —OH group is substituted by a metal, insoluble compounds result, but their solubility varies according to temperature, concentration and other conditions, thus making it possible to use the differences for analysis.

oxo (*Chem.*). The radical O= in organic compounds, e.g., ketones $R_2C{=}O$.

oxonium salts (*Chem.*). Derivatives of a hypothetical oxonium hydroxide, $H_3O{\cdot}OH$, a base with a tetravalent oxygen atom. Such substances are readily produced from several heterocyclic compounds containing oxygen, e.g., dimethylpyrone, which forms salts with hydrochloric acid, etc., by direct addition to the oxygen atom.

oxyacanthine (*Chem.*). $C_{19}H_{21}O_3N$, an alkaloid of the *iso*quinoline group, obtained from the root barks of the *Berberis* species. It crystallizes in needles, m.p. 208°–214°C.

oxyacetylene cutting (*Eng.*). See under **oxyacetylene welding.**

oxyacetylene welding (*Eng.*). Welding with a flame resulting from the combustion of oxygen and acetylene. Many ferrous metals can also be cut by the same process. The cut is started by introducing a jet of oxygen on to the metal after it has been preheated.

oxyacids (*Chem.*). May be regarded as hydroxides capable of donating a proton, e.g., sulphuric(VI)acid $O_2S(OH)_2$, nitric(V)acid O_2NOH.

oxyazo dyes (*Chem.*). *Azo dyes* (q.v.) which contain a hydroxyl group (OH) within their structure.

oxycelluloses (*Chem.*). Products formed by the action of oxidizing agents on cellulose. They dissolve in dilute alkaline solution and have strong reducing properties. When boiled with hydrochloric acid they yield furfuraldehyde quantitatively. This reaction serves for the analytical estimation of oxycelluloses.

oxycephaly (*Med.*). Steeple-head. A deformity of the skull characterized by a high forehead and a pointed vertex, associated with exophthalmos and impairment of vision.

oxychloride cement (*Build.*). A strong, extremely hard-setting cement used in making composition floors; composed of an oxide and a chloride of magnesia chemically combined.

oxychlorocruorin (*Zool.*). The compound formed by the combination of oxygen and *chlorocruorin* (q.v.).

oxychromatin (*Cyt.*). A form of chromatin which stains comparatively lightly, and contains little nucleic acid. Cf. *basichromatin.*

oxydactylous (*Zool.*). Having narrow-pointed digits.

oxygen (*Chem.*). A nonmetallic element, symbol O, at. no. 8, r.a.m. 15·9994, valency 2. It is a colourless, odourless gas which supports combustion and is essential for the respiration of most forms of life. M.p. −218·4°C, b.p. −183°C, density 1·42904 g/dm³ at s.t.p., formula O_2. An unstable form is ozone, O_3. Oxygen is the most abundant element, forming 21% by volume of the atmosphere, 89% by weight of water, and nearly 50% by weight of the rocks of the earth's crust. It is manufactured from liquid air, for use in hot welding flames, in steel manufacture, in medical practice, and in anaesthesia; liquid oxygen is much used in rocket fuels; mixed with charcoal, sawdust, etc., it forms a powerful explosive. See also **Lurgi process.**

oxygen lancing (*Eng.*). A process used principally for cutting heavy sections of steel or cast iron, in which oxygen is fed to the cutting zone through a length of steel tubing which is consumed as the cutting action proceeds. The cutting zone or the end of the tube has to be preheated to commence cutting.

oxygenotaxis or **oxytaxis** (*Biol.*). Response by locomotory movement or reaction of an organism to the stimulus of oxygen. *adjs.* **oxygenotactic, oxytactic.**

oxygenotropism or **oxytropism** (*Biol.*). Synonym for *oxygenotaxis.* In plants, response is by growth curvature.

oxygen scavenger (*Eng.*). An additive to boiler feed water which removes traces of oxygen and helps to prevent corrosion, e.g., hydrazine or tannin.

oxyhaemocyanin (*Zool.*). The compound formed by the combination of oxygen and haemocyanin.

oxyhaemoglobin (*Chem., Zool.*). The product obtained by the action of oxygen upon haemoglobin. The oxygen is readily given up when oxygen-tension is low in surrounding medium.

oxyhornblende (*Min.*). See **basaltic hornblende.**

oxyhydrogen welding (*Eng.*). A method of welding in which the heat is produced by the combustion of a mixture of oxygen and hydrogen.

oxyluciferin (*Biol.*). Substance produced by the action of the enzyme luciferase on luciferin and causing luminescence.

oxyntic (*Zool.*). Acid-secreting.

oxyphile (*Geol.*). Descriptive of elements which have an affinity for oxygen, and therefore occur in the oxide and silicate minerals of rocks rather than in the sulphide minerals or as native elements. Also **lithophile.**

oxyphilic (*Cyt.*). Having an affinity for, staining readily with acidic dyes; *acidophil* (q.v.).

oxyphobic (*Bot., Zool.*). Unable to withstand soil acidity. (*Cyt.*) Not staining with acid dyes.

oxyproline (*Chem.*). 4-Hydroxypyrrolidine-2-carboxylic acid, obtained by cleavage of gelatine.

oxytaxis (*Biol.*). See **oxygenotaxis.**

oxytetracycline (*Pharm.*). A broad spectrum antibiotic used in a wide variety of infections, including whooping-cough.

oxytocin (*Physiol.*). An octapeptide hormone secreted by the posterior lobe of the pituitary body (*neurohypophysis*) which stimulates the uterine muscle to contract, and causes milk ejection from lactating mammary glands.

oxytropism (*Biol.*). See **oxygenotaxis.**

Ozarkian (*Geol.*). The name applied by Ulrich to the equivalents, in N. America (Ozark, Missouri), of the Tremadocian of Great Britain; deposited during a period of transition between the Cambrian and Ordovician periods.

ozocerite (*Min.*). A mineral paraffin wax, of dark yellow, brown, or black colour, m.p. 55°–110°C, rel. d. 0·85–0·95, soluble in petrol, benzene, turpentine; it is found in Galicia and near the Caspian Sea. When bleached, it forms *ceresine wax* (q.v.). Also **ozokerite.**

ozoena, ozena (*Med.*). Chronic atrophic rhinitis. Chronic inflammation of the nasal mucous membrane, which atrophies, with the formation of crusts in the nose and a foul smell. (*Vet.*) Catarrh of the frontal and maxillary sinuses of the horse.

ozone (*Chem.*). O_3. Produced by the action of ultraviolet radiation or electrical corona discharge on oxygen or air. Powerful oxidizing agent.

ozonides (*Chem.*). Explosive organic compounds formed by the addition of an ozone molecule to a double bond. They thus contain the characteristic grouping:

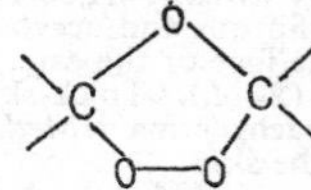

ozonizer (*Chem.*). An apparatus in which oxygen is converted into ozone by being subjected to an electric brush discharge. Ozone was formerly thought to be beneficial in air-conditioning, but is now known to be toxic.

ozonosphere (*Met.*). A layer of the atmosphere at a height of between 15 and 30 km, where ozone is found in greatest quantity.

P

p Symbol for *pico-*; *proton*.
p Symbol for: *electric dipole moment*; *impulse*; *momentum*; *pressure*.
***p*-** (*Chem.*). An abbrev. for: (1) *para-*, i.e., containing a benzene nucleus substituted in the 1.4 positions; (2) *primary*, i.e., containing the functional group attached to a $—CH_2$ group.
φ (*Chem.*). A symbol for: (1) the phenyl radical C_6H_5; (2) *amphi-*, i.e., containing a condensed double aromatic nucleus substituted in the 2.6 positions. (*Elec., Mag.*) Symbol for *phase displacement*.
π See under **pi.**
P (*Gen.*). Symbol for *parental generation*.—P_1, the first parental generation.—P_2, the grandparental generation. See **Mendel's laws.** (*Chem.*) The symbol for *phosphorus*. (*Phys.*) Symbol for *poise*.
P Symbol for: *electric polarization*; *power*; *pressure*.
[*P*] (*Chem.*). The symbol for *parachor*.
Φ Symbol for: *heat flow rate*; *luminous flux*; *magnetic flux*; *work function*.
Π (*Chem.*). A symbol for *pressure*, especially *osmotic pressure*. (*Elec. Eng.*) Symbol for *Peltier coefficient*.
Ψ (*Chem.*). A symbol for *pseudo-*. (*Elec., Mag.*) Symbol for *electric flux*; *magnetic field strength*.
Pa Abbrev. for *pascal*. (*Chem.*) The symbol for *protactinium*.
PABX (*Teleph.*). Abbrev. for *private automatic branch exchange*.
Pacchionian glands (*Zool.*). Villous processes of the arachnoid which penetrate into the veins and venous sinuses of the dura mater and beyond to become embedded in the skull; they serve to drain the cerebrospinal fluid into the venous system.
pace (*Build.*). An area of floor which is raised above the general level of the surrounding floor.
pacemaker (*Med., Surg.*). Electronic device implanted in the chest wall to regulate abnormal heart rhythms by means of electrical pulses. Used when the sino-atrial or atrio-ventricular nodes, which usually regulate heart rhythm, are deficient or diseased. See also **nuclear pacemaker.** (*Physiol.*) A region of the body which determines the activity of other parts of the body; e.g., the sino-atrial node from which originate the impulses causing the heart beat.
Pachuca tank (*Min. Proc.*). Brown tank. Large vertically-set cylindrical vessel used in chemical treatment of ores, in which pulp is reacted with suitable solvents for long periods, while the contents are agitated by compressed air.
pachycarpous (*Bot.*). Having a thick pericarp.
pachydermatocele (*Med.*). A soft flabby tumour, composed of fibrous and nervous tissue, which hangs over the face or the ears.
pachydermatous (*Zool.*). Thick-skinned.
pachydermia, pachyderma (*Med.*). Abnormal thickness of the skin.
pachydermia laryngis (*Med.*). A rare variety of chronic laryngitis in which there is hyperplasia and thickening of the mucous membrane of the vocal cords.
pachymeningitis (*Med.*). Inflammation of the *dura mater* (q.v.).
pachyphyllous (*Bot.*). Having thick leaves.
pachytene (*Cyt.*). The third stage (*bouquet stage*) of meiotic prophase, intervening between zygotene and diplotene, in which condensation of chromosomes commences.
Pacific converter (*Textiles*). Machine for transforming a tow (rope) of man-made fibres into a top of staple fibres. See **tow-to-top.** There are also the *Tubo-Stapler* and *Halle-Seyde converter*.
Pacific Province (*Geol.*). The name applied to those regions bordering the Pacific Ocean within which rocks of Cambrian age occur; characterized by a different faunal assemblage from that occurring in rocks of the same age in the Atlantic Province.
Pacinian corpuscles (*Zool.*). In Vertebrates, sensory nerve-endings in which the nerve-ending is surrounded by many concentric layers of connective tissue. Sensitive to pressure. Also **Vater's corpuscles.**
pack (*Mining*). 'Fill'. Waste rock, mill tailings, etc., used to support excavated stopes. (*Textiles*) The weight-measure by which wool or wool tops are sold, viz. 240 lb (109 kg).
package (*Textiles*). Trade term for a cop, cheese, cone, etc., of yarn, indicating that it is in convenient form for transport, or further processing, e.g., rewinding, warping, etc.
packaged (*Elec. Eng.*). Said of an equipment complete for use, particularly a magnetron, with magnet and coupling unit; also of circuit units having their components sealed or encapsulated together. (*Nuc. Eng.*) Said of a reactor of limited power which can be packaged and erected easily on a remote site.
packed column (*Chem. Eng.*). A column used for distillation or absorption, which consists of a shell filled either with random material such as coke or other broken inert material, or with one of the proprietary rings or similar, usually ceramic or stainless steel (*Raschig, Lessing*, or *Pall rings* (qq.v.)
packer (*Mining*). A man employed in a coal-mine, to build pack walls along roadways, etc., to support the roof.
pack-hardening (*Met.*). *Case-hardening* (q.v.), using a solid carburizing medium, followed by a hardening treatment.
packing (*Build.*). The operation of filling in a double or hollow wall. (*Eng.*) Material inserted (*a*) in *stuffing boxes* (q.v.) to make engine and pump rods pressure-tight; it may consist either of compressible material such as hemp, or of metal rings (see **metallic packing**) or (*b*) between clamped surfaces to space them apart. (*Mining*) See **stowing.** (*Print.*) (1) On a letterpress machine, the prescribed thickness of sheets of paper on the platen or the cylinder, required to apply impression, and used as the basis for the *making-ready* (q.v.). (2) On an offset press, the sheets of paper beneath both the plate and the blanket which can be added to or taken away in either case to alter the 'print length' when obtaining register.
packing density (*Comp.*). Spacing of data in a store, e.g., bits/cm of magnetic tape.
packing fraction (*Nuc.*). $(M\text{-}A)/A$ for a nuclide, where M is its mass in at. mass units and A is the mass number. $M\text{-}A$ is variously known as the *mass defect* and the *mass decrement*.
pad (*Telecomm.*). (1) Small preset adjustable capacitor, to regulate the exact frequency of oscillation of an oscillator, or a tuned circuit

in an amplifier or filter; also **padder, trimmer.** (2) Fixed attenuator inserted in a transmission line, or in a waveguide to obviate reflections.

paddle hole (*Hyd. Eng.*). The opening in a lock-gate through which water flows from the high-level pond to the lock-chamber, or from the lock-chamber to the low-level pond.

paddle plane (*Aero.*). See **cyclogyro.**

paddle-wheel fan (*Eng.*). See **centrifugal fan.**

paddle-wheel hopper (*Eng.*). A hopperfeed for small articles, in which a continuously rotating paddle pushes the articles along an inclined track towards the delivery chute.

paddles (*Print.*). Rotating curved or shaped fingers of the fly which transfer copies from the folder to the delivery belt of web-fed rotary presses.

paddling (*Leather*). A process in tanning light skins by which the skins and the liquor are kept in movement by a revolving paddle.

pad roller (*Cinema.*). Nonsprocket roller for pressing edges of cinematograph film on to sprockets, so that a sufficient number of teeth are engaged by sprocket holes.

padsaw (*Tools*). See **keyhole saw.**

pad stone (*Build.*). A stone *template* (q.v.).

paediatric, pediatric (*Med.*). Pertaining to *paediatrics* (*pediatrics*), i.e., that branch of medical science which deals with the study of childhood and the diseases of children.

paediatrician, pediatrician, paediatrist (*Med.*). A medical man who specializes in the study of childhood and the diseases of children.

paedogamy (*Zool.*). A variety of autogamy in which both nucleus and cytoplasm divide and reunite.

paedogenesis (*Bot.*). Markedly precocious flowering. (*Zool.*) Reproduction by larval or immature forms.

Paenungulata (*Zool.*). A superorder of Mammals containing a loosely related assembly of animals only slightly more advanced than the *Protoungulata* (q.v.), having several digits and no well-marked hooves, and often large tusks and grinding molars. Hyrax, Elephants, Manatees, and extinct forms.

page cord (*Typog.*). Thin strong cord used to secure type pages.

page cut-off (*Print.*). A control on the ink duct of sheet-fed rotary presses which allows ink feed to be cut off from one page.

Page effect (*Elec. Eng.*). A click heard when a bar of iron is magnetized or demagnetized.

page gauge (*Typog.*). An accurately-sized piece of furniture or a length of 6-point or 12-point strip cut to size, used to gauge the depth of type pages when they are made up.

page proofs (*Typog.*). Proofs taken when the type is made up into pages and before imposition.

Paget's disease of bone (*Med.*). Osteitis deformans. A chronic disease characterized by progressive enlargement and softening of bones, especially of the skull and of the lower limbs.

Paget's disease of the nipple (*Med.*). A condition in which chronic eczema of the nipple is associated with the subsequent development of cancer of the breast.

pagina (*Bot.*). A synonym for **lamina.**

pagination (*Typog.*). The allotting of numbers (folios) to the pages of a book, roman numerals being traditionally but not invariably used for the preliminary matter and arabic numerals thereafter, beginning again with the numeral '1' on the first text page, but sometimes continuing the sequence where the preliminary folios left off.

paging (*Bind.*). Applying numbers to the pages of an account book or to sets of stationery with a hand-numbering tool or a pedal-operated paging machine.

pagodite (*Min.*). This is like ordinary massive *pinite* (q.v.) in its amorphous compact texture and other physical characters, but contains more silica. The Chinese carve the soft stone into miniature pagodas and images. Also called **agalmatolite.**

pahoehoe lava (*Geol.*). A Hawaiian term for viscous lava which consists of wrinkled flows, free from large scoriaceous masses.

paint. A suspension of a solid or solids in a liquid which, when applied to a surface, dries by evaporation and/or oxidation to a more or less opaque, adhering solid film. The solids are colour- or opacity-imparting, finely-ground pigments together with extenders; the liquid consists of suitable oils, solvents, resins, aqueous colloidal solutions, or dispersions, etc., together with other lesser ingredients. Used for protective and decorative purposes. See also **oil paints.**

paint harling (*Build.*). Rough-casting for protecting domestic steel walls, the adhesive medium being a special paint.

paint remover or **stripper** (*Paint.*). A liquid (usually containing non-inflammable organic solvents) applied to paint to facilitate removal.

pair (*Build., etc.*). To match two similar objects on opposite hands. (*Electronics*) Two electrons forming a nonpolar valency bond between atoms.

paired cable (*Elec. Eng.*). One in which multiple conductors are arranged as twisted pairs but not quadded.

pairing (*Gen.*). The process by which homologous chromosomes are attracted to one another during the zygotene stage of cell division. (*TV*) Deviation from exactness of interlacing of horizontal lines in a reproduced image, reducing vertical definition. Also **twinning.**

pairing energy (*Nuc.*). A component of the binding energy of a nucleus which represents a decrease in stability for an unpaired proton or neutron.

pair production (*Electronics*). Creation of a *positron* and an *electron* when a *photon* passes into the electric field of an atom. See **Compton effect, photoelectric effect.**

paisanite (*Geol.*). A sodic microgranite containing *riebeckite* as the principal coloured mineral.

PAL (*TV*). The development of the *NTSC* colour television system adopted by most Western European broadcasting systems, the major exception being France.

palae-, palaeo-. Prefix from Gk. *palaios*, ancient.

Palaearctic region (*Zool.*). One of the subrealms into which the Holarctic region is divided; it includes Europe and northern Asia, together with Africa north of the Sahara.

palaeobotany. The study of fossil plants.

Palaeocene (*Geol.*). The name given to the oldest epoch of the Tertiary period, included by some authors in the Eocene.

Palaeogene (*Geol.*). The system of rocks lying above the *Cretaceous* and below the *Neogene*, containing the *Palaeocene*, *Eocene* and *Oligocene* series.

Palaeognathae (*Zool.*). A superorder of *Neornithes*, usually large and flightless birds, with no barbules on the feathers, retrices absent or irregular, wings reduced, no sternal keel. Possibly not a natural group, but deriving instead from several lines. Ostriches, Emu, Rhea, Kiwi.

Palaeolithic Period (*Geol.*). The older stone age, characterized by successive 'cultures' of stone implements, made by extinct types of men. Cf. *Neolithic Period.*

palaeomagnetism (*Geophys.*). A phenomenon exhibited in rocks whose remanent magnetization is in a direction which is different from the present earth's field. Intensive studies of palaeomagnetism have led to the conclusion that both the magnetic pole and continental positions have changed considerably during geological time.

Palaeonemertini (*Zool.*). An order of the class *Anopla* of the phylum *Nemertea*. The body wall muscles form two or three layers, the innermost being of circular muscle. Cf. *Heteronemertini.*

palaeoniscoid scale (*Zool.*). See **ganoid.**

palaeontology (*Geol.*). That branch of geological science which is essentially the study of animal life in past geological periods. It deals with the successive faunas which have peopled the earth since earliest times; with the structure, relationships, evolution, and environment of the individual creatures.

Palaeoptera (*Zool.*). A division of the *Pterygota* which do not fold their wings and which have a prominent network of crossveins. Malpighian tubules are numerous, metamorphosis incomplete. Includes the superorders *Ephemeropteroidea* and *Odonatopteroidea.*

Palaeozoic (*Geol.*). A major division of geological time comprising the Cambrian, Ordovician, Silurian, Devonian, Carboniferous, and Permian Periods.

palaeozoology. The zoology of fossil animals.

palagonite (*Geol.*). A term applied to altered basaltic glass. It occurs as infillings in rocks, and is a soft-brown or greenish-black cryptocrystalline substance. Named from Palagonia, Sicily.

palama (*Zool.*). The webbing of the feet in Birds of aquatic habit.

palate (*Bot.*). The prominent part of the lower lip of a ringent corolla which closes the opening of the corolla. (*Zool.*) In Vertebrates, the roof of the mouth; in Insects, the epipharynx. *adjs.* **palatal, palatine.**

palatine (*Zool.*). Pertaining to the palate; a paired membrane bone of the Vertebrate skull which forms part of the roof of the mouth.

palatine glands (*Histol.*). Mucous secreting glands situated in the roof of the mouth.

palatoplegia (*Med.*). Paralysis of the palate.

pale-, paleo-. Prefix. See **palae-, palaeo-.**

pale, palea (*pl.* **paleae**), **palet** (*Bot.*). (1) The inner bracteole, thin and membranous, which, with the flowering glume, encloses a grass flower. (2) A general name for the glumes associated with the grass flower. (3) The scales which form the ramentum in ferns.

paleaceous (*Bot.*). Chaffy in texture.

pale ale (*Brew.*). Ale of a light colour and with a distinct flavour of hops; usually bottled or canned. See **beer.**

paleocerebellum (*Anat.*). Phylogenetically, the older part of the cerebellum, comprising such parts as pyramis, uvula and paraflocculus.

paleoencephalon (*Anat.*). Phylogenetically, the ancient brain; the brain excluding the cerebral cortex and appendages.

paleogenetic. Originating in the past.

paleola (*Bot.*). See **lodicule.**

palet (*Bot.*). See **pale.**

paliform (*Bot.*). Having the form of a stake.

palingenesis (*Geol.*). The production of granitic magma by the complete or partial melting of previously existing rocks. See **granitization.** (*Zool.*) The reproduction of truly ancestral characters during ontogeny. *adj.* **palingenetic.**

palisade cell (*Bot.*). (1) One cell of the palisade layer of a leaf. (2) One of the terminal cells of the hyphae forming the cortex in a lichen thallus.

palisade layer (*Bot.*). A layer of elongated cells, set at right angles to the surface of a leaf, underlying the upper epidermis, containing numerous chloroplasts, and concerned with photosynthesis.

Palisades sill (*Geol.*). A massive intrusive sheet of dolerite (diabase) intruded during the Triassic period of vulcanicity in the Hudson region.

palisade stereide (*Bot.*). A rod-shaped, thick-walled cell, elongated at right angles to the surface of the seed, and occurring in the testa.

palisade tissue (*Bot.*). One or more layers of palisade cells beneath the epidermis of a leaf.

Palladian (*Arch.*). Style of classic architecture based upon work of Venetian Andrea Palladio, in the 16th century. It involves groupings and combinations of superimposed columns and arches.

Palladian window (*Arch.*). Type of window comprising main window with arched head and, on either side, a long narrow window with square head.

palladinized asbestos (*Chem.*). Asbestos fibres saturated with a solution of a palladium compound, which is subsequently decomposed to give finely divided palladium dispersed throughout the asbestos.

palladious (palladium(II)) iodide (*Chem.*). PdI_2. Precipitated when potassium iodide is added to solutions of palladious chloride. As the other halogen salts of palladium are soluble, this reaction is used to distinguish iodine from the other halogens.

palladium (*Chem.*). A metallic element, symbol Pd, at. no. 46, r.a.m. 106·4. The metal is white, m.p. 1549°C, b.p. 2500°C, rel. d. 11·4. Used as a catalyst in hydrogenation. Native palladium is mostly in grains and is frequently alloyed with platinum and iridium. See also **gold.**

palladium leak (*Electronics*). A method of controlling the entry of gas into a vacuum system by temperature control of a porous palladium block, giving superior control to a needle valve. Widely used with ion sources.

pallaesthesia (*Med.*). Insensibility of bone to vibratory stimuli.

pallasite (*Min.*). A group name for stony meteorites which contain fractured or rounded crystals of olivine in a network of nickel-iron.

pallescent (*Bot.*). Becoming lighter in colour with age.

pallet (*Acous.*). The flap of wood, faced with felt or leather, which is raised to permit the flow of air to wind-chests, etc., in the mechanism of an organ. (*Build.*) A thin strip of wood built into the mortar joint of a wall, to provide a substance to which joinery may be nailed. (*For.*) A portable (wooden) platform used to facilitate the handling of stacked goods by giving sufficient ground clearance. (*Horol.*) The surface or part upon which the teeth of the escape wheel act to give impulse to the pendulum or balance. The first pallet acted upon by a tooth of the escape wheel is known as the *entering pallet*, and the other pallet as the *exit pallet*. See **circular-, equidistant locking-, exposed-.** (*Print.*) A platform on which sheets of paper (or other goods) can be stacked before printing and between printings, and designed for transporting by fork-lift truck. Also called **stillage.**

pallet brick (*Build.*). A purpose-made brick with a groove in one edge to receive a fixing strip.
pallet jewel (or stone) (*Horol.*). The jewel in the pallets upon which the escape-wheel teeth act.
pallet truck. See **fork-lift truck.**
palli-, pallio-. Prefix from L. *pallium*, mantle.
pallial groove (*Zool.*). The groove between the mantle and the body in Bivalves.
pallial line (*Zool.*). The line of attachment of the mantle to the shell in Bivalves.
palliative (*Med.*). Affording temporary relief from pain or discomfort; a medicinal remedy which does this.
pallium (*Zool.*). The mantle in *Brachiopoda* or *Mollusca*, a fold of integument which secretes the shell. In the Vertebrate brain, that part of the wall of the cerebral hemispheres excluding the corpus striatum and rhinencephalon. *adjs.* **pallial, palliate.**
Pall rings (*Chem. Eng.*). *Packed column* (q.v.) filling, consisting of open hollow cylinders of diameter equal to length with several portions of cylinder wall turned into interior as small spurs.
palmar (*Zool.*). See **thenal.**
palmate (*Bot.*). Having several (often 5–7) lobes, segments, or leaflets spreading from the same point, like the fingers from the palm; applied particularly to leaves. (*Zool.*) Having webbed feet.
palmatifid (*Bot.*). Having the leaf blade cut about half-way in, to form a number of diverging lobes.
palmatine (*Chem.*). An alkaloid of the *iso*quinoline group, obtained from calumba root, *Jatrorrhiza palmata*, generally isolated in the form of the sparingly soluble iodide, $C_{21}H_{22}O_4NI{\cdot}2H_2O$, which crystallizes in yellow needles, m.p. 240°C.
palmatisect (*Bot.*). Having the leaf blade cut nearly to the base, so forming a number of diverging lobes.
Palm Beach (*Textiles*). A plain-weave fabric made from cotton warp and lustre worsted weft, or entirely of cotton; warp yarn provides a coloured stripe effect.
palmella stage (*Zool.*). In some holophytic *Mastigophora*, a pseudocolonial stage simulating the lower *Algae*, produced by division taking place when the flagella are withdrawn and the organism is in a resting phase.
palmitic acid (*Chem.*). $C_{15}H_{31}{\cdot}COOH$, a normal fatty acid, m.p. 63°C, b.p. 269°C. It occurs as glycerides in vegetable oils and fats.
palmitins (*Chem.*). The glycerine esters of palmitic acid.
palm-kernel oil (*Chem.*). A yellowish oil from the nuts of *Elaeis guineensis*, m.p. 26°–30°C, rel. d. 0·95, saponification value 247, iodine value 13·5, acid value 8·4.
palm oil (*Chem.*). A reddish-yellow fatty mass from the fruit of *Elaeis guineensis*, m.p. 27°–43°C, rel. d. 0·90–0·95, saponification value 196–205, iodine value 51–57, acid value 24–200.
palp (*Zool.*). See **palpus.** *adj.* **palpal.**
palpacle (*Zool.*). In *Siphonophora*, the tentacle of a dactylozooid.
palpebra (*Anat., Zool.*). An eyelid.
palpebral fissure (*Anat.*). The space between the upper and lower eyelids.
palpifer (*Zool.*). In Insects, a lateral sclerite of the stipes which bears the maxillary palp. Also **palpiger.**
Palpigrada (*Zool.*). An order of *Arachnida* with an elongated and well segmented body, and a long tail. Very small animals, found living under stones.
palpitation (*Med.*). Awareness on the part of a person that his heart is beating against the chest wall, usually associated with increase in the frequency of the heart beat; loosely, increased frequency of the beat of the heart.
palpocil (*Zool.*). In *Coelenterata*, a sense hairlet attached to a sense cell.
palpon (*Zool.*). See **dactylozooid.**
palpus (*Zool.*). In *Crustacea* and Insects, a jointed sensory appendage associated with the mouthparts and representing the exopodite of the gnathobase to which it is attached; in *Polychaeta*, sensory appendage of prostomium. Also **palp.**
paludicolous (*Ecol.*). Living in ponds, streams, and marshes.
paludose (*Ecol.*). Inhabiting wet places.
Paludrine (*Pharm.*). TN for proguanil hydrochloride, 1-4-Chlorophenyl-5-*iso*propyldiguanide, a synthetic quinine substitute for the prophylaxis and treatment of *paludism*, i.e., malaria.
Palynology (*Geol.*). The study of fossil spores. Spores are very resistant to destruction and in many sedimentary rocks are the only fossils that can be used for stratigraphical correlation.
pampero (*Meteor.*). A line-squall (S. Amer.).
pampinody (*Bot.*). The change of parts of a leaf, or of leaves, into tendrils.
pan (*Bot.*). (1) A compact layer of soil particles, lying some distance beneath the surface, cemented together by organic material, or by iron and other compounds, and relatively impermeable to water. (2) A depression in the surface of a salt marsh, in which salt water stands for lengthy periods. (*Build.*) A panel of brickwork, or lath and plaster, in half-timbered work. (*Cinema.*) Panoramic motion of a camera. (*Photog.*) (1) Abbrev. for *panchromatic*. (2) To move a camera in bearing.
panache (*Civ. Eng.*). See **pendentive.**
Panadol (*Pharm.*). TN for a proprietary form of *paracetamol* (q.v.).
panala (*Chem.*). See **naptalam.**
pan-and-tilt head (*Photog.*). Tripod head with handle for moving camera in both bearing and elevation, used chiefly in cinematography.
pan breeze (*Build.*). A mixture of coke breeze and clinker collected in the pan below a furnace consuming coke breeze. It is used as an aggregate in the manufacture of concrete.
pancake coil (*Radio, etc.*). Inductor in which the conductor is wound spirally. Much used in early radio, later as part of a printed circuit.
pancaking (*Aero.*). The alighting of an aircraft at a relatively steep angle, with low forward speed. Also World War II radio control term for landing.
pancarditis (*Med.*). Concurrent inflammation of the three main structures of the heart—the pericardium, the myocardium, and the endocardium.
panchromatic (*Photog.*). Said of emulsions which are reasonably sensitive photographically to all visible colours, particularly towards the red.
pancreas (*Zool.*). A moderately compact though somewhat amorphous structure found in Vertebrates, the larger part of which consists of exocrine glandular tissue with one or more ducts opening to the small intestine, and also containing scattered islets of endocrine tissue. The former produces six or more enzymes involved in the digestion of proteins, carbohydrates, and fats, while the latter secrete the hormones insulin and glucagon.
pancreatectomy (*Surg.*). Surgical removal of the pancreas.

pancreatic enzymes (*Chem.*). The enzymes of the pancreatic juice are trypsin, steapsin, amylopsin, maltase, a lactase, pancreatic erepsin, rennin, and possibly invertase.

pancreatitis (*Med.*). Inflammation of the pancreas.

pancreatolith (*Med.*). A calculus in the pancreas.

pancreatotomy (*Surg.*). Incision of the pancreas.

pancreozymin (*Biochem.*, *Physiol.*). A hormone liberated from the duodenal mucosa by the action of gastric juice, and which causes the discharge of pancreatic enzymes.

pandemic (*Med.*). Of an epidemic, occurring over a wide area such as a country or a continent; an epidemic so widespread.

pandiculation (*Med.*). The combined action of stretching the body and the limbs and yawning.

pandurate, panduriform (*Bot.*). Shaped like the body of a fiddle.

pane (*Build.*). (1) A panel. (2) A sheet of glass cut to size for use as a window light. (*Tools*) The end of a hammer-head opposite to that carrying the hammering face; made to various shapes for particular operations such as riveting, etc. Also **pean, peen, pein.** See **ball-pane hammer, cross-pane hammer, straight-pane hammer.**

panel (*Elec. Eng.*, *etc.*). A sheet of metal, plastic or other material upon which instruments, switches, relays, etc., are mounted. Also called **switchboard panel.** (*Bind.*) A piece of material (leather, paper) fixed to the spine of a book, containing the title and author's name.

panel absorber (*Acous.*). Light panels mounted at some distance in front of a rigid wall in order to absorb sound waves incident on them. Perforated panels are in widespread use for this purpose.

panel back (*Bind.*). A style of *finishing* the spine of a book by decorated panels between raised bands.

panel beating (*Eng.*). A sheet-metal working craft, now mainly used in automobile body repairing, in which complex shapes are produced by stretching and gathering the sheet locally, with subsequent finishing by planishing and wheeling.

panel gauge (*Carp.*). Large form of *marking gauge* (q.v.) for use on wide boards.

panel heating (*Build.*). A system of heating a building in which heating units or coils of pipes are concealed in special panels, or built in wall or ceiling plaster. Also called **concealed heating.**

panelled framing (*Join.*). Doors and frames formed of stiles, rails, and muntins framed together with mortise-and-tenon joints, and having panels fitted into the spaces.

panel mounting (*Elec. Eng.*, *etc.*). The normal method of accommodating a collection of non-portable apparatus. Each piece or unit is constructed separately on its standard panel, which is mounted with others on a standard vertical rack, the different panels being provided with terminal blocks so that the units can be wired together after assembly.

panel pins (*Join.*). Light, narrow-headed nails of small diameter used chiefly for fixing plywood or hardboard to supports.

panels (*Eng.*). In a truss or open-web girder, the framed units of which the truss is composed; the divisions separated by the vertical members.

panel saw (*Join.*). A hand-saw used for panelling, having 7 teeth to the inch.

panel switch (*Elec. Eng.*). See **flush switch.**

panendoscope (*Surg.*). An instrument which can be used to view, or operate upon, the urinary tract, from urethra to bladder, and which is passed via the urethra.

Paneth's cells (*Zool.*). Conspicuous cells, containing oxyphil granules, occurring at the base of the crypts of Lieberkühn.

pangamic (*Zool.*). Of indiscriminate mating.

panhead rivet (*Eng.*). One with a head shaped like a truncated cone.

panhysterectomy (*Surg.*). Complete surgical removal of the uterus.

panic bolt (*Build.*). A special form of door-bolt which is released by pressure at the middle of the door; often used on fire-escape doors.

panicle (*Bot.*). Strictly, a branched raceme with each branch bearing a raceme of flowers, e.g., oat. Loosely, any branched inflorescence of some degree of complexity. *adj.* **paniculate.**

panidiomorphic (*Geol.*). A term applied to igneous rocks which have almost all their constituents of perfect crystalline shape.

Panizza's foramen (*Zool.*). In *Loricata*, a small foramen allowing interchange of blood between the right and left systemic arches.

pan-mill mixer (*Build.*). See **mortar mill.**

panmixia (*Zool.*). Cessation of natural selection, as in an organ which through change of habits or environment of a species is no longer useful.

panne velvet (*Textiles*). A warp pile fabric with silk pile; used for dresses and furnishings.

panniculitis (*Med.*). Inflammation of the subcutaneous fat in any part of the body.

panniculus carnosus (*Zool.*). In some Mammals, an extensive system of dermal musculature covering the trunk and part of the limbs, by means of which the animal can shake itself.

pannier (*Civ. Eng.*). A kind of gabion.

panniform (*Bot.*). Looking like felt.

panning (*Min. Proc.*). Use by prospector or plant worker of gold pan, batea, plaque, or dulong to concentrate heavier minerals in a crushed sample by washing away the lighter ones.

panning head (*Cinema.*). Head or platform for a camera, permitting panoramic motion.

pannose (*Bot.*). Felted.

pannus (*Med.*). The appearance of a curtain of blood vessels round the margin of the cornea; e.g., in trachoma, or in phlyctenular keratitis.

panoistic (*Zool.*). In Insects, said of ovarioles in which nutritive cells are wanting Cf. *polytrophic*, *acrotrophic*.

panophthalmitis, panophthalmia (*Med.*). Inflammation of all the structures of the eye.

panoramic attenuator (*Acous.*). Re-recording device by which a 1-channel recording is reproduced and distributed to, say, three tracks, generally magnetic, to give an illusion, when these are reproduced, of stereophonic sound reproduction. Used with wide cinema screens.

panoramic camera (*Photog.*). A camera intended to take very wide angle views, generally by rotation about an axis and by exposing a roll of film through a vertical slit.

panoramic receiver (*Radio*). One in which tuning sweeps over wide ranges, with synchronized display on a CRT of output signals.

panotrope (*Acous.*). A purely electrical reproducer of records, using a pick-up, electronic amplifier, and one or more loudspeakers.

panphotometric (*Bot.*). Said of a narrow leaf which stands nearly or quite erect.

panradiometer (*Heat*). Instrument for measuring radiant heat irrespective of the wavelength.

pan shot (*Cinema.*). A panoramic shot. See **pan.**

pansinusitis (*Med.*). Inflammation of all the air-containing sinuses which communicate with the nasal cavity.

pansporoblast (*Zool.*). In *Neosporidia*, a cell-complex producing both sporoblasts and spores.

panthenol (*Chem.*). A vitamin of the B complex, affecting the growth of hair.

panting (*Ships*). Pulsating movement of the shell plating of the ship developing in the bows and stern while the ship is under way.

pantobase (*Aero.*). The fitment of a landplane with *hydroskis* (q.v.), enabling it to taxi, take-off from, and alight on water or snow.

pantograph (*Elec. Eng.*). A sliding type of current-collector for use on traction systems employing an overhead contact wire. The contact-strip of the collector is mounted on a hinged diamond-shaped structure, so that it can move vertically to follow variations in the contact wire height. (*Instr.*) A mechanism by which a point is constrained to copy, to any required scale, the path traced by another point. It is based on the geometry of a parallelogram; used in engraving machines, etc. (*Textiles*) That part of an embroidery frame which determines the position of the ground net in relation to the needles.

pantophagous (*Zool.*). Omnivorous.

pantophobia (*Psychiat.*). Generalized morbid fear of unknown or imprecisely-defined future event.

Pantopoda (*Zool.*). An order of *Arachnida*, with no opisthosoma, except the pregenital segment-bearing legs on which the genital pore opens. All are marine and semisedentary, having very small bodies and enormously elongated legs.

pantothenic acid (*Chem.*). See **vitamin B complex.**

papain (*Chem.*). A protein-digesting enzyme present in the juice from the fruits and leaves of the papaya tree (*Carica papaya*); commercially produced as a meat tenderizer.

papaverine (*Chem.*). An alkaloid occurring in opium, colourless prisms, m.p. 147°C, optically inactive. It is 3,4-dimethoxybenzyl-4′,5′-dimethoxy-*iso*quinoline and has the formula:

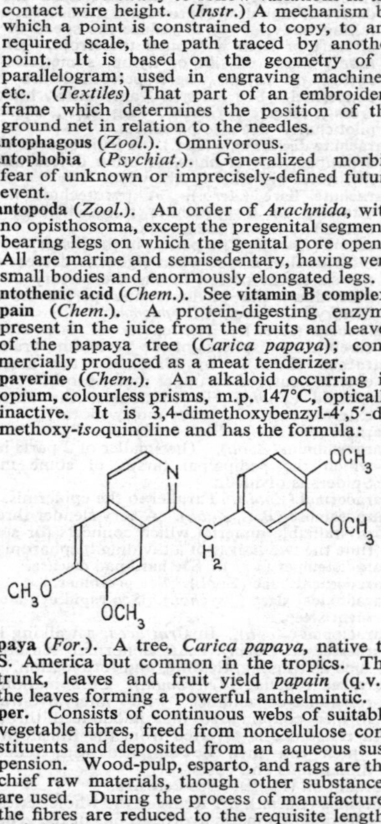

papaya (*For.*). A tree, *Carica papaya*, native to S. America but common in the tropics. The trunk, leaves and fruit yield *papain* (q.v.), the leaves forming a powerful anthelmintic.

paper. Consists of continuous webs of suitable vegetable fibres, freed from noncellulose constituents and deposited from an aqueous suspension. Wood-pulp, esparto, and rags are the chief raw materials, though other substances are used. During the process of manufacture, the fibres are reduced to the requisite length, and their physical properties are modified by mechanical or chemical treatment. See **cellulose.**

paper capacitor (*Elec. Eng.*). One which has thin paper as the dielectric separating aluminium foil electrodes, these being wound together and waxed. See **self-sealing capacitor.**

paper chromatography (*Chem.*). *Chromatography* using a sheet of special grade filter paper as the adsorbent. Advantages: microgram quantities, bands can be formed in two dimensions and cut with scissors.

paper foils (*Bind.*). *Blocking foils* (q.v.) with a glassine substrate. These are the cheapest foils and the most widely used.

paper negatives (*Photog.*). Negatives made with emulsions on paper supports, instead of the more usual film or glass. They have the advantages of lightness, cheapness, ease of retouching, and possibility of large dimensions.

paper sizes. *ISO* (q.v.) recommended paper sizes are based on metric dimensions. The principal series (ISO-A) for printing, stationery, etc., has a basic size of 1 m^2. Each size is got by halving the one above along the longer side, e.g., A0 = 841 × 1189 mm, A1 = 594 × 841 mm, etc.

paper tape (*Comp.*). See **punched tape.**

papilionaceous (*Bot.*). Having some likeness in form to a butterfly; said of flowers like those of the pea.

papilla (*Zool.*). A small conical projection of soft tissue, especially on the skin or lining of the alimentary canal; the conical mass of soft tissue or pulp projecting into the base of a developing feather or tooth. *adjs.* **papillary, papillate.**

papillae foliatae (*Zool.*). In some Mammals, two small oval areas at the back of the tongue, marked by a series of alternating transverse ridges and richly provided with taste buds.

papillitis (*Med.*). Optic neuritis. Inflammation of the disk or head of the optic nerve within the globe of the eye.

papilloedema, papilledema (*Med.*). Choked disk. Swelling and congestion of the disk or head of the optic nerve within the globe of the eye, as a result of increase of pressure within the skull.

papilloma (*Med.*). A tumour (usually innocent) resulting from the new growth of the cells of the skin or of the mucous membrane.

pappose, pappous (*Bot.*). Having a pappus.

pappus (*Bot.*). A ring of fine hairs, variously arranged, developed from the calyx and attached to the top of the fruit of some flowering plants, e.g., dandelion. They assist in wind dispersal.

Pappus' theorem (*Maths.*). The surface area and the volume respectively swept out by revolving a plane area about a nonintersecting coplanar axis is the product of the distance moved by its centroid and either its perimeter or its area. Frequently attributed to Guldin who rediscovered it in the 17th century.

papulae (*Zool.*). The dermal gills of *Echinodermata*, small finger-shaped, thin-walled respiratory projections of the body wall.

papule (*Med.*). A small, circumscribed, solid elevation above the skin, as in chicken-pox. *adj.* **papular.**

papulopustular (*Med.*). Of papules and pustules.

papyrus (*Paper*). An early form of paper prepared from the water reed of the same name and used for records up to the 9th century A.D.

PAR (*Aero.*). Abbrev. for *precision-approach radar*.

para-. Prefix from Gk. *para*, beside. (*Chem.*) (1) Derived from an acid anhydride by combination with an unusually large number of water molecules. (2) A polymer of. . . . (3) A compound related to. . . .

para- (*Chem.*). (1) Containing a benzene nucleus substituted in the 1.4 positions. (2) Consisting of diatomic molecules with antiparallel nuclear spins and an even rotational quantum number.

***para(4)*-aminophenol** (*Photog.*). Developing agent with characteristics similar to *metol*.

***para(4)*-aminosalicylic acid** (*Pharm.*). See PAS.

parabanic acid (*Chem.*). Oxalyl urea. M.p. 242°–244°C. One of the ureides. Can serve as a source of hydantoin by electrolytic reduction:

$$\begin{array}{l} CO - NH \\ | \qquad\quad \rangle CO \\ CO - NH \end{array}$$

parabasal apparatus (*Zool.*). A peculiar striated body surrounding the ring of blepharoplasts in certain *Mastigophora* with a tuft of flagella.

parabasal body (*Zool.*). In certain parasitic *Mastigophora*, a body of unknown function associated with the base of the flagella.

parabiont (*Cyt.*). One of the embryos united in *parabiosis* (q.v.).

parabiosis (*Cyt.*). The union of similar embryos between which a functional connexion exists, e.g., Siamese twins. *adj.* **parabiotic.**

parablastic (*Zool.*). Said of: (*a*) certain large nuclei occurring in the proximal nutritive syncytial layer of tissue which lies on the yolk in the early stages of development of certain Fish; (*b*) certain similar nuclei of yolk-laden cells in the development of some Mammals.

parabola (*Maths.*). See **conic.**

parabolic (*Bot.*). Having a broad base and gradually narrowing by curved sides to a blunt apex.

parabolic catenary (*Maths.*). See **catenary.**

parabolic microphone (*Acous.*). One provided with a parabolic reflector to give enhanced directivity for high audiofrequencies and hence greater range for wanted sounds amid ambient noise.

parabolic mirror or reflector (*Optics, Radio*). One shaped as a paraboloid of revolution. Theoretically produces a perfectly parallel beam of radiation if a source is placed at the focus (or vice versa). Such mirrors are used in reflecting telescopes, car headlamps, etc. See also **dish for parabolic reflector.**

parabolic (or German) nozzle (*Eng.*). A nozzle of parabolic section with a high coefficient of discharge, placed in a pipe to measure the flow of a gas. The pressure drop is measured by a manometer.

parabolic point on a surface (*Maths.*). One at which one of the principal curvatures, and therefore also the Gaussian curvature, is zero. Cf. *elliptical, hyperbolic, umbilical points on a surface.*

parabolic reflector (*Radio*). See **dish, parabolic mirror.**

parabolic spiral (*Maths.*). A spiral with polar equation $r^2 = a^2\theta$. Also called **Fermat's spiral.**

parabolic velocity (*Astron.*). The velocity which a body at a given point would require to describe a parabola about the centre of attraction; smaller values give an ellipse, larger values a hyperbola. Also called **escape velocity**, since it is the upper limit of velocity in a closed curve.

paraboloid (*Maths.*). See **quadric.**

paraboloid of revolution (*Maths.*). A solid figure formed by the revolution of a parabola about its axis. Cf. *quadric.*

parabrake (*Aero.*). A *brake parachute* (q.v.).

parabronchi (*Zool.*). In Birds, the branches into which the secondary bronchi divide; the air-pipes.

paracasein (*Chem.*). A term for the coagulated casein or insoluble curd of milk.

paracentesis (*Med.*). Tapping. The puncture of body cavities with a hollow needle, for the removal of inflammatory fluids.

paracetamol (*Pharm.*). 4-Acetamidohydroxybenzene. Antipyretic and analgesic drug. See **Panadol.**

parachor (*Chem.*). A quantity which may be regarded as the molecular volume of a substance when its surface tension is unity; in most cases it is practically independent of temperature. Its value is given by the expression $[P] = \dfrac{M\gamma^{\frac{1}{4}}}{\rho_L - \rho_V}$, where M is the molecular mass, γ the surface tension, ρ_L and ρ_V are the densities of the liquid and vapour respectively.

parachordals (*Zool.*). In the developing cranium, a pair of flat curved plates of cartilage lying on either side of the notochord.

parachute (*Aero.*). An umbrella-shaped fabric device of high drag (1) to retard the descent of a falling body, or (2) to reduce the speed of an aircraft or item jettisoned therefrom. Commonly made of silk or nylon, sometimes of cotton or rayon where personnel are not concerned. See **antispin-, automatic-, brake-, cross-, reserve-, ribbon-, ring slot-, square-, pilot chute.** See also **parasheet.**

parachute disseminule (*Bot.*). A fruit or seed with a pappus, tuft of hairs, or other device which facilitates dispersal by the wind.

parachute flare (*Aero.*). A pyrotechnic flare, attached to a parachute released from an aircraft to illuminate a region.

paracme (*Zool.*). The period in the history of an individual or a race when vigour is decreasing; the senescent period in the life history of an individual; the phylogerontic period in the history of a race. Cf. *epacme.*

paracoele (*Zool.*). The cavity of one of the cerebral hemispheres of the Vertebrate brain.

paracone (*Zool.*). In Mammals, the antero-external cusp of an upper molar tooth.

paraconid (*Zool.*). In Mammals, the antero-external cusp of a lower molar tooth.

paracorolla (*Bot.*). See **corona.**

paracymbium (*Zool.*). The smaller of 2 parts into which the pedipalpal tarsus of some male Spiders is divided.

paradermal (*Zool.*). Parallel to the epidermis.

paradesmose (*Bot., Zool.*). A very slender thread of stainable material which connects for some time the two halves of a dividing blepharoplast.

paradistemper (*Vet.*). See **hard-pad disease.**

paradoxical bone (*Zool.*). See **prevomer.**

paradoxical sleep (*Psychol.*). See **rapid eye movement sleep.**

paraeiopod (*Zool.*). In *Crustacea*, a walking leg.

paraesthesia (*Med.*). An abnormal sensation, such as tingling, tickling, and formication.

parafeed coupling (*Telecomm.*). Combination of resistance-capacitance with intervalve-transformer coupling so that anode-feed through a resistance is diverted from the transformer, which can then be designed with reduced dimensions and cost.

paraffin oil (*Fuels*). See **kerosine.**

paraffins (*Chem.*). A term for the whole series of saturated aliphatic hydrocarbons of the general formula C_nH_{2n+2}. The term **alkane hydrocarbons** is also used. They are indifferent to oxidizing agents, and not reactive, hence the name *paraffin* (L. *parum affinis*, little allied).

paraffin wax (*Chem.*). Higher homologues of alkanes, waxlike substances obtained as a residue from the distillation of petroleum; m.p. 45°–65°C, rel. d. 0·9, resistivity 10^{13} to 10^{17} ohm metres, permittivity 2–2·3.

parafibula (*Zool.*). A bony element external to the proximal end of the fibula in some *Lacertilia* and young *Metatheria*.

paraformaldehyde (paramethanal) (*Chem.*). $(H\cdot CHO)_2$, a condensation product of formaldehyde (methanal) obtained by the evaporation of an aqueous solution of methanal. It is a white crystalline mass, soluble in water.

parafrons (*Zool.*). In certain Insects, the area between the frontal suture and the eyes.

paraganglia (*Zool.*). In higher Vertebrates, small glandular bodies, occurring in the posterior part of the abdomen, which show a chromaphil reaction and are believed to secrete adrenaline.

paraganglioma (*Med.*). Chromaffinoma; phaeochromacytoma. A tumour composed of chromaffine cells, especially those of the adrenal medulla; usually associated with high blood pressure.

paragaster (*Zool.*). The internal cavity of a Sponge, lined by Choanocytes.

paragastric canals (*Zool.*). In *Porifera*, the canal-system; in *Ctenophora*, the paired blind canals running from the infundibulum to the oral cone.

paragastrula (*Zool.*). A stage in the development of *Porifera* in which the flagellate cells of the amphiblastula are invaginated.

paraglossa (*Zool.*). In Insects, the outer pair of lobes arising from the margin of the prementum.

paragnatha (*Zool.*). In some *Crustacea*, a pair of lobes formed by subdivision of the metastoma.

paragnathous (*Zool.*). Having jaws of equal length; as Birds which have upper and lower beak of equal length.

paragneiss (*Geol.*). A term given to gneissose rocks which have been derived from detrital sedimentary rocks. Cf. *orthogneiss*.

paragonimiasis (*Med.*). Invasion of the lungs by the lung fluke *Paragonimus westermanii*; the condition is endemic in the Far East, and infection results from eating fresh-water crustaceans.

paragonite (*Min.*). A hydrous silicate of sodium and aluminium. It is a sodium mica, has a yellowish or greenish colour, and is usually associated with metamorphic rocks. Differs from *muscovite* (q.v.) chiefly in containing sodium rather than potassium.

paragraphia (*Med.*). Faulty spelling, misplacement of letters and words, and use of wrong words in writing, as a result of a lesion in the brain.

paragula (*Zool.*). A region of the Insect head, beside the gula.

paragutta (*Cables*). A substance containing deproteinized rubber and wax, used for insulating submarine telephone cables.

paragynous (*Bot.*). Said of an antheridium applied to the side of an oögonium.

parahormone (*Zool.*). A substance which, like a hormone, has a controlling effect on distant parts of the body, but which is an ordinary by-product of metabolic activity, not produced for a specific purpose.

***para*hydrogen** (*Chem.*). A hydrogen molecule in which the two nuclear spins are antiparallel, forming a singlet state.

parahypophysis (*Zool.*). In Vertebrates, a vestigial structure in the region of the pituitary body.

parakeratosis (*Med.*). Faulty formation of the horny layer of the skin, with scaling of the skin.

parakinesis (*Psychol.*). The alleged movement of physical objects by nonphysical or supernormal forces.

paralalia (*Med.*). A form of speech disturbance, particularly that in which a different sound or syllable is produced from the one which is intended.

paraldehyde (*Chem.*). $(CH_3\cdot CHO)_3$, a condensation product of ethanal, obtained by the action of concentrated sulphuric acid upon ethanal. It is a colourless liquid, b.p. 124°C, and can be converted again into ethanal by distillation with dilute sulphuric acid. A common hypnotic. Used as a chemical intermediate in the synthesis of *pentaerythritol*.

paralexia (*Med.*). A defect in the power of seeing and interpreting written language, with meaningless transposition of words and syllables, due to a lesion in the brain.

paralimnion (*Ecol.*). The littoral zone of a lake floor, embracing the area lying between the water's edge or shoreline, and the lakeward margin of rooted vegetation.

parallactic angle (*Astron.*). The name given to that angle in the astronomical triangle formed at the heavenly body by the intersection of the arcs drawn to the zenith and to the celestial pole. See **astronomical triangle**.

parallactic ellipse (*Astron.*). The small ellipse on the celestial sphere apparently described by every star in a year about its mean position owing to the earth's orbital motion; the semi-major axis of the ellipse is the star's *annual parallax*.

parallactic inequality (*Astron.*). A periodic term in the mathematical expression of the moon's motion, so called because it is due to the finite distance of the sun and therefore depends on the solar parallax; it amounts to about 2′ 7″ at a maximum, and has for period the synodic month.

parallax. Generally, the apparent change in the position of an object seen against a more distant background when the viewpoint is changed. Absence of parallax is often used to adjust two objects, or two images, at equal distances from the observer. (*Astron.*) The apparent displacement of a heavenly body on the celestial sphere due to a change of position of the observer. See **annual-, geocentric-, horizontal-, secular-, spectroscopic-**. (*Surv.*) A condition set up when the image formed by the object-glass of a surveying telescope is not in the same plane as the cross-hairs so that, on moving his eye from side to side or up and down, the observer can see relative movement between them.

parallax stereogram (*Photog.*). The use of a line screen in front of a positive transparency of alternate strips of two views of an object, made by exposing an emulsion in a similar arrangement with a large lens and two apertures representing the two eyes.

parallel (*Elec. Eng., etc.*). Two circuits are said to be *in parallel* when they are connected so that any current flowing divides between the 2. Two machines, transformers, or batteries are said to be *in parallel* when the terminals of the same polarity are connected together. (*Geog., Surv.*) A *parallel of latitude* is an imaginary line around the earth's surface connecting points of equal terrestrial latitude. (*Maths.*) Applied to pairs of lines or curves which are spaced from each other by a constant distance, the distance being measured along common normals. May be said to meet at infinity.

parallel arithmetic unit (*Comp.*). One in which the digits are operated on simultaneously.

parallel-axis theorem (*Maths.*). A theorem expressing the moment of inertia I of a body about any axis, in terms of its moment of inertia I_a

about a parallel axis through G, its centre of gravity. Thus

$$I = I_a + Md^2,$$

where d is the perpendicular distance between the axes and M the mass of the body.

parallel body (*Ships*). That portion of a ship's form wherein the fullest transverse shape is maintained constant.

parallel circuit (*Elec. Eng.*). See **shunt circuit.**

parallel descent (*Bot., Zool.*). The appearance of similar characteristics in groups of animals or plants which are not directly related in evolutionary descent. Also **parallel evolution, parallelism.**

parallelepiped (*Maths.*). A solid figure bounded by six parallelograms, opposite pairs being identical and parallel.

parallel feed (*Electronics*). Connexion between anode-supply and anode through a high resistance or inductance, while a.c. from the anode is delivered through a capacitance, so that a.c. and d.c. anode currents are separated. Called **RC** or **LC coupling** depending on whether resistive or inductive anode load. Also called **shunt feed.**

parallel-feeder protection (*Elec. Eng.*). A type of balanced protective equipment relying for its action on the fact that the current in two parallel feeders will normally be equal, this balance being upset if a fault occurs on one of the feeders.

parallel folding (*Bind.*). When two or more successive folds in a sheet are in the same direction they are described as *parallel*. Cf. *right-angled folding*.

parallel gutter (*Build.*). A rectangular roof-gutter, e.g., one bounded at the sides by a parapet wall and a pole plate.

parallelism (*Bot., Zool.*). See **parallel descent.**

parallel memory or **store** (*Comp.*). One designed to give as nearly as possible uniform access time.

parallel motion (*Eng.*). (1) A system of links by which the reciprocating motion of one point is copied to an enlarged scale by another—a modified pantograph mechanism. Used on piston-type engine indicators, etc. (2) A mechanism controlling the movement of a straight edge across a drawing-board.

parallelodromous (*Bot.*). Having parallel veins.

parallelogram (*Maths.*). A quadrilateral with two pairs of opposite sides parallel.

parallelogram of forces (*Maths.*). Alternative name for the *parallelogram rule for addition of vectors* when the vectors are forces.

parallelogram rule for addition of vectors (*Maths.*). The sum of two vectors is the diagonal through O of the parallelogram of which the two vectors, when represented by lines through O, are adjacent sides.

parallelotropic (*Bot.*). Said of a plant member set along the direction of stimulus.

parallel-plate capacitor (*Elec.*). Device for storage of electric charge, consisting of two plane parallel conducting electrodes of area A m², separated by a dielectric of *absolute permittivity* ε F/m, thickness d m. Its capacity is given by $C = \varepsilon A/d$ farads; charge stored when p.d. between plates is V volts is CV coulombs.

parallel-plate chamber (*Nuc. Eng.*). Ionization chamber with plane-parallel electrodes.

parallel-plate waveguide (*Radio*). One formed by two parallel conducting or dielectric planes. Often realized in atmospheric propagation under suitable meteorological conditions. See **surface duct.**

parallel resonance (*Elec. Eng.*). See **shunt resonance.**

Parallel Roads (*Geol.*). The strandlines of a glacial lake which occupied Glen Roy (Inverness-shire, Scotland) during the Pleistocene Period, when the lower part of the valley was blocked by ice.

parallel ruler (*Instr.*). A drawing instrument consisting of two straight edges so linked together by connecting pieces that their edges are always parallel, although the distance between them may be varied.

parallel slot (*Elec. Eng.*). The most usual shape of slot for the armature windings of electric machines, the slot having parallel sides. Cf. *taper slots*.

parallel tracking (*Acous.*). Mechanism whereby the pick-up arm is kept exactly tangential to the groove during reproduction from a disk.

parallel-wire line (*Telecomm.*). Form of transmission line widely used at low frequencies where radiation losses will be small.

parallel wire resonator (*Elec. Eng.*). See **Lecher wires.**

paralogia (*Med.*). Impairment of reasoning power characterized by difficulty in expressing logical ideas in speech.

paralogy (*Zool.*). Anatomical similarity that has no phylogenetic or functional implication.

paralutein (*Histol.*). Applied to cells of corpus luteum derived from theca interna. Also called **theca lutein.**

paralyser (*Chem.*). See **catalytic poison.**

paralysis (*Med.*). 'Palsy.' The loss in any part of the body of the power of movement, or of the capacity to respond to sensory stimuli. See **diplegia, hemiplegia, monoplegia, paraplegia; Bell's palsy, Little's disease, general paresis.**

paralysis agitans (*Med.*). Parkinson's disease; 'shaking palsy'. A progressive disease due to degeneration of certain nerve cells at the base of the brain; characterized by rigidity of muscles (the body being fixed in a posture of slight flexion), mask-like expression of the face, and a coarse tremor, especially of the hands. See **retropulsion.**

paralysis time (*Nuc. Eng.*). The time for which a radiation detector is rendered inoperative by an electronic switch in the control circuit.

paralytic ileus (*Med.*). A condition in which, from various causes, there is extensive paralysis of the intestines, leading to persistent vomiting, pain being absent.

paramagnetism (*Elec. Eng., Mag.*). Phenomenon involving a material whose relative permeability is slightly greater than unity. Used to attain very low temperatures by adiabatic demagnetization.

paramastigote (*Zool.*). Having one long principal flagellum and a short accessory flagellum.

paramatta (*Textiles*). Twilled fabrics with a botany worsted or cotton warp and a worsted weft; made into showerproof garments.

paramere (*Zool.*). Half of a bilaterally symmetrical structure; one of the inner pair of gonapophyses in a male Insect.

parameter (*Crystal.*). The *parameters* of a plane consist of a series of numbers which express the relative intercepts of that plane upon the crystallographic axes. Given in terms of the established unit lengths of those axes. (*Maths.*) Generally, a variable in terms of which it is convenient to express other interrelated variables which may then be regarded as being dependent upon the parameter. (*Telecomm.*) A derived constant of a transmission circuit or network, which is more convenient for expressing performance or for use in calculations.

parametric amplifier (*Telecomm.*). Circuit in

which energy is fed from 'pumping' oscillator into signal through a varying reactance, e.g., a germanium diode, made to act as a negative resistance by the pump signal. Also **mavar.**

parametric diode (*Electronics*). One whose series capacitance can be varied by the biasing voltage. Can be a valve or solid-state device.

parametric equations (*Maths.*). Of a curve or surface: equations which express the co-ordinates of points on the curve or surface explicitly in terms of other variables (parameters), which are, for the present, regarded as independent variables.

parametric resonance (*Telecomm.*). The condition of a parametric amplifier when energy transfer from the pump circuit through any resonant cavities to the signal is a maximum.

parametritis (*Med.*). Pelvic cellulitis. Inflammation of the pelvic cellular connective tissue in the region of the uterus, e.g., in the puerperium.

parametrium (*Med.*). The subperitoneal connective tissue surrounding the uterus, especially that in the region of the cervix.

parametron (*Comp.*). Device for digital computers, using parametric oscillation.

paramitosis (*Cyt.*). A form of mitosis occurring in *Protozoa*, in which the longitudinal halves of the chromosomes remain attached at one end after division so that they appear to divide transversely.

paramorph (*Biol.*). A general term for any taxonomic variant within a species, usually used when more accurate definition is not possible. (*Min.*) The name given to a mineral species which can change its molecular constitution without any change of chemical substance. Cf. *pseudomorph.*

paramphistomiasis (*Med.*). Invasion of the human intestine by trematode parasites of the family *Paramphistomidae.*

paramudras (*Geol.*). Flint nodules of exceptionally large size and doubtful significance occurring in the Chalk exposed on the east coast of England.

paramylon (*Bot.*). A polysaccharide storage product allied to starch, but not giving a blue coloration with iodine. Restricted in occurrence to *Euglena* and related organisms.

paramyosinogen (*Chem.*). One of the chief proteins contained in living muscle.

parana pine (*For.*). A tree, *Araucaria brasiliana*, of S. Brazil yielding a softish wood used for internal joinery.

paranema (*Bot.*). A sterile hair. *pl.* **paranemata.**

Paraneoptera (*Zool.*). A section of *Pterygota*, with wings folded in repose; Malpighian tubules are few and metamorphosis incomplete. Includes the superorders *Psocopteroidea, Thysanopteroidea* and *Rhynchota.*

paranephric (*Zool.*). Situated beside the kidney.

paranephros (*Zool.*). See **suprarenal body.**

paranoia (*Psychiat.*). A psychosis characterized by the insidious development of a permanent and unshakeable delusional system, resulting from internal causes, and accompanied by the preservation of clear and orderly thought, action and will. Hallucination is not usually present. Attributed by Freudians to unconscious homosexuality. *adj.* and *n.* **paranoiac.**

paranoid schizophrenia (*Psychiat.*). A type of schizophrenia showing certain symptoms of paranoia, including delusions of persecution and jealousy, but with less systematization of thought and with much personality disintegration.

paranoid state (*Psychiat.*). A state in which a person shows certain characteristics of paranoia, but in a lesser degree and without disintegration of the personality.

paranotum (*Zool.*). In Insects, a lateral expansion of a thoracic tergum. *adj.* **paranotal.**

paranucleus (*Zool.*). A term which has been loosely used to indicate any structure lying beside the nucleus; as a micronucleus of *Ciliophora*, or a mass of mitochondria.

paraophoron (*Zool.*). A rudimentary structure in the ovary of Vertebrates homologous with the paradidymis of the male.

paraphase amplifier (*Telecomm.*). Push-pull stage incorporating *paraphase coupling.* Also **see-saw amplifier.**

paraphase coupling (*Telecomm.*). A push-pull stage, or series of stages, in which reversed phase is obtained by taking a fraction of the output voltage of the first valve of the amplifier and applying it to a similar balancing first valve, which operates succeeding stages in normal push-pull, the number of transformers being minimized thereby.

paraphasia (*Med.*). A defect of speech in which words are misplaced and wrong words substituted for right ones; due to a lesion in the brain.

***para(4)*-phenylene diamine (1,4-diamino-benzene)** (*Photog.*). Fine-grain developing agent, slow in action. Mixed with *catechol* (1,2-dihydroxy-benzene) to increase its speed, it is known as *Meritol*:

paraphilia (*Psychiat.*). Generic term for sexual perversion.

paraphimosis (*Med.*). Persistent retraction of the inner lining of the prepuce behind the glans penis in a case of phimosis.

paraphobia (*Psychiat.*). A mild type of *phobia.*

paraphonia (*Med.*). Alteration of the voice as a result of disease.

paraphototropic (*Bot.*). See **diaphototropic.**

paraphraxia (*Med.*). The performance by a person of an act different from the one required of him; due to a lesion in the brain.

paraphrenia (*Psychiat.*). A type of schizophrenia characterized by the extremely insidious development of ideas of persecution and, later, of exultation, without disintegration of personality. Gradually, ideas of grandeur and of the patient's own importance may replace ideas of persecution, and auditory and other hallucinations may develop.

paraphyll (*Bot.*). A very small leaflike, or much branched, structure found among the leaves of mosses.

paraphysis (*Bot.*). A sterile filament which may be simple or branched and may consist of one or more cells, occurring among reproductive structures in many lower plants. (*Zool.*) A thin-walled sac developed as an outgrowth from the non-nervous roof of the telencephalon, represented in Mammals by the pineal organ. *pl.* **paraphyses.** *adj.* **paraphysate.**

paraphysoid (*Bot.*). A plate of cells occurring between the asci in some *Ascomycetes.*

parapineal organ (*Zool.*). See **parietal organ.**

paraplasm (*Cyt.*). The active, vegetative portion of the cytoplasm.

paraplectenchyma (*Bot.*). Pseudoparenchyma.

paraplegia (*Med.*). Paralysis of the lower part of the body and of the legs.

parapodium (*Zool.*). In *Mollusca*, the epipodia; in *Polychaeta*, a paired fleshy projection of the body wall of each somite used in locomotion. *adj.* **parapodial.**

parapophyses (*Zool.*). A pair of ventrolateral processes of a vertebra arising from the sides of the centrum.

parapsid (*Zool.*). In the skull of Reptiles, the condition when there is one temporal vacuity, this being high behind the eye, usually with the postfrontal and supratemporal meeting below. Found in Mesosaurs and Ichthyosaurs. Cf. *anapsid, diapsid, euryapsid, synapsid.*

parapsychology (*Psychol.*). A division of psychology concerned with investigation of the so-called paranormal phenomena, e.g. telepathy, extrasensory perception, premonitions, etc.

parapteron (*Zool.*). See **tegula.**

parapterum (*Zool.*). In Birds, a group of small feathers attached to the humerus.

paraquadrate (*Zool.*). See **squamosal.**

paraquat (*Chem.*). 1,1′-dimethyl-4,4′-dipyridylium salts, used as a weed-killer. Extremely toxic to human beings, and there is no known antidote. See **alveolitis.**

***para*quinones** (*Chem.*). Quinones in which the two quinone oxygen atoms are in *para* position.

para reds (*Paint.*). Pigments obtained by diazotizing paranitraniline, and coupling it to β-naphthol. Good general-purpose red pigments except for their tendency to *bleeding* in oils.

pararosaniline (*Chem.*). A triamino-triphenylmethanol of the formula:

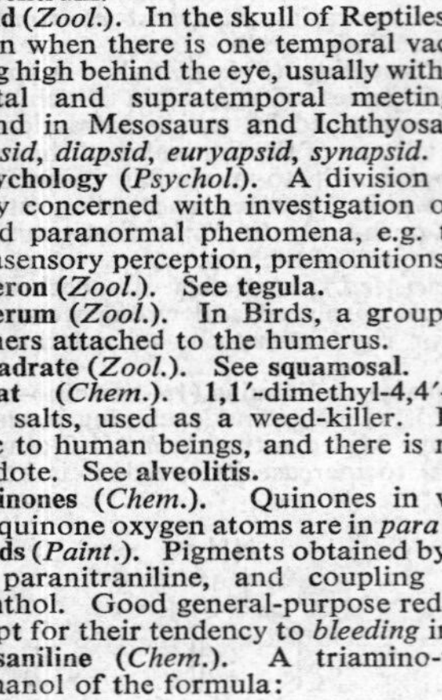

Obtained by oxidation of a mixture of 4-toluidine (1 mol) and aminobenzene (2 mol). As oxidizing agents, arsenic acid or nitrobenzene are used. Acids effect the elimination of water and the formation of a dyestuff with a quinonoid structure, e.g., the hydrochloride of pararosaniline or parafuchsine has the formula:

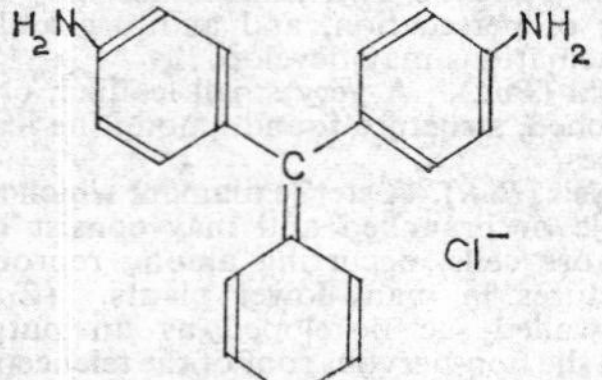

pararosolic acid (*Chem.*). See **aurine.**

paraselenae (*Meteor.*). See **mock moons.**

parasheet (*Aero.*). A simplified form of parachute for dropping supplies, made from one or more pieces of fabric with parallel warp in the form of a polygon, to the apices of which the rigging lines are attached.

parasite (*Ecol.*). An organism which lives in or on another organism and derives subsistence from it without rendering it any service in return. See **parasitism.**

parasite chain (*Ecol.*). A type of food chain which passes from larger to smaller organisms. Cf. *predator chain, saprophyte chain.*

parasitic antenna or **aerial** (*Radio*). Unfed dipole element which acts as a *director* or *reflector.*

parasitic bronchitis (*Vet.*). See **husk.**

parasitic capture (*Nuc.*). Neutron capture in reactor not followed by fission.

parasitic castration (*Bot.*). The condition when a plant is unable to fruit owing to damage to the reproductive organs by a parasite. (*Zool.*) Castration brought about by the presence of a parasite, as in the case of a Crab parasitized by *Sacculina.*

parasitic drag (*Aero.*). An early term for the *drag* of an aircraft other than the drag associated with the production of lift.

parasitic loss (*Elec. Eng.*). A term sometimes used to denote loss in electric machines due to eddy currents occurring in any part of the machine, i.e., in the conductors themselves or in parts of the metal framework. The term does not include losses due to the main flux in the magnetic circuit, which come under the heading of *iron loss* (q.v.).

parasitic male (*Zool.*). A dwarf male in which all but the sexual organs are reduced, and which is entirely dependent on the female for nourishment; as in some deep-sea Angler-fish (Ceratioids).

parasitic oscillation (*Telecomm.*). Unwanted oscillation of an amplifier, or oscillation of an oscillator at some frequency other than that of the main resonant circuit. Generally of high frequency, it may occur during a portion of each cycle of the main oscillation. Also **spurious oscillation.**

parasitic stopper (*Telecomm.*). Components which attenuate a feedback path which otherwise maintains unwanted oscillations.

parasitism (*Ecol.*). A close internal or external partnership between two organisms which is detrimental to one partner (the *host*) and beneficial to the other partner (the *parasite*); the latter obtains its nourishment at the expense of the nutritive fluids of the host.

parasitoid (*Zool.*). An animal which is parasitic in one stage of the life history and subsequently free-living in the adult stage, as the parasitic *Hymenoptera.*

parasitology. The study of parasites and their habits (usually confined to animal parasites).

parasphenoid (*Zool.*). In some of the lower Vertebrates, a membrane bone of the skull, which forms part of the cranial floor.

parastas (*Arch.*). See **pilaster.**

parastat (*Acous.*). A device attached to the tone arm of a record player to clean the record grooves as it tracks and simultaneously to deposit a thin antistatic film on the surface of the record.

parasternum (*Zool.*). The gastralia of certain Reptiles.

parastichy (*Bot.*). A spiral line passing once round a stem through the bases of successive leaves. *pl.* **parastichies.**

parastipes (*Zool.*). See **subgalea.**

parasymbiosis (*Biol.*). The condition when two organisms grow together but neither assist nor harm one another.

parasympathetic nervous system (*Zool.*). In Vertebrates, a subdivision of the autonomic nervous system, also known as the *craniosacral system*. The action of these nerves tends to slow down activity in the glands and smooth muscles which they supply, but promotes digestion, and acts antagonistically to that of the *sympathetic nervous system* (q.v.). Parasympathetic nerves are cholinergic, the preganglionic axons are long and the postganglionic axons short, and they emerge from the central nervous system anteriorly, in association with certain cranial nerves (especially the vagus) and from a few spinal nerves in the sacral region. Cf. *sympathetic nervous system.*

parasympatheticotonia (*Med.*). State of abnormal excitability of the parasympathetic nervous system, characterized by skin flushing, sweating, abdominal cramplike pains, constipation and bradycardia.

parasynapsis, parasyndesis (*Cyt.*). Side-by-side union of the elements of a pair of chromosomes; cf. *syndesis, telosynapsis*. *adjs.* **parasynaptic, parasyndetic.**

paratenic (*Ecol.*). Used of the intermediate host of a parasite that is not essential and neither hinders nor hastens the completion of the life cycle.

parathecium (*Bot.*). A layer of hyphae around the apothecium of a lichen.

parathion (*Chem.*). 2,2-diethyl-2-*p*-nitrophenylthionphosphate:

$$O_2N\text{-}C_6H_4\text{-}O\text{-}P(=S)(OC_2H_5)(OC_2H_5)$$

Used as an insecticide.

parathormone (*Zool.*). The hormone secreted by the *parathyroid* (q.v.).

parathyroid (*Zool.*). An endocrine gland of Vertebrates found near the thyroid or embedded in it. Two pairs are usually present, deriving from the third and fourth pairs of gill pouches. Probably only secrete one hormone (*parathormone*) which controls the metabolism of calcium and phosphorus.

paratomy (*Zool.*). In *Annelida*, a form of reproduction by fission in which regeneration takes place before separation.

paratonic movements (*Bot.*). Plant movements induced by an external stimulus, e.g., tropisms.

paratracheal (*Bot.*). Said of xylem parenchyma occurring at the edge of the annual ring and round the vessels, but nowhere else.

paratuberculosis (*Vet.*). See **Johne's disease.**

paratyphoid (*Med.*). Enteric fever due to infection by *Bacterium paratyphosum A, B*, or *C*; similar to, but more mild than, typhoid fever. A mixed preparation of killed bacteria for immunization against typhoid, paratyphoid A and paratyphoid B is known as **TAB.**

paraxial (*Photog.*). The path of a ray which is parallel to the axis of an optical system.

paraxial focus (*Optics*). The point at which a narrow pencil of rays along the axis of an optical system comes to a focus.

Paraxonia (*Zool.*). A superorder of the cohort *Ferungulata* (*Eutheria*; Mammals) containing the one order *Artiodactyla* (q.v.).

paraxonic foot (*Zool.*). A foot in which the skeletal axis passes between the third and fourth digits, as in *Artiodactyla*. Cf. *mesaxonic foot.*

Parazoa (*Zool.*). A subkingdom of the animals. Multicellular organisms, though their cells are less specialized and interdependent than in the *Metazoa*. Invariably sessile and aquatic, having a single body cavity lined in part or almost completely by collared flagellate cells, numerous pores in the body wall for the intake of water and one or more larger ones for passing it out. Generally have a calcareous, siliceous or horny skeleton. Have no nervous system. Larval forms free swimming and flagellate. Sponges. Cf. *Protozoa, Metazoa.*

Par. C. (*Build.*). Abbrev. for *Parian cement.*

parcel (*For.*). Any quantity of standing, felled, or converted timber forming a unit or item for purposes of trade.

parcel plating (*Elec. Eng.*). The electrodeposition of a metal over a selected area of an article, the remainder being covered with a nonconductor to prevent deposition.

parchmentizing (*Paper.*). The process of passing paper through sulphuric acid, or zinc chloride, which causes the fibres to swell and the paper to become translucent.

parencephalon (*Zool.*). A cerebral hemisphere.

parenchyma (*Bot.*). A tissue composed of blunt-ended cells, having thin walls consisting of cellulose, and often forming a general packing among conducting and mechanical tissues. (*Zool.*) Soft spongy tissue of indeterminate form, consisting usually of cells separated by spaces filled with fluid or by a gelatinous matrix, and generally of mesodermal origin. The specific tissue of a gland or organ as distinct from the interstitial (connective) tissue or stroma. *adj.* **parenchymatous.**

parenchymula (*Zool.*). In *Porifera*, a flagellate larval form, the internal cavity of which is occluded by a form of parenchyma.

parent (*Nuc.*). In radioactive particle decay of *A* into *B*, *A* is the parent and *B* the daughter product.

parental generation (*Gen.*). See **P.**

parenteral (*Med.*). Said of the administration of therapeutic agents by any way other than through the alimentary tract.

parent exchange (*Teleph.*). Of an automatic exchange, the auto-manual or manual exchange which handles the assistance traffic.

parentheses (*Typog.*). Marks of punctuation ( ) used to enclose a definition, explanation, reference, etc., or interpolations and remarks made by the writer of the text himself. *Sing.* **parenthesis.**

parent metal (*Eng.*). A term used in welding to denote the metal of the parts to be welded.

parent peak (*Nuc.*). The component of a mass spectrum resulting from the undissociated molecule.

paresis (*Med.*). Slight or incomplete paralysis. See also **general paresis** (general paralysis of the insane, GPI).

paresis juvenilis (*Med.*). General paresis occurring in a child or young person.

paresthesia (*Med.*). See **paraesthesia.**

pargasite (*Min.*). A monoclinic amphibole of the hornblende group, particularly rich in magnesium, sodium, calcium, and aluminium. Named from Pargas, Finland. It occurs chiefly in metamorphic rocks.

pargeting (*Build.*). The operation of rendering

the interior of a flue with mortar. Also **pergeting.**

parge-work (*Build.*). An ancient form of external plastering with a mixture similar to that used in *pargeting* (q.v.) chimneys. Also called **parging.**

parging (*Build.*). See **parge-work.**

parhelia (*Meteor.*). *Sing.* **parhelion.** See **mock suns.**

Parian cement (*Build.*). A hard plaster made from an intimate mixture of gypsum and borax which has been calcined and then ground to powder.

parichnos (*Bot.*). In certain lower vascular plants, a pair of scars, one on each side of the leaf base; each scar marks the end of a strand of parenchyma passing into the stem.

parietal (*Bot.*). Attached to, or lying near to and more or less parallel with, a wall. Applied particularly to the placenta when this arises from the peripheral wall of a carpel, and to chloroplasts of *Algae* lying not far inside the cell wall. (*Zool.*) A paired dorsal membrane bone of the Vertebrate skull, situated between the auditory capsules; pertaining to, or forming part of, the wall of a structure; in Insects, said of the part of the epicranium between the vertex and the frons.

parietal cells (*Zool.*). The oxyntic cells of a gastric gland in Vertebrates.

parietal foramen (*Zool.*). Small rounded aperture in the middle of the united parietals of the skull. Site of the pineal eye.

parietal organ (*Zool.*). The anterior diverticula of the pineal apparatus. When present may be developed as an eyelike organ, the pineal eye.

parietes (*Anat.*, *Zool.*). The walls of an organ or a cavity. *sing.* **paries.**

paring chisel (*Carp.*, *Join.*). A long chisel with a thinner blade than a firmer tool, used for finishing off work by hand. It is not intended to be struck with a mallet.

paring gouge (*Carp.*, *etc.*). A gouge having the bevel ground upon the inside or concave face of the cutting edge.

paripinnate (*Bot.*). Said of a compound pinnate leaf which has no terminal leaflet.

Paris green (*Chem.*, *Paint.*). Copper ethanoatoarsenate, a bright blue-green pigment once used in anti-fouling compositions but now of use mainly in wood preservative preparations, and as an agricultural chemical. Also **emerald green, Schweinfurt green.**

parison (*Glass*). The intermediate shape produced in the manufacture of a glass article in more than one stage. In the automatic and semiautomatic methods, the parison is formed in one mould, being then transferred to another for the final forming operation.

Paris white (*Paint.*). A fine, size-classified form of *whiting* (q.v.).

parity (*Maths.*). The property of being odd or even, e.g., the parity of 4 is even. (*Med.*) The condition or fact of having borne children. (*Phys.*) The conservation of parity or space-reflection symmetry (*mirror symmetry*) states that no fundamental difference can be made between right and left, or in other words the laws of physics are identical in a right- or in a left-handed system of coordinates. The law is obeyed for all phenomena described by classical physics but was recently shown to be violated by the weak interactions between elementary particles.

parity check (*Comp.*). Addition of a redundant bit to a word to make the total number of 1s even or odd in order to detect simple bit errors.

Parker board (*Print.*). A superior style of mounting board for printing plates, made up of small interlocked pieces of hardwood within a steel confining edge, and ruled off in squares to facilitate spacing out and squaring-up the plates.

Parker's cement (*Civ. Eng.*). See **Roman cement.**

Parke's process (*Met.*). A process used for desilverizing lead. It depends on the fact that when zinc is added to molten lead it combines with any gold or silver present to form compounds that have a very slight solubility in lead. The bullion-rich zinc is then skimmed off.

parkinsonism (*Med.*). The gradual development, as a sequel to encephalitis lethargica, of a state closely similar to that found in *paralysis agitans.*

Parkinson's disease (*Med.*). See **paralysis agitans.**

parliamentary candle (*Light*). See **standard candle.**

parliament hinge (*Join.*). See **H-hinge.**

paroccipital (*Zool.*). A ventrally-directed process of the exoccipital in Mammals.

paroicous (*Bot.*). Said of those *Bryophyta* in which the antheridia and archegonia occur on the same branches but are not mixed, the antheridia being borne lower on the stems than the archegonia.

paronychia (*Med.*). A felon or whitlow. Purulent inflammation of the tissues in the immediate region of the finger nail. (*Zool.*) In Insects, bristles on the pulvillus of the foot.

parosmia (*Med.*). Abnormality of the sense of smell.

parotic process (*Zool.*). In some Vertebrates, a bony process formed by the fusion of the opisthotic with the exoccipital.

parotid gland (*Zool.*). In some *Anura*, an aggregation of poison-producing skin glands on the neck; in Mammals, a salivary gland situated at the angle of the lower jaw.

parotitis (*Med.*). Inflammation of the parotid gland. See **mumps.**

parovarial (*Zool.*). Beside the ovary.

parovarium (*Zool.*). In some female Mammals, a small collection of tubules, near the anterior end of the ovary, which represents the mesonephros.

parpoint work (*Build.*). Stone-wall construction in which the squared stones are laid as stretchers, with occasional courses of headers.

parquet (*Build.*). A floor-covering of hardwood blocks glued and pinned to the ordinary floor boarding and finally wax-polished.

parrot coal (*Min.*). See **boghead coal.**

parrot disease (*Med.*, *Vet.*). See **psittacosis.**

pars (*Zool.*). A part of an organ. *pl.* **partes.**

pars anterior (*Zool.*). See **pars distalis.**

Parscope (*Radar*). Oscilloscope display for weather radars.

pars distalis (*Zool.*). The anterior part of the adenohypophysis of the pituitary.

parsec (*Astron.*). The chief unit used in measuring stellar distances, being the distance at which the semimajor axis of the earth's orbit would subtend 1 second of arc; hence, a star's distance in *parsecs* is the reciprocal of its annual parallax in seconds of arc; 1 parsec $= 206265$ astronomical units or $19{\cdot}17 \times 10^{12}$ miles ($30{\cdot}86 \times 10^{12}$ km), or $3{\cdot}26$ light-years; contraction of *parallax second.* Abbrev. **pc.**

pars intermedia (*Zool.*). In higher Vertebrates, part of the posterior lobe of the pituitary body, which is derived from the hypophysis at first but tends to become spread over the surface of the *pars nervosa* (q.v.) as development proceeds.

pars nervosa (*Zool.*). In higher Vertebrates, part

of the posterior lobe of the pituitary body, developed from the infundibulum.

Parson's steam turbine (*Eng.*). A *reaction turbine* (q.v.) in which rings of moving blades of increasing size are arranged along the periphery of a drum of increasing diameter. Fixed blades in the casing alternate with the blade rings. Steam expands gradually through the blading, from inlet pressure at the smallest section to condenser pressure at the end.

part (*Arch.*). The one-thirtieth portion of a module.

parted (*Bot.*). Cleft nearly to the base.

parthen-, partheno-. Prefix from Gk. *parthenos*, virgin.

parthenapogamy (*Bot.*). A fusion of vegetative nuclei.

parthenocarpy (*Bot.*). The production of a fruit without a preliminary act of fertilization, and without any development of seeds within the fruit. Seedless fruits may be induced artificially in some unfertilized flowers by spraying with a dilute solution of synthetic auxin.

parthenogamy (*Bot.*). The union of 2 female gametes or of structures equivalent to them.

parthenogenetic (*Bot., Zool.*). Reproducing without fertilization by male elements. *n.* **parthenogenesis.**

parthenogonidia (*Zool.*). In some *Mastigophora*, agamonts which multiply by fission to form daughter colonies.

parthenospore (*Bot.*). A spore formed without a previous sexual act.

parthenote (*Bot., Zool.*). An individual developed from an egg containing only 1 nucleus, which is haploid.

partial (*Acous.*). Any one of the single-frequency components of a complex tone; in most musical complex tones, the partials have harmonic frequencies with respect to a fundamental.

partial capacitance (*Elec.*). Where there are a number of conductors (including earth) there is partial capacitance between all pairs, the effective capacitance under specified conditions being calculated from this network. See **capacitance coefficients.**

partial common (*Teleph.*). A trunk which is common to some of the groups in a grading unit.

partial compatibility (*TV*). System of TV transmission for which both monochrome and colour reception are possible but in which the chrominance information is outside the black-and-white video band of frequencies.

partial differential coefficient (*Maths.*). If $z=f(x, y, \text{etc.})$, then the partial differential coefficient of z with respect to x is the limit

$$\underset{h\to o}{Lt}\ \frac{f(x+h, y, \text{etc.})-f(x, y, \text{etc.})}{h},$$

if it exists: written $\partial z/\partial x$, or $f'_x(x, y, \text{etc.})$.

partial drought (*Meteor.*). See **drought.**

partial earth (*Elec. Eng.*). An earth fault having an appreciable resistance.

partial habitat (*Bot.*). The habitat occupied by a plant during one stage in its life history.

partial larva (*Zool.*). A parthenogenetic larval Amphibian produced by artificial stimulation of the ovum without the agency of a spermatozoon.

partial parasite (*Bot.*). A plant which has at least some power of photosynthesis, but obtains some material, mainly mineral salts and carbohydrates, from a host.

partial pressures (*Chem.*). The pressure exerted by each component in a gas mixture. See **Dalton's law of partial pressures.**

partial pressure suit (*Aero.*). A laced airtight overall for aircrew members in very high-flying aircraft. It has inflatable cells to provide the wearer with an atmosphere and external body pressure in the event of cabin pressure failure. Essential for survival above 50 000 ft (15 000 m).

partial pyritic smelting (*Met.*). Blast-furnace smelting of copper ores in which some of the heat is provided by oxidation of iron sulphide and some by combustion of coke. See **pyritic smelting.**

partial roasting (*Met.*). Roasting carried out to eliminate some but not all of the sulphur in an ore or sulphidic concentrate.

partial umbel (*Bot.*). One of the smaller group of flowers which altogether make up a compound umbel.

partial valencies (*Chem.*). The residual affinity which, according to Thiele's theory, still prevails in double bonds.

partial veil (*Bot.*). A hyphal weft or membrane joining the edge of the pileus to the stipe, and on rupture leaving an annulus.

particle (*Min. Proc.*). Single piece of solid material, usually defined (when small) by its mesh, or size passing through a specified size of sieve. (*Phys.*) Volume of air or fluid which has dimensions very small in comparison with the wavelength of a propagating sound wave, but large in comparison with molecular dimensions. See also **elementary particle.**

particle mean size (*Powder Tech.*). The dimension of a hypothetical particle such that, if a powder were wholly composed of such particles, such a powder would have the same value as the actual powder in respect of some stated property.

particle physics (*Phys.*). Study of the properties of elementary particles some of which have very high energy and very brief existence.

particle porosity (*Powder Tech.*). The ratio of the volume of open pores to the total volume of the particle.

particle scattering (*Phys.*). See **scattering.**

particle size (*Powder Tech.*). The magnitude of some physical dimension of a particle of powder. When the particle is a sphere, it is possible to define the size uniquely by a unit of length. For a nonspherical particle, it is not possible to define a size without specifying the method of measurement.

particle-size analysis (*Powder Tech.*). The study of methods for determining the physical structure of powdered materials.

particle velocity (*Acous.*). In a progressive or standing sound wave, the longitudinal alternating velocity of the air or fluid particles, taken either as the maximum or r.m.s. velocity.

particular average (*Ships*). Loss or damage to marine property whose cost is borne by the owner (unless insured against it).

particular integral (*Maths.*). Generally, a solution of a differential equation formed by assigning values to the arbitrary constants in the complete primitive. Nonsingular solution of a differential equation containing no arbitrary constants.

particulate inheritance (*Gen.*). Inheritance, in one individual, of distinctive characteristics of both parents.

parting (*Met.*). Quartation; the process of removing silver from gold-silver bullion. The composition is adjusted so that the silver content is at least 3 times the gold content, the silver being then dissolved by nitric acid. Silver can also be dissolved away in concentrated sulphuric acid, or converted to chloride by bubbling chlorine through molten bullion during ingot refining.

parting bead (*Join.*). A bead fixed to the cased frame of a double-hung window to separate the inner and outer sashes.

parting sand (*Build.*). A layer of dry sand separating 2 layers of damp sand, which are thereby prevented from adhering to one another. (*Foundry*) Dry sand sprinkled on the parting face of a mould to prevent adhesion of the 2 surfaces at the joint when the cope is rammed.

parting slip (*Join.*). A thin lath of wood or zinc which keeps the sash-weights apart within the cased frame of a double-hung window. Also called **mid-feather, wagtail.**

partite (*Zool.*). Split almost to the base.

partition (*Bot.*). See **dissepiment (1).**

partition chromatography (*Chem.*). See **chromatography.**

partition coefficient (*Chem.*). The ratio of the equilibrium concentrations of a substance dissolved in two immiscible solvents. If no chemical interaction occurs, it is independent of the actual values of the concentrations. Also **distribution coefficient.**

partition noise (*Electronics*). That arising when electrons are abstracted from a stream by a number of successive electrodes, as in a travelling-wave tube.

partition plate (*Carp.*). The upper horizontal member of a wooden partition, capping the studding and providing support for joists, etc.

partridge disease (*Vet.*). A popular term for infection of the bowels of partridges by the nematode worm *Trichostrongylus tenuis.*

parturient (*Med.*). Of or pertaining to parturition.

parturient fever, parturient eclampsia, parturient paresis (*Vet.*). See **milk fever.**

parturition (*Zool.*). In viviparous animals, the act of bringing forth young.

parvifoliate (*Bot.*). Having leaves which are small in relation to the size of the stem.

PAS (*Pharm.*). 4-aminosalicylic acid. Used medically, usually in the form of the sodium salt, in the treatment of tuberculosis. Usually administered in conjunction with *isoniazid* (q.v.).

pascal. The SI derived unit of pressure or stress, equals 1 newton per square metre, abbrev. **Pa.**

Pascal's theorem (*Maths.*). That the intersections of pairs of opposite sides of a hexagon inscribed in a conic are collinear. Cf. *Brianchon's theorem.*

Pascal's triangle (*Maths.*). An easily remembered summary of the coefficients in the expansion of $(1+x)^n$ for $n=0, 1, 2, 3 \ldots$ Each row relates to a particular value of *n*, and each digit is the sum of the two above and on either side of it. The triangle, up to $n=5$, is as follows:

```
          1
        1   1
      1   2   1
    1   3   3   1
  1   4   6   4   1
1   5  10  10   5   1
```

Paschen-Back effect (*Phys.*). Extension of *Zeeman effect* (q.v.) obtained with stronger fields which decouple the nuclear and orbital angular momenta. It leads to greater complexity of hyperfine structure.

Paschen circle (*Optics*). A type of mounting in a concave-grating spectrograph which employs only a small part of the Rowland circle, the slit, grating, and plate being fixed in position.

Paschen series (*Phys.*). A series of lines in the infrared spectrum of hydrogen. Their wave numbers are given by the same expression as that for the Brackett series, but with $n_1=3$.

Paschen's law (*Elec.*). That breakdown voltage, at constant temperature, is a function only of the product of the gas pressure and the distance between parallel-plate electrodes.

passage beds (*Geol.*). The general name given to strata laid down during a period of transition from one set of geographical conditions to another; e.g., the Downtonian Stage consists of strata intermediate in character (and in position) between the marine Silurian rocks below and the continental Old Red Sandstone above.

passage cell (*Bot.*). A thin-walled nonsuberized cell in an endodermis or an exodermis, through which solutions can diffuse in a transverse direction. Also **transfusion cell.**

pass band (*Telecomm.*). See **filter transmission band.**

Passeriformes (*Zool.*). An order of *Neognathae* containing the perching birds which comprise about half the known species of bird. Mostly small, living near the ground and having 4 toes arranged to allow gripping of the perch. Young helpless at hatching. Rooks, Finches, Sparrows, Tits, Warblers, Robins, Wrens, Swallows, and many others.

passing hollow (*Horol.*). See **crescent.**

passing place (*Civ. Eng.*). (1) On single-track roads a lateral extension of the roadway to accommodate a vehicle while other traffic passes. (2) A railway siding.

passings (*Plumb.*). The overlap of one sheet of lead past another in flashings, etc.

passivate (*Met.*). To mask the normal electro-potential of a metal. A treatment to give greater resistance to corrosion in which the protection is afforded by surface coatings of films of oxides, phosphates, etc. In the case of oils, passivation results from the stabilizing action of additives, preventing the oil from becoming more acidic.

passive (*Telecomm.*). Said of transducers or filter sections without an effective source e.m.f.

passive electrode (*Elec. Eng.*). The earthed electrode of an electrical precipitation apparatus, being that upon which the particles are deposited. Also called **collecting electrode.**

passive homing guidance (*Aero.*). A missile guidance system which homes on to radiation (e.g., infra-red) from the target.

passive immunity (*Bacteriol.*). Natural passive immunity is the temporary acquired immunity which a young animal acquires by passage of antibodies across the placenta or in the colostrum. This short-term immunity can be artificially induced by administering antibodies.

passive network (*Telecomm.*). One of electrical elements in which there is no e.m.f. or other source of energy.

passive radar (*Radar*). That using microwaves or infrared radiation emitted from source, and hence not revealing the presence or position of the detecting system. Military use.

passive satellite (*Space*). See **active satellite.**

passivity (*Met., Min. Proc.*). Lack of response of metal or mineral surface to chemical attack such as would take place with a clean, newly exposed surface. Due to various causes, including insoluble film produced by ageing, oxidation, or contamination; run-down of surface energy at discontinuity lattices; adsorbed layers. Phenomenon prevents use of cyanide process to dissolve large particles of gold, but is sometimes used to aid froth-flotation by rendering specific minerals passive to collector agents.

pass-over offset (*Plumb.*). The local bend which is given to a pipe so that the latter may pass over another pipe in cases where the axes of the pipes are in the same plane.
pass sheet (*Print.*). The sheet which has been signed by overseer and proofreader and serves as a standard, particularly for colour and register.
paste (*Plastics*). Usually applied to fluid dispersion of PVC in a plasticizer used for a variety of dipping processes such as glove making, leather cloth, and soft toys. (*Print.*) (1) Operating an automatic paster to replace the old reel with a new. (2) Securing two webs together during the run of the press. See **join.**
pasteboards (*Paper*). See **cardboards.**
pasted filament (*Elec. Eng.*). An electric-lamp filament prepared by squirting through a die a paste formed of powdered metal, usually tungsten, together with a binding material, the latter being subsequently removed by heat treatment.
pasted plate (*Elec. Eng.*). A plate in which the active material of the plate is applied in the form of a paste; used for lead-acid accumulators.
paste mould (*Glass*). A mould for blowing light-walled hollow-ware. As a good finish is needed, the moulds are coated with adherent carbon ('paste'), which is wetted before each blowing operation.
paster tab (*Print.*). A strip of gummed paper with a weakened section or other device used to secure the prepared end of the new reel prior to joining it to the old reel on an *automatic reel change.*
pasteurellosis (*Vet.*). The group name for diseases caused by organisms of the genus *Pasteurella. Pasteurella multocida* causes an acute septicaemic and pneumonic disease in several species of animals. See **haemorrhagic septicaemia** (cattle and sheep), **swine plague** (pig), **fowl cholera** (fowl), **rabbit septicaemia** (rabbit).
Pasteur filter (*Chem.*). See **ceramic filter.**
pasteurization (*Med.*). Reduction of the number of microorganisms in milk by maintaining it in a holder at a temperature of from 62·8°–65·5°C for 30 minutes.
patagium (*Zool.*). A lobelike structure at the side of the pronotum in some *Lepidoptera*: in Bats and some other flying Mammals, a stretch of webbing between the fore limb and the hind limb: in Birds, a membranous expansion of the wing. *adj.* **patagial.**
patand (*Build.*). A sill resting on the ground as a support for a post.
Patapsco formation (*Geol.*). The highest division of the Comanchean of the eastern U.S., consisting of brightly coloured, often sandy, clays with fossil plants.
patch (*Telecomm.*). To join together units of apparatus, such as amplifiers, equalizers, etc., by flexible cords terminated on plugs, which are inserted into break-jacks, bridged across terminations of each unit.
patch bay (*Telecomm.*). Section of rack-mounted equipment which includes all break-jacks which terminate units of equipment.
patch board (*Comp.*). One on which problems can be set up and modified as required in analogue computing. See **patch.**
patch cord (*Comp.*). Flexible conductor used in setting up problems on a patch board.
patch in or out (*Telecomm.*). Temporary insertion of spare apparatus, or removal of defective apparatus, in a circuit by patch cords, usually in a patch bay or field.
patching up (*Print.*). (1) In letterpress, as part of *making-ready* (q.v.), applying patches of thin paper which will be included in the *packing* (q.v.) to improve the impression where required. (2) In lithography, the arranging of transfers, retransfers, positives, negatives, in required position on a suitable base, preliminary to making a plate.
patch test (*Med.*). A test for supersensitivity, consisting of the application to the skin of small pads soaked with the allergy-producing substance; if supersensitivity exists inflammation develops at the places where the substance was applied.
patella (*Bot.*). A sessile apothecium, saucerlike, and with a distinct margin. (*Zool.*) (1) Name of a genus of gastropod mollusc, including the limpet. (2) In higher Vertebrates, a sesamoid bone of the knee joint or elbow joint.
patellate, patelliform (*Bot.*). Shaped like a saucer or dish.
patent (*Bot.*). Said of leaves and branches which spread out widely from the stem.
patent glazing (*Build.*). The name applied to various marketed devices for securing together glass sheets (for roof coverings, etc.) without using putty in sashes, the connexion being made usually with special metal sections or flashings.
patent leather (*Leather*). An enamelled leather produced by spraying the leather with a cellulose lacquer.
patent log (*Ships*). See **nautical log,**
patent satin or mitcheline (*Textiles*). A jacquard woven quilting cloth composed of two very firmly interwoven plain cloths featuring an embossed pattern.
patera (*Arch.*). A circular ornament in relief on friezes. *pl.* **paterae.**
paternoster (*Eng.*). A lift for goods or passengers which consists of a series of floored but doorless compartments moving slowly and continuously up and down on an endless chain.
path (*Telecomm.*). Channel through which signals can be sent, particularly *forward, through*, and *feedback* paths of servo systems.
path attenuation (*Radio*). Fall-off in amplitude of a radio wave with distance from the transmitter.
pathetic muscle (*Zool.*). The superior oblique muscle of the Vertebrate eye.
pathetic nerve (*Zool.*). See **trochlear nerve.**
pathogen (*Biol.*). A parasite producing damage in its host. (*Med.*) Any disease-producing microorganism or substance.
pathogenesis (*Med.*). The development or production of a disease-process. *adj.* **pathogenic.**
pathognomonic (*Med.*). Specially indicating a particular disease.
pathological. Concerning pathology: morbid, diseased.
pathology. That part of medical science which deals with the causes and nature of disease, and with the bodily changes wrought by disease.
patin (*Build.*). See **patand.**
patina (*Chem.*). The thin, often multicoloured, film of atmospheric corrosion products formed on the surface of a metal or mineral.
patobionts (*Ecol.*). Animals which normally live all their lives on the forest floor. Cf. *patocoles, patoxenes.*
patocoles (*Ecol.*). Animals found on forest floors, but spending a regular part of their life elsewhere. Cf. *patobionts, patoxenes.*
patoxenes (*Ecol.*). Animals occurring accidentally on forest floors. Cf. *patobionts, patocoles.*
Patra Test Bench (*Print.*). See **Pira Test Bench.**

patroclinous (*Bot.*, *Zool.*). Exhibiting the characteristics of the male parent more prominently than those of the female parent. Cf. *matroclinous.*

patromorphic (*Bot.*). Resembling the father.

patten (*Arch.*). The base of a column or pillar.

patter (*Build.*). A kind of float, made of thick wood, used to consolidate and level cement surfaces.

pattern (*Electronics*). The luminous trace on the screen of a CRO as traced out by the electron beam. (*Foundry*) A wood, metal, or plaster copy, in one piece or in sections, of an object to be made by casting. Made slightly larger than the finished casting to allow for contraction of casting while cooling; and suitably tapered to facilitate withdrawal from the mould. See **double contraction, metal pattern, plate moulding.** (*Photog.*, *Print.*) A regular texture effect (*moiré*) formed by superimposing two or more sets of lines or dots of different pitch or at certain angles; a defect to be avoided, especially in half-tone reproduction of steel-engravings and in half-tone 4-colour work. (*Telecomm.*) Pattern of the radiation field from an aerial system as shown by a polar diagram of field strength and bearing.

pattern-maker's hammer (*Tools*). Light type of cross-pane hammer with long handle.

pattern-maker's rule (*Met.*). One with graduations lengthened so as to compensate for the cooling contraction which must be allowed for in the cast object.

pattern weaver (*Textiles*). A hand-loom weaver engaged in the work of producing section ranges for new-season styles.

Patterson-Cawood graticule (*Powder Tech.*). Eye-piece graticule used in microscopic methods of particle-size analysis. It has a rectangular grid and a series of graded circles for use in sizing the particles.

Pattinson's process (*Met.*). Obsolescent process used for the separation of small quantities of silver from lead by partially solidifying a molten bath of the two metals and separating the remaining liquid. This process is repeated several times and the silver is concentrated in the liquid.

Pattinson's white lead (*Chem.*). A pigment of a lead(II)oxychloride type.

patulin (*Pharm.*). Antibiotic obtained from *penicillium* moulds. White, crystalline powder, m.p. 110°C, also known as **clavacin.**

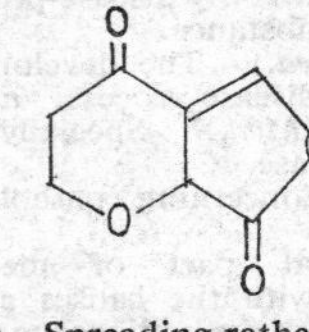

patulous (*Bot.*). Spreading rather widely.

paturon (*Zool.*). The first or basal segment of the chelicerae in *Arachnida.*

Pauli exclusion principle (*Nuc.*). Fundamental law of quantum mechanics that no two fermions can exist in identical quantum states.

paulopost alteration (*Geol.*). Alteration of igneous rocks by residual solutions derived from the magma. Synonymous with deuteric alteration.

paunch (*Zool.*). See **rumen.**

paurometabolic (*Zool.*). Among Insects, those undergoing slight metamorphosis, the young resembling the adults both in form and in mode of life; as in *Collembola.* Cf. *hemimetabolic, holometabolic.*

Pauropoda (*Zool.*). A class of small *Arthropoda*, the adults having 12 segments, 9 pairs of walking legs, branched antennae, no trachea, no eyes and a 1-segmented mandible, and being progoneate. Live under decaying leaves and logs, often in considerable numbers. Regarded by some as an order of *Myriapoda.*

pave (*Civ. Eng.*). To construct a *pavement* (q.v.).

pavement (*Civ. Eng.*). (1) Any surface that has been *paved*, i.e., systematically treated, according to soil circumstances, by bitumen, asphalt or concrete, to withstand expected traffic loads and weathering. (2) In Britain, a footway for pedestrians, whether paved, as above, or surfaced with *flags, pavings*, or *tiles*—or even if unpaved. U.S. sidewalk.

pavement epithelium (*Zool.*). A variety of epithelium consisting of layer of flattened cells.

pavement light (*Build.*). A panel formed of glass blocks framed in concrete, iron, or steel, built into a pavement surface over an opening to the basement of a building, into which it admits light.

pavilion (*Arch.*). An ornamental, detached structure which has a roof but is usually not entirely enclosed by walls. Used on sports fields, or as a place for entertainments.

pavilion roof (*Arch.*). A roof which in plan usually forms a figure of more than 4 straight sides.

paving flags or **stones** (*Civ. Eng.*). Square or rectangular slabs of natural stone or concrete used for the surfacing of *pavements.* They are usually about 600 mm (2 ft) square and 50 mm (2 in.) thick.

pavings (*Build.*). Very hard purpose-made bricks, usually of the dark blue Staffordshire variety, having a surface chequered by grooves to make it less slippery.

pavior (*Build.*). (1) A specially hard brick used in the construction of pavement surfaces. (2) A worker who lays bricks or setts to form pavement surfaces.

Pavy's solution (*Chem.*). A modified Fehling's solution containing sufficient ammonia solution to redissolve the copper(II)oxide.

pawl (*Eng.*). A pivoted catch, usually spring-controlled, engaging with a ratchet wheel or rack to prevent reverse motion, or to convert its own reciprocating motion into an intermittent rotary or linear motion.

Pawlovskys glands (*Zool.*). In *Anoplura*, a pair of glands opening into the stylet sac; believed to secrete a lubricating fluid.

Pawsey stub (*Telecomm.*). Quarter-wavelength coaxial line, added in parallel to end of coaxial line, to match this to a dipole element.

PAX (*Teleph.*). Abbrev. for *private automatic exchange.*

Paxboard or **Paxfelt** (*Build.*). TN for an insulating material.

paxilla (*Zool.*). An upright spine bearing 2 or 3 circles of horizontal spinelets, occurring in some *Asteroidea. pl.* **paxillae.**

Paxolin (*Plastics*). TN for a laminated plastic usually of the phenolic class: used in the manufacture of sheets, tubes, cylinders, and laminated mouldings.

pay (*Mining*). Pay dirt, ore, rock, streak. Any mineral deposit which will repay efficient exploitation.

pay-as-you-view (*TV*). See **toll television.**

payload (*Aero.*). That part of an aeroplane's load from which revenue is obtained. (*Space*) The useful or net weight (as distinct from the weight of the carrier rocket) which is placed into orbit in a space mission.

Pb (*Chem.*). The symbol for *lead.*

PBI (*Med.*). Abbrev. for *protein-bound iodine.*

PBX (*Teleph.*). Abbrev. for *private branch exchange.*

PCE (*Chem.*). See **pyrometric cone equivalent.**

PCNB (*Chem.*). See **quintozene.**

PCPBS (*Chem.*). See **fenson.**

pd or **p.d.** (*Elec.*). Abbrev. for *potential difference.*

Pd (*Chem.*). The symbol for *palladium.*

P display (*Radar*). That of a *PPI* unit. Map display produced by intensity modulation of a rotating radial sweep.

peacock ore (*Min.*). A name given to *bornite* or sometimes *chalcopyrite* (qq.v.), because they rapidly become iridescent from tarnish.

Pea Grit (*Geol.*). A bed of pisolitic limestone, containing shell fragments, found in the Lower Jurassic rocks of Gloucestershire. It is not a true grit.

peak (*Phys., etc.*). The *instantaneous* value of the local maximum of a varying quantity. It is √2 times the r.m.s. value for a sinusoid.

peak clipper (*Telecomm.*). Diode or similar device with sharp cut-off at predetermined level and no effect below this. Used, e.g., in ignition interference suppression and for removing unwanted amplitude modulation from signal using pulse-width modulation.

peak dose (*Radiol.*). Maximum absorbed radiation dose at any point in an irradiated body, usually at a small depth below the surface, due to secondary radiation effects.

peak envelope power (*Radio*). See **peak power.**

peak factor (*Elec. Eng.*). Ratio of peak value of any alternating current or voltage to r.m.s. value. Also **crest factor.**

peak forward voltage (*Elec. Eng.*). The maximum instantaneous voltage in the forward flow direction of anode current as measured between the anode and cathode of a thermionic rectifier, gas-filled tube, or semiconductor diode.

peaking (*Telecomm.*). Inclusion of series or shunt resonant units in, e.g., TV circuits, to maintain response up to a maximum frequency.

peaking circuit (*Telecomm.*). One which produces a sharply peaked output from any input wave. See also **peaking transformer.** Not to be confused with *peaking network.*

peaking network (*Telecomm.*). An interstage coupling circuit which gives a peak response at the upper end of the frequency range that is handled. This is achieved with a resonant circuit and minimizes the fall-off in the frequency response produced by stray capacitances. Not to be confused with *peaking circuit.*

peaking transformer (*Telecomm.*). One in which the core is highly saturated by current in the primary, thus providing a peaky e.m.f. in the secondary as the flux in the core suddenly changes over.

peak inverse voltage (*Elec. Eng.*). **See inverse voltage.**

peak joint (*Build.*). The joint between the members of a roof truss at its ridge.

peak limiter (*Telecomm.*). Circuit for avoiding overload of a system by reducing gain when the peak input signal reaches a certain value.

peak load (*Elec. Eng.*). The maximum instantaneous rate of power consumption in the load circuit. In a power-supply system, the peak load corresponds to the maximum power production of the generator(s).

peak power (*Radio*). Average radiofrequency power at maximum modulation, i.e., of envelope of transmission.

peak programme meter (*Radio*). One bridged across a transmission circuit to indicate the changes in *volume* of the ultimate reproduction of sound, averaging peaks over 1 millisecond.

peak sideband power (*Radio*). The average sideband power of a transmitter over one radiofrequency cycle at the highest peak of the modulation envelope.

peak-to-peak amplitude (*Elec.*). See **double amplitude.**

peak value (*Elec., etc.*). The maximum positive or negative value of an alternating quantity. Also called **amplitude, crest value, maximum value.**

peak voltmeter (*Elec. Eng.*). One for measuring the peak value of an alternating voltage, e.g., a diode which is biased so that it just conducts. Also called **crest voltmeter.**

peak white (*TV*). The level in a TV signal corresponding to white.

pean (*Tools*). See **pane.**

pearl (*Typog.*). An old type size, approximately 5-point. (*Zool.*) An abnormal concretion of nacre formed inside a mollusc shell round a foreign body such as a sand particle or a parasite.

pearlite (*Met.*). A microconstituent of steel and cast iron. It is produced at the eutectoid point by the simultaneous formation of ferrite and cementite from austenite, and normally consists of alternate plates of these constituents (see, however, **granular pearlite**). A carbon steel containing 0·9% of carbon consists entirely of pearlite. See **eutectoid steel, hypo-eutectoid steel, hyper-eutectoid steel.**

pearlitic iron, Perlite Iron (*Met.*). In general, *pearlitic iron* is grey cast-iron consisting of graphite in a matrix of pearlite, i.e., without free ferrite. In particular, *Perlite Iron* is a German proprietary name denoting an iron of low silicon content, which is caused to solidify grey by the use of heated moulds.

pearl lamp (*Light*). See **inside-frosted lamp.**

pearl spar (*Min.*). See **dolomite.**

pearl white (*Chem.*). See **bismuth trichloride.**

pear oil (*Chem.*). See under **amyl acetate.**

pear-push (*Elec. Eng.*). See **pendant push.**

Pearson's coefficient (*Stats.*). See **correlation coefficient.**

pear-switch (*Elec. Eng.*). See **pendant switch.**

peas (*Mining*). See **coal sizes.**

peat (*Geol.*). The name given to the layers of dead vegetation, in varying degrees of alteration, resulting from the accumulation of the remains of marsh vegetation in swampy hollows in cold and temperate regions. Geologically, peat may be regarded as the youngest member of the series of coals of different rank, including brown coal, lignite, and bituminous coal, which link peat with anthracite. Peat is very widely used as a fuel, after being air-dried, in districts where other fuels are scarce and in some areas, e.g., in Russia and Ireland, it is used to fire power stations. It is low in ash, but contains a high percentage of moisture, and is bulky; specific energy content about 16 MJ/kg or 7000 Btu/lb.

peav(e)y or **peavie** (*For.*). A stout wooden lever fitted into a metal socket terminating in a spike, carrying towards its upper end a hinged steel hook; used in handling and turning logs.

pea viner (*Agric.*). A stationary (or mobile) machine to shell green peas on the vine.

pebble-bed reactor (*Nuc. Eng.*). One with cylindrical core into which fuel pellets are introduced at the top and extracted at the base.

pebble-dashing (*Build.*). A rough finish given to a wall by coating it with plaster, on to which, while it is still soft, small stones and liquid lime are thrown.

pebble mill (*Min. Proc.*). *Ball mill* (q.v). in which selected pebbles or large pieces of ore are used as grinding media.

Pebidian (*Geol.*). The name given to a series of igneous rocks, of Pre-Cambrian age, found in Pembrokeshire.

PEC (*Electronics*). Abbrev. for *photoelectric cell.*

pecilokont (*Cyt.*). The flagellum and all homologous organelles of higher organisms.

peckings (*Build.*). Under-burnt, badly-shaped bricks, used only for temporary work or for the inside of walls.

peck order (*An. Behav.*). A *social hierarchy* (q.v.) among small groups of some birds, in which the top-ranking bird will peck, or show aggressive behaviour towards, all the others, the second will behave thus to all but the first, and the last is pecked by all the others.

pecky (*For.*). Showing signs of decay.

Péclet number (*Chem. Eng., etc.*). A dimensionless ratio used for assessing the similarity of motion in heat flow, given by

$$\rho\frac{cul}{\lambda}$$

where ρ = density, c = sp. heat cap., u = velocity, l = length, λ = thermal conductivity. Péclet number = *Reynolds' number* (q.v.) × *Prandtl number* (q.v.).

pectase (*Biochem., Bot.*). An enzyme found in plants which forms fruit and vegetable gels.

pecten (*Zool.*). Any comblike structure; in some Vertebrates (Reptiles and Birds), a vascular process of the inner surface of the retina, possibly concerned with the detection of the movement of small objects, or the adjustment of pressure inside the eyeball when the lens changes shape during accommodation: in *Scorpionidea*, tactile sensory organs under the mesosoma.

pectinate (*Bot., Zool.*). Comblike; said especially of a pinnatifid leaf having many narrow lateral lobes.

pectineal (*Zool.*). Comblike; said: (1) of a process of the pubis in Birds; (2) of a ridge on the femur to which is attached the *pectineus muscle*, one of the protractors of the hind limb.

pectinellae (*Zool.*). In some *Mastigophora*, transversely planted comblike membranellae representing the adoral spiral of cilia.

pectines (*Zool.*). Comblike chitinous structures of tactile function attached to the ventral surface of the second somite of the mesosoma in *Scorpionidea.*

pectins (*Chem.*). Calcium-magnesium salts of polygalacturonic acid, partially joined to methanol residues by ether linkage. They occur in the cell walls of fruits and vegetables, especially in the middle lamella. They are soluble in water and can be precipitated from aqueous solutions by excess alcohol. Acid solutions gel with 65–70% of sucrose—the basis of their use in jam making.

pectization (*Chem.*). The formation of a jelly.

pectolite (*Min.*). A silicate of calcium and sodium, with a variable amount of water, which crystallizes in the triclinic system. It occurs in aggregations like the zeolites in the cavities of basic eruptive rocks, and as a primary mineral in some alkaline igneous rocks.

pectorales (*Zool.*). In Vertebrates, muscles connecting the upper part of the fore-limb with the ventral part of the pectoral girdle. *sing.* **pectoralis.**

pectoral fins (*Zool.*). In Fish, the anterior pair of fins.

pectoral girdle (*Zool.*). In Vertebrates, the skeletal framework with which the anterior pair of locomotor appendages articulate.

pectoriloquy (*Med.*). Conduction, to the chest wall, of the sound of words spoken or whispered by the patient and clearly heard through the stethoscope; indicative of consolidation or cavitation of the lung.

pectose (*Biochem., Bot.*). A carbohydrate constituent of plants which is converted into pectin by the enzyme *pectosinase.*

pectus (*Zool.*). A thoracic sclerite of Insects, formed by the fusion of the pleuron with the sternum; in Vertebrates, the breast region. *adj.* **pectoral.**

peculiar (*Typog.*). A term describing any unusual type character, such as certain accents.

pedal curve (*Maths.*). Of a given curve with respect to a given point: the locus of the point of intersection of a tangent to the curve and the perpendicular to the tangent from the given point.

pedal feed motion (*Textiles*). Motion (linked pedals and cone pulleys and belts) which regulates feeding of cotton sheet to the porcupine cylinder in an opener or the bladed beater on a scutcher.

Pedaliaceae (*Bot.*). A family of tubifloral dicotyledons akin to the bignonias.

pedate leaf (*Bot.*). A palmately divided compound leaf having three main divisions, and having the two outer divisions forked one or more times.

pedatifid (*Bot.*). Having the lamina deeply cut in pedate fashion.

pedes (*Zool.*). See **pes.**

pedesis (*Phys.*). *Brownian movement* (q.v.).

pedestal (*TV*). Level between peak of the synchronizing signals and minimum of video signal, in the total TV signal.

pedestal sit (*TV*). Colloquialism for separation in TV picture signal level between black level (in a positive signal) and blanking level.

pediatric, etc. (*Med.*). See **paediatric, etc.**

pedicel (*Bot.*). The stalk which bears a single flower or a single fruit. (*Zool.*) (1) The second joint of the antennae in Insects; more generally, the stalk of a sedentary organism; the stalk of a free organ, as the *optic pedicel* in some *Crustacea.* (2) A tubular prolongation of some Insect eggs. (3) The narrow necklike zone formed by the wholly or partly constricted second abdominal segment in some *Hymenoptera.*

pedicellaria (*Zool.*). In *Echinodermata*, a small pincerlike calcareous structure, consisting of a basal plate with which are articulated 2 or 3 jaws provided with special muscles and capable of executing snapping movements; it may be stalked or sessile.

pedicellate (*Bot.*). Said of a flower or a fruit which has a stalk. (*Zool.*) That which is provided with a pedicel.

pedicle (*Zool.*). (1) In the vertebrae of the Frog, a pillarlike process springing from the centrum and extending vertically upwards to join the flat, nearly horizontal lamina, which forms the roof of the neural canal. Intervertebral foramina, for the passage of the spinal nerves, occur between successive pedicles. (2) In the skull of the Frog, one branch of the forked proximal end of the posterior quadrate region of the palatoquadrate, which is fused with the trabecular region just in front of the auditory capsule. (3) In some *Arachnida*, a narrow stalk uniting the prosoma and opisthosoma.

pediculosis (*Med.*). Infestation of the body with lice.

pediment (*Arch.*). A triangular or segmental part surmounting the portico in the front of a building.

pedion (*Crystal.*). A crystal form consisting of a single plane; well shown by some crystals of tourmaline which may be terminated by a pedion, with or without pyramid faces.

pedipalps (*Zool.*). In *Arachnida*, the appendages, borne by the first postoral somite, of which the gnathobases function as jaws; they may be tactile organs or chelate weapons.

pedology. The study of soil.

pedometer (*Surv.*). An instrument for counting the number of paces (and hence the approximate distance) walked by its wearer.

peduncle (*Bot.*). The main stalk or stalks of an inflorescence. (*Zool.*) In *Brachiopoda* and *Cirripedia*, the stalk by which the body of the animal is attached to the substratum; in some *Arthropoda*, the narrow portion joining the thorax and abdomen or the prosoma and opisthosoma; in Vertebrates, a tract of white fibres in the brain.

pedunculate (*Bot.*). Having, or borne on, a peduncle.

peeling (*Chem.*). Undesired detachment of an electrodeposition; cf. *stripping*. (*Typog.*) The operation of preparing *overlays* (q.v.) by thinning the hard edges of an illustration.

peen (*Tools*). See **pane.**

peening (*Met.*). Using hammer blows or shot blasting to cold-work metal.

Peep (*Aero.*). TN for electronic air navigation device which projects flight data on to the pilot's windscreen or other screen at eye level. (*P*ilot's *e*lectronic *e*ye-level *p*resentation). Now superseded by *head-up display* (q.v.).

peg (*Bot.*). An outgrowth from the hypocotyl of seedlings of cucumbers and related plants. It plays some part in assisting the seedling to emerge from the testa. (*Teleph.*) A device without external connexions for conductors, which is designed to be placed in a jack to prevent the insertion of a plug.

peg-and-cup dowels (*Eng.*). Metal pegs and sleeves inserted in adjoining parts of a split pattern to hold them in register while ramming the mould.

peggies (*Build.*). Slates 10–14 in. (254–356 mm) in length.

pegmatite (*Geol.*). A term originally applied to granitic rocks characterized by intergrowths of feldspar and quartz, as in graphic granite; now applied to igneous rocks of any composition but of particularly coarse grain, occurring as offshoots from, or veins in, larger intrusive rock bodies, representing a flux-rich residuum of the original magma.

pegmatite phase (*Geol.*). That chapter in the cooling history of a body of magma dealing with the production of liquid residua in which fluxes are concentrated, giving a low viscosity to this late fraction of the magma, and hence great penetrative power and abnormally large crystals.

peg plan (*Weaving*). Plan on squared paper indicating the order in which the healds are to be lifted during weaving. Also called **lifting plan.**

pegtop paving (*Civ. Eng.*). Paving of setts each of which is small in visible area.

pegwood (*Horol.*). Sticks of close-grained wood used for cleaning out the pivot holes of clocks and watches.

pein (*Tools*). See **pane.**

pelagic (*Geol.*). A term applied to any accumulation of sediments under deep-water conditions. (*Zool.*) Living in the middle depths and surface waters of the sea.

Pelagothurida (*Zool.*). An order of *Holothuroidea* having shield-shaped buccal tube feet with enormous ampullae but no retractor muscles, and no tube feet on the trunk; respiratory trees are lacking; the madreporite is external; pelagic free-swimming forms.

pelargonic acid (*Chem.*). Nonan1-oic acid. $CH_3\cdot(CH_2)_7\cdot COOH$. M.p. 12·5°C. Oxidation product of oleic acid. Occurs naturally in fusel oil from beet and potatoes.

Pelecaniformes (*Zool.*). A varied order of *Neognathae*, mainly fish-eating and colonial nesters, with 4-toed webbed feet, bodies adapted for diving, and long beaks with wide gapes and sometimes with a pouch. Pelicans, Cormorants, Gannets.

P-electron (*Nuc.*). One in the P-shell surrounding a nucleus. There is one such electron in the caesium atom and ten in thorium, but the shell is not completed in existing elements.

Pelecypoda (*Zool.*). A synonym for **Lamellibranchiata.**

Pélé's hair (*Min.*). Long threads of volcanic glass which result from jets of lava being blown aside by the wind in the volcano of Kilauea, Hawaii.

pelidisi index (*Nut.*). Index obtained by dividing the sitting height in cm into the cube root of 10 times the weight in grams.

peliosis rheumatica (*Med.*). Purpura rheumatica; Schönlein's disease. Characterized by swelling of the joints and the appearance of purpuric spots and of weals (urticaria) in the skin.

pelitic gneiss (*Geol.*). A gneissose rock derived from the metamorphism of argillaceous sediments.

pelitic schist (*Geol.*). A schist of sedimentary origin, formed by the dynamothermal metamorphism of argillaceous sediments such as clay and shale.

pellagra (*Med.*). Maidismus. A chronic disease due to general dietary deficiency; it is characterized by gastro-intestinal disturbances, a symmetrical erythema of the skin, mental depression, and paralysis. Treatment is by the administration of vitamins, mainly B group.

pellet (*Build.*). A term applied to a moulding characterized by a series of spherical protuberances. (*Plastics*) A compressed mass of moulding material of prescribed form and weight.

pelletization (*Met., Min. Proc.*). Treatment of finely divided ore, concentrate or coal to form aggregates some 1–1·5 cm in diameter, for furnace feed, transport, storage or use (e.g., as coal briquettes). Powder, with suitable additives, is rolled into aggregates (green balls) in pelletizing drum and then hardened in a furnace by specialized baking methods.

pellicle (*Bot.*). The outer layer of the upper surface of a pileus, when it can be stripped off as a delicate membrane. (*Nuc. Eng.*) A strippable photographic emulsion used to form a *stack* in nuclear emulsion techniques. (*Photog.*) A thin film, particularly of an emulsion when dried. (*Zool.*) A thin cuticular investment, as in some *Protozoa*. *adj.* **pelliculate.**

Pellin-Broca prism (*Optics*). A 4-sided prism, which can be imagined to be formed by placing together two 30° prisms and one 45° prism. It is used in wavelength spectrometers where the wavelength of the light received in the telescope is altered by rotation of the prism.

pelma (*Zool.*). See **planta.**

Pelmatozoa (*Zool.*). A subphylum of *Echinodermata* comprising stalked forms of sessile habit; the *Crinoidea* are the only living representatives of the group. Cf. *Eleutherozoa.*

pelophile (*Bot.*). A plant which occurs on clayey soils.

peloria, pelory (*Bot.*). An abnormal condition in which the flowers of a species normally producing irregular flowers are regular.

pelorus (*Nav., etc.*). A pivoted dial, marked in compass points or degrees which is on a stand with a *lubber line* (q.v.) and fitted with sight vanes. It is set manually to match the compass card and used to take bearings.

peloton (*Bot.*). A specialized coiled hypha occurring in mycorrhizal species of fungi, in the cortical cells of the root of the host.

peltate (*Bot.*). Said of any plant member which is more or less flattened, and has its stalk attached to the middle of the lower surface.

Peltier coefficient (*Elec. Eng.*). Energy absorbed or given out per second, due to the Peltier effect, when unit current is passed through a junction of 2 dissimilar metals. Symbol Π.

Peltier effect (*Elec. Eng.*). Phenomenon whereby heat is liberated or absorbed at a junction when current passes from one metal to another.

Pelton wheel (*Eng.*). An impulse water turbine in which specially shaped buckets attached to the periphery of a wheel are struck by a jet of water, the nozzle being either deflected or valve-controlled by a governor.

pelt wool (*Textiles*). Wool removed by the fellmonger from the pelts of slaughtered sheep. See *skin wool.*

pelvic fins (*Zool.*). In Fish, the posterior pair of fins.

pelvic girdle (*Zool.*). In Vertebrates, the skeletal framework with which the posterior pair of locomotor appendages articulate.

pelvimeter (*Med.*). A specially designed calliper for measuring distances among various bony points of the female pelvis.

pelvimetry (*Med.*). Estimation, by the use of pelvimeters or of X-rays, of the size and shape of the female pelvis.

pelvis (*Zool.*). The pelvic girdle or posterior limb girdle of Vertebrates, a skeletal frame with which the hind limbs or fins articulate; in Mammals, a cavity, just inside the hilum of the kidney, into which the uriniferous tubules discharge and which is drained by the ureter. *adj.* **pelvic.**

pelvisternum (*Zool.*). In many *Urodela* and *Xenopus*, a rod of cartilage forked anteriorly and attached to the anterior border of the pubic region. Named thus because it develops independently of the pelvis, and bears the same relation to it as the sternum does to the pectoral girdle. Also known as the **epipubis.**

pemphigus (*Med.*). Inflammatory condition of the skin characterized by the eruption of crops of blisters, the mucous membranes at times also being involved.

pen (*Zool.*). In *Cephalopoda*, the shell or cuttle bone.

pencatite (*Geol.*). A crystalline limestone which contains brucite and calcite in approximately equal molecular proportions. See also **predazzite.**

pencil (*Light*). A narrow beam of light, having a small angle of convergence or divergence. (*Maths.*) Of lines: a number of lines passing through a fixed point (*vertex*). Of planes: a number of planes having a common line.

pencilling (*Paint.*). The operation of painting the mortar joints of a brick wall with white paint so as to accentuate the contrast of colours.

pencil mark (*For.*). Fungal discoloration that appears on the end surface of certain timbers as streaks, not unlike short pencil lines.

pencil stone (*Min.*). The name given to the compact variety of pyrophyllite, used for slate-pencils. The term *pencil ore* has been used for the broken splinters of radiating massive haematite, as they give a red streak.

pendant (*Build.*). Ornamentation suspended below an object or surface to be decorated, as from a ceiling. (*Horol.*) The neck of a watch-band, to which the bow is attached. (*Light*) A lighting fitting suspended by means of a flexible support.

pendant push (*Elec. Eng.*). A pushbutton arranged for attachment to a flexible cord. Also called **pear-push, suspension-push.**

pendant socket-outlet (*Elec. Eng.*). A socket-outlet arranged for attachment to a flexible cord.

pendant switch (*Elec. Eng.*). A switch arranged for attachment to a flexible cord. Also called **pressel-switch, pear-switch, suspension-switch.**

pendentive. A spherical triangle formed by a dome springing from a square base.

pendentive dome. A dome covering a square area to which it is linked at the corners by pendentives.

Pendleside Series (*Geol.*). A series of dark shales and limestones found in North Lancashire; they belong to the uppermost Lower Carboniferous Series; well seen in Pendle Hill.

pendulous (*Bot.*). Said of an ovule which is suspended from a point at or near the top of the ovary.

pendulum. The simple pendulum consists of a small, heavy bob suspended from a fixed point by a thread of negligible weight. Such a pendulum, when swinging freely with small amplitude, has a periodic time given by $T=2\pi\sqrt{\frac{l}{g}}$, where l is the length of the thread and g is the acceleration due to gravity. See **centre of oscillation, compound-.** (*Horol.*) The time-controlling element of a pendulum clock. The theoretical length of a pendulum, in mm, is given by $L=993{\cdot}6\times t^2$, where t is the time of swing in seconds and 993·6 mm is the length of a pendulum beating seconds in London. For household clocks, the pendulum beats 0·5 s or less. For long-case clocks and regulators, a seconds pendulum is used; for tower clocks, it may be up to 2 seconds.

pendulum apparatus (*Geophys.*). Instrument used to determine the absolute value of gravity at a point by measuring the period of a swinging pendulum. It is used for establishing datum stations rather than for survey work.

pendulum bob (*Horol.*). The weighted mass at the end of a pendulum.

pendulum damper (*Aero.*). A short heavy pendulum, in the form of pivoted balance weights, attached to the crank of a radial aero-engine in order to neutralize the fundamental torque impulses and so eliminate the associated critical speed.

pendulum governor (*Eng.*). An engine governor, the many forms of which involve the principle of the conical pendulum. Heavy balls swing outwards under centrifugal force, so lifting a weighted sleeve and progressively closing the engine throttle valve. See **Porter governor, Watt governor.**

pendulum rod (*Horol.*). The rod of a pendulum which supports the bob.

pendulum spring (*Horol.*). The thin ribbon of spring steel used for suspending the pendulum.

pendulum therapy (*Radiol.*). Form of *moving-field therapy* in which the source of radiation swings to and fro in arc concave to patient.

penecontemporaneous (*Geol.*). Describes any process occurring in a sedimentary rock very soon after its deposition, for example the dolomitization of a carbonate sediment.

peneplain (*Geol.*). A gently rolling lowland, produced after long-continued denudation.

penetrance (*Gen.*). The index, expressed as a percentage, of the frequency with which a gene exhibits an effect.

penetrant (*Chem.*). (1) Substance which increases the penetration of a liquid into porous material or between contiguous surfaces, e.g., alkyl-aryl sulphonate. (2) A **wetting agent.**

penetrating shower (*Nuc.*). Cosmic-ray shower containing mesons and/or other penetrating particles. See **cascade shower.**

penetration (*Chem.*). A term used in testing bituminous material. Penetration is expressed as the distance that a standard needle vertically penetrates a sample of the material under known conditions of loading, time, and temperature. (*Elec. Eng.*) Measure of depth of *skin effect* of eddy currents in induction heating, or depth of magnetic field in superconducting metals. Usually to $1/e$ (0·37) of surface value.

penetration depth (*Radio*). That thickness of hollow conductor of the same dimension which, if the current were uniformly distributed throughout the cross-section, would have the same effective resistance as the solid conductor.

penetration factor (*Electronics*). The reciprocal of the amplification factor. Given by the ratio of change of grid voltage to change of anode voltage. (*Nuc.*) Probability of incident particle passing through nuclear potential barrier.

penetration theory (*Chem. Eng.*). Theory that mass transfer across an interphase into a stirred liquid takes place by diffusive penetration of the solute into the liquid surface, which is continually being renewed, hence the rate of mass transfer is proportional to the square root of the *diffusion coefficient*.

penetration twins (*Min.*). See **interpenetration twins,** and cf. *juxtaposition twins*.

penetrometer (*Radiol.*). A device for the measurement of the penetrating power of radiation by comparison of the transmission through various absorbers.

penial setae (*Zool.*). See **copulatory spicules.**

penicilli (*Zool.*). See **Ruysch's penicilli.**

penicillin (*Pharm.*). Group of ether-soluble substances produced by a species of the mould *Penicillium* (*P. notatum*), having an intense growth-inhibiting action against various micro-organisms (including *streptococcus, gonococcus, spirochaeta pallida*). At least 5 members have been distinguished, with slightly different antibacterial activities. They have the following structure, in which R is an open-chain or cyclic radical with 6 or 7 carbon atoms:

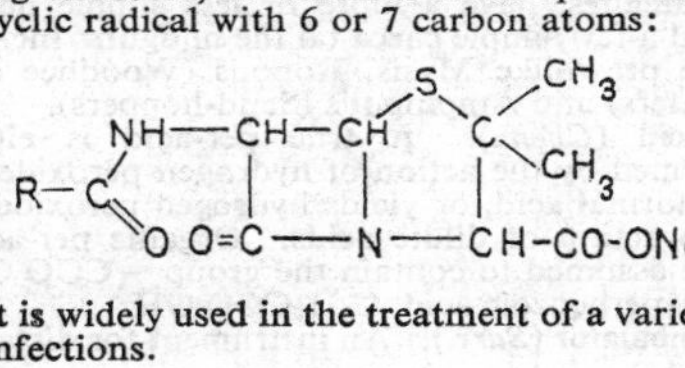

```
                     S     CH3
                   /   \  /
       NH——CH——CH       C
      /     |    |      | \
  R—C       |    |      |  CH3
      \\    |    |      |
       O  O=C————N—————CH-CO-ONa
```

It is widely used in the treatment of a variety of infections.

penis (*Zool.*). The male copulatory organ in most higher Vertebrates: a form of male copulatory organ in various Invertebrates, as *Platyhelminthes, Gastropoda. adj.* **penial.**

pennae (*Zool.*). See **plumae.**

Pennant Series (*Geol.*). A series of coarse sandstones and grits found in the Coal Measures of South Wales and in the Bristol Coalfield, between two series of productive measures.

pennate. Generally, winged. (*Bot.*) Said of diatoms which have an elongated cell which is isobilateral, zygomorphic, or dorsiventral in structure. Sometimes used for *pinnate*.

pennine or **penninite** (*Min.*). A silicate of magnesium and aluminium with chemically combined water. It crystallizes in the monoclinic system and is a member of the chlorite group.

penning (*Civ. Eng.*). The same as **pitching.**

Pennsylvanian System (*Geol.*). The American equivalent of the upper part of the Carboniferous system in Europe.

pennyweight. Troy weight of 24 gr or 1·5552 g, a unit widely used in valuation of gold ores and sale of bullion. Abbrev. dwt.

pen recorder (*Instr.*). A device, which may take various forms, in which the movements of a pen produce on a roll of paper a continuous record of the variations in an electric current.

pen ruling (*Bind.*). The ruling of paper on a pen-ruling machine as distinct from *disk ruling*.

Pensky-Martens test (*Chem.*). Standard test for determining the flash and fire points of oils. Based on closed or open cups depending on the nature of the oil under test.

penstock (*Eng.*). The valve-controlled water conduit between the intake and the turbine in a hydroelectric or similar plant.

pent-, penta- (Gk. *pente*, five). A prefix used in the construction of compound terms, e.g., *pentactinal*, 5-rayed. (*Chem.*) Containing 5 atoms, groups, etc.

pentachlorophenol (*Chem.*). C_6Cl_5OH. Widely used fungicidal and bactericidal compound, used particularly for timber protection. M.p. 189°C.

pentacrinoid (*Zool.*). A larval form of some *Crinoidea*, which resembles *Pentacrinus*.

pentacyclic (*Bot.*). Having the parts arranged in five whorls.

pentad. The period of 5 days; being an exact fraction of a normal year, it is useful for meteorological records.

pentadactyl (*Zool.*). Having 5 digits.

pentadactyl limb (*Zool.*). The characteristic free appendage of *Tetrapoda* with 5 digits.

pentaerythritol (*Chem.*). $C(CH_2OH)_4$. Condensation derivative of ethanal and methanal. M.p. 260°C. Used in the production of surface finishes.

pentaerythritol tetranitrate (*Chem.*). $C(CH_2ONO_2)_4$. A detonating explosive, abbrev. **PETN.**

pentagon (*Maths.*). A five-sided polygon.

pentagonal (*Bot.*). Having 5 angles with convex surfaces among them.

pentagonal dodecahedron (*Crystal.*). A form of the cubic system comprising 12 identical pentagonal faces. Pyrite frequently crystallizes in this form, hence the synonym *pyritohedron*.

pentagram (*Maths.*). A five-pointed star formed by the diagonals of a regular pentagon. The mystic symbol of the Pythagoreans (ca. 500 B.C.).

pentagrid (*Electronics*). A frequency-converting tube for supersonic heterodyne receivers; cathode and first two grids form an oscillator and the modulated electron stream is mixed with incoming signal by the other grids and anode which act as a pentode. Also **heptode.**

pentahydric alcohols (*Chem.*). Alcohols containing 5 hydroxyl groups, e.g., arabitol, $HO{\cdot}CH_2{\cdot}(CHOH)_3{\cdot}CH_2OH$, xylitol (stereoisomeric) and rhamnitol, $HO{\cdot}CH_2{\cdot}(CHOH)_4{\cdot}CH_3$.

pentamerous (*Bot.*). Having 5 members in a whorl.

pentamethylene or ***cyclo*pentane** (*Chem.*). A saturated cyclic hydrocarbon oil of the formula:

CH_2—CH_2
CH_2
CH_2—CH_2

It belongs to the group of naphthenes and occurs in crude petroleum.

pentamethylene-diamine (*Chem.*). See **cadaverine.**

pentamethylene glycol (*Chem.*). $CH_2OH{\cdot}(CH_2)_3{\cdot}CH_2OH{\cdot}$ pentan-1,5-diol. Organic solvent used in syntheses. B.p. 239°C, rel. d. 0·994.

pentane lamp (*Light*). An early photometric standard lamp. Also called **Vernon-Harcourt pentane lamp.**

pentanes (*Chem.*). C_5H_{12}. Low-boiling paraffin hydrocarbons. Pentane has a b.p. 36°C, rel. d. 0·63.

pentangular (*Bot.*). Having 5 angles with flat or concave surfaces among them.

Pentaphane (*Plastics*). TN for the film form of *Penton* (q.v.).

pentaprism (*Photog.*). Five-sided prism which corrects lateral inversion, used on reflex cameras to allow eye-level viewing.

Pentastomida (*Zool.*). An order of *Arthropoda*, generally, but doubtfully, placed in the *Arachnida*. Elongate vermiform parasites of carnivorous mammals, living in the nasal sinuses, which have 2 pairs of claws at the sides of the mouth, and no respiratory or circulatory system.

pentavalent (*Chem.*). Capable of combining with 5 atoms of hydrogen or their equivalent.

pentazocine (*Pharm.*). A powerful pain-killing drug, similar in action to morphine, believed to be non-addictive.

penthouse (*Arch.*). An individual dwelling situated on the roof of a building but forming an integral part of the building. (*Mining*) Pentice. Protective covering for workmen at bottom of shaft.

pentlandite (*Min.*). A sulphide of iron and nickel which crystallizes in the cubic system. It commonly occurs intergrown with pyrrhotite, from which it can be distinguished by its octahedral cleavage. Also called **nicopyrite.**

pentobarbitone (*Pharm.*). See **Nembutal.**

pentode valve (*Electronics*). Five-electrode thermionic tube, comprising an emitting cathode, control grid, a screen (or auxiliary grid) maintained at a positive potential with respect to that of the cathode, a suppressor grid maintained at about cathode potential, and an anode. It has characteristics similar to those of a screened-grid valve, except that secondary emission effects are suppressed.

Penton (*Plastics*). TN for a chlorinated polyether widely used as a coating material for vessels, etc., where very good chemical resistance is required. Relatively expensive but cheaper than the fluorinated plastics. (See **polytetrafluoroethylene**).

pentosans (*Chem.*). $(C_5H_8O_4)x$, polysaccharides, comprising arabinans and xylans.

pentoses (*Chem.*). A group of monosaccharides containing 5 oxygen atoms in the molecule, and having the formula $HO{\cdot}CH_2{\cdot}(CHOH)_3{\cdot}CHO$ and $CH_3{\cdot}(CHOH)_4{\cdot}CHO$. Important pentoses are L-arabinose, L-xylose, rhamnose, and fucose. Pentoses cannot be fermented. They are characterized by the fact that they yield furfuraldehyde or its homologues on boiling with dilute acids. A qualitative test for pentoses is the occurrence of a bright-red colour when they are boiled with HCl and phloroglucinol.

Pentothal (*Chem.*). TN for **thiopentone.**

pent roof (*Arch.*). A roof having only a single slope, often built against another building.

penultimate cell (*Bot.*). The last cell but one at the tip of an ascogenous hypha; it is commonly binucleate, and later becomes the ascus.

penumbra (*Astron.*). See **umbra.**

peotillomania (*Psychiat.*). Pseudomasturbation.

pepo (*Bot.*). A fleshy or succulent fruit, often of large size, formed from an inferior syncarpous ovary, and containing many seeds; it is a particular type of a berry. The cucumber is a familiar example.

Pepper's ghost (*Optics*). An illusion used to introduce a 'ghost' into a stage play. A plane sheet of glass is placed vertically at an angle of 45° to the line of vision of the audience. Thus actors in the wings can be superimposed by reflection on actors on stage.

pepsin (*Zool.*). An enzyme secreted by tubular glands in the fundus of the stomach of Vertebrates. It is an endopeptidase, catalysing the splitting of peptide linkages between the α-carboxyl group of a dicarboxylic amino acid, and the α-amino group of an aromatic amino acid, though inhibited by nearby amino-groups.

peptic ulcer (*Med.*). An ulcer of the stomach or of the duodenum.

peptides (*Chem.*). Substances resulting from the breakdown of proteins, characterized by the occurrence of the —CO—NH— structure in the molecule joining adjacent pairs of amino acids together.

peptization (*Chem.*). The production of a colloidal solution of a substance, especially the formation of a sol from a gel. Deflocculation.

peptonephridia (*Zool.*). In some *Oligochaeta*, modified nephridia which open into the pharynx, and may have a digestive function.

peptones (*Chem.*). Products obtained by the progressive action of enzymes on albuminous matter. They do not coagulate and cannot be precipitated by ammonium sulphate. They still show the xanthoprotein and biuret reactions.

per- (*Chem.*). (1) A prefix which properly should be restricted to compounds which are closely related to hydrogen peroxide, and thus contain 2 oxygen atoms linked together, e.g., *persulphates, percarbonates, perchromates.* (2) A prefix which is loosely used to denote that the central atom of a compound is in a higher state of oxidation than the usual, e.g., *perchlorates, permanganates.*

Peracarida (*Zool.*). Superorder of *Malacostraca* whose carapace, if present, does not fuse with more than four thoracic somites, whose eyes may be stalked or sessile, which possess oöstegites, have a more or less simple heart, and a few simple caeca on the midgut. Include the prawnlike Mysis, Isopods (Woodlice and Slaters) and Amphipods (Sand-hoppers).

per-acid (*Chem.*). A true per-acid is either formed by the action of hydrogen peroxide on a normal acid, or yields hydrogen peroxide by the action of dilute acids. Organic per-acids are assumed to contain the group $-COO{\cdot}OH$, e.g. perbenzoic acid, $C_6H_5COO{\cdot}OH$.

perambulator (*Surv.*). An instrument for distance

measurement consisting of a large wheel (often 6 ft or 2 m in circumference) attached at its axis to a long handle, so that it may be wheeled along the distance to be measured. A recording mechanism records the number of revolutions of the wheel and is calibrated in order to give distance traversed directly. See also **ambulator, odometer.**

percentage articulation (*Telecomm.*). That of elementary speech-sounds received correctly when logatoms are called over a telephone circuit in the standard manner.

percentage depth dose (*Nuc.*). In radiotherapy, the absorbed dose in soft tissue at a specified depth relative to that at the surface.

percentage differential relay (*Elec. Eng.*). A differential relay which operates at a current which, instead of being fixed, is a fixed percentage of the current in the operating coils.

percentage hearing loss (*Acous.*). See **hearing loss.**

percentage modulation (*Radio*). See **depth of modulation.**

percentage registration (*Elec. Eng.*). The registration of an integrating meter expressed as a percentage of the true value.

percentage tachometer (*Aero.*). An instrument indicating the rev/min of a turbojet engine as a percentage, 100 per cent corresponding to a pre-set optimum engine speed. Provides better readability and greater accuracy and also enables various types of engine to be operated on the same basis of comparison.

percentile (*Stats.*). The value below which fall a specified percentage (as 25, 50, 75) of a large number of statistical units.

perception (*Bot.*). The first changes which must be assumed to occur when a plant is stimulated; they lead to the appropriate reaction in due course. (*Psychol.*) The process of becoming aware of something: usually applied to sense perception. Work on visual perception indicates that it is largely a learned ability with a hierarchy of information-processing stages which are achieved at different ages in early life. See **agnosia, illusion and hallucination.**

perching (*Leather*). A process for stretching and softening a skin by working over it with a crutch stake, on the flesh side, while it is fixed to a horizontal pole (*perch*). (*Textiles*) Inspection of cloth after weaving, and at various stages of finishing, for possible defects; the cloth is drawn over the *perch*—two spars or rollers a distance apart, overhead—and the cloth hanging from the front roller is examined in a good light. See **mending.**

perchlorates (*Chem.*). Chlorates(VII). Salts of perchloric acid, $HClO_4$. They are all soluble in water, though the potassium and rubidium salts are only slightly soluble. The alkali perchlorates are isomorphous with the corresponding permanganates (manganates(VII)).

perchlor-ether (*Chem.*). $(C_2Cl_5)_2O$, a solid mass, of camphorlike odour. It is the final substitution product of ether by chlorine, and can be obtained by the action of chlorine upon ether in the dark and in the cold.

perchloric (chloric(VII)) acid (*Chem.*). $HClO_4$. A colourless fuming liquid, m.p. $-112°C$, b.p. 110°C, a powerful oxidizing agent; harmful to the skin; monobasic, and forms perchlorates (VII).

perchloroethene (*Chem.*). $CCl_2{\cdot}CCl_2$. Tetrachloroethene. Organic solvent used for dry-cleaning. B.p. 121°C, rel. d. (20°C) 1·623.

perchromates (*Chem.*). See **peroxychromates.**

percnosome (*Bot.*). An inclusion of obscure nature found in the cytoplasm of sperm mother cells in mosses.

percolating filter (*San. Eng.*). A bed of filtering material, such as broken stone or slag, used in the final or oxidizing stage in sewage treatment. This stage consists in sprinkling the liquid sewage over the filter, through which it percolates. Also called a **continuous filter.** See **contact bed.**

percrystallization (*Chem.*). The crystallization of a substance from a solution being dialysed.

percurrent (*Bot.*). Said of a vein which runs through the whole length of a leaf but does not project beyond the tip. (*Zool.*) Running from one end to the other of the body: extending from base to apex.

percussion (*Med.*). The act of striking with one finger lightly and sharply against another finger placed on the surface of the body, so as to determine, by the sound produced, the physical state of the part beneath.

percussion drill (*Mining*). See **cable drill.**

percussion drilling (*Mining*). System in which string of tools falls freely on rock being penetrated. Also, pneumatic drilling in which hammer blows are struck on drill shank.

percussion figure (*Min.*). A figure produced on the basal pinacoid or cleavage face of mica when it is sharply tapped with a centre punch. It consists of a 6-rayed star, 2 rays more prominent than the others, lying in the unique plane of symmetry.

percussive boring (*Civ. Eng.*). The process of sinking a borehole in the earth by repeatedly dropping on the same spot, from a suitable height, a heavy tool which pulverizes the earth and gradually penetrates.

percussive welding (*Elec. Eng.*). See **resistance percussive welding.**

perdistillation (*Chem.*). Distillation through a dialysing membrane.

perdurens (*Chem.*). Plastics obtained by vulcanizing thioplasts with or without sulphur, e.g., *Thiokol* (q.v.). Resistant to attack by oils.

pereion (*Zool.*). In *Crustacea*, the thoracic region. Cf. *pleon.*

pereiopods (*Zool.*). In higher *Crustacea*, the thoracic appendages modified as walking-legs. Cf. *pleopod.*

perennation (*Bot.*). Survival from season to season, with generally a period of reduced activity between each season.

perennial (*Bot.*). A plant which lives for 3 or more years, and normally flowers and fruits at least in its second and subsequent years.

perennibranchiate (*Zool.*). Retaining the gills throughout life, as certain Amphibians.

perfect (*Bot., Zool.*). (1) Having all organs or parts present. (2) *Monoclinous* (q.v.).

Perfectape Merging Unit (*Typog.*). A high-speed machine to merge an original 6-level tape with a correction tape to produce correct output.

Perfect binding (*Bind.*). A term used as a synonym for *unsewn binding* (q.v.), being the name of the first machine designed for the purpose.

perfect combustion (*Heat*). That of which the products contain neither unburnt gas nor excess air.

perfect crystal (*Crystal.*). A single crystal in which the arrangement of the atoms is uniform throughout.

perfect dielectric (*Elec.*). One in which all the energy required to establish an electric field is returned when the field is removed. In practice, only a vacuum conforms, other dielectrics dissipating heat to varying extent.

perfect flower (*Bot.*). See **perfect.**

perfect fluid (*Hyd.*). An ideal fluid which is incompressible, has a uniform density, and offers no resistance at all to distorting forces.

perfect frame (*Eng.*). A frame which has just sufficient members to keep it stable in equilibrium under the intended load.

perfect gas (*Chem.*). See **ideal gas.**

perfecting engine (*Paper*). See **refiner.**

perfect number (*Maths.*). An integer which is equal to the sum of all its factors including unity. The first four are 6, 28, 496 and 8128, after which they become enormous. Cf. *friendly numbers.*

perfector (*Print.*). A type of machine which prints both sides of the paper before delivery.

perfect paper (*Print.*). Reams of paper made up to a printer's ream, i.e., 516 sheets. Since the standardization of paper quantities the term has been largely discarded.

perfect stage (*Bot.*). Of fungi-stage of life cycle when spores resulting from sexual reproduction are produced.

perfect up (*Print.*). To print the second side of a sheet of paper, and so complete it.

perfoliate (*Bot.*). Said of a leaf base which surrounds a stem completely so that the latter appears to pass through it.

perforate (*Bot.*). (1) Having holes. (2) Containing small rounded transparent dots which give the appearance of holes. (*Zool.*) Having apertures, said especially of shells: of gastropod shells, having a hollow columella.

perforated bars (*Textiles*). Flexible steel strips in the well of a lace machine which guide the warp threads.

perforated brick (*Build.*). A clay brick manufactured with vertical perforations of varying shapes and sizes. So made to reduce weight, improve insulating properties, also to reduce capillary attraction through a wall.

perforating fibres (*Zool.*). See **Sharpey's fibres.**

perforating press (*Eng.*). A power press suitable for removing patterns of small areas of metal by punching.

perforation (*Bot.*). An interruption in the continuity of a stele not due to a leaf gap.

performance (*Aero.*). Those flying qualities of an aircraft capable of quantitative definition, usually speeds, rates of climb, ranges, ceiling, take-off, landing—which require to be specified in conjunction with the flying weights and disposable load.

performance tests (*Psychol.*). Types of psychological tests, usually used in the assessment of intelligence, which apart from any instructions there may be, are nonverbal and involve the manipulation of the test materials, e.g., the construction of set patterns from small coloured blocks.

pergeting (*Build.*). See **pargeting.**

Perhydrol (*Chem.*). TN for a 30% solution of hydrogen peroxide.

perhydrous coal (*Fuels*). Coal of hydrogen content above average for the rank of coal, e.g., *cannel coal* (q.v.).

peri-. Prefix from Gk. *peri*, around.

peri- (*Chem.*). Term for the 1.8-positions in naphthalene derivatives.

perianal (*Anat.*). The region around the anus.

periandra (*Bot.*). The leaves surrounding a group of antheridia in mosses.

perianth (*Bot.*). (1) A general term for calyx and corolla together. (2) The outer whorl of a flower when it is not distinctly composed of sepals and petals. (3) A cup-shaped or tubular sheath surrounding the archegonia of some liverworts.

periarteritis nodosa (*Med.*). A rare and usually fatal disease characterized by inflammation of the arteries, in the walls of which small inflammatory nodules appear.

periarticular (*Anat.*). Said of the tissues immediately around a joint.

periastron (*Astron.*). That point in an orbit about a star in which the body describing the orbit is nearest to the star; applied to the relative orbit of a double star.

periblast (*Zool.*). In meroblastic eggs, the margin of the blastoderm merging with the surrounding yolk. See also **periplasm.**

periblastic (*Zool.*). Of cleavage, superficial.

periblem (*Bot.*). That portion of an apical meristem from which the cortex is ultimately formed.

peribranchial (*Zool.*). Surrounding a gill or gills, as part of the atrium in *Urochorda.*

pericardiomediastinitis (*Med.*). Inflammation both of the pericardium and of the mediastinum.

pericardioperitoneal canal (*Zool.*). In *Selachii*, a small aperture which puts the pericardial and peritoneal cavities into communication.

pericardiotomy (*Surg.*). Incision of the pericardium.

pericarditis (*Med.*). Inflammation of the pericardium.

pericardium (*Zool.*). Space surrounding the heart: the membrane enveloping the heart. *adj.* **pericardial.**

pericarp (*Bot.*). Wall of a fruit, if derived from the wall of the ovary.

pericellular (*Histol.*). Surrounding a cell.

pericentral siphon (*Bot.*). One of the tubular elements surrounding the central siphon in the thallus of certain red algae.

perichaetial bract (*Bot.*). One of the leaves composing the perichaetium in a moss.

perichaetine (*Zool.*). In *Oligochaeta*, a complete ring of chaetae surrounding the body.

perichaetium (*Bot.*). (1) A cuplike sheath surrounding the archegonia in some liverworts. (2) The group of involucral leaves around the archegonia of a moss.

perichondritis (*Med.*). Inflammation of the perichondrium, especially of the perichondrium of the cartilages of the larynx.

perichondrium (*Zool.*). The envelope of areolar connective tissue surrounding cartilage.

perichordal (*Zool.*). Encircling or ensheathing the notocord.

perichordal centra (*Zool.*). Centra which arise by chondrification or ossification of tissues surrounding the notochord without invasion of the notochord sheath.

perichordal sheath (*Zool.*). The mesenchymatous layer surrounding the notochord in adult Cyclostomes and the very young forms of all other Vertebrates.

perichylous (*Bot.*). Having water-storage tissue surrounding the chlorophyll-containing tissue.

periclase (*Min.*). Native magnesia. Oxide of magnesium, which crystallizes in the cubic system. It is commonly found in metamorphosed magnesian limestones, but readily hydrates to the much commoner brucite.

periclinal chimaera (*Bot.*). A chimaera in which one component is completely enclosed by the other.

periclinal wall (*Bot.*). A cell wall which is parallel to the surface of an apical meristem or other part of a plant.

pericline (*Min.*). A variety of *albite* (q.v.) which usually occurs as elongated crystals in alpine type veins. The name is also used for a type of twinning in feldspars (the *pericline law*).

pericolitis (*Med.*). Inflammation of the peritoneum covering the colon.
pericranium (*Zool.*). The fibrous tissue layer which surrounds the bony or cartilaginous cranium in Vertebrates.
pericycle (*Bot.*). A layer, one or more cells in thickness, of nonconducting cells at the periphery of a stele.
pericyclic fibre (*Bot.*). A strand of sclerenchyma in the pericycle.
pericycloid (*Maths.*). See **roulette.**
pericyte (*Histol.*). A cell, often of macrophagic type, found in adventitia of small blood vessels; in some species, also called **Rouget cell.**
periderm (*Bot.*). A protective layer which develops on those parts of plants which last for some time; it consists of cork, the cork cambium, and usually some phelloderm. (*Zool.*) See **perisarc.**
peridesmium (*Zool.*). The coat of connective tissue which ensheathes a ligament.
perididymis (*Zool.*). The fibrous coat which encapsules the testis in higher Vertebrates.
peridium (*Bot.*). A general term for the outer wall of the fruit body of a fungus, when the wall is organized as a distinct layer or envelope surrounding the spore-bearing organs partially or completely.
peridot (*Min.*). See **olivine, Brazilian peridot, Ceylon peridot.**
peridotite (*Geol.*). A coarse-grained ultramafic igneous rock consisting essentially of olivine, with other mafic minerals such as hypersthene, augite, biotite, and hornblende, but free from plagioclase. See **dunite, kimberlite, scyelite.**
peridural (*Zool.*). Surrounding the dura mater.
perigastritis (*Med.*). Inflammation of the external surface of the stomach (as a result usually of gastric ulcer), with the formation of adhesions between it and other abdominal viscera.
perigee (*Astron.*). The point nearest to the earth on the apse line of a central orbit having the earth as a focus.
perigon (*Maths.*). See **round angle.**
perigone (*Bot.*). A perianth which is not clearly differentiated into calyx and corolla.
perigonial bract (*Bot.*). One leaf of the perigonium in mosses.
perigonium (*Bot.*). A group of leaves, often forming a flat rosette, around the base of the group of antheridia in mosses.
perigynium (*Bot.*). A group of leaves around the group of archegonia in mosses: a tubular sheath surrounding the archegonia in hepatics.
perigynous (*Bot.*). Said of a flower in which the receptacle is developed into a flange or into a concave to deeply concave structure, on which the sepals, petals, and stamens are borne; the receptacle remains distinct from the carpels. *n.* **perigyny.**
perihaemal (*Zool.*). Said of: (1) an anterior paired prolongation of the trunk coelom lying beside the dorsal blood vessel in *Enteropneusta*; (2) a system of coelomic cavities in *Echinodermata*, which derives its name from the fact that in *Asteroidea* its radial branches surround the radial blood vessels.
perihaemal system (*Zool.*). The perihaemal coelomic cavities of *Echinodermata*. See **perihaemal.**
perihelion (*Astron.*). That point in the orbit of any heavenly body moving about the sun at which it is nearest to the sun; applied to all the planets and also to comets, meteors, etc.
perihepatitis (*Med.*). Inflammation of the peritoneum covering the liver.
perikaryon (*Histol.*). The cytoplasm of the cell body of a neurone.
perikinetic (*Phys.*). Pertaining to the Brownian movement.
perilla oil (*Paint.*). A useful drying oil for paint-making. It dries faster than linseed oil.
perilymph (*Zool.*). The fluid which fills the space between the membranous labyrinth and the bony labyrinth of the internal ear in Vertebrates. *adj.* **perilymphatic.** Cf. *endolymph.*
perimedullary zone (*Bot.*). See **medullary sheath.**
perimeter (*Med.*). An instrument, in the form of an arc, for measuring a person's field of vision.
perimeter track (*Aero.*). A *taxi track* (q.v.) round the edge of an aerodrome.
perimetritis (*Med.*). Inflammation of the peritoneum covering the uterus.
perimetrium (*Anat.*). The peritoneum covering the uterus.
perimysium (*Zool.*). The connective tissue which binds muscle fibres into bundles and muscles.
perinatal (*Med.*). Said of the period from the seventh month of pregnancy to the first week of life.
perineal glands (*Zool.*). In some Mammals, a pair of small glands beside the anus which secrete a substance with a characteristic odour.
perineoplasty (*Surg.*). Repair of the perineum by plastic surgery.
perineorrhaphy (*Surg.*). Stitching of the perineum torn during childbirth.
perinephric (*Anat.*). Said of the tissues round the kidney, e.g., *perinephric* abscess.
perinephritis (*Med.*). Inflammation of the tissues round the kidney.
perineum, perinaeum (*Zool.*). The tissue wall between the rectum and the urinogenital ducts in Mammals. *adj.* **perineal, perinaeal.**
perineurium (*Zool.*). The coat of connective tissue which ensheathes a funiculus of nerve fibres.
perinium (*Bot.*). See **epispore.**
period (*Chem.*). The elements between an alkali metal and the rare gas of next highest atomic number, inclusive, occupying 1 (*short p.*) or 2 (*long p.*) horizontal rows in the periodic system. (*Cinema.*) Division of time effected by the intermittent mechanism in a projector or camera. (*Elec.*) Time taken for 1 complete cycle of an alternating quantity. Reciprocal of frequency. (*Maths.*) See **periodic function.** (*Nuc.*) Of a radioactive isotope, synonym for U.K., **mean life,** U.S., **half-life.** (*Nuc. Eng.*) In a reactor, time in which the neutron flux changes by a factor of *e*.
periodates (*Chem.*). Iodates(VII). Formed by the oxidation of iodates(V). Periodates form heteropolybasic compounds of the types $M_5[I(WO_4)_6]$ and $M_5[I(MoO_4)_6]$.
periodic acid (*Chem.*). Iodic(VII)acid. H_5IO_6. A weaker acid and a stronger oxidizing agent than iodic acid. May be regarded as orthoperiodic acid. Exists in deliquescent crystals. Resembles phosphoric acid in furnishing partially dehydrated acids.
periodic antenna (*Radio*). One depending on resonance in its elements, thereby presenting a periodic change in input impedance as the frequency of the drive is varied.
periodic chain (*Chem.*). The arrangement of the periodic system in the form of a simple vertical list of elements in order of atomic numbers.
periodic current (*Telecomm.*). An oscillating current whose values recur at equal intervals.
periodic damping (*Telecomm.*). That of a system which shows an oscillatory response to any disturbance. See also **underdamping.**

periodic function (*Maths.*). A function f is periodic with period ω if for all $x, f(x) = f(x+\omega)$ and if ω is the smallest value for which this is true, e.g., sin x has a period of 2π and tan x a period of π. A function of a complex variable may have two or more independent complex periods (i.e., none being a real rational multiple of any other), in which case it is referred to as **doubly-** or **multiply-periodic.**

periodicity (*Biol.*). Rhythmic activity. (*Chem.*) Location of element in periodic table. (*Elec.*) See **frequency.**

periodic law (*Chem.*). See **periodic system.**

periodic ophthalmia (*Vet.*). Moon blindness. Recurrent *iridocyclitis* (q.v.) in the horse; believed to be a form of *leptospirosis* (q.v.).

periodic precipitation (*Chem.*). See **Liesegang phenomenon.**

periodic rating (*Elec. Eng.*). Rating of a component for continuous use with a specified periodically varying load.

periodic system (*Chem.*). Classification of chemical elements into periods (corresponding to the filling of successive electron shells) and groups (corresponding to the number of valence electrons). Original classification by relative atomic mass (Mendeleev, 1869).

periodic table (*Chem.*). The most common arrangement of the *periodic system.*

period-luminosity law (*Astron.*). A relationship between the period and absolute magnitude, discovered by Miss Leavitt to hold for all Cepheid variables; it enables the distance of any observable Cepheid to be found from observation of its light curve and apparent magnitude, this indirectly deduced distance being called the *Cepheid parallax.*

period meter (*Nuc. Eng.*). Instrument for measurement of reactor period.

period of decay (*Nuc.*). See **half-life.**

period of revolution (*Astron.*). The mean value, derived from observations, of one complete revolution of a planet or comet about the sun, or of a satellite about a planet. The mean distance (or semimajor axis of the orbit) is derived from the period by Kepler's third law.

periodontitis (*Med.*). Inflammation of the membrane investing that part of the tooth seated in the jaw.

period range (*Nuc. Eng.*). See **start-up procedure.**

perioesophageal (*Zool.*). Encircling the oesophagus.

peri-oöphoritis (*Med.*). Inflammation of the peritoneum investing the ovary and of the cortex of the ovary.

periopticon (*Zool.*). The ganglionic plate or outer zone of the optic lobe of the brain in Insects.

periosteum (*Zool.*). The covering of areolar connective tissue on bone.

periostitis (*Med.*). Inflammation of the periosteum.

periostracum (*Zool.*). The horny outer layer of a Molluscan shell.

periotic (*Zool.*). In higher Vertebrates, a bone enclosing the inner ear and formed by the fusion of the otic bones: petrosal.

peripharyngeal (*Zool.*). Encircling the pharynx.

peripheral. Situated or produced around the edge. (*Comp.*) Said of units forming part of a computing system but not directly under the control of the main computer. The term applies both to input-output devices and to, e.g., magnetic tape units.

periphlebitis (*Med.*). Inflammation of the outer coat of a vein.

periphysis (*Bot.*). A hairlike extension of the end of a hypha, forming with many others of the same kind a pilelike lining in the ostiole of a perithecium.

periphyton (*Ecol.*). Plants or animals in an aquatic habitat which are attached to, or cling to, stems and leaves of rooted plants or other surfaces projecting from the bottom. Also called **aufwuchs.**

periplasm (*Bot.*). The plasma lying just within the oögonial wall in some *Oömycetes*; it contains degenerating nuclei and contributes to the formation of the wall of the oöspore. (*Zool.*) A bounding layer of protoplasm surrounding an egg just beneath the vitelline membrane, as in Insects.

periplasmodium (*Bot.*). The material produced by the breakdown of the tapetum in the sporangia of *Pteridophyta* and *Spermatophyta*; it helps in nutrition of developing spores.

periplast (*Zool.*). The cuticle of *Mastigophora*, formed by the conversion of the ectoplasm and containing myonemes.

peripneustic (*Zool.*). Of Insects, having the spiracles arranged in a row along each side of the body. Cf. *holopneustic, oligopneustic, apneustic.*

peripodial (*Zool.*). Surrounding an appendage.

peripolesis (*Cyt.*). Movement of lymphocytes around but not into macrophages in tissue culture. Cf. *emperipolesis.*

periproct (*Zool.*). The area surrounding the anus.

periproctitis (*Med.*). Inflammation of the cellular connective tissue round the rectum.

perisalpingitis (*Med.*). Inflammation of the peritoneum covering the Fallopian tube.

perisarc (*Zool.*). In some *Hydrozoa*, the chitinous layer covering the polyps, etc. Cf. *coenosarc.*

periscope (*Light*). An optical instrument comprising an arrangement of reflecting surfaces whose purpose is to enable an observer to view along an axis deflected or displaced with respect to the axis of the observer's eye.

perisomatic (*Zool.*). Surrounding the body, as the *perisomatic cavity* of a developing *Sacculina.*

perisperm (*Bot.*). A nutritive tissue present in some seeds, derived from the nucellus of the ovule.

perisphere (*Phys.*). Theoretical sphere surrounding a body, in which significant electromagnetic effects occur.

perisplenitis (*Med.*). Inflammation of the capsule of the spleen and of the peritoneum investing it.

Perisporales (*Bot.*). See **Erysiphales.**

perispore (*Bot.*). Remains of the contents of the cells of the tapetum, forming a deposit on the outside of the walls of the spores of ferns.

perissodactyl (*Zool.*). Having an odd number of digits. Cf. *artiodactyl.*

Perissodactyla (*Zool.*). An order of Mammals containing the 'odd-toed' hooved animals, i.e., those with a mesaxonic foot with the skeletal axis passing down the third digit. Horses, Tapirs, Rhinoceros, and extinct forms.

peristaltic (*Zool., etc.*). Compressive: contracting in successive circles; said of waves of contraction passing from mouth to anus along the alimentary canal: cf. *antiperistaltic, systaltic. n.* **peristalsis.**

peristaltic pump (*Eng.*). One in which the flow is produced by peristaltic action, e.g., by rollers passing in succession over a length of flexible tube.

peristerite (*Min.*). A whitish variety of albite, or oligoclase, which is beautifully iridescent.

peristomate (*Bot.*). Possessing a peristome.

peristome (*Bot.*). (1) A fringe of elongated teeth around the mouth of the capsule of a moss; the teeth are formed from persistent remains of unevenly thickened cell walls. (2) A fringe of hyphae around the opening of the fruit body of some *Gasteromycetes*. (3) A kind of lip arising as an outgrowth in some *Bryophyta*, assisting in the ingestion of solid food. (*Zool.*) (1) The margin of the aperture of a gastropod shell. (2) In some *Ciliophora*, a specialized food-collecting, frequently funnel-shaped, structure surrounding the cell-mouth. (3) More generally, the area surrounding the mouth. *adj.* **peristomial.**

peristomium (*Zool.*). In *Chaetopoda*, the somite in which the mouth is situated: in some forms (as *Nereis*), 2 somites have been fused to form the apparent peristomium.

peristyle (*Arch.*). A colonnade encircling a building.

perisystole (*Zool.*). The period between diastole and systole in cardiac contraction.

perithecium (*Bot.*). A globose or flask-shaped structure with a sterile wall enclosing asci and paraphyses, the characteristic fruit body of the *Pyrenomycetes*. A true perithecium has an ostiole, but the term is often used to include the *cleistocarp*, which has no ostiole. Also **pyrenocarp.**

perithelioma (*Med.*). A tumour, the cells of which are arranged in sheathlike fashion around thin-walled blood vessels and are believed to arise from endothelium.

peritoneal cavity (*Zool.*). In Vertebrates, that part of the coelom containing the viscera; the abdominal body cavity.

peritoneal dialysis (*Med.*). Passage of fluids through the peritoneal cavity to lower the blood urea in certain cases of renal failure.

peritoneum (*Zool.*). In Vertebrates, a serous membrane which lines the peritoneal cavity and extends over the mesenteries and viscera. *adj.* **peritoneal.**

peritonitis (*Med.*). Inflammation of peritoneum.

peritreme (*Zool.*). In Insects, the annular sclerite which surrounds the external opening of a trachea: the margin of a shell aperture.

Peritricha (*Zool.*). An order of *Ciliata*, the members of which are usually of sedentary habit; they possess a permanent gullet with undulating membrane; cilia are either absent or reduced to a single ring, e.g., *Vorticella.*

peritrichous (*Bot.*). Said of bacteria when there are flagella distributed over the whole surface of the cell. (*Zool.*) Having an adoral wreath of cilia surrounding the cytopharynx.

peritrochoid (*Maths.*). See **roulette.**

peritron (*Electronics*). Cathode-ray tube with facility for axial displacement of screen—used when 3-dimensional display is necessary.

peritrophic (*Zool.*). Surrounding the gut; as the *peritrophic membrane* of Insects, a membranous tube lining the stomach and partially separated from the stomach epithelium by the peritrophic space.

perivascular sheath (*Zool.*). A sheath of connective tissue around a blood or lymph vessel.

perivitelline (*Histol.*). Surrounding an egg yolk.

perivitelline membrane (*Zool.*). See **vitelline membrane.**

Perkin's mauve (*Chem.*). See **mauveine.**

Perkin's phenomenon (*Elec. Eng.*). The decrease which is observed to occur in the conductivity of a rod of graphite when a negative electric charge is imparted to it.

Perkin's synthesis (*Chem.*). The synthesis of unsaturated aromatic acids by the action of aromatic aldehydes upon the sodium salts of fatty acids in the presence of a condensing agent, e.g., acetic (ethanoic) anhydride.

perknite (*Geol.*). A family of coarse-grained ultramafic igneous rocks which consist essentially of pyroxenes and amphiboles, but contain no feldspar.

perlite (*Geol.*). An acid and glassy igneous rock which exhibits perlitic structure.

Perlite Iron (*Met.*). See **pearlitic iron.**

perlitic structure (*Geol.*). A structure found in glassy igneous rocks, which consists of systems of spheroidal concentric cracks produced during cooling.

Perlon (*Chem.*). TN for *polycaprolactam*, nylon 6, a synthetic fibre.

permafrost (*Geol.*). In arctic and sub-arctic regions, the thickness of permanently frozen soil.

Permali (*Elec.*). Beech plywood impregnated with bakelite. Electrical strength is 60–90 kV/cm across the grain and 20–30 kV/cm along the grain.

permalloy (*Met.*). An alloy with high magnetic permeability at low field strength, and low hysteresis loss. Original composition nickel 78·5%, iron 21·5%, but the term is now used generally to cover numerous alloys produced by adding other elements, e.g., copper, molybdenum, chromium, cobalt, manganese, etc.

permanent collenchyma (*Bot.*). Functional collenchyma present in petioles and in the stems of herbaceous plants.

permanent dentition (*Zool.*). In Mammals, the second set of teeth, which replaces the milk dentition.

permanent finish (*Textiles*). A cloth finish which will maintain its appearance; the cloth will not shrink or develop dull spots after rain.

permanent hardness (*Chem.*). Of water, the hardness which remains after prolonged boiling is *permanent hardness*. Due to the presence of calcium and magnesium chlorides or sulphates.

permanent implant (*Radiol.*). An implant with radioactive material of short half-life, e.g., radon, arranged so that the prescribed dose is delivered by the time that the radioactive material has decayed, thus rendering removal of the sources unnecessary.

permanent load (*Eng.*). The dead loading on a structure, consisting of the weight of the structure itself and the fixed loading carried by it, as distinct from any other loads.

permanent magnet (*Elec. Eng.*). Ferromagnetic body which retains an appreciable magnetization after excitation (electric currents, or stroking with another magnet) has ceased. Cobalt steel and various ferritic alloys, e.g. *Alnico* (q.v.), are the usual materials. Used on a large scale in loudspeakers, relays, small motors, magnetrons, etc.

permanent mould (*Met.*). A metal mould (other than an ingot mould) used for the production of castings, e.g., in die-casting.

permanent set (*Eng.*). (1) An extension remaining after load has been removed from a test piece, when the elastic limit has been exceeded. (2) Permanent deflection of any structure after being subjected to a load, which causes the elastic limit to be exceeded.

permanent store (*Comp.*). Storage unit which is not cleared when power is interrupted.

permanent tissue (*Bot.*). Tissue consisting of fully differentiated elements.

permanent way (*Civ. Eng.*). The ballast, sleepers, and rails forming the finished track for a railway, as distinct from a *temporary way.*

permanent white (*Paint.*). Precipitated barium sulphate (*blanc fixe*). High density, low oil absorption extender pigment.

permanganates (*Chem.*). Manganates(VII). Oxidizing agents, the best known being *potassium permanganate* (manganate(VII)), the commonest salt of manganese.

permanganic acid (*Chem.*). Manganic(VII)acid. $HMnO_4$. Powerful oxidizing agent. Decomposes in the presence of organic matter.

permatron (*Electronics*). Hot-cathode gas-discharge diode, gated by applied magnetic field.

permeability (*Elec. Eng.*). *Absolutely*, μ is ratio of magnetic flux density produced in a medium to magnetizing force producing it. *Relatively*, μ_r the ratio of magnetic flux density produced in a medium to that which would be produced in free space by the same magnetizing force. With CGS electromagnetic units, absolute and relative permeabilities are the same. In SI units, the former is equal to the product of the latter and the permeability of free space, $\mu_0 = 4\pi \times 10^{-7}$ H/m, i.e. $\mu = \mu_0\mu_r$. See **differential-, incremental-, initial-**. (*Phys.*) The rate of diffusion of gas or liquid under a pressure gradient through a porous material. Expressed, for thin material, as the rate per unit area, and for thicker material, per unit area of unit thickness. (*Powder Tech.*) The facility with which a porous mass, e.g., a powder bed or compact, permits passage of a fluid.

permeability bridge (*Elec. Eng.*). Device for measuring magnetic properties of a sample of magnetic material, fluxes in different branches of a divided magnetic circuit being balanced against each other. See **Ewing-, Holden-**.

permeability coefficient (*Powder Tech.*). The volume of incompressible fluid per sec. which will flow through a unit cube of a porous mass or a packed powder across which unit pressure difference is maintained. Using cgs units, a flow rate of 1 cm^3/s through a 1 cm cube under 1 $dyne/cm^2$ pressure gives a p.c. of 1 darcy. Using SI units, a flow rate of 1 m^3/s through a 1 m cube under 1 N/m^2 pressure gives a p.c. of 1 **kilodarcy** (1000 darcys).

permeability equations (*Powder Tech.*). Mathematical equations derived to describe the flow of fluids through packed powder beds and porous media in general.

permeability equipment (*Powder Tech.*). See **permeameter.**

permeability surface area (*Powder Tech.*). The surface area of a particle calculated from the permeability of a powder bed under stated conditions.

permeability tuning (*Radio*). Adjusting a tuned circuit by varying the inductance of a coil (by altering position on axis of a sintered iron core, or by sliding a copper spade over the winding).

permeameter (*Mag.*). Instrument for measuring static magnetic properties of ferromagnetic sample, in terms of magnetizing force and consequent magnetic flux. (*Powder Tech.*) Instrument for measuring the specific surface area of a powder by measuring the resistance offered to a flowing fluid by a packed powder bed.

permeameter cell (*Powder Tech.*). That portion of a permeameter in which the plug or bed of the powder under test is assembled. Also **permeability cell.**

permeance (*Elec. Eng.*). Reciprocal of the *reluctance* of a magnetic circuit.

permeants (*Ecol.*). In a terrestrial community above the ground, the highly motile animals, e.g., Birds, Reptiles, Mammals, and Flying Insects, which move freely from one community to another, though not necessarily being indiscriminate in habitat selection.

Permendur (*Met.*). TN for alloy of equal parts iron and cobalt, characterized by high permeability and very high saturation flux-density, ca. 2·35 tesla.

Permian System (*Geol.*). That system of rocks deposited between the Carboniferous and Triassic Periods of geological time, the type area being Perm in Russia. Permian rocks in England include the Magnesian Limestone of the north-east, the Brockram and the Penrith Sandstone, as well as important gypsum and salt deposits, the whole assemblage bearing evidence of accumulation under arid, desert conditions.

Permic (*Geol.*). Equivalent to ***Permian.***

Perminvar (*Met.*). TN for alloy of nickel, cobalt and iron with low hysteresis and high permeability.

permissible dose (*Radiol.*). See **maximum permissible dose.**

permissive waste (*Build.*). Dilapidations in a building which are the result of neglect on the part of a tenant.

permitted explosives (*Mining*). Those which may be used in mines, under specified conditions, where a danger of explosion from inflammable gas exists.

permittivity (*Elec. Eng.*). *Absolutely*, ε is ratio of electric displacement in a medium to electric field intensity producing it. *Relatively*, ε_r is ratio of electric displacement in a medium to that which would be produced in free space by the same field. With CGS electrostatic units, absolute and relative permittivities are the same. In SI units, the former is equal to the product of the latter and the permittivity of free space, $\varepsilon_0 = 8{\cdot}854 \times 10^{-12}$ F/m, i.e. $\varepsilon = \varepsilon_0\varepsilon_r$. Relative permittivity is also termed **dielectric constant or specific inductive capacity.**

permo (*Textiles*). A lustre dress fabric with a crêpelike appearance; the weft is hard-spun botany worsted and the warp lightly spun mohair and cotton. The cotton is removed, after weaving, by carbonizing. Also known as **resilda, mohlaine.**

permonosulphuric acid (*Chem.*). H_2SO_5, a powerful oxidizing reagent; prepared by anodic oxidation of concentrated sulphuric acid.

Permo-Trias (*Geol.*). The Permian and Triassic systems considered together, as is commonly done in areas such as the British Isles where the rocks are of similar type.

permutations (*Maths.*). The different arrangements that can be made of a given number of items. For n items, all different, there are $n(n-1)(n-2)\ldots 2.1$ permutations; taken r at a time, there are $n(n-1)(n-2)\ldots(n-r+1)$.

Permutit (*Chem.*). TN for natural or synthetic zeolite used as ion-exchanger to substitute sodium for calcium or magnesium ions in hard water, to obtain softer water, the spent (i.e., calcium and magnesium) zeolite being regenerated (i.e., to sodium zeolite) periodically by treatment with concentrated sodium chloride solution.

permutoid (*Chem.*). Involving a double decomposition between a soluble substance and an insoluble one.

pernicious anaemia (*Med.*). Addison's anaemia. A disease characterized by anaemia, abnormalities in the size and shape of red blood corpuscles, *achylia gastrica* (q.v.), and changes in the nervous system, due to the absence from the body of an antianaemic factor which is normally present in liver. See **vitamin B_{12}.**

perolene (*Chem.*). Heat exchange organic fluid based on a mixture of diphenyl and diphenyl oxide.

peronate (*Bot.*). Having the stipe, particularly at the base, covered by a thick felted sheath.

peroneus, peronaeus (*Zool.*). In Vertebrates, the fibula; in Birds, a leg muscle. *adj.* **peroneal.**

Peronosporales (*Bot.*). An order of fungi, belonging to the *Phycomycetes*, with biflagellate zoospores; sexual reproduction gives rise to the formation of thick walled oöspores. Includes many plant parasites, some of economic importance, e.g., *Phytophthora infestans* (potato blight).

peroral (*Zool.*). Surrounding the mouth; as the *peroral membrane* of *Ciliophora*, which surrounds the cytopharynx.

perosis (*Vet.*). Slipped tendon. A disease of chickens, turkeys and other birds characterized by swelling and deformity of the hock joint leading to dislocation of the gastrocnemius tendon; caused by a dietary deficiency of manganese or choline.

perovskite (*Min.*). Titanate of calcium and the rare earths which crystallizes in the monoclinic system but is very close to being cubic. An accessory mineral in melilite-basalt, and in contact metamorphosed impure limestones.

peroxidases (*Chem.*). Enzymes which activate hydrogen peroxide and induce reactions which hydrogen peroxide alone would not effect.

peroxides (*Chem.*). (1) Oxides whose molecules contain 2 atoms of oxygen linked together and which yield hydrogen peroxide with acids, e.g., barium peroxide, $Ba^{++}O_2^{--}$. Organic peroxides are known, e.g., diethyl peroxide, $C_2H_5{\cdot}O{\cdot}O{\cdot}C_2H_5$. (2) The term is used loosely for certain oxides in which the valency of the central atom is greater than the usual value, e.g., lead (IV)oxide or peroxide, $O{=}Pb{=}O$.

peroxychromates (*Chem.*). Formed by the action of hydrogen peroxide on chromates(VI), and containing a peroxide group —O—O—. In alkaline solution it forms M_3CrO_8, red salts. In acid: $MCrO_6$, blue salts. In ammonia or KCN, brown coordination compounds are formed, of type $CrO_4{\cdot}3NH_3$.

perpend, perpend-stone (*Build.*). See **through-stone.**

perpetual motion (*Phys.*). Two main kinds are distinguished: (1) the continual operation of a machine which creates its own energy, thereby violating the 1st law of thermodynamics; (2) any device which converts heat completely into work, thereby contravening the 2nd law of thermodynamics. See also **thermodynamics.**

perradius (*Zool.*). In *Coelenterata*, one of the 4 primary radii.

perrhenates (*Chem.*). Rhenates(VII). See **rhenium oxides.**

perrhenic acid (*Chem.*). Rhenic(VII)acid. See **rhenium oxides.**

perron (*Build.*). An external staircase to a building, from ground-level to the first floor.

per-salts (*Chem.*). Salts corresponding to *per-acids*. A solid per-salt reacts immediately and quantitatively with neutral potassium iodide to form iodine.

Perseids (*Astron.*). A swarm of meteors whose orbit round the sun is crossed by the earth on 12 August, on or about which date a shower with its radiant point in the constellation Perseus may be expected.

perseveration (*Med.*). Meaningless repetition of an action or utterance. (*Psychol.*) The continuation or continued influence of an idea, attitude, percept, etc. This varies for different people and may be related to other personality characteristics.

Persian (*Leather*). The terms *Persian goat* and *Persian sheep* are used in the trade for goat and sheep skins tanned in India.

persistence (*Electronics*). Continued visibility or detection of luminous radiation from a gas-discharge tube or luminous screen when the exciting agency has been removed. Long-persistence cathode-ray tubes are used for retaining a display which has been momentarily excited. Also called **afterglow.**

persistence characteristic (*Electronics*). Graph showing decay with time of the luminous emission of a phosphor, after excitation is cut off.

persistence of vision (*Optics*). Brief retention of image on retina of eye after optical excitation has ended. An essential factor in cinematography and TV.

persistent (*Zool.*). Continuing to grow or develop after the normal period for the cessation of growth or development, as teeth (cf. *deciduous*); said also of structures present in the adult which normally disappear in the young stages.

persistent perianth (*Bot.*). A perianth which remains unwithered, and often enlarged, around the fruit.

persistor (*Comp.*). Memory element consisting of bimetallic loop operating in cryostat at superconducting temperature.

persistron (*Electronics*). Electroluminescent photoconducting device, which gives a steady display on being impulsed.

person (*Zool.*). An individual organism.

persona (*Psychol.*). Jung's term for a person's system of adaptation to, or manner assumed when dealing with, the world. It contrasts with the *anima* or *animus* and there is a danger that the individual may become identified, both by himself and by others, with this public face; also called **mask.**

personal dosimeter (*Nuc., Radiol.*). Sensitive tubular electroscope, using a metallized quartz fibre which is viewed against a scale calibrated in röntgens; it is charged by a generator and radiation discharges it. Used by Civil Defence and workers with radio-isotopes.

personal equation. The correction which is to be applied to the reading of an instrument on account of the tendency of the observer to read too high or too low. For a given observer and instrument, the equation is usually constant.

personality (*Psychol.*). The integrated organization of all the psychological, intellectual, emotional, and physical characteristics of an individual, especially as they are presented to other people.

personate (*Bot.*). Said of a 2-lipped corolla which has some likeness to a mask or to the face of an animal, as in the snapdragon.

personnel monitoring (*Radiol.*). Monitoring for radioactive contamination of any part of an individual, his breath or excretions, or any part of his clothing.

persorption (*Chem.*). The extremely effective absorption of a gas by a solid, with the formation of an almost molecular mixture of the two substances.

perspective (*Acous.*). See **acoustic-.**

Perspex (*Plastics*). Proprietary thermoplastic resin of polymethyl methacrylate of exceptional transparency and freedom from colour in the unpigmented state. Available also in wide range of colours.

persuador (*Electronics*). The electrode in the image

orthicon camera tube which deflects electrons returned from the target into the electron multiplier.

PERT (*Work Study*). Abbrev. for *programme evaluation and review technique*.

Perthe's disease (*Med.*). Osteochondritis deformans juvenilis. A deformed condition of the epiphysis of the head of the femur in young children, associated with a painful limp.

perthite (*Min.*). The general name for megascopic intergrowths of potassium and sodium feldspars, both components having been miscible to form a homogeneous compound at high temperatures, but the one having been thrown out of solution at a lower temperature, thus appearing as inclusions in the other. Perthite may also be formed by the replacement of potassium feldspar by sodium feldspar. See also **microperthite.**

perthosite (*Geol.*). A type of soda-syenite consisting to a very large extent of perthitic feldspars, occurring at Ben Loyal and Loch Ailsh in Scotland.

perturbations (*Astron.*). Those departures from regularity in the orbital motion of a planet, comet, or other body, caused by the disturbing gravitational force of another body.

perturbation theory (*Phys.*). Method involving solution of simplified equation representing behaviour of system and substituting this in more exact equation including second-order terms to determine the effect of these second-order perturbations. Widely used in quantum mechanics and in allowing for localized neutron-flux variations in large reactors.

pertusate (*Bot.*). Perforated; pierced by slits.

pertussis (*Med.*). Whooping cough. An acute infectious disease, due to infection with the bacillus *Haemophilus pertussis*; characterized by catarrh of the respiratory tract, also by periodic, recurring spasms of the larynx, which end in the prolonged crowing inspiration known as the 'whoop'.

pervaporation (*Chem.*). The evaporation of a solvent from the outside of a semipermeable or dialysing membrane.

perveance (*Electronics*). The constant G in Langmuir's equation $i = GV^{3/2}$, for the anode current i of a space-charge limited diode operating with anode voltage V.

pes (*Zool.*). The podium of the hind limb in land Vertebrates. *pl.* **pedes.**

pes cavus, pes arcuatus (*Med.*). Claw foot. A condition of the foot in which the balls of the toes approximate to the heel, so that the foot is shortened and the instep abnormally high.

pes planus, pes valgus (*Med.*). Flat foot. A condition of the foot in which the longitudinal arch is lost, so that the foot is flattened and turned outwards.

pessary (*Med.*). An instrument worn in the vagina to remedy displacements of the uterus: any occlusive or medicated appliance for insertion into the vagina to prevent conception.

pessulus (*Zool.*). In Birds, an osseous band which traverses the trachea horizontally at the point of origin of the bronchi, forming a vertical septum between the 2 bronchial apertures.

pestle (*Chem.*). A club-shaped instrument, of glass or porcelain, for grinding and pounding solids in a mortar.

petal (*Bot.*). One of the leaves (often coloured) composing the corolla.

Petalichthyida (*Zool.*). See **Macropetalichthyida.**

petalite (*Min.*). A silicate of lithium and aluminium which crystallizes in the monoclinic system. It typically occurs in granite pegmatites.

petalody (*Bot.*). The transformation of stamens into petals.

petaloid (*Bot., Zool.*). Looking like a petal; petal-shaped, as the dorsal parts of the ambulacra in certain *Echinoidea.*

petalomania (*Bot.*). An abnormal increase in the number of petals.

petasma (*Zool.*). A curtainlike structure of certain Prawns (*Penaeidae, Sergestidae*) formed by the union of lobes projecting inwards from the base of the first pair of pleopods in the male.

pet cock (*Eng.*). A small plug-cock for draining condensed steam from steam-engine cylinders, or for testing the water-level in a boiler.

petechia (*Med.*). A small red spot due to minute haemorrhage into the skin.

Peterfi's method (*Micros.*). See **celloidin-paraffin embedding.**

Petersen coil (*Elec. Eng.*). A reactor placed between the neutral point of an electric-power system and earth. The value of the reactance is such that, when an earth fault occurs, the current through the reactor exactly balances the capacitance current flowing through the fault, so that any tendency to arcing is suppressed. Also called **arc-suppression coil.**

pethidine (*Pharm.*). A synthetic analgesic and hypnotic, having action similar to that of morphine.

petiolate (*Bot.*). Having a leaf stalk.

petiole (*Bot.*). The stalk of a leaf. (*Zool.*) The narrow stalklike zone formed by the constriction of the second abdominal segment in certain *Hymenoptera* (*Apocrita*); cf. *propodeum.*

petiolule (*Bot.*). The stalk of a leaflet of a compound leaf.

petit mal (*Med.*). A form of epileptic attack in which convulsions are absent and certain transient phenomena, e.g., brief loss of consciousness, occur.

PETN (*Chem.*). See **pentaerythritol tetranitrate.**

petoscope (*Electronics*). Development of *aniseikon*, for detecting movement in a wide angle of view, especially in outdoor operations.

petrail (*Carp.*). A heavy beam in the framing of a timber building.

petrifaction, petrified (*Geol.*). Terms applied to any organic remains which have been changed in composition by molecular replacement but whose original structure is in large measure retained. *Petrified wood* is wood which has had its structure replaced by calcium carbonate or silica. Many of the original minute structures are preserved.

petrifying liquid (*Paint.*). (1) A thin, waterproofing liquid for brickwork. (2) Emulsions of oils and resins in size water, etc., containing linseed oil emulsion or a diluted polyvinyl acetate emulsion, added to distempers to improve their weather resistance for outside use.

petrissage (*Med.*). A kneading manipulation.

petrochemicals (*Chem.*). Those derived from crude oil or natural gas. They include light hydrocarbons such as butene, ethene and propene, obtained by fractional distillation or catalytic cracking.

petrocoles (*Ecol.*). Animals living in rock crevices or among rocks.

petrographic province (*Geol.*). See **comagmatic assemblage.**

petrography (*Geol.*). Systematic description of rocks, based on observations in the field, on hand specimens, and on thin microscopic sections. Cf. *petrology.*

petrol (*Fuels*). Originally a TN in Britain for a

particular *motor spirit* (q.v.) but now in general use. Also called **gasoline** or (esp. U.S.A. and Canada) simply **gas**.

petrol-electric generating set (*Elec. Eng.*). A small generating plant using a petrol engine as the prime-mover.

petrol engine (*I.C. Engs.*). A reciprocating engine, working on the Otto 4-stroke or the 2-stroke cycle, in which the air charge is carburetted by a petrol spray from a carburettor, or alternatively by *fuel injection* (q.v.). In 4-stroke engines, inlet and exhaust valves control the entry of charge and the exit of exhaust gases; in 2-stroke engines, the piston is usually made to act both as inlet and exhaust valve. Ignition of the combustible mixture is effected by sparking-plug, operated either by coil and battery or by magneto.

petroleum (*Chem.*). Naturally-occurring green to black coloured mixtures of crude hydrocarbon oils, found as earth seepages or obtained by boring. Petroleum is widespread in the Earth's crust, notably in the U.S., U.S.S.R. and the Middle East. In addition to hydrocarbons of every chemical type and boiling range, petroleum often contains compounds of sulphur, vanadium, etc. Commercial petroleum products are obtained from crude petroleum by distillation, cracking, chemical treatment, etc.

petroleum jelly (*Chem.*). A mixture of petroleum hydrocarbons; used for making emollients, for impregnating the paper covering of electric cables, and as a lubricant. Also **petrolatum.** See **liquid paraffin.**

petrol injection (*I.C. Engs.*). See **fuel injection.**

petrology (*Geol.*). That study of rocks which includes consideration of their mode of origin, present conditions, chemical and mineral composition, their alteration and decay.

petrol pump (*Autos.*). A small pump of the diaphragm type, operated either mechanically from the camshaft, or electrically. It draws petrol from the tank and delivers it to the carburettor.

Petromyzontia (*Zool.*). A subclass and order of the *Cyclostomata*, with a well-developed dorsal fin, a complete branchial basket, and the naso-hypophysial sac ending posteriorly in a blind sac. Lampreys.

petrosal (*Zool.*). In higher Vertebrates, a bone formed by the fusion of the various otic bones.

petrous (*Zool.*). Stony, hard (as a portion of the temporal bone in higher Vertebrates): situated in the region of the petrous portion of the temporal bone.

petticoat (*Elec. Eng.*). One of the umbrella-shaped shields commonly provided on pin-type insulators in order to increase the length of the leakage path which will remain dry under rain conditions.

petzite (*Min.*). A telluride of silver and gold. It is steel-grey to black and often shows tarnish.

Petzval curvature (*Optics*). The curvature of the image surface of a lens system in which spherical aberration, coma, and astigmatism have been corrected.

Petzval lens (*Photog.*). Early type of camera lens comprising two widely separated pairs of achromatic elements. Developed by Josef Petzval in 1840.

pewter (*Met.*). An alloy containing 80–90% of tin and 10–20% of lead. In modern pewter, antimony may replace tin, and from 1–3% copper is usual.

Peyer's patches (*Zool.*). In higher Vertebrates, oval aggregations of lymph follicles occurring on the walls of the intestine and consisting of masses of reticular tissue containing numerous lymphocytes.

Pezizales (*Bot.*). An order of *Ascomycete fungi*, with the asci in a well-developed hymenium and the apothecium expanded, flattened, or cup-shaped. The fruit body is characteristically leathery, gelatinous, or fleshy. Includes saprophytic and parasitic species, e.g., *Sclerotinia fructigena* (brown rot of apples).

pezizoid (*Bot.*). Resembling a cup-shaped apothecium.

pf (*Elec. Eng.*). Abbrev. for *power factor.*

pF (*Elec. Eng.*). Abbrev. for *picofarad.*

PF (*Build.*). Abbrev. for *plain face.*

Pfannkuch protection (*Cables*). A protective system for use with the cables of an electric-power system; some of the strands of the cable are lightly insulated from the others and have an e.m.f. applied between them, this e.m.f. causing a current to flow and operate relays if the insulation is destroyed by a fault.

Pferdestärke, PS (*Eng.*). See **cheval-vapeur.**

PFI & R (*Build.*). Abbrev. for *part fill in and ram.*

pfleidering (*Textiles*). Method and machine for shredding pressed alkali-cellulose for rayon manufacture.

Pflüger's egg-tubes (*Zool.*). In higher Vertebrates, rounded strings of epithelium cells growing inwards into the stroma of the ovary from the germinal epithelium.

PF resins (*Plastics*). *Phenol formaldehyde (methanal) resins.* See **phenolic resins.**

Pfund series (*Phys.*). A series of lines in the far infrared spectrum of hydrogen. Their wave numbers are given by the same expression as that for the Brackett series but with $n_1 = 5$.

P gas (*Chem.*). One based on argon, used for gas-flow counting.

Ph (*Chem.*). A symbol for the phenyl radical C_6H_5—.

pH (*Chem.*). See **pH-value.**

phacella (*Zool.*). In some *Coelenterata*, a gastral filament bearing nematocysts.

Phacidiales (*Bot.*). An order of *Ascomycete fungi* with resemblances to the *Pezizales*; in all plant parasites, the apothecium is embedded in the tissue of the host, the hymenium only being exposed at maturity.

phacoidal structure (*Geol.*). A rock structure in which mineral or rock-fragments of lenslike form are included. The term is applicable to igneous rocks containing softened and drawn-out inclusions; also to metamorphic rocks such as crush-breccias and crush-conglomerates; and to certain gneisses. (Gk. *phakos*, lentil.)

phacolite (*Min.*). A variety of *chabazite* (q.v.), so named from its lenslike crystal form.

phacolith (*Geol.*). A minor intrusion of igneous rock occupying the crest of an anticlinal fold. Its form is due to the folding, hence it is not the cause of the uparching of the roof (cf. *laccolith*). The type example is the Corndon phacolith in Shropshire, England.

phacomalacia (*Med.*). Pathological softening of the lens of the eye.

phaeic, phaeochrous (*Zool.*). Dusky. *n.* **phaeism.**

phaeism (*Zool.*). In butterflies, incomplete *melanism* (q.v.).

phaenogamous (*Bot.*). Relating to a flowering plant.

phaeo-, pheo-. Prefix from Gk. *phaios*, dusky.

phaeochromacytoma (*Med.*). See **paraganglioma.**

phaeochrous (*Zool.*). See **phaeic.**

phaeodium (*Zool.*). In some *Radiolaria*, a greenish or brownish mass, situated at the main opening of the central capsule; it consists of food-material and excretory substances.

Phaeophyta, Phaeophyceae (*Bot.*). The brown seaweeds. A phylum or class of algae, mostly marine, in which the green chlorophyll is more or less masked by a yellow pigment, fucoxanthin, so that the plants look brownish. The thallus may be of simple construction, but it is often very complicated; some members of the group are the largest algae known, and may reach several metres in length.

phaeoplast (*Bot.*). The chromatophore of the *Phaeophyceae*.

phaeosporous (*Bot.*). Having dark-coloured 1-celled spores.

phag-, phago-, -phage, -phagy. Prefix and suffix from Gk. *phagein*, to eat.

phage (*Bacteriol.*). An abbrev. for *bacteriophage*.

phagedaena, phagedena (*Med.*). Rapidly spreading and destructive ulceration.

phagocyte (*Cyt.*). A cell which exhibits amoeboid phenomena, i.e., spontaneous change of form, ability to throw out pseudopodia (and so to engulf foreign bodies), and to migrate from place to place. See **bacteriotropin.**

phagocytosis (*Cyt.*). Destruction of tissue-cells or microparasites by the action of phagocytes.

phagolysis (*Histol.*). The breakdown of phagocytes, usually by bacteria.

phagosome (*Cyt.*). Membrane-bounded cytoplasmic vesicle in which recently pinocytosed material is segregated.

phalanges (*Zool.*). In Vertebrates, the bones supporting the segments of the digits: fiddle-shaped rings composing the reticular lamina of the organ of Corti. *sing.* **phalanx.**

Phalangida (*Zool.*). An order of *Arachnida*, whose prosoma is covered by a single tergal shield and united to the invariably segmented opisthosoma by its whole breadth. Chelicerae 3-jointed and chelate, pedipalps leglike, 2 simple eyes. Harvestmen.

phalanx (*Zool.*). See **phalanges.**

Phalaris staggers (*Vet.*). A nervous disease of sheep in Australia caused by eating the grass *Phalaris tuberosa*; believed to be caused by a toxic substance in the grass when eaten by cobalt-deficient sheep.

phallus (*Zool.*). The penis of Mammals: the primordium of the penis or clitoris of Mammals. *adj.* **phallic.**

phanero-. Prefix from Gk. *phaneros*, visible.

phanerocodonic (*Zool.*). Detached and free-swimming; said of a medusoid person.

phanerocrystalline (*Min.*). Said of an igneous rock in which the crystals of all the essential minerals can be discerned by the naked eye.

Phanerogamae, Phanerogamia (*Bot.*). Old term for plants reproducing by seeds. Now replaced by *Spermatophyta*; subdivided into *Gymnospermae* and *Angiospermae* (qq.v.).

phanerogamous (*Bot.*). Relating to a flowering plant.

phanerophyte (*Bot.*). A tree or shrub with the perennating buds borne more than 25 cm above soil level.

phanotron (*Electronics*). Large hot-cathode gas diode for industrial use.

phantastron (*Electronics*). Valve circuit for generating pulses which are delayed from trigger pulses.

phantom antenna (*Radio*). Same as **artificial antenna.**

phantom circuit (*Telecomm.*). An additional circuit using a pair of transmission lines in parallel for the *go* and another similar pair for *return*, being fed at the midpoint of bridging or superposing transformers.

phantom material (*Radiol.*). That producing absorption and back scatter of radiation very similar to human tissue, and hence used in models to study appropriate doses, radiation scattering, etc. A *phantom* is a reproduction of (part of) the body in this material. Also termed **tissue equivalent material.**

phantom ring (*Light*). The coloured ring of light which appears to surround an observer's shadow when the shadow falls on an extended bank of fog or cloud.

pharate (*Zool.*). In Insects, refers to a phase of development when the old cuticle of one stage is separate from the hypodermis of the next stage, but has not yet been ruptured and cast off. The so-called *pupa* of many Insects actually represents a pharate adult, and the so-called *prepupa* is a pharate pupa.

pharmacodynamics (*Pharm.*). The science of the action of drugs; pharmacology.

pharmacolite (*Min.*). A hydrated arsenate of calcium which crystallizes in the monoclinic system. It is a product of the late alteration of mineral deposits which carry arsenopyrite and the arsenical ores of cobalt and silver.

pharmacology. The scientific study of drugs (Gk. *pharmakon*, drug) and their action. *n.* **pharmacologist.**

pharmacosiderite (*Min.*). Hydrated arsenate(v) of iron. It crystallizes in the cubic system, and is a product of the alteration of arsenical ores.

pharotaxis (*An. Behav.*). A *taxis* (q.v.) in which an animal directs its movements towards one definite spot. This differs from a *telotaxis* (q.v.) in that the stimulus situation acquires its significance by conditioning.

pharyngeal sclerite (*Zool.*). See **basal sclerite.**

pharyngismus (*Med.*). Spasm of the muscles of the pharynx.

pharyngitis (*Med.*). Inflammation of the pharynx.

pharyngobranchial (*Zool.*). The uppermost element of a branchial arch.

pharyngocutaneous duct (*Zool.*). In *Myxinoidea*, a tube running from the pharynx either to an aperture behind, and close to, the last gill slit on the left side, or opening into the left external branchial aperture, and allowing the expulsion of large inhaled particles.

pharyngohyal (*Zool.*). In *Holocephali*, a small cartilage forming part of the hyoid arch, lying above the epihyal. Evidently serially homologous with the pharyngobranchials, it represents the hyomandibular of *Selachii*, but plays no part in the support of the jaws.

pharyngoplasty (*Surg.*). The surgical alteration, or reconstruction, of the pharynx in congenital and acquired disease.

pharyngoplegia (*Med.*). Paralysis of the muscles of the pharynx.

pharyngoscope (*Med.*). An instrument for viewing the pharynx.

pharyngotomy (*Surg.*). Incision into the pharynx.

pharynx (*Zool.*). In Vertebrates, that portion of the alimentary canal which intervenes between the mouth cavity and the oesophagus and serves both for the passage of food and the performance of respiratory functions: in Invertebrates, the corresponding portion of the alimentary canal lying immediately posterior to the buccal cavity, usually having a highly muscular wall. *adj.* **pharyngeal.**

phase (*Astron.*). Of the moon, the name given to the changing shape of the visible illuminated surface of the moon due to the varying relative positions at the earth, sun, and moon during the synodic month. Starting from *new moon*, the phase increases through *crescent*, *first quarter*, *gibbous*, to *full moon*, and then decreases through *gibbous*, *third quarter*, waning to *new moon* again. The inferior planets show the same phases, but in the reverse order; the superior planets can show a gibbous phase, but not a crescent. (*Chem.*) The sum of all those portions of a material system which are identical in chemical composition and physical state, and are separated from the rest of the system by a distinct interface called the *phase boundary*. (*Elec. Eng.*, *Telecomm.*) Fraction of a cycle of periodic disturbance which has been completed at specific reference time, e.g., at the start of a cycle of a second disturbance of the same frequency. Expressed as an angle, one cycle corresponding to 2π radians or 360°. The terms 'in phase', 'in quadrature' and 'out of phase' correspond to phase angles between two disturbances of 0 (or 2π), $\pi/2$ and π respectively. See also **polyphase, single phase, two-phase and three-phase.**

phase advancer (*Elec. Eng.*). A machine connected in the secondary circuit of an induction motor to improve its power factor.

phase angle (*Elec. Eng.*, *Telecomm.*). See under **phase.** Also given by δ, where $\tan \delta$ = (reactance/resistance) for an alternating-current circuit or an acoustic system.

phase compensation (*Telecomm.*). See **line equalizer.**

phase constant (*Telecomm.*). Imaginary part of propagation constant of a line, or of the transfer constant of a filter section. It is expressed in radians per unit length or section. See **image-.**

phase-contrast microscopy (*Micros.*). A method of particular value in examining living unstained specimens, whereby phase differences are converted to differences of light intensity by the phase-contrast microscope. Contrasts in the image are produced wherever differences in refractive index or thickness occur.

phase converter (*Elec. Eng.*). On applying a single-phase voltage to one phase of a rotating 3-phase induction motor, it will be found that 3-phase voltages, which are approximately balanced, may be derived from the 3 terminals of the machine.

phase corrector (*Telecomm.*). A circuit for correcting phase distortion.

phased (*Cinema.*). Said of camera drives when they are switched on to speed-controlled mains, ready to be brought up to speed by starting the generator which supplies these mains.

phased array (*Radar*). A *linear array* which has a facility for steering the beam electronically by varying the progressive phase shift between the elements.

phase defect (*Elec. Eng.*, *Telecomm.*). The phase difference between the actual current in a capacitor and that which would flow in an equivalent ideal (loss-free) capacitor.

phase delay (*Telecomm.*). Delay, in radians or seconds, for transmission of wave of a single frequency through whole or part of a communication system. Also **phase retardation.**

phase delay distortion (*Telecomm.*). That of a signal transmitted over a long line, so that difference in time of arrival of components of a complex wave is significant.

phase deviation (*Telecomm.*). In phase modulation, maximum difference between the phase angle of the modulated wave and the angle of the nonmodulated carrier.

phase diagram (*Met.*). See **constitutional diagram.**

phase difference (*Elec. Eng.*, *Telecomm.*). The phase angle by which one periodic disturbance *lags* or *leads* on another.

phase discriminator (*Telecomm.*). Circuit preceding demodulator in phase-modulation receiver. It converts the carrier to an amplitude-modulated form.

phase distortion (*Telecomm.*). Found in the waveform of transmitted signal on account of nonlinear relation of wavelength constant of a line, or image phase constant of amplifier or network, with frequency.

phase equalization (*Telecomm.*). See **line equalizer.**

phase equalizer (*Elec. Eng.*). See **interphase transformer.**

phase focusing (*Electronics*). Effect used in electron bunching whereby those lagging in phase tend to gain energy from the field and those leading tend to surrender it. A similar effect occurs in synchrotron charged-particle accelerators.

phase intercept (*Telecomm.*). Phase delay, in radians for zero frequency, for the whole or part of a transmission system; obtained by extrapolating the measured or calculated curve for phase delay with respect to frequency.

phase-intercept distortion (*Telecomm.*). Distortion in a received signal waveform solely because phase delay for zero frequency is not an exact multiple of π.

phase inversion (*Telecomm.*). Any arrangement for obtaining an additional voltage of reverse phase for driving both sides of a push-pull amplifier stage.

phase inverter (*Telecomm.*). See **grid neutralization.**

phase lag, phase lead (*Elec. Eng.*, *Telecomm.*). If a periodic disturbance *A* completes a cycle when a second disturbance *B* has completed $\varphi/2\pi$ of a cycle, *B* is said to *lead A* by a phase angle φ. Conversely *A lags* on *B* by φ or *leads B* by $(2\pi-\varphi)$.

phase-locked loop (*Telecomm.*). Control loop incorporating a voltage-controlled oscillator and phase detector in order to lock a signal to a stable reference frequency. Used in frequency synthesis and threshold extension demodulators.

phase margin (*Telecomm.*). Variation in phase of unity-gain from that which would give instability. Readily measured from a Nyquist diagram.

phase meter (*Telecomm.*). One for measuring the phase difference between 2 signals of the same frequency.

phase modifier (*Elec. Eng.*). A term sometimes used to denote a synchronous capacitor when this is used for varying the power factor of the current in a transmission line to effect voltage regulation.

phase modulation (*Telecomm.*). Periodic variation in phase of a high-frequency current or voltage in accordance with a lower impressed modulating frequency. Occurs as an unwanted by-product of amplitude modulation, but can be independently produced.

phaseoliform (*Bot.*). Shaped like a bean.

phase plate (*Optics*). A quarter wave retarding plate used in phase-contrast microscopy.

phase reaction (*Chem.*). A reaction involving a change in the proportions of the phases present in a system. It takes place at constant temperature and involves the absorption or evolution of heat.

phase resonance (*Telecomm.*). That in which the induced oscillation differs in phase by $\pi/2$ from the forcing disturbance. Also termed **velocity resonance.**

phase retardation (*Telecomm.*). See **phase delay.**

phase reversal (*Chem.*). An interchange of the components of an emulsion; e.g., under certain conditions an emulsion of an oil in water may become an emulsion of water in the oil.

phase rule (*Chem.*). A generalization of great value in the study of equilibria between phases. In any system, $P+F=C+2$, where P is the number of phases, F the number of *degrees of freedom* (1, q.v.), C the number of components.

phase sensitive detector (*Telecomm.*). One in which the output signal changes sign when the phase of the input signal is reversed.

phase sequence (*Elec. Eng.*). The order in which the phase voltages of a polyphase system reach their maximum values. If the phases of a three-phase system are given the standard colourings of red, yellow, blue, this phase sequence is said to be a *positive phase sequence*. See **negative-, zero-.**

phase-shift control (*Electronics*). Control of initial discharge of a gas tube, e.g., thyratron, by the phase of voltage applied to a grid.

phase shifter (*Telecomm.*). Waveguide or coaxial line component which produces any selected phase delay in signal transmitted.

phase-shifting circuit (*Telecomm.*). One in which the relative phases of components in an a.c. waveform or signal can be continuously adjusted.

phase-shifting transformer (*Elec. Eng.*). One for which the secondary voltage can be varied continuously in phase. Usually with rotatable secondary winding and polyphase primaries. Also **phasing transformer.**

phase shift keying (*Telecomm.*). Transmission of coded data using phase modulation with several discrete phase angles. Abbrev. **PSK.**

phase-shift oscillator (*Telecomm.*). Same as **line oscillator.**

phase space (*Phys.*). Six-dimensional space, three coordinates of position and three of momentum; used for study of particle systems. See **microscopic state.**

phase splitter (*Elec. Eng.*). A means of producing two or more waves which differ in phase from a single input wave.

phase stability (*Nuc. Eng.*). The automatic synchronization of the time of rotation of the charged particle in a frequency-modulated cyclotron.

phase swinging (*Elec. Eng.*). Periodic variations in the phase angle between 2 synchronous machines running in parallel. (*TV*) Lack of synchronism throughout individual cycles of frame-frequency generators at the transmitting and receiving ends of a TV system, causing the received picture to wander over the screen.

phase velocity (*Phys.*). Velocity of propagation of any one phase state, such as a point of zero instantaneous field, in a steady train of sinusoidal waves. It may differ from the velocity of propagation of the disturbance, or *group velocity*, and, in transmission through ionized air, may exceed that of light. Also called **wave velocity.**

phase voltage (*Elec. Eng.*). See **voltage to neutral.**

phasing (*TV*). In facsimile transmission and reception, adjustment of reproduced picture to correspond exactly with the original, achieved by a phasing signal.

phasing transformer (*Elec. Eng.*). See **phase-shifting transformer.**

phasitron (*Electronics*). Thermionic valve with split concentric electrodes, using 3-phase voltages and fan beams of electrons. It is employed to modulate a high-frequency electron current.

phasmajector (*TV*). Device for providing standard video signal for testing TV circuits, the signal being electronically generated.

Phasmidia (*Zool.*). A class of *Nematoda* having phasmids (caudal sensory organs) and a well-developed excretory system. Males usually have two copulatory spicules.

phasor (*Elec. Eng.*). See **complexor.**

phellandral (*Chem.*). A terpene aldehyde, found in eucalyptus and water fennel oils.

phellandrene (*Chem.*). $C_{10}H_{16}$, a terpene existing in a *d*- and *l*-form, which is Δ-1,5-terpadiene, b.p. 62°C (12 mm).

phellem (*Bot.*). The tissue formed externally to the phellogen; cork.

phelloderm (*Bot.*). A layer or layers of thin-walled cells, with cellulose walls, often containing starch and sometimes chloroplasts, formed internally to the phellogen.

phellogen (*Bot.*). The layer of meristematic cells lying a little inside the surface of a root or stem, forming cork on its outer surface and phelloderm internally; the cork cambium.

phelloid (*Bot.*). A crust of nonsuberized or weakly suberized cells present in the surface of some plants, replacing true cork.

phenacaine (*Pharm.*). Para-phenetidylphenacetin hydrochloride. Used as a local anaesthetic in dilute solution in ophthalmology. Also known as **Holocaine.**

phenacetin (*Pharm.*). $C_2H_5O{\cdot}C_6H_4{\cdot}NH{\cdot}CO{\cdot}CH_3$, aceto-4-phenetidine, colourless crystals, m.p. 135°C, used as an antipyretic.

phenakite (*Min.*). An orthosilicate of beryllium, crystallizing in the rhombohedral system. It is commonly found as a product of pneumatolysis. Sometimes cut as a gemstone, having great brilliance of lustre but lacking fire. The name (Gk. 'the deceiver') refers to the frequency with which it has been confused with quartz.

phenanthraquinone (*Chem.*). A diketone obtained from phenanthrene and chromic(VI)acid:

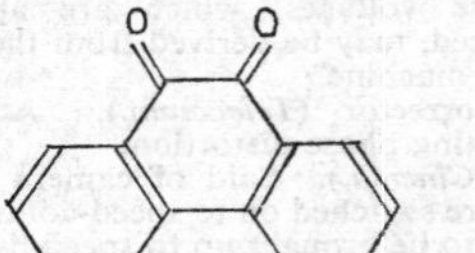

O O

It crystallizes in orange needles: m.p. 200°C. Useful for the detection of diamines with an ortho grouping.

phenanthrene (*Chem.*). $C_{14}H_{10}$, white, glistening plates, m.p. 99°C, b.p. 340°C; its solutions show a blue fluorescence. It has the formula:

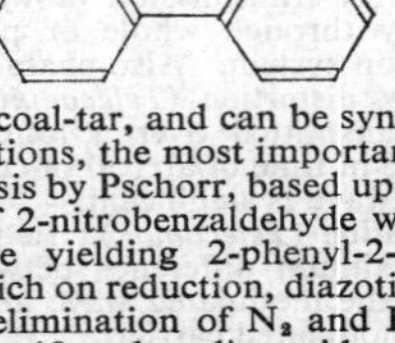

It occurs in coal-tar, and can be synthesized by various reactions, the most important of which is the synthesis by Pschorr, based upon the condensation of 2-nitrobenzaldehyde with sodium phenylacetate yielding 2-phenyl-2-nitrocinnamic acid, which on reduction, diazotization, and subsequent elimination of N_2 and H_2O, yields phenanthrene-10-carboxylic acid. This acid yields phenanthrene on distillation. The 9-10-

phenanthrene bridge is readily attacked by reagents, yielding diphenyl derivatives.

phenates (*Chem.*). Phenolates, phenoxides. Salts formed by phenols, e.g., C_6H_5ONa, sodium phenate (phenolate, phenoxide).

phenazine (*Chem.*). A heterocyclic compound, which can be considered as anthracene in which two CH groups have been replaced by two N atoms. It has the formula:

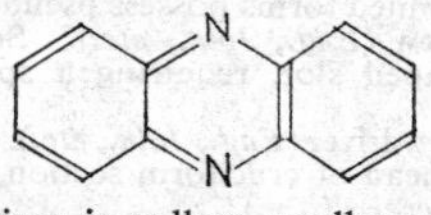

It crystallizes in yellow needles, m.p. 171°C, and can be readily sublimed.

phenazone (*Pharm.*). See **antipyrine.**

4-phenetidine (*Chem.*). $H_2N{\cdot}C_6H_4{\cdot}OC_2H_5$, the ethyl ether of 4-aminophenol, basis for a number of pharmaceutical preparations, e.g., phenacetin, $H_3C{\cdot}CO{\cdot}NH{\cdot}C_6H_4{\cdot}OC_2H_5$.

phenetole (*Pharm.*). $C_6H_5{\cdot}O{\cdot}C_2H_5$, phenyl ethyl ether, a colourless liquid of pleasant odour, b.p. 172°C.

phengite (*Min.*). An end-member variety of muscovite mica, in which Si : Al > 3 : 1 and some aluminium is replaced by magnesium or iron.

Phenidone (*Photog.*). TN for 1-phenyl-3-pyrazolidinone, a developing agent with characteristics similar to *metol* (q.v.):

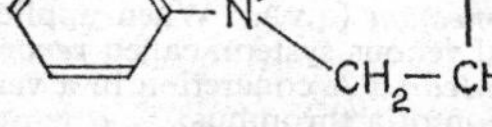

phenobarbitone (*Pharm.*). Phenylethyl-barbituric acid. White crystalline powder; m.p. 174°C. Used as a hypnotic and sedative, particularly in epilepsy. *Luminal* is one of numerous proprietary forms.

phenocrysts (*Geol.*). Large (megascopic) crystals, usually of perfect crystalline shape, found in a fine-grained matrix in igneous rocks. See **porphyritic texture.**

phenol (*Chem.*). C_6H_5OH, carbolic acid, colourless hygroscopic needles, m.p. 43°C, b.p. 183°C, chief constituent of the coal-tar fraction boiling between 170° and 230°C, soluble in caustic soda and potash, forming Na and K phenates or phenolates. It forms with bromine 2,4,6-tribromophenol. Sodium phenolate reacts with CO_2 under certain conditions, yielding sodium salicylate. It is a strong disinfectant.

phenolic acids (*Chem.*). A group of aromatic acids containing one or more hydroxyl groups attached to the benzene nucleus. The 2-hydroxy acids are volatile in steam, soluble in cold chloroform, and give a violet or blue coloration with iron(III)chloride. The 3-hydroxy acids are the most stable acids. Important phenolic acids are *salicylic acid, gallic acid, tannin.*

phenolic alcohols (*Chem.*). A group of compounds possessing phenolic properties in addition to those of an alcohol. Some of them occur as glucosides in nature. Important representatives of this group are saligenin or 2-hydroxybenzyl alcohol, and coniferyl alcohol.

phenolic resins (*Plastics*). A large group of plastics, made from a phenol (phenol, 3-cresol, 4-cresol, catechol, resorcinol, or quinol) and an aldehyde (methanal, ethanal, benzaldehyde, or furfuraldehyde). An acid catalyst produces a soluble and fusible resin (*modified resin*) used in varnishes and lacquers; while an alkaline catalyst results in the formation, after moulding, of an insoluble and infusible resin. Phenolics may be moulded, laminated, or cast.

phenology (*Biol.*). The study of organisms in relation to climate.

phenolphthalein (*Chem.*). A triphenylmethane derivative, obtained by the condensation of phthalic anhydride and phenol by the action of concentrated sulphuric acid. It has the formula:

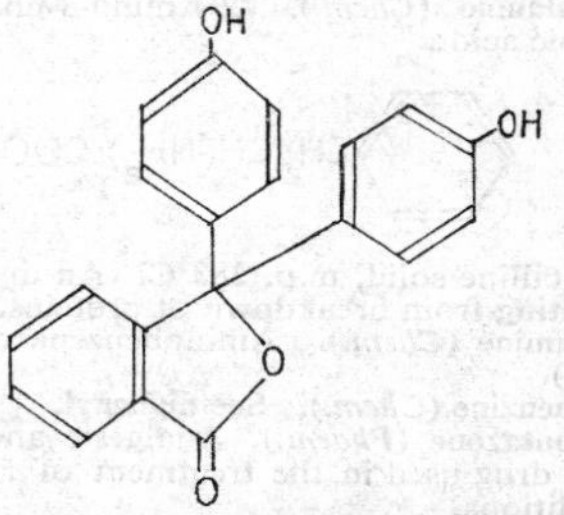

It forms colourless crystals which dissolve with a red colour in alkalis. It is used as an indicator in volumetric analysis. The colourless substance has the lactone formula, whereas its coloured salts have a quinonoid structure. Used as a laxative.

phenol red (*Chem.*). Phenolsulphonphthalein:

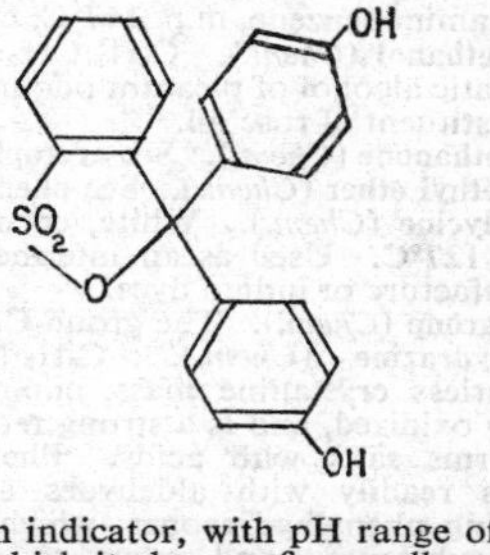

Used as an indicator, with pH range of 6·8 to 8·4, over which it changes from yellow to red; also used for testing the functioning of the kidney.

phenols (*Chem.*). Hydroxybenzenes. A group of aromatic compounds having the hydroxyl group directly attached to the benzene nucleus. They give the reactions of alcohols, forming esters, ethers, thiocompounds, but also have feeble acidic properties and form salts or phenolates by the action of NaOH or KOH. Phenols are divided into mono-, di-, tri-, tetra-, and polyhydric phenols. Phenols are more reactive than the benzene hydrocarbons.

phenomenology (*Psychol.*). The study of conscious experience (not just sensations) in its own right and the insistence on the intuitive foundation and verification of formal concepts.

phenothiazine (*Chem.*). A heterocyclic compound of the formula:

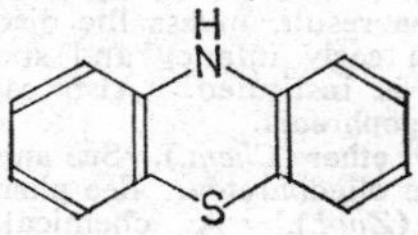

phenotype (*Gen.*). The observable characteristics of an organism produced by the interaction of genes and environment. Cf. *genotype*.

phenoxy resins (*Plastics*). Range of thermoplastic resins based on polyhydroxy ethers, produced from bisphenol A and epichlorhydrin. Used for injection and extrusion mouldings, blow mouldings, and solution coatings.

phenylacetic (ethanoic) acid (*Chem.*). $C_6H_5{\cdot}CH_2{\cdot}COOH$, colourless crystals, m.p. 76°C.

phenylalanine (*Chem.*). 2-Amino-3-phenyl-propanoic acid:

$$C_6H_5{\cdot}CH_2.CH(NH_2){\cdot}COOH$$

Crystalline solid, m.p. 283°C. An amino acid resulting from breakdown of proteins.

phenylamine (*Chem.*). Aminobenzene or *aniline* (q.v.).

phenylbenzine (*Chem.*). See **diphenyl.**

phenylbutazone (*Pharm.*). Analgesic and antipyretic drug used in the treatment of rheumatic conditions.

phenyl cyanide (*Chem.*). See **benzonitrile.**

phenylenediamines (*Chem.*). $C_6H_4(NH_2)_2$, obtained by the reduction of the dinitro, the nitro-amino, or the aminoazo compounds. The diamines are crystalline substances with strong basic properties. There are three isomers, viz., 1,2 diaminobenzene, m.p. 102°C, b.p. 252°C; 1,3 diaminobenzene, m.p. 63°C, b.p. 287°C; 1,4 diaminobenzene, m.p. 147°C, b.p. 267°C.

phenyl ethanol (*Chem.*). $C_6H_5{\cdot}CH_2{\cdot}CH_2OH$, an aromatic alcohol of pleasant odour, b.p. 220°C, a constituent of rose oil.

phenyl ethanone (*Chem.*). See **acetophenone.**

phenyl ethyl ether (*Chem.*). See **phenetole.**

phenylglycine (*Chem.*). White, crystalline solid; m.p. 127°C. Used as an intermediate in the manufacture of indigo dyes.

phenyl group (*Chem.*). The group C_6H_5—.

phenylhydrazine (*Chem.*). $C_6H_5{\cdot}NH{\cdot}NH_2$, a colourless crystalline mass, m.p. 23°C. It is easily oxidized, and is a strong reducing agent. It forms salts with acids. Phenylhydrazine reacts readily with aldehydes and ketones, forming phenylhydrazones, which are crystalline substances and serve to identify the respective aldehydes and ketones.

phenylhydrazones (*Chem.*). The reaction products of *phenylhydrazine* (q.v.) with aldehydes and ketones, formed by the elimination of two hydrogen atoms of the amino group and the oxygen atom of the aldehyde or ketone group as water.

phenylhydroxylamine (*Chem.*). $C_6H_5{\cdot}NH{\cdot}OH$, a colourless crystalline substance, m.p. 81°C, obtained by the reduction of nitrobenzene with zinc dust and water in the presence of a mineral salt. It is very unstable and easily oxidized. Mineral acids cause intramolecular rearrangement into *p*-aminophenol, $H_2N{\cdot}C_6H_4{\cdot}OH$.

phenylketonuria (*Med.*). A rare condition in which, owing to an inborn error of metabolism, the amino acid phenylalanine cannot be converted to tyrosine and accumulates in the blood, to be ultimately excreted in the urine as phenylpyruvic acid. Severe mental retardation occurs as a result, unless the disorder is discovered in early infancy and special dietary treatment is instituted. Also called **phenylpyruvic oligophrenia.**

phenyl methyl ether (*Chem.*). See **anisole.**

phenylpyruvic oligophrenia. See **phenylketonuria.**

pheromone (*Zool.*). A chemical substance secreted by an animal, which influences the behaviour of other animals (cf. *hormone*) generally of the same species, e.g., *gyplure*, and the *queen substance* of honey bees.

phialide (*Bot.*). A short flask-shaped sterigma.

phialiform (*Bot.*). Shaped like a saucer or cup.

phillipsite (*Min.*). A fibrous zeolite; hydrated silicate of potassium, calcium, and aluminium, usually grouped in the orthorhombic system. Some twinned forms possess pseudosymmetry.

Phillips screw (*Eng., Join., etc.*). Screw with a cross shaped slot, requiring a special screwdriver.

Phillips screwdriver (*Eng., Join., etc.*). One with a pointed head of cruciform section, for turning *Phillips screws* (q.v.).

phimosis (*Med.*). Narrowness of the prepuce (foreskin) so that it cannot be drawn back over the glans penis.

phleb-, phlebo-. Prefix from Gk. *phleps*, gen. *phlebos*, vein.

phlebectomy (*Surg.*). Surgical removal of a vein or part of a vein.

phlebenteric (*Zool.*). Having branches of the alimentary canal extending into the appendages or limbs, as in *Pycnogonida*.

phlebitis (*Med.*). Inflammation of the coats of a vein.

phleboedesis (*Zool.*). The condition of having a haemocoelic body cavity, the coelom being reduced by the enlargement of the blood vessels, as in *Mollusca* and *Arthropoda*.

phlebography (*Radiol.*). The radiological examination of veins following the injection of a *contrast medium* (q.v.). When applied to the peripheral venous system, called *venography*.

phlebolith (*Med.*). A concretion in a vein due to calcification of a thrombus.

phlebosclerosis (*Med.*). Thickening of a vein due chiefly to a pathological increase in the connective tissue of the middle coat.

phlebotomus fever (*Med.*). See **sandfly fever.**

phlebotomy (*Surg.*). The cutting of a vein for the purpose of letting blood.

phlegmasia alba dolens (*Med.*). White-leg, milk-leg. Painful swelling of the leg, the skin of which is shiny and white, occurring in women after childbirth; due to thrombosis of veins and obstruction of lymphatics.

phlegmon (*Med.*). Purulent inflammation, with necrosis of tissue. *adj.* **phlegmonous.**

phlobaphene (*Bot.*). Yellow-to-brown substances occurring inside cork cells, probably derived by the decomposition of tannins.

phloem (*Bot.*). The conducting tissue present in vascular plants, chiefly concerned with the transport of elaborated food materials about the plant. When fully developed, phloem consists of sieve tubes, companion cells and parenchyma, but companion cells may not be present.

phloem fibre (*Bot.*). A sclerenchymatous element (or a strand of such elements) present in phloem, affording support to the delicate sieve tubes.

phloem island (*Bot.*). A patch of phloem surrounded by secondary wood.

phloem ray (*Bot.*). That portion of a vascular ray which traverses the phloem.

phloeoterma (*Bot.*). Endodermis in which the radial walls and the inner tangential walls are heavily thickened.

phlogisticated air (*Chem.*). An old name for nitrogen.

phlogopite (*Min.*). Hydrous silicate of potassium, magnesium, iron, and aluminium, crystallizing in the monoclinic system. It is a magnesium mica, and is usually found in metamorphosed limestones or in ultrabasic igneous rocks. Not so good an electrical insulator as *muscovite* (q.v.) at low temperatures, but keeps its water of composition until 950°C.

phloridzin (*Chem.*). A *glucoside* (q.v.) found in the roots of apple, pear, cherry, plum, and other fruit trees; phlorizin; phlorrhizin.

phloroglucinol (*Chem.*). 1,3,5-Trihydroxybenzene, $C_6H_3(OH)_3$, large prisms, m.p. 218°C, which sublime without decomposition. It can be obtained by the fusion of resorcinol with caustic soda. It possesses reducing properties, and gives a violet coloration with ferric chloride. See also **pyrogallol**.

phlycten, phlyctena, phlyctenule, phlyctenula (*Med.*). A small round, grey or yellow nodule, occurring on the conjunctiva where it covers the sclera and cornea of the eye. (Gk. *phlyctaina*, a bleb.)

phlyctenular conjunctivitis (*Med.*). Eczematous conjunctivitis. An inflammation of the conjunctiva covering the sclera and cornea of the eye and giving rise to phlyctens.

pH meter (*Chem., Elec., Phys.*). Specialized millivolt meter which measures potential difference between reference electrodes in terms of pH value of the solution in which they are immersed.

phobia (*Psychiat.*). An abnormal fear of some object or situation which may interfere with normal activities. It is more successfully treated by *behaviour therapy* than by *psychotherapy* and may be the result of maladaptive learning or conditioning.

phobotaxis (*An. Behav.*). Former name for the avoidance reaction of some *Protozoa* (e.g., *Paramecium*), now regarded as *klinokinesis* (q.v.).

phocomelia (*Med.*). A congenital malformation in which the proximal portion of the limb is absent; **phocomelus**. One so affected.

pholidosis (*Zool.*). The order and arrangement of scales, e.g., of Fish and Reptiles.

Pholidota (*Zool.*). An order of Mammals having imbricating horny scales, interspersed with hairs, over the head, body, and tail. They have a long snout, no teeth, and a long thin tongue. The hind feet are plantigrade, and the fore feet have long curved claws. Nocturnal animals, feeding on ants or termites. Resemblances to Edentate ant-eaters probably convergent. Pangolins.

phon (*Acous.*). Unit of the objective loudness on sound-level scale; the decibel unit of the 1 kHz intensity-level scale which is used for deciding the apparent loudness of an unknown sound or noise, when a measure of loudness is required. This is effected either by subjective comparison by the ear, or by objective comparison with a microphone-amplifier and a weighting network. The reference sound pressure level at 1 kHz is 0·0002 microbar = 20 $\mu N/m^2$.

phonation (*Zool.*). Sound-production.

phonendoscope (*Med.*). A stethoscope fitted with a diaphragm for the amplification of the sounds heard through it.

phonetics. Study of speech and vocal acoustics. Used also to describe the system of symbols which uniquely represent the spoken word of any language in writing, enabling the reader to pronounce words accurately in spite of spelling irregularities.

phonic wheel or drum (*Telecomm.*). Elementary synchronous motor capable of being driven with low power from electronic oscillators, so that frequency of the latter can be measured by a revolution counter.

phonmeter (*Acous.*). Apparatus for the estimation of loudness level of a sound on the phon scale by subjective comparison. Also **phonometer.**

phonochemistry (*Chem.*). The study of the effect of sound and ultrasonic waves on chemical reactions.

phonoelectrocardioscope (*Med.*). An instrument embodying a double-beam CRO, whereby the waveforms of two different functions of the heart are shown simultaneously.

phonograph (*Acous.*). See **gramophone.**

phonolite (*Geol.*). A fine-grained igneous rock of intermediate composition, consisting essentially of nepheline, subordinate alkali feldspar (sanidine), and sodium-rich coloured silicates. Termed also **clink-stone**, because it rings under the hammer when struck.

phonometer (*Acous.*). See **phonmeter.**

phonon (*Phys.*). Quantum of thermal energy in an acoustic mode for lattice vibrations of a crystal. In the limit, the interaction of acoustic waves with matter reduces to that between phonons and atoms, although for audible frequencies (ν), phonon energies ($h\nu$) are negligible (h is *Planck's constant*).

phonoplug (*Telecomm.*). A connector used on screen cables in signal circuits. It has a single central pin which is 'live' and an outer earth shield.

-phore. Suffix from Gk. *pherein*, carry. (*Chem.*) A suffix which denotes a group of atoms responsible for the corresponding property, e.g. *chromophore*.

phoresis (*Med.*). Electrical passage of ions through a membrane.

phoresy (*Zool.*). Transport or dispersal achieved by clinging to another animal; e.g., certain Mites which achieve dispersal by attaching themselves to various Insects.

phorone (*Chem.*). $(CH_3)_2C{=}CH{\cdot}CO{\cdot}CH{=}C(CH_3)_2$, yellow crystals, m.p. 28°C, b.p. 196°C, a condensation product of acetone obtained by treating it with HCl gas or $ZnCl_2$. Its activated double bonds make it a useful organic intermediate.

Phoronidea (*Zool.*). A small phylum of hermaphrodite unsegmented *Coelomata*, of tubicolous habit, having a U-shaped gut, a dorsal anus, and a lophophore in the form of a double horizontal spiral; marine forms occurring in the sand and mud of the sea bottom.

phorozooid (*Zool.*). In *Cyclomyaria*, a zooid of the media row, never sexually mature, acting as a nurse, usually bearing a gonozooid with which it is later liberated.

phorozoon (*Zool.*). An asexual form preceding, in the life-history, the sexual form.

phosgene (*Chem.*). $COCl_2$. A very poisonous, colourless, heavy gas with a nauseating, choking smell; b.p. 8°C. It is manufactured by passing carbon monoxide and chlorine over a charcoal catalyst. It is used for the manufacture of intermediates in the dyestuff industry, and as a *war gas*. Also known as **carbonyl chloride.**

phosgenite (*Min.*). A chlorocarbonate of lead, crystallizing in the tetragonal system. It is a rare mineral found in association with cerussite.

phosphagen. See **phospho-creatine.**

phosphatase (*Chem.*). A nonproteolytic protein enzyme occurring in the kidneys, intestines, and bones. Hydrolyses orthophosphoric esters to phosphoric acid and alcohol.

phosphate coatings (*Paint.*). Coatings of iron or zinc phosphate produced by reaction between the metal being coated and an acid solution. This may be simply dilute phosphoric acid, as in the original Coslett process, or may contain a variety of metal phosphates and other additives. The coatings provide a key for subsequent painting or may be filled with oil and dyed, as a decorative and protective film. Commercial processes include *Bonderizing* and *sherardizing* (qq.v.).

phosphates(V) (*Chem.*). Salts of phosphoric acid. There are three series of orthophosphates, MH_2PO_4, M_2HPO_4, and M_3PO_4; the first yield acid, the second are practically neutral, and the third alkaline, aqueous solutions. Metaphosphates, MPO_3, and pyrophosphates, $M_4P_2O_7$, are also known. All phosphates give a yellow precipitate on heating with ammonium molybdate in nitric acid.

phosphatic deposits (*Geol.*). Beds containing calcium phosphate which are formed especially in areas of low rainfall, and which may be exploited as sources of phosphate. See also **guano, phosphatic nodules.**

phosphatic nodules (*Geol.*). Rounded masses containing calcium phosphate, which are found forming under marine conditions at or near the surface of the sea-bed. They have been taken in geological sections to represent a time of inhibited sedimentation.

phosphatides (*Chem.*). Fatlike substances containing phosphoric acid. In some, glycerophosphoric acid is combined with 2 mols of fatty acid and 1 mol of a hydroxy base which may be choline (in *lecithins*), aminoethanol (in *kephalins*) or serine. In others, the base sphingosine is combined with fatty acid, phosphoric acid and choline. Also called **phospholipins.**

phosphating (*Met.*). Treatment of metal surface with hot phosphoric acid before painting in order to inhibit corrosion.

phosphaturia (*Med.*). The presence of an excess of phosphates in the urine.

phosphene (*Physiol.*). Area of luminosity in the visual field. It is caused by pressure on the eye-ball.

phosphides (*Chem.*). Binary compounds of metals with phosphorus, e.g., calcium phosphide Ca_3P_2.

phosphine (*Chem.*). Phosphorus(III)hydride. PH_3. A colourless, evil-smelling gas which usually burns spontaneously in air to form phosphorus (V)oxide. It has reducing properties and precipitates phosphides from solutions of many metallic salts.

phosphines (*Chem.*). Derivatives of PH_3, obtained by the exchange of hydrogen for alkyl radicals; classified according to the extent of substitution into *primary*, *secondary*, and *tertiary phosphines*. They correspond closely to the amines, except that they are easily oxidized even in the air, that they are only feebly basic, and that the P atom has a tendency to pass from the tri- to the quinquevalent state.

phosphites (*Chem.*). Phosphates(III). Salts of phosphorous acid. Soluble in water.

phospho-creatine (*Chem.*). Phosphagen. A compound of creatine and phosphoric acid found in, e.g., muscle. It appears to regulate the balance between adenosine tri-phosphate and adenosine di- and mono-phosphates.

phospholipins (*Chem.*). See **phosphatides.**

phosphomolybdic acid stain (*Micros.*). A 1% solution of phosphomolybdic acid, buffered at pH 5·4, and used to increase the contrast of specimens in electron microscopy.

phosphenium bases (*Chem.*). Compounds formed by the combination of a tertiary phosphine with an alkyl halide.

phosphonium salts (*Chem.*). The so-called phosphonium salts are formed when phosphine is brought into contact with hydrogen chloride, hydrogen bromide, or hydrogen iodide. Formed in a similar way to ammonium compounds.

phosphoproteins (*Chem.*). Compounds formed by a protein with a substance containing phosphorus, other than a nucleic acid or lecithin, e.g., caseinogen, vitellin.

phosphor. Generic name for the class of substances which exhibit *luminescence*. (*Electronics*) More specifically the fluorescent coating used on the screen of CRTs or image intensifiers.

phosphor-bronze (*Met.*). A term sometimes applied to alpha (low tin) bronze deoxidized with phosphorus, but generally it means a bronze containing 10–14% of tin and 0·1–0·3% of phosphorus, with or without additions of lead and nickel. Used, in cast condition, where resistance to corrosion and wear is required, e.g., gears, bearings, boiler fittings, parts exposed to sea water, etc.

phosphorescence (*Chem.*). Greenish glow observed during slow oxidation of white phosphorus in air. (*Light*) Luminescence which persists for more than 0·1 nanosecs after excitation. See **fluorescence, persistence.** (*Zool.*) Luminosity; production of light, usually (in animals) with little production of heat, as in Glow-worms. *adj.* **phosphorescent.**

Phosphoria formation (*Geol.*). Important phosphate-bearing beds occurring in the Permian of Idaho, Wyoming and N. Utah.

phosphoric acid (*Chem.*). H_3PO_4. See **orthophosphoric acid.**

phosphorite or **rock-phosphate** (*Min.*). The fibrous concretionary variety of *apatite*.

phosphorized copper (*Met.*). Copper deoxidized with phosphorus. Contains a small amount (about 0·02%) of residual phosphorus, which lowers the conductivity.

phosphorofluoric acid (*Chem.*). See **hexafluorophosphoric acid.**

phosphorous (*Chem.*). Relating to trivalent phosphorus, e.g., *phosphorus*(III)*oxide* P_4O_6 which with cold water forms *phosphorus*(III)*acid*.

phosphorous (phosphorus(III)) acid (*Chem.*). H_3PO_3. Formed by the action of cold water on phosphorous oxide; decomposes on heating; forms phosphites; reducing agent.

phosphorus (*Chem.*). A nonmetallic element in the fifth group of the periodic system. Symbol P, at. no. 15, r.a.m. 30·9738, valencies 3, 5. White phosphorus is a waxy, poisonous, spontaneously inflammable solid, m.p. 44°C, b.p. 282°C, rel. d. 1·8–2·3. Red phosphorus is nonpoisonous and ignites in air only when heated above about 300°C, m.p. 500–600°C, rel. d. 2·20. Violet phosphorus has rel. d. 2·34, m.p. 590°C. Black has rel. d. 2·7, and is a metallic substance obtained at high temperature and pressure. Phosphorus occurs widely and abundantly in minerals (as phosphates) and in all living matter. Manufactured by heating calcium phosphate with sand and carbon in an electric furnace. It is used mainly in the manufacture of phosphoric acid for phosphate fertilizers; also used in matches and organic synthesis.

phosphorus chloronitrides (*Chem.*). Formed by the interaction of phosphorus(IV)chloride and ammonium chloride.

phosphorus oxychloride (*Chem.*). $POCl_3$. Liquid; fumes in air; slowly hydrolysed by water, forming phosphoric and hydrochloric acids. It is formed when compounds containing a hydroxyl group are treated with phosphorus (v)chloride. Also called **phosphoryl chloride.**

phosphorus pentahalides (*Chem.*). Phosphorus (v)halides. These, *phosphorus*(v)*chloride* (PCl_5), *phosphorus*(v)*bromide* (PBr_5), and *phosphorus*(v) *fluoride* (PF_5), are formed by the action of the dry halogen on the trihalide. The properties of the (v)halides are similar. They transform hydroxyl compounds into the corresponding halides.

phosphorus pentoxide (*Chem.*). Phosphorus(v) oxide. P_2O_5. Powerfully hygroscopic white solid obtained by burning phosphorus in air. Chiefly used to manufacture orthophosphoric acid by reaction with water; also as a drying agent.

phosphorylase (*Biochem.*). An enzyme playing a part in *phosphorylation*, the process of converting a sugar into a phosphate, and of splitting the latter to yield energy for vital processes.

phosphoryl bromide (*Chem.*). $POBr_3$. Formed in a similar manner to *phosphorus oxychloride* (q.v.).

phosphoryl chloride (*Chem.*). See **phosphorus oxychloride.**

phosphoryl fluoride (*Chem.*). POF_3. May be made by the action of hydrofluoric acid on phosphorus(v)oxide; similar in properties to the other phosphoryl compounds.

phosphotungstic acid stain (*Micros.*). A stain used as a 1% or 2% acid or neutral solution in electron microscopy, to increase the contrast by locally increasing the density to electrons.

phot (*Light*). CGS luminance unit equalling one lumen/cm^2 or 10^4 lumen/m^2.

photic zone (*Ecol.*). The zone of the sea where light penetration is sufficient for photosynthesis, corresponding to the limnetic zone of freshwater habitats. Cf. *aphotic zone.*

photobiology (*Biol.*). The study of light as it affects living organisms.

photocatalysis (*Chem.*). The acceleration or retardation of rate of chemical reaction by light.

photocathode (*Electronics*). Electrode from which electrons are emitted on the incidence of radiation in a photocell. It is *semitransparent* when there is photoemission on one side, arising from radiation on the other, as in the signal plate of a TV camera tube.

photocell (*Electronics*). See **photoelectric cell.**

photocell amplifier (*Cinema.*). That located in close proximity to the photocell which receives the light beam modulated by the sound track, when the latter passes through the sound gate in a projector. (*Electronics*) Galvanometer current amplifier in which movement of reflected light beam from galvanometer across the boundary between two photovoltaic cells produces greatly amplified output current. Also PEC or **pec amplifier, photoelectric galvanometer.**

photocell pick-off (*Electronics*). One which is operated by changes in the intensity or magnitude of a light beam.

photocell sensitivity (*Electronics*). Ratio of output current to level of illumination. Expressed in milliampere/lumen.

photochemical cell (*Electronics*). Photocell comprising two similar electrodes, e.g., of silver, in an electrolyte; illumination of one electrode results in a voltage between them; also called **Becquerel cell, photoelectrolytic cell.** See also **backwall cell, barrier-layer cell, frontwall cell and photovoltaic cell.**

photochemical equivalence (*Chem.*). **See Einstein law of photochemical equivalence.**

photochemical induction (*Chem.*). The lapse of an appreciable time between the absorption of light by a system and the occurrence of the resulting chemical reaction.

photochemistry (*Chem.*). The study of the chemical effects of radiation, chiefly visible and ultraviolet, and of the direct production of radiation by chemical change.

photochondria (*Zool.*). Granules occurring in the cytoplasm of a luminous cell or animalcule.

photochromic (*Glass*). Said of special glass which changes colour and hence the amount of light transmitted when the light falling on it increases or decreases. Also **photosensitive.**

photochromics (*Phys.*). Light-sensitive materials which with the advent of the laser make promising storage and read-out media. In holograph form, hundreds of books could be stored in a cm cube of a photochromic material.

photocleistogamy (*Bot.*). A condition in which flowers do not open, because of the more rapid growth of the outer side of the petals.

photocomposing camera (*Typog.*). One which makes negatives of a typeface by operation of a keyboard, for photographic type-setting.

photocomposition (*Typog.*). See **filmsetting.**

photoconducting camera tube (*TV*). One in which the optical image is focused on to a surface, the electrical resistance of which is dependent on illumination.

photoconductive cell (*Electronics*). Photoelectric cell using photoconductive element (often cadmium sulphide) between electrodes.

photoconductivity (*Electronics*). Property possessed by certain materials, such as selenium, of varying their electrical conductivity under the influence of light.

photocurrent (*Electronics*). That released from sensitized surface of photocell on the incidence of light, electrons which form the current being attracted to an anode, polarized positively with reference to the surface. The true photocurrent is augmented, in the presence of gas, through ionization by collision.

photodiode (*Electronics*). Combination of photoconducting cell and junction diode into a two-electrode semiconducting device. Used as optical sensing device for data processing.

photodisintegration (*Nuc.*). Nuclear reaction initiated by photon. Also **nuclear photoeffect, photonuclear reaction.**

photodissociation (*Chem.*). Dissociation produced by the absorption of radiant energy.

photoelasticity (*Light*). Phenomenon whereby strain in certain materials, crystal or foil, results in coloured fringes, when transmitting polarized light between crossed Nicol prisms.

photoelectric absorption (*Nuc.*). That part of the absorption of a beam of radiation associated with the emission of photoelectrons. In general, it increases with increasing atomic number but decreases with increasing quantum energy when *Compton absorption* becomes more significant.

photoelectric cell (*Electronics*). Transducer which has an electric output corresponding to the incident light. Also **photocell.** Abbrev. PEC.

photoelectric constant (*Phys.*). The ratio of Planck's constant to the electronic charge. This quantity is readily measured by experiments on photoelectric emission and forms one of the principal methods by which the value of Planck's constant may be determined.

photoelectric effect (*Electronics*). Phenomena resulting from absorption of photons by electrons, resulting in their ejection with kinetic energy equal to the difference between the energy of photon and the surface *work function* or an atomic binding energy (*Einstein photoelectric equation* (q.v.)). Under these phenomena are photoconductive, photovoltaic, photoemissive and photoelectromagnetic effects. Emission of X-rays or the impact of high-energy electrons on a surface is an inverse photoelectric effect.

photoelectric exposure meter (*Photog.*). A meter which incorporates a photoelectric cell as the essential device for indicating an intensity of illumination from which photographic exposures can be estimated.

photoelectric galvanometer (*Electronics*). See **photocell amplifier.**

photoelectricity (*Electronics*). Emission of electrons from the surface of certain materials by quanta exceeding certain energy (Einstein).

photoelectric multiplier (*Electronics*). See **photomultiplier.**

photoelectric photometer (*Light*). One in which the light from the lamp under test is measured by the current from a photoelectric cell.

photoelectric photometry (*Astron.*). The determination of stellar magnitudes and colour indexes by means of photoelectric devices used at the focus of a large telescope.

photoelectric threshold (*Electronics*). The limiting frequency for which the quantum energy is just sufficient to produce photoelectric emission. Given by equating the quantum energy to the work function of the cathode.

photoelectric tube (*Electronics*). Gas or vacuum type of emissive cell.

photoelectric work function (*Electronics*). The energy required to release a photoelectron from a cathode. It should correlate closely with the thermionic work function.

photoelectric yield (*Electronics*). The proportion of incident quanta on a photocathode which liberate electrons.

photoelectroluminescence (*Electronics*). The enhancement of luminescence from a fluorescent screen during excitation by ultraviolet light or X-rays when an electric field is applied, i.e., the field enhancement of luminescence.

photoelectrolytic cell (*Electronics*). See **photochemical cell.**

photoelectromagnetic effect (*Electronics*). See **photomagnetoelectric effect.**

photoelectromotive force (*Electronics*). That which apparently arises in a photovoltaic cell, considered as a constant-current source.

photoelectron (*Electronics*). One released from a surface by a photon, with or without kinetic energy.

photoemission (*Electronics*). Emission of electrons from surface of a body (usually an electropositive metal) by incidence of light.

photoemissive camera tube (*TV*). Tube operating on the photoemissive principle, the image falling on its photocathode, causing this to emit electrons in proportion to the intensity of the light in the picture elements.

photoemissive cell (*Electronics*). An electronic valve containing a photocathode, the photocurrent flowing from the anode to the cathode. Greater sensitivity at the expense of linearity of response is obtained by the use of gas in the envelope and by operating at the ionization potential of the gas.

photo-engraving (*Print.*). See **process-engraving.**

photoextinction method (*Chem.*). See **nephelometric analysis.**

photofission (*Nuc.*). Nuclear fission induced by gamma-rays.

Photoflood lamp (*Photog.*). A tungsten lamp run at excess voltage, with correspondingly reduced life, to raise its colour temperature.

photoformer (*Telecomm.*). Function generator using optical projection of shadow mask representing required waveform.

photogen (*Zool.*). A light-producing or phosphorescent organ; as in some *Polychaeta.*

photogenic (*Bot., Zool.*). Emitting light, light-producing, e.g., *photogenic bacteria.*

photogenin (*Zool.*). See **luciferase.**

photoglow tube (*Electronics*). One in which the sensitivity is increased by radiation incident on the cathode.

photogrammetry (*Photog.*). The making of photographs (e.g., from the air) for survey work.

photographic borehole survey (*Mining, Surv.*). Check on orientation and angle of long borehole by insertion of a special camera which photographs a magnetic needle and a clinometer at known distance down.

photographic efficiency (*Light*). Of a light source, the fraction of the light energy in the emitted spectrum which is usefully registered on a photographic emulsion.

photographic-emulsion technique (*Nuc.*). Study of the tracks of ionizing particles as recorded by their passage through a nuclear emulsion.

photographic memory (*Psychol.*). See **eidetic imagery.**

photographic photometry (*Phys.*). Measurement of intensity of radiation by comparing a photographic image of the source with that of a standard source.

photographic recording (*Acous.*). The registering of a modulated track on photographic film, so that it can be scanned by a constant beam of light, fluctuations of which, after being converted into corresponding electric currents by a photocell, can be amplified and reproduced as sound by loudspeakers.

photographic surveying (*Surv.*). A method of surveying employing the principles of *intersection* (q.v.) by means of a special instrument called a phototheodolite, with which a series of photographs is taken of the points whose positions are required, each point appearing in at least two different photographs. Also **phototopography.**

photographic zenith tube (*Astron.*). A modern instrument for the exact determination of time; it consists of a fixed vertical telescope which photographs stars as they cross the zenith; instrumental and observational errors are thus eliminated, the instrument being entirely automatic. Abbrev. PZT.

photogravure (*Print.*). An intaglio printing process using copper cylinders or plates etched through *carbon tissue* (q.v.). The Helio-Klischograph is used to scan coloured copy and translate corrected colour separations into depressions cut in the cylinder, thus dispensing with process camera separation and carbon tissue.

photohalide (*Chem.*). A halogen salt which is sensitive to light, e.g., silver bromide.

photokinesis (*An. Behav.*). A *kinesis* (q.v.) occurring in response to variations of light intensity.

photolithography (*Print.*). The normal modern lithographic process in which a photographic image is deposited on a metal plate, of which there are several kinds, including deep etch, albumen (or surface), and a variety of bimetallic and polymetallic plates. When printing down, the deep etch process requires a positive, the albumen process a negative, while a positive is prescribed for some polymetallic plates and a negative for others.

photoluminescence (*Light*). Light emitted by visible, infrared or ultraviolet rays after irradiation.

photolysis (*Bot.*). The grouping of the chloroplasts in relation to the amount of light falling on the plant. (*Chem.*) The decomposition or dissociation of a molecule as the result of the absorption of light.

photomagnetoelectric effect (*Electronics*). Generation of electric current by absorption of light on surface of semiconductor placed parallel to magnetic field. Due to transverse forces acting on electrons and holes diffusing into the semiconductor from the surface. Also termed **photoelectromagnetic effect.**

Photomaton (*Photog.*). Automatic machine which takes a number of photographs in succession and delivers a strip or sheet of finished prints in a few minutes. Prints are produced by automatic reversal of the image on a celluloid and paper base.

photomechanical (*Print.*). Any process by which a printing surface is prepared mechanically with the aid of photography.

photomeson (*Nuc.*). Meson resulting from interaction of photon with nucleus—usually a *pion.*

photometer (*Light*). An instrument for comparing the luminous intensities of two sources of light. Most photometers employ the principle that, if equal illumination is produced on similar surfaces illuminated normally by two light sources, the ratio of their intensities equals the square of the ratio of their distances from the surfaces.

photometer bench (*Light*). A bench upon which is mounted the apparatus for carrying out photometric tests by comparison with a standard lamp. The apparatus consists of a mounting for the standard lamp, the photometer itself, a mounting for the lamp under test, and equipment for moving any or all of these and determining their position.

photometer head (*Light*). The unit in an optical measuring system which contains the device for making the actual balance or measurement.

photometric integrator (*Light*). That part of an integrating photometer which actually sums up the light flux, e.g., the globe of the *Ulbricht sphere photometer*.

photometric surface (*Light*). A surface used for photometric comparisons.

photometry (*Chem.*). Volumetric analysis in which the end-point of a reaction is determined by means of turbidity measurements made with the aid of photoelectric cells. (*Light*) The measurement of the luminous intensities of light sources and of luminous flux and illumination. See also **photometer.**

photomicrography (*Photog.*). Photography through a microscope. Not to be confused with *microphotography* (q.v.).

photomontage. See **montage photograph.**

photomorphosis (*Bot.*). A change in the structure of a plant following exposure to strong light.

photomosaic (*TV*). Sheet of material, usually mica, which is covered by a large number of minute photocells formed by depositing a film of silver, breaking it up and activating it by deposition of caesium, which is also evaporated away from intervening areas. Used in *iconoscope* and *image orthicon* camera tubes.

photomultiplier (*Electronics, etc.*). Photocell with series of dynodes used to amplify emission current by electron multiplication.

photon (*Nuc.*). A quantum of light or of electromagnetic radiation equal to *Planck's constant* (h) multiplied by frequency ν in Hz.

Photon (*Typog.*). See **Lumitype-Photon.**

photonasty (*Bot., Zool.*). Response to variation in the intensity of illumination, or to the stimulus of diffuse light.

photonegative (*Elec.*). Material of which electrical conductivity decreases with increasing illumination. See **photopositive.**

photoneutron (*Nuc.*). Neutron resulting from interaction of photon with nucleus.

photon noise (*Electronics*). That occurring in photocells as a result of the fluctuations in the rate of arrival of light quanta at the photocathode.

photonuclear reaction. See **photodisintegration.**

photopathy (*Biol.*). Negative phototaxis.

photoperceptor (*Biol.*). The part of an eye spot which is sensitive to light.

photoperiodicity (*Ecol.*). The controlling effects of the length of day on such phenomena as the flowering of plants, the reproductive cycles of Mammals, migration and diapause of Insects, and seasonal changes in the feathers of Birds and the hair of Mammals.

photophelein (*Zool.*). See **luciferin.**

photophilous (*Biol.*). Light-seeking, light-loving; said of plants which inhabit sunny places.

photophobia (*Med.*). Intolerance of the eye to light with spasm of the eyelids.

photophore (*Zool.*). A luminous organ of Fish.

photophoresis (*Chem.*). The migration of suspended particles under the influence of light.

photophthalmia (*Med.*). Electric-light ophthalmia. Burning pain in the eyes, lacrimation, photophobia, and swelling and spasm of the eyelids as a result of exposure to an intensely bright light (e.g., a naked arc light); due to the action of ultraviolet rays.

photophygous (*Zool.*). Shunning strong light.

photopic luminosity curve (*Light*). Curve giving the relative brightness of the radiations in an equal-energy spectrum when seen under ordinary intensity levels. See **scotopic luminosity curve.**

photopic vision (*Optics*). That based on *cones* (q.v.), and therefore sensitive to colour. Possible only with adequate ambient illumination. Cf. *scotopic vision.*

photoplasm (*Zool.*). The photochondria-containing cytoplasm of a luminous cell or animalcule.

photopolymer plates (*Print.*). Relief printing plates made of a synthetic material, on a flexible or rigid base as required. After printing down with ultraviolet light through a negative, during which the exposed parts are polymerized, the plate is completed by washing away the unexposed areas with a caustic solution.

photopositive (*Elec.*). Material of which conductivity increases with increasing illumination. See **photonegative.**

photoproton (*Nuc.*). Proton resulting from interaction of photon with nucleus.

photopsy, photopsia (*Med.*). The appearance of flashes of light in front of the eyes, due to irritability of the retina.

photoreceptor (*Zool.*). A sensory nerve-ending receiving light stimuli.

photoresist process (*Electronics*). Process removing selectively the oxidized surface of a silicon slice semiconductor. The photoresist material is an organic substance polymerizing on exposure to ultraviolet light and in that form resisting attack by acids and solvents.

photosedimentation (*Powder Tech.*). Method of particle-size analysis in which concentration changes within a suspension are determined by measuring the attenuation of a beam of light passing through it.

photosensitive (*Glass*). See **photochromic.** (*Photog., Phys.*) Property of being sensitive to action of visible or invisible light, whether

subsequent development is required to exhibit sensitivity or not.
photosetting. See **filmsetting.**
photosphere (*Astron.*). The name given to the visible surface of the sun on which sun-spots and other physical markings appear; it is the limit of the distance into the sun that we can see. (*Zool.*) A light-producing or phosphorescent organ, as in some *Crustacea.*
photostage (*Bot.*). An early stage in the development of a seedling, during which it needs a supply of light.
Photostat (*Photog.*). TN for photographic apparatus (also for any print made by it) designed for rapidly copying flat originals on sensitized paper, giving a negative image.
photosynthesis (*Bot.*). The synthesis of simple carbohydrates from carbon dioxide and water, with the liberation of oxygen, using the energy of light, in green plants chlorophyll being the energy transformer. In photosynthetic bacteria, bacteriochlorophyll and 'chlorobium chlorophyll' are found, oxygen is not produced, but various inorganic and organic compounds are oxidized.
photosynthetic bacteria (*Bot.*). Bacteria capable of photosynthesis, e.g., purple and green sulphur bacteria. The photosynthetic pigments are present in chromatophores visible only by the electron microscope. Frequently inorganic sulphur compounds are the hydrogen donors and usually sulphur or sulphur compounds are liberated or deposited.
photosynthetic number (*Bot.*). The ratio between the number of grams of carbon dioxide absorbed per hour by a unit of leaf and the number of grams of chlorophyll that unit contains.
photosynthetic quotient, photosynthetic ratio (*Bot.*). The ratio between the volume of carbon dioxide absorbed to the volume of oxygen set free, during a given time, by plant material occupied in photosynthesis.
phototaxis (*Biol.*). Locomotory response or reaction of an organism or cell to the stimulus of light. *adj.* **phototactic.**
phototelegraphy (*Telecomm.*). General term for high-grade facsimile transmission of half-tone pictures.
phototheodolite (*Surv.*). A photographic camera of fixed known focal length, with horizontal and vertical cross-wires pressing tightly against the sensitive plate, on which they are photographed. It is mounted on a tripod and fitted with levelling screws, a graduated horizontal circle, and a telescope. See **photographic surveying.**
phototonus (*Bot.*). The condition of a leaf which is able to respond to a stimulus, because it has received an adequate amount of light.
phototopography. See **photographic surveying.**
phototransistor (*Electronics*). A 3-electrode photosensitive semiconductor device. The emitter junction forms a photodiode and the signal current induced by the incident light is amplified by transistor action.
phototrophic (*Biol.*). Said of organisms obtaining their energy from sunlight.
phototropism (*Biol.*). Tendency of growing plants to turn towards the greatest light.
phototropy (*Chem.*). (1) The property possessed by some substances of changing colour according to the wavelength of the incident light. (2) The reversible loss of the colour of a dyestuff under the influence of light of a definite wavelength.
phototube (*Electronics*). See **photoelectric tube.**
photovaristor (*Electronics*). Material, e.g., cadmium sulphide or lead telluride, in which the varistor effect, i.e., nonlinearity of current-voltage relation, is dependent on illumination.
photovoltaic cell (*Electronics*). A class of photoelectric cell which acts as a source of e.m.f. and so does not need a separate battery. See **backwall cell, barrier-layer cell, frontwall cell.**
photovoltaic effect (*Electronics*). The production of an e.m.f. across the junction between dissimilar materials when it is exposed to light or ultraviolet radiation.
phragma (*Zool.*). A septum or partition: an apodeme of the endothorax formed by the infolding of a portion of the tergal region of a somite: an endotergite.
phragmobasidium (*Bot.*). A basidium which becomes septate, and is then divided into four cells.
phragmocone (*Zool.*). In *Belemnoidea*, the chambered part of the internal shell, protected by a thickened guard posteriorly.
phragmosis (*Ecol.*). The habit, found in several Ants and Termites, of closing the openings to their nests with their heads, or, in some Spiders, with their abdomens, these being truncated and circular.
phreatic gases (*Geol.*). Those vapours and gases of atmospheric or oceanic origin which, coming into contact with ascending magma, may provide the motive force for volcanic eruptions.
phrenicectomy (*Surg.*). Excision of a part of the phrenic nerve in order to paralyse the diaphragm on the same side.
phrenicotomy (*Surg.*). The cutting of the phrenic nerve in order to paralyse the diaphragm on one side; done in the treatment of lung disease.
phrenosin (*Chem.*). A *cerebroside* obtained from the brain substance. On hydrolysis it yields a fatty acid (phrenosinic acid), galactose, and a base (sphingosine).
phthaleins (*Chem.*). Triphenylmethane derivatives obtained by the action of phenols upon phthalic (benzene 1,2-dicarboxylic) anhydride.
phthalic acid (*Chem.*). $C_6H_4(COOH)_2$, benzene-*o*-dicarboxylic acid, colourless prisms or plates, m.p. 213°C, soluble in water, alcohol, ether. When heated above the melting-point it yields its anhydride. It is prepared by catalytic air oxidation via the anhydride.
phthalic anhydride (*Chem.*).

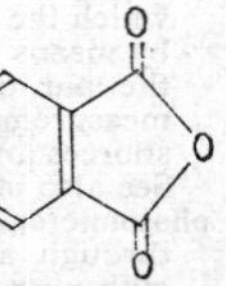

Phthalic (benzene 1,2-dicarboxylic) anhydride. Long prisms which can be sublimed, m.p. 128°C, b.p. 284°C. Used in the production of plasticizers, paints, and polyester resins. Prepared by air oxidation of naphthalene in the presence of a catalyst such as vanadium(v) oxide.
phthalic glyceride resins (*Plastics*). See **alkyd resins.**
phthalimide (*Chem.*).

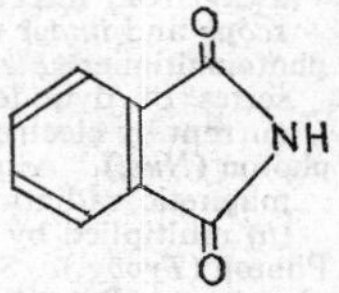

Benzene 1,2-dicarboximide. Colourless crystals, m.p. 238°C, obtainable by passing ammonia over heated phthalic anhydride. The imide hydrogen is replaceable by Na or K.

phthalocyanines (*Chem.*). A group of green and blue organic colouring matters, used mainly as pigments and formed by combining four *phthalonitrile* (q.v.) groups with one atom of a metal, such as copper. **Blue**: Consists of α or β copper phthalocyanine. Widely used in almost all pigmented products. **Green**: Consists of halogenated copper phthalocyanine, and available in a number of shades from deep blue-green to a bright green with only a slight blue tone.

phthalonitrile (*Chem.*). Crystalline solid; m.p. 140°C:

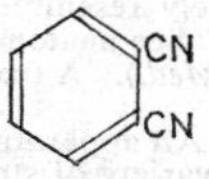

CN
CN

used as an intermediate in the manufacture of *phthalocyanine* (q.v.) colours which are made by heating it with metallic compounds.

phthiriasis (*Med.*). Infestation with lice. See also **pediculosis.**

phthisaner (*Zool.*). An abortive male ant which fails to complete its metamorphosis as a result of its being parasitized by *Orasema.*

phthisergate (*Zool.*). An abortive worker ant which fails to complete its metamorphosis as a result of its being parasitized by *Orasema.*

phthisis (*Med.*). (1) Wasting of the body. (2) Pulmonary tuberculosis. *adj.* **phthisical.**

phthisogyne (*Zool.*). An abortive queen ant which fails to complete its metamorphosis as a result of its being parasitized by *Orasema.*

phugoid oscillation (*Aero.*). A longitudinal periodic fluctuation in speed, i.e., a velocity-modulation, in the motion of an aircraft, accompanied by rising and falling of the nose.

pH value (*Chem.*). A logarithmic index for the hydrogen ion concentration in an aqueous solution. Used as a measure of acidity of a solution; given by $pH = \log_{10}(1/[H^+])$, where $[H^+]$ is the hydrogen-ion concentration. A pH below 7 indicates acidity, and one above 7 alkalinity.

phyc-, phyco-. Prefix denoting association with algae or seaweed (Gk. *phykos*, seaweed),

phycobiont (*Bot.*). The algal partner in a Lichen.

phycocyanin (*Bot.*). Photosynthetic blue pigment occurring with chlorophyll in *Rhodophyta* (red algae) and *Myxophyceae* (blue-green algae).

phycology (*Bot.*). The study of *Algae.*

Phycomycetes (*Bot.*). The lower Fungi, including 1000 species. The thallus ranges from a rounded sac to an ordinary mycelium, in which regular septation is unusual. Many species are aquatic, reproducing freely by zoospores, and the sexual organs, when present, are either isogametes, heterogametes, or well-defined antheridia and oögonia, or else gametangia which unite without individualization of the gametes.

phylactocarp (*Zoo .*). A hydrocladium bearing gonangia.

Phylactolaemata (*Zool.*). A class of fresh-water *Ectoprocta* in which the lophophore is U-shaped and an epistome is present.

phyletic classification (*Biol.*). A scheme of plant classification based on the presumed evolutionary descent of organisms.

phyllary (*Bot.*). One of the involucral bracts on the outside of a capitulum.

phyllidium (*Zool.*). See **bothridium.**

phyllite (*Geol., Min.*). A name which has been used in several different senses: (1) for the pseudohexagonal platy minerals (*phyllosilicates*) including mica, chlorite, and talc (by some French authors); (2) for argillaceous rocks in a condition of metamorphism between slate and mica-schist (by most English authors). Phyllite in the latter (usual) sense is characterized by a silky lustre due to the minute flakes of white mica which, however, are individually too small to be seen with the naked eye.

phyllobranchia (*Zool.*). A gill composed of numerous thin plate-like lamellae.

phyllocarpic movement (*Bot.*). A curvature of the fruit stalk bringing the young fruit under the shelter of the leaves.

phylloclade (*Bot.*). A branch which is so flattened that it looks like a leaf, and which functions as a leaf.

phyllocyst (*Zool.*). In some *Siphonophora*, a dilatation at the apex of a hydrophyllium.

phyllode (*Bot.*). A petiole which is flattened and leaflike and functions as a leaf; phyllodes are often set on the plant so that they turn their edges to the direction of the brightest light.

phyllody (*Bot.*). The transformation of parts of a flower into leaves.

phyllome (*Bot.*). A general term for all leaves and leaflike organs.

phyllopodium (*Bot.*). The petiole and rachis of the leaf of a fern. (*Zool.*) The thin leaflike swimming foot characteristic of *Branchiopoda*, with broad flat exopodite and endopodite. Cf. *stenopodium.*

phylloquinone. See **vitamin K.**

phyllosiphonic (*Bot.*). Said of a siphonostele which has both leaf gaps and branch gaps.

phyllotaxis, phyllotaxy (*Bot.*). The arrangement of leaves on a shoot.

phyllozoid (*Zool.*). See **hydrophyllium.**

phylo-. Prefix from Gk. *phylon*, race.

phyloephebic (*Biol.*). Pertaining to the period of maximum vigour of a race or species.

phylogenetic series (*Biol.*). The complete range of organisms, extant and extinct, which shows the evolution of the higher from the lower forms. Cf. *taxonomic series.*

phylogeny, phylogenesis (*Bot., Zool.*). The history of the development of a race. Cf. *ontogeny. adj.* **phylogenetic.**

phylogerontic (*Biol.*). Said of that period in the history of a race or species which corresponds to the senescent period in the life-history of an individual.

phylon (*Biol.*). A line of descent.

phyloneanic (*Biol.*). Said of that period in the history of a race or species which corresponds to the adolescent period in the life-history of an individual.

phylonepionic (*Biol.*). Said of that period in the history of a race or species which corresponds to the embryonic period in the life-history of an individual.

phylum (*Zool.*). A category or group of related forms constituting one of the major subdivisions of the animal kingdom.

physical astronomy. See **astrophysics.**

physical chemistry (*Chem.*). The study of the dependence of physical properties on chemical composition, and of the physical changes accompanying chemical reactions.

physical mass unit (*Chem.*). See **atomic mass unit.**

physical optics (*Optics*). That branch of the study of light dealing with phenomena, such as diffraction and interference, which are best considered from the standpoint of the wave theory of light. See **geometrical optics.**

physics. Study of electrical, luminescent, mechanical, magnetic, radioactive and thermal phenomena with respect to changes in energy states without change of chemical composition.

physio-. Prefix from Gk. *physis*, nature.

physiognomy (*Bot.*). The characteristic appearance of a plant community, by which it can often be recognized at a distance. It is usually determined by the dominant plants of the community.

physiographic climax (*Bot.*). A plant community maintained at a certain stage of development by some natural feature of the habitat, such as active erosion or slow but persistent movement of the ground.

physiography. The science of the surface of the earth and the inter-relations of air, water and land.

physiological. Relating to the functions of plant or animal as a living organism.

physiological anatomy (*Biol.*). The study of the relation between structure and function.

physiological drought (*Bot.*). Condition in which a plant is unable to take in water because of low temperature, or because the water available to it holds substances in solution which hinder absorption by the plant.

physiological monitor (*Surg.*). One used in operating theatre to record changes in patient's physiological condition.

physiological race (*Biol.*). A group of individuals within the morphological limits of a species but differing from other members of the species in habits (as host, larval food, etc.).

physiological variety or strain (*Bot.*). See **biological form.**

physiological zero (*Biol.*). The threshold temperature below which the metabolism of a cell, organ, or organism ceases.

physiology. The study of the manner in which organisms carry on their life processes.

physiotherapy (*Med.*). The treatment of disease by physical means, other than radiation.

physoclistous (*Zool.*). Of fish, having no pneumatic duct connecting the air-bladder with the alimentary canal. Cf. *physostomous.*

physogastry (*Ecol.*). The condition, found in some termitophilous Insects, of having swollen soft whitish bodies.

physostigmine (*Chem.*). Eserine, $C_{15}H_{21}O_2N_3$, an alkaloid obtained from the seeds of Calabar beans (*Physostigma venenosum*).

physostomous (*Zool.*). Of Fish, not *physoclistous* (q.v.), i.e., having a pneumatic duct.

phytic acid (*Chem.*). *Myo*inositol hexaphosphoric acid, present in cereals, especially the outer parts of grain. It forms insoluble salts with calcium, magnesium, and iron, and so tends to hinder the absorption of these substances from the intestine.

phyto-, -phyte. Prefix and suffix from Gk. *phy̆ton*, plant.

phytobenthon (*Ecol.*). Plant benthon.

phytobiology (*Bot.*). Study of the life history of plants.

phytochemistry (*Bot.*). Study of the chemical processes and composition of plant life.

phytochrome (*Bot.*). A reversible plant pigment which is involved in the responses of green plants to light of red and far-red wavelengths.

phytohormones (*Bot.*). Plant hormones, e.g., auxins, traumatins.

phytology. Synonym for *botany*.

Phytomastigina (*Zool.*). A subclass of *Mastigophora* comprising mainly holophytic forms possessing chromatophores, rarely having more than 2 flagella and frequently having starch reserves; a few members are colourless and practise holozoic or saprophytic nutrition.

phytomitogens (*Biochem.*). Vegetable substances which can induce cell-division.

phyton (*Bot.*). A hypothetical plant unit composed of leaf, blade and stalk.

phytopathology (*Bot.*). The study of plant diseases, and especially of plants in relation to parasites.

phytophagous, phytophilous (*Zool.*). Plant-feeding.

phytophysiology (*Bot.*). Study of the physiology of plants.

phytoplankton blooms (*Ecol.*). Very high densities of plankton which occur quickly, and persist for short times, usually at regular times of the year.

phytosterol (*Chem.*). A terpene alcohol, m.p. 137°C. It closely resembles *cholesterol* (q.v.).

phytotomy (*Bot.*). See **anatomy.**

phytotoxin (*Bot., Med.*). A toxin obtained from a plant.

phytotron (*Bot.*). An apparatus for the growth of plants under a variety of strictly controlled environmental conditions.

pi (π) (*Maths.*). The ratio of the circumference of a circle to its diameter. $\pi = 3{\cdot}141\,592\,653\,6$ to 10 places.

pia-arachnoid membrane (*Histol.*). The connective tissue membrane, derived from the primitive mesenchyme, which lies adjacent to the dura mater, and covers the brain and spinal cord, continuing for a short distance along the cranial and spinal nerves.

pia mater (*Zool.*). In Vertebrates, the innermost of the three membranes surrounding the brain and spinal cord, a thin vascular layer.

pian (*Med.*). See **yaws.**

Pianotron (*Acous.*). Pianoforte in which the normal vibration of the strings is used to modulate the potential applied to electrostatic screw pick-ups, with amplification and loudspeakers.

piano whine (*Acous.*). A defect in gramophone recording caused by variations in turntable speed due to the large amplitudes occurring in piano music.

pi-attenuator or **π-attenuator** (*Telecomm.*). An attenuator network of resistors arranged as in the pi-network.

piazza (*Arch.*). (1) An enclosed court in a building. (2) A colonnade or arcade.

PIB (*Plastics*). Poly*iso*butylene or *polybutene* (q.v.).

pica (*Med.*). Unnatural craving for unusual food. (*Typog.*) An old type size, approximately 12-point, still in use in a few printing offices, and a synonym for 12-point in others. See **em, point.**

picaroon (*For.*). A short pole about one metre long, with a recurved spike or hook, used to pull small products down slopes or for short distances, as in a sawmill.

picene (*Chem.*). $C_{22}H_{14}$, a complicated hydrocarbon, obtained from the coal-tar fraction boiling above 360°C. It has the formula:

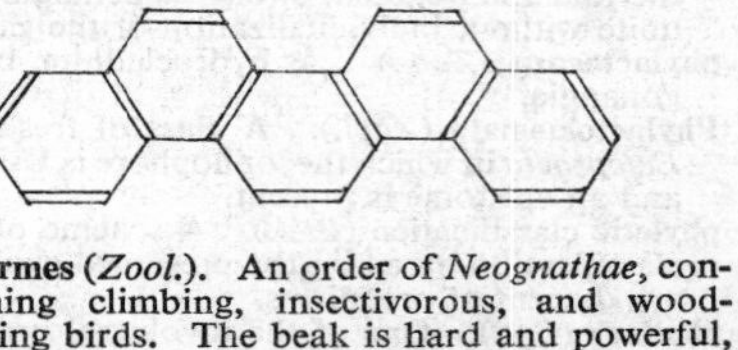

Piciformes (*Zool.*). An order of *Neognathae*, containing climbing, insectivorous, and wood-boring birds. The beak is hard and powerful, the tongue long and protrusible, and the feet zygodactylous. Woodpeckers, Toucans.

pick (*Tools*). A double-headed tool, pointed at both ends, having the handle fastened into the middle of the head (as in a hammer); used for rough digging by quarrymen, road builders, etc. (*Weaving*) One traverse of the shuttle through the warp shed in a loom to lay a weft thread for beating up into cloth. Picks per inch (ppi) denotes the total number of weft threads in every inch of cloth. Also called **shot, shoot.**

pick-and-pick (*Weaving*). A term indicating alternate picks of yarns of two different colours or kinds.

pick-at-will (*Weaving*). A mechanism which enables the shuttle from any box to be propelled through the warp shed as desired.

pick-axe (*Tools*). A tool similar to a pick but having one end edged so that it may cut.

picker (*Paper*). A form of cleaning equipment consisting of screen plates, either perforated or cut with long fine slits, through which the pulp stock passes, any large contraries being held back and removed by water sprays. Also called **strainer.** (*Textiles*) (1) Small tool manipulated by hand for removing laps caused by broken ends on spinning frames. (2) Power-driven fine spindle for doing same job. (3) Buffalo hide, rubber, or plastic component that strikes the shuttle tip to propel weft across loom.

pickeringite (*Min.*). Magnesia alum. Hydrated sulphate of aluminium and magnesium, crystallizing in the monoclinic system. It usually occurs in fibrous masses, and is formed by the weathering of pyrite-bearing schists.

picker leather (*Leather*). Tough, specially impregnated leather used in making tuggers and picking bands for looms. See **picker.**

picket (*Build.*). A narrow upright board in a fence; frequently pointed at the top. (*Surv.*) A short ranging rod about 2 m (6 ft) long.

picket fence (*Elec. Eng.*). Arrangement for stabilizing plasma in a torus, consisting of bands of coils round the tube, with currents alternating in direction, giving an interior field with periodic cusps.

picking (*Build.*). Chiselling or picking small indentations on a wall surface as a key for the finish. Also called **stugging, wasting.** (*Print.*) Removal of part of the paper surface because of faults in the paper or in the printing. (*Textiles*) (1) Removing extraneous matter (outstanding hairs, kemps, slubs, etc.) from the face of fabrics by a burling iron. (2) Propelling the shuttle carrying the weft from side to side, through the warp shed on a loom.

picking belt (*Mining, Min. Proc.*). Sorting belt, on which run-of-mine ore is displayed so that pickers can remove waste rock, debris or a special mineral constituent, which is not to be sent to the mill for treatment.

pickling (*Chem. Eng.*). The removal of mill scale, lime scale, or salt water deposits, e.g., in a ship's engine or evaporators, by circulating a suitable acid containing inhibitors. The commonest acid used is sulphuric, at up to ten times dilution. Inhibitors include glue and acidic long-chain molecular compounds such as HBF_4. (*Leather*) The treatment of light skins with a dilute solution of sulphuric acid and salt, after dehairing. (*Mining*) Treatment of mine timbers before use with wet- or dry-rot preventatives.

picklock (*Textiles*). The term applied by woolsorters to the best sort of wool from a fleece.

pick-off (*Automation*). Any device actuated by relative motion and giving an output, electrical or pneumatic, for operating other mechanisms.

Pick's bundle (*Zool.*). A tract of fibres in the medulla oblongata of higher Vertebrates.

Pick's disease (*Med.*). Concato's disease.

pick-up (*Acous.*). See **pick-up head.** (*Elec. Eng., etc.*) In general, any transducer used to 'pick-up' a signal required for any equipment. (*Teleph.*) In automatic telephony using d.c. signalling, the operation of line-terminating apparatus resulting from a change in line-terminating conditions.

pick-up baler (*Agric.*). Machine for gathering up mown crops, compressing, and tying them with twine or wire into bales of equal size.

pick-up head (*Acous.*). Mechanical-electrical transducer, often piezo, actuated by a sapphire or diamond stylus which rests on the sides of the groove on a gramophone record, and by tracking this groove generates a corresponding voltage (or, for stereo reproduction, two voltages) for driving an audio amplifier. Also called **pick-up.**

pick-up limit (*Teleph.*). The limiting line resistance at which false circuit operation may occur during pick-up.

pick-up needle (*Acous.*). Loosely applied to any needle in a gramophone pick-up. (*Elec. Eng.*) One in which the moving part is a magnetic needle, which by its motion diverts magnetic flux and induces a voltage in coils on the magnetic circuit.

pick-up plate (*Electronics*). One concerned with electrostatic signals from a number of capacitors as in a TV camera mosaic, cathode-ray tube store, electronic musical instrument.

pick-up reaction (*Nuc.*). Nuclear reaction in which incident particle collects nucleon from a target atom and proceeds with it.

pick-up tube (*TV*). See **camera tube.**

pick-up well (*Autos.*). A small petrol reservoir arranged between the metering jet and the spraying tube in some carburettors; it provides a temporarily enriched mixture during acceleration.

pico-. SI prefix for 1-millionth-millionth, or 10^{-12}. Formerly, *micromicro-*. Abbrev. **p.**

picofarad (*Elec. Eng.*). 10^{-12} farad. Same as **micromicrofarad.** Abbrev. **pF.**

picolines (*Chem.*). Methyl-pyridines, $CH_3{\cdot}C_5H_4N$. The three isomers are 2-picoline, b.p. 129°C; 3-picoline, b.p. 142°–143°C; 4-picoline, b.p. 144°–145°C.

picosecond. The million millionth part of a second, 10^{-12} s. Abbrev. **ps.**

picotite (*Min.*). A dark-coloured spinel containing iron, magnesium, aluminium, and chromium; a chromium-bearing hercynite.

picramic acid (*Chem.*). Red crystalline solid; m.p. 168°C, obtained by reduction of *picric acid* (trinitrophenol):

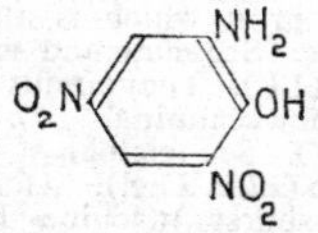

NH$_2$ / O$_2$N / OH / NO$_2$

Used in manufacture of azo dyes.

picric acid (*Chem.*). $C_6H_2(NO_2)_3{\cdot}OH$, 2,4,6-trinitrophenol, yellow plates or prisms, m.p. 122°C, made by nitrating phenol; slightly soluble in water. It is a strong acid and dyes wool and silk yellow. Used for the preparation of explosives; lyddite or melinite is compressed or fused picric acid.

picrite (*Geol.*). A general name for ultramafic coarse-grained igneous rocks, consisting essentially of olivine and other ferromagnesian minerals, together with a small amount of plagioclase.

picro-carmine stain (*Micros.*). A combined fixative and stain used for staining tissues in bulk and for general staining. It contains carmine, ammonia solution, and picric acid.

picrolite (*Min.*). A fibrous antigorite.

picro-Mallory stain (*Micros.*). A microanatomical polychromatic stain, giving sharp differentiation of tissues. A modification of Mallory with a picro-haematoxylin basis.

picture (*TV*). Complete TV image, constructed from one, or two interlaced, fields. In U.S., a **frame.**

picture-chasing (*TV*). Circuit used in conjunction with intermediate film-scanning system, with continuously moving film.

picture element (*TV*). Same as **picture point.**

picture frequency (*TV*). Same as **frame frequency.**

picture head (*Cinema.*). That part of the projector mechanism which includes the intermittent motion for projecting the picture frames on the screen.

picture inversion (*TV*). Conversion of negative to positive image (or vice versa) when carried out electronically. In facsimile transmission, it will correspond to the reversal of the black and white shades of the recorded copy.

picture monitor (*TV*). CRT for exhibiting TV picture or related waveform for purposes of control.

picture noise (*Radar*). See **grass.**

picture point (*TV*). Any one of the large number of minute illuminated areas which go to make up a TV image. Also called **picture element.**

picture ratio (*Cinema., TV*). Ratio of the width to height of reproduced image, e.g., 5/4. Sound film has a number of picture ratios, original being 4/3. Also **aspect ratio.**

picture signal (*TV*). That portion of a TV signal which carries information relative to the picture itself, as distinct from synchronizing portions. See **pedestal.**

picture slip (*TV*). Vertical displacement of received picture due to bad synchronism between field frequency of the frame time base and that of the signal.

picture telegraphy (*Teleg.*). Same as **facsimile telegraphy.**

picture tone (*Teleg.*). The frequency of the carrier used in amplitude-modulated facsimile transmission.

picture-to-synchronization ratio (*TV*). The ratio of the total amplitude of the television waveform assigned to picture information to that which is assigned to the synchronizing pulses and flyback times.

picture track (*Cinema.*). The part of the width of a sound-film print which is allocated to the picture frames. See **mute** and **sound-track.**

picture traverse (*TV*). That circuit which controls frame-frequency scanning.

picture tube (*TV*). See **teletube.**

Pidgeon machine (*Elec. Eng.*). An improved form of the Wimshurst machine having special features, such as embedded sectors.

pie (*Typog.*). To upset type-matter accidentally.

piece cart (*Textiles*). A piece of apparatus in which lengths of woollen and worsted fabrics are placed, in the process of finishing, and subjected to pressure.

pieces (*Textiles*). (1) Staples, from different types of fleece, accumulated during sorting and sold as mixed lots or otherwise. (2) Lengths of woven or knitted cloths in processing stages after weaving.

pie characters (*Typog.*). The *outside sorts* on line-composing machines which cannot be accommodated in a magazine but are stored in small cases beside the machine and inserted in the line by hand. Also called **side sorts.**

piedmont alluvial plain (*Geog.*). See **alluvial cone.**

piedmont glacier (*Geol.*). A glacier of the 'expanded foot' type; one which, after being restricted within a valley, spreads out on reaching the flat ground into which the latter opens.

piedmont gravels (*Geol.*). Accumulations of coarse breccia, gravel, and pebbles brought down from high ground by mountain torrents and spread out on the flat ground where the velocity of the water is checked. Literally, mountain-foot gravels, typical of the outer zone of arid areas of inland drainage such as the Lop Nor Basin in Chinese Turkestan.

piedmontite (*Min.*). A hydrous silicate of calcium, aluminium, manganese, and iron, crystallizing in the monoclinic system. Also called **manganepidote, piemontite.**

piedroit (*Build.*). A pier projecting from a wall but having neither cap nor base. Cf. *pilaster.*

piend (*Build.*). See **arris.**

piend check (*Build.*). The rebate cut along a lower corner of a stone step to enable it to sit upon the step below.

piend rafter (*Carp.*). An *angle rafter.*

Pieper system (*Elec. Eng.*). See **automixte system.**

pier (*Build.*). (1) The part of a wall between doors and windows. (2) See **buttress.**

pierced (*Typog.*). Said of a block with an internal portion removed to accommodate type.

Pierce oscillator (*Electronics*). Original crystal oscillator in which positive feedback from anode to grid in a triode valve is controlled by piezoelectric mechanical resonance of a suitably cut quartz crystal.

piercing (*Eng.*). (1) Making a hole in a cylindrical billet of steel by passing it over a fixed point or mandrel, to produce a seamless tube. (2) A presswork operation for punching holes in strip or sheet. (*Met.*) See **Mannesmann process.**

pierre perdue (*Civ. Eng.*). Work consisting of stone blocks deposited at random to form a foundation, e.g., in some breakwater construction.

pier template (*Civ. Eng.*). A stone slab laid on a brickwork pier to ensure that the load is distributed over the full area of the pier.

pie shute (*Typog.*). The *pie characters* (q.v.) have no accommodation in the magazine and, having a complete set of teeth, travel the full length of the distributor bar to slide down the pie shute.

pieze. Unit of pressure in the metre-tonne-second system, equivalent to 10^3 N/m² or 1 kN/m².

piezo-. Prefix from Gk. *piezein*, to press.

piezochemistry (*Chem.*). The study of the effect of high pressures on chemical reactions.

piezoelectric crystal (*Telecomm.*). One showing piezoelectric properties which may be shaped and used as a resonant circuit element or transducer (microphone, pick-up, loudspeaker, depth-finder, etc.).

piezoelectric effect or piezoelectricity (*Elec.*). Electric polarization arising in some anisotropic (i.e., not possessing a centre of symmetry) crystals (quartz, Rochelle salt, barium titanate) when subject to mechanical strain. An applied electric field will likewise produce mechanical deformation.

piezoelectric loudspeaker (*Acous.*). A piezoelectric crystal (quartz or Rochelle salt) used to generate mechanical waves (or conversely electric potentials due to incident sound waves, i.e., to act as a microphone). Such devices may also be designed to operate under water. Also **subaqueous loudspeaker.**

piezoelectric microphone (*Acous.*). See **crystal microphone.**

piezoelectric resonator (*Radio*). A crystal used as a standard of frequency, controlling an electronic oscillator.

piezoid (*Elec. Eng.*). Blank of piezo crystal, adjusted to a required resonance, with or without relevant electrodes.

piezomagnetism (*Phys.*). An effect analogous to *piezoelectricity* in which mechanical strain in certain materials (e.g. Ferroxcube, TN) produces a magnetic field, and vice versa.

piezo receiver (*Acous.*). A head-telephone receiver, usually grouped in pairs, in which the diaphragm is driven electrostatically through the medium of a Rochelle salt or other suitable crystal, as in a piezo microphone.

pig (*Glass*). An iron block laid against the pot mouth as a support for the blowing iron. (*Met.*) A mass of metal (e.g., cast-iron, copper, or lead) cast in a simple shape for transportation or storage, and subsequently remelted for purification, alloying, casting into final shapes, or into ingots for rolling.

pig bed (*Met.*). A series of moulds for iron pigs, made in a bed of sand. Connected to each other and to the tap hole of the blast-furnace by channels, along which the molten metal runs.

pigeon-holed (*Build.*). Said of a wall built with regular gaps in it (e.g., a *honeycomb wall*, q.v.).

pigeonite (*Min.*). One of the monoclinic pyroxenes, intermediate in composition between clinoenstatite and diopside. It is poor in calcium, has a small optic axial angle and occurs in quickly chilled lavas and minor intrusions.

pigeon's milk (*Zool.*). In Pigeons and Doves, a white slimy secretion of the epithelium of the crop in both sexes. It contains protein and fat, and is produced during the breeding season under the influence of *prolactin* (q.v.), being regurgitated to feed the young.

pig iron (*Met.*). The crude iron produced in the blast furnace and cast into pigs which are used for making steel, cast-iron, or wrought-iron. Principal impurities are carbon, silicon, manganese, sulphur, and phosphorus. Composition varies according to the ores used, the smelting practice, and the intended usage.

pigment (*Chem.*, *Paint.*). Insoluble, natural, or artificial, black, white, or coloured materials reduced to powder form which, when dispersed in a suitable medium, are able to impart colour and/or opacity. Ideally, pigments should maintain their colour under the most unfavourable conditions; in practice, may tend to fade or discolour under the influence of acids, alkalis, other chemicals, sunlight, heat, and other conditions. They may also *bleed*. They are used in many industries, e.g., in paint, printing ink, plastics, rubber, paper, etc. (*Biol.*) Any of various substances which impart colour to the tissues or cells of animals and plants. See **carotenes, chlorophyll, xanthophyll.**

pigmentary colours (*Zool.*). Colours produced by the presence of drops or granules of pigment in the integument, as in most Fish. Cf. *structural colours*.

pigment cell (*Cyt.*). See **chromatophore.**

pigment foils (*Bind.*). *Blocking foils*, which leave a black, white, or coloured matt transfer.

pigmentosa (*Bot.*). The pigmented part of an eye-spot.

pigment process (*Photog.*). Any process which involves pigment suspended in a tissue, which may be partially dissolved away to form the required image.

Pignet's index (*Nut.*). The height in centimetres, less the weight in kilograms, less the chest girth in centimetres; an index which gives an inverse qualitative indication of corpulence.

pigsty timbering (*Mining*). Roof support formed of timbers laid in stope to form rectangular frames chocked at corners, and usually filled with waste rock.

pigtail (*Elec. Eng.*). The short length of flexible conductor connecting the brush of an electric machine to the brush-holder.

pila (*Zool.*). A pillar-shaped structure; as the *pilae antoticae*, pillars of cartilage in the developing chondrocranium.

pilaster (*Build.*). A square pier projecting from a wall, having both a cap and a base. Cf. *piedroit*.

pilaster strip (*Build.*). A pilaster without a cap.

Pilat process (*Chem.*). Method of separating the fractions of asphalt oils without distillation, by dissolving out the asphalt with propane and saturating the residual oil with methane under pressure, with the result that the lubricating oil fractions separate out in the order of decreasing viscosity.

pile (*Civ. Eng.*). A column or sheeting which is sunk into the ground to support vertical loading or to resist lateral pressures. (*Elec. Eng.*) Abbrev. for *thermopile*, a close packing of thermocouples in series, so that alternate junctions are exposed for receiving radiant heat, thus adding together the e.m.fs. due to pairs of junctions. (*Light*) The light reflected from a glass plate incident at the Brewster angle will be plane-polarized if the transmitted light is partially-polarized normal to the plane of the reflected polarized light. By using a pile of glass plates the transmitted light becomes increasingly plane-polarized. (*Med.*) See **haemorrhoid.** (*Met.*) A number of wrought-iron bars arranged in an orderly pile which is to be heated to a welding heat and rolled into a single bar. (*Nuc. Eng.*) Original name for pile of graphite blocks which formed the moderator of the first nuclear reactor. (*Textiles*) A covering on the surface of a fabric, formed by threads that stand out from it. Pile in loop form is termed *loop pile* or *terry*; if the loops are cut, it is termed *cut pile*. The latter is produced by weaving cloths face to face, and cutting them apart. Some carpets and moquettes are made this way.

pileate (*Bot.*). (1) Shaped like a cap. (2) Having a pileus. (*Zool.*) Crested.

pile cap (*Civ. Eng.*). (1) A horizontal beam connecting the heads of piles. (2) See **dolly.**

pile delivery (*Print.*). On a printing press, a delivery arrangement which can accommodate a large pile of sheets, the delivery board lowering automatically to floor level as the sheets are delivered.

pile-drawer (*Civ. Eng.*). An appliance for extracting driven or partly driven piles from the ground.

pile-driver (*Civ. Eng.*). A framed construction erected above the spot where a pile is to be driven into the ground. It is provided with a heavy weight which runs in upright guides and is so arranged that it may fall by gravity on to the head of the pile and drive it in. Also called **pile frame.**

pile hoop (*Civ. Eng.*). An iron or steel band fitted around the head of a pile to prevent brooming.

pile oscillator (*Nuc. Eng.*). The means of maintaining a neutron-absorbing body in periodic motion in a nuclear reactor.

piles. See **haemorrhoid.**
pile shoe (*Civ. Eng.*). The iron or steel point fitted to the foot of a pile to give it strength to pierce the earth and so assist driving.
pileum (*Zool.*). In Birds, the top of the head.
pile-up (*Teleph.*). Set of spring contacts operated in a relay. Also **stack.**
pileus (*Bot.*). The widened caplike portion of an agaric: by extension, the corresponding part of the fruit body of other fungi. (*Meteor.*) See **cap.**
pilidium (*Zool.*). A helmet-shaped larval form of some *Nemertea*, having a prominent apical tuft of cilia and a pair of lateral ciliated lappets.
pilifer (*Zool.*). One of a pair of lateral projections of the labrum, found in some families of *Lepidoptera*.
piliferous (*Bot.*). Bearing or producing hairs; ending in a delicate hair-like point.
piliferous layer (*Bot.*). The outermost cell layer of a young root, corresponding to the epidermis of a stem. It bears the root hairs.
piliform (*Bot.*). Resembling a long, zigzag hair.
piling-up (*I.C. Engs.*). Deposition of liquid petrol in the induction manifold of a petrol-engine, due to the engine's being motored round at a small throttle opening, thus leading to an over-rich mixture on opening up.
Pilizocarpeae (*Bot.*). See **Hydropteridineae.**
Pilkington twin process. See **twin-plate process.**
pill. See **oral contraception.**
pillar (*Elec. Eng.*). A structure of pillar form for containing switch or protective gear. Also called a **switchgear pillar.** (*Horol.*) Any cylindrical pieces of brass or steel which act as distance pieces to hold the plates of a clock or watch in their correct relative position. (*Mining*) Column of unserved ore left as roof support in stope.
pillar-and-stall (*Mining*). See **bord-and-pillar.**
pillar drill (*Eng.*). A drilling machine in which the spindle and table are supported by brackets carried by a pillar, the table bracket usually being slidable thereon.
pillar plate (*Horol.*). In a clock or watch, the plate to which the pillars are fixed.
pillow distortion (*Optics, Photog., TV*). See **pincushion distortion.**
pillow lava (*Geol.*). A lava-flow exhibiting pillow structure.
pillowphone (*Acous.*). Telephone receiver, concealed in a pillow, for reproducing, without sound radiation, programmes for a listener.
pillow structure (*Geol.*). A term applied to lavas consisting of ellipsoidal and pillow-like masses which have cooled under submarine conditions. The interspaces between the pillows consist, in different cases, of chert, limestones, or volcanic ash.
pilocarpidine (*Chem.*). $C_{10}H_{14}N_2O_2$, an alkaloid obtained from *pilocarpus* (q.v.). Viscid oil; soluble in water and alcohol.
pilocarpine (*Chem.*). $C_{11}H_{16}O_2N_2$, an alkaloid of the glyoxaline group, obtained from *pilocarpus* (q.v.). It is a colourless oil, b.p. 260°C (5 mm), soluble in water, alcohol, chloroform, but it is almost insoluble in ether and light petroleum.
pilocarpus (*Chem.*). The dried leaves of *Pilocarpus jaborandi*; source of *pilocarpidine* and *pilocarpine* (qq.v.).
pilomotor (*Anat.*). Causing movements of hair.
pilot (*Elec. Eng.*). In power systems, a conductor used for auxiliary purposes, not for the transmission of energy. Also called **pilot-wire.**
pilotaxitic texture (*Geol.*). The term applied to the groundmass of certain holocrystalline andesitic lavas in which there is a feltlike interweaving of feldspar microlites. Cf. *hyalopilitic texture*.
pilot balloon (*Meteor.*). A small rubber balloon, filled with hydrogen, used for determining the direction and speed of air currents at high altitudes. The balloon is observed by means of a theodolite after being released from the ground.
pilot carrier (*Telecomm.*). In a suppressed carrier system (as in single sideband working) a small portion of original carrier wave transmitted to provide a reference frequency with which local oscillator at the receiving end may be synchronized.
pilot cell (*Elec. Eng.*). A cell of a battery upon which readings are taken in order to give an indication of the state of the whole battery.
pilot chute (*Aero.*). A small parachute which extracts the main canopy from its pack.
pilot cloth (*Textiles*). Heavily milled, blue dyed, raised fabric, made from fine merino wools; used for uniforms, overcoatings, etc.; imitations are mixtures of wool and cotton.
pilot controller (*Elec. Eng.*). See **master controller.**
piloted head (*Heat.*). Gas burner head or nozzle having a by-pass by which low-pressure feeder flames are produced around the main flame, to secure positive retention when the velocity of the combustible mixture exceeds the flame speed.
pilot electrode (*Electronics*). Additional electrode or spark gap which triggers and makes certain main discharge, e.g., in a discharge spark gap, for creating ions, or in a TR switch, or a mercury-arc rectifier. Also **ignitor, keep-alive electrode, starter, trigger electrode.**
pilot engine (*Rail.*). A separate locomotive preceding a train as a precaution against accidents to the latter.
pilot gauge (*Eng.*). A plug gauge which has a circumferential groove near its free end to facilitate entry into the hole to be gauged.
pilotherm (*Heat*). A thermostat in which the temperature control is brought about by the deflection of a bimetallic strip, thereby switching on and off the electric heating current.
pilot lamp (*Elec. Eng.*). One giving visual indication of the closing of a circuit.
pilot nail (*Carp.*). A temporary nail used in fixing shuttering.
pilot pin (*Eng.*). A pin with rounded or tapered end, fixed to a tool or machine element and projecting beyond it so as to enter a hole in a mating part before engagement, thus ensuring alignment. Used in press tools to position the strip material.
pilot pins (*Cinema.*). (1) Pins which enter the perforations of the film while it is in the film gate of a camera and hold it stationary during exposure. (2) In the manufacture of film-stock, pins for positioning the film during the process of perforation. (3) In a film printer, the detachable pins which are changed over when a double-coated positive is being exposed for the second time.
pilot plant (*Chem. Eng.*). Smaller version of a projected industrial plant, used to gain experience and data for the design and operation of the final plant.
pilot spark (*Electronics*). See **pilot electrode.**
pilot (or relay) valve (*Eng.*). A small balanced valve, operated by a governor or by hand, which controls a supply of oil under pressure to the piston of a *servomotor* (q.v.) or relay connected to a large control valve, which it is desired to operate.
pilot voltmeter (*Elec. Eng.*). A voltmeter used in a

power station or substation to indicate the voltage at the remote end of a feeder to which it is connected by means of a pilot.

pilot wave (*Telecomm.*). A carrier oscillatory current or voltage which is amplified in a high-efficiency amplifier independently of the side-frequencies, which are added subsequently.

pilot wire (*Telecomm.*). In a multi-core transmission cable, a wire which is solely concerned with detecting deterioration of the main insulation of the cable. (*Elec. Eng.*) See **pilot.**

pilot-wire regulator (*Telecomm.*). An automatic device in transmission circuits, e.g., for compensating changes in transmission arising from temperature variations.

pimaricin (*Chem.*). Antibiotic effective against many fungi and yeasts. Also **natamycin.**

pimelic acid (*Chem.*). $HOOC{\cdot}(CH_2)_5{\cdot}COOH$, a saturated dibasic acid of the oxalic acid series, crystals, m.p. 105°C.

pi-meson or **π-meson** (*Nuc.*). See **meson.**

pi-mode (*Electronics*). Operation of a multicavity magnetron, whereby voltages on adjacent segments of the anode differ by π radians (half-cycle).

pimple (*Acous.*). A defect in a gramophone record caused by a dent in the stamper, the dent being due to some hard impurity in the record stock.

pimpling (*Nuc. Eng.*). Small swellings on the surface of a reactor fuel can, usually caused by swelling of fuel during burn up.

pin (*Brew.*). Cask holding 4½ galls. (*Carp., Join., etc.*) (1) A small wooden peg or nail. (2) The male part of a dovetail joint. (*Eng.*) A cylinder or tube, often heat-treated, used to connect members in a structure or machine, usually when freedom of angular movement at the joint is required.

pinacocytes (*Zool.*). The flattened epithelial cells forming the outer part of the dermal layer in Sponges.

pinacoid (*Crystal.*). An open crystal form which includes two precisely parallel faces.

pinacol (*Chem.*). $(CH_3)_2{=}C(OH){\cdot}C(OH){=}(CH_3)_2$ tetramethyl-ethan 1,2-diol, crystallizes with 6 H_2O. The anhydrous substance has a m.p. 38°C, b.p. 172°C, and is obtained by the reduction and condensation of acetone by the action of metallic sodium. Pinacol forms *pinacolone* (q.v.) by the elimination of water and intramolecular transformation in the presence of dilute acids. It is the simplest member of a series of tetra alkyl glycols known as *pinacols.* Also **pinacone.**

pinacolone (*Chem.*). $CH_3{\cdot}CO{\cdot}C(CH_3)_3$, 3,3-dimethylbutan-2-one, produced by the action of dilute sulphuric acid upon *pinacol* (q.v.). A colourless liquid, b.p. 106°C, rel. d. 0·800 at 16°C.

pinakryptol (*Photog.*). Group of dyes used as desensitizers.

Pinatype (*Photog.*). A process involving the differential dyeing of soft portions of the gelatin portions of the image.

pin barrel (*Horol.*). A cylindrical piece on the periphery of which are short vertical pins for lifting the hammers in a chiming clock, or for lifting the comb in a musical clock.

pincers (*Zool.*). Claws adapted for grasping; as chelae, chelicerae.

pinch (*Acous.*). An effect due to the variation in the width of the groove cut by a stylus in gramophone recording; the tip of the round reproducing needle rises in the groove when it crosses the mean track. (*Electronics*) An airtight glass seal through which pass the electrode connections in a thermionic valve.

pinchbeck alloy (*Met.*). Red brass, with 6–12% zinc.

pinched post (*Paper*). A size of writing paper, 14½ × 18½ in. (368 × 470 mm).

pinch effect (*Nuc. Eng.*). Because of magnetic attraction of parallel currents, tendency for fast streams of electrons or plasma, e.g. that in the evacuated torus in *Zeta* and *stellarator*, to contract, with consequent instability, reduced by a longitudinal magnetic field. Also called **rheostriction.**

pinch-off (*Electronics*). Cut-off of the channel current by the gate signal in a field-effect transistor.

pincushion distortion (*Optics, Photog., TV*). A type of *curvilinear distortion* (q.v.) in which the lines are convex, when the stop is in front of the lens, giving a shape similar to a pincushion. Also called **pillow distortion.** See also **barrel distortion.**

PIN or ***p-i-n*** **diode** (*Electronics*). A semiconductor *p-n* diode with a layer of an intrinsic semiconductor incorporated between the *p* and *n* junctions. Used as an r.f. switching element.

pine (*For.*). See **pitch-, Scots-.** (*Vet.*) Bush sickness; enzootic marasmus; vinquish. A disease of sheep and cattle caused by a deficiency of cobalt in the diet, due to low levels of cobalt in the soil; characterized by debility, emaciation, and anaemia.

pineal apparatus (*Zool.*). In some Vertebrates, two median outgrowths from the roof of the diencephalon, one (originally the left) giving rise to the *parietal organ* (q.v.) and the other (originally the right) to the *pineal organ* (q.v.).

pineal body (or **gland**) (*Zool.*). See **pineal organ.**

pinealectomy (*Surg.*). Surgical removal of the pineal gland.

pineal eye (*Zool.*). An anatomically imprecise term referring to eyelike structures formed by one or other of the two outgrowths of the pineal apparatus. In *Cyclostomata* it derives from the pineal organ, but in Lizards it derives from the parietal organ.

pinealoma (*Med.*). A tumour of the pineal gland.

pineal organ (*Zool.*). One of the outgrowths of the pineal apparatus. In *Cyclostomata* it forms an eyelike functional photosensitive structure involved in a diurnal rhythm of colour change, and may be sensitive to light in some fish. It persists in higher Vertebrates, and may function as an endocrine gland.

pinene (*Chem.*). There are four terpenes known as pinenes. α-Pinene, $C_{10}H_{16}$, is the chief constituent of turpentine, eucalyptus, juniper oil, etc.; b.p. 155°–156°C. It has the formula:

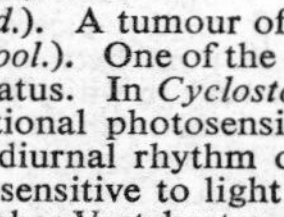

CH_3

H_3C

CH_3

It forms a hydrochloride, $C_{10}H_{17}Cl$, a white crystalline mass of camphorlike odour, m.p. 131°C. As it contains a double bond it forms a dibromide which can be converted into a glycol.

pine oil (*Min. Proc.*). Commercial frothing agent widely used in flotation of ores. Distillate of wood, varying somewhat in chemical composition according to timber used and scale of heating.

pinetree antenna (*Radio*). Vertical array of horizontal dipoles driven by twisted vertical transmission lines, so that they all radiate in phase. Unit in *Kooman's array* (q.v.).

pi-network (*Telecomm.*). Section of circuit with one series arm preceded and followed by shunt arms to return leg of circuit. Also **π-network.**

pin-eyed (*Bot.*). Having the throat of the corolla more or less closed by a stigma shaped like a pin head; applied to the primrose and its relatives. Cf. *thrum-eyed*.

ping (*Acous.*). Brief pulse of medium frequency sound, reproduced from the subaqueous reflection of asdic ultrasonic signals. Its length in space will be equal to the product of ping duration time and the velocity of sound.

pinguecula (*Med.*). A yellow, triangular patch on the conjunctiva covering the sclera.

pin hinge (*Join.*). A form of butt hinge which has a removable pin connecting the two leaves.

pinholes (*Paper*). Small holes through the paper caused by defects in manufacture. (*Photog.*) A photographic defect during development, whereby air-bells or dust particles prevent access of developer to the emulsion, causing lack of density. (*Textiles*) In knitted fabrics, defects, such as irregular texture, which may be due to using too dry or inelastic yarns.

pinholing (*Paint.*). A painting or varnishing defect in which the surface becomes pitted with small holes.

pin insulator (*Elec. Eng.*). An insulator which is supported from the cross-arm by a pin. Suitable up to 33kV.

pinion (*Eng.*). The smaller of a pair of high-ratio toothed spur-wheels. (*Horol.*) A small-toothed wheel, which normally has fewer than 12 teeth (leaves).

pinion leaf (*Horol.*). A tooth of a pinion.

pinion wire (*Horol.*). Steel or brass wire drawn to the section of a pinion.

pinite (*Min.*). A hydrous silicate of aluminium and potassium which is usually amorphous. It is an alteration product of cordierite, spodumene, feldspar, etc., approximating to muscovite in composition. See **pagodite.**

pin joint (*Eng.*). A joint between members in a structural framework in which moments are not transmitted from one member to another.

pink disease (*Med.*). See **erythroedema.**

pink-eye (*Med.*). Acute mucopurulent conjunctivitis (the inflammation of the conjunctiva making it red) due to infection with various bacteria. (*Vet.*) See **equine influenza.**

pinking (*I.C. Eng.*). See **knocking.**

pink-noise generator (*Elec. Eng.*). A random-noise generator providing a frequency spectrum with higher amplitudes at the low frequency end of the audible spectrum. See **white noise.**

Pinkus' organ (*Zool.*). In some *Dipnoi* (e.g., *Protopterus* and *Lepidosiren*), an organ of unknown function developing from the spiracular rudiment.

pin mark (*Typog.*). A mark near the top of the type shank, made in casting.

pinna (*Bot.*). (1) A leaflet, when part of a pinnate compound leaf. (2) A branch of a thallus, when these are arranged in opposite rows. (*Zool.*) In Fish, a fin; in Mammals, the outer ear; in Birds, a feather or wing.

pinnate (*Bot.*). (1) Said of a compound leaf having leaflets arranged in two ranks, one on each side of the rachis. (2) Said of a thallus having branches arranged on each side of a middle axis. (*Zool.*) Feather-like; bearing lateral processes.

pinnatifid (*Bot.*). Said of a leaf-blade which is cut, about half-way towards the midrib, into a number of pinnately arranged lobes.

pinnatiped, pinniped (*Zool.*). Having the digits of the feet united by flesh or membrane. Cf. *fissiped*.

pinnatisect (*Bot.*). Pinnatifid but with the cuts reaching nearly to the midrib.

pinning-in (*Build.*). The operation of inserting small splinters of stone in the joints of coarse masonry.

pinnings (*Build.*). A Scottish term for different varieties of stones introduced into a wall for decorative purposes.

pinnule (*Bot.*). One of the lobes or segments when a leaflet of a pinnate leaf is itself more or less divided into parts in a pinnate manner. (*Zool.*) One of the branchlets borne by the arms in *Crinoidea*.

pinocamphone (*Chem.*). Dicyclic ketone,

CH_3 H_3C =O H_3C

The laevorotatory isomer constitutes the main constituent of oil of hyssop.

pinocytosis (*Cyt.*). The absorption of liquids by cells, by the process of ingestion of droplets from the cell membrane into the cytoplasm.

pin pallet escapement (*Horol.*). An escapement in which the pallets are vertical pins of steel or jewels. The impulse is derived entirely from the teeth of the escape wheel. Used extensively for inexpensive watches, alarm clocks, and small drum movements.

pinpoint (*Aero.*). An aircraft's ground position as fixed by direct observation. Cf. *fix*.

pint. A unit of capacity or volume in the Imperial system. Equal to $\frac{1}{8}$ Imperial gallon or 0·5682 dm^3. In U.S. equal to 0·473 dm^3 (liquid) or 0·551 dm^3 (dry). 20 fluid ounces (U.S. 16 fl. oz.).

pinta (*Med.*). Caraate. Mal de los pintos. A contagious skin disease, characterized by patches of coloured pigmentation; probably due to infection with various fungi; occurs in tropical America.

pintle (*Eng., etc.*). (1) The pin of a hinge. (2) The king pin of a wagon. (3) An iron bolt on which a chassis turns. (4) One of the metal braces on which a rudder swings, supported by a *dumb-pintle* at its heel. (5) The plunger or needle of an oil-engine injection valve, opened by oil pressure on an annular face, and closed by a spring.

pintle chain (*Eng., etc.*). A sprocket chain.

pin tongs (*Horol.*). Small hand-vices with a split draw-in chuck.

pin wheel (*Horol.*). A wheel having pins, fixed at right angles to the plane of the wheel, which lift the hammer of a striking clock.

pin wheel escapement (*Horol.*). An escapement, used in turret clocks, in which semicircular or D-shaped pins standing at right angles to the plane of the whee lgive impulse to the pallets. The action of the escapement is similar to that of the dead-beat.

pion (*Nuc.*). See under **meson.**

Pioneer (*Space*). A series of U.S. space probes which have been used successfully for studying interplanetary matter and solar-terrestrial relationships.

pioneer community (*Bot.*). The first plant community to become prominent on a piece of ground which has been stripped of its vegetation and is being reoccupied by plants.

pioneer species (*Bot.*). A species of which the members tend to be among the first to occupy bared ground; these plants are often intolerant of competition, and especially of shading, and may be crowded out as the community develops.

pionnate (*Bot.*). A continuous layer of fungal spores, often slimy.

pioscope (*Chem.*). An instrument in which the fat content of milk is estimated colorimetrically.

pip (*Radar*). Significant deflection or intensification of the spot on a CRT giving a display for identification or calibration. Particularly applied to the peaked pattern of a radar signal. See **blip.**

pipe (*Acous.*). A musical instrument, mainly used in organs, in which the note is produced by longitudinal resonance of an air column when excited by an edge-tone at one end. The timbre of the emitted note depends on the scale, taper and material of the pipe, since these determine the relative responses to overtones. See **closed-, open-, pitch-, reed-, stopped-.** (*Horol.*) A tubular boss of extension. (*Met.*) A conical cavity formed in the top central portion of ingots. It arises because the solid occupies less volume than the molten metal and solidification proceeds from the sides and bottom towards the top and centre, where the effects of the liquid-solid contraction are concentrated.

pipeclay (*Geol.*). A white clay, nearly pure and free from iron, used in the pottery industry.

pipe coupling (*Plumb.*). A short collar with female threads at both ends into which screw the ends of successive lengths of piping which are to be connected co-axially.

pipe factor (*Mining*). Compensating factor used when samples are taken from casings which go into or through running sands or gravels, so that the amount of material raised from the section traversed by the drill does not correspond with the volume enclosed by the pipe.

pipe fitting (*Eng.*). Any piece of the wide variety of pipe connecting-pieces used to make turns, junctions, and reductions in piping systems.

pipe moulding (*Foundry*). The production of cast-iron pipes either by moulding in green sand, using split patterns, or by the process of *centrifugal casting* (q.v.).

piperazine (*Chem.*). Diethene-diamine, a cyclic compound of the formula:

$$HN\begin{matrix}\diagup CH_2-CH_2\diagdown \\ \diagdown CH_2-CH_2\diagup\end{matrix}NH$$

colourless crystals, m.p. 104°C, b.p. 145°C. It is a strong base and has the property of forming salts with uric acid which are easily soluble in water; it is therefore used in medicine.

pipe resonance (*Acous.*). Acoustic resonance of a pipe when the length, allowing for a constant end-correction, is an integral number of half-wavelengths when it is open at both ends, and odd multiples of a quarter-wavelength when it is closed at one end.

pipe resonator (*Acous.*). An acoustic resonator, in the form of a pipe, which may be open at one or both ends, as in organ pipes, or entirely closed. Resonance arises from the stationary waves set up by a plane-progressive wave being reflected at the ends, open or closed.

piperidine (*Chem.*). $C_5H_{11}N$, a heterocyclic reduction product of pyridine, having the formula:

$$CH_2\begin{matrix}\diagup CH_2-CH_2\diagdown \\ \diagdown CH_2-CH_2\diagup\end{matrix}NH.$$

It is a colourless liquid, of peculiar odour, b.p. 106°C. It is soluble in water and ethanol, has strong basic properties, and forms salts. It is a secondary amine, and the imino hydrogen is replaceable by alkyl or acyl radicals.

piperine (*Chem.*). An optically inactive alkaloid occurring in pepper, $C_5H_{10}N{\cdot}C_{12}H_9O_3$, piperyl-piperidine, which crystallizes in prisms, m.p. 129°C.

piperitone (*Chem.*). A terpene ketone:

$$CH_3{\cdot}C\begin{matrix}\diagup CH_2-CH_2\diagdown \\ \diagdown CH-CO\diagup\end{matrix}CH-CH(CH_3)_2$$

The laevo-rotatory isomer is found in eucalyptus oils, whilst the dextro isomer occurs in Japanese peppermint oil. Colourless oil with peppermint odour.

pipe roller (*Print.*). See **idler.**

piperonal (*Chem.*). $CH_2O_2{=}C_6H_3{\cdot}CHO$, methylene-protocatechuic aldehyde, a phenolic aldehyde of very pleasant odour, used as a perfume under the name of **heliotropin.**

pipe sample (*Mining, etc.*). One obtained by driving open-ended pipe into heap of material and withdrawing the core it collects.

pipe stopper (*San. Eng.*). An expanding form of drain plug for closing the outlet of drain pipes which are to be tested.

pipette method (*Powder Tech.*). A sedimentation technique in which concentration changes within a sedimenting *suspension* (q.v.) are measured by pipetting samples out of the suspension and determining the concentration of the samples.

pipe-ventilated (*Elec. Eng.*). Said of an electric machine so constructed that a supply of ventilating air can be drawn from and returned to a source at some distance from the machine itself; used where the surrounding air is too dirty for ventilating purposes.

pipe wrench (*Plumb.*). A tool adapted to turning a pipe or rod about its axis. Also called a **cylinder wrench.**

pipless lamp (*Elec. Eng.*). An electric lamp in which the sealing-off tip of the bulb is inside the cap.

pip-pip tone (*Teleph.*). A voice-frequency signal comprising two pulses of tone, indicating that a particular switching stage has been reached.

piqué (*Textiles*). Good quality, dobby woven cotton fabric with a plain weave and cords running across from selvedge to selvedge. Used for dress waistcoats, dress ties, etc.

Pirani gauge (*Vac. Tech.*). Used to measure moderate vacuums to about 0·01 N/m². Employs the principle that the heat lost by a hot body depends on the local gas pressure. A wire is heated electrically and its temperature is indicated by its electrical resistance.

Pira test bench (*Print.*). A small compact laboratory designed by the Printing Division of the Research Association for the Paper, Board, Printing and Packaging Industries, equipped to test paper, ink and printing surfaces before printing and to diagnose troubles encountered during and after printing.

pirn (*Weaving*). Small wooden bobbin which fits the loom shuttle and carries weft rewound from cones, cheeses or hanks.

piroplasmosis (*Vet.*). See **babesiosis.**

Pirquet's index (*Nut.*). See **pelidisi index.**

Pirquet's reaction (*Med.*). The reaction of scarified skin to the presence on it of tuberculin; a local inflammatory swelling—positive reaction—indicates that a tuberculous lesion, not necessarily active, exists in the body.

piscicolous (*Zool.*). Living within a Fish, as certain parasites.

piscivorous (*Zool.*). Fish-eating.

pi-section filter (*Telecomm.*). Unbalanced filter in which one series reactance is preceded and followed by shunt reactances.

pisé de terre (*Build.*). A kind of cob wall used sometimes in cottage construction, the cob usually being moulded between forms.

pisiform (*Zool.*). Pea-shaped; as one of the carpal bones of Man.

pisolite (*Geol.*). A type of limestone built of rounded bodies similar to oöliths, but of less regular form and 2 mm or more in diameter. See **Pea Grit.**

pisolitic (*Geol.*). A term descriptive of the structure of certain sedimentary rocks containing pisoliths (see **pisolite** above). Calcite-limestones, dolomitic limestones, laterites, iron-ores, and bauxites may be pisolitic.

pistacite (*Min.*). Synonym for **epidote.**

pistil (*Bot.*). The ovary of a flower with its style and stigma. See **gynaeceum.**

pistillate (*Bot.*). Said of a flower which has a gynaeceum, but in which the stamens are lacking or non-functional.

pistillidium (*Bot.*). Archegonium.

pistillode (*Bot.*). An abortive or non-functional pistil.

pistillum (*Zool.*). In *Siphonophora*, the central tube of an aurophore.

piston (*Acous.*). In brass musical instruments, a valve mechanism which alters the effective length of the resonating air-column and thereby permits the voicing of extra notes in the scale. (*Elec. Eng.*) Closely fitting sliding short-circuit in a waveguide. (*Eng.*) A cylindrical metal piece which reciprocates in a cylinder, either under fluid pressure, as in engines, or to displace or compress a fluid, as in pumps and compressors. Leakage is prevented by spring rings, leather packing, hat leather, etc.

piston attenuator (*Elec. Eng.*). Attenuator employing a waveguide system beyond cut-off. The attenuation in dB is linearly proportional to length, but the minimum attenuation is high.

piston mechanism (*Bot.*). A device, found in some flowers, in which the pollen is shed into a tube from which it is pushed by the style, thus coming into contact with an insect visitor.

pistonphone (*Acous.*). Device in which a rigid piston is vibrated, so that, by measurement of its motion, acoustic pressures and velocities can be calculated.

piston pin (*I.C. Engs.*). See **gudgeon pin.**

piston ring (*Eng.*). A ring, of rectangular section, fitted in a circumferential groove in a piston, and springing outward against the cylinder wall to prevent leakage. It is cut through at one point to increase its springiness and allow of fitting. See **junk ring, mitre-cut piston ring, scraper ring.**

piston rod (*Eng.*). The rod connecting the piston of a reciprocating engine with the crosshead.

piston rod gland (*Eng.*). The gland in the stuffing box in the cylinder end of an engine through which the piston rod passes.

piston slap (*I.C. Engs.*). The light knock caused by a worn or loose piston slapping against the cylinder wall when the connecting rod thrust is reversed.

piston valve (*Eng.*). A steam-engine slide valve in which the sealing or sliding surfaces of the valve are formed by two short pistons attached to the valve rod, working over cylindrical port faces in the steam chest; commonly used on steam locomotives.

pit (*Bot.*). (1) A thin localized area in the wall of a cell or other element of plant structure. (2) The two opposite thin areas in the walls of two cells or vessels in contact. (3) A local thin spot in the wall of the oögonium of some *Oömycetes*. (*Eng.*) See **engine pit.** (*Mining, etc.*) (1) A place whence minerals are dug. (2) The shaft of a mine. The *pit eye* is the bottom, whence daylight is visible; the *pit frame* is the superstructure carrying poppet head and sheaves. The *pit head* is the surface landing-stage.

pit cavity (*Bot.*). The excavation in the wall where the thinning is apparent.

pitch (*Acous.*). The subjective property of a simple or complex tone which enables the ear to allocate its position on a frequency scale. If the fundamental of a complex tone is absent the pitch of this fundamental is still recognized because of subjective difference tones amongst the partials. See **concert pitch.** (*Aero.*) The distance forward in a straight line travelled by an airscrew in one revolution; often colloquially though wrongly applied to the blade angle of incidence. See also **pitching.** (*Build.*) The ratio between the rise and the span of a roof, or the angular slope to the horizontal of an embankment. (*Chem.*) A dark-coloured, fusible, more or less solid material containing bituminous or resinous substances, insoluble in water, soluble in several organic solvents. Usually obtained as the distillation residue of tars. (*Cinema.*) The distance between perforations (sprocket holes) on the edges of cinematograph film. (*Elec. Eng.*) A term used in connexion with electrical machines to denote the distance measured along the armature periphery between various parts. See **pole-, slot-, winding-.** (*Eng.*) The linear distance or interval between similar features arranged in a pattern, as between rivets along a seam or the turns of a screw thread. (*Join.*) Of a plane, the angle between the blade and the sole, normally 45°, but in moulding planes up to 55° and in some smoothing planes 50° (York pitch). In planes with reversed bevel it is less, e.g., block planes ca. 20° and low-angle planes ca. 12°. (*Nuc. Eng.*) Distance between centres of adjacent fuel channels in a reactor. (*Paper*) The resinous substance deposited on parts of the machine and showing as defects in the paper. Originating from the wood pulp being used. (*Ships*) See **pitching period.**

pitchblende (*Min.*). A variety of uraninite. Radium was first discovered in this mineral. This and helium are due to the disintegration of uranium.

pitchboard (*Build.*). A triangular board used as a templet for setting out stairs, the sides of which correspond to the *rise*, the *going*, and the *pitch*.

pitch circle (*Eng.*). In a toothed wheel, an imaginary circle along which the tooth pitch is measured, and with respect to which tooth proportions are given. For two wheels in mesh, the pitch circles roll in contact.

pitch cone (*Eng.*). A conical surface through the teeth of a bevel wheel, corresponding to the pitch circle of a spur gear. For two bevels in mesh, the pitch cones roll together.

pitch control (*Aero.*). The *collective* and *cyclic pitch* (i.e. blade incidence) *controls* (qq.v.) of a helicopter's main rotor(s). (*TV*) Control of number of lines per unit length in a television image.

pitch cylinder (*Eng.*). Cylinder coaxial with a screw thread and which intersects the flanks of the thread symmetrically.

pitch diameter (*Eng.*). The diameter of the pitch circle of a gear wheel:

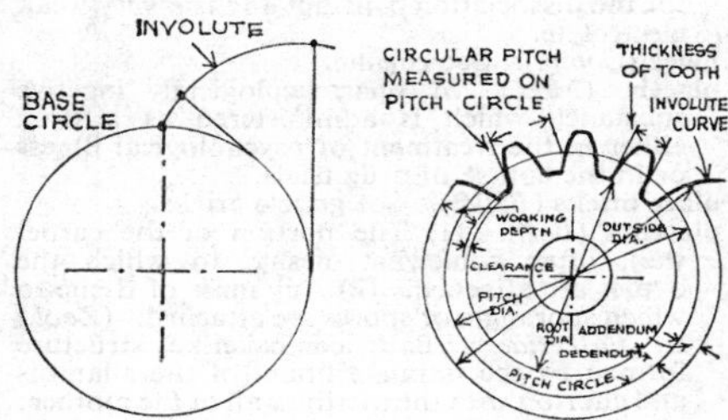

pitch edge (*Print.*). The leading edge of the paper as it enters the printing, or some other, machine.

pitched roof (*Build., Civ. Eng.*). A roof having a sloping surface or surfaces.

pitched work (*Build.*). Stone facing work for the slopes of jetties, breakwaters, etc., executed by pitching the stones into place with some regularity. Cf. *pierre perdue, coursed masonry.*

pitcher (*Bot.*). An urn-shaped or vase-shaped modification of a leaf, or part of a leaf, developed by certain plants; it serves as a means of trapping insects and other small animals, which are killed and digested. (*Build.*) Thick-edged mason's chisel for rough dressing stone. (*Civ. Eng.*) A term applied to a granite sett used in paving, or facing a dam or sea wall.

pitch face (*Build.*). A stone surface left with a rough finish produced by the hammer.

pitching. The angular motion of a ship or aircraft in a vertical plane about a lateral axis. (*Brew.*) The process of mixing the yeast intended to ferment a brew of beer with a small quantity of wort, in the fermenting vat, previous to admitting the remainder of the wort. This induces rapid fermentation. (*Civ. Eng.*) The foundation layer of well-rammed and consolidated broken stone upon which a road surfacing is built.

pitching fold (*Geol.*). A fold whose crest or trough line is not horizontal when traced in the general direction of strike.

pitching moment (*Aero.*). The component of the couple about the lateral axis, acting on an aircraft in flight.

pitching period (*Ships*). The movement of a ship in waves about a transverse horizontal axis is known as *pitching* as the bow goes down and *scending* as the bow rises. The time taken for the complete movement, down and up, is the *pitching period.*

pitching-piece (*Join.*). See **apron piece.**

pitching tool (*Build.*). A chisel with a very blunt edge, used to knock off superfluous stone.

pitch line (*Eng.*). The line along which the pitch of a rack is marked out, corresponding to the pitch circle of a spur wheel. (*Print.*) See **crash line.**

pitch of organ pipe stops (*Acous.*). The pitch of a stop, which potentially brings into action a rank of pipes, is known as the *footage* (q.v.). The footages are related, in terms of frequency, as follows:

Relative Frequency	Name of Relative Pitch	Name of Manual Stop Pitch
½ f	*sub-octave, double, contra* .	16 ft
1 f	*unison, prime* .	8 ft
1½ f	*sub-third octave, quint* .	5⅓ ft
2 f	*octave, principal* .	4 ft
3 f	*nazard* or *12th* .	2⅔ ft
4 f	*double octave* or *15th* . .	2 ft
5 f	*tierce* or *17th* .	1⅖ ft
6 f	*larigot* or *19th* .	1⅓ ft
7 f	*septième* or *20th* .	1⅐ ft
8 f	*3rd octave* or *22nd* .	1 ft
9 f	*nonième* . .	ft

For pedals the unison is 16 ft, the other pitches of pedal stops being in accordance.

pitch of propeller (*Ships, etc.*). The theoretical distance the propeller would advance through a solid in one revolution.

pitch of rivets (*Ships*). Distance between rivets measured centre to centre.

pitch pine (*For.*). *Pinus rigida,* yielding very resinous softwood timber; also low-grade tar and pitch. Commonly used for heavy framing and for piles, as well as for internal joinery.

pitch pipe (*Acous.*). A pipe, somewhat remote from an organ, which is tuned to a standard frequency and then used for adjusting pipes in the organ, its remoteness being necessary to obviate acoustic coupling and the pulling of the frequency of emitted sounds from pipes which are slightly mistuned.

pitch setting (*Aero.*). The blade angle of adjustable- or variable-pitch *airscrews* (q.v.).

pitchstone (*Geol.*). A glassy igneous rock which has a pitch-like (resinous) lustre and contains crystallites and microlites. It is usually of acid to subacid composition, and contains a notable amount of water (4% or more).

pith (*Bot.*). A cylinder of cells, chiefly parenchymatous, lying centrally in an axis and surrounded by vascular tissue. (*Horol.*) The pith of elderwood used in cleaning watches and clocks.

pith-ball electroscope (*Elec. Eng.*). Primitive detector of charges, whereby two pith-balls, suspended by silk threads, attract or repel each other, according to inverse-square law.

pithed (*Zool.*). Having the central nervous system (spinal cord and brain) destroyed.

pithiatism (*Med.*). Those phenomena of hysteria which can be produced by suggestion and removed by persuasion.

pith ray (*Bot.*). See **vascular ray.**

pith-ray fleck (*Bot.*). A dark spot in timber, composed of cells which have filled a cavity resulting from the attacks of insects on the cambium.

pit membrane (*Bot.*). The thin sheet of unbroken wall between two opposite pit cavities.

pitmen (*Mining*). Men employed in shaft sinking or shaft inspection and repair.

pit moulding (*Eng.*). Process by which extremely large castings are moulded in a pit, with brick-lined sides and cinder-lined bottom carrying connecting vent pipes to floor level.

Pitocin (*Pharm.*). TN for *oxytocin* (q.v.).

Pitot tube (*Aero., Civ. Eng., Mining, etc.*). Tube inserted parallel to flow stream. It has two orifices, one facing flow and hence receiving total pressure, and the other registering the static pressure at one side. Connected to a

U-tube from the exit point, it is used to measure pressure manometrically. There are numerous varieties and modifications, e.g., *pitometer* (with one orifice facing upstream, the other downstream) and *pitot-static tube* (see **pressure head**).

Pitressin (*Physiol.*). TN for **vasopressin.**

pit sampling (*Mining*). In alluvial prospecting, sinking of small untimbered pits in shallow ground or old ore dumps, to gain access to underlying strata.

pit-saw (*Tools*). A large two-handled rip-saw used for cutting logs. Also called a **cleaving-saw.**

pitted (*Bot., etc.*). (1) Having pits in the walls. (2) Having the surface marked by small excavations.

pitting (*Build.*). See **blowing.** (*Eng.*) (1) Corrosion of metal surfaces, as boiler plates, due to local chemical action. (2) A form of failure of gear teeth, due to imperfect lubrication under heavy tooth pressure.

pitting factor (*Met.*). In assessment of metal corrosion, the depth of penetration of the deepest pit divided by average loss of thickness as calculated from loss of weight.

Pittsburgh process (*Glass*). Method of making sheet glass by drawing glass vertically upward from a bath of molten glass, the edges of the sheet being formed by rollers and the draw rated by a submerged drawbar.

pituicyte (*Histol.*). Glial cell of the *pars nervosa* of the pituitary gland and other parts of the neurohypophysis.

pituitary gland (*Zool.*). The major endocrine gland of Vertebrates, formed by the fusion of a downgrowth from the floor of the diencephalon (the *infundibulum*) and an upgrowth of ectoderm from the roof of the mouth (*hypophyseal pouch*). The former becomes the *neurohypophysis*, comprising the neural lobe and infundibulum, and the latter becomes the *adenohypophysis*, comprising the *pars distalis*, the *pars tuberalis* and the *pars intermedia*. The *pars intermedia* and the neural lobe together form the posterior lobe. The *adenohypophysis* produces several hormones affecting growth, adrenal cortex activity, thyroid activity, reproduction (*gonadotropic hormones*) and melanophore cells. The *neurohypophysis* secretes two hormones, *vasopressin* and *oxytocin*.

Pituitrin (*Pharm.*). TN for hormone extract from pituitary gland, containing the hormones oxytocin and vasopressin, and used medically as an antidiuretic for diabetes insipidus.

pit work (*Mining*). The moving beams and balance bobs actuated in a shaft by a Cornish pump or beam engine.

pityriasis (*Med.*). A term common to various skin diseases in which branny scales appear.

pivot (*Horol.*). The reduced end of an arbor or staff which runs in a hole, jewel, or screw. It may be parallel, shouldered, or conical.

pivot bridge (*Eng.*). A form of swing bridge in which the vertical pivot is located at the middle of the length of the bridge.

pivoted brace (*Horol.*). A form of hooking for the mainspring to the barrel. The brace or post is pivoted into holes in the barrel and barrel cap, the end of the spring being looped round the centre portion of the brace.

pivoted detent (*Horol.*). A detent which is carried on pivots, as distinct from a spring detent such as is used in English chronometers.

pivot factor (*Elec. Eng.*). In an electrical indicating instrument, the (full-scale torque)/(mass of movement), a measure of freedom from error due to friction in the bearings.

pivot jaw (*Elec. Eng.*). A fixed jaw to which the blade of a switch is pivoted.

pivot joint (*Zool.*). An articulation permitting rotary movements only.

pK (*Chem.*). Value which is the logarithm to the base 10 of the reciprocal of the dissociation constant of a weak electrolyte. A high pK value indicates the substance has a small value for the dissociation constant and is a very weak electrolyte.

place (*Comp.*). See **column.**

placebo (*Med.*). A pharmacologically inactive substance which is administered as a drug either in the treatment of psychological illness or in the course of drug trials.

place bricks (*Build.*). See **grizzle bricks.**

placenta (*Bot.*). (1) The portion of the carpel wall, often somewhat fleshy, to which the ovules are attached. (2) Any mass of tissue to which sporangia or spores are attached. (*Zool.*) In *Eutheria*, a flattened cakelike structure formed by the intimate union of the allantois and chorion with the uterine wall of the mother; it serves for the respiration and nutrition of the growing young. *adjs.* **placental, placentate, placentiferous, placentigerous.**

Placentalia (*Zool.*). See **Eutheria.**

placental scale (*Bot.*). See **ovuliferous scale.**

placentation (*Bot.*). The arrangement of the placentae in an ovary, and of the ovules on the placentae. (*Zool.*) The method of union of the foetal and maternal tissues in a placenta.

placenta vera (*Zool.*). A deciduate placenta in which both maternal and foetal parts are thrown off at birth. Cf. *semiplacenta*.

placentiform (*Bot.*). Like a flat cake or cushion.

placers, placer deposits (*Geol.*). Superficial deposits, chiefly of fluviatile origin, rich in heavy ore minerals such as cassiterite, native gold, platinum, which have become concentrated in the course of time by long-continued disintegration and removal from the neighbourhood of the lighter associated minerals. See also **auriferous deposit.**

placode (*Zool.*). Any platelike structure; in Vertebrates, an ectodermal thickening contributing to a dorsal nerve-ganglion in the head region.

Placodermi (*Zool.*). A class of *Gnathostomata* comprising the earliest jawed Vertebrates, known from fossils from the Silurian to Permian. All had a heavy defensive armour of bony plates, and hyoid gill slits with no spiracle. The hyoid arch did not support the jaw, i.e., the aphethyoidean condition was retained. Most possessed paired fins, but with no general pattern.

placodioid (*Bot.*). Said of the thallus of a lichen which is rounded in outline and has that outline edged by small scales.

placodium (*Bot.*). A hardened hyphal layer surrounding the openings of the ostioles of perithecia embedded in a stroma.

placoid (*Zool.*). Plate-shaped; as the scales and teeth of *Selachii*.

placula (*Zool.*). A flattened blastula with reduced segmentation cavity occurring in the development of some *Urochorda*.

plafond (*Arch.*). The under surface of the corona in an entablature.

plages (*Astron.*). Dark or bright areas on calcium or hydrogen spectroheliograms, identified as areas of cool gas or heated gas respectively on the sun's surface; cf. *flocculi*, which refers to small patches only.

plagio-. Prefix from Gk. *plagios*, slanting, oblique.

plagiocephaly, plagiocephalism (*Med.*). An asymmetrical and twisted condition of the head.
plagioclase feldspars (*Min.*). An isomorphous series of triclinic silicate minerals which consist of albite and anorthite combined in all proportions. They are essential constituents of the majority of igneous rocks.
plagiophototaxy (*Bot.*). The arrangement of chlorophyll granules at an oblique angle to incident light.
plagiotropic, plagiotropous (*Bot.*). Said of members which become orientated at a constant angle to the direction in which the stimulus is acting.
plague (*Med.*). A disease of rodents due to infection with the *Bacillus pestis*, transmitted to man by rat-fleas, epizootics in rats invariably preceding epidemics. In Man the disease is characterized by enlargement of lymphatic glands (*bubonic plague*), severe prostration, a tendency to septicaemia, and occasional involvement of the lungs.
plagula (*Zool.*). A chitinous plate of the prosoma of *Solifugae*.
plaid (*Textiles*). (1) A woollen shawl or wrap, usually with a kind of check pattern. (2) Term used loosely to describe other types of simple check design cloth.
plain antenna system (*Radio*). An early form of spark transmission system in which the spark gap was included in the antenna circuit itself.
plain conduit (*Elec. Eng.*). See **plain steel conduit.**
plain coupler (*Elec. Eng.*). A short length of tubing serving to connect the end of two adjacent pieces of plain steel conduit in line with each other in an electrical installation. Also called a **sleeve.**
plain flap (*Aero.*). A wing flap in which the whole trailing edge (apart from the ailerons) is lowered so as to increase the camber. Also, occasionally, **camber flap.**
plain loom (*Weaving*). Machine for high-speed production of cloth of plain weave for printing, etc.
plain muscle (*Zool.*). See **unstriated muscle.**
plain-sawn (*For.*). See **back-sawn.**
plain steel conduit (*Elec. Eng.*). Conduit consisting of light-gauge steel tubing not having the ends screwed; used for containing the conductors in electrical installations. Cf. *screwed steel conduit*. Also called **plain conduit.**
plain tile (*Build.*). The ordinary flat tile, with two nibs for hanging from the battens.
plain weave (*Weaving*). Simplest interlacing of warp and weft. Each warp thread is alternately over and under the weft, while adjacent warp threads work opposite to each other. Also called **calico weave, tabby weave.**
plaiting (*Textiles*). See **cuttling** (1).
planar diode (*Electronics*). (1) Vacuum tube with plane parallel electrodes. A photoelectric cathode is usually used. (2) See **planar semiconductor.**
planar process (*Electronics*). A basic part of the technology of silicon transistors, being a combination of oxidation, selective oxide removal and then heating to introduce dope materials by diffusion.
planar semiconductor (diode, transistor or integrated circuit) (*Electronics*). One prepared on the surface of a semiconductor crystal, usually silicon, by the *planar process*.
planceer or **plancier** (*Build.*). A soffit, especially the under surface of the corona in a cornice.
planceer piece (*Build.*). A horizontal timber to which the soffit boards of an overhanging eave are fastened.
Planckian colour (*Phys.*). The colour or wavelength-intensity distribution of the light emitted by a black body at a given temperature.
Planckian locus (*Phys.*). Line on *chromaticity diagram* joining points with coordinates corresponding to black body radiators.
Planck's law (*Phys.*). Basis of quantum theory, that the energy of electromagnetic waves is confined in indivisible packets or quanta, each of which has to be radiated or absorbed as a whole, the magnitude being proportional to frequency. If E is the value of the quantum expressed in energy units and ν is the frequency of the radiation, then $E=h\nu$, where h is known as *Planck's constant* and has dimensions of energy × time, i.e., action. Present accepted value is $6{\cdot}626\times10^{-34}$ J s. See **photon.**
Planck's radiation formula (*Phys.*). Expression for the electromagnetic energy density E_ν of frequency ν radiated per second from unit area of a black body at absolute temperature T is:

$$E_\nu=\frac{h\nu^5}{c^3(e^{h\nu/kT}-1)},$$

where c is the velocity of light, h is Planck's constant, k is Boltzmann's constant, and e is the base of natural logarithms.
plançon (*For.*). A log of hardwood timber roughly sawn or hewn to an octagonal shape, with a minimum of 10 in. (250 mm) between opposite faces.
plane (*Carp., Join.*). A wood-working tool used for the purpose of smoothing surfaces, reducing the size of wood and, in specialized forms, for grooving, rebating, and other purposes. (*Maths.*) A flat surface; one whose radius of curvature is infinite at all points.
plane baffle (*Acous.*). Plane board, with a hole, at or near the centre, for mounting and loading an open-diaphragm loudspeaker unit.
plane earth factor (*Phys.*). Electromagnetic wave propagation, the ratio of the electric field strength which would result from propagation over an imperfectly conducting earth to that resulting from propagation over a perfectly conducting plane.
plane-iron (*Carp., etc.*). The cutting part of a plane, which actually shapes the work.
plane of collimation (*Surv.*). The imaginary surface swept out by the *line of collimation* of a levelling instrument, when its telescope is rotated about its vertical axis.
plane of polarization (*Light*). The plane containing the incident and reflected light rays and the normal to the reflecting surface. The magnetic vector of plane-polarized light lies in this plane. The electric vector lies in the plane of vibration, which is that containing the plane-polarized reflected ray and the normal to the plane of polarization.
plane of saturation (*Civ. Eng.*). The natural level of the *ground water* (q.v.).
plane of section (*Bot.*). The direction in which a plant member is cut, or assumed to be cut, for purposes of elucidating its structure. In radially symmetrical members, the plane may be *transverse* (i.e., at right angles to the longitudinal axis), *longitudinally radial* (i.e., along one of the radii) or *longitudinally tangential* (i.e., parallel to a tangent to the surface). In a flower, the plane may be *median* or *anterior-posterior* (i.e., passing through the bract and axis), *lateral* (i.e., parallel to the ground), or neither of these, when it is *oblique*.
plane of symmetry (*Crystal.*). In a crystal, an imaginary plane on opposite sides of which faces, edges, or solid angles are found in similar

positions. One half of the crystal is hence a mirror image of the other. (*Maths.*) See **symmetry (1).**

plane polarization (*Micros.*). Restriction of light waves to one wave direction on emergence from a polarizing system. (*Phys.*) That of an electromagnetic wave when the electric (or magnetic) field at any point does not vary in direction (except for reversal) over a cycle. A light beam striking a glass plate at the Brewster angle gives rise to a reflected beam which is plane-polarized.

planer (*Typog.*). A flat piece of wood or rubber which is placed on a forme of type and tapped with a mallet to level the surface.

planer tools (*Eng.*). Planing machine cutting-tools, similar to those used for turning, clamped vertically in a block pivoted in the *clapper box* (q.v.) on the head.

plane stock (*Carp., etc.*). The body of a plane, holding the plane-iron in position.

plane surveying (*Surv.*). Surveying which makes no correction for curvature of earth's surface.

planet (*Astron.*). (Gk. *planētēs*, wandering.) The name given in antiquity to the seven heavenly bodies, including the sun and moon, which were thought to travel among the fixed stars. The term is now restricted to those bodies, including the Earth, which revolve in elliptic orbits about the sun; in the order of distance they are: Mercury, Venus, Earth, Mars, Jupiter, Saturn, Uranus, Neptune, and Pluto. The two planets, Mercury and Venus, which revolve within the Earth's orbit are designated *inferior planets*, the planets Mars to Pluto are *superior planets*. Planets reflect the sun's light and do not generate light or heat.

plane table (*Surv.*). A drawing-board mounted on a tripod so that the board can be levelled and also rotated about a vertical axis and clamped in position. An alidade completes the essential parts of a plane table. It is set up at ends of a suitable baseline where required survey points can be seen from these ends. By intersecting sights a rough plan can then be produced.

planetarium (*Astron.*). A building in which an optical device displays the apparent motions of the heavenly bodies on the interior of a dome which forms the ceiling of the auditorium.

planetary electrons (*Phys.*). Those which orbit round the nucleus of an atom.

planetary gear (*Eng.*). Any gear-wheel whose axis describes a circular path round that of another wheel, e.g., the bevel wheels carried by the crown wheel of *differential gear* (q.v.).

planetary nebulae (*Astron.*). Small regular nebulae showing a disk which resembles a planet. They consist of shells of gas surrounding a central star; about 500 are known, mostly within our Galaxy, but a few have been detected in other galaxies.

planetary probe (*Space*). A space vehicle intended to explore the neighbourhood of a planet, or to land instruments on its surface, and send information back to Earth. See **Mariner, Zond.**

planetesimal hypothesis (*Astron.*). A theory of the evolution of our solar system which postulates a non-rotating spherical mass of gas, so affected by the tidal action of a passing star as to emit two opposite jets of matter while also beginning to rotate, the spiral arms so formed condensing into planets by the larger condensations annexing the smaller (called *planetesimals*).

plane-tile (*Build.*). See **crown-tile.**

planetoid (*Astron.*). See **asteroid.**

plane wave (*Phys.*). One for which equiphase surfaces are planes.

planidium (*Zool.*). A larval type of certain parasitic *Hymenoptera*, of active habit, possessing chitinized segmental plates and locomotor spines, and developing from an egg laid away from the host.

planigraphy (*Radiol.*). See **tomography.**

planimeter (*Eng., etc.*). A form of integrator for measuring mechanically the area of a plane surface, e.g., the area of an indicator diagram. A tracing point on an arm is moved round the closed curve, whose area is then given to scale by the revolutions of a small wheel supporting the arm.

planing bottom (*Aero.*). The part of the under surface of a flying-boat hull which provides hydrodynamic lift.

planing machine (*Eng.*). A machine for producing large flat surfaces. It consists of a gear-driven reciprocating work-table sliding on a heavy bed, the stationary tool being carried above it by a saddle, which can be traversed across a horizontal rail carried by uprights. See **clapper box.**

planisher (*Met.*). (1) Hammer or tool for planishing. (2) A *rolling mill* (q.v.).

planishing (*Met.*). Giving a finish to metal surfaces by hammering.

plank root (*Bot.*). A root which is very markedly flattened so that it stands out from the base of the stem like a plank set edgeways to the surface of the ground; plank roots give additional support to the plant.

planktohyponeuston (*Ecol.*). Organisms of the main water mass that accumulate near the surface at night.

plankton (*Ecol.*). Animals and plants floating in the waters of seas, rivers, ponds, and lakes, as distinct from animals which are attached to, or crawl upon, the bottom; especially minute organisms and forms, possessing weak locomotor powers.

plankton pulse (*Ecol.*). Periodic variation in the abundance of plankton in any particular area, due to various ecological factors.

plank truss (*Carp.*). A roof or bridge truss constructed of planks.

planning grid (*Arch.*). Squared grid scaled in *modules* (q.v.), used in designing for modular construction.

Plannipennia (*Zool.*). An order of the *Neuropteroidea*, with a prognathous head, biting mouthparts, and two pairs of long membranous wings with veins bifurcated at the ends. The larvae are terrestrial with long, curved, grooved mandibles for piercing and sucking. Lace Wings and Ant-lions.

planoblast (*Zool.*). A free-swimming Medusa.

plano-convex (*Optics, Photog.*). Said of a lens with one surface flat and the other curved.

planogamete (*Bot., Zool.*). A motile or wandering gamete; a zoogamete.

planogamic (*Bot.*). Having motile gametes.

planographic process (*Print.*). Process in which the printing image is on a level with the plate, which is specially treated to accept ink while the surrounding areas reject it. See **collotype, lithography.**

planont (*Zool.*). In some *Neosporidia*, an initial amoebula phase liberated from a spore within the body of the host.

planosome (*Cyt.*). An odd (supernumerary) chromosome resulting from non-disjunction of a pair during meiosis.

planospore (*Bot.*). See **zoospore.**

planozygote (*Bot.*). A motile zygote.

plan-position indicator (*Radar*). Screen of a CRT with an intensity-modulated and persistent radial display, which rotates in synchronism with a highly directional antenna. The surrounding terrain is thus painted with relevant reflecting objects, such as ships, aircraft, and physical features. Abbrev. **PPI.** See **azimuth stabilized PPI.**

plant (*Bot.*). An organism which has little or no power of dealing with solid food, and which therefore takes in all or most of the material used in nutrition in solution in water. (*Eng., etc.*) (1) The machines, tools, and other appliances requisite for carrying on a mechanical or constructional business; the term sometimes includes also the buildings and the site and, in the case of a railway, the rolling stock. (2) The permanent appliances needed for the equipment of an institution.

planta (*Zool.*). The sole of the foot in land Vertebrates; the flat apex of a proleg in Insects. *adj.* **plantar.**

plantation (*Build.*). A slate size, 330 × 280 mm.

planted moulding (*Join.*). A moulding cut out of a separate strip of wood of the required section and secured to the surface to be decorated.

Planté plate (*Elec. Eng.*). See **formed plate.**

plant formation (*Bot.*). See **association.**

plantigrade (*Zool.*). Walking on the soles of the feet, as Man. Cf. *digitigrade, unguligrade.*

plant indicator (*Bot.*). See **indicator.**

planting (*Join.*). The operation of forming a *planted moulding* (q.v.).

plant load factor (*Elec. Eng.*). The ratio of the total number of kWh supplied by a generator or generating station to the total number of kWh which would have been supplied if the generator or generating station had been operated continuously at its maximum continuous rating.

plant pathology (*Bot.*). See **phytopathology.**

plantula (*Zool.*). See **pulvillus.**

planula (*Zool.*). A larval form of some Invertebrates, especially *Coelenterata*; it consists of an outer layer of ciliated ectoderm and an inner mass of endoderm cells.

plaque (*Bot.*). Areas of cell destruction in a bacterial colony grown on a solid medium, or tissue culture monolayer preparation due to infection with a virus. (*Med.*) A layer of amorphous material adhering to the surfaces of teeth. (*Min. Proc.*) White-enamelled saucer-shaped disk used in spot checking of products made during ore treatment. It has taken the place of the old vanning shovel. A sample is gently manipulated on it with added water, to separate the light from the heavy constituents.

plashing (*Build., Civ. Eng.*). The process of intertwining branches in forming hurdles, etc.

plasm (*Biol.*). Protoplasm, especially in compound terms, as *germ plasm.*

plasma (*Min.*). A bright-green translucent variety of cryptocrystalline silica (*chalcedony*). It is used as a semiprecious gem. (*Electronics*) Synonym for the positive column in a gas discharge. (*Phys.*) Ionized gaseous discharge in which there is no resultant charge, the number of positive and negative ions being equal, in addition to un-ionized molecules or atoms. (*Physiol.*) The watery fluid containing salts, protein and other organic compounds, in which the cells of the blood are suspended. When blood coagulates it loses certain constituents (for example, fibrinogen) and becomes serum.

plasma-, plasmo, -plasm. Prefix and suffix from Gk. *plasma*, gen. *plasmatos*, anything moulded.

plasma-arc cutting (*Eng.*). Cutting metal at temperatures approaching 35 000°C by means of a gas stream heated by a tungsten arc to such a high temperature that it becomes ionized and acts as a conductor of electricity. Heat is not obtained by a chemical reaction; the process can therefore be used to cut any metal.

plasmacytoma (*Med.*). A tumour appearing in bone and composed of cells closely resembling plasma cells. Also **plasmoma.**

plasma display panels (*Electronics*). An array of gas cells which can be selectively ignited to form any desired letter, number or diagram.

plasmagene (*Gen.*). A cytoplasmic hereditary factor; cytogene.

plasmalemma (*Zool.*). See **plasma membrane.**

plasma membrane (*Bot.*). A very thin layer of specialized protoplasm, forming the outer boundary of the protoplast where that is in contact with the cell wall. (*Zool.*) An extremely thin membrane enclosing the apparently naked protoplasm of forms like *Amoeba.* See **cell wall.** Also **plasmalemma.**

plasmatoparous (*Bot.*). Said of a spore which, in the earliest stage of germination, emits its contents as a naked protoplast, which then forms a wall and puts out a germ tube.

plasma torch (*Phys.*). One in which solids, liquids, or gases are forced through an arc within a water-cooled tube, with consequent ionization; de-ionization on impact results in very high temperatures. Used for cutting and depositing carbides.

plasmatron (*Electronics*). Discharge tube in which anode current can be regulated by control of plasma, either by a grid or through the electron stream originating the plasma.

plasmids (*Biol.*). Structures in the cell cytoplasm which are able to reproduce autonomously, e.g., mitochondria.

plasmin (*Med.*). A substance in blood capable of destroying fibrin as it is formed.

plasminogen (*Med.*). The precursor of plasmin in the blood.

plasmocyte (*Zool.*). See **leucocyte.**

plasmoderma (*Bot.*). A very thin specialized layer of protoplasm around a vacuole.

plasmodesm (*Bot.*). An extremely delicate strand of protoplasm passing through a fine perforation in a cell wall, and, with many other plasmodesms, providing a connexion between the protoplasts of contiguous cells.

plasmodiocarp (*Bot.*). A sporangium, formed by some *Myxomycetes*, which is of irregular or sinuous form.

Plasmodiophorales (*Bot.*). An order of obligate plant parasites belonging to the lower fungi. Sometimes classified with the Myxomycetes because of the plasmodial vegetative phase and absence of a mycelium. Includes *Plasmodiophora brassicae* (club root of cruciferous plants) and *Spongospora solani* (corky scab of potato).

plasmoditrophoblast (*Zool.*). In the Mammalian placenta, a syncytium formed by the thickening of the upper part of the trophoblast which is in contact with the uterine wall.

plasmodium (*Bot.*). A thallus having the form of a naked multinucleate mass of protoplasm which can creep in amoeboid manner and take in solid food material. (*Zool.*) A syncytium formed by the union of uninucleate individuals without fusion of their nuclei. *adj.* **plasmodial.**

Plasmodium (*Zool.*). A genus of sporozoan *Protozoa* which includes the causative organisms of malaria.

plasmogamy (*Biol.*). Fusion of cytoplasm as distinct from fusion of nucleoplasm; plastogamy.

plasmoid (*Phys.*). Any individual section of a plasma with a characteristic shape.

plasmolysis (*Biol.*). Removal of water from a cell by osmotic methods, with resultant shrinking.

plasmoma (*Med.*). See **plasmacytoma.**

plasmon (*Gen.*). The sum total of the cytoplasmic hereditary factors.

plasmosome (*Cyt.*). A small cytoplasmic granule; a type of *nucleolus* (q.v.) which stains with acid dyes and disappears during mitosis without mingling with the chromosomes.

plasmotomy (*Zool.*). Fission, by division of cytoplasm only, of multinucleate *Protozoa* to form multinucleate offspring.

plaster. A general name for plastic substances which are used for coating wall surfaces, and which set hard after application. See also **acoustic plaster.**

plaster board (*Build.*). A building-board made of plaster with paper facings.

plasterer's putty (*Build.*). A preparation similar to *fine stuff* (q.v.), made by dissolving pure lime in water and passing it through a fine sieve.

plaster mould casting (*Eng.*). Small, precision parts of non-ferrous alloys are cast in plaster moulds which are destroyed when the casting is removed.

plaster of Paris (*Chem., etc.*). Partly dehydrated gypsum, $2CaSO_4$, H_2O (hemihydrate). When mixed with water, it evolves heat and quickly solidifies, expanding slightly; used for making casts.

plaster slab (*Build.*). A block, frequently perforated, made from plaster of Paris and coarse sand; used in the construction of partitions.

plastic (*TV*). Said of a TV image when distortion results in apparent moulding or relief.

plastic bronze (*Met.*). Bronze containing a high proportion of lead; used for bearings. Composition: 72–84% copper, 5–10% tin, and 8–20% lead plus zinc, nickel, and phosphorus.

plastic clay (*Build.*). See **foul clay.**

plastic deformation (*Met.*). Permanent change in the shape of a piece of metal, or in the constituent crystals, brought about by the application of mechanical force. (*Geol.*) Also said of certain minerals, e.g., *calcite*, so affected during metamorphism.

plasticity (*Eng.*). A property of certain materials by which the deformation due to a stress is largely retained after removal of the stress.

plasticizers (*Chem.*). High-boiling liquids used as ingredients in lacquers and certain plastics, e.g., PVC; they do not evaporate but preserve the flexibility and adhesive power of the cellulose lacquer films or the flexibility of plastic sheet and film. Well-known plasticizers are triphenyl phosphate, tricresyl phosphate, high-boiling glycol esters, etc.

plastic limit (*Agric.*). Of a soil, the moisture-content corresponding to a departure from a plastic condition, as indicated by a standardized apparatus.

plastic material (*Bot.*). Any substance which is used up in growth processes.

plastic paint (*Paint.*). Thick-texture paint which can be worked to a patterned finish.

plastic rail-bond (*Elec. Eng.*). A rail-bond made by inserting plastic conducting material between the rail itself and the fishplate.

plastics. A generic name for certain organic substances, mostly synthetic (see **synthetic resins**) or semi-synthetic (casein and cellulose derivatives) condensation or polymerization products, also for certain natural substances (shellac, bitumen, but excluding natural rubber), which under heat and pressure become plastic, and can then be shaped or cast in a mould, extruded as rod, tube, etc., or used in the formation of laminated products, paints, lacquers, glues, etc. Plastics are *thermoplastic* (q.v.) or *thermosetting* (q.v.). Adaptability, uniformity of composition, lightness, and good electrical properties make plastic substances of wide application, though relatively low resistance to heat, strain, and weather are, in general, limiting factors of consequence. See articles on the different classes of plastics.

plastic setting (*Textiles*). Setting woollen yarn, previously placed under tension, by the influence of moisture at high temperature, to eliminate curl which tends to develop in certain types.

plastics moulding (*Eng.*). A process for manufacturing articles of plastics materials, using either injection moulding machines for the rapid production of small articles or hydraulic presses for compression moulding of relatively large articles.

plastic sulphur (*Chem.*). Formed when sulphur is distilled into water. Unstable and changes to the rhombic form.

plastic surgery (*Surg.*). That branch of surgery which deals with the repair and restoration of damaged or lost parts of the body.

plastid (*Cyt.*). Any small dense protoplasmic inclusion in a cell. Plastids are probably special centres of chemical activity, and many of them, when exposed to light, become pigmented and become chloroplasts.

plastidome (*Bot.*). The total outfit of plastids in a cell.

plastiponics (*Agric.*). Technique of improving arid soils by the use of plastic foam, which forms a plastic soil, retaining the soil water which plant roots need.

plastochrone (*Bot.*). The period of time that elapses between the formation of one leaf primordium and the next, on the growing point of a shoot in which there is a stable spiral phyllotaxis.

plastocont (*Cyt.*). See **chondriocont.**

plastogamy (*Zool.*). Union of individual *Protozoa* without fusion of their nuclei.

plastonema (*Bot.*). Deeply staining peripheral cytoplasm in sporogenous tissue in mosses.

plastron (*Anat.*). The sternum and the costal cartilages. (*Zool.*) (1) The ventral part of the bony exoskeleton in *Chelonia*; any similar structure. (2) In some aquatic Insects (e.g., *Coleoptera* and *Hemiptera*) a thin air film over certain parts of the body held by minute hydrofuge hairs which may be as numerous as 2 million per sq. mm. This serves as a physical gill rather than as a store of air. *adj.* **plastral.**

platband (*Build.*). (1) An *impost* (q.v.). (2) A flat projecting moulding, which projects from the general wall surface by an amount less than its own breadth. (3) A door or window lintel.

plate (*Carp.*). The top horizontal timber of a wall, supporting parts of the structure. Also **platt.** (*Chem. Eng.*) In an industrial fractionating column, a tray over which the liquid flows, with nozzles or bubble caps so that the vapour can bubble through the liquid. See **Murphree efficiency, still.** (*Elec.*) (1) Each of the two extended conducting electrodes which, with a dielectric between, constitutes a capacitor. (2) U.S. for *anode* in valves. (*Eng.*) A large, flat body of steel, thicker than sheet, which is produced by the working of ingots, billets, or slabs, in a rolling mill. (*Photog.*) Glass used as a support for sensitive emulsions during exposure and processing. See **collodion process.** (*Print.*) (1) Any original and duplicate for letterpress printing; also plates for lithographing, and the several kinds of intaglio plate. (2) An illustration, especially one that is printed separately from the text which it illustrates. (*Zool.*) See **plax.**

plate amalgamation (*Min. Proc.*). Trapping of metallic gold on an inclined plate made of copper or an alloy, which has been coated with a pasty film of mercury. Method largely superseded by use of *strake* (q.v.).

plateau (*Nuc. Eng.*). See under **Geiger region.**

plateau-basalts (*Geol.*). Basic lavas of basaltic composition occurring as thin, widespread flows, forming extensive plateaux (e.g., the Deccan in India).

plateau-building movements (*Geol.*). See **epeirogenic earth movements.**

plateau eruptions (*Geol.*). Volcanic eruptions by which extensive lava-flows are spread in successive sheets over a wide area and eventually build a plateau; as in Idaho. See **fissure eruptions.**

plateau gravel (*Geol.*). Deposits of sandy gravel occurring on hill-tops and plateaux at heights above those normally occupied by river terrace gravels. Originally deposited as continuous sheets, plateau gravel has been raised by earth movements to its present level and deeply dissected. Of Pliocene or early Pleistocene age in the main.

plateau length (*Nuc. Eng.*). The voltage range which corresponds to the plateau of a Geiger counter tube.

plateau slope (*Nuc. Eng.*). The ratio of the percentage change in count rate for a constant source, to the change of operating voltage. Measured for a median voltage corresponding to the centre of the Geiger plateau.

plate battery (*Telecomm.*). B-battery or high-tension battery, especially for portable equipment.

plate-bend detector (*Electronics*). One which applies incoming signal to grid of triode valve with considerable negative bias; the demodulated signal is taken from anode circuit. Also **anode-bend detector.**

plate cam (*Eng.*). A flat, open cam for sliding movement used, for example, in automatics.

plate clutch (*Eng.*). See **disk clutch.**

plate columns (*Chem. Eng.*). Distillation or absorption columns which contain plates of various types, e.g., *bubble cap, Glitsch trays, Kuhm, sieve, Turbogrid* (qq.v.), spaced at regular intervals with nothing between the plates.

plate cylinder (*Print.*). The cylinder on a printing press to which the printing plate is attached.

plated carbon (*Elec. Eng.*). An arc-lamp carbon upon which a layer of copper has been deposited by electroplating, in order to improve its conductivity and ensure good contact with the holder.

plated fabrics (*Textiles*). Fabrics produced by arranging two threads in a knitting machine so that one appears on the face of the fabric and the other on the back; e.g., cotton on wool.

plate dissipation (*Electronics*). The energy dissipated by the electrons striking the plate electrode of a valve at high velocities.

plate efficiency (*Electronics*). Same as **anode efficiency.**

plate exchanger (*Chem. Eng.*). Heat exchanger comprising either a series of alternating flat and ribbed plates forming the flow channel, or flat plates which are hollow with internal flow passages immersed in or forming part of the wall of vessels.

plate frame (*Elec. Eng.*). The nickel-plated framework for supporting the perforated steel tubes of the electrode of a nickel-iron accumulator.

plate gauge (*Eng.*). A limit gauge or single external gauge formed by cutting slots of the required gauge width in a steel plate, the surfaces of which are hardened. See **limit gauge.**

plate girder (*Eng.*). A built-up steel girder consisting of a single web-plate along each edge of which is riveted, as a flange, a pair of angles. In the larger sizes, flange-plates are riveted to the angles.

plate glass (*Glass*). Glass of superior quality, originally cast on an iron bed and rolled into sheet form, and afterwards ground and polished. Modern methods have to a large extent superseded this, save for special kinds of glass. Also **polished plate.**

plate glazing (*Paper*). The operation of passing paper and polished metal sheets between heavy rolls, the process being repeated until a sufficiently smooth surface has been imparted to the paper.

plate group (*Elec. Eng.*). The complete unit, consisting of an accumulator plate or plates, terminal bar and terminal lug, forming the electrode of an accumulator cell. Also called a **plate section.**

plate link chain (*Eng.*). A chain comprising pairs of flat links connected by pins.

plate load impedance (*Electronics*). The total impedance between the plate (anode) and cathode of a thermionic valve, excluding the electron stream.

plate-lug (*Elec. Eng.*). A projection on an accumulator plate used for connecting it to a terminal bar.

plate modulation (*Electronics*). See **anode modulation.**

plate moulding (*Foundry*). A method of mounting the halves of a split pattern on opposite sides of a wood or metal plate, placed between the cope and drag, thus eliminating the making of the joint faces.

platen (*Eng.*). The work table of a machine tool, usually slotted for clamping bolts.

plate neutralization (*Elec. Eng.*). A method of neutralizing an amplifier (i.e., preventing its oscillation) by shifting through π radians a part of the plate-cathode a.c. voltage and applying it to the cathode-grid circuit via a neutralizing capacitor.

plate nip (or **cling**) (*Print.*). The tight fit of the metal plate on the cylinder.

platen machine (*Print.*). A printing machine in which the impression is taken with a flat surface, not a cylinder. Small commercial work is usually printed on platen machines.

plate proof (*Print.*). A proof taken from a plate as distinct from one taken direct from type.

plate rectifier (*Elec. Eng.*). One of large area for large output currents, e.g., for electrolytic bath supply, or electric traction.

plate resistance (*Electronics*). More strictly the *dynamic plate resistance of a thermionic valve*, defined by

$$R_P = \left(\frac{\Delta E_P}{\Delta I_p}\right)_{E_G},$$

where ΔE_P and ΔI_p represent very small changes of plate voltage and current for a constant grid voltage E_G.

plates (*Horol.*). The circular or rectangular plates of brass which form the framework of a watch or clock and which are drilled to receive the pivots of the train, etc.

plate screws (*Surv.*). The screws directly connecting the head of a theodolite or level with its tripod, and serving as a means of adjustment for bringing the head to a level position. Also called **foot screws.**

plate section (*Elec. Eng.*). See **plate group.**

plate slap (*Print.*). The noise resulting from the impact of loose or badly fitting plates against the cylinder.

plate support (*Elec. Eng.*). A support from which the plates of an accumulator are suspended or upon which they rest.

plate tectonics (*Geol.*). The interpretation of the Earth's major structures in terms of the movements of large plates of lithosphere acting as rigid slabs.

platform escapement (*Horol.*). An escapement mounted on an independent plate.

platform gantry (*Build.*). A gantry formed to support a platform on which is erected a scaffold used for the handling of materials.

platforming (*Chem. Eng.*). Process for reforming low-grade into high-grade petrol, using a platinum catalyst.

platinammines (*Chem.*). Compounds of platinum and ammonia of the form $Pt(NH_3X)_4$.

platinates (IV) (*Chem.*). See **platinic hydroxide.**

platinectomy (*Surg.*). The operative removal of the stapedial footplate in the middle ear.

plating (*Bind.*). The operation of inserting *plates* (see **plate** (*Print.*) (2)) in a volume, e.g., by attaching them to guards. (*Print.*) Making stereos, electros, or other duplicates.

plating sequence (*Print.*). The order in which printing plates are secured to the press.

plating up (*Print.*). The securing of plates to the printing cylinders.

platinic hydroxide (*Chem.*). Platinum (IV) hydroxide. $Pt(OH)_4$. Dissolves in acids to form platinic salts and in bases a series of salts called *platinates* (IV). A type of compound formed by the other members of the platinum group of metals.

platinic oxide (*Chem.*). Platinum (IV) oxide. PtO_2. Dark grey powder formed when platinic (platinum (IV)) hydroxide is heated. Also called **platinum dioxide.**

platinite (*Met.*). Alloy containing iron 54–58%, and nickel 42–46%, with a trace of carbon. Has the same coefficient of expansion as platinum, and is used to replace it in some light bulbs and measuring standards.

platinized asbestos (*Chem.*). Asbestos permeated with finely divided platinum. Used as a catalyst.

platinoid (*Met.*). Alloy containing copper 62%, zinc 22%, nickel 15%. Has high electrical resistance and is used for resistances and thermocouples.

Platinotron (*Radar*). TN for a high-power microwave tube for radar, similar to a *magnetron*.

platinotype (*Photog.*). Early type of printing process involving the reduction of a platinum salt to platinum by ferrous oxalate, after the latter has been reduced from ferric oxalate during exposure.

platinous hydroxide (*Chem.*). Platinum (II) hydroxide. $Pt(OH)_2$. Soluble in the haloid acids (hydrochloric acid, etc.), forming platinous (platinum (II)) salts.

platinous oxide (*Chem.*). PtO. Formed when platinous hydroxide is gently heated.

platinum (*Met.*). A metallic element, symbol Pt, at. no. 78, r.a.m. 195·09, rel. d. at 20°C 21·45, electrical resistivity at 20°C $9{\cdot}97 \times 10^{-8}$ ohm metres, m.p. 1773·5°C, b.p. 3910°C, Brinell hardness 47. Platinum is the most important of a group of six closely related rare metals, the others being osmium, iridium, palladium, rhodium, and ruthenium. It is heavy, soft, and ductile, immune to attack by most chemical reagents and to oxidation at high temperatures. Used for making jewellery, special scientific apparatus, electrical contacts for high temperatures and for electrodes subjected to possible chemical attack. Also used as a basic metal for resistance thermometry over a wide temperature range. Native platinum is usually alloyed with iron, iridium, rhodium, palladium, or osmium, and crystallizes in the cubic system.

platinum black (*Chem.*). Platinum precipitated from a solution of the (IV) chloride by reducing agents. A velvety-black powder. Uses: catalyst and gas absorber, e.g., in the *hydrogen electrode*.

platinum dioxide (*Chem.*). Platinum (IV) oxide. See **platinic oxide.**

platinum sponge (*Chem.*). See **spongy platinum.**

platinum tetrachloride (*Chem.*). Platinum (IV) chloride. $PtCl_4$. Formed by dissolving platinum in aqua regia. Similar chlorides are formed with the other platinum metals.

platinum thermometer (*Elec. Eng.*). See **resistance thermometer.**

platt (*Carp.*). See **plate.**

platting (*Build.*). The top course of a brick clamp.

platybasic (*Zool.*). Said of the chondrocranium of developing Vertebrates which has the trabeculae wide apart.

platycephalic, platycephalus (*Anat.*). Having a flattened or broad head, with a breadth-height index of less than 70.

Platyctenea (*Zool.*). A group of *Tentaculata* which includes a number of aberrant forms of flattened creeping habit, having much reduced tentacles.

platydactyl (*Zool.*). Having the tips of the digits flattened.

Platyhelminthes (*Zool.*). A phylum of bilaterally symmetrical, triploblastic *Metazoa*; usually dorsoventrally flattened; the space between the gut and the integument is occluded by parenchyma; the excretory system consists of ramified canals containing flame-cells; there is no anus, coelom, or haemocoele; the genitalia are usually complex and hermaphrodite. Flat Worms.

platykurtosis (*Stats.*). A distribution curve which is flat-topped and short-tailed compared with the normal.

platyphylous (*Bot.*). (1) Broadly lobed. (2) Having wide leaves.

platysma (*Zool.*). A broad sheet of dermal musculature in the neck region of Mammals.

platysperm (*Bot.*). A seed which is flattened in transverse section.

platytrabic (*Zool.*). Platybasic.

plauenite (*Geol.*). A name sometimes applied to the well-known Dresden syenite, consisting essentially of orthoclase, hornblende, some oligoclase, and a little quartz.

plax (*Zool.*). A flat platelike structure, as a lamella or scale.

play (*An. Behav.*). A rather diffuse category of animal behaviour concerned with the practising of skills by young animals using trial and error, and also allowing a thorough, and potentially useful, exploration of the environment, possible when conditions of life are relatively easy, and appetitive behaviour is to some extent freed from the constraints of primary needs. (*Eng.*) Limited movement between mating parts of a mechanism, due either initially to the type of fit and dimensional allowance specified or subsequently to wear.

playback (*Acous.*). Immediate reproduction from a disk or tape recording; used for testing the quality of the reproduced sound before actual records are made, or as a check on other types of recording.

playback equalizer (*Acous.*). A resistance-capacitance network introduced into an inter-stage coupling so that all frequencies are reproduced with equal intensity in the recording of music.

pleasure principle (*Psychol.*). The original urge of the *id* to seek gratification for its impulses regardless of any other considerations. It rules human behaviour at the start of life and remains the guiding principle of the *unconscious*.

pleated-diaphragm loudspeaker (*Acous.*). A loud-speaker in which the radiating element is a pleated diaphragm, the pleats being radial and the rim clamped. It is driven by a pin at the centre.

Plecoptera (*Zool.*). An order of the *Orthopteroidea* (Insects; *Polyneoptera*). Prognathous, with biting mouthparts, long antennae, two pairs of well-developed wings with, some say, primitive venation (but weak fliers), 3-jointed tarsi, long cerci. The nymphs are aquatic, usually in swift-flowing streams, with gill tufts in varied positions. Widely distributed. Stoneflies.

Plectascales (*Bot.*). An order of *Ascomycetes* belonging to the *Plectomycetes*, with the asci in a cleistocarp; saprophytic and weakly parasitic members. Includes the genera *Penicillium* and *Aspergillus*, both of which are sources of penicillin, and which may also be classified with the *Fungi Imperfecti*, because of the absence of a perfect stage.

plectenchyma (*Bot.*). Pseudoparenchyma formed by the interweaving of hyphae. *adjs.* **plectenchymatous, plectenchymic, plectenchymoid.**

Plectomycetes (*Bot.*). Members of the *Ascomycetes* with the asci borne on ascogenous hyphae and usually enclosed within a closed spherical cleistocarp. Includes the orders *Plectascales* and *Erysiphales*.

plei-, pleio-, pleo-, plio-. Prefix from Gk. *pleiōn*, more.

Pleiades, The (*Astron.*). The name given to the open cluster in the constellation *Taurus*, of which the seven principal stars, forming a well-known group visible to the naked eye, each have a separate name.

pleiandrous (*Bot.*). Having a large and indefinite number of stamens.

pleiochasium (*Bot.*). A cymose inflorescence in which each branch bears more than two lateral branches.

pleiomerous (*Bot.*). Having a large number of parts or organs.

pleion (*Meteor.*). An area over which some weather element, such as temperature, is above the normal. The reverse is *antipleion*.

pleiosporous (*Bot.*). Many-spored.

pleiotaxy (*Bot.*). An increase in the number of whorls in a flower.

pleiotomy (*Bot.*). Multiple apical division with the formation of multiplets.

pleiotropism (*Gen.*). The condition when one gene has an effect simultaneously on more than one character in the offspring.

Pleistocene Period (*Geol.*). The period of geological time which followed the Pliocene. It was during this period that an ice-sheet covered the greater part of N. Europe and N. America; hence it has been called the *Great Ice Age*.

plenum chamber (*Aero.*). A sealed chamber pressurized from a *ram intake* (q.v.). Centrifugal flow turbojets having double-entry impellers (see **double-entry compressor**) have to be mounted in plenum chambers to ensure even air pressure on both impeller faces.

plenum system (*Build.*). An air-conditioning system in which the air propelled into the building is maintained at a higher pressure than the atmosphere. The conditioned air is usually admitted to rooms from 2·5–3 m above floor-level, while the vitiated air is extracted at floor-level on the same side of the room.

pleo-. Prefix. See **plei-**.

pleocholia (*Med.*). Excessive formation of bile pigment.

pleochroic haloes (*Min.*). Dark-coloured zones around small inclusions of radioactive minerals which are found in certain crystals, notably biotite. The colour and pleochroism of the zones are stronger than those of the surrounding mineral, and result from radioactive emanations during the conversion of uranium or thorium into lead. Their characters have been closely studied as providing evidence of the age of the rocks containing them.

pleochroism (*Min.*). The property of a mineral by which it exhibits different colours in different crystallographic directions on account of the selective absorption of transmitted light.

pleochromatic (*Biol.*). Presenting different colours according to changes in the environment or with different physiological conditions.

pleocytosis (*Med.*). An increase in the number of white-blood cells, especially in the cerebrospinal fluid.

pleomorphism (*Zool.*). See **polymorphism.**

pleomorphous (*Zool.*). Polymorphic.

pleon (*Zool.*). In *Crustacea*, the abdominal region. Cf. *pereion*.

pleonaste (*Min.*). Oxide of magnesium, iron, and aluminium, crystallizing in the cubic system. It is a member of the spinel group and may be dark-green, brown, or black in colour. Also called **ceylonite.**

pleonexia (*Psychiat.*). Obsessive greed; plutomania.

pleophagous (*Bot.*). Said of a parasite which attacks several species of host plant.

pleopod (*Zool.*). In *Arthropoda* (especially *Crustacea*), an abdominal appendage adapted for swimming.

plerergate (*Zool.*). A worker ant which stores, in its distended gaster, liquid food for the community.

plerocercoid (*Zool.*). A solid elongate metacestode in which the scolex is directly derived from the onchosphere.

plerome (*Bot.*). The central region of an apical meristem, from which the stele is ultimately formed.

plesiometacarpal (*Zool.*). In most Old World Deer (*Artiodactyla*), only retaining the proximal extremities of the metacarpals corresponding to the reduced lateral toes. Cf. *telemetacarpal*.

plessite (*Min.*). A eutectic intergrowth of kamacite and taenite occurring in some iron meteorites; appears as dark areas on the polished surface.

plethora (*Med.*). An increase, above normal, in the volume of the blood, with or without an increase in the total number of red cells.

plethysmograph (*Med.*). An apparatus for measuring variations in the size of bodily parts and in the flow of blood through them. See **electroarteriograph.**

pleur-, pleuro-. Prefix from Gk. *pleura*, side.

pleura (*Zool.*). The serous membrane lining the pulmonary cavity in Mammals and Birds.

pleuracrogenous (*Bot.*). Produced at the tip and also laterally.

pleural membrane (*Zool.*). In *Arthropoda*, the lateral wall of a somite when it is membranous

pleurapophysis (*Zool.*). A lateral vertebral process; usually applied to the true ribs.

pleurethmoid (*Zool.*). A bone of some Fish,

representing the fused ectethmoid and prefrontal.

pleurisy (*Med.*). Inflammation of the pleura, which may be either dry or accompanied by effusion of fluid into the pleural cavity.

pleurites (*Zool.*). Chitinous plates forming the lateral wall of a somite, especially a thoracic somite in Insects.

pleuritis (*Med.*). Pleurisy.

pleurobranchiae (*Zool.*). In *Arthropoda* (especially the higher *Crustacea*), gills which arise from the pleura.

pleurocarpous (*Bot.*). Having the fruit in a lateral position.

pleuroccipital (*Zool.*). Exoccipital.

pleurocentrum (*Zool.*). A lateral element of the centrum in some of the lower Vertebrates.

pleurodont (*Zool.*). Having the teeth fastened to the side of the bone which bears them, as in some Lizards.

pleurodynia (*Med.*). Fibrositis of the muscles between the ribs, or of other muscles attached to the ribs, with pain in the chest on breathing or coughing.

pleurogenous (*Bot.*). Borne in a lateral position. (*Med.*) Having origin in the pleura, e.g., *pleurogenous* cirrhosis of the lung. Also **pleurogenic.**

pleurography (*Radiol.*). Examination of the pleural cavity following an injection of air or other contrast media.

pleuron (*Zool.*). In some *Crustacea*, a lateral expansion of the tergite; more generally, in *Arthropoda*, the lateral wall of a somite. *pl.* **pleura.** *adj.* **pleural.**

pleuropericarditis (*Med.*). Concurrent inflammation of the pleura and of the pericardium.

pleuropneumonia (*Med.*). Combined inflammation of the pleura and of the lung. (*Vet.*) A contagious disease of cattle due to infection by *Mycoplasma mycoides*; characterized by an exudative fibrinous pneumonia and pleurisy.

pleuropneumonia-like organisms (*Bacteriol.*). A group of nonpathogenic organisms, closely resembling the pleuropneumonia organisms, isolated from the throat and vagina. See **Mycoplasmatales.**

pleuropodite (*Zool.*). A basal joint preceding the coxa in some *Crustacea*.

pleurorhizal (*Bot.*). Said of an embryo when the radicle is placed against the edges of the cotyledons.

pleurosphenoid (*Zool.*). Sphenolateral.

pleurosporous (*Bot.*). Having the spores borne in a lateral position.

pleurothotonos, pleurothotonus (*Med.*). Forced bending of the body to one side as a result of muscular spasm, as in tetanus.

p-levels (*Light*). See **principal series.**

plexitis (*Med.*). Inflammation of the components of a nerve plexus.

plexus (*Zool.*). A network; a mass of interwoven fibres, as a *nerve plexus*. *adj.* **plexiform.**

plica (*Zool.*). A fold of tissue; a foldlike structure. *adjs.* **plicate, pliciform.**

plicate aestivation (*Bot.*). A type of valvate aestivation in which the perianth segments are plicate.

plimsoll mark (*Ships*). See **load lines.**

plinth (*Build.*). (1) The projecting course or courses at the base of a building. (2) The cuboidal base of a column or pedestal. (3) The base of a bookcase, wardrobe, etc.

plinth block (*Carp.*). See **architrave block.**

plinth course (*Build.*). A projecting course lai d at the base of a wall.

plio-. Prefix. See **plei-.**

Pliocene Period (*Geol.*). The period of geological time which followed the Miocene and preceded the Pleistocene. Rocks of this age are found in East Anglia and Southern England.

pliodynatron (*Electronics*). Multielectrode tube for obtaining negative resistance by secondary emission; an oscillator circuit using such.

pliotron (*Electronics*). Large hot-cathode vacuum tube for industrial use, with one or more grids. Applied at one time to any vacuum tube.

Plombière's douche (*Med.*). Lavage of the colon by the slow injection of, e.g., salt solution, through a rubber tube inserted into the rectum.

plotter (*Comp., etc.*). Instrument which prepares graph showing relationship between two voltage signals applied to X and Y inputs. Widely used as output unit of analogue computer to represent the relationship between two physically different variables.

plough or **plow** (*Agric.*). See **coulter, landside, mouldboard, share, skim coulter, slade.** (*Bind.*) A hand tool for cutting the edges of books, now only occasionally used for handbound books. (*Carp., Join.*) (1) A form of grooving plane which has an adjustable fence and is capable of being fitted with various irons. (2) To cut a groove. (*Elec. Eng.*) A current-collector used on the conduit system of electric street traction. Also called **underground collector.**

plough carrier (*Elec. Eng.*). The frame under a tram, which carries the plough used in the conduit system. The arrangement is such that the plough can slide laterally, so that it may follow any variations in the relative positions of the conduit and the track rails.

ploughed-and-tongued joint (*Join.*). A joint formed between the square butting edges of two boards, each having a plough groove into which a common tongue is inserted.

plow. A variant spelling of *plough.*

plucked (*Textiles*). Term used to denote irregularity in a top, roving, or yarn; generally caused by excessive draft or speed.

plucked wool (*Textiles*). Wool obtained from a sheep that has been dead a few days; occasionally, wool from slaughtered sheep. See **pelt wool.**

plug (*Build.*). A wooden or plastic piece driven into a hole cut in a wall or partition and finished off flush, so as to provide a material to which joinery or fittings may be nailed. (*Elec. Eng.*) Male termination on wire or cable—makes electrical connexion with a (female) socket. (*Geol.*) A roughly cylindrical orifice through which igneous rock is injected. It is frequently filled with igneous rock which constitutes the plug proper. (*Nuc. Eng.*) Piece of absorbing material used to close the aperture of a channel through a reactor core or other source of ionizing radiation.

plug adapter (*Elec. Eng.*). See **lampholder plug.**

plugboard (*Comp.*). Detachable unit with an area of sockets, which by *jumpers* establishes sets of circuits quickly for the solution of specific analogue programs or problems. See also **patch board.**

plug centre bit (*Carp.*). A form of centre bit in which the projecting central point is replaced by a plug of metal, adapting the bit for use in holes already drilled.

plug cock (*Eng., Plumb.*). A simple valve in which the fluid passage is a hole in a rotatable plug fitted in the valve body. Rotation of the plug through a right angle stops the flow by opposing to it the undrilled diameter of the plug.

plug flow (*Chem. Eng.*). Idealized flow pattern in which successive particles of fluid flow through an apparatus always in the same order as they entered it, i.e., without back mixing.

plug fuse (*Elec. Eng.*). A form of fuse in which the fuse-link is contained in a plug which can be inserted into a suitable socket.

plug gauge (*Eng.*). A gauge, made in the form of a plug, used for testing the diameter of a hole; in a *plug limit gauge* two plugs are provided, a 'go' and a 'not go'. See **limit gauge.**

plugging (*Build.*). The operation of drilling a hole in a wall or partition, and driving in a wall plug. (*Elec. Eng.*) A method which provides for the braking of an electric motor by arranging the connexions so that it tends to run in the reverse direction.

plugging-up (*Teleph.*). The transference of a faulty line from its normal connexions to the test-desk, at which position the test-clerk can apply suitable tests to ascertain the fault and issue instructions for its removal.

plug-in (*Electronics*). Originally, inductor with contact pins for insertion into sockets, for preselecting a range of tuning in radio circuits. Now applied to most valves, some capacitors, and many printed circuit boards. See **wired-in.**

plug-in unit (*Comp.*). Any panel or component which can be inserted and interchanged in a computing system, esp. a *cross-connexion panel*, which can be set up independently.

plug tap (*Eng.*). (1) The final tap required to finish an internal thread in a blind hole. Also **bottoming tap, third tap.** (2) A *plug cock* (q.v.).

plug tenon (*Carp.*). See **stub tenon.**

plug welding (*Eng.*). A method of welding, akin to *spot welding*, in which plugs projecting from one part through apertures in another are welded to the latter.

plum (*Civ. Eng.*). A large undressed stone embedded with others in mass concrete on large work, such as dams, in order to save concrete.

plumae (*Zool.*). Feathers having a stiff shaft and a firm vexillum, and usually possessing hamuli; they appear on the surface of the plumage and determine the contours of the body in addition to forming the remiges and rectrices. *adjs.* **plumate, plumous, plumose, plumigerous.**

plumasite (*Geol.*). A very rare rock type consisting esssentially of oligoclase and corundum only.

plumb (*Build., Civ. Eng.*). Vertical.

plumb- (*Chem.*). From the Latin, *plumbum*, lead; e.g., plumbic chloride ($PbCl_4$), plumbous chloride ($PbCl_2$).

plumbago (*Min.*). *Graphite* (q.v.); used for the making of crucibles because of its refractory qualities.

plumb-bob (*Surv.*). A small weight or 'bob', hanging at the end of a cord, used to centre survey instrument over signal mark. Also **plummet.**

plumber's solder (*Plumb.*). A lead-tin alloy of varying ratio from 1 : 1 to 3 : 1 for different classes of work, the melting-points being always considerably lower than that of lead itself.

plumbing (*Build.*). (1) The craft of working lead for structural purposes, or for the installation of domestic water-supply systems, sanitary fittings, etc. (2) The operation of arranging vertically. (*Telecomm.*) Colloquialism for waveguides and their jointing in establishing microwave systems. Also for piped vacuum systems.

plumbing fork (*Surv.*). An accessory for the plane-table when large-scale work is being done, enabling the point on the paper representing the plane-table station to be located exactly above the corresponding point on the ground. It consists of a U-shaped piece having two long equal limbs, one of which rests on the board to mark, with its end, the point on the paper, while the other passes under the board and carries at its end a plumb line.

plumbing unit (*Build., Plumb.*). A prefabricated assembly of pipes and fittings, generally of storey height, and now commonly used in multistorey flats. In *system building* (q.v.), it may also include the complete bathroom assembly.

plumbism (*Med.*). Lead poisoning.

plumbites (plumbates (II)) (*Chem.*). See **lead (II) hydroxide.**

plumb level (*Surv.*). A *level tube* for showing plumb direction, with a small bubble at right angles to the main one. When the latter is held vertical the small bubble is central.

plumb line (*Surv.*). A cord with a *plumb-bob* (q.v.) attached to the end.

plumbojarosite (*Min.*). A basic hydrous sulphate of lead and iron, crystallizing in the trigonal system. See **jarosite.**

plumb rule (*Build.*). A narrow board used for determining verticals; it has at one end a point of suspension for a plumb-bob, which is free to swing in an egg-shaped hole at the other end of the board.

plume (*Bot.*). A light, hairy or feathery appendage on a fruit or seed serving in wind dispersal. (*Meteor.*) Snow blown over the ridge of a mountain. (*Zool.*) A feather; any featherlike structure.

plumed disseminule (*Bot.*). A fruit or seed bearing a plume.

plumiped (*Zool.*). A Bird having feathered feet.

plummer block (*Eng.*). A journal bearing for line shafting, etc., consisting of a box-form casting holding the bearing brasses, split horizontally to take up wear.

Plummer-Vinson syndrome (*Med.*). The association of difficulty in swallowing, chronic inflammation of the tongue, achlorhydria, and anaemia.

plummet (*Surv.*). See **plumb-bob.**

plumose (*Bot., Zool.*). Hairy; feathered.

plumping (*Leather*). The swelling of a pelt, or leather, during the process of manufacture. See **liming.**

plumulae (*Zool.*). Feathers having a soft shaft and vane and lacking hamuli; in some cases the shaft is entirely lacking; they form the deep layer of the plumage. *adjs.* **plumulate, plumulaceous.**

plumule (*Bot.*). The first apical bud on the embryo in the seed; it is the rudimentary shoot. (*Zool.*) Down feather. Form covering of nestling, sometimes persisting in adult, between the contour feathers. Barbules and hooks little developed.

plunge angle (*Surv.*). See **angle of depression.**

plunge-cut milling (*Eng.*). Milling without transverse movement of the workpiece relative to the cutter, resulting in a groove or slot with a curved bottom.

plunge grinding (*Eng.*). See **profile grinding.**

plunger (*Elec. Eng.*). Device for altering length of a short-circuited coaxial line, taking the effective form of an annular disk. For very high frequencies there need not be a contact, because a small clearance gives sufficient capacitance for an effective short-circuit. (*Eng.*) The ram or solid piston of a force-pump.

plunger key (*Teleph.*). A telephone key with spring contacts which are opened or closed by pressing a small plunger in line with the springs. Used in keysenders.

plunging fold (*Geol.*). One whose axis is not horizontal. The angle between the axis and the horizontal is called the *plunge* or the *pitch*.

plunging shot (*Surv.*). Downward theodolite sight. To *plunge* is to transit the instrument.

plural gel (*Chem.*). A gel formed from two or more sols.

pluriglandular (*Med.*). Pertaining to, affected by, or affecting, several (ductless) glands.

plurilocular (*Bot.*). Said of a sporangium or an ovary which is divided by septa into several compartments.

plurisporous (*Bot.*). Having two or more spores.

plurivalent (*Cyt.*). In certain types of cell-divisions, said of compound chromatin rods formed of more than two chromosomes.

plush (*Texti es*). (1) Fabric with cut pile, less dense than velvet, on one side. Warp pile is generally made by weaving two cloths together, with a pile warp common to both, which is afterwards cut. (2) One type of knitted fabric has the loops on the back of the cloth.

plush copper ore (*Min.*). Chalcotrichite.

plus strain (*Bot.*). Often written (+) *strain.* One of the two strains of a heterothallic mould, often distinguished from the corresponding (−) *strain* by its stronger growth.

pluteus (*Zool.*). In *Echinoidea* and *Ophiuroidea*, a pelagic larval form, being a modification of the *dipleurula* (q.v.), in which the ciliated band remains continuous, forming only a small preoral lobe, the postanal region is greatly developed and the arms are supported by calcareous rods.

Pluto (*Astron.*). The outermost planet, discovered in 1930. It is a small body, rather less than half of the earth's diameter, and rotating in 6·4 days. It travels in an eccentric orbit inclined at 17° to the ecliptic in a period of 248 years. Its mean distance from the Sun is 39·5 A.U., but its eccentric orbit allows it to come to perihelion (about 1989) at a distance of 29·6 A.U., which is less than the mean distance of Neptune. Pluto is so different from the giant planets that it has been suggested that it was originally a satellite of Neptune.

plutomania (*Psychiat.*). Delusion of sufferer that he has great wealth. A pathological pre-occupation with money.

plutonic intrusions (*Geol.*). A term applied to large intrusions which have cooled at great depth beneath the surface of the earth. Also **major intrusions.** Cf. *minor intrusions.*

plutonites (*Geol.*). All rocks occurring in plutonic intrusions. More precisely, igneous rocks of the coarse grain-size group.

plutonium (*Chem.*). Element, symbol **Pu**, at. no. 94, product of radioactive decay of *neptunium*. Has many isotopes, the fissile isotope ^{239}Pu, produced from ^{238}U by neutron absorption in a reactor, being the most important for the production of nuclear power.

plutonium reactor (*Nuc. Eng.*). One in which ^{239}Pu is used as the fuel.

pluviometer (*Meteor.*). See **rain gauge.**

pluviometric coefficient (*Meteor.*). The mean rainfall at a place for a given period, expressed as a percentage of the amount that would have fallen, if the rainfall were evenly distributed throughout the year.

Plyglass (*Arch., Build.*). A proprietary form of window or underceiling light covering. Composed of two sheets of glass enclosing glass fibres or plastics to diffuse entering light.

plywood (*Carp.*). A board consisting of a number of thin layers of wood glued together so that the grain of each layer is at right angles to the grain of its neighbour. See also **multi-ply.**

Pm (*Chem.*). The symbol for *promethium.*

PM (*Build.*). Abbrev. for *purpose-made.*

PMBX (*Teleph.*). Abbrev. for *private manual branch exchange.*

PMC (*Build.*). Abbrev. for *plaster-moulded cornice.*

P.M.T.S. (*Work Study*). Abbrev. for *predetermined motion time system.*

PMX (*Teleph.*). Abbrev. for *private manual exchange.*

p-n **boundary** (*Electronics*). The surface on the transition region between *p*-type and *n*-type semiconductor material, at which the donor and acceptor concentrations are equal.

pneum-, pneumo-, pneumat-, pneumato-. Prefix from Gk. *pneuma*, gen. *pneumatos*, breath.

pneumathode (*Bot.*). A more or less open outlet of the ventilating system of a plant, usually some loosely packed cells on the surface of the plant; through it exchange of gases between the air and the interior of the plant is facilitated.

pneumatic (*Eng., etc.*). Operated by, or relying on, air-pressure or the force of compressed air. (*Zool.*) Containing air, as, in physostomous Fish, the *pneumatic duct* leading from the gullet to the air-bladder, and, in Birds, those bones which contain air-cavities.

pneumatically-operated switch (or circuit-breaker) (*Elec. Eng.*). A switch (or circuit-breaker) in which the force for closing is obtained from a piston operated by compressed air. Cf. *motor-operated switch, solenoid-operated switch.*

pneumatic brake (*Eng.*). A continuous braking system, used on some railway trains, in which air-pressure is applied to brake cylinders throughout the train. See **air brake (1), continuous brake, electropneumatic brake.**

pneumatic conveyer (*Eng.*). A system by which loose material is conveyed through tubes by air in motion, the air velocity being created by the expansion of compressed air through nozzles.

pneumatic drill (*Eng.*). A hard-rock drill in which compressed air is arranged to reciprocate a loose piston which hammers the shank of the bit or an intermediate piece, or in which the bit is clamped to a piston rod.

pneumatic flotation cell (*Min. Proc.*). One in which low-pressure air is blown in and diffused upward through the cell.

pneumatic gauge (*Instr.*). Instrument for making accurate measurements of the dimensions of solid objects based on the comparison of pressures from two airstreams, one from a fixed, the other from a variable orifice, when the object which is to be measured is placed in the path o f the latter.

pneumaticity (*Zool.*). The condition of containing air-spaces, as the bones of Birds.

pneumatic keys (*Teleg.*). The keys which control high-frequency currents in a high-power telegraph radio transmitter; operated by compressed air to ensure rapidity of operation of contacts in making and breaking heavy currents.

pneumatic lighting (*Mining*). Use of small turbo-motor driven by compressed air to drive a small dynamo connected to an electric lamp, thus avoiding wiring extensions underground where a compressed air service exists.

pneumatic loudspeaker (*Acous.*). One in which a jet of high-pressure air is modulated by a transducer driven by audio currents, e.g., stentorphone.

pneumatic motor (*Acous.*). The small bellows which, when air is admitted or exhausted, acts as a driving force against a spring, for the operation of air paths in organs.

pneumatic pick (*Eng.*). A road contractor's tool in which, by mechanism similar to that of a *pneumatic drill* (q.v.), a straight pick is hammered rapidly by a reciprocating piston driven by compressed air.

pneumatic riveter (*Eng.*). A high-speed riveting machine similar in arrangement to a *hydraulic riveter* (q.v.) but in which a rapidly reciprocating piston driven by compressed air delivers 1000–2000 blows per minute.

pneumatic tools (*Eng.*). Hand tools, such as riveters, scaling and chipping hammers, and drills, driven by compressed air. See **pneumatic drill, pneumatic pick, pneumatic riveter.**

pneumatic trough (*Chem.*). A vessel used, in chemical laboratories, for the collection of gases.

pneumatic tube conveyor (*Eng.*). A system in which small objects enclosed in suitable containers are transported along tubes, the container acting as a moving piston which is impelled either by means of pressure or vacuum.

pneumato-. Prefix. See **pneum-.**

pneumatocele (*Med.*). (1) A hernial protrusion of lung through some defect in the chest wall. (2) Any air-containing swelling.

pneumatocyst (*Zool.*). (1) Any air-cavity used as a float. (2) In Fish, the *air bladder* (q.v.). (3) The cavity of a pneumatophore.

pneumatolysis (*Geol.*). The after-action of the concentrated volatile constituents of a magma, effected after the consolidation of the main body of magma. See **greisenization, kaolinization, tourmalinization.**

pneumatophore (*Bot.*). A specialized root which grows vertically upwards into the air from roots embedded in mud, and, being of loose construction, makes possible the access of air to the buried roots. (*Zool.*) In *Siphonophora*, an apical float containing gas; it possibly represents a modified medusoid.

pneumatopyle (*Zool.*). In certain *Siphonophora*, the aperture by which the cavity of a pneumatophore communicates with the exterior.

pneumaturia (*Med.*). The passing of urine containing gas or air.

pneumectomy (*Surg.*). See **pneumonectomy.**

pneumococcus (*Bacteriol.*). A Gram-positive diplococcus, a causative agent of pneumonia, though it may occur normally in throat and mouth secretions. Four main types, distinguishable by their agglutination reactions, have been recognized.

pneumoconiosis (*Med.*). See **pneumonoconiosis.**

pneumoencephalitis (*Vet.*). See **Newcastle disease.**

pneumogastric (*Zool.*). A term applied to the tenth cranial or *vagus* (q.v.) nerve in Vertebrates because it sends branches to the lungs and stomach.

pneumohaemo-pericardium (-thorax) (*Med.*). The presence of air and blood in the pericardial sac (or in the pleural cavity).

pneumohydro-pericardium (-thorax) (*Med.*). The presence of air and a clear effusion in the pericardial sac (or in the pleural cavity).

pneumolith (*Med.*). A concretion in the lung, formed usually as a result of calcification of a chronic tuberculous focus.

pneumolysis (*Surg.*). The operation of freeing the outer layer of the pleura from the chest wall (*external pneumolysis*), or of dividing adhesions between the outer layer of the pleura and that covering the lung (*internal pneumolysis*); both measures used for producing or increasing collapse of the lung.

pneumon-, pneumono-. Prefix from Gk. *pneumōn*, gen. *pneumonos*, lung.

pneumonectomy (*Surg.*). Removal of lung tissue.

pneumonia (*Med.*). A term generally applied to any inflammatory condition of the lung accompanied by consolidation of the lung tissue; more especially, *lobar pneumonia*, in which the consolidation affects one or more lobes of the lung.

pneumonitis (*Med.*). Pneumonia.

pneumonoconiosis (*Med.*). A disease of the lungs due to inhalation of dust in excessive quantities, usually occupational, the chief being silicosis, asbestosis and (coal-miner's) anthrocosis.

pneumonomycosis (*Med.*). A term applied to disease of the lung caused by any one of a number of various fungi.

pneumo-oil switch (or circuit-breaker) (*Elec. Eng.*). A switch or circuit-breaker in which the operation is carried out partly by pneumatic means, and partly by hydraulic means using oil as the medium.

pneumopericardium (*Med.*). The presence of air in the pericardial sac.

pneumoperitoneum (*Med.*). (1) The presence of air or gas in the peritoneal cavity. (2) The injection of air into the peritoneal cavity for radiographic purposes.

pneumopyopericardium (*Med.*). The presence of air and pus in the pericardial sac.

pneumopyothorax (*Med.*). The presence of air and pus in the pleural cavity.

pneumostome (*Zool.*). In *Arachnida*, the opening to the exterior of the lung books. In *Pulmonata*, the opening to the exterior of the lung formed by the mantle cavity.

pneumotaxis (*Biol.*). (1) Response or reaction of an organism to the stimulus of carbon dioxide in solution. (2) Response to the stimulus of gases generally. *adj.* **pneumotactic.**

pneumothorax (*Med.*). (1) The presence of air or gas in the pleural cavity. (2) The therapeutic injection of air or gas into the pleural cavity for the purpose of collapsing diseased lung (*artificial pneumothorax*).

pneumotropism (*Biol.*). Pneumotaxis.

***p-n-i-p* transistor** (*Electronics*). Similar to *p-n-p* transistor with a layer of an instrinsic semiconductor (germanium of high purity) between the base and the collector to extend the high-frequency range.

***p-n* junction** (*Electronics*). Boundary between *p*- and *n*-type semiconductors, having marked rectifying characteristics; used in diodes, photocells, transistors, etc.

***p-n-p* transistor** (*Electronics*). A junction transistor in which a thin slice of *n*-type is sandwiched between slices of *p*-type, and amplification arises from hole conduction controlled by the electric field in the *n*-type slice.

Po (*Chem.*). The symbol for *polonium*.

P.O. box (or **bridge**) (*Elec. Eng.*). Abbrev. for *Post Office box* (or *bridge*).

pock (*Med.*). A pustule; any small elevation of the skin, containing pus, occurring in an eruptive disease (especially smallpox).

Pockel's effect (*Elec. Eng.*). The *electro-optical effect* (q.v.) in a piezoelectric material.

pocket (*Join.*). The hole in a pulley stile through which the counterpoise weights are passed into the box of a sash and frame.

pocket chamber (*Radiol.*). A small ionization chamber used by individuals to monitor their exposure to radiation. It depends upon the loss of a given initial electric charge being a measure of the radiation received.

pocket chisel (*Join.*). See **sash pocket chisel.**

pocket chronometer (*Horol.*). A pocket watch fitted with the chronometer escapement. On the continent of Europe the term is used for any high-precision pocket watch.

poculiform (*Bot.*). Cup-shaped.

pod (*Aero.*). See **engine pod.** (*Bot.*) A dry fruit formed from a single carpel, having a single loculus containing one (rarely) to several seeds, and usually opening at maturity by splitting along both ventral and dorsal sutures. The pea-pod is a good example.

pod-, podo-, -pod. Prefix and suffix from Gk. *pous*, gen. *podos*, foot.

podagra (*Med.*). Gout.

podal (*Zool.*). Pedal.

podauger (*Tools.*). An auger having a straight groove cut in its length to hold the chips.

podeon (*Zool.*). See **petiole.**

podetium (*Bot.*). A stalklike, cuplike, or much-branched erect thallus which is formed by some lichens.

podex (*Zool.*). The anal region. *adj.* **podical.**

podical plates (*Zool.*). In Insects, a pair of small sclerites in the anal region, representing the sternum of the eleventh abdominal segment.

Podicipitiformes (*Zool.*). An order of *Neognathae*, containing compact-bodied birds of cosmopolitan distribution. Almost completely aquatic, and build floating nests. Toes lobate, and feet placed far back. Grebes.

podite (*Zool.*). A walking leg of *Crustacea.*

podium (*Arch.*). A continuous low wall under a row of columns. (*Zool.*) (1) In land Vertebrates, the third or distal region of the limb; manus or pes; hand or foot. (2) Any footlike structure, as the locomotor processes or tube-feet of *Echinodermata*. *pl.* **podia.** *adj.* **podial.**

podobranchiae (*Zool.*). In *Crustacea*, gills arising from the coxopodites of the legs.

podocarp (*Bot.*). A stalk to a carpel.

podocarpus (*For.*). Conifer yielding a commercial softwood, found chiefly in South Africa and Australasia.

pododerm (*Zool.*). The dermal layer of a hoof lying within the keratinous layers.

podomere (*Zool.*). In *Arthropoda*, a limb segment.

podophthalmite (*Zool.*). In *Crustacea*, the distal segment of the eye-stalk; the eye-stalk itself.

podophyllin (*Chem.*). Extract from the mandrake *Podophyllum peltatum* (America) or *Emodi* (India), containing resins which act as strong purgatives.

podosoma (*Zool.*). In *Acarina*, the region of the body composed of leg-bearing segments.

podzol (*Agric.*). Typical ashlike soil of regions having subpolar climate. Also **podsol.**

poecilitic (*Geol.*). See **poikilitic.**

poecilogeny (*Zool.*). Larval polymorphism, as in some *Diptera* (*Oligarces, Miastor*).

Poetsch process (*Mining*). Freezing of water-logged strata by circulation of refrigerated brine through boreholes surrounding the section through which a shaft or tunnel is to be driven.

Poggendorff cell (*Elec. Eng.*). A single-fluid form of the bichromate cell.

Poggendorff compensation method (*Chem.*). A method of measuring an unknown e.m.f. by finding the point at which it just opposes the steady fall of potential along a wire.

Pogonophora (*Zool.*). A subphylum of Chordate animals, without tail, alimentary canal or bony tissue, and having a simple nervous system and no sense organs. There is a muscular heart, and the blood contains haemoglobin. The extremely long bodies are divided into three sections, the posterior one serving to anchor the animals in the chitinous tubes in which they live, and the anterior part bearing well-developed tentacles which probably serve to gather food, digest it, and absorb the products of digestion.

poikil-, poikilo-. Prefix from Gk. *poikilos*, many-coloured.

poikilitic texture (*Geol.*). Texture in igneous rocks in which small crystals of one mineral are irregularly scattered in larger crystals of another, e.g., small olivines embedded in larger pyroxenes, as in some peridotites.

poikiloblastic (*Geol.*). A textural term applicable to metamorphic rocks in which small crystals of one mineral are embedded in large crystals of another. The texture is comparable with the *poikilitic* of igneous rocks.

poikilochlorophyllous (*Bot.*). Used of *Angiospermae* that completely lose and regain their chlorophyll during changes in environmental conditions.

poikilocyte (*Med.*). A malformed red blood cell.

poikilocytosis (*Med.*). Presence of malformed red cells in the blood, e.g., in severe anaemia.

poikilohydrous (*Bot.*). Used of *Angiospermae* that become dormant in the dry season by losing most of their water.

poikilosmotic (*Ecol.*). Of an aquatic animal, being in osmotic equilibrium with its environment, the concentration of its body fluids changing if the environment becomes more dilute or more concentrated. Marine Invertebrates are frequently poikilosmotic. Cf. *homoiosmotic.*

poikilothermal (*Zool.*). See **cold-blooded.**

point (*Comp.*). Decimal point in numbers using 10 as radix; similarly with 2 as radix. *Fixed* and *floating* points imply omission of point or its movement among the digits by multiplication or division respectively by the radix. (*Elec. Eng.*) In electric-wiring installations, a termination of the wiring for attachment to a lighting fitting socket-outlet or other current-using device. (*Maths.*) See **double-, multiple-, singular-.** (*Typog.*) The unit of measurement for type and materials. The 12-point em (the foundation of the system) measures 0·1660 in. (approx. $\frac{1}{6}$ in.) (4·21 mm); 1 point measures 0·0138 in. (0·351 mm); 72 points or 6 ems measure 0·996 in. (25·3 mm). The old type sizes such as *nonpareil*, *brevier*, *pica* (approx. 6-point, 8-point, 12-point) have largely been discarded. See also **points,** page 911.

pointal (*Carp.*). A king-post.

point bar (*Textiles*). Horizontal bar supporting the points at the back and front of a lace machine; has a motion related to the swing of the carriages.

point brilliance (*Optics*). The effect produced by a point source of light expressed in terms of the illumination at the iris of the eye, a constant pupillary aperture being assumed.

point contact (*Electronics*). Condition where current flow to a semiconductor is through a point of metal, e.g., use of end of metal wire as in 'cat's whisker'.

point-contact rectifier (*Electronics*). One comprising a metal point pressing on to a crystal of semiconductor and delivering *holes* when made positive.

point-contact transistor (*Electronics*). One in which the base electrode is joined by two or more point-contact electrodes. An early construction of the transistor, not very reliable and of low power-handling capability.

point counter tube (*Electronics*). One using gas amplification in which the central electrode is a point or a small sphere.

point defect (*Crystal.*). See **defect.**

point d'esprit (*Textiles*). A narrow spotted traverse net.

point draft (*Textiles*). An order for the drawing-in of warp threads, e.g., from loom heald shafts 1 to 8 and then from 8 to 1 (front to back and conversely).

pointed arch (*Build.*). An arch which rises on each side from the springing to a central apex.

pointed ashlar (*Build.*). A block of stone whose face-markings have been done with a pointed tool.

pointer (*Build.*). A tool used for raking out old mortar from brickwork joints prior to pointing.

Pointers (*Astron.*). The name used in popular language for the two stars of the Great Bear, α and β Ursae Majoris; they are roughly in line with the Pole Star and so help to identify it.

point gamma (*TV*). The contrast gamma for a specified level of brightness. In TV reception the instantaneous slope of the curve connecting log (input voltage) and log (intensity of light output).

pointing (*Build., Civ. Eng.*). The process of raking out the exposed jointing of brickwork and refilling with, normally, cement mortar.

point mutation (*Gen.*). A heritable change taking place at a single gene position; a mutation proper. Also **transgenation.**

point of inflexion on a curve (*Maths.*). A point at which a curve changes from convex to concave. For the curve $y=f(x)$, the point where $x=a$ is a point of inflexion if $f''(a)=0$, and if the first nonzero higher order derivative at $x=a$ is of odd order.

point of osculation (*Maths.*). A multiple point on a curve through which two branches, having a common tangent at the point, pass.

points (*Print.*). General term for all punctuation marks, particularly the full stop. (*Rail.*) Movable tapered blades or tongues of metal for setting alternative routes of running rails. Each such blade is pivoted at the *heel*, its *toe* being locked against the stock rail, *facing points* if the train approaches the toe, *trailing points* if the train approaches the heel. (*Textiles*) The tapered metal pins on a lace machine which carry the twists of bobbin and warp threads to the position where the lace is actually made.

point tie (*Textiles*). An arrangement of the harness in a loom jacquard, in which one section is tied up left to right, or right to left, and the next section conversely.

poise (*Chem.*). To maintain the oxidation reduction potential of a solution constant by the addition of a suitable compound. (*Horol.*) Equilibrium. A balance is said to be *in poise* when, supported horizontally by its pivots on knife edges, it has no tendency to rotate, or if rotated, no tendency to take up any set position. (*Phys.*) The CGS unit of dynamic viscosity of a fluid, applies when a force of one dyne maintains unit rate of shear of a film of unit thickness between surfaces of unit area. Otherwise one dyne sec cm^{-2} (10^{-1} N s m^{-2}). (Named from the physicist Poiseuille.) Symbol P. See **viscosity.**

Poiseuille's formula (*Phys.*). An expression for the volume of liquid per second Q, which flows through a capillary tube of length L and radius R, under a pressure P, the viscosity of the liquid being η:

$$Q=\frac{\pi PR^4}{8L\eta}.$$

poising callipers (*Horol.*). A form of callipers between the jaws of which a balance may be mounted and rotated, as a test for truth and poise.

poison. Any substance or matter which, introduced into the body in any way, is capable of destroying or seriously impairing life. Poisons include products of decomposition or of bacterial organisms, and in some cases the organisms themselves; also very numerous chemical substances forming the residue of industrial, agricultural and other processes (see **pollution**). Poisons are generally classified as *irritants* (e.g. cantharides, arsenic) and *corrosives* (e.g. strong mineral acids, caustic alkalis); *systemics* and *narcotics* (e.g. prussic acid, opium, barbiturates, henbane); *narcotic-irritants* (e.g. nux vomica, hemlock). Among gases, carbonic acid, carbonic oxide, sulphuretted hydrogen, sulphur dioxide, sulphide of ammonium and numerous others (e.g. fumes of leaded petrol) are of significance in industry and daily life. See also **war gases.** (*Phys.*) Any contaminating material which, because of high-absorption cross-section, degrades intended performance, e.g., fission in a nuclear reactor, radiation from a phosphor. See also **catalytic poison.**

poison computer (*Comp.*). Reactor control instrument consisting of analogue computer operating in real time and providing continuous record of level of xenon poisoning in a reactor.

poisoning (*Chem. Eng., Min. Proc., etc.*). Loading of resin sites in ion exchange with ions of unwanted species which therefore prevent the capture of those required by the process. In liquid-liquid ion exchange, fouling of the organic solvent in similar manner.

Poisson distribution (*Stats.*). Statistical distribution, characterized by a small probability of a specific event occurring during observations over a continuous interval (e.g., of time or distance). A limiting form of *binomial distribution* expressible as

$$y=\frac{m^{\gamma}e^{-m}}{m!}.$$

Applicable, e.g., to most nuclear processes, such as radioactive decay.

Poisson's equation (*Maths.*). The equation $\nabla^2 V=-4\pi\rho$, where ∇ is the vector operator del, V is the electric or gravitational potential function, and ρ is the charge or mass density respectively. Cf. *Laplace's equation.*

Poisson's ratio (*Phys.*). One of the elastic constants of a material. It is defined as the ratio of the lateral contraction per unit breadth to the longitudinal extension per unit length, when a piece of the material is stretched. For most substances its value lies between 0·2 and 0·4. The relationship between Poisson's ratio ν, Young's modulus E, and the modulus of rigidity G, is given by:

$$\nu=\frac{E}{2G}-1.$$

polar axis (*Astron.*). (1) That diameter of a sphere which passes through the poles. (2) In an equatorial telescope, the axis, parallel to the earth's axis, about which the whole instrument revolves in order to keep a celestial object in the field. (*Crystal.*) A crystal or symmetry axis to which no two- or four-fold axes are normal; thus the arrangements of faces at the two ends of such an axis may be dissimilar. The principal axis of tourmaline is a polar axis of three-fold symmetry, the top of the crystal being terminated by pyramid faces, the bottom end by a single plane in some cases.

polar body (*Biol.*). One of two small cells detached from the ovum during the maturation divisions.

polar bond (*Chem.*). See **electrovalence.**

polar control (*Aero.*). See **twist and steer.**

polar coordinates (*Maths.*). (1) Of a point *P* in a plane: the distance of the point, *OP*, from a fixed point *O* in the plane, called the *pole*, and the angle made by *OP* with a fixed direction. (2) Of a point *P* in space: the distance of the point, *OP*, from a fixed point *O*, called the *pole*, and (*a*) the angles made by *OP* with two non-parallel (usually perpendicular) fixed planes (**spherical coordinates**) or (*b*) the angle made by the projection of *OP* on to a fixed plane with a fixed angle in that plane, and the perpendicular distance of *P* from that plane (**cylindrical coordinates.**)

polar crystal (*Crystal.*). A crystal, such as sodium chloride, with ionic bonding between atoms.

polar curve or diagram (*Phys. etc.*). A curve drawn in polar coordinates showing the field distribution around a radiator or receiver of radiated energy, to show its directional efficacy. E.g., a contour of equal field-strength around a transmitting antenna; or, in a receiving antenna, contour path of a mobile transmitter producing a constant signal at the receiver. Similar diagrams are prepared for sound fields round electro-acoustical transducers, sensitivity curves around photocells, scintillation counters, etc.; and for all other energy detectors or radiators with directional sensitivity.

polar distance (*Astron.*). A term (generally preceded by *North* or *South*, and written initially —N.P.D., S.P.D.) denoting the angular distance of a heavenly body from the pole of the celestial sphere; hence it is equal to the complement of the body's declination.

polarfield (*Zool.*). See **polar plate**.

polar guide (*Telecomm.*). Waveguide filter with circular components.

polarilocular spore (*Bot.*). A two-celled spore with a very thick median septum traversed by a canal.

polarimeter (*Chem.*). An instrument in which the optical activity of a liquid is determined by inserting Nicol prisms in the path of a ray of light before and after traversing the liquid.

polarimetry (*Chem.*). The measurement of optical activity, especially in the analysis of sugar solutions.

Polaris (*Astron.*). The name given to the Pole star, α Ursae Minoris, near enough to the north celestial pole to mark it for rough observations, its North Polar Distance being less than 1°.

polariscope (*Light*). Instrument for studying the effect of a medium on polarized light. Interference patterns enable elastic strains in doubly refracting materials to be analysed. It may consist of a polarizer and an analyser, with facilities for placing transparent specimens between them. The analyser is usually a Nicol prism. The polarizer may be also a Nicol prism or a *pile* (q.v.) of plates. Modern polariscopes use light-polarizing films, e.g., Polaroids, instead. See **photoelasticity.**

polarity (*Elec.*). (1) Distinction between positive and negative electric charges (Franklin). (2) General term for difference between two points in a system which differ in one respect, e.g., potentials of terminals of a cell or electrolytic capacitor, windings of a transformer, video signal, legs of a balanced circuit, phase of an alternating current. (*Mag.*) Distinction between positive (north) and negative (south) magnetic poles of an electro- or permanent magnet; these poles do not exist, but describe locations from or towards which magnetic flux leaves or enters the magnetic material. (*Zool.*) Existence of a definite axis.

polarity of sunspots (*Astron.*). Of the magnetic polarity of sunspot groups, the following spots have the reverse sign to that of the leading spot, all groups in the same hemisphere behaving in the same way. In the other hemisphere the signs are reversed, and they also reverse at the beginning of each new sunspot cycle.

polarization (*Chem.*). The separation of the positive and negative charges of a molecule. (*Elec.*) (1) Nonrandom orientation of electric and magnetic fields of electromagnetic wave. (2) Change in a dielectric as a result of sustaining a steady electric field, with a similar vector character; measured by density of dipole moment induced. (*Mag.*) Same as **intensity of magnetization.** (*Phys.*) For a system of particles possessing spin, a measure of the extent of unbalance between parallel and antiparallel spin vectors.

polarization current (*Elec. Eng.*). That causing or caused by polarization (*soakage*) in a dielectric, with possible late discharge.

polarization error (*Radio*). Error in determining the direction of arrival of radio waves by a direction-finder when the desired wave is accompanied by downward components which are out of phase. Formerly termed **night error.**

polarization method (*Micros.*). A technique using the polarization of the fluorescence of submicroscopic particles to determine their size and shape.

polarized beam (*Phys.*). (1) In *electromagnetic radiation* (q.v.), a beam in which the vibrations are partially or completely suppressed in certain directions. (2) In *corpuscular radiation* (q.v.), a beam in which the individual particles have non-zero spin and in which the distribution of the values of the spin component varies with the direction in which they are measured.

polarized capacitor (*Elec. Eng.*). Electrolytic capacitor designed for operation only with fixed polarity. The dielectric film is formed only near one electrode and thus the impedance is not the same for both directions of current flow.

polarized diversity (*Radio*). See **diversity reception.**

polarized plug (*Elec. Eng.*). One which can be inserted into a socket in only one position.

polarized relay (*Elec. Eng.*). See **relay types.**

polarizer (*Light*). Prism of doubly refracting material, or Polaroid plate, which passes only plane-polarized light, or produces it through reflection. See **Nicol prism, pile, Polaroid.**

polarizing angle (*Elec. Eng.*). See **Brewster angle.**

polarizing filter (*Photog.*). One made of Polaroid sheet or other polarizing material, used to reduce unwanted reflections.

polarizing monochromator (*Astron.*). A filter consisting of a succession of quartz crystals and calcite or Polaroid sheets; the light passing through is restricted to a narrow band, useful in observing the solar chromosphere.

polar line and plane (*Maths.*). Of a point (the *pole*) with respect to a conic or quadric respectively: the locus of harmonic conjugates of the pole with respect to the pairs of points intersected on the conic or quadric by lines through it. When the pole is outside the conic or quadric the polar is the line or plane determined by the points of contact of tangents from the pole to the conic or quadric.

polar molecule (*Phys.*). One with unbalanced electric charges, usually valency electrons, resulting in a dipole moment and orientation.

polar nuclei (*Bot.*). Two nuclei in the embryo sac which unite to give the polar fusion nucleus.

polarogram (*Elec. Eng.*). A current-voltage curve

obtained with a polarograph. The height of a current 'wave' is proportional to the concentration of a substance, and the corresponding voltage indicates its nature.

polarograph (*Elec. Eng.*). Instrument recording current-voltage characteristic for polarized electrode. Used in solution chemical analysis. See **polarogram.**

Polaroid (*Optics, Photog.*). TN for a variety of photographic and optical products, including a transparent, light-polarizing plastic sheet which transmits plane-polarized light and is used in spectacles and filters for minimizing the effect of reflections and glare.

Polaroid camera (*Photog.*). TN for a range of cameras, called *Land cameras* after the American inventor, which can produce a processed print 10 sec (50 sec for colour) after exposure, by a variant of the diffusion-transfer reversal process.

polaron (*Electronics*). Electron in substance, trapped in potential well produced by polarization charges on surrounding molecules; analogous to *exciton* in semiconductor.

polar plate (*Zool.*). In *Ctenophora,* one of two ciliated areas forming part of the aboral sense-organ.

polar pyrenoid (*Bot.*). A pyrenoid which is not wholly enveloped in a sheath of starch grains.

polar reciprocation (*Maths.*). Transformation in which lines are replaced by points and points by lines, usually by replacing a line by its pole and a point by its polar with respect to a conic. Also called **dualizing.**

polar response curve (*Acous.*). The curve which indicates the distribution of the radiated energy from a sound reproducer for a specified frequency. Also the relative response curve of a microphone for various angles of incidence of a sound wave for a given frequency. Generally plotted on a radial decibel scale.

polar sequence (*Astron.*). An adopted scale for determining photographic stellar magnitudes It consists of a number of stars near the North Pole which are used as a standard of comparison; they range from Polaris to the faintest observable.

polder (*Civ. Eng.*). A piece of low-lying land reclaimed from the water or artificially protected therefrom.

pole (Gk. *polos*, hinge, axis). Generally, the axis or pivot on which anything turns; one of the ends of the axis of a sphere, especially of the earth. (*Astron.*) See **celestial-, terrestrial-.** (*Bot.*) One end of an elongated spore. (*Carp.*) A long piece of timber of circular section and small in diameter. (*Cyt.*) One end of the achromatic spindle, where the spindle fibres come together. (*Elec. Eng., etc.*) (1) That part of an electric machine which carries one of the exciting windings. (2) A term sometimes used to denote one of the terminals of a d.c. generator, battery, or electric circuit, e.g., *negative pole, positive pole.* (3) A term sometimes used in connexion with an electric arc to denote the extremity of either of the electrodes between which the arc burns. (4) A wooden, steel, or concrete column for supporting the conductors of an overhead transmission or telephone line. (*Electronics*) That part of the anode between adjacent cavities in a multiple-cavity magnetron. (*Mag.*) The part of a magnet, usually near the end, towards which the lines of magnetic flux apparently converge or from which they diverge, the former being called a *south pole* and the latter a *north pole.* (*Maths.*) (1) Of a function: if a function $f(z)$ is expanded in *Laurent's expansion* (q.v.), in an annular domain surrounding a singularity $z=a$,

viz. $f(z)=\sum_{1}^{\infty} b_n(z-a)^n+\sum_{1}^{\infty} c_n(z-a)^{-n}$, the second term is called the *principal part* of $f(z)$ at $z=a$. If the principal part has a finite number of terms, m say, then $f(z)$ is said to have a pole of order m at $z=a$. If the principal part is an infinite series, $f(z)$ is said to have an *isolated essential singularity* at $z=a$. (2) Of a line or plane with respect to a conic or quadric, see **polar line and plane.** (3) Of a circle (or part thereof) on a sphere, the ends of a diameter of the sphere normal to the circle. (*Zool.*) Point; apex; an opposite point (as *aboral pole*); axis.

pole arc (*Elec. Eng.*). The length of the pole face of an electric machine measured circumferentially around the armature surface.

pole bevel (*Elec. Eng.*). A portion of the pole face of an electric machine, near the pole tip, which is made to slope away from the armature surface instead of being concentric with it, the object being to obtain a more satisfactory shape of flux wave.

polecat (*Zool.*). Carnivorous mammals, widely found, with long bodies and short legs.

pole-changing control (*Elec. Eng.*). A method of obtaining two or more speeds from an induction motor, the connexions of the stator winding being altered so that it sets up different numbers of poles.

pole core (*Elec. Eng.*). See **pole shank.**

pole end-plate (*Elec. Eng.*). A thick plate placed at each end of the laminations of a laminated pole.

pole face (*Elec. Eng.*). That surface of the pole piece of an electric machine which faces the armature.

pole face loss (*Elec. Eng.*). Iron losses which occur in the iron of the pole face of an electric machine on account of the periodic flux variations caused by the armature teeth.

pole-finding paper (*Elec. Eng.*). Paper prepared with a chemical solution, which, when placed across two poles of an electric circuit, causes a red mark to be made where it touches positive.

pole horn (*Elec. Eng.*). The portion of the pole shoe of an electric machine which projects circumferentially beyond the pole shank.

pole piece (*Elec. Eng.*). Specially-shaped magnetic material forming an extension to a magnet, e.g., the salient poles of a generator, motor or relay, for controlling flux.

pole pitch (*Elec. Eng.*). The distance between the centre-lines of two adjacent poles on an electric machine; it is measured circumferentially around the surface of the armature of the machine.

pole plate (*Carp.*). A horizontal member supporting the feet of the common rafters and carried upon the tie-beams of the trusses.

pole shading (*Elec. Eng.*). See **shaded pole.**

pole shank (*Elec. Eng.*). The part of a pole piece around which the exciting winding is placed. Cf. *pole shoe.*

pole shim (*Mag.*). See **shim.**

pole shoe (*Elec. Eng.*). That portion of the pole piece of an electric machine which faces the armature; it is frequently detachable from the pole shank.

pole star (*Astron., Surv.*). The star Polaris, readily identified by means of the *Pointers.* Its altitude (when suitably corrected) is equal to the local latitude, and is so used in the northern hemisphere in navigation and surveying.

pole strength (*Mag.*). The force exerted by a particular magnet pole upon a unit pole supposed situated at unit distance from it.

pole tip (*Elec. Eng.*). The edge of the pole face of an electric machine which runs parallel to the axis of the machine. Hence *leading pole tip, trailing pole tip*.

pole top switch (*Elec. Eng.*). A switch which may be mounted at the top of a transmission-line pole; arranged for hand-operation by some mechanical device.

polianite (*Min.*). (IV)oxide of manganese, crystallizing in the tetragonal system. A variety of *pyrolusite* from which it is distinguished by greater hardness and crystallinity.

Polian vesicle (*Zool.*). In some *Echinodermata*, a small stalked sac attached to the water-vascular ring.

policis muscles (*Zool.*). In Birds, muscles enabling independent movement of the first digit, or bastard wing.

poling (*Met.*). In the fire-refining of copper, the impurities are eliminated by oxidation and the oxygen is in turn removed by the reducing gases produced when green logs (*poles*) are burned in the molten metal. If the final oxygen content is too high the metal is *underpoled*, if too low *overpoled*, and if just right *tough pitch* (q.v.). (*Telecomm.*) See **turnover**.

poling boards (*Civ. Eng.*). Rough vertical planks used to support the sides of narrow trenches after excavation; placed in pairs on opposite sides of the trench at intervals along its length, each pair being wedged apart by wooden struts. (*Mining*) Fore-poling boards are used in tunnelling through loose (running) rock, and are driven horizontally ahead to support roof.

polio-. Prefix from Gk. *polios*, grey.

polioencephalitis (*Med.*). Inflammation of the grey matter of the brain.

polioencephalomyelitis (*Med.*). Inflammation of the grey matter both of the brain and of the spinal cord.

poliomyelitis (*Med.*). Inflammation of the grey matter of the spinal cord. Often used as a synonym for *acute anterior poliomyelitis*, or *infantile paralysis*, due to infection, chiefly of the motor cells of the spinal cord, with a virus; characterized by fever and by variable paralysis and wasting of muscles.

polioplasm (*Biol.*). Granular protoplasm.

polished face (*Civ. Eng.*). A fine surface finish produced on granite, or other natural stone, by rubbing down a sawn face with iron sand, fine grit, and polishing powder in turn.

polished foil (*Print.*). Blocking foil, which leaves a glossy transfer. These are available from eggshell to full gloss and will mark surfaces ranging from cheap paper to polypropylene.

polished plate (*Build.*). A superseded sheet-glass over $\frac{3}{16}$ in. thick, used for shop-windows.

polished specimen (*Met., Min. Proc.*). Characteristic hand specimen of ore, metal, alloy, compacted powder, etc., one face of which is ground plane and mirror-smooth by abrasive powder and/or polishing laps, or electrolytic methods.

polishing lathe (or head) (*Eng.*). A headstock and spindle carrying a *polishing wheel* or mop rotated at high speed by belt drive or built-in motor; used for buffing and polishing.

polishing stake (*Horol.*). A flat polished piece of steel on which materials for polishing are mixed. It should be kept in a container to exclude dust.

polishing stick (*Eng.*). A stick of wood, one end of which is charged with emery or rouge, used for finishing small surfaces. It is twisted in the hands or held in the chuck of a drilling machine.

politzerization (*Med.*). Inflation of the middle ear by means of a special bag devised by Politzer.

polje (*Geol.*). A large depression found in some limestone areas, due in part to subsidence following underground solution.

poll-adze (*Tools*). An adze having a blunt head (*poll*) opposite to the cutting edge.

pollard (*For.*). A tree that has been topped to obtain a head of shoots, usually above the height to which browsing animals can reach.

pollen (*Bot.*). The dusty or sticky material produced in anthers; it consists of many pollen grains each of which ultimately contains 2 male nuclei which are equivalent to male gametes.

pollen analysis (*Bot.*). A method of investigating the occurrence and abundance of plant species in the past, by a study of pollen grains preserved in peat and sedimentary deposits.

pollen chamber (*Bot.*). The cavity formed in the apex of the nucellus in *Gymnospermae*, in which pollen grains lodge after pollination has occurred. The pollen grains slowly develop there and ultimately bring about fertilization.

pollen count (*Med.*). A graded assessment of the level of plant pollen in the atmosphere, of importance to sufferers from *hay fever* (q.v.).

pollen flower (*Bot.*). A flower which produces no nectar, but liberates large amounts of pollen, which attracts insects.

pollenosis (*Med.*). See **hay fever**.

pollen sac (*Bot.*). A cavity in an anther in which pollen is formed.

pollen tube (*Bot.*). A tubular outgrowth from the pollen grain, which grows to and into the embryo sac and conveys the male nuclei to the neighbourhood of the egg nucleus.

poll-evil (*Vet.*). Inflammation of the bursa of the ligamentum nuchae of the horse.

pollex (*Zool.*). The innermost digit of the anterior limb in *Tetrapoda*.

pollination (*Bot.*). The transfer of pollen from an anther to a stigma, preliminary to fertilization.

pollinium (*Bot.*). A mass of pollen grains united by a sticky or other substance and transported as a whole in pollination.

pollinodium (*Bot.*). Old name for an *antheridium*.

pollucite (*Min.*). A rare hydrated alumino-silicate of caesium, occurring as clear colourless or white crystals with cubic symmetry. It occurs in granite pegmatites.

pollution (*Ecol.*). Term applied to any environmental state or manifestation which is harmful or unpleasant to life, resulting from man's failure to achieve or maintain control over the chemical, physical or biological consequences or side-effects of his scientific, industrial and social habits.

pollution carpet (*Ecol.*). In stagnant and polluted waters, a slimy layer occurring on the bottom; it consists mainly of bacteria, detritus-feeding protozoa and fungi.

polocyte (*Zool.*). Polar body.

polonium (*Chem.*). Radioactive element, symbol Po, at no. 84. Important as an α-ray source relatively free from γ-emission.

poly-. Prefix from Gk. *polys*, many. (*Chem.*) (1) Containing several atoms, groups, etc. (2) A prefix denoting a polymer, e.g., polyethene.

polyacetals (*Plastics*). *Polyformaldheydes* (q.v.).

polyacrylate (*Chem.*). A polymer of an ester of acrylic acid or its chemical derivatives.

polyadelphous (*Bot.*). Said of an androecium in which the stamens are joined by their filaments into several separate bundles.

polyalkane (*Chem.*). A hydrocarbon polymer essentially of long chain molecules with only saturated carbon atoms in the main chain.

polyandry (*Bot.*). The condition of having a large and indefinite number of stamens. (*Zool.*) The practice of a female animal consorting with more than one male.
polyarch (*Bot.*). Said of a stele having many protoxylem strands.
polyarthritis (*Med.*). Inflammation affecting several joints at the same time.
polyaster (*Zool.*). A complex mitotic figure formed in an ovum after polyspermy.
polyaxon (*Zool.*). Having many axes; said of Sponge spicules.
polybasic acids (*Chem.*). Acids with two or more replaceable hydrogen atoms in the molecule.
polybasite (*Min.*). Sulphide of silver and antimony, often with some copper, crystallizing in the monoclinic system.
polyblast (*Zool.*). A type of phagocytic cell.
polybutadienes (*Plastics*). Range of rubber polymers produced by the stereospecific polymerization of butadiene (butan-1,3-diene) using *Ziegler catalysts*. Some of the rubbers have high resilience, a high abrasion resistance and good low-temperature properties.
polybutenes (*Plastics*). Polymers of 2-methylpropene, used as rubber substitutes on account of excellent electrical and moisture-resisting properties. See **Vistanex.**
polycarbonates (*Plastics*). Range of thermoplastics based on the condensation products of phosgene (carbonyl chloride) with dihydroxy organic compounds, such as diphenylol propane. Relatively expensive materials used for 'tough' applications, e.g., shatterproof lamp bowls.
polycarp (*Zool.*). In some *Urochorda*, a form of gonad on the inner surface of the mantle.
polycarpellary (*Bot.*). Consisting of many carpels.
polycarpic (*Bot.*). Able to fruit many times in succession.
polycarpous (*Bot.*). Having an apocarpous gynaeceum.
polycercoid (*Zool.*). See **echinococcus.**
polycercous (*Zool.*). Having many tails or scolices, as an *echinococcus* (q.v.).
Polychaeta (*Zool.*). A class of *Annelida* of marine habit, having locomotor appendages (parapodia) bearing numerous setae; there is usually a distinct head; the perivisceral cavity is subdivided by septa; the sexes are generally separate, with numerous gonads; and development after external fertilization involves metamorphosis, with a free swimming trochosphere larva. Marine Bristle-worms.
polychasium (*Bot.*). A cymose inflorescence in which the branches arise in sets of three or more at each node.
polychlamydeous chimaera (*Bot.*). A periclinal chimaera in which the skin is more than 2 layers of cells in thickness.
polychloroprene (*Chem.*). See **Neoprene.**
polychromasia (*Med.*). The diffuse bluish staining of the young immature red blood corpuscles with eosin and methylene blue. Also called **polychromatophilia.**
Polycladida (*Zool.*). An order of marine *Turbellaria* in which the gut has many branches which may ramify or anastomose.
polyclinal chimaera (*Bot.*). A chimaera which is made up of more than two components.
polyconic projection (*Geog.*). A type of *conical projection* (q.v.) in which the scale is made true along every parallel and the central meridian. It can only conveniently be divided into strips, each with its own central meridian, and is thus useful for large scale topographical maps.
polycormic (*Bot.*). Said of a woody plant having several strong vertical trunks.
polycotyledonous (*Bot.*). Having more than two cotyledons.
polycyclic (*Bot.*). Said of a stele in which there are two or more concentric rings of vascular strands. (*Chem.*) Containing more than one ring of atoms in the molecule. (*Zool.*) Said of shells having numerous whorls.
polycystic (*Med.*). Containing many cysts, e.g., a *polycystic* kidney.
polycythaemia, polycythemia (*Med.*). An increase in the number of red corpuscles per unit of circulating blood. See also **erythraemia.**
polydactylism, polydactyly (*Med., Zool.*). Having more than the normal number of digits. *adj.* **polydactylous.**
polydipsia (*Med.*). Excessive thirst.
polydisc strobilization (*Zool.*). The production of more than one ephyra at a time by a Scyphozoan hydratuba.
polyembryony (*Bot., Zool.*). The presence of more than one embryo in one ovule or fertilized ovum.
polyester amide (*Plastics*). A polymer in which the structural units are linked by ester and amide (and/or thio-amide) groupings.
polyesters (*Plastics*). Range of polymers formed by the condensation polymerization of polyhydric alcohols and polycarboxylic acids or anhydrides. Maleic and fumaric (i.e. ethene 1,2-dicarboxylic) acids and ethylene and propyl alcohol are the usual starting materials, which are combined with e.g. styrene and glass fibre to give important *reinforced plastics* for very wide use. See also **Terylene.**
polyester wax (*Micros.*). A low melting-point infiltration medium for histological use. Based upon 400 polyethene glycol distearate and cetyl alcohol. M.p. 38°C. Soluble in hydrocarbons, alcohols, esters and ethers.
polyethers (*Plastics*). Long-chain glycols made from alkene oxides such as epoxyethane (ethene oxide). Used as intermediates in the production of *polyurethanes*, antistatic agents and emulsifying agents.
polyeth(yl)ene (*Plastics*). Polythene. A thermoplastic polymer of ethene; an elastomer of superior electrical properties at very high radio-frequencies; many other uses include weather-proof sheeting.
polyeth(yl)ene terephthalate (*Plastics*). The chemical name for the polyester forming the basis of Terylene and Melinex.
polyformaldehydes (*Plastics*). Polymethanal. Acetal resins. A range of thermoplastics based on polymers of formaldehyde. High melting-point, crystalline polymers used in engineering because of good dimensional stability and abrasion resistance. Typified by American plastic, Delrin, and British plastic, Kematal.
polygalacturonase (*Chem.*). Enzyme which destroys pectin and causes softening in pickled vegetables, etc.
polygamous (*Bot.*). Having staminate, pistillate and hermaphrodite flowers on the same and on distinct individual plants. (*Zool.*) Mating with more than one of the opposite sex during the same breeding season. *n.* **polygamy.**
polygastric (*Zool.*). Having several gastric cavities, as a Siphonophoran colony.
polygenes (*Gen.*). Genes which control a single continuous character (e.g. height).
polygenetic (*Chem.*). Producing two or more shades with different mordants.
polygon (*Maths.*). A many-sided plane figure.
polygonal roof (*Build.*). A roof which in plan

forms a figure bounded by more than four straight lines.

polygonal rubble (*Build.*). A form of ragwork in which the rubble wall is built up of stones having polygonal faces.

polygoneutic (*Zool.*). Having several broods in a year.

polygon of forces (*Mech.*). A polygon whose sides are parallel to and proportional to forces acting at a point, the directions of the forces being cyclic around the polygon. The polygon is closed if the forces are in equilibrium, otherwise the closing side of the polygon is parallel and proportional to the equilibrant of the forces.

polygynous (*Bot.*). Said of a flower having several distinct styles. (*Zool.*) Said of a male animal which consorts with more than one female.

polyhalite (*Min.*). A hydrated sulphate of potassium, magnesium, and calcium, $K_2MgCa_2(SO_4)_4{\cdot}2H_2O$. Found in salt deposits.

polyhedral viruses (*Bot.*). A group of insect viruses characterized by the formation of polyhedral protein structures, containing the virus, inside the cytoplasm of nucleus of the host cells; e.g., the 'jaundice' disease of silkworms.

polyhedron (*Maths.*). A solid rectilineal figure. Cf. *regular convex solids*.

polyhydric (*Chem.*). Containing a number of hydroxyl groups in the molecule, e.g., *polyhydric alcohols*, alcohols with three, four, or more hydroxyl groups; *polyhydric acids*, acids containing a number of hydroxyl groups in the molecule, apart from the carboxyl group.

polyimide (*Chem.*). Product of the polycondensation reaction between pyromellitic dianhydride and an aromatic diamine, giving a high-temperature dielectric with strong mechanical stability and electrical properties over a wide temperature range.

polymastia, polymastism (*Med., Zool.*). The presence of supernumerary breasts.

Polymastigina (*Zool.*). An order of *Zoomastigina* the members of which have usually more than three flagella, and an extranuclear division centre.

polymastigote (*Zool.*). Having a tuft of flagella.

polymastism (*Med., Zool.*). See **polymastia.**

polymelus (*Med.*). A foetal monster with more than the normal number of limbs.

polymer (*Plastics*). A material built up from a series of smaller units (*monomers*), which may be relatively simple, such as ethene (the unit of polythene), or relatively complex, such as methyl methacrylate (see Perspex). The molecular size of the polymer helps to determine the mechanical properties of the plastic material and ranges from a few hundred of the basic units to perhaps hundreds of thousands.

polymerization (*Chem.*). The combination of several molecules to form a more complex molecule having the same empirical formula as the simpler ones. It is often a reversible process.

polymery (*Bot.*). The condition when a whorl consists of many members.

polymethylene derivatives (*Chem.*). Cyclic compounds containing three or more methylene groups in the ring.

polymethyl methacrylate (*Plastics*). The chemical name for *Perspex* (q.v.).

polymorph (*Aero.*). The term applied by Sir Barnes Wallis to his supersonic aeroplane designs incorporating *variable sweep* (q.v.) wings.

polymorphic (*Zool.*). Showing a tendency to division of labour among the members of a colony. See also **polymorphism.**

polymorphic transformation (*Met.*). A change in a pure metal from one form to another, e.g., the change from γ- to α-iron.

polymorphism (*Min.*). The property possessed by certain chemical compounds of crystallizing in several forms which are structurally distinct; thus TiO_2 (titanium (IV) oxide) occurs as the mineral species *anatase*, *brookite*, and *rutile*. (*Zool.*) The occurrence of different structural forms at different stages in the life-cycle of an individual; the occurrence of morphologically different types of adult individuals within the same species. *adjs.* **polymorphic, polymorphous.**

polymorphonuclear leucocyte (*Physiol.*). See **neutrophil** (2).

polymyositis haemorrhagica (*Med.*). A form of *dermatomyositis* (q.v.) in which haemorrhages occur in and among the muscles.

polymyxins (*Chem.*). Generic name for a series of antibiotic polypeptides used for treating Gram-negative infections, derived from *B. polymyxa.*

Polyneoptera (*Zool.*). A section of the *Pterygota*, which fold the wings in repose and have a rich wing venation. Malpighian tubes are numerous and metamorphosis incomplete. Includes the superorders *Blattopteroidea*, *Orthopteroidea*, and *Dermapteroidea*.

polyneuritis (*Med.*). Multiple neuritis. A widespread affection of many peripheral nerves with flaccid paralysis of muscles and/or loss of skin sensibility, due to infection or poisoning with various agents, such as lead, alcohol, arsenic, diphtheria toxin, etc.

polynomial (*Maths.*). An algebraic expression of the form

$$A_n x^n + A_{n-1} x^{n-1} + \ldots + A_1 x + a.$$

polynosic fibres (*Plastics*). Modified rayon fibres with improved properties, such as good dimensional stability on washing, arising from better uniformity of fibres.

polynucleate (*Biol.*). Having many nuclei. Also **multinucleate.**

polyoestrous (*Zool.*). Exhibiting several oestrous cycles during the breeding season. Cf. *monoestrous*.

polyoicous (*Bot.*). Said of *Bryophyta* in which antheridia and archegonia may occur on the same plant or on separate plants.

polyoma virus (*Med.*). A virus that can induce a wide variety of cancers in mice, hamsters, and also in other species.

polyorrhomenitis (*Med.*). Concato's disease.

polyoses (*Chem.*). Polysaccharides.

polyoxybiontic (*Ecol.*). Requiring a copious oxygen supply.

polyoxymethylene resins (*Plastics*). Alternative name for **polyformaldehydes.**

polyp (*Med.*). See also **polypus.** (*Zool.*) An individual of a colonial animal.

polypeptides (*Chem.*). Protein derivatives obtained by the condensation of amino acids.

Polypetalae (*Bot.*). A group of Dicotyledons in which the corolla consists of separate petals.

polypetalous (*Bot.*). Said of a corolla made up of distinct petals.

polyphagous (*Bot.*). Said of a fungal parasite which attacks several host cells at the same time. (*Zool.*) Feeding on many different kinds of food, as *Sporozoa* which exist in several different cells during one life-cycle, or phytophagous Insects with many food plants.

polyphase (*Elec. Eng.*). Said of a.c. power supply circuits (usually 3) carrying currents of equal frequency with uniformly spaced phase differences. Normally using common return conductor.

polyphase motor (*Elec. Eng.*). An electric motor designed to operate from a polyphase supply.
polyphyletic (*Gen.*). Descended from diverse ancestors, showing characters derived from several ancestral types; convergent.
polyphyllous (*Bot.*). Consisting of members which are separate from one another.
polyphyodont (*Zool.*). Having more than 2 successive dentitions. Cf. *diphyodont, monophyodont*.
polypide (*Zool.*). A polyp; an individual of a colonial animal; in *Polyzoa*, that part of an individual excluding the zooecium or body wall. Also **polypite**.
Polyplacophora (*Zool.*). An order of *Amphineura* in which the foot occupies the whole ventral surface of the body, the shell consists of 8 transverse dorsal plates, and the branchiae form a row on each side between the mantle and the foot; sluggish marine animals living among rocks and feeding on algae. Chitons, Coat-of-Mail Shells.
polyplexer (*Radar*). Device acting as duplexer and lobe switcher.
polyploid (*Cyt.*). Having more than twice the normal haploid number of chromosomes. The condition is known as **polyploidy.** *Artificial polyploidy*, which can be induced by the use of chemicals (notably *colchicine*, q.v.), is of immense potential economic importance in producing hybrids with desired characteristics.
polypod (*Zool.*). In the development of larval Insects, having the abdomen completely segmented and bearing the full number of appendages. Cf. *oligopod, protopod.*
Polyporales (*Bot.*). See **Aphyllophorales.**
polyposis (*Med.*). The development of many polypi in a part, as in the intestine.
polypropene (*Chem.*). *poly*merized *propene*, a plastic with properties similar to polyethylene but with greater resistance to heat, organic solvents; numerous uses include tufted carpet backings.
polyprotic (*Nuc.*). Having more than one proton.
polyprotodont (*Zool.*). Having numerous pairs of small subequal incisor teeth; pertaining to the *Polyprotodontia*. Cf. *diprotodont*.
polypus (*Med.*). A smooth, soft, pedunculated tumour growing from mucous membrane.
polyribosome (*Biochem.*). Cluster of two or more *ribosomes*.
polyrod antenna (*Radio*). One comprising a number of tapered dielectric rods emerging from a waveguide.
polysaccharides (*Chem.*). Polyoses; a group of complex carbohydrates such as starch, cellulose, etc. They may be regarded as derived from x monosaccharide molecules by the elimination of $x-1$ molecules of water. Polysaccharides can be hydrolysed step to step, ultimately yielding a monosaccharide.
polysaprobe (*Biol.*). An organism able to live in heavily contaminated water.
polysepalous (*Bot.*). Said of a calyx consisting of separate sepals.
polyserositis (*Med.*). Concato's disease.
polysiloxanes (*Chem.*). The basis of silicone chemistry. Polymers of the form:

$$\diagup O \diagdown \underset{R}{\overset{R}{Si}} \diagup O \diagdown \underset{R}{\overset{R}{Si}} \diagup O \diagdown .$$

Prepared by the hydrolysis of chlorosilanes R_3SiCl, R_2SiCl_2, ethers $R_3SOR, R_2S(OR)_2$, or mixtures.

polysiphonous (*Bot.*). Said of an algal thallus consisting of a central row of elongated cells surrounded by layers of peripheral cells.
polysomy (*Cyt.*). A chromosome complement in which some of the chromosomes are present in more than the normal diploid number, e.g. a *trisomic* (q.v.).
polyspermous (*Bot.*). Containing many seeds.
polyspermy (*Zool.*). Penetration of an ovum by several sperms.
polyspondyly (*Zool.*). The condition of having more than two vertebral centra corresponding to a single myotome. *adjs.* **polyspondylic, polyspondylous.**
polyspore (*Bot.*). A multicellular spore.
polyspory (*Bot.*). The formation of more than the normal number of spores.
polystele (*Bot.*). See **dictyostele.**
polystemonous (*Bot.*). Polyandrous.
polystichous (*Bot.*). Arranged in several rows.
polystomatous (*Zool.*). Having many apertures, as *Porifera*; having more than one mouth-opening.
polystyrenes (*Plastics*). See **styrene** (phenyl ethene) resins.
polytetrafluoroeth(yl)ene (*Plastics*). PTFE. $(CF_4)_n$. The principal fluoropolymer noted for its very marked chemical inertness and heat resistance. Widely used for a variety of engineering and chemical purposes, and as a coating for 'non-stick' kitchen equipment. TNs **Fluon, Teflon.** Cf. *Penton*.
polythalamous (*Zool.*). In *Protozoa*, said of a shell or test which has two or more chambers.
polythelia (*Med.*). The occurrence of supernumerary nipples.
polythene (*Plastics*). See **polyeth(yl)ene.**
polytokous (*Zool.*). Bringing forth many young at a birth; prolific; fecund. *n.* **polytoky.**
polytomous (*Bot.*). Having several branches arising at the same level.
polytrifluorochloroeth(yl)ene (*Plastics*). PTFCE. A plastic material with many of the good properties of PTFE, but somewhat easier to fabricate as it is more thermoplastic.
polytrophic (*Zool.*). In Insects, said of ovarioles in which nutritive cells alternate with the oöcytes; more generally, obtaining food from several sources. Cf. *acrotrophic, panoistic.*
polyurethanes (*Plastics*). Range of resins, both thermoplastic and thermosetting, based on the reaction of di-isocyanates with dihydric alcohols, polyesters, or polyethers. Widely used in the production of foamed materials.
polyuria (*Med.*). Excessive secretion of urine.
polyvalent (*Chem.*). Having a valency greater than unity.
polyvinyl acetal (*Chem.*). (1) Group comprising polyvinyl acetal, polyvinyl formal, polyvinyl butyral and other compounds of similar structure. (2) A thermoplastic material derived from a polyvinyl ester in which some or all of the ester groups have been replaced by hydroxyl groups and some or all of these hydroxyl groups replaced by acetal groups. Generally a colourless solid of high tensile strength.
polyvinyl acetate (*Plastics*). PVA. Polymerized vinyl acetate (ethanoate) used chiefly in adhesive and coating compositions. See **vinyl polymers.**
polyvinyl alcohol (*Plastics*). PVA. Polymer produced by the hydrolysis of polyvinyl acetate. Chiefly water soluble and used in adhesives and greaseproofing, but a non-water-soluble form is used as fibre-forming material.
polyvinyl-alcohol enamel (*Print.*). An all-purpose *resist* suitable for use when making line, half-tone, or combined blocks.
polyvinyl butyral (*Plastics*). A polyvinyl resin

made by reacting butanal with a hydrolysed polyvinyl ester such as the acetate (ethanoate). Used chiefly as the interlayer in safety-glass windscreens.

polyvinyl chloride (*Plastics*). PVC. Best known and most widely used of the vinyl plastics. Used in unplasticized form (*rigid PVC*) for pipework, ducts, etc., where high chemical resistance is called for. In plasticized form used for sheeting, cable covering, mouldings and as a fabric for clothing and furnishings. See **vinyl polymers.**

polyvinyl formal (*Plastics*). A polyvinyl resin made by reacting formaldehyde with a hydrolysed polyvinyl ester. Used in lacquers and other coatings, varnishes and some moulding materials.

polyvinylidene chloride (*Plastics*). Based on vinylidene chloride (asymmetrical dichloroethylene). Usually used as copolymer with small amounts of vinyl chloride or acrylonitrile. Basis of fibres, films for packaging, and chemically resistant piping.

polyvinyl polymers (*Plastics*). See **vinyl polymers.**

Polyzoa (*Zool.*). Animals now separated into 2 phyla: *Entoprocta* and *Ectoprocta* (qq.v.).

polyzoic (*Zool.*). Having many zooids, as a hydroid colony; containing many sporozoites, as a spore.

pome (*Bot.*). A term for a fleshy fruit mainly developed from the receptacle, the seeds being contained inside a papery core formed from the inner walls of the united carpels, e.g., apple.

pomiform (*Bot.*). Apple-shaped.

pommel (*Build.*). (1) An ornament in the shape of a ball, e.g., a ball finial. (2) See **punner.**

Ponceaux (*Chem.*). A group of dyestuffs prepared by the interaction of various diazo-salts with naphthol-sulphonic acids.

Poncelet wheel (*Eng.*). An undershot water-wheel with curved vanes; of higher efficiency than the flat-vane type. See **undershot wheel.**

pond (*Ecol.*). Small body of fresh water in which the littoral zone is relatively large and the limnetic and profundal zones are small or absent. (*Hyd. Eng.*) A reach or level stretch of water between canal locks. Also **pound.**

ponderable (*Nuc.*). U.S. term for quantity of radioactive isotope sufficient to be weighed.

pondermotive force (*Min. Proc.*). In high-tension separation, the electrostatic force exerted on a particle as it passes through the field of a corona-type separator. In magnetic separation, the flux field intensity together with density of particle determines its deflection from a straight path.

pongee (*Textiles*). Originally made from real silk in Northern China and in the handloom industry of India, the Lancashire type is made from very fine cotton yarns, bleached and mercerized to imitate silk.

pons (*Zool.*). A bridgelike or connecting structure; a junction. *pl.* **pontes.** *adj.* **pontal.**

pons Varolii (*Zool.*). In Mammals, a mass of transversely coursing fibres joining the cerebellar hemispheres.

pontal flexure (*Zool.*). The flexure of the brain occurring in the same plane as the cerebellum; it bends in the reverse direction to the primary and nuchal flexures and tends to counteract them.

Pontian (*Geol.*). The uppermost stage of the Miocene series in Europe.

pontie, pontil (*Glass*). See **punty.**

pontoon (*Civ. Eng.*). A flat-bottomed floating vessel for the support or transport of plant, materials, or men.

pontoon bridge (*Civ. Eng.*). A temporary bridge carried on *pontoons.*

pontoon dock (*Civ. Eng.*). A floating *dry dock* (q.v.) made up of a number of sections or pontoons.

pony girder (*Eng.*). A secondary girder carried across side-by-side cantilevers.

pony motor (*Elec. Eng.*). An auxiliary motor used to bring synchronous machinery up to speed before synchronizing.

pool cathode (*Electronics*). An emissive cathode which consists of a liquid conductor, e.g., mercury. See **pool tube.**

pool reactor (*Nuc. Eng.*). See **swimming-pool reactor.**

pools zone (*Ecol.*). In small streams, zones of deeper water where current velocity is reduced so that silt and other loose materials tend to settle to bottom, providing conditions unfavourable for surface benthos, but favouring burrowing forms, nekton, and, in some cases, plankton. Cf. *rapids zone.*

pool tube (*Electronics*). One where a mercury cathode provides electrons by spot impact of the mercury positive ions formed in the vapour. Large sizes are used for power or traction rectification. Also **tank tube.** See **excitron, ignitron, mercury-arc rectifier.**

poor lime (*Build.*). Lime in which the proportion of impurities insoluble in acids is in excess of 15%.

POP (*Photog.*). Abbrev. for *printing-out paper*, i.e., a photographic paper which produces an image on printing which needs only fixing, no development. Now obsolete for general purposes, but sometimes used for proofing portraits.

poplar (*For.*). A common tree *Populus* of the temperate regions of the Northern hemisphere. Yields *salicin* (q.v.).

poplin (*Textiles*). A cotton fabric of plain weave, with fine lines or cords running across the piece (due to the ends per inch greatly exceeding the picks per inch); usually mercerized. Used for shirtings, pyjamas, and dresses. Formerly, poplin fabrics were made from a combination of silk, worsted, and cotton.

popliteal (*Zool.*). In *Tetrapoda*, pertaining to the region above the cruro-femoral articulation.

poppet head (*Mining*). Headframe of hoisting shaft. The poppet is the bearing in which the winding pulley is set.

poppets (*Ships*). Temporary structures erected beneath a ship's hull to transfer the weight to the sliding ways, before and during launching.

poppet valve (*I.C. Engs.*). The mushroom or tulip-shaped valve, made of heat-resisting steel, commonly used for inlet and exhaust valves. It consists of a circular head with a conical face which registers with a corresponding seating round the port, and a guided stem by which it is lifted from its seating by the rocker or tappet. Also called **mushroom valve.** See **stellifed valves, valve inserts.**

popping (*Build.*). A defect in plasterwork resulting from the use of a lime which has not been properly slaked. (*For.*) The audible splitting of the bole as a result of the release of internal stresses when a tree is being felled.

popping-back (*I.C. Engs.*). An explosion through the inlet pipe and carburettor of a petrol engine, due to a weak, slow-burning mixture.

Pop rivet (*Eng.*). A proprietary design of hollow rivet, requiring work access from one side only for placing and clinching. Similar to *Chobert rivet* (q.v.).

population inversion (*Phys.*). The reversal of the

normal ratio of populations of two different energy states, i.e., that normally fewer and fewer atoms occupy states of successively higher energies. It forms the basis of laser action.

population types (*Astron.*). The 2 broad types of stellar population. Population I includes hot blue stars such as those in the sun's neighbourhood; they are found in the arms of spiral galaxies, and share in the galactic rotation. Population II stars are found in the central regions of galaxies and in globular clusters, where dust and gas are absent; they are red stars, having high velocities and do not share in the regular rotation of the system.

pop valve (*Eng.*). A boiler safety-valve in which the head of the wing valve is so shaped as to cause the steam to accelerate the rate of lift when a small lift occurs, giving rapid pressure release.

poradenitis venerea (*Med.*). See **lymphogranuloma inguinale.**

porandrous (*Bot.*). Said of stamens which open by pores, not by slits.

porcelain clay (*Geol.*). See **china clay.**

porcelain insulator (*Elec. Eng.*). An insulator for supporting high-voltage electric conductors; made of a hard-quality porcelain.

porch (*TV*). In a complete TV picture signal, the *front porch* is that between the front of the horizontal blanking signal and the front of synchronizing signal. The *back porch* is that between the end of the horizontal synchronizing signal and the end of blanking signal.

porcupine drawing (*Textiles*). A method of producing worsted roving in a rounded form without twisting; effected by passing the sliver between oscillating rubbing leathers. Also known as **Continental** (or **French**) **system of drawing.**

porcupine roller (*Spinning*). Bladed cylinder in opening machine which beats dust, leaf, etc., out of matted cotton. (*Textiles*) A roller, covered with wire carding, which draws the lace fabric over the facing bar on to the work roller.

pore (*Bot.*). (1) The aperture of a stoma. (2) The ostiole in *Pyrenomycetes*. (3) One of the tubular cavities lined by basidia in the pore-bearing fungi. (*Min. Proc., Powder Tech.*) Cavity in particle. If communicating with surface, open; if internal, closed. (*Zool.*) A small aperture. *adjs.* **poriferous, porous, poriform.**

pore distribution (*Powder Tech.*). (1) The size and shape distribution of pores in a consolidated body such as a sintered metal compact. (2) In a loose powder compact, the space confined by adjacent particles can be considered as a system of interconnecting pores. The term *pore distribution* is used to describe the size variation in these so-called pores.

porencephaly, porencephalia (*Med.*). The presence, in the substance of the brain, of cysts or cavities containing colourless fluid; due to a defect in development.

pore plate (*Zool.*). The perforated portion of the central capsule in certain *Radiolaria*; the madreporite of *Echinodermata*.

poricidal (*Bot.*). Said of anthers which open by pores.

Porifera (*Zool.*). The single phylum of the subkingdom *Parazoa*, having the characteristics of the subkingdom.

porocyte (*Zool.*). In *Porifera*, a conical cell extending through from the dermal layer to the paragaster and pierced from base to apex by a tube through which water passes to the paragaster.

porogamy (*Bot.*). The entry of the pollen tube through the micropyle in early stages of fertilization.

poroid (*Bot.*). Having more or less obvious pores.

poromeric (*Chem.*). Permeable to water-vapour; said of a polyurethane-base synthetic leather used in the manufacture of shoe 'uppers'.

porometer (*Bot.*). An instrument for measuring the rate at which air can be drawn through a portion of a leaf; it is a means of measuring the degree to which the stomata are open.

porose (*Bot.*). Said of cell walls which are pierced by pores.

porosity (*Build.*). The percentage of space in a material. (*Geol.*) Of rocks, the ratio, usually expressed as a percentage, of the volume of the pore space to the total volume of the rock. (*Met., etc.*) In castings, unsoundness caused by shrinkage during cooling, or blowholes. In compaction of powders, the percentage of voids in a given volume under specified packing conditions.

porous bearing (*Eng.*). A bearing produced by powder metallurgy which, after sintering and sizing, is impregnated with oil by a vacuum treatment. Bearing porosity can be readily controlled and may be as high as 40% of the volume.

porous dehiscence (*Bot.*). The liberation of pollen from anthers, and of seeds from fruits through pores in the wall of the containing structure.

porous pot (*Elec. Eng.*). An unglazed earthenware pot serving as a diaphragm in a 2-fluid cell.

porpezite (*Min.*). A variety of gold which contains up to 10% of palladium.

porphin (*Chem.*). Group of 4 pyrrole nuclei linked by methene groups, having a complete system of conjugated double bonds, which accounts for the (reddish) colour of its derivatives:

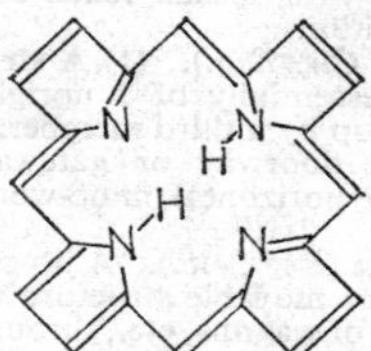

porphyria (*Med.*). An inborn error of metabolism resulting in the excretion of an abnormal pigment in the urine which turns a dark red colour on standing. It is characterized by bouts of abdominal pain and vomiting, abnormal skin pigmentation and photosensitivity.

porphyrin (*Chem.*). A substituted porphin free from metal.

porphyrite (*Geol.*). A fine-grained igneous rock containing phenocrysts, and equivalent in composition to diorite, i.e. porphyritic microdiorite. The term is usually only applied to dyke rocks.

porphyritic texture (*Geol.*). The term applied to the texture of igneous rocks which contain isolated euhedral crystals larger than those which constitute the groundmass in which they are set.

porphyroblastic (*Geol.*). A textural term applicable to metamorphic rocks containing conspicuous crystals in a finer groundmass, the former being analogous with the phenocrysts in a normal igneous rock, but having developed in the solid.

porphyry (*Geol.*). A general term used rather loosely for igneous rocks which contain relatively large isolated crystals set in a fine-grained groundmass, e.g., *granite-porphyry*. It is better used as a textural qualifier combined with a specific rock name, e.g., *porphyritic micro-granite*, a medium- to fine-grained rock of granitic composition with porphyritic texture.

porpoising (*Aero.*). Recurring movements of a seaplane, flying-boat, or amphibian, when taxiing: instability on the water, as distinct from instability under airborne conditions.

porrect (*Bot.*). Extending forwards.

port. Place of access to a system, used for introduction or removal of energy or material, e.g., *glove-box port*. (*Eng.*) An opening, generally valve-controlled, by which a fluid enters or leaves the cylinder of an engine, pump, etc. See **piston valve, poppet valve, sleeve valve, slide valve.**

porta (*Zool.*). Any gatelike structure. *adj.* **portal.**

portable colour duct (*Print.*). An interchangeable ink duct which can be attached to the inking system of a printing unit and used to provide an additional colour.

portable electrometer (*Elec. Eng.*). A portable form of the absolute attracted-disk type of electrometer.

portable engine (*Eng.*). A steam or I.C. engine carried on road wheels but not self-propelled thereby.

portable ink pump (*Print.*). A pump-operated *portable colour duct*.

portable instrument (*Elec. Eng.*). An electrical measuring instrument specially designed for carrying about for testing purposes. Cf. *switch-board instrument, substandard instrument*.

portable substation (*Elec. Eng.*). A substation comprising the converting or transforming plant and the necessary switch and protective gear, mounted on a railway truck or other vehicle in order that it can be quickly moved to any site for dealing with special loads or other emergency conditions.

portal (*Build., Civ. Eng.*). (1) A structural frame consisting essentially of 2 uprights connected across the top by a third member. (2) An arch spanning a doorway or gateway. (*Mining*) Entrance to horizontal mine-working on hill-side.

portal jib crane (*Eng., etc.*). A jib crane mounted on a fixed or movable structure which permits the passage of wagons, etc., through an opening directly under the crane.

portal system (or **circulation**) (*Zool.*). A vein which breaks up at both ends into sinusoids or capillaries; as (in Vertebrates) the *hepatic portal system*, in which the hepatic portal vein collects from the capillaries of the alimentary canal and passes the blood into the sinusoids and capillaries of the liver, and the renal portal system.

ported baffle (*Acous.*). *Box baffle* (q.v.) with parts adjacent to the radiating loudspeaker unit which radiate in antiphase at low frequencies, thus unifying the acoustic response.

ported loudspeaker (*Acous.*). Open-diaphragm loudspeaker mounted on a face of a box with adjacent ports.

porte-lumière (*Surv.*). A simpler form of *heliostat* (q.v.), worked by hand.

porter (*Brew.*). A beer partaking of the character of mild ale and stout; usually it has a sweet flavour. (*Textiles*) A basic measure in linen weaving. The porter consists of 20 splits or openings, and the loom reed may be gauged by the number of porters in a width of 37 in. Also called **portie.** Cf. *beer*.

Porter governor (*Eng.*). A pendulum-type governor in which, usually, the ends of 2 arms are pivoted to the spindle and sleeve respectively, and carry heavy balls at their pivoted joints. The sleeve carries an additional weight. See **pendulum governor.**

portico (*Arch.*). A covered colonnade at one side of a building (usually the entrance side).

Portland Beds, Portlandian Stage (*Geol.*). A group of sands (below) and limestones (above) found in the Upper Jurassic rocks of southern England. It includes the famous Portland Stone (an oölitic freestone), much used for building purposes.

Portland blast-furnace cement (*Build., Civ. Eng.*). A cement differing from a Portland cement in that it contains a proportion of blast-furnace slag.

Portland cement (*Build., Civ. Eng.*). A much-used cement made by intimately mixing clay and lime, and afterwards burning the mixture in a kiln.

portlandite (*Min.*). Calcium hydroxide, $Ca(OH)_2$, occurring as hexagonal plates in the Chalk-dolerite contact-zone at Scawt Hill, Co. Antrim. Occurs also in Portland cement, hence the name.

Portmann's operation (*Surg.*). Removal of the stapes, followed by replacement with a polythene prosthesis.

position (*Radio*). In radio-navigation, a set of coordinates used to specify location and elevation.

positional astronomy (*Astron.*). See **spherical astronomy.**

positional notation (*Maths.*). That system of representation of a number by symbols whose position relative to other symbols in the number affects their absolute value. It is used in the conventional Arabic numerals (as compared with the Roman system where each symbol has an absolute value, the number being represented by the sum of those values modified by the convention of subtracting smaller values from succeeding larger values). Thus in the number 37, the 3 represents ten times as much as it does in the number 53; in the number 437 246 the 3 represents 10^4 times as much: each position back from the right multiplying its value by 10, hence decimal notation. Any base can be used in place of 10.

position angle (*Astron.*). A measure of the orientation of one point on the celestial sphere with respect to another. The position angle of any line with reference to a given point is the inclination of the line to the hour circle passing through the point; it is measured from 0° to 360° from the north point round through east.

position error (*Aero.*). That part of the difference between the *equivalent* and *indicated air speeds* (qq.v.) due to the location of the *pressure head* or *static vent* (qq.v.). Position error is not a constant factor, but varies with airspeed due to the variations in the airflow around an aircraft at different *angles of attack*.

position-finding (*Radio*). Determination of location of transmitting station (e.g., an aeroplane) by taking a number of bearings by direction-finders which receive a signal from the transmitter.

position habit (*An. Behav.*). A *fixation* (q.v.) in which an animal always chooses the alternative in the same position, i.e., always right or always left. Frequently shown in a frustrating or insoluble problem situation.

position head (*Hyd.*). See **elevation head.**

positive (*Elec.*). Said of point in circuit which is higher in electric potential than earth. (*Photog.*) An image in the same scale of contrast as the original, i.e., black for black and white for white, as contrasted with a *negative* (q.v.), where the scale is reversed, i.e., white for black and black for white, with intervening gradations. (*Weaving*) Term describing movement in any part of a loom that is due to mechanical or electrical means; *negative* movement is that achieved by using springs or weights.

positive after-image (*Optics*). The continued image perceived after visual fatigue in the retina, when the object is replaced by a dark surface or the eye is closed.

positive amplitude modulation (*TV*). That in which the amplitude of a video signal increases with an increase in luminance of the picture elements.

positive coarse pitch (*Aero.*). An extreme blade angle which is reached and locked after engine failure to reduce the drag of a nonfeathering airscrew.

positive column (*Electronics*). Luminous plasma region in gas discharge, adjacent to positive electrode.

positive coupling (*Elec. Eng.*). When two coils are inductively coupled so that the magnetic flux associated with the mutual inductance is in the same direction as that associated with the self-inductance in each coil, the mutual inductance and the coupling are both termed *positive*. If these fluxes oppose the mutual inductance and coupling they are known as *negative*.

positive cyanotype (*Photog.*). A positive blue-printing process in which the iron is included in a viscous sensitizing liquid, while the ferro-cyanide is provided in the developer.

positive electricity (*Elec.*). Phenomenon in a body when it gives rise to effects associated with deficiency of electrons, e.g., positive electricity appears on glass rubbed with silk. See **negative electricity**.

positive electrode (*Electronics*). (1) That connected to a positive supply line. (2) The *anode* in a voltameter. (3) The *cathode* in a primary cell.

positive electron (*Nuc.*). See **positron**.

positive emulsion (*Photog.*). Fine-grain emulsion of the ortho-type used for printing positives and for registering photographic sound-tracks; not required to be specially sensitive to a variety of coloured rays.

positive feedback (*Telecomm.*). Interconnexion between output and input circuits of an amplifier to facilitate the voltage, current, or power drive of input by addition of voltage, current, or power from output. This reduces resistance of the source of amplified power and, if feedback is sufficient, sets up sustained oscillations independently of input drive. See also **regeneration**.

positive feeder (*Elec. Eng.*). A feeder connected to the positive terminal of a d.c. supply.

positive film stock (*Cinema.*). The unexposed film on which positive rush and release prints are made. It is also used in sound cameras for exposing for the sound-track negative, because of its fine grain as compared with the pan stock used in the picture cameras.

positive *g* (*Space*). The force exerted on the human body in a gravitational field or during acceleration so that the force of inertia acts in a head-to- oot direction, the reaction varying with the individual. See **negative *g***.

positive grid oscillator (*Telecomm.*). See **Barkhausen-Kurz** and **Gill-Morrell oscillators**.

positive *g* tolerance (*Space*). At 5*g* positive, the downward pull on the blood equals the ability of the heart to push the blood towards the brain. If 6*g* is continued for a few minutes, death results due to circulation failure. Cf. *negative-g tolerance*.

positive ion (*Phys.*). An atom (or a group of atoms which are molecularly bound) which has lost one or more electrons, e.g., α-particle is a helium atom less its 2 electrons. In an electrolyte, the positive ions (*cations*) produced by dissolving ionic solids in a polar liquid like water, have an independent existence and are attracted to the anode. Negative ions likewise are those which have gained one or more electrons.

positive (or negative) light modulation (*TV*). Increase from low to high modulation (or decrease from high to low modulation) of radiated carrier with increase of light in transmitted image.

positive magnetostriction (*Mag.*). See **Joule magnetostriction**.

positive mineral (*Min.*). A mineral in which the ordinary ray velocity is greater than that of the extraordinary ray, that is, ω is less than ε. Quartz is a positive mineral for which ω = 1·544 and ε = 1·553. See **optic sign**.

positive ore (*Mining*). Ore blocked out on four sides in panels sufficiently small to warrant assumption that the exposed mineral continues right through the block as a calculable tonnage.

positive phase sequence (*Elec. Eng.*). See under **phase sequence**.

positive-ray parabolas (*Phys.*). Parabolic traces on the photographic plate obtained by Sir J. J. Thomson's original method of positive-ray analysis, which consisted in subjecting a narrow beam of the rays to the action of mutually perpendicular magnetic and electrostatic fields. See **positive rays, mass spectrograph**.

positive rays (*Phys.*). Streams of positively charged atoms or molecules which take part in electrical discharge in a rarefied gas; they are accelerated through a perforated cathode on to a photographic plate, being deflected by magnetic and electrostatic fields. Used in Thomson's parabolic traces and Aston's mass spectrometer. Also called **anode rays, canal rays**.

positive reaction (*Bot.*). A tactism or tropism in which the plant moves, or the plant member grows, from a region where the stimulus is weaker to one where it is stronger.

positive reinforcing stimulus (*An. Behav.*). A *reinforcing stimulus* (q.v.) which generally increases the intensity of a consummatory act. Also called a **reward**.

positive stagger (*Aero.*). See **stagger**.

positive taxis (*Biol.*). See **taxis**.

positive transference (*Psychiat.*). The type of *transference* (q.v.) in which the infantile emotions displaced on to the analyst are those of love.

positive video signal (*TV*). One in which increasing amplitude corresponds to increasing light value in the transmitted image. White is 100%, and the black level makes about 30% of the maximum amplitude of signal.

positive wire (*Elec. Eng., etc.*). Any wire in a circuit which is connected to the positive or other d.c. end of the battery power supply.

positron (*Nuc.*). *Positive electron*, of the same mass as and charge of opposite sign to the normal (negative) electron. Produced in the decay of radioisotopes, and in *pair production* by X-rays of energy much greater than 1 MeV.

positronium (*Nuc.*). Short-lived particle, combination of a positron and an electron, its mean life being less than 10^{-7} seconds. The energy levels of a positronium are similar to those of the hydrogen atom but because of the different *reduced mass* the spectral line of frequencies is only approximately half of the corresponding lines of hydrogen.

possible ore (*Mining*). Ore probably existing, as indicated by the apparent extension of proved deposits, but not yet entered and sampled.

post (*Glass*). The formed *gather* prior to drawing by the hand process. (*Paper*) A standard size of printing paper, $15\frac{1}{4} \times 19$ in. (388×483 mm).

post-. Prefix from L. *post*, after.

postabdomen (*Zool.*). In Scorpions, the narrow posterior part of the abdomen; metasoma.

postal (*Paper*). A standard board size, $22\frac{1}{2} \times 28\frac{1}{2}$ in. (572×725 mm).

postanal gut (*Zool.*). In some embryonic *Craniata*, a continuation of the intestine backwards into the haemal canal, perhaps indicating that the anus has shifted forwards during evolution.

post and pan (or pane), post and petrail (*Arch.*). A term applied to half-timbering formed with brickwork or lath and plaster panels.

postcard (*Paper*). A standard size of cut card, $3\frac{1}{2} \times 5\frac{1}{2}$ in. (89×140 mm).

postcardinal (*Zool.*). Posterior to the heart; as the *postcardinal* sinus of *Selachii*.

postcaval vein (*Zool.*). In higher Vertebrates, the posterior vena cava conveying blood from the hind parts of the body and viscera to the heart. Called **inferior vena cava** in Man.

postclavicle, postclavicula (*Zool.*). See **postcleithrum.**

postcleithrum (*Zool.*). A bone of the pectoral girdle in Fish, situated behind the cleithrum. Also called **postclavicle, postclavicula.**

postclimax (*Ecol.*). A relict of a former climax held under edaphic control in an area where the climate is no longer favourable for its development.

postdeflection acceleration (*Electronics*). In a CRT, acceleration of the beam electrons after deflection, so reducing the power required for this. Abbrev. **PDA.**

post-dipping lameness (*Vet.*). Fever and lameness in sheep due to cellulitis of the limbs caused by the bacterium *Erysipelothrix insidiosa* (*E. rhusiopathiae*); occurs after sheep have been dipped in baths contaminated by the organism.

post-emphasis (*Acous.*). See **de-emphasis.**

poster (*Print.*). A large sign printed on paper, for advertisement or propaganda purposes. Posters are usually printed by lithography, offset or direct, very large ones being executed in convenient sections. In Britain, the 1-sheet poster measures 30×20 in. (double crown) and larger sizes are 2-sheet, 4-sheet, etc.

posterior (*Bot.*). (1) Inserted on the back of another organ. (2) The part of the flower nearest to the axis. (3) The rear. (*Zool.*) In a bilaterally symmetrical animal, further away from the head region; behind. Cf. *anterior*.

postern (*Arch.*). A private door or gate, generally at the back or side of a building.

poster stick (*Typog.*). A long wooden composing stick suitable for setting up large type, usually wooden.

postfertilization stages (*Bot.*). The developmental processes which go on between the union of the gametic nuclei in the embryo sac and the maturity of the seed.

postfloral movement (*Bot.*). A change in position of the flower stalk or inflorescence stalk after fertilization has occurred, bringing the young fruits into a more favourable position for development, or placing the seeds in good conditions for germination.

post-forming sheet (*Plastics*). A grade of laminated sheet, of the thermosetting type, suitable for drawing and forming to shape when heated.

postfrontal (*Zool.*). In some Vertebrates, a paired lateral membrane bone of the skull.

postganglionic (*Physiol.*). See **preganglionic.**

postganglionic fibre (*Physiol.*). An *excitor* (q.v.) neurone.

post head (*Elec. Eng.*). A post or pillar at which cables supplying a third-rail traction system may be terminated and connexion made to the conductor rail.

postheterokinesis (*Cyt.*). A form of reduction division in which the sex-chromosome passes undivided to 1 pole in the second spermatocyte division.

posthitis (*Med.*). Inflammation of the prepuce.

post-hole digger (*Mining, etc.*). Large auger worked by hand or mounted on truck, used to dig shallow holes in unconsolidated earth for sampling or for constructional work.

postical (*Bot.*). Relating to or belonging to the back or lower part of a leaf or stem.

posticous (*Bot.*). Outward or behind. (*Zool.*) Extrorse.

posting plumb (*For.*). An adjustable gauge, operating on the principle of either a plumb-bob or a spirit level, by means of which frame-saw blades can be posted with the required degree of hang.

post insulator (*Elec. Eng.*). A porcelain insulator built in the form of a post; used for supporting bus-bars, etc., in a high-voltage outdoor substation.

postminimus (*Zool.*). A rudimentary extra digit of some Vertebrates.

postmortem (*Comp.*). Program designed to locate and diagnose in an electronic computer a fault which has caused faulty operation.

postnotum (*Zool.*). In the wing-bearing segments of adult Insects, a narrow sclerite behind the notum arising in the intersegmental membrane.

Post Office box (or bridge) (*Elec. Eng.*). A Wheatstone bridge in which the resistances making up the arms are contained in a box and varied by means of plugs. Frequently abbrev. to **P.O. box (or bridge).**

postorbital (*Zool.*). In some Vertebrates, a paired lateral membrane bone of the orbital region of the skull.

post pallet (*Build.*). A *pallet* (q.v.) with corner posts, used when the unit load either cannot withstand the weight of another to be placed direct above it or has an irregular surface on which another palletized unit load could not be stacked direct.

postparietal (*Zool.*). In some Vertebrates, a paired membrane bone of the skull, lying between the parietal and the interparietal.

postpartum (*Med.*). Occurring after childbirth; e.g., *postpartum* haemorrhage.

postpatagium (*Zool.*). In Birds, a cutaneous expansion between the upper arm and the trunk. *adj.* **postpatagial.**

postscoring (*Cinema.*). In motion-picture production, the recording of sound, such as a musical accompaniment or special noise effects, after the cinematograph pictures have been made.

postscutellum (*Zool.*). See **postnotum.**

postsynchronize (*Cinema.*). To add a sound-recording to an independently shot picture.

post-tensioned concrete (*Civ. Eng.*). A method whereby concrete beams or other structural members are compressed, after casting and reaching the requisite strength, to enable them to act in the same way as *prestressed concrete*.

Achieved by the use of high-tensile wires or rods threaded through ducts in the member(s).

Post-Tertiary (*Geol.*). Name assigned to geological events which occurred after the close of the Tertiary era, i.e., during Pleistocene and Recent times.

post-trematic (*Zool.*). Posterior to an aperture; as, in *Selachii*, that branch of the ninth cranial nerve which passes posterior to the first gill-cleft. Cf. *pretrematic*.

postulate (*Maths.*). An assumption. Almost synonymous with *axiom*. An axiom generally relates to an assumption which by its nature is unlikely ever to be proved whereas a postulate is generally rather less basic and may admit, in certain circumstances, of being proved. Since the end of the 19th c. *axiom* has assumed both meanings and *postulate* is now rarely used.

postventitious (*Bot., Zool.*). Delayed in development.

postzygapophysis (*Zool.*). A facet or process on the posterior face of the neurapophysis of a vertebra, for articulation with the vertebra next behind. Cf. *prezygapophysis*.

pot (*Elec. Eng.*). A shortened form of **potentiometer (2)**. (*Glass*) A vessel of fireclay holding from a few kg to about 2 tonnes, according to the type of manufacture; used to contain the glass during melting in the pot furnace. Such pots may be open or closed (i.e., provided with a hood to prevent furnace gases from acting on the glass).

potamoplankton (*Ecol.*). The plankton of rivers and streams.

potamous (*Ecol.*). Living in rivers and streams.

pot annealing (*Met.*). See **close annealing.**

potash-syenite (*Geol.*). A syenitic rock characterized by a large excess of potash-feldspar or feldspathoid over soda-feldspar.

potassium (*Chem.*). A very reactive alkali metal, soft and silvery white, symbol K, at. no. 19, r.a.m. 39·102, m.p. 63°C, b.p. 762°C, rel. d. 0·87. In the form of the element, it has little practical use, although its salts are used extensively. In combination with other elements it is found widely in nature. It shows slight natural radioactivity due to ^{40}K (half-life $1{\cdot}30 \times 10^9$ years). May be used as coolant in liquid-metal reactors, usually as an alloy with sodium. See **NaK.**

potassium alum (*Min.*). A hydrous sulphate of aluminium and potassium, crystallizing in the cubic system. It is found in connexion with volcanoes and also as a result of the action of ascending acid waters on argillaceous rocks containing disseminated sulphides.

potassium antimonyl tartrate (*Chem.*). See **tartar emetic.**

potassium bichromate (*Chem.*). See **potassium dichromate.**

potassium bromide (*Chem.*). KBr. Used in medicine as a sedative, and in photography to produce bromide papers.

potassium carbonate (*Chem.*). K_2CO_3. Potash. White deliquescent powder, soluble in water. Basic. Manufactured by extraction from Stassfurt and other potassium salt beds, or from the electrolysis of potassium chloride. Uses: washing, glass flux, potassium cyanide, soft soaps, pharmacy.

potassium chlorate(V) (*Chem.*). $KClO_3$. Detonates with heat; used in the manufacture of matches, fireworks, and explosives, and in the laboratory as a source of oxygen.

potassium chloride (*Chem.*). KCl. Occurs extensively in nature. With sodium chloride, it is extracted on a commercial scale from the waters of the Dead Sea. Uses: fertilizers, mineral waters, pharmacy, photography.

potassium cyanide (*Chem.*). KCN. White, deliquescent solid, smelling of bitter almonds. Extremely poisonous. In the fused condition, is a powerful reducing agent. Used in chemical analysis and in metallurgy, e.g., in the cyanide process for extracting gold. Also in HT salts, fumigants, pharmacy, photography. On the industrial scale, its use has largely been superseded by that of the cheaper sodium cyanide.

potassium dichromate(VI) (*Chem.*). $K_2Cr_2O_7$. Used in analytical chemistry. Mixed with sulphuric acid, it is used as a cleanser of laboratory vessels, particularly after contamination with organic matter.

potassium feldspar (*Min.*). Silicate of aluminium and potassium, $KAlSi_3O_8$, occurring in 2 principal crystalline forms—*orthoclase* (monoclinic) and *microcline* (triclinic). Both are widely distributed in acid and intermediate rocks, especially in granites and syenites and the fine-grained equivalents. See **adularia, feldspar, sanidine.**

potassium ferricyanide (*Chem.*). Potassium hexacyano ferrate(III). $K_3Fe(CN)_6$. Used in chemical analysis. Gives characteristic colour reactions. Also used in dyeing, etching, blue print paper, and as a fertilizer.

potassium hexachloroplatinate (*Chem.*). K_2PtCl_6. Results from the reaction of chloroplatinic acid and potassium chloride.

potassium hydride (*Chem.*). KH. Formed when potassium is heated in hydrogen.

potassium hydrogen carbonate (*Chem.*). $KHCO_3$.

potassium hydrogen fluoride (*Chem.*). KHF_2. A double salt. Formed when potassium fluoride is dissolved in hydrofluoric acid and the solution evaporated.

potassium iodate(V) (*Chem.*). KIO_3. Potassium salt of iodic acid.

potassium iodide (*Chem.*). KI. Used in chemical analysis, medicine, photography.

potassium mica (*Min.*). See **muscovite, sericite.**

potassium nitrate (*Min.*). KNO_3. Salt of potassium and nitric acid. Strong oxidizing agent. Used in pyrotechnics, explosives, HT salts, glass manufacture and as a fertilizer. Also known as **nitre, saltpetre.**

potassium oxalate (*Chem.*). Potassium ethandioate. The normal salt, $K_2C_2O_4{\cdot}H_2O$, is soluble in water. The acid salt, KHC_2O_4, is less soluble and occurs in many plants. A compound of these two, potassium quadroxalate, $K_3HC_4O_8{\cdot}2H_2O$, known as *salts of sorrel*, is used for bleaching and removing iron stains.

potassium permanganate (*Chem.*). Potassium manganate (VII). $KMnO_4$. Dark purple-brown crystals and solution, with characteristic sweet-bitter taste. Strong oxidizing agent. Used in analytical chemistry and as a disinfectant.

potassium propionate (*Chem.*). Potassium propanoate. $CH_3{\cdot}CH_2{\cdot}COOK$. Compound used as a fungicide in edible products such as baked goods.

potassium sorbate (*Chem.*). $CH_3{\cdot}(CH{=}CH)_2COOK$. Solid compound used as an antifungal in edible products in place of sorbic acid when high water solubility is needed.

potassium titanate(IV) (*Chem.*). K_2TiO_3. Formed by the fusion of potassium hydroxide and titanium (IV) oxide.

potato scab (*Agric.*). A fungal disease of potatoes caused by *Streptomyces scabies*, characterized by scab-like surface blemishes.

potcher (*Paper*). A machine similar to a Hollander

beater in which rags, esparto, or wood pulp are washed, bleached, and reduced to pulp in readiness for the next process of beating.

potential (*Elec.*). Scalar magnitude, negative integration of the electric (or magnetic) field intensity over a distance. Hence all potentials are relative, there being no absolute *zero potential*, other than a convention, e.g., earth, or at infinity distance from a charge. (*Nuc.*) The *potential energy* of a nucleon expressed in terms of its position in the nuclear field. A central potential, e.g., is one that is spherically symmetric, i.e., the potential energy V is a function only of the distance $\boldsymbol{r}$ from the centre of the field. (*Zool.*) Latent.

potential attenuator (*Elec. Eng.*). Same as *potentiometer* (1), as contrasted with a normal attenuator which adjusts power.

potential barrier (*Phys.*). Maximum in the curve covering two regions of potential energy, e.g., at the surface of a metal, where there are no external nuclei to balance the effect of those just inside the surface. Passage of a charged particle across the boundary should be prevented unless it has energy greater than that corresponding to the barrier. Wave-mechanical considerations, however, indicate that there is a definite probability for a particle with less energy to pass through the barrier, a process known as **tunnelling.** Also called **potential hill.**

potential coefficients (*Elec.*). Parts of total potential of a conductor produced by charges on other conductors, treated individually.

potential-determining ions (*Chem., Min. Proc.*). Those which leave the surface of a solid immersed in an aqueous liquid before saturation point (equilibrium) has been reached.

potential difference (*Elec.*). That between two points when maintained by an e.m.f., or by a current flowing through a resistance; unit is one *volt* when one *joule* of work is done (or received) by one *coulomb* moving between the two points. Abbrev. **pd.** (*Mag.*) Line integral of a magnetic field intensity between two points by any path.

potential divider (*Elec. Eng.*). **See voltage divider.**

potential drop (*Elec.*). Difference of potential along a circuit because of current flow through the finite resistance of the circuit.

potential energy (*Phys.*). Universal concept of *energy* stored by virtue of position in a field, without any observable change, e.g., after a mass has been raised against the pull of gravity. A body of mass m at a height h above the ground possesses potential energy mgh, since this is the amount of work it would do in falling to the ground. In electricity, potential energy is stored in an electric charge when it is taken to a place of higher potential through any route. A body in a state of tension or compression (e.g., a coiled spring) also possesses potential energy.

potential fuse (*Elec. Eng.*). A fuse used to protect the voltage circuit of a measuring instrument or similar device.

potential galvanometer (*Elec. Eng.*). A galvanometer having a resistance sufficiently high to enable it to be used as a voltmeter.

potential gradient (*Elec. Eng.*). Potential difference per unit length along a conductor or through a dielectric; equal to slope of curve relating potential and distance. Also **electromotive intensity.**

potential hill (*Phys.*). See **potential barrier.**

potential indicator (*Elec. Eng.*). An instrument which shows whether a conductor is alive. Also called a **charge indicator.**

potential mediator (*Chem.*). A substance which is added to an oxidation-reduction system to accelerate establishment of definite potential.

potential temperature (*Heat*). The temperature which a specimen of gas would have if it were brought to standard pressure adiabatically. The potential temperature is given by the expression:

$$\theta = T\left(\frac{P_0}{P}\right)^{\frac{\gamma-1}{\gamma}}$$

where T is the absolute temperature, P the pressure of the gas, P_0 is the standard pressure, and γ is the ratio of the specific heat capacities (=1·40 for air).

potential theory. That branch of applied mathematics which studies the properties of a potential function without reference to the particular subject (e.g., hydrodynamics, electricity and magnetism, gravitational attraction) in which the function is defined.

potential transformer (*Elec. Eng.*). An undesirable synonym for **voltage transformer.**

potential trough (*Phys.*). Region of an energy diagram between two neighbouring hills, e.g., arising from the inner electron shells of an atom.

potential well (*Nuc.*). Localized region in which the energy of a particle is appreciably lower than outside. Such a well forms a trap for any incident particle.

potentiometer (*Elec. Eng.*). (1) Precision-measuring instrument in which an unknown potential difference or e.m.f. is balanced against an adjusted potential provided by a current from a steady source. (2) Three-terminal voltage divider, often shortened to **pot.** The resistance change with shaft rotation or slider position may follow various laws. Thus one may have, e.g., linear, logarithmic, cosine, etc., potentiometers. A form of *rheostat* (q.v.).

potentiometer braking (*Elec. Eng.*). A braking method used for series motors; the series field and a rheostat are connected in series across the supply, and the armature is connected across the field and a variable proportion of the rheostat.

potentiometer braking controller (*Elec. Eng.*). A controller making the necessary connexions for potentiometer braking of a series motor.

potentiometer card generator (*Elec. Eng.*). Potentiometer wound with length of turns dependent on rate of change of a function, which determines shape of card.

pot(entiometer) cut (*Acous.*). Colloquialism for introducing a *cut* (interruption of reproduction) by bringing down a potentiometer to zero for a short time.

potentiometer for alternating voltage (*Elec. Eng.*). In this instrument, there are two variable alternating voltages in quadrature which are together used to balance the unknown a.c. potential difference. See also **coordinate potentiometer.**

potentiometer function generator (*Elec. Eng.*). One in which functional values of voltage are applied to points on a potentiometer, which becomes an interpolator.

potentiometer pick-off (*Elec. Eng.*). One in which output voltage depends on position of a slide contact on a resistor, thereby transmitting a position.

potentiometer-type field rheostat (*Elec. Eng.*). A field rheostat which is connected across the supply, the field winding being connected between one pole of the supply and a variable tapping on the rheostat. See **reversible-.**

potentiometric titration (*Chem.*). Method of quantitative analysis which measures changes in electromotive force, useful where colour changes cannot be observed.

potette (*Glass*). A hood shaped like a pot, but with no bottom, which is placed in a tank furnace so that it reaches below the glass level. It protects the man gathering glass on his pipe or iron from furnace gases; also, the glass here is somewhat cooler than that in the main part of the furnace, where melting is taking place. Also **boot, hood.**

pot furnaces (*Glass*). Furnaces in which are set a number of *pots*. They may be: (*a*) direct-fired from below; (*b*) gas-fired from below through a central opening in the circular siege, using the recuperative principle; (*c*) fired through ports in the siege or in the walls, the waste gases escaping through similar openings. In the last-named process, which holds generally for noncircular furnaces, the regenerative principle may be used.

Potier construction (*Elec. Eng.*). A graphical construction for determining the reactance, armature reaction, and regulation of a synchronous generator from the open-circuit, short-circuit, and zero-power factor characteristics.

Potier reactance (*Elec. Eng.*). The reactance of a synchronous machine as determined by the Potier construction.

pot magnet (*Elec. Eng.*). One embracing a coil or similar space, excited by current in the coil or a permanent magnet in the central core. Main use is with a circular gap at an end of the core for a moving coil. Miniature split-sintered pot magnets are also used to contain high-frequency coils.

Potsdamian (*Geol.*). An obsolete stage name for the Upper Cambrian in North America.

pot still (*Chem.*). A still consisting of a boiling vessel with condenser attached. The use of a fractionating column is optional. Used for batch distillation.

potstone (*Min.*). A massive variety of *steatite* (q.v.), more or less impure.

Potter-Bucky grid (*Radiol.*). Type of lead grid designed to avoid exposure of film to scattered X-radiation in diagnostic radiography. Mechanical oscillation of the grid eliminates reproduction of the grid pattern on the negative.

potters' clay (*Geol.*). See **ball clay.**

Pott's disease (*Med.*). Spinal caries. Tuberculous infection of the spinal column.

Pott's fracture (*Surg.*). Fracture-dislocation of the ankle-joint, the lower parts of the tibia and of the fibula being broken.

Pottsville Series (*Geol.*). A thick formation of sandstone and quartz-conglomerate, of torrential origin and containing coal-seams, which occurs at the base of the Pennsylvanian in the Appalachian region and in Virginia.

pouce (*Textiles*). Name applied in linen trade to particles of cortex formed in roughing flax.

pouch (*Zool.*). Any saclike or pouchlike structure; as the abdominal *brood-pouch* of marsupial Mammals.

poughite (*Min.*). A hydrated sulphate and tellurite of iron, crystallizing in the orthorhombic system.

Poulsen arc (*Radio*). An arc discharge maintained between electrodes of carbon and water-cooled copper enclosed in an atmosphere of hydrogen and situated in a powerful magnetic field. Used for the generation of continuous waves.

Poulsen-arc converter (*Radio*). An arc oscillator which can oscillate up to a frequency of the order of 100 kHz.

pound. The unit of mass in the British system of units established by the Weights and Measures Act, 1856, and until 1963 defined as the mass of the Imperial Standard Pound, a platinum cylinder kept at the Board of Trade. In 1963 it was redefined as 0·453 592 37 kg. The U.S. pound is defined as 0·453 592 427 7 kg.

pound (*Civ. Eng.*). See **tamp.** (*Hyd. Eng.*) See **pond.**

poundal (*Phys.*). Unit of force in the foot-pound-second system of units. It is that force which will produce an acceleration of 1 ft/sec^2 in a mass of 1 lb on which it acts. 32·2 poundals = 1 pound-weight (lbwt) or pound-force (lbf). 1 poundal = 0·138 255 N.

pound-calorie (*Eng.*). An engineering heat unit, often called *Centigrade heat unit* (C.H.U.). Defined as 1/100 part of the heat required to raise the temperature of 1 lb of water from 0°C to 100°C (= approx. 1·9 kJ). Abbrev. **lb-cal.** Obsolete.

pounding (*Ships*). The heavy falling of the fore end of the ship into the sea when it has been lifted clear of the water by wave action. Also striking the ground under the ship due to wave action.

Poupart's ligament (*Anat.*). The inguinal ligament.

pouring basin (*Eng.*). Part of the passage system for bringing molten metal to the mould cavity in metal casting. It is provided on large moulds, next to the sprue hole top, to simplify pouring and to prevent slag from entering the mould.

pour point (*Chem.*). The lowest temperature at which a petroleum-based oil, chilled under test conditions, will flow.

povidone (*Pharm.*). Polyvinylpyrrolidone; PVP; mixture of polymers of 1-vinyl-pyrrolid-2-one, $(C_6H_9NO)_n$. A suspending and dispersing agent, used also for binding tablets and as a granulating agent. Strongly hygroscopic.

powder (*Powder Tech.*). Discrete particles of dry material with a maximum dimension of less than 1000 μm (1 mm).

powder core (*Elec. Eng.*). Core of powdered magnetic material with an electrically insulating binding material to minimize the effects of eddy currents, thus permitting use for high-frequency transformers and inductors with low loss in power.

powder density (*Plastics*). The mass in grams of 1 cm^3 of loose moulding powder.

powder down-feather (*Zool.*). A down-feather the end of which readily breaks off, forming a fine dust.

powderless etching (*Print.*). The etching of line blocks in one stage without recourse to *dragon's blood* between a series of etches, using specially-designed etching baths with several features including close temperature regulation, planetary movement of the plate, and controlled application of the etching fluid. A measured quantity of special inhibitor is added to the etchant and this forms a protecting film over the metal to control the progress of the etch, by protecting the sides of the lines from under-cutting, and stopping the etch when a suitable depth is reached. Originally introduced for line work only, on magnesium alloy, it has been adapted for half-tone and is particularly suitable for combined work, using micrograin zinc or copper.

powder metallurgy (*Met.*). The working of metals and certain carbides in powder form by pressing and sintering. Used to produce self-lubricated bearings, tungsten filaments and shaped cutting inserts (carbide), etc. See **cemented carbides.**

powder method (*Min.*). A method of X-ray determination of minerals in which powder is used instead of a single crystal. It is a useful method for any type of crystalline substance.

powder pattern (*Mag.*). See **Bitter pattern.**

powder photography (*Met., Min., Min. Proc.*). Method of identification of minerals or crystals, in which a small rod is coated with the powdered preparation, mounted vertically in a special camera in which it rotates, and subjected to a suitably modified beam of X-rays. The pattern, characterized by a set of concentric rings produced by rays diffracted at the Bragg angle relative to the incident beam, is diffracted on to a surrounding strip of film to give positive identification.

powder porosity (*Powder Tech.*). The ratio of the volume of voids plus the volume of open pores to the total volume of the powder.

powder technology. That covering the production and handling of powders, particle-size analysis, and the properties of powder aggregates.

power (*Maths.*). The product of a number of equal factors, generalized to include negative and fractional numbers. (*Mech.*) Rate of doing work. Measured in joules per second and expressed in *watts* (1 W = 1 J/s). The foot-pound-second unit of power is the *horsepower*, which is a rate of working equal to 550 ft-lbf per second. 1 horsepower is equivalent to 745·7 watts. 1 watt is equal to 10^7 CGS units, that is 10^7 erg/second.

power amplification (*Elec. Eng.*). The amplification of a signal so that it can deliver worthwhile energy; the difference between output and input power-levels of an amplifier, expressed in *decibels*. See **applied power.**

power amplifier (*Elec. Eng.*). Stage designed to deliver output power of an amplifier. It may be separate from other parts of the same amplifier, and may contain its own power supply. Intended to give the required power output with specified degree of nonlinear distortion, gain not being considered. In some cases, the voltage gain is fractional (or, in dB, negative). Also **power unit.**

power-assisted controls (*Aero.*). Primary flying controls wherein the pilot is aided by electric motors or double-acting hydraulic jacks.

power-assisted steering (*Autos.*). In this system, a hydraulic ram is fitted to the steering drag arm. The ram is powered by a small hydraulic pump driven off the engine.

power breeder (*Nuc. Eng.*). A nuclear reactor which is designed to produce both useful power and fuel.

power circuit (*Elec. Eng.*). That portion of the wiring of an electrical installation which is used to supply apparatus other than fixed lighting, usually through socket outlets.

power coefficient (*Nuc. Eng.*). The change of reactivity of a reactor with increase in power. In a heterogeneous reactor, due to temperature differences of the fuel, moderator and coolant, it differs from the temperature coefficient.

power component (*Elec. Eng.*). See **active component.**

power control rod (*Nuc. Eng.*). One used for control of the power level of a nuclear reactor. May be a full rod or a part of the moderator. In a thermal reactor it is generally a neutron absorber such as cadmium or boron steel.

power controls (*Aero.*). A primary flying control system where movement of the surfaces is done entirely by a power system, commonly hydraulic, but sometimes electrohydraulic or electric. The system is always at least duplicated (power source, supply lines, and operating rams) and both circuits are usually running continuously. Reversion to manual control was common in early systems, but the trend is toward a multiplication of reserve circuits and supply sources. See **power-assisted controls, *q*-feel.**

power density (*Nuc. Eng.*). Energy released per second per unit volume of a reactor core. Expressed normally in watts/cm³.

power detector or demodulator (*Elec. Eng.*). Rectifier which accepts a relatively large carrier voltage and thereby achieves low nonlinear distortion in the demodulation process.

power drag line (*Eng.*). An excavator comprising a large scraper pan or bucket which is dragged through the material towards the machine and below its level.

power-driven system (*Teleph.*). A system in which the wipers of the selectors are operated by a motor which is common to a number of selectors.

powered supports (*Mining*). Pit props held to the roof of a coal seam by hydraulic pressure. In fully mechanized collieries they form part of a mechanically operated system.

power efficiency (*Eng.*). The ratio of power delivered by a transducer (optical, mechanical, acoustical, electrical, etc.) to the power supplied; usually quoted as a percentage.

power factor (*Aero.*). See **height-.** (*Elec. Eng.*) Ratio of the total power (in watts) dissipated in an electric circuit to the total equivalent volt-amperes applied to that circuit. In single- and balanced 3-phase systems, it is equal to cos φ, where φ is the phase angle between the applied voltage and the applied current in a single-phase circuit, or between the phase voltage and phase current in a balanced 3-phase circuit. In normal dielectrics, it is exactly equal to $G(G^2+\omega^2C^2)^{-1/2}$, where C=capacitance, G= shunt conductance, $\omega=2\pi\times$frequency, and thus nearly equals $G/\omega C$. Abbrev. **pf.**

power-factor meter (or indicator) (*Elec. Eng.*). An instrument which reads the power factor of a circuit directly.

power feed (*Eng.*). The power-operated feed motion of tables, heads, etc., of machine tools, used when the duration of feed, loads, and type of cut justify it.

power frequency (*Elec. Eng.*). That of the a.c. supply mains, viz., 50 Hz in the U.K., 60 Hz in the U.S. Also called **mains frequency.**

power gain (*Telecomm.*). Power delivered to a load divided by that which would be delivered if the load were matched to the source, expressed in *decibels*. If the ratio is less than unity, there is *loss*, a negative number of decibels.

power gas (*Fuels*). See **producer gas.**

power grid detector (*Elec. Eng.*). A valve which is operated at a fairly high plate voltage with low values of grid leak and grid capacitor.

power hammer (*Eng.*). Any type of hammer which is operated, either continuously or intermittently, by power, e.g., by directly coupling the hammer to a steam or pneumatic cylinder.

power-level (*Telecomm.*). See **transmission level.**

power-level diagram (*Telecomm.*). Diagram indicating how maximum power levels vary at different points of a transmission channel, thereby indicating how various losses are neutralized by appropriate amplifier gains.

power-level indicator (*Telecomm.*). See **level indicator.**

power line (*Elec.*). U.S. for **mains.** (*Elec. Eng.*) See **bus-line.**

power loading (*Aero.*). The gross weight of an

aircraft divided by the take-off power of its engine(s). See **thrust loading.**

power loss (*Elec. Eng.*). (1) The ratio of the power absorbed by a transducer to that delivered to the load. (2) The energy dissipated in a passive network or system.

power navvy (*Eng.*). See **power shovel.**

power of lens (*Photog.*). The relative focusing power of a lens, measured in dioptres, which is the reciprocal of the focal length in metres.

power output (*Eng.*). The total power delivered to the load of a system. Occasionally (according to context) the useful power if an unwanted component is also supplied.

power-pack (*Elec. Eng.*). Power-supply unit for an amplifier, e.g., in a radio or television receiver, wherein the requisite steady voltages are obtained by rectifiers from mains. Also the last or power stage of an amplifier when this is integral with the power supply proper.

power range (*Nuc. Eng.*). See **start-up procedure.**

power rating (*Aero.*). The power, authorized by current regulations, of an aero-engine under specified conditions; e.g., maximum take-off rating, combat rating, maximum continuous rating, weak-mixture cruising rating, etc. The conditions are specified by r.p.m. and, for piston engines, *manifold pressure* (q.v.) and torque (in large engines), for turboprops *jet-pipe temperature* and torque, for turbojets J.P.T., for *rocket motors* combustion-chamber pressure.

power reactor (*Elec. Eng.*). One designed to produce useful power.

power relay (*Elec. Eng.*). One which operates at a specified power level.

power series (*Maths.*). A series of the form $\Sigma a_n x^n$.

power shovel (or navvy) (*Eng.*). An excavator consisting of a jib carrying a radial arm to the end of which a large bucket or scoop is attached. The bucket makes a radial cut, digging above the level of the excavator.

Powerstat (*Elec. Eng.*). TN for class of adjustable autotransformers. Similar to **Variac.**

power supply (*Elec.*). (1) Arrangement for delivering available power from a source, e.g., public mains, in a form suitable for valves or transistors, etc., generally involving a transformer, rectifier, smoothing filter, circuit-breaker or other protection and frequently incorporating electronic regulation. In a full-wave supply, use is made of a full-wave rectifier and filter. (2) U.S. for mains.

power-supply vibrator (*Elec. Eng.*). A vibrator used to provide an a.c. supply from a d.c. source.

power transformer (*Electronics*). Term for the mains transformer providing the high tension and filament supplies in electronic equipment.

power transistor (*Elec. Eng.*). One capable of being used at power rating of greater than about 10 watts, and generally requiring some means of cooling.

power unit (*Aero.*). An engine (or assembly of engines) complete with any extension shafts, reduction gears, or airscrews. (*Elec. Eng.*) See **power amplifier.**

power valve (*Electronics*). See **output valve.**

pox (*Med.*). *Pl.* of *pock* (q.v.); hence popular names for diseases characterized by pustules, e.g., *chickenpox, smallpox*; specifically (vulgar), syphilis.

pox viruses (*Bot.*). A group of fairly large viruses, brick-shaped in shadow-cast electron micrographs, usually characterized by the formation of cytoplasmic inclusion bodies in the cells they invade; usually cause skin lesions, e.g., smallpox virus.

Poynting-Robertson effect (*Astron.*). The combined effect of solar radiation and of relativity in causing small particles to move into smaller orbits and spiral into the sun; the dust in the region of the sun is thereby removed in the course of time and must be constantly replenished, from the débris of comets or of asteroid collisions.

Poynting's theorem (*Elec. Eng., Radio*). That which shows that the rate of flow of energy through a surface is equal to the surface integral of the Poynting vector formed by the components of field lying in the plane of the surface. Used for calculating the power radiated from an antenna.

Poynting vector (*Elec. Eng.*). One whose flux, through a surface, represents the instantaneous electromagnetic power transmitted through the surface. Equal to vector cross product of the electric and magnetic fields at any point. In electromagnetic wave propagation, where these fields have complex amplitudes, half the vector product of the electric field intensity and the complex conjugate of the magnetic field intensity is termed the *complex Poynting vector*, and the real part of this gives the time average of the power flux.

pozzuolana, pozzolana (*Civ. Eng., Geol.*). A volcanic dust, first discovered at Pozzuoli in Italy, which has the effect, when mixed with mortar, of enabling the latter to harden either in air or under water.

PPC (*Build.*). Abbrev. for *plain plaster cornice.*

ppi (*Weaving*). Abbrev. for *picks per inch.* See **pick.**

PPI (*Radar*). Abbrev. for *plan-position indicator.*

***p-p* junction** (*Electronics*). One between **p-type** crystals having different electrical properties.

ppm (*Chem.*). An abbrev. for *parts per million.*

PPQ bar (*Zool.*). See **pterygopalatoquadrate bar.**

P.Q. developer (*Photog.*). One compounded of pyrazolidone and hydroquinone, with characteristics similar to M.Q.

Pr (*Chem.*). (1) The symbol for *praseodymium.* (2) The propyl radical C_3H_7—.

practical units (*Elec.*). Obsolete system of electrical units, whereby the ohm, ampere, and volt were defined by physical magnitudes. Replaced now by the SI system, in which these units are defined in terms of arbitrarily fixed units of length, mass, time and electric current.

prae-. Prefix. See **pre-.**

praecoces (*Zool.*). Birds which when hatched have a complete covering of down and are able at once to follow the mother on land or into water to seek their own food. Cf. *altrices.*

Praesepe (*Astron.*). A well-known open cluster in Cancer.

praetarsus (*Zool.*). A terminal outgrowth of the tarsus in some Insects.

Prandtl number (*Chem. Eng.*). A dimensionless group much used in heat exchange calculations. It is

$$\frac{\text{specific heat capacity at constant pressure} \times \text{dynamic viscosity}}{\text{thermal conductivity}}.$$

prase (*Min.*). A translucent and dull leek-green variety of *chalcedony* (q.v.).

praseodymium (*Chem.*). A metallic element, a member of the rare earth group. Symbol Pr, at. no. 59, r.a.m. 140·9077, rel. d. 6·48, m.p. 940°C. It closely resembles neodymium and occurs in the same minerals.

Pratt truss (*Eng.*). See **Whipple-Murphy truss.**

PRBS. Abbrev. for *pseudo random binary sequence*.

preabdomen (*Zool.*). In Scorpions, the broad anterior part of the abdomen. Cf. *postabdomen*.

preadaptation (*Biol.*). The property of an organism to adjust to a different environment. (*Zool.*) Change of structure preceding appropriate change of habit.

preamplifier (*Acous.*). Amplifier with at least a stage of valve or transistor gain following a high-impedance source from which the level is too low for line transmission and clearance above noise level.

prebaratic chart (*Meteor.*). A chart of the *meteorological elements* (q.v.) which are expected to exist in the near future, probably in 3, 6, or 12 hr time. A forecast weather chart.

Pre-Cambrian (*Geol.*). The era of geological time which preceded the Cambrian age, being the oldest yet defined.

precast (*Civ. Eng.*). Said of concrete blocks, etc., which are cast separately before they are fixed in position.

precast stone (*Civ. Eng.*). See **reconstructed stone.**

precaval vein (*Zool.*). The anterior vena cava conveying blood from the head and neck to the right auricle. Called **superior vena cava** in Man.

precession (*Eng.*). An effect exhibited by a rotating body, such as a gyroscope, when a torque is applied to it in such a way as to tend to change the direction of its axis of rotation. If the speed of rotation and the magnitude of the applied torque are constant, the axis slowly generates a cone as its precessional motion. (*Maths.*) A regular cyclic motion of a dynamical system in which, with suitably chosen coordinates, all except one remain constant, e.g., the regular motion of the inclined axis of a top around the vertical.

precession of the equinoxes (*Astron.*). The westward motion of the equinoxes caused mainly by the attraction of the sun and moon on the equatorial bulge of the earth. This *luni-solar precession* together with the smaller planetary precession combine to give the general precession amounting to 50″·27 per annum. The equinoxes thus make one complete revolution of the ecliptic in 25 800 years, and the earth's pole turns in a small circle of radius 23°27′ about the pole of the ecliptic, thus changing the coordinates of the stars.

prechordal (*Zool.*). Anterior to the notochord or to the spinal cord.

precipitation (*Chem.*). The formation of an insoluble solid by a reaction which occurs in solution. It is widely used for the separation and identification of substances in chemical processes and analyses. *n.* and *v.* **precipitate.** (*Meteor.*) Moisture falling on the earth's surface from clouds; it may be in the form of rain, hail, or snow. (*San. Eng.*) The process of assisting the settlement of suspended matters in sewage by the addition of chemicals to the sewage before admission to the sedimentation tanks.

precipitation hardening (*Met.*). In a metal or alloy, precipitation of one phase in the lattice of another of different ionic diameter. This keys the structure against 'creep' or slip under stress. Produced by precipitation during cooling of a super-saturated solution. See **ageing, temper-hardening.**

precipitin (*Chem., Med.*). An antibody substance analogous to *agglutinin* (q.v.), but its action characterized by clouding and precipitation.

precipitinogen (*Chem., Med.*). The substance which, introduced into the blood plasma, calls forth the specific precipitin.

precise levelling (*Surv.*). Particularly accurate levelling in which the allowable discrepancy between two determinations of the level difference between two bench marks M km apart is very low—of the order of $0{\cdot}003\ \sqrt{M}$ m or less.

precision aids (*Typog.*). A variety of specially-designed equipment to ensure precision and expedite the work throughout the preparation of letterpress printing surfaces.

precision-approach radar (*Aero.*). A primary radar system which shows the exact position of an aircraft during its approach for landing. Abbrev. **PAR.**

preclimax (*Ecol.*). A seral stage which just precedes the climax.

pre-combustion chamber (or **antechamber**) (*I.C. Engs.*). A small chamber formed in the cylinder-head of some compression-ignition engines into which the oil fuel is injected at the end of the compression stroke. The high pressure caused by the partial combustion of the fuel expels the rich mixture through a neck or perforated throat plate into the engine cylinder, where combustion is completed. Derived from the oil engine of Akroyd Stuart (1892).

pre-conduction current (*Telecomm.*). Low anode current when discharge is not self-sustaining.

preconscious (*Psychol.*). (1) That part of the mind containing mental processes of which the individual is not actually aware at a given moment, but which can be brought into consciousness by mental effort. (2) That part of the mind which is in contact with reality and obeys the reality principle.

precoracoid (*Zool.*). An anterior ventral bone of the pectoral girdle in Amphibians and Reptiles, corresponding to the epicoracoid of Monotremes.

precordial (*Anat.*). Situated or occurring in front of the heart.

precoxa (*Zool.*). See **pleuropodite.**

predation (*Ecol.*). A form of species interaction in which an individual of one species of animal (the *predator*) directly attacks, kills, and eats, one of another species (the *prey*). The predator is usually larger than its prey. Cf. *parasite*.

predator (*Ecol.*). See **predation.**

predator chain (*Ecol.*). A type of food chain starting from a plant base and then proceeding from smaller to larger animals. Cf. *parasite chain, saprophyte chain*.

predazzite, pencatite, brucite-marble (*Geol.*). Mixtures of brucite, calcite, periclase, and hydromagnesite, found originally at Predazzo, Italy; formed from magnesian limestone by contact-metamorphism. See **brucite-marble, pencatite.**

predentin(e) (*Histol., Physiol.*). The soft primitive dentin, composed of reticular Korff's fibres, which later becomes calcified to form dentin(e).

predetermined motion time system (*Work Study*). A work measurement technique whereby times established for basic human motions are used to build up the time for a job at a defined level of performance. Abbrev. **P.M.T.S.**

pre-distorting network (*Telecomm.*). A network which anticipates subsequent frequency distortion in the transmission path, as along a line, so that the line distortion has not entirely to be compensated at the receiving end.

pre-distortion (*Telecomm.*). The principle of altering the response of a circuit to compensate, fully or partially, anticipated distortion; the aim is to make transmission as high as practicable above the anticipated noise level.

Preece's formula (*Elec. Eng.*). A formula stating that the fusing current of a wire is proportional to the 1·5th power of the rated current.

pre-echo (*Acous.*). The weak prior impression of an oncoming loud sound which results from a slight deformation of an adjacent groove on a record, or transfer of magnetic signal between layers of tape on a reel (see **print through**).

pre-eclampsia (*Med.*). See **pregnancy toxaemia.**

pre-emphasis (*Acous.*). Marked alteration of response at the beginning of a part of a transmission system (as in magnetic recording and reproduction, frequency modulation and demodulation, disk-recording and reproduction), for a technical reason, such as noise level, restriction of amplitude. See **de-emphasis.**

preen gland (*Zool.*). See **oil gland.**

prefabricated building (*Build.*). Building for which walls, roofs, and floors are constructed off the site. See **system building.**

prefading (*Acous.*). Listening to programme material and adjusting its level before it is faded up for transmission or recording.

preferential mating (*Zool.*). The theory, postulated by Darwin, that if there are two or more rival males unequally endowed, the female will exercise the prerogative of selection.

preferred numbers, sizes, values (*Eng., etc.*). Where a range of components (e.g., screws, resistors) is made for use in manufactured articles, and they are identical except for one variable (e.g., diameter, resistance), their values usually follow a geometric progression, rounded at each stage to a convenient *preferred* value, giving a roughly constant proportional increment of variable from one component to the next. E.g., resistors, accurate to $\pm 10\%$, are valued 10, 12, 15, 18, 22, . . ., 68, 82, 100 ohms, each value being approx. 20% higher than its predecessor.

preferred orientation (*Met.*). During slip, metal crystals change their orientation; when a sufficient amount of deformation has been performed, the random orientation of the original crystals is converted into an arrangement in which a certain direction in all the crystals is parallel to the direction of deformation.

prefloration (*Bot.*). Aestivation.

prefoliation (*Bot.*). Vernation.

preformed precipitate (*Nuc. Eng.*). A precipitate which is used for the co-separation of a tracer element and has been formed before mixing with the tracer to be adsorbed.

preformed rope (*Cables*). Wire rope, each strand of which is correctly bent before being laid up, thus minimizing risk of kinking.

prefrontal (*Zool.*). In some Vertebrates, a paired lateral membrane bone of the orbital region of the skull. (*Surg.*) *Prefrontal lobotomy.* See **leucotomy.**

preganglionic (*Zool.*). Said of neurones and nerve fibres of the autonomic nervous system. The first (preganglionic) neurone originates in the central nervous system (brain stem and spinal cord) and its medullated axon, the *preganglionic fibre*, terminates at a synapse in a ganglion, either part of the paravertebral chain of ganglia or situated close to or within the innervated organ (peripheral ganglia). The ganglion cell with its unmyelinated axon constitutes the second or postganglionic neurone, and its axon (*postganglionic fibre*) innervates the effector organ.

pregnancy (*Med.*). Gestation. The state of being with child.

pregnancy test (*Med.*). Any physiological test diagnosing pregnancy at an early stage and with a very high degree of certainty. Earlier methods are described at *Aschheim-Zondek test* (q.v.); subsequently male and female toads have been used. More recently haemagglutination inhibition tests have been employed; these afford immediate results.

pregnancy toxaemia (*Med.*). Pre-eclampsia. A condition which may occur late in pregnancy, characterized by swollen ankles, increased weight, albuminuria, and heightened blood pressure. Unless treated promptly, may develop into the more serious *eclampsia.* (*Vet.*) Twin lamb disease. An acute, usually fatal, metabolic disease of unknown cause affecting ewes towards the end of pregnancy, in which ketone bodies accumulate in the tissues and are excreted in the urine and breath. Characterized by depression and a variety of nervous symptoms, and associated with a low caloric intake in late pregnancy, especially when twin lambs are present.

pregnanediol (*Chem.*). Steroid, apparently the end product of the metabolism of progesterone, found in the urine during pregnancy and at one stage of the menstrual cycle. It is excreted partly free and partly combined with glycuronic acid.

pregs (*Min. Proc.*). See **pregnant solution.**

prehallux (*Zool.*). In Amphibians and Mammals, a rudimentary additional digit of the hind limb.

prehalteres (*Zool.*). The squamae of *Diptera.*

preheating time (*Electronics*). Minimum time for heating a cathode before other voltages are applied, thus ensuring full emission. Automatic delays are often incorporated in large amplifiers and radio transmitters.

prehensile (*Zool.*). Adapted for grasping.

preheterokinesis (*Cyt.*). A form of reduction division in which the sex chromosome passes undivided to 1 pole in the first spermatocyte division. Cf. *postheterokinesis.*

prehnite (*Min.*). A hydrous pale-green and usually fibrous silicate of calcium and aluminium, crystallizing in the orthorhombic system. It occurs, with zeolites, in geodes in altered igneous rocks and in calcareous metamorphic rocks.

preignition (*I.C. Engs.*). The ignition of the charge in a petrol-engine cylinder before normal ignition by the spark; caused by overheated plug-points, the presence of incandescent carbon, etc.

preimaginal conditioning (*An. Behav.*). A conditioning effect, such as a response to a particular odour, carried over from the larval stage of an insect to the adult stage.

preischiopodite (*Zool.*). The distal 1 of 2 joints into which the basipodite of *Crustacea* is sometimes divided.

prelacteal (*Zool.*). In Mammals, said of teeth developed prior to the formation of the milk dentition.

preliminary matter (*Typog.*). The pages of a book preceding the actual text. The order should be half-title, frontispiece or 'advertisement', title (at the back of this, number of editions, imprint, etc.), dedication, preface, contents, list of illustrations, introduction. Frequently abbrev. to **prelims.**

premaxillary (*Zool.*). A paired membrane bone of the Vertebrate skull which forms the anterior part of the upper jaw: anterior to the maxilla. Also **premaxilla.**

premeiotic mitosis (*Cyt.*). The nuclear division immediately preceding the organization of nuclei which will divide by meiosis.

premolars (*Zool.*). In Mammals, the anterior grinding or cheek teeth, which are represented in the milk dentition.

premorse (*Bot.*). Looking as if the end had been bitten off.

prenasal (*Zool.*). A small bone at the tip of the nose in Pigs and a few other forms.

preoperculum (*Zool.*). In Fish, an anterior membrane bone forming part of the gill cover.

preoral pit (*Zool.*). In embryonic *Cephalochorda*, a depression opening to the exterior which originates from the left of a pair of pouches which grow out from the anterior end of the alimentary canal. It afterwards forms the groove of Hatschek and the wheel organ.

preorbital (*Zool.*). In some Fish, a membrane bone situated in front of the orbit: anterior to the orbit.

préparation mécanique (*Min. Proc.*). See **mineral processing.**

prepared (*Textiles*). In lace manufacturing, bobbin yarn made smooth by special calendering between rollers in hank form.

prepatagium (*Zool.*). In Birds, a cutaneous expansion between the forearm and the upper arm. *adj.* **prepatagial.**

prepayment meter (*Elec. Eng., etc.*). A meter used in connexion with electricity or gas supplies. It is designed so that the insertion of a coin or coins operates a switch or other device, and allows a certain predetermined quantity of electrical energy or gas to be used.

prepollex (*Zool.*). In some Vertebrates (Amphibians, Mammals), a rudimentary extra digit of the forelimb.

prepotency (*Bot.*). The ability of some pollen to bring about fertilization more readily than other pollen. (*Zool.*) The capacity of one parent to transmit more characteristics to the offspring than the other parent. *adj.* **prepotent.**

prepubic (*Zool.*). Pertaining to the anterior part of the pubis: in front of the pubis; as bony processes in some Marsupials and Rodents.

prepubis (*Zool.*). In some Fish, an anterolateral process of the pubis.

prepuce (*Zool.*). In Mammals, the loose flap of skin which protects the glans penis. *adj.* **preputial.**

prepupa (*Zool.*). In the development of Insects, a stage just before the last larval ecdysis, when the larval cuticle has separated from the hypodermis, and the developing pupa lies within the persistent larval cuticle, often deforming it and changing the external appearance of the larva, which is generally quiescent and does not feed. Not, however, a distinct instar.

presbyope (*Med.*). One suffering from presbyopia.

presbyophrenia (*Psychiat.*). Mental illness of old age characterized by disorientation, amnesia, and confabulation.

presbyopia (*Med.*). Long-sightedness and impairment of vision due to loss of accommodation of the eye in advancing years.

prescoring (*Cinema.*). In motion-picture production, the use of previously recorded sound, such as a musical sequence, for cueing the action of dancers or singers.

prescutum (*Zool.*). In Insects, the most anterior of 3 sclerites into which the notum is typically subdivided.

Presdwood (*Build.*). TN for a strong building board having water-resisting properties.

preselection (*Radio*). Use of selective circuits in early stages of a supersonic heterodyne receiver, prior to frequency changer, to reduce second channel interference and cross-modulation, and increase the overall sensitivity and selectivity. (*Teleph.*) Type of automatic switching used in telephone exchanges.

preselector gearbox (*Autos.*). A gearbox, generally epicyclic, in which the gear ratio is selected, before it is actually required, by a small lever, being afterwards engaged by pressure on a pedal.

presentation (*Med.*). The relation which the long axis of the foetus bears to that of the mother; various presentations are defined in terms of the presenting part, as *shoulder presentation, breech presentation*, etc.

presentation time (*Biol.*). The shortest period during which an organism must be exposed to a stimulus in order that response may subsequently occur.

preset (*Electronics*). Same as **quiescent.**

preset guidance (*Aero.*). The guidance of controlled missiles by a mechanism which is set before launching and is subsequently unalterable.

preset iris (*Photog.*). Device which enables the aperture to be selected before focusing at maximum aperture, thus obviating the necessity for resetting visually.

preset oscillator (*Radio*). One in which frequency is determined by one or more control knobs acting on a resistance-capacitance or a tuned circuit. Slight drifts arise from mains voltage or temperature variations, but can be made negligible in effect.

presphenoid (*Zool.*). A medium cartilage bone of the Vertebrate skull forming part of the floor of the brain case in front of the basisphenoid.

press (*Eng.*). A machine used for applying pressure to a workpiece, via a tool, usually for the purpose of carrying out cold-working operations like cutting, bending, drawing, and squeezing. (*Print.*) (1) Any hand-operated machine used for proofing or printing small runs. (2) A general term for the printing stage of a job; exemplified in such phrases as 'going to press', 'in the press'. (3) A general term for the printing and publishing industry, particularly newspapers.

press-and-blow machines (*Glass*). Machines in which the *parison* (q.v.) is formed by the pressing action of a plunger forced into a mass of plastic glass dropped in a parison mould; the parison is then blown to the shape of the finished ware in another mould.

pressboard (*Elec. Eng.*). Compressed paper in thick sheets. See **Elephantide pressboard.**

press brake (*Eng.*). A power press akin to a guillotine, used mainly for producing long bends, as in corrugating or seaming, but also for embossing, trimming, and punching.

pressed amber (*Min.*). See **ambroid.**

pressed brick (*Build.*). A high-quality brick moulded under pressure, as a result of which it has sharp arrises and a smooth face, making it especially suitable for exposed surface work.

pressel switch (*Elec. Eng.*). See **pendant switch.**

presse-pâte (*Paper*). A machine which converts *half-stuff* into a web of pulp, so that the material can be handled more conveniently.

press fit (*Eng.*). A class of fit for mating parts, tighter than a sliding fit and used when the parts do not normally have to move relative to each other.

press forging (*Eng.*). A die-forging process using a vertical press to apply a slow squeezing action to deform the plastic metal, as distinct from the rapid impact employed in drop forging.

pressing (*Acous.*). Disk record formed by pressure, with or without heat; the negative of the recording on a stamper is transferred to a large number of pressings for distribution. (*Textiles*) Final finishing of woollen, worsted, or blended fabrics. Flat bed presses—steam or electrically heated—set the yarns, stabilize stresses and give lustre to the cloth. In lace-making, the process of compression and pressure-steaming to which brass bobbins are subjected as soon as they are wound.

pressing boards (*Print.*). Glazed boards used for removing the impression from printed sheets.

pressor (*Zool.*). Causing a rise of arterial pressure.

presspahn (*Elec. Eng.*). A fibrous insulating material made from wood pulp; used, in thin sheets, for the insulation of electrical equipment; an early variety of pressboard.

press proof (*Print.*). The last proof checked over before going to press.

press rolls (*Paper*). Pairs of heavy rolls on the paper-making machine which press out moisture from the wet web. Before the last pressing, the web is reversed in order to remove felt marks. The top roll is usually of iron, brass or granite, the bottom one is rubber covered. Cf. *suction roll*.

press tool (*Eng.*). A tool for cutting, usually comprising at least a punch and a die, or for forming, or for assembling, used in a manual or in a power press.

pressure. See **stress, atmospheric pressure.**

pressure altitude (*Aero.*). Apparent altitude of the local ambient pressure related to the International Standard Atmosphere.

pressure angle (*Eng.*). In toothed gearing, the angle between the common tangent to the pitch circles of two teeth at their point of contact and the common normal at that point. Two angles are in common use: 14½° and 20°.

pressure cabin (*Aero.*). An airtight cabin which is maintained at greater than atmospheric pressure for the comfort and safety of the occupants. Above 20 000 ft (6000 m) a differential of 6½ lbf/in² (45 kN/m²) is usual, and above 40 000 ft (12 000 m) one of 8¼ lbf/in² (57 kN/m²). Pressurization can be either by a shaft-driven *cabin blower*, or by air bled from the compressor of turbine main engines.

pressure cable (*Cables*). A paper-insulated power cable operated under a hydrostatic pressure (up to 1·5 MN/m²) by means of gas (nitrogen) contained in an outer steel pipe or, in more modern forms, an outer reinforced lead sheath; this minimizes ionization. See also **Oilostatic cable.**

pressure capsule (*Eng.*). See **sylphon bellows.**

pressure circuit (*Elec. Eng.*). See **voltage circuit.**

pressure diecasting (*Met.*). A process by means of which precision castings of various alloys are made by squirting liquid metal under pressure into a metal die. See **diecasting.**

pressure drag (*Aero.*). The resolved component of the pressure due to drag normal to the surface; sum of *form drag* and *induced drag* (qq.v.).

pressure forging (*Eng.*). See **drop forging.**

pressure gauge (*Eng.*). (1) A flattened tube bent to a curve, which tends to straighten under internal pressure, thus indicating, by the movement of an indicator over a circular scale, the fluid pressure applied to it. Also called **Bourdon gauge.** (2) A liquid manometer.

pressure governor (*Gas*). A gas governor which operates to supply gas at constant pressure.

pressure gradient (*Meteor.*). (1) The rate of change of the atmospheric pressure horizontally in a certain direction on the earth's surface as shown by isobars on a weather chart. (2) The rate of change of pressure with distance over the ground, normal to the isobars. The force acting on the air is the *pressure-gradient force*.

pressure-gradient microphone (*Acous.*). One which offers so little obstruction to the passage of a sound wave that the diaphragm, in practice a ribbon, is acted on by the difference in the excess pressures on the two sides, and therefore tends to move with the particle velocity in the wave. Also **velocity microphone.**

pressure head (*Aero.*). A combination of a *static-pressure* and a *Pitot tube* which is connected to opposite sides of a differential pressure gauge, for giving a visual reading corresponding to the speed of an airflow. Sometimes called **pitot-static tube.** (*Hyd.*) A distance representing the height above a datum which would give unit mass of a fluid in a conduit a potential energy equal to its pressure energy. If, at a given point, its pressure is p N/m² and its density is ρ kg/m³, its pressure head is $p/\rho g$ m.

pressure helmet (*Aero., Space*). A flying helmet for the crew of high-altitude aircraft or space craft for use with a *pressure*, or *partial pressure* suit. Usually of plastic, with a transparent facepiece, which may be in the form of a visor, the helmet incorporates headphones, microphone, and oxygen supply, and there is usually a feeding trap near the mouth.

pressure in bubbles (*Phys.*). A spherical bubble of radius r, formed in a liquid for which the surface tension is T, contains air (or some other gas or vapour) at a pressure which exceeds that in the liquid in its immediate vicinity by $\frac{2T}{r}$. The excess pressure within a soap bubble in air is $\frac{4T}{r}$ since the soap film has two surfaces.

pressure jet (*Aero.*). A type of small jet-propulsion unit fitted to the tips of helicopter rotor blades, in which small size (to give low drag for *autorotation* (q.v.)) is of greater importance than the losses due to ejecting the efflux at pressures as high as 2 or 3 atmospheres.

pressure leaching (*Met., Min. Proc.*). Chemical extraction of values from ore pulp in autoclaves, perhaps followed by precipitation as metal or refined salt.

pressure microphone (*Acous.*). Any type which is operated by the excess pressure in a sound wave, as contrasted with a *ribbon* or *pressure-gradient* type. See **capacitor microphone.**

pressure of atmosphere (*Phys.*). See **atmospheric pressure, barometric pressure.**

pressure pad (*Photog.*). The device which keeps the film in a gate so that it remains exactly in focus.

pressure-pattern flying (*Aero.*). The use of barometric pressure altitude to obtain the most favourable winds for long-distance, high-altitude aerial navigation.

pressure ratio (*Aero.*). The absolute air pressure, prior to combustion, in a *gas turbine, pulse-jet*, or *ramjet* divided by the ambient pressure: analogous to the *compression ratio* of a reciprocating engine.

pressure roller (*Eng.*). A roller sometimes used in centreless grinding of short, heavy workpieces to assist rotation of the workpiece.

pressure suit (*Aero., Space*). An airtight fabric suit, similar to that of a diver, for very high altitude and space flight. It differs from the *partial pressure suit* (q.v.) in being loose-fitting, with bellows or other form of pressure-tight joint, to permit limited movement by the wearer. Essential for long flights above 60 000 ft (18 000 m).

pressure-type capacitor (*Elec. Eng.*). One in which the dielectric is an inert gas under high pressure. Useful for high voltage work since it is self-healing in the event of breakdown.

pressure unit (*Acous.*). A metal or plastic dome forming the diaphragm of a small moving-coil loudspeaker unit situated in the throat of a horn, for use at intense acoustic pressures.

pressure vessel (*Eng.*). A container, often made of steel or aluminium, for fluids subjected to pressure above or below atmospheric. Classified as *fired* or *unfired*. Usually cylindrical with dished ends, for ease of construction and support, or spherical, for very high pressures.

pressure waistcoat (*Aero.*). A double-skinned garment, covering the thorax and abdomen, through which oxygen is passed under pressure on its way to the wearer's lungs, to aid breathing at great heights, i.e., above 40 000 ft (12 000 m).

pressure welding (*Eng.*). Welding while the parts to be united are pressed against each other, as in forge welding or in various electrical-resistance welding processes or in cold welding.

pressurized. Fitted with a device that maintains nearly normal atmospheric pressure, e.g., in an aircraft.

pressurized water reactor (*Nuc. Eng.*). One using water cooling at a pressure such that its boiling point is above the highest temperature reached.

presswork (*Print.*). Printing of the job, quality of the result depending on care and attention given; also called **machining**, but not in the case of hand press printing.

presternum (*Zool.*). In *Anura*, an anterior element of the sternum, of paired origin and doubtful homologies; the reduced sternum of whalebone whales; the anterior part of the sternum.

prestoring (*Comp.*). Supplying of data to a computer store in advance of its requirement for a particular program.

prestressed concrete (*Civ. Eng.*). Concrete which is pre-compressed, using high-tensile wires, thereby enabling it to withstand tensile stresses which it would otherwise be unable to do. Cf. *reinforced concrete*.

presynchronization (*Cinema.*). The recording of a sound track for use in cueing artists, motion pictures of whom are to be fitted to the sound track during the process of editing.

presystolic (*Med.*). Pertaining to, or occurring just before, the beginning of the systole of the heart, e.g., *presystolic* murmur.

pretarsus (*Zool.*). In some Insects, a terminal outgrowth of the tarsus.

pre-TR cell (*Radar*). A gas-filled RF switching valve which protects the *TR tube* in a radar receiver from excessive power. Also acts as a block to receiver frequencies other than the fundamental.

pretrematic (*Zool.*). Anterior to an aperture as (in *Selachii*) that branch of the 9th cranial nerve which passes anterior to the first gill cleft. Cf. *post-trematic*.

pretzel (*Nuc. Eng.*). Shape of the stellarator, in which a torus is folded into a figure-8, the end loops being flat and in parallel planes.

preventitious bud (*Bot.*). A dormant bud which may, if conditions allow, produce an epicormic branch.

preventive choke-coil (*Elec. Eng.*). A choking coil connected between the two halves of the moving-contact used in varying the tapping of a transformer or battery; its purpose is to reduce the circulating current which flows as a result of the short-circuiting of the turns or cells between adjacent tappings by the moving-contact, as it travels from one fixed-contact to the next.

preventive resistance (*Elec. Eng.*). A resistance connected between the two halves of the moving-contact used in varying the tapping of a transformer or battery; its purpose is to reduce the circulating current which flows as a result of the short-circuiting of the turns or cells between the adjacent tappings, as the moving-contact travels from one fixed-contact to the next.

prevernal (*Bot.*). Flowering early in the year. (*Ecol.*) Pertaining to the early part of spring, when it is found convenient to divide this season into two. Cf. *vernal*.

prevernal aspect (*Bot.*). The condition of the vegetation of a community early in the year.

prevomer (*Zool.*). In Monotremes, a median dumb-bell-shaped membrane bone of uncertain homologies lying in the floor of the nasal cavities. Also called **dumb-bell bone, paradoxical bone.**

prey (*Ecol.*). See **predation.**

prezygapophysis (*Zool.*). A facet or process on the anterior face of the neurapophysis of a vertebra, for articulation with the vertebra next in front. Cf. *postzygapophysis*.

PRF (*Telecomm.*). See **pulse repetition frequency.**

priapism (*Med.*). Abnormally, and often painfully, persistent erection of the penis unaccompanied by sexual desire; due either to a lesion of the penis or to nervous disorder.

Priapulida (*Zool.*). A phylum of unsegmented wormlike animals, living in mud. They have a large haemocoelic body cavity, a straight gut with an anterior mouth and a posterior anus. The posterior end of the body bears a series of caudal appendages, the nervous system is not separated from the epidermis and the urinogenital system is simple, with solenocytes.

Price's guard-wire (*Elec. Eng.*) A conductor placed around the edge of a piece of insulating material under test; it is arranged to be at the same potential as the surface of the material to prevent leakage current from the surface to earth.

pricking-up (*Build.*). The operation of scoring the surface of the first coat of plaster to provide a key for the next: the whole operation of laying and scoring such a coat.

prickle (*Bot.*). A hard epidermal appendage resembling a thorn but not containing woody tissue.

prickle cells (*Zool.*). In stratified epithelium, the cells of the deeper layers, which are connected to one another by intercellular fibrils giving them a prickly appearance.

prickly heat (*Med.*). See **miliaria.**

prick punch (*Horol.*). A punch used to locate the centre of holes in a plate prior to drilling.

prill (*Met.*). (1) To make granular or crystalline solids fluid, e.g., for extraction from slag. (2) To turn into pellet form by melting and letting the drops solidify in falling. (3) Bullion bead produced by cupellation of lead button during fire assay of gold or silver.

prilocaine (*Med.*). Local anaesthetic, α-propylaminopropio-*o*-toluidide hydrochloride, similar in potency to, but longer-lasting than, *lignocaine*.

prima (*Print.*). The first word of the page (or sheet) following the one being read. [Rare.]

primacord fuse (*Mining, etc.*). Fuse based on *pentaerythritol tetranitrate*, with detonating effect. Speed of detonation, 7000 m/s.

primaquine (*Med.*). Anti-malarial drug, 8-(4-amino-1-methylbutamino)-6-methoxyquinoline diphosphate, used prophylactically and therapeutically.

primaries (*Zool.*). In Birds, the remiges attached to the manus.

primary (*Bot., Zool.*). Original, first formed; as *primary meristem, primary body-cavity*: principal, most important; as *primary feathers, primary axis*. (*Chem.*) A substance which is obtained directly, by extraction and purification, from natural raw material; **e.g., benzene,**

phenol, anthracene are coal-tar *primaries*. Cf. *intermediate*. (*Elec. Eng.*) See **primary winding**.

Primary (*Geol.*). An obsolete synonym for *Palaeozoic*.

primary acids (*Chem.*). Acids in which the carboxyl group is attached to the end carbon atom of a chain, i.e., to a $—CH_2$ group.

primary additive colours (*Light*). Minimum number of spectral colours (red, green, blue) which can be adjusted in intensity and, when mixed, visually make a match with a given colour. This match with real colours cannot be perfect, since one primary may have to be negative in intensity. A typical set of primary additive spectral colours are, in nanometres: red (640), green (537), blue (464). For colour TV and photography, original colour has to be separated into such arbitrary components.

primary alcohols (*Chem.*). Alcohols containing the group $—CH_2{\cdot}OH$. On oxidation, they form aldehydes and then acids containing the same number of carbon atoms as the alcohol.

primary amines (*Chem.*). Amines containing the amino group $—NH_2$. Primary amines are converted into the corresponding alcohol by the action of nitrous acid, nitrogen being eliminated.

primary axis (*Bot.*). (1) The main shoot of a plant. (2) The main stalk of an inflorescence.

primary body (*Bot.*). That part of the plant body formed directly from cells cut off from the apical meristems.

primary body cavity (*Zool.*). The blastocoel or segmentation cavity formed during cleavage, or that part of it which is not subsequently obliterated by mesenchyme.

primary bow (*Meteor.*). See **rainbow**.

primary cell (*Elec. Eng.*). Voltaic cell in which chemical energy of constituents is changed to electrical energy when current is permitted to flow. Such a cell cannot be recharged electrically because of the irreversibility of chemical reaction occurring therein. It may be used as a sub-grade standard cell for calibration purposes.

primary cell wall (*Bot.*). The cell wall that surrounds the protoplast until that is approximately mature. It is thin, usually nonstratified, contains much pectin material and not so much cellulose, and later persists as the middle lamella.

primary coil (*Elec. Eng.*). A coil, forming part of an electrical machine or piece of apparatus, in which flows a current setting up the magnetic flux necessary for the operation of the machine or apparatus.

primary colours (*Paint.*). For pigments these are red, yellow, and blue. See **primary additive colours, primary subtractive colours.**

primary constants (*Elec. Eng.*). Those of capacitance, inductance, resistance, and leakance of a conductor to earth (coaxial or concentric) or to return (balanced) conductor, per unit length of line.

primary crushing (*Min. Proc.*). Reduction of run-of-mine ore as severed to somewhere below 6-in. (15 cm) diameter lump, performed in jaw or gyratory breakers.

primary current (*Electronics*). That formed by *primary electrons*, as contrasted with secondary electrons, which reduce or even reverse it.

primary electrons (*Electronics*). (1) Those incident on a surface whereby secondary electrons are released (see also **primary ionization**). (2) Those released from atoms by internal forces and not by external radiation as with secondary electrons.

primary emission (*Electronics*). Electron emission arising from the irradiation (including thermal heating) or the application of a strong electric field to a surface.

primary flexure (*Zool.*). The flexure of the mid-brain by which, in Vertebrates, the forebrain and its derivatives are bent at a right angle to the axis of the rest of the brain.

primary flow (*Electronics*). That of *carriers* when they determine the main properties of a device.

primary gneissic banding (*Geol.*). This is exhibited by certain igneous rocks of heterogeneous composition. Possible causes are admixture of two magmas only partly miscible, injection of magma along bedding or foliation planes in the country rocks, or selective mobilization of a rock under metamorphism.

primary increase (*Bot.*). Increase in the size of a stem or root not brought about by the addition of cells from a cambium.

primary ionization (*Nuc.*). (1) In collision theory, the ionization produced by the primary particles, in contrast to total ionization, which includes the *secondary ionization* produced by delta-rays. (2) In counter tubes, the total ionization produced by incident radiation without gas amplification.

primary lamella (*Bot.*). The first-formed layer of the wall of a spore.

primary luminous standard (*Light*). A standard of luminous intensity which is reproducible from a given specification.

primary medullary ray (*Bot.*). A vascular ray passing radially from the pith to the cortex.

primary meristem (*Bot.*). A meristem derived immediately from a promeristem and giving rise to cells which build up the primary body of the plant.

primary metal (*Met.*). Ingot metal produced from newly smelted ore, with no addition of re-circulated scrap.

primary nitro-compounds (*Chem.*). Nitro-compounds containing the group $—CH_2{\cdot}NO_2$.

primary node (*Bot.*). The node at which the cotyledons are inserted.

primary orientation (*An. Behav.*). The normal stance of an animal, i.e., the position it adopts when inactive, and from which active movements generally start. Cf. *secondary orientation.*

primary phloem (*Bot.*). The phloem formed from a procambial strand and present in a primary vascular bundle. It consists of protophloem and metaphloem.

primary productivity (*Ecol.*). In any ecological system, the rate at which energy is stored by photosynthetic and chemosynthetic activity of the producer organisms.

primary radar (*Radar*). Basic radar in which target reflects some of transmitted energy, which is detected at the transmitter.

primary radiation (*Phys.*). Radiation which is incident on the absorber, or which continues unaltered in photon energy and direction after passing through the absorber. Also **direct radiation.**

primary sere (*Bot.*). A plant succession beginning on land which has never borne vegetation in recent geological time.

primary service area (*Radio*). That in which groundwave reception of a broadcast transmission is sufficient, in relation to interference and ionospheric reflections, to give consistently good reproduction of programmes without fading.

primary solid solution (*Met.*). A constituent of

alloys that is formed when atoms of an element *B* are incorporated in the crystals of a metal *A*. In most cases solution involves the substitution of *B* atoms for some *A* atoms in the crystal structure of *A*, but in a few instances the *B* atoms are situated in the interstices between the *A* atoms.

primary standard. A standard agreed upon as representing some unit (e.g., length, mass, e.m.f.) and carefully preserved at a national laboratory. Cf. *secondary standard*.

primary store (*Comp.*). (1) The main store built into a computer, not necessarily the fast access store. (2) Relatively small immediate or very rapid access store incorporated in some computers for which the main memory is a slower secondary store.

primary stress (*Eng.*). An axial or direct tensile or compressive stress, as distinct from a bending stress resulting from deflection.

primary structure (*Aero.*). All components of an aircraft structure, the failure of which would seriously endanger safety, e.g., wing or tail plane spars, main fuselage frames, engine bearers, portions of the skin which are highly stressed.

primary subtractive colours (*Light, etc.*). Minimum number of spectral colours (cyan, magenta, yellow), which when subtracted in the right intensity from a given white, result in a match with a given colour. These are complementary to the primary additive colours and are used in printing colour films, e.g., Technicolor.

primary succession (*Ecol.*). A succession beginning on an area not previously occupied by a community, e.g., a newly exposed rock or sand surface. Cf. *secondary succession*.

primary thickening (*Bot.*). The first layers of wall material to be laid down on the very young cell wall, often rich in pectin materials.

primary tissue (*Bot.*). Tissue formed from cells derived from primary meristems.

primary trisomic (*Bot.*). A plant which has the ordinary diploid chromosome complement, together with one extra chromosome.

primary vascular bundle (*Bot.*). A vascular bundle formed from a procambial strand.

primary voltage (*Elec. Eng.*). The voltage applied to the primary winding of a transformer or induction motor.

primary winding (*Elec. Eng.*). (1) The winding of a transformer which is on the input side. (2) The input winding, usually on the stator, of an induction motor. Sometimes called **primary**.

primary wood (or **xylem**) (*Bot.*). The xylem formed from a procambial strand and present in a primary vascular bundle. It consists of protoxylem and metaxylem.

Primates (*Zool.*). An order of Mammals. Dentition complete, but unspecialized; brain, especially the neopallium, large and complex; pentadactyl; eyes well developed and directed forwards, the orbit being closed behind by the union of the frontal and jugal bones; basically arboreal animals; uterus a single chamber; few young produced, parental care lasting a long time after birth. Lemurs, Tarsiers, Monkeys, Apes, and Man.

Primaton overlay (*Print.*). A mechanical overlay produced by dusting a proof with special powder and heating the sheet to fuse the powder.

prime mover (*Eng.*). An engine or other device by which a natural source of energy is converted into mechanical power. See **gas engine, internal-combustion engine, oil engines, petrol engine, steam engine, steam turbine, water turbine.**

prime number (*Maths.*). A natural number other than 1, divisible only by itself and 1. Or, any natural number which has precisely two divisors. The following numbers are prime: 2, 3, 5, 7, 11, 13, ... 37, ... 5521. Every natural number greater than 1 may be resolved uniquely into a product of prime numbers; e.g., $8316 = 2^2 \times 3^3 \times 7 \times 11$. In the case of a prime number *p*, the *product* has to be interpreted as *p* itself.

primer (*Mining*). Cartridge in which detonator is placed in order to initiate explosion of string of high-explosive charges in bore hole. (*Paint.*) The first coat of paint on a bare surface, formulated to provide a suitable surface for the next coats. For wood, the absorption of the material must be allowed for, for metals, their tendency to corrode, and the high alkalinity in the case of concrete. Special primers may have to be formulated for unusual surfaces.

prime vertical (*Astron.*). The great circle passing through the zenith and cutting the observer's horizon in the east and west points.

primigravida (*Med.*). A woman who is pregnant for the first time.

primine (*Bot.*). The outer integument of an ovule.

priming (*Eng.*). (1) The delivery by a boiler of steam containing water in suspension, due to violent ebullition or frothing. (2) The operation of filling a pump intake with fluid to expel the air. (3) The operation of injecting petrol into an engine cylinder to assist starting.

priming illumination (*Electronics*). A steady illumination applied to a photocell to bring it to the most sensitive condition.

priming pump (*Aero.*). A manual, or electric, fuel pump which supplies the engine during starting where an injection carburettor or fuel injection pump is fitted.

priming valve (*Eng.*). A valve fitted on the suction side of a pump to assist in priming.

primipara (*Med.*). A woman who gives, or has given, birth to a child for the first time.

primitive (*Bot., Zool.*). Original, first-formed, of early origin; as the *primitive streak*. (*Maths.*) See **complete primitive.**

primitive groove (*Zool.*). A shallow groove along the median line of the primitive streak.

primitive knot (*Zool.*). A small elevation in front of the primitive pit representing the dorsal lip of the blastopore.

primitive pit (*Zool.*). A small depression at the front end of the primitive streak.

primitive streak (*Zool.*). In developing Birds and Reptiles, a thickening of the upper layer of the blastoderm along the axis of the future embryo; represents the fused lateral lips of the blastopore.

primordial cell (*Bot.*). A cell which has not yet formed a cell wall.

primordial germ cells (*Zool.*). In the early embryo, cells which will later give rise to the gonads.

primordial leaf (*Bot.*). (1) The next leaf formed after the cotyledons. (2) The very small protuberance from which a leaf starts its development.

primordial meristem (*Bot.*). See **promeristem.**

primordial utricle (*Bot.*). In a mature cell, the peripheral layer of protoplasm, and the vacuole which it encloses.

primordium (*Bot., Zool.*). The earliest recognizable rudiment of an organ or structure in development; also **anlage.** *adj.* **primordial.**

primrose chrome (*Paint.*). A chrome yellow pigment consisting of lead chromate with as much as 40% of lead sulphate and lead carbonate, and consequently a clean but pale yellow colour of high opacity and good light fastness, etc. See **chrome yellows.**

primuline (*Chem.*). A thiazole colouring-matter, obtained by heating sulphur with 4-toluidine, which dyes cotton directly in primrose-yellow shades. Its formula is:

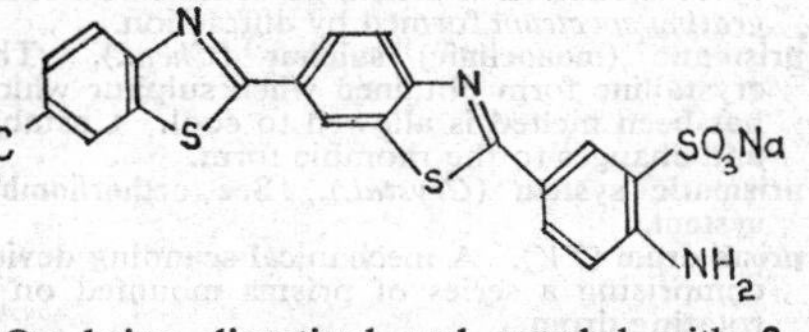

On being diazotized and coupled with 2-naphthol, it yields primuline red.

primuline process (*Photog.*). See **diazo process.**

princess posts (*Carp.*). Additional vertical ties on each side of the queen posts, introduced to give extra support to the tie beam in roofs of greater span than about 45 ft (14 m).

principal (*Build., Civ. Eng.*). See **roof truss.**

principal axes of a body (*Maths.*). At any given point O: three mutually perpendicular axes $Oxyz$ such that the three products of inertia about the coordinate planes are all zero. The principal axes at the centre of gravity are the axes of symmetry if any exist.

principal axis (*Optics.*). See **optic axis.** (*Radio*) Direction of maximum sensitivity or response for a transducer or antenna.

principal direction (*Maths.*). See **curvature (3).**

principal normal (*Maths.*). See **moving trihedral.**

principal part (*Maths.*). Of a function $f(z)$ at a singularity: see **pole.**

principal planes of a lens (*Light*). Planes drawn through the principal points, at right angles to the principal axis.

principal points of a lens (*Light*). Two points on the principal axis of a lens or system of lenses such that, if the object distance u be measured from one of them, and the image distance v from the other, the simple lens formula $\frac{1}{f}=\frac{1}{v}-\frac{1}{u}$ may be used to give the equivalent focal-length f of the lens. See **nodal points.**

principal quantum number (*Nuc.*). That which specifies the shell the electron occupies.

principal radius of curvature (*Maths.*). See **curvature (3).**

principal rafter (*Build., Civ. Eng.*). A rafter forming part of the roof truss proper and supporting the purlins.

principal ray (*Optics*). From an object point lying off the axis, the ray passing through the centre of the entrance pupil of the system.

principal series (*Light*). Series of optical spectrum lines observed in the spectra of alkali metals. Energy levels for which the orbital quantum number is unity are designated **p-levels.**

principal stress (*Eng.*). The component of a stress which acts at right angles to a surface, occurring at a point at which the shearing stress is zero.

principal values of the inverse trigonometrical functions (*Maths.*). Those usually taken are those which contain the smallest numerical value irrespective of its sign.

principle of expectancy (*An. Behav.*). See **expectancy, principle of.**

principle of least action (*Phys.*). Principle stating that the actual motion of a conservative dynamical system between two points takes place in such a way that the action has a minimum value with reference to all other paths between the points which correspond to the same energy.

principle of least constraint (*Mech.*). The motions of any number of interconnected masses under the action of forces deviate as little as possible from the motions of the same masses if disconnected and under the action of the same forces. The motions are such that the constraints are a minimum, the constraint being the sum of the products of each mass and the square of its deviation from the position it would occupy if free.

principle of least time (*Phys.*). See **Fermat's principle of least time.**

principle of relativity (*Phys.*). A universal law of nature which states that the laws of mechanics are not affected by a uniform rectilinear motion of the system of coordinates to which they are referred. Einstein's relativity theory is based on this principle, and on the postulate that the observed value of the velocity of light is constant and is independent of the motion of the observer. See **relativity.**

principle of the equipartition of energy (*Chem.*). The total energy of a molecule in the normal state is divided up equally between its different capacities for holding energy, or degrees of freedom.

print (*Foundry*). See **core prints.** (*Photog.*) The positive obtained by exposing a negative image —either by contact or by projection in an enlarger—on film, or paper, or glass (for preparing slides), with subsequent development.

printed circuit (*Electronics*). Circuit of an equipment, formed into a unit which can be realized by copper conductors laminated on a phenol base. They may be deposited as a powder photographed on a resist and etched, or deposited by electrodeposition. Circuit components are added by hand or dip soldering.

printer (*Comp.*). Apparatus for printing results of computation. *Line* (or *page*) prints whole of line or page at a time; *fly* prints from non-stopping typewheel; *wheel* stops and prints; *stylus, matrix,* or *wire,* synthesizes characters directly or through inked tape (*Samastronic*), or electric typewriter (*IBM*). (*Weaving*) A plain grey cotton cloth made in various qualities, subsequently printed and shipped to Eastern markets. Also called **Burnley printer.**

printergram (*Teleg.*). A telegram sent by teleprinter.

printer plate (*Photog.*). The support for an emulsion which is exposed, and differentially hardened with special developers, while it is being formed into a matrix for 3-colour printing.

printer's ream (*Paper*). 516 sheets, the extra allowance being for spoilage and make-ready. The printer is thus enabled to deliver 500 perfect sheets of printed matter to the ream.

printing. Any process of producing copies of designs or lettering by transferring ink to paper (or other material) from a printing surface. There are 3 main classes according to the method of application of the ink to the printing surface: (*a*) *relief,* or *letterpress,* printing surfaces have the ink-carrying parts in relief, so that rollers deposit ink on these parts only, as in printer's type; (*b*) *planographic* printing surfaces are prepared so that parts accept the ink from the rollers although there is no difference in level; the ink-accepting parts may be greasy, the remainder being moist and ink-rejecting, as in *lithography* (q.v.); (*c*) *intaglio* printing surfaces have the ink-carrying portions hollowed out; the whole surface is covered with ink and then cleaned off, leaving the hollows filled with ink, which is lifted out when the paper is pressed into contact. All classes can be adapted for use with a cylindrical printing surface which can be printed at high speed by continuous rotation against another cylinder, with the paper to be printed running between them. Flat letterpress surfaces are the most adaptable and are widely used, but they necessitate a reciprocating motion, which limits the speed of the output. (*Photog.*) The production of a positive (or negative) picture by exposing a negative (or positive) image on to sensitive emulsion.

printing diameter (*Print.*). The correct diameters of printing cylinder and impression surface.

printing down (*Print.*). A stage in the making of printing surfaces, for any of the main processes, in which the surface, after being made light-sensitive, is exposed to suitable light through a negative (or, in some cases, a positive).

printing ink (*Print.*). A mixture of carbon black, or other pigments, in a vehicle of mineral oil, linseed oil, etc. Inks are formulated to dry by penetration, evaporation, oxidation, or by a combination of these, and also can be *heat-set*, *cold-set*, or *moisture-set*.

printing-out paper (*Photog.*). See **POP**.

printing telegraph (*Teleg.*). A telegraph system in which the received signals are translated and operate a printing machine, giving a readable message. See **multiplex printing**.

print-out (*Comp.*). Automatic reproduction by *printer* of data in store or recorded on punched card or tape.

print-through (*Acous., Comp.*). In magnetic tape recording, the transfer of a recording from one layer to another when the tape is spooled or reeled giving rise to a form of distortion; also called **transfer**.

priority indicators (*Comp.*). Code signals which form a queue of data awaiting processing so that this is handled in order of importance.

prisere (*Bot.*). A primary succession.

prism (*Crystal.*). A hollow (open) crystal form consisting of three or more faces parallel to a crystal axis. (*Maths.*) A solid of which the ends are similar, equal, and parallel polygons, and of which the sides are parallelograms. (*Optics, etc.*) Triangular prisms made of glass and other transparent materials are used in a number of optical instruments. Equilateral prisms are used at minimum deviation in spectroscopes for forming spectra, 90° prisms are used for totally reflecting a ray through a right angle in binoculars, periscopes, and range-finders.

prismatic. Prism-shaped; composed of prisms.

prismatic astrolabe (*Surv.*). An instrument for observing stars at an altitude of 60° (in some instruments, 45°) at different azimuths around the horizon, these observations being used for the computation of latitude and local time.

prismatic binoculars (*Optics*). Binoculars in which the tubes are not straight and parallel, but formed thus: ┗┳┛ and are fitted with total reflecting prisms at the angles.

prismatic coefficient (*Ships*). The ratio between the immersed volume of the vessel and the volume of an enclosing prism with a constant transverse section identical with the maximum immersed cross-section area of the vessel.

prismatic compass (*Surv.*). A handheld form of surveyor's compass in which the eye vane carries a prism reflecting, to an eye placed opposite the sighting slit, a view of a graduated ring, attached to and moving round with the compass needle.

prismatic layer (*Zool.*). In the shell of *Mollusca*, a layer consisting of calcite or aragonite lying between the periostracum and the nacreous layer. In the shell of *Brachiopoda*, the inner layer of the shell, composed mainly of calcareous, but partly of organic material.

prismatic spectrum (*Optics*). A spectrum formed by refraction in a prism, as contrasted with a *grating spectrum* formed by diffraction.

prismatic (monoclinic) sulphur (*Chem.*). The crystalline form obtained when sulphur which has been melted is allowed to cool. Unstable and changes to the rhombic form.

prismatic system (*Crystal.*). See **orthorhombic system.**

prism drum (*TV*). A mechanical scanning device comprising a series of prisms mounted on a rotating drum.

prism light (*Build.*). A pavement light in which glass prisms internally reflect light.

prismoid. A body which has plane parallel ends and is bounded by plane sides.

prismoidal formula (*Civ. Eng.*). A formula used in the calculation of earthwork quantities. It states that the volume of any prismoid is equal to one-sixth its length multiplied by the sum of the two end-areas plus four times the mid-area.

prism square (*Surv.*). A form of *optical square* (q.v.) in which the fixed angle of 90° between the lines of sight is obtained by reflection from the surfaces of a suitably shaped prism.

prison ashlar (*Build.*). A block of stone dressed so that the faces are wrought into holes.

privacy system (*Telecomm.*). See **scrambler, inverter.**

private automatic branch exchange (*Teleph.*). A small automatic exchange on a subscriber's premises, for internal telephone connexions, with extensions over the public telephone system through lines to the local exchange. Abbrev. **PABX.**

private automatic exchange (*Teleph.*). An automatic exchange on private premises; not connectable with the public telephone system. Abbrev. **PAX.**

private branch exchange (*Teleph.*). An automatic or manual exchange on a subscriber's premises which is used for internal connexions, with extension through the local exchange to the public telephone system. Abbrev. **PBX.**

private exchange (*Teleph.*). An exchange in a private establishment, which is not connected in any way with the public telephone service. Abbrev. **PX.**

private manual branch exchange (*Teleph.*). A small manually-operated exchange on a subscriber's premises for establishing internal connexions and extensions, and external connexions over lines to the local exchange. Abbrev. **PMBX.**

private manual exchange (*Teleph.*). A manually-operated exchange, not connected with the public telephone service. Abbrev. **PMX.**

private wire (*Teleph.*). See **P-wire.**

pro-. Prefix from Gk. and L. *pro*, before in time or place, used in the sense of 'earlier' or 'more primitive', or 'placed before'.

proamnion (*Zool.*). In the embryos of higher Vertebrates, an area of blastoderm in front of the head, formed of ectoderm and endoderm only.

proatlas (*Zool.*). In *Sphenodon*, Crocodiles, and Chameleons, a median bone intercalated between the occipital region of the skull and the atlas vertebra.

probable ore (*Mining*). Inferred ore, partly exposed and sampled but not fully *blocked-out* in panels.

probability density (*Phys.*). Quantum mechanics suggests that electrons must not be regarded as being located at a defined point in space, but as forming a cloud of charge surrounding the nucleus, the cloud density being a measure of the probability of the electron being located at this point. See **uncertainty principle.**

probability density function (*Stats.*). See **distribution function.**

probability distribution function (*Stats.*). See **distribution function.**

proban (*Chem.*). Flameproof finish for textile fabrics based on a phosphorus compound, tetrakis (hydroxymethyl) phosphonium chloride.

probang (*Vet.*). A flexible tube which may be passed into the oesophagus and stomach of animals; used for relieving obstructions and administering fluids.

probasipodite (*Zool.*). The proximal 1 of 2 joints into which the basipodite of *Crustacea* is sometimes divided.

probe (*Acous.*). Device using narrow tube to transmit acoustic pressure to measuring microphone. (*Electronics*) (1) Electrode of small dimensions compared with the gas volume, placed in gas-discharge tube to determine the space potential. (2) Magnetic or conducting device to extract power from a waveguide. (3) Coil or semiconductor sensing element associated with a fluxmeter. (*Nuc.*) Portable radiation-detector unit, cable-connected to counting or monitory equipment. (*Space*) A space vehicle, especially one unmanned, sent to explore outer space (moon, planets, etc.), to collect and transmit data back to earth. (*Surg.*) A surgical instrument with a blunt end, used for exploring wounds, sinuses, and cavities.

probertite (*Min.*). Hydrated oxide of sodium, calcium, and boron, crystallizing in the monoclinic system. Occurs in borate deposits in California.

probiotic (*Biol.*). The conditions existing on earth before the appearance of life, which provided an environment in which the first living organisms could evolve from nonliving molecules.

problem-orientated language (*Comp.*). One directed towards the development of a suitable algorithm which could be used to solve a problem, but not to the specific program required for processing the solution on a particular computer.

problem-solving behaviour (*An. Behav.*). The nonrandom behaviour of an animal in a situation such as a *maze* or a *puzzle box* (qq.v.), which may often represent attempts to cope with the problem situation.

Proboscidea (*Zool.*). An order of Mammals having a long prehensile proboscis with the nostrils at the tip, large lophodont molars, and a pair of incisors of the upper jaw enormously developed as tusks; semiplantigrade; forest-living herbivorous forms of Africa and India. It includes Elephants and the extinct Mammoths.

proboscis (*Zool.*). An anterior trunklike process: in *Turbellaria* and *Polychaeta*, the protrusible pharynx: in *Nemertinea*, a long protrusible muscular organ lying above the mouth: in some Insects, the suctorial mouth-parts: in *Hemichorda*, a hollow club-shaped or shield-shaped structure in front of the mouth: in *Proboscidea*, the long flexible prehensile nose.

probuds (*Zool.*). In *Cyclomyaria*, reproductive bodies (blastozooids) which break free from the stolon and migrate across the body of the parent to become attached to the cadophore.

procaine (*Pharm.*). Di-ethylaminoethyl-4-aminobenzoate:

$$H_2N\text{-}C_6H_4\text{-}COOC_2H_4{\cdot}N(C_2H_5)_2$$

Crystalline solid. Its hydrochloride, Novocain(e), is used as a local anaesthetic.

procambial strand (*Bot.*). An elongated group of meristematic cells, derived from an apical meristem, and by division giving rise to the xylem and phloem of a vascular bundle; if the whole of the strand is used up in this way, the bundle contains no cambium, and is closed; if a thin strip of cambium remains between xylem and phloem, the bundle is open.

procarp (*Bot.*). The multicellular female organ of the *Rhodophyta*, consisting of an archicarp and a trichogyne.

procartilage (*Zool.*). An early stage in the formation of cartilage in which the cells are still angular in form and undergoing constant division; embryonic cartilage.

Procellariiformes (*Zool.*). An order of *Neognathae*. Wandering ocean birds, often very large, with long narrow wings. Lay one white egg, often in a burrow. Petrels, Shearwaters, Albatrosses.

procercoid (*Zool.*). A larval stage of some *Cestoda*, occurring in the secondary host.

process (*Bot., Zool.*). An extension or projection.

process annealing (*Eng.*). Heating steel sheet or wire between cold-working operations to slightly below the critical temperature and then cooling it slowly. The process is similar to tempering but will not give as much softness and ductility as full annealing.

process camera (*Photog.*). A camera specially designed for the reproduction processes, there being a large variety to suit particular needs, e.g., the traditional gallery camera, darkroom camera, etc.

process chart (*Eng.*). Chart in which a sequence of events is portrayed diagrammatically by means of conventional symbols.

process control (*Electronics*). In a complicated industrial or chemical process, control of various sections of the plant by electronic, hydraulic or pneumatic means, taking rates of flow, accelerations of flow, changes of law, temperatures and pressures into account automatically.

process-engraving (*Print.*). A relief printing plate made with the aid of the process camera followed by etching; or by means of an *electronic engraving* machine.

process factor (*Nuc.*). See **separation factor.**

processing (*Photog.*). The chemical sequences involved from the exposing of the negative in the camera to the obtaining of the final positive for use.

processing data (*Comp.*). The various operations of adding, subtracting, etc., performed by an electronic computer on data supplied, either as punched cards or tape, or on magnetic tape, according to a program.

processing routes (*Min. Proc.*). Term used in considering alternative methods of treating a specified ore or concentrate. Main 'routes' are physical, chemical and pyro-metallurgical.

process pump (*Vac. Tech.*). A pump which maintains the system at the operating pressure.

prochlorite (*Min.*). See **ripidolite.**

procidentia (*Med.*). A falling down or prolapse, esp. severe prolapse of the uterus.

Procion (*Chem.*). TN for series of *reactive dyes* (q.v.) used for cellulosic fabrics.

proclimax (*Ecol.*). A community which appears

to have the extent and permanence of a climax, but is not controlled by climate.

procoelous (*Zool.*). Concave anteriorly and convex posteriorly; said of vertebral centra.

procoracoid (*Zool.*). In the pectoral girdle of the Frog, a bar of cartilage anterior to the epicoracoid, having the clavicle closely applied to it.

proct-, procto-. Prefix from Gk. *proktos*, anus.

proctal (*Zool.*). Anal.

proctalgia (*Med.*). Neuralgic pain in the rectum.

proctectomy (*Surg.*). Surgical removal of the rectum.

proctitis (*Med.*). Inflammation of the rectum.

proctoclysis (*Med.*). The slow injection of large amounts of fluid into the rectum (the Murphy drip method).

proctodaeum (*Zool.*). That part of the alimentary canal which arises in the embryo as a posterior invagination of ectoderm. Cf. *stomodaeum, mid-gut*. *adj*. **proctodaeal.**

proctodynia (*Med.*). Pain in or around the anus.

proctoscope (*Med.*). An instrument which is used for inspecting the mucous membrane of the rectum.

proctosigmoiditis (*Med.*). Inflammation of the rectum and of the sigmoid flexure of the colon.

proctotomy (*Surg.*). Surgical incision of the anus or rectum for the relief of stricture.

procumbent (*Bot.*). Said of a stem which lies on the ground for all or most of its length.

procuticle (*Zool.*). In the cuticle of Insects, a multilaminar layer initially present below the *epicuticle* (q.v.) and pierced by pore canals running perpendicularly to it. In soft transparent areas this undergoes no apparent change after formation, but in other cases the outer part becomes hard, dark sclerotized *exocuticle* (q.v.), the inner unchanged part then being called the *endocuticle* (q.v.). See **tanning, sclerotin.**

Prodorite (*Build.*). TN for material and method for rendering various kinds of industrial structures acid-proof.

prodromal (*Med.*). Premonitory of disease.

producer gas (*Fuels*). A low-grade gas made by the partial combustion of coal, coke, or anthracite, in a mixed air-steam blast. Combustible constituents are (per cent) CO 24 to 30, H_2 10 to 14, CH_4 0·2 to 3·5; balance, CO_2 3 to 5, N_2 50 to 53. Energy density 5 to 6 MJ/m^3. Used chiefly for furnaces and production of power.

producers (*Ecol.*). In an ecosystem, autotrophic organisms, largely green plants, which are able to manufacture complex organic substances from simple inorganic compounds. Cf. *consumers, decomposers*.

product (*Maths.*). See **multiplication.**

production reactor (*Nuc. Eng.*). One designed for large-scale production of transmutation products such as plutonium.

products of inertia (*Maths.*). Of a body about two planes, the sum Σmxy taken over all particles of the body where m is the mass of a particle and x and y the perpendicular distances of the particle from the specified planes.

proecdysis (*Zool.*). The period of preparation for a moult in *Arthropoda* during which the new cuticle is laid down and the old one ultimately detached from it.

proembryo (*Bot.*). The first cells formed from the zygote in *Spermatophyta*, which by further development form the suspensor and the embryo.

proenzyme (*Biochem.*). See **zymogen.**

proepipodite (*Zool.*). In *Crustacea*, an epipodite borne upon the pleuropodite.

profile (*Build.*). A temporary guide set out at corners, normally with small timbers, to act as a guide for the foundations of a building. (*Surv.*) A longitudinal section, usually along the centre line of a proposed work such as a railway.

profile drag (*Aero.*). The 2-dimensional drag of a body, excluding that due to lift; the sum of the surface-friction and form drag.

profile grinding (*Eng.*). The grinding of cylindrical work without traversing the wheel whose periphery is profiled to the form required and extends over the full length of the work. Also **plunge grinding.**

profile paper (*Civ. Eng.*). A drawing paper ruled with horizontal and vertical lines spaced according to certain scales; used for plotting profiles of proposed engineering works.

profile position (*Bot.*). A position assumed by chloroplasts and by leaves when the edge of the structure is turned towards the position from which the brightest light is coming.

profiling (*Eng.*). Producing the profile of a die or other workpiece, (1) with a modified milling machine, incorporating a tracing mechanism, or (2) with a grinding machine using a wheel dressed to correspond to the required profile.

profilometer (*Civ. Eng.*). See **roughness integrator.**

proflavine (*Pharm.*). 2,8-Aminoacridine sulphate,

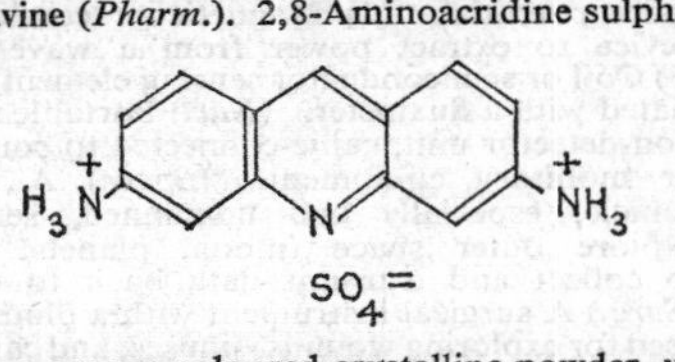

Deep orange-coloured crystalline powder, used in dilute solution as an antiseptic.

profundal zone (*Ecol.*). In freshwater habitats, the bottom and deep water area, which is beyond the depth of effective light penetration. Often absent in ponds.

profundus (*Zool.*). In Vertebrates, a cranial nerve representing the dorsal root of the first somite, and corresponding to the oculomotor ventrally.

progametangium (*Bot.*). A fungal hypha from which a gametangium is subsequently cut off by a transverse septum.

progeria (*Med.*). Premature old age. Occurring in children, the condition is characterized by dwarfism, falling out of hair, wrinkling of the skin and senile appearance, due, it is believed, to a destructive lesion of the pituitary gland.

progestational (*Physiol.*). Used of the luteal phase of the oestrous or menstrual cycle during which the endometrium is prepared for nidation; of the proliferative reaction of the endometrium towards the fertilized ovum or an irritant foreign body mimicking the ovum; of the hormones which bring about these effects.

progesterone (*Biochem.*). One of the female sex hormones. A steroid:

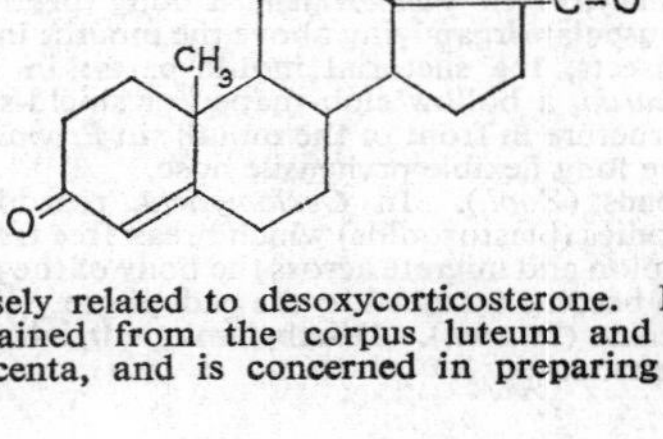

closely related to desoxycorticosterone. It is obtained from the corpus luteum and the placenta, and is concerned in preparing the uterus to receive the fertilized ovum, and in maintaining the continuance of pregnancy. It is used therapeutically in cases of uterine bleeding.

progestogens (*Chem.*). Generic name for a range of steroid hormones of the progesterone type; several synthetic progestogens (e.g., norethynodrel, megestrol lynoestrenol, megestrol) are used in *oral contraceptives*. See **oral contraception.**

proglottis (*Zool.*). One of the reproductive segments forming the body in *Cestoda*; produced by strobilization from the back of the scolex. *pl.* **proglottides.**

prognathous (*Zool.*). Having protruding jaws; having the mouth parts directed forwards. Cf. *orthognathous.*

prognosis (*Med.*). A forecast of the probable course of an illness.

progonal (*Zool.*). Said of that anterior portion of the genital ridge which does not contribute to the gonad.

progoneate (*Zool.*). Having the genital opening at the anterior end of the body, as in the *Symphyla*. Cf. *opisthogoneate.*

program (*Comp.*). Sequence of events to be performed by an electronic computer, as laid down by a programmer, in processing a given class of data, e.g., payroll, insurance instalments, etc. Realized on a punched card or paper tape which becomes a unit in the *library* for a specific computer.

program control (*Automation*). System in which the *index* (or *desired*) *value* is changed according to a plan of operations.

program evaluation and review technique (*Work Study*). A management control tool for the organization of a large-scale project by computer so that continuous assessment and evaluation of progress may be obtained. Abbrev. **PERT.**

programmable read only memory (*Comp.*). A fast access store similar to a *ROM*, but with some provision for setting the stored data after manufacture. Abbrev. **PROM.**

programmed learning (*Psychol.*). The process of learning from a systematic presentation of data constructed so that each step leads to the next. The learner has no need of recourse to material other than the program and progresses by answering questions at each stage which are necessary for the understanding of the rest. Such programs can be presented on a *teaching machine.*

programme level (*Telecomm.*). The level, related to datum power level of 1 milliwatt in 600 ohms, as indicated by a volume-unit (VU) meter, as defined for this purpose. Measured in decibels.

programme line (*Telecomm.*). A transmission line, of superior propagation characteristics, for relaying broadcasting programmes between a studio and transmitters, or between control points of broadcasting administrations.

programmer (*Comp.*). Person who analyses a problem and sets out a program for its solution by a specific electronic computer. Also **compiler.**

programme repeater (*Telecomm.*). Amplifier which is of sufficiently high-grade performance for insertion into transmission lines for relaying broadcasting programmes, with or without automatic means for reversing its direction of operation.

programme signal (*Telecomm.*). The electric waves in audio systems which correspond to the various sounds that are to be reproduced.

programming (*Comp.*). Working out detailed sequence of steps which determine operations of an electronic digital computer. The program is realized in holes punched in cards or tapes which *set* the machine prior to the insertion of data to be processed.

program-orientated language (*Comp.*). One used in preparing a source program (source deck) and suitable for direct compilation into machine language. This may be a transcription of a more fundamental algorithm suitable for solving the problem expressed in a *problem-orientated language*.

program tape (*Comp.*). That in computing and data processing which contains the sequence of instructions.

progressive heating (*Elec. Eng.*). Same as **scanning heating.**

progressive interlace (*TV*). Normal scanning of a television image, whereby all the *odd* lines are scanned (one *field* in U.S.) and then the *even* lines (another *field* in U.S.), and so on.

progressive metamorphism (*Geol.*). The progressive changes in mineral composition and texture observed in rocks within the aureole of contact metamorphism round igneous intrusions; also in rocks which have experienced regional metamorphism of varying degrees of intensity. The particular degree of metamorphism in the latter case is indicated by the 'metamorphic grade' of the rock.

progressive press tool (*Eng.*). A *press tool* (q.v.) which performs several operations simultaneously but at successive stations.

progressive proofs (*Print.*). In colour printing, a set of proofs supplied to the printer as a guide to colour and registration, each colour being shown both separately and imposed on the preceding ones.

progressive provisioning (*An. Behav.*). In Insects and Spiders, the supply of food by the parent to the larva or young form at regular intervals throughout its development. Cf. *mass provisioning, nest provisioning.*

progressive scanning (*TV*). Earlier system than *progressive interlace* in which all the lines of a TV image are scanned sequentially, without *interlace.*

progressive stain (*Micros.*). A stain which cannot be modified by a differentiating reagent and with which, therefore, care must be taken not to overstain. Cf. *regressive stain.*

projected area (*Eng.*). The product of the diameter of a bearing or rivet and the length along which it is in contact with another element.

projected diameter (*Powder Tech.*). The diameter of a circle which has the same area as the projected profile of the particle.

projected-scale instrument (*Elec. Eng.*). An instrument in which an image of the scale or the pointer is projected on to a screen.

projection (*Geog.*). Representation on a plane surface (e.g., a map), of details existing on a spherical surface (e.g., the earth). See 3 main types of map projections: **conical-, cylindrical-, zenithal-.** (*Maths.*) One figure is the projection of another when there is a *(1, 1)* correspondence between the lines and points in the two figures. (*Psychol.*) The process whereby we ascribe to other people and to the outside world mental factors and attributes really in ourselves.

projection balance (*Chem.*). Balance showing swing of beam on lighted graduated scale.

projection booth (*Cinema.*). Enclosure in a cinema from which the projectors are operated.

projection distance (*Cinema.*). The distance between the projector and the screen.

projection lamp (*Cinema., Photog.*). The source of illumination for projecting the image of a

cinematographic film or of a transparency on to a screen.
projection lantern (*Light*). See **projector**.
projection lens (*Cinema.*). The objective lens in a cinema projector, which projects the image on the film in the gate on to the screen.
projection-room(-box) (*Cinema.*). The enclosure containing cinematograph projectors in a cinema or studio; from it motion-pictures are projected through windows. See also **projection booth.**
projection television (*TV*). Optical reproduction, on a large open screen, of a TV image established on the phosphor of a small CRT.
projection welding (*Eng.*). A process similar to spot welding but allowing a number of welds to be made simultaneously. One of the workpieces has projections at all points or lines to be welded, the other is flat. Both are held under pressure between suitable electrodes, the current flowing from one to the other at the contact points of the projections.
projective properties (*Maths.*). Of a figure, properties unaltered by projection, e.g., the class or order of a curve.
projective technique (*Psychol.*). A method of psychological testing, used in personality assessment, in which the subject projects his own interpretation on to the test material, e.g., pictures, inkblots (see **Rorschach test**). The aim is to gain insight into the subject's total personality by sampling his way of seeing the world in which he lives.
projective transformation (*Maths.*). A transformation projecting one figure into another.
projector (*Cinema.*). Apparatus for projecting motion-picture film on to a screen, with or without sound reproduction. (*Light*) A '2-lens' optical system for projecting on to a screen a magnified image of a transparency or 'lantern slide'. The condenser, a lens whose function is to illuminate the slide evenly, forms an image of the source of light on the projection lens, the slide being placed in the converging beam of light between the 2 lenses in such a position that the projection lens forms on the screen a real inverted image of it. See **floodlight-**, **searchlight-**. (*Optics*) **Overhead projector**, i.e. one which throws an enlarged image of a transparency on a screen or wall behind and above the operator. Used for teaching purposes.
projector-type filament-lamp (*Elec. Eng.*). An electric filament-lamp in which the filament is arranged in a concentrated form, so that it can be focused for projection purposes.
prokaryon (*Biol.*). The nucleus of a blue-green alga or of a bacterium.
prolactin (*Biochem., Physiol.*). Lactogenic hormone; a mammotrophic protein hormone, secreted by cells of the pars anterior of the pituitary gland, which stimulates the developed mammary gland to secrete milk; has a similar effect on the pigeon crop gland; also called **luteotrophin** in respect of its other, gonadotrophic, properties.
prolamines (*Chem.*). A group of alcohol-soluble simple proteins which are insoluble in water and in strong alcohol but soluble in 70–90% alcohol.
prolan (*Biochem., Physiol.*). Formerly general name for gonadotrophic hormones of mammals found in various tissues and body fluids during pregnancy; prolan A is equivalent to *follicle-stimulating hormone* (q.v.) and prolan B to *luteinizing hormone* (q.v.).
prolapse (*Med.*). The falling out of place or sinking of an organ or part of the body.
prolate cycloid (*Maths.*). See **roulette.**
prolate ellipsoid (*Maths.*). An ellipsoid obtained by rotating an ellipse about its major axis, i.e., like a rugger ball. Cf. *oblate ellipsoid*.
proleg (*Zool.*). One of several pairs of fleshy conical retractile projections borne by the abdomen in larvae of most *Lepidoptera*, Saw-flies and Scorpion-flies; used in locomotion.
proletarian (*Bot.*). A plant having little or no reserves of food material.
proleucocytes (*Zool.*). Cells which will develop into leucocytes.
proliferation, prolification (*Bot.*). (1) A renewal of growth in a mature organ after a period of inactivity. (2) The production of vegetative shoots from a reproductive structure. (3) The formation of a sporangium inside the empty walls of a previously discharged sporangium. (4) The production of offshoots, which may become detached and established as new plants. (*Med.*) Growth or extension by the multiplication of cells. *adj.* **proliferous.**
proliferous (*Bot.*). (1) Bearing offshoots. (2) Producing abnormal or supernumerary outgrowths. (3) Producing progeny by means of offshoots. (4) Showing excessive development in some respect.
prolification (*Bot.*). Development of buds in the axils of sepals and petals. See **proliferation.**
proline (*Chem.*). Pyrrolidine-2-carboxylic acid, a cleavage product of certain proteins. A non-essential amino acid, with respect to growth effect in rats.
prolonged (*Bot.*). Drawn out into a long point which is not hollowed along its sides.
PROM (*Comp.*). Abbrev. for **programmable read only memory.**
promenade deck (*Ships*). On a passenger ship, the upper deck on which passengers walk.
promenade tile (*Build.*). See **quarry tile.**
promeristem (*Bot.*). The meristem in an embryo and at the growing points.
prometal (*Met.*). Cast-iron resistant to heat.
prometaphase (*Cyt.*). The stage in cell division, mitosis and meiosis, between prophase and metaphase.
promethazine (*Pharm.*). 10-(2-Dimethylaminopropyl) phenothiazine hydrochloride. An antihistamine drug of slow but prolonged action.
promethium (*Chem.*). A radioactive element of the rare earth series, symbol Pm, at. no. 61, having no known stable isotopes in nature. Its most stable isotope, $^{145}_{61}$Pm, has a half-life of over 20 years.
promine (*Med.*). Substance found in thymus, muscle, and big blood vessels, which makes cancer cells grow faster. See also **retine.**
prominences (*Astron.*). Tongues of glowing gas standing out from the sun's disk, sometimes to a height of many thousands of kilometres, and displaying a great variety of form and motion. First discovered during total solar eclipses, but now observable at any time with the spectro-helioscope.
promitosis (*Zool.*). A primitive type of *mitosis*.
promontory (*Zool.*). A projecting structure: a small ridge or eminence.
promoter (*Chem.*). A substance which increases the activity of a catalyst.
prompt critical (*Nuc.*). Condition in which a reactor could become critical solely with prompt neutrons. For a reactor which is super-critical, the reaction rate will rise rapidly and in this condition the reactor is difficult to control. See **delayed neutrons.**
prompt gamma (*Nuc.*). The γ-radiation emitted at the time of fission of a nucleus.

prompt neutrons (*Nuc.*). Those released at a primary fission with practically zero delay.

promycelium (*Bot.*). A short germ tube, put out by some fungal spores, on which other spores of different type are developed.

promyelocyte (*Histol.*). A uninuclear bone-marrow cell which develops into a myelocyte or granulocyte; a large uninuclear myeloid cell seen in leukaemia.

pronation (*Zool.*). In some higher Vertebrates, movement of the hand and forearm by which the palm of the hand is turned downwards and the radius and ulna brought into a crossed position; cf. *supination*. *adj.* **pronate.**

pronator (*Zool.*). A muscle effecting pronation.

pronephros (*Zool.*). Archinephros; in *Craniata*, the anterior portion of the kidney, functional in the embryo but functionless and often absent in the adult. Also called **fore-kidney, head-kidney.** *adj.* **pronephric.** Cf. *mesonephros*; *metanephros*.

prong test (*For.*). A seasoning test to show by the movement of stress sections whether the timber is in compression, tension, or neutral.

pronotum (*Zool.*). The notum of the prothorax in Insects. *adj.* **pronotal.**

Prontosil (*Pharm.*). TN for *sulphanilamide*; now little used.

pronucleus (*Zool.*). The nucleus of a germ cell after the maturation divisions.

Prony brake (*Eng.*). An absorption dynamometer consisting of a pair of friction blocks bolted together across a brake drum, the torque absorbed being balanced by weights at the end of a torque arm attached to the blocks.

pro-oestrus (*Zool.*). In Mammals, the coming on of heat in the oestrus cycle

proof. In alcoholometry, a designation (*proof-spirit*) for spirituous liquid containing 49·28% of real alcohol by weight, 57·10% by volume, with rel. d. of 0·920 at 15·6°C. Proof spirit is taken as the standard strength of alcoholic liquids for fiscal purposes. A spirituous liquid which is x% *overproof* contains as much alcohol in 100 vol. as in 100+x vol. of proof spirit; x% *underproof* signifies the opposite condition. (*Typog.*) Impression taken from type matter or blocks for checking and correction only, not as representative of the finished appearance and quality of the work.

proof by contradiction (*Maths.*). See **reductio ad absurdum proof.**

proof corrections (*Typog.*). Additions or emendations to a proof. They should be made in ink, and clearly indicated in the margin in accordance with B.S. 1219: 1958. A certain amount of correcting is usually allowed for in the estimated price, corrections in excess of this being charged separately.

proofed tape (*Cables*). Cotton cloth coated with a rubber compound, wrapped round rubber-insulated cables.

proofing press (*Print.*). Models are available for all three main printing processes. In letterpress printing there is a wide variety, ranging from simple entirely hand-operated models for proofing galleys to power-driven models capable of impression and register comparable to a printing press, but without the feeding and delivery mechanisms required for a long run. Multicolour proofing presses are also made.

proof load (*Aero.*). The load which a structure must be able to withstand, while remaining serviceable.

proof plane (*Elec. Eng.*). A piece of conducting material, mounted on an insulating handle, which may be used for receiving or removing charges in electrostatic experiments.

proof stress (*Met.*). The stress required to produce a certain amount of *permanent set* (q.v.) in metals which do not exhibit a sudden yield point. Usually it is the stress producing an extension of 0·1 or 0·5%. Often *yield point*.

pro-otic (*Zool.*). An anterior bone of the auditory capsule of the Vertebrate skull.

prop (*Build.*). A post, usually relatively short and made of timber, used as a strut. (*Mining*) Sturdy supporting post set across a lode or seam underground to hold up roof after excavation. See **powered supports.**

propagation (*Bot.*). Increase in the number of plants by vegetative means. (*Phys., etc.*) Transmission of energy in the form of waves in the direction normal to a wavefront, which is generally spherical, or part of a sphere, or plane. Applies to acoustic, electromagnetic, water, etc. waves.

propagation constant (*Phys., etc.*). Measure of diminution in magnitude and retardation in phase experienced by current of specified frequency in passing along unit length of a transmission line or waveguide, or through one section of a periodic lattice structure. Given by the natural logarithm of the ratio of output to input current or of acoustic particle velocity.

propagation loss (*Phys., etc.*). The transmission loss for radiated energy traversing a given path. Equal to the sum of the *spreading loss* (due to increase of the area of the wavefront) and the *attenuation loss* (due to absorption and scattering).

propagation of light (*Light*). Consists of transverse electromagnetic waves propagated through free space with a velocity of $2{\cdot}9979 \times 10^8$ m/s and wavelengths about 400 to 800 nm. The ratio (velocity of light in free space/velocity in medium) is the refractive index of the medium. According to the special theory of relativity, the velocity of light is absolute and no body can move at a greater speed.

propagulum (*Bot.*). A small vegetative outgrowth which becomes detached from the parent and grows into a new plant.

propane (*Chem.*). C_3H_8, an alkane hydrocarbon, a colourless gas at atmospheric pressure and temperature. B.p. −45°C; found in crude petroleum. In liquid form it constitutes *liquefied petroleum gas* (LPG), a clean-burning fuel, used by some as a petrol substitute.

propanoic acid (*Chem.*). See **propionic acid.**

propanol (*Chem.*). See **propyl alcohol.**

propanone (*Chem.*). See **acetone.**

propargyl alcohol (*Chem.*). $HC{\equiv}C{\cdot}CH_2{\cdot}OH$, propinyl alcohol, a monohydric unsaturated alcohol, a mobile liquid of agreeable odour, lighter than water, b.p. 114°C.

propargylic acid (*Chem.*). See **propiolic acid.**

propellant (*Aero., Space*). Comprehensive name for the combustible materials for a chemical *rocket* motor, comprising the fuel (hydrocarbons such as kerosene or hydrazine) and *oxydant* (q.v.) in the case of liquid rockets; powder or cordite in dry rockets.

propeller (*Eng.*). See **airscrew, marine screw propeller.**

propeller brake (*Aero.*). A shaft brake to stop, or prevent windmilling of, a turboprop, principally to avoid inconvenience on the ground.

propeller efficiency (*Aero.*). The ratio of the actual thrust power of an airscrew to the torque power supplied by the engine shaft; 80–85% is the usual value.

propeller fan (*Eng.*). A fan consisting of an impeller or rotor carrying several blades of airscrew form, working in a cylindrical casing sometimes provided with fixed blades; usually driven by a direct-coupled motor.

propeller hub (*Aero.*). The detachable fitting by which a propeller is attached to the power-driven shaft.

propeller post (*Ships*). See **stern frame.**

propellers (*Print.*). Individually mounted rollers, draw rollers or bosses, with peripheries of rubber, synthetic material or knurled metal, usually operating in conjunction with a driven roller to assist in controlling the reels on web-fed presses.

propeller shaft (*Autos.*). The driving shaft which conveys the engine power from the gearbox to the rear axle of a motor vehicle. It is usually connected through universal joints to permit displacement of the rear axle on the springs.

propeller turbine engine (*Aero.*). See **turboprop.**

propeller-type water turbine (*Eng.*). A water turbine having a runner similar to a four-bladed ship's propeller. It gives a high specific speed under low heads, thus reducing the size of a direct-coupled generator. See **Kaplan water turbine.**

propelling nozzle (*Aero.*). The constricting nozzle at the outlet of a turbojet *exhaust cone* (q.v.) or *jet pipe* which reduces the gases to slightly more than ambient atmospheric pressure and accelerates them to raise their kinetic energy, thereby increasing the thrust.

propeltidium (*Zool.*). The conspicuously swollen head of *Solifugae*.

propene (*Chem.*). See **propylene.**

propenol (*Chem.*). See **allyl alcohol.**

proper exciple (*Bot.*). A hyphal layer around the fructification of a lichen which contains no algal cells.

proper fraction (*Maths.*). See **division.**

proper motion (*Astron.*). That component of a star's own motion in space which is at right angles to the line of sight, so that it constitutes a real change in the position of the star relative to its neighbouring stars.

prop-free front (*Mining*). Long wall working of coal seam in which roof is supported by beams cantilevered from behind, so as to leave an unobstructed working face.

prophage (*Bot.*). A form of bacteriophage known as a temperate phage which exists in the bacterial host cell in an inactive form, i.e., without causing lysis.

propham (*Chem.*). Isopropyl *N*-phenylcarbamate:

used as a herbicide. Abbrevs. **IPC, IPPC.**

prophase (*Cyt.*). The preliminary stages of mitosis or meiosis leading up to the formation of the astroid.

prophylactic (*Med.*). Tending to prevent or to protect against disease, especially infectious disease; any agent which does this.

prophylaxis (*Med.*). The preventive treatment of disease.

prophyll (*Bot.*). A bracteole.

propiolic acid (*Chem.*). CH ⋮ C·COOH, acetylene-carboxylic acid; silky crystals, m.p. 6°C, b.p. 144°C, soluble in water and ethanol; it forms an explosive silver salt. Also called **propargylic acid.**

Propionibacteriaceae (*Bacteriol.*). A family of bacteria belonging to the order *Eubacteriales*. Gram-positive rods. Important as fermenters of carbohydrates, organic acids, and alcohols, e.g., in rumen and in production of cheeses.

propionic (propanoic) acid (*Chem.*). $CH_3{\cdot}CH_2{\cdot}COOH$, a monobasic fatty acid, a colourless liquid, m.p. −36°C, b.p. 141°C. A constituent of pyroligneous acid; it is formed in certain fermentations. A more up-to-date process involves the use of the 'oxo' process with ethylene as the starting material.

propionyl group (*Chem.*). The monovalent radical $CH_3{\cdot}CH_2{\cdot}CO$—.

proplastid (*Bot.*). A minute inclusion in the cytoplasm from which a plastid may develop.

propneustic (*Zool.*). Among insect larvae, having the prothoracic spiracles only; unusual, but found in the pupae of some *Diptera*. Cf. *amphipneustic, metapneustic*. See **oligopneustic.**

propodeum, propodeon (*Zool.*). In certain *Hymenoptera*, the first somite of the abdomen which becomes fused with the metathorax; epinotum.

propodite (*Zool.*). In some *Crustacea*, the 4th joint of the endopodite of the walking legs or maxillipeds.

propodium (*Zool.*). In *Gastropoda*, the anterior part of the foot projecting beneath the head.

propodosoma (*Zool.*). In *Acarina*, the segments of the first and second pairs of legs.

proportional control (*Automation*). Feedback signal in a control system which is proportional to the discrepancy between the actual and desired values of the controlled quantity.

proportional counter (*Nuc. Eng.*). One which uses the *proportional region* in a tube characteristic, where the gas amplification in the tube exceeds unity but the output pulse remains proportional to initial ionization.

proportional ionization chamber (*Nuc. Eng.*). One in which the initial ionization current is amplified by electron multiplication in a region of high electric field strength, as in a proportional counter. This device is not for counting but measures ionization currents over a period of time.

proportional navigation (*Nav.*). Navigation in which rate of turn is proportional to rate of turn of line of sight.

proportional region (*Nuc. Eng.*). The range of operating voltages for a counter tube (or ionization chamber) in which the gas amplification is greater than unity and is not dependent on the primary ionization.

proprioceptor (*Zool.*). A sensory nerve-ending receptive to internal stimuli; interoceptor. *adj.* **proprioceptive.**

propriospinal (*Zool.*). Arising within and confined to the spinal cord; said of certain endogenous fibres of the dorsal and ventrolateral columns of the spinal cord which are derived from cells in the grey matter of the cord itself.

prop root (*Bot.*). A root formed from the stem, usually close to the ground; it helps to hold the stem erect.

propterygium (*Zool.*). In *Selachii*, the outermost or most anterior of the 3 basal cartilages of the pectoral fin. Cf. *mesopterygium, metapterygium*.

proptosis (*Med.*). Displacement forwards or protrusion of a part of the body, esp. of the eye.

propulsion reactor (*Eng.*). One designed to supply energy for the propulsion of a vehicle—at present invariably a ship.

propulsive ducti(*Aero.*). Generic term for the simplest form of reaction-propulsion aero-engine having no compressor/turbine rotor. Thrust is generated by initial compression due to forward motion, the form of the duct converting kinetic energy into pressure, the addition and combustion of fuel, and subsequent ejection of the hot gases at high velocity. See **pulse-jet, ramjet.**

propulsive efficiency (*Aero.*). (1) The propulsive horsepower divided by the torque horsepower. (2) In a turbojet, the net thrust divided by the gross thrust.

propupa (*Zool.*). See **prepupa.**

propygidium (*Zool.*). In some Insects, a sclerite anterior to the pygidium.

propyl alcohol (*Chem.*). $C_3H_7 \cdot OH$, a monohydric aliphatic alcohol, existing in 2 isomers: (*a*) *n*-propyl alcohol (propan-1-ol), $CH_3 \cdot CH_2 \cdot CH_2 \cdot OH$, b.p. 97°C, rel. d. 0·804, obtained from fusel-oil, miscible with water in all proportions; (*b*) *iso*propyl alcohol (propan-2-ol), $CH_3 \cdot CH(OH) \cdot CH_3$, b.p. 81°C, rel. d. 0·789, which can be prepared by the reduction of acetone with sodium amalgam, or by the hydrolysis of propylene sulphate obtained by the absorption of propylene in sulphuric acid. It also results as a by-product in the synthesis of methanol by the Fischer-Tropsch process.

propylene (*Chem.*). Propane. C_3H_6, $CH_2{=}CH \cdot CH_3$, an alkene hydrocarbon, a gas, b.p. −48°C. An important by-product of oil refining used for the manufacture of many organic chemicals including propanone, glycerine, and polypropene.

propylene dichloride (*Chem.*). 1,2-Dichloropropane. $CH_3 \cdot CHCl \cdot CH_2Cl$. B.p. 94°C. Solvent for oils, fats, waxes, and resins.

propylene glycol (*Chem.*). Propan-1,2-diol. $CH_3CH(OH)CH_2OH$. B.p. 188°C. Used as a solvent, humectant and plasticizer. Also used as a chemical intermediate.

propyne (*Chem.*). Allylene, $CH_3 \cdot C{\equiv}CH$. Methyl acetylene.

pros- (*Chem.*). Containing a condensed double aromatic nucleus substituted in the 2,3 positions. Symbol ρ-.

proscapula (*Zool.*). See **clavicle.**

proscenium (*Build.*). The stage frame in a theatre, fitted with curtains and a fire-proof safety curtain to cut off the stage from the auditorium.

proscenium lights (*Light*). Rows of incandescent lamps around the back of the proscenium arch for illuminating the stage of a theatre.

proscolex (*Zool.*). See **cysticercus.**

prosector (*Med.*). One who dissects dead bodies for anatomical demonstration and teaching.

prosencephalon (*Zool.*). In *Craniata*, the part of the forebrain which gives rise to the cerebral hemispheres and the olfactory lobes.

prosenchyma (*Bot.*). A tissue composed of elongated cells with pointed ends; the cells are often empty, and are concerned with affording support and with conducting material.

prosiphon (*Zool.*). In some decapod *Mollusca* with a spiral chambered shell, a tube passing through the initial chamber.

Prosobranchiata (*Zool.*). An order of *Gastropoda*. The adults show torsion, nearly always have a shell and an operculum, have the visceral loop of the nervous system twisted into a figure-eight, the mantle cavity opening anteriorly, the ctenidia in front of the heart and separate sexes. Previously known as *Streptoneura*. Whelks, Limpets, Periwinkles.

prosocoele (*Zool.*). In *Craniata*, the cavity of the forebrain or first brain vesicle in the embryo; fore ventricle.

prosodetic (*Zool.*). In Bivalves, anterior to the beak.

prosodus (*Zool.*). In *Porifera*, a canal leading from an incurrent canal to the flagellated chamber.

prosoma (*Zool.*). In *Arachnida*, the region of the body comprising all the segments in front of the segment bearing the genital pore; in *Acarina*, the gnathosoma together with the podosoma.

prosoplectenchyma (*Bot.*). A false tissue of elongated fungal hyphae.

prosopothoracopagus (*Med.*). Twin foetuses which are joined at the faces, necks, and chests.

prosopyle (*Zool.*). In *Porifera*, a pore or aperture opening from an incurrent canal into a flagellated chamber.

prospect (*Geol., Mining*). Area which shows sufficient promise of mineral wealth to warrant exploration. Methods of search include aerial survey, magnetometry, geophysical and geochemical tests, seismic probe, electroresistivity measurement, pitting, trenching and drilling.

prospory (*Bot.*). The formation of sporangia on very young plants.

prostaglandins (*Biochem.*). A series of complex organic compounds derived naturally from the prostate gland but also synthetically. Certain members of the group have been applied in the induction of labour and abortion.

prostate (*Zool.*). In *Oligochaeta*, said of glands of unknown function associated with the male genitalia; in *Cephalopoda*, said of a gland of the male genital system associated with the formation of spermatophores; in eutherian Mammals (except *Edentata* and *Cetacea*), including Man, said of a gland associated with the male urogenital canal.

prostatectomy (*Surg.*). Removal of the prostate gland.

prostatism (*Med.*). The concurrence of mental irritability, eroticism, and frequent micturition in elderly men, usually indicative of abnormal enlargement of the prostate gland.

prostatitis (*Med.*). Inflammation of the prostate gland.

prostatorrhoea, prostatorrhea (*Med.*). Chronic gleety or mucous discharge from the prostate.

prosternum (*Zool.*). The sternum of the *prothorax*.

prosthesis (*Surg.*). The supplying of an artificial part of the body to replace one which is deficient or absent; the artificial part supplied. *adj.* **prosthetic.** See also **dentistry.**

prosthetic group (*Biochem.*). The non-protein component of an enzyme consisting of a protein and a non-protein portion.

prosthetics (*Surg.*). That branch of surgical science concerned with prosthesis.

prosthomere (*Zool.*). A preoral somite.

prostomium (*Zool.*). In annelid Worms, that part of the head region anterior to the mouth.

protactinium (*Chem.*). A radioactive element. Symbol Pa, at. no. 91, half-life of 2×10^4 years. One radioactive isotope is ^{233}Pa which is an intermediate in the preparation of the fissionable ^{233}U from thorium.

protamines (*Chem.*). The simplest proteins, e.g., salmine, sturine, etc. They are found, in combination with nucleic acid, only in fish testicles. Strong bases, they can be precipitated by alcohol from their sulphuric acid solution. They are not coagulated by heat, but can be precipitated with NaCl or with ammonium sulphate. They are laevorotatory.

protandrous (*Bot., Zool.*). Said of organisms in which the male germ cells ripen before the female germ cells. Cf. *protogynous*.

protanopic (*Med.*). Colour blind to red.

protease (*Chem.*). A term for any protein-splitting enzyme. Proteases include pepsin, trypsin, erepsin, rennin, and several plant proteases, e.g., papain and bromelin.

protected-type (*Elec. Eng.*). Said of electrical

machinery or other apparatus in which any internal rotating or live parts are protected against accidental mechanical contact in such a way as not to impede ventilation. See **screen-protected motor.**

protection cap (*Elec. Eng.*). See **fender.**

protection ratio (*Telecomm.*). The lowest ratio of the signal strength of the required signal to that of the interfering signals necessary to obtain satisfactory and consistent reception.

protective coating (*Chem.*). A layer of a relatively inert substance, on the surface of another, which diminishes chemical attack of the latter.

protective colloid (*Chem.*). A lyophilic colloid which is adsorbed on the dispersed phase of a lyophobic colloid, thus decreasing its tendency to coagulate, e.g., albumen in mayonnaise.

protective furnace atmosphere (*Heat*). Inert gas produced from the products of combustion of gas and air in predetermined proportions, the atmosphere being first cooled, cleaned, dehydrated, and desulphurized before delivery to heating chambers of furnaces operated for bright-annealing of nonferrous metals and bright treatment of steels. See also **furnace atmosphere.**

protective gap (*Elec. Eng.*). A spark gap arranged between an electric circuit and earth, or across a piece of apparatus, so adjusted that, should the voltage across the gap exceed a certain safe value, the gap breaks down, thereby limiting the voltage appearing across the part of the circuit being protected to the breakdown voltage of the gap.

protective gear (*Elec. Eng.*). The apparatus associated with a protective system, e.g., relays, instrument transformers, pilots, etc.

protective layer (*Bot.*). A layer of suberized cells lying across the place where a leaf comes away at leaf fall; it checks both loss of water from within the plant and also the entry of parasites.

protective system (*Elec. Eng.*). An arrangement of apparatus designed to isolate a piece of electrical apparatus should a fault occur on it.

protector (*Electronics*). Tube in which glow discharge from a cold cathode prevents high voltage across a circuit.

protegulum (*Zool.*). Embryonic shell (*Brachiopoda*).

protein-bound iodine (*Med.*). Iodine bound to the plasma proteins (serum albumin fraction) and used as an indication of thyroid function. Abbrev. **PBI.**

proteins (*Chem.*). Complex nitrogenous substances composed mainly of α-amino acid residues joined by the 'peptide linkage' which is formed when water is eliminated from the carboxyl group of 1 amino acid molecule and the amino group of another. Proteins have molecular weights which tend to be approximate multiples of 17 600. Most form colloidal solutions in water or dilute salt solutions. They are odourless, tasteless, difficult to crystallize, decomposed by heat without definite melting points, hydrolysable by acids, alkalis, or certain enzymes. Some are conjugated—i.e., contain a relatively simple group such as haematin in haemoglobin, or nucleic acid in nucleoprotein. They are essential constituents of the living cell, and must be provided in the food to make good tissue wastage and allow of growth.

proteolytic, proteoclastic (*Zool.*). Said of enzymes which catalyse the breakdown of proteins into simpler substances, e.g., trypsin.

proteoses (*Chem.*). Protein derivatives, soluble in water, not coagulated by heat, but precipitated by saturation with ammonium or zinc sulphate.

proter-, protero-. Prefix from Gk. *proteros*, before, former.

proterandrous (*Bot., Zool.*). See **protandrous.**

proteroglyph (*Zool.*). Having specialized canine teeth in the upper jaw.

proterogyny, proterogynous (*Bot.*). See **protogynous.**

proterokont (*Bacteriol.*). A bacterial flagellum, not homologous with the flagella of higher organisms.

proterosoma (*Zool.*). See **prosoma.**

Proterozoic (*Geol.*). A division of the Pre-Cambrian comprising the less ancient rocks of that system, and lying above the Archaeozoic.

prothallial cell (*Bot.*). A small cell present in pollen grain of Gymnosperms, representing a last reminiscence of a male prothallus.

prothallus, prothallium (*Bot.*). (1) A small plant derived from a spore, and bearing the antheridia and archegonia, occurring in the life cycles of ferns and related plants. The zygote and young spore-bearing plant develop at the expense of the prothallus, which dies as the spore-bearing plant enlarges. The term prothallus is extended to cover homologous stages in the life cycle of Gymnosperms, where the relations are less clear. (2) The very earliest stages in the development of the lichen thallus.

protheca (*Zool.*). The calyx rudiment in coral formation.

prothecium (*Bot.*). A primitive or rudimentary perithecium.

prothetely (*Zool.*). In some Insects, the condition, thought to result from abnormal metamorphosis, when forms intermediate between nymph and adult, or larva and pupa (e.g., larvalike forms with wing lobes) are produced. Cf. *metathetely*.

prothorax (*Zool.*). The first or most anterior of the three thoracic somites in Insects. Cf. *mesothorax, metathorax*.

prothrombin (*Biochem.*). A proteinlike substance present in blood plasma. In shed blood it is converted to *thrombin* (q.v.) by the action of thromboplastin (thrombokinase), which is liberated from the blood platelets or is derived from tissue fluid. Calcium is concerned in the process. Also **serozyme.**

Protista (*Bot., Zool.*). A large group of simple organisms, which show at least some characters common to plants and animals. The more plantlike are placed in the *Protophyta*, the more animal-like in the *Protozoa*. Two classes of *Protista* (*Flagellata* and *Mycetozoa* or *Myxomycetes*) are claimed by the zoologists as animals and by the botanists as plants. That these forms fit so neatly into both zoological and botanical classifications seems to point to the fact that they are among the most primitive forms of life, evolved before the animal and plant kingdoms began to diverge. However, while this is probably true for the *Flagellata*, there are reasons for considering the *Myxomycetes* to be a somewhat specialized group.

protium (*Chem.*). Lightest isotope of hydrogen, of mass unity (^{1}H), most prevalent naturally. The other isotopes are deuterium (^{2}H) and tritium (^{3}H).

proto-. Prefix from Gk. *prōtos*, first.

Protobranchiata (*Zool.*). A small order of *Lamellibranchiata* in which the gills are simple bipectinate structures with free, flat, nonreflected filaments; the foot is flat and the byssus gland only slightly developed; ciliary feeders.

protocatechuic acid (*Chem.*). 2,3-Dihydroxybenzoic acid:

Crystalline solid; m.p. 199°C. Found in onions and various other related plants.

protocercal (*Zool.*). See **diphycercal.**

protocerebron, protocerebrum (*Zool.*). In higher *Arthropoda* (as Insects and *Crustacea*), the fused ganglia of the first somite of the head, forming part of the 'brain'.

Protochordata (*Zool.*). A division of *Chordata* comprising the subphyla *Hemichorda, Urochorda,* and *Cephalochorda,* which are distinguished by the absence of a cranium, vertebral column and of specialized anterior sense organs. Cf. *Vertebrata.*

Protociliata (*Zool.*). In some classifications, a subclass of *Ciliophora,* having no mouth (astomatous) or peristome; parasitic forms.

protocnemes (*Zool.*). In *Zoantharia,* the 6 primary pairs of mesenteries.

protoconch (*Zool.*). In *Mollusca,* the larval shell.

protocone (*Zool.*). In Mammals, the inner cusp of an upper molar tooth.

protoconid (*Zool.*). In Mammals, the inner cusp of a lower molar tooth.

protocooperation (*Ecol.*). See **cooperation.**

protocorm (*Bot.*). A tubelike structure formed in the early stages of club mosses and some other plants which appear to live in close relations with fungi in youth.

protoderm (*Bot.*). See **dermatogen.**

protogenic (*Chem.*). Capable of supplying a hydrogen ion (proton).

protogynous (*Bot., Zool.*). Said of organisms in which the female germ cells ripen before the male germ cells; cf. *protandrous. n.* **protogyny.**

protolysis (*Bot.*). The decompositio of chlorophyll by light.

protomala (*Zool.*). In *Myriapoda,* the mandible.

protomerite (*Zool.*). In some *Gregarinidea,* the part of the body intervening between the epimerite and the nucleus.

Protomonadina (*Zool.*). An order of *Zoomastigina* the members of which have one or two flagella, never show active amoeboid movement over the whole surface of the body, and have no extranuclear division centre. Includes the parasitic *Trypanosomes.*

protomorphic (*Zool.*). Primordial; primitive.

proton (*Nuc.*). Elementary particle, of positive charge and unit atomic mass, atom of lightest isotope of hydrogen without its electron. Appears to be joined with neutrons in building up nucleus of atoms, there being Z protons in the nucleus, where Z is the atomic number. (Each chemical element can have a variable number of neutrons to form isotopes.) It is the lightest baryon (rest mass = 1·007276 u), believed to be the only one completely stable in isolation.

protonema (*Bot.*). (1) A branched filamentous plantlet, looking like an alga, produced when a moss spore germinates; from it the leafy moss plants arise by the development of lateral buds. (2) The early filamentous stages of some algae, which differ in form from the adult plants.

protonephridial system (*Zool.*). The excretory system of *Platyhelminthes,* consisting of flame cells and ducts.

protonephridium (*Zool.*). A larval nephridium, usually of the flame cell type.

protonic solvent (*Chem.*). One that yields a proton, H^+, as the cation in *self-dissociation.*

proton microscope (*Micros.*). A microscope employing a beam of protons in place of the beam of electrons used in the electron microscope. This gives greater resolving power and also better contrast in thin sections.

proton precessional magnetometer (*Nuc. Eng.*). Precision magnetometer based on measurement of *Larmor frequency* (q.v.) of protons in a sample of water. See **nuclear magnetic resonance** and **proton resonance.**

proton-proton chain (*Nuc.*). A series of thermonuclear reactions which are initiated by a reaction between two protons. It is thought that the proton-proton cycle is more important than the carbon cycle in the cooler stars.

proton resonance (*Nuc.*). A special case of nuclear magnetic resonance. Since the nuclear magnetic moment of protons is now well known, that of other nuclei is found by comparing their resonant frequency with that of the proton.

proton-synchrotron (*Nuc. Eng.*). A synchrotron which is modified to allow the acceleration of protons by frequency modulation of the RF acceleration voltage.

protopathic (*Med.*). (1) Relating to first symptom or lesion; primary. (2) Relating to certain type of nerve which is only affected by the coarser stimuli, e.g., pain, and not by the more sensitive sensory influences, e.g., localization stimuli.

protophilic (*Chem.*). Able to combine with a hydrogen ion.

protophloem (*Bot.*). The first phloem to be formed from the procambial strands; it may be generalized in structure, with poorly formed sieve tubes and no companion cells.

protophyll (*Bot.*). A sterile leaf.

protophyte (*Bot.*). A simple unicellular plant.

protoplasm (*Biol.*). The material basis of all living matter, a greyish semitransparent semifluid substance, of complex chemical composition, within which physical, chemical, and electrical changes are constantly taking place. *adj.* **protoplasmic.** Also **bioplasm.**

protoplasmic circulation (*Biol.*). The streaming motion that may be seen in the protoplasm of a living cell.

protoplasmic inclusions (*Cyt.*). See **inclusion.**

protoplasmic respiration (*Bot.*). Respiration going on at the expense of protein materials in a starved plant.

protoplast (*Biol.*). See **energid.**

protopod (*Zool.*). Said of a phase in the development of larval Insects in which the abdomen is imperfectly segmented and bears no appendages. Cf. *oligopod, polypod.*

protopodite (*Zool.*). The basal portion of a typical Arthropod limb.

protoporphyrin (*Chem.*). 1,3,5,8-Tetramethyl-2,4-divinyl-6,7-dipropanoic acid-porphin. It combines with iron (II) to give reduced haematin, the prosthetic group of haemoglobin.

protoporphyrinuria (*Med.*). The presence of protoporphyrin in the urine.

protopsis (*Med.*). Protrusion of the eye.

Protoselachii (*Zool.*). An order of *Elasmobranchii* having two types of tooth, pointed ones in front, and flattened ones for crushing molluscs, behind them. Many post-Devonian forms, and one living genus, the Port Jackson Sharks.

proto set (*Mining*). Mine rescue equipment, weighing some 18 kg and incorporating an oxygen cylinder, breathing bag, and face mask, used for work in foul underground air. Other types include the lighter Salvus.

protostele (*Bot.*). A stele in which the vascular

tissue forms a solid core, with centrally placed xylem surrounded by phloem.

protostigmata (*Zool.*). In a Craniate embryo, the primary pair of gill slits.

protothallus (*Bot.*). The first stages in the formation of the thallus of a lichen, often before the fungus and alga have become associated. The name is also applied to fringes of hyphae growing out from the edges of a mature thallus.

Prototheria (*Zool.*). A subclass of primitive Mammals, which probably left the main stock in the Mesozoic, and are found only in Australasia. The adults have no teeth, the cervical vertebrae bear ribs, the limbs are held laterally, the shoulder girdle has precoracoids and an interclavicle, and large yolky eggs are laid. The young are fed after hatching on milk produced by specialized sweat glands, whose ducts do not unite to open on nipples. One order, the *Monotremata*. Duck-billed Platypus, Spiny Anteater.

Prototracheata (*Zool.*). See **Onychophora.**

prototroch (*Zool.*). In a trochophore larva, the preoral circlet of cilia.

prototropic change (*Chem.*). Term applied in the form of isomerism known as *tautomerism* (q.v.) when the 'movement' of a proton is involved.

prototropy (*Chem.*). The reversible conversion of tautomeric forms by migration of hydrogen as a proton.

prototype Generally, the first or original type or model from which anything is developed. (*Zool.*) An ancestral form; an original type or specimen.

prototype filter (*Telecomm.*). Basic type which has the specified nominal cut-off frequencies, but which must be developed into derived forms to obtain further desirable characteristics, such as constancy of image impedance with frequency.

Protoungulata (*Zool.*). A superorder of the cohort **Ferungulata** (*Eutheria*; Mammals) containing the fossil order *Condylarthra,* order *Tubulidentata* (including the contemporary aardvark) and three small fossil orders known from S. America. First known from early Tertiary, i.e. Palaeocene era.

protovertebra (*Zool.*). In a Craniate embryo, the dorsal portion of a mesoderm band or mesoblastic somite, bordering the central nervous system and notochord.

protoxide (*Chem.*). Of a metal which forms more than one oxide, that containing the least combined oxygen.

protoxylem (*Bot.*). The first xylem to be formed from the procambial strand; it has annular, spiral, and loose scalariform vessels or tracheids, and these are usually of much smaller diameter than the corresponding elements of metaxylem and secondary xylem.

Protozoa (*Zool.*). A subkingdom and phylum of animals. Small animals of varying shape and complex microstructure, with no nucleus ever having sole charge of a specialized part of the cytoplasm, i.e., they are unicellular or acellular. Nutrition holophytic, holozoic or saprophytic; reproduction by fission or conjugation; locomotion by cilia, flagella, or pseudopodia; free-living or parasitic. *sing.* **protozoon.** Cf. *Metazoa, Parazoa.*

protozoaea (*Zool.*). In decapod *Crustacea,* a larval stage preceding the *zoaea* (q.v.), in which the abdominal region shows as yet no trace of segmentation.

protractor (*Instr.*). An instrument used by the draughtsman for measuring or setting out angles on paper, etc. (*Zool.*) A muscle which by its contraction draws a limb or a part of the body forward or away from the body. Cf. *retractor.*

Protura (*Zool.*). An order of minute *Apterygota* having 12 abdominal somites; antennae and compound eyes are lacking; 3 pairs of abdominal appendages usually occur; mouthparts piercing; usually found under bark, leaves, and stones, or in moss.

Protype (*Print.*). A minor filmsetting system, for display lines, using an ultraviolet light source and operated in ordinary daylight.

proud (*Eng., Join., etc.*). A part or portion of a part projecting above another or above its surroundings; standing proud.

proud flesh (*Med.*). The popular name for *granulation tissue* (q.v.) formed during wound healing.

proustite (*Min.*). Sulphide of silver and arsenic which crystallizes in the trigonal system. It is commonly associated with other silver-bearing minerals. Also called **light-red silver ore, ruby silver ore.** Cf. *pyrargyrite.*

proved reserves (*Mining*). Tonnages of economically valuable ore which have been tested adequately by being blocked out into panels and sampled at close intervals.

proventriculus (*Zool.*). (1) In Birds, the anterior thin-walled part of the stomach, containing the gastric glands. (2) In *Oligochaeta* and Insects, the gizzard—a muscular thick-walled chamber of the gut posterior to the crop. (3) In *Crustacea,* the stomach or gastric mill.

provisional sum (*Civ. Eng.*). A sum of money fixed by the engineer, and included in the bill of quantities, to provide for work not otherwise included therein or for any unforeseen contingency arising out of the contract.

provitamin (*Chem.*). A substance, not a vitamin, which is readily transformed into a vitamin within an organism; thus, β-carotene results in vitamin A, and ergosterol, after irradiation, in vitamin D_2.

proxan (*Chem.*). Sodium isopropylxanthate,

$$S{=}C\begin{cases}OCH(CH_3)_2\\SNa\end{cases}$$

used as a herbicide.

Proxima Centauri (*Astron.*). The nearest star to the sun; a faint companion to the double star *Alpha Centauri* in the constellation Centaurus, its distance being 4·3 light-years.

proximal (*Biol.*). Pertaining to or situated at the inner end, nearest to the point of attachment. Cf. *distal.*

proximity effect (*Elec.*). Increase in effective high-frequency resistance of a conductor when it is brought into proximity with other conductors, owing to eddy currents induced in the latter. It is especially prominent in the adjacent turns of an inductance coil.

proximity fuse (*Radar*). Miniature radar carried in shells or other missiles so that they explode within a preset distance of the target.

prozymogen (*Zool.*). Chromidial substance; a substance which will give rise to zymogen.

pruina (*Bot.*). A powdery bloom or secretion on the surface of a plant.

pruinose (*Bot.*). (1) Covered with a waxy or powdery bloom. (2) Covered with minute points which give a frosted appearance to the surface.

pruniform (*Bot.*). Shaped like a plum.

prunt (*Glass*). An applied mass of glass bearing a monogram or badge, e.g., of the owner or vintner.

pruriginous (*Med.*). Of the nature of prurigo.

prurigo (*Med.*). A term common to various skin diseases the chief characteristic of which is a papular eruption and intense itching.

pruritus (*Med.*). Severe persistent itching, a characteristic of numerous skin diseases.

Prussian blue (*Paint.*). Iron (III) hexacyanoferrate (II), $Fe_4[Fe(CN)_6]_3$. A blue pigment obtained through reactions between solutions of iron salts and alkaline hexacyanoferrates (II and III). Known by a variety of other names, e.g., **Berlin blue.**

prussic acid (*Chem.*). A solution of *hydrogen cyanide* (q.v.) in water.

Pryor's index (*Nut.*). An index, basiliac diameter to height, which, used in human growth studies, numerically expresses constitutional types, e.g., *amplosome*, *leptosome*, and *mesosome* (qq.v.).

ps. English transliteration of Gk. letter ψ. Pronounced *S*.

psalterium (*Zool.*). In ruminant Mammals, the third division of the stomach; also **omasum, manyplies.**

psammitic gneiss (*Geol.*). A gneissose rock which has been produced by the metamorphism of arenaceous sediments.

psammitic schists (*Geol.*). Schists formed from arenaceous sedimentary rocks. Cf. *pelitic gneiss, pelitic schist.*

psammoma (*Med.*). A tumour arising from the meninges, composed of fibrous tissue and endothelial cells and *brain-sand.*

psammophile (*Bot.*). A plant which inhabits sandy soils.

psammophyte (*Bot.*). A plant which occurs only on sand.

pseud-, pseudo-. Prefix from Gk. *pseudēs*, false.

pseudapogamy (*Bot.*). A replacement of a normal fusion of sexual nuclei by a fusion of two female nuclei, or of a female nucleus and a vegetative nucleus, or of two vegetative nuclei.

pseudapospory (*Bot.*). The formation of a spore without meiosis, the spore being diploid.

pseudautostyly (*Zool.*). A type of jaw suspension in which the upper jaw is fused with the ethmoidal, orbital, and otic regions of the cranium. Cf. *autostyly. adj.* **pseudautostylic.**

pseudaxis (*Bot.*). See **monochasium.**

pseudepisematic (*Zool.*). Alluring by false resemblance.

pseudhaemal (*Zool.*). Said of the so-called vascular system of *Echinodermata.*

pseudholoptic (*Zool.*). Of *Diptera*, having the condition of the compound eyes between holoptic and dichoptic.

pseudo- (*Chem.*). A prefix which is sometimes used to indicate a tautomeric, isomeric, or closely related compound. Symbol, ψ.

pseudoacid (*Chem.*). A substance which can exist in two tautomeric forms, one of which functions as an acid.

pseudoalums (*Chem.*). A name sometimes given to double sulphates of the alum type, where there is a bivalent element in place of the univalent element of ordinary alums.

pseudoamitosis (*Bot.*). An irregular nuclear division caused by treating cells with poisons.

pseudoangina (*Med.*). Angina innocens. The occurrence of symptoms of angina pectoris in the absence of organic disease of the heart.

pseudoaposematic (*Zool.*). Warning or aposematic coloration borne by animals which are not dangerous or distasteful, but show *Batesian mimicry* of animals which are.

pseudoasymmetry (*Chem.*). The asymmetry of an atom which results from the attachment thereto of two enantiomeric groups.

pseudobase (*Chem.*). A substance which can exist in two tautomeric forms, one of which functions as a base.

pseudoberry (*Bot.*). A fleshy fruit which looks like a berry, but in which some of the succulent material is derived from the enlarged, persistent perianth.

pseudobrachium (*Zool.*). In some Fish, an appendage used for propulsion along a substratum or on dry land; formed by modification of the pectoral fin.

pseudobranch (*Zool.*). A nonfunctional or vestigial gill, as the spiracular gill of *Selachii.*

pseudobulb (*Bot.*). A swollen stem internode formed by some orchids.

pseudocarp (*Bot.*). See **false fruit.**

pseudocaudal (*Zool.*). Said of a type of tail-fin found in some bony Fish which is contributed to by the dorsal and ventral median fins.

pseudocelli (*Zool.*). In *Collembola* and *Protura*, sense organs of unknown function distributed over various parts of the body.

pseudocilium (*Bot.*). A very thin, motionless, elongated outgrowth from an algal cell.

pseudocode (*Comp.*). Instructions written in symbolic language which must be translated into an acceptable program language or direct into machine language before they can be executed.

pseudocoele (*Zool.*). In higher Vertebrates, a space enclosed by the inner walls of the closely opposed cerebral hemispheres; the 5th ventricle.

pseudocolony (*Zool.*). In *Protozoa*, a collection of individuals united only by nonliving material.

pseudoconditioning (*An. Behav.*). The process occurring when a neutral (*unconditioned*) *stimulus* is presented several times, and then a new stimulus is similarly presented, and the normal (*unconditioned*) *response* to the *unconditioned stimulus* is given to the new stimulus.

pseudocone (*Zool.*). In Insects, said of compound eyes in which there is no true crystalline cone in each ommatidium, its place being taken by a transparent viscous fluid.

pseudoconjugation (*Zool.*). In some *Sporozoa*, temporary union of individuals without fusion or exchange of nuclear material.

pseudocowpox (*Vet.*). A papular and vesicular eruption affecting the bovine teat, believed to be due to an infectious agent unrelated to the virus of *vaccinia* (q.v.).

pseudocubic, pseudotetragonal, pseudohexagonal, etc. (*Min.*). See **pseudosymmetry.**

pseudocyesis (*Med.*). Spurious pregnancy. The condition in which women desiring offspring imagine themselves to be pregnant, the abdomen often being considerably enlarged owing to rapid accumulation of fat or to gas in the intestines. (*Zool.*) See **pseudopregnancy.**

pseudocyst (*Zool.*). An animal cell in which there is no definite nucleus but chromatin is scattered through the cytoplasm.

pseudodeltidium (*Zool.*). In some *Brachiopoda*, a plate partially or entirely closing the deltidial fissure of the ventral valve.

pseudoderm (*Zool.*). In some Sponges, a compact outer layer of tissue resembling skin.

pseudodont (*Zool.*). Having horny pads or ridges in place of true teeth, as Monotremes.

pseudo fowl plague (*Vet.*). See **Newcastle disease.**

pseudogamy (*Bot., Zool.*). Development without fertilization; male gamete stimulates development of ovum without fusion and therefore without contributing to its hereditary constitution. A union between 2 vegetative cells which are not closely related.

pseudogaster (*Zool.*). In some Sponges, a false gastric cavity into which true oscula open.

pseudogastrula (*Zool.*). In experimental embryology, an abnormal type of gastrula, produced artificially, in which there is invagination of the ectoderm.

pseudogyne (*Zool.*). An ant which has the characters of the worker combined with those of the female.

pseudohaemophilia hepatica (*Med.*). A haemorrhagic state resembling that of haemophilia, occurring in disease of the liver.

pseudoheart (*Zool.*). In *Oligochaeta*, one of a number of paired contractile anterior vessels by which blood is pumped from the dorsal to the ventral vessel; in *Echinodermata*, the axial organ.

pseudohypertrophic paralysis (*Med.*). A grave form of paralysis in children, in which progressive muscular weakness is associated with enlargement of certain groups of muscles (such as the calves, the buttocks, the muscles of the shoulder-blade), and wasting of others.

pseudoleucite (*Min.*). An aggregate showing the crystal shape of *leucite* (q.v.) but consisting mainly of potassium feldspar and nepheline.

pseudomalachite (*Min.*). Phosphate and hydroxide of copper which resembles malachite and crystallizes in the monoclinic system.

pseudomarine (*Ecol.*). Applied to freshwater forms bearing a superficial resemblance to marine types, but not necessarily closely related to them.

pseudomasturbation (*Psychiat.*). Nervous involuntary pulling of the penis. Also called **peotillomania.**

pseudomerism (*Chem.*). A form of tautomerism in which only one tautomer is known, although derivatives corresponding to both forms can be prepared.

pseudometamerism (*Zool.*). The condition of repetition of parts, found in some *Cestoda*, which bears a superficial resemblance to metamerism.

pseudomixis (*Biol.*). Fusion between two vegetative cells, or between cells which are not differentiated as gametes.

Pseudomonadaceae (*Bacteriol.*). A family of bacteria belonging to the order *Pseudomonadales*. Gram-negative rods, occurring in water and soil and including also animal and plant pathogens, e.g., *Pseudomonas aeruginosa* (blue pus), *Xanthomonas hyacinthi* (yellow rot of hyacinth). Acetobacter species are used in production of vinegar.

Pseudomonadales (*Bacteriol.*). One of the two main orders of true bacteria, distinguished by the polar flagella of motile forms. Gram-negative. Spiral, spherical or rod-shaped cells. See also **Eubacteriales.**

pseudomonotropic (*Chem.*). Existing in two forms stable under all conditions, but such that the transition between them takes place in only one direction.

pseudomorph (*Min.*). A mineral whose external form is not the one usually assumed by its particular species, the original mineral having been replaced by another substance or substances.

pseudomycorrhiza (*Bot.*). An association between a fungus and a higher plant in which the fungus is distinctly parasitic.

pseudonavicella (*Zool.*). In some *Sporozoa*, a small boat-shaped spore containing sporozoites.

pseudonotum (*Zool.*). See **postnotum.**

pseudonucleolus (*Cyt.*). A net knot in nuclear reticulum, not a true nucleolus.

pseudoparenchyma (*Bot.*). A mass of closely interwoven hyphae, appearing very like parenchyma, in prepared sections.

pseudoperianth (*Bot.*). A cuplike envelope surrounding the archegonia in some *Hepaticae*.

pseudoperidium (*Bot.*). A sheath of sterile hyphae surrounding the aecidium of the *Uredinales*.

pseudopod (*Zool.*). A footlike process of the body wall, characteristic of some Insect larvae.

pseudopodiospore (*Zool.*). See **amoebula.**

pseudopodium (*Bot.*). (1) A leafless branch formed by some mosses, bearing gemmae. (2) The stalk of the capsule in the bog mosses. (*Zool.*) In *Sarcodina* and phagocytic cells of *Parazoa* and *Metazoa*, a temporary protrusion of cytoplasm serving for locomotion or prehension.

pseudopregnancy (*Med.*). See **pseudocyesis.** (*Zool.*) In some Mammals, uterine changes following oestrus and resembling those characteristic of pregnancy.

pseudopterygium (*Med.*). The adherence of a tip of a fold of oedematous conjunctiva to a corneal ulcer, thus simulating a pterygium.

pseudorabies (*Vet.*). See **Aujesky's disease.**

pseudoracemic (*Chem.*). Consisting of mixed crystals of the dextrorotatory and laevorotatory forms of a compound.

pseudo random binary sequence (*Comp., Telecomm.*). A fixed length binary sequence which satisfies many of the tests for a true random sequence. Normally generated using a shift register with logical feedback. An *n*-stage register provides a sequence $p < 2^n$ bits. The maximum length sequence is $(2^n - 1)$. Abbrev. **PRBS.**

Pseudoscorpionidea (*Zool.*). An order of *Arachnida*, resembling *Scorpionidea*, but with no tail, i.e., the opisthosoma is not divided into mesosoma and metasoma; there is no telson; the pedipalps are large, chelate, and contain poison glands; respiration is by tracheae; the forelegs of the male are sometimes modified as sexual organs; small carnivorous forms found under stones, leaves, bark, and moss; occasionally found in houses; they are distributed by flying insects, to which they cling by their pedipalps. Also called **Chelonethida, False Scorpions.**

pseudoseptum (*Bot.*). A septum which is perforated by one or more pores; found in some lower Fungi.

pseudosessile (*Zool.*). In certain hymenopterous Insects, having a very short petiole so that the abdomen and thorax appear directly joined.

pseudosolution (*Chem.*). A colloidal solution or suspension.

pseudosperm (*Bot.*). A small indehiscent fruit which looks like a seed.

pseudostipule (*Bot.*). An appendage at the base of a leaf stalk, which looks like a stipule but is really part of the lamina.

pseudosymmetry (*Min.*). A term applied to minerals whose symmetry elements place them on the borderline between two crystal systems, e.g., a mineral with the *c*-axis very nearly equal to the *b*- and *a*-axes might, on casual inspection, appear cubic, though actually tetragonal. It would be described as possessing *pseudocubic symmetry*. The phenomenon is due to slight displacement of the atoms from the positions which they would occupy in the class of higher symmetry.

pseudotachylite (*Geol.*). Flinty crush-rock, resulting from the vitrification of rock powder produced during faulting under conditions involving the development of considerable heat by friction, as in the Glencoe *cauldron subsidence*.

pseudovelum (*Zool.*). In *Scyphozoa*, an internal flange occupying the same position as the true velum of *Hydrozoa* but lacking muscles and a nerve ring.

pseudovilli (*Zool.*). Projections from the surface of the trophoblast in some Mammals, as distinct from the true villi, which are definite out-growths.

pseudovitellus (*Zool.*). In some hemipterous Insects, an abdominal mass of cells which contains symbiotic microorganisms. See also **mycetocytes, mycetome.**

pseudovum (*Zool.*). A parthenogenetic ovum, capable of development without fertilization.

pseudozoaea (*Zool.*). In stomatopod *Crustacea*, a larval stage characterized by suppressed development of the thoracic appendages.

pseudozygospore (*Bot.*). See **azygospore.**

P-shell (*Nuc.*). The outermost electron shell for most of the heavy stable elements. Electrons in this shell have a principal quantum number 6.

psilomelane (*Min.*). A massive hydrous oxide of manganese which contains varying amounts of barium, potassium, and sodium. It is a secondary mineral formed by alteration of manganese carbonates and silicates, and is used as an ore of manganese.

Psilophytales (*Bot.*). An extinct order of palaeozoic *Pteridophyta*. Includes the oldest known land plants.

psilosis (*Med.*). See **sprue.**

Psilotales (*Bot.*). A mostly tropical and subtropical order of *Pteridophyta* related to the oldest known land plants, the *Psilophytales*. Contains the genera *Psilotum* and *Tmesipteris*.

Psittaciformes (*Zool.*). An order of *Neognathae* containing one family. Mainly vegetarian, with powerful hooked beaks; feet typically zygodactylous. The birds are often vividly coloured, and capable of mimicry. Parrots, Cockatoos.

psittacinite (*Min.*). See **mottramite.**

psittacosis (*Med., Vet.*). Parrot disease. A contagious disease of parrots and other birds due to a virus and communicable to man, in whom the symptoms of the disease resemble those of typhoid fever, accompanied by inflammation of the lungs.

psittacosis-lymphogranuloma viruses (*Bot.*). A group of Rickettsiae-like organisms, sometimes classified with the animal viruses, but which may be more closely related to bacteria. Large antigenically related organisms, which, unlike the small viruses, contain ribonucleic acid and deoxyribonucleic acid, muramic acid in the cell wall, and are susceptible to sulphonamides and certain antibiotics. Includes causative agents of trachoma, inclusion conjunctivitis, lymphogranuloma venereum, and psittacosis.

PSK (*Telecomm.*). Abbrev. for **phase shift keying.**

Psocoptera (*Zool.*). An order of the *Psocopteroidea* (Insects; *Paraneoptera*) with orthognathous head, biting mouthparts, long antennae, small wings with reduced or no venation, and 2- or 3-segmented tarsi. Booklice.

Psocopteroidea (*Zool.*). A superorder of the *Paraneoptera*, including the orders *Psocoptera*, *Mallophaga* and *Anoplura*.

psophometer (*Telecomm.*). An instrument, incorporating a suitable weighting network, which gives a visual indication that is equivalent to the aural effect of disturbing voltages of various frequencies.

psophometric voltage (*Telecomm.*). That which measures, by reference of the random spectrum to 800 Hz, the noise in a communication circuit arising from interference of any kind.

psoriasis (*Med.*). A chronic disease of the skin in which red scaly papules and patches appear, especially on the outer aspects of the limbs.

psoroptic mange (*Vet.*). Mange of animals due to mites of the genus *Psoroptes*. Sheep scab is caused by *Psoroptes communis ovis*.

pst (*Aero.*). Abbrev. for *static thrust*.

p-state (*Electronics*). That of an orbital electron when the orbit has angular momentum of one Bohr unit.

psych-, psycho-. Prefix from Gk. *psychē*, soul, mind.

psychalgia (*Med.*). (1) Hysterical pain or pain orginating in the mind of the sufferer. (2) Pain following brain surgery.

psychasthenia (*Psychiat.*). A functional mental disorder, characterized by fixed ideas, ruminative states, and hypochondriacal conditions.

psychedelic (*Psychol.*). Literally 'mind-opening'. Said of an agent, a drug such as *lysergic acid* (q.v.), which gives the consumer the impression of gaining access to new worlds of perception and knowledge. Some psychiatrists say that such agents bring unconscious materials to conscious levels. The word (and its better form *psychodelic*) is often applied to the blatantly bizarre manifestations of such mentality, whether real or imagined.

psychiatry. That branch of medical science which deals with disorders and diseases of the mind; in particular, with such as arise from physical and organic causes.

psychical research (*Psychol.*). See **parapsychology.**

psychic determinism (*Psychol.*). The Freudian theory that mental processes are determined by unconscious motivations.

psychism (*Biol.*). The doctrine that living matter possesses attributes not recognized in nonliving matter; the distinctive attributes or 'mentality' of living things.

psychoanalgesia (*Psychol.*). Relief of pain by psychological means, e.g., hypnotism.

psychoanalysis. The method of treatment of functional nervous disorder introduced by Freud. It consists in bringing unconscious conflicts into consciousness by free association, dream analysis, and use of the transference situation.

psychobiology (*Psychol.*). The study of the formation and function of the personality and of its psychosomatic effects.

psychogalvanic reflex (*Zool.*). The decrease in the electrical resistance of the skin under the stimulation of various emotional states.

psychogenic (*Med.*). Having a mental origin.

psychology. The study of human behaviour; the science of the mind. The general term is used to include all the observations, investigations and recordings of the mind and its functions.

psychometrics (*Psychol.*). (1) The measurement of intelligence. (2) The quantitative approach to psychological measurement which attempts to avoid the effects of subjective judgment.

psychoneurosis (*Psychiat.*). Functional disorder of the mind in a legally sane person who shows insight into his condition; the term includes such conditions as hysteria, obsessional states, and anxiety states. Freud distinguished psychoneurosis from the real or actual *neuroses* (q.v.). In both there is a derangement of the normal ways of gratification of the *libido* (q.v.), but in psychoneurosis this is always accompanied by unconscious mental conflict between the ego and the id, giving rise to symptoms and pathological states which are capable of being relieved by psychotherapy.

psychopath (*Psychol.*). An individual who shows a marked degree of emotional instability, but with no specific mental disorder. There is inability or unwillingness to conform to social mores but intellectual functions are not generally affected.

psychopathia sexualis (*Psychiat.*). Mental illness characterized by abnormal sexual behaviour.

psychopathology. That branch of psychology

which deals with the abnormal working of the mind.

psychophilae (*Bot.*). Plants fertilized by diurnal *Lepidoptera*; characterized by brightly coloured flowers and the provision of nectar.

psychophysics (*Psychol.*). The experimental study of the relationships between physical stimuli (or sets of stimuli) and sensation, perception and the responses to them.

psychophysiology (*Psychol.*). Study of the physical function of the brain.

psychoprophylaxis (*Med.*). A method of psychological and physical training for childbirth, aimed at making labour painless.

psychosexual development (*Psychol.*). The development of sexuality in the widest sense, including associated attitudes, objects, etc. Especially important in psychoanalysis which claims that the training and satisfaction obtained during the stages of infantile sexuality have a permanent effect on the personality.

psychosis (*Psychiat.*). A grave disorder of the mind, characterized by such phenomena as illusions, delusions, hallucinations, mental confusion, etc., with absence of insight into his condition on the part of the patient; the term includes such conditions as manic depressive psychosis, schizophrenia, paranoia, etc.

psychosocial (*Biol.*). Describes an organism which is effective, to some extent, in determining its own evolution, e.g., man.

psychosomatic medicine (*Med.*). Branch of medicine that stresses the relationship of bodily and mental happenings, and combines physical and psychological techniques of investigation. Particular attention is paid to the possibility of physical disease (i.e., duodenal ulcer, asthma) being induced by mental states.

psychosurgery (*Surg.*). Brain surgery for the relief of the symptoms of either mental or physical disease.

psychotherapist. An individual, usually a physician, but not always so at present, who practises *psychotherapy* (q.v.).

psychotherapy. The treatment of functional (and even so-called organic) psychic disorder. Methods: the different forms of analysis, as psychoanalysis (Freud) and its modifications, known as Jungian, Adlerian, Stekelian, and direct reductive analysis (Hadfield); explanation, persuasion, re-education, progressive relaxation, suggestion, hypnosis; occupational therapy.

psychotomimetic (*Chem.*). Said of drugs which cause bizarre psychic effects in humans as well as marked behavioural changes in animals, e.g., esters of *N*-methyl-3-hydroxypiperidine.

psychrometer (*Meteor.*). The *wet and dry bulb hygrometer* (q.v.).

psychrophilic (*Bot.*). Growing best below 20°C.

Pt (*Chem.*). The symbol for *platinum*.

Ptd.A. (*Build.*). Abbrev. for *pointed arch*.

pteralia (*Zool.*). See **axillary.**

pterergate (*Zool.*). A worker or soldier ant with vestigial wings, but otherwise unmodified.

Pteridophyta (*Bot.*). (Gk. *pteris*, gen. *pteridos*, fern.) The ferns, horsetails and club mosses, together with a few smaller groups. Vascular plants showing well-marked alternation of generations, with independent gametophytes bearing antheridia and/or archegonia, from which the spore-bearing plants develop. A very old group with many ancient fossil representatives.

Pteridospermae (*Bot.*). An ancient group of plants known from fossils but having no living representatives. They were fernlike in some respects, but produced seeds, often apparently on the edges of their leaves.

Pterobranchiata (*Zool.*). A class of *Hemichorda.* Sessile tubicolous animals with gills reduced or absent and a U-shaped gut. The collar bears two or several hollow branched ciliated arms.

pteropaedes (*Zool.*). Young Birds which are able to fly as soon as they are hatched.

pteropegum (*Zool.*). In insects, the socket for the insertion of the wing.

pteropleuron (*Zool.*). In *Diptera*, a thoracic sclerite lying between the root of the wing, the sternopleuron, and the mesopleuron.

pteropod ooze (*Geol.*). A calcareous deep-sea deposit which contains a large number of pteropod remains.

Pteropsida (*Bot.*). A comprehensive name for the *Filicineae*, *Gymnospermae*, and *Angiospermae*, plants having in general large leaves, and invariably having leaf gaps in the stele.

pterosphenoids (*Zool.*). In the skull of *Teleostei*, large paired bones lying above the single basisphenoid, and forming the outer parts of the side walls of the cranium.

pterostigma (*Zool.*). In insects, an opaque cell on the wing.

pterotheca (*Zool.*). In the pupal stage of some Insects, the wing case.

pterotic (*Zool.*). A bone of the lateral wall of the auditory capsule of the skull in some Vertebrates.

pteroyl glutamic acid (*Chem.*). See **vitamin B complex.**

pterygial (*Zool.*). In Fish, an element of the fin skeleton; pertaining to a fin; pertaining to a wing.

pterygium (*Med.*). The encroachment on to the cornea from the side of a thickened, vascular, wing-shaped area of the conjunctiva. (*Zool.*) In Vertebrates, a limb; in some *Coleoptera*, a process of the prothorax.

pterygoda (*Zool.*). See **tegula.**

pterygoid (*Zool.*). A paired cartilage bone of the Vertebrate skull, formed by the ossification of the front part of the PPQ bar; a membrane bone which replaces the original pterygoid in some Vertebrates; more generally, wing-shaped.

pterygopalatoquadrate bar (*Zool.*). In Fish with a cartilaginous skeleton, the rod of cartilage forming the upper jaw and known as the **PPQ bar.**

pterygophore, pterygiophore (*Zool.*). In *Elasmobranchii* and *Osteichthyes*, cartilaginous rays, which may or may not ossify, supporting the median and paired fins.

Pterygota (*Zool.*). A subclass of Insects (cf. *Apterygota*). The young initially have the full complement of adult segments, there are no abdominal locomotory appendages, Malpighian tubules are present, and the mouthparts are free. The adults normally possess wings. Contains the divisions *Palaeoptera* and *Neoptera*.

pterylae (*Zool.*). The tracts of contour feathers on the body of a Bird.

pterylosis (*Zool.*). In Birds, the arrangement of the feathers in distinct feather tracts or pterylae, whose form and arrangement are important in classification.

PTFCE (*Plastics*). See **polytrifluorochloroethene.**

PTFE (*Plastics*). See **polytetrafluoroethene.**

pthanite (*Geol.*). A fine-grained sedimentary rock composed of silica; a *chert*.

ptilinum (*Zool.*). In certain *Diptera* (*Cyclorrhapha*), an expansible membranous cephalic sac by which the anterior end of the puparium is thrust off at emergence.

Ptolemaic system (*Astron.*). The final form of

Greek planetary theory as described in Claudius Ptolemy's treatise. In this the earth was the centre of the world, the planets, including the sun and moon, being supposed to revolve round it in motions compounded of eccentric circles and epicycles; the fixed stars were supposed to be attached to an outer sphere concentric with the earth.

Ptolemy's theorem (*Maths.*). That the product of the diagonals of a cyclic quadrilateral is equal to the sum of the products of the opposite sides.

ptomaines (*Chem.*). Poisonous amino compounds produced by the decomposition of proteins, especially in dead animal matter. The ptomaines include substances such as putrescine, cadaverine, choline, muscarine, neurine. Few of the ptomaines are known to be poisonous by mouth, food poisoning being caused by specific bacteria, e.g., *Bacillus botulinus*.

ptosis (*Med.*). (1) Paralytic dropping of the upper eyelid. (2) Downward displacement of any bodily organ.

P-trap (*San. Eng.*). A type of trap shaped like the letter P, and having a nearly horizontal outlet; used in sanitary pipes.

ptyalin (*Zool.*). An enzyme, found in saliva, which catalyses the conversion of starch into sugar.

ptyalism (*Med.*). Excessive secretion of saliva.

***p*-type conduction** (*Electronics*). That arising in a semiconductor containing acceptor impurities, with conduction by (positive) holes.

***p*-type conductivity** (*Electronics*). That apparently resulting from movement of positive charges; actually *holes* in impurity component in a semiconductor.

***p*-type semiconductor** (*Electronics*). One in which the hole conduction (absence of electrons) exceeds the electron conduction, the acceptor impurity predominating.

ptyxis (*Bot.*). The manner in which an individual leaf is folded in bud.

Pu (*Chem.*). Symbol for ***plutonium***.

puberty (*Physiol.*). Sexual maturity.

puberulent (*Bot.*). Feebly pubescent.

pubescence (*Bot.*). A covering of downy hairs closely pressed on to the surface bearing them. (*Zool.*) A covering of fine hairs or down. *adj.* **pubescent.**

pubiotomy (*Surg.*). The operation of cutting the pubic bone to one side of the midline, so as to facilitate childbirth in difficult labour.

pubis (*Zool.*). In *Craniata*, an element of the pelvic girdle (contr. of *os pubis*). *adj.* **pubic.**

public-address system (*Acous.*). Sound-reproducing system for large space and outdoor use, usually with high-powered horn radiators, or columns of open diaphragm units which concentrate radiation horizontally.

publisher's binding (*Bind.*). See **edition binding.**

pucella (*Glass*). An implement for opening out the top of a wine glass in the off-hand process.

pucherite (*Min.*). Vanadate of bismuth, crystallizing in the orthorhombic system.

pucking cutter (*Mining*). A man employed in a coal mine to cut the floor in cases of creep or upheaval towards the roof.

Puckle time base (*Telecomm.*). The means of generating a sawtooth waveform by charging a capacitor via a pentode and discharging through a multivibrator.

puddingstone (*Geol.*). A popular term for *conglomerate*. Hertfordshire Puddingstone, consisting of rounded flint pebbles set in a siliceous sandy matrix, is a good example.

puddle (*Civ. Eng.*). See **clay puddle.**

puddled ball (*Met.*). The mass of iron intimately mixed with slag which is formed by the process of puddling pig iron. See **puddling.**

puddled bars (*Met.*). Bars of wrought iron which have been rolled from the *puddled ball* (q.v. after squeezing the ball to compact it and eliminate some of the slag.

puddling (*Met.*). The agitation of a bath of molten pig iron, by hand or by mechanical means, in an oxidizing atmosphere to oxidize most of the carbon, silicon, and manganese and thus produce wrought iron. (*Min. Proc.*) Concentration of diamond from 'blue ground' clays (weathered kimberlite) by forming an aqueous slurry in a mechanized stirring pan in which the heavier fraction is retained for periodic retrieval, while the lighter fraction overflows.

puddling furnace (*Met.*). A small reverberatory furnace in which iron is puddled.

Pudlo (*Build.*). TN for a substance, sold in powder form, which is used as a waterproofing agent for concrete surfaces.

puerilism (*Psychiat.*). Reversion to a childlike state of mind.

puering (*Leather*). The process of steeping skins in a warm fermenting solution containing pancreatic enzymes to soften them before tanning.

puerperal (*Med.*). Pertaining to or ensuing upon childbirth, e.g., **puerperal** fever, a condition caused by infection of the genital tract in the course of childbirth.

puerperium (*Med.*). Strictly, the period between the onset of labour and the return to normal of the generative organs: usually, the first 5 or 6 weeks after the completion of labour.

puffing (*Bot.*). The simultaneous and violent discharge of ascospores from many asci.

pug (*Hyd. Eng.*). To prevent leakage by packing cracks with clay; the material so used. (*Mining*) In metalliferous mining, the parting of soft clay which sometimes occurs between the walls of a vein and the country rock.

pugging (*Build.*). A special mixture carried on boards between the floor joists, serving to insulate the room against sounds from below. (*Paint.*) Mixture of pigment with minimum oil to form a very thick paste.

pugioniform (*Bot.*). Shaped like a dagger.

pug lifter (*Mining*). One who removes coal left adhering to the floor by a coal-cutting machine.

pug mill (*Civ. Eng.*). Mixing machine for wet materials, as used in the making of mortar.

pug piles (*Civ. Eng.*). Dovetailed piles.

pulaskite (*Geol.*). An alkali syenite containing a wide variety of sodic ferromagnesian minerals and often a small amount of nepheline.

pulegone (*Chem.*). $C_{10}H_{16}O$, occurs in oil of pennyroyal, a ketone of the terpene series.

Pulfrich refractometer (*Phys.*). Instrument for measuring the refractive index of oils and fats.

pull (*Typog.*). An impression taken for checking.

pulled coil (*Elec. Eng.*). An armature coil wound with parallel sides on a suitable former and then pulled out to the correct coil span.

pullet disease (*Vet.*). See **avian monocytosis.**

pulley (*Eng.*). A wheel on a shaft, sometimes having a crowned or cambered rim, for carrying an endless belt, or grooved for carrying a rope or chain. A *fast pulley* is one that is keyed to the shaft and revolves with it; a *loose pulley* is not attached to the shaft. The term 'pulley' is also applied to a small grooved wheel over which a sash cord, etc., runs.

pulley mortise (*Join.*). A form of joint between the end of a ceiling joist, which is tenoned, and the binding joist, which is mortised, so as to let in the ceiling joist in a position such that the lower faces of both are in the same plane.

pulley stile (*Join.*). One of the upright sides of the frame of a double-hung window, to which is secured the pulley over which the sash cord passes.

pulling (*Telecomm.*). Variation in frequency of an oscillator when the load on it changes.

pulling by crystal (*Crystal.*). Growing both metal and nonmetal crystal by slowly withdrawing a crystal from a molten surface. See **zone refining**.

pulling figure (*Telecomm.*). The stability of an oscillator, measured by the maximum frequency change when the phase angle of the complex reflection coefficient at the load varies through 360° and its modulus is constant and equal to 0·2.

pulling focus (*Cinema.*). The alteration of focus during a shot, e.g., during a pan or tracking shot, so that the same or different objects remain in focus in spite of varying distance.

pull-off (*Elec. Eng.*). A fitting used in connexion with the overhead contact-wire of an electric traction system for retaining the contact-wire in the correct position above the track on curves.

pullorum disease (*Vet.*). Bacillary white diarrhoea (BWD). A disease of young chicks caused by *Salmonella pullorum*; affected chicks appear drowsy and huddled, may develop diarrhoea, and often die.

pull-out (*Acous.*). A common defect in the pressing of gramophone records, a small portion of the pressing being pulled away from the disk and remaining on the stamper. (*Aero.*) The transition from a dive or spin to substantially normal flight. (*Bind.*) See **fold-out**. (*Telecomm.*) Sudden release of an oscillator, which has been pulled off its own frequency of oscillation by a variable-frequency independent oscillator.

pull-out distance (*Aero.*). A naval term for the distance travelled by an aircraft while arresting on the deck of a carrier.

pull-out torque (*Elec. Eng.*). (1) The value of the torque at which a synchronous motor falls out of synchronism. (2) The maximum torque of an induction motor.

pull-over mill (*Met.*). A rolling-mill using a single pair of rolls. The metal, after passing through the rolls, is pulled back over the top roll to be fed through the mill a second time.

pull switch (*Elec. Eng.*). See **ceiling switch**.

pullulation (*Bot.*). Budding, sprouting.

pulmobranch (*Zool.*). See **lung book**.

pulmonary (*Zool.*). In land Vertebrates, pertaining to the lungs; in pulmonate *Mollusca*, pertaining to the respiratory cavity.

pulmonary adenomatosis (*Vet.*). Jaagziekte. A disease of sheep characterized by chronic pneumonia, believed to be caused by an infective agent.

pulmonary valvotomy (*Surg.*). The surgical restoration of pulmonary valve function, when the valve will not open properly.

Pulmonata (*Zool.*). An order of *Gastropoda*. Hermaphrodite; exhibit torsion; have a shell but no operculum; nervous system symmetrical; mantle cavity forming a lung with no ctenidium but a vascular roof and a small aperture (the pneumostome); one kidney; direct development. Land and freshwater snails, land slugs.

pulmonate (*Zool.*). Possessing lungs or lung-books; air-breathing.

pulmonectomy (*Surg.*). See **pneumonectomy**.

pulmones (*Zool.*). Lungs. *sing.* **pulmo.** *adj.* **pulmonary.**

pulmonitis (*Med.*). Inflammation of lung tissue; pneumonia.

pulp (*Min. Proc.*). Finely ground ore freely suspended in water at a consistency which permits flow, pumping or settlement in quiet conditions. (*Paper*) See **wood pulp**. (*Zool.*) A mass of soft spongy tissue situated in the interior of an organ, as *spleen pulp*, *dental pulp*.

pulp boards (*Paper*). See **cardboards**.

pulping (*For.*). Any process, mechanical or chemical, that disintegrates the fibres in wood. Also called **defibration**.

pulp-saver (*Paper*). A machine through which water from the paper-making machine passes to avoid wastage of fibrous materials.

pulpwood (*Join.*). A board composed of wood-pulp and adhesive pressed into sheets; used for panelling, partitions, etc.

pulpy kidney disease (*Vet.*). An acute and fatal toxaemia of lambs due to enteric infection by the bacterium *Clostridium perfringens* (*Cl. welchii*), type D; the kidneys of affected lambs frequently show a characteristic degenerative change after death.

pulsar (*Astron.*). A peculiar radio source, emitting regular but extremely rapid pulses of radio energy, generally with a period of only a fraction of a second. First detected in 1967, more than sixty pulsars are now known. The effect is probably due to the rapid rotation of a *neutron star*.

pulsatance (*Phys.*). See **angular frequency**.

pulsating current (*Elec. Eng.*). One taking the form of a succession of isolated pulses, and usually unidirectional, which changes in magnitude in a regularly recurring manner.

pulsating stars (*Astron.*). Variables of the Cepheid type, the pulsation theory assuming that the stars contract and expand, being brightest when most contracted. The theory is inadequate to explain all the features of Cepheid variability.

pulsator (*Agric.*). A piece of apparatus attached to a milking machine which causes alternations of suction and release on the cow's teats when the machine is in operation; the pulsations range from 50–60 a minute. (*Min. Proc.*) Mineral jig of Harz type.

pulse (*Elec. Eng.*, *Telecomm.*). One *step* followed by a reverse *step*, after a finite interval. A unidirectional flow of current of nonrepeated waveform, i.e., consisting of a transient and a zero-frequency component greater than zero. Measured either by peak value, duration, or integration of magnitude over time. Obsoletely, **impulse.** (*Med.*) The periodic expansion and elongation of the arterial walls caused by the pressure wave which follows each contraction of the heart.

pulse amplifier (*Telecomm.*). One with very wide frequency response which can amplify pulses without distortion of the very short rise time of the leading edge.

pulse amplitude (*Telecomm.*). That of the crest relative to the quiescent signal level. Sometimes a mean taken over the pulse duration.

pulse amplitude modulation (*Telecomm.*). That which is impressed upon a pulse carrier as variations of amplitude. They may be either unidirectional or bidirectional according to the system employed.

pulse bandwidth (*Telecomm.*). The frequency band occupied by Fourier components of the pulse which have appreciable amplitude and which make an appreciable contribution to the actual pulse shape.

pulse code (*Telecomm.*). Coding of information

by pulses, either in amplitude, length, or absence or presence in a given time interval.

pulse-code modulation (*Telecomm.*). Modulation in accordance with a *pulse code* (q.v.).

pulse crest factor (*Telecomm.*). The ratio of the peak and r.m.s. amplitudes.

pulsed columns (*Chem. Eng.*). Columns used for *liquid-liquid extraction* (q.v.) in which there is a series of horizontal plates containing many small holes. Liquids flow continuously through the column but a reciprocating pump is attached to cause pulsations in the flow which leads to high velocity mixing as the liquids are pulsed through the perforated plates. Can also be applied to packed or even simple empty columns.

pulsed Doppler (*Radar*). Doppler system in which direction and range can be measured.

pulse decay time (*Telecomm.*). Time for decay between the arbitrary limits of 0·90 and 0·10 of the maximum amplitude. Also **pulse fall time.**

pulse delay circuit (*Telecomm.*). One through which the propagation of a signal takes a known time.

pulse discriminator (*Telecomm.*). Any circuit capable of discriminating between pulses varying in some specific respect (e.g., duration, amplitude or interval).

pulsed radar system (*Radar*). One transmitting short pulses at regular intervals and displaying the reflected signals from the target on the screen of a CRO.

pulse droop (*Telecomm.*). The exponential decay of amplitude which is often experienced with nominally rectangular pulses of appreciable duration.

pulse duration (*Telecomm.*). (1) Time interval for which the amplitude exceeds a specified proportion (usually $1/\sqrt{2}$) of its maximum value. Also called **pulse length, pulse width.** (2) The duration of a rectangular pulse of the same maximum amplitude and carrying the same total energy.

pulse duty factor (*Telecomm.*). The ratio of pulse duration to spacing.

pulse extra-high-tension generator (*TV*). The means of generating extra-high-tension supply for a TV tube by the rectification of the voltage pulses occurring in the anode circuit of a valve, when the anode current is suddenly 'cut-off'.

pulse fall time (*Telecomm.*). **Pulse decay time.**

pulse-forming line (*Radar*). An artificial line which generates short high-voltage pulses for radar.

pulse frequency modulation (*Telecomm.*). Frequency modulation of a pulse carrier wave by an input signal. Also **pulse interval modulation.**

pulse generator (*Telecomm.*). One supplying single or multiple pulses, usually adjustable for *pulse repetition frequency* amplitude and width. May be self-contained or require sine wave input signal.

pulse height analyser (*Electronics*). (1) A single or multichannel pulse height selector followed by equipment to count or record the pulses received in each channel. The multichannel units are known as **kicksorters.** (2) One which analyses statistically the magnitudes of pulses in a signal using pulse code modulation.

pulse height discriminator (*Electronics*). See **discriminator** (2).

pulse height selector (*Electronics*). Circuit which accepts pulses with amplitudes between two adjacent levels and rejects all others. An output pulse of constant amplitude and profile is produced for each such pulse accepted. The interval between the two reference amplitudes is termed the *window* or *channel width* and the lower level the *threshold.*

pulse interleaving or **interlacing** (*Telecomm.*). Adding independent pulse trains on the basis of time division multiplex along a common path.

pulse interval modulation (*Telecomm.*). See **pulse frequency modulation.**

pulse ionization chamber (*Nuc. Eng.*). One for detection of individual particles by their primary ionization. Must be followed by a very high gain stable amplifier but has a much shorter dead time than proportional or Geiger counters.

pulse jet (*Aero.*). A propulsive duct with automatic air intake valves, or a frequency-tuned jet pipe, so that pressure builds up between 'firings', thus achieving thrust at reasonable economy at moderate air speeds, e.g., 200–400 m.p.h. (350–650 km/h).

pulse jitter (*Telecomm.*). Irregularities in pulse spacing.

pulse method of measuring sound velocity (*Telecomm.*). The velocity is measured by comparing the time of transit of an acoustic pulse through the medium with that of a radar signal over the same distance.

pulse pile-up (*Telecomm.*). Failure to separate adjacent pulses due to finite resolving time of detection circuit. In nuclear counting it occurs where pulses have random distribution, and in communication circuits when the pulse repetition frequency or pulse duration are variable.

pulse position modulation (*Telecomm.*). A form of **pulse-time modulation.**

pulser (*Telecomm.*). A simple *pulse generator* (q.v.) used to drive pulse-controlled circuits.

pulse regeneration or **restoration** (*Telecomm.*). Correction of pulse to its original shape after phase or amplitude distortion.

pulse repeater (*Telecomm.*). See **repeater.**

pulse repetition frequency or **rate** (*Telecomm.*). The average number of pulses in unit time.

pulse rise time (*Telecomm.*). Time required for amplitude to rise from 0·10 to 0·90 of its maximum value.

pulse shaping and re-shaping (*Telecomm.*). Adjustment of a pulse to square-wave or other form by electronic means.

pulse spectrum (*Telecomm.*). The distribution, as a function of frequency, of the magnitudes of the Fourier components of a pulse.

pulse spike (*Telecomm.*). A subsidiary pulse superimposed upon a main pulse.

pulse stretcher (*Telecomm.*). An electronic unit used to increase the time duration of a pulse.

pulse-time modulation (*Telecomm.*). Modulation by pulses, the duration or relative position of these depending on a signal to be transmitted.

pulse time multiplex (*Telecomm.*). See **time division multiplex.**

pulse transformer (*Telecomm.*). One designed to accept the very wide range of frequencies required to transmit pulse signals without serious distortion.

pulse wave (*Med., Physiol.*). A wave of increased pressure travelling along the arterial system and generated by the discharge of blood from the heart into the aorta during ventricular systole.

pulse-width modulation (*Telecomm.*). One form of **pulse-time modulation.**

pulsing (*Telecomm.*). Said of an oscillatory valve when it alternates between, or sustains oscillations of, two frequencies.

pulsing signal (*Teleph.*). One which carries information to steer the call to the desired point.

pulsometer pump (*Eng.*). A steam pump in which an automatic ball valve, the only moving part, admits steam alternately to a pair of chambers, so forcing out water which has been sucked in by condensation of the steam after the previous stroke.

pulverized-coal burners (*Eng.*). See **fantail burner, turbulent burner.**

pulverized coal or fuel (*Fuels*). Finely ground coal which is burnt as it issues from a nozzle when blown through by compressed air.

pulvillus (*Zool.*). One of a pair of pads situated beneath the tarsal claws in Insects. *adj.* **pulvillar, pulvilliform.**

pulvinate (*Bot.*). Shaped like a cushion.

pulvinated (*Carp.*). Said of a frieze which presents a bulging face.

pulvinule (*Bot.*). The small *pulvinus* (q.v.) of a leaflet.

pulvinus (*Bot.*). A swollen leaf base, often capable of changes of form, bringing about movement of the leaf.

pulviplume (*Zool.*). See **powder down-feather.**

pumice (*Geol.*). An 'acid' vesicular glass, formed from the froth on the surface of some particularly gaseous lavas. The sharp edges of the disrupted gas vesicles enable pumice to be used as an abrasive.

pummel (*Build.*). See **punner.**

pump. A machine driven by some prime mover, and used for raising fluids from a lower to a higher level, or for imparting energy to fluids.

pumped storage (*Hyd. Eng.*). Pumping of water to high-level storage during off-peak periods of electric power generation, for return to source via turbines at periods of high demand.

pumped tube (*Electronics*). Transmitting valve, X-ray tube or other electronic device which is continuously evacuated during operation.

pumped vacuum system (*Vac. Tech.*). A vacuum system from which gas or vapour is exhausted continuously or intermittently.

pumpellyite (*Min.*). A complex greenish hydrous orthosilicate of calcium, magnesium, iron, and aluminium crystallizing in the monoclinic system. Found in low grade metamorphic rocks and in amygdales in some basaltic rocks.

pump frequency (*Electronics*). Frequency of oscillator used in *maser* or *parametric amplifier* to provide the stored energy released by the input signal.

pumping (*Phys.*). The motion of mercury in a barometer arising from the movement of a ship or from fluctuations of air pressure in a varying wind.

pumping speed (*Phys.*). Rate at which a pump removes gas in creating a near-vacuum; measured in dm^3 or cu. ft per minute, s.t.p., against a specified pressure.

pump-line (*Elec. Eng.*). A cable extending throughout the length of an electric train for the control of auxiliary apparatus such as air compressors.

punch (*Civ. Eng.*). A follower placed on the head of a pile in any case where the pile has been driven in beyond the stroke of the monkey. (*Eng.*) (1) A tool for making holes by shearing out a piece of material corresponding in outline to the shape of the punch; a machine incorporating such a tool. (2) A hand tool, struck with a hammer, for marking a surface, or for displacing metal as in riveting. See **centre punch.** (*Typog.*) The first stage in the making of type, the letter being cut, nowadays usually by machine, in mild steel which, after hardening, is struck into copper or one of its alloys, to make a *matrix* from which the type is cast.

punched card (*Comp.*). One providing stored information or instructions for use in computer or data processing system.

punched screens (*Min. Proc.*). Suitably perforated robust plates which act as industrial screens.

punched tape (*Comp.*). Tape into which coded patterns of holes are punched, constituting the source of input and output data; originated in telegraphy. Also **paper tape.**

puncheon (*Carp.*). A short post giving intermediate support to a beam carried between principals, especially in trussed partitions.

punching (*Comp.*). Process of preparing punched cards or tapes. (*Elec. Eng.*) See **lamination.**

punching machine (*Eng.*). A machine for punching holes in plates, the punch being driven either mechanically by a crank and reciprocating block, or by fluid power.

punch-through (*Electronics*). Collector-emitter voltage breakdown in transistors.

punctate basophilia (*Med.*). See under **basophilia.**

punctum (*Zool.*). A minute aperture; a dot or spot in marking. *adj.* **punctate.**

punctured (*Build.*). A term applied to a variety of rusticated work distinguished by holes picked in the faces of the stones, either in lines or irregularly.

puncture test (*Paper*). Test measuring the energy required to force a standard pyramidal puncture head operated by the fall of a loaded sector-shaped pendulum through a sample of container board.

pungent (*Bot.*). Ending in a sharp, hard point.

punishment (*An. Behav.*). See **negative reinforcing stimulus.**

punner (*Build., Civ. Eng.*). A heavy-headed tool, with a long upright handle, used in the operation of punning, the tool being repeatedly lifted and dropped on to the surface to be consolidated. Also applied to rods used for thrusting into and ensuring the thorough consolidation of concrete.

punning (*Build., Civ. Eng.*). (1) The operation of ramming or consolidating the surface of hardcore, concrete, earth, etc., with repeated blows from a heavy-headed tool, such as a *punner*. (2) The operation of consolidating concrete by means of constant driving in of a metal rod to distribute the mortar and aggregate evenly within the mix.

punt (*Glass*). The bottom of a container.

punt codes (*Glass*). Identification marks on the base of a glass container.

punty, puntee, pontie, pontil (*Glass*). A short iron rod, at one end of which is either a button of hot glass or a suitably shaped piece of metal, which is applied hot to the end of a partially formed glass article in order that (*a*) it may be cracked off the blowpipe and manipulated on the punty, or (*b*) in the case of tube drawing, the mass of glass may be drawn out between punty and blowpipe.

pupa (*Zool.*). An inactive stage in the life history of an Insect during which it does not feed and reorganization is taking place to transform the larval body into that of the imago. *adj.* **pupal.**

puparium (*Zool.*). The hardened and separated last larval skin which is retained to form a covering for the pupa in some *Diptera*.

pupil (*Zool.*). The central opening of the iris of the eye. *adj.* **pupillary.**

pupilometer (*Optics*). An instrument for measuring the size and shape of the pupil of the eye and its position with respect to the iris.

Pupin cable (*Elec. Eng.*). A cable in which the conductors are coil-loaded at intervals, with resulting attenuation, which is uniform up to a cut-off frequency and then increases rapidly.

Pupin coil (*Elec. Eng.*). See **loading coil.**

pupiparous (*Zool.*). Giving birth to offspring which have already reached the pupa stage, as some two-winged flies.

Purbeck Beds, Purbeckian Stage (*Geol.*). A series of limestones, largely nonmarine, dirt-beds, and 'marbles', found in the Upper Jurassic rocks of Southern England, following directly upon the Portlandian Stage.

purchase (*Eng.*). (1) Mechanical advantage or leverage. (2) Mechanical appliance for gaining mechanical advantage.

Purdy's solution (*Chem.*). A modified Fehling's solution containing sufficient ammonia to re-dissolve the copper (I) oxide as it is formed. Similar to Pavy's solution.

pure clay (*Build.*). See **foul clay.**

pure colour (*Light*). One with *CIE coordinates* lying on the *spectrum locus* or on the *purple boundary*.

pure continuous waves (*Radio*). Waves which are not modulated or broken up into trains except by keying, as distinguished from interrupted continuous waves.

pure culture (*Bot.*). A culture containing a pure stock of one species of plant, especially of lower plants.

pure line (*Zool.*). (1) A homogenous collection of individuals within a species, resulting from autogamous reproduction. (2) A population consisting of individuals whose descent can be traced to a single ancestor.

pure metal crystals (*Met.*). The crystals of which a solid pure metal is composed. Each crystal in a given metal has a similar structure consisting of the same atoms arranged in the same way, though one crystal differs from another in orientation.

pure tone (*Telecomm.*). See **simple tone.**

purgative (*Med.*). An evacuant, e.g., senna, cascara, etc., used to treat functional constipation.

purine group (*Chem.*). A group of cyclic diureides derived from 1 molecule of a dibasic hydroxy acid and 2 molecules of urea. The simplest member of this group is *purine*, of the formula:

Adenine, caffeine, guanine, hypoxanthine, theobromine, theophylline and xanthine are other members of the purine group.

Purkinje cells (*Zool.*). Large flask-shaped cells lying between the two layers of grey matter in the cortex of the cerebellum in higher Vertebrates.

Purkinje effect (*Optics*). Shift of maximum sensitivity of human eye towards blue end of spectrum at very low illumination levels.

Purkinje fibres (*Zool.*). A network of large beaded trabeculae lying under the endocardium in some animals.

purl (or pearl) fabrics (*Textiles*). Knitted fabrics in which the reverse side stitches are brought to the surface for effect; used extensively for pull-overs, etc.

purlin (*Build., Carp.*). A member laid horizontally on the principal rafters and supporting the common rafters.

purlin post (*Carp.*). A post placed beneath a purlin to support it against sagging.

purple boundary (*Light*). Straight line joining the ends of the *spectrum locus* on a *chromaticity diagram*. The coordinates of all real colours fall within the loop formed by these 2 lines.

purple brown (*Paint.*). A precipitated iron oxide pigment calcined at a high temperature to give a very deep red-purple colour.

purple copper ore (*Min.*). Bornite.

purpleheart (*For.*). Heavy and resilient hardwood from the genus *Peltogyne*, used for fancy veneers, etc. Also called **amaranth, violetwood.**

purple of Cassius (*Chem.*). Produced by adding a mixture of tin (IV) and tin (II) chlorides to a very dilute solution of gold (III) chloride; hydrated tin (IV) oxide is precipitated and the gold (III) chloride reduced to metal. The red-to-violet colour is due to the precipitation of finely divided gold on the tin (IV) hydroxide. Used in the making of ruby glass.

purpose-made brick (*Build.*). A brick which has been specially moulded to a shape suiting it for use in a particular position, e.g., an arch brick shaped like the voussoir of an arch.

purposive behaviour (*An. Behav.*). Behaviour considered with respect to its goals, though not implying awareness of the goal on the part of infra-human animals.

purposiveness (*Zool.*). Correlation of individual reactions to a definite end.

purpura (*Med.*). The condition in which spontaneous haemorrhages appear beneath the skin and the mucous membranes, forming purple patches; these may occur as a result of infection, or the cause may be unknown. See also **Henoch's purpura and peliosis rheumatica.** *adj.* **purpuric.**

purpura haemorrhagica (*Med.*). Haemogenia. Essential thrombocytopenia. Werlhof's disease. Morbus maculosus. A disease characterized by purpuric haemorrhages in the skin, bleeding from mucous membranes, and abnormal diminution of the platelets in the blood (*thrombocytopenia*). (*Vet.*) An acute, noncontagious disease occurring mainly in horses, characterized by mild fever, subcutaneous oedema and haemorrhages in the mucous membranes. In the horse usually follows an infectious disease, such as strangles or equine influenza, and is thought to be allergic in origin.

purpuric acid (*Chem.*). Barbituryl iminoalloxan, of the formula:

The ammonium salt is *murexide* (see **murexide test**).

pursuit (*Nav.*). Navigation of a missile by electronic means whereby the guided vehicle is always on line of sight.

purulent (*Med.*). Forming or consisting of pus; resembling or accompanied by the formation of pus; of the nature of pus.

pus (*Med.*). Matter. The yellowish fluid formed by suppuration, consisting of serum, pus cells (white blood cells), bacteria, and the debris of tissue destruction.

pushbutton (*Elec. Eng.*). A device, carrying a small current, which closes or opens an electric circuit by means of pressure on a small button.

pushbutton tuning (*Radio, TV*). Selection by push-button of any one of a number of preset tuned circuits in a receiver, to change stations or channels quickly.

pushed punt (*Glass*). A concave bottom to a container. Also **push-up.**

pusher airscrew (*Aero.*). See **airscrew.**

pusher-on (*Mining*). The man in charge of haulage hands in a coal mine. Also called **master haulier.**

push-off stacker (*Agric.*). A stacker, usually tractor mounted, to discharge the crop from the raised tines by a forward thrust of the back of the frame.

push piece (*Horol.*). (1) A small cylindrical plunger which projects just beyond the band of a watch case and is pressed in when it is required to set the hands. (2) The button and stem of a hunting watch, which, on being pressed, causes the opening of the case.

push-pull (*Acous.*). Term applied to sound-tracks which carry sound recordings in antiphase. They are *class-A* when each carries the whole waveform and *class-B* when each carries half the waveform, both halves being united optically or in a push-pull photocell. (*Telecomm.*) General principle of transmission, whereby one leg of a balanced circuit is driven by a periodic waveform, while the other leg is driven by the same waveform with phase reversed.

push-pull amplifier (*Electronics*). One in which two matching amplifying devices (e.g. transistors) each conduct alternate halves of the input signal, their outputs being recombined in antiphase. Used for the reduction of harmonic distortion in amplifiers, and in short-wave oscillators and amplifiers, etc.

push-pull microphone (*Acous.*). Carbon microphone in which two carbon-granule cells are mounted on either side of a stretched diaphragm, so that amplitude distortion arising in one is partially balanced by the opposite-phase amplitude distortion in the other.

push-pull oscillator (*Telecomm.*). One depending on a pair of push-pull amplifiers, positive feedback being taken from each output to the opposite input, thus maintaining symmetry and balance.

push-pull sound-track (*Cinema.*). A system of sound recording on film in which the two halves of recorded waves are made transparent and displaced, there being no transmitted light with zero modulation.

push-push amplifier (*Telecomm.*). One which uses two similar valves (or transistors) with their grids in phase opposition but their anodes parallel-connected to a common load. By this means, even-order harmonics are emphasized.

push rod (*I.C. Engs.*). A rod through which the tappet of an overhead-valve engine operates the rocker arm, when the camshaft is located in the crankcase.

push-up (*Glass*). See **pushed punt.**

pustular stomatitis (*Vet.*). See **horse pox.**

pustule (*Bot.*). A mass of fungal spores and the hyphae bearing them. (*Med.*) A small elevation of the skin containing pus. *adjs.*, **pustular, pustulous.**

pusule (*Bot.*). A small vacuole present in the protoplast of some lower plants, which is able to expand and contract. (*Zool.*) In *Dinoflagellata*, a noncontractile vacuole discharging to the exterior by a duct.

putamen (*Bot.*). The hard endocarp of a drupe, such as the stone of a plum. (*Zool.*) In Birds, the shell membrane of the egg: in higher Vertebrates, the lateral part of the lentiform nucleus of the cerebrum.

putlog (*Build.*). A transverse bearer which, in the case of a bricklayer's scaffold, is lashed at one end to the ledger and at the other end is wedged into a hole left by the bricklayer in the wall; used to support the scaffold boards.

putrefaction (*Chem.*). The chemical breaking down or decomposition of plants and animals after death. This is caused by the action of anaerobic bacteria and results in the production of obnoxious or offensive substances.

putrescine (*Chem.*). $H_2N{\cdot}(CH_2)_4{\cdot}NH_2$, 1,4-diaminobutane, crystals, m.p. 27°C, formed during the putrefaction of flesh.

putter (*Mining*). A man employed in a coal mine to take empty tubs from a flat or siding to the working face, and to bring back full ones. Also called **kibbler.**

putter and filler (*Mining*). A man employed in a coal mine to fill the tubs, and move them to the main underground haulage.

putty (*Build.*). See **glazier's-, painter's-, plasterer's-.**

putty and plaster (*Build.*). See **gauged stuff.**

putty powder (*Build.*). Tin oxide, used for polishing glass.

puy (*Geol.*). The name given to a small volcanic cone, especially in the Auvergne, France.

puzzle box (*An. Behav.*). A box in which an animal is confined, and from which it can escape only by performing a particular series of manipulations which it must discover by trial and error or problem-solving behaviour.

puzzolano (*Build., Civ. Eng.*). See **pozzuolana.**

PVA (*Plastics*). Abbrev. for *polyvinyl acetate*, and for *polyvinyl alcohol.*

PVC (*Plastics*). Abbrev. for *polyvinyl chloride.*

P-wire (*Teleph.*). The third, guard, or private wire of the three which constitute a channel through an exchange; it holds the circuit when established, and, by retaining an earth, guards the circuit at every point from an extraneous connexion.

PX (*Teleph.*). Abbrev. for *private exchange.*

Py (*Chem.*). A symbol for the *pyridine* (q.v.) a nucleus.

py-, pyo-. Prefix from Gk. *pyon*, pus.

pyaemia (*Med.*). The condition in which infection of the blood with bacteria, from a septic focus, is associated with the development of abscesses in different parts of the body. Also **pyemia.** *adj.* **pyaemic.**

pycastyle (*Build.*). See **pycnostyle.**

pycn-, pycno-, pykn-, pykno-. Prefix from Gk. *pyknos*, compact, dense.

pycnidiospore (*Bot.*). A spore formed within a pycnidium.

pycnidium (*Bot.*). A roundish fructification formed by many species of fungi, usually with an opening, and having a general resemblance to a perithecium. It contains fertile hyphae and pycnidiospores, but no asci, and has apparently no connexion with any sexual act, real or modified.

pycniospore (*Bot.*). A term used, chiefly in America, to denote the spermatium in *Uredinales.*

pycnium (*Bot.*). A term used, chiefly in America, to denote the spermogonium of the *Uredinales.*

pycnoconidium (*Bot.*). A conidium formed inside a pycnidium.

Pycnogonida (*Zool.*). See **Pantopoda.**

pycnometer (*Chem.*). A small, graduated glass vessel, of accurately defined volume, used for determining the relative density of liquids. (*Civ. Eng.*) A bottle device for measuring the relative density of soil particles smaller than about 5 mm.

pycnosis (*Bot.*). The formation of a perithecium, under the cover of the tissue of a stroma. (*Cyt.*) The shrinkage of the stainable material of a nucleus into a deeply staining knot, usually a feature of cell degeneration.

pycnospore (*Bot.*). (1) A spore formed inside a pycnidium. (2) See **spermatium** (*Uredinales*).

pycnostyle (*Build.*). A colonnade in which the space between the columns is equal to 1½ times the lower diameter of the columns. Also **pycastyle.**

pycnoxylic wood (*Bot.*). The compact wood characteristic of pine trees, with little or no parenchyma.

pyelitis (*Med.*). Inflammation of the pelvis of the kidney. (Gk. *pyelos*, trough.)

pyelocystitis (*Med.*). Inflammation of the pelvis of the kidney and of the bladder.

pyelography (*Radiol.*). Radiography of the pelvis, the kidney and the ureter, after these have been filled with a substance opaque to X-rays.

pyelolithotomy (*Surg.*). The operation for removal of a stone from the pelvis of the kidney.

pyelonephritis (*Med.*). Inflammation of the kidney and of its pelvis.

pyemia (*Med.*). See **pyaemia.**

pygal (*Zool.*). Pertaining to the posterior dorsal extremity of an animal; in *Chelonia*, a posterior median plate of the carapace. (Gk. *pygē*, rump.)

pygidium (*Zool.*). In Insects, the tergum of the last abdominal somite; the suranal plate.

pygochord (*Zool.*). In some *Hemichorda*, a median ventral ridgelike outgrowth of the intestinal epithelium.

pygostyle (*Zool.*). In Birds, a bone at the end of the vertebral column formed by the fusion of some of the caudal vertebrae.

pyinkado (*For.*). Durable timber from *Xylia dolabriformis*, found in India and Burma. Typical uses include dock work, heavy duty flooring, railway sleepers, piling, etc.

pyknic type (*Psychol.*). One of Kretschmer's three types of individual, characterized by short squat stature, small hands and feet, domed abdomen, round face, and short neck, the limbs being short in relation to the trunk. People in this group are extrovert individuals with happy jovial temperaments, tending to develop manic-depressive psychosis if any breakdown occurs.

pyknolepsy (*Med.*). A form of epilepsy in which there are sudden attacks of momentary loss of consciousness; eventually the attacks disappear.

pyknometer (*Chem.*). See **pycnometer.**

pyknosome (*Nut.*). See **amplosome.**

pylephlebitis (*Med.*). Inflammation of the portal vein (the vein formed by veins running from the spleen, stomach, and intestines, and entering the liver) with or without thrombosis; in suppurative pylephlebitis abscesses form in the liver.

pylocyte (*Zool.*). See **porocyte.**

pylome (*Zool.*). In some *Sarcodina*, an opening through which pseudopodia can be protruded and food particles taken in.

pylon (*Eng.*). A slender vertical structure which is self-supporting. Cf. *mast.* (*Elec. Eng.*) See **tower.**

pylorectomy (*Surg.*). Excision of the pylorus.

pyloric gland (*Histol.*). Branched mucous secreting glands, found embedded in the stroma, in the pyloric region of the stomach.

pyloric stenosis (*Med.*). See **hypertrophic pyloric stenosis.**

pyloroplasty (*Surg.*). An operation for widening the lumen of the pylorus when this has been pathologically narrowed.

pylorospasm (*Med.*). Spasm of the circular muscle of the pyloric part of the stomach.

pylorus (*Zool.*). In Vertebrates, the point at which the stomach passes into the intestine. *adj.* **pyloric.**

pyocolpos (*Med.*). A collection of pus in the vagina, the result of infection of a *haematocolpos* (q.v.) which has been inadequately treated.

pyogenic (*Med.*). Having the power to produce pus.

pyometra (*Med.*). A collection of pus in the cavity of the uterus.

pyonephrosis (*Med.*). Accumulation of pus in the pelvis of the kidney.

pyopneumothorax (*Med.*). The presence of pus and air or gas in the pleural cavity.

pyorrhoea, pyorrhea (*Med.*). (Lit., a flow of pus.) The term now used as a synonym for *pyorrhoea alveolaris*, a purulent inflammation of the periosteum round a tooth.

pyosalpingitis (*Med.*). Purulent inflammation of a Fallopian tube.

pyosalpinx (*Med.*). Accumulation of pus in a Fallopian tube.

pyralspite (*Min.*). A group name for the *pyr*ope, *al*mandine, and *sp*essartine garnets.

pyramid (*Crystal.*). A crystal form with three or more inclined faces which cut all three axes of a crystal. See also **bipyramid.** (*Ecol.*) See **biomass, energy, numbers—pyramids of.** (*Maths.*) A polyhedron formed of a multiplicity of triangular faces joined to each other around a common point (the *vertex*) and to a polygonal face (the *base*). (*Zool.*) A conical structure, as part of the medulla oblongata in Vertebrates. *adj.* **pyramidal.**

pyramidal disease (*Vet.*). Exostosis affecting the extensor (pyramidal) process of the third phalanx of the foot of the horse.

pyramidal horn (*Acous.*). One with linear flare-out in both planes. Cf. *sectoral horn.*

pyramidal system (*Crystal.*). See **tetragonal system.**

pyramidal tract (*Zool.*). In the brain of Mammals, a large bundle of motor axons carrying voluntary impulses from particular areas of the cerebral cortex.

pyranometer (*Phys*). An instrument designed to measure solar radiation by its heating action on two blackened metallic strips of different thickness which thereby assume different temperatures. Also called **solarimeter.**

pyrargyrite (*Min.*). Sulphide of silver (I) and antimony (III) which crystallizes in the trigonal system. It is commonly associated with other silver-bearing minerals; cf. *proustite.* Also called **dark-red silver ore, ruby silver ore.**

pyrazinamide (*Pharm.*). Pyrazinecarboxyamide, $C_5H_5N_3O$, used in treatment of tuberculosis.

pyrazines (*Chem.*). Six-membered heterocyclic rings containing 2 nitrogen atoms in the 1,4 positions.

pyrazole (*Chem.*). $C_3H_4N_2$, long needles, m.p. 70°C, b.p. 185°C. It is a weak secondary base. Fuming sulphuric acid forms a sulphonic acid. Pyrazole and its derivatives can be halogenated, nitrated, diazotized, and generally treated in a similar way to benzene or pyridine.

pyrazoles (*Chem.*). Heterocyclic compounds containing a 5-membered ring consisting of 3 carbon and 2 nitrogen atoms arranged thus:

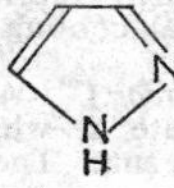

Pyrazole derivatives are formed by the condensation of hydrazines with compounds containing 2 CO groups, or a CO and a COOH group, in the *beta* position, or which contain a CO or COOH group attached to a doubly linked carbon atom. Pyrazoles have a similar chemical character to benzene and pyridine.

pyrazolone dyes (*Chem.*). A series of dyes made by direct condensation of pyrazolone with diazonium salts.

pyrene (*Bot.*). A small hard body containing a single seed, comparable with the stone of a plum except that pyrenes often occur several together in one fruit. (*Chem.*) A tetracyclic hydrocarbon obtained from the coal-tar fraction boiling above 360°C:

$= C_{16}H_{10}$

Colourless, monoclinic crystals, m.p. 148°C; soluble in ether, slightly soluble in ethanol and insoluble in water. Has carcinogenic properties.

pyrenin (*Biochem.*, *Cyt.*). A chromatinlike material of nucleoli; paranuclein.

pyrenocarp (*Bot.*). See **perithecium.**

pyrenoid (*Bot.*). A small mass of refractive protein occurring singly or in numbers in or on the chromatophores of some algae and *Bryophyta*, and concerned in the formation of carbohydrates. (*Zool.*) A centre of starch formation in some *Mastigophora*.

Pyrenomycetes (*Bot.*). One of the major subdivisions of the *Ascomycetes*, including some 12 000 species, mostly of small fungi. The characteristic fructification is a perithecium, either formed singly on the hyphae or developed in groups in a mass of hyphae known as a stroma, which assumes many forms. In general the perithecia are dark brown or black, though some are brightly coloured. Includes many plant parasites.

pyrethrins (*Chem.*). Active constituents of pyrethrum flowers used as standard contact insecticide in fly-sprays, etc.; remarkable for the very rapid paralysis ('knock-down' effect) produced on flies, mosquitoes, etc. Pyrethrum root is the source of a similar substance used as a sialagogue.

pyretic (*Med.*). Pertaining to fever.

pyretotherapy (*Med.*). The treatment of disease by artificially increasing body temperature. This is generally accomplished by inoculation with a fever-producing organism (e.g. that concerned with malaria); it has been used in certain cases of arthritis, syphilis, and cancer.

Pyrex glass. TN designating a proprietary range of glasses used as heat-resisting glasses and as line insulators. The glass is essentially a soda-alumina-borosilicate glass, having very little alkali and a high silica content.

pyrexia (*Med.*). An increase above normal of the temperature of the body; fever. *adj.* **pyrexial.**

pyrgeometer (*Heat*). An instrument consisting of a number of blackened and polished surfaces, designed to measure the loss of heat by radiation from the earth's surface. The surfaces exhibit differential cooling depending on the radiation loss.

pyrheliometer (*Meteor.*). An instrument for measuring the rate at which heat energy is received from the sun. The earlier forms used mercury thermometers; more modern types embody a bolometer or a thermocouple to measure the rate of rise of temperature of a black surface heated by the sun. See **solar constant.**

pyridazines (*Chem.*). Six-membered heterocyclic rings containing two nitrogen atoms in the 1,2 positions.

pyridine (*Chem.*). A heterocyclic compound containing a ring of 5 carbon atoms and 1 nitrogen atom, having the formula C_5H_5N:

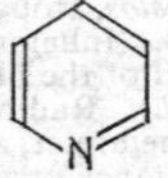

It occurs in the coal-tar fraction with a boiling range between 80°C and 170°C; a colourless liquid of pungent, characteristic odour, b.p. 114°C; a very stable compound and resists oxidation strongly.

pyridoxal phosphate (*Chem.*). See **vitamin B complex.**

pyridoxin (*Chem.*). See **vitamin B complex**

pyriform (*Bot.*, *Zool.*). Pear-shaped, as the *pyriform organ* of a *Cyphonautes* larva.

pyriform organ (*Zool.*). In the larval stage of some *Ectoprocta*, a mass of cells in close relation to the ectodermal groove, believed to be of sensory function.

pyrimidine (*Chem.*). A six-membered heterocyclic compound containing two nitrogen atoms in the 1,3 positions.

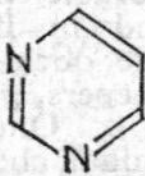

pyrite, pyrites (*Min.*). Sulphide of iron (FeS_2) crystallizing in the cubic system. It is brassy yellow and is the commonest sulphide mineral of widespread occurrence. Also known as **fool's gold, iron pyrite, mundic.**

pyritic smelting (*Met.*). Blast furnace smelting of sulphide copper ores, in which heat is partly supplied by oxidation of iron sulphide.

pyritohedron (*Crystal.*). See **pentagonal dodecahedron.**

pyro- (*Chem.*). A prefix used to denote an acid (and the corresponding salts) which is obtained by heating a normal acid, and thus contains relatively less water, e.g., *pyrosulphuric acid*, $H_2S_2O_7$. (*Photog.*) Abbrev. for *pyrogallol.*

pyroborates (*Chem.*). Generally known as *borates.*

pyroboric acid (*Chem.*). See **boric acid.**

pyrocatechin, pyrocatechol (*Chem.*). See **catechol.**

Pyrochlor (*Chem.*). TN for a non-inflammable transformer oil. Mixture of 60% hexachlorodiphenyl and 40% trichlorobenzene.

pyrochlore (*Min.*). Essentially a niobate of sodium, calcium and other bases, with iron, uranium, manganese, zirconium, titanium, thorium, and fluorine; crystallizes in the cubic system. It is found in nepheline-syenites, and in alkaline pegmatites.

pyrochroite (*Min.*). Hydroxide of manganese, crystallizing in the trigonal system.

pyroclastic rocks (*Geol.*). A name given to fragmental deposits of volcanic origin.

pyrocondensation (*Chem.*). A molecular condensation caused by heating to a high temperature, e.g., the formation of biuret from urea.

pyroelectric effects (*Min. Proc.*). In high-tension (electrostatic) separation, the electrical charging of particles by heating.

pyroelectricity (*Min.*). Polarization developed in some hemihedral crystals by an inequality of temperature.

pyrogallol (*Chem.*). Pyrogallic acid, 1,2,3-trihydroxy-benzene, $C_6H_3(OH)_3$, white plates, m.p. 132°C:

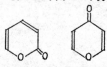

It sublimes without decomposition, is soluble in water, and is a strong reducing agent. Used as an absorbing agent for oxygen in gas analysis. Also as a developing agent with tanning action, now little used owing to its staining propensities. See also **phloroglucinol.**

pyrogenic (*Chem.*). Resulting from the application of a high temperature.

pyrogens (*Med.*). Bacterial polysaccharides which produce febrile reactions.

pyroligneous acid (*Chem.*). An aqueous distillate obtained by the destructive distillation of wood, which contains ethanoic acid, methanol, acetone, and other products.

pyrolusite (*Min.*). (IV) oxide of manganese, commonly containing a little water, and crystallizing in the tetragonal system. It typically occurs massive and as a pseudomorph after manganite, and is used as an ore of manganese, as an oxidizer, and as a decolorizer. Cf. *polianite.*

pyrolysis (*Chem.*). The decomposition of a substance by heat.

pyromeride (*Geol.*). An anglicized French term for nodular rhyolite. It is a quartz-felsite or devitrified rhyolite containing spherulites up to several centimetres in diameter which impart a nodular appearance to the rock.

pyrometallurgy (*Met.*). Treatment of ores, concentrates, or metals dry, at high temperatures. Techniques include smelting, refining, roasting, distilling, alloying, and heat treatment.

pyrometer (*Heat*). Instrument similar to thermometer used particularly for measuring high temperatures. See **Seger cones.**

pyrometric cone equivalent (*Chem.*). A measure of the m.p. of a refractory, carried out by means of *Seger cones* (q.v.). Abbrev. **PCE.**

pyrometric cones (*Heat*). See **Seger cones.**

pyromorphite (*Min.*). Phosphate and chloride of lead, crystallizing in the hexagonal system. It is a mineral of secondary origin, frequently found in lead deposits; a minor ore of lead.

pyrones (*Chem.*). 6-membered heterocyclic compounds containing a ring of 5 carbon atoms and 1 oxygen atom, 1 of the former being oxidized to a CO group. According to the position of the CO and the O in the molecule, there are α-pyrones (1,2) and γ-pyrones (1,4):

α-pyrone γ-pyrone

pyrope (*Min.*). The fiery-red garnet; silicate of magnesium and aluminium, crystallizing in the cubic system. It is often perfectly transparent and then prized as a gem, being ruby-red in colour. It occurs in some ultrabasic rocks and in eclogites.

pyrophilous (*Bot.*). Growing on ground which has been recently burnt over.

pyrophoric metals (*Met.*). Those liable to spontaneous combustion under conditions which may arise in a nuclear reactor. The nuclear fuels U, Th, and Pu are all pyrophoric.

pyrophoric powders (*Chem.*). Finely divided powders which take fire or oxidize extremely rapidly when exposed to the air; usually a metal, or a mixture of a metal and its oxide.

pyrophosphoric acid (*Chem.*). HPO_3, obtained by the loss, through heating, of one H_2O molecule from orthophosphoric acid, H_3PO_4.

pyrophyllite (*Min.*). A soft hydrous silicate of aluminium crystallizing in the monoclinic system. It occurs in metamorphic rocks; often resembles talc.

pyrophyte (*Bot.*). A plant which is protected by a thick bark from permanent damage by fires.

pyrosis (*Med.*). See **waterbrash.**

Pyrosomatidae (*Zool.*). An order of *Thaliacea* in which the larval stage is lacking; the oözooid is degenerate and retained within the parent; a cylindrical colony is formed from the blastozooids by budding. Also **Luciae.**

pyrostibite (*Min.*). See **kermesite.**

pyrotechny (*Chem.*). The study and manufacture of fireworks.

Pyrotenax (*Cables*). TN of a type of *mineral-insulated cable* (q.v.); used for a magnesia insulation for low-voltage cables. It is very tough, non-inflammable, and heat-resisting.

pyrotron (*Nuc. Eng.*). Thermonuclear device based on cylindrical discharge chamber with magnetic mirrors at the ends.

pyroxene group (*Min.*). A number of mineral species which, although falling into different systems (orthorhombic, monoclinic), are closely related in form, composition, and structure. They are metasilicates of calcium, magnesium, and iron, sometimes with manganese, titanium, sodium, or lithium. See **aegirine, augite, diallage, diopside, enstatite, hypersthene, orthopyroxene.**

pyroxenite (*Geol.*). A coarse-grained, holocrystalline igneous rock, consisting chiefly of pyroxenes. It may contain biotite, hornblende, or olivine as accessories. See **enstatite, hypersthenite.**

pyroxilins (*Chem.*). Nitrocelluloses with a low nitrogen content, containing from 2 to 4 nitrate groups in the molecule. Used in an ethanol-ethoxyethane solution to form collodion. Pyroxilin is a synonym for *guncotton* (q.v.)

pyroxmangite (*Min.*). A triclinic metasilicate of manganese and iron.

pyrrhotite (*Min.*). Ferrous sulphide which contains variable amounts of dissolved sulphur Fe_7S_8 to FeS; it crystallizes in the monoclinic system but is pseudohexagonal. Also called **magnetic pyrite.**

pyrrole (*Chem.*). A heterocyclic compound having a ring of 4 carbon atoms and 1 nitrogen:

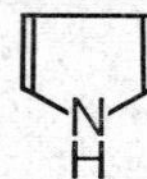

a colourless liquid of chloroform-like odour b.p. 131°C, rel. d. 0·984. Pyrrole is a secondary base, and is found in coal tar and in bone-oil (*Dippel's oil* (q.v.)). Numerous natural colouring matters are derivatives of pyrrole, e.g., chlorophyll and haemoglobin.

pyrrolidine (*Chem.*). The final reduction product of pyrrole, a colourless, strongly alkaline base, b.p. 86°C, having the formula:

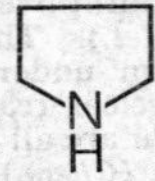

pyrroline (*Chem.*). A reduction product of pyrrole obtained by treating it with zinc and glacial acetic acid. It is a colourless liquid, b.p. 91°C, and is a strong secondary base. It has the formula:

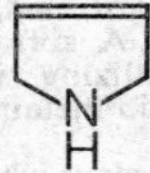

Pyruma (*Build.*). TN for a fireclay cement used in forming heat-resistant joints.

pyruvic acid (*Biochem.*). A 2-keto acid $CH_3COCOOH$. It occurs in several metabolic reaction pathways and can undergo reduction to lactic acid, be converted to carbohydrate, be used for the formation of oxalacetic or malic acid, be transaminated to form alanine or be oxidized to CO_2 and acetyl coenzyme A.

Pythagoras's theorem (*Maths.*). That the square on the hypotenuse of a right-angled triangle is equal to the sum of the squares on the other two sides.

pyuria (*Med.*). The presence of pus in the urine.

pyx, trial of the (*Met.*). Periodic official testing of sterling coinage.

pyxidate (*Bot.*). Having a lid.

pyxidium (*Bot.*). A capsule which dehisces by means of a transverse circular split, so that the upper part of the pericarp forms a lid.

PZT (*Astron.*). Abbrev. for *photographic zenith tube*.

q (*Chem.*). A symbol for the quantity of heat which enters a system.

Q (*Chem. Eng., etc.*). Symbol for *throughput* (q.v.) (*Hyd. Eng.*). Symbol for quantity of water discharged, usually in m³/s.

Q (*Elec*). (1) A symbol for *charge*. (2) Symbol of merit for an energy-storing device, resonant system or tuned circuit. Parameter of a tuned circuit such that

$$Q = \omega L/R, \quad \text{or} \quad 1/\omega CR,$$

where L=inductance, C=capacitance and R=resistance, considered to be concentrated in either inductor or capacitor. Q is the ratio of shunt voltage to injected e.m.f. at the resonant frequency $\omega/2\pi$. $Q=f_r/(f_1-f_2)$, where f_r is the resonant frequency and (f_1-f_2) is the bandwidth at the half-power points. For a single component forming part of a resonant system it equals 2π times the ratio of the peak energy to the energy dissipated per cycle. For a dielectric it is given by the ratio of displacement to conduction current. Also called **magnification factor, *Q*-factor, quality factor, storage factor.**

***Q*₁₀** (*Biol.*). See **temperature coefficient.**

***Q*-band** (*Radio*). Frequency band used in radar, 36 to 46 GHz.

Q code (*Aero.*). An internationally-understood code established by *ICAO*, consisting of groups of letters which represent interrogatory phrases used in various aircraft flight operations.

***Q*-electron** (*Nuc.*). One of two in the Q-shell for radioactive elements of at. no. 88 and above.

***Q*-factor** (*Elec.*). See *Q*.

***q*-feel** (*Aero.*). A term given (because of the use of q in a relative equation) to a device which applies an artificial force on the control column of a power-controlled aircraft proportional to the aerodynamic loads on the control surfaces, thereby simulating the natural 'feel' of the aircraft throughout its speed range. Colloquially *q-pot* because of its barometric bellows.

q.g. (*For.*). Abbrev. for *quarter girth*.

***Q*-gas** (*Chem.*). One based on helium (98·2% He, 1·8% butane), widely used in gas-flow counting.

***Q*-meter** (*Phys.*). Laboratory instrument which measures the *Q*-factor of a component.

***Q*-point** (*Electronics*). The static operating point for a valve or transistor.

***q*-pot** (*Aero.*). See ***q*-feel.**

QPP amplifier (*Telecomm.*). Abbrev. for *quiescent push-pull amplifier*.

Q.S. (*Build.*). *Quick sweep; quantity surveyor.*

***Q*-shell** (*Nuc.*). The outermost electron shell ever normally occupied by electrons.

***Q*-signal** (*Teleg.*). First of 3-letter code for standard messages in international telegraphy. (*TV*) In the NTSC colour system, that corresponding to the narrow-band axis of the chrominance signal.

***Q*-switching** (*Phys.*). The optical pumping of a laser having a low *Q*-factor and then suddenly increasing the *Q* of the system. In this way, a very short output pulse of high intensity is obtained.

***Q*-value** (*Nuc. Eng.*). See p. 967.

quad (*Elec. Eng.*). Either four insulated conductors twisted together (*star-quad*) or two twisted pairs (*twin-quad*). Normally a single structural unit of a multiconductor cable. (*Paper*) Prefix often used in premetrication paper sizes to denote a size 4 times single and twice double size, e.g., demy, 17½ × 22½ in.; double-demy, 22½ × 35 in.; quad demy, 35 × 45 in. Double quad demy would be 45 × 70 in. Similarly with other basic sizes, e.g., quad crown, quad royal, etc. (*Typog.*) See **quadrat.**

quadr-, quadri-. Prefix from L. *quattuor*, four.

quadra (*Arch.*). A plinth at the base of a podium.

quadrant (*Civ. Eng.*). A special granite sett, usually 18 × 18 × 8 in. deep, shaped like a sector of a circle enclosing 90°. (*Eng.*) A slotted segmental guide through which an adjusting lever (e.g., a reversing lever) works. It is provided with means for locating the lever in any desired angular position. See **link motion.** (*Maths.*) (1) Of a circle, a quarter of the circle. (2) Rectangular cartesian axes divide the plane into four quadrants numbered one to four anticlockwise around the origin from the first in which both variables (x and y) are positive. (*Surv.*) An angle-measuring instrument of the sextant type, but embracing an angle of 90° or a little more. (*Zool.*) A section of a segmenting ovum originating from 1 of the 4 primary blastomeres.

quadrantal deviation (*Ships, etc.*). Those parts of the *deviation* (q.v.) which vary as sine and cosine of twice the *compass course* (q.v.), thus changing their sign quadrantally with change in direction of the ship's head.

quadrantal points (*Nav.*). Points of the compass which in moving from north correspond to the headings NE(45°), SE(135°), SW(225°), and NW(315°). Cf. *cardinal points.*

quadrant dividers (*Carp.*). A form of dividers in which one limb moves over an arc fixed rigidly to the second limb, and may be secured to it by tightening a binding screw.

quadrant electrometer (*Elec. Eng.*). See **Dolezalek quadrant electrometer.**

quadrat (*Bot.*). A square area of vegetation marked off for study. (*Typog.*) A piece of metal less than type height, for spacing. Also **quad.**

quadrate (*Bot.*). Square to squarish in cross-section or in face view. (*Zool.*) A paired cartilage bone of the Vertebrate skull formed by ossification of the posterior part of the PPQ bar, or the corresponding cartilage element prior to its ossification; except in Mammals, it forms part of the jaw-articulation.

quadratic equation (*Maths.*). An algebraic equation of the second degree, i.e.,

$$ax^2+bx+c=0,$$

whose solution is

$$x=\frac{-b\pm\sqrt{b^2-4ac}}{2a}.$$

quadratic system (*Crystal.*). See **tetragonal system.**

quadrature (*Astron.*). Position of the moon or a superior planet in elongation 90° or 270°, i.e., when the lines drawn from the earth to the sun and the body in question are at right angles. (*Elec. Eng., Telecomm.*) See under **phase.**

quadrature component (*Elec. Eng.*). See **reactive component.**

quadrature reactance (*Elec. Eng.*). A term used in the 2-reaction theory of synchronous machines to denote the ratio which the synchronous reactance drop produced by the quadrature component of the armature current bears to actual value of quadrature component.

quadrature transformer (*Elec. Eng.*). A transformer designed so that secondary e.m.f. is 90° displaced from primary e.m.f.

quadratus (*Zool.*). A muscle of rectangular appearance, e.g., *quadratus femoris.*

quadric (*Maths.*). The three-dimensional surface represented by a general second-degree equation in three variables. By a suitable choice of coordinates such an equation can be reduced to one of the following standard equations:

(1) $\pm\frac{x^2}{a^2}\pm\frac{y^2}{b^2}\pm\frac{z^2}{c^2}=1$,
an ellipsoid (+ + +),
a hyperboloid of one sheet (one minus),
a hyperboloid of two sheets (two minuses),
an imaginary (virtual) quadric (− − −).
(2) $ax^2+by^2=2cz$,
an elliptic paraboloid (a and b of same sign),
a hyperbolic paraboloid (a and b of opposite sign).
(3) $ax^2+by^2+cz^2=0$, a cone.
(4) $ax^2+2hxy+by^2=1$, a cylinder.

Plane sections of quadrics are conics. For a cylinder, sections parallel to the plane $z=0$ determine its type, which is elliptic, parabolic or hyperbolic. For an ellipsoid a, b and c are the lengths of its principal semi-axes.

quadriceps (*Zool.*). A muscle having 4 insertions, as one of the thigh muscles of Primates.

quadricorrelator (*TV*). Abbrev. for *quadrature information correlator*. Term applied to certain forms of automatic frequency and phase control systems in TV which use the correlation existing between a pair of measurements of a synchronizing signal at quadrature phases to derive additional information.

quadrigeminal bodies (*Zool.*). See **corpora quadrigemina.**

quadrilateral (*Maths.*). A four-sided polygon.

quadrilateral speed-time curve (*Rail., Eng.*). A simplified form of speed-time curve used in making preliminary calculations regarding energy consumption and average speed of railway trains. The acceleration and coasting portions of the curve are sloping straight lines and the braking portion is neglected, so that the curve becomes a quadrilateral. Cf. *trapezoidal speed-time curve.*

quadripartition (*Bot.*). Division of a spore mother cell to yield 4 spores.

quadripennate (*Zool.*). See **tetrapterous.**

quadriplegia (*Med.*). Paralysis of both arms and both legs.

quadripolar spindle (*Bot.*). An achromatic spindle with 4 poles, seen in preparations of spore mother cells in course of division.

quadripole (*Telecomm.*). A network with 2 input and 2 output terminals. A balanced wave-filter section.

quadrituberculate (*Zool.*). Of tuberculate cheek-teeth, having 4 cusps.

quadrivalent (*Cyt.*). A nucleus having 2 pairs of homologous chromosomes: an organism plant containing such nuclei. (*Electronics*) An atom with 4 electrons in its valency shell.

quadrumanous (*Zool.*). Of Vertebrates, having all 4 podia constructed like hands, as in Apes and Monkeys.

quadruped (*Zool.*). Of Vertebrates, having all 4 podia constructed like feet, as Cattle.

quadruple-expansion engine (*Eng.*). A steam engine in which the steam is expanded successively in 4 cylinders of increasing size, all working on the same crankshaft. See **multiple-expansion engine, triple-expansion engine.**

quadruple point (*Chem.*). A point on a concentration-pressure-temperature diagram at which a 2-component system can exist in 4 phases.

quadruplex (*Gen.*). Having four dominant genes, in polyploidy (see **polyploid**).

quadruplex system (*Teleg.*). A system of Morse telegraphy arranged for simultaneous independent transmission of two messages in each direction over a single circuit.

quadrupole (*Phys.*). A collection of charges such that the potential at a point distant from their centre of mass may be expressed by an infinite series of terms in inverse powers of r. The inverse third power term is the *quadrupole potential.*

quadrupole moment (*Phys.*). That derived from the series expansion (see **quadrupole**) of charges multiplied by space coordinates. The sum of the quadratic terms is the *quadrupole moment,* which is possessed by most metals.

quagginess (*For.*). A term used to indicate the defective condition of timber having shakes at the heart of the log.

qualitative analysis (*Chem.*). Identification of constituents of a material irrespective of their amount. In inorganic mixtures, metals and acid radicals are tested for separately. In the case of organic mixtures, a preliminary determination of the elements present is made. The components of a mixture may often be separated by distillation, extraction, chromatography, etc., and tests applied for the detection of certain groups of atoms.

quality (*Acous.*). (1) In sound reproduction, the degree to which a sample of reproduced sound resembles a sample of the original sound. The general description of freedom from various types of acoustic distortion in sound-reproducing systems. See **high-fidelity.** (2) The timbre or quality of a note which depends upon the number and magnitude of harmonics of the fundamental. (*Eng.*) The condition of a saturated vapour, particularly steam, expressed as the ratio per cent of the vaporized portion to the total weight of liquid and vapour. (*Radiol.*) In radiography, it indicates approximate penetrating power. Higher voltages produce higher quality X-rays of shorter wavelength and greater penetration. (This term dates from a period before the nature of X-rays was completely understood.)

quality control (*Eng.*). A form of inspection involving sampling of parts in a mathematical manner to determine whether or not the entire production run is acceptable, a specified number of defective parts being permissible.

quality factor (*Elec.*). See *Q*. (*Radiol.*) See **relative biological effectiveness.**

quality number or **spinning number** (*Textiles*). Number assigned to worsted tops to indicate the counts for which they are suitable, e.g., 58s and downwards are crossbred quantities, 60s and upwards botany. In yarns and noils, this number indicates the material from which they are derived. In cotton spinning, numbers and letters are used to indicate a blend of various staples for a particular range of counts; varies in every mill (e.g., AM/MID60). See also **quality terms.**

quality terms (*Textiles*). Terms applied to worsted and woollen fabrics to indicate the type of wool and the system of yarn spinning employed in their manufacture. *Botany* indicates merino wool, spun on the worsted system. *Crossbred* denotes a worsted cloth made from crossbred wool. *Saxony* means composed of merino wool, spun on the woollen system. *Cheviot* is applied to cloth made from Cheviot quality wool, spun on the woollen system. In cotton spinning, T indicates twist yarns; D.W., doubling weft; R.Y., reel yarn, etc.

quantitative analysis (*Chem.*). The determination of amounts in which various constituents of a material are present. Apart from purely chemical methods of analysis (see **gas analysis, gravimetric analysis, volumetric analysis**), the measurement of physical properties, such as ultraviolet, infrared and nuclear magnetic resonance spectra is of great value. Properties such as refractive index can be used for binary mixtures.

quantity of electricity (*Elec. Eng.*). Product of flow of electricity (current) and time during which it flows. The term may also refer to a charge of electricity. See **coulomb.**

quantity of light (*Light*). Product of luminous flux and time during which it is maintained; usually stated in lumen-hours.

quantity of radiation (*Radiol.*). Product of intensity and time of X-ray radiation. Not measured by energy, but by energy density and a coefficient depending on ability to cause ionization.

quantity sensitivity (*Elec. Eng.*). The throw, in mm, on a scale 1 m from an instrument designed to measure the quantity of electricity.

quantity surveyor (*Build., Civ. Eng.*). One who measures up from drawings and prepares a bill (or schedule) of quantities showing the content of each item. This is then used by contractors for estimating. The quantity surveyor is also concerned in measuring and assessing at intervals the value of work already carried out.

quantization (*Comp.*). See **digitization.** (*Nuc.*) In quantum theory, the division of energy of a system into discrete units (*quanta*), so that continuous infinitesimal changes are excluded. (*Telecomm.*) In information theory, division of the amplitude of a wave into a restricted number of finite amplitudes (subranges).

quantized computer (*Comp.*). Computer which converts continuous amplitudes into digital amplitudes, processes these, and converts back to continuous magnitudes for reading-out.

quantometer (*Met.*). An instrument showing by spectrographical analysis the percentages of the various metals present in a metallic sample.

quantum (*Nuc., etc.*). (1) General term for the indivisible unit of any form of physical energy; in particular the *photon*, unit of electromagnetic radiation energy, its magnitude depending only on frequency. See also **graviton, magnon, phonon, roton.** (2) An interval on a measuring scale, fractions of which are considered insignificant.

quantum efficiency (*Nuc.*). Number of electrons released in a photocell per photon of incident radiation of specified wavelength.

quantum electronics (*Phys.*). Those concerned with the amplification or generation of microwave power in solid crystals, governed by quantum mechanical laws.

quantum limit (*Phys.*). See **boundary wavelength** and **minimum wavelength.**

quantum mechanics (*Phys.*). See **statistical mechanics.**

quantum number (*Nuc.*). One of a set, describing possible states of a magnitude when quantized, e.g., nuclear spin. See **spin** (3).

quantum statistics (*Nuc.*). Statistics of the distribution of particles of a specified type in relation to their energies, the latter being quantized. See **Bose-Einstein statistics** and **Fermi-Dirac statistics.**

quantum theory (*Phys.*). That developed from *Planck's law*. There is a quantum theory for most branches of classical physics. See **theories of light.**

quantum voltage (*Nuc.*). Voltage through which an electron must be accelerated to acquire the energy corresponding to a particular quantum.

quantum yield (*Nuc.*). Ratio of the number of photon-induced reactions occurring, to total number of incident photons.

quaquaversal fold (*Geol.*). A domelike structure of folded sedimentary rocks which dip uniformly outwards from a central point. See **dome.**

quarantine (*Med.*). Isolation or restrictions placed on the movements of individuals associated with a case of a communicable disease; place or period of detention of ships coming from infected or suspected ports, or of animals on importation.

quarl (*Heat.*). See **burner firing block.**

quarrel (*Build.*). The diamond-shaped pane of glass used in *fret-work* (q.v.).

quarries (*Build.*). *Quarry tiles* (q.v.).

quarry (*Mining*). (1) An open working or pit for granite, building-stone, slate, or other rock. (2) An underground working in a coal-mine for stone to fill the goaf. Distinction between quarry and mine somewhat blurred in law, but usage implies surface workings.

quarry-faced (*Build.*). A term applied to a building-stone whose face is hammer-dressed before leaving the quarry. See **rock face.**

quarry-pitched (*Build.*). A term applied to stones which are roughly squared before leaving the quarry.

quarry sap (*Civ. Eng., etc.*). The moisture naturally contained in building-stone freshly cut from the quarry.

quarry-stone bond (*Build.*). A term applied to the arrangement of stones in rubble masonry.

quarry tile (*Build.*). The common unglazed, machine-made paving tile not less than ¾ in (20 mm) in thickness. Also **promenade tile.**

quartan (*Med.*). Term denoting *quartan malaria*, in which the febrile paroxysm recurs every fourth day (i.e., at an interval of 72 hr). Quartan malaria is due to infection with the parasite *Plasmodium malariae*.

quartation (*Met.*). See **inquartation.**

quarter (*Astron.*). The term applied to the phase of the moon at quadrature. The first quarter occurs when the longitude of the moon exceeds that of the sun by 90°, the last quarter when the excess is 270°. The two other quarters are the *new moon* and *full moon* (qq.v.). (*Carp.*) A rough vertical timber in the framework of a stud partition.

quarter bend (*Plumb.*). A bend, as in a piece of pipe, connecting two parts whose directions are mutually at right angles.

quarter bond (*Build.*). The ordinary brickwork bond obtained by using a 2¼-in (57mm) closer.

quarter-bound (*Bind.*). A term applied to a book having its back and part of its sides covered in one material and the rest of its sides in another.

quarter-chord point (*Aero.*). The point on the *chord line* (q.v.) at one quarter of the chord length behind the leading edge. It is of importance because *sweepback* is usually quoted by the angle between the line of the quarter-chord points and the normal to the aeroplane fore-and-aft centre-line.

quartered partition (*Carp.*). A partition built with quarterings.

quarter girth (*For.*). The girth of a log or tree divided by four. A measure commonly used in countries where volumes are reckoned in Hoppus feet. Abbrev. **q.g.**

quarter ill, quarter evil (*Vet.*). See **blackleg.**

quartering (*Civ. Eng., Min. Proc.*). A method of obtaining a representative sample of an aggregate with occasional shovelfuls, of which a heap or cone is formed. This is flattened out and two opposite quarter parts are rejected. Another cone is formed from the remainder which is again quartered, the process being repeated until a sample of the required size is left. (*For.*) A piece of timber of square section between 2 and 6 in. (50 and 150 mm) side.

quarter lines (*Ships*). The aggregation of waterlines, buttocklines, sections, and diagonals indicative of a ship's form, drawn on a scale of ¼ in. = 1 ft. See **fairing.**

quarter page folder (*Print.*). A supplementary device to give a third fold in line with the run of the paper on web-fed presses.

quarter-phase systems (*Elec. Eng.*). See **two-phase systems.**

quarter rack (*Horol.*). The rack of the striking work of a clock or repeater-watch which regulates the striking of the quarters.

quarters (*Build., Civ. Eng.*). See **flanks.**

quarter sawing (*For.*). A mode of converting timber, adopted when it is desired that the growth rings shall all be at no less than 45° to the cut faces. Also called **rift sawing.**

quarter screws (*Horol.*). Four screws in the rim of a compensating balance, one placed at either end of the arms, and the other two at right angles to the arm. Used for rating the watch but not for compensation purposes.

quarter snail (*Horol.*). The snail in a chiming clock or repeater-watch which controls the number of teeth picked up on the quarter rack.

quarter-space landing (*Build.*). A landing extending across only half the width of a staircase.

quarter stuff (*Carp.*). A board ¼ in. in thickness.

quarter turn (*Join.*). A wreath subtending an angle of 90°.

quarter-wave antenna (*Radio*). One whose overall length is approximately a quarter of free-space wavelength corresponding to frequency of operation. Under these conditions it is oscillating in its first natural mode, and is half a dipole.

quarter-wave bar (*Telecomm.*). See **quarter-wave line.**

quarter-wavelength stub (*Telecomm.*). Resonating two-wire or coaxial line, approximately one quarter-wavelength long, of high impedance at resonance. Used in antennae, as insulating support for another line, and as a coupling element.

quarter-wave line (*Telecomm.*). Quarter-wavelength section of balanced transmission line (Z_0) designed to operate as a matching device between lines of different impedance levels (Z_1, Z_2), such that $Z_0^2 = Z_1Z_2$. Also called **quarter-wave bar.**

quarter-wave plate (*Light*). A plate of quartz, cut parallel to the optic axis, of such thickness that a retardation of a quarter of a period is produced between ordinary and extraordinary rays travelling normally through the plate. By using a quarter-wave plate, with its axis at 45° with the principal plane of a Nicol prism, circularly polarized light is obtained.

quartet, quartette (*Bot.*). The group of 4 related nuclei or cells formed as a result of meiosis. (*Zool.*) A set of 4 related cells in a segmenting ovum.

quartic equation (*Maths.*). An algebraic equation of the fourth degree, i.e.,

$$ax^4 + bx^3 + cx^2 + dx + e = 0.$$

Its resolution into a pair of quadratic equations, and hence its solution, depends upon the solution of a subsidiary cubic equation.

quarto (*Print.*). The quarter of a sheet, or a sheet folded twice to make 4 leaves or 8 pages; written 4to.

quarto galley (*Typog.*). The widely used galley for jobbing work, approximately 13 in. × 8¾ in.

quartz (*Min.*). Crystalline silica occurring either in prisms capped by rhombohedra (low-temperature quartz, stable up to 573°C) or in hexagonal bipyramidal crystals (high-temperature quartz, stable above 573°C). Widely distributed in rocks of all kinds; igneous, metamorphic, and sedimentary; usually colourless and transparent (rock crystal), but often coloured by minute quantities of impurities as in citrine, cairngorm, etc.; also finely crystalline in the several forms of chalcedony, jasper, etc. See also **cristobalite, tridymite, twinning.** Quartz retains its high resistivity at high temperatures and is used for mercury and halogen vapour lamps, radio transmitting-valves, etc.

quartz crystal (*Radio*). A disk or rod cut in the appropriate directions from a specimen of piezoelectric quartz, and accurately ground so that its natural resonance shall occur at a particular frequency.

quartz-crystal clock (*Horol.*). A synchronous electric clock of great accuracy, having a quartz crystal to control the frequency of the a.c. supply to the motor unit.

quartz-diorite (*Geol.*). A coarse-grained holocrystalline igneous rock of intermediate composition, composed of quartz, plagioclase feldspar, hornblende, and biotite, and thus intermediate in mineral composition between typical diorite and granite. Also called *tonalite* (q.v.).

quartz-dolerite (*Geol.*). A variety of dolerite which contains interstitial quartz usually intergrown graphically with feldspar, forming patches of micropegmatite. A dyke-rock of worldwide distribution, well represented by the Whin Sill rock in N. England.

quartz-fibre balance (*Chem.*). A very sensitive spring balance, the spring being a quartz fibre.

quartz-fibre dosimeter (*Nuc., Radiol.*). See **personal dosimeter.**

quartz glass (*Glass*). See **vitreous silica.**

quartz-iodine lamp (*Elec. Eng.*). Compact high-intensity light source, consisting of a bulb with a tungsten filament, filled with an inert gas containing iodine (sometimes bromine) vapour. The bulb is of quartz, glass being unable to withstand the high operating temperature (600°C). Used for car-lamps, cine projectors, etc. Also **quartz-halogen lamp, tungsten-halogen lamp.**

quartzite (*Geol.*). The characteristic product of the metamorphism of a siliceous sandstone or grit. The term is also used to denote sandstones and grits which have been cemented by silica.

quartz-keratophyre (*Geol.*). A type of soda-trachyte carrying accessory quartz.

quartz lamp (*Radiol.*). One which contains a mercury arc under pressure, a powerful source of ultraviolet radiation.

quartz oscillator (*Radio*). One whose oscillation frequency is controlled by a piezoelectric quartz crystal.

quartz-porphyrite (*Geol.*). A porphyrite carrying quartz as an accessory constituent; the representative in the medium grain-size group of the fine-grained dacite.

quartz porphyry (*Geol.*). A medium-grained igneous rock of granitic composition occurring normally as minor intrusions, and carrying prominent phenocrysts of quartz.

quartz resonator (*Radio*). A standard of frequency comparison making use of the sharply resonant properties of a piezoelectric quartz crystal.

quartz-syenite (*Geol.*). A potash- or soda-syenite carrying quartz as an accessory constituent, and hence on the borderline between true syenite and granite.

quartz thermometer (*Electronics*). One with digital read-out obtained by beating a temperature-dependent quartz oscillator against a temperature-invariant one.

quartz topaz (*Min.*). See **citrine.**

quartz-trachyte (*Geol.*). A soda- or potash-trachyte carrying quartz as an accessory constituent only. Increase in quartz content would convert the rock into rhyolite.

quartz wedge (*Micros.*). A thin wedge of quartz which provides a means of superposing any required thickness of quartz on a mineral section being viewed under a polarizing microscope, the wedge being cut parallel to the optic axis of a prism of quartz crystal. It enables the sign of the birefringence of biaxial minerals to be determined from their interference figure in convergent light.

quasar (*Astron.*). From *quasi*-stell*ar* object. Star-like objects first identified by radio, but now recognized by the spectrum, which shows sharp emission lines with a very large red shift. Quasars are the most distant and most luminous bodies known, producing enormous energy from a comparatively small source. More than 200 have now been optically identified, but the source of their intense radiation remains unknown.

Quasi-arc welding (*Elec. Eng.*). A system of arc welding in which covered iron electrodes are used, the covering consisting of asbestos yarn impregnated with a fluxing compound. The covering protects the deposited metal from oxidation.

quasi-bistable circuit (*Elec. Eng.*). An astable circuit which is triggered at a high rate as compared with its natural frequency.

quasi-duplex (*Teleph.*). A circuit which operates apparently duplex, but actually functions in one direction only at a time, e.g., a long-distance telephone or a radio link, which is automatically switched by speech.

quasi-Fermi levels (*Electronics*). Energy levels in a semiconductor from which the number of electrons or holes contributing to conduction under non-equilibrium conditions can be calculated in the same way as from the true Fermi level under equilibrium conditions.

quasi-instruction (*Comp.*). A symbolic representation of a computer instruction in a compiler.

quasi-optical waves (*Phys.*). Electromagnetic waves of such short wavelength that their laws of propagation are similar to those of visible light.

quasi-stationary front (*Meteor.*). A *front* (q.v.) which is moving slowly and irregularly so that it cannot be described as either a *cold front* (q.v.) or a *warm front* (q.v.).

quaternary (*Chem.*). Consisting of 4 components, etc.; also, connected to 4 like atoms.

quaternary ammonium bases (*Chem.*). Bases derived from the hypothetical ammonium hydroxide $NH_4{\cdot}OH$, in which the four hydrogen atoms attached to the nitrogen are replaced by alkyl radicals, e.g., $(C_2H_5)_4N{\cdot}OH$, tetraethylammonium hydroxide.

quaternary diagram (*Met.*). *Constitutional diagram* of 4-phase alloy.

Quaternary Era (*Geol.*). The era of geological time which succeeded the Tertiary Era. It includes the Pleistocene and Recent Periods.

quebracho extract (*Chem.*). An aqueous extract of the wood of *Quebracho Colorado*, growing in South America. The wood contains up to 20% of tannin, and the extract is used for tanning.

Queckenstedt's sign (*Med.*). Increase in the pressure of the cerebrospinal fluid when the jugular veins in the neck are compressed; if this manoeuvre causes no rise of pressure in the spinal fluid, an obstruction is present at a higher level.

queen (*Build.*). A slate, 36×24 in. (914×610 mm). (*Zool.*) In social Insects, a sexually perfect female.

Queen Anne arch (*Arch.*). A combination of a central semicircular arch with side camber arches carried from the same springings.

queen bee substance (*Zool.*). See **queen substance.**

queen bolt (*Carp.*). A long iron or steel bolt serving in place of a timber queen-post.

queen closer (*Build.*). A half-brick made by cutting the brick lengthwise.

queen-post (*Carp.*). For roofs of more than about 30 ft. (ca 10 m) span, the central support of the tie-beam by the king-post is insufficient, and two other vertical ties, one on each side of the king-post, are required. These are called *queen-posts.*

queen-post roof (*Carp.*). A timber roof having 2 queen-posts but no king-post.

queen substance (*Zool.*). A pheromone produced by queen honey-bees (*Apis mellifera*; *Hymenoptera*) consisting of 9-ketodecanoic acid. Its effects include the suppression of egg-laying and of the building of queen cells by workers.

quench (*Elec. Eng.*). Resistor or resistor-capacitor shunting a contact, to reduce high-frequency sparking when a current is broken in an inductive circuit.

quenched cullet (*Glass*). *Cullet* (q.v.) made by running molten glass into water. Also **dragaded cullet, dragladled cullet.**

quenched spark gap (*Elec. Eng.*). Spark gap in which the discharge takes place between cooled or rapidly moving electrodes, e.g., studs on a disk.

quenched spark converter (*Elec. Eng.*). A generator which utilizes the oscillatory discharge of a capacitor through an inductor and spark gap as a source of RF power.

quencher (*Phys.*). That which is introduced into a luminescent material to reduce the duration of phosphorescence.

quench frequency (*Telecomm.*). The lower frequency signal used to quench intermittently a high-frequency oscillator, e.g., in a super-regenerative receiver. See **squegging.**

quenching (*Met.*). Generally means cooling steel from above the critical range by immersing in oil or water, in order to harden it. Also applied to cooling in salt and molten-metal baths or by means of an air blast, and to the rapid cooling of other alloys after solution treatment. See **oil-hardening steel, tempering.** (*Nuc. Eng.*) Process of inhibiting continuous discharge, by choice of gas and/or an external valve circuit, so that discharge can occur again on the incidence of a further photon in a counting tube. Essential in a *Geiger-Müller counter* (q.v.). (*Telecomm.*) Suppression of oscillation, particularly periodically, as in a super-regenerative receiver.

quenching oscillator (*Telecomm.*). One with a frequency slightly above the audible limit, and which generates the voltage necessary to quench the high-frequency oscillations in a super-regenerative receiver.

quench time (*Nuc. Eng.*). That required to quench the discharge of a Geiger tube. *Dead time* for internal quenching, *paralysis time* for electronic quenching.

quercitol (*Chem.*). $C_6H_7(OH)_5$, a polyhydroxy derivative of cyclohexane, found in oak-bark. Colourless crystals, m.p. 235°C, dextrorotatory.

queuing (*Comp., etc.*). Retardation and waiting of data for acceptance, e.g., for computer processing, or objects in an automation line.

Quevenne scale (*Chem.*). A scale indicating the relationship between the percentages of fat and total solids and the specific gravity. Used in milk analysis.

quick-break switch (*Elec. Eng.*). A switch having a spring or other device to produce a quick break, independently of the operator.

quick gouge (*Carp., etc.*). A gouge having a cutting edge shaped to a small radius of curvature. See **flat gouge, middle gouge.**

quicking (*Elec. Eng.*). Electrodeposition of mercury on a surface before regular plating.

quick-levelling head (*Surv.*). A fitting provided on some levels to facilitate setting up the instrument approximately level, the instrument head being secured to a ball-and-socket joint for adjustment to desired position, where it can be clamped.

quicklime (*Chem.*). See **caustic lime, lime.**

quick make-and-break switch (*Elec. Eng.*). See **snap switch.**

quick return (*Eng.*). A reciprocating motion, for operating the tool of a shaping machine, etc., in which the return is made more rapidly than the cutting stroke, so as to reduce the 'idling' time.

quick-revolution engines (*Eng.*). A term formerly applied to high-speed steam-engines direct-coupled to generators.

quick S (*Typog.*). A device for securing pages as an alternative to and improvement on page cord, and in which they can be imposed, consisting of four spring-loaded and adjustable side pieces joined together to form a tight frame.

quicksand (*Civ. Eng.*). Loose sand mixed with such a high proportion of water that its bearing-pressure is very low. Also **running sand.**

quick-setting inks (*Print.*). A general term for inks formulated to set quickly, allowing handling of the stock soon after printing and before drying.

quick-setting level (*Surv.*). A level fitted with a *quick-levelling head* (q.v.).

quicksilver (*Chem.*). See **mercury.**

quick sweep (*Build.*). A term applied to circular work in which the radius is small.

quidding (*Vet.*). A condition in the horse in which food is expelled from the mouth after being chewed; usually due to disease of the mouth.

quiescent (*Electronics*). General term for a system waiting to be operated, as a valve ready to amplify or a gas-discharge tube to fire. Also **preset.**

quiescent carrier transmission (*Telecomm.*). One for which the carrier is suppressed in the absence of modulation.

quiescent current (*Telecomm.*). Current in a valve or transistor, in the absence of a driving or modulating signal. Also called **standing current.**

quiescent operating point (*Telecomm.*). The steady-state operating conditions of a valve or transistor in its working circuit but in the absence of any input signal.

quiescent period (*Telecomm.*). That between pulses in a pulse transmission.

quiescent push-pull amplifier (*Telecomm.*). Thermionic valve or transistor amplifier, in which one side alone passes current for one phase, the other side passing current for the other phase. Abbrev. QPP **amplifier.**

quiescent tank (*San. Eng.*). A form of sedimentation tank in which sewage is allowed to rest for a certain time without flow taking place.

quiet automatic volume control (*Radio*). Same as **delayed automatic gain control.** The application of this is known as **quieting.**

quieting sensitivity (*Radio*). The minimum input signal required by a frequency-modulation radio receiver to give a specified signal/noise ratio at the output.

quiet-littoral fauna (*Ecol.*). Animals living at the edges of lakes with gently sloping shores.

Quiktran (*Comp.*). A program-orientated computing language based on *Fortran* (q.v.), but designed specifically for users employing a central computing service from a remote terminal, usually via the regular telephone service, although private lines and microwave radio links can also be used.

quill or **quill drive** (*Elec. Eng.*). A form of drive used for electric locomotives in which the armature of the driving motor is mounted on a quill surrounding the driving axle, but connected to it only by a flexible connexion. This enables a small amount of relative motion to take place between the motor and the driving axle. See **geared quill drive.** (*Eng.*) A hollow shaft revolving on a solid spindle. They may be connected by a clutch, or by a flange at one end to a flexible drive. (*Zool.*) See **calamus.**

quill feathers (*Zool.*). In Birds, the remiges and rectrices.

quilo. Name, proposed by M. Fulton in 1968, for the fundamental SI unit of mass (the kilogram).

quilt (*Acous.*). See **Euphon quilt, Chance's.**

quilted (*For.*). Ridges on sawn wood surfaces, usually due to the uneven cutting of reciprocating saws.

quinaldine (*Chem.*). $C_{10}H_9N$, 2-methylquinoline, a colourless refractive liquid, b.p. 246° C, which occurs to the extent of 25% in quinoline obtained from coal-tar.

quinapyramine (*Pharm., Vet.*). Same as *Antrycide.*

Quincke's disease. See **angioneurotic oedema.**

Quincke's method (*Mag.*). One for determining the magnetic susceptibility of a substance in solution by measuring the force acting on it in terms of the change of height of the free surface of the solution when placed in a suitable magnetic field.

Quincke tube (*Telecomm.*). Parallel adjustable sound transmission tube system forming interference filter.

quincuncial aestivation (*Bot.*). A particular type of imbricate aestivation in a 5-petalled corolla. Two petals overlap their neighbours by both edges, 2 are overlapped on both edges, and 1 overlaps 1 neighbour and is overlapped by the other.

quinhydrone (*Chem.*). $C_6H_4O_2 + C_6H_4(OH)_2$, an additive compound of 1 molecule of 1,4-quinone and 1 molecule of 1,4-dihydroxybenzene. It crystallizes in green prisms with a metallic lustre.

quinhydrone electrode (*Chem.*). A system consisting of a clean, polished, gold or platinum electrode dipping into a solution containing a little quinhydrone, for determining pH-values.

quinine (*Chem., Pharm.*). $C_{20}H_{24}O_2N_2 3H_2O$, an alkaloid of the quinoline group, present in Cinchona bark. It is a diacid base of very bitter taste and alkaline reaction. It crystallizes in prisms or silky needles, m.p. 177°C; the hydrochloride and sulphate are used as a febrifuge but have been largely superseded as a remedy for malaria. Its constitution is:

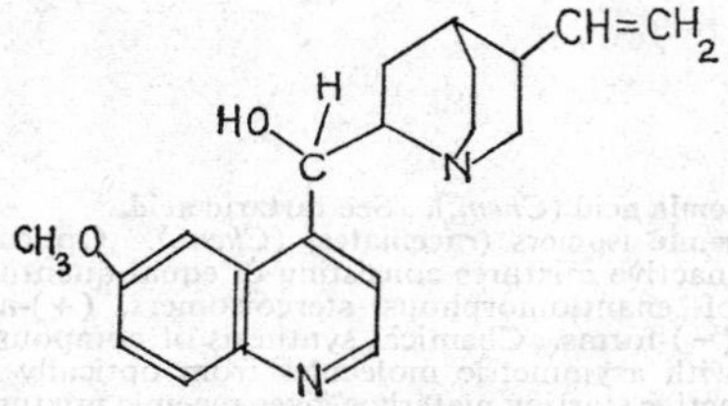

quinizarine (*Chem.*). A synonym for 1,4-dihydroxy-anthraquinone.

quinol (*Chem., Photog.*). See **hydroquinone.**

quinoline (*Chem.*). A heterocyclic compound consisting of a benzene ring condensed with a pyridine ring:

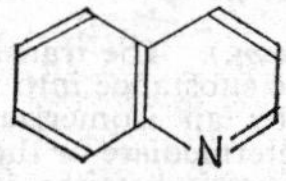

It is a colourless, oily liquid, m.p. −19·5°C, b.p. 240°C, rel. d. 1·08, of characteristic odour, insoluble in water, soluble in most organic solvents. It is found in coal-tar, in bone oil and in the products of the destructive distillation of many alkaloids. It can be synthesized by heating a mixture of aniline, glycerine and nitrobenzene with concentrated sulphuric acid. See **Skraup's synthesis.**

quinoline blue (*Photog.*). See **cyanine.**

quinones (*Chem.*). Compounds derived from benzene and its homologues by the replacement of 2 atoms of hydrogen with 2 atoms of oxygen, and characterized by their yellow colour and by being readily reduced to dihydric phenols. According to their configuration they are divided into 1,2-quinones and 1,4-quinones, e.g.:

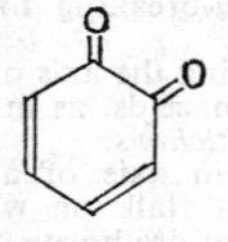

1,2-quinones 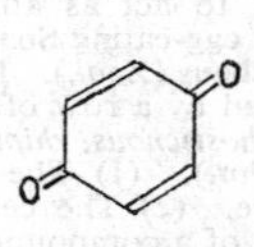 1,4-quinones

quinonoid formula (*Chem.*). A formula based upon the diketone configuration of 1,4-quinone (*benzoquinone* (q.v.)), involving the rearrangement of the double bonds in a benzene nucleus; adopted to explain the formation of dyestuffs, e.g., coloured salts of compounds of the triphenyl-methane series.

quinoxalines (*Chem.*). A group of heterocyclic compounds consisting of a benzene ring condensed with a diazine ring:

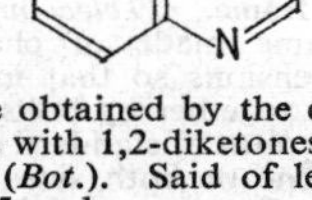

They can be obtained by the condensation of 1,2-diamines with 1,2-diketones.

quinquefarious (*Bot.*). Said of leaves which are arranged in 5 ranks.

quinquefoliolate (*Bot.*). Having 5 leaflets.

quinquituberculate (*Zool.*). Of tuberculate cheek-teeth, having 5 cusps.

quinsy (*Med.*). Acute suppurative tonsillitis; peritonsillar abscess. Acute inflammation of the tonsil with the formation of pus around it.

quintal. Unit of mass in the metric system, equal to 100 kg. Abbrev. **q.**

quintic equation (*Maths.*). An algebraic equation of the fifth degree. Unlike equations of lower degree, its general solution (and that of equations of higher degree) cannot be expressed in terms of a finite number of root extraction.

quintocubital (*Zool.*). In Birds, a term indicating the presence of the fifth secondary remex or fifth flight feather carried by the ulna. Cf. *diastataxy*.

quintozene (*Chem.*). Pentachloronitrobenzene, used as a fungicide. Also called **PCNB.**

quintuple point (*Chem.*). A point on a concentration-pressure-temperature diagram at which a 3-component system can exist in 5 phases.

quintuplinerved (*Bot.*). Having 5 main veins or ribs in the leaf.

quire (*Paper*). One-twentieth of a *ream* (q.v.).

quired paper (*Paper*). Reams folded in quires.

quire spacing (*Print.*). On a rotary printing press, as the product is delivered, it is separated into quires or batches by the *kicker* which delivers a *kick copy* at the required interval.

quirewise (*Bind.*). Sections which after printing are folded and *inserted* one in the other. This method allows the booklet to be stitched instead of stabbed.

quirk (*Join.*). The narrow groove alongside a bead which is sunk into the face of the work so as to be flush with it.

quirk-bead (*Join.*). See **bead-and-quirk.**

quirk float (*Tools*). A plasterer's trowel specially shaped for finishing mouldings.

quirk moulding (*Join.*). A moulding having a small groove in it.

quirk-router (*Join.*). A form of plane for shaping quirks.

quitclaim (*Mining*). A deed of relinquishment of a claim or portion of mining ground.

quittor (*Vet.*). A chronic suppuration of the lateral cartilage and its surrounding tissues within a horse's foot.

quoin (*Build.*). An exterior angle of a building, especially one formed of large squared corner-stones projecting beyond the general faces of the meeting wall surfaces. (*Typog.*) A wooden wedge or a metal device used to lock up formes.

quoin header (*Build.*). A brick which is so laid at the external angle of a building that it is a header in respect of the face of the wall proper, and a stretcher in respect of the return wall.

quoin key (*Typog.*). Each pattern of mechanical quoin is operated by a key, some being square in shape, others ribbed.

quoin-post (*Hyd. Eng.*). See **heel-post.**

quotation marks, quotes (*Typog.*). If double quotes (" ") are used to indicate a quotation, then single quotes (' ') are used for a quotation within the passage quoted; the reverse procedure is becoming more popular. In hand-set matter the opening quotes are inverted commas, the closing quotes being apostrophes; in machine-set matter the opening quote is a special sort.

quotations (*Typog.*). Metal spaces, not less than 4 ems by 2 ems used for filling blanks in pages or formes.

quotient (*Maths.*). See **division.**

***Q*-value** (*Nuc. Eng.*). Quantity of energy released in a given nuclear reaction. Normally expressed in MeV, but occasionally in atomic mass units.

R

r **(*Chem.*). With subscript, a symbol for *specific refraction*.**

r- (*Chem.*). An abbrev. for *racemic*.

ρ. See under rho.

ρ- (*Chem.*). A symbol for *pros-*, i.e., containing a condensed double aromatic nucleus substituted in the 2,3 positions.

R (*Build.*). Abbrev. for *render*. (*Chem.*) A general symbol for an organic hydrocarbon radical, especially an alkyl radical. (*Heat*) Symbol used to indicate *Rankine scale* (q.v.) of temperature. (*Radiol.*) Symbol for *röntgen* unit in X-ray dosage.

R (*Chem., Phys.*). A symbol for (1) the *gas constant*; (2) the *Rydberg constant*.

[*R*] (*Chem.*). With subscript, a symbol for molecular refraction.

ra- (*Chem.*). A symbol for *radio-*, i.e., a radioactive isotope of an element, e.g., *ra*Na, *radio-*sodium.

Ra (*Chem.*). The symbol for *radium*.

R.A. (*Astron.*). See **right ascension.**

Raabe's convergence test (*Maths.*). If, for the series of positive terms Σu_n, the product $n\left(\frac{u_n}{u_{n+1}} - 1\right) \to \sigma$, then Σu_n is convergent if $\sigma > 1$ and divergent if $\sigma < 1$. This is a special case of *Kummer's convergence test* (q.v.).

rab (*Build.*). A stick for mixing hair with mortar.

rabbet (*Carp.*). A corruption of *rebate*.

rabbeted lock (*Join.*). A lock which is fitted into a recess cut in the edge of a door.

rabbit (*Nuc. Eng.*). See **shuttle, single-stage recycle.**

rabbit septicaemia (*Vet.*). An acute septicaemic and pneumonic disease of rabbits caused by the bacterium *Pasteurella multocida*; a mild form of the disease, in which the infection is confined to the upper respiratory tract, is called **snuffles.**

rabble (*Met., Mining*). Mechanized **rake** used to loosen sluice bed or move ore or concentrate through a kiln or furnace.

rabies (*Med.*). An acute disease of dogs, wolves, and other animals, due to infection with a virus and communicable to man by the bite of the infected animals. In man the disease is characterized by intense restlessness, mental excitement muscular spasms (especially of the mouth and throat), and paralysis. Also called **hydrophobia.**

race (*Build., Geol.*). Fragments of limestone sometimes found in certain brick earths of a hard marly character. (*Eng.*) The inner or outer steel rings of a *ball-bearing* or *roller-bearing* (qq.v.). (*Hyd. Eng.*) A channel conveying water to or away from a hydraulically operated machine. (*Zool.*) A category of variant individuals occurring within a species and differing slightly in characteristics from the typical members of the species; a breed of domesticated animals; a microspecies.

race board (*Weaving*). Thin, polished, accurately levelled hardwood board nailed to the top of the loom stay. Each pick sends the shuttle across it to the opposite box before beat-up of weft. Also called **shuttle race.**

racemates (*Chem.*). See **racemic isomers.**

raceme (*Bot.*). (1) An indefinite inflorescence, in which stalked flowers are borne in acropetal succession on an unbranched main stalk. (2) A group of sporangia similarly arranged.

racemic acid (*Chem.*). See tartaric acid.

racemic isomers (racemates) (*Chem.*). Optically inactive mixtures consisting of equal quantities of enantiomorphous stereoisomers, (+)-and (−)-forms. Chemical synthesis of compounds with asymmetric molecules from optically inactive starting materials gives racemic mixtures. These can be resolved into the optically active components by various methods, e.g., by coupling with an optically active substance, such as an alkaloid, and subsequent fractional crystalization, by the action of lower plant organisms, e.g., bacteria, moulds, yeasts, etc., which attack only one of the isomers, leaving the other one intact.

racemization (*Chem.*). The transformation of an optically active substance into racemic inactive form, either by an isomerization through a symmetrical intermediate or through a reaction by which a new substance is formed via a similar intermediate or transition state.

racemose (*Bot.*). (1) Bearing racemes. (2) Said of an inflorescence which is a raceme. (*Zool.*) Shaped like a bunch of grapes; said especially of glands.

racemule (*Bot.*). A small raceme.

race-track (*Nuc. Eng.*). Discharge tube or ion-beam chamber where particles are constrained to an oval path.

rachi-, rachio-. Prefix from Gk. *rhachis*, spine.

rachianaesthesia (*Med.*). Anaesthesia produced by the injection of anaesthetic agents into the space round the spinal cord.

rachilla (*Bot.*). The axis in the centre of a spikelet of a grass.

rachiodont (*Zool.*). Having some of the anterior thoracic vertebrae with the hypapophysis enlarged, forwardly directed, and capped with enamel to act as an egg-breaking tooth, as certain egg-eating Snakes.

rachiostichous (*Zool.*). Having the axis of the fin occupied by a row of somactids, as in *Dipnoi*. Cf. *orthostichous, rhipidostichous*.

rachis (*Bot.*). (1) The main axis of an inflorescence. (2) The central stalk on which the leaflets of a compound leaf are borne. (*Zool.*) The shaft or axis; the shaft of a feather; the vertebral column. *adj.* **rachidial.**

rachischisis (*Med.*). A developmental defect of the spinal column. See **spina bifida.**

rachitic (*Med.*). Affected with or pertaining to rickets.

rachitis (*Med.*). *Rickets* (q.v.).

rachitomous (*Zool.*). Temnospondylous.

R-acid (*Chem.*). 2-Naphthol-3,6-disulphonic acid; used in preparation of azo-dyes for wool.

rack (*Horol.*). A toothed segment; used in the striking and chiming mechanism of a clock or a repeater-watch. The number of blows struck by the hammer is controlled by the number of teeth gathered on the rack. (*Min. Proc.*) Reck. See **ragging frame.** (*Telecomm.*) A vertical mounting frame made of channel iron of standard dimensions so that individual racks can be joined together by bolts to form bays. Panels of apparatus of standard dimensions can be fixed on one or both sides of each rack. (*Textiles*) (1) Component in spinning mule headstock which operates the short copping rail governing size of yarn package. (2) Term in lace manufacture to signify number of motions concerned in making a piece of lace or net; varies with type of machine.

rack-and-pin (*Cinema.*). Processing method in which film is wound spirally around vertical pillars mounted on a rack which can be immersed in shallow tanks of solution. Suitable only for lengths under 50 ft (ca 15 m).

rack-and-pinion (*Eng.*). A method of transforming rotary into linear motion, or vice versa; accomplished by a pinion or small gear-wheel which engages a straight, toothed rack. See **pitch line.**

rack-and-pinion steering-gear (*Autos.*). One in which a pinion carried by the steering column engages with a rack attached to the divided track rod.

rack clock (*Horol.*). One mounted on a vertical toothed rack and operated by gravity.

rack development (*Cinema.*). See **frame development.**

racked (*Carp.*). Said of temporary timbering which is braced so as to stiffen it against deformation. (*Textiles*) Term applied to knitted patterns of the type used extensively for machine knitted outerwear. Known also as **shaker patterns, shogged-rib.**

rack hook (*Horol.*). Part of the rack-striking work. Just before the hour, the lifting piece lifts the *rack hook*, which allows the rack to fall.

racking (*Brew.*). The operation of conveying the beer by hose from the fermenting vat to casks, after removal of the yeast. (*Mining*) The operation of separating ore by washing on an inclined plane. (*Ships*) Distortion of the ship's transverse shape. (*Surv.*) See **fixed-needle surveying.**

racking back (*Build.*). The procedure adopted when the full length of a wall is not built at once, the unfinished end being stepped or *racked back* at an angle, so that when the remainder of the wall is built there is not a vertical line of junction, which might cause cracking, owing to uneven settlement.

rack mounting (*Telecomm.*). The use of standard racks, of varying height but otherwise of standard dimensions, for mounting panels carrying apparatus, such as repeaters or filters, with a uniform scheme of wiring; such mounting gives both accessibility and compactness.

rack railway (*Rail.*). A device for overcoming adhesion difficulties, as met on mountain railways. A *pinion* on the locomotive axle engages with a *rack* laid parallel with, usually in the centre of, the normal rails, which continue to carry the weight. Also **mountain railway.**

rack saw (*Carp.*). A saw having wide teeth.

racon (*Nav.*). *Ra*dar bea*con* used by navigators for identifying objects at a distance.

rad (*Maths.*). Abbrev. for *radian.* (*Radiol.*) Unit of radiation dose which is absorbed, equal to 0·01 J/kg of the absorbing (often tissue) medium.

radar. In general, a system using pulsed radio waves, in which reflected (*primary radar*) or regenerated (*secondary radar*) pulses lead to measurement of distance and direction of target (*ra*dio *d*etection *a*nd *r*anging). In an alternative radar system continuous RF waves are sent out by the transmitter.

radar beacon (*Radar*). A fixed radio transmitter whose radiations enable a craft to determine its own direction or position relative to the beacon by means of its own radar equipment.

radar indicator (*Radar*). Display on a CRT of a radar system output, either as a radial line or a coordinate system for range and direction. The echo signal gives a brightening of the luminous spot, which remains for some seconds because of afterglow. Also **radar screen.** See **A, B, C, etc., display, plan-position indicator.**

radar performance figure (*Radar*). The ratio of peak transmitter power to minimum signal required by receiver.

radar pulse (*Radar*). Repeated pulse on very high frequency, usually a few microseconds in duration repeated, e.g., 1000 times per second, frequency of the radiation being up to 40 GHz.

radar range (*Radar*). Usually given as that at which a specified object can be detected with 50% reliability.

radar scan (*Radar*). The path followed by a radar beam in space.

radar screen (*Radar*). See **radar indicator.**

radar V-beam (*Radar*). One using 2 fan-shaped beams to produce 3-dimensional position data on a number of targets.

raddle or wraithe (*Weaving*). Steel comb or half-reed used at the front of the beamer to spread the threads to width of the warp beam.

radechon (*Electronics*). A storage tube in which the controlling element is a fine mesh grid.

radiac (*Instr.*). Class of instruments for radiation survey, in mineral prospecting, Civil Defence, handling radioactive materials, and working with nuclear reactors.

radial. Radiating from a common centre; pertaining to a *radius* (q.v.). (*Zool.*) In *Crinoidea*, a whorl of ossicles supporting the oral disk.

radial commutator (*Elec. Eng.*). A commutator for a d.c. machine in which the commutator bars are arranged radially from the axis to form a disk instead of a cylinder.

radial drill (*Eng.*). A large drilling machine in which the drilling head is capable of radial adjustment along a rigid horizontal arm carried by a pillar.

radial ducts (*Elec. Eng.*). In an electric machine, ventilating ducts which run radially from the shaft.

radiale (*Zool.*). A bone of the proximal row of the carpus in line with the radius. *pl.* **radiala.** See **somactids.**

radial engine (*Aero., Eng.*). An aircraft or other engine having the cylinders arranged radially at equal angular intervals round the crankshaft. See **double-row-, single-row-, master connecting-rod.**

radial feeder (*Elec. Eng.*). See **independent feeder.**

radial fission (*Zool.*). Repeated longitudinal fission in which the daughter individuals remain in position, as in some *Protozoa.*

radial grating (*Electronics*). One consisting of radial wires inserted in a circular waveguide to suppress E-modes.

radial longitudinal section (*Bot.*). A section cut in the radius of the longitudinal axis of an organ.

radial-ply (*Autos.*). Term applied to tyres which have a semi-rigid breaker strip under the tread and relatively little stiffening in the walls. Also **braced-tread.** Cf. *cross-ply.*

radial recording (*Acous.*). Same as *lateral recording* (q.v.).

radial symmetry (*Biol.*). The condition in which the body of an animal can be divided into two similar halves by any one of several vertical planes passing through the centre, as in *Echinodermata* and *Coelenterata.* Cf. *bilateral symmetry.*

radial system (*Cables*). A distribution system in which the cables radiate out from a generating or supply station. If a ault occurs, all consumers beyond the fault are cut off.

radial valve gear (*Eng.*). A steam-engine valve gear in which the slide-valve is given independent component motions proportional to the sine and cosine of the crank angle respectively.

radial vascular bundle (*Bot.*). A vascular strand having the xylem and phloem on different radii, as is usual in roots.

radial velocity (*Astron.*). See **line-of-sight velocity.**

radial wall (*Bot.*). An anticlinal wall placed in or across a radius of an organ.

radian (*Maths.*). The SI unit of plane angular measure, defined as the angle subtended at the centre of a circle by an arc equal in length to the radius. Abbrev. **rad.**

$$2\pi \text{ radians} = 360°$$
$$1 \text{ radian} = 57°{\cdot}295\,779\,513\,1\ldots$$
$$1° = 0{\cdot}017\,453\,292\,5\ldots \text{ radian.}$$

radiance (*Phys.*). Of a surface, the luminous flux radiated per unit area.

radian frequency (*Phys.*). Same as **angular frequency.**

radiant (*Astron.*). Point on the celestial sphere from which a series of parallel tracks in space, such as those followed by the individual meteors in a shower, appear to originate.

radiant flux (*Phys.*). The time rate of flow of radiant electromagnetic energy.

radiant-flux density (*Phys.*). A measure of the radiant power per unit area that flows across a surface. Also called **irradiance.**

radiant heat (*Phys.*). Heat transmitted through space by *infra-red radiation.*

radiant intensity (*Phys.*). The energy emitted per second per unit solid angle about a given direction.

radiant-tube furnace (*Heat.*). Modified form of muffle furnace, the heating chamber having a series of steel alloy tubes in which the fuel is burned, thus excluding products of combustion from the heating chamber.

radiant-type boiler (*Eng.*). A water-tube boiler having one or more drums and a circulation system consisting of vertically or horizontally inclined banks of tubes, heating surfaces of bare or protected water-tubes forming the walls of the combustion chamber; firing is generally by pulverized fuel.

radiant umbel (*Bot.*). One in which the outermost flowers are larger than the flowers in the middle.

radiate (*Bot.*). (1) Said of a capitulum which has ray florets. (2) Said of a flower which spreads like a ray from the periphery of any densely packed inflorescence. (3) Said of a stigma in which the receptive surfaces radiate outwards from a centre.

radiated power (*Radio*). The actual power level of the radio signals transmitted by an antenna. See also **effective radiated power.**

radiating brick (*Build.*). See **compass brick.**

radiating circuit (*Telecomm.*). Any circuit capable of sending out power, in the form of electromagnetic waves, into space; especially the antenna circuit of a radio transmitter.

radiating surface (*Heat*). The effective area of a radiator available for the transmission of heat by radiation.

radiation (*Heat*). A process by which heat may be transferred from a source to a receiver without heating of the intervening medium or without the existence of a material medium, e.g., heat received by the earth from the sun. (see **solar constant**). More generally (*Phys.*), the dissemination of energy from a source. Its intensity falls off as the inverse square of the distance from the source in the absence of absorption. Applied to electromagnetic waves (radio waves, light, X-rays, γ-rays, etc.), and acoustic waves and emitted particles (α and β, protons, neutrons, mesons, etc.). See articles on particular types of radiation (*black-body, electromagnetic, infrared, ultraviolet, visible, etc.*); also **Planck's radiation formula, Stefan-Boltzmann law, Wien's laws.** (*Surv.*) A method of plane-table surveying in which a point is located on the board by marking its direction with the alidade and measuring off its distance to scale from the instrument station. See **plane table.**

radiation belts (*Astron.*). See **Van Allen radiation belts.**

radiation burn (*Med.*). A burn caused by over-exposure to radiant energy.

radiation chemistry (*Chem.*). That of radiation-induced chemical effects (e.g., decomposition, polymerization, etc.). Not to be confused with *radiochemistry.*

radiation counter (*Nuc. Eng.*). One used to detect individual particles or photons in nuclear physics.

radiation danger zone (*Radiol.*). A zone within which the *maximum permissible dose rate* or *concentration* is exceeded.

radiation diagram (*Phys.*). See **radiation pattern.**

radiation efficiency (*Radio*). Ratio of actual power radiated by antenna to that provided by the drive. Also called **aerial efficiency.**

radiation field (*Phys.*). See **field.**

radiation flux density (*Phys.*). Rate of flow of radiated energy through unit area of surface normal to the beam (for particles this is frequently expressed in number rather than energy). Also called **radiation intensity.**

radiation hazard (*Radiol.*). The danger to health arising from exposure to ionizing radiation, either due to external irradiation or to radiation from radioactive materials within the body.

radiation impedance (*Phys.*). Impedance per unit area. Measured, e.g., by the complex ratio of the sound pressure to the velocity at the surface of a vibrating body which is generating sound waves, or by the corresponding electromagnetic quantities.

radiation intensity. See **radiation flux density.**

radiation length (*Nuc.*). The path length in which relativistic charged particles lose e^{-1} of their energy by radiative collisions. See **bremsstrahlung.**

radiation loss (*Telecomm.*). That power radiated from a nonshielded radiofrequency transmission line.

radiation pattern (*Phys.*). Polar or Cartesian representation of distribution of radiation in space from any source and, in reverse, effectiveness of reception. Also called **radiation diagram.**

radiation potential (*Phys.*). See **ionization potential.**

radiation pressure (*Phys.*). Minute pressure exerted on a surface normal to the direction of propagation of an electromagnetic wave. Electromagnetic theory leads to the result that, for a perfect reflector, this pressure should be equal to the total energy density in the medium. This has been confirmed experimentally in spite of the extreme smallness of the pressure (about 10^{-5} N m^{-2} for sunlight). In the case of a sound wave in a fluid, the pressure gives rise to 'streaming', i.e., a flow of the fluid medium.

radiation pyrometer (*Heat*). A device for ascertaining the temperature of a distant source of heat, such as a furnace, by allowing radiation from the source to face, or be focused on, a thermojunction connected to a sensitive galvanometer, the deflection of the latter giving, after suitable calibration, the required temperature. For temperatures *ca.* 500°–1500°C.

radiation resistance (*Telecomm.*). That part of impedance of an antenna system related to power radiated; the power radiated divided by the square of the current at a specified point, e.g., at the junction with the feeder.

radiation sickness (*Med.*). Illness, marked by internal bleeding, decrease in blood cells, etc., due to excessive absorption of radiation.

radiation therapy (*Med.*). The use of any form of radiation, e.g., electromagnetic, electron or neutron beam, or ultrasonic, for treating disease.

radiation trap (*Nuc. Eng.*). (1) Beam trap for absorbing intense radiation beam with a minimum of scatter. (2) Maze or labyrinth formed by entry corridor with several right-angle bends, used for approach to multicurie radiation sources or some accelerating machines.

radiative collision (*Phys.*). One in which kinetic energy is converted into electromagnetic radiation. See **bremsstrahlung.**

radiative equilibrium (*Astron.*). An ideal state of a star, postulated in astrophysical researches, implying a control of temperature by the transfer of heat from one part of the star to another by radiation instead of by convection.

radiator (*Autos.*). A device for dissipating heat created in an engine. It consists of thin-walled tubes, or narrow passages of honeycomb form, through which the water is conducted, and across which an airstream is induced either by the motion of the vehicle or by a fan. (*Nuc.*) In radioactivity, the origin of α-, β-, and/or γ-rays; also called a **source.** (*Radio*) Element of an antenna which is effective in radiating power as electromagnetic waves; usually a resonating quarter- or half-wavelength wire, or a multiple of this, with suppression or neutralization of elements, or of their radiation in the opposite phase or direction.

radiator flaps (*Aero.*). See **gills.**

radical (*Bot.*). Appearing as if springing from the root at soil-level. (*Chem.*) (1) A molecule or atom which possesses an odd number of electrons, e.g., $Br.CH_3$. It is often very short-lived, reacting rapidly with other radicals or other molecules. (2) A group of atoms which passes unchanged through a series of reactions, but is normally incapable of separate existence. This is now more usually called a *group*.

radical axis (*Maths.*). The straight line from each point of which the tangents to two circles are equal.

radication (*Bot.*). The general characters of the root system of a plant.

radicicolous (*Bot.*). Said of a parasite which attacks roots.

radicivorous (*Zool.*). Root-eating.

radicle (*Bot.*). (1) The root of the embryo of a flowering plant. (2) A rhizoid of a moss. (3) Any very small root. *adj.* radiculose.

radiculectomy (*Surg.*). The operation of cutting the roots of spinal nerves.

radiculitis (*Med.*). Inflammation of the root of a spinal nerve.

radio. Generic term applied to methods of signalling through space, without connecting wires, by means of electromagnetic waves generated by high-frequency alternating currents.

radio- (*Chem.*). A prefix denoting an artificially prepared radioactive isotope of an element.

radioactinium (*Chem.*). Symbol RdAc. A thorium isotope of mass number 227 which is formed naturally from the β-decay of actinium-227. Radioactinium emits α-particles to produce radium-223.

radioactivation analysis (*Nuc.*). Method based on measurement of radionuclides formed by neutron irradiation or, less frequently, proton or deuteron irradiation by a particle accelerator.

radioactive atom (*Nuc.*). One which decays into another species by emission of an α- or β-ray (or by electron capture). Activity may be natural or induced.

radioactive chain (*Nuc.*). Chart of natural radioactive isotopes, showing how one is related to others through radiation and decay, finally to an isotope of lead. Three natural radioactive series exist, their members having mass numbers: (*a*) $4n$ (thorium series); (*b*) $4n+2$ (uranium series); (*c*) $4n+3$ (actinium series). Members of the $4n+1$ (neptunium series) can be produced artificially. (*N.B.*—Radioactive decay leads to a decrease of one in the value of n above whenever an α-particle is emitted.) Also called **radioactive series.**

radioactive dating (*Geol., etc.*). General term for age measurements made in a natural system when these are based on the laws of radioactive decay.

radioactive decay (*Nuc.*). See **disintegration constant, half-life, radioactive atom.**

radioactive equilibrium (*Nuc.*). Eventual stability of products of radioactivity if contained, i.e., rate of formation (quantitative) equals rate of decay. Particularly important between radium and radon.

radioactive isotope (*Nuc.*). Natural or artificial isotope exhibiting radioactivity, used as a source for medical or industrial purposes. Also **radioisotope.**

radioactive series (*Nuc.*). See **radioactive chain.**

radioactive standard (*Nuc.*). A radiation source for calibrating radiation measurement equipment. The source has usually a long half-life and during its decay the number and type of radioactive atoms at a given reference time is known.

radioactive tracer (*Nuc.*). Small quantity of radioactive preparation added to corresponding nonactive material to *label* or *tag* it so that its movements can be followed by tracing the activity. (The chemical behaviour of radioactive elements and their nonactive isotopes is identical.)

radioactive tube (*Electronics*). Any electron tube in which the cathode is activated by a radioactive substance; about 1 μCi of radium bromide has been commonly used.

radioactivity (*Nuc.*). Spontaneous disintegration of certain natural heavy elements (radium, actinium, uranium, thorium) accompanied by the emission of α-rays, which are positively charged helium nuclei; β-rays, which are fast electrons; and γ-rays, which are short X-rays. The ultimate end-product of radioactive disintegration is an isotope of lead. See also **artificial radioactivity, induced radioactivity.**

radio altimeter (*Aero.*). Device for determining height, particularly of aircraft in flight, by electronic means, generally by detecting the delay in reception of reflected signals, or change in frequency, using the Doppler effect; also called **terrain-clearance indicator.**

radioastronomy. The study of radio emissions from discrete radio sources, interstellar hydrogen, the Sun and planets, etc.; also includes the use of radio-echo methods in the measurement of meteor streams and planetary distances.

radio beacon (*Radio*). Stationary radio transmitter which transmits steady beams of radiation along certain directions for guidance of ships or aircraft, or one which transmits from an omnidirectional antenna and is used for the taking of bearings, using an identifying code. Also **aerophare.**

radio beam (*Radio*). Concentration of electromagnetic radiation within narrow angular limits, such as is emitted from a highly directional antenna, in the form of a curtain or bowl.

radio bearing (*Radio*). Direction of arrival of a radio signal, as indicated by a loop or goniometer, for navigational purposes.

radiobiology (*Biol.*). Branch of science involving study of effect of radiation and radioactive materials on living matter.

radio broadcasting (*Radio*). The transmission by means of electromagnetic (i.e., radio) waves, of a programme of sound or picture for general reception. The separation of the frequency channels, usually about 10 kHz, is decided by international agreement.

radiocaesium (*Chem.*). Cs-137, a radioactive isotope recovered from the waste of nuclear reactors in nuclear power plants. Usefu o mass-radiation and sterilization of foodstuffs. Also for high-intensity X-ray radiation of surface tumours in place of much more expensive radium. Half-life 37 years.

radiocarbon (*Chem.*). C-14, a weakly radioactive isotope much used in biological and agricultural tracer studies, e.g., of the products of plants grown in an atmosphere containing CO_2 with C-14, which has a half-life of 5740 years.

radiochemical purity (*Chem.*). The proportion of a given radioactive compound in the stated chemical form. Cf. *radioisotopic purity*.

radiochemistry (*Chem.*). Study of science and techniques of producing and using radioactive isotopes or their compounds to study chemical compounds. Cf. *radiation chemistry*.

radio circuit (*Radio*). Communication system including a radio link, comprising a transmitter and antenna, the radio transmission path with possible reflections or scatter from ionized regions, and a receiving antenna and receiver.

radiocolloids (*Nuc.*). Radioactive atoms in colloidal aggregates.

radio communication (*Radio*). Transmission of information by channels including radio links, fixed or mobile, either coded (telegraph), analysed (video and facsimile), or plain (broadcasting and telephony). Postwar extensions include radar, hyperbolic navigation, and telemetry from space vehicles (see **communications satellite**).

radio compass (*Radio*). Originally a rotating loop, later rendered more sensitive by a goniometer system, and by display on a cathode-ray tube. Any device, depending on radio, which gives a bearing. See **Adcock antenna, Bellini-Tosi antenna.**

radio direction-finding (*Radio*). Formerly used for all methods of direction-finding by radio techniques, now principally for passive reception of direction-finding signals from radio beacons or navigational transmitters as distinct from active radar. Abbrev. **RDF.**

radio echo methods (*Astron.*). Measurement of position and velocities by the reception of radio signals reflected from meteor trails, aurorae, the moon and planets, etc. See **radio telescope, meteor streams.**

radioelement (*Nuc.*). One exhibiting natural radioactivity.

radio engineering (*Eng.*). The science which deals with the design, construction, and maintenance of apparatus used for radio communication.

radio exchange (*Radio*). Radio receiving station, with multiple power amplifiers, for distributing to subscribers radio programmes on a relay basis, via overhead wires, telephone lines, or electric power mains; also termed **rediffusion, relay system.** See **carrier telegraphy.**

radio frequency (*Telecomm.*). One suitable for radio transmission, above 10^4 Hz and below 3×10^{12} Hz approx. Abbrev. **RF.** Also **radio spectrum.**

radio-frequency heating (*Elec. Eng.*). See **dielectric heating, induction heating.**

radio-frequency spectrometer (*Nuc. Eng.*). A valve containing cathode, several grids (seven in the double stage spectrometer for observing positive and negative ions), and a plate; of considerable application in the study of negative atomic ions. Ions formed in the cathode are passed through the fields of the grids and, by applying to the fourth grid a blocking potential of the same order as the energy of ions leaving the radio-frequency field, and observing the current to the plate, rate of ion formation may be ascertained.

radio-frequency spectroscopy (*Nuc.*). Study of absorption spectrum of nuclei spinning in strong magnetic field by nuclear paramagnetic resonance; or of molecular electron spin resonances.

radiogenic (*Nuc.*). Said of stable or radioactive products arising from radioactive disintegration.

radiogoniometer (*Radio*). Rotating coil within crossed field established by crossed loops (Bellini-Tosi) for direction-finding or interference-reduction on long-distance reception.

radiogram (*Radio*). (1) Message regularly transmitted by radio telegraphy. Once termed **marconigram,** after its inventor, Marconi. (2) Combination of gramophone and radio broadcast receiver. (*Radiol.*) See under **radiography.**

radiography (*Radiol.*). Registration of images in photographic material by X-rays or γ-rays, the result being properly termed a **radiogram.** Such images depend on differential absorption of the substances, including human bone or tissue, passed through.

radio heating (*Heat*). See **high-frequency heating.**

radio horizon (*Radio*). In the propagation of electromagnetic waves over the earth, the line which includes the part of the earth's surface which is reached by direct rays.

radioiodine (*Chem.*). I-125, I-131 and I-132, radioactive isotopes useful in diagnosis and treatment of thyroid gland disorders. I-125 and 131 are used in organ-function and blood volume studies.

radioisotope (*Nuc.*). See **radioactive isotope.**

radioisotopic purity (*Chem.*). The proportion of the activity of a given compound which is due to material in the stated chemical form. Cf. *radiochemical purity*.

Radiolaria (*Zool.*). An order of marine planktonic *Sarcodina*, the members of which have numerous fine radial pseudopodia which do not anastomose; the ectoplasm is vacuolated; there is a central capsule and usually a skeleton of siliceous spicules.

radiolarian chert, radiolarite (*Geol.*). A cryptocrystalline siliceous rock in part composed of the remains of Radiolaria. Most described examples seem to be of shallow-water origin, such as that which reaches a thickness of 9000 ft (3000 m) in New South Wales and contains 20 million Radiolaria per cubic centimetre.

radiolarian ooze (*Geol.*). A variety of non-calcareous deep-sea ooze, deposited at such depth that the minute calcareous skeletons of such organisms as *Foraminifera* pass into solution, causing a preponderance of the less soluble siliceous skeletons of *Radiolaria*. Confined to the Indian and Pacific Oceans, and passes laterally into red clay.

radiolarite (*Geol.*). See **radiolarian chert.**

radio link (*Radio*). Self-contained radio circuit capable of working in both directions, for insertion between two landline circuits.

radiolocation (*Radar*). Former term for **radar.**

radiology. The science and application of X-rays, gamma-rays and other penetrating ionizing radiations.

radioluminescence (*Nuc.*). Luminous radiation arising from rays from radioactive elements, particularly in mineral form.

radiolysis (*Nuc.*). Chemical decomposition induced by ionizing radiation.

Radiometal (*Met.*). An alloy of permalloy type. Contains iron 50%, nickel 45%, and copper 5%. Used because of high magnetic permeability and low hysteresis loss.

radiometer (*Acous., Phys.*). Instrument devised for the detection (e.g., Dicke's radiometer, two-receiver radiometer, subtraction-type radiometer) and measurement (e.g., thermopile, bolometer, microradiometer) of electromagnetic radiant energy. Also for the measurement of acoustic energy. See **Crookes radiometer, Rayleigh disk.**

radio microphone (*Radio*). One with a miniature radio transmitter, which allows freedom for the speaker. It transmits a signal that is picked up by a receiver nearby and demodulated for applying to an audio reproducing system or relayed over a broadcasting system.

radiomimetic (*Med.*). Said of drugs which imitate the physiological action of X-rays, notably in causing mutations. They act upon the nucleic acid content of biological material.

radionavigation (*Nav.*). The use of radio signals for obtaining position fixes or indicating departures from a planned course line.

radionuclide (*Nuc.*). Any nuclide (isotope of an element) which is unstable and undergoes natural radioactive decay.

radiopaque, radioparent (*Radiol.*). Opaque, transparent to radiation (especially X-rays).

radiophare (*Radio*). Same as **radio beacon.**

radiophone (*Radio*). Telephone system using a radio link, e.g., over oceans or between ships.

radiophotoluminescence (*Phys.*). Luminescence produced by irradiation followed by exposure to light.

radio pill (*Med.*). A capsule, which may be swallowed, containing a tiny radio transmitter that will send out information about body processes.

radio range (*Nav.*). Specific system of radio homing for aircraft, in which crossed loops are separately modulated with complementary signals, which coalesce on reception when the aircraft is *on course.*

radio receiver (*Radio*). One for receiving and interpreting radio transmissions. Normally *either* a radio-frequency tuned amplifier, demodulator, and a low-frequency amplifier, *or* an oscillator (first detector or remodulator), a band-pass intermediate frequency amplifier, a second detector or demodulator, followed by a low-frequency amplifier.

radioreceptor (*Biol.*). A specialized structure sensitive to the stimulus of light or temperature.

radio relay (*Radio*). See **wire broadcasting.**

radio relay station (*Radio*). An intermediate station receiving a signal from the primary transmitter and reradiating it to its destination.

radioresistant (*Radiol.*). Able to withstand considerable radiation doses without injury.

radiosensitive (*Radiol.*). Quickly injured or changed by irradiation. The gonads, the blood-forming organs, and the cornea of the eye are biologically most radiosensitive in man.

radiosonde (*Radio*). Sounding balloon which ascends to high altitudes, carrying meteorological equipment which modulates (usually frequency) radio signals transmitted back to equipment on earth.

radio spectrum (*Radio*). See **radio frequency.**

radiosperm (*Bot.*). A seed which is approximately circular in cross-section; by extension, a plant bearing such seeds, especially some fossil plants.

radio stars (*Astron.*). *Discrete radio sources*; only a few have been identified with visual objects, and these appear to contain gases in violent motion (see **Crab nebula, Cygnus A, Cassiopeia A**).

radio telegraph (*Teleg.*). Using a radio channel for telegraph purposes, e.g., by interrupted carrier, change of frequency, or modulation with interrupted audio tone.

radio telephony (*Teleph.*). Use of a radio channel for transmission of telephonic speech. Methods include simple modulation, suppressed carrier and one sideband, inverted sidebands and carrier, scrambling before modulation, one in a group modulation, or pulse modulation.

radio telescope (*Astron.*). Any type of receiver of cosmic radio signals. It may consist of a fixed interferometer arrangement or of a large parabolic reflector which can be directed to any part of the sky, and which focuses the incoming signal on to a small aerial.

radiotherapy (*Radiol.*). Theory and practice of medical treatment of disease, particularly any of the forms of cancer, with large doses of X-rays or other ionizing radiations.

radiothermoluminescence (*Phys.*). Luminescence produced by irradiation followed by exposure to heat. Now used for personal dosimetry.

radiothorium (*Chem.*). Symbol RdTh. A disintegration product and isotope of thorium, with a half-life of 1·90 years.

radiotropospheric duct (*Meteor.*). Stratum in which, because of a negative gradient of refractive modulus, there is an abnormal concentration of radiated energy.

radio-ulna (*Zool.*). The shaft-bone of the forearm in some Amphibians, in which the two elements are fused.

radiovision (*TV*). See **closed circuit.**

radium (*Chem.*). Radioactive metallic element, one of the alkaline earth metals. Symbol Ra, at. no. 88, r.a.m. 226, half-life 1620 years. The metal is white and resembles barium in its chemical properties; m.p. 700°C. It occurs in bröggerite, cleveite, carnotite, pitchblende, in certain mineral springs, and in sea water. Pitchblende and carnotite are the chief sources of supply.

radium bomb (*Radiol.*). An apparatus containing radium, emitting gamma rays for medical treatment.

radium cell (*Radiol.*). A sealed container in shape of thin-walled tube (usually metal), normally loaded into the larger containers, e.g., *radium needle, radium tube* (qq.v.).

radium emanation (*Chem.*). See **radon.**

radium needle (*Radiol.*). A container in form of needle, usually platinum-iridium or gold alloy, designed primarily for insertion into tissue. See **radium cell.**

radium plaque (*Radiol.*). Container in which radium is distributed over a surface, usually with small screenage in one direction to allow transmission of β- as well as γ-rays.

radium therapy (*Radiol.*). Radiotherapy by the use of the radiations from radium.

radium tube (*Radiol.*). *Radium cell* container in form of blunt-ended tube.

radius (*Bot.*). The group of ray flowers in a capitulum. (*Maths.*) A straight line joining the centre of a conic (particularly a circle) to the curve; the length of such a line. (*Zool.*) In land Vertebrates, the pre-axial bone of the antebrachium; one of the veins of the wing in Insects; in *Echinodermata* and *Coelenterata*, one of the primary axes of symmetry. *adj.* **radial.** See **per-radius.**

radius brick (*Build.*). See **compass brick.**

radius gauge (*Eng.*). A sheet metal strip with accurately formed external and internal radii of specified size, used as a profile gauge to determine size of an unknown radius by comparison. Usually one of a set of gauges of graded sizes.

radius mounting (*Optics*). The mounting of a grating at one end of a bar, normal to the bar. The bar is of length equal to half the radius of curvature of the grating and the other end is attached to a fixed bearing about which the bar and grating rotate.

radius of action (*Aero.*). Half the *range* (q.v.) in still air; the total range is out and home again.

radius of atom (*Chem.*). See **atomic diameters.**

radius of convergence (*Maths.*). See **circle of convergence.**

radius of curvature (*Maths.*). See **curvature.**

radius of gyration (*Maths.*). See **moment of inertia.**

radius of inversion (*Maths.*). See **inversion.**

radius of spherical curvature (*Maths.*). See **osculating sphere.**

radius of torsion (*Maths.*). See **curvature.**

radius rod (*Build.*). A rod pivoted at one end and carrying a marking point at the other end so that, as the rod is swung around, a circle or part of a circle, of radius equal to the length of the rod, may be marked out. Also **gig stick** (*Eng.*) A rod attached to the die or block of a *Walschaert's valve gear* (q.v.) for transmitting its motion to the end of the combination lever pivoted to the valve rod.

radius vector (*Astron.*). The line joining the focus to the body which moves about it in an orbit, as the line from the sun to any of the planets, comets, etc.

radix (*Comp., Maths.*). Basis of a denominational number system. A radix of 10 is normally used except for digital computers which usually use 2 because of the ease with which two-state electronic circuits can be constructed. Also called **base.** (*Zool.*) The root or point of origin of a structure, as the *radix aortae*.

radix point (*Comp., Maths.*). That separating positive and negative indices of the radix; the decimal point when the radix is 10.

radome (*Radar*). Housing for radar equipment, transparent to the signals, e.g., a plastic shell on aircraft or a balloon on the ground; also **blister, raydome.**

radon (*Chem.*). A zero-valent, radioactive element, the heaviest of the noble gases. Symbol Rn, at. no. 86, r.a.m. 222, half-life 3·82 days, b.p. −65°C, m.p. −150°C. It is formed by the disintegration of radium. Isotopes are actinon (at. no. 219, half-life 4 seconds, from actinium) and thoron (at. no. 220, half-life 54 seconds, from thorium). Also **radium emanation.** See **emanations.**

radon seeds (*Radiol.*). Short lengths of gold capillary tubing containing radon used in treatment of malignant and nonmalignant neoplasms.

Radstockian Series (*Geol.*). The highest series in the Westphalian stage of the Carboniferous System which comprises the upper coal measures of Somerset (Radstock) and the Keele Series of North Staffordshire.

radula (*Zool.*). In *Mollusca*, a strip of horny basement membrane bearing numerous rows of horny or chitinous teeth. *adjs.* **radular, radulate, raduliform.**

RaEm (*Chem.*). A symbol for *radium emanation* or *radon*.

raffinate (*Chem. Eng., Min. Proc.*). Liquid layer in solvent extraction system from which required solute has been extracted, e.g., in the chemical extraction of uranium from its ores, the liquid left after the uranium has been extracted by contact with an immiscible solvent.

raffinose (*Chem.*). Melitriose, $C_{18}H_{32}O_{16}+5H_2O$, a non-reducing trisaccharide found in the sugar-beet, in molasses, in cotton-seed cake, etc. On hydrolysis it gives D-glucose, D-fructose and D-galactose.

raft bridge (*Civ. Eng.*). One somewhat similar to a *pontoon bridge* but supported on rafts.

raft foundation (*Build.*). A layer of concrete, usually reinforced, extending under the whole area of a building and projecting outside the line of its walls; used to provide a foundation in cases where the ground is unduly soft or the loading to be put upon it is unduly heavy.

rag-bolt (*Eng.*). A foundation bolt with a long tapered head of increasing size towards its end, and having jagged points on its surface to prevent withdrawal.

rag content papers (*Paper*). Papers containing 25% or more of rag (cotton, etc.) fibres.

rag duster (*Paper*). A revolving long hollow cylinder covered with coarse wire through which rags are given a preliminary dusting.

ragging (*Min. Proc.*). Rough concentration or washing, for a low ratio of concentration. In mineral processing, grooves cut on surface of roll to improve grip on feed. In jigging, the bed of heavy mineral or metal shot maintained on the jig screen.

ragging frame (*Min. Proc.*). **Reck. Tilting table,** which may be worked automatically, on which finely ground ore is treated by sluicing.

raglan, raglin (*Carp.*). A slender ceiling joist.

raglet (*Plumb.*). A narrow groove cut into a masonry or brickwork surface, for receiving the edge of a flashing which is to be fixed to it.

rag stone (*Build.*). A general term for coarse-grained sandstone, often with a calcareous cement, e.g., Kentish rag.

raguinite (*Min.*). Thallium iron sulphide, crystallizing in the orthorhombic system.

rag-work (*Build.*). Term applied to wall construction in which undressed flat stones of about the thickness of a brick are built up into a wall the outer faces of which are left rough.

rail (*Civ. Eng.*). A steel bar, usually of special section, laid across sleepers to provide a track for the passage of rolling stock with flanged wheels. The standard section for main line rails in Gt. Britain is the 110 lb (per yard) flat-bottomed rail of 60 ft length, made in accordance with specification of the B.S.I. See also **continuous welded rail.** (*Join*). (1) A horizontal member in framing or panelling. (2) The upper member in a balustrade.

rail bender (*Eng.*). A short stiff steel girder with claws at the ends and a central boss carrying a heavy screw.

rail bond (*Elec. Eng.*). An electrical connexion between two adjacent lengths of track or conductor rail on a railway.

railbus (*Rail.*). A lightweight four-wheel railway coach powered by one or two bus-type engines.

railcar (*Rail.*). A self-propelled bogie railway coach (power car), usually powered by two bus-type diesel engines; also, its trailer.
rail chair (*Rail.*). See **chair.**
rail gauge. See **gauge, standard gauge, broad gauge, narrow gauge.**
rail guard (*Rail.*). See **check rail.**
rail post (*Build.*). A newel post.
railroad disease (*Vet.*). See **transit tetany.**
railway curve (*Instr.*). A drawing instrument similar to a French curve but cut at the edge to an arc of large radius. Used for drawing arcs when these are too large for beam compasses. Sets are supplied to cover a wide range of radii.
railway transit (*Surv.*). A transit theodolite unequipped with means for measuring vertical angles.
rain (*Meteor.*). Result of condensation of excess water vapour when moist air is cooled below its dew-point. Rain falls when droplets increase in size until they form drops whose weight is equivalent to the frictional air resistance. The greater proportion of raindrops have a diameter of 0·2 cm or less; in torrential rain a small proportion may reach 0·4 cm. Rain effects important geological work by assisting in the mechanical disintegration of rocks; also chemically, in bringing about solution of carbonates, etc.; and, through the agency of running water, in redistributing the products of erosion and disintegration.
rain band (*Meteor.*). An absorption band in the solar spectrum on the red side of the D lines, produced by water vapour in the earth's atmosphere.
rainbow (*Meteor., Phys.*). A rainbow is formed by sunlight which is refracted and internally reflected by raindrops, the concentration of light in the bow corresponding to the position of minimum deviation of the light. The angular radius of the primary bow is 42°, this being equal to 360° minus the angle of minimum deviation for a spherical drop. The colours, ranging from red outside to violet inside, are due to dispersion in the water. See **secondary bow.**
rainbow quartz (*Min.*). See **iris.**
rain chamber (*Met., Min. Proc.*). Washing tower or other space where rising dust and fumes are brought into contact with descending sprays of water.
rain cloud (*Meteor.*). See **nimbus.**
rain day (*Meteor.*). A period of 24 hr from 9 a.m. in which 0·01 in or 0·2 mm or more of rain is recorded. See **rain gauge.**
Rainey's corpuscles (*Zool.*). In certain *Sarcosporidia*, sickle-shaped spores occurring encysted in muscle.
Rainey's tubes (*Zool.*). See **Miescher's tubes.**
rain gauge (*Meteor.*). An instrument for measuring the amount of rainfall over a given period, usually 24 hr. The usual form consists of a sharp-rimmed funnel, 5 in. in diameter, leading into a narrow-necked graduated collecting vessel. The *Dines tilting siphon* is of the self-recording type, noting both the time and the amount of rainfall.
rainmaking (*Meteor.*). Artificial stimulation of precipitation by scattering solid carbon dioxide on supercooled clouds, or by silver iodide nucleation.
rain prints (*Geol.*). More or less circular, vertical, or slanting pits occurring on the bedding planes of certain strata; believed to be the impressions of heavy raindrops falling on silt or clay, hard enough to retain the impression before being covered by a further layer of sediment.
rain shadow (*Ecol.*). A dry area, often a desert, on the sides of mountains away from the sea, due to the deposition of most of the moisture from the winds blowing off the ocean on the slopes facing the ocean. The higher the mountain, the greater the effect.
rain spell (*Meteor.*). 15 consecutive *rain days* (q.v.).
rain stage (*Meteor.*). That part of the condensation process taking place at temperatures above 0°C so that water vapour condenses to water liquid.
rain-wash (*Geol.*). The creep of soil and superficial rocks under the influence of gravity and the lubricating action of rain.
rainwater pipe (*Build.*). See **downpipe.**
raise (*Mining*). A shaft or winze excavated upwards.
raised bands (*Bind.*). Bands which show on the back of a book when bound. This indicates that the book has been sewn *flexible* (q.v.).
raised beach (*Geol.*). Beach deposits which are found above the present high-water mark; due to the relative uplift of the land or to a falling sea-level. See **eustatic movements.**
raised oil (*Leather*). See **raising.**
raised panel (*Join.*). A panel whose surface stands *proud* (q.v.) of the general surface of the framing members.
raiser (*Build.*). See **riser.**
raising (*Leather*). The process by which oil is recovered from the wash waters in the manufacture of chamois leather. (*Textiles*) Machine operation in which wire covered rollers or teazles on cylinders revolve over the cloth surface to produce a nap, e.g., on woollen and worsted cloths, mixture fabrics, cotton blankets.
raising-plate (*Carp.*). A horizontal timber resting on part of a structure and supporting a superstructure. Also called **reason-piece.** See **pole plate, wall plate (2).**
rake. An angle of inclination. (*Arch.*) In a theatre, the upward slope, from the horizontal, of both the stage and the auditorium. (*Eng., etc.*) Angular relief, e.g., *top-rake, side-rake*, given to the faces of cutting tools to obtain the most efficient cutting angle. The face of a cutting tool with negative *rake* has no angular relief but, on the contrary, meets the workpiece with the cutting edge trailing, the tool being stronger in shear, impact, and abrasion than one having positive *rake*. (*Mining*) (1) A forked tool for loading coal underground. (2) An irregular vein of ironstone; in Derbyshire, any transverse fissure vein. (3) Train or 'journey' of mineral trucks. (*Ships*) A term used in shipbuilding to denote *not perpendicular to the datum line*. (*Tools*) A long-handled tool with projecting teeth at one end, used for mixing plaster.
rake classifier (*Min. Proc.*). Inclined tank into which ore pulp is fed continuously, the slow settling portion overflowing and the coarser material gravitating down, to be gathered and raked up to a top discharge.
raker (*Build.*). See **pointer.** (*Carp.*) See **raking shore.**
raker set (*Tools*). A pattern of saw teeth in which one straight tooth alternates with two teeth set in opposite directions.
raking bond (*Build.*). A form of bond sometimes used for very thick walls, or for strengthening the bond in footings carrying heavy loads. The courses are built diagonally across the wall, successive courses crossing one another in respect of rake; triangular *bats* are added to enable square facework to be completed. Also

diagonal bond. See also **herring-bone bond.**

raking cornice (*Arch.*). A cornice decorating the slant sides of a pediment.

raking flashing (*Plumb.*). A *flashing* (q.v.) much used where a masonry chimney projects from a sloping roof.

raking-out (*Build.*). The operation of preparing mortar joints in brickwork for pointing.

raking pile (*Civ. Eng.*). A pile which is not driven in vertically.

raking prop (*Mining*). An inclined timber support.

raking shore (*Carp.*). An inclined baulk of timber, one end of which rests upon the ground while the other presses against the wall to which temporary support is to be given.

râle (*Med.*). A bubbling or crackling sound produced in a diseased lung by the passage of air over or through secretions in it.

r.a.m. (*Chem.*). Abbrev. for *relative atomic mass.*

ram (*Civ. Eng.*). (1) The monkey of a pile-driver. (2) To consolidate the surface of loose material by punning. (*Eng.*) (1) The reciprocating head of a press. (2) The falling weight of a pile-driver.

RAM (*Comp.*). Random access memory. See **access to store.**

ram-air turbine (*Aero.*). A small turbine motivated by ram (i.e., free stream) air; used (*a*) to drive fuel pumps, hydraulic pumps, or electrical generators in guided weapons because of the absence of shaft drives with rockets and ramjets, (*b*) as an emergency power source for driving hydraulic pumps or electrical generators for high-speed aeroplanes, particularly those with power controls. Abbrev. **RAT.**

Raman scattering (*Phys.*). See **scattering.**

ramark (*Nav.*). A nondirectional radio beacon, usually pulsed, used as a radar beacon to enable a craft to determine its direction relative to it.

ramentaceous (*Bot.*). Covered with ramenta.

ramentum (*Bot.*). A thin brownish scale, one cell-layer in thickness, occurring on the stems, petioles, and leaves of ferns. *pl.* **ramenta.**

rami (*Zool.*). In *Collembola*, the free ends of a minute pair of appendages on the third abdominal segment. They form the organ known as the *retinaculum* (q.v.).

ramicole (*Bot.*). Living on twigs.

rami communicantes (*Physiol.*). The preganglionic myelinated nerve fibres of sympathetic nervous system connecting spinal nerves and sympathetic chain of ganglia.

ramicorn (*Zool.*). Of Insects, having branched antennae.

ramie (*Textiles*). *Bast fibre*, strong and very white, grown in Japan, India, China, and other tropical countries. Once widely used for gas mantles, also for bank-note paper. Fibre length varies enormously and poor elasticity is its serious disadvantage.

ram intake (*Aero.*). A forward-facing engine (or accessory) air intake which taps the kinetic energy in the airflow and converts it into pressure energy by diffusion; in supersonic flight very high pressure ratios can be obtained.

ramisection, ramisectomy (*Surg.*). The operation of cutting the sympathetic nerves between the spinal cord and the sympathetic ganglia.

ramjet (*Aero.*). The simplest *propulsive duct* (q.v.) deriving its thrust by the addition and combustion of fuel with air compressed solely as a result of forward speed. In subsonic flight kinetic energy is converted into pressure by a *diffuser* (q.v.), or widening duct, which also slows it sufficiently to permit combustion to be maintained; about Mach 1 the shock wave generated by the air-intake lip improves the compression when it decelerates the air to subsonic velocity prior to diffusion. At high Mach numbers, 1·5 and upward, two shock waves are required for the dual purpose of raising the pressure and slowing the air for combustion—pressure ratios of 6 : 1 are attainable at Mach 2, 36 : 1 at Mach 3. In supersonic flight the jet efflux of a ramjet has to be accelerated to high velocity by a *venturi* or *convergent-divergent nozzle* (qq.v.).

rammelsbergite (*Min.*). Essentially composed of diarsenide of nickel; crystallizes in the orthorhombic system.

rammer (*Build., Civ. Eng.*). A *punner* (q.v.). (*Foundry*) Hand tool, which takes various forms, for packing the sand of a mould evenly round the pattern. The process of *ramming* is one in which the moulding sand is firmly consolidated before the pattern is removed from the moulding box.

Rammstedt's operation (*Surg.*). Incision of the pylorus down to the mucous membrane, done in the treatment of congenital hypertrophy of the pylorus.

ramp (*Civ. Eng.*). An inclined surface, often in place of steps. (*Join.*) A sudden rise in a handrail when it is concave upwards. Cf. *knee.* (*San. Eng.*) A short length of sewer laid locally at a much steeper gradient than normal.

rampant arch (*Civ. Eng.*). An arch whose abutments are not in the same horizontal line.

rampant centre (*Carp.*). A centre for a rampant arch.

ramp voltage (*Elec. Eng.*). Steadily rising voltage, as in a sawtooth waveform.

Ramsauer effect (*Phys.*). Sharp decrease to zero of scattering cross-section of atoms of inert gases, for electrons of energy below a certain critical value.

Ramsay and Young's rule (*Chem.*). The ratio of the boiling-points of two liquids of similar chemical character is approximately constant, independent of the pressure at which they are measured.

Ramsden circle (*Optics*). The *exit pupil* of a telescope, found as a ring of light at the eyepiece of a telescope focused on a diffuse source at infinity. The magnification is the *objective* diameter divided by that of this circle.

Ramsden eyepiece (*Light, Surv.*). An eyepiece often used in an optical instrument in which crosswire measurements are to be made. It consists of two similar plano-convex lenses separated by a distance equal to two-thirds the focal length of each, and having their convex faces towards each other. The focal plane is just outside the system.

ramulus (*Bot.*). A very small branch of a stem or of a leaf. (*Zool.*) See **hydrocladia.** *pl.* **ramuli.**

ramus (*Zool.*). The barb of a feather; in Vertebrates, one lateral half of the lower jaw, the mandible; in *Rotifera*, part of the trophi; any branchlike structure; a ramification.

ramus communicans (*Zool.*). In *Craniata*, a branch of the spinal nerves, connecting them with ganglia of the sympathetic nervous system, and containing visceral motor and visceral sensory nerve fibres.

rance (*Build.*). A shore.

random (*Build.*). Said of rubble masonry in which the stones are of irregular shape and the work is not coursed. (*Typog.*) The sloping top on a composing frame.

random access (*Comp.*). See **access to store.**

random coincidence (*Comp.*). Simultaneous operation of two or more coincidence counters as a result of their discharge by separate incident particles arriving together (instead of common discharge by a single particle as is normally assumed in interpreting the readings).

random error (*Stats.*). Error unlikely to be repeated in any series of measurements.

random event (*Stats.*). Event which has no effect on the probability of occurrence of subsequent events.

random noise (*Acous.*). Noise due to the aggregate of a large number of elementary disturbances with random occurrence in time.

random number series (*Comp., Maths.*). One that exhibits no regular pattern, e.g., the numbers drawn in a lottery. No completely satisfactory definition has yet been proposed.

random response (*Acous.*). See **reverberation response.**

random searching (*Ecol.*). A process of completely unorganized 'search' by which some ecologists suggest that animal populations find food, mates, and suitable places to live.

random-tooled ashlar (*Build.*). A block of stone finished with groovings irregularly cut.

random winding (*Elec. Eng.*). See **mush winding.**

Raney nickel (*Chem.*). A nickel catalyst, used for hydrogenation, produced by the action of alkali on a nickel-aluminium alloy.

range (*Acous.*). Maximum distance for effective acoustical reception. (*Aero.*) Maximum horizontal distance covered by projectile. (*Nav.*) The distance from a navigation point. (*Nuc.*) (1) Length of track along which ionization is produced in a nuclear particle. (2) Distance of effective operation of nuclear forces. (*Radio*) (1) Maximum distance of radio transmitter at which effective reception is possible (not normally constant). (2) The optical short-wave radio range calculated from straight-line propagation. (*Surv.*). To fix points, either by eye or with the aid of an instrument, to be in the same straight line.

rangefinder (*Photog.*). Device for assisting focusing by measuring distance of object from camera. Most common method is by correcting parallax of image seen simultaneously from two slightly separated viewpoints. The *split-image type* (q.v.) works on a different principle.

range height indicator (*Radar*). A display used in conjunction with a *plan-position indicator* for airport control. It displays a vertical plane on which the elevation and bearing of the target can be seen.

range of stress (*Met.*). The range between the upper and lower limits of a cycle of stress in a fatigue test. The midpoint is the *mean stress.*

range tracking (*Radar*). System in which an electronic device rejects signals other than those from a moving target.

ranging figures (*Typog.*). See **lining figures.**

ranging rod (*Surv.*). A wooden pole used to mark stations conspicuously, or to assist in *ranging.*

rangs (*Textiles*). See **tier.**

Ranikhet disease (*Vet.*). See **Newcastle disease.**

ranine (*Zool.*). Pertaining to, or situated on, the under surface of the tongue.

ranivorous (*Zool.*). Feeding on frogs.

rank (*Maths.*). Of a matrix: a matrix is of rank r if it contains at least one determinant of order r ($\neq 0$) and all higher-order determinants are zero.

Rankine cycle (*Eng.*). A composite steam plant cycle used as a standard of efficiency, comprising introduction of water by a pump to boiler pressure, evaporation, adiabatic expansion to condenser pressure, and condensation to the initial point.

Rankine efficiency (*Eng.*). The efficiency of an ideal engine working on the *Rankine cycle* (q.v.), under given conditions of steam pressure and temperature.

Rankine scale (*Heat*). Absolute scale of temperature, based on degrees Fahrenheit. See **Fahrenheit scale.**

Rankine's formula (*Civ. Eng.*). An empirical formula giving the collapsing load for a given column. It states that

$$P=\frac{f_c \, . \, A}{1+a\left(\frac{l}{k}\right)^2},$$

where P = the collapsing load, f_c = safe compressive stress for very short lengths of the material, A = area of cross-section, l = the length of the pin-jointed column, k = the least radius of gyration of the section, a = a constant for the material $=\frac{f_c}{\pi^2 \, . \, E}$, where E is Young's modulus for the material.

rankinite (*Min.*). A monoclinic calcium disilicate, $Ca_3Si_2O_7$, found in highly metamorphosed siliceous limestones.

rank of coal (*Geol.*). This is determined by the extent to which original 'mother substance' of the coal has been modified by heat, pressure, and chemical change after burial. Thus brown coal is of *low rank*, anthracite of *high rank.*

rank of selectors (*Teleph.*). The whole set of selectors concerned with a specified stage in setting up a call through an exchange.

ranula (*Med.*). A cystic tumour formed on the lower surface of the tongue or on the bottom of the mouth, caused by blocking or dilatation of mucous gland.

ranunculaceous (*Bot.*). Like a buttercup.

Ranvier's nodes (*Zool.*). Constrictions of the neurolemma occurring at regular intervals along peripheral medullated nerve fibres.

Raoult's law (*Chem.*). The vapour pressure of an ideal solution at any temperature is the sum of the vapour pressure of each component, P_i, multiplied by the mole fraction of that component, x_i: $P=\Sigma P_i x_i$.

Rapakivi Granite (*Geol.*). A type described from a locality in Finland, characterized by the occurrence of rounded pink crystals of orthoclase surrounded by a mantle of whitish sodic plagioclase. A widely-used textural term.

raphe (*Bot.*). (1) An elongated mass of tissue, containing a vascular strand, and lying on the side of an anatropous ovule, between the chalaza and the attachment to the placenta. (2) A line running longitudinally on the valve of a diatom indicating the position of a narrow slit in the wall; it bears a nodule at each end and one in the middle. (*Zool.*) A broad junction, as between the halves of the Vertebrate brain.

raphide (*Bot.*). A long needle-shaped crystal, usually of calcium oxalate, occurring singly, in rounded masses, or in sheaves, in plant cells.

Raphidioptera (*Zool.*). An order of *Neuropteroidea*, with a prognathous head, biting mouthparts, a long prothorax, two pairs of membranous wings with pterostigma, and terrestrial larvae. Snake-flies.

rapid access memory (*Comp.*). One in which time of access is of the order of microseconds.

rapid eye movement sleep (*Psychol.*). Part of the sleep cycle, characterized by rapid movement of the eyes, loss of muscle tone and a low amplitude encephalogram record. It coincides with the periods of dreaming. Also called **paradoxical sleep.** Abbrev. **REM sleep.**

rapids zone (*Ecol.*). In small streams, the zone where current velocity is great enough to keep the bottom clear of silt and other loose materials, thus providing a firm bottom. Cf. *pools zone*.

rapping (*Foundry*). The process of loosening a pattern in a mould to facilitate its withdrawal. A spike or lifting screw is inserted in the pattern and tapped smartly in every direction.

rapport (*Psychol.*). The emotional bond or atmosphere existing between analyst, or hypnotist, and patient, which is conducive to *suggestion* (q.v.).

Rap-rig (*Build.*). TN for a form of light scaffolding capable of being speedily erected; it is especially convenient for interior use.

raptatory, raptorial (*Zool.*). Adapted for snatching or robbing, as birds of prey.

rare earth elements (*Chem.*). A group of metallic elements possessing closely similar chemical properties. They are trivalent, but otherwise similar to the alkaline earth metals. The group consists of the lanthanide elements 57 to 71, plus scandium (21) and yttrium (39). Extracted from monazite, and separated by repeated fractional crystallization, liquid extraction, or ion exchange.

rare earths (*Chem.*). The oxides (M_2O_3) of the rare earth elements.

rarefaction (*Med.*). Abnormal decrease in the density of bone as a result of absorption from it of calcium salts, as in infection of bone. (*Phys.*) Decrease of density due to expansion of the low-pressure wavefront of a sound wave.

rare gases (*Chem.*). See **inert gases.**

Raritan Sandstones (*Geol.*). Lignitic sands of aeolian origin occurring in the Cretaceous of the Atlantic Plain, succeeding the Comanchean; equivalent to the Dakota Sandstone of the Great Plains.

Raschel (*Textiles*). 2-bar warp loom, fitted with latch needles.

Raschig rings (*Chem. Eng.*). *Packed column* (q.v.) fillers which are hollow open cylinders of diameter equal to length.

rashing (*Mining*). A thin layer of shale or inferior coal, sometimes found between the coal-seam and the roof or floor.

rasorial (*Zool.*). Adapted for scratching.

Rasorite (*Min.*). A brand name for sodium borates originating from California.

rasp (*Tools*). Coarse type of file with teeth in the form of raised points.

raspail test (*Paper*). A deep red colour is formed if resin is present when a drop of concentrated cane sugar is placed on the paper, blotted off, and a drop of concentrated sulphuric acid added.

rastellus (*Zool.*). In Spiders which dig burrows, a rakelike row of stout teeth on the outside lower edge of the first joint of the chelicerae.

raster (*TV*). The area of the screen of a TV picture tube which is illuminated by the pattern of horizontal lines produced during a complete frame scanning period. Also called **scanning field.**

raster burn (*TV*). Deterioration of the scanned area of the screen of a TV picture or camera tube as a result of use.

raster stereoscopy (*Cinema.*). A form of cinematography in which a 3-dimensional image is obtained by using a radial grid of conical plastic lens elements in front of the screen.

Rastinon (*Pharm.*). See **tolbutamide.**

RAT (*Aero.*). Abbrev. for **ram-air turbine.**

rat-bite fever (*Med.*). See **sodoku.**

ratch (*Spinning*). The distance between the nip of the front and back rollers in a drawing machine.

ratchet brace (or **drill**) (*Eng.*). A drilling brace in which the drill spindle is rotated intermittently by a ratchet wheel engaged by a pawl on a hand-lever; used in confined spaces, repair work.

ratchet mechanism (*Eng.*). A mechanism comprising a *ratchet wheel* (q.v.) and a *pawl* (q.v.) with which it engages. It is used to convert reciprocating motion of the pawl pivot into intermittent rotary or linear motion in one direction only.

ratchet screwdriver (*Tools*). One with a ratchet mechanism to make it operate only in one direction.

ratchetting (*Nuc.*). Periodic movement of reactor fuel elements due to thermal cycling and differential thermal expansion effects.

ratchet-toothed escape wheel (*Horol.*). An escape wheel with fine-pointed teeth; used in English lever watches.

ratchet wheel (*Eng.*). A wheel with inclined teeth used in a *ratchet mechanism* (q.v.).

ratching (*Spinning*). After completing its outward run the mule carriage is moved a little farther forward with the front rollers stopped. This tightens the twisting yarns to take out snarls or slubs, etc., before winding commences.

rate action (*Automation*). Same as **booster response.**

rate constant (*Chem.*). See **velocity constant.**

rated altitude (*Aero.*). The height, measured in the *International Standard Atmosphere*, at which a piston aero-engine delivers its maximum power. Cf. *power rating*.

rated blowing-current (*Elec. Eng.*). The current at which a fuse-link is specified by the maker to melt and break the circuit.

rated breaking-capacity (*Elec. Eng.*). The r.m.s. current, or the kVA at the rated voltage, which a circuit-breaker is specified by the maker to interrupt without damage.

rated capacity (*Eng.*). General term for the output of an equipment, which can continue indefinitely in conformity with a criterion, e.g., heating, distortion of signals or of waveform. See **continuous rating, intermittent rating** or **periodic rating.**

rate-determining step (*Chem. Eng.*). Where a process consists of a series of consecutive steps, the overall rate of the process is largely determined by the step with the slowest rate, so that efforts to speed up the process must chiefly be directed to this step.

rated impedance (*Elec. Eng.*). Particularly applied to a loudspeaker, in which impedance rises with frequency, with an added sharp rise at frequency of bass resonance. That resistance, equal in magnitude to the modulus of the minimum impedance above this resonant frequency, which replaces the loudspeaker when measuring the power applied to the loudspeaker during testing.

rated making-capacity (*Elec. Eng.*). The maximum asymmetrical current which a circuit-breaker can make at the rated voltage.

rate fixing (*Eng., etc.*). The determination of a time allocation for carrying out a specified task of work, usually as a basis for remuneration.

rate gene (*Gen.*). A gene influencing the rate of a developmental process.

rate gyro (*Instr.*). Gyroscope with single gimbal which produces a couple proportional to the rate of rotation.

ratemeter (*Nuc. Eng.*). See **count ratemeter.**

rate of climb (*Aero.*). Generally, rate of ascent from the earth. In performance testing, the vertical component of the air path of an ascending aircraft, corrected for standard atmosphere. See **International Standard Atmosphere.**

rate-of-climb indicator (*Aero*). See **vertical speed indicator.**

rate time (*Automation*). Period during which control depends on a rate of change or derivative of position.

Rathke's pouch (*Zool.*). In developing Vertebrates, the diverticulum formed from the dorsal aspect of the buccal cavity ectoderm which gives rise to the adenohypophysis.

ratine (*Textiles*). A dress fabric with a rough surface, and either plain, fancy, or stripe effects; made from worsted or cotton yarns.

rating (*Elec. Eng.*). Specified limit to operating conditions, e.g., current rating, etc. (*Work Study*) Numerical value or symbol used to denote rate of working.

rating nut (*Horol.*). A milled nut which supports the pendulum bob. By rotating it the bob is raised or lowered, thus altering the time of vibration of the pendulum.

ratio (*Maths.*). See **cross-, harmonic-.**

ratio arms (*Elec. Eng.*). Two adjacent arms of a Wheatstone bridge, the resistances in which can be made to have one of several fixed ratios.

ratio detector (*Radio*). Detector circuit used for frequency-modulated carriers.

ratio error (*Elec. Eng.*). A departure of the ratio between the primary and secondary voltages or currents of a voltage or current transformer from the rated value.

rational horizon (*Surv.*). See **true horizon.**

rationalization (*Psychol.*). Justifying by (spurious) reasoning after the event. Frequently an *ego-defence mechanism* used to avoid the recognition of the true motives for an action and consequent feelings of guilt.

rationalized units (*Elec.*). Systems of electrical units for which the factor 4π is introduced in Coulomb's laws so that it shall be absent from more widely used relationships. In *Heaviside-Lorentz units* (rationalized Gaussian units) this is done directly, thus modifying values of the unit charge and unit pole. In *MKSA units* it is done indirectly by modifying the values of the permittivity and permeability of free space.

rational number (*Maths.*). A number which may be expressed as the ratio of two integers, e.g., $\frac{3}{4}$.

ratio of compression (*Eng.*). See **compression ratio.**

ratio of slenderness (*Build.*). The ratio between length or height of a pillar and its least radius of gyration.

ratio of specific heat capacities (*Phys.*). The ratio of specific heat capacity of a gas at constant pressure to that at constant volume has a constant value of about 1·67 for monatomic gases, 1·4 for diatomic gases, and approaches unity for polyatomic gases. This ratio, denoted by γ, enters into the *adiabatic equation.*

RATOG (*Aero.*). *R*ocket *a*ssisted *t*ake *o*ff *g*ear.

rat-race (*Telecomm.*). See **hybrid junction.**

rat-tail file (*Tools*). Small round file; used for enlarging holes, etc.

rattle (*Paper*). The crackling noise when paper is handled. It indicates the degree to which the pulp or fibre has been hydrated in the process of beating, and can be augmented by the addition of starch and other additives. (*Zool.*) The series of horny rings representing the modified tail-tip scale in Rattlesnakes (*Colubridae*).

rattle echo (*Meteor.*). Unmusical multiple echo, generally associated with thunder, which is formed by nearby flashes giving rise to a sharp acoustic impulse which is reflected between mountains or strata of air.

rat-trap bond (*Build.*). A form of bond in which a 9-in wall is built up of bricks on edge, so arranged as to enclose a 9×3 inch cavity.

Rauber's cells (*Zool.*). In Mammals, cells of the trophoblast situated immediately over the embryonic plate.

Raunkiaer's life forms (*Bot.*). A classification system for vegetation based on the position of the perennating buds with respect to soil level. See **megaphanerophyte, mesophanerophyte, microphanerophyte, nanophanerophyte, phanerophyte.**

Rauwolfia serpentina (*Pharm.*). One of a number of sedative and antihypertensive drugs, also including *R. canescens* and *R. vomitoria*, obtained from the dried roots of trees of the *Rauwolfia* genus. See **reserpine, rescinnamine.**

rawhide (*Leather*). A hide which has been dried or treated with a preservative to prevent putrefaction prior to tanning.

rawhide hammer (*Eng.*). A hammer the head of which consists of a close roll of hide projecting from a short steel tube; used by fitters to avoid injuring a finished surface.

Rawlplug (*Build.*). TN for a small tube of tough compressed fibre or plastic for insertion in a hole, to provide a fixing plug into which a screw may be turned.

raw silk (*Textiles*). The natural material from which silk yarns are made. The filaments from 4, 5, 6, or more cocoons are reeled to form 'single' threads, which are converted into yarns by throwing.

ray (*Phys.*). General term for the geometrical path of the radiation of wave energy, always in a direction normal to the wavefront, but with possible reflection, refraction, diffraction, divergence, convergence, and diffusion. By extension, also the geometrical path followed by a beam of particles in an evacuated chamber. This may be curved in electric or magnetic field. See **particle.** (*Zool.*) A skeletal element supporting a fin; a sector of a radially symmetrical animal.

raydome (*Radar*). See **radome.**

ray floret (*Bot.*). One of the small flowers radiating out from the margin of a capitulum or other dense inflorescence.

ray initial (*Bot.*). One of the cells of the cambium which takes part in the formation of a vascular ray.

Rayleigh disk (*Acous.*). Small light mica disk (in water a lead disk is used) pivoted about a vertical diameter, hung by a fine thread of glass or quartz. If placed at an angle to a progressive sound wave, the disk experiences a torque which depends on the square of the velocity of the molecules in the medium. It provides a useful method of measurement of sound intensity, calculated from the measured torque using a formula due to König.

Rayleigh distillation (*Chem.*). A simple distillation in which the composition of the residue changes continuously during the course of the distillation.

Rayleigh distribution (*Phys.*). Spectral distribution of noise energy which has a skew distribution.

Rayleigh-Jeans law (*Phys.*). Distribution of energy E in the spectrum of a black body as a function of temperature T and wavelength λ.

$$E d\lambda = \frac{8\pi kT}{\lambda^4}\, d\lambda,$$

where k = Boltzmann's constant and E = energy density in the range $\lambda \rightarrow \lambda + d\lambda$. The formula applies only to large values of λ; for small λ, *Planck's radiation formula* is used.

Rayleigh ilmit (*Phys.*). One-quarter of a wavelength, the maximum difference in optical paths between rays from an object point to the corresponding image point for perfect definition in a lens system.

Rayleigh refractometer (*Phys.*). An instrument for measuring the refractive index of a gas by an optical interference method. Each of two interfering light beams passes through a tube which may contain air or gas or be evacuated. By observing the shift of the interference fringes when one of the tubes is evacuated and the other contains gas, the refractive index of the gas may be calculated from the expression $n=1+\frac{s\lambda}{l}$ where l is the length of tube, λ the wavelength used, and s the number of fringes shifted.

Rayleigh scattering (*Phys.*). See **scattering.**

Rayleigh's theorem (*Phys.*). Relation between energy of an impulse function in terms of time and the same energy as the sum of energies of frequency components in a frequency spectrum.

Raynaud's disease (*Med.*). A paroxysmal disorder of the arteries of the fingers and toes characterized by attacks of pain in them, the fingers (or toes) going white and then blue; gangrene may supervene.

rayon (*Textiles*). Generic name for a group of regenerated cellulose filaments, yarns, etc. Filaments formed from solutions of modified cellulose, made by various processes. After passing through the spinneret or spinning jet, the filaments are combined to form yarn. In the viscose and copper processes of making rayon, the filaments pass through a bath which coagulates them. In acetate manufacture, spinning is a dry process.

ray tracheide (*Bot.*). A somewhat thick-walled cell which, with many other similar cells, occurs in the vascular rays of pine trees; it has bordered pits, and conducts aqueous solutions horizontally.

Rb (*Chem.*). The symbol for *rubidium.*

rbe (*Radiol.*). Abbrev. for *relative biological effectiveness.* See also **dose.**

R.C. (*Build.*). Abbrev. for *rough cutting*; *reinforced concrete.*

RC coupling (*Electronics*). Abbrev. for *resistance-capacitance coupling.* See **parallel feed.**

R-cells (*Zool.*). In the larvae of some *Diptera,* cells in the ring gland (see **Weismann's ring**) which are probably homologous with the prothoracic glands of other Insects.

rd (*Nuc.*). Abbrev. for *rutherford.*

R.D. (*Med.*). Reaction of degeneration. Term used in physiotherapy for abnormal responses to the electrical reactions of muscles.

RdAc (*Chem.*). The symbol for *radioactinium.*

RDF (*Rad o*). Abbrev. for *radio direction-finding.*

RdTh (*Chem.*). The symbol for *radiothorium.*

Re (*Chem.*). The symbol for *rhenium.*

reach (*Hyd. Eng.*). A clear uninterrupted stretch of water.

reactance (*Elec.*). The imaginary part of the impedance. Reactances are characterized by the storage of energy rather than by its dissipation as in resistance.

reactance chart (*Elec. Eng.*). A chart of logarithmic scales so arranged that it is possible to read directly the reactance of a given inductor or capacitance at any frequency.

reactance coupling (*Elec. Eng.*). Coupling between two circuits by a reactance common to both, e.g., a capacitor or inductor.

reactance drop (or rise) (*Elec. Eng.*). The decrease (or increase) in the available voltage at the terminals of a circuit caused by the reactance voltage set up within that circuit.

reactance modulation (*Elec. Eng.*). Use of a variable reactance, e.g., capacitor or inductor, or a reactance valve, to effect frequency modulation.

reactance relay (*Elec. Eng.*). An impedance relay which operates as soon as reactances of the circuit to which it is connected fall below a predetermined value.

reactance theorem (*Elec.*). See **Foster's reactance theorem.**

reactance valve (*Elec. Eng.*). One in which, by alteration of a voltage on a grid, anode impedance appears as a changing capacitance or inductance. This can alter resonance frequency of oscillatory circuit and hence effect frequency modulation.

reactance voltage (*Elec. Eng.*). The voltage produced by current flowing through the reactance of a circuit; equal to the product of the current (amps) and the reactance (ohms).

reactants (*Chem.*). The substances taking part in a chemical reaction.

reactatron (*Electronics*). A low-noise microwave amplifier with a semiconductor diode.

reaction (*Bot., Zool.*). Any change in behaviour of an organism in response to a stimulus. (*Chem.*) (1) See **chemical-.** (2) The acidity or alkalinity of a solution. (*Mech.*) The equal and opposite force arising when a force is applied to a material system; in particular the force exerted by the supports or bearings on a loaded mechanical system. (*Telecomm.*) *Positive feedback* by coupling between an anode and a previous grid, whereby a small voltage applied to the latter is reinforced by voltage or current of the former. Original use was to reduce effective loss in a tuned circuit and hence improve its gain and selectivity. Excess of reaction leads to oscillation. Also **retroaction.** See **regeneration.**

reaction chain (*Chem.*). See **chain reaction.**

reaction chamber (*Aero.*). The chamber, usually cylindrical but sometimes spherical, in which the reaction, or combustion, of a rocket's fuel and oxidant occur.

reaction coil (*Electronics*). One included in the anode circuit of a thermionic valve and inductively coupled to the grid circuit.

reaction formation (*Psychol.*). The development of a character trait which is usually the exact opposite of the original simple character trait. Formed as a reaction against the environment or from endogenous causes, or both, e.g., the character trait of submissiveness in place of the original one of aggression.

reaction generator (*Elec. Eng.*). A special form of synchronous generator excited by alternating current.

reaction isochore (*Chem.*). See **van't Hoff's reaction isochore.**

reaction isotherm (*Chem.*). See **van't Hoff's reaction isotherm.**

reaction order (*Chem.*). See **order of reaction.**

reaction pair (*Geol.*). Two minerals of different composition which exhibit the reaction relationship (see **reaction principle**). Thus forsterite at high temperature is converted into enstatite at a lower temperature, by a change in the atomic structure involving the addition of silica from the magma containing it. Forsterite and enstatite form a *reaction pair.*

reaction principle (*Geol.*). The conversion of one mineral species stable at high temperature into a different one at lower temperatures, by reaction between crystal phase and liquid magma containing it. The change may be continuous over a wide temperature range (*continuous reaction*), or may occur at certain fixed temperatures only (*discontinuous reaction*).

reaction products (*Chem.*). The substances formed in a chemical reaction. Also **resultants.**

reaction propulsion (*Aero.*). The scientifically correct expression for all forms of jet and rocket propulsion; they act by ejection of a high-velocity mass of gas, from which the vehicle reacts with an equal and opposite momentum, according to Newton's Third Law of Motion.

reaction rate (*Nuc.*). The rate of fission in a nuclear reactor.

reaction rim (*Geol.*). The peripheral zone of mineral aggregates formed round a mineral or rock fragment by reaction with magma during consolidation of the latter. Thus, quartz caught up by basaltic magma is partially resorbed, at the same time being surrounded by a reaction rim of granular pyroxene.

reaction series (*Geol.*). One in which the minerals of igneous rocks are arranged in the order of temperature at which they crystallize from magmas.

reaction time (*Bot., Zool.*). The time between a stimulus and the appropriate reaction.

reaction turbine (*Eng.*). A turbine in which the fluid expands progressively in passing alternate rows of fixed and moving blades, the kinetic energy continuously developed being absorbed by the latter.

reactivation (*Electronics*). When a thoriated tungsten filament loses its emission, the raising of temperature for a time (without anode voltage) to bring fresh thorium to the surface is called reactivation.

reactive (*Chem.*). Readily susceptible to chemical change.

reactive anode (*Ships, etc.*). See **sacrificial anode.**

reactive component of current (or **voltage**) (*Elec. Eng.*). Preferred terms for component of vectors representing an alternating current (voltage) which is in quadrature (90°) with the voltage (current) vector. Also called **idle component, inactive component, quadrature component, wattless component.**

reactive dyes (*Chem.*). Classes of dyestuffs which react chemically with textile fibres. Many dyes of this type are based on cyanuric chloride and the chemical linkage between dye and fibre depends on a triazine nucleus.

reactive factor (*Elec. Eng.*). The ratio of reactive volt-amperes to total supply volt-amperes.

reactive iron (*Elec. Eng.*). Iron inserted in the leakage-flux paths of a transformer to increase its leakage reactance.

reactive load (*Elec. Eng.*). A load in which current lags behind or leads on the voltage applied to its terminals.

reactive power (*Elec. Eng.*). The reactive volt-amperes, i.e., the product of the voltage of a circuit and the reactive component of the current.

reactive voltage (*Elec. Eng.*). That component of the vector representing voltage of an a.c. circuit which is in quadrature (at 90°) with the current.

reactive volt-ampere-hour (*Elec. Eng.*). A unit used in measuring the product of reactive volt-amperes in a circuit and the time during which they have been passing.

reactive volt-ampere-hour meter (*Elec. Eng.*). An integrating meter which measures and records the total number of reactive volt-ampere-hours which have passed in the circuit to which it is connected.

reactive volt-amperes (*Elec. Eng.*). Product of the reactive voltage and the amperes in a circuit, or the reactive current (amperes) and voltage of the circuit; measure of the wattless power in the circuit. Abbrev. VAr (*volt-amperes-reactive*).

reactivity (*Nuc. Eng.*). The departure of the multiplication constant of a reactor from unity, measured in different ways. See **cent, dollar nile, inhour.**

reactor (*Elec. Eng.*). Electric circuit component which stores energy, i.e. a capacitor or an inductor. (*Nuc. Eng.*) Assembly in which the fission of a greater-than-critical mass of a fissile material is regulated by a moderator. Controlling factor is the neutron flux, regulated by absorbing rods (boron-steel, cadmium). Used for neutron irradiation and for releasing nuclear energy. Reactors are classified according to their *fuel, moderator*, and *coolant* (or less frequently according to *size, power output*, or *function*). Several hundred types of reactor have been tested or suggested based on the following alternatives: *fuel*: uranium-235; plutonium-239; uranium-233; *moderator*: light water; heavy water; graphite; beryllium; organic liquid (or none in fast reactors); *coolant*: gas; light water; heavy water; organic liquid; liquid metal. Fuels are classed as *natural, slightly enriched*, or *heavily enriched* according to the extent to which the proportion of fissile material has been increased beyond its normal isotopic abundance.

reactor noise (*Nuc. Eng.*). Random statistical variations of neutron flux in reactor.

reactor oscillator (*Nuc. Eng.*). Device for producing mechanical oscillation of neutron absorbing sample in reactor core.

reactor simulator (*Nuc. Eng.*). Analogue computer which simulates variations in reactor neutron flux produced by changes in any operating parameter.

reactor trip (*Nuc. Eng.*). Rapid reduction of reactor power to zero by emergency insertion of control members.

reader (*Typog.*). One who reads and corrects printers' proofs, comparing them with the original copy. Also called **proofreader.**

read in (*Comp.*). To insert data signals into a computer, from punched tape or cards, or magnetic tape.

reading (*Comp.*). See **sensing.**

reading (or **shunted**) **capacitor** (*Teleg.*). A resistance shunted with a capacitor; inserted in a telegraph line to improve the definition of received signals over great distances.

reading microscope (or **telescope**) (*Phys., etc.*). See **cathetometer.**

reading-off (*Textiles*). (1) The operation of converting a lace pattern draft into a figure sheet or other form from which jacquard cards, etc., can be punched. (2) In weaving, translating design on squared paper into punched cards for jacquard.

reading speed (*Comp.*). (1) Rate at which words or bits can be recovered from memory. (2) Rate at which input of computer reads these from program card or tape.

read only memory (*Comp.*). A fast access store containing fixed data. Abbrev. **ROM.**

read-out (*Comp.*). (1) Output unit of a computer. (2) Data from above in form of mechanical or xerographic printing, registering on magnetic tape or for punching paper tape.

read-out pulse (*Comp.*). Pulse applied to binary cells to extract the bit of information stored.

ready-mixed concrete (*Build., Civ. Eng.*). A method whereby concrete constituents are loaded into a vehicle carrying a mixing plant and mixed with water to the right consistency while the vehicle is travelling to the place of

deposit. Frequently used on congested sites where room is not available for traditional plant for manufacture of concrete.
reagent (*Chem.*). A substance or solution used to produce a characteristic reaction in chemical analysis.
reagent feeder (*Min. Proc.*). Appliance which dispenses chemicals into a continuously moving flow of ore or pulp at a controlled rate.
real absorption coefficient (*Nuc.*). See **true absorption coefficient.**
real axis (*Maths.*). See **Argand diagram.**
realgar (*Min.*). A bright-red monosulphide of arsenic; monoclinic. Occurs associated with orpiment.
real image (*Optics*). See **image.**
realised natality (*Ecol.*). See **natality.**
reality principle (*Psychol.*). The conditions imposed by the social environment on the demands of the id and satisfaction of the *pleasure principle*. Normally the requirements of society are incorporated in the maturing personality.
real number (*Maths.*). Any rational or irrational number.
real part (*Maths.*). See **complex number.**
real-time computer (*Comp.*). Process of calculation by analogue electronic computers in which operations are synchronized in time to the effects being studied or calculated, e.g., computer which controls a radio telescope while tracking satellites.
real-time processing (*Comp.*). Automatic handling of data at the same speed as it is received from transducers or instrumentation of any kind.
ream (*Glass*). A nonhomogeneous layer in flat glass. (*Paper*) Twenty quires; now usually 500 sheets. See also **printer's ream.**
reamer (*Eng.*). A hand- or machine-operated tool for finishing drilled holes. It consists of a cylindrical or conical shank on which cutting edges are formed by longitudinal or spiral flutes, or in which separate teeth are inserted.
reany (*Textiles*). Novelty yarn made by doubling a single and a twofold thread together.
rearranged twills (*Textiles*). Fancy weaves made by rearranging the ends of a regular twill.
reason-piece (*Carp.*). See **raising-plate.**
Réaumur malleable cast-iron (*Met.*). See **malleable cast-iron.**
Réaumur scale (*Phys.*). A temperature scale ranging from 0°R to 80°R (freezing-point and boiling-point of pure water at normal pressure).
rebalsaming (*Photog., etc.*). The process of separating and recementing lens elements with Canada balsam, after deterioration of the latter.
rebate (*Build.*). A shallow recess, in which door or window frames are to be fitted, formed with small projections in the jambs of external walls. (*Carp.*) A groove cut into the edge of a piece of timber. Also **rabbet.**
rebate plane (*Join.*). A plane specially adapted to cutting a groove in the corner of a board.
rebecca-eureka (*Radar*). Radar system on aircraft carrying low-power interrogator transmitters (rebecca), working with fixed beacon responders (eureka), sending coded signals when triggered by interrogator pulses.
reboiler (*Chem. Eng.*). The vessel at the bottom of a still in which reflux is reboiled to produce vapour for the fractionating column. Usually a shell and tube heat exchanger.
rebore (*Autos.*). The treatment of a worn cylinder by boring it out and replacing the piston by a (necessarily) larger one.
Reboul effect (*Electronics*). The emission of electrons, soft X-rays, and ultraviolet light from the surface of certain semiconductors when very intense electric fields are applied. See also **electroradiescence.**
recalescence (*Met.*). Release of heat in ferromagnetic material as it cools through a temperature at which a change of crystal structure occurs, normally associated with change in magnetic properties.
recapitulation theory (*Biol.*). States that stages in the life-history of the race are reproduced during the developmental stages of the individual, i.e., ontogeny tends to recapitulate phylogeny. Also called **biogenetic law, Haeckel's law.**
receiver (*Mining*). Vessel which stores compressed air intermediately between air compressor and distributing system. (*Telecomm.*) Final unit in transmission system where received energy is stored, recorded or converted into preferred form, e.g., Magslip receiver, telephone receiver.
receiver response (*Acous.*). The response of a telephone receiver operating into a real or artificial ear; expressed as the ratio of the square of the excess-pressure in the specified cavity to the electrical power applied to the receiver.
receptacle (*Bot.*). A term of wide application and diverse meaning. It may mean: (1) in fungi, a spore-bearing structure, especially if more or less concave; (2) in algae, a swollen end of a branch containing reproductive organs; (3) in liverworts, a cup containing gemmae; (4) in mosses, a group of sexual organs surrounded by leaves; (5) in ferns, the cushion of tissue bearing the sporangia; (6) in flowering plants: (*a*) the more or less enlarged end of the flower stalk bearing the parts of the flower (also known as **torus, thalamus**); (*b*) the enlarged end of the peduncle, bearing the flowers of a crowded inflorescence.
receptaculum (*Zool.*). (1) A receptacle; a sac or cavity used for storage. (2) A sac in which ova are stored, as in some *Oligochaeta.*
receptaculum seminis (*Zool.*). A sac in which spermatozoa are stored, as in many Invertebrates; a spermatheca.
reception wall (*Build.*). See **retention wall.**
receptiveness (*Bot.*). The condition of the stigma when effective pollination is possible.
receptive papilla (*Bot.*). In some *Phycomycetes*, a small outgrowth, from the oögonium into the antheridium, to which the antheridium becomes attached.
receptive spot (*Bot.*). A clear area in the eggs of some fungi and algae through which the sperm enters. (*Zool.*) The point on the surface of an ovum at which the sperm enters.
receptor (*Biol.*). Theoretical sites in cells and tissues to which drugs or hormones are believed to become attached in the process of exerting their particular action. See **side-chain theory.** (*Zool.*) An element of the nervous system specially adapted for the reception of stimuli; for example, a sense-organ or sensory nerve-ending.
recess (*Zool.*). A small cleft or depression, as the *optic recess.*
recessed arch (*Arch.*). See **compound arch.**
recessed pointing (*Build.*). A method of pointing designed to prevent any peeling off; the mortar at all joints both vertical and horizontal is pressed back about ¼ in (6 mm) from the face of the wall.
recessed switch (*Elec. Eng.*). See **flush switch.**
recessive character (*Gen.*). Of a pair of allelomorphic characters, the one which will not be manifested if both are present since it is masked

by the corresponding alternative or dominant character. Also **recessive**.

reciprocal (*Maths.*). The reciprocal of the number a is the number $\frac{1}{a}$, i.e., a^{-1}.

reciprocal cross (*Bot.*). A cross between two plants in which each plant receives pollen from the other.

reciprocal diagram (*Eng.*). See **force diagram.**

reciprocal hybrids (*Zool.*). A pair of hybrids obtained by crossing the same two species, in which the male parent of one belongs to the same species as the female parent of the other, e.g., mule and hinny.

reciprocal innervation (*Zool.*). Innervation of an organ by two sets of nerves having opposite effects, e.g., the innervation of the arteries by vasoconstrictor and vasodilator fibres.

reciprocal networks (*Elec. Eng.*). Those the product of whose impedances remains a constant at all frequencies; thus an inductance is reciprocal to a capacitance.

reciprocal polar triangles (*Maths.*). See **conjugate triangles.**

reciprocal proportions (*Chem.*). See **law of equivalent (or reciprocal) proportions.**

reciprocal theorem (*Eng.*). The statement, enunciated by Clerk Maxwell, that on any elastic structure, if a load W applied at a point A causes a deflection y at another point B, then, if the loading be taken off A and applied at B, it will cause a deflection y at A, provided that W acts at B along the line in which y was measured, and that the deflection at A is measured along the original line of action of W at A.

reciprocal translocation (*Cyt.*). A mutual interchange of portions between two chromosomes.

reciprocating compressor (*Eng.*). Any compressor which employs a piston working in a cylinder, the piston causing the periodic compression of the working fluid.

reciprocating engine (*Eng.*). Any engine which employs a piston working in a cylinder, the piston being caused to oscillate by the periodic pressure of the working fluid.

reciprocating pump (*Eng., Mining, Min. Proc.*). One which uses displacing action of a plunger, piston or diaphragm to move water in a pulsated stream. Also, *pulsometer pump* (q.v.), using steam or compressed air in a valved system for similar pumping.

reciprocation (*Maths.*). See **polar-.** (*Telecomm.*) Operation of finding a reciprocal network to a given network. Used in electric wave filters.

reciprocity (*Eng.*). The principle enunciated in the *reciprocal theorem* (q.v.).

reciprocity calibration (*Elec. Eng.*). Absolute calibration of microphone by use of reversible microphone-loudspeaker. See **reciprocity theorem.**

reciprocity constant (*Elec. Eng.*). See **reciprocity theorem.**

reciprocity principle (*Nuc.*). That the interchange of radiation source and detector will not change the level of radiation at the latter, whatever the shielding arrangement placed between them.

reciprocity theorem (*Elec. Eng.*). The interchange of electromotive force at one point in a network and the current produced at any other point results in the same current for the same electromotive force. In an electrical network comprising 2-way passive linear impedances, the so-called transfer impedance is given by the ratio of the e.m.f. introduced in a branch of the network to the current measured in any other branch. By the reciprocity theorem this ratio is equal in phase and magnitude to that observed if positions of current and e.m.f. were interchanged. In its application to calibration of transducers the reciprocity theorem concerns the quotient of the value of the ratio of open circuit voltage at output terminals of the transducer (when used as a sound receiver) to the value of the free-field sound pressure (referred to some arbitrarily selected point of reference near the transducer) divided by the value of the sound pressure at a distance d from the point of reference to the current flowing at input terminals of the transducer (used as a sound transmitter). The value of this quotient, termed the *reciprocity constant*, is independent of the constructional nature of the transducer.

recirculating ball-type steering (*Autos.*). Steering gear resembling the screw-and-nut type, but using a half-nut with an eccentric ball-race which is operated by a spiral cam.

recirculating loop memory (*Comp.*). Small element of magnetic memory in which stored information recirculates continuously for rapid access.

recirculation heating system (*Heat.*). Heating industrial ovens and low-temperature furnaces with the atmosphere of the working chamber under constant recirculation throughout the complete system.

reck (*Min. Proc.*). See **ragging frame.**

Recklinghausen's disease (or syndrome) (*Med.*). See **molluscum fibrosum, fibrocystic disease.**

reclaimed rubber (*Chem.*). Rubber recovered from waste or used rubber goods. The most extensively used method of reclamation is that of heating ground scrap in a dilute solution of caustic soda at about 180°C for 12 to 20 hr.

recoil atom (*Nuc.*). One which experiences a sudden change of direction or reversal, after the emission from it of a particle or radiation. Also **recoil nucleus.**

recoil escapement (*Horol.*). A clock escapement (invented by Hooke, 1635–1703) in which the acting faces of the pallets are arcs of circles, and at the end of each swing of the pendulum the pallets push the escape wheel backwards a small amount, causing the recoil. This escapement is used largely for domestic clocks. Although departing from the requirements of the ideal escapement, it gives very satisfactory results, as it tends to be self-correcting, i.e., if there is any tendency for the arc of vibration to increase, there is a proportionally greater recoil, which reduces the arc. Also **anchor escapement.**

recoil nucleus (*Nuc.*). See **recoil atom.**

recoil particles (*Nuc.*). Those arising through collision or ejection, e.g., Compton recoil electrons.

recombination (*Cyt., Gen.*). Regrouping of linked characters caused by crossing-over. (*Electronics*) Neutralization of free electron and hole in semiconductor, thus eliminating two current carriers. The energy released in this process must appear as a light photon or less probably as several phonons. See **exciton, phonon, photon.** (*Nuc.*) Neutralization of ions in gas, by combination of charges or transfer of electrons. Important for ions arising from the passage of high-energy particles.

recombination coefficient (*Electronics*). Ratio of the rate of recombination per unit volume to the product of the densities of positive and negative current carriers.

recombination-rate surface (or volume) (*Electronics*). Rate of recombination per unit volume, near the surface (or in the interior) of a semiconductor.

re-combing (*Textiles*). A second combing of worsted tops, to remove additional noil and ensure a better yarn. Known as **double combing** in fine cotton spinning.

reconcentration (*Min. Proc.*). Additional treatment of a mineral product to raise its grade or to separate out one constituent.

reconditioned carrier (*Electronics*). Isolation of a pilot carrier for re-insertion of a carrier adequate for demodulation.

reconstructed stone (*Civ. Eng.*). Artificial stone made of concrete blocks faced to resemble natural stone. Also called **precast stone.**

reconstruction (*Zool.*). The reconstitution of the structure of an organ or organism from a series of sections.

record (*Acous.*). See **recording.** (1) Data preserved in a suitable store, print, punched card or tape, etc. (2) To enter data in such a store.

recorder (*Acous.*). A machine for registering a sound, either magnetically, photographically, or on lacquer. (*Instr.*) An instrument which measures, and records the quantity measured. The mechanism may be actuated by clockwork, by air pressure, electrically or electronically, and frequently incorporates a servo device. Recording is on chart or photosensitive paper moved at a speed to give a time scale.

recording (*Acous.*). (1) The process of making a record of a received signal. (2) A disk, tape, or film on which a sound record is stored. (Disk recordings are commonly known simply as *records*).

recording altimeter (*Aero.*). A barographic type of instrument which traces height against time.

recording amplifier (*Acous.*). That preceding the recording heads of any type of recorder.

recording drum (*Cinema.*). The smooth wheel which carries the unexposed film as it is being subjected to the modulated light-beam in a sound-film recorder.

recording head (*Acous.*). The transducer (magnetic, electric, mechanical or electro-optical) used to record sound on tape, disk, or film.

recording stylus (*Acous.*). The instrument that cuts the groove in an original disk recording.

recordist (*Acous.*). Operator of the controls which determine the amplitude of electric currents which control a sound-recording device.

recovery (*Met.*). Ability of stressed metal, etc., to return to original state after distortion under load. (*Mining*) Percentage of schedule tonnage actually mined. (*Min. Proc.*) Percentage of desired product actually concentrated as value.

recovery pegs (*Surv.*). Special reference pegs established in known survey relation to the working setting-out pegs, so that the location of these can be re-established if disturbed.

recovery rate (*Radiol.*). That at which recovery occurs after radiation injury. It may proceed at different rates for different tissues.

recovery time (*Electronics*). Time required for control electrode of gas tube to regain control. (*Nuc.*) (1) For Geiger tube, the period between the end of the dead time and the restoration of full normal sensitivity. (2) For counting system, the minimum time interval between two events recorded separately. (*Radar*) That required by TR tube in radar system to operate (usually measured to point where receiving sensitivity is 6 dB below maximum).

recovery voltage (*Elec. Eng.*). The normal frequency or d.c. voltage which appears across the contacts of a switch, circuit-breaker, or fuse after it has interrupted the circuit.

recrystallization (*Chem.*). The process of reforming crystals, usually by dissolving them, concentrating the solution, and thus permitting the crystals to reform. Frequently performed in the process of purification of a substance. (*Met.*) The replacement of deformed crystals by a new generation of crystals, which begin to grow at certain points in the deformed metal and eventually absorb the deformed crystals. This process leads to elimination of strain-hardening.

recrystallization temperature (*Met.*). That at which solidification from the molten state begins. That marking a change in crystal form. The range of temperature through which strain-hardening disappears. Lead, tin, and zinc can recrystallize at air temperature; iron, copper, aluminium, and nickel have to be heated.

rectal gills (*Zool.*). In the larvae of some *Odonata*, tracheal gills in the form of an elaborate system of folds in the wall of the rectum, the latter being known as the branchial basket. Water is alternately taken into them and then expelled, so as to ventilate the gills.

rectal gland (*Zool.*). In *Selachii*, a small dorsal glandular diverticulum of the rectum; of unknown function.

rectal papillae (*Zool.*). In the rectum of many Insects, a variable number of inward projections which are often richly tracheated and are possibly concerned with the resorption of water and inorganic ions from the faeces.

rectangle (*Maths.*). A parallelogram whose angles are right angles.

rectangular axes (*Maths.*). Coordinate axes which are mutually at right angles.

rectangular loop hysteresis (*Elec. Eng.*). Colloquial expression for hysteresis curve of ferromagnetic or ferroelectric materials suitable for use in bistable or switching circuits. Characterized by very steep slope followed by unusually sharp onset of saturation.

rectangular notch (*Civ. Eng.*). A notch plate having a rectangular notch cut in it; used for the measurement of large discharges over weirs.

rectangular pulse (*Telecomm.*). Idealized pulse with infinitely short rise and fall times and constant amplitude. Pulse amplitudes and durations are often specified in terms of those of the nearest equivalent rectangular pulse carrying the same energy (subtending the same area under the curve).

rectangular scan (*Electronics*). Any scanning system producing a rectangular field.

recti-. Prefix from *rectus*, straight.

rectification (*Chem.*). Purification of a liquid by distillation. (*Elec. Eng.*) The conversion of a.c. into d.c. using some form of rectifier or rectifying apparatus.

rectification by leaky grid (*Radio*). One form of demodulation.

rectification efficiency (*Elec. Eng.*). Ratio of d.c. output power to a.c. input. Often expressed as percentage.

rectified airspeed (*Aero.*). See **calibrated airspeed.**

rectified spirit (*Chem.*). Distilled ethanol containing only 4·43% water. This is a constant-boiling (azeotropic) mixture with a boiling-point of 78°C.

rectifier (*Elec. Eng.*). Component for converting an a.c. into a d.c. by inversion or suppression of alternate half-waves. Examples: semiconductor or thermionic diode, mercury-arc, rotary (motor and dynamo on one shaft). (*Nuc. Eng.*) In isotope separation, section of a cascade between feed point and withdrawal point.

rectifier instrument (*Elec. Eng.*). An a.c. instrument in which the current to be measured is rectified and measured on a d.c. instrument.

rectifier leakage current (*Elec. Eng.*). That passing through a rectifier without rectification (e.g., due to parallel capacitance and finite reverse conduction).

rectifier ripple factor (*Elec. Eng.*). The amount of a.c. voltage in the output of a rectifier after d.c. rectification. It is measured, in per cent, by the ratio of the r.m.s. value of the a.c. component to the algebraic average of the total voltage across the load. See also **ripple.**

rectifier stack (*Elec. Eng.*). A pile of rectifying elements (usually semiconductor) series-connected for higher voltage operation.

rectifier voltmeter (*Elec. Eng.*). One in which applied voltage is rectified in a bridge circuit before measurement with a d.c. meter.

rectifying detector (*Elec. Eng.*). Detector of electromagnetic waves which depends for its action on rectification of high-frequency currents, as opposed to one using thermal, electrolytic-breakdown, or similar effects.

rectifying valve (*Electronics*). Any thermionic valve in which direct use is made of unilateral or asymmetrical conductivity effects, as opposed to one used primarily for amplification, e.g., a diode used as a rectifier for anode voltage supply to a receiver, or a triode used as a detector. Also **Fleming valve.**

rectigon (*Electronics*). Thermionic gas diode formerly used as low-voltage rectifier.

rectilinear lens (*Photog.*). A lens which provides images with no distortion, as far as parallel lines are concerned.

rectilinear scan (*TV*). A *raster* (q.v.) in which a rectangular area is scanned by a series of parallel lines.

rectinerved (*Bot.*). Having straight or parallel veins.

rectipetaly (*Bot.*). The tendency of a plant member to grow in a straight line.

rectirostral (*Zool.*). Having a straight beak.

rectiserial (*Bot.*). Arranged in straight rows.

recto (*Print., Typog.*). A right-hand page of a book, bearing an odd page number. Cf. *verso.*

recto- (*Med.*). Prefix used in terms pertaining to the rectum.

rectocele (*Med.*). A prolapse or protrusion of the lower part of the posterior vaginal wall, carrying with it the anterior wall of the rectum.

rectrices (*Zool.*). In Birds, the stiff tail feathers used in steering. *sing.* **rectrix.** *adj.* **rectricial.**

rectum (*Zool.*). (L. *rectum intestinum*, straight intestine.) The posterior terminal portion of the alimentary canal leading to the anus. *adj.* **rectal.**

rectus (*Zool.*). A name used for various muscles which are of equal width or depth throughout their length, e.g., the *rectus abdominis* in Vertebrates.

recuperative air heater (*Eng.*). An air heater in which heat is transmitted from hot gases to the air through conducting walls, the flows of gas and air being continuous and unidirectional.

recuperator (*Met.*). An arrangement of flues which enables the hot gases leaving a furnace to be utilized in heating the incoming air (and sometimes gas). Outgoing hot gases and incoming cold gases pass in opposite directions through parallel flues and heat is transferred through the dividing walls.

recurrence (*Nuc.*). See **Regge trajectory.**

recurrent (*Bot.*). Said of the smaller veins of a leaf when they bend back towards the midrib. (*Zool.*) Returning towards point of origin.

recurrent novae (*Astron.*). A small group of novae which have shown more than one outburst of light, as T Coronae Borealis in 1866, 1898, and 1933; they show smaller ranges of brightness than most novae.

recurrent sensibility (*Zool.*). In Vertebrates, sensibility shown by ventral roots of spinal nerves (motor) due to sensory fibres of dorsal roots.

recurrent surge oscillograph (*Elec. Eng.*). An instrument comprising a surge generator, a cathode-ray oscillograph, and the necessary auxiliary circuits so arranged that the phenomenon to be recorded is repeated at such a frequency that a steady picture is provided for visual or photographic record.

recurrent vision (*Optics*). The perception of repeated images of brightly illuminated objects when the source of illumination is suddenly removed.

recurring decimal (*Maths.*). One in which, after a certain point, a number or set of numbers is repeated indefinitely. The figures which recur are indicated by dots placed above them, e.g., $0{\cdot}\dot{3} \equiv 0{\cdot}333\,333\ldots$, $0{\cdot}5\dot{2}4\dot{3} \equiv 0{\cdot}524\,324\,324\,3\ldots$ Also called **repeater, repeating decimal.**

recurvirostral (*Zool.*). Having the beak bent upwards.

recursion (*Maths.*). Of a formula, enabling a term in a sequence to be computed from one or more of the preceding terms.

recursive (*Maths.*). Of a definition, consisting of rules which allow values or meaning to be determined with certainty.

recycle (*Phys.*). Repetition of fixed series of operations, e.g., biodegradation of organic material followed by regrowth; applied also to isotope separation (see **single-state-**).

red algae (*Bot.*). See **Rhodophyta.**

Redalon (*Build.*). TN for a retarder applied to concrete surfaces which are to be plastered. Also used to slow the setting process of cement to enable the surface to be brushed or scrubbed or treated otherwise.

red blood corpuscle (*Zool.*). See **erythrocyte.**

red body (*Zool.*). See **red gland.**

red brass (*Met.*). Copper-zinc alloy containing approximately 15% zinc; used for plumbing pipe, hardware, condenser tubes, etc. Red casting brass may contain up to 5% of lead and/or tin in place of the zinc.

Red Chalk (*Geol.*). A thin bed of brick-red chalk occurring in the Cretaceous rocks of Lincolnshire and Norfolk. It is equivalent in age to the Gault Clay of Southern England, and separates the Carstone from the White Chalk above.

red clay (*Geol.*). A widespread deep-sea deposit restricted to the oceanic abysses; essentially a soft, plastic clay consisting dominantly of insoluble substances which have settled down from the surface waters; these substances are partly of volcanic, partly of cosmic origin, and include nodules of manganese and phosphorus, crystals of zeolites, and rare organic remains such as shark's teeth.

red-conscious (*Photog.*). Said of an electron camera which is unduly sensitive to light of long wavelengths. This results in inartistic enhancement of relative brightness of image areas which are red.

red corpuscle (*Zool.*). See **erythrocyte.**

Red Crag (*Geol.*). A local group of richly fossiliferous sands which accumulated as shell banks in land-locked bays in the Pliocene Sea; restricted in distribution to parts of Norfolk and Suffolk.

red deal (*For.*). A light-yellow soft wood obtained from the Scots fir; commonly used for timbering trenches, heavy framing, piles, joinery, etc.

reddle (*Eng.*). A mixture of red lead and oil wiped

over one of two surfaces to be bedded together to indicate high spots to be removed by scraping. (*Min.*) A red and earthy variety of haematite, often with a certain admixture of clay.

red gland (or body) (*Zool.*). In some Fish, a structure found on the wall of the air-bladder, consisting of a network of blood capillaries and a number of tubular glands, which open into the cavity of the bladder and are believed to be responsible for secretion or absorption of gas.

red hardness (*Met.*). Tool materials such as high-speed steel and cemented carbides which, unlike most metals, retain a considerable part of their hardness at red heat are said to have a high red hardness.

red heat (*Met.*). As judged visually, a temperature between 500°C and 1000°C.

redia (*Zool.*). The secondary larval stage of *Trematoda*, possessing a pair of locomotor papillae and a rudimentary pharynx and intestine, and capable of paedogenetic reproduction.

rediffusion (*Telecomm.*). See **radio exchange.**

redingtonite (*Min.*). A hydrated chromium sulphate, occurring as mauve fibrous crystals at the Redington mercury mine in Knoxville, California.

redistilled zinc (*Met.*). Zinc from which impurities have been eliminated by selective distillation. The process takes advantage of different boiling-points of zinc (907°C) and the impurities lead (1750°C) and cadmium (767°C). A purity of over 99·9% zinc is obtainable.

red lead (oxide) (*Chem.*). Dilead (II) lead (IV) oxide, Pb_3O_4. Formed by heating lead (II) oxide in air at approximately 450°C. It occurs as red and yellow crystalline scales. Commercial varieties contain up to approximately 35% PbO_2. Widely used as an anticorrosive pigment in iron and steel primers.

Redler conveyor (*Eng.*). An enclosed continuous drag chain conveyor used, for example, in the disposal of dust recovered in a cyclone.

red marls (*Geol.*). A lithological term applied to red silts and calcareous clays, which in some cases may have accumulated in the same way as loess, i.e., as wind-driven desert dust, but in other cases were water-deposited. They occur as the most extensive deposit in the Old Red Sandstone of parts of England and Wales, and in the Keuper Series.

red marrow (*Zool.*). The reddish vasoformative tissue occupying cavities of some bones in Mammals.

red mud (*Met.*). In the Bayer process, iron-rich residue from digestion of bauxite with caustic soda.

red muscles (*Zool.*). In Vertebrates, muscles which perform long-continued actions and are therefore rich in sarcoplasm and haemoglobin, and are of a red colour.

red nucleus (*Zool.*). See **tegmentum** (2).

Redonda phosphate (*Chem.*). Mineral composed chiefly of aluminium phosphate, $AlPO_4$; found in the West Indies and used in manufacture of phosphorus.

redox (*Chem.*). Abbrev. for *oxidation-reduction.*

red oxide of copper (*Min.*). See **cuprite.**

red oxide of zinc (*Min.*). See **zincite.**

red pine (*For.*). See **Douglas fir.**

redruthite (*Min.*). A name frequently applied to the mineral *chalcocite* because of its occurrence, among other Cornish localities, at Redruth.

red shift (*Astron.*). The displacement of nebular spectral lines towards the red end of the spectrum. Interpreted as a Doppler effect, this leads to *Hubble's law,* that velocity is proportional to distance, and velocities up to 80% of the velocity of light have been measured.

red-short (*Met.*). See **hot-short.**

red silver ore (*Min.*). For *dark-red silver ore,* see **pyrargyrite**; for *light-red silver ore,* see **proustite.**

red snow (*Bot.*). Snow stained by a surface growth of unicellular algae rich in haematochrome; sometimes seen on mountains.

red spot (*Astron.*). A large marking on the planet Jupiter which appears to be of a permanent nature. Probably first seen in 1831, it was very prominent in 1878 when it was bright red, but the colour and visibility vary, and the spot shows a definite drift in longitude.

reduced apogamy (*Bot.*). The development of a sporophyte from a cell or cells of a gametophyte, without any fusion of gametes, giving a plant whose nuclei have the gametic number of chromosomes. Also **meiotic euapogamy.**

reduced fertilization (*Biol.*). The substitution for a normal sexual fusion between male and female gametes or nuclei of some other union, as, for instance, union of two female nuclei.

reduced level (*Surv.*). The elevation of a point above or below a specified datum.

reduced mass (*Phys.*). Quantity $mM/(M+m)$ used in studying the relative motion of two particles, masses m and M, about their common centre of gravity. This is used in place of the smaller mass; movement of the larger is then ignored.

reduced width (*Nuc.*). For neutron capture, width of nuclear energy level divided by square root of resonance energy.

reducer (*Photog.*). A solution which acts on the silver image and dissolves it away by chemical or abrasive action, thus reducing contrast and/or density. (*Plumb.*) See **reducing socket.**

reducing (*Textiles*). In worsted yarn trade, an operation carried out previous to spinning to attenuate sliver and roving. Known as *drawing* (q.v.) in cotton spinning; performed on draw-frames and speedframes.

reducing agent (*Chem.*). A substance which is capable of bringing about the chemical change known as *reduction.*

reducing atmosphere (*Met.*). One deficient in free oxygen, and perhaps containing such reactive gases as hydrogen and/or carbon monoxide; used in reducing furnace to lower oxygen content of mineral.

reducing flame (*Chem.*). One containing excess of fuel over oxygen, hence capable of acting as a *reducing agent.*

reducing screen (*Elec. Eng.*). A transparent screen used in photometry to absorb a certain predetermined fraction of the flux falling on it.

reducing socket (*Plumb.*). A pipe-fitting used to connect together pipes of differing sizes. For screwed pipes it will be threaded internally to required standard size at each end (e.g. 1″ and ¾″ BSP); as a compression or capillary fitting, the smaller end will accept the smaller size pipe, the larger end will either do likewise for the larger pipe, or be designed to simulate the larger pipe for insertion into another fitting. Also **reducer, reducing pipe-joint.**

reducing surface (*Phys.*). A prepared surface, used in photometry, which reflects only a certain predetermined proportion of the luminous flux falling on it.

reductases (*Chem.*). Enzymes which bring about the reduction of organic compounds.

reductio ad absurdum proof (*Maths.*). One which proves a proposition by proving that its inverse leads to contradictions, e.g., to prove that $A=B$, one proves that the assumption that

$A \neq B$ leads to contradictions, i.e., to absurdities.

reduction (*Chem.*). Any process in which an electron is added to an atom or an ion. Four common types of reduction are removal of oxygen from a molecule, addition of hydrogen to a molecule, the liberation of a metal from its compounds, and diminution of positive valency of an atom or ion. (*Mining*) The extraction of gold from ore. The *reduction officer* is the official in charge of mill, extraction plant, or reduction works. [S. Africa.]

reduction division (*Cyt.*). See **meiosis.**

reduction factor (*Light*). The ratio of the mean spherical luminous intensity of a light source to its mean horizontal luminous intensity.

reduction intensity (*Chem.*). See **rH-value.**

reduction of levels (*Surv.*). The process of computing reduced levels from staff readings booked when levelling.

reduction roasting (*Met.*). Reducing roast. Use of heat in a controlled atmosphere to lower oxygen content of a mineral, e.g., iron (II) to iron (III) oxide by reaction with CO.

reduction to soundings (*Ships, etc.*). Correction of depths observed in tidal waters for comparison with charted depths referred to chart datum.

reductive analysis (*Psychol.*). See **direct reductive analysis.**

redundancy (*Comp.*). Inclusion of symbols in the coding, which are not strictly necessary for conveying the desired information, but may be used for verification or safeguarding. (*Electronics*) Provision of extra equipment, e.g., valves or capacitors in parallel, so as to retain operation after component failure. (*Teleg.*) The provision of greater-than-minimum number of codings so as to ensure accuracy of interpretation (decoding) after transmission through adverse conditions.

redundancy rate (*Comp.*). Percentage of the maximum bits for the message which is above that which is sufficient to convey the essential information.

redundant (*Eng.*). Term applied to a structural framework having more members than it requires to be perfect. Also **over-rigid.**

reduplicate aestivation (*Bot.*). Valvate aestivation in which the edges of the segments are turned outwards.

Redux bonding (*Aero.*). A proprietary method of joining primary sheet metal aircraft structures with a two-component adhesive under controlled heat and pressure. It is widely used for the making of *honeycomb* sandwich, for doubling sheet metal and for attaching *stringers* or skin stiffeners.

red variables (*Astron.*). See **Mira stars.**

redwater (*Vet.*). (1) Babesiosis; blackwater; moor ill; piroplasmosis. A disease of cattle caused by infection of the erythrocytes by the protozoon *Babesia bovis*, which is transmitted principally by the tick *Ixodes ricinus*; characterized by fever, diarrhoea, anaemia, and haemoglobinuria. A similar tick-borne disease, called *Texas fever* (q.v.), is caused by *Babesia bigemina.* (2) Haemorrhagic disease; infectious icterohaemoglobinuria. A disease of cattle, occurring in America, caused by *Clostridium haemolyticum.*

redwood (*For.*). (1) A conifer (*Sequoia*) of the Pacific coastal area of the U.S., of rapid growth (sometimes reaching over 300 ft (90 m)), yielding a softwood used for commercial purposes. (2) Name given to *red deal* (q.v.) in north of England.

Redwood second (*Eng.*). Unit used in viscometry. See **viscosity.**

Redwood viscometer (*Eng.*). One of several designs of standard *viscometer* (q.v.) in which viscosity is determined in terms of number of seconds required for the efflux of a certain quantity of liquid through an orifice under specified conditions.

reed (*Acous.*). Vibrating tongue of wood or metal, for generating air vibrations in musical instruments. Cane wood reeds are used for tongue action, as in the clarinet and saxophone, and the metal reed, used in organ reed pipes, may also be used as one of the plates of a capacitor for electronic amplification. (*Weaving*) A comblike arrangement of flattened steel wires or dents fixed in a frame. Held in the loom slay it separates the warp threads, forms a guide for the shuttle, and beats the weft to the fell of the cloth. The reed determines the number of ends per inch; a 90 reed usually means 90 threads per inch—an indication of the quality of a fabric. (*Zool.*) See **abomasum.**

reed-counting systems (*Textiles*). (1) In one system the count shows number of splits in a stipulated space. (2) Another method gives count as number of groups of dents in stipulated spaces. See **sett systems.**

reed loudspeaker (*Acous.*). Small loudspeaker with a driving mechanism in which the essential element is a magnetic reed, which is drawn into the gap between pole-pieces on a permanent magnet by the currents in the driving coils.

reed marks (*Weaving*). Marks running lengthwise in a piece of cloth; a fault due to the tendency of the warp threads to run in groups, especially when there are 3 or 4 ends in a dent. *adj.* **reedy.**

reed pipe (*Acous.*). Organ pipe in which pitch of the note is determined by vibration of a reed, the associated pipe reinforcing the generated note by resonance.

reed relay (*Elec. Eng.*). One with contacts on springs in vacuum, operated by external coils.

reeds (*Join.*). Moulding in form of several side-by-side beads sunk below general surface.

reef (*Geol.*). See **coral-.** (*Mining*) Originally an Australian term for a *lode* (q.v.). Now used for a gold-bearing tabular deposit or flattish lode.

reef knolls (*Geol.*). Large masses of limestone formed by reef-building organisms; found typically in the Craven district of Yorkshire where they have weathered out as rounded hills above the lower ground on the shales. These are of Carboniferous age.

reef picking (*Min. Proc.*). On Rand, removal of gold-bearing banket ore from barren waste rock, a reversal of the more usual hand sorting.

reel (*Cinema.*). The standard length of cinematograph film supplied for exposure or for projection in a theatre. The 35-mm length is for 1000-ft cameras; modern projectors can take 2000 ft, the greatest length allowed by British fire regulations, which takes 22 min to run through. Standard sizes for 16 mm substandard apparatus are: camera, 50 or 100 ft; projector, 400 or 1600 ft. (*Textiles*) (1) Power-driven cylindrical frame (140 cm circum.), which winds yarn from cops, bobbins, or other packages into *hanks* (Lancs) or *skeins* (Yorks). *Reeling* is the winding of yarn on to a reel. (2) In the silk trade refers to the silk filaments from a number of cocoons which are thrown to form a silk thread.

reel-and-tray (*Cinema.*). Film-processing apparatus in which the film is wound on a large rotating drum taking up to 200 ft, which is dipped automatically into a succession of semi-cylindrical trays containing the solutions.

reel bogie or truck (*Print.*). A special truck used for transporting a reel of paper to a rotary press.

Reema (*Build., Civ. Eng.*). A proprietary form of system building, one of the first developed in the U.K. It consists of precast storey-height hollow concrete cross-wall panels with recesses providing for a stabilizing frame *in situ*. Floors are of precast hollow reinforced concrete panels and the cross-wall panels are fitted with concrete *in situ* for multistorey buildings. Cladding varies according to circumstances. See **system building.**

re-entrant angle (*Maths., Surv.*). Of a closed figure, e.g., a polygon, an angle which points inward, being above 180° as viewed from the interior. Cf. *salient angle* (2).

re-entrant horn (*Acous.*). A horn for coupling a sound-reproducing diaphragm with the outer air. To conserve space, the horn divides at a distance from the throat and, after convolutions, unites before expanding to the flare.

re-entrant polygon (*Maths.*). A polygon in which one or more internal angles are greater than 180°. Such an angle is called a *re-entrant angle*.

re-entrant winding (*Elec. Eng.*). A term used in connexion with armature windings for d.c. machines; a *singly* (or *doubly*) *re-entrant winding* is one containing one (or two) independent closed circuits. The majority of windings are singly re-entrant.

refection (*Zool.*). In Rabbits, Hares, Shrews, and probably other herbivores, the habit of eating freshly-passed faeces. They probably obtain in this way vitamins of the B-complex which have been produced by bacteria in their large intestines.

reference address (*Comp.*). One that is used as a locating point for a group of related addresses.

reference direction (*Nav.*). One relative to which bearings are given in a navigational system.

reference equivalent (*Teleph.*). The number of decibels by which a given piece of telephonic apparatus differs from the standardized piece of apparatus in the master transmission reference system.

reference frequency (*Telecomm.*). Alternation of 1000 Hz, widely used in acoustics and circuit testing, being approximately in the middle of useful ranges of audio frequencies. In radio testing, a modulating frequency of 400 Hz. Once the frequency $5000/2\pi$ Hz was in use for telephony. Also called **reference tone.**

reference mark (*Surv.*). A distant point from which angular distances to other marks may be taken at a station. Also **reference object.** (*Typog.*) See **mark of reference.**

reference noise (*Telecomm.*). Circuit noise level corresponding to that produced by 10^{-12} watt of power at a frequency of 1 kHz.

reference power (*Telecomm.*). That used as standard level when signal inputs or outputs are expressed in decibels—commonly 1 mW in 600 ohms.

reference system (*Teleph.*). See **master telephone transmission reference system.**

reference tone (*Telecomm.*). See **reference frequency.**

reference tube (*Electronics*). See **voltage regulator tube.**

reference voltage (*Elec. Eng.*). (1) Closely controlled d.c. signal obtained from regulator tube, Zener diode or standard cell and used for calibration, e.g., in analogue computer. (2) A.c. voltage used as phase reference, e.g., in synchro-type servo systems.

reference volume (*Elec. Eng.*). That transmission voltage which gives zero recording level on the standard volume unit meter.

refined iron (*Met.*). Wrought-iron made by puddling pig-iron.

refiner (*Paper*). Machine for increasing wetness of pulp stock by fixed and rotating metal bars. These may be conical or disk-shaped. Formerly called **perfecting engine.**

refinery (*Met.*). Plant where impure metals or mixtures are treated by electrolysis, distillation, liquation, pyrometallurgy, chemical, or other methods to produce metals of a higher purity or specified composition.

refining (*Glass*). See **founding.**

refining of metals (*Met.*). Operations performed after crude metals have been extracted from their ores, to obtain them in a condition of higher purity. See **electrolytic copper, Hoopes process.**

reflectance (*Light*). See **reflection factor.**

reflected (*Zool.*). Said of a structure, especially a membrane, which is folded back on itself.

reflected wave or **ray** (*Phys.*). One turned back from a discontinuity in a continuous medium. (*Radio*) See **ionospheric ray or wave.** (*Telecomm.*) One propagated back along a waveguide or transmission line system as a result of a mismatching at the termination.

reflecting galvanometer (*Elec. Eng.*). See **mirror galvanometer.**

reflecting level (*Surv.*). An instrument, used for levelling, which employs the principle that a ray of light which strikes a reflecting plane at right angles is reflected back in the same direction. In its practical forms, it usually consists of a hanging mirror which takes up a position in the vertical plane, and has an unsilvered part through which a distant staff may be seen and also a reference horizontal line upon it. When the eye is in such a position that the image of the pupil is bisected by the horizontal line, the line of sight to staff is horizontal.

reflecting telescope (*Astron.*). A form of telescope invented by Newton to overcome difficulties of chromatic aberration; in it light is reflected from a polished and figured surface, forming an image which is then magnified by an eyepiece.

reflection (*Nuc.*). Return of neutrons to reactor core after a change of direction experienced in the shield surrounding the core. (*Phys.*) (1) Phenomenon which occurs when a wave meets a surface of discontinuity between two media and part of it has its direction changed so as not to cross this surface and in accordance with the *reflection laws*. See also **scattering.** (2) Change in direction of a beam of charged particles produced by electric and/or magnetic fields. See **reflex klystron, magnetic mirror.** (*Telecomm.*) Reduction of power from the maximum possible, because a load is not *matched* to the source and part of the energy transmitted is returned to the source. Also, reduction in power transmitted by a wave filter due to iterative impedance becoming highly reactive outside the pass bands. In all instances, loss of power (**reflection** or **return loss**) is measured in dB below maximum, i.e., when *properly matched*.

reflection coefficient (*Telecomm.*). Ratio of electric voltage or field amplitude for reflected wave to that for incident wave. Given by $(Z_2-Z_1)/(Z_2+Z_1)$, where Z_1 and Z_2 are the impedances of the medium (or line) and the load respectively. For acoustic reflection Z is the acoustic impedance.

reflection factor (*Light*). The ratio which the luminous flux reflected from a surface bears to that falling upon it. Also **coefficient of reflection, reflectance.** (*Telecomm.*) (1) Ratio of current actually delivered to load, to that which would be delivered to a perfectly matched load. (2) See **reflection coefficient.**

reflection gain (*Telecomm.*). Gain in power received in a load from a source because of introduction of a matching network, such as a transformer; measured in dB.

reflection laws (*Phys.*). For wave propagation: (1) incident beam, reflected beam and normal to surface are coplanar; (2) the beams make equal angles with the normal.

reflection layers (*Phys.*). Layers of very low-pressure atmosphere partly ionized by particles from the sun. Each layer progressively refracts electromagnetic waves, so that (below a definite frequency) they are effectively reflected downwards. See **E-layer, F-layer.**

reflection loss (*Telecomm.*). See **reflection.**

reflection point (*Telecomm.*). Point at which there is a discontinuity in a transmission line, and at which partial reflection of a transmitted electric wave occurs.

reflection tube (*Electronics*). CRT in which electron beam is reflected backwards on to screen.

reflectivity (*Phys.*). Proportion of incident energy returned by a surface of discontinuity.

reflectometer (*Phys.*). Instrument measuring ratio of energy of reflected wave to that of incident wave in any physical system.

reflector (*Electronics*). Electrode in reflex klystron connected to negative potential and used to reverse direction of electron beam. U.S. term **repeller.** (*Light*) A device consisting of a bright metal surface shaped so that it reflects in a desired direction light or heat falling upon it. (*Nuc.*) Layer of material surrounding core of reactor which turns back some of the neutrons leaving the core. (*Radio*) Part of an antenna array which reflects energy that would otherwise be radiated in a direction opposite to that intended.

reflector arc (*Cinema.*). See **mirror arc.**

reflector sight (*Aero.*). Mirror sight which projects the aiming reticule and computed correcting information for speed and deflexion on to a transparent glass screen. Used first in World War II, modern sights incorporate elaborate radar information and firing instructions which allow an attack and breakaway under completely blind conditions.

reflex (*Zool.*). Involuntary.

reflex action (*Zool.*). An automatic or involuntary response to a stimulus.

reflex angle (*Maths.*). An angle greater than 180°.

reflex arc (*Zool.*). The simplest functional unit of the nervous system, consisting of an afferent sensory neuron conveying nerve impulses from a receptor to the C.N.S., generally the spinal cord, where they are passed, either directly or via an internuncial or association neuron, to a motor neuron which conveys them to a peripheral effector, such as a muscle.

reflex camera (*Photog.*). One which incorporates a negative-sized glass screen on to which the image is reflected for composing and focusing, either through camera lens (*single-lens reflex*) or through a separate lens of the same focal length (*twin-lens reflex*). In single-lens type the reflecting mirror is retracted when the shutter is released.

reflexed (*Bot.*). Turned back abruptly.

reflexive (*Maths.*). A relation in a set S is reflexive if it applies from x to x for all elements x in S.

reflex klystron (*Electronics*). One in which the electron beam, after passing through one rhumbatron, is reversed by a negative reflector and, arriving at the same rhumbatron in antiphase, returns energy at a possible oscillation frequency; also **klystron reflection.**

reflorescence (*Bot.*). See **double flowering.**

reflux (*Chem.*). Boiling a liquid in a flask, with a condenser attached so that the vapour condenses and flows back into the flask, thus providing a means of keeping the liquid at its b.p. without loss by evaporation.

reflux ratio (*Chem. Eng.*). In distillation design, the ratio of liquid reflux to vapour of any point in the column. In distillation plant operation, the reflux returned per unit quantity of condensate removed as product.

reflux valve (*Civ. Eng.*). A nonreturn type of valve used in pipelines at rising gradients to prevent water which is ascending the gradient from flowing back in the event of a burst lower down.

reforming processes (*Chem. Eng.*). A group of proprietary processes in which low-grade or low molecular weight hydrocarbons are catalytically reformed to higher grade or higher molecular weight materials. (*Platforming* (q.v.) is one such process.)

refracted (*Bot.*). Bent abruptly backwards from the base.

refracting telescope (*Astron.*). Original form of telescope, invented in 17th century, in which light from a distant object passes through a converging lens (now always compound) and is brought to a focus on the principal axis, the image then being magnified by an eyepiece.

refraction (*Phys.*). Phenomenon which occurs when a wave crosses a boundary between two media in which its phase velocity differs. This leads to a change in the direction of propagation of the wavefront in accordance with *Snell's law*.

refraction correction (*Astron.*). That small amount which is subtracted from the observed altitude of a heavenly body to allow for refraction of light by earth's atmosphere, which so bends the rays that all bodies appear higher than they are by an amount which is a maximum at the horizon and zero at the zenith.

refractive index (*Phys.*). The absolute refractive index of a transparent medium is the ratio of the phase velocity of electromagnetic waves in free space to that in the medium. It is given by the square root of the product of complex relative permittivity and complex relative permeability. See **refraction, Snell's law.** Symbol n.

refractive modulus (*Meteor.*). One million times the excess of modified refractive index above unity in M units.

refractivity (*Chem.*). *Specific refraction* (q.v.).

refractometer (*Phys.*). Instrument for measuring refractive indices. Refractometers used for liquids, such as the *Pulfrich refractometer* (q.v.), usually measure the critical angle at surface between liquid and a prism of known refractive index. See **Rayleigh refractometer.**

refractor (*Light*). A device by which the direction of a beam of light is changed by causing it to pass through the boundary between two substances of different densities; the principle is used in certain types of lighting fittings.

refractories (*Met.*). Materials used in lining furnaces, etc. They must resist high temperatures, changes of temperature, the action of molten metals, and slags and hot gases carrying solid particles. China clay, ball clay, and fireclay are all highly refractory, the best qualities fusing at above 1700°C. Other materials are silica, magnesite, dolomite, alumina, and chromite. See **silica.**

refractory cement (*Build.*). A form of cement capable of withstanding very high temperatures.

refractory clay (*Met.*). See **refractories.**

refractory concrete (*Build., Civ. Eng.*). Concrete used for resistance to high temperatures such as in construction or lining of chimneys. Constituents are normally high alumina cement and crushed fire brick but other specially prepared aggregates may be used.

refractory metals (*Met.*). Term applied to transition group elements in periodic table which have high melting-points. They include chromium, titanium, platinum, tantalum, wolfram, and zirconium.

refractory ore (*Min. Proc.*). Gold ore non-responsive to amalgamation process.

refractory period (*Zool.*). Time interval during which an excitable tissue is incapable of response to a second stimulus, applied after a previous one.

refrigerants. Substances suitable for use as working agents in a refrigerator, e.g., ammonia, carbon dioxide, sulphur dioxide, chloromethane, etc.

refrigeration cycle (*Eng.*). Any thermodynamic cycle which takes heat at a low temperature and rejects it at a higher temperature. The cycle must receive power from an external source.

refrigerator (*Eng.*). A machine or plant by which mechanical or heat energy is utilized to produce and maintain a low temperature.

refringent. Refractive.

reftone (*Telecomm.*). Abbrev. for *reference tone*. See **reference frequency.**

refusal (*Civ. Eng.*). A term applied to the resistance offered by a pile to continued driving.

regain (*Textiles*). (1) The increase in the length of a thread taken from a cloth, e.g., if in 100 metres of cloth the warp length is 105 metres owing to its interlacement with the weft, the *regain* is 5%. (2) The weight of moisture which is present in fibres, yarns, or fabrics, expressed as a percentage of the oven-dry weight.

regattas (*Textiles*). Usually 2×1 twill weave all-cotton cloth featuring blue/white, black/white or red/white warp stripes in fast-to-washing dyes; used for aprons, summer dresses, etc.

regelation (*Heat*). Process by which ice melts when subjected to pressure and freezes again when pressure is removed. Regelation operates in the forming of a snowball by pressure, in the flow of glaciers, and in the slow passage through a block of ice by a weighted loop of wire.

regeneration (*Comp.*). Replacement or reforming of stored data, e.g., in computer register or charge storage tube. (*Min. Proc.*) Reconstitution of liquid used in chemical treatment of ores before returning it to head of attacking process (e.g., in cyanide process). Freshening of 'poisoned' ion-exchanger resins. (*Nuc. Eng.*) (1) Reprocessing of nuclear fuel by removal of fission products. (2) Recovery of reactor from effect of xenon or other poisoning. (*Telecomm.*) Same as *positive feedback* (q.v.), but particularly applied to a super-regenerative receiving valve, which oscillates periodically through self-quenching. (*Zool.*) Renewal or replacement of an organ or structure which has been lost or damaged.

regenerative air heater (*Eng.*). An air heater in which heat-transmitting surfaces of metallic plates or bricks are exposed alternately to the heat-surrendering gases and to the air.

regenerative braking (*Elec. Eng.*). A method of braking for electric motors in which the motors are operated as generators by momentum of the equipment being braked, and return energy to the supply.

regenerative detector (*Elec. Eng.*). One in which high-frequency components in the output are fed back (*reaction, retroaction*) to the input, thus increasing gain and selectivity.

regenerative furnace (*Eng., Met.*). A furnace in which the hot gases pass through chambers containing fire-brick structures, to which the sensible heat is given up. The direction of gas flow is reversed periodically, and cold incoming gas is preheated in the chambers. See **heat economizer.**

regenerative receiver (*Telecomm.*). One with positive feedback for the carrier, enhancing efficiency of amplification and demodulation.

regenerator (*Met.*). Labyrinth which transfers heat of exit gases to air entering furnace, or feed-water to boiler. (*Teleph.*) Device arranged to accept and store trains of pulses having characteristics (speed and ratio) within a specified range and to transmit corresponding trains of pulses having characteristics within a more limited range.

Regge trajectory (*Nuc.*). A graph relating spin angular momentum to energy for a nuclear particle. Possible quantized values of spin correspond to large discrete energy increments on the graph. This enables recurrences of nuclear particles to be predicted—the extra energy corresponding to the greater rest mass expected to be associated with such particles. A *recurrence* is a particle identical in all respects, except energy (or mass) and spin momentum, with a known particle, and is regarded as being a higher energy equivalent of the normal particle.

region (*Maths.*). See **domain** (1).

regional metamorphism (*Geol.*). All those changes in mineral composition and texture of rocks due to compressional and shearing stresses, and to rise of temperature occasioned by intense earth movements. The characteristic products are the crystalline schists and gneisses.

region of limited proportionality (*Nuc.*). Range of operating voltages for a counter tube in which the gas amplification depends upon the number of ions produced in the initial ionizing event as well as on the voltage. For larger initial events the counter saturates.

register (*Build.*). A metal damper to close a chimney. (*Comp.*) (1) Device which stores small amount of information, usually one word, with fast access. (2) Mechanical, electrical, or electromechanical device which stores and displays data; usually one decimal number in a measuring system. (*Photog., Print.*) Exact correspondence of superimposed work, e.g., when the separate colours in colour photography are printed or projected together to reproduce the original picture. (*Telecomm.*) Device which records pulses received over a code signalling system.

register-controlled system (*Teleph.*). System of automatic switching in which the selectors are positioned by signals supplied by registers in response to information provided by dialling or other means.

registered breadth (*Ships*). The breadth measured over the shell plating at widest part.

registered depth (*Ships*). Depth measured from top of *ceiling* (q.v.) to top of deck beam at mid-length in the centre line of the vessel. Deck to which it is measured is usually stated.

registered dimensions (*Ships*). Dimensions appearing on the Certificate of Registry. Their main purpose is to identify the ship and they are also called the **identification dimensions.** They are *registered length, registered breadth,* and *registered depth* (qq.v.).

registered length (*Ships*). Length from the fore side of the stem at the top to after side of stern post or, in a vessel without a stern post, to the centre of the rudder stock.

register(ed) tonnage (*Ships*). See **net register tonnage.**

register galley (*Typog.*). A galley with a fourth, removable, side in which the page make-up can be checked under lock-up pressure. Supplied with three plastic sheets, one with 12-pt. ruling, one clear on which a key proof can be taken, and one with a ground surface on which a layout can be traced.

register length (*Comp.*). The number of bits stored in a computer register.

register lock-up (*Print.*). Mechanism allowing fine positioning of plates on the cylinders of web-fed presses.

register rollers (*Print.*). Adjustable rollers that provide a means of varying the web length between one unit of a web-fed press and another.

register sender (*Teleph.*). A device, common to a number of input circuits, which accepts and stores information relating to a called number or service. It is thereafter capable of controlling the setting up of a part or all of the wanted connexion.

Register Sets (*Print.*). A combination of *Mixed Forme Base* and *honeycomb base*, each supplied in a variety of accurately sized units, to be assembled with type to the size required for a particular plate, for which it provides both a mount and a means of attaining register.

register sheet (*Print.*). The sheet used in obtaining correct register or position.

register-translator (*Teleph.*). Device in which the functions of a register and translator are combined, e.g., a *director* (q.v.).

reglet (*Arch.*). (1) A flat narrow rectangular moulding. (2) A *facette* (q.v.). (*Typog.*) A thin strip of wood used for spacing; usually 6 or 12 points in thickness (known as *nonpareil reglet* and *pica reglet* respectively).

reglette (*Surv.*). The short graduated scale attached at each end of the special measuring tape or wire used in baseline measurement.

Regnault's hygrometer (*Meteor.*). A type of hygrometer in which the silvered bottom of a vessel contains ethoxyethane, through which air is bubbled to cool it, its temperature being indicated by a thermometer.

regrating (*Build.*). Operation of redressing the faces of old hewn stone work.

regression (*Biol.*). A tendency to return from an extreme to an average condition, as when a tall parent gives rise to plants of average stature. (*Psychol.*) A return to an earlier stage of development, whereby the libido takes up earlier modes of expression and gratification in accordance with the particular phase to which it has regressed, e.g., a regression from adolescence to childhood.

regression of nodes (*Astron.*). An effect due to planetary perturbations by which nodes of an inclined orbit regress, or move in reverse direction to motion of the planet or satellite.

regressive stain (*Micros.*). A stain with which the best results are obtained by overstaining and subsequent modification, or partial removal of the stain by a differentiating reagent. Cf. *progressive stain.*

regular (*Bot.*). Said of a flower which has its parts so arranged that it can be divided into halves by several longitudinal planes passing through the centre; symmetrical, actinomorphic.

regular convex solids (*Maths.*). Solids having congruent all faces bounded by plane surfaces and all corners. They are (1) *tetrahedron*, 4 equilateral triangular faces, (2) *hexahedron* or *cube*, 6 equal squares as faces, (3) *dodecahedron*, 12 regular pentagons as faces, (4) *octahedron*, 8 equilateral triangles as faces, (5) *icosahedron*, 20 equilateral triangles as faces.

regular-coursed (*Build.*). Said of rubble walling built up in courses of the same height.

regular function (*Maths.*). See **analytic function.**

regular polygon (*Maths.*). One with all its sides equal and all its angles equal; e.g., a regular polygon of three sides is an equilateral triangle.

regular reflection factor (*Light*). The ratio which the luminous flux regularly reflected from a surface bears to the total flux falling on that surface.

regular transmission (*Light*). Transmission of light through a surface in such a way that the beam of light, after transmission, appears to proceed from the light source.

regular transmission factor (*Light*). The ratio which the luminous flux regularly transmitted through a surface bears to the total luminous flux falling on the surface.

regulating rod (*Nuc.*). Fine control rod of reactor.

regulation (*Automation*). Control of plant or process at required activity level. (*Electronics*) (1) Fractional change in voltage level when load is connected to supply. (2) Anode voltage range for voltage regulator tube between maximum and minimum anode currents.

regulator (*Electronics*). Glow tube which provides a constant potential difference across its electrodes when fed through a resistor; used for supplying small currents at constant voltage at a point in a circuit for control and servo purposes. (*Horol.*) (1) A precision long-case clock with a seconds pendulum. The dial has independent hands and zones for the hours, minutes and seconds. (2) See **index.** (*Mining*) Gauge floor. One which regulates the quantity of air passed through underground workings.

regulator cell (*Elec. Eng.*). One of several cells which are arranged at the end of a battery of accumulator cells and are connected to a regulating switch so that they can be cut in or out of circuit in order to adjust the voltage of the battery as a whole. Also called **end cell.**

regulator gene (*Gen.*). A gene which controls the rate of synthesis of gene product by another gene.

reguline deposit (*Elec. Eng.*). Good electrodeposited metal, as opposed to a *burnt deposit.*

regulon (*Gen.*). A group of genes regulating the production of an enzyme but not necessarily genetically linked, as in an *operon* (q.v.).

regulus (*Maths.*). One of the sets of lines forming a *ruled surface* (q.v.).

regulus of antimony (*Met.*). Commercially pure metallic antimony. Also, the impure metallic mixture which is produced during smelting.

regurgitation. The bringing back into the mouth of (undigested) food. (*Med.*). The flowing of blood in the reverse direction to the circulation in the heart as a result of valvular disease, e.g., *aortic regurgitation.*

reheat (*Aero.*). Injection of fuel into the jet pipe of a turbojet for the purpose of obtaining supplementary thrust by combustion with the unburnt air in the turbine efflux. Reheat is the British, and original, term, but is gradually being superseded by the American term *afterburning*, with *afterburner* for the device itself.

reheating, resuperheating (*Eng.*). The process of passing steam, which has been partially expanded in a steam turbine, back to a superheater before subjecting it to further expansion. Reheating is also used, sometimes repeatedly, in heat-treating processes, like annealing, and in pneumatic systems for operating power tools.

reheating furnace (*Met.*). The furnace in which metal ingots, billets, blooms, etc., are heated to temperature required for hot-working.

Rehfuss test (*Med.*). Fractional test meal. The analysis of contents of stomach at 15-min intervals after the swallowing of a pint of oatmeal mixture, the contents being removed by a small stomach tube.

Reichert-Meissl number (*Chem.*). A standard used in butter analysis. A Reichert-Meissl number of *n* means that the soluble volatile fatty acids liberated from 5 g of butter fat under specified conditions require *n* cm^3 of 0·05 M barium hydroxide solution for their neutralization.

re-ignition (*Electronics*). (1) Re-establishment of conduction by ionization in a discharge tube, after conduction has been cut off, but during the de-ionization period. (2) A process by which multiple counts are generated in a counting tube by excitation arising from one ionization.

Reil's island (*Zool.*). In Mammals, a small lobe of the cerebrum situated at the bottom of the Sylvian fissure.

Reimer-Tiemann reaction (*Chem.*). The synthesis of phenolic aldehydes by heating a phenol with trichloromethane in the presence of conc. KOH. The intermediate dichloro derivative is hydrolysed to an aldehyde. The CH=O group takes up the 2- or 4-position with respect to the hydroxyl group.

re-imposition (*Typog.*). (1) Transferring the page from one chase to another, the latter being perhaps a machine chase or a foundry chase. (2) Altering position of pages in a forme to suit a size of paper or the requirements of printing and binding equipment.

rein (*Civ. Eng.*). See **springer.**

Reinartz circuit (*Radio*). One using reaction controlled by a capacitor, suitable for short-wave reception.

reinforced concrete (*Civ. Eng.*). Concrete work in which steel bars or wires (*reinforcement*) are embedded to provide increased strength.

reinforced masonry (*Build.*). That in which metal rods or mesh are embedded to increase strength.

reinforced plastics (*Plastics*). Range of plastic materials in which the basic plastic has been strengthened by incorporating a fibrous reinforcing agent, e.g. asbestos, paper, cloth, glass fibre.

reinforcement (*Acous.*). Sound reproduction in which the received enhanced level appears to come from the actual source, as is required, e.g., in theatres. In *public-address* the received level comes from sources not necessarily coalesced with the actual source. See **Haas effect.**

reinforcing stimulus (*An. Behav.*). An environmental event which, when presented after an animal has made the desired response during operant conditioning, increases or alters the response intensity. See also **positive reinforcing stimulus, negative reinforcing stimulus.**

reinsertion (*TV*). See **d.c. restoration.**

Reiss microphone (*Acous.*). Carbon transmitter in which a large quantity of carbon granules between a cloth or mica diaphragm and a solid backing, such as a block of marble, is subjected to the applied sound wave. Characterized by high damping of the applied vibrational forces, and freedom from carbon noise by virtue of packing amongst the granules.

Reissner's fibre (*Zool.*). In Vertebrates, a wirelike fibre of unknown function running from its attachment at the posterior end of the nerve-tube to the posterior commissure in the roof of the midbrain, via the cavity of the nerve-tube.

Reissner's membrane (*Zool.*). In Mammals, a delicate connective-tissue membrane which cuts off the outer lower portion of the *scala* vestibuli.

reiteration (*Surv.*). Method of checking angular measurements made with a theodolite (and of securing greater accuracy) by repeating the observations after reversing face (turning the sighting telescope through 180°). Cf. *repetition.*

rejection band (*Telecomm.*). Frequency band over which signals are highly attenuated when passed through a wave filter, the iterative impedance becoming highly reactive, causing reflections.

rejector circuit (*Telecomm.*). Parallel combination of inductance and capacitance, tuned to the frequency of an unwanted signal, to which it offers a high impedance when placed in series with a signal channel.

rejointing (*Build., Civ. Eng.*). *Pointing* (q.v.).

rejuvenation (*Geol.*). A term applied to the action of a river system which, following uplift of the area drained by it, can resume down-cutting in the manner of a younger stream.

rejuvenescence (*Biol.*). (1) Conversion of the contents of a cell into one or more cells of a different and usually more active character. (2) Renewal of growth from old or injured parts.

relapsing fever (*Med.*). Spirochaetosis. A term applied to a number of diseases which are transmitted by lice or by ticks and which are due to infection with various spirochaetes; characterized by recurrent attacks of fever and by enlargement of the liver and spleen.

relation (*Maths.*). Let S and T be sets. Then $S \times T$ denotes the set of all ordered pairs (x, y) with x in S and y in T. Any subset p of $S \times T$ is called a *relation from S to T.*

relative abundance (*Ecol.*). A rough measure of population density, relative, e.g., to time (e.g., the number of birds seen per hour) or percentages of sample plots occupied by a species of plant. (*Nuc.*) See **abundance.**

relative address (*Comp.*). One which identifies the position of a word in a particular routine; cf. *absolute address.*

relative atomic mass (*Chem.*). Mass of atoms of an element formerly in *atomic weight units* but now more correctly given on the *unified scale* where 1u is $1{\cdot}660 \times 10^{-27}$ kg. See **atomic weight.**

relative bearing (*Ships, etc.*). Angle between direction of ship's head and of an object.

relative biological effectiveness (*Radiol.*). Inverse ratio of an absorbed dose of ionizing radiation to the absorbed dose of 200 kV X-rays (or sometimes radium γ-rays), which would produce an equivalent biological damage. (*N.B.* Absorbed doses measure the radiation energy absorbed regardless of physical or biological consequences—see **rad.**) Rounded-off values of rbe used in radiological protection are known as *quality factors.* Abbrev. **rbe.**

relative density (*Phys.*). The ratio of the mass of a given volume of a substance to the mass of an equal volume of water at a temperature of 4°C. Originally **specific gravity.**

relative efficiency (*Eng.*). In an internal-combustion engine, ratio of actual indicated thermal efficiency to efficiency of some ideal cycle, such as *air standard cycle* (q.v.), at the same compression ratio.

relative humidity (*Meteor.*). Existing *water vapour pressure* (q.v.) of the atmosphere expressed as a percentage of the *saturated water vapour pressure* (q.v.) at the same temperature.
relative molecular mass (*Chem.*). Preferred term for *molecular weight*.
relative permeability, permittivity (*Elec. Eng.*). See **permeability, permittivity.**
relative plateau slope (*Nuc. Eng.*). See **plateau slope.** *N.B.* Often expressed as percentage change in count rate for 100 volt change in potential.
relative sexuality (*Bot.*). Occurrence in a species of strains giving gametes able to fuse with those produced by either of the normal strains.
relative stopping power (*Nuc.*). See **stopping power.**
relative visibility factor (*Light*). Ratio of apparent brightness of a monochromatic source to that of a source of wavelength 550 nm having the same energy.
relativistic (*Phys.*). Said of any deviation from classical physics and mechanics based on relativity theory.
relativistic mass equation (*Phys.*). When a particle is accelerated up to a velocity (v) which is more than a small fraction of the phase velocity of propagation of light in vacuum (c), it is said to be *relativistic* with mass increased according to the formula

$$m=m_0/\sqrt{1-v^2/c^2},$$

where m_0=rest mass (at low velocities). Required to be considered in cyclotron, betatron and linear-accelerator design.
relativistic particle (*Phys.*). One having a speed comparable with that of light.
relativistic velocity (*Phys.*). That of a particle when an appreciable part of imparted energy passes temporarily to accession of mass, e.g., for an electron above 10^4 eV.
relativity (*Phys.*). Theory based on equivalence of observation of the same phenomena from differing frames of reference having different velocities and accelerations. Einstein's *special theory*, announced in 1905, was generalized a decade later and verified by line-reddening, bending of light in strong gravitational fields, and the explanation of precession of the perihelion of the planet Mercury. Two important results of this restricted theory are the *relativistic mass equation* and the principle of *mass-energy equivalence*. See **field theory.**
relaxation (*Eng., etc.*). Exponential return of system to equilibrium after a sudden disturbance. Time constant of exponential function is *relaxation time*.
relaxation method (*Civ. Eng.*). Method of solving structural equations by making an initial estimate of the solution and then systematically reducing the errors in estimation.
relaxation oscillation (*Telecomm.*). One of irregular waveform and frequency, produced, e.g., by charging and discharging a capacitor. Other examples are the flapping of a flag in the wind, the heart-beat, the scratching of a knife on a plate, etc. See **multivibrator.**
relaxation oscillator (*Telecomm.*). One, usually electric, which generates relaxation oscillations. Characterized by peaky or rectangular waveforms, and the possibility of pulling into step (locking) by an independent source of impulses of nearly the same frequency.
relaxation time (*Chem.*). The time constant of *relaxation*, e.g., from the higher to the lower energy level in nuclear magnetic resonance. (*Eng.*) See **relaxation.** (*Nuc.*) In isotope separation, the time required in a separating plant under given conditions for the mole fraction of the top stage in the process to attain $(1-e^{-1})$ of the steady state value. (*Physiol.*) In excitable tissues, the period during which activation subsides after cessation of a stimulus.
relay (*Acous.*). Broadcast of a performance, e.g., concert-hall or theatre, not specially staged for broadcasting, taken over a radio link, a music line or by recording. (*Elec. Eng.*) An electrically-operated switch employed to effect changes in an independent circuit. Examples are alternating-current, balanced-armature, gas-tube, moving-coil, moving-iron, solenoid, telephone-type, thyratron and transistor relays. All function as single or multiple non-manual switches. See **relay types.**
relay cell (*Anat., Physiol.*). An internuncial neurone or interneurone of the central nervous system, particularly one forming a link between afferent and efferent neurones of a reflex arc.
relay mechanism (*Automation*). System in which an appropriate monitoring device signals to a valve, motor, or other controlling mechanism to vary its setting when the work undergoing scrutiny varies beyond preset limits.
relay spring (*Elec. Eng.*). Flexible part of a relay which keeps it in an unoperated condition. It is stressed on operation, and restores the relay to normal on cessation of the operating current, thereby operating contacts.
relay system (*Radio*). See **radio exchange.**
relay types (*Elec. Eng.*). *Allström*—a sensitive form using a light beam and photocell. *Control*—one operated by permitting the next step in a control circuit. *Differential*—one operating on the difference between, e.g., two currents. *Frequency*—one operating with a selected change of the supply frequency. *Polarized*—one in which the movement of the armature depends on the current direction on the armature control circuit.
relay valve (*Eng.*). See **pilot valve.**
release (*Photog.*). The trigger arrangement for releasing the shutter and effecting exposure in a camera. (*Print.*) A term used in hot press stamping to describe tendency of foil substrate to adhere to surface being marked. A foil with 'good release' is suitable for use with high-speed automatic machines. Release properties depend also on correct choice of temperature and pressure. (*Teleph.*) In automatic telephony, the release of apparatus which has been seized for establishing a connexion. In manual telephony, the positive disengagement of apparatus on cessation of a conversation.
release mesh (*Min. Proc.*). That at and below which screen size mineral is released from a closed crushing or grinding circuit and passed to next stage of treatment.
release papers (*Paper*). Papers treated so that an adhesive surface will become easily detached from them without rupture of the paper surface. Used as protective backing to self-adhesive materials.
release print (*Cinema.*). A print of a cinematograph film for public use in cinemas.
releaser (*An. Behav.*). A stimulus which releases or sets off a species-specific response. See also **sign stimulus.**
release wire (*Teleph.*). An extra wire in the exchange circuits which is sometimes used solely for releasing selectors and switches when a connexion is to be broken down.
releasing key (*Civ. Eng.*). A tapered piece used to ease shuttering away from concrete after it has set.

reliability (*Elec. Eng.*). Probability that an equipment or component will continue to function when required. Expressed as average percentage of failure per 1000 hrs of availability. Also **fault rate.**

relic (*Mining*). Block of ore temporarily or permanently left close to a drive through solid rock, forming a wall between this and the stoped-out portion of the deposit. Cf. *remnant*.

relict (*Ecol.*). A species, whether terrestrial, marine, or freshwater, which occurs at the present time in circumstances different from those in which it originated.

relief block (*Print.*). A printing block (e.g., line, half-tone) which can be used with printing type.

relief holes (*Mining*). Relieving cut. In tunnelling or shaft sinking, the drill-holes blasted after the initial cut holes.

relief map (*Surv.*). One with contour lines, shading, or colouring used to indicate changes in surface configuration of the area mapped.

relief process (*Photog.*). Any colour process using matrices. (*Print.*) See **printing.**

relieving (*Eng.*). (1) Interrupting a bearing surface, such as a machine tool slide or a plain journal bearing, so as to improve alignment or lubrication conditions. (2) In cutting tools, removing material adjacent to a cutting edge so as to improve the flow of chips or cooling and lubrication conditions.

relieving arch (*Build., Civ. Eng.*). One built on the spandrel of a main arch, to distribute the load, or over a lintel, to relieve it of the weight of wall above.

relieving gear (*Ships*). Any gear attached to the rudder, directly or indirectly, in such a way as to reduce stresses on the steering gear due to wave action on the rudder.

relight (*Aero.*). Term used for igniting an aircraft gas turbine in flight.

relish (*Join.*). Projection of shoulder from flanks of tenon.

reluctance (*Elec. Eng.*). Magnetomotive force applied to whole or part of a magnetic circuit divided by the flux in it. It is the reciprocal of *permeance*.

reluctance pick-up (*Elec. Eng.*). Transducer for detecting vibrations or reproducing records. The signal causes a change in the reluctance of a magnetic circuit, which induces an e.m.f. linked to it.

reluctivity (*Elec. Eng.*). The reciprocal of *permeability*.

rem (*Radiol.*). See **röntgen equivalent man.**

remainder (*Maths.*). See **division.**

Remak's fibres (*Zool.*). In Vertebrates, amyelinate fibres occurring in peripheral nerves.

remanence (*Mag.*). See **residual magnetization.**

remiges (*Zool.*). In Birds, the large contour feathers of the wing. *sing.* **remex.**

remiped (*Zool.*). Having the feet adapted for paddling, as many aquatic Birds.

remission (*Med.*). An abatement (often temporary) of the severity of a disease; the period of such abatement.

Remitron (*Electronics*). TN for gas tube used in counting systems.

remittent (*Med.*). Of a fever, characterized by remissions in which the temperature falls, but not to normal; as in malaria.

remnant (*Mining*). Block of ore or stope pillar left well clear of the underground travelling ways on completion of stoping. See also **relic.**

remodulation (*Radio*). Transferring modulation from one carrier to another carrier, as in the frequency changer in a supersonic heterodyne radio receiver.

Remos (*Acous.*). An acoustic-absorbing material, consisting of porous boards made from a certain type of dried moss.

remote (*Bot.*). Said of the gills of agarics which do not reach the stipe but leave a free space around it.

remote control. Control, usually by electric or radio signals, carried out from a distance in response to information provided by monitoring instruments.

remote cut-off tube (*Electronics*). U.S. term for variable mu(tual) conductance valve, which requires a large increasing negative grid bias to reduce the anode current to zero; also called **extended tube, long-tail tube, supercontrol tube.**

remote handling equipment (*Nuc. Eng.*). Apparatus developed to enable an operator to manipulate highly radioactive materials from behind a suitable shield, or from a safe distance.

remote mass-balance weight (*Aero.*). A *mass balance* (q.v.) weight which, usually because of limitations of space, is mounted away from the control surface, to which it is connected by a mechanical linkage.

remould (*Autos.*). Used tyre which has had a new tread vulcanized to the casing, including a coating of rubber on the walls. Cf. *retread*.

removable isolated singularity (*Maths.*). If a function $f(z)$, having a singularity at $z=a$, has no negative powers of $(z-z_0)$ in *Laurent's expansion* (q.v.), then the singularity may be removed by defining $f(a)=a_0$, where a_0 is the first coefficient of $(z-z_0)$ in the expansion.

removes (*Typog.*). Quotations, etc., set in smaller type than the main text. The difference in size is usually 2 points. Thus a book the text of which is set in 12-point or pica should have its quotations set in 10-point.

Remscope (*Electronics*). TN for an oscilloscope in which a trace on the phosphor is maintained indefinitely.

REM sleep (*Psychol.*). See **rapid eye movement sleep.**

renal (*Zool.*). Pertaining to kidneys.

renal portal system (*Zool.*). In some lower Vertebrates, that part of venous system which brings blood from capillaries of posterior part of body and passes it into capillaries of the kidneys.

render and set (*Build.*). Two-coat plaster work on walls.

rendered (*Build.*). A term applied to laths which are split rather than sawn, so as to conserve the maximum strength.

render, float, and set (*Build.*). Three-coat plaster work on walls.

rendering (*Build.*). Operation of covering brick or stonework with a coat of coarse stuff; the coating itself.

reniform (*Bot.*). Kidney-shaped, either solid or flat, and having the outline of a kidney cut longitudinally. Also **nephroid.**

renin (*Physiol.*). Protein enzyme, liberated from the ischaemic kidney into the bloodstream, where it reacts with hypertensinogen to produce *hypertensin* (q.v.).

rennet (*Chem.*). A commercial preparation for making junket (clotted milk) prepared from the mucous membrane of the stomach of calves, and containing the enzyme *rennin*. See **abomasum.**

rennin (*Chem.*). Enzyme found in gastric juice, causing clotting of milk.

rep (*Radiol.*). See **röntgen equivalent physical.** (*Textiles*) Fabric with a corded surface of cotton, silk, wool, or of silk and wool, spun rayon, etc., using coarse and fine yarns in alternate order in both warp and weft. Sometimes two warp beams are used. Also called **repp.**

repagula (*Zool.*). Rodlike bodies, circles of which are used by the females of some *Plannipennia* (*Neuropteroidea*) to protect their eggs from predators.

repeated emergence (*Bot.*). A condition in Fungi in which zoospores, after swimming for a time, encyst and then emerge from the cysts without any change in morphology.

repeater (*Horol.*). A watch which 'repeats the time' by striking a sequence of blows on gongs when a slide that projects from the band of the case is pushed. In a *quarter repeater* the last hour (as shown by the watch) is struck, followed by the number of quarters; a *minute-repeater* strikes, in addition, the number of minutes since the last quarter struck. (*Maths.*) See **recurring decimal.** (*Telecomm.*) Device which transmits what it receives, but amplified or augmented. Applied to telegraph relay, later to one-way or two-way telephonic amplifier, with or without echo suppression. A *pulse repeater* is one which receives pulses from one circuit and transmits corresponding pulses to another circuit. If regenerative, it acts as a pulse regenerator.

repeater balance (*Telecomm.*). The balancing network associated with the hybrid coil in a two-way repeater. See **line balance.**

repeater distribution frame (*Teleph.*). A frame providing for the interconnection in repeater stations of amplifiers, transformers, and signalling units.

repeater gain (*Teleph.*). Power delivered by a repeater divided by the power which would be delivered in the absence of the repeater; expressed in decibels.

repeater test rack (*Teleph.*). Rack in a repeater station at which measurements can be made of input and output levels of audio-frequency amplifiers and signalling units.

repeating back (*Photog.*). Sliding back for a camera, in which images for colour separation can be taken successively and side-by-side.

repeating coil (*Teleph.*). Unity-ratio transformer for separating telephonic circuits, with windings balanced to earth.

repeating decimal (*Maths.*). See **recurring decimal.**

repeating selector (*Teleph.*). A selector which is operated by the first train of impulses received, and also repeats all received impulses for operating further selectors.

repeating work (*Horol.*). Chiming mechanism of a repeater watch.

repeller (*Electronics*). U.S. term for **reflector,** as in a klystron.

repent (*Bot.*). Lying on the soil and rooting.

reperforator (*Teleg.*). Instrument on which received signals cause the code of the corresponding characters or functions to be punched on a tape.

repetition (*Surv.*). Method of checking angular measurements made with a theodolite by repeating the observation after unclamping the lower plate and sighting on the back station so that the vernier reading is unaltered, and then sighting forward to get a new reading on the vernier, which should be double the previous reading. Cf. *reiteration.*

repetition compulsion (*Psychol.*). Factor in mental life which compels early patterns of behaviour to be repeated, irrespective of pleasure/displeasure thereby experienced by the individual.

repetition rate (*Telecomm.*). Rate at which recurrent signals, usually pulses, are repeated. (*Teleph.*) Number of times repetition is demanded in a telephone conversation, used as a measure of effective transmission of the circuit. Usually expressed as repetitions per 100 seconds of conversation.

replaceable hydrogen (*Chem.*). Those hydrogen atoms in the molecule of an acid which can be replaced by atoms of a metal on neutralization with a base.

replacement (*Geol.*). The process by which one type of rock occupies the space previously occupied by another rock; also applies to minerals.

replete (*Zool.*). See **plerergate.**

replicate septum (*Bot.*). Septum in some algae which bears a collarlike appendage projecting into cavity of the cell.

replicatile (*Zool.*). Said of a wing which folds over on itself in the resting position.

replum (*Bot.*). Thin wall dividing fruit into two chambers, formed by an ingrowth from the placentas; not a true part of the carpellary walls.

report call (*Teleph.*). Call made to ascertain whether a desired subscriber is available for connexion.

repp (*Textiles*). See **rep.**

Reppe synthesis (*Chem.*). German wartime development in synthetic organic chemistry based on acetylene as the starting material. Important features were production of butadiene for synthetic rubbers via butynediol (from acetylene and formaldehyde) and production of acrylic esters by reaction of acetylene with carbon monoxide and water, using a nickel catalyst.

representation (*Comp.*). System of representing data by numbers, their position in relation to the radix, and their significance.

representative sample (*Min. Proc.*). One cut from the bulk of material or ore deposit in such a way as to make it reasonably representative of the whole body.

repression (*Psychol.*). An unconscious mental mechanism which excludes unacceptable thoughts and desires from consciousness without preventing them from affecting psychic activity. As a result, the true origin of emotional conflict may be hidden.

reproducer (*Acous.*). (1) Complete sound reproduction system. (2) Loudspeaker.

reproducibility. Precision with which a measured value can be repeated in a process or component.

reproduction (*Acous.*). Replay of recorded sounds, or re-creation of sounds from electric waves produced by microphone. (*Biol.*) The process of generation of new individuals whereby the species is perpetuated. *adj.* **reproductive.**

reproduction constant (*Nuc.*). Deprecated term for **multiplication constant.**

reproduction proof (*Print.*). Print taken of a letterpress forme, on *art paper* or *baryta paper*, from which a plate is to be made photomechanically.

reproduction ratio (*Teleg.*). Ratio of linear size between received and transmitted documents.

Reptilia (*Zool.*). A class of *Craniata.* They are pentadactyl and have shelled amniote eggs. The vertebrae are gastrocentrous, the kidney metanephric and the skin completely covered by ectodermal scales, or, sometimes, by bony plates. They are poikilothermous, breathe by lungs, and retain both aortic arches. Known as fossils from late Carboniferous, they were dominant and varied in the Mesozoic (*Dinosaurs*) but became less numerous in the Cretaceous. Living forms include Lizards, Snakes, Turtles, Tortoises, Crocodiles, and Alligators.

repugnatorial glands (*Zool.*). In *Arthropoda*, glands, usually abdominal in position, which produce a repellent secretion of an odoriferous pungent, or corrosive nature which can be used in self-defence.

repulsion-induction motor (*Elec. Eng.*). A single-phase induction motor having, in addition to the squirrel-cage winding on the rotor, a commutator winding with its brushes short-circuited, so that the motor starts as a repulsion motor with a high starting torque and runs with the characteristics of an induction motor.

repulsion motor (*Elec. Eng.*). A type of single-phase commutator motor in which power is supplied to the stator winding, and the armature winding is short-circuited through the brushes.

repulsion-start induction motor (*Elec. Eng.*). A repulsion motor having a centrifugal device which short-circuits all the commutator bars when the motor reaches a certain speed, so that it runs as a single-phase induction motor, and starts as a repulsion motor with a high starting torque.

reradiation (*Radio*). (1) Radiation from resonating elements, such as masts, antennae, telephone lines, giving errors in bearings or displaced television images. (2) Phenomenon which occurs when a receiver employing reaction on to the antenna circuit is adjusted to the point of oscillation. The signal strength for nearby receivers is thereby increased.

rere arch (*Arch.*). A flat soffit arch laid over splayed jambs.

re-recording (*Acous.*). Recording acoustic waveforms immediately upon reproduction from the same, or any other type of, recording medium as that in use.

re-reeler (*Print.*). An auxiliary unit to rewind the reel on web-fed presses for subsequent operations.

re-run (*Comp.*). Repeat of part of program for a computer. Most programs incorporate a re-run point for every few minutes running time. In the event of an error only the part subsequent to the previous re-run point is repeated.

resaw (*For.*). Circular or band saw used to saw boards, cants, planks, slabs, and other material from the head saws.

rescinnamine (*Pharm.*). A synthetic sedative drug of which the natural base is *Rauwolfia serpentina*.

réseau (*Astron.*). A network of parallel lines photographed on to the plates used in certain branches of stellar photography, to facilitate subsequent measurement of star positions. (*Photog.*) Mosaic resulting from ruling coloured lines, when making a screen on an emulsion.

resection (*Surg.*). Cutting off of a part of a bodily organ, especially the ends of bones and other structures forming a joint. (*Surv.*) Positional fix of a point which is not necessarily to be occupied by sighting it from two or more known stations.

reserpine (*Pharm.*). An antihypertensive, sedative drug obtained from *Rauwolfia serpentina*.

reserve buoyancy (*Aero.*). Potential buoyancy of a seaplane or amphibian which is in excess of that required for normal floating. The downward force required for complete immersion. (*Ships*) Watertight volume above the load water line.

reserve factor (*Aero.*). Ratio of actual strength of an aircraft structure to estimated minimum strength for a specified load condition.

reserves (*Mining*). Block of ore proved by development (normally by driving levels and winzes or raises so as to expose its four sides in a rectangular panel, and sufficiently sampled and tested) to warrant exploitation.

reservoir (*Nuc. Eng.*). Any volume in an isotope separation plant which is for the purpose of storing material or to ensure smooth operation. (*Zool.*) In some *Mastigophora*, a noncontractile vacuole which opens into the gullet.

reset (*Comp.*). After (or before) completion of an operation by an electronic computer, restoration of all circuits and mechanisms so that the system is ready to perform its function again, from the beginning of its cycle of operations, according to its *instruction* or *program*. (*Elec. Eng.*) General term for preparation of any circuit or apparatus for a fresh performance of its duty. Amplifiers require no resetting, apart from switching on; timers or counters do require resetting. Resetting can be automatic, or initiated by an external signal, arbitrary in time.

reset circuit (*Elec. Eng.*). One which, when operated, resets a functional circuit, i.e., establishes it in a ready condition for operating.

reshabar (*Meteor.*). A dry, squally, northeast wind blowing down some mountain ranges of southern Kurdistan.

reshaping (*Elec. Eng.*). Restoration to intended shape, in amplitude and time, of pulses which have become distorted.

residual activity (*Nuc. Eng.*). In a nuclear reactor the remaining activity after the reactor is shut down following a period of operation.

residual affinity (*Chem.*). Chemical attractive forces which remain after saturation of normal valencies of the atoms in a molecule. They are responsible for the formation of molecular compounds and for chemisorption.

residual air (*Physiol.*). Volume of air left in the lungs after the strongest possible forced expiration, usually of the order of 1500 cm^3 in man.

residual current (*Electronics*). That in a thermionic valve when the anode is at the same potential as the cathode (due to the finite velocity of emission of electrons at the cathode).

residual deposits (*Geol.*). Accumulations of rock waste resulting from disintegration *in situ*. They cover the whole range of grain size, from residual boulder beds to residual clays.

residual errors (*Eng.*). Errors which remain in an observation despite all attempts to eliminate them.

residual field (*Mag.*). Magnetic field remaining in a magnetic circuit after removal of the magnetizing force.

residual flux density (*Mag.*). Flux density remaining after exciting magnetic field has been removed. This depends on the geometry and nature of the whole magnetic circuit.

residual gas (*Chem. Eng.*). Small amount of gas which inevitably remains in a 'vacuum' tube after pumping. If present to excess, it causes erratic operation of the tube, which is said to be *soft*.

residual induction (*Mag.*). Same as *residual magnetization*, but could be applicable to an entire magnetic circuit.

residual magnetization (*Mag.*). Magnetization persisting in ferromagnetic material when the exciting magnetizing force is removed. Also termed **remanence**.

residual resistance (*Elec.*). That persisting at temperatures near zero on the absolute scale, arising from crystal irregularities and impurities, and in alloys.

residual volume (*Physiol.*). See **residual air**.

residue (*Maths.*). (1) The *n*th *power residues* of p are the remainders when r^n, for $r=1, 2, 3 \ldots$, are divided by p. Thus, the quadratic residues of 13 are 1, 3, 4, 9, 10 and 12. (2) Of an analytic function $f(z)$ at an isolated singularity,

$$\frac{1}{2\pi c}\int_c f(z)dz,$$

where c is a simple closed curve around the singularity, and the singularity is deleted from the region contained therein.

resilda (*Textiles*). See **permo.**

resilience (*Eng.*). Stored energy of a strained material, or the work done per unit volume of an elastic material by a bending moment, force, torque, or shear force, in producing strain.

resilient escapement (*Horol.*). An escapement in which the banking pins yield to any excess pressure due to overbanking, allowing the impulse pin to pass the lever, which has no horns; or one in which the teeth of the escape wheel are so formed as to provide a recoil.

resilium (*Zool.*). Elastic hinge joining halves of a shell in a Bivalve.

resin (*Chem.*). The product from the secretion of the sap of certain plants and trees. Resins are hard, fusible, and more or less brittle, insoluble in water, soluble in certain organic solvents. They consist of resinous matter, i.e., certain highly polymerized acids and neutral substances mixed with terpene derivatives. See also **rosin.** (*Plastics*) The term *resin* is widely but loosely used to describe any synthetic plastics material. More precisely, it is applied to a polymeric compound prior to curing. See **synthetic resins.**

resinates (*Chem.*). Calcium, magnesium, aluminium, iron, nickel, cobalt, zinc, tin, manganese, and lead salts of rosin, obtained by fusion of *rosin* (q.v.) with metal oxides.

resin-bonded plywood (*Build.*). Plywood in which the wood veneers are held together with synthetic resin, glues, or glue-impregnated paper and finally formed with pressure and heat.

resin canal (*Bot.*). Intercellular space, often bordered by secreting cells, containing resin or turpentine.

resin esters (*Paint.*). *Ester gums* (q.v.).

resin flux (*Bot.*). Abnormal escape of resin from a plant due to parasitic attack. Also resinosis.

resin-in-pulp (R.I.P.) (*Min. Proc.*). Ion-exchange method of continuously treating ores by acid leaching. Baskets containing *ion-exchange resins* (q.v.) are jigged through tanks containing finely ground ore pulp as it flows from vessel to vessel. Abbrev. **R.I.P.**

resinophore groups (*Chem.*). Groups occurring in the molecule of certain substances which make them readily susceptible to polymerization and resin formation. Resinophore groups are, e.g.,

$$-N=C=N- \qquad -N=\underset{|}{P}=N-$$
$$-CO \cdot CH=CH-$$

resinosis (*Bot.*). See **resin flux.**

resinous substances (*Chem.*). Term applied to (*a*) true resins; (*b*) substances resembling true resins in their physical properties.

resin pocket (*For.*). A well-defined intercellular cavity in timber, often more or less lens-shaped, containing a resinous substance.

resin poisons (*Min. Proc.*). In ion-exchange processes, substances which reduce efficiency of resin loading by masking activated resin sites.

resin soaps (*Chem.*). See **soaps (2).**

resist (*Photog.*). A coating of chemically neutral substance placed over a surface when the latter has to be protected at some stage in processing, as in etching or selective dyeing. (*Print.*) When making printing surfaces, the protection over the lines and dots of the image which preserve it from the action of etching fluid.

resistance (*Elec.*). In electrical and acoustic fields, real part of the impedance, characterized by the dissipation of energy as opposed to its storage. *N.B.*—Electrical resistance may vary with temperature, polarity, field illumination, purity of material, etc. See **impedance, ohm, reactance.** (*Phys.*) Opposition to motion leading to dissipation of energy. (*Psychol.*) An unconscious barrier in the mind against making unconscious processes conscious. Manifested in psychoanalysis by reluctance on the part of an individual to accept interpretations from the analyst concerning his unconscious processes.

resistance box (*Elec. Eng.*). One containing carefully constructed and adjusted resistors, which can be introduced into a circuit by switches or keys. At higher frequencies there are disturbing inductive and capacitive effects which complicate measurements using resistance boxes, but which are mitigated by suitable design. The boxes are then described as *nonreactive*.

resistance butt-seam welding (*Elec. Eng.*). Resistance welding process in which coaxial roller electrodes conduct current to the edges of the seam to be joined, the mechanical pressure being applied independently. Extensively used in tube-making from strip.

resistance butt-welding (*Elec. Eng.*). A resistance welding process in which the two parts to be joined are butted together. See **resistance butt-seam welding, resistance flash-welding, resistance upset-butt welding.**

resistance-capacitance coupling (*Elec. Eng.*). That in which signal voltages developed across a load resistance are passed to the subsequent stage through a d.c. blocking capacitor.

resistance-capacitance oscillator (*Elec. Eng.*). One producing a sine waveform of frequency determined by phase shift in a resistance-capacitance section of an artificial line.

resistance coupling (*Elec. Eng.*). That between successive stages of an amplifier using thermionic valves, by which changes in anode potential across a resistance are impressed on grid of succeeding valve, correct grid bias being obtained separately. See **direct coupling.**

resistance drop (*Elec. Eng.*). Voltage drop produced by a current flowing through the resistance of a circuit; equal to product of current and effective resistance.

resistance-flash welding (*Elec. Eng.*). A resistance welding process in which an arc is struck and maintained between the parts until the correct temperature is attained, after which the current is cut off and the parts are forced together by mechanical pressure. Also called **flash-butt welding.**

resistance frame (*Elec. Eng.*). A frame containing a number of resistors connected to a multiple-contact switch at the top, so that any desired number of them can be included in the circuit in which the frame is connected.

resistance furnace (*Elec. Eng.*). See **resistance oven.**

resistance grid (*Elec. Eng.*). A resistance unit generally used for heavy currents. Made up of a cast-iron grid designed so that current enters one end and passes through all the sections in series, before leaving at the other end.

resistance lamp (*Elec. Eng.*). One used to limit the current in a circuit.

resistance lap-welding (*Elec. Eng.*). A resistance welding process in which the two parts to be joined overlap one another. See **resistance roller-spot-welding, resistance seam-welding, resistance spot-welding,** and **resistance stitch-welding.**

resistance noise. Same as **thermal noise.**

resistance oven (*Elec. Eng.*). An oven in which the heating is carried out by means of heating resistors. Also called a **resistance furnace.**

resistance percussive-welding (*Elec. Eng.*). A resistance welding process in which a heavy electric current is discharged momentarily across the electrodes, and a momentary mechanical force is applied simultaneously.

resistance projection-welding (*Elec. Eng.*). A variant of resistance spot-welding in which current is concentrated at the desired points by projections on one of the parts.

resistance pyrometer (*Elec. Eng.*). See **resistance thermometer.**

resistance roller-spot-welding (*Elec. Eng.*). A variant of resistance seam-welding but with current pulses so separated that the spots produced do not overlap.

resistance seam-welding (*Elec. Eng.*). A resistance welding process in which the welding electrodes consist of two rollers having mechanical pressure between them through which the work passes while the current flows continuously or intermittently, producing a line of overlapping welds.

resistance spot-welding (*Elec. Eng.*). A resistance welding process in which the electrodes consist of two points and cause welding at one spot.

resistance stitch-welding (*Elec. Eng.*). A form of resistance spot-welding consisting of a series of overlapped spot welds to form a seam-weld.

resistance strain gauge (*Elec. Eng.*). Foil, wire or thin film resistor which has a value which varies with mechanical strain. Normally used in a bridge circuit. See **strain gauge.**

resistance thermometer (*Elec. Eng.*). One using resistance changes for temperature measurement. Resistance element may be platinum wire for extreme precision or semiconductor (*thermistor*) for high sensitivity. Also **resistance pyrometer** for higher temperatures.

resistance to extinction (*An. Behav.*). The number of times a conditioned response to a particular stimulus occurs during *experimental extinction* (q.v.) before reaching some low intensity.

resistance upset-butt welding (*Elec. Eng.*). A resistance welding process in which mechanical pressure is first applied to the joint and then current is passed until welding temperature is reached and the weld is 'upset'. Also called **slow-butt welding.**

resistance welding (*Elec. Eng.*). Pressure welding, in which the heat to cause fusion of the metals is produced by the welding current flowing through the contact resistance between the two surfaces to be welded, these being held together under mechanical pressure. See **resistance butt-welding, resistance flash-welding, resistance percussive-welding, resistance seam-welding, resistance spot-welding.**

resistant (*Biol., Med.*). Not readily attacked by a parasite or disease.

resisticon (*Photog.*). High-velocity camera tube.

resistive component (*Elec. Eng.*). That part of the impedance of an electrical system which leads to the absorption and dissipation of energy as heat.

resistive load (*Elec. Eng.*). Terminating impedance which is nonreactive, so that the load current is in phase with the source e.m.f. Reactive loads are made entirely resistive by adding inductors or capacitors in series or shunt (tuning).

resistivity (*Elec. Eng.*). Intrinsic property of a conductor, which gives the resistance in terms of its dimensions. If R is the resistance in ohms, of a wire l m long, of uniform cross-section a m², then $R = \rho . l/a$, where the resistivity ρ is in ohm metres (*not* ohm m^{-3}). Also called, erroneously, **specific resistance.**

resistor (*Elec. Eng.*). Electric component designed to introduce known resistance into a circuit and to dissipate accompanying loss of power. Types are wirewound, composition, metal film, etc.

resnatron (*Elec. Eng.*). High-power, high-frequency tetrode in which concentric cylinders form bypass capacitors and cavities from which output energy is taken.

resol (*Chem.*). A synthetic resin produced from a phenol and an aldehyde, having reactive methylol groups in the molecule.

resolution (*Chem.*). The separation of an optically inactive mixture or compound into its optically active components. (*Med.*) Retrogression of the phenomena of inflammation; the subsidence of inflammation. *v.* **resolve.** (*Nuc.*) The smallest time interval between ionizing particles recorded separately by any detector. (*Photog.*) The minimum image detail recorded on a photographic emulsion (measured in lines per millimetre). (*Phys.*) (1) The smallest measurable difference in wavelength, frequency, or energy for a light, sound, or particle beam spectrum. (2) The smallest mass difference detectable in a mass spectrograph. (3) The smallest separation of two points recorded separately by an electron or optical microscope —also the smallest angular separation possible with an optical or radio telescope. (*TV*) The definition of a picture in TV or facsimile (measured by the number of lines used to scan the image of the picture).

resolution of forces (*Mech.*). The process of substituting two forces in different directions for a single force, the latter being equal to the resultant of the two components. If these are at right angles to each other, the one which makes an angle θ with the original force P is equal to $P \cos \theta$, the other being $P \sin \theta$.

resolution time (*Nuc.*). Minimum time between two events recorded separately. The maximum time between two events recorded as coinciding.

resolution-time correction (*Nuc.*). Correction applied to observed counting rate for random events, which allows for those not recorded because of the finite resolution time.

resolvant equation (*Maths.*). An equation which is used in the solution of a higher-order equation.

resolver (*Comp.*). Apparatus which converts polar coordinates into cartesian coordinates.

resolving power of the eye (*Optics*). The angle subtended by a small object which can just be determined visually.

resolving time (*Nuc.*). See **resolution time.**

resonance (*Aero.*). See **ground-.** (*Chem.*) A state of a molecule which is intermediate between those represented by two bond formulae which are theoretically possible, e.g., the benzene molecule shows resonance. (*Elec.*) Balance between positive (inductive) and negative (capacitive) reactance of a circuit, accompanied by large currents and voltages resulting from relatively small applied e.m.f. at resonant frequency. (*Nuc.*) Increased probability of nuclear reaction when energy of incident particle or photon would raise compound nucleus to natural energy level. Shown by increase in effective cross-section to particles of this energy. (*Phys.*) Phenomenon of minimum mechanical or acoustical impedance as the frequency of the applied disturbing force is varied, resulting in a maximum velocity of motion. Rods or plates are potentially vibrating systems with several modes of vibration, the frequencies of resonance generally not being exactly harmonic. *Sharpness of resonance* (q.v.) is measured by the ratio of the dissipation to the inertia of the system, which also measures rate of decay of motion of the vibrating system when it is impulsed. See **decay factor.**

resonance bridge (*Elec. Eng.*). One for which balance depends upon adjustment for resonance.

resonance curve (*Elec.*). One showing variation of current in a resonant circuit in series with an e.m.f. as the ratio of the resonance frequency to the frequency of the generator is varied through unity.

resonance escape probability (*Nuc.*). In a reactor, the probability of a fission neutron slowing down to thermal energy without experiencing resonance.

resonance heating (*Phys.*). See **magnetic pumping.**

resonance integral (*Nuc.*). One used in reactor theory which is expressed in terms of the logarithm of the resonance escape probability multiplied by the slowing-down power of the absorber.

resonance lamp (*Elec. Eng.*). One which depends on the absorption and reradiation of a prominent line from a mercury arc, excited in mercury vapour.

resonance level (*Nuc.*). An excited level of the compound system which is capable of being formed in a collision between two systems, such as between a nucleon and a nucleus.

resonance potential (*Electronics*). See **excitation potential.**

resonance radiation (*Phys*). Emission of radiation from gas or vapour when excited by photons of higher frequency.

resonances (*Nuc.*). During nuclear reactions very unstable mesons or hyperons are frequently created. These decay through the strong interaction, with a half-life of the order of 10^{-23} s. Consequently such particles are undetectable and their formation as an intermediate step in the reaction can only be inferred from indirect measurements. These temporary states are known as *resonances* to distinguish them from metastable particles with half-lives of the order of 10^{-10} s, which are detectable.

resonance scattering (*Nuc.*). See **scattering.**

resonance step-up (*Elec. Eng.*). Ratio of the voltage appearing across a parallel tuned circuit to the e.m.f. acting in the circuit (usually induced in the coil) when the circuit is resonant at the applied frequency. See *Q*.

resonance test (*Aero.*). A test in which an aircraft, while suspended by cables or supported on inflated bags, is excited by forced oscillations over a range of frequencies, so as to establish the natural frequencies and modes of oscillation of the structure.

resonant cavity or **chamber** (*Telecomm.*). One in which resonant effects result from the possibility of a modal pattern of electric and magnetic fields, as in magnetrons, klystrons, waveguide couplers. Also applies in acoustics.

resonant circuit (*Electronics*). One consisting of an inductor and a capacitor in series or parallel. The series circuit has an impedance which falls to a very low value at the resonant frequency; that of the parallel circuit rises to a very high value.

resonant frequency (*Chem.*). See **nuclear magnetic resonance.** (*Telecomm.*) That at which reactances of a series resonant circuit, or susceptances of a parallel resonant circuit, balance out; numerically equal to $1/2\pi\sqrt{LC}$ Hz, where L is inductance in henries and C capacitance in farads.

resonant gap (*Electronics*). The interior volume of the resonant structure of a transmit-receive tube in which the electric field is concentrated.

resonant line (*Telecomm.*). Parallel wire or coaxial transmission line open or short-circuited at the ends and an integral number of quarter-wavelengths long. Used for stabilizing the frequency of short-wave oscillators and in antenna systems.

resonant mode (*Electronics*). Field configuration in a tuned cavity. In general, resonance occurs at several related frequencies corresponding to different configurations.

resonator (*Electronics*). Any device exhibiting a sharply defined electric, mechanical, or acoustic resonance effect, e.g., a stub, piezoelectric crystal, or Helmholtz resonator.

resonator grid (*Electronics*). Electrode traversed by an electron beam and which provides a coupling to a resonator.

resorcin brown (*Chem.*). The sodium salt of xylidine-azo-sulphanilic-azo-resorcin; used as a primary diazo dye.

resorcinol or **resorcin** (*Chem.*). $C_6H_4(OH)_2$, 1,3-dihydroxy-benzene, a dihydric phenol, colourless crystals, m.p. 111°C, b.p. 276°C. Used as a lotion in certain skin diseases. With formaldehyde used in the preparation of cold-setting adhesives.

resorption (*Geol.*). The partial or complete solution of a mineral or rock fragment by a magma, as a result of changes in temperature, pressure, or composition of the latter.

respiration (*Biol.*). A term originally used to describe the process in which many living organisms take in oxygen from their surroundings and give out carbon dioxide. In many terrestrial animals this is connected with breathing and it is still used popularly in this sense (e.g., artificial respiration). Nowadays this activity is called *external respiration* (see **respiration, external**), in contrast to the intracellular process in which the oxygen is used in the breakdown of *respiratory substrates* (q.v.) to release energy, and carbon dioxide is produced, this being called *internal respiration* (see **respiration, internal**). Thus in its widest sense respiration includes, (*a*) external respiration or gaseous exchange, (*b*) *gas transport* (q.v.) (to the cells), and (*c*) internal or tissue respiration.

respiration, aerobic (*Biol.*). A type of *internal respiration* found in most animals and plants, in which the respiratory substrate is completely broken down to water and carbon dioxide with the use of oxygen (in animals, often transported to the site of respiration by the blood) and a relatively large amount of energy is released, the three main stages (if glucose is the substrate) being glycolysis, oxidative decarboxylation and oxidative phosphorylation (the latter two occurring largely in the cell mitochondria).

respiration, anaerobic (*Biol.*). A type of *internal respiration* found, e.g., in yeast, parasitic flatworms and, on occasions, muscle, in which respiratory substrates are only partly broken down to simpler organic compounds (e.g., ethyl alcohol, lactic acid), and, sometimes, carbon dioxide, no oxygen is used, and relatively less energy is released. If glucose is the substrate, glycolysis is followed only by one or two further reactions.

respiration, external (*Biol.*). The phase of respiration in which a living organism takes oxygen in from its surroundings and gives up carbon dioxide (also known as **gaseous exchange**). It occurs in both plants and animals, but in the latter occurs on a larger scale, smaller and/or less active animals using all or part of their external surface. Many larger and/or more active animals have specialized organs with a relatively large moist surface (the *respiratory surface*) with a good blood supply, at which gaseous exchange takes place (e.g., lungs, gills), this generally being internal in terrestrial animals to avoid excessive loss of water. In many animals there is some way of passing the external medium (either water or air) rapidly over the respiratory surface, this activity being known as *breathing*.

respiration, internal (*Biol.*). The metabolic process occurring within the cells of living organisms which results in the release of energy in a form in which the organism can make use of it, this being achieved by gradually breaking down complex organic compounds either into simpler organic compounds or into simple inorganic compounds. In animals the *respiratory substrates* (q.v.) are usually carbohydrates or fats (see **respiratory quotient**) or deaminated amino-acids, and they are often transported to respiring cells by the bloodstream; in plants they derive from the products of photosynthesis and reach the respiring cells by translocation. Internal respiration may be *aerobic* or *anaerobic*. In both cases the energy released at certain stages of the breakdown is used to convert ADP (adenosine diphosphate) to ATP (adenosine triphosphate), and this can then be transported intracellularly to be used in processes needing energy. Also called **tissue respiration.**

respiratory cavity, respiratory chamber (*Bot.*). A large intercellular space lying immediately beneath a stoma.

respiratory centre (*Zool.*). In Vertebrates, a nerve-centre of the hind-brain which regulates the respiratory movements.

respiratory chromogen (*Bot.*). A colourless substance which gives rise to a coloured substance on oxidation or reduction and may play a part in respiration.

respiratory heart (*Zool.*). In Birds and Mammals, the auricle and ventricle of the right side of the heart, which supply blood to the lungs. Cf. *systemic heart*.

respiratory index (*Bot.*). The number of milligrams of carbon dioxide set free from one gram of plant material (weighed as dry) when the temperature is 10°C, when the amount of respirable material is unlimited, and when oxygen is present in the same proportions as in the ordinary atmosphere.

respiratory movements (*Zool.*). The muscular movements associated with the supply of air or water to the respiratory organs.

respiratory organs (*Biol.*). The specialized structures which enable an organism to obtain a sufficient supply of oxygen for its needs.

respiratory pigments (*Zool.*). In the blood of many animals, including most, if not all, *Craniata*, coloured compounds which combine readily, but reversibly, with oxygen if the tension or partial pressure of the latter is high (as in the blood capillaries of an animal's respiratory surface), but give up the oxygen at low tensions or partial pressures (as in the capillaries of the body tissues), generally changing their colour and absorption spectrum as they do so. More oxygen can be combined with a respiratory pigment than could be dissolved in blood plasma (especially if a relatively high concentration of the pigment is achieved by having it in discrete corpuscles, such as erythrocytes), and complete combination with oxygen can be achieved at relatively low oxygen tensions. The relation between the amount of oxygen combined and the oxygen tension (the *dissociation curve*) is not linear, but sigmoid. Examples of such pigments are haemoglobin, haemocyanin, chlorocruorin.

respiratory quotient, respiratory ratio (*Biol.*). The ratio between the volume of carbon dioxide given off and that of oxygen taken in, during a given time. It is approximately unity when an organism is respiring at the expense of carbohydrates, but is less than unity when fatty or protein material is being utilized.

respiratory substrate (*Biol.*). Any substance broken down to release the energy which it contains in the process of internal respiration.

respiratory surface (*Biol.*). See **respiration, external.**

respiratory system (*Biol.*). See **respiratory organs.**

respiratory trees (*Zool.*). In *Holothuroidea*, a pair of elaborate tubular diverticula of the cloaca, the ultimate branches of which end in small spherical ampullae; believed to have a respiratory and hydrostatic function.

respiratory trumpets (*Zool.*). In the pupae of *Culicidae* (Mosquitoes, etc.), a pair of tubes projecting from the thorax which allow air to pass to the spiracles.

respiratory tube (*Zool.*). In some *Cyclostomata*, a median ventral tube by which water passes from the gullet to the gills.

respiratory valve (*Zool.*). In some Fish, e.g., Trout, a pair of transverse membranous folds, one attached to the floor, the other to the roof of the mouth, which prevent water from escaping through the mouth during expiration.

respond (*Build.*). (1) A pilaster which forms a pair with another. (2) A *reveal* (q.v.).

responder (*Telecomm.*). That part of a transponder which replies automatically to the correct interrogation signal.

response (*An. Behav.*). (1) A part of behaviour or a change in part of an animal's behaviour which can be observed to vary systematically as a function of time or other changes in the animal's environment. The amount of variance as the result of experimental manipulation is the *response intensity*. (2) Any activity of an animal resulting from stimulation, such as muscular activity, glandular secretion, etc. (*Telecomm.*) That of a transmission system at any particular frequency is given by the ratio of the output to input level. If these levels are defined on a logarithmic scale, e.g., in dB, the response of the complete system is the sum of responses of the separate parts. See also **polar diagram.**

response curve (*Telecomm.*). That which exhibits the trend of the response of a communication system or a part thereof, for the range of frequency over which the system or part is intended to operate. Usually plotted in dB against a logarithmic frequency scale.

response latency (*An. Behav.*). The time elapsing between the onset of a stimulus and the beginning of an animal's response to it.

response time (*Electronics*). Time constant of change in output of an electronic circuit after a sudden change in input, or of indication given by any instrument after change in signal level.

responsor (*Radar*). Receiver of secondary radar signal from transponder. See **identification, friend or foe.**

rest (*Eng.*). In lathes, the part which supports or imparts movement to the tool or tool post;

sometimes that which supports the workpiece against the cutting pressure.

rest bend (*San. Eng.*). A right-angle bend off a horizontal drain-pipe, fitted with a flat seating for connexion to a vertical pipe. Also called **duckfoot bend.**

restiform (*Zool.*). Ropelike.

restiform bodies (*Zool.*). In some Fish, a pair of lateral extensions of the medulla oblongata which lie behind the cerebellum.

resting nucleus (*Cyt.*). A nucleus which is not dividing.

resting spore (*Bot.*). A thick-walled spore able to endure drought or other unfavourable conditions, and normally remaining quiescent for some time before it germinates.

restitution (*Teleg.*). The series of conditions assumed as a consequence of a telegraph modulation by the appropriate device of a receiving apparatus, each condition being associated.

rest mass (*Phys.*). Newton mass of a particle at zero or low velocities, i.e., not augmented by relativistic mass (Einstein). Associated energy on annihilation is the rest mass multiplied by the square of the velocity of light. See **relativistic mass equation.**

restorative (*Med., etc.*). Capable of restoring to health, consciousness or good condition; any remedy which does this.

restore (*Comp.*). Return of a variable address or word or cycle index to its initial value.

restoring moment (*Aero.*). A moment which, after any rotational displacement, and dependent upon that displacement, tends to restore an aircraft to its normal attitude.

restrainer (*Photog.*). Ingredient of a developer which checks the development of unexposed silver halide, thus reducing tendency to fog. Usually potassium bromide.

restriking voltage (*Elec. Eng.*). The high-frequency transient voltage which appears across the contacts of a switch, circuit-breaker, or fuse immediately after it has interrupted a circuit, and which is superimposed on the recovery voltage.

resuing (*Mining*). In mining narrow seams or lodes, separate prior detachment of adjacent barren rock to facilitate attack on ore.

resultant (*Maths.*). (1) Of two forces: that force, obtained by the parallelogram of forces, which is equivalent to the two forces. (2) Of two polynomials $f(x)$ and $g(x)$: the result of eliminating x from the equations $f(x)=0$ and $g(x)=0$. If the resultant is zero the two equations have at least one common root. Also called **eliminant.**

resultants (*Chem.*). See **reaction products.**

resuperheating (*Eng.*). See **reheating.**

resupinate (*Bot.*). Reversal in position, usually through 180°, so that orientation is abnormal. The fruit bodies of some fungi with the hymenium on the upper surface are regarded as being resupinate, and in some flowers, those of orchids for example, the flower stalks or the ovaries twist during development, so that the open flower is really upside down.

resurgent gases (*Geol.*). Superheated steam and other volatiles which play an active rôle in volcanic action, and which were derived from the water included in sedimentary rocks at the time of their accumulation.

retaining mesh (*Min. Proc.*). In sizing ore before further treatment, the screen aperture above which size the material is arrested.

retaining wall (*Civ. Eng.*). A wall built to support earth at a higher level on one side than on the other. Also **revetment.**

retardation coil (*Elec. Eng.*). Inductor for separating d.c. from a.c., particularly from a rectifier or supply with ripple.

retardation test (*Elec. Eng.*). A method of determining the iron, friction, and windage losses of electrical machinery by determining the rate at which it retards under the influence of these losses after being run up to speed and then disconnected from the supply.

retarded field and **retarded potential** (*Elec. Eng.*). Those at a point which arise later than at some other point because of finite speed of propagation of waves in the medium.

retarded hemihydrate plaster (*Build.*). A plaster based on calcium sulphate hemihydrate (*plaster of Paris* (q.v.)) but with the addition of a retarder in varying quantities. Particularly suitable for use as an undercoat on plasterboard, fibre board, or metal lath, because of low expansion.

retarder (*Build.*). A substance which delays or prevents the setting of cement. (*Chem.*) A negative catalyst which is added to a reacting system to prevent the reaction from being too vigorous. (*Rail.*) An arrangement of braking surfaces placed alongside, and parallel with, the running rails in a shunting yard; operated from a signal-box by electric, pneumatic, hydraulic, or mechanical means. Also **wagon retarder.**

retarding field (*Electronics*). Electric field such as between a positively charged grid and a lower potential outer grid in a valve, so that electrons entering this region lose energy to the field. Also called **brake field.**

retarding-field detector (*Electronics*). Use of a retarding-field valve for demodulation of ultra-high-frequency signals.

retarding-field oscillator (*Electronics*). One which depends on the electron-transit time of a positive grid oscillator valve.

rete (*Zool.*). A netlike structure. *pl.* **retia.**

rete Malpighii (or mucosum) (*Zool.*). See **Malpighian layer.**

rete mirabile (*Zool.*). (1) A network of small blood vessels, as in the so-called *red gland* (q.v.) of Fish. (2) In *Holothuroidea*, a vascular plexus which connects the dorsal vessel with the intestine.

retene (*Chem.*). $C_{18}H_{18}$, methyl-isopropyl-phenanthrene; it occurs in the coal-tar fraction boiling above 300°C and crystallizes in white plates which sublime without decomposition.

retention (*An. Behav.*). The persistence of a learned response over an interval during which neither trials nor training have been given; cf. *experimental extinction.* (*Med.*) The abnormal keeping back in the body of matter (e.g., urine) normally evacuated. (*Nuc.*) The fraction of radioactive atoms which are not separable from the target compounds after the production of these atoms by nuclear reaction or radioactive decay.

retention wall (*Build.*). A thin wall built alongside an external wall of a building leaving a ½ in. to 1 in. (ca. 12–25 mm) cavity between, which is later filled with waterproofing material to form a vertical damp-proof course.

retentivity (*Mag.*). Residual magnetic induction after removal of field producing saturation induction.

rete ovarii (*Zool.*). In female Mammals, a structure homologous with the rete testis of the male, joining the epoöphoron to the medullary cords.

rete testis (*Zool.*). In Mammals, a network of intercommunicating vessels within the mediastinum.

reticul-, reticulo-. Prefix from L. *reticulum*, net.

reticular (*Med., Zool.*). Resembling a net; of or pertaining to the reticuloendothelial system.
reticular activating system (*Psychol.*). An area of the brain stem having an *arousal* function, the destruction of which results in coma. Afferent nerves going to the cortex have collateral connexions with the reticular activating system which also has direct connexions to and from the cortex. See **arousal.**
reticular tissue (*Zool.*). A form of connective tissue in which the intercellular matrix is replaced by lymph; it derives its name from the network of collagenous fibres which it shows.
reticulated (*Build.*). A term applied to a variety of rusticated work distinguished by irregularly shaped sinkings separated by narrow margins of regular width.
reticule (*Surv.*). A cell carrying cross-hairs and fitting into the diaphragm of a surveying telescope. Also called a **graticule.**
reticulin (*Chem., Zool.*). A collagen, containing phosphorus, occurring in reticular fibrous tissues.
reticulitis (*Vet.*). Inflammation of the reticulum.
reticulocyte (*Zool.*). A variety of erythrocyte having a granular or reticular appearance; usually occurring in small numbers, but more numerous in young animals and in adults after haemorrhage.
reticulocytosis (*Med.*). An increase in the number of reticulocytes in the blood, as in pernicious anaemia.
reticuloendothelial system (*Zool.*). A system of special phagocytic cells which show a special affinity for certain colloidal dyes (such as pyrrhol blue), fine suspended particles, and lipoid matter introduced into the circulation.
reticulum (*Zool.*). In ruminant Mammals, the second division of the stomach, or *honeycomb bag*; any netlike structure. *adj.* **reticular.**
retiform tissue (*Zool.*). See **reticular tissue.**
retina (*Zool.*). The light-sensitive layer of the eye in all animals. Human retina contains 2 types of sensitive element: *rod* and *cone* (qq.v.). *adj.* **retinal.**
retinaculum (*Zool.*). In *Cirripedia*, a small hooked process which retains the egg-sac; in *Collembola*, the partially fused, rudimentary appendages of the third abdominal somite which retain the furcula in position; in *Lepidoptera*, a locking mechanism of the fore-wing which retains the frenulum in position.
retinal fatigue (*Physiol.*). Retention of images after removal of excitation, due to chemical changes in the retina.
retinal illumination (*Optics*). The luminous flux received by unit area of the retina. It equals

$$KLS/l^2,$$

where K=a transmission factor of the eye, L=the luminance of a uniformly diffusing surface, S=pupil area of eye, l=distance of retina from the second nodal point.
retine (*Med.*). Substance found in thymus, muscles, and big blood vessels, which tends to stop the growth of cancer cells, and may make a developed cancer regress. See also **promine.**
retinene (*Chem.*). Vitamin A_1-aldehyde, a component of rhodopsin. The retinene liberated by bleaching rhodopsin is the all-*trans* form, but the form which recondenses with opsin has the *cis* configuration about the 11,12 bond.
retinerved (*Bot.*). Net-veined.
retinite (*Min.*). A large group of resins, characterized by the absence of succinic acid.
retinitis (*Med.*). Inflammation of the retina.
retinitis pigmentosa (*Med.*). A familial and hereditary disease in which chronic and progressive degeneration of the choroid occurs in both eyes, with progressive loss of vision.
retinoblastoma (*Med.*). A tumour of the retina composed of small round cells, arising from embryonic retinal cells; it is locally destructive and forms metastases.
retinochoroiditis (*Med.*). Inflammation of the retina and of the choroid.
retinophore (*Zool.*). One of the crystal cells or vitrellae of an ommatidium.
retinoscopy (*Med.*). Skiascopy; shadow test. A method of estimating the refractive state of the eye by reflecting light on to it from a mirror and observing the movement of the shadow across the pupil.
retinulae (*Zool.*). In *Arthropoda*, the visual cells of the compound eye, forming the base of each ommatidium.
retort (*Fuels*). A closed chamber of refractory material, heated by producer gas on the outside, in which charges of coal are carbonized. Retorts are mostly of the continuous vertical type, the other types being horizontal retorts and intermittent vertical retorts. (*Met.*) A metallic or refractory vessel used in the distillation of metals, e.g., in extracting zinc from its ores, and in eliminating mercury from gold amalgams.
retort bench (*Fuels*). See **bench.**
retrace (*Radar, TV*). See **flyback.**
retractable landing gear (*Aero.*). Strictly, an alighting gear unit which can be withdrawn from its operative position so as to reduce drag, but in practice all the units collectively.
retractable radiator (*Aero.*). A liquid cooler for an aero-engine, capable of being withdrawn out of the airstream, for reducing drag and controlling the temperature of the cooling liquid.
retractile (*Zool.*). Capable of being withdrawn, as the claws of most *Felidae*.
retraction lock (*Aero.*). A device preventing inadvertent retraction of the landing gear while an aircraft is on the ground. Also **ground safety lock.**
retractor (*Zool.*). A muscle which by its contraction draws a limb or a part of the body towards the body. Cf. *protractor*.
retransfer (*Print.*). See **transfer.**
retread (*Autos.*). A reconditioned tyre, on which only the tread has been renewed. Cf. **remould.**
retreating systems (*Mining*). Systems in which the removal of ore or coal is commenced from the boundary of the deposit, which is then worked towards the entry through undisturbed rock.
retree (*Paper*). Slightly damaged paper from reams. It is marked XX in the U.K. and R in the U.S., and is invoiced lower than *good*.
retro-. Prefix from L. *retro*, backwards, behind.
retroaction (*Telecomm.*). See **reaction.**
retrobulbar neuritis (*Med.*). Inflammation of that part of the optic nerve behind the eyeball.
retrocaecal, retrocecal (*Med.*). Behind the caecum.
retrocerebral glands (*Zool.*). In Insects, a collective name applied to a number of endocrine glands in the head, behind the brain, which are concerned with the coordination of postembryonic development and metamorphosis. See **corpora cardiaca, corpora allata.**
retroflexed (*Med.*). Said of the uterus when its body is bent back on the cervix. Cf. **retroversion.** *n.* **retroflexion.**
retrograde metamorphism (*Geol.*). See **retrogressive metamorphism.**
retrograde motion (*Astron.*). (1) Motion of a comet (or satellite) whose orbit is inclined more than 90° to the ecliptic (or to the planet's equatorial plane). (2) Apparent motion of a planet from east to west among the stars,

caused by a combination of its true motion with that of the earth.

retrograde vernier (*Surv.*). A vernier in which n divisions on the vernier plate correspond in length to $(n+1)$ divisions on the main scale.

retrogression (*Zool.*). Degeneration; the assumption of features characteristic of lower forms.

retrogressive (or retrograde) metamorphism (*Geol.*). A term descriptive of those changes which are involved in the conversion of a rock of high metamorphic grade to one of lower grade, through the advent of metamorphic processes less intense than those which determined the original mineral content and texture of the rock.

retromorphosis (*Zool.*). Tendency to degeneration during development.

retroperitoneal (*Med.*). Situated or occurring behind the peritoneum.

retropharyngeal (*Med.*). Situated or occurring in the tissues behind the pharynx.

retropulsion (*Med.*). The running backwards of a patient with paralysis agitans or Parkinsonism; the patient's centre of gravity is displaced backwards, the rigidity of his posture making it difficult for him to recover his balance.

retrorocket (*Space*). A small rocket motor used for reducing the velocity of a space vehicle in landing, or in any manoeuvre calling for a thrust in the direction opposite to the motion.

retroserrate (*Bot.*). Having marginal teeth strongly directed backwards. (*Zool.*) Having backwardly directed teeth.

retroversion (*Med.*). The abnormal displacement backwards of the uterus, with or without *retroflexion.*

retting or **rotting** (*Textiles*). Soaking flax straw or jute in ponds, canals, tanks, etc., for bacteria to soften the woody tissue to enable the fibres to be extracted by scutching (beating).

return (*Radar*). Refers to radar reflections, e.g., land (or ground) return, sea return.

return airway or **aircourse** (*Mining*). One leading foul air away from the mine workings to the upcast shaft.

return bead (*Join.*). A double-quirk bead formed on the exterior angle of a timber.

return crank (*Eng.*). A short crank which replaces an eccentric in the *Walschaert's valve gear* (q.v.) on outside cylinder locomotives. It is fixed to the outer end of the main crank pin.

return electrons (*Electronics*). Those electrons which, having impinged on the fluorescent screen, are on their way back to the anode.

return feeder (*Elec. Eng.*). See **negative feeder.**

return-flow system (*Aero.*). A gas-turbine combustion system in which the air is turned through 180° so that it emerges in the opposite direction to that in which it entered; sometimes **reverse-flow system.**

return-flow wind tunnel (*Aero.*). One in which the air is circulated round a closed loop to preserve its momentum and so reduce the power requirement.

returning charge (*Met., Min. Proc.*). In custom smelting, that imposed by the smelter per unit of mineral treated. It may be modified by penalties or premiums if the ore or concentrate varies from a specified composition.

return line flyback (*Electronics*). Faint trace formed on the screen of a cathode-ray tube by the beam during the flyback period. Usually suppressed. Also **return trace.**

return loss (*Telecomm.*). See under **reflection.**

return trace (*Electronics*). See **return line flyback.**

return wall (*Civ. Eng.*). A short length of wall built perpendicularly to one end of a longer wall.

retuse (*Bot.*). Having a bluntly rounded apex with a central notch.

Retzius' fibre cells (*Zool.*). Thin rigid nucleated cells occurring between the hair cells of the cochlea.

Reuben-Mallory cell (*Chem.*). Small robust primary cell, of very level discharge characteristic, having a zinc anode and a (red) mercuric oxide cathode, which also depolarizes. Made in minute sizes for hearing-aids, internal radio transmitters, watches.

Reuleaux valve diagram (*Eng.*). See **valve diagram.**

revalé (*Build.*) Said of a cornice, moulding, etc., finished when the work is in position.

reveal (*Build.*). The depth of wall revealed, beyond the frame, in the sides of a door or window opening. Also called **respond,** and (Scottish) **ingo** or **ingoing.**

revehent (*Zool.*). Carrying back.

reverberation absorption coefficient (*Phys.*). That of a large plane uniform surface when the incident sound wave is of random intensity and direction, as in the reverberant field in an enclosure.

reverberation bridge (*Acous.*). Method of measuring the reverberation time in an enclosure; the rate of decay of the sound intensity is balanced against the adjusted and known decay of the discharge of a capacitance through a resistance.

reverberation chamber (*Acous.*). One with the minimum acoustic absorption for acoustic absorption measurements, or for adding echo effects in sound reproduction. Also **echo chamber.**

reverberation period (*Acous.*). See **reverberation time.**

reverberation response (*Acous.*). That of a microphone for reverberant sound, i.e., for the simultaneous arrival of sound waves of random phase, magnitude and direction. Substantially equal to the mean-spherical response at each frequency of interest.

reverberation response curve (*Acous.*). That of a microphone to reverberant sound waves. Plotted with the response in decibels as ordinates, on a logarithmic frequency base.

reverberation time (or period) (*Acous.*). The time, in seconds, required for the decay of the average sound intensity in an auditorium over an amplitude range of 1 million, or 60 dB, there being no emission of sound power during this decay.

reverberatory furnace (*Met.*). A furnace in which the charge is melted on a shallow hearth by flame passing above the charge and heating a low roof. Firing may be with coal, pulverized coal, oil, or gas. Much of the heating is done by radiation from the roof. Has many applications.

reversal colour film (*Photog.*). Film in which the negative image in the respective colour layers is reversed in processing to give a positive transparency.

reversal of control (*Aero.*). Reversal of a control moment (or couple) which occurs when displacement of the control surface results in such high forces that distortion of the main structure counteracts the effect of the surface. This overloading is a function of air speed, since control forces increase proportionately to the square of the velocity, and *reversal speed* is the lowest *E.A.S.* (q.v.) at which reversal occurs.

reversal of spectrum lines (*Light*). The appearance of a line as a broad, diffuse bright line with a narrow dark line down the centre. The effect is caused by cool vapour surrounding a hot source such as an electric arc, which produces a narrow absorption line on the short range of continuous spectrum given by the same vapour, at a high temperature, at the centre of the arc. Only certain lines are thus affected.

reversal speed (*Aero.*). See **reversal of control.**

reverse (*Build.*). A *template* (q.v.); so called from the fact that it is cut to the reverse profile of the work for which it is to be used as a reference.

reverse compatibility (*TV*). Exhibition of an acceptable black-and-white (monochrome) picture on a TV receiver designed to accept and exhibit colour pictures.

reverse coupling (*Telecomm.*). See **negative feedback.**

reverse (or inverse) current (*Electronics*). That in an electrode opposite to that intended, e.g., that in a triode grid because of ionization.

reverse-current relay (*Elec. Eng.*). A relay for use in electric circuits, arranged to operate when the current is in the opposite direction to normal.

reverse curve (*Surv.*). A curve composed of 2 arcs, of the same or different radii, having their centres on opposite sides of the curve. Also **S-curve.**

reversed (*Print.*). Said of any printed image (lettering, line-drawing, etc.) which appears, e.g., white on black. (*Zool.*) Inverted, as a spiral shell with a sinistral coil instead of a dextral coil, or vice versa.

reversed arch (*Civ. Eng.*). Same as *inverted arch* (q.v.).

reversed cleavage (*Zool.*). See **spiral cleavage.**

reversed drainage (*Geol.*). A phenomenon associated with river-capture, manifested by the flowing of a stream in a direction contrary to that which would be normally consistent with the existing geological structure.

reversed fault (*Geol.*). A type of *fault* (q.v.) in which compression has forced the strata on the side towards which the fracture is inclined to over-ride the strata on the downthrow side. Cf. *normal fault*.

reversed image (*TV*). See **lateral inversion, negative image.**

reversed jeanette (*Textiles*). See **jeanette.**

reversed ogee (*Arch.*). A *cyma reversa* (q.v.).

reverse-flow system (*Aero.*). See **return-flow system.**

reverse grid current (*Electronics*). That which flows away from the grid of a thermionic tube through the external circuit to the cathode.

reverse indention (*Typog.*). See **hanging indention.**

reverse mutation (*Gen.*). Change of a mutant gene back to its original form.

reverse-pitch airscrew (*Aero.*). See **airscrew.**

reverse plating (*Textiles*). Arrangements on flat or circular machines to effect the transfer of threads of different kinds or colours from back to front of the fabric to form the desired pattern or effect. See **plated fabrics.**

Reverse Polish Notation (*Maths.*). On some electronic calculators, the procedure whereby a number is entered, followed by an instruction for operating it on the number already held in the machine. Abbrev. **RPN.**

reverser (*Elec. Eng.*). On traction vehicles, the switch used for altering the connexions of the traction motors in order to reverse the direction of running.

reverse reaction (*Elec. Eng.*). That which opposes self-oscillation; used to neutralize the effects of interelectrode capacitance coupling.

reverse recovery time (*Electronics*). That for the reverse current or voltage in a semiconductor diode to reach a specified value after the instantaneous switching from a steady forward current to a reverse bias in a particular circuit.

reverse voltage (*Electronics*). That applied to a valve or semiconductor device in a normally nonconducting direction.

reversible absorption current (*Elec. Eng.*). That which decreases with time much less rapidly than the 'geometrical' capacitor charging current, normally exponential; returned on short-circuiting the plates.

reversible cell (*Elec. Eng.*). See **accumulator.**

reversible colloid (*Chem.*). See **lyophilic colloid.**

reversible potentiometer-type field rheostat (*Elec. Eng.*). A potentiometer-type rheostat used for controlling the field current of an electric machine and arranged so that the field current may be reversed.

reversible reaction (*Chem.*). A chemical reaction which can occur in both directions, and which is therefore incomplete, a mixture of reactants and reaction products being obtained, unless the equilibrium is disturbed by removing one of the products as rapidly as it is formed. Examples of reversible reactions are the formation of an ester and water from an alcohol and an acid, the dissociation of vapours, e.g., ammonium chloride, and the ionic dissociation of electrolytes. Also **equilibrium reaction.**

reversible saturation-adiabatic process (*Meteor.*). An idealized, alternating condensation-evaporation process occurring in the atmosphere, assuming that none of the condensation products are removed by precipitation.

reversible transducer (*Telecomm.*). One for which the loss is independent of the direction of transmission.

reversible unit (*Print.*). A printing unit on a web-fed press which can print in either direction of rotation, having reversible drives.

reversing commutator (*Elec. Eng.*). Any form of reversing switch, but more particularly the type in which the contacts form two halves of a cylinder upon which bear two brushes.

reversing face (*Surv.*). The process of transiting a theodolite telescope, thereby changing its position from face left to face right, or vice versa.

reversing field (*Elec. Eng.*). In a commutator machine, a field of opposite polarity to that in which an armature coil had previously been moving; designed to produce a reversed e.m.f. to assist commutation. The field may be produced by a compole, or by shifting the brushes from the neutral axis.

reversing gear (*Eng.*). See **Joy's valve-gear, link motion, Walschaert's valve-gear.**

reversing layer (*Astron.*). The name given to the lower part of the sun's chromosphere where the absorption lines of the solar spectrum are formed by 'reversal' from bright emission lines to dark absorption lines.

reversing mill (*Met.*). A type of rolling-mill in which the stock being rolled passes backwards and forwards between the same pair of rolls, which are reversed between passes. See **continuous mill, pull-over mill, three-high mill.**

reversing switch (*Elec. Eng.*). A switch used for reversing the connexions in an electrical circuit.

reversion (*Gen.*). See **atavism.**

revertive control system (*Teleph.*). A register-controlled system of automatic switching in which a selector, when set in motion by a positioning signal from a register, transmits progress signals back to the register to enable it to determine when the selector has reached the desired position.

revertose (*Chem.*). A disaccharide obtained by the action of maltase on D-glucose.

revetment (*Civ. Eng.*). A retaining wall.

revise (*Typog.*). A second or third proof supplied

in order that corrections made on the preceding proof may be checked over; to prepare and submit such a proof.

reviver (*Typog.*). A type alloy rich in tin and antimony added to the remelting pot in a prescribed quantity to maintain the formulation of the type metal by making good the loss incurred during each use.

revolute (*Bot.*). Rolled backwards and, usually, downwards, as the apex or margins of a leaf.

revolution (*Astron.*). The term generally reserved for orbital motion, as of the earth about the sun, as distinct from *rotation* (q.v.) about an axis. (*Geol.*) A period (usually regarded as of relatively short duration) of intense change in the disposition of sea and land, and of the surface configuration. The chief revolutions in the geological history of Britain were the *Caledonian, Armorican*, and *Alpine* (qq.v.).

revolving boilers (*Paper*). Large vessels in which rags, wood pulp, etc., are digested. Noncellulose portions are loosened and removed.

revolving centre (*Eng.*). A *centre* (q.v.), which is mounted on rolling bearings and revolves with the workpiece so as to obviate relative motion and thus friction between centre and workpiece.

revolving flats (*Spinning*). Endless chains on either side of a flat card. These hold metal bars over the top of the main cylinder. Equipped with fine steel wire teeth and set to extremely fine limits, these exert a combing action on the fibres on the cylinder teeth.

reward (*An. Behav.*). See **positive reinforcing stimulus.**

rewrite (*Comp.*). To return data to a store when it has been erased during reading.

Rexilite (*Build.*). TN for a bituminous roofing felt.

Reynolds number (*Chem. Eng., etc.*). The dimensionless group

$$\frac{\text{density} \times \text{velocity} \times \text{a linear measure}}{\text{viscosity}}$$

all values being in the same system of units, e.g. FPS or SI. If used in pipe or tube flow linear dimension is internal diameter and this has special application in heat exchange calculations. In other applications, e.g., mixers in vessels, the linear dimension may be, for example, the diameter of moving part.

R_F (*Chem.*). Term used in chromatography. The ratio of the distance moved by a particular solute to that moved by the solvent front.

RF (*Radio*). Abbrev. for *radio frequency*.

R.F.S. (*Build.*). Abbrev. for *render, float, and set.*

R_H, r_H, rH (*Chem.*). See **rH value.**

Rh (*Chem.*). The symbol for *rhodium*. (*Med.*) See **Rh (Rhesus) factor.**

Rhabditata (*Zool.*). An order of *Phasmidia* (*Nematoda*). Small, mainly free-living worms, the oesophagus having one or two bulbs, and the mouth either having six papillae or being simple.

rhabdites (*Zool.*). Small rodlike bodies, of doubtful function, secreted by certain cells of the epidermis or parenchyma in *Turbellaria.*

rhabditiform (*Zool.*). Of *Nematoda*, having a short straight oesophagus with a double bulb.

Rhabditis-form (*Zool.*). A free-living sexual stage in the life history of certain parasitic *Nematoda*, e.g., *Rhabdonema.*

rhabditoid (*Zool.*). Resembling *Rhabditis*; said of a type of nematode larva in which the buccal cavity resembles that of *Rhabditis* (i.e., is narrow with parallel sides).

Rhabdocoela (*Zool.*). An order of *Turbellaria* in which the gut, which may or may not have lateral pouches, is straight and the mouth is near the anterior end. Occur in fresh and salt water, the marine forms being very small.

rhabdolith (*Zool.*). In some *Protozoa*, a small calcareous skeletal rod.

rhabdom (*Zool.*). In the compound eyes of *Arthropoda*, a rodlike refractive body secreted by the retinulae and forming part of each ommatidium.

rhabdomere (*Zool.*). One of the constituent portions of a rhabdom, secreted by a single visual cell.

rhabdomyoma (*Med.*). A tumour composed of voluntary or striped muscle fibres.

rhabdomyosarcoma (*Med.*). A malignant rhabdomyoma.

rhachi-, rhachio-. See **rachi-.**

rhachis (*Bot., Zool.*). See **rachis.**

Rhachitomi (*Zool.*). An order of the *Labyrinthodontia* (Amphibians), being the most plentiful and typical members of this group. Found in the Permian and Triassic, but may have arisen as early as Lower Carboniferous. Vertebrae with a semi-lunar wedge-like intercentrum, one or two posterior pleurocentra, and a vertical neural arch. The large depressed skull had large interpterygoid cavities and an otic notch. Superficially crocodilian, but probably mainly terrestrial, e.g., *Eryops.*

rhachitomy (*Zool.*). See **temnospondyly.**

Rhaetic (*Geol.*). The uppermost stage of the Triassic system. In Britain, Rhaetic sediments pass upwards into Jurassic rocks and differ in character from the underlying Triassic, and have therefore been treated as Jurassic by some.

rhagades (*Med.*). Ulcerated fissures or cracks in the skin, especially those at the angles of the mouth occurring in congenital syphilis.

rhagadiose (*Bot.*). Deeply marked by cracks or fissures.

rhagon (*Zool.*). A form of Sponge colony in the shape of a flattened pyramid attached by its base, opening at the apex by an osculum, and having flagellated chambers in the upper wall only.

rhamnazin (*Chem.*). 3,5,4′-Trihydroxy-7,3′-dimethoxyflavone, a dye occurring in nature in the form of glucosides.

rhamnetin (*Chem.*). 3,5,3′,4′-Tetrahydroxy-7-methoxyflavone, a dye occurring in nature in the form of glucosides.

rhamnose (*Chem.*). 6-deoxy mannose $CH_3 \cdot (CHOH)_4 \cdot CHO$, a methylpentose obtained from several glucosides; crystallizes with 1 H_2O, m.p. 93°C. On distillation with sulphuric acid it yields 5-methyl-furfuraldehyde.

rhamphotheca (*Zool.*). In Birds, the horny coverings ensheathing the upper and lower jaws.

rhaphe. See **raphe.**

Rheiformes (*Zool.*). An order of *Palaeognathae*, containing two species of large running bird found on the South American pampas, occupying an ecological niche approximately similar to that of the emu. Have 3 toes.

rhenic(VII)acid (*Chem.*). See **rhenium oxides.**

rhenium (*Chem.*). A metallic element in the subgroup manganese, rhenium, technetium. Symbol Re, at. no. 75, r.a.m. 186·2, rel. d. 21, m.p. 3000°C. Valencies 2, 3, 4, 6, 7. A very rare element, occurring in molybdenum ores. A small percentage increases the electrical resistance of tungsten. Used in high-temperature thermocouples.

rhenium oxides (*Chem.*). Re_2O_7, ReO_3, ReO_2, and Re_2O_3. The volatile (VII)oxide Re_2O_7 is formed when the metal or its compounds are heated in air. Rhenium (VI)oxide is the anhydride of rhenic acid H_2ReO_4. The (VII)oxide dissolves in water to form perrhenic acid, which forms metallic perrhenates (rhenates(VII)). Cf. *permanganates*.

Rhenish bricks (*Build.*). Very light bricks made of calcareous material bound together with dolomitic lime. Also called **floating bricks**.

rheo-. Prefix from Gk., meaning current, flow.

rheobase (*Physiol.*). The minimal electrical stimulus (volts) that will produce a physiological response and below which no stimulus will excite however long it is maintained; it is half the strength of that for *chronaxie*.

rheokinesis (*An. Behav.*). A *kinesis* (q.v.) occurring in response to the intensity of water currents.

rheolaveur (*Min. Proc.*). Coal-cleaning plant where raw coal is sluiced through a series of troughs, the high-ash fraction gravitating down to a separate discharge.

rheology (*Phys.*). The science of flow of matter. The critical study of elasticity, viscosity, and plasticity.

rheomorphism (*Geol.*). Flowage of rocks resulting from severe deformation, especially applied to those undergoing partial melting as a result of heating to a high temperature.

rheophily (*Ecol.*). A preference for situations where there is a water current. See also **rheotaxis**.

rheoreceptors (*Zool.*). Receptors of Fish and certain Amphibians which respond to stimulus of water current, e.g., lateral line system.

rheostat (*Elec. Eng.*). Electric component in which resistance introduced into a circuit is readily variable by a knob or handle, or by mechanical means such as an electric motor.

rheostriction (*Nuc. Eng.*). See **pinch effect**.

rheotaxis (*An. Behav.*). A *taxis* (q.v.) occurring in response to the direction of water currents.

rheotron (*Nuc. Eng.*). Same as **betatron**.

Rhesus factor (*Med.*). See **Rh factor**.

Rheum (*Bot.*). The rhubarb genus.

rheumatism (*Med.*). A general term for a wide range of diseases characterized by painful inflammation and degeneration particularly of joints and muscles. Acute rheumatic fever is characterized by multiple arthritis and heart complications both immediate and long-delayed. It is related to streptococcal infection.

rheumatoid arthritis (*Med.*). A disease characterized by inflammation and swelling of several joints (especially the small joints of the extremities) affecting the periarticular tissues, synovial membranes, and joint cartilages. Its cause is obscure, but complex immunological disturbances are present.

rhexigenic (*Bot., Zool.*). See **lysigenic**.

rhexis (*Med.*). Rupture of a bodily structure, especially of a blood vessel.

Rh (Rhesus) factor (*Med.*). So-called because first discovered by the use of serum from guinea-pigs immunized to the blood of Rhesus monkeys. A group of weakly antigenic agglutinogens in human red blood cells, inherited according to Mendelian laws. About 85% of British and American people are classed by tests against sera containing the appropriate agglutinins as 'Rh-positive' and 15% as 'Rh-negative', but each of these groups can be subdivided into at least four subgroups. An Rh-negative person, by repeated treatment with Rh-positive blood, may develop antibodies, so that further injection of Rh-positive blood produces agglutination. The Rh factors are important in transfusion. An Rh-negative woman, carrying an Rh-positive foetus, may become 'sensitized' and the mother's antibodies reacting on the child may cause *haemolytic disease* (q.v.) or even miscarriage.

rhesus monkey (*Zool.*). One of a species, *Macaca mulatta*, of the macaque monkeys, native to SE Asia. Robust and intelligent, they are widely used in medical (see, e.g., **Rh factor**) and space research.

rhin-, rhino-. Prefix from Gk. *rhis*, gen. *rhinos*, nose.

rhinal (*Zool.*). Pertaining to the nose.

rhinarium (*Zool.*). In Mammals, the moist skin around the nostrils, also known as the muzzle, which is lacking in Anthropoids.

rhinencephalon (*Zool.*). The olfactory lobes of the brain in Vertebrates.

rhinitis (*Med.*). Inflammation of nasal mucous membrane. For *allergic rhinitis*, see **hay fever**.

rhinocoele (*Zool.*). The cavity of the rhinencephalon; olfactory ventricle of the Craniate brain.

rhinopharyngitis (*Med.*). Inflammation of the nose and of the pharynx.

rhinophore (*Zool.*). In certain *Mollusca*, an olfactory organ, usually borne on the tentacles, consisting of a patch of sensory epithelium which is sometimes developed into a pit with folded walls.

rhinophyma (*Med.*). Overgrowth of the subcutaneous tissue and the skin of the nose as a result of chronic vascular congestion, associated with dyspepsia.

rhinoplasty (*Surg.*). The repair of a deformed, diseased, or wounded nose by plastic surgery.

rhinorrhoea, rhinorrhea (*Med.*). Discharge of mucus from the nose.

rhinoscleroma (*Med.*). A disease, occurring in eastern Europe, in which hard granulomatous swellings appear in the nose, pharynx, and larynx, the characteristic feature of the swellings being the presence of large, round, clear cells, often containing bacilli.

rhinoscope (*Med.*). A speculum for viewing the interior of the nose.

rhinosporidiosis (*Med.*). A chronic fungus disease of the nasal and ocular mucous membranes of man, cattle, and horses, caused by *Rhinosporidium seeberi*.

rhinotheca (*Zool.*). The horny sheath enclosing the upper part of the beak of a Bird.

rhinotomy (*Surg.*). Incision into the nose.

rhipidate (*Zool.*). Fan-shaped.

Rhipidistia (*Zool.*). An order of *Crossopterygii*, thought to have become extinct some 50 000 000 years ago, until the discovery of the Coelacanth, *Latimeria*, off the east coast of S. Africa, in 1938.

rhipidium (*Bot.*). A fan-shaped cymose inflorescence.

rhipidostichous (*Zool.*). Having the peripheral somactids of the fin arranged in a fanlike manner. Cf. *orthostichous*, *rachiostichous*.

rhizanthous (*Bot.*). Apparently forming flowers from the root.

rhizine (*Bot.*). The rhizoid of a lichen; it may be single or it may be a strand of rhizoids.

Rhizobiaceae (*Bacteriol.*). A family of bacteria belonging to the order *Eubacteriales* (Bergey classification). Includes the symbiont *Rhizobium* which is important in nitrogen fixation by leguminous plants.

rhizocaline (*Bot.*). A hypothetical root-forming substance.

rhizocarpic, rhizocarpous (*Bot.*). Producing flowers underground as well as in the normal position.

rhizocaul (*Zool.*). In hydroid colonies, the hydrorhizae or creeping stolons.

Rhizocephala (*Zool.*). An order of *Cirripedia* containing forms which are parasitic, almost always on decapod *Crustacea*, and sometimes causing parasitic castration. They never have an alimentary canal, and the adults have neither appendages nor segmentation. After a free living nauplius stage the *cypris larvae* initially attach to their host by an antennule, and the adults are attached by a stalk from which roots proceed into the host's tissue, e.g., *Sacculina*.

rhizocorm (*Bot.*). A stout fleshy rhizome.

rhizodermis (*Bot.*). See **piliferous layer.**

rhizogenic, rhizogenetic (*Bot.*). Producing roots.

rhizoid (*Bot.*). (1) A short fungal hypha serving to attach the fungus to the substratum, and to collect nutritive material. (2) A short hairlike organ formed by liverworts, mosses, and fern prothalli, serving the same functions as above.

Rhizomastigina (*Zool.*). An order of *Zoomastigina*, the members of which have one or two flagella and numerous pseudopodia.

rhizome (*Bot.*). An underground stem consisting of more than 1 year's growth, usually lying horizontally in the soil, having a superficial resemblance to a root, but bearing scale leaves and one or more buds.

rhizomorph (*Bot.*). A densely packed strand of fungal hyphae, looking like a root.

rhizophagous (*Zool.*). Root-eating.

rhizophilous (*Bot.*). Growing on roots.

rhizoplast (*Bot.*). A very delicate thread running between the centrosome and the blepharoplast. (*Zool.*) In *Mastigophora*, a thread, or bunch of threads, connecting the basal granule to the nucleus or to the parabasal body.

Rhizopoda (*Zool.*). See **Sarcodina.**

rhizopodia (*Zool.*). Rootlike pseudopodia which branch and anastomose.

rhizosphere (*Bot.*). The region in the soil surrounding the root system of a plant, affected by its excretions and characterized by considerable microbiological activity.

Rhizostomeae (*Zool.*). An order of *Discomedusae* (*Scyphozoa*) whose medusae have no tentacles, and whose manubrium is branched, forming sucking mouths.

r.h.m. (*Radiol.*). Abbrev. for *röntgen-hour-metre*.

rhod-, rhodo-. Prefix from Gk. *rhodon*, rose.

rhodamines (*Chem.*). Dyestuffs of the triphenylmethane group, closely related to fluorescein. They are obtained by the condensation of phthalic anhydride with *N*-alkylated *m*-aminophenols in the presence of sulphuric acid.

rhodanizing (*Met.*). The process of electroplating with rhodium, especially on silver, to prevent tarnishing.

rhodeose (*Chem.*). $C_6H_{12}O_5$, a methylpentose sugar, an isomer of rhamnose.

rhodinal (*Chem.*). See **citronellal.**

rhodium (*Chem.*). A metallic element of the platinum group. Symbol Rh, at. no. 45, r.a.m. 102·9055, rel. d. at 20°C 12·1, m.p. ca. 2000°C; electrical resistivity approx. $5{\cdot}1 \times 10^{-8}$ ohm metres. A noble silvery-white metal, it resembles platinum and is alloyed with the latter to form positive wire of the platinum-rhodium-platinum thermocouple. Used for plating silver and silverplate to prevent tarnishing, and in catalysts and alloys for high-temperature thermocouples.

rhodochrosite (*Min.*). Carbonate of manganese which crystallizes in the trigonal system, occurring as rose-pink rhombohedral crystals. It is a minor ore of manganese. Also called **manganese spar or dialogite.**

rhodonite (*Min.*). Metasilicate of manganese, generally with some iron and calcium, crystallizing in the triclinic system. It is rose-coloured, and is sometimes used as an ornamental stone.

rhodophane (*Zool.*). A coloured oily substance, globules of which are found in the cones of Birds and in parts of the retina in some other forms.

Rhodophyta, Rhodophyceae (*Bot.*). A phylum or class of (marine) algae, distinguished by the prevalence of a red colour masking the chlorophyll. None is known to have motile reproductive organs. Simpler forms are filamentous; larger forms have complicated **thalli.** Red seaweeds.

rhodoplast (*Bot.*). Chromatophore of red algae.

rhodopsin (*Chem.*). A purple pigment found in the rods of the retina and required for the appreciation of low intensities of light, e.g., for night vision. It is a conjugated protein consisting of opsin and retinene, and these components are liberated when rhodopsin is bleached by bright light. Lack of rhodopsin, due to some congenital defect or to vitamin A deficiency, causes night blindness (*nyctalopia*).

rhodosporous (*Bot.*). Having pink spores.

rhombencephalon (*Zool.*). See **hind-brain.**

rhombic antenna (*Radio*). Directional antenna comprising an equilateral parallelogram of conductors, each several quarter-wavelengths long, usually arranged in a horizontal plane. The wires are connected through a resistance at one apex at the end of the longer diagonal, and to the transmitting or receiving apparatus at the other end. The maximum directive effect is along the longer diagonal; used for *musa* (q.v.).

rhombic dodecahedron (*Crystal.*). A crystal form of the cubic system, consisting of 12 exactly similar faces, each of which is a regular rhombus. Does not occur in the orthorhombic system, in spite of its name. See **pyritohedron.**

rhombic system (*Crystal.*). See **orthorhombic-.**

rhombohedral class (*Crystal.*). A class of the trigonal system, a characteristic form being the *rhombohedron* (q.v.), which is exhibited by crystals of quartz, calcite, dolomite, etc.

rhombohedron (*Crystal.*). A crystal form of the trigonal system, bounded by 6 similar faces, each a rhombus or parallelogram.

rhomboideus (*Zool.*). In higher Vertebrates, a muscle connecting the scapula with the spinal column.

rhomb-porphyry (*Geol.*). A medium-grained rock of intermediate composition, usually occurring in dykes and other minor intrusions; characterized by numerous phenocrysts of anorthoclase which are rhomb-shaped in cross-section, set in a finer-grained groundmass. Related to laurvikite among the coarse-grained, and to kenyte among the fine-grained rocks.

rhom brick (*Build.*). A brick made in the shape of a parallelogram to facilitate bonding.

rhomb-spar (*Min.*). An old-fashioned synonym for *dolomite*.

rhombus (*Maths.*). A quadrilateral with all its sides equal. Loosely, a diamond shape.

rho meson (*Nuc.*). Term formerly used to denote a meson which stopped in a nuclear emulsion without apparently any concomitant decay event or nuclear interaction.

Rhometal (*Met.*). An alloy of permalloy type. Contains 64% of iron and 36% of nickel. Used in high-frequency electrical circuits.

rhometer (*Met.*). One for the measurement of impurity content of molten metals by means of the variation in electrical conductivity.

rhonchus (*Med.*). A harsh, prolonged sound, heard on auscultation, produced by air passing over secretions in the bronchial tubes; indicative of bronchitis. *adj.* **rhonchal.**

rho-theta (*Nav.*). Navigational system which gives distance and bearing from a known point radio source.

rhumbatron (*Electronics*). Type of cavity resonator used, e.g., in klystron. It acts as a tuned circuit comprising a parallel-disk capacitor surrounded by a single-turn toroidal inductance, and is used to velocity-modulate an electron beam passing through holes in the capacitor disks. See **buncher.**

rhumb line (*Nav.*). Navigational line of constant direction, crossing successive meridians at the same angle.

rH value (*Chem.*). Logarithm, to the base 10, of the reciprocal of hydrogen pressure which would produce same electrode potential as that of a given oxidation-reduction system, in a solution of same *pH value* (q.v.). The greater the oxidizing power of a system, the greater the rH value.

rhyacolite (*Min.*). See **sanidine.**

rhyncho-. Prefix from Gk. *rhynchos*, beak, snout, proboscis.

Rhynchobdellida (*Zool.*). An order of *Hirudinea*, the members of which are all marine or freshwater forms parasitic on Snails, Fish, Reptiles and aquatic Birds; they lack botryoidal tissue and possess a protrusible proboscis without jaws; the blood is colourless.

Rhynchocephalia (*Zool.*). An order of the *Lepidosauria* with two temporal vacuities, a large parietal foramen which, in the one living form, *Sphenodon*, contains a nonfunctional median eye. Known as fossils from Middle Trias, *Sphenodon* survives in coastal islands off New Zealand. Vertebrae are amphicoelous and intercentra present, ribs single-headed with uncinate processes. Sternum and abdominal ribs are present, and teeth are acrodont. Lungs, heart and brain are like those of *Squamata.*

rhynchocoel (*Zool.*). In *Nemertea*, the cavity in which the proboscis lies.

rhynchodaeum (*Zool.*). In *Nemertea*, that part of the proboscis in front of the brain.

rhynchodont (*Zool.*). Having a toothed beak.

rhynchophorous (*Zool.*). Having a beak.

rhynchostome (*Zool.*). In *Nemertea*, the anterior terminal pore via which the proboscis is everted.

Rhynchota (*Zool.*). A superorder of the *Paraneoptera*, including the two orders *Homoptera* and *Heteroptera* (q.v.), previously included in the single order *Hemiptera.*

Rhynie Chert (*Geol.*). A silicified peaty bed containing well-preserved plant remains as well as spiders, scorpions, and insects; discovered at Rhynie, Aberdeenshire, in the Middle Old Red Sandstone (i.e., Devonian Age).

rhyolite (*Geol.*). General name for fine-grained igneous rocks having a similar chemical composition to granite, commonly occurring as lava flows, although occasionally as minor intrusions, and generally containing small phenocrysts of quartz and alkali-feldspar set in a glassy or cryptocrystalline groundmass. Sometimes called **liparite.** See also **nevadite, obsidian, pitchstone, pumice, pyromeride** (nodular rhyolite).

rhythmic crystallization (*Geol.*). A phenomenon exhibited by rocks of widely different composition but characterized by development of orbicular structure.

rhythmic sedimentation (*Geol.*). A regular interbanding of two or more types of sediment or sedimentary rocks due to a seasonal change in conditions of sedimentation, such as alternation of wet and dry periods. See, e.g., **varve clays.**

rhytidome (*Bot.*). An external covering to a plant member made up of alternating sheets of cork and dead cortex or dead phloem.

ria (*Geol.*). A normal valley drowned by a rise of sea-level relative to the land. Cf. *fiords*, in the production of which glacial action plays an essential part. A good example of a ria type of coastline is S.W. Ireland, the *rias* being long synclinal valleys lying between anticlinal ridges.

rib (*Aero.*). A fore-and-aft structural member of an *aerofoil* which has the primary purpose of maintaining the correct contour of the covering, but is usually also a stress-bearing component of the main structure. Ribs are usually set either parallel with the longitudinal axis or at right angles to the front spar; cf. *nose ribs.* (*Bot.*) One of the larger veins of a leaf. (*Build., Civ. Eng.*) (1) A curved member of a centre or ribbed arch. (2) A moulding projecting for purposes of ornamentation from a ceiling or vault surface. (3) The vertical portion of a T-beam. (*Textiles*) A prominent line running lengthwise across or diagonally in a fabric, and forming a cord effect, depending on the weave, e.g., weft rib, warp rib, and the arrangement of coarse yarns to fine. (*Zool.*) A small ridge or riblike structure; in Vertebrates, an element of the skeleton in the form of a curved rod connected at one end with a vertebra; it serves to support the body walls enclosing the viscera.

riband (*Carp.*). A flat rail fixed across posts in a palisade.

rib and panel (*Build.*). A term applied to a vault formed of separate ribs and panels, the latter being supported upon the former.

ribbed arch (*Civ. Eng.*). An arch composed of side-by-side ribs spanning the distance between the springings.

ribbed flutings (*Build.*). Flutings separated by a flat or slightly convex listel.

ribbon (*Cinema.*). Loop of stretched Duralumin in the light valve, which is opened by the passage of modulation currents.

ribbon-flame burner (*Heat*). A tubular gas-burner on which a ribbon of flame is produced by alternating corrugated and plain steel strips inserted in a milled slot, thus forming honeycombed flame ports, or by the tube being drilled with lines of very fine holes in close formation.

ribbonite (*Plumb.*). Soft lead in the form of slender ribbons, used for caulking pipe-joints.

ribbon microphone (*Acous.*). Same as **moving-conductor microphone.**

ribbon parachute (*Aero.*). A parachute in which the canopy is made from light webbing with spaces between, instead of conjoined fabric gores, so as to give greater strength against ripping for deployment at high speed. Commonly used for *brake parachutes* (q.v.).

ribbon ring (*I.C. Engs.*). A ring fitted round a piston, to retain a floating gudgeon pin without any other locking means.

ribbon saw (*Carp.*). A thin, narrow band-saw having a width of not more than 2 in. (50 mm).

ribbon strip (*Carp.*). A horizontal timber attached to vertical timbers as a support for joists. Also called **girt strip, ledger board.**

rib fabric (*Textiles*). A type of fabric in which the reverse-side stitches alternate in a vertical direction with the face stitches, forming vertical furrows and rows. This results in contraction of the fabric, but also renders it elastic and suitable for the extremities of garment parts.

rib mesh (*Build., Civ. Eng.*). See **expanded metal.**

Ribmet (*Build.*). A registered form of expanded

steel reinforcement used for concrete flooring, roofing, plaster ceilings, partitions, etc.

riboflavin (*Chem.*). See **vitamin B complex.**

ribonuclease (*Biochem.*). A family of enzymes in the pancreas, etc., which hydrolyse ribonucleic acids. They were the first enzymes to be synthesized.

ribose (*Chem.*). $C_5H_{10}O_5$, a pentose, a stereoisomer of arabinose. D-Ribose occurs in certain nucleic acids.

ribosome (*Biochem.*). Cytoplasmic particle containing ribonucleic acid. Most of the cell ribonucleic acid occurs in the ribosomes. They are regarded as the assembly units upon which polypeptide chains are synthesized to the specifications contained on the messenger RNA. See also **cell-free system, genetic code, nucleic acids, transcription.**

Riccati equation (*Maths.*). A differential equation of the form

$$\frac{dy}{dx} = py^2 + qy + r,$$

where *p, q, r* are functions of *x* alone.

Ricco's law (*Light*). That with steady illumination on the macular region of the retina, the condition that a luminous sensation may be just perceived is that the total quantity of light must reach a definite value.

rice paper (*Paper*). The finely cut pith of *Fatsia papyrifera*; not a true paper.

rice weaves (*Textiles*). Fancy weaves formed by breaking certain regular twills, e.g., the 6- or 8-end twills. These are broken in the middle and reversed.

Richardson-Dushman equation (*Electronics*). Original Richardson formula, as modified by Dushman, for the emission of electrons from a heated surface, current density being

$$I = AT^2e^{-\phi/kT}.$$

T is the absolute temperature, *A* is a material constant, *k* is Boltzmann's constant, with φ the work function of the surface and *e* the base of natural logarithms.

Richardson effect (*Electronics*). Same as **Edison effect.**

rich lime (*Build.*). *Fat lime* (q.v.).

rich mixture (*I.C. Engs.*). A combustible mixture in which the fuel is in excess of that physically correct for the air.

richmondite (*Min.*). A hydrated phosphate o aluminium, $AlPO_4.4H_2O$.

richterite (*Min.*). Hydrous metasilicate of sodium, calcium, magnesium, manganese, and iron occurring as monoclinic crystals in alkaline igneous rocks and in thermally metamorphosed limestones and skarns. A member of the amphibole group.

ricin (*Chem.*). An albumin occurring in the castor bean, with highly toxic effects due to agglutination of the erythrocytes.

ricinoleic acid (*Chem.*). An oily liquid, $CH_3(CH_2)_5 \cdot CHOH \cdot CH_2 \cdot CH{=}CH(CH_2)_7 \cdot COOH$, which, in glyceride form, is the chief constituent of castor oil.

rickers (*For.*). Round timber of less than 2½ in. (65 mm) diameter in the middle.

rickets (*Med.*). A nutritional disease of childhood characterized by defective ossification and softening of bones: due to deficiency of vitamin D and failure to absorb and utilize calcium salts. Also **rachitis.**

rickettsiae (*Med.*). Micro-organisms, intermediate between bacteria and viruses, found in the tissues of lice, ticks, mites and fleas and which can cause disease, e.g., typhus, when transmitted to man and other animals.

rictus (*Zool.*). Of Birds, the mouth aperture *adj.* **rictal.**

riddle. Strong coarse sieve used to size gravel, furnace clinker, etc. Large pieces removed by hand in riddling are called *knockings*, remaining on-screen material *middlings* and through passing particles *fells, undersize*, or *smalls.*

Rideal-Walker test (*Chem.*). Test for germicidal power of a disinfectant, carbolic acid (phenol) being taken as the standard. A series of dilutions of disinfectant is tested with a typhoid broth culture, samples being taken at short intervals and subjected to incubation.

rider (*Chem.*). A small piece of platinum wire used on a chemical balance as a final adjustment. (*Mining*) (1) A *horse*, i.e., mass of country rock occurring in a mineral deposit. (2) A thin seam of coal above a thick one. (3) A guide for a *bowk*, in sinking.

rider rollers (*Print.*). Steel rollers in the inking system of a printing machine which secure the inkers (and distributors), act as a necessary link between their tacky surfaces, and contribute to efficient distribution and inking.

rider's bone (*Med.*). Ossification at either end of one of the adductor muscles of the thigh, particularly following a riding injury.

rider shore (*Carp.*). An inclined baulk of timber used in a system of raking shores for a high building. It abuts at its lower end against a length of timber laid along the back of the outer raking shore, instead of against the ground.

ridge (*Build., Civ. Eng.*). The summit-line of a roof; the line on which the rafters meet. (*Meteor.*) An outward extension of the isobars from a centre of high pressure.

ridge-board (*Carp.*). A horizontal timber at the upper ends of the common rafters, which are nailed to it.

ridge capping (or covering) (*Build.*). The covering applied over a ridge to protect the intersection of the sloping roof surfaces.

ridge course (*Build.*). The last (i.e., the top) course of slates or tiles on a roof, cut to length as required.

ridge pole, ridge-piece (*Build., Civ. Eng.*). A timber member laid horizontally along the ridge of a roof.

ridger (*Agric.*). Mouldboards shaped to produce low, broad-based ridges. Primarily intended for shallow planting of potatoes.

ridge roll (*Build.*). A ridge piece, of rounded section, over which a lead flashing is secured as a covering.

ridge roof (*Build.*). A pitched roof whose sloping surfaces meet to form an apex or ridge.

ridge stop (*Build.*). A piece of sheet-lead shaped over the junction between a roof ridge and a wall; used in cases where the one runs into the other and a watertight joint has to be made.

ridge tile (*Build.*). A purpose-made tile specially shaped for use as a covering over the ridge of a roof.

ridging (*Agric.*). The process of forming ridges in surface soil so that as much as possible of the latter is exposed to weathering. (*Build.*) The operation of covering the ridge of a roof with specially shaped ridge tiles or other material.

riding lamps (*Aero.*). Lamps displayed at night by a float plane or flying boat when moored or at anchor. Colours and positions as in the Maritime code.

riding shore (*Carp.*). See **rider shore.**

riebeckite (*Min.*). Hydrous metasilicate of sodium and iron(II) found in alkaline igneous rocks as black monoclinic prismatic crystals. A member of the amphibole group: the blue asbestos *crocidolite* is a variety occurring in metamorphosed ironstones.

Riedel's disease (*Med.*). Chronic thyroiditis. A chronic inflammation of the thyroid gland, which becomes enlarged and hard as the result of excessive formation of dense fibrous tissue.

Riedel's lobe (*Med.*). An anomalous downward prolongation of the right lobe of the liver.

Rieke diagram (*Electronics*). Polar form of load impedance diagram representing the components of the complex reflection coefficient of the oscillator load in a microwave oscillator.

Riemann surface (*Maths.*). An extension of the complex plane whereby a multivalued function can be regarded as a one-valued function. The surface has to be designed for the particular function concerned, e.g., for the two-valued function $w = \pm\sqrt{z}$, the complex plane is extended by superimposing a similar plane above it, cutting both planes from the origin along the real axis to infinity, and joining the opposite edges of the two cuts crosswise. The result is that regaining a starting point by tracing a circle around the origin now takes two revolutions (argument increase of 4π) because both the original and the additional planes have to be covered. Thus effectively the complex plane has been doubled and the two-valued function $w = \pm\sqrt{z}$ can be laid out as a one-valued function of position. That portion of a multivalued function which lies on a single sheet of a Riemann surface is called a *branch of a function*, and the connecting point of the several sheets of the surface a *branch point*.

Riemann zeta function (*Maths.*). The function $\xi(z)$ of a complex variable defined by

$$\xi(z) = \sum_{n=1}^{\infty} \frac{1}{n^z}.$$

Riffel loudspeaker (*Acous.*). One in which the radiating element takes the form of two sheets curved to an edge, on which is located the driving element, the latter being a current-carrying conductor in the longitudinal gap of a magnet. The arrangement provides for radiation in a horizontal plane with restricted vertical radiation.

riffle (*Mining*). A groove in the bottom of a sluice, or a strip of wood fixed across a sluice, or on a Wilfley table, to catch heavy mineral.

riffler (*Eng.*). A file bent so as to be capable of operating in a shallow depression. (*Mining, Powder Tech.*) A device for dividing a stream of crushed material, e.g., coal, into truly representative samples. (*Paper*) See **sand trap**.

rift and grain (*Geol.*). The two directions, approximately at right angles to one another, along which granite and other massive igneous rocks can be split; rift being the easier of the two.

rift sawing (*For.*). See **quarter sawing**.

rift valley (*Geol.*). See **graben**.

Rift Valley fever (*Med.*). An infectious disease of cattle and sheep in Africa, characterized by high fever and hepatitis, and caused by a virus; probably transmitted by mosquitoes. Man is also susceptible.

rig (*Mining, etc.*). A well-boring plant, e.g., for oil. (*Textiles*) A woolsorting term—fleece divided up the middle to help in sorting. In wool cloth finishing, a machine in which the piece can be arranged selvedge to selvedge, for accurate folding and/or cutting.

riga last (*For.*). A unit of timber measure containing 80 ft^3 of squared or 65 ft^3 of round timber.

rigaree (*Glass*). A name given to a broken design produced by a small ridged metal wheel; applied to raised bands or collars of glass as decoration.

Rigden's permeameter (*Powder Tech.*). One in which the powder bed under test is connected across the two arms of a U-tube manometer when the liquid columns of the manometer are displaced with respect to each other. Air is allowed to flow through the powder bed to equalize the pressure. The time required for a specified movement of the manometer is used to calculate the surface area.

rigging (*Aero.*). (1) The operation of adjusting and aligning the various components, notably flight and engine control, of an aircraft. (2) In airships and balloons, the system of wires by which the weight to be lifted is distributed over the envelope or gas-bag.

rigging angle of incidence (*Aero.*). See **angle of incidence**.

rigging diagram (*Aero.*). The drawing giving the manufacturer's instructions as to the positioning and aligning of the components and control systems of an aeroplane.

rigging line (*Aero.*). See **shroud line**.

rigging position (*Aero.*). The position in which an aeroplane is set up in order to effect the adjustment and alignment of the various parts, i.e., with the lateral axis and an arbitrarily chosen longitudinal datum line horizontal.

right angle (*Maths.*). One quarter of a complete rotation. 90° or $\pi/2$ radians.

right-angled folding (*Bind.*). Each fold is at right angles to the preceding one. Cf. *parallel folding*.

right ascension (*Astron.*). The angle at the pole between the equinoctial colure and the hour circle through the body whose coordinates are being measured; also expressed as the arc of the equator from the first point of Aries to the foot of the body's hour circle, and measured in time from 0 to 24 hr in the eastwards direction. Abbrev. **R.A.**

right circular cone (*Maths.*). See **cone (2)**.

right circular cylinder (*Maths.*). See **cylinder (2)**.

right-handed engine (*Aero.*). An aero-engine in which the airscrew shaft rotates clockwise with the engine between the observer and the airscrew.

right-hand rule (*Elec. Eng.*). See **Fleming's rule**.

right-hand twine (*Textiles*). See **openband twine**.

right helicoid (*Maths.*). The surface generated by a straight line moving so that it always intersects a helix and cuts its axis perpendicularly. A right circular helicoid is like a spiral staircase.

right-reading, wrong-reading (*Print.*). When making plates the various printing processes have their particular requirements from the process camera or filmsetting system, e.g., right-reading negative for direct letterpress, wrong-reading positive for deep-etch offset; the direction is read from the emulsion side.

rigid arch (*Civ. Eng.*). A continuous arch without hinges or joints, the arch being rigidly fixed at the abutments.

Rigidex (*Plastics*). British TN for high density Ziegler-type polyeth(yl)ene.

rigid expanded polyurethane (*Plastics*). Material used in thermal insulation and in providing light structural reinforcement. Reaction products of diisocyanates with polyesters or polyethers in the presence of water.

rigidity (*Phys.*). See **elasticity of shear**.

rigidity, test for (*Paper*). A sample strip of paper or board is clamped vertically at one end and the amount of bending from the vertical measured.

rigid PVC (*Plastics*). PVC in its unplasticized form

used chiefly for corrosion-resistant applications such as chemical pipework.

rigid support (*Elec. Eng.*). A support for an overhead transmission line designed to withstand, without appreciable bending, a longitudinal load as well as transverse and vertical loads.

rigor (*Bot.*). An inert condition assumed by a plant when conditions for growth are unfavourable. (*Med.*) A sudden chill of the body, accompanied by a fit of shivering, which heralds the onset of fever. (*Zool.*) A state of rigidity and irresponsiveness into which some animals pass on being subjected to a sudden shock; it is the result of *reflex action* (q.v.) and is known popularly as 'shamming dead'.

rigor mortis (*Med.*). Stiffening of the body following upon death.

rill stoping (*Mining*). Overhand or upward stoping, in which ore is detached from above the miner so as to form an inverted stoped pyramid spreading from a winze at its apex, through which broken ore is withdrawn.

rim (*Bot.*). The overhanging part of a wall about a bordered pit.

rima (*Zool.*). A narrow cleft. *adjs.* **rimate, rimose, rimiform.**

rima glottidis (*Med.*). The gap in the larynx between vocal cords in front and arytenoid cartilages of the larynx behind.

rime (*Build.*). A rung of a ladder.

rim lock (*Join.*). A form of lock distinct from *mortise lock* (q.v.) in that, in its metal case, it is screwed to the face of the door.

rimming (or rimmed) steel (*Met.*). Steel that has not been completely deoxidized before casting. Gas is evolved during solidification and bubbles are entrapped. Ingots contain blowholes but no pipe. See also **killed steel.**

rim wheel (or pulley) (*Spinning*). Large pulley on the rim shaft of the mule. It transmits power to the tin roller driving the spindles, then reverses them to back-off the yarn and next revolves them at a very slow speed for winding-on during the inward run of the carriage.

Rinco process (*Print.*). Reproduction proofs taken with white ink on black paper and photographed to yield positives suitable for printing down, particularly for photogravure.

rind (*Bot.*). (1) The outer layers of the fruit body, or the sclerotium of a fungus. (2) The outer layers of the bark of a tree.

rinderpest (*Vet.*). Cattle plague. An acute, highly contagious and often fatal viral infection of cattle, sheep, and goats; characterized by fever and ulceration of the mucous membranes, especially of the alimentary tract, causing severe diarrhoea and discharges from the mouth, nose, and eyes.

ring (*Bot.*). See **annulus.** (*Maths.*) Let S be a set with two operations called *addition* and *multiplication.* This system is called a ring if (1) for all elements x, y in S, $x+y$ and xy belong to S; (2) the set S is an Abelian group with respect to the addition operation; (3) the multiplication is (i) associative, (ii) distributive with respect to the addition operation: e.g., the positive and negative multiples of 3 with 0, i.e. the set $\{\ldots -9, -6, -3, 0, 3, 6, 9, \ldots\}$, with the usual addition and multiplication of numbers. A ring which is such that all its nonzero elements form an Abelian group with respect to the multiplication operation is called a **field**: e.g., the rational numbers with the usual addition and multiplication of numbers. (*Teleph.*) The centre one of the three contacts on the terminating plugs of the flexible cords of an operator's cord circuit. See **R-wire.**

ring armature (*Elec. Eng.*). An electric-machine armature having a ring winding.

ring bark (*Bot.*). Bark which splits off in more or less complete rings.

ringbone (*Vet.*). An exostosis on the phalangeal bones of the horse's foot. *High ringbone* involves the first interphalangeal joint and *low ringbone* involves the second interphalangeal joint. *False ringbone* is an exostosis affecting the first or second phalanx but not involving a joint.

ringbound (*Vet.*). See **ringwomb.**

ring canal (*Zool.*). (1) The circular canal of a medusa which runs around the periphery of the umbrella. (2) A circular vessel of the water-vascular system in *Echinodermata.*

ring complex (*Geol.*). See **ring dyke.**

ring counter (*Telecomm.*). A number of counting circuits in complete series, for sequence operation in counting impulses.

ring course (*Build.*). The course farthest from the intrados of an arch.

ring current (*Astron.*). A term used for the intense electric current that is believed to flow in a westerly direction at a height of about 30 000 miles (50 000 km) above the earth; associated with the *Van Allen radiation belts* (q.v.), its existence is postulated to explain the anomalies in the earth's magnetic field.

ring doffer (*Spinning*). Junior operative who removes full bobbins from a ring frame, and places empty bobbins or tubes on the spindle so that spinning may begin on a new cycle.

ringdown (*Teleph.*). Method of signalling operator in wire telephony.

Ring drier (*Chem. Eng.*). A proprietary device for drying solids in which the wet solid is entrained in a hot air stream which travels in a circular duct itself arranged in a large diameter circle.

ring dyke (*Geol.*). An almost vertical intrusion of igneous rock which rose along a more or less cylindrical fault which had an approximately circular outcrop. In certain Tertiary instances (e.g., Ardnamurchan) several successive ring dykes, separated by 'screens' of *country rock* and approximately concentric, form *ring complexes.*

Ringelmann smoke chart. One of a series of six charts, numbered 0 to 5 and shaded from white to black, indicating a shade of grey against which density of smoke may be gauged.

ringent (*Bot.*). Said of a corolla consisting of two distinct, widely gaping lips.

ringers (*Geol.*). See under **Lingula Flags.**

ring filament (*Light*). The usual form taken by the filament of an electric filament lamp; the ring lies in a plane at right angles to the lamp axis.

ring fire (*Elec. Eng.*). Thin streaks of fire appearing round the commutator of an electric machine; due to small particles of copper or carbon which have become embedded in the mica between the commutator and are raised to incandescence by the current. Ring fire indicates that the commutator needs to be cleaned.

ring frame (*Spinning, etc.*). Continuous spinning machine which imparts twist to drafted roving by a tiny traveller mounted on a ring on the lifter rail. Used for cotton, woollen, and worsted yarns, the modern frame is a high-speed productive machine which has steadily increased in favour. Produces a very wide range of yarns for weaving or knitting purposes.

ring gauge (*Eng.*). A hardened steel ring having an internal diameter of specified size within very small limits of error; used to check the diameter of finished cylindrical work.

ring gland (*Zool.*). See **Weismann's ring.**
ringing (*Telecomm.*). Extended oscillation in a tuned circuit, at its natural frequency, continuing after an applied voltage or current has been shut off, dying away according to its *decay constant*, but running into the next oscillation. (*Teleph.*) The application of alternating current to a telephone circuit, causing the distant operation of apparatus which attracts the attention of operators or subscribers—in the former instance by operating a relay and lighting a lamp, and in the latter by ringing a bell. (*TV*) Multiple displaced black and white TV images arising from free oscillations in the system.
ringing key (*Teleph.*). The nonlocking lever key used by a telephone operator to put ringing current on a subscriber's line.
ring latch (*Join.*). A latch in which the fall bar is operated by a handle in the shape of a ring, pivoted at the top so that it always falls into the vertical position.
ring main (*Chem.*). Closed loop of piping through which chemicals in solution, or such finely divided materials as powdered coal, are circulated in suspension past suitable draw-off points. (*Elec. Eng.*) A domestic a.c. wiring system in which a number of outlet sockets are connected in parallel to a ring circuit which starts and finishes at a mains supply point. All plugs used in the power outlets are fitted with individual fuses.
ring modulator (*Elec. Eng.*). Four rectifying elements in complete series, which act as a switch, being fed with appropriate currents at the corners.
ring oscillator (*Elec. Eng.*). One in which a number of valves feed each other in a circle or circus and in which the frequency is determined by a ring cut from a quartz crystal, suspended at its nodes to minimize damping; used, e.g., as a standard time-keeper (*quartz-crystal clock*) at 10^5 Hz.
ring-porous (*Bot.*). Said of wood which contains more vessels, or larger vessels, in the spring wood than elsewhere, so that it is marked in cross-section by rings, or portions of rings, of small holes.
rings and brushes (*Crystal.*). Name applied to the patterns produced when convergent or divergent plane-polarized light, after passing through a doubly refracting crystal cut perpendicular to the optic axis, is examined by an analyser. See **interference figure.**
ring scaler (*Electronics*). One consisting of a ring of counting elements triggered in turn, the output pulse being from one of them to another ring. See **Dekatron.**
ring shake (*For.*). See **cup shake.**
ring shift (*Comp.*). See **end-around shift.**
ring size (*Min. Proc., Mining*). Description of rock too large for handling by screening, in accordance with the diameter of a ring which can be slipped over it.
ring slot parachute (*Aero.*). A parachute the canopy of which has slots all round its circumference, to give it greater stability.
rings of Saturn (*Astron.*). See **Saturn.**
ring spanner (*Tools*). One in the form of a notched ring in which diametrically opposite notches fit over the nut.
ring stress (*Mining*). That adjacent to the walls of an unsupported underground excavation.
ring traveller (*Spinning*). See **traveller.**
ring vessel (*Zool.*). A vessel in the form of a ring, as the vessel in the scolex of *Cestoda* which connects the longitudinal excretory trunks.
ring winding (*Elec. Eng.*). (1) A helical winding arranged on a ring of iron or other material. Also called a **toroidal winding.** (2) A form of armature winding in which the armature core is a hollow cylinder with each turn of the winding threaded through the centre.
ring-wire (*Teleph.*). See **R-wire.**
ringwomb (*Vet.*). Ringbound. Incomplete dilatation of the cervix at parturition in the ewe.
ringworm (*Med.*). Tinea. A contagious disease characterized by formation of ring-shaped patches on the skin; due to infection with moulds, especially of the three genera *Microsporon*, *Trichophyton*, and *Epidermophyton.*
Rinmann's green (*Chem.*). A pigment prepared from a cobalt(II) salt, a zinc compound and alumina by fusion. The green has a yellow tone and is much duller than cobalt blue.
riometer (*Phys.*). Apparatus for continuously measuring ionospheric absorption.
rip (*Carp.*). To saw timber along the direction of the grain.
R.I.P. (*Min. Proc.*). Abbrev. for **resin-in-pulp.**
Ripart and Petit's fluid (*Micros.*). A combined fixative and preservative for delicate organisms. It contains camphor water, glacial ethanoic acid, copper(II) ethanoate, and copper(II) chloride.
rip-bit (*Mining*). See **jackbit.**
ripcord (*Aero.*). (1) A cable used for opening the pack of a personnel parachute. (2) An emergency release for gas in an aerostat envelope.
ripening (*Photog.*). Process in the manufacture of emulsions in which the silver halide grains attain their ultimate size after being precipitated, either by coalescence or by Ostwald **ripening**, in which the larger grains grow at the expense of the smaller owing to difference in solubility.
ripidolite (*Min.*). A species of the chlorite group of minerals, crystallizing in the monoclinic system. It is essentially a hydrous silicate of magnesium and aluminium with iron(II). Also called **prochlorite.**
ripper (*Build.*). Slater's tool with a cranked handle and a long flat blade ending in an arrow-shaped head. Used for removing slates by cutting the fixing-nails.
ripping saw (*Carp.*). See **rip-saw.**
ripple (*Elec. Eng.*). The a.c. component in the output of a rectifier delivering d.c. reduced by series choke and shunting (smoothing) capacitor, or Zener diode. Measured as a percentage of the steady (average) current. (*Phys.*) Small wave on the surface of a liquid for which the controlling force is not gravity, as for large waves, but surface tension. The velocity of ripples diminishes with increasing wavelength, to a minimum value which for water is 23 cm/s for a wavelength of 1·7 cm.
ripple cloths (*Textiles*). Plain weave good quality all-wool, wool/viscose or cotton fabrics, piece-dyed, with the nap raised in a special wavy form. Used for dressing-gowns.
ripple control (*Elec. Eng.*). A method of controlling street lighting or other equipment from some central point by means of a high-frequency ripple superimposed on the current-carrying conductors of an electric power system.
ripple filter (*Elec. Eng.*). A low-pass filter which is designed to reduce the ripple current but at the same time permits the free passage of the d.c. current, e.g., from a rectifier. Also **smoothing circuit.**
ripple finish (*Paint.*). See **wrinkle finish.**
ripple frequency (*Elec. Eng.*). The frequency of the ripple current in rectifiers, etc. Usually double the supply frequency in a full-wave rectifier.
ripple generator (*Nuc.*). Thermoelectric generator

powered by heat from a radioactive source and suitable for long periods of maintenance-free operation on remote sites, e.g., for lighting marine navigational buoys.

ripple marks (*Geol.*). Undulating ridges and furrows found on the bedding planes of certain sedimentary rocks, due to the action of waves or currents of air or water on the sediments before they were consolidated. Such ripple and rill marks can be seen in the process of formation today on most sandy beaches, on sand dunes, and in deserts.

ripple trays (*Chem. Eng.*). Distillation column trays (or plates) which consist of thin metal formed into a series of parallel channels and perforated to provide space for rising vapour.

rippling (*Textiles*). Removal of seed, leaves, shive, and other impurities from flax by iron combs.

rip-rap (*Civ. Eng.*). A foundation formed in water, or on a bed of soft material, by depositing broken stones loosely.

rip-saw, ripping saw (*Carp.*). A saw designed for cutting timber along the grain.

rise (*Build., Civ. Eng.*). (1) The vertical distance from the centre of span of an arch in the line of the springings to the centre of the intrados. Also called the **versed sine.** (2) The vertical height from end supports to ridge of a roof. (3) The height of a step in a staircase. (*Mining*) See **raise.**

rise-and-fall pendant (*Elec. Eng.*). A pendant *luminaire* the height of which can be adjusted by a pulley and counter-weight or spring.

rise and fall system (*Surv.*). A system of reduction of levels in which the staff reading at each successive point after the first is compared with that preceding it, and the difference of level entered as a *rise* or a *fall*. See **collimation system.**

rise and run (*Carp.*). Term applied to the amount of any given slope quoted as a given *rise* (vertical distance) in a given *run* (horizontal distance).

risen moulding (*Join.*). A moulding decorating a panel and projecting beyond the general surface of the surrounding framing.

rise of floor line (*Ships*). A tangent to the curve of the *bilge* (q.v.) to meet the extremity of the flat of the keel.

rise of tide (*Ocean.*). The vertical distance of high-water level, at a given place, above a fixed datum (usually low water of ordinary spring tides at the place).

riser (*Build.*). The vertical part of a step. (*Elec. Eng.*) See **commutator lug.** (*Foundry*) In a mould, a passage up which the metal flows after filling the mould cavity. It allows dirt to escape, indicates that the mould is full, and supplies metal to compensate for contraction on solidification. Also called **out-gate.**

rise time (*Telecomm.*). Time for pulse signal in an amplifier or filter to rise to from 10 to 90% of the maximum amplitude. Also **build-up time.**

rising and falling saw (*Carp.*). A circular saw whose spindle can be moved in relation to the position of the working table, so that more or less of the saw can be made to project for cutting grooves of different depths.

rising and setting (*Astron.*). The positions of a heavenly body when it is exactly on the great circle of the observer's horizon, east or west of the meridian respectively.

rising arch (*Civ. Eng.*). An arch whose springing line is not horizontal.

rising butt hinge (*Join.*). A butt hinge with a loose leaf which, when opened, rises on the centre pin due to helical bearing surfaces on the two leaves. This enables a door, on opening, to rise above a carpet, and to close automatically.

rising front (*Photog.*). A sliding panel for carrying the lens in a field or technical camera; used to diminish foreground and avoid distortion of perspective when photographing high buildings, trees, etc.

rising mains (*Elec. Eng.*). A mains circuit which runs from one floor of a building to another. (*Plumb.*) A mains water supply where it enters premises.

rising shaft (*Mining*). A shaft which is excavated from below upward. Cf. *sinking*.

rising-sun magnetron (*Elec. Eng.*). One in which the resonances of cavities are of two magnitudes alternately.

rising type, rising spaces (*Typog.*). Faulty chases, furniture, typesetting, and make-up can cause the type to rise when locked up, and careless lock-up can cause it to rise no matter how good the equipment and make-up. This results in a vertical movement of the pages concerned at each impression and a tendency for spacing material to rise to surface level and print as a blemish. Precision equipment and careful workmanship reduce this trouble to a minimum; printing from plates eliminates it.

risus sardonicus (*Med.*). Wrinkling of the forehead and retraction of the angles of the mouth due to spasm of the facial muscles (as in tetanus), giving appearance of a grin.

Ritchie wedge (*Optics*). A photometer head in the form of a wedge with two white diffusing surfaces set 90° apart.

Ritter's disease (*Med.*). Dermatitis exfoliativa infantum. Severe pemphigus of infants, in which the horny layer of the skin separates from the underlying layer over wide areas.

Rittinger's law (*Chem. Eng.*). States that energy required in a crushing operation is directly proportional to the area of fresh surface produced, i.e., $E=k_r(1/d_2-1/d_1)$, where E is the energy used in crushing, k_r is a constant, depending on the characteristics of the material and on the type and method of operation of the crusher, and d_1 and d_2 are the average initial and final linear dimensions of the material crushed.

ritualization (*An. Behav.*). The evolutionary process by which behaviour patterns which originated as displacement reactions secondarily become *social releasers* (q.v.).

river capture (*Geol.*). The beheading of a stream by a neighbouring stream which has greater power of erosion.

rivers (*Typog.*). In widely spaced text matter the spaces in successive lines can form channels of space running down the page, a tendency which is reduced by close spacing. Also called **streets.**

rivers, geological work of (*Geol.*). This involves (*a*) corrasion (wearing away) of their banks and beds, largely through the sediment suspended in the water and pushed along their beds; (*b*) transportation of immense quantities of rock waste produced by the agents of denudation in the high ground drained by the rivers, and of salts carried in solution.

river wall (*Civ. Eng.*). A wall built as a side boundary to the flow of a river, thereby confining it to a definite path.

rivet (*Eng.*). A headed shank for making a permanent joint between two or more pieces. It is inserted in a hole which is made through the pieces, and 'closed' by forming a head on the projecting part of the shank by hammering or other means. The head may be rounded, flat, pan-shaped, or countersunk.

riveted joint (*Eng.*). A joint between plates, sheets,

or strips of materials (usually metallic) secured by rivets. See also **butt joint, lap joint.**

riveting machine (*Eng.*). A machine for closing, clinching, or setting rivets. See also **hydraulic riveter, pneumatic riveter.**

riving knife (*For.*). A knife used for splitting shingles, barrel staves, hurdle members, etc.

rivulose (*Bot.*). Marked with lines, giving an appearance of rivers as shown on a map.

R-meter (*Radiol.*). One calibrated in röntgens, incorporating an ionization chamber and indicating meter for radiation survey.

r.m.m. Abbrev. for **relative molecular mass.**

r.m.s. power (*Elec.*). The effective mean power level of an alternating electric supply. Abbrev. for **root-mean-square power.**

r.m.s. value (*Phys.*). Measure of any alternating waveform, the square root of the mean of the squares of continuous ordinates (e.g., voltage or current) through one complete cycle. If there are harmonics of the fundamental, the total r.m.s. value is the sum of the r.m.s. values of the fundamental and the harmonics taken separately. For a simple sinusoid, it is $1/\sqrt{2}$ times the peak value. Employed because the energy associated with any wave depends on its intensity, i.e., upon the square of the amplitude. Also **effective value.** Abbrev. for **root-mean-square value.**

Rn (*Chem.*). The symbol for *radon*.

RNA (*Biochem.*). See **nucleic acids.**

Roach Beds (*Geol.*). Beds of cavernous limestone in the Portland Stone; full of fossil casts.

road (*Civ. Eng.*). See **arterial-, by-pass-, county-, district-, trunk-; dual carriageway, motorway.**

road bed (*Rail.*). Foundation carrying the sleepers, rails, chairs, points, and crossings, etc., of a railway track.

road line composition (*Paint.*). A thermoplastic composition consisting of rosin, or a derivative, and mineral oil, filled with coarse and fine extenders and pigmented with up to 10% titanium dioxide. Used for marking white lines and Zebra crossings on roads.

road metal (*Civ. Eng.*). Broken stone for forming the surfaces of macadamized roads. Also called **metal, metalling.**

road studs (*Civ. Eng.*). Rubber or metal pads, the former generally incorporating small reflectors (cat's eyes), built in to the surface of a road to define traffic lanes at night or in fog. Generally laid at 30 ft (10 m) intervals on straight roads and at 12 ft (4 m) intervals on curves.

road surface (*Civ. Eng.*). See **wearing course.**

road tar (*Civ. Eng.*). See **gas tar.**

roak (*Met.*). A *seam* (q.v.).

roan (*Leather*). Sheepskin, dyed and finished with a smooth grain; used in bookbinding.

roaring (*Vet.*). A respiratory affection of horses in which an abnormal sound is produced in the larynx during inspiration; caused by paralysis of the laryngeal muscles or by inflammatory conditions affecting the larynx. *Whistling* is a similar condition in which a sound of higher pitch is produced.

roasting (*Met.*). The operation of heating sulphide ores in air to convert to oxide. Sometimes the sulphur-bearing gases produced are used to make sulphuric acid. See **chloridizing-, sulphating-, sweet-.**

roasting furnace (*Met.*). A furnace in which finely ground ores and concentrates are roasted to eliminate sulphur. Part or all of the necessary heat may be provided by the burning sulphur. The essential feature is free access of air to the charge. This is done by having a shallow bed which is continually rabbled. Many types have been devised; *multiple-hearth* is most widely used.

robalo (*Zool.*). An American pike-like fish (Centropomus), related to sea-perches.

robbings (*Textiles*). Wool fibres, longer than noil, removed during combing and re-used in blends.

Robinson bridge (*Elec. Eng.*). An a.c. bridge used for the measurement and control of frequency. Also **Robinson-Wien bridge.**

Robinson cement (*Build.*). A slow- but hard-setting plaster used for interior work. It is fire-resisting and does not expand or contract in setting.

Robinson direction-finder (*Nav.*). Rotating-loop direction-finding system having an auxiliary loop, which can be reversed in series with the main loop and so permits an audible signal to be balanced by switching, instead of nodalizing signal and noise.

Roboting (*Electronics*). TN for system operating driverless trolleys, which follow a cable energized by a route setter.

Rob Roy (*Typog.*). See **Multifont type tray.**

roburite (*Chem.*). Chlorinated dinitrobenzine with ammonium nitrate. A flameless explosive.

Rochelle salt (*Chem.*). Crystal of sodium potassium tartrate, having strong piezoelectric properties, but high damping. Used as a *bimorph* in microphones, loudspeakers, and pick-ups. Its disadvantage is the limited range of piezoelectric property (−18°C to 23°C) and high temperature coefficient.

roche moutonnée (*Geol.*). A mound of bare rock which is usually smoothed on the upstream side and roughened by plucking on the downstream side, as a result of a moving ice-sheet.

Roche's limit (*Astron.*). The least distance from a planet at which a liquid satellite could revolve without being disrupted by tidal forces. Roche's value is 2·44 times the planet's radius.

rock (*Geol.*). Any of the solid materials making up the Earth. As used by geologists, the term also includes unconsolidated material such as sand, mud, clay and peat, in addition to the harder materials described as rock in conventional usage.

rockallite (*Geol.*). An alkaline igneous rock composed of aegirine, quartz, albite and microcline, with traces of the rare zirconium silicate, elpidite. Named after the Atlantic island of Rockall.

rock burst (*Mining*). Sudden failure of stope pillars, walls or other rock buttresses adjacent to underground works, with explosively violent disintegration.

rock crystal (*Min.*). The name given to colourless quartz whether in distinct crystals or not; particularly applicable to quartz of the quality formerly used in making lenses.

rock drill (*Civ. Eng.*). A tool specially adapted to the boring of holes through rock.

rocker (*Mining*). A short and easily portable rocking trough or cradle for washing concentrates, gold-bearing sand, etc.

rocker arms (*I.C. Engs.*). Pivoted levers operated by pushrods or overhead camshaft, which carry the tappets of an overhead valve system. Also called **valve rockers.**

rocker gear (*Elec. Eng.*). The hand-wheel or other device for moving a brush-rocker.

rocket (*Electronics*). Slotted-line voltage standing-wave ratio measuring system with end cones, to match slotted line to external circuits. (*Space*) (1) A missile projected by a rocket system. (2) A system, or a vehicle, powered by *reaction* or *rocket propulsion* (qq.v.).

rocket propulsion (*Aero.*). *Reaction propulsion* (q.v.) using internally stored, instead of atmospheric, oxygen for combustion. Used primarily where there is insufficient oxygen, e.g., above 70 000 ft (20 km), or where the lightness and compactness of the motor offset the high *propellant* (q.v.) weight, e.g., for assisted take-off and missiles.

rocket tester (*San. Eng.*). A rocket giving off dense smoke; used to test a drain for leaks.

rock face (*Build.*). The form of face given to a building-stone which has been *quarry-faced* (q.v.).

Rock fever (*Med.*). *Undulant fever* (1, q.v.).

rock flour (*Geol.*). A term used for finely comminuted rock material found at the base of glaciers and ice-sheets. It is mudlike and is composed largely of unweathered mineral particles.

rock-forming minerals (*Geol.*). The minerals which occur as dominant constituents of igneous rocks, including quartz, feldspars, feldspathoids, micas, amphiboles, pyroxenes, and olivine.

rocking bar (*Horol.*). A pivoted bar carrying the intermediate wheels in a keyless mechanism.

rock meal (*Min.*). A white and light variety of calcium carbonate, resembling cotton; it becomes a powder on the slightest pressure.

rock milk (*Min.*). A very soft white variety of calcium carbonate which breaks easily in the fingers; it is sometimes deposited in caverns or about sources holding lime in solution.

rockoon (*Space*). A technique for exploring the upper atmosphere by firing small rockets at great heights from balloons of the *Skyhook* type; serious air drag in the lower atmosphere is thus avoided.

rock phosphate (*Min.*). See **phosphorite.**

rock roses (*Min.*). See **desert rose.**

rock salt (*Min.*). Halite.

rock soap (*Min.*). Steatite.

rock wall failure (*Civ. Eng., Mining*). Collapse due to one or more of the following: rock fall, simple dropping; rock flow, slope failure; plane shear, failure along weakness plane; rotational shear, stress where soil has dropped leaving a rounded cavity.

Rockwell hardness test (*Met.*). A method of determining the hardness of metals by indenting them with a hard steel ball or a diamond cone under a specified load and measuring the depth of penetration. See **Brinell hardness test.**

rock wool (*Met.*). Fibrous insulating material made by blowing steam through molten slag. Also called **mineral wool, slag wool.**

Rocky Mountain fever (*Med.*). Blue disease; black fever; spotted fever. A disease, more or less limited to certain states of the U.S., characterized by fever, headache, muscular pains, enlargement of the spleen, and a macular eruption on the skin; the disease is spread by a tick and is associated with infection by a *rickettsia* organism.

Rocky Point effect (*Electronics*). Of obscure origin, occasionally occurring in high-voltage transmitting valves; characterized by a transient but violent discharge between the electrodes, not always damaging the valve. First observed at Rocky Point radio station. Also called **flash arc.**

rococo (*Arch.*). Style of exaggerated ornamentation prevalent in Europe in 17th and 18th centuries; florid and ornate decoration characterized by curved lines imitating shells, foliage, and scrolls together.

rod (*Build.*). A unit of brickwork which measures $16\frac{1}{2}$ ft (= 1 lineal rod ≈ 5 m) by $16\frac{1}{2}$ ft by $1\frac{1}{2}$ bricks thick = 306 ft³ or $11\frac{1}{3}$ yd³. It contains about 4500 bricks plus about 75 ft³ of mortar. Since 1 rod ≈ 5 m, 1 rod of brickwork ≈ 25 m² $1\frac{1}{2}$ bricks thick, which is approx. $8\frac{2}{3}$ m³. (*Nuc. Eng.*) A rod-shaped reactor fuel element or sample, intended for irradiation in reactor. (*Zool.*) One of the noncolour-sensitive light perceptive elements on periphery of retina. Rods respond to lower illumination levels than *cones* (see cone).

rodding (*Build.*). Operation of clearing a stoppage in a pipe by inserting a rod to break down the obstruction and remove it; descaling of encrusted pipework with scrapers attached to jointed rods.

rodding eye (*Build.*). The removable cover on a small opening in a vertical or horizontal drain pipe which enables any obstruction to be removed by drain rods.

Rodentia (*Zool.*). An order of the cohort *Glires*, numerically the largest of all the Mammalian groups. Generally small animals with never more than a single pair of chisel-shaped upper incisors which have open roots. Lower incisors can move like scissors as there is no anterior symphysis between the mandibles. Canines are never present, but a wide diastema between the gnawing incisors and the grinding cheek teeth which vary in numbers and frequently have persistently open roots. Glenoid cavities are elongated antero-posteriorly, the lower jaw being moved forwards for gnawing and backwards for grinding. Jaw muscles are greatly enlarged. Mostly unguiculate, pentadactyl, and plantigrade, clavicle usually present, herbivorous with a large caecum. Brain little, if at all, convoluted. They are almost universally distributed, and are of considerable economic and medical importance as pests of stored food and carriers of plague fleas. Adaptive, radiation wide, terrestrial, amphibian, burrowing, arboreal, gliding, saltatorial. Divided into three ill-defined suborders: *Sciuromorpha*—Squirrels, Beavers, and Gophers; *Myomorpha*—Jerboas, Voles, Rats, Mice, and Hamsters; and *Hystricomorpha*—Porcupines, Guinea Pigs, and Agoutis.

rodent ulcer (*Med.*). A slow-growing ulcerating cancer of the skin which usually affects the upper part of the face; arises from the basal cells of the skin and is of low malignancy.

rod epithelium (*Zool.*). A form of epithelial tissue in which cells appear to be striated.

rod fibre (*Zool.*). The fibre with which a retinal rod is connected internally.

rod gap (*Elec. Eng.*). A spark gap in which electrodes are two coaxial rods whose ends (between which discharge occurs) are cut perpendicular to the axis.

Rodinal (*Photog.*). Alkaline solution of 4-aminophenol containing a little sodium sulphite; used as a photographic developer.

rodman (*Surv.*). A staffman (U.S.).

rod mill (*Min. Proc.*). Sturdily constructed cylindrical grinding mill in which heavy metal rods are used instead of steel for the purpose of pulverizing rock.

roe number (*Paper*). Bleachability of the raw fibres indicated by the amount of bleach required to produce a good colour and measured by the amount of chlorine gas consumed by the sample.

roentgen (*Radiol.*). See **röntgen.**

rogue (*Gen.*). A *sport* (q.v.).

Rok (*Build.*). A weatherproof roof-covering material, made from wool-felt saturated with an elastic waterproofing compound and coated with a natural bitumen.

Rokalba (*Build.*). Asbestos-surfaced Rok.

roke (*Met.*). A *seam* (q.v.).

Rokitansky's tumour (*Med.*). Compound hydrops folliculorum. An ovarian tumour formed by the aggregation of a number of Graafian follicles distended with fluid (see **hydrops folliculi**).

Rolf (*Mining*). Automatic electronically controlled coal-cutting machine.

roll (*Aero.*). Aerobatic manoeuvre consisting of a complete revolution about the longitudinal axis. In a *slow roll* the centreline of the aircraft follows closely along a horizontal straight line; an *upward roll* is similar, but considerable height is gained; a *hesitation roll* is one where the pilot brings his aircraft momentarily to rest in its rolling motion. A vertical upward or downward roll is usually called an *aileron turn* because these are the only control surfaces involved. A *flick roll* is an entirely different, very rapid and violent manoeuvre in which the aircraft makes its revolution along a helical path; high structural stresses are imposed and many countries ban this aerobatic. A *half-roll* is lateral rotation through 180°. (*Bind.*) Tool used to impress designs on leather book covers. Usually brass, set in a long handle, which is necessary because of the pressure required in its operation. (*Plumb.*) A joint between the edges of two lead sheets on the flat, the edges being overlapped over a 2-in. (50 mm) diameter wood roll fastened to the surface to be covered with lead. (*Ships*) The phenomenon of a ship's behaviour in waves, wherein she changes her angle of heel. See **rolling period.**

roll boiling (*Textiles*). An old process for producing a permanent lustre on fine woollen fabrics, after raising, by steeping them vertically, wound on a wooden or perforated metal cylinder, in a tank of cold water, to which steam is admitted. See **crabbing.**

roll-capped (*Build.*). Said of ridge tiles which are finished with a roll or cylindrical projection along the apex.

roll damper (*Aero.*). See **damper.**

rolled gold (*Met.*). Composite sheet made by soldering or welding a sheet of gold on to both sides of a thicker sheet of silver, or other suitable metal, and rolling the whole down to the thickness required.

rolled laminated tube (*Plastics*). Tube produced by winding, under heat, pressure, and tension, a synthetic resin impregnated or coated fabric or paper on to a former.

rolled steel joist (*Eng.*). Abbrev. **R.S.J.** See **H-beam.**

rolled-steel sections (*Eng.*). Steel bars rolled into I-shaped, channel, T, angle, cruciform, or similar cross-sections for different applications in structural work, each section being made in graded standardized sizes. See also **H-beam.**

roller (*Acous.*). In organ mechanism, a cylindrical rod for transmitting motion by rotation, the forces being applied at the ends of radial pegs. (*Cinema.*) See **drag-, pad-.** (*Paint.*) A hollow cylinder covered with absorbent material (e.g., lambswool, foamed rubber), revolving on an axle and fitted with a handle; used to apply paint to flat surfaces. (*Print.*) See **rollers.**

roller-bearing (*Eng.*). A shaft-bearing consisting of inner and outer steel races between which a number of parallel or tapered steel rollers are located by a cage; suits heavier loads than the *ball-bearing* (q.v.).

roller chain (*Eng.*). A driving or transmission chain in which the links consist of rollers and sideplates, the rollers being mounted on pins which connect the sideplates. See **driving chain.**

roller coating (*Paint.*). An industrial process for coating endless strips of metal with paint, etc., on a coating machine. A cylindrical roller picks up the paint from a trough, and distributes it over one or more rollers, from the last of which it is picked up by the strip.

roller conveyor (*Eng.*). A *conveyor* (q.v.) comprising a series of closely spaced rollers set so that their crests can support articles to be conveyed. They may be power-driven, or freely revolving. Tapered ones are used to form bends in the conveyor line.

roller delivery motion (*Spinning*). Mechanism for driving the rollers of a mule so as to deliver a short length of drafted roving which takes up some of the twist as the carriage makes its inward run.

roller mill (*Eng.*). In its simplest form, consists of two rolls of suitable material, mounted with their axes horizontal, running in opposite directions; used for crushing and mixing operations. (*Min. Proc.*) *Chilian mill* (q.v.).

rollers (*Print.*). Means by which ink on a printing machine is both distributed into a thin film and applied to the printing surface. In letterpress printing various modern plastic materials are used in addition to the long-established *composition rollers* (q.v.).

roll feed (*Eng.*). A mechanism incorporating nipping rollers for feeding strip material to power presses. The rollers turn intermittently by a set distance and are timed in relation to the press so as to advance the required length of strip, correctly timed, for each press stroke.

roll film (*Photog.*). Film, wound on a spool, with opaque paper fitting closely to the end-cheeks of the spool, excluding light sufficiently to allow loading and unloading in daylight.

roll forging (*Eng.*). A hot forging process for reducing or tapering short lengths of bar stock in the manufacture of axles, chisels, knife blades, tapered tubes, etc. Parts are placed by tongues into profiled grooves in pairs of forging rollers.

roll forming (*Eng.*). A cold forming process in which a series of mating rolls progressively forms strip metal into tube or section, usually of complex shape, at high rolling speeds. Often associated with a continuous resistance welding machine.

rolling (*Aero.*). The angular motion of an aircraft tending to set up a rotation about a longitudinal axis. One complete revolution is called a *roll* (q.v.). (*Ships*) See **rolling period.**

rolling bearing (*Eng.*). A ball-bearing, roller-bearing, needle bearing, or other type of bearing in which elements roll between load-bearing surfaces.

rolling instability (*Aero.*). See **lateral instability.**

rolling lift bridge (*Eng.*). A type of bascule bridge in which the bascule or cantilever part has, at the shore end, a surface of segmental profile rolling on a flat bearing.

rolling load (*Eng.*). See **moving load.**

rolling locker (*Textiles*). A traverse net machine having a double tier of carriages which swing in an arc by means of reciprocating rollers with which they engage.

rolling mills (*Met.*). Sets of rolls used in rolling metals into numerous intermediate and final shapes, e.g., blooms, billets, slabs, rails, bars, rods, sections, plates, sheets, and strip.

rolling moment (*Aero.*). The component of the couple about the longitudinal axis acting on an aircraft in flight.

rolling period (*Ships*). The movement of a ship in waves about a longitudinal horizontal axis is known as *rolling*. The rolling period is the time taken for a complete roll from the upright position first to one side then to the other and back to upright again. Cf. *pitching period*.

rolling resistance (*Eng.*). When a wheel rolls on a surface, distortion of the contact surfaces due to the normal force between them destroys the ideal line contact. This introduces a force with a component, called the rolling resistance, opposing the motion.

rolling stock (*Rail.*). A general and collective term for all coaches, trucks, etc., which run along a railway track.

roll-off frequency (*Radio*). Frequency where response of an amplifier or filter is 3 dB below maximum.

roll out (*Comp.*). Method of reading number stored in register of computer, by adding digits to each column in turn until this reverts to zero.

rolls (*Min. Proc.*). Crushing rolls are pairs of horizontally mounted cylinders, faced with manganese steel, between which ore is crushed as they rotate inward.

ROM (*Comp.*). Abbrev. for *read only memory*.

roman (*Typog.*). Ordinary upright type, as distinct from *italic*, or sloping.

Roman cement (*Civ. Eng.*). A natural hydraulic cement made by calcining nodules found in London Clay which contain about 66% chalk, 25% clay, and some iron(II)oxide. It was much used for underwater work but is now largely superseded by Portland cement, whose ultimate strength is 2 or 3 times as great. Also called **Parker's cement.**

Roman mosaic (*Build.*). See **tessellated pavement.**

Romberg's sign (*Med.*). Sign present when a patient, standing with feet close together and eyes closed, sways more than when his eyes are open; it indicates disease of the sensory tracts in the spinal cord (as in tabes dorsalis).

roméite (*Min.*). Naturally occurring antimonite of calcium, sometimes with manganese and iron, crystallizes in the cubic system, often as brownish octahedra.

rone (*Plumb.*). In Scotland, an *eaves gutter* (q.v.). Also **rhone.**

ronnel (*Chem.*). See **fenchlorphos.**

röntgen (*Radiol.*). Unit of X-ray or gamma dose, for which the resulting ionization liberates a charge of each sign of $2{\cdot}58 \times 10^{-4}$ coulombs per kilogram of air. Symbol R.

röntgen equivalent man (*Radiol.*). Unit of biological dose given by the product of the absorbed dose in rads and the relative biological efficiency of the radiation. Abbrev. **rem.**

röntgen equivalent physical (*Radiol.*). Obsolete unit of absorbed dose equal to that received by water from an exposure dose of 1 röntgen. This was taken as 93 erg/gm. Hence, 1 rep $= 0{\cdot}93$ rad. Abbrev. **rep.**

röntgen-hour-metre (*Radiol.*). Unit of γ-ray source strength, producing 1 röntgen/hr at 1 m distance. Abbrev. **r.h.m.**

röntgenology (*Radiol.*). Same as **radiology.**

röntgen rays (*Phys.*). Electromagnetic rays of wavelengths of the order of 0·01 to 1 nanometres, between much longer (ultraviolet) and much shorter (γ-rays) wavelengths. Generated by high-voltage electrons hitting a target (usually tungsten) in an evacuated tube. *X-rays.*

roof antenna (*Radio*). See **flat-top antenna.**

roof boards (*Carp.*). Boards laid on a roof to provide a foundation and undercovering to the covering materials proper, e.g., slates, tiles.

roof bolting (*Mining*). Method of roof support in which steel bolts are inserted in drill holes so as to pin supporting steel beams under the roof of a stope.

roofers (*Carp.*). *Roof boards* (q.v.).

roof guard (*Build.*). A device fitted to a roof to prevent snow from sliding off it.

roofing slate (*Geol.*). A term widely applied to rocks of fine grain in which regional metamorphism has developed a good slaty cleavage. Cf. *Collyweston Slate, Stonesfield Slate.*

roof pendant (*Geol.*). A mass of country rock projecting downwards, below the general level of the roof, into an intrusive rock body.

roof tree (*Carp.*). A *ridge-board* (q.v.).

roof truss or **roof** (*Build., Civ. Eng.*). The structural framework built to support the roof covering for a building.

room-and-pillar (*Mining*). **See bord-and-pillar.**

room index (*Light*). The coefficient of utilization of lamps in a room depends on the shape of the room. This shape is expressed by the room index, k. For rectangular rooms

$$k = \frac{0{\cdot}9w + 0{\cdot}1 0l}{h}$$

where w = width of room, l = length of room, h = height of room.

room noise (*Acous.*). See **ambient noise.**

root cap (*Bot.*). A hollow cone of cells covering the growing tip of a root and protecting the meristematic cells from damage as the tip is pushed through the soil.

root diameter (*Eng.*). Of a gear-wheel, the diameter at the bottom of the tooth spaces.

rooter (*Electronics*). Arrangement of electronics designed to obtain an output amplitude which is proportional to the square root of the input amplitude. Required in compressors for reducing contrast in sound reproduction.

root fork (*Agric.*). Implement used with trip-operated loaders. The tines have rounded tips to avoid damage to crops such as sugar beets, turnips, and mangolds.

root hair (*Bot.*). A tubular outgrowth from a superficial cell of a young root, serving for absorption of water and mineral salts from soil. Its cavity is continuous with that of the cell from which the root hair springs. See **piliferous layer.**

root locus (*Elec. Eng.*). Locus for the roots of the closed-loop response for a system, derived by plotting the poles and zeros of the open-loop response in the complex plane. Used in the study of system stability.

root-mean-square (r.m.s.). Square root of sum of squares of individual observations divided by total number of observations.

root-mean-square power (*Elec.*). See **r.m.s. power.**

root-mean-square value (*Phys.*). See **r.m.s. value.**

root pressure (*Bot.*). A pressure sometimes demonstrable in roots, and shown by the exudation of fluid that occurs when the stem is cut just above ground level. Also known as **bleeding pressure.**

Roots blower (*Eng.*). An air compressor for delivering large volumes at relatively low pressure ratios; in it a pair of hour-glass-shaped members rotate with a small clearance within a casing, no valves being required.

rootstock (*Bot.*). See **rhizome.**

root tuber (*Bot.*). A swollen root containing reserve food material.

root tubercle (*Bot.*). A swelling on a root, inhabited by symbiotic bacteria.

rope (*Radar*). See **window.**

rope brake (*Eng.*). An absorption dynamometer consisting of a rope encircling a brake drum or flywheel, one end of the rope being loaded by weights and the other supported by a spring balance. The effective torque absorbed is obtained by multiplying the drum radius by the difference of the tensions.

rope-scouring machine (*Textiles*). Machine for scouring woollen, worsted, or cotton, etc., cloths in which the ends of the pieces are sewn together for washing to form an endless band and the material twisted like a rope.

ropy lava (*Geol.*). A lava flow, particularly basalt, the surface appearance of which resembles loose coils of rope. This is the characteristic surface appearance of flows of fluid lava. Cf. *block lava*, which results from consolidation of a very viscous lava from which the included gases had been boiled off before extrusion.

Rorschach test (*Psychol.*). A perception test, designed to gauge intelligence, personality, and mental state, in which the subject interprets ten inkblots of standard form, his responses being scored and analysed in terms of prescribed norms. See **projective technique.** Also called the **inkblot test.**

rosacea (*Med.*). See **acne rosacea.**

rosaniline (*Chem.*). Triamino - diphenyl - tolyl-hydroxy-methane, a base of the fuchsine dyes.

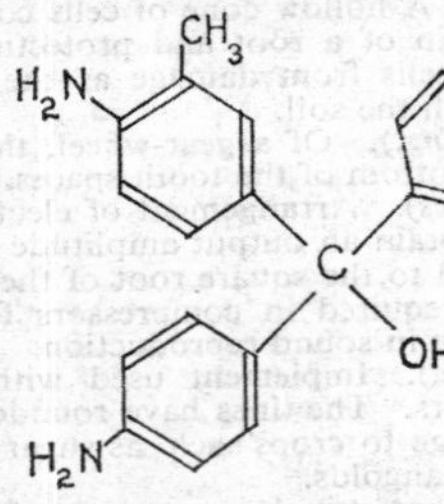

It is obtained by oxidation of an equi-molecular mixture of aniline, 2-toluidine and 4-toluidine.

rosasite (*Min.*). A hydrous carbonate of copper and zinc, occurring in fibrous masses as a secondary mineral in oxidized zinc-copper-lead ores.

roscoelite (*Min.*). This mineral is essentially *muscovite* (q.v.) in which vanadium has partly replaced the aluminium. Its colour is clove-brown to greenish-brown.

rose (*Arch.*). A decorative circular escutcheon through which, e.g., the spindle of a door-handle or the flexible cord of a pendant *luminaire* may pass. See also **ceiling rose,** a more practical extension of the latter use.

rose bengal stain (*Med.*). Dichlorotetraiodofluorescin, $C_{20}H_4I_4Cl_2O_5K_2$, a dye used as a bacterial stain. Its rate of disappearance from the blood stream is used to test liver function.

Rose crucible (*Chem.*). A crucible the lid of which is fitted with an inlet tube. It is used for igniting substances in a current of gas.

rose cutter (*Horol.*). A small hollow milling cutter used on the lathe for rapidly producing pivots, screws, etc.

Rosenmüllers organ (*Zool.*). See **epididymis.**

roseola (*Med.*). Any rose-coloured rash.

rose opal (*Min.*). A variety of opaque common opal having a fine red colour.

rose quartz (*Min.*). Quartz of a pretty rose-pink colour, due probably to titanium in minute quantity. The colour is apt to be destroyed by exposure to strong sunlight. See also **Bohemian gem-stones.**

rose topaz (*Min.*). The yellow-brown variety of topaz changed to rose-pink by heating. These crystals often contain inclusions of liquid carbon dioxide.

rosette (*Bot.*). (1) A group of leaves lying close together on or near the surface of the soil, and radiating out from the apex of a short stem. (2) Four cells in the embryo of a pine tree, lying just above the suspensor. (*Zool.*) Any rosette-shaped structure; in some *Oligochaeta*, a large ciliated funnel by which the contents of the vesiculae seminales pass to the exterior; in some *Crinoidea*, a thin calcareous plate (*rosette plate, rosette ossicle*) formed by the coalescence of the basal plates.

rose window (*Arch.*). A circular window with radial bars. Also called **Catherine wheel, marigold window.**

rosewood (*For.*). Decorative cabinet-wood from the sapwood of species of *Dalbergia* exported mainly from Brazil, Honduras, Jamaica, India, and Africa. Also called **jacaranda.**

rosin (*Chem.*). Colophony. The residue from the distillation of turpentine. The colour varies from colourless, to yellow, red, brown, and black. Rel. d. 1·08, m.p. 100°C–140°C. Wood-rosin is obtained by the extraction of long-leaf pine wood; chief sources, U.S. and France. Used as a soldering flux, in varnish, soap, and size manufacture, and (in the form of resinates) as a drier in paint. See **resinates.**

rosinates (*Chem.*). *Resinates* (q.v.).

Rosiwal intercept method (*Geol., etc.*). Particle-size analysis technique based on measurement of the intercepts made by a line drawn through a selection of particle images on a photomicrograph or in the field of view of a projection microscope.

Roslin Sandstone (*Geol.*). A yellow sandstone found in the Midland Valley of Scotland; of Carboniferous age, falling partly in the Lower, and partly in the Upper Carboniferous; it is the Scottish equivalent of the English *Millstone Grit.*

rosolic acid (*Chem.*). Quinonoid triarylcarbinol anhydride:

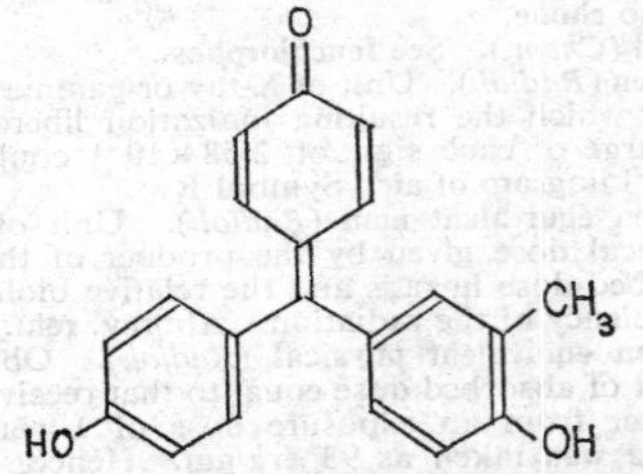

An acidic dyestuff of the triphenylmethane series, obtained by oxidizing a mixture of phenol and 4-cresol with arsenic(V) acid and sulphuric acid. Green glistening crystals, insoluble in water, dissolving in alkalis with a red colour.

Rossi counter (*Radiol.*). An important type of spherical tissue-equivalent proportional counter widely used in radiobiology.

Rossi-Forell scale (*Geol.*). A scale of apparent intensity of earthquake movements, now replaced by the *Mercalli scale.*

rostellum. A beak-shaped process. (*Bot.*) A beaklike outgrowth from the column in the flower of an orchid. (*Zool.*) A small rostrum; in *Cestoda*, the hook-bearing projection at the top of the scolex; in *Cephalopoda*, a process of the shell. *adjs.* **rostellar, rostellate.**

rostrum (*Zool.*). In Birds, the beak; a beak-shaped process; in *Cirripedia*, a ventral plate of the carapace; in some *Crustacea*, a median anterior projection of the carapace. *adjs.* **rostral, rostrate.**

rosula (*Bot.*). A *rosette* (q.v.) of leaves.

rotachute (*Aero.*). A 'parachute', usually for stores or the recovery of missiles, in which the normal retarding canopy is replaced by rotor blades, which are freely-revolving and act like a *rotor*.

Rotameter (*Phys.*). TN for rate-of-flow indicator in which a tapered float moves vertically in a transparent tube in accordance with velocity of rising liquid. Other types include magnetic and spinning Rotameters.

rotaplane (*Aero.*). A heavier-than-air aircraft which derives its lift or support from the aero-dynamical reaction of freely rotating rotors. See **rotorcraft.**

rotary amplifier (*Elec. Eng.*). Rotary generator, output of which is field-controlled by another generator or amplifier.

rotary arc (*Light*). An arc lamp in which one or both carbon electrodes are rotated automatically by a motor, as well as being fed, so that the carbons are consumed regularly and the arc is steady.

rotary combustion engine (*I.C. Engs.*). See **epitrochoidal engine, Wankel engine.**

rotary converter (*Elec. Eng.*). Combination of electric motor and dynamo used to change form of electric energy, e.g., d.c. to a.c. or one d.c. voltage to another.

rotary drier (*Met., Min. Proc.*). Tubular furnace sloped gently from feed to discharge end, through which moist material is tumbled as it rotates slowly, while rising hot gases remove moisture.

rotary engine (*Aero.*). An early type of aero engine in which the crankcase and radially disposed cylinders revolved round a fixed crank; not to be confused with the modern radial engine.

rotary exhaust machine (*Vac. Tech.*). A mass-production unit, for the manufacture of lamps or valves, in which operations such as loading, pumping, degassing, cathode activation, gettering, and sealing-off are performed at successive positions on a multihead rotary machine.

rotary field (*Elec. Eng.*). See **rotating field.**

rotary indexing machine (*Eng.*). A transfer machine, used for cutting, assembling, or other production work, in which workpieces or tools are carried on a circular turntable which rotates intermittently.

rotary machine (*Print.*). A machine in which the printing surface is a cylinder. Used for letterpress, photogravure, and lithography, and may be sheet-fed or web-fed.

rotary pump (*Eng.*). A pump, similar in principle to a *gear pump* (q.v.), in which two specially shaped members rotate in contact; suited to large deliveries at low pressure. See **Roots blower.**

rotary regenerative heater (*Eng.*). An air heater consisting of a slowly revolving rotor made up of concentric rings of corrugated and flat plates, which pass alternately and continuously through the hot gases and the air drawn across opposite halves of the rotor.

rotary shutter (*Cinema.*). The rotating vanes which cut off the light from the screen while the frames are being moved and located in the picture gate of a projector.

rotary strainer (*Paper*). A machine which removes foreign material from the pulp before it travels on to the paper-making machine.

rotary switch (*Elec. Eng.*). A switch operated by a rotating handle capable of rotation in one direction only.

rotary valve (*I.C. Engs.*). A combined inlet and exhaust valve in the form of a ported cylinder rotating on cylindrical faces in the cylinder head, usually parallel with the crankshaft.

rotate (*Bot.*). Said of a sympetalous corolla having a very short tube, the petals spreading like the spokes of a small wheel or like a star.

rotating amplifier (*Elec. Eng.*). A form of d.c. generator in which the electrical power output can be accurately and rapidly controlled by a small electrical signal applied to the control field of the machine. Mainly used in industrial closed-loop systems. See **cross-field generator.**

rotating anode (*Electronics*). Tube in which the anode is rotated about an axis normal to but not through the target area, so that heat from the impact of electrons is dispersed.

rotating crystal method (*Crystal.*). A widely used method of X-ray analysis of the atomic structure of crystals. A small crystal, less than 1 mm in maximum size, is rotated about an axis at right angles to a narrow incident beam of X-rays. The diffraction of the beam by the crystal is recorded photographically on a flat plate or on a film bent into a cylinder round the axis.

rotating (or rotary) disk contactor (*Chem. Eng.*). A liquid-liquid extraction device which consists of a column with annular stators fitted to it and a central shaft carrying rotors of diameter nearly equal to the holes in the stators. Liquids flow counter-current through the column and under the effect of high speed of rotation of the shaft and rotors improved contact is obtained.

rotating field (*Elec. Eng.*). One in which the magnitude is constant at a point but whose direction is rotating about a point in a fixed reference system.

rotating-field magnet (*Elec. Eng.*). The rotating portion of an electric machine, usually a synchronous motor or generator, in which the field poles rotate and the armature is stationary.

rotating joint (*Telecomm.*). Short length of cylindrical waveguide, constructed so that one end can rotate relative to the other; it is used to couple two other waveguide systems normally of rectangular cross-section.

rotating reflector rangefinder (*Photog.*). One in which two images from reflectors separated by a fixed distance are made to coincide by rotating one of the reflectors. The range varies with the amount of rotation necessary.

rotating scanner (*TV*). Any form of mechanical scanner in which the moving parts revolve, as in a *Nipkow disk* (q.v.). Cf. *oscillatory scanning.*

rotating-wedge rangefinder (*Photog.*). Similar to *swinging-wedge rangefinder* (q.v.), but the variable-angle effect is produced by two prisms of equal angle rotating in their own planes in opposite directions.

rotation (*Astron.*). The term generally confined to the turning of a body about an axis passing through itself, e.g., *rotation* of the earth about its polar axis in 1 sidereal day. (*Elec. Eng.*) See **curl.**

rotational field (*Elec. Eng.*). A field in which the *circulation* (q.v.) is, in some parts, not always zero.

rotational moulding (*Plastics*). A *moulding* (q.v.) process in which the material (e.g., PVC) in the mould is rotated while in the heating oven and during cooling; used for some hollow articles.

rotational therapy (*Radiol.*). Form of *moving-field therapy* in which there is relative rotational movement between patient and beam of radiation about an axis passing through the patient.

rotation axes of symmetry (*Crystal.*). Symmetrically placed lines, rotation about which causes every atom in a crystal structure, as revealed by X-ray analysis, to occupy identical positions a given number (2, 3, 4, 6) of times. Cf. *screw axes*.

rotation of a vector (*Maths.*). See **curl.**

rotation of the plane of polarization (*Chem., Phys.*). A property possessed by optically active substances. See **optical activity.**

rotation speed (*Aero.*). The speed during take-off at which the nosewheel of an aeroplane is raised from the ground prior to *lift-off* (q.v.).

rotator (*Phys.*) A device for rotating the plane of a wave in a waveguide.

rotatory dispersion (*Chem., Phys.*). Variation of rotation of the plane of polarized light with wavelength for an optically active substance.

rotatory evaporator (*Chem.*). A device for facilitating the evaporation of a liquid, generally under reduced pressure, by continuously rotating the flask in which it is contained.

rotatory power (*Chem., Phys.*). See **specific rotation.**

rotatrol (*Elec. Eng.*). Electrodynamic amplifier used as component of some servo systems.

rotenone (*Chem.*). See **derris.**

Rothera test (*Chem.*). A test for the presence of acetone or acetoacetic acid, based on the appearance of a dark violet colour in the presence of an ammoniacal solution of sodium pentacyanonitrosylferrate(II).

Rotifera (*Zool.*). A phylum of the *Metazoa* containing minute acoelomate unsegmented animals typically having a ciliated trochal disk for locomotion and food collection. Alimentary canal complete with an anterior mouth, a muscular pharynx (or mastax) with chitinous jaws (trophi) and a posterior anus. The excretory system has flame cells, no circulatory system or respiratory organs, nervous system simple, sexes separate and usually markedly dimorphic. There are two types of egg, one developing immediately without fertilization, and the other, which is fertilized, being thick-walled and developing only after a resting period. Freshwater or marine.

rotogravure (*Print.*). Photogravure printing on a *rotary machine* (q.v.).

roton (*Phys.*). Quantum of rotational energy analogous to the *phonon.*

rotor (*Aero.*). A system of revolving aerofoils producing lift. (*Autos.*) The revolving arm of a distributor. (*Elec. Eng.*) See **armature** (3). (*Horol.*) A revolving eccentric weight which winds a self-winding watch. (*Zool.*) A muscle which by its contraction turns a limb or a part of the body on its axis.

rotor core (*Elec. Eng.*). That portion of the magnetic circuit of an electric machine which lies in the rotor.

rotorcraft (*Aero.*). Any aerodyne which derives its lift from a rotor, or rotors.

rotor head (*Aero.*). The structure at the top of the rotor pylon, including the hub member to which the blades of a rotorcraft are attached.

rotor hinge (*Aero.*). A hinge for the blades of a rotorcraft. See **drag hinge, feathering hinge, flapping hinge.**

rotor hub (*Aero.*). The rotating portion of the rotor head of a rotorcraft to which the rotor blades are attached.

rotor starter (*Elec. Eng.*). A motor starter used for slip-ring induction motors; it cuts out resistance previously inserted in the rotor circuit.

rotor-tip jets (*Aero.*). Propulsive jets in the tips of a rotorcraft's blades that are used to obtain a drive with minimum torque reaction; they may be *pulse jets, ramjets*, combustion units fed with air and fuel from the fuselage, *pressure jets*, or small *rocket* units.

rotor wind (*Meteor.*). See **orographic ascent.**

Rototrol (*Elec. Eng.*). TN of a range of rotating amplifiers.

rottenstone (*Geol.*). A material used commercially for polishing metals; formed by the weathering of impure siliceous limestones, the calcareous material being removed in solution by percolating waters.

rotting (*Textiles*). See **retting.**

rotula (*Zool.*). In higher Vertebrates, the kneecap; in *Echinoidea*, one of five radially directed bars running inwards from the junctions of the epiphyses and forming part of *Aristotle's lantern.*

rotunda (*Arch.*). A building or room which is circular in plan and is covered by a dome.

rot v (*Maths.*). See **curl.**

rouge (*Chem.*). Hydrated oxide of iron in a finely divided state; used as a polish for metals.

Rouget cells (*Histol.*). In Vertebrates, cells found lying against the walls of the capillaries; the power of contraction of the capillaries was formerly ascribed to these. See also **pericyte.**

rough (*Paper*). The unglazed surface of machine-made drawing papers specially induced in drying by a coarse felt. See **hot-pressed** and **not.**

rough ashlar (*Build.*). A block of freestone as taken from the quarry.

rough brackets (*Carp.*). Pieces of wood nailed to the sides of the *carriage* (q.v.), to provide intermediate support for treads of a wooden stair.

rough-cast (*Build.*). A rough finish given to a wall by coating it with a plaster containing gravel or small stones.

rough coat (*Build.*). The first coat of plaster applied to a wall surface.

rough colony (*Bacteriol.*). A bacterial colony produced by mutation from a smooth colony. The morphological change is frequently accompanied by physiological changes, e.g., altered virulence.

rough grounds (*Join.*). Unplaned strips of wood used as *grounds* (q.v.) when the attached joinery will entirely cover them.

roughing (*Min. Proc.*). Production of an impure concentrate as an early stage in ore processing, thus reducing bulk for more thorough treatment. (*Textiles*) Straightening and cleaning scutched flax by drawing it in handfuls through a series of pins in a board.

roughing-in (*Build.*). The first coat of 3-coat plaster work.

roughing pump (*Vac. Tech.*). A pump used for evacuation of a system from atmospheric pressure to the pressure at which the process pump can operate.

roughing tool (*Eng.*). A lathe or planer tool, generally having a round-nosed or obtuse-angled cutting edge, used for roughing cuts.

roughness integrator (*Civ. Eng.*). An instrument for measuring roughness of a road surface. Also called a **corrugmeter** or **profilometer.**

rough proof (*Print.*). A print of the work in hand, much below the finished quality to be expected, submitted for checking purposes only.

rough-string (*Carp.*). See **carriage.**

rough trimmed (*Bind., Print.*). A design feature in many well-produced books where the tail edges are merely cleaned up by cutting the farthest projecting leaves, the pages having been deliberately positioned to ensure a variation in the tail margin. See **cut edges, trimmed size.**

rouleaux (*Zool.*). Aggregations of erythrocytes resembling piles of plates.

roulette (*Maths.*). The locus of any point, or envelope of any line, moving with a first curve which rolls without slipping on a second curve. Referred to respectively as *point* or *line roulette*. The locus of a point on a circle rolling on a straight line is called a *curtate cycloid*, a *cycloid* or a *prolate cycloid* respectively according as the point is inside, on the circumference or outside the rolling circle. The locus of a point on a circle rolling on another circle is called an *epicycloid* or *hypocycloid* respectively according as the rolling circle is outside or inside the fixed circle, if the point is on the circumference, or an *epitrochoid* or *hypotrochoid* respectively if the point is not on the circumference of the rolling circle. Epicycloids and epitrochoids in which the rolling circle completely encloses the fixed circle are sometimes called *pericycloids* and *peritrochoids* respectively. Curtate and prolate cycloids are sometimes called *trochoids*. (*Photog.*) A steel wheel, having pointed projections, used for retouching copper plates which have been etched.

roumanite, rumänite (*Min.*). A variety of amber obtained from Roumania.

round (*Build.*). A rung of a ladder.

round angle (*Maths.*). A complete rotation, an angle of 360° or 2π radians. Also called a **perigon.**

round dance (*An. Behav.*). Bee dance consisting of rapid locomotion in a tight circle, and indicating the existence of a food source 50-100 metres from the hive.

round heart disease (*Vet.*). A disease of chickens characterized by enlargement and muscular degeneration of the heart and sudden death; the cause is unknown.

rounding (*Bind.*). The process of giving the back of a book a convex shape before casing; usually performed along with *backing* (q.v.). (*Leather*) Removing from the hide, before or after tanning, the shoulder, belly, and neck parts.

rounding-off (*Maths.*). The reduction of the number of digits in a number to a given number of significant figures (accuracy) or places of decimals counting from the point in either direction (precision). Only the first digit to be deleted is considered (the others are ignored); if it is 0, 1, 2, 3 or 4 the last remaining digit is unaltered; if it is 5, 6, 7, 8 or 9 it is increased by 1. (Other rules have been applied in the past to 5, but this is now generally accepted.) Thus 1·2345 reduced to 3 sig. figs. is 1·23, 9·8765 becomes 9·88; 135 791 reduced to the nearest 1000 becomes 136 000, and 8·6429 reduced to the nearest 1/100 (2 places of decimals) becomes 8·64. Not the same as **truncation.**

rounding plane (*Join.*). See **rounds.**

rounding-up tool (*Horol.*). A tool for correcting the size and shape of the teeth of a toothed wheel.

round key (*Eng.*). A circular bar or pin fitted in a hole drilled half in the shaft and half in the boss, parallel to the shaft axis; used for light work to avoid fitting. See **key.**

roundness (*Powder Tech.*). A shape factor defined by the relationship

$$R_w = \frac{\Sigma r/R}{N},$$

where r is the radius of curvature of a corner of the particle surface, R is the radius of the maximum inscribed circle in the longitudinal section of the particle, and N is the number of corners.

round of beam (*Ships*). See **camber.**

rounds (*Join.*). The general name for planes having a concave sole and cutting iron, used for forming rounded surfaces. Cf. *hollows*.

round step (*Build.*). A step finished with a semicircular end.

Round valve (*Electronics*). Early grid detector valve (named after inventor), in which the effectiveness of demodulation was increased by heating a pellet which released small quantities of gas.

roundworm (*Med., Zool.*). Name applied to a number of parasitic Nematodes, especially those of the genus *Ascaris*, including the large intestinal roundworm of man (*A. lumbricoides*).

roup (*Vet.*). A term applied to symptoms of oculonasal discharge and swelling of the face and wattles of fowl, which occur in infections of the upper respiratory tract such as infectious coryza, fowl pox, and infectious laryngotracheitis. Also synonym for *infectious coryza*.

Rousay Flags (*Geol.*). A group of blue and calcareous flags with sandy beds which belong to the Middle Old Red Sandstone of the Devonian System; found in the North of Scotland and in Orkney.

rousing (*Brew.*). Mixing thoroughly.

Rous' sarcoma (*Vet.*). A tumour, occurring in fowls, which can be transmitted to other fowls by inoculation of a cell-free filtrate of the tumour, which is therefore thought to be the result of infection with a virus.

Rousseau diagram (*Elec. Eng.*). A diagram by the use of which the total output (in lumens) of a light source can be obtained, if the polar curve of the lamp about the vertical axis is known.

Rousselet's solution (*Micros.*). A narcotic solution used in the examination of small living organisms, or to prevent their distortion on fixation. It contains cocaine and ethanol.

Roussin's salts (*Chem.*). Formed when sodium trisulphide is added to a solution of iron(II) chloride saturated at −2°C with nitrogen(II) oxide, and converted by sodium sulphide into so-called *Roussin's red salt*; by treatment with dilute acids this is converted into *Roussin's black salt*.

rout (*Carp., Join.*). To cut out wood from the bottom of a sinking with a router plane.

router (*Join.*). (1) A plane adapted to work on circular sashes; operated in the manner of a spokeshave. (2) The side wing of a *centre-bit* (q.v.), which removes the material in forming the hole.

router plane (*Carp., Join.*). A plane having a central projecting cutting iron, adapted to smoothing the bottom of a recess.

Routh's rule (*Maths.*). A rule summarizing the values of the radius of gyration of a rectangular lamina, an elliptic lamina and an ellipsoid about a principal axis through its centre of gravity. According to the rule, k^2 equals the sum of the squares of the other semi-axes divided by 3, 4 or 5 respectively.

routine (*Comp.*). Prescribed series of operations which can be repeated, in conjunction with others, at any time without special planning. Kept in a *library* in the form of punched paper tape or cards. See **subroutine.**

routiner (*Teleph.*). Apparatus which tests, as a routine, all machine-switching apparatus in an exchange, so that faults may be rapidly detected and rectified, and contacts kept clean.

routing machine (*Print.*). A machine having a revolving tool which removes unwanted metal from printing plates.
roving or rove (*Textiles*). A continuous strand of fibres sufficiently drawn by speedframe operations to a diameter suitable (and in bobbin form) for drafting and twisting into yarn on mules or ringframes.
roving frames (*Spinning*). See **fly frames.**
row crop thinner (*Agric.*). Implement with tines of various widths to give a choice in the length of gap made between plants.
Rowland circle (*Light*). A circle having the radius of curvature of a concave diffraction grating as diameter. It has the property that, if the slit is placed anywhere on the circumference of the circle, the spectra of various orders are formed in exact focus also round the circumference of the circle. This fact is used in designing mountings for the concave grating.
rowlock (*Arch.*). A term applied to a course of bricks laid on edge.
rowlock-back (*Arch.*). A term applied to a wall whose external face is formed of bricks laid flat in the ordinary manner, while the back is formed of bricks laid on edge.
row vector (*Maths.*). A single-row matrix.
royal (*Paper*). A standard size of printing and writing paper, 20×25 in. (508×635 mm).
royalette (*Textiles*). Dress cloth made with a 5-shaft sateen weave from cotton warp and fine botany worsted weft.
royal octavo (*Typog.*). A book-size $9\frac{3}{4} \times 6\frac{1}{8}$ in. (metric, 234×156 mm).
royal pair (*Zool.*). The pair of sexually perfect reproductory individuals which are the founders of a termite colony. See **Isoptera.**
royal quarto (*Typog.*). A book-size $12\frac{1}{4} \times 9\frac{7}{8}$ in. (metric, 312×234 mm).
royals (*Min. Proc.*). See **pregnant solution.**
royer (*Foundry*). Short fast-moving belt used in conditioning of foundry sand by throwing it into air.
R.P.N. See **Reverse Polish Notation.**
R.R. alloys (*Met.*). Properly *Hiduminium R.R. alloys.* A series of aluminium alloys of the Duralumin type (rel. d. approx. 2·75) composed of aluminium with the following elements in percentages varying between the limits shown: copper 0·8–3·0, magnesium up to 4·0, silicon 0·6–2·8, nickel 0·5–2·0, iron 0·6–1·5, titanium 0·02–0·3.
RR Lyrae variables (*Astron.*). Variable stars with periods of less than 1 day; common in globular clusters, and used, like the cepheids, to measure galactic distances.
R.S.F. (*Build.*). Abbrev. for *rough sunk face.*
R.S.J. (*Build.*). Abbrev. for *rolled steel joist.*
Ru (*Chem.*). Symbol for *ruthenium.*
Ruabon (*Build.*). A term applied to hard, smooth-surfaced, impervious facing-bricks resembling red terra-cotta.
Rubarth's disease (*Vet.*). See **infectious canine hepatitis.**
rubber (*Chem.*). The main source of natural rubber or caoutchouc is the tree, *Hevea brasiliensis*, originally a native of Central and South America, but since the end of the 19th century widely grown in plantations in S.E. Asia. Commercial rubber consists of caoutchouc, a polymerization product of isoprene, of resinlike substances, nitrogenous substances, inorganic matter, and carbohydrates. The caoutchouc portion is soluble in CS_2, CCl_4, trichloromethane or benzene, forming a viscous colloidal solution. When heated, rubber softens at 160°C, and melts at about 220°C. Rubber easily absorbs a large quantity of sulphur either by heating or in the cold by contacting with S_2Cl_2, etc. This process is called *vulcanization* (q.v.). Carbon black, in a fine state of division, is used as a reinforcing filler; other substances, produced by the condensation of aldehydes with amines, retard the oxidation of vulcanized rubber. The uses of rubber are innumerable. See also **crêpe rubber, synthetic rubber.** For *foamed rubber*, see **expanded plastics.**
rubber blanket (*Print.*). See **offset printing.**
rubber feed rolls (*Agric.*). Attachment to a seed drill for extra gentle handling of brittle seeds; especially large seeds such as peas and beans.
rubber jaw (*Vet.*). See **osteodystrophia fibrosa.**
rubber line (*Acous.*). In gramophone disk-record manufacture, the terminating mechanical resistance in the sound-recording mechanism used for driving the stylus in making the original record. By using concentric tubes of rubber, a substantially pure mechanical resistance over the required range of frequency is obtained.
rubber plates (*Print.*). Extensively used for all grades of work. Flexible, they can be affixed easily to rotary cylinders; have good inking qualities; and are durable; require precision grinding to exact thickness and *kiss impression* (q.v.); hand-cut rubber plates are extensively used for bold designs, particularly on packages.
rubbers (*Build.*). See **cutters.**
rubbing leathers (*Textiles*). Endless leather or oil-resistant synthetic rubber aprons, one above another, which travel with an oscillating movement and convert the ribbon-shaped slivers in the condenser, or from a card ring-doffer, into round twistless rovings wound on to a large bobbin for mule or ringframe spinning. Also known as **condenser leathers.**
rubbing stone (*Build.*). An abrasive stone with which the bricklayer rubs smooth the bricks which he has cut to a special shape. See **gauged arch.**
rubble (*Build.*). Rough uncut stones, of no particular size or shape, used for rough work, for filling between facing walls, etc.
rubble concrete (*Civ. Eng.*). A form of masonry often used on massive works such as solid masonry dams; composed of very large blocks of stone set about 6 in (15 cm) apart in fine cement concrete and faced with squared rubble or ashlar.
rubeanic acid (*Chem.*). Dithio-oxamide,

$$\begin{array}{l} S{=}C{-}NH_2 \\ \quad\;\; | \\ S{=}C{-}NH_2 \end{array}$$

an orange crystalline powder, sparingly soluble in water but soluble in alcohol; used as a reagent to detect small amounts of copper, with which it forms a black precipitate.
rubefacient (*Med.*). Producing reddening of the skin; any agent which does this, a counter-irritant.
rubella (*Med.*). German measles. An acute infectious disease which is distinct from measles; characterized by slight fever, enlargement of glands in the neck and at the back of the head, and a pink papular-macular rash.
rubellite (*Min.*). The pink or red variety of tourmaline, sometimes used as a semiprecious gemstone.
rubescent (*Bot.*). Turning pink or red.
rubicelle (*Min.*). A yellow or orange-red variety of spinel; an aluminate of magnesium.
rubidium (*Chem.*). A metallic element in the first group of the periodic system, one of the alkali metals. Symbol Rb, at. no. 37, r.a.m. 85·47,

m.p. 38·5°C, b.p. 690°C, rel. d. 1·532. The element is widely distributed in nature, but occurs only in small amounts; the chief source is carnallite. The metal is slightly radioactive.

rubric (*Typog.*). A heading or passage printed in red, the main text being in black. Also the marginal headings of minutes, etc., although printed in black.

ruby (*Min.*). The blood-red variety of the mineral corundum, the oxide of aluminium (Al_2O_3), which crystallizes in the trigonal system. Also called *true ruby* (to distinguish it from the various types of *false ruby*, q.v.) and *Oriental ruby*, though the adjective *Oriental* is quite unnecessary, since it merely stresses the fact that rubies come from the East (Burma, Thailand, Ceylon, Afghanistan). See also **balas ruby, ruby spinel.** (*Typog.*) An old type size approx. 5½-point.

ruby pin (*Horol.*). The impulse pin of a lever escapement.

ruby silver ore (*Min.*). See **proustite, pyrargyrite.**

ruby spinel (*Min.*). That variety of magnesian spinel, $MgAl_2O_4$, which has the colour, but none of the other attributes, of true ruby. Also **almandine spinel. Spinel ruby** is a deceptive misnomer.

rudder (*Aero.*). A movable surface in a vertical plane for control of an aeroplane in angles of yaw (i.e., movement in a horizontal plane about a vertical axis). Usually located at the rear end of the body and controlled by the pilot through a system of rods and/or cables. (*Ships*) Broad, flat device, varying in form, hinged vertically to, or behind, the sternpost of a vessel; the rudder serves to change the vessel's course when it is moved from a position in line with the keel.

rudder bar or pedals (*Aero.*). A mechanism consisting of differential foot-operated levers by which the pilot actuates the rudder of a glider or an aeroplane, or controls the pitch of a helicopter tail rotor through mechanical or hydraulic relaying devices.

rudder post (*Ships*). See **stern frame.**

rudenture (*Arch.*). A cylindrical moulding carved in imitation of a rope.

ruderal (*Bot.*). A plant which grows usually on rubbish heaps or waste places.

rudiment (*Bot., Zool.*). The earliest recognizable stage of a member or organ.

Rudistes (*Geol.*). A group of heavily-built lamellibranchs which are characteristic of the Cretaceous rocks formed in the southern ocean (the Tethys) of the period; it includes the genera *Hippurites, Requienia,* and *Monopleura.* Rudistids also occur in the Cretaceous Trinity Series of Texas and Mexico.

Ruffini's organs (*Zool.*). In Vertebrates, a type of sensory nerve ending in which the nerve gives rise, within a cylindrical capsule, to small ramifications which end in flattened expansions.

rufous (*Bot.*). Red-brown.

rugose (*Biol.*). Having a wrinkled surface. *dim.* **rugulose.**

Ruhmkorff coil (*Elec. Eng.*). Electromagnetic high-voltage generator consisting essentially of a transformer with interrupted d.c. primary supply. Source of energy for early wireless telegraphy spark transmitters.

rule (*Build.*). See **floating rule.** (*Typog.*) Type-high brass or metal strip of various thicknesses and designs; a dash or score (see **em rule, en rule**).

rule border (*Typog.*). A frame of rules fitted around an advertisement or other displayed matter.

ruled surface (*Maths.*). One generated by the motion of a straight line with one degree of freedom, e.g., a cone.

ruling (*Print.*). The operation of making lines on writing, account-book, and ledger paper, etc.; the paper is conveyed on an endless belt and makes contact with suitably adjusted disks or pens.

ruling gradient (*Civ. Eng.*). The maximum gradient permissible for any given section of road or railway.

rumble (*Acous.*). Low-frequency noise produced in disk recording when turntable is not dynamically balanced.

rumen (*Zool.*). The first division of the stomach in Ruminants and *Cetacea,* being an expansion of the lower end of the oesophagus used for storage of food; the paunch.

rumenotomy (*Vet.*). Operation of cutting into the rumen.

Rumford's photometer (*Light*). A *photometer* (q.v.) consisting of a rod standing vertically in front of a white screen on which are cast shadows of it by the two light sources whose intensities are to be compared. When the shadows are of equal darkness, ratio of intensities of the sources equals the square of the ratio of their distances from the screen.

ruminant (*Zool.*). See **rumination.**

ruminate (*Bot.*). Mottled, as if composed of a mixture of two or more differently coloured parts; said of endosperm.

rumination (*Med.*). See **merycism.** (*Zool.*) The regurgitation of food that has already been swallowed, and its further mastication before reswallowing. *adj.* and *n.* **ruminant.**

rummel (*San. Eng.*). See **soakaway.**

run (*Comp.*). Complete set of operations in sequence, e.g., running a magnetic tape store of open accounts for *updating,* changes of address, etc. (*Horol.*) The movement of the lever, in a lever escapement, to the banking pins, due to draw. (*Plumb.*) That part of a pipe or fitting which is in the same straight line as the direction of flow in the pipe to which it is connected. (*Print.*) The number of copies to be printed. (*Surv.*) In a level tube, the movement of a bubble with change of inclination. (*Textiles*) (1) Batches of the same type of material run through the same set of machines. (2) The American unit of length in counting woollen yarns, viz., 100 yd.

run around (*Typog.*). Type set alongside an illustration to less than the full measure.

runaway electron (*Electronics*). One under an applied electric field in an ionized gas which acquires energy from the field at a greater rate than it loses through particle collision.

runcinate (*Bot.*). Said of a leaf having a lamina composed of lobes with their points backwardly directed.

run down (*Telecomm.*). That part of a repeated waveform which is a decay, particularly if a large fraction of the cycle, as in a linear time-base generator, e.g., of the Miller type.

runic texture (*Geol.*). An alternative term suggested by A. Johannsen for *graphic texture,* because intergrown quartz and feldspar resemble runic characters.

run in (*Typog.*). U.S. term for **run on.**

runite (*Geol.*). See **graphic granite.**

runner (*Bot.*). A prostrate shoot which roots at the end and there gives rise to a new plant. (*Eng.*) The rotor or vaned member of a *water turbine* (q.v.). (*Foundry*) The vertical passage into the interior of a mould through which the metal is poured. Also called **runner-gate.** See **ingate.**

runner bush (*Foundry*). A small cast-iron box, without a bottom, lined with sand; placed over a runner gate to act as a funnel and reservoir for the metal during pouring.

runners (*Civ. Eng.*). A form of sheet pile much used for timbering wide excavations. It consists of short planks shaped to a chisel point at one end and usually shod with thin steel strip, so that as each runner is driven in, it wedges up against its neighbour. (*Horol.*) The cylindrical sliding pieces which support the work in a pair of turns. (*Typog.*) Marginal figures for reference purposes indicating the number of each line in a poem or play.

runner stick (*Foundry*). See **gate stick.**

running (*Build.*). The operation of forming a plaster moulding, cornice, etc., *in situ*, by running a horsed mould along the material while it is still plastic.

running bond (*Build.*). The same as **stretching bond.**

running ground (*Mining*). Weak walls of excavation requiring special support to prevent collapse.

running heads (*Typog.*). The headings at the top of the page, the usual arrangement being title of book on left-hand page and title of chapter on right-hand page.

running on (*Print.*). The actual printing of an edition after the *make ready* operations have been completed.

running rule (*Build.*). A wood strip fixed temporarily to serve the same purpose as a running screed.

running sand. See **quicksand.**

running screed (*Build.*). A band of plaster laid on the surface of a wall as a guide to the movement of a horsed mould in the process of running a moulding.

running shoe (*Build.*). The zinc part of a horsed mould, giving protection to the wood and facilitating running.

running tapes (*Print.*). Tapes travelling at the press speed for the purpose of leading or conveying the web of paper.

running trap (*San. Eng.*). See **siphon trap.**

runoff (*Civ. Eng., Geol.*). The resultant discharge of a river from a catchment area; surface water as distinct from that rising from deep-seated springs.

run on (*Typog.*). An indication that a new paragraph is not to be made. Marked in copy and proof by a line running from the end of one piece of matter to the beginning of the next.

run-on chapters (*Typog.*). Chapters in a book which do not commence on a new page but after a few lines of space.

run-out (*Cinema.*). The end-trailer of a print, i.e., the length of film between the last effective frame and the end. (*Foundry*) See **break-out.**

run-through ruling (*Bind.*). A term which indicates that the ruling continues from edge to edge, horizontally or vertically, without interruption. See **stopped heading.**

runway threshold (*Aero.*). The usable limit of a runway; in practice it is usually the current downwind end which is intended.

runway visual markers (*Aero.*). See **aerodrome markers.**

runway visual range (*Aero.*). In bad weather, the horizontal distance at which black-and-white markers of standard size are visible, the figure being transmitted to pilots approaching by *air traffic control*. Abbrev. **RVR.**

rupia (*Med.*). A syphilitic ulcer of the skin which is covered by a layer of crusts formed by dried secretion and dead tissue.

rupicolous (*Bot., Zool.*). Living or growing on or among rocks.

rupture (*Med.*). (1) Forcible breaking or tearing of a bodily organ or structure. (2) To break or to burst (said of a blood vessel or viscus). (3) *Hernia* (q.v.).

rupture disk (*Chem. Eng.*). See **bursting disk.**

rupturing capacity (*Elec. Eng.*). See **breaking capacity.**

rushes (*Cinema.*). First prints from the exposed cinematograph film in motion-picture production. These prints are usually made overnight, but it is possible to complete them in a few hours. All editing is done on rushes, and the negative is not cut until final sequences with rushes are approved.

russel cord (*Textiles*). A dress fabric of plain weave, with a cord effect; made from cotton warp and worsted or mohair weft, the warp being in tapes.

Russell effect (*Photog.*). Fogging on development, arising from previous prolonged action on the emulsion by (possibly) hydrogen peroxide, arising from oxidation of, e.g., clean zinc, resins, oils, etc.

Russell-Saunders coupling (*Nuc.*). Extreme form of coupling between orbital electrons of atoms. The angular and spin momenta of the electrons combine and the combined momenta then interact. Also called **l-s coupling.**

Russell's test (*Elec. Eng.*). A method of determining the insulation resistance of a 3-wire d.c. distribution network. The value is obtained by calculation from readings of an electrostatic voltmeter connected between the neutral wire and earth, both with and without a known resistance in parallel.

Russia leather (*Leather*). Formerly a special product of Russia made from calf's skin; now made by tanning skins in the usual way and imparting the familiar smell, by means of birch bark oil, in the finishing processes.

Russian cord (*Textiles*). A cloth with a plain ground which has, at intervals, prominent cords running the length of the piece. These special cords entail the use of a leno harness in the loom.

Russian cotton (*Textiles*). A strong cotton of about 1 in (25 mm) staple, but rather harsh. Used chiefly in Russia. Quality and staple are variable but have greatly improved recently.

rust (*Met., etc.*). Product of oxidation of iron or its alloys, due either to atmospheric attack or electrolytic effect of cell action round impurities.

rust fungi (*Bot.*). See **Uredinales.**

rusticated ashlar (*Build.*). Ashlar work in which the face stands out from the joints, at which the arrises are bevelled. The face may be finished rough or smooth or tooled in various ways.

rustic joint (*Build.*). A sunken joint between adjacent building-stones.

rustics (*Build.*). Bricks having a rough-textured surface, often multicoloured. Also **texture bricks.**

rust joint (*Build.*). A watertight joint between adjoining lengths of guttering or pipes.

rustle (*Cinema.*). See **valve rustle.**

rusty gold (*Min. Proc.*). Native gold which has become surface-filmed by adherent staining substances and is non-amalgamable and non-treatable by cyanide process in consequence.

rut (*Zool.*). The noise made by certain animals, as Deer, when sexually excited; oestrus; to be sexually excited, i.e., to be in the oestrous period; to copulate.

ruthenium (*Chem.*). A metallic element. Symbol

Ru, at. no. 44, r.a.m. 101·07, m.p. 2400°C, rel. d. 12·26. The metal is silvery-white, hard and brittle. It occurs with the platinum metals in osmiridium, and is used in certain platinum alloys.

rutherford (*Nuc.*). Unit of radioactive decay rate equal to 10^6 disintegrating atoms/sec, or alternatively that amount of radioactive material undergoing 10^6 atomic disintegrations/sec. Abbrev. **rd.** Superseded by curie. 1 Ci = $3{\cdot}7 \times 10^4$ rd.

Rutherford atom (*Nuc.*). Earliest modern concept of atomic structure, in which nearly all the mass of the atom is in the nucleus, the electrons, equal in number to the atomic number, occupying the rest of the atomic volume in a fixed pattern.

rutherfordium (*Chem.*). Proposed U.S. name for transuranic element at. no. 104. Discovery of this element is also claimed by the Russians who gave it the name *kurchatovium*.

Rutherford scattering (*Phys.*). See **scattering.**

Ruths accumulator (*Eng.*). A steam accumulator of variable pressure type for smoothing fluctuations of demand on a boiler. It consists of a large insulated vessel containing water into which steam is injected at a high pressure, to be liberated later when the pressure is reduced.

rutilant (*Bot.*). Brightly coloured in red, orange, or yellow.

rutilated quartz (*Min.*). See **needle stone.**

rutile (*Min.*). (IV)oxide of titanium which crystallizes as reddish brown prismatic crystals in the tetragonal system. It is found in igneous and metamorphic rocks, and in sediments derived from these, also in quartz (see **flèches d'amour**), and it is a source of titanium.

rutin (*Pharm.*). Drug, obtained from buckwheat, some eucalyptus leaves, etc., used in the treatment of capillary bleeding.

Ruysch's penicilli (*Zool.*). In Vertebrates, tufts of capillary arterioles opening into the interstices of the spleen pulp.

R-value (*Nuc. Eng.*). Percentage decrease in density of reactor fuel for 1% burn-up. See also **S-value.**

R-wire (*Teleph.*). In the cord circuit on a telephone switchboard, the ring wire connected to ring contacts on terminating plugs, and eventually to the B-wire of subscriber's line.

R.W.P. (*Build.*). Abbrev. for ***rainwater pipe.***

rybat (*Build.*). An inband or outband.

Rydberg constant (*Phys.*). The frequencies of spectrum lines in the same series are less than those of the spectrum limit by an amount $R/(n+k)$ where n can have any value, k is a constant for each series, R is a universal constant, termed the Rydberg constant. It was first determined empirically from spectrographic data, but has since been shown to be given by:

$$R = \frac{2\pi^2 e^4}{c\,h^3}\,M_r$$

where M_r is the reduced mass of the radiating electron, e=electronic charge, c=velocity of light, h=Planck's constant.

Rydberg formula (*Phys.*). A formula, similar to that of Balmer, for expressing the wave numbers (ν) of the lines in a spectral series:

$$\nu = R\left[\frac{1}{(n+a)^2} - \frac{1}{(m+b)^2}\right],$$

where n and m are integers and $m > n$, a and b are constants for a particular series, and R is the *Rydberg constant.*

R-Y signal (*TV*). Component of colour TV chrominance signal. Combined with luminance (Y) signal it gives primary red component.

S

s Symbol for *second* (time).

s Symbol for: distance along a path; solubility; specific entropy.

s- (*Chem.*). An abbrev. for: (1) symmetrically substituted (also *sym-*); (2) *secondary*, i.e., substituted on a carbon atom which is linked to two other carbon atoms; (3) *syn-*, i.e., containing the corresponding radicals on the same side of the plane of the double bond between a carbon and a nitrogen atom or between two nitrogen atoms.

σ (*Chem.*). A symbol for the diameter of a molecule.

σ Symbol for: conductivity; normal stress; Stefan-Boltzmann constant; surface charge density; surface tension; wave number.

S (*Chem.*). (1) In names of dyestuffs, a symbol for *black*. (2) The symbol for *sulphur*. (*Elec.*) Symbol for *siemens*.

S Symbol for: area; entropy; Poynting vector.

Σ Symbol for *sum of*.

Sabatier effect (*Photog.*). An image reversal phenomenon in photography which appears to be connected with the image development.

sabin (*Acous.*). Unit of acoustic absorption; equal to the absorption, considered complete, offered by 1 ft^2 of open window to low-frequency reverberant sound waves in an enclosure.

sabinene (*Chem.*). A terpene derivative occurring in marjoram oil; b.p. 163°–165° C, rel. d. 0·848, refractive index 1·4675.

Sabine reverberation formula (*Acous.*). Earliest formula (named after investigator) for connecting the period of reverberation of an enclosure, T seconds, with the volume, V in cubic metres, and the total acoustic absorption in the enclosure, ΣaS, where a is the absorption coefficient of a surface of S square metres. The formula is $T = 0{\cdot}17\,V/\Sigma aS$.

Sabin vaccine (*Med.*). A live, attenuated vaccine, taken orally, which immunizes the host against poliomyelitis. Cf. *Salk vaccine*.

Sabouraud pastille (*Radiol.*). A means for measuring the dose of X-rays to which a surface has been exposed.

sabulose, sabuline (*Bot.*). Growing in sandy places.

saccate fruit (*Bot.*). A fruit having a baglike envelope around it.

saccharase (*Chem.*). See **invertase**.

saccharides (*Chem.*). *Carbohydrates* (q.v.), which according to their complexity are usually divided into *mono-*, *di-*, *tri-*, and *polysaccharides*.

saccharimeter (*Chem.*). A special type of polarimeter adapted for use with white light; used in sugar analysis.

saccharimetry (*Chem.*). The estimation of the percentage of sugar present in solutions of unknown strength, especially by measurements of optical activity.

saccharin (*Chem.*). 2-Sulphobenzimide, a white crystalline powder, 300 times as sweet as sugar, not very soluble in water. The imido-hydrogen is replaceable by Na, forming a salt which is readily soluble in water. It is used in medicine in cases where sugar is harmful, e.g., in diabetes.

saccharobiose (*Chem.*). Cane sugar or sucrose.

saccharoidal textures (*Geol.*). Granular textures which resemble loaf sugar; found especially in limestones and marbles.

saccharolytic (*Bacteriol.*). Said of bacteria which use simple carbohydrates and starches as sources of energy.

saccharometer (*Chem.*). A hydrometer which is used to determine the concentration of sugar in a solution.

Saccharomycetaceae (*Bot.*). A family of *Endomycetales*, characterized by single asci. Includes the yeasts (*Saccharomyces*) used in alcohol production and in baking, and a few parasitic species, e.g. *Candida albicans* (thrush).

saccule, sacculus (*Zool.*). A small sac; the lower chamber of the auditory vesicle in Vertebrates. *adj.* **sacculate.**

sacculiform (*Biol.*). Shaped like a little bag.

saccus (*Zool.*). A sac or pouchlike structure; in male Insects, the 9th abdominal sternite.

S-acid (*Chem.*). 1-Amino-8-naphthol-4-sulphonic acid, an intermediate for dyestuffs.

2S-acid (*Chem.*). 1-Amino-8-naphthol-2,4-disulphonic acid, an intermediate for dyestuffs.

sack-pusule (*Bot.*). See **pusule**.

sacralgia (*Med.*). Pain in the sacral region.

sacralization (*Med.*). A developmental anomaly in which one or both transverse processes of the 5th lumbar vertebra become abnormally large and strong, appearing to form part of the sacrum.

sacral ribs (*Zool.*). Bony processes uniting the sacral vertebrae to the pelvis, distinct in Reptiles but fused to the transverse processes in other *Tetrapoda*.

sacral vertebrae (*Zool.*). In higher Craniata, those vertebrae which articulate with the ilia of the pelvis via sacral ribs, there being one in the frog and two in the lizard, coming between the lumbar vertebrae and the caudal vertebrae (if any). In Birds and Mammals they are fused with other vertebrae to form the *sacrum* (q.v.).

sacrificial anode (*Ships, etc.*). An anode, used in *cathodic protection* (q.v.), commonly magnesium alloyed with about 6% aluminium and 3% zinc. When attached to steel in sea water the natural potential difference is such as to make the steel cathodic. The anode dissolves and requires renewal.

sacroiliac joint (*Zool.*). In some Craniata, the almost immovable joints between the *sacrum* (q.v.) and the two ilia of the pelvis. The articular surfaces of the bones are partly covered with cartilage and partly roughened for the attachment of the *sacroiliac ligament*.

sacroiliac ligament (*Zool.*). In the *sacroiliac joints* (q.v.) of some Craniata, a strong interosseus ligament which, together with bundles of fibres above and below the joint, holds the parts together. Shortly before parturition in Mammals, these ligaments become loosened under the influence of hormones circulating at the time, and more movement becomes possible at the joints.

sacrum (*Zool.*). In the skeleton of some Craniata, part of the vertebral column which articulates immovably with the ilium of the pelvis at the *sacroiliac joint* (q.v.), being composed of several fused vertebrae, including the *sacral vertebrae* (q.v.). In Birds it consists of one thoracic vertebra, five or six lumbar vertebrae, the two sacral vertebrae and the anterior five caudal vertebrae (and is sometimes called the **synsacrum**). In Mammals it comprises varying numbers of vertebrae in different orders (e.g., four in the rabbit and five in man), the first one or two being regarded as sacral, and the others as caudal. The former have low spines and expanded ventral surfaces for the attachment of muscles.

saddle (*Civ. Eng.*). A block surmounting one of the towers of a suspension bridge, providing bearing or fixing for the suspension cables. (*Elec. Eng.*) A U-shaped cleat for securing screwed or other lighting conduit to a flat surface. (*Eng.*) The part of a lathe which slides on the bed, between headstock and tailstock.

saddle back (*Geol.*). See **anticline.**

saddle-back board (*Join.*). A narrow board, chamfered along each of the upper edges, which is fixed on the floor across the threshold of a doorway so that the gap beneath the door will be small when the latter is shut and large enough when it opens to accommodate a carpet.

saddle-back coping (*Build.*). A copingstone whose upper surface slopes away on both sides from the middle.

saddleback harrow (*Agric.*). A toothed harrow with arched frames for use on rigid land.

saddle bar (*Build.*). A metal bar fixed across a window to support glazing held in lead cames.

saddle boiler (*Build.*). An inverted U-shaped boiler fitted in a domestic range to supply the hot-water system.

saddle coils (*Electronics*). Rectangularly formed coils which are bent around the neck of a cathode-ray tube; used for magnetic deflection of the beam.

saddle key (*Eng.*). A key sunk in a key-way in the boss, but having a concave face which bears on the surface of the shaft, which it grips by friction only. See **key.**

saddle point (*Meteor.*). See **neutral point.** (*Nuc. Eng.*) Point on plot of potential energy against distortion for nucleus at which fission will occur, instead of return to equilibrium.

saddle scaffold (*Build.*). A scaffold erected over a roof from standards on both sides of the building; used for repair work on, e.g., a chimney at the middle of the roof.

saddle-stitching (*Bind.*). A method of wire-stitching in which the book is placed astride a saddle-shaped support and stitched through the back.

saddle stone (*Build.*). An *apex stone* (q.v.).

S.A.E. (*Aero., Autos., etc.*). Abbrev. for *S*ociety of *A*utomotive *E*ngineers (U.S.). Gives name to a widely-used viscosity scale for classifying motor oils.

safe (*Print.*). Condition of press when locked. See **lock.**

safe edge (*Tools*). The edge of a file on which no teeth are cut.

safeguard (*Rail.*). See **check rail.**

safe-lights (*Photog.*). Special light filters used for lighting dark-rooms, the colour passed being adjusted to the types of emulsions used. The filters may be of glass, gelatine, or dye solutions in glass boxes.

safe load (*Civ. Eng.*). See **factor of safety.**

Safeticurb (*Civ. Eng.*). A proprietary form of gutter consisting of lengths of concrete blocks, containing a continuous bore along their length and with a continuous or discontinuous slot along the top surface, to allow access for water draining off a road or area. Some forms also have a slight kerb formed on top to warn traffic of the definition of the edges of the road.

safety action (*Horol.*). The action in a lever escapement that ensures that the notch in the lever is always in its correct position for the reception of the impulse pin.

safety arch (*Build.*). See **discharging arch.**

safety barrier (*Aero.*). A net which is erected on the forward part of the deck of an aircraft carrier to stop any aircraft which misses the *arrester gear* (q.v.). A cable and/or nylon net which can be quickly raised to prevent an aeroplane from over-running the end of a runway. The barrier is held by friction brakes or weights so that it imposes a 1 or 2 g deceleration on the aircraft.

safety cage (*Mining*). A cage fitted with a 'safety catch' to prevent it from falling if the hoisting rope breaks.

safety coupling (*Eng.*). A friction coupling adjusted to slip at a predetermined torque, to protect the rest of the system from overload.

safety cut-out (*Elec. Eng.*). An overload protective device in an electric circuit.

safety factor (*Civ. Eng.*). See **factor of safety.**

safety (or nonflam) film (*Cinema.*). Negative or positive film with cellulose acetate or polyester base. So called because it cannot flare up when ignited.

safety finger (*Horol.*). The pin or finger attached to the end of the lever adjacent to the notch. It butts against the edge of the safety roller if the escapement is subject to a jerk, or when setting the hand back, in the case of a watch. Also called **guard pin** and sometimes **dart.**

safety fuse (*Elec. Eng.*). A protective fuse in part of an electric circuit.

safety glass (*Glass*). (1) Laminated glass, formed of a sandwich of a plastic material such as polyvinyl butyral, (0·4-0·75 mm) between two glass sheets, certain intermediate layers being used to facilitate adhesion. (2) Toughened glass, formed by heating a sheet of glass to the point of incipient softening and then chilling it rapidly to a certain extent, but not sufficiently to cause fracture. The stresses have the effect of imparting a considerably greater resistance to shock. (3) Glass reinforced with wire mesh incorporated in the body.

safety height (*Aero.*). The height below which it is unsafe to fly on instruments because of high ground.

safety lamp (*Mining*). Oil-burning miners' lamp which will not immediately ignite firedamp or gas in a coal mine, e.g., a *Davy lamp.* Also used for detecting gas.

safety lintel (*Arch.*). A lintel doing the work of a relieving arch, and serving to protect another more decorative lintel used for architectural reasons.

safety plug (*Eng.*). See **fusible plug.**

safety rail (*Rail.*). See **check rail.**

safety rods (*Nuc. Eng.*). Rods of neutron-absorbing material used for shutting down a nuclear reactor in case of emergency.

safety roller (*Horol.*). The roller mounted on the balance staff of a lever escapement, against the edge of which the guard pin acts when the safety action is brought into play.

safety speed (*Aero.*). The lowest speed above stalling at which the pilot can maintain full control about all three axes. It is particularly applicable to multi-engined aeroplanes, where it is taken to be the minimum speed at which control can be maintained after complete failure of the engine most critical to directional control.

safety switch (*Elec. Eng.*). See **emergency stop.**

safety valve (*Eng., etc.*). A valve, spring or dead-weight loaded, fitted to a boiler or other pressure vessel, to allow fluid to escape to the atmosphere when the pressure exceeds the maximum safe value.

safflorite (*Min.*). Diarsenide of cobalt and iron, sometimes with a small amount of nickel. It crystallizes in the orthorhombic system.

safflower (*Bot.*). Carthamus tinctorius. A thistle-like composite cultivated in India. The dried petals are used for making a red dye and rouge.

safranine (*Photog.*). A basic dye used for desensitizing emulsions, so that they can be developed with appreciable light.

safranines (*Chem.*). A group of azine dyestuffs. They are 2,8-diamino derivatives and have also a phenyl or a substituted phenyl group attached to the nitrogen in position 10. Phenosafranine has the formula:

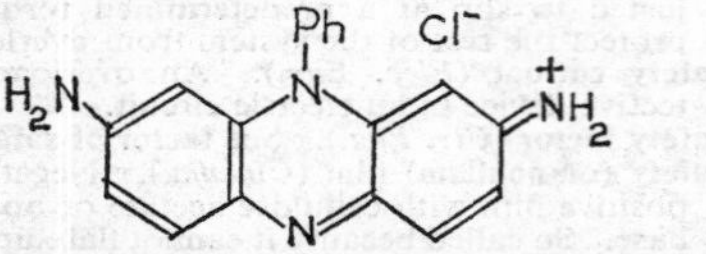

sag correction (*Surv.*). A correction applied to the observed length of a base line, to correct for the sag of the measuring tape.

sage (*Aero.*). Air defence system whereby information is received from radar and other sources and is processed at a central station to give an evaluation of a situation. (*S*emi-*a*utomatic *g*round *e*nvironment.)

saggar (spelling variable) (*Eng.*). A clay box in which pottery is packed for baking.

sagging (*Glass*). Forming glass by reheating until it conforms with the mould or form on which it rests. (*Ships*) This occurs when the ends of the ship are supported on wave crests while the middle is in a trough or when the middle is more heavily loaded than the ends. If the ship actually bends she is said to be *sagged.* Cf. *hogging*.

sagitta (*Civ. Eng.*). (1) See **keystone.** (2) See **rise** (1).

sagittal (*Zool.*). Elongate in the median vertical longitudinal plane of an animal, as the *sagittal* suture between the parietals, the *sagittal* crest of the skull; used also of sections.

sagittal field (*Optics*). The image surface formed by the sagittal foci of a series of object points lying in a plane at right angles to the axis.

sagittal focus (*Optics*). The focus of an object point lying off the axis of an optical system in which the image is drawn out by the astigmatism of the system into a line radial to the optical axis.

sagittate (*Bot., Zool.*). Shaped like an arrowhead with the barbs pointing backwards.

sahlite, salite (*Min.*). A mineral of the clinopyroxene group, intermediate in composition between diopside and hedenbergite.

sailing courses (*Build.*). See **oversailing courses.**

sail-over (*Build.*). To project over. See **oversailing courses.**

sailplane (*Aero.*). A glider designed for sustained motorless flight by the use of air currents. The most advanced methods of streamlining and very high *aspect ratio* are used to reduce *drag* to the barest minimum.

sainfoin (*Bot.*). Onobrychis viciaefolia. A leguminous fodder-plant.

St Andrew's Cross bond (*Build.*). See **Dutch bond.**

St Anthony's fire (*Med.*). See **erysipelas.**

Saint Elmo's fire (*Elec.*). Brush discharge from isolated points above the ground, e.g., ship's mast and aircraft.

St Vitus dance (*Med.*). See **chorea.**

saixe (*Tools*). See **sax.**

saki (*Zool.*). A South American monkey of the genus Pithecia, with long bushy non-prehensile tail.

Sakmarian (*Geol.*). The lowest stage of the Permian system in eastern Europe and the U.S.S.R.

salamander (*Zool.*). A genus (Salamandra) of tailed amphibians, closely related to the newts, harmless but long thought to be poisonous.

Salamander (*Build.*). TN for a fire-resisting decorative building-board made from pulped asbestos.

sal-ammoniac (*Min.*). Chloride of ammonia, which crystallizes in the cubic system. It is found as a white encrustation around volcanoes, as at Etna and Vesuvius. It is used in chemical analysis, in medicine, in dry batteries, as a soldering flux, and in textile printing.

salbutamol(*Pharm.*). A *sympathomimetic* broncho-dilator, used in the treatment of bronchial asthma.

salempores (*Textiles*). Cotton fabrics, with bars of colour across the piece; made for S. American, Indian, and African markets.

salicin (*Chem.*). $C_{13}H_{18}O_7$. A glucoside found in varieties of *Salix* and *Populus*. It is hydrolysed to saligenin and dextrose.

salic minerals (*Geol.*). Those minerals of the *norm* which are rich in *s*ilicon and *al*uminium, including quartz, feldspars, nepheline, lencite and kalsilite.

salicylic acid (*Chem., Pharm.*). $HO{\cdot}C_6H_4{\cdot}COOH$, 2-hydroxybenzene-1-carboxylic acid, colourless monoclinic prisms, m.p. 157°C–159°C, insoluble in cold water, soluble in hot. Iron(III) chloride colours its aqueous solution violet. It is an antiseptic and an important intermediate for a number of derivatives, e.g., aspirin. It can be prepared by heating sodium phenolate to 180°C–220°C in a stream of CO_2.

salient (*Surv.*). A jutting-out piece of land.

salient angle (*Maths., Surv.*). One in a closed figure which points outward, being below 180°. Cf. *re-entrant angle*.

Salientia (*Zool.*). A superorder of Aspidospondyli. Adults four-legged, the hind limbs being especially well-developed, and short-bodied, with no tail. Toads and Frogs. Also known as **Anura, Batrachia.**

salient junction (*Build.*). See **external angle.**

salient pole (*Elec. Eng.*). A type of field pole protruding beyond the periphery of the circular yoke in the case of a stator field system, or the circular core in the case of a rotor field system.

salient-pole generator (*Elec. Eng.*). An alternating-current generator whose rotor field system is of the salient-pole type, e.g., in the case of slow-speed water-turbine-driven generators.

saligenin (*Chem.*). See **phenolic alcohols.**

salina, saline lake (*Geog., Geol.*). A salt lake, lagoon, marsh, spring, etc. See **salt lakes.**

salinometer (*Phys.*). A *hydrometer* (q.v.) for measuring the density of sea water, the stem being scaled in arbitrary units; used by engineers for estimating the amount of dissolved solids in feed water.

salite (*Min.*). See **sahlite.**

saliva (*Zool.*). The watery secretion produced by the salivary glands, whose function is to lubricate the passage of food and, sometimes, to carry out part of its digestion. In Insects saliva may contain amylase, invertase, protease, and lipase, according to the usual diet, and in some blood-sucking insects it contains anticoagulants. In Mammals it contains water, mucin and, in man and some herbivores, the amylase ptyalin, which catalyses the breakdown of starch to maltose.

salivarium (*Zool.*). In some Insects, the posterior or ventral portion of the space between the mouthparts and the labrum, not strictly part of the intestine. Cf. *cibarium.*

salivary glands (*Zool.*). Glands present in many land animals, the ducts of which open into or near the mouth.

Salkowski's test (*Chem.*). A colour test for cholesterol, based on the appearance of a red colour in trichloromethane and of a green in concentrated sulphuric acid when these liquids are shaken up in the presence of cholesterol.

Salk vaccine (*Med.*). Antipolio vaccine, first cultivated on monkey kidney tissue by Dr Jonas Salk of New York in 1955. It consists of killed virus. Cf. *Sabin vaccine.*

salle (*Paper*). A large well-lighted room being the department which sorts, counts, and packs the paper. Also called the **finishing house.**

sallenders (*Vet.*). See **mallenders and-.**

sally (*Carp.*). A re-entrant angle cut into the end of a timber, so as to allow it to rest over the arris of a cross-timber.

salmine (*Chem.*). A protamine isolated from fish testicles.

Salmo (*Zool.*). The salmon and trout genus of fishes, giving name to the family **Salmonidae.**

Salmonella (*Bacteriol.*). A group of Gram-negative, carbohydrate fermenting, nonsporing, bacilli, all pathogenic to animals. Associated with food poisoning in man. Includes *S. typhi* and *S. paratyphi.*

salmonellosis (*Med., Vet.*). Disease of man and warm-blooded animals, due to infection by *Salmonella* bacilli. Characterized by gastroenteritis, sometimes with more serious complications.

Salopian (*Geol.*). Name sometimes applied to the Middle and Upper Series of the Silurian System. This is now subdivided into the Wenlockian and Ludlovian Series.

Salpidae (*Zool.*). An order of *Thaliacea,* with no larval stage, a well-formed free oozooid, no lateral walls to the pharynx, and blastozooids incapable of budding but adhering as a chain and eventually breaking free in groups.

salping-. A prefix from the Gk. *salpinx* (gen. *salpingos*), trumpet, referring esp. to the Fallopian tubes. See **salpinx.**

salpingectomy (*Surg.*). Removal of a Fallopian tube.

salpingitis (*Med.*). Inflammation of a Fallopian tube.

salpingo-oöphorectomy (*Surg.*). Removal of a Fallopian tube and of the ovary on the same side.

salpingo-oöphoritis (*Med.*). Inflammation of both the Fallopian tube and the ovary.

salpingorrhaphy (*Surg.*). The suturing of a Fallopian tube to the ovary on the same side, after a part of the latter has been removed.

salpingostomy (*Surg.*). The operative formation of an opening into a Fallopian tube whose natural opening has been closed by disease.

salpingotomy (*Surg.*). Incision into a Fallopian tube.

salpinx (*Anat., Zool.*). The Eustachian tube; the Fallopian tube; a trumpet-shaped structure. *adj.* **salpingian.**

SALR (*Meteor.*). Abbrev. for *saturated adiabatic lapse rate.*

salsuginous (*Bot.*). Growing on a salt marsh.

salt (*Chem.*). A compound which results from the replacement of one or more hydrogen atoms of an acid by metal atoms or electropositive radicals. Salts are generally crystalline at ordinary temperatures, and form positive and negative ions on dissolution in water, e.g., chlorides, nitrates, carbonates, sulphates, silicates, and phosphates. For *common* or *rock salt,* see **halite.**

saltant (*Biol.*). Variant of a species, developed suddenly, and differing from the original in morphological or physiological characters; term frequently used when genetical basis of change unknown.

saltation (*Biol.*). A sudden variation, often irreversible and inheritable, into a *saltant.*

saltatorial, saltatory (*Zool.*). Used in, or adapted for, jumping, as the third pair of legs in Grasshoppers.

saltatory (*Zool.*). (1) Process of transmission of the nerve impulse in myelinated nerve fibres due to the current flow reaching a maximum at each node of Ranvier, the action currents so developed at each node acting as stimulating currents to the adjacent nodes. (2) See **saltatorial.**

salt bath (*Met.*). A bath of molten salts used for heating steel, for hardening or tempering. Salt baths give uniform heating and some protect against oxidation. Certain salts are employed only to transmit heat to the immersed material, and different salts are used for different temperatures. For *tempering* baths, sodium and potassium nitrate are used. For *hardening* baths, sodium cyanide, and sodium, potassium, barium, and calcium chlorides are used. An electric salt-bath furnace is a conductor-type electric furnace in which the salt is melted by the passage of the current.

salt chuck (*For.*). A salt water log pond.

salt gland (*Bot.*). A hydathode from which a saline exudation oozes, the salts drying on the outside of the leaf. (*Histol.*) Groups of cells found on the gills of certain fishes and presumed to be chloride-secreting. The mechanism of secretion is unknown.

saltigrade (*Zool.*). Progressing by jumps, as Grasshoppers.

salting (*Mining*). Fraudulent enrichment of ore samples, made to increase apparent value of a mine. Originally, to sprinkle salt in dry mines to allay dust. (*Photog.*) When printing papers are being prepared before sensitization, the preliminary soaking in a solution of a chloride, with added size or colloid, to regulate the subsequent contrast.

salting-out (*Chem.*). The removal of an organic compound from an aqueous solution by the addition of a salt.

salt (or saline) lakes (*Geol.*). Enclosed bodies of water in areas of inland drainage, whose concentration of salts in solution is much higher than in ordinary river water. See **soda lakes.**

saltpetre (*Min.*). See **potassium nitrate.**

salts (*Glass*). See **gall.**

salts of sorrel (*Chem.*). See **potassium oxalate (ethandioate).**

salt water (*Glass*). See **gall.**

Salvarsan (*Pharm.*). 3,3′-Diamino-4,4′-dihydroxyarsenobenzene hydrochloride, a yellow, crystalline powder, easily soluble in water, methanol and glycerine. Formerly widely used as a treatment for syphilis; now replaced by the penicillins.

Salviniales. See **Hydropteridineae.**

sal volatile (*Chem.*). Ammonium carbonate, the main constituent of smelling salts.

samara (*Bot.*). A single-seeded, dry, indehiscent fruit, bearing a winglike extension of the pericarp; ash keys provide a familiar example.

samariform (*Bot.*). Winged, like an ash key.

samarium (*Chem.*). A metallic element. Symbol Sm, at. no. 62, r.a.m. 150·35, m.p. 1350°C, b.p. 1600°C, hard and brittle, rel. d. 7·7. Found in allanite, cerite, gadolinite and samarskite. Feeble, naturally radioactive, can be produced by decay of fission fragments, and forms reactor poison.

Samastronic (*Comp.*). See **printer.**

samel bricks (*Build.*). See **grizzle bricks.**

sampling (*Phys.*). Selection of an irregular signal over stated fractions of time or amplitude (pulse height). (*Ecol.*) A fundamental activity of ecologists, consisting of the collection, by appropriate techniques in particular cases, of a small but representative section of a population, from which parameters of the whole population, such as, for example, density, abundance or distribution, can be deduced with a greater or lesser degree of uncertainty. Frequently uses, or interacts with, statistical sampling techniques.

sampling action (*Automation*). Type of control system in which error signal only operates controller intermittently.

sampling circuit (*Automation*). One with output of a form suitable for the error signal in a controller using *sampling action*. See **sampling gate.**

sampling gate (*Automation*). A circuit with an output only when the gate is opened by an activating pulse.

SAN (*Plastics*). A copolymer of *styrene* and *acrylonitrile*.

sanatron (*Electronics*). Valve circuit for fast time bases.

sand (*Foundry*). See **moulding sands.** (*Geol.*) A term popularly applied to loose, unconsolidated accumulations of detrital sediment, consisting essentially of rounded grains of quartz. Restricted in sedimentary petrology to sediments whose grains lie between 1 mm and 0·1 mm diameter; cf. *grit*. In the mechanical analysis of soil, sand, according to international classification, has a size between 0·02 and 2·0 mm. See silt. In coral sand the term implies a grade of sediment the individual particles of which are fragments of coral, not quartz.

sandal bricks (*Build.*). A local term for *grizzle bricks*.

sandalwood (*For.*). A scented wood from several trees of the family *Santalaceae*, found in the East Indies. Important for its essential oil and used for joss-sticks and small ornaments. One species is found in Australia.

sandarach (*Bot.*). Realgar, the resin of the Moroccan sandarach tree.

sand-blasting. A method of cleaning metal or stone surfaces by sand, steel shot, or grit blown from a nozzle at high velocity; also used for forming a key on the surface of various materials requiring a finish, such as enamel.

sand-calcites (*Geol.*). See **Fontainebleau Sands.**

sand casting (*Foundry*). Formation of shapes by pouring molten metal into a shaped cavity in a moulding flask.

sand colic (*Vet.*). Colic caused by the collection of sand in the intestines.

sandcrack (*Vet.*). A fissure of the horse's hoof.

sand cushion (*Civ. Eng.*). A bag of sand placed beneath a helmet to protect the top of the pile from damage due to impact of the monkey when it is being driven.

sand dunes (*Geol.*). Rounded or crescentic mounds of loose sand which have been piled up by wind action on seacoasts or in deserts. See also **barkhan.**

sand fill (*Mining*). Underground support of worked-out stopes by return of ore tailings from mill, usually by hydraulic flow. See **stowing.**

sand filter (*Civ. Eng.*). A bed of sand arranged in layers of decreasing texture, i.e., the fine sand at the top and the coarser layers underneath. Used widely in purification of public water supplies and in lime soda water softeners.

sandfly fever (*Med.*). Phlebotomus fever. An acute disease caused by infection with a virus conveyed by the bite of a sand-fly *Phlebotomus papatasii*; characterized by a 3-day fever, pains in the joints and the back, diarrhoea, and a slow pulse.

sand lime bricks (*Build.*). Bricks made by mixing suitable sand with approx. 6% of hydrated lime and water, moulding under high pressure and then curing in steam at high pressure.

Sandmeyer's reaction (*Chem.*). The replacement of the diazonium group, $-N_2^+$, in a diazonium compound by chlorine, bromine, or the cyanogen radical, which is effected by heating a solution of the diazonium compound with, e.g., a concentrated solution of cuprous chloride in hydrochloric acid. In this case the diazonium group is replaced by Cl, with evolution of gaseous N_2.

sandpaper (*Carp., etc.*). Stout paper or cloth with a thin coating of fine sand glued on to one side, for use as an abrading material.

sand pump (*Civ. Eng.*). See **sludger.**

sand-pump dredger (*Civ. Eng.*). A long pipe reaching down from a vessel into the sand, the latter being raised under the suction of a centrifugal pump and discharged into the vessel itself or an attendant barge. Also called a **suction dredger.**

sands (*Met., Min. Proc.*). Particles of crushed ore of such a size that they settle readily in water and may be leached by allowing the solution to percolate. See also **slimes.**

sandslinger (*Eng.*). A machine for reproducing the action of a moulder in filling a mould by hand. Sand is delivered in wads at high speed by centrifugal force, and directed by the operator into the mould as required.

sandstones (*Geol.*). Compacted and cemented sedimentary rocks, which consist essentially of rounded grains of quartz, between the diameters of 1 mm and 0·1 mm, with a variable content of 'heavy mineral' grains. According to the nature of the cementing materials the varieties *calcareous sandstone, ferruginous sandstone, siliceous sandstone* may be distinguished; *glauconitic sandstone, micaceous sandstone*, etc., are so termed from the presence in quantity of the mineral named.

sand trap (*Paper*). An inclined trough across which bars are set at intervals. During the passage of the pulp to strainers, any heavy particles such as sand sink to the bottom and are retained by the bars. Also called a **riffler.**

sand volcano (*Geol.*). A structure formed by sand flowing upwards through an overlying bed of sediment and spilling out on to the surface; not related to any type of volcanic activity or volcanic rock.

sandwich (*Nuc. Eng.*). Photographic nuclear

research emulsion forming series of thin layers with intervening layers of a material in which some event or process is to be studied.

sandwich beam (*Build.*). See **flitch beam.**

sandwich compounds (*Chem.*). Compounds in which a metal atom is 'sandwiched' between two rings, e.g., dibenzene chromium:

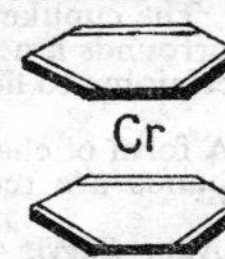

Ferrocene (dicyclopentadienyl iron) is another example.

andwich construction (*Aero.*). Structural material, mainly used for skin or flooring, possessing exceptionally good stiffness for weight characteristics. It consists of two approximately parallel thin skins with a thick core having different mechanical properties, so that the tensile and compressive stresses develop in the skin, and the core both stabilizes these surfaces and gives great strength in bending; core materials range from balsa wood through metal-foil honeycomb (light-alloy or steel) to corrugated sheet.

sandwich irradiation (*Radiol.*). The irradiation of tissues from opposite sides.

sandy clay (*Build.*). See **loam.**

sanguicolous (*Zool.*). Living in blood.

sanguiferous (*Zool.*). Blood-carrying.

sanguivorous (*Zool.*). Blood-feeding, as Fleas.

sanidine (*Min.*). A form of potassium feldspar identical in chemical composition with orthoclase, but physically different, formed under different conditions and occurring in different rock types. It is the high-temperature form of orthoclase, to which it inverts at below 900°C. Occurs in lavas and dyke-rocks.

sanies (*Med.*). A thin, offensively smelling discharge of pus, mixed with blood or with serum.

Sanio's band (or beam) (*Bot.*). See **trabecula.**

Sankey diagram (*Chem. Eng.*). See **heat balance.**

sanserif (*Typog.*). A type face without serifs, e.g., Gill Sans.

Sanson-Flamsteed projection (*Geog.*). See **sinusoidal projection.**

santonin (*Pharm.*). Anthelmintic drug, largely superseded because of its toxic effects.

Santorin (*Build., Civ. Eng.*). A natural pozzuolana or volcanic ash from the Greek island of Santorin (Thera).

Santorini, cartilages of (*Zool.*). In the larynx of Mammals, two small cartilaginous nodules situated at the apices of the arytenoids.

Santorini's duct (*Zool.*). The dorsal or accessory pancreatic duct of Mammals. Cf. *Wirsung's duct.*

sap (*Bot.*). An aqueous solution of mineral salts, sugars, and other organic substances, present in the xylem of plants.

sap cavity (*Bot.*). The large vacuole filled with fluid, which occupies the middle of an adult cell.

sapele (*For.*). An African tree, *Entandrophragma cylindricum*, yielding a mahogany-like silky-grained hard-wood used for furniture, etc.

saphir d'eau (*Min.*). French 'water sapphire'. An intense-blue variety of the mineral cordierite, occurring in water-worn masses in the river gravels of Ceylon; used as a gemstone.

sapogenins (*Chem.*). Hydroxy-derivatives by hydrolysis of *saponins* (q.v.), e.g., *digitogenin.*

saponification (*Chem.*). The hydrolysis of esters into acids and alcohols by the action of alkalis or acids, or by boiling with water, or by the action of superheated steam. It is the reverse process to *esterification* (q.v.) if acids are used, but, when alkalis are used then soaps result, hence the term.

saponification number (*Chem.*). The number of milligrams of potassium hydroxide required to saponify 1 g of a fat or oil.

saponins (*Chem.*). Steroid vegetable glycosides that act as emulsifiers of oils. They dissolve the red corpuscles, irritate the eyes and organs of taste and are toxic to lower animals, e.g., *digitonin*, found in *Digitalis purpurea.*

saponite (*Min.*). A hydrous hydrated silicate of magnesium, with some aluminium. A clay mineral of the smectite (montmorillonite) group, occurring as white soapy masses in serpentinite. Also called **bowlingite.**

sapphire (*Min.*). The fine blue transparent variety of crystalline corundum, of gemstone quality; obtained chiefly from Ceylon, Kashmir, Thailand and Australia.

sapphire needle (*Acous.*). A gramophone record reproducing stylus ground from natural sapphire; by virtue of its hardness in comparison with that of the record surface, it is relatively hard-wearing although inferior in this respect to the diamond stylus.

sapphire quartz (*Min.*). A very rare indigo blue variety of silicified crocidolite occurring at Salzburg; used as a semiprecious gemstone. Also known as **azure quartz.**

sapphirine (*Min.*). A silicate of magnesium and aluminium with lesser iron, crystallizing in the monoclinic system. It occurs as blue grains in metamorphosed, aluminous, silica-poor rocks.

sapphism (*Psychiat.*). Lesbianism.

sappy wool (*Textiles*). Raw wool containing an excessive amount of natural grease.

sapr-, sapro-. Prefix from Gk. *sapros*, rotten, rancid.

sapraemia, sapremia (*Med.*). The presence in the blood of toxic products resulting from the putrefactive action of saprophytes on dead tissue (e.g., on retained discharges and placental tissue in the uterus); the condition (intoxication) that results from this.

saprobe (*Bot.*). A plant growing in foul water.

saprobiotic (*Biol.*). Feeding on dead or decaying animals or plants.

saprogenous (*Bot.*). Growing on decaying matter.

sapropel (*Geol.*). Slimy sediment laid down in stagnant water, largely consisting of decomposed organic material.

sapropelic (*Zool.*). Said of aquatic organisms which live in the decaying organic matter of a muddy bottom.

sapropelite (*Geol.*). A term applied by H. Potonié to coals derived from algal materials. Cf. *humite.*

saprophile (*Ecol.*). An animal not properly characteristic of stagnant and polluted waters, but capable of breeding therein by reason of its ability to resist or overcome oxygen-scarcity.

saprophilous (*Bot.*). Saprogenous.

saprophyte (*Bacteriol.*). Any bacterium which breaks up dead animal and vegetable matter and does not produce disease in the animal or plant which it inhabits. More generally, an organism which obtains its food from dead organic material.

saprophyte chain (*Ecol.*). A type of food chain which goes from dead organic matter to micro-organisms. Cf. *parasite chain, predator chain.*

saprophytic (*Biol.*). Feeding on dead or decaying organic materal. *n.* **saprophytism.**

saproplankton (*Bot.*). Plankton growing in foul water.
saprozoic (*Biol.*). Feeding on dead or decaying organic material.
sapstain (*For.*). A discoloration produced in the sapwood of felled timber by the growth of certain fungi.
sapwood (*Bot.*). The layer of recently formed secondary wood forming a sheath over the whole of the surface of the xylem of a woody plant; it contains living cells, conducts sap, and is usually light-coloured. Also **alburnum.**
Sarah (*Radar*). TN (*s*earch-*a*nd-*r*escue-*a*nd-*h*oming) of a small radar transmitter attachable to an airman's life jacket.
Saran (*Plastics*). Synthetic fibre or film based on a copolymer of vinylidene chloride and vinyl chloride.
sarc-, sarco-. Prefix from Gk. *sarx*, gen. *sarkos*, flesh.
sarcenchyma (*Zool.*). A form of parenchyma occurring in Sponges, characterized by closely packed granular cells in a reduced gelatinous matrix.
sarciniform (*Bot.*). In the form of small packets.
sarcoblast (*Zool.*). An elongate multinucleate muscle cell.
sarcocarp (*Bot.*). The fleshy part of the pericarp of a drupe.
sarcocyte (*Zool.*). A middle layer of the ectoplasm in some *Protozoa*.
sarcodic, sarcodous, sarcoid (*Zool.*). Pertaining to or resembling flesh.
Sarcodina (*Zool.*). A class of *Protozoa* which in their principal phase are amoeboid, without flagella, are usually not parasitic, and, though they may have a phase of sporulation, do not form large numbers of spores after syngamy. Also called **Rhizopoda.**
sarcody (*Bot.*). Conversion into something of fleshy texture.
sarcoidosis (*Med.*). A disease, cause unknown, characterized by tubercle-like nodules; affects the lungs particularly, and other organs.
sarcolemma (*Zool.*). The extensible sheath of a muscle fibre enclosing the contractile substance.
sarcolytes (*Zool.*). In the histolysis which accompanies metamorphosis in some Insects, fragments of muscle fibres.
sarcoma (*Med.*). A malignant tumour of connective tissue origin (e.g., of fibrous tissue, bone, cartilage); the tumour invades adjacent tissue and organs, and metastases are formed via the blood stream. *pl.* **sarcomata** or **sarcomas.** Cf. *carcinoma.* (*Zool.*) The fleshy portion of an animal body as opposed to the skeletal portion.
sarcomatosis (*Med.*). The presence of many sarcomata in the body.
sarcomatous (*Med.*). Pertaining to, of the nature of, or resembling sarcoma.
sarcomeres (*Zool.*). The serial portions into which a muscle fibril is divided by the membranes of Krause.
sarcophagous (*Zool.*). Feeding on flesh.
sarcoplasm (*Zool.*). The interfibrillar protoplasm of muscle fibres.
sarcoplasmic reticulum (*Zool.*). In striated muscle of Vertebrates, a system of submicroscopic tubules which transmit action potentials and allow the circulation of energy, in the form of ATP and other substances.
sarcoptic mange (*Vet.*). Mange of animals due to mites of the genus *Sarcoptes*.
sarcosine (*Chem.*). Monomethyl-glycine, $H_3C\cdot HN\cdot CH_2\cdot COOH$. It is obtained by the decomposition of creatine or caffeine. Crystals, m.p. 212°C, readily soluble in water. It may be synthesized from chloroacetic ester and aminomethane, or by hydrolysis with barium hydroxide of methylaminoethanonitrile.
sarcosporidiosis (*Vet.*). Infection of the muscles of swine, sheep, horses, cattle, goats, and birds by the parasitic *Sarcosporidia.*
sarcotesta (*Bot.*). A fleshy layer in a testa.
sarcotheca (*Zool.*). The cuplike expansion of the perisarc which surrounds a nematophore.
sarcous (*Zool.*). Pertaining to flesh; pertaining to muscle tissue.
sardonyx (*Min.*). A form of chalcedony in which the alternating bands are reddish-brown and white. Cf. *onyx.*
Sargent diagram (*Nuc. Eng.*). Log-log plot of radioactive decay constant against maximum β-ray energy, for various β-emitters.
sari (*Textiles*). Plain woven light cotton fabric distinguished by fancy coloured borders woven or printed and deep fancy heading at each end. Sari resembles *dhootie* (q.v.) and is worn by Indian women.
sarking felt (*Build.*). A bituminous underlining placed beneath slates or tiles.
sarmentose (*Bot.*). Having a stem which arises in a small arch from the root, and then becomes prostrate.
sarong (*Textiles*). Plain woven, warp striped cotton fabric with deep headings, fancy borders, and white selvedges. Used as dress material in the Far East.
saros (*Astron.*). A cycle of 18 years 11 days, which is equal to 223 synodic months, 19 eclipse years and 239 anomalistic months. After this period the relative positions of the sun, moon, and node recur; known to the ancient Babylonians, it was used to predict eclipses.
sarothrum (*Zool.*). The pollen-brush of a honey bee, i.e., the first tarsal joint of the third leg, bearing stiff straight bristles.
sarsen (*Geol.*). Irregular masses of hard sandstones which are found in the Reading and Bagshot Beds of the Tertiary System in Southern England. They often persist as residual masses after the softer sands have been denuded away.
sartorius (*Zool.*). A thigh muscle of *Tetrapoda* which by its contraction causes the leg to bend inwards.
Sartorius sedimentation balance (*Powder Tech.*). One in which the pan in the suspension is compensated and the weight of particles sedimented on to the pan is recorded by an electronic pen recorder.
sash (*Join.*). A framing for window panes.
sash and frame (*Join.*). A cased frame in which counterweighted sashes slide vertically.
sash bar (*Join.*). A *transom* or a *mullion.*
sash centres (*Join.*). The points about which a pivoted sash is moved.
sash cramp (*Join.*). A contrivance for holding parts of a frame in place during construction. It usually consists of a steel bar along which slide two brackets between which the work is fixed, one of the brackets being pegged into a hole in the bar while the other is adjustable for position by means of a screw.
sash door (*Join.*). A door which has its upper part glazed.
sash fastener (or lock) (*Join.*). A fastening device secured to the meeting rails of the sashes of a double-hung window, serving to fix both sashes in the shut position.
sash fillister (*Join.*). A special plane for cutting grooves in stuff for sash bars.
sash mortise chisel (*Join.*). One somewhat lighter than a mortise chisel, used for mortising softwood.

sash pocket chisel (*Join.*). A strong-bladed chisel with a narrow edge, used for cutting the pocket in the pulley stile of a sash and frame.

sash rail (*Join.*). See **transom.**

sash saw (*Join.*). A saw similar to the tenon saw but slightly smaller and finer; used for making window sashes.

sash stuff (*Join.*). The timber prepared for use in the making of sashes.

sash weights (*Join.*). Weights which are used as counter-poises in balancing the sashes of windows.

sassafras (*For.*). Tree of the family *Lamaceae*, yielding from the bark an essential oil used in perfumery and cosmetics, and from the root an extract used for flavouring.

sassolite (*Chem.*). See **boric acid.**

sateen (*Textiles*). Cotton or spun rayon fabric with a weft surface, with only one interlacing of each warp and weft in the weave repeat. Made in numerous qualities on dobby looms for linings and dress goods. Complementary to a satin which has a warp surface.

satellite (*Astron.*). The name given to a smaller body revolving round another, generally a planet, e.g., the moon, which is the earth's satellite. (*Cyt.*) A small part of a chromosome, attached to one end of the main body of the chromosome by a fine threadlike connexion. Also called **trabant.** (*Telecomm.*) A vehicle put into orbit around the earth or other body in space and used to transmit back information (e.g., *Explorer*, q.v.) or as a means of communication. A *communications satellite* (q.v.) orbits the earth and reflects or transmits back radio waves using highly directive antennae for transmission and reception, orientated by computer calculation of orbit (e.g., *Echo*, *Telstar*, *Early Bird*).

satellite computer (*Comp.*). One used to relieve a central processing device of relatively simple but time-consuming operations, such as compiling, editing and controlling input and output devices.

satellite exchange (*Teleph.*). A small automatic-telephone exchange which is dependent on a main automatic exchange for completion of its calls to subscribers other than those connected to it.

satellite station (*Surv.*). In triangulation by theodolite, one resected and thus fixed for reference purposes, but not occupied. (*Telecomm.*) (1) One which re-broadcasts a transmission received directly, but on another wavelength. (2) One designed for transmission to, and reception from, earth satellites. See **satellite.**

satin (*Textiles*). A silk or rayon (acetate or viscose) fabric with a smooth lustrous surface devoid of pattern; four-fifths of the warp usually appears on the face.

satin drill (*Textiles*). A cotton cloth, bleached or dyed, which is frequently worn in tropical countries. Cloths of this type, dyed brown, blue or khaki are also used as overalls everywhere.

satinet, satinette (*Textiles*). Similar to satin and sateen, but with weaves which repeat on only 4 ends and 4 picks. Formed by breaking and reversing the 3-and-1 twill every 2 ends.

satin leather (*Leather*). Leather with a perfectly smooth finish and without grain marks.

satin spar (*Min.*). Name given to fine fibrous varieties of calcite, aragonite, and gypsum, the gypsum variety being distinguished from the others by its softness (it can be scratched by a finger-nail).

satin walnut (*For.*). Heartwood of *American red gum* (q.v.). Varies in colour from brown to reddish-brown. Chiefly used for furniture, fittings, cooperage, and panelling.

satinwood (*For.*). See **East Indian-.**

saturable reactor (*Elec. Eng.*). Inductor in which the core is saturable by turns carrying d.c., which controls the inductance. Used for modulation, control of lighting, and developed in the magnetic amplifier.

saturated adiabatic lapse rate (*Meteor.*). The *adiabatic lapse rate* (q.v.) which occurs while water vapour condenses within the parcel of air. About 6°C/km at ordinary temperatures but increases at lower temperatures. Cf. *dry adiabatic lapse rate*.

saturated calomel electrode (*Chem.*). A calomel electrode containing saturated potassium chloride solution.

saturated compounds (*Chem.*). Compounds which do not contain any free valencies and to which no hydrogen atoms or their equivalent can be added, i.e., which contain neither a double nor a triple bond.

saturated diode (*Electronics*). A diode valve having a tungsten filament giving a limited emission; used for charging or discharging a capacitor at a constant rate in some forms of linear time base.

saturated humidity mixing ratio (*Meteor.*). The *humidity mixing ratio* (q.v.) which will produce *saturation of the air* (q.v.) at the existing temperature.

saturated solution (*Chem.*). A solution which can exist in equilibrium with excess of the dissolved substance as a second phase.

saturated steam (*Eng.*). Steam at the same temperature as the water from which it was formed, as distinct from steam subsequently heated. See **dry steam.**

saturated vapour (*Phys.*). A vapour which is sufficiently concentrated to exist in equilibrium with the liquid form of the same substance.

saturated water vapour pressure (*Meteor.*). The maximum *water vapour pressure* (q.v.), which can occur when the water vapour is in contact with a free water surface at a particular temperature. The water vapour pressure existing when effective evaporation ceases.

saturation (*Acous.*). See **acoustic-.** (*Electronics*) (1) Condition obtaining when all the electrons emitted from the cathode are swept away to the anode or other electrodes, so that further increase in anode potential produces no corresponding increase in anode current. (2) A similar condition in a screened-grid or pentode valve when all the electrons which pass the screen grid go on to the anode, although the cathode emission is not the limiting factor. (*Mag.*) Application of a sufficiently intense magnetizing force to result in maximum temporary or permanent magnetization in magnetic material. Usually applied by large currents from an impulse current transformer. (*Nuc. Eng.*) Condition where field applied across ionization chamber is sufficient to collect all ions produced by incident radiation. (*Optics, Photog.*) Degree to which a colour departs from white and approaches the pure colour of a spectral line. *Desaturation* is the inverse of this.

saturation activity (*Nuc. Eng.*). The limiting artificial radioactivity induced in a given sample by a specific level of irradiation.

saturation coefficient (*Build.*). The ratio between the natural capacity of a material (such as a building-stone) to absorb moisture and its porosity.

saturation current (*Elec. Eng.*). The steady

current in a winding of an iron-cored transformer which causes the inductance of the winding to be seriously reduced. (*Electronics*) That passed by a valve, phototube, or ion chamber, when operated under saturation conditions. Also **total emission.**

saturation curve (*Mag.*). The characteristic curve relating magnetic flux density to the strength of the magnetic field.

saturation factor (*Elec. Eng.*). The ratio of the increase of field excitation to the increase of generated voltage which it produces.

saturation limit (*Elec. Eng.*). The maximum flux density economically attainable.

saturation of the air (*Meteor.*). The air is said to be saturated when the water vapour contained in it is at *saturated water vapour pressure*.

saturation scale (*Optics*). Minimum visual steps of saturation in scale of spectrum colours, varying with wavelength.

saturation signal (*Radar*). A signal whose amplitude is larger than the dynamic range of the receiver.

saturation voltage (*Nuc. Eng.*). Voltage applied to a device to operate under saturation conditions.

Saturn (*Astron.*). One of the four giant planets of the solar system, remarkable for the system of flat rings which surrounds it. There are two bright rings, separated by a gap (*Cassini's division*) and an inner dusky ring. The whole system is 272 000 km in external diameter, and is 61 500 km wide, but is very thin; it consists of individual particles, possibly of frozen gases. The planet revolves in 29·46 years at a distance of 9·539 A.U. from the Sun; compared with the earth its mass is 95·2 and volume 743·6, so that its density is only 0·71. The atmosphere is probably similar to that of Jupiter. There are ten satellites, of which Titan, the largest, is known to possess an atmosphere.

satyriasis or satyromania (*Psychiat.*). See **gynaecomania.**

saucer (*Eng.*). See **cup wheel.** (*Hyd. Eng.*) A flat form of camel, used for raising a vessel in shallow waters.

saucisse, saucisson (*Civ. Eng.*). Similar to a *fascine*, but of stouter and longer material.

sauconite (*Min.*). One of the montmorillonite group of clay minerals. See **smectites.**

Saurischia (*Zool.*). An order of the *Archosauria*, including extinct 'dinosaurs' with a triradiate pelvis and two or one antorbital vacuities in the skull. Included such forms as *Tyrannosaurus* and *Brontosaurus*. Cf. *Ornithischia*.

saussurite (*Geol., Min.*). Once thought to be one mineral, *saussurite* consists of an aggregate of albite and zoisite or epidote, occasionally together with prehnite and calcite. It results from the alteration of feldspars.

savanna (*Ecol.*). A biome consisting of grassland with scattered trees or clumps of trees, found in Africa, South America and Australia. The few plant species must be resistant to drought and fire, and there are many Mammalian herbivores.

save-all (*Paper*). A trough to collect water draining from the stock under the paper machine wire; or a system for separating the solids from the water in the backwater.

Savi's ampulla (*Zool.*). In *Selachii*, a long tube expanding internally into an ampulla and opening externally on the surface of the head, containing mucus and innervated by a branch of the lateralis.

SAW (*Radar*). Abbrev. for *surface acoustic wave*.

saw arbor (*For.*). The shaft on which a circular saw is mounted.

sawney (*Textiles*). All the ends (yarns) broken at once on a mule or ringframe, usually by mechanical failure.

sawtooth oscillator (*Telecomm.*). See **relaxation oscillator, time-base generator.**

sawtooth roof (*Build.*). A roof formed of a number of north light trusses, presenting a serrated profile when viewed from the end elevation.

sawtooth truss (*Eng.*). A truss used for small-span roofs of sawtooth form, braced by vertical and diagonal members.

sawtooth wave (*Electronics*). One generated by a *time-base generator* for scanning in a CRT, for uniform sweep and high-speed return. See **flyback.**

sax (*Tools*). An axe used for shaping slates; it has a pointed peen for piercing the nail-holes. Also called **saixe, slate axe.**

saxicavous (*Zool.*). Rock-boring.

saxicole, saxicolous (*Bot.*). Growing on rocks.

saxonite (*Geol.*). A coarse-grained, ultramafic igneous rock, consisting essentially of olivine and orthopyroxene, usually hypersthene. A hypersthene-peridotite.

saxony (*Textiles*). Woollen yarn, or cloth, made from wool of good quality. It may have a milled, clear-cut finish, be raised, or finished with a rough surface.

Saybolt colour (*Chem.*). Test result for the colour of white oils.

Saybolt Universal second (*Eng.*). U.S. unit of measurement used in viscometry. See **viscosity.**

say-cast (*Textiles*). Coarse part of wool fleece, at the tail end.

Sayers winding (*Elec. Eng.*). A type of armature winding in which commutator sparking is prevented by means of additional coils influenced by auxiliary commutating poles.

Sb (*Chem.*). The symbol for *antimony*.

S.B. (*Radio*). Abbrev. for *simultaneous broadcasting*.

SBA (*Aero.*). See **standard beam approach.**

S.B.A.C. (*Aero.*). Abbrev. for *Society of British Aerospace Companies*.

SB alloy (*Elec. Eng.*). A resistance material having a low-temperature coefficient of resistance.

S.B.C. (*Elec. Eng.*). Abbrev. for *small bayonet cap*.

SBP spirit (*Chem.*). Special boiling-point spirit.

Sc (*Chem.*). The symbol for *scandium*.

scab (*Vet.*). See **mange.**

scabbling (*Build.*). The operation of rough-dressing a stone face with an axe, prior to smoothing it.

scabbling hammer (*Build.*). The pointed hammer used in the rough-dressing of a nigged ashlar.

scabellum (*Zool.*). In *Diptera*, the dilated basal portion of a haltere.

scabies (*Med.*). A contagious skin disease caused by the acarine parasite *Sarcoptes scabiei*, the female of which burrows in the horny layer of the skin. (*Vet.*) See **mange.**

scabrid (*Bot.*). Having a surface rough like a file.

scabrous (*Bot.*). Having a surface roughened by small wartlike upgrowths. *dim.* **scaberulous.**

scaffold, scaffolding (*Build.*). A temporary erection of timber or steelwork, used in the construction, alteration, or demolition of a building, to support or to allow of the hoisting, lowering, or standing of men, materials, etc.

scaffold poles (*Build.*). Round timber from 6 to 15 cm (2½ to 6 in) diameter in the middle.

scala (*Zool.*). A ladderlike structure, as the canals in the cochlea of the Mammalian ear known as the *scala vestibuli*, the *scala tympani*, and the *scala media*. *pl.* **scalae.**

scalar (*Maths.*). A real number or element of a field. The term *scalar* is used in vector geometry and in vector and matrix algebra to contrast numbers with vectors and matrices. See also **vector space.**

scalar matrix (*Maths.*). A device by which an ordinary number (or scalar), λ, may be expressed as a matrix. It is defined as λI, where I is the unit matrix, and has λs in the principal diagonal and zeros elsewhere.

scalar product (*Maths.*). Of two vectors, a scalar (i.e., real number) equal to the product of the magnitudes of the two vectors and the cosine of the angle between them. Denoted by a dot, e.g., **A**. **B**.

scalariform cell or tracheide or vessel (*Bot.*). A cell, tracheide, or vessel having wall-thickenings which give a ladderlike (**scalariform**) pattern.

scale (*Acous.*). (1) Ratio of length to diameter of organ pipe, a factor which, among others, determines the timbre of a note. (2) Set of (usually) 12 notes forming an octave or frequency range of 2 to 1. (*Bot.*) (1) A thin, flat, semitransparent plant member, usually of small size, and green only when very young, if then. (2) A hardened, usually nongreen bract of a catkin. (*Instr.*) Numerical factor relating measured quantity to indication of instrument. (*Maths.*) The ratio between the linear dimensions of a representation or model and those of the object represented, as on a map or technical drawing. (*Met.*) Mill, roll, hammer scale are oxide coats formed on surface of metal. May require removal by pickling or de-scaling. (*Photog.*) The range of densities over which a constant gamma is obtained when an emulsion is normally processed; it measures the acoustic contrast obtainable when the emulsion is used for photographic recording. (*Zool.*) A small exoskeletal outgrowth of tegumentary origin, of chitin, bone, or some horny material, usually flat and platelike.

scale bark, scaly bark (*Bot.*). (1) Bark which becomes detached in irregular patches. (2) *Rhytidome* (q.v.).

scale factor (*Comp.*). Any numerical which alters a numeric attached to the magnitude of a quantity, especially in calculations with an analogue computer, or that related to the *radix*, which shifts the significant figures in a digital computer.

scale hair (*Bot.*). A multicellular flattened hair.

scale leaf (*Bot.*). A leaf, usually reduced in size, membranous, of tough texture, and ordinarily protective in function.

scalene (*Zool.*). In higher Vertebrates, applied to a pair of muscles of the body wall passing from the ribs along the side of the neck.

scalene triangle (*Maths.*). A triangle in which no two sides are equal.

scale-of-ten (*Telecomm.*). Ring, or other, system of counting elements which divides counts by 10.

scale-of-two (*Telecomm.*). Any bistable circuit which can divide counts by 2, when operated by pulses. See **flip-flop.**

scaler (*Telecomm.*). Instrument, incorporating one or more scaling circuits, used to register a count.

scaling (*Acous.*). The adjustment of the notes of a musical instrument to a specified scale, e.g., natural scale or tempered scale. See **aliquot scaling.** (*Mining*) Removal of rock loosened by blast, by use of *scaling bar*, a long light rod which prizes off dangerous fragments.

scaling circuit (*Telecomm.*). One which divides counts of pulses by an integer, so that they are more readily indicated, to a required degree of accuracy. If scalers of 10 are used in cascade, indications of counts in decimal numbers are possible, e.g., by *Dekatrons.*

scaling hammer (*Eng.*). See **boilermaker's hammer.**

scalloping (*Electronics*). The axial variation of the focusing field in a travelling-wave tube which gives rise to a corresponding variation of the cross-section of the electron beam.

scallops (*Nuc. Eng.*). Alternating sections, with reverse curvature, in a stellarator, to avoid drift in the plasma.

scalpella (*Zool.*). In *Diptera*, a pair of pointed processes belonging to the mouth parts. *sing.* **scalpellum.**

scalpriform (*Zool.*). Chisel-shaped, as the incisor teeth of Rodents.

scaly leg (*Vet.*). A form of mange affecting the feet and legs of fowl, due to the mite *Cnemidocoptes mutans* burrowing in the skin.

scan (*Radar*). Systematic variation of a radar beam direction for search or angle tracking. See **A, B, etc., display.** (*TV*) See **scanning.**

Scan-a-Colour (*Print.*). An *electronic engraving* (q.v.) machine for making sets of colour plates, developed from the *Scan-a-Graver.*

Scan-a-Graver (*Print.*). An *electronic engraving* (q.v.) machine for making same-size plates on flexible plastic sheet.

scan area (*TV*). The area swept out on the screen by the scanning beam.

Scan-a-Sizer (*Print.*). An *electronic engraving* (q.v.) machine for making plates on flexible plastic sheet to enlarged or reduced sizes, developed from the *Scan-a-Graver.*

Scanatron (*Print.*). Electronic equipment for producing sets of colour-corrected separations.

scandium (*Chem.*). A metallic element, classed with the rare earth metals. Symbol Sc, at. no. 21, r.a.m. 44·956. It has been found in cerite, orthite, thortveitite, wolframite, and euxenite; discovered in the last named. Scandium is the least basic of the rare earth metals.

scanner (*Telecomm.*). Mechanical arrangement for covering a solid angle in space, for the transmission or reception of signals, usually by parallel lines or *scans.*

scanning (*Radar*). Coverage of a prescribed area by a directional radar antenna or sonar beam. (*TV*) Coverage of original or reproduced TV image, now generally by interlaced horizontal lines controlled electronically, but originally by mechanically controlled disks, drums or mirrors. (*Comp.*) The checking of recorded data (punched cards, letters, figures, etc.) by means of an electronic device.

scanning beam (*TV*). Any beam of light or electrons which scans a TV image.

scanning coils (*TV*). Coils mounted in CRT, and carrying suitable currents for deflecting electron beam, so as to sweep the picture area.

scanning disk (*TV*). (1) Disk with colour-filter sectors, used in colour-sequence TV. (2) Rotating disk carrying a series of apertures, lenses, mirrors, or other optical devices, used for mechanical scanning in primitive TV systems. See **Nipkow disk.**

scanning field (*TV*). Same as **raster.**

scanning frequency (*TV*). Number of times an image is scanned each second, e.g. 25 in Europe and 30 in North America. The scan is synchronized with the supply mains so that any interference produced is static.

scanning heating (*Heat*). *Induction heating* where the workpiece is moved continuously through the heating region, as in *zone refining* of germanium. Also called **progressive heating.**

scanning line (*TV*). Trace of single traverse of the picture by the scanning spot from side to side in horizontal scanning, or vertically in vertical scanning.

scanning linearity (*TV*). Uniformity of scanning speed for a CRO or TV receiver. This is necessary to avoid waveform or picture distortion.

scanning loss (*Radar*). That which arises from relative motion of a scanning beam across a target, as compared with zero relative motion.

scanning microscope (*Phys.*). An instrument which enables the surface structure of a specimen to be examined. The specimen is scanned by a small-diameter electron probe which gives rise to secondary-electron emission from the specimen. The secondary-emission current is detected and causes a picture to be projected on a cathode-ray screen.

scanning slit (*Cinema.*). In a sound-film projector, a slit in front of the photocell on to which an image of the sound-track is projected. The dimensions are accurately calculated since the frequency response varies with the effective width of the slit.

scanning speech (*Med.*). A disturbance of speech in which the utterance is slow and halting, the words being broken into syllables; a sign of a lesion in the nervous system, as in disseminated sclerosis.

scanning speed (*TV*). That of a scanning spot across the screen of a CRT. Usually accurately specified for a CRO to facilitate time measurements.

scanning spot (*TV*). Spot of light formed by scanning beam on the screen of a reproducer, or, in some early forms of transmission, on the object being televised.

scan plates (*Print.*). A general term for plates made by *electronic engraving*.

scansion (*TV*). Same as **scanning.**

scansorial (*Zool.*). Adapted for climbing trees.

scantling (*Build.*). Stones more than 2 m (6 ft) long. (*For.*) A piece of timber 5–10 cm (2–4 in) thick and 5–11 cm wide.

scapal organ (*Zool.*). A sensory structure situated in the expanded base of a haltere in *Diptera*.

scape (*Bot.*). A peduncle, quite or nearly leafless, arising from the middle of a rosette of leaves, and bearing a flower, several flowers, or a crowded inflorescence; the dandelion furnishes a familiar example. *adj.* **scapigerous.** (*Build.*) See **apophyge.** (*Zool.*) The basal joint of the antenna in Insects.

'scape pinion, 'scape wheel (*Horol.*). See **escape pinion, escape wheel.**

scaph-, scapho-. Prefix from Gk. *skaphē*, boat.

scaphium (*Zool.*). (1) In *Ostariophysi*, one of the Weberian ossicles. (2) In male *Lepidoptera*, part of the genital apparatus. (3) More generally, any boat-shaped structure.

scaphocephalic (*Med.*). Having a keel-shaped head.

scaphocerite (*Zool.*). In decapod *Crustacea*, the boat-shaped exopodite of the second antenna.

scaphognathite (*Zool.*). In decapod *Crustacea*, the exopodite of the second maxilla, the movements of which cause water to flow through the branchial chamber.

scaphoid (*Zool.*). See **navicular bone.**

Scaphopoda (*Zool.*). A class of *Mollusca*, being bilaterally symmetrical with a tubular shell open at both ends, a reduced foot used for burrowing, a head with many prehensile processes, a radula, separate cerebral and pleural ganglia, no ctenidia, circulatory system rudimentary, larva a trochosphere.

scapolite or **wernerite** (*Min.*). A group of minerals forming an isomorphous series, varying from meionite, a silicate of aluminium and calcium with calcium carbonate, to marialite, a silicate of aluminium and sodium with sodium chloride. Common scapolite is intermediate in composition between these two end-member minerals; see also **dipyre** and **mizzonite.** The scapolites crystallize in the tetragonal system and are associated with altered lime-rich igneous and metamorphic rocks. A transparent honey yellow variety is cut as a gemstone.

scappling (*Build.*). A variant of *scabbling*.

scapula (*Zool.*). In Vertebrates, the dorsal portion of the pectoral girdle; the shoulder blade; any structure resembling the shoulder blade. *adj.* **scapular.**

scapulars (*Zool.*). In Birds, small feathers attached to the humerus, and lying along the side of the back.

scapulodynia (*Med.*). Pain in the region of the shoulder blade.

scapulo-humerals (*Zool.*). In Birds, relatively small muscles extending from the scapula to the humerus, and concerned, together with the coraco-humerals, in rotational movements of the wing.

scapus (*Build.*). See **apophyge.** (*Zool.*) The stiff axial rod of a contour feather. Also **stem.**

scarabeiform (*Zool.*). In Insects, a type of oligopod larva, being stout, subcylindrical and C-shaped, with short thoracic legs and no caudal processes. Typical of the super-family *Scarabaeoidea* (Stag Beetles, Chafers, etc.) of the *Coleoptera*, but also found in other families of *Coleoptera*. Probably derived from campodeiform larvae leading an inactive life with ample food supplies. Cf. *campodeiform*.

scarcement (*Build.*). A ledge formed at a place where part of a wall is set back from the general face of the wall.

scarf (*Carp.*). A joint between timbers placed end to end, notched and lapped, and secured together with bolts or straps.

scarfed joint (*Elec. Eng.*). A cable joint in which the conductor ends are bevelled off so that, after soldering, there is no appreciable increase in conductor diameter at the joint.

scarfing (*Eng.*). (1) Tapering the ends of materials for a lap joint, so that the thickness at the joint is substantially the same as that on either side of it. (2) Preparing metal edges for forge welding.

scarifier (*Civ. Eng.*). A spiked mechanical picking appliance for breaking up road surfaces as a preliminary to remetalling.

scariose, scarious (*Bot.*). Dry, thin, more or less transparent, and usually brownish as if scorched, especially at the tip and along the edges.

scarlatina (*Med.*). See **scarlet fever.**

scarlatiniform (*Med.*). Resembling or having the form of (the rash of) scarlet fever.

scarlet fever (*Med.*). An acute infectious fever due to infection of the throat with a haemolytic streptococcus; characterized by sore throat, headache, raised temperature, and a punctate erythema of the skin, which subsequently peels. Also called scarlatina.

Scarpa's ganglion (*Zool.*). In higher Vertebrates, a ganglion of the vestibular division of the eighth nerve.

scarp face (*Geol.*). See **escarpment.**

scattering (*Phys., etc.*). General term for irregular reflection or dispersal of waves or particles, e.g., in acoustic waves in an enclosure leading to diffuse reverberant sound, *Compton effect* (q.v.)

on electrons, light in passing through material, electrons, protons, and neutrons in solids, radio waves by ionization. See **forward** and **backward scatter.** Particle scattering is termed *elastic* when no energy is surrendered during scattering process—otherwise it is *inelastic.* If due to electrostatic forces it is termed *coulomb scattering* (q.v.), if short-range nuclear forces are involved it becomes *anomalous.* Long-wave electromagnetic wave scattering is *classical* or *Thomson* (q.v.), while for higher frequencies *resonance* or *potential* scattering occurs according to whether the incident photon does or does not penetrate the scattering nucleus. Coulomb scattering of α-particles is *Rutherford scattering.* For light, scattering by fine dust or suspensions of particles is *Rayleigh scattering*, while that in which the photon energy has changed slightly, due to interaction with vibrational or rotational energy of the molecules, is *Raman scattering. Shadow scattering* results from interference between scattered and incident waves of the same frequency. See **acoustic-, atomic-.**

scattering amplitude (*Nuc.*). Ratio of amplitude of scattered wave at unit distance from scattering nucleus to that of incident wave.

scattering cross-section (*Nuc.*). Effective impenetrable cross-section of scattering nucleus for incident particles of low energy. The radius of this cross-section is the *scattering length.*

scattering mean free path (*Phys., etc.*). The average distance travelled by a particle between successive scattering interactions. It depends upon the medium traversed and the type and energy of the particle.

scavenging (*Met.*). Addition made to molten metal to counteract an undesired substance. (*Min. Proc.*) Final stage of froth flotation in which a low-grade concentrate or middling is removed.

scavenging (or scavenger) pump (*I.C. Engs.*). An oil-suction pump used to return used oil to the oil tank from the crank-case of an engine using the dry sump system of lubrication.

scavenging stroke (*Eng.*). See **exhaust stroke.**

Scawfell Tuff (*Geol.*). A volcanic tuff found near Scawfell in the English Lake District; it forms part of the Borrowdale Volcanic Series of Ordovician age.

scawtite (*Min.*). A monoclinic silicate and carbonate of calcium occurring in vesicles at the contact of dolerite and limestone, as at Scawt Hill, Co. Antrim.

s.c.c. (*Elec. Eng.*). Abbrev. for *single cotton-covered (wire).*

scending (*Ships*). See **pitching period.**

scene-slating attachment (*Cinema.*). Device attached to a motion-picture camera for identifying individual takes on the film. Used in place of *clappers* (q.v.).

scenograph (*Build.*). A drawing showing a general view of a building or part of a building.

scent scales (*Zool.*). See **androconia.**

Schafer's method (*Med.*). A method of artificial respiration in which the patient lies prone, the head supported on one forearm, and the operator, his knees on either side of the patient's hips, exerts pressure with each hand over the lower ribs at the back at intervals of from 3 to 5 seconds. Cf. *kiss of life.*

Schäffer's acid (*Chem.*). Sulphonated 2-naphthol:

OH

HO_3S

used as an intermediate in manufacture of dyes, such as the pigment naphthol green.

schappe (*Textiles*). Name given to waste silk thread from which a certain proportion of natural gum (sericin) has been removed by the process of schapping, i.e., fermentation by steaming.

Schauddin fixative (*Micros.*). A fixative, used particularly for *Protozoa*, containing mercury(II) chloride, ethanol, and glacial ethanoic acid.

Scheele's green (*Chem.*). Copper(II) hydrogen arsenite ($CuHAsO_3$). Poisonous. Formerly used in wallpapers. [From K. W. Scheele (1742-86), a Swedish chemist.]

scheelite (*Min.*). An ore of tungsten formed under pneumatolytic conditions. It occurs in association with granites and pegmatites, has the composition calcium tungstate and crystallizes in the tetragonal system.

schefferite (*Min.*). A manganese-bearing variety of diopside-hedenbergite found in metamorphic manganese deposits.

Scheiner film-speed (*Photog.*). See **Weston-Scheiner film-speed.**

Schellbach tubing (*Glass*). Enamel back tubing having a cobalt blue line plied centrally between the other plies.

scheme arch (*Civ. Eng.*). See **skene arch.**

Scherbius advancer (*Elec. Eng.*). An expedor type of phase advancer, comprising an a.c. motor coupled to a d.c. exciter, for use with slip-ring induction motors.

Schering bridge (*Elec. Eng.*). An a.c. bridge of the 4-arm (or Wheatstone) type usually used to measure capacitance and power factor of small capacitors.

Schick's test (or reaction) (*Med.*). (Test for) the reaction of skin to intradermal injection of a measured amount of diphtheria toxin of known potency. If the skin reacts by redness and swelling (as compared with a control injection of heated toxin) the subject is susceptible to diphtheria.

schiffle machine (*Textiles*). See **Swiss machine.**

Schiff's bases (*Chem.*). A term for benzylidene anilines, e.g., $C_6H_5{\cdot}CH{=}N{\cdot}C_6H_5$.

Schiff's reagent (*Chem.*). A reagent, consisting of a solution of fuchsine decolorized by sulphurous acid, for testing the presence of aldehydes. Their presence is shown by a red-violet colour.

Schilder's disease (*Med.*). See **encephalitis periaxialis.**

schillerization (*Geol.*). A play of colour (in some cases resembling iridescence due to tarnish) produced by the diffraction of light in the surface layers of certain minerals. The diffraction is due to regularly orientated rods of opaque iron ores in labradorite and bronzite, and to a microscopic interlamination of two different mineral species in diallage.

Schimmelbusch's disease (*Med.*). A condition characterized by the formation of cysts in the breast, and by hyperplasia of the epithelium of the glandular tissue of the breast.

schindylesis (*Zool.*). A form of articulation in which a thin plate of skeletal material fits into a narrow cleft.

schirmerite (*Min.*). A silver, lead, bismuth sulphide, which crystallizes in the orthorhombic system.

schist (*Geol.*). The name given to a group of metamorphic rocks which have a tendency to split on account of the presence of folia of flaky and elongated minerals, such as mica, talc and chlorite. These rocks are formed from original sedimentary or igneous rocks by the action of combined heat and pressure.

schistaceous, schistose (*Bot.*). Slate-coloured; easily split.

schistosity (*Geol.*). The tendency in certain rocks to split easily along weak planes produced by regional metamorphism and due to the abundance of mica or other cleavable minerals lying with their cleavage planes parallel.

schistosomiasis (*Med.*). Infestation by parasitic worms of the *Schistome* group, transmitted via water by a complex cycle involving snails as intermediate hosts; common in many parts of the tropics. The chief varieties affect mainly the bladder (*Sch. haematobium*, see **bilharziasis**) and rectum (*Sch. mansonii*).

schizo-. Prefix from Gk. *schizein*, to cleave.

schizocarp (*Bot.*). A dry fruit, formed from more than one carpel, and separating when ripe into a number of 1-seeded parts which do not dehisce.

schizocoel (*Zool.*). Coelom produced within the mass of mesoderm by splitting or cleavage; cf. *enterocoel*. *adj.* **schizocoelic.**

schizocotyly (*Bot.*). The forking of cotyledons.

schizogamy (*Zool.*). In *Polychaeta*, a method of reproduction in which a sexual form is produced by fission or germination from a sexless form.

schizogenesis (*Zool.*). Reproduction by fission.

schizogenetic (or schizogenic) space (*Bot.*). An intercellular space formed by the splitting of cell walls.

schizogenous gland (*Bot.*). A glandular cavity formed by the separation of cells by splitting of the common middle lamellae.

schizognathous (*Zool.*). In Birds, said of a type of palate in which the basipterygoid processes are either absent, or spring from the base of the rostrum; the vomer is small and pointed, or may be absent; the palatines articulate posteriorly with the rostrum; and the maxillopalatines do not unite with one another. Found in the pigeon and the fowl. Cf. *aegithognathous, desmognathous, dromaeognathous.*

schizogony (*Zool.*). In *Protozoa*, vegetative reproduction by fission.

schizoid (*Psychiat.*). Showing qualities of a schizophrenic personality, such as asocial behaviour, introversion, tendency to phantasy.

schizolysigenous cavity (*Bot.*). An intercellular space formed in part by the separation, in part by the breakdown, of cells.

Schizomycetes (*Bot.*). Literally 'fission fungi'. Term sometimes used for *bacteria* (q.v.), and some bacteria-like organisms, e.g., *Rickettsiae.*

schizont (*Zool.*). In *Protozoa*, a mature trophozoite about to reproduce by schizogony.

schizontocyte (*Zool.*). In some *Sporozoa*, a cytomere or stage in the life cycle produced by division of the schizont and itself giving rise by further fission to merozoites.

schizopelmous (*Zool.*). Having the toes provided with two separate flexor tendons, as in some Birds.

schizophrenia (*Psychiat.*). A psychosis associated with withdrawal, dissociation and inability to distinguish reality from unreality, delusions, etc.

Schizophyta (*Bot.*). A group, doubtless artificial, of plants, including bacteria and *Myxophyceae*; these multiply by fission, have no well-defined nucleus, do not appear to have a sexual process.

schizopod (*Zool.*). Having the limbs split, i.e., having each limb provided with an endopodite and an exopodite, as in the mysis larva of higher *Crustacea.*

schizorhinal (*Zool.*). Having the posterior margin of the nares slitlike.

schizothymia (*Psychiat.*). Manifestation of schizoid traits within socially acceptable limits.

schizotrypanosomiasis (*Med.*). Chagas' disease. A disease, occurring in parts of South America, due to infection of the muscles, heart, and brain of Man with the protozoal parasite *Trypanosoma cruzi*, the infection being conveyed by the bite of an insect.

schizozoite (*Zool.*). A merozoite; a stage in the life cycle of a *Sporozoa* produced by schizogony.

Schlemm canal (*Histol.*). The circular venous sinus of the eye, which forms a link between the aqueous humour of the anterior chamber and the ciliary veins.

Schlenke loudspeaker (*Acous.*). One with a large stretched Duralumin diaphragm driven eccentrically by a current-carrying coil located in the circular gap of a pot magnet.

schlieren photography (*Aero., Phys.*). Technique by which the flow of air or other gas may be photographed, the change of refractive index with density being made apparent under a special type of illumination. The method is used in studying the behaviour of models in transonic and supersonic wind tunnels.

Schlippe's salt (*Photog.*). Sodium sulphantimoniate; used for redevelopment in sulphide toning and for intensification with mercury.

Schmidt lines (*Nuc. Eng.*). Lines in plot of nuclear magnetic moment against nuclear spin, as a result of the spin of an odd unpaired proton or neutron. Experimental magnetic moments for the majority of such nuclides lie between these lines. Also known as **Schmidt limits.**

Schmidt optical system (*Optics*). A projection optics system used in telescopes and for projection TV, employing a spherical concave mirror in place of the more perfect parabolic type. Resulting spherical aberration is avoided by use of a moulded correction plate.

Schmitt limiter (*Telecomm.*). A bistable pulse generator which gives a constant amplitude output pulse provided the input voltage is greater than a predetermined value.

Schmitt trigger (*Telecomm.*). Monostable circuit giving accurately-shaped constant amplitude rectangular pulse output for any input pulse above the triggering level. Widely used as pulse shaper after radiation counter tubes.

Schnee bath (*Med.*). An electrotherapeutic method in which the four limbs are immersed in four separate baths, when currents such as sinusoidal, galvanic, or faradic, are used.

Schneider furnace (*Elec. Eng.*). A type of high-frequency induction furnace.

Schneiderian membrane (*Zool.*). In Vertebrates, the olfactory mucous membrane.

schnorkel. A retractable tube or tubes containing pipes for discharging gases from, or for taking air into, a submerged submarine or other underwater vessel; a tube for bringing air to a submerged swimmer. Also **snorkel, snort.**

Schone elutriator (*Powder Tech.*). A vertical liquid elutriator. A single fractionating chamber is used and the cut size varied by using different fluid velocities.

Schönherr process (*Elec. Eng.*). A process for direct fixation of atmospheric nitrogen, employing a single-phase high-pressure arc furnace.

Schönlein's disease (*Med.*). See **peliosis rheumatica.**

Schön-Punga motor (*Elec. Eng.*). A compensated single-phase induction motor having a double rotor, one being an auxiliary running concentrically with the main rotor between the latter and the stator.

Schopper-Riegler test (*Paper*). A control test for

beating, depending on the measurement of drainage through the pulp under set conditions.

schorl (*Min.*). Black, iron-bearing end-member of tourmaline series. Common *tourmaline* (q.v.).

schorlomite (*Min.*). A black variety of andradite garnet even richer in titania (5–20% TiO_2) than melanite.

schorl-rock (*Geol.*). A rock composed essentially of aggregates of black tourmaline (schorl) and quartz. A Cornish term for the end product of tourmalinization. See also **luxulianite.**

Schotten-Baumann reaction (*Chem.*). The introduction of a benzoyl group into hydroxyl, amino, or imino compounds, by shaking with benzoyl chloride and excess aqueous alkali.

Schottky defect (*Chem.*). Deviation from the ideal crystal lattice by removal of some of the molecules to the surface, leaving within the lattice equivalent numbers of randomly spaced anion and cation vacancies.

Schottky effect (*Electronics*). Increase of saturation with increasing potential gradient near cathode-emitting surface of a triode.

Schottky noise (*Electronics*). Strictly, noise in the anode current of a thermionic valve due to random variations in the surface condition of the cathode. Frequently extended to include **shot noise.**

Schott treatment (*Med.*). Treatment of heart disease by means of baths and regulated exercises.

schradan (*Chem.*). *bis*(Dimethylamino)phosphonous anhydride:

$(CH_3)_2N$ O O $N(CH_3)_2$
P P
$(CH_3)_2N$ O $N(CH_3)_2$

used as an insecticide.

Schrage motor (*Elec. Eng.*). A variable-speed induction motor employing a commutator winding on the rotor from which an e.m.f. is collected and injected into the stator winding.

Schreiner finish (*Textiles*). A lustrous finish imparted to cotton and rayon satins and sateens by rollers engraved with fine lines which, by means of heat and pressure, are imprinted on the fabrics.

Schrödinger equation (*Phys.*). General equation of particle waves forming basic law of wave mechanics. Its solution is a wave function for which the square of the amplitude expresses the probability of the particle being located at the corresponding point.

Schroteffekt (*Electronics*). See **shot noise.**

Schuler pendulum (*Instr.*). A theoretically ideal pendulum which will not be affected by the earth's rotation, having a length equal to the earth's radius and hence a period of 84 min. Practically, such a pendulum is not physically realizable but a stable platform servomechanism can be constructed to simulate pendular motion of the above period, and is then said to be *Schuler-tuned.* Conversely a gyroscopically stabilized platform, constrained to move parallel to the earth's surface in its motion over the earth, will possess a period of 84 min and will be conditionally stable. Damping to reduce the maximum error may be introduced by rate feedback.

Schultz-Charlton reaction (*Med.*). The local blanching of the scarlet-fever rash when scarlatinal antitoxin is injected into the skin.

Schumann plates (*Phys.*). Photographic emulsion with low gelatine content for use in ultraviolet light, where gelatine absorption is serious.

schwannoma (*Med.*). A tumour growing from the sheath of a nerve (neurofibroma) and containing cells resembling those of the neurolemma.

Schwann's sheath (*Zool.*). See **neurolemma.**

Schwartz board (*Textiles*). Instrument for measuring the stress/strain properties of elastomeric yarns and rubber thread.

Schwartzschild antenna (*Radio*). System of bent plates reflecting a radar wave and achieving a very narrow fan beam pattern of radiation.

Schwarz's inequality (*Maths.*).

$$\{\int f(x)\,.\,g(x)dx\}^2 < \{\int f(x)dx\}\{\int g(x)dx\}$$

unless $f(x) = kg(x)$, where k is independent of x. This is the integral analogue of *Cauchy's inequality.*

Schweinfurt green (*Chem., Paint.*). See **Paris green.**

Schweitzer's reagent (*Chem.*). A reagent for cellulose. It consists of a 0·3% solution of precipitated copper(II) hydroxide in a 20% ammonium hydroxide solution. This mixture is a solvent for cellulose, which can be re-precipitated by the addition to the solution of mineral acids.

Schwinger coupling (*Elec. Eng.*). Directional coupler which links a longitudinal magnetic field and a transverse magnetic field in separate guides.

sciatic (*Zool.*). Situated in, or pertaining to, the ischial or hip region.

sciatica (*Med.*). Inflammation of the fibrous elements of the sciatic nerve, resulting in pain and tenderness along the course of the nerve in the buttock and the back of the leg; less strictly, any pain along the course of the sciatic nerve.

science. The ordered arrangement of ascertained knowledge, including the methods by which such knowledge is extended and the criteria by which its truth is tested. The older term *natural philosophy* implied the contemplation of natural processes *per se*, but modern science includes such study and control of nature as is, or might be, useful to mankind, and even proposes control of the destiny of man himself. *Speculative science* is that branch of science which suggests hypotheses and theories, and deduces critical tests whereby uncoordinated observations and properly ascertained facts may be brought into the body of science proper.

scientific alexandrite (*Min.*). Synthetic corundum coloured with vanadium oxide and resembling true alexandrite in some of its optical characters.

scientific emerald (*Min.*). Beryl glass coloured with chromic oxide, resembling true emerald in colour.

scintillation (*Astron.*). The twinkling of stars, a phenomenon due to the deflection, by the strata of the earth's atmosphere, of the light rays from what are virtually point sources. (*Nuc.*) Minute light flash caused when α-, β- or γ-rays strike certain phosphors, known as **scintillators.** The latter are classed as liquid, inorganic, organic or plastic according to their chemical composition. (*Telecomm.*) Undesired transient changes in carrier frequency, arising from the modulation process.

scintillation counter (*Nuc. Eng.*). Counter in which a photomultiplier generates pulses that are used to count nuclear events causing scintillations in a phosphor or crystal.

scintillation spectrometer (*Nuc. Eng.*). Scintillation counter for which all the energy of the incident particle is normally absorbed in the scintillator, so that the amplitude of the resulting light flash (or corresponding signal pulse from photomultiplier) is a measure of the energy of the incident particle or photon. These signal pulses are analysed simultaneously or sequentially by a kicksorter or single channel pulse height analyser. Different types of events counted can then be distinguished by their differing energies. See **alpha-, beta-,** and **gamma-ray spectrometers.**

scintling (*Build.*). Placing half-dry raw bricks diagonally and a little distance apart, so as to admit air between them.

sciograph (*Build.*). A drawing showing a sectional view of a building.

scion (*Bot.*). (1) A portion of a plant, usually a piece of young stem, which is inserted into a rooted stock in grafting. (2) A young plant formed at the end of, or along the course of, a runner. (3) A stolon.

sciophyllous (*Bot.*). Having leaves which can endure shading.

sciophyte (*Bot.*). A plant which grows in shade.

scirocco (*Meteor.*). See **sirocco.**

scirrhous carcinoma (*Med.*). A hard cancer, in which there is an abundance of connective tissue and few cells.

scirrhous cord (*Vet.*). See **botryomycosis.**

scirrhus (*Med.*). A scirrhous carcinoma.

scissors truss (*Carp.*). A type of truss used for a pitched roof, consisting of two principal rafters braced by two other members, each of which connects the foot of a rafter to an intermediate point in the length of the other rafter.

sclera (*Zool.*). The tough fibrous outer coat of the Vertebrate eye. *adj.* **sclerotic.**

scleranthium (*Bot.*). Dry one-seeded fruit enclosed in the hardened remains of the calyx.

scleratogenous (*Zool.*). Skeleton-forming.

sclere (*Zool.*). A skeletal structure; a sponge spicule.

sclereide (*Bot.*). (1) A general term for a cell with a thick, lignified wall, i.e., any sclerenchymatous cell. (2) A thick-walled cell mixed with the photosynthetic cells of a leaf, giving them mechanical support. (3) A stone cell.

sclerema neonatorum (*Med.*). A disease of newborn infants in which there is symmetrical hardening of the subcutaneous fat in certain parts of the body.

sclerencephalia (*Med.*). Hardening of the brain.

sclerenchyma (*Bot.*). A tissue composed of cells with thick lignified walls and with little or no living contents. The cells are elongated, with pointed ends which often interlock, or less often short and blunt-ended. Sclerenchyma supports and protects the softer tissues of the plant. (*Zool.*) Hard skeletal tissue, as of Corals.

sclerite (*Zool.*). A hard skeletal plate or spicule.

scleritis (*Med.*). Inflammation of the sclera of the eye. See also **episcleritis.**

scleroblast (*Zool.*). In *Porifera*, a spicule-forming cell.

scleroblastic (*Zool.*). See **scleratogenous.**

sclerocaulous (*Bot.*). Having a hard, dry stem.

sclerocoel (*Zool.*). The cavity of a sclerotome.

sclerocorneal (*Zool.*). Pertaining to the sclerotic and the cornea.

sclerodactylia (*Med.*). Sclerodermia of the hands, the skin being drawn tightly over the fingers.

scleroderm (*Zool.*). A hard integument.

sclerodermatous (*Zool.*). Possessing exoskeletal structures.

sclerodermia, scleroderma (*Med.*). A condition of hardness and rigidity of the skin as a result of overgrowth of fibrous tissue in the dermis and subcutaneous tissue, the fat of which is replaced by the fibrous tissue.

sclerodermite (*Zool.*). The exoskeleton of an Arthropodan somite.

scleroma (*Med.*). A condition in which hard nodules of granulomatous tissue appear in the nose, or occasionally in the trachea.

scleronychia (*Med.*). Thickening and dryness of the nails.

Scleroparei (*Zool.*). An order of *Teleostei* in which one of the suborbital bones extends across the cheek towards the pre-operculum; carnivorous forms with pointed teeth; marine. Mail-cheeked Fishes, Scorpion-fishes, Rockperches, Gurnards, Sticklebacks, Greenlings.

sclerophyll (*Bot.*). A hard, stiff, and tough leaf, with a strongly cutinized epidermis.

scleroproteins (*Chem.*). Insoluble proteins forming the skeletal parts of tissues, e.g., keratin from hoofs, nails, hair, etc., chondrin and elastin from ligaments. Also **albuminoids.**

scleroscope hardness test (*Met.*). The determination of the hardness of metals by measuring the rebound of a diamond-tipped hammer dropped from a given height.

sclerosed, sclerotic (*Bot.*). Having hard, usually lignified walls. (*Med.*) Affected with sclerosis.

sclerosis (*Bot.*). The hardening of cell walls or of tissues by thickening and lignification. (*Med.*) An induration or hardening, as of the arteries. See also **disseminated sclerosis.**

sclerotesta (*Bot.*). A hard layer in the testa of a seed in which a sarcotesta is also present.

sclerotic (*Zool.*). The sclera of the eye; pertaining to the sclera.

sclerotic cell (*Bot.*). See **sclereide.**

sclerotic ossicles (or **bones**) (*Zool.*). A ring of small membrane bones, derived from the sclera of the eye in some Reptiles and Birds.

sclerotiform, sclerotioid (*Bot.*). Resembling a sclerotium.

sclerotin (*Zool.*). A substance consisting of protein molecules linked or tanned by quinones which fills the spaces between the chitin micelles of insect cuticle during the process of tanning.

sclerotium (*Bot.*). A hard mass of fungal hyphae, usually black on the outside, crustlike to globular, and serving as a resting stage from which fructifications are formed later. (*Zool.*) The resting or encysted stage in the life cycle of *Mycetozoa*, in which the organism is represented by a number of multinucleate cellulose cysts. *adj.* **sclerotial.**

sclerotization (*Zool.*). In Insects, the process by which most of the *procuticle* (q.v.) becomes hardened and darkened to form tough, rigid, discrete sclerites of *exocuticle* (q.v.). See also **tanning.**

sclerotome (*Zool.*). One of the metamerically arranged masses of mesenchyme which give rise in part to the axial skeleton in developing Vertebrates.

sclerotomy (*Surg.*). Operative incision of the sclera.

scobicular, scobiform (*Bot.*). Looking like sawdust.

scobinate (*Bot.*). Said of a surface which feels as if it has been roughened with a coarse file.

scoinson arch (*Build.*). See **squinch.**

scolecite (*Bot.*). An old name for an *archicarp.* (*Min.*) A member of the zeolite group of minerals; a hydrated silicate of calcium and aluminium, occurring usually in fibrous or acicular groups of crystals.

scolecosporous (*Bot.*). Having thread-shaped or worm-shaped spores.

scolex (*Zool.*). The terminal organ of attachment of a tapeworm (Cestode). *pl.* **scoleces,** *erron.* **scolices.** *adj.* **scolecid, scoleciform.**

scoliokyphosis (*Med.*). Abnormal curvature of the

spine in which scoliosis is combined with kyphosis.

scoliosis (*Med.*). Abnormal curvature of the spine laterally.

scolopale (*Zool.*). A hollow rodlike structure forming part of a scolophore in Insects.

scolophore (*Zool.*). A spindle-shaped nerve-ending in Insects, auditory in function and consisting essentially of a bipolar nerve cell. Also called **scolopidium.**

scolus (*Zool.*). A thornlike process of the body wall, characteristic of some Insect larvae.

scontion (*Build.*). An inside quoin, as laid in a splayed jamb.

scoop (*Cinema.*). One or more suspended *broads* (q.v.). (*Surg.*) A spoonlike instrument for clearing out cavities.

scopa (*Zool.*). The pollen brush of Bees, consisting of short stiff spines on the posterior metatarsus.

scope (*Electronics*). Colloquial term for *oscilloscope*. (*Radar*) A radarscope which produces a mode of information display. See **A, B,** etc., **display.**

Scophony television (*TV*). Electromechanical system in which a light beam is modulated by a Kerr cell and scanning is carried out by rotating mirrors.

scopolamine (*Pharm.*). Hyoscine. A coca-base alkaloid, present in *Datura meteloides.* It is the scopoline ester of tropic acid, i.e.,

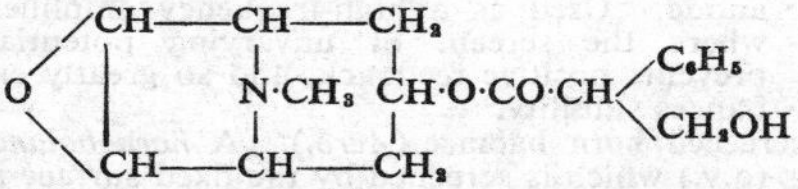

```
   /CH——CH——CH2
  /   |       |               /C6H5
O     N·CH3  CH·O·CO·CH<
  \   |       |               \CH2OH
   \CH——CH——CH2
```

Has a sedative effect on the central nervous system and the hydrobromide is used to induce *twilight sleep* (q.v.); once used to obtain criminal confessions.

scopula (*Zool.*). A small tuft of hairs; as in some Spiders (*Tengellidae, Pycnothelidae,* etc.), a small tuft of tarsal hairs of use in climbing; a rodlike sponge spicule with a number of rays at one end.

scorbutic (*Med.*). Pertaining to scorbutus (scurvy).

score (*Typog.*). See **em rule, en rule.**

scoria (*Geol.*). A cavernous mass of volcanic rock which simulates a clinker.

scorification (*Met.*). Assay in which impurities are slagged while bullion metals dissolve in molten lead, from which they are later separated by *cupellation* (q.v.).

scorifier (*Met.*). A crucible of bone ash or fireclay used in assaying and in the metallurgical treatment of precious metals. See also **scorification.**

scoring system (*Cinema.*). One used for sound recording in cinematography where music has to be synchronized with the film.

scorodite (*Min.*). An orthorhombic hydrated arsenate of iron and aluminium.

scorpioid cyme (*Bot.*). A cymose inflorescence in which the branches develop alternately right and left, but do not all lie in one plane; in bud, the axis of the inflorescence is coiled.

Scorpionidea (*Zool.*). An order of *Arachnida* with the prosoma covered by a dorsal carapace, the opisthosoma being divided into a distinct mesosoma and metasoma, and consisting of 12 segments and a telson. The chelicera and pedipalps are chelate, there are 4 pairs of walking legs, the mesosomatic segments carry the genital operculum, the pectines and 4 pairs of lung books, the metasoma forming a flexible tail wielding the terminal sting. Viviparous and terrestrial. Known as fossils from the Silurian. Scorpions.

scotch (*Build.*). See **scutch.**

Scotch beaming (*Textiles*). See **Scotch dressing** (2).

Scotch block (*Rail.*). Attachment to running rails, to prevent the passage of rolling stock.

Scotch boiler (*Eng.*). See **marine boiler.**

Scotch bond (*Build.*). See **English garden-wall bond.**

Scotch carpet (*Textiles*). May be made of two distinct plain weave cloths interlaced with each other; jute or cotton. Known also as **Kidderminster** or **ingrain carpet.**

Scotch crank (or yoke) (*Eng.*). A form of crank, used on a *direct-acting pump* (q.v.), in which a square block, pivoted on the overhung crank pin, works in a slotted crosshead carried by the common piston rod and ram.

Scotch dressing (*Textiles*). (1) Method of sizing cotton yarns of extra fine count; they are passed from the warper's beam through a size box, then brushed, dried by hot air, and again brushed. Also **dresser sizing.** (2) Method of producing a striped warp by running yarns of different colours from a number of warper's beams on to the weaver's beam, according to the colour design. Also called **Scotch beaming, dry taping.**

Scotchprint (*Print.*). A *conversion system* using a film on which an impression of a relief printing surface is taken, to provide a positive for litho platemaking.

Scotch yoke (*Elec. Eng.*). A triangular framework used in certain types of electric tractor for coupling two traction motors to the driving-wheel system. (*Eng.*) See **Scotch crank.**

scotia (*Arch.*). See **cavetto.**

scotoma (*Med.*). (1) A blind or partially blind area in the visual field, the result of disease of, or damage to, the retina or optic nerve or visual cortex. (2) The appearance of a black spot in front of the eye, as in choroiditis. *pl.* **scotomata.**

scotometer (*Med.*). An instrument for detecting and measuring *scotomata* (1).

scotophor (*Electronics*). Material which darkens under electron bombardment, used for screen of CRT in storage oscilloscopes. Recovers upon heating. Usually potassium chloride.

scotophyte (*Bot.*). A plant which lives in the dark.

scotopic luminosity curve (*Light*). The curve giving relative brightness of the radiations in an equal-energy spectrum when seen at a very low intensity level. See **photopic luminosity curve.**

scotopic vision (*Optics*). That which occurs at low illumination levels through the medium of the retinal rods.

Scots pine (*For.*). *Pinus sylvestris,* native to many parts of Europe and the best known home-grown species for producing commercial timbers. Also called **Baltic redwood, Scots fir.**

Scott-Bentley discriminator (*Elec. Eng.*). In series motors requiring a wide range of speed, a field divertor arrangement for obtaining light-load speeds above normal.

Scott connexion (*Elec. Eng.*). A method of connecting two single-phase transformers so as to convert a 3-phase 3-wire a.c. supply to a 2-phase 3-wire supply, and vice versa.

Scottish topaz (*Min.*). A term applied in the gemstone trade to yellow transparent quartzes, resembling Brazilian topaz in colour, used for ornamental purposes. Not a true topaz. See **cairngorm, citrine.**

scourer (*Eng.*). A flour-milling machine in which the wheat, for cleaning purposes, is subjected to the action of revolving beaters in a ventilated casing.

scouring (*Hyd. Eng.*). Said of the eroding action of water flowing at high velocity. (*Textiles*) (1) Process to which raw wool and manufactured woollen fabrics are subjected, raw wool to remove natural grease and impurities, yarns and cloth to remove oil and dirt. (2) Process by which the natural gum is removed from silk yarns prior to dyeing. (3) Cleaning and polishing steel rollers on textile machines.

scram (*Nuc. Eng.*). General term for emergency shutdown of a plant, especially of a reactor when the safety rods are automatically shot in to stop the fission process.

scrambler (*Telecomm.*). Multiple modulating and demodulating system which interchanges and/or inverts bands of speech, so that speech in transmission cannot be intelligible, the reverse process restoring normal speech at receiving end.

scram rod (*Nuc. Eng.*). An emergency safety rod used in a reactor.

scraper (*Carp.*). A thin flat steel blade with a square straight edge on which a burr is raised; used to pare wood from a surface which is being finally dressed. (*Civ. Eng.*) A device, towed behind a tractor, which collects earth, with the object of reducing the ground level to that required.

scraper board (*Print.*). A coated cardboard, the coating being either white on white or black on white; it can be readily scraped away to expose the underlayer. Used by commercial artists to obtain the effect of wood engraving, the result being made into a printing plate photomechanically.

scraper plane (*Join.*). Tool shaped like a *spokeshave* (q.v.) but with a thin blade set at a slight angle from the perpendicular. Used for cleaning up hardwood and removing tears left after ordinary planing.

scraper ring (*I.C. Engs.*). A ring usually fitted on the skirt of a petrol- or oil-engine piston, to prevent excessive oil consumption. It may have a bevelled upper edge or a slotted groove, the oil being scraped off the cylinder wall and led back to the sump through holes in the piston wall.

scrapie (*Vet.*). A chronic nervous disease of sheep, caused by a virus, and characterized by nervous symptoms, especially pruritus and an uncoordinated gait.

scratch (*Join.*). A tool with an upright shaped blade fixed in a wooden body; used for working small mouldings.

scratch-coat (*Build.*). The first of three coats applied in plastering. It consists of coarse stuff.

scratched figure (*Typog.*). A figure cast with a stroke through it to indicate a cancel, thus: 6̸.

scratcher (*Build.*). A tool used to make scratch marks in a cement surface to provide a grip for a subsequent coat.

scratch filter (*Acous.*). A low-pass filter network to eliminate noise due to traverse of the stylus in the grooves of a gramophone record.

scratch work (*Paint.*). See **sgraffito.**

scree (*Geol.*). A tumbled mass of angular rock debris strewn on a hillside or at a mountain foot, resulting from frost action or, in arid regions, exfoliation. Also called **talus.**

screeching (*Aero.*). A cacophonous form of unstable combustion that can occur with rockets, and occasionally in turbine engines, causing very rapid damage due to resonance stresses on the jet pipe or nozzle. Also called **howl.**

screed (*Build.*). A band of plaster laid on the surface of a wall as a guide to the thickness of a coat of plaster to be applied subsequently. (*Civ. Eng.*) A strip of wood or metal temporarily inserted in a road surface to form a guide for the template for forming the final surface of the road.

screed-coat (*Build.*). A coat laid level with the screeds.

screen (*Build., Civ. Eng.*). A large sieve used for grading fine or coarse aggregates. (*Elec. Eng.*) Electrode consisting of a relatively fine mesh network of wires interposed between two other electrodes, to reduce the electrostatic capacitance between them. It is usually maintained at positive potential, and connected to earth through a capacitor. See **electrostatic shield, Faraday cage.** (*Electronics*) End of tube on which is exhibited a display through fluorescent excitation by electron beam; a phosphor. (*Photog., Print.*) The meshwork of lines at right angles, ruled on glass, used to translate the subject of a *half-tone* (q.v.) illustration into dots. See **coarse-, medium-, fine-, very fine-.**

screen analysis (*Powder Tech.*). See **sieve analysis.**

screen burning (*Electronics*). Gradual falling off in luminosity, sometimes accompanied by discoloration, in the fluorescent screen of a CRT, particularly if operated under adverse conditions.

screened-grid valve (*Electronics*). Four-electrode valve, with cathode, control grid, screen and anode. Used as a high-frequency amplifier, where the screen, of unvarying potential, prevents positive feedback, and so greatly enhances stability.

screened horn balance (*Aero.*). A *horn balance* (q.v.) which is screened by the fixed surface in front of it.

screened pentode (*Electronics*). Pentode valve having a fine-mesh auxiliary grid and consequently small grid-to-anode capacitance, for use at high frequencies.

screened wiring (*Elec. Eng.*). Insulated conductors enclosed in earthed and continuously conducting metal tubes or conduits, mainly for mechanical protection, but frequently for preventing induction to or from conductors.

screen factor (*Electronics*). Ratio of actual area of grid structure to total area of surface containing the grid.

screen grid (*Electronics*). One placed between control grid and anode in a valve, having invariant potential. Originally of slats (Hall), but now of gauze or spirals, with the intention of removing effect of anode circuit on the input control grid circuit, thus eliminating positive feedback and instability through anode/grid capacitance. See **Miller effect.**

screening (*Elec. Eng., Nuc.*). Use of a screen, in the form of a metal or gauze can, normally earthed, so that electrostatic effects inside are not evident outside, or vice versa. Similarly, nucleus of an atom is screened by its surrounding electrons. (*Min. Proc.*) See **screens.** (*Radiol.*) See **fluoroscopy.**

screening constant (*Phys.*). Of an element, the atomic number minus the apparent atomic number which is effective for, e.g., X-ray emission.

screening protector (*Elec. Eng.*). See **line choking coil.**

screenings (*Build., Civ. Eng.*). The residue from a sieving operation.

screen memory (*Psychol.*). Early childhood impressions and ideas which break through into consciousness, but are distorted into something unrecognizable to the individual.

screen modulation (*Radio*). That in which the potential of the screen in a multi-electrode valve is varied in accordance with impressed modulating currents.
screen process printing (*Print.*). See **silk-screen printing.**
screen-protected motor (*Elec. Eng.*). A protected type of electric motor in which the openings for ventilation are covered with wire-mesh screens.
screens, screening (*Min. Proc.*). Perforated or woven cloths (metal, fibre, rods, bars), used to size ore or products as part of treatment required to regulate concentration. Include *grizzlys, trommels* (qq.v.), and mechanically and electrically vibrated screens.
screw-and-nut steering-gear (*Autos.*). One in which a square-threaded screw formed on the lower end of the steering-column engages with a nut provided with trunnions, which work in blocks sliding in a short slotted arm, connecting with the remainder of the steering system.
screw-auger (*Tools*). An auger having a helical groove cut in its surface so as to carry away the chips from the cutting edge.
screw axes (*Min.*). Axes of symmetry about which the atoms in a mineral are symmetrically disposed. Rotation about a 4-fold screw axis, for example, will carry an atom 1 into the positions successively occupied by similar atoms 2, 3, and 4, after rotations of 90°, 180°, 270°, and 360°. Cf. *rotation axes of symmetry.*
screw box and tap (*Tools*). Device for making wooden screws on furniture legs, wooden cramps, etc.
screw chases (*Typog.*). Chases used in newspaper work. They are tightened by screws, which obviate the use of separate quoins.
screw chasing (*Eng.*). See **chaser.**
screw composing-stick (*Typog.*). An old type of composing-stick fastened with a thumbscrew. The modern style is fastened at the correct measure by means of a lever.
screw conveyor (*Eng.*). See **worm conveyor.**
screw-cutting lathe (*Eng.*). A metal turning-lathe provided with a lead screw driven by *change wheels* (q.v.), for traversing the pointed tool used in screw-cutting.
screwdriver (*Tools*). Tool with shank terminating in a blade of size and shape to fit the slot in screws. See also **Phillips-, ratchet-.**
screwed steel conduit (*Elec. Eng.*). Light steel tubing, having screwed ends for connecting up in lengths by means of sockets, in which electrical installation wiring is run. Cf. *plain steel conduit.*
screwing die (*Eng.*). An internally threaded hardened steel block, sometimes split in halves, on which cutting edges are formed by longitudinal slots. Held in a stock, lathe, or screwing machine for cutting external threads.
screwing machine (*Eng.*). A form of lathe adapted for the continuous production of screws, or screwed pieces, by means of dies.
screw jack (*Eng.*). See **jack.**
screw micrometer (*Eng.*). See **micrometer gauge.**
screw nail (*Eng.*). A nail in whose surface shallow helical depressions are formed, so that as it is driven in place with blows from a hammer it turns like a screw.
screw pile (*Civ. Eng.*). A pile having a wide projecting helix or screw at the foot, useful in alluvial ground.
screw plate (*Eng.*). A hardened steel plate in which a number of screwing dies of different sizes are formed.
screw plug (*San. Eng.*). A drain plug consisting of a rubber ring held between two steel disks which, on being screwed together, force the ring out to close the drain pipe in which the plug is placed.
screw press (*Eng.*). See **fly press.**
screw propeller (*Eng.*). See **airscrew, marine screw propeller.**
screws (*Med.*). See **caisson disease.**
screw shackle (*Eng.*). A long nut screwed internally with a right-hand thread at one end and a left-hand thread at the other, serving to connect the ends of two rods which are to be joined together, and providing a means of adjusting the total length. See **coupling.**
screw thread (*Eng.*). A helical ridge of approximately triangular (or V), square, or rounded section, formed on a cylindrical core, the pitch and core diameter being standardized under various systems. See **British Association-, British Standard Whitworth-, Sellers-, Unified-,** etc.
scribbler (*Spinning*). Woollen carding set, comprising a combination of two cylinders, feed rollers, working and stripping rollers, fancy, doffer; all are covered with wire fillet.
scriber (*Tools*). A pointed steel tool used for making an incised mark on timber or metal, to guide a subsequent cutting operation.
scribing block (*Eng.*). A tool for gauging the height of some point on a piece of work, above a surface plate or machine table. It consists of a base supporting a pivoted column, to which a scriber is slidably clamped. Also called **surface gauge.**
scribing gouge (*Tools*). One sharpened with the bevel on the inside, used for work where an upright cut is necessary.
scrieve board (*Ships*). A formation of portable portions of flat wooden boards whereon are scrieved (or scribed) the ship's transverse frame sections and lines indicative of shell seams, decks, stringers, etc. Scrieve boards are used for setting the soft iron, to which the frames, etc., are turned.
scrim (*Build.*). An open weave fabric used to cover the joints between fibre boards before applying a finish. (*Textiles*) A fabric generally made from low-quality linen tow, in an open weave. Used as a reinforcement in bookbinding, upholstery, plaster work, etc. A low muslin cotton cloth is also given this name.
script (*Typog.*). A style of type which imitates handwriting.
scrobe (*Zool.*). In some *Arthropoda*, especially Insects, a groove for the reception of an antenna.
scrobiculate (*Bot., Zool.*). Having the surface dotted all over with small rounded depressions; pitted.
scrobicule, scrobicula (*Zool.*). In *Echinoidea*, the smooth area of the test around the boss of a spine.
scrobiculus (*Bot., Zool.*). A small pit or rounded depression.
scrofula (*Med.*). Caseating tuberculosis of the lymphatic glands. *adj.* **scrofulous.**
scrofulodermia, scrofuloderma (*Med.*). Tuberculous infection of the skin from the bursting of a deep-seated tuberculous abscess; a subcutaneous tuberculous abscess.
scroll (*Textiles*). Large steel cam to which is fastened powerful ropes. These draw in and check the inward run of mule carriages after twisting and winding-on.
scroll chuck (*Eng.*). A self-centring chuck, having jaws slotted to engage with a raised spiral or scroll on a plate which is rotated by a key, so as to advance the jaws while maintaining their concentricity. See also **three-jaw chuck.**

scrotiform (*Bot.*). Like a bladder.

scrotum (*Anat., Zool.*). In Mammals, a muscular sac forming part of the ventral body wall into which the testes descend. *adj.* **scrotal.**

scrubber (*Agric.*). An implement using overlapping metal bars, or the metal-shod edges of overlapping heavy timbers, to break clods. (*Chem. Eng.*) One kind of *dampener* (q.v.), which depends for its action on changing the velocity of gas flowing by altering direction and flow area for the gas stream. The alterations are provided by baffles in the scrubber.

scrubbers (*Gas*). In a gasworks plant, the apparatus in which the gas is freed from tar, ammonia, and sulphuretted hydrogen.

scrub typhus (*Med.*). See **typhus.**

scudding (*Leather*). The operation of removing the epidermis hair roots, pigment cells, and lime salts from the grain side of a hide, before tanning.

scull (*Glass*). The glass remaining in a ladle after most of the molten glass has been poured out.

scum (*Build.*). A surface formation of lime crystals appearing on new cement work.

scumble (*Paint.*). A light-coloured, low-opacity paint allowing a darker background colour to *grin through* deliberately.

S-curve (*Surv.*). See **reverse curve.**

scurvy (*Med.*). Scorbutus. A nutritional disease due to deficiency in the diet of vitamin C (ascorbic acid), characterized by anaemia, debility, apathy, sponginess of the gums, ulceration of the mouth, and haemorrhages from various parts of the body.

scutch (*Build.*). The bricklayer's cutting tool for dressing bricks to special shapes. Also called **scotch.**

scute (*Zool.*). An exoskeleton scale or plate. *adj.* **scutate.**

scutellar epithelium (*Bot.*). A layer of elongated cells covering the surface of the scutellum, lying against the endosperm, and producing enzymes which assist in the utilization of the latter.

scutellation (*Zool.*). Scale arrangement; pholidosis.

scutellum (*Bot.*). A flattened portion of the embryo of a grass, probably the expanded cotyledon; it is applied to the endosperm and serves as an absorptive organ. (*Zool.*) The posterior of three sclerites into which the notum is typically divided in Insects; in Birds, a tarsal scale; more generally, a shield-shaped structure.

scutum (*Zool.*). In Insects, the middle sclerite of three into which the notum is typically divided. *adj.* **scutal.**

scybalum (*Med.*). A round, hard, and dry faecal mass in the intestine.

scyelite (*Geol.*). A coarse-grained ultramafic igneous rock, named from the original locality at Loch Scye in Sutherland, Scotland; it consists essentially of mafic minerals including serpentine pseudomorphs after olivine set poikilitically in large amphibole crystals associated with large bronze mica crystals. A mica-hornblende-peridotite.

scyphi (*Bot.*). See **scyphus.**

scyphistoma (*Zool.*). In *Scyphozoa*, the segmenting polyp stage. Cf. *hydratuba.*

Scyphomedusae (*Zool.*). See **Scyphozoa.**

Scyphozoa (*Zool.*). A class of *Cnidaria* in which the polyp stage is inconspicuous and may be completely absent. Where present it is known as a *scyphistoma*, and gives rise to the ephyra larvae which grow into medusae by transverse fission or strobilation. There are vertical partitions (mesenteries) in the enteron of some forms, and the larval enteron of others. Velum and nerve ring generally absent, gonads endodermal, no skeleton. Jellyfish.

scyphula (*Zool.*). See **scyphistoma.**

scyphus (*Bot.*). A cuplike widening of the distal end of the podetium in some lichens. *pl.* **scyphi,** *adj.* **scyphiferous.**

Se (*Chem.*). Symbol for *selenium.*

S.E. (*Build.*). Abbrev. for *stopped end.*

sea. An expanse of salt water on the face of the globe; strictly, one of lesser extent than an ocean. Sea water contains the majority of the common elements in small amounts, the principal ions being chlorine and sodium; others are calcium, magnesium, and potassium; dissolved gases, e.g., oxygen, nitrogen, carbon dioxide. The salinity holds a constant relation to the chlorine content and varies between 32 and 37·4 parts per thousand in the open sea; that of the Red Sea is about 40%. See also **ocean.**

sea anchor (*Ships*). A float to which a ship may be attached by a hawser to ride out a gale. Used in small boats in the form of a canvas drogue.

sea breeze (*Meteor.*). See **land and sea breezes.**

sea cell (*Elec.*). Primary electrolytic cell which functions as a source of electric power when immersed in sea water. A battery of such cells is possible, power being available although cells are partially short-circuited by the sea water. Fitted to life-belts, etc., so that an indicating light is produced automatically in the event of use at night.

sea earth (*Teleg.*). A sea earth is effected by the removal of the main earth connexion of the termination of a submarine cable to some miles out to sea, instead of making it near the shore, to avoid the earth coupling of a number of such submarine cables. The earth connexion takes the form of an insulated conductor inside the cable, ultimately terminated on the armouring, which is in contact with the sea.

Sea-Island cotton (*Textiles*). A long staple, silky cotton produced in the West Indies. Always bagged but never densely compressed into bales, it is still made into the finest quality plain and striped shirtings.

seal (*Brew.*). The final head of yeast on a vat when fermentation has finished. (*Electronics*) In a vacuum tube, the point at which the tube is closed after pumping, and the act of closing off. (*Print.*) A small printing plate usually used to indicate the edition of a newspaper. See **seal cylinder, seal unit.** (*San. Eng.*) The water contained in a trap, which prevents the flow of air or gases from one side to the other. (*Telecomm.*) In a relay, colloquialism for *lock*, whether by holding contact or residual magnetism.

seal cylinder (*Print.*). The auxiliary cylinder to which the *seal* is attached.

sealed cover (*San. Eng.*). An air-tight cast-iron or precast concrete cover fitting into a frame and used to cover a manhole.

sealed pressure balance (*Aero.*). An *aerodynamic balance*, used mainly on ailerons, consisting of a continuous projection forward of the hinge line within a cavity formed by close fitting *shrouds* projecting rearward from the main surface, the gap between the balance and the main surface being sealed to prevent communication of pressure between lower and upper surfaces. Sometimes called a Westland-Irving balance after its inventor and the company which developed it.

sealed pumped vacuum system (*Vac. Tech.*). A vacuum system in which, after preliminary evacuation, the pumping means is included in the system and does not eject gas or vapour from it.

sealed pumpless vacuum system (*Vac. Tech.*). A vacuum system which, after evacuation, is sealed and contains no pumping means.
sealed tube (*Electronics*). Electron tube which is permanently evacuated and not pumped during operation. Cf. *demountable* tube.
sea level (*Surv.*). The datum line from which heights are measured in surveying; the term is also used more loosely to denote the mean surface level of the oceans or the surface of the geoid.
sea level static thrust (*Aero.*). See **static thrust.**
sea lily (*Zool.*). See **Crinoidea.**
sealing box (*Cables*). A box in which the end of a paper-insulated cable is hermetically sealed.
sealing-in (*Elec. Eng.*). The making of an air-tight joint between the filament wires and the glass envelope of an incandescent lamp or vacuum tube.
sealing-in burner (*Heat*). Gas or oxy-gas burner for sealing glass containers such as the envelopes of electric lamps and radio transmitting and receiving valves.
sealing-off (*Elec. Eng.*). The final sealing of the exit to the evacuating pumps of an incandescent lamp bulb or vacuum tube.
Sealocrete (*Build.*). TN of a waterproofing agent for cement surfaces.
seal unit (*Print.*). The small auxiliary printing unit used for printing the *seal*, usually in a second colour on front page.
seam (*Met.*). A surface defect in worked metal, the result of a blowhole being closed but not welded; it remains as a fine crack. Ridge in casting, effect of enclosure of impurity or scale in working. (*Mining*) (1) A tabular, generally flat deposit of coal or mineral; a stratum or bed. (2) A joint or fissure in a coal bed. (*Plumb.*) See **welt.**
sea marker (*Aero.*). Any device dropped from an aircraft on to water to make an observable patch from which the drift of the aircraft may be determined. Usually filled with fluorine for use during the day, and with a flame-producing device for night use.
seaming machine (*Eng.*). A press for forming and closing longitudinal and circumferential interlocking joints, used in the manufacture of containers from sheet metal.
seamless tube (*Met.*). Tube other than that made by bending over and welding the edges of flat strip. May be made by extrusion (nonferrous metals), or by piercing a hole through a billet and then rolling down over a mandrel to form a tube of the required dimensions.
seam roll (*Plumb.*). See **hollow roll.**
seam welding (*Elec. Eng.*). (1) See **resistance seam-welding.** (2) Uniting sheet plastic by heat arising from dielectric loss, the electric field being applied by electrodes carrying a high-frequency displacement current. Also **high-frequency welding, jig welding.**
seaplane (*Aero.*). An aeroplane fitted with means for taking off from and alighting on water. See **float seaplane, flying-boat, hydroskis.**
seaplane tank (*Aero.*). A long, narrow water tank with a powered carriage carrying equipment by which the water performance of a seaplane can be observed and precisely measured.
search (*Comp.*). Continuous examination and rejection of data until a desired value is found.
search coil (*Elec. Eng.*). See **exploring coil.** (*Nav.*) Rotating coil in a radiogoniometer or phase adjuster.
search gas (*Vac. Tech.*). A gas used in leak detection tests.
searchlight projector (*Light*). A projector embodying a parabolic reflector; capable of housing a high-intensity lamp for producing a parallel beam of light for picking out objects at night at considerable distances.
search radar (*Radar*). One designed to cover a large volume of space and to give a rapid indication of any target which enters it.
season (*Astron.*). One of the four divisions of the tropical year taken from the passage of the sun through the equinoctial and solstitial points: *spring*, reckoned from the equinox on March 21, is 92 days 19 hr; *summer*, from the solstice on June 21, is 93 days 15 hr; *autumn*, from the equinox on Sept. 23, is 89 days 20 hr; *winter*, from the solstice on Dec. 22, is 89 days 0 hr.
seasonal polymorphism (*Ecol.*). The occurrence of different forms of the same species at different seasons, a phenomenon especially characteristic of lake faunas.
seasonal succession (*Ecol.*). The changes in the distribution and abundance of plants and animals which occur cyclically in one habitat as the seasons change. Strictly speaking not a *succession* (q.v.) since it is not a steady unidirectional sequence of changes but a repeated cycle.
season cracking (*Met.*). Spontaneous cracking of brass and other metals. Intergranular cracks arise from residual internal stresses due to cold-working operations, in conjunction with surface corrosion.
seasoning (*For.*). The process in which the moisture content of timber is brought down to an amount suitable for the purpose for which the timber is to be used. (*Leather*) The process of coating leather, after dyeing, with some form of liquid albumen, preparatory to glazing and polishing.
seat board (*Horol.*). The board or platform that carries the movement of a long case clock.
seat earth (*Geol.*). See **underclay.**
seating (*Eng., etc.*). A surface for the support of another piece, e.g., the end of a girder, or a masonry block.
sebaceous (*Bot.*). Looking like lumps of tallow. (*Zool.*) Producing or containing fatty material, as the *sebaceous* glands of the scalp in Man.
sebaceous cyst (*Med.*). A cyst formed as a result of blockage of the duct of a sebaceous gland, often present on the face, scalp, or neck.
sebacic acid (*Chem.*). Obtained by heating castor oil with sodium hydroxide. A white crystalline solid, m.p. 129°C, with the structure:

$$(CH_2)_8\begin{cases}COOH\\COOH\end{cases}$$

Its esters are used in the production of resins and plasticizers.
sebiferous (*Zool.*). Conveying fatty material.
sebiparous (*Zool.*). Sebaceous.
seborrhoea, seborrhea (*Med.*). Overactivity of the sebaceous glands, resulting in an abnormally greasy skin.
seborrhoeic (or **seborrheic**) **dermatitis** (*Med.*). An inflammatory disease of the skin characterized by the presence of reddish patches covered with greasy scales; especially of the scalp, causing 'scurfy head'.
sebum (*Zool.*). The fatty secretion produced by the sebaceous glands, which protects and lubricates hair and skin.
sec, secant. See **trigonometrical functions.**
Secatron (*Print.*). Electronically controlled equipment for registering pre-printed webs into sheeters, die cutters, folders, perforators, and other paper converters. Cf. *Webatron.*

Secchi's classification (*Astron.*). The earliest classification of about 4000 stars into spectral types; made by Father Secchi (1818-78), who divided them into four groups designated by Roman figures, starting with the white, or helium, stars and ending with the reddest. Superseded by *Harvard classification* (q.v.).

secodont (*Zool.*). Having teeth adapted for cutting.

Seconal (*Pharm.*). TN for *quinalbarbitone sodium*, a sedative and hypnotic drug.

second. (1) 1/60 of a minute of time, or 1/86 400 of the mean solar day; once defined as the fraction 1/31 556 925·974 7 of the tropical year for the epoch 1900 January 0 at 12 h ET. Since 1964 defined, in terms of the resonance vibration of the caesium-133 atom, as the interval occupied by 9 192 631 770 cycles. This was adopted in 1967 as the SI unit of time-interval. Abbrev. s. (2) Unit of angular measure, equal to 1/60 of a minute of arc; indicated by the symbol ″. (3) In duodecimal notation, 1/12 of an inch; indicated by ‴. (4) Unit for expressing flow times in capillary viscometers (Redwood, Saybolt Universal, or Engler), e.g., an Engler second is that viscosity which allows 200 cm^3 of fluid through an Engler viscometer in one second.

secondary (*Zool.*). Arising later; of subsidiary importance; in Insects, the hind-wing; in Birds, a quill feather attached to the forearm, and also called the **cubital.**

secondary alcohols (*Chem.*). Alcohols containing the group >CH(OH). When oxidized they yield ketones (alkanones).

secondary amines (*Chem.*). Amines containing the imino group >NH. They yield nitrosamines with nitrous (nitric(III)) acid.

secondary association (*Bot.*). The coming together of bivalent chromosomes during meiosis.

secondary battery (*Elec.*). A number of secondary cells connected to give a larger voltage or a larger current than a single cell.

secondary beam (*Build.*). In floor construction, a beam carried by *main beams* (q.v.) and transmitting loads to them.

secondary body cavity (*Zool.*). See **coelom.**

secondary bow (*Meteor.*). A *rainbow* having an angular radius of 52°, the red being inside and the blue outside, usually fainter than the primary bow. It is produced in a manner similar to the primary bow except that two internal reflections occur in the raindrops.

secondary cell (*Elec. Eng.*). See **accumulator.**

secondary cell wall (*Bot.*). The layers of wall material deposited on the primary wall as the cell ages; it usually contains more cellulose and less pectin than the primary wall, and is often pitted.

secondary coil (*Elec. Eng.*). A coil which links the flux produced by a current flowing in another coil (*primary coil*).

secondary colours (*Paint.*). Colours produced by mixing primary colours.

secondary constants (*Elec. Eng.*). Those for a transmission line which are derived from the *primary constants*. They are the *characteristic impedance* (*impedance level*), as of an infinite line, and the *propagation constant* (*attenuation* and *phase delay constant*).

secondary cortex (*Bot.*). See **phelloderm.**

secondary depression (*Meteor.*). A *depression* (q.v.) embedded in the circulation of a larger primary depression.

secondary electrode (*Elec. Eng.*). See **bipolar electrode.**

secondary electrons (*Electronics*). Those which are emitted from a surface by electronic bombardment, as distinct from the primary bombarding electrons, or photoelectrons.

secondary emission (*Electronics*). Emission of electrons from a surface (usually conducting) by the bombardment of the surface by electrons from another source. The number may greatly exceed that of the primaries, depending on the velocity of the latter and the nature of the surface.

secondary emission multiplier (*Electronics*). See **electron multiplier.**

secondary enrichment (*Geol.*). The name given to the addition of minerals to, or the change in the composition of the original minerals in, an ore body, either by precipitation from downward-percolating waters or upward-moving gases and solutions. The net result of the changes is an increase in the amount of metal present in the ore at the level of secondary enrichment.

secondary gneissic banding (*Geol.*). A prominent mineral banding exhibited by coarse-grained crystalline rocks which have been subjected to intense regional metamorphism, involving rock-flowage. Often it is difficult to distinguish from *primary gneissic banding* (q.v.).

secondary grid emission (*Electronics*). Electrons released by bombarding electrons, depending on the surface of the grid material.

secondary growth. See **secondary thickening.**

secondary hardness (*Met.*). Further increase in hardness produced on tempering high-speed steel after quenching.

secondary leakage (*Elec. Eng.*). The magnetic leakage associated with the secondary winding of a transformer.

secondary memory (*Comp.*). (1) Storage unit associated with, but outside, a computer. (2) Part of computer store for data less often required, which has a longer access time than the *primary store* (q.v.). Also called **auxiliary** or **backing** or **secondary store.**

secondary meristem (*Bot.*). A meristem formed from permanent tissue.

secondary metal (*Met.*). That retrieved from scrap, and worked up for return to industry.

secondary mineral (*Geol.*). One resulting from alteration of primary mineral (deposited during rock formation). (*Min. Proc.*) One of minor interest in ore body undergoing exploitation.

secondary mycelium (*Bot.*). (1) Hyphae growing down from the developing fruit body of a fungus and taking up food material for its nutrition. (2) The mycelium of binucleate segments, bearing clamp connexions, formed by many *Basidiomycetes*.

secondary nitro-compounds (*Chem.*). Nitro-compounds containing the group >$CH(NO_2)$.

secondary nucleus (*Bot.*). The nucleus formed in embryo sac by union of the two polar nuclei.

secondary orientation (*An. Behav.*). The normal spatial relationships of an animal with various features of its usual habitat or environment. Cf. *primary orientation*.

secondary petiole (*Bot.*). The petiole of a leaflet of a compound leaf.

secondary phloem (*Bot.*). Phloem formed by the activity of a cambium.

secondary radar (*Radar*). One in which the received pulses have been transmitted by a responder which has been triggered by a *primary radar*.

secondary radiation (*Nuc.*). That produced by interaction of primary radiation and an absorbing medium.

secondary service area (*Radio*). Area in which

reasonably consistent broadcast programmes are received at night via an ionospheric reflected wave.

secondary spectrum (*Optics*). The residual longitudinal chromatic aberration in a lens corrected to bring two wavelengths to the same focus.

secondary standard. A copy of a *primary standard* for general use in a standardizing laboratory.

secondary store (*Comp.*). See **secondary memory** (2).

secondary stress (*Eng.*). A bending stress, resulting from deflection, as distinct from a direct tensile or compressive stress.

secondary succession (*Ecol.*). A succession proceeding in an area from which a previous community has been removed, e.g., a ploughed field. Cf. *primary succession.*

secondary thickening (*Bot.*). The increase in diameter of a stem or root when elongation has ceased and all primary tissues have been differentiated.

secondary voltage (*Elec. Eng.*). The voltage at the terminals of the secondary winding of a transformer.

secondary wall layer (*Bot.*). See **secondary cell wall.**

secondary wave (*Radio*). A wave deriving from the main or desired wave forming a communication link but arising when this wave is partially reflected, refracted, or scattered.

secondary winding (*Elec. Eng.*). A winding which links the flux produced by a current flowing in another winding (*primary winding*, q.v.).

secondary wood (or xylem) (*Bot.*). Wood formed by the activity of a cambium.

secondary yeasts (*Brew.*). See **wild yeasts.**

second-channel interference (*Radio*). In reception by supersonic heterodyne receivers, the interference from signals which are not desired, but whose frequency differs from local oscillator frequency by the same amount as the wanted signal. Both these signals produce an intermediate frequency output acceptable to the receiver. Discriminated against by pre-oscillator tuning.

second detector (*Radio*). In supersonic heterodyne receivers, detector which demodulates the received signal after passing through the intermediate-frequency amplifier.

second development (*Photog.*). The second development in a reversal process, after the first image has been removed by bleaching and the remaining silver halide rendered developable by further exposure to light.

second moment of area (*Eng.*). Of a plane area about an axis XX in the same plane: $I_{XX} = \Sigma ar^2$, where I is the second moment of area, a is an element of the total area and r is the perpendicular distance from axis XX. I_{XX} can also be obtained from the second moment of area about a parallel axis through the centroid of the area (I_{GG}), by the equation $I_{XX} = I_{GG} + Ah^2$, where A is the total area and h the perpendicular distance between the two axes. For a rectangle of breadth b (parallel to the axis) and depth d, $I_{GG} = \frac{bd^3}{12}$. For a triangle of base b (parallel to the axis) and height d, $I_{GG} = \frac{bd^3}{36}$. For a circle about a diameter (d), $I_{GG} = \frac{\pi d^4}{64}$. Symbol I.

second-operation work (*Eng.*). Machining work carried out after cutting from bar or other material and for which a second clamping operation is necessary.

seconds (*Build.*). Bricks similar to *cutters* (q v.) but of a slightly uneven colour. (*Textiles*) (1) Term applied by wool-sorters to the rather coarse wool from the edges of the front of a fleece; also to coarse skirtings from merino fleeces. (2) Goods slightly damaged in manufacturing or finishing, e.g., setting-on places in cloths, lost-pile or pulled stitches in knitted fabrics, etc.

second tap (*Eng.*). A tap used, after a taper tap, to carry the full thread diameter further down the hole, or to give the finished size of thread in a through hole.

second ventricle (*Zool.*). In Vertebrates, the cavity of the right lobe of the cerebrum.

secrecy (or privacy) system (*Telecomm.*). See **scrambler, inverter.**

secretagogue (*Med.*). Substance, e.g., hormone, which stimulates secretion.

secret dovetail (*Join.*). An angle joint between two members in which neither shows end grain, the visible external parts being mitred, while the dovetails are kept back from both faces. Also called **dovetail mitre, mitre dovetail.**

secretin (*Biochem.*). In Vertebrates, a hormone produced by certain cells forming part of the lining of the intestine when stimulated by hydrochloric acid from the stomach; it passes into the bloodstream and stimulates the pancreas to secretion and apparently also causes a flow of bile into the intestine. The first hormone to be experimentally demonstrated and described.

secretion (*Biochem.*). A substance elaborated, collected, and discharged by a gland or gland-cell; the process of elaboration of such a substance.

secretory (*Zool.*). Secretion-forming.

secretory cell (*Bot.*). A cell in which oils, resins, nectary, etc., are formed; the secretions may be retained in the cell or may exude from it.

secretory duct (or passage) (*Bot.*). An elongated intercellular space in which secretions accumulate.

secretory tissue (*Bot.*). A group of secretory cells.

secret switch (*Elec. Eng.*). See **locked-cover switch.**

section (*Biol.*). A taxonomic group, esp. a subdivision of a genus, but sometimes of a higher rank. (*Bind.*) A folded *sheet* of a book. Also **signature.** (*Micros.*) A thin slice of plant, animal or rock material, sufficiently transparent to be capable of investigation with the compound microscope. (*Surv.*) The representation to scale of the variations in level of the ground surface along any particular line. Drawing which shows a plane through a solid object, succession of geological strata, mine, etc. (*Telecomm.*) Unit of a ladder network, derived through design techniques to give specified transmission performance with respect to frequency. (*Typog.*) A reference mark (§) directing the reader's attention to a footnote.

sectional pontoon dock (*Civ. Eng.*). A form of *self-docking dock* (q.v.) built up of a number of separate pontoons lying transversely to the length of the dock and carrying its two side walls, which are sufficiently far apart to permit any one of the pontoons to be unbolted and supported by the dock.

section gap (*Elec. Eng.*). An arrangement for dividing the overhead contact wire of an electric traction system into sections, both electrically and mechanically, without interfering with the smooth passage of the current collector. Also called **overlap span.**

section insulator (*Elec. Eng.*). An insulator

forming the joint between two sections of an overhead contact wire.

section modulus (*Eng.*). Ratio of *second moment of area* to distance of the farthest stressed element from the *neutral axis.* It is an important property of structural members, for calculating bending stresses. Symbol *Z.* Also called, loosely, **elastic modulus.** Units cm^3.

section mould (*Carp., etc.*). A templet whose profile corresponds to the shape of the section of a required member. This shape is marked on the ends of a timber and used as a reference in making the member.

section pillar (*Elec. Eng.*). A cable box, in the form of a cast-iron pillar or small kiosk, in which different sections of feeder cables are joined together by removable links.

section switch (*Elec. Eng.*). A switch whose function is to connect or disconnect two sections of an electric circuit, generally two bus-bar sections.

section warping (*Weaving*). Winding from yarn cheeses or cones to form sections side by side on a beam. The yarns, after sizing, are wound on to the weaver's beam.

sector (*Maths.*). A plane figure enclosed by two radii of a circle (or of an ellipse) and the arc cut off by them.

sectoral horn (*Acous.*). Waveguide horn with two surfaces parallel to side of guide and the other two flared out—classified according to whether the flaring is in the plane of the electric or magnetic field.

sector disk (*Phys.*). Rotating disk with angular sector removed, interposed in path of beam of radiation. Used to produce known attenuation or to chop or modulate intensity of transmitted beam.

sector display (*Radar*). A form of radar display used with continuously rotating antennae, *not* with sector scanning. (So termed because the display uses a long persistence CRT excited only when the antenna is directed into a sector from which a reflected signal is received.) Cf. *sector scan.*

sectorial (*Zool.*). (1) Adapted for cutting. (2) In the wings of Insects, a cross vein joining certain branches of the radius vein.

sectorial chimaera (*Bot.*). A *chimaera* in which two or more distinct types of tissue are arranged in sectors which come to the surface.

sector-pattern instruments (*Elec. Eng.*). Switchboard instruments contained in cases having a sector shape instead of the usual circular shape; used to save space on the switchboard.

sector regulator (*Civ. Eng.*). A form of drum weir. It consists of a hollow reinforced concrete sector of a cylinder placed transversely across the direction of flow and capable of rotation about a horizontal axis on the downstream side, with accommodation for the sector in a special pit in the bed of the stream.

sector scan (*Radar*). Scan in which the antenna rotates only through a limited sector. (Not to be confused with a *sector display.*)

Sectra (*Build., Civ. Eng.*). A method of system building, developed in France. In this method, precision-made steel forms are used to precast rectilinear tunnel-shaped units in room widths and ceiling heights. Internal heating of the form work is used to accelerate the curing of the units. This method is one of the most important examples of onsite prefabrication but relies much on dimensional accuracy.

sectroid (*Build.*). The curved surface between adjacent groins on a vault surface.

sectrometer (*Chem.*). A potentiometer for electrometric titrations in which the microammeter is replaced by a cathode ray tube.

secular acceleration (*Astron.*). A nonperiodic term in the mathematical expression for the moon's motion, by which the mean motion increases ca. 11″ per century; caused by perturbations and by tidal friction in shallow seas.

secular changes (*Geol., etc.*). Changes which are extremely slow and take many centuries to accomplish; they may apply to climate, levels of land and sea, or, as in geomagnetism, to long period changes in the magnetic field at any place.

secular equilibrium (*Nuc.*). Radioactive equilibrium where parent element has such long life that activities remain effectively constant for long periods.

secular parallax (*Astron.*). An effect, so slow as to be undetectable, by which, owing to the motion of the solar system as a whole through space, the apparent places of the stars (and hence the shape of the constellations) will in the course of time entirely change.

secular variation (*Geophys.*). Changes of the earth's magnetic field as observed at a fixed station, which are progressive for decades or centuries.

secund (*Bot.*). Having the lateral members all turned to one side.

secundine (*Bot.*). The inner integument when two are present.

security paper (*Paper*). Paper for cheques and other similar documents made under security conditions, specially watermarked and/or treated with chemicals or animal fibres to prevent the possibility of forgery.

sedentary (*Zool.*). Said of animals which remain attached to a substratum.

Sedgewick-Rafter cell (*Powder Tech.*). A shallow glass trough used in microscope methods of particle-size analysis to contain drops of suspension which are to be examined.

sedimentary rocks (*Geol.*). All those rocks which result from the wastage of pre-existing rocks. They include the fragmental rocks deposited as sheets of sediment on the floors of seas, lakes, and rivers and on land; also deposits formed of the hard parts of organisms, and salts deposited from solution, in some cases with the aid of lowly organisms. Igneous and metamorphic rocks are excluded.

sedimentation (*Chem., Powder Tech.*). Method of analysis of suspensions, by measuring the rate of settling of the particles under gravity or centrifugal force and calculating a particle parameter from the measured settling velocities.

sedimentation balance (*Powder Tech.*). A device widely used in particle-size analysis by sedimentation techniques. The accumulation of particles at the bottom of a suspension vessel is automatically recorded, the particle size distribution being derived from the rate of accumulation of the particles at various times. See **Bostock-, Sharples micromerigraph, Shimadzu sedimentograph.**

sedimentation potential (*Chem.*). The converse of *electrophoresis* (q.v.); a difference of potential which occurs when particles suspended in a liquid migrate under the influence of mechanical forces, e.g., gravity.

sedimentation tank (*San. Eng.*). A tank into which sewage from the detritus pit is passed so that suspended matters may sink to the bottom, from which they can be removed.

sedimentation techniques (*Powder Tech.*). Group of methods of particle-size analysis in which concentration changes occur within a suspension, the changes being apparently caused by differential rates of settling of various sizes of particle.

sedimentation test (*Med.*). The measurement of the rate of sinking of red blood cells (erythrocytes) in drawn blood placed in a tube; the rate is increased in disease and in pregnancy.

sediment transporters (*Ecol.*). Aquatic animals which stir up bottom deposits and mix up organic and mineral constituents.

Se-duct (*Build., Heat*). A proprietary form of rectangular precast duct used for transferring gas-fired heated air throughout a building. Manufactured with aluminous cement and refractory aggregate to resist sulphate attack.

Seebeck effect (*Elec. Eng.*). Phenomenon by which an (thermoelectric) e.m.f. is set up in a circuit in which there are junctions between different bodies, metals or alloys, the junctions being at different temperatures. Also **thermoelectric effect.**

seed (*Bot.*). A multicellular structure containing the embryo of a higher plant, with stored food, the whole protected by a seed coat or *testa* (q.v.). (*Glass*) Small bubbles.

seed crystal (*Chem.*). A crystal introduced into a supersaturated solution or a supercooled liquid in order to initiate crystallization.

seed leaf, seed lobe (*Bot.*). A cotyledon in a flowering plant.

seed stalk (*Bot.*). See **funicle.**

seed vessel (*Bot.*). A dry fruit.

seedy (*Textiles*). Of wools, contaminated by grass seeds, difficult to remove.

seedy toe (*Vet.*). An affection of the horse's hoof in which the wall of the hoof becomes separated from the subcorneal tissue, forming a space which becomes filled with abnormal, crumbly, horn; causes lameness when severe.

seersucker (*Textiles*). A lightweight fabric distinguished by a crinkly appearance in which two or more broad colour effects sometimes alternate lengthways in the piece. Can be made with one or two warp beams and in some cases is also woven as a check effect.

see-saw amplifier (*Telecomm.*). Same as **paraphase amplifier.**

Seessel's pocket (*Zool.*). A pit of unknown significance occurring just behind *Rathke's pouch* on the dorsal side of the oral cavity in developing Vertebrates.

Seewer governor (*Elec. Eng.*). A hydraulic turbine governor for controlling the speed of high-pressure Pelton wheels; a needle valve varies the divergence of the conical pressure jet issuing from the nozzle.

Seger cones (*Heat.*). Small cones of clay and oxide mixtures, calibrated within defined temperature ranges at which the cones soften and bend over. Used in furnaces to indicate, within fairly close limits, the temperature reached at the position where the cones are placed. Also **fusion cones, pyrometric cones.**

segment (*Bot.*). (1) A multinucleate portion of a hypha or filament, delimited by transverse walls. (2) A daughter cell cut off by the division of a single apical cell. (3) A portion of the lamina of a leaf when deeply lobed but not divided into true leaflets. (*Elec. Eng.*) One of many elements, insulated from one another, which collectively form a commutator. (*Maths.*) A plane figure enclosed by the chord of a circle (or of an ellipse) and the arc cut off by it. The segment of a sphere or of an ellipsoid is the portion cut off by a plane. (*Zool.*) One of the joints of an articulate appendage: one of the divisions of the body in a metameric animal: a cell or group of cells produced by cleavage of an ovum. *adj.* **segmental.**

segmental (*Zool.*). In metameric animals, repeated in each somite; as *segmental arteries, segmental papillae.*

segmental apparatus (*Zool.*). The brain stem of Vertebrates, i.e., that part of the brain which shows the same type of organization as the spinal cord. Cf. *suprasegmental structures.*

segmental arch (*Civ. Eng.*). An arch having the shape of a circular arc struck from a point below the springings.

segmental core disk (*Elec. Eng.*). An armature core disk made up in segments; used when a disk in a single piece would be so large as to be unwieldy.

segmental duct (*Zool.*). The archinephric or pronephric duct of Vertebrates.

segmental Gothic arch (*Arch.*). An arch whose outline is formed by two segments of circles meeting obtusely.

segmental interchange (*Cyt.*). The exchange of portions between two chromosomes which are not homologous.

segmental organ (*Zool.*). An embryonic excretory organ of a metameric animal.

segmentation (*Zool.*). Meristic repetition of organs or of parts of the body: the early karyokinetic divisions of a fertilized ovum, leading to the formation of a blastula or analogous stage.

segmentation cavity (*Zool.*). See **blastocoele.**

segmentation nucleus (*Zool.*). The nucleus of a fertilized ovum formed by the union of the male pronucleus with the female pronucleus.

Segrè chart (*Nuc.*). A chart on which all known nuclides are represented by plotting the number of protons vertically against the number of neutrons horizontally. Stable nuclides lie close to a line which rises from the origin at 45° and gradually flattens at high atomic masses. Nuclides below this line tend to be β-emitters whilst those above tend to decay by positron emission or electron capture. Data for half-life, cross-section, disintegration energy, etc., are frequently added.

segregation (*Gen.*). (1) The process by which a pair of allelomorphic characters become separated out in the pure dominants and the pure recessives of the second and subsequent generations. (2) The separation of hereditary factors from one another during spore formation. (*Met.*) Nonuniform distribution of impurities, inclusions, and alloying constituents in metals. Arises from the process of freezing, and usually persists throughout subsequent heating and working operations. See **inverse-, normal-.**

seiche (*Meteor.*). An apparent tide in a lake (originally observed on Lake Geneva) due to the pendulous motion of the water when excited by wind, earth tremors or atmospheric oscillations.

Seidlitz powder (*Chem.*). Effervescent powder. A mixture of sodium hydrogen carbonate with tartaric acid, acid sodium tartrate, or some similar acid or salt. Purgative.

Seignette salt (*Chem.*). Same as **Rochelle salt.**

seine-net (*Ocean.*). A long shallow net with a buoyed head-rope and a weighted bottom rope; used to surround a certain area of water or ground so that anything within that area may be captured; it may be worked from the shore or from a boat.

seismaesthesia (*Biol.*). Perception of physical vibrations.

seismic prospecting (*Geophys.*). A method of geophysical exploration in which the travel times of reflected and refracted artificial shock waves are used to give information about the depth and character of subsurface rock formations.

seismograph (*Geophys.*). An instrument which registers earthquake shocks and concussions.

seismology (*Geophys.*). The study of earthquake phenomena, in particular of the shock waves which they create. Studies of the velocities and refraction of seismic waves enable the deeper structure of the earth to be investigated.

seismonasty (*Bot.*). Movement by a plant in response to a stimulus provided by mechanical shock.

seismotherapy (*Med.*). Treatment of disease by means of mechanical vibration.

seistan (*Meteor.*). The 120-day summer north wind in E. Persia.

seizing signal (*Teleph.*). One sent from the outgoing end of a circuit at the start of a call and having the primary function of preparing the apparatus at the incoming end of the circuit for the reception of subsequent signals.

seizure or seizing-up (*Eng.*). The locking or partial welding together of sliding metallic surfaces normally lubricated, e.g., a journal or bearing.

selachine (*Zool.*). A neurohumor of selachians which causes blanching of the skin.

Selbornian (*Geol.*). The stage of rocks in the Cretaceous System which includes the Gault and Upper Greensand of Southern England; approximately equivalent to the Albian Stage of the Cretaceous. Obsolete.

selcal (*Aero.*). An automatic signalling system used to notify the pilot that his aircraft is receiving a call. It makes constant monitoring of the receiving equipment unnecessary. A contraction of *sel*ective *cal*ling.

selectance (*Radio*). Ratio of sensitivities of a receiver to two specified channels.

selected areas (*Astron.*). Two hundred and six areas distributed nearly uniformly over the whole sky, and forty-six special areas mostly near the Galaxy, to which intensive research under international cooperation is being applied to extend and complete the statistical work on stellar motions done by Kapteyn.

selection fauna (*Zool.*). Immigrants from one type of habitat which succeed in surviving in a different one, e.g., forms typical of running fresh waters surviving in stagnant waters.

selection rules (*Phys.*). See **nuclear selection rules.**

selective absorption (*Bot.*). The power sometimes said to be possessed by a plant of taking in some substances and rejecting others. (*Light*) Absorption of light, limited to certain definite wavelengths, which produces so-called absorption lines or bands in the spectrum of an incandescent source, seen through the absorbing medium. See **Kirchhoff's laws** (*Phys.*), **Fraunhofer lines.** (*Radiol.*) Concentration of a compound—usually labelled—in a specific organ as a result of natural processes. See **differential absorption ratio.**

selective assembly (*Eng.*). The assembly of mating parts selected by trial for their accuracy so as to obtain the required precision of fit.

selective dump (*Comp.*). Recording on tape or printing-out the full contents of a specific part of a computer store.

selective emission (*Elec. Eng.*). The property of an incandescent body whereby it emits radiation, predominantly of one frequency.

selective fading (*Telecomm.*). That affecting some parts of a composite signal more than others, e.g., sound and not vision in TV reception.

selective freezing (*Met.*). A process involved in the solidification of alloys, as a result of which the crystals formed differ in composition from the melt. Thus, in alloys in a *eutectic system* (q.v.) (except the eutectic alloy), crystals of one metal are formed from a melt containing two, and this continues until the melt reaches the *eutectic point* (q.v.).

selective mating (*Zool.*). See **preferential mating.**

selective network (*Telecomm.*). One for which the loss and/or phase shift are a function of frequency.

selective protection (*Elec. Eng.*). A term applied to methods of protecting power transmission networks in which an automatic disconnexion of the faulty section occurs without disturbance of the remainder of the network.

selective resonance (*Radio*). Resonance which occurs at one or more discrete frequencies, instead of extending over a band of frequencies as in some forms of filter.

selectivity (*Radio*). Ability of a receiver to distinguish by tuning between specified wanted and unwanted signals. Measured by frequency difference for the half-power points of the pass band of the receiver. Often aided by directive reception.

selector (*Teleph.*). The unit device in automatic telephone switching, operated either by the dialled impulses originated by the subscriber, or by self-generated or machine-generated impulses arising in the exchange.

selector forks (*Autos.*). In a gearbox, forked members whose prongs engage with grooves cut in bosses which they move along a splined shaft for changing gear. They are secured to cranks that are operated through rods by the gear lever.

selector plug (*Teleph.*). A plug for making connexion to a selector, so that the latter can be taken out of service temporarily for testing, without removal from its shelf.

selector shelf (or **panel**) (*Teleph.*). A group of selector switches so arranged in a row that their bank wiring is taken out to one terminal strip.

selector switch (*Elec. Eng.*). A switch which prepares the closing of a circuit on any one of two alternative paths by means of a main switch or circuit-breaker.

selector valve (*Aero.*). A valve used to direct the flow of the hydraulic fluid or compressed air in a system into the desired actuating circuit.

selenite (*Chem.*). The salt of the hypothetical selenous acid $HSeO_2$, e.g., sodium selenite, $NaSeO_2$. (*Min.*) The name given to the colourless and transparent variety of *gypsum* (q.v.) which occurs as distinct monoclinic crystals, especially in clay rocks.

selenitic cement (or **lime**) (*Build.*). A mixture of a feebly hydraulic lime with approximately 5% of plaster of Paris ground together to suppress the slaking action of the lime; used for plastering or rendering.

selenium (*Chem.*). A nonmetallic element, symbol Se, at. no. 34; r.a.m. 78·96; valencies 2, 4, 6. A number of allotropic forms are known. *Red selenium* is monoclinic; m.p. 180°C; rel. d. 4·45. *Grey* (*metallic*) *selenium*, formed when the other varieties are heated at 200°C, is a conductor of electricity when illuminated; m.p. 220°C; b.p. 688°C; relative density 4·80; electrical resistivity 12×10^{-8} ohm metres. Selenium is widely distributed in small quantities, usually as selenides of heavy metals. It is obtained from the flue dusts of processes in which sulphide ores are used, and from the anode slimes in copper refining. It is used as a decolorizer for glass, in red glasses and enamels, and in photoelectric cells and rectifiers. Selenium is similar to sulphur in chemical properties, but resembles tellurium more closely still.

selenium cell (*Electronics*). (1) Photoconductive cell which depends on change in electrical resistance on incidence of light. (2) Photovoltaic cell which depends on the illumination of a barrier layer between selenium and other materials.

selenium glass (*Photog.*). A red orange glass filter used in colour cinematography.

selenium halides (*Chem.*). Selenium has a greater affinity for the halogens than sulphur. Selenium(VI)fluoride, SeF_6, is the only (VI)halide. (IV)halides known. No compounds with iodine.

selenium rectifier (*Electronics*). One depending on a barrier layer of crystalline selenium on an iron base. Widely used in power supplies for small electronic apparatus at one time.

selenium red (*Paint.*). See **cadmium red.**

selenodont (*Zool.*). Having cheek teeth with crescentic ridges on the grinding surface.

selenography (*Astron.*). The description and delineation of the moon's surface.

selenophone (*Acous.*). Original system of photographically recording sound on paper, the track being reproduced by scanning with a focused slit, the modulated reflected light being received into a photocell.

self-absorption (*Nuc.*). See **self-shielding.**

self-actor mule (*Spinning*). An automatic, intermittent machine for spinning cotton or woollen yarns and dry-spun worsted yarns. It incorporates drafting rollers, a bobbin creel, and travelling carriage with spindles.

self-aligning ball-bearing (*Eng.*). A *ball-bearing* (q.v.) in which the two rows of balls roll between an inner race and a spherical surface in the outer race, thus allowing considerable shaft deviation from the normal.

self-annealing (*Met.*). A term applied to metals such as lead, tin, and zinc, which recrystallize at air temperature and in which little strain-hardening is produced by cold-working.

self-baking electrode (*Elec. Eng.*). An arc-furnace electrode in the form of a hollow tube, into which a pastelike electrode material is continuously fed as it becomes hard-baked and burns away in the furnace.

self-balance protection (*Elec. Eng.*). A method of protecting transformers and a.c. generators from internal faults, based on the fact that the instantaneous sum of the phase currents in a symmetrical 3-phase system is always zero.

self-bias (*Electronics*). In a class-A amplifier, grid bias obtained from p.d. across a resistor in cathode circuit taking anode current.

self-capacitance (*Elec.*). See **capacitance.**

self-centring (or universal) chuck (*Eng.*). A lathe-chuck for cylindrical work in which the jaws are always maintained concentric by a scroll, or by radial screws driven by a ring gear operated by a key. See **scroll chuck.**

self-centring lathing (*Build.*). Expanded metal specially manufactured with raised ribs, greatly stiffening the sheet and enabling it to be used for lathing purposes with the minimum of framing. Also called **stiffened expanded metal.**

self-cleansing (*San. Eng.*). A term applied to a velocity of flow of sewage material sufficient to prevent deposition of solid matters.

self-compatible (*Bot.*). Said of a plant forming reproductive organs which will function together; self-fertile.

self-conjugate directions (*Maths.*). See **conjugate directions.**

self-conjugate triangle (*Maths.*). See **self-polar triangle.**

self-discharge (*Elec. Eng.*). (1) Loss of capacity of primary cell or accumulator as a result of internal leakage. (2) Loss of charge from capacitor due to finite insulation resistance between plates.

self-dissociation (*Chem.*). The weak tendency of water and some other liquids (e.g., liquid ammonia) with strongly polar molecules (see **associated liquid**) to break up into their component ions, such as H^+ (proton) and HO^- (hydroxyl), the former of which usually attaches itself to a complete water molecule H_2O, forming a hydronium ion, $H_2O{\cdot}H^+$.

self-docking dock (*Civ. Eng.*). A floating dock built up in sections, so that any section can be unbolted and lifted up on to the remainder for repair or maintenance purposes.

selfedge (*Mining*). See **selvage.**

self end-papers (*Bind., Print.*). Instead of having separate end-papers, the first two and last two leaves of the book are left blank, the first and last leaves being pasted down on the cover; also called **own ends.**

self-excitation (*Elec. Eng.*). A form of machine excitation in which the supply to the field system is obtained either from the machine itself or from an auxiliary machine which is coupled to it.

self-excited oscillator (*Radio*). Normal form of oscillator, in which excitation of the grid circuit is derived from a.c. in the anode circuit.

self-faced (*Build.*). A term applied to stone, e.g., flagstone, which splits along natural cleavage planes, leaving faces which do not have to be dressed.

self-fertilization (*Zool.*). The situation in a hermaphrodite animal when its female gametes are fertilized by its own male gametes.

self-fluxing ore (*Met.*). Mineral charged to smelter which contains its own slag-forming constituents.

self-hardening steel (*Met.*). Steel which hardens on cooling in air, i.e., does not require to be quenched in oil or water. The effect is produced by adding alloying elements which lower and retard the normal transformation from austenite to pearlite.

self-heterodyne (*Radio*). See **autodyne.**

self-incompatible (*Bot.*). Said of a plant producing reproductive organs which will not function together; self-sterile.

self-induced e.m.f. (*Elec. Eng.*). The e.m.f. induced in an electric circuit as a result of a change in the current flowing in it.

self-inductance (*Elec.*). Realization, in a current-carrying coil, of self-induction. See **mutual inductance.**

self-inductance coefficient (*Elec. Eng.*). See **inductance coefficient.**

self-induction (*Elec.*). Property of an electric circuit by which it resists any change in the current flowing (Lenz's law).

self-injection (*An. Behav.*). An experimental procedure by which rats can obtain the opportunity to inject themselves with various chemicals by learning to perform instrumental acts, this often acting as a positive reinforcement.

self-levelling level (*Surv.*). Instrument which levels automatically by means of a pendulum-operated system of prisms.

self-lubricating bearing (*Eng.*). Plain, powdered-metal bearing having a porous structure impregnated with lubricant which is released gradually during relative movement of the bearing.

self-polar triangle (*Maths.*). A triangle whose sides are the polars of the opposite vertices with respect to a conic. Also called **self-conjugate triangle.** Cf. *conjugate triangles.*

self-pollination (*Bot.*). The transfer of pollen from the anthers to the stigmas of the same flower, or to the stigmas of another flower on the same plant, or to those of a flower on another plant of the same clone.

self-quenching (*Nuc.*). Said of counter tubes which do not depend on an external circuit for quenching, the residual gas providing sufficient resetting for the next operation of detecting a further photon or particle.

self-quenching oscillator (*Telecomm.*). Same as **squegging oscillator.**

self-rectifying (*Radiol.*). Said of an X-ray tube when an alternating voltage is applied directly between target and cathode.

self-regulating (*Nuc. Eng., etc.*). Said of a system when departures from the required operating level tend to be self-correcting, and in particular of a nuclear reactor where changes of power level produce a compensating change of reactivity, e.g., through negative temperature coefficient of reactivity.

self-restoring coherer (*Radio*). A coherer in which the cohering contact reverts automatically to its original condition after the finish of a signal, without aid of a decoherer, e.g., a mercury-carbon contact.

self-scattering (*Nuc.*). The scattering of radioactive radiations by the body of the material which is emitting the radiation.

self-sealing capacitor (*Elec. Eng.*). One in which a puncture of dielectric by excessive field-strength causes oxidation of the neighbouring metal electrode through heat, with consequent restoration of insulation and negligible loss of capacitance. Also called **Mansbridge capacitor.**

self-sealing paper (*Paper*). See **gummed paper.**

self-shielding (*Nuc.*). In large radioactive sources, the absorption in one part of the radiation arising in another part. (*Telecomm.*) A coaxial line is self-shielding in that the return transmission current is in the inside surface of the outer conductor, while the interfering currents, if of sufficiently high frequency, are on the outside surface of the outer conductor.

self-starting rotary converter (*Elec. Eng.*). A synchronous converter designed to start up from the a.c. supply as an induction motor, thus requiring no separate starting motor.

self-sterility (*Bot., Zool.*). In a hermaphrodite animal or plant, the condition in which self-fertilization is impossible or ineffective.

self-synchronizing (*Elec. Eng.*). A term applied to a synchronous machine that can be switched on to the a.c. supply without being in exact synchronism with it.

self-tapping screw (*Eng.*). One made of hard metal which cuts its own thread when driven in.

self-toning paper (*Photog.*). Printing-out paper having inherent toning qualities which render separate toning normally unnecessary.

self-winding clock (*Horol.*). Clock rewound automatically from electricity supply after a definite period of running.

self-winding watch (*Horol.*). Watch which winds itself while being worn through movements of wearer turning a *rotor* (q.v.); the winding may also be performed by the opening or shutting of the case.

sellaite (*Min.*). Magnesium fluoride, crystallizing in the tetragonal system, occurring in volcanic *fumaroles* such as those of Vesuvius.

sella turcica (*Zool.*). In Vertebrates, a pocket in the floor of the chondrocranium in which the pituitary body comes to lie: in some *Crustacea*, part of the endoskeleton of certain posterior somites.

Sellers (or U.S.S.) screw thread (*Eng.*). The U.S. standard thread, having a profile angle of 60°, and a flat crest formed by cutting off $\frac{1}{8}$ of the thread height.

Sellmeier's dispersion formula (*Light*).

$$n^2 = 1 + C_1 \cdot \frac{\lambda^2}{\lambda^2 - \lambda_1^2} + C_2 \cdot \frac{\lambda^2}{\lambda^2 - \lambda_2^2} + \ldots$$

An expression giving the refractive index n of a medium for light wavelength λ; λ_1, λ_2, etc., being the wavelengths of absorption bands in the medium, C_1 and C_2 being constants for a given medium. See **anomalous dispersion.**

selsyn (*Elec. Eng., Telecomm.*) (*Self-syn*chronous). See **synchro, differential-.**

selsyn motor (*Elec. Eng.*). A small self-synchronizing motor (used for transmitting signals), which indicates the position of a switch or reproduces instrument indications at a distance.

selvage, selfedge (*Mining*). A layer or parting of clay or pug occurring on wall of a vein.

selvedge (*Textiles*). The frame or strong edge of a woven cloth, generally $\frac{1}{4}$ in (6 mm) wide on both sides, but wider in expensive materials. It sometimes bears a woven trade-mark or name. It is the part of the cloth securely gripped by pins or clips when drying to width after finishing. Also known as **lists.**

Semaeostomeae (*Zool.*). An order of *Discomedusae*, the adults having neither septa nor gastric pouches, e.g., *Aurelia.*

semantic error (*Comp.*). One which results in ambiguous or erroneous meaning of a computer program. Most programs have to be *debugged* to eliminate these errors before use.

semantide (*Biol.*). A molecule thought to carry information (as in a gene) or a transcript thereof.

semantophoric (*Biol.*). Descriptive of a molecule that carries information (as in a gene) or a transcript thereof.

sematic (*Zool.*). Warning; signalling; serving for warning or recognition, as *sematic colours.*

semeiology (*Med.*). The branch of medical science which is concerned with the symptoms of disease. (Gk. *sēmeion*, sign.)

semeiotic (*Med.*). Pertaining to, or relating to, the symptoms and signs of disease.

semen (*Zool.*). The fluid formed by the male reproductive organs in which the spermatozoa are suspended. *adj.* **seminal.**

semi-. Prefix from L. *semi*, half.

semi-apogamy (*Bot.*). A union of cells for reproductive purposes, when the cells are not of opposite sexes but when one at least of them is more or less gametic in nature.

semi-automatic (*Elec. Eng.*). Said of an electric control in which the initiation of an operating sequence is manually performed and the sequence of operations subsequently proceeds automatically.

semi-automatic exchange (*Teleph.*). An exchange in which the operators set up the desired connexions by means of remotely controlled switches.

semi-automatic exposure control (*Photog.*). A photoelectric device coupled to the lens of a cine camera which operates indicators, e.g., pointers, which can be aligned by moving the iris ring, thus obtaining the correct aperture setting for the available light.

semicarbazide (*Chem.*). $H_2N{\cdot}CO{\cdot}NH{\cdot}NH_2$, a base forming salts, e.g., hydrochloride. M.p. 96°C; may be prepared from potassium cyanate and hydrazine hydrate. It reacts with aldehydes and ketones, forming *semicarbazones* (q.v.).

semicarbazones (*Chem.*). The reaction products of aldehydes or ketones with semicarbazide. The two amino hydrogen atoms of the semicarbazide react with the carbonyl oxygen of the aldehydes or ketones, forming water, and the two molecular groups then combine to form the semicarbazone.

semichemical pulp (*Paper*). A brown pulp produced by grinding wood chips which have been softened by steam without the addition of chemicals. Cf. *mechanical* and *chemical wood pulps*.

semicircular canals (*Zool.*). Structures forming part of the labyrinth of the inner ear of most Vertebrates, there usually being three at right angles to each other, two being vertical and one horizontal. Movements of the head cause movement of the endolymph in the canals, which moves the gelatinous cupula attached to the sensory hairs of a neuromast sense organ in the swellings (called *ampullae*) at the base of the canals, initiating nerve impulses which travel to the brain via the 8th cranial nerve. They thus serve as organs of dynamic equilibrium.

semicircular deviation (*Ships*). Those components of the *deviation* (q.v.) which vary as the sine and cosine of the *compass course* (q.v.). That is, deviations which have the same sign in one semicircle of courses and the opposite sign in the other semicircle.

semiclosed slot (*Elec. Eng.*). A slot whose width narrows sharply at the top, necessitating the insertion of the conductors from the ends of the slot.

semicoke (*Fuels*). See **Coalite.**

semiconductor (*Electronics*). Poor conductor (at room temperature resistivity is between 10^{-5} and 10^{-1} ohm m), showing limited electrical conductivity due to either or both the following effects: (*a*) *intrinsic semiconductor*—the crossing of forbidden energy bands through thermal energy; effect increases with temperature and produces equal numbers of electron and hole (or *n* and *p*) carriers; (*b*) *extrinsic semiconductor*—the introduction of current carriers by donor and acceptor impurities with localized energy levels near top or bottom of forbidden band; so producing *n*- or *p*-type conductivity. Important semiconducting materials are Ge, Si, Se, Cu_2O, SiC, PbS, PbTe, etc., being used as rectifiers, transistors, photocells, thermistors, etc.

semiconductor diode (*Electronics*). Two-electrode, point contact or junction, semiconducting device with asymmetrical conductivity.

semiconductor junction (*Electronics*). One between *donor* and *acceptor* impurity semiconducting regions in a continuous crystal; produced by one of several techniques, e.g., alloying, diffusing, doping, drifting, fusing, growing, etc.

semiconductor radiation detector (*Electronics*). Semiconductor diodes, e.g., silicon junction, are sensitive under reverse voltage conditions to ionization in the junction depletion layer, and can be used as radiation counters or monitors.

semiconductor trap (*Electronics*). Lattice defects in a semiconductor crystal that produce potential wells in which electrons or holes can be captured.

semidiameter (*Astron.*). Half the angular diameter of a celestial body; the measure always used in expressing the apparent size of bodies with disks, such as the sun and the moon.

semi-elliptic spring (*Autos.*). A *carriage spring* (q.v.), so called because when a pair is used, one inverted and attached by its ends to the other, the arrangement resembles an ellipse.

semi-enclosed (*Elec. Eng.*). Said of electric motors in which ventilation is provided but access to live parts necessitates opening the case.

semigroup (*Maths.*). A set S together with an operation $*$ in S is said to be a semigroup if (1) the set S is closed with respect to $*$; i.e. for all elements x, y in S, $x{*}y$ is an element of S; (2) the operation $*$ in S is associative; i.e., for all elements x, y, z in S, $(x{*}y){*}z = x{*}(y{*}z)$. E.g. the set of natural numbers with addition and multiplication.

semi-immersed liquid-quenched fuse (*Elec. Eng.*). A liquid-quenched fuse in which the fuse link is above the liquid before operation but drawn down into it during or after fusion.

semi-indirect fitting (*Elec. Eng.*). A lighting fitting used for a semi-indirect lighting scheme; more than 60% and less than 90% of the light flux is emitted in the upper hemisphere.

semi-indirect lighting (*Light*). A method of illumination by means of translucent bowls surrounding lamps that have no reflectors.

semimetallic foils (*Print.*). Coloured *blocking foils* which give a metallic effect when viewed from certain angles. Used for stamping the reverse side of moulded acrylate badges.

semimonocoque construction (*Aero.*). See **monocoque** and **stressed-skin construction.**

semimuffle-type furnace (*Heat*). See **oven-type furnace.**

seminal receptacle (*Zool.*). See **vesicula seminalis**.

seminatural vegetation (*Bot.*). Vegetation which is influenced in its occurrence and persistence by human agency, though not planted by man.

seminiferous (*Bot.*). Seed-bearing. (*Zool.*) Semen-producing or semen-carrying.

seminoma (*Med.*). A malignant tumour of the testis arising from the germinal cells.

Semionotoidea (*Zool.*). An order of *Holostei*, which arose in the late *Palaeozoic*, and faded in the *Cretaceous*, but with one surviving genus, *Lepidosteus*, the Gar-pike. Also **Ginglymodi, Lepisosteiformes.**

semioviparous (*Zool.*). Giving birth to imperfectly developed young, as marsupial Mammals.

semipalmate (*Zool.*). Having the toes partially webbed.

semipermeable membrane (*Chem.*). A membrane which permits the passage of solvent but is impermeable to dissolved substances.

semiplacenta (*Zool.*). A nondeciduate placenta in which only the foetal part is thrown off at birth. Cf. *placenta vera.*

semirigid airship (*Aero.*). See **airship.**

semisteel (*Met.*). Cast-iron with low carbon content incorporating a proportion of melted down scrap steel.

semistreptostyly (*Zool.*). In Vertebrates, the condition of having a slightly movable articulation between the quadrate and the squamosal. Cf. *streptostyly, monimostyly.*

semiterete (*Bot.*). Half cylindric.

semitone (*Acous.*). Difference of pitch between two sounds with a frequency ratio equal to the 12th root of 2 on the even-tempered scale.

semitransparent mirror (*Photog.*). A mirror which partially reflects and partially transmits light rays without appreciable diffusion.

semitransparent photocathode (*Electronics*). One where the electrons are released from the opposite side to the incident radiation.

semitubular rivet (*Eng.*). A rivet which has been drilled hollow part-way up the shank to provide a thin shank wall for easy setting.

semiwater gas (*Chem.*). A mixture of carbon monoxide, carbon dioxide, hydrogen, and nitrogen obtained by passing a mixture of air and steam continuously through incandescent coke. Its calorific value is low, about 45 MJ/m^3.

Semper's rib (*Zool.*). In *Lepidoptera*, a degenerate trachea accompanying the ordinary trachea within the cavity of some of the wing-nervures.

sempervirent (*Bot.*). Evergreen.

senarmontite (*Min.*). The (III)oxide of antimony, crystallizing in the cubic system. A secondary mineral resulting from the oxidation of stibnite and other antimony ores.

sender (*Teleph.*). The same as *keysender*.

sender transmitting station (*Radio*). Location of radio transmitters and associated aerials (antennae). It can be *(semi-)attended, automatic, booster, master, remotely controlled, satellite* or *unattended*.

Senegal gum (*Chem.*). See **gum arabic.**

senescent (*Biol.*). Said of that period in the life history of an individual when its powers are declining prior to death.

senility (*Biol.*). Condition of vital exhaustion or degeneration due to racial or to individual old age.

Senonian (*Geol.*). One of the stages of the Cretaceous system, corresponding in England to the Upper Chalk.

sensation (*Psychol.*). The result of the stimulation of sense receptors. Formerly thought to consist of elements out of which percepts were constructed.

sensation curves (*Photog.*). Curves which give the relative response of the eye to different colours having the same intensity.

sensation level (*Acous.*). Difference in level of a single-frequency sound, as applied to the ear, from the level which is just audible at the same frequency. The number of decibels a single frequency note has to be attenuated before it becomes just inaudible. See also **phon** and **sone.**

sensation unit (*Acous.*). Original name of the *decibel*; so called because it was erroneously thought that the subjective loudness scale of the ear is approximately logarithmic.

sense (*Radio*). Relative progression along a line, e.g., up or down, left or right, the line not in itself giving this information. In radio direction-finding, in which the direction of arrival of a radio wave is ascertained by directive antennae, the sense of the direction as determined by a simple loop, frame, or Adcock antenna is indeterminate, but is readily determined by injecting an additional e.m.f. depending on the arriving wave and derived from a simple elevated aerial.

sense dome (*Zool.*). In Insects, a campaniform sensillum.

sense-hillocks (*Zool.*). See **neuromasts.**

sensibility (*Bot.*). The condition of a plant of being liable to parasitic attack.

sensible heat (*Heat*). That heat which effects a change in a body which is detectable by the senses; i.e. it causes the temperature of the body to change. Measured by the product of the specific heat capacity, the mass of the body and the change of temperature.

sensible horizon (*Surv.*). See **visible horizon.**

sensiferous, sensigerous (*Zool.*). Sensitive.

sensillum (*Zool.*). In Insects, a small sense organ on the integument, typically comprising a cuticular and/or hypodermal structure through which a stimulus is amplified or directed or transduced into other mechanical or chemical changes, and one or more sensory neurones respond by nervous activity. *pl.* **sensilla.**

sensing (*Comp.*). In scanning punched cards or punched tape, process of mechanical, electrical, or photoelectric examination of presence or absence of a hole in a set. Also **reading.** (*Radar*) Removal of 180° ambiguity in bearing as given by simple vertical loop antenna, by adding signal from open aerial.

sensitive (*Zool.*). Capable of receiving stimuli.

sensitive drill (*Eng.*). A small drilling machine in which the drill is fed into the work by a hand-lever attached directly to the drilling spindle, the operator being thus given sensitive control of the rate of drilling.

sensitive flame (*Phys.*). A gas flame which changes its height or shape when sound waves fall on it. The simplest type, in which the gas, issuing from a fine orifice, burns with a tall narrow flame, is sensitive to very high-pitched sounds, which cause the flame to shorten and roar.

sensitive periods (*Psychol.*). Periods of development during which the individual is particularly susceptible to certain types of learning. There is more agreement on their existence in language and perceptual learning than in social learning.

sensitive time (*Nuc.*). Period for which conditions of supersaturation in a cloud chamber or bubble chamber are suitable for formation of tracks.

sensitive tint plate (*Min.*). A thin, optically-orientated plate of a crystal, usually gypsum, used to measure the optical properties of minerals and other crystalline substances with a polarizing microscope.

sensitive volume (*Nuc.*). The portion of an ionization chamber or counter tube across which the electric field is sufficiently intense for incident radiation to be detected. The portion of living cells believed to be susceptible to ionization damage. See **target theory.**

sensitivity (*Phys. etc.*). General term for ratio of response (in time and/or magnitude) to a driving force or stimulus, e.g., galvanometer response to a current, minimum signal required by a radio receiver for a definite output power, ratio of output level to illumination in a camera tube or photocell.

sensitization (*Chem.*). The process by which a sol of a lyophilic colloid becomes lyophobic in character, with the result that it may readily be coagulated by electrolytes.

sensitized paper (*Paper*). Paper usually used for photographic printing, being treated with an emulsion or passed through sensitive salts; also applied to papers chemically treated to expose erasures, etc.

sensitizer (*Chem.*). A substance, other than the catalyst, whose presence facilitates the start of a catalytic reaction. (*Photog.*). Chemical, usually dye, used to increase the sensitivity of photographic emulsions, generally or to specific colours.

sensitometer (*Photog.*). Wedge photometer, specially arranged for measuring densities of photographic sound-track on film. See **gamma.**

sensitometry (*Photog.*). Measurement of *gamma*, e.g., of photographic film used for optical film recording.

sensor (*Instr.*). General name for detecting device used to locate (or detect) presence of matter (or energy, e.g., sound, light, radio, or radar waves).

sensori-motor phase (*Psychol.*). Earliest phase of development (*Piaget*), 0–18 months, in which reflex movements lead to a system of movements and displacements leading to a concept of permanence.

sensorium (*Zool.*). The seat of sensation; the nervous system. *adj.* **sensorial.**

sensory (*Zool.*). Directly connected with the sensorium: pertaining to, or serving, the senses.

sensory adaptation (*An. Behav.*). A decrease in the intensity or strength of a response due solely to changes in the state of a receptor organ, produced by protracted or repetitive stimulation, or a corresponding increase due to recovery of a receptor from such stimulation.

sensory deprivation (*Psychol.*). The reduction to a minimum of all external stimulation reaching the body. After an extended period of sensory deprivation mental ability usually deteriorates, there is inability to concentrate and there may be hallucinations.

sentinel (*Comp.*). A symbol used to indicate the end of a specific block of information in a data-processing system. Also called **marker, tag.**

sentinel pile (*Med.*). An oedematous mass of rolled-up anal mucosa situated at the margin of the anus at the lower end of an anal fissure.

sepal (*Bot.*). One of the leaflike members forming the calyx of a flower.

sepalody (*Bot.*). An abnormal condition shown by the conversion of some other member of a flower into a sepal.

separated lift (*Aero.*). Lift generated by very low aspect ratio wings (usually of *delta* or *gothic* plan form) at high angles of incidence (ca. 20°) through separated vortices causing large suction forces. The *jet flap* (q.v.) is also a separated lift system.

separate excitation (*Elec. Eng.*). A form of machine excitation in which the supply to the field system is obtained from a separate direct-current source.

separate-lead-type cable (*Cables*). See **S.L.-type cable.**

separates (*Print.*). Same as **offprints.**

separate system (*San. Eng.*). A system of sewerage in which two sewers are provided in every street, one for the sewage proper, and the other for the rain water. Cf. *combined system.*

separating calorimeter (*Eng.*). A device for mechanically separating and measuring the water associated with very wet steam; used in conjunction with the *throttling calorimeter* (q.v.) in determining dryness fractions.

separating drum (*Eng.*). An auxiliary steam-collecting drum attached by tubes to the upper drum of some water-tube boilers to avoid priming or foaming.

separating funnel (*Chem.*). A funnel with a tap at the bottom and a stoppered top, in which two immiscible liquids can be dispersed by shaking, then separated by settling and drawing off the lower layer; used in *liquid-liquid extraction.*

separating weir (*Civ. Eng.*). See **leaping weir.**

separation (*Aero.*). The spacing of aircraft arranged by *air-traffic control* (q.v.) to ensure safety, which may be vertical, lateral, longitudinal, or a combination of the three.

separation disk (*Bot.*). A biconcave disk of intercellular material, found between the cells of a filament of a blue-green alga, and assisting in the break-up of the filament to form hormogones.

separation energy (*Nuc.*). That required to separate 1 nucleon from a complete nucleus.

separation factor (*Nuc.*). The abundance ratio of material taken from the product end of an isotope separation system or unit, divided by that of material taken from reject end. $(S-1)$ is known in U.K. as *separation* or *process factor*; in U.S. as *enrichment factor*. S is given by $(n'_1/n'_2)\div(n_1/n_2)$ where n_1 and n_2 are the mol fractions of two isotopes of mass numbers m_1 and m_2 respectively and n'_1 and n'_2 are the corresponding quantities after processing.

separation filters (*Photog.*). Green, red, and blue filters used for *colour separation* (q.v.).

separation layer (*Bot.*). See **absciss layer.**

separation of losses (*Elec. Eng.*). The itemizing of the individual losses from the combined losses obtained when testing an electrical machine.

separation point (*Phys.*). The point at which streamline flow, *laminar* or *turbulent*, separates from the surface of a body.

separation potential (*Nuc. Eng.*). Separative work content per mol.

separative efficiency (*Nuc. Eng.*). Ratio of change in separative work content produced by a separation plant in unit time to the integrated separative powers of its elements.

separative element (*Nuc. Eng.*). One unit of a cascade forming a complete isotope separation plant.

separative power (*Nuc. Eng.*). Change in separative work content produced by an isolated separative element in unit time.

separative work content (*Nuc. Eng.*). The quantity:

$$Q\left\{2(C_2-1)\ln\frac{C_2(1-C_1)}{C_1(1-C_2)}+\frac{(C_2-C_1)(1-2C_1)}{C_1(1-C_1)}\right\}$$

for an isotope separation plant. Q is the molar quantity of the product, and C_1 and C_2 are the abundances of the required isotope in the raw material and the product. In U.S. the term **value** is used.

separator (*Civ. Eng.*). A distance piece, usually of steel or cast-iron, bolted between the webs of parallel side-by-side steel joists to give rigidity and ensure unity of action. (*Comp.*) A *flag* used to separate items of data. (*Elec. Eng.*) A thin sheet of wood, perforated ebonite, or porous polyvinyl chloride separating the plates of a secondary cell. (*Eng.*) A trap in a pipe containing a gas with condensed vapours. Removes, e.g., water to produce dry steam. (*Mag.*) An electromagnet used to select iron and steel from mixed scrap. (*Min. Proc.*) Concentrating machine, used to separate constituent minerals of mixed ore from one another. (*Radio*) Same as **buffer stage.**

separator circuit (*TV*). One for separating line and frame synchronizing signals from the picture (video) signal and from each other.

sepdumag (*Cinema.*). International code for motion picture with two magnetic sound-tracks on a separate film.

sepduopt (*Cinema.*). International code for a motion picture having two optical sound-tracks on a separate film.

Sephadex (*Biochem.*). TN for a cross-linked *dextran* which acts as a molecular sieve and is used for the separation of proteins.

sepiolite (*Min.*). See **meerschaum.**

sepmag (*Cinema.*). International code name for a motion picture having a separate magnetic sound-track.

sepopt (*Cinema.*). International code name for a motion picture having an optical sound-track on a separate film.

sepsis (*Med.*). The invasion of bodily tissue by nonspecific pathogenic bacteria. *adj.* **septic.**

sept-, septi-, septo-. Prefixes used in the construction of compound terms. (1) L. *septum,* partition. (2) L. *septem*, seven. (3) Gk. *septos*, rotten.

septal neck (*Zool.*). In some *Cephalopoda* having a chambered shell, a tubular extension of a septum supporting the siphuncle.

septaria, septarian nodules (*Geol.*). Concretionary nodules containing irregular cracks which have been filled with calcite or other minerals.

septate (*Bot.*). (1) Divided into cells or compartments by walls (fungal spores and hyphae, algal filaments). (2) Divided into two or more chambers by partitions (ovaries of flowering plants).

septate fibre (*Bot.*). A fibre of which the lumen is divided into several compartments by transverse septa.

septation (*Bot.*). The division of a plant member into separate parts by walls (usually transverse).

septavalent (*Chem.*). See **heptavalent.**

septechlorites (*Min.*). A group of sheet silicates closely related chemically to the chlorites, and structurally to the serpentines and kandites. Includes *chamosite* and *greenalite* (qq.v.).

septemfid (*Bot.*). Deeply divided into 7 parts.

septenate (*Bot.*). Having parts in sevens.

Septibranchiata (*Zool.*). An order of *Lamellibranchiata* in which the gills form a horizontal muscular septum, which surrounds and is continuous with the foot and cuts off an upper respiratory chamber from the mantle-cavity; this septum is pierced by paired orifices, and by its contraction pumps water through the respiratory chamber; the byssus gland is rudimentary or absent. Carnivorous or carrion-feeding forms living in deep water.

septic tank (*San. Eng.*). A tank in which sewage is left for about 24 hr, during which time a scum forms on the surface and the sewage below is to some extent purified by the action of the anaerobic bacteria functioning in the absence of oxygen.

septicaemia, septicemia (*Med.*). The invasion of the blood stream by bacteria and their multiplication therein; associated with high fever, chills, and petechial haemorrhages into the skin. *adj.* **septicaemic.**

septicidal (*Bot.*). Said of the manner of dehiscence of a fruit composed of several compartments when the split forms through the middle of the partitions.

septicopyaemia, septicopyemia (*Med.*). Combined septicaemia and pyaemia.

septifragal (*Bot.*). Said of the manner of dehiscence of a fruit when the outer walls of the carpels break away from the partitions.

septum (*Bot.*). (1) A transverse wall in a fungal hypha, an algal filament, or a spore. (2) A wall between one cell and another. (3) The partition between two neighbouring chambers of an ovary. (*Electronics*) Dividing partition in a waveguide. (*Photog.*) A dividing piece in a beam-splitting camera, to avoid interference between the two beams because of contiguity of the lenses. (*Zool.*) A partition separating two cavities. *adj.* **septal.**

septum transversum (*Zool.*). See **diaphragm.**

sequence (*Cinema.*). The unit of the scenario, involving one general idea or happening and a number of scenes, each of which may include a number of shots. (*Elec. Eng.*) The order in which the several phases of a polyphase alternating-current supply undergo their cyclic variation of e.m.f. (*Maths.*) A set of numbers derived according to a rule, each member being determined either directly or from the preceding terms.

sequencer (*Comp.*). Instrument used for sorting information into the required order for data processing.

sequence register (*Comp.*). A register which is reset following the execution of an instruction and which then holds the memory address of the next instruction.

sequence switch (*Teleph.*). A rotary switch for making complicated circuit changes in a prescribed order.

sequence valve (*Aero.*). A type of automatic selector valve in a hydraulic, or pneumatic, system, much used in aircraft, whereby the action of one component is dependent upon that of another.

sequential colour systems (*TV*). Colour TV systems in which colour information for each channel is transmitted sequentially. Systems may be *field-*, *line-* or *dot-sequential.*

sequential memory (*Comp.*). See **access to store.**

sequential monitoring (*TV*). Viewing in a cycle a number of transmissions.

sequential operation (*Comp.*). Operation of digital computer when all instructions are carried out sequentially.

sequential scanning (*TV*). Scanning in which the spot traverses each line in the same direction, returning rapidly from the end of one line to the beginning of the next. Cf. *oscillatory scanning.*

sequential transmission (*TV*). A technique of transmitting pictures so that the picture elements are selected at regular times and are then delivered to the communication channel in the correct sequence.

sequestering agent (*Chem.*). One which removes an ion or renders it ineffective, by forming a complex with the ion. See **complexones.**

sequestrectomy (*Surg.*). The surgical removal of a sequestrum.

sequestrum (*Med.*). A piece of bone, dead as a result of infection and separated off from healthy bone.

sequoia (*For.*). See **redwood.**

seral community (*Bot.*). Any plant community which is not stabilized, but represents a stage in a succession.

Serber force (*Nuc.*). Close range force between nucleons arising from both ordinary *Wigner force* and *Majorana force* (qq.v.).

sere (*Bot.*). A series of plant communities making up a succession.

serein (*Meteor.*). The rare phenomenon of rainfall out of an apparently clear sky.

serge (*Textiles*). Dress and suiting fabrics of 2×2, 3×3, or 4×4 twill weave, made from worsted spun crossbred yarns, also with worsted warp and woollen weft, and with cotton warp and worsted weft.

serial arithmetic unit (*Comp.*). One in which the digits of a number are operated on sequentially.

serial bud (*Bot.*). A supernumerary bud lying to the side of the axillary bud.

serial computer (*Comp.*). Computer which operates successively on each *bit* of a word in contradistinction to a parallel computer where all bits are operated on simultaneously.

serial radiography (*Radiol.*). A technique for making a number of radiographs of the same subject in succession.

serial store (*Comp.*). A memory for which stored data can only be read out sequentially, with access time depending upon location.

sericin (*Zool.*). The water-soluble, gelatinous, protein forming the outer layer of the *silk* fibres.

sericite (*Min.*). A fine-grained white potassium mica, like muscovite in chemical composition and general characters but occurring as a secondary mineral, often as a decomposition product of orthoclase.

series (*Elec., etc.*). Said of electric components when a common current flows through them. (*Maths.*) The sum where each term of a *sequence* is added to the previous one. (*Phys.*) Lines in a spectrum described by a formula related to the possible energy levels of the electrons in outer shells of atoms.

series arm (*Telecomm.*). In a wave filter, that which is in series with one leg of the transmission line.

series capacitor (*Elec. Eng.*). A capacitor connected in series with a transmission line or distribution circuit to compensate for the inductive reactance drop and thereby improve the regulation.

series characteristic (*Elec. Eng.*). The characteristic graph relating terminal voltage and load current in the case of a series-wound direct-current machine.

series-characteristic motor (*Elec. Eng.*). An electric motor having a speed torque characteristic similar to that of a d.c. series motor, i.e., one in which the speed falls with an increase of torque. Also **inverse-speed motor.**

series feed (*Electronics*). The application of a direct voltage to an electrode of an amplifying device via the same impedance through which the signal current flows.

series field (*Elec. Eng.*). The main field winding of a motor when series connected.

series-gap capacitor (*Elec. Eng.*). Two variable vane capacitors, usually with air dielectric, with the moving vanes on the same rotating shaft; used in high-frequency circuits with the two capacitances in series, to obviate taking the current through a rubbing contact or through a pigtail, the latter being inductive.

series modulation (*Radio*). Anode modulation in which modulator and modulated amplifier are connected directly in series, to eliminate the necessity for a modulation transformer or choke coupling.

series motor (*Elec. Eng.*). An electric motor whose main excitation is derived from a field winding in series with the armature.

series-parallel controller (*Elec. Eng.*). A method of controlling the speed and tractive effort of an electric tractor having one or more pairs of series motors, whereby the motors can be connected either in series or in parallel.

series-parallel network (*Elec. Eng.*). One in which the electrical components are composed of branches which are successively connected in series and/or in parallel.

series-parallel winding (*Elec. Eng.*). See **multiplex winding.**

series regulator (*Elec. Eng.*). A regulating resistance in the main circuit of a motor.

series resonance (*Elec. Eng.*). The condition of a tuned circuit when it offers minimum impedance to an a.c. voltage supply connected in series with it (due to the circuit reactances neutralizing each other). The term **tunance** is sometimes used in place of resonance if this condition is attained by adjustment of a component value and not of frequency.

series stabilization (*Telecomm.*). A technique of stabilization using amplifier feedback in which the feedback and amplifier circuits are in series at each end of the amplifier.

series system (*Elec. Eng.*). The constant current system of d.c. distribution developed by Thury, in which generators and motors are all connected in series to form a single d.c. circuit. Also called **Thury system.**

series transformer (*Elec. Eng.*). A power transformer operating under constant-current instead of constant-voltage conditions. See also **current transformer.**

series winding (*Elec. Eng.*). A field winding connected in series with the armature of the motor.

serif (*Typog.*). The short strokes of a letter, at the extremities of the main strokes and hair lines.

serine (*Chem.*). $CH_2OH{\cdot}CH(NH_2){\cdot}COOH$, 2-amino-3-hydroxypropanoic acid, obtained by the hydrolysis of various proteins.

sero-amniotic connexion (*Zool.*). In a developing embryo of a higher Vertebrate, the point of union between the serous membrane, or chorion, and the amnion.

serological typing (*Bacteriol.*). A technique used for the identification of pathogenic organisms, e.g., bacteria, particularly strains within a species, when morphological differentiation is difficult or impossible. It is based on antibody-antigen reactions, specific proteins of the organism acting as antigens.

serology (*Med.*). The study of sera.

serophyte (*Med.*). Any microorganism which will grow in the presence of fresh serum exuding into a wound, such as the *streptococcus* and the *staphylococcus* (qq.v.).

seropurulent (*Med.*). Said of a discharge or effusion which is both serous and purulent.

serosa (*Zool.*). See **serous membrane.**

serositis (*Med.*). Inflammation of a serous membrane.

serotherapy (*Med.*). The curative or preventive treatment of disease by the injection into the body of animal or human serums which contain antibodies to the bacteria or toxins causing the disease.

serotinal (*Ecol.*). Designating late summer, when it is found convenient to divide this season into two parts. Cf. *aestival*.

serotonin (*Biochem.*). 5-Hydroxytryptamine. A potent vasoconstrictor found particularly in brain and intestinal tissue, blood platelets and mast cells.

serous (*Zool.*). Watery: pertaining to, producing, or containing a watery fluid or serum.

serous membrane or **serosa** (*Zool.*). One of the delicate membranes of connective tissue which line the internal cavities of the body in *Craniata*; the chorion.

serozyme (*Biochem.*). See **prothrombin.**

serpentine (*Min.*). A hydrous silicate of magnesium which crystallizes in the monoclinic system. The three chief polymorphic forms are *antigorite*, *chrysotile*, and *lizardite*. The serpentine minerals occur mainly in altered ultrabasic rocks, where they are derived from olivine or from enstatite. Usually dark green, streaked and blotched with red iron oxide, whitish talc, etc. The translucent varieties are used for ornamental purposes; those with a fibrous habit form one type of asbestos.

serpentine-jade (*Min.*). A variety of serpentine, resembling bowenite, occurring in China.

serpentinization (*Geol.*). A type of metamorphism effected by water, which results in the replacement of the original mafic silicates in peridotites by the mineral serpentine (bastite, chrysotile) and secondary fibrous amphibole (tremolite, actinolite).

Serpollet boiler (*Eng.*). A *flash boiler* (q.v.) used in the early Serpollet steam cars.

Serpulite Grit (*Geol.*). A minor subdivision of the Eriboll Quartzite of Cambrian age in the North-West Highlands of Scotland, characterized by the occurrence of the organism *Serpulites*, by some geologists regarded as a very primitive nautiloid cephalopod, by others as merely a segmented worm.

serrated roller (*Print.*). A knurled or serrated roller on web-fed presses which gives a pull to the paper and minimises ink pick-up.

serratulate, serrulate (*Bot.*). Minutely serrate.

serratus magnus (*Zool.*). In higher Vertebrates, a muscle connecting the scapula with the anterior ribs.

Serret-Frenet formulae (*Maths.*). See **Frenet's formulae.**

serricorn (*Zool.*). Having serrate antennae.

serrodyne (*Elec. Eng.*). Frequency translator using linear sawtooth modulation of transit times.

serrula (*Zool.*). A comblike ridge on the chelicerae of *Chelonethida.*

Sertoli cells (*Zool.*). Cells found in the epithelium of the seminiferous tubules of Mammals. They are closely associated with the developing germ cells. Metabolic products probably pass from them to the developing spermatids, and they phagocytose any residual protoplasm, and probably effect the correlation of the cycle of activity in the seminiferous epithelium.

serum (*Med., Physiol.*). (1) The watery fluid which separates from blood in coagulation. (2) Blood serum containing antibodies, taken from an animal that has been inoculated with bacteria or their toxins, used to immunize people or animals. *adj.* **serous.**

serum albumin (*Biochem.*). An albumin obtained from blood and nutritive fluids. A crystalline, water-soluble substance, which is not precipitated by NaCl, but coagulates at 70°–75°C.

serum globulins (*Biochem., Physiol.*). Proteins of blood serum, separable into three main groups, α, β, and γ.

serum hepatitis (*Med.*). See **infectious hepatitis.**

serum sickness (*Med.*). The reaction that sometimes occurs about eight days after the therapeutic injection of serum, viz. slight temperature, urticaria, pain and swelling in joints.

service area (*Radio*). That surrounding a broadcasting station where the signal-strength is above a stated minimum and not subject to fading.

service band (*Radio*). That allocated in the frequency spectrum and specified for a definite class of radio service, for which there may be a number of channels.

service capacity (*Elec. Eng.*). The power output of an electric motor, as specified on the maker's nameplate.

service ceiling (*Aero.*). The height at which the rate of climb of an aircraft has fallen to a certain agreed amount (in British practice, originally, 100 ft/min, but for jet aeroplanes 500 ft/min).

service ell (*Plumb.*). An ell having a male thread at one end.

service mains (*Elec. Eng.*). Cables of small conductor cross-section which lead the current from a distributor to the consumer's premises.

service pipe (*Civ. Eng.*). A branch pipe drawing supplies from a main. Also called **supply pipe.**

service quality (*Automation*). Proportion of time during which there is correct functioning on demand of an equipment.

service reservoir (*Hyd. Eng.*). A small reservoir supplying a given district, and capable of storing the water which is filtered during the hours of small demand for use when the requirements become greater. Also called **distribution reservoir, clear water reservoir.**

service tanks (*Aero.*). See under **fuel tanks.**

service tee (*Plumb.*). A tee having a female thread on the branch and one end of the run and a male thread on the other end of the run.

serving (*Cables*). (1) A layer of jute, tape, or yarn, impregnated with bitumen or similar substance, to prevent the steel-wire armouring biting into the lead sheath of a cable. Also **bedding.** (2) The process of covering a cable with some form of mechanically strong insulating and binding tape.

servo (*Automation*). General term for system in which the response is determined by a drive which is actuated by the difference (*error*) between a set target and the actual response. A servo system aids or replaces human action, by force, time of operation, or location. Error usually requires power amplification by electrical, hydraulic, etc., means.

servo-amplidyne system (*Elec. Eng.*). One in which an *amplidyne*, together with a control amplifier, is used in order to amplify mechanical power.

servo amplifier (*Elec. Eng.*). One designed to form the part of a servomechanism from which output energy can be drawn.

servo brakes (*Autos.*). Power-assisted brakes worked either by a *vacuum servo* (q.v.) or mechanically from the transmission.

servocontrol (*Automation*). Control of any system by means of a *servomechanism* (q.v.). The term usually infers control systems powered, or power assisted, by hydraulic jacks, or electric motors, or a combination of the two. (*Aero.*) A reinforcing mechanism for the pilot's effort. It may consist of *servo tabs* (q.v.).

servo link (*Elec. Eng.*). A mechanical power amplifier which permits low strength signals to operate control mechanisms that require fairly large powers.

servomechanism (*Automation*). Closed-cycle control system in which a small input power controls a much larger output power in a strictly proportionate manner; e.g., movement of a gun turret may be accurately controlled by movement of a small knob or wheel.

servomechanism types (*Automation*). Type-0 has no integrating term and requires an offset signal. Type-1 has one integrating circuit in its closed loop and a velocity error. Type-2 has two integrating circuits and no errors.

servomotor (*Elec. Eng.*). A small motor (electric, hydraulic, etc.) for use in an automatic control system. (*Eng.*) A device for magnifying a relatively small effort, usually by hydraulic means; e.g., for providing a large force for operating a governor-controlled valve by a governor of small power. See **pilot valve.**

servo tab (*Aero.*). A control surface *tab* (q.v.) moved directly by the pilot, the moment from which operates the main surface, the latter having no direct connexion with the pilot.

SES (*Chem.*). See **2,4-DES.**

sesamoid (*Bot.*). Granular. (*Zool.*) A small rounded ossification forming part of a tendon, usually at, or near, a joint; as the patella.

sesin (*Chem.*). See **2,4-DES.**

sesone (*Chem.*). See **2,4-DES.**

sesqui- (*Chem.*). Containing two kinds of atom, radical, etc., in the proportions of 2 : 3.

sesquiterpenes (*Chem.*). A group of terpene derivatives of the empirical formula $C_{15}H_{24}$.

sessile (*Bot.*). (1) Having no stalk. (2) Fixed and stationary.

sessile benthos (*Bot.*). Plants growing attached at the bottom of water.

seston (*Biol., Ocean.*). Very small plankton organisms retained only by the finest nets.

set (*Build.*). See **setting coat.** (*Carp.*) See **nail**

punch. (*Comp.*) To prepare circuits so that they operate when required or triggered. Binary circuits are *set* when in '1' configuration, *reset* or *cleared* when restored to '0'. (*Eng.*) See **cold sett.** (*Hyd. Eng.*) The direction of a current of water. (*Maths.*) Any collection of entities ('elements') defined by specifying the elements. See **subset, universal-, Venn diagram.** (*Mining*) (1) A frame of timber used in a shaft or tunnel. (2) A section of a leased mining area in Cornwall. (*Psychol.*) An attitude or predisposition which renders a person liable to particular types of thought, action, or perception. Also called **einstellung.** (*Typog.*) (1) Width of a type character. (2) To *compose* (q.v.) type-matter. (*Weaving*) See **sett.**

seta (*Bot.*). (1) A single elongated cell or row of cells, with scanty colourless contents, found in some algae. (2) A long hollow outgrowth from the cell wall. (3) A thick-walled unicellular structure found among the asci in some *Ascomycetes.* (4) The multicellular stalk which bears the capsule of mosses and liverworts. (5) A slender, straight prickle. (6) A bristle. (*Zool.*) A small bristle-like structure; a chaeta. *adjs.* **setaceous, setiferous, setigerous, setiform, setose, setulose.**

set flush (*Typog.*). A typographic instruction to set all lines without indention.

set-hands dial (*Horol.*). A small dial on a turret clock movement, the hands of which read the same as the main dial. Used in regulating the clock.

set-hands square (*Horol.*). The square for setting the hands of a key-wound watch.

setigerous sac (*Zool.*). See **chaeta sac.**

set noise (*Electronics*). Interfering noise which arises in the receiver itself, such as *thermal noise* and *shot noise.*

set of chromosomes (*Cyt.*). A group of chromosomes consisting of one each of the various kinds of chromosomes contained in the nucleus of the gamete.

set-off (*Build.*). See **offset.** (*Print.*) Smudging of ink from one sheet to another before ink has dried. Obviated by interleaving (see **slip sheets**) or by using *anti-set-off spray* (q.v.).

set point (*Elec. Eng.*). See **control point.**

set screw (*Eng.*). A screw, usually threaded along the entire shank length, which is used to prevent relative motion by exerting pressure with its point.

set solid (*Typog.*). A typographic instruction to use no leading between lines.

sett (*Build.*). A small rectangular block of stone 6 in. deep by 3–4 in. wide, and from 6 to 9 in. in length; formerly used for surfacing roads where traffic was heavy. Best setts were of either Scottish or Welsh granite. (*Civ. Eng.*) Kind of *follower* (q.v.), used where a pile has been driven in beyond immediate reach of monkey. (*Eng.*) See **cold sett.** (*For.*) Two or more linked vehicles used for transporting long timber. (*Weaving*) The count or number of the loom reed. It largely determines the ends (warp yarns) per inch in a fabric. Also **set.**

setting (*Astron.*). See **rising and-.** (*Build.*) The hardening of a lime, cement, mortar, or concrete mixture, or a plaster.

setting and lustring (*Textiles*). Different processes of finishing by which woollen and worsted fabrics are set and their appearance enhanced. The processes, which may be applied at various stages of finishing, all involve the application of moisture and heat while the fabrics are under tension. See **crabbing, decatizing, dry blowing, roll boiling.**

setting coat or **set** (*Build.*). The finishing coat of plaster—a thin layer, about $\frac{1}{8}$ in. (3 mm) thick, of fine stuff. Also called **skimming coat.**

setting-out (*Leather*). A mechanical process for removing creases and marks from leather, which is stretched out in a wet condition.

setting point (*Chem.*). The temperature at which a melted wax, when allowed to cool under definite specified conditions, first shows the minimum rate of temperature change.

setting-point test (*Oils*). A test made to ascertain the temperature at which an oil will cease to flow; used in order to determine suitability for use at low temperatures, e.g., in refrigerating machines.

setting rule (*Typog.*). See **composing rule.**

setting stick (*Typog.*). See **composing stick.**

settle mark (*Glass*). Slight variation in thickness of the glass in the walls of a container.

settlement (*Build., Civ. Eng.*). The subsidence of a wall, structure, etc.

settling (*Powder Tech.*). Classification effected by the rate of fall in a fluid which may have a horizontal component of velocity.

settling tank (*San. Eng.*). See **sedimentation tank.**

sett paving (*Civ. Eng.*). Pavement constructed with setts on a suitable foundation. A **causeway** in Scotland.

sett systems (*Textiles*). Systems for arranging, or for expressing, the fineness of cloths, i.e., the number of ends or threads per inch. Systems vary, but fundamentally they are all concerned with the number of ends per inch in a fabric. Also known as **reed-counting systems.**

set-up (*Surv.*). Location of theodolite above a station point. (*TV*) Ratio between black and white reference levels measured from blanking level for facsimile transmission.

set-up instrument, set-up-scale instrument, set-up-zero instrument (*Elec. Eng.*). See **suppressed-zero instrument.**

set-work (*Build.*). Two-coat plasterwork on lath.

severy (*Arch.*). See **civery.**

sewage farm (*San. Eng.*). A farm on which sewage (especially sewage conveyed from a town) is used as a manure. See also **land treatment.**

sewage (or **sludge**) **gas** (*San. Eng.*). A self-generated combustible gas collected from the digesting tanks of sewage sludge. General composition: 66% CH_4 and 33% CO_2, with energy density in the region of 25 MJ/m^3. The gas has a very slow rate of flame propagation.

sewing (*Bind.*). The operation of joining the gathered sections of a book by sewing.

sex (*Biol.*). The sum total of the characteristics, structural and functional, which distinguish male and female organisms, especially with regard to the part played in reproduction: to determine the sex of a specimen. *adj.* **sexual.**

sex cells (*Biol.*). See **gametes.**

sex chromosome (*Cyt.*). The chromosome which is responsible for the initial determination of sex. Also *X, Y* or *W chromosome* (qq.v.). Cf. *autosome.*

sex determination (*Zool.*). The phenomena occurring prior to, and during, the development of an individual which lead to the establishment of its sex.

sexfarious (*Bot.*). In 6 rows.

sex gland (*Zool.*). See **gonad.**

sex hormones (*Biochem.*). Hormones produced by the gonads, e.g., male sex hormone, *testosterone*; female sex hormones, *oestrogens, progesterone.*

sex-intergrade (*Bot.*). A plant bearing staminate and pistillate flowers, but belonging to a species of which the members are normally dioecious.

sex-limited character (*Bot., Zool.*). A character developed only by individuals which belong to a particular sex.

sex-linked (*Med.*). Of genes, characters, or diseases, carried on the X chromosome. Diseases of this type characteristically affect males only but are transmitted by females only.

sex mosaic (*Zool.*). An individual showing characteristics of both sexes; an intersex.

sex reversal (*Zool.*). The gradual change of the sexual characters of an individual, during its lifetime, from male to female or vice versa.

sextant (*Surv., etc.*). A reflecting instrument in the form of a quadrant, for measuring angles up to about 120°. It consists essentially of two mirrors: a fixed *horizon glass*, half silvered and half plain glass, and a movable *index glass*, to which is attached an arm moving over a scale graduated to read degrees directly. The index glass reflects an image of one signal, or body, into the silvered part of the horizon glass, and this image is brought into coincidence with the other signal or body as seen through the plain part of the same glass. The sextant is used chiefly for measuring the altitude of the sun at sea, the reflected image of the sun being made to touch the visible horizon. See **air-**.

sextol (*Chem.*). See **methylcyclohexanol (IV)**.

sex transformation (*Zool.*). See **sex reversal**.

sexual cell (*Biol.*). A male or female germ cell.

sexual coloration (*Zool.*). Characteristic colour difference between the sexes, especially marked at the breeding season. See also **epigamic**.

sexual dimorphism (*Bot., Zool.*). Marked differences between the males and females of a species, especially differences in superficial characters, such as colour, shape, size, etc.

sexuales (*Zool.*). In some polymorphic *Homoptera* (e.g. *Aphidae*, *Phylloxeridae* and *Adelgidae*), those male and female individuals which reproduce sexually (cf. the various parthenogenetic stages). Female sexuales are usually oviparous and apterous, and their eggs develop into fundatrices.

sexual inversion (*Psychiat.*). Choice by a person of another of the same sex as a love object; may not become a full homosexual relationship.

sexual organs (*Zool.*). The gonads and their accessory structures; reproductive system.

sexual reproduction (*Bot., Zool.*). The union of gametes or of gametic nuclei, preceding the formation of a new individual.

sexual selection (*Zool.*). A phase of natural selection, based on the struggle for mating, by which some authorities have attempted to explain the existence of the secondary sexual characteristics.

sexupara (*Zool.*). In some polymorphic *Homoptera* (e.g., *Aphidae*, *Phylloxeridae* and *Adelgidae*), parthenogenetic females that lay eggs from which hatch *sexuales* (q.v.).

Seyfert galaxies (*Astron.*). Spiral galaxies in which the arms are inconspicuous but the nuclei extremely bright, giving spectra of highly ionized gases in violent motion. They radiate most strongly in the infra-red region, but some are also X-ray sources.

Seyler's classification (*Chem.*). System of classifying coal, based on its hydrogen and carbon contents; used particularly for Welsh coals.

sferics (*Radio*). U.S. for **atmospherics**.

sgraffito (*Paint.*). A mode of surface decoration in which two finishing coats of contrasting colours are applied, one on top of the other. Before the upper one has set, parts are removed according to some design, thus exposing the coat below. Also **graffito**.

shackle (*Eng.*). U-shaped machine element for connecting two parts which may have some relative movement. The ends are usually connected by a retaining bolt or pin.

shackle insulator (*Elec. Eng.*). A porcelain insulator whose ends are secured to metal shackles.

shade (*Paint.*). A gradation of colour. (*Surv.*) A disk of coloured glass used in telescope of theodolite when making sun observations.

shade chomophyte (*Bot.*). A plant inhabiting shaded rock crevices.

shaded pole (*Elec. Eng.*). A pole having a short-circuited ring around one section, thus altering the phase of the flux over that section. Sometimes used to make small single-phase motors self-starting.

shading (*Telecomm.*). The technique of varying the directivity of a transducer by controlling the amplitude and phase distribution over the active area of the transducer. (*TV*) Variations in brightness in a televised image because of local defects in the signal plate of a camera tube, arising from inadequate discharge. Corrected by injected waveforms in the output signal. See **tilt-and-bend-**.

shadow (*Light*). The shadow of an obstacle cast by a point source of light is the geometrical projection of the obstacle, except for small-scale diffraction effects at the edge. See **umbra.** (*Radio, TV*) Ineffectiveness of reception because of an obstacle, e.g., due to the topography of the terrain, between the transmitter and the receiver.

shadow bands (*Astron.*). A phenomenon sometimes occurring just before totality in a solar eclipse, in which parallel striations of light and shade are seen moving rapidly perpendicularly to their length across the ground; it is an atmospheric effect resulting from irregular refraction in the earth's atmosphere of the light from the thin crescent.

shadow casting technique (*Powder Tech.*). Method, carried out in high vacuum equipment, of determining the thickness of a particle of powder on a microscope slide. A beam of vaporized metal (usually gold or chromium) is directed towards the specimen slide at an oblique angle. The thickness of the particles can be measured from the angle of approach of the beam and the dimensions of the shadow cast by the particles.

shadow fringe test (*Optics*). A technique for examining the optical quality of glass. The shadows formed on transmission of a beam of light limited laterally are examined, since inequalities in the refractive index appear as fringes in the shadow.

shadowing technique (*Micros.*). A technique of shadow-casting used in electron microscopy, in which a very thin nongranular film of a metal, e.g., chromium, gold, uranium, is deposited obliquely on to the surface of the specimen prior to examination. This gives a three-dimensional effect and improves the clarity of the surface contours.

shadow-mark (*Paper*). A defect of paper showing as a faint reproduction of the holes of the suction couch or press rolls in the look-through.

shadow-mask tube (*TV*). Type of directly viewed 3-gun cathode-ray tube for colour TV display, in which beams from 3 electron guns converge on holes in a shadow mask placed behind a tricolour phosphor-dot screen.

shadow photography (*Photog.*). High-speed technique using an electric spark or similar light-source to photograph the shadow of a fast-moving object such as a projectile. Exposures of 10^{-6} s have been attained.

shadow photometer (*Light*). See **Rumford's photometer.**

shadow scattering (*Phys.*). See **scattering.**

shadow scratch (*Cinema.*). Defect in a recorded sound track, similar in appearance to optical scratch.

shadow stripes (*Textiles*). Cotton or man-made fibre cloths, of plain or satin weave, in which stripes are produced by using warp yarns of different directions of twist, i.e., 'S' or 'Z' twist. The shadow effect is due to light being reflected in different directions by the different twists.

shadow test (*Med.*). See **retinoscopy.**

shaft (*Arch.*). The principal portion of a column, between the capital and the base. (*Civ. Eng., Mining*) A passage, usually vertical, leading from ground-level into an underground excavation, for purposes of ventilation, access, etc. (*Zool.*) The part of a hair distal to the root: the straight cylindrical part of a long limb bone: the rachis, or distal solid part of the scapus of a feather.

shaft furnace (*Met.*). One in which ore and fuel are charged into the top and gravitate vertically, reacting as they proceed to bottom discharge.

shaft governor (*Eng.*). A compact type of spring-loaded governor used for controlling the speed of small oil engines, etc.; it is arranged to rotate about the crankshaft axis, and is sometimes housed in the flywheel. See **spring-loaded governor.**

shafting (*Eng.*). See **line shafting.**

shaft pillar (*Mining*). Solid block of coal or ore left unworked round the bottom of a shaft or pit for support.

shaft station (*Mining*). Room excavated underground adjacent to shaft, to accommodate special equipment such as pumps, crushing machine, truck tipples, ore sorting equipment, and surge storage bins.

shaft turbine (*Aero.*). Any gas turbine aero-engine wherein the major part of the energy in the combustion gases is extracted by a turbine and delivered, through appropriate gearing, by a shaft; **see free turbine** and **turboprop.**

shafty wool (*Textiles*). Wool with a good staple length of excellent spinning nature.

shagreen (*Leather*). Leather made from belly part of shark skin.

shake (*For.*). A partial or complete separation between adjacent layers of fibres. (*Mining*) (1) A cave in limestone. (2) In a coal-mine, a vertical crack in the seam and roof.

shaker patterns (*Textiles*). See **racked.**

shaking grate (*Eng.*). A grate for a hand-fired boiler furnace in which the pivotally supported fire-bars can be rocked by hand levers in order to break up clinker.

shaking palsy (*Med.*). See **paralysis agitans.**

shaking table (*Min. Proc.*). See **Wilfley table.**

shale (*Geol.*). A consolidated clay rock which possesses definite lamination. Cf. *mudstone*. See also **oil-shale.**

shale oils (*Fuels*). Oils obtained by the pyrolysis of oil-shale at ca. 550°C. Shale oils are usually characterized by the presence of a large proportion of unsaturated hydrocarbons, alkenes and di-alkenes. While shale oils are generally paraffinaceous in nature, some, such as that obtained from Narke (Swedish) shales, are predominantly aromatic in nature.

shalloon (*Textiles*). A lightweight woollen fabric of 2×2 twill weave, usually dyed a dark colour; used for coat lining.

shallow well (*Civ. Eng.*). A shaft sunk nearly to the bottom of a superficial permeable stratum to tap the waters in it.

shamal (*Meteor.*). The summer north-west wind in the plain of Mesopotamia.

shank (*Build.*). (1) Shaft of column, pillar, etc. (2) Shaft of tool, connecting head and handle. (*Eng.*) (1) Part of a twist drill which fits into chuck. (2) Portion of a screw between the head and the beginning of the thread. (3) Stem of rivet. (*Met.*) A ladle for molten metal. (*Typog.*) See **body.**

Shannon equation (*Telecomm.*). One in information theory which gives theoretical limit to rate of transmission of binary digits with a given bandwidth and signal/noise ratio.

shantung (*Textiles*). (1) Plain silk cloth of light brown colour, with a rough surface; made from tussah, the silk produced by the wild silkworm. (2) A cotton cloth woven with a slub weft and a crisp finish.

shaped-beam tube (*Electronics*). One in which the cross-section of the beam of electrons is formed to the shape of various characters.

shaped-conductor cable (*Elec. Eng.*). A three-phase cable in which the conducting cores are specially shaped so as to give the best utilization of the total available cross-section of the cable.

shape factor (*Powder Tech.*). A parameter descriptive of the shape of a particle of a powder or the ratio of two average particle sizes determined by techniques in which the shape of the particles of a powder influence the measured parameter in different ways.

shaper rail (*Spinning*). See **copping rail.**

shaper tools (*Eng.*). Cutting tools similar to those used on planing machines, and similarly supported in a clapper box.

Shap Granite (*Geol.*). A well-known granite occurring on Shap Fell in the Lake District of England. Much quarried as a valuable building-stone, distinctive in appearance by reason of the abundant pink phenocrysts of alkali-feldspar which it contains. A valuable indicator of the directions of ice movement in northern England.

shaping machine (*Eng.*). A machine tool for producing small flat surfaces, slots, etc. It consists of a reciprocating ram carrying the tool horizontally in guide ways, and driven by a *quick return* (q.v.) mechanism. Either the tool or the table may be capable of traverse.

shaping network (*Telecomm., etc.*). One which determines or restores the shape of a pulse, especially in radar and computing.

shard (*Geol.*). A fragment of volcanic glass, often with curved edges, representing the margins of disrupted gas bubbles. Glass shards are important constituents of some pyroclastic rocks.

share (*Agric.*). A pointed wedge-shaped implement of cast-iron or steel which is fixed on the front part of the breast of a plough. It makes the horizontal cut that separates the furrow slice from the undersoil. See **coulter.**

shared-channel broadcasting (*Radio*). See **common-frequency broadcasting.**

shared store (*Comp.*). One to which two independent computers both have access.

sharp (*Build., Civ. Eng.*). Said of sand the grains of which are angular, not rounded. (*Paint.*) Said of oil paint which dries rapidly to give a flat surface. It generally contains much pigment and little medium together with solvent, and is used as primer or sealer.

sharp-bend (*Elec. Eng.*). In electrical installation work, a bend of short radius for joining two lengths of steel conduit which are at 90° to each other.

sharp-edged orifice (*Eng.*). A circular orifice cut in a thin plate; placed in a pipe, or the wall of an air-box, to measure the flow of air or gas.

Sharpey's fibres (*Zool.*). Calcified bundles of white fibres and elastic fibres, prolonged from the periosteum into the periosteal lamellae of bone.

sharp flutings (*Arch.*). Flutings which are so close together as to form sharp arrises.

sharp gas (*Mining*). Mine air so contaminated with methane as to burn inside the Davy-type lamp and therefore to be dangerous.

Sharples micromerigraph (*Powder Tech.*). A special form of sedimentation balance in which particles sedimented through a gas column are collected in a balance pan. A servomechanism keeps the balance pan in equilibrium and continuously records the weight of particles which have sedimented out of the gas column.

sharp mouth (*Vet.*). Overgrowth of a part of one or more teeth of a horse through loss of wear.

sharpness (*Radio*). Equivalent to *selectivity*, but referring more directly to the change in circuit adjustment necessary to alter signal strength from its maximum to a negligible value.

sharpness of resonance (*Phys.*). The rapidity with which resonance phenomena are exhibited as the frequency of excitation of a constant driving force is varied.

sharp paint (*Paint.*). See **sharp.**

sharp series (*Light*). Series of optical spectrum lines observed in the spectra of alkali metals. Has led to energy levels for which the orbital quantum number is zero being designated s-levels.

shaving (*Acous.*). Machining surface of master disk recording to give fresh surface for further use. (*Leather*) A process by which skins and hides are brought to an even thickness; formerly done by hand, now by the bank knife splitting machine.

shear (*Eng., Phys., etc.*). A type of deformation in which parallel planes in a body remain parallel but are relatively displaced in a direction parallel to themselves; in fact, there is a tendency for adjacent planes to slide over each other. A rectangle, if subjected to a shearing force parallel to one side, becomes a parallelogram. See **elasticity of shear, strain, torsion.**

shearer loader (*Mining*). Machine which cuts coal from seam and loads it in the same operation to a conveyor belt working parallel to the face. In fully mechanized mining the assembly, together with roof support props, is moved hydraulically and can be remotely controlled.

shearing of rocks (*Geol.*). Shear zones, which are common in metamorphic rocks, are indicated by bands of crushed rock (cataclasite, etc.) and by the development of such minerals as chlorite. See **strain-slip cleavage.**

shear-legs (*Eng.*). See **sheers.**

shear mouth (*Vet.*). An increase in the obliquity of the wearing surfaces of the molar teeth of horses.

shear pin (*Eng.*). Pin used as a safety device to connect elements in a power transmission system, which is strong enough to transmit permissible loads but will fail by shearing when these loads are exceeded.

shears (*Eng.*). See **ways.**

shear stress (*Eng.*). The intensity of shear force per unit area of cross-section, varying in some definite way across the section, e.g., as the radius in a twisted shaft, and parabolically across a beam. See also **stress.**

Shea's operation (*Surg.*). The surgical removal of the stapes, leaving part of the crura behind to articulate with vein graft over the oval window.

sheath (*Bot.*). The leaf base when it forms a vertical coating surrounding the stem. (*Elec. Eng.*) The covering on a cable. (*Electronics*) Excess of positive or negative ions in a plasma, giving a shielding or *space-charge* effect. (*Nuc. Eng.*) The can protecting a nuclear fuel element. (*Zool.*) An enclosing or protective structure, e.g., elytron of some Insects.

sheath-circuit eddies (*Elec. Eng.*). The paths of currents in the sheaths of separate cables which flow only when the sheaths are bonded. See also **sheath eddies.**

sheath current (*Elec. Eng.*). The eddy current flowing in the metallic sheath of an alternating-current cable.

sheath eddies (*Elec. Eng.*). Currents which are induced in the sheath of a single cable, and which flow even when the sheaths are isolated from each other. Cf. *sheath-circuit eddies.*

sheathed-pilot system (*Elec. Eng.*). A system of selective protection using multicore pilot cables in which each pilot wire is provided, outside its main insulation, with a thin metallic sheath.

sheath effects (*Elec. Eng.*). The phenomena associated with the metallic sheaths of cables carrying alternating currents.

sheathing (*Carp.*). Close boarding nailed to the framework of a building to form the walls or the roof.

sheathing paper (*Build.*). A flexible waterproof lining material made from bitumen reinforced with fibre, and faced with stout kraft paper.

sheath of Schwann (*Zool.*). See **neurolemma.**

sheave (*Eng.*). Grooved pulley for use with vee-belts, ropes, or round belts.

shed (*Nuc.*). Minute unit of nuclear cross-section, 10^{-52} m² or 10^{-24} barn. (*Weaving*) Opening created by dividing the warp threads so that the shuttle, rapier, gripper, or pneumatic, etc., means can take the weft through to be beaten up into cloth. See **closed shed, open shed.**

shedding (*Weaving*). Dividing the warp threads in a loom, by tappet, dobby, or jacquard mechanism, to form a passage for the shuttle. After a pick has been inserted the shed is changed.

shed roof (*Build.*). See **lean-to roof.**

sheep ked (*Vet.*). A blood-sucking, wingless fly, *Melophagus ovinus*, which lives on the wool and skin of sheep.

sheep pox (*Vet.*). A highly contagious disease of sheep caused by a virus and characterized by a papulo-vesicular eruption of the skin and mucous membranes of the respiratory and alimentary tracts.

sheep scab (*Vet.*). See **psoroptic mange.**

sheer lawns (*Textiles*). Name given to very fine linens, used for ladies' handkerchiefs, etc.

sheers (*Eng.*). A large lifting device used in shipyards, etc., resembling a crane in which a pair of inclined struts take the place of a jib. Also called **sheer-legs, shear-legs.**

sheerstrake (*Ships*). The top strake or line of plating below, but extending a little above, the freeboard deck.

sheet (*Aero.*). The general term for aircraft structural material under 0·25 in (6 mm) thick; above that it is usually called plate. (*Bind., Print.*) A term applied to any one piece of printing paper printed or plain.

sheet anchor (*Ships*). A third bower anchor carried abaft the starboard bower for use in emergency. Formerly called **waist anchor.**

sheet-fed (*Print.*). A term applied to a rotary machine indicating that it prints separate sheets and not a reel or web.

sheet furnace (*Met.*). One in which metal sheet is

heated before further size reduction in rolling mill.

sheet glass (*Glass*). Glass used for common glazing purposes; produced by drawing a continuous thin film of glass from a molten bath and, after a suitable time interval for cooling, cutting up the product into sheets. It is not of such good quality, nor so flat, as plate glass, which is ground and polished.

sheeting (*Civ. Eng.*). (1) Rough horizontal boards used to support the sides of narrow trenches during excavation in very loose soils, each pair of boards on opposite sides of the trench being wedged apart with struts. (2) See **wearing course.**

sheetings (*Textiles*). Cotton or linen cloths used for bed coverings; most are woven grey, then bleached, also woven checks, all-over tints and floral prints, etc. The weave used is *2-and-2 twill* (q.v.)

sheet lead (*Build.*). Lead in a form in which it is commonly used in building construction, viz. in the form of sheets. It is the trade practice to refer to it in terms of the mass of unit area, e.g., 7-lb sheet lead, meaning 7lb/ft^2.

sheet lightning (*Meteor.*). Diffuse illumination of clouds by distant lightning of which the actual path of the discharge is not seen.

sheet pavement (*Civ. Eng.*). A road surfacing formed of continuous material such as concrete, and free therefore from the frequent joints associated with *block pavements* (q.v.). See **contraction joints.**

sheet piling (*Civ. Eng.*). Timber or steel sheeting supported in a vertical position by guide piles, and serving to resist lateral pressures. Prestressed concrete piles are also used.

sheet severer (*Print.*). An automatic device for cutting the web to prevent wrap around.

sheet wander (*Print.*). Undesirable lateral movement of the running web of paper.

sheet-work (*Print.*). Work in which two formes are used, one for each side of the paper, to give one complete copy of the job or section per sheet. Cf. *work-and-turn.*

shelf-back (*Bind.*). See **spine.**

shelf nog (*Carp.*). A wooden piece built into a brick wall, so as to leave a projecting part which will serve as a support for a shelf.

Shelford's law of tolerance (*Ecol.*). See **tolerance, law of.**

shell (*Mag.*). (1) Shell-like magnet in which magnetization is always normal to the surface and inversely proportional to the thickness. The *strength of the shell* is the product of magnetization and thickness of shell. (2) Theoretical concept of a double layer of poles, i.e., multitudinous magnetic dipoles, which is, in general, not plane. (*Nuc.*) Pattern of orbital electrons surrounding nucleus of atom, characterized by principal quantum numbers. See **Bohr atom.** (*Zool.*) A hard outer case or exoskeleton of inorganic material, chitin, lime, silica, etc.

shellac (*Chem.*). The purified product of *lac* (q.v.); thin yellow or brown leaflets, which can be bleached. Rel. d. 1·08–1·13, saponification value 194–213, acid value 48–64, ester value 137–163. Electric strength (1-min value for a film 2·5 mm thick) is 15–25 kV/mm. Shellac is a thermoplastic moulding material, and forms the basis of photographic and other varnishes. The normal gramophone disk record is based on PVC or shellac, the black colour being obtained by the addition of carbon black as a filler. Also used in polishes, rubber compounds, binders, inks, etc.

shell and tube exchanger (*Chem. Eng.*). Heat exchanger in which a large number of small bore tubes fixed between plates are surrounded by a single shell. One fluid flows through the tubes and the other flows over the outside of tubes in the shell. The flow of the fluids is thus approximating to right angle. There are two types of exchanger: the *fixed-tube sheet* type, in which, due to the conditions, no provision is made for differential expansion, and the *floating-head* type, in which provision for expansion is made by attaching the sheets carrying the tubes in such a way that one is fixed with reference to the shell and the other is free to move.

shell bark (*Bot.*). See **rhytidome.**

shell bit (*Tools*). Bit shaped like a narrow gouge, used for screw holes. Also **gouge-bit.**

shell gimlet (*Carp.*). A gimlet having a parallel shank.

shell gland (*Zool.*). In some Invertebrates, a glandular organ which secretes the materials for the formation of the shell. Also **shell sac.**

shell ligament (*Zool.*). The dorsal ligament joining the valves of the shell in lamellibranch *Mollusca.*

shell model (*Nuc.*). A model of the nucleus of an atom, with protons and neutrons in shells, by analogy with electrons in shells outside the nucleus.

shell pump (*Civ. Eng.*). See **sludger.**

shell reamer (*Eng.*). Reamer in the form of a hollow cylinder, end-mounted on an arbor. This construction is used for economy, for relatively large reamers are made of expensive materials.

shell sac (*Zool.*). See **shell gland.**

shell shake (*For.*). Part of a cup shake, as exhibited on the surface of converted timber.

shell-shock (*Psychol.*). See **war neurosis.**

shell star (*Astron.*). One of a number of stars of spectral type O, B, or A, which is surrounded by a shell of luminous gas giving bright emission lines.

shell-type transformer (*Elec. Eng.*). A transformer in which the magnetic circuit surrounds the windings more or less completely.

shelly (*For.*). Said of timber exhibiting shell shake.

shelter deck (*Ships*). A term correctly interchangeable with *awning deck*. Sometimes used in a more casual way to identify a deck above a weather deck.

shepherd's check (*Textiles*). Small check 2×2 twill pattern of black and white, or other contrasting colours, arranged 2 (or 4) warp, 2 (or 4) weft.

sherardizing (*Met.*). The process of coating steel or iron with a corrosion-resistant layer of zinc, by heating the object to be coated to a temperature of approximately 300°C in a closed box containing a powder consisting of zinc dust with some zinc oxide.

sheridanite (*Min.*). A mineral in the chlorite group poor in iron and relatively low in silica.

Sherringham daylight (*Light*). An artificial-daylight illumination obtained by means of a special lighting fitting in which an opaque reflector is sprayed with zones of different tints so as to give the correct resultant illumination.

shide (*Build.*). See **shingle.**

shield (*Civ. Eng.*). A kind of *curb* (q.v.) adapted for use at the working face in driving a tunnel through loose or water-bearing ground. It is driven forward as excavation proceeds by means of hydraulic jacks around its edge bearing on preceding lines of tunnel. (*Elec. Eng., Nuc. Eng., etc.*) Screen used to protect persons or equipment from electric or magnetic fields, X-rays, heat, neutrons, etc. In a nuclear reactor the shield surrounds it to prevent the escape of neutrons and radiation into a protected area. See **Faraday cage, biological-, magnetic-, neutron-, thermal-.** (*Electronics*) Electrode which controls intensity of cathode ray beam. Usually negatively charged with respect to cathode, which it surrounds or encloses.

shielded box (*Nuc. Eng.*). *Glove box* protected by lead walls, and with facilities for manipulation of contents by remote handling equipment.

shielded line (*Elec. Eng.*). Line or circuit which is specially shielded from external electric or magnetic induction by shields of highly conducting or magnetic material. (*Telecomm.*) Transmission line enclosed within a conducting sheath, so that the transmitted energy is enclosed within the sheath and not radiated.

shielded nuclide (*Nuc.*). One which when found among fission fragments, is assumed to have been a direct product because it is known not to be formed as a result of beta decay.

shielded pair (*Telecomm.*). Balanced pair of transmission lines within a screen, to mitigate interference from outside.

shielded-pole instrument (*Elec. Eng.*). An induction instrument having a shaded pole.

shield grid (*Elec. Eng.*). Auxiliary grid which protects control grid from heat and deposition of cathode material.

shielding (*Elec. Eng.*). (1) Prevention of interfering currents in a circuit, due to external electric fields. Any complete metallic shield earthed at one point is adequate. (2) Use of high permeability material, e.g., Mumetal, for shielding devices susceptible to a magnetic field, e.g., cathode-ray beam; the field, direct or alternating, is shunted away from spaces where it would cause interference. (*Nuc. Eng.*) Protective use of very light materials to thermalize strong neutron beams. (*Radiol.*) Use of dense matter to attenuate radiation when it might be harmful to operator or measuring system. The most common material used for large areas is concrete, and for smaller areas lead.

shielding pond (*Nuc. Eng.*). Deep tank of water used to shield operators from highly radioactive materials stored and manipulated at the bottom.

shielding windows (*Nuc. Eng.*). Dense glass blocks or liquid-filled tanks used as windows for inspecting the interior of shielded boxes.

shift (*Build.*). See **breaking joint.** (*Comp.*) Movement of a set of figures in a computer, to the right or left, as if multiplied or divided by the *radix*. (*Electronics*) Movement of a pattern on a CRT phosphor, by imposition of steady voltages, e.g., X-shift, Y-shift. (*Phys.*) (1) Change in wavelength of spectrum line due, e.g., to Doppler or Zeeman effects or to Raman scattering. (2) Change in value of energy level (*level shift*) for quantum mechanical system arising from interaction or perturbation. (*Teleg.*) Double use of code, using one code for changing over, as in Telex, teleprinter, teletype, analogous with typewriter keyboard. In teleprinters, one shift is capital letters, the other figures and special signs (*case shift*).

shifting of brushes (*Elec. Eng.*). The displacement of the brushes of a commutator motor from the neutral position.

shift-phase recording (*Cinema.*). In a light-valve, the separation of currents for the two ribbons, so that the retarded currents in one ribbon neutralize amplitude distortion arising from the comparable velocity of the ribbons and the steady velocity of the film passing normally to them.

shift pulse (*Comp.*). One which initiates and controls a shift in a digital computer.

shift register (*Comp.*). A temporary storage or delay register in which the output is released a fixed number of clock pulses after the input.

shilling-stroke (or **mark**) (*Typog.*). See **solidus.**

shim (*Eng.*). A thin strip of material, used singly or multiplied, to take up space between clamped parts. (*Mag.*) A packing piece consisting of a thin sheet of magnetic material for placing behind a pole piece in a magnetic circuit to adjust an air-gap. (*Print.*) A sheet of metal or plastic on which flexible plates can be mounted and then secured to the plate cylinder, or to the bed. Sometimes called *draw sheet*, thus risking confusion with the top sheet of the impression cylinder.

Shimadzu sedimentograph (*Powder Tech.*). A sedimentation balance for use with liquid suspensions. The accumulation of particles on the balance pan is recorded intermittently, the balance being counterpoised when necessary by the automatic addition of small steel balls.

shimamushi fever (*Med.*). Tsutsugamushi fever; flood fever; Japanese river fever. An acute febrile disease associated with infection by rickettsiae, transmitted by the bite of a larval mite; it is characterized by fever, enlargement of lymphatic glands in the neck, axilla, and groin, conjunctivitis, and a dark-red macular rash.

shimming (*Mag.*). Adjustment of magnetic field with soft iron shims or, by extension, small compensating coils.

shimmy (*Aero.*). The violent oscillation of a castoring wheel (in practice *nose* or *tail wheel* of an aeroplane) about its castor axis, which occurs when the coefficient of friction between the surface and the tyre exceeds a critical value. It is usually suppressed by a friction, spring, or hydraulic device called a *shimmy damper* (see **damper**). (*Autos.*) See **wheel wobble.**

shim rod (*Nuc. Eng.*). Coarse control rod of reactor. It is usually positioned so that the reactor will be just critical when the rod is near the centre of its travel path.

shin (*Eng.*). See **fish-plate.**

shin-bone (*Zool.*). The *tibia*.

shiner (*Build.*). A thin flat stone laid on edge in a rubble wall, the width of the stone being equal to the depth of at least two courses of the other stones. (*Print.*) See **light table.**

shiners (*Paper*). Particles, usually from mica in the china clay, or undissolved alum, showing on the surface of the paper.

shingle (*Build.*). A thin, flat, rectangular piece of wood laid in the manner of a slate or tile, as a roof covering or for the sides of buildings. Normally of red cedar. Also called **shide.** (*Geol.*) Loose detritus, generally of coarser grade than gravel though finer than boulder beds, occurring typically on the higher parts of beaches on rocky coasts.

shingles (*Med.*). *Herpes zoster*. An infection by a virus apparently related to that of chickenpox.

shingle trap (*Civ. Eng.*). A low barrier wall built out into flowing water or into the sea to catch the shingle and sand and minimize scour.

ship caisson (*Hyd. Eng.*). A floating caisson shaped like a ship, and capable of being floated into position across the entrance to a basin, lock, or graving dock, and then sunk into grooves in the sides and bottom of the entrance.

shiplap (*Carp.*). A term applied to parallel boards having a rebate cut in each edge, the two rebates being on opposite faces. They are especially adapted for use as sheathing.

shippers (*Build.*). Bricks which are sound and hard-burned, but not of good shape.

shipping fever (*Vet.*). See **equine influenza** and **haemorrhagic septicaemia.**

shipping pneumonia (*Vet.*). See **haemorrhagic septicaemia.**

ship plane (*Aero.*). Any aeroplane designed or adapted for operating from an aircraft carrier: special modifications are strengthened landing gear, strong points for catapulting, arrester hook and, if large, folding wings.

shivering (*Vet.*). A disease of horses, of unknown cause; characterized by involuntary spasmodic contractions of the muscles of one or both hind limbs and tail.

shives (*Paper*). Undigested particles of wood showing as pale yellow to brown splinters in the wood pulp or paper. (*Textiles*) (1) Vegetable matter found in wool fleece, exclusive of burrs. (2) Small particles of woody tissue adhering to flax fibres after scutching, and removed during hackling and combing.

s.h.m. (*Phys.*). Abbrev. for **simple harmonic motion.**

shoad or **shode** (*Mining*). Float-ore. Water-worn fragments of vein minerals found on the surface away from the outcrop.

shoal (*Hyd. Eng.*). A submerged sand bank.

shock (*Acous.*). See **acoustic-.** (*Eng., etc.*) The sudden application of load to a member. (*Med.*) Acute peripheral circulatory failure due to diminution in the volume of circulating blood and usually characterized by a low blood pressure and a weak thready pulse. It occurs (often associated with congestive cardiac failure) in, e.g., coronary thrombosis, pulmonary embolism, acute infection, or following trauma.

shock absorber (*Aero.*). See **oleo.** (*Autos.*) A frictional or hydraulic damper, having links attached to the chassis frame and axle respectively, to prevent spring rebound and damp out oscillation.

shock excitation (*Telecomm.*). Excitation of transient currents in an oscillatory circuit (ringing) at its natural resonant frequency by the sudden application or removal of an e.m.f. having some other frequency. The cause of interference by keying clicks from a continuous-wave telegraph transmitter.

shock heating (*Nuc.*). Heating, especially of a plasma, by the passage of a shock wave.

shockproof switch (*Elec. Eng.*). A switch having all its external metallic parts covered, or protected by insulating material, in order to guard against the possibility of electric shock. Also called **all-insulated switch, Home Office switch.**

shockproof watch (*Horol.*). A watch provided with a flexible mounting for the balance staff, to avoid damage to the pivots when the watch is subjected to severe shock.

shock shoe (*Agric.*). Attachment to a plough which absorbs shock when plough is lowered at the head furrow. Prevents risk of damage to shares on hard, dry ground.

shock therapy (*Psychiat.*). See **electroconvulsive therapy.**

shock tube (*Aero.*). A laboratory device for the simulation of hypersonic gas flow conditions by firing a shock wave down a long open-ended tube.

shock wave (*Aero.*). A region of infinitesimal width in which the airflow changes abruptly from *subsonic* to *supersonic*, i.e., from viscous to compressible fluid conditions, thus causing an increase in entropy—an abrupt rise in pressure and temperature. When a shock wave is caused by the passage of a supersonic body the airflow will decelerate to subsonic conditions through a second shock wave. A supersonic body normally sets up a conical shock wave with its nose (the angle becomes increasingly acute the higher the speed), subsidiary shock waves from projections on the body, and a decelerating shock wave from its tail. The nose and tail shock waves, either attached or travelling on after the passage of an aircraft, are the source of the pressure waves causing *sonic booms* (q.v.). Abbrev. **shock.** (*Phys.*) Wave of high amplitude, in which the group velocity is higher than the phase velocity, leading to a steep wavefront. This happens, for example, when an explosive is detonated. The speed at which the chemical reaction travels through the material is higher than the speed of sound in the material, hence a shock wave occurs. Strong shocks cause luminosity in gases, and so are useful for spectroscopic work. Also called **blast wave.**

shoddy (*Textiles*). Short, fibrous material obtained from old, cleaned, woven or knitted wool cloth after treatment in a rag-tearing machine. Sprayed with oil, it is used again in low woollen blends for cheap suits and coatings. The oil is removed in wet scouring. The lowest shoddy is also used as a fertilizer.

shoe (*Build.*). The short bent part at the foot of a downpipe, directing the water away from the wall. (*Civ. Eng.*) (1) The iron point fitted over the driving end of a pile. (2) The block, plate, or piece giving support to the foot of a rafter or post. (*Elec. Eng.*) The device by which an electric tractor collects current from a live rail. (*Glass*) A crucible-shaped clay vessel laid horizontally at the side of the pot mouth for heating the ends of blowing irons. (*Min. Proc.*) The replaceable steel wearing part of the head of a stamp or muller of a grinding pan.

shogged-rib (*Textiles*). See **racked.**

Shone ejector (*San. Eng.*). An apparatus employed to force sewage from a low-level sewer into a near-by high-level sewer, under the pressure of compressed air from an air compressor.

shonkinite (*Geol.*). A coarse-grained basic igneous rock (named originally from the Shonkin Sag Laccolith in Montana, U.S.), gabbroic in composition, sodic in character, and, as defined by Rosenbusch, containing nepheline as an essential constituent.

shoo flies (*Print.*). A feature of the *two-revolution* press; they direct the gripper edge of the printed sheet clear of the opened grippers and the strippers with which they work in conjunction.

shoot (*Hyd.*). See **chute.** (*Weaving*) A thread of weft. See **pick.**

shooting (*Join.*). The operation of truing with a *jointing plane* (q.v.) the edges of timbers which are to be accurately fitted together.

shooting board (*Join.*). A prepared board used to steady a piece of timber whilst shooting the edges. It has a stop against which the piece of timber abuts endwise, and a guide surface against which the jointing plane runs.

shooting plane (*Join.*). See **jointing plane.**

shooting star (*Astron.*). See **meteor.**

shooting stick (*Typog.*). A short tapered length of hard material used with a mallet to lock-up and unlock formes by tapping the wooden quoins; originally of wood, occasionally of metal, but now usually of hard plastic.

shop rivet (*Eng.*). A rivet which is put in when the work is being erected on the floor of the assembly shop prior to delivery to the site.

shop traveller (*Eng.*). See **overhead travelling crane.**

shoran (*Radar*). Short-range navigation, using H-radar (*shor*t *ran*ge).

shore effect (*Radio*). Horizontal refraction as a radio wave crosses a shoreline at an angle, because of different retardations of the ground wave. This causes direction-finding errors. Also **coastline effect.**

shoring (*Build., etc.*). The method of temporarily supporting by **shores**, i.e., props of timber or other material in compression, the sides of excavations and, especially, unsafe buildings.

short (*Glass*). Fast-setting. (*Mining*) Brittle.

short-chord winding (*Elec. Eng.*). An armature winding employing coils whose span is less than the pole pitch.

short-circuit (*Elec. Eng.*). Reduction of p.d. between two points in a circuit to zero by connexion of a conductor of zero impedance, in which no power is dissipated. If the short-circuit is not intended, damage may result if the circuit is not opened quickly elsewhere.

short-circuit calculator (*Elec. Eng.*). An assembly of variable impedances or resistances which can be connected to represent in miniature the circuits of a power system. If a low voltage is applied and a short-circuit put on the system, the currents which flow represent to scale the short-circuit currents which would flow in the actual system under similar conditions. Cf. *network calculator.*

short-circuit characteristic (*Elec. Eng.*). The characteristic graph relating e.m.f. or excitation to load current in the case of a machine operating under short-circuit conditions.

short-circuited rotor (*Elec. Eng.*). The same as *cage rotor*.

short-circuit impedance (*Elec. Eng.*). Input impedance of a network when the output is short-circuited or grounded.

short-circuiting device (*Elec. Eng.*). A switching device on the rotor of a slip-ring induction motor; operated by a mechanical clutch, which short-circuits the rotor windings when the motor has gained speed.

short-circuit protector (*Elec. Eng.*). A device for preventing damage from excessive currents caused by a short-circuit. See **overcurrent release, overload protective system.**

short-circuit ratio (*Elec. Eng.*). The ratio of the field ampere turns of a synchronous generator at normal voltage and no load to the field ampere turns on short-circuit with full-load stator current flowing. The value is important in evaluating and comparing the regulation and stability of machines.

short-circuit test (*Elec. Eng.*). A low-voltage test carried out on an electrical machine with its output terminals short-circuited and full-load current flowing.

short-circuit voltage (*Elec. Eng.*). The e.m.f. necessary to cause full-load current to flow under short-circuit conditions.

short coal (*Mining*). See **short.**

short column (*Eng.*). A column the diameter of which is so large that bending under load may be neglected, and in which failure would occur by crushing; commonly assumed as a column of height less than 20 diameters.

short-day plant (*Bot.*). One in which the onset of flowering is hastened by giving the plant alternating periods of relatively short illumination and relatively long darkness.

short delay line (*Comp.*). Any part of an electronic computer which adjusts phases of signals by transmission through coaxial lines, extended coils, or as magnetostriction pulses through nickel wires.

short descenders (*Typog.*). The length of the descenders (g, j, etc.) is a feature of type design; for bookwork long descenders are usually considered more elegant, but many successful display types have very short descenders.

shortening capacitor (*Radio*). One inserted in series with an antenna to reduce its natural wavelength.

short inks (*Print.*). See **long inks.**

short-oil (*Paint.*). Term applied to varnishes, etc., with a low proportion, i.e., less than about 40%, of oil content. Cf. *long-oil.*

short-period comets (*Astron.*). Those comets moving in elliptical orbits of such a size that the comets' returns to the neighbourhood of the sun are repeated at intervals comparable to the sidereal periods of the planets, one such group having periods between 3 and 8 yr.

short-period variables (*Astron.*). See **variable stars.**

short-range forces (*Nuc.*). Noncoulomb forces which act between nucleons when very close together, and are responsible for the stability of the nucleons. They are of two kinds: ordinary forces and exchange forces. See **Bartlett force, Heisenberg force, Majorana force, Serber force, Wigner force.**

shorts (*Mining*). (1) The contents of a wagon of coal containing very much dirt. (2) The shortage in production under a royalty lease.

short shoot (*Bot.*). A short branch borne in the axil of a scale leaf and bearing the true foliage leaves of the plant.

short-sightedness (*Med.*). See **myopia.**

short stick (*Textiles*). A cotton trade term which indicates a yard length of exactly 36 in. 120ss means 120 warps of 36 in. each and a length of 120 yards exactly.

short-time breakdown voltage (*Cables*). The voltage required to break down a cable in a short time (minutes).

short-time rating (*Elec. Eng.*). The output which an electrical machine can deliver for a specified short period (½ hr or 1 hr) without exceeding a specified safe temperature.

short wave (*Radio*). Rather vague designation of radio transmission with wavelengths between about 15 m and 100 m.

short-wave converter (*Radio*). A combination of a local oscillator and a mixer to produce a beat signal in the appropriate frequency band to convert short-wave signals to the region of standard broadcast frequencies.

short-wave therapy (*Med.*). Treatment by short-wave generators in the electrical capacitor field with high-frequency energy of from 6 m to 30 m wavelength. Therapeutic results of short-wave therapy are due, principally, to the heat produced in body tissues. See also **diathermy.**

shot (*Cinema.*). The unit in motion-picture production, comprising the sequence of events while the cameras are in operation. For each shot there may be several takes, to ensure a selection of good registrations. (*Weaving*) See **pick.**

shot-drilling (*Mining*). Boring deep holes by means of hard steel shot fed down a rotating hollow cylinder.

shot effects (*Textiles*). Changeable nonpatterning effects produced in fabrics in which the weft is a different colour from the warp.

shot firer (*Mining*). Miner, qualified under Coal Mines Act, 1911, who tests for gas and then fires explosive charges in colliery.

shot hole (*Civ. Eng.*). A hole bored in rock for

the reception of a blasting charge. (*For.*) A small hole made in timber by a wood-boring insect.

shot-hole disease (*Bot.*). A disease of leaves of various plants, caused by fungi which destroy small patches of tissue, which drop out.

shot noise (*Electronics*). The *Schroteffekt* (small-shot effect) of Schottky, which arises inevitably in the anode circuit of a thermionic valve, because electron emission is not strictly continuous, but consists of a series of random pulses from emitting surfaces.

shot peening (*Met.*). See **peening.**

shoulder (*Civ. Eng.*). Either of the verges of a road. A **hard shoulder** is part of the verge designed as a *pavement* (1) to take the weight of a vehicle. (*Electronics*) That part of a characteristic curve at which the response tends to fall off, as at the upper limit of thermionic characteristics, the γ-curves of photographic emulsions, or peak response of a transformer. (*Typog.*) The space from the foot of the *bevel* to the edge of the type body.

shouldered arch (*Arch.*). A lintel supported over a door opening upon corbels.

shoulder girdle (*Zool.*). See **pectoral girdle.**

shoulder heads (*Typog.*). Subheadings set flush to the left.

shoulder nipple (*Plumb., etc.*). A nipple which is not threaded over its full length but only at the two ends.

shoulder notes (*Typog.*). Notes which are printed, only one per page, in the outer margin level with the first line of text.

shoulder plane (*Tools*). Metal type of rebate plane with reversed cutting-bevel; used for fine trimming, especially of wide shoulders.

shoulders (*Acous.*). The alteration of the shape of the needle tip during the reproduction from gramophone records causes ridges (*shoulders*) to be formed on the side of the needle, which then slides over the flat surfaces of the disk rather than in the groove. (*Build.*) Of an arch, see **flanks.** (*Carp.*) The abutting surfaces left on each side of a tenon; they abut against the cheeks of the mortise. (*Civ. Eng.*) See **shoulder.**

shovel beak (*Vet.*). See **mandibular disease.**

shower (*Phys.*). Result of impact of a high-energy cosmic ray particle, consisting of ionizing particles and photons, directed downwards.

showerproofing (*Textiles*). The rainproofing of woollen and worsted cloths by treating them with metallic salts, insoluble soap, or silicone-based preparations. The thermal and ventilating properties and the general appearance of a fabric are not affected by these treatments.

shower unit (*Phys.*). The mean path length for the reduction of 50% of the energy of cosmic rays as they pass through matter.

show-through (*Print.*). Appearance of print through another sheet placed on top of printed sheet; paper should be sufficiently opaque to prevent this. Cf. *strike-through.*

shread head (*Build.*). See **jerkin head.**

shrinkage (*Civ. Eng.*). The difference in the spaces occupied by material before excavation and after settlement in embankment. Also, the contraction of concrete after placing. (*Textiles*) Term indicating the losses in weight which occur in wool materials during the different processes of scouring and finishing, by heating, wetting, etc. Also called **sinkage.**

shrinkage allowance (*Eng.*). The difference in diameter, when both are cold, of two parts to be united by shrinking. See **shrinking-on.**

shrinkage stoping (*Mining*). While excavating (stoping) ore from a vein or lode, the leaving of sufficient broken ore in the stope to keep the walls from falling in. This broken ore is subsequently withdrawn.

shrinking-on (*Eng., etc.*). The process of fastening together two parts by heating the outer member so that it expands sufficiently to pass over the inner and on cooling grips it tightly, e.g., in the attachment of steel tyres to locomotive wheels.

shrinking processes (*Textiles*). Various methods for rendering wool cloths unshrinkable during tailoring and wear. Cloths are shrunk effectively by the manufacturer before sale. The oldest method is *London shrinking*. See **London-shrunk.**

shrink-ring commutator (*Elec. Eng.*). A high-speed type of commutator in which the segments are held together by a steel ring shrunk on over a layer of insulation.

shroud (*Aero.*). (1) Rearward extension of the skin of a fixed aerofoil surface to cover the whole or part of the leading edge of a movable surface, e.g. flap, elevator, hinged to it. (2) See **jet pipe-.** (*Eng.*) (1) Circular webs used to stiffen the sides of gear-teeth. See **full-, half-.** (2) An outer or peripheral strip used to strengthen turbine blading. (3) A semicircular deflecting wall formed at one side of an inlet port in some I.C. engines to promote air swirl in the cylinder. Also **shrouding.** (*Radio, etc.*) Extension of metal parts in valves, and other electrical devices subject to high voltages, so that parts of the insulating dielectric are not excessively stressed.

shrouded balance (*Aero.*). An *aerodynamic balance* in which the area ahead of the hinge line moves within a space formed by shrouds projecting aft from the fixed surface.

shroud line (*Aero.*). Any one of the cords attaching a parachute's load to the canopy. Also **rigging line.**

Shuckneckt's operation (*Surg.*). Stapedectomy, followed by replacement with a prosthesis of wire and fat between the incus and oval window.

shudder (*For.*). Vibration of the wood in cutting operations, causing corrugation of the surface and irregularity in thickness.

shuffs (*Build.*). See **chuffs.**

shunt (*Elec.*). Addition of a component to divert current in a known way, e.g., from a galvanometer, to reduce temporarily its effective sensitivity. (*Mag.*) Diversion of some flux from the gap in a magnetic circuit by a magnetic slide or screw in a moving-coil indicating instrument. (*Rail.*) To divert a train from one track to another, especially to allow another train to pass along the principal track.

shunt characteristic (*Elec. Eng.*). The characteristic graph relating terminal voltage and load current in the case of a shunt-wound d.c. machine.

shunt circuit (*Elec. Eng.*). (1) Electric or magnetic circuit in which current or flux divides into two or more paths before joining to complete the circuit. Also called **parallel circuit.** (2) See **voltage circuit.**

shunted capacitor (*Teleg.*). See **reading capacitor.**

shunt-excited antenna (*Radio*). Antenna consisting of a vertical radiator (frequently the mast itself) directly earthed at the base, and connected to the transmitter through a lead attached to it a short way above ground.

shunt feed (*Electronics*). See **parallel feed.**

shunt field (*Elec. Eng.*). The main field winding of a motor when shunt connected.

shunt-field relay (*Telecomm.*). One with two coils on opposite sides of a closed magnetic circuit, so that a bridging magnetic circuit takes no flux while the currents in the two coils magnetize the circuit in the same direction. Flux passes in this bridging circuit when one current is reversed.

shunt-field rheostat (*Elec. Eng.*). A rheostat for insertion in the shunt field of a d.c. shunt machine; used to vary the speed of a shunt motor, or the voltage of a shunt generator.

shunt motor (*Elec. Eng.*). One whose main excitation is derived from a shunt-field winding.

shunt resonance (*Elec. Eng.*). The condition of a parallel tuned circuit connected across an a.c. voltage supply when maximum impedance is offered to the supply, and the circulating loop current is also a maximum. The term **tunance** is sometimes used in place of resonance if this condition is attained by adjustment of a component value and not of frequency. Also called **parallel resonance.**

shunt trip (*Elec. Eng.*). A solenoid-type tripping device, connected in shunt across either the main or an auxiliary supply, by which a circuit-breaker may be tripped by a suitable relay.

shunt voltage regulation (*Elec. Eng.*). That performed by control of a variable impedance in parallel with the output.

shunt winding (*Elec. Eng.*). A field winding connected in shunt across the armature circuit of a motor.

shut-down (*Nuc. Eng.*). Reduction of power level in nuclear reactor to lowest possible value by maintaining core in subcritical condition.

shut-down amplifier (*Nuc. Eng.*). See **trip amplifier.**

shute (*Hyd.*). See **chute.**

shuting (*Build.*). See **eaves gutter.**

shutter (*Cinema.*). (1) The rotating device in a camera which exposes when the film is stationary and closes while it is moving. (2) In a projector, the device which cuts off the light from the screen while the film is advanced. Each frame may be flashed on to the screen two or three times to reduce the flicker. See **automatic-, flicker-.** (*Photog.*) The device in a camera for exposing the sensitized surface of the film to the image of the object during a known time and at will.

shuttering (*Civ. Eng.*). The general term for temporary works for the support of reinforced concrete while it is setting. Also **formwork.**

shutter weir (*Civ. Eng.*). A type of movable weir consisting of a row of large panels hinged at the bottom and inclined slightly downstream towards the top when the weir is closed.

shutting stile (*Join.*). The stile of a door further from the hinges. Also **meeting stile.**

shuttle (*Nuc. Eng.*). Container for samples to be inserted in, and withdrawn from, nuclear reactors, when they are made radioactive by irradiation with neutrons; also **rabbit.** (*Space*) Rocket-powered airplane-like vehicle launched by rocket or from a high-flying aircraft to transfer crew and supplies to and from orbiting spacecraft, whence it returns and lands on earth like an ordinary airplane. (*Weaving, etc.*) (1) Loom accessory which carries the weft (in the form of a cop or pirn) across a loom, through the upper and lower warp threads. Made of boxwood, cornel, or persimmon, it is boat-shaped, with a metal tip at each end. One end has a porcelain eye through which the weft passes. (2) In a sewing machine, a sliding or rotating device that carries the lower thread to form a lock-stitch.

shuttle armature (*Elec. Eng.*). A simple form of armature; used on small d.c. machines in which there are only two slots, so that the armature stampings assume an H-shape; the winding is a single coil connected to a 2-part commutator. Also called **H-armature.**

shuttle box (*Weaving*). Box-like extension at each end of race board, from which shuttle is thrown to and fro when loom is working.

shuttle guard (*Weaving*). Robustly constructed metal guard fixed to loom slay to deflect and keep low a shuttle which by accident flies out of loom.

shuttle machine (*Textiles*). See **Swiss machine.**

shuttle race (*Weaving*). See **race board.**

shuttling (*Weaving*). Inserting a new cop or weft pirn in the shuttle. On automatic looms, this is carried out mechanically by a transfer hammer operating with a weft magazine at one side of the loom.

Si (*Chem.*). The symbol for *silicon.*

SI. Abbrev. for *Système International* (d'Unités). See **SI units.**

sial (*Geol.*). The discontinuous earth shell of granitic composition which forms the foundation of the continental masses and which is in turn underlain by the *sima.* So called because it is essentially composed of *si*liceous and *al*uminous minerals.

sialagogue, sialogogue (*Med.*). Stimulating the flow of saliva (Gk. *sialon*); any medicine which does this.

sialic acids (*Biochem.*). Naturally-occurring N- and O-acyl derivatives of neuraminic acid.

sialo-adenitis, sialadenitis (*Med.*). Inflammation of a salivary gland.

sialoid (*Zool.*). Resembling saliva.

sialolith (*Med.*). A calculus in a salivary gland.

sialorrhoea (*Med.*). Excessive secretion of saliva.

sibilus (*Med.*). A sibilant or whistling *rhonchus* (q.v.) due to the passage of air over secretions in the smaller bronchial tubes.

Sicilian (*Textiles*). A plain cloth, made from fine cotton warp and coarse mohair weft; used for dress goods. Another type is made from silk (or rayon) warp and woollen weft.

sickle-cell anaemia (*Med.*). See **drepanocytosis.**

sickle spanner (*Tools*). One for turning large, narrow nuts, having a sickle-shaped end with a projection which fits into a notch in the nut.

side-and-face cutter (*Eng.*). A milling cutter with plain or staggered teeth on the sides as well as on the periphery, widely used for cutting slots.

sidebands (*Radio*). Those added to a carrier in the process of any modulation, the carrier remaining unchanged in the ideal case. Either sideband, without the carrier, contains the transmitted information. See **carrier.**

side bones (*Vet.*). Ossification of the lateral cartilages of a horse's foot.

side chains (*Chem.*). Alkyl groups which replace hydrogen in ring compounds.

side-chain theory (*Biol.*). A theory propounded by Ehrlich in 1897 to explain the phenomena of the poisoning and immunity of living cells. It assumes that the molecules of protoplasm possess a cyclic structure and side chains similar to those found in derivatives of benzene. Molecules of foodstuffs or toxins are incorporated into the cell by attachment to these side chains or *receptors.* Interaction of toxin molecules with a number of receptors in a cell tends to stimulate overproduction of receptors, which are liberated into the blood, where they act as antibodies. Variants of this theory led to the development of modern chemotherapy, in particular the sulphonamides.

side circuit (*Teleph.*). The telephone circuit loop which is used as one leg of a *phantom circuit*, the two wires being effectively in parallel for the phantom circuit.

side cutting (*Civ. Eng.*). Excavation taken from the side of a railway or canal when the amount of the normal cutting is less than the fill.
side draught (*I.C. Engs.*). Said of a carburettor in which the mixture is drawn in at right angles to the force of gravity.
side drift (*Civ. Eng.*). An *adit* (q.v.).
side frequency (*Radio*). Any frequency of a sideband.
side hook (*Carp.*). See **bench hook.**
side keelson (*Ships*). See under **keelson.**
side lobe (*Electronics*). Any lobe of the radiation pattern of an acoustic or radio transmitter other than that containing the direction of maximum radiation.
side pond (*Hyd. Eng.*). A storage space at the side of a canal lock-chamber, the two being interconnected by a sluice, so that the normal loss of water occurring in the process of passing a vessel through the lock may be reduced.
side posts (*Carp.*). See **princess posts.**
side rail (*Rail.*). See **check rail.**
sidereal day (*Astron.*). The interval between two successive transits of the First Point of Aries over the meridian; also, neglecting the very small effect of precession, the interval between two successive transits of the same fixed star; the period of the earth's rotation on its axis.
sidereal month (*Astron.*). The interval (amounting to 27·321 66 days) between two successive passages of the moon through the same point in her orbit, relative to the fixed stars.
sidereal noon (*Astron.*). The instant when the First Point of Aries is on the meridian at upper culmination, indicated by 0^h local sidereal time.
sidereal period (*Astron.*). The interval between two successive positions of a celestial body in the same point with reference to the fixed stars; applied to the moon and planets to indicate their complete revolution of the heavens as against their synodic revolution relative to the line joining the earth and sun.
sidereal time (*Astron.*). A method of reckoning intervals based on the rotation of the earth on its axis as the fundamental period; sidereal time at any moment is the hour angle of the First Point of Aries at the moment, and increases from 0 to 24^h, beginning with sidereal noon.
sidereal year (*Astron.*). The interval between two successive passages of the sun in its apparent annual motion through the same point relative to the fixed stars; it amounts to 365·256 36 days, slightly longer than the tropical year, owing to the annual precessional motion of the equinox.
side-rebate plane (*Join.*). A rebate plane with its cutting edge on the side, not sole, of the tool.
siderite (*Min.*). (1) Ferrous carbonate, crystallizing in the trigonal system and occurring in sedimentary iron ores and in mineralized veins. Sometimes called **chalybite.** (2) A name for iron meteorites as a class.
sideropenia (*Med.*). Deficiency of iron.
siderophile (*Geol.*). Descriptive of elements which have an affinity for iron, and whose geochemical distribution is influenced by this property.
siderophyllite (*Min.*). The iron and aluminium-rich end-member of the biotite micas.
siderosis (*Med.*). (1) *Pneumonoconiosis* (q.v.) due to the inhalation of metallic particles by workers in tin, copper, lead, and iron mines, and by steel grinders. (2) Excessive deposit of iron in the body tissues.
siderostat (*Astron.*). An instrument designed on the same principle as the coelostat to reflect a portion of the sky in a fixed direction; applied specially to a form of telescope called the *polar siderostat*, in which the observer looks down the polar axis on to a mirror.
sideslip (*Aero.*). The component of the motion of an aircraft in the plane of its lateral axis; generally a piloting, or stability, error, but also used intentionally to obtain a steep glide descent without gaining speed: *angle of sideslip* is that between the plane of symmetry and the direction of motion.
side sorts (*Typog.*). See **pie characters.**
side-stable relay (*Teleg.*). One polarized, but without central position, free from *mark* or *space* contacts.
side-stick (*Typog.*). A tapering piece of wood placed at the sides of pages when locking them up. Quoins are wedged between them and the sides of the chase. (*Vet.*) A cylindrical stick fixed to the head-collar and to the surcingle to limit the movement of a horse's head.
side stitching (*Bind.*). See **flat stitching.**
side thrust (*Acous.*). Radial force on pick-up arm caused by stylus drag.
side timber (*Carp.*). A roof purlin supporting the common rafters.
side tone (*Teleph.*). A signal reaching the receiver of a radio-telephone station from its own transmitter.
side tool (*Eng.*). A cutting tool in which the cutting face is at the side, and which is fed laterally along the work.
side valve (*I.C. Engs.*). Said of an engine having the valves situated at the side of the cylinder block with their posts on the same side of the combustion chamber as the piston and their tappets operated directly by the camshaft.
sidewalk (*Civ. Eng.*). See **pavement.**
side wave (*Radio*). Isolated frequency component in the sideband. Analysis shows that a sinusoidally modulated wave may be physically resolved into a carrier wave of frequency ω and two side waves of frequencies $\omega+\rho$ and $\omega-\rho$ respectively, where ρ is the modulating frequency.
side waver (*Carp.*). See **side timber.**
siding (*Rail.*). A short length of side line on to which one train from the main line may be shunted to allow the passage of another train on the main line.
Sieber number (*Paper*). A measure of the bleachability of pulps, defined as the amount of available chlorine consumed by 100 parts of pulp in 1 hr at 20°C under specified conditions.
Siebold's organ (*Zool.*). See **crista acustica.**
siegbahn (*Radiol.*). X-unit, 10^{-13} m, obsolete.
siege (*Build.*). A mason's or bricklayer's *banker* (q.v.). (*Glass*) The floor of a tank furnace or pot furnace.
Siegwart (*Build.*). TN for a form of fire-resisting floor, built of reinforced concrete tubes laid close together and grouted in between the joints in the several tubes to complete the floor, ready for any top finish.
siemens (*Elec.*). SI unit of electrical conductance; reciprocal of *ohm*. Abbrev. S.
Siemens dynamometer (*Elec. Eng.*). A dynamometer-type of instrument arranged for measuring current or power.
Siemens furnace (*Met.*). Gas-heated reverberatory furnace.
Siemens-Halske process (*Min. Proc.*). Chemical extraction of sulphidic copper from its ores or concentrates with sulphuric acid and ferrous sulphate.
Siemens-Martin process (*Met.*). See **open-hearth process.**

Siemens ozone tube (*Chem.*). Apparatus used in the preparation of ozone by the corona discharge of electricity.

sienna (*Paint.*). Yellow-brown pigment consisting of naturally-occurring hydrated iron oxide, manganese salts, clays, etc. Used 'raw' or 'burnt'; when 'burnt' (calcined), it becomes brownish-red.

sieve analysis (*Build., Civ. Eng.*). A simple method of assessing the suitability of sand or other aggregate for concrete by passing a representative sample through progessively finer sieves and measuring the quantity passed on each occasion. (*Powder Tech.*) The measurement of the size distribution of a powder by using a series of sieves of decreasing mesh aperture.

sieve area (*Bot.*). A limited area on the longitudinal wall of a sieve tube, perforated by numerous fine pores through which material may pass.

sieverfield (*Bot.*). One of the perforated areas into which a sieve plate may be divided by a network of thick strands of wall material.

sieve mesh number (*Powder Tech.*). The number of apertures occurring in the surface of a sieve per linear inch. Unless the size of the wire used in weaving the mesh is specified, the mesh number does not uniquely specify the aperture size.

sieve plate (*Bot.*). A perforated area in the lateral or end wall of a sieve tube. (*Chem. Eng.*) Distillation column plate (or tray) with large number of small holes through which the vapour rises, and fitted with weirs and downcomers to retain liquid through which the vapour rises.

Sievert unit (*Radiol.*). A unit of γ-ray dose, being the dose of radiation delivered in 1 hr at a distance of 1 cm from a point source of 1 mg of radium element enclosed in platinum 0·5 mm in thickness. Numerically equal to ca. 8·4 röntgens or 21·6 C/kg.

sieve tube (*Bot.*). A long tubular element, enclosed by a thin wall of cellulose and containing living contents but no nucleus, occurring in the phloem of vascular plants. Sieve tubes conduct elaborated food material about the plant, and communicate with one another by means of sieve areas or sieve plates.

sigatoka (*Bot.*). A fungal disease of bananas, causing wilting of the leaves, premature ripening, and reduction in crop size.

sig. fig. (*Maths.*). *Significant figures.*

sight (*Optics*). Sensation produced when light waves impinge on the photosensitive cells of the eye. (*Surv.*) Bearing from instrument at known point on a distant signal or topographical feature. See **back observation, fore sight, intermediate sight.**

sight-feed lubricator (*Eng.*). A small glass tube through which oil-drops from a reservoir can be seen, or which is filled with water so that oil from the pump rises in visible drops on its way to the oil-pipe.

sight lines (*Cinema.*). The extreme angles from which the screen can be seen in a cinema.

sight rail (*Surv.*). An above-ground horizontal wooden rail fixed to two upright posts, one on each side of a trench excavation for a sewer, drain, etc. Used with others to establish a reference line from which the sewer, etc., may be laid at the required gradient.

sight rule (*Surv.*). See **alidade.**

sigma meson (*Nuc.*). An old term denoting a meson which gives rise to a 'star'. Such mesons are actually *pions* (see **mesons**).

sigma-particle (*Nuc.*). Hyperon triplet, rest mass equivalent to 1190 MeV, hypercharge 0, isotopic spin 1.

sigma pile (*Nuc. Eng.*). One comprising neutron source and moderating materials without any fissile element. Used in study of neutron properties of moderator.

sigmatron (*Nuc.*). A cyclotron and a betatron operating in tandem to produce very high energy X-rays.

sigmoid curve (*Radiol.*). An S-shaped curve which is often obtained in dose-effect curves in radiobiological studies.

sigmoidectomy (*Surg.*). Excision of part of the sigmoid flexure of the colon.

sigmoid flexure (*Zool.*). An S-bend.

sigmoidoscope (*Med.*). A tube fitted with a lamp for viewing the mucous membrane of the rectum and pelvic colon.

sigmoidostomy (*Surg.*). The surgical formation of an opening (artificial anus) in the sigmoid flexure of the colon.

sign (*Med.*). Any objective evidence of disease or bodily disorder, as opposed to a *symptom* (q.v.), which is a subjective complaint of a patient.

signal (*Surv.*). A device, such as a ranging rod, heliostat, etc., used to mark a survey station. (*Telecomm.*) General term referring to a conveyor of information, e.g., an audio waveform, a video waveform, series of pulses in a computer. Colloquially, the message itself. In radio, the signal modulates a carrier, and is recovered during reception by demodulation.

signal code (*Teleph.*). In voice-frequency signalling, the plan for representing each of the required signalling functions as a voice-frequency signal.

signal coloration (*Zool.*). Distinctive markings for the recognition of other members of the same species.

signal component (*Teleph.*). That part of a signal which continues uniform in character throughout its duration. In a multi-component signal with spaces between current pulses a space may be regarded as a signal component.

signal distortion (*Phys.*). Modification of the information content of a signal, sometimes irreversibly; e.g. the suppression or introduction of *harmonics*.

signal electrode or plate (*TV*). In a TV camera tube, the mosaic from which a signal arises during scanning. See **image dissector, image iconoscope, orthicon.**

signal element (*Telecomm.*). The portion of a signal occupying the smallest interval of the signal code.

signal frequency shift (*Telecomm.*). The bandwidth between white and black signal levels in frequency-modulation facsimile transmission systems.

signal generator (*Radio*). Oscillator designed to provide known voltages (usually from 1 volt to less than 1 microvolt) over a range of frequencies. Used for testing or ascertaining performance of radio-receiving equipment. It may be amplitude-, frequency-, or pulse-modulated.

signal imitation (*Teleph.*). In voice-frequency signalling, an unwanted response of a voice-frequency receiver to any condition other than a true signalling condition.

signal lamp (*Elec. Eng.*). An indicating lamp on a switch- or control-board.

signal level (*Radio*). The level at any point in a transmission system, as measured by a voltmeter (volume-unit meter) across the circuit when properly terminated; expressed in dB or VU in relation to a reference level, now 1 mW in 600 ohms.

signalling capacitors (*Teleg.*). Large capacitors placed in series with the ends of a submarine cable, to improve the definition of arriving signals. See **double block, single block.**

signal/noise ratio (*Acous.*). See **noise ratio.**

signal output current (*Electronics*). The absolute difference between the output current and the dark current of a phototube or a camera tube.

signal plate (*TV*). Electrode which carries mosaic in an iconoscope.

signal shaping (*Telecomm.*). Use of specially designed electric network to correct distortion produced during transmission or propagation of signals.

signal-to-cross-talk ratio (*Teleph.*). In line telephony, the ratio of the test level in the disturbed circuit to the level of the cross-talk at the same point which is caused by the disturbing circuit operating at the test level.

signal wave (*Telecomm.*). One which allows intelligence to be conveyed.

signal windings (*Elec. Eng.*). U.S. term for **control turns (or windings)** of a saturable reactor.

signature (*Bind., Print.*). See **section.**

signature mark (*Bind., Print.*). A number or letter of the alphabet placed on the first page of a section as a guide to the binder in *gathering*.

sign digit (*Comp.*). A digit, generally preceding a number, indicating its sign.

signed minor (*Maths.*). See **cofactor.**

significant figures (*Maths.*). Of a number: those digits which make a contribution to its value, e.g., in the number 00·1230, the first two zeros are insignificant and the digits 1, 2, 3 are significant. The last zero also should be significant, indicating that the number is accurate to four places of decimals. Care should be exercised before making this assumption, however, because some writers add such zeros indiscriminately. Also **significant digits.**

sign stimulus (*An. Behav.*). A part, or change in a part, of an animal's environment correlated with the occurrence of a species-specific response which is not a reflex action, i.e., it releases the behaviour. See also **supernormal-, releaser.**

Sikes hydrometer (*Chem.*). A hydrometer used for determining the strengths of mixtures of alcohol and water.

sikyotic (*Bot.*). Parasitic by fusion of the plasma of host and parasite.

silage. See **ensilage.**

silal (*Met.*). High-silica (up to 5%) cast iron.

silanes (*Chem.*). A term given to the **silicon hydrides**: silane, SiH_4, disilane H_3Si—SiH_3, trisilane, $H_3Si(SiH_2)SiH_3$, etc.

Silastic (*Plastics*). TN for a range of silicone rubbers. Noted for very good heat resistance and a wide temperature range of application. Very good chemical resistance and electrical properties.

silencer (*I.C. Engs.*). An expansion-chamber fitted to the exhaust pipe of an I.C. engine to dampen the noise of combustion. U.S. **muffler.**

silent period (*Radio*). Stated period within each hour during which all marine transmissions must close down and listen on the international distress frequency of 500 kHz.

silex (*Min. Proc.*). Silica brick used to line grinding mills when contamination by abraded steel must be avoided.

silica (*Chem.*). Dioxide [(IV)oxide] of silicon, SiO_2, occurring in crystalline forms as quartz, cristobalite, tridymite; as cryptocrystalline chalcedony; as amorphous opal; and as an essential constituent of the silicate groups of minerals. Used in the manufacture of glass and refractory materials. Refractory materials containing a high proportion of silica (over 90%) are known as *acid refractories* (e.g., gannister), and are used in open-hearth and other metallurgical furnaces to resist high temperatures and attack by acid slags.

silica gel (*Chem.*). Hard amorphous granular form of hydrated silica, chemically inert but very hygroscopic. Used for absorbing water and vapours of solvents, especially in enclosed electronic equipment. When saturated, it may be regenerated by heat.

silica glass (*Glass*). See **vitreous silica.** (*Min.*) Fused quartz, occurring in shapeless masses on the surface of the Libyan Desert, in Moravia, in parts of Australia, and elsewhere; believed to be of meteoritic origin. See **tektites.**

silica poisoning (*Min. Proc.*). Loading of resins used in ion-exchange process with silica, thus reducing the efficiency of reaction with desired ions.

silicates (*Min.*). The salts of the silicic acids, the largest group among minerals; of widely different, and in some cases extremely complex, composition, but all containing silica as an essential component. The micas, amphiboles, pyroxenes, feldspars, and garnets are examples of groups of rock-forming silicates.

silica valve (*Elec. Eng.*). One in which the envelope is made of fused silica to withstand high temperatures. Originally developed for use in warships.

siliceous clay (*Geol.*). A clay rock with an admixture of silica in a finely divided state.

siliceous deposits (*Geol.*). Those sediments, incrustations, or deposits which contain a large percentage of silica in one or more of its modes of occurrence. They may be chemically or mechanically formed, or may consist of the siliceous skeletons of organisms such as diatoms and Radiolaria. See also **silicification.**

siliceous sinter (*Geol.*). Cellular quartz or translucent to opaque opal, found as incrustations or fibrous growths and deposited from thermal waters containing silica or silicates in solution.

silicic acid (*Chem.*). An acid formed when alkaline silicates are treated with acids. Amorphous, gelatinous mass. Dissociates readily into water and silica.

silicides (*Chem.*). Compounds formed by the combination of silicon with other elements, chiefly metals.

silicification (*Geol.*). The process by which silica is introduced as a cement into rocks after their deposition, or as an infiltration or replacement of organic tissues or of other minerals such as calcite. See also **novaculite.**

silicle, silicula, silicule (*Bot.*). A capsule derived from two united carpels, divided internally by a replum; contains many seeds, and is short (never more than four times as long as broad, and often not longer than broad), e.g., honesty. Cf. *siliqua*.

silicole (*Bot.*). A plant which grows on soils rich in silica, and usually acid in reaction.

silicolous (*Bot.*). Growing on rocks containing much silica.

silico-manganese steel (*Met.*). See **manganese alloys.**

silicon (*Chem.*). A nonmetallic element, symbol Si, at. no. 14, r.a.m. 28·086, valency 4. Amorphous silicon is a brown powder; rel. d. 2·42. Crystalline silicon is grey; rel. d. 2·42, m.p. 1420°C, b.p. 2600°C. This element is the second most abundant, silicates being the chief constituents of many rocks, clays, and soils. Silicon is manufactured by reducing silica with carbon in an electric furnace, and is used in glass and in making certain alloys (see **ferro-silicon**). It has semiconducting properties, being used for transistors and certain crystal diodes.

silicon bronze (*Met.*). A noncorroding alloy based on copper and tin.

silicon carbide (*Chem.*). SiC. Formed by fusing a mixture of carbon and sand or silica in an electric furnace (see **Acheson furnace**). Used as an abrasive and refractory.

silicon-controlled rectifier (*Electronics*). A 3-junction semiconductor device which is the solid state equivalent of a thyratron.

silicon copper (*Met.*). An alloy (20–30% Si), a 'getter' used to remove oxygen from molten copper alloys.

silicon detector (*Radio*). Stable silicon crystal diode for demodulation.

silicon dioxide (*Chem.*). Silicon(IV)oxide. See **silica.**

silicone rubbers (*Chem.*). An important group of synthetic rubbers (dimethylsiloxene polymers), having both high and low temperature resistances which are better than those for natural rubbers.

silicones (*Chem.*). Open-chain and cyclic organo-silicon compounds containing $—SiR_2O—$ groups, prepared mainly by hydrolysing alkyl or aryl silicon dichlorides, R_2SiCl_2, which are themselves made by the Grignard reaction. The simpler substances are oils of very low melting-point, the viscosity of which changes little with temperature, used as lubricants, shock-absorber fluids, constituents of polishes, etc. More complex solid products, stable to heat and cold, and chemically inert, are exceptionally good electrical insulators for electric motors, etc. Also used in gaskets and a wide variety of special applications.

silicon hydrides (*Chem.*). See **silanes.**

silicon iron (*Met.*). Iron or low carbon steel to which 0·75–4·0% silicon has been added. Has low magnetic hysteresis, and is resistant to mild acids. Used for sheets for transformer cores. Typical composition: silicon 4%, manganese under 0·1%, phosphorus 0·02%, sulphur 0·02%, carbon 0·05%.

silicon rectifier (*Electronics*). A semiconductor diode rectifier usually based on *p-n* junction in silicon crystal.

silicon resistor (*Electronics*). A resistor of special silicon material which has a fairly constant positive temperature coefficient, making it suitable as a temperature-sensing element.

silicon tetrachloride (*Chem.*). Tetrachlorosilane, $SiCl_4$. Formed by the action of chlorine on a mixture of silica and carbon, or silicon. Liquid.

silicon tetrafluoride (*Chem.*). Tetrafluorosilane, SiF_4. A gaseous compound formed by the action of hydrofluoric acid on silica. Readily hydrolyses into silica and hydrofluoric acid.

silicosis (*Med.*). *Pneumonoconiosis* (q.v.), due to the inhalation of particles of silica by masons and by miners who work in the presence of silica; tuberculosis of the lung is a common complication.

siliqua (*Bot.*). A capsule with the general characters of a *silicle*, but at least 4 times as long as it is broad. Also **silique.** (*Zool.*) A pod-shaped structure, as the *siliqua* olivae of the Mammalian brain, a tract of fibres investing the olive.

silit resistor (*Elec. Eng.*). Tubular element made from a mixture of silicon carbide and silicon.

silk (*Min.*). A sheen resembling that of silk, exhibited by some corundums, including ruby, and due to minute tubular cavities, or to rutile needles, in parallel orientation. The colour of such stones is paler than normal by reason of the inclusions. (*Textiles*) A fabric made from the natural product of the silkworm, of which there are many varieties, both wild and cultivated. The best-known of the cultivated type is the genus *Bombyx.* Of the wild varieties, the worm of the moth *Antheraea mylitta* produces the thread from which tussah or tussur silk is made. See **throwing, spun silk, schappe, rayon.** Natural silk is specially useful for the insulation of very fine wires not immersed in oil. Electric breakdown strength about 16 kV/mm. (*Zool.*) A fluid substance secreted by various *Arthropoda.* It is composed mainly of fibroin, together with sericin and other substances, and hardens on exposure to air in the form of a thread. Used for spinning cocoons, webs, egg cases, etc.

silk-screen printing (*Print.*). A mesh-stencil process in which ink is squeezed through the open parts of the mesh on to the surface to be printed, which may be of any material, smooth or rough, flat or curved. The stencil may be prepared by hand, or by photography, or by a mixture of both. Screens of metal and of fibres other than silk are also used, and **screen process printing** has become the preferred name.

sill (*Build.*). The horizontal timber, stone, etc., at the foot of an opening, as for a door (also **threshold**), window, embrasure, etc. (*Geol.*) A minor intrusion of igneous rock injected as a tabular sheet between, and more or less parallel to, the bedding planes of rocks. (*Hyd. Eng.*) The top level of a weir, or the lowest level of a notch.

sillénite (*Min.*). Cubic bismuth(III)oxide, polymorphous with bismite. Occurs as a secondary mineral at Durango, Mexico.

sillimanite (*Min.*). An orthorhombic orthosilicate of aluminium. The high-temperature polymorph of Al_2SiO_5; cf. *kyanite* and *andalusite.* It occurs in high-grade metamorphosed argillaceous rocks. *Fibrolite* is a fibrous variety.

silo (*Civ. Eng.*). A tall construction, usually of reinforced concrete, serving as a container for the storage of loose materials, especially grain. (*Space*) Underground chamber housing a guided missile which is ready to be fired.

siloxen (*Chem.*). A crystalline substance resembling graphitic acid. Polymerized silicon analogue. Obtained by the action of hydrochloric acid on calcium silicide in the absence of air.

Siloxicon (*Chem.*). TN for a material obtained by heating a mixture of silica with carbon at about 2500°C; used as a refractory.

Silsbee rule (*Elec. Eng.*). A wire of radius r cannot carry a superconducting current greater than $\frac{1}{2}rH_c$, where H_c is the critical field. The self-magnetic field would then destroy superconductivity.

silt (*Geol.*). Material of an earthy character deposited in a finely divided form by flowing water. See **silt grade.**

silt box (*San. Eng.*). A removable iron box placed at the bottom of a gulley; serves to accumulate the deposited silt for periodic removal.

silt grade (*Geol.*). Size classification of diameters of sedimentary particles which under British standard defines *coarse silt*, 0·06-0·02 mm; *medium silt*, 0·02-0·006 mm; *fine silt*, 0·006-0·002 mm.

Siluminite (*Elec. Eng.*). TN designating materials composed principally of asbestos or mica for electrical insulating materials.

Silurian (*Paper*). A mottled effect in special papers produced by adding deeply dyed fibres to the main furnish. Under certain circumstances this can be a paper defect.

Silurian System (*Geol.*). The rocks which succeed

the Ordovician System and precede the Devonian System. They consist essentially in Great Britain of dark shales with interbedded fossiliferous limestones. Found in Wales, the Lake District, Shropshire, and the Southern Uplands of Scotland. Named by Sir R. I. Murchison after the ancient British tribe, the Silures. Sometimes called **Gothlandian.** Also **Siluric.**

Siluric (*Geol.*). Equivalent to *Silurian*

silver (*Chem.*). A pure white metallic element, symbol Ag, at. no. 47, r.a.m. 107·868, rel. d. at 20°C 10·5, m.p. 960°C, b.p. 1955°C, casting temp. 1030–1090°C, Brinell hardness 37, electrical resistivity approx. $1{\cdot}62 \times 10^{-8}$ ohm metres. The metal is not oxidized in air. Occurs massive, or assumes arborescent or filiform shapes. The best electrical conductor and the main constituent of photographic emulsions. Native silver often has variable admixture of other metals—gold, copper, or sometimes platinum. Used for ornaments, mirrors, cutlery, jewellery, etc., and for certain components in food and chemical industry where cheaper metals fail to withstand corrosion.

silver amalgam (*Min.*). Arquerite; a solid solution of mercury and silver, which crystallizes in the cubic system. Of rare occurrence, it is found scattered in mercury or silver deposits.

silver frost (*Meteor.*). A deposit of frozen dew.

silver glance (*Min.*). See **argentite.**

silver grain (*For.*). The light greyish, shining flecking seen in oak timber, caused by vascular rays exposed in preparing the timber when it is cut radially through the centre of a log. Also seen in beech.

silver(I)halides (*Chem.*). Silver fluoride, AgF; silver iodide, AgI; silver chloride, AgCl; and silver bromide, AgBr. The last three are sensitive to light and are of basic importance in photography.

silvering (*Glass*). This process is carried out on a perfectly clean surface of glass by pouring on to it an ammoniacal silver solution, mixed with Rochelle salt or with a nitric-acid/cane-sugar/alcohol mixture. The silver film so formed is then washed, backed with varnish and painted.

silver lead ore (*Min.*). The name given to galena containing silver. When 1% or more of silver is present it becomes a valuable ore of silver. Also called **argentiferous galena.**

silverlock bond (*Build.*). See **rat-trap bond.**

silver oxide (*Chem.*). See **argentic [silver(II)] oxide, argentous [silver(I)] oxide.**

silver solder (*Met.*). See **brazing solders.**

Silverstat (*Automation*). TN for control device in which a series of contacts are successively closed by an increasing applied force.

silver steel (*Eng., Met.*). A bright-drawn carbon steel containing up to 0·3% silica, 0·45% manganese, 0·5% chromium and 1·25% carbon; low in phosphorus and sulphur.

silver voltameter (*Elec. Eng.*). An electrolytic cell used for determining accurately the average value of a current from the quantity of silver deposited from the silver nitrate solution forming the electrode.

silvex (*Chem.*). See **fenoprop.**

sima (*Geol.*). A layer of the earth with the composition of a basic or ultrabasic igneous rock. Such rocks contain *si*licon and *ma*gnesium as their principal constituents, hence the name. Applied by some authors to the earth's mantle as opposed to the crust. More commonly applied to the lower rather than the upper part of the crust, although it is not certain that its composition justifies this name.

similar polygons (*Maths.*). Polygons whose corresponding angles are equal and in which the sides about the corresponding angles are in direct proportion.

Simmonds' disease (*Med.*). Hypophyseal cachexia. A rare disease due to destruction of the pituitary gland; characterized by cachexia, atrophy of the skin and the bones, premature senility, loss of hair, and loss of sexual function. *Progeria* is thought to be an example of Simmonds' disease.

simo chart (*Work Study*). **See simultaneous motion cycle chart.**

simoom (*Meteor.*). A hot dry wind or whirlwind of brief duration, occurring in the African and Arabian deserts. It usually carries sand.

simple (*Bot.*). (1) Consisting of one piece. (2) Unbranched.

simple curve (*Surv.*). A curve composed of a single arc connecting two straights.

simple eye (*Zool.*). See **ocellus.**

simple fruit (*Bot.*). A fruit formed from one pistil.

simple harmonic motion (*Phys.*). Motion which satisfies the equation $\ddot{x} = -\omega^2 x$. The majority of small amplitude oscillatory motions are simple harmonic motions, e.g., the oscillation of a weight on a spring, the vibration of a violin string, the swing of a pendulum. When such motion takes place in a resistive medium, e.g., air, the oscillations die away with time; the motion is then called *damped simple harmonic motion*. Abbrev. **s.h.m.**

simple leaf (*Bot.*). A leaf in which the lamina consists of one piece, which, if lobed, is not cut into separate parts reaching down to the midrib.

simple press tool (*Eng.*). *Press tool* (q.v.) which performs only one operation at each stroke of the press, as distinct from a *compound press tool* (q.v.) and a *progressive press tool* (q.v.).

simple process (*Nuc. Eng.*). The physical mechanism on which each stage of an isotope separation plant depends.

simple process factor (*Nuc. Eng.*). The separation factor obtainable with a simple process. Its theoretical value is the *ideal SPF*, that attained in a single stage of a cascade is the *effective SPF* or *stage separation factor*.

simple sorus (*Bot.*). A sorus of one sporangium.

simple steam-engine (*Eng.*). An engine with one or more cylinders in which the steam expands from the initial pressure to the exhaust pressure in a single stage.

simple tissue (*Bot.*). A tissue made up of cells all of the same kind.

simple (or pure) tone (*Telecomm.*). Single-frequency tone for testing telephone circuits, for voice-frequency telegraphy and for facsimile carrier.

simple umbel (*Bot.*). An umbel in which the flower stalks arise directly from the apex of the main stalk.

simple venation (*Bot.*). A type of venation in which the midrib alone is clearly visible.

simplex (*Teleg.*). General term for simple one-way transmission and reception over a channel. **Duplex** is use of the same channel in both directions at the same time, with different transmissions. **Diplex** is use of one channel for two different transmissions in the same direction.

simplex dialling (*Teleph.*). In subscriber dialling, use of the two line-wires in parallel, with earth return, to get the maximum change in current for impulsing.

Simplex piling (*Civ. Eng.*). A system of piling in which a cast-iron point is driven to the required depth by means of a steel pipe. The pipe is then filled with concrete and afterwards slowly withdrawn, the concrete filling adapting itself to the irregularities in the ground.

simplex winding (*Elec. Eng.*). An armature winding through which there is only one electrical path per pole.

simply-connected domain (*Maths.*). See **connected domain.**

Simpson's rule (*Maths.*). The area under the curve $y=f(x)$ from $x=x_0$ to $x=x_2$ is approximately $\frac{1}{6}\{f(x_0)+4f(x_1)+f(x_2)\}\{x_2-x_0\}$, where x_0, x_1 and x_2 are equally spaced.

Sims' speculum (*Med.*). A speculum, shaped like a duck's bill, for viewing the lining of the vagina and the cervix uteri.

simulated line (*Telecomm.*). Same as **artificial line.**

simulation (*Electronics*). Setting up of an electronic circuit, especially with analogue computers, which will obey the same laws as a physical system. (*Psychol.*) The investigation of thought processes by the use of computers programmed to imitate them. (*Zool.*) Mimicry; assumption of the external characters of another species in order to facilitate the capture of prey or escape from enemies. *v.* **simulate.**

simultaneity (*Phys.*). Basic issue in relativity theory. Two outside events are simultaneous to an observer if a clock with him indicates the same time for both. For observers in other frames of reference, this is not so.

simultaneous broadcasting (*Radio*). Transmission of one programme from two or more transmitters on different wavelengths.

simultaneous motion cycle chart (*Work Study*). A chart, often based on film analysis, used to record simultaneously on a common time scale the *therbligs* (q.v.) or groups of therbligs performed by different parts of the body of one or more workers. Also **simo chart.**

sin, sine (*Maths.*). See **trigonometrical functions.**

sinapism (*Med.*). A mustard plaster.

Sindanyo (*Elec. Eng.*). TN designating materials, composed principally of asbestos, for the mounting of switchgear of all types for electrical insulation work generally, and for arc shields, barriers, furnace linings, and other purposes.

sine bar (*Eng.*). A hardened steel bar carrying two plugs of standard diameter accurately spaced to some standard distance; used in setting out angles to close limits.

sine condition (*Optics*). A condition which must be satisfied by a lens if it is to form an image free from aberrations (other than chromatic). It may be stated $n_1l_1 \sin \alpha_1 = n_2l_2 \sin \alpha_2$, where n_1 and n_2 are the refractive indices of the media on the object and image sides of the lens respectively, l_1 and l_2 are the linear dimensions of the object and image, and α_1 and α_2 are the angles made with the principal axis by the conjugate portions of a ray passing between object and image.

sine galvanometer (*Elec. Eng.*). A galvanometer in which the coil and scale are rotated to keep the needle at zero. The current is then proportional to the sine of the angle of rotation. The arrangement can be made more sensitive than the *tangent galvanometer*.

sine potentiometer (*Elec. Eng.*). Voltage divider in which the output of an applied direct voltage is proportional to the sine of the angular displacement of a shaft.

sine wave (*Phys.*). Waveform of a single frequency, indefinitely repeated in time, the only waveform whose integral and differential has the same wave-form as itself. Its displacement can be expressed as the sine (or cosine) of a linear function of time or distance, or both. In practice there must be a transient at the start and finish of such a wave.

singeing (*Textiles*). Removing outstanding fibres from the face of a cloth to increase the lustre; effected by passing the cloth rapidly over gas flames or incandescent plates or rollers.

singing (*Telecomm.*). Self-oscillation in a transmission system caused by feedback across a source of gain because of unbalance in the circuit.

singing arc (*Elec. Eng.*). See **Duddell arc.**

singing point (*Telecomm.*). Gain of amplifier under specified conditions just below oscillation.

singing tube (*Acous.*). An acoustic generator consisting of a piece of gauze fitted in the lower end of a vertical open pipe. When the gauze is heated a loud noise is produced.

single-acting cylinder (*Eng.*). Fluid-power cylinder in which the piston is displaced in one direction by the fluid and returned mechanically, usually by a spring.

single-acting engine (*Eng.*). A reciprocating en gine in which the working fluid acts on one side of the piston only, as in most I.C. engines.

single address (*Comp.*). Said of the coding accompanying a *word* to guide it to a position in the store.

single-beat escapement (*Horol.*). An escapement in which the balance receives impulse only at every alternate vibration, e.g., the chronometer and duplex escapements.

single block (*Teleg.*). A large capacitor at the receiving end only of a submarine cable, to improve the definition of the received fluctuation of current. See **double block.**

single-break switch (*Elec. Eng.*). A switch having only one pair of main contacts.

single bridging (*Carp.*). Bridging in which a pair of diagonal braces are used to connect adjacent floor-joists at their middle points.

single-catenary suspension (*Elec. Eng.*). A catenary suspension system in which the conductor wire is hung from a single catenary or bearer wire.

single-channel pulse height analyser (*Telecomm.*). See **pulse height analyser.**

single circuit (*Telecomm.*). One which accepts a particular signal from a random collection of signals. Also called **one-shot circuit.**

single-commutation direct-current signalling (*Teleph.*). That which uses double current signals and in which the change in the direction of a current is controlled from a single change-over contact commutating a single battery.

single-core cable (*Elec. Eng.*). A cable having only one conductor.

single current (*Teleg.*). In a telegraph system, the use of line currents which are of one polarity for mark, and zero for spacing.

single-electrode system (*Elec. Eng.*). Electrode of an electrolytic cell and the electrolyte with which it is in contact. Also called **half-cell, half-element.**

single-ended (*Elec. Eng.*). (1) Unit or system designed for use with unbalanced signal, having one input and one output terminal permanently earthed. (2) Valve with all electrodes connected to pins at the same end.

single-entry compressor (*Aero.*). A centrifugal compressor which has vanes on one face only.

single Flemish bond (*Build.*). A form of bond combining English bond for the body of the wall with Flemish bond for the face-work.

single floor (*Carp.*). A floor in which the bridging

joists span the distance from wall to wall without intermediate support.

single-frequency signalling (*Teleph.*). A system of automatic signalling using a.c. signals within the audio range and in which all of the components of the signal code contain the same signalling frequency for a given direction of signalling.

single-hung window (*Build.*). A window having top and bottom sashes, of which only one (usually the bottom sash) is balanced by sash cord and weights so as to be capable of vertical movement.

single laths (*Build.*). Wood laths 1 in (25 mm) by $\frac{1}{8}$ to $\frac{3}{16}$ in (3 to 5 mm) thick in section.

single-layer winding (*Elec. Eng.*). A type of armature winding having only one coil-side per slot.

single measure (*Join.*). Joinery work which is square on both sides.

single-parity check (*Comp.*). A system for detecting errors similar to a binary code.

single phase (*Elec. Eng.*). Of electrical power transmission; two conductors, one of which may be the earth or at earth potential, between which is a sinusoidally alternating potential difference, as for domestic a.c. supply.

single-phase induction regulator (*Elec. Eng.*). An induction regulator for use on a single-phase circuit; the arrangement is such that the voltage on the secondary side is always in phase with that on the primary side. Cf. *three-phase induction regulator*.

single-pivot instrument (*Elec. Eng.*). An instrument in which the moving element is supported at its centre of gravity on a single pivot.

single-plate clutch (*Eng.*). A *friction clutch* (q.v.) in which the disk-shaped or annular driven member, fabric-faced, is pressed against a similar face on the driving member by springs, being withdrawn against them through a thrust collar; used in automobiles.

single pole (*Elec. Eng.*). Of a switch or relay contact which makes, breaks or changes over connection on one *pole* (2) only, a.c. or d.c., of a circuit.

single quotes (*Typog.*). See **quotation marks.**

single-rate prepayment meter (*Elec. Eng.*). A prepayment meter in which the circuit is broken and the supply cut off after a predetermined number of units have been consumed.

single-retort underfeed stoker (*Eng.*). An *underfeed stoker* (q.v.) consisting of a retort along the bottom of which coal is fed by a steam-driven ram or a screw conveyor, air being supplied through tuyères round the upper edge of the retort and into the sealed ash pit below.

single-revolution (*Print.*). A design of printing machine in which the cylinder rotates continuously while the bed reciprocates, and prints one impression during each revolution. Cf. *stop-cylinder, two-revolution.*

single-row ball-bearing (*Eng.*). *Ball-bearing* (q.v.) comprising a set of balls arranged in a single plane, as distinct from a double-row bearing.

single-row disk harrow (*Agric.*). A disk harrow with two gangs of disks, working side by side, one right hand and one left hand.

single-row radial engine (*Eng.*). A *radial engine* (q.v.) in which the cylinders are disposed in one plane, operating on a common crank pin.

singles (*Build.*). Roofing slates about 12×18 in. (305×457 mm). (*Mining*) See **coal sizes.** (*Textiles*) (1) Threads of raw silk, as reeled from the cocoons, which consist of a small number of strands, slightly twisted. (2) Weak yarn where one of two rovings has broken or, in doubling, one of the two components has broken.

single-sideband suppressed carrier (*Radio*). See **vestigial-sideband transmitter.**

single-sideband system (*Radio*). One in which, after modulation, one sideband and carrier are removed by push-pull modulator and filters. This minimizes the bandwidth required for transmission and improves signal/noise ratio. On reception an identical carrier is inserted before demodulation. Cf. *double-sideband transmitter*.

single-spindle automatic (*Eng.*). *Automatic screw machine* or lathe with a tool turret and one chucking spindle, producing one part per operating cycle.

single-state recycle (*Nuc. Eng.*). Isotope separation operating on the principle used in gaseous diffusion whereby the diffusing gas is divided into 2 fractions. One of these portions is recirculated within the first stage while the other proceeds to the next stage. Also **rabbit.**

single switch call (*Teleph.*). In international telephony, a transit call which involves two international circuits. See **double switch call.**

singlet (*Chem.*). A state in which there are no unpaired electrons.

single-turn coil (*Elec. Eng.*). An armature coil consisting of a single turn of copper bar.

single-turn transformer (*Elec. Eng.*). A current transformer in which the primary winding takes the form of a single straight conductor of heavy cross-section, to which the cable or bus-bar is connected.

single-valued (*Maths.*). Said of a function which can have only one value for each set of values of the independent variables.

single-wave rectification (*Elec. Eng., Radio*). See **half-wave rectification.**

single-wire circuit (*Elec. Eng.*). One with a single live wire, including coaxial or concentric, with sheath, earth, or frame return.

single-wire feeder (*Radio*). One for an antenna, similar to an ordinary downlead, but connected to the antenna in such a manner that it is terminated in its characteristic impedance, so that no standing waves are formed on it.

singularity (*Maths.*). Of a function: a point where the function ceases to be analytic.

singular point (on a curve) (*Maths.*). A point on the curve $f(x, y)=0$ at which

$$\frac{\partial f}{\partial x}=\frac{\partial f}{\partial y}=0.$$

One at which there is either no real tangent or two or more tangents. Cf. *double point*.

singular solution (*Maths.*). Of a differential equation: a solution which cannot be derived as a particular integral from a complete primitive.

sinigrin (*Chem.*). Chief constituent of black mustard seed, a glucoside consisting of glucose combined with the sulphur in the following structure:

$$C_3H_5N{=}C\begin{matrix} \diagup S-(\text{glucose}) \\ \diagdown O-SO_2-O^-K^+ \end{matrix}$$

sinistral fault (*Geol.*). See **dextral fault.**

sinistrorse (*Bot., Zool.*). Twisting in a spiral from right to left. Cf. *dextrorse*.

sink (*Cinema.*). See **synchronization.** (*Electronics*) Unstable operating region on Rieke diagram. (*Horol.*) A recess; the spherical depression around a pivot hole for holding the oil. (*Print.*) A depression in the printing surface of a plate.

sinkage (*Textiles*). See **shrinkage.**

sinker (*Bot.*). An absorbing organ formed by mistletoe, which penetrates the wood of the host. (*Textiles*) The mechanism in a knitting machine that pushes a length of thread over the spring needles to form a new course of loops.

sinker bar (*Mining*). A heavy bar attached to the cable above the drilling tools used in percussive drilling.

sink-float process (*Min. Proc.*). See **heavy media separation.**

sinking (*Civ. Eng., etc.*). The operation of excavating for a shaft, pit, or well. (*Join.*) A recess cut below the general surface of the work.

sinking pump (*Mining*). One of light construction and not needing fixed foundations or rigid power connexions, used to de-water shafts during sinking.

sino-atrial (or **sino-auricular**) **node** (*Zool.*). See **Keith and Flack's node.**

sino- (or **sinu-**) **auricular** (*Anat., Physiol.*). Applied to structures located in the right atrium near the opening of the venae cavae, corresponding to the sinus venosus, e.g., sinu-atrial (or sino-auricular) node.

sinography (*Med.*). The radiological examination of the large intracranial venous sinuses following carotid or other *angiography* (q.v.).

sinor (*Elec. Eng.*). See **complexor.**

sinter (*Chem.*). To coalesce into a single mass under the influence of heat, without actually liquefying. (*Min.*) A concretionary deposit of opaline silica which is porous, incrusting, or stalactitic in habit; found near geysers, as at Yellowstone National Park (U.S.). Also called **geyserite.**

sintered carbides (*Met.*). See **cemented carbides, sintering.**

sintered crucible (*Chem.*). A crucible with a permeable sintered base, used as a combination filter and crucible.

sintering (*Elec. Eng.*). The process of consolidating the filament of an electric lamp by passing a relatively high current through it when in a vacuum. (*Met.*) The fritting together of small particles to form larger particles, cakes, or masses; in case of ores and concentrates, it is accomplished by fusion of certain constituents. As used in powder metallurgy, sintering consists in mixing metal powders having different melting-points, and then heating the mixture to a temperature approximating the lowest m.p. of any metal included. In sintered carbides, powdered cobalt, having the lowest m.p., acts as the binder holding together the unmelted particles of the hard carbides.

sinuate (*Bot.*). (1) Said of gills of agarics which show a sudden curvature as they reach the stipe. (2) In general, having a margin divided into wide irregular teeth or lobes, separated by shallow notches.

sinuitis (*Med.*). See **sinusitis.**

sinuous flow (*Phys.*). See **turbulent flow.**

sinupalliate (*Zool.*). Of *Pelecypoda*, having a distinct indentation of the pallial line posteriorly, owing to the presence of a siphon.

sinus (*Bot.*). A depression or notch in a margin between two lobes. (*Zool.*) A cavity or depression of irregular shape.

sinusitis (*Med.*). Nasal sinusitis. Inflammation of any one of the air-containing cavities of the skull which communicate with the nose. The condition is called ethmoid, frontal, maxillary or sphenoid sinusitis, according to the site affected. It may be acute or chronic, purulent or nonpurulent.

sinusoid (*Zool.*). In Vertebrates, a sinuslike blood space connected usually with the venous system and lying between the cells of the surrounding tissue or organ.

sinusoidal (*Elec. Eng.*). An alternating quantity is said to be *sinusoidal* when its trace, plotted to a linear time base, is a sine wave.

sinusoidal current (*Elec. Eng.*). One which varies sinusoidally with time, having a frequency, amplitude and phase. It flows in each direction alternately for equal periods.

sinusoidal projection (*Geog.*). A type of *Bonne's projection* (q.v.), in which the standard parallel is the equator and the projected meridians appear as sine curves. Area is correct, but shape is distorted away from the centre, so that it is chiefly suitable for showing 'slices' of the globe. When a world map is required, this projection can be recentred or 'interrupted', i.e., more central meridians may be used for different parts of the map. Also called **Sanson-Flamsteed projection.**

sinusoidal spirals (*Maths.*). Curves with polar equation $r^n = a^n \cos n\theta$. Not true spirals since the tracing point does not recede continuously. For certain values of n the curves have specific names, e.g., $n=-2$ (hyperbola), $n=-1$ (straight line), $n=-\frac{1}{2}$ (parabola), $n=\frac{1}{2}$ (cardioid) and $n=2$ (lemniscate).

sinus pocularis (*Zool.*). See **uterus masculinus.**

sinus rhomboidalis (*Zool.*). In an early Vertebrate embryo, the posterior part of the neural groove before the medullary folds have closed over it; the wide diamond-shaped cavity of the lumbar region of the spinal cord into which this develops later.

sinus venosus (*Zool.*). In a Vertebrate embryo, the most posterior chamber of the developing heart; in lower Vertebrates, the tubular chamber into which this develops and which receives blood from the veins, or sinuses and passes it into the auricle.

sipho-, siphono-. Prefix from Gk. *siphōn*, gen. *siphōnos*, tube.

siphon (*Bot.*). An elongated cell which extends the whole length of a joint in some red algae. (*Civ. Eng.*) (1) A pipe system comprising a rising leg and a falling leg, typically in the shape of an inverted 'U'; pressure at the inlet to the rising leg is atmospheric pressure; as liquid rises in the rising leg the pressure falls below atmospheric pressure and reaches a minimum at or near the apex, recovering to atmospheric pressure in the down leg. The lift of a siphon can equal or exceed the equivalent of atmospheric pressure expressed in *head* of the fluid which enters the siphon; this results from entrained gases emerging throughout the rising leg. (2) A pipe or aqueduct crossing a valley and rising again to somewhat less than its inlet level, so as to have the necessary hydraulic gradient. More correctly called an **inverted siphon.** (*Glass*) A refractory block used in the construction of a tank furnace bath. (*Zool.*) A tubular organ serving for the intake or output of fluid, as the pallial siphons of many bivalve *Mollusca*. *adj.* **siphonate.**

Siphonaptera (*Zool.*). The only order of the superorder *Siphonapteroidea*. Wingless insects ectoparasitic on warmblooded animals. Laterally compressed, short antennae lying in grooves, mouthparts for piercing and sucking, coxae large, tarsi 5-jointed with prominent claws. Metamorphosis complete, larvae legless, pupae exarate and enclosed in a cocoon. Serve as vectors for some diseases, e.g., myxomatosis, plague. Fleas. Also **Aphaniptera.**

Siphonapteroidea (*Zool.*). A superorder of the *Oligoneoptera*, with one order *Siphonaptera*.

siphoneous (*Bot.*). Tubular.

siphonet (*Zool.*). The cornicle or honeydew tube of an Aphid. Also called **siphuncle.**

siphonium (*Zool.*). In *Crocodilia*, a membranous tube connecting the air-passages of the quadrate with an air space in the os articulare of the mandible. *adj.* **siphonial.**

siphonogam (*Bot.*). A plant in which the contents of the pollen grain pass into the embryo sac through a pollen tube.

siphonoglyph (*Zool.*). In *Anthozoa*, a longitudinal ciliated groove of the stomodaeum.

Siphonophora (*Zool.*). An order of *Hydrozoa* consisting of pelagic or drifting colonies of markedly polymorphic individuals. These originate from a planula larva which metamorphoses to form a single individual which is either medusiform or else forms an apical float or pneumatophore, whose epithelium secretes gas. In the former case the colonies bud off similar assemblages of individuals (cormidia) along their length, but in the latter these are all crowded into a compact colony, e.g., Portuguese Man o' War (*Physalia*).

siphonostele (*Bot.*). A hollow tube of vascular tissue, enclosing a pith and embedded in ground tissue.

siphonostomatous (*Zool.*). Having a tubular mouth; of *Gastropoda*, having the anterior end of the shell-aperture produced into a spout.

siphonozooid (*Zool.*). In *Alcyonaria* (*Anthozoa*), a type of polyp lacking tentacles and gonads and having filaments only on the dorsal mesenteries. Cf. *autozooid*.

siphon recorder (*Teleg.*). Early paper-tape recorder in which a fine siphon is fed with ink and, under the control of coil currents, records signals on a continuously moving paper.

siphon spillway (*Civ. Eng.*). A siphon connecting the upstream and the downstream sides of a reservoir dam, thus enabling flood waters to pass, as in the case of a *bye-channel* (q.v.).

siphon trap (*San. Eng.*). A trap having a double bend like an S on its side, the lower bend containing the water seal preventing reflux of foul gases. Fitted to closets and sinks.

siphorhinal (*Zool.*). Having tubular nostrils.

siphosome (*Zool.*). The lower part of a Siphonophoran colony, to which the nutritive and reproductive individuals are attached.

siphuncle (*Zool.*). In Aphids, same as *siphonet*; in *Nautiloidea*, a narrow vascular tube extending from the visceral region of the body through all the chambers of the shell to its apex. *adj.* **siphunculate.**

Siporex (*Build.*). A material of light density and high insulation value manufactured with sand, cement, and a catalyst and cured under high-pressure steam conditions. TN.

Sippy treatment (*Med.*). The treatment of gastric ulcer by diet and by administering alkalis to neutralize the hydrochloric acid of the gastric juice.

Sipunculoidea (*Zool.*). A class of *Annelida*, of doubtful affinities, having a spacious uninterrupted coelom, a single pair of nephridia and no chaetae. In the adult there is no sign of segmentation, but the trochosphere larva shows three pairs of somites which soon disappear. Intestine coiled and anus dorsal and anterior. Sessile sand dwellers.

siren (*Acous.*). A powerful source of noise of a more or less pure tone; the noise is usually generated by the periodic escape of compressed air through a rotary shutter.

Sirenia (*Zool.*). An order of large aquatic *Paenungulata* of herbivorous habit; the fore limbs are finlike, the hind limbs lacking; there is a horizontally flattened tail-fin; the skin is thick with little hair and underlying blubber, there are two pectoral mammae, no external ears, and the neck is very short. Manatees, Dugongs.

sirenomelus (*Med.*). A foetal monster with fused legs but no feet.

sirocco, scirocco (*Meteor.*). A warm moist wind from the south or southeast, which blows before the eastward passage of a depression in Mediterranean regions.

sisal hemp (*Bot.*). A fibrous material used extensively for cordage and binder-twine. The chief source is *Agave sisalana*, an acaulescent herbaceous plant, native to Yucatan, Mexico. Grown in the tropics.

sister cell (*Biol.*). One of the two cells formed by the division of a pre-existing cell.

sister nucleus (*Biol.*). One of the two nuclei formed by the division of a pre-existing nucleus.

S.I.T. (*Eng.*). Abbrev. for *spontaneous ignition temperature*.

site error (*Radio*). That in a radio bearing which depends only on site surroundings and which can be used as a correction to readings.

site rivet (*Eng.*). See **field rivet.**

sitfast (*Vet.*). A small hard lump on the skin of a horse's back, due to necrosis of the skin caused by pressure of the saddle or harness.

sitology. The study of diet and nutrition.

sitophore sclerite (*Zool.*). In *Psocoptera* and *Mallophaga*, a sclerite forming the basal part of the cibarial surface of the hypopharynx. Sometimes called the **oesophageal sclerite.**

sitosterol (*Chem.*). The β form $C_{29}H_{50}O$, a sterol derivative, found in corn oil, which closely resembles cholesterol. Also occurs in other forms: α_1-, α_2-, α_3- and γ-sitosterol.

sitotoxin (*Med., Biol.*). A food poison.

SI units. System of coherent metric units (*S*ystème *I*nternational d'Unités) proposed for international acceptance in 1960 and now increasingly used. It was developed from the Giorgi/MKSA system (see **MKSA**) by the addition of the *kelvin* and *candela*, and later the *mole*, as base units. There are numerous *derived units* (*newton, joule*, etc.) and scales of decimal multiples and submultiples, all with agreed symbols.

six-phase (*Elec. Eng.*). A term applied to circuits or systems of supply making use of six alternating voltage phases, vectorially displaced from each other by $\pi/3$ radians.

Six's thermometer (*Meteor.*). A form of *maximum and minimum thermometer* (q.v.) consisting of a bulb containing alcohol joined to a capillary stem bent twice through 180°. A long thread of mercury is in contact with the alcohol in the stem, and this mercury moves as the alcohol in the bulb expands and contracts. Each end of the mercury thread pushes a small steel index in front of it, one of which registers the maximum temperature and the other the minimum.

sixteenmo (*Print.*). The 16th of a sheet or a sheet folded 4 times to make 16 leaves or 32 pages. Also called **sextodecimo,** and written 16*mo*.

six-tenths rule (*Chem. Eng.*). A rule, generally regarded as empirical but with a theoretical basis, which is used in plant cost estimating and says for similar plants of different capacities:

$$\frac{\text{Cost of Plant } A}{\text{Cost of Plant } B}=\left(\frac{\text{Capacity of Plant } A}{\text{Capacity of Plant } B}\right)0{\cdot}66.$$

six-twelve potential (*Chem.*). See **Lennard-Jones potential.**

sixty-fourmo (*Print.*). The 64th part of a sheet or a sheet folded 6 times to make 64 leaves or 128 pages. More than 4 folds are not practicable and the sheet would be cut into 4 parts, each being folded 4 times. Written 64*mo*.

size (*Paint., etc.*). An animal glue in powder or jelly form, used in cheap work as an undercoating for distempers and wallpapers to prevent the wall from absorbing moisture from them.

size distribution (*Powder Tech.*). Proportions of each size of particles in a powder or colloidal system. Expressed either as cumulative or frequency distribution.

size fraction (*Powder Tech.*). A portion of a powder composed of particles between two given size limits, expressed in terms of the weight, volume, surface area, or number of particles.

size-grading (*Chem.*). The process of determining the frequency distribution of particles of different sizes in a material.

size water (*Paint., etc.*). A solution of size in hot water. It is added to plaster to make it more slow-setting, or used as an undercoating on walls. See **size.**

sizing (*Textiles*). Embraces several yarn or cloth processes, e.g., application of binding materials such as flours, starches, animal sizes; softeners, such as lubricants; weighting materials. All have some special significance in protecting or strengthening the product.

S.J. (*Plumb.*). Abbrev. for *soldered joint*.

skarn (*Geol.*). A rock containing calcium silicate minerals produced by metasomatic alteration of limestone close to the contact of an igneous intrusion. Skarns are sometimes enriched in ore minerals such as magnetite or scheelite. *Tactite* is a synonymous term which is used in America.

Skarne (*Build.*). A method of system building, developed in Sweden. In this construction first the lift shaft(s) and stair well surround(s) are constructed of *in situ* reinforced concrete. Storey height precast wall elements are then progressively erected in conjunction with *in situ* reinforced concrete floors. See **system building.**

skate (*Elec. Eng.*). The special type of sliding contact piece by which an electric tractor collects current in the surface-contact system of electric traction supply. (*Rail.*) See **retarder.**

skatole (*Chem.*).

3-methylindole, colourless plates, of faecal odour, m.p. 95°C.

skein (*Cyt.*). The nuclear reticulum. (*Textiles*) (1) Fixed quantity of yarn of any textile fibre wound on a frame to a fixed length, and then hanked-up and secured by a knot for convenience in handling. (2) Unit of length in the counting of woollen yarns and silk yarns. See **count of yarn.**

skeletal muscle (*Zool.*). See **voluntary muscle.**

skeletogenous (*Zool.*). Forming or taking part in the formation of the skeleton.

skeleton (*Anat., Zool.*). The rigid or elastic, internal or external, framework, usually of inorganic material, which gives support and protection to the soft tissues of the body and provides a basis of attachment for the muscles, forming a system of jointed levers which they can move. *adjs.* **skeletal, skeletogenous.** (*Surv.*) The network of survey lines providing a figure from which the shape and salient features of the survey may be determined.

skeleton crystals (*Geol.*). Imperfect crystals of minute size occurring in glassy igneous rocks; often merely 3-dimensional frameworks, the interstices in which would have been filled in under conditions of slower cooling. See also **dendrite.**

skeleton drums (*Paper*). Drums over which paper is dried after being animal-sized. Fans revolve in a direction opposite to the travel of the paper and expedite the drying process.

skeleton flashing (*Plumb.*). See **stepped flashing.**

skeleton movement (*Horol.*). A clock or watch movement the plates of which have been pierced or cut away to show the mechanism.

skeleton steps (*Build.*). Steps in a stair, of a construction such that there are no risers but only treads fixed at suitable positions above one another between side supporting pieces.

skeleton-type switchboard (*Elec. Eng.*). A switchboard consisting of a metal framework upon which the switches and other apparatus are mounted. Also called **frame-type switchboard.**

skelp (*Met.*). Mild steel strip from which tubes are made by drawing through a bell at welding temperature, to produce lap-welded or butt-welded tubes.

skene arch (*Civ. Eng.*). An arch having the shape of a circular arc subtending less than 180°. Also called **scheme arch.**

skep (*Mining*). See **skip.**

skew arch (*Arch., Civ. Eng.*). An arch which has its axis or line of direction oblique to its face.

skewback (*Civ. Eng.*). The courses of stones from which an arch springs (on the top of a pier) where the upper and lower beds are oblique to each other.

skewbacks (*Glass*). Specially shaped blocks, supporting a furnace crown or other arch. Also **springers.**

skew bevel gear (*Eng.*). See **hypoid bevel gear.**

skew butt (*Build.*). See **skew corbel.**

skew coil (*Elec. Eng.*). An unsymmetrical coil inserted in the armature winding of an alternator having an odd number of pole pairs.

skew corbel (*Arch.*). The projecting masonry or brickwork supporting the foot of a gable coping.

skewed pole (*Elec. Eng.*). A field pole whose cross-section is a parallelogram instead of the usual rectangle.

skewed slot (*Elec. Eng.*). A slot whose diameter is not parallel to the axis of rotation.

skew fillet (*Build.*). See **tilting fillet.**

skew flashing (*Plumb.*). A flashing fixed down a gable wall.

skew lines (*Maths.*). Nonparallel finite straight lines which do not intersect.

skew nailing (*Carp.*). The operation of driving nails in obliquely.

skew rebate plane (*Join.*). A rebate plane with its cutting edge arranged obliquely across the sole.

skew-table (*Build.*). A stone which is bonded in with a gable wall, as a support for the foot of the coping.

skew wall (*Acous.*). In a studio, a wall which does not form a face of a parallelepiped; the walls are so arranged that continuous reflections between opposite walls are obviated.

skiagram, skiagraph (*Radiol.*). Same as *radiogram*.

skiascopy (*Med.*). See **retinoscopy.**

skiatron (*Electronics*). Cathode-ray tube in which an electron beam varies the transparency of a phosphor, which is illuminated from behind so that its image is projected on to a screen, as in a photographic slide projector.

Skiddavian (*Geol.*). An alternative name given to the lowest division (Arenig Series) of the Ordovician System in the British Isles, and applied particularly to the graptolitic shale facies of that formation in the Lake District.

Skiddaw Slates (*Geol.*). A thick group of highly cleaved slates, found in the core of the Lake District of England, and of Lower Ordovician (Arenigian to Llanvirnian) age.

skids (*Build.*). Small pieces of timber packed under a surface to bring it to the plane.

skid-senser (*Autos.*). A servo device which automatically relieves brake pressure when the road wheels start to lock.

skillet (*Met.*). Mould for casting bullion.

skim coulter (*Agric.*). A form of coulter sometimes fitted to a plough, in advance of the coulter; it pares the edge of the furrow slice.

Skimmia (*Bot.*). An Asiatic genus of shrubs, cultivated for its holly-like leaves and drupes.

skimming coat (*Build.*). See **setting coat.**

skin (*Aero.*). The outer surface other than fabric of an aircraft structure; the outer surface in a *sandwich construction.* (*Bot.*) Plant material forming the outer part of a periclinal chimaera. (*Eng.*) The hard surface layer found on iron castings due to the rapid cooling effect of the mould, or on steel plates, strip and sheet, due to rolling, or on other materials or products due to the surface-hardening effect of the finishing process. (*Zool.*) The protective tissue layers of the body wall of an animal, external to the musculature. See **epidermis.**

skin depth (*Elec.*). The distance into a conductor at which the amplitude of an electromagnetic wave falls to $1/e = 0{\cdot}3679$ of its value at the surface, given by $\delta = \sqrt{\rho/\pi f \mu}$. If the magnetic field at the surface is H, the power dissipated is $\frac{1}{2}H^2R_s$ W/m², where the effective surface resistance $R_s = \rho/\delta$, ρ is the resistivity and μ the absolute permeability of the metal, and f is the frequency of the wave. See **depth of penetration, skin effect.**

skin dose (*Radiol.*). Absorbed or exposure radiation dose received by or at the skin of a person exposed to sources of ionization. Cf. *tissue dose.*

skin effect (*Elec. Eng.*). When a conductor carries a high-frequency current it tends to be confined to a thin surface layer. Thus the high-frequency resistance is greater than the low-frequency value. See **depth of penetration, litz wire, skin depth.**

skin friction (*Aero.*). See **surface-friction drag.**

skin gills (*Zool.*). See **papulae.**

skinner (*Elec. Eng.*). The length of insulated wire between the point of connexion to a solder tag and the cable form from which it emerges.

Skinner box (*An. Behav.*). A box from which an animal cannot escape, provided with objects whose manipulation by the animal is followed, often automatically, by a positive reinforcing stimulus. Manipulations may include pressing a bar, pecking a key, pulling a string, etc.

skintled (*Build.*). Said of brickwork in which the bricks are laid irregularly, so as to leave an uneven surface on the wall; also of a similar effect produced by protruding mortar squeezed from the joints.

skin tuberculosis of cattle (*Vet.*). A benign lymphangitis producing nodules in the skin of cattle, caused by bacteria which resemble in some respects the causative organism of true tuberculosis.

skin wool (*Textiles*). Wool removed from the fleeces of slaughtered sheep by a liming, sulphide, or sweating process. In the first two, the skins are painted on the flesh side with lime or sodium sulphide; in the last they are steeped in water and hung in a closed chamber. Also known as **slipe wool.**

skiophyte (*Bot.*). See **sciophyte.**

skip (*Civ. Eng.*). A bucket or large container used for movement of building, etc., materials or refuse, e.g., from demolition. (*Mining*) Rectangular metal box, sliding between guides in a vertical shaft or running on skip track in inclined shaft. Usually has capacity of several tons of ore or coal. Term sometimes used to describe *bowks* or *kibbles* (q.v.), which differ in that they hang freely in the shaft. Also called **skep.** (*Radio*) See **hop.**

skip distance (*Radio*). Region of no-signal between the limit of reception of the direct (*ground*) wave and the first downcoming reflection from an ionized layer; prominent with short waves.

skip draft (*Textiles*). A method of drawing warp threads through the eyes of loom healds, a number of healds being *skipped*, or missed, at intervals, e.g., 1, 3, 2, 4.

skipping printer (*Cinema.*). A contact printer of cinematograph film which prints selected frames rather than each one in turn.

skip slitter (*Print.*). A cam-operated slitter giving intermittent cuts, whereby both broadsheet and tabloid sections can be incorporated in a single copy by means of cylinder collection.

skip-tooth saw (*Tools*). A saw from which alternate teeth are cut away.

skirt (*Chem. Eng.*). A cylindrical support on large vessels and fractionating columns welded to the main part of the cylindrical shell and enclosing the bottom of the vessel or column. (*Telecomm.*) Lower side portions of a resonance curve, which should be symmetrical.

skirt dipole (*Telecomm.*). Quarter-wavelength wire, emerging coaxially from the conductor of a coaxial transmission feeder. Also called **sleeve dipole.**

skirting board (*Build.*). A board covering the plaster wall where it meets the floor. Also called **baseboard, mopboard, washboard.** (*Mining, Min. Proc.*) At delivery on to belt conveyor, boards which direct falling rock toward centre of belt, or prevent spill-over.

skirtings (*Textiles*). (1) Stained wool removed from skirt of a fleece after shearing. (2) Term used in classifying certain rags for recovery.

skittle pot (*Glass*). A small pot, in shape resembling a skittle, which can be set in a furnace in some small corner to melt a special glass, e.g., a colour. Some small firms use a furnace holding only 4 or 6 of these, fired by coke.

skiver (*Leather*). The grain split from a sheepskin, used in the finished state for bags, purses, etc.

skot (*Light*). A deprecated unit of illumination of low intensity. A light flux emission of 1 skot corresponds to that of a source radiating 10^{-3} apostilb (10^{-3} lm m^{-2}) at a colour temperature of 2045 K.

skotograph (*Photog.*). A developable image produced in a photographic emulsion by radiation from organic tissue in the dark.

skototaxis (*An. Behav.*). A *taxis* (q.v.) in which there is a positive response to a dark area as opposed to a negative response to light. Cf. *phototaxis.*

Skraup's synthesis (*Chem.*). The synthesis of quinoline by heating aminobenzene with glycerine and sulphuric acid, nitrobenzene or arsenic(V)acid acting as oxidizing agent.

skua (*Zool.*). A genus (*Stercorarius*) of large predatory gulls.

skull (*Zool.*). In Vertebrates, the brain case and sense-capsules, together with the jaws and the branchial arches.
skutterudite (*Min.*). Grey or whitish arsenide of cobalt, which crystallizes in the cubic system and sometimes assumes a massive granular habit. Often contains appreciable nickel and iron substituting for cobalt.
sky (*Meteor.*). See **blue of the sky.**
Skyhook (*Space*). A large nondilatable plastic balloon which can carry a heavy load to a height of about 30 000 m, and is used in the exploration of the upper atmosphere. See **rockoon.**
Skylab (*Space*). A U.S. experimental space station, placed in orbit in May 1973, a 'floating workshop' for three separate manned missions, ferried to it by spacecraft.
skyline (*For.*). Systems requiring a cable (skyline) elevated between spars (head and tail) and supporting the load through a moving carriage and a (fall) block.
sky ray or wave (*Radio*). See **ionospheric ray, space wave.**
skyteens (*Textiles*). Satin weave shirting with a light blue ground and stripes of other colours.
S.L. (*Build.*). Abbrev. for *short lengths.*
slab (*Carp.*). An outer piece of a log cut away in the process of slabbing. (*Civ. Eng.*) (1) Normally a reinforced concrete floor supported, at intervals, on beams and/or columns. (2) A thin flat piece of concrete or stone (not necessarily flat in the case of concrete).
slab and girder floor (*Civ. Eng.*). A reinforced concrete floor in which the beams are designed to be homogeneous with the slab and to provide continuity and reduce the amount of concrete and steel.
slabbing (*Carp.*). The operation of squaring a log. (*Print.*) Finishing operations on electrotypes, low areas being made level by punching up from the back.
slab coil (*Elec. Eng.*). A coil in the form of a flat spiral; the term is normally applied to inductance coils.
slab resolver (*Telecomm.*). Four contacts on a square, giving ($\pm R\cos\theta$) and ($\pm R\sin\theta$), concentric with a slab potentiometer fed with $\pm R$ at its ends, θ being angular displacement.
slab serif (*Typog.*). A type style with strokes of uniform thickness and with straight serifs of the same thickness as the strokes. Synonymous with **Egyptian.**
slab tail (*Aero.*). A 1-piece horizontal tail surface, pivoted and power operated so as to serve as a stabilizing tail plane, elevator and, through a lower gearing, trimming tab.
slack (*Mining*). Small coal, coal dirt, as in slack heap, a tip or dump.
slack blocks (*Civ. Eng.*). A pair of wedge-shaped blocks of hard wood packed beneath each end of a centre, and placed in contact with their thin ends pointing in opposite directions, so that by moving them relatively the centre may be gradually lowered when the work has been completed.
slackline (*For.*). System requiring a 4-drum power unit carrying a skyline, mainline, haulback, and strawline, and a 1- or 2-sheave carriage, to which mainline, butt hooks, and haulback line are all attached.
slack sheet (*Print.*). Insufficient tension of the web between units of the press.
slack-water navigation (*Civ. Eng.*). River or canal navigation rendered possible by the construction of dams across the stream at intervals, dividing it into separate reaches, communication being maintained by the use of locks. Also called **still-water navigation.**
slade (*Agric.*). A flat plate attached to the body of a plough to take up the vertical reactions of the ground when the plough is working. (*Build.*) An inclined pathway.
slag (*Met.*). The top layer of the 2-layer melt formed during smelting and refining operations. In smelting it contains the gangue minerals and the flux; in refining, the oxidized impurities.
slag cement (*Civ. Eng.*). An artificial cement made by granulating slag from blast furnaces by chilling it in water and then grinding it with lime, to which it imparts hydraulic properties.
slag hole (*Met.*). In smelting furnace, aperture above that through which molten metal is withdrawn, used to tap off accumulated slag.
slag wool (*Met.*). See **rock wool.**
slaked lime (*Build., Chem., etc.*). See under **caustic lime.**
slaking (*Build.*). The process of combining quicklime with water.
slamming stile (*Join.*). The upright member of a door case against which the door shuts and into which the bolt of a rim lock engages.
slant range (*Surv.*). Distance along sloping sight between two points as measured by tacheometer or tellurometer.
slap dash (*Build.*). A rough finish given to a wall by coating it with a plaster containing gravel or small stones.
slasher (*Textiles*). (1) A machine which coats warp yarn with size, drys it, and then rewinds it on to a weaver's beam. (2) The operative in charge of the slasher sizing machine.
slash-sawn (*For.*). See **back-sawn.**
slat (*Aero.*). An auxiliary aerofoil which constitutes the forward portion of a slotted aerofoil, the space between it and the main portion of the structure forming the slot. (*Carp.*) A thin, flat strip of wood.
slat conveyor (*Eng.*). See **apron conveyor.**
slate (*Geol.*). A sedimentary rock of the clay or silt grade which, because of regional metamorphism, has developed a good *slaty cleavage.* (*Elec. Eng.*) Slate is used for switchboard panels. Care must be taken to avoid the use of pieces with conducting veins of pyrites, magnetite, etc. The electrical strength is 2 to 3 kV/cm for centimetre thicknesses.
slate axe (*Tools*). See **sax.**
slate boarding (*Build.*). Close boarding laid as an underlining to, and a support for, roofing slates.
slate cramp (*Build.*). A piece of slate about $7\times2\frac{1}{2}\times1$ in. cut to a narrow waist at the middle, and fitted flush into mortises in adjacent stones to bind them together.
slate hanging (*Build.*). Similar to *weather tiling* (q.v.), slates being used instead of tiles.
slating and tiling battens (*Build.*). Any pieces of square-sawn converted timber between $\frac{1}{2}$ and $1\frac{1}{2}$ in. in thickness and from 1 to $3\frac{1}{2}$ in. in width; commonly used as a basis for slating and tiling.
slatted lens (*Radio*). One with shaped metal slats, parallel to E or H vector in wave from waveguide. Also used for low-frequency acoustic waves. Also called **egg-box lens.**
slaty cleavage (*Geol.*). The property of splitting easily along regular, closely spaced planes of fissility, produced by pressure in fine-grained rocks, the cleavage planes lying in the direction of maximum elongation of the mass.
slave mechanism (*Automation*). Device which responds to signals in an automatically controlled continuous production line, initiated by a monitoring device. Used to vary valve settings, feed rates, etc.

slavery (*Zool.*). See **dulosis.**

slay (*Weaving, etc.*). See **sley.**

sledge-amphibian (*Ships, etc.*). An amphibious vehicle designed to move over water and swamp as a boat and over ice and snow as a sledge. Propelled by air propeller and steered by two rudders, one in the air and one in the snow.

sledge-hammer (*Tools*). A heavy double-faced or straight-pane hammer, weighing up to 100 lb (45 kg), swung by both hands.

sledger (*Civ. Eng.*). A machine for the first stage of crushing of rock in quarrying. Also **scalper.**

sleekers (*Foundry*). Moulders' tools having a smoothing face made to various shapes, for smoothing over small irregularities in the sand of the mould. Also called **smoothers.** See **corner tool.**

sleeper (*Carp.*). A horizontal timber supporting a vertical shore or post, and distributing the load over the ground. (*Rail.*) A timber or pre-stressed concrete beam passing transversely beneath the rails to support them and prevent them from spreading apart. Also called **cross-tie, cross-sill.** Also, a sleeping-car.

sleeper plate (*Build.*). A wall plate resting upon a sleeper wall.

sleeper wall (*Build.*). A low wall built under the ground storey of buildings having no basement, as a support for the floor joists. When in brick, the wall is built honeycombed to leave spaces for ventilation, and when in stone, small piers at intervals provide the support required.

sleeping sickness (*Med.*). See **trypanosomiasis.**

sleep movement (*Bot.*). The folding together of the leaflets of a compound leaf at night, so that surfaces bearing numerous stomata are brought together; the significance of the movement is not yet understood.

sleepy sickness (*Med.*). See **epidemic encephalitis, Von Economo's disease.**

sleet (*Meteor.*). A mixture of rain and snow, or partially melted snow.

sleeve (*Elec. Eng.*). See **plain coupler.** (*Eng.*) A tubular piece, usually one machined externally and internally. (*Radio*) Quarter-wavelength coaxial line for coupling a coaxial line to a dipole at its centre. See **balun.** (*Teleph.*) (1) The thin slug which is placed over the core of the line relay in a subscriber's line circuit. Made of nickel-iron, so that the inductance of the relay, which also acts as a feeding bridge, is enhanced, with a reduction of the number of turns and consequent lower **resistance.** (2) The outermost contact on 3-conductor flexible cords, terminating in conductor plugs which contact the cord circuit. It is connected to the S-wire of the circuit. See **S-wire.**

sleeve control switchboard (*Teleph.*). One in which the supervisory signals are controlled over the cord circuit sleeve connexions by the line terminating equipment. The cord circuits contain no transmission bridges.

sleeve dipole (*Telecomm.*). Same as **skirt dipole.**

sleeve joint (*Elec. Eng.*). A conductor joint formed by a sleeve fitting over the conductor ends. It is either pinned or soldered to the conductors.

sleeve piece (*Plumb.*). A short length of brass or copper pipe, used in forming a joint between a lead pipe and one of some other material. Also called **thimble.**

sleeve valve (*I.C. Engs.*). A thin steel sleeve fitted between the cylinder and the piston of a petrol or oil engine. It is given a reciprocating and rotary oscillating motion, thus causing ports cut in it to register alternately with corresponding inlet and exhaust ports in the cylinder wall.

sleeve-wire (*Teleph.*). See **S-wire.**

sleeving (*Elec. Eng.*). Tubular flexible insulation for threading over bare conductors.

slenderness ratio (*Build.*). See **ratio of slenderness.**

s-levels (*Light*). See **sharp series.**

slewing rollers (*Print.*). Rollers whose angle can be adjusted to alter the run of the web to maintain sidelay register, manual control being superseded by electronic equipment such as *Autotron.*

slew rate (*Electronics*). The maximum rate of change of output voltage for an amplifier when a voltage step is applied at the input. Normally measured in V/μs.

sley or slay (*Weaving, etc.*). (1) To pass the warp threads through a reed, for dressing or weaving. (2) Heavy metal or wooden beam carrying the shuttle race board, reed, and shuttle-boxes. As it moves backwards the shuttle is picked from one box to the other carrying the weft. When the sley comes forward the reed beats-up the weft to form cloth. (3) A guideway in a knitting machine. (4) The part of a lace machine, between the beams and the thread guides, which functions in keeping the threads properly arranged.

sley sword (*Weaving*). One of the two metal arms, attached to a rocking shaft, supporting the sley in a loom and providing adjustment points.

slice (*Paper*). The arrangement at the wet end of the machine which distributes the flow of the paper stock evenly on to the machine wire. Many different types are used, the simplest being a flat plate across the width of the machine.

slicer (*Telecomm.*). Combination of limiters such that only a range of signal between two limiting values is passed on.

slicing (*Mining*). Removal of a layer from a massive ore body. In top slicing this is horizontal, a mat of timber separating it from the overburden. Side and bottom slicing are also practised.

slick (*Join.*). A chisel with a wide cutting edge, used for paring the sides of tenons and mortises.

slick bit (*Tools*). Type of interchangeable scraping bit attached to a shank for use in electric drills.

slickensides (*Geol.*). Smooth, grooved, polished surfaces produced by friction on fault-planes and joint faces of rocks which have been involved in faulting.

slicker (*Foundry*). A small implement used by a moulder for smoothing the surface of a mould. (*Leather*) A metal, vulcanite, stone, or glass implement for smoothing or stretching leather.

slide (*Mining*). A vertical crack in a vein, along which movement has taken place; the clay filling of such a crack; a fault. (*Photog.*) A photographic image on a transparent support, usually glass, for use in a projector.

slide-back (*Elec. Eng.*). Original method for measuring high-frequency voltage applied to grid of triode; the voltage peak is measured by increase of negative bias which just cuts off grid current.

slide-back voltmeter (*Elec. Eng.*). Traditional circuit for alternating voltage measurement in which an adjusted and known steady voltage just balances the peak of the unknown voltage, as indicated by flow of anode or grid current. See **valve voltmeter.**

slide bars (*Eng.*). See **guide bars.**

slide-rails (*Elec. Eng.*). Slotted rails, secured to the floor, on which the base plate of an electric motor is fixed. As the driving belt stretches, the motor can be moved along the rails to take up the slack.

slide resistance (*Elec. Eng.*). A rheostat whose

ohmic value is adjusted by sliding a contact over the resistance wire.

slide rest (*Eng.*). A slotted table carrying the tool post of a lathe. It is mounted on the saddle or carriage, and is capable of longitudinal and cross traverse.

slide rule. A device for performing mechanically arithmetical processes, as multiplication, division, etc. It consists of one rule sliding within another, so that their adjacent similar logarithmic scales permit of the addition and subtraction, corresponding to the multiplication and division, of the numbers engraved thereon.

slide-valve (*Acous.*). The slide, containing holes, which is drawn across the supply of air to a rank of organ pipes, to stop the pipes speaking when a key is depressed. Operated by draw-stops through trackers and stickers. (*Eng.*) A steam-engine inlet and exhaust valve shaped like a rectangular lid. It is reciprocated inside the steam chest, over a face in which steam ports are cut, so as alternately to admit steam to the cylinder and connect the ports to exhaust through the valve cavity. See **D slide-valve, piston valve.**

slide-valve lead (*Eng.*). The amount by which the steam port of a steam-engine is already uncovered by the valve when the piston is at the beginning of its working stroke. See also **inside lead.**

slide wire (*Elec. Eng.*). One for potential division, by sliding a contact along a wire. With a concentric tube with a slot and a probe contact, this becomes a coaxial slotted line for voltage standing-wave ratio measurements at high frequencies.

sliding bevel (*Tools*). See **bevel.**

sliding caisson (*Hyd. Eng.*). A floating body used to open or shut the entrance to a dock or basin, and capable of being drawn for the former purpose into a recess at right angles to the channel.

sliding contact (*Elec. Eng.*). Tangential movement between contacting metal surfaces, to remove film and establish conduction contact. Wear of contact is proportional to total use.

sliding growth (*Bot.*). The rearrangements which are presumed to take place in xylem as the tracheids and vessels elongate and fit in among one another.

sliding-mesh gear-box (*Autos.*). One in which the ratio is changed by sliding one pair of wheels out of engagement and sliding another pair in.

sliding-panel weir (*Civ. Eng.*). A form of frame weir in which the wooden barrier consists of wooden panels sliding in grooves between each pair of frames.

sliding sash (*Join.*). A sash that moves horizontally on runners, as distinct from a *balanced sash* sliding vertically.

sliding ways (*Ships*). The portion of a ship's launching ways which move with the ship on launching.

slime flux (*Bot.*). An exudation of a watery solution of sugars and other substances from trees which have been wounded or are attacked by various parasites.

slime mould (*Bot.*). The popular name for a *Myxomycete*.

slime plug (*Bot.*). A mass of slimy material filling the pores in a sieve tube.

slimes (*Met., Min. Proc.*). Particles of crushed ore which are of such a size that they settle very slowly in water and through a bed which water does not readily percolate. Such particles must be leached by agitation. By convention these particles are regarded as less than 1/400 in (0·0635 mm) in diameter (mesh number 200). *Primary slimes* are naturally weathered ore, or associated clays. *Secondary slimes* are produced during comminution. See **anode slime.**

slime string (*Bot.*). A viscous mass of food material passing through a pore in a sieve plate.

sling fruit (*Bot.*). A fruit from which the seeds are projected by elastic tissue.

slip (*Build.*). A long narrow piece of wood of the full thickness of a mortar joint, built into surface brickwork to provide material to which joinery may be nailed. (*Elec. Eng.*) The fraction by which the rotor speed of an induction motor is less than the speed of rotation of the stator field, i.e., the ratio of the speed at which the rotor slips back from exact synchronism with the stator field to the latter's speed of rotation. (*Eng., etc.*) (1) See **belt slip.** (2) In a screw propeller, the amount by which the product of the pitch and r.p.m. exceeds the actual forward velocity, expressed as a percentage of the former. (*Met.*) The process involved in the plastic deformation of metal crystals in which the change in shape is produced by parts of the crystals moving with respect to each other along crystallographic planes. (*Ships*) (1) The difference between the distance travelled by the ship through the water and the distance the propeller would have moved in a solid. Expressed as a percentage of the latter. (2) A sloping masonry or concrete surface for the support of a vessel in process of being built or repaired. (*Teleg.*) The continuous narrow strip of paper which is perforated with holes representing telegraphic signals intended for subsequent accurate and speedy transmission in a transmitter such as the Wheatstone (Morse code), or Siemens multiplex (Baudot code). (*TV*) Vertical shift of TV image because of imperfect *field* synchronization. Similarly for *line*, but the shift is horizontal.

slip angle (*Autos.*). Slight angle of the tyres to the actual track of the wheels when cornering, due to deformation under centrifugal load.

slip bands (*Met.*). Steps or terraces produced on the polished surface of metal crystals as a result of the parts moving with respect to each other during slip.

slip correction (*Powder Tech.*). Correction for *Knudsen flow* (q.v.) in permeability equations.

slip dock (*Ships*). A dock from which the water can be discharged, and which is equipped with a *slip* (q.v.).

slipe (*For.*). A small sledge.

slipe wool (*Textiles*). See **skin wool.**

slip feather (*Join.*). A wooden tongue for a *ploughed-and-tongued joint* (q.v.), distinguished as a *straight tongue*, *feather tongue*, or *cross-tongue* (qq.v.) according to grain direction.

slip flow (*Aero.*). The molecular shearing which replaces normal gas flow conditions at *hypersonic* velocities above *Mach 10*. See **Mach number.**

slip form (*Civ. Eng.*). A *form* that can be moved slowly as work progresses.

slip gauge (*Eng.*). Accurate rectangular block or plate used singly or in combination with others, the distance between end faces forming a gauging length.

slip meter (*Elec. Eng.*). A device for measuring the slip of an induction motor.

slip mortise (*Join.*). A *slot mortise* or a *chase mortise*.

slipped bank (or **multiple**) (*Teleph.*). A bank of outgoing trunks so connected that they are tested in the same order by all switches but starting from different points. Cf. *straight bank*.

slipped tendon (*Vet.*). See **perosis.**

slipper (*Rail.*). See **retarder.**

slipper brake (*Elec. Eng.*). An electromechanical brake acting directly on the rails of a tramway.

slipper piston (*I.C. Engs.*). A light piston having the lower part or skirt cut away between the thrust *faces*, to save weight and reduce friction.

slipper tank (*Aero.*). An *auxiliary tank* mounted externally, close up under wing or fuselage.

slip planes (*Crystal.*). The particular set or sets of crystallographic planes along which slip takes place in metal and other crystals. These are usually the most widely spaced set or sets of planes in the crystals concerned. See **gliding planes.**

slip proof (*Typog.*). A proof taken from a galley of type matter before it is made up into pages.

slip regulator (*Elec. Eng.*). A regulating resistance connected in series with the rotor of a slip-ring type of induction motor to alter the slip, and thus vary the speed of the machine.

slip-ring motor (*Elec. Eng.*). An induction motor having a wound rotor, the connexions to which are brought out to *slip-rings*.

slip-ring rotor (*Elec. Eng.*). The rotor of a slip-ring induction motor; it has a 2- or 3-phase winding brought out to slip-rings. Cf. *cage rotor*.

slip-rings (*Elec. Eng.*). The rings mounted on, and insulated from, the rotor shaft of an a.c. machine, which form the means of leading the current into or away from the rotor winding.

slips (*Bind.*). Name given to ends of tapes or cords after sewing and before covering.

slip sheets (*Print.*). Sheets of paper, oiled manilla being very suitable, placed between sheets as they are printed to avoid *set-off*; largely superseded by *anti-set-off spray* (q.v.).

slip sill (*Build.*). A sill of length equal to the distance between the jambs of the opening, so that it can be placed in position after the shell of the building has been completed.

slip stone (*Carp., etc.*). See **gouge slip.**

slipstream (*Aero.*). The helical airflow from an airscrew, faster than the aircraft.

slip tank (*Aero.*). Alternative to *drop tank* (q.v.).

slipway (*Ships*). See **slip.**

slip winding (*Textiles*). The process of transferring yarn from a hank to flanged bobbins, for lace manufacture.

slit (*Cinema.*). See **scanning-.**

slitless spectroscope (*Optics*). See **objective prism.**

slitter marks (*Bind., Print.*). Two marks in the centre gutter of a printed sheet as a guide to slitting either on the printing machine or on the guillotine.

slitters (*Print.*). Circular rotating knives used (1) to separate printed sheets on sheet-fed printing or folding machines, (2) to divide the printed web on reel-fed rotaries.

slit test (*Cinema.*). The routine test in which a photograph of the focused slit is made, either in sound-camera or projector, to ascertain whether it is of the correct dimensions and located exactly 90° with respect to the edge of the film.

sliver (*Spinning*). Continuous strand of fibres formed after carding, combing, or drawing. From then onwards it is known as *slubbing*, then *roving*.

slope (*Build.*). See **splay brick.** (*Civ. Eng.*) (1) The inclined side of a cutting or embankment. (2) The angle of inclination of the above side. (*Electronics*) See **mutua conductance.** Also called **slope conductance.** (*Maths.*) A measure of the inclination of a line with respect to a fixed line, usually horizontal. On a graph, the tangent of the angle between a line and the *x*-axis. Also called **gradient.**

slope correction (*Surv.*). A correction applied to the observed length of a baseline to correct for differences of level between the ends of the measuring tape.

slope deflection method (*Civ. Eng.*). A method of structural analysis normally used where only the bending moment at every point is evaluated in terms of the loads applied. The reactions are then determined and the analysis can be completed by the application of normal statics.

sloped roman (*Typog.*). Italic founts in which the letters are not cursive but have the same form as the roman.

slope resistance (*Elec. Eng.*). See **differential anode resistance.**

slope staking (*Surv.*). The locating and pegging of points at which proposed earth slopes in cutting or bank will meet the original ground surface.

slot (*Elec. Eng.*). An axially-cut trench, cut out of the periphery of the stator or rotor of an electrical machine, into which the current-carrying conductors forming the winding are embedded.

slot antenna (*Radio*). Radiating element formed by metal surrounding a slot.

slot-fed dipole (*Elec. Eng.*). One normal to a coaxial line, and coupled to it by adjacent longitudinal slots.

slot leakage (*Elec. Eng.*). In an electrical machine, the leakage flux that passes across the slots.

slot link (*Eng.*). See **link motion, radial valve gear.**

slot mortise (*Join.*). A mortise made in the end of a member.

slot permeance (*Elec. Eng.*). The total permeance of the several parallel portions of the slot-leakage flux path.

slot pitch (*Elec. Eng.*). The distance between successive slots around an armature.

slot ripple (*Elec. Eng.*). The harmonic ripple in the e.m.f. wave of an electrical machine. It arises from the regular variation in the permeability of the magnetic path between stator and rotor, caused by the repeated change in air-gap length as the rotor and stator slots pass one another and the intervening teeth.

slotted aerofoil (*Aero.*). Any aerofoil having an air passage (or slot) directing the air from the lower to the upper surface in a rearward direction. Slots may be permanently open, closable, automatic, or manually operated.

slotted core (*Elec. Eng.*). The usual type of armature core, in which slots are provided for the windings. Cf. *smooth core*.

slotted flap (*Aero.*). A trailing-edge *flap* which opens a slot between itself and the main aerofoil as it is lowered or extended.

slotted line (*Telecomm.*). Rigid coaxial line, with slot access for contact with central conductor. Used for impedance and voltage standing-wave ratio measurements at wavelengths comparable with length of slotted line.

slotting machine (*Eng.*). A machine tool resembling a *shaping machine* (q.v.) but in which the ram has a vertical motion and is balanced by a counterweight, the tool cutting on the down stroke, towards the table.

slotting tools (*Eng.*). Cutting tools used for keyway cutting, etc., in a *slotting machine* (q.v.); they are of narrow edge and deep, stiff section, with top and side clearance but little rake.

slot wedge (*Elec. Eng.*). A wood or metal wedge driven into the opening of an armature slot so as to keep the conductors in place against the action of centrifugal forces.

slough (*Med.*). A mass of dead, soft, bodily tissue in a wound or infected area; to form dead tissue (said of the soft parts of the body); to come away as a slough. (*Zool.*) The cast-off outer skin of a Snake.

sloven (*For.*). The torn, splintered portion of timber left on a stump or the end of a butt, where the key finally broke when the tree fell.

slow-acting relay (*Telecomm.*). A relay designed to operate at an appreciable time after the application of voltage. A copper sleeve or slug is placed over the core, or a short-circuited winding is used.

slow-break switch (*Elec. Eng.*). A knife-switch with a single rigid blade forming the moving part of each pole.

slow-butt welding (*Elec. Eng.*). See **resistance upset-butt welding.**

slowing-down area (*Nuc.*). One-sixth of the mean square distance from the neutron source to the point where the neutrons have the required average energy. See **Fermi age.**

slowing-down density (*Nuc.*). Rate at which neutrons fall below a given energy level, per unit volume.

slowing-down length (*Nuc.*). The square root of the slowing-down area.

slowing-down power (*Nuc.*). Increase in neutron lethargy per unit distance travelled in medium. The lethargy is approximately given by the negative of the natural logarithm of the energy of the neutron.

slow neutron (*Nuc.*). See **neutron.**

slow reactor (*Nuc. Eng.*). Obsolete synonym for **thermal reactor.**

slow-running cut-out (*Aero.*). See **fuel cut-off.**

slow-wave structure (*Elec. Eng.*). One in which the apparent velocity of a high-frequency wave is markedly retarded, e.g., in the helix in a travelling-wave tube or in a spiral cable.

SLR (*Photog.*). Abbrev. for *single-lens reflex.*

S.L.-type cable (*Cables*). *S*eparate-*l*ead-type cable. Each core has its own lead sheath, and a fourth lead sheath encloses the whole.

slub (*Textiles*). A fault in any drafted, twisted yarn, which appears as a thicker part, with little twist. Slubs can also be deliberately made in yarn at regular or random intervals for ornamentation and special weave effects.

slubbing (*Spinning*). Drawframe sliver drafted by the rollers in a slubbing frame and twisted and wound on to a bobbin by a flyer and presser.

sludge (*Mining*). Soft mud produced in drilling or boring, or by settling out of water in a pump sump. (*San. Eng.*) A slime produced by the precipitation of solid matters from liquid sewage in sedimentation tanks. See **sewage gas.**

sludger (*Civ. Eng.*). A long cylindrical tube, fitted with a valve at the bottom and open at the top, used for raising the mud which accumulates in the bottom of a boring during the sinking process. Also called **sand pump, shell pump.** (*Mining*) (1) A scraper for clearing mud out of a shot hole. (2) A hand-operated sand pump, consisting of a tube closed by a foot valve, pumped up and down in drill holes to collect and remove samples or disturbed sands and muds.

sludging (*Elec. Eng.*). The formation of a brown deposit in transformer and switch oil, the rate of deposition determining the 'grade' or 'class' of the oil. (*Hyd. Eng.*) (1) Free-running mud. (2) The process of filling the crevices left in the dried clay of an embankment formed by the method of *flood flanking* (q.v.).

slug (*Glass*). Any nonfibrous glass in a glass-fibre product. (*Nuc. Eng.*) Unit of fuel in nuclear reactor, either rod or slab of fissile material encased in a hermetic can of Al, Be, Zr, or stainless steel. Also **cartridge.** See **fuel rod.** (*Phys.*) Unit of mass in the ft-lb-sec system (=14·593 kg). The mass of a body which when acted upon by a force of 1 lbf will accelerate at 1 ft/s². (*Telecomm.*) Thick copper band, comparable with a portion of a winding on a telephone-type relay which, through induced eddy currents, retards the operation and fall-off of the relay. (*Typog.*) A solid line of type as cast by the Linotype, Intertype, or Ludlow.

slug tuning (*Radio*). Alteration of inductance in radiofrequency tuning circuits, by inserting a magnetic core (original 'musical' choke) or a copper disk or cylinder (*spade tuning*).

sluice (*Hyd. Eng.*). A water channel equipped with means of controlling the flow, enabling a sudden rush of water to be used at harbours or canal-locks for the purpose of cleaning out silt, mud, etc., obstructing navigation. (*Mining*) A long trough for washing gold-bearing sand, clay, or gravel. Also called **sluice box** and *launder* (q.v.).

sluice gate (*Hyd. Eng.*). A barrier plate free to slide vertically across a water or sewage channel or an opening in a lock gate, so controlling flow and enabling a sudden rush of water to be used.

sluicing (*Hyd. Eng.*). The process of deepening a navigation channel by discharging into it through an open sluice impounded water from a reservoir.

slump test (*Civ. Eng.*). A test for the consistence of concrete, made with a metal mould in the form of a frustum of a cone with the following internal dimensions: bottom diameter 8 in., top diameter 4 in., height 12 in. This is filled with the concrete, deposited and punned in layers 3 or 4 in. thick, and then the mould is removed and the height of the specimen measured when it has finished subsiding.

slur (*Print.*). A printing fault in which the image lacks sharpness through improper inking, impression, or movement of paper.

slurgalls (*Textiles*). Faults in knitted fabrics which occur in a horizontal direction; due to trapping or tightening of the thread while it is being fed to the needles.

slurry (*Met., Min. Proc., etc.*). A thin paste produced by mixing some materials, especially Portland cement, with water, sufficiently fluid to flow viscously. Used, e.g., to repair (*fettle*) slag-eroded brickwork in smelting furnace, etc.

slurry reactor (*Nuc. Eng.*). One in which fuel exists as a slurry carried by the coolant fluid.

slushed-up (*Build.*). A term applied to brickwork the joints of which are filled with mortar.

slushing compound (*Met.*). A rust-inhibiting liquid composition consisting of mineral oil and anticorrosive additives, such as barium petroleum sulphonates.

slush moulding (*Plastics*). Method based on using certain plastics, particularly P.V.C., in sol form, i.e., in the form of a thin paste. This is placed in a hollow heated mould which is rotated until the paste forms the solid replica of the mould configuration. Used, e.g., for dolls' heads.

slush pulp (*Paper*). Pulp which is pumped direct from the pulp mill to the paper mill for use without passing through the pulp drying stage.

slush pump (*Mining*). See **mud pump.**

Sm (*Chem.*). The symbol for *samarium.*

small bayonet cap (*Elec. Eng.*). A bayonet cap of about 16 mm (⅝ in) diameter; used for small lamps, e.g., automobile head and side lamps.

small-bore (*Heat*). Term applied to pump-assisted hot water central heating systems with ½-in or 15 mm copper or stainless steel pipes.

small calorie (*Heat*). See **calorie.**

small capital (*Typog.*). A letter having the form of a capital but the height of a lower-case letter, e.g., C; indicated in manuscript or proof by two lines under the letter. See **even small caps.**

small circle (*Maths.*). A section of a sphere by a plane not passing through its centre.

small Edison screw-cap (*Elec. Eng.*). An Edison screw-cap having a screw-thread of about ½ in. diameter and about 9 threads per in.

small pica (*Typog.*). An old type size, approximately 11-point.

smallpox (*Med.*). Variola. An acute viral infection characterized by fever, severe headache, pain in the loins, and a rash which is successively macular, papular, vesicular, and pustular, affecting chiefly the peripheral parts of the body.

smalls (*Min. Proc.*). See **riddle.**

small-signal parameters (*Electronics*). See **transistor parameters.**

smallwares (*Textiles*). General name given to tapes, ribbons, and other narrow goods, woven on special narrow looms or braiding machines.

smalt (*Glass*). A glass made by fusing cobalt oxide and silica.

smaltite, smaltine (*Min.*). Smaltite-chloanthite; essentially the diarsenide of cobalt, crystallizing in the cubic system. Chloanthite (diarsenide of nickel) is also invariably present, and the two species graduate into each other; occur in veins; ores of cobalt and nickel.

smaragdite (*Min.*). A fibrous green amphibole, replacive after the pyroxene, omphacite, in such rocks as eclogite.

smashing (*Bind.*). The pressing of a book in a machine after sewing, thereby crushing and expelling air. Also called **crushing, nipping.**

SMC (*Aero.*). Abbrev. for *standard mean chord.*

smearer (*Telecomm.*). A particular circuit used to eliminate the overshoot of a pulse.

smear metal (*Eng.*). Particles and serrated edges produced by metal cutting and caused, by the cutting heat generated, to be welded or combined into an amorphous metal substance.

smear test (*Med.*). Diagnostic test for cancer by laboratory examination of the cells of the blood and other body, e.g., cervical, fluids. Cytodiagnosis. (*Nuc. Eng.*) A method of estimating the loose, i.e., easily removed, radioactive contamination upon a surface. Made by wiping the surface and monitoring the swab.

smectic (*Phys.*). Said of a mesomorphous substance whose atoms or molecules are oriented in parallel planes. Cf. *nematic.*

smectites (*Min.*). A group of clay minerals including montmorillonite, beidellite, nontronite, saponite, sauconite, and hectorite. They are 'swelling' clay minerals and can take up water or organic liquids between their layers, and they show cation exchange properties.

smegma (*Med.*). A thick greasy secretion of the sebaceous glands of the glans penis.

smell (*Physiol.*). The sensation produced by stimulation of the mucous membrane of the olfactory organs.

smelting (*Met.*). Fusion of an ore or concentrate with suitable fluxes, to produce a melt consisting of two layers—on top a slag of the flux and gangue minerals, molten impure metal below.

Smith chart (*Elec. Eng.*). Polar chart with circles for constant resistance and reactance, lines for vector angles, and standing-wave ratio circles. Used for impedance calculations, especially with data from slotted lines for very high frequencies and from waveguides. See **voltage standing-wave ratio.**

Smith's coupling (*Autos.*). Type of electromagnetic clutch comprising a coupling on the gear box mainshaft fitted into the flywheel. Between the two is a small gap filled with metallic powder which is magnetized by a square section coil in a stationary field member in the flywheel casing, and connected to the dynamo, which is modified to ensure that the coupling becomes 'solid' at a maximum engine torque.

Smith's fixative (*Micros.*). See **formol-dichromate fixative.**

smithsonite (*Min.*). Carbonate of zinc, crystallizing in the trigonal system. It occurs in veins and beds and in calcareous rocks, and is commonly associated with hemimorphite. The honeycombed variety is known as *dry bone ore.* In Britain sometimes *calamine.*

smoke (*Powder Tech.*). Visible cloud of airborne particles derived from combustion, or from chemical reaction; the particles are mainly smaller than 1 μm.

smoke control area. An area, statutorily defined, in which emission of smoke from chimneys is prohibited.

smokeless fuel (*Fuels*). A fuel that burns without emitting smoke and is authorized by statute for use in *smoke control areas* or 'smokeless zones'.

smoke point (*Chem.*). In the testing of kerosene, the maximum flame height (in millimetres) at which a kerosene will burn without smoking, under prescribed conditions.

smoke test (*San. Eng.*). A test sometimes applied to drain-pipes suspected of leakage. The ends are plugged and dense smoke is introduced into the pipe, which is watched for any escape.

smoky quartz (*Min.*). See **cairngorm.**

smooth ashlar (*Build.*). A block of stone dressed ready for use, usually for the stone-facing of walls.

smooth colony (*Bacteriol.*). A bacterial colony of a typical regular, glistening appearance; sometimes developed from a rough colony by mutation and frequently differing from it physiologically, e.g., altered sensitivity to bacteriophages.

smooth core (*Elec. Eng.*). An old form of armature core in which the windings were laid on the smooth cylindrical surface of the core and secured by binding wires or small pegs.

smooth-core rotor (*Elec. Eng.*). A rotor carrying a field winding embedded in tunnels, used in high-speed steam-turbine-driven generators.

smoother (*Elec. Eng.*). Combination of capacitors and inductors for removing the ripple from rectified power supplies.

smoothers (*Foundry*). See **sleekers.**

smoothing choke (*Elec. Eng.*). Inductor in a filter circuit which attenuates ripple from a rectifier in supplying d.c. to transistor or valve circuits.

smoothing circuit (*Elec. Eng.*). See **ripple filter.**

smoothing plane (*Carp.*). A bench plane, about 8 in. (20 cm) in length, which is not provided with a handle; used after the trying plane to give a smooth, even finish to the work.

smooth mouth (*Vet.*). Smooth and polished grinding surface of the molar teeth of horses.

smooth muscle (*Zool.*). See **unstriated muscle.**

SMPE (*Cinema.*). *S*ociety of *M*otion *P*icture *E*ngineers standard for sound-tracks on sub-standard cinematograph film in which the sound-track is located on the left-hand side of the picture as normally projected on the screen, as contrasted with the *DIN* standard, at one time in considerable use in Europe, which places the sound-track on the right-hand side. In each type the sound-track replaces one set of sprocket holes on the side of the film.

smudge (*Plumb.*). A lampblack and glue size mixture which is painted over lead surfaces so that solder will not adhere. Also called **soil.**

smut (*Bot.*). See **Ustilaginales.** Also **smut fungus.** (*Mining*) (1) Bad soft coal containing earthy matter. (2) Worthless outcrop material of a coal seam.

smythite (*Min.*). A trigonal sulphide of iron, Fe_3S_4.

sn (*Maths.*). See **elliptic functions.**

Sn (*Chem.*). The symbol for *tin* (stannum).

snail (*Horol.*). A cam the contour of which may be a smooth curve or stepped. Used for the gradual lifting or discharge of a lever. The snail in the rack striking work controls the number of teeth picked up on the rack, e.g., at 12 o'clock sufficient motion is allowed to the rack for the gathering pallet to pick up 12 teeth.

snailed (*Horol.*). Said of (*a*) a surface finished with eccentric curves; (*b*) a barrel arbor so shaped that the inner coil of the mainspring passes over the hook without forming a kink.

snail plant (*Bot.*). A plant which is pollinated by snails, i.e., malacophilous.

snakewood (*For.*). (1) Hardwood from a tree of the genus *Piratinera.* Very heavy, it is dark red in colour, with darker markings resembling a snake skin. Also called **letterwood, leopard wood, tortoiseshell wood.** (2) The *Strychnus nux vomica*, a tree species yielding *strychnine*.

snaking (*Aero.*). An uncontrolled oscillation in yaw, usually at high speed, of approximately constant amplitude.

snap (*Elec. Eng.*). Sudden action in magnetic amplifiers with excessive positive feedback, arising from hysteresis in the core. (*Eng.*) (1) A form of punch with a hemispherically recessed end, used to form rivet heads. (2) A limit gauge of plate or calliper type. (*Textiles*) The unit of length used in the West of England for woollen yarn, viz., 320 yd. See **count of yarn.**

snapped header (*Build.*). A half-length brick, sometimes used in Flemish bond.

snap switch (*Elec. Eng.*). A switch which makes and breaks the circuit with a quick snap; it comprises blades whose rate of motion is controlled by a spring. Also **quick make-and-break switch.**

snap the line (*Build., Civ. Eng.*). To pluck a well-chalked string, held taut in position, against work to mark a straight line.

snare (*Surg.*). A wire loop for removing soft tumours such as nasal polypi.

snarls (*Spinning*). Small double thicknesses of yarn where the twist has run back on itself. They are caused by excessive twist or inadequate tension on the yarn, or excessive speed.

S/N curve (*Met.*). *Stress-number curve.*

sneak circuit (*Telecomm.*). In developing the design of a circuit, a part of it which obviates the required performance and has to be circumvented.

sneck (*Build.*). One of the smaller stones used to fill in between the larger ones in a snecked masonry wall. (*Join.*) The lifting lever which passes through a slot in a door and actuates the fall bar. See **Norfolk latch.**

snecked (*Build.*). Said of rubble walls in which the stones are roughly squared but of irregular size and uncoursed.

sneezewood (*For.*). South African tree *Ptaeroxylon utile*, giving a hard and strong timber. Its principal uses are for fencing posts, piling, poles, etc. It has a smell akin to pepper.

Snell's law (*Phys.*). A wave refracted at a surface makes angles relative to the normal to the surface, which are related by the law $n_1 \sin \theta_1 = n_2 \sin \theta_2$. n_1 and n_2 are the refractive indices on each side of the surface, and θ_1 and θ_2 are the corresponding angles. The two rays and the normal at the point of incidence on the boundary lie in the same plane.

Snettisham Clay (*Geol.*). A clay rock of marine origin found in the Lower Greensand series of the Cretaceous System in Norfolk.

snore piece (*Mining*). The lower part of the pipe which admits the water to a mine pump; the nose, wind bore, or suction pipe.

snorkel, snort. See **schnorkel.**

snow (*Meteor.*). Precipitation in the form of small ice crystals, which may fall singly or in flakes, i.e., tangled masses of snow crystals. The crystals are formed in the cloud from water vapour. (*Telecomm.*) Effect of electrical noise on display of intensity-modulation signals. The effect is random and resembles falling snow and may be seen, e.g., on a TV screen in the absence of a signal, or when the signal is weak.

snow boards (*Build.*). Horizontal boards, about 20 cm high, fixed over roof gutters to prevent snow from sliding off in mass. The snow, on melting, drops into the gutter through gaps left between the boards. Also called **gutter boards, snow guards.**

Snowdonian Lavas (*Geol.*). A thick series of rhyolitic lavas and ashes exposed on Snowdon in N. Wales; they belong to the Caradocian stage of the Ordovician System.

snow guards (*Build.*). See **snow boards.**

snow load (*Build.*). The unit loading assumed in the design of a roof to allow for the probable maximum amount of snow lying upon it.

snow stage (*Meteor.*). That part of the condensation process taking place at temperatures below 0°C so that water vapour condenses directly to ice.

S/N ratio (*Acous.*). Abbrev. for *signal/noise ratio, speech/noise ratio.*

snuffles (*Med.*). Symptoms resulting from nasal discharge and obstruction of the nose in infants, due usually to congenital syphilis. (*Vet.*) A term popularly applied to several diseases of animals in which the nasal cavities are affected. See **atrophic rhinitis** (*pig*), **osteodystrophia fibrosa** (*pig*), **rabbit septicaemia** (*rabbit*).

soakaway (*Build., San. Eng.*). A pit excavated to receive excess surface water which can duly seep into the surrounding ground.

soakers (*Plumb.*). Small pieces of sheet-lead or zinc bonded in for watertightness with the slates or tiles of a roof at joints with walls or at valleys and hips.

soaking (*Glass, Met., etc.*). A phase of a heating operation during which metal or glass is maintained at the requisite temperature until uniformly heated, and/or until any required phase transformation has occurred.

soaps (*Build.*). Bricks, $9 \times 2\frac{1}{4} \times 2\frac{1}{4}$ in. in size, which are often pierced for use as air bricks. (*Chem.*) (1) The alkaline salts of fatty acids, chiefly palmitic, stearic, or oleic acids. *Soft soaps* contain the potassium salts, whereas the sodium salts are *hard soaps*. *Metallic soaps*, water-insoluble compounds of fatty acids with bases of copper, aluminium, lithium, calcium, etc., are used as waterproofing agents and as the bases of many greases. (2) The alkali salts of resins, so-called *resin soaps*.

soapstone (*Min., etc.*). See **steatite.**

soaring (*Aero.*). The art of sustained motorless flight by the use of thermal upcurrents and other favourable air streams.

sobole (*Bot.*). A creeping underground stem which develops leaf buds and roots at intervals. *adj.* **soboliferous.**

social (*An. Behav.*). Living together; gregarious; living in organized colonies.

social facilitation (*An. Behav.*). A type of behaviour, found especially in Birds, in which the performance of a more or less instinctive behaviour pattern by one individual tends to act as a releaser for the same type of behaviour in other individuals of the same species. Also used, more generally, to describe any influence exerted by the nearby presence of another organism on the rate of performance or even the character of a physiological process or a behaviour pattern.

social hierarchy (*An. Behav.*). A state of affairs within a group in which some individuals are superior, some intermediate, and some inferior, this being shown by their behaviour to other individuals differently located in this hierarchy. Also **dominance hierarchy.** See **peck order.**

social parasitism (*An. Behav.*). A form of association between a colony of social animals and another species, in which the latter lives in the colony and either preys on the members or makes use of their food, without conferring any benefit in turn. *n.* **social parasite.**

social plants (*Bot.*). Species which grow many together and occupy wide areas.

social psychology (*Psychol.*). The study of the individual in society, including the study of the formation and structure of human groups, communication within them and their effects on the individual.

social releasers (*An. Behav.*). Stimuli provided in some way by an individual of a particular species which act as releasers for a social response by another animal of the same species. They are usually simple, conspicuous, and specific.

social symbiosis (*An. Behav.*). A form of association between a colony of social animals and another species which is mutually beneficial.

society (*Bot.*). A plant community of minor rank forming part of a consociation.

sociopathic personality disturbances (*Psychiat.*). Personality disorders involving inability or unwillingness to conform to the sociocultural norms and mores. The term is used in place of 'psychopathic personality' and refers to the disorders of antisocial reaction, dyssocial (amoral) reaction, and sexual deviation.

socket (*Elec. Eng.*). The female portion of a plug-and-socket connexion in an electric circuit. (*Mining*) (1) Portion of drill hole left undisturbed after blasting, and liable to contain unexploded dynamite. (2) Capping eye on end of steel winding rope. (*Plumb.*) (1) Enlarged end of a pipe, into which the end of another pipe of the same diameter can be fitted to form a joint between the two lengths. U.S., **bell** or **hub.** (2) Pipe fitting used to join two pipes in line.

socket chisel (*Join.*). A robust type of chisel used for mortising; it has a hollow tapering end to the steel shank into which the handle fits.

socket head screw (*Eng.*). One which has a hexagonal recess in its head, so that it can be turned by means of a hexagonal bar formed into a key.

socket spanner (*Eng.*). One with a recessed head, the sides of the recess having six or more flats to fit a hexagonal nut in various positions.

socle (*Arch.*). A plain projecting block or plinth at the base of a pedestal, wall, or pier.

soda-ash (*Min. Proc.*). Impure (commercial) sodium carbonate. Widely used in pH control of flotation process.

soda lakes (*Geol.*). Salt lakes the water of which has a high content of sodium salts (chiefly chloride, sulphate, and acid carbonate). These salts also occur as an efflorescence around the lakes.

soda-lime-silica glass (*Glass*). The commonest type of glass, used for windows, containers and electric lamp bulbs. Contains silica, soda, and lime in approx. proportions SiO_2 70%, Na_2O 15%, CaO 10%. *Crown glass* is of this type, but in *optical crown* barium oxide is now normally substituted for lime.

sodalite (*Min.*). A cubic feldspathoid mineral, essentially silicate of sodium and aluminium with sodium chloride, occurring in certain alkali-rich syenitic rocks.

sodamide (*Chem.*). $Na^+NH_2^-$; an ionic compound formed when ammonia gas is passed over hot sodium.

soda nitre (*Min.*). Nitrate of sodium, crystallizing in the trigonal system. It is found in great quantities in northern Chile, where beds of it are exposed at the surface and are known as *caliche*. Also called **Chile saltpetre, nitratine.**

soda recovery (*Paper*). Liquor resulting from the digestion of raw materials with caustic soda is concentrated, and organic matter burnt off in a furnace. The soda is recovered as soda-ash and causticized for further use.

soda-syenite (*Geol.*). A syenitic igneous rock containing an excess of sodium feldspar or feldspathoid. Cf. *potash-syenite.*

Soderberg electrode (*Elec. Eng.*). An amorphous carbon electrode manufactured by a continuous electric arc furnace process.

sodium (*Chem.*). Metallic element, one of the alkali metals. Symbol Na, at. no. 11, r.a.m. 22·9898, rel. d. 0·978, m.p. 97·5°C, b.p. 883°C. Sodium does not occur in nature in the free state, owing to its reactivity, but it is widely distributed combined as the chloride, nitrate, etc. Metallic sodium is made by electrolysis of fused caustic soda. It is a soft, silver-white metal, very active, with a moderate thermal neutron cross-section. Used in reactors as a liquid metal heat-transfer fluid.

sodium aluminate (*Chem.*). See **aluminate.**

Sodium Amytal (*Pharm.*). See **amylobarbitone.**

sodium chloride (*Chem.*). NaCl. See **halite.**

sodium-cooled reactor (*Nuc. Eng.*). One in which liquid sodium is used as the primary coolant.

sodium-cooled valves (*I.C. Engs.*). High-duty engine exhaust valves cooled by filling the hollow stem (and head) with sodium to about 60% of the volume.

sodium cyclamate (*Chem.*). The sodium salt of *N*-cyclohexylsulphamic acid (see **cyclamates**). A white, crystalline powder readily soluble in water. Used as an artificial sweetening material, e.g., in soft drinks and diabetic sweetening tablets, but banned in some countries because of possible health hazards in long-term use.

sodium hydroxide (*Chem.*). See **caustic soda.**

sodium nitrate (*Min.*). See **soda nitre.**

sodium thiosulphate (*Chem., Photog.*). See **hypo.**

sodium vapour lamp (*Light*). An electric lamp of the gaseous-discharge type whose electrodes operate in an atmosphere of sodium vapour.

sod oil (*Chem*). See **degras** (2).

sodoku (*Med.*). Rat-bite fever. A disease due to infection with the microorganism *Spirillum minus*, conveyed by the bite of a rat; it is characterized by inflammation of the skin around the bite, relapsing fever, swelling of the lymphatic glands, and a red, patchy, rash.

sofar (*Nav.*). System of underwater navigation depending on reception of pulses from distant explosions. (*So*und *f*ixing *a*nd *r*anging.)

soffit (*Build., Civ. Eng.*). (1) A term often used for *intrados* (q.v.), but more particularly applied to that part of the intrados in the immediate vicinity of the keystone. (2) The under surface of a stair or floor or of the head of an opening such as a door or window opening.

soft (*Electronics*). Said of valves and tubes when there is appreciable gas pressure within the envelope, such as gas-discharge tubes and photocells. Particularly said of valves when gas is released from the envelope or electrodes. See **hard, outgassing, Round valve.** (*Glass*) Having a relatively low softening point.

soft commissure (*Zool.*). In Mammals, the point at which the thickened sides of the thalamencephalon touch one another across the constricted third ventricle.

softening (*Met.*). A process for removing arsenic, antimony and tin from lead, after drossing. A bath of molten metal is oxidized by furnace gases and the addition of lead oxide. Impurities are oxidized and form a dross. Heating of metal towards its *arrest point* (q.v.), followed by slow cooling. Also called **improving.**

softening-point test (*Build., Civ. Eng.*). See **melting-point test.**

soft-focus lens (*Photog.*). A lens for producing slightly diffused outlines.

soft iron (*Met.*). Iron low in carbon and unable to retain magnetism; useful as solenoid cores, etc.

soft-iron armature (*Elec. Eng.*). The attracted part of an electromagnet retaining little residual magnetism.

soft-iron instrument (*Elec. Eng.*). An undesirable synonym for *moving-iron instrument*.

soft knitting yarn (*Textiles*). (1) A four-ply worsted yarn for knitting stockings and socks. (2) Cotton and/or mixture yarns with low turns per inch, intended for hosiery fabrics, e.g., vests.

softness (*Met.*). Tendency to deform easily. It is indicated in tensile test by low ultimate tensile stress and large reduction in cross-section before fracture. Usually the elongation is also high. In notched bar test, specimens bend instead of fracturing, and energy absorbed is relatively small. See **brittleness, toughness.**

soft palate (*Zool.*). In Mammals, the posterior part of the roof of the buccal cavity which is composed of soft tissues only.

soft radiation (*Radiol.*). General term used to describe radiation whose penetrating power is very limited, e.g., low frequency X-rays.

soft-shelled egg (*Vet.*). A bird's egg in which lime salts are deficient or absent.

soft soaps (*Chem.*). See **soaps.**

soft solder (*Met.*). Alloys of lead and tin used in soldering. Tin content varies from 63% to 31%. The remainder is mainly lead, but some types contain about 2% antimony and others contain cadmium. The best-known types are *tinman's solder* and *plumber's solder* (qq.v.).

software (*Comp.*). General term for programming or compiling accessories used for computing or data-processing systems. Cf. *hardware*.

soft water (*Chem.*). Generally, water free from calcium and magnesium salts (cf. *hard water*), naturally or artificially. See **water softening.**

soil (*Agric.*). The uppermost layer of the earth, which supports the growth of plant life. Soils may be heavy clay, loam (natural mixture of sand and clay), peaty (excess of decaying leafy matter), light or sandy (chiefly silica). The loamy type is most generally useful for agriculture, and other types may be corrected by the admixture of clay, manure, or fertilizer with the sandy types, and by breaking up the clay with the subsoil for weathering. Sour soil (excess acidity) is corrected by lime dressings. (*Plumb.*) See **smudge.**

soil flora (*Bot.*). Plants, chiefly fungi, living in the soil.

soil mechanics (*Civ. Eng.*). The process of determining the properties of any soil, e.g., water content, bulk, density, permeability, shear strength, etc.

soil pipe (*Build., San. Eng.*). A vertical cast-iron or plastic pipe conveying waste matter from W.C.s, etc., to the drains. Abbrev. **S.P.**

soil sampler (*Civ. Eng.*). A hollow circular tool, with a sharp edge, for extracting specimens of soil for examination or analysis. Also **soil borer, soil pencil.**

sol (*Chem.*). A colloidal solution, i.e., a suspension of solid particles of colloidal dimensions in a liquid.

solanine bases (*Chem.*). Alkaloid bases derived from the Solanum genus, e.g., *Atropa belladonna.*

solar (*Zool.*). Having branches or filaments radially arranged.

solar antapex (*Astron.*). The point on the celestial sphere diametrically opposite to the *solar apex*.

solar apex (*Astron.*). The point on the celestial sphere towards which the solar system as a whole is moving at the rate of 20 km/s. It is located in the constellation Hercules in equatorial coordinates R.A. 271° and declination +30° approximately.

solar attachment (*Surv.*). See **shade.**

solar cell (*Electronics*). Photoelectric cell using silicon, which collects photons from the sun's radiation and converts the radiant energy into electrical energy with reasonable efficiency. Used in artificial satellites and for remote locations lacking power supplies, e.g., for telephone amplifiers in the desert.

solar constant (*Phys.*). The rate of reception, per unit area, of energy normal to the earth's surface, corrected for loss by absorption in the earth's atmosphere. The value (which is not constant) is about 1340 watts per metre squared.

solar corona (*Astron.*). A halo, pearly white in colour, surrounding the sun; its full extent can be seen during a total solar eclipse; the inner parts can be studied at any time with a *coronagraph* (q.v.).

solar day (*Astron.*). See **apparent-, mean-.**

solar eclipse (*Astron.*). See **eclipse.**

solar flare (*Astron.*). Bright eruption of the sun's surface associated with sun spots, causing intense radio and particle emission.

solar flocculi (*Astron.*). The name given to small bright and dark markings on the solar chromosphere as seen on calcium or hydrogen spectroheliograms.

solar gain (*Build.*). Heat gain in a building due entirely to the sun.

solar granulation (*Astron.*). The mottled appearance of the sun's photosphere, small bright granules (about 1000 km in diameter) being seen in rapid change against the darker background.

solarimeter (*Phys.*). See **pyranometer.**

solarization (*Bot.*). The temporary stoppage of photosynthesis in a leaf when it is exposed for a long time to bright light. (*Glass*) Changes of the light transmission properties of glass as a result of exposure to sunlight or other radiation in or near the visible spectrum. (*Photog.*) The reversal of an image because of excessive exposure to light, i.e., a supposed negative appears to be a positive.

solar oil (*Fuels*). See **gas oil.**

solar parallax (*Astron.*). The mean value of the angle subtended at the sun by the earth's equatorial radius. It cannot be measured directly, but is derived from Kepler's laws when the distance between the earth and any other planet is known. The value of 8″·80 for the solar parallax has been in use for half a century, but recent measures of the distance of Venus by radar in 1961 gave the value 8″·794 05, corresponding to a value of 92 957 130 miles ($149 \cdot 6 \times 10^6$ km) for the astronomical unit.

solar plexus (*Zool.*). In higher Mammals, a ganglionic centre of the autonomic nervous system—situated in the anterior dorsal part of the abdominal cavity—from which nerves radiate in all directions.

solar radio noise (*Astron.*). A hissing noise on short-wave radio, due to solar radiation; an almost constant background noise is greatly increased at times of solar disturbances such as sunspots and flares. The origin of the radiation lies in the chromosphere and the corona.

solar rotation (*Astron.*). The rotation of the sun takes place in the same direction as the orbital motion of the planets. The rotation is non-uniform, taking 24·65 days at the sun's equator, but increasing to about 34 days near the poles. Because of the earth's motion, the equatorial solar rotation has a synodic period of 27 days, and this interval is apparent in the recurrence of magnetic storms, aurorae, etc.

Solar System (*Astron.*). The term designating the sun and the attendant bodies moving about it under gravitational attraction; comprises nine major planets, and a vast number of asteroids, comets, and meteors.

solar wind (*Astron.*) A constant plasma stream of protons moving radially away from the sun at a speed of 250–800 km/s. It affects the earth's magnetic field and also causes acceleration in the tails of comets.

solation (*Chem.*). The liquefaction of a gel.

solder (*Met.*). A general term for alloys used for joining metals together. The principal types are *soft solder* (lead-tin alloys) and *brazing solders* (alloys of copper and zinc).

solder-covered wire (*Elec. Eng.*). Copper wire coated with solder instead of tin, to facilitate connections between components.

soldered dot (*Plumb.*). A device for fixing sheet-lead to woodwork. A hollow formed in vertical boarding which is covered with sheet-lead, secured in the hollow by splayed screws, the hollow being then filled with solder.

soldering (*Eng.*). Hot joining of metals by adhesion using, as a thin film between the parts to be joined, a metallic bonding alloy having a relatively low melting point.

soldering iron (*Plumb.*). See **copper bit.**

solder paint (*Eng.*). Mixture of powdered solder and flux, applied by brush to surfaces to be joined by *soldering* to form the bonding film.

soldier (*Build.*). A term applied to a course of bricks laid so that they are all standing on end. (*Civ. Eng.*) Vertical member of timber or steel inserted to prop walings when struts have to be removed to facilitate working. (*Zool.*) In some social Insects, a form with especially large head and mandibles adapted for defending the community, for fighting, and for crushing hard food particles.

sole (*Carp., Join.*). The lower surface of the body of a plane. (*Eng.*) (1) The bed plate of a marine engine; secured through bearers to the hull of the ship. Also **sole plate.** (2) A timber base for supporting the feet of *raking shores* (q.v.). (3) The plate supporting a leg of a process vessel.

soleil (*Textiles*). A worsted dress fabric, with a fine cord of the Ottoman rib type formed from twist-way and weft-way yarns.

Solenhofen stone (*Geol.*). An exceedingly fine and even-bedded limestone, thinly stratified, of Upper Jurassic age, occurring in S.E. Bavaria; formerly widely used in lithography.

solenia (*Zool.*). In some hydroid colonies, diverticula of the enteron formed by hollow strands of endoderm.

Solenicthyes (*Zool.*). An order of *Teleostei* in which the mouth is situated at the tip of a long tubular snout; marine forms occurring in all tropical and temperate waters. Shrimp-fishes, Cornet-fishes, Snipe-fishes, Pipe-fishes, Sea-horses.

solenocyte (*Zool.*). In Invertebrates and lower *Chordata*, an excretory organ consisting of a hollow cell with branched processes, in the lumen of which occurs a bunch of cilia which by their movements maintain a downward current.

solenoid (*Elec. Eng.*). Current-carrying coil, of one or more layers. Usually a spiral of closely wound insulating wire, in the form of a cylinder, not necessarily circular. Generally used in conjunction with an iron core, which is pulled into the cylinder by the magnetic field set up when current is passed through the coil.

solenoidal field (*Elec. Eng.*). One in which divergence is zero, and the vector is constant over any section of a tube of force.

solenoidal magnetization (*Mag.*). The distribution of the magnetization on a piece of magnetic material when the poles are at the ends. Cf. *lamellar magnetization.* Also called **circuital magnetization.**

solenoid brake (*Elec. Eng.*). An electromechanical brake in which the brake toggle is operated by the plunger of a solenoid.

solenoid-operated switch (or circuit-breaker) (*Elec. Eng.*). A switch in which the closing force is provided by a solenoid. Cf. *pneumatically-operated switch, motor-operated switch.*

solenoid relay (*Elec. Eng.*). A relay in which the contacts are closed by the action of a solenoid-operated plunger.

solenostele (*Bot.*). A form of siphonostele in which there is an endodermis both outside and inside the tube of vascular tissues.

sole piece (*Build.*). The plate to which the feet of the shores, in a system of raking shores, are secured, and which forms an abutment for them at their lower ends. (*Ships*) See **stern frame.**

sole plate (*Eng.*). See **sole.**

soleus (*Zool.*). A muscle of the lower part of the hind limb in *Tetrapoda.*

solfatara (*Geol.*). The name applied to a volcanic orifice which is in a dormant or decadent stage and from which gases (especially sulphur dioxide) and volatile substances are emitted.

solid (*Chem.*). A state of matter in which the constituent molecules or ions possess no translational motion, but can only vibrate about fixed mean positions. A solid has a definite shape and offers resistance to a deforming force. (*Maths.*) A figure having 3 dimensions.

solid angle (*Maths.*). Of an area subtended at a point, the area intercepted on a unit sphere, centred at the point, by a cone having the given area as its base and the point as its vertex. See **steradian.**

solid carbon (*Light.*). A carbon electrode in which a core of softer material is absent.

solid circuit (*Electronics*). (1) Modification of properties of a material, e.g., silicon, so that components can be realized in one mass, e.g., resistors, capacitors, transistors, diodes. (2) Subminiature realization of a circuit in three dimensions, e.g., as built up as parts of a semiconductor crystal or by etching or deposition on a substrate.

solid diffusion (*Met.*). Movement of atoms through the crystals of a solid metal, as when carbon diffuses into or out of steel during carburizing or decarburizing respectively.

solid floor (*Build.*). A floor made of wood blocks laid on a concrete subfloor.

solid head (*I.C. Engs.*). A cylinder or cylinder block cast in one piece, as distinct from one with a detachable head.

solidification range (*Chem.*). The range of temperature in which solidification occurs in alloys and silicate melts, etc., other than those which freeze at constant temperature. It extends from a point on the liquidus to one on the solidus.

solid injection (*Eng.*). See **airless injection.**

solidium (*Arch.*). The body of a pedestal.

solid-laid cable (*Cables*). See **solid system of cable-laying.**

solid matter (*Typog.*). Type-matter set up without leads between the lines; or, in mechanical typesetting, type-matter cast on its own body size rather than on a larger one.

solid newel (*Build.*). The centre post of a winding stair, as distinct from a *hollow newel* (q.v.).

solid panel (*Join.*). A panel whose surface is in line with the faces of the stiles.

solid pole (*Elec. Eng.*). A field pole of an electrical machine which is not built up from laminations.

solid propellant (*Space*). Rocket propellant in solid state, usually in caked, plasticlike form, comprising a fuel-burning compound of fuel and oxidizer.

solid solubility (*Met.*). The extent to which one metal is capable of forming solid solutions with another.

solid solution (*Chem., Phys.*). Usually a *primary solid solution*, but the term may also be applied to the case when an intermediate constituent dissolves one of its components.

solid-state capacitor (*Electronics*). That presented by a depletion layer at a *p-n* junction, in which the carriers from each side substantially neutralize each other at low voltages. Effective capacitance depends on applied voltage.

solid-state maser (*Electronics*). One made from ruby at a few degrees above absolute zero in intense magnetic fields. A pulse of microwaves raises energy of electrons to a high level so that the crystal acts as an amplifier.

solid-state physics (*Phys.*). Branch of physics which covers all properties of solid materials, including electrical conduction in crystals of semiconductors and metals, superconductivity and photoconductivity.

solid system of cable-laying (*Cables*). A system of cable-laying in which the cables are laid in troughing in an open trench. The troughing is of stoneware, cast-iron, asphalt, or treated wood, and is filled with a bituminous or asphaltic compound.

solid-type cable (*Cables*). See **straight-type cable.**

solidus (*Chem.*). A line in a constitutional diagram indicating the temperatures at which solidification is completed, or melting begins, in alloys and other melts of different composition. See **liquidus, solidification range.** (*Typog.*) The diagonal or oblique stroke.

solifluction, solifluxion (*Geol.*). Soil-creep on sloping ground, characteristic of, though not restricted to, regions subjected to periods of alternating freezing and thawing.

Solifuga (*Zool.*). An order of *Arachnida* with the body divided into a prosoma and opisthosoma, but no pedicel. The body and limbs are very hairy, there is no tail, the pedipalps have a small sucker and there is a well developed tracheal system. Tropical and subtropical.

solitaria phase (*Zool.*). One of the two main phases of the Locust (*Orthoptera*) (see **Uvarov's phase theory**), which occurs when nymphs are reared in isolation. They adjust their colour to match their background and lack the higher activity and gregarious tendencies of the *gregaria phase* (q.v.).

sollar (*Build.*). A loft which is open to the sun. (*Mining*) Landing stage between two ladders in a shaft. Also **solar, soller.**

Sollas' centre (*Zool.*). The centre or morphological centre of form in the skull of *Primates*.

solochrome black (*Chem.*). Indicator for the complexometric titration of both calcium and magnesium ions in hard water (red colour) with *EDTA* (q.v.). Often used in parallel with *murexide* (q.v.), which responds to calcium only.

solstices (*Astron.*). The two moments in the year when the sun in its apparent motion attains its maximum distance from the celestial equator; *summer solstice* (about June 21st) occurs when the sun's apparent longitude is 90°; *winter solstice* (about December 22nd) occurs when the sun's apparent longitude is 270°.

solubility (*Chem.*). The mass of a dissolved substance which will saturate 100 g of a solvent under stated conditions.

solubility curve (*Chem.*). The curve showing the variation of the solubility of a substance with temperature.

solubility product (*Chem.*). The value, at saturation, of the product of the activities (concentrations in mol/dm^3) of the ions into which a dissolved substance dissociates.

soluble oil (*Eng.*). Cutting fluid consisting of oil and an emulsifier to which water is added.

soluble-RNA (*Biochem.*). See **s-RNA.**

soluble starch (*Chem.*). A product of the hydrolysis of starch obtained by treating starch with dilute acids, or by boiling with glycerine, or by the action of diastase.

solute (*Chem.*). A substance which is dissolved in another.

solution (*Bot.*). The abnormal separation of parts normally united. (*Chem.*) Homogeneous mixture of molecules, e.g., of gas in liquid, solid in liquid, gas in solid, etc. Simple chemical compounds ionize when dissolved in water, radicals or atoms separating with numbers of electrons which differ from their neutral complement, depending on their valency. Such ions bunch together and regulate the conductivity and acidity. See **pH value.**

solution mining (*Mining*). Winning of soluble salts by use of percolating liquor introduced through shafts, drives and/or bores. Resulting saturated solution is pumped to surface for further treatment.

solution pressure (*Chem.*). The tendency of a substance to pass into solution.

solution treatment (*Met.*). The operation of heating suitable alloys (e.g., duralumin) in order to take the hardening constituent into solution. This is followed by quenching, to retain the solid solution, and the alloy is then age-hardened at atmospheric or elevated temperature.

solutizer process (*Chem. Eng.*). Process for removing mercaptans from petroleum fractions by 2-stage treatment with caustic soda and sodium cresylate solution.

solvation (*Chem.*). The association or combination of molecules of solvent with solute ions or molecules.

Solvay's ammonia soda process (*Chem.*). A process based on the fact that when a concentrated solution of sodium chloride is saturated with ammonia, and carbon dioxide is passed through, sodium hydrogen carbonate is precipitated and ammonium chloride remains in solution. Used for the manufacture of sodium carbonate from chloride.

solvent (*Chem.*). That component of a solution which is present in excess, or whose physical state is the same as that of the solution. (*Paint.*) A liquid capable of dissolving the *binder* (q.v.) and added to paint to make it work more freely.

solvent extraction (*Min. Proc.*). In chemical extraction of values from ores or concentrates, selective transfer of desired metal salt from aqueous liquor into an immiscible organic liquid after intimate stirring together followed by phase separation.

solventless coatings (*Paint.*). Compositions of brushing viscosity which 'dry' by the reaction of two parts, mixed immediately before use. The usual system is a pigmented *epoxy resin* cured with a mixture of phenols and amines. Flow is controlled by including bentones in the pigmented part to impart structure.

solvent naphtha (*Chem.*). Middle and high-boiling benzene hydrocarbons chiefly consisting of toluene and xylene, obtained from the fractionation of light tar oils after the benzene fractions have been distilled off.

solvent tanning (*Leather*). Tanning with a quick-working solvent such as acetone.

solvolysis (*Chem.*). See **lyolysis.**

solvsbergite (*Geol.*). An alkaline igneous rock; a porphyritic microsyenite.

soma (*Zool.*). The body of an animal, as distinct from the germ cells; cf. *germen. pl.* **somata.** *adj.* **somatic.**

somactids (*Zool.*). In Fish, cartilaginous or bony rods placed in the basal part of the median fin folds and supporting the dermotrichia.

somascope (*Med.*). An instrument using ultrasonic waves converted into a television image to show the character of diseased internal tissues of the body.

somatic apogamy (*Bot.*). The development of a sporophyte, having nuclei with the zygotic number of chromosomes, from a cell or cells of the gametophyte, the fusion of gametes being omitted.

somatic cell (*Zool.*). One of the nonreproductive cells of the parent body, as distinct from the reproductive or germ cells.

somatic doubling (*Cyt.*). A doubling of the number of chromosomes in the nuclei of somatic cells.

somatic mitosis (*Cyt.*). Division of the metabolic nucleus.

somatic mutation (*Gen.*). A mutation arising in a somatic cell and not in a reproductive structure.

somatic segregation (*Bot., Gen.*). A change in nuclear or hereditary constitution during vegetative growth.

somatoblast (*Zool.*). In development, a cell which will give rise to somatic cells: in developing *Chaetopoda*, a micromere (d_2) which divides after gastrulation to form the ventral plate.

somatocyst (*Zool.*). In some *Siphonophora*, a dilatation at the upper end of the central canal of a nectocalyx, sometimes with an oil drop.

somatogenic (*Zool.*). Arising as the result of external stimuli: developing from somatic cells, as opposed to germ cells.

somatopleure (*Zool.*). The outer body wall of coelomate animals: the outer layer of the mesoblast which contributes to the outer body-wall; cf. *splanchnopleure. adj.* **somatopleural.**

somatotropin (*Biochem.*). See **growth hormone.**

somatotropism (*Bot.*). Directed growth movements in plants so that the members come to be placed in a definite position in relation to the substratum.

somatotypes (*Psychol.*). Types of physique (Sheldon), based on anthropometric measurements, associated with various types of temperament. There are three types, each measured on a seven-point scale, all used in the description of any individual. See **ectomorphy, endomorphy, mesomorphy.**

somite (*Zool.*). One of the divisions or segments of the body in a metameric animal: a mesoblastic segment in a developing embryo.

sommaite (*Geol.*). An alkaline igneous rock, similar to essexite but with leucite in place of nepheline.

Sommerfeld atom (*Nuc.*). Atomic model developed from Bohr atom, but allowing for elliptic orbits with radial, azimuthal, magnetic, and spin quantum numbers. Modern theories modify this by regarding the electrons as forming a cloud, the density of which is described in terms of their wave function.

sommering lines (*Civ. Eng.*). The radiating lines giving the direction of the bed-joints of the voussoirs of an arch.

somnambulism (*Med.*). (1) The fact or habit of walking in the sleep. (2) A hysterical state of automatism in which the patient performs acts of which he is unaware at the time or when he comes out of the state.

sonar (*Acous.*). See **asdic, echo sounding.**

sonde (*Meteor.*). Small telemetering system in satellite, rocket, or balloon.

sone (*Acous.*). Unit of loudness equal to a tone of 1 kHz at a level of 40 dB above the threshold of the listener.

Soneryl (*Pharm.*). See **butobarbitone.**

sonic boom (*Aero.*). Noise phenomenon due to the shock-waves projected outwards and backwards through the atmosphere from leading and trailing edges of an aircraft travelling at *supersonic speed* (q.v.). The waves are discontinuities of atmospheric pressure and are heard as a characteristic double report which may be of sufficient intensity to cause damage to buildings, etc.

sonic line (*Phys.*). The locus of field points in 2-dimensional flow where the medium attains the velocity of sound under local conditions.

sonics (*Phys.*). General term for study of mechanical vibrations in matter.

sonims (*Met.*). Solid nonmetallic inclusions in metal.

sonne (*Nav.*). German navigational system, forerunner of *consol* and *consolan* (qq.v.).

Sonne dysentery (*Med.*). Dysentery caused by bacteria described by Carl Sonne and differing from the usual dysentery bacilli described by Shiga and by Flexner.

sonobuoy (*Radio*). Equipment dropped and floated on the sea, to pick up aqueous noise and transmit a bearing of it to aircraft; three of such bearings enable the aircraft to 'fix' the source of underwater noise, e.g., from submarines.

sonometer (*Acous.*). See **monochord.**

soralium (*Bot.*). A group of soredia surrounded

by a distinct margin formed from the thallus of the lichen.

soral membrane (*Bot.*). The wall surrounding the sorus in some lower fungi.

sorbic acid (*Chem.*). $CH_3\cdot(CH:CH)_2\cdot COOH$. Hexa-2,4-dienoic acid. Solid, m.p. 133–135°C. Used as a preservative for pharmaceuticals, cosmetics, etc. Anti-fungal for edible products.

sorbite (*Met.*). Decomposition product of hardenite (cementite and ferrite in solid solution), of heterogeneous structure, thus distinguishing it from *pearlite* (q.v.).

sorbitol (*Chem.*). A hexahydric alcohol, isomeric with mannitol.

sorbose (*Chem.*). A ketohexose, an isomer of fructose.

sordes (*Med.*). Foul, dark brown crusts which collect on the lips and the teeth in prolonged fever (e.g., in typhoid fever).

soredial branch (*Bot.*). A branch of a lichen thallus formed by a soredium beginning to develop while still attached to the parent thallus.

sorediate (*Bot.*). Having small patches on the surface.

soredium (*Bot.*). One or more algal cells enclosed in hyphae, forming a tiny mass which separates from the lichen thallus and gives rise to a fresh thallus if transported to a suitable place for growth.

sore-heels (*Vet.*). See **grease.**

Sorel's cement (*Build.*). Calcined magnesite (MgO) mixed with a solution of magnesium chloride of a concentration of about 20° Baumé. It sets within a few hours to a hard mass. The basis of artificial flooring cements.

Sørensen's formol titration (*Chem.*). See **formol titration.**

sore-shins (*Vet.*). An inflammation of the periosteum of the large metacarpus (shin bone), or occasionally of the metatarsus, of young horses.

Soret effect (*Heat*). See **thermal diffusion.**

soriferous (*Bot.*). Bearing sori.

sorosilicate (*Min.*). A silicate mineral whose atomic structure contains paired silicon-oxygen tetrahedra (Si_2O_7 groups), e.g. melilite.

sorosis (*Bot.*). A fleshy fruit formed from a number of crowded flowers, such as a pineapple.

sorosphere (*Bot.*). A hollow ball of spores formed by some lower plants.

sorption (*Chem.*). A general term for the processes of absorption, adsorption, chemisorption, and persorption.

sorting (*Build.*). The process adopted when a roof is to be covered by slates of different sizes; the largest slates are nailed at the eaves and the smallest at the ridge. (*Maths.*) The process of selecting the elements of a set with a common attribute from a random collection. (*Textiles*) Selecting yarns for doubling, winding, warping, and weaving, so that there may be no appreciable difference in types or shade.

sort mechanics (*Civ. Eng.*). The process of determining properties of any sort, e.g., water content, bulk, density, permeability, shear strength, etc.

sorts (*Typog.*). Particular type letters as distinct from complete founts. A case is 'out of sorts' when one or more of the boxes is empty. 'Outside sorts' are required for mathematical and foreign language setting.

sorus (*Bot.*). (1) In Lichens, a powdery mass of soredia lying on the surface of the thallus. (2) In Fungi, Ferns, etc., a group of sporangia usually accompanied by some protective structures.

Sothic cycle (*Astron.*). A period of 1460 years, familiar in ancient Egyptian chronology as the time in which the Egyptian calendar year of 365 days precessed through the seasons; so called because the Egyptian new year began with the heliacal rising of the star Sirius, whose Egyptian name was Sothis.

souffle (*Med.*). A murmuring, blowing sound, especially a murmur heard over the heart.

sough (*Civ. Eng.*). A drain at the foot of a slope, e.g., an embankment, to receive and carry away surface waters from it.

sound (*Acous.*). (1) The periodic mechanical vibrations of a medium. (2) The sensation felt when the eardrum is acted upon by air vibrations if within a limited frequency range, i.e., 20 Hz to 20 kHz or less. (*Med.*) A solid rod used for exploring hollow viscera (e.g., the bladder, the uterus) or for dilating stenosed passages.

sound absorption factor (*Acous.*). See **acoustic absorption factor.**

sound analyser (*Acous.*). One which measures each frequency, or a small band of frequencies, in the spectral distribution of energy. Particularly useful for tracing sources of vibration or noise in rotating equipment.

sound articulation (*Acous.*). Percentage of all elementary speech sounds received correctly, when logatoms are called over a circuit or in an auditorium, in a standard manner.

sound boarding (*Carp.*). Boards fitted in between the joists of a floor, to carry the *pugging* (q.v.) which is to insulate the room from sound and smell from the room below.

sound camera (*Cinema.*). The machine in which photographic film is exposed for registration of the sound record in motion-picture production. Except in newsreel cameras, this machine is entirely distinct from the camera taking the scene.

sound carrier (*Telecomm.*). Radiofrequency electromagnetic wave, modulated at a sound frequency and carrying the information, to be reconverted into sound vibrations at the receiver.

sound channel (*TV*). The carrier frequency with its associated sidebands which are involved in the transmission of the sound in TV.

sound energy density (*Acous.*). Density of energy in a diffused sound field, in ergs/cm³ or J/m³, averaged over a volume, in reverberation time calculations.

sounder (*Ocean.*). Any instrument used for determining the depth of the sea, e.g., *Lucas sounder* (by line and weight), *Kelvin sounder* (q.v.). See **echo sounding.** (*Teleg.*) A telegraph-receiving instrument in which Morse signals are translated into sound signals which are determined by the time intervals between two sounds.

sound field (*Acous.*). (1) Region through which sound waves, standing or progressive, propagate from a source. Such fields diverge from point, cylindrical, or plane sources. (2) Enclosed space in which the diffused sound waves are random in magnitude, phase, and direction, constituting reverberant sound.

sound-gate (*Cinema.*). The precise location in a sound-head where the sound-track is scanned by the constant focused beam of light from the exciter lamp, the transmitted light being received into the photoelectric cell.

sound-head (*Cinema.*). That unit in a projector which reproduces the sound-track on the edge of the film.

sounding (*Surv.*). The depth of an underwater

point below some chosen reference datum. Cf. *reduced level.*

sounding balloon (*Meteor.*). A small free balloon carrying a meteorograph, used for obtaining records of temperature, pressure, and humidity in the upper atmosphere.

sounding line (*Surv.*). A stout cord, divided into fathoms and feet and weighted at one extremity with a lead weight; used in finding soundings.

sound insulation (*Build.*). The property, possessed in varying degrees by different materials, of blocking the transmission of sound.

sound intensity (*Acous.*). Flux of sound power through unit area normal to the direction of propagation. If p=r.m.s. excess pressure, ρ= density, c=velocity of propagation, then the intensity is given by

$$I=p^2/\rho c.$$

sound intensity level (*Acous.*). At any audio-frequency, the intensity of a sound, expressed in decibels above an arbitrary level, 10^{-12} W/m², which is equivalent, in air, to a pressure of 20 μPa. Also **sound level.**

sound interval (*Acous.*). The interval between two sounds is the ratio of their fundamental frequencies or its logarithm.

sound level (*Acous.*). Same as **sound intensity level.**

sound-level meter (*Acous.*). Microphone-amplifier-indicator assembly which indicates total intensity in decibels above an arbitrary zero. With suitable weighting networks, the indicator gives approximately the level on the *phon scale*, at least for the comparison of sounds of a similar class. See **audiogram.**

sound locator (*Acous.*). Apparatus for determining the direction of arrival of sound waves, particularly the noise from aircraft and submarines.

sound pick-up (*Acous.*). Used loosely for part of a sound-reproducing system, such as a microphone, the sound-head in a projector, or a reproducer of gramophone recordings.

sound-picture spoilation (*TV*). The leakage of the sound signal of a TV transmission into the vision circuit of a receiver, giving rise to alternate dark and light horizontal bands in the screen picture.

sound pressure (*Acous.*). R.m.s. value of the instantaneous pressure exerted by the sound wave over an integral number of periods. The unit is the pascal, microbar, or N/m². Also **excess pressure.**

sound pressure level (*Acous.*). Sound pressure expressed in decibels relative to a reference pressure which is taken as 20 μPa or 2×10^{-5} N/m².

sound probe (*Acous.*). Usually a very small microphone to minimize the disturbance of the sound field it is being used to measure. It is often equipped with a fine tube which is alone inserted in the field.

sound ranging (*Acous.*). Determination of locality of a source of sound, e.g., from guns, by simultaneously recording through spaced microphones and making deductions from the differences of times of arrival.

sound recording (*Acous.*). The practice of registering sound so that it can be reproduced at some subsequent time. See **recording, recorder.**

sound-reflection factor (*Acous.*). The percentage of energy reflected from a large plane surface of uniform material on the incidence of a sound wave at a specified angle.

sound reinforcement system (*Acous.*). One used to increase the uniform sound intensity in large halls by employing public address systems. Also by the control of reverberation time using an electronic method, involving microphones, resonators, and loudspeakers.

sound-reproducing system (*Acous.*). That comprising sound recording, transducers and equipment for sound reproduction. For *monaural* reproduction, the system comprises one or more microphones with pre-amplifiers, mixers, amplifiers and loudspeakers. In *binaural* reproduction, two microphones are used to simulate positions of the ears of a human and reproduction of each component is through a separate earphone heard by one ear only.

sound spectrograph (*Acous.*). An electronic instrument which makes a graphic representation (a *sound spectrogram*) of qualities of a sound, e.g., frequency, intensity, etc.

sound stage (*Cinema.*). The main floor of a motion-picture studio on which sets are built and the artists perform during shooting.

sound-track (*Cinema., etc.*). Track on magnetic tape or ciné film on which sound signals have been or can be recorded. *Optical tracks* may use variable density or variable area modulation.

sound velocity (*Phys.*). See **velocity of sound.**

souple (*Textiles*). Silk yarns or fabrics from which the sericin has not been removed; they are not so lustrous as scoured silks. See **boiling-off.**

source (*Elec. Eng.*). An active pair of terminals, which can deliver power to a load. (*Nuc.*) See **radiator.**

source computer (*Comp.*). That which is used to compile the source program, if this is different from the one which processes the object program. Cf. *object computer.*

source deck (*Comp.*). Stack of program cards ready to insert into compiler of some computers operated by punched cards.

source impedance (*Elec. Eng.*). See **output impedance.**

source range (*Nuc. Eng.*). See **start-up procedure.**

source resistance (*Elec. Eng.*). See **internal resistance.**

sources of neutrons (*Nuc.*). Fast neutrons are obtained by (i) nuclear transformations and (ii) fission in nuclear reactors. To obtain slow neutrons, a moderator such as paraffin wax is used. For laboratory sources, see **neutron source.**

source strength (*Radiol.*). Activity of radioactive source expressed in disintegrations per second, curies, or, less usually, röntgen-hour-metre value.

souring (*Textiles*). The treatment of cotton yarn or cloth with dilute sulphuric or hydrochloric acid as a preliminary to bleaching.

Southern Cross (*Astron.*). A striking constellation of the southern hemisphere, visible only in latitudes below 30° N. It is a cruciform group of 4 stars, having the 2 bright stars α and β Centauri some way to the east, which makes it easy to identify.

Southerndown Beds (*Geol.*). A local series of massive and in part conglomeratic limestones found in S. Wales and forming part of the Lower Lias; deposited as coastal deposits near islands of Carboniferous Limestone rising through the Liassic Seas.

Southey's tubes (*Med.*). Small cannulas which, inserted into oedematous tissues, drain off the excess fluid.

southing (*Astron.*). The word used of a star crossing the meridian at upper culmination; applicable only in the northern hemisphere, where, unless the star be north of the zenith, it will be due south of the observer at the moment of transit. (*Surv.*) Measured difference southward from a reference latitude.

south pole (*Geog.*). See **terrestrial poles.** (*Mag.*) See **pole.**

Soxhlet apparatus (*Chem.*). A laboratory apparatus for the continuous extraction of a solid substance with a solvent, consisting of a distillation flask, a reflux condenser, and a cylindrical vessel fitted between them, to which a syphon system is attached.

soya-bean, soybean (*Nut.*). The seed of *Glycine soja* and *G. hispida*, and numerous other species. The oil (ca. 18%) is used in soaps, paints, varnishes, and rubber substitutes, while the high protein content (ca. 40%) makes it a valuable foodstuff (said to be capable of still greater development).

Sp. (*Chem.*). An abbrev. for *spirit*.

S.P. (*San. Eng.*). Abbrev. for *soil pipe*. (*Elec. Eng.*) Abbrev. for *single pole*.

space. Continuous and boundless extension in which extended things may exist and move: the distance between points or objects, whether regarded as filled or unfilled. (*Teleg.*) The period of time in transmission during which the key is open, i.e., not in contact. (*Typog.*) A type less than type height, and thinner than a quadrat; used to separate words, etc.

space bands (*Typog.*). On the Linotype and Intertype machines, the steel wedges between each word, which are pushed upwards until they expand the line of matrices to the full measure.

space box (*Typog.*). A small box with a few divisions for different spaces, used when making corrections to type matter.

space charge (*Electronics*). Collection or cloud of electrons near a source, e.g., heated cathode, which stabilizes emission by repelling those not attracted to another electrode, the field in its region being very low.

space-charge grid (*Electronics*). One between control grid and cathode of a valve (*bigrid*), to reduce the space charge, so reducing voltages necessary on the anode to operate the valve.

space-charge limitation (*Electronics*). Condition in a thermionic valve when electron current leaving a cathode is limited by balance between attractive electric forces from other electrodes and repulsion within space charge. See **filament limitation.**

space-charge pentode (*Electronics*). See **beam-power valve.**

space contact (*Teleg.*). The contact which is made and which retains the telegraphist's key when it is not depressed; by extension, the contact which is made by a machine when sending spacing current to line. Cf. *mark contact*.

space current (*Electronics*). See **thermionic current.**

spaced antennae (*Radio*). Those used with diversity systems, or to enhance directivity.

spaced-loop direction-finder (*Radio*). One including two loops spaced sufficiently in terms of the wavelength to enhance their normal directivity, as exhibited by the polar diagram of response.

space draft (*Weaving*). Entering the warp threads in the healds, harness, and reed, in the order necessary to ensure that the pattern is right in the cloth.

spaced slating (*Build.*). Slating laid with gaps between adjacent slates in any course.

spaced trials (*An. Behav.*). An experimental procedure in which animals are given several consecutive trials, there being a relatively long interval between the end of one trial and the beginning of the next; cf. *massed trials*.

space factor (*Elec. Eng.*). The ratio of the active cross-sectional area of an insulated conductor to the total area occupied by it.

space group (*Crystal.*). Classification of crystal lattice structures into groups with corresponding symmetry elements.

space lattice (*Crystal.*). Three-dimensional regular arrangement of atoms characteristic of a particular crystal structure. There are 14 such simple symmetrical arrangements, known as *Bravais lattices*. See also **symmetry class.**

space parallax (*Acous.*). The difference in bearing between a moving object, such as a machine in flight, and the direction of arrival of the sound waves emitted by it. This arises from the comparable velocity of flight with that of the propagation of sound waves.

space parasite (*Bot.*). A plant which inhabits intercellular spaces in another plant, obtaining shelter but possibly taking nothing else.

space probe (*Space*). Space vehicle for exploration between the planets, sending back data by radio telemetry.

space-reflection symmetry (*Phys.*). See **parity.**

spaces of Fontana (*Histol.*). Meshes in the ligament at the junction of the cornea and sclerotic of the eye, which communicate with the anterior chamber and allow fluid to pass into the canal of Schlemm and hence into the ciliary veins.

space suit (*Space*). A suit for use in free fall, or on the moon or other planets. The design always includes the provision of a pressurized oxygen supply, temperature control and the purification of exhaled gases.

space-time (*Phys.*). Normal 3-dimensional space plus dimension of time, modified by gravity in relativity theory.

space velocity (*Astron.*). The rate and direction of a star's motion in space of three dimensions, as deduced from its observable components (*a*) in the line of sight by the spectroscope, and (*b*) perpendicular to the line of sight by proper motions.

space wave (*Radio*). Wave from an antenna which is not a ground wave, but which travels rectilinearly in space, apart from reflection (negative *refraction* and *bending*) when it enters an ionized region; also called **sky wave (or ray).**

spacing current (*Teleg.*). The current in the circuit which corresponds to the nondepression of the telegraphist's key, and which also, when there is no mark on the slip, operates the sending machine.

spacing material (*Typog.*). Lower than type height; there are spaces for each size of type and a wide variety made to 12-pt sizes for making up pages and spacing them apart. See **clumps, furniture, leads, quadrat, quotations, reglet.**

spacing ratio (*Light*). The ratio of the distance between equally spaced electric lights to their vertical distance above the plane to be illuminated.

spacing wave (*Telecomm.*). Emitted wave corresponding to *spacing* impulses in a code, e.g., Morse code; also called **back wave.**

spacistor (*Electronics*). A form of *depletion-layer transistor* (q.v.).

spade tuning (*Radio*). A crude form of tuning in which the inductance of a coil is varied by moving a spade-shaped metal disk across the face of the coil.

spadiceous (*Bot.*). (1) Having the colour of a date. (2) Bearing a spadix; spadixlike. (*Zool.*) Shaped like a palm branch. Also **spadiciform, spadicose.**

spadix (*Bot.*). A spike with a swollen fleshy axis, enclosed in a spathe. (*Zool.*) In male *Nauti-*

loidea, a cone-shaped structure formed by the modification of four of the tentacles and believed to be homologous with the hectocotylized arm of male Squids: in some Coelenterate embryos, the endodermal anlage of the manubrium.

spall (*Build.*). Fragment detached by weather action, by internal movement, or by the process of chiselling; also called **galet**.

spallation (*Nuc.*). Any nuclear reaction when several particles result from a collision, e.g., cosmic rays with atoms of the atmosphere; chain-reaction in a nuclear reactor or weapon.

spalling (*Build., Civ. Eng.*). (1) The breaking off of fragments on a concrete surface through weathering. Generally caused by infiltration of water through hair cracks and subsequent freezing. (2) The operation of breaking off splinters of stone from a block by slanting blows with a chisel, when dressing to shape.

spalt (*For.*). The material left after conversion of shingle bolts to shingles.

span (*Aero.*). The distance between the wing tips of an aeroplane. (*Civ. Eng., etc.*) Horizontal distance between supports of a bridge, arch, etc. (*Elec. Eng.*) (1) The distance between two transmission-line towers. (2) The number of slots separating the two sides of an armature coil. Also called **throw**.

spanaemia, spanemia (*Med.*). Poorness of the blood; deficiency of red cells in the blood; anaemia.

spandrel (*Civ. Eng.*). The space between the haunches and the underside of an arch.

spandrel step (*Arch.*). An individual step in a stair, which consists of a solid block, triangular in section, arranged so that one face is parallel to the slope of the stair.

spandrel wall (*Arch.*). A wall constructed upon the extrados of an arch.

Spanish topaz (*Min.*). Not a true topaz but an orange-brown quartz, the colour resembling that of the honey-brown Brazilian topaz. It is often amethyst which has been heat treated. See **citrine**.

span loading (*Aero.*). The gross weight of an aeroplane or glider divided by the square of the span.

spanner (*Tools*). A tool used for turning a nut. See **box-, ring-, sickle-**.

span piece (*Carp.*). The horizontal beam connecting the rafters of a collar-beam roof.

span pole (*Elec. Eng.*). The pole to which the span wires are attached.

span roof (*Arch.*). A pitched roof with two sloping sides having the same inclination.

span saw (*Tools*). See **frame saw**.

span wire (*Elec. Eng.*). One of several wires by which the trolley wire of a tramway or trolleybus system is suspended from street poles or buildings.

spar (*Aero.*). A main spanwise member of an aerofoil, or control, surface. The term can be applied either to individual beam(s) designed to resist bending, or to the box structure of spanwise vertical webs, transverse ribs, and skin which form a torsion box in most modern designs. (*For.*) A round timber more than 6 in. in diameter in the middle. (*Min.*) Transparent to translucent crystalline mineral with vitreous lustre and clean cleavage planes, e.g., fluorspar. See **Iceland-**.

Sparagmite (*Geol.*). A comprehensive term which includes the late Pre-Cambrian rocks of Scandinavia. These, like the Torridonian Sandstone of Northern Scotland, consist of conglomerates and red feldspathic grits and arkoses.

spar drying (*Paper*). Air drying of paper after tub-sizing is carried out by passing the paper around slatted drums through which hot air moves.

spar frame (*Aero.*). A specially strong transverse fuselage, or hull, frame to which a wing spar is attached.

sparganosis (*Med.*). Infestation of bodily tissues with the larvae (spargana) of various tapeworms, the adult stage of which may be unknown.

sparging (*Brew.*). A process in which the grains in the mash tun are treated with water at a temperature of about 180°F (82°C), which is sprinkled on them after the sweet wort has been run off.

spark (*Elec. Eng.*). Breakdown of insulation between two conductors, such that the field is sufficient to cause ionization and rapid discharge. See **arc, field discharge, lightning**.

spark absorber (*Elec. Eng.*). Resistance and/or capacitor placed across a break in an electrical circuit to damp any possible oscillatory circuit which would tend to maintain an arc or spark when a current is interrupted.

spark chamber (*Nuc. Eng.*). Radiation detector in which tracks of ionizing particles can be studied by photographing a spark following the ionization. Consists of a stack of parallel metal plates with the electric field between them raised nearly to the breakdown point.

spark coil (*Elec. Eng.*). Induction or Ruhmkorff coil used as the source of high voltage in a spark transmitter. Obsolete for radio, but used in motor-cars.

spark erosion (*Eng.*). Electrochemical metal machining process in which an electrode of male form is maintained very close to the workpiece, both being submerged in a dielectric liquid. High local temperatures are produced by passing a current through the gap, detaching and repelling particles from the workpiece.

spark frequency (*Radio*). Frequency of repetition of discharge in a spark transmitter.

spark gap (*Elec. Eng.*). In simplest form, two shaped electrodes separated by a dielectric which breaks down at a quasi-constant voltage gradient; this may be triggered by an externally applied electric field. Main uses: (1) voltage limiting safety device (e.g. *lightning arrestor*); (2) low-reactance switch (e.g. *Marx generator*); (3) generator of electromagnetic waves (e.g. *spark system*); (4) concentrated energy deposition (e.g. *spark erosion*).

spark-gap generator (*Elec. Eng.*). Radiofrequency generator for induction heating in which a capacitor is charged from a high-tension transformer and discharged through an oscillatory circuit when a spark gap breaks down.

spark-gap modulation (*Radio*). A method of pulse modulation in which a pulse-forming line discharges across a spark gap in the transmitter circuit.

sparking (*Elec. Eng.*). The occurrence of a spark discharge between the brushes and the surface of a commutator.

sparking contact (*Elec. Eng.*). An auxiliary contact used on circuit-breakers; designed to make circuit before, and to break circuit after, the main contact, so that any sparking takes place on the auxiliary contact. It has removable contact tips, usually of carbon.

sparking limit (*Elec. Eng.*). The limiting output of a d.c. machine as determined by considerations of commutator sparking.

sparking plug (*I.C. Engs.*). A plug screwed into the cylinder head of a petrol engine for ignition purposes, a spark gap being provided between an insulated central electrode, connected to the

H.T. distributor, and one or more earthed points.

sparking potential (*Elec. Eng.*). Potential difference between the ends of an insulator, sufficient to cause a spark discharge through or over the insulator. Also called **sparkover potential.**

sparkless commutation (*Elec. Eng.*). A term applied to methods of current commutation in which the reactance voltage is neutralized before actual commutation occurs, so that the formation of a commutation spark or arc is avoided.

spark machining (*Eng.*). See **spark erosion.**

sparkover potential (*Elec. Eng.*). See **sparking potential.**

sparkover test (*Elec. Eng.*). See **flashover test.**

spark photography (*Photog.*). High-speed technique employing a high-intensity electric spark for illumination. Used in ballistics and in *schlieren photography* (q.v.).

spark recorder (*Instr.*). A recorder using an electric spark to mark the paper, thus avoiding pen friction.

spark resistance (*Elec. Eng.*). That between electrodes after a discharge has commenced; if excessive and in an oscillatory circuit, it causes loss of power and a high decrement.

spark spectra (*Phys.*). The most important way of exciting spectra is by means of an electric spark. The high temperature reached will generate the spectrum lines of multiply ionized atoms as well as of uncharged and singly ionized ones. (As distinct from *arc spectrum* (q.v.).) Also evaporation of metal from the electrodes leads to additional lines not associated with the gas through which the discharge takes place.

spark system (*Radio*). Oldest form of radio telegraphy, using a spark transmitter, in which high-frequency currents are generated by charging a capacitor from an induction coil, or other source of high voltage, and then discharging it through an inductance in series with a spark gap. The inductance is coupled to the antenna, which may also form part, or all, of the capacitor.

spark test (*Chem. Eng.*). Test used for examining rubber and plastic linings in steel process vessels. A very high voltage low-energy current is applied to a strip of conductor held near to the surface and the circuit closed to the base metal. In sound linings a very small discharge of sparks takes place uniformly along the length of the strip; but a defect such as a pinhole is indicated by a single large spark.

spark transmitter (*Radio*). See **spark system.**

spar piece (*Carp.*). See **span piece.**

spartalite (*Min.*). An old name for *zincite.*

sparteine (*Chem.*). $C_{15}H_{26}N_2$, an alkaloid of the quinuclidine group, obtained from the branches of the common broom, *Cytisus scoparius*; a colourless oil, b.p. 188°C (18 mm), sparingly soluble in water, soluble in ethanol, trichloromethane or ethoxyethane. It resembles coniine in its physiological action.

spasm (*Zool.*). Involuntary contraction of muscle fibres. *adj.* **spasmodic.**

spasmodic torticollis (*Med.*). A nervous disorder in which the muscles of either side, or both sides, of the neck are in a state of continuous or of intermittent spasm.

spasmoneme (*Zool.*). In *Ciliophora* with contractile stalks, the powerful stalk muscle formed by the union of the longitudinal myonemes.

spasmophilia (*Med.*). A term used to indicate the hypothetical heightened irritability of the nervous system of patients who have a tendency to spasms, convulsions, or tetany.

spasmus nutans (*Med.*). Nodding spasm. Rhythmic nodding of the head seen in babies in the first year of life.

spastic (*Med.*). Of the nature of spasm (sudden contraction) of muscle: characterized or affected by muscular spasm: rigid, or in a state of continuous spasm; e.g., *spastic paralysis.* One affected by spastic paralysis.

spastic paralysis (*Med.*). Paralysis of the voluntary movements caused by prenatal brain damage; characterized by spasm of muscles.

Spatangoida (*Zool.*). An order of *Echinoidea* in which Aristotle's lantern is lacking, the anus and often the mouth are eccentric, the body is more or less heart-shaped, and the ambulacral areas are petal-shaped; sand-living forms. Heart Urchins.

spathe (*Bot.*). A large, usually coloured foliar organ which subtends and more or less encloses a spadix; the white part of the arum lily is a familiar example.

spathic iron (*Min.*). See **siderite (1).**

spatula (*Zool.*). Any spoon-shaped structure.

spatulate (*Bot.*). Having a broad, short lamina and a long petiole. (*Zool.*) Shaped like a spoon.

spavin (*Vet.*). Chronic arthritis of the hock joint of a horse.

spawn (*Bot.*). The mycelium of a mushroom. (*Zool.*) To deposit eggs or discharge spermatozoa: a collection of eggs, such as that deposited by many Fish.

spay (*Vet.*). To remove or destroy the ovaries.

S.P.D. (*Astron.*). See **polar distance.**

speak-back circuit (*TV*). See **talk-back circuit.**

speaker (*Acous.*). Abbrev. for *loudspeaker.*

speaking pair (*Teleph.*). When wires are grouped for trunking through automatic switching, the pair carrying the speech currents is termed the *speaking pair,* as contrasted with the guard wire, private wire, or meter wire.

speaking stop (*Acous.*). A stop key on an organ console which controls a rank of pipes for potential operation by the keyboard, as contrasted with *nonspeaking stops,* which are couplers or other devices not directly controlling pipes.

Spearfast cross-cut saw (*Tools*). One with the teeth divided into groups of 4 by a series of U-shaped cuts. Coarser than the normal cross-cut saw but quicker-working.

spear pyrites (*Min.*). The name given to twin crystals of marcasite which show re-entrant angles, in form somewhat like the head of a spear. Cf. *cockscomb pyrite.*

special character (*Comp.*). Any character recognized by a particular computing system which is neither a letter nor a numeral.

specialization (*Bot.*). The tendency of a parasite to attack only one species or variety of host plant.

special modifying factor (*Radiol.*). See **modifying factor.**

special rules zone (*Aero.*). A 3-dimensional space, under *air traffic control,* wherein aircraft must obey special instructions.

special steel (*Met.*). See **alloy steel.**

special unitary groups (*Nuc.*). See **Lie algebra.**

speciation (*Bot., Zool.*). Formation of new biological species including formation of polyploids.

species (*Bot., Zool.*). A term used in classification to denote a group of closely allied, mutually fertile individuals, showing constant differences from allied groups. *adj.* **specific.** In the system of *binomial nomenclature* (q.v.) of plants and animals, the second name (i.e., the name by which the species is distinguished from other species of the same genus) is termed the *specific* *epithet* or *specific name*. The latter, however, correctly refers to the full name, e.g., *Lilium candidum* (Madonna Lily), where *candidum* is the *specific epithet*.

species-specific behaviour (*An. Behav.*). The behaviour patterns shown by a great majority of individuals of a species in the same environment under the same conditions. Also termed **inborn, innate** or **instinctive behaviour.**

specific. Term used generally to indicate that the property described relates to unit mass of the substance involved, e.g. specific entropy is entropy per kilogram, but see footnote †. (*Med.*)† A treatment or medicine effective against a particular disease. (*Zool.*)† Of a parasite, restricted to a particular host. See also **specific strain.** *n.* **specificity.**

specific action potential (*An. Behav.*).† A tension, specific for a particular innate behaviour pattern, which tends to build up with time, and which, in the absence of a suitable releaser, becomes 'dammed up', resulting in a lowering of the threshold for releasing stimuli, and which may, eventually, accumulate to a point where the action pattern occurs with no external stimulation, this being known as *vacuum activity* or *overflow activity*. It includes the earlier idea of *action specific energy*.

specific activity (*Radiol.*). See **activity.**

specification (*Eng.*). (1) A detailed description, including dimensions and other quantities, of the function, construction, materials, and quality of a manufactured article or an engineering project. (2) A description, by an applicant for a patent, of the operation and purpose of the invention.

specific characters (*Biol.*).† The constant characteristics by which a species is distinguished.

specific charge (*Phys.*). Charge/mass ratio of elementary particle, e.g., the ratio e/m_e of the electronic charge to the rest mass of the electron $=1{\cdot}759\times10^{11}$ coulomb per kilogram.

specific conductivity (*Elec. Eng.*).† The inverse of **resistivity.**

specific conductivity of wood (*Bot.*).† The rate at which water flows through a piece of wood of standard area and length in a given time.

specific damping (*Elec. Eng.*).† The attenuation constant per kilometre of a cable.

specific depression (*Heat*).† See **depression of freezing point.**

specific dielectric strength (*Elec. Eng.*).† The *dielectric strength* of an insulating material, expressed in V/mm.

specific dynamic action (*Biochem., Physiol.*).† The special calorigenic property of foodstuffs, and particularly of proteins, of raising the metabolic rate after ingestion by an amount in excess of their calorific value. May be expressed as the ratio of the calories in excess of the basal to urinary nitrogen in excess of the basal.

specific electric loading (*Elec. Eng.*).† The electric loading, in ampere-conductors, of the armature of a machine per cm of circumference.

specific epithet (*Bot., Zool.*).† See **species.**

specific exhaustibility (*An. Behav.*).† The falling off of a particular response with time, not due to physiological exhaustion, and providing evidence for *specific action potential*.

specific fuel consumption (*I.C. Engs.*).† The mass of fuel used by an engine per unit energy delivered; generally expressed in pounds/B.H.P.-hour or kg/MJ.

specific gamma-ray emission (*Nuc.*).† Exposure dose from unit point gamma-ray source at unit distance. Usually expressed in röntgen/hour at 1 cm for a 1 mCi source. Formerly **k-factor.**

specific gravity (*Phys.*).† See **relative density.**

specific gravity bottle (*Phys.*). See **density bottle.**

specific heat capacities of gases (*Phys.*). Gases have two values of specific heat capacity: c_p, the s.h.c. when the gas is heated and allowed to expand against a constant pressure, and c_v, the s.h.c. when the gas is heated while enclosed within a constant volume. See **ratio of specific heat capacities.**

specific heat capacity (*Phys.*). The quantity of heat which unit mass of a substance requires to raise its temperature by one degree. This definition is true for any system of units, including SI, but whereas in all earlier systems a unit of heat was defined by putting the s.h.c. of water equal to unity, SI employs a single unit, the joule, for all forms of energy including heat, which makes the s.h.c. of water 4·1868 kJ/kg K. See **mechanical equivalent of heat.**

specific humidity (*Meteor.*). The mass of water vapour/unit mass of moist air.

specific impulse (*Aero.*).† The thrust available from the ideal combustion of 1 unit of a stoichiometric mixture of rocket propellants (fuel plus oxidant) and expanded perfectly to atmospheric pressure; symbol I_{sp}.

specific inductive capacity (*Elec. Eng.*).† See **permittivity.**

specific ionization (*Nuc.*).† Number of ion pairs formed by ionizing particle per cm of path. Also called the *total specific ionization* to avoid confusion with the *primary specific ionization* which is defined as the number of ion clusters produced for unit length of track.

specific latent heat (*Heat.*). See **latent heat.**

specific magnetic loading (*Elec. Eng.*).† The average flux density (i.e., the total magnetic loading divided by the peripheral area) in the armature of a machine.

specific name (*Bot., Zool.*).† See under **species.**

specific output (*Elec. Eng.*).† The ratio of the electrical output of a machine to its weight, its volume, or some other function of its dimensions. Cf. *specific torque coefficient*.

specific permeability (*Elec. Eng.*).† Same as **relative permeability.** See **permeability.**

specific power (*Nuc. Eng.*). U.S. for **fuel rating.**

specific reaction rate (*Chem.*).† See **velocity constant.**

specific refraction (*Chem.*).† The molecular refraction of a compound, defined by the Lorentz-Lorenz equation, divided by the molecular weight. Symbol *r*.

specific resistance (*Elec. Eng.*).† See **resistivity.**

specific rotation (*Chem., Phys.*).† The angle through which the plane of polarization of a ray of sodium D light would be rotated by a column of liquid 1 decimetre in length, containing 1 g of an optically active substance per cm³.

specific stain (*Micros.,*).† A stain which will pick out certain structures or tissues in contrasting colours or shades. Cf. *general stain*.

specific surface (*Powder Tech.*).† The surface area of the particles in a unit mass of the powder determined under stated conditions. Also sometimes used to describe the surface area of the particles in unit volume of the powder.

specific temperature rise (*Elec. Eng.*).† The temperature rise of an electrical machine per unit of radiating surface.

specific torque coefficient (*Elec. Eng.*).† A co-

† The word 'specific' in this entry derives from historical usage, and does not imply a relationship to unit mass; see *specific*.

efficient used in the design of electrical machines, giving a figure representing the torque per unit of volume enclosed by the air-gap periphery. Also known as **output coefficient, Esson coefficient.**

specific volume (*Phys.*). The volume of unit mass; the reciprocal of density.

specpure (*Chem.*). A trade term for *spectroscopically pure* (see **spectrum analysis**).

spectacle crown (*Glass*). White crown glass of refractive index 1·523, used for ophthalmic purposes.

spectacle flint (*Glass*). A glass of high refractive index which is fused to *spectacle crown* (q.v.) in the manufacture of bifocal spectacle lenses.

spectral characteristic (*Phys.*). Graph of photocell sensitivity, as related to wavelength of radiation.

spectral colour (*Light*). Colour with degrees of saturation between no-hue and a pure spectral colour on the rim of the chromaticity diagram.

spectral distribution curve (*Light*). The curve showing the relation between the radiant energy and the wavelength of the radiation from a light source.

spectral line (*Phys.*). Component consisting of a very narrow band of frequencies isolated in a spectrum. These are due to similar quanta produced by corresponding electron transitions in atoms. The lines are broadened into bands when the equivalent process takes place in molecules.

spectral sensitivity (*Photog.*). The comparative response of an emulsion to exposures of light of different wavelengths but constant intensity.

spectral series (*Phys.*). Group of related spectrum lines produced by electron transitions from different initial energy levels to the same final one. The recognition and measurement of series has been of great importance in atomic and quantum theories.

spectral transmission (*Photog.*). The relative transmission, opacity, or density of a filter in respect of light of different wavelengths.

spectral types (*Astron.*). The Harvard classification of stars according to their spectra, giving a graded list represented by the letters

(W)OBAFGKMS(RN),

which represents a sequence (called the *main sequence*) of descending temperature, the O type stars being hot, white, and gaseous, while the cooler M type show molecular band spectra.

spectre of the Brocken (*Meteor.*). The shadow of an observer cast by the sun on to a bank of mist. The phenomenon, often seen from a hilltop, may present the illusion that the shadow is a gigantic form seen through the mist. Also **Brockenspectre.**

spectrograph (*Phys.*). Normally used of spectroscope designed for use over wide range of frequencies (well beyond visible spectrum) and recording the spectrum photographically. The *mass spectrograph* (q.v.) separates particles of different specific charge in an analogous manner to the separation of spectrum lines of an optical spectrum.

spectroheliogram (*Astron.*). The recorded result of an exposure on the sun by the spectroheliograph.

spectroheliograph (*Astron.*). An instrument for photographing the sun in monochromatic light. It consists essentially of a direct-vision spectroscope, with a second slit instead of an eyepiece, which can be set so that only light of a desired wavelength passes through it on to a photographic plate.

spectrohelioscope (*Astron.*). An instrument in principle the same as the spectroheliograph, but adapted for visual use by the employment of a rapidly oscillating slit which, by the persistency of vision, enables an image of the whole solar disk to be viewed in light of one wavelength; it also detects the velocities of moving gases in the solar atmosphere by an adjustment called the 'line-shifter'.

spectrometer (*Phys.*). Instrument used for measurements of wavelength or energy distribution in a heterogeneous beam of radiation.

spectrophotometer (*Phys.*). Instrument for measuring photometric intensity of each colour or wavelength present in an optical spectrum.

spectroradiometer (*Phys.*). A spectrometer for measurements in the infrared.

spectroscope (*Phys.*). General term for instrument (spectrograph, spectrometer, etc.) used in spectroscopy. The basic features are a slit and collimator for producing a parallel beam of radiation, a prism or grating for 'dispersing' different wavelengths through differing angles of deviation, and a telescope, camera, or counter tube for observing the dispersed radiation.

spectroscopic analysis (*Chem.*). See **spectrum analysis.**

spectroscopic binary (*Astron.*). A binary whose components are too close to be resolved visually, but which is detected by the mutual shift of their spectral lines owing to their varying velocity in the line of sight.

spectroscopic parallax (*Astron.*). The name given to the indirect method of deducing the distances of stars too far away to have detectable annual parallaxes; it involves the inferring of their absolute magnitudes from spectroscopic evidence which then, combined with the observed apparent magnitudes, gives their distances.

spectroscopy (*Phys*). The practical side of the study of spectra, including the excitation of the spectrum, its visual or photographic observation, and the precise determination of wavelengths.

spectrum (*Phys.*). Arrangement of components of a complex colour or sound in order of frequency or energy, thereby showing distribution of energy or stimulus among the components. A mass spectrum is one showing the distribution in mass, or in mass-to-charge ratio of ionized atoms or molecules. The mass spectrum of an element will show the relative abundances of the isotopes of the element.

spectrum analyser (*Electronics*). (1) Electronic spectrometer usually working at microwave frequencies and displaying energy distribution in spectrum visually on a cathode-ray tube. (2) Pulse-height analyser for use with radiation detector.

spectrum analysis (*Chem.*). The analysis of a substance by observation of its spectrum. It is a valuable method for the detection of traces of impurities.

spectrum colours (*Light*). When split up into a spectrum, white light is shown to be composed of a continuous range of merging colours; red, orange, yellow, green, blue, indigo, violet.

spectrum line (*Phys.*). Isolated component of a spectrum formed by radiation of almost uniform frequency. Due to photons of fixed energy radiated as the result of a definite electron transition in an atom of a particular element.

spectrum locus (*Light*). Locus of points on chromaticity diagram representing visual stimulus produced by light beams of uniform wavelength (*monochromatic radiations*).

spectrum-luminosity law (*Astron.*). See **Hertzsprung-Russell diagram.**
spectrum stripping (*Phys.*). Process of separating a required gamma-ray spectrum from background, scattered radiation and other spectra. Can be carried out analytically or automatically, e.g., by an analogue technique.
specular density (*Photog.*). The photographic density in an image measured with parallel light, as contrasted with diffuse density, when the total light passed is measured, including that dispersed.
specular iron (*Min.*). The name given to a crystalline rhombohedral variety of haematite which possesses a splendent metallic lustre often showing iridescence.
specular reflectance (*Phys.*). Quotient of reflected to incident luminous flux for a polished surface.
specular reflection (*Phys.*). General conception of wave motion in which the wavefront is diverted from a polished surface, so that the angle of the incident wave to the normal at the point of reflection is the same as that of the reflected wave. Applicable to heat, light, radio, and acoustic waves. See **reflection laws.**
specular transmittance (*Phys.*). See **transmittance.**
speculum (*Med.*). A hollow or curved instrument for viewing a passage or cavity of the body. (*Met.*) Alloy of 1-tin to 2-copper, providing wide spectral reflection from a grating ruled on a highly polished surface.
speech (*Acous.*). The fundamental method of communicating thoughts, which consists in regulating the pitch and intensity of voiced sounds, and the intensity of unvoiced sounds, by the larynx, and in modifying the spectral content of these elementary sounds by posturing the cavities of the mouth (assisted by the nasal cavities), which form double or triple Helmholtz resonators. See **articulation.**
speech clipping (*Telecomm.*). Removal of high peaks in speech, to get higher loading of transmitting valves, with some change in intelligibility.
speech frequency (*Telecomm.*). See **voice frequency.**
speech inverter (*Telecomm.*). See **inverter.**
speech/noise ratio (*Acous.*). See **noise ratio.**
speech power (*Acous.*). See **average speech power.**
speech scrambler (*Telecomm.*). See **scrambler.**
speech-sounds (*Acous.*). The least distinctive elements in speech, such as vowels and consonants. For telephonic purposes, about 30 English speech-sounds are sufficient for recognition in telephonic work, but phoneticians recognize about 60. See **logatom.**
speed. (1) Ratio of the distance covered by a moving body to the time taken, either in a straight line or in a continuous curve. Uniform speed is a theoretical conception approached only in astronomical bodies. A practical uniform speed is a matter of approximate inference, and is always an average speed. Units of speed are feet or metres/second, miles or kilometres/hour, knots, etc. (2) Angular velocity, usually expressed in rev/min or rad/s. (*Photog.*) (1) The measure of the rate of exposure required by an emulsion. See **A.S.A., DIN, H and D number.** (2) Of a lens, the ratio of the working aperture diameter to the focal length. See ***f*-number.** (*Teleg.*) Number of standard words/minute or *bands.* (*Vac. Tech.*) The speed of a pump for a given gas is the volume passing through the pump per second measured at the pressure at the mouth of the pump. Unit is litres per second.
speed-adjusting rheostat (*Elec. Eng.*). A rheostat arranged in the field or armature circuit of an electric motor for varying the motor speed.
speed bulges (*Aero.*). Streamlined bulges on the fuselage, or nacelles near the trailing edge of the wing, which meet the requirements of *area rule* (q.v.) for a smooth *nose-to-tail drag curve* (q.v.) where it is impractical to give the fuselage a wasp waist to reduce transonic *wave drag* (q.v.).
speed control (*Elec. Eng.*). The method by which the speed of an electric motor may be varied.
speed-distance curve (*Eng., etc.*). The curve showing the relation between the speed of a moving object or vehicle and the distance it has travelled.
speed frames (*Spinning*). See **fly frames.**
speed-frequency (*Elec. Eng.*). The product of rotor speed and the number of pole-pairs in an induction motor.
speed governing (*Elec. Eng.*). The method of keeping the speed of a prime mover independent of the load which it is driving.
speed indicator (*Eng.*). See **speedometer.**
speed of light (*Phys.*). See **velocity of light.**
speed of rotation (*Phys.*). In a rotating body, the number of rotations about the axis of rotation divided by the time (see **speed**). Units are revolutions/second, minute, or hour, or radians/second, minute, or hour. The axis of rotation may have a translatory speed of its own. See **moment of inertia.**
speed of sound (*Phys.*). See **velocity of sound.**
speedometer (*Autos.*). A *tachometer* (q.v.) fitted to the gearbox or propeller shaft of a road vehicle, so graduated as to indicate the speed in m.p.h., km/h, or both. It may be centrifugal, magnetic, air-vane, chronometric, or electrical.
speed-time curve (*Eng.*). A curve of vehicle speed plotted against running time. The area beneath it represents the distance covered between any two given instants.
speed-torque characteristic (*Elec. Eng.*). The curve showing the relation between the speed of a motor and the torque developed. Also known as **mechanical characteristic.**
speedy-cut (*Vet.*). Injury of the foreleg of a horse near the knee, made by the shoe of the opposite foot.
Speeton Clay (*Geol.*). A series of clays and marls, with beds of phosphatic nodules, occurring near Speeton on the Yorkshire coast; equivalent in age to the Gault, Lower Greensand, and Wealden of Southern England.
speise, speiss (*Met.*). Metallic arsenides and antimonides produced in the smelting of cobalt and lead ores.
spelaeology, speleology (*Zool.*). The study of the fauna and flora of caves.
speleothems (*Geol.*). Calcium carbonate encrustations deposited in caves by running water.
spelter (*Met.*). Zinc of about 97% purity, containing lead and other impurities.
Spence Shale (*Geol.*). A subdivision of the Middle Cambrian at Mt Stephen in the Canadian Rockies, famous for its remarkable fossils.
spend safe (*Brew.*). Small vessel into which specimens of wort from taps in mash tun are drawn off for temperature testing, etc.
spent fuel (*Nuc. Eng.*). Reactor fuel element which must be replaced due to (*a*) swelling and/or bursting, (*b*) burn-up or depletion, (*c*) poisoning by fission fragments. The fissile material is not exhausted and so-called spent fuel is normally subsequently reprocessed.
Spergen Limestone (*Geol.*). See **Salem Limestone.**
sperm (*Zool.*). See **spermatozoon.**
sperm-, sperma-, spermi-, spermo-, spermato-. Prefix from Gk. *sperma*, gen. *spermatos*, seed.
spermaceti (*Chem.*). A glistening white wax from

the head of the sperm whale, consisting mainly of cetyl palmitate, $C_{15}H_{31}\cdot COO\cdot C_{16}H_{33}$; m.p. 41°–52°C, saponification number 120–135, iodine number nil. Used in the manufacture of cosmetics and ointments.

spermagonium (*Bot.*). See **spermogonium.**

spermaphytic (*Bot.*). Seed-bearing.

spermary (*Zool.*). See **testis.**

spermateleosis (*Zool.*). The process by which a mature spermatozoon is developed from a spermatid; used also by some authors in the sense of *spermatogenesis* (q.v.).

spermatheca (*Zool.*). A sac or cavity used for the reception and storage of spermatozoa in many Invertebrates; receptaculum seminis.

spermatic (*Zool.*). Pertaining to spermatozoa: pertaining to the testis.

spermatid (*Zool.*). A cell formed by division of a secondary spermatocyte, and developing into a spermatozoon without further division.

spermatiophore (*Bot.*). A hypha which bears a spermatium.

spermatium (*Bot.*). (1) The nonmotile male gamete in the red algae, which is carried by water to the trichogyne. (2) A sporelike structure which is formed by some lichens and some fungi, and which may have sexual functions.

spermatoblast (*Zool.*). A spermatid.

spermatocele (*Med.*). A cyst of the epididymis or of the tubules of the testis as a result of blocking of the ducts of the epididymis; contains a clear fluid and spermatozoa.

spermatocide (*Med.*). Any agent (especially chemical) which kills spermatozoa. *adj.* **spermatocidal.**

spermatocyte (*Zool.*). A stage in the development of the male germ cells, arising by growth from a spermatogonium or by division from another spermatocyte, and giving rise to the spermatids.

spermatogenesis (*Zool.*). Sperm formation; the maturation divisions of the male germ cells by which spermatozoa are produced from spermatogonia. See **spermateleosis.**

spermatogonium (*Zool.*). A sperm mother cell; a primordial male germ cell. *adj.* **spermatogonial.**

spermatogons (*Zool.*). Clear cubical epithelium cells lying on the basement membrane of the seminiferous tubules in higher Vertebrates.

spermatophore (*Zool.*). A packet of spermatozoa enclosed within a capsule.

Spermatophyta (*Bot.*). Seed-bearing plants. See **Phanerogamae.**

spermatoplast (*Bot.*). A male gamete.

spermatorrhoea, spermatorrhea (*Med.*). Involuntary, frequent discharge of seminal fluid in the absence of sexual excitement or intercourse.

spermatozeugma (*Zool.*). Union, by conjugation, of two or more spermatozoa.

spermatozoid (*Bot.*). The motile male gamete of many lower plants.

spermatozoon (*Zool.*). The male gamete, typically consisting of a head containing the nucleus, a middle piece containing a mitochondrion, and a tail whose structure is similar to that of a flagellum; pl. **spermatozoa**; abbrev. **sperm.**

spermaturia (*Med.*). The presence of spermatozoa in the urine.

spermidual glands (*Zool.*). In many Vertebrates, glands opening into or near the spermiducts.

spermiduct, spermaduct (*Zool.*). A duct by which sperms are carried from the testis to the external genital opening; the vas deferens. *adj.* **spermiducal.**

spermiogenesis (*Bot.*). The conversion of a spermatid into a spermatozoid.

sperm morula (*Zool.*). A spherical mass consisting of a protoplasmic core with an investment of developing sperms, such as occurs in the vesiculae seminales of the Earthworm.

spermogonium, spermogone (*Bot.*). A flask-shaped structure in which spermatia are formed.

spermology (*Bot.*). The study of seeds.

sperm pronucleus (*Zool.*). A male pronucleus.

Sperry arc lamp (*Light*). An arc lamp used in searchlights, which has a positive carbon with a fast-burning impregnated core, thus giving rise to a deep crater of very high brilliancy.

sperrylite (*Min.*). Diarsenide of platinum, crystallizing in the cubic system. It has a brilliant metallic lustre and is tin-white in colour.

spes phthisica (*Med.*). Euphoria and hopefulness of recovery in patients with advanced pulmonary tuberculosis.

spessartine, spessartite (*Min.*). Manganese garnet; silicate of manganese and aluminium, crystallizing in the cubic system. Usually contains a certain amount of either ferrous or ferric iron. The colour is dark, orange-red, sometimes having a tinge of violet or brown. Strictly, *spessartite* is the name for a lamprophyre rock. See **garnet.**

spew (*Acous.*). The superfluous irregular rim of wax which has to be removed from a gramophone record after it has been pressed between two stampers in the hot-pressing machine.

SPF (*Nuc. Eng.*). *Simple process factor* (q.v.).

sp. gr. (*Chem.*). An abbrev. for *specific gravity.*

sphacelate (*Bot.*). Dark and shrunken.

sphaeridia (*Zool.*). Small rounded bodies containing ganglion cells, situated on the surface of the test in some *Echinodermata*; believed to be sense organs. *sing.* **sphaeridium.**

sphaerocarpous (*Bot.*). Having a globular fruit.

sphaerocephalous (*Bot.*). Having the flowers crowded in a rounded head.

sphaerocrystal (*Bot.*). A rounded crystalline mass of calcium oxalate in the cells of some plants.

Sphaeropsidales (*Bot.*). An order of the *Fungi Imperfecti*, characterized by conidia borne in rounded fruit bodies or in cavities within the host, plant parasites, e.g., *Septoria apii* (leaf spot of celery).

sphaerraphide (*Bot.*). A rounded spiky mass of calcium oxalate found, usually singly, in the cells of many plants.

sphagnicolous (*Ecol.*). Living in peat-moss.

sphagnophilous (*Ecol.*). Living in peaty waters.

sphalerite (*Min.*). Zinc sulphide which crystallizes in the cubic system as resinous black or brown crystals. The commonest zinc mineral, occurring chiefly in vein deposits with fluorite, galena, etc. Also known as **blende or zinc blende.**

sphen-, spheno-. Prefix from Gk. *sphēn*, a wedge.

sphene (*Min.*). Orthosilicate of calcium and titanium, with varying amounts of iron, manganese, and the rare earths. It crystallizes in the monoclinic system as lozenge-shaped black or brown crystals and occurs as an accessory mineral in many igneous and metamorphic rocks. Also called **titanite.**

sphenethmoid (*Zool.*). In Amphibians, a bone of the interorbital region which extends forward to the nasal region and replaces the orbitosphenoids.

Sphenisciformes (*Zool.*). The only order of the *Impennae*; flight feathers are lacking; the wings are stiff and used as paddles in swimming; the feet are webbed; the bones are solid and there are no air sacs; flightless marine forms, with streamlined bodies, powerful swimmers and divers; confined to the southern hemisphere. Penguins.

Sphenodon (*Zool.*). See **Rhynchocephalia.**

sphenoid (*Crystal.*). A wedge-shaped crystal-form consisting of 4 triangular faces. The tetragonal and orthorhombic analogue of the cubic tetrahedron. (*Zool.*) A bone of the Vertebrate skull, one of the *sphenoidalia* (q.v.).

sphenoidal (*Bot., Zool.*). Wedge-shaped.

sphenoidalia (*Zool.*). A set of cartilage bones forming the walls of the middle part of the brain-case in the Vertebrate skull.

sphenoiditis (*Med.*). Inflammation of the air-containing sinus in the sphenoid bone.

sphenolateral (*Zool.*). In the developing chondrocranium, a pair of dorsal bars which are situated in front of the basal plate and parallel to the trabeculae.

sphenotic (*Zool.*). A bone of the lateral wall of the auditory capsule of the skull in some Vertebrates.

sphere-crystal (*Bot.*). A sphaerraphide.

sphere gap (*Elec. Eng.*). *Spark gap* between spherical electrodes.

spherical aberration (*Optics*). Loss of image definition arising from the geometry of a spherical surface. Parabolic mirrors are normally used in astronomical telescopes in order to avoid this defect. See also **Schmidt optical system.**

spherical astronomy (*Astron.*). The branch of astronomy concerned with the position of heavenly bodies regarded as points on the observer's celestial sphere. It comprises all diurnal and seasonal phenomena and the precise assignment of coordinates to the heavenly bodies. Also called **positional astronomy.** See also **astrometry.**

spherical candle-power (*Light*). The illumination on a sphere of unit radius having the source of light at its centre.

spherical curvature (*Maths.*). See **osculating sphere.**

spherical excess (*Surv.*). The amount by which the sum of the three angles of a spherical triangle exceeds 180°. It is equal to the area of the triangle divided by the square of the sphere's radius.

spherical polar coordinates (*Maths.*). See **polar coordinates** (2).

spherical radiator (*Radio*). Same as **omnidirectional radiator.**

spherical roller-bearing (*Eng.*). A roller-bearing having two rows of barrel-shaped rollers of opposite inclination, working in a spherical outer race, thus providing a measure of self-alignment.

spherical triangle (*Maths.*). A triangle formed by three lines (see **line**) on the surface of a sphere.

sphericity (*Powder Tech.*). The ratio of the surface area of a particle to the surface area of the sphere having the same volume as the particle.

spherics (*Radio*). See **atmospherics.**

spherocytosis (*Med.*). Acholuric jaundice. A congenital and familial disease in which the red cells of the blood are smaller than normal, biconvex instead of biconcave, and abnormally fragile; it is characterized by jaundice and splenomegaly.

spheroid (*Maths.*). The surface obtained by rotating an ellipse about one of its principal axes. It is therefore an ellipsoid in which two of its three principal axes are equal. If the axis of rotation is the major axis of the ellipse it is called a *prolate spheroid*, otherwise it is an *oblate spheroid*.

spheroidal jointing (*Geol.*). Spheroidal cracks found in both igneous and sedimentary rocks. Some are due to cooling and resultant contraction in the igneous rock body; others are due to a shell-like type of weathering.

spheroidal state (*Phys.*). Water dropped upon a clean, horizontal, red-hot metal plate gathers into spheroidal drops which roll about, rather like mercury drops, without boiling. This is prevented by a cushion of steam on which the drop rides, the rapid evaporation of the drop showing that its temperature is near the boiling-point.

spheroidal structure (*Geol.*). A structure exhibited by certain igneous rocks, which appear to consist of large rounded masses, surrounded by concentric shells of the same material. Presumably a cooling phenomenon, comparable with perlitic structure, but on a much bigger scale, and exhibited by crystalline, not glassy, rocks.

spheroidizing (*Eng.*). Process of producing, by heat treating, a structure in which the cementite in steel is in a spheroidal distribution, giving improved machinability.

spherome (*Bot.*). A cell inclusion which gives rise to globules of fat and oil.

spherometer (*Optics*). An instrument for measuring the curvature of a lens surface.

spheroplasts (*Cyt.*). Mitochondria.

spherulite (*Geol.*). A crystalline spherical body built of exceedingly thin fibres radiating outwards from a centre and terminating on the surface of the sphere, which may vary in diameter in different cases from a fraction of a millimetre to that of a large apple.

spherulitic texture (*Geol.*). A type of rock fabric consisting of spherulites, which may be closely packed or embedded in an originally glassy groundmass. Commonly exhibited by rhyolitic rocks.

sphincter (*Zool.*). A muscle which by its contraction closes or narrows an orifice. Cf. *dilator*.

sphincter of Oddi (*Anat.*). Fibres of muscle surrounding duodenal end of bile-duct.

sphingomyelin (*Chem.*). A phosphatide of complex structure found in the brain. On breakdown by hydrolysis it forms *choline* (q.v.), *sphingosine*, fatty acids and phosphoric acid.

sphingosine (*Chem.*). See **kerasin, phrenosin.**

sphygmogram (*Med.*). A tracing of the movements of the pulse made by a sphygmograph.

sphygmograph (*Med.*). An instrument for recording the movements of the arterial pulse by means of tracings.

sphygmomanometer (*Med.*). An instrument for measuring the arterial blood-pressure (in millimetres of mercury), an inflatable bag being applied to the arm and attached to a manometer.

sphygmus (*Zool.*). The pulse; the beat of the heart and the corresponding beat of the arteries.

spica (*Med.*). A figure-of-eight bandage with turns that cross one another.

spicula (*Bot.*). A small spike.

spicular cell (*Bot.*). A hard, thick-walled cell, spindle-shaped or branched, occurring among thin-walled soft tissues.

spiculate (*Bot.*). Said of a surface covered by fine points.

spicule (*Zool.*). A small pointed process: one of the small calcareous or siliceous bodies which form the skeleton in many *Porifera* and *Coelenterata*. *adj.* **spicular, spiculate, spiculiferous, spiculiform.**

spiculum (*Bot.*). A little spine. (*Zool.*) Any spiculelike structure: in Snails, the dart.

spider (*Acous.*). Mount with flexible arms used for centring coil of dynamic loudspeaker. (*Cinema.*) In a motion-picture studio, the local distributing box for lighting cables. It consists of a box with a number of paralleled substantial jacks, into which the ends of the lamp cables are plugged. (*Elec. Eng.*) The centre part of an armature core, upon which the core stampings are built up. (*Eng.*) See **cathead.**

spider wrench (or spanner) (*Tools*). A *box spanner* (q.v.) having heads of different sizes at the ends of radial arms.

spiegeleisen (*Met.*). Pig-iron containing 15–30% manganese and 4–5% carbon. Added to steel as a deoxidizing agent and to raise the manganese content of the steel.

Spigelian lobe (*Zool.*). A division of the right lateral lobe of the Mammalian liver.

spigot (*Brew.*). A wooden pin used to stop vent-hole in cask. (*Electronics*) A projection in the base of a thermionic valve which serves to locate this in its correct position in the holder. (*Eng.*) Male part of joint between two tubes.

spigot-and-socket joint (*Civ. Eng.*). A common type of joint for cast-iron pipes, the plain or spigot end of one length fitting into the enlarged or socket end of the next length, the joint being made tight by caulking.

spike. A stout nail more than 4 in (10 cm) long. (*Bot.*) An indefinite inflorescence with sessile flowers. (*Elec. Eng.*) Short pulse of voltage or current. (*Nuc.*) Zone surrounding track of charged particles in which atoms have been displaced (or heating has occurred, i.e., *thermal spike*). (*Radar*) Initial rise in excess of the main pulse in transmission. (*Telecomm.*) See **pulse-.**

spiked chain harrow (*Agric.*). A chain harrow in which the links carry spikes or knives.

spike knot (*For.*). A knot which has been cut lengthwise. Also called a **splay knot.**

spikelet (*Bot.*). One of the units of the inflorescence of a grass. It consists of a central rachis bearing one or more sterile glumes at the base, followed by one or more flowers, each enclosed between a flowering glume and a pale; all the parts are crowded together.

spile (*Civ. Eng.*). A timber pile.

spiling (*Mining*). See **spilling.**

spilite (*Geol.*). A fine-grained igneous rock of basaltic composition, generally highly vesicular and containing the sodium feldspar, albite. The pyroxenes or amphiboles are usually altered. These rocks are frequently developed as submarine lava-flows and exhibit pillow structure.

Spillane permeameter (*Powder Tech.*). A modified form of *Rigden's permeameter* (q.v.), the manometer movement shown on a dial which can be calibrated to read directly in surface area units.

spill burner (*Aero.*). A gas turbine burner wherein a portion of the fuel is recirculated instead of being injected into the combustion chamber.

spilling or spiling (*Mining*). A method of tunnelling through loose ground, by driving *spills* (sharp-edged thick planks) ahead of and around the timber frames.

spill-over (*Teleph.*). In V.F. signalling on multi-link connections, that part of a signal which passes from one section to another before the connection between the sections is split.

spillway (*Civ. Eng.*). See **bye-channel.**

spillway dam (*Civ. Eng.*). A reservoir dam over which flood water is allowed to flow to a downstream escape channel situated at the foot of the dam.

spilosites (*Geol.*). Spotted slates formed by contact metamorphism near the junctions of basic igneous rock and shales or slates; frequently associated with adinoles.

spin (*Aero.*). A continuous, but not necessarily even, spiral descent with the mean *angle of incidence* to the relative airflow above the *stalling* angle. In a *flat spin* the mean angle of incidence is nearer the horizontal than the vertical, while in an *inverted spin* the aircraft is actually upside down. (*Nuc.*) (1) Of an elementary particle, its quantized angular momentum in absence of orbital motion. (2) Of a nucleus, its quantized angular momentum, including contributions from the orbital motion of nucleons. (3) Of an electron, the quantized angular momentum which adds to the orbital angular momentum, thus producing fine structure in line spectra. (*N.B.* Quantized spin is always an integral multiple of half the Dirac unit $\hbar$).

spina (*Zool.*). A small sharp-pointed process; a spine: in Insects, a median apodeme arising from the poststernellum.

spina bifida (*Med.*). Developmental malformation of the bony spinal canal, spinal segments failing to meet. Often associated with defect at brain base preventing circulation of cerebrospinal fluid and causing brain damage and *hydrocephalus* (q.v.). There may also be leg paralysis, deformity of the feet, or other abnormalities (see **syringomyelocele**). Also called **rachischisis.**

spinacene (*Chem.*). See **squalene.**

spinal (*Zool.*). Pertaining to the vertebral column, or to the spinal cord.

spinal anaesthesia (*Med.*). A form of regional anaesthesia. It is produced by injecting a solution of a suitable drug within the spinal theca (intrathecally), causing temporary paralysis of the nerves with which it comes into contact.

spinal canal (*Zool.*). The tubular cavity of the vertebral column which houses the spinal cord.

spinal caries (*Med.*). See **Pott's disease.**

spinal cord (*Zool.*). In *Craniata*, that part of the dorsal tubular nerve cord posterior to the brain.

spinales (*Zool.*). Muscles connecting the vertebrae. *sing.* **spinalis.**

spinal reflex (*Zool.*). Reflex situated in spinal cord in which higher nerve centres play no part.

spinal shock (*Zool.*). The state of diminished excitability of the spinal cord, due to injury.

spina ventosa (*Med.*). A condition, resulting from tuberculous infection of a small bone (e.g., of the hand or foot), in which the centre part of the bone is destroyed and new bone forms under the periosteum, the bone thus appearing to expand.

spin bath (*Chem. Eng.*). The acid bath with various additives into which the spinning solution is injected to form the thread in viscose rayon production.

spin chute (*Aero.*). See **antispin parachute.**

spindle (*Cyt., Zool.*). Any spindle-shaped structure; especially the framework of achromatin fibres, formed between the centrioles during nuclear division by meiosis or mitosis; also used for special sensory receptor in muscle—*muscle spindle*. (*Eng.*) (1) The tubular member revolving with the headstock of a lathe, through which bar material may be introduced into the chuck or collet. (2) Generally, a machine element acting as a revolving axis or a pin on which another element revolves.

spindle fibre (*Cyt.*). A delicate line seen in a preparation of a dividing nucleus and, with many others, making up the spindle. It is doubtful whether it is a material fibre, or merely a line of strain in the material of the nucleus.

spindleless reel stand (*Print.*). One supporting the reel on free-running cones at each side.

spine (*Bind.*). The back of a book, i.e., the edge where the gathered sections are sewn together. The spine faces outwards when the book is placed on a shelf, hence the name often used, *shelf-back*. (*Bot.*) The end of a branch or leaf which has become rounded in section, hard and sharply pointed. *adj.* **spiniferous, spiniform.** (*Zool.*) A small sharp-pointed process: the backbone or vertebral column: a pointed process of a vertebra, as the neural spine: a fin-ray: the scapular ridge. *adjs.* **spinate, spiniform, spinose, spinous.**

spinel (*Min.*). A group of closely related oxide minerals crystallizing in the cubic system, usually in octahedra. Chemically, spinels are aluminates, chromates, or ferrates of magnesium, iron, zinc, etc., and are distinguished as *iron spinel* (hercynite), *zinc spinel* (gahnite), *chrome spinel* (picotite), and *magnesian spinel*. See also **ruby-, synthetic-, balas ruby, chromite, franklinite, magnetite.**

spinel ruby (*Min.*). See **ruby spinel.**

spinescent (*Bot., Zool.*). Spiny, tapering. (*Zool.*) Showing a tendency to become spinous, as some animals during the declining period of the racial history. *n.* **spinescence.**

spinicarpous (*Bot.*). Having spiny fruit.

spiniger, spinigerous (*Bot.*). Producing thorns.

spinner (*Aero.*). A streamlined fairing covering the hub of an *airscrew* and rotating with it.

spinneret (*Textiles*). In the spinning of man-made fibres, a plate containing fine holes through which filaments of plastic material are extruded. Also **spinnerette.** (*Zool.*) In Spiders, one of the spinning organs, consisting of a mobile projection bearing at the tip a large number of minute pores by which the silk issues.

spinner-gate or **whirl-gate** (*Foundry*). An ingate incorporating a small whirl-chamber into and from which the metal flows tangentially, so releasing any dirt, which rises to the top of the chamber or up a riser. See **ingate.**

spinnerule (*Zool.*). In Spiders, a duct by which the fluid silk is discharged.

spinning (*Eng.*). Forming a metal disk into a hollow shape by rotating it in a lathe over a former or chuck, by pressing a spinning tool against it. (*Textiles*) (1) Twisting together short fibres of animal, vegetable or mineral origin to form a yarn. (2) The processes of *wet spinning, dry spinning*, or *melt spinning*, by which most of the man-made fibres are formed.

spinning glands (*Zool.*). The silk-producing glands of *Arthropoda*.

spinning jenny (*Textiles*). First spinning machine employing vertical spindles, invented by James Hargreaves, Blackburn, Lancs, in 1767.

spinning number (*Textiles*). See **quality number.**

spinning riffler (*Powder Tech.*). Device used primarily for sample reduction but also for systematic bulk sampling. It consists of a rotating series of containers passing under a stream of powder, flowing from a hopper. For the device to be efficient, the ratio T/t, where T is the time for the bulk supply to flow through the hopper, and t is the time for one revolution of the turntable, has to be a large number.

spinning tunnel (*Aero.*). See **vertical wind tunnel.**

spinode (*Maths.*). See **double point.**

spino-occipital (*Zool.*). Arising in the trunk region, and later becoming incorporated in the occipital region of the skull.

spinose, spinous (*Bot.*). Bearing sharp, spiny teeth. (*Zool.*) Covered with spines.

spinous process (*Zool.*). A process of (*a*) the proximal end of the tibia, (*b*) the sphenoid bone: the neural process or spine of a vertebra.

spin polarization (*Nuc.*). Orientation of the spin axis of a nuclear particle.

spin quantum number (*Electronics*). Contribution to the total angular momentum of the electron of that due to the rotation of the electron about its own axis.

spinthariscope (*Radiol.*). Eyepiece system through which individual scintillations arising from impact of alpha-particles on zinc sulphide can be seen.

spinule (*Bot., Zool.*). A very small spine or prickle.

spiracle (*Zool.*). In Insects and some *Arachnida*, one of the external openings of the tracheal system: in Fish, the first visceral cleft, opening from the pharynx to the exterior between the mandibular and the hyoid arches: in amphibian larvae, the external respiratory aperture: in *Cetacea*, the external nasal opening. *adjs.* **spiracular, spiraculate, spiraculiform.**

spiral (*Maths.*). A plane curve traced by a point which winds about a fixed pole from which it continually recedes. See **Archimedes'-, Cornu's-, equiangular-, hyperbolic-, parabolic-, sinusoidal-.**

spiral binding (*Bind.*). A speciality binding in which holes are punched near edge of book and a spiral of wire is fed through them.

spiral cell, spiral tracheide, spiral vessel (*Bot.*). A cell, tracheide, or vessel in which the secondary wall is laid down in the form of spirally arranged thickenings.

spiral cleavage (*Zool.*). A type of segmentation of the ovum occurring in many *Turbellaria*, most *Mollusca*, and all *Annelida*; the early micromeres rotate with respect to the macromeres, so that the micromeres lie opposite to the furrows between the macromeres; the direction of rotation (viewed from above) is normally clockwise (*dexiotropic*) but in 'reversed cleavage' it is anti-clockwise (*laeotropic*). Also **alternating cleavage.**

spiral galaxies (*Astron.*). The largest class of extra-galactic nebulae, having a symmetrical structure, two spiral arms emerging from a bright central nucleus about which they rotate. Each is a stellar system comparable with our own Galaxy, and many millions are within the reach of our telescopes.

spiral ganglion (*Zool.*). In Mammals, a continuous ganglionic cord lying at the base of the spiral lamina and connected with the cochlear branch of the auditory nerve.

spiral gear (*Eng.*). A toothed gear for connecting two shafts whose axes are at any angle and do not intersect. The teeth are of spiral form (i.e., parts of a multiple-threaded screw) and engage as in a *worm gear* (q.v.).

spiral instability (*Aero.*). That form of lateral instability which causes an aircraft to develop a combination of side-slipping and banking, the latter being increasingly too great for the turn. This causes the machine to follow a spiral path.

spiral lamina (*Zool.*). In Mammals, the bony shelf which partially subdivides the bony labyrinth of the cochlea.

spiral lead-in and **spiral throw-out** (*Acous.*). Grooves in disk recording used to guide stylus into and out of the modulated track.

spiral ligament (*Zool.*). A projection of reticular connective tissue, by which the basilar membrane is attached to the outer wall of the cochlea in the Mammalian ear.

spiral ratchet screwdriver (*Tools*). *Ratchet screwdriver* (q.v.), which can be made to turn in either direction by downward pressure on a sleeve which applies torque to helical grooves in the shank.

spiral reel (*Photog.*). In a light-tight developing tank, a reel on which film has been wound in a spaced spiral so that, on rotation, the developer has free access to the emulsion.

spiral roller (*Print.*). **(1) An ink roller with spiral channels or grooves to improve ink supply and distribution. (2) An idling roller with a spiral**

groove to smooth out unevenness in the web of paper.

spiral scanning (*TV*). Scanning in which the spot traverses the scan area in a spiral path from the centre to the outside, whence it returns rapidly to the centre, or vice versa.

spiral spring (*Eng.*). A spring formed by coiling a steel ribbon into an elongated spiral or a helix of increasing diameter. When compressed completely it forms a true spiral.

spiral throw-out (*Acous.*). See **spiral lead-in.**

spiral stairs (*Build.*). *Circular stairs* (q.v.) of small diameter and usually open.

spiral time-base (*Electronics*). Arrangement for causing the fluorescent spot to rotate in a spiral path at a constant angular velocity, to obtain a much longer base-line than is possible with linear deflection. Used for detailed delineation of events relatively widely spaced in time, with or without a memory through long-glow or photography.

spiral tracheide (*Bot.*). See **spiral cell.**

spiral valve (*Zool.*). In Lampreys, Selachii, and some Lung fish, part of the intestinal canal, which is provided with an internal spiral fold to increase its absorptive surface.

spiral vessel (*Bot.*). See **spiral cell.**

spiral yarn (*Textiles*). Fancy yarn with a spiral effect, produced by doubling together, the opposite way, two threads of different counts, one fine and one coarse. The finer thread is usually subjected to more tension.

spiratron (*Electronics*). Travelling-wave tube, with radial electrostatic beam focusing.

spire (*Arch.*). A slender tower tapering to a point.

spireme (*Cyt.*). A stage in *prophase* (q.v.) in which the nuclear chromatin takes the form of a long thread.

spirillum (*Bacteriol.*). A bacterium which is a variation of the rod form, i.e., a curved or corkscrew shaped organism. Varies from 0·2–50 μm in length. One or more flagella may be present. Also the name of a particular genus, which includes *Sp. minus* (rat-bite fever).

spirit (*Chem.*). An aqueous solution of ethanol, especially one obtained by distillation.

spirit duplicating (*Print.*). The matter to be duplicated is typed, written or drawn on the under side of a smooth-surfaced paper with a special carbon paper in contact with the smooth surface, several colours of carbon being available, and multicoloured arrangements easily made. The image transferred from the carbon paper is sufficient for about 250 copies, being moistened by a volatile fluid between each impression on the duplicating machine.

spirit level (*Surv.*). See **level tube.**

spirits of salts (*Chem.*). Hydrochloric acid.

spirits of wine (*Chem.*). Ethanol.

spirit stain (*Build.*). A stain for wood—colouring matter dissolved in methanol.

spirit varnish (*Paint.*). A varnish made by dissolving certain alcohol-soluble gums or resins like shellac in industrial alcohol.

Spirochaetes or **Spirochaetales** (*Bacteriol.*). Filamentous flexible bacteria showing helical spirals: without true flagella. Divided into two families, the *Spirochaetaceae* and the *Treponemataceae*. Saprophytic and parasitic members. Pathogenic species include the causative agents of syphilis (*Treponema pallidum*), spirochaetosis (relapsing fever) in Man (*Borrelio recurrentis*), infectious jaundice (*Leptospira icterohaemorrhagiae*), and yaws (*Treponema pertenue*).

spirochaetosis (*Med.*). See **relapsing fever.**

spirochaetosis icterohaemorrhagica (*Med.*). Weil's disease; infectious jaundice. An acute disease characterized by fever, jaundice, haemorrhages from the mucous membranes, enlargement of the liver, and nephritis; due to infection with *Leptospira icterohaemorrhagiae*, conveyed to Man by rats, which excrete the organism in their urine.

spirolobous (*Bot.*). Said of an embryo having spirally rolled cotyledons.

spirometer (*Med.*). An instrument for measuring the air inhaled and exhaled during respiration.

spironeme (*Zool.*). In certain *Ciliophora*, the contractile spiral thread of the stalk.

Spirotricha (*Zool.*). An order of *Ciliata*, having a permanently open gullet with an undulating membrane. Adoral wreath curves clockwise, rest of body generally covered by cilia, body not usually flattened. Some suborders include parasitic or symbiotic forms, e.g., *Balantidium, Stentor, Stylonichia, Vorticella.*

Spirurata (*Zool.*). An order of *Phasmidia*, with a cylindrical oesophagus, often partly muscular, partly glandular, the males having two copulatory spicules with well developed alae and papillae, vagina of females elongate and tubular. Need an intermediate host, e.g., *Wucheria.*

spit (*Agric.*). The unit of depth of digging—the length of a fork or spade.

spitzkasten (*Min. Proc.*). Crude *classifiers* (q.v.), consisting of one or more pyramid-shaped boxes with regulated holes in down-pointing apexes. Ore pulp streaming across either settles (*coarse particles*) to bottom discharge or overflows (*fires*). In the **spitzlutten**, efficiency is improved by adding upflow of low-pressure hydraulic water.

splanchnic (*Zool.*). Visceral.

splanchno-. Prefix from Gk. *splanchnon*, the inward parts.

splanchnocoel (*Zool.*). In Vertebrates, the larger posterior portion of the coelom which encloses the viscera, as opposed to the pericardium.

splanchnocranium (*Zool.*). See **viscerocranium.**

splanchnomegaly (*Med.*). Enlargement of bodily organs.

splanchnopleure (*Zool.*). The wall of the alimentary canal in coelomate animals; the inner layer of the mesoblast which contributes to the wall of the alimentary canal; cf. *somatopleure. adj.* **splanchnopleural.**

splanchnoptosis (*Med.*). Glenard's disease. General displacement downwards, or dropping, of the abdominal viscera.

splanchnotome (*Zool.*). In an early Vertebrate embryo, the ventral division of a mesoblastic somite.

splash baffle (*Electronics*). See **arc baffle.**

splashproof fitting (*Light*). See **weatherproof fitting.**

splat (*Build.*). A cover strip for joints between adjacent sheets of building-board.

splay brick (*Build.*). A purpose-made brick bevelled off on one side. Also **cant brick, slope.**

splayed coping (*Build.*). See **feather-edged coping.**

splayed grounds (*Join.*). Grounds with splayed or rebated edges, providing a key for holding the plaster to the wall in cases where the grounds also serve as screeds.

splayed jambs (*Build.*). Internal jambs or sides of a door or window opening which are not built at right angles to the wall but slope away from it, to admit more light or to increase the width.

splayed skirting (*Build.*). A skirting board having its top edge chamfered.

splay end (*Build.*). That end of a brick opposite to the square end.

splaying arch (*Build., Civ. Eng.*). An arch in which the opening at one end is less than that at the other end, so that the arch is funnel-shaped. Also called a **fluing arch.**

splay knot (*For.*). See **spike knot.**

spleen (*Zool.*). In *Craniata*, a large mass of reticular tissue involved in the formation, storage, and destruction of blood corpuscles, and a defence against disease. In the embryos of all *Craniata*, both red and white corpuscles are formed here, as also in the adults of all groups except Mammals, in which it only forms lymphocytes. Red corpuscles are stored here and, when worn out or diseased, are destroyed here, by macrophages which also destroy infectious materials in the bloodstream.

splenectomy (*Surg.*). Removal of the spleen.

splenial (*Zool.*). In some Vertebrates, a membrane bone on the inner side of the lower jaw behind the dentary.

splenial teeth (*Zool.*). A pair of large compound teeth in the lower jaw of some Lung fish.

splenic anaemia (*Med.*). A chronic disease characterized by splenomegaly, anaemia, leucopenia, and a tendency to bleeding; when there is also cirrhosis of the liver the condition is known as *Banti's disease* (q.v.).

splenic organs (*Zool.*). In some Insects, certain bilateral groups of phagocytic cells, believed to give rise to fresh leucocytes.

splenitis (*Med.*). Inflammation of the substance of the spleen.

splenium (*Zool.*). A posterior bend of a commissure; as in the Vertebrate brain, that part of the corpus callosum which passes downwards and forwards to unite with the fornix.

splenius (*Zool.*). An anterior dorsal trunk muscle of higher Vertebrates.

splenocyte (*Histol.*). A large mononuclear leucocyte thought to be produced by the spleen.

splenomegaly (*Med.*). Abnormal enlargement of the spleen.

splenopexy (*Surg.*). Fixation, by suture, of the spleen to the abdominal wall.

splenotomy (*Surg.*). Incision of the spleen.

splice (*Carp.*). See **scarf.** (*Cinema.*) A joint in a cinematograph film made by removing emulsion and cementing the two bases together, either straight across at the edge of a frame or diagonally across a frame.

splice bump (*Cinema.*). See **bloop.**

spliced joint (*Elec. Eng.*). A cable joint in which the conductor strands are spliced, in the manner of a rope.

splice piece (*Rail.*). See **fish-plate.**

splicer (*Photog.*). Device for joining lengths of film.

splines (*Eng.*). A number of relatively narrow keys formed integral with a shaft, somewhat resembling long gear-teeth; produced by milling longitudinal grooves in the shaft (*external splines*); similarly, the grooved ways formed in a hole into which the splined shaft is to fit (*internal splines*). Used, instead of keys, for maximum strength.

splint bones (*Zool.*). In *Equidae*, the reduced second and fourth metacarpals and metatarsals; so-called on account of their position on either side of the third metacarpal or metatarsal.

splints (*Vet.*). Exostoses on the small metacarpal or metatarsal bones of the horse.

splint wood (*Bot.*). Wood in which living cells are scattered throughout.

split (*Leather*). A hide that has been split into two or more layers, parallel to the surface. (*Mining*) (1) Separating system for underground ventilating air. (2) The upper or lower portion of a divided coal-seam. (*Weaving*) The name applied to (1) a wire in a reed; (2) the space between two adjacent wires; (3) cloth woven with a centre selvedge which, when divided, produces two cloths.

split-anode magnetron (*Electronics*). Early type with split anode, to give a push-pull output from electrons from the filament cathode, when gyrating in a (nearly) coaxial magnetic field.

split-beam CRT (*Electronics*). A tube containing only one electron gun, but with the beam subdivided so that two traces are obtained on the screen.

split bearing (*Eng.*). A shaft bearing in which the housing is split, the bearing bush or brasses being clamped between the two parts.

split boards (*Bind.*). Two layers of board between which the tapes are securely held.

split bump (*Cinema.*). See **bloop.**

split compressor (*Aero.*). An axial-flow *gas turbine* (q.v.) compressor in which front and rear sections are mounted on separate concentric shafts (being powered by separate turbines) as a means of increasing the pressure ratio without incurring difficulties with *surge*. Also called a **two-spool compressor.**

split-conductor cable (*Cables*). A cable in which each conductor is divided into two sections lightly insulated from each other and connected in parallel at the ends. Used with special schemes of protection.

split-conductor protection (*Elec. Eng.*). A current-balance system of feeder protection avoiding the use of pilot wires by splitting each phase conductor into two parallel sections lightly insulated from each other.

split course (*Build.*). A course of bricks which have been cut lengthwise, so that the depth of the course is less than that of a brick.

split crankcase (*I.C. Engs.*). An engine crankcase split horizontally at about the centre line of the crankshaft. Cf. *barrel-type crankcase.*

split duct (*Print.*). Process in which the forme or plate can be printed in two or more colours by separating each from the other across the inking system.

split fitting (*Elec. Eng.*). A bend, elbow, or tee used in electrical installation work, which is split longitudinally so that it can be placed in position after the wires are in the conduit.

split flap (*Aero.*). A trailing edge flap in which only the lower surface of the aerofoil is lowered.

split-flow reactor (*Nuc. Eng.*). One in which the coolant enters at the central section and flows outwards at both ends (or vice versa).

split fractions (*Typog.*). Founts of figures and spaces which can be assembled to form any required fraction, the horizontal rule being cast, sometimes on the numerator, sometimes on the denominator; a fount of 10-pt split fractions would be cast on a 5-pt body.

split-image rangefinder (*Photog.*). Optical device generally used in conjunction with the camera lens to assist reflex focusing. Two very acute-angled prisms are juxtaposed so that their vertices lie together in the focal plane. When the image is out of focus, it is laterally displaced across the line where the vertices meet.

split pastern (*Vet.*). Fracture of the first phalanx of the foot of the horse.

split personality (*Psychol.*). See **multiple personality.**

split-phase (*Elec. Eng.*). A term denoting a circuit arrangement for changing a single-phase to a 2-phase supply.

split-pin (*Eng.*). One formed of wire with hemispherical section, bent round till the flat faces meet. See **cotter pin.**

split-pole converter (*Elec. Eng.*). A synchronous

converter in which the flux distribution under the poles can be varied by means of auxiliary windings on the individual pole limits.

split pulley (*Eng.*). A belt pulley split diametrically, the halves being bolted together on the shaft; used when a solid pulley cannot be fitted.

split-ring clutch (*Eng.*). A small friction clutch commonly used in machine tools. It consists of a split ring which is expanded into a sleeve by a cam or lever mechanism. See **friction clutch.**

splits (*Build.*). Bricks of the same length and breadth as ordinary bricks but of smaller thickness. (*Weaving*) See **split.**

split-seconds chronograph (*Horol.*). A chronograph with two independent centre seconds hands, one underneath the other. On pressing the button, the two hands travel together, but if a push piece in the band of the case is pressed, one hand remains stationary, the other continuing until stopped by a second pressing of the button. A third pressing of the button causes both hands to fly back to zero.

splitting limits (*Mining*). Divergence between assay of ore, concentrate or metal made by vendor and purchaser, inside which mutual adjustment will be made without going to arbitration.

splitting ratio (*Nuc. Eng.*). See **cut.**

splitting time (*Teleph.*). The period measured from the arrival of current at the V.F. signal receiver until the time when the through circuit is opened to avoid possible interference.

spodogram (*Bot.*). A preparation of the ash of a portion of a plant, especially a woody portion, used in investigating structure.

spodumene or **triphane** (*Min.*). A silicate of aluminium and lithium which crystallizes in the monoclinic system; a pyroxene. It usually occurs in granite-pegmatites, often in very large crystals. The rare emerald-green variety *hiddenite* and the clear lilac coloured variety *kunzite* are used as gems.

spoil (*Civ. Eng.*). The excess of cutting over filling on any given construction. Also called waste.

spoilage or **spoils** (*Print.*). (1) Accidental spoilage, the cost of reprinting being budgeted for in the costing system. (2) Ordinary spoilage, allowed for on each job.

spoil bank (*Civ. Eng.*). An earthwork bank formed by depositing spoil.

spoiler (*Aero.*). A device for changing the airflow round an aerofoil to reduce, or destroy, the lift. There are 3 principal types: (1) a small fixed spanwise ridge on the wing-root leading edge along the line of the *stagnation point* (q.v.) which improves lateral stability at the stall by ensuring that it starts at the root; (2) controllable devices at, or near, both wingtips which, by destroying lift on the side raised, impart a rolling moment to the aircraft; (3) small-chord spanwise flaps on top of the wing of a sailplane which can be raised to destroy a large part of the lift so as to make landing of the lightly-loaded aircraft more positive. Similar devices are often now fitted to jet aircraft so that by destroying lift at touchdown better braking can be achieved. See **thrust-.**

spokeshave (*Carp., Join.*). A form of double-handled plane which is used in shaping convex surfaces.

spondyl (*Zool.*). A *vertebra* (q.v.). *adj.* **spondylous.**

spondylarthritis (*Med.*). Inflammation of a vertebral joint.

spondylitis (*Med.*). Inflammation of the vertebrae.

spondylitis deformans (*Med.*). A condition in which the ligaments of the spinal column become ossified and the vertebrae fused together, so that the spine is bent, rigid, and immobile.

spondylolisthesis (*Med.*). A forward displacement of the 5th lumbar vertebra (carrying the vertebral column with it) on the sacrum.

sponge (*Elec. Eng.*). Loose, fluffy cathode deposit in electrolysis, contrasted with reguline.

Sponge (*Zool.*). See **Porifera.**

sponge beds (*Geol.*). Deposits, either calcareous or siliceous, which contain a large proportion of the remains of sponge organisms belonging to the phylum *Porifera.*

sponge cloth (*Textiles*). (1) Coarse cloth of plain (or irregular) weave, made from cotton yarns of about 4^s count. (2) Coarse open cloth of gauze structure.

spongicolous (*Zool.*). Living in association with a Sponge, usually within the chambers of the water-vascular system.

spongin (*Zool.*). A horny skeletal substance, occurring usually in the form of fibres, in various groups of *Porifera.*

sponginblast, spongoblast (*Zool.*). In *Porifera*, a cell which secretes spongin.

spongioblastoma (*Med.*). A soft, rapidly growing, malignant tumour occurring in the brain (or in the spinal cord) and derived from cells of the supporting structure of the brain.

spongioblasts (*Zool.*). Columnar cells of the neural canal giving rise to neuroglia cells.

spongophare (*Zool.*). In the *rhagon*-type of Sponge colony, the upper wall containing the flagellated chambers. Cf. *hypophare.*

spongy layer, -mesophyll, -parenchyma, -tissue (*Bot.*). A loosely constructed layer of irregularly shaped cells, separated by large intercellular spaces. It lies just above the lower epidermis of a dorsiventral leaf, and contains many chloroplasts.

spongy platinum (*Chem.*). The spongy mass resulting from the calcination of ammonium chloroplatinate(IV).

sponson (*Aero.*). A short, winglike projection from a flying-boat hull to give lateral stability on the water.

spontaneous alternation (*An. Behav.*). The behaviour sometimes shown by rats in multiple T- or Y-mazes, comprising regular alternation between left and right turns. An example of a *hypothesis* (q.v.).

spontaneous behaviour (*An. Behav.*). Behaviour occurring in the apparent absence of any stimuli.

spontaneous emission (*Phys.*). In contrast to *stimulated emission* (basis of laser action), process involving the loss of energy in an atomic system without any external stimulation. See **fluorescence.**

spontaneous fission (*Nuc.*). Nuclear fission occurring without absorption of energy. The probability of this increases with increasing values of the fission parameter Z^2/A (Z=atomic number, A = relative atomic mass) for the fissile nucleus.

spontaneous generation (*Biol.*). The production of living matter or organisms from nonliving matter. Also **autogeny.**

spontaneous ignition temperature (*Eng.*). The temperature at which a liquid or gaseous fuel will ignite in the presence of air or oxygen, measured, for liquid fuels, by allowing a drop to fall into a heated pot. Abbrev. **S.I.T.**

spontaneous remission (*Psychol.*). The disappearance of a neurosis without therapy, more easily explained in terms of *extinction* than by *psychoanalysis*.

spool (*Elec.*). The support of a coil. (*Weaving*) A pirn or bobbin (usually a bobbin of weft).

spoon bit (*Carp.*). A wood-boring bit of semi-cylindrical form, with the cutting edge ground to a conoidal shape. Also called **dowel bit, duck's-bill bit.**

sporadic E-layer (*Phys.*). Ionization in *E-layer* (q.v.) of ionosphere occurring erratically during the day as a result of particulate emission from sun, and leading to **sporadic reflections.**

sporadic lymphangitis (*Vet.*). Monday morning disease; weed. An acute disease of horses characterized by fever and lameness due to lymphangitis affecting the limbs. The disease is often recurrent and occurs especially in working horses that have been rested for a day or two on full working rations; the cause is not known.

sporangial sac (or **vesicle**) (*Bot.*). A very thin-walled and often evanescent outgrowth from the sporangium of many lower fungi, in which the zoospores complete their development, and from which they are set free.

sporangiocyst (*Bot.*). A thick-walled sporangium which is able to remain alive but inert under unfavourable conditions.

sporangiole, sporangiolum (*Bot.*). A sporangium which contains one spore or very few spores.

sporangioliferous head (*Bot.*). A rounded group of sporangiola.

sporangiophore (*Bot.*). A hypha or filament on which a sporangium is borne.

sporangiospore (*Bot.*). A spore—especially a non-motile spore—formed within a sporangium.

sporangium (*Bot.*). A walled structure in which spores are formed. (*Zool.*) In certain *Protozoa*, a capsule containing spores.

spore (*Bot.*). A reproductive body characteristic of plants. It consists of one cell or of a few cells, never contains an embryo, and when set free may, if conditions are favourable, give rise to a new plant. (*Zool.*) In *Protozoa*, a minute body formed by multiple fission: more strictly, a seedlike stage in the life cycle of *Protozoa* arising as a result of sporulation, and contained in a tough resistant envelope.

spore group (*Bot.*). A multicellular spore, each cell of which is capable of independent germination.

sporeling (*Bot.*). (1) The early stages of a plant developing from a spore. (2) A young fern plant.

spore mother cell (*Bot.*). A cell which divides to give four spores, meiosis occurring during the process. (*Zool.*) In *Protozoa*, a stage in the life history that will give rise to spores.

spore print (*Bot.*). The pattern obtained by placing the cap of an agaric, gills downwards, on paper, and allowing the sticky spores to fall and adhere to the paper.

spore sac (*Bot.*). The layer of cells in immediate contact with the sporogenous cells in the young capsule of a moss.

spore tetrad (*Bot.*). The group of four spores formed from a spore mother cell.

sporetia (*Cyt.*). See **idiochromidia.**

spori-, sporo-. Prefix from Gk. *sporos*, seed.

sporidesm (*Bot.*). See **spore group.**

sporidiiferous (*Bot.*). Bearing sporidia.

sporidium (*Bot.*). (1) A spore formed from a promycelium. (2) An old name for an *ascospore*.

sporoblast (*Bot.*). One cell or segment of a spore group. (*Zool.*) In *Protozoa*, a spore mother cell from which a spore will arise.

sporobola (*Bot.*). The path followed by a spore which is shot off horizontally from a basidium and then falls under the influence of gravity.

sporocarp (*Bot.*). (1) A multicellular structure in which spores are formed. (2) The spore-containing structure of some water ferns.

sporocyst (*Zool.*). The tough resistant envelope secreted by, and surrounding, a Protozoan spore.

sporocyte (*Zool.*). A spore mother cell.

sporodochium (*Bot.*). A cushion-shaped mass consisting of many conidiophores crowded together.

sporoducts (*Zool.*). In some *Gregarinidea*, ducts in the cyst through which the spores are ejected by the swelling of the residual protoplasm derived from the sporonts.

sporogenesis (*Bot., Zool.*). Spore formation.

sporogenous (*Bot.*). Producing or bearing spores.

sporogenous cell (*Bot.*). A spore mother cell.

sporogenous layer (*Bot.*). See **hymenium.**

sporogenous tissue (*Bot.*). A layer or group of cells from which spore mother cells are formed.

sporogonium (*Bot.*). The spore-bearing plant in the *Bryophyta*. It develops from the fertilized egg and lives as an almost complete parasite on the gametophyte.

sporogony (*Zool.*). In *Protozoa*, propagative reproduction, usually involving sexual processes and always ending in the formation of spores.

sporont (*Zool.*). A stage in the life history of some *Protozoa* which, as a gametocyte, gives rise to gametes, which in turn, after a process of syngamy, may give rise to spores.

sporophore (*Bot.*). The spore-bearing structure in Fungi.

sporophyll (*Bot.*). A leaf, more or less modified, which bears one or more sporangia, or subtends a sporangium.

sporophyte (*Bot.*). The spore-bearing plant.

sporoplasm (*Zool.*). (1) A binucleate amoebula stage in the life cycle of some *Protozoa*. (2) In the spores of certain *Sporozoa*, the body of protoplasm.

sporotrichosis (*Med.*). An infection of the skin (and rarely of muscles and bones) with the fungi of the genus *Sporotrichum*, causing granulomatous lesions.

Sporozoa (*Zool.*). A class of *Protozoa* comprising forms which are always parasitic, usually being at some stage intracellular, in the principal phase have no external organs of locomotion or are amoeboid, lack a meganucleus, and form large numbers of spores after syngamy, these being either sporozoites or undivided zygotes.

sporozoite (*Zool.*). In *Protozoa*, an infective stage developed within a spore.

sport (*Gen.*). Any individual differing markedly from the normal by reason of genetical factors; it may be due to a *mutation* (q.v.) or to other causes, such as a rare recombination of factors. Also called **rogue.**

sporulation (*Bot.*). The production of spores. (*Zool.*) In *Protozoa*, a form of multiple gemmation in which the parent organism breaks up almost completely into buds, leaving a little mass of residual protoplasm.

spot (*Electronics*). Point on a phosphor which becomes visible through impact of electrons in a beam. See **ion burn.**

spot board (*Build.*). The square wooden board on which the plasterer works up the coarse or fine stuff prior to applying it to the walls.

spot face (*Eng.*). Machined surface produced in the vicinity of a hole in a casting or other rough part, to permit a bolt head, washer, or adjacent part to seat evenly.

spot-knocking (*Electronics*). The removal of local

imperfections on an electrode by operating the tube concerned at very high voltages so that spark erosion of discharge points takes place.

spot level (*Surv.*). The reduced level of a point (usually on ground surface), not necessarily lying along a traverse or survey line.

spot punch (*Comp.*) Hand punch for correcting isolated errors on punched cards or tape.

spot speed (*Telecomm.*). In facsimile recording, the speed of the recording or scanning spot within the allotted time. In TV, the product of the number of spots in a scanning line multiplied by the number of scanning lines/second.

spotted fever (*Med.*). Meningococcal meningitis, cerebrospinal fever. See also **Rocky Mountain fever.**

spotted gum (*For.*). *Eucalyptus maculata*, from New South Wales and Queensland, growing up to 45 m in height. The sapwood is said to be susceptible to powder-post beetle and termite attack. It bends well after steam treatment. It is used for ship building, sleepers and wagon making.

spotted slate (*Geol.*). Argillaceous rock which has been altered by low or moderate grade metamorphism so that the homogeneous colouring matter of the rock becomes concentrated at some points and an early stage of crystallization of other minerals may be found in other points, giving a spotted appearance.

spotting (*Eng.*). The operation of turning a short length of a bar or forging to form a journal, by which the work is to be supported by the jaws of a steady rest. See **steady.** (*Horol., etc.*) A method of finishing plates or other flat surfaces with a regular pattern of circular patches.

spotting drill (*Eng.*). A flat drill having a point so shaped as to centre and face the end of a bar at one operation. See also **centre drill.**

spot welding (*Civ. Eng.*). The process of welding when the joint formed is required to resist only light and temporary stresses, e.g., many steel meshes are spot welded purely to resist stresses arising from handling and not for resisting applied structural loads. (*Elec. Eng.*) See **resistance spot-welding.**

spot wobble (*TV*). Oscillatory movement normal to scanning lines, to smooth them and make them less obvious. The amplitude of the wobble is slightly greater than the width of a line and about 200–400 vertical cycles occur during each line period. Now seldom used.

sprag (*Mining*). (1) Timber prop; short piece of wood, used to prevent the wheels of a train from revolving. (2) Slanting prop used to support coal face. Also **gib.**

sprag clutch (*Eng.*). One incorporating balls and wedging surfaces, which allows power to be transmitted from one rotating shaft to another in one direction only.

sprain (*Med.*). A wrenching of a joint with tearing or stretching of its ligaments, damage to the synovial membrane, effusion into the joint, occasionally rupture of muscles or tendons attached to the joint, but without dislocation.

spray arrester (*Elec. Eng.*). (1) A lightning arrester designed for removing accumulation of static charge; it consists of a spray of water from an earthed pipe impinging on a plate connected to the live circuit. (2) A sheet of glass placed over an open accumulator cell to prevent splashing of acid spray.

spray column (*Min. Proc.*). Tower packed with coarse material, e.g., coke, or set with grids or trays down which a liquid trickles countercurrent to rising gas so as to facilitate interaction. See **Glover tower.**

spray discharge (*Min. Proc.*). In high-tension separation the corona radiated from the electric discharge wire (18 000 volts or more) on to a passing stream of finely divided mineral particles. In classification, a spray-shaped discharge from the hydrocyclone.

spray drying (*Chem. Eng.*). Rapid drying of a solution or suspension by spraying into a flow of hot gas, the resultant powder being separated by a cyclone. Used to prepare powdered milk, detergent, fertilizer, etc.

spray gate (*Foundry*). An ingate consisting of a number of small separate gates, fed from the runner; used for shallow castings where there is insufficient depth for a single large gate. See **ingate.**

spray gun (*Civ. Eng., Paint., etc.*). An apparatus for forming by pneumatic pressure a fine spray or mist of paint or cement mortar, or asbestos, which can be directed on to the work.

spraying (*Met.*). The process of coating the surface of an article with metal by projecting on to it a spray of molten metal.

spray-shielded valve (*Electronics*). See **metallized valve.**

spray tower (*Chem. Eng.*). A plant for purifying gases, which pass up through a tower into which a suitable liquid is sprayed from the top.

spread (*Ecol.*). The establishment of a species in a new area. (*Print.*) An illustration or diagram occupying each side of the fold. (*Radio*) Angle, in degrees, within which fall a number of bearings, ostensibly of a distant radio transmitter, when corrected for site and other errors.

spreadboard (*Textiles*). A short conveyor on which cleaned and graded flax fibre is laid in handfuls for conversion into sliver.

spreader (*Radio*). Wooden or metal spar for keeping the wires of a multiwire antenna spaced apart.

spread factor (*Phys.*). See **distribution coefficients.**

Sprengel pump (*Vac. Tech.*). A vacuum pump which functions by entrainment of gas between successive short columns formed by a liquid (normally mercury) falling into a vertical tube.

sprig (*Build.*). A small nail with little or no head. (*Foundry*). Small nail pushed into a weak edge of sand in a mould to reinforce it during pouring. See **brad.**

sprig bit (*Join.*). See **bradawl.**

spring (*Eng.*). A device capable of deflecting so as to store energy, used to absorb shock or as a source of power or to maintain pressure between contacting surfaces or to measure force. See **carriage-, contact-, helical-, relay-, spiral-.**

spring assembly (*Telecomm.*). The collection of contact springs in a relay. The moving springs are pushed by the movement of the armature, on operation of the relay, so that the appropriate circuit changes are made.

spring-back (*Bind.*). A style of stationery binding in which the book opens quite flat. (*Eng.*) Elastic recovery of a work-piece after completion of a bending operation.

spring-balance (*Chem.*). Balance in which the weight of the sample is balanced by the extension of a spring.

spring bows (*Instr.*). Small compasses whose two limbs are not hinged together but are connected by a bow of spring steel, the distance apart of the marking points being adjusted by means of a screw.

spring constant (*Eng.*). See **spring rate.**

spring control (*Elec. Eng.*). A method of controlling the movement of an indicating instrument by means of a spring.

spring cramp (*Tools*). One in the shape of a broken ring made of tensile steel, used for light work.

Springdale Time-Life Scanner (*Print.*). Electronic equipment whose product is a set of colour-corrected continuous tone separations for making colour printing plates.

springer or **springing** (*Civ. Eng.*). The lowest voussoir on each side of an arch. Also **rein.**

springers (*Glass*). See **skewbacks.**

springing line (*Arch.*). The line joining the springings on both sides of an arch.

springings (*Arch.*). The lines where the intrados of an arch meets the abutments or piers.

spring-loaded governor (*Eng.*). An engine governor consisting of rotary masses which move outwards under centrifugal force and are controlled by a spring. See **Hartnell governor.**

spring-loaded idler (*Print.*). See **jockey roller.**

spring needle (*Textiles*). One with hook or beard which is flexed by the pressers on a knitting machine. Used to produce close and even texture in knitted fabrics. Also called **bearded needle.**

spring pawl (*Eng.*). See **pawl.**

spring points (*Rail.*). Points which are normally held closed by springs in determining the route when they are *facing points*, but can be passed through as *trailing points* (see **points**).

spring rate (*Eng.*). Ratio of load to deflection of a spring, measured, e.g., in newtons per millimetre. Also called spring **constant.**

spring safety valve (*Eng.*). See **safety valve.**

spring tab (*Aero.*). A *balance tab* connected so that its angular movement is geared to the compression or extension of a spring incorporated in the main control circuit. Its primary purpose is to reduce the effort required by the pilot to overcome the air loads on the main control surface resulting from high airspeeds. Cf. **servo-**, and **trimming-.**

spring tides (*Astron.*). Those high tides occurring when the moon is new or full, at which times the sun and moon are acting together to produce a maximum tide.

spring-tine harrow (*Agric.*). A harrow with rows of spring tines carried on bars, which can be partially rotated to adjust the working depth.

spring wood (*Bot.*). Secondary wood formed during the spring and early summer, often distinguished by the larger and thinner walled elements of which it is composed. Cf. *summer wood.*

sprinkler (*Build.*). A pipe system installed in a building, having at frequent intervals spray nozzles protected by connexions made of a fusible alloy; these, in the event of a fire, melt and release water for automatic fire fighting.

sprocket (*Carp.*). A small wedge-shaped piece of wood nailed to the upper surface of the lower end of a common rafter in cases where the latter is carried beyond the pole plate to form a projecting eave. It provides a break in the slope of the roof near the eaves. (*Cinema.*) A cylindrical wheel, with protruding pins on one rim or both rims for pulling film by means of perforations. (*Eng.*) See **sprocket wheel.**

sprocket holes (*Cinema.*). The accurately punched perforations on the edges of cinematograph film, for pulling it through cameras and projectors at the correct speed.

sprocket noise (or **hum**) (*Cinema.*). The extraneous noise, of 96 Hz fundamental, arising from a displacement of the film in a sound gate, so that the light from the exciter lamp passes through the adjacent perforations and is thereby modulated.

sprocket wheel (*Eng., etc.*). A toothed wheel used for chain drives, as on the pedal shaft and rear hub of a bicycle.

sprout cell (*Bot.*). A cell formed as a bud from a mother cell.

sprouting or **sprout germination** (*Bot.*). See **budding** (1).

spruce (*For.*). See **Canadian-.**

sprue (*Foundry*). See **gate.** (*Med.*) Psilosis. A disease affecting the gastrointestinal tract. Characterized by loss of energy, loss of weight, anaemia, inflammation of the tongue and the mouth, and by the frequent passage of pale, bulky, acid, frothy stools, there being inability to absorb adequately fat, glucose, and calcium. (*Plastics*) Waste plastic formed during injection moulding processes, being the material setting in the main inlet passages of the mould.

spun-dyed (*Textiles*). Man-made fibres formed from substances to which the colouring matter has been added before the filaments are formed.

spun silk (*Textiles*). Yarn made from silk waste which is spun in a manner very similar to the woollen and condenser systems.

spur (*Bot.*). (1) A tubular prolongation of the base of a petal or of a gamopetalous corolla. (2) An extension of the base of the leaf beyond its point of attachment. (3) See **short shoot.** (4) A short branch in many trees on which flowers and fruit are borne. (*Build.*) A strut. (*Geol.*) A hilly projection extending from the flanks of a valley. (*Zool.*) See **calcar.**

spur gear (*Eng.*). A system of gear-wheels connecting two parallel shafts. The pitch surfaces are cylinders with straight, helical, or double-helical teeth, generally of involute form.

spuriae (*Zool.*). In Birds, the feathers of the bastard wing.

spurious coincidences (*Nuc. Eng.*). Those recorded by a coincidence counting system, when a single particle has not passed through both or all the counters in the system. They usually result from the almost simultaneous discharge of two counters by different particles.

spurious counts (*Nuc. Eng.*). Those arising in counter tubes from leakages and defects in external quenching circuits.

spurious dissepiment (*Bot.*). A partition in a fruit which is not an ingrowth from the edges of the carpels nor an upgrowth from the receptacle. Also called **false septum.**

spurious fruit (*Bot.*). A group of fruits having the appearance of a single fruit; a fig is a familiar example.

spurious oscillation (*Telecomm.*). See **parasitic oscillation.**

spurious pregnancy (*Med.*). See **pseudocyesis.**

spurious pulse (*Nuc. Eng.*). One arising from self-discharge of particle counter leading to erroneous counting.

spurious radiation (*Telecomm.*). Undesired transmission, e.g., harmonics of carrier or modulation, outside specified band, causing interference with reception of other transmissions.

spurious response ratio (*Telecomm.*). Ratio of field strengths of signals producing spurious and required response in telecommunication receiving equipment, e.g., of image frequency or intermediate frequency relative to required signal frequency.

spur pelory (*Bot.*). An abnormal condition in which all the petals of an irregular flower develop spurs, so that the flower becomes regular.

spurrite (*Min.*). A carbonate and silicate of

calcium, $2Ca_2SiO_4{\cdot}CaCO_3$, occurring somewhat rarely in limestones containing silica in the metamorphic aureoles round basic igneous intrusions, as at Scawt Hill, Antrim.

Sputnik (*Space*). A series of Russian artificial satellites, first launched in October 1957.

sputtering (*Electronics*). In a gas discharge, the removal of atoms from the cathode by positive ion bombardment, like a cold evaporation. These unchanged atoms deposit on any surface and are used to coat dielectrics with thin films of various metals.

sputum (*Med.*). Matter composed of secretions from the nose, throat, bronchi, or lungs, which is spat out.

spyndle (*Textiles*). A unit of length (14 400 yd) used in counting jute and dry spun flax yarns.

Spy Wood Grit (*Geol.*). A sandstone-grit bed which forms the basal member of the Caradocian stage in the Ordovician System in the Shelve district of Shropshire.

S.Q. (*Build.*). Abbrev. for *squint quoin*.

squalene (*Chem.*). A symmetrical triterpene, formerly known as *spinacene*, originally detected in shark oil but also occurring in mammalian and plant tissues. $(CH_3)_2C{=}CH{\cdot}CH_2{\cdot}(CH_2{\cdot}C{\cdot}CH_3{=}CH{\cdot}CH_2)_2(CH_2{\cdot}CH{=}C{\cdot}CH_3{\cdot}CH_2)_2CH_2{\cdot}CH{=}C(CH_3)_2$. It is the biological precursor of lanosterol and related tetracyclic triterpenes and of pentacyclic triterpenes.

squall (*Meteor.*). A temporary increase in the mean wind speed, lasting for some minutes.

squama (*Bot., Zool.*). A scale: a scalelike structure. (*Zool.*) In some *Crustacea*, a scalelike body arising from the second joint of the antenna and believed to represent the exopodite: in *Diptera*, the *antisquama* (q.v.). *adjs.* **squamate, squamose, squamous.**

Squamata (*Zool.*). An order of Reptiles in which the skull has lost either one or both temporal vacuities, the vertebrae are nearly always procoelous, the sacrum, when present, consists of two vertebrae, the quadrate is movably articulated with the skull, and there is no inferior temporal arch. Snakes and Lizards.

squamation (*Zool.*). See **pholidosis.**

squamiform (*Zool.*). Scalelike.

squamosal (*Zool.*). A paired membrane bone of the Vertebrate skull which overlies the quadrate.

squamous epithelium (*Zool.*). Epithelium consisting of one or more layers of flattened scalelike cells; pavement epithelium.

squamula (*Zool.*). A small scale: in Insects, the tegula: in some *Gymnophiona*, one of the small rounded areas of the pouch scales.

squamule (*Bot.*). A small scale. *adj.* **squamulose.**

square (*For.*). A piece of square-section timber of side up to 6 in (15 cm). (*Maths.*) A rectangle whose adjacent sides are equal.

squared rubble (*Build.*). Walling in which the stones are roughly squared to rectangular faces but are of irregular size.

square folding (*Bind.*). See **right-angled folding.**

square joint (*Carp., Join.*). A joint formed between the squared ends of two jointing pieces which come together but do not overlap.

square law (*Phys.*). The law of inverse squares expressing the relation between the amount of radiation falling upon unit area of a surface and the distance of the surface from the source. (*Telecomm.*) Said of any device, such as a rectifier or (de)modulator, in which the output is proportional to the square of the input amplitude.

square-law capacitor (*Elec. Eng.*). Variable vane capacitor, used for tuning, in which capacitance is proportional to the square of the scale reading, so that wavelength of circuit which it tunes becomes directly proportional to it.

square-law demodulator (*Elec. Eng.*). See **detector.**

square-law rectifier (*Elec. Eng.*). One in which the rectified output current is proportional to the square of the applied alternating voltage.

square-mile foot (*Elec. Eng.*). A hydraulic constant defined as the quantity of water that will cover an area of a square mile to a depth of 1 ft. It is equal to 27·88 million ft^3 (789 000 m^3) or 174 million gal. (789 million litres).

square parachute (*Aero.*). A parachute of which the canopy is approximately square when laid out flat.

square rebate plane (*Join.*). A rebate plane with its cutting edge square across the sole.

square roof (*Arch.*). A roof in which the principal rafters enclose an angle of 90°.

square root (*Maths.*). The square root of a number is that quantity which when multiplied by itself gives the number.

squares (*Bind.*). The protrusion of the covers of a book beyond the leaves.

square sets (*Mining*). See **girt.**

square staff (*Build.*). An angle staff of square-section material, as distinct from an angle bead.

square step (*Build.*). An individual stone step in a stair which consists of a solid block, rectangular in section, either lapping over the back edge of the step below or rebated to fit over it.

square thread (*Eng.*). A screw thread of substantially square profile.

square wave (*Telecomm.*). Pulse wave with very rapid (theoretically zero) rise and fall times; and pulse duration equal to half period of repetition. Mark-space ratio of unity.

squaring bands (*Spinning*). See **steadying bands.**

squaring shears (*Eng.*). Manual or power-operated press used for shearing sheets of steel.

squaring-up (*Build., Civ. Eng.*). A process following *taking-off* in drawing up a bill of quantities, superficial areas of items being calculated by multiplying the relevant dimensions entered on the dimensions paper.

squarrose (*Bot.*). Rough, with many scales or hairs standing out at right angles.

squeaking (*Telecomm.*). Using a gliding tone, either from a beat (heterodyne) oscillator or from a record, for testing frequency-response curve of lines, amplifiers, recorders, etc.

squeegee (*Photog.*). A rubber roller or brush for squeezing out surplus water, for good contact between photographic surfaces or supports.

squeeze (*Print.*). The amount of impression between plate and impression cylinders.

squeezer (*Eng.*). A moulding machine, operated by hand, compressed air, hydraulic power, or magnetic means, in which the sand is squeezed or compressed into the box and round the pattern by a ram.

squeeze track (*Cinema.*). This indicates closing the light-valve ribbons in variable-density recording, or cutting off superfluous light or deflecting the modulating mirror in variable-area recording, so that the negative sound-track is as transparent as possible. It ensures that the positive print, allowing for the desired modulation, is as dense as possible, so that noise arising from scratches and grain is minimized.

squegging (*Telecomm.*). Mode of oscillation of an oscillator when operated under certain conditions, e.g., with excessive resistance in the grid circuit. The oscillations build up to a certain value and then abruptly stop, the process being repeated at a rate determined by the time constant of the capacitance and resistance of the grid circuit. Used also in a CRT time-base generator.

squegging oscillator (*Telecomm.*). One in which normal oscillation is periodically shut off because of grid current blocking and release. Used in super-regenerative receivers. Also **self-quenching oscillator.**

squelch (*Telecomm.*). Colloquialism for automatically reducing the gain of a receiver to cope with a strong signal.

squid (*Aero.*). A dynamically stable condition of a fully-deployed parachute canopy which will not fully distend.

squinch (*Arch.*). A small arch running diagonally across the corner of a square tower or room, to support a side of an octagonal tower or spire above. Also called a **scoinson arch.**

squint (*Build.*). A purpose-made brick of shape suiting it for use as a squint quoin. (*Med.*) See **strabismus.** (*Radio*) Difference between the geometrical axis of an aerial array and the axis of the radiation pattern; particularly applicable to very-high-frequency Yagi arrays.

squint quoin (*Build.*). A quoin enclosing an angle which is not a right angle. Abbrev. **S.Q.**

squirrel-cage motor (*Elec. Eng.*). An induction motor whose rotor winding consists of a number of copper bars distributed in slots round the periphery, with the ends solidly connected to two heavy copper end-rings, the whole forming a rigid cage embedded in the rotor. See also **cage rotor, cage winding.**

squirrel-cage rotor (*Elec. Eng.*). See **cage rotor.**

squirrel-cage winding (*Elec. Eng.*). See **cage winding.**

squirted filament (*Elec. Eng.*). An electric-lamp filament prepared by squirting a composition through a die and subsequently heat treating.

sr (*Maths.*). Symbol for *steradian.*

Sr (*Chem.*). The symbol for *strontium.*

s-RNA (*Biochem.*). Soluble-RNA, a low molecular weight sequence of nucleic acids which serves as the adaptor between the m-RNA and the ribosome. It is visualized as having two ends, one for attaching to a specific amino-acid, and the other for attaching to its appropriate codon on the m-RNA. There are at least as many different species of s-RNA as there are different amino-acids. Also called **t-RNA.**

S.R.V. (*Electronics*). Abbrev. for *surface recombination velocity.*

s/s (*Print.*). An instruction to the process worker to make the block or plate the same size as the original copy.

s.s. and s.c. lathe (*Eng.*). A sliding, surfacing, and screw-cutting lathe, i.e., one suitable for working on the periphery and on the end faces of workpieces, and capable of cutting a screw thread using a single-point tool.

SST (*Aero.*). Abbrev. for *s*uper*s*onic *t*ransport aeroplane.

s state (*Electronics*). State of zero orbital angular momentum.

S.T. (*Build.*). Abbrev. for *surface trench.*

stab (*Typog.*). See **establishment.**

stabbing (*Bind.*). A special kind of *flat stitching* required for very thick books, pairs of wires being necessary (driven from the front page and from the back) clenching not being possible; often (wrongly) used as a synonym for *flat stitching.* (*Build.*) Making a brick-work surface rough to provide a key for plasterwork.

stabilate (*Biol.*). A population of an organism preserved in viable condition on a unique occasion. Distinguished from a strain because of its assured stability.

stabilator (*Aero.*). See **all-moving tail.**

Stabilit (*Elec. Eng.*). A vulcanized rubber insulating material similar to ebonite and vulcanite.

stability. A general property of mechanical, electrical, or aerodynamical systems, whereby the system returns to a state of equilibrium after disturbance, any consequent oscillation dying away through dissipation of energy. (*Aero.*) An aircraft has three axes about which to be stable and three degrees of freedom (angular, normal displacement, and change of velocity) about each. See **lateral-, longitudinal-, static-.** (*Elec. Eng.*) The property of a transmission system whereby there is a general tendency for changes in load demand to be met without any falling out of step on the part of synchronous machines.

stability derivatives (*Aero.*). Quantitative expressions for the variation of forces and moments on an aircraft due to disturbances of steady motion.

stability factor (*Radio.*) Ratio $\frac{\Delta I_c}{\Delta I_{c0}}$ for transistor where ΔI_c is change in collector current for a particular circuit, which results from a change in temperature, and ΔI_{c0} is the corresponding change measured under conditions of zero emitter current. The larger this factor the greater the sensitivity of the circuit to temperature changes.

stability test (*Cables*). In this test, the cable is subjected to working voltage (or a higher voltage) whilst it is alternately heated and allowed to cool. The power factor is measured during each heating and cooling period. If the power factor increases steadily during the test, the cable is said to be unstable.

stabilivolt (*Electronics*). Tube containing a gaseous discharge through a series of electrodes, potentials between which tend to a greater constancy than the voltage applied through a resistance to end electrodes.

stabilization (*Bot.*). Establishment of an equilibrium between vegetation of a locality and external conditions, especially climatic. (*Electronics*) Maintenance of voltage at a point in a circuit, against variations of load or supply voltage. Series pentode valves or shunt *Zener diodes* may be used. (*Radio*) Maintenance of frequency of a transmitter within limits prescribed internationally for its type of service. Often effected by quartz-crystal drive in a thermostatically-controlled oven.

stabilized feedback amplifier (*Telecomm.*). One in which amplification is stabilized against changes in supply voltages, etc., by the application of negative feedback.

stabilized glass (*Glass*). (1) Glass heat-treated so that it is in an equilibrium state corresponding to some particular temperature. (2) Glass heat-treated so as to suffer no permanent change of dimensions or properties over a particular range of temperature. (3) Glass resistant to darkening by high-energy, short-wave radiation.

stabilizer (*Aero.*). In U.S., *horizontal stabilizer* is *tail plane*, and *vertical stabilizer* is *fin.* See **automatic-.** (*Chem.*) (1) A negative catalyst. (2) A substance which makes a solution stable.

stabilizer tube (*Electronics*). Gas-discharge tube, the voltage across which is much more stable than a voltage applied to it in series with a resistor.

stabilizing choke (*Elec. Eng.*). A reactive choke coil inserted in series with an electric discharge lamp to compensate its negative resistance characteristic.

stable (*Chem., etc.*). Said of systems not exhibiting sudden changes, particularly atoms which are not radioactive. (*Nuc.*) Used to indicate the incapability of following a stated mode of spontaneous change, e.g., *beta stable* means incapable of ordinary beta disintegration but capable of isomeric transition or alpha disintegration, etc. (*Automation*). State of an amplification system when it satisfies the *Nyquist criterion*, either conditionally or unconditionally.

stable equilibrium (*Mech.*). The state of equilibrium of a body when any slight displacement increases its potential energy. A body in stable equilibrium will return to its original position after a slight displacement.

stable orbit (*Nuc. Eng.*). The circle of constant radius described by accelerated particles; cf. *betatron* and *synchrotron*.

stable oscillation (*Telecomm.*). One for which amplitude and/or frequency will remain constant indefinitely. A statically stable system may be dynamically unstable and follow a divergent oscillation when subjected to a disturbance. In this sense, the term *dynamically stable* means that an induced oscillation will be convergent, i.e., of decreasing amplitude.

stable platform (*Telecomm.*). Structure which can be controlled in position with great precision, e.g., by gyroscopes, and which forms base for other information to be measured and transmitted by telemetry, e.g., from satellites.

stable pneumonia (*Vet.*). See **equine influenza.**

stachyose (*Chem.*). $C_{24}H_{42}O_{21}+4\frac{1}{2}H_2O$, a tetrasaccharide, found in the roots of *Stachys tuberifera* and of several *Labiatae*; m.p. (anhydrous) 170°C.

stack (*Nuc.*). Pile of photographic plates exposed to radiation together, and used to study tracks of ionizing particles. (*Teleph.*) See **pile-up.**

stacked array (*Radio*). See **tiered array.**

stacked ceramic tube (*Electronics*). A development of the *megatron* or lighthouse tube with electrodes separated by ceramic discs. These tubes operate at such elevated temperatures that additional cathode heating is superfluous.

stacking elevator (*Agric.*). A machine to lift crops to stack by means of a powered endless conveyor.

stack mixing (*Textiles*). A method of mixing cotton, spun rayon or wool of different types by piling it in horizontal layers, to produce raw material suitable for spinning perfectly uniform yarns. See **direct mixing.**

stadia hairs (*Surv.*). The two additional horizontal hairs, one on each side of the central hair, fitted to the diaphragm of a telescope to be used in *tacheometry* (q.v.).

stadia rod (*Surv.*). A special form of levelling staff bearing bold graduations suitable for the long sights usual in stadia tacheometry.

stadium (*Zool.*). An interval in the life history of an animal between two consecutive ecdyses.

staff (*Build.*). See **angle staff.** (*Horol.*) An arbor or axis, especially that of the balance or pallets. (*Surg.*) A grooved rod introduced into the urethra as a guide for cutting a stricture. (*Surv.*) See **levelling staff.**

staff angle (*Build.*). See **angle staff.**

staff bead (*Build.*). See **angle bead.**

staffman (*Surv.*). The surveyor's assistant whose duty it is to hold the levelling staff while the instrument is sighted upon it and readings are being taken.

Staffordian Series (*Geol.*). The so-called Transition Group of the British Coal Measures, between the Middle and Upper Coal Measures, in the Carboniferous System. They include the Newcastle-under-Lyme Group and the Etruria Marl, and the Blackband Group in N. Staffordshire.

Staffordshire blues (*Build.*). Hard, dense, and almost impervious bricks of the *engineering brick* (q.v.) class, dark blue in colour owing to an oxide of iron content.

stage (*Build., Civ. Eng.*). A ledge or working platform associated with scaffolding. (*Geol.*) A succession of ricks which were deposited during an age of geological time. A subdivision of a geological series. (*Nuc.*) Unit of cascade in isotope separation plant, consisting of single separative element, or group of these elements, operating in parallel on material of same concentration.

stage efficiency (*Elec. Eng.*). Ratio of a.c. output power to d.c. input power for any stage of an electronic amplifier.

stage micrometer (*Micros.*). A device for measuring the magnification achieved with a given microscope or for calibrating an eyepiece graticule. It usually consists of a small accurate scale mounted on a microscope slide.

stage separation factor (*Nuc. Eng.*). See **simple process factor.**

stagger (*Aero.*). The horizontal distance between the leading edges of the wings of a multiplane as projected vertically. If the upper plane is ahead of the lower, stagger is positive, if behind it is negative.

staggered tuning (*Radio*). Attempt to get a wide band response by a number of tuned circuits, having slightly different frequencies of resonance.

staggering (*Elec. Eng.*). A term signifying the displacement of the brushes of a commutator motor from the neutral zone.

staggers (*Vet.*). See **megrims.**

stagger-tuned amplifier (*Radio*). One with couplings tuned to different high frequencies to give a band-pass response.

stag-headed (*Bot.*). Said of a tree which, owing to a diseased condition, is devoid of twigs and leaves at the top, some dead main branches standing up like antlers.

staging (*Build.*). See **builder's staging.**

stagnation point (*Aero.*). The point, at or near the nose of a body in motion in a fluid, where the flow divides and where, in a viscous fluid pressure is at a maximum, and in an inviscid one the fluid is at rest.

stagnation temperature (*Aero.*). The temperature which would be reached if a flowing fluid were brought to rest adiabatically, which is almost applicable in supersonic flight for the leading edges and air intakes, where the air in the boundary layer of a body is drastically and rapidly decelerated. Also **total temperature.**

stagnicolous (*Ecol.*). Living in stagnant water.

stainer (*Paint.*). A pigment added to paint when a final colour is required which is different from that of the base used. Also called **colouring pigment.**

staining power (*Chem.*). The degree of intensity of colour which a coloured pigment will impart when mixed with a standard white pigment under standardized conditions.

stainless steel (*Met.*). Corrosion-resistant steel of a wide variety of compositions, but always containing a high percentage of chromium (8–25%). The following are typical compositions: 13% Cr, 0·35% C; 18% Cr, 0·15% C, 8% Ni; 20% Cr, 0·25% C, 10% Ni, 1·2% Nb. These are highly resistant to corrosive attack by organic acids, weak mineral acids, atmospheric oxidation, etc. Used for cutlery, furnace parts, chemical plant equipment, stills, valves, turbine blades, ball-bearings, etc.

staking (*Eng.*). Fastening operation similar to rivet setting, in which a projection on one part is upset by means of a punch so as to fit tightly against a mating feature in another part.

staking machine (*Leather*). A machine in which leather undergoes a process of stretching and softening, to prevent cohesion of the fibres while drying.

stalactite (*Geol.*). A concretionary deposit of calcium carbonate which is formed by percolating solutions and hangs icicle-like from the roofs of limestone caverns.

stalactited (*Build.*). A term applied to a variety of rusticated work distinguished by having ornaments resembling icicles on the faces of the stone.

stalagmite (*Geol.*). A concretionary deposit of calcium carbonate, precipitated from dripping solutions on the floors and walls of limestone caverns. Stalagmites are often complementary to stalactites, and may grow so that they eventually join with these.

stalagmometer (*Chem.*, *Phys.*). Apparatus which measures surface tension of liquid in terms of mass of a drop leaving a specified orifice.

staling (*Bot.*). The accumulation in the substratum of waste metabolic products from bacteria or fungi, rendering the substratum unfit to sustain further growth of the organisms. (*Zool.*) Applied to cultures of cells or micro-organisms in general, when growth is retarded or stopped by the accumulation of toxic waste products.

stalk (*Civ. Eng.*). The upright part of a reinforced concrete retaining wall, springing from the horizontal base. (*Horol.*) The thin rod or wire which carries the hammer of a striking or chiming clock.

stall (*Aero.*). The progressive breakdown of the lift-producing airflow over an aerofoil, which occurs just after the angle of maximum lift. (*Eng.*) Of an engine, to stop owing to the too sudden application of a load or brake. (*Mining*) The working compartment or room in the *bord-and-pillar* (q.v.) method of working coal; a coal-miner's working place.

stallboard (*Join.*). A substantial and often wide sill at the foot of the window sash of a shop front.

stalling speed (*Aero.*). The airspeed of an aeroplane at which the wing airflow breaks down.

stalling torque (*Elec. Eng.*). The overload torque which is sufficient to slow down to zero the speed of an electric motor operating under load.

stalloy (*Met.*). Steel with 3·5% silicon content, which has low hysteresis loss, and is used for transformer stampings.

stall riser (*Build.*). The upright part, of wood, marble, etc., between the pavement and the stallboard of a shop front.

stall-warning indicator (*Aero.*). A device fitted to aeroplanes which do not give any positive warning of the approach of the stall by *buffeting* (q.v.). Usually operated by the change of pressure and movement of the *stagnation point* (q.v.) near the stall, warning may be audible, visual, or by a stick-shaking (or forward-pushing) electric motor. See **stick pusher**.

stamen (*Bot.*). One of the members of the flower, which produces pollen. It usually consists of a slender filament surmounted by an anther in which the pollen develops.

staminal (*Bot.*). Pertaining to a stamen: derived from a stamen.

staminate (*Bot.*). Said of a flower possessing stamens but not pistils, and, by extension, of an inflorescence consisting of such flowers.

staminode (*Bot.*). An imperfectly developed or vestigial stamen.

staminose (*Bot.*). Said of a flower in which the stamens are very obvious.

stamp (*Mining*). (1) To crush. (2) A freely falling weight, attached to a long rod and lifted by means of a cam; once widely used for crushing ores.

stamping (*Elec. Eng.*). See **lamination.**

stamping press (*Bind.*). See **blocking press.**

stanchion (*Civ. Eng.*). A pillar, usually of steel, for the support of a superstructure.

standard. Established unit of measurement, or reference instrument or component, suitable for use in calibration of other instruments. Basic standards are those possessed or laid down by national laboratories or institutes (N.P.L., B.S.I., etc.). (*Bot.*) The large petal which stands up at the back of the flower of the pea and related plants, and does much to make the flower conspicuous. Also called **banner, vexillum.** (*Build.*) (1) One of the upright poles forming a principal part of a scaffold. (2) A measure of the volume of timber in bulk, viz., 165 cu. ft. (*For.*) See **board foot, Petersburg standard.**

standard atmosphere (*Aero.*). See **International Standard Atmosphere.** (*Phys., etc.*) Unit of pressure, defined as 101 325 N/m^2; equivalent to that exerted by a column of mercury 760 mm high at 0°C. Abbrev. **atm.**

standard beam approach system (*Aero.*). A system of radio navigation which provides an aircraft with lateral guidance and marker-beacon indications at specific points during its approach. Abbrev. **SBA.**

standard book number. A number allotted to a book by agreement of (international) publishers which shows area, publisher and individual title, plus a check digit. Abbrevs. **SBN, ISBN.**

standard cable (*Telecomm.*). See **mile of standard cable.**

standard calomel electrode (*Chem.*). A half-element consisting of mercury, a paste of mercury and calomel (mercury (I) chloride), and a standard solution of potassium chloride saturated with calomel; used as a standard potential difference in e.m.f. measurements.

standard (or parliamentary) candle (*Light*). A source of light of standard intensity once used in photometry but now replaced by more reliable standards such as the pentane lamp and the Hefner lamp.

standard cell (*Elec.*). See **Weston standard cadmium cell.**

standard chamber (*Nuc. Eng.*). Ionization chamber used for calibration of radioactive sources, or of absolute values of exposure doses.

standard deviation (*Stats.*). The root of the average of the squares of the differences from their mean, $\bar{x}$, of a number, n, of observations, x. Usually denoted by s or σ. Thus

$$\sigma^2 = \frac{1}{n} \Sigma (x-\bar{x})^2.$$

The most frequently employed measure of the spread of a series of values from their mean. Cf. *variance.*

standard electrode potential (*Chem.*). The potential of a chemical element dipping into a solution containing its ions at unit activity, referred to that of hydrogen under a pressure of one atmosphere as zero.

standard film stock (*Cinema.*). The normal size of

film used in theatres and film studios, i.e., 35 mm in width (specialized wide-screen apparatus sometimes uses broader stock (e.g., CinemaScope 55·625 mm, ToddAO 65 mm)).

standard filter (*Photog.*). A filter which, when placed in front of a specified source, e.g., a tungsten lamp, gives a standard light—white of black-body temperature 4800 K.

standard form (*Maths.*). See **canonical form.**

standard frequency (*Elec. Eng.*). Fifty or sixty Hz (cycles/second), the standards of *power frequency* in most countries of the world.

standard gauge (*Rail.*). That employed in most countries of the world, viz. 4 ft $8\frac{1}{2}$ in (1·435 m).

standard hour (*Work Study*). Unit representing the amount of unrestricted work carried out in 1 hr at standard performance, i.e., at 100 B.S. scale.

standard illuminant (*Light*). One used for accurate colour measurements. The C.I.E. specify 3 alternative standards: A, B, & C with corresponding colour temperatures: 2848 K, 4800 K, 6500 K (the latter sources being realized by use of specified liquid filters).

standard knot (*For.*). A knot which is $1\frac{1}{2}$ in (40 mm) or less in diameter.

standard man (*Radiol.*). Averaged characteristics of adult human body as specified in the recommendations of the International Commission on Radiological Protection.

standard mean chord (*Aero.*). The average chord, i.e., gross wing area divided by the span.

standard measurement (*Build.*). The method recommended by the Chartered Surveyors' Institution for measurement of building works.

standard M-gradient (*Meteor.*). The constant rate of change of excess modified refractive index with height regarded as equivalent to normal propagation conditions at a given place on the earth. (The excess modified refractive index is the amount by which the modified refractive index exceeds unity.)

standard oxidation-reduction potential (*Chem.*). The potential established at an inert electrode dipping into a solution containing equimolecular amounts of an ion or molecule in two states of oxidation.

standard page (*Print.*). The largest size of page on any particular rotary press, there being no standard size.

standard performance (*Work Study*). The rate of output which qualified workers will naturally achieve without over-exertion as an average over the working day or shift when adhering to the specified method.

standard propagation (*Radio*). With standard refraction, the propagation of radio waves over a perfectly smooth earth with uniform electrical characteristics.

standard radio atmosphere (*Radio*). Radio atmosphere having the *standard M-gradient* (q.v.).

standard rating (*Work Study*). The *rating* (q.v.) corresponding to the average rate at which qualified workers will naturally work at a job.

standard reflector (*Light*). A reflector which conforms with the British Standard Specification for industrial reflectors.

standard refraction (*Radio*). Refraction arising in a *standard radio atmosphere.*

standard refractive modulus gradient (*Meteor.*). Reference based on 0·12 M units/m (3·6 units/100 ft). See **standard M-gradient.**

standard signal generator (*Radio*). An oscillator whose output is calibrated as regards frequency and amplitude, and sometimes depth of modulation; used for the testing of radio equipment, receivers, etc.

standard solenoid (*Elec. Eng.*). A laboratory standard of inductance consisting of an air-cored solenoid with a secondary coil located at its centre. The dimensions are such that a value of the mutual inductance between the windings can be calculated from them.

standard solution (*Chem.*). A solution whose strength is known. Such solutions are used in *volumetric analysis.*

standard specification (*Eng., Elec. Eng.*). A *specification* (q.v.) incorporating standard quantities and features, usually of a restricted range, drawn up and issued by, for example, the British Standards Institution.

standard temperature and pressure. See **s.t.p.**

standard time (*Astron.*). The system of time-reckoning with reference to selected meridians (usually 15° apart) and changing by one hour from one zone to the next, thus avoiding the use of many differing *local times* (q.v.). For navigational and astronomical purposes *Greenwich Mean Time* (*universal time*) is the accepted standard. *Central European Time* (1 hour in advance of G.M.T.) is the standard for, e.g., Germany, Italy, Switzerland. In the U.S. and Canada five zones are commonly used, viz. *Atlantic* (4 hours behind G.M.T.), *Eastern* (5 hours), *Central* (6 hours), *Mountain* (7 hours) and *Pacific* (8 hours) standard times. In some countries the adopted civil time differs from the true standard time for that area; thus in France and Spain the clocks are permanently advanced by one hour, giving Central European Time. Also **zone time.** See **date line.**

standard volume indicator (*Telecomm.*). One specified for indicating power levels on a transmission line, particularly when the level is fluctuating, as with speech.

Standard Wire Gauge (*Eng.*). One of the systems of *gauge numbers* (q.v.) legally recognized (until 1964) in the U.K., in which numbers from 0,000,000 (=0·5 in.) to 50 (=0·001 in.) represent wire diameters or sheet thickness. Abbrev. **S.W.G.** Also called **Imperial** or **British Standard Wire Gauge.** See BS 3737 and ISO/R 388.

stand-by losses (*Elec. Eng.*). That part of the power expended in a generating station to maintain plant in instant readiness to take a sudden load.

standing crop biomass (*Ecol.*). The total dry weight of organisms present at any one time, this being larger for larger organisms, since smaller organisms have greater metabolism per gram of biomass.

standing current (*Telecomm.*). See **quiescent current.**

standing formes (*Typog.*). Type kept locked up in formes in the expectation of a reprint.

standing-off dose (*Radiol.*). Absorbed dose after which occupationally exposed radiation workers must be temporarily or permanently transferred to duties not involving further exposure. (Doses are normally averaged over 13-week periods and standing-off would then continue for the remainder of the corresponding period.)

standing panel (*Join.*). A door panel whose height is greater than its width.

standing wave (*Telecomm.*). Pattern of maxima and minima when two sets of oppositely travelling waves of the same frequency interfere with each other. They can be produced, e.g., by a source of sound held over the open end of a pipe of suitable length closed with a rigid bung; in this case it is the incident and reflected waves which interfere.

standing-wave indicator (*Telecomm.*). Voltage detector, germanium or silicon diode, for sliding along a *slotted line*, detecting maxima and minima of voltages in an electromagnetic standing-wave system. Shunt for sliding along open transmission wire, to detect maxima and minima of its currents. In acoustic standing wave, the maxima and minima may be detected by means of a probe microphone.

standing-wave meter (*Telecomm.*). One designed to measure voltage standing-wave ratio in wave-guide.

standing-wave ratio (*Telecomm.*). Where standing and progressive waves are superimposed, the SWR is the ratio of the amplitudes at nodes and antinodes. For a transmission line or wave-guide, it is equal to $\frac{1-r}{1+r}$, where r is the co-efficient of reflection at the termination. It may alternatively be defined by the reciprocal of this value as shown by its value being numerically greater than unity.

standing ways (*Ships*). The portion of a ship's launching ways which are fixed to the ground. The sliding ways move on these ways and are positioned by an upstanding rib integral with the fixed ways.

stand-insulator (*Elec. Eng.*). An insulator on which stands the structure used for supporting an accumulator or battery.

stand-off bomb (*Aero.*). A small, fast powered aircraft containing a nuclear warhead, which is released from a bomber to fly a hundred or more miles to the target. It is automatically piloted and navigated, usually by a *Doppler* and/or *inertial system*.

stand oil (*Paint.*). A lithographic varnish, also used in printer's inks, paints, and enamels. Obtained by heating, from old clarified tanked oil. It may be prepared from any *drying* or *semidrying oil* (q.v.).

stand pipe (*Eng.*). An open vertical pipe connected to a pipeline, to ensure that the pressure head at that point cannot exceed the length of the stand pipe.

stand sheet (*Build.*). A window having no frame. Also called **fast sheet, fixed sash.**

standstill (*Elec. Eng.*). A term pertaining to the electrical behaviour of a machine when it is at rest.

standstill torque (*Elec. Eng.*). The load torque which would bring an electric motor to a standstill.

Stanley Kent's fibres (*Zool.*). The fibres which constitute the auriculoventricular bundle of the heart and maintain connexion between the muscle of the auricles and the muscle of the ventricles. Also **His's bundle.**

stannane (*Chem.*). Tin hydride, SnH_4.

stannates(II) (*Chem.*). See **stannites.**

stannates(IV) (*Chem.*). Analogous to the carbonates. Formed by heating solutions of, say, tin(IV)chloride with alkaline carbonates.

stannic(IV) acid (*Chem.*). Acids of two types, formed by action of alkalis on solutions of tin(IV) chloride and by the action of nitric acid on the metal; called respectively α-stannic(IV) acid and metastannic acid or β-stannic(IV) acid.

stannic oxide (*Chem.*). Tin(IV)oxide. SnO_2. Formed (1) by combustion of tin, (2) when stannic acids are calcined. Forms alkali stannates(IV) when fused with alkali carbonates.

stannite (*Min.*). A sulphostannate of copper and iron, which crystallizes in the tetragonal system. It usually occurs in tin-bearing veins, having been deposited from hot ascending solutions. It is also called **bell-metal ore, tin pyrites.**

stannites(II) (*Chem.*). Stannates(II). Salts of stannous acid. Formed when stannous hydroxide is dissolved in alkaline solutions.

Stannius' corpuscles (*Zool.*). In Fish, inter-renal glandular organs scattered in or on the ventral surface of the mesonephros.

stannotype (*Photog.*). A photomechanical process in which an exposed and developed bichromated film is coated with tinfoil and used directly for pressure printing.

stannous hydroxide (*Chem.*). Tin(II)hydroxide. $Sn(OH)_2$. Precipitated when sodium hydroxide is added to a solution of tin(II) chloride. When heated in carbon dioxide, forms black tin(II) oxide, SnO, which, heated in air, forms tin(IV) oxide, SnO_2.

Stanton number (*Chem. Eng.*). A dimensionless parameter equal to the reciprocal of the *Prandtl number*.

stapedectomy (*Surg.*). Excision of the stapes.

stapes (*Zool.*). In Amphibians, a small nodule of cartilage in connexion with the fenestra ovalis of the ear: in Mammals, the stirrup-shaped innermost auditory ossicle. *adj.* **stapedial.**

staphylococcus (*Bacteriol.*). A Gram-positive coccus of which the individuals tend to form irregular clusters. The commonest types, associated with various acute inflammatory and suppurative conditions, including mastitis in animals, are *S. aureus* (golden yellow colonies) and *S.* (*pyogenes*) *albus* (white colonies).

staphyloma (*Med.*). Local bulging of the weakened sclera of the eye (as in glaucoma or myopia); bulging of a corneal scar in which the iris of the eye has become fixed.

staphylorrhaphy (*Surg.*). The operation of closing a cleft in the soft palate.

staple. (1) A U-shaped piece of bar or wire with ends pointed to be driven part-way into a softer part to take a padlock or hook or to secure a wire. (2) A wire fastener in the form of three sides of a rectangle, the free ends being pointed for driving through sheets of paper or other materials and subsequently bent to secure. (*Mining*) An internal shaft connecting two coal-seams. Also **staple pit.** Cf. *winze.* (*Textiles*) Quality of any short fibre, estimated by length, strength and colour.

star (*Astron.*). Celestial body similar to the sun, an incandescent mass undergoing thermonuclear reactions, e.g. the *carbon cycle* or *proton-proton chain.* See **magnitudes, nova, proper motion, spectral types, variable stars.** (*Photog.*) Radiating tracks in photographic emulsion, arising from particle disintegration on collision, and dispersal of energy to other particles.

starch (*Chem.*). Amylum $(C_6H_{10}O_5)_x$; polysaccharide found in all assimilating (green) plants. A white hygroscopic powder which can be hydrolysed to dextrin and finally to D-glucose. Diastase converts starch into maltose. Starch does not reduce Fehling's reagent and does not react with phenylhydrazine. It forms a blue compound with iodine.

star charts (*Astron.*). The name given to systematic and accurately made maps of the heavens in which the star positions are generally plotted according to equatorial coordinates.

starch crescent (*Bot.*). A strand of cells, crescentic in cross-section, containing starch grains which are presumed to act as statoliths.

starch grain (*Bot.*). A rounded or irregularly shaped inclusion in a cell consisting of a series of layers of starch, giving a stratified appearance, surrounding a central hilum.

starch gum (*Chem.*). See **dextrin.**

starch plant (*Bot.*). A plant in which the carbohydrate formed in excess of immediate requirements is stored in the cells of the leaf as temporary starch.

starch sheath (*Bot.*). (1) A 1-layered cylinder of cells lying on the inner boundary of the cortex of a young stem, with prominent starch grains in the cells. It is homologous with an endodermis. (2) A layer of starch grains around a pyrenoid in an algal cell.

star clusters (*Astron.*). See **galactic clusters, globular clusters, open clusters.**

star connection (*Elec. Eng.*). That of a source or load to a 3-phase transmission system so that each of three windings connects a common neutral point to one conductor, as opposed to delta connection where each winding connects two conductors. See **star point, voltage to neutral.**

star-delta starter (*Elec. Eng.*). A starting switch for an induction motor which, in one position, connects the stator windings in star for starting and, in the other position, reconnects the windings in delta when the motor has gained speed.

Stark effect (*Phys.*). Splitting of atomic energy levels, and of corresponding emission spectrum lines, by placing source in region of strong electric field; cf. *Zeeman effect*.

Stark-Einstein equation (*Chem.*). The energy absorbed per mole for a photochemical reaction is $E = Nh\nu$, where N is Avogadro's constant, h is Planck's constant and ν is the frequency of the absorbed light.

starling (*Civ. Eng.*). Piling driven around a bridge pier to afford it some protection.

starlite (*Min.*). A name suggested (from a fancied resemblance to starlight) for the blue zircons which are heat-treated and used as gem-stones.

star lots (*Textiles*). Wool offered for sale at the London sales in lots not usually exceeding three bales.

star magnitudes (*Astron.*). See **magnitudes.**

star-mesh transformation (*Telecomm.*). Technique for simplifying network, whereby any number of branches meeting at a point can be replaced by an equivalent mesh, thereby reducing the number of connexions.

star network (*Telecomm.*). One with many branches connected at a point; a *T*- or *Y-network* has 3 branches.

star observation (*Surv.*). Use of theodolite to locate or orient a ground station by sighting on a star.

Star Peak Group (*Geol.*). A group which comprises three great limestones with three intervening quartzites in Nevada.

star point (*Elec. Eng.*). The common junction of the several phases of a star-connected 3-phase system.

star-quad (*Elec. Eng.*). See **quad.**

starred signature (*Bind., Print.*). When a section is to be inserted in another it is given the same signature mark followed by an asterisk (*) or the figure 2.

star ruby, -sapphire, -quartz (*Min.*). The prefix 'star' has reference to the narrow-rayed star of light exhibited by varieties of the minerals named. The star is seen to best advantage when they are cut *en cabochon*. It is caused by reflections from exceedingly fine inclusions lying in certain planes. See also asterism.

star shake (*For.*). A number of shakes radiating from the heart of a log.

star-streaming (*Astron.*). A phenomenon, discovered from analysis of observed stellar motions (after removing the effects of the observer's own motions), by which the stars are found to have two preferential directions of motion, one towards the point R.A. 90°, declination 15° south, and the other towards R.A. 285°, declination 64° south; the first stream contains about 60% of the observed stars. The effect is due to the rotation of the Galaxy.

star target (*Print.*). A circular symbol with numerous radii, tapering to the centre but not meeting, leaving a small clear inner circle, used particularly on lithographic plates as a guide to maintenance of colour and quality of printing.

starter (*Elec. Eng.*). A device for starting an electric motor and accelerating it to normal speed. Also **motor starter.** (*Electronics*) See **pilot electrode.**

starter gap (*Electronics*). The conducting path between the pilot electrode and the electrode to which the starting voltage is applied in a glow-discharge tube.

starter voltage (*Electronics*). That applied to the pilot electrode of a cold-cathode discharge tube or mercury-arc rectifier; cf. *starting voltage*.

starting box (*An. Behav.*). An enclosure at the beginning of a maze, or a simple straight runway, in which an animal is placed before being released and allowed to run through the maze.

starting current (*Elec. Eng.*). The current drawn by a motor from the mains when starting up.

starting resistance (*Elec. Eng.*). A fixed resistance connected in series with the main circuit of a motor when starting up.

starting sheet (*Met.*). A sheet of pure metal used as the initial cathode on which the metal being refined is deposited during electrolytic refining.

starting torque (*Elec. Eng.*). The torque developed by a motor at starting.

starting-up time (*Nuc. Eng., etc.*). That required by instrument or system (e.g., nuclear reactor, chemical plant, etc.) to reach equilibrium operating conditions.

starting voltage (*Electronics*). That which initiates current passing in a gas-discharge tube after nonconduction; much greater than that required to maintain conduction. Also known as **striking voltage** or **ionizing voltage.** (*Nuc.*) See **threshold voltage.**

starting winding (*Elec. Eng.*). An auxiliary winding on the armature of a single-phase motor (enabling it to start up as a 2-phase machine) or of a synchronous converter.

start-stop (*Teleg.*). In machine telegraphy, the principle by which depression of a key on the keyboard of a transmitting machine sends the corresponding code to line, together with start and stop signals, the former to trip the printing machine so that it scans the sent code correctly, the latter to restore the scanning mechanism to the condition required for its operation by the next signal to arrive.

start-up procedure (*Nuc. Eng.*). That followed when bringing a nuclear reactor into operation. It involves four successive stages: (*a*) *source range*, where a neutron source is introduced to generate the required neutron flux; (*b*) *counter range*, where reactor is just critical but counters are required to monitor neutron-flux changes; (*c*) *period range*, where changes in reactivity are monitored on period meter; (*d*) *power range*, where reactor is operating within its designed power ratings.

star voltage (*Elec. Eng.*). See **voltage to neutral.**

star wheel (*Eng.*). (1) Toothed wheel moved one

tooth per revolution of an adjacent shaft by a pin attached to that shaft; cf. *Geneva movement*. (2) A continuously or intermittently rotating disk with scalloped circumference, used to guide bottles or other containers on to a conveyor at correct intervals. (*Horol.*) A wheel with pointed triangular teeth. (*Typog.*) On *line-casting machines* a four-toothed wheel which stops rotating, to indicate to the operator that a line of matrices is too long.

stasimorphy (*Zool.*). Structural modification due to arrested development.

stasis (*Bot.*). Stoppage of growth. (*Med.*) (1) Complete stoppage of the circulation of blood through the capillaries and smallest blood vessels in a part. (2) Arrest of the contents of the bowel at any point from obstruction or weakness of the bowel wall.

Stassfurt Deposits (*Geol.*). A series of saline deposits found in the Triassic rocks at Stassfurt, Saxony, which include halite, anhydrite, kieserite, gypsum, and boracite.

stassfurtite (*Min.*). A massive variety of boracite which sometimes has a subcolumnar structure and resembles a fine-grained white marble or granular limestone. See **boracite.**

stat- (*Elec.*). Prefix to name of unit, indicating derivation in obsolete electrostatic system of units, e.g., statampere, statohm, etc.

state (*Nuc.*). The energy level of a particle as specified by the appropriate quantum numbers.

statement number (*Comp.*). See **instruction number.**

statenchyma (*Bot.*). A tissue consisting of cells containing statoliths.

state of matter (*Chem.*). Traditionally all matter was in one of three states: solid (fixed volume and shape), liquid (fixed volume, shape that of container), gaseous (filling the containing vessel). Now is added plasma, the state of ionized nuclei.

stathmokinesis (*Cyt.*). Inhibition of cell division.

static (*Elec. Eng., etc.*). Nonmovable or nonrotating, e.g., a transformer or rectifier is a static converter. (*Radio*) Said of all electrical disturbances to a radio system which arise through electrostatic induction, particularly from lightning flashes.

statical stability (*Ships*). The ability of a ship to return to her initial position when forcibly inclined.

static balancer (*Elec. Eng.*). See **a.c. balancer.**

static bending (*For.*). The bending of timber specimen or member, supported at points near the ends, when it is subjected to a load at one or more intermediate points, applied at a low steady rate to the dimensions.

static breeze (*Med.*). Electric brush discharge used in therapy.

static capacitor (*Elec. Eng.*). A static piece of electrical apparatus having the characteristic property of drawing a leading current from an a.c. supply, in consequence of which it is widely used for power-factor correction.

static characteristic (*Elec. Eng.*). Curve, or set of curves, which describes relation between specified voltages and currents of electrodes under unvarying conditions, as compared with *dynamic characteristic* (q.v.), which implies operation under normal load conditions.

static convergence (*TV*). The adjustment to a colour TV whereby the three coloured *rasters* are made to coincide in the centre of the screen.

static discharge wicks (*Aero.*). Wicks, usually of cotton impregnated with metal silver, or of nichrome wire, fitted at the trailing edges of an airplane's flight control surfaces, by which static electricity is discharged into the atmosphere.

static electricity (*Elec.*). See **dynamic electricity.**

static frequency changer (*Radio*). A transformer having a magnetically saturated iron core arranged to accentuate the harmonic content of the secondary current; formerly used to produce high-frequency currents for radio transmission.

static impedance (*Elec. Eng., etc.*). The electrical impedance of a machine or transducer when it is stopped from moving. In loudspeakers, it has the same meaning as *blocked impedance*.

static inverter (*Aero.*). A nonrotating device for converting d.c. current to a.c. supply, usually of high voltage, for radio and instrument services.

static jet thrust (*Aero.*). See **static thrust.**

static line (*Aero.*). A cable joining a parachute pack to the aircraft, so that when the wearer jumps, the parachute is automatically deployed.

static machine (*Elec. Eng.*). See **electrostatic generator.**

static marks (*Cinema.*). Marks on film caused by exposure to sparks of static electricity induced by friction.

Staticon (*TV*). TN for a *photoconducting* TV camera.

static pressure (*Aero.*). The pressure at any point on a body moving freely with a fluid in motion; in practice, the pressure normal to the surface of a body moving through a fluid.

static-pressure tube (*Aero.*). A tube with openings placed so that when the air is moving past it the pressure inside is that of still air. See **Pitot tube, static vent.**

statics. That branch of applied mathematics which studies the way in which forces combine with each other usually so as to produce equilibrium. Until the early part of the 20th century the term also embraced the study of gravitational attractions, but this is now normally regarded as a separate subject.

static stability (*Aero.*). Positive static stability in an aircraft means that if it is disturbed from a trimmed speed there will be no tendency for the disturbance to increase. (*Elec. Eng.*) The stability of a transmission system with reference to gradual changes in load demand.

static thrust (*Aero.*). The net thrust of a jet engine at *International Standard Atmosphere* sea level and without translational motion; quoted in pounds in the British system; 2·6 lb thrust is equivalent to 1 h.p.; abbrevs. **lb.s.t.** or **pst.**

static vent (*Aero.*). An opening, usually about half way along the fuselage, found by experiment, where there is minimum *position error* and which is used instead of the *static-pressure tube*.

station (*Elec. Eng.*). In general, a generating station. Specifically, a key point on an electricity supply system. (*Eng.*) Location, on a transfer machine or on an assembly line, at which a workpiece or an assembly is halted for the placing of a component or for the execution of a machining or fastening operation. (*Radio*) Location of radio transmitters and/or receivers with antennae, for sending or receiving radio signals on one or more wavelengths. (*Surv.*) (1) A point at an apex of a triangle in a skeleton, or otherwise situated in a line of the skeleton. (2) A point whose reduced level is to be found.

stationary baler (*Agric.*). A ram or press baler for use whilst stationary.

stationary dredger (*Civ. Eng.*). The type of bucket-ladder dredger which discharges the dredged materials into attendant vessels, and consequently dredges more or less continuously.

stationary gap (*Electronics*). A device used in conjunction with a pulse-forming line to form short period high-voltage pulses for modulating microwave oscillators.

stationary orbit (*Astron.*, *Space*). That circular orbit of a satellite which holds it fixed above a point on its parent body's equator. For the earth, such an orbit is about 35 800 km above the equator. Very useful for communications relays since the surface antennae do not require to be steered. Also **synchronous orbit.**

stationary period (*Cinema.*). That fraction of the complete time cycle, expressed in degrees, during which the mechanism holds the frame stationary in the gate. See **period.**

stationary phase (*Chem.*). See **chromatography.**

stationary point on a curve (*Maths.*). A point at which the tangent is parallel to the x-axis. For the curve $y=f(x)$, the point where $x=a$ is a stationary point if $f'(a)=0$. All turning points are stationary points but not all stationary points are turning points.

stationary points (*Astron.*). Those points in the apparent path of a planet where its direct motion in right ascension changes to retrograde motion, or vice versa.

stationary wave (*Radio*). Earlier alternative to **standing wave.**

station pointer (*Surv.*). Instrument for obtaining a mechanical solution of the *three-point problem* (q.v.). It consists of a full-circle protractor with one fixed radial arm and two movable radial arms, which can be set to the correct mutual directions of the 3 points.

station roof (*Build.*). A roof which is cantilevered out to one side or to both sides of a single line of stanchions; used for roofing railway platforms or loading bays for vehicles.

statistical diameters (*Powder Tech.*). Statistical average of a specified parameter on microscopic methods of particle-size analysis.

statistical error (*Nuc.*). That arising in measurements of average count rate for random events, as a result of statistical fluctuations in the rate.

statistical mechanics (*Phys.*). Theoretical predictions of the behaviour of a macroscopic system by applying statistical laws to the behaviour of component particles. *Quantum mechanics* is an extension of classical statistical mechanics introducing the concepts of the quantum theory, especially the Pauli exclusion principle. *Wave mechanics* is a further extension based on the Schrödinger equation, and the concept of particle waves.

statistical weight (*Nuc.*). In nuclear reactors, the effect on reactivity of introducing an absorber at a given site. Also called **weighting factor.** (*Phys.*) In statistical mechanics, the number of possible microscopic states for the system.

statoblast (*Zool.*). In some *Ectoprocta*, an internal bud arising on the funiculus and becoming surrounded by a chitinous capsule which, unlike the colony, can survive the rigours of winter and germinate to produce a new colony in the spring.

statocone (*Zool.*). One of a number of small calcareous granules occurring in the statocyst in various animals.

statocyst (*Bot.*). A cell containing starch grains or other solid inclusions which act as statoliths. The contents of the cell are rather fluid, so that the statoliths move readily to the lower face of the cell if the position of the latter is altered by a displacement of the plant. Also called **statocyte.** (*Zool.*) An organ for the perception of the position of the body in space, consisting usually of a sac lined by sensory cells and containing a free hard body or bodies, either introduced or secreted; an otocyst.

statolith (*Bot.*). A solid inclusion in a cell, such as a starch grain, which moves readily in the somewhat fluid contents of the cell, comes to rest on the portion of the protoplast lining the lower wall of the cell, and, it is said, plays some part in the perception of gravity by plants. (*Zool.*) A secreted calcareous body contained in a statocyst.

stator (*Aero.*). The row of fixed, radially disposed aerofoils which forms an essential part of the dynamics of an axial compressor or axial turbine. (*Elec. Eng.*) Stationary part of machine, especially a dynamo or motor; cf. *armature* (2).

stator blade (*Aero.*). A small aerofoil, usually of thin highly-cambered section, and of approximately parallel chord, mounted in the outer case of an axial compressor or turbine. See **exhaust-.**

stator core (*Elec. Eng.*). The assembly of laminations forming the magnetic circuit of the stator of an a.c. machine.

statorhab (*Zool.*). In some *Trachomedusae*, a tentacular process bearing the statolith and projecting into the cavity of the statocyst.

stator-rotor starter (*Elec. Eng.*). A combined stator circuit switch and rotor-circuit regulating resistance for use with slip-ring induction motors.

stator winding (*Elec. Eng.*). That part of the electrical winding of a machine accommodated in the stator.

status epilepticus (*Med.*). A succession of severe epileptic convulsions with no recovery of consciousness between each convulsion.

status lymphaticus (*Med.*). An ill-defined condition in which there are hyperplasia of lymphatic tissue and enlargement of the thymus; patients in this state are liable to sudden death during anaesthesia.

staurolite (*Min.*). Silicate of aluminium, iron and magnesium, with chemically combined water, sometimes occurring as brown cruciform twins, and crystallizing in the orthorhombic system. It is typically found in medium grade regionally metamorphosed argillaceous sediments.

Stauromedusae (*Zool.*). A subclass of *Scyphozoa*, including sessile forms superficially more like a polyp than a medusa, but usually supposed to be a medusa. The egg develops into an individual exactly like the parent.

stave (*Textiles*). See **heald.**

stay (*Eng., etc.*). See **brace.**

stay pile (*Civ. Eng.*). A pile which is driven into the earth as an anchorage for a *land tie* (q.v.).

stay tap (*Eng.*). A long tap for threading the holes for stays connecting adjacent plates in boilers, thus ensuring that the two holes are threaded in correct pitch relation.

stay tubes (*Eng.*). Boiler fire tubes acting as stays to the flat surfaces which they join; sometimes threaded and nutted to the plates for extra strength.

stay wire (*Elec. Eng.*). One of several steel cables by which a transmission-line pole is secured to the ground.

S.T.D. (*Teleph.*). Abbrev. for *subscriber trunk dialling.*

steady (*Eng.*). A support for backing up slender work in the lathe, attached either to the bed or the carriage. It consists of 3 slotted radial jaws, adjusted to bear on the rough-turned work. Also called **back rest, back stay.**

steady flow (*Phys.*). See **viscous flow.**

steady head (*Min. Proc.*). Water supply system from storage at fixed height above draw-off points, ensuring constant pressure and volume under set conditions of delivery.

steadying bands (*Spinning*). Ropes running round pulley underneath a spinning mule carriage, which ensure that the ends make the same movement as the centre of the carriage during its outward and inward run. Also called **squaring bands.**

steadying resistance (*Elec. Eng.*). The ballast resistance placed in series with a d.c. arc lamp, to counteract negative resistance in the arc.

steady pin. A pin which permits mechanical parts to be fitted together accurately with one fixing screw. (*Horol.*) A pin used where two parts have to be fixed accurately relative to one another; a steady pin is fixed to one part and is a close fit in a hole in the other part. Used for the location of bridges, cocks, etc.

steady state (*Phys.*). One in dynamic equilibrium, with entropy at its maximum. (*Telecomm.*) Said of any oscillation system which continues unchanged indefinitely.

steady-state theory (*Astron.*). See **continuous creation.**

steam (*Phys.*). Water in the vapour state; formed when specific latent heat of vaporization is supplied to water at boiling point. The specific latent heat varies with the pressure of formation, being approximately 2257 kJ/kg at atmospheric pressure. See **dry-, saturated-, super-heated-.**

steam accumulator (*Eng.*). A large pressure vessel, partly filled with water, into which surplus high-pressure steam is blown and condensed. A supply of saturated steam is thus available by lowering the pressure at the outlet valve, thus causing evaporation of the water stored. See **Ruth's accumulator.**

steam car (*Eng.*). An automobile propelled by steam. Oil-fired *flash boilers* (q.v.) are generally used, no gearbox is necessary, and control is simple. Water is recovered by condensing the exhaust steam in a radiator.

steam chest (*Eng.*). The chamber in which the slide-valve of a steam engine works, and to which the steam pipe is connected.

steam coal (*Mining*). Two varieties classified by the National Coal Board are *dry steam coal* (also called *semianthracite*) and *coking steam coal* (rank 202, 203, 204).

steam distillation (*Chem.*). The distillation of a substance by bubbling steam through the heated liquid. It is a useful method of separation for substances which are practically insoluble in water. The rapidity with which a substance distils in steam depends on its vapour pressure and on its vapour density.

steam dome (*Eng.*). See **dome.**

steam economizer (*Eng.*). See **economizer.**

steam-electric generating set (*Elec. Eng.*). A generating set in which the prime mover is a steam engine, e.g., a steam turbine or reciprocating steam engine.

steam engine (*Eng.*). An external combustion engine whose working fluid is steam.

steam-generating heavy water reactor (*Nuc.*). A class of high-temperature, high-flux, compact deuterium-moderated nuclear reactors. A U.K. prototype with a rating of 100 MWe has been built at Winfrith and is cooled by light water boiling in pressure tubes in the core, the outlet steam driving the turbines directly.

steam generating station (*Elec. Eng.*). A generating station in which the prime movers driving the electric generators are operated by steam, e.g., steam turbines or reciprocating steam engines.

steam generator (*Eng.*). A steam boiler.

steaming up (*Vet.*). The practice of increasing the nutritional plane of dairy cattle a few weeks before calving.

steam injector (*Eng.*). See **injector.**

steam jacket (*Eng.*). A jacket formed round a steam-engine cylinder; supplied with live steam to prevent excessive condensation of the working steam in the cylinder.

steam lap (*Eng.*). See **outside lap.**

steam locomotive (*Eng., Rail.*). A steam engine and boiler integrally mounted on a frame which is fitted with road wheels driven by the engine. The term is usually restricted to locomotives used to haul passenger or goods traffic on a railway. See also **traction engine.**

steam navvy (*Civ. Eng.*). A mechanical excavator having a single large bucket ($\frac{1}{2}$ to 8 yd^3, 0·4 to 6 m^3) at the end of a long beam carried in a revolving jib. Also called **steam shovel.**

steam nozzle (*Eng.*). See **nozzle, convergent-divergent nozzle.**

steam ports (*Eng.*). Passages leading from the valve face to the cylinder of a steam engine; through them the steam is supplied and exhausted.

steam reversing gear (*Eng.*). A power reversing gear, used in steam locomotives, by which movement of the driver's reversing lever admits steam to an auxiliary cylinder, whose piston operates the reversing links of the valve gear.

steam shovel (*Civ. Eng.*). See **steam navvy.**

steam tables (*Eng.*). List of figures giving the properties of steam over a pressure range.

steam trap (*Eng.*). A device into which condensed steam from steam pipes, etc., is allowed to drain, and which automatically ejects it without permitting the escape of steam.

steam turbine (*Eng.*). A machine in which steam is made to do work by expanding so as to create kinetic energy, which is then partly absorbed by causing the steam to act on moving blades attached to a disk or drum. See **back-pressure turbine, disk-and-drum turbine, extraction turbine, impulse turbine, mixed-pressure turbine, reaction turbine.**

steaning (*Civ. Eng.*). See **steining.**

steapsin (*Zool.*). A fat-digesting enzyme occurring in the digestive juices of various animals, as the pancreatic juice of Vertebrates.

stearic acid (*Chem.*). $C_{18}H_{36}O_2$, a monobasic fatty acid; m.p. 69°C, b.p. 287°C; obtained from mutton suet, or by reducing oleic acid. It occurs free in a few plants, as glycerides in many fats and oils, and as esters with the higher alcohols in certain waxes.

stearin (*Chem.*). A term for the glyceryl ester of stearic acid. The name is also applied to a mixture of stearic acid and palmitic acid.

steatite or **soapstone** (*Min., etc.*). A coarse, massive, or granular variety of talc, greasy to the touch. On account of its softness it is readily carved into ornamental objects. Fired at 800 to 1000°C, it becomes very hard and strong. Used in the making of sparking-plug insulators and giant high-voltage insulators.

steatopygia (*Med.*). Gross accumulation of fat in the buttocks. A Hottentot deformity.

steatorrhoea, steatorrhea (*Med.*). The presence of an excess of fat in the stools, due either to failure of absorption, or to deficiency of the fat-splitting enzymes in the digestive juices, as a result of disease of the pancreas.

steel (*Met.*). Essentially an alloy of iron and carbon. Contains less than 2% carbon, less than 1% manganese, and small amounts of silicon, phosphorus, sulphur, and oxygen. Mechanical properties can be varied over a wide range by changes in composition and heat treatment. Mild steel contains less than 0·15% C, medium 0·15-0·3%C, hard more than 0·3%C. If used as a structural material in reactors as a neutron or thermal shield, it must be outside the active section. This restriction is due to some of its constituents becoming strongly radioactive, with long lives, on capturing neutrons. See also **alloy steel, stainless steel.**

steel-cored aluminium (*Elec. Eng.*). An electrical conductor consisting of a layer or layers of aluminium wire surrounding a core of galvanized steel strands.

steel-cored copper conductor (*Elec. Eng.*). A conductor made in the same way as steel-cored aluminium, except that the steel core is covered by a layer of insulating tape, to prevent corrosion of the surrounding copper.

steel gauge (*Eng.*). A *gauge* (q.v.) for indicating or recording steam pressure in a boiler or other part of a steam system.

steel-making (*Met.*). The process of making steel from pig-iron, with or without admixture with steel scrap. The processes used include *Bessemer, open-hearth, crucible, electric-arc, high-frequency induction*, and *duplex.*

steel-pipe pressure cable (*Cables*). See **pressure cable.**

steel-tank rectifier (*Elec. Eng.*). A mercury-arc rectifier in which the arc chamber is of steel. Cf. *glass-bulb rectifier.*

steel tape (*Surv.*). See **band chain.**

steel tower (*Elec. Eng.*). The framed steel structure carrying a high-voltage transmission line. Also called **pylon.**

steely (*Brew.*). Term applied to barley grains which are glassy and translucent, making them unsuitable for malting.

steening (*Civ. Eng.*). See **steining.**

steeping (*Brew.*). Soaking barley for a period of about 50 hr in water at a temperature of 50-55°F (10-13°C), to induce germination: a preliminary stage in malting.

steeple-head (*Med.*). See **oxycephaly.**

steerable antenna (*Radio*). Fixed multiple antenna, major lobe of sensitivity being adjustable by altering the phase of the separate contributions of the elements. See **musa.**

steering (*Radio*). Alteration by mechanical or electrical means of the direction of maximum sensitivity of a directional antenna, e.g., a radar or radio telescope.

steering arm (*Autos.*). An arm rigidly attached to a stub axle, to which it transmits angular movement from the motion of the steering rod, and attached to it by a ball joint.

steering box (*Autos.*). The housing which encloses the steering gear and provides an oil-bath for the working surfaces. It is rigidly attached to a side-member of the chassis frame. It contains gears which transmit the action of the steering column to the steering rod.

steering gear (*Autos.*). The two geared members attached to the steering column and the drop-arm spindle respectively. They transmit motion from the steering wheel to the stub axles through the drop arm, steering rod or drag link, steering arms, and track rod. See **cam-type-, rack-and-pinion-, screw-and-nut-, worm-and-wheel-.**

steering rod (*Autos.*). See **drag link.**

Stefan-Boltzmann law (*Phys.*). Total radiated energy from a black body per unit area per unit time is proportional to the 4th power of its absolute temperature, i.e., $E=\sigma T^4$, where σ (Stefan-Boltzmann constant) is equal to $5{\cdot}69\times 10^{-8}$ W m^{-2} K^{-4}.

stegma (*Bot.*). A small elongated cell nearly filled with silica.

stegocarpous (*Bot.*). Said of the capsule of a moss when it is provided with a lid.

stegocrotaphy (*Zool.*). The condition of having the temporal region of the skull without fossae. *adj.* **stegocrotaphic.**

Steinach's operation (*Surg.*). Ligature of the vas deferens (the excretory duct of the testis) for the purpose of increasing sexual vigour in the male.

Steiner's tricusp (*Maths.*). A hypocycloid in which the radius of the rolling circle is one-third or two-thirds that of the fixed circle. It has three cusps.

Steinhall mix (*Paper*). A paste consisting of raw starch and starch granules which have been burst, together with urea-formaldehyde resin, for use as a heat-seal adhesive.

steining (*Civ. Eng.*). The process of lining a well with bricks, stone, timber, or metal, so as to prevent the sides from caving in.

Steinmetz coefficient (*Mag.*). The constant of proportionality in the *Steinmetz Law.* Also called the **hysteresis coefficient.**

Steinmetz law (*Mag.*). Empirical statement that within the range 0·1 to 1 tesla the energy loss in a ferromagnetic material per unit volume during each cycle of a hysteresis loop is proportional to $B^{1{\cdot}6}_{max}$ where B_{max} is the maximum absolute value of the flux density attained.

Stekelian analysis (*Psychol.*). The analytical method introduced by Stekel of Vienna which stresses mainly the importance of the active and intuitive approach, as contrasted with the more passive method of Freudian analysis. It uses the *shock factor*, intentionally producing profound emotional reaction, in lessening the duration of analytic treatment.

stele (*Bot.*). The central region of a stem or root, containing the vascular tissues, often with a central pith and an external parenchymatous sheath, the pericycle, in which some sclerenchyma may be present.

stellarator (*Nuc. Eng.*). Twisted torus for nuclear fusion experiments. The reverse curvatures are to balance out first-order drift, which is serious in the simple torus because of variation of magnetic field across the section of the plasma. (*stellar* gener*ator.*)

stellar energy (*Astron.*). See **carbon cycle, proton-proton chain.**

stellar interferometer (*Astron., Phys.*). A device, developed by Michelson, by means of which, when fitted to a telescope, it is possible to measure the angular diameters of certain giant stars (all of which are below the limit of resolution of even the largest telescopes) by observations of interference fringes at the focus of the telescope.

stellar magnitudes (*Astron.*). See **magnitudes.**

stellar populations (*Astron.*). See **population types.**

stellar rotation (*Astron.*). The rotation of certain stars may be detected by *line broadening* (q.v.); the actual circumstances can be deduced in some eclipsing binaries, but not in the case of a single star.

stellate (*Bot., Zool.*). Radiating from a centre, like a star.

stellate hair (*Bot.*). A hair which has several radiating branches.

stellate reticulum (*Histol.*). The epithelial cells of the enamel pulp of the developing dental germ, which have a reticular connective tissue-like appearance.

Stellite (*Met.*). TN for a series of alloys with cobalt, chromium, tungsten, and molybdenum in various proportions. The range is chromium 10–40%, cobalt 35–80%, tungsten 0–25%, and molybdenum 0–10%. Very hard. Used for cutting-tools and for protecting surfaces subjected to heavy wear.

stellulate (*Bot.*). Resembling a small star.

Stelvetite (*Plastics*). TN for PVC-coated steel sheet used for a variety of purposes including cabinets, wall sections, domestic equipment.

stem (*Bot.*). The ascending axis of a plant. Sometimes subterranean, e.g., rhizome, but distinguished from a root by the presence of leaves or scale leaves and characteristic arrangement of the vascular bundles. (*Horol.*) In a keyless watch, the shaft to which the button is attached. (*Typog.*) See **body.** (*Zool.*) See **scapus.**

stem bar (*Ships*). The portion of material forming the extreme forward end of a ship. The hull proper is secured thereto; sometimes known as **stem post.**

stem cell (*Histol.*). A primordial germ cell; a primitive cell (haemocytoblast), giving rise to the myeloid series of blood cells and perhaps to lymphoid cells; certain primitive cells appearing in the blood in leukaemia.

stem correction (*Heat*). If only the bulb of a thermometer is immersed, the stem may be at a different temperature from the bulb, and the reading may have to be corrected accordingly.

stemma (*Zool.*). See **ocellus.**

stemming (*Mining*). The stopping material (e.g., clay) used to tamp the explosive in a shot-hole. See **tamp** (1).

stem post (*Ships*). See **stem bar.**

stem succulent (*Bot.*). A plant with a succulent stem and with very small leaves, often reduced to spines.

stem tendril (*Bot.*). A tendril which is a modified stem.

stem xerophyte (*Bot.*). A plant characteristic of very dry places, with ephemeral or much reduced leaves, and with the photosynthetic tissue located in the peripheral cells of the stem.

stench trap (*San. Eng.*). See **air trap.**

stencil (*Print., etc.*). A plate of fine metal, plastic, waxed paper, etc., used for design and lettering, perforated in the required pattern so that ink, paint or other colouring substance may be passed through according to that pattern.

steno-. Prefix from Gk. *stenos*, narrow.

stenode (*Radio*). Supersonic heterodyne receiver in which there is very sharp tuning in intermediate-frequency circuits, using piezoelectric quartz crystals, with frequency correction of audio signal after demodulation.

stenoecious (*Ecol.*). Tolerating a narrow range of habitats. Cf. *euryoecious*.

stenohaline (*Zool.*). Capable of existence within a narrow range of salinity only. Cf. *euryhaline*.

stenohydric (*Ecol.*). Tolerating a narrow range of soil water content. Cf. *euryhydric*.

Stenonian duct (*Zool.*). In Mammals, the duct of the parotid gland opening into the mouth near the molars of the upper jaw.

stenonotal (*Zool.*). Of Insects, having a small or narrow thorax.

stenopetalous (*Bot.*). Having narrow petals.

stenophagic (*Ecol.*). Tolerating a narrow range of foods. Cf. *euryphagic*.

stenophyllous (*Bot.*). Having narrow leaves.

stenopodium (*Zool.*). The typical biramous limb of *Crustacea*, having slender exopodite and endopodite. Cf. *phyllopodium*.

stenosis (*Med.*). Narrowing or constriction of any duct, orifice, or tubular passage as a result of disease. *adj.* **stenosed.**

stenothermal (*Ecol.*). Tolerating only a very narrow range of temperature. Cf. *eurythermal*.

Stenson's duct (*Zool.*). See **Stenonian duct.**

Stenson's gland (*Zool.*). In *Amniota*, a large gland in the lateral ventral wall of the nasal cavity opening into the vestibule.

stenter (*Textiles*). Machine on which fabrics, after leaving the drying chamber, are stretched to full width between travelling chains equipped with strong pins or clips which grip the cloth selvedges. Also known as **tenter.**

stentorphone (*Acous.*). Apparatus for the reproduction of sound at high level by the electromagnetic control of the flow of high-pressure air through a grid orifice.

step (*Aero.*). The discontinuity in the form of a hydrofoil, as in the bottom of a flying-boat, to facilitate take-off from the water surface. (*Elec. Eng.*) Synchronous machines are said to keep *in step* when they remain in synchronism with each other. (*Nuc. Eng.*) The off-setting of a hole in a reactor shield to form a zig-zag joint and so avoid leaving a straight path for the possible escape of any radiation. (*Telecomm.*) See **transient.**

step-and-repeat (*Print.*). A printing-down machine used in graphic industries to repeat a design as often as necessary and at accurate intervals on the printing surface.

step-by-step method (*Elec. Eng.*). A method of determining the hysteresis curve of a magnetic material, in which the field strength is increased and reversed in steps.

step-down amplifier (*Telecomm.*). One in which a high negative voltage is applied to the anode, current in a positive grid being taken as a measurement of the voltage.

step-down transformer (*Elec. Eng.*). Reverse of the *step-up transformer*, i.e., in an electrical transformer, the transfer of energy from a high to a low voltage.

step faults (*Geol.*). A series of tensional or normal faults which have a parallel arrangement, throw in the same direction, and hence progressively 'step down' a particular bed.

step function (*Maths.*). One which makes an instantaneous change in value from one constant value to another. Its *Fourier analysis* shows an infinite number of harmonics present. (*Telecomm.*) A function which is zero for all time preceding a certain instant, and has a constant infinite value thereafter.

Stephanian (*Geol.*). The uppermost stage of the Carboniferous system, corresponding in Britain to the beds above the Coal Measures.

stephanite (*Min.*). A sulphide of silver and antimony which crystallizes in the orthorhombic system. It is usually associated with other silver-bearing minerals and is deposited from ascending solutions. Also **brittle silver ore.**

stephanokontan (*Bot.*). Bearing a crown of cilia.

Stephenson's link motion (*Eng.*). See **link motion.**

step irons (*Build.*). See **foot irons.**

stepped (*Typog.*). When type is to be printed close to an irregularly shaped block, the mount must be cut in steps to allow this.

stepped bowl centrifuge (*Powder Tech.*). A disk centrifuge in which sedimenting particles are collected on metal strips mounted on the steps of the bowl and the remaining suspension withdrawn from the centrifuge whilst it is running. Also known as **Donoghue and Bostock centrifuge.**

stepped flashing (*Plumb.*). A *flashing* (q.v.) much used where a brick chimney projects from a sloping roof, the lead being cut in steps so that the horizontal edges of the 'steps' may be secured into raglets cut in the joints of the brickwork. Also called **skeleton flashing.**

stepping (*Civ. Eng.*). Laying foundations in horizontal steps on sloping ground. See

benched foundation. (*Surv.*) The process of chaining over sloping ground by making the measurement in horizontal lengths with the chain always held horizontally, one end in the air.

stepping tube (*Electronics*). One in which an electron beam can be moved between a few stable positions by applied voltages and so act as a switch.

step principle (*Space*). The method of constructing multistage rockets or space vehicles by mounting two or more rockets nose-to-tail. As each booster rocket uses up its fuel, it is dropped away, thus increasing the mass ratio and therefore the efficiency of the whole system.

step printer (*Cinema.*). An intermittent printer which prints one frame at a time.

step-rate prepayment meter (*Elec. Eng.*). A prepayment meter in which a high charge per unit is made until a given number of units have been consumed, when the gear ratio is automatically changed to give a lower charge per unit until the mechanism is reset by hand.

step turner (*Plumb.*). A wooden tool for bending sheet-lead through a right angle, as at the horizontal edges of the steps of a stepped flashing.

step-up instrument (*Elec. Eng.*). See **suppressed-zero instrument.**

step-up transformer (*Elec. Eng.*). A two-winding transformer with more secondary than primary turns. It thus transforms low to high voltages, and matches low to high impedances.

step wedge (*Photog.*). A strip of processed sensitized material which has been subjected to a series of exposures increased in geometrical progression or other regular succession.

steradian (*Maths.*). The SI unit of solid angular measure. It is defined as the solid angle subtended at the centre of a sphere by an area on its surface numerically equal to the square of the radius. Symbol **sr.**

sterba antenna (*Radio*). Vertical panel, with reflector, formed by folding a single wire so that normal radiation arises from current of one phase in the vertical sections, while the other phase current balances out in space in the crossovers.

stercobilin (*Biochem., Physiol.*). A brownish pigment found in faeces and formed in the bowel from the bile pigment, *bilirubin.*

stercolith, stercorolith (*Med.*). A hard faecal concretion, impregnated with calcium salts, in the intestine. (Latin *stercus*, gen. *stercoris*, dung.)

stercomarium (*Zool.*). In some *Sarcodina*, the system of stercome-containing tubes.

stercome (*Zool.*). In some *Sarcodina*, faecal matter in the form of masses of brown granules.

stercoraceous (*Med.*). Consisting of, or pertaining to, faeces.

stercoral (*Med.*). Of, pertaining to, or caused by, faeces.

stere (*For.*). A stacked cubic metre of timber, which equals 35·3 stacked cubic feet.

stereide (*Bot.*). Stone cell.

stereo (*Acous.*). Abbrev. for *stereophonic*.

stereoblastula (*Zool.*). An abnormal form of echinoderm larva.

stereocamera (*Photog.*). One equipped with two matched lenses or a beam-splitting device for taking photos in stereo pairs. Also called **binocular camera, stereoscopic camera.**

stereochemistry (*Chem.*). Branch of chemistry dealing with arrangement in space of the atoms within a molecule.

stereognosis (*Med.*). The ability to recognize similarities and differences in the size, weight, form, and texture of objects brought into contact with the surface of the body.

stereogram, stereograph (*Photog.*). General terms for photographs intended to be viewed in an apparatus to give a 3-dimensional appearance. See **parallax stereogram.**

stereographic projection (*Geog.*). A type of *zenithal projection* (q.v.), in which the viewpoint is on the opposite diameter of the sphere.

stereoisomerism (*Chem.*). The existence of different substances whose molecules possess an identical structure but different arrangements of their atoms in space. Also **alloisomerism.**

stereokinesis (*Biol.*). Movement of an organism in response to contact stimuli.

stereome (*Bot.*). A general term for the mechanical tissue of the plant.

stereome cylinder (*Bot.*). A cylinder of strengthening tissue lying in a stem, usually just outside the phloem.

stereomicrophone system (*Acous.*). Dual microphone with, e.g., interlocking figure-of-eight polar diagrams; used to provide signals for both channels of a stereophonic sound-reproduction system.

stereophonic recording (*Acous.*). (1) Use of adjacent tracks on magnetic tape, with multiple recording and reproducing channels. (2) Use of a spiral cut on disks, two channels being represented by stylus motions at right angles, each at 45° to the surface; reproduction is by one stylus operating a double transducer.

stereophony (*Acous.*). Method of sound reproduction which attempts to give the hearer an effect of auditory perspective similar to that created by the original sound. Methods used require two loudspeakers and three microphones or a pair of microphones placed close together. *adj.* **stereophonic.** Also called **auditory perspective, localization.**

stereoplasm (*Zool.*). The viscous part of protoplasm: a substance filling the spaces between the septa in some Corals.

stereoscope (*Optics, Photog.*). Device for producing apparently binocular (3D) image by presenting differing plane images to the two eyes.

stereoscopic camera (*Photog.*). See **stereocamera.**

stereoscopic high-power microscope (*Micros.*). A modified light microscope, with a single objective lens, which enables specimens to be examined stereoscopically at high magnifications. The method is based on splitting the light rays from the objective into two beams, thus producing superimposed images from two different depths of the specimen.

stereoscopy (*Optics*). Sensation of depth obtainable with binocular vision due to small differences in parallax producing slightly-differing images on the two retinas.

stereospondyly (*Zool.*). The condition of having the parts of the vertebrae fused to form one solid piece; cf. *temnospondyly*. *adj.* **stereospondylous.**

stereotaxia (*Med.*). The electrical destruction of a small area of brain tissue, aimed at alleviating unmanageable emotional reactions in people with such disorders as parkinsonism and epilepsy and in irritable mentally-retarded people.

stereotaxis (*Biol.*). Response or reaction of an organism to the stimulus of contact with a solid body; as the tendency of some animals to insert themselves into holes or crannies, or to attach themselves to solid objects. *adj.* **stereotactic.**

stereotropism (*Biol.*). See **stereotaxis.**

stereotype (*Print.*). A *duplicate plate* (q.v.) made from the original surface (type and/or blocks) or from an existing plate; for metal plates, the mould is made in flong; for rubber or plastic plates, a thermosetting plastic sheet is used.
stereotype alloys (*Met.*). Lead-based alloys with 5–10% tin and 10–15% antimony.
stereotyped behaviour (*An. Behav.*). See **fixation.**
stereotypy (*Med.*). The repetition of senseless movements, actions, or words by the insane.
steric hindrance (*Chem.*). The retarding influence of neighbouring groups on reactions in organic molecules, e.g., *ortho*(2) substitution in aromatic acids considerably retards esterification.
sterigma (*Bot.*). A short, and often somewhat swollen, hypha on which a fungal spore is borne. *pl.* **sterigmata.**
sterile (*Biol.*). Unable to breed; freed from bacteria and moulds by treatment with heat or with antiseptics. (*Bot.*) Unable to produce spores or seeds. *n.* **sterility.**
sterile cell (*Bot.*). The terminal cell of a chain of aecidiospores.
sterile flower (*Bot.*). A staminate flower.
sterile glume (*Bot.*). One of the glumes at the base of the spikelet of a grass, which does not subtend a flower.
sterile line (*Elec. Eng.*). One that has no direct connexion with adjoining circuits and is therefore unaffected by them.
sterile vein (*Bot.*). A strand or sheet of interwoven hyphae occurring with the spore-bearing hyphae in the fruit bodies of some fungi.
sterilization (*Bot., Zool.*). (1) Loss of sexual function. (2) The preparation, usually by heating, of a substratum free from any living organism, on which fungi, bacteria, or protozoa may subsequently be grown in pure culture. (*Med.*) A method of destroying the reproductive powers, either by inhibition of the functions of the sex organs, or by their removal.
sterlingite (*Min.*). See **zincite.**
sternal (*Zool.*). See **sternite, sternum.**
sternal canal (*Zool.*). In some *Crustacea*, a median ventral cavity of the skeleton in which the thoracic part of the nerve-cord lies.
sternebrae (*Zool.*). In Mammals, a median ventral series of bones which alternate with the ribs.
sternellum (*Zool.*). In Insects, a ventral thoracic sclerite situated behind the eusternum. See also **basisternum.**
stern frame (*Ships*). In a twin-screw ship a heavy bar attached to the keel and to a strong transverse plate at the top, with gudgeons to carry the rudder. In a single-screw ship it surrounds the propeller aperture and is almost rectangular. The forward vertical part is the *propeller post*, the after part is the *rudder post*, (sometimes called the *stern post*), the lower horizontal part is the *sole piece* and the upper part is the *arch piece.*
Stern-Garlach experiment (*Nuc.*). Atomic beam experiment which provided fundamental proof of quantum theory prediction that magnetic moment of atoms can only be orientated in certain fixed directions relative to an external magnetic field.
sternite (*Zool.*). The sternum in Arthropods when it forms a chitinous plate. *adj.* **sternal.**
sternohyoid (*Zool.*). In some Vertebrates, a muscle running from the hyoid to the sternum.
sternopleuron, sternopleurite (*Zool.*). In Insects, a compound thoracic sclerite formed by the fusion of the episternum with the sternum.
stern post (*Ships*). See **stern frame.**
sternum (*Zool.*). The ventral part of a somite in Arthropods; the breast bone of Vertebrates, forming part of the pectoral girdle, to which, in higher forms, are attached the ventral ends of the ribs. *adj.* **sternal.**
sternutation (*Med.*). The act of sneezing; a sneeze.
steroids (*Chem.*). Compounds containing the perhydrocyclopentenophenanthrene nucleus. They include the sterols, bile acids, sex hormones, adrenocortical hormones, cardiac glycosides, sapogenins and some alkaloids.
sterols (*Chem.*). A group of steroid alcohols obtained originally from the non-saponifiable portions of the lipid extracts of tissues. The best known is *cholesterol.*
Sterry process (*Photog.*). Method of softening gradation of image by the use of potassium dichromate and ammonia.
stet (*Typog.*). A reader's mark in the margin of a proof indicating that the correction marked is to be ignored; in the text matter, a dotted line (. . . . . .) is drawn under the original wording.
stethoscope (*Med.*). Instrument for study of sounds generated inside the human body.
Stevenson screen (*Meteor.*). A form of housing for meteorological instruments consisting of a wooden cupboard having a double roof and louvred walls, these serving to protect the instruments from the sun and wind while permitting free ventilation. The base of the screen should be about 1 metre above the ground.
stew (*Cinema.*). Any undesired sound accompanying sound-film reproduction, particularly one arising from defective or badly adjusted apparatus.
Stewart's organs (*Zool.*). In some *Echinoidea*, internal gills associated with the lantern coelom.
sthene. Unit of force in the metre-tonne-second system, equivalent to 10^3 newtons.
stib- (*Chem.*). Root word denoting *antimony*, from the L. *stibium*. As in *stibine, stibnite.*
stibialism (*Med.*). Antimony poisoning.
stibic, stibious (*Chem.*). Referring to (V) and (III) oxidation state of antimony respectively.
stibine (*Chem.*). SbH_3. Antimony (III) hydride A poisonous gas. Less stable than arsine.
stibnite (*Min.*). (III)sulphide of antimony, that crystallizes in grey metallic prisms in the orthorhombic system. It is sometimes auriferous and also argentiferous. It is widely distributed but not in large quantity, and is the chief source of antimony. Formerly called **antimony glance.** Also called **antimonite.**
stichidium (*Bot.*). A special branch of the thallus in red algae on or within which the tetraspores are formed.
stichobasidium (*Bot.*). A basidium, usually elongated and cylindrical, in which the spindles of the dividing nuclei lie obliquely or longitudinally.
stichtite (*Min.*). A lilac or pink trigonal hydrated and hydrous carbonate of magnesium and chromium. Found at Dundas, Tasmania.
stick-and-rag work (*Build.*). Plasterwork formed of canvas stretched across a wooden frame and coated with a thin layer of gypsum plaster.
stick force (*Aero.*). The force exerted on the *control column* by the pilot when applying aileron or elevator control.
stick-force recorder (*Aero.*). A device attached to the control column of an aircraft by which the pilot's effort is measured and transmitted to a recording instrument.
sticking (*Join.*). Shaping a *stuck moulding.*
sticking board (*Join.*). A prepared board used to steady a piece of timber whilst shaping a moulding from it out of the solid.
sticking probability (*Nuc.*). Probability of an incident particle, which reaches the surface of a

nucleus, being absorbed and forming a compound nucleus.

sticking relay (*Telecomm.*). One which *makes* when d.c. current passes through the operating coil, remains operated on cessation of current (thereby saving current), and falls off or *releases* on a reversed current.

sticking voltage (*Electronics*). Potential in electron-beam tube above which electrons collected at screen cannot all be dispersed, leading to negative charge accumulating and neutralizing the excess voltage. In CRT, that accelerating voltage which fails to increase brightness of spot on a phosphor because of insufficient secondary-electron emission or conduction for dispersal of incident electrons.

stick pusher (*Aero.*). A device fitted to the control column of some high-performance aeroplanes with swept-back wings which moves the column sharply forward to prevent a stall. See **stall-warning indicator.**

stick shaker (*Aero.*). See **stall-warning indicator.**

stick-slip motion (*Eng.*). The motion of sliding surfaces of various materials, in which the force to start them moving is greater than the force to keep them moving.

stiction (*Mech.*). *St*atic fr*iction* (see **friction**).

stiffened expanded metal (*Build.*). See **self-centring lathing.**

stiffened suspension bridge (*Eng.*). A suspension bridge in which the tendency for the suspension cables to change shape under different load systems is counteracted by the provision of stiffening girders which are supported on the bridge piers and connected to the cables by suspension rods.

stiffener (*Aero.*). A member attached to a sheet for the purpose of restraining movement normal to the surface. Usually of thin drawn or extruded light-alloy, L, Z, or U section, attached by riveting, metal bonding or spot welding. See **integral-, stringer.** (*Eng.*) A steel angle or bar riveted or welded across the web of a built-up girder to stiffen it.

stiff lamb disease (*Vet.*). (1) See **white muscle disease.** (2) Arthritis of lambs due to infection by *Erysipelothrix insidiosa* (*E. rhusiopathiae*).

stiff neck (*Med.*). See **torticollis.**

stiffness control (*Acous.*). In a mechanically vibrating system, the condition in which the motion is mainly determined by the stiffness of the retaining springs and negligibly by the resistance and mass of the system. The resonant frequency of such a system is then much higher than the frequencies of the driving forces.

stiffness criterion (*Aero.*). The relationship between the stiffness, strength, and other structural properties which will prevent *flutter* or dangerous aeroelastic effects.

stiff ship (*Ships*). A ship with a large transverse metacentric height and short rolling period. Cf. *tender ship*.

stiff sickness (*Vet.*). See **osteomalacia.**

stifle (*Vet.*). The femorotibial joint of animals.

stigma (*Bot.*). (1) See **eye spot.** (2) The distal end of the style, more or less enlarged, on which pollen alights and germinates. (*Zool.*) In *Protozoa*, an eye spot; in *Arthropoda*, one of the external apertures of the tracheal system; in *Urochorda*, a gill slit; generally, a spot or mark of distinctive colour, as on the wings of many Butterflies. *pl.* **stigmata.**

stigmata (*Med.*). Physical characteristics of a disease process, or syndrome.

stilb (*Light*). Unit of luminance, equal to 1 cd/cm^2 or 10^4 cd/m^2 of a surface.

stilbene (*Chem.*). $C_6H_5{\cdot}CH{=}CH{\cdot}C_6H_5$, *s*-diphenylethene; monoclinic plates or prisms; m.p. 125°C, b.p. 306°C. Can occur as two isomers, the *trans* solid form and the *cis* liquid form.

stilbite (*Min.*). A zeolite; silicate of sodium, calcium, and aluminium with chemically combined water; crystallizes in the monoclinic system, the crystals frequently being grouped in sheaflike aggregates. Found both in igneous rock cavities and in fissures in metamorphic rocks. Also called **desmine.**

stilboestrol (*Pharm.*). 4,4′-Dihydroxy αβdiethylstilbene:

OH
$CH_3\ CH_2$
HO
$CH_2\ CH_3$

White crystalline solid; m.p. about 170°C. Used medically as an *oestrogen* (q.v.).

stilboid (*Bot.*). Said of a fungus having a stalked head of spores or other reproductive structures.

stile (*Join.*). An upright member in framing or panelling. Often incorrectly spelt **style.**

Stiles-Crawford effect (*Optics*). Light entering the eye near to the margin of the pupil is less effective in producing a sensation of brightness than the same amount of light entering through the centre of the pupil.

still (*Chem.*). Apparatus for the distillation of liquids, consisting of a reboiler, a fractionating column, and arrangement for reflux.

stillage (*Brew., etc.*). A raised platform for storing casks, etc., off the ground. (*Elec. Eng.*) A stand for accommodating accumulator cells. (*Print.*) See **pallet.**

still air range (*Aero.*). The theoretical ultimate range of an aircraft without wind and with allowances only for take-off, climb to cruising altitude, descent, and alighting.

Stilling's nucleus (*Zool.*). In higher Vertebrates, a group of large nerve cells at the base of the dorsal horn of the spinal cord; isolated grey nuclei lying in the white matter of the middle lobe of the cerebellum over the roof of the fourth ventricle.

still-picture transmission (*Telecomm.*). See **facsimile telegraphy** and cf. *television*.

Still's disease (*Med.*). Acute polyarthritis of children (resembling rheumatoid arthritis), with fever and enlargement of the spleen and of the lymphatic glands.

still-water navigation (*Civ. Eng.*). See **slack-water navigation.**

stilpnomelane (*Min.*). Monoclinic hydrous iron(II), magnesium, potassium, aluminium silicate occurring in regionally metamorphosed sediments and in veins and druses. Resembles biotite but has different optical qualities.

stilted arch (*Arch.*). An arch rising from points below its centre, and having the form of a circular arc above its centre.

stilt root (*Bot.*). An adventitious root formed by a stem from a point above ground-level, passing downwards into the soil and affording support to the stem.

stimulated emission (*Phys.*). See **laser** and **maser.**

stimulation. The application of stimuli.

stimulose (*Bot.*). Bearing stinging hairs.

stimulus (*An. Behav.*). A physical event impinging on the receptors of an animal, exciting them, and possibly being correlated in an orderly manner with the occurrence of a particular response by the animal. (*Bot., Zool.*) An agent which will provoke active response or reaction in a living organism. Especially (*Zool.*), an agent which will cause propagation of a nerve impulse in a nerve fibre. *pl.* **stimuli.**

stimulus satiation (*An. Behav.*). A phenomenon postulated to underlie any more or less temporary decrease in responsiveness to a specific stimulus. Distinct from *habituation* (q.v.).

stimulus threshold (*An. Behav.*). The value of a quantified stimulus which elicits a particular response at a definite intensity.

sting (*Zool.*). A sharp-pointed organ by means of which poison can be injected into an enemy or a victim; as the poisonous fin-spines of some Fishes, the ovipositor of a worker Wasp. See **urticaria.**

stinging hair (*Bot.*). A multicellular hair with a brittle tip, which breaks off on contact with an animal, leaving a sharp edge which penetrates the skin and injects an irritant fluid.

stink damp (*Mining*). Underground ventilation tainted by sulphuretted hydrogen.

stink-trap (*San. Eng.*). See **interceptor.**

stinkwood (*For.*). A strong wood from the S. African tree *Ocotea bullata* (Cape walnut), known for its beautiful and distinctive figuring. It has an unpleasant smell when green.

stipate (*Bot.*). Crowded.

stipe (*Bot.*). (1) A general term for the stalk of the fruit body of a fungus when it consists of a large number of more or less interwoven hyphae; the stalk of a mushroom is a familiar example. (2) The petiole of a fern up to the lowermost leaflet. (3) Stalk of thallus of seaweeds. (*Geol.*) One of the branches of a fossil *graptolite.*

stipel (*Bot.*). One of the two small leaflike appendages present at the base of a leaflet in some compound leaves. *adj.* **stipellate.**

stipes (*Zool.*). A stalklike structure: an eyestalk: in Insects, the second joint of the maxilla, articulating with the distal border of the cardo. *pl.* **stipites.** *adjs.* **stipitate, stipiform.**

stipple (*Print.*). See **mechanical stipple.**

stippling (*Paint., Build.*). The operation of breaking up the smoothness of a paint, distemper, plaster, or cement surface by dabbing it repeatedly with the point of a special brush.

stipular trace (*Bot.*). The vascular tissue running into a stipule.

stipule (*Bot.*). One of the two appendages, usually leaflike, often present at the base of the petiole of a leaf. *adj.* **stipulate.**

Stirling boiler (*Eng.*). A water-tube boiler in which two upper drums and one lower drum are connected by highly inclined banks of water tubes curved so as to enter the drums radially, the upper drums being also connected by horizontal steam and water tubes.

Stirling engine (*Eng.*). An *external* combustion reciprocating engine, patented by a Scottish clergyman, Robert Stirling, in 1827. It consists essentially of a cylinder in which two pistons (a working piston and a displacer) operate. When the air (or a suitable gas) in the cylinder is heated it expands, driving the working piston. The second piston transfers the air to a cold region for cooling; it is then recompressed by the working piston and transferred by the displacer to the hot region to start the cycle again. Such an engine is very much quieter and cleaner than a petrol or diesel engine. A modern version has been developed in Holland in which helium under pressure is used as the working medium.

Stirling's approximation (*Maths.*).

$$n! = \sqrt{2\pi}\,.\,n^{(n+\frac{1}{2})}\,.\exp\left(-n+\frac{1}{12n}-\frac{1}{360n^3}+\cdots\right).$$

stirps (*Bot.*). Any well-established variety which keeps its characters in cultivation. *pl.* **stirpes.**

stirrup (*Civ. Eng.*). A vertical steel rod which loops together the top and bottom reinforcing bars of a reinforced-concrete beam and helps to resist the shear. (*Horol.*) A support; such as the bottom of the rod of a mercurial pendulum on which rests the container for the mercury.

stishovite (*Min.*). A high-density form of silica. Synthesized at $1{\cdot}6\times10^{10}$ N/m^2 and 1200°C, found in Meteor Crater, Arizona, in shock-loaded sandstone.

stitching (*Bind.*). Joining the sections of an *inserted book* (see **inset** (2)) along the back by means of thread or wire.

stitch welding (*Elec. Eng.*). *Seam welding* (q.v.), using small mechanically-operated electrodes, similar to a sewing machine.

stoa (*Arch.*). A covered colonnade or portico.

Stobie furnace (*Elec. Eng.*). An electric furnace of the coreless induction type in which a definite path for the magnetic flux is provided in the form of an external laminated yoke.

stochastic (*Stats., etc.*). Random; having an element of chance.

stochastic noise (*Acous.*). That which maintains a statistically random distribution.

stock (*Acous.*). The material from which pressed records are made before it is actually heated and placed between two stampers in the press. (*Bot.*) (1) The rooted stem into which the scion is inserted in grafting. (2) The perennial portion of a herbaceous perennial. (3) A race. (*Carp., Join.*) The principal part of a tool, e.g., the body of a plane, in which the cutting iron is held: the stouter arm of a bevel, in which the blade is fastened. (*Gen., Zool.*) A direct line of descent: an individual originating a line of descent: an asexual zooid giving rise by budding to sexual zooids of one sex only. (*Geol.*) Similar to **boss.** (*Print.*) The general term for the material being printed: paper, board, foil, etc. See also **stocks.**

stock board (*Build.*). A bottom made to fit the mould used in the handmoulding of bricks.

Stockbridge damper (*Elec. Eng.*). A vibration damper used to prevent the mechanical vibration of overhead lines.

stock brush (*Build.*). A brush used to moisten surfaces with water, prior to plastering, so that the surface will not absorb moisture from the plaster.

Stockholm tar (*Chem.*). The tarry product obtained in carbonization of wood.

stockless anchor (*Ships*). A form of anchor in which there is no crosspiece on the shank and the arms are pivoted so that both of them can engage at the same time; the shank can be drawn into the hawsepipe of the ship.

stock lock (*Join.*). A form of rim lock contained in a heavy metal-bound wooden case.

stock lumber (*For.*). Lumber which is sawn to stock market sizes.

stock pile (*Mining, Min. Proc.*). Temporarily stored tonnage of ore, middlings, concentrates or saleable products.

stock rail (*Rail.*). The outer fixed rail against which the point (see **points**) works at a turn-out.

stocks (*Build.*). Bricks which are fairly sound and hard-burned but are more uneven in colour than *shippers* (q.v.); the bricks most used for ordinary building purposes. (*Ships*) The massive timbers supporting a ship in course of construction.

stockwork (*Geol.*). An irregular mass of interlacing veins of ore; good examples occur among the tin ores of Cornwall and in the Erzgebirge. (Ger. *Stockwerk.*)

stoichiometry (*Chem.*). Determination of exact proportion of elements to make pure chemical compounds, alloys such as semiconductors, or ceramic crystals.

stokes (*Phys.*). The CGS unit of kinematic viscosity (10^{-4} m^2/s). Abbrev. St.

Stokes's Law (*Phys.*). (1) The resisting force offered by a fluid of dynamic viscosity η to a sphere of radius r, moving through it at steady velocity v is given by

$$R = 6\pi\eta rv,$$

whence it can be shown that the *terminal velocity* of a sphere of density ρ, falling under gravitational acceleration g through fluid of density ρ_0 is given by

$$v = \frac{2gr^2}{9\eta}(\rho - \rho_0).$$

Applies only for viscous flow with Reynolds number less than 0·2. (2) Incident radiation is at a higher frequency and shorter wavelength than the re-radiation emitted by an absorber of that incident radiation.

Stokes's line (*Phys.*). A spectrum satisfying *Stokes's law*, i.e., line seen in Raman spectrum on the long wave side of the Rayleigh line when monochromatic light is scattered.

Stokes's theorem (*Maths.*). The surface integral of the curl of a vector function equals the line integral of that function around a closed curve bounding the surface, i.e.,

$$\int_L E\,dl = \int_S \text{curl}\,E\,ds.$$

STOL (*Aero.*). *S*hort *T*ake-*O*ff and *L*anding, a term applied to aircraft with high-lift devices enabling them to operate from small air strips, 1000 ft (300 m) or less being the criterion.

stolon (*Bot.*). (1) A weak stem, growing more or less horizontally from the main stem of the plant, bearing scale leaves and adventitious roots, and giving rise to one or more buds which develop into new plants, or end in a tuber, as in the potato. The term *stolon* is sometimes confined to underground stems having the characters given above, while similar stems lying on the surface of the soil are called *runners*. (2) A long hypha produced by some fungi, which lies on the substratum, forming tufts of rhizoids and of sporangiophores at intervals. *adj.* **stoloniferous.** (*Zool.*) A cylindrical stemlike structure: a tubular outgrowth in hydroid colonies of *Coelenterata* and *Entoprocta* from which new individuals or colonies may arise: in *Urochorda*, the cadophore. *adj.* **stolonate.**

Stolonoidea (*Zool.*). An order of *Graptolita* containing very irregular branching forms, the branches being made up of complete rings and not half rings, as in the other orders.

-stoma. Suffix from Gk. *stōma*, mouth, applied esp. in zoological nomenclature. *pl.* -stomata.

stoma. A small aperture. Specifically (*Bot.*), a minute perforation in the epidermis of a leaf or a young stem, together with two guard cells, one on each side of the pore, and sometimes certain accessory cells. The term is sometimes used to denote the pore only. A stoma allows of the diffusion of gases and vapours from the interior of the leaf to the external atmosphere, and of gases from without into the leaf. *pl.* **stomata.** Also **stomate.**

stomach (*Zool.*). In Vertebrates, the saclike portion of the alimentary canal intervening between the oesophagus and the intestine. The term is loosely applied in Invertebrates to any saclike expansion of the gut behind the oesophagus. *adj.* **stomachic.**

stomach insecticide (*Chem.*). One acting on ingestion and applicable only to insects which eat as distinct from sucking insects which draw food in liquid form from host plant or animal; may be used on foliage against leaf-eating insects, or as poison-bait ingredient against locusts, etc. Examples: lead arsenate, DDT, Gammexane.

stomatal, stomate, stomatiferous, stomatose stomatous. *adjs.* from *stoma.*

stomatitis (*Med.*). Inflammation of mucous membrane of the mouth. (*Vet.*) See **horse pox.**

stomatogastric (*Zool.*). Pertaining to the mouth and stomach; said especially of that portion of the autonomic nervous system which controls the anterior part of the alimentary canal.

Stomatopoda (*Zool.*). An order of *Hoplocarida.*

stomidia (*Zool.*). In some *Actinaria*, apertures on the disk which represent the terminal pores of tentacles which have disappeared.

stomium (*Bot.*). A part of the wall of a fern sporangium composed of thin-walled cells; splitting begins here when the sporangium dehisces.

stomodaeum (*Zool.*). That part of the alimentary canal which arises in the embryo as an anterior invagination of ectoderm; cf. *proctodaeum, mid-gut*. *adj.* **stomodaeal.**

-stomy. Suffix from Gk. *stōma*, mouth, referring esp. to the formation of an opening by surgery.

stone (*Horol.*). A jewel; used especially of the pallet jewels. (*Typog.*) The smooth, milled cast-iron surface on which formes are locked up.

stone canal (*Zool.*). In *Echinodermata*, a vertical tube leading upwards or downwards from the madreporite, having calcified walls and forming part of the water-vascular system.

stone cell (*Bot.*). A thick-walled cell, not much longer than broad, with lignified walls.

stone-hand (*Typog.*). A compositor who specializes in *imposition* (q.v.).

stone head (*Mining*). (1) The solid rock first met with in sinking a shaft. (2) A heading or tunnel in stone.

stone insulator (*Elec. Eng.*). A low-voltage type of insulator made from stoneware.

stone saw (*Build.*). A smooth-faced blade which in use is fed with an abrasive such as sand, carborundum, or diamond powder, as it cuts its way through stone.

Stonesfield Slate (*Geol.*). A stratigraphical subdivision of the Jurassic System occurring as thinly bedded sandy limestones; formerly exploited for roofing material in the Cotswolds.

stone square (*Brew.*). See **Yorkshire square.**

stone tongs (*Build.*). An accessory used in hoisting blocks of stone. It resembles a large pair of scissors with the points curved inwards. These clip into the sides of the block, while chains connect the loops of the tongs to the hoisting ring. Also called **nippers.**

stoneware (*San. Eng.*). A material used for some sanitary fittings, etc.; made from plastic clays of the Lias formation, with a small amount of sharp sand, etc., added to reduce shrinkage.

stoneworts (*Bot.*). See **Charales.**

stony meteorites (*Geol.*). Those meteorites which consist essentially of rock-forming silicates. See **achondrite, aerolites, chondrite.**

stoop (*Build.*). A low platform outside the entrance door of a house. Also called **stoep.**

stop (*Carp., Join.*). (1) A projecting piece set in

the top of a bench at one end and adjustable for height. It is used to steady work which is being planed. (2) An ornamental termination to a stuck moulding. (3) See **door stop.** (*Optics, Photog.*) (1) Circular opening which sets the effective aperture of a lens, e.g., the iris of a camera or the rim of the objective in a telescope. (2) An *f-number* (q.v.), esp. as marked on the iris scale of a camera lens.

stop-bath (*Photog.*). Acid bath, e.g., acetic acid, chrome alum, used between development and fixing to arrest development and prevent staining by contamination of fixer by traces of developer.

stop-cock (*Plumb.*). A short pipe opened or stopped by turning a key or handle.

stop-cylinder (*Print.*). A printing machine in which the cylinder makes one revolution and prints a sheet as the bed travels in the printing direction, and then remains stationary while the bed returns to its starting position. Cf. *single-revolution, two-revolution.*

stop down (*Photog.*). To reduce the working aperture of a lens in order to increase depth of field and/or exposure time.

stope (*Mining*). (1) To excavate ore from a reef, vein, or lode. (2) Space formed during extraction of ore underground. Types are flat, open, overhand, rill, shrinkage, and underhand, with variations to suit shape, geology, and size of deposit.

stope pillar (*Mining*). Supporting column of ore left during mining to support roof of workings and possibly removed ('robbed') before abandonment of area.

stoping (*Geol.*). A mining term applied by R. A. Daly to a process in the emplacement of some igneous rock bodies, by which blocks of the overlying country rock are wedged off and sink into the advancing magma.

stoping width (*Mining*). Distance between upper (hanging) and lower (foot) walls excavated during stoping, irrespective of the width of the valuable lode or vein.

stop key (*Acous.*). A finger key for operating the stop mechanism and so bringing into potential action groups of pipes in an organ.

stop moulding (*Join.*). A *stuck moulding* (q.v.) terminating in a stop.

stopped end (*Build.*). A square end to a wall.

stopped heading (*Bind.*). In paper ruling, the heading for account books and other work may require the ruling to stop at more than one stage both vertically and horizontally.

stopped mortise (*Join.*). See **blind mortise.**

stopped pipe (*Acous.*). An organ pipe which is closed at the outer end, so that the wavelength of its fundamental resonance is four times the length of the air column.

stopper (*Electronics*). (1) Resistance next to the grid of a valve to reduce high-frequency potentials on the grid and consequent build-up of parasitic oscillations. (2) Resistor-capacitor combination for decoupling anode or grid supply circuits, so as to obviate oscillation or motor-boating in thermionic amplifiers.

stopping (*Paint.*). Plastic material used to fill holes and cracks in timber, e.g., before painting.

stopping equivalent (*Nuc.*). Thickness of a standard substance which would produce the same energy loss as the absorber under consideration. The standard substance is usually air at s.t.p. but can be Al, Pb, H_2O, etc. See **air equivalent.**

stopping motions (*Textiles*). Electrical or mechanical devices employed on many textile machines when a fault develops in raw material feeding arrangements (openers, scutchers, drawframes, etc.) or a yarn breaks in winding, warping, or weaving.

stopping-off (*Elec. Eng.*). Coating a conducting surface with a resist to prevent electrodeposition.

stopping-out (*Photog.*). See **blocking-out.** (*Print.*) Painting out with a protecting varnish the darker tones of a half-tone block, in stages during the etching, to obtain the best rendering of the subject; also painting out unwanted background or detail on a negative before printing down.

stopping potential (*Electronics*). Reverse difference of potential required to bring electrons to rest against their initial velocity from either thermal or photoelectric emission.

stopping power (*Nuc.*). Energy loss, resulting from a particle traversing a material. The *linear stopping power* S_L is the energy loss per unit distance and is given by $S_L = -dE/dx$, where x is path distance and E is the kinetic energy of the particle. The *mass stopping power* S_M is the energy lost per unit surface density traversed and is given by $S_M = S_L/\rho$, where ρ is the density of the substance. If A is taken as the rel. at. mass of an element and n the number of atoms per unit volume, then the *atomic stopping power* S_A of the element is defined as the energy loss per atom per unit area normal to the motion of the particle, and is given by $S_A = S_L/n = S_M A/N$, where N is Avogadro number. The *relative stopping power* is the ratio of the stopping power of a given substance to that of a standard substance, e.g., air or aluminium.

stop press (*Typog.*). See **fudge.**

stop slide (*Horol.*). See **all-or-nothing piece.**

stop valve (*Eng.*). The main steam valve fitted to a boiler to control the steam supply and to allow isolation of the boiler from the main steam pipe.

stop watch (*Horol.*). A watch, usually having seconds and minutes hands only, which is started and stopped by pressure of the winding knob. The normal type reads to 1/5 s, special types read to 1/50 s. See also **chronograph.**

storage battery (*Elec. Eng.*). See **accumulator.**

storage bin (*Civ. Eng.*). See **silo.**

storage capacity (*Comp.*). Maximum number of *bits* which can be stored, located, and recovered for a main or fast store of a computer.

storage element (*Comp.*). One unit in a memory, capable of retaining one *bit* of information. Also the smallest area of the surface of a charge-storage tube which retains information different from that of neighbouring areas.

storage excretion (*Zool.*). A form of excretion in which certain waste substances are not actually passed to the exterior, but are stored in an inert form in specialized areas, e.g., the fat body of some Insects.

storage factor (*Elec.*). See ***Q***.

storage heater (*Elec. Eng.*). A heater with large thermal capacity, used to store heat during off-peak periods and release it over a longer period. A fan may be used to increase the rate of heat output.

storage oscilloscope (*Electronics*). One in which trace is retained indefinitely or until deliberately wiped off. See **Remscope, scotophor, Storascope.**

storage pith (*Bot.*). Pith in which starch or water is stored by the plant.

storage tracheide (*Bot.*). A thick-walled cell resembling a tracheide, without living contents, in which water is stored.

storage tube (*Electronics*). Tube which stores charges deposited on a plate or screen in a cathode-ray tube, a subsequent scanning by the electron beam detecting, reinforcing, or abolishing the charge. (*TV*) Camera tube in which charge produced by an image can be retained in an element of capacitor. See **iconoscope.**

Storascope (*Electronics*). TN for infinite-persistence cathode-ray oscilloscope.

storax (*Pharm.*). Balsam from the bark of *Liquidambar orientalis*, rel. d. 1·12, used in medicine and microscopy, an ingredient of friar's balsam. Also **styrax.**

store (*Comp.*). The section of a computer in which the *program* and *data* are held until required and in which the results are held until *read out* to a display or printer. Fast stores may use MOS transistors or magnetic cores, while slow stores may use, e.g., magnetic drums or magnetic tape. Also called **memory.** (*Electronics*) Another name for **storage tube.**

storey rod (*Build.*). A pole on which is marked the level of the courses and which gives a guide to the ultimate level.

storied cork (*Bot.*). A type of cork composed of short cells arranged in somewhat irregular radial groups.

storm (*Radio*). Large persistent ionospheric disturbance to transmission on band 3 to 30 MHz, or one coincident with magnetic disturbances lasting several days. Caused by particle emission from the sun and hence delayed after *SID* (sudden ionospheric disturbance) and *SPA* (sudden phase anomaly).

storm-centre (*Meteor.*). The position of lowest pressure in a cyclonic storm.

storm pavement (*Civ. Eng.*). See **breakwater-glacis.**

storm-water tanks (*Civ. Eng.*). Tanks into which, in times of storm, when the amount of surface water increases, sewage in excess of three times the mean dry weather flow passes, for the removal of road grit and other solid matters, before final discharge into a river.

storm window (*Build.*). (1) A window arranged with double sashes enclosing air, which acts as a sound and heat insulator. (2) A small upright window set in a sloping roof surface so as not to project beyond it. Cf. *dormer*.

stout (*Brew.*). A dark brown beer with a strong flavouring of hops; the colour is due to the amount of black malt in the grist. See **beer, malt.**

Stovaine (*Pharm.*). Amylocaine hydrochloride. Proprietary synthetic alkaloid, used as an anaesthetic, especially for spinal purposes.

stoving (*Paint.*). An industrial process of quickly drying specially formulated paints by heat (radiant or convected). Temperatures generally exceed 180°F (82°C) (for lower temperatures, see **force drying**). (*Textiles*) A bleaching process in which wool or woollen fabric, in a moist condition, is brought in contact with sulphur fumes. The term is also used in printing, where cotton cloth is subjected to dry heat. Also **stove drying.**

stowage factor (*Ships, etc.*). The space required to contain unit weight of a commodity allowing for all packing, dunnage and unavoidable lost space between units. Usually, measured in cubic feet per ton or cubic metres per tonne.

stowing (*Mining*). Back fill, packing. Use of waste rock, mill tailings, etc., as roof support in worked-out areas of mine. See also **gob.**

s.t.p., STP. *S*tandard *t*emperature and *p*ressure; a temperature of 0°C and a pressure of 101 325 N/m^2. See **standard atmosphere.**

strabismus (*Med.*). Squint. A condition in which the visual axes of the eyes assume a position relative to each other which is abnormal.

strabotomy (*Surg.*). The surgical operation of curing *strabismus* by dividing one or more muscles of the eye.

straddle milling (*Eng.*). The use of two or more side-cutting milling-cutters on one arbor so as to machine, for example, both side-faces of a piece of work at one operation.

straddle scaffold (*Build.*). See **saddle scaffold.**

straggling (*Nuc.*). Variation of range or energy of particles in a beam passed through absorbing material, arising from random nature of interactions experienced. Additional straggling may arise from instrumental effects such as noise, source thickness and gain instability.

straight angle (*Maths.*). An angle of 180° or π radians.

straight arch (*Build.*). See **flat arch.**

straight bank (*Teleph.*). A bank of outgoing trunks connected in a regular way, so that they are tested in order, starting from the same one each time, by each switch having access to them. Also called **straight multiple.** Cf. *slipped bank*.

straight eight (*I.C. Engs.*). An 8-cylinder-in-line engine, as distinct from an 8-cylinder-V-type engine.

straighteners (*Aero.*). See **honeycomb.**

straight-flow system (*Aero.*). A gas-turbine combustion system in which air enters and the gases leave in the same direction; the more common system. See also **return-flow system.**

straight-flute drill (*Eng.*). A conical pointed drill having backed-off cutting edges, formed by cutting straight longitudinal flutes in the shank; more rigid than a twist drill and often used for soft metals.

straightforward-junction working (*Teleph.*). The system of manual operation in which the attention of a B-operator is obtained by an A-operator connecting to any one of a number of junctions leading to the B-operator.

straight joint (*Build.*). A continuity of vertical joints in brickwork.

straight-line capacitor (*Radio*). Variable capacitor whose value varies linearly with scale reading.

straight-line frequency capacitor (*Radio*). Variable capacitor whose value is inversely proportional to the square of the scale reading, so that the frequency of the circuit which it tunes is directly proportional thereto.

straight-line lever escapement (*Horol.*). A lever escapement in which the balance staff, pallet staff, and escape wheel are planted in the same straight line. The majority of watches with the club-tooth escapement are of this form.

straight-line wavelength capacitor (*Radio*). Variable capacitor whose value is proportional to square of scale reading, so that it can tune a circuit with a linear relationship between scale reading and wavelength.

straight-pane hammer (*Tools*). A fitter's hammer the head of which has a flat striking face at one end and a blunt chisel-like edge, parallel with the shaft, at the other.

straight receiver (*Radio*). One in which all high-frequency amplification, if any, is at the same frequency as that of the original signal. See **supersonic heterodyne receiver.**

straight run (*Print.*). Running a web-fed press without cylinder collection and sometimes without turner bars to give the maximum number of copies to the cylinder revolution.

straight tie (*Weaving*). See **Norwich tie.**

straight tongue (*Join.*). A wooden tongue for a *ploughed-and-tongued joint* (q.v.), cut so that the grain is parallel to the grooves.

straight- (or solid-) type cable (*Cables*). A cable which has oil-impregnated paper as the dielectric. Used up to 66 kV in the form of single-core H-type cable.

straight-up-and-down filament (*Elec. Eng.*). A filament made in the form of long zigzags between an upper and lower spider support; used in the ordinary vacuum filament lamp.

strain (*Bot., Zool.*). A variety of a species, with distinct morphological and/or physiological characters. (*Eng., Phys., etc.*) When a material is distorted by the forces acting on it, it is said to be in a state of strain, or strained. As a quantity, strain is the ratio

$$\frac{\text{deflection}}{\text{dimension of material}}$$

Strain is thus a number which has no units, since the units of numerator and denominator cancel out. The main types of strain are: direct (*tensile* or *compressive*) strain: E (or *e*)

$$= \frac{\text{elongation or contraction}}{\text{original length}}$$

shear strain: φ

$$= \frac{\text{deflection in direction of shear force}}{\text{distance between shear forces}}$$

$$\text{volumetric strain} = \frac{\text{change of volume.}}{\text{original volume}}$$

See **modulus of elasticity, elasticity of shear.**

strain-ageing (*Met.*). An increase in metal strength and hardness that proceeds with time, after cold-working. It takes place slowly at air temperature and is accelerated by heating. It is most pronounced in iron and steel, but also occurs in other metals.

strain disk (*Glass*). A glass disk of calibrated birefringence, used as a comparative measure of the degree of annealing of glass.

strainer (*Paper*). See **picker.**

strain gauge (*Eng.*). Metal or semiconductor filament on a backing sheet by which it can be attached to a body to be subjected to strain, so that the filament is correspondingly strained. The strain alters the electrical properties of the filament, this alteration forming a basis of measurement.

strain-hardening (*Met.*). Increase in resistance to deformation (i.e., in hardness) produced by deformation. See **cold-working, work-hardening.**

straining (*Leather*). The operation of fastening hides or skins, in a damp state, on wooden frames to maintain their full size while drying.

straining beam (*Build.*). The horizontal beam secured between the heads of the queen-posts in a timber roof having no king-post to prevent them from being forced outwards.

straining sill (*Build.*). A piece of scantling lying on the tie-beam of a timber roof and butting against the feet of the queen-posts, or between the feet of the queens and princesses to keep them apart.

strain insulator (*Elec. Eng.*). An insulator inserted in the span wire of an overhead contact-wire system.

strain-slip cleavage (*Geol.*). A variety of cleavage in which the cleavage planes are parallel shear planes; between each pair the rocks are puckered into small sigmoidal folds.

strain viewer (*Optics*). Eyepiece or projection unit of polariscope.

Straits tin (*Met.*). See **Banka tin.**

strait work (*Mining*). (1) Narrow headings in coal. (2) A method of working coal by driving parallel headings and then removing the coal between them.

strake (*Agric., For.*). A form of metal lug, attached and usually hinged to rim of a pneumatic-tyred wheel to give it additional purchase. (*Min. Proc.*) Gently-sloped, flat table used for catching grains of heavy water-borne **mineral.** See **blanket-, tye.** (*Ships*) A row of plates positioned end to end.

stramineous (*Bot.*). Straw-coloured.

strand (*Elec. Eng.*). One of several wires which together constitute a stranded conductor.

stranded cable (*Elec. Eng.*). One whose core (or cores) consists of stranded conductor.

stranded caisson (*Civ. Eng.*). A watertight box, having a solid floor, which is floated over the site where a bridge pier is to be constructed. Construction goes on in the dry on the floor of the box, which sinks finally to a previously levelled bed under water, the sides of the box being kept always above water. Also called an **American caisson.**

stranded conductor (*Elec. Eng.*). One woven from individual wires or strands, as in the case of a rope.

stranding effect (*Cables*). An increase (20–30%) of the stress at the surface of the conductor, caused by stranding. A usual increase is 25%. Stranding effect is overcome by sector-shaped conductors or by lead sheathing the conductor.

strand plant (*Bot.*). A plant growing in a place where it is not submerged at high tide, but receives salt spray.

strand-space waxing (*Cables*). Waxing which occurs on the conductor side of a conductor paper between two strands. There is no waxing at the crest of the strand because of the good contact between paper and crest.

Stranducer (*Telecomm.*). TN for transducer incorporating strain gauges.

strand waxing (*Cables*). Waxing which occurs near the conductor because of the latter's stranded shape. It is at a maximum at the crest of the strand.

strangeness (*Nuc.*). Quantum number which represents unexplained delay found in strong interactions between certain elementary particles. The *strangeness number* is zero or integral (positive or negative) and is conserved during interactions but not during decay. K-mesons and hyperons which have a non-zero strangeness number are termed *strange particles.* Strangeness is associated with the displacement of the charge centre of particle multiplets from the expected value. An alternative quantity expressing the same concept and more widely used today is *hypercharge.* The hypercharge of a particle is equal to its strangeness plus its baryon number (1 for baryon, −1 for anti-baryon, 0 for meson). Since leptons do not participate in strong interactions they are not assigned strangeness or hypercharge numbers.

strangler (*I.C. Engs.*). See **choke** (2).

strangles (*Vet.*). A contagious disease of horses, due to infection by *Streptococcus equi*, characterized by rhinitis and suppurative adenitis.

strangury (*Med.*). Slow and painful micturition.

strap (*Carp.*). A metal plate or band securing timbers together at a joint.

strap hinge (*Join.*). A hinge having one long leaf for securing to a heavy door or gate.

strapping (*Build.*). A general term for battens fixed to the internal faces of walls as a support for laths and plaster or for plaster board. (*Electronics*) Alternate connexion of segments in a magnetron, to stabilize phases and mode of resonance in the cavities.

strapping motion (*Spinning*). See **governor motion.**

strapping wires (*Elec. Eng.*). Parallel single-wire connexions between a pair of 2-way electric light switches for dual control of a lighting point.

strass (*Chem.*). A very dense glass of high

refractive power; used largely in making artificial jewellery.

Strassburger-Flemming solution (*Micros.*). A preserving solution, containing alcohol and glycerol, which also prevents hardening of tissues.

stratification (*Bot.*). (1) Banding seen in thick cell walls, due to presence of wall layers differing in water content, chemical composition and physical structure. (2) Grouping of vegetation of a wood into two or more well-defined layers differing in height, as trees, shrubs, and ground vegetation. (3) Method of breaking dormancy period of seeds by storage in moist sand at approximately 41°F (4°C). (*Geol.*) The layering in sedimentary rocks due to changes in the rate of deposition, or in the nature of the sediment, or to contemporaneous recurrent pauses in the process of sedimentation. See also **lamination.**

stratification, principle of (*Ecol.*). An important general principle of community organization, embracing all objectively delimitable vertical or horizontal layers of organisms, their by-products, or the results of their activities on the environment.

stratified cambium (*Bot.*). A cambium in which the cells, seen in tangential section, appear arranged in fairly regular horizontal rows.

stratified epithelium (*Zool.*). A type of epithelium consisting of several layers of cells, the outer ones flattened and horny, the inner ones polygonal and protoplasmic.

stratified thallus (*Bot.*). A lichen thallus composed of a layer of algal cells between layers of fungal hyphae.

stratiform, stratose (*Zool.*). Arranged in layers.

stratigraphical break (*Geol.*). The geological record is incomplete, the succession of strata being broken by unconformities and non-consequences, these representing longer or shorter periods of time during which no sediment was deposited or erosion predominated.

stratigraphical level (*Geol.*). See **horizon.**

stratigraphic column (*Geol.*). The succession of stratified rocks in order of age from the oldest at the bottom to the youngest at the top. The term is used in an abstract sense, and also to describe a written table in which the succession of rocks of a particular area is listed.

stratigraphy (*Geol.*). The historical study of the rocks of the earth's crust, their relations and structure, their arrangement into chronological groups, their lithology and the conditions of their formation, and their fossil contents.

strato-cumulus (*Meteor.*). Cloud in large globular or rolled masses, in groups or lines, usually below about 8000 ft (2400 m).

stratose (*Bot.*). Made up of well-defined layers; said particularly of the thallus of a heteromerous lichen.

stratosphere (*Meteor.*). The earth's atmosphere, above the *troposphere*, about 5 miles (8 km) high at the poles and about 10 miles (16 km) high at the equator, extending about 50 miles (80 km) up to where the *ionosphere* begins; has little moisture, and is isothermal in character.

stratum (*Geol.*). A single bed of rock bounded above and below by divisional planes originally almost horizontal—the planes of *stratification* (q.v.). A stratum differs from a lamination only in thickness. *pl.* **strata.** *adj.* **stratified.** (*Zool.*) A layer of cells: a tissue layer.

stratum compactum, -spongiosum (*Zool.*). Layers of the decidua vera in Mammals.

stratum contours (*Geol.*). Contours drawn on the surface of a bed of rock. The position of the outcrop of the bed can be predicted from the intersection of these contours with the surface of the ground.

stratum corneum, -granulosum, -lucidum, -Malpighii (*Zool.*). Layers of the skin in Vertebrates.

stratum germinativum (*Zool.*). See **Malpighian layer.**

stratum society (*Bot.*). A plant society occurring as a well-defined layer in a plant community, as shrubs in a wood.

stratus (*Meteor.*). Cloud in low horizontal sheets.

strauss reaction (*Vet.*). Orchitis and inflammation of the scrotal sac in the guinea-pig following the inoculation of *Actinobacillus mallei* (*Malleomyces mallei*); used in the diagnosis of *glanders.*

strawberry footrot (*Vet.*). A proliferative dermatitis of sheep affecting particularly the lower parts of the legs; caused by the fungus *Dermatophilus pedis.*

straw-rope (*Foundry*). Rough rope of twisted straw or hay wound on a bar or barrel to serve as a foundation for the loam in making a *struck core* (q.v.).

stray capacitance (*Electronics*). Any occurring within a circuit other than that intentionally inserted by capacitors, e.g., capacitance of connecting wires, giving rise to *parasitic oscillation.*

stray field (*Elec. Eng.*). A magnetic field set up in the neighbourhood of electric machines or current-carrying conductors, which serves no useful purpose and which may interfere with the operation of measuring instruments, etc.

stray flux (*Elec. Eng.*). The leakage flux in an a.c. machine or transformer.

stray induction (*Elec. Eng.*). The equivalent induction of the leakage flux effective in producing a reactive voltage drop.

stray losses (*Elec. Eng.*). The stray load losses of an electrical machine, due to stray fields and harmonic flux pulsations in the iron circuit and eddy currents in the windings.

stray radiation (*Nuc.*). Direct and secondary radiation from irradiated objects which is not serving a useful purpose.

stray resonance (*Elec. Eng.*). That arising from unwanted inductance and capacitance, e.g., in leads between conductors, in leads inside canned capacitors, between turns of inductors.

strays (*Radio*). Same as **atmospherics.**

streak (*Bot.*). (1) A line of tissue differing in colour or structure from the tissues on each side of it. (2) A furrow. (3) A kind of viral disease in which necrotic streaks develop in the plant. Also called **stripe.** (*Min.*) The name given to the colour of the powder obtained by scratching a mineral with a knife or file or by rubbing the mineral on paper or an unglazed porcelain surface (*streak plate*). For some minerals, this differs from the body colour.

streaking (*TV*). Extension of image element because of relative phase delay of frequency components in the video signal.

stream anchor (*Ships*). An anchor of lighter weight than a bower anchor; used as a stern anchor.

stream factor (*Chem. Eng.*). That proportion of the time, during which a complete plant is in operation, that any individual item of the plant is working.

stream feeder (*Print.*). An automatic feeder in which the front edge of each sheet moves forward after the previous sheet has travelled a few inches; it is succeeded by the next sheet, and they all travel slowly down the feedboard, each beneath the sheets in front. When each sheet approaches the lays it has only a short final distance to travel, and at a moderate speed; a necessity for good performance on large-sized machines.

streaming effect (*Nuc.*). See **channelling effect.**

streaming potential (*Chem.*). The difference of electrical potential induced between the two ends of a capillary by forcing a liquid through it.

streamline (*Phys.*). A line in a fluid such that the tangent at any point follows the direction of the velocity of the fluid particle at the point, at a given instant. When the streamlines follow closely the contours of a solid object in a moving fluid, the object is said to be of streamline form.

streamline burner (*Eng.*). See **fantail burner.**

streamline flow (*Phys.*). Path taken by fluid molecules or minute suspended particles. Usually qualified as *laminar flow* or *turbulent flow* (qq.v.). See also **viscous flow.**

streamline motion (*Phys.*). The steady motion of a fluid in *laminar flow* (q.v.) past a body with neither abrupt changes in direction nor close curves.

streamlines (*Meteor.*). A set of lines on a chart showing the direction of the horizontal wind at some particular level.

streamline wire (*Aero.*). High-tensile steel wire of elliptical, not true streamline, cross-section, used to reduce the drag of external bracing wires. Fitted principally to biplanes and some early types of monoplane.

stream tin (*Min.*). Cassiterite occurring as derived grains in sands and gravels in the beds of rivers.

streets (*Typog.*). See **rivers.**

strengths of acids (*Chem.*). The relative hydrogen ion *activities* (2), or concentrations (mol dm^{-3}), in solutions containing equivalent quantities of acid.

strepsinema (*Cyt.*). A stage in division reduction; chromosome thread in the strepsitene stage; strepsitene.

Strepsiptera (*Zool.*). An order of the superorder *Hymenopteroidea.* Males have only one pair of wings, the anterior pair being modified to form balancers. Females are larviform, viviparous and endoparasitic in other insects, mainly *Hymenoptera* and *Homoptera*, often causing parasitic castration. Larva initially a triungulin, then becomes parasitic, undergoing hypermetamorphosis. Stylops.

strepsitene (*Cyt.*). A stage in meiotic prophase, where the diplotene threads appear to be twisted, and during which *crossing-over* (q.v.) occurs.

streptococcus (*Bacteriol.*). A Gram-positive coccus of which the individuals tend to be grouped in chains. Many forms possess species-specific capsular polysaccharides by which they can be divided into groups, which include the causative agents of scarlet fever, erysipelas and one form of mastitis. Other streptococci include *Diplococcus pneumoniae* (one cause of pneumonia). Some types occur normally in the mouth, throat, and intestine.

streptomycin (*Pharm.*). $C_{21}H_{39}N_7O_{12}$. An antibiotic produced by the actinomycete soil-fungus *Streptomyces griseus*, an organic base of very complicated molecular constitution. Many infections (some not affected by penicillin) respond to its therapeutic use, particularly tularaemia, influenzal meningitis, and tuberculosis.

Streptoneura (*Zool.*). See **Prosobranchiata.**

streptoneural, streptoneurous (*Zool.*). Having an asymmetrical nervous system; said especially of certain *Gastropoda* in which the visceral nerve loop is twisted into a figure of eight.

streptostyly (*Zool.*). In Vertebrates, the condition of having the quadrate movably articulated with the squamosal; cf. *monimostyly.*

streptotrichosis (*Med.*). A rare infection (especially of the lungs) with a branching fungus.

stress (*Eng., etc.*). The force per unit area acting on a material and tending to change its dimensions, i.e. cause a *strain.* The *stress* in the material is the ratio of force applied to the area of material resisting the force; i.e.:

$$\text{stress} = \frac{\text{force}}{\text{area}}$$

The two main types of stress are *direct* or *normal* (i.e., *tensile* or *compressive*) *stress* (symbol σ or f) and *shear stress* (symbol τ or f_s or q). The usual units are kPa, MPa, lbf/in^2, tonf/in^2, kN/m^2, MN/m^2, bar or hbar. (See **newton, bar.**) *To stress* may also mean 'to calculate stresses'; thus a structure has been *stressed* when its strength has been verified by calculation.

stress cone (*Cables*). A cone formed by wrapping paper tapes on the core of a cable from the place where the lead sheath ends. The purpose is to prevent flashover or tracking across the surface of the core. The stress cone is covered on its sloping face by a lead flare.

stress diagram (*Eng.*). See **force diagram.**

stressed-skin construction (*Aero.*). The general term for aircraft structures in which the skin (usually light alloy, formerly plywood, occasionally plastic such as Fibreglass, and in supersonic aircraft titanium alloy or steel) carries a large proportion of the loads. In the more elementary forms, the framework may take bending and shear, with a thin skin transmitting torsion, but when of developed form, the skin is thick enough to support bending loads in the form of tension and compression in the respective surfaces. Since 1950 the principle of thick skin has been developed in the form of 'sculpturing' to vary the thickness to suit the local loads and to incorporate integral stiffeners, either by machining or by acid etching. See **monocoque.**

stress marks (*Photog.*). Markings on finished prints due to abrasion of the film surface. These can be removed with methanol, possibly assisted by thiocarbamide.

stress minerals (*Geol.*). A term used for minerals occurring in metamorphic rocks whose formation is favoured by shearing-stresses. Such minerals include kyanite, chlorite, chloritoid, epidote, albite.

stress-number (or S/N) **curve** (*Met.*). A curve obtained in fatigue tests by subjecting a series of specimens of a given material to different ranges of stress and plotting the range of stress against the number of cycles required to produce failure. In steel and many other metals, there is a limiting range of stress below which failure will not be produced even by an indefinite number of cycles.

stress relief (*Met.*). Annealing of metals at moderate temperature to remove the internal stresses caused by work-hardening or quenching.

stress relieving (*Chem. Eng.*). A process used in the manufacture of welded pressure vessels for severe duties, particularly involving strong caustic solutions or very thick walls, in which the completed vessel is raised to a high temperature, usually red hot, held for a time and then allowed to cool in still air. The exact temperatures and times are determined by *codes of construction* (q.v.). (*Met.*) Generally applied to steel, by heating to a temperature below that liable to alter the crystalline structure and with the object of reducing or eliminating any harmful residual stresses arising from other processes.

stress-strain curve (*Met.*). A curve similar to a *load-extension curve*, except that the load is divided by the original cross-sectional area of the test piece and expressed in units of *stress*, while the extension is divided by the length over which it is measured and expressed as a ratio. In the elastic range, it follows *Hooke's law* (q.v.), after which further stress produces plastic flow, work-hardening and residual strain (*permanent set*).

stress zone (*Mining*). Depth of rock surrounding an underground excavation, e.g., a stope, which is now bearing the transferred stress originally supported by the removed ore.

stretch (*Cinema.*). (1) The relative increase in length of film during processing or exposure to moist conditions. (2) The introduction of extra frames in a print by holding the negative in an intermittent printer, to obtain a slower motion of the images on projection, or to permit an increase in speed of projection without increasing motion of images.

stretched diaphragm (*Acous.*). A diaphragm in a microphone or loudspeaker which has its rigidity increased by radial stretching, frequently by screwing on to it a rim near its edge. Resonance then becomes marked, and the tension is adjusted so that the major resonant frequency nears the upper limit of the desired transmission frequency band. Trend of the response curve is then adjusted by altering the damping.

stretcher (*Build.*). Brick laid parallel to the course. (*Mining*) A bar fixed across a narrow working place or tunnel to support a rock drill.

stretch forming (*Eng.*). Process for forming large sheets of thin metal into symmetrical shapes by gripping the sheet edges in horizontally sliding stretcher jaws and moving a forming punch, without a die, vertically between them against the sheet.

stretching bond (*Build.*). The form of bond, used largely for building internal partition walls $4\frac{1}{2}$ in (114 mm) thick, in which every brick is laid as a stretcher, each vertical joint lying between the centres of the stretchers above and below, so that angle closers are not required. Also **stretcher bond**. See **chimney bond**.

stria, striation. A faint ridge or furrow; a streak; a linear mark.

striae (*Geol.*). Parallel lines or grooves occurring on glaciated pavements, roches moutonnées, etc.; produced by hard stones frozen into the base of a moving ice-sheet; also seen on slicken-sided rock surfaces along which movement has taken place during faulting. (*Min.*) Parallel lines occurring on the faces of some crystals; caused by oscillation between two crystal forms. The striated cubes of pyrite are good examples.

striae atrophicae (*Med.*). Greyish-white bands of atrophied skin in areas where the skin has been unduly stretched, as in pregnancy (*striae gravidarum*).

stria medullaris (*Zool.*). See **habenula.**

striated muscle (*Zool.*). A form of contractile tissue composed of multinucleate unbranched fibres enclosed by a sarcolemma, showing marked transverse striations and having the nuclei at the periphery. Also striped muscle. Cf. *cardiac muscle*, *unstriated muscle*, and see **voluntary muscle.**

striation (*Electronics*). Phenomenon in low-pressure gas discharge, which forms luminous bands (*striae*) across the line between electrodes.

stria vascularis (*Zool.*). A vascular pigmented membrane of the organ of Corti.

strickle board (*Foundry*). A board profiled along one edge to the required shape of the surface of a loam mould or core; used to sweep or strike the loam to the correct section. See **loam.**

stricture (*Med.*). Any abnormal narrowing of a duct or passage in the body, especially the narrowing of the urethra due to gonorrhoeal inflammation. See also **stenosis.**

striding level (*Surv.*). Sensitive spirit level which can be placed astride a theodolite by resting the V-shaped ends of its legs on the trunnion axis, enabling the latter to be accurately levelled.

stridor (*Med.*). A harsh vibrating noise produced by any obstruction in the respiratory tubes, e.g., in diphtheria of the larynx.

stridulating organs (*Zool.*). The parts of the body concerned in sound production by stridulation.

stridulation (*Zool.*). Sound production by friction of one part of the body against another, as in some Insects.

striga (*Zool.*). Synonym for **stria.** *adjs.* **strigate, strigose.**

Strigiformes (*Zool.*). An order of *Neognathae* containing nocturnal birds of prey with hawk-like beaks and claws. Plumage adapted for silent flight, retina containing mainly rods, eyes large and immovable, probably hunt mainly by sound. Owls.

strigil, strigilis (*Zool.*). In certain *Hemiptera*, a curious asymmetrical organ consisting of rows of black comblike plates situated on the dorsal surface of the abdomen: in some Bees, a mechanism for cleaning the antennae situated at the junction of the tibia and the tarsus of the first leg.

strigose (*Bot.*). Bearing hairs which are usually rough and all pointing in the same general direction. *dim.* **strigillose.**

strike (*Geol.*). The horizontal direction which is at right angles to the dip of a rock. (*Typog.*) To drive a hardened steel punch into a brass or copper bar, so producing a matrix from which types are cast. The term is also used to describe the impression itself. (*Vet.*) See **blowfly myiasis.**

strike fault (*Geol.*). A fault aligned parallel to the strike of the strata which it cuts. Cf. *dip fault*.

strike lines (*Geol.*). Synonymous with *stratum contours* and commonly used if the beds are dipping evenly.

strike note (*Acous.*). The note, largely subjective, which is initially prominent when a bell is struck. It rapidly attenuates, leaving the hum note and some overtones.

strike-through (*Print.*). Ink percolating through from the other side of a paper, due to lack of opacity or sizing; often seen in newsprint. Cf. *show-through*.

striking (*Build., Civ. Eng.*). The operation of removing temporary supports or shuttering from a structure. (*Foundry*) See **striking-up.**

striking plate (*Carp., Join.*). A metal plate screwed to the jamb of a door case in such a position that when the door is being shut the bolt of the lock strikes against, and rubs along, the plate, finally engaging in a hole in the latter.

striking potential (*Electronics*). That sufficiently large to break down a gap and cause an arc, or start discharge in a cold-cathode tube.

striking-up or striking (*Foundry*). The process of generating a loam mould surface by means of a *strickle board* (q.v.).

striking voltage (*Electronics*). See **starting voltage.**

striking wedges (*Civ. Eng.*). A pair of wedge-shaped blocks of hard wood packed beneath each end of a centre and placed in contact, with their thin ends pointing in opposite directions, so that, by moving them relatively, the centre may be gradually lowered on completion of the work. Also **easing wedges, lowering wedges.**

string (*Carp.*). A sloping wooden joist supporting the steps in wooden stairs. See cut-. (*Cinema.*) The phosphor-bronze flattened wires in a recording vibrator, and the Duralumin strip in a light valve. (*Comp.*) Series of numbers in order but not related, i.e., at random. (*Elec. Eng.*) The series of insulator units combining to form a suspension insulator.

string chart (*Elec. Eng.*). A diagram from which the relation between the sag of an overhead line and the temperature may be rapidly obtained.

string course (*Build.*). A projecting course in a wall.

string diagram (*Work Study*). A scale plan or model on which a thread is used to trace and measure the path of workers, materials or equipment, during a specified sequence of events.

string efficiency (*Elec. Eng.*). The ratio of the flashover voltage of a suspension-insulator string to the product of the flashover voltage of each unit and the number of units forming the string.

string electrometer (*Elec. Eng.*). An electrometer consisting of two metal plates oppositely charged between which a conducting fibre is displaced from a middle position in proportion to the voltage between the plates.

stringer. A long horizontal member in a structural framework. (*Aero.*) A light auxiliary member parallel with the main structural members of a wing, fuselage, float, or hull, mainly for bracing the transverse frames and stabilizing the skin material. See **stiffener.** (*Nuc. Eng.*) Group of reactor fuel elements strung together for insertion into one channel of the core.

string galvanometer (*Cinema.*). A vibrator used in variable-area sound-film recording, which deflects the image of the slit across the track or along the track. (*Elec. Eng.*) See **Einthoven galvanometer.**

stringhalt (*Vet.*). A disease of horses, characterized by involuntary sudden and excessive flexion of one hind limb or both.

string-pulling behaviour (*An. Behav.*). The behaviour of some birds in reaching food suspended on a string by pulling up the string and holding the pulled-in loop with one foot while reaching with the beak for the next pull. Possibly an example of *insight learning* (q.v.).

string warp machines (*Textiles*). Lace machines which are supplied by spools as well as by beams in making warp laces.

striolate (*Bot.*). Finely striate.

stripe (*Bot.*). See **streak** (3). (*Cinema.*) Magnetic sound-track(s) on cinematograph film for sound-film reproduction, adjacent to the printed picture track. Can be added to final print.

striped muscle (*Zool.*). See **striated muscle.**

strip heating (*Join.*). Method of heating glued joints by bringing the wood into contact with an electrically heated bare metal strip.

stripline (*Electronics*). Waveguide formed of strips of copper on dielectrics, formed by etching a printed circuit.

strip mining (*Mining*). Form of opencast work, in which the overburden is removed (stripped) after which the valuable ore is excavated, the work usually being done in series of benches, steps, or terraces.

strippable coatings (*Paint.*). Coatings for the inside of spray booths, etc., which can be removed mechanically, bringing away all overspray. These are formulated from well-plasticized ethyl cellulose or similar systems.

stripped atom (*Phys.*). Ionized atom from which at least one electron has been removed.

stripper (*Nuc.*). The section of an isotope-separation system which strips the selected isotope from the waste stream.

stripper and grinder (*Textiles*). A skilled operative whose duty is to sharpen the doffers, cylinders, flats or rollers on cards by revolving carborundum rollers, then reset the machine.

stripper and worker (*Textiles*). A pair of wire-covered rollers forming part of the regulating mechanism on a roller and clearer card, i.e., wool or waste card.

strippers (*Print.*). A feature of delivery arrangement of some printing machines, including the 2-revolution, by which the printed sheet is led away after being released from the grippers.

stripping (*Chem.*). Removal of an electrodeposit by any means, i.e., by chemical agent or by reversed electrodeposition. (*Mining*) Removal of barren overburden in opencast work. (*Nuc.*) A phenomenon observed in deuteron (or heavier nuclei) bombardment in which only a portion of the incident particle merges with the target nucleus, the remainder proceeding with most of its original momentum practically unchanged in direction. (*Photog.*) The process of removing the negative emulsion film from its glass support for transfer to another glass or other support (as in process-block making).

stripping agent (*Acous.*). The substance which is coated on the surface of a copper matrix when a copy of this is made by electroplating. When copper is copperplated, a silver iodide film is used, while the bichromate salt is used in nickel-plating. For effective plating, the stripping agent must be adequately conducting.

strip-wound armature (*Elec. Eng.*). An armature whose winding consists of conductors in the form of copper strip.

strobe (*Electronics*). (1) General term for detailed examination of a designated phase or epoch of a recurring waveform or phenomenon. (2) Enlargement or intensification of a part of a waveform as exhibited on a CRT. Also called **linearity control.** (3) Process of viewing vibrations with a *stroboscope*; colloquially, the stroboscope itself.

strobe lighting (*Aero.*). An anti-collision lighting system based on the principle of a capacitor-discharge flash tube. A capacitor is charged to a very high voltage which is then discharged in a controlled sequence as a high-intensity flash of light, usually blue-white, through xenon-filled tubes located at wing tips and tail of an aeroplane.

strobe pulse (*Electronics*). Pulse much shorter than a repeated waveform, for examination of a display.

strobila (*Zool.*). In *Scyphozoa*, a scyphistoma in process of production of medusoids by transverse fission: in *Cestoda*, a chain of proglottides. Also **strobile.** *adjs.* **strobilate, strobilaceous, strobiliferous, strobiloid.**

strobilaceous (*Bot.*). Of, or resembling, a cone.

strobilate (*Bot.*). Of the nature of a cone.

strobile. See **strobila, strobilus.**

strobilization (*Zool.*). Production of strobilae: in *Scyphozoa*, transverse fission of a scyphistoma to form medusoids: in *Cestoda*, production of proglottides by budding from the back of the scolex: in some *Polychaeta*, reproduction by gemmation.

strobilus (*Bot.*). A group of sporophylls with their

sporangia, more or less tightly packed around a central axis, forming a well-defined group; a cone. Also **strobile.**

stroboscope (*Electronics*). A flashing lamp, of precisely variable periodicity, which can be synchronized with the frequency of rotating machinery or other periodic phenomena, so that, when viewed by the light of the stroboscope, they appear to be stationary.

strobotron (*Electronics*). Triggered gas-discharge tube used as pulsed light source in stroboscope.

stroke (*Electronics*). The time-base sweep moving with uniform velocity across the screen of a CRO. See **flyback.** (*Eng.*) The travel or excursion of a piston, press ram, or other (reciprocating) part of a machine. (*Med.*) An apoplectic seizure; a sudden attack of paralysis. See also **apoplexy.**

stroked (*Build.*). A term applied to the face of an ashlar which has been so tooled as to present a regular series of small flutings.

stroma (*Bot.*). (1) The denser part of a chloroplast; it is colourless. (2) A dense mass of interwoven hyphae, fleshy to horny in texture, cushion-like, columnar, club-shaped, or branched, in which many fungi develop their fructifications. (*Zool.*) A supporting framework, as the connective tissue framework of the ovary or testis in Mammals. *pl.* **stromata.** *adjs.* **stromate, stromatic, stromatiform, stromatoid, stromoid, stromatous.**

stroma starch (*Bot.*). Starch formed in the stroma of a chloroplast at times when photosynthesis is active.

stromatoporoid limestone (*Geol.*). A calcareous sedimentary rock type, rich in the remains of the reef builder *Stromatopora*, important from Palaeozoic times onwards.

Strombolian eruption (*Geol.*). A type of volcanic eruption characterized by frequent small explosions as trapped gases break through overlying viscous lava.

strombuliferous (*Zool.*). Having the organs coiled in a spiral fashion; bearing spirally coiled structures.

strombuliform (*Bot.*). Said of a spirally twisted fruit.

strombus (*Bot.*). A spirally coiled pod.

strong clay (*Build.*). See **foul clay.**

strong electrolyte (*Chem.*). An *electrolyte* (q.v.) which is completely ionized even in fairly concentrated solutions.

strong interaction (*Nuc.*). One produced by short-range nuclear forces associated with mesons and completed in a time of the order of 10^{-23} s. Baryons and mesons participate in strong interactions.

Strongylata (*Zool.*). An order of *Phasmidia* having a simple mouth with no papillae, the males having two spicules and a true bursa, and the oesophagus being club-shaped or cylindrical. Includes *Ancylostoma spp.* (Hookworms).

strongyloid (*Zool.*). Said of a type of nematode larva: said of a biradiate monaxonic sponge spicule with rounded ends.

strongyloidiasis, strongyloidosis (*Med.*). Infestation of Man with the nematode worm *Strongyloides stercoralis*, the worm living in the intestines and causing diarrhoea; common in the Tropics.

strontianite (*Min.*). Carbonate of strontium, crystallizing in the orthorhombic system. Its colour varies from pale green or grey to brown, and it occurs in hydrothermal veins in limestones and is found occasionally in igneous rocks. It is named for the locality at Strontian, Argyll: the element *strontium* was first discovered in this mineral.

strontium (*Chem.*). Metallic element, symbol Sr, at. no. 38, r.a.m. 87·62, rel. d. 2·54, m.p. 800°C, b.p. 1300°C. Silvery-white in colour, it is found naturally in *celestine* and in *strontianite*; it also occurs in mineral springs. Similar chemical quantities to calcium. Compounds give crimson colour to flame and are used in fireworks. The radioactive isotope, *strontium-90*, is produced in fission of uranium, and has a long life, hence its presence in 'fall-out' after a nuclear explosion.

strontium age (*Geol.*). Age estimated from the relative numbers of atoms of stable strontium and radioactive rubidium present.

strontium dating (*Min.*). The geological method based on the *beta decay* of rubidium-87 to strontium (decay half-life 5×10^{10} years, hence rate of strontium-87 production is 3·7 μg per g of rubidium per million years).

strontium unit (*Nuc.*). One used to measure concentration of radioactive strontium-90 in calcium; an SU is 10^{-22} Ci/g.

strophiolate (*Bot.*). See **carunculate.**

strophiole (*Bot.*). See **caruncle.**

strophotron (*Electronics*). Multi-reflection oscillator for microwave frequencies, capable of a small range of tuning. See **Barkhausen-Kurz oscillator.**

struck (*Build.*). (1) Taken away, dismantled; said, e.g., of scaffolding or shuttering. (2) Joints on an exposed face of a wall are said to be *struck* when the mortar is recessed. (*Vet.*) (1) A form of *enterotoxaemia of sheep* (q.v.) caused by *Clostridium perfringens* (*Cl. welchii*), type C. (2) Said of sheep affected with *blowfly myiasis* (q.v.).

struck core (*Foundry*). A loam core formed by revolving the built-up core, loam-covered, against a *strickle board* (q.v.). See also **straw-rope.**

struck-joint pointing (*Build.*). See **weathered pointing.**

structural colours (*Zool.*). Colour effects produced by some structural modification of the surface of the integument, as the iridescent colours of some Beetles. Cf. *pigmentary colours.*

structural damping (*Aero.*). See **damping.**

structural hybrid (*Gen.*). An organism in which the two sets of chromosomes in the diploid complement are different in composition.

structural timber (*Build.*). The Canadian name for *carcassing timber*.

structure (*Chem.*). See **molecular structure.**

structure of the atom (*Chem.*). See **atomic structure.**

struma (*Bot.*). A swelling on one side at base of capsule of a moss. (*Med.*) (1) See **scrofula.** (2) See goitre.

strumose (*Bot.*). (1) Bearing a swelling at one side of the base. (2) Bearing cushionlike swellings.

strut (*Civ. Eng.*). A timber or adjustable metal member used for bracing purposes during excavation or construction works. (*Eng.*) Any light structural member or long column which sustains an axial compressive load. Failure occurs by bending before the material reaches its ultimate compressive stress. See **column.**

Struthioniformes (*Zool.*). An order of *Palaeognathae* retaining only two toes and whose feathers lack an aftershaft. They extend their rudimentary wings when running. Known from the Pliocene onwards. Ostriches.

strutting (*Build.*). The process of using props to give temporary support between two surfaces.

strychnine (*Chem.*). $C_{21}H_{22}N_2O_2$, a monoacidic alkaloid base, rhombic prisms, m.p. 265°C, b.p. 270°C at 666 Pa; very poisonous, causing tetanic spasms. It occurs in the seeds of *Strychnos Ignatii* and *Strychnos nux vomica*, in *Upas Tieuté*, and in *Lignum colubrinum*. Strychnine is almost insoluble in water, but is readily soluble in chloroform and benzene.

strychnine bases (*Chem.*). A group of alkaloids obtained from *Strychnos nux vomica*. They include *strychnine* (q.v.) and *brucine* (q.v.).

stub (*Build.*). A small projection on the under surface at the top edge of a tile, enabling it to be hung on a batten. (*Telecomm.*) An auxiliary section of a waveguide or transmission line connected at some angle with the main section. See **coaxial-, quarter-wavelength-.**

stub aerial (*Radio*). Straight wire which resonates at a wavelength ca. 4 times its length.

stub axle (*Autos.*). A short dead axle. If carrying a steered wheel it is capable of limited angular movement about a swivel-pin carried by the end of the axle beam.

stub plane (*Aero.*). A short length of wing projecting from the fuselage, or hull, of some types of aircraft to which the main planes are attached.

stub tenon (*Carp.*). A very short tenon for fitting into a blind mortise. Also **joggle.**

stub-tooth gear (*Eng.*). A gear tooth of smaller height and of more robust form than that normally employed; used in the manufacture of automobile gears.

stub tuning (*Elec. Eng.*). Use of shunt stubs connected to short-circuited sections of line or waveguide in order to produce matched conditions. In the single-stub tuner the stub susceptance is $-jb$ and it is connected to the main transmission line at a point where the transformed load admittance is $y = 1 + jb$ (normalized).

stuc (*Build.*). Plasterwork resembling stone.

stucco (*Build.*). A smooth-surfaced plaster or cement rendering applied to external walls.

stuck moulding (*Join.*). A moulding shaped out of the solid of a member.

stud (*Build.*). The vertical bracing members in a hollow partition framework. (*Carp.*) An upright scantling in a timber framework or partition. (*Eng.*) A shank, or headless bolt, generally screwed from both ends and plain in the middle. It is permanently screwed into one piece, to which another is then secured by a nut.

student's t-test (*Stats.*). See **t-distribution.**

stud partition (*Carp.*). A wooden partition based on rough timber framing.

stuff (*Build.*). (1) See **coarse stuff, fine stuff.** (2) Timber sawn or manufactured from logs.

stuff chest (*Paper*). A large cylindrical vessel in which pulp is stored before passing forward to the strainers, etc. The fibres are kept in suspension by revolving paddles.

stuffed (*Bot.*). Said of the stipe of an agaric when the interior is occupied by a cottony or spongy mass different in texture from the peripheral parts.

stuffing (*Leather*). A process, similar to currying, by which leather is impregnated with grease, generally in a stuffing drum.

stuffing-box (*Eng.*). A cylindrical recess provided in, for example, a cylinder cover, at the point at which the piston rod emerges; it is filled with packing which is compressed by a *gland* (q.v.) to make a pressure-tight joint.

stuffing-drum (*Leather*). A heated drum in which leather is impregnated with grease.

stugging (*Build.*). See **picking.**

stuke (*Build.*). See **stucco.**

stull (*Mining*). A timber prop between the walls of a stope.

stull covering (*Mining*). A platform in a stope, to carry men or minerals.

stump tenon (*Carp.*). A tenon differing from a *stub tenon* (q.v.) in that it tapers so as to have a greater thickness at the root.

stupe (*Med.*). A piece of cloth, flannel, or the like, soaked in hot water, wrung out dry, and medicated for external application.

stupor (*Med.*). A state of mental and physical inertia; inhibition of instinctive activity and indifference to social environment.

stupose (*Bot.*). Towlike.

sturdy (*Vet.*). See **coenuriasis.**

Sturm's theorem (*Maths.*). A theorem by which the number of real roots of an algebraic equation which lie in any given interval can be determined. It utilizes sign changes in the partial remainders that occur in calculating the H.C.F. of $f(x)$ and $f'(x)$, where $f(x) = 0$ is the given equation.

Stuttgart disease (*Vet.*). Canine typhus. The name formerly given to a disease of the dog characterized by apathy, stomatitis, and gastro-enteritis, but which probably was, in most cases, the uraemic stage of nephritis caused by *Leptospira canicola.*

sty or **stye** (*Med.*). Hordeolum. Staphylococcal infection of a sebaceous gland of the eyelid.

stylar canal (*Bot.*). A tube or space occupied by loose tissue, through which the pollen tubes pass as they grow through the style.

style (*Bot.*). The portion of the pistil between the ovary and the stigma; it is often elongated and threadlike. (*Join.*) See **stile.** (*Zool.*) In *Hydrocorallinae*, a calcareous projection arising from the tabula at the bottom of each cup; in some of the lower orders of Insects, the unmodified outer gonapophysis of the ninth segment in the male; in some *Diptera*, an appendage of the antenna. Also **stylus.**

style of the house (*Typog.*). The customary style of spelling, punctuation, capitalization, etc., used in a printing establishment. It is followed in the absence of contrary instructions.

stylet (*Zool.*). A small pointed bristlelike process.

stylidium (*Bot.*). The upper portion of an archegonium.

stylifer (*Zool.*). A process or sclerite from which a style arises.

styliform (*Zool.*). Bristle-shaped.

stylo-. Prefix from Gk. *stylos*, pillar.

stylobate (*Arch.*). A continuous pedestal supporting a row of columns.

styloconic (*Zool.*). Said of a type of sensilla in Insects which consists of one or more basiconic pegs elevated on a cone.

styloglossus (*Zool.*). In higher Vertebrates, a muscle connecting the styloid process and the tongue. *adj.* **styloglossal.**

stylohyal (*Zool.*). In the hyoid apparatus of Mammals, one of the three bones forming each of the two anterior cornua. In some Mammals, including Man, it, together with the tympanohyal, becomes fused to the periotic and tympanic bones, forming the styloid process.

stylohyoid (*Zool.*). Pertaining to, or connecting, the styloid process and the hyoid; as a ligament in higher Vertebrates.

styloid (*Zool.*). Pillar-shaped; as a process of the otic capsule in the Mammalian skull.

stylomastoid (*Zool.*). In the skull of Mammals, a foramen between the tympanic and periotic bones, through which passes the seventh cranial nerve.

stylopized (*Zool.*). Of Bees, parasitized by *Strepsiptera* (after the principal genus *Stylops*).
stylopodium (*Bot.*). A swelling at the base of a style. (*Zool.*) The proximal segment of a typical pentadactyl limb; brachium or femur; upper arm or thigh.
stylospore (*Bot.*). See **pycnidiospore, pycnospore.**
stylus (*Acous.*). Needle for cutting or replaying a disk recording. With the introduction of lightweight pickups, the more durable sapphire- or diamond-tipped reproducer styli replaced the earlier steel or chrome-plated needles. (*Zool.*) In primitive Insects, a small appendage attached to the coxae of the middle and hind pairs of legs; in Mammals, a molar cusp. See also **style.**
S-type cathode (*Electronics*). One which conforms to a rigid specification, with Ni or Pt core heater wire, and Ba and Sr oxides co-precipitated, applied, and processed.
styphnic acid (*Chem.*). $(NO_2)_3 \cdot C_6H \cdot (OH)_2$, the trinitro derivative of 1,3-dihydroxybenzene; formed by the action of nitric acid upon many gum resins.
stypsis (*Med.*). The application, or use, of styptics.
styptic (*Med.*). Astringent; tending to stop bleeding by coagulation.
styrax (*Pharm.*). See **storax.**
styrene (*Chem.*). Phenylethene, $C_6H_5 \cdot CH : CH_2$, a constituent of essential oils and coal-tar. It is a colourless aromatic liquid, b.p. 145°C, soluble in ethanol and ethoxyethane.
styrene-butadiene rubber (*Plastics*). See **Buna, GR-S.**
styrene joint (*Cables*). A joint filled with hot liquid styrene, which polymerizes on cooling into a very hard solid and prevents displacement of the cores.
styrene resins (*Plastics*). Compounds, polystyrenes, formed by the polymerization of styrene, $C_6H_5 \cdot CH{=}CH_2$. The power factor and dielectric constant are lower than for other plastics, and they do not disintegrate at ultrashort-wave and TV frequencies. They have excellent mechanical properties, and are resistant to moisture, concentrated sulphuric acid, strong alkalis, and alcohol. Foamed polystyrene is an important heat insulation material.
styrol resins (*Plastics*). See **styrene resins.**
sub-. Prefix from L. *sub*, under, used in the following senses: (1) deviating slightly from, e.g., *subtypical*, not quite typical; (2) below, e.g., *subvertebral*, below the vertebral column; (3) somewhat, e.g., *subspatulate*, somewhat spatulate; (4) almost, e.g., *subthoracic*, almost thoracic in position.
subacute (*Med.*). Said of a disease whose symptoms are less pronounced than those of the acute form; between acute and chronic.
subacute combined degeneration (*Med.*). Anaemic spinal disease. A condition in which there is degeneration of motor and sensory nerve tracts in the spinal cord, giving rise to paraplegia and loss of sensibility of the skin, the disease being associated with pernicious anaemia.
sub-additive function (*Maths.*). See **additive function.**
subaqueous loudspeaker (*Acous.*). See **piezoelectric loudspeaker.**
subaqueous microphone (*Acous.*). See **hydrophone.**
subarachnoid haemorrhage (*Med.*). Haemorrhage into the space between the arachnoid and the pia mater, especially as a result of rupture of an aneurysm of one of the arteries.
subassembly (*Eng.*). Assembly which can be handled and stored as a unit and subsequently incorporated in a more complex assembly.
subatomic (*Chem.*). Said of particles or processes at less than atomic level, e.g., radioactivity, production of X-rays, nuclear shells, etc.
subatomics (*Chem.*). The study of processes which involve changes inside an atom.
subatrial ridge (*Zool.*). In embryonic *Amphioxus* two longitudinal ridges which form on the inner faces of the metapleural folds and eventually meet and coalesce, forming the ventral wall of the atrium.
subaudio frequency (*Acous.*). One below those usefully reproduced through a sound-reproducing system or part of such system.
subaudio telegraphy (*Teleg.*). That which transmits by a.c. of frequency below 300 Hz.
subcarrier (*Telecomm.*). One frequency which is modulated over a narrow range by a measured quantity, and then used to modulate (with others) a carrier that will be finally demodulated on reception.
subcellular (*Cyt.*). Used of individual structures, units, and organelles within a cell, e.g., mitochondria, chromosomes.
subcentric oösphere (*Bot.*). A fungal oösphere with the protoplasm surrounded by one layer of fatty globules and with two or three additional layers on one side only.
subchelate (*Zool.*). In *Arthropoda*, having the distal joint of an appendage modified so that it will bend back and oppose the penultimate joint, like the blade and handle of a penknife, to form a prehensile weapon; cf. *chelate*.
subchord (*Surv.*). The chord length from a tangent point on a railway or highway curve to the adjacent chainage peg around the curve when this is less than the full chord distance employed in setting out the chainage pegs.
subcircuit (*Elec. Eng.*). One of several lighting circuits supplied from a common branch distribution fuse board.
subclavian (*Zool.*). Passing beneath, or situated under, the clavicle; as the *subclavian artery*.
subclavius (*Zool.*). A wing muscle of Birds.
subclimax (*Bot.*). A community which has not attained the full development possible under the prevailing climatic conditions because of some limitation by an edaphic or biotic factor.
subconscious (*Psychol.*). See **preconscious.**
subcooled water (*Meteor.*). See **supercooled water.**
subcortical (*Psychol.*). Pertaining to activity of the brain in which the cortex is not involved. (*Zool.*) Below the cortex or cortical layer; as certain cavities in Sponges.
subcosta (*Zool.*). In Insects, one of the primary veins of the wing. *adj.* **subcostal.**
subcostal (*Zool.*). Below the costae.
subcoxa (*Zool.*). The true basal segment of the primitive leg in Insects, usually reduced or much modified; probably represented in the *Pterygota* by the trochantin and some pleural sclerites of the thoracic wall. Replaced as the functional base of the leg by the *coxa* (q.v.).
subcritical (*Nuc.*). Assembly of fissile material, for which the multiplication factor is less than unity.
subcrust (*Civ. Eng.*). A cushioning layer between the pavement and the foundation of a carriageway, or the base formed on the natural foundation. See **cushion course.**
subculture (*Bot.*). A culture of bacteria or fungi prepared from a pre-existing culture.
subcutaneous (*Zool.*). Situated just below the skin.
subdorsal (*Zool.*). Situated just below the dorsal surface.
subdural (*Med., Zool.*). Situated beneath the dura mater, e.g., *subdural abscess*.

subentire (*Bot.*). Said of a margin which is very faintly indented.
subepidermal tissue (*Bot.*). Hypodermis.
suberect (*Bot.*). Upright below, nodding at the top.
suberic acid (*Chem.*). Hexan-1,6-dicarboxylic acid or octan-1,8-di-oic acid. $HOOC \cdot (CH_2)_6 \cdot COOH$, a saturated dibasic acid; m.p. 140°C.
suberification (*Bot.*). See **suberization.**
suberin (*Bot.*). A complex mixture of fatty substances present in the cell walls of corky tissue making them waterproof and decay-resistant.
suberin lamella (*Bot.*). A layer of wall material impregnated with suberin.
suberization (*Bot.*). The impregnation of cell walls with suberin, with consequent formation of cork.
subfactorial *n* (*Maths.*). The number of different ways of arranging *n* objects so that no object occupies its original position. Written $n¡$ or $\underline{\parallel n}$. Cf. *factorial n.*
subgalea (*Zool.*). In Insects, an inner sclerite of the stipes; parastipes.
subgenital (*Zool.*). Below the genital organs, as the subgenital pouches of *Aurelia.*
subgenital portico (*Zool.*). In some *Scyphozoa*, a spacious chamber lying below the stomach and formed by the fusion of the subgenital pouches.
subgenual organ (*Zool.*). In many Insects, chordotonal organs adapted for perceiving vibrations of the leg.
subglacial drainage (*Geol.*). The system of streams beneath a glacier or ice sheet; formed chiefly of melt-waters. Cf. *englacial streams.*
subgroup (*Maths.*). Let $(G; *)$ be a *group*. Let H be a subset of G which forms a group with the operation $*$. Then $(H; *)$ is said to be a *subgroup* of the group $(G; *)$; e.g., integers with addition constitute a subgroup of the rational numbers with addition.
subharmonic (*Acous.*). Having a frequency which is a fraction of a fundamental. Subharmonics appear in some forms of nonlinear distortion.
subhedral (*Geol.*). See **hypidiomorphic.**
subhydrous coal (*Fuels*). Coal of hydrogen content below average for the rank of coal, e.g., coals containing a high proportion of fusain.
subhymenial layer (*Bot.*). A layer of hyphae immediately beneath a hymenium.
subiculum, subicle (*Bot.*). A felted or cottony mass of fungal hyphae underlying the fruit bodies of some fungi.
subimago (*Zool.*). In *Ephemeroptera* (Mayflies) the stage in the life history emerging from the last aquatic nymph. It is usually dull in appearance and has somewhat translucent wings which have hair fringes. This stage then moults to give the true imago, this *ecdysis*, which is unique among Insects, involving the casting of a delicate pellicle from the whole body, including the wings. *adj.* **subimaginal.**
subinvolution (*Med.*). Partial or complete failure of the uterus to return to the normal state after childbirth.
subjective noise meter (*Acous.*). A noise meter for assessing noise levels on the phon scale, the loudness of the noise level being measured by ear with the adjusted reference tone, 1000 Hz. See **objective noise meter.**
sublevel caving (*Mining*). Method of mining massive ore deposit in which ore is drawn down to a delivery road under-running the deposit, and overburden is allowed to cave in, the process being repeated.
sublevel stoping (*Mining*). Method in which ore is blasted in stopes and drawn down to a sublevel in the footwall through ore passes.
sublimate (*Chem.*). The product of sublimation.
sublimation (*Chem.*). The vaporization of a solid (especially when followed by the reverse change) without the intermediate formation of a liquid. (*Psychol.*) An unconscious mechanism whereby the energy attaching to an instinct finds indirect gratification by being diverted into socially and ethically useful channels.
sublimation from the frozen state (*Vac. Tech.*). See **freeze-drying.**
sublimation pump (*Vac. Tech.*). Type of vacuum pump in which gas is effectively adsorbed on to a pad—usually of titanium. It cannot be used to attain a high vacuum unaided but greatly increases the pumping speed obtained when used with, e.g., an ion-getter pump.
sublimed white lead (*Chem.*). Basic lead(II) sulphate fume.
subliminal (*Physiol.*). Said of stimuli which are below threshold, i.e., not strong enough to produce a perceptible response.
sublingua (*Zool.*). In Marsupials and Lemurs, a fleshy fold beneath the tongue.
sublittoral plant (*Ecol.*). A plant which grows near the sea, but not on the shore.
sublittoral zone (*Ecol.*). In a lake, the lake bottom below the paralimnion, extending from the lakeward limit of rooted vegetation to the upper limit of the hypolimnion.
sublunar point (*Astron.*). See **substellar point.**
subluxation (*Med.*). Partial, incomplete dislocation of a joint.
submarginal ore (*Mining*). Developed ore which, at current market price of extracted values, cannot be profitably treated.
submarine telegraph cable (*Cables*). Deep-sea cable is of gutta-percha or polyethylene with steel-wire armouring. Shallow sea cables have been made with dry cellulose or polyethylene, lead-sheathed, steel-wire armoured.
submaxillary (*Zool.*). Situated beneath the lower jaw.
submentum (*Zool.*). The proximal sclerite of the basal part of the labium in Insects.
submerged arc welding (*Elec. Eng.*). Automatic *arc welding* (q.v.) process, using a single, bare electrode which passes through a blanket of granular, fusible flux laid along the seam to be welded, so that the entire welding action takes place beneath this blanket which eliminates spatter losses and protects the joint from oxidation.
submerged combustion (*Heat*). A method of heating liquids by submerged burner equipment of special design, which maintains the flame in direct contact with the liquid.
submerged heating (*Elec. Eng.*). Induction heating of workpiece submerged in quenching liquid.
submerged repeater (*Cables*). One used at intervals along a submarine cable, to compensate losses. Built into the cable, or in a steel enclosure to withstand pressure, it has components of long life, e.g., 20 yr.
submerged resonance (*TV*). See **tailing hangover.**
submicron (*Phys.*). A particle of diameter less than a *micron*. Visible by ultra-microscope.
subminiature camera (*Photog.*). One which takes very small negatives, i.e., smaller than the normal 35-mm miniature format.
subminiature valve. See **miniature valve.**
submodulator (*Radio*). Low-frequency amplifier which immediately precedes the modulator in a radio-telephony transmitter.
subnormal (*Maths.*). Of a curve: the projection on to the *y*-axis of that part of a normal to the curve lying between its point of normalcy with the curve and its intersection with the *y*-axis.

subnotochord (*Zool.*). In some *Chordata*, a skeletal rod lying beneath the true notochord; hypochord.
subocular (*Zool.*). In the skull of *Cyclostomata*, an inverted cartilaginous arch lying below, and affording support to, the eyeball; below the eyeball.
suboperculum (*Zool.*). In Fish, a membrane bone of the gill cover.
suborbital (*Zool.*). One of a series of membrane bones surrounding, or lying below, the eye in some Fish.
suboutcrop (*Mining*). See **blind apex.**
subpetiolate (*Bot.*). Said of a bud which grows concealed by the petiole.
subpress (*Eng.*). Unit comprising a carrier plate for the punch and another for the die of a press tool, connected by pillars sliding in bushes so that the alignment of punch and die is maintained accurately during working of the press tool. Also called **die set.**
subproportional (*Photog.*). Said of reducers or intensifiers which produce their maximum effect on the weaker parts of the image.
sub-radius (*Zool.*). In radially symmetrical forms, a radius of the fourth order, lying between an ad-radius and a per-radius or an inter-radius.
subramose (*Bot.*). (1) Not branching freely. (2) Having few branches.
subrefraction (*Meteor.*). Refraction less than standard refraction.
subroutine (*Comp.*). Short routine, ready-made for a given computer, which is used as a unit to shorten the total program peculiar to a desired calculation.
subscriber's extension station (*Teleph.*). The subset to which an incoming call can be extended from a main subset on a subscriber's premises.
subscriber's line (*Teleph.*). The line connecting the subscriber's main telephone instrument to the exchange, as contrasted with an extension line from this instrument.
subscriber's station (*Teleph.*). See **substation.**
subscriber trunk dialling (*Teleph.*). System whereby subscribers can make trunk calls by dialling code numbers for the exchange required, followed by the normal exchange numbers. Abbrev. **STD. International subscriber dialling (ISD)** operates between many towns in W. Europe and extends to many countries worldwide; a code is dialled for the country required, followed by the STD number.
subscription television (*TV*). See **toll television.**
subsequent pick-up (*Teleph.*). The pick-up occurring at the end of a pulse train due to the change in line-relay holding conditions.
subset (*Maths.*). Intuitively, a part of a set; e.g., the positive even integers constitute a subset of the set of positive integers. However, in maths, it is also accepted that (i) each set is a subset of itself; (ii) the empty set is a subset of every set. These two extreme cases and the above intuitive notion are incorporated in this mathematical definition of subset; A set X is a subset of a set Y if and only if each element of X is also an element of Y.
subsidence (*Build., Civ. Eng.*). (1) The sinking or caving-in of the ground. (2) The settling down of a structure, etc., to a lower level.
subsidiary cell (*Bot.*). See **accessory cell.**
subsoil (*Geol.*). Residual deposits lying between the soil above and the bed-rock below, the three grading into one another.
subsoil drain (*Civ. Eng.*). A drain laid just below ground-level to carry off waters from saturated ground. It consists usually of unsocketed earthenware pipes laid end to end at the bottom of a trench, which is covered in with broken stones.
subsolar point (*Astron.*). See **substellar point.**
subsonic (*Acous.*). Deprecated term for **infrasonic.**
subsonic speed. Any speed below the speed of sound in a fluid. (*Aero.*) Any speed of an aircraft where the airflow round it is everywhere below Mach 1. See **Mach number.**
subspecies (*Zool.*). A category of individuals within a species distinguished by certain common characteristics from typical members of the species; a variety.
subspontaneous (*Bot.*). Said of a plant which has been introduced but maintains itself successfully by its ordinary means of reproduction.
subsporangial vesicle (*Bot.*). A swelling on a sporangiophore immediately beneath the terminal sporangium.
substance (*Chem.*). A kind of matter, with characteristic properties, and generally with a definite composition independent of its origin. (*Glass*) The thickness of flat glass; in the case of sheet glass, usually expressed as mass per unit area. (*Paper*) See **basis weight.**
substandard film stock (*Cinema.*). A type of film stock which is narrower than the standard. Substandard sizes in use are 17·5, 16, 9·5, and 8 mm, as contrasted with the standard 35 mm.
substandard instrument (*Elec. Eng.*). A laboratory instrument whose accuracy is very great and which has been calibrated against an international standard of measurement.
substantia (*Zool.*). Substance; matter.
substantia gelatinosa of Rolando (*Histol.*). A network of neuroglial and nerve cells found in the posterior horn of grey matter towards the surface of the spinal cord.
substantive dyes (*Chem.*). Dyestuffs which can dye cotton and other fibres direct without the aid of a mordant. Many are derived from benzidine and its derivatives.
substantive variation (*Bot., Zool.*). Variation in the constitution of an organ or organism, as opposed to variation in the number of parts.
substation (*Elec. Eng.*). A switching, transforming or converting station intermediate between the generating station and the low-tension distribution network. (*Surv.*) Apex of a subsidiary triangle in a skeleton. (*Teleph.*) A subscriber's telephone located on his premises. Abbrev. for *subscriber's station.*
substellar point (*Astron.*). The point on the earth's surface, regarded as spherical, where it is cut by a line from the centre of the earth to a given star; hence the point where the star would be vertically overhead, the point whose latitude is equal to the star's declination. Applied also to the sun and moon as *subsolar point* and *sublunar point* respectively.
subtertian malaria (*Med.*). See **tertian.**
substitution (*Chem.*). The replacement of hydrogen by other groups, e.g., halogen, alkyl, hydroxyl, etc.
substitutional resistance (*Elec. Eng.*). A resistance, equal to the normal resistance of an arc lamp, which is automatically cut into circuit upon the failure of any one of several arc lamps connected in series.
substrate (*Biochem.*). A chemical substance, or substances, being changed in an enzyme-controlled reaction, e.g., respiratory substrate. (*Bot.*) The substances used by a plant in respiration. (*Electronics*) Single crystal of semiconductor used as a basis for integrated circuit or transistor. (*Glass*) Material of panel on which subminiaturized components are mounted or deposited, e.g., glass or ceramic.

substratum (*Bot.*). Nonliving material to which a plant is attached and from which it obtains substances used in its nutrition. (*Zool.*) The solid underlying material or basis on which an animal moves, or to which it is attached; as the sea-bottom.

subsultus tendinum (*Med.*). Involuntary twitching of muscles or groups of muscles in patients whose vitality is lowered by prolonged fever, as in typhoid fever.

subsynchronous (*Elec. Eng.*). Below *synchronism*.

subtangent (*Maths.*). Of a curve, the projection on to the *x*-axis of that part of a tangent lying between its point of tangency with the curve and its intersection with the *x*-axis.

subtectal (*Zool.*). Lying beneath the roof, as of the skull; in some Fish, a cranial bone.

subtend (*Bot.*). To have a bud, or something developed from a bud, or a sporangium in its axil. (*Maths.*) (1) The portion of a straight line or arc intercepted by and between the arms of an angle is said to subtend the angle. (2) The chord joining the points at the end of an arc is said to subtend the arc.

subtense bar (*Surv.*). A horizontal bar, bearing two targets fixed at a known distance apart, used as the distant base in one system of tacheometry.

subterranean (*Bot.*). Growing beneath the surface of the soil.

subthalamus (*Zool.*). A mass of grey matter lying below the thalamus and representing the prolongation of the tegmentum of the crus cerebri in the Mammalian brain; hypothalamus. *adj.* **subthalamic.**

subtraction (*Bot.*). The loss of an hereditary factor. (*Maths.*) The inverse operation to addition. Denoted by the minus sign $-$. Thus in the statement $m-s=d$, the terms m, s, and d are referred to as the *minuend*, *subtrahend* and *difference* respectively.

subtractive-coloured light (*Light*). The monochromatic illumination obtained from a polychromatic light source by the aid of an appropriate absorption screen.

subtractive process (*Photog.*). The printing of images, corresponding to the three primary colours, in their subtractive or complementary colours, so that transmitted light from a white source, or reflected light from a white support, loses in turn the minus colour of the original primary.

subtractor (*Photog.*). A filter which stops the transmission of a specified primary colour in 3-colour subtractive printing.

subtrahend (*Maths.*). See **subtraction.**

subtransient reactance (*Elec. Eng.*). The reactance of the armature winding of a synchronous machine corresponding to the leakage flux which occurs in the initial stage of a short-circuit. This flux is smaller than that corresponding to the transient reactance on account of eddy currents which may be set up in the rotor during the first one or two half-cycles of a short-circuit.

subula (*Bot.*). A delicate sharp-pointed prolongation of an organ. *adj.* **subulate.**

subumbrella (*Zool.*). The concave inner or lower surface of the umbrella of a medusa.

subunguis (*Zool.*). In Vertebrates, the ventral scale contributing to a nail or claw.

succession (*Cyt.*). The tendency sometimes shown by the sex chromosomes to pass to the poles of the meiotic spindle after the autosomes. (*Ecol.*) The sequence of communities (i.e., a sere) which replace one another in a given area, until a relatively stable community (i.e., the climax) is reached, which is in equilibrium with local conditions.

succinamic acid (*Chem.*). $H_2N{\cdot}CO{\cdot}CH_2{\cdot}CH_2{\cdot}COOH$, monoamidobutandioic (succinic) acid.

succinic acid (*Chem.*). Butandioic acid. $HOOC{\cdot}CH_2{\cdot}CH_2{\cdot}COOH$, dibasic acid; monoclinic prisms; m.p. 185°C, b.p. 235°C, with partial decomposition into its anhydride. It occurs in the juice of sugar cane, in the castor-oil plant, and in various animal tissues, where it plays an important part in metabolism (see **citric acid cycle**). Succinic anhydride is used in the manufacture of alkyd resins and in organic synthesis.

succinite (*Min.*). (1) A variety of *amber* (q.v.), separated mineralogically because it yields succinic acid. (2) A name given to an amber-coloured garnet of the grossular species.

succinyl (*Chem.*). The bivalent acid residue $-CO{\cdot}CH_2{\cdot}CH_2{\cdot}CO-$.

succinylcholine chloride (*Med.*). A chemical agent used in electric shock therapy to induce complete muscular relaxation in order to avoid bone fractures.

succinylsulphathiazole (*Pharm.*). See **sulphonamides.**

succise (*Bot.*). Ending below abruptly, as if cut off.

succubous (*Bot.*). Having the lower edge of the leaf in front of the stem and overlapping the upper edge of the next leaf below it on the same side of the stem.

succulent (*Bot.*). Juicy, soft, and thick. (*Succulent* is often used when *fleshy*—which is not specially juicy—should be employed.) A xerophytic plant with water-storing tissue.

succus (*Zool.*). A juice secreted by a gland.

succus entericus (*Zool.*). A collective name for the enzymes secreted by glandular cells (*Brunner's glands* and *Lieberkühn's crypts*) in the walls of the duodenum, these including erepsin (itself a mixture), invertase, maltase, lactase, nucleotidase, nuclease, lipase, and enterokinase.

succussion (*Med.*). The act of shaking a patient to detect the presence of fluid in a pleural cavity already containing air (*pneumothorax*).

sucker (*Bot.*). A strongly growing shoot arising from the base of the stem or from a root. (*Elec. Eng.*) A time-delay device of the *dash-pot* type, employing a disk immersed in glycerine. (*Zool.*) A suctorial organ adapted for adhesion or imbibition, as one of the muscular sucking disks on the tentacles of *Cephalopoda*: the suctorial mouth of animals like the Leech and the Lamprey: a newly born Whale: one of a large number of Fishes having a suctorial mouth or other suctorial structure, as the Remora (*Echeneis*), members of the genus *Lepadogaster*, etc.

sucking booster (*Elec. Eng.*). A booster whose function is to overcome voltage drop occurring in a feeder.

sucking stomach (*Zool.*). In many Arthropods, a muscular dilatation of the alimentary canal which can produce imbibition of fluid by the mouth.

sucrase (*Chem.*). See **invertase.**

sucrol (*Chem.*). See **dulcin.**

sucrose (*Chem.*). See **cane-sugar.**

suction box (*Paper*). A compartment removing water by suction from the moist paper web carried by the machine wire. Also called **vacuum box.**

suction-cutter dredger (*Civ. Eng.*). A dredger in which rotary blades dislodge the material to be excavated, which is then removed by suction as in a *sand-pump dredger* (q.v.).

suction dredger (*Civ. Eng.*). See **sand-pump dredger.**
suction pressure (*Bot.*). The avidity with which the cell takes in water; it is equivalent to the difference between the osmotic pressure of the cell sap, which tends to bring water into the cell, and the pressure exerted by the elastic cell wall, which tends to force water out of the cell.
suction roll (*Paper*). A perforated press roll removing water by suction as well as by pressure. Cf. *press rolls.*
suction valve (*Eng.*). See **foot valve** (1).
Suctoria (*Zool.*). A subclass of *Ciliophora* in which only the young forms possess cilia, the adults being sedentary organisms devoid of cilia and capturing their food by means of suctorial tentacles. They usually have a cuticular stalk, the end of which may be expanded, forming a cup.
suctorial (*Zool.*). Drawing in: imbibing: tending to adhere by producing a vacuum: pertaining to a sucker.
suctorial mouth-parts (*Zool.*). Tubular mouth-parts adapted for the imbibition of fluid nourishment; found in some Insects and many ectoparasites.
sudamina (*Med.*). Whitish vesicles on the skin, due to retention of sweat in the sweat glands. *sing.* **sudamen.** *adj.* **sudaminal.**
Sudan stains (*Micros.*). A group of aniline dyes, e.g., Sudan black, used for staining lipoids.
sudation (*Bot.*). The exudation of a dilute watery solution of various substances from a plant.
Sudbury Series or **Sudburian** (*Geol.*). See **Timiskaming Group.**
sudd (*Bot.*). Dense mats of aquatic vegetation formed usually in tropical and subtropical lakes and swamps and causing damage to fisheries, navigation, etc.
sudetic phase (*Geol.*). A phase of the Variscan (Hercynian) orogeny marked by folding in several parts of Britain during Carboniferous times. Named after the Sudeten mountains.
sudor (*Med.*). Sweat or perspiration.
sudoriferous, sudoriparous (*Zool.*). Sweat-producing: sweat-carrying.
sudorific (*Med.*). Connected with the secretion of sweat: stimulating the secretion of sweat: a drug which does this.
Suffolk coulter (*Agric.*). A coulter shaped something like the prow of a boat; used on a drill to make a shallow trench for the seed.
Suffolk latch (*Join.*). A variant of the *Norfolk latch* (q.v.).
Suffolk whites (*Build.*). See **gaults.**
suffrutescent, suffruticose (*Bot.*). Said of a plant in which many of the branches die after flowering, leaving a persistent woody base.
suffuse (*Bot.*). Spread out on the substratum.
sugar (*Chem.*). (1) A water-soluble, crystalline mono- or polysaccharide. (2) The common term for *sucrose*, or *cane-sugar*, $C_{12}H_{22}O_{11}$.
sugar-beet harvester (*Agric.*). A machine to top, lift, clean and either load, windrow or carry sugar-beet roots.
sugar charcoal (*Chem.*). Highly pure form of charcoal derived from sucrose.
sugar-dye (*Chem.*). See **caramel.**
sugar plant (*Bot.*). A plant which forms little or no temporary starch, the carbohydrate formed in photosynthesis remaining as sugar.
sugar soap (*Paint.*). An alkaline cleansing or stripping preparation for paint surfaces.
sugent, sugescent (*Zool.*). See **suctorial.**
suggestion (*Psychol.*). The adoption of ideas or attitudes as a result of the activity of another person, through processes other than rational thought. Sometimes used in psychotherapy either alone or with hypnosis.
Suhl effect (*Electronics*). The reduction in lifetime of holes injected into an *n*-type semiconducting filament, by deflecting them to the surface using a powerful transverse magnetic field. Reverse of *Hall effect* (q.v.).
suite (*Teleph.*). A row of racks or switch-boards.
sulculus (*Zool.*). The 'dorsal' siphonoglyph of *Anthozoa.*
sulcus (*Zool.*). A groove or furrow, as one of the grooves on the surface of the cerebrum in Mammals: in *Dinoflagellata*, a longitudinal groove in which a flagellum lies: in *Anthozoa*, the 'ventral' siphonoglyph.
sulfate, sulfur, etc. See **sulphate, sulphur, etc.**
sulfotep (*Chem.*). Bis-*OO*-diethylphosphorothionic anhydride, used as an insecticide. Also **dithio, thiotep, dithioTEPP.**
sullage (*Civ. Eng.*). The mud and silt deposited by flowing waters.
Sully Beds (*Geol.*). A series of fossiliferous grey and greenish marls which are found exposed in Somerset and Glamorganshire. They belong to the lower division of the Rhaetic Stage in the Jurassic System.
sulpha drugs (*Pharm.*). See **sulphonamides.**
sulphaemoglobinaemia, sulphemoglobinemia (*Med.*). The presence in the blood of sulphaemoglobin, due to the combination of haemoglobin with hydrogen sulphide absorbed from the intestine. See also **enterogenous cyanosis.**
sulphamic acid (*Chem.*). The monoamide of sulphuric acid: $HO—SO_2—NH_2$. Prepared by the ammonolysis of sulphuric acid; commercially, by the reaction of urea (carbamide) with fuming sulphuric acid.
sulphamide (*Chem.*). The diamide of sulphuric acid: $SO_2(NH_2)_2$. Prepared by the ammonolysis of sulphuryl chloride.
sulphanilamide (*Pharm.*). See **sulphonamides.**
sulphanilic acid (*Chem.*). $H_2N{\cdot}C_6H_4{\cdot}SO_3H$, 4-aminobenzene-sulphonic acid; rhombic plates; crystallizes with $2H_2O$; sparingly soluble in water.
sulphate of ammonia (*Chem.*). $(NH_4)_2SO_4$. Commercially the most important of the ammonium salts, particularly for use as fertilizer. Produced partly as a by-product of gas works, coke ovens, etc., but now largely by direct synthesis.
sulphate of iron (*Min.*). See **melanterite.**
sulphate of lead (*Min.*). See **anglesite.**
sulphate of lime (*Min.*). See **anhydrite, gypsum.**
sulphate of strontium (*Min.*). See **celestine.**
sulphate-resisting cement (*Civ. Eng.*). Cement manufactured for use in concrete to resist normal concentrations of sulphates as in most flues; also for underwater work.
sulphates (*Chem.*). Salts of sulphuric acid. Produced when the acid acts on certain metals, metallic oxides, hydroxides, and carbonates. The acid is dibasic, forming two salts—normal and acid sulphates.
sulphating roasting (*Met.*). Roasting carried out under conditions designed to convert part of the contained sulphur (sulphide mineral) to sulphate.
sulphation (*Elec. Eng.*). The formation of the insoluble white sulphate of lead ($PbSO_4$) in the plates of a lead-acid type of secondary cell, a process which diminishes the efficiency and capacity of the cell.
sulphide dyestuffs (*Chem.*). Dyestuffs of unknown constitution, containing sulphur, e.g., vidal black, obtained by fusing 4-aminophenol with

sulphur, or fast black B by fusing 1,8-dinitronaphthalene with sulphur.

sulphide fog (*Photog.*). Fog caused by stale or contaminated developer in which the sodium sulphite has been reduced to sulphide by fungi or other organisms.

sulphides (*Chem.*). Salts of hydrogen sulphide. Many sulphides are formed by direct combination of sulphur with the metal.

sulphide toning (*Photog.*). A process of toning photographic prints in which the silver is converted to silver sulphide via silver bromide.

sulphide zone (*Mining*). Primary (unaltered) zone of sulphide-mineral lode, underlying leached (superficial) zone and that of secondary enrichment in which there has been redeposition of values oxidized from leached zone by penetrating water.

sulphinic acids (*Chem.*). Acids containing the monovalent sulphinic acid group —SO·OH.

sulphite process (*Paper*). See **acid process**.

sulphites (*Chem.*). Sulphates(IV). Salts of sulphurous (sulphuric(IV)) acid. The acid forms 2 series of salts, acid sulphites or bisulphites and normal sulphites.

sulphite wood pulp (*Paper*). Chipped and crushed wood which, by boiling with bisulphite liquor at high pressure, has been reduced to its constituent fibres.

sulphocyanides (*Chem.*). See **thiocyanates**.

sulphonamides (*Pharm.*). A group of drugs with a powerful antibacterial action, used in the treatment of various infections. They are more effective in the body than in the test tube, and are believed to act by interfering with the metabolic processes of bacteria, which are thereby rendered more susceptible to the natural defences, the leucocytes and immune bodies. The earlier drugs were most effective against haemolytic streptococci, but later members of the group are potent against other organisms. In chemotherapeutic practice the sulphonamides have been to a considerable extent displaced by antibiotics, but they continue to occupy an important place, partly because of cheapness and ease in administration, partly because of their use in certain cases of antibiotic resistance, or in conjunction with antibiotics. As a class they are subject to a wide range of toxic reactions, which, however, are usually (though not entirely) minor and responsive to adjustment. Many compounds of this group have been tested and used. A comprehensive list cannot be accommodated here, but the following are representative members:

Prontosil, the first to be used, a reddish powder, the hydrochloride of 2′·4′-diaminoazobenzene-4-sulphonamide:

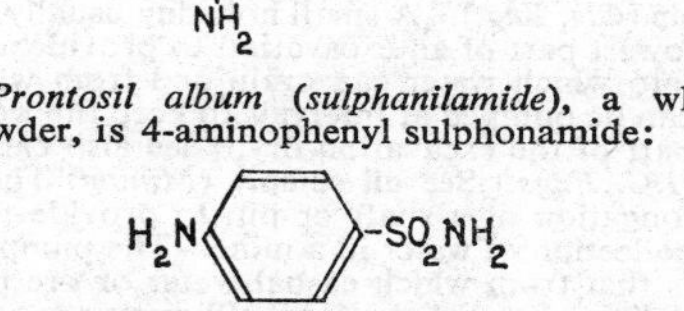

Prontosil album (*sulphanilamide*), a white powder, is 4-aminophenyl sulphonamide:

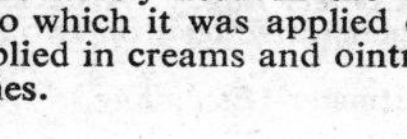

It was once widely used in the treatment of wounds, to which it was applied direct, but it is now applied in creams and ointment, also as dental cones.

Sulphapyridine (*M. and B. 693*), 2-(4-aminobenzene sulphonamide) pyridine,

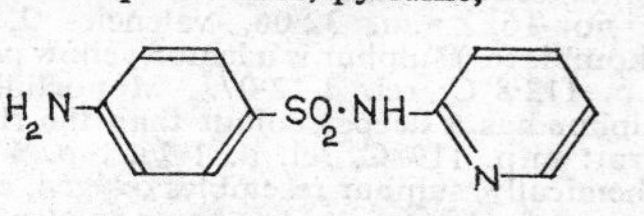

is a white crystalline powder, sparingly soluble in water. It is usefully effective against certain forms of dermatitis.

Sulphathiazole (*Thiazamide*) is the thiazole analogue of sulphapyridine (which it resembles in physical properties):

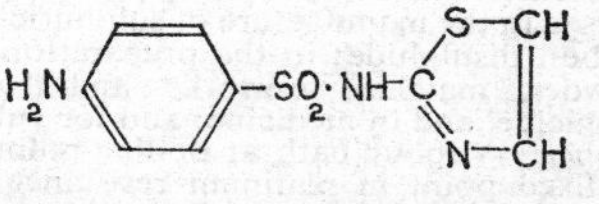

It is more active than sulphapyridine against staphylococci, but less so against pneumococci.

Sulphaguanidine is the guanidine analogue of sulphapyridine. It is almost insoluble in water and is therefore useful in infections of the alimentary tract, e.g., dysentery.

Sulphacetamide (*Albucid*), $H_2N{\cdot}C_6H_4{\cdot}SO_2{\cdot}NH{\cdot}COCH_3$, is particularly suitable for local application to delicate tissues and therefore used for eye-infections, burns, etc.

Sulphadiazine

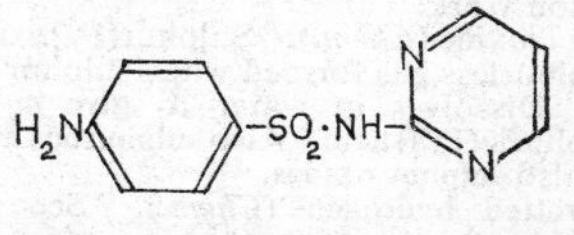

and its homologues *sulphamerazine* and *sulphadimethylpyrimidine* (*sulphadimidine*) are particularly suitable for maintaining a persistent high blood level, being easily absorbed and slowly excreted.

Succinylsulphathiazole and *phthalylsulphathiazole* are similar to sulphaguanidine in physical properties and therapeutic uses, both being used for infections of the gastric intestinal tract, bacillary dysentery, etc.

Other important members of the group include *sulphamethoxydiazine* and *sulphamethoxypyridazine* (urethritis, etc.), *sulphadimethoxine* (bacillary dysentery, respiratory infections), *sulphaphenazole* (streptococcal and urinary tract infections), *sulphafurazole* (meningococcal and urinary tract infections), *sulphamethizole* (urinary tract infections), *sulphasalazine* (ulcerative colitis). For certain purposes preparations of mixed sulphonamides are preferable.

sulphonation (*Chem.*). The reversible process of forming sulphonic acids by the action of concentrated sulphuric acid on aliphatic or aromatic compounds.

sulphones (*Chem.*). Compounds of the formula $RR'{=}SO_2$. The sulphur is hexavalent. Used in the treatment of leprosy (see **dapsone**).

sulphonic acids (*Chem.*). Acids containing the monovalent sulphonic acid group $—SO_2{\cdot}OH$.

sulphosol (*Chem.*). A colloidal solution in concentrated sulphuric acid.

sulphoxides (*Chem.*). Compounds of the formula RR′SO. The sulphur is in the (IV) oxidation state.

sulphoxylic acid (*Chem.*). The hypothetical oxv acid of sulphur, $S(OH)_2$.

sulphur (*Chem.*). A nonmetallic element occurring in several allotropic forms. Symbol S, at. no. 16, r.a.m. 32·06, valencies 2, 4, 6. Rhombic (α-) sulphur is a lemon yellow powder; m.p. 112·8°C, rel. d. 2·07. Monoclinic (β-) sulphur has a deeper colour than the rhombic form; m.p. 119°C, rel. d. 1·96, b.p. 444·6°C. Chemically, sulphur resembles oxygen, and can replace the latter in many compounds, organic and inorganic. It is abundantly and widely distributed in nature, in the free state and combined as sulphides and sulphates, occurring around volcanoes and hot springs and in sulphate deposits as a result of the reduction of sulphates. It is manufactured by purifying the native material or by heating pyrites. Sulphur is used in the manufacture of sulphuric acid and carbon disulphide; in the preparation of gunpowder, matches, fireworks, and dyes; as a fungicide, and in medicine; and for vulcanizing rubber. Vapour bath at boiling-point is used as fixed point in platinum resistance thermometry.

sulphur bacteria (*Bacteriol.*). Bacteria which live in situations where oxygen is scarce or absent, and which act upon compounds containing sulphur, liberating the element. Occur in two families of true bacteria, the *Thiorhodaceae* (purple sulphur bacteria) and the *Thiobacteriaceae* (colourless sulphur bacteria). See also **Beggiatoales.**

sulphur cement (*Build.*). A cement made of sulphur and pitch mixed in equal parts; used to fix iron work.

sulphur dioxide (*Chem.*). Sulphur(IV)oxide; SO_2. A colourless gas formed when sulphur burns in air. Dissolves in water to give sulphurous (sulphuric(IV)) acid. Also **sulphurous anyhdride.** See also **sulphur oxides.**

sulphuretted hydrogen (*Chem.*). See **hydrogen sulphide.**

sulphur hydrides (*Chem.*). Four well-defined hydrides—H_2S, H_2S_2, H_2S_3, and H_2S_5.

sulphuric acid (*Chem.*). A strong dibasic acid, H_2SO_4. The concentrated acid is a colourless oily liquid; rel. d. 1·85, b.p. 338°C; it dissolves in water with the evolution of heat, and is very corrosive, largely owing to its dehydrating action. It is manufactured from sulphur dioxide, obtained by burning either pyrites or sulphur, by the *contact process* (q.v.) or the *chamber process* (q.v.). It is an important heavy chemical, used extensively in the dyestuffs and explosives industries; as a drying agent in chemical process; in the manufacture of other acids, e.g., HCl, HF, phosphoric acid; in fertilizers, pickling liquors and leaching solutions; in petroleum treatment and the manufacture of alkyl sulphonate detergents; in rayon production, etc. The salts of sulphuric acid are called *sulphates* (VI).

sulphuric anhydride (*Chem.*). Sulphur(VI)oxide; sulphur trioxide; SO_3. Dissolves in water to give sulphuric acid.

sulphurous acid (*Chem.*). An aqueous solution of sulphur dioxide, which contains the hypothetical compound H_2SO_3. The corresponding salts, sulphites, are well known.

sulphurous anhydride (*Chem.*). See **sulphur dioxide.**

sulphur oxides (*Chem.*). A series of oxides—SO, S_2O_3, SO_2, SO_3, S_2O_7, and SO_4.

sulphur print (*Met.*). See **Baumann print.**

sulphur trioxide (*Chem.*). See **sulphuric anhydride.**

sumatra (*Meteor.*). A southwesterly summer squall, with thunder and lightning, blowing in the Straits of Malacca.

summation (*Physiol.*). The production of an effect by repetition of a causal factor which would be insufficient in a single application, as *summation of contractions*, the production of a state of tetanic contraction by a series of stimuli.

summation check (*Comp.*). Figure added to a *word*, indicating a summation of the digits so that accuracy of processing can be verified. Totals arrived at by two methods for verifying processing of data.

summation instrument (*Elec. Eng.*). An instrument for indicating, integrating or recording the sum total of the energy, power or current in two or more circuits.

summation metering (*Elec. Eng.*). A system of metering electrical energy in which the consumption of several distinct load circuits is summated and indicated by one instrument.

summation of losses (*Elec. Eng.*). The process of adding together the individual losses, after allowing for any corrections, in order to obtain the guaranteed efficiency of an electrical machine.

summation panel (*Elec. Eng.*). A switchboard panel on which are mounted the instruments for measuring and recording the total output of a number of generators.

summation tone (*Acous.*). One of the combination tones produced when two or more pure tones are subjectively perceived by the ear, or partially rectified through amplitude distortion during reproduction.

summer or summer tree (*Carp.*). A large beam or lintel for the support of dead load only.

summer annual (*Bot.*). A plant which lives for a short period in summer, setting seed at the end of its growth and then dying.

summer draught (*Ships*). The *draught* (q.v.) when loaded to the *summer-load waterline* (q.v.).

summer egg (*Zool.*). In many fresh-water animals, a thin-shelled, rapidly developing egg laid during the warm season. Cf. *winter egg.*

summer-load waterline (*Ships*). The waterline to which a ship may be loaded in summer. It is indicated in the freeboard markings.

summer mastitis (*Vet.*). Mastitis of nonlactating cows due to infection by *Corynebacterium pyogenes*, or occasionally other organisms; possibly transmitted by flies.

summer solstice (*Astron.*). See **solstices.**

summer wood (*Bot.*). Secondary wood formed in summer as secondary thickening comes to an end for the season; the elements of summer wood are often thick walled, and smaller than those of *spring wood* (q.v.). Sometimes, but less accurately, called **autumn wood.**

summing point (*Telecomm.*). See **mixing point.**

summit canal (*Hyd. Eng.*). A canal crossing a summit; one, therefore, to which water must be supplied.

sum of infinite series (*Maths.*). More precisely, sum to infinity. The limit, as n tends to infinity, of the sum of the first n terms.

sump (*Civ. Eng.*). A small hole dug usually at the lowest part of an excavation to provide a place into which water can drain and from which it can be pumped at intervals to keep the working part of the excavation dry. See also **catch pit.** (*I.C. Engs.*) See **oil sump.** (*Mining*) The prolongation of a shaft or pit, to provide for the collection of water in a mine. The pump sump is that from which casual water or ore pulp is delivered to the mine or mill pumps.

Sumpner test (*Elec. Eng.*). A back-to-back load test, or regenerative test, on two similar transformers.

Sumpner wattmeter (*Elec. Eng.*). An iron-cored

type of dynamometer wattmeter for use on a.c circuits.

Sun (*Astron.*). The central body of the solar system, containing all but one-thousandth of its mass. It is an incandescent gaseous sphere, in which the commonest elements are hydrogen and helium; its diameter is 865 000 miles (1 392 000 km), and its mass 333 000 times that of the Earth (2×10^{30} kg). The mean distance of the Sun from the Earth is 93 000 000 miles (149 600 000 km). It is a star of type G and absolute magnitude 5·0. See also astronomical articles at **solar.**

sun arc (*Cinema.*). An arc lamp of sufficient capacity to imitate the sun in studio or outdoor locations in motion-picture production.

sunburner (*Glass*). An excessive local thickness of material in a mouth-blown glass article.

sun cracks (*Geol.*). Polygonal cracks, usually in a fine-grained sedimentary rock, indicative of desiccation at the time of formation. Common in beds laid down under arid conditions.

Sundance Series (*Geol.*). Marine Jurassic clays and sands exposed in the Front Range district of Colorado. Similar, though thicker, strata with some limestones occur also in the Big Horn Mts, where the series is 300 ft (90 m) thick.

sun gear (*Eng.*). See **epicyclic gear.**

sunk coak (*Carp.*). The mortiselike recess in one of the mating surfaces of a scarfed joint, into which the *coak* (q.v.) fits.

sunk face (*Build.*). A term applied to a stone in whose face a panel is sunk by cutting into the solid material. Abbrev. **S.F.**

sunk fence (*Build.*). See **ha-ha.**

sunk key (*Eng.*). A key which is sunk into key ways in both shaft and hub. See **key.**

sun observation (*Surv.*). Use of theodolite to fix latitude and/or longitude of a station, or to orient a survey line by direct sighting and calculation of sun's position.

sun pillar (*Meteor.*). A vertical column of light passing through the sun, seen at sunset or sunrise. It is caused by reflection of sunlight by horizontal ice crystals.

sun-ray therapy (*Med.*). See **ultraviolet therapy.**

sunseeker (*Space*). A photoelectric device mounted in rockets or space vehicles, in which an instrument such as a spectrograph can be directed constantly to the sun.

sunshine recorder (*Meteor.*). The *Campbell-Stokes recorder* consists of a glass sphere arranged to focus the sun's image on to a bent strip of card, on which the hours are marked. The focused heat burns through the card, and the duration of sunshine is read off from the length of the burnt track.

sunshine unit (*Nuc.*). U.S. term for **strontium unit.**

sunspot (*Astron.*). A disturbance of the solar surface which appears as a relatively dark centre (*umbra*), surrounded by a less dark area (*penumbra*); spots occur generally in groups, are relatively shortlived, and with few exceptions are found in regions between 30° N. and S. latitude; their frequency shows a marked period of about 11 yr (*sunspot cycle*), they have intense magnetic fields and are sometimes associated with magnetic storms on the earth. See also **polarity of sunspots.** (*Cinema.*) A large incandescent lamp, provided with a glass filter, to imitate the light from the sun in colour cinematography.

sunstone (*Min.*). See **aventurine feldspar.**

sunstroke (*Med.*). Heat hyperpyrexia. A condition produced by exposure to high atmospheric temperature and characterized by a rapid rise of bodily temperature, convulsions, and coma. See also **insolation.**

sun wheel (*Eng.*). A gear-wheel round which one or more planet wheels or planetary pinions rotate in mesh.

sup (*Electronics*). See **suppressor grid.**

super-, supra. Prefix from L. *super*, over, above.

superacidity (*Chem.*). The enhanced acidity which results from the dissolution of certain usually weak acids in certain non-aqueous solvents.

super-additive function (*Maths.*). See **additive function.**

superalloy (*Met.*). High-temperature alloy, based on iron, chromium, cobalt, manganese and/or nickel, having good stability in the 600°C–1000°C range.

superaudio frequency (*Acous.*). One above those usefully transmitted through an audio-frequency reproducing system, or part of such.

superaudio telegraphy (*Teleg.*). That which transmits by a.c. of frequency above 3400 Hz.

superaxillary (*Bot.*). Developing from above the axil.

supercalendered paper (*Paper*). Paper which has been given an extra smooth surface by means of a *calender* (q.v.).

supercharger (*I.C. Engs.*). A compressor, commonly of the rotary vane or the centrifugal type, used to supply air or combustible mixture to an I.C. engine at a pressure greater than atmospheric; driven either directly by the engine or by an exhaust gas turbine.

supercharging (*Aero., Eng.*). (1) In aero engines, maintenance of ground-level pressure in the inlet pipe up to the rated altitude by means of a centrifugal or other blower. Necessary for flying at heights at which the air pressure is low and normal aspiration would be insufficient. (2) In other I.C. engines, the term is used synonymously with boosting (see **boost**).

superciliary (*Zool.*). Pertaining to, or situated near, the eyebrows; above the orbit.

supercirculation (*Aero.*). A form of boundary layer control for high lift in which air is blown supersonically over the leading edge of a plain flap so that it carries the main airflow downward below the actual surface as an invisible extension. See also **jet flap.**

supercompression engine (*Eng.*). An unsupercharged engine of above normal compression ratio, to be run at full throttle only at high altitudes, when the normal power decreases owing to the fall in atmospheric pressure. Below these altitudes the throttle opening is limited by the use of a gate.

superconducting amplifier (*Electronics*). One using superconductivity to give noise-free amplification.

superconducting gyroscope (*Electronics*). Frictionless gyroscope supported in vacuum through magnetic field produced by currents in superconductor.

superconducting magnet (*Electronics*). Very powerful electromagnet, made possible by the exceptionally large current-carrying capacity of superconductors.

superconductivity (*Elec.*). Property of many pure metals within a few degrees of *absolute zero* of having negligible resistance (dissipation) to a flow of electric current. When the current is established it continues nearly indefinitely. Property removed by a magnetic field; hence the phenomenon can be used for memory or store for computers. Such a device is known as a *cryotron* (q.v.). The superconducting characteristics of a material are the critical transition temperature and the critical magnetic field curve. Metals in this state are called **superconductors.**

supercontrol tube (*Electronics*). See **remote cut-off tube.**

supercooled (*Chem.*). Cooled below normal boiling- or freezing-point without change of phase.

supercooled (or subcooled) water (*Meteor.*). Water which continues to exist as a liquid at temperatures below 0°C.

supercritical (*Nuc.*). Assembly of fissile material for which the multiplication factor is greater than unity.

superego (*Psychol.*). The strong unconscious inhibitory mechanism which criticizes the ego and causes it pain and distress when it accepts unworthy impulses from the id. The superego is the main constituent of conscience, and the term was originally used by Freud to mean the development of a moral standard in the child as a result of the resolution of the Oedipus complex. Other authorities claim that the superego is formed at an even earlier stage, by the necessity of giving up the earliest love choices of the id, e.g., the breast. It is influenced in its formation by the parents' moral and ethical standards, many of which are taken over by *introjection* into the child's superego.

superelevation (*Surv.*). The amount by which the outer rail of a railway curve is elevated above the inner rail to counteract the effect of the centrifugal force of the moving train. Superelevation is also applied in the construction of highway curves. See also **cant.**

supererogation (*Zool.*). A supplementary process, as the developmental processes associated with regeneration of a part, as opposed to the primary developmental processes involved during the first formation of the part.

superficial deposits (*Geol.*). See **drift.**

superficial radiation therapy (*Radiol.*). X-ray therapy (usually by soft radiation produced at less than 140 kVp) of the skin or of any surface of the body made accessible.

superfinishing (*Eng.*). An abrasive process, resembling honing and lapping, for removing *smear metal* (q.v.), and scratches and ridges produced by grinding and machining operations, from bearing surfaces.

superfluid (*Chem.*). Condensed degenerate gas in which a significant proportion of the atoms are in their lowest permitted energy state. In practice this affects only *liquid helium II*.

superfoetation, superfetation (*Med.*). Fertilization of an ovum in a woman already pregnant, some time after fertilization of the first ovum.

supergiant stars (*Astron.*). Stars of late type and abnormal luminosity, such as Betelgeuse and Antares; they are of enormous size and low density, and are indicated by their position on the Hertzsprung-Russell diagram well above the main sequence.

supergrid (*Elec. Eng.*). The national electric power network at voltages of 275 kV and 400 kV. See **grid.**

superhardboard (*Build.*). Hardboard of high density specially treated.

superheat (*Aero.*). The increase (positive) or decrease of the temperature of the gas in a gas-bag as compared with the temperature of the surrounding air. Similarly, *superpressure*.

superheated steam (*Eng.*). Steam heated at constant pressure out of contact with the water from which it was formed, i.e., at a higher temperature than that of saturation.

superhet (*Radio*). *Super*sonic *het*erodyne.

superimposed drainage (*Geol.*). A river system unrelated to the geological structure of the area, as it was established on a surface since removed. Cf. *consequent drainage*.

superimposition (*Print.*). Coloured blocking foils are frequently used as a base for over-stamping by gold, silver, or other coloured foil. This 'superimposition' is of particular importance to bookbinders and display-card printers.

superior. Placed above something else; higher, upper (as the *superior* rectus muscle of the eyeball). (*Bot.*) Posterior, as applied to the petal of a corolla. Cf. *inferior*.

superior conjunction (*Astron.*). See **conjunction.**

superior figures (or letters) (*Typog.*). Small figures or letters printed above the general level of the line. They are used instead of *marks of reference*, and in mathematical work, etc.; thus: x^2, e^x, 10^6.

superior vena cava (*Zool.*). See **precaval vein.**

superlingual (*Zool.*). In primitive Insects, a pair of small lobes associated with the hypopharynx.

Supermalloy (*Elec. Eng.*). TN for magnetic material similar to permalloy but with higher permeability.

Supermendur (*Elec. Eng.*). TN for rectangular hysteresis loop magnetic material, containing iron, cobalt, and vanadium.

Supermiser (*Eng.*). A combination of air preheater and economizer for boilers; the flue gases pass through annular spaces between the water tubes and the flue tubes surrounded by the air stream.

supernatant liquid (*Chem.*). The clear liquid above a precipitate which has just settled out.

supernormal sign stimulus (*An. Behav.*). A sign stimulus discovered experimentally, and sometimes constructed artificially, which renders it more effective in eliciting a particular species-specific response.

supernovae (*Astron.*). Novae of absolute magnitude -14 to -16; 3 have been recorded in our own Galaxy, and about 50 more in spiral nebulae. The violent outburst results from the gravitational collapse of a massive star, the outer layers being ejected, while the core is left as a *neutron star*.

superoctave (*Acous.*). A stop on an organ which sounds two octaves higher than the nominal pitch of the key or pedal which actuates it.

superovulation (*Zool.*). Hormone-induced excess *ovulation* (q.v.). See **insemination.**

superoxides (*Chem.*). Compounds of the alkali and alkaline earth metals containing the O_2^- group, e.g., $K^+(O_2^-)$. Differ from peroxides in yielding oxygen as well as hydrogen peroxide on hydrolysis.

superparasitism (*Zool.*). In Insect parasitoids, the situation when there are too many eggs or larvae in one host, so that none of them reaches maturity. Occurs less often than would be expected if the eggs were placed in the hosts at random by the females, thus implying a degree of host selection. Should not be confused with **hyperparasitism.**

superphosphate (*Chem.*). Superphosphate of lime, an agricultural fertilizer; a mixture of calcium sulphate and dihydrogen calcium phosphate; made by treating bone ash or basic slag (calcium phosphate) with sulphuric acid.

superphosphate-triple (*Chem. Eng.*). See **triple superphosphate.**

superposed, superposition (*Telecomm.*). Said of that added to a normal circuit or circuits, e.g., a phantom on telephone circuits or d.c. telegraphy on a telephone circuit.

superposition image (*Zool.*). The type of image

formed in a compound eye by overlapping of the separate images formed by the various facets. Cf. *mosaic image*.

superposition theorem (*Elec. Eng.*). That any voltage/current pattern in a linear network is additive to any other voltage/current pattern.

superpressure (*Aero.*). See under **superheat.**

superproportional (*Photog.*). Said of reducers and intensifiers which produce their maximum effect on the denser parts of the image.

super-refraction (*Meteor.*). Refraction greater than standard refraction.

super-regeneration (*Radio*). Reaction in a receiver to a degree that normally causes self-oscillation. This is prevented by the application of a *quenching* voltage to the reacting valve, which is thereby intermittently paralysed at a frequency which is high enough to be inaudible. Also **super-reaction.** (*Zool.*) The formation of an organ a second time by regeneration, without removal; the formation of an organ in multiple form by regenerative processes.

super-regenerative receiver (*Radio*). One with sufficient positive feedback to result in a quenched supersonic oscillation (*squegging*), with consequent increase in sensitivity but also increase in distortion of demodulated signals.

super-royal (*Paper*). A pre-metric size of printing paper, $20\frac{1}{2} \times 27\frac{1}{2}$ in.

supersaturation (*Chem.*). Solution containing solute in excess of equilibrium. Condensation can take place on nuclei, particularly ions, e.g., those produced by high-speed charged particles, exhibiting a track of minute but visible water drops, as in a *Wilson chamber* (q.v.).

supersonic (*Phys.*). Faster than the speed of sound in that medium. Erroneously used for ultrasonic. See **Mach number, ultrasonic.**

supersonic amplification (*Radio*). That at an ultrasonic frequency, i.e., following the frequency changer in a supersonic heterodyne receiver.

supersonic frequency (*Radio*). The ultrasonic frequency used for the postfrequency-changer amplification in a supersonic heterodyne receiver, viz., from 100 to 450 kHz.

supersonic heterodyne receiver (*Radio*). One in which received signal has the frequency of its carrier wave changed, by means of the heterodyne principle, to some predetermined frequency above the audible limit, after which it is amplified and finally rectified and demodulated. Also **beat receiver, double detection receiver.**

supersonic speed. Any speed above the speed of sound in a medium. (*Aero.*) Any speed of an aircraft where the airflow round it is everywhere above Mach 1.

supersonic wind tunnel (*Aero.*). A wind tunnel in which the stream velocity is greater than the local speed of sound in the tunnel.

super stall (*Aero.*). This phenomenon appeared with the adoption of high tail planes for swept-wing jet aeroplanes. When the disturbed airflow from a stalled wing renders the tail controls inoperative, the aircraft will remain in a stable, substantially level attitude, while descending very rapidly. Recovery is by releasing the tail parachute to raise the tail so that the elevators again become operative.

superstition (*An. Behav.*). A response which, due to a fortuitous association with positive reinforcement (e.g., food), increases in intensity.

superstructure (*Civ. Eng.*). The part of a structure carried upon any main supporting level. (*Ships*) A decked structure on the *freeboard deck* (q.v.) extending from side to side of the ship. An integral part of the hull.

supersulphated cement (*Civ. Eng.*). Cement manufactured using a high proportion of granulated blast-furnace slag.

supervisory control (*Elec. Eng.*). A method of remote control of electrical plant from a distant centre in which back-indication of the several control operations is given to the control centre.

supervoltage therapy (*Radiol.*). Application of voltage, over a million volts, to X-ray tubes or accelerators in therapy.

supination (*Zool.*). In some higher Vertebrates, movement of the hand and forearm by which the palm of the hand is turned upwards and the radius and ulna are brought parallel to one another; cf. *pronation. adj.* **supinate.**

supinator (*Zool.*). A muscle effecting supination.

supplemental, supplementary (*Zool.*). Additional; extra; supernumerary, as (in some *Foraminifera*) *supplemental* skeleton, a deposit of calcium carbonate outside the primary shell.

supplemental air (*Physiol.*). The volume of air which can be expelled from the lungs by a forced expiration after a normal expiration, which is approximately 1500 cm^3 in man; also known as **expiratory reserve volume.**

supplemental chords (*Maths.*). Of a circle, ellipse or hyperbola: the chords joining any point on the curve to the ends of any diameter. Diameters parallel to supplemental chords are conjugate (perpendicular for a circle).

supplementary angles (*Maths.*). Two angles whose sum is 180°. Cf. *complementary angles.*

supplementary lens (*Photog.*). A so-called magnifying lens for temporary attachment in front of a normal camera lens to alter the focal length.

supply frequency (*Elec. Eng.*). The electrical frequency, in Hz, of an a.c. supply.

supply meter (*Elec. Eng.*). An instrument for measuring the total quantity of electrical energy supplied to a consumer during a certain period.

supply pipe (*Civ. Eng.*). See **service pipe.**

supply point (*Elec. Eng.*). A point on an electric power system from which electrical energy may be drawn.

supply station (*Elec. Eng.*). See **generating station.**

supply terminals (*Elec. Eng.*). Those at which connexion may be made to a supply point.

supply voltage (*Elec. Eng.*). The voltage across a pair of supply terminals.

suppository (*Med.*). A conical or cylindrical plug of a medicated mass for insertion into the rectum, vagina, or urethra.

suppressed carrier system (*Radio*). One in which the carrier wave is not radiated but is supplied by an oscillator at the receiving end; used with single sideband working because of the phase distortion arising when a carrier is inserted between two sidebands. Cf. *transmitted carrier system.*

suppressed-zero instrument (*Elec. Eng.*). An indicating or graphic instrument in which the zero position first scale reading is off scale, i.e., beyond the range of travel of the pointer. Also called **inferred-zero instrument, set-up instrument, set-up-scale instrument, set-up-zero instrument, step-up instrument.**

suppression (*Bot.*). Failure to develop; said of a member normally present. (*Comp.*) Elimination of specified data or digits, e.g., initial zeros. (*Med., Zool.*) Stoppage of discharge, as by obstruction of a duct; absence of some organ or structure normally present. *adj.* **suppressed.**

suppressor (*Elec. Eng.*). (1) Component, usually a low resistance adjacent to an electrode of a valve, to obviate conditions which promote parasitic oscillations. (2) Component, such as a capacitor or resistor, or both, which damps high-frequency oscillations liable to arise on breaking a current at a contact, causing radio interference.

suppressor grid (*Electronics*). That between anode and screen in pentode valves, to repel secondary electrons back to the anode. Abbrev. **sup.**

suppressor-grid modulation (*Electronics*). Insertion of the signal voltage into the (suppressor) grid circuit of a valve which rectifies and is driven by the carrier. Also **grid modulation.**

suppressor mutation (*Gen.*). One which suppresses the effects of another existing mutation by introducing a correction.

suppuration (*Med.*). The softening and liquefaction of inflamed tissue, with the production of pus. *adj.* **suppurative.**

supra-angulare (*Zool.*). See **surangulare.**

supraclavicle, supraclavicula (*Zool.*). See **supracleithrum.**

supracleithrum (*Zool.*). A dorsal bone of the pectoral girdle in Fish.

supradorsal (*Zool.*). On the back; above the dorsal surface; a dorsal intercalary element of the vertebral column.

supra-epimeron (*Zool.*). In Insects, the upper part of the epimeron when that sclerite is subdivided.

supra-episternum (*Zool.*). In Insects, the upper half of the episternum when that sclerite is subdivided.

supraneuston (*Ecol.*). Aquatic animals associated with the upper side of the surface film, e.g., Water Skaters. Cf. *infraneuston.*

supra-occipital (*Zool.*). A median dorsal cartilage bone of the Vertebrate skull forming the roof of the brain case posteriorly.

supra-orbital (*Zool.*). In some Vertebrates, a paired lateral membrane bone of the orbital region of the skull.

suprapharyngeal (*Zool.*). A bone lying above the pharynx in some Fish, formed by the fusion of the pharyngobranchials; situated above the pharynx.

suprarenal (*Zool.*). Situated above the kidneys.

suprarenal body (or **gland**) (*Zool.*). In higher Vertebrates, one of the endocrine glands lying close to the kidney and pouring into the blood secretions having important effects on the metabolism of the body.

suprasegmental structures (*Zool.*). The cerebellum, and, in higher forms, the cerebral cortex, as opposed to the *segmental apparatus* (q.v.).

suprasphenoid (*Zool.*). In some Fish, a membrane bone dorsal to the sphenoid cartilage and within the cranial wall.

supratemporal (*Zool.*). In the skull of some Vertebrates, a paired membrane bone occurring lateral to each parietal.

supravital staining (*Micros.*). The use of non-poisonous dyes (vital stains) to stain living material.

suprethmoid (*Zool.*). In the skull of Fish, an unpaired membrane bone lying between the nasals.

sural (*Zool.*). Of or pertaining to the calf of the leg.

suranal (*Zool.*). Above the anus, as the *suranal* plate or pygidium of Insects.

suramin (*Pharm.*). Germanin, Bayer 205. Non-arsenical drug mainly used to treat certain forms of trypanosomiasis.

surangulare (*Zool.*). In some lower Vertebrates, a membrane bone on the outer posterior part of the lower jaw.

surbase (*Join.*). See **dado rail.**

surbased arch (*Arch.*). An arch with a rise that is less than one-half its span.

surcharge (*Civ. Eng.*). A term applied to the earth supported by a retaining wall at a level above the top of the wall.

surculus (*Bot.*). A sucker.

surd (*Maths.*). An imprecise term usually applied to irrational roots or to the sum of such roots, e.g., $\sqrt{2}$ or $\sqrt{3}+3\sqrt{5}$.

surface absorption coefficient (*Nuc.*). See **absorption coefficient.**

surface acoustic wave (*Radar*). Acoustic wave propagated along the surface of a piezoelectric substrate. Used in microwave components and amplifiers. Abbrev. **SAW.**

surface activity (*Chem.*). The influence of certain substances on the surface tension of liquids.

surface barrier (*Electronics*). Potential barrier across surface of semiconductor junction due to diffusion of charge carriers.

surface-barrier transistor (*Electronics*). One in which very thin barriers (by means of etching techniques) are used, so permitting the frequency range to be extended to 100 MHz.

surface charge (*Elec. Eng.*). See **bound charge.**

surface chemistry (*Chem., Min. Proc.*). That of the interface or interphase between two systems in which the substrate of one or both has become ionically unbalanced.

surface combustion (*Heat*). Bringing a combustible mixture of gas and air into contact with a suitable refractory material so as to produce flameless or nearly flameless combustion, the surface of the refractory material being maintained in a state of incandescence.

surface compressibility (*Chem.*). The compressibility of the layer of adsorbed molecules in response to surface tension forces.

surface concentration excess (*Chem.*). The excess concentration (may be negative) of a solute per unit area in the surface layer of a solution.

surface condenser (*Eng.*). A steam condenser for maintaining a vacuum at the exhaust pipe of a steam-engine or turbine. It consists of a chamber in which cooling water is circulated through tubes, and which is evacuated by an air pump. See **condenser, condenser tubes.**

surface conductivity (*Phys.*). See **surface resistivity.**

surface-contact system (*Elec. Eng.*). A system of electric traction supply employing insulated fixed contacts or studs, placed at intervals between the running rails. The studs are normally 'dead', but become 'alive' as soon as the electric tractor is over them, by connexion to a conductor running in a closed conduit underneath.

surface density (*Elec.*). The amount of electric charge/unit surface area. (*Nuc.*) In nuclear physics, mass/unit area, a common method of indicating the thickness of an absorber, etc.

surface duct (*Radio*). Atmospheric propagation duct for which the earth's surface forms the lower boundary.

surface energy (*Phys., etc.*). Free energy per unit area. Surface tension (N/m) multiplied by surface area in m^2.

surface friction drag (*Aero.*). That part of the drag represented by the components of the pressures at points on the surface of an aerofoil, resolved tangential to the surface.

surface gauge (*Eng.*). See **scribing block.**

surface-grinding machine (*Eng.*). A grinding machine for finishing flat surfaces. It consists of a high-speed abrasive wheel, mounted above a reciprocating or rotating work-table on which flat work is held, often by a *magnetic chuck.*

surface hardening (*Met.*). *Cementation* (q.v.) of low-carbon steels. See **nitriding.**

surface irradiation (*Radiol.*). Irradiation of a part

of the body by applying a mould or applicator loaded with radioactive material to the surface of the body.

surface leakage (*Elec. Eng.*). That along the surface of a nonconducting material or device. May vary widely with contamination, humidity, etc. It sets a practical limit to the value of high resistors for use with electrometers, etc.

surface lifetime (*Electronics*). The lifetime of current carriers in the surface layer of a semiconductor (where recombination occurs most readily). Cf. *volume lifetime*.

surface loading (*Aero.*). The average force per unit area, normal to the surface, on an aerofoil under specified aerodynamic conditions.

surface measure (*For.*). A method of measuring timber in quantity, by the area of one face, irrespective of thickness. Cf. *board measure*.

surface noise (*Acous.*). See **needle scratch.**

surface of operation (*Build.*). A surface which is dressed to a plane as a reference from which the rest of the work can be set out and executed.

surface plate (*Eng.*). A rigid cast-iron plate whose surface is accurately scraped flat; used to test the flatness of other surfaces or to provide a truly plane datum surface in marking off work for machining.

surface plates (*Print.*). A general name for litho plates which are not *deep etch* (q.v.), including plates made by the *albumen process* (q.v.); a negative is usually used when printing down.

surface pressure (*Chem.*). The 2-dimensional analogue of gas pressure. Defined as the difference between the surface tension of a pure liquid and that of a surface active solution, it represents the tendency of the adsorbed surfactant molecules to spread over the clean liquid surface. See **Gibbs' adsorption theorem.**

surface recombination velocity (*Electronics*). Electron-hole recombination on surface of semiconductor occurs more readily than in the interior, hence the carriers in the interior drift towards the surface with a mean speed termed the *surface recombination velocity*. It is defined as the ratio of the normal component of the impurity current to the volume charge density near the surface. Abbrev. **S.R.V.**

surface resistivity (*Phys.*). That between opposite sides of a unit square inscribed on the surface. Its reciprocal is *surface conductivity*.

surface sterilization (*Nuc.*). Radiation with low-energy rays which penetrate thin surface layers only, e.g., with ultraviolet rays.

surface tension (*Phys.*). A property possessed by liquid surfaces whereby they appear to be covered by a thin elastic membrane in a state of tension, the surface tension being measured by the force acting normally across unit length in the surface. The phenomenon is due to unbalanced molecular cohesive forces near the surface. Units of measurement are dyne cm^{-1}, $N\ m^{-1}$. See **capillarity, liquid-drop model, pressure in bubbles.**

surface wave (*Phys.*). See **wave.**

surface wind (*Meteor.*). The wind at a standard height of 10 m (33 ft) above ground. Differs from the *geostrophic wind* (q.v.) and the *gradient wind* (q.v.) because of friction with the earth's surface.

surface wiring (*Elec. Eng.*). A wiring installation in which the insulated conductors are attached to the surfaces of a building, either enclosed in conduit or secured by cleats.

surfactant (*Chem.*). An abbreviated form of *surface active agent*, i.e., a substance which has the effect of altering the interfacial tension of water and other liquids or solids, e.g., a detergent or soap.

surge (*Aero.*). Unstable airflow condition in the compressor of a gas turbine due to a sudden increase (or decrease) in mass airflow without a compensating change in pressure ratio. (*Elec. Eng.*) A large but momentary increase in the voltage of an electric circuit.

surge absorber (*Elec. Eng.*). A circuit device which diverts, and may partly dissipate, the energy of a surge, thus preventing possible damage to apparatus or machines connected to a transmission line. Also **surge modifier.**

surge arrester (*Elec. Eng.*). See **lightning arrester.**

surge bin or **tank** (*Min. Proc., Mining*). Hopper (dry material) or reservoir with means of agitation (ore pulps), used to iron out irregularities in process delivery and flow.

surge-crest ammeter (*Elec. Eng.*). An instrument for recording a surge on a transmission line by measurement of the residual magnetism in a piece of magnetic material which has been magnetized by the surge current.

surge generator (*Elec. Eng.*). See **impulse generator.** (*Radiol.*) See **Marx generator.**

surge impedance (*Elec.*). See **characteristic impedance** (1).

surge modifier (*Elec. Eng.*). See **surge absorber.**

surge point (*I.C. Engs.*). Of a centrifugal supercharger, the value of the mass airflow at which, during throttling of the delivery, surging occurs. See **surging** (1).

surge tank (*Eng.*). One used to absorb irregularities in flow.

surgical spirit (*Chem.*). Methanol, to which is added small amounts of oil of wintergreen and castor oil; used chiefly for sterilizing the skin in surgical operations.

surging (*I.C. Engs.*). (1) In centrifugal superchargers, an abrupt decrease or severe fluctuation of the delivery pressure as the weight of air delivered is reduced. See **surge point.** (2) In valve springs, the coincidence of some harmonic of the cam lift curve with the spring's natural frequency of vibration, leading to irregular action and failure.

surmounted (*Arch.*). A term applied to a vault springing from points below its centre and having the form of a circular arc above its centre.

surra (*Vet.*). A form of trypanosomiasis affecting horses, dogs, cattle, elephants, and camels, occurring in Asia and the Sudan, caused by *Trypanosoma evansi*; symptoms include emaciation and subcutaneous oedema, usually fatal. Transmitted by biting flies.

surround (*Acous.*). The flexible support of the large conical diaphragm in open-diaphragm electrodynamic loudspeakers. It can contribute, by resonance, to the radiation of sound-power of very low frequency.

surveillance radar (*Aero., Radar*). A plan position indicator radar showing the position of aircraft within an air traffic control area or zone.

surveying. Measurement of the relative positions of points on the surface of the earth and/or in space, to enable natural and artificial features to be depicted in their true horizontal and vertical relationship by drawing them to scale on paper.

survival curve (*Radiol.*). One showing the percentage of organisms surviving at different times after they have been subjected to large radiation dose. Less often, one showing percentage of survivals at given time against size of dose.

survivorship curve (*Ecol.*). The number or percentage of an original population surviving, plotted against time, giving an indication of the mortality rate at different ages.

Susa fixative (*Micros.*). A fixative used for general purposes, and containing mercuric chloride, sodium chloride, trichloroacetic acid, glacial acetic acid, and formalin.

susceptance (*Elec.*). Imaginary part of *admittance* (q.v.), equal to $\frac{-X}{R^2+X^2}$ in a circuit of impedance $R+jX$.

susceptibility (*Elec. Eng.*). See **electric** or **magnetic susceptibility.**

susceptibility curves (*Elec. Eng.*). Curves of susceptibility plotted to a base of magnetic field strength.

susceptor phase advancer (*Elec. Eng.*). A phase advancer which injects into the secondary circuit of an induction motor an e.m.f. which is a function of the open-circuit secondary e.m.f. Cf. *expedor phase advancer*.

suspended scaffold (*Build.*). A form of scaffold used in the construction, repair, cleaning, etc., of buildings. It consists of working platforms, on light frameworks, slung from fixed higher points in the building.

suspended span (*Civ. Eng.*). The middle length of a bridge span connecting, and carried upon, the cantilever arms, in cases where these are not built out until they meet. See **cantilever bridge.**

suspension (*Autos.*). A system, primarily of springs (leaf and coil) and dampers (usually hydraulic), designed to support the body of a vehicle and to protect it, and hence its occupants, from road shocks. See **Hydrolastic.** (*Chem.*) A system in which denser particles, which are at least microscopically visible, are distributed throughout a less dense liquid or gas, settling being hindered either by the viscosity of the fluid or the impacts of its molecules on the particles.

suspension bridge (*Civ. Eng.*). A bridge in which the deck is suspended from a flexible connexion between the two sides. For large span bridges, experience has shown that the possibility of harmonic action must be taken into account and that wind tunnel experiments are necessary to determine the degree of lateral stiffness required or even necessary.

suspension cable anchor (*Civ. Eng.*). The anchorage, which may have various forms, of the cables of a suspension bridge.

suspension insulator (*Elec. Eng.*). A freely hanging insulator made of units connected in series, by which an overhead line is suspended from the arm of a transmission-line tower; it cannot withstand any other force than a tension.

suspension push (*Elec. Eng.*). See **pendant push.**

suspension spring (*Horol.*). The thin ribbon of spring steel which supports a pendulum.

suspension switch (*Elec. Eng.*). See **pendant switch.**

suspensoid (*Chem.*). See **lyophobic colloid.**

suspensor (*Bot.*). (1) A hypha which forms the stalk of a gametangium in the *Zygomycetes.* (2) A small mass of cells which forces the developing embryo of a higher plant down into nutritive tissue.

suspensotium (*Zool.*). In Vertebrates, the apparatus by which the jaws are attached to the cranium.

suspensory (*Zool.*). Pertaining to the suspensorium; serving for support or suspension.

Sussex garden-wall bond (*Build.*). The form of *garden-wall bond* (q.v.) in which one header and three stretchers are laid in each course.

sustained oscillations (*Telecomm.*). Externally-maintained oscillations of a system at or very near its natural resonant frequency. Cf. *forced oscillations*, *free oscillations.*

sustentacular (*Zool.*). Supporting; as the *sustentacular* cells surrounding the gustatory cells in a taste-bud.

sustentator (*Zool.*). The cremaster of *Lepidoptera.*

sutural bones (*Zool.*). Small supernumerary bones occurring in the sutures between the other bones of the skull in some Primates. Also called **Wormian bones.**

suture (*Bot.*). (1) The line of union between two members of a whorl or the united edges of one member. (2) A line of weakness along which dehiscence may occur. (*Med.*) Surgical stitch, or group or row of such stitches. (*Zool.*) A line of junction of two structures, as the line of junction of adjacent chambers of a Nautiloid shell; a synarthrosis or immovable articulation between bones, as between the bones of the cranium; junctions of exoskeletal cuticular plates in Insects. *adj.* **sutural.**

sutured (*Geol.*). A textural term descriptive of the sinuous interlocking grain boundaries of rocks which have undergone extensive recrystallization, e.g. quartzites.

S-value (*Nuc. Eng.*). Percentage increase in volume of nuclear fuel after 1% burn up. See also **R-value.**

SVD (*Vet.*). See **swine vesicular disease.**

swab (*Foundry*). See **bosh.** (*Med.*) (1) Any small mass of cotton-wool or gauze used for mopping up blood or discharges, or for applying antiseptics to the body, or for cleansing surfaces (e.g., the lips, the mouth). (2) A specimen of a secretion taken on a swab for bacteriological examination.

swage (*Tools*). Smith's tool (dolly) used in shaping drill bits.

swaging (*Met.*). Reducing of cross-section of metal rod or tube by forcing it through a tapered aperture between two grooved dies.

swallowtail (*Carp., Join.*). See **dovetail.**

Swammerdam's glands (*Zool.*). See **calcigerous glands.**

swamp fever (*Vet.*). See **equine infectious anaemia.**

swamping resistance (*Elec. Eng.*). A high resistance connected in series with the coil of a voltmeter; made from material having a negligible temperature coefficient, so as effectively to 'swamp' the variation with temperature of the resistance of the copper coil.

Swan cube (*Optics*). The prism system used in the Lummer-Brodhum photometer. Consists of two 45° prisms placed with their hypotenuse faces together, but with one face ground so that the prisms are touching only over the central part of the faces.

swan-neck (*Join.*). The bend formed in a handrail when a knee and a ramp are joined together without any intermediate straight length.

swan-neck insulator (*Elec. Eng.*). A pin-type insulator with a bent pin, arranged so as to bring the insulator into approximately the same horizontal plane as that of the support.

swan-neck lock mortise chisel (*Tools*). A hook-shaped chisel for chopping lock mortises in doors.

swansdown (*Textiles*). A cotton fabric of the fustian type, bleached and with a raised surface, or piece dyed; used for box linings, covering loom shuttle raceboards, etc.

swansdown twill (*Weaving*). See **crow twill.**

swarf (*Acous.*). In cutting original gramophone disks, the thin thread of waste which is removed from the lacquer surface by a cutting stylus.

It is often electrified and inflammable. Also called **chip.** (*Eng.*) Metal or plastic chips and turnings removed by cutting tools during machining operations.

swarm (*Zool.*). A large number of small animals in movement together; especially a number of Bees emigrating from one colony to establish another under the guidance of a queen.

swarm cell (*Bot.*). See **zoospore.**

swarm sporangium (*Bot.*). See **zoosporangium.**

swarm spore (*Zool.*). A ciliated planula larva. See **zoospore.**

swash bank (*Hyd. Eng.*). The upper part of the slope of a sea embankment. See **footing, outburst bank.**

swash letters (*Typog.*). Ornamental italic letters with tails and flourishes, such as $\mathscr{A}$, $\mathscr{B}$, etc. They should be used only at the beginning or end of a word.

swash plate (*Eng.*). A circular plate mounted obliquely on a shaft; sometimes used in conjunction with working cylinders mounted axially concentric with the shaft, as a substitute for an engine or pump crank mechanism.

swath aerator (*Agric.*). A machine with a pick-up mechanism to lift and loosen *swaths* (rows of cut grass or grain) or windrows without inverting them.

swayback (*Vet.*). Enzootic ataxia. A nervous disease of newborn and young lambs characterized by degenerative changes in the cerebrum of the brain, causing inability or difficulty in standing or walking. Associated with a low copper content of tissues in the lamb and its ewe and preventable by administering copper to the pregnant ewe.

sway rod (*Build.*). A member inserted in a structural framework to resist wind forces.

SWD (*Civ. Eng.*). Abbrev. for *stoneware drain.*

sweat cooling (*Aero., Chem. Eng.*). Cooling of a component by the evaporation of a fluid through a porous surface layer; used for high-performance gas turbine blades.

sweated joint (*Eng.*). See under **sweating.**

sweating (*Brew.*). The removal of surplus moisture from barley by slow heating in a kiln. (*Build.*) A term applied to a surface showing traces of moisture due either to formation of condensate or to water having got through a porous material of which the surface is part. (*Eng.*) The operation of soldering pieces together by 'tinning' the surfaces and heating them while pressed into contact.

Swedish iron (*Met.*). Wrought iron of high purity, with some 300 MN/m^2 ultimate tensile strength, and 33% elongation before fracture.

sweep (*Aero.*). The angle, in plan, between the normal to the plane of symmetry and a specified spanwise line on an aerofoil. Most commonly, the quarter-chord line is used, but leading and trailing edges are sometimes stipulated. Sweep increases longitudinal stability by extending the centre of pressure and delays compressibility drag by reducing the chordwise component of the airflow. *Sweepback*, the more usual, is the aft displacement of the wings and *forward sweep* the opposite.

sweep circuit (*Electronics*). That which supplies deflecting voltage to one pair of plates or coils of a cathode-ray tube, the other pair being connected to the source of current or voltage under examination. See **linear scan.** (*TV*) In TV and facsimile transmission the circuits which produce the required rectangular scan by the electron beam.

sweeper (*Electronics*). Frequency swept oscillator, particularly at microwave frequencies.

sweeps (*Met.*). Dust and debris in jeweller's workshops, gold refineries, bullion assay offices, etc., collected and treated periodically to recover valuable contents.

sweep-saw (*Tools*). A thin-bladed saw which is held taut in a special frame and may be used for making curved cuts. Also called **turning-saw.**

sweet (*Glass*). Easily workable.

sweet clover disease (*Vet.*). A fatal, haemorrhagic disease of cattle and other animals caused by feeding sweet clover (genus *Melilotus*), which has been damaged during haymaking or ensiling, causing the formation of a toxic substance, *dicoumarin*, which interferes with blood coagulation.

sweet nitre (*Chem.*). Solution of ethyl nitrite in ethanol, of about 2½% strength and used as a heart stimulant.

sweet roasting (*Met.*). Ignition of metal sulphides to remove bulk of sulphur and arsenic as gaseous fume, leaving the mineral as its oxide.

sweet water (*Chem.*). Term applied in soap-making to the aqueous solution of glycerol arising from the hydrolysis of the fat charge.

swell (*Acous.*). (1) Mechanism for altering the volume of sound in an organ by opening or closing shutters. Operated by the *swell pedal.* (2) In an electronic organ, volume control operated by potentiometer. (*Min. Proc., Mining*) Volumetric increase due to crushing, which creates more void space in a given weight of rock. (*Weaving*) Spring or pneumatic device at either front or back of the loom shuttle box. It brings the shuttle to rest quickly without shock and holds it firmly for picking through the warp shed.

swelled head (*Vet.*). See **big head disease of sheep.**

swelled rules (*Typog.*). Decorative rules in a variety of sizes with a central swelling, occasionally tooled; sometimes called **Bodoni rules.**

swelling (*Arch.*). The slight bulge given to the profile of a column near the middle of its length, to correct for the apparent concavity which it would have if it were a straight taper. (*Nuc.*) Of fissile materials, a change in volume without necessarily any change in shape, which may occur during radiation. See **R-value, S-value.**

swell pedal (*Acous.*). The foot-operated lever for regulating the loudness of stops drawn on the swell manual. See **grand swell.**

swept volume (*Vac. Tech.*). See **volumetric displacement.**

SWG (*Eng.*). Abbrev. for *Standard Wire Gauge.*

swifts or cylinders (*Textiles*). (1) The large, wire-covered rollers on roller or flat cards. Working with either workers and strippers or flats, they clean and comb wool, cotton, or flax fibres. (2) Light revolving frames of wood or metal which carry the hanks during unwinding.

swim bladder (*Zool.*). See **air bladder.**

swimmerets (*Zool.*). In some *Crustacea*, paired biramous abdominal appendages used in part for swimming.

swimming bell (*Zool.*). See **nectocalyx.**

swimming funnel (*Zool.*). In dibranchiate *Cephalopoda*, the funnel-shaped tube through which water is forcibly expelled from the mantle-cavity, producing a jerky backward movement of the animal.

swimming ovaries (*Zool.*). Clumps of ripe ova floating freely in the fluid of the body cavity, as in *Acanthocephala.*

swimming plate (*Zool.*). See **ctene.**

swimming-pool reactor (*Nuc. Eng.*). One which uses uranium-235 in aluminium cans immersed in water. Also **aquarium reactor, pool reactor.**

Swinburne test (*Elec. Eng.*). An indirect test

applied to shunt and compound d.c. motors for determining the load losses.

swine erysipelas (*Vet.*). Diamond skin disease. A disease of pigs caused by infection by *Erysipelothrix insidiosa* (*E. rhusiopathiae*). The acute form of the disease is a septicaemia characterized by fever, urticaria-like patches on the skin and sometimes lameness due to arthritis; in the chronic form endocarditis occurs.

swine fever (*Vet.*). Hog cholera. A highly contagious disease of pigs, caused by a virus but usually complicated by secondary bacterial infection; symptoms may include fever, diarrhoea, pneumonia, and nervous symptoms.

swine influenza (*Vet.*). An acute viral infection of pigs characterized by fever, coughing, and respiratory distress.

swine paratyphoid (*Vet.*). A disease of pigs caused by *Salmonella cholerae suis*, and characterized by septicaemia in the acute form and necrotic enteritis in the chronic form.

swine plague (*Vet.*). A septicaemic and pneumonic disease of pigs due to infection by the bacterium *Pasteurella multocida*.

swine pox (*Vet.*). Variola. A disease of pigs characterized by the formation of papules, vesicles, and pustules on the skin, caused by a virus related to that of *vaccinia* (q.v.). A similar disease, in which papules and scabs develop, is caused by an unrelated virus.

swine vesicular disease (*Vet.*). A virus disease, confined (unlike *foot-and-mouth disease*, which it closely resembles clinically) to pigs. It is characterized by vesicular lesions, and is transmitted by contact, or through infected feed. Abbrev. SVD.

swing (*Aero.*). The involuntary deviation from a straight course of an aircraft while taxiing, taking-off, or alighting. (*Eng.*) Lathe dimension, equal to the radial size of the largest workpiece that can be rotated or swung in it. (*Radio*) Angle, expressed as plus or minus half the excursion or spread, over which the dial of a goniometer or rotating aerial must be swung to estimate the reading of a bearing. (*Telecomm.*) Extreme excursion from positive peak to negative peak in an alternating voltage or current waveform.

swing back (*Photog.*). The back of a camera which can tilt upwards or sideways, or both, so that distortion of objects (such as the vertical lines of buildings) may be minimized, or objects at different distances may be brought into focus.

swing bridge (*Civ. Eng.*). A type of movable bridge which is capable of swinging about a vertical pivot, to allow the passage of a vessel.

swing front (*Photog.*). The provision for tilting the front of a camera, with the lens, so that distortion of an object due to its receding along the axis can be minimized.

swinging choke (*Elec. Eng.*). Iron-cored inductor with saturable core, used in smoothing circuits where decreasing impedance with increasing current improves regulation.

swinging grippers (*Print.*). Found on sheet-fed rotaries and on high-speed single- and 2-revolution machines, the sheets being transferred at speed from the feedboard to grippers on the continuously rotating impression cylinder.

swinging post (*Build.*). See **hingeing post.**

swinging-wedge rangefinder (*Photog.*). One in which the two images are reflected from the end surfaces of a rhomboid prism, coincidence being effected by displacing a convex lens about its centre of curvature in front of one of the reflectors.

swing-wing (*Aero.*). See **variable sweep.**

S-wire (*Teleph.*). The sleeve-wire which is added to a subscriber's pair in its passage through an exchange; it is connected to the sleeves of the plugs of the cord circuit.

swirl chamber (*I.C. Engs.*). Type of diesel-engine combustion chamber, separated from the cylinder by a short vent and shaped so as to set up an eddy in the air drawn in through the intake valve.

swirl sprayers (*Aero.*). Fuel injectors in a gas turbine which impart a swirling motion to the fuel.

swirl vanes (*Aero.*). Vanes which impart a swirling motion to the air entering the flame tube of a gas-turbine combustion chamber.

Swiss lapis (*Min.*). A fraudulent imitation of lapis lazuli, obtained by staining pale-coloured jasper or ironstone with hexacyano ferrate(II). Also known as **German lapis.**

Swiss (or schiffle or shuttle) machine (*Textiles*). A type of embroidery machine in which the shuttles are placed diagonally.

Swiss (or Thury) screw-thread (*Eng.*). A metric thread having a profile angle of 47½°, the crest of the thread being formed by cutting off ⅙ of the thread height, and rounding.

Swiss-type automatic (*Eng.*). *Single-spindle automatic* (q.v.), in which several single-point tools are arranged radially around the stock, used primarily for small precision parts for watches and instruments.

switch (*Comp.*). A point in a program at which a branch is possible. (*Elec. Eng.*) Device for opening and closing an electric circuit. (*Electronics*) Electronic circuit for switching between two independent inputs, e.g., by valves, tubes, or transistors. (*Mag.*) Reversal by saturation of the residual flux in a ferrite core. (*Eng.*) A mechanism for shifting a moving body in another direction. (*Rail.*) A device for moving a small section of a railway track so that rolling stock may pass from one line of track to another. (*Teleph.*) See **double switch call, single switch call.**

switch base (*Elec. Eng.*). The insulating base on which a switch is mounted.

switch blades (*Rail.*). See **points.**

switchboard (*Elec. Eng., etc.*). An assembly of switch panels.

switchboard instrument (*Elec. Eng.*). An electric measuring instrument arranged for mounting on a switchboard. See **sector-pattern instruments.**

switchboard panel (*Elec. Eng., etc.*). See **panel.**

switch-box (*Elec. Eng.*). An enclosure housing one or more switches operated by means of an external handle.

switch-desk (*Elec. Eng.*). A control desk on which a number of miniature switches are mounted, each of which serves to initiate some control operation.

switch-fuse (*Elec. Eng.*). A knife-switch carrying a fuse in each blade.

switchgear (*Elec. Eng.*). The generic name for that class of electrical apparatus whose sole function is to open and close electric circuits.

switchgear pillar (*Elec. Eng.*). See **pillar.**

switching (*Telecomm.*). Alternative to *clamp* (q.v.); connexion of a circuit point to a known potential for a definite period of time.

switching constant (*Elec. Eng.*). The ratio of *switching time* and *magnetic-field intensity*.

switching time (*Elec. Eng.*). Time required for complete reversal of flux in ferroelectric or ferromagnetic core.

switch panel (*Elec. Eng.*). An insulating panel on which a switch is mounted.

switch plant (*Bot.*). A plant with small leaves, often reduced to nonfunctional scales, and long thin stems and branches with photosynthetic tissue in their cortical regions.

switch plate (*Elec. Eng.*). A plate for covering one or more flush switches. Also called **flush plate.**

switch-starter (*Elec. Eng.*). A combination of knife-switch and starting regulator, in which the circuit is closed and the resistance progressively cut out in one continuous movement.

swivelling airscrew (*Aero.*). See **airscrew.**

swivel-pin (*Autos.*). See **king-pin.**

swivel weaving (*Weaving*). A method of weaving figures in a fabric by means of small shuttles, which are located above the reed and contain the extra material required.

swivel wheel (*Agric.*). A supporting wheel for the forward part of a frame.

sycamore (*For.*). A product of the largest hardwood trees (*Acer*), native to Europe and W. Asia; a hard and dense wood, used largely for internal work. See also **London plane.**

sycon grade (*Zool.*). The second of the three types of canal system in *Porifera*; the flagellated chambers open in radial fashion from a central paragaster, which itself opens to the exterior by a single osculum. Cf. *ascon grade*, *leucon grade*.

syconium, syconus (*Bot.*). The fruit of the fig, consisting mainly of the much enlarged receptacle of the inflorescence.

sycosis barbae (*Med.*). Inflammation of the hair follicles of the beard region, due to infection with the staphylococcus.

Sydenham's chorea (*Med.*). See **chorea.**

syenite (*Geol.*). A coarse-grained igneous rock of intermediate composition, composed essentially of alkali-feldspar to the extent of at least two-thirds of the total, with a variable content of mafic minerals, of which common hornblende is characteristic.

syenite-porphyry (*Geol.*). An igneous rock of syenitic composition and medium grain size, commonly occurring in minor intrusions; it consists of phenocrysts of feldspar and/or coloured silicates set in a microcrystalline groundmass. See also **microsyenite.**

syenodiorite (*Geol.*). An alternative name for *monzonite*, a rock type intermediate between syenite and diorite.

syeno-gabbro (*Geol.*). A term introduced independently by J. W. Evans and A. Johannsen for coarse-grained igneous rocks of intermediate composition, and intermediate between syenite and diorite in feldspar content.

Sykes microphone (*Acous.*). One in which a limp coil in the radial field of a pot magnet generates an e.m.f. when subjected to an acoustic pressure.

syllable articulation (*Teleph.*). See **articulation.**

sylphon bellows (*Eng.*). A thin-walled cylindrical metal bellows consisting of a number of elements arranged concertina-fashion, responding to external or internal fluid pressure; used in pressure-governing systems. See also **bellows.**

sylvanite (*Min.*). Telluride of gold and silver, which crystallizes in the monoclinic system and is usually associated with igneous rocks and, in veins, with native gold. It is an ore of gold.

sylvestrene (*Chem.*). Δ-1,8-*m*-Terpadiene, the chief constituent of Swedish and Russian turpentine. It is dextrorotatory, b.p. 175°C.

Sylvian aqueduct (*Zool.*). In Vertebrates, the cavity of the mesencephalon.

Sylvian fissure (*Zool.*). In Mammals, a deep lateral fissure of the cerebrum.

sylvic acid (*Chem.*). See **abietic acid.**

sylvine, sylvite (*Min.*). Chloride of potassium, which crystallizes in the cubic system. It occurs in bedded salt deposits (*evaporites*), and as a sublimation product near volcanoes; it is a source of potassium compounds, used as fertilizers.

sylvinite (*Min.*). A general name for mixtures of the two salts sylvine and halite, the latter predominating, occurring at Stassfurt and in Alsace. Also used as a commercial name for *sylvine*.

sym-. See **syn-.**

symbatic (*Biol.*). The condition, in morphometric analysis, where two distinguishable types of polymorphism share a common dimension.

symbiosis (*Biol.*). An internal partnership between two organisms (*symbionts*), in which the mutual advantages normally outweigh the disadvantages. See also **mutualism.**

symblepharon (*Med.*). Adhesion of the eyelid to the globe of the eye.

symbol (*Chem.*). See **chemical symbol.**

symbolic address (*Comp.*). See **floating address.**

symbolic language (*Comp.*). That of a computer program prepared in any coding other than the specific machine language, and so requiring to be assembled or compiled before it can be carried out.

symbolic logic. See **mathematical logic.**

symbolic method (*Elec. Eng.*). A powerful method of a.c. circuit analysis utilizing the vector operator $j = \sqrt{-1}$, a multiplier signifying rotation of a vector through a right angle in the positive direction.

Symbranchii (*Zool.*). A small order of eel-like *Teleostei*, lacking scales and paired fins; found in rivers and swamps of tropical Asia, Australasia, and South America.

symmetrical (*Bot., Zool.*). See **actinomorphic.**

symmetrical components (*Elec. Eng.*). A term applied to a method of calculating short-circuit currents in a.c. systems, making use of symmetrical phase-sequence components of the currents and voltage.

symmetrical deflection (*Electronics*). The application of a voltage to a pair of deflection plates in a CRO such that they vary symmetrically above and below an average value, which is equal to the final anode potential of the tube. This procedure minimizes the possibility of trapezoidal distortion of the screen image.

symmetrical flutter (*Aero.*). See **flutter.**

symmetrical grading (*Teleph.*). Grading in which all groups of selectors are equally favoured in seeking outlets.

symmetrical network (*Elec. Eng.*). One which can be divided into two mirror half-sections.

symmetrical short-circuit (*Elec. Eng.*). An a.c. short-circuit in which each phase carries the same current.

symmetrical winding (*Elec. Eng.*). A term applied to an armature winding which fulfils certain conditions of electrical symmetry.

symmetric dyadic (*Maths.*). See **conjugate dyadics.**

symmetric relation (*Maths.*). A relation is symmetric if when it applies from x to y, then it applies from y to x; e.g., 'is parallel to', 'is the complement of'. 'Is greater than' is not a symmetric relation.

symmetry (*Crystal.*). The quality possessed by crystalline substances by virtue of which they exhibit a repetitive arrangement of similar faces. This is a result of their peculiar internal atomic structure, and the feature is used as a basis of crystal classification. (*Maths.*) (1) A geometric configuration is said to be symmetrical about a point, line or plane if any line through the point or any line perpendicular to the line or plane cuts the configuration in pairs of points equally spaced from the point, line or plane, which are then referred to respectively as the *centre of symmetry*, the *axis of symmetry* and the *plane of symmetry*. (2) A function of several variables is symmetric if it is unchanged when any two of the variables are interchanged. (*Zool.*) (1) The method of arrangement of the constituent parts of the animal body. (2) In higher animals, the disposition of such organs as show bilateral or radial symmetry.

symmetry class (*Crystal.*). Crystal lattice structures can show 32 combinations of symmetry elements—each combination forming a possible symmetry class.

Symons's earth thermometer (*Meteor.*). See **earth thermometer.**

sympathectomy, sympatheticectomy (*Surg.*). Excision or cutting of a part of a sympathetic nerve.

sympathetic image (*Photog.*). An image formed by rendering insoluble gelatine which has penetrated into a support of unsized paper.

sympathetic nervous system (*Zool.*). In some Invertebrates (Crustaceans and Insects), a part of the nervous system supplying the alimentary system, heart, and reproductive organs and spiracles. In Vertebrates, a subdivision of the autonomic nervous system, also known as the **thoracicolumbar system.** The action of these nerves tends to increase activity, speed the heart and circulation, and slow digestive processes. Sympathetic nerves produce noradrenaline and, to a lesser extent, adrenaline. The preganglionic axons are long and the postganglionic axons short, the ganglia forming prominent chains. The axons emerge from the C.N.S. in the thoracic and lumbar region of the spinal cord. Cf. *parasympathetic nervous system.*

sympathetic reaction (*Chem.*). See **induced reaction.**

sympathin (*Physiol.*). Substance released from endings of noradrenergic postganglionic sympathetic nerves; *noradrenaline* (q.v.). It has two forms: excitatory (SE) which causes blood vessels to constrict, and inhibitory (SI) which causes them to dilate.

sympathomimetic (*Physiol.*). Mimicking the sympathetic nervous system, i.e., used of drugs which produce effects like those produced by stimulation of the sympathetic nervous system, particularly adrenergic and noradrenergic effects.

Sympetalae (*Bot.*). A large subdivision of Dicotyledons in which the corollas consist of united petals. Also **Metachlamydeae, Gamopetalae.**

sympetalous (*Bot.*). Said of a corolla composed of a number of united petals; gamopetalous.

symphile (*Zool.*). In a community of social Insects, especially Ants or Termites, a true guest species which is fed and reared by the members of the community.

Symphyla (*Zool.*). A class of *Arthropoda*, having long antennae, a single pair of legs on most of the body segments, the gonads opening on the fourth postcephalic segment, and a single pair of spiracles situated on the head. Thought by some to derive from the same forms which gave rise to the ancestral Insects.

symphysiotomy, symphyseotomy (*Surg.*). The operation of cutting through the pubic joint to facilitate the birth of a child.

symphysis (*Zool.*). Union of bones in the middle line of the body, by fusion, ligament, or cartilage, as the mandibular *symphysis*, the pubic *symphysis*; growing together or coalescence of parts, as acrodont teeth with the jaw; the point of junction of two structures; chiasma; commissure. *adj.* **symphysial.**

symplast (*Cyt.*). A multinucleate cell formed by fragmentation of the nucleus within a single energid.

symplectic (*Zool.*). In some Fish, a bone supporting the quadrate, formed by the ossification of the lower part of the hyomandibular cartilage.

sympodite (*Zool.*). See **protopodite.**

sympodium (*Bot.*). A branch system in which the main axis ceases to elongate after a time, and one or more lateral branches grow on; these cease to grow and give laterals which repeat the process.

symptom (*Med.*). Evidence of disease or disorder as experienced by the patient (e.g., pain, weakness, dizziness); any abnormal sensation or emotional expression or thought accompanying disease or disorder of the body or the mind; less accurately, any objective evidence of disease or bodily disorder (see **sign**).

symptomatology (*Med.*). The study of symptoms; a discourse or treatise on symptoms; the branch of medical science concerning symptoms of disease.

syn-, sym-. Prefix from Gk. *syn*, with, generally signifying fusion or combination. (*Chem.*). See s- (1), (3).

synaldoximes (*Chem.*). The stereoisomeric forms of aldoximes in which the H and OH groups are on the same side of the plane of the double bond.

synandrium (*Bot.*). A mass of united anthers.

synandrous (*Bot.*). Having several united stamens.

synangium (*Bot.*). A group of sporangia united side by side. (*Zool.*) An arterial trunk from which several arteries arise, as the terminal portion of the truncus arteriosus in lower Vertebrates. *adj.* **synangial.**

synaposematic (*Zool.*). Said of a form of protective mimicry in which there is resemblance to a powerful or dangerous species.

synapse (*Zool.*). The region where one nerve cell makes functional contact with another, and where nerve impulses can pass from one to the other by the diffusion across the small gap between the pre- and post-synaptic neurons of small quantities of specific transmitter substances (e.g., acetylcholine, noradrenaline) which are released from the presynaptic nerve terminals and act on the postsynaptic membrane.

synapsid (*Zool.*). In the skull of Reptiles, the condition when there is one temporal vacuity, this being low behind the eye, with the post-orbital and squamosal meeting above. Found in *Pelycosauria* and mammal-like Reptiles. Cf. *diapsid.*

Synapsida (*Zool.*). A subclass of Reptiles, often mammal-like, with a single lateral temporal vacuity, primitively lying below the post-orbital and the squamosal. The brain case was high, and the inner ear low down. Teeth in more advanced forms were heterodont, the lower jaw was flattened from side to side, and the dentary was relatively large. There were both a coracoid and a pre-coracoid. Permian to Jurassic.

synapsis (*Cyt.*). See **syndesis.**

synaptene (*Cyt.*). See **zygotene.**

synaptic mates (*Cyt.*). The pairs of homologous chromosomes which form a close association in synapsis.

synapticula (*Zool.*). Transverse rodlike connecting structures, as skeletal rods connecting gill bars in *Amphioxus* or septa in Corals.

synaptic vesicles (*Zool.*). Structures about 50 nanometres in diameter found in the bulb-shaped endings of presynaptic nerves at synapses, and probably concerned with the storage of the chemical transmitters.

Synaptida (*Zool.*). An order of *Holothuroidea*, having pinnate buccal tube-feet without ampullae but with retractor muscles; there are no tube-feet on the trunk; respiratory trees are lacking; the madreporite is internal; burrowing forms.

synapton (*Biol.*). A mechanical model of an activated complex capable of self-reproduction; a model constructed to imitate some of the properties of living material.

Synaptosauria (*Zool.*). A subclass of Reptiles containing a variety of extinct aquatic forms with a single temporal vacuity high in the skull (*parapsid* condition). Included the *Plesiosaurs*, some of which were over 12 m long.

synarthrosis (*Zool.*). An immovable articulation, especially an immovable junction between bones. Cf. *amphiarthrosis, diarthrosis*.

sync (*Cinema.*). See **synchronization.**

Syncarida (*Zool.*). A superorder of *Malacostraca* having no carapace, eyes stalked, sessile or absent, most of the thoracic limbs having exopodites and none being chelate or subchelate, no oöstegites, a tail fan, and simple caeca on the mid gut.

syncarp (*Bot.*). A multiple fleshy fruit.

syncarpous (*Bot.*). Said of a gynaeceum consisting of two or more united carpels.

syncaryon (*Biol.*). See **synkaryon.**

syncerebrum (*Zool.*). In *Arthropoda*, a cerebral ganglionic mass formed by the fusion of one or more of the ventral ganglia with the true primary cerebron consisting of supraoesophageal ganglia only.

synchondrosis (*Zool.*). Connexion of two bones by cartilage, usually with little possibility of relative movement.

synchro (*Telecomm.*). A general term used for a family of self-synchronous angle data transmitters and receivers. Also called **selsyn.**

synchrocyclotron (*Nuc. Eng.*). A *cyclotron* in which frequency modulated voltage is used to accelerate particles to relativistic energies at the expense of a pulsed output instead of a continuous output.

synchromesh gear (*Autos.*). A gear in which the speeds of the driving and of the driven members which it is desired to couple are first automatically synchronized by small *cone clutches* (q.v.) before engagement of the dogs or splines, thus avoiding shock and noise in gear-changing.

synchronism (*Telecomm.*). Said of two oscillations of the same frequency when the phase angle between them is zero.

synchronization (*Cinema.*). Linear adjustment of sound-track and mute in cutting rush prints; also of the negatives, when the separate prints are required to be reproduced together, the reproduced sound fitting exactly with the accompanying motion pictures. Colloquial abbrevs. **sync, sink.** *v.* **synchronize.** (*TV*) Adjustment of line and frame frequencies in a TV receiver so as to coincide with those at the transmitter, and the keeping of them so adjusted and locked.

synchronization of oscillators (*Telecomm.*). Phenomenon when two oscillators, having nearly equal frequencies, are coupled together. When the degree of coupling reaches a certain point, the two suddenly pull into step.

synchronized clock (*Horol.*). One driven by any means but having its accuracy corrected electrically at given intervals.

synchronizer (*An. Behav.*). See **zeitgeber.** (*Comp.*) A unit used to maintain synchronism when transmitting information between two devices. It may merely control the speed of one (e.g., by clutch); or if the speeds are very different, may include buffer storage. (*Elec. Eng.*) See **synchroscope.**

synchronizing (*Elec. Eng.*). The operation of bringing a machine into synchronism with an a.c. supply.

synchronizing modulation (*TV*). The range of modulation depth reserved for the synchronizing pulses, as distinct from that for picture (video) signals.

synchronizing power (*Elec. Eng.*). The power developed in a synchronous machine that keeps it in synchronism with the a.c. supply system to which it is connected.

synchronizing pulse (*TV*). That transmitted at the beginning and/or end of each frame and scanning line, to regulate synchronization.

synchronizing signals (*TV*). Those added between the video scanning signals, to trigger line and field scanning pulses in the receiver.

synchronizing torque (*Elec. Eng.*). The torque power that keeps a synchronous machine in synchronism with the a.c. supply system to which it is connected.

synchronizing valve (*TV*). One used for injecting the synchronizing pulses into a time-base circuit.

synchronizing wheel (*TV*). Rotating wheel, having a large moment of inertia, used for maintaining synchronism over the individual cycles of frame and line frequency oscillators.

synchronometer (*Elec. Eng.*). A device which counts the number of cycles in a given time. If the time interval is unity the device becomes a digital *frequency meter* (q.v.).

synchronous-asynchronous motor (*Elec. Eng.*). A slip-ring type of induction motor whose rotor is fed from a d.c. exciter coupled to it. The machine operates asynchronously during starting-up, and runs on load as a synchronous motor.

synchronous booster (*Elec. Eng.*). An a.c. generator coupled to a synchronous converter and having its armature connected in series with that of the converter.

synchronous camera (*Cinema.*). A motion-picture camera which is driven by a synchronous motor. In one sound-film system all such cameras, including sound-cameras, are switched on to the normal a.c. mains. In another sound-film system they are switched to special mains excited by a generator; as this generator is started from rest, it brings all cameras up to their synchronous speed, while keeping them interlocked.

synchronous capacitor (*Elec. Eng.*). A lightly loaded synchronous motor supplying a leading current for power-factor correction.

synchronous capacity (*Elec. Eng.*). The synchronizing power of an interconnector linking two a.c. power systems. It is defined as the change of kilowatts transmitted over the interconnector per radian change of angular displacement of the voltages of the two systems.

synchronous carrier system (*Radio*). Simultaneous broadcasting by two or more transmitters having the same carrier frequency, the various drive circuits being interlocked so as to avoid heterodyne beats between them.

synchronous clock (*Horol.*). One operated from the electric mains, the motor being in step with the mains frequency.

synchronous communications satellite (*Space*). See **communications satellite.**

synchronous computer (*Comp.*). One in which all operations are timed by a master clock, which

may be a separate pulse source or timed pulses reproduced from a track on a magnetic drum.
synchronous converter (*Elec. Eng.*). A synchronous machine for converting polyphase alternating current to direct current. It comprises a double-purpose armature, rotating within a salient-pole direct-current field system.
synchronous demodulator (*Radio*). Demodulator capable of deriving modulation components in phase synchronism with a local phase reference signal. Also called **synchronous detector.**
synchronous gate (*Comp.*). Time gate controlled by clock pulses and used to synchronize various operations in, e.g., data-processing systems.
synchronous generator (*Elec. Eng.*). See **alternator.**
synchronous homodyne (*Radio*). Reception in which the incoming modulated carrier has added to it a local oscillation of correct phase, with possible locking.
synchronous impedance (*Elec. Eng.*). The ratio of the open-circuit e.m.f. to the short-circuit current of a synchronous machine, both values having reference to the same field excitation.
synchronous induction motor (*Elec. Eng.*). An induction motor in which a direct current is passed into the rotor winding after it has run up to speed, so that, after starting as an induction motor (with a high starting torque), it runs as a synchronous motor.
synchronous machine (*Elec. Eng.*). An a.c. machine whose electrical frequency is independent of the load.
synchronous motor (*Elec. Eng.*). A.c. electric motor designed to run in synchronism with supply voltage.
synchronous orbit. See **stationary orbit.**
synchronous phase modifier (*Elec. Eng.*). A large synchronous machine used solely for varying the power factor at the receiving end of a transmission line to maintain the voltage constant under all conditions of loading.
synchronous reactance (*Elec. Eng.*). The vector difference between the synchronous impedance and the effective armature resistance of a synchronous machine.
synchronous spark gap (*Radio*). *Rotary spark gap* (q.v.) driven by a synchronous motor running from the same supply as that for the transformer furnishing the high voltage.
synchronous vibrator (*Elec. Eng.*). See **vibrator.**
synchronous watt (*Elec. Eng.*). A unit of torque used loosely in connexion with a.c. machines. It is defined as the torque which, at the synchronous speed of the machine, would develop a power of one watt.
synchroscope (*Elec. Eng.*). An instrument indicating the difference in frequency between two a.c. supplies. Also **synchronizer.**
synchrotron (*Nuc. Eng.*). Machine for accelerating charged particles in a vacuum; they are guided by an alternating magnetic field, as in a *betatron*, with acceleration by a high-frequency electric field at one point in their orbit.
synchrotron radiation (*Phys.*). A theory of the origin of cosmic radio waves; it is suggested that electrons moving in an orbit in a magnetic field are accelerated as in a synchrotron, but on a vastly larger scale.
synchysis (*Med.*). Abnormal softening and fluidity of the vitreous body of the eye.
syncline (*Geol.*). A flexure found in sedimentary rocks, the form of which is concave upwards.
synclitism (*Med.*). The compensatory difference in the rates of descent of the anterior and posterior portions of the presenting foetal part in the pelvis during labour.
syncope (*Med.*). A fainting attack or sudden loss of consciousness due to sudden anaemia of the brain, as a result of an unstable nervous system or of heart failure.
syncotyly (*Bot.*). Union of cotyledons.
syncraniate (*Zool.*). Having vertebral elements fused with the skull.
syncranterian (*Zool.*). Having the teeth in a continuous row, with no gap between anterior and posterior sets, as some Snakes. Cf. *diacranterian.*
syncryptic (*Zool.*). Showing superficial resemblance through protective mimicry of surrounding objects, without any more fundamental relationship.
syncytium (*Zool.*). A multinucleate cell; a tissue complex in which there are many nuclei but no distinguishable cell walls, arising from division or concrescence of cells. *adj.* **syncytial.**
syndactyl (*Zool.*). Showing fusion of two or more digits, as some Birds. *n.* **syndactylism.**
syndesis (*Cyt.*). In meiotic nuclear division, fusion of homologous chromosomes. Also **synapsis.**
syndesmochorial placenta (*Zool.*). A chorio-allantoic placenta in which the uterine epithelium disappears so that the chorion is in contact with the endometrium or glandular epithelium of the uterus. Usually cotyledonary, e.g., sheep.
syndesmosis (*Zool.*). Connexion of two bones by a ligament, usually with little possibility of relative movement.
syndiazo compounds (*Chem.*). The stereoisomeric forms of diazo compounds in which the groups attached to the nitrogen atoms are on the same side of the plane of the double bond.
syndiotactic (*Plastics*). Term describing very long chain hydrocarbon polymers in which the substituent groups lie alternately on each side of the main chain. See also **atactic, isotactic.**
syndrome (*Med.*). A concurrence of several symptoms or signs in a disease which are characteristic of it; a set of concurrent symptoms or signs.
syndyotaxy (*Chem.*). Polymerization exhibiting regular alternation of differences in stereochemical structure. *adj.* **syndyotactic.** Cf. *isotaxy.*
synechia (*Med.*). A morbid adhesion of the iris of the eye to the cornea or to the lens.
synechthran (*Ecol.*). In a community of social Insects (especially Ants or Termites), a scavenging or predatory species living within the community and arousing marked hostility.
synecology (*Ecol.*). The study of the ecology of a group of organisms associated together as a unit. Cf. *autecology.*
synema (*Bot.*). A group of united filaments when the stamens in a flower are monadelphous.
Synentognathi (*Zool.*). An order of *Teleostei* having a distinct lateral line, median fins far back, lower pharyngeals united into a single bone; some forms have the jaws elongated into a beak; mainly marine forms. Skippers, Garfishes, Flying Fishes.
syneresis (*Chem.*). Spontaneous expulsion of liquid from a gel.
synergic, synergetic (*Zool.*). Working together; said of muscles which cooperate to produce a particular kind of movement.
synergidae, synergids (*Bot.*). Two naked cells which lie in the embryo sac at the end toward the micropyle, and appear to play some part in guiding the tip of the pollen tube towards the egg nucleus.
synergism (*Ecol.*). A type of *social facilitation*

(q.v.) in which the nearby presence of another organism enhances the efficiency or intensity of a physiological process or behaviour pattern in an individual.

synergist (*Chem.*). Substance which increases the effect of another.

syngameon (*Zool.*). A category of individuals based on the type of reproduction.

syngamiasis (*Vet.*). Infection of the trachea, bronchi, and lungs of Birds by the nematode worm *Syngamus trachea*.

syngamy (*Biol.*). Sexual reproduction; fusion of gametes.

syngenesious (*Bot.*). Said of anthers which are united laterally so that they form a hollow tube around the style.

syngenesis (*Biol.*). Sexual reproduction; reproduction distinguished by the fusion of male and female elements. *adj.* **syngenetic.**

syngenetic (*Min.*). A category of ore bodies comprising all those which were formed contemporaneously with the enclosing rock. Cf. *epigenetic*, i.e., formed at some subsequent time.

syngnathous (*Zool.*). Of certain Fish, having the jaws fused to form a tubular structure.

syngynous (*Bot.*). See **epigynous.**

synizesis (*Cyt.*). In meiotic prophase, contraction of the chromatin towards one side of the nucleus, obscuring the individual loops.

synkaryon, syncaryon (*Bot.*). A pair of nuclei in close association in a hyphal segment, and dividing at the same time to give daughter nuclei which associate in similar manner. (*Zool.*) A zygote nucleus resulting from the fusion of two pronuclei.

synkinesia (*Med.*). The occurrence simultaneously of both voluntary and involuntary movements, e.g., in the movement of a partly-paralysed muscle in conjunction with the voluntary movement of healthy muscle.

synnema (*Bot.*). An erect bunch of hyphae.

synodic month (*Astron.*). The interval (amounting to 29·530 59 days) between two successive passages of the moon through conjunction or opposition respectively; therefore, the period of the phases.

synodic period (*Astron.*). An interval of time between two similar positions of the moon or a planet, relative to the line joining the earth and sun; hence the length of time from one conjunction or opposition to another, and the period of the phases of the moon or a planet.

synoekete (*Ecol.*). In a community of social Insects (especially Ants or Termites), an indifferently tolerated guest species living in the community without attracting notice.

synoicous (*Bot.*). Having antheridia and archegonia together in the same group, surrounded by an involucre.

synoptic chart (*Meteor.*). See **weather map.**

synoptic meteorology (*Meteor.*). That part of the science of meteorology which deals with the preparation of a *synoptic chart* of the observed *meteorological elements* (q.v.) and, from consideration of this chart, the production of a weather forecast.

synosteosis (*Zool.*). See **ankylosis.**

synostosis (*Zool.*). A union between two bones which is so perfect that no trace of original separableness remains.

synotic tectum (*Zool.*). That part of the chondrocranium which forms a roof over the hind-brain between the otic capsules, and later ossifies in higher forms to form the supra-occipitals and exoccipitals.

synovia (*Zool.*). In Vertebrates, *glair*-like lubricating fluid, occurring typically within tendon sheaths and the capsular ligaments surrounding movable joints.

synovial membrane (*Zool.*). The delicate connective tissue layer which lines a tendon sheath or a capsular ligament, and is responsible for the secretion of the *synovia* (q.v.).

synoviparous (*Zool.*). Secreting synovia.

synovitis (*Med.*). Inflammation of the synovial membrane of a joint.

synsacrum (*Zool.*). In Birds, part of the pelvic girdle formed by the fusion of some of the dorsal and caudal vertebrae with the sacral vertebrae.

syntannins (*Chem., Leather*). Synthetic tannins compounded usually from formaldehydes and phenolsulphuric acids.

syntechnic (*Zool.*). Said of unrelated forms showing resemblance due to environmental factors; convergent.

syntenosis (*Zool.*). Union of bones by means of tendons, as in the phalanges of the digits.

synthetic papers (*Paper*). A term denoting papers made from manmade fibres, such as nylon, etc. Usually they have good strength and chemical resistance.

synthetic-resin adhesive (*Plastics*). One made from a thermosetting resin, e.g., urea or phenol formaldehyde, and used with an accelerator to regulate setting conditions, or from a thermoplastic resin, e.g., polymethyl methacrylate or polyvinyl acetate.

synthetic resins (*Chem.*). Resinous compounds made from synthetic materials, as by the condensation or polymerization of phenol and formaldehyde, formaldehyde and urea, glycerol and phthalic anhydride, polyamides, vinyl derivatives, etc. See **phenolic resins, thermosetting compositions, vinyl polymers.**

synthetic rubber (*Plastics*). Any of numerous artificial rubberlike compounds, e.g., polymers of isoprene or its derivatives, copolymers of vinyl acetate and vinyl chloride. In some respects they surpass natural rubber. See **Buna, butyl rubber, Neoprene, Thiokol, vinyl polymers.**

synthetic ruby, synthetic sapphire (*Min.*). In chemical composition and in all their physical characters, including optical properties, these stones are true crystalline ruby or sapphire; but they are produced in quantity in the laboratory by fusing pure precipitated alumina with the predetermined amount of pigmentary material. They can be distinguished from natural stones only by the most careful expert examination.

synthetic sands (*Foundry*). Sands deficient in clay which have been blended with bentonite or other claylike material to make them suitable for moulding.

synthetic spinel (*Min.*). This is produced, in a wide variety of fine colours, by the *Verneuil process*; in chemical and optical characters identical with natural magnesian spinel, it is widely used as a gemstone.

synthetic stop (*Acous.*). In an organ, the production of low-pitched notes by sounding two low notes together with a musical interval of a fifth, the impression of the desired difference tone being subjective.

syntonic jars (*Elec. Eng.*). Two similar Leyden jars each fitted with spark gap and equal lengths of conductor for connecting thereto. When one jar is charged and then discharged across the gap, a spark appears at the gap connected to the other jar. Used by Lodge to demonstrate the principle of *tuning* or *syntony.*

syntony (*Elec. Eng.*). See **current (or voltage) resonance.**

synusia (*Bot.*). An ecological unit based on the life-forms of the plants growing in company.
syphilide, syphilid (*Med.*). Any skin affection caused by syphilis. Also **syphiloderm, syphiloderma.**
syphilis (*Med.*). A contagious venereal disease due to infection with the microorganism *Spirochaeta pallida* (*Treponema pallidum*); contracted in sexual intercourse, by accidental contact, or (by the foetus) from an infected mother.
syphiloderm, syphiloderma (*Med.*). See **syphilide.**
syphiloma (*Med.*). A syphilitic tumour. See **gumma.**
syphon. Same as *siphon.*
Syrian garnet (*Min.*). A name for *almandine* (q.v.) of gemstone quality.
syringitis (*Med.*). Inflammation of the Eustachian tube.
syringium (*Zool.*). In some Insects, a tubular structure for the ejection of repellent liquid. *adj.* **syringial.**
syringobulbia (*Med.*). A disease characterized by increase of neuroglia and the presence of cavities in the medulla oblongata, giving rise to such nervous phenomena as paralysis of the palate, pharynx, and larynx. See also **syringomyelia.**
syringomyelia (*Med.*). A chronic, progressive disease of the spinal cord in which increase of neuroglia and the formation of irregular cavities cause paralysis and wasting of muscles and loss of skin sensibility to pain and to temperature. See also **syringobulbia.**
syringomyelocele (*Med.*). A form of spina bifida in which the part protruding through the defective spinal column consists of the greatly distended central canal of the spinal cord.
syrinx (*Med.*). A fistula, or a fistulous opening. (*Zool.*) (1) The vocal organs in Birds, situated at the posterior end of the trachea. (2) A portion of the internal skeleton of certain *Brachiopoda*. *pl.* **syringes.** *adj.* **syringeal.**
systaltic (*Zool.*). Alternately contracting and dilating; pulsatory, as the movements of the heart. Cf. *peristaltic*. *n.* **systalsis.**
system. Generally, anything formed of parts placed together or adjusted into a regular or connected whole. (*Biol.*) (1) Tissues of the same histological structure, e.g., the osseous system. (2) Tissues and organs uniting in the performance of the same function, e.g., the digestive system. (3) In *Urochorda* (*Tunicata*), a colony of individuals possessing a common cloaca, e.g., *Botryllidae*. (4) A method or scheme of classification, e.g., the *Linnaean system*. (5) A systematic treatise on the animal or plant kingdom or any part of either. *adj.* **systematic.** (*Chem., etc.*). A portion of matter, or a group or set of things that forms a complex or connected whole. (*Elec. Eng.*) General term covering entire complex of apparatus involved in the transmission and distribution of electric power. (*Geol.*) (1) Name given to the succession of rocks which were formed during a certain period of geological time, e.g., *Jurassic System*. (2) Term applied to the sum of the phases which can be formed from one or more components of minerals under different conditions of temperature, pressure, and composition.
systematic errors (*Civ. Eng., Maths., etc.*). Errors which are always in the same direction, i.e., errors which are always positive or always negative. Sometimes known as **cumulative errors.** In calculations such errors can arise, e.g., by always rounding fives upwards.
systematics (*Biol.*). The branch of biology which deals with classification and nomenclature.
system building (*Build., Civ. Eng.*). Methods of construction aimed at increasing speed of construction by preparing component parts of the building in advance of commencing to piece it together. Sometimes called **industrialized building.**
system engineering (*Automation*). The approach to automation in which the design is based on general consideration of the required process and the available control elements, rather than on replacing manual operators by automatic devices which function in a similar manner.
Système International d'Unités. See **SI Units.**
systemic (*Zool.*). Pertaining to the body as a whole, not localized, as the *systemic* circulation.
systemic arch (*Zool.*). In Vertebrates, the main vessel or vessels carrying blood from the heart to the body as a whole.
systemic heart (*Zool.*). In Birds and Mammals, the auricle and ventricle of the left side of the heart, which supply blood to the body generally. Cf. *respiratory heart*.
systemic infection (*Bot.*). The condition when a parasitic fungus perennates in the perennial parts of its host, and passes from them into the new shoots developed during each season of growth.
systems analysis. Complete analysis of all phases of activity of an organization, and development of a detailed procedure for all collection, manipulation and evaluation of data associated with the operation of all parts of it.
systems of crystals (*Crystal.*). The 7 large divisions into which all crystallizing substances can be placed, viz., cubic, tetragonal, hexagonal, trigonal, orthorhombic, monoclinic, triclinic. This classification is based on the degree of *symmetry* (q.v.) displayed by the crystals.
systole (*Bot., Zool.*). Rhythmical contraction, as of the heart, or of a contractile vacuole. Cf. *diastole*. (*Cyt.*) Collapse of the nucleus and outflow of nuclear material into the cytoplasm during mitosis, due to dissolution of the nuclear membrane.
systrophe (*Bot.*). Clumping of chloroplasts when exposed to very bright light.
systyle (*Arch.*). A colonnade in which the space between the columns is equal to twice the lower diameter of the columns.
syzygy (*Astron.*). A word (derived from the Greek *syzygia*, close union) applied to the moon when in conjunction or opposition. (*Zool.*) An association of gregarine individuals adhering in strings; in *Crinoidea*, a close suture of two adjacent brachia. *adj.* **syzygial.**
Szerelmey's Stone Liquid (*Build.*). TN for a preserving agent for building-stones.
Szilard-Chalmers process (*Nuc.*). One in which a nuclear transformation occurs with no change of atomic number, but with breakdown of chemical bond. This leads to formation of free active radicals from which material of high specific activity can be separated chemically.

T

t. Symbol for tonne (metric *ton*).
t*- (*Chem.*).** An abbrev. for (1) ***trans*-**, **i.e.**, containing the two radicals on opposite sides of the plane of a double bond or alicyclic ring; (2) ***tertiary, i.e., substituted on a carbon atom which is linked to three other carbon atoms.
T. Symbol for *tera*, 10^{12}. (*Elec.*). Symbol for *tesla*.
T* (*Chem.*).** With subscript, a symbol for ***transport number.
***T*- (*Chem.*).** A symbol indicating the presence of a triple bond which begins on the corresponding carbon atom.
2, 4, 5-T (*Chem.*). 2,4,5-trichlorophenoxyethanoic acid:

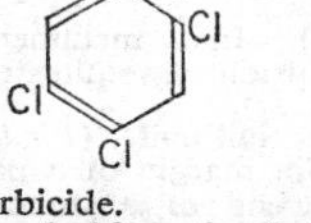

$O.CH_2.COOH$ Cl Cl Cl

Used as a herbicide.
τ (*Chem.*). A symbol for a time interval, especially half-life or mean life.
Ta (*Chem.*). The symbol for *tantalum*.
tab (*Aero.*). Hinged rear portion of a flight control surface. See **balance-, servo-, spring-,** and **trimming-.**
TAB (*Med.*). See **paratyphoid.**
tabashir, tabasheer (*Bot.*). A mass of silica found in the stems of bamboos.
tabby weave (*Weaving*). See **plain weave.**
tabescent (*Bot.*). Shrivelling.
tabes dorsalis (*Med.*). Locomotor ataxia (ataxy). A disease of the nervous system marked by attacks of pain in the legs, anaesthesia of certain areas of the skin, ataxia, loss of the pupil reflex to light, and other nervous affections; due to degenerative changes in the nerves, especially of the sensory roots of the spinal cord, as a late result of syphilis.
tabes mesenterica (*Med.*). Tuberculous infection of the lymphatic glands in the abdomen, causing enlargement of the abdomen, diarrhoea, anaemia, wasting of the body, and weakness.
tabetic (*Med.*). Pertaining to, affected by, or caused by tabes dorsalis; an affected person.
table (*Carp.*). See **coak.** (*Eng.*) The horizontal portion of a drilling, milling, shaping or other machine, which supports the workpiece and is adjustable relative to the tool. (*Maths., etc.*) A collection of data (e.g., square-root values) laid out in rows and columns for reference, or stored in a computer memory as an array.
table matter (*Typog.*). See **tabular matter.**
table of strata (*Geol.*). A column which depicts a series of rocks arranged in chronological order, the oldest being at the bottom. It is usual to draw this to scale so that the average thicknesses of the beds are also shown.
table rolls (*Paper*). Rolls from 5 to 30 cm (2–12 in) in diameter, which carry the paper machine wire, and also assist in the removal of water by capillary action.
tablet (*Acous.*). A tilting key for operating stops on the console of an organ, with or without illuminated indication of position. (*Plastics*) A piece of moulding composition of the correct weight and density, and of suitable diameter and thickness to fit the mould; not preformed to the approximate shape of the moulding.
tabling (*Carp.*). Operation of shaping a *coak* (q.v.).
tabloid newspaper (*Print.*). A newspaper with a smaller page size, usually half the normal, using a cross fold to form the spine.
taboparesis, taboparalysis (*Med.*). General paresis and tabes dorsalis affecting the same person.
tabula (*Zool.*). In *Hydrocorallinae* and certain Corals, a horizontal calcareous partition completely shutting off the lower part of the polyp cup. Cf. *dissepiment*.
tabular (*Bot., etc.*) Horizontally flattened.
tabulare (*Zool.*). A paired membrane bone of the skull in some Vertebrates, lying above the otic capsule.
tabular matter (*Typog.*). Type arranged in accurately spaced columns; if there are rules between the columns, it is *table matter*. Can be set in one measure by hand, by Monotype or by filmsetters, employing a 'set' system, but not easily by linecasting systems.
tabular spar (*Min.*). See **wollastonite.**
tabulator (*Comp.*). Machine which prints a line up to 100 characters at a time, on continuous paper, at a speed of, e.g., 900 lines a minute, controlled by punched cards or by the output signals of an electronic computer.
tacan (*Nav.*). *Tac*tical *a*ir *n*avigation system (U.S.) using ultra-high radio frequencies. The airborne equipment directly indicates both bearing and range for any station to which it may be tuned.
tacheometer or **tachymeter (*Surv.*).** Theodolite fitted with telescope which measures distances by sighting on levelling staff, the space between two crosshairs in the eyepiece being an index (after correction) of the distance concerned. See **additive constant, multiplying constant.**
tacheometry (*Surv.*). The process of surveying and levelling by means of angular measurements from a known station, combined with determination of distances from the station.
tachometer (*Eng.*). An instrument for measuring speed of rotation.
tachometric electrometer (*Elec. Eng.*). An electrometer for measuring very small currents; it is kept at a fixed reading by compensating for its loss of charge by periodic charges from a capacitor, these being controlled by a variable-speed motor, the speed of which is observed.
tachyauxesis (*Biol.*). Growth of a part at an accelerated rate in relation to the whole.
tachycardia (*Med.*). Frequent action of the heart, beyond the normal rate.
tachygenesis (*Zool.*). Accelerated development with elimination of certain embryonic stages, as in some Caecilians, in which the free-living tadpole stage is suppressed. *adj.* **tachygenetic.**
tachylite, tachylyte (*Geol.*). A black glassy igneous rock of basaltic composition, which occurs as a chilled margin of dykes and sills. In Hawaii it forms the bulk of certain lava-flows.
tachymeter (*Surv.*). See **tacheometer.**
tachyon (*Nuc.*). A theoretical particle moving faster than the speed of light.
tachypnoea, tachypnea (*Med.*). Excessive frequency of respiration.
tachysporous (*Bot.*). Said of a plant which liberates its seeds quickly.
tachytely (*Biol.*). Abnormally fast evolution.

tacitron (*Electronics*). Thyratron in which extinction is effected by small negative bias.

tack (*Build.*). A small clout nail. (*Print.*) A quality in printing inks and the rollers of inking systems, required for good ink coverage, which can be measured empirically and chosen in both by formulation; in some rollers it deteriorates with age.

tack marks (*Bind.*, *Print.*). (1) Small dots printed in the centre of a back margin to indicate to the binder the lay edges of the sheet, two dots indicating the second side of a work-and-turn sheet; only seen occasionally. (2) A dot printed to overlap the side lay edge of the sheet when the first printing is being run off, so that it shows on the edge of the pile of sheets, to indicate the lay required for subsequent printings and which will be cut off when the job is trimmed. When wood furniture is used the mark may be made literally by a tack.

tack welding (*Eng.*). Making short, provisional welds along a joint to hold it in position and prevent distortion during subsequent continuous welding.

tacky (*Paint.*). Said of paint or varnish which has not quite dried and is in a sticky condition.

Taconic revolution (*Geol.*). A period of intense folding which affected the eastern parts of N. America at the end of the Ordovician period. The effects are best seen in the Taconic Mts on the borders of New York State and Massachusetts.

tacticity (*Chem.*). The stereochemical arrangement of units in the main chain of a polymer.

tactile (*Zool.*). Pertaining to the sense of touch.

tactile bristle (*Bot.*). A stiff hair which transmits a contact stimulus.

tactile perception (*Physiol.*). The perception of vibration by the sense of touch; developed particularly in deaf persons, who can be trained to detect and interpret vibrations in another person's larynx, or to interpret vibrations applied selectively to their fingers by vibrators operated through filters by a microphone.

tactile pit (*Bot.*). A sharply defined area of thin cell wall in an epidermal cell of a tendril, which appears to be concerned in the perception of pressure.

tactite (*Geol.*). See **skarn.**

tactoid (*Chem.*, *Phys.*). A droplet of non-spherical rod-shaped or flat particles, exhibiting double refraction, capable of parallel orientation and appearing in colloidal solutions.

tactosol (*Chem.*, *Phys.*). Sol containing *tactoids.*

taenia (*Zool.*). A ribbon-shaped structure, such as the *taenia pontis*, a bundle of nerve fibres in the hind-brain of Mammals.

taeniasis (*Med.*). The state of infestation of the human body with tapeworms (*Taenia*), which as adults may inhabit the intestine and as larvae the muscles and other parts of the body.

taenidia (*Zool.*). Thickenings of the *endotrachea* (q.v.) which keep the trachea distended and generally take the form of a spiral thread. *sing.* **taenidium.**

taenioles (*Zool.*). The ribbon-shaped longitudinal gastric ridges of a scyphula.

taenite (*Min.*). A solid solution of iron and nickel occurring in iron meteorites; it appears as bright white areas on a polished surface. It crystallizes in the cubic system and has 27% to 65% Ni.

taffeta (*Textiles*). A lightweight plain weave fabric produced from silk, rayon, or nylon yarns; also in cotton and worsted. Usually stripe or check pattern, used for blouses, etc.

taffrail log (*Ships*). See **nautical log.**

taft joint (*Plumb.*). A joint between the ends of two lengths of lead pipe, one of which is rasped to a smaller size so as to fit into the opened-out end of the other, the whole being soldered with a blowlamp.

tag (*Comp.*). See **sentinel.**

tagatose (*Chem.*). A *ketohexose.*

tag block (*Elec. Eng.*). Terminal block, holding varying numbers of double-ended solder tags, which is fitted to every panel of apparatus supported on standard apparatus racks. External wiring to a unit can then be connected without interference with the internal wiring, which is completed during manufacture. External connexions to bays of apparatus are also made to tag blocks mounted at the top of the racks by cable forms.

tagged atom (*Nuc.*). See **labelled atom**; also **radioactive tracer.**

tagger (*Met.*). Thin sheet-iron or tin-plate.

tagma (*Zool.*). A distinct region of the body of a metameric animal, formed by the grouping or fusion of somites; as the thorax of an Insect. *pl.* **tagmata.**

tagmosis (*Zool.*). In a metameric animal, the grouping or fusion of somites to form definite regions (*tagmata*).

tail (*Aero.*). See **tail unit.** (*Bind.*, *Typog.*) The bottom or foot margin of a page or volume. (*Build.*) That end of a stone step which is built into a wall. (*Min. Proc.*) U.S. term for *tailings* (q.v.). Also **tails.** (*Zool.*) See **cauda.**

tailband (*Bind.*). See **headband.**

tail-bay (*Hyd. Eng.*). The part of a canal lock immediately below the tail gates.

tail beam (*Carp.*). A floor joist which at one end is framed into a trimmer.

tail boom (*Aero.*). One or more horizontal beams which support the tail unit where the fuselage is truncated; commonly used for cargo aeroplanes to facilitate loading trucks and bulky freight through full-width rear doors.

tail cap (*Bind.*). See **head cap.**

tail chute (*Aero.*). A parachute mounted in an aircraft tail; cf. *antispin parachute* and *brake parachute.*

tail cone (*Aero.*). The tapered streamline *fairing* (q.v.) which completes a fuselage or *tail boom* (q.v.).

tailed (*Bot.*). Bearing a long slender point.

tail-first aeroplane (*Aero.*). An aeroplane in which the horizontal stabilizer (i.e., tailplane) is mounted ahead of the main plane; common on pioneer aeroplanes and re-introduced for supersonic flight; sometimes **canard.**

tail gates (*Hyd. Eng.*). The gates at the low-level end of a lock.

tail heaviness (*Aero.*). The state in which the combination of forces acting upon an aircraft in flight is such that it tends to pitch up.

tailing (*Build.*). Operation of building in and fixing the end of a member which projects from a wall. Also **tailing in, tailing down.** (*TV*) Lag in response from black to white in facsimile reproduction.

tailing hangover (*TV*). Blurring of reproduced picture because of slow decay in electronic circuits. Also **submerged resonance.**

tailing iron (*Build.*). A steel section built into a wall, across the top of the encastered end of a projecting member.

tailings (*Min. Proc.*). (1) Rejected portion of an ore; waste. (2) Portion washed away in water concentration. May be impounded in a tailings dam or pond, or stacked dry on a dump.

tailings dam (*Min. Proc.*). One used to hold mill residues after treatment. These arrive as fluent slurries. Dam may include arrangements for run-off or return of water after the slow-settling solids have been deposited.

tail joist (*Carp.*). See **tail beam.**

tailless aircraft (*Aero.*). An aeroplane, or glider, in which longitudinal stability and control in flight is achieved without a separate balancing horizontal aerofoil. This balance is achieved by *sweep* (q.v.), and many *delta-wing* (q.v.) aeroplanes are tailless because their sharp angle of sweepback renders a tail plane unnecessary.

tailpiece (*Print.*). An engraving, design, etc., occupying the bottom of a page, as at the end of a chapter.

tail plane (*Aero.*). A horizontal surface, fixed or adjustable, providing longitudinal stability of an aeroplane or glider. See **stabilizer.**

tail race (*Hyd. Eng.*). A channel conveying water away from a hydraulically-operated machine. (*Mining*) The launder or trough for the discharge of water-borne tailings.

tail rotor (*Aero.*). See **auxiliary rotor.**

tail slide (*Aero.*). A difficult aerobatic in which an aeroplane is pulled up into a zoom and allowed to slide backward along its longitudinal axis after the vertical speed drops to zero.

tail-stock (*Eng.*). A casting mounted slidably on a lathe bed. It carries a spindle in true alignment with the centre of the head-stock, longitudinally adjustable, and coned internally to receive a centre. See **lathe.**

tail trimmer (*Carp.*). A trimmer close to a wall, used in cases where it is not desired to build the joists into the wall.

tail unit (*Aero.*). Hindmost parts of an aircraft, usually the horizontal *tailplane* of an aeroplane. Also **empennage.**

tail wheel landing gear (*Aero.*). That part of the alighting gear taking the weight of the rear of the plane when on the ground. It consists of a shock-absorber carrying a wheel (*tail wheel*) or a shoe (*tail skid*).

take (*Cinema., TV*). The unit of registration technique in shooting sound-film or transmitting television images. A number of takes are separate samples of the same shot, the best take being selected for the final sequence in sound-film production. Each take therefore corresponds to a continuous sequence in the rushes. (*Typog.*) A suitable portion of the copy which has been divided up for sharing among the hand or machine compositors.

take-off rocket (*Aero.*). A rocket, usually jettisonable, used to assist the acceleration of an aeroplane. Cordite rockets were introduced for naval aeroplanes during the Second World War and replenishable liquid-fuel rockets, some with controllable thrust, thereafter. Sometimes referred to as a **booster rocket,** although strictly this is for the acceleration of missiles. Abbrev. **RATOG** (*r*ocket-*a*ssisted *t*ake-*o*ff *g*ear) and U.S. abbrev. **JATO** (*j*et-*a*ssisted *t*ake-*o*ff).

take-off sprocket wheel (*Cinema.*). The sprocket which accepts the film after it has come off the constant-speed drive sprocket, or impedance wheel. This is driven synchronously, and protects the constant-speed drive from the pull on the film exerted by the take-up reel.

take-up, take-up reel (*Cinema.*). The drive and the reel which is necessary to accept the cinematograph film after exposure in the gate of a camera or projector. It is driven through a friction drive, so that adequate tension is maintained in pulling the film from the take-off sprocket wheel.

taking-off (*Build., Civ. Eng.*). The first process involved in drawing up a bill of quantities. It consists of measuring the various dimensions of each item on the drawing, and entering them in a systematic manner on sheets specially ruled for the purpose.

taking out turns (*Typog.*). Inserting the correct characters for those substituted and turned upside down (*turns*) when the required sorts were unavailable during the original setting.

talandic (*Cyt.*). Applied to rhythmic changes in a quantity or state within a cell, postulated as part of the mechanism of metabolic and morphological control. In medical usage applied to periodic fevers.

talbe (*Radio*). Air sea rescue system using V.H.F. continuous waves. (*T*alk *a*nd *l*isten *be*acon.)

talbot (*Light*). Unit of luminous energy, such that 1 lumen is a flux of 1 talbot per second.

Talbot process (*San. Eng.*). An anticorrosion process applied to cast-iron pipes; they are coated internally with a mixture of bitumen and a hard siliceous material, which is distributed over the surface by centrifugal action during rapid rotation.

talc (*Min.*). A monoclinic hydrous silicate of magnesium, $Mg_6Si_8O_{20}(OH)_4$. It is usually massive and foliated and is a common mineral of secondary origin associated with serpentine and schistose rocks; also found in metamorphosed siliceous dolomites. Purified talc is used medically and in toilet preparations. See **steatite.**

talipes (*Med.*). Club-foot. A general term for a number of deformities of the foot: *talipes calcaneus*, in which the toes are drawn up from the ground and the patient walks on the heel; *talipes equinus*, in which the heel is drawn up and the toes point downwards; *talipes equinovarus*, in which the foot is inverted and turned inwards, with the toes pointing down; *talipes valgus*, in which the foot is abducted and everted, so that the patient walks on its inner side.

talk-back (or **speak-back**) **circuit** (*TV*). One which enables the controller of a programme to give directions to those originating a performance or rehearsal in a studio or location.

talk-down (*Radar*). See **ground-controlled approach.**

tall-boy (*Build.*). A fitting added to the top of a chimney to prevent down-draught.

tallow wood (*For.*). *Eucalyptus microcorys*, a native of New South Wales, Fraser Island and Queensland; it is durable, impermeable to preservative fluids, and difficult to work. It will not glue well because of its greasy nature.

tally (*Surv.*). A brass tag attached to a chain at every tenth link, and so marked or shaped as to enable the position of the tally along the chain to be immediately read. Also called a **teller.**

talon (*Arch.*). An ogee moulding. (*Zool.*) A sharp-hooked claw, as that of a bird of prey.

talose (*Chem.*). An *aldohexose*.

talus (*Civ. Eng.*). An earthwork or batter wall slope. (*Geol.*) See **scree.** (*Zool.*) A synonym for **astragalus.** *pl.* **tali.**

talus wall (*Build.*). A wall the face of which is built on a batter.

tamarugite (*Min.*). A hydrated sulphate of sodium and aluminium, crystallizing in the monoclinic system.

tambour lace (*Textiles*). An embroidery on net or muslin; a variety of Limerick lace.

Tamman's temperature (*Heat*). The temperature at which the mobility and reactivity of the molecules in a solid become appreciable. It is approximately half the melting-point in kelvins.

tamp (*Civ. Eng., Mining*). (1) To fill a charged shot-hole with clay or other stemming material to confine the force of the explosion. See **stemming**. (2) To ram or pound down ballast on a railway track, or road-metal. See also **punning**.

tamper (*Nuc. Eng.*). Heavy casing placed round core of nuclear weapon to delay expansion and act as a neutron reflector.

tampin (*Plumb.*). A conical plug of boxwood used in opening out the end of a lead pipe.

tampon (*Surg.*). A plug or packing made of gauze, cotton-wool, and the like, for insertion into orifices or cavities (especially the vagina and uterus) for the control of haemorrhage, the removal of secretions, or the dilating of passages.

tamponade (*Surg.*). The surgical use of the tampon.

tandem (*Telecomm.*). Connexion of output of one four-terminal network to input of a second.

tandem dialling (*Teleph.*). The dialling, by subscriber or operator, of the required subscriber's number, preceded by two or more code digits to route the call via one or more automatic exchanges.

tandem disk harrow (*Agric.*). A disk harrow comprising four gangs of two rows in tandem, in which each gang can be angled in opposite directions.

tandem engine (*Eng.*). An engine in which the cylinders are arranged axially, or end to end, with a common piston rod.

tandem exchange (*Teleph.*). One used primarily as a switching point for traffic between other exchanges.

tandem generator (*Elec. Eng.*). One equivalent to two Van de Graaff generators in series so that twice the normal voltage is obtained.

tandem knife-switch (*Elec. Eng.*). A switch in which two or more blades are mechanically coupled in order to operate simultaneously as a multiple-pole switch.

tandem selection (*Teleph.*). Selection of outlets by two uniselectors in series, so that the maximum possible availability of outlets is obtained.

tandem working (*Teleph.*). The using of an intermediate exchange during the transition period when a manual system is being converted to automatic working. Working is effected by trains of impulses which are set up on key senders on instructions from A-operators in originating exchanges.

tang (*Carp., Join., etc.*). The end of a tool which is driven into or about its haft. See **socket chisel**.

tangent (*Maths.*). (1) See **trigonometrical functions**. (2) A line or plane which touches a curve or surface. See **moving trihedral**. The problem of finding tangents was first considered by the Greeks and is now solved by the differential calculus.

tangent distance (*Surv.*). The distance between the intersection point and one of the tangent points of a railway or highway curve.

tangent galvanometer (*Elec. Eng.*). A vertical circular coil with its plane parallel to the meridian. If I is the current flowing, and r the effective coil radius in cm, then

$$I=\frac{rH}{50n}\tan\theta \text{ amperes,}$$

where θ is the angle of deflection of a magnetometer needle placed at the centre of the coil, n is the number of turns in the coil, and H the horizontal component of the earth's magnetic field in amperes per metre.

tangential field (*Optics*). The image surface formed by the tangential foci of a series of object points lying in a plane at right angles to the axis.

tangential focus (*Optics*). The focus of an object point lying off the axis of an optical system, in which the image is drawn out by the astigmatism of the system into a line tangential to a circle centred on the optic axis.

tangential keys (*Eng.*). A pair of keys fitted in such a way that one side is entirely sunk in the shaft and the other in the hub, the two keys facing in opposite directions; used for heavy drives. See **key**.

tangential longitudinal section (*Bot.*). A section taken along the length of a plant member, parallel to a tangent to its surface.

tangential wave path (*Meteor.*). That of a direct wave, tangential to the surface of the earth and which is curved by atmospheric refraction.

tangent point (*Surv.*). The point of commencement, or of termination, of a railway or highway curve.

tangent scale (*Elec. Eng.*). The scale of an electrical instrument in which the measured quantity varies as the tangent of the angle of deflection.

tangent screw (*Surv.*). A screw by which a fine adjustment may be made to the setting of a theodolite about its axis, either in order to bring the line of sight into coincidence with a signal, or to adjust the vernier reading to a given value.

tangoreceptor (*Biol.*). A specialized structure sensitive to slight differences of pressure.

tanh (*Maths.*). See **hyperbolic functions**.

tank (*Cinema.*). (1) Colloquialism for a portable camera booth used in motion-picture production. (2) (also *Photog.*) A receptacle for solutions in the development of films. (*Telecomm.*) Said of any storage, e.g., signals in mercury delay line, oscillations in tuned circuit.

tank circuit (*Telecomm.*). Section of a resonating coaxial transmission line or a tuned circuit, which accepts power from an oscillator and delivers it, harmonic-free, to a load.

tanker (*Ships*). A term covering all types of ships carrying liquid in bulk, from light acids and oils to molasses and latex.

tank furnace (*Glass*). Essentially a large 'box' of refractory material holding from 6 to 200 tons of glass, through the sides of which are cut 'ports' fed with a combustible mixture (producer gas and air, coke-oven gas and air, or oil spray and air), so that flame sweeps over the glass surface. With the furnace is associated a regenerative or recuperative system for the purpose of recovering part of the heat from the waste gases.

tanking (*Build., Civ. Eng.*). Waterproof material included in an underground structure to prevent infiltration of subsoil water.

tank line (*Radio*). Quarter-wavelength line used as a frequency stabilizer in an ultra-short-wave radio transmitter.

tank reactor (*Nuc. Eng.*). Covered type of *swimming-pool reactor* (q.v.).

tank rectifier (*Elec. Eng.*). Mercury-arc rectifier enclosed in a metal tank with vitreous seals for the conductors.

tank tube (*Electronics*). Same as **pool tube**.

tank vent pipe (*Aero.*). The pipe leading from the air space in an aircraft fuel, or oil, tank to atmosphere, for equalizing changes in pressure due to alterations in altitude; in *aerobatic* aircraft a nonreturn valve is fitted to prevent liquid escaping when inverted.

tan liquor (*Leather*). A solution of tanning materials, prepared for the tanyard by grinding the materials and macerating them with water. The liquor may be strengthened by the addition of tanning extracts. See **leaching, tannin.**

tannic acid (*Chem.*). See **tannin.**

tannin (*Chem.*). A mixture of derivatives of polyhydroxy-benzoic acids. When pure it forms a colourless amorphous mass, easily soluble in water, of bitter taste and astringent properties. Occurs in many trees, e.g., *Quebracho Colorado*. The best source for the production of pure tannic acid is the gall formed on certain oak trees. Tannin is used for tanning leather because it has the property of precipitating gelatin and forming an insoluble compound. For fancy leathers, sumach is extensively used. Also called **tannic acid.** See **gallotannins.**

tanning (*Leather*). Any of the processes by which skins and hides are converted into leather, viz. (*a*) the use of vegetable tanning materials; (*b*) impregnation with various mineral salts; (*c*) treatment with oils and fats. (*Zool.*) In newly-formed cuticle of Insects, the process in which the spaces between the chitin micelles are filled with sclerotin, which consists of protein molecules linked together or tanned by quinones. This makes the cuticle tougher and darker.

tanning developer (*Photog.*). One which causes differential hardening or insolubility in emulsions, e.g., *catechol, hydroquinone, pyrogallol* (qq.v.). Used mainly in carbon and imbibition dye printing.

tannin sac (*Bot.*). A cell containing much tannin.

Tannoy (*Acous.*). TN for a sound-reproducing and amplifying system.

tantalite (*Min.*). Tantalate of iron and manganese, crystallizing in the orthorhombic system. It usually has an admixture of the niobate of iron and manganese and the mineral passes from the pure tantalate (*tantalite*) to the pure niobate (*columbite*). In some varieties (*manganotantalite*) the iron is almost entirely replaced by manganese. It commonly occurs in pegmatite veins, and is a source of tantalum.

tantalum (*Chem.*). A metallic element, symbol Ta, at. no. 73, r.a.m. 180·948, rel. density at 20°C 16·6, m.p. 2850°C, electrical resistivity $15{\cdot}5 \times 10^{-8}$ ohm metres, Brinell hardness 46. It occurs in crystals and grains (usually containing, in addition, small amounts of niobium) in the Ural and Altai Mts. It is used as a substitute for platinum for corrosion-resisting laboratory apparatus, as acid-resisting metal in chemical industry, and in the form of carbide in cemented carbides. Used in surgical insertions because of its lack of reaction to body fluids. Also in the manufacture of electronic equipment—plates and grids for high-frequency valves, rectifiers, capacitors, etc., and for lamp filaments. It has a high absorption rate for residual gases in vacuum valves, high melting-point, good ductility and is easily welded. Also used in making nonreactive alloys, e.g., with nickel and molybdenum.

tantalum capacitor (*Electronics*). Miniature electrolytic capacitor employing tantalum foil.

tantalum rectifier (*Electronics*). One employing a tantalum electrode.

T-antenna (*Radio*). One comprising a top conductor with a vertical downlead attached at the centre; much used for long waves.

tap (*Eng.*). A screwed plug of accurate thread, form, and size, on which cutting edges are formed along longitudinal grooves; screwed into a hole, by hand or power, to cut an internal thread.

tap changing (*Elec. Eng.*). A method of varying the voltage ratio of a transformer by tapping the windings in such a way that the *turns ratio* (q.v.) is altered.

tap density (*Powder Tech.*). The apparent powder density of a powder bed formed in a container of stated dimensions when a stated amount of the powder is vibrated or tapped under stated conditions.

tape (*Build., Surv.*). A long flexible measuring scale of thin strip steel, linen, or linen in which wire is interwoven to increase its strength, coiled up in a circular leather case fitted with a handle for winding purposes. (*Electronics*) Tape can be used: (1) as an insulator, e.g., round conductors in multiple cables; (2) for instruction and control, e.g., perforated paper tape as input for control system or computer; and (3) see **magnetic tape.**

tape condenser (*Spinning*). The delivery end of a woollen or cotton waste card. The doffer web is converted into ribbons of fibres which are rubbed between travelling aprons into sliver form for spinning.

tape deck (*Acous.*). Platform incorporating essentials for magnetic recording—motor(s), spooling, recording and erasing heads—for adding to amplifier, microphone, loudspeaker, to form a complete recording and reproducing equipment (**tape recorder**). (*Comp.*) A *main store* in the form of a long magnetic tape on which computer data is stored as in a file, with facilities for rewinding and searching (*random access*) and replacement of data.

tape joint (*Build.*). Paper or gauze tape used to cover the joints between wall or centring boards before a decorative surface is applied.

taper (*Eng.*). In conical parts, the difference in diameter per unit length.

tape reader (*Comp.*). Apparatus for providing signals related to holes punched in paper tape as it is moved by a tape transport.

tape recording (*Acous.*). Longitudinal recording on magnetic particles dispersed in a medium carried on plastic tape. There is a residual magnetization of a high-frequency biasing current, modulated by the signal current. The residual m.m.f. in the particles allows the modulation to be reproduced by induction in a magnetic circuit. See also **tape deck.**

tapered roller bearing (*Eng.*). A roller bearing rendered capable of sustaining end thrust by the use of tapered rollers, in conjunction with internally and externally coned races.

tapering (*Bot.*). Said of a leaf base which becomes gradually narrowed towards the petiole.

tapering gutter (*Plumb.*). A parapet gutter having an increasing width in the direction of flow, so as to secure the necessary fall.

taper key (*Eng.*). A rectangular key having parallel sides, but slightly tapered in thickness along its depth. See **key.**

taper-loaded cable (*Cables*). Submarine cable in which the continuous loading is effectively reduced towards the ends of the cable but is uniform for the centre length.

taper pin (*Eng.*). A pin, used as a fastener, very slightly tapered to act as a wedge.

taper pipe (*Plumb., San. Eng., etc.*). A pipe of diminishing diameter, serving to connect pipes of different diameters.

taper slots (*Elec. Eng.*). Slots in an armature core in which the intervening teeth have parallel sides.

taper tap (*Eng.*). The first tap used in threading a hole. The first few threads are ground down to the core diameter to provide a guide, gradually increasing to the full thread size. See **tap, plug tap, second tap.**

taper-turning attachment (*Eng.*). An attachment bolted to the back of a lathe, with a guide bar which may be set at the angle or taper it is desired to impart to the turned part, the lathe tool being guided by the bar via a slide.

tapes (*Bind.*). See **bands.** (*Print.*) Strips of leather, plastic or fabric for carrying paper from one part of press to another.

tapestry (*Textiles*). Handwoven furnishing fabric in which a design was produced by stitches across the warp. Now employed for jacquard figured design on power-loom woven cloths.

tapestry brick (*Build.*). See **rustics.**

tapetal plasmodium (*Bot.*). A multinucleate mass formed by the breakdown of the cell walls between the cells of a tapetum.

tapetum (*Bot.*). A layer of cells surrounding a mass of spore mother cells, which finally breaks down and contributes material for nourishing the developing spores. *adj.* **tapetal.** (*Zool.*) In the eyes of certain nightflying Insects, a reflecting structure; in some Vertebrates, a reflecting layer of the retinal side of the choroid; in the Vertebrate brain, a tract of fibres in the corpus callosum.

tapeworms (*Med., Zool.*). Parasitic worms of the class *Cestoda*, generally taking the form of a scolex with hooks and/or suckers for attachment to the host and a chain of individual proglottids in successive stages of development. Species infecting man are *Hymenolepis nana* or Dwarf Tapeworm, *Taenia solium* from infected pork, *Diphyllobothrium latum* from infected fish, and *Echinococcus granulosus* which spends the larval stage only in man, causing hydatid cysts, the adult worm inhabiting dogs.

tap-field control (*Elec. Eng.*). A method of controlling the speed of a series motor; the field excitation is varied by means of tappings on the field windings.

tap-field motor (*Elec. Eng.*). A series motor whose field windings are arranged for tap-field control.

taphephobia (*Med.*). Morbid fear of being buried alive.

tap hole (*Foundry*). Opening in side of smelting or refining furnace through which molten metal or slag is withdrawn. Closed by *bod* (q.v.), and opened by pointed bar or tie tap bar.

Taphrinales (*Bot.*). An order of *Fungi* of which the systematic position is uncertain. All plant parasites—includes *Taphrina deformans* (peach leaf curl), filamentous in the host, but yeastlike on culture media.

taphrogenesis (*Geol.*). Vertical movements of the Earth's crust, resulting in the formation of major faults.

tapiolite (*Min.*). Tantalate and niobate of iron and manganese. This really forms a variable series of minerals that may be considered as dimorphous with the columbite-tantalite series. The various molecules have been named *tapiolite* (tantalate of iron), *mossite* (niobate of iron), and *ixiolite* (tantalate of manganese). The minerals crystallize in the tetragonal system.

tapper (*Radio*). Electromechanically-operated hammer for decohering a coherer after actuation by high-frequency signals.

tappet (*Eng.*). A sliding member working in a guide; interposed between a cam and the push rod or valve system which it operates, to eliminate side thrust. (*Weaving*) See **wiper.**

tapping (*Elec. Eng.*). An intermediate connexion on a circuit element such as a resistor, often used to vary the potential applied to another electrical system. (*Eng.*) The operation of forming a screw-thread in a hole by means of taps, the tapping size of the hole corresponding to the core diameter of the thread. (*Met.*) The operation of running molten metal from a furnace into a lade. (*Surg.*) See **paracentesis.**

T.A.P.P.I. standard methods (*Paper*). Laboratory test methods conforming to the *T*echnical *A*ssociation of the *P*ulp and *P*aper *I*ndustry of the U.S. Widely used internationally.

taproot (*Bot.*). A strongly developed main root which grows vertically downwards and normally bears lateral roots much smaller than itself.

tar (*Chem., etc.*). See **coal-tar, gas tar.**

Tarantula (*Nuc. Eng.*). An experiment on shock-waving heating of a plasma, carried out at the Culham Laboratories.

tarbuttite (*Min.*). The hydrous phosphate of zinc, which crystallizes in the triclinic system. The crystals are often found in sheaflike aggregates.

Tardigrada (*Zool.*). An order of minute cryptozoic *Arachnida* with a reduced head and no specialized respiratory system; the mouth-parts are suctorial, and the only appendages are four pairs of stumpy clawed legs; common forms, of wide distribution, are found among moss and débris in ditches and gutters and on tree trunks, and can survive desiccation.

tare. The weight of a vessel, wrapping, or container, which subtracted from the gross weight gives the net weight.

target (*Electronics*). Any electrode or surface upon which electrons impinge at high velocity, e.g., fluorescent screen of a cathode-ray tube, or any intermediate electrode in an electron multiplier, or anode or anti-cathode in an X-ray tube. (*Nuc. Eng.*) Material irradiated by beam from accelerator. (*Radar*) Reflecting object which returns a minute portion of radiated pulse energy to the receiver of a radar system. (*TV*) Plate in a TV camera tube on which external scenes are focused and scanned by an electron beam.

target capacitance (*Electronics*). That between a camera tube target and backplate.

target diagram (*Elec. Eng.*). A diagram for estimating the uniformity of a batch of electric lamps; obtained by plotting luminous intensity against input power for each lamp. The more uniform the lamps, the more closely together will the plotted points lie in the diagram.

target rod (*Surv.*). A type of levelling staff provided with a sliding target, which can be moved by the staffman, under direction from the leveller, to a position in which it is in line with the line of sight of the level, the staff reading being recorded by the staffman.

target strength (*Acous.*). *T*. Defined in dB by $T=E-S+2H$, where E=echo level, S=source level and $2H$=transmission loss.

target theory (*Radiol.*). Proposed explanation of radiobiological effects, in which only a small sensitive region of each cell is susceptible to ionization damage.

tarmacadam (*Civ. Eng.*). A road or runway surfacing of broken stone which has been covered with tar; spread in a layer of uniform thickness and well rolled. Two layers are usually applied, the upper one being of stone of smaller size. In the U.S., **Tarmac** is a widely-used proprietary mixture. See also **macadamized road.**

tarn (*Geog.*). A small lake occupying a depression in a *cirque*.

tarnish (*Chem.*). The discoloration produced on the surface of an exposed metal or mineral, generally as the result of the formation of an oxide or a sulphide film.

tarsal (*Zool.*). One of the bones composing the *tarsus* (q.v.) in Vertebrates. Also **tarsale.**

tarsalgia (*Med.*). Pain in the instep of the foot.
tarsal glands (*Zool.*). See **Meibomian glands.**
tarsia (*Build.*). Wood inlay, of geometric or architectural patterns, in which comparatively large pieces of wood are used.
tarsometatarsus (*Zool.*). In Birds, a bone formed by the fusion of the distal row of tarsals with the metatarsals.
tarsus (*Zool.*). (1) In Vertebrates, an elongate plate of dense connective tissue which supports the eyelid. (2) In Insects, *Myriapoda*, and some *Arachnida* (as Mites), the terminal part of the leg, consisting typically of five joints. (3) In land Vertebrates, the basal podial region of the hind limb; the ankle. *adj.* **tarsal.**
tartar emetic (*Chem.*). Potassium antimonyl tartrate, $2[K{\cdot}SbO{\cdot}C_4H_4O_6]{\cdot}H_2O$.
tartareous (*Bot.*). Said of the surface of a lichen when it is rough and crumbly.
tartaric acid (*Chem.*). $HOOC{\cdot}CH(OH){\cdot}CH(OH){\cdot}COOH$, dihydroxy-succinic acid. It exists in four modifications, viz., (+)-tartaric acid, m.p. 170°C; (−)-tartaric acid, m.p. 170°C; *racemic* or (±)-tartaric acid, m.p. 206°C; *meso*tartaric acid, m.p. 143°C. The (+)-tartaric acid is found in nature, free and as salts of potassium, calcium, and magnesium. It occurs in a large number of plants and fruits; the acid potassium salt is deposited from wine (see **argol**).
tartrates (*Chem.*). The salts of tartaric acid. Extensively used in medicine, the potassium salts and Rochelle salt as saline purges.
tarus (*Build.*). A cylindrical projection along the intersection between the two sloping roof surfaces on one side of the ridge of a mansard roof.
tarviated (*Civ. Eng.*). A term applied to macadam road surfacings in which the stone is bound together with tar.
T.A.S. (*Aero.*). Abbrev. for *true airspeed.*
tasmanite (*Geol.*). A type of practically pure spore coal; a variety of *cannel coal* (q.v.). See **boghead coal.**
tassement polaire (*Bot.*). The formation of a dense mass by the chromosomes at the poles of the spindle, as telophase comes on.
taste-bud (*Zool.*). In Vertebrates, an aggregation of superficial sensory cells subserving the sense of taste; in higher forms, usually on the tongue.
T-attenuator (*Elec. Eng.*). One comprising three resistors, one end of each being connected together. The free ends of two are connected respectively to an input and an output terminal while the free end of the third is connected to the common input and output terminals.
taurine (*Chem.*). $H_2N{\cdot}CH_2{\cdot}CH_2{\cdot}SO_2OH$, amino-ethylsulphonic acid; m.p. 240°C, with decomposition. Found in the animal body, in ox-gall, etc.
taurocholic acid (*Chem.*). *Cholyl-taurine*, found in bile.
tautomerism (*Chem.*). The existence of a substance as an equilibrium mixture of two inter-convertible forms, usually because of the mobility of a hydrogen atom. Thus tautomeric compounds can give rise to two series of derivatives. See **ethyl aceto-acetate.**
tautonym (*Bot., Zool.*). A name in which the specific epithet repeats the generic name.
tawa (*For.*). New Zealand native hardwood from the genus *Beilschmedia*, used for furniture, veneering, fittings, and flooring.
Tawara's node (*Zool.*). A plexiform mass of very small muscle fibres on the septal wall of the right auricle in Mammals.
tawing (*Leather*). The process by which lamb, kid, and deer skins are tanned. After soaking, liming, and bating, the skins are drummed with a paste consisting of alum, flour, salt, and egg yolk (or olive oil). Tawed leathers are white and are used for gloves, etc.; they are brush-dyed on the grain side, if necessary.
taxeopodous (*Zool.*). Having the proximal and distal tarsal bones in straight lines parallel with the axis of the limb, as some *Ungulata.*
taxi-channel markers (*Aero.*). See **aerodrome markers.**
taxines (*Chem.*). A mixture of alkaloids constructed from nitrogen-free polyhydroxylic compounds called *taxicins* partially esterified with 3-dimethylamino-3-phenyl propanoic acid and ethanoic acid.
taxis (*An. Behav.*). Term applied to a type of *orientation behaviour* (q.v.), divided into *klinotaxis, tropotaxis*, and *telotaxis* (qq.v.) on the basis of the exact orientation mechanism used, and into *phototaxis, geotaxis*, etc., according to nature of the stimulus. Cf. *kinesis.* (*Biol.*) Movement of a whole organism directly towards a source of stimulation (*positive taxis*) or directly away from it (*negative taxis*). It is confined to motile plants of microscopic dimensions, and to motile reproductive bodies.
taxi track (*Aero.*). A specially prepared track on an aerodrome used for the ground movement of aircraft. See **perimeter track.**
taxi-track lights (*Aero.*). Lights, usually violet, so placed as to define manoeuvring areas and tracks.
taxonomic series (*Biol.*). The range of extant living organisms, ranging from the simplest to the most complex forms.
taxonomy (*Biol.*). The science of classification as applied to living organisms, including study of means of formation of species, etc.
Taylor's series (*Maths.*). Series expansion for a continuous function, giving the value of the function for one value of the independent variable in terms of that for another value. Under specified conditions the series is

$$f(a+h) = f(a)+hf'(a) + \frac{h^2f''(a)}{2!} \ldots \text{etc.}$$

Tay-Sachs' disease (*Med.*). Amaurotic (see **amaurosis**) family idiocy. A rare neurological disease of infants, characterized by idiocy, progressive paralysis of the body, blindness, and death before the age of two years, as a result of the progressive degeneration of the nerve cells of the brain and spinal cord.
Tb (*Chem.*). The symbol for *terbium.*
T-beam (*Civ. Eng.*). A beam forming part of the construction of a reinforced concrete floor; regarded as being composed of the beam part projecting below the floor slab, and portions of the floor slab on both sides, the whole having the form of a letter T.
TBO or **tbo** (*Aero.*). See **time between overhauls.**
T-bolt (*Eng.*). A bolt having a head in the form of a short cross-piece; used to hold work on a machine table in conjunction with a corresponding T-slot into which the head is dropped and secured by turning through 90°. Also **tee bolt.**
Tc (*Chem.*). The symbol for *technetium.*
Tchebychef. See under **Chebyshev.**
TCNB (*Chem.*). See **tecnazene.**
T.D.I. (*Chem.*). Abbrev. for *tolylene di-isocyanate.*
t-distribution (*Stats.*). A test of the statistical significance of experimental data, i.e., of the probability of their being due to chance alone. Also called **student's t-test.**
Te (*Chem.*). The symbol for *tellurium.*
teaching machine (*Psychol.*). General term for the

instruments used in presenting material for *programmed learning* (q.v.).

teak (*For.*). Valuable wood from *Tectona grandis*, a tree of the verbena family, found in India, Burma, Vietnam, and other eastern countries, renowned for its durability; used for shipbuilding, piers, furniture, etc.

teaming (*Civ. Eng.*). The removing of excavated material from cutting to bank.

tear factor (*Paper*). The ratio of the tear strength of a paper to the *substance*.

tear fault (*Geol.*). A horizontal displacement of a series of rocks along a more or less vertical plane, as a result of differential stresses acting upon the bed. Cf. *normal fault*, *thrust plane*.

tear gases (*Chem.*). Volatile compounds which even in low concentration make vision impossible by their irritant action on the eyes. They are halogenated organic compounds, e.g., *xylyl bromide*, $CH_3{\cdot}C_6H_4{\cdot}CH_2Br$, and *ethyl iodoacetate*, $CH_2I{\cdot}COOC_2H_5$.

tear gland (*Zool.*). See **lacrimal gland.**

tearing (*TV*). Break-up of image in facsimile transmission due to faulty synchronization.

teaser transformer (*Elec. Eng.*). The smaller of the windings in a *Scott connexion* (q.v.).

teats (*Zool.*). In female Mammals, paired projections from the skin on which the lactiferous tubules of the mammary glands open.

technetium (*Chem.*). Radioactive element not found in ores. First produced as a result of deuteron and neutron bombardments of molybdenum. Symbol Tc, at. no. 43, the most common isotope ^{99}Tc has half-life of $2{\cdot}1 \times 10^6$ years. Found among fission products of uranium, and (unexplained) in the spectra of some stars.

Technetron (*Electronics*). TN for a French transistor which employs centripetal striction to vary the conductance of the semiconductor.

technical jack plane (*Tools*). Small type of *jack plane* (q.v.), with low-cut rear section.

Technicolor (*Cinema.*). TN for colour films made by various techniques, and printed by imbibition or dye-transfer processes.

Technitron (*Electronics*). TN for a high-frequency triode.

technology. The practice, description, and terminology of any or all of the applied sciences which have practical value and/or industrial use.

tecnazene (*Chem.*). 1,2,4,5-tetrachloro-3-nitrobenzene,

Used as a fungicide. Also **TCNB.**

tectogene (*Geol.*). A downward extension of the Earth's crust, postulated to exist below an *orogenic belt*.

tectology (*Biol.*). The doctrine of structure; morphology in which the organism is considered as being composed of morphological rather than physiological units, each organ being regarded as an individual unit.

tectonic (*Geol.*). Said of rock structures which are directly attributable to earth movements involved in folding and faulting.

tectonics (*Geol.*). The study of the major structural features of the Earth's crust.

tectorial (*Zool.*). Covering; as the tectorial membrane (*membrana tectoria*) of Corti's organ.

tectosilicates (*Min.*). Those silicates having an atomic structure in which atoms of silicon and oxygen are linked in a continuous frame-work.

tectospondylous (*Zool.*). Having cartilaginous vertebrae which are calcified in several concentric rings.

tectotype (*Biol.*). Description of a species from the microscopical examination of a section.

tectrices (*Zool.*). In Birds, small feathers covering the bases of the remiges and filling up the gaps between them. Also called **auriculars.**

tectum (*Zool.*). A covering or roofing structure; as the *tectum synoticum*, part of the roof of the cartilaginous skull which connects the two auditory capsules.

tedder (*Agric.*). A machine to loosen swaths or windrows.

tee (*Plumb.*). A short pipe fitting used to connect two pipe lengths with a short offset at right angles for the connection of a branch pipe.

tee bolt (*Eng.*). See **T-bolt.**

tee hinge (*Join.*). A large strap hinge shaped like the letter T, the long arm, corresponding to the upright part of the T, being secured to the door, and the crosspiece to the hingeing post.

tee joint (*Elec. Eng.*). A joint in a cable formed by tapping off a branch circuit, without cutting the main cable.

teem (*Glass*). To pour molten glass from a pot in the rolling process.

teeming (*Met.*). The operation of filling ingot moulds from a ladle of molten metal.

Teepol (*Chem.*). TN of liquid anionic detergent based on mixed sodium alkyl sulphates of long-chain alcohols, such as lauryl alcohol (*n*-dodecanol, $CH_3(CH_2)_{11}OH$).

tee rest (*Eng.*). See **T-rest.**

Teflon (*Plastics*). TN for **polytetrafluoroethene.**

tegmen (*Bot.*). The inner leaves of a testa. (*Zool.*) In some *Crinoidea*, the leathery membrane covering the top of the calyx; in *Icthyopsida*, the roof of the chondrocranium, in *Orthoptera*, the hardened leathery fore-wing. *pl.* **tegmina.**

tegmentum (*Zool.*). (1) Upper layer of a shell plate in *Amphineura*. Cf. *articulamentum*. (2) A reticular mass of fibres with much grey matter in the mesencephalon of higher Vertebrates. Also known as **red nucleus.**

tegula (*Build.*). A roofing tile. (*Zool.*) One of the articular sclerites of the wing in Insects, a small tile-shaped structure. *pl.* **tegulae.** *adj.* **tegular.**

tegulated (*Zool.*). Composed of or covered by plates overlapping like tiles.

tegumen (*Zool.*). In male *Lepidoptera* (Insects) the ninth segment of the abdomen, a narrow ring encircling the apex of the body. The sternal region is known as the *vinculum* (q.v.) and the *uncus* (q.v.) is attached to the hind margin of the tergal region.

tegumentary system (*Bot.*). The layer or layers of cells which cover the surface of a plant.

tegumentum (*Zool.*). See **tegmentum.**

tegumere (*Zool.*). In metameric animals, the portion of the integument within one somite.

t.e.h.p. (*Aero.*). Abbrev. for *total equivalent brake horsepower*.

teichoic acids (*Biochem.*). A group of polymers of ribitol or glycerol phosphate in which adjacent polyol residues are linked by a phosphodiester bond. The polymers contain D-alanine residues in ester linkage with hydroxyl groups, and the glycerol phosphate polymers contain glycosidically-bound sugar residues. They occur in bacterial cell walls but are also found intracellularly.

teichopsia (*Med.*). Temporary loss of sight in part of the visual field, and the appearance before the eye of a spot of light which enlarges and

becomes zigzag in shape and many-coloured; a symptom of *migraine* (q.v.).
tier (*Textiles*). See **tier**.
Teklan (*Plastics*). TN for synthetic textile fibre based on a modified acrylic (modacrylic) base. Chiefly noted for its inherent flameproof properties.
tektites (*Min.*). A group term suggested by Suess in 1900 to cover moldavites, billitonites, australites. They are natural glasses of non-volcanic origin and may be of extra-terrestrial origin.
tektosilicates (*Min.*). See **tectosilicates.**
tela (*Zool.*). A weblike tissue.
telangiectasis (*Med.*). Morbid dilatation of capillaries and arteries. *adj.* **telangiectatic.**
tele-. Prefix from Gk. *tēle*, afar, at a distance.
telearchics (*Telecomm.*). See **telecontrol.**
teleblem (*Bot.*). A membrane of closely interwoven hyphae covering the entire fructifications of some agarics.
telecentric stop (*Optics*). A stop placed in the second focal plane of a positive lens, forming a viewing system by which a scale can be read without parallax errors.
teleceptor, telereceptor (*Biol.*). A sense organ which responds to stimuli of remote origin.
telecine (*TV*). Projector of cinematograph films, specially designed for television scanning and video transmission.
telecommunication. Any communication of information in verbal, written, coded, or pictorial form by electric means, whether by wire or by radio.
telecontrol (*Telecomm.*). Control of mechanical devices remotely, either by radio (as ships and aircraft), by sound waves, or by beams of light. Also called **telearchics.**
teledeltos (*Electronics*). Sensitive paper used in direct electric recording (electrography).
Telediphone (*TV*). TN of device for monitoring, recording, and reproducing for written record any programme material.
telefilm (*TV*). Regular sound-film made specially for subsequent transmission to television audiences.
telegamic (*Biol.*). Attracting mates from a distance.
telegenesis (*Biol.*). Artificial insemination.
telegony (*Zool.*). The supposed influence of a male with which a female has previously been mated, as evinced in offspring subsequently borne by that female to another male.
telegraph (*Teleg.*). Combination of apparatus for conveying messages over a distance by electrical pulses sent along special overhead wires or underground cables. Such pulses are often interrupted audio-frequency currents.
telegraph distortion set (*Teleg.*). An instrument which generates selected telegraph signals, in Morse or five-unit code, for transmission through circuits or apparatus to be tested, and which also indicates the resulting distortion of the signal, either on a dial or a cathode-ray oscillograph.
telegraph speed (*Teleg.*). Reciprocal of the unit interval measured in seconds. Expressed in *bauds* (q.v.).
telegraph word (*Teleg.*). Standard word for calculations is five letters and letter space, e.g., PARIS.
telegraphy. The branch of telecommunication which is concerned in any process providing reproduction at a distance of documentary matter such as written, printed or pictorial matter, or the reproduction at a distance of any kind of information in such a form. See also **carrier telegraphy, facsimile telegraphy, printing telegraph, start-stop.**
teleguided missile (*Aero.*). A small subsonic missile for attacking surface targets, e.g., tanks and ships, controlled by command guidance from an operator, or automatic device, by signals transmitted through fine wires connected to the control box and uncoiled in flight from the missile. Also **wire-guided.**
telemetacarpal (*Zool.*). In some deer (*Artiodactyla*), retaining the distal extremities of the metacarpals corresponding to the reduced lateral toes. Cf. *plesiometacarpal.*
telemeter (*Elec. Eng.*). An instrument for the remote indication of electrical quantities, such as voltage, current, power, etc. (*Surv.*) General name for instrument which acts as a distance measurer, without the use of a chain or other direct-measuring apparatus.
telemetry (*Surv.*). Measurement of linear distances by use of tellurometer. (*Telecomm.*) Transmission to a distance of measured magnitudes by radio or telephony, with suitably coded modulation, e.g., amplitude, frequency, phase, pulse.
telencephalon (*Zool.*). One of the two divisions of the Vertebrate fore-brain or prosencephalon (the other being the *diencephalon*), comprising the cerebral hemispheres (with the cerebral cortex or pallium and the corpus striatum), the olfactory lobes, and the olfactory bulbs.
teleo-. Prefix from Gk. *teleios*, perfect.
teleology (*Biol.*). The interpretation of animal or plant structures in terms of purpose and utility. *adj.* **teleological.**
teleonomy (*Biol.*). Impression of purpose arising from adaptation through natural selection.
teleophore (*Zool.*). See **gonotheca.**
teleoptiles (*Zool.*). The types of feathers characteristic of an adult Bird, as filoplumes, plumulae, and pennae. Cf. *neossoptiles.*
teleosis (*Biol.*). Purposive development.
Teleostei (*Zool.*). A superorder of *Actinopterygii*, including fish with a wide diversity of form and psycho-physiological adaptations. Gills fully filamentous, tail externally (and in many cases internally) homocercal, endoskeleton completely ossified, fins completely fanlike with no trace of an axis. Bony fishes.
telepantoscope (*TV*). Device similar to an iconoscope, except that scanning motion of the beam is in one direction only (i.e., line-scanning direction), the frame scanning being accomplished by mechanical means.
telepathy (*Psychol.*). Direct communication between minds without recourse to known sensory data.
telephone (*Teleph.*). (1) Combination of apparatus for conveying speech over a distance by audio-frequency current sent along special overhead wires or underground cables; (2) a sound reproducer held to the ear, driven by signals derived from any source.
telephone capacitor (*Teleph.*). Fixed capacitor, having a capacitance ca. 0·001 μF in parallel with telephones in a crystal receiver to bypass radio-frequency currents.
telephone interference (or **influence**) **factor** (*Teleph.*). The weighting-factor required for determining the total interference of induced electromotive forces arising from harmonic induction in telephone lines from adjacent power lines. The factor takes into account the average relative sensitivity of the ear for varying frequency, and also the average response curves of telephone receivers. Abbrev. **TIF.**
telephone-telegram circuit (*Teleg.*). A circuit for

transmitting telegrams verbally between telegraph offices.

telephony. The conversion of a sound signal into corresponding variations of electric current (or potential), which is then transmitted by wire or radio to a distant point where it is reconverted into sound.

telephotography (*Telecomm.*). See **phototelegraphy.**

telephoto-lens (*Photog.*). A combination of a convex and a concave lens to increase the effective focal length (so magnifying the image) without altering distance between the lens and film of the camera.

teleprinter (*Teleg.*). Telegraph transmitter, having a typewriter keyboard and a type-printing telegraph receiver; widely used in commercial offices and for public and news services. U.S. term *teletypewriter*. See **telex.**

Teleprompter (*TV*). TN for a visual prompter for displaying a script inconspicuously to a speaker in front of a TV camera.

teleradiography (*Radiol.*). A technique to minimize distortion in taking X-ray photographs by placing X-ray tube some distance from the body.

telerecording (*TV*). Recording television programme material on film, for editing and subsequent transmission. See **Ampex, Vera.**

telescope (*Optics*). An optical instrument for making distant objects appear nearer; it consists of arrangements of lenses or mirrors by which the light is brought to a focus, the image there formed being magnified. See **astronomical-, meniscus-, reflecting-, refracting-, terrestrial-.**

telescope tube (*Electronics*). One in which an infrared object is focused on a photoemitting surface, electrons from which form an enlarged image, exhibited on a fluorescent screen. See **image tube.**

telescopic shaft (*Eng.*). An assembly of two or more tubes sliding within each other to provide a shaft of infinitely variable length.

telescopic stars (*Astron.*). Those stars whose apparent magnitudes are numerically greater than the 6th and which are too faint to be seen with the naked eye. The largest telescopes can photograph stars as faint as apparent magnitude 23.

teletherapy (*Radiol.*). Treatment by X-rays from a powerful source at a distance, i.e., by high-voltage X-ray tubes, or radioactive sources, such as cobalt-60 or caesium, up to 2000 curies.

telethmoid (*Zool.*). See **prenasal.**

teletorium, telestudio (*TV*). An enclosure, sound-proofed and treated acoustically, used for live TV or broadcasting programmes.

teletron (*TV*). CRT specially designed for synthesizing TV test images, either for direct viewing or for projection.

teletube (*TV*). Abbrev. for *tele*vision *tube*. A cathode-ray tube specifically designed for reproduction of TV images. Also **picture tube,** and in U.S. **kinescope.**

teletypesetting (*Typog.*). A method of operating *line-casting machines* by using a six-unit punched tape produced on a separate keyboard; can be used within the same office where increased output results through specialization and the optimum use of equipment and skill; or the keyboard output can be transmitted by wire or radio to be converted to tape at the receiving end.

teletypewriter (*Teleg.*). U.S. for type of *teleprinter*, start-stop telegraph system using normal keyboards.

teleutosorus (*Bot.*). A group of teleutospores, together with their supporting hyphae, forming a pustule on the surface of the host. See **telium.**

teleutospore, teleutogonidium (*Bot.*). A thick-walled spore, consisting of two or more cells, formed by rust fungi towards the end of the season; capable of remaining quiescent for some time, and then germinating to give one or more promycelia, on which the basidiospores are developed. See **teliospore.**

teleutostage (*Bot.*). The stage in the life-history of a rust fungus when *teleutospores* (q.v.) are formed. See **telial stage.**

television (*Telecomm.*). Electric transmission of visual scenes and images by wire or radio, in such rapid succession as to produce in the observer at the receiving end illusion of being able to witness events as they occur at the transmitting end. Effected by scanning a *raster* which covers the direct image or a picture on a standard film. In broadcasting, the transmission is by radio waves, in closed circuit, by line. In the British system, the picture is scanned in 405 (V.H.F.) or 625 (U.H.F.) lines both at 25 times/sec., and with interlaced scanning, whereby the odd number lines are scanned first, followed by even lines, so giving two frames for a complete picture. See **Ampex, compatible colour television, line frequency, monochrome receiver, Vera.**

television amplifier (*TV*). One operating uniformly in gain and phase delay, between zero and several MHz, depending on the definition in the system.

television cable (*TV*). One capable of transmitting frequencies sufficiently high to accommodate TV signals without undue attenuation or relative phase delay; usually coaxial, with as much air insulation as possible.

television camera (*TV*). Converter of an external scene into a video signal for transmission. Prominent are *emitron, iconoscope, image orthicon, orthicon, Vidicon* (qq.v.).

television channel (*TV*). One with a sufficiently wide frequency band ($\sim$4 MHz) to be used for TV transmission.

television field frame (*TV*). In interlaced scanning a *frame* consists of the full sequence of scanning lines, and is divided into two or more equal fields scanned alternately.

television microscope (*TV*). A device which gives a much enlarged image of a very small object using TV techniques.

television receiver (*TV*). That part of a TV system in which the picture and associated sounds are reproduced from the input signal. Also **television set.**

television transmitter (*TV*). One which radiates video, aural and synchronizing components of a television signal as a modulated RF wave.

televisor (*TV*). TV receiver.

telex (*Teleph.*). An audio-frequency teleprinter system for use over telephone lines (provided in Britain by the Post Office). (Automatic *Tel*etypewriter *Ex*change Service.)

Telfener rack (*Civ. Eng.*). A form of rack railway in which the rack is centrally located, is formed of two angles (placed back-to-back, with teeth cut in them), and is sometimes strengthened with flat bars between the angles.

telial stage (*Bot.*). An American term for *teleutostage.*

teliospore (*Bot.*). An American term for *teleutospore.*

telium (*Bot.*). An American term for *teleutosorus.*

teller (*Surv.*). See **tally.**

telltale clock (*Horol.*). A portable clock which

gives, on a chart, a record of the time a watchman visits certain fixed points in his round of inspection.

telluric bismuth (*Min.*). An intermetallic compound, Bi_2Te_3, crystallizing in the trigonal system. The name has also been used as a synonym for *tetradymite* (q.v.).

telluric current (*Elec. Eng.*). Current in, or put into, the earth, which is used in exploration of strata.

telluric lines (*Astron.*). Absorption lines or bands in stellar and planetary spectra, caused by absorption in the earth's atmosphere, mainly by water vapour and oxygen.

tellurides (*Chem.*). Compounds of divalent tellurium [Te(II)], analogous to *sulphides* (q.v.).

tellurite (*Min.*). Orthorhombic tellurium dioxide.

tellurium (*Chem.*). A semimetallic element, tin-white in colour. Symbol Te, at. no. 52, r.a.m. 127·60, rel. density at 20°C 6·24, m.p. 452°C, valencies 2, 4, 6, electrical resistivity 2×10^{-3} ohm metres. Used in the electrolytic refining of zinc in order to eliminate cobalt; alloyed with lead to increase the strength of pipes and cable sheaths. The chief sources are the slimes from copper and lead refineries, and the fine dusts from telluride gold ores.

tellurobismuth (*Min.*). See **telluric bismuth.**

Tellurometer (*Surv.*). Electronic instrument used to measure survey lines for distances up to 40 miles by measurement of time required for a radar signal to echo back, accuracy being of order of 1:100 000 in good weather. See **trilateration.**

telo-. Prefix from Gk. *telos*, end.

teloblast (*Zool.*). A large cell from which many smaller cells are produced by budding, as one of the primary mesoderm cells in developing *Polychaeta.*

telodendra (*Zool.*). The terminal twigs into which an efferent axon breaks up at a synapse; cf. *dendron. sing.* **telodendron.**

telokinesis (*Cyt.*). See **telophase.**

telolecithal (*Zool.*). A type of egg which is large in size, with yolk constituting most of the volume of the cell, and with the relatively small amount of cytoplasm concentrated at one pole. Found in Sharks, Skates, Reptiles, and Birds. Cf. *mesolecithal, oligolecithal.*

telolemma (*Zool.*). The connective tissue sheath of a muscle-spindle.

telomitic (*Cyt.*). In cell-division, having the chromosomes attached to the fibres of the spindle by their ends.

telophase (*Cyt.*). Final phase of mitosis with cytoplasmic division; the period of reconstruction of nuclei which follows the separation of the daughter chromosomes in mitosis; telokinesis.

Telosporidia (*Zool.*). A subclass of *Sporozoa* in which the trophozoite is uninucleate and ceases to exist when spores are formed; the spore cases are simple and usually contain several sporozoites.

telosynapsis, telosyndesis (*Cyt.*). End-to-end union of the chromosome halves in meiosis; cf. *parasynapsis, syndesis.*

telotaxis (*An. Behav.*). A type of *taxis* (q.v.) in which an animal fixates a source of stimulation with its sense organ and advances towards it so that a certain region of the receptor apparatus is always acted on by the chosen stimulus, and other stimuli are disregarded. Cf. *klinotaxis, tropotaxis.*

telotroch (*Zool.*). The abapical tuft of cilia in a trochophore.

telotrocha (*Zool.*). See **trochophore.**

telpher line (*Civ. Eng.*). A form of monorail in which an electrically driven truck runs along a single rail, the load being suspended below the truck and rail.

telson (*Zool.*). The post-segmental region of the abdomen in *Crustacea* and *Arachnida.*

Telstar (*Telecomm.*). First Transatlantic telecommunications satellite, launched 1962.

telum (*Zool.*). In Insects, the last abdominal somite.

Temnocephalea (*Zool.*). An order of *Turbellaria*, in some ways linking them to *Trematoda.* They are ectocommensals on freshwater Crustaceans, having reduced ciliation and tentacle-like prolongations on the anterior end, with a large sucker on the posterior end, e.g., *Temnocephala.*

temnospondyly (*Zool.*). The condition of having the vertebrae in articulated parts; cf. *stereospondyly. adj.* **temnospondylous.**

témoin (*Civ. Eng.*). An undisturbed column of earth left on an excavated site, as an indication of the depth of the excavation.

temper (*Glass*). The amount of residual stress in annealed ware measured by comparison with strain disks.

temperature (*Heat*). The degree of hotness or coldness measured with respect to an arbitrary zero (e.g., melting point of ice at *standard pressure*, as for the Celsius scale) or *absolute zero* (e.g., kelvin scale). See **degree, fixed points, fundamental interval, thermometer.** (*Met.*) See **tempering.** (*Med., Vet.*) Human body temperature in health is about 37°C. Normal values for animals: cattle, 38·6°C; goats, sheep, 39·4°C; horses, 38·1°C; pigs, 39·2°C; cats, dogs, 38·6°C.

temperature coefficient (*Biol.*). The ratio of the rate of progress of any reaction or process, at a given temperature, to the rate at a temperature 10°C lower. Also Q_{10}. (*Eng., Phys.*) The fractional change in any particular physical quantity per degree rise of temperature.

temperature coefficient of resistance (*Elec. Eng.*). In any conductor, if

$$R = R_0[1 + \alpha(T - T_0)],$$

R is the resistance at temperature T, compared with the resistance R_0 at temperature T_0, the mean coefficient in the range $T_0 \rightarrow T$ being α. This is useful only for *linear* resistances, i.e., pure metals.

temperature coefficient of voltage drop (*Electronics*). The ratio of the change in voltage drop across a glow-discharge tube to the change in operating temperature.

temperature coefficient of voltage regulator (*Electronics*). The rate of change of striking voltage of a glow-discharge tube with ambient temperature.

temperature correction (*Surv.*). A correction applied to the observed length of a base line to correct for any difference between the temperature of the tape during the measurement and that at which it was calibrated.

temperature cycle (*Nuc. Eng.*). Method of processing thick photographic nuclear research emulsions to ensure uniform development. These must be soaked in the solutions at refrigerated temperatures, and then warmed for the required processing period.

temperature inversion (*Meteor.*). Anomalous increase in temperature with height in the troposphere.

temperature lapse (*Meteor.*). A decrease of temperature with height.

temperature-limited (*Electronics*). Said of a thermionic device operated under saturation conditions, i.e., with the electrode currents limited by the cathode temperature.

temperature rise (*Elec. Eng.*). The difference in temperature between an electrical machine, after it has been on load for some time, and the surrounding air. It is a measure of the machine's capacity for dissipating the heat generated by the electrical power losses in the machine, and thus forms part of the electrical specification.

temperature screws (*Horol.*). The screws in the rim of a compensation balance, excluding the quarter screws.

temper brittleness (*Met.*). A type of brittleness that is shown by the notched bar test, but not by the tensile test, in certain types of steel after tempering; influenced to a marked extent by the composition of the steel, the tempering temperature, and the subsequent rate of cooling. Caused by precipitation of carbides or unequal segregation.

temper colour (*Met.*). In tempering hardened steel cutting tools, etc., the colour of the oxide layer which forms on reheating and which indicates approximately the correct quenching temperature for a particular purpose. See **tempering.**

tempered scale (*Acous.*). The musical scale of keyboard instruments, and, by implication, any other instruments or voices which are concerted with them, in which all semi-tones have frequencies of the same ratio, so that 12 semitones amount to one octave. Also called **equitempered scale, equal-tempered scale.** See **natural scale.**

temper-hardening (*Met.*). A term applied to alloys that increase in hardness when heated after rapid cooling; also to the operation of producing this. Also called **artificial ageing**; distinguished from *ageing* (q.v.), which occurs at atmospheric temperature. Both processes are covered by the term *precipitation hardening* (q.v.).

tempering (*Met.*). The reheating of hardened steel at any temperature below the critical range, in order to decrease the hardness. Also called **drawing.** Sometimes applied to reheating after rapid cooling, even when this results in increased hardness, e.g., in the case of steels that exhibit secondary hardening. See also **austempering.**

template (*Build.*). (1) A long flat stone supporting the end of a beam, to spread the load over several joints in the brickwork. (2) A framework of timber or steel used for the final setting out of girders, roof trusses, etc. (*Eng., etc.*) A thin plate, cut to the shape or profile required on a finished surface, by which the surface is marked off or gauged during machining or other operation. Sometimes called **templet.**

temple (*Weaving*). Twin spiked or fluted rollers fitted to a rod secured to the breast-beam of a loom. Their purpose is to keep the cloth at warp width and suitably tensioned from side to side as it is woven.

templet (*Build., Eng.*). See **template.**

tempolabile (*Chem.*). Tending to change with time.

temporal (*Zool.*). A cartilage bone of the Mammalian skull formed by the fusion of the petrosal with the squamosal.

temporalis (*Zool.*). In Vertebrates, one of the adductor muscles which by their contraction raise the lower jaw.

temporal vacuities or openings (*Zool.*). In Reptiles, openings in the skull, varying in number (none, one or two) and position, and used in classification. The various conditions found in different groups are known as *anapsid, synapsid, parapsid, euryapsid* and *diapsid* (qq.v.).

temporary collenchyma (*Bot.*). Collenchyma present in a young organ and disappearing as secondary thickening progresses.

temporary film (*Met.*). A soluble removable protective coating for metals, usually prepared from a lanolin derivative combined with suitable resins and dyed to a distinctive colour. The whole is dispersed in petrol or other solvent.

temporary hardness (*Chem.*). Hardness of water caused by the presence of the hydrogen carbonates of calcium and magnesium and therefore removable by boiling to precipitate the carbonate.

temporary memory (*Comp.*). One for holding data during a stage in processing. Also called **buffer memory.**

temporary star (*Astron.*). See **nova.**

temporary starch (*Bot.*). Starch which is stored for a time in the chloroplasts, when the plant is forming carbohydrates more rapidly than they are being used or removed from the leaf.

temporary way (*Civ. Eng.*). The ballast, sleepers, and rails laid temporarily by a contractor for his use on constructional works in transporting material.

temse (*Build., etc.*). See **sieve.**

tenacity (*Met.*). See **ultimate tensile stress.**

tenaculum (*Zool.*). In *Neopterygii*, a fibrous band extending from the eyeball to the skull.

tender ship (*Ships*). A ship with a small transverse metacentric height and long rolling period. Cf. *stiff ship.*

tendinous (*Zool.*). See **tendon.**

tendo calcaneus (*Zool.*). See **Achilles tendon.**

tendon (*Zool.*). A cord, band, or sheet of fibrous tissue by which a muscle is attached to a skeletal structure, or to another muscle. *adj.* **tendinous.**

tendril (*Bot.*). A slender, simple or branched, elongated organ used in climbing, at first soft and flexible, later becoming stiff and hard. May be a modified stem, leaf, leaflet, or inflorescence.

tenebrescence (*Min.*). The reversible bleaching observed in the hackmanite variety of sodalite. This mineral has a pink tinge when freshly fractured; the colour fades on exposure to light, but returns when the mineral is kept in the dark for a few weeks or is bombarded by X-rays.

tenent (*Zool.*). Used for clinging or attachment.

tenesmus (*Med.*). Painful and ineffectual straining at stool.

tenia, teniasis, etc. See **taenia, etc.**

tennantite (*Min.*). The sulphide of copper and arsenic, which crystallizes in the cubic system. This mineral is isomorphous with *tetrahedrite* (q.v.). The crystals are frequently dodecahedral and contain antimony, and hence grade into tetrahedrite. It may also contain minor amounts of iron, zinc, silver, and bismuth. Also called **fahlerz.**

tenon (*Carp.*). A tongue formed on the end of a member by cutting away from both sides one-third of the thickness of the member. The projecting part fits into a mortise in a second member in order to make a joint between them. See also **tusk tenon.**

tenon-and-slot mortise (*Join.*). A joint, such as that made between the posts and heads of solid door frames, in which a tenon cut on the end of the head fits into a *slot mortise* (q.v.) on the end of the post.

tenon saw (*Join.*). A saw with a very thin parallel blade, having fine teeth (4 to 6 per centimetre) and a stiffened back along its upper edge. Also called a **mitre saw.**

tenoreceptor (*Histol.*). A sense organ of tendons responding to contraction.
tenorite (*Min.*). Oxide of copper(II), crystallizing in the triclinic system. Occurs in minute black scales as a sublimation product in volcanic regions or associated with copper veins. *Melaconite* is a massive variety.
tenosynovitis, tenovaginitis (*Med.*). Inflammation of the sheath of a tendon.
tenotomy (*Surg.*). The cutting of a tendon for the correction of deformity.
tenovaginitis (*Med.*). See **tenosynovitis.**
tensegrity system (*Eng.*). Structural system in which compression and tension forces are separated out and handled by different members so that the structures have no inherent size limitations.
tensile stress (*Met.*). See **stress, ultimate tensile stress.**
tensile test (*Met.*). Two main forms are (*a*) those in which a static increasing pull is applied until fracture results. From this a stress-strain curve may be plotted and the proof stress, yield point, ultimate tensile stress, and elongation determined; and (*b*) those in which a dynamic load is applied giving data on fatigue and impact.
tensile testing machine (*Eng.*). A machine for applying a tensile or compressive load to a test piece, by means of hand- or power-driven screws, or by a hydraulic ram. The load is usually measured by a poise weight and calibrated lever.
tensimeter (*Chem.*). An apparatus for the determination of transition points by observation of the temperature at which the vapour pressures of the two modifications become equal.
tension (*Elec. Eng.*). Former term used to designate a potential difference, e.g., low-tension (such as an accumulator), or high-tension supplies (such as a power-distribution cable).
tension flange (*Bot.*). Mechanical tissue developed on the concave side of a coiled tendril.
tension insulator (*Elec. Eng.*). A suspension insulator for overhead transmission lines, which is designed to withstand the pull of the conductors; it is used, therefore, at terminal, anchor, or angle towers.
tension pin (*Eng.*). A hollow dowel pin of elastic material, slotted so as to be deformed on assembling, to take up misalignment or shape irregularity, or to exert a force.
tension plate lock-up (*Print.*). A method of locking plates to the cylinder by fingers which grip in slots on the underside of the plate. Cf. *compression plate lock-up*.
tension rod (*Eng.*). A structural member subject to tensile stress only. Also **tie rod.**
tension sleeve (*Eng.*). See **screw shackle.**
tensometer (*Eng.*). A versatile, portable testing machine used for a variety of mechanical tests, including tensile tests.
tensor (*Maths.*). The magnitude of a vector. A mathematical entity specifiable by a set of components with respect to a system of coordinates and such that the transformation that has to be applied to the components to obtain components with respect to a new system of coordinates is related in a certain way to the transformation that had to be applied to the system of coordinates. (*Zool.*) A muscle which stretches or tightens a part of the body without changing the relative position or direction of the axis of the part. Cf. *laxator*.
tensor force (*Nuc.*). A noncentral force in nuclear physics whose direction depends in part on the spin orientation of the nucleons.
tent (*Surg.*). A roll or plug of soft absorbent material, or of expansible material, for keeping open a wound or dilating an orifice. See also **laminaria tent.**
tentacle (*Bot.*). One of the hairs on the leaf of sundew, which helps in capturing small insects and produces enzymes which digest the prey. (*Zool.*) An elongate, slender, flexible organ, usually anterior, fulfilling a variety of functions in different forms, as exploring, feeling, grasping, holding, and sometimes locomotion. In *Protozoa*, a pseudopod; in *Coelenterata*, a prehensile appendage bearing stinging cells; in *Polychaeta*, a tactile process of the prostomium or peristomium; in *Arthropoda* generally, an antenna (popular); in *Cirripedia*, a thoracic appendage or cirrus; in Insect larvae, any slender fleshy process of the body; in *Gastropoda*, a sensory cephalic process or horn; in some *Pelecypoda*, a sensory process of the mantle-edge; in *Cephalopoda*, an arm; in *Crinoidea*, an arm or ray; in some *Hemichorda*, a process of the collar; in *Cephalochorda*, a tactile process of the velum or the buccal hood; in Fish, a barbel. Also **tentaculum.** *adjs.* **tentacular, tentaculiferous, tentaculiform.**
Tentaculata (*Zool.*). A class of *Ctenophora* distinguished by the possession of tentacles.
tentaculocyst (*Zool.*). In some coelenterate medusoids, a sense organ situated at the margin of the umbrella and consisting of a modified vesicular tentacle, frequently containing lithites.
tenter (*Textiles*). (1) Female operative watching cardroom processes or ringframes. (2) See **stenter.**
tenterhook willow (*Textiles*). See **fearnought.**
Tentest (*Build.*). TN of an insulation board made of fibre.
tenth-value thickness (*Nuc. Eng.*). Thickness of absorbing sheet which attenuates intensity of beam of radiation by a factor of ten.
tentilla (*Zool.*). Branches of a tentacle.
tentorium (*Zool.*). (1) In the Mammalian brain, a strong transverse fold of the dura mater, lying between the cerebrum and the cerebellum. (2) In Insects, the endoskeleton of the head, composed of two or three pairs of apodemes which coalesce at their bases. Its functions are to provide a basis for the attachment of cephalic muscles and give rigidity to the head, to give support to the brain and fore-intestine, and to strengthen the articulatory points of some of the mouthparts.
tepal (*Bot.*). One of the members of a perianth which is not clearly differentiated into a calyx and a corolla.
tepee buttes (*Geol.*). Conical hills of Cretaceous shale, with steep, smooth slopes of talus and a core of shell-limestone, formed *in situ* by the growth of successive generations of lamellibranchs (*Lucina*). Found in the Great Plains of the U.S.A.
tephigram (*Meteor.*). A printed diagram with curves of *DALR* (q.v.) and *SALR* (q.v.) for selected dry-bulb temperatures. When the *ELR* (q.v.) is plotted thereon, the stability conditions at all heights are displayed.
tephrite (*Geol.*). A fine-grained igneous rock resembling basalt and normally occurring in lava flows; characterized by the presence of a feldspathoid mineral in addition to, or in place of, feldspar. According to the particular feldspathoid mineral present, *leucite-tephrite*, *nepheline-tephrite*, and *analcite-tephrite* may be distinguished.
tephroite (*Min.*). An orthosilicate of manganese, which crystallizes in the orthorhombic system. It forms a member of the olivine isomorphous

group, and occurs with zinc and manganese minerals in New Jersey and Sweden.

TEPP (*Chem.*). Bis-*O,O*-diethylphosphoric anhydride, used as an insecticide. Also **ethylpyrophosphate.**

tera-. Prefix, symbol T, denoting 10^{12} times, e.g., a *terawatt-hour* is 10^{12} watt-hours.

teratogen (*Gen.*). An agent that raises the incidence of congenital malformations.

teratology (*Biol.*). The study of monstrosities. (Gk. *teras*, gen. *teratos*, a wonder).

teratoma (*Med.*). A tumourlike mass in the body consisting of tissues derived from the three germ layers (ectoderm, mesoderm, and endoderm) as a result of some abnormality in development.

terbium (*Chem.*). A metallic element, a member of the rare earth group. Symbol Tb, at. no. 65, r.a.m. 158·9254. It occurs in the same minerals as dysprosium, europium, and gadolinium.

terebene, terebine, terebanthenes (*Paint.*). Volatile solvents and thinners derived from heavy petroleum and rosin oils, or petroleum and rosin oils mixed with turpentine.

terebra (*Zool.*). The actual sting or modified ovipositor of female *Hymenoptera*.

terebrate (*Bot.*). Having scattered perforations. (*Zool.*) Possessing a boring organ, or a sting.

terebrator (*Bot.*). The trichogyne of a lichen.

terephthalic acid (*Chem.*). $C_6H_4(COOH)_2$, benzene-1,4-dicarboxylic acid, a powder, hardly soluble in water or ethanol, which sublimes unchanged. It is prepared by the oxidation of 4-toluic acid, or industrially by the catalytic oxidation of 4-xylene. Important material in the manufacture of *Terylene* (q.v.).

terete (*Bot.*). Elongated cylindrical-conical, tapering to a point.

terfas (*Bot.*). The edible subterranean fruit-bodies of the fungus *Terfezia*.

tergite (*Zool.*). The tergum in *Arthropoda* when it forms a chitinous plate.

tergum (*Zool.*). The dorsal part of a somite in *Arthropoda*; one of the plates of the carapace in *Cirripedia*. *adj.* **tergal.**

term diagram (*Nuc.*). Energy-level diagram for isolated atom in which levels are usually represented by corresponding quantum numbers.

terminal (*Bot.*). (1) Situated at the tip of anything. (2) Said of secondary wood parenchyma when this develops only at the end of the growing season, and therefore at the limit between one annual ring and the next. (*Elec. Eng.*) A point in an electrical circuit at which any electrical element may be connected.

terminal bar (*Elec. Eng.*). A bar to which a group of plates of an accumulator is attached. Also called **connector bar, terminal yoke.**

terminal curvature (*Geol.*). A sudden local change in the dip of stratified rocks in the near neighbourhood of a fault, due to the drag of the downward displaced block against the fault plane.

terminal curve (*Horol.*). The curve which connects the inner end of a balance spring to the collet or the outer end to the stud.

terminal equipment (*Teleph.*). The special apparatus required for connecting the normal telephone exchange pairs to special transmission systems (such as radio-telephone links, carrier systems) or to trunk-lines.

terminal impedance (*Elec.*). End or load impedance.

terminal lug (*Elec. Eng.*). A projection on a group of accumulator plates for connexion to an external circuit.

terminal pillar (*Elec. Eng.*). See **post head.**

terminal pole (*Elec. Eng.*). A pole at the end of a power-transmission or telephone line so designed as to withstand the longitudinal load of the conductors as well as the vertical load.

terminals (*Bot.*). The fine end-branches of the veins of a leaf.

terminal screw (*Elec. Eng.*). See **clamping screw.**

terminal tower (*Elec. Eng.*). The transmission line tower at the end of an overhead transmission line; arrangements must be made for taking the pull of the conductors, and for connecting them to the substation or other apparatus at the end of the line.

terminal velocity (*Aero.*). The maximum *limiting velocity* attainable by an aircraft as determined by its total *drag*. (*Phys.*) The constant velocity acquired by a body falling through a fluid when the frictional resistance is equal to the gravitational pull.

terminal-velocity dive (*Aero.*). A nose dive to the greatest obtainable velocity of the machine at that altitude.

terminal voltage (*Elec. Eng.*). The voltage at the supply terminals of an electrical machine.

terminal yoke (*Elec. Eng.*). See **terminal bar.**

terminated level (*Teleph.*). The reading of a level-measuring set at a point in a system when terminated at that point by a resistance equal to the nominal impedance of the system.

termination (*Elec. Eng., etc.*). A device placed at the end of a transmission system. Approximately equivalent to load but implies greater interest in the physical nature of the device, i.e., the load is often 3 ohms when the termination is a loudspeaker. See **matched load.**

terminator (*Astron.*). The border between the illuminated and dark hemispheres of the moon or planets. Its apparent shape is an ellipse and it marks the regions where the sun is rising or setting.

termitarium (*Zool.*). A mound of earth built and inhabited by termites and containing an elaborate system of passages and chambers.

termite shield (*Build.*). A sheet of copper or other noncorroding metal inserted between the foundation and woodwork in buildings, that serves as a barrier to the movement of subterranean termites from the ground to the woodwork of the structure. Also called **antproof course.**

termitocole (*Ecol.*). An organism living in the nests of termites, either in the galleries with the termites (*termitophile*) or not.

termolecular (*Chem.*). Pertaining to three molecules.

ternary (*Chem.*). Consisting of three components, etc. (*Maths.*) Counting with *radix* 3.

ternary fission (*Nuc.*). The splitting of the nucleus into three nuclear fragments. The term is not of universal acceptance in its general application.

ternary system, ternary diagram (*Met.*). The alloys formed by three metals constitute a ternary alloy system, which is represented by the ternary constitutional diagram for the system.

ternate (*Bot.*). (1) Said of a compound leaf with three leaflets. (2) Arranged in threes, as branches arising at about the same point from a stem.

terne metal (*Met.*). Lead-based alloy with 18% tin and 1½–2% antimony.

terne plate (*Eng.*). Iron or steel sheet coated by hot-dipping with terne metal which acts as a carrier for lubricant used in subsequent drawing operations, as rust protection, as a paint base or to facilitate soldering.

terpadienes (*Chem.*). $C_{10}H_{16}$, isocyclic compounds, containing two double bonds.

Numerous compounds in the terpene series are terpadienes. The carbon atoms are numbered as follows:

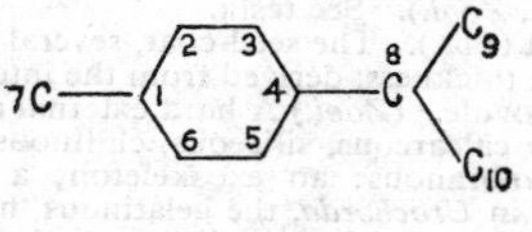

terpenes (*Chem.*). Compounds of the formula $(C_5H_8)_n$, the majority of which occur in plants. The value of *n* is used as a basis for classification. (i) Monoterpenes $C_{10}H_{16}$; (ii) sesquiterpenes $C_{15}H_{24}$; (iii) diterpenes $C_{20}H_{32}$; (iv) triterpenes $C_{30}H_{48}$; (v) tetraterpenes $C_{40}H_{64}$. All naturally occurring terpenes can be built up of isoprene units. In practice the name terpene is often used for monoterpenes.

terpinenes (*Chem.*). A series of three monocyclic, isomeric terpenes, of which two occur naturally in various vegetable oils and the other is a synthetic product. Their structures are:

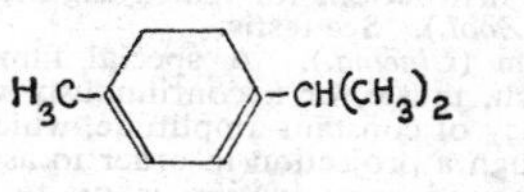

α-terpinene

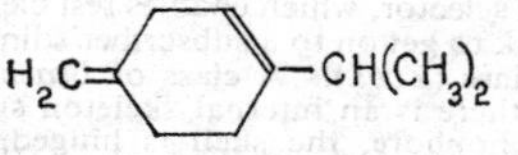

β-terpinene (synthetic)

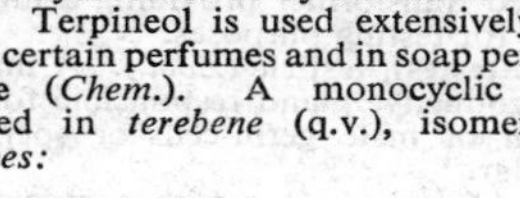

γ-terpinene

terpineol (*Chem.*). $C_{10}H_{17}OH$, colourless crystals; m.p. 37°C, b.p. 218°C; it can be obtained from limonene hydrochloride by the action of caustic potash. Terpineol is used extensively as the basis of certain perfumes and in soap perfumery.

terpinolene (*Chem.*). A monocyclic terpene contained in *terebene* (q.v.), isomeric with *terpinenes:*

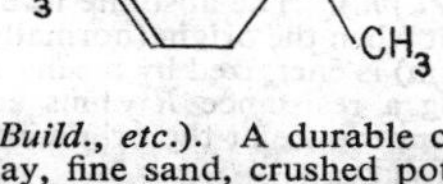

terracotta (*Build., etc.*). A durable composition of fine clay, fine sand, crushed pottery waste, etc., easily manipulable before hardening (by fire, like bricks). Generally unglazed, used for sculpture, pottery, and architecturally in the form of building blocks, decorative tiles, etc.

terrain-clearance indicator (*Aero.*). Same as **radio altimeter.**

Terramycin (*Pharm.*). Oxytetracycline. TN for antibiotic produced by *Streptomyces rimosus*. The name was derived from the grey, earthy appearance of the mould. This antibiotic is effective against both Gram-positive and Gram-negative bacteria. It has the following structure:

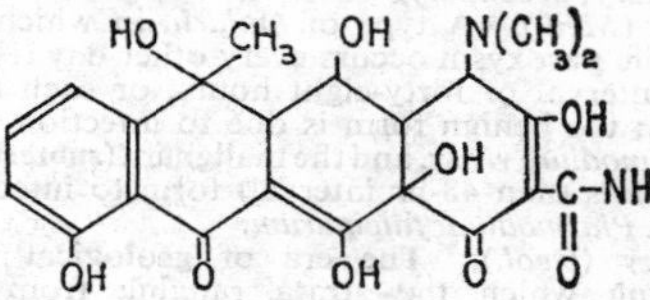

terrazzo (*Build.*). A rendering of cement (white or coloured) and marble or granite chippings, used as a covering for concrete floors, on which it is floated and finally polished with abrasive blocks and fine grit stones; frequently precast (particularly for sills). Also **Venetian mosaic.**

terrestrial equator (*Geog.*). An imaginary circle on the surface of the earth, the latter being regarded as cut by the plane through the centre of the earth perpendicular to the polar axis; it divides the earth into the northern and southern hemispheres, and is the primary circle from which terrestrial latitudes are measured.

terrestrial magnetism (*Geog., Phys.*). The magnetic properties exhibited within, on, and outside the earth's surface. There is a nominal (magnetic) North pole in Canada and a nominal South pole opposite, the positions varying cyclically with time. The direction indicated by a compass needle at any one point is that of the horizontal component of the field at the point. Having the characteristics of flux from a permanent magnet, the earth's magnetic field probably depends on currents within the earth and also on those arising from ionization in the upper atmosphere, interaction being exhibited by the Aurora Borealis.

terrestrial poles (*Geog.*). The two diametrically opposite points in which the earth's axis cuts the earth's surface are the geographical poles. The magnetic poles (see **terrestrial magnetism**), the positions to which the compass needle will point, are unstable and differ from the geographical. (N magnetic pole: ca. 76°N, 101°W; S magnetic pole: ca. 66°S, 139°E).

terrestrial radiation (*Meteor.*). At night the earth loses heat by radiation to the sky, the maximum cooling occurring when the sky is cloudless and the air dry. Dew and hoar-frost are the result of such cooling.

terrestrial telescope (*Optics*). Telescope consisting of an objective and a four-lens eyepiece (*terrestrial eyepiece*), giving an erect image (see **erecting prism**) of a distant object.

terre verte (*Paint.*). A green iron silicate pigment of low opacity and tinting strength, now little used. Also **Verona (or Veronese) green.**

terricolous (*Bot.*). Living in or on the soil.

terrigenous sediments (*Geol.*). Those sediments that are deposited on the shallower parts of the sea-floors; they consist of detritus derived from the land areas. Cf. *deep-sea deposits*.

territory (*An. Behav.*). A *home range* (q.v.), which is partly or wholly defended by an individual, thus ensuring even spacing and adequate food and shelter. Territories are held by some species of most classes of Vertebrates, but especially by Birds, and also by some *Arthropoda*.

terro-metallic clinkers (*Build.*). Bricks similar to *Dutch clinkers* (q.v.) but made from a clay which burns to a nearly black colour.

terry (*Textiles*). Special cloth, e.g., towelling or bedspreads, using an additional warp beam for looped pile on one or both sides.

tertial (*Zool.*). In Birds, a flight feather of the third row, attached to the upper arm. Cf. *primary, secondary*.

tertian (*Med.*). A type of *malaria* in which the febrile paroxysm occurs every other day (i.e. at an interval of forty-eight hours, or each *third* day); the benign form is due to infection with *Plasmodium vivax*, and the malignant (**subtertian**; i.e., less than 48-hr interval) form to infection with *Plasmodium falciparum*.

Tertiary (*Geol.*). The era of geological time during which the strata ranging from the Eocene to the Pliocene were deposited.

tertiary alcohols (*Chem.*). Alcohols containing the group $\geqslant C{\cdot}OH$. When oxidized, the carbon chain is broken up, resulting in the formation of two or more oxidation products containing a smaller number of carbon atoms in the molecule than the original compound.

tertiary amines (*Chem.*). Amines containing the nitrogen atom attached to three groups. Tertiary aliphatic-aromatic amines yield 4-nitroso compounds with nitrous (nitric(III)) acid.

tertiary cell wall, tertiary layer, tertiary thickening (*Bot.*). A deposit of wall-thickening on the inner surface of the secondary wall of a cell, tracheid, or vessel, usually in the form of rings or of a loose spiral band.

Tertiary igneous rocks (*Geol.*). The various types of igneous rocks which were intruded or extruded during early Tertiary times, especially over a region stretching from Britain to Iceland; e.g., in the Inner Hebrides and northeast Ireland (the Thulean Province).

tertiary layer (*Bot.*). See **tertiary cell wall.**

tertiary nitro compounds (*Chem.*). Nitro compounds containing the group $\geqslant C{\cdot}NO_2$. They contain no hydrogen atom attached to the carbon atom next to the nitro group, and they have no acidic properties.

tertiary thickening (*Bot.*). See **tertiary cell wall.**

tertiary winding (*Elec.*). A third winding on a transformer core, linking the same flux as the primary and secondary windings. The functions, and the names, of all three are interchangeable. May be used for additional output or monitoring.

tervalent (*Chem.*). See **trivalent.**

Terylene (*Chem.*). TN for straight-chain polyester fibres derived from condensation of the diglycol ester of *terephthalic acid* (q.v.) with elimination of ethan-1,2-diol, used widely in the manufacture of fabrics, clothing materials, and other textiles.

Teschen disease (*Vet.*). Infectious pig paralysis; virus encephalomyelitis of swine, characterized by mild fever, nervous excitement, and paralysis.

teschenite (*Geol.*). A coarse-grained basic (gabbroic) igneous rock consisting essentially of plagioclase, near labradorite in composition, titanaugite, ilmenite, and olivine (or its decomposition products); primary analcite occurs in wedges between the plagioclase crystals, which it also veins.

tesla (*Elec.*). SI unit of magnetic flux density equal to 1 weber per square metre. Symbol T.

Tesla coil (*Elec. Eng.*). Simple source of high-voltage oscillations for rough testing of vacua and gas (by discharge colour) in vacuum systems.

Tesla current (*Elec.*). High-frequency current of moderately high voltage for therapeutic use.

tessella (*Arch.*). See **tessera.**

tessellate (*Bot.*). Said of a surface marked out in little squarish areas, like a tiled pavement.

tessellated pavement (*Build.*). A pavement formed of small pieces of stone, marble, etc., in the manner of a mosaic. Also **Roman mosaic.**

tessera, tessella (*Arch.*). One of the small pieces of stone, marble, etc., used in the mosaic of a tessellated pavement. *pls.* **tesserae, tessellae.**

tesseralkies (*Min.*). See **skutterudite.**

test (*Bot., Zool.*). See **testa.**

testa, test (*Bot.*). The seed coat, several layers of cells in thickness, derived from the integuments of the ovule. (*Zool.*) A hard external covering, usually calcareous, siliceous, chitinous, fibrous or membranous; an exoskeleton; a shell; a lorica; in *Urochorda*, the gelatinous 'house'; in *Echinoidea*, the hard calcareous shell. *adjs.* **testaceous, testacean.**

testaceous (*Bot.*). Of the colour of old red bricks.

test bed (*Eng.*). Area with full monitoring facilities for testing new or repaired machinery under full working conditions.

test board (*Elec. Eng.*). A switchboard carrying instruments and switches for connecting up to apparatus to be tested.

test card, test chart (*TV*). See **test pattern.**

test desk (*Teleph.*). The special position, away from the operators, at which are situated test clerks who can apply special tests to faulty lines to ascertain the nature of such faults, and who issue instructions for remedying them.

testes (*Zool.*). See **testis.**

test film (*Cinema.*). A special film, either of speech, music, or a continuously varying frequency of constant amplitude, which is passed through a projection in order to ascertain that the sound reproduction is up to a specified standard.

test final selector (*Teleph.*). The selector following the test selector, which enables test clerks at the test desk to get on to a subscriber's line.

Testicardines (*Zool.*). A class of *Brachiopoda* in which there is an internal skeleton supporting the lophophore, the shell is hinged, and the intestine ends blindly, there being no anus.

testicle (*Zool.*). See **testis.**

testing machine (*Eng.*). A machine for applying accurately measured loads to a test piece, to determine the suitability of the latter for a particular purpose.

testing position (*Teleph.*). A position equipped for testing purposes and forming part of a test desk or test rack.

testing set (*Elec. Eng.*). A self-contained set of apparatus, including switches, instruments, etc., for carrying out certain special tests.

testing transformer (*Elec. Eng.*). A specially designed transformer providing a high-voltage supply for testing purposes.

testis (*pl.* **testes**), **testicle** (*Zool.*). A male gonad or reproductive gland responsible for the production of male germ-cells or sperms. *adj.* **testicular.**

test jack (*Teleph.*). One with contacts in series with a circuit, so that a testing device can be immediately introduced for locating faults.

test level (*Teleph.*). The absolute level at a point in a circuit when the origin (normally the 2-wire termination) is energized by means of a generator having a resistance R ohms equal to the nominal impedance at the origin and an e.m.f. of $2\sqrt{R/1000}$ volts.

testosterone (*Biochem.*). A steroid:

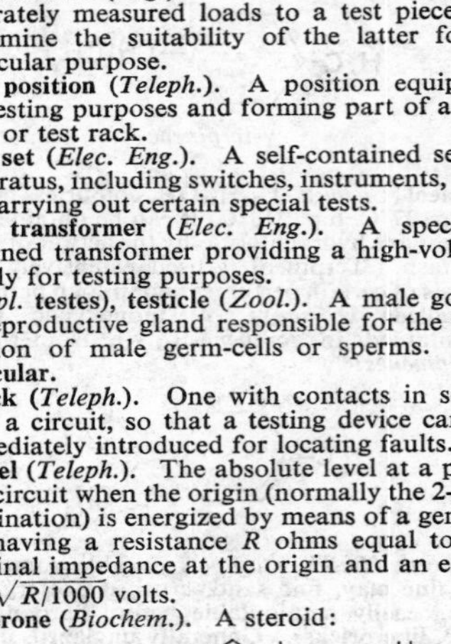

H
OH
O

obtained from the testis. It is the most active androgenic substance so far isolated and is the chief male sex hormone. See also **androsterone.**

testosterone propanoate (*Med.*). A hormonal extract obtained from the testes of bulls. It can also be prepared synthetically. Useful for children with mental retardation resulting from malnutrition.

test pattern (*TV*). A transmitted chart with lines and details to indicate particular characteristics of a transmission system, used in TV for general testing purposes. Also **test card, test chart.**

test piece (*Eng.*). A piece of material accurately turned or shaped, often to specified standard dimensions, for subjecting to a tensile test, shock test, etc., in a testing machine.

test point (*Elec. Eng.*). Designated junction between components in an equipment, where the voltage can be stated (as a minimum or with tolerance) for a quick verification of correct operation.

test program (*Comp.*). See **check program.**

test record (*Acous.*). A gramophone record specially made for the testing of reproducing equipment, having on its surface constant-frequency or gliding tones, or selected recordings of speech or music, to emphasize particular faults in the subsequent reproduction.

testriol (*Chem.*). See **chimyl alcohol.**

test selector (*Teleph.*). A selector operated by a test clerk in an automatic exchange; by means of it, through a *test final selector* (q.v.), he is able to get on to any line in the exchange.

test strip (*Cinema.*). The specially exposed unmodulated sound-track which is made to ascertain the current in the exciter lamp of a recording machine which gives the requisite density on the negative, after normal development. See **slit test.**

test terminals (*Elec. Eng.*). Circuit terminals to which a connexion is made for purposes of testing.

test vehicles (*Aero.*). Aircraft for aerodynamic, control and other tests in guided weapon development. They may simply be for gathering basic information, or they may be actual missiles without a warhead. They are known by their initials: **CTV**, command (control) test vehicle; **GPV**, general-purpose vehicle; **MTV**, missile test vehicle; **RJTV**, ramjet test vehicle; **RTV**, rocket test vehicle.

tetanic contraction (*Zool.*). See **tetanus.**

tetanus (*Med.*). Lockjaw. A disease due to infection with the tetanus bacillus, *Clostridium tetani*, the toxin secreted by which causes the symptoms and signs of the disease, viz., painful tonic spasms of the muscles, which usually begin in the jaw and then spread to other parts. (*Zool.*) The state of prolonged contraction which can be induced in a voluntary muscle by a rapid succession of stimuli.

tetany (*Med.*). A condition characterized by heightened excitability of the motor nerves and intermittent painful muscular cramps, occurring in many abnormal states, especially those associated with hypocalcaemia.

tetartohedral (*Crystal.*). Containing a quarter of the number of faces required for the full symmetry of the crystal system.

tête-beche (*Radio*). Single sideband transmitters placed far apart but operating over same frequency range. One uses the upper, the other the lower sideband for transmission.

tetra-. From Gk. *tetra-*, a prefix indicating four, e.g., *tetracyte.*

tetraboric acid (*Chem.*). See **boric acid.**

Tetrabranchiata (*Zool.*). An order of *Cephalopoda* with well-developed calcareous shells, the living form having two pairs of ctenidia and kidneys, numerous arms without suckers, simple eyes, and no chromatophores. Includes the fossil *Ammonoidea* first found in the mid-Palaeozoic, and the *Nautiloidea* including the one living form, the Pearly Nautilus.

tetracerous (*Zool.*). Having four horns.

tetrachloroethane (*Chem.*). $C_2H_2Cl_4$, used as a solvent and also as a vermifuge. B.p. 146°C. Prepared by the chlorination of ethyne in the presence of antimony(III)chloride. Intermediate in the manufacture of *trichloroethene.*

tetrachloromethane (*Chem.*). See **carbon tetrachloride.**

tetrachlorosilane (*Chem.*). See **silicon tetrachloride.**

tetracoccus (*Zool.*). A micro-organism occurring in groups of four.

Tetractinellida (*Zool.*). An order of *Demospongiae* in which the skeleton is composed of tetraxial and monaxial spicules.

tetracyclic (*Bot.*). Said of a flower which has four whorls or members.

tetracycline (*Pharm.*). A yellow crystalline antibiotic used esp. for treatment of the respiratory and urinary tracts, but additionally for a very wide range of other infections. Used also as a preservative in certain foods, e.g., fish and poultry. Originates from a soil organism *Streptomyces viridifaciens.* TNs **Achromycin, Tetracyn, Panmycin,** etc.

tetracyte (*Bot.*). One of the four cells formed after a meiotic division.

tetrad (*Bot.*). A group of four spores, formed by meiosis, remaining together until they are nearly or quite mature. (*Cyt.*) A bivalent chromosome, formed during the latter part of meiotic prophase, which shows signs of division into four longitudinal threads (*chromatids*).

tetradactyl (*Zool.*). Having four digits.

tetrad division (*Bot.*). The nuclear and cell divisions occurring when a spore mother cell divides to give four spores.

tetradidymous (*Bot.*). Fourfold.

tetradymite (*Min.*). An ore of tellurium, of composition Bi_2Te_2S, sometimes also containing selenium; crystallizes in the trigonal system. Bismuth tellurides are commonly found in gold-quartz veins or in contact metamorphic rocks.

tetradynamous (*Bot.*). Said of a flower which has six stamens, four longer than the others.

tetrafluorosilane (*Chem.*). See **silicon tetrafluoride.**

tetragonal system (*Crystal.*). The crystallographic system in which all the forms are referred to three axes at right angles; two are equal and are taken as the horizontal axes, whilst the vertical axis is either longer or shorter than these. It includes such minerals as zircon and cassiterite. Also **pyramidal system.**

tetragonous (*Bot.*). Having four angles and four convex faces.

tetrahedrite (*Min.*). A sulphide of copper and antimony, $(Cu, Fe)_3 (Sb, As)S_3$, crystallizing in the tetrahedral division of the cubic system; frequently contains other metals such as bismuth, mercury, silver (as in the old silver-mines of Devon and Cornwall), zinc, and iron. It is used as an ore of copper and, in some cases, of these other metals. Also called **fahlerz, fahl ore, grey copper ore.** See also **tennantite.**

tetrahedron (*Maths.*). A polyhedron having four triangular faces. *adj.* **tetrahedral.**

tetrahydrofuran (*Chem.*). Tetramethene oxide. Organic solvent used in refining lubricating oils and as a solvent for a wide range of plastics and resins. B.p. 64–66°C, rel. d. (20°C) 0·888:

H_2C—CH_2
H_2C CH_2
O

tetrahydronaphthalene (*Chem.*). See **tetralin.**

tetrakontan (*Bot.*). Having 4 flagella.

tetralin (*Chem.*). Organic solvent with a high boiling-point (206°C), used in a wide variety of products, such as floor polishes, paints, and varnishes. Also called **tetrahydronaphthalene.**

tetramerous (*Bot., Zool.*). Having 4 parts; arranged in 4s; arranged in multiples of 4.

tetramethene oxide (*Chem.*). See **tetrahydrofuran.**

tetramorphous (*Chem.*). Existing in 4 different crystalline forms.

tetraoses (*Chem.*). See **tetroses.**

tetraploid (*Cyt.*). Possessing twice the normal number of chromosomes. Cf. *haploid, diploid.*

tetrapneumonous (*Zool.*). Having 4 lung-books.

tetrapod (*Zool.*). Having 4 feet.

tetrapterous (*Zool.*). Having 4 wings.

tetrarch (*Bot.*). Having 4 strands of xylem.

tetrasomic (*Cyt.*). A tetraploid nucleus (or organism) having 1 chromosome 4 times over, the others in duplicate.

tetrasporangium (*Bot.*). A unilocular sporangium containing tetraspores.

tetraspore (*Bot.*). One of the asexual spores of the red algae; these spores are formed in groups of four.

tetrasporous (*Bot.*). Having or containing 4 spores.

tetraster (*Cyt.*). A complex mitotic figure formed in an ovum after polyspermy.

tetrastichous (*Bot.*). Arranged in 4 rows.

tetrathognathous (*Zool.*). Having the jaws borne on the 4th somite of the head.

tetravalent (*Chem.*). Of an atom, having 4 electrons in the outer or valency shell.

tetraxon (*Zool.*). Of sponge spicules, having 4 axes.

tetrazo dyes (*Chem.*). See **disazo dyes.**

tetrazole (*Chem.*). A 5-membered heterocyclic compound containing 4 nitrogen atoms and 1 carbon atom in the ring.

tetrazooid (*Zool.*). In some *Urochorda*, one of 4 zooids developed from the stolon.

tetrode (*Electronics*). 4-electrode thermionic valve, e.g., *screened-grid valve* (q.v.).

tetrode transistor (*Electronics*). Transistor with additional base contact to improve high-frequency performance.

tetronal (*Pharm.*). A crystalline compound, 3,3-bis-ethylsulphonyl pentane ($C_9H_{20}S_2O_4$), got by the action of hydrochloric acid on ethanethiol hydrate and pentan-3-one; a hypnotic and sedative drug, highly toxic, so rarely used.

tetroses (*Chem.*). Tetraoses. Monosaccharides containing 4 carbon atoms in the molecule, e.g., $HO{\cdot}CH_2{\cdot}(CHOH)_2{\cdot}CHO$.

tetryl (*Chem.*). N,2,4,6-tetranitromethylaniline, a yellow crystalline compound used as a detonator.

TE-wave (*Telecomm.*). Abbrev. for *transverse electric* wave, having no component of electric force in the direction of transmission of electromagnetic waves along a waveguide. Also known as **H-wave** (since it must have magnetic field component in direction of transmission).

tex (*Textiles*). A unit to measure the thickness of thread. Thread of 1 tex has a mass of 1 gram per kilometre length, or 1 milligram per metre length. Equals 9 denier.

Texas fever (*Vet.*). A disease of cattle caused by the protozoon *Babesia bigemina.* See **redwater** (1).

text (*Typog.*). The body of matter in a printed book, exclusive of preliminary matter, end matter, notes, and illustrations; words set to music, as distinct from the accompanying music.

textile fibres (*Textiles*). Filaments of vegetable, animal, mineral, or synthetic origin used for making yarns and fabrics. Cotton, ramie and flax are vegetable fibres; wool, mohair, and silk are of animal origin; asbestos is a mineral. Rayon is a regenerated cellulose. Nylon, Terylene, Acrilan and many others are true manmade fibres, being totally synthetic types derived from polymerization, etc., processes.

textiles. Term used loosely to describe any fibres, yarns or cloths, woven, felted or bonded.

texture. The mode of union or disposition, in regard to each other, of the elementary constituent parts in the structure of any body or material. (*Geol.*) That quality of a rock which is determined by the relative sizes, disposition, and arrangement of the component minerals. The nomenclature and classification of rocks are governed by mineral composition and texture. See, e.g., **graphic texture, ophitic texture, poikilitic texture.** (*Photog.*) The quality of the surface of a photograph.

texture brick (*Build.*). See **rustics.**

T-gauge (*Join.*). Type of marking gauge with long marker pin and large flat fence; used for marking jobs where mouldings, beading, or other surface items have to be cleared.

Tg point (*Glass*). See **transformation points.**

TGT (*Aero.*). See **gas temperature.**

Th (*Chem.*). The symbol for *thorium*.

thalamencephalon (*Zool.*). See **diencephalon.**

thalamifloral (*Bot.*). Said of a flower which has all its members inserted separately on the receptacle, with the gynaeceum superior.

thalamium (*Bot.*). The hymenium of an apothecium.

thalamus (*Bot.*). The *receptacle* of a flower. (*Zool.*) In the Vertebrate brain, the larger, more ventral part of the dorsal zone of the thalamencephalon.

thalasso-. Prefix from Gk. *thalassa*, sea.

thalassoid (*Ecol.*). See **pseudomarine.**

thalassophyte (*Bot.*). A seaweed.

Thaliacea (*Zool.*). A class of *Urochorda* in which the adult is pelagic and tailless, with a degenerate nervous sytem and a posterior atriopore; the gill-clefts are not divided by external longitudinal bars.

thalidomide (*Pharm.*). α-Phthalimidoglutarimide, $C_{13}H_{10}N_2O_4$. A non-barbiturate sedative drug, used especially in early pregnancy until withdrawn (UK) in 1961 because of teratogenic effects.

thalline exciple (*Bot.*). An exciple which contains algal cells.

thalline margin (*Bot.*). The margin of an apothecium in a lichen when it is of the same structure as the thallus, and usually coloured like it.

thallium (*Chem.*). Symbol Tl, at. no. 81, r.a.m. 204·37, rel. d. 11·85, m.p. 303·5°C, b.p. 1650°C. White malleable metal like lead. Several thallium isotopes are members of the uranium, actinium, neptunium, and thorium radioactive series. Thallium isotopes are used in scintillation crystals.

thallodic (*Bot.*). Belonging to a thallus.

Thallophyta (*Bot.*). A major division of the plant kingdom, including fungi, algae, and a number of smaller groups, e.g., lichens. The *Myxomycetes* (slime moulds) and *Bacteria* are frequently also included. The plant body is a thallus, but varies a good deal in complexity, and the plants do not produce flowers or seeds.

thallous carbonate (*Chem.*). Thallium(I). Tl_2CO_3. Has marked fungicidal potency against mildew growth in textile fabrics.

thallus (*Bot.*). A plant body which is not differentiated into root and shoot; it may be a single cell, a filament of cells, or a complicated branching multicellular structure, in which case it has no true root.

thalofide cell (*Electronics*). A photoconducting cell employing thallium oxy-sulphide as the light-sensitive agent.

thalweg (*Geol.*). (German 'valley way'.) The name frequently used for the longitudinal profile of a river, i.e., from source to mouth.

thanatocoenosis (*Geol.*). Assemblage of fossil remains of organisms which were not associated during their life but were brought together after death.

thanatoid (*Zool.*). Poisonous, deadly, lethal; as some venomous animals.

thatching (*Build.*). A form of roof-covering composed of courses of reeds, straw, or heather, laced together.

thawing (*Heat*). A term used to describe the beginning of the fusion process of a solid. The corresponding temperature is the *thaw-point* of the solid.

theatre main (*Elec. Eng.*). A special service main providing an alternative electric supply to a theatre if the normal supply fails.

thebaine (*Pharm.*). $C_{19}H_{21}NO_3$; glistening plates; m.p. 193°C; contained in opium; it is a morphine in which both hydroxyls are methylated.

Thebesian valve (*Zool.*). An auricular valve of the Mammalian heart.

theca (*Bot.*). (1) An ascus. (2) The capsule of a moss. (3) A pollen sac. (4) An anther. (*Zool.*) A case or sheath covering or enclosing an organ, as the *theca vertebralis* or dura mater enclosing the spinal cord; a tendon sheath; the wall of a coral cup. *adjs.* **thecal, thecate.**

theca lutein (*Histol.*). See **paralutein.**

thecaspore (*Bot.*). An ascospore.

theciferous (*Bot.*). Containing asci.

thecium (*Bot.*). The hymenium of the apothecium of a lichen.

thecodont (*Zool.*). Having the teeth implanted in sockets in the bone which bears them.

Theelin (*Biochem.*). TN for brand of *oestrone.*

Theelol (*Biochem.*). TN for brand of *oestriol.*

theine (*Pharm.*). See **caffeine.**

thelygenic (*Biol.*). Producing mainly female offspring.

thelykaryotic (*Zool.*). Possessing a female pronucleus only.

thelytoky (*Zool.*). Parthenogenesis resulting in the production of females only.

thenal (*Zool.*). Pertaining to the palm of the hand.

Thénard's blue (*Paint.*). See **cobalt blue.**

theobromine (*Pharm.*). 3,7-Dimethyl-2,6-dihydroxypurine. Occurs in cacao beans. A crystalline powder with a bitter taste. It is used as a diuretic, a cardiac stimulant, and for arterial dilatation.

theodolite (*Surv.*). An instrument for measuring horizontal and vertical angles by means of a telescope mounted on an axis made vertical by levelling screws, and rotated both horizontally on this axis and in horizontal bearings. Circular graduated plates are used to measure amount of rotatory motion when the telescope is sighted on successive signal stations.

theophylline (*Pharm.*). 1,3-Dimethyl-2,6-dihydroxypurine. Occurs in tea and is a strong diuretic and muscle relaxant. Little used, its more soluble derivatives, e.g., aminophylline, being preferred. Colourless crystals; m.p. 264°C.

theorem (*Maths.*). A statement of a mathematical truth together with any qualifying conditions.

theorem of the equipartition of energy (*Chem.*). See **principle of the equipartition of energy.**

theoretical plate (*Chem. Eng.*). A concept, used in distillation design, of a plate in which the vapour and liquid leaving the plate are in equilibrium with each other.

theories of light (*Phys.*). Interference and diffraction phenomena are explained by the *wave theory*, but when light interacts with matter the energy of the light appears to be concentrated in *quanta*, called photons. The *quantum* and *wave theories* (qq.v.) are supplementary to each other.

theory of indicators (*Chem.*). See **Ostwald's—.**

theralite (*Geol.*). A coarse-grained, holocrystalline igneous rock composed essentially of the minerals labradorite, nepheline, purple titanaugite, and often with soda-amphiboles, biotite, analcite, or olivine.

therapeutic (*Med.*). Pertaining to the medical treatment of disease (**therapy**); remedial; curative; preventive. Hence **therapeutics**, that part of medical science which deals with the treatment of disease; the art of healing.

therapeutic ratio (*Radiol.*). The ratio between tumour lethal dose and tissue tolerance. In radio-resistant tumours, the tumour lethal dose equals, or is greater than, the dose required to destroy normal tissues.

therblig (*Work Study*). Name given to each of the specific divisions of movement, according to the purpose for which it is made. [Backward spelling of *Gilbreth*, name of the inventor.]

Theria (*Zool.*). A subclass of the Mammals, containing the extinct *Pantotheria* or *Trituberculata*, the *Metatheria* (marsupials) and the *Eutheria* (placentals).

therm (*Heat*). A unit of energy used mainly for the sale of gas; equals 10^5 Btu or 105·5 MJ.

thermal (*Aero., Meteor.*). An ascending current due to local heating of air, e.g., by reflection of the sun's rays from a beach.

thermal agitation noise (*Electronics*). See **thermal noise.**

thermal ammeter (*Elec. Eng.*). One in which the deflection of the pointer depends on the sag of a fine wire, carrying the current to be measured, due to thermal expansion.

thermal analysis (*Met.*). The use of cooling or heating curves in the study of changes in metals and alloys. The freezing-points and the temperatures of any polymorphic changes occurring in pure metals may be determined. The freezing ranges and temperatures of changes in solid alloys may also be studied. The data obtained are used in constructing constitutional diagrams.

thermal capacity (*Heat*). See **water equivalent.**

thermal catastrophe (*Electronics*). See **thermal runaway.**

thermal circuit-breaker (*Elec. Eng.*). A miniature-type circuit-breaker whose overload device operates by virtue of thermal expansion.

thermal column (*Nuc. Eng.*). Column or block of moderator in reactor which guides large thermal neutron flux to given experimental region.

thermal comparator (*Heat*). A device which enables comparative measurements of the thermal conductivities of solids to be made rapidly. The instrument is also used to make comparative measurements of foil thickness, surface deposits, etc.

thermal conductivity (*Heat*). A measure of the rate at which heat flows through a given material. Measured by the amount of heat, Q, that flows in time, t, through temperature difference, θ, across thickness, d, of cross-section area, A, of the material, where no heat is gained or lost during transit. Given by

$$k=\frac{Qd}{At\theta}.$$

Units are watts per metre kelvin (W/m K). Materials with high electrical conductivities (most metals) tend to have high thermal conductivities, and vice versa, although the range of values is more restricted for the latter.

thermal converter (*Elec. Eng.*). The combination of a thermoelectric device, e.g., thermocouple, and an electrical heater thus converting an electrical quantity into heat and then into a voltage. Used in telemetering systems; see **thermocouple meter**. Also called **thermo-element**.

thermal cross-section (*Nuc.*). Effective nuclear cross-section for neutrons of thermal energy.

thermal cut-out (*Elec. Eng.*). A thermal circuit-breaker designed to screw into a standard barrel-type fuse-holder.

thermal cycle (*Heat*). An operating cycle by which heat is transferred from one part to another. In reactors, separate heat transfer and power circuits are usual to prevent the fluid flowing through the former, which becomes radioactive, from contaminating the power circuit.

thermal cycling (*Heat*). The subjection of a substance to a number of temperature and pressure cycles in succession. Used in petrol refining (*cracking*).

thermal death-point (*Biol.*). The temperature at which an organism is killed or a plant virus inactivated.

thermal detector (*Elec. Eng.*). Any detector of high-frequency currents which operates by virtue of their heating effect when passed through a resistance.

thermal diffusion (*Heat*). Process in which a temperature gradient in a mixture of fluids tends to establish a concentration gradient. Has been used for isotope separation. Also known as **Soret effect**.

thermal diffusivity (*Heat*). Thermal conductivity divided by the product of specific heat capacity and density; more generally applicable than thermal conductivity in most heat transfer problems. Unit is metre squared per second.

thermal dissociation (*Chem.*). The dissociation of certain molecules under the influence of heat.

thermal efficiency (*Eng.*). Of a heat engine, the ratio of the work done by the engine to the mechanical equivalent of the heat supplied in the steam or fuel.

thermal effusion (*Phys.*). The leaking of a gas through a small orifice, the gas being at a low pressure so that the mean free path of the molecules is large compared with the dimensions of the orifice.

thermal electromotive force (*Elec. Eng.*). That which arises at the junction of different metals because of a temperature different from the rest of the circuit. Widely used for measuring temperatures relative to that of the cold junction, e.g., ice in a vacuum flask.

thermal emissivity (*Bot.*). The loss of heat from a leaf by radiation, conduction, and convection.

thermal excitation (*Nuc.*). Collision processes between particles by which atoms and molecules can acquire extra energy.

thermal fission factor (*Nuc. Eng.*). In reactors, the ratio of the fast neutrons emitted from, to the thermal neutrons absorbed by, the fuel elements.

thermal flasher (*Heat*). A flasher the operation of which depends upon the heating effect of the current which it is controlling. The movement of a wire or bimetallic strip, when heated by the current, causes it to interrupt the circuit, which is not remade until it has cooled.

thermal inertia (*Nuc. Eng.*). See **thermal response**.

thermal instability (*Elec. Eng.*). The condition when, the power factor rising so rapidly with temperature that the losses increase more rapidly than the dissipation, the dielectric becomes progressively hotter and burns out.

thermal instrument (*Elec. Eng.*). An instrument the operation of which depends upon the heating effect of a current. See **hot-wire, thermocouple instrument**.

Thermalite (*Build.*). TN for a manufactured material for building blocks of light density and high insulation value.

thermalization (*Nuc.*). Process of slowing fast neutrons to thermal energies. In reactors normally the function of the moderator.

thermal leakage factor (*Nuc. Eng.*). Ratio of number of thermal neutrons lost from, to the number absorbed in, reactor core.

thermal limit (*Elec. Eng.*). The maximum permissible output of an electrical machine as governed by considerations of safe temperature rise.

thermal metamorphism (*Geol.*). The process by which the atoms forming the constituent parts of rocks are regrouped as a result of induced temperature changes, due to contact with, or proximity to, molten magma. Cf. *regional metamorphism*.

thermal microphone (*Acous.*). A microphone which depends for its operation on the variation of resistance of a fine wire or foil on the passage of a sound wave, the velocity of the particles in the sound wave cooling the heated wire.

thermal neutrons (*Nuc.*). See **neutron**.

thermal noise (*Electronics*). That arising from random (Brownian) movements of electrons in a resistor, and which limits the sensitivity of electronic apparatus dealing with transmission frequency bands. The noise voltage V is given by

$$V=\sqrt{4RkT\,.\,\delta f},$$

where δf = frequency bandwidth,
R = resistance of source,
k = Boltzmann's constant,
T = absolute temperature.

Also known as **circuit noise, Johnson noise, output noise, resistance noise**. See **Nyquist noise theorem**.

thermal ohm (*Heat*). See **thermal resistance**.

thermal precipitator (*Powder Tech.*). A device for sampling aerosol particles. The gas stream is drawn through a chamber containing a hot wire. The particles near the wire are bombarded by gas molecules of high kinetic energy, then driven away from it and collected on microscope slides located on opposite sides of the hot wire.

thermal radiation (*Med.*). Analysis of the heat produced by the human body, by the use of infrared radiation to penetrate surface tissue, in order to diagnose certain diseases affecting deeper tissue and to locate superficial tumours.

thermal reactor (*Nuc. Eng.*). One for which fission is principally due to thermal neutrons. Formerly called **slow reactor**.

thermal receiver (*Acous.*). See **thermophone.**

thermal relay (*Elec. Eng.*). A relay the operation of which depends upon the heating effect of an electric current.

thermal resistance (*Heat*). Resistance to the flow of heat. The unit of resistance is the *thermal ohm*, which requires a temperature difference of 1°C to drive heat at the rate of 1 watt. If the temperature difference is θ°C, the resistance S thermal ohms, and the rate of driving heat W watts, then $\theta = SW$.

thermal resistor (*Electronics*). See **thermistor.**

thermal response (*Nuc. Eng.*). Of a reactor, its rate of temperature rise if no heat is withdrawn by cooling. Its reciprocal is the *thermal inertia.*

thermal runaway (*Electronics*). The effect arising when the current through a semiconductor creates sufficient heat for its temperature to rise above a critical value. The semiconductor has a negative temperature coefficient of resistance, so the current increases and the temperature increases again, resulting in ultimate destruction of the device. Also **thermal catastrophe.**

thermal shield (*Nuc. Eng.*). Inner shield of a reactor, used to protect biological shield from excess heating.

thermal shock (*Heat*). That resulting when a body is subjected to sudden changes in temperature. The ability of the material to withstand its effects is called its *thermal shock resistance.*

thermal siphon (*Heat*). The system causing flow round a vertical loop of fluid when the bottom of one column is heated, the fluid rises, cools at the top and falls down the other column. Used, e.g., for heating buildings, isotope separators.

thermal spike (*Electronics*). Surface hot spot(s) due to friction. (*Nuc.*) See **spike.**

thermal station (*Elec. Eng.*). An electric generating station in which the prime movers are steam-turbines or internal-combustion engines.

thermal transpiration (*Vac. Tech.*). The passage, under molecular flow conditions, of gas or vapour through a connexion between two vessels, due to a difference in the temperatures of the vessels, which results in a pressure gradient when gas (or vapour) transfer equilibrium is reached.

thermal trip (*Elec. Eng.*). A tripping relay for a large circuit-breaker, which operates by thermal expansion.

thermal tuning (*Electronics*). Change of resonant frequency in oscillator or amplifier produced by controlled temperature change. Used, e.g., with crystal (see **quartz thermometer**) or resonant cavity in microwave tube.

thermal unit (*Heat*). See **British-, calorie, joule.**

thermal utilization factor (*Nuc.*). Probability of thermal neutron being absorbed by fissile material (whether causing fission or not) in infinite reactor core.

thermal wind (*Meteor.*). The vector difference between the wind at some level in the upper air and the wind at some lower level.

thermanaesthesia (*Med.*). Loss of skin sensibility to heat and cold.

thermie (*Heat*). A unit of heat used originally in France. The quantity of heat required to raise the temperature of 1 tonne of pure water from 14·5°C to 15·5°C at standard atmospheric pressure. Equals 4·186 MJ.

thermion (*Phys.*). A positive or negative ion emitted from incandescent material.

thermionic amplifier (*Electronics*). Any device employing thermionic vacuum tubes for amplification of electric currents and/or voltages.

thermionic cathode (*Electronics*). One from which electrons are liberated as a result of thermal energy, due to high temperature.

thermionic conduction (*Electronics*). That which arises through electrons being liberated from hot bodies.

thermionic current (*Electronics*). One represented by electrons leaving a heated cathode and flowing to other electrodes, as distinguished from current which flows through the cathode for heating it. Also called **space current.**

thermionic detector (*Radio*). Valve used for detection of radio-frequency alternating currents. See **Fleming diode.**

thermionic emission (*Electronics*). That from a thermionic cathode in accordance with the Richardson-Dushman equation. See **Edison effect.**

thermionic generator (*Electronics*). Producer of high-frequency a.c., using thermionic valves for converting d.c. to a.c. in an oscillator.

thermionic oscillator (*Electronics*). A generator of electrical oscillations using high-vacuum thermionic valves.

thermionic rectifier (*Electronics*). Thermionic valve used for rectification or demodulation.

thermionic relay (*Electronics*). Three-(or more) electrode valve in which the on-off potential applied to one electrode controls the current flowing to another, usually without expenditure of energy at the control electrode.

thermionics. Strictly, science dealing with the emission of electrons from hot bodies. Applied to the broader subject of subsequent behaviour and control of such electrons, especially *in vacuo.*

thermionic valve (*Electronics*). One containing a heated cathode from which electrons are emitted, an anode for collecting some or all of these electrons, and generally additional electrodes for controlling flow to the anode. Normally the glass or metal envelope is evacuated but a gas at low pressure is introduced for special purposes. Also called **thermionic tube, thermionic vacuum tube.**

thermionic work function (*Electronics*). Thermal energy surrendered by an electron liberated through thermionic emission from a hot surface.

thermistor (*Electronics*). Abbrev. for *therm*al res*istor*. Semiconductor, a mixture of cobalt, nickel, and manganese oxides with finely divided copper, of which the resistance is very sensitive to temperature. Sometimes incorporated in a waveguide system to absorb and measure all the transmitted power. Also used for temperature compensation and measurement.

thermistor bridge (*Electronics*). One used for measuring microwave power absorbed by a thermistor, in terms of the resulting change of resistance.

thermite (*Met.*). A mixture of aluminium powder and half an equivalent amount of iron oxide (or other metal oxides) which gives out a large amount of heat on igniting with magnesium ribbon. The molten metal forms the medium for welding iron and steel (Thermit welding). See **aluminothermic process.**

thermo-. Prefix. See **therm-.**

thermoammeter (*Elec. Eng.*). A.c. type in which current is measured in terms of its heating effect, usually with a thermocouple.

thermochemistry (*Chem.*). The study of the heat changes accompanying chemical reactions and their relation to other physicochemical phenomena.

thermocline (*Ecol.*). In lakes, a region of rapidly

changing temperature, found between the epilimnion and the hypolimnion.

thermocouple (*Elec.*). If two wires of different metals are joined at their ends to form a loop, and a temperature difference between the two junctions unbalances the *contact potentials*, a current will flow round the loop. If the temperature of one junction is kept constant, that of the other is indicated by measuring the current. Conversely, a current driven round the loop will cause a temperature difference between the junctions. See **Peltier** and **Seebeck effects**.

thermocouple instrument (*Elec. Eng.*). An instrument the operation of which depends upon the heating of a thermocouple by an electric current.

thermocouple meter (*Elec. Eng.*). A combination of a thermocouple and an ammeter or voltmeter. The current in the external circuit passes through a coil of suitable gauge wire which is electrically insulated from the thermojunction but in very close thermal contact; sometimes couple and heater are enclosed in the evacuated quartz bulb. The rise in temperature of the junction causes the thermoelectric current to flow in the connected meter. If the heating current is alternating (such devices may be used for radio-frequency currents), then the meter indications are *root-mean-square values*.

thermocouple wattmeter (*Elec. Eng.*). One which uses thermoelements in suitable bridge to measure average a.c. power.

thermoduric (*Phys.*). Resistant to heat.

thermodynamic concentration (*Chem.*). See **activity (2)**.

thermodynamic potential (*Chem.*). See **free energy**.

thermodynamics (*Phys.*). The mathematical treatment of the relation of heat to mechanical and other forms of energy. Its chief applications are with respect to heat engines (steam-engines and I.C. engines; see **Carnot cycle**) and to chemical reactions (see **thermochemistry**). *Laws of thermodynamics. 1st law*: Energy can be transformed but not destroyed (conservation of energy). *2nd law*: Heat can never pass spontaneously from a colder to a hotter body; a temperature difference can never appear spontaneously in a body originally at uniform temperature (Clausius). *3rd law*: The entropy of a substance approaches zero as its temperature approaches the absolute zero.

thermodynamic scale of temperature (*Heat*). See **Kelvin thermodynamic scale of temperature**.

thermoelectric cooling (*Heat*). Abstraction of heat from electronic components by Peltier effect, greatly improved and made practicable with solid-state materials, e.g., Bi_2Te_3. Devices utilizing this effect, e.g., frigistors, are used for automatic temperature control, etc., and are energized by d.c.

thermoelectric effect (*Elec. Eng.*). See **Seebeck effect**.

thermoelectricity (*Elec., Eng.*). Interchange of heat and electric energy. See **Peltier effect, Seebeck effect**.

thermoelectric materials (*Met.*). Any set of materials (metals) which constitute a thermoelectric system, e.g., *binary* (bismuth or lead telluride), *ternary* (silver, antimony and telluride), *quaternary* (bismuth, tellurium, selenium, and antimony, termed *Neelium*).

thermoelectric power (*Elec. Eng.*). Defined as dE/dT, i.e., the rate of change with temperature of the thermo e.m.f. of a thermocouple.

thermoelectric pyrometer (*Elec. Eng.*). The combination of apparatus forming a temperature-indicating instrument whose action derives from a thermoelectric current; for high temperatures.

thermoelectroluminescence (*Phys.*). See under **electrothermoluminescence**.

thermoelement (*Elec. Eng.*). See **thermal converter**.

thermogalvanometer (*Elec. Eng.*). A very sensitive thermocouple meter.

thermogenesis (*Zool.*). Production of heat within the body.

thermograph (*Meteor.*). A continuously recording thermometer. In the commonest forms the record is made by the movement of a bimetallic spiral, or by means of the out-of-balance current in a Wheatstone bridge containing a resistance thermometer in one of its arms.

thermographic (*Print.*). Of a process of writing, involving the use of heat.

thermojunction (*Elec. Eng.*). See **thermocouple**.

thermolabile (*Chem.*). Tending to decompose on being heated.

thermoluminescence (*Light*). Release of light by previously irradiated phosphors upon subsequent heating.

thermoluminescent dosemeter (*Radiol.*). One which registers integrated radiation dose, the read-out being obtained by heating the element and observing the thermoluminescent output with a photomultiplier. It has the advantage of showing very little fading if read-out is delayed for a considerable period, and forms an alternative to the conventional film badge for personnel monitoring.

thermolysis (*Chem.*). The dissociation or decomposition of a molecule by heat. (*Zool.*) Loss of body heat.

thermomagnetic effect (*Heat*). See **magnetocaloric effect**.

thermometer (*Heat*). An instrument for measuring temperature. A thermometer can be based on any property of a substance which varies predictably with change of temperature. For instance, the *constant volume gas thermometer* is based on the pressure change of a fixed mass of gas with temperature, while the *platinum resistance thermometer* is based on a change of electrical resistance. The commonest form relies on the expansion of mercury or other suitable fluid with increase in temperature.

thermometric scales (*Heat*). See **Centigrade scale, Celsius scale, Fahrenheit scale, international practical temperature scale, Kelvin thermodynamic scale, Rankine scale, Réaumur scale**; also **fixed points**.

thermometry (*Heat*). The measurement of temperature.

thermomolecular pressure (*Phys.*). The pressure difference arising between two vessels containing a gas at different absolute temperatures (T_1 and T_2), which are separated by a thermally insulating partition containing an orifice whose dimensions are small compared with the mean free path of the gas molecules. Elementary kinetic theory gives $P_1/P_2 = \sqrt{T_1/T_2}$ as the pressures in the respective vessels. The effect is important in low-temperature gas systems.

thermonasty (*Bot.*). A nastic movement in a plant in relation to heat.

thermonuclear bomb (*Nuc.*). See **hydrogen bomb**.

thermonuclear energy (*Nuc.*). Energy released by fusion process, which occurs between particles as a result of their thermal energy and not because of electric acceleration. The rate of reaction increases rapidly with temperature and, in the hydrogen bomb, a fission bomb is used initially to obtain the required temperature ($>20\times10^6$°C) to make use of thermonuclear reactions. The energy of most stars is believed to be acquired from exothermic thermonuclear reactions. For laboratory experiments designed to release thermonuclear energy, see, e.g., **mirror machine, stellarator, Zeta.**

thermonuclear reaction (*Nuc.*). One involving the release of thermonuclear energy.

thermoperiodism (*Bot.*). The relationship existing in some plants between the diurnal temperature cycle and the rate of growth and development.

thermophile, thermophilous, thermophilic (*Bot.*). Said of a plant which requires a high temperature for growth, or which can tolerate exposure to high temperatures.

thermophilic bacteria (*Bot.*). Bacteria which need a temperature of from 45°C to 65°C for their development.

thermophone (*Acous.*). Electroacoustic transducer in which heating effect of current produces a calculable sound pressure wave that can be used for calibration of microphones.

thermophyte (*Bot.*). A plant growing in warm situations.

thermopile (*Elec. Eng.*). See **pile.**

thermoplastic (*Chem.*). Becoming plastic on being heated. Specifically (*Plastics*), any resin which can be melted by heat and then cooled, the process being able to be repeated any number of times without appreciable change in properties, e.g., cellulose derivatives, vinyl polymers, polystyrenes, polyamides, acrylic resins.

thermoplastic binding (*Bind.*). A synonym for *unsewn binding* (q.v.), whether the adhesive used is thermoplastic or not.

thermoplastic plates (*Print.*). Duplicate printing plates made from thermoplastic material, usually flexible and suitable for rotary printing; normally recoverable for subsequent plates in contrast to *thermosetting plates* (q.v.).

thermoplastic putty (*Build.*). Glazier's putty which is rendered pliable by the addition of tallow and so is able to give and take with the expansion and contraction of large sheets of glass.

thermoprinting machine (*Print.*). A type of *blocking press*, used in conjunction with automatic food packaging machines using heat-sealing wraps. These are over-printed with details of the contents.

thermoregulator (*Heat*). Type of thermostat which keeps a bath at a constant temperature by regulating its supply of heat.

thermoscopic (*Heat*). Perceptive of change of temperature.

thermosetting compositions (*Plastics*). Compositions in which a chemical reaction takes place while they are being moulded under heat and pressure; the appearance and chemical and physical properties are entirely changed, and the product is resistant to further applications of heat (up to charring point), e.g., phenol formaldehyde, urea formaldehyde, aniline formaldehyde, glycerol-phthalic anhydride.

thermosetting plates (*Print.*). Duplicate printing plates made from thermosetting plastic material, not flexible and must be used flat or be pre-curved; not recoverable like *thermoplastic plates* (q.v.). Both have certain advantages over metal plates, such as good inking qualities, long life, light weight, easy storage.

thermosiphon (*Eng.*). The method of establishing circulation of a cooling-liquid by utilizing the slight difference in density of the hot and the cool portions of the liquid. At one time, the water cooling system of many automobile engines relied on this method entirely.

thermosphere (*Astron.*). The region of the earth's atmosphere, above the *mesosphere* (q.v.), in which the temperature rises steadily with height.

thermostable (*Chem.*). Not decomposed by heating.

thermostage (*Bot.*). A stage in the life-history of a flowering plant, when, at the outset of development from the embryo in the seed, low temperatures are needed to ensure further normal development.

thermostat (*Heat*). An apparatus which maintains a system at a constant temperature which may be preselected. Frequently incorporates a *bimetallic strip.*

thermotaxis (*An. Behav.*). A *taxis* (q.v.) occurring in response to the direction of a source of radiant heat. Also **thermotropism.**

thermotolerant (*Bot.*). Able to endure high temperatures, but not growing well under such conditions.

therophyllous (*Bot.*). Having leaves only in the warmer part of the year; deciduous.

therophyte (*Bot.*). An annual plant, passing the winter or dry season in the form of seeds. One of the subdivisions into which plants can be divided in the system known as *Raunkiaer's life forms* (q.v.).

Thévenin's theorem (*Elec. Eng.*). The current produced in a load, when connected to a source, is that produced by an e.m.f. equal to the open-circuit voltage of the source divided by the load impedance plus the apparent internal impedance of the source, i.e., the conjugate impedance to that which extracts maximum power from the source. Also known as **Helmholtz's theorem.** See **Norton's theorem.**

thi-, thio- (*Chem.*). A prefix denoting a compound in which a sulphur atom occupies a position normally filled by an oxygen atom.

thiamides (*Chem.*). A group of compounds derived from amides by the exchange of oxygen for sulphur, e.g., $CH_3 \cdot CS \cdot NH_2$.

thiamin (*Chem.*). See **vitamin B complex.**

Thiazamide (*Pharm.*). TN for *sulphathiazole.* See **sulphonamides.**

thiazines (*Chem.*). Six-membered heterocyclic compounds, containing in the ring four carbon, one sulphur, and one nitrogen atoms.

thiazole (*Chem.*). A colourless, very volatile liquid which closely resembles pyridine; b.p. 117°C. It forms salts, but is hardly affected by concentrated sulphuric acid. Its formula is:

thiazole dyestuffs (*Chem.*). Thiazole derivatives, e.g., primulines, obtained by heating a mixture of 4-toluidine with sulphur at ca. 200°C. Other dyestuffs of this group are the **thioflavines.**

thickener (*Met., Min. Proc.*). Apparatus in which water is removed from ore pulp by allowing solids to settle. To obtain continuous working the solids are worked towards a central hole in the bottom by means of revolving rakes.

thickening fibre (*Bot.*). One of the spiral bands of thickening on the wall of a cell, tracheid, or vessel.

thick-film lubrication (*Eng.*). The state of fluid-medium lubrication which exists when the lubricant separates the journal and bearing surfaces to prevent metal-to-metal contact. The bearing friction is then dependent largely on the force necessary to shear the lubricant film, and perfect lubrication is said to prevail.

thick leg disease (*Vet.*). See **osteopetrosis gallinarum.**

thick lens (*Light*). Any lens, or system of lenses, in which the distance between the outer faces is not small compared with the focal length.
thickness chart (*Meteor.*). A chart of upper air showing the difference in height between two particular pressure levels. Centres of low thickness are *cold pools*, and of high thickness, *warm pools*. Elongations of these are *cold troughs* and *warm ridges* respectively. See **constant-pressure chart.**
thickness dummy (*Bind., Print.*). A book made up of blank leaves to show size and physical appearance in advance.
thickness moulding (*Build.*). A moulding serving to fill up the bare space beneath a projecting cornice.
thickness ratio (*Aero.*). The ratio of the maximum depth of an *aerofoil*, measured perpendicular to the *chord line*, to the *chord length*; usually expressed as a percentage.
thick source (*Nuc.*). Radioactive source with appreciable self-absorption.
thick space (*Typog.*). The normal standard space between words set in lower case, one-third of the type size; reduced if possible, rather than increased, when justifying text matter.
thick target (*Nuc.*). One which is not penetrated by primary or secondary radiation beam.
thick-wall chamber (*Nuc. Eng.*). Ionization chamber in which build-up of ion current is produced by contribution of knock-on particles arising from wall material.
thief sampler (*Powder Tech.*). Instrument for sampling powders, comprising two tubes, one fitting closely inside the other, rectangular slits in corresponding positions. The instrument is thrust into the powder with the holes closed; the inner tube is rotated to open the holes to collect the sample, which is withdrawn with the holes closed.
thigmo-. Prefix from Gk. *thigma*, touch.
thigmocyte (*Zool.*). See **thrombocyte.**
thigmokinesis (*An. Behav.*). A *kinesis* (q.v.) in response to contact, a low intensity of stimulation by contact usually inducing a high level of activity, and vice versa.
thigmotaxis (*An. Behav.*). A *taxis* (q.v.) occurring in response to contact, the balanced action of equal zones of contact on either side resulting in the animal pursuing a straight course.
thigmotropism (*Biol.*). Turning of an organism (or of part of it) towards or away from object providing touch stimulus.
thill (*Mining*). The floor in a coal-mine.
thimble (*Plumb.*). See **sleeve piece.**
thimble ionization chamber (*Nuc. Eng.*). A small cylindrical, spherical or thimble-shaped ionization chamber, volume less than 5 cm^3 with air-wall construction. Used in radiobiology.
thimble-tube boiler (*Eng.*). A heat-recovery boiler, consisting of an annular water-drum from which short thimble-like tubes or pockets project into the central flue, through which the hot exhaust products are passed.
thin film (*Phys.*). Said of a layer which is of (or approximates to) monomolecular thickness and is laid down by vacuum deposition. Many types of electronic components and complete microcircuits can be produced in this way.
thin-film capacitor (*Elec. Eng.*). One constructed by evaporation of two conducting layers and an intermediary dielectric film (e.g., silicon monoxide) on an insulating substrate.
thin-film lubrication (*Eng.*). When metal-to-metal contact exists in a journal bearing for all or part of the time, thin-film or imperfect lubrication is said to prevail and the materials and surface characteristics of journal and bearing affect the bearing friction significantly.
thin-film memory (*Comp.*). The use of an evaporated thin film of magnetic material on glass as an element of a computer memory when a d.c. magnetic field is applied parallel to the surface. A large capacity memory will contain thousands of these elements which can be produced in one operation. Also called **magnetic-film memory.** See **magnetic memory.**
thin-film resistor (*Elec. Eng.*). Modern high-stability type formed by a conducting layer a few tens of nanometres thick on an insulating substrate.
thin-layer chromatography (*Chem.*). A form of chromatography in which compounds are separated by a suitable solvent or solvent mixture on a thin layer of adsorbent material coated on a glass plate. It is rapid, and excellent separations can be obtained. Abbrev. **TLC.**
thin source (*Nuc.*). Radioactive source with negligible self-absorption.
thin space (*Typog.*). The narrowest of the justifying spaces, one-fifth of the type size; the *hair space* (q.v.) is not a justifying space.
thin target (*Nuc.*). One penetrated by primary radiation beam so that detecting instrument(s) may be used on opposite side of target to source.
thin-wall bearings (*I.C. Engs.*). Bearings made of steel sheet coated with a thin layer of soft alloy.
thin-wall chamber (*Nuc. Eng.*). Ionization chamber in which the number of knock-on particles arising from wall material is negligible.
thio-acids (*Chem.*). Acids in which the hydroxyl of the carboxyl group has been replaced by SH, thus forming the group —CO·SH.
thio-alcohols (*Chem.*). See **mercaptans.**
thiocarbamide (*Chem.*). See **thiourea.**
thiocyanates (*Chem.*). Compounds formed when alkaline cyanides are fused with sulphur.
thio-ethers (*Chem.*). Compounds in which the ether oxygen has been replaced by sulphur; general formula R.S.R′. They form additive crystalline compounds with metallic salts; they are capable of combining with halogen or oxygen, which becomes attached to the sulphur atom, thereby converting the latter from the bivalent to the tetravalent state; and they form additive crystalline compounds with alkyl halides, e.g., $(CH_3)_3SI$.
thioflavines (*Chem.*). See **thiazole dyestuffs.**
thiogenic (*Chem.*). Sulphur-producing.
thioglycollic acid (*Chem.*). $CH_2(SH)·COOH$. A colourless liquid, with a slight odour when pure but extremely unpleasant in impure form. A strong reducing agent, used as a reagent for detecting iron, with which it gives a violet colour in ammoniacal solution. Used in cold-waving treatment of hair, and in certain textile treatments for crease-resistant finishes.
Thiokol (*Plastics*). TN for a synthetic rubber of the polysulphide group, derived from sodium tetrasulphide and organic dichlorides, with sulphur as an accelerator.
thiopentone (*Pharm.*). A yellow powder, sodium ethyl (1-methyl-butyl) thiobarbiturate, used intravenously in solution to give general anaesthesia of short duration. TN **Pentothal.**
thiophen (*Chem.*). A five-membered heterocyclic compound of the formula:

a colourless liquid, b.p. 84°C, which closely resembles benzene and occurs in coal-tar.

thiophil (*Chem.*). Having an affinity for sulphur and its compounds, e.g., bacteria.

thiospinels (*Min.*). A group of minerals which have the same type of crystal structure as *spinel*, but in which sulphur atoms occupy the positions occupied by oxygen in spinel.

thiosulphuric acid (*Chem.*). $H_2S_2O_3$, formerly called **hyposulphurous acid** (q.v.).

thiotep (*Chem.*). See **sulfotep.**

thio-uracil (*Chem., Pharm.*). A pyrimidine derivative with the structure:

It appears to have some action in inhibiting the production of the thyroid hormone.

thiourea (*Chem.*). Thiocarbamide, $NH_2{\cdot}CS{\cdot}NH_2$, colourless prisms, m.p. 180°C; it is slightly soluble in water, ethanol, and ethoxyethane. Used in organic synthesis and as a reagent for bismuth.

thiourea resins (*Plastics*). Resins made from thiourea and an aldehyde. They are more water-resistant and more stable than the urea resins, but they cure more slowly, and the sulphur in them causes trouble with the steel moulds and the dyestuffs used to colour them.

thiram (*Chem.*). *bis*(dimethylthiocarbamyl)disulphide:

$$(CH_3)_2{\cdot}N\text{—}\underset{\underset{S}{\|}}{C}\text{—}S\text{—}S\text{—}\underset{\underset{S}{\|}}{C}\text{—}N(CH_3)_2$$

Used as a fungicide; also **TMT, TMTD.**

third-angle projection (*Eng.*). A system of projection used in engineering drawing, in which each view shows what would be seen by looking on the near side of an adjacent view.

third pinion (*Horol.*). The pinion on the same axis as the third wheel, engaging with the centre wheel.

third-rail insulator (*Elec. Eng.*). See **conductor-rail insulator.**

third-rail system (*Elec. Eng.*). The system of electric traction supply by which current is fed to the electric tractor from an insulated conductor rail running parallel with the track.

thirds (*Paper*). A premetrication size of cut card, $1\frac{1}{2}\times 3$ in (38×76 mm).

third tap (*Eng.*). See **plug tap (1).**

third ventricle (*Zool.*). In Vertebrates, the cavity of the diencephalon, joining the two lateral ventricles of the cerebral hemispheres via the foramen of Monro, and the fourth ventricle in the medulla oblongata by the cerebral aqueduct.

third wheel (*Horol.*). The wheel between the centre and the fourth wheel of a watch train.

thirl or **thurl** (*Mining*). To cut through from one working into another.

thirty-twomo (*Print.*). Written 32mo. The thirty-second part of a sheet, or a sheet folded five times to make thirty-two leaves or sixty-four pages. More than four folds are not in fact practicable.

Thistle board (*Build.*). A form of building-board made from gypsum faced with sheets of brown paper on both sides.

thistle funnel (*Chem.*). A glass funnel with a thistle-shaped head and a long narrow tube.

thistle microphone (*Acous.*). See **apple-and-biscuit microphone.**

thixotrope (*Chem.*). A colloid whose properties are affected by mechanical treatment.

thixotropy (*Chem., Paint.*). Rheological property of fluids and plastic solids, characterized by a high viscosity at low stress, but a decreased viscosity when an increased stress is applied. A useful property in paints, because it makes for a thick film which is nevertheless easily worked.

tholeiite (*Geol.*). A fine-grained basic igneous rock essentially basaltic in composition, intersertal in texture, with wedges of mesostasis or glass between laths of plagioclase. *Quartz-dolerite* is the medium-grained equivalent.

Thomas bar (*Med.*). A leather bar placed externally on the sole of the shoe in the treatment of *hallux rigidus* (q.v.).

Thomas-Gilchrist process (*Met.*). Use of basic-lined Bessemer converters to remove phosphorus during steel production.

Thomas resistor (*Elec. Eng.*). A standard manganin resistor which has been annealed in an inert atmosphere and sealed into an envelope.

Thomas's splint (*Surg.*). A skeleton splint consisting of two parallel metal rods and a padded leather ring, used for maintaining the hip and the knee-joint in fixed extension.

Thomsen's disease (*Med.*). See **myotonia congenita.**

Thomson compass (*Ships*). See **Kelvin compass.**

Thomson effect (*Elec. Eng.*). The e.m.f. produced by temperature differences in a single conductor, and the heat change associated with current flow between temperature differences.

thomsonite (*Min.*). An orthorhombic zeolite; a hydrated silicate of calcium, aluminium and sodium, found in amygdales and crevices in basic igneous rocks.

Thomson scattering (*Phys.*). The scattering of electromagnetic radiation by electrons in which, on a *classical* interpretation, some of the energy of the primary radiation is lost because the electrons radiate when accelerated in the transverse electric field of the radiation. The scattering cross-section of an electron according to Thomson is given by

$$b=\frac{8}{3}\pi\left(\frac{e^2}{mc^2}\right)^2=0{\cdot}66\text{ barn},$$

where e is electronic charge, m is mass of electron, and c is velocity of light.

thoracentesis, thoracocentesis (*Surg.*). The operation of drawing off a morbid collection of fluid in the pleural cavity through a hollow needle stuck through the wall of the chest.

Thoracica (*Zool.*). An order of *Cirripedia*, having an alimentary canal, six pairs of biramous thoracic limbs, and no abdominal somites, being permanently attached by the pre-oral region. Barnacles.

thoracic glands (*Zool.*). In Insects, endocrine glands of ectodermal origin which produce the *moulting hormone* (q.v.) when stimulated by the brain hormone.

thoracicolumbar system (*Zool.*). See **sympathetic nervous system.**

thoracopagus (*Med.*). A foetal monstrosity in which twins are joined together at the thorax.

thoracoplasty (*Surg.*). The operation for collapsing a diseased lung by removal of portions of the ribs.

thoracoscope (*Surg.*). An instrument for viewing the pleura covering the lung and the chest wall. It is inserted through the chest wall into a pleural cavity previously filled with air.

thoracotomy (*Surg.*). Incision of the wall of the

chest, for draining pus from the pleural cavity or from the lung.

Thoraeus filter (*Radiol.*). A combination of tin, copper and aluminium metal filters used to harden radiation in the 200–400 kV region.

thorax (*Zool.*). In *Crustacea* and *Arachnida*, a region of the body lying between the head and the abdomen and usually fused with the former, distinguished by the nature of its appendages; in Insects, one of the three primary regions of the body, lying between the head and the abdomen, and bearing in the adult three pairs of legs and the wings (if present); in some tubicolous *Polychaeta*, a region of the body behind the head, distinguished by the form of its segments and the nature of its appendages; in land Vertebrates, the region of the trunk between the head or neck and the abdomen which contains heart and lungs and bears the fore-limbs, especially in the higher forms, in which it is enclosed by ribs and separated from the abdomen by the diaphragm. *adj.* **thoracic.**

thorianite (*Min.*). Thorium oxide; crystallizes in the cubic system and is found in pegmatites and gem gravel washings, as in Ceylon. An important source of thorium and uranium.

thoriated filament (*Electronics*). Tungsten filament containing a small proportion of thorium to reduce the temperature at which copious electronic emission takes place. With heat, the thorium diffuses to the surface, forming a tenuous emitting layer.

thorides (*Nuc.*). Naturally occurring radioactive isotopes in the radioactive series containing thorium.

thorite (*Min.*). Tetragonal thorium orthosilicate, found in syenites and syenitic pegmatites.

thorium (*Chem.*). A metallic radioactive element, dark-grey in colour. Symbol Th, at. no. 90, r.a.m. 232·0381, rel. d. 11·2, m.p. 1845°C. It occurs widely in beach sands (*monazite*). The thorium radioactive series starts with thorium of mass 232. It is fissile on capture of fast neutrons and is a fertile material, ^{233}U (fissile with slow neutrons) being formed from ^{232}Th by neutron capture and subsequent beta decay.

thorium reactor (*Nuc. Eng.*). Breeder reactor in which ^{233}U is bred in a blanket of ^{232}Th.

thorium series (*Nuc.*). The series of nuclides which result from the decay of ^{232}Th. The mass numbers of all the members of the series are given by $4n$, n being an integer.

thorn (*Bot.*). A leaf, part of a leaf, or a shoot which contains vascular tissue and ends in a hard, sharp point.

thoron (*Nuc.*). Thorium emanation, an isotope of radon (^{220}Rn); half-life 54·5 s. Symbol Tn.

thoroughpin (*Vet.*). Through-pin. A swelling on the hock of the horse caused by distension of the synovial sheath of the flexor perforans tendon.

thortveitite (*Min.*). A relatively rare monoclinic disilicate of scandium and yttrium, found in pegmatites.

thousand (*Build.*). Formerly a trade term for 1200 slates; now for 1000, so also called **thousand actual.**

thread (*Eng.*). See **screw thread.**

thread-cells (*Zool.*). See **cnidoblast.**

thread-dial indicator (*Eng.*). A device which forms part of a screw-cutting lathe. It indicates, while the lathe is running, the correct time for engaging the half-nut so that the tool will repeatedly follow the thread cut.

thread grinding (*Eng.*). The accurate production of screw threads by a form grinding wheel, profiled to the thread section and automatically traversed along the revolving work.

threading or **threading-up** (*Cinema.*). The operation of inserting the start of the film into the mechanism of camera, or projector, as it leaves the feed reel, and attaching it to the take-up reel.

thread rolling (*Eng.*). Producing a screw thread by rolling between flat or cylindrical dies a blank of material sufficiently plastic to withstand the cold working forces without disintegrating.

thread sewn (*Bind.*). The traditional method of securing and joining to each other the sections of a book, by hand and on tapes or cords for the best work, by machine for *edition binding* (q.v.).

thread-stitching (*Bind.*). The traditional method, by hand or machine, of securing the leaves of insetted (or quirewise) books or pamphlets; better and more expensive than *wire-stitching* (q.v.).

three-ammeter method (*Elec. Eng.*). A method of measuring the power carried by a single-phase circuit making use of three ammeters. Cf. *three-voltmeter method.*

three-body problem (*Astron.*). The problem of the behaviour of three bodies which mutually attract each other; no general solution is possible but certain particular solutions are known. See **Trojan group.**

three-centred arch (*Arch.*). An arch having the form of a false ellipse struck from three centres.

three-coat work (*Build.*). Plastering in three successive coats. See **floating, pricking-up, roughing-in, setting.**

three-colour process (*Photog.*). Any system of colour photography which analyses the colours by three colour filters each giving a black-and-white record. These records are used in the synthesis of the final positive coloured image, either by the additive process, in which lights approximating to the filters are added together, or by the subtractive process, in which the colours approximating to filters are subtracted from white light, or from a white reflecting surface, by the use of colours complementary to those of the filters. (*Print.*) The subtractive process applied to printing. The yellow is, as a rule, printed first, followed by magenta (red), then cyan (blue). A fourth printing, in black or grey, may be added in high-class work.

three-core cable (*Elec. Eng.*). A cable having three conducting cores arranged symmetrically about the axis of the cable and insulated.

three-day sickness (*Vet.*). Ephemeral fever. A benign, noncontagious virus disease of cattle in tropical countries characterized by fever, lameness, and stiffness; transmitted by mosquitoes.

three-eighths rule (*Maths.*). The area under the curve $y=f(x)$ from $x=x_0$ to $x=x_3$ is approximately

$$\tfrac{3}{8}\{f(x_0)+3f(x_1)+3f(x_2)+f(x_3)\}(x_3-x_0),$$

where x_0, x_1, x_2 and x_3 are equally spaced.

three-electrode valve (*Electronics*). See **triode valve.**

three-electron bond (*Chem.*). Resonance structure involving an unpaired shared electron, similar to the *one-electron bond* (q.v.).

three-high mill (*Met.*). A rolling mill with three rolls, which are rotated in such a way that the metal is passed in one direction through the bottom pair of rolls and in the opposite direction through the top pair.

three-hinged arch (*Arch.*). An arch which is hinged at the crown and the abutments.

three-jaw chuck (*Eng.*). A *scroll chuck* (q.v.) with three jaws for holding cylindrical workpieces, materials, or tools, particularly useful on the lathe or drilling machine.

three-level maser (*Electronics*). Solid-state maser involving 3 energy levels.

three-light window (*Join.*). A window having 2 mullions dividing the window space into 3 compartments.

three-phase (*Elec. Eng.*). An electric supply system in which the alternating potentials on the 3 wires differ in phase from each other by 120°.

three-phase four-wire system (*Elec. Eng.*). A system of 3-phase a.c. distribution making use of 3 outgoing conductors (lines) and a common return conductor (neutral), the voltage between lines being $\sqrt{3}$ times the voltage between any line and the neutral.

three-phase induction regulator (*Elec. Eng.*). An induction regulator for use on 3-phase circuits, in which the e.m.f. induced in the secondary winding is constant in magnitude but variable in phase, so that the total e.m.f. on the secondary side bears a small phase displacement to the primary voltage.

three-phase six-wire system (*Elec. Eng.*). A system of 3-phase a.c. distribution in which each phase has separate outgoing and return conductors.

three-pinned arch (*Arch.*). See **three-hinged arch.**

three-pin plug (*Elec. Eng.*). A plug with 3 contact pins, 2 for the main circuit and 1 for the earth connexion.

three-point landing (*Aero.*). The landing of an aeroplane so equipped on the 2 wheels and tail skid (or wheel) simultaneously; the normal 'perfect landing'.

three-point problem (*Surv.*). A field problem, arising in plane table and hydrographical surveying, in which it is required to locate on the plan the position of the instrument station, given that only 3 points represented on the plan are in fact visible from the station.

three-point switch (*Elec. Eng.*). See **three-way switch.**

three-quarter bat (*Build.*). A brick made or cut equal to ¾ the full length of a brick.

three-quarter bound (*Bind.*). Similar to *quarter bound* (q.v.), but having the material used for the back covering a large part of the sides too.

three-quarter plate watch (*Horol.*). A watch with the upper plate cut away so that the balance may be in the same plane as the plate.

three-start thread (*Eng.*). See **multiple-threaded screw.**

three-to-two folder (*Print.*). On a web-fed press, a type of folder in which the folding cylinder has a circumference of 3 cut-offs and the cutting cylinder 2 cut-offs, giving a ratio of cylinder sizes 3 : 2. Cf. *two-to-one folder*.

three-voltmeter method (*Elec. Eng.*). A method of measuring the power in a single-phase circuit by means of 3 voltmeters and a nonreactive resistance. Cf. *three-ammeter method*.

three-wattmeter method (*Elec. Eng.*). A method of measuring the power carried by a 3-phase 4-wire circuit, making use of 3 wattmeters whose current coils are connected in the lines and whose voltage coils are connected between the lines and the neutral.

three-way switch (*Elec. Eng.*). A rotary-type single-pole switch having 3 independent contact positions.

three-wire meter (*Elec. Eng.*). An electricity supply meter performing the simultaneous integration of the energy supplied by the 2 sides of a *three-wire system*.

three-wire mooring (*Aero.*). A 3-wire system of cables attached to points in the ground in such positions that the dynamic lift due to the prevailing wind is counterbalanced, and the airship rides at a constant safe height above the ground.

three-wire system (*Elec. Eng.*). A supply system in which, e.g., 220 volt is the p.d. between 2 of the wires, while ca. 110 volt exists between the other 2-wire combinations.

thremmatology (*Gen.*). The science of breeding animals and plants under domestic conditions.

threonine (*Chem.*). 2-Amino-3-hydroxy-butanoic acid. Has the same configuration as the tetrose, threose, and is one of the 'essential' amino acids.

threshold. Generally, the lowest intensity of an effect which is detectable, e.g. (*Physiol.*) of visibility, below which neither the cones nor the rods in the retina of the eye respond to a light stimulus. (*Psychol.*) The smallest intensity of a stimulus, or the smallest difference between two stimuli, which can be perceived. (*Telecomm.*) See **pulse height selector.**

threshold amplitude (*Electronics*). The lowest amplitude level which a pulse height selector or window discriminator will accept.

threshold current (*Electronics*). That at which a gas discharge becomes self-sustaining.

threshold dose (*Radiol.*). The smallest dose of radiation that will produce a specified result.

threshold effect (*Electronics*). The marked increase in background noise which occurs in a valve circuit when on the verge of oscillation.

threshold energy (*Nuc.*). Minimum energy for a particle which can just initiate a given endoergic reaction. Exoergic reactions may also have threshold energies.

threshold frequency (*Electronics*). Minimum frequency in a photon which can just release an electron from a surface.

threshold lights (*Aero.*). A line of lights across the ends of a runway, strip, or landing area to indicate the usable limits.

threshold of feeling (*Acous.*). Minimum intensity or pressure of sound wave which causes sensation of discomfort or pain in average normal human listener.

threshold of sound (or **of audibility**) (*Acous.*). Minimum intensity or pressure of sound wave which average normal human listener can just detect at any given frequency. Commonly expressed in decibels relative to 2×10^{-5} Pa.

threshold treatment (*Chem.*). Addition of minute quantities of dehydrated phosphates to water, to inhibit furring and corrosion in pipes, containers, etc.

threshold voltage (*Nuc.*). That which must be applied to a radiation counter before pulses can be observed. Also **starting voltage,** esp. in U.S.

thrill (*Med.*). A tremor or vibration palpable at the surface of the body, especially in valvular disease of the heart.

throat (*Bot.*). The aperture of a gamopetalous corolla or of a gamosepalous calyx. (*Build., Civ. Eng.*) See **drip.** (*Eng.*) (1) The C-shaped aperture of a gap press. (2) The root portion of a saw tooth. (*Glass*) The submerged channel through which glass passes from the melting end to the working end of a tank furnace.

throatless chamber (*Aero.*). A rocket without a restriction between the combustion chamber and the expansion nozzle.

throat microphone (*Acous.*). One worn against the throat and actuated by contact pressure against the larynx. Used, e.g., by air pilots and deep sea divers. Also called **laryngophone.**

thromb-, thrombo-. Prefix from Gk. *thrombos*, lump, clot.

thrombectomy (*Surg.*). The operation of removing a venous thrombus.

thrombin (*Biochem.*). A proteinlike substance formed in shed blood from *prothrombin* (q.v.). It reacts with the soluble protein fibrinogen, converting it to insoluble fibrin, and so causes the blood to clot.

thrombo-angiitis obliterans (*Med.*). A disease characterized by inflammation and thrombosis of the larger arteries and veins (especially those of the leg); it gives rise to muscular cramps on walking and, later, to ulceration and gangrene of the foot. Also called **Buerger's disease.**

thrombocyte (*Zool.*). A minute greyish circular or oval body found in the blood of higher Vertebrates in numbers varying from 20 000 to 300 000 per cubic millimetre; it plays an important role in coagulation; a blood-platelet.

thrombocytopenia (*Med.*). Abnormal decrease in the number of platelets (thrombocytes) in the blood. See **purpura haemorrhagica.**

thrombokinase (*Biochem.*). A *kinase* (q.v.) which converts prothrombin into active thrombin in the presence of soluble calcium salts. Also called **thromboplastin, thrombozyme.**

thrombopenia (*Med.*). See **thrombocytopenia.**

thrombophilia (*Med.*). A tendency to the formation of thrombi in the blood vessels.

thrombophlebitis (*Med.*). Combined inflammation and thrombosis of a vein.

thromboplastin (*Biochem.*). See **thrombokinase.**

thrombosis (*Med.*). The formation of a clot in a blood-vessel during life. (*Zool.*) Coagulation; clotting.

thrombozyme (*Biochem.*). See **thrombokinase.**

thrombus (*Med.*). A clot formed in a blood vessel during life and composed of thrombocytes (platelets), fibrin, and blood cells.

throttle valve (*Eng.*). (1) In steam engines and turbines, a governor-controlled steam valve, usually a *double beat valve* (q.v.). (2) In petrol engines, the *butterfly valve* (q.v.). (3) In refrigerators, the regulating valve controlling the pressure and temperature range of the working agent.

throttling (*Eng.*). The process of reducing the pressure of a fluid by causing it to pass through minute or tortuous passages so that no kinetic energy is developed and the total heat remains constant. See **refrigerator, throttling calorimeter.**

throttling calorimeter (*Eng.*). A device for measuring the *dryness fraction* (q.v.) of wet steam by throttling to a measured lower pressure and measuring the resulting superheat.

through bridge (*Civ. Eng.*). A bridge in which the track is carried by the lower stringers. Cf. *deck bridge.*

through level (*Teleph.*). The reading of a high impedance level-measuring set at a point in a system, no correction being made for any difference between the actual impedance of the system and the impedance with respect to which the set is calibrated.

through path (*Telecomm.*). The forward path from loop input to loop output in a feedback circuit.

through-pin (*Vet.*). See **thoroughpin.**

throughput (*Chem. Eng., etc.*). A measure of the rate of production, in terms of mass of product per unit time per unit volume of plant. In vacuum technology, the quantity of gas or vapour passing a given section of a pump or pipe line in unit time; measured as the product of the pressure and the volume per second at the pressure at that section. Units: litre torr per second or litre pascal per second. Symbol Q.

through-stone (*Build.*). A bondstone whose length is equal to the full thickness of the wall in which it is laid as a header. Also **perpend.**

throw (*Cinema.*). The distance between the projector and the screen. (*Elec. Eng.*) See **span** (2). (*Eng.*) The total travel of a crank or similar element, being twice the radius of eccentricity. Sometimes half this distance is called the throw. (*Geol.*) The amount of vertical displacement (*upthrow* or *downthrow*) of a particular rock, vein or stratum, due to faulting. See **fault,** and cf. *lateral shift.* (*Horol.*) A hand-driven, dead-centre lathe, used by clockmakers. (*Mining*) (1) See **throw** (*Geol.*). (2) The amplitude of shake of a concentrating table. (3) Deviation of a deep borehole from the planned path.

throw-away (*Print.*). A term for any simple and cheap advertising leaflet.

throwback (*Acous.*). In a public-address system, when the microphone is near the reproducers, the *throwback* is the sound intensity which is applied to the microphone by the reproducers. If this is excessive, the system becomes paralysed with self-sustained oscillations.

throwing (*Textiles*). In silk manufacture, the processes of reeling, doubling, twisting, scouring, etc., to bring the raw silk filaments into the form of a silk thread.

throwing power (*Chem.*). The property of a solution in virtue of which a relatively uniform layer of metal may be electrodeposited on a relatively irregular surface.

throw-off trip (*Print.*). An attachment on a printing machine which allows the impression to be suspended without stopping the machine.

throw-out (*Bind.*). See **fold-out.**

thrum-eyed (*Bot.*). Having the throat of the corolla more or less closed by the anthers; said of the primrose and its relatives.

thrums (*Textiles*). See **beating.**

thruput (*Chem. Eng.*). See **throughput.**

thrush (*Med.*). Infection of the mouth with the fungus *Candida albicans*, characterized by the appearance of white patches on the mucous membrane and the tongue. (*Vet.*). The name is sometimes applied to *horse pox* (q.v.). See also **equine thrush.**

thrust (*Aero.*). Propulsive force developed by a jet- or rocket-motor. (*Arch.*) The equal horizontal forces acting upon the abutments of an arch, due to the loading carried by the arch. (*Eng.*) The reaction to a compressive force on a rod.

thrust bearing or **thrust block** (*Eng.*). A shaft bearing designed to take an axial load. It consists either of a plain bearing pad, a *Michell bearing* (q.v.), or a ball-bearing provided with lateral races.

thrust chamber (*Aero.*). The compartment in a rocket where the propulsive forces are developed before ejection; usually, but not necessarily, the *reaction chamber* (q.v.).

thrust deflector (*Aero.*). A device, usually a combination of doors closing the jet pipe and a cascade of guide vanes, for deflecting the efflux of a turbojet downward to provide upward thrust for *STOL* or *VTOL*.

thrust deflexion (*Aero.*). The direction of the efflux from a *turbojet*, *ramjet*, or *rocket* in a direction other than along its axis, for the purpose of obtaining a thrust component normal to this axis. Generally used for the guidance of rockets and for *STOL* and *VTOL* aircraft. See **jet deflexion.**

thrust loading (*Aero.*). The gross weight of a jet-propelled aeroplane divided by the sea level static thrust of its engine(s).

thrustor (*Eng.*). An assembly comprising a motor,

a hydraulic pump, and a piston, which forms an alternative to a solenoid for controlling switchgear, operating hydraulic valves, releasing brakes, and similar operations.

thrust plane (*Geol.*). A thrust plane, or thrust, is a *reversed fault* which dips at a low angle. The rocks overlying the thrust plane are described as having been thrust over the underlying rocks.

thrust reverser (*Aero.*). A device for deflecting the efflux of a turbojet forward in order to apply a positive braking thrust after landing. There are two basic types: mechanical ones in which the jet is blocked by hinged doors, which also direct the gases forward; and aerodynamic ones wherein high-pressure air injected into the centre of the jet causes it to impinge upon peripheral louvres that turn it forward.

thrust spoiler (*Aero.*). A controllable device mounted on, or just behind, the nozzle of a jet-propulsion engine to deflect and thus negative the thrust. See also **thrust reverser.**

thrust/weight ratio (*Aero.*). The thrust of an aircraft's power plant(s) divided by the gross weight at take-off.

thudichium speculum (*Surg.*). An instrument used to inspect the nasal interior, from the facial aspect.

thulite (*Min.*). A variety of zoisite, pink in colour due to small amounts of manganese.

thulium (*Chem.*). A metallic element, a member of the rare earth group. Symbol Tm, at. no. 69, r.a.m. 168·9342. One of the rarest elements, occurring in small quantities in euxenite, gadolinite, xenotime, etc. Radioactive isotope emits 84 keV gamma-rays, frequently used in radiography.

thumbat (*Build.*). A *wall-hook* (q.v.) intended for attaching sheet-lead to a wall surface.

thumb latch (*Join.*). A latch which is operated by the pressure of the thumb. See **Norfolk latch.**

thumb plane (*Tools*). Rebate plane with curved sole; used for rounded rebates.

thumbscrew (*Tools*). Small type of G cramp used for light work.

thunder (*Meteor.*). The crackling, booming, or rumbling noise which accompanies a flash of lightning. The noise has its origin in the violent thermal changes accompanying the discharge, which cause nonperiodic wave disturbances in the air. Its reverberatory characteristic arises mainly from the continuous arrival of the brief noise from sections of the discharge at increasingly remote locations, since the spark may be many kilometres long. Claps of thunder occur when the spark is, roughly, normal to the line of observation. The time interval between lightning and the corresponding thunder (seconds divided by three) gives the distance of the storm centre (in kilometres).

thundercloud (*Meteor.*). See **cumulo-nimbus.**

thunderstorm (*Meteor.*). A storm in which *lightning* and *thunder* (qq.v.) occur, usually associated with *cumulo-nimbus* (q.v.) cloud. The mechanism by which the cloud becomes electrically charged is not fully understood but it appears that this only occurs when large quantities of liquid water and ice are present together in the cloud at temperatures well below −20°C.

thuringite (*Min.*). An oxidized chlorite relatively poor in silica.

thurl (*Mining*). See **thirl.**

Thury regulator (*Elec. Eng.*). An automatic voltage regulator in which the rheostat arm is moved by a pawl-and-ratchet mechanism actuated by solenoids.

Thury screw-thread (*Eng.*). See **Swiss screw-thread.**

Thury system (*Elec. Eng.*). See **series system.**

thyme (*Bot.*). Any member of the labiate genus *Thymus*, low shrubby plants with two-lipped corolla and calyx and four diverging stamens.

thymectomy (*Surg.*). Surgical removal of the thymus.

thymine (*Biochem.*). 5-methyl-2,6-dioxytetrahydropyrimidine,

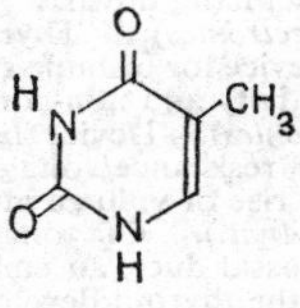

contained in the *nucleic acid* (q.v.) of animal nucleoprotein. Pairs with adenine in the *genetic code* (q.v.) of DNA.

thymitis (*Med.*). Inflammation of the thymus.

thymocyte (*Histol., Physiol.*). A small lymphocyte-like cell occurring in the thymus.

thymol (*Chem.*). $C_{10}H_{14}O$, 1-methyl-4-methylethyl-3-hydroxybenzene, of the formula:

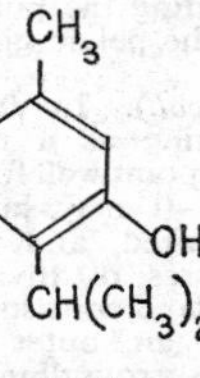

Large crystals; m.p. 51°C, b.p. 230°C; it occurs in thyme oil, and can be synthesized from propan-2-ol and 1-methyl-3-hydroxybenzene. It is used as a disinfectant, and in mouthwashes and in dentistry. Isomeric with *carvacrol.*

thymol blue (*Chem.*). Thymol-sulphon-phthalein, used as an indicator with two pH ranges, 1·2 to 2·8 (red→yellow), and 8·0 to 9·6 (yellow→blue).

thymoleptic (*Pharm.*). Any drug or other agent stimulating the emotions or mental activities.

thymolphthalein (*Chem.*). Indicator obtained by reaction between thymol and phthalic anhydride, having a pH range of 9·3 to 10·5, over which it changes from a colourless to a blue solution.

thymoma (*Med.*). A tumour arising in the thymus, usually highly malignant.

thymus (*Zool.*). In Vertebrates, a glandular structure originating as a series of outgrowths from the gill pouches and recently discovered to be the source of lymphocytes necessary for immune responses and also of a hormone enabling these cells to make antibodies.

thyratron (*Electronics*). Originally a TN for a gas-filled triode operating in an atmosphere of mercury-vapour. Now applied to any gas-filled triode, other common gas fillings being argon, helium, hydrogen, and neon. Ionization starts with sufficient positive swing of the negative grid potential, and anode and grid potentials lose control.

thyratron characteristic (*Electronics*). Graph relating grid striking voltage to anode voltage for a thyratron.

thyratron firing angle (*Electronics*). Phase angle of a.c. anode voltage supply to thyratron (measured relative to zero) at instant when it strikes.

thyreohyal (*Zool.*). In some Fish, a vertical plate connected with the posterior side of the basihyal and serving for the attachment of the muscles of the hyoid apparatus.

thyridium (*Zool.*). In some Insects (as *Trichoptera*), a semitransparent hairless whitish spot on the wing.

thyriothecium (*Bot.*). An inverted perithecium in which the asci hang down.

thyristor (*Electronics*). Thyratronlike semiconductor device for bistable switching between high conductivity and non-conductive modes.

thyrite (*Electronics*). Device having an inverse exponential resistance/voltage characteristic, which limits rise of voltage in circuits.

thyroglossal (*Anat.*). Of tongue and thyroid, e.g., thyroglossal duct, an embryonic structure from which the thyroid develops.

thyroid cartilage (*Zool.*). An unpaired cartilage forming part of the laryngeal skeleton, and lying between the body of the hyoid and the cricoid cartilage.

thyroidectomy (*Surg.*). The surgical removal of part of the thyroid gland.

thyroid fenestra (*Zool.*). In Turtles, *Sphenodon* and Lizards, an opening covered by a membrane developing under the region of the obturator muscle, separating the pubis and ischium and thus making the pelvic girdle into a tripartite structure.

thyroid gland (*Zool.*). In Vertebrates, a ductless gland originating as a median ventral outgrowth from a point well forward on the floor of the pharynx. It may be a single structure, bilobed, or paired, and there may be small accessory masses of thyroid tissue in other places. The gland consists of spherical follicles composed of an outer layer of cuboidal secretory cells surrounding and discharging into a central cavity. In this are found the hormones thyroxine and tri-iodothyronine, which are concerned with the rate of tissue metabolism, and the development of the nervous system and behaviour (deficiency causing *cretinism*), and, in *Amphibia*, with the control of metamorphosis. Evolutionarily the thyroid originates from the endostyle of amphioxus (of the *Cephalochorda*) and *Tunicata*, and the ammocoete larva of Lampreys.

thyroiditis (*Med.*). Inflammation of the thyroid gland. See also **Riedel's disease.**

thyrotomy (*Surg.*). See **laryngofissure.**

thyrotoxicosis (*Med.*). The condition resulting from overactivity of the thyroid gland (hyperthyroidism), the secretion of which is probably abnormal, as in Basedow's disease.

thyrotrophic (*Physiol.*). Maintaining or nourishing the thyroid; used of the hormone of the pars anterior of the pituitary gland, which stimulates growth and function of the thyroid gland.

thyroxine (*Biochem.*). One of the active principles of the thyroid gland (see also **tri-iodothyronine**) which controls the metabolic rate, i.e., the rate at which oxidation occurs:

thyrsus (*Bot.*). (1) A densely branched inflorescence, with main branching racemose but lateral branches cymose. (2) Any closely branched inflorescence with many small stalked flowers.

Thysanoptera (*Zool.*). The only order of the superorder *Thysanopteroidea*; minute insects with asymmetrical piercing mouthparts, prothorax large and free, tarsi have a protrusible adhesive terminal vesicle, two pairs of wings fringed with hairs. Some are serious pests causing malformation of plants and sometimes inhibiting the development of fruit. Thrips.

Thysanura (*Zool.*). The only order of the *Ectotropha*, having biting mouthparts, an eleven-segmented abdomen, some or all segments bearing styliform appendages which probably represent the coxites of limbs no longer present. There are paired segmented cerci and a segmented terminal filament. Tracheal system and Malpighian tubules present, metamorphosis slight or wanting, and moulting may continue throughout life. Silverfish, Bristle-tails.

thysanuriform (*Zool.*). See **campodeiform.**

Ti (*Chem.*). The symbol for *titanium.*

tibia (*Zool.*). In land Vertebrates the pre-axial bone of the crus; in Insects, *Myriapoda*, and some *Arachnida*, the fourth joint of the leg.

tibiale (*Zool.*). A bone of the proximal row of the tarsus in line with the tibia.

tibiofibula (*Zool.*). In some *Tetrapoda*, a bone of the leg formed by the fusion of tibia and fibula.

tibiotarsus (*Zool.*). In Birds, a leg bone formed by the fusion of the tibia with the astragalus.

tic (*Med.*). See **habit spasm.**

tic douloureux (*Med.*). Trigeminal neuralgia. An affection of the fifth cranial nerve characterized by paroxysmal attacks of pain in the face and the forehead.

tick-borne fever (*Vet.*). A febrile disease of sheep and cattle caused by infection by *Rickettsia phagocytophila*; transmitted by ticks.

ticker (*Telecomm.*). A printing tape machine, operated by start-stop signals.

tick fever (*Vet.*). Texas fever. See **redwater.**

tickler coil (*Electronics*). Inductance coil included in the anode current of a valve, magnetically coupled to the grid circuit for reaction.

tick pyaemia (*Vet.*). A pyaemic disease of lambs caused by bacterial infection by *Staphylococcus aureus*, which causes abscesses in various parts of the body, especially the joints and muscles; the infection is believed to enter via tick bites.

ticks (*Zool.*). See **Acarina.**

tic-tac escapement (*Horol.*). Type of anchor escapement covering 2 teeth of the escape wheel.

tidal air (*Physiol.*). The volume of air moving in and out of the lungs of Vertebrates (and of the tracheal system of Insects) during normal (unforced) breathing; in man about 500 cm^3.

tidal dock (*Civ. Eng.*). A dock within which the level is the same as outside.

tidal friction (*Astron.*). The friction caused by the ebb and flow of the tides, especially in narrow channels; it is responsible for a reduction of the earth's rate of rotation, and for an acceleration of the moon's motion, so that eventually the day and the month will be equal in length. This effect in the past has caused the moon to present the same face to the earth (rotation and revolution being equal).

tide (*Astron.*). The effect of the gravitational attraction of the moon, and in a lesser degree of the sun, on the waters of the earth, by which they tend to become heaped up at the point below the moon, and at the opposite point to this, so that twice in each lunar day there is an alternate inflow and outflow on the shores, modified by local configurations.

tide gauge (*Surv.*). An apparatus for determining the variation of sea-level with time.

tie (*Eng.*). A frame member sustaining only a tensile load.

tie-beam (*Eng.*). A structural member connecting the lower ends of a pair of principal rafters to prevent them from moving apart. Also applied to a member connecting parts of a structure which may otherwise move apart under load.

Tiedemann's bodies (or vesicles) (*Zool.*). In some *Echinodermata*, small glandlike structures of unknown function borne by the water-vascular ring.

tied letters (*Typog.*). A synonym for *ligature*.

tie line (*Surv.*). A survey line forming part of a *skeleton* (q.v.) and serving to fix its shape, e.g., a diagonal of a four-sided skeleton. (*Teleph.*) A line which may pass through exchanges, but which is used solely for connecting private branch exchanges, and over which incoming calls cannot be extended. Also called **inter-switchboard line.**

Tiemann-Reimer reaction (*Chem.*). See **Reimer-Tiemann reaction.**

tier or teir or rangs (*Textiles*). Terms for the carriages in a lace machine, arranged in working order.

tiered array (*Radio*). Antenna comprising a number of radiating elements, one above the other. Also called **stacked array.**

tie rod (*Eng.*). See **tension rod.**

tie-rod stator frame (*Elec. Eng.*). A form of stator frame for large electrical machines in which several frame sections are laterally secured by means of tie rods parallel with the axis of the machine.

tie wall (*Civ. Eng.*). A cross-wall built upon the extrados of an arch at right angles to the spandrel wall or walls.

tie wire (*Elec. Eng.*). A wire used to attach a transmission or telephone line conductor to a supporting insulator. Also called a **binding wire.**

T.I.F. (*Teleph.*). Abbrev. for *telephone interference* (or *influence*) *factor*.

tige (*Arch.*). The principal part of a column, between the capital and the base.

tiger's eye (*Min.*). A form of silicified crocidolite stained yellow or brown by iron oxide.

tiger-snake (*Zool.*). The most deadly Australian snake (*Notechis scutatus*), brown with black cross-bands.

tight coupling (*Elec. Eng.*). That between 2 circuits which causes alteration of the current in either to affect materially the current in the other. In mutual reactance coupling, coupling is said to be *tight* when the ratio of mutual reactance to the geometric mean of the individual reactances (of the same sign) of the 2 circuits approaches unity. Also called **close coupling.**

tight-edged (*Print.*). Said of a reel of paper which has dried out at the edges, resulting in a web which is slack in the middle.

tiglic acid (*Chem.*). $CH_3{\cdot}CH : C(CH_3)COOH$. Trans-form of dimethylacrylic acid. More stable than *angelic acid* (q.v.), the cis-form. M.p. 64·5°C, b.p. 198°C. Occurs in croton oil.

tigroid (*Cyt.*). Applied to the cytoplasmic chromophil granules (*Nissl granules*) of nerve cells.

tigrolysis (*Cyt.*). Chromatolysis of tigroid (*Nissl*) granules.

tile (*Build.*). A thin slab, often highly ornamental, of baked clay, terracotta, glass, cement, or asbestos-cement, used for roofing or for covering walls or floors.

tile-and-a-half tile (*Build.*). A purpose-made tile of extra width.

tile creasing (*Build.*). A course formed of 2 or 3 thicknesses of plain roofing tiles set in mortar and breaking joint. Laid immediately below a brick-on-edge coping and projecting about 2 in over each side of the wall, with the top surface sloped in cement, in order to prevent the percolation of water into the wall below the coping.

tilefish (*Zool.*). A brightly coloured American Atlantic fish, *Lopholatilus chamaeleonticeps*, noted for the sudden changes in its numbers.

tile hanging (*Build.*). See **weather tiling.**

tile lintel floor (*Build.*). A type of fire-resisting floor having a steel framework similar to the *filler joist floor* (q.v.), but with hollow tile, terracotta, or fireclay lintels filling in the panels between filler joists, thus reducing the amount of concrete required for encasement.

tile ore (*Min.*). The earthy brick-red variety of cuprite; often mixed with red oxide of iron.

Tilia (*Bot.*). The lime or linden genus, giving name to the family Tiliaceae.

tiling batten (*Build.*). See **slating and tiling battens.**

till (*Geol.*). A poorly sorted mixture of debris produced by the erosion of rocks by moving ice. Synonymous with *boulder clay*.

tiller (*Agric.*). Implement for penetration of soil to renovate old pasture land. (*Bot.*) A branch produced from the base of the stem, especially in corn and grasses.

tillering (*Agric.*). Shoots springing from base of original stalks; suckering as in elms, etc.

tilleyite (*Min.*). A monoclinic silicate and carbonate of calcium. Found in highly metamorphosed siliceous limestones.

tillite (*Geol.*). Consolidated and lithified till.

tilt-and-bend shading (*TV*). Adjusted compensation for inequality of effectiveness of a mosaic in translating an image of an external screen into a video signal.

tilting (*Cinema.*, *Photog.*). Swinging the camera vertically instead of sideways. See **pan.**

tilting fillet (*Build.*). A strip of wood laid beneath a *doubling course* (q.v.) to tilt it up slightly, so that the slates may rest properly on the roof. Also called **eave-board, skew fillet.**

tilting level (*Surv.*). A type of level whose essential characteristic is that the telescope and attached level tube may be levelled without the necessity for setting the rotation axis truly vertical.

tilt roof (*Build.*). A roof having the form of a circular arc in which the rise is small compared with the span.

timber (*For.*). Felled trees or logs suitable for conversion by sawing or otherwise.

timber brick (*Build.*). See **wood brick.**

timbering (*Build.*, *Civ. Eng.*). Temporary timbers arranged for the support of the earth in excavations, to prevent collapse of the sides.

timbre (*Acous.*). The characteristic tone or quality of a sound, which arises from the presence of various harmonics or overtones of the fundamental frequency.

time (*Astron.*). Originally measured by the *hour angle* of a selected point of reference on the celestial sphere with respect to the observer's meridian. The fundamental unit of time measurement now is the *second* based on an atomic oscillation. See also **apparent solar time, ephemeris time, Greenwich Mean Time, local time, mean solar time, sidereal time, standard time, universal time.**

time base (*Electronics*). (1) In a graph of the variation of a parameter with time, the horizontal scale is the time base. (2) In a cathode-ray tube, it records the voltages (or currents) producing a deflexion of the cathode beam which is uniform with time.

time-base generator (*Electronics*). Any circuit for deflecting the spot in the horizontal or vertical direction (or in some cases in a circular path) in a known manner (usually linearly) with time and at adjustable frequency. Same as *scanner* and *scanning* in television. Two time bases of sawtooth waveform are required in a television receiver, one to operate at line frequency, the other to deflect the beam vertically at frame frequency, thus ensuring that each line is traced below the previous one.

time between overhauls (*Aero.*). The period in hours of running time between complete dismantling of an aero-engine. Abbrev. **TBO** or **tbo.** Also **overhaul period.**

time constant (*Elec. Eng.*). When any quantity varies exponentially with time, the time required for a fractional change of amplitude equal to:

$$100\left(1-\frac{1}{e}\right)=63\%,$$

where e is the exponential constant (the base of natural logarithms). For a capacitance C to be charged from a constant voltage through a resistance R, the time constant is RC. For current in an inductance L being fed from a constant voltage through a resistance R, the time constant is L/R. See **response time.** (*Vac. Tech.*) Of a vacuum system, the ratio of the volume of the system to the pumping speed.

time-delay relay (*Elec. Eng.*). One which closes contacts in one circuit a specified time after those in a second circuit have been closed. Widely used to delay application of high-tension voltage until valve cathodes are hot.

time discriminator (*Telecomm.*). Circuit which gives an output proportional to the time difference between two pulses, its polarity reversing if the pulses are interchanged.

time-division multiplex (*Telecomm.*). System of multiplex transmission which allocates a physical channel in sequence to a number of communication channels, each using timed pulses.

timed spark system (*Radio*). A spark system of radiotelegraphy employing a rotary spark gap, with the result that the discharges occur at regular intervals.

time element (*Elec. Eng.*). The time-delay feature in the action of a circuit-breaker.

time exposures (*Photog.*). Exposures in cameras for periods which are long in comparison with so-called instantaneous exposure; generally operated by hand with the assistance of a stop-watch.

time-lag device (*Elec. Eng.*). The apparatus providing the time element in a circuit-breaker. See **time-limit attachment.**

time-limit attachment (*Elec. Eng.*). The mechanical device whereby a circuit-breaker opens only after a predetermined time delay. Cf. *time-limit relay.*

time-limit relay (*Elec. Eng.*). An electric relay which comes into action some time after it has received the electrical operating impulse.

time-meter (*Elec. Eng.*). An instrument for measuring the time during which current flows in a circuit. Also called **hour-counter, hour-meter.**

time-of-flight spectrometer (*Nuc. Eng.*). Spectrometer used with beams of particles, especially neutrons. A chopper admits the particles to a flight tube in short bursts, and another at the far end of the tube allows through only those for which the time of flight corresponds to the interval between the choppers opening. See **sector disk.**

time of operation (*Nuc.*). Time between the occurrence of a primary ionizing event and the occurrence of the count. (*Telecomm.*) In relays, time between application of current or voltage and occurrence of a definite change in circuits controlled by its contacts.

time of oscillation (*Horol.*). The time of oscillation of a pendulum or balance is twice that of the single vibration.

timepiece (*Horol.*). A general trade term for any clock that shows the time but does not strike.

timer (*Elec. Eng.*). Device, operated by electric motor, clockwork, or an electronic or resistor-capacitor circuit, which opens or closes at specified times, with or without delay, control circuits for lighting lamps, operating motors or valves, etc., in a process controller.

time response (*Automation*). Same as **booster response.**

time scale (*Comp.*). When time is a variable in analogue computing, time scale can be *real* (tracking satellites or aircraft), *fast, slow,* or *extended.*

time-shared amplifier (*Telecomm.*). Form of multiplex system in which one amplifier handles several signals simultaneously using successive short intervals of time for each.

time-sharing (*Comp.*). Technique by which more than one terminal device can use the input, processing and output facilities of a central computer simultaneously. A large computer may control many peripheral units operating at their own speed with maximum efficiency.

time signal (*Telecomm.*). One indicating standardized time, radiated by radio or over telegraph lines for exact calibration. Now determined by atomic clock.

time study (*Work Study*). A work measurement for recording the times and rates of working for the elements of a specified job carried out under specified conditions, and for analysing the data so as to obtain the time necessary for carrying out the job at a defined level of performance.

time switch (*Elec. Eng.*). A switch arranged to open or close a circuit at a predetermined time, operating by some form of electrical or clock-work mechanism.

time valve (*Photog.*). A small valve on the operating bulb of the release of a camera shutter, fitted to enable the actual time of operation of the shutter to be delayed arbitrarily.

timing (*I.C. Engs.*). The process of setting the valve-operating mechanism of an engine so that the valves open and close in correct relation to the crank during the cycle: a similar adjustment of the magneto or distributor drive; the actual valve or magneto setting, called **valve timing, ignition timing.** (*Horol.*) The process of: (*a*) setting a clock or watch to time; (*b*) observing the rate of a clock or watch.

timing chain (*I.C. Engs.*). Chain which drives the camshaft from the crankshaft in some types of timing gear.

timing gear (*I.C. Engs.*). The drive between the crankshaft and the camshaft, by direct gearing, bevel-shaft, or chain and sprocket-wheels, giving a reduction ratio of 1 : 2.

timing nuts (*Horol.*). The 2 nuts on the rim of a chronometer balance, one at each end of the arm, used for timing purposes.

timing screws (*Horol.*). The 4 diametrically opposite screws in the rim of a compensating balance used for bringing the watch to time.

timing washers (*Horol.*). Thin washers placed under the heads of the screws of a balance to produce slight alteration in the moment of inertia of the balance and so modify the time of vibration.

Timiskaming Group (*Geol.*). An important member of the Pre-Cambrian succession in the Canadian Shield, of post-Laurentian, pre-Huronian age, and consisting essentially of quartzites and arkoses about 29 000 ft (ca. 9000 m) thick. Equivalent to the Sudbury Series (or Sudburian) of Coleman, the Hastings Series of S.W. Ontario, and the Pontiac Series around Lake Quebec.

tin (*Chem.*). A soft, silvery-white metallic element, ductile and malleable, existing in three allotropic forms. Symbol Sn (Latin *stannum*, tin), at. no. 50, r.a.m. 118·69, rel. d. at 20°C 7·3, m.p. 231·85°C. Not affected by air or water at ordinary temperatures. Electrical resistivity is $11{\cdot}5 \times 10^{-8}$ ohm metres at 20°C. The principal use is as a coating on steel in tinplate; also used as a constituent in alloys and with lead in low melting-point solders for electrical connexions. Occurs as tin-oxide, SnO_2, or *cassiterite*. See **tin alloys.**

tin alloys (*Met.*). Tin is an essential constituent in soft solders, type metals, fusible alloys, and certain bearing metals. These last contain 50–92% of tin alloyed with copper and antimony, and sometimes lead. Tin is also a constituent of bronze and pewter.

Tinamiformes (*Zool.*). An order of *Palaeognathae*, containing small superficially partridge-like, almost tailless birds which are essentially cursorial, but can fly clumsily for short distances, and have a keeled sternum. May in fact be nearer the *Galliformes* of the *Neognathae*. Tinamus.

tincal (*Min.*). The name given since early times to crude borax obtained from salt lakes, e.g., in Kashmir and Tibet. See **borax.**

tinctorial power (*Photog.*). The measure of the depth of colour produced by a dye. Precisely, it is the reciprocal of the concentration required to yield a given density in a given thickness of emulsion.

tinea (*Med.*). See **ringworm.**

tin fusion gas analysis (*Vac. Tech.*). A form of hot extraction gas analysis in which the sample is dissolved in molten tin.

tingle (*Plumb.*). A flat strip of lead or copper used as clip between jointing sheets of lead.

ting-tang (*Horol.*). A clock that strikes the quarters on two notes only.

tinguaite (*Geol.*). A fine-grained, usually porphyritic igneous rock which normally occurs in dykes. It has the composition of phonolite, with aegirine-augite and nepheline as essential constituents.

tinman's solder (*Met.*). A tin-lead solder melting below a red heat, used for tinning. The most fusible solder contains 65% tin.

tin-nickel (*Met.*). Metal finish which results from the simultaneous electrodeposition of tin and nickel on a polished surface of brass in a carefully controlled bath, resulting in a non-tarnishable and noncorrodible polished surface of low friction.

tinnitus, tinnitus aurium (*Med.*). Persistent sensation of ringing noises in the ear.

tin-plate (*Met.*). Thin sheet-steel covered with an adherent layer of tin formed by passing the steel through a bath of molten tin or by electrodeposition. Resists atmospheric oxidation and attack by many organic acids. Used for food-containers, etc.

tin pyrites (*Min.*). See **stannite.**

tin-stone (*Min.*). See **cassiterite.**

tint (*Light*). An unsaturated colour. (*Paint.*) A colour softened by the addition of white. *Tinting* signifies the addition of pigment to white. (*Print.*) A *mechanical stipple* (q.v.), especially when used for line-colour purposes.

tintype (*Photog.*). See **ferrotype.**

tin-zinc (*Met.*). Metal finish which results from the simultaneous electrodeposition of tin and zinc on a clean steel surface, giving a non-corrodible finish to chassis for electronic apparatus.

tip (*Eng.*). The part of a cutting tool containing the cutting edges, if made of a material of superior quality to that of the remainder of the tool and securely fastened to the latter by brazing or otherwise. (*Mining*) See **dump.** (*Teleph.*) The outermost contact on the 3-way plugs which terminate the flexible cords of an operator's cord circuit in a manual exchange.

tip-cat folder (*Print.*). A type of folder in which the folding blade shaft is actuated by a specially shaped cam and forms the transverse fold by pushing the copy through the folding rollers.

tip-path plane (*Aero.*). The plane of rotation of the tips of a rotorcraft's blades, which is higher than the rotor hub in flight. See **coning angle.**

tipping-in (*Bind.*). See **plating.**

tipple (*Mining*). Frame into which ore trucks are run, gripped, and rotated to discharge contents.

tip wire (*Teleph.*). See **T-wire.**

Tirill regulator (*Elec. Eng.*). An automatic voltage regulator in which a vibrating contact device short-circuits the regulating rheostat to an extent controlled by voltage relays.

T-iron (*Civ. Eng.*). A structural member of wrought-iron or rolled mild steel having a T-shaped cross-section.

Tiros (*Meteor.*). Name given to one of a series of satellites giving meteorological information by transmitting television pictures of sun-illuminated cloud-systems and associated areas on the earth's surface.

tissue (*Biol.*). An aggregate of similar cells forming a definite and continuous fabric, and usually having a comparable function; as *epithelial tissue, nervous tissue, vascular tissue.*

tissue culture (*Bot., Zool.*). The growth of detached pieces of tissue in nutritive fluids under conditions which exclude bacteria and fungi.

tissue dose (*Radiol.*). Absorbed *depth dose* of radiation received by specified tissue. Cf. *skin dose.*

tissue equivalent material (*Radiol.*). See **phantom material.**

tissue respiration (*Biol.*). See **respiration, internal.**

tissue system (*Bot.*). The whole of the tissues present in a plant which have the same function, whether or not they are in continuity throughout the plant and whatever their position.

tissue tension (*Bot.*). The mutual compressions and stretchings exerted by the tissues of a living plant.

Titan (*Astron.*). A satellite of *Saturn.*

titanates (*Chem.*). Compounds found in minerals, or formed when titanium(IV)oxide is fused with alkalis; containing the anion TiO_4^{4-} or TiO_3^{2-}.

titanaugite (*Min.*). A titaniferous variety of the monoclinic pyroxene *augite.*

titania (*Chem.*). See **titanium(IV)oxide.**

titaniferous iron ore (*Min.*). See **ilmenite.**

titanite (*Min.*). See **sphene.**

titanium (*Chem.*). A metallic element resembling iron. Symbol Ti, at. no. 22, r.a.m. 47·90, rel. d. (20°C) 4·5, m.p. 1850°C, b.p. above 2800°C. Manufactured commercially since 1948, it is characterized by strength, lightness, and corrosion resistance. Widely used in aircraft manufacture, for corrosion resistance in some wet extraction processes and as a deoxidizer for special types of steel, in stainless steel to diminish susceptibility to intercrystalline corrosion, and as a carbide in cemented carbides. Sometimes used for the solid horns of magnetostrictive generators. It occurs only in combination, and is widely distributed on the earth's crust (0·6% by weight). (See **ilmenite, ilmenorutile.**)

titanium(IV)oxide or **titanium dioxide** (*Chem.*). TiO_2. It is a pure white pigment of great opacity. Widely used industrially in paints, plastics, etc., and see **titanizing.** Forms titanates when fused with alkalis. Also called **titania.**

titanium yellow (*Paint.*). A lemon-yellow pigment of high opacity but low tinting strength, consisting of titanium dioxide with antimony oxide and a small proportion of nickel oxide, prepared by calcination at high temp.; heat- and chemical-resistant and light-fast.

titanizing (*Glass*). Use of the titanium dioxide TiO_2 in the manufacture of heat-resisting and durable glass by incorporating it to replace certain proportions of soda, the viscosity increasing proportionately.

title block (*Eng.*). A framed and subdivided space on an engineering drawing, containing the drawing number and other information.

titler (*Cinema.*). Specially constructed screen on which lettering may be set up and photographed against suitable backgrounds.

title signature (*Print.*). The first *section* (or *signature*) of a book, containing the title page and other prelims, and not normally requiring a *signature mark* (q.v.).

titling fount (*Typog.*). Consists of capitals, figures, and punctuation marks only, no lowercase, a narrow beard being sufficient; used for dropped initials and when close spacing between lines is required.

titration (*Chem.*). The addition of a solution from a graduated vessel (burette) to a known volume of a second solution, until the chemical reaction between the two is just completed. A knowledge of the volume of liquid added and of the strength of one of the solutions enables that of the other (the *titre*) to be calculated.

titre (*Bacteriol.*). A term used in immunology, of antibody, to mean the quantity of antibody present in an organism. (*Chem.*) See **titration.**

titrimeter (*Chem.*). Apparatus for electrometric titrations in which potential changes are followed continuously and automatically.

titubation (*Med.*). Staggering and reeling movements of the body, due to disease of the nervous system.

Tl (*Chem.*). The symbol for *thallium.*

TLC (*Chem.*). **Thin-layer chromatography.**

Tm (*Chem.*). The symbol for *thulium.*

T-maze (*An. Behav.*). A T-shaped pathway with a starting box at the base of the T and goal boxes containing reinforcing stimuli at the ends of either or both arms. Discriminative stimuli may be placed in the arms of the T near the *choice point* (q.v.). See also **multiple-.**

TMT and **TMTD** (*Chem.*). **See thiram.**

TM-wave (*Telecomm.*). Abbrev. for *transverse magnetic wave*, having no component of magnetic force in the direction of transmission of electromagnetic waves along a waveguide. Also known as **E-wave** (it must have electric field component in direction of transmission).

Tn (*Nuc.*). The symbol for *thoron.*

T-network (*Telecomm.*). One formed of two equal series arms with a shunt arm between.

T.N.T. (*Chem.*). An abbrev. for *trinitrotoluene.*

toad's-eye tin (*Min.*). A variety of *cassiterite* (q.v.) occurring in botryoidal or reniform shapes which show an internal concentric and fibrous structure. It is brownish in colour.

toadstone (*Geol.*). An old and local name for the basalts found in the Carboniferous Limestone of Derbyshire. The name may be derived from the rock's resemblance in appearance to a toad's skin, or from the fact that it weathers into shapes like a toad, or from the German word *todstein* ('dead stone') in reference to the absence of lead.

to-and-fro aerial ropeway (*Civ. Eng.*). See **jig-back.**

toat (*Carp., Join.*). The handle of a bench plane.

Tobin bronze (*Met.*). A type of alpha-beta brass or Muntz metal containing tin. It contains 59–62% copper, 0·5–1·5% tin, the remainder being zinc. Used when resistance to sea water is required. Also **Admiralty or naval brass.**

tocopherol (*Chem.*). See **vitamin E.**

Todd-AO (*Cinema.*). Wide-screen cinematography using 65–70 mm film run at 30 frame/s.

toe (*Civ. Eng.*). The part of the base of a reinforced concrete retaining wall projecting in front of the face of the *stalk* (q.v.). (*Photog.*) The lower curvature of the gamma curve of an emulsion, where densities are obtained which are greater than those expected from a given exposure.

toed (*Carp.*). Said of an upright or inclined timber which is fastened to a horizontal timber by nails driven in obliquely through its foot.

toe-in (*Autos.*). A slight forward convergence given to the planes of the front wheels to promote steering stability and equalize tyre-wear. Generally ⅛ in (3 mm).

toe of the brush (*Elec. Eng.*). See **entering edge.**

toe-picking (*Vet.*). The vice, acquired by individual budgerigars, of biting the feet of birds of other species, particularly finches, within the same aviary.

toe piston (*Acous.*). A large pushbutton, arranged with others at the side of the balanced pedals in an organ, for operating groups of stops with the feet.

Toepler machine (*Elec. Eng.*). An early form of electrostatic generator. Also **Voss machine.**

Toepler pump (*Vac. Tech.*). A vacuum pump which functions by means of the alternate raising and lowering of a column of liquid (normally mercury).

toe wall (*Civ. Eng.*). A dwarf retaining-wall built at the foot of an embankment slope as a safeguard against the earth spreading.

toggle (*Telecomm.*). Bistable trigger circuit, a multivibrator with coupling capacitors omitted, which switches between 2 stable states depending on which valve (or transistor) is triggered. Once **flip-flop** in U.K.

toggle joint (*Eng.*). A mechanism comprising two levers hinged to each other, the opposite end of one being hinged on a fixed point, the opposite end of the other being hinged on a press ram or other load point. If the levers form an obtuse angle and an effort tending to increase this angle is applied at their common hinge point, a considerable force is produced at the load point.

toggle press (*Eng.*). A power press, often double-acting, incorporating a toggle joint, used for deep drawing operations.

toilet (*Med.*). The cleaning and dressing of a wound or injured part.

tolbutamide (*Pharm.*). *N*-Butyl-*N'*-toluene-*p*-sulphonylurea, $C_{12}H_{18}N_2O_3S$, used for oral treatment of diabetes. TN **Rastinon.**

tolerance (*Bot.*). (1) The ability of a plant to endure adverse environmental conditions, especially drought and shading. (2) The ability of the plant to withstand the development within it of a parasite without showing signs of serious disease. (*Eng.*) The range between the permissible maximum and minimum limits of size of a workpiece or of distance between features (e.g., hole centres) on a workpiece.

tolerance dose (*Radiol.*). Maximum dose which can be permitted to a specific tissue during radiotherapy involving irradiation of any other adjacent tissue.

tolerance, 'law' of (*Ecol.*). Organisms have an ecological minimum and maximum, with a range between, which represents the limits of tolerance. Absence or failure of an organism can be controlled by any factor approaching the limits of tolerance, i.e., a *limiting factor* (q.v.).

toll call (*Teleph.*). A short-distance *trunk call* covering exchanges in the neighbourhood of a major centre. Now obsolete. In America, any long-distance call.

toll television (*TV*). Programme service which, through technical scrambling devices, is available only by *ad hoc* payment. Also called **pay as you view, subscription television.**

Tolu balsam (*Chem.*). See **balsam of Tolu.**

toluene (*Chem.*). $C_6H_5{\cdot}CH_3$, a colourless liquid, m.p. −94°C, b.p. 110°C. It occurs in coal- and wood-tar; insoluble in water; miscible with ethanol, ethoxyethene, trichloromethane. Used as a solvent and as an intermediate for its derivatives. Also called **methylbenzene, toluol.**

toluene di-isocyanate (*Chem.*). T.D.I. $CH_3{\cdot}C_6H_3{\cdot}(NCO)_2$. The most commonly used isocyanate in the production of polyurethane foams. The isocyanate is reacted with a high molecular weight glycol to form a liquid pre-polymer. The pre-polymer reacts with water to give the polyurethane and carbon dioxide, which causes the foam.

toluidine blue stain (*Micros.*). An alkaline aniline dye, used particularly for mast cells and nerve cells in smear preparations or sections.

toluidines (*Chem.*). Methyl(aminobenzenes). $H_3C{\cdot}C_6H_4{\cdot}NH_2$, homologues of aniline. There are three isomers, viz., 2-toluidine, a liquid, b.p. 197°C; 4-toluidine, crystals, m.p. 43°C, b.p. 198°C; 3-toluidine, a liquid, b.p. 199°C.

tomentum (*Bot.*). Covering of felted cottony hairs. *adj.* **tomentose.**

Tomes's fibres (*Histol.*). Fibrelike processes of the odontoblasts which pass through the dentinal tubules.

tommy bar (*Tools*). See **box spanner.**

tomography (*Radiol.*). Movement of source and photoplate in diagnostic radiography, so that the ratio of their distances from a chosen plane in the body remains constant, this plane only giving rise to a defined image. Also **body-section radiography, laminography, planigraphy.** See also **transaxial tomography.**

Tomosvary's organs (*Zool.*). In some Centipedes, transparent chitinous projections, covered with fine hairs, placed on the head near the base of the antennae; possibly auditory in function.

-tomy. Suffix from Gk. *tomē*, a cut.

ton. A unit of mass for large quantities. The *long ton*, commonly used in Britain, is 2240 lb. The *short ton*, commonly used in America, is 2000 lb. The *metric ton* or *tonne* (1000 kilograms; symbol t) is 2204·6 lb. In Britain the *short ton* is used in metalliferous mining, the *long ton* in coal-mining. (*Ships*) See **tonnage.**

tonalite (*Geol.*). A coarse-grained igneous rock of dioritic composition carrying quartz as an essential constituent, i.e., quartz-mica-diorite. Two varieties are distinguished: *soda-tonalite*, with albite in excess of anorthite, and *lime-tonalite*, with anorthite in excess of albite.

tone (*Acous.*). Single frequency for testing; but a tone is now often used to specify a complex note having a constant fundamental frequency. A sinusoidal sound wave is termed a pure tone. (*Telecomm.*) That frequency of interruption of the scanning light in facsimile which becomes the carrier in transmission. (*TV*) Brightness of an area in a picture, described as *high-key* or *low-key*. (*Zool.*) The condition of elasticity or tension proper to the living tissues of the animal body, especially muscles. Also **tonicity.** *adj.* **tonic.** See also **tonus.**

tone control (*Telecomm.*). One for altering characteristic response of audiofrequency amplifier for more pleasing output quality.

tone control circuit (*Telecomm.*). A circuit network or element to vary the frequency response of an audiofrequency circuit, thus varying the quality of the sound reproduction.

tone-control transformer (*Telecomm.*). One in which the leakage and/or self-capacitance can be altered in such a way as to regulate its response over the operating frequency range.

tongs (*Build.*). See **stone tongs.**

tong-test ammeter (*Elec. Eng.*). An a.c. ammeter and current transformer combination whose iron core can be opened and closed round a cable, thus forming the single-turn primary winding of the transformer.

tongue (*Join.*). A slip feather. (*Zool.*) In Vertebrates, the movable muscular organ lying on, and attached to, the floor of the buccal cavity; it has important functions in connexion with tasting, mastication, swallowing, and (in higher forms) sound production; in Invertebrates, especially Insects, any conformation of the mouth-parts which resembles the tongue in structure, appearance, or function—proboscis: antlia; haustellum; radula; ligula; any structure which resembles the tongue.

tongue-and-groove joint (*Join.*). A joint formed between the butting edges of two boards, one of which has, along the middle of its length, a projecting fin cut to fit into a corresponding plough groove in the other.

tongue bars (*Zool.*). In *Cephalochorda*, the secondary gill-bars.

tongue-worm (*Vet.*). *Linguatula serrata.* An aberrant Arthropod belonging to the class *Pentastomida*, which occurs in the nasal passages of the dog, fox, and wolf, and more rarely other animals; the larval and nymphal stages occur in herbivorous animals.

tonicity (*Zool.*). See **tone.**

tonic train (*Radio*). Same as **interrupted continuous waves.**

toning (*Photog.*). See **chemical toning, dye toning.**

tonnage (*Ships*). A measurement assigned by the relevant authority for assessing dues, etc. One ton equals 100 cubic feet. See **gross tonnage, net register tonnage.**

tonnage breadth (*Ships*). A number of tonnage breadths are measured horizontally inside the frames or *spar ceiling* (q.v.) if fitted. Internal cross-section areas are calculated from these by *Simpson's rule* (q.v.). These areas are then used in the calculation of *tonnage* (q.v.).

tonnage depth (*Ships*). A number of tonnage depths are measured at certain points throughout the length of the ship from the top of the inner bottom or *ceiling* (q.v.), if fitted, to the top of the deck, deducting one-third of the *camber* (q.v.). Used, together with *tonnage breadths* (q.v.), in the calculation of cross-section areas.

tonnage dimensions (*Ships*). The internal dimensions used in the calculation of *tonnage* (q.v.). These are *tonnage length*, *tonnage breadth* and *tonnage depth* (qq.v.).

tonnage length (*Ships*). The length measured along the uppermost continuous deck in ships having less than three decks and the second continuous deck from below in others. It is measured from a point where the line of the inside of the frames or *spar ceiling* (q.v.) cuts the centre line forward to a similar point aft.

tonne. Metric ton, 1000 kg. See **ton.**

tonofibrillae (*Zool.*). In Insects, fibrillar non-striated structures which, in some cases where muscles are attached to cuticle, extend from myofibrillae through or among the hypodermal cells, and may traverse the endocuticle.

tonometer (*Med.*). An instrument for measuring (1) hydrostatic pressure within the eye; (2) blood pressure, etc. (*Acous.*) A device consisting of a series of tuning-forks for determining the frequencies of tones. (*Phys.*) An instrument for measuring vapour pressure.

tonoplast (*Bot.*). The cytoplasmic membrane surrounding a vacuole in the protoplast.

tonotaxis (*Biol.*). The property of responding to a change in density of the surrounding medium.

tonsillectomy (*Surg.*). The surgical removal of the tonsils.

tonsillitis (*Med.*). Inflammation of the tonsils.

tonsillotomy (*Surg.*). The surgical removal of part of a hypertrophied tonsil.

tonsils (*Zool.*). In Vertebrates, lymphoid bodies of disputed function situated at the junction of the buccal cavity and the pharynx.

tonus (*Zool.*). A state of persistent excitation; in plain or involuntary muscle, a prolonged state of contraction, independent of continued excitation by nervous impulses, in striated or voluntary muscle, a similar state of prolonged contraction dependent on continued impulses from the nerve-centres; in certain nerve-centres, as the respiratory centre, the state in which impulses are constantly given out without any corresponding afferent impulses from the receptors. Also **tonic spasm.**

tooled ashlar (*Build.*). A block of stone finished with parallel vertical flutes.

tooling (*Bind.*). Decorating by hand a book cover, usually leather.

toolmaker (*Eng.*). A highly skilled engineering craftsman employed to make press tools, cutters, and other precision equipment.

tool post (*Eng.*). The clamp by which a lathe or shaping-machine tool is held in the slide rest or ram. In its simplest form it consists of a slotted post, the end of which carries a clamping screw.

tool-post grinder (*Eng.*). A small grinding machine held on the tool post of a lathe and fed across the work by means of the regular longitudinal or compound rest. It is used for grinding miscellaneous small workpieces in the lathe, and for truing lathe centres.

tool steel (*Met.*). Steel suitable for use in tools, usually for cutting or shaping wood or metals. The main qualities required are hardness, toughness, ability to retain a cutting edge, etc. Contains 0·6–1·6% carbon. Many tool steels contain high percentages of alloying metals—tungsten, chromium, molybdenum, etc. (see **high-speed steel**). Usually quenched and tempered, to obtain the required properties.

tool-using (*An. Behav.*). The manipulation by an animal of an inanimate object to achieve some end which it could not otherwise accomplish.

tooth (*Bot.*). (1) Any small irregularity on the margin of a leaf. (2) The free tip of one petal of a gamopetalous corolla. (*Elec. Eng.*) The projection formed in the core-plate of an armature between adjacent slots. (*Zool.*) A hard projecting body with a masticatory function. In Vertebrates, a hard calcareous or horny body attached to the skeletal framework of the mouth or pharynx, used for trituration or fragmentation of food; in Invertebrates, any similar projection of chitinous or calcareous material used for mastication or trituration.

toothed wheels (*Eng.*). See **bevel gear, helical gears, spiral gear, spur gear, worm.**

toothing plane (*Tools*). One with a serrated blade set perpendicularly; used for providing a key for gluing veneers, etc.

toothings (*Build.*). The recesses left in alternate courses of a wall when later extension is expected, so that the extension can be properly bonded in.

tooth ratio (*Elec. Eng.*). The ratio of slot width to tooth width as measured at the circumference of the armature.

tooth ripple (*Elec. Eng.*). See **slot ripple.**

tooth thickness (*Eng.*). Of a gear-wheel, the length of arc of the pitch circle between opposite faces of the same tooth. See **pitch diameter (Fig. II).**

top (*Acous.*). A colloquialism for the higher range of audiofrequencies in sound reproduction. (*Spinning*) Combed wool sliver which has been prepared for spinning into worsted yarn.

topaz (*Min.*). Silicate of aluminium and fluorine, usually containing hydroxyl, which crystallizes in the orthorhombic system. It usually occurs in veins and druses in granites and granite-pegmatites. It is colourless, pale blue, or pale yellow in colour, and is used as a gemstone. Cf. *citrine*, *Oriental topaz*, *Scottish topaz*, *Spanish topaz*.

topazolite (*Min.*). A variety of the calcium-iron garnet *andradite* (q.v.), which has the honey-yellow colour and transparency of topaz.

top beam (*Carp.*). The horizontal beam connecting the rafters of a collar-beam roof.

top blanket (*Print.*). On web-fed relief presses the outside dressing of an impression cylinder. Also **top sheet.** Cf. *underblanket*.

top-contact rail (*Elec. Eng.*). A type of contact rail, as used in electric traction, where contact between the collecting-shoe and the rail is made at the upper surface of the rail.

top dead-centre (*Eng.*). See **inner dead-centre.**

topectomy (*Surg.*). Removal by surgery of a portion of cerebral cortex, especially in the treatment of certain psychoses.

top girth (*For.*). In standing trees, the girth at the top of the merchantable length of the bole.

tophaceous (*Med.*). Sandy, gritty, of the nature of tophus, e.g., *tophaceous gout*.

top hamper (*Ships*). That part of the structure of the ship which is above the *superstructure* (q.v.). Also called **deck houses or erections.**

top-hung (*Build.*). Said of a window-sash arranged to open outwards about hinges on its upper edge.

tophus (*Med.*). A hard nodule composed of crystals of sodium biurate which are deposited in bodily tissues in gout.

topochemistry (*Chem.*). The study of reactions which occur only at definite regions in a system.

topography (*Geog., Surv.*). The delineation of the natural and artificial features of an area.

topology (*Maths.*). The study of those properties of shapes and figures which remain invariant under one-to-one continuous transformations (i.e., topological transformations). If the shapes are imagined as drawn on rubber then the totality of properties which are preserved if the rubber is arbitrarily distorted, without being torn, constitutes topology. Topology is often said to be a study of continuity. Formerly called **analysis situs.**

topotaxis (*Biol.*). Response or reaction of an organism to a stimulus, in which the organism orientates itself in relation to the stimulus and moves towards or away from it.

topotype (*Zool.*). A specimen collected in the same locality as the original type specimen of the same species.

topping (*Civ. Eng.*). See **wearing course.** (*Electronics*) Operation such that the anode voltage remains constant during a cycle, achieved by a catching diode leading to a definite potential.

top rake (*Eng.*). In cutting tools, the angle which that part of the cutting face which lies immediately behind the tool point makes with the horizontal.

topset beds (*Geol.*). Gently inclined strata deposited on the subaerial plain or the just-submerged part of a delta. They are succeeded seawards by the *foreset beds* and, in deep water, by the *bottomset beds.*

top sheet (*Print.*). See **top blanket.**

top shot (*Photog.*). One taken with axis of camera nearly vertical.

top yeast (*Bot.*). The yeast which vegetates at the surface of a fluid in which fermentation is proceeding.

torbanite (*Min.*). A variety of oil shale containing 70–80% of carbonaceous matter, including an abundance of spores. It is dark-brown in colour, and is found at Torbane Hill near Bathgate (Scotland). See **boghead coal.**

torbernite (*Min.*). A beautiful rich-green hydrated phosphate of uranium and copper which crystallizes in the tetragonal system. It occurs associated with autunite and frequently in parallel growth with it, and also with other uranium minerals. Also called **copper** (or **cupro-) uranite.**

torch brazing (*Eng.*). Brazing by means similar to oxyacetylene welding. Heat is applied locally by a gas torch, and filler metal in wire form is melted into the joint. Flux is applied by immersing the wire.

torch igniter (*Aero.*). A combination igniter plug and fuel atomizer for lighting-up gas turbines.

torching (*Build.*). The operation, sometimes performed on slates which have been laid on battens only and not on boarding, of pointing the horizontal joints from the inside with hair mortar or cement.

torfschlamm (*Ecol.*). Synonym for **dy.**

tormae (*Zool.*). Two small chitinous processes of the pharyngeal surface of the labrum in Insects.

tormentor (*Cinema.*). In motion-picture production, a portable panel for absorbing sound waves and regulating the reflection of sound waves in a set.

tormogen cell (*Zool.*). In the setae of Insects, the hypodermal cell which produces the articular membrane. Cf. *trichogen cell.*

tornado (*Meteor.*). An intensely destructive, advancing whirlwind formed from strongly ascending currents. When over the sea, the apparent drawing up of water arises from the condensing of water vapour in the vacuous core; also, in W. Africa, the squall following thunderstorms between the wet and dry seasons.

tornaria (*Zool.*). The ciliated pelagic larval form of *Balanoglossida.*

toroid or torus (*Elec. Eng.*). Said of a coil or transformer which corresponds in shape to an anchor ring. Adopted because of the ease of making windings exactly balanced to the circuit in which they are inserted and to earth. Also, on account of the entire enclosure of magnetic field, there is no interaction with similar adjacent coils; thus, loading coils for transmission lines are threaded on rods in a pot. See also **doughnut.** (*Maths.*) A solid generated by rotating a circle about an external point in its plane.

toroidal-intake guide vanes (*Aero.*). The flared annular guide vanes which guide the air evenly into the intake of a centrifugal *impeller.*

toroidal surface (*Optics*). A lens surface in which the curvature in one plane differs from that in a plane at right angles.

toroidal winding (*Elec. Eng.*). **See ring winding (1).**

torpedo divider (*Agric.*). A cone-shaped divider to assist the cutting of tangled crops.

torque (*Aero.*). Airscrew torque is the measure of the total air forces on the airscrew blades, expressed as a moment about its axis. (*Mech.*) The turning moment exerted by a tangential force acting at a distance from the axis of rotation or twist. Measured by the product of the force and the distance, unit: newton metre.

torque amplifier (*Elec. Eng.*). (1) Mechanical amplifier in which torque (not angle) is varied by differential friction of belts on drums. (2) Electrical servo system performing same function.

torque converter (*Eng.*). A device which acts as an infinitely variable gear, but generally at varying efficiency, e.g., a centrifugal pump in circuit with an inward-flow turbine.

torque limiter (*Aero.*). Any device which prevents a safe torque value from being exceeded, but specifically one which is used on a constant-speed *turboprop* (q.v.) to prevent it from delivering excess power to its airscrew.

torque link (*Aero.*). A mechanical linkage, usually of simple scissor form, which prevents relative rotation of the telescopic members of an aircraft *landing gear* shock absorber.

torque-loaded quoin key (*Typog.*). One in which the lock-up pressure is consistent no matter who uses the key, which is adjusted to apply the optimum pressure and no more.

torquemeter (*Aero.*). A device for measuring the torque of a reciprocating aero-engine or *turboprop,* the indication of which is used by the pilot, together with r.p.m. and other readings, to establish any required power rating. (*Eng.*) A torsion meter attached to a rotating shaft, the angle of twist of a known length of shaft between the gauge points of the meter being indicated by optical or electrical means, thus enabling the power transmitted to be calculated. A form of transmission dynamometer.

torque motor (*Elec. Eng.*). One exerting high torques at low speeds.

torque spanner (*Civ. Eng., Mech.*). A spanner with a special attachment whereby a prescribed force can be applied to a bolt through the medium of the nut.

torr (*Phys.*). Unit of low pressure equal to head of 1 mm of mercury or 133·3 N/m^2.

torrenticoles (*Ecol.*). Animals living in swiftly running waters, such as mountain streams.

Torricellian vacuum (*Meteor., Phys.*). See **mercury barometer.**

Torridonian (*Geol.*). A succession of conglomerates and red sandstones and arkoses forming part of the Pre-Cambrian System in the north-west highlands of Scotland. They were deposited in mountain-girt basins under desertlike conditions, and rest on the Lewisian schists and gneisses.

Torrid Zone (*Geog.*). A name sometimes given to the region of the earth bounded by the two Tropics and bisected by the equator. So called because, owing to the height of the sun in the sky, the climate is always hot and has almost no seasonal temperature variations.

torsiograph (*Eng.*). An instrument for measuring and recording the frequency and amplitude of torsional vibrations in a shaft.

torsion (*Eng.*). State of strain set up in a part by twisting. The external twisting effort is opposed by the shear stresses induced in the material. (*Bot.*) Twisting without marked displacement. (*Maths.*) See **curvature (2).** (*Zool.*) The preliminary twisting of the visceral hump in Gastropod larvae which results in the transfer of the pallial cavity from the posterior to the anterior face, as distinct from the secondary or spiral twisting of the hump exemplified by the spiral form of the shell.

torsional pendulum (*Eng.*). A *torsion pendulum* (q.v.) in which the bob is a cylindrical mass suspended co-axially from a rod or wire.

torsion balance (*Phys.*). A delicate device for measuring small forces such as those due to gravitation, magnetism, or electric charges. The force is caused to act at one end of a small horizontal rod, which is suspended at the end of a fine vertical fibre. The rod turns until the turning moment of the force is balanced by the torsional reaction of the twisted fibre, the deflection being measured by a lamp and scale using a small mirror fixed to the suspended rod.

torsion bar suspension (*Autos.*). A springing system, used in some independent suspension designs, in which straight bars, anchored at one end, are subjected to torsion by the weight of the car, thereby acting as springs.

torsion galvanometer (*Elec. Eng.*). A galvanometer in which the controlling torque is measured by the angle through which the suspension head must be rotated in order to bring the pointer back to zero.

torsion meter (*Eng.*). See **torquemeter.**

torsion pendulum (*Horol*). A pendulum in which the bob rotates, in a plane perpendicular to the line of suspension, by the twisting of the suspension ribbon. Used where a long time of vibration is required, as with 400-day clocks.

torticollis (*Med.*). Wry neck. Stiff neck. A twisted position of the head on the neck, due to disease of the cervical vertebrae or to affections (especially rheumatic) of the muscles of the neck. See also **spasmodic torticollis.**

torticone (*Zool.*). A spirally twisted shell.

tortoise-shell (*Zool.*). A translucent mottled material. See **turtle-shell.**

tortoiseshell wood (*For.*). See **snakewood.**

tortuose, tortuous (*Bot., etc.*). Pursuing a somewhat zigzag line.

toruloid, torulose (*Bot.*). Elongated and cylindrical, with fairly evenly spaced swellings; necklace-like.

torulosis (*Med.*). A disease due to infection with the yeastlike micro-organism *Torula histolytica*; affects especially the central nervous system.

torulus (*Zool.*). The socket for the insertion of the antenna in Insects.

torus (*Bot.*). (1) The *receptacle* of a flower. (2) The tiny thickening on the middle of the closing membrane of a bordered pit. (*Elec. Eng.*) See **toroid.** (*Maths.*) A surface of revolution shaped like an anchor ring. It is generated by rotating a circle about a nonintersecting coplanar line as axis. Also called **anchor ring.** (*Zool.*) A ridge or fold, as in *Polychaeta*, a ridge bearing uncini.

toscanite (*Geol.*). A quartz-bearing trachyandesite, i.e., a fine-grained acid igneous rock, commonly occurring as lava-flows and characterized by the presence of orthoclase and plagioclase feldspars.

toss-bombing (*Aero.*). A manoeuvre for the release of a bomb (usually with a nuclear warhead) which allows the pilot to evade the blast. A special computing sight enables the pilot to loop before reaching the target, when the bomb is lobbed forward; or the pilot can overfly the target and loop to toss the bomb 'over-the-shoulder'.

tossing or kieving (*Min. Proc.*). The operation of raising the grade or purity of a concentrate by violent stirring, followed by packing in a *kieve.*

tosyl (*Chem.*). The toluene-4-sulphonyl group:

$$H_3C-\langle\bigcirc\rangle-SO_2-$$

total absorption coefficient (*Nuc.*). The coefficient which expresses energy losses due to both absorption and scattering. Relevant to narrow beam conditions. Preferred term **attenuation coefficient.**

total air-gas mixture (*Heat*). Air-gas mixture in which the proportion of air is the amount needed for perfect combustion.

total body burden (*Radiol.*). (1) The summation of all radioactive materials contained in any person. (2) The maximum total amount of radioactive material any person may be permitted to contain.

total cross-section (*Nuc.*). The sum of the separate cross-sections for all processes by which an incident particle can be removed from a beam. If all the atoms of an absorber have the same total cross-section, then it is identical with the *atomic absorption coefficient* (q.v.).

total curvature (*Maths.*). See **curvature (3)**

total differential (*Maths.*). See **complete differential.**

total electron binding energy (*Electronics*). That required to remove all the electrons surrounding a nucleus to an infinite distance from it and from one another.

total emission (*Electronics*). See **saturation current.**

total equivalent brake horsepower (*Aero.*). The brake horsepower at the airscrew shaft plus the b.h.p. equivalent of the residual jet thrust of a turboprop; abbrev. **t.e.h.p. or e.h.p.**

total head (*Aero.*). In fluid flow, the algebraic sum of the *dynamic pressure* and the *static pressure.*

total impulse (*Aero., Space*). The total thrust available from a self-contained rocket, such as a missile or a *take-off rocket*, expressed as the product of the mean thrust, in lbf or N, and the firing time, in s, expressed as lbf s or N s.

total internal reflection (*Phys.*). Complete reflection of incident wave at boundary with medium in which it travels faster, under conditions where Snell's law of refraction cannot be satisfied. The angle of incidence at which this occurs (corresponding to an angle of refraction of 90°) is known as the *critical angle.*

total losses (*Elec. Eng.*). The totality of power loss in an electrical machine, equal to the difference between the input and the output powers.

totally enclosed motor (*Elec. Eng.*). A motor with no provision for ventilation but not necessarily water- or gas-tight.

total modulation (*Radio*). Amplitude modulation to a depth of 100%.

total normal curvature (*Maths.*). See **curvature (3)**.

total output panel (*Elec. Eng.*). A panel of a generating station switchboard carrying instruments for measuring and recording the total power output of the station.

total response (*Acous.*). The response of a loudspeaker expressed as the relative ratio of the output sound power to the applied electrical power for each frequency over the operating frequency range. Substantially equal to the mean-spherical response of the device.

total temperature (*Aero.*). See **stagnation temperature.**

toti-. Prefix from L. *totus*, all, whole.

totipotent (*Zoöl.*). Capable of development into a complete organ or embryo; capable of self-differentiation.

touch (*Plumb.*). The plumber's term for tallow.

touchstone (*Min.*). See **Lydian stone.**

touchwood (*Bot.*). Wood much decayed as a result of fungal attack; it crumbles readily, and when dry is easily ignited by a spark.

toughened glass (*Glass*). See **safety glass.**

toughness (*Met.*). A term denoting a condition intermediate between brittleness and softness. It is indicated in tensile tests by high ultimate tensile stress and low to moderate elongation and reduction in area. It is also associated with high values in notched bar tests. (*Min. Proc.*) Ability of mineral to withstand disruption, assessed empirically by comparison with standard minerals under controlled test conditions.

tough pitch (*Met.*). A term applied to copper in which the oxygen content has been correctly adjusted at about 0·03% by poling. Distinguished from overpoled and underpoled copper.

tourbillon watch (*Horol.*). A watch fitted with a revolving carriage which carries the balance and escapement round the fourth wheel, for the purpose of eliminating the positional errors.

tourmaline (*Min.*). A complex silicate of boron and aluminium, with, in addition, magnesium (*dravite*), iron (*schorl*), or lithium (*elbaite*), and fluorine in small amounts, which crystallizes in the trigonal system. It is usually found in granites or gneisses. The variously coloured and transparent varieties are used as gemstones, under the names *achroite* (colourless), *indicolite* (blue), *rubellite* (pink). The common black variety is *schorl.*

tourmalinization (*Geol.*). The process whereby minerals or rocks are replaced wholly or in part by tourmaline. These processes result from the invasion by mineralizing fluxes and gases. See **pneumatolysis**

tourniquet (*Surg.*). Any instrument or appliance which, by means of a constricting band, a pad to lie over the artery, and a device for tightening it, exerts pressure on an artery so as to control bleeding from it.

tower (*Elec. Eng.*). The lattice-type steel structure used to carry the several conductors of a transmission line at a considerable height above the ground. Also called **pylon.** (*Eng.*) See **pylon.**

tower beater (*Paper*). A beating machine which has a smaller roll than the Hollander, but the circulation is in the vertical, being assisted by a pump at the base.

tower bolt (*Join.*). Large type of *barrel bolt* (q.v.).

tower crane (*Eng.*). A rotatable cantilever pivoted to the top of a steelwork tower, either fixed or carried on rails. The load is balanced by the lifting machinery carried on the opposite side of the pivot. Commonly used in shipbuilding.

Townend ring (*Aero.*). A cowling for radial engines consisting of an aerofoil section ring, which ducts the air on to the engine cylinders and directs a streamline flow on to the fuselage or nacelle, thus reducing drag; now obsolete.

tow-net (*Ocean.*). A conical net having the mouth kept open by a frame and at the apex a glass or metal vessel in which the catch accumulates; it is towed behind a boat in order to obtain samples of the fauna of the surface waters.

town or town's gas (*Fuels*). Gas made and supplied for domestic or trade use. Usually a mixture of coal gas and carburetted water gas; the energy density is about 500 Btu/ft^3 or 20 MJ/m^3, which is about half that of *natural gas.*

Townsend avalanche (*Nuc.*). Multiplication process, whereby a single charged particle, accelerated by a strong field, causes, through collision, a considerable increase in ionized particles.

Townsend coefficient (*Nuc.*). Number of ionizing collisions per cm of path in the direction of an applied electric field.

Townsend discharge (*Nuc.*). A *Townsend avalanche* initiated by an external ionizing agent.

tow-to-top (*Textiles*). Untwisted rope of man-made continuous filaments, cut or broken into staple by a machine to enable it to be drafted finer. See **Pacific converter.**

toxaemia, toxemia (*Med.*). The condition of a patient caused by the absorption into the tissues and into the blood of toxins formed by micro-organisms at the site of infection.

toxaphene (*Chem.*). Chlorinated *camphene* (67–69% Cl); used as an insecticide.

toxicology (*Med.*). The branch of medical science dealing with the nature and effects of poisons.

toxin (*Biol., Med.*). A poisonous substance of plant or animal origin.

toxoid (*Med.*). A toxin which has been so treated as to remove its poisonous (toxic) properties but leave its antigenic properties.

toxoplasmosis (*Vet.*). Infection of animals and man by protozoa of the genus *Toxoplasma.*

T.P. (*Surv.*). *Turning point* (see **change point**).

2,4,5-TP (*Chem.*). See **fenoprop.**

T.P.I. (*Eng.*). Abbrev. for threads per inch, the inverse of the pitch of a screw thread. (*Ships*) Abbrev., *tons per inch*: weight required to increase the *mean draught* 1 in.

TPN (*Biochem.*). See **NADP.**

trabant (*Cyt.*). See **satellite.**

trabecula (*Bot.*). A rodlike structure, or a rod-like cell, running across a cavity. (*Zool.*) A row of cells bridging a cavity; a band or plate of fibrous tissue forming part of the internal supporting framework of an organ; one of a pair of cartilaginous bars lying just anterior to the parachordals in the developing cranium. *adjs.* **trabecular, trabeculate.**

trabeculate (*Bot.*). (1) Said of peristome teeth marked by transverse bars. (2) Having trabeculae.

TRACE (*Aero.*). Abbrev. for *T*est-equipment for *R*apid *A*utomatic *C*heck-out and *E*valuation. A computerized general purpose testing rig for aircraft electrical and electronic systems.

trace (*Electronics*). Image on phosphor on electron beam impact, forming the display.

trace chemistry (*Chem.*). Small traces of chemical reagents, e.g., radioactive materials which are free of nonradioactive isotopes, behave quite differently from the same material in ordinary amounts and concentrations. A knowledge of trace chemistry behaviour is therefore of considerable importance in the handling of radioactive materials. Cf. *tracer chemistry.*

trace elements (*Ecol.*). See **micronutrients.**

tracer atom (*Nuc.*). Labelled atom introduced into a system to study structure or progress of a process. Traced by variation in isotopic mass or as a source of weak radioactivity which can be detected, there being no change in the chemistry. See **radioactive isotope.**

tracer chemistry (*Chem.*). The use of isotopic tracers in chemical studies. Cf. *trace chemistry.*

tracer compound (*Nuc.*). One in which a small proportion of its molecules is labelled with a radioactive isotope.

tracer element (*Nuc.*). One of the *radioelements* (q.v.), e.g., radiophosphorus, used for experiments in which its radioactive properties enable its location to be determined and followed. Tracer technique may be applied to physiological, biological, pathological, and technological experiments. For some purposes stable isotopes, e.g., ^{13}C and heavy hydrogen (see **deuterium**), are more conveniently used than radioisotopes.

trace routine (*Comp.*). One which verifies operations by a *program*, or reveals faults.

trachea (*Bot.*). See **vessel.** (*Zool.*) An air-tube of the respiratory system in certain *Arthropoda*, as Insects; in air-breathing Vertebrates, the windpipe leading from the glottis to the lungs.

tracheal gills (*Zool.*). In some aquatic Insect larvae, filiform or lamellate respiratory outgrowths of the abdomen richly supplied with tracheae and tracheoles.

tracheal system (*Zool.*). In certain *Arthropoda*, as Insects and Myriapods, a system of respiratory tubules containing air and passing to all parts of the body.

tracheid(e) (*Bot.*). An elongated element with pointed ends, occurring in wood. It is derived from a single cell, which lengthens and develops thickened pitted walls, losing its living contents. Tracheides conduct water.

tracheidal (*Bot.*). Of the nature of a tracheide.

tracheitis (*Med.*). Inflammation of the mucous membrane of the trachea.

trachelate (*Zool.*). Necklike. (Gk. *trachēlos*, neck.)

trachelectomy (*Surg.*). Surgical excision of the cervix uteri.

trachelorrhaphy (*Surg.*). The surgical repair of lacerations of the cervix uteri.

tracheoblasts (*Zool.*). In Insects, large stellate end cells in which tracheoles develop intracellularly before later joining a developing trachea.

tracheobronchial (*Zool.*). Formed from, or pertaining to, the trachea and bronchi; as a particular type of syrinx.

tracheobronchitis (*Med.*). Inflammation of the mucous membrane of both the trachea and the bronchi. (*Vet.*) An inflammation of the trachea and bronchi of dogs due to infection by the nematode worm *Oslerus osleri*.

tracheocele (*Med.*). An air-containing swelling in the neck due to the bulging of the wall of the trachea between the cartilages of the trachea.

tracheole (*Zool.*). The ultimate branches of the tracheal system.

tracheoscopy (*Surg.*). Inspection of the interior of the trachea by means of a tubular instrument fitted with a lamp.

tracheotomy (*Surg.*). The operation of cutting into the trachea, usually for the relief of respiratory obstruction.

trachoma (*Med.*). A highly contagious infection of the conjunctiva covering the eyelids, characterized by the presence of small elevations on the inner side of the lid and leading often to blindness.

trachyandesite (*Geol.*). Fine-grained igneous rock, commonly occurring as lava flows, intermediate in composition between trachyte and andesite, that is, containing both orthoclase and plagioclase in approximately equal amounts. See also **toscanite.**

trachybasalt (*Geol.*). A fine-grained igneous rock commonly occurring in lava flows and sharing the mineralogical characters of trachyte and basalt. The rock contains sanidine (characteristic of trachyte) and calcic plagioclase (characteristic of basalt).

trachyglossate (*Zool.*). Having a rough tongue, adapted for rasping.

Trachylina (*Zool.*). An order of *Hydrozoa* in which the medusoid develops directly from the egg, the polyp being either reduced to a minute fixed individual or being represented only by the planula larva which metamorphoses into a medusa. They have sense tentacles with endodermal concretions.

trachyspermous (*Bot.*). Having seeds with rough surfaces.

trachyte (*Geol.*). A fine-grained igneous rock-type, of intermediate composition, in most specimens with little or no quartz, consisting largely of alkali-feldspars (sanidine or oligoclase) together with a small amount of coloured silicates such as diopside, hornblende, or mica.

tracing (*Eng.*). An engineering drawing transferred to transparent tracing paper or cloth, in Indian ink for permanence and for making good dye-line prints. A *tracer's* work carries no design responsibility. See **draughtsman.**

track (*Acous.*). (1) The groove which is cut on a blank during disk recording and is then carried through the processing to the finished disk. (2) Single circumferential area on a magnetic drum, or longitudinal area on a magnetic tape, alongside other tracks, allocated to specific recording and reproduction channel. (3) Space on a disk or soundfilm allocated to one channel of sound-recording and hence reproduction. See **stripe.** (*Aero.*) 1) The distance between the outer points of contact of port and starboard main wheels. (2) The distance between the vertical centre-lines of port and starboard undercarriages where the wheels are paired. (3) The projection of the flight path upon the earth's surface. (*Nav.*) Course followed by ship or aircraft. (*Nuc.*) Path followed by particle especially when rendered visible in photographic emulsion by cloud chamber, bubble chamber or spark chamber. (*TV*) To move a camera on a dolly in a defined path while taking a shot.

track brake (*Elec. Eng.*). A solenoid-operated tram-car brake in which a brake-shoe acts directly on the track rail.

track chamber (*Nuc. Eng.*). A bubble, cloud, or spark chamber which can be used to render the tracks of ionizing particles visible.

track circuit (*Elec. Eng.*). The electric circuit formed by the two running rails of a traction system, as used for track-circuit signalling.

track-circuit signalling (*Elec. Eng.*). An electric signalling system making use of the change in resistance of a track circuit when an electric train passes over a section of the railway track.

tracker (*Acous.*). The feed-motion arrangement for traversing the cutting-head across the blank during the making of a record.

tracker wires (*Elec. Eng.*). Wires forming a mechanical connexion between a switch and its operating mechanism situated some distance away.

track guide (*Aero*). A radio beacon system

providing one or more tracks for the guidance of craft.

tracking (*Acous.*). The action of a reproducing needle in following the recorded groove on a gramophone record; also, the guidance given by the mechanism to the cutting head, so that the stylus cuts along an accurately spaced spiral. (*Automation*) Following, by an inverse feedback or servo loop, variation of a quantity. (*Cables*) Tracks along the surface of oil-impregnated paper caused by a surface stress. Tracking causes waxing and carbon formation. (*Elec.*) Excessive leakage current between two insulated points, e.g., due to moisture. (*Radar*) Automatic holding of radar beam on to target through operation of return signals. (*Space*) The determination of the trajectory of a space vehicle from optical or radio observations. (*Telecomm.*) The accuracy of adjustment of a set of ganged tuned circuits.

tracking shot (*Cinema.*). A shot in motion-picture production which involves the longitudinal or sideways motion of the camera, which, with its operators, is mounted on a dolly.

track rail bond (*Elec. Eng.*). A rail bond for preserving the electrical continuity of the track rails when these are used for carrying traction or other currents.

track relay (*Elec. Eng.*). A relay used in track-circuit signalling for controlling the electrically operated signals.

track rod (*Autos.*). A transverse link which, through ball-joints, connects arms carried by the stub axles, in order to convey angular motion from the directly-steered axle to the other.

track-sectioning cabin (*Elec. Eng.*). A cabin housing switchgear by means of which the supply to different sections of an electrified railway line may be disconnected.

track switch (*Elec. Eng.*). A switch controlling the supply of current to a section of an electrified railway line.

Tracoba (*Build., Civ. Eng.*). A method of system building, developed in France. It consists basically of large area cross wall precast carrying *in situ* floors which incorporate lightening material and services. Cladding walls are carried on the ends of cross walls and are of sandwich construction, incorporating an insulating material and with an external finish to suit individual circumstances.

tract (*Zool.*). The extent of an organ or system, as the *alimentary tract*; an area or expanse, as the *ciliated tracts* of some *Ctenophora*; a band of nerve-fibres, as the *optic tract*: see **pterylae**.

tractellum (*Zool.*). An anterior flagellum of *Mastigophora*, which draws the organism after it.

tractile fibre (*Cyt.*). A spindle fibre which begins to develop from an attachment to a chromosome and extends to the pole of the spindle.

traction (*Med.*). Treatment involving tension on affected parts (e.g. fractures) by means of suitably applied weights or otherwise.

traction battery (*Elec. Eng.*). See **vehicle battery**.

traction engine (*Eng.*). A road locomotive in which large road wheels are gear-driven from a simple or compound engine mounted on top of the boiler, a rope drum being provided for haulage purposes.

traction generator (*Elec. Eng.*). A d.c. generator used solely for supplying power to an electric traction system.

traction lamp (*Elec. Eng.*). An electric lamp having a specially robust filament to withstand vibration: used on trains or road vehicles.

traction load (*Elec. Eng.*). That part of the load carried by a d.c. generating station which is formed by the traction system which it supplies. (*Geol.*) That part of the load of solid material carried by a river which is rolled along the bed; also applied to the load carried by a glacier.

traction motor (*Elec. Eng.*). An electric motor specially designed for traction service.

traction permeameter (*Elec. Eng.*). An apparatus for measuring the permeability of a sample of iron by weighing the mechanical force between the end of the sample and a surface forming part of the yoke of the apparatus.

traction rope (*Civ. Eng.*). The endless rope employed in an aerial ropeway system to effect movement of the carriers transporting the loads. Also called a **hauling rope**.

tractive force (or effort) (*Elec. Eng.*). The pull necessary to detach the armature from an excited electromagnet. (*Eng.*) Of a locomotive, the pull which the engine is capable of exerting at the draw-bar, the limiting value of which is given by the product of the weight on the coupled wheels and the coefficient of friction between wheels and rails.

tractor (*Aero.*). A propeller or airscrew which is in front of the engine and the structure of the aircraft, as contrasted with a *pusher*, which is behind the engine and pushes the aircraft forward. See under **airscrew**.

tractor mower (*Agric.*). A machine similar in design to the *horse mower* (q.v.) but with longer cutter-bars, which are operated directly from a shaft on the tractor.

tractor plough (*Agric.*). A plough consisting of a frame to which is attached the requisite number of coulters, shares, and breasts, usually not exceeding five of each. Mechanism enables all the working parts to be lifted by the pull of a cord, when the plough is being turned. See **mouldboard plough**.

tractrix (*Maths.*). The involute of a catenary.

tractrix horn (*Acous.*). A horn which is so shaped that the area at a distance from the throat is dependent on the tractrix curve, as contrasted with the exponential horn.

trade effluent (*San. Eng.*). The liquid discharge from a manufacturing process.

trade-winds (*Meteor.*). Persistent winds blowing from the N.E. in the northern hemisphere and from the S.E. in the southern hemisphere between the *horse latitudes* (calm belts at 30° N. and S. of the equator) towards the *doldrums* (q.v.).

traffic flow (*Teleph.*). The number of calls which an exchange, or a set of switches, is carrying at any instant.

traffic meter (*Teleph.*). A meter which, inserted at any part of an automatic telephone exchange, totals the number of calls passing through.

traffic unit (*Teleph.*). The measure of the occupancy of telephonic apparatus during conversation. One traffic unit equals the use of one circuit for one minute or for one hour. Abbrev. **T.U.** See **congestion traffic-unit meter.**

tragacanth (*Bot.*). A gum obtained from several species of the genus *Astralagus*. Used as an excipient, etc., in pharmacy.

tragus (*Zool.*). In the ear of some Mammals, including the *Microchiroptera* (bats), an inner lobe to the pinna.

trail (*Autos.*). The distance by which the point of contact of a steered wheel with the ground lies behind the intersection of the swivel-pin axis and the ground. See **caster action.**

trailer (*Cinema.*). (1) The blank piece of film, generally black, which is attached or printed at the end of every reel, so that when the projected

picture finishes there is time to close the picture shutter before the threaded film runs out. (2) Short film advertising future programme item(s).
trailing action (*Autos.*). See **caster action.**
trailing axle, -springs, -wheels, etc. (*Eng.*). In a locomotive, the parts belonging to the rearmost axle.
trailing cable (*Cables*). A flexible cable carrying the current to a transportable piece of electric apparatus.
trailing edge (*Aero.*). The rear edge of an aerofoil, or of a strut, wire, etc. (*Elec. Eng.*) See **leaving edge.** (*Telecomm.*) The falling portion of a pulse signal. See also **leading edge.**
trailing flap (*Aero.*). A *flap* which is mounted below and behind the wing trailing edge so that it normally trails at neutral incidence and is rotated to various positive angles of incidence to increase lift, there always being a gap between the wing undersurface and the flap leading edge.
trailing points (*Rail.*). See **points.**
trailing pole tip (or horn) (*Elec. Eng.*). The edge of a field pole which is passed last by an armature conductor, irrespective of the direction of rotation of the armature.
trailing vortex (*Aero.*). The vortex passing from the tips of the main surfaces of an aeroplane and extending over the tail and behind it.
trails (*Astron.*). (1) Long flashes of brightness seen in the wake of some large meteors in the sky. (2) The lines left on a photographic plate when a star image, instead of being kept stationary by moving the plate, is allowed to 'trail' across it during the exposure.
train (*Eng.*). Similar or identical parts in a machine, arranged in series, such as a simple or compound train of gears. (*Horol.*) the interconnected wheels and pinions of a watch, clock, or similar mechanism. A watch is said to have an 18 000 train when the train is suitable for a balance making 18 000 vibrations per hour; a 21 600 train is used for small wristwatches, and a 14 400 train for chronometers.
train brake (*Eng.*). See **vacuum brake, air brake.**
train control (*Elec. Eng.*). The method by which the mechanical control operations carried out by the driver of an electric or diesel-electric train are translated into the appropriate changes in the electric circuit conditions pertaining to the supply of the traction motors.
train describer (*Elec. Eng.*). An automatic or semi-automatic device for giving information regarding the destination of trains. Also called **destination indicator.**
training (*An. Behav.*). The procedure of subjecting an animal to processes which result in conditioning or the establishment of a discrimination. See **discrimination-.**
training works (*Hyd. Eng.*). Works undertaken to remedy instability and eccentricity of flow in channels. See **dyke, levee, groynes, ground sills.**
train-line (*Elec. Eng.*). An electric cable extending over the length of a railway coach, and terminating in sockets at each end so that couplers may maintain continuity over the whole length of the train. See **bus-line, control-line, pump-line.**
train of waves (*Radio*). A group of waves of limited duration, such as that resulting from a single spark discharge in an oscillatory circuit.
trait (*Psychol.*). In terms of personality theory, an individual characteristic, frequently thought of as being made up of a group of habits. Likewise a group of traits are thought of as characterizing a personality type.
trajectory (*Meteor.*). The actual path along which a small quantity, or parcel, of air travels during a definite time interval. (*Space*) The path of a rocket or space probe. The word *trajectory* is used rather than *orbit* when the path is limited in length, e.g., a *trajectory* to the moon.
Trakatron (*Print.*). Electronically-controlled equipment for lateral register of printed reels on slitting and re-reeling machines.
tram (*Mining*). A small wagon, tub, cocoa-pan, corve, corf, or hutch, for carrying mineral. (*Spinning*) The term for silk yarn intended for weft; formed from two or more single threads, slightly twisted. (*Elec. Eng.*) See **tram(car).**
trama (*Bot.*). The loosely packed hyphae which occupy the middle of the gill of an agaric.
tram(car) (*Eng.*). A passenger vehicle, running on rails, powered by electricity, horse, steam etc., power. A contemporary electric vehicle can carry up to 200 passengers at high speeds to city centres.
trammel-net (*Ocean.*). A form of *gill-net* (q.v.) consisting of two taut outer nets of large mesh and a larger slack middle net of finer mesh, all three being attached to each other at the head, foot, and ends.
trammels (*Instr.*). See **beam compasses.**
tramontana (*Meteor.*). A northerly mountain wind blowing over Italy.
tramp iron (*Min. Proc.*). Stray pieces of iron or steel mixed with ore from mine, which must be removed before crushing.
tramp metal (*Eng.*). Stray metal pieces which are accidentally entrained in food or other processed materials and must be removed before the material leaves the process.
Trancor (*Met.*). TN for grain-orientated silicon-iron magnetic alloy.
tranquilliser (*Med.*). A natural or synthetic substance used to quieten a disturbed person or to alleviate a mood of depression. Tranquillisers are more specific in their actions than sedatives, which have a general depressive effect on mental and physical activity.
trans- (*Chem.*). That geometrical isomer in which the two radicals are situated on opposite sides of the plane of a double bond or alicyclic ring.
Transactor (*Comp.*). TN for interrogator/reply device for inputs to a computer.
transadmittance (*Elec.*). Output a.c. current divided by input a.c. voltage for electronic device when other electrode potentials are constant.
transaxial tomography (*Radiol.*). A process whereby serial radiographs are taken transverse to the vertical axis of the body.
transceiver (*Telecomm.*). Equipment, e.g., 'walkie-talkie', in which circuitry is common to transmission or reception. Also **transreceiver.**
transcendental function (*Maths.*). Any non-*algebraic function* (q.v.), e.g., trigonometric, exponential, logarithmic, Bessel or gamma functions.
transconductance (*Electronics*). Reciprocal of *transfer impedance*. Ratio of current in one part of a circuit to e.m.f. or p.d. in another part, not necessarily with the same frequency, e.g., in a demodulator. U.S. term for **mutual conductance.**
transcriber (*Comp.*). Apparatus for transferring input or output data from a register of information in one program language to the medium and language employed in a different computer, or the reverse process.
transcription (*Biochem.*). Starting from the linear genetic determinants on a DNA strand, the flow of cellular information is assumed to descend to specify the structure of proteins via the transcription into messenger RNA molecules having

complementary base sequences to the parent DNA strands. The codons on the m-RNA are then translated into amino-acid sequences of the polypeptide chains formed by the ribosomes. (*Radio*) The recording of a broadcast performance for subsequent rebroadcast or other use.

transcrystalline failure (*Met.*). The normal type of failure observed in metals. The line of fracture passes through the crystals, and not round the boundaries as in intercrystalline failure.

transducer (*Elec. Eng.*). General term for any device which converts a physical magnitude of one form of energy into another form, generally on a one-to-one correspondence, or according to a specified formula, e.g., an electroacoustic transducer is one receiving an alternating current (or e.m.f.) and supplying an output of acoustic waves as in a loudspeaker. The microphone operates in the reverse way. An electric motor is an electromechanical transducer.

transducer translating device (*Elec. Eng.*). One for converting error of the controlled member of a servomechanism into an electrical signal that can be used for correcting the error.

transduction (*Bot.*). The transfer of genetic material from the host bacterial cell to an uninfected cell by bacteriophage. A similar process occurs in the transfer of the genetic factor controlling colicin production from a colicogenic to a non-colicogenic strain in a mixed culture of bacterial cells.

transductor (*Elec. Eng.*). Arrangement of windings on a laminated core, which, when excited, permits current amplification. Part of a *magnetic amplifier*.

transect (*Bot.*). A line or belt of vegetation marked off for study.

transept (*Arch.*). Part of a church at right angles to the *nave*, or of another building to the body: either wing of such a part where it runs right across.

transeptate (*Bot.*). Having all the septa placed transversely.

transfer (*An. Behav.*). The situation where the strengths of a set of responses initially given by an animal to one set of stimuli tally with the strengths of the same set of responses given in the presence of other sets of stimuli. (*Comp.*) See **print-through.** (*Print.*) An impression taken of a nonlithographic surface (intaglio, letterpress), on one of several specially coated papers, for transferring to a lithographic surface; a transfer from an existing lithographic surface is called a *retransfer*. (*Psychol.*) The facilitation (positive) or hindrance (negative) of performance on a learning or training task as a result of previous activity. The positive and negative transfer effects appear to be a function of the similarity of the tasks.

transfer admittance (*Elec.*). Ratio of current in any part of a mesh to e.m.f. producing it and applied in another part of the mesh, all other e.m.f.s being reduced to zero.

transfer characteristic (*Electronics*). See **transistor characteristics.** (*TV*) Relationship between TV camera tube illumination and corresponding signal current.

transfer circuit (*Teleph.*). A circuit between operators' positions in an exchange, allowing the functions of one operator to be taken over by another during slack periods.

transfer constant (*Telecomm.*). The natural logarithm of the ratio of the output and input currents for a network.

transfer current (*Telecomm.*). Current in a control electrode to initiate ionization and gas discharge.

transfer ellipse (*Space*). The trajectory (part of an ellipse) by which a space vehicle may transfer from an orbit about one body (e.g., the earth) into an orbit about another (e.g., the sun).

transference (*Psychiat.*). The displacement of affect, positive or negative, from the person to whom it was originally directed, on to another. Used specifically in relation to the psychoanalytical situation where all the earlier emotional attitudes and reactions of the patient to his parents are displaced on to the analyst, and relived in relation to him. See **counter-, negative-, positive-.**

transfer function (*Telecomm.*). A mathematical expression relating the output of a closed loop servo system to its input.

transfer impedance (*Elec.*). In any network or transducer, complex ratio of r.m.s. voltage applied at any pair of terminals to the r.m.s. current at some other pair. Also **mutual impedance.**

transfer instrument (*Elec. Eng.*). Instrument which gives d.c. indication independent of frequency, including zero frequency, so that when calibrated for d.c. it can be used for calibrating a.c. instruments, as with electrostatic wattmeters.

transfer lettering systems (*Typog.*). Alphabets, symbols, rules, and tints which can be stripped into position on layouts or film for photoreproduction, e.g., Artype, Chart-Pak, Craf Type, Letraset, Letter-on, Prestype, Zip-a-Tone.

transfer line (*Eng.*). A long series of machines, operating on a succession of similar parts, e.g., car cylinder blocks, automatically or semi-automatically.

transfer machine (*Eng.*). Machine in which a workpiece or an assembly passes automatically through a number of stations, at each of which it undergoes one or more production processes.

transfer moulding (*Plastics*). Injection moulding with thermosetting compositions.

transfer orbit (*Space*). An elliptical trajectory tangent to the orbits of both departure planet and target planet.

transfer port (*Nuc. Eng.*). The aperture through which items are inserted into or removed from a dry box, glove box, or shielded box (usually by sealing into plastic sac attached to the rim of the port).

transfer process (*Photog.*). Any means whereby an image, dyed or pigmented, is transferred to a new emulsion.

transfer-RNA (*Biochem.*). See **s-RNA.**

transfinite numbers (*Maths.*). A system of cardinal and ordinal numbers, invented by G. Cantor (1845-1918), relating to infinite sets. In effect Cantor classifies different types of infinity. For example, the infinite associated with the set of positive integers is different from that associated with the set of all real numbers.

transfixion (*Surg.*). A cutting through, as in amputation.

Transfluxor (*Comp., etc.*). TN for arrangement with apertured magnetic cores for memory or switching.

transform (*Comp.*). In digital computers, a transform changes the form of the information without a significant alteration in meaning. (*Maths.*) A process or rule for deriving from a given mathematical entity (e.g., point, line, function, etc.) a corresponding entity. The word 'transform' is also sometimes used in respect of the corresponding entity itself (e.g., Fourier and Laplace transform). Certain problems (e.g., differential equations) can be transformed so as to obtain a simpler problem whose solution can then be transformed back to give the solution of the original problem. Also called **transformation.** See **affine transformation, conjugate elements of a group, Fourier transform, Laplace transform, linear transformation** and **isogonal transformation.**

transformation (*Bot.*). The inheritable change from rough to smooth cultures, of certain bacteria, which occurs when the rough colonies are grown with filtrates of smooth colonies of the same or different serological types. (*Gen.*) Specifically, the transformation of one species into another related species by incorporation of DNA from the related species into the receptor genes. Common agents of transformation are viruses, and common subjects are bacteria. (*Maths.*) See **transform.** (*Met.*) A constitutional change in a solid metal, e.g., the change from gamma to alpha iron, or the formation of pearlite from austenite. (*Nuc.*) (1) Change of phase, e.g., by reactor fuel. (2) See **atomic transmutation.**

transformation constant (*Nuc.*). See **disintegration constant.**

transformation points (*Glass*). In the measurement of thermal expansion coefficient, the two temperatures (Mg point and Tg point) at which the slope of the graph of expansion against temperature changes fairly sharply.

transformation ratio (*Elec. Eng.*). See **turns ratio.**

transformation theory (*Eng.*). A theory in which the factor relating to yield is assumed to be the shear-strain energy stored in unit volume.

transformer (*Elec. Eng.*). An electrical device without any moving parts, which transfers energy of an alternating current in the primary winding to that in one or more secondary windings, through electromagnetic induction. Except in the case of the *autotransformer* there is no electrical connexion between the two windings and usually (excepting the isolating transformer) a change of voltage is involved in the transformation. (*Mech.*) Mechanical transformers involving lever arms of different lengths are used to vary *mechanical advantage*.

transformer booster (*Elec. Eng.*). A transformer connected with its secondary in series with the line, so that its voltage is added to that of the circuit; used to compensate the voltage drop in a feeder or distributor.

transformer core (*Elec. Eng.*). The structure, usually of laminated iron or ferrite, forming the magnetic circuit of a transformer.

transformer coupling (*Elec. Eng.*). Transference, in either direction, of electrical energy from one circuit to another by a transformer, of any degree of coupling. Also **mutual coupling.**

transformer oil (*Elec. Eng.*). A mineral oil of high dielectric strength, forming the cooling and insulating medium of electric power transformers.

transformer plate (*Elec. Eng.*). Sheet-iron of low magnetic loss, for transformer core laminations.

transformer ratio (*Elec. Eng.*). See **turns ratio.**

transformer-ratio bridge (*Elec. Eng.*). An a.c. bridge similar to a Wheatstone bridge but with two transformer windings used for the two ratio arms.

transformer stampings (*Elec. Eng.*). The laminations, stamped out of transformer plate, which are assembled to form the transformer core.

transformer switch (*Elec. Eng.*). A switch or circuit-breaker for disconnecting a transformer from the supply.

transformer tank (*Elec. Eng.*). The steel tank encasing the core and windings of a transformer and holding the transformer oil.

transformer tapping (*Elec. Eng.*). A means of varying the voltage ratio of a transformer by making a connexion to a point on one winding intermediate between the ends.

transformer tube (*Elec. Eng.*). One of a number of steel tubes on the outside of a transformer tank to provide a vertical path of circulation for the transformer oil.

transformer winding (*Elec. Eng.*). The electrically active part of a transformer, which surrounds the magnetically active transformer core.

transforming station (*Elec. Eng.*). A point on an electricity supply system where a change of supply voltage occurs.

transfusion (*Med.*). Blood transfusion. The operation of transferring the blood (or any required constituent of it) of one person into the veins of another, either to make good loss or counteract deficiency.

transfusion cell (*Bot.*). See **passage cell.**

transfusion tissue (*Bot.*). A group of short tracheids lying by the side of the xylem in a leaf of a pine (and related plants), by means of which material passes to or from the vascular strand and the rest of the leaf.

transgenation (*Gen.*). See **point mutation.**

transient (*Acous.*). A sound of short period and irregular nonrepeating waveform, which implies a continuous spectrum of sound-energy contributions, the frequency distribution of which determines whether the sound is correctly recognizable or not. (*Elec. Eng.*) A short surge of voltage or current. The voltage or current before steady-state conditions have become established. (*Telecomm.*) Any noncyclic change in a part of a communication system. The most general transient is the *step*, while the steady state is represented by any number of sinusoidal variations. See **Heaviside unit function.** (*Zool.*) A distinct individual mode in a developmental series, such as a line of descent, corresponding in the time-character concept to a species in the taxonomic concept.

transient analyser (*Telecomm.*). Test instrument which generates repeating transients and displays their waveform (which is usually adjustable) at different points in the system under investigation, on a CRT screen.

transient distortion (*Telecomm.*). Distortion arising only when there is a rapid fluctuation in frequency and/or amplitude of the stimulus.

transient equilibrium (*Nuc.*). Radioactive equilibrium between daughter product(s) and parent element of which activity is decaying at an appreciable rate. Characterized by ratios of activity, but not magnitudes, being constant.

transient flow permeability (*Powder Tech.*). A permeametry technique in which sample of the powder under test, contained in a long vertical column, is evacuated. Then gas at a known pressure is introduced at the foot of the column. The time required for the gas to diffuse through the column is used to calculate the average pore diameter of the powder bed and hence the surface area.

transient reactance (*Elec. Eng.*). The reactance of the armature winding of a synchronous machine which is caused by the leakage flux. Cf. *synchronous reactance*.

transient stability (*Elec. Eng.*). That of a power system under transient current conditions.

transient state (*Telecomm.*). Transition period and associated phenomena between steady states in the repetition of a waveform.

transient wave (*Telecomm.*). One set up in transmission circuits or filters because of changes in the current amplitude and/or frequency, the effects dying out in local circuits but propagated along lines.

transillumination (*Med.*). The passing of a strong light through the walls of a cavity so that its outlines may become visible to the observer and any abnormalities in density may be detected.

transistor (*Electronics*). Three-electrode semiconductor device with thin layer of *n*- (or *p*-) type semiconductor sandwiched between two regions of *p*- (or *n*-) type, thus forming two *p-n* junctions back to back. The emitter junction is given a forward bias and the collector junction a reverse bias. Current carriers entering the emitter diffuse through the base to the collector, so i_c is of the order of 0·98 i_e. Due to the low forward resistance of the emitter junction and the high reverse resistance of the collector junction considerable power gain is possible for signals in the emitter or base leads. The latter arrangement also gives current gain. Amplification in the *p-n-p* transistor is due to hole conduction, that in an *n-p-n* transistor to electron conduction. See also **transistor constructions.**

transistor amplifier (*Electronics*). One which uses transistors as the source of current amplification. Depending on impedance considerations, there are three types, with *base*, *emitter*, or *collector* grounded.

transistor characteristics (*Electronics*). General name for graphs relating d.c. electrode currents and/or voltages in a manner similar to that adopted for thermionic valves. Many such characteristics are employed, especially: (1) *feedback characteristic*—input voltage *v.* output voltage for fixed input current; (2) *input characteristic*—input voltage *v.* input current for fixed output voltage; (3) *output characteristic*—output current *v.* output voltage for fixed input current; (4) *transfer characteristic*—output current *v.* input current for fixed output voltage.

transistor constructions (*Electronics*). The sandwich forms described under *transistor* (q.v.) are known as *junction transistors*, e.g., *n-p-n*, *p-n-p*, etc., and they can be grown by carefully controlled addition of impurities to a single crystal of the base material, or they may be 'fabricated' by fusing gallium or indium to either side of a thin slice of the base crystal (the 'dice'). In the modern so-called 'mesa' technique the dice are vacuum-coated by evaporating the desired metal on to the surfaces. The earliest type of semiconductor detector was the *cat's whisker* and crystal of crystal broadcast receivers, and *point-contact transistors* are still used.

transistor current gain (*Electronics*). The slope of the output current against input current characteristic for constant output voltage. In a common-base circuit it is inherently a little less than unity but in a common-emitter circuit it may be relatively large. The current gain is the hybrid parameter h_{21}.

transistor equivalent circuit (*Electronics*). For the purpose of circuit analysis, a transistor may be represented by a four-terminal network having two common terminals. In this way the transistor characteristics may be expressed in terms of four independent variables, the input and output voltages and currents of the equivalent circuit. Such possible circuits are called the *common base*, the *common emitter* and the *common collector*.

transistor ignition (*Autos.*). Ignition system fitted with transistor unit which amplifies the battery current, thus enabling a very small control current to be used, reducing wear on the contact-breaker while increasing spark efficiency.

transistor parameters (*Electronics*). In circuit analysis the performance of transistors is calculated from parameters obtained from the slope of the various characteristic curves (as with thermionic valves). Many such sets of parameters have been used, the most widely adopted probably being the hybrid parameters, h_{11}, h_{12}, h_{21}, and h_{22} (also called **h-parameters, small-signal parameters**). These are given by the slopes of the input, feedback, transfer and output characteristics respectively, at the selected working point.

transistor pentode (*Electronics*). A point-contact transistor having 1 collector but 3 emitters, used for switching, mixing, or modulation.

transistor power-pack (*Electronics*). One in which high-tension supply of low power is obtained by rectifying transformed high-voltage current from a transistor oscillator fed at low voltage.

transistor tetrode (*Electronics*). Transistor operating at higher frequencies than the *pentode*, having an emitter, collector and two base connexions.

transit (*Astron.*). (1) The apparent passage of a heavenly body across the meridian of a place, due to the earth's diurnal rotation. See **culmination.** (2) The passage of a smaller body across the disk of a larger body as seen by an observer on the earth, e.g., of Venus or Mercury across the sun's disk, or of a satellite across the disk of its parent planet. (*Surv.*) Rotation of the telescope of a theodolite about its trunnion axis, so that the positions of the ends of the telescope are reversed. See **change face, transit theodolite.**

transit angle (*Electronics*). Product of delay or transit time and angular frequency of operation. In a velocity-modulated valve the transient time corresponds to the time taken for an electron to pass through a drift space.

transit call (*Teleph.*). In international telephony, a call which is established over more than one international telephone circuit, the transit country not being concerned with the switching, except in emergency. See **single switch call.**

transit circle (*Astron.*). See **meridian circle.**

transition (*Aero.*). In VTOL aircraft flight, the action of changing to, or from, the vertical lift mode (jet, fan, or rotor) to forward flight with wing lift. (*Electronics*) In electrons, change of energy level associated with emission of quantum of radiation. (*Gen.*) A mutation which is caused by the replacement in the DNA of one pyrimidine base by another pyrimidine or one purine by another purine. (*Nuc.*) Change in configuration associated with α-, β-, or γ-emission, and having probability given by relevant selection rule.

transitional (*Bot.*). Said of an inflorescence with some racemose and some cymose characters.

transitional epithelium (*Zool.*). A stratified epithelium consisting of only three or four layers of cells; especially that found lining the ureters, the bladder, and the pelvis of the kidney in Vertebrates.

transition cell (*Bot.*). A thin-walled cell at the end of a vein in a leaf, the last of the phloem.

transition curve (*Surv.*). A curve of special form connecting a straight and a circular arc on a railway or road. Designed to eliminate sudden change of curvature between the two, and to allow of superelevation being applied gradually to the outer rail or outer part of the curve. Also called an **easement curve.**

transition element. See **transition metal.**

transition energy (*Electronics*). That at which phase focusing changes to defocusing in a synchrotron accelerator. This necessitates a sharp arbitrary change of phase in the radio-frequency field.

transition fit (*Eng.*). A class of fit intermediate between a clearance fit and an interference fit.

transition frequency (*Acous.*). That at which frequency response of disk-recording system changes from constant amplitude to constant velocity characteristic.

transition metal (*Phys.*). One of the group which have an incomplete inner electron shell. They are characterized by large atomic magnetic moments. Also **transition element.**

transition point (*Aero.*). The point where the flow in a *boundary layer* changes abruptly from *laminar* to *turbulent*. (*Chem.*) The temperature at which one crystalline form of a substance is converted into another solid modification, i.e., that at which they can both exist in equilibrium.

transition probability (*Nuc.*). The probability per unit time that a system in state k will undergo a transition to state l. In the case of radioactive transitions the total transition probability is termed the **disintegration constant.**

transition region (*Bot.*). The portion of the axis of a young plant in which the change from root structure to shoot structure occurs. (*Electronics*) That over which the impurity concentration in a doped semiconductor varies.

transition resistor (*Elec. Eng.*). A resistor connected across that part of a transformer winding short-circuited during onload *tap changing*.

transition state (*Chem.*). The atomic arrangement of highest energy in the course of a one-step chemical reaction.

transition stops (*Elec. Eng.*). In a traction-motor controller, intermediate electrical positions inserted between the main circuit positions in order to avoid breaking the circuit.

transition temperature (*Electronics*). (1) That corresponding to a change of phase. (2) That at which a metal becomes superconducting.

transitive (*Maths.*). A relationship is transitive if when it applies from A to B and from B to C it also applies from A to C, e.g., 'is less than', 'is similar to', 'is a subset of', 'belongs to the same family as'. 'Is the father of' is not a transitive relation.

transitman (*Surv.*). U.S. term for a man operating a *transit theodolite*.

transitory starch (*Bot.*). Starch formed temporarily in a leaf in which photosynthesis is proceeding faster than the removal or consumption of carbohydrates.

transitron (*Electronics*). Pentode in which suppressor grid acts as control grid, characterized by negative mutual conductance between suppressor and screen grids.

transitron oscillator (*Electronics*). One in which a retarding field is established by holding a screen grid negative in a screened grid valve.

transit tetany (*Vet.*). Railroad disease. A form of hypocalcaemia occurring in pregnant cattle during or soon after a journey by road or rail in hot, crowded conditions; characterized by restlessness, incoordination and coma. A similar condition occurs in horses, especially lactating mares. Also called **fleeting tetanus.**

transit theodolite (*Surv.*). A theodolite whose telescope is capable of being completely rotated about its horizontal axis. See **Everest theodolite, wye theodolite.**

transit time (*Electronics*). (1) Time taken by an electron to go from cathode to anode of a thermionic valve; important factor in the operation of valves at very high frequencies. (2) In a transistor, time necessary for injected charge carriers to diffuse across the barrier region.

translation (*Teleph.*). Alteration of the number and composition of the last two coded trains of impulses which are dialled by a subscriber and represent a desired exchange. The translation is effected in the director, and is for the purpose of routing the call over a multiplicity of junctions. (*Zool.*) Change of position of an organ. Also **translocation.**

translation field (*Teleph.*). The frame of terminal tags by means of which the coded impulses dialled by a subscriber are translated.

translation of heterosis (*Zool.*). An apparent change in the position of a merome from one somite to another, due to the expansion or contraction of tagmata, as the lateral fins of Fish.

translator device (*Comp.*). One for converting information from one form into another. In computers, a network such that input signals in a certain code are changed to output signals in another code without altering the significance of the information.

translocated injury (*Biol.*). Injury occurring in an area remote from the original directly affected part of an animal or plant, but associated with it in type and extent.

translocation (*Bot.*). The movement of material in solution inside the body of the plant. (*Cyt.*) The transfer of a portion of a chromosome, either to another part of the same chromosome or to a different chromosome. (*Zool.*) See **translation.**

translucent or translucid (*Bot., Min.*). More or less transparent.

transmission (*Autos.*). The means by which power is transmitted from the engine of an automobile to the *live axle*. Includes the change gear, propeller shaft, clutch, differential gear, etc. See also **automatic transmission.** (*Phys.*) See **transmittance.** (*Telecomm.*) Conveying electrical energy over a distance by wires, either to operate controls or indications (*telemetering*), acoustic information (*telephony, broadcasting*) or pictorial information (*facsimile, television*). Also denotes radio, optical and acoustic wave propagation.

transmission band (*Telecomm.*). Section of a frequency spectrum over which minimum attenuation is desired, depending on the type and speed of transmission of desired signals.

transmission bridge (*Teleph.*). Device for separating a connexion into incoming and outgoing sections for the purpose of signalling, at the same time permitting the through transmission of voice frequencies.

transmission chain (*Eng.*). Roller or inverted-tooth chain designed for transmitting power.

transmission coefficient (*Nuc.*). Probability of penetration of a nucleus by a particle striking it. (*Phys.*) Ratio of intensities of transmitted (refracted) and incident waves at the interface between two media. Also **transmissivity.**

transmission dynamometer (*Eng.*). A device for measuring the torque in a shaft, and hence the power transmitted, either (*a*) by inference from the measured twist over a given length of shaft, obtained by a torsion meter, or (*b*) by direct measurement of the torque acting on the cage carrying the planetary pinions of an interposed differential gear.

transmission experiment (*Nuc.*). One in which radiation transmitted by a *thin target* (q.v.) is measured, to investigate the interaction which takes place. Such experiments are used in the measurement of total cross-sections for neutrons.

transmission gain (*Telecomm.*). The increase of power (usually expressed in dB) in a transmission from one point to another.

transmission level (*Telecomm.*). Electric power in a transmission circuit, stated as the decibels or nepers by which it exceeds a reference level. Also called **power level.**

transmission line (*Elec.*). General name for any conductor used to transmit electric or electromagnetic energy, e.g., power line, telephone line, coaxial feeder, G-string, waveguide, etc. Also for the acoustic equivalent.

transmission-line amplifier (*Telecomm.*). One in which grids of valves are driven by, and anodes feed, appropriate points on an artificial line, giving wideband amplification. Also called **distributed amplifier.**

transmission-line control (*Radio*). Control of frequency of an oscillator by means of a resonant line in the form of a tapped quarter-wavelength stub.

transmission loss (*Telecomm.*). Difference between the output power level and the input power level of the whole, or part, of a transmission system in decibels or nepers.

transmission measuring set (*Telecomm.*). Apparatus consisting essentially of a sending circuit and a level measuring set. The sending circuit has a specified impedance (e.g., 600 ohms) and its output power is known.

transmission modes (*Telecomm.*). Field configurations by which electromagnetic or acoustic energy may be propagated by transmission lines, especially waveguides.

transmission pressure (*Elec. Eng.*). The nominal voltage at which electric power is transmitted from one place to another.

transmission primaries (*TV*). In colour TV, the set of three primaries chosen so that each one corresponds in magnitude to one of the independent signals which comprise the colour signal.

transmission ratio (*Light, Photog.*). The ratio of the transmitted luminous flux to that incident upon a transparent medium. Reciprocal of *opacity*.

transmission reference system (*Teleph.*). See **master telephone transmission reference system.**

transmission speed (*Comp.*). The number of bits or elements of information transmitted in unit time.

transmission tower (*Elec. Eng.*). The steel structure that carries a high-voltage transmission line.

transmission unit (*Acous.*). Former name for *decibel* (q.v.). Abbrev. **T.U.**

transmissivity (*Phys.*). See **transmission coefficient.**

transmit-receive tube (*Electronics*). Switch which does or does not (*ATR tube*) permit flow of high-energy radar pulses. It is a vacuum tube containing argon for low striking, and water vapour to assist recovery after the passage of a pulse. Used to protect a radar receiver from direct connexion to the output of the transmitter when both are used with the same scanning aerial through a common waveguide system. Abbrev. **TR tube.**

transmittance (*Phys.*). Ratio of energy transmitted by a body to that incident on it. If scattered emergent energy is included in the ratio, it is termed *diffuse transmittance*—otherwise *specular transmittance*. Also called **transmission.**

transmitted carrier system (*Radio*). One in which the carrier wave is radiated. Cf. *suppressed carrier system*.

transmitter (*Radio*). Strictly, complete assemblage of apparatus necessary for production and modulation of radiofrequency current, together with associated antenna system; but frequently restricted to that part concerned with the conversion of d.c. or mains a.c. into modulated RF current. (*Teleg.*) The mechanical device which sends accurate signals at a uniform speed over a telegraph circuit; operated by punched holes on a slip.

transmitter frequency tolerance (*Telecomm.*). Maximum permitted *frequency departure*.

transmitting valve (*Radio*). One which handles output power of a radio transmitter; may be in parallel or push-pull with others.

transmittivity (*Phys.*). Transmittance of unit thickness of nonscattering medium.

transmutation (*Nuc.*). See **atomic transmutation.**

transom or **transome** (*Join.*). An intermediate horizontal member of a window frame, separating adjacent panes.

transonic range (*Aero.*). The range of air speed, in which both *subsonic* and *supersonic* airflow conditions exist round a body. Largely dependent upon body shape, curvature and *thickness ratio*, it can be broadly taken as Mach 0·8 to Mach 1·4.

transpalatine (*Zool.*). In *Crocodilia*, a cranial bone connecting the jugal and maxilla with the pterygoid.

transparency (*Electronics*). Ratio of free space (window), between grid or screen wires in a thermionic valve, to their total area. (*Nuc.*) Proportion of energy or number of incident photons or particles which pass through window of ionization chamber or Geiger counter.

transparent parchment (*Paper*). See **glassine.**

transpiration (*Aero.*). The flow of gas along relatively long passages, the flow being determined by the pressure difference and the viscosity of the gas, surface friction being negligible. (*Bot.*) The loss of water vapour from a plant, mainly through the stomata.

transpiration current (or **stream**) (*Bot.*). The stream of water which passes through the plant from the roots to the leaves, whence it escapes chiefly as water vapour.

transplant (*Surg., Zool.*). (1) In surgery and experimental zoology, the process of transferring a part or organ from its normal position to another position in the same individual or to a position in another individual. Also **transplantation.** (2) The part or organ transferred in this way.

transpolarizer (*Elec. Eng.*). Ferroelectric dielectric impedance, controlled electrostatically.

transponder (*Telecomm.*). A form of transmitter-receiver which transmits signals automatically when the correct interrogation is received. An example is a radar beacon mounted on a flight vehicle (or missile), which comprises a receiver tuned to the radar frequency and a transmitter which radiates the received signal at an intensity appreciably higher than that of the reflected signal. The radiated signal may be coded for identification.

transport (*Comp.*). Mechanism for moving tape during recording or reproducing. (*Nuc.*) Rate at which desired material is carried through any section of processing plant, e.g., isotopes in isotope separation.

transport cross-section (*Nuc. Eng.*). Reciprocal of *transport mean free path*.

transporter bridge (*Civ. Eng.*). A bridge consisting of two tall towers, one on each side of the river, connected at the top by a supporting girder along which a carriage runs. A small platform at the ordinary road-level is suspended from the carriage, and this system can be made to travel along the girder across the river. Such bridges, inadequate for modern traffic, are obsolescent.

transport mean free path (*Nuc.*). If Fick's law of diffusion is applicable to the conditions in a nuclear reactor, then the mean free path is three times the diffusion coefficient of neutron flux. In practice the theory usually has to be modified to account for anisotropy of scattering and persistence of velocities.

transport number (*Chem.*). The fraction of the total current flowing in an electrolyte which is carried by a particular ion.

transport theory (*Nuc.*). Rigorous theoretical treatment of neutron migration which must be used under conditions where *Fick's law* (q.v.) does not apply. See **diffusion theory.**

transpose of a matrix (*Maths.*). The matrix whose rows are the columns of the given matrix. Sometimes, usually earlier, **conjugate matrix.**

transposition (*Elec. Eng.*). Ordered interchange of position of the lines on a pole route, and also of phases in an open power line, so that effects of mutual capacitance and inductance, with consequent interference, are minimized or balanced. See **barrel.**

transposition insulator (*Elec. Eng.*). A special type of insulator used at transposition points on a transmission line.

transposition tower (*Elec. Eng.*). A transmission tower specially designed to allow of the transposing of the conductors.

transreceiver (*Telecomm.*). See **transceiver.**

transrectification factor (*Elec. Eng.*). Ratio of change in average output current to change in alternating voltage applied to a rectifier.

transudate (*Med.*). A passive effusion of fluid from blood-vessels due to obstruction of the circulation, the fluid containing little protein, few cells, and not clotting outside the body.

transuranic elements (*Chem.*). The artificial elements 93 and upwards, which possess heavier and more complex nuclei than uranium, and which can be produced by the neutron bombardment of uranium. More than ten of these have been produced, including neptunium, plutonium, curium, lawrencium, etc.

Transvaal jade (*Min.*). Massive light green hydrogrossular garnet.

transvar (*Telecomm.*). Variable directional coupler for transferring power between waveguides.

transverse or transversal (*Bot., Zool., etc.*). Broader than long: lying across the long axis of the body or of an organ: lying crosswise between two structures: connecting two structures in crosswise fashion. See also **transversely.**

transverse architrave (*Carp.*). The moulding across the top of a door or window opening.

transverse-beam travelling-wave tube (*Electronics*). One in which the directions of propagation of the electron beam and the electromagnetic wave carrying the signal are mutually perpendicular. The cavity magnetron is an example.

transverse-current microphone (*Acous.*). A type of Reiss microphone; sometimes modified by having two cells of carbon granules in series, so that, by varying the size of the granules, a combination of characteristics is possible.

transverse electric wave (*Telecomm.*). See **TE-wave.**

transverse-field travelling-wave tube (*Electronics*). One in which the electric fields associated with the signal wave are normal to the direction of motion of the electron beam.

transverse frame (*Aero.*). The outer-ring members of a rigid airship frame. It may be of a stiff-jointed type, or braced with taut radial members to a central fitting. It connects the main longitudinal girders together. (*Ships*) A stiffening member of a ship's hull, disposed transversely to the longitudinal axis. In double bottom construction, it is that portion above the tank margin.

transverse heating (*Elec. Eng.*). Dielectric heating in which electrodes impose a high-frequency electric field normal to layers of laminations.

transverse joint (*Build.*). Any joint in a brick wall which cuts across the bed from the front to the back surface, such joint in the best practice being always a continuous one in order to avoid the setting up of *straight joint* (q.v.).

transversely (*Bot.*). Descriptive of a member which is longer one way than the other, when it is attached by one of its longer sides; thus, *transversely oval.*

transverse magnetic wave (*Telecomm.*). See **TM-wave.**

transverse metacentre (*Ships*). The *metacentre* (q.v.) obtained by inclining the vessel through a small angle about a longitudinal axis. Cf. *longitudinal metacentre.*

transverse springs (*Autos.*). Laminated springs arranged transversely across the car, parallel to the axles, instead of longitudinally; usually *semi-elliptic* (q.v.) and anchored centrally to the chassis.

transverse wave (*Phys.*). A wave motion in which the disturbance of the medium occurs at right angles to the direction of wave propagation, e.g., the vibration of a violin string.

transversum (*Zool.*). In *Crocodilia*, a cranial bone joining the palatine and the pterygoid to the maxilla.

transverter (*Elec. Eng.*). Apparatus for converting alternating to direct current, and vice versa. It makes use of a multiphase transformer, in conjunction with a stationary commutator and synchronously rotating brushgear. Cf. *commutator rectifier.*

transvestism (*Psychol.*). Pursuit of sexual excitement by dressing in the attire of the opposite sex.

transwitch (*Comp.*). A *p-n-p-n* silicon device for on-off switching in computers.

trap (*Electronics*). Crystal lattice defect at which current carriers may be trapped in a semiconductor. This trap can increase recombination and generation or may reduce the mobility of the charge carriers. (*Geol.*) See **trappean rocks.** (*San. Eng.*) A bend in a pipe so arranged as to be always full of water, in order to imprison air within the pipe. Also **air trap,** under which see other synonyms.

trap amplifier (*Telecomm.*). Parallel amplifier with grid of the amplifying valve connected to grid of an amplifying valve in a main amplifier; used to ensure that a short-circuit on any part of the monitoring equipment, such as headtelephones, does not affect transmission in the main current.

trapezium (*Maths.*). A four-sided plane figure with two opposite sides parallel. (*Zool.*) In the Mammalian brain, a part of the medulla oblongata consisting of transverse fibres running behind the pyramid bundles of the pons varolii. In Mammals, one of the carpal bones, being the inner member of the distal row.

trapezium diagram (*Electronics*). Pattern on the screen of a cathode-ray oscillograph when an amplitude-modulated radiofrequency voltage is applied to one pair of plates and the modulating voltage is applied to the other pair.

trapezium distortion (*Electronics*). That associated with *trapezium effect*.

trapezium effect (*Electronics*). Phenomenon in which the deflecting voltage applied to the deflector plates of a cathode-ray tube is unbalanced with respect to the anode. If equal alternating voltages, of different frequencies, are applied to the two sets of plates, resulting pattern on the screen is trapezoidal instead of square.

trapezius (*Zool.*). In land Vertebrates, one of the levators of the forelimb.

trapezohedron (*Crystal.*). A form in the quartz class of the trigonal system. See also **icositetrahedron**

trapezoidal rule (*Surv.*). A rule for the estimation of the area of an irregular figure. For this purpose it is divided into a number of parallel strips of equal width. The lengths of the boundary ordinates of the strips are measured, and the area is calculated from the rule stating that the area is equal to the common width of the strips multiplied by the sum of half the first and half the last ordinates plus all the others.

trapezoidal speed-time curve (*Eng.*, *Maths.*). A simplified form of speed-time curve used in making preliminary calculations regarding the energy consumption and average speed of moving vehicles. The acceleration and braking portions of the curve are sloping straight lines, while the coasting portion is a horizontal straight line, so that the complete curve becomes a trapezium. Cf. *quadrilateral speed-time curve*.

trappean rocks (*Geol.*). An obsolete term (from Swedish *trappa*, a stair) applied originally to dolerites and basalts whose outcrops gave rise to a terraced type of scenery; later widened to include a large variety of igneous rocks excluding granite, e.g., mica-trap.

trapped mode (*Phys.*). Propagation in which the radiated energy is substantially confined within a tropospheric duct.

trapping (*Comp.*). Feature of some computers by which they make an unscheduled jump to some specified location if an abnormal arithmetic situation arises.

trapping region (*Phys.*). Three-dimensional space in which particles from the sun are guided into paths towards the magnetic poles, giving rise to *aurora*, and otherwise forming ionized shells high above the ionosphere. Also called **magnetic tube.**

trappoid breccias (*Geol.*). A succession of breccias found near Nuneaton, Charnwood, and Malvern, consisting of angular blocks of rhyolite and feldspathic tuffs; of Permian age. They probably represent fossil scree material.

trap points (*Rail.*). Points placed in running rails to prevent unauthorized switching of trains.

trass (*Build.*, *Geol.*). A material similar to pozzuolana, found in the Eifel district of Germany; used to give additional strength to lime mortars and plasters.

trass mortar (*Build.*, *Civ. Eng.*). A mortar composed of lime, sand, and trass or brick-dust, or of lime and trass without sand, the trass making the mortar more suited for use in structures exposed to water.

trauma (*Med.*). A wound or body injury. (*Psychiat.*) An emotional shock that may be the origin of neurosis or a psychosis.

traumatic (*Bot.*). Relating to wounds.

traumatic response (*Bot.*). A reaction of the plant to wounding.

traumatic ring (*For.*). A zone of traumatic tissue produced by a cambium that has been injured.

traumatonasty (*Bot.*). A nastic movement following wounding.

traumotaxis (*Biol.*). Movements of protoplasts and nuclei after wounding.

traumotropism (*Bot.*). The development of curvatures following wounding.

travel (*Eng.*). The distance between the extreme positions reached by a mechanism executing a reciprocating or other reversing motion.

traveller (*Spinning*). Small C- or E-shaped metal or nylon device running round the flange of the ring on a ringframe. It acts as a yarn guide, inserts twist and winds the yarn at proper tension on to the bobbin.

traveller gantry (*Build.*). A gantry of the platform gantry type, but having a movable carriage on rails in place of the platform; the carriage, on which is fixed a crab or winch, is capable of movement along or across the gantry.

travelling matt (*Cinema.*). Method of trick photography involving the registration of more than one film during photography and/or printing to produce a single composite image.

travelling wave (*Phys.*). A wave carrying energy continuously away from the source.

travelling-wave amplifier (*Telecomm.*). One using a travelling-wave tube, or one in which simple valves inject gain at repeated points in a wave filter.

travelling-wave antenna (*Radio*). One in which radiating and nonradiating elements are formed into the edges and diameters of a series of adjacent square boxes. Also **TW antenna, box antenna.**

travelling-wave magnetron (*Electronics*). Multiple-cavity magnetron in which cavities are coupled by travelling-wave systems.

travelling-wave maser (*Electronics*). One in which the signal interacts with the electrons in paramagnetic material while propagated through a travelling-wave system.

travelling-wave tube (*Electronics*). One in which energy is interchanged between a helix delay line and an electron beam, which can be at an angle. Used to amplify ultra-high and microwave frequencies. See also **transverse-beam-, transverse-field-.**

traverse (*Surv.*). A survey consisting of a continuous series of lines whose lengths and bearings are measured.

traverse tables (*Surv.*). Tables from which the differences of latitude and departure of a line of any length and bearing may be read off.

traversing (*Eng.*). The sliding motion in a self-acting lathe or, more generally, the sideways movement of part of a machine.

traversing bridge (*Eng.*). A type of movable bridge which is capable of rolling backwards and forwards across an opening, such as a dock entrance, to allow of the passage of a vessel.

travertine (*Geol.*). A variety of calcareous tufa of light colour, often concretionary and compact, but varying considerably in structure; some varieties are porous.

trawl (*Ocean.*). A sack-like net the mouth of which is kept open by some kind of framework; used for fishing and for obtaining samples of the fauna of the sea-bottom. There are many different types, e.g., the Agassiz trawl, beam trawl, otter trawl.

tray (*Electronics*). See **card box.**

TR tube (*Electronics*). See **transmit-receive tube.**

tread (*Build.*). The horizontal part of a step. (*Eng.*) In the wheels of a vehicle, that part of the tyre in contact with the road or rail. (*Vet.*) An injury of the coronet of a horse's hoof due to striking with the shoe of the opposite foot.

trebles (*Mining*). See **coal sizes.**
trechmannite (*Min.*). A silver arsenic sulphide, crystallizing in the trigonal system.
tree (*Bot.*). A tall, woody perennial plant having a well-marked trunk and few or no branches persisting from the basal parts. (*Chem.*) Crystal growth structure. See **dendrite, lead tree.** (*Telecomm.*) A number of connected circuit branches which do not include meshes.
treeing (*Acous.*). Irregular build-up of the metal which is electrodeposited on the surface of a lacquer record, because of excessive potential drop across the plating solution, resulting in a coarse metal surface.
trega-. A prefix signifying 10^{12} times. It is replaced in the SI system of unit notation by the prefix *tera-* (q.v.).
tregohm (*Elec. Eng.*). A million *megohms* (q.v.).
Tremadocian (*Geol.*). A stratigraphic name for the lowest part of the Ordovician in Europe and the uppermost part of the Cambrian in Britain.
trematic (*Zool.*). Pertaining to the gill-clefts.
Trematoda (*Zool.*). A class of *Platyhelminthes* all the members of which are either ectoparasites or endoparasites, and have a tough cuticle, a muscular non-protrusible pharynx, and a forked intestine; eye-spots never occur; a ventral sucker for attachment is usually present, and a sucker surrounding the mouth. Liver Flukes.
trembler bell (*Elec. Eng.*). A bell with a self-interrupting armature; actuated by d.c. or a.c.
trembling (*Vet.*). See **myoclonia congenita.**
tremelloid, tremellose (*Bot.*). Jellylike.
tremie (*Civ. Eng.*). A large metal funnel used for the distribution of freshly mixed concrete over a site which is below water.
tremolite (*Min.*). A hydrous silicate of calcium and magnesium which crystallizes in the monoclinic system. It is usually grey or white, and occurs in bladed crystals or fibrous aggregates associated with metamorphic rocks; with the entry of iron replacing magnesium, it grades into *actinolite* (q.v.). See **amphiboles, asbestos.**
tremor (*Med.*). Involuntary agitation of the muscles of the body, or of a limb, due to emotional disturbance, old age, or disease of the nervous system.
tremulant (*Acous.*). The stop-key in an organ console which actuates a vibrating piston in the wind-chest supporting and operating ranks of pipes, so that the sound emitted has a pulsating or tremolo effect.
trenail (*Carp.*). A hardwood pin driven transversely through a mortise and tenon to secure the joint. Also called **trunnel.** See **draw-bore.**
trench drain (*Civ. Eng.*). See **French drain.**
trench fever (*Med.*). A disease common among troops in World War I; symptoms were relapsing fever, headache, pains in the back and in the limbs, often a rose-red eruption; due to infection with a virus conveyed by lice.
trenching plane (*Tools*). See **dado plane.**
Trenton Limestone (*Geol.*). An important member of the Ordovician succession in the region of the Adirondacks, lying between the Black River Limestone below and the Utica Shales above. Usually regarded as the highest member of the Champlainian Series, it is typically exposed at Trenton Falls, Utica, and is an important oil-bearing formation in the central States.
trepan (*Surg.*). (1) To trephine. (2) A form of trephine no longer in use.
trepanation (*Surg.*). An operation with the trepan; trephining.
trepanning (*Eng.*). Producing a hole by removing a ring of material, as opposed to disintegrating all the material corresponding to the hole.
trephination (*Surg.*). The operation of trephine.
trephine (*Surg.*). (1) To operate with the trephine; to remove by surgical means a part of the skull; to remove by surgical means a disk from any part, e.g., from the globe of the eye in the treatment of glaucoma. (2) A crown saw which is designed to remove a circular area of bone from the skull.
trephone (*Biol.*). A substance formed by cells, perhaps during their breakdown, and used by other cells in the synthesis of their protoplasm; it may stimulate cell division.
Treponemataceae (*Bacteriol.*). A family of mainly parasitic, small spirochaetes, many of which are pathogenic, e.g., *Treponema pallidum* (syphilis), *Treponema pertenue* (yaws).
treption (*Cyt.*). An adaptive response of the cell to a change in the internal or external environment.
Trescatyl (*Pharm.*). TN for *ethionamide.*
T-rest (*Eng.*). A T-shaped rest clamped to the bed of a wood-turning lathe for supporting the tool. Also **tee rest.** See **L-rest.**
trevette or trivet (*Textiles*). A knife used to release the wire in a row of loop pile, by drawing it along a groove on the top; this produces cut pile in moquette, carpet, etc., fabrics.
trevorite (*Min.*). A rare nickel ferric iron spinel found in the Transvaal.
tri-. Prefix from L. *trēs*, Greek *tria*, three.
triacetin (*Chem.*). Glycerol triacetate. Chiefly used as a plasticizer for cellulose acetate and cellulose ether plastics.
triacetonamine (*Chem.*). Condensation product of acetone (3 molecules) and ammonia, following initial *aldol condensation* (q.v.), to form mesityl oxide and then diacetonamine.

$$\begin{array}{ccc} (CH_3)_2C & \text{---} & CH_2 \\ | & & | \\ NH & & CO \\ | & & | \\ (CH_3)_2C & \text{---} & CH_2 \end{array}$$

triacid (*Chem.*). Containing three hydroxyl groups replaceable by acid radicals on neutralization.
triad (*Chem.*). A trivalent atom. (*Nav.*) Group of three transmitters used to obtain navigational position fix. (*Nuc.*) Group of three neighbouring isobaric nuclides. (*TV*) A colour cell of a screen, with three colour dots on the phosphor.
triaene (*Zool.*). In *Porifera*, a spicule in the form of a trident.
trial (*An. Behav.*). A single occasion in an experimental situation on which a specific response either occurred or could have occurred.
trial-and-error learning (*An. Behav.*). The development of an association by an animal, as a result of reinforcement during appetitive behaviour, between a stimulus or situation and an independent voluntary response in that behaviour, when both stimulus and response precede the reinforcement, and the response is not the inherited response to the reinforcement. *Operant conditioning* (q.v.) is the core of trial-and-error learning, but it also contains an element of *classical conditioning* (q.v.).
trial pit (*Civ. Eng.*). A pit sunk into the ground to obtain information as to nature, thickness, and position of strata.
triamterene (*Pharm.*). 2,4,7-Triamino-6-phenylpteridine, a diuretic drug which may cause slight blue coloration of the urine.
triandrous (*Bot.*). Having three stamens.
triangle (*Maths.*). A three-sided rectilineal plane figure. See **equilateral-, isosceles-, scalene-.**

triangle of error (*Surv.*). The triangle formed in the trial-and-error solution of the three-point problem when, on drawing back rays through the three known points on plan, they form a small triangle instead of intersecting at a single point, as a result of the positioning of the survey instrument or the measurement of one or more angles being incorrect.

triangle of forces (*Mech.*). A particular case of the polygon of forces drawn for three forces in equilibrium at a point. See **polygon of forces.**

triangular. Having three angles. (*Bot.*) Having three angles and three flat faces.

triangular notch (*Civ. Eng.*). See **vee notch.**

triangulation (*Surv.*). The process of dividing up a large area for survey purposes into a number of connected triangles with their apexes (triangulation or 'trig' stations) mutually visible, measuring one side of one of the triangles (the 'base line') and all the angles. See **intersection, trilateration.**

triarch (*Bot.*). Having three strands of xylem in the stele.

Trias (*Geol.*). The geological system of rocks which succeeds the Permian and precedes the Jurassic System. It was named by von Alberti from the three-fold division in Germany. The rocks in Britain consist of red sandstones and marls, and were deposited under desert conditions. The Trias in N. America is largely continental, but marine sands, shales, and sandstones occur in Idaho, Wyoming, and Utah.

triaster (*Cyt.*). A complex mitotic figure resulting from triple mitosis, as in the ovum after polyspermy, or in cancer cells.

triaxon (*Zool.*). Having three axes.

triazole (*Chem.*). A heterocyclic compound consisting of a five-membered ring containing three nitrogen atoms, i.e:

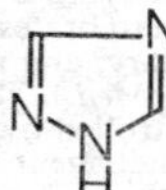

tribasic (*Chem.*). Containing three replaceable hydrogen atoms in a molecule.

tribe (*Bot.*). A section of a family consisting of a number of related genera.

tribo-electrification (*Phys.*). Separation of charges through surface friction. If glass is rubbed with silk, the glass becomes *positive* and the silk *negative*, i.e., the silk takes electrons. The phenomenon is that of contact potential, made more evident in insulators by rubbing.

triboluminescence (*Phys.*). Luminescence generated by friction.

tribology. The science and technology of interacting surfaces in relative motion (and the practices related thereto), including the subjects of friction, lubrication and wear.

tribromoethanol (*Pharm.*). See **bromal.**

tributyl tin compound (*Chem.*). Compound consisting of three carbon atoms bonded to one of tin, joined to an oxide group. Effective pesticide, which decomposes when exposed to light and air, leaving only inorganic tin.

tricarballylic acid (*Chem.*). $HOOC{\cdot}CH_2CH(COOH){\cdot}CH_2{\cdot}COOH$, a saturated tribasic acid which crystallizes in rhombic prisms; m.p. 166°C.

tricarboxylic acid cycle (*Chem.*). Synonym for *citric acid cycle* (q.v.).

tricarpellary (*Bot.*). Consisting of three carpels.

triceps (*Zool.*). A muscle with three insertions.

trich-, tricho-. Prefix from Gk. *thrix*, gen. *trichos*, hair.

trichiasis (*Med.*). Distortion of the eyelashes so that they rub against the eye.

trichiniasis, trichinosis (*Med.*). Infestation of the human intestine, as a result of eating raw or underdone pork, with the nematode worm *Trichinella* (or *Trichina*) *spiralis*, the larvae of which migrate to, and become encysted in, the muscles of the body.

trichite (*Bot.*). A hypothetical crystal, very thin and elongated, presumed to be present in very large numbers in a starch grain. (*Geol.*) Thin filament- or hair-like crystallite which occurs in volcanic rocks in irregular or radiating groups. (*Zool.*) A type of Sponge spicule; a rodlike element of the shell in some *Ciliophora.*

trichlorethene (*Chem.*). An ethyne derivative, C_2HCl_3, used as a solvent in dry-cleaning, in the extraction of fat from wool, and in the manufacture of paints and varnishes. Used in surgery to give general analgesia, and, with nitrogen(I) oxide, light general anaesthesia. TN **Trilene.**

trichloroacetic(-ethanoic) acid (*Chem.*). $CCl_3{\cdot}COOH$. Organic acid prepared by oxidizing chloral ($CCl_3{\cdot}CHO$) with nitric(V)acid. Acid with a high *dissociation constant*.

trichlorocarbanilide (*Chem.*). 3 : 4 : 4′trichlorocarbanilide. $Cl{\cdot}C_6H_4{\cdot}NH{\cdot}CO{\cdot}NH{\cdot}C_6H_3Cl_2$. Used as a germicide in toilet soaps. Low toxicity.

1,1,1-trichloroethane (*Chem.*). $CH_3{\cdot}CCl_3$. Methyl chloroform. R.m.m. 133·5; b.p. 74°C. Chlorinated solvent with low toxicity (much safer in use than tetrachloromethane). Non-inflammable. Widely used industrially for cleaning electrical equipment. TNs, Chlorothene NU, Genklene.

trichloromethane. See **chloroform.**

trichocephaliasis (*Med.*). See **trichuriasis.**

trichocyst (*Zool.*). In some *Ciliophora*, a minute hairlike body lying in the subcuticular layer of protoplasm; it is capable of being shot out, and is believed to an organ of attachment.

trichogen cell (*Zool.*). In the setae of Insects, the hypodermal cell producing the actual hollow cuticular hair. Cf. *tormogen cell.*

trichogyne (*Bot.*). A threadlike extension of the female organ in some Fungi, red and green Algae, and Lichens, which appears to function, at least sometimes, in receiving the male organ.

trichoid (*Zool.*). Hairlike.

trichome (*Bot.*). (1) A plant hair, i.e., a superficial outgrowth consisting of one or more cells. (2) The thread of cells which, together with the sheath, makes up the filament in *Myxophyceae* (*Cyanophyceae*).

trichome hydathode (*Bot.*). A multicellular hair which secretes water.

trichophore (*Zool.*). A chaeta sac.

trichophyllous (*Bot.*). Said of a plant of dry places which has the young stems and leaves protected from desiccation by a thick coating of hairs.

Trichoptera (*Zool.*). An order of *Mecopteroidea* with vestigial or absent mandibles, well developed maxillary and labial palps, and membranous hairy wings held rooflike over the body when at rest. Larvae generally eruciform, with the last body segment bearing two hooked appendages. Caddis flies.

trichosiderin (*Biochem., Physiol.*). An iron-containing red pigment which is extractable from hair.

trichosis (*Zool.*). Arrangement or distribution of hair.

trichothallic growth (*Bot.*). A type of growth of an algal filament in which cell division occurs only in a few cells located towards, or at the base of, the filament.

trichotillomania (*Med.*). An obsessional impulse to pull one's own hairs.

trichotomous (*Bot.*). Having 3 equal or nearly equal branches arising from the same part of the stem.

trichromatic coefficients (*Light*). The relative intensities of 3 primaries of a given trichromatic system of colour specification required to match a colour sample. Generally add to unity.

trichromatic filter (*Photog.*). In colour photography, a set of 3 filters arranged to suit a specified emulsion.

trichromatic process (*Photog.*). See **three-colour process.**

trichrome carbro (*Photog.*). Variant of *carbro process* (q.v.) for colour prints, using pigmented gelatine tissues on to which colour separation bromide prints are reproduced in yellow, cyan, and magenta and transferred in register on to a final support.

Trichurata (*Zool.*). Order of *Nematoda* having no phasmids (caudal sensory organs) and poorly developed excretory system: oesophagus cylindrical; males have one spicule.

trichuriasis (*Med.*). Infestation of the human intestine with the nematode whip-worm *Trichuris trichiura* (also known as *Trichocephalus dispar*).

tricipital (*Zool.*). *Adj.* from **triceps.**

trick valve (*Eng.*). See **Allan valve.**

Tricladida (*Zool.*). An order of *Turbellaria*, including marine, freshwater, and terrestrial forms; the gut has three branches, one directed forwards and two backwards.

triclinic system (*Crystal.*). The lowest system of crystal symmetry containing crystals which possess only a centre of symmetry. Also called **anorthic system.**

tricolour chromatron (*TV*). Directly-viewed single-gun cathode-ray tube for colour TV display, in which the screen is composed of successive horizontal stripes of the three colour phosphors, the electron beams being deflected on to the correct phosphors by an electrostatic field produced by a grid of horizontal wire electrodes. Also called **Lawrence tube.**

tricolour filter (*Photog.*). One of a set of filters for 3-colour photography or a composite filter having the colours on adjacent areas.

tricolour ratio (*Photog.*). The relative inertias of an emulsion for specified primary colours, i.e., deep-blue, green-yellow, and red.

tricon (*Nav.*). System based on pulsed signals from a transmitting triad, which arrive simultaneously when an aircraft is on course.

tricotine (*Textiles*). A dress fabric made from fine botany worsted yarns, featuring a whipcord effect. Used also to describe knitted dress material, flat, circular, or warp-knitted from fine silk or man-made filament yarns.

tricotylous (*Bot.*). Having 3 cotyledons.

tricrotic (*Physiol.*). Having a triple beat in the arterial pulse. *n.* **tricrotism.**

tricusp (*Maths.*). See **Steiner's tricusp.**

tricuspid (*Zool.*). Having 3 points, as the right auriculo-ventricular valve of the Mammalian heart.

tricycle landing gear (*Aero.*). A landing gear with a nose-wheel unit.

tridactyle (*Zool.*). Said of large pedicellariae having a partially flexible stalk and 3 toothed jaws, broad at the base and narrow distally, usually without poison-glands; found in *Echinoidea.*

tridymite (*Min.*). A high-temperature form of silica, SiO_2, crystallizing in the orthorhombic system, but possessing pseudohexagonal symmetry. The stable form of silica from 870°C to 1470°C. An α and β form are recognized. Typically occurs in acid volcanic rocks.

triennial (*Bot.*). Lasting for 3 years.

triethanolamine (*Chem.*). $(HOC_2H_4)_3N$. Strongly alkaline organic solvent used in some paint strippers. Also used as a stabilizer for chlorinated hydrocarbon solvents.

trifacial (*Zool.*). The 5th cranial or trigeminal nerve of Vertebrates.

trifarious (*Bot.*). Arranged in 3 rows.

trifid (*Bot.*). Divided about half-way down into 3 parts.

trifoliate (*Zool.*). Said of very small pedicellariae found in *Echinoidea*, having a very flexible stalk and 3 broad leaflike blades without teeth or poison-glands.

trifoliolate (*Bot.*). Said of a compound leaf having 3 leaflets.

trifoveolate (*Bot.*). Marked by 3 hollows.

trifurcate (*Bot.*). Bearing 3 prongs. (*Zool.*) Having 3 branches.

trifurcating box (*Elec. Eng.*). A cable dividing box for enclosing the joints between a 3-arc or triple concentric cable and 3 single-core cables or conductor terminals.

trigamma (*Zool.*). In *Lepidoptera*, a characteristic feature of wing venation, consisting of a 3-pronged fork formed by the veins, M_3, Cu_{1a} Cu_{1b}.

trigatron (*Electronics*). An envelope with an anode, cathode and trigger electrode, containing a mixture of argon and oxygen. The device operates as an electronic switch, in which a low energy pulse ionizes the gas in the switch, and permits discharge of a much higher energy pulse across the main electrodes.

trigeminal (*Zool.*). Having 3 branches; the 5th cranial nerve of Vertebrates, dividing into the ophthalmic, maxillary, and mandibular nerves.

trigeminal neuralgia (*Med.*). See **tic douloureux.**

trigenic (*Gen.*). Controlled by 3 genes.

trigger (*Chem.*). The agent which causes the initial decomposition of a chain reaction. (*Comp.*) Manual or automatic signal for an operation to start.

trigger circuit (*Telecomm.*). A circuit having a number of states of electrical condition which are either stable (or quasi-stable) or unstable with at least one stable state, and so designed that desired transition can be initiated by the application of suitable trigger excitation.

trigger electrode (*Electronics*). See **pilot electrode.**

trigger hair (*Zool.*). See **cnidocil.**

trigger level (*Telecomm.*). The minimum input level at which a trigger circuit will respond.

trigger pulse (*Telecomm.*). One which operates a trigger circuit.

trigger relay (*Electronics*). Gas-filled triode or certain combinations of high-vacuum thermionic valves, in which a disturbance of sufficient magnitude can initiate or terminate a discharge but has no subsequent control thereof. (*Telecomm.*) Relay which, when operated, remains in its operated condition when the operating current or other control is removed, because of residual magnetism or a mechanical latch.

trigger valve (*Electronics*). A thermionic or gas discharge valve used as a trigger relay. A typical example is the discharging valve in a linear time-base circuit.

triglyceride (*Chem.*). Term applied to a fatty acid ester of glycerol in which all 3 hydroxyl groups are substituted.

triglyph (*Arch.*). A group of 3 glyphs, or of 2 glyphs and 2 half-glyphs, used as a decoration for a flat surface.
trigonal (*Bot.*). Triangular in section.
trigonal system (*Crystal.*). A style of crystal architecture characterized essentially by a principal axis of threefold symmetry; otherwise resembling the hexagonal system. Such important minerals as calcite, quartz, and tourmaline crystallize in this system.
trigone (*Anat.*). Triangular area of interior of urinary bladder between the openings of the ureters and of the urethra. (*Bot.*) A thickened angle of a cell. (*Zool.*) A triangular space or area. Also **trigonum.** *adjs.* **trigonal, trigonate.**
trigonitis (*Med.*). Inflammation of the trigone.
trigonometrical functions (*Maths.*). If θ is any

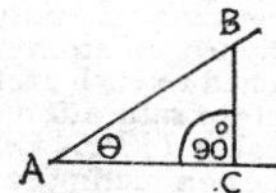

angle, and ABC is the right-angled triangle formed by dropping a perpendicular BC from a point B in one of the lines enclosing the angle to the other, the trigonometrical functions, or ratios, are as follows:

$$\text{sine } \theta = \frac{BC}{AB}; \text{ cosecant } \theta = \frac{1}{\text{sine } \theta}$$

$$\text{cosine } \theta = \frac{AC}{AB}; \text{ secant } \theta = \frac{1}{\text{cosine } \theta}$$

$$\text{tangent } \theta = \frac{BC}{AC}; \text{ cotangent } \theta = \frac{1}{\text{tangent } \theta}$$

Usually abbreviated to *sin, cosec, cos, sec, tan, cot*. Independent arithmetic definitions of *sin, cos,* and *tan* are as follows:

$$\sin x = \sum_{0}^{\infty} (-1)^r \frac{x^{2r+1}}{(2r+1)!}$$

$$\cos x = \sum_{0}^{\infty} (-1)^r \frac{x^{2r}}{(2r)!}$$

$$\tan x = \frac{\sin x}{\cos x}.$$

See also **inverse trigonometrical function.**
trigonometrical station (*Surv.*). A survey station used in a triangulation.
trigonometrical survey (*Surv.*). A survey based on a triangulation.
trigonous (*Bot.*). Having 3 obtuse angles.
trigonum (*Zool.*). See **trigone.**
trihydric alcohols (*Chem.*). Alcohols containing 3 hydroxyl groups attached to 3 different carbon atoms, e.g., *glycerine* (q.v.).
tri-iodothyronine (*Chem.*). A hormone stored, as is thyroxine, as part of the thyroglobulin protein molecule within the thyroid gland. See **thyroid gland, thyroxine.**
trilateration (*Surv.*). Land survey, by triangulation, in which distances are measured direct by a *tellurometer* (q.v.). In the *shoran* system, up to 800 kilometres can be thus measured by aid of airborne magnetometer flown between 2 ground stations.
Trilene (*Chem.*). See **trichlorethene.**
trilling (*Crystal.*). A twinned crystal made up of three individuals.
trillion (*Maths.*). The cube of a million; (U.S.) the cube of ten thousand. Colloquial only.
Trilobita (*Zool.*). A class of *Arthropoda* with the body moulded longitudinally in three lobes, one pair of antennae and, on all other somites, similar appendages which are biramous and have gnathobases. Numerous in the Cambrian and Silurian but became extinct by the Secondary Era. Trilobites.
trim (*Aero.*). Adjustment of an aircraft's controls to achieve stability in a desired condition of flight; cf. *trimming strip* and *-tab*. (*Build.*) Architraves and other finishings around a door or window opening. (*Ships*) The difference between the draughts measured at the forward and after perpendiculars. May be expressed as an angle.
Trimask (*Print.*). An all-purpose film used for *colour masking* (q.v.); similar to *Multi-mask*.
trimer (*Chem.*). Substance in which molecules are formed from 3 molecules of a monomer.
trimeric (*Chem.*). Having the same empirical formula but a relative molecular mass three times as great.
trimerous (*Bot.*). Arranged in 3s or in multiples of 3.
trimethadione (*Pharm.*). A drug, 3,5,5-trimethyloxazolidine-2, 4-dione, $C_6H_9NO_3$, whose anticonvulsive properties make it effective in the treatment of *petit mal* epilepsy.
trimethyl-aminoethanoic acid (trimethylglycine) (*Chem.*). See **betaine.**
trimethylene glycol (*Chem.*). $CH_2OH{\cdot}CH_2{\cdot}CH_2OH$. propan-1,3-diol. Organic solvent. B.p. 214°C (with decomposition); rel. d. (20°C) 1·060.
trimethylglycine (*Chem.*). See **betaine.**
trimetric system (*Geol.*). An obsolete name for the *orthorhombic system*.
trimmed edges (*Bind.*). See **cut edges.**
trimmed size (*Bind., Print.*). A necessary specification for books and for any subdivision of a sheet, the minimum trim being ⅛″ from each edge that requires it. *International Paper Sizes* are always given as trimmed.
trimmer (*Aero.*). See **trimming tab.** (*Carp.*) The cross-member which is framed between the full-length members to afford intermediate support to the shortened joists in a trimming. (*Elec. Eng.*) See **trimming capacitor.** (*Telecomm.*) See **pad** (1).
trimmer arch (*Build.*). A somewhat flat arch turned from the wall to the trimmer to support a hearthstone. Also called a **brick-trimmer.**
trimmer joint (*Carp.*). A joint formed with a *tusk tenon* (q.v.).
trimming (*Carp.*). The operation by which bridging joists or rafters are shortened and given intermediate support around a fireplace or chimney. (*Eng.*) The removal of flash from rubber or plastics mouldings or from castings, or of material from the edge of a workpiece.
trimming capacitor (*Elec. Eng.*). Variable capacitor of small capacitance used in conjunction with ganging for taking up the discrepancies between self and stray capacitances of individual ganged circuits, so that they remain in step for all settings of the main tuning control. Also called **trimmer.**
trimming joist (*Carp.*). One of the 2 full-length members between which the trimmer is framed. As these members have to carry more load than the other bridging joists, they are wider.
trimming strip (*Aero.*). A metal strip, or a cord or wire doped in place with fabric, on the trailing edge of a control surface to modify its balance or trim; it is adjustable only on the ground.

trimming tab (*Aero.*). A *tab*, which can be adjusted in flight by the pilot, for trimming out control forces; coll. **trim tab or trimmer.**

trimonoecious (*Bot.*). Having perfect, pistillate and staminate flowers on the same plant.

trimorphic (*Bot.*). Said of a species which has 3 kinds of flowers, differing in the relative lengths and positions of the filaments, anthers, and stigmata. *n.* **trimorphism.**

trimorphous (*Chem.*). Having 3 crystalline forms.

trim tab (*Aero.*). See **trimming tab.**

trimyarian (*Zool.*). Having 3 muscular layers.

trinacriform (*Bot.*). Having 3 prongs.

trinitrides (*Chem.*). Salts of hydrazoic acid. Also called **azides and hydrazoates.**

trinitroglycerine (*Chem.*). See **nitroglycerine.**

Trinitron (*TV*). TN for a 3-colour TV tube using a vertical grating and vertical phosphor stripes. A common electron gun assembly and deflexion system is used with three cathodes to provide the electron beams.

trinitrophenol (*Chem.*). 2,4,6-trinitro-1-hydroxy-benzene. See **picric acid.**

trinitrotoluene (*Chem.*). The symmetrical isomer, 2,4,6-trinitrotoluene, $C_6H_2(CH_3)(NO_2)_3$, is a solid, melting at 82°C. It is manufactured by slowly adding toluene to a mixture of nitric and sulphuric acids containing oleum. It is used as a high explosive, and is known as **T.N.T.**

trinoscope (*TV*). Cathode-ray tube with 3 electron guns, designed for the reproduction of TV images in colour.

triode-heptode (*Electronics*). Thermionic valve used as a frequency changer in a similar manner to a triode-hexode.

triode-hexode (*Electronics*). Combination of triode and hexode in the same envelope, used as a frequency converter in a supersonic heterodyne receiver. The triode section is used as oscillator and the hexode as demodulator.

triode valve (*Electronics*). Thermionic vacuum tube containing an emitting cathode, an anode, and a control electrode or grid, whose potential controls the flow of electrons from cathode to anode. Also called **three-electrode valve.**

trioecious (*Bot.*). Having perfect, staminate and pistillate flowers on distinct plants of the same species.

triolein (*Chem.*). Naturally-occurring triglyceride in which all three fatty acid chains are oleic acid.

trioses (*Chem.*). The simplest monosaccharides. They contain 3 carbon atoms in the molecule, e.g., $HO{\cdot}CH_2{\cdot}CO{\cdot}CH_2{\cdot}OH$.

trip (*Nuc. Eng.*). Automatic reduction of reactor power initiated by signal from one of the safety circuits. This may be a gradual power setback or an emergency trip.

tripack (*Photog.*). A process involving the use of 3 emulsions on separate bases, so that exposure is effected by light passing through them when they are all in contact. See also **integral-.**

trip amplifier (*Nuc. Eng.*). One operating the trip mechanism of a nuclear reactor. Also termed a **shut-down amplifier.**

tripartite (*Bot., etc.*). Divided nearly to the base into 3 parts.

trip circuit (*Elec. Eng.*). The electric circuit operating the tripping mechanism of a circuit-breaker. Cf. *shunt trip.*

trip coil (*Elec. Eng.*). Any magnet coil which operates some other circuit or mechanism by motion of an armature; more particularly, a coil which operates a circuit-breaker, or the release mechanism of a telegraph machine.

trip gear (*Eng.*). A valve-actuating gear, used for drop valves and rocking (Corliss) valves of large steam-engines, in which the valve is opened by a trigger mechanism, which is then tripped out of engagement to allow the valve to close under a heavy spring. See **Corliss valve, drop valve.**

triphane (*Min.*). See **spodumene.**

triphenylmethane dyes (*Chem.*). A group of dyestuffs derived from triphenylmethane. They comprise the malachite green group derived from diaminotriphenylmethane, the rosaniline group derived from triaminotriphenylmethane, the aurine group derived from trihydroxytriphenylmethane, the phthalein group derived from triphenylmethane-carboxylic acid.

triphosphopyridine (*Biochem.*). See **NADP.**

triphylite (*Min.*). An orthorhombic lithium iron phosphate isomorphous with *lithiophilite.*

tripinnate (*Bot.*). Said of a pinnate compound leaf with pinnately divided leaflets, themselves pinnately divided.

triple bond (*Chem.*). An indication of a state of unsaturation between 2 polyvalent atoms, showing that 2 hydrogen atoms or their equivalent can be attached to each atom connected by a triple bond before saturation is reached.

triple-concentric cable (*Elec. Eng.*). A 3-core cable in which the conducting cores are arranged concentrically about the axis of the cable.

triple-expansion engine (*Eng.*). An engine in which the steam expands, successively, in a high pressure, intermediate pressure, and low pressure cylinder, working on the same crankshaft. See **multiple-expansion engine.**

triple fusion (*Bot.*). The nuclear union in the embryo sac between the 2 polar nuclei and a male nucleus; it provides the starting-point for the development of the endosperm.

triplener (*TV*). A filter device to enable three TV bands to use a common VHF downlead from the aerial system.

triple point (*Phys.*). The temperature and pressure at which 3 phases of a substance can coexist (see **phase rule**). The *triple point of water* is the equilibrium point (273·16 K at 610 N/m^2) between pure ice, air-free water and water vapour, obtained in a sealed vacuum flask. It is one of the six fundamental fixed points of the *international practical temperature scale.*

triple-pole switch (*Elec. Eng.*). A switch for simultaneously making or breaking a 3-wire electric circuit.

triple superphosphate (*Chem. Eng.*). The product obtained by reacting phosphate rock with phosphoric acid giving a higher concentration of soluble calcium phosphate than obtains in ordinary or 'single' *superphosphate* (q.v.).

triplet (*Chem.*). A state in which there are two unpaired electrons.

triplets (*Bot.*). Individuals resulting from the division of the ovum into 3 parts, each then developing. (*Zool.*) In Mammals, 3 individuals produced at the same birth.

triplex (*Bot.*). A tetraploid zygote which has 3 doses of any given dominant.

triplex boards (*Paper*). See **cardboards.**

Triplex glass (*Glass*). A patented form of laminated glass. See **safety glass.**

triplex winding (*Elec. Eng.*). A d.c. armature winding having 3 parallel paths per pole between positive and negative terminals.

triplinerved (*Bot.*). With 3 main veins in the leaf.

triplite (*Min.*). Manganese fluorophosphate, crystallizing in the monoclinic system.

triploblastic (*Zool.*). Having three types of tissue in the body, there being mesoderm between the ectoderm and endoderm, which gives rise to connective, skeletal and muscular tissues, etc. Cf. *diploblastic.*

triplocaulescent (*Bot.*). Having a main stem

bearing branches, being themselves branched.

triploid (*Cyt.*). Having 3 times the haploid number of chromosomes for the species.

tripod (*Surv., etc.*). Device by which some surveying and other instruments are supported firmly off the ground. It consists of 3 legs hinged to a common head on which the instrument is secured. In photography, used for steadying a camera for long exposures, telephoto shots, etc.

tripod bush (*Photog.*). That part of the tripod head to which the camera is screwed. Found with both continental and British thread, adaptors being available for interchangeability.

tripod drill (*Mining*). Rock drill on heavy tripod.

tripod head (*Photog.*). See **ball-and-socket head.**

Tripoli powder (*Min.*). See **tripolite.**

tripolite (*Min.*). A variety of opaline silica which is formed from the siliceous frustules of diatoms. It looks like earthy chalk or clay, but is harsh to the feel and scratches glass. When finely divided it is sometimes called **earthy tripolite.** Also called **diatomite, infusorial earth.**

tripping (*Horol.*). An escapement is said to *trip* when a tooth of the escape wheel runs past the locking face.

tripping battery (*Elec. Eng.*). The secondary battery which provides the supply for the trip-coil circuits of a number of circuit-breakers.

trip relay (*Elec. Eng.*). A relay controlling the electromagnetic tripping mechanism of a circuit-breaker.

triprosthomerous (*Zool.*). Having 3 somites in front of the mouth.

trip switch (*Elec. Eng.*). A control switch for closing the tripping circuit of a circuit-breaker.

triptane (*Fuels*). Trimethyl butane, a powerful additive for aviation fuel.

tripus (*Zool.*). In *Ostariophysi*, one of the Weberian ossicles.

trip value (*Elec. Eng.*). The current or voltage required to operate a relay.

triquetrous (*Bot.*). Having 3 angles and 3 concave faces.

trisaccharides (*Chem.*). Carbohydrates consisting of molecules composed of 3 monosaccharide anhydrides; they result from the elimination of 2 molecules of water from 3 molecules of a monosaccharide, e.g. maltotriose is $G\alpha1 \rightarrow 4G\alpha1 \rightarrow 4G$ (G = glucose).

trismus (*Med.*). Lockjaw; tonic spasm of the muscles of the jaw, causing the jaws to be clenched, as in tetanus.

trisomic (*Cyt.*). Said of an otherwise normal diploid organism in which one chromosome type is represented thrice instead of twice.

tristearin (*Chem.*). Naturally-occurring triglyceride in which all 3 fatty acid chains arise from stearic acid.

tristichous (*Bot.*). Having leaves or branches arranged one above another in 3 rows. (*Zool.*) Arranged in 3 rows.

tristimulus values (*Optics*). Amounts of each of 3 colour primaries that must be combined to form an objective colour match with a sample.

tritanopic (*Med.*). Colour blind to blue.

tritium (*Nuc.*). Symbol T, r.a.m. 3·0221, mass no. 3. The radioactive isotope of hydrogen, of half-life 12·5 years. It is very rare, the abundance in natural hydrogen being one atom in 10^{17} but tritium can be produced artificially by neutron absorption in lithium. Can be used to label any aqueous compound and consequently is of great importance in radiobiology.

tritium unit (*Chem.*). A proportion of tritium in hydrogen of one part in 10^{18}. This represents 7 disintegrations per minute in 1 litre of water.

tritocerebron, tritocerebrum (*Zool.*). In higher *Arthropoda*, as Insects and *Crustacea*, the fused ganglia of the 3rd somite of the head, forming part of the 'brain'.

tritognathous (*Zool.*). Having the jaws borne on the 3rd somite of the head.

triton (*Nuc.*). The tritium nucleus, consisting of 1 proton combined with 2 neutrons.

tritor (*Zool.*). The masticatory surface of a tooth.

tritubercular (*Zool.*). Of teeth, possessing 3 cusps.

triturate (*Chem.*). To grind to a fine powder, especially beneath the surface of a liquid.

trityl (*Chem.*). The triphenylmethyl group, $C(C_6H_5)_3$; the triphenylmethyl radical is generally accepted as being the first organic radical to be obtained in a free state.

triungulin (*Zool.*). In *Meloidae* (Oil-Beetles), the small active hard-skinned campodeiform larva: the similar larva of *Strepsiptera.*

trivalent (*Chem.*). Capable of combining with 3 atoms of hydrogen, or their equivalent. (*Cyt.*) Said of association of 3 chromosomes in meiosis.

trivet (*Textiles*). See **trevette.**

trivium (*Zool.*). In *Echinodermata* generally, the 3 radii farthest from the madreporite; in *Holothuroidea*, the 3 rays which form the 'ventral' surface of the body.

t-RNA (*Biochem.*). *Transfer-RNA*, another name for *s-RNA* (q.v.).

trocar (*Surg.*). A sharp-pointed perforator which, inserted into a cannula, enables this to be introduced into the body.

trochal (*Zool.*). Wheel-shaped. See also **trochus.**

trochal disk (*Zool.*). In *Rotifera*, the flattened anterior end.

trochanter (*Zool.*). The second joint of the leg in Insects; a prominence for muscle attachment near the head of the femur in Vertebrates.

trochantin (*Zool.*). In primitive Insects, an articular sclerite situated at the base of the coxa.

trochilus (*Arch.*). A hollow moulding whose profile is formed of 2 circular arcs of different radii.

trochlea (*Zool.*). Any structure shaped like a pulley, especially any foramen through which a tendon passes. *adj.* **trochlear.**

trochlear nerve (*Zool.*). The 4th cranial nerve of Vertebrates, running to the superior oblique muscle.

trochoblasts (*Zool.*). Those cells of a segmenting ovum destined to become a trochophore which will give rise to the prototroch.

trochoid (*Maths.*). See **roulette.**

trochoidal mass analyser (*Nuc. Eng.*). A form of mass spectrometer in which the ion beams traverse trochoidal paths within electric and magnetic fields mutually perpendicular.

trochophore, trochosphere (*Zool.*). A free-swimming pelagic larval form of *Annelida, Mollusca*, and *Polyzoa*, possessing a prominent pre-oral ring of cilia, an apical tuft of cilia, a ventrally curved gut, and a blastocoelic body cavity containing the primitive mesoblasts.

trochotron (*Nuc. Eng.*). Abbrev. for *trocho*idal magne*tron*. High-frequency counting tube, which uses crossed electric and magnetic fields to deflect a beam on to radially disposed electrodes.

trochus (*Zool.*). In *Rotifera*, the inner pre-oral circlet of cilia; cf. *cingulum*. *adj.* **trochal.**

troctolite (*Geol.*). A coarse-grained basic igneous rock, consisting essentially of olivine and plagioclase only. The former mineral occurs as dark spots on a light ground of feldspar, giving the rock a characteristic spotted appearance, whence the name *troutstone.*

Trogoniformes (*Zool.*). An order of *Neognathae*,

containing one family of tropical birds whose size is between that of a thrush and a crow, and which have a loose and beautiful plumage.

troilite (*Min.*). A nonmagnetic iron sulphide, FeS, which occurs mainly in meteorites.

Trojan group (*Astron.*). A number of minor planets, named after the heroes of the Trojan war, which have the same mean motion as Jupiter and travel in the same orbit. They are divided into 2 clusters, one of which is 60° of longitude ahead of Jupiter, the other 60° behind; each planet oscillates about a point which forms an equilateral triangle with Jupiter and the sun. These are 2 particular solutions of the *three-body problem* (q.v.).

troland (*Light, Optics*). A unit of illuminance used in retinal work. If the apparent area of the entrance pupil of the eye is 1 mm^2, then the troland is the visual stimulation obtained from an illuminance of 1 candela/m^2.

trolleybus (*Elec. Eng.*). A rail-less passenger vehicle, powered by current transmitted from two overhead trolley wires through two roof-mounted *trolley poles*.

trolley bush (*Elec. Eng.*). The graphite bushing between the steel axle-pin and the trolley wheel.

trolley cord (*Elec. Eng.*). The rope by which the trolley head is drawn down to disengage the trolley wheel from the overhead contact wire.

trolley exhaust (*Vac. Tech.*). A mass production method of manufacture of large lamps and valves in which they are mounted on individual pumping units capable of movement to fixed stations where the successive operations of manufacture are performed.

trolley-frog (*Elec. Eng.*). A device used at a junction of two overhead contact wires on a traction system to permit the passage of the current-collector along either wire as desired.

trolley head (*Elec. Eng.*). The complete fitting housing the trolley wheel.

trolley pole (*Elec. Eng.*). The steel tube or pole carrying the trolley head, which is insulated from it, and down which runs the cable connecting the trolley wheel to the traction-motor circuit.

trolley system (*Elec. Eng.*). The overhead current-collecting system used on tramcars and trolley-buses, in which a small grooved wheel or skid runs under the contact wire.

trolley wheel (*Elec. Eng.*). The small grooved wheel by means of which current is collected from the overhead contact wire in the trolley system.

trolley wire (*Elec. Eng.*). The overhead contact wire in the trolley system of current collection.

trombone (*Telecomm.*). A U-shaped length of waveguide which is of adjustable length for use in a waveguide circuit.

trommel (*Min. Proc.*). A cylindrical revolving sieve for sizing crushed ore or rock.

trondhjemite (*Geol.*). A coarse-grained igneous rock consisting essentially of plagioclase (ranging from oligoclase to andesine), quartz, and small quantities of biotite, in some instances accompanied, or replaced, by amphibole and pyroxene.

troostitic structures (*Met.*). Structures in steel which consist of very fine aggregates of ferrite and cementite, which cannot be resolved under the microscope, which etch rapidly, and appear very dark. Such structures may be very fine pearlite, or the product of tempering martensite at temperatures lower than those which give sorbite.

tropacocaine (*Pharm.*). An alkaloid found in the cocoa plant, with the structure:

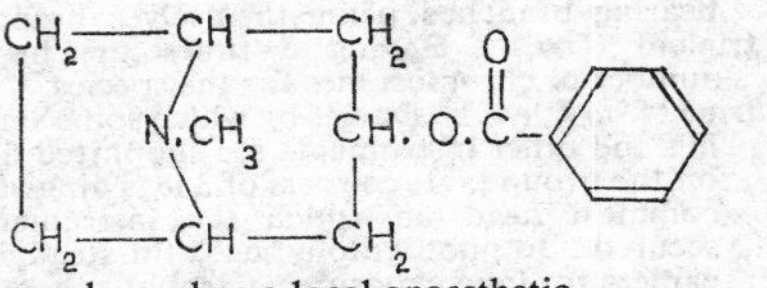

```
CH2——CH——CH2   O
 |    /    |    ||
 |  N.CH3  CH.O.C—C6H5
 |    \    |
CH2——CH——CH2
```

formerly used as a local anaesthetic.

tropaeolines (*Chem.*). A group of dyes derived from 4-hydroxyazobenzene.

tropeic (*Zool.*). See **carinate**.

troph-, tropho-. Prefix from Gk. *trophē*, nourishment.

trophallaxis (*Zool.*). Mutual exchange of food between imagines and their larvae, as in some social Insects.

trophamnion (*Zool.*). In the eggs of certain parasitic *Hymenoptera*, a protoplasmic sheath surrounding a central embryonic mass during early development.

trophi (*Zool.*). In Insects, the mouth-parts; in *Rotifera*, the masticatory mechanism of the mastax.

trophic (*Bot., Zool.*). Pertaining to nutrition.

trophic level (*Ecol.*). Organisms whose food is obtained from plants by the same number of intermediate steps (or links in a food chain) belong to the same *trophic level*. Thus green plants occupy the first trophic level, herbivores the second, carnivores which eat herbivores the third, etc.

trophic race (*Zool.*). A collection of individuals within the limits of a species, but differing from the typical members of the species in their choice of food.

trophic structure (*Ecol.*). A characteristic feature of any ecosystem, measured and described either in terms of the standing crop per unit area, or energy fixed per unit area per unit time, at successive trophic levels. It can be shown graphically by the various *ecological pyramids* (q.v.).

trophifer, trophiger (*Zool.*). That part of the head of an Insect with which the mouth-parts articulate.

trophoblast (*Zool.*). The differentiated outer layer of epiblast in a segmenting Mammalian ovum.

trophochondrioma (*Cyt.*). Mitochondria as concerned in nutrition.

trophochromatin (*Cyt.*). A substance within the nucleus which controls the metabolism of the cell.

trophochromidia (*Cyt.*). Vegetative chromidia; chromidia concerned with nutritive processes.

trophocyte (*Zool.*). In Insects, a cell of the larval fat-body which accumulates albuminoid reserve material and plays a part in tissue-building during histogenesis.

trophology. The study of nutrition.

trophonemata (*Zool.*). Finger-shaped projections of the wall of the uterus in some viviparous lower Vertebrates.

trophoneurosis (*Med.*). Any functional disorder of the body due to derangement of the trophic action of the nerves.

trophonucleus (*Zool.*). In some *Mastigophora*, the large vegetative nucleus which regulates metabolism and growth. Cf. *kinetonucleus*.

trophophore (*Zool.*). In *Porifera*, an aggregation of cells destined to become a gemmule.

trophophyll (*Bot.*). A vegetative leaf.

trophoplasm (*Cyt.*). Protoplasm which is mainly concerned with nutrition.

trophoplast (*Bot.*). A plastid.

trophosome (*Zool.*). All the zooids of a hydroid colony which are concerned with nutrition.

trophospongium (*Zool.*). Holmgren's canaliculi of nerve cells occupied by branching processes of neuroglia cells.
trophothylax (*Zool.*). In certain Ant larvae (*Pseudomyrminae*), a food-pouch on the first abdominal somite.
trophotropism (*Bot.*). A reaction in a growing organ induced by the chemical nature of the environment.
trophozoite (*Zool.*). In *Protozoa*, the trophic phase of the adult, which generally reproduces by schizogony.
trophozooid (*Zool.*). In some *Urochorda* colonies, a nutritive zooid.
tropibasic (*Zool.*). Said of a skull in which the trabeculae are near together and fuse in the middle line. Also **tropitrabic.**
tropic acid (*Chem.*). $CH_2OH{\cdot}CH(C_6H_5){\cdot}COOH$, 2-phenyl-3-hydroxypropanoic acid; crystallizes in fine prisms; m.p. 117°C. Obtained from atropine by hydrolysis together with *tropine*, $C_8H_{15}ON$, the latter being oxidized by chromic (VI) acid to a ketone, *tropinone*, $C_8H_{13}ON$.
tropical month (*Astron.*). The period of lunar revolution with respect to the equinox (27·321 58 days).
tropical revolving storm (*Meteor.*). A small intense cyclonic depression originating over tropical oceans. See **cyclone, hurricane, typhoon, willy-willy.**
tropical switch (*Elec. Eng.*). A switch mounted on feet or bosses; it thus guards against the effect of excessively damp climates by having an air space between its base and mounting surface. Also called **feet-switch.**
tropical year (*Astron.*). The interval between 2 successive passages of the sun in its apparent motion through the First Point of Aries; hence the interval between 2 similar equinoxes or solstices and the period of the seasons; its length is 365·242 194 mean solar days.
tropic curvature (*Bot.*). A curvature of a plant organ caused by one-sided growth due to a stimulus falling on the plant from one side.
Tropics (*Geog.*). The name given to those 2 parallels of celestial latitude which pass through the *solstices* (q.v.), and which therefore represent the limits of the sun's extreme north and south *declinations* (q.v.). The old terms *Tropic of Cancer* and *Tropic of Capricorn* are still applied to the northern and southern parallels respectively. The terrestrial counterparts are the 2 parallels of terrestrial latitude on either side of the equator and each distant from it by about 23° 27′, the value of the *obliquity of the ecliptic* (q.v.). In popular language, the term is sometimes used of the zone of the earth, bounded by these parallels, in which the sun can be vertically overhead.
tropine, tropinone (*Chem.*). See **tropic acid.**
tropism (*Biol.*). A reflex response of a cell or organism to an external stimulus; growth movement or other directed reaction of sessile plants and animals. See **geotropism, phototropism.** Cf. *taxis.* In *An. Behav.* term formerly used to include directed movements of animals, on the basis of Loeb's theory which held that they were similar in their basic mechanism to the tropisms of plants.
tropitrabic (*Zool.*). See **tropibasic.**
troposphere (*Meteor.*). Atmosphere up to region where temperature ceases to decrease with height. This upper boundary is termed the **tropopause.** See also **stratosphere.**
tropospheric (*Radio*). Said of reflection, absorption or scattering of a radio wave when encountering variations in the troposphere.
tropospheric wave (*Radio*). A radio wave whose path between two points at or near the earth's surface lies wholly within the troposphere and will be governed by meteorological conditions.
tropotaxis (*An. Behav.*). A type of *taxis* (q.v.) in which an animal directs itself towards or away from a source of stimulation by simultaneously comparing the intensity of stimulation on either side by symmetrically placed receptor organs. Cf. *klinotaxis, telotaxis.*
tropotron (*Electronics*). Form of magnetron valve.
Trotter photometer (*Light*). A portable photometer in which the brightness of the comparison screen is varied by tilting.
trouble-shooting (*Electronics*). U.S. for **fault-finding.**
trough (*Meteor.*). An outward extension of the isobars from a centre of low pressure.
troughed belt conveyor (*Eng.*). A *conveyor* (q.v.) comprising an endless belt which is made hollow or troughed on the carrying run to increase the carrying capacity and to obviate spillage.
trough fault (*Geol.*). Actually a pair of parallel, normal faults hading towards one another and throwing in opposite directions.
trough gutter (*Build.*). A gutter used along roof valleys or parapets.
Trousseau's phenomenon (*Med.*). Spasm of the muscles of a limb whose blood-vessels or nerves are compressed, occurring in tetany.
Trouton's rule (*Chem.*). For most nonassociated liquids, the ratio of the latent heat of vaporization per mole, measured in joules, to the boiling-point, on the absolute scale of temperature, is approx. equal to 88 at atmospheric pressure.
troutstone (*Geol.*). See **troctolite.**
trowel (*Build.*). A flat steel tool used for spreading and smoothing mortar or plaster.
TR tube (*Electronics*). Abbrev. for **transmit-receive tube.**
truck-type switchgear (*Elec. Eng.*). Switchgear in which each circuit-breaker, with its associated equipment, is mounted on a truck capable of withdrawal, so that it may be completely removed from the gear for maintenance and repair. Also called **carriage-type switchgear.**
true (or **real**) **absorption coefficient** (*Nuc.*). The absorption coefficient applicable when scattered energy is not regarded as absorbed. Applicable to broad-beam conditions.
true airspeed (*Aero.*). The actual speed of an aircraft through the air, computed by correcting the indicated airspeed for altitude, temperature, position error, and compressibility effect. Abbrev. **T.A.S.**
true altitude (*Astron., Surv.*). *Altitude* (q.v.) of a heavenly body as deduced from the *apparent altitude* (q.v.) by applying corrections for atmospheric refraction, for instrumental errors, and where necessary for geocentric parallax, sun's semi-diameter, and dip of horizon.
true angle of incidence (*Aero.*). See **angle of incidence.**
true azimuth (*Surv.*). That measured relative to true geographical north.
true bearing (*Surv.*). That measured clockwise from true geographical north.
true coincidences (*Nuc.*). Those produced by a single particle discharging both or all counters; cf. *spurious coincidences.*
true course (*Ships, etc.*). The angle between the true meridian and the direction of the ship's head.
true density (*Powder Tech.*). The mass of the particle divided by its volume, excluding *open* and *closed pores* (qq.v.).
true horizon (*Surv.*). A great circle of the celestial

sphere parallel to the horizon and passing through the earth's centre. Also called the **rational horizon.**

true north (*Surv., etc.*). The direction of the geographical north pole.

true ohm (*Elec.*). The actual realization of the practical unit of resistance, exactly equal to 10^9 electromagnetic units of resistance. Cf. *international ohm, B.A. ohm, legal ohm.*

true resistance (*Elec. Eng.*). See **d.c. resistance.**

true section (*Surv.*). A section which has been drawn, with the same scales, horizontally and vertically.

true watts (*Elec.*). Power dissipated in a.c. circuit.

trumpet arch (*Build., Civ. Eng.*). See **splaying arch.**

trumpet hypha (*Bot.*). A filament inside the thallus of a brown alga which is markedly enlarged at each transverse septum.

truncate (*Bot.*). Blunt-ended, as if cut off abruptly.

truncation (*Comp.*). (1) Ending of a computational procedure in accordance with some program rule as soon as a specified accuracy has been reached. (2) Rejection of final digits in a number, thus lessening precision (but not necessarily accuracy).

truncus (*Zool.*). A main blood vessel; as the *truncus transversus* or *Cuvierian duct* (q.v.), and the *truncus arteriosus* or great vessel, through which blood passes from the ventricle. Also **trunk.**

trunk (*Anat., Zool.*). The body, apart from the limbs; the proboscis of an elephant. See also **truncus.** (*Arch.*) Shaft of a column. (*Bot.*) Upright, massive main stem of a tree. (*Cables*) See **trunk feeder.** (*Comp.*) See **highway.** (*Teleph.*) In U.S., a *link* (q.v.).

trunk call (*Teleph.*). In Britain, a telephone call from one telephone area to another, involving cable connexions between two trunk centres each dealing with all calls passing in or out of its area. See **long-distance call, toll call.**

trunk circuit (*Teleph.*). In Britain, a 2- or 4-wire connexion between trunk centres, for establishing trunk calls between telephone areas. In America, a circuit between exchanges in the same telephone area. Also **trunk line.**

trunk conveyor (*Mining*). In colliery, belt conveyor in main road.

trunk dialling (*Teleph.*). The dialling of trunk telephone calls directly, connexions not being made by an operator. See also **subscriber trunk dialling.**

trunk distribution frame (*Teleph.*). A frame carrying the terminals for connecting the trunks between ranks of selectors.

trunk exchange (*Teleph.*). An exchange in a telephone area which is connected, by trunk or long-distance lines, to other trunk exchanges, and to subscribers through local exchanges.

trunk feeder (*Cables*). A feeder connecting 2 generating stations, or a generating station and a large sub-station. Also **trunk main.**

trunk frame terminal assembly (*Teleph.*). A cross-connexion terminal frame for connecting trunks between ranks of selectors.

trunking (*Teleph.*). The cables which contain the links between one rank of selectors and others in the sequence of operation, the cables taking a common route through the exchange building.

trunking diagram (*Teleph.*). A diagram which indicates the cable routes between the various groups of telephone switching apparatus in an automatic telephone exchange.

trunk junction circuit (*Teleph.*). The junction between an exchange and the trunk exchange for routing subscribers to the trunk exchange system.

trunk line (*Teleph.*). See **trunk circuit.**

trunk main (*Cables*). See **trunk feeder.**

trunk-offering final selector (*Teleph.*). The final selector which enables a trunk operator to offer a trunk call to a subscriber although he may be engaged on another call.

trunk-offering selector (*Teleph.*). The first selector operated by impulse trains coming from a trunk operator who wishes to break into a conversation to offer a trunk connexion.

trunk piston (*Eng.*). A piston, long in relation to its diameter, used where there is no piston-rod or crosshead, the piston having to take the connecting-rod thrust; most I.C. engine pistons are of this type.

trunk position (*Teleph.*). A position in a trunk exchange for handling delayed trunk calls.

trunk-record circuit (*Teleph.*). The circuit to a trunk operator who records the long-distance connexion requirements of a subscriber, so that the latter may be called when the trunk line is available, and may be properly charged, when delay-working is in operation.

trunk-record position (*Teleph.*). A position in a trunk exchange at which particulars of a trunk call are recorded on a card which is routed by a trunk operator, who rings back the calling subscriber when a line is free. The record is made by any trunk operator when no-delay is in operation.

trunk road (*Civ. Eng.*). A main arterial road; legally one for which the responsibility for control and maintenance has been transferred from the local highway authorities to the central government.

trunnion axis (*Surv.*). The horizontal axis about which the telescope of a theodolite or tacheometer may be rotated on its trunnion bearings.

trunnion mounting (*Eng.*). A pair of short journals, supported in bearings, projecting coaxially from opposite sides of a vessel or cylinder required to pivot about their axis.

Truscon floor (*Build.*). A type of floor formed in monolithic reinforced or precast concrete.

truss (*Civ. Eng.*). A framed structure built up entirely from tension and compression members, arranged in panels so as to be stable under load; used for supporting loads over long spans. (*Surg.*) A surgical appliance consisting of a pad incorporated in a spring or belt for retaining a reduced hernia in place.

truss-beam (*Build.*). A framework acting as a beam.

trussed partition (*Carp.*). A partition which is framed so as to be self-supporting between its ends; used in cases where the floor is not strong enough to carry it.

trying plane (*Carp.*). A tool similar to the jack plane but about 22 in. (56 cm) long; used after the jack plane to obtain a straight and true surface.

trypaflavine (*Chem., Med.*). 3,6-Diaminomethylacridinium chloride, a dye used for the treatment of wounds, on account of its antiseptic and nontoxic properties. The formula is

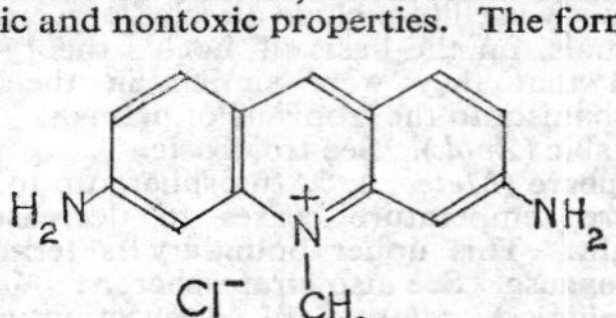

trypan blue stain (*Micros.*). A vital stain, which is

selectively absorbed by the macrophages of the reticulo-endothelial system.

trypano-. Prefix from Gk. *trypanon*, borer.

trypanomonad (*Zool.*). See **crithidial.**

trypanorhynchus (*Zool.*). In *Cestoda*, a protractile proboscis armed with hooks.

trypanosomes (*Zool.*). A group of flagellate *Protozoa* including many causing disease of man (see **trypanosomiasis**) and animals.

trypanosomiasis (*Med.*). Infection with trypanosomes. In man there are two African forms, Gambian (*Tr. zambiense*) and Rhodesian (*Tr. rhodesiense*) both spread by tsetse flies and both chiefly affecting the brain ('sleeping sickness'). For the American form, see **schizotrypanosomiasis.**

tryparsamide (*Chem., Pharm.*). Sodium N-phenylglycineamide-4-arsonate:

O
HO—As—⟨C₆H₄⟩—NH·CH$_2$·CONH$_2$
ONa

used in treatment of sleeping sickness.

trypsin (*Zool.*). A protein-digesting enzyme, an endopeptidase, of the alimentary canal of Vertebrates, secreted by the pancreas. *adj.* **tryptic.**

trypsinogen (*Zool.*). In Vertebrates, an inactive substance produced by the pancreas from which *trypsin* (q.v.) is formed by the action of the enterokinase of the succus entericus.

L-tryptophan (*Chem.*). $C_{11}H_{12}N_2O_2$, β-indole-2-aminopropanoic acid, obtained by the cleavage of certain proteins, e.g., of casein by pancreatic enzymes.

try square (*Carp., etc.*). A tool similar to the bevel but having the blade fixed at 90°.

T.S. (*Chem.*). Abbrev. for *test solution.*

Tschebycheff. See under **Chebyshev.**

tschermakite (*Min.*). An end-member subspecies in the hornblende group of amphiboles, rich in aluminium and calcium.

T-section cramp (*Join.*). Strong form of *sash cramp* (q.v.) having a T-section steel bar for the sliding members.

T-section filter (*Telecomm.*). T-network ideally formed of non-dissipative reactances, having frequency pass band over which attenuation is theoretically zero or very low. See **Butterworth filter, Chebyshev filter.**

tsetse fly disease (*Vet.*). See **nagana.**

tsutsugamushi fever (*Med.*). See **shimamushi fever.**

T-tail (*Aero.*). A *tail unit* characterized by positioning the horizontal stabilizer at or near the top of the vertical stabilizer. The unit is employed principally on high-performance aircraft, notably those having rear-mounted engines, and has variable incidence. See **all-moving tail.**

T.U. (*Teleph.*). Abbrev. for *traffic unit* and for *transmission unit.*

tub (*Mining*). A tram, wagon, corf, or corve.

tubbing (*Mining*). The lining of a circular shaft, formed of timber or by steel segments.

tubby (*Acous.*). An acoustically defective set in motion-picture or other recording; characterized by reverberant booming for frequencies which are familiar when barrels are struck.

tube (*Bot.*). The lower, cylindrical portion of a gamopetalous corolla. (*Civ. Eng.*) Abbrev. for *tube railway.* (*Electronics*) (1) Enclosed device with gas at low pressure, depending for its operation on ionization originated by electrons accelerated from a cathode by a field applied by an anode. (2) U.S. for all vacuum and gas-discharge devices. The term is now widely used in the U.K. where formerly *valve* was almost universally used.

tube-drawing (*Eng.*). The production of seamless tubes by drawing a large, roughly formed tubular piece of material through dies of progressively decreasing size.

tube extrusion (*Eng.*). A method of producing tubes by direct extrusion from a billet, using a mandrel to shape the inside of the tube.

tube-feet (*Zool.*). See **podium** (2).

tube fuse (*Elec. Eng.*). A fuse in which the fuse wire is enclosed in an insulating tube. Cf. *cartridge fuse.*

tube germination (*Bot.*). Germination of a spore by the formation of a hypha (germ tube).

tubeless tyre (*Autos.*). One in which the air-seal is provided by adhesion between the beads and the wheel-rim.

tube mill (*Min. Proc.*). Horizontal mill in which diameter/length ratio is usually high compared with that of standard ball mill, and which has high discharge.

tube nucleus (*Bot.*). A nongametic nucleus in a pollen tube which probably plays a part in regulating the development and behaviour of that organ.

tube of force (*Phys.*). Space enclosed by all the lines of force passing through a closed contour. Of unit magnitude when it contains unit flux.

tube plates (*Eng.*). The end walls of a surface condenser, between which the water tubes are carried; they are bolted between the casing and water-chamber covers. See **condenser tubes.**

tuber (*Bot.*). A swollen, underground stem, or less often a root, consisting mainly of parenchymatous cells containing much stored food material.

tube railway (*Civ. Eng.*). An underground electric railway running in tunnels of circular cross-section, and lined with cast-iron precast, or precast reinforced concrete, facing segments, each tunnel normally accommodating one track.

Tuberales (*Bot.*). An order of higher *Fungi* belonging to the *Ascomycetes* and characterized by a closed subterranean fruit body. Includes saprophytic, e.g., truffles, and mycorrhizal species.

tuber cinereum (*Zool.*). A nerve-centre of the diencephalon.

tubercle (*Bot.*). (1) A general name for a small swelling. (2) A small swelling on the roots of beans and other plants, inhabited by symbiotic bacteria. (*Med.*) (1) Any small rounded projection on a bone or other part of the body. (2) A solid elevation of the skin larger than a papule. (3) A small mass or nodule of cells resulting from infection with the bacillus of tuberculosis. (4) Loosely, tuberculosis; the tubercle bacillus. (*Zool.*) A small rounded projection; the dorsal articulator process of a rib; a cusp of a tooth. Also called **tuberculum.** *adjs.* **tubercled, tubercular, tuberculate, tuberculose.**

tubercular (*Med.*). Of, pertaining to, resembling, or affected with, nodules (tubercles); less correctly, affected with tuberculosis (i.e., *tuberculous*).

tubercularoid (*Bot.*). Having a warted surface.

tuberculate (*Bot.*). Provided with tubercles.

tuberculide, tuberculid (*Med.*). Any skin lesion due to infection with bacillus of tuberculosis.

tuberculiform (*Bot.*). Wartlike.

tuberculin (*Bacteriol.*). The endotoxin produced by *Mycobacterium tuberculosis*. Preparations of tuberculin are used in the diagnosis of tuberculosis in animals (*tuberculin test*) and man (*Mantoux test*).

tuberculoma (*Med.*). A slow-growing, circumscribed tuberculous lesion, sometimes present in the brain.

tuberculosis (*Med.*). Infection by *Mycobacterium tuberculosis*, especially of the lungs; characterized by the development of tubercles in the bodily tissues and by fever, anorexia, and loss of weight. Spread by air droplets and raw milk.

tuberculous (*Med.*). Pertaining to, affected with, or caused by, tuberculosis.

tuberculum olfactorium (*Zool.*). In the brain of the rabbit, slight rounded elevations on the ventral surfaces of the cerebral hemispheres behind the olfactory bulbs, connected to them by the olfactory tracts.

tuberiform, tuberous (*Bot.*). Having the form of a tuber.

tube ring (*Telecomm.*). Undesired ringing noise, sustained in an amplifying system because of continual mechanical impulsing of a microphonic valve from some external source.

tuberose (or **tuberous**) **sclerosis** (*Med.*). A condition in which hyperplasia of the neuroglia gives rise to hard, tumourlike masses in the brain, associated with epilepsy and mental deficiency; the disease is part of the developmental defect known as *epiloia* (q.v.).

tuberosity (*Zool.*). A prominence on a bone, generally for muscle attachment, especially prominences near the head of the humerus.

tuberous (*Bot.*). Thickened and forming tubers.

tubicolous (*Zool.*). Living in a tube.

tubifacient (*Zool.*). Tube-building, as certain *Polychaeta*.

tubiparous (*Zool.*). Tube-producing; said of certain glands in some tubicolous animals which secrete the material with which the tube is built.

tubocurarine (*Med.*). See **curarine.**

tub-sizing (*Paper*). The impregnation of paper with gelatine by passing the moving web through a bath and drying with hot air over spar drums.

tubular lamp (*Light*). An electric lamp in the form of a tube, the connexion to which is generally made at either end.

tubular rivet (*Eng.*). A rivet with a shank from which the centre has been removed to leave a thin wall, so that the rivet can be used to punch its own hole in thin, soft materials, and to facilitate setting.

tubular scaffold (*Build.*). A form of scaffold constructed of steel tubes which can be clamped together in any desired manner by special steel collar-pieces with screw fixings.

tubule, tubulus (*Bot.*). (1) The neck of a perithecium. (2) A pore lined by a hymenium-bearing basidia. (*Zool.*) Any small tubular structure. *adjs.* **tubulate, tubuliferous, tubuliform, tubulose.**

Tubulidentata (*Zool.*). An order of *Protoungulata*, the only surviving form having a long snout, round mouth, and long tongue, with peglike teeth with no enamel but hexagonal columns of dentine separated by pulp. Limbs specialized for digging, and olfactory turbinals more developed than in any other mammal. Feeds on termites. Aardvark.

tubuliform (*Bot.*). Composed of cylindrical, pipe-like filaments.

tucker, tucking blade (*Print.*). See **folding blade.**

tuck pointing (*Build.*). Pointing finished by cutting a groove in the surface at the joints and tucking into the groove a narrow projecting artificial joint of putty.

Tudsbury machine (*Elec. Eng.*). A form of electrostatic generator which operates inside a vessel containing compressed air, and is therefore able to employ smaller clearances for a given voltage.

tufa (*Geol.*). A porous, concretionary, or compact form of calcium carbonate which is deposited from solution around springs.

tuff (*Geol.*). A rock formed of compacted volcanic fragments, some of which can be distinguished by the naked eye. If the fragments are larger, then the rock grades into an agglomerate.

Tufnol (*Plastics*). A proprietary laminated plastic; lightweight, tensile strength approx. 55-110 MN/m²; strong insulation qualities. Widely used as an engineering plastic for bearings, gear wheels, and pulleys.

tufted (*Bot.*). Having many short crowded branches all arising at about the same level.

tularaemia, tularemia (*Med., Vet.*). A disease of rodents due to infection with *Bacterium tularense*, transmitted to Man by blood-sucking flies, fleas, or bugs, or directly from the infected rodent. In Man it is characterized by prolonged fever, enlargement of the lymphatic glands, or by a condition which resembles that of typhoid fever.

tulipwood (*For.*). Yellowish wood with reddish grain from several tree species including the Australian *Harpullia pendula*. Used for cabinet-work, moulding, etc.

Tullgren funnel (*Ecol.*). A device similar to the *Berlese funnel* (q.v.).

tumble-home (*Ships*). A term defining the narrowing of a ship's breadth. It is the measure of the inward fall when the deck breadth is less than the maximum breadth.

tumbler gear (*Eng.*). A gear in a train, mounted on a pivot arm so that it can be swung into and out of engagement with an adjacent gear.

tumbler switch (*Elec. Eng.*). A small single-pole switch having a quick-break action, universally used in electric-lighting installations for controlling individual lamp circuits.

tumbling bay (*Civ. Eng.*). A form of weir used to measure the rate of flow of water passing over it, or to act as an overflow dam diverting water in excess of a given discharge.

tumbling-in (*Build.*). A term applied to the brickwork forming the top surface of a pier and sloping in towards the general face of the wall.

tumbu disease (*Med.*). A disease, common in Central and West Africa, due to invasion of the surface of the body by the larvae of the tumbu fly *Cordylobia anthropophaga*; characterized by the formation of a boil or a warble in the skin.

tumefaction (*Med.*). The process or act of swelling; the state of being swollen. Also **tumescence**; *adj.* **tumescent.**

tumid (*Bot.*). Swollen; inflated.

tumour, tumor (*Med.*). Any swelling or morbid enlargement. The term now usually denotes neoplasm, a non-inflammatory mass formed by the growth of new cells in the body and having no physiological function. An *innocent tumour* is encapsulated and usually solitary, pressing upon, but not invading, adjacent tissues; a *malignant tumour* (*carcinoma, sarcoma*) invades tissues, tends to recur, and spreads to other parts of the body.

tunance (*Elec. Eng.*). See **shunt resonance.**

Tunbridge Wells Sand (*Geol.*). A succession of ferruginous sandstones which are almost unfossiliferous. It occurs in Southern England,

and forms the upper division of the Hastings Sand Group of the Wealden Beds, which are usually grouped in the Cretaceous System.

tundra (*Ecol.*). A biome which is essentially an Arctic grassland. The vegetation consists of lichens, grasses, sedges, and dwarf woody plants. It covers two large areas, one in the Palearctic, and one in the Nearctic region.

tune (*Acous.*). To adjust for resonance or syntony, especially musical instruments or radio receivers. See also **tuning.**

tuned amplifier (*Telecomm.*). One containing tuned circuits, and therefore sharply responsive to particular frequencies.

tuned anode (*Electronics*). An inductor shunted by a capacitor (either or both of which may be variable) in series with the lead to the anode of a thermionic valve.

tuned-anode coupling (*Electronics*). That between stages of a high-frequency thermionic valve amplifier, in which the coupling impedance is a tuned anode circuit.

tuned antenna (*Radio*). One operating at its natural resonant frequency.

tuned-base oscillator (*Electronics*). One in which the *tuned circuits* are in series with the base of a transistor (equivalent to the conventional triode-tuned grid oscillator).

tuned cell (*Telecomm.*). Adjustable cavity in a wave-guide structure, particularly in a wave-filter section.

tuned circuit (*Electronics*). One comprising an inductor (L henries) and a capacitor (C farads) in series (parallel) which offers a low (high) impedance to alternating current at the *resonant frequency* given by $f = 1/2\pi\sqrt{LC}$ Hz.

tuned-emitter oscillator (*Radio*). One in which the *tuned circuits* are in series with the emitter of a transistor.

tuned-grid circuit (*Electronics*). Parallel tuned circuit included between grid and cathode of a thermionic valve, to provide high step-up at the resonant frequency. See *Q*.

tuned magnetron (*Radio*). One capable of tuning to a range of frequencies.

tuned oscillator (*Radio*). Valve or transistor oscillator with frequency controlled by tuned circuit. Classified according to the electrode tuned, e.g., tuned anode (tuned plate), tuned anode tuned grid (tuned plate tuned grid), tuned base, tuned collector, etc. The transistor tuned base oscillator is equivalent to the tuned grid oscillator.

tuned radiofrequency (TRF) receiver (*Radio*). Receiver which does not use frequency changing before detection.

tuned relay (*Radio*). One which responds only at a resonant frequency.

tuned transformer (*Elec. Eng.*). Interstage coupling transformer in which one or more, usually both, windings are tuned to resonate with the signal frequency. A higher secondary voltage can be built up than would be the case without resonance.

tuner (*Telecomm.*). Assemblage of one or more resonant circuits, used for accepting a wanted signal and rejecting others.

Tungar rectifier (*Elec. Eng.*). A rectifier of the gas-discharge type, employing a thermionic cathode and operating in an atmosphere of inert gas, e.g., argon.

tung oil or **China wood oil** (*Chem.*). A yellow drying oil, obtained from the seeds of *Aleurites cordata*. M.p. 31°–44°C, rel. d. 0·936–0·943, saponification value 193, iodine value 150–165. Used in paints, varnishes, enamels.

tungsten (*Chem.*). Symbol W, at. no. 74, r.a.m. 183·85, rel. d. 19·1, m.p. 3370°C. A hard, grey metal, resistant to corrosion, used in cemented carbides for drills and grinding tools, and as wire in incandescent electric lamps. Also **wolfram.**

tungsten alloy (*Met.*). A protective material containing tungsten, copper, and nickel, and having a density about 50% greater than that of lead, and thus providing better protection from ionizing radiation.

tungsten arc (*Elec. Eng.*). A high-intensity arc of small dimensions, obtained between tungsten electrodes enclosed in a glass bulb.

tungsten blues or **tungsten bronzes** (*Chem.*). Colloidal pigments obtained by reduction of tungstates in aqueous solution at various temperatures with hydrogen, zinc, electrolysis, etc.

tungsten-halogen lamp (*Elec. Eng.*). See **quartz-iodine lamp.**

tungsten lamp (*Elec. Eng.*). An electric lamp employing an incandescent tungsten filament.

tungstic acid (*Chem.*). WO_3. The starting-point for the preparation of tungsten metal. Also called **tungstic oxide** and **tungsten (VI)oxide.**

tungstic ochre or **tungstite** (*Min.*). Hydrous (VI)oxide of tungsten, which probably crystallizes in the orthorhombic system. It is usually earthy and yellow or greenish in colour, and is a mineral of secondary origin, usually associated with wolframite.

tunic (*Zool.*). An investing layer. *adj.* **tunicate.**

tunica albuginea (*Cyt.*). See **albuginea.**

Tunicata (*Zool.*). See **Urochorda.**

tunicate, tunicated (*Bot.*). Having a coat or covering. (*Zool.*) Enclosed by a nonliving test or mantle.

tunicate bulb (*Bot.*). A bulb composed of a number of swollen leaf bases, each of which completely encloses all parts of the bulb inside it; an onion is a familiar example.

tunica vaginalis (*Zool.*). The serous layer covering the tunica albuginea of the testis.

tunicin (*Zool.*). A gelatinous substance, allied to cellulose, found in the test of *Urochorda.*

tuning (*Acous.*). (1) The pitch adjustment of one note to another, or to a specified frequency of oscillation. (2) The adjustment of tension in the strings of a piano, harp, or violin, so that the specified notes emitted coincide in frequency with a standard scale, e.g., concert pitch. (3) The adjustment of the length of pipes in organs to obtain the correct emitted pitch. See **aliquot.** (*Elec. Eng.*) See **current (or voltage) resonance.** (*Radio*) (1) Operation of adjusting circuit settings of a radio receiver so as to produce maximum response to a particular signal, generally by varying one or more capacitors and/or inductors. (2) Carrying out a similar process by electronic or thermal means. Also **tuning-in.**

tuning capacitor (*Radio*). Variable capacitor for tuning purposes, generally consisting of air-spaced vanes; several can be ganged.

tuning coil (*Radio*). See **tuning inductance.**

tuning control (*Radio*). Mechanical means for tuning a resonant circuit.

tuning curve (*Radio*). That relating the resonant frequency of a tuned circuit to the setting of the variable element, e.g., a capacitor.

tuning error (*Radio*). See **loop tuning error.**

tuning fork (*Acous.*). A fork with two tines and heavy cross-section, generally made of steel. Expressly designed to retain a constant frequency of oscillation when struck. See **maintained tuning fork.**

tuning-fork control (*Radio*). Control of the frequency of the wave emitted from a radio transmitter by a tuning-fork oscillator, fundamental of which can be multiplied by valve circuits.

tuning-fork oscillator (*Radio*). See **maintained tuning fork.**

tuning-in (*Radio*). See **tuning.**

tuning indicator (*Elec. Eng., Radio*). A simple cathode-ray tube in which a metal fin is employed as a control electrode. The opening or closing of a fluorescent pattern is the indication of balance. Used as a detector for a.c. bridges and as a tuning indicator in radio receivers. See **indicator tube.**

tuning inductance (*Radio*). Fixed or variable inductor used for tuning. Also called **tuning coil.**

tuning note (*Radio*). Steady musical note radiated from a broadcasting transmitter before a programme, to facilitate the tuning of receivers and for lining-up studio and land-line circuits.

tuning-out (*Radio*). Opposite of *tuning-in*, i.e., adjustment for minimum response to a signal, consistent with acceptance of another.

tuning screw (*Electronics*). A screw used to provide a variable reflection coefficient in a waveguide matching system. An alternative system to *stub tuning*.

tunnel burners (*Heat*). Industrial gas burners using a refractory tunnel at the burner exit for the main purpose of positive flame retention. The tunnel serves as an ignition zone, and accelerates the rate of flame propagation through turbulence and temperature rise, to a point where it is in equilibrium with the relatively high air-gas mixture velocity employed.

tunnel diode (*Electronics*). Junction diode with such a thin depletion layer that electrons bypass the potential barrier. Negative differential resistance is exhibited. It may be used for low noise amplification up to 1000 MHz. Also called **Esaki diode.**

tunnel effect (*Electronics*). Piercing of a narrow potential barrier by a current carrier which cannot do so classically, but, according to wave-mechanics, has a finite probability of penetrating.

tunnel furnace (*Met.*). Kiln through which material moves slowly on cars, racks, or suspending gear.

Tunnelite (*Civ. Eng.*). A form of rapid-hardening cement.

tunnelling (*Phys.*). See **potential barrier.**

tunnel slots (*Elec. Eng.*). See **closed slots.**

tunnel vault (*Build.*). See **barrel vault.**

tunnel windings (*Elec. Eng.*). A term sometimes applied to armature windings in which the conductors are inserted, end-on, into closed slots.

tup (*Civ. Eng.*). See **monkey.**

turacin (*Zool.*). In some Birds (*Turacos*, order *Cuculiformes*), a red pigment containing copper and an isomer of uroporphyrin. Soluble in dilute alkali.

turanose (*Chem.*). A disaccharide, formed from one molecule of fructose and one molecule of glucose, obtained by the hydrolysis of the trisaccharide melezitose.

Turbellaria (*Zool.*). A class of *Platyhelminthes* comprising forms of free-living habit, marine, freshwater, or terrestrial; with a ciliated ectoderm containing rhabdites; usually with a muscular protrusible pharynx and a pair of eye-spots; rarely have suckers. Planarians.

turbidimeter (*Powder Tech.*). Equipment for determining the surface area of a powder by measuring the light scattering properties of a fluid suspension.

turbidimetric analysis (*Chem.*). See **nephelometric analysis.**

turbidity (*Photog.*). Property of an emulsion whereby light is diffused into areas receiving no direct illumination.

turbinal (*Zool.*). Coiled in a spiral; one of certain bones of the nose in Vertebrates which support the folds of the olfactory mucous membrane.

turbinate (*Bot.*). Shaped like a top and attached by the point. (*Zool.*) In the form of a whorl or an inverted cone; as certain Gastropod shells.

turbinate bone. See **turbinal.**

turbine (*Aero.*). See **axial-flow turbine.**

turbine aero-engine (*Aero.*). See **by-pass turbojet, ducted fan, turbojet, turboprop.**

turbine blades (*Aero.*). The small aerofoils of heat-resisting alloy which extract pressure energy from the combustion products of a gas turbine. See **turbine buckets.**

turbine buckets (*Aero.*). (1) U.S. term for *turbine blades*. (2) The energy-extracting members of an impulse turbine.

turbinectomy (*Surg.*). Removal of a turbinal.

turbine disk (*Aero.*). The rotating member upon which the blades of a turbine are mounted.

turbine entry duct (*Aero.*). The duct which leads the combustion products into the turbine.

turbine shroud ring (*Aero.*). A peripheral ring which prevents gas escaping outward past the tips of turbine blades.

turbine wheel (*Aero.*). The assembly of the *turbine disk* and its blades.

turbo-compound aero-engine (*Aero.*). A reciprocating engine with an *exhaust-driven supercharger* (q.v.) of which any surplus turbine power is fed into the airscrew by a fluid drive or infinitely-variable gear.

turbo-convertor (*Elec. Eng.*). A combination of turbine-driven induction generator and rotary convertor, in which the rotor of the generator revolves at turbine speed; the stator, being in this case free to revolve also, is coupled to the armature of a rotary convertor.

turbo-dynamo (*Elec. Eng.*). A specially designed d.c. generator for direct coupling to a high-speed steam turbine.

turbo-electric propulsion (*Elec. Eng.*). A form of electric drive, used in marine and locomotive work, in which turbine-driven generators supply electric power to motors coupled to the propeller or axle shafts.

turbofan (*Aero.*). See **ducted fan.**

turbo-generator (*Elec. Eng.*). The arrangement of a steam turbine coupled to an electric generator for electric power production.

Turbogrid plates (*Chem. Eng.*). Proprietary plates for distillation columns which consist of a series of parallel narrow slots with or without downcomers.

turbojet (*Aero.*). An internal-combustion aero-engine comprising compressor(s) and turbine(s), of which the net gas energy is used solely for reaction propulsion through propelling nozzle(s). See also **by-pass turbojet, ducted fan, split compressor.**

turboprop (*Aero.*). A *shaft turbine* (q.v.) where the torque output is transmitted to an *airscrew* through a reduction gearbox; it may be of *single shaft*, *twin shaft*, or *free turbine* form. A constant-power, or supercharged, turboprop has an oversize compressor/turbine assembly which enables it to maintain full power up to a considerable altitude.

turbopump (*Aero.*). A combination *ram-air turbine* (q.v.) and hydraulic, or fuel, pump for a guided weapon or aircraft (emergency).

turboramjet (*Aero.*). An engine consisting of a

turbojet mounted within a *ramjet* duct, so that the efficiency of the former in subsonic flight is combined with the advantages of the latter at high supersonic speeds.

turborocket (*Aero.*). A composite engine in which a rocket propellant (an example would be high-test peroxide catalysed to super-heated steam and oxygen) is used to energize a turbine, which in turn drives a compressor, its air delivery joining the products from the turbine for combustion with a fuel to produce a propulsive jet. The object is to obtain a high ceiling, say 100 000 ft (30 000 m), without the enormous propellant consumption of a rocket.

turbo-starter (*Aero.*). An aero-engine starter in which rotation is imparted by a turbine motivated either by compressed air, a gas source, or the decomposition by catalysis of an unstable chemical, such as hydrogen peroxide.

turbo-supercharger (*Aero.*). See **exhaust-driven supercharger.**

turbulence (*Phys.*). See **turbulent flow.**

turbulent burner (*Eng.*). A pulverized-coal burner in which the coal-bearing primary air and secondary air pass through the burner, resulting in a short flaring flame.

turbulent flow (*Phys.*). Fluid flow in which the particle motion at any point varies rapidly in magnitude and direction. This irregular eddying motion is characteristic of fluid motion at high Reynolds' numbers. Gives rise to high drag, particularly in the *boundary layer* (q.v.) of aircraft. Also **turbulence.** See **streamline flow.**

turgescence (*Bot.*). The condition of cells or tissues which are distended with water. (*Med.*) The act or condition of swelling up: the state of being swollen. *adj.* **turgescent.**

turgid (*Bot.*). (1) Said of a cell which is distended and tense, well supplied with water. (2) Said of a young or soft plant member which is stiff and rigid owing to internal pressure arising from a plentiful supply of water.

turgidity (*Bot.*). The condition of rigidity when the cells of a plant member are distended and press against one another, owing to turgor pressure.

turgite (*Min.*). See **hydrohaematite.**

turgor (*Bot.*). The balance between the osmotic pressure of the cell sap and the elasticity of the cell wall.

turgor pressure (*Bot.*). The hydrostatic pressure set up within the cell by the water present acting against the elasticity of the wall.

turion (*Bot.*). A swollen perennating bud, containing much stored food, formed by a number of water plants; it comes away from the parent, remains inert during winter, and gives rise to a fresh plant in the following spring.

Turkey red (*Paint.*). A lighter red pigment obtained by calcining precipitated iron oxide.

Turkey-red oil (*Chem.*). Sulphonated castor oil, rel. d. 0·95, acid value 174, iodine value 82, saponification value 189. Used in dyeing.

turn (*Glass*). A work shift in which a definite number of articles, usually two *moves* (q.v.), is produced. (*Typog.*) See **taking out turns.**

turn-and-slip indicator (*Aero.*). A pilot's instrument for blind flying which indicates the rate of turn and correctness, or error, in banking; also **turn-and-bank indicator.**

turn bridge (*Civ. Eng.*). A *swing bridge* (q.v.) or *pivot bridge* (q.v.).

turnbuckle (*Eng.*). See **screw shackle.**

turn-down ratio (*Chem. Eng.*). The fraction or % of full throughput to which the actual throughput of a plant item can be reduced without loss of continuity. Particularly applied to gas and oil burners. The 'ratio' to 1 is always understood (1 representing 100% throughput).

turned sorts (*Typog.*). Characters purposely turned face-downwards so that the feet print prominent black marks in a proof, thus ensuring that missing letters shall be inserted later.

turner bars (*Print.*). See **angle bars.**

Turner's syndrome (*Cyt.*). A condition in humans in which a person looks superficially like a female but has only one X-chromosome.

turn indicator (*Aero.*). Any instrument that indicates the departure of an aircraft from its set course in a horizontal plane. Necessary for flying in clouds or at night.

turning (*Build.*). A term applied to the process of building an arch. (*Eng.*) Producing cylindrical or tapered workpieces in a lathe.

turning-bar (*Build.*). An iron bar supporting the arch over a fireplace opening.

turning-on (*Textiles*). See **beaming.**

turning-piece (*Build.*). A simple form of centring, consisting of a single solid wooden piece shaped to the form of an intrados, and supported in its temporary position by wooden struts.

turning-pin (*Plumb.*). See **tampin.**

turning point (*Surv.*). (1) The point which consecutive straight lines of a traverse meet at an angle. (2) See **change point.**

turning point on a curve (*Maths.*). A peak (maximum) or a trough (minimum) on a curve. For the curve $y=f(x)$, the point where $x=a$ is a turning point if $f(a+h)-f(a)$ is of constant sign for all values of h sufficiently small, i.e., $f'(a)=0$ and the first nonzero higher order derivative at $x=a$ must be of even order.

turnings (*Eng.*). Chips or swarf produced as waste in turning.

turning-saw (*Tools*). See **sweep-saw.**

turning tools (*Eng.*). See **lathe tools.**

turnout (*Rail., etc.*). The movable tapered rails or points by which a train or tram is directed from one set of rails to another.

turnover (*Nuc.*). In isotope separation, the total flow of material entering a given stage in a cascade. (*Radiol.*) Rate of renewal of a particular chemical substance in a given tissue. (*Telecomm.*) Reversing the legs of a balanced transmission circuit. This test is very important in all transmission measurements with balanced circuits, because if the same results are not obtained when any legs of the balanced system are interchanged, the presence of longitudinal currents is indicated, and no measurement can be accurate unless such currents are eliminated. Also called **poling.** (*Typog.*) The part of a divided word turned over into the next line, or a short line at the end of a paragraph.

turnover board (*Eng.*). A smooth square board on which an inverted bottom-half box is placed and rammed up round a pattern having a flat joint, thus saving the labour of making the facing joint. After turning over, removing the board, and adding facing sand, the top half may be rammed up at once.

turnover frequency (*Acous.*). In disk recording, the frequency, generally between 200 and 500 Hz, where the change from constant-amplitude to constant-velocity recording takes place. Also called **cross-over frequency.**

turns (*Horol.*). A small dead-centre lathe used by watchmakers. Usually held in a vice, and driven by a hand wheel or a bow. Used for pivoting, polishing, and turning small parts.

turnsick (*Vet.*). See **coenuriasis.**

turns ratio (*Elec. Eng.*). Ratio N of the turns in any pair of windings on a transformer. Power passing between windings changes its impedance level inversely as N^2, because the e.m.f. is proportional to the number of turns. Also called **transformer, transformation** or **voltage ratio.**

turnstile antenna (*Radio*). Two normal dipoles, crossed over at their centre, driven with equal currents in quadrature.

turntable (*Acous.*). The rotating table which supports the lacquer-blank during cutting and the processed record while being reproduced. It is of relatively high inertia, to keep down fluctuations of speed. (*Rail.*) A circular platform capable of rotation about its centre; used to reverse locomotives, which are driven on, turned through a half-circle, and driven off pointing the opposite way. In general, any such rotating platform, or system of rings rotating one inside the other.

turn tread (*Build.*). A tread, generally triangular in plan, to form a step at a change of direction of the stair.

Turonian (*Geol.*). A stratigraphical stage in the Upper Cretaceous rocks of Europe.

turpentine (*Chem.*). An essential oil, $C_{10}H_{16}$, obtained by the steam distillation of rosin. It is a colourless liquid, of aromatic pine-like odour; b.p. 155°–165°C, rel. d. 0·85–0·91; the chief constituent is pinene. American turpentine is dextrorotatory, others are usually laevorotatory. An important solvent for lacquers, polishes, etc.

turquoise (*Min.*). A hydrous phosphate of aluminium and copper which crystallizes in the triclinic system. It is a mineral of secondary origin, found in thin veins or small masses in rocks of various types, and used as a gem. The typical sky-blue colour often disappears when the mineral is dried. Much of the gem turquoise of old was fossil bone of organic origin and not true turquoise.

turret (*Eng.*). A turntable or wheel for carrying a number of alternative tools, e.g., in a turret lathe or a turret press.

turret clock (*Horol.*). A tower clock; a large clock in which the movement is quite separate from the dials.

turret lathe (*Eng.*). A high-production lathe for long workpieces, using a large number of tools carried on the revolving tool-holder or turret and on the cross slide. The turret is mounted on a saddle which slides on the lathe bed.

turret lenses (*Photog.*). Series of lenses, usually of different focal lengths, mounted on a rotatable turret so that any one of them may be instantly brought into use.

turret press (*Eng.*). A power press in which pairs of punches and dies of various sizes are held in upper and lower turrets, the turrets being geared together to bring corresponding punches and dies into positions of exact alignment in which they are then locked.

turret tuner (*TV*). As used in TV receivers, a drum carrying preset tuned circuits which can be rotated for selection.

turtle-shell (*Zool.*). The horny plates of the hawk's bill turtle. Commonly *tortoise-shell* (q.v.).

Tuscan red (*Paint.*). A purple shade of synthetic iron oxide, containing some lake pigment.

tusk tenon (*Carp.*). A form of tenon used for framing one horizontal piece into another, e.g., a trimmer into a trimming joist. The tenon is strengthened by a short projection underneath, and by a bevelled shoulder above, both fitting into a suitably cut mortise in the other piece.

tusks (*Build.*). See **tusses.**

tussah, tussore, or **tussur silk** (*Textiles*). Yarn or fabric produced from the silk of the wild silkworm, of which the tussah moth of India (*Antheraea mylitta*) is an example. The fabric is light brown and of rather irregular texture.

tusses (*Build.*). Stones left projecting from the face of a wall, when later extension is allowed for. Also called **tusks.**

tussive (*Med.*). Pertaining to, or caused by, a cough.

tussores (*Textiles*). Plain cotton dress cloth composed of fine warp and coarse weft. Woven from dyed yarns, or from mercerized yarns, then piece-dyed; made for Eastern and African markets.

tut-work (*Mining*). Work paid for according to the amount excavated, e.g., per fathom.

tuyère or **twyere** (*Met.*). A nozzle through which air is blown into a blast furnace. May be kept cool by circulating water.

Twaddell or **Twaddle scale** (*Chem.*). One for measuring the relative density of acids, etc., used in the trade. It is abbreviated °Tw, and the calculation in relation to relative density is x°Tw = (rel. d. − 1) × 200. The measurement is carried out at 60°F.

TW antenna (*Radio*). See **travelling-wave antenna.**

tweel blocks (*Glass*). (1) Refractory blocks placed in front of a newly-set pot. (2) Refractory blocks for controlling the flow of glass in forming machine canals. (3) Refractory blocks in a counterweighted vertical damper or door of a furnace.

tweeter (*Acous.*). A loudspeaker used in high-fidelity sound reproduction for the higher frequencies (> 5 kHz). See **cross-over frequency, woofer.**

twelvemo (*Print.*). Written *12mo*. The twelfth of a sheet or a sheet folded to make twelve leaves or twenty-four pages. Also **duodecimo.**

21 centimetre line (*Astron., Phys.*). A line in the radio spectrum of neutral hydrogen at 21·105 cm. It is caused by the spontaneous reversal of direction of spin of the electron in the magnetic field of the hydrogen nucleus, but it may be detected only in the vast hydrogen clouds of the Galaxy.

24-circuit combining equipment (*Teleph.*). In carrier telephony, equipment for combining a basic group (60–108 kHz) with a group in the range 12–60 kHz for transmission on a 24-circuit line link and for performing the reverse process.

twenty-four hour rhythm (*An. Behav.*). See **circadian rhythm.**

twenty-fourmo (*Print.*). Written *24mo*. The twenty-fourth part of a sheet, or a sheet folded four times to make twenty-four leaves or forty-eight pages.

twilight (*Astron.*). The period after sunset, or before sunrise, when the sky is not completely dark. Astronomical twilight is defined as beginning (or ending) when the sun is 18° below the horizon; hence twilight will last all night for a period in the summer months in all latitudes greater than about 48½°. See **civil-, nautical-.**

twilight sleep (*Med.*). A state of semi-consciousness produced by the administration of morphine and scopolamine; used for diminishing pain in labour and for producing forgetfulness of the event after it is over.

twilled mats (*Textiles*). See **Celtic twills.**

twills (*Textiles*). Fabrics with diagonal lines on the face. Regular twills have continuous lines; zigzag twills have the lines reversed at intervals.

twin. One of a pair of two and related entities similar in structure or function; often synonymous with *double*. See also **twins.**

twin cable (*Elec. Eng., etc.*). A lead-sheathed cable comprising two individually insulated conductors twisted together. A twin cable for telecommunication may have a large number of such pairs, e.g., up to 2400 pairs for telephone connexions between large exchanges.

twin-carbon arc lamp (*Elec. Eng.*). An arc lamp having 2 pairs of carbons, so arranged that the second pair comes into operation automatically as soon as the first pair has burnt away. Also called **double-carbon arc lamp.**

twin check (*Comp.*). Continuous check achieved by duplication of *hardware* and comparison of results.

twin columns (*Arch.*). Two columns springing from one base.

twin-concentric cable (*Elec. Eng.*). A 2-core cable in which the conducting cores are concentrically arranged about the axis of the cable.

twin crystal, twinned crystal (*Crystal., Min.*). A crystal composed of two or more individuals, either in contact or intergrown, in a systematic crystallographic orientation with respect to one another. See *interpenetration twins*, *juxtaposition twins*. Twinned crystals of quartz are unsuitable for use in radio oscillators.

twin diode (*Electronics*). Two independent thermionic *diodes* in one envelope. Cf. *binode*.

twiner (*Bot.*). A plant that climbs by winding around a support. (*Textiles*) Mule machine designed to double or cable yarns. The creel moves to and fro from the spindles.

twin feeder (*Radio*). Twin-wire transmission line, balanced to earth, on a pole-route.

twin flexible cord (*Elec. Eng.*). Two flexible cords plaited together to form *go* and *return* leads for pendant or portable electric fittings.

twin lamb disease (*Vet.*). See **pregnancy toxaemia.**

twin lens reflex camera (*Photog.*). A camera with matched lenses, one for exposing, the other for focusing, generally with a reflex mirror.

twinned grooves (*Acous.*). A defect in cutting a lacquer record, in which the grooves are not uniformly spaced, generally because of the flexing of part of the traversing mechanism.

twinning (*Crystal.*). Intergrowth of crystals of near-symmetry, such that (in quartz) the piezoelectric effect is not sufficiently determinate. See **twin crystal.** (*TV*) See **pairing.**

twin-plate process (*Glass*). A process for making polished plate glass in which rolling, annealing, and grinding are carried out on a continuously produced ribbon of glass without first cutting it into sections in which top and bottom surfaces are ground simultaneously. Also **Pilkington twin process.**

twinplex (*Teleg.*). In a radio-telegraph system, a method of diplex operation in which 1 of 4 conditions is transmitted, each condition representing a permutation of the 2 conditions of each channel.

twin-quad (*Elec. Eng.*). See **quad.**

twins (*Biol.*). (1) Individuals arising from the division into 2 of the fertilized egg, each part proceeding to develop. (2) In Mammals, 2 individuals produced at the same birth.

twin-shaft turbine (*Aero.*). See **split compressor.**

twin T-network (*Telecomm.*). One consisting of two T-networks which have their input terminal pairs connected in parallel and their output terminal pairs connected in parallel.

twin triode (*Electronics*). A combination of 2 triode valves within the same envelope.

T-wire (*Teleph.*). The tip-wire connected to the tips of the plugs which terminate the cords of an operator's cord-circuit, and which eventually connect with the A-wire of the subscriber's line.

twist (*Textiles*). (1) A cotton trade term for warp yarn. (2) The number of turns per inch inserted into a yarn by the revolving, spinning, or doubling spindle. (3) Yarn in which the twist conforms to 'Z' direction.

twist and steer (*Aero.*). Control of a guided weapon or *drone*, about the pitch and roll axes only, turns being achieved by rolling into a 90° bank so that the elevator can provide the required yawing moment. The system simplifies the autopilot and power requirements and is sometimes used with differentially-mounted variable-incidence wings, e.g., Bloodhound.

twist bit (*Tools*). Bit with long spiral cutting section, used for deep holes for dowels, etc.

twist cop (*Textiles*). A large mule-spun cop of warp and doubling weft yarn.

twist drill (*Eng.*). A hardened steel drill in which cutting edges, of specific rake, are formed by the intersection of helical flutes with the conical point which is backed off to give clearance; of universal application.

twisted aestivation (*Bot.*). See **contorted aestivation.**

twister (*Radar*). Plate with slats giving double reflection of a radar wave, one being half-wave retarded, to give a twist in direction of polarization of electric component of wave.

twist gimlet (*Carp.*). A gimlet having a tapered shank for easy withdrawal.

twisting frame (*Spinning*). See **doubling frame.**

twisting paper (*Paper*). A long fibred paper suitable for waxing and intended for wrapping around sweets, toffee, etc.

twist lace (*Textiles*). A lace formed simply by twisting together bobbin and warp threads.

twistor (*Comp.*). Unit of a store, consisting of a set of copper wires crossing magnetic wires. At the point where pulses in dissimilar wires are coincident, helical magnetization is locally produced. For detection, a single pulse in a copper wire which over-rides reverse magnetic field if present as a stored *bit*.

twitch (*Vet.*). A noose for compressing the lip of a horse as a means of restraint.

twitty (*Textiles*). Cotton or woollen yarns and slubbings characterized by irregular twisted or uneven portions.

two-address program (*Comp.*). That which is used in nonsequential computers where each instruction must include the address of two registers—one for the operand and one for the result of the operation.

two-and-two (2-and-2) twill (*Textiles*). A weave in which all ends work 2-up and 2-down, the ends being lifted in consecutive order, thus producing a diagonal line in the cloth. Known in the cotton trade as *Harvard twill* or *sheeting twill*; in the woollen and worsted trades as the *cassimere twill*.

two-body force (*Phys.*). A type of interaction between 2 particles which is unmodified by the presence of other particles.

two-circuit prepayment meter (*Elec. Eng.*). A prepayment meter for use when the load is connected to 2 separate circuits, energy being charged at a different rate in each.

two-circuit tuner (*Radio*). One, formerly much used for receivers, in which the antenna circuit is tuned by an inductance and variable capacitor inductively coupled to a closed resonant circuit to which the detector is connected.

two-circuit winding (*Elec. Eng.*). An alternative name for **wave winding.**

two-coat work (*Build.*). Plastering in 2 coats—a first coat of coarse stuff, and a second coat of fine stuff.

two-colour process (*Photog.*). Any additive or

subtractive colour photographic process for obtaining colour positives with 2 records only, corresponding to 2 arbitrarily selected colour-bands. (*Print.*) The application of the subtractive process to printing for the reproduction of a 2-colour original.

two-core cable (*Elec. Eng.*). A cable containing 2 individually insulated conducting cores.

two-dimensional gas (*Chem.*). A unimolecular film whose behaviour in 2 dimensions is analogous, qualitatively and quantitatively, to that of an ordinary gas in 3 dimensions.

two-edged (*Bot.*). Flattened and having 2 sharp edges.

two-electrode valve (*Electronics*). See **diode.**

two-fluid theory (*Elec.*). The obsolete theory of electrostatics which regarded positive and negative electricity as 2 separate fluids which neutralized each other when present in equal quantities. Also called **double-fluid theory.**

two-group theory (*Nuc.*). Simplified theoretical treatment of neutron diffusion in which only 2 energy groups are considered, i.e., partly-thermalized neutrons are neglected.

two-hinged arch (*Build.*). An otherwise continuous rigid arch, hinged at the abutments.

two-light frame (*Join.*). A window frame having 1 mullion dividing the window space into 2 compartments.

two-lipped (*Bot.*). See **bilabiate.**

two-magazine mixer (*Typog.*). Certain models of *Linotype* and *Intertype* and the *Fotosetter* are able to mix in 1 line matrices from 2 adjacent magazines, thus making available up to 9 alphabets.

two-motion selector (*Teleph.*). An electromechanical selector having vertical and rotary motions.

two-part prepayment meter (*Elec. Eng.*). A prepayment meter in which the supply is cut off by the combined action of a time element (to collect the fixed charge) and an energy element (to collect the running charge).

two-part step-rate prepayment meter (*Elec. Eng.*). A prepayment meter combining the functions of a 2-part prepayment meter and a step-rate prepayment meter.

two-part tariff (*Elec. Eng.*). An electricity tariff divided into a fixed charge, calculated per annum, and a variable charge for the actual amount of electrical energy consumed in a given period.

two-phase (*Elec. Eng.*). A term applied to a.c. systems employing 2 phases, whose voltages are displaced from one another by 90 electrical degrees.

two-phase four-wire system (*Elec. Eng.*). A system of 2-phase a.c. distribution employing 2 conductors per phase.

two-phase three-wire system (*Elec. Eng.*). A system of 2-phase a.c. distribution in which 2 conductors (lines) belong 1 to each phase, and the 3rd (neutral) is common to both phases.

two-pin plug (*Elec. Eng.*). A plug for connecting a twin-flex lead to a socket forming the terminal of a 2-wire electric circuit.

two-position action (*Automation*). Control action of *bang-bang* type in which control element always takes one of two fixed positions.

two-rate meter (*Elec. Eng.*). A meter for use with a 2-part tariff.

two-rate two-part prepayment meter (*Elec. Eng.*). A 2-part prepayment meter in which the running charge collected by the energy element is automatically changed to a lower rate per unit during certain hours of the day.

two-reaction theory (*Elec. Eng.*). A theory used in calculations on salient-pole synchronous machines; the m.m.fs. in the machine are assumed to be divided into 2 components, one acting along the axis of the main poles, the other at 90° to this.

two-revolution (*Print.*). Type of letterpress machine in which the cylinder revolves continuously, making 2 revolutions while the carriage reciprocates once; the cylinder is pulled down on the bearers for the printing revolution, and rises clear of the forme during a second revolution, while the forme returns to the printing position. Cf. *single-revolution, stop-cylinder.*

two set (*Print.*). Plating a rotary press with two sets of plates to produce two copies for each cylinder revolution.

two-spool compressor (*Aero.*). See **split compressor.**

two-stage pressure-gas burner (*Heat.*). Natural-draught type designed for operating with gas under pressure, normally about 35 kN/m², and having primary and secondary air inspirating stages in the injector.

two-start thread (*Eng.*). See **double-threaded screw.**

two-step relay (*Teleph.*). Telephone relay which is partially operated by a weak current, and so makes an *x-contact* or *fly-contact*, thereby closing a winding in a local circuit. This passes sufficient current for full operation of the remaining contacts of the relay, which *locks.*

two-stroke cycle (*I.C. Engs.*). An engine cycle completed in 2 piston strokes, i.e., in 1 crankshaft revolution, the charge being introduced by a blower or other means, compressed, expanded, and exhausted through ports in the cylinder wall, before and during the entry of the fresh charge. See **Otto cycle, diesel cycle.**

two-terminal pair network (*Telecomm.*). See **quadripole.**

two-tone (*Textiles*). Term given to any doubled or cabled yarn, or fabric in which the different yarns have differing dyeing properties.

two-tone keying (*Teleg.*). Keying of modulated continuous wave through a circuit which changes the modulation frequency only.

two-to-one folder (*Print.*). On a web-fed press a type of folder in which the folding cylinder has a circumference of two cut-offs and the cutting cylinder 1 cut-off giving a ratio of cylinder sizes 2 : 1. Cf. *three-to-two folder.*

two-up (*Print.*). A printing surface made up to print 2 copies at 1 impression; can also be arranged for 3-up, 4-up or any suitable number.

two-way circuit (*Telecomm.*). Bidirectional channel, which operates stably in either direction; used when remote sources are contributing a montage to a programme.

two-wire circuit (*Telecomm.*). A circuit in which go and return wires take equal currents, with potentials balanced with respect to earth.

two-wire repeater (*Teleph.*). A repeater for insertion into a 2-wire telephone circuit, in which the 2 amplifiers, 1 for amplifying the telephonic current in each direction, are connected into the 2-wire line by hybrid coils.

two-wire system (*Elec. Eng.*). A system of d.c. transmission and distribution making use of 2 conductors.

twyere (*Met.*). See **tuyère.**

tychopotamous (*Ecol.*). Occurring in a river fauna, but derived from a pond fauna.

tye (*Min. Proc.*). *Strake* in which a considerable thickness of low-grade concentrate is collected.

tyler (*For.*). A tight-line system designed for swinging across deep narrow gorges, operated with a special carriage and fall block, and a 4-drum power-unit carrying 2 main lines, one of which is used solely for raising and lowering the fall block.

Tyler sieves (*Min. Proc.*). Widely used series of laboratory screens in which mesh sizes are in $\sqrt{2}$ progression with respect to linear distance between wires.

tylosis (*Bot.*). A bladderlike growth of a parenchymatous cell through a pit into the lumen of a neighbouring vessel or tracheide; the vessel or tracheide becomes blocked by the tylosis, and ceases to function as a conducting element. Also **tylose.** (*Zool.*) The development of irregular cells in a cavity.

tympan (*Print.*). In a hand-press, the frame on which the paper is placed when printing. In printing machines the sheets of paper used to adjust impression, the outermost being tympan paper (a strong sulphite or manilla), oiled or plain.

tympan hooks (*Print.*). In a hand-press, thumb-hooks used for locking the outer and inner tympans together.

tympanic bulla (*Zool.*). In some Mammals, a bony vesicle surrounding the outer part of the tympanic cavity and external auditory meatus formed by the expansion of the tympanic bone.

tympaniform (*Bot.*). Drumlike.

tympanites (*Med.*). Distension of the abdomen by accumulation of gas in the intestines or in the peritoneal cavity. (*Vet.*) Rapid distension of the rumen and reticulum of cattle, due to the formation of gases.

tympanohyal (*Zool.*). In the hyoid apparatus of some Mammals, including Man, a small bone which, together with the stylohyal, becomes fused to the periotic and tympanic bones, forming the styloid process.

tympanum (*Arch.*). The triangular or segmental space forming the central panel of a pediment. (*Zool.*) A drumlike structure; in some Insects, the external vibratory membrane of a chordotonal organ; in some members of the Grouse family (*Tetraonidae*), an inflatable air-sac of the neck-region; in Vertebrates, the middle ear, or the resonating membrane of the middle ear; in Birds, the resonating sac of the syrinx. *adjs.* **tympanic, tympanal.**

Tyndall effect (*Light*). Scattering of light by very small particles of matter in the path of the light, the scattered light being mainly blue.

tyndallimetry (*Chem.*). The determination of the concentration of suspended material in a liquid by measurement of the amount of light scattered from a Tyndall cone.

type (*Biol.*). The individual specimen on which the description of a new species or genus is based; the sum-total of the characteristics of a group. *adj.* **typical.** (*Psychol.*) In terms of personality theory, a major classification of *traits* and characteristics by which it is allegedly possible to typify certain individuals. (*Typog.*) A rectangular piece of metal on top of which is cast the character. There are several styles, each further subdivided, including: Roman, *Italic*, **Slab Serif**, Sans Serif, Black Letter.

type-0, 1, and **2** (*Automation*). See **servomechanism types.**

type face (*Typog.*). A particular family or fount of type in which the characters have distinctive features. Type faces used in modern English bookwork include Aldine Bembo, Baskerville, Caslon, Ehrhardt, Fournier, Garamond, Gill Sans, Imprint, Perpetua, Plantin, Times Roman, Univers. Each type face has its own special characteristics.

type family (*Typog.*). A range of variations on the basic type design—light, medium, bold, extra bold, condensed, etc., a notable example being Gill Sans.

type-high (*Typog.*). When a printing plate or block is mounted on wood or metal and brought to the proper height for printing it is said to be *type-high*, the British/American standard being 0·918 in (23·32 mm). *n.* **type-height.**

type holder (*Bind.*). A hand tool for holding the letters to be impressed on the cover of a book by the finisher.

type locality (*Geol.*). The locality from which a rock, formation, etc., has been named and described, usually because of its characteristic occurrence there.

type metal (*Met.*). A series of alloys of lead, antimony, and tin, used for type. One composition is antimony 10–20%, tin 2–12%, and the remainder lead. In another, tin may reach 26%, and up to 1% of copper may be added.

type number (*Zool.*). The most frequently occurring chromosome number in a taxonomic group.

typesetting machines (*Typog.*). See **composing machines.**

type specimen (*Biol.*). The actual specimen from which a given species was first described.

typewriter composition (*Typog.*). The use of electric typewriters to produce typesetting for reproduction by printing processes. The product has lower typographic standards than normal typesetting or the major filmsetting systems but usually has justified lines and a difference of set between wide and narrow letters. Among systems and accessories in use are Justi-Gage, Justowriter, Marginator, NHM (No Hot Metal), Optype, VariTyper.

typhlitis (*Med.*). Inflammation of the caecum.

typhlosole (*Zool.*). In some Invertebrates, a longitudinal dorsal inwardly projecting fold of the wall of the intestine, by which the absorptive surface is increased.

typhoid, typhoid fever (*Med.*). The most serious form of enteric fever (which includes *paratyphoid*) due to infection with the bacillus *Eberthella typhi*. Prolonged fever, a rose rash and inflammation of the small intestine with ulceration occur. Infection is by faecal contamination of food or water.

typhoon (*Meteor.*). A *tropical revolving storm* (q.v.), in the China Sea and Western North Pacific.

typhus, typhus fever (*Med.*). Infection with the rickettsa *Rickettsia prowazeki* derived from the bite of lice, an acute epidemic fever with high mortality. There are varieties of lesser severity (scrub t., murine t., etc.) due to related organisms.

typographer (*Typog.*). A specialist in typographic design who may also be a compositor.

typography. The art of arranging the printed page, including choice of type, illustration, and method of printing.

tyramine (*Chem.*, *Med.*). 1-hydroxy-4 (aminoethyl) benzene:

$$HO-\langle\text{benzene ring}\rangle-CH_2.CH_2\,NH_2$$

formed in decaying animal products, including cheeses, by the action of bacteria on *tyrosine*. Used as a diagnostic agent and also has a mild action in increasing blood pressure.

tyre (*Eng.*). (1) A renewable, forged steel, flanged ring, shrunk on the rim of a locomotive wheel.

(2) A steel or rubber band or air-filled tube of rubber (*pneumatic tyre*) surrounding a wheel to strengthen it or to absorb shock. See **cross-ply, radial-ply,** and **tubeless tyre.**

tyre fabrics (*Textiles*). Strongly constructed cloths of cotton or manmade fibres used in the construction of pneumatic tyres.

tyrosinase (*Chem.*). A group of enzymes found in various plants, which oxidize tyrosine to the dark pigment melanin. A similar group in animals produces the melanin of hair and skin.

L-tyrosine (*Chem.*). An amino-acid, β- (4-hydroxyphenyl) alanine: $HO{\cdot}C_6H_4{\cdot}CH_2{\cdot}CH(NH_2){\cdot}COOH$, found in many proteins and also formed *in vivo* by the oxidation of phenylalanine. It is used in the animal body for the production of adrenalin, thyroxin, and melanin.

Tzaneen disease (*Vet.*). A febrile disease, usually mild, of cattle and buffalo in Africa, caused by the protozoon *Theileria mutans*; transmitted by ticks.

U

u Symbol for unit of *unified scale* (q.v.).

u (*Chem.*). (1) With subscript, a symbol for velocity of ions. (2) Symbol for *specific internal energy*.

U (*Chem.*). The symbol for *uranium*. (*Vac. Tech.*) See **conductance.**

U (*Chem.*). A symbol for *internal energy*. (*Elec.*) Symbol for *potential difference, tension*.

U-bend (*San. Eng.*). See **air trap.**

ubiquinones (*Biochem.*). Compounds containing polyisoprenoid chains of varying lengths attached to a substituted 1, 4-quinone nucleus; they act as electron carriers in the electron transport chain.

U-bolts (*Autos.*). Bars bent into U-shape and threaded at each end; used for anchoring a semi-elliptic spring to an axle beam, a plate being threaded over the ends and secured by nuts.

U-centre (*TV*). See **colour centre.**

UCP/21 (*Chem.*). See **etem.**

UCR (*An. Behav.*). Abbrev. for *unconditioned reflex*. See also **classical conditioning.**

UCS (*An. Behav.*). Abbrev. for *unconditioned stimulus*.

udder (*Vet.*). The popular name for the mammary glands of certain animals, e.g., of the cow, mare, sow, and ewe.

udometer (*Meteor.*). A *rain gauge* (q.v.).

Uehling force (*Phys.*). Interaction arising from vacuum polarization.

U.F. (*Plastics*). Abbrev. for *urea-formaldehyde* plastics (see **urea resins**).

uferflucht (*Ecol.*). A concentration of limnetic plankton towards the centre of a lake.

Uganda mahogany (*For.*). Hardwood from *Khaya anthotheca*. The timber seasons well and is used for cabinet-making, veneering, plywood manufacture, high-class joinery, and furniture.

ugandite (*Geol.*). An alkaline, ultramafic, volcanic rock composed of olivine with subordinate augite, leucite, perovskite, biotite and iron oxide, in a matrix of dark glass.

ugrandite (*Min.*). A group name for the *u*varovite, *gr*ossular and *and*radite garnets.

UHF (*Radio*). Abbrev. for *ultra-high frequencies*.

uintaite (*Min.*). A variety of natural asphalt occurring in the Uinta Valley, Utah, as rounded masses of brilliant black solid hydrocarbon. Also called **gilsonite.**

Ulbricht sphere photometer (*Light*). A photometer for measuring directly a lamp's mean-spherical candle-power. It comprises a hollow sphere, whitened inside, with the lamp under test at the centre. Owing to the internal reflection, the illumination on any part of the sphere's inside surface is proportional to the lamp's total light output, measured through a small window.

ulcer (*Med.*). A localized destruction of an epithelial surface (e.g., of the skin or of the gastric mucous membrane), forming an open sore; it is usually a result of infection.

ulceration (*Med.*). The process of forming an ulcer; the state of being ulcerated.

ulcerative (*Med.*). Of the nature of, or pertaining to, ulcers; causing ulceration; associated with ulceration (e.g., *ulcerative colitis*).

ulcerative cellulitis (*Vet.*). See **ulcerative lymphangitis.**

ulcerative dermal necrosis (*Vet.*). A fungus disease of salmon.

ulcerative lymphangitis (*Vet.*). Ulcerative cellulitis. A chronic contagious lymphangitis of the horse, due to infection by *Corynebacterium pseudotuberculosis* (*C. ovis*).

U-leather, U-packing (*Eng.*). A flexible ring of U-section, used to pack the glands of fluid power pistons or rams. The open side is the pressure side, and the fluid in expanding the inner face against the ram, effectively seals it against leakage.

ulexite (*Min.*). A hydrated borate of sodium and calcium occurring in borate deposits in arid regions, as in Chile and Nevada, where it forms rounded masses of extremely fine acicular white crystals. Also called **cotton ball.**

uliginose, uliginous (*Bot.*). Growing in places which are wet.

U-links (*Telecomm.*). The spring links used to join isolated parts of telegraph or other communication channels, the ends of which are brought to a special link-board. Removal of a link opens the circuit, so that testing apparatus can be rapidly inserted to detect and locate faults.

ullage (*Brew.*). The dregs left in a cask after the saleable liquor has been drawn off. (*Ships, etc.*) The quantity that a container (e.g., a tank for liquid cargo or fuel) lacks of being full.

ullmanite (*Min.*). Nickel(II)antimony(III)sulphide. It crystallizes in the cubic system and occurs in hydrothermal veins.

ulna (*Zool.*). The post-axial bone of the antebrachium in land Vertebrates. *adj.* **ulnar.**

ulnare (*Zool.*). A bone of the proximal row of the carpus in line with the ulna. *pl.* **ulnaria.**

ulotrichous (*Zool.*). Having woolly or curly hair.

Ulsterian (*Geol.*). The lower of the two divisions of the Middle Devonian rocks in N. America; it is succeeded by the Erian stage.

ultimate analysis (*Chem.*). *Quantitative analysis* (q.v.) of the elements in the materials being examined.

ultimate limit switch (*Elec. Eng.*). See **final limit switch.**

ultimate load (*Aero.*). The maximum load which a structure is designed to withstand without a failure; see also **load factor, limit load** and **proof load.**

ultimate pressure (*Vac. Tech.*). The limiting pressure approached in a vacuum system after pumping for sufficient time to establish that further reductions in pressure will be negligible. The term *blanked-off pressure* is sometimes used for the special case of a pump with the intake isolated.

ultimate tensile stress (*Met.*). The highest load applied to a metal in the course of a tensile test, divided by the original cross-sectional area. In brittle or very tough metals it coincides with the point of fracture, but usually extension continues under a decreasing stress, after the ultimate stress has been passed. Also called **tenacity.**

ultimobranchial bodies (*Zool.*). In Vertebrates, a pair of glandular bodies, arising from the last pair of gill-pouches, and possibly functioning in Fish like the parathyroid.

ultor (*Electronics*). Anode, especially in a cathode-ray tube, which has highest potential with respect to cathode.

ultra-. Prefix from L. *ultra*, beyond.
ultra-abyssal zone (*Ocean.*). See **hadal zone.**
ultra-audion oscillator (*Telecomm.*). A Colpitts oscillator in which the resonant circuit is a section of transmission line.
ultra-basic rocks (*Geol.*). Igneous rocks containing less silica than the basic rocks (i.e., less than 45%), and characterized by a high content of mafic constituents, particularly olivine (in the peridotites) and amphiboles and pyroxenes (in the perknites and picrites).
ultra-centrifuge (*Chem.*). A high-speed centrifuge for the separation of submicroscopic particles.
ultradextral (*Zool.*). Said of sinistral Gastropod shells when the organization of the animal is dextral.
Ultrafax (*Telecomm.*). TN of system for very high speed transmission of printed information, which utilizes radio, TV facsimile and film recording.
ultra-filtration (*Chem.*). The separation of colloidal or molecular particles by filtration, under suction or pressure, through a colloidal filter or semi-permeable membrane.
ultra-high frequencies (*Radio*). Those frequencies between 3×10^8 Hz and 3×10^9 Hz. Abbrev. UHF.
ultralinear (*Elec. Eng.*). Said of a power amplifier when nonlinear distortion is reduced to a very low value, e.g., when a pentode valve has negative feedback applied to its screen.
ultramafic rocks (*Geol.*). Igneous rocks in which there is an abnormally high content of ferromagnesian silicates, but which contain no feldspar; subdivided into picrites (with accessory plagioclase), pyroxenites, and peridotites.
ultramarine or **ultramarine blue** (*Paint.*). A brilliant blue pigment, of poor opacity and low tinting strength, derived either from the natural mineral *lapis lazuli* (q.v.), or manufactured artificially by calcining together silica, china clay, sulphur, soda ash, and other minor ingredients.
ultramicroscope (*Optics*). An instrument for viewing particles too small to be seen by an ordinary microscope, e.g., fog or smoke particles. An intense light projected from the side shows, against a dark background, the light scattered by the particles.
ultramicrotome (*Micros.*). A modified microtome developed for cutting ultra-thin sections for examination with the electron microscope. The cutting surface may be of steel, but is more usually glass or diamond, and the movement of the specimen block towards the knife is very delicately controlled, e.g., by the thermal expansion of a rod.
ultramicrowaves (*Radio*). Submillimetre waves having frequencies between 300 GHz and 3000 GHz.
ultra-short waves (*Radio*). Electromagnetic waves of wavelength less than 10 metres.
ultrasonic (*Acous.*). Said of frequencies above the upper limit of the normal range of hearing, at or about 20 kHz. **Ultrasonics** is the general term for the study and application of ultrasonic sound and vibrations.
ultrasonic cleaning (*Eng.*). A cleaning process effective for small crevices, blind holes, fine screw threads, etc., in which the part is immersed in water or solvent vibrated at ultrasonic frequency. (*Phys.*) The separation of dirt and other foreign particles from a substance by subjecting it to ultrasonic irradiation. Solid bodies such as glass lenses can be polished in this way.
ultrasonic coagulation (*Phys.*). Under suitable conditions, coalescence of particles into large aggregates by ultrasonic irradiation.
ultrasonic delay line (*Telecomm.*). A device which utilizes the finite time for the propagation of sound in liquids or solids to produce variable time delays. Such systems may also be used for storage in digital computers. Mercury or quartz are used as transmitting media, and nickel wire for magnetostrictive delay lines.
ultrasonic depth finder (*Instr.*). Instrument for measuring or displaying the depth of water under a ship by measuring the time of propagation of a pulse of ultrasonic waves to the sea bed and back.
ultrasonic detector (*Telecomm.*). Electro-acoustic transducer for the detection of ultrasonic radiation.
ultrasonic disintegrator (*Biochem.*). An apparatus for cell disintegration, e.g., the disruption of micro-organisms, red blood cells, and the preparation of emulsions and cell fractions. The basis of the apparatus is either a ring-shaped transducer surrounding a beaker containing the cell suspension, or a probe, emitting the ultrasonic radiation, dipping into it.
ultrasonic dispersion (*Chem.*). High-intensity ultrasonic waves can produce a dispersion of one medium in another, e.g., mercury in water.
ultrasonic generator (*Elec. Eng.*). One for the generation of ultrasonic waves, e.g., quartz crystal, ceramic transducer, supersonic air jet, magnetostrictive vibrator.
ultrasonic grating (*Phys.*). The presence of acoustic waves in a medium leads to a periodic spatial variation of density which produces a corresponding variation of refractive index. Diffraction spectra can be obtained in passing a light beam through such a sound field.
ultrasonic machining (*Eng.*). Drilling, tapping, and other operations carried out, primarily on hard and brittle materials, by the use of abrasive grains carried in a liquid between the relatively soft tool and the workpiece. The tool, which must match the surface to be machined, changes its normal length by contraction and expansion; the tool holder is attached to a transducer and oscillates at ultrasonic frequencies to drive the abrasive grains into the workpiece.
ultrasonics (*Acous.*). See **ultrasonic.**
ultrasonic soldering (*Eng.*). Form of soldering in which a specially designed soldering bit replaces the machine tool. Soldering is particularly difficult with aluminium and the application of ultrasonics is supposed to break up the aluminium oxide layer.
ultrasonic sound apparatus (*Surg.*). Apparatus which produces very-high-frequency vibrations, used to destroy the semicircular canals in *Ménière's disease*.
ultrasonic stroboscope (*Eng.*). One in which an ultrasonic field is applied to the modulation of a light beam to obtain stroboscopic illumination. See **stroboscope.**
ultrasonic testing (*Eng.*). A method of testing in which an ultrasonic source is pressed against the part to be tested and the vibrations are reflected from a discontinuity or another surface of the part, to generate a signal in the receiver. The time lag of the echo is measured to determine the distance to the discontinuity or the workpiece thickness.
ultrasonic underwater communication (*Telecomm.*). Communication carried out underwater using a modified sonar system.
ultrasonic welding (*Eng.*). A solid-state process for bonding sheets of similar or dissimilar materials, usually with a lap joint. No heat is

applied but vibratory energy at ultrasonic frequencies is applied in a plane parallel to the surface of the weldment.

ultrastructure (*Biol.*). The submiscroscopic, ultimate structure of protoplasm.

ultraudion (*Radio*). Earliest form of detector circuit, using reaction from three-electrode valve.

ultraviolet cell (*Phys.*). A cell having a maximum response to light in the ultraviolet end of the spectrum.

ultraviolet microscope (*Micro.*). Instrument using ultraviolet light for illuminating the object. Its resolving power is therefore considerably increased as the resolution varies inversely with the wavelength of the radiation.

ultraviolet photography (*Photog.*). Photography using (*a*) an ultraviolet filter or (*b*) ultraviolet illumination and an ultraviolet-absorbing filter. Applied to the detection of fingerprints, invisible ink, alterations in documents, etc.

ultraviolet radiation (*Phys.*). Electromagnetic radiation in the wavelength range 4×10^{-7} m to 5×10^{-9} m approximately, covering the gap between X-rays and visible light rays.

ultraviolet sound recording (*Cinema.*). Use of ultraviolet light beam in order to reduce chromatic aberration and diffraction in photographic sound tracks.

ultraviolet spectrometer (*Phys.*). An instrument similar to an optical spectrometer but employing nonvisual detection and designed for use with ultraviolet radiation.

ultraviolet therapy (*Med.*). Treatment of disease by ultraviolet rays. The therapeutic rays (300–400 nm) are generated by two types of apparatus: carbon and tungsten arc lamps, and quartz mercury-vapour lamps.

ulvöspinel (*Min.*). An end-member species in the magnetite series, with composition ferrous and titanium oxide. First recognized in ore from Södra Ulvön, in northern Sweden.

umbel (*Bot.*). An inflorescence consisting of numerous small flowers in flat-topped groups, borne on stalks all arising from about the same point on the main stem; in most umbels this sort of branching is repeated, the stalks which bear the groups of flowers themselves arising at about the same point on a main axis.

umbellate (*Bot.*). Having the characters of an umbel; producing umbels.

umbellifer (*Bot.*). A plant which has its flowers in umbels.

umber (*Paint.*). Naturally occurring brown pigment, consisting principally of hydrated iron oxide, manganese dioxide, silica, carbon, and calcium carbonate.

umbilectomy (*Surg.*). Removal of the umbilicus.

umbilic (*Maths.*). The limit point of circular sections of a quadric as the radius tends to zero. Every quadric has four real and eight imaginary umbilics.

umbilical cord (*Anat., Zool.*). In eutherian Mammals, the vascular cord connecting the foetus with the placenta. (*Bot.*) See **funicle.** (*Space*) Term (frequently *umbilical*) applied to any flexible and easily disconnectable cable, e.g. for conveying information, power or oxygen to a missile or spacecraft before launching, for connecting an operational spacecraft with an external astronaut, or an astronaut's spacesuit with his backpack.

umbilical point on a surface (*Maths.*). One at which the curvatures of all normal sections are positive and constant. Cf. *elliptical*, *hyperbolic* and *parabolic point on a surface*.

umbilicate, umbilicated (*Bot.*). Having a small central depression. (*Med.*) Having a depression which resembles the umbilicus.

umbilicus (*Anat., Zool.*). In Gastropod shells, the cavity of a hollow columella; in Birds, a groove or slit in the quill of a feather; in Mammals, an abdominal depression marking the position of former attachment of the umbilical cord. *pl.* **umbilici.**

umbo (*Bot.*). (1) A small central hump in the middle of the top of the pileus of an agaric. (2) A small projection on the apophysis of the scale of a pine cone. (*Zool.*) A boss or protuberance; the beaklike prominence which represents the oldest part of a Bivalve shell. *pl.* **umbones.**

umbonate (*Bot., Zool.*). Bearing, resembling, or pertaining to an umbo.

umbra (*Astron.*). Region of complete shadow of an illuminated object, e.g., dark central portion of the shadow of the earth or moon. Generally applied to eclipses of the moon or of the sun, the term is also applied to the dark central portion of a sunspot. The outer, less dark, shadow is known as the **penumbra.**

umbraculiform (*Bot.*). Shaped like an umbrella.

umbrella (*Zool.*). A flat cone-shaped structure, especially the contractile disk of a medusa.

umbrella antenna (*Radio*). An antenna comprising a vertical uplead from the top of which a number of wires extend radially towards the ground.

umbrella roof (*Build.*). See **station roof.**

umbrella-type alternator (*Elec. Eng.*). A vertical-shaft alternator, driven by a water turbine, in which the field system is overhung and revolves around the stationary armature.

umbrine (*Bot.*). Dull darkish-brown.

Umklapp process (*Phys.*). A type of collision between phonons, or between phonons and electrons, in which crystal momentum is not conserved. Such processes provide the greater part of the thermal resistance in solid dielectrics.

Unaflow (or **Uniflow**) **engine** (*Eng.*). A steam-engine in which the steam enters through drop valves at the ends of the cylinder and exhausts through a piston-controlled belt of ports at the centre.

unarmed (*Bot.*). Without thorns or prickles.

unarmoured cable (*Cables*). A cable in which the outer covering of steel wire (armouring) is absent.

unary (*Chem.*). Consisting of one component, etc.

unavailable energy (*Eng.*). The energy which becomes unavailable to do work, in the course of an irreversible process.

unbalanced (*Elec. Eng.*). (1) Of a bridge circuit in which detector signal is not zero. (2) Of a pair of conductors in which magnitudes of voltage or current are not symmetrical with reference to earth.

unbalanced circuit (*Elec. Eng.*). One whose two sides are inherently unlike. See **balanced circuit.**

unbalanced load (*Elec. Eng.*). A load which is unequal on the two sides of a three-wire d.c. system, or on the three phases of a symmetrical three-phase a.c. system.

unbalanced network (*Elec. Eng.*). One arranged for insertion into an unbalanced circuit, the earthy terminal of the input being directly connected to the earthy terminal of the output.

unbalanced system (*Elec. Eng.*). A three-phase a.c. system carrying an unbalanced load.

uncate, unciform, uncinate (*Bot., Zool.*). Hooked at the tip; hooklike.

uncertainty principle (*Phys.*). Pairs of quantities with the dimensions of action (e.g., time and energy) cannot be determined to an accuracy such that the product of the errors is less than Dirac's constant. Also known as **Heisenberg principle, indeterminacy principle.**

uncinate fit (*Med.*). A hallucination of smell or of taste, due to a cerebral tumour or to epilepsy.

uncinus (*Zool.*). A hook, or hooklike structure, e.g., a hooklike chaeta of *Annelida*; in *Gastropoda*, one of the marginal radula-teeth.

unconditional jump (*Comp.*). Jump involving rigorous instruction to alter the normal sequence of operation. See **conditional jump.**

unconditionally stable (*Telecomm.*). Of amplification in a system which continues to satisfy the Nyquist criterion when the gain is reduced.

unconditioned reflex (*An Behav.*). Term used to refer to the original reflex action involved in *classical conditioning* as a result of which it has become a *conditioned reflex*. Abbrev. **UCR.**

unconditioned stimulus (*An Behav.*). A stimulus normally eliciting a reflex action which has been associated with some other, previously neutral, stimulus in the process of *classical conditioning* (q.v.). Abbrev. **UCS.**

unconformity (*Geol.*). A break in the succession of stratified sedimentary rocks, following a period when deposition was not taking place. If the rocks below the break were folded or tilted before deposition was resumed, their angle of dip will differ from that of the overlying rocks, in which case the break is an **angular unconformity.**

unconscious (*Psychol.*). A general term used to include all processes which cannot be made conscious by direct effort of will. These processes are present in a dynamic state, and require a powerful force to maintain them in the unconscious. This force is supplied by the *super-ego* (q.v.), and by the *censor* (q.v.) present in dreams, causing distortion, displacement, and condensation, which are forms of *resistance* (q.v.). The unconscious contains the reservoir of instinct, real desires, and all experiences in the past history of the individual which have never been made completely conscious and resolved. Jung also includes the racial history of the individual, forming a racial or collective unconscious.

uncoursed (*Build.*). See **random.**

uncus (*Zool.*). In *Rotifera*, the head of the malleus; in some male *Lepidoptera*, a hooklike dorsal process of the ninth abdominal somite; any hook-shaped structure.

uncut (*Bind.*). Said of a book whose edges have been left untrimmed, the bolts therefore remaining *uncut*.

undamped oscillations (*Radio*). Same as **continuous oscillations.**

underback (*Brew.*). Vessel in which wort from the mash tun is collected.

under blanket (*Print.*). On web-fed relief presses the first dressing on the impression cylinder. Cf. *top blanket*.

underbunching (*Electronics*). Less than optimum efficiency in a velocity-modulation system.

undercarriage (*Aero.*). Each of the units (consisting of wheel(s), shock-absorber(s), and supporting struts) of an aircraft's alighting gear is an undercarriage, i.e., two main and either tail or nose undercarriages. Colloquially, the whole *landing gear*.

undercast (*Mining*). A duct to carry ventilating air under a roadway at a crossing.

underclay (*Geol.*). A bed of clay, in some cases highly siliceous, in many others highly aluminous, occurring immediately beneath a coal-seam, and representing the soil in which the trees of the Carboniferous swamp-forests were rooted. Stigmarian roots commonly occur as fossils in underclays, many of which are used as fire-clays. Also called **seat earth.**

undercloak (*Plumb.*). The first or lower sheet of lead in a *roll* (q.v.). Cf. *overcloak*.

undercoat (*Paint.*). Paint applied before the finishing coat. It is usually highly pigmented in order to provide good hiding power.

under-compensated meter (*Elec. Eng.*). An induction-type meter provided with insufficient phase compensation, as the result of which it reads high with leading currents and low with lagging currents.

undercontact rail (*Elec. Eng.*). A conductor-rail supported and protected from above, in order that the collecting-shoe of the electric tractor may make contact with the bottom surface.

undercut (*Eng.*). (1) A mould or pattern is undercut when it has a re-entrant portion and thus cannot be opened or withdrawn in a straight motion. (2) A short portion of a bolt, stepped shaft, or similar part, with a diameter smaller than the peripheries or diameters of the adjacent features. It obviates difficulties of machining sharp corners and facilitates subsequent assembling.

undercutting (*Print.*). (1) A fault to be guarded against in the etching of line blocks, traditionally by the use of dragon's blood, but nowadays by powderless etching. (2) Cutting away part of the edge of a mount to allow the block to overhang a crossbar.

underdamping (*Telecomm.*). Sometimes synonymous with **periodic damping,** but often restricted to cases where a critically-damped response would be preferable.

underdone bell (*Elec. Eng.*). A *trembler bell*, in which the mechanism is enclosed by the gong.

under-exposure (*Photog.*). Insufficient exposure of the sensitive surface of a plate, paper, or film to the incident image, resulting in lack of detail and reduced contrast.

underfeed stoker (*Eng.*). A *mechanical stoker* (q.v.) in which the fuel is fed automatically and progressively from below the fire, and gradually forced up into the active zone, air being injected into the fuel bed just below the combustion-level. See **single-retort-, multiple-retort-.**

underflow (*Comp.*). A number below the accepted minimum which cannot be stored (in its existing form) with adequate accuracy by a particular computer register. See also **overflow.**

underfold (*Print.*). See **overfold.**

undergettings (*Mining*). See **shorts** (2).

underground collector (*Elec. Eng.*). See **plough.**

underground conduit (*Cables*). A cement or stoneware underground channel in which cables are laid.

underhand stopes (*Mining*). Stopes in which excavation is carried downslope from access level.

underlay (*Mining*). The departure of a vein or thin tabular deposit from the vertical; it may be measured in horizontal feet per fathom of inclined depth. Also **underlie.** (*Print.*) To paste paper or card under the mount of a printing plate in order to bring it to type-height.

underlay shaft (*Mining*). A shaft sunk in the rock below an inclined reef or lode, and at the average dip or inclination of the lode.

underleaf (*Bot.*). One of a row of leaves on the underside of the stem of a liverwort.

underlie (*Mining*). See **underlay.**

underlining felt (*Build.*). See **sarking felt.**

undermodulation (*Cinema.*). That which is unnecessarily low in relation to the possible level of modulation that can be accommodated on the sound-track. (*Radio*) That state of adjustment of a radio-telephone transmitter at which the peaks of speech or music do not produce 100% modulation, so that carrier power is not used to full advantage.

underpick (*Weaving*). Propelling the shuttle from its box across the loom slay by a picking stick, pivoted and actuated below the loom, striking in a short powerful arc.

underpinning (*Build.*, *Civ. Eng.*). The operation of propping part of a building to avoid damaging or weakening the superstructure.

underpitch groin (*Build.*). See **Welsh groin.**

underpoled copper (*Met.*). See **poling.**

undersaturated exciter (*Elec. Eng.*). An auxiliary exciter operating on the straight part of its magnetization curve; used to compensate a rotary convertor for field weakening due to lagging currents.

Underseal (*Autos.*). TN for coat of bituminous or other corrosion-resisting substance, applied to the exposed portions of the underside of an automobile.

undershoot (*Telecomm.*). See **overshoot.**

undershot wheel (*Eng.*). A water-wheel used for low heads, in which the power is obtained almost entirely from the impulse of the water on the vanes. See **Poncelet wheel.**

undershrub (*Bot.*). A small woody perennial in which the flowering branches die back after flowering, leaving a persistent basal system of woody branches from which new growth arises in the next season.

undersize (*Min. Proc.*). See **riddle.**

understeer (*Autos.*). Tendency of a vehicle to preserve directional stability by reacting against external forces applied to the steering mechanism. Cf. *oversteer.*

under-voltage no-close release (*Elec. Eng.*). A device which acts upon the trip coil of a circuit-breaker in such a way as to prevent the circuit-breaker being closed if the voltage is below a certain predetermined value.

under-voltage release (*Elec. Eng.*). A device to trip an electrical circuit should the voltage fall below a certain predetermined value. Also called a **low-volt release.**

underwater cutting (*Eng.*). A method of cutting iron or steel under water, utilizing a combusted mixture of oxygen and hydrogen (cf. *oxy-hydrogen welding*), the flame being protected by an airshield formed by a hood.

underwater transducer (*Acous.*). One (e.g., a microphone or loudspeaker) designed for operation under water.

undistorted output (*Telecomm.*). That of a valve amplifier free from nonlinear distortion. The *maximum undistorted output* is defined as that obtained with a specified degree of nonlinear distortion.

undistorted transmission (*Telecomm.*). That of any type of line for which the velocity of propagation and coefficient of attenuation are both independent of frequency.

undulant fever (*Med.*). Brucellosis. (1) Malta fever, Mediterranean fever, Gibraltar fever. A disease characterized by alternating febrile and afebrile periods, splenomegaly, transient painful swelling of joints, neuralgia, and anaemia; due to infection with *Brucella melitensis*, conveyed to man by infected goats or their milk. (2) A disease with the same *signs* as (1), due to infection by *Brucella abortus*, a micro-organism which causes abortion in cows and is conveyed to man in cow's milk or by contact with infected animals.

undulating membrane (*Zool.*). A film of protoplasm connecting the body of some *Protozoa* (e.g., *Trypanosoma*) with a flagellum running parallel to the body surface for some distance. Should not be confused with *vibratile membrane.*

undulator (*Teleg.*). In Morse systems, a high-speed tape-recorder in which a two-position stylus feeds into a continuously moving paper tape in conformity with two conditions of the incoming signal.

unequal (*Bot.*). (1) Having the two sides not symmetrical. (2) Not all of the same length.

unequally pinnate (*Bot.*). See **imparipinnate.**

unessential element (*Bot.*). A chemical element which may be found in plants, but which appears to play no part in nutrition and seems not to be necessary to the welfare of the plant.

uneven working (*Print.*). See **even working.**

unexcited (*Nuc.*). Said of an atom in its ground state. See **excitation.**

unfired (*Electronics*). Said of any gas discharge device when in an un-ionized state.

ungual (*Med.*). Pertaining to or affecting the nails. See **unguis.**

Unguiculata (*Zool.*). A cohort of the *Eutheria*, containing both extinct and living orders which retain archaic mammalian characters. Living forms include Hedgehogs, Bats, Armadillos, Pangolins, and Primates.

unguiculate (*Bot.*, *Zool.*). Provided with claws. Specifically (*Bot.*) applied to a petal with an expanded limb supported on a long narrow stalklike base.

unguis (*Zool.*). In Insects, one of the tarsal claws; in Vertebrates, the dorsal scale contributing to a nail or claw (cf. *subunguis*); more generally, a nail or claw. *pl.* **ungues.** *adjs.* **ungual, unguinal.**

unguitractor plate (*Zool.*). In most Insects, a median structure between the paired claws of the tarsus, and ventral to them, serving as an attachment for the flexor muscle of the claws.

ungula (*Zool.*). A hoof. *adj.* **ungulate.**

Ungulata (*Zool.*). A term not now used, but previously used to cover a multitude of hoofed grazing animals whose common features (e.g., grinding teeth, muscular, articular and alimentary adaptations, long necks, and social tendencies related to self preservation) probably arose independently in response to the same conditions. The group in fact comprised a variety of widely divergent types, and thus is now divided into 4 superorders, *Protoungulata*, *Paenungulata*, *Mesaxonia* and *Paraxonia*, all being included with the superorder *Ferae* (one order, *Carnivora*) in the cohort *Ferungulata.*

unguligrade (*Zool.*). Walking on the tips of enlarged nails of one or more toes, i.e., hoofs, as in horses, etc. Cf. *digitigrade*, *plantigrade.*

uniaxial (*Bot.*, *Zool.*). Having an unbranched main axis; haplocaulescent. (*Min.*) A term embracing all those crystalline minerals in which there is only one direction of single refraction (parallel to the principal crystal axis and known as the optic axis). All minerals which crystallize in the tetragonal, trigonal, and hexagonal systems are *uniaxial*. Cf. *biaxial.*

uniaxial construction (*Bot.*). The structure of an algal thallus, consisting of a main filament with its branches packed closely round it.

uniaxial crystal (*Min.*). See under **uniaxial.**

unicellular (*Biol.*). Consisting of a single cell.

unicostate (*Zool.*). Having one rib.

unidactyl (*Zool.*). Having one digit.

unidirectional antenna (*Radio*). One in which the radiating or receiving properties are largely concentrated in one direction.

unidirectional current (*Elec. Eng.*). One which, although its amplitude may vary, never changes sign.

unified field theory (*Phys.*). See **field theory.**

unified model of nucleus (*Nuc.*). Modern model of nucleus combining as many valuable features from older models as possible.

unified scale. The scale of atomic and molecular weights which is based on the mass of the ^{12}C isotope of carbon being taken as 12 exactly, hence the atomic mass unit equals $1{\cdot}660 \times 10^{-27}$ kg. This scale was adopted in 1960 by the International Unions of both Pure and Applied Physics and Pure and Applied Chemistry, hence the name *unified scale.* See **atomic mass unit, atomic weight.**

unified screw thread (*Eng.*). A screw thread form adopted by Great Britain, Canada and the U.S.A. It has a 60° angle and is in many respects similar to the U.S. *standard screw thread.* See **Sellers (or U.S.S.) screw thread.**

Uniflow engine (*Eng.*). See **Unaflow engine.**

unifoliolate (*Bot.*). Said of a compound leaf which has one leaflet only.

uniform convergence (*Maths.*). (1) A sequence $\alpha_1(x), \alpha_2(x) \ldots \alpha_n(x) \ldots$, converges uniformly to the limit $\alpha(x)$ in the interval (a, b) if, given any $\varepsilon > 0$, there exists, for all x in (a, b), N such that $|\alpha_n(x) - \alpha(x)| < \varepsilon$ when $n \geq N$. (2) A series converges uniformly if the sequence of partial sums converges uniformly. (3) A product converges uniformly if the sequence of partial products converges uniformly. See **convergence, M-test of Weierstrass for-.**

uniform extension (*Met.*). The extension produced in a tensile test before the ultimate tensile stress is reached, and uniform over the gauge length.

uniform field (*Maths.*). One which is described by the same vector at all points. A constant field.

uniform line (*Elec. Eng.*). One for which the electric properties are identical throughout its length.

uniform system (*Photog.*). A method of marking the stop adjustment of lenses so that equal unit-steps correspond to doubling or halving the exposure, unity on the scale corresponding to *f*/4. Abbrev. U.S. See ***f*-number.**

unijugate (*Bot.*). Said of a compound leaf having one pair of leaflets.

unilateral (*Bot.*). Said (*a*) of members which are all inserted on one side of the axis, or are all turned to one side of the axis; (*b*) of a raceme with all the flowers turned to one side; (*c*) of a stimulus falling on the plant from one side, as when a plant is placed so that light reaches it through a small aperture.

unilateral conductivity (*Elec.*). The property of unipolarity by which current can flow in one direction only; exhibited by a perfect rectifier.

unilateral impedance (*Elec.*). Any electrical or electromechanical device in which power can be transmitted in one direction only, e.g., a thermionic valve or carbon microphone.

unilateralization (*Elec.*). Neutralization of feedback so that transducer or circuit has unilateral response, i.e., there is no response at the input if the signal is applied to the output terminals. While many valve circuits are inherently unilateral, equivalent transistor ones require external neutralization.

unilateral tolerance (*Eng.*). A tolerance with dimensional limits either entirely above or entirely below the basic size.

unilateral transducer (*Elec. Eng.*). One for which energy can be transmitted only forward.

unilocular (*Bot.*). Having a single compartment.

unimolecular layer, unimolecular reaction (*Chem.*). See **monomolecular layer, monomolecular reaction.**

uninsulated conductor (*Elec. Eng.*). A conductor at earth potential, such that no care need be taken to insulate it from earth.

uninucleate (*Biol.*). Containing one nucleus.

union (*Maths.*). The union of sets A and B is the set of elements which are in A or in B or in both A and B. (*Med.*) In the process of healing, the growing together of parts separated by injury (e.g., the two ends of a broken bone, the edges of a wound). (*Plumb.*) A connexion for pipes. (*Textiles*) A mixture of two or more materials in a yarn or cloth; generally the presence of the less expensive material is not apparent, e.g., wool and cotton.

union blanket (*Textiles*). A blanket made from cotton warp and woollen or shoddy weft; or one made from angola yarns.

union paper or **union kraft** (*Paper*). Wrapping paper consisting of two layers of kraft paper, between which is a layer of bitumen.

uniparous (*Zool.*). Giving birth to one offspring at a time.

Unipivot instrument (*Elec. Eng.*). An instrument whose moving-coil system is balanced on a single pivot passing through its centre of gravity.

unipolar (*Zool.*). Said of nerve cells having only one process. Cf. *bipolar, multipolar.*

unipolar transistor (*Electronics*). One with one polarity of carrier.

unipole antenna (*Radio*). Isotropic antenna conceived as radiating uniformly in phase in all directions. Theoretically useful, but not realizable in practice.

unipotential cathode (*Electronics*). Same as **indirectly-heated cathode.**

uniramous (*Zool.*). Having only one branch; as some Crustacean appendages. Cf. *biramous.*

uniselector (*Teleph.*). A selector switch which only rotates its wipers about an axis, in contrast with the *two-motion selector*, in which wipers are raised to a specified level in the rows of contacts by the impulse trains, and then enter the bank of contacts, either by hunting or by a further train of impulses. See **selector.**

uniselector distribution frame (*Teleph.*). A frame carrying terminals, so that the extent of the multiple outlets from uniselectors can be varied to suit the demand on outlets in the multiple.

uniserial (*Zool.*). Said of a type of fish fin in which the radial elements occur on one side only of the basalia.

uniseriate (*Bot.*). (1) Arranged in a single row, series, or layer. (2) Said of a vascular ray which is one cell wide in cross-section.

uniseriate sorus (*Bot.*). A sorus having a single series of sporangia, forming a rosette around a central cushion.

Unisexales (*Bot.*). See **Casuarinales.**

unisexual (*Bot., Zool.*). Showing the characters of one sex or the other; distinctly male or female. Cf. *hermaphrodite.*

unison (*Acous.*). In an organ, the production of notes which have the same pitch as the keys which actuate them; in singing or orchestral music, the use of octaves only in the harmony.

unistratose (*Bot.*). Forming a single layer.

unit. A dimension or quantity which is taken as a standard of measurement. (*Mining*) 1% of a specified element or compound in a parcel of ore, concentrates or metal being sold.

unit arch (*Print.*). On a web-fed press a perfecting unit arranged in an arch or inverted U design.

unitary ratio (*Elec. Eng.*). The ratio between the electromagnetic and electrostatic absolute units; equal to the velocity of propagation of radiated energy, viz., 3×10^{10} cm/sec. Disused.

unit cell (*Chem.*). The smallest group of atoms,

ions, or molecules, whose repetition at regular intervals, in three dimensions, produces the lattice of a given crystal.

unit characters (*Gen.*). Independent characteristics, which act as units, are traceable in each generation, and are assorted and distributed by the laws of chance.

unit charge or quantity of electricity (*Elec.*). An electric charge which experiences a repulsive force of 1 dyn when 1 cm, or 1 N when 1 m, from a like charge *in vacuo*.

unit-distance code (*Comp.*). The arrangement of a pattern to avoid ambiguity errors in a computer.

Uniterm system (*Comp.*). Data-recording system used by libraries, based on classifying keywords in a coordinate indexing system.

unit heater (*Eng.*). A combination of air heater and circulator, often in the form of a heated cellular core or finned tube over which air is blown by a fan.

unit interval (*Teleg.*). In a system using an equal-length code or in a system using an isochronous modulation, the interval of time such that the theoretical durations of the significant intervals of a telegraph modulation (or restitution) are whole multiples of this interval.

unitized (*Autos.*). See **chassis**.

unit matrix (*Maths.*). A unit matrix is a square matrix satisfying the following conditions: (i) the leading diagonal entries, i.e. the entries on the diagonal from top left to bottom right, are all unity; (ii) the entries not on the leading diagonal are all zero. The unit matrix of order *n* is an identity element for multiplication in the set of all square matrices of order *n*. A unit matrix is usually denoted by *I*.

unit of attenuation (*Telecomm.*). See **decibel, neper.**

unit of bond (*Build.*). That part of a brickwork course which, by being constantly repeated throughout the length of the wall, forms a particular bond.

unit of illumination (*Light, etc.*). See **lux, phot.**

unitor (*Elec. Eng.*). Multiple plug and socket device, suitable for temporary multichannel connexions, especially of interchangeable chassis.

unit pole (*Mag.*). A magnetic pole which experiences a repulsive force of 1 dyn when 1 cm, or 1 N when 1 m, from a like pole *in vacuo*. A mathematical concept, useful for establishing magnetic and electric units.

unit sampling (*Telecomm.*). Gating in *pulse code modulation* (q.v.).

unit system (*Typog.*). A feature of the Monotype system and of certain major filmsetting systems. See **Monotype unit system.**

unit type press (*Print.*). A web-fed press with one or more printing units in line on the bed plate.

unit weight (*Eng.*). In a steam boiler, the ratio of the mass of the complete boiler to that of the hourly maximum steam production.

univalent (*Chem.*). *Monovalent* (q.v.). (*Cyt.*) One of the single chromosomes which separate in the first meiotic division.

univalve (*Zool.*). In one piece; said of some Molluscan shells. Cf. *bivalve*.

univariant (*Chem.*). Having one *degree of freedom*.

universal chuck (*Eng.*). See **self-centring chuck.**

universal combustion burner (*Heat.*). Natural-draught gas burner having one injector for the entrainment of primary air prior to combustion, and a secondary injector through which the flow of additional air into the combustion chamber can be regulated.

universal controller (*Elec. Eng.*). A crane controller, similar in action to the 'joystick' of an aeroplane, in which a vertical motion of the control handle operates the hoist motor, whilst a lateral motion controls the travel motor.

universal grinder (*Eng.*). A machine tool similar in layout to a centre lathe in which the toolpost is replaced by a grinding wheel with drive, used for grinding cylindrical and tapered work.

universal indicator (*Chem.*). A mixture of *indicators* (q.v.) which gives a definite colour change for each integral change of pH-value over a wide range.

universal joint (*Autos., etc.*), A joint at the ends of the propeller shaft to allow for movement of the rear axle relative to the gear box; usually of modified Hooke's type, enclosed in a grease-cover. Universal joint couplings are used generally where angular drive is required (e.g., machine tools).

universal milling machine (*Eng.*). A *milling machine* (q.v.) similar to a plane milling machine but with the additional feature that the table swivels horizontally, and is provided with a dividing head as standard equipment.

universal motor (*Elec. Eng.*). A fractional-horsepower commutator motor suitable for use with both direct current and single-phase alternating current.

universal plane (*Tools*). Multi-purpose plane adaptable for rebating, grooving, trenching, cutting mouldings, etc.

universal planer (*Eng.*). A planer that will cut on the forward and on the reverse strokes.

universal set (*Maths.*). A term used to denote the set containing all the elements relevant to a particular mathematical study. In logic, sometimes called the **universe of discourse.**

universal shunt (*Instr.*). A series of high stability resistors used to shunt galvanometers to provide different ranges of measurement.

universal time (*Astron.*). A name for *Greenwich Mean Time* recommended in 1928 by the International Astronomical Union to avoid confusion with the pre-1925 GMT which began at noon, not midnight. Abbrev. **UT.** See also **ephemeris time.**

universal veil (*Bot.*). A coating of hyphae which surrounds the young fruit body of a fungus.

universal viewfinder (*Photog.*). One with objectives for lenses of varying focal length.

univibrator (*Telecomm.*). Term for monostable multivibrator circuit. See **flip-flop.**

univoltine (*Zool.*). Producing only one set of offspring during the breeding season or year. Cf. *multivoltine*.

unloaded antenna (*Radio*). One with no inductance coils to increase its natural wave length.

unmarried print (*Cinema.*). Pair of *reels*, one carrying the optical, the other a photographic or magnetic sound-track, of a film, the reels being synchronized during projection.

unmasked hearing (*Acous.*). In deaf-aids, the principle of applying to one ear the higher-frequency components of wanted sounds, in order that their presence may be appreciated; such frequencies are swamped by amplitude-distortion, and consequent alien tones, when all the desired component frequencies are applied at high level to the other ear.

unmodulated track (*Cinema.*). Those parts of the finished sound-track on which the modulation is zero, the transparent part of the track being responsible for the residual noise-level.

unmodulated waves (*Radio*). Waves which do not vary in amplitude with time, such as those radiated from a radio-telephone transmitter when no sound enters the microphone.

unpitched sound (*Acous.*). Any sound or noise

which does not exhibit a definite pitch, but consists of components spread more or less continuously over the frequency spectrum.

unreduced apogamy (*Bot.*). See **euapogamy.**

unsaturated (*Chem.*). (1) Less concentrated than a saturated solution or vapour. (2) Containing a double or a triple bond, especially between two carbon atoms; unsaturated molecules can thus add on other atoms or radicals before saturation is reached.

unseptate (*Bot.*). Aseptate.

unsewn binding (*Bind.*). The sections of the book are gathered and fed to a machine (of which there are several designs); part of the back is cut off and the edge roughened, an adhesive applied, and, when it sets, the leaves compacted together. Used mainly for paperbacks. Often called **Perfect binding, thermoplastic binding.**

unsoundness (*Met.*). The condition of a solid metal which contains blowholes or pinholes due to gases, or cavities resulting from its shrinkage during contraction from liquid to solid state, i.e., *contraction cavities* (q.v.).

unstability (*Cables*). See under **stability test.**

unstable (*Build.*). A term applied to a structural framework having fewer members than it would require to be perfect. Also **deficient.** (*Chem.*) Subject to spontaneous change.

unstable community (*Bot.*). A plant community which does not remain constant over a period of years.

unstable equilibrium (*Mech.*). The state of equilibrium of a body when any slight displacement decreases its potential energy. The instability is shown by the fact that, having been slightly displaced, the body moves farther away from its position of equilibrium.

unstable oscillation. Any oscillation, in a mechanical body, electrical circuit, etc., which increases in amplitude with time.

unstick (*Aero.*). See **lift-off.**

unstriated muscle (*Zool.*). A form of contractile tissue composed of spindle-shaped fibrillar uninucleate cells, occurring principally in the walls of the hollow viscera. Also **smooth muscle.** Cf. *striated muscle*, and see **voluntary muscle.**

unsymmetrical (*Bot.*). Said of a flower in which all the parts are not regular.

unsymmetrical grading (*Teleph.*). *Grading* (q.v.) in which subscribers originating higher-than-average traffic are given access to a greater proportion of individual trunks.

unsymmetrical oscillations (*Radio*). Oscillations in which the positive and negative parts of the waveform are unequal and of different shape.

untrimmed floor (*Carp.*). A floor consisting of bridging joists only.

untuned antenna (*Radio*). One not separately tuned to the operating frequency, although effectively tuned by coupling to one or more resonant circuits.

untuned circuit (*Telecomm.*). One not sharply resonant to any particular frequency.

unvoiced sound (*Acous.*). In speech, any elemental sound which has not discrete harmonic frequencies but consists of noise frequencies generated by air rushing through the mouth and nasal cavities; these are modified in spectral energy distribution by the posture of the cavities of the mouth and the resultant broad resonances.

up (*Mining*). A working place which has reached the boundary, working limit, or line of fault.

up and/or down locks (*Aero.*). Safety locks which hold the units of a retractable landing gear up in flight and down on the ground.

upcast (*Geol.*). In faulted rocks, *upcast* refers to the strata in which the relative displacement has been upwards, i.e., away from the earth's centre, along the fault plane. Cf. *downthrow.* See **fault.** (*Mining*) The lifting of a seam or bed by a dyke. Also **uptake.**

upcast shaft (*Civ. Eng.*). A ventilating shaft through which the vitiated air passes in an upward direction.

update (*Comp.*). (1) To modify a computer instruction so that the address specified is increased every time the instruction is carried out. (2) To correct entries in a file of data in accordance with a specified procedure.

updraught (*I.C. Engs.*). Said of a carburettor in which the mixture is drawn upwards against the force of gravity.

upholsterer's hammer (*Tools*). Lightly built claw-hammer for use with small pins and tacks.

upmake (*Typog.*). (1) To arrange lines of print into columns or pages. (2) Print so arranged.

upper (*Bot.*). Posterior.

upper atmosphere (*Astron.*). A term used somewhat loosely for the region of the earth's atmosphere above about 20 miles (30 km), which is not normally explored by sounding balloons, but can be studied by rockets and artificial satellites.

upper bound (*Maths.*). See **bounds of a function.**

upper case (*Typog.*). Frequently used as a synonym for capitals, the name deriving from the arrangement (now largely superseded) of the capitals in an upper case and the small letters in a lower *case* (q.v.).

upper culmination (*Astron.*). See **culmination.**

upper deck (*Ships*). The term correctly denotes the main strength deck of a ship. From this deck all scantlings are determined, freeboard assigned, and subdivision arranged, according to type of vessel.

Upper Greensand (*Geol.*). So named to distinguish it from Lower Greensand, it is the arenaceous facies of the Albian stage of the English Cretaceous rocks. Best developed in the more westerly outcrops, it passes eastwards laterally into the Gault Clay at Maidstone, Kent, to the east of which town no Upper Greensand occurs.

upper leathers (*Leather*). The name for leathers, more lightly tanned than heavy hides, used extensively for the uppers of boots and shoes, harness, straps, etc. They are produced from a wide variety of skins. See also **poromeric.**

upper mean-hemispherical candle-power (*Light*). See **mean-hemispherical candle-power.**

upper transit (*Astron.*). Another name for *upper culmination.*

upright (*Build.*). A vertical member in a structure.

uprighting (*Horol.*). The process of correcting the pivot holes in a plate so that the hole in one plate is exactly in line with the corresponding hole in the other plate, and a line through the two holes is at right angles to the plates.

upset (*For.*). A timber defect, the fibres being torn across the grain.

upsetting (*Met.*). See **jumping-up.**

upstream injection (*Aero.*). A gas turbine fuel system in which the fuel is injected towards the compressor in order to achieve maximum vaporization and turbulence.

uptake (*Eng.*). The flue or duct which leads the flue gases of a marine boiler to the base of the funnel. (*Mining*) See **upcast.** (*Radiol.*) In radiobiology, the quantity (or proportion) of administered substance subsequently to be found in a particular organ or tissue.

up-time (*Comp.*). Period during which a computer is available for intended operation.

upturn (*Plumb.*). The part of a lead flashing which is dressed up against a wall face.

uracil (*Biochem.*). 2,6-Dioxypyrimidine having the following constitution:

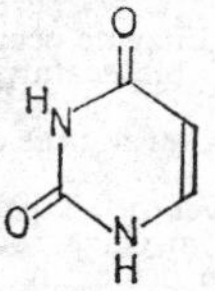

In the *genetic code* (q.v.) uracil in the RNA codons replaces thymine in the equivalent DNA codons, pairing with adenine.

uraemia, uremia (*Med.*). The state resulting from failure of a diseased kidney to perform its normal functions; associated with retention of urea in the blood, and characterized by varied symptoms, among which are headache, foul breath, diarrhoea and vomiting, visual disturbances, lethargy, convulsions, and coma.

Uralian emerald (*Min.*). Not an emerald; green variety of andradite garnet, occurring as nodules in ultra-basic rocks in the Nizhni-Tagilsk district of the Urals; used as a semiprecious gemstone, though rather soft for this purpose. Known also as **Bobrovska garnet**.

uralite (*Min.*). A bluish-green monoclinic amphibole, generally actinolitic in composition, resulting from the alteration of pyroxene.

uralitization (*Geol.*). A type of alteration of pyroxene-bearing rocks, involving the replacement of the original pyroxenes by fibrous amphiboles, as in some epidiorites.

uranic, uranous (*Chem.*). Referring to (VI) and (IV) uranium respectively.

uranides (*Chem.*). Name for elements beyond protactinium in the periodic system.

uraninite (*Min.*). Uranium(IV)oxide, UO_2, often more or less hydrated and containing also lead, thorium, and the metals of the lanthanum and yttrium groups. It occurs as brownish to black cubic crystals and is an accessory mineral in granite rocks and in metallic veins. When massive, and apparently amorphous, known as *pitchblende*.

uranite (*Min.*). For *copper uranite*, see **torbernite**; for *lime uranite*, see **autunite.**

uranium (*Chem.*). A hard grey metal with a number of isotopes. Symbol U, at. no. 92, r.a.m. 238·03, rel. d. 18·68, m.p. 1150°C. ^{235}U is the only naturally occurring readily fissile isotope and exists as one part in 140 of natural uranium. ^{238}U is a fertile material (half-life $4{\cdot}5 \times 10^9$ yr) which has a small fission cross-section for fast neutrons. ^{233}U is a fissile material which can be produced by the neutron irradiation of thorium-232; ^{235}U has great value in nuclear reactors and nuclear weapons. Metallic uranium exists in three phases known respectively as *alpha, beta*, and *gamma uranium* (all these show normal radioactive alpha-emission) with transition temperatures of 660°C and 770°C. The large expansion associated with the transition from alpha to beta uranium precludes the operation of nuclear reactors with metallic uranium fuel elements at temperatures above 660°C.

uranium acetate (ethanoate) stain (*Micros.*). A 0·5% uranium acetate solution buffered at pH 6·1, used to increase contrast by forming electron-dense deposits in specimens prepared for examination with the electron microscope.

uranium hexafluoride (*Chem.*). Uranium(VI) fluoride. A gaseous compound of uranium with fluorine, used in the gaseous diffusion process for separating the uranium isotopes. Very corrosive.

uranium reactor (*Nuc. Eng.*). One in which the fissile fuel material is uranium. Such reactors are classified according to whether they use natural or enriched uranium fuel elements, the proportion of the fissile ^{235}U isotope in the latter having been artificially increased by gaseous diffusion.

uranoplasty (*Surg.*). Plastic operation for closing a cleft in the hard palate.

Uranus (*Astron.*). Planet discovered by W. Herschel in 1781. Its mean distance from the Sun is 19·18 A.U. and its sidereal period 84 years. It rotates in 10·8 hours, its axis being inclined at 98° to the orbit, so that its rotation and the revolution of its five satellites are all retrograde. Recent measures of the diameter give 50 800 km, making the planet larger than Neptune. The atmosphere is probably mainly hydrogen, but methane has been detected.

uranyl (*Chem.*). The radical UO_2^{++}, e.g., uranyl nitrate, $UO_2(NO_3)_2$.

urate (*Chem.*). Salt of uric acid.

urceolus (*Bot., Zool.*). An urn-shaped or pitcher-shaped structure, as the tube in which some *Rotifera* live. *adj.* **urceolate.**

urea (*Chem.*). Carbamide, $H_2N{\cdot}CO{\cdot}NH_2$; long rhombic prisms or needles; m.p. 132°C, very soluble in water, insoluble in ether. Found in the urine of Mammals. Wöhler synthesized (1828) urea from ammonium isocyanate, which undergoes an intramolecular transformation when its aqueous solution is heated, forming urea. Manufactured by heating carbon dioxide and ammonia under high pressure, it is highly nitrogenous, and widely used for fertilizer and animal feed additive.

urea resins (*Plastics*). Thermosetting resins manufactured by heating together urea and an aldehyde, generally formaldehyde (methanal). Pale-coloured or water-white and translucent, they can therefore take delicate dyes and tints. They are non-inflammable, and are resistant to weathering, weak acids and alkalis (pH between 6 and 8·4), ethanol, propanone, greases and oils. Also **aminoaldehydic resins.**

urease (*Chem.*). An enzyme which occurs in several plants, e.g., in the soya bean and in *Micrococcus ureae*, hydrolysing urea quantitatively with the formation of ammonium carbonate.

ured-, uredo-. Prefix from L. *uredo*, a blight.

uredicole (*Bot.*). Growing as a parasite upon a rust fungus.

Uredinales (*Bot.*). A group of parasitic *Basidiomycetes*, including about 2000 species. The mycelium lives inside the host plant, spores being formed in pustules on the surface of the host. Some species pass part of their life on one host and part on another. Some have a complicated life-history, forming *uredospores, teleutospores, basidiospores, spermatia*, and *aecidiospores*; others omit one or more of these kinds of spores. The *Uredinales* are popularly known as *rust fungi*; e.g., the destructive *wheat rust*.

uredinial stage (*Bot.*). U.S. term for **uredostage.**

urediniospore (*Bot.*). U.S. term for **uredospore.**

uredinium (*Bot.*). U.S. term for **uredosorus.**

urediospore (*Bot.*). U.S. term for **uredospore.**

uredogonidium (*Bot.*). Alternative term for **uredospore.**

uredosorus (*Bot.*). A pustule consisting of uredospores, with their supporting hypha, and some sterile hyphae.

uredospore (*Bot.*). An orange or brownish spore formed by rust fungi when growth is vigorous, and serving as a means of rapid propagation; it gives rise to a mycelium which may produce more uredospores, or, later in the year, uredospores and teleutospores.

uredostage (*Bot.*). The phase in the life-history of a rust fungus when uredospores are formed.

ureides (*Chem.*). The acid derivatives of urea. They correspond to amides or anilides. The cyclic ureides are known as the *purine group*.

ureido acids (*Chem.*). The mono-substituted products of urea and a dibasic acid, in which only one hydroxyl of the carboxyl group has been substituted by the urea group with elimination of water.

uremia (*Med.*). See **uraemia.**

ureotelic (*Chem.*). Excreting nitrogen in the form of urea.

ureter (*Zool.*). The duct by which the urine is conveyed from the kidney to the bladder or cloaca.

ureteralgia (*Med.*). Pain in the ureter.

ureteritis (*Med.*). Inflammation of the ureter.

ureterocele (*Med.*). Cystic dilatation of that part of the ureter which lies within the wall of the urinary bladder, due to congenital narrowing at its point of entry into the bladder.

ureterocolostomy (*Surg.*). The operation of implanting the ureter into the colon so that it may drain into it.

ureterography (*Radiol.*). Radiography of a ureter filled with a medium opaque to X-rays.

ureterolithotomy (*Surg.*). The operation of cutting into the ureter to remove a stone from it.

ureteropyelitis (*Med.*). Inflammation both of a ureter and of the pelvis of the kidney on the same side.

ureterotomy (*Surg.*). Surgical incision of a ureter.

urethane (*Pharm.*). $H_2N{\cdot}CO{\cdot}OC_2H_5$, ethyl carbamate; crystallizes in large plates; m.p. 50°C, b.p. 184°C, soluble in water. Used in some cases of leukaemia. See also **polyurethanes.**

urethra (*Zool.*). The duct by which the urine is conveyed from the bladder to the exterior, and which in male Vertebrates serves also for the passage of the semen. *adj.* **urethral.**

urethritis (*Med.*). Inflammation of the urethra. **Non-specific urethritis (NSU)** is thought to be of venereal nature.

urethrocele (*Med.*). Prolapse of the floor of the female urethra; usually associated with cystocele.

urethrocystitis (*Med.*). Inflammation of both the urethra and the urinary bladder.

urethrospasm (*Med.*). Spasmodic contraction of the muscular tissue of the urethra.

urethrotomy (*Surg.*). The operation of cutting a stricture of the urethra.

uretidine (*Chem.*). A four-membered heterocyclic compound having two nitrogen atoms:

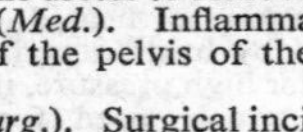

$$H_2C\langle\begin{matrix}NH\\NH\end{matrix}\rangle CH_2$$

uric acid (*Chem.*). An acid of the purine group, $C_5H_4N_4O_3$, 2,6,8-trihydroxypurine. It has the constitution:

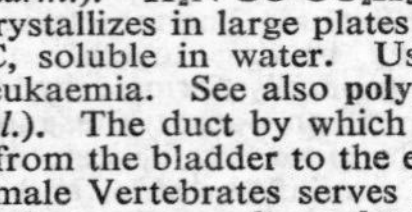

O, H, HN, N, =O, O, N, N, H, H

It is a white crystalline powder, insoluble in cold, hardly soluble in hot, water. Uric acid deposits in the organism are the cause of gout and rheumatism. It forms soluble lithium and piperazine salts. It can be recognized by the *murexide test* (q.v).

uricase (*Biochem.*). Enzyme occurring in the liver and kidneys, and which catalyses the oxidation of *uric acid* (q.v.).

uricolytic (*Chem.*). Causing the decomposition of uric acid.

Uriconian Rocks (*Geol.*). Lavas (largely rhyolites), tuffs, and intrusive rocks of presumed Pre-Cambrian age occurring in Shropshire in the hills east of the Longmynd, and in the Wrekin.

uricotelic (*Chem.*). Excreting nitrogen in the form of uric acid.

uridrosis (*Med.*). Condition in which the constituents normally found in urine are excreted by the sweat glands.

urine (*Zool.*). In Vertebrates, the excretory product elaborated by the kidneys, usually of a more or less fluid nature. *adj.* **urinary.**

uriniferous, uriniparous (*Zool.*). Urine-secreting, urine-producing; as the glandular tubules of the kidney.

urinogenital (*Zool.*). Pertaining to the urinary and genital systems.

urinogenital system (*Zool.*). The organs of the urinary and genital systems when there is a direct functional connexion between them, as in male Vertebrates.

urinometer (*Med.*). An instrument for measuring the density of urine.

urite, uromere (*Zool.*). In *Arthropoda*, one of the somites of the abdomen.

urn (*Bot.*). The capsule of a moss.

uro-. Prefix from (*a*) Gk. *ouron*, urine, (*b*) Gk. *oura*, tail.

urobilin (*Zool.*). A brownish pigment of urine, chiefly in the form of a chromogen, *urobilinogen*; *stercobilin* (q.v.), reabsorbed from the intestine and excreted in the urine.

urobilinaemia, urobilinemia (*Med.*). The presence of urobilin in the blood.

urobilinogen (*Zool.*). See **urobilin.**

urobilinuria (*Med.*). The presence of (an excess of) urobilin in the urine.

urocardiac (*Zool.*). In higher *Crustacea*, said of one of the ossicles of the gastric mill.

urochord (*Zool.*). Having the notochord confined to the tail region.

Urochorda (*Zool.*). A subphylum of *Protochordata*, in which only the larvae have a hollow dorsal nerve cord and a notochord, the adults being without coelom, segmentation and bony tissue, and having a dorsal atrium, a reduced nervous system, and a test composed of tunicin, a substance closely related to cellulose. Sea Squirts. Also called **Tunicata.**

urochrome (*Zool.*). A yellow pigment found in urine.

urodaeum (*Zool.*). The division of the cloaca which receives the urinary ducts.

Urodela (*Zool.*). An order of Amphibians, the adults having four similar pentadactyl limbs and a prominent tail. The larvae have external gills which persist in the adults of neotenous forms, and in some others gill slits persist. Newts and Salamanders. Also known as **Caudata.**

urodelous (*Zool.*). Having a persistent tail; as Salamanders.

urography (*Radiol.*). The radiological examination of the urinary tract following: (*a*) the intravenous injection and subsequent excretion of a *contrast medium* (q.v.), or (*b*) the introduction of a suitable *catheter* (q.v.).

urogomphi (*Zool.*). In the larvae of some *Coleoptera* (Insects), fixed and unjointed abdominal cerci.
urohyal (*Zool.*). A posterior element of the hyoid copula.
urology (*Med.*). That part of medical science which deals with diseases and abnormalities of the urinary tract and their treatment. Hence **urologist.**
uromere (*Zool.*). See **urite.**
uropatagium (*Zool.*). In Insects, the podical plate; in *Cheiroptera*, the membrane of the hind limbs.
uropod (*Zool.*). In *Malacostraca*, an appendage of the abdominal somite preceding the telson.
uropterine (*Chem.*). See **xanthopterine.**
uropygial (*Zool.*). Pertaining to or situated on the uropygium; as the oil gland.
uropygial gland (*Zool.*). See **oil gland.**
uropygium (*Zool.*). In Birds, the short caudal stump into which the body is prolonged posteriorly.
uroscopy (*Med.*). The inspection and scientific examination of urine for diagnostic purposes.
urosome (*Zool.*). In aquatic Vertebrates, the tail region; in *Crustacea*, the hinder part of the abdomen.
urostege, urostegite (*Zool.*). In *Ophidia*, a ventral tail-plate.
urosternite (*Zool.*). In *Arthropoda*, the sternite of an abdominal somite.
urosthenic (*Zool.*). Having the tail adapted for propulsion.
urostyle (*Zool.*). In Fish, the hypural bone; in *Anura*, a rodlike bone formed by the fusion of the caudal vertebrae.
urotropine (*Chem.*). See **hexamethylene-tetramine** (methanal and ammonia).
Ursa Major cluster (*Astron.*). An open cluster which includes most of the stars of the Plough, Sirius, and a number of other bright stars in the region near the sun; the sun is not a member.
urticant, urticating (*Zool.*). Irritating; stinging.
urticaria (*Med.*). A condition in which smooth, elevated, whitish patches (weals) appear on the skin and itch intensely, as a result of taking drugs or certain foods (e.g., shellfish), or as a reaction to the injection of serum, insect bites, or the stings of plants (nettle-rash).
urtite (*Geol.*). An intrusive igneous rock composed mainly of nepheline.
U.S. (*Photog.*). Abbrev. for *uniform system.*
useful life (*Elec. Eng.*). The life which can be expected from a component before the chance of failure begins to rise.
useful load (*Aero.*). The gross weight of an aircraft, less the tare weight. Usually includes fuel, oil, crew, equipment not necessary for flight (such as parachutes), and paying load.
use inheritance (*Zool.*). Inheritance of acquired characters.
U.S.S. thread (*Eng.*). See **Sellers screw thread.**
Ustilaginales (*Bot.*). A group of parasitic *Basidiomycetes*, including about 400 species. The mycelium lives inside the host, and frequently the flowers of the host are destroyed, and their ovaries replaced by a mass of dark-coloured spores looking like small masses of soot. Several species cause a good deal of damage to cereals. The *Ustilaginales* are popularly known as *smuts* or *smut fungi.*
UT (*Astron.*). Abbrev. for *universal time.*
UT chart (*Radio*). A contour chart giving, for a stated time, the world-wide values of an ionospheric characteristic such as the critical frequency.
uterus (*Bot.*). The peridium in some fungi, especially in the *Gasteromycetes.* (*Zool.*) In female Mammals, the muscular posterior part of the oviduct in which the foetus is lodged during the prenatal period; in lower Vertebrates and Invertebrates, a term loosely used to indicate the lower part of the female genital duct, or in certain cases (as in *Platyhelminthes*) a special duct in which eggs are stored or young developed. *adj.* **uterine.**
uterus masculinus (*Zool.*). In male Mammals, a small blind tube embedded in the prostate gland, opening dorsally into the urethra; believed to represent the vestige of the Müllerian ducts.
utilization factor (*Elec. Eng.*). A transformer feeding a rectifier system cannot supply the same maximum current as one feeding an a.c. circuit. The ratio of the maximum permitted currents in these two cases is the utilization factor. (*Light*) The ratio of the luminous flux reaching a specified plane to the total flux emanating from an electric lamp.
utricle (*Bot.*). A more or less inflated, membranous, bladderlike envelope surrounding the fruits of various plants. (*Zool.*) A small sac; in Vertebrates, the upper chamber of the inner ear from which arise the semicircular canals; the *uterus masculinus* (q.v.). Also **utriculus.**
utricular, utriculiform (*Bot., Zool.*). Like a bladder; pertaining to a utricle.
utriculoplasty (*Surg.*). The operation of excising a portion of the body of the uterus; done for the treatment of uterine haemorrhage.
uva (*Bot.*). A berry formed from a superior ovary.
uvarovite (*Min.*). A variety of garnet, of an attractive green colour; essentially silicate of calcium and chromium. Named after Uvarov, a Russian statesman.
Uvarov's phase theory (*Zool.*). The theory that each species of locust can exist in two main forms or phases, *solitaria* and *gregaria* (qq.v.), which differ structurally and biologically. Only the gregaria phase forms destructive swarms of *nymphs* (q.v.) and adults, and while nymphs reared in isolation develop into the solitaria phase, crowded nymphs develop into the gregaria phase.
uvea (*Zool.*). (1) In Vertebrates, the posterior pigment-bearing layer of the iris of the eye. (2) The iris, the ciliary body, and the choroid considered as one structure; also called the **uveal tract.**
uveitis (*Med.*). Inflammation affecting the iris, the ciliary body, and the choroid.
uveoparotid fever (*Med.*). A condition characterized by inflammation of the parotid glands and bilateral iridocyclitis, often with paralysis of the seventh cranial nerve.
uvula (*Anat.*) (1) A small conical process hanging from the middle of the lower border of the soft palate; part of the inferior vermis of the cerebellum. (2) A slight elevation of the mucous membrane of the urinary bladder in the male caused by the median lobe of the prostate.

V

Articles prefixed by V may be found either under V- (e.g., **V-band**) or under vee (e.g., **vee gutter**).

v- (*Psychol.*). Symbol representing verbal ability.

ν (*Geol.*). See **outcrop.** (*Phys.*) (1) The symbol for *velocity*. (2) Symbol for the specific volume of a gas.

ν- (*Chem.*). See **vicinal.**

V (*Chem.*). The symbol for *vanadium*. (*Elec.*) Symbol for *volt*.

V (*Elec.*). Symbol for *potential, potential difference, electromotive force*. (*Phys.*) Symbol for *volume*.

V_1 (*Aero.*). Abbrev. for *critical speed*.

V_{lo} (*Aero.*). See **lift-off.**

V_{ne} (*Aero.*). Abbrev. for the maximum permissible *indicated air speed*: a safety limitation (the suffix means '*n*ever *e*xceed') because of strength or handling considerations. The symbol is used mainly in operational instructions.

V_{no} (*Aero.*). *Normal operating speed*, usually of an airliner or other civil aircraft; this term is used mainly in flight operation documents and may be quoted in *E.A.S.*, *I.A.S.*, or *T.A.S.*

V_r (*Aero.*). Abbrev. for *rotation speed*.

vacancy (*Nuc.*). Unoccupied site for ion or atom in crystal.

vaccinal (*Med.*). Of, pertaining to, or caused by vaccine or vaccination.

vaccination (*Med.*). (1) Inoculation into the skin of the virus of vaccinia in order to immunize the person so treated against smallpox. (2) The therapeutic application of a vaccine made from any micro-organism.

vaccine (*Med.*). (1) The lymph (containing virus) taken from the cow-pox vesicle. (2) Of, or pertaining to, vaccinia. (3) Of, pertaining to, or connected with, the cow. (4) A preparation of any micro-organism or virus, either killed or so treated as to lose its virulence, for introduction into the body in order to stimulate the production of antibodies to the micro-organism(s) introduced, in order to confer immunity against any subsequent infection by the same type of micro-organism(s).

vaccinia (*Med., Vet.*). Cow-pox. A disease of cows characterized by the eruption of papules and pustules on the udder and teats. Communicable to man, and caused by a virus related to that causing smallpox in man.

vaccinial (*Med.*). Of, pertaining to, or caused by, vaccinia.

vacciniin (*Chem.*). $C_6H_{11}O_6COC_6H_5$. Glucoside, rel. mol. mass 284·16, m.p. 104°C, soluble in water and ethanol.

vacuolar membrane (*Biol.*). The protoplasmic membrane which bounds a vacuole, separating it from the surrounding cytoplasm.

vacuolate (*Biol.*). Vesicular; having vacuoles.

vacuole (*Biol.*). A small space or cavity in cytoplasm, generally containing fluid.

vacuum (*Vac.*). Literally, a space totally devoid of any matter. Does not exist, but is approached in inter-stellar regions. On earth, the best vacuums produced have a pressure of about 10^{-8} N/m². Used loosely for any pressure lower than atmospheric, e.g., train braking systems, 'vacuum' cleaners, etc.

vacuum arc furnace (*Chem.*). One in which a small specimen is heated by a high-voltage arc in an inert gas, e.g. argon, at low pressure.

vacuum augmenter (*Eng.*). An *air ejector* (q.v.) placed in a steam condenser to produce a higher degree of vacuum than is obtainable by the use of an air-pump alone.

vacuum blasting (*Eng.*). Blasting process, utilizing compressed air to apply the abrasive, which with the resulting debris is then withdrawn by vacuum, cleaned, and re-used.

vacuum brake (*Eng.*). A brake system used on railway trains and goods vehicles, in which a vacuum, maintained in reservoirs by exhausters, and under the control of the driver, operates all brake cylinders. See **continuous brake.**

vacuum concrete (*Civ. Eng.*). Concrete enclosed in specially prepared shuttering which incorporates fine filters and air ducts. After pouring, strong suction is applied to the ducts by means of pumps and the excess water is extracted. This enables the shuttering to be removed and re-used much sooner than by traditional methods.

vacuum crystallization (*Chem.*). Crystallization of a solution in vacuum at a temperature lower than its b.p. at ordinary pressure; used in sugar refineries to separate sugar from syrups.

vacuum distillation (*Chem.*). Distillation under reduced pressure. As a reduction of pressure effects a lowering of the boiling-point, many thermolabile substances can be distilled.

vacuum evaporation (*Phys., Space*). Under normal conditions, there is an equilibrium of molecular exchange at the surface of a solid body—molecules leaving the surface and others captured by it. Under vacuum conditions and in space, these molecules are lost.

vacuum-filament lamp (*Elec. Eng.*). An incandescent electric lamp in which the filament is enclosed in a highly evacuated bulb.

vacuum filtration (*Chem. Eng.*). A process of filtration where a partial vacuum is applied to increase the rate of filtration by causing the liquid to be sucked through the filter.

vacuum forming (*Plastics*). Shaping process applied to heated sheets of thermoplastics by clamping the plastic sheet in a holder, heating the sheet (usually electrically) then applying suction when the sheet is in the pliable, or 'rubbery', state.

vacuum furnace (*Met.*). One in which material is outgassed by radiant treating *in vacuo*.

vacuum-fusion gas analysis (*Vac. Tech.*). Determination of total oxygen, nitrogen or hydrogen in a sample by melting it in a vacuum in contact with carbon or graphite and analysing the gas evolved. A molten metal bath may or may not be used to dissolve the sample.

vacuum impregnation (*Elec. Eng.*). The process of treating armature and transformer windings by applying moisture-resisting varnish to the insulation, under vacuum, thereby ensuring that the varnish penetrates the pores of the insulating material when normal atmospheric conditions are restored.

vacuum melting (*Met.*). Steel refining process in which selected raw materials are remelted in an induction furnace surrounded by a vacuum tank.

vacuum oven (*Elec. Eng.*). An oven for heating armature and transformer windings under vacuum, so as to drive off all moisture from the insulation prior to impregnation.

vacuum photocell (*Electronics*). High-vacuum photoemissive cell in which anode current equals total photoemission currents, so that strict proportionality between current and incident illumination is obtained.

vacuum printing frame (*Print.*). A frame from which air can be exhausted to ensure close contact when printing down for any of the printing processes.

vacuum pump (*Eng.*). General term for apparatus which displaces gas against a pressure.

vacuum servo (*Autos.*). Servo mechanism operated by a vacuum provided by the induction pipe of the engine; used in power-assisted brake systems.

vacuum solidification (*Vac. Tech.*). An assessment of the volume of gas contained in aluminium, copper and their alloys by observation of the effect of gas evolution upon the surface appearance of a liquid sample solidifying in a vacuum.

vacuum switch (*Elec. Eng.*). One operating in a vacuum to avoid arcing.

vacuum tube (*Electronics*). See **valve.**

vacuum tube rectifier (*Elec. Eng.*). One which exploits the unidirectional movement of electrons flowing from heated electrode to gathering electrode.

vacuum tube voltmeter (*Elec. Eng.*). U.S. for **valve voltmeter.**

vagility (*Ecol.*). The inherent power of movement of individuals of a species or its disseminules, especially their power to cross barriers.

vagina (*Bot.*). A sheathing leaf-base. (*Zool.*) Any sheathlike structure; the terminal portion of the female genital duct leading from the uterus to the external genital opening. *adjs.* **vaginal, vaginant, vaginate, vaginiferous.**

vaginal plug (*Zool.*). In female Rodents and Insectivores, the coagulated secretion of Cowper's glands which blocks the vagina and prevents premature escape of seminal fluid.

vaginicolous (*Zool.*). Living in a sheath, as some *Protozoa.*

vaginismus (*Med.*). Painful spasmodic contraction of the muscles of the vagina and/or of the muscles forming the pelvic floor; can prevent conception and is usually psychogenic in origin.

vaginitis (*Med.*). Inflammation of the vagina.

vaginula (*Bot.*). A minute sheath surrounding the base of the seta in *Bryophyta.*

vagitus uterinus (*Med.*). The crying of a child while still *in utero*, just before birth.

vagotonia, vagotony (*Med.*). The condition of heightened activity of the vagus nerve.

vagus (*Zool.*). (1) The tenth cranial nerve of Vertebrates, supplying the viscera and heart (see also **pneumogastric**), and, in lower forms, the gills and lateral line system. (2) In Mammals, the larynx.

valence band (*Chem.*). Range of energy levels of electrons which bind atoms of a crystal together.

valence electrons (*Chem.*). Those in the outer shell of an *atom*, which, by gaining, losing, or sharing such electrons, may combine with other atoms to form *molecules.*

valency (*Chem.*). The combining power of an atom or group in terms of hydrogen atoms (or equivalent). The valency of an ion is equal to its charge. For *valency bond, valence*, see **chemical bond.** (*Cyt.*) The numerical arrangement of the chromosomes in a nucleus, i.e., single, paired, etc. See **univalent, bivalent.**

Valentian Series (*Geol.*). The lowest series of rocks in the Silurian System in England and Wales. Equivalent to the Llandovery Series.

valentinite (*Min.*). (III)oxide of antimony, Sb_2O_3, occurring as orthorhombic crystals or radiating aggregates; snow-white when pure: it is formed by the decomposition of other ores of antimony.

valerian (*Pharm.*). A drug obtained from the plant *Valeriana officinalis*, used as a *depressant* of the central nervous system.

valeric acids (*Chem.*). $C_4H_9 \cdot COOH$, monobasic fatty acids, of which isomers are known, viz., n-valeric acid (pentanoic acid), $CH_3 \cdot (CH_2)_3 \cdot COOH$, b.p. 185°C; isovaleric acid (3-methylbutanoic acid, $(CH_3)_2{=}CH \cdot CH_2COOH$, b.p. 175°C; methylethylacetic acid (2-methylbutanoic acid), $(CH_3)(C_2H_5)CH \cdot COOH$, b.p. 177°C; pivalic acid (2,2-dimethylpropanoic acid), $(CH_3)_3C \cdot COOH$, b.p. 164°C.

valine (*Chem.*). $(CH_3)_2CH \cdot CH(NH_2) \cdot COOH$, α-amino-isovaleric acid, an amino acid obtained by the hydrolysis of certain proteins.

valium (*Pharm.*). TN for diazepam, a tranquillizer of wide use, e.g. as premedication, anticonvulsant, etc.

valley (*Build.*). The re-entrant angle formed between two intersecting roof slopes. (*Geol.*) Any hollow or low-lying tract of ground between hills or mountains, usually traversed by streams or rivers, which receive the natural drainage from the surrounding high ground. Usually valleys are developed by stream erosion; but in special cases, faulting may also have contributed, as in rift valleys.

valley board (*Build.*). A board nailed along the top of the valley rafter as a support for a *laced valley* (q.v.).

valonia (*Bot.*). The large unripe acorn-cup of the Valonia oak (*Quercus aegilops*) of Asiatic Turkey; used extensively in tanning.

valsoid (*Bot.*). Having the perithecia in a circle in the stroma.

value (*Nuc. Eng.*). **Separative work content (U.S.).**

valvate aestivation (*Bot.*). The condition when the perianth segments touch but do not overlap.

valvate dehiscence (*Bot.*). Liberation of pollen from anthers, or of seeds from dry fruits, by means of little flaps of upraised wall material.

valve (*Bot.*). (1) A part of the wall of the fruit-wall which separates at dehiscence. (2) One of the two silicified halves of the cell wall of a diatom. (*Electronics*) Simple vacuum device for amplification by an electron stream. There are many types. Also called **bottle.** See **tube** (2). (*Eng.*) Any device which controls the passage of a fluid through a pipe. (*Zool.*) Any structure which controls the passage of material through a tube, duct, or aperture, usually in the form of membranous folds, as the auriculo-ventricular valves of the heart: any membranous fold resembling a valve, as the valve of Vieussens: in *Mollusca, Cirripedia, Brachiopoda*, one of several separate pieces composing the shell: in Insects, a covering plate or sheath, especially one of a pair which can be opposed to form a tubular structure, as the valves of the ovipositor. Also **valva.**

valve adaptor (*Electronics*). Device which enables a valve to be fitted to a socket for which it was not originally designed.

valve admittance (*Electronics*). Control grid admittance under working or specified conditions, including that arising from base sockets, but not from leads thereto.

valve base (*Electronics*). Insulating cap cemented to the envelope of a valve and fitted with contacts connected to the electrodes. It enables the valve to be plugged readily into the circuit.

valve bounce (*Eng.*). The unintended secondary opening of an engine valve due to inadequate rigidity of various parts in the valve gear. The deflected parts allow the valve to seat too early and when their strain energy is released, the valve opens again. This bounce causes valve breakage, seat wear, and irregular functioning.

valve box (or chest) (*Eng.*). In a force-pump or steam engine, the chamber which contains the valves or valve: the *steam chest* (q.v.) of a steam engine.

valve characteristic (*Elec. Eng.*). Graphical relation between voltage and current for specified electrodes, all other potentials being maintained constant.

valve coupling (*Elec. Eng.*). Use of a valve, which permits power to pass in only one direction in a transmission channel.

valve diagram (*Eng.*). For a steam-engine slide-valve, a graphical method of correlating the throw and angle of advance of the eccentric, the lead and laps of the valve, and the points of admission, cut-off, compression, and release. The 'Bilgram', the 'Reuleaux', and the 'Zeuner' valve diagrams are examples of the method.

valve effect (*Elec. Eng.*). The unilateral conductivity of certain electrodes (notably aluminium) in suitable solutions. As anodes, they may withstand several hundred volts, although current will pass freely in the opposite direction.

valve face (*Eng.*). The sealing surface of a valve which slides over, or beds on to, the seating.

valve gear (*Eng.*). The linkage by which the valves of an engine derive their motion and timing from the crankshaft rotation.

valve hiss (*Telecomm.*). See **valve noise.**

valve inserts (*I.C. Engs.*). Valve seatings of special heat- and lead-resisting steel which are pressed into the alloy heads of high-duty petrol engines.

valve noise or hiss (*Telecomm.*). That present in telephones connected in the anode circuit of a valve, in the absence of signals applied to the grid, due to shot effect, microphonic action of electrodes, thermal agitation voltage, etc.

valve nomenclature (*Electronics*). Description of valves in terms of the number of electrodes therein, e.g., *diode*, *triode*, etc.

valve-opening diagram (*I.C. Engs.*). A diagram showing the lift or the opening area of a valve to a base of engine crank angle or piston displacement.

valve parameters (*Electronics*). Numerical quantities obtained from the characteristic curves and used in circuit analysis. See **amplification factor, differential anode resistance, mutual conductance, transistor parameters.**

valve rectifier (*Elec. Eng.*). A rectifier of the vacuum or the gas-discharge type.

valve relay (*Elec. Eng.*). A thermionic valve arranged to operate as a synchronous voltage relay in high-frequency a.c. circuits.

valve rockers (*I.C. Engs.*). See **rocker arms.**

valve rustle (*Cinema.*). Rustling noise accompanying soundfilm reproduction; caused by the clashing of the ribbons in the light valve in the recorder through over-modulation or incorrect adjustment.

valve socket (*Electronics*). Arrangement of contact springs or pins into which a valve can be plugged for connexion to the rest of the circuit; moulded into low-loss materials.

valve spring (*Eng.*). The helical spring (or springs) used to close a poppet valve after it has been lifted by the cam; generally, any spring which closes a valve after it has been lifted mechanically or by fluid pressure.

valve timing (*I.C. Engs*). See **timing.**

valve tower (*Civ. Eng.*). A hollow concrete or masonry tower built within a reservoir; equipped with draw-off pipes at different levels for taking off water for supply purposes.

valve tube (*Electronics*). See **kenotron.**

valve voltmeter (*Elec. Eng.*). Valve used for measuring voltages, rectified output current being dependent on voltage applied to the input. It takes negligible power from measured circuit; can be calibrated at low frequencies for use at very high frequencies, when other means are impossible. See **diode voltmeter, slide-back voltmeter**; U.S. term **vacuum tube voltmeter.**

valvula (*Zool.*). In bony Fish, a process of the cerebellum projecting downwards and forwards beneath the roof of the mid-brain.

valvulae conniventes (*Zool.*). Folds of the mucous membrane of the small intestine.

valvular (*Bot.*). Opening by means of valves.

valvulate, valvate (*Zool.*). Said of pedicellariae in which the jaws are broader than they are long; such pedicellariae are always sessile.

valvulitis (*Med.*). Inflammation of a valve of the heart.

vanadinite (*Min.*). Vanadate(V) and chloride of lead(II), typically forming brilliant reddish hexagonal crystals or globular masses encrusting other minerals in lead-mines.

vanadium (*Chem.*). A very hard, whitish metallic element. Symbol V, at. no. 23, r.a.m. 50·941, rel. d. at 20°C 5·5, m.p. 1710°C, electrical resistivity 22×10^{-8} ohm metres at 20°C. Its principal use is as a constituent of alloy steel, e.g., in chromium-vanadium, manganese-vanadium, and high-speed steels.

vanadium steel (*Met.*). Typical constitution is carbon 0·4–0·5%, chromium 1·1–1·5%, vanadium 0·15–0·2%, balance being iron. The vanadium removes occluded oxygen and nitrogen thus giving improved tensile and elastic quality. Up to 0·5% V may be used.

vanadocytes (*Zool.*). In some *Ascidiacea*, blood corpuscles carrying a green pigment containing vanadium, which has been claimed, but not proved, to be a respiratory pigment.

vanadous, vanadic (*Chem.*). Referring to divalent and trivalent vanadium respectively.

vanadyl (*Chem.*). The cation VO^{2+}.

Van Allen radiation belts (*Astron.*). Zones of charged particles trapped in and accelerated by the earth's magnetic field. At least two zones between 1200 and 20 000 miles from the earth's surface have been detected by artificial satellites.

vancometer (*Paper*). An instrument for measuring the penetration of oil or ink into paper by timing the disappearance of a standard film of the oil formed on the paper by a roller on an inclined plane.

Van de Graaff generator (*Elec. Eng.*). Very-high-voltage electrostatic machine, using a high-speed belt to accumulate charge in a large Faraday cage, which takes the form of a metal globe. Recent models use Freon or nitrogen gas under high pressure. Used as voltage source for accelerator tubes, e.g., in neutron sources.

Van de Pol oscillator (*Electronics*). Class of pentode relaxation oscillator.

van der Hoeven's organ (*Zool.*). In male *Nautiloidea*, a median body lying between the groups of tentacles and having its surface divided into folds.

van der Waals' equation (*Chem.*). An equation of state which takes into account the effect of intermolecular attraction at high densities and the reduction in effective volume due to the actual volume of the molecules: $(P+a/v^2)(v-b)=RT$, a and b being constant for a particular gas. See **gas laws.**

van der Waals' forces (*Chem.*). Weak attractive forces between molecules or crystals, represented by the coefficient *a* in *van der Waals' equation.* They vary inversely as the seventh power of the interatomic distance, and are due to momentary dipoles caused by fluctuations in the electronic configuration of the molecules.

Vandyke brown (*Paint.*). A naturally occurring brown artists' pigment, consisting of iron oxide with vegetable decomposition products and earth. A mixture of iron oxide pigments and lampblack, or water-soluble brown dyestuff obtained by extraction of the pigment. Also **Cassel brown.**

Vandyke pieces (*Plumb.*). The lead scraps remaining after the cutting out of a stepped flashing.

vane (*Eng.*). (1) One of the elements which variably divide the fluid space in a *vane pump* (q.v.). (2) An alternative name for blade in turbines, flow meters and similar rotary devices. (*Surv.*) A disk attachment to a levelling staff; it provides a sliding target which the staffman can move into the line of sight of the level. See **target rod.** (*Zool.*) The web of a feather, composed of the barbs and barbules; also called **vexillum.**

vane pump (*Eng.*). A type of pump used, e.g., as a vacuum or oil pump or as a compressor, in which a slotted rotor is mounted eccentrically in a circular stator (or a similar geometrical arrangement), and vanes sliding in the rotor slots divide the crescent-shaped fluid space into variable volumes.

vane relay (*Elec. Eng.*). Relay in which the moving element carrying the contacts comprises a disk or vane, which is propelled by the reaction of eddy currents induced therein by a coil-excited magnet.

vanes (*Aero.*). See **inlet guide —, nozzle guide —, swirl —.**

vane wattmeter (*Elec.*). Instrument for measuring power transmitted in waveguide, depending on mechanical forces induced in a vane.

Van Gieson stain (*Micros.*). A general stain, containing picric acid and acid fuchsin and usually used in conjunction with iron haematoxylin.

vanillin (*Chem.*). 3-Methoxy-4-hydroxy-benzene carbaldehyde, found in vanilla pods and some other plants. It crystallizes in white needles, m.p. 80°C, and has the formula:

CHO

OCH$_3$

OH

vanner (*Min. Proc.*). See **frue vanner.**

vanning (*Min. Proc.*). Rough estimate of cassiterite or other heavy mineral, made by washing finely ground sample on a flat shovel, or in a vanning plaque.

van't Hoff factor (*Chem.*). The ratio of the number of dissolved particles (ions and undissociated molecules) actually present in a solution to the number there would be if no dissociation occurred. Symbol *i*.

van't Hoff's law (*Chem.*). The osmotic pressure of a dilute solution is equal to the pressure which the dissolved substance would exert if it were in the gaseous state and occupied the same volume as the solution at the same temperature.

van't Hoff's reaction isochore (*Chem.*). For a reversible reaction taking place at constant volume,

$$\frac{d \log_e K}{dT} = \frac{\Delta U}{RT^2},$$

where K is the equilibrium constant, T is the absolute temperature, R is the gas constant, ΔU is the heat absorbed in the complete reaction.

van't Hoff's reaction isotherm (*Chem.*). For a reversible reaction taking place at constant temperature,

$$-\Delta A = RT \log_e K - RT\Sigma n \log_e c,$$

where $-\Delta A$ is the decrease in free energy, R the gas constant, T the absolute temperature, K the equilibrium constant, and $\Sigma n \log_e c$ is of the same form as $\log_e K$, but with the equilibrium concentrations replaced by the initial values.

vapography (*Photog.*). An effect obtained by the action of vapours on sensitive photographic materials.

vaporimeter (*Chem.*). An apparatus in which the volatility of oils is estimated by heating them in a current of air.

vaporization (*Chem.*). The conversion of a liquid or a solid into a vapour. See **entropy of fusion.**

vaporizing oil (*Fuels*). See **kerosine.**

Vapotron (*Electronics*). TN for high-frequency transmitting valve, the anode of which is cooled by boiling water.

vapour (*Phys.*). A gas which is at a temperature below its critical temperature and can therefore be liquefied by a suitable increase in pressure.

vapour barrier (*Chem.*). Covering which prevents water condensing within the insulation around a cold surface. Often a pigmented vinyl polymer solution applied by brush.

vapour compression cycle (*Eng.*). A reversed Carnot cycle modified by replacing the isothermal compression stage by removal of heat at constant pressure, involving the condensation of fluid, replacing the adiabatic expansion stage by expansion through a regulating valve, and replacing the isothermal expansion stage by absorption of heat at constant pressure, involving evaporation of the fluid.

vapour concentration (*Meteor.*). Same as **absolute humidity.**

vapour-liquid-solid mechanism (*Chem.*). Method, applicable to most crystalline substances, of growing different near-perfect crystalline forms.

vapour lock (*Eng.*). Of a volatile fluid in a pipe, the formation of vapour in a petrol feed-pipe to a carburettor caused by undue heating of the pipe, resulting in an interruption of flow.

vapour permeability (*Paper*). The rate of passage of a vapour, e.g., water vapour, through the paper.

vapour phase inhibitor (*Chem.*). Stable organic chemicals coated on paper or board which slowly evaporate, so preventing air reaching the metallic articles enclosed in the package and preventing corrosion, e.g., ethanolamine benzoate, cyclohexylamine nitrite (nitrate(III)). Abbrev. **V.P.I.**

vapour pressure (*Phys.*). The pressure exerted by a vapour, either by itself or in a mixture of gases. The term is often taken to mean saturated vapour pressure, which is the vapour pressure of a vapour in contact with its liquid form. The saturated vapour pressure increases with rise of temperature. See **saturation of the air.**

vapour pump (*Vac. Tech.*). A pump employing vapour jet(s) as stage(s) in pumping process.

Vaqueros formation (*Geol.*). Strata of shallow-water origin and of Lower Miocene age. The formation includes the chief oil-bearing sands of the Coalinga district, California.

VAr (*Elec. Eng.*). Abbrev. for *v*olt-*a*mperes *r*eactive. Unit of reactive power. See **reactive volt-amperes.**

var (*Nav.*). (*V*isual-*a*ural *r*ange). Navigational system giving mutually perpendicular courses, one of which is displayed visually and the other aurally.

varactor (*Electronics*). Two-electrode semiconductor with a non-linear capacitance instantaneously dependent on voltage; cf. *varistor*.

varec (*Bot.*). The French name for **kelp.**

variable (*Maths.*). A term used in algebra to describe symbols like x, y, suggesting that the symbol can be replaced by an element from a specified set to give a meaningful statement. If x can be replaced by elements of a set S, then x is said to be a variable in the set S.

variability (*Chem.*). The number of *degrees of freedom* (1, q.v.) of a system.

variable-aperture shutter (*Cinema.*). One which can be closed or opened gradually while filming a scene, producing a fade-out or fade-in. A lap-dissolve is obtained by rewinding to the commencement of the operative before taking the second scene.

variable area (*Cinema.*). Pertaining to a sound-track, which is divided laterally into transparent and opaque areas. The line which divides these areas corresponds to the waveform of the recorded signal.

variable-area propelling nozzle (*Aero.*). A turbojet *propelling nozzle* (q.v.) which can be varied in effective outlet area, either mechanically or aerodynamically, to match it to the optimum engine operating conditions (principally thrust), thereby improving fuel economy: essential for the efficient use of an *afterburner* (see **reheat**) and in supersonic flight.

variable attenuator (*Telecomm.*). See **attenuator.**

variable-contrast enlarger (*Photog.*). One which, by means of polarizing filters, controls the ratio of scattered light to transmitted light, thus varying the contrast of the projected image.

variable coupling (*Elec. Eng.*). An electromagnetic coupling between two a.c. circuits in which the mutual inductance is continuously variable between wide limits.

variable density (*Cinema.*). Pertaining to a sound-track, whose average light transmission varies in proportion to some characteristic of the applied signal.

variable-density wind tunnel (*Aero.*). A closed-circuit wind tunnel wherein the air may be compressed to increase the *Reynolds number* (q.v.); also **compressed-air wind tunnel.**

variable elevation beam antenna (*Radio*). One comprising a large number of dipoles, the maximum lobe of radiation being controlled by adjusting the phases of separate contributions.

variable-focus lens (*TV*). See **zoom lens.**

variable geometry (*Aero.*). See **variable sweep.**

variable inductor (*Elec. Eng.*). An *inductor* whose self-inductance is continuously variable.

variable-inlet guide vanes (*Aero.*). See **inlet guide vanes.**

variable mu(tual) conductance valve (*Electronics*). See **remote cut-off tube.**

variable phase (*Nav.*). See **vor.**

variable-pitch airscrew (*Aero.*). See **airscrew.**

variable-ratio transformer (*Elec. Eng.*). A transformer whose voltage ratio can be varied by altering the number of active turns in either the primary or the secondary winding.

variable-reluctance pick-up (*Mag.*). One in which the reluctance of a magnetic circuit is varied, the consequent modulation of flux from a permanent magnet generating an e.m.f., as in a transducer or gramophone-record reproducer.

variable resistance (*Elec. Eng.*). See **rheostat.**

variable-speed drive (*Elec. Eng.*). An electric drive whose speed is continuously variable between wide limits.

variable speed indicator (*Aero.*). A sensitive form of differential pressure gauge which measures variations in pressure sensed at the static pressure tube and indicates them in terms of rates of climb and descent.

variable-speed motor (*Elec. Eng.*). An electric motor whose speed is continuously variable between wide limits.

variable-speed scanning (*TV*). System of scanning employed in a velocity-modulation system, a constant-intensity beam going slow for highlights and fast for dark areas.

variable stars (*Astron.*). Those stars whose apparent magnitudes are not constant but vary over a range. There are three main classes, distinguished by their light-curves: (1) *eclipsing binaries*; (2) pulsating stars, including the short-period variables (the *cepheids* and *RR Lyrae stars*), and the long-period *Mira variables*; (3) the eruptive variables, including the *novae*, *supernovae* and *flare stars*. More than 20 000 variable stars have been catalogued.

variable sweep (*Aero.*). An aeroplane with wings so hinged that they can be moved backward and forward in flight to give sharp *sweepback* for low *drag* in supersonic flight and high *aspect ratio*, with good lifting properties, for take-off and landing. Colloq. **swing-wing.**

variable-voltage control (*Elec. Eng.*). A system of controlling speed by varying the voltage applied at the motor terminals.

Variac (*Telecomm.*). TN of autotransformer in the form of a toroid winding on ring laminations, the output voltage being varied by a rotating brush contact on the turns.

variance (*Stats.*). The average of the squares of the deviations of a number of observations of a quantity from their mean value, the quantity being termed a *variate*. Cf. *standard deviation.*

variant (*Biol.*). A specimen differing in its characteristics from the type and produced either by changed environmental conditions and/or by mutation.

variate (*Automation*). Measurement of machined part as compared with that intended, in machine-tool control. (*Stats.*) The variable quantity being studied. See also **variance.**

variation (*Astron.*). The name given to the fourth principal periodic term in the mathematical expression of the moon's motion, caused by the variation of the residual attraction of the sun on the earth-moon system during a synodic month; it has a maximum value of 39′ and a period of 14·77 days. (*Biol.*) The differences between the offspring of a single mating; the differences between the individuals of a race, subspecies, or species; the differences between analogous groups of higher rank. (*Surv., etc.*) See **magnetic declination.**

variation factor (*Light*). The ratio between the maximum and the minimum illumination along a street or roadway illuminated at intervals by overhead lamps.

variation of latitude (*Astron.*). A phenomenon, first detected in 1888 by Küstner, who showed that, owing to the spheroidal form and non-rigid consistency of the earth, its axis of rotation does not remain constant in direction but varies in a regular manner about a mean position, so that the latitude of a place also undergoes periodic variations.

variation order (*Civ. Eng.*). A document giving authority for some alteration in work being done under contract.

variations (*Build., Civ. Eng.*). See **extras.**

varicella (*Med.*). See **chickenpox.**

varicocele (*Med.*). A varicose condition of the plexus of veins which leave the testis to form the spermatic vein, forming at the upper part of the testis a swelling which feels like a mass of spaghetti.

varicose (*Bot.*). Dilated. (*Med.*) Of the nature of, pertaining to, or affected by, a varix or varices; (of veins) abnormally dilated, lengthened, and tortuous.

variegated (*Bot., etc.*). Marked irregularly with patches of diverse colour.

variegated copper ore (*Min.*). A popular name for *bornite* (q.v.). So named from the characteristic tarnish that soon appears on the freshly fractured surface.

variety (*Biol.*). A race; a stock or strain; a sport or mutant; a breed; a subspecies; a category of individuals within a species which differ in constant transmissible characteristics from the type but which can be traced back to the type by a complete series of gradations; a geographical or biological race.

Varigroove (*Acous.*). Arrangement of variable groove spacing on microgroove records, which allows more recording time per radial inch of record to be accommodated.

varimeter, varmeter, or **varometer** (*Elec. Eng.*). Equivalent terms for instrument measuring reactive volt-amperes in circuit.

variocoupler (*Elec. Eng.*). Device comprising two inductors whose mutual inductance can be varied for variable inductive coupling between two circuits.

Vario-Klischograph (*Print.*). A development of the *Klischograph* (q.v.), designed to produce plates enlarged or reduced from the copy, single colour or colour-corrected sets.

variola (*Med.*). See **smallpox.** (*Vet.*) See **swine pox.**

variola minor (*Med.*). See **alastrim.**

variolarioid, variolose (*Bot.*). Covered with a powdery coating or with very small tubercles.

variolite (*Geol.*). A fine-grained igneous rock of basic composition containing small more or less spherical bodies (*varioles*) consisting of minute radiating fibres of feldspar, comparable with the more perfect spherulites in acid igneous rocks. The term would be better used as an adjective, i.e., *variolitic* basalt, etc.

varioloid (*Med.*). A mild type of smallpox occurring in one who has been vaccinated.

variometer (*Elec. Eng.*). Variable inductor comprising two coils connected in series and arranged one inside the other, the inner coil rotated so as to vary the mutual inductance between them and hence the total inductance.

varioplex (*Teleg.*). Device used in conjunction with *time-division multiplex* (q.v.), which enables the multiplexed channels to be distributed automatically between the users in a variable manner according to the number of users who are transmitting at any given time.

variscite (*Min.*). A greenish hydrated phosphate of aluminium ($AlPO_4 \cdot 2H_2O$) occurring as nodular masses in Utah, U.S.

varistor (*Electronics*). Two-electrode semi-conductor with a nonlinear resistance dependent on instantaneous voltage. Used to short-circuit transient high voltages in delicate electronic devices.

VariTyper (*Typog.*). A justifying typewriter producing copy suitable for reproduction by a printing process, and requiring a second typing for line justification. See **typewriter composition.**

varix (*Med.*). Varicose vein; a vein that is abnormally dilated, lengthened, and tortuous. *pl.* varices.

Varley loop test (*Elec. Eng.*). A method of determining the position of a cable fault, in which resistance measurements are made with a resistance bridge, first, so that the fault forms one junction of the bridge, and, secondly, so that the conductor resistance of the cable is measured directly.

varnish (*Paint*). A solution of a resin or resinous gum in spirits or oil applied as an extra glossy coat on a painted surface, as protection.

varnished (*Bot.*). Said of a surface which appears to be covered by a thin shining film.

varnished insulation (*Elec.*). Varnished cotton, silk, and cambric which were much used for insulating wires and flexes.

varved clays (*Geol.*). Distinctly and finely stratified clays of glacial origin, deposited in lakes during the retreat stage of glaciation. The stratification is thought to be a seasonal banding, and its study enabled Baron de Geer to work out the chronology of the Pleistocene Ice Age.

vas (*Zool.*). A vessel, duct, or tube carrying fluid. *pl.* vasa. *adj.* vasal. See **vas deferens.**

vasa efferentia (*Zool.*). A series of small ducts by which the semen is conveyed from the testis to the vas deferens.

vasa recta (*Zool.*). In the Mammalian kidney, groups of small arteries and veins alternating with groups of uriniferous tubules in the part of the medulla nearest to the cortex.

vasa vasorum (*Zool.*). In Vertebrates, small blood vessels ramifying in the external coats of the larger arteries and veins.

vascular (*Bot., Zool.*). Pertaining to vessels which convey fluids or provide for the circulation of fluids: provided with vessels for the circulation of fluids.

vascular anastomosis (*Bot.*). A small transversely directed vascular bundle, acting as a link between the main vascular bundles of a stem or root.

vascular area (*Zool.*). See **area vasculosa.**

vascular bundle, vascular strand (*Bot.*). A strand of conducting tissue, consisting of xylem and phloem, sometimes separated by cambium; sclerenchymatous supporting tissue may also be present.

vascular ray (*Bot.*). A sheet of cells, usually mostly parenchymatous, lying radially in a stem or root, appearing in cross-section as a narrow radial streak, in radial longitudinal section as a plate of cells; the ray lies partly in the xylem and partly in the phloem, and serves to conduct solutions horizontally in the stem or root. Often called a **medullary ray.**

vascular-ray initial (*Bot.*). A cell of the cambium which divides to give daughter cells which are converted into the cells forming a vascular ray.

vascular system (*Zool.*). The organs responsible for the circulation of blood and lymph, collectively.

vascular tissue (*Bot.*). Xylem and phloem.

vasculiform (*Bot.*). Shaped like a little pot.

vasculum (*Bot.*). A receptacle for collecting botanical specimens.

vas deferens (*Zool.*). A duct leading from the testis to the ejaculatory organ, the urino-genital canal, the cloaca, or the exterior. *pl.* vasa.

vasectomy (*Surg.*). Excision of the vas deferens, or of part of it, either therapeutically or for sterilization.

Vaseline (*Chem.*). TN for high-boiling residues obtained from the distillation of petroleum; a *petroleum jelly* (q.v.).

VASI (*Aero.*). Abbrev. for *visual approach slope indicator*.

vasicentric (*Bot.*). See **paratracheal.**

vasifactive (*Zool.*). See **vasoformative.**

vasiform orifice (*Zool.*). In certain *Homoptera* ('white-flies', family *Aleyrodidae*) a conspicuous opening on the dorsal surface of the last abdominal segment. The anus opens within the orifice, and honeydew excreted through the anus accumulates on the *lingula* (q.v.) inside it. It is present in both larval and adult stages.

vasiform tracheide (*Bot.*). A wide tracheide capable of conducting water.

vaso-. Prefix from L. *vas*, vessel.

vasochorial placenta (*Zool.*). A chorioallantoic placenta in which the epithelium and the endometrium of the uterus disappear, and the chorion is in intimate contact with the endothelial wall of the maternal capillaries. Also known as **endotheliochorial placenta.**

vasoconstrictor (*Zool.*). Of certain autonomic nerves, causing constriction of the arteries.

vasodilator (*Pharm.*). A drug, e.g., glyceryl nitrate, which effects expansion of the arteries. (*Zool.*) Of certain autonomic nerves, causing expansion of the arteries.

vaso-epididymostomy (*Surg.*). The operation of anastomosing the vas deferens to the upper part of the epididymis, forming a communication between the two; performed for the treatment of sterility in the male.

vasoformative (*Zool.*). Pertaining to the formation of blood or blood vessels.

vasoformative cells (*Zool.*). See **angioblast.**

vasoganglion (*Zool.*). See **red gland.**

vasohypertonic (*Zool.*). See **vasoconstrictor.**

vasohypotonic (*Zool.*). See **vasodilator.**

vasoinhibitory (*Zool.*). See **vasodilator.**

vasoligature (*Surg.*). Ligature of the vas deferens, as in *Steinach's operation* (q.v.).

vasomotor (*Zool.*). Causing constriction or expansion of the arteries; as certain nerves of the autonomic nervous system.

vasopressin (*Physiol.*). A hormone, produced by the posterior lobe of the pituitary, which is antidiuretic, and also causes contraction of involuntary muscles, especially of the arterioles. Used in the diagnosis and treatment of *diabetes insipidus*. **TN Pitressin.**

vasopressor (*Med.*). Substance which causes a rise of blood pressure.

vat dyestuffs (*Chem.*). A series of insoluble dyestuffs that can be reduced to their water-soluble leuco-compounds, which are oxidized by exposure to the air, thus producing the dyestuff direct on the fibre.

vaterite (*Min.*). A less common polymorph of calcium carbonate, crystallizing in the hexagonal system. Cf. *calcite, aragonite*.

Vater's corpuscles (*Zool.*). See **Pacinian corpuscles.**

vat machine (*Paper*). See **cylinder mould machine.**

vault (*Build.*). (1) An arched roof or ceiling. (2) A room or passage covered by an arched ceiling. (3) An underground room.

vault light (*Build.*). A form of *pavement light*.

V-band (*Radar*). Frequency band from 4·6 to $5{\cdot}6 \times 10^{10}$ Hz used in radar.

V-beam (*Radar*). Scanning by 'fan' beams, one vertical, the other inclined. Interval between reflections depends on target elevation.

V-chromosomes (*Gen.*). Those having two arms.

VCO (*Electronics*). Abbrev. for *voltage controlled oscillator*.

V-connexion (*Elec. Eng.*). An alternative name for the open delta connexion of two phases of a three-phase a.c. system.

V-curve (*Cables*). The power-factor/temperature curve of a cable with moisture, which shows a pronounced minimum at about 40°C.

V-cut (*Mining*). Centre, wedge or burn cut. In blasting a solid rock face, removal of a wedge by means of several pairs of shots drilled towards each other, preparatory to subsequent blasting.

VD (*Med.*). Abbrev. for *venereal disease*.

VDU (*Comp.*). Abbrev. for *video display unit*.

vector (*Aero.*). The course of an aircraft, missile, etc. (*Biol.*) An agent (usually an insect) which transmits a viral, bacterial, or fungal disease from one host to another. (*Elec. Eng.*) The representation of single-frequency alternating current or voltage, which with similar vectors in a plane can be manipulated, in accordance with relevant vector laws, on what is strictly an Argand diagram. Vectors, which are properly used to represent electric and magnetic fields in space, are normally distinguished by clarendon type, e.g. **i, v.** (*Maths.*) A vector or vector quantity is one which has magnitude and direction, e.g., force or velocity; two such quantities of the same kind obey the parallelogram law of addition. A *localized vector* is one in which the line of action is fixed, as contrasted with a *free vector*, in which only the direction is fixed.

vector addition (*Maths.*). Compounding of two vector quantities according to parallelogram law.

vector algebra (*Maths.*). Manipulation of symbols representing vector quantities according to laws of addition, subtraction, and multiplication, which these quantities obey.

vectored thrust (*Aero.*). The deflexion of the thrust from turbojet(s) to provide a **jet-lift** component. Particularly applied to a system using a ducted fan engine with bifurcated nozzles on the fan and the jet pipe so that there are four sources of thrust, thereby contributing to the balance of the system. The swivelling nozzles which deflect the thrust at any angle from horizontally aft to several degrees forward of the vertical are under the control of the pilot.

vector potential (*Elec. Eng.*). Potential postulated in electromagnetic field theory. Space differentiation (*curl*) of the vector potential yields the field. Magnetic vector potential is due to electric currents, while electric vector potential is assumed to be due to a flow of magnetic charges.

vector product (*Maths.*). Of two vectors, the vector perpendicular (right-hand screw convention) to both the given vectors, of magnitude equal to the product of the magnitudes of the two given vectors, multiplied by the sine of the angle between them. Vector products are usually denoted by $\mathbf{a} \times \mathbf{b}$, or $\mathbf{a} \wedge \mathbf{b}$, while scalar products are denoted by **a.b.** Scalar products alone are used in electrical engineering.

vector ratio (*Elec. Eng., etc.*). That between two alternating quantities, e.g., currents, in which both relative amplitudes and phases are expressed as vectors.

vectorscope (*TV*). Instrument which displays phase and amplitude of an applied signal, e.g., of chrominance signal in a colour TV system.

vee antenna (*Radio*). Line radiator folded in a V in the horizontal plane.

vee belt (*Eng.*). A power transmission belt having a cross-section of truncated vee form, usually running in corresponding vee grooves in pulleys.

vee gutter (*Build.*). A gutter of V-shape, as required, for example, along the valley between two roofs sloping towards each other.

vee joint (*Carp.*). A joint between *matched boards* (q.v.) which have been chamfered along their edges on the same side to present a vee depression at their junction.

vee notch (*Civ. Eng.*). A notch plate having a triangular notch cut in it, used for the measurement of small discharges.

veering (*Meteor.*). A clockwise change in the direction from which the wind comes. Cf. *backing*.

vee roof (*Build.*). A roof formed by two lean-to roofs meeting to enclose a valley.

vee-tail (*Aero.*). An aeroplane tail unit consisting of two surfaces on each side of the centre line, usually at about 45° to the horizontal, which serve both as tailplane and fin. The associated hinged control surfaces are so actuated that they move in unison up/down as elevators and left/right as rudders, following conventional movements of the control column and rudder bar respectively. Also **butterfly tail.**

vee thread (*Eng., etc.*). A screw thread in which the thread profile is V-shaped (as for the common Whitworth thread), as distinct from other forms, e.g., square thread.

vegetable black (*Chem.*). Originally soot produced by burning vegetable oils in the presence of insufficient air. Now usually another name for *lampblack* (q.v.). It is used as a pigment.

vegetable cork (*Heat.*). Granules of cork, compounded with resin binder and moulded to shapes of use in thermal insulation.

vegetable oils (*Chem.*). Oils obtained from plants, seeds, etc. Cf. *mineral oils.*

vegetable parchment (*Paper*). Produced by immersing a web of unsized paper through cold 70% sulphuric acid whereby the paper is parchmentized, and finally washing, neutralizing, and passing through dilute glycerine to soften it.

vegetable pole (*Zool.*). The lower portion or pole of an ovum in which cleavage is slow owing to the presence of yolk. Cf. *animal pole.*

vegetation (*Bot.*). All plants in a given area.

vegetative cone (*Bot.*). The apical meristem of a stem.

vegetative functions (*Zool.*). The autonomic or involuntary functions, as digestion, circulation.

vegetative multiplication (*Bot.*). Increase in the number of plants by the production of portions of the plant body which root and become detached from the parent.

vegetative nervous system (*Med.*). Sympathetic nervous system.

vegetative nucleus (*Zool.*). See **macronucleus.**

vegetative reproduction (*Zool.*). Propagation by budding.

vehicle (*Paint.*). The liquid substance which, when mixed with a pigment, forms a paint.

vehicle battery (*Elec. Eng.*). A battery of heavy-duty-type secondary cells, which forms the source of electrical energy in self-contained electric road vehicles.

veil (*Bot.*). (1) An evanescent membrane over an apothecium. (2) A sheath of hypha forming a complete membrane over a young fruit body of an agaric. (3) The calyptra of a moss. Also **velum.** (*Zool.*) See **velum.**

vein (*Bot.*). One of the smaller strands of conducting tissue in a leaf. (*Geol.*) A narrow planar body of rock, penetrating a mass of a different type of rock. Sometimes applied to particularly narrow igneous intrusions (dykes and sills), the term is more often applied to material deposited by solutions, such as quartz veins or calcite veins. Many ore deposits consist of veins in which the ore mineral is one of several constituents. (*Zool.*) A vessel conveying blood back to the heart from the various organs of the body: a wing nervure. *adj.* **venous.**

vein islet (*Bot.*). A very small patch of photosynthetic cells in a leaf, more or less surrounded by a small vein.

vein stuff (*Mining*). The minerals occurring in veins of fissures.

velamen (*Bot.*). A multilayered sheath of dead empty cells occurring on the surfaces of some aerial roots; the air-filled cells give it a silvery appearance. *pl.* **velamina.**

velarium (*Zool.*). In scyphozoan medusae, the thin flexible marginal region of the umbrella with its fringe of tentacles; it is distinguished from a true *velum* (q.v.) of *Hydrozoa* by its containing endoderm canals.

veldt sore (*Med.*). Barcoo rot. A chronic ulcerated sore of the exposed hairy parts of the body occurring in those living in hot, dry, sandy, or desert country; thought to be due to a deficiency in diet.

veliger (*Zool.*). The secondary larval stage of most *Mollusca*, developing from the trochophore and characterized by the possession of a velum.

Vello process (*Glass*). A method of continuously drawing glass tubing or cane, in which glass is fed downwards through an annular orifice to a horizontal draw back.

vellum (*Paper*). An early form of writing surface made from the skins of calves, lambs, or goats. In modern usage, a thick writing paper.

vellus (*Zool.*). In Man, the widespread short downy hair which replaces the fine lanugo which almost covers the foetus from the fifth or sixth month until shortly before birth.

Velocitron (*Electronics*). TN for U.S. external cavity type of reflex klystron.

velocity (*Maths.*). Of a point: the rate of change with respect to time of the position of the point. In cartesian coordinates, the velocity of the point (x, y) is $(\dot{x}, \dot{y})$, where $\dot{x}$ represents $\frac{dx}{d}$ and similarly for $\dot{y}$. (*Phys.*) In a wave, distance travelled by a given phase of a wave divided by time taken. It is a vector quantity (cf. *speed*, which is scalar) so that it has a magnitude and a direction expressed relative to some frame of reference. Symbol $\boldsymbol{v}$.

velocity amplitude (*Acous.*). On a gramophone record, the r.m.s. or peak lateral velocity of the sinuous track when the disk is rotating at its correct speed. This equals the product of the frequency, in radians per second, of the note recorded on the track and the amplitude of the excursion of the track as measured along a radius of the disk.

velocity budget (*Space*). The sum of the *characteristic velocities* involved in a complete space mission.

velocity constant (*Chem.*). See **velocity rate constant.**

velocity head (*Hyd.*). A distance representing the height above a datum which would give unit mass of a fluid in a conduit a potential energy equal to its velocity energy or kinetic energy. If, at a given point its velocity is v m/s, its velocity head is $\frac{v^2}{2g}$ metres. Also known as **kinetic head.**

velocity microphone (*Acous.*). See **pressure-gradient microphone.**

velocity-modulated oscillator (*Electronics*). One in which an electron beam is velocity-modulated (*bunched*) by passing through a toroidal cavity resonator (*rhumbatron*), the energy exciting a further cavity (*collecting*) and feeding back into the first.

velocity modulation (*Electronics*). Modulation in a klystron in which the velocities of the electrons, and hence their *bunching*, is related to radio signals to be transmitted. (*TV*) System in which gradation of light and shade in the reproduced picture is effected by variation of velocity of the scanning spot across the phosphor, the intensity of illumination of the spot being constant.

velocity of approach (*Hyd.*). The velocity of the liquid at the point from which the head above the sill is measured, in the case of flow over a weir or notch-plate. It may be appreciable if the cross-sectional area of the liquid in the approach channel is small.

velocity of energy (*Phys.*). Energy flux per unit area divided by the energy density.

velocity of light (*Phys.*). $2{\cdot}997\,925 \times 10^8$ m/s *in vacuo*. This is a natural constant equal to the maximum speed with which energy can be propagated and normally represented by *c*. In a dispersive medium, the phase and group velocities are not the same, both normally being less than *c*. In waveguides and ionized media, the phase velocity may be greater—the geometric mean of the phase and group velocities then being equal to *c*. Also **speed of light.**

velocity of propagation (*Elec. Eng.*). The velocity of an electromagnetic wave, or the velocity with which a wave travels along a transmission line. For free space, or an air-spaced cable, it equals the *velocity of light*.

velocity of sound (*Phys.*). In dry air at s.t.p. 331·4 m/s (750 mi/h). In fresh water, 1410 m/s, and in sea water, 1540 m/s. The above values are used for sonar ranging but do not apply to explosive shock waves. They must be corrected for variations of temperature, humidity, etc. (*Aero.*) A key factor in aircraft design and operation, *Mach* (q.v.) 1·0. Under *ISA* (q.v.) sea level conditions the speed of sound in air is 761·6 mi/h (1229 km/h), reducing with temperature, to 660·3 mi/h (1062 km/h) at the tropopause (ISA 36 090 ft 11 000 m), where the temperature is −56·46°C (−69·63°F), above which height it remains constant. Also **speed of sound.**

velocity rate constant (*Chem.*). The speed of a chemical reaction, in moles of change per cubic metre per second, when the active masses of all the reactants are unity. If Rate $= k[\mathrm{A}]^x[\mathrm{B}]^y$, for $\mathrm{A}+\mathrm{B}\rightarrow$products, and x, y are partial orders, then k is the rate constant.

velocity ratio (*Mech.*). The ratio of the distance moved through by the point of application of the effort to the corresponding distance for the load in a machine. The ratio of the *mechanical advantage* (q.v.) to the velocity ratio is termed the efficiency of the machine.

velocity resonance (*Telecomm.*). See **phase-.**

velodyne (*Elec. Eng.*). Tachogenerator in which rotational speed of output shaft is proportional to applied voltage through feedback.

velour (*Textiles*). A woollen cloth with a fine dense pile, cropped level. The name is often applied to woollen cloths finished with a soft thick pile laid in one direction.

velour paper (*Paper*). Paper made by depositing short wool fibres on an adhesive coated paper.

Velox boiler (*Eng.*). A forced-circulation boiler in which very high rates of heat release and heat transfer are obtained by the use of a supercharged furnace and high gas velocities, resulting in exceptionally low *unit weight* (q.v.) and high efficiency.

velum (*Bot.*). See **veil.** (*Zool.*) A veil-like structure, as the *velum pendulum* or posterior part of the soft palate in higher Mammals: in some *Ciliophora*, a delicate membrane bordering the oral cavity: in *Porifera*, a membrane constricting the lumen of an incurrent or excurrent canal: in hydrozoan medusae, an annular shelf projecting inwards from the margin of the umbrella: in *Rotifera*, the trochal disk: in some Bees, a membrane attached to the inner side of the cubital spur: in *Mollusca*, the ciliated locomotor organ of the veliger larva: in *Cephalochorda*, the perforated membrane separating the buccal cavity from the pharynx.

velum interpositum (*Zool.*). In the Mammalian brain, a delicate vascular membrane forming the roof of the diacoele or third ventricle, and containing a choroid plexus.

velum partiale (*Bot.*). See **partial veil.**

velum universale (*Bot.*). See **universal veil.**

velutinous (*Bot.*). Having a velvety surface.

velvet (*Textiles*). A silk or rayon fabric with a dense short pile, formed in loops which are cut with a trevette. Most velvets are now made on power looms with cotton or worsted ground. In one method, two cloths are woven face to face. The pile warp weaves through both, and is cut while weaving to produce two separate cloths. (*Zool.*) The tissue layers covering a growing antler, consisting of periosteum, skin, and hair.

velveteen (*Textiles*). Imitation velvet.

vena contracta (*Hyd.*). Of a jet of fluid discharged by an orifice, the point of minimum cross-sectional area at which the converging streamlines become parallel.

venae cavae (*Zool.*). The caval veins; in higher Vertebrates, three large main veins conveying blood to the right auricle of the heart.

vena spuria (*Zool.*). In one family of *Diptera* ('hover-flies', *Syrphidae*) a characteristic feature of the wing, rarely found in other *Diptera*, and consisting of a veinlike thickening of the wing membrane which can be distinguished from true veins, as it is fainter and ends without associating with any of the true veins.

venation (*Bot., Zool.*). The arrangement of the veins or nervures; by extension, the veins themselves considered as a whole.

V-end connexions (*Elec. Eng.*). V-shaped conductors connecting the ends of corresponding pairs of bars in a bar-wound armature.

veneer (*Civ. Eng.*). See **wearing course.** (*For.*) Timber in the form of a thin layer of uniform thickness; it is generally cut from timber of fine appearance, the veneer being glued to a less expensive material and appearing on the surface.

veneered construction (*Build.*). A mode of construction in which a thin external layer of facing material is applied to the steel or reinforced concrete framework.

veneering hammer (*Tools*). One with a flat wooden head in which a brass strip is inserted. Used for smoothing veneers and forcing out the glue.

veneer saw (*Tools*). Small saw with curved cutting edge, unset teeth, and a wooden grip running along the back. Used for cutting veneers.

venereal disease (*Med.*). One of a number of contagious diseases (see **chancroid, gonorrhoea, granuloma inguinale, lymphogranuloma inguinale, syphilis**) usually contracted in sexual intercourse. Abbrev. **VD.**

Venetian (*Typog.*). A style of type based on the fifteenth-century original of Nicolas Jenson, and characterized by strong colour, prominent serifs and oblique bar to 'e'.

Venetian arch (*Arch.*). (1) A *Queen Anne arch* (q.v.). (2) A pointed arch in which the extrados and the intrados are not parallel.

Venetian mosaic (*Build.*). See **terrazzo.**
Venetian red (*Paint.*). A bright red oxide pigment, containing 40% of ferric oxide and 60% calcium sulphate, prepared from ferrous sulphate and lime. One of the cleaner shades obtainable from iron oxide, with excellent durability.
Venetian shutters (*Build.*). See **jalousies.**
Venetian white (*Paint.*). A pigment consisting of equal parts of white lead and barium sulphate.
Venetian window (*Join.*). A window having two mullions dividing the window space into three compartments, usually a large centre light and two narrow side lights.
Venn diagram. In logic and mathematics, a diagram consisting of shapes (rectangles, circles, etc.) that show by their inclusion, exclusion or intersection, the relationships between classes or sets.
venography (*Med.*). See **phlebography.**
venomous (*Zool.*). Having poison-secreting glands.
venosclerosis (*Med.*). Hardening of a vein due to thickening of its walls.
venose (*Bot.*). Having raised anastomosing ridges on the surface, looking like veins.
venous system (*Zool.*). That part of the circulatory system responsible for the conveyance of blood from the organs of the body to the heart.
vent (*Aero.*). The opening (usually at the centre) in a parachute canopy which stabilizes it by allowing the air to escape at a controlled rate. (*Eng.*) To allow air to enter, or escape from, a confined space to facilitate movement (of liquid, a piston, etc.) within the space. (*Geol.*). See **volcanic vent.** (*Zool.*) The aperture of the anus or cloaca in Vertebrates.
venter (*Bot.*). The dilated basal part of an archegonium, containing one egg. (*Zool.*) A protuberance; a median swelling; the abdomen in Vertebrates; the ventral surface of the abdomen.
vent gleet (*Vet.*). An infectious disease of fowls, characterized by inflammation of the cloaca.
ventifacts (*Geol.*). A general name for pebbles shaped by wind under (usually) desert conditions. See also **dreikanter.**
ventil (*Acous.*). A key for disabling a section of an organ which has become faulty during playing, so that it is not possible to draw the included stops.
ventilated wind tunnel (*Aero.*). A wind tunnel for *transonic* testing in which part of the walls in the working section are perforated, or porous, to prevent choking by the presence of the model, which would otherwise render measurements unreliable in the range from Mach 0·9 to 1·4.
ventilating bead (*Join.*). See **deep bead.**
ventilating fan (*Elec. Eng.*). An electrically driven fan whose function is to force cooling air through the ventilating ducts of an electrical machine.
ventilating plant (*Elec. Eng.*). Electrically driven fans and their attendant apparatus, designed for specific ventilating purposes.
ventilating tissue (*Bot.*). The sum total of the intercellular spaces in a plant, by means of which air circulates in the plant body.
venting (*Foundry*). The process of making holes through the rammed sand of a mould or core in order to allow gases to escape during pouring and so avoid blown castings. See **vent wires, wax vent.**
vent pipe (*San. Eng.*). A small escape pipe which carries off foul gases from a sanitary fixture and leads into the vent stack. Abbrev. V.P.
ventral (*Bot.*). (1) Anterior or in front. (2) Uppermost (of the upper face of a leaf). (3) Nearest to the axis. Synonymous with **adaxial.** (*Zool.*) Pertaining to that aspect of a bilaterally symmetrical animal which is normally turned towards the ground.
ventral canal cell (*Bot.*). An unwalled cell, nonfunctional, which lies in the venter of the archegonium above the ovum, of which it is the sister cell.
ventral fins (*Aero.*). Fins mounted under the rear fuselage to increase directional stability, usually under high incidence conditions when the main fin may be blanketed.
ventral light response (*An. Behav.*). The response of keeping the ventral surface perpendicular to rays of light which normally fall from above, used to maintain *primary orientation* (q.v.) in some of the few animals which normally have their ventral surface uppermost. Cf. *dorsal light response.*
ventral suture (*Bot.*). The presumed line of junction of the edges of the infolded carpel.
ventral tank (*Aero.*). An auxiliary *fuel tank*, fixed or jettisonable, mounted externally under the fuselage; sometimes **belly tank.**
ventral trace (*Bot.*). One of the two laterally placed vascular strands often present in the wall of a carpel.
ventricle, ventriculus (*Zool.*). A chamber or cavity, especially the cavities of the Vertebrate brain and the main contractile chamber or chambers of the heart (in Vertebrates or Invertebrates). *adj.* **ventricular.**
ventricose (*Bot., Zool.*). (1) Swollen in the middle. (2) Having an inflated bulge to one side.
ventriculography (*Radiol.*). Radiography of the brain after the cerebrospinal fluid in the lateral ventricles has been replaced by air.
ventrifixation (*Surg.*). The operation of stitching the uterus to the anterior wall of the abdomen, for the treatment of retroversion of the uterus.
ventrisuspension (*Surg.*). An operation for replacing the retroverted uterus by transplanting the round ligaments of the uterus into the anterior abdominal wall in such a way that they exert a strong pull on the uterus.
ventrofixation (*Surg.*). See **ventrifixation.**
ventrosuspension (*Surg.*). See **ventrisuspension.**
vent stack (*San. Eng.*). A vertical pipe carried up from the highest point in a system of house drains to a level clear of all windows and opening skylights; it provides an escape for foul gases from drains and sanitary fixtures.
Ventube (*Mining*). A flexible ventilating duct some distance away from the source of fresh air.
venturi (*Aero.*). A convergent-divergent duct in which the pressure energy of an air stream is converted into kinetic energy by the acceleration through the narrow part of the wasp-waisted passage. It is a common method of accelerating the airflow at the working section of a supersonic wind tunnel. Small venturis are used on some aircraft to provide a suction source for vacuum-operated instruments, which are connected to the low-pressure neck of the duct.
venturi flume (*Civ. Eng.*). A flume which is constricted at one section with convergent upstream and divergent downstream walls, the difference in water-level at the constriction and at a point in the full channel upstream affording a means of measuring the rate of flow. See **flume.**
venturi meter (*Eng.*). One in which flow rate is measured in terms of pressure drop across a venturi (or tapered throat) in a pipe.
vent wires (*Foundry*). Wires ranging from $\frac{1}{16}$ to $\frac{3}{8}$ in. diameter, used for making vent holes in the rammed sand of a mould or core.

Ventzke scale (*Chem.*). Scale used in graduating polarimeters in the sugar industry, so that, if 26 g of a sample containing sugar is dissolved in 100 cm^3 water and the polarization read in a 20 cm tube at ordinary temperature, the reading gives the direct percentage of sugar contained in the sample (provided that no other optically active substance than sucrose is present).

venule (*Zool.*). In *Chordata*, small blood vessels which receive blood from the capillaries and unite to form veins. Their thin nonelastic walls are strengthened with a definitive sheath of connective tissue.

Venus (*Astron.*). The brightest of the planets, revolving about the Sun at a distance of 0·723 A.U. in a period of 224·7 days; its volume is 0·857 and its mass 0·815 that of the earth. The rotation is now known to be retrograde in a period of 243 days. Space probes have shown a surface temperature over 400°C, and a very dense cloudy atmosphere consisting mainly of carbon dioxide.

Venus' hair stone (*Min.*). Also **Veneris crinis.** See **flèches d'amour.**

Vera (*TV*). A machine for recording television pictures and sound on magnetic tape for almost immediate playback. (*V*ision *e*lectronic *r*ecording *a*pparatus.)

veranillo (*Meteor.*). The short period of fine weather which ends the rainy reason in the tropical countries of America.

verano (*Meteor.*). The dry season in the tropical countries of America.

veratrine (*Chem.*). (1) Crystalline veratrine is the same as *cevadine* (q.v.). (2) A mixture of alkaloid bases obtained from the *Veratrum* genus.

verbigeration (*Med.*). The persistent repetition of meaningless words and phrases.

verde antico (*Chem.*). A green patina formed on old bronze by oxidation; it is imitated artificially by pickling.

Verdet's constant (*Light*). The angle, per unit thickness traversed, of a transparent medium through which a ray of polarized light rotates in unit magnetic field according to the Faraday effect. For light of 589 nm in water, rotation is 0·000 477 rad/A.

verdigris (*Chem.*). The green basic copper(II) carbonate (i.e. $CuCO_3{\cdot}Cu(OH)_2$) formed on copper exposed to moist air.

verdite (*Min.*). A green rock, consisting chiefly of green mica (fuchsite) and clayey matter, occurring as large boulders in the North Kaap River, South Africa; used as an ornamental stone.

verge (*Build.*). The edge of the roof covering projecting beyond the gable of a roof. (*Horol.*) The axis of a clock pallet, especially that of the verge escapement.

verge board (*Build.*). See **barge board.**

verge escapement (*Horol.*). One of the earliest known escapements, now obsolete. The pallets are set at right angles to the escape wheel (crown wheel) axis, and its action has a recoil.

verge tile (*Build.*). A tile which is purpose-made to a wider size than the normal, to assist in forming the bond at the end of a roof.

verifier (*Comp.*). Apparatus which verifies the accuracy of punched cards or tapes and indicates errors, or which compares two punched cards or tapes.

veristron (*Electronics*). Device based on electron spin, analogous to *maser* (q.v.).

vermiculation (*Build.*). A variety of rustication, distinguished by worm-shaped sinkings.

vermicule (*Zool.*). A small wormlike structure or organism, as the motile phase of certain *Sporozoa.*

vermiculites (*Min.*). A group of hydrous silicates, closely related chemically to the chlorites, and structurally to talc. They occur as decomposition products of biotite mica. When slowly heated, they exfoliate and open into long wormlike threads, forming a very lightweight water-absorbent aggregate. They are used in seed planting and, in building, as an insulating material.

vermiform (*Zool.*). Wormlike, as the *vermiform appendix.*

vermiform larva (*Zool.*). In most, if not all, *Orthoptera*, the short-lasting first instar, which has a loose cuticle which envelops the appendages so that they are pressed to the sides of the body, and their segmentation is indistinct.

vermifuge (*Med.*). Having the power to expel worms from the intestines; any drug which has this power. See also **anthelminthic.**

vermilion (*Paint.*). Mercuric sulphide. Orange to blue-red pigment, once made by grinding cinnabar, having good opacity combined with low oil absorption. Use limited by its poisonous nature.

vermis (*Zool.*). In lower Vertebrates, the main portion of the cerebellum; in Mammals, the central lobe of the cerebellum.

vernal (*Ecol.*). Of, or belonging to, spring.

vernal aspect (*Bot.*). The condition of a plant community in spring, characterized by some species which displays particular activity at that time.

vernal equinox (*Astron.*). The instant at which the sun's apparent longitude is 0°. It then crosses the celestial equator from south to north, the date being about March 21st. See also **equinox.**

vernalization (*Agric.*, *Bot.*). Treatment of seeds before sowing by wetting at low temperature or by freezing for a definite period after germination has begun, which speeds up the plant's development. Principally used on cereals, especially winter wheats, winter seed being converted to spring seed.

vernation (*Bot.*). The manner in which the leaves are packed in a bud.

Verneuil process (*Min.*). The technique invented by the French chemist Verneuil for the manufacture of synthetic corundum and spinel by fusing pure precipitated alumina, to which has been added a predetermined amount of the appropriate oxide for colouring, in a vertical, inverted blow-pipe type of furnace.

vernier (*Instr.*). A small movable auxiliary scale, attached to, and sliding in contact with, a scale of graduation. It enables readings on the latter to be made to a fraction (usually a tenth) of a division.

vernier arm (*Surv.*). The part of an instrument carrying the vernier or verniers.

vernier capacitor (*Elec. Eng.*). A variable capacitor of small capacitance, connected in parallel with a larger fixed one, used for fine adjustment of total capacity.

vernier potentiometer (*Elec. Eng.*). Precision pattern based on the *Kelvin-Varley slide* (q.v.). Balance can be attained purely by the operation of switches so that the possibility of wear associated with sliding contacts is avoided.

Verona (or **Veronese**) **green** (*Paint.*). See **terre verte.**

veronal (*Chem.*). See **barbitone.**

verruca (*Bot.*). A granular or wartlike outgrowth on a thallus. (*Zool.*) A wartlike process; especially one of a number of wartlike processes situated around the base of certain kinds of alcyonarian polyp.

verruciform (*Bot., Zool.*). Resembling a wart.

verrucose (*Bot., Zool.*). Said of a surface covered with wartlike upgrowths. *dim.* **verruculose.**

verrucous (*Bot., Zool.*). Studded with wartlike tubercles.

versatile (*Bot.*). Of an anther, attached to the tip of the filament by a small area on its dorsal side, so that it turns freely in the wind, facilitating the dispersal of the pollen. (*Zool.*) Capable of free movement, as the toes of birds when they may be turned forwards or backwards.

versed sine (*Civ. Eng.*). See rise (1).

versene (*Chem.*) Sodium versenate, the sodium salt of *EDTA* (q.v.), used for the *complexometric titration* (q.v.) of calcium ion.

versicolorous (*Bot., Zool.*). Not all of the same colour: changing in colour with age.

versiform (*Bot.*). Said of organs of the same kind which are not all alike in shape.

versine (*Maths.*). A trigonometrical **function** of an angle, required for the solution of spherical triangles. It is given by

$$\text{vers(ine)}\ \theta = 1 - \cos\text{(ine)}\ \theta.$$

version (*Med.*). The act of turning manually the foetus *in utero* in order to facilitate delivery.

verso (*Typog.*). A left-hand page of a book, bearing an even number. Cf. *recto.*

Verson's cells (*Zool.*). In the testis of *Orthoptera, Dictyoptera*, and some *Lepidoptera, Diptera* and *Homoptera*, large cells, also known as **apical cells**, which have abundant mitochondria which are transferred to the spermatogonial cytoplasm during spermatogenesis.

verst. Russian measure of length, 0·6629 mile (1·065 km).

vertebra (*Zool.*). One of the bony or cartilaginous skeletal elements of mesodermal origin which arise around the notochord and compose the backbone: in *Ophiuroidea*, one of the brachial ossicles. *pl.* **vertebrae.** *adjs.* **vertebral, vertebrate.**

vertebra prominens (*Zool.*). The seventh cervical vertebra in some Mammals in which the neural spine is very large.

Vertebrata (*Zool.*). A subphylum of *Chordata* in which the notochord stops beneath the fore-brain, and a skull is always present. There are usually paired limbs. The brain is complex and associated with specialized sense organs, and there are at least ten pairs of cranial nerves. The pharynx is small, and there are rarely more than seven gill slits. The heart has at least three chambers, and the blood has corpuscles containing haemoglobin.

vertebraterial canals (*Zool.*). In Vertebrates, small canals found one on each side of all or most of the cervical vertebrae. They are formed by the articulation or fusion of the two heads of the small or vestigial cervical ribs to the centra and transverse processes, and the vertebral arteries run through them.

vertex (*Build., Civ. Eng.*). See **crown.** (*Maths.*) Of a polygon or polyhedron: one of the points in which the sides or faces intersect. Of a conic, the points in which it is cut by its axes. See also **cone, pencil.** (*Zool.*) In higher Vertebrates, the top of the head, the highest point of the skull; in Insects, the dorsal area of the head behind the epicranial suture.

vertical aerial photograph (*Surv.*). A photograph taken from the air, for purposes of aerial survey work, with the camera pointing directly at the ground so that the optical axis is vertical or nearly so.

vertical blanking (*TV*). The elimination of the vertical trace on a CRT during frame flyback.

vertical boiler (*Eng.*). A steam boiler having a vertical cylindrical shell and domed or spheroidal firebox, from which (generally) vertical flue tubes lead to the smoke box and chimney.

vertical circle (*Astron.*). A great circle which passes through the observer's zenith and cuts the horizon at right angles; hence the term is applied to the great circles on which altitudes of heavenly bodies are measured. (*Surv.*) The graduated circular plate used for the measurement of vertical angles by theodolite.

vertical component (*Elec. Eng.*). The vertical component of the force experienced by unit magnetic pole as the result of the action of the earth's magnetic field. Cf. *horizontal component.*

vertical curve (*Surv.*). The curve, generally parabolic, which is introduced between two railway or highway gradients in order to provide a gradual change from one to the other.

vertical-draw-out metal-clad switchgear (*Elec. Eng.*). A form of metal-clad switchgear in which the circuit-breaker can be withdrawn for inspection or repair by being lowered away from the bus-bar chambers.

vertical effect (*Radio*). Original name for **antenna effect** (1).

vertical engine (*Eng.*). Any engine in which the cylinders are arranged vertically above the crankshaft.

vertical escapement (*Horol.*). An escapement in which the axis of the balance is at right angles to that of the escape wheel. In a verge watch, the balance staff is vertical when the watch is in the laying position.

vertical force instrument (*Ships, etc.*). An instrument used in adjusting magnetic compasses, particularly in the correction of *heeling error* (q.v.).

vertical frequency (*TV*). That of frame-scanning voltage in a normal TV system in which the line scanning is horizontal.

vertical gust (*Aero.*). A vertical air current, which can be of dangerous intensity, particularly when met by aircraft flying at high speed.

vertical-gust recorder (*Aero.*). An *accelerometer* (q.v.) which records graphically the intensity of accelerations due to vertical gusts and, simultaneously, the airspeed; used in assessment of aircraft fatigue life: abbrev. **v.g. recorder.**

vertical hold control (*TV*). The control in a TV receiver which varies the free-running period of the oscillator providing the vertical deflection.

vertical-lift bridge (*Civ. Eng.*). A bridge consisting of two towers connected by a span which can be raised and lowered vertically, maintaining its horizontal position between the towers.

vertical milling machine (*Eng.*). A *milling machine* (q.v.) in which the cutter spindle is vertical, the table movements being the same as in plane machines. It uses end mills.

vertical polarization (*Elec. Eng.*). State of electromagnetic wave when electric component lies in the vertical plane and magnetic component in the horizontal plane, as in the wave emitted from a vertical aerial.

vertical recording (*Accous.*) Same as **hill-and-dale** recording.

vertical scanning (*TV*) That in which individual lines are vertical, not, as normal, horizontal.

vertical separation (*Aero.*) See **separation.**

vertical shaft alternator *Elec. Eng.*. A water-turbine driven alternator designed to operate with its shaft vertically above, and directly coupled to, the turbine shaft.

vertical spindle motor (*Elec. Eng.*) An electric

motor specially designed to operate with its spindle in a vertical position.

vertical take-off and landing (*Aero.*). See **VTOL.**

vertical tiling (*Build.*). See **weather tiling.**

vertical wind tunnel (*Aero.*). A wind tunnel wherein the air flow is upward and which is used principally for testing freely spinning models. Also **spinning tunnel.**

verticil (*Bot.*). See **whorl.**

verticillaster (*Bot.*). A kind of inflorescence found in dead nettles and related plants. It looks like a dense whorl of flowers, but is really a combination of two crowded dichasial cymes, one on each side of the stem.

verticillate (*Bot.*). Arranged in whorls.

vertigo (*Med.*). Dizziness: a condition in which the person has the sensation of turning or falling, or of surrounding objects turning about himself. See also **Ménière's disease.**

very fine screen (*Photog., Print.*). Term for half-tones used only for specially fine detail, from 175-400 lines/in (7-16 lines/mm); 1st figure is limit for half-tone blocks, 2nd is rarely used.

very high frequencies (*Radio*). Those between 30 and 300 MHz. Abbrev. **VHF.**

very low frequencies (*Radio*). Those between 10 and 30 kHz. Abbrev. **VLF.**

vesica (*Zool.*). The urinary bladder.

vesicant (*Med.*). Causing blisters; any agent which does this. See **war gas.**

vesicle (*Bot.*). A thin-walled globular swelling, usually at the end of a hypha. (*Geol.*) See under **vesicular structure.** (*Med., Zool.*) A small cavity containing fluid; a small saclike space containing gas; one of the three primary cavities of the Vertebrate brain; a small bladder-like sac. Also **vesicula.**

vesicular (*Bot., Zool.*). Like, or pertaining to, a vesicle; like a bladder.

vesicular exanthema (*Vet.*). A febrile, virus disease of swine in which vesicles develop on the snout, lips, tongue, and feet.

vesicular stomatitis (*Vet.*). A virus disease of horses, and occasionally cattle, characterized by vesicle formation on the tongue and mucosa of the mouth.

vesicular structure (*Geol.*). A character exhibited by many extrusive igneous rocks, in which the expansion of gases has given rise to more or less spherical cavities (vesicles). The latter may become filled with such minerals as silica (chalcedony, agate, quartz), zeolites, chlorite, calcite, etc.

vesicula seminalis (*Zool.*). In many animals, including Man, a sac in which spermatozoa are stored during the completion of their development.

vesiculitis (*Med.*). Inflammation of the vesiculae seminales.

vesiculose (*Bot.*). (1) Swollen like a bladder. (2) Appearing as if made up of small bladders.

vespoid (*Zool.*). Wasplike.

vessel (*Bot.*). A long water-conducting tube in the xylem, formed from a vertical row of cells by the more or less complete breakdown of the horizontal walls between the individual cells. When mature, the vessel has no living contents, and has thick lignified walls which are pitted in various ways. Also **trachea.** (*Zool.*) A channel or duct with definitive walls, as one of the principal vessels through which blood flows.

vessel element, -segment, -unit (*Bot.*). One of the cells which, with many others above and below it, forms a vessel.

vestibule (*Build.*). A small antechamber at the entrance to a building, or serving as an entrance room to a larger room. (*Zool.*) A passage leading from one cavity to another or leading into a cavity from the exterior: in *Protozoa*, a depression in the ectoplasm at the base of which is the mouth; in *Entoprocta*, the space within the ring of tentacles; in a female Mammal, the space between the vulva and the junction of the vagina and the urethra (urinogenital sinus); in Birds, the posterior chamber of the cloaca; in Vertebrates generally, the cavity of the internal ear. *adjs.* **vestibular, vestibulate.**

vestibulectomy (*Surg.*). Surgical removal of the membranous labyrinth of the inner ear.

vestigial (*Zool.*). Of small or reduced structure: of a functionless structure representing a useful organ of a lower form. *n.* **vestige.**

vestigial sideband (*TV*). Reduction of total bandwidth associated with normal amplitude modulation by markedly reducing one sideband, especially for TV transmission.

vestigial-sideband transmitter (*Radio*). Station which radiates either a carrier with only one full sideband (for bandwidth reduction); or the one full sideband with no carrier (for maximum economy of radiated power). The latter system is termed **single-sideband suppressed carrier.**

vestiture (*Bot., Zool.*). A covering, e.g., of hairs, feathers, fur, or scales.

vesuvianite (*Min.*). A hydrous silicate of calcium and aluminium, with lesser magnesium and iron, crystallizing in the tetragonal system. It occurs in metamorphosed limestones. Also known as **idocrase.**

veterinary (*Vet.*). Relating to, concerned with, the diseases of domestic animals.

vexillar (*Bot.*). Relating to the vexillum.

vexillum (*Bot.*). See **standard.** (*Zool.*) See **vane.**

VFR (*Aero.*). Abbrev. for *visual flight rules*.

V.F. (or v-f) telegraphy. Abbrev. for *voice-frequency telegraphy*.

v.g. recorder (*Aero.*). Abbrev. for *vertical-gust recorder*.

VHF (*Radio*). Abbrev. for *very high frequencies*.

VHF rotating talking beacon (*Radio*). An automatic very-high-frequency radio-telephone beacon system, having a radiation pattern which is continuously rotated.

V.I. (*Acous.*). Abbrev. for *volume indicator*.

viable (*Bot., Zool.*). Capable of living and developing normally.

viaduct (*Civ. Eng.*). A structure which carries a road or railway across a wide and deep valley; it consists of a series of small-span bridges in line, supported on intermediate piers.

vial (*Surv., etc.*). The glass tube containing the liquid in a *level tube*.

viameter (*Surv.*). See **perambulator.**

vibracula, vibracularia (*Zool.*). Long, bristlelike structures, found in *Ectoprocta*, which represent modified zooids.

vibratile membrane (*Zool.*). In some *Protozoa* (*Ciliata*), a thin membrane which vibrates rhythmically, is composed of fused cilia, and may be used to draw food particles into the gullet. Not to be confused with *undulating membrane*.

vibrating capacitor (*Elec. Eng.*). One in which the potential on the electrode is varied by mechanical oscillation, so that the steady applied potential is converted to an alternating potential, which can be more easily amplified. Also **oscillating capacitor.** See **Vibron.**

vibrating conveyor (*Eng.*). A tubular or flat trough with sides, to which vibrators are attached. Via the trough, the vibrators impart an upward and forward conveying movement to granular materials in the trough.

vibrating-reed amplifier (*Elec. Eng.*). D.c. amplifier in which vibrating-reed electrometer converts signal to a.c. This is amplified and fed to phase-sensitive detector to convert back to d.c. output. Also known as **chopper amplifier** and **contact-modulated amplifier.**

vibrating-reed electrometer (*Elec. Eng.*). One which utilizes a *vibrating capacitor* (q.v.) to measure small charges. Also **dynamic-capacity electrometer.**

vibrating-reed instrument (*Elec. Eng.*). An instrument for measuring frequency, consisting of a row of steel reeds each having a different natural frequency; these are in the field of an electromagnet excited by the current whose frequency it is desired to measure. The reed which is in resonance with the field vibrates strongly and is easily distinguished.

vibrating-reed rectifier (*Elec. Eng.*). A rectifier in which a vibrating steel reed acts as a commutator, the reed being made to vibrate synchronously with the a.c. supply by means of an electromagnet.

vibrational energy (*Chem.*). Energy due to the relative oscillation of two contiguous atoms in a molecule.

vibration dampers (*Eng.*). Devices fitted to an engine crankshaft in order to suppress or minimize stresses resulting from torsional vibration at critical speeds. See **dynamic damper, frictional damper.**

vibration galvanometer (*Elec. Eng.*). Moving-coil taut-suspension galvanometer with natural frequency of vibration of coil, tunable usually over range 40 Hz to 1000 Hz. Small a.c. currents at the resonant frequency excite a large response, hence these instruments form sensitive detectors for circuits such as a.c. bridges or potentiometers.

vibration pick-up (*Elec. Eng.*). One which uses some form of microphone or transducer, e.g., crystal, capacitance, electromagnetic, to transform the oscillatory motion of a surface, e.g., of machinery, into an electrical voltage or current.

vibration-rotation spectrum (*Phys.*). The infrared end of the electromagnetic spectrum which arises from vibrational and rotational transitions within a molecule.

vibrator (*Cinema.*). An electromechanical device, consisting of two stretched wires in a magnetic field and carrying a mirror, which deflects a beam of light from the exciter lamp in making the photographic sound-record on the location of the sound-track. Essentially the same as the *Duddell oscillograph* (q.v.). (*Civ. Eng.*) Equipment which, by vibration, consolidates loose soil or compacts cohesive soil for better bearing resistance, etc. **Vibroflotation** incorporates water in the vibratory process. (*Elec. Eng.*) Electro-mechanical interrupter used to supply intermittent d.c. current to primary of transformer when voltage conversion is required. Formerly widely used for radio equipment in vehicles. The **synchronous vibrator** also uses contacts operating simultaneously to rectify the a.c. voltage from the transformer secondary. (*Eng.*) An oscillating mechanism, usually electromagnetically excited, which imparts vibrations to hoppers, conveyors, or other parts of machines, usually for the purposes of dislodging, loosening, or propelling materials or workpieces. (*Horol.*) An instrument for checking the time of vibration of a balance and its spring. It consists of a vertical arm to which the free end of the spring is clipped and a master balance and spring in a container with a glass cover in the base. The master balance is vibrated, and the vibrations of the balance to be tested are compared with it.

vibratory bowl feeder (*Eng.*). A small parts feeder, comprising a bowl, a spring suspension system, and an electromagnetic or hydraulic exciter, used for separating, orienting, and feeding workpieces to an assembling machine, machine tool, etc.

vibrionic abortion (*Vet.*). A contagious form of abortion in cattle and sheep caused by infection of the uterus and placenta by the bacterium *Vibrio foetus.*

vibrissa (*Zool.*). (1) In Mammals, one of the stiff tactile hairs borne on the sides of the snout and about the eyes. (2) In *Diptera*, one of the stout cephalic bristles placed close to the sides of the epistoma. (3) One of the vaneless rictal feathers of certain Birds, e.g., Flycatchers. *pl.* **vibrissae.**

Vibroflotation (*Civ. Eng.*). See **vibrator.**

vibrometer (*Elec. Eng.*). An instrument used for the measurement of the displacement, velocity or acceleration of a vibrating body.

vibromill (*Chem. Eng.*). A *ball mill* (q.v.) in which the impacts between the balls and the material to be ground are achieved by vibrating the mill at high frequency.

Vibron (*Elec. Eng.*). TN for a *vibrating capacitor* (q.v.) which can modulate an applied potential and so facilitate its amplification and measurement.

Vibrophore (*Met.*). A machine for testing for metal fatigue under various stresses and for testing damping properties and the modulus of elasticity.

vibrotaxis (*An. Behav.*). A *taxis* (q.v.) occurring in response to vibrations occurring in water, air, the threads of a spider's web, etc.

vibrotron (*Electronics*). A special form of triode valve in which the anode can be vibrated by a force external to the envelope.

vicarious trial and error (*An. Behav.*). Movements of the head of an animal back and forth at a *choice point* (q.v.). Possibly a form of *intention movement* (q.v.).

Vicat needle (*Civ. Eng.*). An apparatus which tests the setting-time of a cement specimen by measuring the effect produced by a specially shaped loaded needle which is pressed against the surface of the specimen.

vice (*Eng., etc.*) A clamping device, usually consisting of two jaws which can be brought together by means of a screw, toggle, or lever, used for holding work that is to be operated on. Generally named after the trade in which it is used. Also **vise.**

vicinal (*Chem.*). Substituted on adjacent carbon atoms, e.g., on the 1,2,3,4 atoms in a naphthalene nucleus.

vicinal faces (*Min.*). Facets modifying normal crystal faces, but themselves abnormal, as their indices cannot be expressed in small whole numbers; they usually lie nearly in the plane of the face they modify.

Vickers diamond pyramid hardness test (*Met.*). A method of determining the hardness of metals by indenting them with a diamond pyramid under a specified load and measuring the size of the impression produced.

Vicq d'Azyr, bundle of (*Anat.*). Mamillo-thalamic tract.

vidal black (*Chem.*). See **sulphide dyestuffs.**

video (*TV*). Relates to the bandwidth (~ megahertz) and spectrum position of the signal arising from TV scanning, or to the signal, or to the resultant image, or to television.

video amplifier (*TV*). Wideband amplifier which, in a TV system, passes the picture signal.

video carrier (*TV*). The carrier wave which is modulated with a video signal.

video display unit (*Comp.*). A cathode-ray tube or similar display system used to display data from a computer. Abbrev. **VDU**.

video frequency (*TV*). Said of circuits or devices which operate over the frequency range required for a video signal—approximately 0–5 MHz.

video integration (*Telecomm.*). In facsimile reception, the integration and averaging of successive video signals to improve the overall signal/noise ratio.

video pulse (*Telecomm.*). Colloquialism for fast rise-time pulse, i.e., one occurring in TV.

video signal (*TV*). That part of a TV signal which conveys all the information (intensity, colour and synchronization) required for establishing the visual image in monochrome or colour TV. See **negative-, positive-**.

video stretching (*Telecomm.*). A method of increasing the duration time of a video pulse.

video tape recording (*TV*). Registering TV signals on a high-speed magnetic tape (transverse or longitudinal) for subsequent transmission. See **Ampex, Vera.**

vidian nerve (*Zool.*). In Vertebrates, a nerve of the head formed by the union of the deep petrosal nerve and the palatine nerve which runs to the spheno-palatine ganglion.

Vidicon (*TV*). TN for a camera tube operating on the photoconducting principle.

Vierendeel girder (*Eng.*). A girder formed of upper and lower booms connected rigidly by upright members but not braced by diagonal members, the rigidity of the girder being secured by rigidity at the joints.

Vieussens' valve (*Zool.*). In the Vertebrate brain, a transverse fold formed by the isthmus together with the anterior wall of the cerebellum which dips into the fourth ventricle: the anterior medullary velum.

viewer (*Cinema.*). Compact projection apparatus with rotating prism and ground-glass screen, used for editing and cutting film. Also **editor, animated viewer.** (*Photog.*) Small device comprising magnifying lens and slide holder with or without artificial illumination, for viewing transparencies.

viewfinder (*Photog.*). An attachment to a camera for ascertaining the field of view about to be registered on the emulsion. See **Albada-, bright-line-, brilliant-, direct-vision-, frame finder, Galilean finder.**

vigia (*Ships*). A reported navigational danger of unknown nature, marked on charts although its existence has not been confirmed.

vigilance (*Psychol.*). The tendency for a person to be especially sensitive to stimuli which have particular significance, good or bad, for him.

vigoureux printing (*Textiles*). See **melange.**

Villard effect (*Photog.*). Partial destruction of an X-ray latent image by exposure to diffuse white light.

Villari effect (*Mag.*). Temporary change in magnetization, arising from longitudinal stretching.

villitis (*Vet.*). Inflammation of the coronet of the horse.

villose, villous (*Bot.*). Covered with loose, long, weak hairs.

villus (*Bot.*). A thin branching outgrowth from the stem of a moss. (*Zool.*) A hairlike or finger-shaped process, such as the absorptive processes of the Vertebrate intestine; one of the vascular processes of the Mammalian placenta which fit into the crypts of the uterine wall. *pl.* **villi.** *adjs.* **villous, villiform.**

Vincent's angina (*Med.*). See **fusospirillosis.**

vinculum (*Zool.*). n male *Lepidoptera* (Insects), the sternal region of the *tegumen* (q.v.) to which the claspers are hinged.

vinegar (*Chem.*). The product of the alcoholic and acetic fermentation of fruit juices, e.g., grape juice, cider, etc., or of malt extracts. Vinegar consists of an aqueous solution of acetic (ethanoic) acid (3–6%), mineral salts, and traces of esters.

vinyl floor coverings. Floor coverings, in roll, tile or rug form, made from PVC, often copolymerized with PVA.

vinquish (*Vet.*). See **pine.**

vinyl foils (*Print.*). Blocking foils which have no final adhesive layer and transfer exclusively to thermoplastics. Also **cello foils.**

vinyl group (*Chem.*). The unsaturated monovalent radical $CH_2{=}CH$—.

vinyl polymers (*Plastics*). Thermoplastic polymers formed by the co-polymerization of vinyl chloride, $CH_2{=}CHCl$, and vinyl ethanoate, $CH_3{\cdot}COOCH{=}CH_2$. They are odourless and tasteless. **PVC** (*polyvinyl chloride*, q.v.); used instead of rubber in electric cables; resists oil and some chemicals, but is slightly inferior to rubber in electrical properties. **PVA** (*polyvinyl acetate*, q.v.) is similarly resistant and of wide application (sheets, hose, belts, etc.).

violetwood (*For.*). See **purpleheart.**

Virchow-Robin space (*Histol.*). A term used for the subarachnoid space in the deeper layers of nervous tissue of the central nervous system. Route for the spread of pathological processes.

virescence (*Bot.*). An abnormal green condition, sometimes accompanied by the development of small, crowded leaflike structures, due to attack by a parasite or other disease.

virgate (*Bot.*). Long, slender, and stiff, and not much branched.

virgin metal (*Met.*). Metal or alloy first produced by smelting, as distinct from secondary metal containing re-circulated scrap.

virgin neutrons (*Nuc.*). Those which have not yet experienced a collision and therefore retain their energy at birth.

virgin state (*Mag.*). Same as **neutral state.**

viridian green (*Paint.*). See **Guignet's green.**

viridine (*Min.*). A green iron(III) and manganese(II)-bearing variety of *andalusite* (q.v.).

virilism (*Med.*). The development of masculine characteristics, physical and mental, in the female, often due to hyperplasia of, or the presence of a tumour in, the cortex of the adrenal gland.

virology. The study of viruses.

virtual cathode (*Electronics*). Region in a space charge where a potential minimum gives the effect of a source of electrons.

virtual earth (*Elec. Eng.*). A point maintained close to ground potential by negative feedback, although not directly connected to ground; e.g., the input terminal of an operational amplifier to which negative shunt voltage feedback is applied.

virtual height (*Radio*). The apparent height of an ionized layer as deduced from the time interval between the transmitted signal and the resulting ionospheric echo at normal incidence.

virtual image (*Phys.*). See **image.**

virtually inert anode (*Ships, etc.*). An anode, used in *cathodic protection* (q.v.), with an impressed direct current from mains or battery. Silicon iron or graphite may be used. Has a limited life but lasts much longer than a *sacrificial anode*.

virtual quantum (*Phys.*). In higher order perturbation theory, a matrix element which connects an initial state with a final state involves intermediate states in which energy is not conserved.

A photon or quantum in one of these states is designated a *virtual quantum*. This concept enables the coulomb energy between two electrons to be regarded as arising from the emission of virtual quanta by one of the electrons and their absorption by the other.

virtual slope (*Hyd.*). The slope showing the rate of loss of head due to resistance at any point in a system of fluid flow.

virtual state (or level) (*Nuc.*). (1) Nuclear state which can only be deduced indirectly or which appears as an intermediate stage when the transition between two other states is computed. See **resonance.** (2) Sometimes the so-called unbound singlet state of the deuteron. This state was 'invented' to calculate a theoretical binding energy of the singlet neutron-proton system that would give the same scattering length as that observed in neutron-proton scattering.

virtual value (*Phys.*). Coll. for *r.m.s. value*.

virulence (*Bot.*). The capacity of a parasite to cause disease.

virus (*Bot., Med.*). A particulate infective agent smaller than accepted bacterial forms, usually invisible by light microscopy, incapable of propagation in inanimate media and multiplying only in susceptible living cells, in which specific cytopathogenic changes frequently occur. Causative agent of many important diseases of man, lower animals, and plants, e.g., poliomyelitis, foot and mouth disease, tobacco mosaic. See also **bacteriophage.**

virus pneumonia of pigs (*Vet.*). A contagious, pneumonic disease of pigs, usually chronic in form, caused by a virus. Abbrev. **VPP.**

visceral (*Med.*). See **viscus.**

visceral arch (*Zool.*). See **gill arch.**

visceral clefts (*Zool.*). The gill clefts, especially the abortive gill clefts of higher Vertebrates.

visceral gout (*Vet.*). See **avian gout.**

visceral lobe (*Zool.*). The nerve centre in the medulla oblongata of the Vertebrate brain, which is the seat of the sense of taste.

viscerocranium (*Zool.*). The jaws and visceral arches of a Vertebrate skull. Also **splanchnocranium.** Cf. *neurocranium*.

visceroptosis (*Med.*). See **enteroptosis.**

viscerotonia (*Psychol.*). A pattern of temperament associated with the type *endomorphy*—sociable, fond of bodily comforts, etc.

viscid disseminule (*Bot.*). A spore or seed which has a sticky surface, or bears sticky hairs, and is dispersed by adhering to the coat of an animal.

viscin (*Bot.*). The sticky substance present in the fruits of the mistletoe.

viscometer (*Phys.*). An instrument for measuring viscosity. Many types of viscometer employ *Poiseuille's formula* (q.v.) for the rate of flow of a viscous fluid through a capillary tube.

viscose (*Chem.*). See **cellulose xanthate.**

viscose rayon (*Textiles*). *Rayon* (q.v.) in which the cellulose constituent is derived from purified wood pulp, sometimes mixed with cotton linters, the agents being caustic soda and carbon disulphide. Used in the trade as continuous filament or staple (cut fibre) yarns. See **acetate rayon, cuprammonium rayon.**

viscosity (*Phys., Eng.*). The resistance of a fluid to shear forces, and hence to flow. Such shear resistance is proportional to the relative velocity between the two surfaces on either side of a layer of fluid, the area in shear, the *coefficient of viscosity* (q.v.) of the fluid and the reciprocal of the thickness of the layer of fluid. For comparing the viscosities of liquids various scales have been devised, e.g., *Redwood No. 1 seconds* (U.K.), *Saybolt Universal seconds* (U.S.), *Engler degrees* (Germany). See also **kinematic viscosity.**

viscountess (*Build.*). A slate size, 18×10 in. (457×254 mm).

viscous damping (*Phys.*). Opposing force, or torque, proportional to velocity, e.g., resulting from viscosity of oil or from eddy currents.

viscous flow (*Phys.*). A type of fluid flow in which fluid particles, considered to be aggregates of molecules, move along streamlines so that at any point in the fluid the velocity is constant or varies in a regular manner with respect to time, random motion being only of a molecular nature. The name is also used to describe *laminar flow* or *streamline flow* (qq.v.).

viscous hysteresis (*Elec. Eng.*). The phenomenon of time-lag between the intensity of magnetization and the magnetizing force producing it.

viscus (*Med., Zool.*). Any one of the organs situated within the chest and the abdomen: heart, lungs, liver, spleen, intestines, etc. *pl.* **viscera,** *adj.* **visceral.**

vise (*Eng., etc.*) See **vice.**

viséite (*Min.*). A zeolite somewhat resembling analcite, but with phosphorus replacing some of the silicon and aluminium.

vishnevite (*Min.*). The sulphate-bearing equivalent of *cancrinite* (q.v.), found in nepheline syenite.

visibility (*Meteor.*). The maximum distance at which an object of sufficient size can be seen and recognized in normal daylight. The international scale of horizontal visibility on land runs from zero with visibility less than 0·1 km (110 yd) through 20 for 2·0 km (2200 yd), 40 for 4·0 km (4400 yd), 60 for 10 km (5·4 n.mls), 80 for 30 km (16·2 n.mls) to 89 for more than 70 km (37·8 n.mls). At sea the scale runs from 90 for less than 0·05 km (55 yd) through 94 for 1 km (1100 yd), 97 for 10 km (5·4 n.mls) to 99 for 50 km (27·0 n.mls) or more. (*Photog.*) The ratio of the luminous flux, in lumens, to the corresponding energy flux, in watts.

visibility curve (*Photog.*). The relation between visibility and wavelength. Owing to varying sensitivity of the eye, this curve indicates a maximum at 555 nm, which is a bright green.

visibility factor (*Telecomm.*). The ratio of the minimum signal input power to a radar, TV, or facsimile receiver for which an ideal instrument can detect the output signal, to the corresponding value when the output signal is detected by an observer watching the CRT.

visibility meter (*Meteor.*). A meter which attenuates visibility to a standardized value, and measures such visibility on a scale.

visible horizon (*Surv.*). Junction of sea or earth with sky as seen from observer's position. Also called **apparent** or **sensible horizon.**

visible radiation (*Phys.*). Electromagnetic radiation which falls within the wavelength range of 780 to 380 nm, over which the normal eye is sensitive.

visible speech (*Acous.*). Display of oscillogram patterns corresponding to characteristic speech sounds; used as an aid to the speech training of the totally deaf.

visiogenic (*TV*). Artistically suitable for visual reproduction by TV.

vision modulation (*TV*). The modulation of the carrier effected by the picture signal, as distinct from that reserved for the synchronizing impulses.

Vistanex (*Plastics*). TN for a *polybutene*.

VistaVision (*Cinema.*). Form of wide-screen cinematography using standard 35 mm film fed horizontally at twice normal speed. Horizontal feed enables the frame size to be approximately doubled, improving picture quality; increased speed improves sound reproduction.

visual acuity (*Optics*). A term used to express the spatial resolving power of the eye. Measured by determining the minimum angle of separation which has to be subtended at the eye between two points before they can be seen as two separate points.

visual approach slope indicator (*Aero.*). A luminous device for day and night use, consisting of red, green, and amber light bars on each side of a runway which, by being directed through restricting visors, show a pilot if he is below, on, or above and in line with the approach path for an accurate touchdown. Developed by the Royal Aircraft Establishment from a World War II night lighting system as the *VGPI* (*visual glide path indicator*). Abbrev. **VASI.**

visual binary (*Astron.*). A *double star* (q.v.), whose two components may be seen as separate in a telescope of sufficient resolving power.

visual cliff (*An. Behav.*). An experimental set-up in which there is a vertical drop, over which an animal is prevented from falling by a sheet of glass. Some animals avoid this area as a result of the visual perception of the drop.

visual exploration (*An. Behav.*). The activity of animals looking out of a box at things and events around them. The opportunity to do this can act as reinforcement in learning experiments with monkeys.

visual exposure meter (*Photog.*). One in which light value is measured (*a*) by viewing the subject through a step wedge of graduated translucence, or (*b*) by attenuating the light reflected from the subject until it is equated with that from a standard test image.

visual fatigue (*Photog.*). Loss in visual sensitivity due to prolonged perception of distorted images, particularly of cinematographic images on a screen.

visual flight rules (*Aero.*). The regulations set out by the controlling authority stating the conditions under which flights may be carried out without radio control and instructions. The regulations usually specify minimum horizontal visibility, cloud base, and precise instructions for the distance to be maintained below and away from cloud. Abbrev. **VFR.**

visual meteorological conditions (*Aero.*). Weather conditions in which an aircraft can fly under *visual flight rules*, which implies general freedom from air-traffic control except in controlled air space. Abbrev. **VMC.**

visual purple (*Optics*). See **rhodopsin.**

visual range (*Nuc. Eng.*). Observable range of ionizing particle in bubble chamber, cloud chamber, or photographic emulsion. (*Telecomm.*) Optical range of transmitter.

visual record (*Comp.*). One which can be read directly without decoding.

visual violet (*Optics*). A photosensitive retinal cone pigment; iodopsin.

vital capacity (*Physiol.*). The volume of air exhaled by a maximal expiration following a maximal inspiration, usually of the order of 4 litres in Man; the sum of the complemental air (*inspiratory reserve volume*), supplemental air (*expiratory reserve volume*) and the tidal air (*tidal volume*).

vitality index (*Nut.*). Index of chest girth (average of inspiration and expiration) divided by the height.

vitallium (*Med.*). A metallic alloy used in orthopaedic surgery because of its nonrust, nonreactive properties. These alloys, based on 62½–65% cobalt, carry such other elements as chromium (27–35%), molybdenum (5–5·6%), manganese (0·5–0·6%), iron (up to 1%), and nickel (up to 2%). They show good resistance to heat and corrosion.

vital stain (*Micros.*). A stain which can be used on living cells without killing them.

vitamins (*Chem., Med.*). Organic substances required, in relatively small amounts, for the proper functioning of the animal organism. They (or in some cases their immediate precursors) must be present in the food. Lack causes certain deficiency diseases, curable by administration of the appropriate vitamin; partial lack may cause minor disturbances or less well-defined ill-health. Some vitamins have a valuable therapeutic action when given in relatively large doses. As they were discovered, vitamins were distinguished by letters (A, B, C, etc.), but were later given names as their chemical nature was determined. There are two main groups: *fat-soluble* (i.e., associated with fats, soluble in fats and the usual fat solvents) which contain only carbon, hydrogen, and oxygen (vitamins A, D, E and K); and *water-soluble*, most of which contain nitrogen (vitamin C and the vitamin B complex). Several of the latter group are associated with enzymes concerned in oxidation-reduction processes.

vitamin A (*axerophthol*). Unsaturated alcohol of structure:

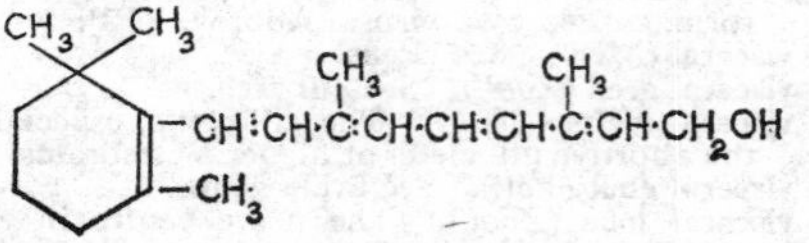

CH_3 CH_3 CH_3 CH_3

$-CH{:}CH{\cdot}C{:}CH{\cdot}CH{:}CH{\cdot}C{:}CH{\cdot}CH_2OH$

$-CH_3$

Animals can obtain it from its precursors, or 'provitamins', the *carotenes* (q.v.). Lack of vitamin A causes night-blindness and, in more severe cases, xerophthalmia, excessive keratinization of other epithelia (and hence, perhaps, increased susceptibility to infection) and, in the young, deficient growth. The richest natural sources are the liver oils of fish (halibut, cod, etc.), but valuable sources of the vitamin and its precursors are milk, butter, eggs, and various vegetables. The daily requirement of the normal adult is not known with certainty but is believed to be at least 1 mg.

vitamin B complex. A group of water-soluble vitamins which includes vitamins B_1, B_2, B_6, B_{12}, niacin, pantothenic acid, biotin and folic acid.

Vitamin B_1 (*aneurin, thiamin*). A somewhat heat-labile water-soluble vitamin, complete deprivation of which causes beri-beri, while partial lack results in peripheral neuritis and cardiac abnormalities. Isolated as its chloride-hydrochloride, with structure:

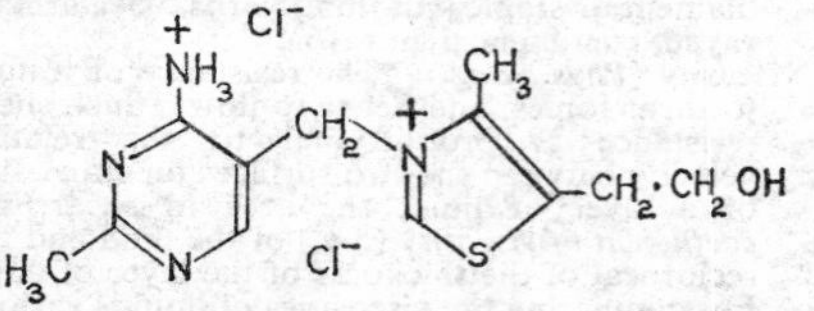

$\overset{+}{N}H_3$ Cl^- CH_3 CH_2 $\overset{+}{N}$ N N H_3C Cl^- S $CH_2{\cdot}CH_2OH$

Vitamin B_1 i s essential for the proper metabolism of carbohydrate (of especial importance to brain and nerve tissue), its pyrophosphate ester forming the co-enzyme of the enzyme carboxylase. The richest natural sources are yeast and the germs of cereals, but it is present in most foods. The minimum daily requirement is about 1 mg.

Vitamin B_2 (*riboflavin*). A yellow crystalline substance with the structure:

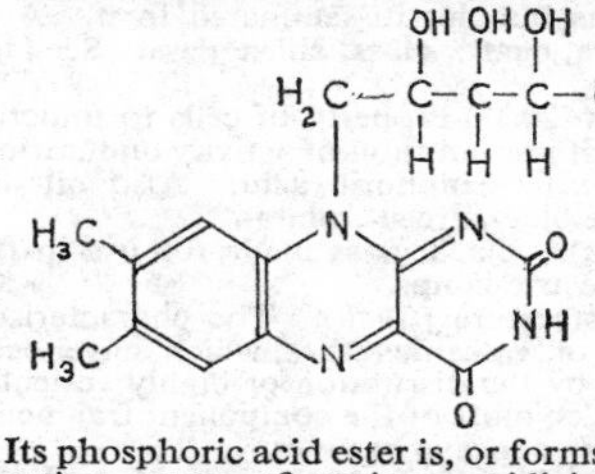

Its phosphoric acid ester is, or forms part of, the active group of various oxidizing enzymes. Lack of it causes failure of growth and, in Man, a syndrome characterized by thickening and cracking of the lips and by corneal lesions. The daily requirement is believed to be about 1 to 3 mg.

Niacin or *nicotinic acid* (pellagra-preventing factor, pyridine-*m*-carboxylic acid) also forms an essential part of various oxidizing enzymes. Lack of it leads to *pellagra* (q.v.), maize being an extremely poor source. The daily requirement is approximately 10 to 20 mg.

Vitamin B_6 (*pyridoxine, pyridoxal phosphate, pyridoxamine*) is also a pyridine derivative, with the structure:

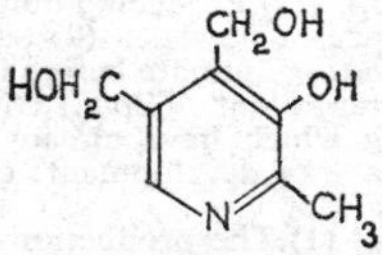

Pyridoxal phosphate plays a central role in reactions by which a cell transforms nutrient amino acids into a mixture of amino acids and other nitrogenous compounds required for its own activity.

Pantothenic acid forms part of the 'filtrate factor' which has been of therapeutic value in the treatment of certain anaemias. It has the structure: $HO{\cdot}CH_2{-}C(CH_3)_2{-}CH\ (OH){-}CO{-}NH{\cdot}CH_2{\cdot}CH_2{\cdot}CO{\cdot}OH$. Deprivation causes a characteristic dermatitis in chickens. It is necessary for rat growth. Human needs are unknown.

Biotin has the structure:

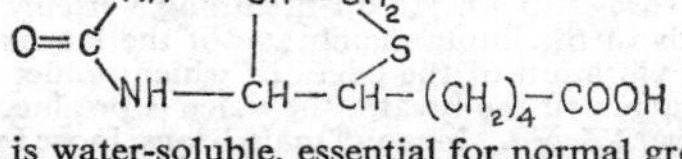

It is water-soluble, essential for normal growth of yeast and protecting rats or chicks against a nutritional injury caused by eating excess of raw egg-white. Human needs, if any, are unknown

Folic acid (*folacin*). Strictly, folic acid is the yellow substance isolated from spinach by Mitchell and found to be a growth factor for *Streptococcus lactis R* (*S. faecalis*) and for *Lactobacillus casei*. It is believed to be pteroic acid, a derivative of pteridine (see **xanthopterine**) and 4-amino-benzoic acid. It has the following structure:

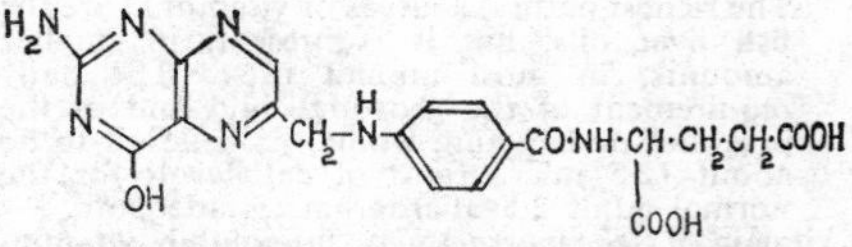

The substance used therapeutically under the name folic acid for the treatment of pernicious anaemia and some other macrocytic anaemias (e.g., sprue) is *pteroyl glutamic acid*, having glutamic acid joined by a peptide linkage to the carboxyl of pteroic acid. It has been synthesized and isolated from liver and yeast as vitamin B_c. Folic acid deficiency is expressed primarily as a failure to make purines and thymine required for DNA synthesis. *Pteroyl diglutamyl glutamic acid* (from fermentation residues of certain bacteria) and *pteroyl hexaglutamyl glutamic acid* (vitamin B_c conjugate from yeast) are also known. It has been suggested that sulphonamides act by preventing bacteria from synthesizing folic acid, which is essential for their growth.

Vitamin B_{12} (*cyanocobalamin*). A red crystalline substance isolated from liver and other natural sources. The most potent and complex of the vitamins, it contains cobalt and phosphorus. The nucleus (*corrin*) was synthesized in 1964. Pernicious anaemia results from defective secretion of *intrinsic factor* (q.v.), without which ingested vitamin B_{12} cannot be absorbed.

vitamin C (*ascorbic acid*). A white crystalline substance, stable when dry but very easily oxidized in solution, especially in neutral or alkaline solution; the oxidation is greatly accelerated by traces of copper. The enolic form of 3-keto-1-gulofuranolactone, with the following structure:

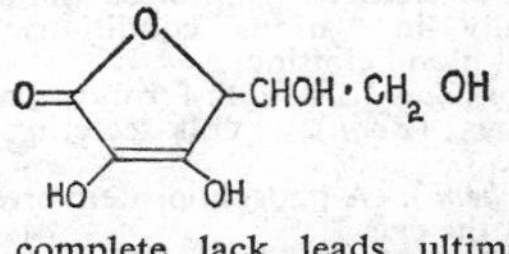

Almost complete lack leads ultimately to scurvy, partial lack to vague ill-health. The daily requirement is at least 15 mg. Certain fruits are very rich in ascorbic acid (rose-hips, blackcurrants, citrus fruits, etc.), as are many green vegetables; potato and turnip are also valuable sources.

vitamin D. The fat-soluble, antirachitic vitamin which probably acts by regulating the absorption of calcium and phosphate from the intestine. Some ten antirachitic substances have been obtained by ultraviolet irradiation of sterol precursors, but only two have been isolated from natural sources. Of these, *calciferol* (D_2) is usually prepared artificially from ergosterol; *cholecalciferol*, D_3, appears to be the commonest natural vitamin. They are white crystalline substances. Vitamin D_3 has the following structure:

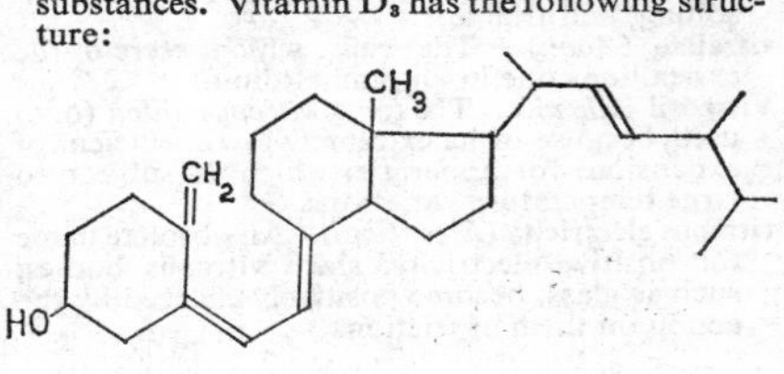

The richest natural sources of vitamin D are the fish liver oils, but it is present, in smaller amounts, in most animal fats. The daily requirement of the growing child and of the pregnant or lactating woman is believed to be about 12·5 micrograms of calciferol; for the normal adult 2·5 micrograms is adequate.

vitamin E (*tocopherol*). A fat-soluble vitamin. In some species at least, deficiency causes, in the females, inability to reproduce, the foetus dying and being resorbed. In the males complete deprivation of the vitamin ultimately produces degenerative changes and permanent sterility. Other symptoms of deficiency, in some species, are muscular dystrophy and necrosis of the liver. Human requirements are not known. Three closely related tocopherols are known; the most active of these, *a*-tocopherol, has the structure:

CH_3, HO, H_3C, CH_3, O, CH_3, $[(CH_2)_3 \cdot CH(CH_3)]_3 \cdot CH_3$

The richest natural sources are the oils of cereal germs.

vitamin K. A fat-soluble vitamin required for the production of prothrombin and therefore for normal blood coagulability. Two active substances have been isolated. Vitamin K_1 (*phylloquinone*), found most abundantly in the green leaves of plants, is 3-phytyl-2-methyl-1,4-naphthoquinone; K_2, formed by putrefactive bacteria, is 3-difarnesyl-2-methyl-1,4-naphthoquinone. Active analogues, all derivatives of 2-methyl-1,4-naphthoquinone (some water-soluble), have been synthesized and used therapeutically in various conditions involving delayed blood clotting.

vitellarium (*Zool.*). A yolk-forming gland.

vitelligenous (*Zool.*). Yolk-secreting or producing.

vitellin (*Chem.*). A phosphoprotein present in the yolk of the egg.

vitelline (*Bot., Zool.*). Egg-yellow; pertaining to yolk.

vitelline membrane (*Zool.*). A protective membrane formed around a fertilized ovum to prevent the entry of further sperms.

vitellophags (*Zool.*). In some *Arthropoda*, isolated yolk-consuming cells which play a part in the formation of the hypoblast.

vitellus (*Zool.*). Yolk of egg.

vitiligo (*Med.*). See **leucodermia.**

Viton (*Plastics*). Synthetic rubber based on a copolymer of vinylidene fluoride and hexafluoropropene. Maintains its rubber-like properties over a very wide temperature range and is chemically very resistant. TN.

vitrain (*Fuels*). A separable constituent of bright coal, of vitreous appearance, glossy, non-soiling, and friable.

vitrellae (*Zool.*). The cells which secrete the crystalline cone in an ommatidium.

Vitreosil (*Chem.*). TN for a *vitreous silica* (q.v.) used, because of its extremely low coefficient of expansion, for apparatus which is subject to large temperature variations.

vitreous electricity (*Elec. Eng.*). An obsolete name for positive electricity, since vitreous bodies, such as glass, become positively charged by the action on them of friction.

vitreous enamel (*Foundry*). Glazed coating fused on to steel surface for protection and/or decoration.

vitreous humour (*Zool.*). The jellylike substance filling the posterior chamber of the Vertebrate eye, between the lens and the retina.

vitreous silica (*Glass*). A vitreous material consisting almost entirely of silica, made in translucent and transparent forms. The former has minute gas bubbles disseminated in it. Also **fused silica, quartz glass, silica glass.** See also **Vitreosil.**

vitrification (*Cyt.*). Property of cells to undergo freezing with resumption of activity on thawing.

vitriol (*Chem.*). Sulphuric acid. Also **oil of-.** (*Min.*) See **blue-, green-, white-.**

vitrite (*Glass*). Black glass made for use in the caps of electric lamps.

vitroclastic structure (*Geol.*). The characteristic structure of volcanic ashes which have been produced by the disruption of highly vesicular glassy rocks, most of the component fragments thus having concave outlines.

vitrodentine (*Zool.*). In the cosmoid scale of *Crossopterygii*, the hard glossy substance forming the outer layer of the outer cosmine layer.

Vitrolite (*Glass*). TN for a type of opaque glass with a fire-finished surface.

vitta (*Bot.*). (1) A stripe. (2) A thin, elongated cavity containing oil, present in the pericarp of some fruits.

vittate (*Bot.*). Bearing longitudinal ridges or stripes.

vivax malaria (*Med.*). See **tertian.**

vivianite (*Min.*). Blue-iron earth. Hydrated iron phosphate ($Fe_3P_2O_8{\cdot}8H_2O$). Monoclinic. Hardness $1\frac{1}{2}$–2 (Mohs' scale), rel. d. 2·66.

viviparous (*Bot.*). (1) Producing bulbils or young plants in place of flowers. (2) Said of a seed which begins to germinate before it is detached from the parent plant. (*Zool.*) Giving birth to living young which have already reached an advanced stage of development; cf. *oviparous*. *n.* **viviparity.**

vivipary (*Bot.*). (1) The production of buds and young plants instead of, and in place of, flowers. (2) The production of seeds which begin to germinate before they are set free from the parent.

VLF (*Radio*). Abbrev. for *very low frequencies.*

VLS mechanism (*Phys.*). See **vapour-liquid-solid mechanism.**

VMC (*Aero.*). Abbrev. for *visual meteorological conditions.*

***V-n* diagram** (*Aero.*). See **flight envelope.**

v notch (*Civ. Eng.*). See **notch plate.**

vocabulary (*Comp.*). Set of *words* for data processing.

vocabulary test (*Psychol.*). A type of mental test used to determine the candidate's store of understood words.

vocal cords (*Zool.*). In air-breathing Vertebrates, folds of the lining membrane of the larynx by the vibration of the edges of which, under the influence of the breath, the voice is produced.

vocal sac (*Zool.*). In many male Frogs, loose folds of skin at each angle of the mouth which can be inflated from within the mouth into a globular form, and act as resonators.

vocoder (*Acous.*). System for synthetic speech using recorded speech elements.

vodas (*Telecomm.*). *V*oice *o*perated *d*evice *a*nti-*s*ing; used for the suppression of echoes in transoceanic radio telephony.

voder (*Acous.*). *V*oice *o*peration *de*monstrato*r*. System for producing synthetic speech through keyboard control of electronic oscillators.

vogad (*Telecomm.*). *V*oice *o*perated *g*ain-*ad*justing *d*evice. Used in telephone systems to give an approximately constant volume output for a wide range of input signals.
vogesite (*Min.*). A hornblende-lamprophyre, the other essential constituent being orthoclase. Cf. *spessartine*.
Vogt loudspeaker (*Acous.*). An electrostatically-driven loudspeaker operating on the push-pull system, the stretched diaphragm being located between two damping grilles which connect the displaced air with the outer air.
voice coil (*Acous.*). The coil attached to the cone of a loudspeaker. The coil currents react with the magnetic field to drive the cone. Also used in microphones to generate the signal.
voiced sound (*Acous.*). In speech, an elemental sound in which the component frequencies are exact multiples of a fundamental frequency which is determined by the tension of the oscillating muscles in the larynx.
voice filter (*Acous.*). Device which deliberately distorts speech for specific purpose, e.g., telephonic imitation.
voice frequency (*Telecomm.*). One in the approximate range 200 to 3500 Hz (that required for the normal human voice). Also called **speech frequency.**
voice-frequency multichannel telegraphy (*Telecomm.*). The use of a large number of voice-frequency channels, e.g., 12 or 18, for the fullest utilization of the transmission properties of normal audio-frequency telephone circuits.
voice-frequency relay (*Telecomm.*). A relay, incorporating a tuned reed, which is selectively operated by voice-frequency current, as in ringing on a long-distance telephone line.
voice-frequency sender (*Telecomm.*). Actuation of a register (sender) at the receiving end of a line by pulses of voice-frequency a.c. transmitted from the sending end by the depression of digit keys.
voice-frequency telegraphy (*Teleg.*). See **carrier telegraphy.**
voicing (*Acous.*). The art of adjusting the volume, pitch, and timbre of organ pipes so that they operate together in an organ.
void (*Powder Tech.*). In a powder compact or in a powder fluid system, the space between particles.
voidage (*Powder Tech.*). In powder compact or powder fluid system, the fractional quantity of voids in the system. For a system containing dense particles, the voidage and the powder porosity are numerically equal.
void waxing (*Cables*). Waxing which occurs at random anywhere in a dielectric. Irregular in size and shape, it is free from carbon and treeing.
Voigt effect (*Phys.*). Double refraction of electromagnetic waves passing through a vapour, when an external magnetic field is applied.
Voigt loudspeaker (*Acous.*). A moving-coil loudspeaker in which a large open diaphragm terminates the throat of a tractrix horn.
volant (*Zool.*). Flying; pertaining to flight.
volatile (*Chem.*). Changing readily to a vapour. (*Comp.*) Said of stored data which is lost if power-supply should fail.
volatility product (*Chem.*). The product of the equilibrium pressures of the gases formed by the dissociation of a solid or a liquid.
volatilization (*Chem.*). See **vaporization.**
volcanic ash (*Geol.*). The typical product of explosive volcanic eruptions, consisting of comminuted rock and *lava* (q.v.), the fragments varying widely in size and in composition, and including deposits of the finest dust, lapilli, and bombs. See also **agglomerate, pyroclastic rocks, tuff.**
volcanic bomb (*Geol.*). A spherical or ovoid mass of lava, in some cases hollow, formed by the disruption of molten lava by explosions in an active volcanic vent. See also **bread-crust bomb.**
volcanic muds and sands (*Geol.*). The products of explosive volcanic eruptions (*volcanic ash*) which have been deposited under water and have consequently been sorted and stratified, thus showing some of the characters of normal sediments, into which they grade.
volcanic neck (*Geol.*). A vertical pluglike body of igneous rock or volcanic ejectamenta, representing the feeding channel of a volcano.
volcanic vent (*Geol.*). The pipe which connects the crater with the source of magma below; it ultimately becomes choked with agglomerate or volcanic ash, or with consolidated lava.
volcano (*Geol.*). A centre of volcanic eruption, having the form typically of a conical hill or mountain, built of ashes and/or lava-flows, penetrated irregularly by dykes and veins of igneous rocks, with a central crater from which a pipe leads downwards to the source of magma beneath. Volcanoes may be active (periodically), dormant, or extinct; the eruptions may involve violent explosions (e.g., Krakatoa) or the relatively quiet outpouring of lava, particularly in those cases where the lava is basaltic (e.g., Hawaii). See also **lava.**
Volkmann's canals (*Zool.*). Canals carrying blood vessels which traverse the periosteal lamellae of a bone to join the Haversian canals.
Volkmann's contracture (*Med.*). A contracture of the flexor muscles of the forearm and leg due to the pressure of splints or tight bandages used in the treatment of fractures. It causes obstruction to the veins, so that the muscles swell, become hard, and then undergo fibrosis.
Volscan (*Comp.*). Analogue computer which can handle all landing approach problems.
volsella forceps (*Surg.*). Forceps whose blades have pronged ends.
volt (*Elec. Eng.*). The SI unit of *pd* (q.v.), electric potential or e.m.f., such that the pd across a conductor is 1 volt when 1 ampere in it dissipates 1 watt of power. Named after Count Alessandro Volta (1745-1827).
Volta effect (*Elec.*). Potential difference which results when two dissimilar metals are brought into contact: the basis of voltage cells and corrosion.
voltage (*Elec.*). The value of an e.m.f. or pd expressed in *volts*.
voltage-amplification factor (*Electronics*). See **amplification factor.**
voltage amplifier (*Telecomm.*). An amplifier whose function is to increase the voltage of the applied signal, without necessarily increasing its power. The output impedance must therefore be high.
voltage between lines (*Elec. Eng.*). The voltage between any two of the line wires in a single- or three-phase system; between the two lines of the same phase in a two-phase system; between any two lines which are consecutive as regards phase sequence in a symmetrical six-phase system. Also called **line voltage, voltage between phases, voltage of the system.**
voltage circuit (*Elec. Eng.*). The circuit of an instrument or relay which is connected across the lines of the circuit under test, and which therefore carries a current proportional to the voltage of this circuit. Also called **pressure circuit.** See **shunt circuit.**
voltage coefficient (*Elec. Eng.*). The constant by which the product of the armature speed in

revolutions per minute, the flux in volt-lines, and the number of armature conductors in series must be multiplied in order to obtain the e.m.f. of a d.c. generator.

voltage controlled oscillator (*Electronics*). Oscillator whose frequency is controlled by a bias signal; a *varactor* diode may be used as the controlling element. Abbrev. **VCO.**

voltage divider (*Elec. Eng.*). Chain of resistors such that the voltage across one or more is an accurately known fraction of that applied to all; used for calibrating voltmeters. Also called **potential divider, volt-box.** See **Kelvin-Varley slide.**

voltage doubler (*Elec. Eng.*). Power supply circuit in which both half-cycles of a.c. supply are rectified, and the resulting d.c. voltages are added in series.

voltage drop (*Elec.*). (1) Diminution of potential along a conductor, or over an apparatus, through which a current is passing. (2) The possible diminution of voltage between two terminals when current is taken from them.

voltage-fed antenna (*Telecomm.*). One which is fed with power from a line at a point of high impedance, where, through resonance, there is a voltage loop in the standing-wave system.

voltage feedback (*Elec. Eng., Electronics*). In amplifier circuits, a feedback voltage directly proportional to the load voltage. It may be applied in series or shunt with the source of the input signal. See also **current feedback, negative feedback, positive feedback.**

voltage gain (*Elec. Eng., Electronics*). The ratio of the change in output voltage (for a network or amplifier) to the change in input voltage which produces it. Often expressed in dB as $A_V = 20 \log (V_o/V_i)$, although this definition applies strictly only to the case when the input and output impedances are equal.

voltage gradient (*Elec. Eng.*). The difference in potential per unit length of a conductor, or per unit thickness of an insulating medium.

voltage level (*Telecomm.*). Peak-to-peak value at any point in a network, expressed relative to a specified reference level. When this is 1 volt, the symbol dBv is used.

voltage multiplier (*Elec. Eng.*). Circuit for obtaining high d.c. potential from low-voltage a.c. supply, effective only when load current is small, e.g., for anode supply to CRT. A ladder of half-wave rectifiers charges successive capacitors connected in series on alternate half-cycles. See **Westeht.**

voltage ratio (*Elec. Eng.*). Same as **turns ratio.**

voltage reference tube (*Electronics*). A glow-discharge tube designed to operate with anode-cathode voltage as nearly as possible constant regardless of the anode current, and hence suitable for use as a standard of pd.

voltage regulation (*Elec. Eng.*). The percentage variation in the output voltage of a power supply for either a specified variation in supply voltage or a specified change of load current.

voltage-regulator tube (*Electronics*). One in which, over a practical range of current, voltage between electrodes in a glow discharge remains substantially constant; also called **reference tube, v.r. tube.** See **stabilivolt.**

voltage resonance (*Elec. Eng.*). See **current resonance.**

voltage ripple (*Elec. Eng.*). The peak-to-peak a.c. component of a nominally d.c. supply voltage.

voltage stabilizer or **voltage-stabilizing tube** (*Electronics*). See **voltage-regulator tube.**

voltage standing-wave ratio (*Telecomm.*). Ratio between a maximum and a minimum in a standing wave, particularly on a transmission line or in a waveguide, arising from inexact impedance terminations. Abbrev. **VSWR.**

voltage to neutral (*Elec. Eng.*). The voltage between any line and neutral of a three- or six-phase system. Also called **phase voltage, star voltage, Y-voltage.**

voltage transformer (*Elec. Eng.*). A small transformer of high insulation for connecting a voltmeter to a high-tension a.c. supply.

voltaic cell (*Chem.*). Any device with electrolyte (ionized chemical compound in water), and two differing electrodes which establish a difference of potential.

voltaic current (*Elec.*). Current (direct) produced by chemical action.

voltaic pile (*Elec. Eng.*). A source of d.c. supply. It comprises a battery of primary cells in series, arranged in the form of a pile of disks, successive disks being of dissimilar metals separated by a pad soaked in the chemical agent.

voltameter (*Elec. Eng.*). An instrument for measuring a current by means of the amount of metal deposited, or gas liberated, from an electrolyte in a given time due to the passage of the current.

volt-ampere-hour (*Elec. Eng.*). Unit of apparent power, equal to watt-hour divided by *power factor.*

volt-amperes (*Elec.*). Product of actual voltage (in volts) and actual current (amperes), both r.m.s., in a circuit. See **active volt-amperes, reactive volt-amperes.**

volt-box (*Elec. Eng.*). See **voltage divider.** Also **volt ratio-box.**

voltinism (*Zool.*). Breeding rhythm; brood frequency. See **univoltine, bivoltine, multivoltine.**

volt-line (*Elec.*). A unit of magnetic flux equal to 10^3 maxwells or 10^{-5} webers.

voltmeter (*Elec. Eng.*). An instrument for measuring potential differences.

volt-ohm-milliammeter (*Elec. Eng.*). An electrical d.c. test instrument measuring voltage, resistance, and current. Usually a.c. volts can also be measured.

volume (*Acous.*). A general term comprehending the general loudness of sounds, or the magnitudes of currents which give rise to them. Volume is measured by the occasional peak values of the amplitude, when integrated over a short period, corresponding to the time constant of the ear. See **volume indicator, volume unit.** (*Phys.*) The amount of space occupied by a body; measured in cubic units. Symbol **V.**

volume bottles (*Chem. Eng.*). *Dampener* (q.v.), in the form of empty pressure vessels (usually steel), depending on the relationship between their volume and volume of gas passing to produce their dampening effect.

volume compression and expansion (*Acous.*). Automatic compression of the volume range in any transmission, particularly in speech for radio-telephone transmission, so that the envelope of the waveform is transmitted at a higher average level with respect to interfering noise levels. After expansion at receiving end, resulting transmission is freer from noise. See **compander.**

volume compressor (*Acous.*). In communication systems depending on amplitude modulation, intelligence transmitted as a modulation is limited to 100%. So that this is not exceeded with very loud sounds in the modulation, original transmission has to be compressed into a relatively small dynamic range to maintain a high signal/noise ratio.

volume control (*Telecomm.*). The manually-operated potentiometer which is used to regulate the communication transmission levels.

volume implant (*Radiol.*). A type of implant used when the volume to be treated approximates to a regular solid form. Sources are distributed both within the volume and on its periphery.

volume indicator (*Acous.*). Voltage-measuring device which, when placed across a communication channel carrying current for later conversion into sounds, gives a relative estimation of the apparent loudness of these reproduced sounds. The original type of volume indicator used a bridging transformer and an anode-bend rectifying thermionic valve, but indicators having a semiconductor rectifier are now used. The indicating meter and associated circuit have in each case a time constant which results in the indication being an integration over a short period, e.g. 0·2 second, of the varying voltage waveform applied to it. In these volume indicators, frequency-weighting networks are not incorporated for programmes. See **volume unit.**

volume ionization (*Nuc.*). The mean ionization density in any given volume without reference to the specific ionization of the particles.

volume lifetime (*Electronics*). That of current carriers in the bulk of a semiconductor; cf. *surface lifetime.*

volume limiter (*Telecomm.*). Circuit which automatically restricts volume of signal to specified maximum level.

volume range (*Acous.*). The difference between the maximum amplitude and the minimum useful amplitude of a waveform in an original sound, e.g., of an orchestra, expressed in decibels. In speech, it is generally taken to be 15–20 dB, and for full orchestra 60–70 dB.

volume resistivity (*Elec. Eng.*). See **resistivity.**

volume shape factor (*Powder Tech.*). The ratio

$$d_v = \frac{W}{Pd_m^3}$$

where W = mass of a particle; P = density of the particle; d = diameter of the particle as measured by a specified technique. Since d_m depends upon the method of measurement there are many shape factors for a given particle. If the value of volume shape factor is averaged over several particles, this value is called the volume shape factor of the powder.

volume sterilization (*Nuc.*). Complete trans-irradiation of material, e.g., foodstuffs, with penetrating radiation, e.g., high-energy gamma-rays from large radioactive sources.

volumetric analysis (*Chem.*). A form of chemical analysis using standard solutions for the estimation of the particular constituent present in solution by titrating the one against a known volume of the other. See **burette, end-point, indicator (1), pipette, titrimeter analysis.**

volumetric displacement (*Vac. Tech.*). The volume of air, calculated from the dimensions of the pump, passed per revolution through a mechanical pump when the pressure at the intake and the exhaust side is equal to the atmospheric pressure. Also called **swept volume.**

volumetric efficiency (*Eng.*). In an I.C. engine or air compressor, the ratio of the weight of air actually induced per unit time to the weight which would fill the swept volume at **s.t.p.**

volumetric heat (*Phys., etc.*). See **molal specific heat capacity.**

volumetric strain (*Eng.*). The algebraic sum of three mutually perpendicular principal strains in a material.

volume unit (*Acous.*). One used in measuring variations of modulation in a communication circuit, e.g., telephone or broadcasting. The unit is the decibel expressed relative to a reference level of 1 mW in 600 ohms, and standard *volume indicators* are calibrated in these units. Abbrev. **VU.**

voluntary muscle (*Zool.*). Any muscle which is controlled by the will, e.g., arm, thigh, neck, etc. Also called **skeletal muscle.** The muscles of the viscera are *involuntary.* In general, voluntary muscles are *striated* and involuntary ones *unstriated*, except for *cardiac muscle* (q.v.), which is slightly striated.

volutine granules (*Cyt.*). Granular cytoplasmic inclusions, of ribonucleic acid, which stain intensely with basic dyes; believed to contribute to the formation of chromatin; metachromatic bodies.

volva (*Bot.*). A sheath of hypha enclosing the whole of the fruit body of some agarics, becoming ruptured as the fruit body enlarges, and sometimes remaining as a cup or pouch round the base of the pileus.

Volvocina (*Zool.*). An order of *Phytomastigina* which usually have a flask-shaped green chromatophore with one or more pyrenoids, but are sometimes colourless, though never holozoic. Form starch reserves, have no gullet, usually have a cellulose cuticle and regularly undergo syngamy. Often form colonies.

volvulus (*Med.*). Torsion of an abdominal viscus, especially of a loop of bowel, causing internal obstruction.

vomer (*Zool.*). A paired membrane bone forming part of the cranial floor in the nasal region of the Vertebrate skull; believed not to be homologous in all groups. *adj.* **vomerine.**

vomerine teeth (*Zool.*). In *Holocephali, Teleostei, Dipnoi* and *Amphibia*, teeth, sometimes atypical, borne on the vomers.

Von Ebner's glands (*Zool.*). The serous glands of the tongue in Mammals.

Von Economo's disease (*Med.*). The disease called by Von Economo **lethargic encephalitis,** popularly known as **sleepy sickness.** See **epidemic encephalitis.**

Von Gudden's atrophy (*Zool.*). Secondary atrophy caused by disuse of nerve cells in the same physiological path as a damaged cell which has failed to regenerate.

Von Gudden's commissure (*Zool.*). A tract of fibres in the Mammalian brain connecting the posterior corpora quadrigemina.

Von Gudden's tract (or bundle) (*Zool.*). A tract of fibres in the Mammalian brain, passing from the corpora mammillaria to the tegmentum of the mid-brain.

Von Recklinghausen's disease (*Med.*). See **fibrocystic disease, molluscum fibrosum.**

vor (*Nav.*). (*V*ery-high-frequency *o*mnidirectional *r*adio-range.) Navigational ranging system in which any bearing can be obtained from the phase difference between an audio reference phase and one which is a function of azimuth, termed the variable phase.

vorobyevite (*Min.*). See **morganite.**

vortex (*Aero.*). An eddy, or intense spiral motion in a limited region; a *vortex sheet* is a thin layer of fluid with intense vorticity; *tip vortices* are a form of *trailing vortex* from aerofoils, caused by shedding of lateral and line-of-flight airflows. (*Hyd.*) The rotary flow of liquids or gases round a depressed central area.

vortex generators (*Aero.*). Small aerofoils,

mounted normal to the surface of a main aerofoil and at a slight angle of incidence to the main airflow, which re-energize the *boundary layer* (q.v.) by creating vortices. Used on the wings and tail surfaces of high-speed aeroplanes to reduce *buffeting* (q.v.) caused by compressibility effects, so raising the critical *Mach number* (q.v.), and sometimes to improve the airflow over control surfaces near the stall, thereby improving controllability. Cf. *wing fence*.

vortex street (*Aero.*). A regular arrangement of vortices in parallel rows; e.g., those springing from the wing-tips of an aeroplane and extending downstream behind it.

Voss machine (*Elec. Eng.*). See **Toepler machine.**

Vostok (*Space*). A series of Russian manned earth satellites, first used successfully in April 1961.

Votator (*Chem. Eng.*). TN for a heat exchanger in which a type of Archimedean screw rotates and nearly scrapes a cylindrical jacket. Used for viscous liquids and pastes which are moved by the screw over the inner wall of the jacket, which is heated or cooled as required.

vough (*Mining*). See **vug.**

voussoir (*Civ. Eng.*). See **arch stone.**

vowel articulation (*Teleph.*). See **articulation.**

V.P. (*San. Eng.*). Abbrev. for *vent pipe.*

V-particle (*Nuc.*). Particle exhibiting V-shaped decay track, and later identified as neutral meson or hyperon.

V.P.I. (*Chem.*). Abbrev. for *vapour phase inhibitor.*

VPP (*Vet.*). Abbrev. for *virus pneumonia of pigs.*

V-rings (*Elec. Eng.*). V-shaped mica rings insulating the segments of a commutator from the end rings.

v.r. tube (*Electronics*). Abbrev. for *voltage-regulator tube.*

V.S. (*Chem.*). Abbrev. for *volumetric solution.*

VSI (*Aero.*). Abbrev. for *Vertical speed indicator.*

VSWR (*Telecomm.*). Abbrev. for *voltage standing-wave ratio.*

V.T.B. curve (*Cables*). *V*oltage/*t*ime-to-*b*reakdown curve, i.e., a curve connecting the time and voltage for breakdown in this time. See **short-time breakdown voltage** and **asymptotic breakdown voltage.**

VTOL (*Aero.*). A general term for aircraft, other than conventional helicopters, capable of vertical take-off and landing: in Britain, the initials *VTO* were originally used, but the U.S. expression *VTOL* has become general.

V-type commutator (*Elec. Eng.*). A commutator whose segments are provided with projecting spigots, which dovetail into the end rings.

VU (*Acous.*). Abbrev. for *volume unit.*

vug or **vough** (*Mining*). A cavity in rock or a lode, usually lined with crystals.

Vulcan (*Astron.*). See **intramercurial planet.**

Vulcan coupling (*Eng.*). A hydraulic shaft coupling, of the Föttinger type, used for connecting marine diesel engines to the propeller shaft in order to avoid torsional vibration troubles. See **Föttinger coupling.**

vulcanite (*Chem.*). Hard vulcanized rubber, in the making of which a relatively high proportion of sulphur is used. *Ebonite* (q.v.) is one form; coloured varieties are obtained by adding various ingredients, such as the sulphides of antimony and mercury. See **vulcanization of rubber.**

vulcanites (*Geol.*). A general name for igneous rocks of fine grain-size, normally occurring as lava flows. Cf. *plutonites.*

vulcanization of rubber (*Chem.*). The treatment of rubber with sulphur or sulphur compounds, resulting in a change in the physical properties of the rubber. Sulphur is absorbed by the rubber, and the process can be carried out either by heating raw rubber with sulphur at a temperature between 135°C and 160°C, or by treating rubber sheets in the cold with a solution of S_2Cl_2. To increase the velocity of vulcanization, *accelerators* (q.v.) may be used.

vulcanized fibre (*Chem.*). A fibre obtained by treating paper pulp with zinc chloride solution. It consists of amyloid 90%, the remainder being water with some slight trace of insoluble salts. Used for low-voltage insulation.

vulpinite (*Min.*). A granular, scaly form of the mineral anhydrite, occurring at Vulpino, Lombardy, where it is cut and polished for ornamental purposes.

vulsinite (*Geol.*). A variety of trachyandesite containing phenocrysts of andesine bordered with sanidine, together with mica, hornblende, and, rarely, olivine, embedded in a groundmass consisting essentially of sanidine microliths.

vultex (*Chem.*). See under **latex.**

vulva (*Zool.*). The external genital opening of a female Mammal. *adj.* **vulviform.**

vulvitis (*Med.*). Inflammation of the vulva.

vulvovaginitis (*Med.*). Inflammation of both the vulva and the vagina.

VU meter (*Acous.*). Instrument calibrated to read intensity of electroacoustic signals directly in *volume units*. See also **volume indicator.**

W

w (*Civ. Eng.*). Symbol for load per foot run or for weight per cubic foot. (*Mech.*) Symbol for *work*. (*Stats.*) Symbol for range.

W (*Chem.*). The symbol for *tungsten*. (*Eng.*, *Phys.*) Unit symbol for *watt*.

W (*Civ. Eng.*). Symbol for total load. (*Elec. Eng.*) Symbol for electrical energy. (*Mech.*) Symbol for *weight*; *work*. (*Nuc.*) Symbol for radiant energy.

wacke (*Geol.*). A sandstone in which the grains are poorly sorted with respect to size.

Wackenroder's solution (*Chem.*). A concentrated solution of sulphurous acid into which hydrogen sulphide is passed at 0°C, to produce the thioacids $H_2S_xO_y$.

wad (*Min.*). Bog manganese, hydrated oxide of manganese(IV). See **asbolane.**

Wadell's sphericity factor (*Powder Tech.*). A shape factor used in particle-size analysis, defined by the equation

$$S=\frac{d_c}{D_c},$$

where S = Wadell's sphericity factor, d_c = the diameter of a circle equal in area to the projected image of the particle when the particle rests on its larger face, and D_c = the diameter of the smallest circle circumscribing the defined projection diameter.

wadi (*Geog.*). A river valley which is dry for most of the year, in desert and semi-desert areas.

Wadsworth mounting (*Optics*). A form of stigmatic mounting used for large concave gratings. The grating is illuminated by parallel light and the spectrum is focused at a distance of approximately one-half the radius of curvature of the grating.

waggle dance (*An. Behav.*). One of the bee dances, consisting of a broad figure of eight during which the bee waggles her abdomen from side to side and emits a pulsed sound while making the relatively straight transverse path across the middle. The number of these dances in a unit time decreases with increasing distance of the food source while the number of sound pulses rises, and, when performed on the vertical surface of a honeycomb in the dark, the direction of the waggle run to gravity indicates the direction of the food source with reference to the sun.

Wagner earth (*Elec. Eng.*). A pair of impedances with their common point earthed, connected across an a.c. bridge network in order to neutralize the effect of stray capacitances. This is done by simultaneously balancing the normal bridge and that formed by the Wagner earth with the ratio arms.

wagon retarder (*Rail.*). See **retarder.**

wagon vault (*Build.*). See **barrel vault.**

wagtail (*Join.*). See **parting slip.** (*Zool.*) Any bird of the family *Motacillidae*, so named from the constant vertical movement of the tail.

wainscot, wainscoting (*Join.*). A wooden lining, usually panelled, applied to interior walls.

wainscoting cap (*Join.*). A moulding surmounting a given piece of wainscoting.

wainscot oak (*Join.*). Selected oak, cut radially to display the silver grain; much used for panelling.

wairakite (*Min.*). The monoclinic, pseudocubic, calcium-bearing equivalent of *analcime* (q.v.), composition $CaAl_2Si_4O_{12}{\cdot}2H_2O$. Found hydrothermally replacing plagioclase.

waist anchor (*Ships*). See **sheet anchor.**

wait time (*Comp.*). The time interval during which a processing unit is waiting for information to be retrieved from a serial access file or for it to be located by a search.

wake (*Aero.*). The region behind an aircraft in which the *total head* (q.v.) of the air has been modified by its passage.

Waldegg valve gear (*Eng.*). See **Walschaert's valve gear.**

Walden inversion (*Chem.*). The transformation of certain optically active substances into their stereoisomeric derivatives by chemical reactions; sometimes a complete cycle is involved, e.g.,

$$\begin{array}{ccc} (-)\text{-chlorosuccinic acid} & \xrightarrow[\text{(moist } Ag_2O)]{} & (-)\text{-malic acid} \\ \uparrow (PCl_5) & & \downarrow (PCl_5) \\ (+)\text{-malic acid} & \xleftarrow[\text{(moist } Ag_2O)]{} & (+)\text{-chlorosuccinic acid} \end{array}$$

wale (*Textiles*). Ridge formed vertically as part of the pattern in flat or circular knitted fabrics.

walings (*Civ. Eng.*). Rough planks which run horizontally in front of the poling boards used in timbering trenches. The struts wedging the timbers apart on both sides are placed between the walings. Also wales.

walk-about disease (*Vet.*). See **Kimberley horse disease.**

Walker phase advancer (*Elec. Eng.*). A type of rotary phase advancer, comprising a generator with three-phase star-connected armature, commutator, and three field poles, the latter being in series with the armature circuit and the slip rings of the induction motor to be compensated.

walkie-talkie (*Radio*). A sender and receiver radio communication set which is carried by one person and may be operated while moving.

walking beam (*Eng.*). A mechanism for conveying solid articles, comprising a fixed horizontal grid of separated bars on which the articles rest, and a similar second grid so positioned as to be capable of rising through the fixed grid and lifting the articles, moving a short, horizontal distance relative to the fixed grid to carry the articles forward, and descending below the surface of the fixed grid to deposit the articles in an advanced position. While below the surface of the fixed grid, the moving grid returns to its original position to repeat the cycle. (*Mining*) Rocking beam used for transmitting power, e.g., for actuating the cable in cable-drilling for oil.

walking dragline (*Mining*). Large power shovel mounted on pads which are mechanically worked to manoeuvre it as required.

walking line (*Build.*). An imaginary line, always 18 in. from the centre line of the handrail, used in setting out winders for a stair, the width of the winder measured on this line being made approx. the same as the going of the risers.

Wallace's line (*Zool.*). An imaginary line passing through the Malay Archipelago and dividing the Oriental faunal region from the Australasian region.

wallboard (*Build.*). *Fibre-board* (q.v.), usually of laminated construction.

wall box (*Build.*). A support built into a wall to carry the end of a timber.

wall effect (*Nuc.*). (1) The contribution of electrons liberated in the walls of an ionization chamber to the recorded current. (2) The reduction in the count rate recorded with a Geiger tube due to ionizing particles not having the energy to penetrate the walls of the tube. (*Phys.*) Reduction of free-fall velocity of body in tube containing liquid due to nearness of wall.

wall energy (*Phys.*). The energy per unit area stored in the domain wall bounding two oppositely magnetized regions of a ferromagnetic material.

Wallerian degeneration (*Zool.*). Degeneration of nerve fibres (axis cylinder and myelin sheath) distal to the site of a lesion, such as a crush.

wall frame (*Build., Civ. Eng.*). A method of system building for high buildings, comprising large precast concrete components.

wall hanger (*Build.*). A support partly built into a wall to carry the end of a structural timber, which itself is not to be built into the wall.

wall hook (*Build.*). An L-shaped nail used as a means of attachment to a wall.

wall insulator (*Elec. Eng.*). An insulator specifically designed to enable a conductor at high potential to earth to pass through a brick or concrete wall.

wall-less ionization chamber (*Nuc. Eng.*). One in which (by the use of a guard ring) the collecting volume is defined by the applied field and the contribution of knock-on particles to the ionization is avoided.

Wallman amplifier (*Telecomm.*). See **cascode amplifier.**

'wallpaper' colour (*Print.*). A web, usually of advertising, preprinted in more than one colour on one or both sides in a repeating design, enabling it to be run through rotary presses of any cut-off.

wall plate (*Build.*). (1) The vertical member in a system of raking shores, held against the wall by hooks driven into the mortar joints, and providing support for the heads of the shores. (2) The timber or rolled-steel member built into or upon a wall as support for the ends of floor joists or other bearers.

wall plug (*Elec. Eng.*). A plug-in device for connecting a flexible conductor to a circuit terminal in the form of a wall socket. See **two-pin plug, three-pin plug.**

wall-sided (*Ships*). A term signifying absence of *tumble-home*, and indicating that the maximum breadth is maintained to deck-level.

wall socket (*Elec. Eng.*). A circuit terminal in the form of a receptacle into which a plug is inserted in order to make a connexion to the circuit. It is insulated and wall-mounted.

wall string (*Carp.*). A string, generally a *housed string*, positioned against a wall and supporting the inner ends of the steps.

wall tie (*Build.*). A galvanized iron piece built into the two parts of a cavity wall, thus serving to bond them together.

walnut (*For.*). Tree of the genus *Juglans*, giving a hardwood popular for high-class furniture, panelling, gun stocks.

Walschaert's valve gear (*Eng.*). A valve gear of the radial type used in some steam locomotives. The valve is driven through a 'combination lever' whose oscillation is the resultant of sine and cosine components of the piston motion, derived from connexions with the engine crosshead and with an eccentric or return crank at 90° to the main crank. Sometimes called **Waldegg valve gear.**

wamoscope (*Electronics*). Wideband display tube incorporating a travelling-wave tube. (*Wave-mo*dulated oscillo*scope.*)

wandering cells (*Zool.*). Migratory amoeboid cells; may be leucocytes or phagocytes.

wane (*For.*). A defect in converted timber; some of the original rounded surface of the tree is left along an edge.

Wankel engine (*I.C. Engs.*). Rotary automobile engine having an approximately triangular central rotor geared epitrochoidally to the central driving shaft and turning in a close-fitting oval-shaped chamber so that the power stroke is applied to each of the three faces of the rotor in turn as they pass a single sparking plug.

warble (*Vet.*). A small skin tumour of horses and cattle caused by bot-fly (*Hypoderma bovis*) which deposits its eggs in the fetlocks of cattle. The larvae reach the back of the animal and form chrysalises immediately beneath the skin, through which they emerge. Skins so damaged are of less value as leather.

warbler (*Telecomm.*). A rotating capacitor or other device for rapidly varying the carrier frequency of a transmitter in the warbling carrier system.

warble tone (*Acous.*). Narrow frequency band current for testing microphones, etc. To minimize errors due to standing waves in the acoustic space, the frequency is varied cyclically, thus the standing-wave pattern is constantly changing, and errors balanced out.

warbling carrier system (*Telecomm.*). Method of increasing degree of secrecy obtainable with a radio-telephone system using inversion.

Ward-Leonard control (*Elec. Eng.*). A method of speed control for large d.c. motors, employing a variable-voltage generator to supply the motor armature, driven by a shunt motor.

Ward-Leonard-Ilgner system (*Elec. Eng.*). A modification of the Ward-Leonard system of speed control, in which a flywheel is included on the motor generator shaft to smooth out peak loads otherwise taken from the supply.

warehouse (*Bind., Print.*). The department of a printing works where cutting, folding, and the simpler methods of binding are undertaken.

warfarin (*Chem.*). 3-(α-acatonyl-benzyl)-4-hydroxycoumarin, used (usually in the form of its sodium derivative) as a blood anticoagulant; also widely used as a selective rodenticide.

war gas (*Chem.*). Any solid, liquid or gaseous chemical substance used in warfare (or in riot control) to produce poisonous or irritant effects upon the human body. War gases are classified according to the length of time they are effective, **non-persistent** or **persistent** (less or more than 10 min in normal atmospheric conditions), and according to their effect: **lung irritants** or choking gases (phosgene, diphosgene); **lachrymators** or tear gases (chloracetophone, CS gas [orthochlorobenzylidene malononitrile], bromobenzylcyanide); *nerve gases* (derivatives of fluorphosphoric acid); **paralysants** or blood gases (hydrocyanic acid); **sternutators** or irritant smokes (diphenylaminechlorarsine); *vesicants* or blister gases (lewisite, mustard gas). The lachrymators and sternutators, being less toxic, are used in riot and crowd control.

warm-blooded (*Zool.*). Said of animals which have the bodily temperature constantly maintained at a point usually above the environmental temperature, of which it is independent; idiothermous; homoiothermous.

warm front (*Meteor.*). The leading edge of a mass of advancing warm air as it rises over colder air. There is usually heavy rain in advance of it.

Warminster Beds (*Geol.*). Usually referred to the Upper Greensand, these consist of about 18 ft of fossiliferous glauconitic sands with chert and cherty sandstone composed largely of sponge spicules, and occur typically just below the Chalk in the Vale of Warminster, Wiltshire.

warm pool and **warm ridge** (*Meteor.*). See **thickness chart.**

warm-up (*An. Behav.*). Increments in response strength sometimes observed to occur over the first few occasions within a short period of time when the response occurs, regardless of any reinforcement. Occurs both in learned and unlearned behaviour.

war neurosis (*Psychol.*). A preferable synonym for *shell-shock.* The term was originally used (World War I) for all types of nervous conditions resulting from war experiences, especially those caused by a bursting shell, which might result in (*a*) a condition of physical shock or concussion to the nervous system, (*b*) the precipitation of a psychoneurosis in a predisposed individual, (*c*) a combination of these conditions.

warning (*Horol.*). In a striking clock, the partial unlocking of the striking train, just before the hour.

warning coloration (*Zool.*). See **aposematic coloration.**

warning piece (*Horol.*). In the striking work, a projection on the lifting piece which projects through a slot in the dial plate, and against which a pin on the warning wheel butts, to hold up the train. Exactly at the hour the warning piece drops clear of the pin and frees the train.

warning pipe (*Plumb.*). An overflow pipe fitted to cisterns, etc., to warn of a defective valve.

warning wheel (*Horol.*). The last wheel in a striking train, which is held up by the warning piece during warning.

warp (*For.*). Permanent distortion of a timber from its true form, due to causes such as exposure to heat or moisture. (*Weaving*) Total number of yarns wound under substantial tension on to the weaver's beam. Passing through the healds and reed they constitute the lengthwise threads in a piece of cloth. In lace manufacture, the warp comes from the warp beam, and from independent beams. See **bobbin, chain warp, shed, shedding.**

warper's brasses (*Textiles*). A series of brass plates in a lace machine, with holes arranged in rows through which the warp threads pass.

warping (*Acous.*). Departure from flatness in a gramophone record, usually caused by excessive temperature during storage; obviated by storing vertically under moderate pressure in rigid racks. (*For.*) See **warp.**

warping mill (*Textiles*). (1) A cylindrical cage in which threads from jack bobbins are wound to fixed lengths, for use in a lace machine. (2) Large wooden or light alloy reel, vertical or horizontal, on to which yarns from cheeses or cones in a creel are wound, in preparation for a weaver's beam.

warp knitted fabric (*Textiles*). Fine, knitted fabric formed wholly from warp yarns wound on a beam the width of the machine. The yarns interloop to form cloth.

warp machine (*Textiles*). A straight-bar lace frame with bearded needles, in which individual threads pass to individual needles to form the fabric.

warp satin (*Textiles*). A term applied to a satin fabric merely to emphasise the fact that it is a true satin (i.e., with a warp face).

warp stop motion (*Weaving*). Mechanical or electrical device employing fine droppers on each yarn; these operate to stop a loom when a warp thread breaks.

warp twist (*Textiles*). See **crossband.**

Warren girder (*Build., Eng.*). A form of girder consisting of horizontal upper and lower members, connected by members inclined alternately in opposite directions.

Warren Hill Series (*Geol.*). A group of volcanic rocks of Pre-Cambrian age occurring in the Malvern Hills, England, and probably to be correlated with the Uriconian of Shropshire, which they closely resemble.

Warrington hammer (*Tools*). Type of cross-pane hammer with slightly convex face.

wart (*Bot.*). A small blunt-topped rounded upgrowth. *adj.* **warted.** (*Med.*) A tumour of the skin formed by overgrowth of the prickle-cell layer, with or without hyperkeratosis; due to infection with a virus. See also **verruca.**

wash (*Civ. Eng., Paint.*). A thin coat of water-colour paint applied to part of a drawing as an indication of the nature of the material to be used for the part represented, particular colours conventionally indicating particular materials.

washable distemper (*Paint.*). Distemper based on slaked lime, casein, borax, and the required pigments.

washboard (*Build.*). See **skirting board.**

wash box (*Min. Proc.*). Box in which raw coal is jigged in coal washery.

washed clay (*Build.*). See **malm.**

washer (*Build., Eng., etc.*). Annular piece, usually flat, used under a nut to distribute pressure, or between jointing surfaces for a tight joint, etc.

wash gravel (*Mining*). Alluvial sands worth exploitation for mineral values contained.

wash-in and wash-out (*Aero.*). Increase (wash-in) or decrease (wash-out) in the *angle of incidence* (q.v.) from the root toward the tip of an aerofoil, principally used on wings to ensure that the wing-tips stall last so as to maintain aileron control.

washing (*Photog.*). The essential process of removing soluble salts from emulsions, particularly hypo after fixing silver images, residual hypo corroding the image after a time.

wash leather (*Leather*). See **chamois leather.**

wash-out (*Acous.*). The elimination of a record on a magnetic tape by saturating it magnetically with a direct or ultrasonic current in an *erase head*, thus permitting further records to be made on the same tape. Same as **wipe-out.** (*Aero.*) See **wash-in.** (*Geol.*) A channel cut through a bed of sediment and filled with later material. Wash-outs are common in sedimentary rocks deposited under shallow water, as in deltas, and are common in Coal Measures.

wash-out valve (*Civ. Eng.*). A valve inserted in a pipeline at the bottom of a valley, in order to enable a particular length of the pipe to be emptied as required.

wash plate (*Ships*). Fitted in tanks to prevent large quantities of water or oil rushing from side to side when the vessel rolls.

Wassermann reaction (*Med.*). A test of the blood-serum (or of the cerebrospinal fluid) to determine whether the person from whom it is drawn is infected with syphilis. This condition is indicated by the presence of syphilitic antibodies in the serum.

waste (*Civ. Eng.*). See **spoil.** (*Mining*) Waste rock, either host (enclosing) rock mined with the true lode, or ore too poor to warrant further treatment. (*Nuc. Eng.*) (1) Depleted material rejected by an isotope separation plant. (2) Unwanted radioactive material for disposal.

waste heat recovery (*Heat*). The recovery of heat from furnaces, kilns, combustion engines, flue gases, etc., for utilization in, e.g., air preheating, feed water heating, waste heat boilers.

waste-light factor (*Light*). A factor used in the design of floodlighting installations to allow for the light which, although emitted along the beam from the projector, does not fall on the area to be illuminated.

waste preventer (*San. Eng.*). A type of cistern used for flushing a water closet; a fixed quantity of water is released for the purpose by pulling a chain or depressing a lever.

waster (*Build.*). A facing-brick having some defect which renders it unsuitable for facing work, although it may be quite suitable for other building purposes. (*Tools*) A mason's chisel, sometimes with claw head.

waste weir (*Civ. Eng.*). The weir provided in reservoir construction to discharge all surplus water flowing into the reservoir in flood-time, so as to prevent the water level from rising above the limit which was allowed for in designing the dam.

wasting (*Build.*). The operation of removing stone from a block by blows with a pick, prior to squaring and dressing.

WAT curves (*Aero.*). Complicated graphs relating the take-off and landing behaviour of an aeroplane to its weight, aerodrome altitude, and ambient temperature. Their preparation and use is mandatory for British public transport aircraft. Many other countries use the *w*eight/*a*ltitude/*t*emperature information in tabular form.

water (*Chem., Phys.*). A colourless, odourless, tasteless liquid, m.p. 0°C, b.p. 100°C. It is hydrogen oxide, H_2O, the liquid probably containing associated molecules, H_4O_2, H_6O_3, etc.; on electrolysis it yields two volumes of hydrogen and one of oxygen. It forms a large proportion of the earth's surface, occurs in all living organisms, and combines with many salts as water of crystallization. Water has its maximum density of 1000 kg/m^3 at a temperature of 4°C. This fact has an important bearing on the freezing of ponds and lakes in winter, since the water at 4°C sinks to the bottom and ice at 0°C forms on the surface. Besides being essential for life, water has a unique combination of solvent power, thermal capacity, chemical stability, permittivity and abundance. See **hard water, soft water, ion exchange**; also **triple point, heavy water.**

water balance (*Bot.*). The ratio between the water taken in by a plant and the water lost by it.

water ballast (*Ships*). Water carried for purposes of stability. Also, water taken into ballast tanks to balance or redress change of draught due to consumption of fuel, etc.

water bar (*Build.*). A galvanized iron (or non-ferrous metal) bar set in the joint between the wood and stone sills of a window, to prevent penetration of water. Also called **weather bar.**

water blast (*Mining*). A sudden escape of confined air due to water pressure, e.g., in rise workings.

water bloom, water flowers (*Bot.*). Large masses of algae, chiefly *Myxophyceae*, which sometimes develop very suddenly in bodies of fresh water.

waterbound macadam (*Civ. Eng.*). A road surfacing formed of broken stone, well rolled and covered with a thin layer of hoggin, which is watered in and binds the stones together.

waterbrash (*Med.*). A sudden gush into the mouth of a watery secretion from the salivary glands combined with acid fluids from the stomach; often sign of duodenal ulcer. Also **pyrosis.**

water calyx (*Bot.*). A calyx, in the form of a closed sac, into which hydathodes secrete much water, so that the other parts of the flower continue their development without risk of damage from dryness.

water-carriage system (*San. Eng.*). The system of disposing of waste matter from buildings by water closets, etc., involving the use of water to carry away the waste matter. Cf. *conservancy system.*

water cell (*Astron.*). A water filter used to eliminate the infra-red radiations, and thus separate these from the reflected sunlight, in determining planetary heat.

water channel (*Aero.*). An open channel in which the behaviour of the surface of water flowing past a stationary body gives a visual simulation of supersonic airflow.

water-checked (*Build.*). Said of a casement, the stiles and mullions of which have grooves cut in the meeting edges to prevent entry of rain.

water-chrysolite (*Min.*). See **bottle-stone.**

water closet (*San. Eng.*). A closet which is connected to a water-supply system so that the excreta may be carried away by flushing. There are two types, the standard 'wash-down' and the more efficient 'siphonic'.

water colours (*Paint.*). Pigments ground up in an aqueous, gummy medium.

water-cooled engine (*I.C. Engs.*). An engine cooled by the circulation of water through jackets, which are usually cast integral with the cylinder block.

water-cooled motor (*Elec. Eng.*). A motor employing water as a cooling medium.

water-cooled resistance (*Elec. Eng.*). A resistance kept cool by immersion in water, which circulates in channels provided for the purpose.

water-cooled transformer (*Elec. Eng.*). A transformer in which the oil is kept cool by means of water circulating in pipes immersed in the oil.

water-cooled valve (*Electronics*). Large thermionic vacuum tube in which the heat generated by the electronic bombardment of the anode is carried away by water circulating around or through it. In the former the anode is made an integral part of the envelope. Cf. *cooled-anode valve.*

water culture (*Bot.*). An experimental means of determining the mineral requirements of a plant; the plant is grown with its roots dipping into solutions of known composition.

water-cushion principle (*Space*). Immersion in water of the human body increases *g*-force tolerance to 13*g* for 4 minutes. See **positive *g* tolerance, negative *g* tolerance.**

water development (*Photog.*). The transference of an emulsion from a normal developing bath to a water bath after partial development, so as to reduce final contrast without increasing the maximum density; the development continues in those parts where the density is low, the high densities having used up all the locally absorbed developer.

water-displacing liquid (*Chem.*). A solvent containing surface active materials capable of removing water from moist surfaces and substituting a thin film of rust-inhibitive chemicals. A typical system would consist of organic fatty acids, non-ionic surfactants and barium petroleum sulphonate, dissolved in kerosine.

water equivalent (*Heat*). The mass of water which would require the same amount of heat as a body to raise its temperature by one degree. It is its **thermal capacity** (the product of its mass and its *specific heat capacity*) divided by the s.h.c. of water (4·186 kJ/kg K).

waterfall siphon (*Civ. Eng.*). A *siphon* (q.v.) in which the flow of liquid is insufficient to keep the pipe full, resulting in water falling freely under gravity in the down pipe.
water-finish (*Paper*). The application of water to a calender bowl to damp the surface of the paper or board in order that a higher finish can be obtained.
water flowers (*Bot.*). See **water bloom.**
water gas (*Fuels*). See **blue water gas, carburetted water gas, semiwater gas.**
water gauge (*Eng.*). A vertical or inclined protected glass tube connected, at its upper and lower ends respectively, to the steam and water spaces of a boiler, for showing the height of the water level. (*Mining*) An instrument (e.g., *Pitot tube*) for measuring the difference in pressure produced by a ventilating fan or air current.
water-glass (*Chem.*). A concentrated and viscous solution of sodium or potassium silicate in water. It is used as an adhesive, as a binder, as a protective coating in waterproofing cement, as a preservative for eggs, and in the bleaching and cleaning of fabrics.
water hammer (*Eng.*). A sharp hammerlike blow from a steep-fronted pressure wave in water caused by the sudden stoppage of flow in a long pipe when a valve is closed sufficiently rapidly.
Waterhouse stops (*Photog.*). Removable metal plates, each with a hole giving the required aperture, which can be inserted in front of the camera lens. Used in process cameras.
water-jet driving (*Civ. Eng.*). A process of pile driving often adopted when the piles have to be sunk into alluvial deposits; a pressure water jet is used to displace the earth around the point of the pile.
water-jet pump (*Eng.*). A simple suction pump, capable of producing a moderate degree of vacuum, in which air is drawn through the branch of a T-pipe by the action of a fast jet of water passing through the straight section. The principle is similar to that of an *ejector* (q.v.) and the pump has no moving mechanical parts.
waterleaf (*Paper*). Paper which has not been sized with rosin prior to the tub-sizing operation.
water lime (*Build., Civ. Eng.*). See **hydraulic cement.**
water lines (*Ships*). The intersection of the various water planes with the ship's form.
water lodge (*Mining*). An underground reservoir or sump.
waterlogged (*Civ. Eng.*). A term applied to ground which is saturated with water.
Waterman's wax (*Micros.*). A modified paraffin-wax embedding medium, used when a hard wax is necessary, e.g., for brittle specimens. It contains paraffin, stearic acid, spermaceti and ceresin.
watermark (*Paper*). Lettering or a design impressed into the paper during manufacture by means of the dandy roll or a projecting wire on the mould. Added as a trademark, or to give distinction to a paper.
water/methanol injection (*Aero.*). (1) The use of the *latent heat of evaporation* of water (the methanol is an antifreeze agent) injected into a piston engine intake to cool the charge, thereby permitting the use of greater power without detonation for take-off; (2) the injection of water into the airflow of the compressor of a *turbojet* or *turboprop* to restore take-off power by cooling the intake air at high ambient temperatures.
water monitor (*Nuc. Eng.*). One for measuring the level of radioactivity in a water supply, similar to, but much more sensitive than, an effluent monitor.
water of capillarity (*Build.*). The moisture drawn up by capillary action from the soil into the walls of a building.
water of hydration (or **crystallization**) (*Chem.*). The water present in hydrated compounds. These compounds when crystallized from solution in water retain a definite amount of water, e.g., copper(II)sulphate, $CuSO_4.5H_2O$.
water paint (*Paint.*). Any paint that can be diluted with water, e.g., emulsion paints, washable and nonwashable distempers.
water plane (*Ships*). A horizontal section through a ship's hull. Usually named by measurement from the base line, but sometimes from the load water plane.
water pore (*Bot.*). See **hydathode.** (*Zool.*) A madreporite.
water power station (*Elec. Eng.*). An alternative name for *hydroelectric generating station.*
waterproof paper (*Paper*). Paper which has been impregnated with pitch or bitumen to make it waterproof; often lined with hessian or other coarse material.
water recovery (*Aero.*). The recovery, principally by condensation, of the water in the exhaust gases of an aero engine. Used in airships for ballast purposes, as a partial set-off against the loss of weight due to the consumption of fuel during flight.
water resistor (*Elec. Eng.*). One made by immersing two electrodes in an aqueous solution. The resistance depends on the strength of the solution, and dimensions of the conducting path.
water rheostat (*Elec. Eng.*). A *water resistor* whose resistance can be varied, usually by moving one electrode relative to the other.
water-rib tile (*Build.*). A purpose-made tile having a projecting rib that serves to prevent entry of rain or snow.
water sapphire (*Min.*). See **saphir d'eau.**
water seal (*San. Eng.*). See **seal.**
watershed (*Civ. Eng.*). The line of separation between adjacent catchment areas.
water softening (*Chem.*). The removal of 'hardness' in the form of calcium and magnesium ions, which form precipitates with soap. See **Calgon, double decomposition, hard water, ion exchange, Permutit.**
waterspout (*Meteor.*). See **tornado.**
water stain (*Build.*). A stain for wood, consisting of colouring matter dissolved in water.
water stoma (*Bot.*). The opening through which water is discharged from a hydathode.
Waterstones (*Geol.*). The term applied by Hull to the higher part of the Keuper Sandstone of the English Midlands, consisting of irregularly bedded red and grey sandstones with curious markings on the bedding planes, resembling 'watered' silk. The name does not refer to the water-bearing qualities of the rocks, which are not exceptional.
water-storage tissue (*Bot.*). A group of large and often thin-walled cells inside a plant, in which water is stored and from which it is withdrawn in times of drought.
water table (*Build.*). See **canting strip.** (*Geol.*) The surface below which fissures and pores in the strata are saturated with water. It roughly conforms to the configuration of the ground, but is smoother. Where the water table rises above ground level a body of standing water exists.
watertight fitting (*Elec. Eng.*). An electric-light fitting designed to exclude water under certain prescribed conditions. Cf. *weatherproof fitting.*

watertight flat (*Ships*). When a watertight bulkhead above a deck is not directly over the bulkhead below, that part of the deck between the two is a *watertight flat*.

water torch (*Tools*). Machine which cuts through concrete and steel without generating heat and with relatively little noise. Uses pumping equipment in order to provide water supply to a pressure of 150 MN/m², and has a specially designed nozzle which produces a fine, penetrating needle of water.

water tower (*Civ. Eng.*). A tower containing tanks in which water is stored, built at or near the summit of an area of high ground in cases where the ordinary water pressure would be inadequate for distribution to consumers in the area.

water-tube boiler (*Eng.*). A boiler consisting of a large number of closely spaced water-tubes connected to one or more drums, which act as water pockets and steam separators, giving rapid water circulation and quick steaming. See **Babcock and Wilcox boiler, forced-circulation boilers, Stirling boiler, Yarrow boiler.**

water tunnel (*Aero.*). A tunnel in which water is circulated instead of air to obtain a visual representation of flow at high Reynolds numbers with low stream velocities. A shallow water tunnel, or channel, can represent compressible flow patterns qualitatively.

water turbine (*Eng.*). A prime mover in which a wheel or runner carrying curved vanes is supplied with water directed by a number of stationary guide vanes; usually direct-coupled to large alternators.

water twist (*Spinning*). Rather more than the usual amount of twist; a ring yarn has usually $4\sqrt{\text{counts}}$ as the turns per inch, but for water twist $4{\cdot}25\sqrt{\text{counts}}$ are usual.

water vapour pressure (*Meteor.*). That part of the atmospheric pressure which is due to the water vapour in the atmosphere.

water-vascular system (*Zool.*). In *Echinodermata*, a system of coelomic canals, associated with the tube-feet, in which water circulates; in *Platyhelminthes*, the excretory system.

water vesicle (*Bot.*). A much enlarged epidermal cell which serves for the storage of water.

water wave (*Acous.*). An optical effect on the surface of a gramophone record, caused by periodic alteration of the recording stylus with reference to the surface of the record, which varies the depth of cut.

water wheels (*Eng.*). Large wheels carrying, round the periphery, buckets or shrouded vanes on which water is caused to act, either by falling under gravity or by virtue of its kinetic energy. See **overshot wheel, undershot wheel.**

Wates (*Build., Civ. Eng.*). A method of *system building* (q.v.), developed in the U.K. The cross walls are of *in situ* reinforced concrete and the cladding is of precast panels with an applied external finish. The floor slabs are of the wide slab type which are precast on site and contain longitudinal cores.

Watling Shales (*Geol.*). The basal member of the Eastern Longmyndian Series, of Pre-Cambrian age, occurring in the eastern part of the Longmynd in South Shropshire; they consist of green shales (with occasional purple mudstones), with rare calcareous bands.

Watson-Crick DNA model (*Gen.*). The basis of modern molecular genetics. The genetic material is shown to consist of a twin-stranded, α-helical structure, of precise specification. Each strand consists of a sequence of nucleic acids, which defines the genetic information, and the base-pairs on each strand are complementary to each other. This concept of base-sequences and complementary base-pairs in a twin-stranded linear structure, of crystalline precision, provides a theoretical explanation for the fundamental genetic phenomena of replication and development.

watt (*Elec.*). SI unit of power equal to 1 joule per second. Thus, 1 horse-power (h.p.) equals 745·70 watts. Symbol W.

wattful loss (*Elec. Eng.*). See **ohmic loss.**

Watt governor (*Eng.*). A simple *pendulum governor* in which a pair of links are pivoted to the vertical spindle and terminate in heavy balls. Shorter links are pivoted to the mid-points of the first, and to the sleeve operating the engine throttle.

watt-hour (*Elec.*). A unit of energy, being the work done by 1 watt acting for 1 hour, and thus equal to 3600 joules.

watt-hour efficiency (*Elec.*). The ratio of the amount of energy available during the discharge of an accumulator to the amount of energy put in during charge. Cf. *ampere-hour efficiency*.

watt-hour meter (*Elec. Eng.*). Integrating meter for measurement of total electric energy consumed in a circuit. The conventional domestic electricity meter is of this type.

wattle and daub (or **dab**) (*Build.*). A type of wall construction in which wicker work is interlaced about a rough timber framework and the whole covered with plaster.

wattless component (*Elec. Eng.*). See **reactive component.**

wattmeter (*Elec. Eng.*). Instrument for measuring active power (active volt-amperes) in a circuit.

wattmeter method (*Elec. Eng.*). A method of testing the electrical quality of iron specimens by measuring the power loss with a.c. magnetization.

Waucobian (*Geol.*). A thick series of strata referred to the Lower Cambrian of western N. America (Waucoba Springs, California). Also known as the **Georgian Series** (Georgia, Vermont) in an eastern area extending from Boston to Newfoundland.

Waulsortian (*Geol.*). A reef-knoll facies of the Lower Carboniferous of Belgium, resembling that of the Mid-Pennine region of England. See **reef knoll.**

wave (*Elec., Phys., etc.*). A time-varying quantity which is also a function of position. A wave proceeding from a radiator will carry energy into the medium. Mechanical waves will involve displacement of the actual medium, while in electromagnetic waves it is the change in the electric and magnetic field intensities from their equilibrium values which represents the wave disturbance. In the case of electromagnetic waves radiated from a relatively low antenna one component is the *ground wave*. A component dependent on the nature of the surface is a *surface wave*.

wave analyser (*Elec. Eng.*). One which separates frequency components in a continuously repeated waveform, usually by a quartz-crystal filter or gate. In the heterodyne type, the heterodyne principle is applied to measure the various frequency components of a wave.

wave angle (*Radio, etc.*). Either angle of elevation or azimuth of arrival or departure of a radio wave with respect to the axis of an antenna array. Applied similarly to *sonar*.

wave antenna (*Radio*). Directional receiving antenna comprising a long wire running horizontally to the direction of arrival of the incoming waves at a small distance above the ground. The receiver is connected to one end, and the other end is connected to earth through a terminating resistance. The induced voltage depends on the forward tilt of the ground wave. Also called **Beverage antenna.**

waveband (*Radio*). Range of wavelengths occupied by transmissions of a particular type, e.g., the *medium waveband* (from 200 to 550 metres) used mostly for broadcasting.

wave clutter (*Radar*). Spurious marine radar echoes from waves on the sea.

wave drag (*Aero.*). The drag caused by the generation of *shock waves*, usually applied to the wing as a whole. See **compressibility drag.**

wave filter (*Elec. Eng.*). Four terminal network, consisting of pure reactances, designed to pass particular bands of frequency and to reject others.

waveform (*Phys.*). Variation of instantaneous amplitude as a function of time. A waveform may be **periodic, transient,** or **random,** for example.

wave-formed mouth (*Vet.*). A variation in height of the molar teeth of horses.

wavefront (*Elec., Phys., etc.*). Imaginary surface joining points of constant phase in a wave propagated through a medium. The propagation of waves may conveniently be considered in terms of the advancing wavefront, which is often of simple shape, such as a plane, sphere, or cylinder.

wave function (*Phys.*). Mathematical equation representing the space and time variations in amplitude for a wave system. The term is used particularly in connection with the Schrödinger equation for particle waves.

waveguide (*Elec., Phys., etc.*). (1) Hollow metal conductor within which very-high-frequency energy can be transmitted efficiently, in one of a number of modes of electromagnetic oscillation. Dielectric guides, consisting of rods or slabs of dielectric, operate similarly, but normally have higher losses. (2) Also exists around earth for electromagnetic waves, the ionized layers (ionosphere) acting as the upper reflecting 'surface'. (3) Also, for acoustic waves, the temperature reversal region in the atmosphere corresponding to the electromagnetic ionized layer. Similar conditions obtain under the sea. The acoustic waveguide corresponding to (1) would have very rigid walls.

waveguide attenuator (*Elec. Eng.*). Conducting film placed transversely to the axis of the waveguide.

waveguide choke flange (*Elec. Eng.*). Coupling flange between waveguide sections which offers zero impedance to signal without requiring metallic continuity.

waveguide coupler (*Elec. Eng.*). Arrangement for transferring part of the signal energy from one waveguide into a second, crossing or branching off from the first, e.g., in a *directional coupler*, the direction of flow of the energy transferred to the second guide reverses when the direction of propagation in the first guide is reversed.

waveguide filter (*Elec. Eng.*). One having distributed properties, giving frequency discrimination in a waveguide in which it is inserted.

waveguide impedance (*Elec.*). Ratio Z derived from $W = V^2/Z$, or I^2Z, where W is power, V a voltage and I a current, the last being defined in relation to type of wave and shape of waveguide.

waveguide iris (*Elec. Eng.*). Diaphragm placed across a waveguide forming a reactance.

waveguide junction (*Elec. Eng.*). Unit joining three or more waveguide branches, e.g., *hybrid junction* (q.v.).

waveguide lens (*Phys.*). An array of short lengths of waveguide which convert an incident plane wavefront into an approximately spherical one by refraction.

waveguide modes (*Elec.*). Modes of propagation in a waveguide. Classified as TE_{mn} (or H_{mn})—transverse electric, and TM_{mn} (or E_{mn})—transverse magnetic. In rectangular guide the subscripts refer to the number of half-cycles of field variation along the axes parallel to the sides. In circular guide they refer to the field variation in the angular direction and in the radial direction. Waves below the cut-off frequency are said to be **evanescent.**

waveguide stub (*Elec.*). A short-circuited length of waveguide used as a reactance for matching.

waveguide switch (*Elec. Eng.*). One which switches power from waveguide A to B or C, with considerable loss between A and C or B, and between B and C.

waveguide tee (*Elec. Eng.*). A T-shaped junction for connecting a branch section of a waveguide in parallel or series with the main waveguide transmission line.

waveguide transformer (*Elec. Eng.*). Unit placed between waveguide sections of different dimensions for impedance matching.

wave impedance (*Elec.*). Complex ratio of transverse electric field to transverse magnetic field at a location in a waveguide. For an acoustic wave, the ratio is pressure/particle velocity.

wave interference (*Phys.*). Relatively or completely stationary patterns of amplitude variation over a region in which waves from the same source (or two different coherent sources) arrive by different paths of propagation; *constructive interference* arises when the two waves are in phase and their amplitudes add; *destructive interference* arises when they are out of phase and their amplitudes partly or totally neutralize each other.

wave-interference error (*Nav.*). See **heiligtag effect.**

wavelength (*Phys.*). Symbol λ. (1) Distance, measured radially from the source, between two successive points in free space at which an electromagnetic or acoustic wave has the same phase; for an electromagnetic wave, it is equal, in metres, to 300 000 ÷ frequency (kHz). (2) Distance between two similar and successive points on an alternating wave, e.g., between successive maxima or minima; equal to the velocity of propagation divided by the frequency of the alternations when travelling along wires. (3) For electrons in motion when considered as a *wave train*, $\lambda = h/p$, h being Planck's constant and p the momentum of the electron. Similar for neutrons and other particles. See **wave mechanics.**

wavelength constant (*Phys., etc.*). The imaginary part of the *propagation constant*.

wavelength minimum (*Phys., etc.*). For an X-ray spectrum, given by

$$\lambda_{min} = \frac{hc}{V_{max}\, e}$$

where V_{max} = highest voltage applied to tube, λ_{min} = wavelength, h = Planck's constant, e = electronic charge, c = velocity of light.

wavelength of light (*Phys.*). In the electromagnetic spectrum (see **spectrum, electromagnetic**

wave) the order of magnitude of the wavelength of visible light may be taken as 300 nm, with a frequency of 10^{15} Hz and quantum energy of 3 eV. On either side lie the *infrared* (3–300 μm; $10^{14}-10^{12}$ Hz) and the *ultraviolet* (30–3 nm; $10^{16}-10^{17}$ Hz).

wavellite (*Min.*). Orthorhombic hydrated phosphate of aluminium, occurring rarely in prismatic crystals, but commonly in flattened globular aggregates, showing a strongly developed internal radiating structure. Named after Dr Wavell, who discovered the mineral in cherty shales near Barnstaple.

wave mechanics (*Phys.*). The modern form of the quantum theory in which all events on an atomic or nuclear scale are explained in terms of interaction between the associated wave systems as expressed by the Schrödinger equation. The energy of a quantum is the same whether considered as a wave train ($h\nu$) or as a particle (mvc), where h is Planck's constant, ν is the frequency of the wave, m is the mass of the particle, v its velocity and c the velocity of light. As a particle, this energy is bound up as a standing wave within a definite length which must contain an integral number of wavelengths, fractions being inadmissible; hence quantization. See also **statistical mechanics.**

wavemeter (*Instr.*). Instrument for measuring or indicating frequency or wavelength of electromagnetic radiation, either for laboratory use, or for monitoring in a radio transmitting station.

wave number (*Phys., etc.*). In an electromagnetic wave, the reciprocal of the wavelength, i.e., the number of waves in unit distance.

wave parameter (*Phys., etc.*). See **wavelength constant.**

wavers (*Print.*). Ink rollers that reciprocate to distribute the ink.

waveshape (*Elec., Phys., etc.*). See **waveform.**

wave tail (*Elec., etc.*). The portion of a waveform that follows the peak or crest.

wave theory (*Phys.*). Macroscopic explanation of diffraction, interference, and optical phenomena as an electromagnetic wave, predicted by Maxwell and verified by Hertz for radio waves. See **theories of light.**

wave tilt (*Elec. Eng.*). The angle between the normal to the ground and the electric vector, in a ground wave polarized in the plane of propagation.

wave train (*Phys., etc.*). Group of waves of limited duration, such as those which result from a single spark discharge occurring in an oscillatory circuit.

wave trap (*Radio, etc.*). A circuit tuned to parallel resonance connected in series with the signal source to reject an unwanted signal, e.g., between a radio receiver and the aerial. See also **wave filter.**

wave velocity (*Phys.*). See **phase velocity.**

wave winding (*Elec. Eng.*). A type of armature winding in which there are only two parallel circuits through the armature, irrespective of the number of poles.

wavicle (*Phys.*). Quantum mechanical entity which shows both particle and wave properties, related by principle of complementarity.

waving groin (*Build.*). A groin which is not straight in plan.

wavy paper (*Paper*). A defect of paper showing as undulations, especially near the edges, due to the moisture content of the paper not being in equilibrium with the surrounding atmosphere.

wawa (*For.*). See **obeche.**

wax (*Chem.*). Esters of monohydric alcohols of the higher homologues; e.g., *beeswax* is the myricyl ester of palmitic acid, $C_{30}H_{61}O{\cdot}CO{\cdot}C_{15}H_{31}$. For properties and uses, see also **beeswax, cable wax, candelilla wax, carnauba wax.**

waxing (*Cinema.*). The thin layer of wax which is placed on the edges of positive release prints, on the emulsion side, to form lubrication when the film is being projected.

wax pocket (*Zool.*). In Bees, a ventral abdominal pouch which secretes wax.

wax vent (*Foundry*). A pliable wax taper with a cotton core, placed in intricate cores during moulding. This wax melts when the core is dried, leaving a clear hole for the escape of gases.

wax wall (*Mining*). A wall of clay built round the gob or goaf, to prevent the entry of air or egress of gas.

waxy flexibility (*Med.*). See **flexibilitas cerea.**

ways (*Eng., etc.*). (1) The machined surfaces of the top of a lathe bed on which the carriage and tailstock slide; sometimes called **shears.** (2) The framework of timbers on which a ship slides when being launched.

way up (*Geol.*). The attitude (e.g., inverted) of a geological structure or formation in an area of severe folding.

W-chromosome (*Cyt.*). The X-chromosome when the female is the heterogametic sex.

weak coupling (*Elec. Eng.*). An inductive coupling in which the mutual inductance between two circuits is small; more generally known as **loose coupling.** Cf. *tight coupling.*

weak electrolyte (*Chem.*). An electrolyte which is only slightly ionized in moderately concentrated solutions.

weak interaction (*Nuc.*). One due to forces of the order of 10^{-13} times as great as in strong interactions. It appears to be due to some unknown coupling between fermions (baryons and leptons) and may be due to an as yet undetected boson (the intermediate charged vector boson) in the same way that strong interactions are believed to be due to meson coupling. Decay produced by weak interactions is not subject to conservation of strangeness, parity or hypercharge. See **strong interaction.**

Wealden Series (*Geol.*). A series some 2300 ft (700 m) thick, comprising the Hastings Beds below and the Weald Clay above, deposited under deltaic conditions in S.E. England; approximately equivalent to the marine Neocomian stage of the Cretaceous System.

weapons system (*Aero.*). The overall planned equipment and backing required to deliver a weapon to its target, including production, storage, transport, launchers, aircraft, etc.

wearing course (*Civ. Eng.*). Uppermost layer in a carriageway construction. Also termed **carpet, coat, crust, road surface, sheeting, topping, veneer, wearing surface.**

wearing depth (*Elec. Eng.*). The permissible amount of radial wear on a commutator, prior to renewing the segments.

wearing surface (*Civ. Eng.*). See **wearing course.**

weather bar (*Build.*). See **water bar.**

weather-board (*Join.*). A board used with others for covering sheds and similar structures. Weather-boards are fixed horizontally and usually overlap each other.

weather check (*Build.*). A *drip* (q.v.). Mastic or special metal strips are frequently incorporated.

weathercock stability (*Aero.*). The tendency for an aircraft to turn into the relative wind, due to the side areas aft of the cg exceeding the value for directional stability (as with aeroplanes designed for flying at low air speeds); excessive weathercock stability causes an oscillating yawing motion when flying in a cross-wind.

weathered pointing (*Build.*). The method of pointing in which, in order to throw the rain off the horizontal joints, the mortar is sloped inwards, either from the lower edge of the upper brick, or from the upper edge of the lower brick, the latter method being preferred by bricklayers. Also called **struck-joint pointing.**

weather fillet (*Build.*). See **cement fillet.**

weather forecast (*Meteor.*). See **forecast, weather map.**

weathering (*Build.*). (1) The deliberate slope at which an approximately horizontal surface is built or laid so that it may be able to throw off the rain. See **coping** (1). (2) The gradual process by which materials on the external faces of a building are affected by natural climatic conditions. (*Geol.*) The processes of disintegration and decomposition effected in minerals and rocks as a consequence of exposure to the atmosphere and to the action of frost, rain, and insolation. These effects are partly mechanical, partly chemical, and for their continuation depend upon the removal, by transportation, of the products of weathering. *Denudation* (q.v.) involves both weathering and transportation. (*Mining*) Long-term change in physical and/or chemical structure of rocks or their constituents due to climatic attack by heat, cold, wind action, oxidation.

weather map (*Meteor.*). A map on which are marked synchronous observations of atmospheric pressure, temperature, strength and direction of the wind, the state of the weather, cloud, and visibility. Weather maps (also known as **synoptic charts**) are used as a basis for forecasting.

weather minima (*Aero.*). The minimum horizontal visibility and cloud base stipulated (*a*) by the air traffic authority and (*b*) by the standing orders of each airline, under which take-off and landing is permitted.

weather moulding (*Build.*). See **dripstone.**

weatherometer (*Paint.*). An instrument used to determine the weather-resisting properties of paints, cycles of artificial conditions approximating closely to those of natural weathering.

weatherproof fitting (*Light*). An electric-light fitting having an enclosure which excludes rain, snow, etc. Also called **splashproof fitting.**

weather satellites (*Space*). Satellites used for the study of cloud formations and other meteorological conditions, e.g., the *Tiros* series (U.S.).

weather slating (*Build.*). See **slate hanging.**

weather strip (*Join.*). See **door strip.**

weather-struck (*Build.*). A term applied to mortar joints finished by the method of *weathered pointing* (q.v.).

weather tiling (*Build.*). Tiles hung vertically to the face of walls, in order to protect them against wet and to help maintain an even temperature within the building. Also called **tile hanging.**

weaver's warp (or web) (*Weaving*). Sheet of warp yarns wound on to a beam. These, heavily tensioned, pass through the healds and reed, to be raised and lowered in a definite order. The shuttle carrying the weft passes between the upper and lower sheets of yarn to form cloth. See **warp.**

weaving (*Textiles*). The interlacing of two or more sets of threads to form a fabric. See **loom, reed, shuttle, warp, weft.**

web (*Build., Eng.*). The relatively slender vertical part or parts of an I-beam or built-up girder (such as a box girder) separating the 2 flanges. (*Paper*) The endless sheet of paper coming along the paper machine. (*Zool.*) The mesh of silk threads produced by Spiders, some Insects, and other forms; the vexillum of a feather; the membrane connecting the toes in aquatic Vertebrates, such as the Otter.

Webatron (*Print.*). Electronically-controlled equipment for registering pre-printed reels into sheeters and bag-machines at high speeds (up to 2000 per minute). Cf. *Secatron.*

webbed (*Zool.*). Having the toes connected by membrane, as in Frogs, Penguins, Otters.

web-break detector (*Print.*). Electronic equipment to stop the press immediately a web breaks.

weber (*Elec.*). The SI unit of magnetic flux. An e.m.f. of 1 volt is induced in a circuit through which the flux is changing at a rate of 1 weber per second. The weber thus equals 1 volt-second, also 1 joule per ampere. Also used as unit of magnetic pole strength. Abbrev. **Wb.**

Weber dynamometer (*Elec. Eng.*). An early form of dynamometer, having a fixed coil within which a small moving-coil is suspended by a bifilar suspension.

Weber-Fechner law (*Physiol.*). That the physiological sensation produced by a stimulus is proportional to the logarithm of the stimulus.

Weberian apparatus (*Zool.*). See **Weberian ossicles.**

Weberian ossicles (*Zool.*). In some *Teleostei* (order *Ostariophysi*, including Carps, Catfish and others) a chain of small bones, derived from processes of the anterior vertebrae, which connect the air bladder to the ear, transmitting vibrations from the former to a perilymphatic sac from which they pass to the endolymph of the inner ear. They correspond functionally to the middle ear ossicles of higher Vertebrates.

Weber photometer (*Elec. Eng.*). A transportable photometer in which a direct comparison is made between the brightness of two screens, one illuminated by an unknown light source and the other by a standard lamp.

web-fed (*Print.*). A term indicating that the printing machine uses a reel of paper and not single sheets.

web frame (*Ships*). An extra strong frame usually made up of a plate with double angle bars on outer and inner edges. Commonly used to resist *panting* (q.v.).

web offset (*Print.*). An offset litho machine using a reel of paper.

websterite (*Geol.*). A coarse-grained ultramafic igneous rock, consisting essentially of both ortho- and clino-pyroxenes; a diallage-hypersthene-pyroxenite.

Weddle's rule (*Maths.*). The area under the curve $y=f(x)$ from $x=x_0$ to $x=x_6$ is approximately

$$\tfrac{3}{10}\{f(x_0)+5f(x_1)+f(x_2)+6f(x_3)+f(x_4)+5f(x_5)+f(x_6)\}\{x_6-x_0\}$$

where x_0, x_1, x_2, x_3, x_4, x_5 and x_6 are equally spaced.

wedge (*Elec. Eng.*). (1) Total attenuator, in the form of a wedge of absorbing material, for terminating a waveguide. (2) Insertion of various lossy materials, put into a section of waveguide, to add fixed or variable attenuation in the circuit. (*Meteor.*) See **ridge.** (*Photog.*) See **wedge filter.**

wedge aerofoil (*Aero.*). A supersonic aerofoil section (much used for missiles) comprising plane, instead of curved, surfaces tapering from a very sharp leading edge at an acute included angle to give a *thickness ratio* (q.v.) of 5% or less; the aerofoil may have a blunt trailing edge, or it may have the section of a very elongated lozenge, or it may have a parallel mid-portion with leading- and trailing-edge wedges, the two latter cases being known as **double-wedge aerofoils.**

wedge bones (*Zool.*). In some Lizards, as Sphenodon and Gecko, small wedge-shaped intercentra formed by the ossification of the ventral portions of the intervertebral disks.

wedge contact (*Elec. Eng.*). A contact consisting of two fingers between which a wedge-shaped contact on the moving element is forced; used for circuit-breakers, etc.

wedge cut (*Mining*). Wedge shot in blasting. See **V-cut.**

wedge filter or **wedge** (*Photog.*). An adjustable light-attenuating device, made of material of a uniform grey translucence, the varying loss in light being obtained by the varying thickness of path. See **neutral wedge filter.**

wedge photometer (*Elec. Eng.*). A type of photometer in which the illumination of the two sides of a wedge is directly compared.

wedge spectrogram (*Photog.*). A spectrogram taken with a neutral wedge with its varying thickness parallel to the slit of the spectrometer. The resultant photographic image indicates, by the height of the density contours, the differential colour sensitivity of the emulsion.

wedging (*Mining*). Use of deflecting wedge near bottom of deep diamond-bore holes, either to restore direction or to obtain further samples; also called **whipstocking.**

wedging crib, curb, or ring (*Mining*). A segmented steel ring on which shaft tubbing is built up and wedged in place.

weed (*Bot.*). A plant growing where it is not wanted by man; a potato growing in a bed of geraniums would be a weed. (*Vet.*) See **sporadic lymphangitis.**

weephole (*Civ. Eng.*). A pipe laid through an earth-retaining wall, with a slope from back to front to allow the escape of collected water.

weft (*Weaving*). The threads across the width of a fabric are called the *weft* (formerly *woof*); a single one of these is called a *pick*, or, less frequently, a *shot* (or *shoot*) of weft.

weft fork (*Weaving*). Mechanism at the side or centre of loom raceboard. It stops the machine when the weft breaks or runs out, and remains inoperative while the weft is being inserted.

Weg rescue apparatus (*Mining*). Portable breathing apparatus with self-contained oxygen supply controlled automatically by wearer's breathing action.

Weierstrass's test for uniform convergence (*Maths.*). See **M-test.**

Weigert-Pal stain (*Micros.*). A method used for the differential staining of the myelin sheath of medullated nerve fibres. The material is overstained with haematoxylin and subsequently differentiated until only the myelin remains stained.

weigh batching (*Civ. Eng.*). A method whereby, using certain plant, the constituents of concrete are measured by weight before mixing, instead of by volume.

weighing bottle (*Chem.*). A thin-walled cylindrical glass container with a tightly fitting lid, used for weighing accurately hygroscopic, etc., materials.

weight (*Aero.*). *Maximum*, or *gross, weight* is the total weight of an aircraft as authorized for flight under the current regulations; *maximum landing weight* is the highest safe weight for landing because of structural strength; *tare weight* is the design weight of an aircraft type in flying condition, without fuel, oil, crew, removable equipment not necessary for flight, and payload; *zero fuel weight*, used in airline load calculations, is the weight of the loaded aircraft after all usable fuel has been consumed. (*Phys.*) The gravitational force acting on a body at the Earth's surface. Units of measurement are the newton, dyne or pound-force. Symbol **W.** Weight = mass × *acceleration due to gravity* (q.v.), and must therefore be distinguished from *mass*, which is determined by quantity of material and measured in pounds, kilograms, etc.

weight coefficient (*Elec. Eng.*). The ratio of the weight of an electrical machine to its rated output.

weighting (*Textiles*). Adding sizing material, such as china clay, or weighting material, such as metallic salts, to yarn or cloth to increase the weight and give more body.

weighting factor (*Nuc.*). See **statistical weight.**

weighting network (*Telecomm.*). One designed to produce unequal attenuation for different frequency components of a signal, thereby weighting these differently in the final output.

weighting observations (*Surv.*). The operation of assigning factors or 'weights' to each of a number of observations to represent their relative liability to error under their individual conditions of measurement.

weight kit (*Agric.*). Set of three weights and attaching bolts, which fit snugly into a plough frame. A valuable aid to penetration in very hard ground.

weightlessness (*Space*). The condition associated with *free fall* (q.v.).

weightometer (*Min. Proc.*). Device which automatically weighs and records the tonnage of ore in transit on a belt conveyor.

weights. Standardized masses used for comparison with unknown masses, balances of various grades of sensitivity and sensibility being employed. For high-grade analysis, weights are often plated with noble metal to ensure that diminution of mass with time, through corrosion, is minimized.

Weigner method (*Powder Tech.*). Original technique devised for the *density change method*.

Weil loudspeaker (*Acous.*). A loudspeaking receiver with an open diaphragm, tuned to operate over a restricted frequency range, so that a number, tuned to different ranges, are essential for normal use, e.g., in a cinema.

Weil's disease (*Med.*). See **spirochaetosis icterohaemorrhagica.**

weir (*Civ. Eng.*). A dam placed across a river to raise its level in dry weather.

Weir-Mitchell treatment (*Med.*). Treatment of neurasthenia by rest, liberal feeding and massage.

Weisbach triangle (*Surv.*). A method used in orienting underground workings, in which the theodolite is deliberately set up off the line of the two hanging wires used to transfer direction from above-ground to below-ground, so that the triangle between the instrument and the wires may be solved to enable the setting-out to proceed.

Weismann's ring (*Zool.*). In the larvae of some *Diptera* (Insects), a small ringlike structure behind the brain and around the aorta, containing three types of glandular cell homologous with the corpora allata, corpora cardiaca and prothoracic glands of other Insects, and controlling metamorphosis in a similar manner. Sometimes known as the **ring gland.**

Weissenberg method (*Crystal.*). A technique of X-ray analysis in which the crystal and photoplate are rotated in the beam of X-rays while the plate is moved parallel to the axis of rotation.

Weiss theory (*Mag.*). Early theory of ferromagnetism based on the concept of independent molecular magnets.

welded joint (*Elec. Eng.*). A joint between two metals, made by electric welding.

welded tuff (*Geol.*). A *tuff* composed of glass fragments which have partially fused together as a result of being deposited while still at a high temperature. Synonymous with *ignimbrite*.

welding (*Eng.*). (1) Joining pieces of suitable metals or plastics, usually by raising the temperature at the joint so that the pieces may be united by fusing or by forging or under pressure. The welding temperature may be attained by external heating, by passing an electric current through the joint, or by friction. (2) Joining pieces of suitable metals by striking an electric arc between an electrode or filler metal rod and the pieces. See **arc-, cold-, resistance-, seam-**.

welding manipulator (*Eng.*). A support or framework with clamps, capable of being rotated or otherwise moved manually or under power, for securely carrying a workpiece and moving it as required relative to welding equipment.

welding regulator (*Elec. Eng.*). A reactance by means of which the welding current may be varied in an a.c. welding set; it is variable by tappings controlled by a handwheel.

welding rod (*Elec. Eng.*). Filler metal in the form of a wire or rod; used in electric welding when the electrode itself does not furnish the filler metal. Also called **filler rod**.

welding set (*Elec. Eng.*). The apparatus for electric arc welding, either a.c. or d.c., comprising a supply unit and a regulator, which may be combined or separate.

welding transformer (*Elec. Eng.*). A transformer specially designed to supply one or more welding regulators.

weldment (*Eng.*). A welded assembly.

Weldon's process (*Chem.*). A process for the manufacture of chlorine, using manganese dioxide and hydrochloric acid.

well (*Nuc.*). Pictorial term used to depict variation of potential energy of nucleon with distance from nucleus.

well-conditioned (*Surv.*). A term used in triangulation to describe triangles of such a shape that the distortion resulting from errors made in measurement and in plotting is, or is nearly, a minimum; achieved in practice by making the triangles equilateral or approximately so.

well counter (*Nuc. Eng.*). One used for measurements of radioactive fluids placed in a cylindrical container surrounded by the detecting element (scintillation crystal or sensitive volume of special Geiger tube).

well foundation (*Civ. Eng.*). A type of foundation formed by sinking *monoliths* (q.v.) to a firm stratum and filling in the open wells with concrete.

well-hole (*Build.*). The vertical opening enclosed between the ends of the flights in a winding stair.

well logging (*Geol.*). The recording of the composition and physical properties of the rocks encountered in a borehole, particularly one drilled during petroleum exploration. Well logging includes a variety of techniques, e.g., resistivity log, gamma-ray log, neutron log, spontaneous or self-potential log, temperature log, calliper log, photoelectric log, acoustic velocity log, etc.

Welsh groin (*Build.*). A groin formed by two intersecting cylindrical vaults of different rises. Also called an **underpitch groin**.

welt (*Plumb.*). A joint made between the edges of two lead sheets on the flat. Made by turning up each edge at right angles to the flat surface, bringing the two turned-up parts together, doubling them over, and dressing them down flat. Also called a **seam**. (*Weaving*) (1) A cord or rib. (2) A piqué fabric which has a transverse cord as its chief characteristic.

wen (*Med.*). See **sebaceous cyst**.

Wenlockian (*Geol.*). One of the major series into which the Silurian System is divided, comprising a considerable thickness of shales, limestones, or grits, lying between the Llandovery Series below and the Ludlow Series above. Named from the Shropshire town of Wenlock, which is situated on the main outcrop.

Wenner winding (*Elec. Eng.*). Form of winding used in wirewound resistors to construct standard resistances of low residual reactance for use at relatively high frequencies.

Werlhof's disease (*Med.*). See **purpura haemorrhagica**.

wernerite (*Min.*). An obsolescent name for *scapolite* (q.v.).

Werner sedimentation techniques (*Powder Tech.*). A 2-layer method of particle-size analysis. Particles sediment through a vertical column of liquid and collect in a narrow capillary at the foot of the column. The height of the column of particles is used as a measure of the particles sedimented out.

Werner's theory (*Chem.*). A method of formulation of complex inorganic compounds based on the assumption that saturated groups are held to the central atom by residual valencies. The total number of such groups and ordinary unsaturated radicals which surround the central atom to form an un-ionizable coordination complex is characteristic of the central atom.

Wertheim's operation (*Surg.*). The operation of removing the uterus, the glandular tissue in the pelvis, and the upper part of the vagina, in the treatment of cancer of the cervix uteri.

Westcott flux (*Nuc.*). A theoretical neutron flux defined as equal to the reaction rate of a detector with cross-section which is unity for thermal neutrons (velocity 2200 m/s) and varies inversely with the neutron velocity.

Westcott convention (*Nuc.*). An approximation widely applied to the neutron flux in thermal reactor design. It is divided into a thermal component with a Maxwellian distribution and a fast component with distribution proportional to dE/E, where E is the neutron energy. Sometimes a third component covering the thermalization region may be considered.

Westeht (*Elec. Eng.*). TN for copper-oxide rectifier voltage multiplier unit used for extra-high-tension supply to CRT.

Western duck sickness (*Vet.*). See **alkali disease (1)**.

West Indian ebony (*For.*). See **cocuswood**.

westing (*Surv.*). A west departure.

Westlake process (*Glass*). An automatic process using vacuum gather for producing articles in paste moulds.

Westmorland slates (*Build.*). Thick slates (6·5–16 mm; $\frac{1}{4}$–$\frac{5}{8}$ in) of varying size.

Weston film-speed (*Photog.*). A practical film-speed scale derived so that optimum development of the emulsion results in a suitable average density. The degrees of the scale are such that 16 Weston is twice as fast as 8 Weston (half-exposure) and half as fast as 32 Weston (double-exposure). See **f-number, Weston-Scheiner film-speed.**

Weston-Scheiner film-speed (*Photog.*). A film-speed scale derived so that an increase of 3 in the scale means doubling the actual speed (halving exposure), the values for particular emulsions being selected to give, under optimum conditions of development, a suitable average density. See **Weston film-speed.**

Weston standard cadmium cell (*Elec.*). Practical portable standard of e.m.f., in which the cathode is 12½% Cd and 87½% Hg by weight, with anode of amalgamated Pt or highly purified Hg. Saturated sol. aq. Cd_2SO_4 as electrolyte. E.m.f. of cell at 20°C is 1·018 636 V, temperature coefficient only 0·000 04 V/°C. Used for calibrating potentiometers and hence all other voltage measuring devices. Cells with unsaturated solutions have a lower temperature coefficient of e.m.f., but do not give an equally high absolute standard of reproducibility. Also **standard cell.**

Westphal balance (*Phys.*). One in which relative density is determined by suspending solid specimen from balance beam in liquid of known density, or *vice versa*.

Westphalian (*Geol.*). The name of a stratigraphical stage in the Carboniferous rocks of Europe, approximately corresponding to the Coal Measures in England and Wales.

wet and dry bulb hygrometer (*Meteor.*). A pair of similar thermometers mounted side by side, one having its bulb wrapped in a damp wick dipping into water. The rate of evaporation of water from the wick and the consequent cooling of the 'wet bulb' is dependent on the relative humidity of the air; the latter can be obtained by means of a table from readings of the two thermometers.

wet assay (*Chem., Met.*). Qualitative or quantitative analysis of ores or their constituents in which dissolution and digestion with suitable solvents plays a part. See also **cupellation, dry assay, scorification.**

wet cell (*Chem.*). A primary cell which contains liquid electrolyte, in contrast to the paste of a dry cell.

wet dock (*Civ. Eng.*). A dock in which water is impounded at a suitable level by means of dock gates; entrance is generally effected by means of locks.

wet drilling (*Mining*). Drilling of shot holes in which water is injected to allay dust as it is generated.

wet electrolytic capacitor (*Elec. Eng.*). One in which the negative electrode is a solution of a salt, e.g., aluminium borate, which is suitable for maintaining the aluminium oxide film without spurious corrosion.

wet end (*Paper*). The part of a paper-making machine where water is extracted (wire, suction boxes, etc.), as distinct from the drying end, where moisture is evaporated by heated cylinders.

wet expansion (*Paper*). The percentage increase in the length of a strip of paper after immersion in water.

wet felt (*Paper*). The continuous felt at the first press section on the machine carrying the paper immediately after the wire.

wet flashover voltage (*Elec. Eng.*). The voltage at which the air surrounding a clean wet insulator completely breaks down. Also called **wet sparkover voltage.**

wet flong (*Print.*). Layers of paper pasted together and used in a very damp condition for jobbing stereotyping.

wet-on-wet (*Print.*). Printing two or more colours in quick succession, particularly coloured illustrations, the tack of the ink being graded in order to prevent pick-up by the succeeding colours.

wet-plate process (*Photog.*). See **collodion process.**

wet raising (*Textiles*). A process of raising which produces a dress-face pile, as on beaver and box-cloth fabrics, the fabrics being kept in a wet state during the operations.

wet rot (*Build.*). A decay of timber which is due to chemical decomposition in the growing tree; it is sometimes set up in wood saturated with water and exposed to alternations of moisture and dryness.

wet sparkover voltage (*Elec. Eng.*). See **wet flashover voltage.**

wet spinning (*Spinning*). Method of flax spinning used for fine counts. The roving passes through a trough of hot water above the frame. This assists drafting by softening the gum, and aids in the production of a level smooth yarn.

wet steam (*Eng.*). A steam-water mixture, such as results from partial condensation of dry saturated steam on cooling.

wet-strength papers (*Paper*). Papers treated with a synthetic resin, such as melamine or urea formaldehydes, showing considerable strength when saturated with water.

wettability (*Chem.*). The extent to which a solid is wetted by a liquid, measured by the force of adhesion between the solid and the liquid phases.

wetted perimeter (*Hyd.*). In a channel or pipe through which flow is taking place, the part of the perimeter with which the liquid is actually in contact is called the *wetted perimeter*.

wetting agent (*Chem.*). Surface active agent which lowers the surface tension of water by a considerable amount, although present only in very low concentration.

wetwood (*For.*). Wood with an abnormally high water content and a translucent or water-soaked glassy appearance. This condition develops only in living trees and not through soaking in water.

w.f. (*Typog.*). The standard mark for *wrong fount*. It is written in the proof margin and the letter is underlined or struck through.

WGTA (*An. Behav.*). Abbrev. for *Wisconsin general testing apparatus.*

whalebone (*Zool.*). See **baleen.**

whaling glass (*Textiles*). Small magnifying glass used to count the ends and picks in woven cloths, or stitches in knitted structures.

wharf (*Civ. Eng.*). See **quay.**

Wharfedale machine (*Print.*). The first successful *stop-cylinder* printing machine, invented in 1858 by William Dawson and David Payne at Otley, in Wharfedale, Yorkshire.

Wharton's duct (*Zool.*). In Mammals, the duct of the submaxillary salivary gland.

Wharton's jelly (*Zool.*). In Mammals, jellylike embryonic connective tissue occurring in the umbilical cord.

wheal (*Mining*). A Cornish name for a mine.

Wheatstone automatic system (*Teleg.*). A Morse system in which signals are transmitted automatically using a previously prepared perforated tape and recorded automatically on perforated or inked tape.

Wheatstone bridge (*Elec. Eng.*). An apparatus for measuring electrical resistance using a *null indicator*, comprising two parallel resistance branches, each branch consisting of two resistances in series. Prototype of most other bridge circuits.

Wheatstone transmitter (*Teleg.*). A high-speed Morse transmitter which is operated by a slip on which are perforated holes which represent the dots and dashes of the coding of the desired message. These holes are scanned by peckers, which, if permitted to pass through the slip, operate contacts to form the transmitted signal.

Wheelabrating (*Met.*). TN for method of cleaning metal surfaces of sand, scale, and foreign materials by steel shot or grit hurled by centrifugal force under control from the rapidly spinning Wheelabrator unit.

wheel base (*Eng.*). The distance between the leading and trailing axles of a vehicle.

wheel girdles (*Agric.*). Traction aids attached to solid-tyred wheels of ploughs for muddy or slippery land.

wheeling (*Eng.*). A sheet-metal working process, using a machine with one flat and one convex wheel, for producing curved panels or for finishing after *panel beating* (q.v.).

wheeling step (*Build.*). See **winder.**

wheel-ore (*Min.*). See **bournonite.**

wheel organ (*Zool.*). In *Rotifera*, the ciliated disk; in *Cephalochorda*, a wheel-shaped ciliated organ on the undersurface of the oral hood, which creates a current of water flowing towards the mouth.

wheel-quartering machine (*Eng.*). A horizontal drilling machine having two opposed spindles at opposite ends of the bed; used to drill the crank-pin holes in both wheels on a locomotive coupled-axle simultaneously, and in precise angular relationship.

wheel stretcher (*Horol.*). A tool used for slightly increasing the diameter of a wheel. The wheel rim is passed between rollers under pressure, which slightly reduce the thickness but increase the width of the rim.

wheel window (*Arch.*). See **rose window.**

wheel wobble (*Autos.*). A periodic angular oscillation of the front wheels, resulting generally from insufficient castor action or from backlash in the steering-gear.

whetstone (*Geol.*). See **honestone.**

whewellite (*Min.*). Hydrated calcium oxalate, crystallizing in the monoclinic system, occurring commonly in coal.

whim (*Mining*). A vertical rope drum revolved by a horse; used for hoisting from shallow shafts.

whin or whinstone (*Geol.*). A popular term applied to doleritic intrusive igneous rock resembling that of the well-known Whin Sill.

whine (*Acous.*). The fluctuation in apparent loudness and pitch of a reproduced sound when the speed of the recording or reproducing machine is varying at a slow rate. See **wow.**

Whin Sill (*Geol.*). A sheet of intrusive quartz-dolerite or quartz-basalt, unique in the British Isles, as it is exposed almost continuously for nearly 200 miles (320 km) from the Farne Islands to Middleton-in-Teesdale.

whipcords (*Textiles*). Fabrics with a bold steep warp twill, used chiefly for dresses, suitings, and coatings, particularly the last.

whiplash (*Med.*). Term descriptive of an extension flexion injury to the soft tissues of the neck including the ligaments and apophyseal joints, causing instability and chronic pain; frequent in car accidents involving sudden collision from behind.

Whipple-Murphy truss (*Eng.*). A bridge truss having horizontal upper and lower chords connected by vertical and diagonal members, so that the panels resemble the letter N. Also called **Linville truss, N-truss, Pratt truss.**

whipstitching or whipping (*Bind.*). See **overcasting.**

whipstocking (*Mining*). See **wedging.**

whirler (*Print.*). Mechanical or hand-operated equipment using centrifugal force to spread an even coating on plates.

whirl-gate (*Foundry*). See **spinner-gate.**

whirling arm (*Aero.*). An apparatus for making certain experiments in aerodynamics, the model or instrument being carried round the circumference of a circle, at the end of an arm rotating in a horizontal plane.

whirlwind (*Meteor.*). A small rotating wind-storm which may extend upwards to a height of many hundred feet; a miniature cyclone.

whisker (*Chem.*). A thin strong filament or fibre made by growing a crystal, e.g. of silicon carbide, sapphire, etc. (*Radio*). See **cat's-.**

whistle box (*Cinema.*). A portable inductance, with or without a shunting capacitor, for filtering the supply ripple in the electrical supply to arc lamps, which would otherwise emit noise.

whistlers (*Acous.*). Atmospheric electric noises which produce relatively musical notes in a communication system.

whistling (*Vet.*). See **roaring.**

Whitby method (*Powder Tech.*). A sedimentation technique in which a bucket centrifuge is used to measure sizes of very fine particles. The particles accumulate in a fine-bore capillary at the bottom of the tube and the weight of particles is estimated from the height of the sediment in the tube.

white arsenic (*Chem.*). See under **arsenic.**

White Bombway (*For.*). Timber from the *Terminalia procera*, a native of the Andaman Islands. Planed surfaces of the wood are mildly silky, the texture moderately open and rather uneven, while the grain is straight. Wood used for joinery, panelling, furniture, and interior fittings.

white bricks (*Build.*). See **gaults.**

white cell (*Zool.*). See **leucocyte.**

white coat (*Build.*). The last or finishing coat of plaster.

white comb (*Vet.*). See **avian favus.**

white copperas (*Min.*). Goslarite.

white corundum (*Min.*). See **white sapphire.**

white damp (*Mining*). Carbon monoxide. Produced by the incomplete combustion of coal in a mine fire or by gas or dust explosions. Invisible; very poisonous.

white deal (*For.*). A whitish soft wood obtained from the spruce fir, and commonly used for interior constructional work. Also called **whitewood.**

White Dwarf (*Astron.*). The name given to a large class of small stars of extremely high density and low luminosity for their spectral type; composed of degenerate gas, they are very common and seem to represent a phase in the life history of all stars.

white fibres (*Zool.*). Unbranched, inelastic fibres of connective tissue occurring in wavy bundles. Cf. *yellow fibres.*

white fibrocartilage (*Zool.*). A form of fibrocartilage in which white fibres predominate.

white frost (*Meteor.*). See **hoar frost.**

white glass (*Glass*). See **opal glass.**

white gold (*Met.*). See **gold.**

white-heart process (*Met.*). See **malleable cast-iron.**

white heat (*Met.*). As judged visually, temperature exceeding 1000°C.

white heifer disease (*Vet.*). A condition in cattle in which the vagina, cervix, and uterus develop abnormally; associated with the gene for white coat colour and occurring mainly in white Shorthorn heifers.

white iron (*Met.*). Pig-iron or cast-iron in which all the carbon is present in the form of cementite (Fe_3C). White iron has a white crystalline fracture, and is hard and brittle.

white iron pyrite (*Min.*). See **marcasite (1).**

white lead (*Chem.*). Basic lead(II)carbonate or hydroxycarbonate. Made by several processes of which the oldest and best known is the *Dutch* (or stack) *process.* Used extensively as a paint pigment and for pottery glazes.

white lead ore (*Min.*). See **cerussite.**

white-leg (*Med.*). **See phlegmasia alba dolens.**

white level (*TV*). That percentage of maximum amplitude in a negative video signal which corresponds to white in a transmitted picture, a lesser amplitude being concerned with synchronizing. Usually between 30% and 40% in a negative video system. Used so that interference causes black spots which are less conspicuous than the white interference spots (*snow*) associated with a system based on a *black level.*

White Lias (*Geol.*). So named to distinguish it from the (true) *Blue Lias*, this is a subdivision of the Rhaetic Series, consisting essentially of massive bedded light-coloured limestones, separated by partings of marl.

white light (*Phys.*). Light containing all wavelengths in the visible range at the same intensity. This is seen by the eye as white. The term is used, however, to cover a wide range of intensity distribution in the spectrum and is applied by extension to continuous spectra in other wavelength bands (e.g. *white noise*, q.v.).

white line (*Civ. Eng.*). A line (usually painted) on a carriageway for guiding, regulating or warning traffic. (*Typog.*) A line of space.

white matter (*Zool.*). An area of the central nervous system, mainly composed of cell processes, and therefore light in colour.

white metal (*Met.*). Usually denotes tin-base alloy (over 50% tin) containing varying amounts of lead, copper, and antimony; used for bearings, domestic articles, and small castings; sometimes also applied to alloys in which lead is the principal metal; also called **antifriction metal, bearing metal.**

white muscle disease (*Vet.*). A disease of calves and lambs (*stiff lamb disease*) in which muscular degeneration occurs due to vitamin-E deficiency; characterized by stiffness and lameness, and symptoms of cardiac and respiratory disorder.

white muscles (*Zool.*). In Vertebrates, muscles which do not perform long-continued actions and which are therefore poor in sarcoplasm and haemoglobin and of a light colour.

white nickel (*Min.*). A popular name for the cubic diarsenide of nickel(IV), $NiAs_2$, the scientific name for which is *chloanthite* (q.v.).

white noise (*Acous.*). Noise which has a uniform frequency spectrum. Obtained by amplifying noise generated by a gas discharge tube or noise diode.

white object (*Photog.*). An object which reflects all wavelengths of light impartially, and therefore, when illuminated with white light, appears to the eye to be hueless.

white-out (*Meteor.*). A situation where the horizon is indistinguishable, when the sky is overcast and the ground is snow-covered. (*Typog.*) To open out composed type-matter with spacing, in order to fill the allotted area or improve the appearance.

white-out lettering (*Print.*). A term indicating that, in the reproduction required, the lettering is to be reversed to white on black.

white radiation (*Phys.*). See **white light.**

white rami (*Zool.*). In Vertebrates, the *rami communicantes* connecting the spinal cord with the sympathetic system.

white reference level (*Telecomm.*). Signal modulation level corresponding to maximum brightness (white) in monochrome facsimile transmission.

whiter than white (*TV*). That part of a negative video signal of even less amplitude than that required for total whiteness. See **white level.**

whites (*Vet.*). Leucorrhoea of cows.

white sapphire (*Min.*). More reasonably called *white corundum*, is the colourless, pure variety of crystallized corundum, Al_2O_3, free from those small amounts of impurities which give colour to the varieties 'ruby' and 'sapphire'; when cut and polished, it makes an attractive gemstone.

white scour (*Vet.*). Calf scour. Diarrhoea affecting calves during the first few weeks of life, caused usually by *Escherichia coli* infection, probably in conjunction with nutritional and environmental factors.

white spirit (*Paint.*). A petroleum distillate used as a substitute for turpentine in mixing paints, and in paint and varnish manufacture.

white vitriol (*Min.*). A popular name for *goslarite*, $ZnSO_4.7H_2O$.

whitewash, whitening (*Paint.*). See **limewash.**

white water (*Paper*). The drainage water from the machine wire containing fine fibres and loading, usually circulated for re-use.

whitewood (*For.*). (1) Softwood from *Liriodendron*, a native of the eastern half of the North American continent. It will not polish satisfactorily but takes stain and paint well. (2) See **white deal.**

whiting (*Paint.*). A filler or extender made from ground, size-classified, pure chalk, and extensively used in putty, distemper, certain paints, plastics, and rubber articles.

whitlockite (*Min.*). Trigonal calcium phosphate, occurring in sedimentary phosphate deposits.

whitlow (*Med.*). See **paronychia.**

whitneyite (*Min.*). An alloy of copper and arsenic, crystallizing in the cubic system.

Whitworth screw-thread (*Eng.*). See **British Standard Whitworth thread.**

whizzer (*Eng.*). A machine consisting of a perforated cylinder inside which loose material to be dried is flung outwards by revolving paddles, thus removing the water by centrifugal force. Also called **hydroextractor.**

whizz-pan (*Cinema.*). Effect due to rapid *pan* (q.v.) from one point of interest to another. Used chiefly in newsreel photography.

whole bound (*Bind.*). See **full bound.**

whole-brick wall (*Build.*). A wall whose thickness is the length of a whole brick.

whole-circle bearing (*Surv.*). The horizontal angle measured from 0° to 360° clockwise, from true north to a given survey line.

whole-coiled winding (*Elec. Eng.*). An armature winding for an alternator having one armature coil per pole, the two sides of the coil being separated by a distance which is equal to the pole pitch.

whole timber (*Carp.*). A roughly squared timber of size greater than 15 × 15 cm (6 × 6 in).

whooping cough (*Med.*). **See pertussis.**

whorl (*Bot.*). A group of similar members arising from the same level on a stem, and forming a circular group around it. *adj.* **whorled.** (*Zool.*) A single turn of a spirally-coiled shell or other spiral structure.

W.I. (*Met.*). Abbrev. for *wrought iron.*

Wickersham quoin (*Typog.*). A mechanical quoin using the cam principle for expanding the sides.

Widal's test (*Med.*). A test for the presence of typhoid fever, blood-serum from a patient with this infection agglutinating (clumping) a suspension of typhoid bacilli (in a tube) towards the end of the first week of the disease.

wide-angle lens (*Photog.*). A lens, of necessarily short focal-length, with a wide angle of view; used for photographing buildings, etc.

wideband amplifier (*Telecomm.*). One which amplifies over a wide range of frequencies, with corresponding low gain.

wide-cut fuel (*Aero.*). Low octane petrol (gasoline) obtained from wide-cut distillation used in turbojets in order to conserve kerosine, the world output of which is limited. Abbrev. avtag.

wide film (*Cinema.*). Film which is greater in width than the standard 35 mm film.

wide screen (*Cinema., TV*). A screen with *picture ratio* (q.v.) above normal. See **CinemaScope, Todd-AO, VistaVision.**

wide-spectrum (*Pharm.*). Of antibiotics, etc., effective against a wide range of micro-organisms. Also **broad-spectrum.**

Widmanstätten structure (*Met.*). Meshlike appearance along preferred crystal planes, produced when steel is cooled rapidly from 1000°C, causing cementite or ferrite to be precipitated.

widow (*Typog.*). A single word or part of a word occupying the last line of a paragraph; to be avoided, particularly as the first line of a page. See **club line.**

width (*Phys.*). The spread or uncertainty in a specified energy level, arising as a result of Heisenberg's *uncertainty principle*, proportional to the instability of the state concerned.

Wiedemann effect (*Elec. Eng.*). Tendency to twist in a rod carrying a current when subject to a magnetizing field.

Wiedemann-Franz law (*Phys.*). That the ratio of thermal to electrical conductivity of any metal equals absolute temperature times $2{\cdot}6\times10^{-8}$ V^2/K^2.

Wien bridge (*Elec. Eng.*). A four-arm a.c. bridge circuit used for measurement of capacitance, inductance and power factor.

Wien-bridge oscillator (*Elec. Eng.*). One in which positive feedback is obtained from a Wien bridge, the variable frequency being determined by a resistance in an arm of the bridge.

Wien effect (*Elec. Eng.*). The increase in the conductivity of an electrolyte observed with very high voltage gradients.

Wien's laws (for radiation from a black body) (*Phys.*). (1) *Displacement law*: $\lambda_m T$=constant (0·0029 metre kelvin). (2) *Emissive power* (E_λ): within the maximum intensity wavelength interval $d\lambda$, $E_\lambda = CT^5 d\lambda$, where $C = 1{\cdot}288\times10^{-5}$ W/m^2K^5 (T is absolute temperature in kelvins). (3) *Emissive power* (dE) in the interval $d\lambda$ is $dE = A\lambda^{-5}\exp(-B/\lambda T)d\lambda$. λ_m is wavelength at E_{max}, $A = 4{\cdot}992$ J m, $B = 0{\cdot}0144$ m K.

wigans (*Weaving*). Plain grey cloths of medium or heavy type, used for boot linings, casements.

Wigner effect (*Nuc.*). The process by which an atom is knocked out of its position in a crystal lattice by direct nuclear impact and the displaced atom may come to rest at an interstitial position or at a lattice edge. A change of chemical or physical properties may result, e.g., graphite bombarded by neutrons changes in dimensions and the shape of the crystal lattice is altered. Also called **discomposition effect.**

Wigner energy (*Nuc.*). Energy stored within a crystalline substance, due to the Wigner effect.

Wigner force (*Nuc.*). Ordinary (non-exchange) short-range force between nucleons.

Wigner nuclides (*Nuc.*). Those isobars of odd mass number in which the atomic number and neutron number differ by one.

Wigner theorem (*Phys.*). A deduction from the quantum theory which states that in a collision of the second kind the angular momentum of the electron spin is conserved.

wig-wag (*Horol.*). A machine which gives an oscillatory motion to a polisher; used for polishing pivots, etc. (*Rail.*) A level-crossing signal which gives its indication, with or without a red light, by swinging about a fixed axis.

Wild-Barfield furnace (*Elec. Eng.*). A resistance furnace comprising wirewound resistance elements; widely used for heat treatment processes.

wildcatting (*Mining*). Prospecting at random, particularly in speculative boring for oil.

wild track (*Cinema.*). A sound-track which is recorded independently of any photographic track or mute, but is destined to be used in editing a sound-film.

wild wall (*Cinema.*). A detachable or floating wall, suitable for inclusion in the camera angle; it is covered with sound-absorbing material.

wild yeasts (*Brew.*). Yeasts of species other than those normally used in brewing. Also **secondary yeasts.**

Wilfley table (*Min. Proc.*). Flat rectangular desk, adjustable about the long axis for tilt, is given rapid but gentle throwing motion along this horizontal axis, while classified sands are washed across and down-tilt, against restraint imposed by horizontal riffles. Heavy minerals work across and progressively lighter ones gravitate down to separate discharge zones. Also called **shaking table.**

wilkeite (*Min.*). A member of the apatite group of minerals in which some phosphorus is replaced by silicon and sulphur.

Willans line (*Eng.*). The graph of total steam consumption against load, for a steam turbine.

willemite (*Min.*). Orthosilicate of zinc, Zn_2SiO_4, occurring as massive, granular, or in trigonal prismatic crystals, white when pure but commonly red, brown, or green through manganese or iron in small quantities. In New Jersey (Franklin) and elsewhere it occurs in sufficient quantity to be mined as an ore of zinc. Noteworthy as exhibiting an intense bright-yellow fluorescence in ultraviolet light.

Willesden paper (*Build.*). Paper treated with cuprammonium hydroxide to make it rotproof; placed under slates in roofing to keep out wet and act as a sound and heat insulator.

willey or **willow** (*Paper, Textiles*). A machine for separating wool or waste. It has a large cylinder with protruding spikes. Similar spikes on the inside of the casing enclosing the cylinder assist the beating action.

Williams tube (*Electronics*). TN for CRT used as electrostatic store, e.g., in computing.

Williot diagram (*Eng.*). A graphical construction for finding the deflexion of a given point in a structural framework under load.

Willis's circle (*Zool.*). In Vertebrates, an arterial ring surrounding the hypophysis.

Willis's law (*Zool.*). See **age and area theory.**

willow (*For.*). A hardwood from the genus *Salix*, used for cricket bats. (*Paper*) See **willey.** (*Textiles*) See **willey.**

willy-willy (*Meteor.*). A *tropical revolving storm* (q.v.) near the coast of N.W. Australia.

Wilson chamber (*Nuc. Eng.*). *Cloud chamber* (q.v.) of expansion type.

Wilson effect (*Elec. Eng.*). Production of electric

polarization when dielectric material is moved through region of magnetic field, due to Faraday induced e.m.f. in the dielectric.

Wilson electroscope (*Elec. Eng.*). Single-leaf electroscope, with the leaf hanging vertically in its most sensitive position.

Wilson's disease (*Med.*). A biochemical condition in man, characterized by the degeneration of certain parts of the nervous system and the liver, caused by abnormal copper metabolism, and inherited as an autosomal recessive.

Wilson seal (*Vac. Tech.*). A type of shaft seal, suitable for vacuum work, which consists of an elastomer disk having a hole of smaller diameter than the shaft, through which the latter is forced.

wilting (*Bot.*). The loss of rigidity in leaves and young stems following on marked loss of water from the plant; growth stops in wilted material.

wilting coefficient (*Bot.*). The percentage of moisture present in a soil when plants growing on that soil begin to wilt.

Wimshurst machine (*Elec. Eng.*). Early type of electrostatic induction generator.

winceyette (*Textiles*). A plain or simple twill cotton cloth of light weight, raised slightly on both sides; used for pyjamas, nightgowns, or underwear. May be colour woven or piece-dyed.

winch (*Eng.*). A mechanism arranged to hoist, manually or by power through the medium of a rope, cable or chain winding on a drum.

winchester (*Glass*). A narrow- or wide-mouthed cylindrical bottle used for the transportation of liquids.

winch launch (*Aero.*). Launching a glider by towing it into the air by a cable on a motorized winch, or pulling through a pulley by a car.

wind (*Elec. Eng.*). A stream of air arising at any sharply pointed electrical conductor charged to a high potential. (*Meteor.*) The horizontal movement of air over the earth's surface. The direction is that from which the wind blows. The speed may be given in metres per second, miles per hour or knots.

windage (*Eng.*). In any machine, the energy dissipated in overcoming air resistance to motion.

wind axes (*Aero.*). Coordinate axes, having their origin within the aircraft, and directionally orientated by the relative airflow.

wind chest (*Acous.*). The box, containing air under pressure, upon which the ranks of pipes in an organ are mounted.

wind dispersal (*Bot.*). The dispersal of spores, seeds, and fruits by the wind.

wind-driven generator (*Elec. Eng.*). A generator driven by a prime-mover of the windmill type, or directly (in the case of aircraft) by an airscrew carried on the generator shaft.

winder (*Build.*). A step, generally triangular in plan, used at a change in direction of the stair. Also **wheeling step**.

wind frost (*Meteor.*). An air frost where the cold air has been propelled by the wind.

wind gag (*Acous.*). A bag of thin cloth or silk placed over a microphone when the latter is used out of doors, to eliminate hissing noises due to wind.

windgall (*Vet.*). Distension of the joint capsule (**articular windgall**) or tendon sheath (**tendinous windgall**) of the fetlock joint of the horse.

wind gauge (*Meteor.*). See **anemometer**.

winding (*Bot.*). A movement of a stem due to unequal rates of growth in successive longitudinal strips of the stem, causing the tip to describe a circle in space; if pronounced, the stem will twine around a support. (*Elec. Eng.*) The system of insulated conductors forming the current-carrying element of a dynamo-electric machine or static transformer. (*Textiles*) Coiling thread on a spindle or bobbin, in spinning, doubling or pirning, used to form a package convenient to handle.

winding coefficient (*Elec. Eng.*). An alternative name for **winding factor**.

winding diagram (*Elec. Eng.*). A diagram showing in schematic form the arrangement and sequence of an armature winding and its circuit connexions.

winding drum (*Eng.*). An engine or motor-driven drum on to which a haulage rope is wound, as the wire rope of a mine cage. The drum may be cylindrical or conical, with a plain or a helically grooved surface for the rope.

winding factor or coefficient (*Elec. Eng.*). A factor which takes account of the difference between the vector and arithmetic sums of the e.m.fs. induced in a series of armature coils occupying successive positions round the periphery of the armature.

winding gear (*Elec. Eng.*). The mechanical gear associated with an electric winder.

winding pitch (*Elec. Eng.*). The distance, measured as the number of slots, separating an armature coil from its successor in the winding sequence.

winding plant (*Elec. Eng.*). The complex of apparatus constituting an electrically driven winder.

winding space (*Elec. Eng.*). The cross-sectional space available in an armature slot for the insertion of the insulated conductors.

winding square (*Horol.*). The square end of a barrel or fusee arbor on which the key fits for winding.

winding stair (*Build.*). A stair formed in a circular, helical, elliptical, or other mathematical plan.

wind load (*Build., Eng.*). The force acting on a structure due to the pressure of the wind upon it.

window (*Elec. Eng.*). (1) The winding space of a transformer, i.e., the cross-sectional spaces between the limbs and yokes of a multicore transformer. (2) Conducting diaphragms inserted into waveguide which act inductively or capacitively according to how they are positioned. (*Geol.*) A closed outcrop of strata lying beneath a thrust plane and exposed by denudation. The strata above the thrust plane surround the 'window' on all sides. (*Nuc. Eng.*) Thin portion of wall of radiation counter through which low energy particles can penetrate. (*Radar*) Strips of metallic foil of dimensions calculated to give radar reflections and hence confuse locations derived therefrom. Also **chaff, rope.** (*Telecomm.*) See **pulse height selector.**

window efficiency ratio (*Light*). See **daylight factor.**

window lock (*Join.*). See **sash fastener.**

wind pollination (*Bot.*). The conveyance of pollen from anthers to stigmas by means of the wind; anemophily.

wind portal (*Civ. Eng.*). A portal which has special bracings to resist wind pressure.

wind pump (*Civ. Eng.*). A pump which is operated by the force of the wind rotating a multibladed propeller.

wind road (or way) (*Mining*). An underground passage used for ventilation.

windrow (*Agric.*). A row of material formed by combining two or more swaths; a row of peats set up for drying.

wind shear (*Meteor.*). A change of wind velocity with height.

wind sock (*Aero.*). A truncated conical fabric sleeve, on a 360° free pivot, which indicates local wind direction; also called **wind stocking.**

windsucking (*Vet.*). A vice acquired by certain horses of swallowing air when 'cribbing' or gripping an object with the incisor teeth.

wind T (*Aero.*). A T-shaped device displayed at aerodromes to indicate the direction of the surface wind. The leg of the T corresponds to the wind direction.

wind trunk (*Acous.*). The metal or wooden piping which distributes the air under pressure to the various wind chests in an organ.

wind tunnel (*Aero.*). Apparatus for producing a steady airstream past a model for aerodynamic investigations. Also **flume.**

wing (*Aero.*). The main supporting surface(s) of an aeroplane or glider. (*Bot.*) (1) One of the lateral petals of a flower of a pea and related plants. (2) A flattened outgrowth from a fruit or a seed, increasing the area without greatly increasing the weight, and serving in wind dispersal. (3) The downwardly continued base of a decurrent leaf. (*Build.*) A section of a building projecting from the principal part of it. (*Zool.*) Any broad flat expansion; an organ used for flight, as the fore-limb in Birds and Bats, the membranous expansions of the mesothorax and metathorax in Insects.

wing area (*Aero.*). See **gross wing area, net wing area.**

wing car (*Aero.*). See **car.**

wing compasses (*Carp.*). A form of *quadrant dividers* (q.v.).

wing coverts (*Zool.*). See **tectrices.**

winged (*Bot.*). Of a stem, bearing the thin flattened bases of decurrent leaves; of a fruit or seed, having a flattened appendage.

wind fence (*Aero.*). A projection extending chordwise along a wing and projecting from its upper surface. It modifies the pressure distribution by preventing a spanwise flow of air which would otherwise cause a breakaway of the flow near the wing tips and lead to tip stalling. Also called a **boundary layer fence.** Cf. *vortex generators.*

wing loading (*Aero.*). The gross weight of an aeroplane or glider divided by its *gross wing area* (q.v.).

wing nut (*Eng.*). A nut having radial lugs or wings to enable it to be turned by thumb and fingers. Also **bow nut, fly nut** or **butterfly nut.**

wing rail (*Rail.*). See **check rail.**

wing shafts (*Ships*). The port and starboard propeller shafts of a triple- or quadruple-screw steamship.

wing-tip float (*Aero.*). A watertight float which gives stability and buoyancy on the water; placed at the extremities of the wings of a seaplane, flying-boat, or amphibian.

wing valve (*Eng.*). A mitre-faced or conical-seated valve guided by three or four radial vanes or wings fitting inside the circular port.

wing wall (*Civ. Eng.*). A lateral wall built on an abutment and serving to retain earth in embankment.

wink (*Work Study, etc.*). Unit denoting 1/2000 minute.

Winkler burette (*Chem.*). A form of burette employed for the analysis of gaseous mixtures containing a constituent soluble in water.

Winkler reagent for oxygen (*Chem.*). Quantitative absorption by a solution of alkaline pyrogallol, with formation of a brown colour.

wink-light (*Photog.*). Type of non-expendable, battery-operated flashlight synchronized with the shutter.

Winner winding (*Elec. Eng.*). Form of winding used in constructing standard resistances of low residual reactance for use at relatively high frequencies.

winsey (*Textiles*). Plain weave, cotton flannelette of good quality; may have a union or woollen weft.

Winslow's foramen (*Zool.*). A small opening by which the cavity of the bursa omentalis communicates with the rest of the abdominal cavity in Mammals.

winter annual (*Bot.*). A plant which lives for a short time, grows and sets seeds during the colder part of the year, dies, and is represented during summer only by the resting seeds.

winter dysentery (*Vet.*). An intestinal infection of cattle caused by the bacterium *Vibrio jejuni*; characterized by fever and diarrhoea.

winter egg (*Zool.*). In some fresh-water animals, a thick-shelled egg laid at the onset of the cold season which does not develop until the following warm season. Cf. *summer egg.*

Winter-Eichberg-Latour motor (*Elec. Eng.*). A single-phase a.c. motor of the compensated repulsion type; mainly used in electric traction.

wintergreen (*Chem.*). The methyl ester of salicylic (2-hydroxybenzoic) acid, $HO{\cdot}C_6H_4CO{\cdot}OMe$. Characteristic smell. Widely used in liniments and other medicinal products.

wintergreen plants (*Bot.*). Small plants, especially woodland plants, which retain green leaves throughout the winter.

winter solstice (*Astron.*). See **solstices.**

winze (*Mining*). Internal shaft, usually between two underground levels in plane of lode, used in exploration and subsequent extraction of valuable ore.

wipe (*Cinema.*). Effect obtained in film editing by gradually removing one scene as the next appears. (*Print.*) Defect in which, instead of an even film, the ink, because of unsuitable make-up, forms a ridge at the edge of the type.

wiped joint (*Plumb.*). A joint formed between two lengths of lead pipe, one of which is opened out with a tampin while the other is tapered to fit into the first. Molten solder in a plastic condition is then wiped around the joint by hand with a pad.

wipe-out (*Acous.*). See **wash-out.** (*Telecomm.*) Interference of such intensity as to render impossible the reception of desired signals.

wipe-out area (*Telecomm.*). Area surrounding a transmitter where wipe-out occurs.

wiper (*Elec. Eng.*). See **brush.** (*Teleph.*) In a uniselector or selector, the conducting arm which is rotated over a row of contacts and comes to rest on an outlet. (*Weaving*) A mechanism used to convert a rotating to a reciprocating motion. Also called **cam, tappet.**

wipe test (*Nuc. Eng.*). See **smear test.**

wiping gland (*Cables*). A projecting sleeve on a junction box, which is connected by a wiped joint to the lead sheath of a cable that enters the box.

wiping solder (*Plumb.*). Lead-based soft solder used in plumbing, containing up to 35% tin.

wire (*Teleph.*). A continuous connexion through a system, particularly a telephone exchange, whether automatic or manual.

wirebar (*Met.*). High-purity copper cast to a tapered ingot shape; used to produce drawn copper wire.

wire broadcasting (*Radio*). Distribution of broadcast programmes to multiple receivers on direct wired circuits. Also called **radio relay.**

wire cloth (*Paper*). A continuous band of wire gauze with a mesh about 60/80 per inch. The moulding unit of a paper-making machine. Also called **machine wire.**

wire comb (*Build.*). A form of *scratcher* (q.v.).

wire-cut bricks (*Build.*). Bricks made by forcing the clay through a rectangular orifice, and cutting suitable lengths off the resulting bar of clay by pressing wires through the plastic mass, before burning.

wired glass (*Glass*). A form of sheet-glass produced by rolling wire mesh into the ribbon of glass so that it acts as a reinforcement and holds the fragments together in the event of the sheet being fractured.

wired-in (*Electronics*). Said of components connected in circuits, particularly subminiature valves and semiconductor devices, which are too small to be plugged safely into a holder.

wire-drawing (*Eng.*). (1) The process of reducing the diameter of rod or wire by pulling it through successively smaller holes. (2) The fall in pressure when a fluid is throttled by passing it through a small orifice or restricted valve-opening.

wired wireless (*Teleg.*). See **carrier telegraphy.**

wire gauge (*Eng.*). Any system of designating the diameter of wires by means of numbers, which originally stood for the number of successive passes through the die-blocks necessary to produce the given diameter. See **Birmingham-, Brown and Sharpe-, Standard-.**

wire guide (*Paper*). Guide located near the couch roll to ensure that the machine wire travels true at all times.

wire-guided missile (*Aero.*). See **teleguided missile.**

wireless. Original term for **radio.**

wireless bearing (*Ships, etc.*). A direction obtained by the use of a directional aerial.

wirephoto (*Telecomm.*). A photograph transmitted over a wire circuit by electrical means.

wire recorder (*Acous.*). Early type of magnetic recorder with recording medium in form of iron wire.

wire rope (*Eng.*). Steel rope made by 'twisting' or 'laying' a number of strands over a central core, the strands themselves being formed by twisting together steel wires. See **Lang lay.**

wiresonde (*Meteor.*). Meteorological data-transmission system operated by cable from captive balloon.

wire stabbing (*Bind.*). A simple method of binding in which one or more wire staples are passed through the back margin of all the leaves or sections.

wire-stitching (*Bind.*). The securing of a booklet by means of wire staples. See **saddle-stitching, stabbing.**

wirewound resistor (*Elec. Eng.*). One with metallic wire element.

wiring point (*Elec. Eng.*). A point in an interior wiring installation where an external connexion can be made to the electric circuit.

Wirsung's duct (*Zool.*). The ventral or main pancreatic duct of Mammals. Cf. *Santorini's duct.*

Wisconsin general testing apparatus (*An. Behav.*). An apparatus used particularly in the study of discrimination and learning by monkeys, in which the animal can choose between two or three stimuli on top of small wells usually containing food as a positive reinforcement, its behaviour being observed by an experimenter through a one-way vision screen. Abbrev. **WGTA.**

wisdom teeth (*Anat.*). The third molars, which do not usually erupt until adulthood.

wishbone (*Autos.*). V-shaped member used in independent suspension systems.

wishful thinking (*Psychol.*). A type of thinking in which the individual substitutes phantasy of fulfilment of a wish for the actual achievement.

witches' broom (*Bot.*). A dense tuft of poorly developed branches formed on a woody plant attacked by a parasite (chiefly fungi and mites).

witch of Agnesi (*Maths.*). The curve whose cartesian equation is $x^2y=4a^2(2a-y)$.

withamite (*Min.*). A mineral belonging to the epidote group, named after Dr Witham of Glencoe where the mineral was discovered. Contains about 1% of manganese(IV)oxide, i.e., it is a manganese-poor piedmontite.

withe (*Build.*). The partition wall between adjacent flues in a chimney stack. Also called **mid-feather.**

withering (*Brew.*). A stage in the process of malting when the malt remains unturned for a period of about 24 hours, so that accumulation of heat dries the grain and arrests growth. See **steeping.**

withering persistent (*Bot.*). Said of a leaf which dies and withers, but remains attached to the stem; a condition found in many herbaceous perennials.

witherite (*Min.*). Barium carbonate, $BaCO_3$, crystallizing in the orthorhombic system as yellowish or greyish-white complex crystals of hexagonal appearance due to twinning; also massive. Occurs with galena in the lead mines on Alston Moor and near Hexham, Northumberland. Exploited as an important source of barium. Hardness 3½ Mohs' scale. Rel. d. 4·3.

withers (*Vet.*). The region of the horse's back above the shoulders.

witness (*Eng.*). A remnant of original surface or scribed line, left during machining or hand working to prove that a minimum quantity of material has been removed or an outline accurately preserved.

Wobbe index (*Chem.*). Relationship between the energy density and relative density of fuel gases,

$$K = \frac{\text{energy density}}{\sqrt{\text{relative density}}}$$

wobble crank (*Eng.*). A short-throw crank in which the pin, machined from and at an angle to the axis of the crankshaft, has been used to give an elliptical motion to a sleeve valve by a short connecting-rod and ball joint.

wobble plate (*Eng.*). See **swash plate.**

wobble-plate engine (*Eng.*). A multicylinder engine in which a wobble-plate or swash-plate mechanism replaces cranks and connecting rods. The cylinders are arranged axially round the shaft, their pistons operating on the wobble plate through sliding blocks. The arrangement is very compact but the mechanical efficiency is usually low.

wobble saw (*Carp.*). See **drunken saw.**

wobbulator (*Telecomm.*). Colloquialism for a signal generator whose frequency is automatically varied periodically over a definite range; used in testing frequency response of systems.

Woburn Sands (*Geol.*). A group of brown to white sands, generally coarse in grain, of Lower Cretaceous age, extending from Leighton Buzzard northeastward to Cambridge, and including the Leighton Buzzard silver sand.

wöhlerite (*Min.*). A silicate of calcium, sodium and zirconium, crystallizing in the monoclinic system.

Wöhler test (*Met.*). A fatigue test in which one end of a specimen is held in a chuck and rotated in a ball-bearing placed on the other end. The ball-bearing carries a weight and, as the specimen rotates, the stress at each point on its surface passes through a cycle from a maximum in tension to a maximum in compression.

Wolfcampian (*Geol.*). The lowest stage of the Permian system in north America.

Wolffian body (*Zool.*). The *mesonephros* (q.v.).

Wolffian duct (*Zool.*). The kidney duct of Vertebrates, which originates in all Vertebrates as a pronephric (or archinephric) duct, and is then taken over by the mesonephros. In adult anamniotes (*Agnatha*, Fish, and *Amphibia*) it serves as a kidney duct and a sperm duct in males, while in adult amniotes (Reptiles, Birds and Mammals) whose metanephric kidney has a separate duct, it is present only in males, forming the *vas deferens*.

Wolffian ridges (*Zool.*). Short longitudinally-running ridges representing the limb-rudiments in many Vertebrates.

wolf note (*Acous.*). Extraneous nonharmonic note made by a bow on a violin string.

wolfram (*Chem.*). See **tungsten.**

wolframite (*Min.*). Tungstate of iron and manganese (FeMn)WO_4, occurring as brownish-black monoclinic crystals, columnar aggregates, or granular masses in association with tin ores, as in Cornwall. It forms a complete series from ferberite ($FeWO_4$) to hübnerite ($MnWO_4$). An important ore of tungsten.

Wolf-Rayet stars (*Astron.*). An abnormal class of stars, with spectra similar to those of novae, broad bright lines predominating, indicating violent motion in the stellar atmosphere.

wolfsbergite (*Min.*). See **chalcostibite.**

Wolf's equation (*Radio*). $R=k(10G+N)$, where R=relative sunspot number, k=a constant, depending on the type of telescope used, G=observed number of sunspot groups, N= observed total number of sunspots, either singly or in groups. Used in forecasting maximum usable frequency.

wolf tooth (*Zool.*). In *Equidae*, a vestigial first upper premolar.

wolgidite (*Geol.*). An alkaline volcanic rock composed of leucite, magnophorite and diopside.

wollastonite (*Min.*). A triclinic silicate of calcium, $CaSiO_3$, occurring as a common mineral in metamorphosed limestones and similar assemblages, resulting from the reaction of quartz and calcite. Also called **tabular spar.**

Wollaston prism (*Optics*). A double-image polarizing device, made of two geometrically similar wedge-shaped prisms of calcite or quartz, cemented with glycerine or castor oil. The cutting of the two prisms is arranged so that their optic axes cause two coloured, oppositely polarized emergent beams. It is useful in determining the proportion of polarization in a partially polarized beam providing, for examination, two images with oscillations in two perpendicular directions. Similar to, but distinct from, Rochon prism.

Wommelsdorf machine (*Elec. Eng.*). A form of electrostatic generator, having alternate fixed and moving plates, and collectors in the form of steel wires dipping into grooves in the rims of the moving plates.

wood (*Bot.*). See **xylem.**

wood alcohol (*Chem.*). Same as **methanol.**

wood brick (*Build.*). A piece of wood the shape of a brick but larger by the amount of the mortar joints. It is bonded to surface brickwork in the course of building and is held in position by friction alone, its function being to provide a substance to which, e.g., **skirtings** may be nailed.

woodcut (*Print.*). An engraving cut on the blank grain of wood; an impression from it.

wooden tongue (*Vet.*). See **actinobacillosis.**

wood fibre (*Bot.*). A thick-walled, elongated, dead element found in wood; it is developed by the elongation and lignification of the wall of a single cell, but differs from a tracheide in its thicker wall and general inability to conduct water.

wood flour. A fine powder made from sawdust and wood waste; used as a filler in many industries, and in the manufacture of guncotton.

wood furniture (*Typog.*). Spacing material, made of wood, as distinct from metal or plastic.

wood letters (*Typog.*). Large type-letters cut in wood; used in some poster work.

wood nog (*Build.*). See **nog.**

wood opal (*Min.*). A form of common opal which has replaced pieces of wood entombed as fossils in sediments, in some cases retaining the original structure.

wood parenchyma (*Bot.*). See **xylem parenchyma.**

wood pulp (*Paper*). Wood reduced to a pulp, either mechanically or by a chemical process. It is used in the manufacture of paper.

wood ray (*Bot.*). See **xylem ray.**

wood-ray parenchyma (*Bot.*). Parenchymatous cells in a xylem ray.

Woodruff key (*Eng.*). A key consisting of a segment of a disk, fitted in a shaft key-way milled by a cutter of the same radius, and a normal key-way in the hub.

Wood's glass (*Glass*). A glass with very low visible and high ultraviolet transmission.

wood spirit (*Chem.*). **Methanol.**

wood sugar (*Chem.*). **Xylose.**

wood tar (*Chem.*). A product of the destructive distillation of wood, containing alkanes, naphthalene, phenols.

wood tin (*Min.*). A botryoidal or colloform variety of cassiterite showing a concentric structure of brown, radiating, woodlike fibres.

wood vessel (*Bot.*). See **vessel.**

wood-wool slabs (*Build.*). Made from long wood shavings with a cementing material; used for linings, partitions, etc.

woody tissues (*Bot.*). Tissues which are hard because of the presence of lignin in the cell walls.

woof (*Weaving*). See **weft.**

woofer (*Acous.*). Large loudspeaker used to reproduce lower part of audiofrequency spectrum only. See **cross-over frequency-**, **tweeter (1).**

wool (*Bot.*). A tangled mass of long, soft whitish hairs on a plant. *adj.* **woolly.** (*Zool.*) A modification of hair in which the fibres are shorter, curled, and possess an imbricated surface. Specifically, the covering of a sheep. The fibres are composed of carbon, nitrogen, hydrogen, sulphur, and oxygen; they are covered with small scales. The length and diameter of the fibres varies according to breed; commercial wools vary in length from 1 in (25 mm) to about 14 in (350 mm), the majority being about 6–10 in (150–250 mm). Diameter about 0·002–0·0005 in (0·04–0·015 mm). See also **fleece wool.**

wool and silk dyeing (*Textiles*). The colouring of fibres, yarns, or fabrics of wool or silk both of which are of animal origin.

woolsorter's disease (*Med.*). An acute disease due to infection with the *Bacillus anthracis*, conveyed to Man by infected wool or hair of animals; characterized by fever, the appearance on the skin of vesicles which become covered with a black scab, and sometimes by infection of the lungs or of the intestines. See also **anthrax.**

word (*Comp.*). Set of digits, having definite meaning, often of a given number, called (*machine*) *word length*.

word time (*Comp.*). That required to read and rewrite one word. (The time quoted is normally that for the immediate access store.)

work (*Eng., Phys., etc.*). One manifestation of energy, usually applied when a mechanical effort is involved either as the input or as the output of a transducer. For example, a pump is said to do work when it raises water to a reservoir; that water is said to do work again in driving a turbine on returning to river level. As for all forms of energy, the SI unit of work is the joule, performed when a force of 1 newton moves its point of application through 1 metre, alternatively when 1 watt of power is expended for 1 second. In electrical work, measurement is often carried out using the 'unit' of 1 kilowatt-hour (kWh), the work done in expending 1 kilowatt of power for one hour, equal to 3·6 megajoules or 3 600 000 joules.

work-and-tumble (*Print.*). A method of turning printed sheets in which, after printing the first side, the opposite edge must be used as the front lay for the second side.

work-and-turn (*Print.*). Process in which one forme is used to print on both sides of the paper which is cut after printing to give two copies of the job or section. Cf. *sheet-work*.

work-and-twist (*Print.*). Two impressions are taken on the same side of the paper using the same forme, the paper being turned 180° before taking the second impression. Useful when printing rule formes.

work bar (*Textiles*). See **facing bar.**

work coil (*Elec. Eng.*). Same as **heating inductor.**

work cycle (*Work Study*). The sequence of *elements* (q.v.) which are required to perform a job or to yield a unit of production.

work decrement (*An. Behav.*). A decrease in response intensity related both to the number of times an animal has made the response, and to the physical work involved in it.

work electrode (*Elec. Eng.*). See **applicator.**

worker (*Textiles*). See **stripper and worker.** (*Zool.*) In social Insects, one of a caste of sterile individuals which do all the work of the colony.

work function (*Electronics*). The minimum energy required by an electron (a few electron-volts) for it to pass through a potential barrier. Important for electrode metals in valves. Also called **electron affinity**.

work-hardening (*Met.*). The increase in strength and hardness (i.e., resistance to deformation) produced by working metals. It is most pronounced in cold-working, and in the case of metals such as iron, copper, aluminium, and nickel. Lead, tin, and zinc are not appreciably hardened by cold-working, because they can recrystallize at room temperature.

workhead transformer (*Elec. Eng.*). One associated with the workpiece in induction heating when the generator is at a distance and feeds power through a cable.

working (*Teleph.*). The technique of routing calls over a telephonic system.

working aperture (*Photog.*). See ***f*-number.**

working chamber (*Civ. Eng.*). The compressed-air chamber at the base of a hollow caisson, being the part in which the work of excavation proceeds. See **air lock.**

working depth (*Bot.*). The average distance reached by the general root system of a plant.

working edge (*Carp., Join.*). An edge of a piece of wood trued square with the working face to assist in truing the other surfaces square.

working face (*Carp., Join.*). That face of a piece of wood which is first trued and then used as a basis for truing the other surfaces. See **face mark.**

working flux (*Elec. Eng.*). That part of the total flux produced by the magnetic system of an electrical machine which links the armature winding; numerically equal to the difference between the total flux and the leakage flux.

working standard (*Elec. Eng.*). A standard for everyday use, calibrated against a *secondary standard* (q.v.).

working stress (*Eng.*). The (safe) working stress is the ultimate strength of a material divided by the applicable factor of safety.

work lead (*Met.*). See **base bullion.**

workpiece (*Elec. Eng.*). The material placed within a *heating inductor* for the purpose of induction heating by high-frequency electric currents.

work study (*Eng.*). A generic term for those techniques, particularly method study and work measurement, which are used in the examination of human work in all its contexts and which lead systematically to the investigation of all the factors which affect the efficiency and economy of the situation being reviewed, in order to effect improvement.

worm (*Eng.*). A gear of high reduction ratio connecting shafts whose axes are at right-angles but do not intersect. It consists of a core carrying a single- or multi-start helical thread of special form (the *worm*), meshing in sliding contact with a concave face gear-wheel (the *worm wheel*). Also **worm gear, worm wheel.** (*Zool.*) An imprecise term applied to elongated Invertebrates with no appendages, as in flatworm (*Platyhelminthes*), roundworm, eelworm (*Nematoda*), earthworm (*Lumbricus spp.*), etc. Also applied to immature forms of some Insects, as in mealworm (*Tenebrio, Coleoptera*), cutworm (some *Lepidoptera*), and wireworm (*Elateridae, Coleoptera*), Click-beetles, and also Millipedes (*Diplopoda*).

worm-and-wheel steering gear (*Autos.*). Steering gear in which the steering column carries a worm, in mesh with a worm wheel or sector, attached to the spindle of the drop arm.

worm (or screw) conveyor (*Eng.*). A conveyor in which loose material such as grain, meal, etc., is continuously propelled along a narrow trough by a revolving worm or helix mounted within it.

worm gear, wheel (*Eng.*). See **worm.**

Wormian bones (*Zool.*). See **sutural bones.**

Woronin's hypha (*Bot.*). A simplified form of archicarp formed by some *Ascomycetes*, consisting of a coiled, somewhat thickened, hypha.

wort (*Bot.*). Malt extract much used as a medium for the culture of micro-organisms. (*Brew.*) The liquor which is run off from the mash tun; the first running is termed *sweet wort*. When hops have been added to the wort, and both boiled, the liquor is termed *hopped wort*. See **beer, hops, malt, mash tun, sparging.**

Woulfe's bottles (*Chem.*). Tubulated glass bottles used for washing gases.

wound cork (*Bot.*). A layer of cork cambium and cork formed below and around wounds, which, if the wound is not too large, heals the damage and prevents the entry of parasites.

wound hormone (*Bot.*). A substance produced in wounded tissues which is able to influence the subsequent development of parts of the plant. It stimulates cell division in neighbouring undamaged mature cells, e.g., traumatic acid.

wound parasite (*Bot.*). A parasite which gains entry to the body of the plant by means of a wound.

wound rotor (*Elec. Eng.*). An alternative term for **slip-ring rotor.**

wound tissue (*Bot.*). A pad of parenchymatous cells formed by the cambium after wounding; it may give rise to groups of meristematic cells from which roots and buds form.

wovenboard (*Build.*). See **interwoven fencing.**

woven steel fabric (*Civ. Eng.*). A mechanically woven fabric of steel wires interlaced and welded at the intersections, used as a reinforcement in reinforced concrete construction.

wove paper (*Paper*). Plain nonwatermarked paper produced by running a plain wove dandy roll on the web.

wow (*Acous.*). Low-frequency modulation introduced in sound reproduction system as a result of speed variation. Similar to, but lower in frequency than, *flutter* (q.v.).

wrack (*For.*). Inferior quality of wood used chiefly for packing-cases.

wraithe (*Weaving*). See **raddle.**

wrapper plate (*Eng.*). (1) In a locomotive boiler, the plate bent round and riveted to the tube plate and back plate, forming the sides and crown of the fire box. (2) The outer casing of the fire box.

wrap round (*Bind.*). See **outset.**

wrap-round plate (*Print.*). A flexible relief printing plate which is clamped round the cylinder.

wreath (*Carp.*). The part of a continuous hand-rail curving in plan around the well-hole of a geometrical stair.

wreathed string (*Carp.*). The continuous curved outer string around the well-hole of a wooden stair.

wreath filament (*Elec. Eng.*). The usual type of filament in large gas-filled electric lamps; the filament wire is festooned from a horizontal supporting spider.

wrench (*Maths.*). A system comprising a force and a couple whose axis is parallel to the force. The force is called the *intensity* of the wrench and the ratio of the moment of the couple to the force its *pitch.* Also called **wrench on a screw.** Any system of forces can be reduced to a wrench. (*Plumb.*) See **pipe wrench.**

wringing (*Eng.*). A method of temporarily combining several *slip gauges* (q.v.) by pressing them together with a slight twisting motion until they adhere, thus ensuring that the combined length equals the sum of the individual lengths.

wrinkle finish (*Paint.*). A paint with a deliberate pattern of small, fairly uniformly distributed wrinkles. Much used on industrial metal articles (cameras, typewriters, etc.) to hide surface imperfections of the metal. Also known as **ripple finish.**

wrinkling (*Met.*). Uneven texture developing on surface of metallic uranium.

Wrisberg's nerve (*Zool.*). In higher Vertebrates, certain sensory fibres of the seventh cranial nerve.

wrist-drop (*Med.*). Paralysis of the extensor muscles of the hand and of the fingers.

wrist pins (*I.C. Engs.*). Pins carried by the big end of the master connecting rod of a radial aero engine, forming the crank pins of the *articulated connecting rods* (q.v.).

write (*Comp.*). (1) To make a permanent record of coded data on, e.g., punched card or tape, magnetic tape, or print by electric typewriter. (2) To feed data into a store or memory.

writer's cramp (*Med.*). A condition in which writing becomes irregular and difficult or even impossible owing to spasm of the muscles of the hand and forearm; an occupational neurosis: not the result of organic disease.

writing speed (*Electronics*). Speed of deflection of trace on phosphor, or rate of registering signals on charge storage device.

wrong fount (*Typog.*). The use in error of a character from a type *fount* other than that currently being used in composition. See also the abbrev. w.f.

wrong lead (*Print.*). See **lead (of the web).**

wrong-reading (*Print.*). See **right-reading.**

Wronskian (*Maths.*). Of n functions u_i of an independent variable x, the determinant whose i, jth element is $\frac{d^{i-1} u_j}{dx^{i-1}}$. It is zero when the functions u_i are linearly dependent.

wrought grounds (*Join.*). Planed strips of wood used as *grounds* (q.v.) when the attached joinery will leave them partly exposed to view.

wrought iron (*Met.*). Iron containing only a very small amount of other elements, but containing 1–3% by weight of slag in the form of particles elongated in one direction, giving the iron a characteristic grain. It is more rust-resistant than steel and welds more easily. Used for chains, hooks, bars, etc., and also widely in decorative iron-work.

wryneck (*Med.*). See **torticollis.**

wulfenite (*Min.*). Molybdate(VI) of lead(II), $PbMoO_4$, occurring as yellow tetragonal crystals in veins with other lead ores. Named after an Austrian mineralogist, von Wülfen.

Wulf string electrometer (*Elec. Eng.*). A type of string electrometer used with an ionization chamber in order to measure short current pulses.

wurtzite (*Min.*). Sulphide of zinc, ZnS, of the same composition as sphalerite, but crystallizing at higher temperatures and in the hexagonal system, in black hemimorphic, pyramidal crystals.

Würtz synthesis (*Chem.*). The reduction of solutions of alkyl halides (in ether) with metallic sodium to yield the corresponding hydrocarbons. If mixtures of different alkyl halides are used, mixtures of hydrocarbons formed by different combinations of the alkyl groups are obtained.

wüstite (*Min.*). Cubic iron(II)oxide, FeO. Known only artificially.

wye (*Plumb.*). A *branch pipe* (q.v.) having only one branch, which is not at right angles to the main run.

wye level (*Surv.*). A type of level whose essential characteristic is the support of the telescope, which is similar to that of the wye theodolite.

wye rectifier (*Elec. Eng.*). Full-wave rectifier system for a three-phase supply.

wye theodolite (*Surv.*). A form of theodolite differing from the transit in that the telescope is not directly mounted on the trunnion axis but is supported on two Y-shaped forks, in which it may be turned end-for-end in order to reverse the line of sight.

wyomingite (*Geol.*). An alkaline volcanic rock, composed of leucite, phlogopite and diopside.

Wyssen (*For.*). A tight-line gravity system invented in Switzerland, primarily designed for the logging of selection-worked mountain forests in Europe. An ingenious device locks the carriage to the skyline while releasing the fall block from the carriage and, vice versa, enables the system to be operated with a single-drum portable power unit.

wythe (*Build.*). See **withe.**

X

x (*Chem.*). A symbol for *mole fraction.*

X (*Chem.*). (1) A general symbol for an electronegative atom or group, especially a halogen.

X (*Elec.*). A symbol for *reactance*.

xalostocite (*Min.*). A pale rose-pink grossular which occurs embedded in white marble at Xalostoc in Mexico. Also called **landerite.**

xanthates (*Chem.*). The salts of xanthic acid, $CS(OC_2H_5)SH$. Potassium xanthate:

$$S=C\begin{matrix}OC_2H_5\\SK\end{matrix}$$

Obtained by the action of potassium ethoxide on carbon disulphide. *Cellulose xanthate* (q.v.) is the basis of the viscose rayon process.

xanthation (*Textiles*). A process in the manufacture of viscose rayon; alkali cellulose is converted into cellulose xanthate by mixing it with carbon bisulphide.

xanthein (*Bot.*). A yellow colouring matter sometimes present in cell sap; it is allied to xanthophyll.

xanthene (*Chem.*). Diphenylene-methane oxide:

H_2 C — O

colourless plates, m.p. 98·5°C.

xanthene dyestuffs (*Chem.*). Dyestuffs which may be regarded as derivatives of xanthene containing the pyrone ring:

$$C\begin{matrix}C=C\\C=C\end{matrix}O$$

They comprise the pyronines, derivatives of diphenylmethane, and the phthaleins, derivatives of triphenylmethane.

xanthine (*Chem.*). 2,6-Dihydroxy-purine, a white amorphous mass which is both basic and acidic, and can be obtained by the action of nitrous acid upon guanine. It has the formula:

O, N, HN, O, N H, N H

xanthine oxidase (*Biochem.*). A flavin enzyme found in liver and milk which can utilize oxygen directly. It catalyses the oxidation of hypoxanthine to xanthine and xanthine to uric acid.

xanthochroism (*Zool.*). A condition in which all skin pigments other than golden and yellow ones disappear; as in the Goldfish.

xanthochromia (*Med.*). Any yellowish discoloration, especially of the cerebrospinal fluid.

xanthodont (*Zool.*). Having yellow teeth, as certain Rodents, the incisors of which are stained a yellowish colour.

xanthoma (*Med.*). A yellow tumour composed of fibrous tissue and of cells containing cholesterol ester, occurring on the skin (e.g., in diabetes) or on the sheaths of tendons, or in any tissue of the body (*xanthoma multiplex*).

xanthophore (*Zool.*). A cell occurring in the integument and containing a yellow pigment; as in Goldfish. Also **guanophore, ochrophore.**

xanthophyll (*Bot., Zool.*). $C_{40}H_{56}O_2$. One of the two yellow pigments present in the normal chlorophyll mixture of green plants; a yellow pigment occurring in some *Phytomastigina.*

xanthophyllite (*Min.*). A brittle mica, crystallizing in the monoclinic system; a hydrous calcium, magnesium, aluminium silicate. Differs from clintonite in its optical properties.

xanthopicrite (*Chem.*). See **berberine.**

xanthoplasts (*Zool.*). Yellow chromatophores found in some *Phytomastigina.*

xanthoprotein reaction (*Chem.*). The reaction between albuminous matter and concentrated nitric acid, resulting in a yellow coloration on heating, often already in the cold. If NaOH is added to the reaction mixture, the solution turns reddish-brown; with ammonia, an orange colour is obtained.

xanthopsia (*Med.*). Yellow vision. The condition in which objects appear yellow to the observer, as in jaundice or after taking santonin.

xanthopsin (*Zool.*). A pigment found in the retinular elements of the eyes of certain night-flying Insects.

xanthopterine (*Chem.*). Yellow pigment originally isolated by Hopkins from the wings of butterflies of the family Pieridae. Later shown to have the structure 2-amino-4,6-dihydroxypteridine:

O, H, N, O, HN, H_2N, N, N

It is identical with *uropterine* which is found, in traces, in normal urine. It is chemically related to folic acid (see **vitamin B complex**), but it is doubtful whether xanthopterine itself has any effect in producing red blood cells.

xanthosiderite (*Min.*). See **goethite.**

x-axis (*Aero.*). The longitudinal, or roll, axis of an aircraft. Cf. *axis*. (*Maths.*) Conventionally, the horizontal axis of any type of graph.

X-back (*Cinema.*). A conducting surface on the back of negative cinematograph film, to eliminate scratches arising from the discharge of electric charges, which are separated by friction on the film. See **squeeze track.**

X-band (*Radar*). Frequency band widely used for 3 cm radar. Now correctly designated **Cx-band.**

X-body (*Bot.*). An amorphous inclusion in a plant cell suffering from a virus disease.

X-chromosome (*Cyt.*). One associated with sex determination, usually occurring paired in the female, and alone in the male, zygote and cell.

X-cut (*Crystal.*). Special cut from a quartz crystal, normal to the electric (X) axis.

X-disease (*Vet.*). See **bovine hyperkeratosis.**

Xe (*Chem.*). The symbol for *xenon.*

xenarthral (*Zool.*). Having additional facets for articulation on the dorsolumbar vertebrae.

xenia (*Bot.*). The effect of the foreign pollen upon the characters of the young plant resulting from pollination.

xeno-. Prefix from Gk. *xenos*, strange, foreign.
xenocryst (*Geol.*). A single crystal or mineral grain which has been incorporated by magma during its uprise and which therefore occurs as an inclusion in igneous rocks, usually surrounded by reaction rims and more or less corroded by the magma. Cf. *xenolith.*
xenogamy (*Bot.*). Pollination of a flower from a flower of the same species but on another plant.
xenolith (*Geol.*). A fragment of rock of extraneous origin which has been incorporated in magma, either intrusive or extrusive, and occurs as an inclusion, often showing definite signs of modification by the magma. Xenoliths in granite used for ornamental purposes detract from the value of the stone.
xenomenia (*Med.*). Vicarious menstruation. A condition in which, in the absence of normal menstruation, bleeding occurs at regular monthly intervals from other parts of the body (e.g., from the nose).
xenomorphic (*Min.*). A textural term implying that the minerals in a rock do not show their own characteristic shapes, but are without regular form by reason of mutual interference. See also **granitoid texture.**
xenon (*Chem.*). A zero-valent element, one of the rare gases, present in the atmosphere in the proportion of 1 : 170 000 000 by volume. Symbol Xe, at. no. 54, r.a.m. 131·30, m.p. −140°C, b.p. −106·9°C, crit. temp. +16·6°C, density at s.t.p. 5·89 g/dm^3. Its isotope ^{135}Xe has the highest known capture cross-section for thermal neutrons ($2{\cdot}7 \times 10^6$ barns). Formed as a result of radioactive decay of uranium fission fragments, this isotope forms the most serious reactor poison, and may delay restart of a reactor after a period of shutdown.
xenon lamp (*Cinema.*). High-intensity lamp developed as an alternative to the carbon-arc for motion-picture projection. Also used in standard fading tests for textiles.
xenophya (*Zool.*). Elements of the shell or skeleton not secreted by the organism itself; foreign particles; as the shell of *Difflugia*. Cf. *autophya.*
xenoplastic (*Zool.*). In experimental Zoology, said of transplantation in which transplant and host belong to the young germs of different species or genera. Cf. *autoplastic*, *heteroplastic*, *homoioplastic.*
Xenopterygii (*Zool.*). A small order of *Teleostei* having a broad sucking disk between the pelvic fins and a smooth scaleless skin; marine forms found clinging to loose stones between tide-marks and feeding on worms and small Crustaceans. Cling-fishes.
xenotime (*Min.*). Yttrium phosphate(V). YPO_4, often containing small quantities of cerium, erbium, and thorium, closely resembling zircon in tetragonal crystal form and general appearance, and occurring in the same types of igneous rock, i.e., in granites and pegmatites as an accessory mineral. An important source of the rare elements named.
xer-, xero-. Prefix from Gk. *xeros*, dry.
xerarch succession (*Bot.*). A succession starting on land where conditions are very dry.
xeric (*Bot.*). Synonym for **xerophytic**; see **xerophyte.**
xeric environment (*Bot.*). An environment in which the soil contains very little water, and where atmospheric conditions favour rapid loss of water from the plants.
xerocoles (*Ecol.*). Animals living in dry situations.
xerodermia, xeroderma (*Med.*). See **ichthyosis.**
xerodermia (xeroderma) pigmentosum (*Med.*). A disease of young children in which prolonged exposure to sunlight produces on the skin erythematous patches which later become pigmented, scaly, wartlike, and finally cancerous.
xerographic printer (*Comp.*). One using photoelectric principles adapted to high-speed *readout* and printing of data on forms. Data is presented by CRT and formed by an image from a negative, both being printed together.
xerography. Non-chemical photographic process in which light discharges a charged dielectric surface. This is dusted with a dielectric powder, which adheres to the charged areas, rendering the image visible. Permanent images can be obtained by transferring particles to a suitable backing surface (e.g., paper or plastic) and fixing, usually by heat. Used for document copying and for making lithographic surfaces, usually on paper, for small-offset printing.
xeromorphic (*Bot.*). Said of parts of a plant protected against excessive loss of water by thick cuticles, coatings of hairs, etc.
xerophilous (*Bot.*). Tolerant of a droughty habitat.
xerophthalmia (*Med.*). A dry lustreless condition of the conjunctiva with or without keratomalacia, due to deficiency of vitamin A.
xerophyte (*Bot.*). A plant able to inhabit places where the water supply is scanty, or where there is physiological drought. *adj.* **xerophytic.**
xeroradiography (*Radiol.*). Radiography in which a xerographic, and not photographic, image is produced. This enables reduced exposure times to be used for low energy radiography.
xerosere (*Bot.*). A succession beginning on dry land. Cf. *hydrosere.*
xerosis (*Med.*). See **xerophthalmia.**
xerostomia (*Med.*). Excessive dryness of the mouth.
X-generation (*Bot.*). The gametes.
X-guide (*Elec. Eng.*). A transmission line with an X-shaped cross-section dielectric, used for guiding surface waves.
x-height (*Typog.*). The height of the lower-case letters exclusive of extenders, varying between the extremes of small and large according to the design of the type face.
Xi particle (*Nuc.*). Hyperon sometimes known as **cascade particle** since it decays to a nucleon through a two-step process apparent as a cascade in a photographic emulsion or bubble chamber. It has hypercharge −1 and isotopic spin ½.
xiphi-, xipho-. Prefix from Gk. *xiphos*, sword.
xiphihumeralis (*Zool.*). In Vertebrates, a muscle leading from the humerus to the xiphoid cartilage.
xiphiplastron (*Zool.*). In *Chelonia*, one of the plates composing the plastron, lying posterior to the hypoplastron.
xiphisternum (*Zool.*). A posterior element of the sternum, usually cartilaginous.
xiphoid (*Zool.*). Sword-shaped.
xiphoid cartilage (*Zool.*). The xiphisternum when it is cartilaginous.
xiphoid process (*Zool.*). The posterior portion of the sternum, the *xiphisternum*; so called after its shape in Man.
Xiphosura (*Zool.*). An order of *Arachnida* which are aquatic, and have a broad prosoma divided by a hinge from the opisthosoma, of which the first six segments are present, being fused together dorsally, and each bearing a pair of biramous appendages, five of which bear gill books. The prosoma bears chelicerae, pedipalps, and four pairs of walking legs. A caudal

spine may represent the lost abdominal segments. One living genus, *Limulus* (King Crab).

xonotlite (*Min.*). A hydrous calcium silicate, of composition $Ca_6Si_6O_{17}(OH)_2$.

X-organ (*Zool.*). A neurosecretory organ in the eye-stalks of certain Crustaceans.

X-plates (*Electronics*). Pair of electrodes in a CRT to which horizontal deflecting voltage is applied in accordance with cartesian coordinate system.

X-ray bath (*Radiol.*). X-ray therapy by means of very large fields irradiating, e.g., the whole abdomen, or the whole trunk.

X-ray crystallography. The study of crystal structures by examination of the diffraction patterns occurring when a fine beam of X-rays is directed on to a crystal specimen.

X-ray diffractometer (*Crystal.*). An instrument containing a radiation detector used to record the X-ray diffraction patterns of crystals or powders.

X-ray fluorescent spectrometry (*Chem.*). A method of chemical analysis in which the sample is bombarded by very hard X-rays or gamma-rays, and secondary radiations, characteristic of the elements present, are studied spectroscopically.

X-ray focal spot (*Electronics*). That small area of the target (anode) of an X-ray tube on which the electron beam is incident, and from which emitted X-rays emerge. High-power tubes frequently have a line focus to minimize localization of the heat dissipated at the anode.

X-ray hardness (*Phys.*). The degree of penetration of X-rays, which is greater the shorter the wavelength.

X-ray microscope (*Micros.*). Modification of *electron microscope* in which a transmission type of X-ray target is also the seal to vacuum system. Hence a specimen can be examined and photographed in air.

X-ray protective glass (*Glass*). Glass containing a high percentage of lead and sometimes also barium, with a high degree of opacity to X-rays.

X-rays (*Phys.*). Electromagnetic waves of short wavelength (ca. 10^{-3} to 1 nm) produced when cathode rays impinge on matter. They may be detected photographically by fluorescence and by ionization produced in gases. They penetrate matter which is opaque to light; this makes them valuable for examining inaccessible regions of the body. When X-rays fall on matter, secondary X-rays (*characteristic X-rays*) are emitted which contain monochromatic radiations that vary in wavelength according to the atoms from which they are scattered. See **Compton effect, K-capture, L-capture, Moseley's law.**

X-ray sources (*Astron.*). The Sun and other bodies which emit X-rays, detected by artificial satellites; more than 100 sources have been found, but only a few (e.g., the *Crab nebula*) have been identified. Most X-ray sources lie near the plane of the Galaxy, but some extragalactic sources are also known (e.g., the Seyfert galaxies).

X-ray spectrography (*Chem.*). See **X-ray fluorescent spectrometry.**

X-ray spectrometer (*Crystal.*). The name originally used for the *X-ray diffractometer*, but now abandoned in order to avoid confusion with *X-ray fluorescent spectrometry*.

X-ray spectrum (*Phys.*). Wavelength distribution of energy radiated from X-ray tube. Consists of a line spectrum characteristic of the target metal superimposed on a continuous background.

X-ray telescope (*Instr.*). One used initially to investigate the X-ray emission from the sun and which showed greater variability than might have been expected from optical measurements in the visible region. To focus the X-rays by mirrors, grazing incidence techniques had to be used. A gas-filled X-ray proportional counter was placed behind an aperture at the mirror focus and the counts were telemetered to earth, the telescope being rigidly fixed to a rocket which itself was scanned.

X-ray television (*Radiol.*). A closed-circuit system used for nondestructive industrial testing of materials, joints, etc., thus avoiding use of photographic film.

X-ray therapy (*Radiol.*). The use of X-rays for medical treatment.

X-ray transformer (*Radiol.*). A special type of high-voltage transformer for use with X-ray tubes.

X-ray tube (*Electronics*). The vacuum tube in which X-rays are produced by a cathode-ray beam incident on an anode (or anticathode). Such tubes may be sealed high vacuum or continuously pumped. Older designs had small residual gas pressure and cold cathodes. See **Coolidge tube.**

X's (*Radio*). Same as **atmospherics.**

X-synchronization (*Photog.*). See **flash-synchronized.**

X-tgd (*Build.*). Abbrev. for *cross-tongued.*

X-unit (*Radiol.*). A unit expressing the wavelength of X-rays or gamma-rays equivalent to approximately 10^{-13} m. ($1\ XU = 1{\cdot}002\,02 \pm 0{\cdot}000\,03 \times 10^{-13}$ m.) Abbrev. **XU.** Earlier, **siegbahn.**

X-wave (*Phys.*). The extraordinary component of an electromagnetic wave.

XX (*Paper*). See **retree.**

xyl-, xylo-. Prefix from Gk. *xylon*, wood.

xylem (*Bot.*). Wood; usually consisting of vessels, fibres, and/or tracheides, all with lignified walls, together with some parenchyma having more or less lignified walls. Xylem is concerned with the conduction of aqueous solutions about the plant body, and with mechanical support.

xylem core (*Bot.*). A solid strand of xylem occupying the middle of the stele.

xylem mother cell (*Bot.*). A daughter cell cut off from a cell of the cambium which is later converted into a component of the xylem.

xylem parenchyma (*Bot.*). Parenchymatous cells occurring in xylem, apart from those present in the vascular rays.

xylem ray (*Bot.*). The portion of a vascular ray traversing the xylem. Also called **wood ray.**

xylenes (*Chem.*). $C_6H_4(CH_3)_2$, dimethylbenzenes. There are three isomers which all occur in coal-tar but cannot be separated by fractional distillation: 1,2-xylene, m.p. −28°C, b.p. 142°C; 1,3-xylene, m.p. −53°C, b.p. 139°C; 1,4-xylene, m.p. +13°C, b.p. 138°C, which is an important starting material for polyester synthetic fibres such as Terylene. Commercial preparation termed xylol. Used, in microscopy, as a clearing agent, in the preparation of specimens for embedding, and also in the preparation of tissue sections, etc., for mounting.

xylenol resin (*Chem.*). Synthetic resin of the phenolic type produced by the condensation of a xylenol with an aldehyde.

xylenols (*Chem.*). $(CH_3)_2{\cdot}C_6H_3{\cdot}OH$, monohydric phenols derived from xylenes: 1,2,3-xylenol (*adj. ortho*), m.p. 73°C, b.p. 213°C; 1,3,4-xylenol (*asym. ortho*), m.p. 65°C, b.p. 222°C;

1,2,5-xylenol (*para*), m.p. 75°C, b.p. 209°C.

xylic gap (*Bot.*). A gap in the xylem opposite a leaf base.

xylochrome (*Bot.*). A mixture of substances to which the colour of heart wood is due. It includes tannins, gums, and resins.

xylogenous, xylophilous (*Bot., Zool.*). Growing on wood; living on or in wood.

xylol (*Chem.*). See xylenes.

xyloma (*Bot.*). A sclerotium-like body which forms spores internally, and does not put out branches which develop into sporophores.

xylometer (*For.*). An apparatus used for determining the volumes of irregular pieces of wood by measuring the amount of water or any other liquid they displace.

Xylonite (*Plastics*). A thermoplastic of the nitrocellulose type.

xylophagous (*Zool.*). Wood-eating.

xylophilous (*Bot., Zool.*). See xylogenous.

xylose (*Chem.*). (−)Xylose, known as *wood sugar*, is a pentose found in many plants. It has the following constitution:

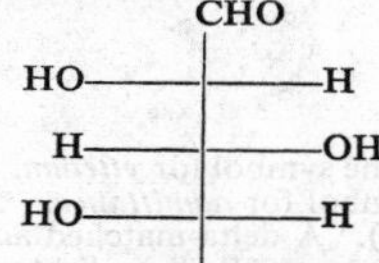

It is a stereoisomer of arabinose.

xylotomous (*Zool.*). Wood-boring; wood-cutting.

x-y recorder (*Instr.*). One which traces on a chart the relation between two variables, not including time. Time may be introduced by moving the chart linearly with time and controlling one of the variables so that it changes proportionally with time.

X-zone (*Histol.*). A transitory region of the inner zones of the adrenal cortex.

Y

Y (*Chem.*). The symbol for *yttrium*.
Y (*Elec.*). Symbol for *admittance*.
y-aerial (*Radio*). A delta-matched antenna.
Yagi antenna or **array** (*Radio*). System of end-fire radiators or receivers, characterized by directors in front of the normal dipole radiator and rear reflector.
Yankee machine (*Paper*). See **MG machine.**
yapp (*Bind.*). A style of binding with overlapping limp covers, much used for pocket bibles.
yard. Unit of length in the foot-pound-second system formerly fixed by a line standard (Weights and Measures Act of 1878), redefined in 1963 as 0·9144 m.
yardage (*Civ. Eng.*). The volume of excavation in cubic yards.
yardangs (*Geog.*). Ridges of soft rock formed by wind action in the deserts of central Asia, often overhanging due to the erosive action of sand blown along the ground.
yard trap (*San. Eng.*). See **gulley trap.**
yarn count (*Textiles*). See **count of yarn, tex.**
Yarrow boiler (*Eng.*). A marine water-tube boiler employing an upper steam drum connected by banks of inclined tubes to three lower water drums, between two of which superheating elements are arranged.
yaryan (*Paper*). The name applied to an evaporation method employed in the recovery of soda from the liquor used in digesting raw materials.
yaw (*Aero., Ships*). Angular rotation of an aircraft or other vessel about a vertical axis. The *yaw angle* is measured between the relative wind and the axis of the vessel.
yaw damper (*Aero.*). See **damper.**
yawing moment (*Aero.*). The component about the normal axis of an aircraft due to the relative airflow.
yaw meter (*Aero.*). An instrument, usually on experimental aircraft or missiles, which detects changes in the direction of airflow by the pressure changes induced thereby.
yaws (*Med.*). Framboesia; pian. A contagious tropical disease affecting dark-skinned races, due to infection with *Treponema pertenue*, and characterized by raspberry-like papules on the skin; as in syphilis, the bones and joints may later become infected.
yaw vane (*Aero.*). A small aerofoil on a pivoted arm at the end of a long boom or probe, attached to the nose of an aircraft or missile, which measures the angle of the relative airflow and transmits it to recording instruments.
y-axis (*Aero.*). The lateral, or pitch, axis of an aircraft. Cf. *axis*. (*Maths.*) Conventionally, the axis perpendicular to and in the horizontal plane through the *x-axis* in any type of graph.
Yb (*Chem.*). The symbol for *ytterbium*.
Y/B ratio (*Optics*). A term used to classify a type of *dichromatism* (q.v.). An observer sees only two colours when examining the solar spectrum, blue and yellow, separated by a white patch. The relative extent of the two colours is the *Y/B ratio*.
Y-chromosome (*Cyt.*). One of a pair of heterochromosomes, in the heterogametic sex, associated with sex determination. See also **X-chromosome.**
Y-class insulation (*Elec. Eng.*). A class of insulating material to which is assigned a temperature of 90°C. See **class -A, -B, -C, etc., insulating materials.**
Y-connexion (*Elec. Eng.*). An alternative name for **star connexion.**
Y-cut (*Crystal.*). Special cut of a quartz crystal, normal to the mechanical (Y) axis.
year (*Astron.*). The civil or calendar year as used in ordinary life, consisting of a whole number of days, 365 in ordinary years, and 366 in leap years, and beginning with January 1. See also **anomalistic-, eclipse-, leap years, sidereal-, tropical-.**
yearling's wool (*Textiles*). See under **fleece wool.**
yeast (*Bot.*). Micro-organisms producing zymase, which induces the alcoholic fermentation of carbohydrates.
yellow body (*Histol.*). See **corpus luteum.**
yellow cells (*Zool.*). In *Chaetopoda*, yellowish cells forming a layer investing the intestine and playing a rôle in connexion with nitrogenous excretion; chloragogen cells; minute symbiotic algae occurring in the extracapsular protoplasm of some *Radiolaria*; *xanthoplasts* (q.v.).
yellow copper ore (*Min.*). A little-used popular name for the sulphide of copper(II) and iron(II), *chalcopyrite*.
yellow enzymes (*Physiol.*). Flavoproteins; combinations of a protein with alloxazine mononucleotide (*riboflavin phosphate*) and alloxazine adenine dinucleotide (*riboflavin phosphate—adenylic acid compound*) as prosthetic groups. Important in cellular respiration due to their oxidation-reduction properties.
yellow fever (*Med.*). Yellow jack. An acute infectious disease caused by a virus, conveyed to man by the bite of the mosquito *Aëdes aegypti* (*Stegomyia fasciata*); characterized by high fever, acute hepatitis, jaundice, and haemorrhages in the skin and from the stomach and bowels; it occurs in tropical America and West Africa.
yellow fibres (*Zool.*). Straight, branched elastic fibres occurring singly in areolar connective tissue. Cf. *white fibres*.
yellow fibrocartilage (*Zool.*). A form of fibrocartilage in which yellow fibres predominate.
Yellow Ground (*Geol.*). The upper, oxidized zone of the Blue Ground, a decomposed volcanic ash occurring in the diamond-bearing vents of South Africa.
yellowing (*Textiles*). Discoloration which sometimes occurs on textile materials during use or storage over a long period.
yellow ochre (*Paint.*). See **ochre.**
yellow pine (*For.*). A very soft, even-grained wood from Canada. Useful for joinery.
yellow quartz (*Min.*). See **citrine.**
yellows (*Vet.*). See **canine leptospirosis.**
yellowses (*Vet.*). Head grit. A form of photosensitization in sheep characterized by dermatitis, subcutaneous oedema affecting particularly the head, and sometimes generalized jaundice; the cause is unknown.
yellow snow (*Bot.*). Snow coloured yellow by the growth on it of certain algae; sometimes observed in the Alps and in the Antarctic.
yellow spot (*Zool.*). Macula lutea; the small area at the centre of the retina in Vertebrates at which day vision is most distinct.

yellow tellurium (*Min.*). A synonym for *sylvanite.*

yellow vision (*Med.*). See **xanthopsia.**

yellow wove (*Paper*). An azure paper with a greenish tint.

yenite (*Min.*). A name rejected by Continental mineralogists, because it commemorated the battle of Jena, 1806. The mineral is now known as *ilvaite* (q.v.).

yew (*For.*). A tree, *Taxus*, indigenous to Europe, Asia, and the western hemisphere, giving a softwood immune to powder-post beetle attack. Durable when used in exposed positions; used for bowmaking.

yield (*Mining*). Tonnage extracted, or ratio of known tonnage, to that recoverable profitably. (*Nuc.*) (1) Ion pairs produced per quantum absorbed or per ionizing particle. (2) See **fission yield.**

yield-G (*Nuc.*). A term in radiation chemistry, being the number of molecules produced or converted per 100 eV of energy absorbed.

yielding attachment (*Horol.*). A method of attaching the outer end of a mainspring to its barrel. It permits of a more concentric uncoiling of the spring.

yielding prop (*Mining*). Support used just behind coal face, which shortens slightly under load, but can be reclaimed and re-used.

yield point (*Met.*). The stress at which a substantial amount of plastic deformation takes place under constant or reduced load. This sudden yielding is a characteristic of iron and annealed steels. In other metals, plastic deformation begins gradually and its incidence is indicated by measuring the proof stress, which, however, is frequently called the yield point.

YIG (*y*ttrium *i*ron *g*arnet) (*Elec. Eng.*). A material which has a lower acoustic attenuation loss than quartz and which has been considered for use in delay lines.

YIG filter (*Elec. Eng.*). One using a YIG crystal which is tuned by varying the current in a surrounding solenoid, a permanent magnet being used to provide the main field strength.

-yl (*Chem.*). A suffix denoting (1) a monovalent organic radical; (2) an electropositive inorganic radical which contains oxygen.

ylem (*Nuc.*). The basic substance from which it has been suggested that all known elements may have been derived through nucleogenesis (fusion of fundamental particles to form nuclei). It would have a density of 10^{16} kg/m^3, and would consist chiefly of neutrons.

Y-level (*Surv.*). See **wye level.**

Y-maze (*An. Behav.*). A maze similar to a *T-maze*, but in which the arms are not at right angles to the stem.

Y-network (*Telecomm.*). Same as **T-network;** three-branch star network.

Yngel trawl. See **young-fish net.**

yoderite (*Min.*). A silicate of aluminium, iron and magnesium, crystallizing in the monoclinic system.

yohimbine (*Chem.*). $C_{22}H_{28}O_3N_2+H_2O$, an alkaloid with a pentacyclic nucleus, obtained from the bark of *Corynanthe johimbe*; colourless needles; m.p. 234°C; soluble in ethanol and trichloromethane; slightly soluble in ethoxyethane. It is poisonous in excess, allegedly acts as an aphrodisiac, and also exerts a local anaesthetic action.

yoke (*Civ. Eng.*). Stout timbers bolted round the shuttering for a column to secure the parts together during the process of pouring and setting. (*Electronics*) Combination of current coils for deflecting the electron beam in a CRT. (*Mag.*) Part or parts of a magnetic circuit not embraced by a current-carrying coil, especially in a generator or motor, or relay.

yoke suspension (*Elec. Eng.*). See **bar suspension.**

yolk (*Zool.*). The nutritive nonliving material contained by an ovum.

yolk duct (*Zool.*). Vitelline duct.

yolk epithelium (*Zool.*). The epithelium surrounding the yolk sac.

yolk gland (*Zool.*). See **vitellarium.**

yolk nucleus (*Zool.*). Vitelline body; a structure which appears in the cytoplasm of oöcytes before the formation of yolk platelets.

yolk plug (*Zool.*). A mass of yolk-containing cells which partially occludes the blastopore in some Amphibians.

yolk sac (*Zool.*). The yolk-containing sac which is attached to the embryo by the yolk stalk in certain forms.

yolk stalk (*Zool.*). A short stalk by which the yolk sac is attached to the embryo and by which yolk substance may pass into the alimentary canal of the embryo.

Yorkshire bond (*Build.*). See **monk bond.**

Yorkshire square (*Brew.*). Cubic fermenting vessel with an upper and lower chamber connected by a vent through which the yeast rises and a valved return pipe ('organ pipe') for running back wort which has been pumped to the upper chamber. Originally made of stone or slate, now generally of stainless steel. Also **stone square.**

young-fish net. A large tow-net the mouth of which is kept open by *otter boards* (q.v.), used for capturing small fishes at the surface or in mid-water. Also **Yngel trawl.**

Young-Helmholtz theory of colour vision (*Optics*). The supposition that the eye contains 3 systems of colour perception, with maximum response to 3 primary colours. It is the theory adopted for the realization of colour photography. There is little medical support for the theory, but the practice is justified by the physical possibility of matching practically every natural colour by the addition of contributions from 3 primary colours. See **colour.**

younging, direction of (*Geol.*). The direction in which a series of inclined sedimentary rocks becomes younger.

Youngman flap (*Aero.*). A trailing-edge flap which is extended below the main aerofoil to form a slot before being traversed rearward to increase the wing area before being deflected downward to increase lift and drag coefficients.

Young's equation (*Min. Proc., Phys.*). Index of surface wettability, used in flotation research on minerals: $\gamma_S=\gamma_{SL}+\gamma_L \cos\theta$, where θ is contact angle (that between water and air bubble adhering to mineral), γ is free energy per unit area, *S* and *L* are solid and liquid phases.

Young's modulus (*Phys.*). A modulus of elasticity applicable to the stretching of a wire (or thin rod), or to the bending of a beam. It is defined as the ratio

$$\frac{\text{tensile (or compressive) stress}}{\text{tensile (or compressive) strain}}.$$

Symbol *E*. Known also as stretch or elongation modulus. Since *strain* is a numeric, its units are those of *stress*, viz. MN/m^2.

Y-parameter (*Electronics*). The short-circuit admittance parameter of a transistor.

y-pipe (*Plumb.*). See **wye.**

Y-plates (*Electronics*). Pair of electrodes to which voltage producing vertical deflection of spot is applied in accordance with cartesian coordinate system.

ypsiloid cartilage (*Zool.*). In Salamanders with functional lungs, a cartilage attached to, but formed independently of, the pubis, furnishing attachment for muscles connected with respiration.

Y rectifier (*Elec. Eng.*). See **wye rectifier.**

yrneh (*Elec.*). Unit of reciprocal inductance (*henry* backwards).

Y signal (*TV*). The monochromatic signal in colour TV which conveys the intelligence of brightness. Combines with the three chrominance components to produce the three colour primary signals. See **B—Y signal, G—Y signal, R—Y signal.**

Y theodolite (*Surv.*). See **wye theodolite.**

ytterbium (*Chem.*). A metallic element, a member of the rare-earth group. Oxide Yb_2O_3, white, giving colourless salts. Symbol Yb, at. no. 70, r.a.m. 173·04, m.p. of metal about 1800°C.

yttrium (*Chem.*). A metallic element usually classed with the rare earths because of its chemical resemblance to them. Oxide, Y_2O_3, white, giving colourless salts. Symbol Y, at. no. 39, r.a.m. 88·9059, m.p. of metal 1250°C.

yttrocerite (*Min.*). A massive, granular or earthy mineral, essentially a cerian fluorite, with the metals of the yttrium and cerium groups, commonly violet-blue in colour, and of rare occurrence.

yttrotantalite (*Min.*). An orthorhombic mineral, the name of which conveys the essential chemical composition, though niobium, cerium, yttrium, uranium, iron, and calcium are present in varying amounts. At the type-locality, Ytterby (Sweden), it occurs with feldspar in pegmatite.

yugawaralite (*Min.*). A rare monoclinic zeolite; a hydrated silicate of calcium and aluminium.

Yukawa potential (*Nuc.*). A potential function of the form

$$\frac{V = V_0 \exp(-kr)}{r},$$

r being distance. Characterizes the meson field surrounding a nucleon. The exponential tail of the Yukawa potential extends with appreciable strength to larger values of *r* than does that of the coulomb potential.

yu-stone (*Min.*). Yu or yu-shih, the Chinese name for the highly prized jade of gemstone quality.

Y-voltage (*Elec. Eng.*). See **voltage to neutral.**

z (*Elec. Eng.*). Symbol for *figure of merit*.
z (*Chem.*). A symbol for the valency of an ion.
Z (*Chem.*). Symbol for: (1) gram-equivalent weight; (2) number of molecular collisions per second; (3) atomic number. (*Elec.*) Symbol for *impedance*. (*Eng.*) Symbol for *section modulus*.
ζ (*Chem.*). A symbol for *electrokinetic potential*.
zaffre or **zaffer** (*Met.*). Impure cobalt oxide remaining when arsenic and sulphur have been removed by roasting.
zalambdodont (*Zool.*). Having molar teeth with V-shaped ridges, as some *Insectivora*.
zaratite or **emerald nickel** (*Min.*). A hydrated basic nickel carbonate, occurring as emerald green stalactitic or mammillary masses encrusting crystals of chromite and magnetite at Unst (Shetland Islands) and elsewhere.
zawn (*Mining*). Underground cavern.
zax (*Tools*). See **sax**.
z-axis (*Aero.*). The normal, or yaw, axis of an aircraft. Cf. *axis*. (*Maths.*) Conventionally, the vertical axis in any 3-dimensional coordinate system.
Z(-axis) modulation (*Electronics*). Variation of intensity of beam, producing varying intensity of brightness of trace.
Z-chromosome (*Cyt.*). The Y-chromosome when the female is the heterogametic sex.
zeatin (*Biochem.*). Plant growth substance, a derivative of adenine, isolated from sweet-corn (*Zea mays*); has strong cell-dividing properties.
zeaxanthin (*Chem.*). $C_{40}H_{56}O_2$. A *xanthophyll* (q.v.) occurring in many plants and in egg yolk. It is isomeric with *lutein* (q.v.).
Zebra (*Comp.*). TN for small digital computer. (*TV*) TN for single-gun colour TV tube.
zebra time (*Telecomm.*). See **zulu time**.
Zechstein (*Geol.*). The higher of the two series into which the Permian System of Germany is divided. The English Permian of Durham and Yorkshire is all probably of Zechstein age.
Zeeman effect (*Phys.*). The splitting up of the individual components of the line spectrum of a substance when in a strong magnetic field.
zein (*Chem.*). A prolamine obtained from maize.
Zeisel's method (*Chem.*). A method for the determination of methoxyl and ethoxyl groups in organic compounds, in which the substance is heated with hydriodic acid; the iodoalkane thus formed is passed into an ethanolic solution of silver(I)nitrate, and the resulting silver(I) iodide weighed.
Zeiss-Endter particle-size analyser (*Powder Tech.*). A device for measuring particle size distribution using a photomicrograph, on which the image of the particle is compared directly with a circular spot of light, the area of which can be varied by an iris diaphragm.
zeitgeber (*An. Behav.*). A stimulus—such as a change in light intensity—occurring at regular 24-hour intervals, thus enabling an endogenous *circadian rhythm* to be precisely synchronized.
Zener breakdown (*Electronics*). Temporary and nondestructive increase of current in diode because of critical field emission of holes and electrons in depletion layer at definite voltage.
Zener cards (*Psychol.*). A pack of 25 cards each bearing one of five different symbols (usually cross, square, circle, star, wavy lines), used in parapsychological experiments.
Zener current (*Electronics*). Current produced in an insulator by electrons which have been raised in energy from the valence bond to the conduction bond through the agency of a strong electric field.
Zener diode (*Electronics*). One with characteristic showing sharp increase of reverse current at a certain negative potential, and suitable for use as voltage reference. The effect is now believed due to Townsend discharge and not to Zener current, and the device is therefore also known as an **avalanche diode** or **breakdown diode**. See **Townsend avalanche, Zener breakdown**.
Zener effect (*Electronics*). Pronounced and stable curvature in the reverse voltage/current characteristic of a semiconductor point-contact diode; predicted by Zener, and widely used in bridges as a control element.
Zener voltage (*Electronics*). The electric field necessary to excite the Zener current, being of the order of 10^7 volt cm^{-1}. It is also used to denote the negative voltage which remains approximately constant over a range of current values in the reverse voltage-ampere characteristic of a semiconductor. See **Zener diode**.
zenith (*Astron.*). The point on the celestial sphere vertically above the observer's head; one of the two poles of the horizon, the other being the *nadir*.
zenithal equal-area projection (*Geog.*). A type of *zenithal projection* (q.v.), in which the areas are in correct proportion, but shape and distance are distorted away from centre. Also called **Lambert's projection**.
zenithal equidistant projection (*Geog.*). A type of *zenithal projection* (q.v.), in which all distances measured from the centre are correct, not distorted.
zenithal projection (*Geog.*). A type of projection in which the plane of projection is tangential to the sphere. Best for representing polar regions.
zenith distance (*Astron.*). The angular distance from the zenith of a heavenly body, measured as the arc of a vertical great circle; hence the complement of the altitude of the body.
zenith sector (*Astron.*). A telescope pivoted at its upper end and allowing only a limited displacement from the zenith. It was used for the accurate determination of the declinations of zenithal stars, and led Bradley to his discovery of aberration and nutation.
zenith telescope (*Astron.*). An instrument similar to the meridian circle, but fitted with an extremely sensitive level and a declination micrometer; used to determine latitude, by observing the difference in zenith distance of two stars whose meridian transit is at a small and equal distance from the zenith, one north and one south.
Zenker's degeneration (*Med.*). Hyaline degeneration of *striated muscle*, occurring, for example, in the abdominal muscles in typhoid fever.
Zenker's fluid (*Micros.*). A micro-anatomical fixative containing mercuric chloride, potassium dichromate(VI), sodium sulphate and glacial acetic acid.
Zeno's paradoxes (*Maths.*). Four paradoxes, of which that about Achilles and the tortoise is well known, apparently designed by Zeno (ca. 450 B.C.) to demonstrate that either of the hypo-

theses that space and time consist of indivisible quanta or that space and time are divisible ad infinitum leads to a dilemma.

zeolite process (*Chem.*). Standard water-softening process, formerly based on using naturally-occurring *zeolites* (q.v.), but nowadays on synthetic zeolites (insoluble synthetic resins). See **ion exchange, Permutit.**

zeolites (*Min.*). A group of alumino-silicates of sodium, potassium, calcium, and barium, containing very loosely held water, which can be removed by heating and regained by exposure to a moist atmosphere, without destroying the crystal structure. They occur in geodes in igneous rocks, and as authigenic minerals in sediments, and include chabazite, natrolite, mesolite, stilbite, heulandite, harmotome, phillipsite, thomsonite, etc.

Zeomorphi (*Zool.*). In some classifications, an order of marine *Teleostei* having a laterally compressed body, spiny fin-rays, large protractile mouth, and truncate or rounded tail-fin; carnivorous. John Dories.

zephyr (*Meteor.*). Properly, a warm westerly wind blowing in the Mediterranean.

zepp antenna (*Radio*). Horizontal half-wavelength antenna fed from a resonant transmission line. It is connected at one end to one wire of the transmission line and the transmitter or receiver is connected between the two wires, the length of the line being critical.

zero (*Maths.*). If a function, analytic in a domain D, is expanded in a *Taylor's series* (q.v.) about any point $z=a$ in D, viz., $f(z)=\sum_{0}^{\infty} b_n(z-a)^n$, and if all values of b_n up to b_{m-1} vanish but b_m does not vanish, then $f(z)$ is said to have a zero of order m at $z=a$.

zero-access store (*Comp.*). Fast store from which information is immediately available, e.g., delay line containing one word.

zero-address instruction (*Comp.*). Operation in computing where the location of the operands is defined by the order code and not specified independently.

zero-beat reception (*Radio*). Of radio telephony, reception in which a locally generated oscillation, having the same frequency as the incoming carrier, is impressed simultaneously on the detector. See **heterodyne reception, homodyne reception.**

zero clearing (*Radio*). Operation of any arrangement for balancing antenna effect to obtain the sharpest minimum signal in determining a radio bearing with a rotating antenna or a radiogoniometer. Also called **minimum clearing.**

zero compression (*Comp.*). A technique of data processing used to eliminate the storing of nonsignificant leading zeros.

zero-cut crystal (*Radio*). Quartz crystal cut at such an angle to the axes as to have a zero frequency/temperature coefficient. Used for accurate frequency standards.

zero-energy reactor (*Nuc. Eng.*). See **zero-power reactor.**

zero error (*Instr.*). (1) Residual time delay which has to be compensated in determining readings of range. (2) Error of any instrument when indicating zero, either by pointer, angle, or display.

zero frequency (*Elec. Eng.*). The component of a complex signal corresponding to the d.c. level. Abbrev. **z.f.**

zero fuel weight (*Aero.*). See **weight.**

zero-g (*Space*). The state of weightlessness or *free fall* (q.v.).

zerograph (*Teleg.*). An early form of *telegraph.*

zero level (*Elec. Eng.*). Any voltage, current or power reference level when other levels are expressed in dB relative to this.

Zerol gear (*Eng.*). A type of *bevel gear*, having curved teeth and a zero helical angle.

zero method (*Elec. Eng.*). Measuring system in which an unknown value can be deduced from other values when a sensitive but not necessarily calibrated instrument indicates zero deflection, as in a Wheatstone bridge or potentiometer. Also called **null method.**

zero order reaction (*Chem.*). One in which the rate is independent of the concentration of the reacting species.

zero pause (*Elec. Eng.*). The momentary cessation of an alternating current when passing through a zero value between successive half-cycles on which the action of a.c. circuit-breakers largely depends.

zero phase sequence (*Elec. Eng.*). A three-phase vector system in which all three vectors are equal in magnitude and are in phase with one another. Cf. *negative phase sequence, positive phase sequence* (under **phase sequence**).

zero phase-sequence component (*Elec. Eng.*). One of three vectors forming a zero phase-sequence system, and one of three components into which any vector forming part of an unbalanced three-phase system can be resolved. See **phase sequence.**

zero-point energy (*Phys.*). Total energy at the absolute zero of temperature. The uncertainty principle does not permit a simple harmonic oscillator particle to be at rest exactly at the origin, and by the quantum theory, the ground state still has one half-quantum of energy, i.e., $h\nu/2$, and the corresponding kinetic energy.

zero-point entropy (*Phys.*). As follows from the third law of thermodynamics, the entropy of a system in equilibrium at the absolute zero must be zero.

zero potential (*Elec. Eng.*). Theoretically, that of a point at infinite distance, used for defining capacitance. Practically, the earth is taken as being of invariant potential. That of any large mass of metal, e.g., equipment chassis.

zero power-factor characteristic (*Elec. Eng.*). A curve obtained by plotting the terminal voltage of a synchronous generator delivering full load current at zero power factor lagging, against the field excitation.

zero power level (*Telecomm.*). Arbitrary power level for referring other power levels, either in decibels or nepers. Zero power level was formerly 5·8 milliwatts in U.S., but is now 1 milliwatt, both at the standard 600 ohms impedance level.

zero-power reactor (*Nuc. Eng.*). An experimental reactor with an extremely low neutron flux so that the small power level involved means that no forced cooling is required.

zero stability (*Telecomm.*). Drift in no-signal output level of amplifier or indicator, either with time or with operating conditions, e.g., mains voltage supply.

zero suppression (*Comp.*). The elimination, in computing and data processing, of non-significant zeros to the left of the integral part of a quantity, especially before the start of printing.

zero-type dynamometer (*Elec. Eng.*). A dynamometer in which the electrical forces are balanced by mechanical forces, in such a manner as to bring the indicating pointer back to zero, before a reading can be taken.

zero-valent (*Chem.*). Incapable of combining with other atoms. Also **nonvalent.**

Zeta (*Nuc. Eng.*). (*Ze*ro-energy *t*hermonuclear *a*pparatus.) An apparatus, toroid in shape, used at Harwell for research into fusion reactions and aspects of plasma physics.

zeta function (*Maths.*). See **Riemann-.**

zeta potential (*Chem.*). See **electrokinetic potential.**

zeugite (*Bot.*). A cell in which nuclear fusion occurs.

zeugopodium (*Zool.*). The second segment of a typical pentadactyl limb, lying between the stylopodium and the autopodium; antebrachium or crus; forearm or shank.

zeunerite (*Min.*). A pseudotetragonal green hydrated copper uranate and arsenate, found on quartz in copper ores. A member of the autunite group.

Zeuner valve diagram (*Eng.*). See **valve diagram.**

Zeus (*Nuc. Eng.*). Zero-energy research reactor used to obtain design data for Dounreay fast reactor.

z.f. (*Elec. Eng.*). Abbrev. for *zero frequency*.

ziberline (*Textiles*). A cloth with a merino warp and a weft of wool and camel hair; raised during finishing to obtain a lustrous pile. Cheaper varieties are made with a worsted or a cotton warp and a cashmere weft.

Ziegler catalyst (*Chem.*). Stereo-orienting catalyst discovered by Ziegler for organic polymerization, notably the low-pressure polyethylene process. Prepared by the reaction of compounds of strongly electropositive transition metals with organometallic compounds, e.g., titanium trichloride, aluminium alkyl.

zigzag connexion (*Elec. Eng.*). A symmetrical three-phase star connexion of six windings, situated in pairs on three cores. Each leg of the star consists of two of the windings in series; these windings, being on different cores, have e.m.fs. in them differing in phase by 120°. Used in transformers for eliminating harmonics, and in reactors to obtain an artificial neutral.

zigzag harrow (*Agric.*). A harrow with a zigzag frame or frames and teeth at the junctions of the frame members.

zigzag leakage (*Elec. Eng.*). Magnetic leakage occurring along the zigzag path between stator and rotor teeth when a stator tooth is opposite to a rotor slot.

zinc (*Chem.*). A hard white metallic element with a bluish tinge. Symbol Zn, at. no. 30, r.a.m. 65·37, rel. d. at 20°C 7·12, m.p. 418°C, electrical resistivity $6{\cdot}0 \times 10^{-8}$ ohm metres. Because of its good resistance to atmospheric corrosion, zinc is used for protecting steel (see **galvanized iron, sherardizing, spelter, spraying**). It is also used in the form of sheet and as a constituent in alloys (see **zinc alloys**). Used as an electrode in a Daniell cell and in dry batteries. It is important nutritionally, trace amounts being present in many foods.

zinc alkyls (*Chem.*). Organometallic compounds such as zinc dimethyl and zinc diethyl prepared by treating the alkyl iodides with zinc-copper couple. Used occasionally as *Grignard reagents* (q.v.), and to replace the halogen attached to a tertiary carbon atom by an alkyl group.

zinc alloys (*Met.*). Zinc base alloys, containing aluminium 3–4%, copper 0–3·5%, and magnesium 0·02–0·1%, are used extensively for die-casting. This metal is also used extensively in brass, of which it is an essential constituent. Light aluminium zinc alloys are also used.

zincate (*Chem.*). The anion ZnO_2^{--} or $Zn(OH)_4^{--}$.

zinc blende (*Min.*). A much-used name for *sphalerite* (q.v.), the common sulphide of zinc.

zinc bloom (*Min.*). A popular name for the massive basic zinc carbonate, *hydrozincite* (q.v.).

zinc chrome (*Paint.*). See **zinc yellow.**

zinc dust (*Min. Proc.*). Finely divided powder produced either by condensation of zinc vapour or by atomization of molten zinc. Widely used in *Merrill-Crowe process* (q.v.) to precipitate gold and silver from pregnant solution in the cyanidation of gold ore.

zincite (*Min.*). Oxide of zinc, crystallizing in the hexagonal system and exhibiting polar symmetry; occurring rarely as crystals, usually as deep-red masses; an important ore of zinc, known also as **red oxide of zinc, spartalite,** and **sterlingite.**

zinckenite (*Min.*). A steel-grey mineral, essentially sulphide of lead and antimony, $PbSb_2S_4$, occurring in antimony-mines at Wolfsberg in the Harz Mountains, in Colorado, and in Arkansas as columnar hexagonal crystals, sometimes exceptionally thin, forming fibrous masses.

zinco (*Print.*). A line block executed in zinc, i.e., the normal line block.

zinc oxide (*Pharm.*). Oxide of zinc used, for its astringent and soothing qualities, as a constituent of creams, baby ointments, etc.

zinc protector (*Ships*). Introduced originally for the *cathodic protection* (q.v.) of copper sheathing on wooden ships. Now used in steel ships near bronze propellers.

zinc-rich paint (*Paint.*). A paint containing an extremely high proportion of metallic zinc dust in the dry film (about 95% by weight), applied to iron and steel as an anticorrosive primer. It may be regarded as a less durable form of *cold galvanizing* (q.v.).

zinc spinel (*Min.*). Same as *gahnite*. See **spinel.**

zinc telluride (*Met.*). A semiconductor capable of high-temperature operation (up to about 750°C) without excessive intrinsic conductivity.

zinc white (*Paint.*). A fine white powder (zinc oxide) used as a non-poisonous, permanent pigment in paint manufacture.

zinc yellow (*Paint.*). A chromate of zinc pigment. Also called **zinc chrome.**

zineb (*Chem.*). Zinc ethylene-1,2-dithiocarbamate:

```
CH₂—NH—CS—S
|           \
|            Zn
|           /
CH₂—NH—CS—S
```

Used as a fungicide.

zingiberine (*Chem.*). The chief constituent of ginger oil.

zinkosite (*Min.*). Anhydrous zinc sulphate, occurring at a mine in the Sierra Almagrera (Spain).

zinnwaldite (*Min.*). A mica related in composition to lepidolite (i.e., containing lithium and potassium) but including iron as an essential constituent; occurring in association with tin-stone ores at Zinnwald in the Erzgebirge, in Cornwall, and elsewhere.

Zip-a-Tone (*Print.*). TN for a range of tone patterns which can be stripped into position when preparing drawings and layouts for reproduction.

ziram (*Chem.*). Zinc dimethyldithiocarbamate:

$$\left[(CH_3)_2\,N\!-\!\underset{\underset{S}{\|}}{C}\!-\!S\!-\right]_2 Zn$$

used as a fungicide.

zircon (*Min.*). A tetragonal accessory mineral widely distributed in igneous, sedimentary, and metamorphic rocks, and occurring in three forms differing in density and optical characters. It varies in colour from brown to green, blue, red, golden-yellow, while colourless zircons make particularly brilliant stones when cut and polished. In composition, it is essentially silicate of zirconium, but often contains yttrium and thorium. A small amount of the rare element hafnium is present.

zirconate(IV) (*Chem.*). The anion ZrO_3^{--}.

zirconia (*Chem.*). Zirconium(IV)oxide, ZrO_2, used as an opacifier in vitreous enamels, as a pigment, and as a refractory.

zirconium (*Chem.*). A metallic element, symbol Zr, at. no. 40, r.a.m. 91·22, rel. d. 4·15, m.p. 2130°C. When purified from hafnium, its low neutron absorption and its retention of mechanical properties at high temperature make it useful for the construction of nuclear reactors. Also used as a refractory, as a lining for jet engines, and as a getter in the manufacture of vacuum tubes. The zirconates are finding application as acoustic transducer materials. Tritium adsorbed in zirconium is a possible target in accelerator neutron sources.

zirconium lamp (*Phys.*). One having a zirconium oxide cathode in an argon-filled bulb. It provides a high-intensity point source with only a small emission of the longer visible wavelengths.

Z-line (*Zool.*). In striated or voluntary muscle of Vertebrates, the line found at either end of a sarcomere. It carries a system of submicroscopic transverse tubules, which are in contact with the *sarcoplasmic reticulum* (q.v.), and are thought to transmit action potentials from the outer sarcolemma of the muscle fibres to the inner fibrils.

Z marker beacon (*Aero.*). A form of marker beacon radiating a narrow conical beam along the vertical axis of the cone of silence of a radio range.

z-modulation (*Radar, TV*). The variations in intensity in the electron beam of a CRT which form the display or picture on a sweep or raster.

Zn (*Chem.*). The symbol for *zinc*.

zoaea (*Zool.*). A larval stage of some *Malacostraca* in which the appendages of the head and the first two thoracic somites are well developed and the abdomen distinctly segmented, while the posterior thoracic region is only partially segmented and bears no appendages.

Zoantharia (*Zool.*). An order of *Actinozoa* with varying numbers of mesenteries typically arranged in pairs, the longitudinal muscles of which usually face each other. Tentacles simple, there usually being six or some multiple of six, mesenteric filaments trefoil-shaped in section, stomodaeum with two ciliated grooves or siphonoglyphs. Includes solitary Sea Anemones (e.g., *Metridium*) and the colonial Corals (e.g., *Madrepora*).

zoarium (*Zool.*). The zooids of a polyzoan colony, collectively.

zocle (*Arch.*). See **socle.**

Zodiac (*Astron.*). A name, of Greek origin, given to the belt of stars, about 18° wide, through which the ecliptic passes centrally. The Zodiac forms the background of the motions of the sun, moon, and planets.

zodiacal light (*Astron.*). A faint illumination of the sky, lenticular in form and elongated in the direction of the ecliptic on either side of the sun, fading away at about 90° from it; best seen after sunset or before sunrise in the tropics, where the ecliptic is steeply inclined to the horizon; it is caused by small particles reflecting sunlight, and appears to be an extension of the solar corona to a distance well beyond the earth's orbit.

zoetrope, zoechrome (*Photog.*). Early processes for colour cinematography, using rapidly repeated images of the selected colours in sequence on a screen, the synthesis arising from persistence of vision in the eye.

zoid (*Bot.*). A zoospore.

zoidiophilous (*Bot.*). Pollinated by animals.

zoidophore (*Zool.*). In *Haemosporidia*, a sporoblast derived from the oöcyte.

zoisite (*Min.*). Hydrous alumino-silicate of calcium, crystallizing in the orthorhombic system and occurring chiefly in metamorphic schists; also a constituent of so-called saussurite. Clinozoisite is of the same composition, but crystallizes in the monoclinic system.

zona (*Med.*). See **herpes zoster.** (*Zool.*) An area, patch, strip, or band; a zone. *adjs.* **zonal, zonary, zonate.**

zona fasciculata (*Zool.*). The middle zone of the adrenal cortex, between the *zona glomerulosa* and *zona reticularis*, in which the cells are arranged in fairly straight cords running at right angles to the surface and with capillaries between them. It is probably the site of production of the *glucocorticoids* (q.v.).

zona glomerulosa (*Zool.*). The outer zone of the adrenal cortex, below the capsule, in which the cells are arranged in rounded irregular clusters; probably site of *aldosterone* (q.v.) production.

zona granulosa (*Zool.*). The mass of membrana granulosa cells of the Graafian follicle around the ovum; **discus proligerus or cumulus oophorus.**

zonal index (*Geol.*). See **zone.**

zona pellucida (*Zool.*). A thick transparent membrane surrounding the fully formed ovum in a Graafian follicle.

zona radiata (*Zool.*). The envelope of the Mammalian egg outside the vitelline membrane.

zona reticularis (*Zool.*). The inner zone of the adrenal cortex, next to the adrenal medulla, in which the cells are arranged in cords running in various directions with sinusoidal capillaries between.

zonary placentation (*Zool.*). The condition in which the villi are on a partial or complete girdle round the embryo, as in *Carnivora* and *Proboscidea.*

zonate (*Bot.*). Said of tetraspores formed in a row of four, and not in a tetrahedral group.

zonation (*Bot.*). (1) The formation of bands of different colour on the surface of a plant. (2) The formation, by fungi in culture, of concentric bands different in colour, texture, abundance of sporulation, and so on. (3) The occurrence of vegetation in well-marked bands, each band having its characteristic dominant species. (4) A stage in the development of an oögonium, when the contents are arranged in two or more well-marked zones.

Zond (*Space*). Russian planetary space probe to Venus.

zone (*Bot.*). A band of colour, or of hairs, warts, or other surface feature. (*Chem.*) A region of oriented molecules. (*Geol.*) A subdivision of a stratigraphical series, comprising a group of strata characterized by a distinctive fauna or flora, and bearing the name of one fossil, called the **zonal index**; e.g., the *Dibunophyllum Zone*, a group of strata of variable lithology occurring high in the Lower Carboniferous Series and characterized by the rugose coral, *Dibunophyllum.* See also the following **zone** articles.

zone law of Weiss (*Crystal.*). Expresses the condition that a crystal face (*hkl*) shall lie in the zone (*UVW*), as the equation:

$$Uh+Vk+Wl=0.$$

zone levelling (*Electronics*). An analogous process to *zone refining* (q.v.), carried out during the processing of semiconductors in order to distribute impurities evenly through the sample.

zone of audibility (*Acous.*). The hearing of explosions at great distances from the source, although, nearer, there is a *zone of silence*.

zone of cementation (*Geol.*). That 'shell' of the earth's crust lying immediately below the zone of weathering, within which loose sediments are cemented by the addition of such minerals as calcite, introduced by percolating meteoric waters.

zone of faces (*Crystal.*). A number of faces, belonging to one or several forms, the normals to which lie in one plane (the *zone plane*) and whose edges of intersection are parallel to a line passing through the centre of the crystal (the *zone axis*).

zone of silence (*Acous.*). Local region where sound or electromagnetic waves from a given source cannot be received at a useful intensity.

zone of vegetation (*Bot.*). A belt of plants having well-marked characters, occurring with other zones of different characters; the condition may often be observed in seaweeds and in the vegetation occupying the side of a mountain.

zone of weathering (*Geol.*). An 'earth shell' comprising the exposed surface and that part which, through porosity, fracturing, and jointing, is subject to the destructive action of the atmosphere, rain, and frost.

zone plate (*Optics*). A transparent plate divided into a series of zones by circles whose radii are in the ratio $1 : \sqrt{2} : \sqrt{3} : \sqrt{4}$, etc. Alternate zones are blacked, a concentration of light forming at a point on the axis if a plane wave is normally incident on the plate, which thus acts as a lens.

zone refining or purification (*Elec. Eng.*). Passage of a melted zone along a trough of metal, particularly germanium. Impurities dissolve into this and do not freeze out. Heating is performed by high-frequency eddy currents from a coil which progresses along the trough.

zone television (*TV*). System in which different parts or zones of the image are scanned by separate devices and separately transmitted to the receiver, where they are recombined.

zone time (*Astron.*). See **standard time.**

zoning (*Aero.*). (1) The specification of areas surrounding an aerodrome in which there is a known clearance above obstruction for the safe landing and taking-off of aeroplanes. (2) The division of an aeroplane's fuselage, wings and engine nacelles into specific areas for precise location of equipment, identification and fire protection purposes. (*Min.*) Concentric layering parallel to the periphery of a crystalline mineral, shown by colour banding in such minerals as tourmaline, and by differences of the optical reactions to polarized light in colourless minerals like feldspars; it is due to the successive deposition of layers of material differing slightly in composition.

zonula ciliaris (*Zool.*). In the Vertebrate eye, a double fenestrated membrane connecting the ciliary process of the choroid with the capsule surrounding the lens.

zonule (*Zool.*). A small belt or zone, such as the zonula ciliaris of the Vertebrate eye.

zoo-. Prefix from Gk. *zōon*, animal.

zoobiotic (*Biol.*). Parasitic on, or living in association with, an animal.

zooblast (*Zool.*). An animal cell.

zoocaulon (*Zool.*). See **zoodendrium.**

zoochlorellae (*Zool.*). Symbiotic green algae found in various animals.

zoochorous (*Bot.*). Said of spores or seeds dispersed by animals.

zoocyst (*Zool.*). See **sporocyst.**

zoodendrium (*Zool.*). The branched stalk connecting the members of the colony in certain colonial *Ciliophora*.

zoodynamics. Animal physiology.

zooecium (*Zool.*). The body-wall or enclosing chamber of a polyzoan individual.

zooerythrin (*Zool.*). A red pigment found in the plumage of certain Birds.

zoofulvin (*Zool.*). A yellow pigment found in the plumage of certain Birds.

zoogamete (*Zool.*). A motile gamete.

zoogamy (*Zool.*). Sexual reproduction of animals.

zoogeography (*Zool.*). The study of animal distribution.

zoogloea (*Bot.*). A mucilaginous mass of bacteria embedded in slimy material derived from the swollen cell walls.

zoogonid (*Bot.*). See **zoospore.**

zoogonous (*Zool.*). See **viviparous.**

zooid (*Zool.*). An individual forming part of a colony in *Protozoa* (*Volvocina*), *Coelenterata*, *Rhabdopleura*, *Urochorda*, and *Polyzoa*; in *Polychaeta*, a posterior sexual region formed by asexual reproduction; polyp; a polypide.

zoolith (*Zool.*). A fossil animal.

zoom (*TV, etc.*). Rapid enlargement of, e.g., a TV picture either by *zoom lens* or electronically.

Zoomastigina (*Zool.*). A subclass of *Mastigophora* comprising forms which lack chromatophores, generally practise holozoic nutrition, often have more than two flagella, and never have starch reserves.

zooming (*Aero.*). Utilizing the kinetic energy of an aircraft in order to gain height. *Zoom-bombing* involves the release of a nuclear bomb during a zooming manoeuvre to give the aircraft time to escape the blast. See **toss-bombing.**

zoom lens (*Cinema., TV*). A lens system, used with a camera, which avoids the interruptions when employing a lens turret by permitting the continuous variation of the focal length of the camera over a considerable range. Also **variable-focus lens.**

zoonomy. Animal physiology.

zoonosis (*Vet.*). A disease of animals communicable to man.

zoop (*Acous.*). A peculiar type of extraneous noise modulation, arising during sound-recording on blanks, reproduced with the desired sounds.

zoophytes (*Zool.*). See **Hydrozoa.**

zooplankton (*Zool.*). Floating and drifting animal life.

zoosperm (*Zool.*). A spermatozooid.

zoosporangium (*Bot.*). A sporangium in which zoospores are formed.

zoospore (*Bot.*). An asexual reproductive cell which can swim by means of flagella. (*Zool.*) In *Protozoa*, an active germ produced by sporulation.

zootaxy. Zoological classification.

zootechnics. Animal husbandry.

zootomy (*Zool.*). See **anatomy.**

Zoraptera (*Zool.*). An order of *Blattopteroidea* containing small Insects with biting mouthparts. Wings, when present, have reduced venation and can be shed by basal fractures, but most species have both winged and wingless forms of both sexes. Prothorax well developed. Gregarious, but not organized into societies, living under bark and sometimes near termite galleries.

zoster (*Med.*). See **herpes zoster.**

z-parameters (*Electronics*). The open-circuit impedance parameters of transistors.

Zr (*Chem.*). The symbol for *zirconium.*

Zuckerkandl organs (*Histol.*). Paraganglia lying in the retroperitoneal tissue in the region of the bifurcation of the abdominal aorta.

zulu time (*Telecomm.*). Used in telecommunications for *GMT*; formerly termed **zebra time.**

zunyite (*Min.*). A rare basic orthosilicate of aluminium, containing fluorine and chlorine; it occurs in minute cubic crystals at the Zuñi mine, Silverton, Colorado, and elsewhere.

zussmanite (*Min.*). A pale green, tabular, trigonal hydrous silicate of iron(II), magnesium and potassium.

zwitterion (*Chem.*). An ion carrying both a positive and a negative charge, e.g., present in solid and liquid amino acids such as glycine (aminoethanoic acid), $N^+H_3CH_2COO^-$.

zyg-, zygo-. Prefix from Gk. *zygon*, yoke.

zygantrum (*Zool.*). In Snakes and some Lizards, an additional vertebral articulation, consisting of a fossa on the posterior surface of the neural arch, into which fits the zygosphene.

zygapophyses (*Zool.*). Articular processes of the vertebrae of higher Vertebrates arising from the anterior and posterior sides of the neurapophyses.

zygobranchiate (*Zool.*). Having paired, symmetrically placed gills.

zygocardiac (*Zool.*). A term used to describe certain paired lateral ossicles in the gastric mill of Crustaceans.

zygodactylous (*Zool.*). Said of Birds which have the first and fourth toes directed backwards, as Parrots.

zygodont (*Zool.*). Having the cusps of the molar teeth united in pairs.

zygogenetic (*Zool.*). A product of fertilization.

zygoma (*Zool.*). The bony arch of the side of the head in Mammals which bounds the lower side of the orbit.

zygomatic (*Zool.*). Pertaining to the zygoma. See also **jugal.**

zygomatic arch (*Zool.*). See **zygoma.**

zygomatic bone (*Zool.*). See **jugal.**

zygomorphic, zygomorphous (*Bot., Zool.*). Divisible into half by one longitudinal plane only; bilaterally symmetrical. *n.* **zygomorphy.**

Zygomycetes (*Bot.*). A subdivision of the *Phycomycetes*, including about 300 species, mostly saprophytes, but some parasitic on Insects or on other Fungi. There are no motile stages in the life-history, and sexual reproduction occurs by the union of multinucleate gametangia which seldom differ much in size and shape.

zygonema (*Cyt.*). The zygotene phase of meiosis.

zygoneury (*Zool.*). In some *Gastropoda*, the condition of having the pallial nerves from the pleural ganglion passing direct to the ganglion of the visceral commissure of the same side, from which the mantle nerves of that side appear to originate. Cf. *dialyneury.*

zygophase (*Biol.*). The diploid portion of the life-history.

zygophore (*Bot.*). A mycelial branch bearing a gametangium in the *Zygomycetes.*

zygopleury (*Zool.*). Bilateral symmetry.

zygopodium (*Zool.*). That part of the forelimb in *Tetrapoda* between the brachium and the basipodium; forearm.

zygosis (*Biol.*). See **conjugation.**

zygosome (*Cyt.*). See **mixochromosome.**

zygosphene (*Zool.*). In Snakes and some Lizards, an additional vertebral articulation, consisting of a process on the anterior surface of the neural arch, which fits into the zygantrum.

zygospore (*Bot.*). A thick-walled resting spore formed after the union of isogametes or of isogametangia. (*Zool.*) See **zygote.**

zygosporophore (*Bot.*). The suspensor in the *Zygomycetes.*

zygotaxis (*Zool.*). Mutual attraction between gametes of the opposite sex.

zygote (*Bot., Zool.*). The product of the union of two gametes; in Botany (by extension), the diploid plant developing from that product. Also **amphiont.**

zygotene (*Cyt.*). The second stage of meiotic prophase, intervening between leptotene and pachytene, in which the chromatin threads approximate in pairs and become loops.

zygotic (*Bot., Zool.*). Relating to, or belonging to, a zygote.

zygotic meiosis (*Cyt.*). Meiosis occurring at the first two divisions of the nucleus resulting from gametic union.

zygotic number (*Cyt.*). The diploid chromosome number of somatic cells.

zygotonucleus (*Zool.*). A nucleus resulting from the union of two gametonuclei.

zygozoospore (*Zool.*). A motile cell produced by the union of two similar type cells.

zym-, zymo- (*Chem.*). Prefixes relating to fermentation of enzymes.

zymase (*Chem.*). An enzyme inducing the alcoholic fermentation of carbohydrates.

zymin (*Zool.*). An ex-enzyme.

zymogen (*Biochem.*). A non-catalytic substance formed by plants and animals as a stage in the development of an enzyme; it is convertible into the active enzyme and a protein by the action of a kinase or zymoexcitor. Also called **proenzyme.**

zymosis (*Med.*). Fermentation. The morbid process, thought to be analogous to fermentation, constituting a zymotic (infectious) disease.

zymotic (*Med.*). Of, pertaining to, or causing, an infectious disease; an infectious disease.

APPENDICES

TABLE OF CHEMICAL ELEMENTS

The Relative Atomic Masses are based on the 1969 international agreed values of the International Union of Pure and Applied Chemistry, the basis being the carbon-12 isotope. Values in brackets indicate the mass numbers of the most stable isotopes. See also the Periodic Table on page 1297, which incorporates the electron distributions within the shells.

Symbol	Name	Derived from	Valence No.	Atomic No.	Atomic Mass (Relative)	Relative Density	Melting or Fusing Pt. °C	Discovered by	Date
Ac	Actinium	Greek, *aktis* = ray	3	89	(227)	..	..	Debierne	1899
Ag	Silver (argentum)	Anglo-Saxon, *seolfor*	1	47	107·868	10·5	960	Prehistoric	..
Al	Aluminium	Latin, *alumen* = alum	3	13	26·9815	2·58	658	Wöhler	1828
Am	Americium	America	..	95	(243)	..	..	Seaborg, James and others	1944
Ar	Argon	Greek, *argos* = inactive	0	18	39·948	Gas	−188	Rayleigh and Ramsay	1894
As	Arsenic	Latin, *arsenicum*	3 or 5	33	74·9216	5·73	Volatile, 450	Schröder	1649
At	Astatine	Greek, *astatos* = unstable	..	85	(210)	..	..	Corson, Mackenzie and Segrè	1940
Au	Gold (aurum)	Anglo-Saxon, *gold*	1 or 3	79	196·9665	19·3	1062	Prehistoric	..
B	Boron	Persian, *būrah*	3	5	10·81	2·5	2300	Davy	1808
Ba	Barium	Greek, *barys* = heavy	2	56	137·34	3·75	850	Davy	1808
Be	Beryllium (glucinum)	Greek, *beryllion* = beryl	2	4	9·0122	1·93	1281	Wöhler	1828
Bi	Bismuth	German (origin unknown)	3 or 5	83	208·9806	9·80	268	Valentine	1450
Bk	Berkelium	Berkeley, California	..	97	(249)	..	..	..	1950
Br	Bromine	Greek, *bromos* = stench	1 or 7	35	79·904	3·19	−7·3	Balard	1826
C	Carbon	Latin, *carbo* = charcoal	4	6	12·011	3·52	Infusible	Prehistoric	..
Ca	Calcium	Latin, *calx* = lime	2	20	40·08	1·58	851	Davy	1808
Cd	Cadmium	Greek, *kadmeia* = calamine	1 or 2	48	112·40	8·64	320	Stromeyer	1817
Ce	Cerium	Planet Ceres	3 or 4	58	140·12	6·68	623	Berzelius	1803
Cf	Californium	California	..	98	(251)	..	..	..	1950
Cl	Chlorine	Greek, *chloros* = green	1, 3, 5, 7	17	35·453	Gas	−102	Scheele	1774
Cm	Curium	Pierre and Marie Curie	..	96	(247)	..	..	Seaborg, James and others	1944
Co	Cobalt	German, *Kobold* = goblin	2 or 3	27	58·9332	8·6	1490	Brandt	1739
Cr	Chromium	Greek, *chroma* = colour	2, 3 or 6	24	51·996	6·5	1510	Vauquelin	1797
Cs	Caesium	Latin, *caesium* = bluish-grey	1	55	132·9055	1·88	26	Bunsen	1860
Cu	Copper (cuprum)	Cyprus	1 or 2	29	63·546	8·9	1083	Prehistoric	..
Dy	Dysprosium	Greek, *dysprositos*	3	66	162·50	..	..	Boisbaudran	1886
Er	Erbium	Ytterby, a Swedish town	3	68	167·26	4·8	..	Mosander	1843
Es	Einsteinium	Einstein	..	99	(254)	..	..	..	1952
Eu	Europium	Europe + *ium*	2 or 3	63	151·96	..	..	Demarcay	1896
F	Fluorine	Latin, *fluo* = flow	1	9	18·9984	Gas	−223	Scheele	1771
Fe	Iron (ferrum)	Anglo-Saxon, *iren*	2 or 3	26	55·847	7·86	1525	Prehistoric	..
Fm	Fermium	Fermi	..	100	(253)	..	..	..	1952

INTERMEDIATE (*contd.*)	DIORITE CLAN		
	Quartz-mica-diorite (=TONALITE) DIORITE (soda-lime feldspar in excess)	MICRODIORITE DIORITE-PORPHYRY (='Porphyrite' of some Authors) Diorite-aplite	ANDESITE
BASIC Silica percentage <55	GABBRO CLAN		
	ALKALI GABBRO (See also Expanded Table, p. 1301)	ALKALI MICROGABBRO e.g., NEPHELINE-DOLERITE ESSEXITE-DOLERITE, etc. TESCHENITE, CRINANITE } in part (See also Expanded Table, p. 1302)	FELDSPATHOIDAL BASALTS TEPHRITES BASANITES, etc. (See also Expanded Table, p. 1301)
	KENTALLENITE	MICROKENTALLENITE	TRACHYBASALT
	CALC-ALKALI (NORMAL) GABBRO including NORITE Gabbro-pegmatite, Norite-pegmatite } ultra-coarse (See also Expanded Table, p. 1302)	MICROGABBRO=DOLERITE (='DIABASE' of some Authors) (See also Expanded Table, p. 1302) Gabbro-aplite	CALC-ALKALI (NORMAL) BASALT including Tachylite =basalt-glass (See also Expanded Table, p. 1302)
	ULTRAMAFIC TYPES (see also Expanded Table, p. 1303)		
	PICRITE PERKNITE PERIDOTITE (See also Expanded Table, p. 1303)		ULTRAMAFIC BASALT (='MAGMA-BASALT') AUGITITE LIMBURGITE Lamprophyres of some types: e.g. { MONCHIQUITE, ALNÖITE }

IGNEOUS ROCKS—B. EXPANDED TABLE

Note.—The DIORITE CLAN is omitted from the table, it being set out as fully as is necessary in TABLE A, p. 1299.

COARSE-GRAINED	MEDIUM-GRAINED	FINE-GRAINED
	GRANITE CLAN	
ALKALI GRANITES (alkali feldspar $> \frac{2}{3}$ of total) POTASH-GRANITES $\left(\begin{matrix}\text{orthoclase}\\ \text{microcline}\end{matrix}\right\rangle$ albite) e.g., Charnockite Potash-leucogranite = aplogranite SODA-GRANITES (albite > orthoclase) e.g., Rockallite (mafic var.) Soda-leucogranite = aplogranite with $<5\%$ dark minerals Textural variants: graphic granite = Runite orbicular granite granite-aplite, fine-grained granite-pegmatite, coarse-grained	0·5 mm grain diameter POTASH-MICROGRANITES POTASH-GRANITE-PORPHYRIES (= 'Quartz-porphyries' of some Authors) SODA-MICROGRANITES SODA-GRANITE-PORPHYRIES some 'Quartz-porphyries' Aegirine-microgranite = Grorudite Riebeckite-microgranite graphic microgranites = potash-granophyres soda-granophyres	0·05 mm grain diameter POTASH-RHYOLITES (= LIPARITES) SODA-RHYOLITES (= LIPARITES) e.g., Pantellerites (with anorthoclase) rhyolitic pumice rhyolitic obsidian rhyolitic pitchstone perlitic pitchstone spherulitic rhyolite felsite (devitrified)
ADAMELLITES	MICRO-ADAMELLITES ADAMELLITE-PORPHYRIES	TOSCANITES
GRANODIORITES (plagioclase $> \frac{2}{3}$ total feldspar) Trondhjemite—orthoclase absent or accessory only	MICRO-GRANODIORITES GRANODIORITE-PORPHYRIES	DACITES

SYENITE CLAN		
POTASH-SYENITES over-saturated: Quartz-syenite, e.g., Plauenite saturated: Potash-orthosyenite under-saturated: Leucite-syenite	POTASH-MICROSYENITES porphyritic = syenite-porphyry Leucite-microsyenite Leucite-syenite-porphyry	POTASH-TRACHYTES Quartz-trachytes Normal (Ortho-)trachyte Leucite-trachyte = Leucitophyre feldspar-free type = Wyomingite
SODA-SYENITES over-saturated: Quartz-syenite, e.g., Nordmarkite saturated: Soda-orthosyenite, e.g., Laurvikite	SODA-MICROSYENITES 'Soda-syenite-porphyry' e.g., microlaurvikite = 'rhomb-porphyry' (in part) Soda-syenite-aplites Bostonite	SODA-TRACHYTES Quartz-soda-trachyte Normal soda-trachyte, e.g., Keratophyre Soda-syenite-aplite, e.g., Bostonite (in part)
under-saturated: Nepheline-syenites { Foyaite with orthoclase; Laurdalite with anorthoclase; Pulaskite with plagioclase; Litchfieldite with albite dominant; Monmouthite, feldspar accessory }	Nepheline-microsyenites e.g., microfoyaite	Nepheline-trachyte = Phonolite Tinguaite (with aegirine) Kenyte (with anorthoclase) Haüyne-nepheline-trachyte = haüynophyre
Sodalite-nepheline syenite } Ditróite Sodalite-syenite Analcite-syenite	Sodalite-microsyenite Analcite-microsyenite	
Syenite-pegmatites (ultra-coarse)		Syenite-aplites
GABBRO CLAN		
Grain area > 1 mm² ALKALI GABBROS POTASH-GABBROS Shonkinite Borolanite Missourite (feldspar-free)	ALKALI DOLERITES	Grain area < 0·05 mm² FELDSPATHOIDAL BASALTS with leucite: Leucite-basalts { Leucite-tephrite; Leucite-basanite; Leucitite; Olivine-leucitite } (Leucitite and Olivine-leucitite without feldspar)

IGNEOUS ROCKS—B. EXPANDED TABLE—*continued*

COARSE-GRAINED	MEDIUM-GRAINED	FINE-GRAINED
	GABBRO CLAN—*continued*	
SODA-GABBROS nepheline essential: Nepheline-gabbros { Theralite (plagioclase-bearing); Covite (orthoclase-bearing); Ijolite, Jacupirangite } feldspar-free types analcite essential: Analcite-gabbros { Teschenite; Crinanite; Lugarite accessory nepheline and analcite: Essexite (Mafic facies = Kylite)	Nepheline-dolerites { Theralite, in part. Logically should be Microtheralite, Micro-ijolite, etc. Analcite-dolerites { Teschenite, etc., in part. Logically should be Microteschenite, Microcrinanite, etc.	with nepheline: Nepheline-basalts { Nepheline-tephrite; Nepheline-basanite; Nephelinite, Olivine-nephelinite } without feldspar with analcite: Analcite-basalts { Analcite-tephrite; Analcite-basanite; Analcitite, Olivine-analcitite } without feldspar
CALC-ALKALI GABBROS clinopyroxene dominant { Gabbro; Quartz-gabbro; Olivine-gabbro; Hypersthene-gabbro; Eucrite (olivine-hypersthene-gabbro) orthopyroxene dominant { Norite; Orthonorite; Quartz-norite; Olivine-norite pyroxene-free types { Troctolite (labradorite plus olivine); Allivalite (anorthite plus olivine) feldspar only: Anorthosite	MICROGABBROS = DOLERITES Dolerite Quartz-dolerite Olivine-dolerite Hypersthene-dolerite Diabase of U.S.A. Geol. Surv. (for pyroxene-free types and anorthosite:) No known medium- and fine-grained equivalents.	CALC-ALKALI BASALTS Quartz-basalt Olivine-basalt Basalts (s.s.) Albite-basalt (= Spilite) Oligoclase-basalt including Mugearite Plateau-basalt Picrite-basalt (= Oceanite) Tachylyte (= tachylite, Basalt-glass) Variolite (variolitic basalt)

ULTRAMAFIC TYPES

PICRITES (olivine essential, feldspar accessory) Hornblende-picrite Enstatite-picrite Augite-picrite PERKNITES (neither feldspar nor olivine) Pyroxenites { essentially ortho-pyroxene: Enstatitite, Hypersthenite, Bronzitite Pyroxenites { essentially clino-pyroxene: Diallagite Pyroxenites { with ortho- and clino-pyroxenes: Websterite Pyroxenites { with mica: Biotite-pyroxenite Hornblendites Eclogites sometimes grouped here PERIDOTITES (olivine essential) with orthopyroxene { Saxonite, Harzburgite with diallage: Lherzolite with pyroxene and hornblende: Cortlandite with hornblende and mica: Scyelite with mica: Kimberlite		Ultramafic lavas. See TABLE A, p. 1299.
	Based on the classification by Hatch and Wells in *Petrology of the Igneous Rocks* (1937).	

SEDIMENTARY ROCKS

	FRAGMENTAL (CLASTIC) DEPOSITS		Indurated Equivalents
RUDACEOUS	blocks ← 250 mm	SCREES (angular debris) BOULDER BEDS	BRECCIAS (in part) BOULDER CONGLOMERATES
	cobbles ← 50 mm pebbles ← 2 mm	GRAVELS (pebbles and/or cobbles plus sand) PEBBLE BEDS {monogenetic, polygenetic}	} CONGLOMERATES
ARENACEOUS	sand grain (rounded) grit grain (angular) {very coarse ← 1·0 mm coarse ← 0·5 mm medium ← 0·25 mm fine} ← 0·10 mm	SAND {very coarse, coarse, medium, fine}	SANDSTONES / GRITSTONES: Quartzites, Chalcedonic sandstones, Calcareous sandstones, Ferruginous sandstones, Sideritic sandstones, Glauconitic sandstones and gritstones ARKOSE (with >10% feldspar)
	silt particle: coarse ← 0·05 mm fine ← 0·005 mm	SILT: coarse, fine	SILTSTONES
ARGILLACEOUS	clay particle	CLAY = rock flour + clay minerals, e.g., kaolinite, dickite, montmorillonite, halloysite, nontronite, etc.	{ MUDSTONES, massive ARGILLITES, hard, with conchoidal fracture SHALES, fissile FLAGS, often sandy or calcareous
		MARINE CLAYS: Blue mud Green mud Red mud Black mud	
		LACUSTRINE CLAY, e.g., Pipe clay of Bovey Tracey	
		AEOLIAN CLAYS: Loess Adobe	
		GLACIAL CLAY: Boulder clay (some) Varve clay	Tillite Varve shale

		CLAYS ultimately of volcanic origin: Bentonite, Fullers' earth, Palagonite muds MARL = calcareous clay	Hydraulic limestone (= 'cement-stone')

CALCAREOUS DEPOSITS > 50% carbonate

1. ORGANIC
   - Shell-fragment limestone
   - Coral limestone—reef limestone (in part)
   - Foraminiferal limestone, e.g., nummulitic limestone
   - Crinoidal limestone
   - Algal limestone: marine—modern and ancient reef limestones; fresh-water—Chara limestone
   - Pellet limestone

2. DETRITAL
   product of disintegration and redeposition of pre-existing limestones
   - Limestone sand
   - Coral sand
   - Coral mud

3. CHEMICALLY PRECIPITATED
   - Aragonite mud (unconsolidated) =drewite (in marine environment) ⟶ { Calcite mudstone (indurated); Oölitic limestone; Pisolitic limestone }
   - Hot spring and lacustrine deposits { Calcareous sinter; Travertine; Tufa { lithoid; thinolithic; dendritic } }
   - Cave deposits { Stalactite; Stalagmite }

4. DOLOMITE- AND MAGNESITE-LIMESTONES
   - Dolomitic limestone } Dolomite rock } primary ⟨ detrital; precipitated
   - secondary ⟨ penecontemporaneous, e.g., laminosa dolomite; subsequent (vein-) dolomite
   - Pseudobreccia (partially dolomitized)
   - Magnesian limestone
   - Magnesite rock

SILICEOUS DEPOSITS

ORGANIC	INDURATED EQUIVALENTS
Radiolarian ooze Radiolarian earth (Barbados earth)	Radiolarian chert 'phtanite' jasper (some) Radiolarite
Diatomaceous ooze Diatomaceous earth Tripoli powder 'Infusorial earth' Kieselguhr	Diatomite Diatomaceous chert
Spicular chert	

CARBONACEOUS DEPOSITS

RECENT	TERTIARY AND EARLIER
Peat { Tundra peat; Moorland peat; Fen peat }	Lignite and jet Brown coal Humic coal Semianthracite Anthracite
Sapropel	Cannel (spore coal) Boghead coal (algal) torbanite tasmanite Oil shale

SEDIMENTARY ROCKS—*continued*

SILICEOUS DEPOSITS—*continued*

INORGANIC
- Siliceous sinter (=geyserite)
- Rhynie chert = silicified peat
- Silicified limestones } metasomatic in part
- Oölitic chert } metasomatic in part
- Spherulitic chert } metasomatic in part

CARBONACEOUS DEPOSITS—*continued*

BITUMENS
- Asphalt, e.g., from Pitch Lake, Trinidad
- Maltha (semi-fluid asphalt)
- Albertite = gilsonite
- Elaterite
- Ozokerite

FERRIFEROUS DEPOSITS

I. PRIMARY, CONTEMPORANEOUS

(A) MARINE

classified as (1) ferrous (without), and (2) ferric (with), free iron oxide:

- Sideritic limestones, e.g., parts of the Marlstone
- Siderite mudstones
- Chamosite-siderite-rocks
  - Cleveland ironstone
  - Northampton ironstone
- Chamosite ironstones (=chamosite mudstone)
  - Marlstone (in part)
  - Pisolitic chamosites of N. Wales
  - Wabana iron ores, in Lower Ordovician of Newfoundland
- Primary haematitic limestones
  - Rhiwbina ore of S. Wales
  - Bryozoa Bed of S.W. Province of the Carboniferous
- Limonitic ironstones
  - Oöliths of limonite in chamosite (Frodingham)
  - Oöliths of limonite in sideritic clay (Claxby)
  - Oöliths of limonite in sandstone (Kellaway's Rock)
- Glauconitic sands, sandstones, and muds = greensands
- Greenalite ironstones

(B) FRESH-WATER

- Bog limonites (Recent) (=bog iron ores)
- Clay ironstones of Coal Measures (siderite-rich nodules)
- Blackband ironstone—free from clay, but with 10–20% coaly material

(C) RESIDUAL

- Pisolitic and oölitic limonites
- Sphaerosiderites in Coal Measures and Wealden
- Laterite

II. SECONDARY REPLACEMENT ORES (included here, though metasomatic)

- Haematite ores in Carboniferous of Lancashire, etc.

THE ANIMAL KINGDOM

Note.—Names in brackets are synonyms in fairly wide use. Extinct groups are denoted by an asterisk.

Subkingdom PROTOZOA

Phylum	Class	Subclass	Superorder	Order
Protozoa	Mastigophora (Flagellata)	Phytomastigina		Chrysomonadina Cryptomonadina Euglenoidina Chloromonadina Dinoflagellata Volvocina
		Zoomastigina		Rhizomastigina Holomastigina Protomonadina Polymastigina Opalina
	Sarcodina (Rhizopoda)			Amoebina Foraminifera Radiolaria Heliozoa Mycetozoa
	Sporozoa	Telosporidia		Coccidiomorpha Gregarinidea
		Neosporidia		Cnidosporidia Acnidosporidia
	Ciliophora	Ciliata		Holotricha Spirotricha Peritricha Chonotricha
		Suctoria		

Subkingdom PARAZOA (Porifera)

Phylum	Class	Subclass	Superorder	Order
Porifera	Calcarea			
	Hexactinellida			
	Demospongiae			Monaxonida Keratosa Myxospongiae Tetractinellida

Subkingdom METAZOA

Phylum	Class	Subclass	Superorder	Order
Cnidaria[1]	Hydrozoa			Calyptoblastea (Leptomedusae) Gymnoblastea (Anthomedusae) Hydrida Trachylina Hydrocorallina Siphonophora
	Scyphozoa (Scyphomedusae)	Stauromedusae Discomedusae		Cubomedusae Coronatae Semaeostomeae Rhizostomeae
	Actinozoa (Anthozoa)			Alcyonaria (Octoradiata) Zoantharia (Hexaradiata)
Ctenophora[1]	Tentaculata Nuda			
Platyhelminthes	Turbellaria			Acoela Rhabdocoela Alloiocoela Tricladida Polycladida Temnocephalea

[1] Cnidaria and Ctenophora are placed by some as subphyla of the phylum Coelenterata.

THE PLANT KINGDOM—MODERN

Subkingdom	Superphylum	Phylum	Subphylum	Class
Thallophyta	Algae (of traditional system)	Chlorophyta		
		Chrysophyta		Heterokontae (Xanthophyceae) Chrysophyceae Bacillariaceae
		Pyrrophyta		Cryptophyceae Dinophyceae (Peridineae) Chloromonadophylaea
		Euglenophyta		Euglenida
		Phaeophyta		
		Rhodophyta		
		Cyanophyta (Myxophyta)		Cyanophyceae (Myxophyceae)
	Fungi (of traditional system)	Schizomycophyta		Bacteriaceae
		Myxomycophyta		Myxomycetes Acrasiae Labyrinthulae Plasmodiophorales
		Eumycophyta		Phycomycetes Ascomycetes Basidiomycetes
Embryophyta		Bryophyta		Hepaticae Musci
		Tracheophyta	Psilopsida	
			Lycopsida	Lycopodineae
			Sphenopsida	Equisetineae
			Pteropsida	Filicineae Gymnospermae Angiospermae

			Dialypetalae	Renales Rhoeodales Sarraceniales Parietales Cucurbitales Guttiferales Malvales Tricoccae Geraniales Rutales Sapindales Celastrales Rhamnales Rosales Myrtiflorae Opuntiales Umbelliflorae
			Sympetalae	Ericales Primulales Plumbaginales Ebenales Oleales Contortae Tubiflorae Plantaginales Rubiales Campanulales

THE PLANT KINGDOM (TRADITIONAL)—*continued*

Subkingdom	Division (Phylum)	Class	Subclass	Order
Pteridophyta (*continued*)		Sphenopsida		Equisetales Sphenophyllales
		Pteropsida		Filicales
Spermatophyta	Pteridospermae (extinct)			
	Gymnospermae			Cycadales Ginkgoales Coniferales Taxales Gnetales
	Angiospermae	Monocotyledones		Pandanales Helobiae Triuridales Glumiflorae Spadiciflorae Farinosae Liliiflorae Scitamineae Microspermae
		Dicotyledones	Monochlamydeae	Salicales Garryales Juglandales Julianiales Fagales Casuarinales Urticiflorae Proteales Santalales Aristolochiales Polygonales Piperales Centrospermae

		Ascomycetes		Plectascales Erysiphales Pezizales Helvellales Phacidiales Saccharomycetales Hypocreales Sphaeriales Laboulbeniales
		Basidiomycetes		Uredinales Ustilaginales Aphyllophorales Agaricales Gasteromycetales
		Fungi Imperfecti		Hyphomycetes (Moniliales) Sphaeropsidales Melanconiales Mycelia Sterilia
	Bacteria			Eubacteriales Actinomycetales Chlamydobacteriales Thiobacteriales Myxobacteriales Spirochaetales
	Lichenes	Ascolichenes		Pyrenocarpeae Gymnocarpeae
		Hymenolichenes		
Charophyta				Charales
Bryophyta		Hepaticae		Jungermanniales Marchantiales Anthocerotales
		Musci		Bryales Sphagnales
Pteridophyta		Psilopsida		Psilotales Psilophytales
		Lycopsida		Lycopodiales Isoetales

	Vertebrata (Craniata)	Agnatha (Marsipobranchii)	*Euphanerida *Heterostraci (Pteraspida) *Anaspida *Osteostraci (Cephalaspida) *Coelolepida Cyclostomata	Petromyzontia Myxinoidea		
		Gnathostomata	*Placodermi (Aphetohyoidea)	Acanthodii Arthrodira Petalichthyida (Macropetalichthyida) Antiarchi (Pterichthyomorphi) Rhenanida (Stegoselachii) Palaeospondylia		
			Elasmobranchii (Chondrichthyes)	Selachii[8]		*Cladoselachii (Pleuropterygii) *Pleuracanthodii (Ichthyotomi) Protoselachii Euselachii
				Bradyodonti[8]		Eubradyodonti Holocephali
			Osteichthyes	Actinopterygii[8]	Chondrostei (Palaeopterygii)	Palaeoniscoidea Acipenseroidea Polypterini (Cladistia)
					Holostei[9]	Semionotoidea (Ginglymodi; Lepisosteiformes) Amioidea (Protospondyli)
					Teleostei[9]	Isospondyli Ostariophysi Apodes Mesichthyes Acanthopterygii

[8] Some raise Selachii, Bradyodonti, Actinopterygii and Crossopterygii (see p. 1322) to classes.
[9] Some group Holostei and Teleostei together as Neopterygii; some divide Teleostei into 30 or more orders.

Phylum	Subphylum	Superclass	Class	Subclass	Superorder	Order
				Crossopterygii[8] (Choanichthyes)		Rhipidistia (Actinistia; Coelacanthini) Dipnoi
			Amphibia	Aspidospondyli	*Labyrinthodontia	Ichthyostegalia Rhachitomi Stereospondyli Embolomeri Seymouriamorpha[10]
					Salientia (Anura; Batrachia)	Anura
				Lepospondyli		*Aistopoda *Nectridia *Microsauria (Adelospondyli) *Phyllospondyli Urodela[11] (Caudata) Apoda[11] (Gymnophiona; Caecilia)
			Reptilia	Anapsida		*Cotylosauria Chelonia (Testudinata)
				Ichthyopterygia		*Ichthyosauria
				Synaptosauria		*Protorosauria *Sauropterygia
				Lepidosauria		*Eosuchia Rhynchocephalia Squamata
				Archosauria		*Thecodontia Crocodilia (Crocodylia; Loricata) *Pterosauria *Saurischia *Ornithischia

[10] Placed by some in the Cotylosauria (reptiles).
[11] Urodela and Apoda are ranked as subclasses by some.

				Synapsida		*Pelycosauria *Therapsida *Ictidosauria *Mesosauria[12]
			Aves	Archaeornithes		*Archaeopterygiformes
				Neornithes	Odontognathae	*Hesperornithiformes *Ichthyornithiformes
					Palaeognathae	Struthioniformes Rheiformes Casuariiformes Apterygiformes *Dinornithiformes *Æpyornithiformes Tinamiformes
					Impennae	Sphenisciformes
					Neognathae	Gaviiformes Podicipitiformes (Colymbiformes) Procellariiformes Pelecaniformes Ciconiiformes Anseriformes Falconiformes Galliformes Gruiformes Charadriiformes *Diatrymiformes Columbiformes Psittaciformes Cuculiformes Strigiformes Caprimulgiformes Micropodiformes (Apodiformes) Coliiformes Trogoniformes Coraciiformes Piciformes Passeriformes
			Mammalia	Eotheria		*Doconodonta
				Prototheria		*Triconodonta Monotremata

[12] Placed by some in Ichthyopterygia.

Phylum	Subphylum	Superclass	Class	Subclass	Infraclass	Cohort	Superorder	Order
				Allotheria				*Multituberculata
				Theria	*Pantotheria (Trituberculata)			Dryolestoidea Symmetrodonta
					Metatheria (Marsupiala; Didelphia)			Marsupiala
					Eutheria (Placentalia)	Unguiculata		Lipotyphla[13] Menotyphla[13] Primates Dermoptera Chiroptera *Taeniodontia *Tillodontia Edentata Pholidota
						Glires		Rodentia Lagomorpha
						Mutica		Cetacea
						Ferungulata	Ferae	Carnivora
							Protoungulata	*Condylarthra *Notoungulata *Litopterna *Astrapotheria Tubulidentata
							Paenungulata	Hyracoidea Proboscidea *Pantodonta *Dinocerata *Pyrotheria *Embrithopoda Sirenia
							Mesaxonia	Perissodactyla
							Paraxonia	Artiodactyla

[13] Some group these 2 orders into the order Insectivora.

PHYSICAL CONCEPTS IN SI UNITS

Concept	Symbol	Name of Unit	Abbreviation of Unit Name	Definition or Defining Equation	Explanations; Equivalent Units; Alternative Definitions; etc.
Length	l	metre	m	1 m = 1 650 763·73 wavelengths in vacuo of radiation ($2p_{10}-5d_5$) of krypton-86.	
Mass	m	kilogram	kg	International Prototype Kilogram.	
Time	t	second	s	1 s = 9 192 631 770 periods of the radiation corresponding to the transition between the two hyperfine levels of the ground state of the caesium-133 atom.	
Electric current	I	ampere	A	An ampere in each of two infinitely long parallel conductors of negligible cross-section 1 metre apart in vacuo will produce on each a force of 2×10^{-7} N/m.	
Thermodynamic temperature	T	kelvin	K	The kelvin is 1/273·16 of the thermodynamic temperature of the triple point of water.	
Luminous intensity	I	candela	cd	The luminous intensity of a black body radiator at the temperature of freezing platinum at a pressure of 1 std. atm. viewed normal to the surface is 6×10^5 cd/m².	
Amount of substance	n	mole	mol	The amount of substance of a system which contains as many elementary units as there are carbon atoms in 0·012 kg of ^{12}C. The elementary unit must be specified (atom, molecule, ion, etc.).	
All the above are internationally agreed base units.					
Plane angle	α, β, θ, etc.	radian	rad	A radian is equal to the angle subtended at the centre of a circle by an arc equal in length to the radius.	
Solid angle	Ω, ω	steradian	sr	A steradian is equal to the angle in three dimensions subtended at the centre of a sphere by an area on the surface equal to the radius squared.	
Area	A, a	square metre	m^2	$a = l^2$	
Volume	V, v	cubic metre	m^3	$V = l^3$	
Velocity	v, u	metre/second	$m\ s^{-1}$	$v = dl/dt$	
Acceleration	a	metre/second²	$m\ s^{-2}$	$a = d^2l/dt^2$	
Density	ρ	kilogram/metre³	$kg\ m^{-3}$	$\rho = m/V$	
Mass rate of flow	$\dot{m}, \dot{M}$	kilogram/second	$kg\ s^{-1}$	dm/dt	
Volume rate of flow	$\dot{V}$	cubic metre/second	$m^3\ s^{-1}$	dV/dt	
Moment of inertia	I	kilogram metre²	$kg\ m^2$	$I = Mk^2$	
Momentum	p	kilogram metre/second	$kg\ m\ s^{-1}$	$p = mv$	
Angular momentum	$I\omega$	kilogram metre²/second	$kg\ m^2\ s^{-1}$	$I\omega$	
Force	F	newton	N	$F = ma$	$kg\ m\ s^{-2}$
Torque (Moment of Force)	$T, (M)$	newton metre	N m	$T = Fl$	
Work (Energy, Heat)	$W, (E)$	joule	J	$W = \int F dl$	1 J = 1 N m = 1 $kg\ m^2\ s^{-2}$ by definition
Potential energy	V	joule		$V = \int F dl$	

PHYSICAL CONCEPTS IN SI UNITS—*Continued*

Concept	Symbol	Name of Unit	Abbreviation of Unit Name	Definition or Defining Equation	Explanations; Equivalent Units; Alternative Definitions; etc.
Kinetic energy	$T, (W)$	joule	J	$T = \frac{1}{2} mv^2$	
Heat (Enthalpy)	$Q, (H)$	joule	J	$H = U + pV$	Definition for a fluid U = Internal energy
Power	P	watt	W	$P = dW/dt$	1 W = 1 J s^{-1} by definition
Pressure (Stress)	$p(\sigma, f)$	pascal	Pa	$p = F/A$	1 Pa = 1 Nm^{-2}
Surface tension	$\gamma(\sigma)$	newton/metre	N m^{-1}	$\gamma = F/l$	Free surface energy
Viscosity, dynamic	η, μ	newton second/metre^2	N s m^{-2}	$\frac{F}{A} = \eta dv/dl$	1 N s m^{-2} = 1000 centipoise (cP)
Viscosity, kinematic	ν	metre^2/second	m^2 s^{-1}	$\nu = \eta/\rho$	1 m^2 s^{-1} = 10^6 centistokes (cSt)
Temperature	T, θ	kelvin, degree Celsius	K, °C	T K = $(\theta + 273{\cdot}15)$°C	International Temperature Scale
Velocity of light	c	metre/second	m s^{-1}	Fundamental, measured, constant	
Permeability of free space	μ_0	henry/metre	H m^{-1}	$\mu_0 = 4\pi \times 10^{-7}$ H/m	Defined value to give coherent rationalized electrical units
Permittivity of free space	ε_0	farad/metre	F m^{-1}	$\varepsilon_0 = 1/\mu_0 c^2$	Derived in Maxwell's theory of e.m. radiation
Electric charge	Q	coulomb	C	$F = (Q_1 Q_2)/(4\pi\varepsilon_0 r^2)$ (Coulomb's Law)	A s Also $q = \int idt$
Electric potential (Potential difference)	V	volt	V	$V = \int_\infty Edl$ $(V_{ab} = -\int_b^a Edl)$	
Electric field strength (Electric force)	E	volt/metre	V m^{-1}	$E = -dV/dl$	N C^{-1} E = Force on unit point charge
Electric resistance	R	ohm	Ω	$R = V/I$	
Conductance	G	siemens	S	$G = \frac{1}{R}$	1 S = 1 Ω^{-1}
Electric flux	Ψ	coulomb	C		
Electric flux density (Displacement)	D	coulomb/metre^2	C m^{-2}	$D = d\Psi/dA$	
Frequency	f	hertz	Hz		s^{-1} or cycles per second
Permittivity	ε	farad/metre	F m^{-1}	$\varepsilon = D/E$	
Relative permittivity	ε_r			$\varepsilon_r = \varepsilon/\varepsilon_0$	a numeric
Magnetic field strength	H	amp. turn/metre	At m^{-1}	$dH = idl \sin\theta/4\pi r^2$	The *turn* is a numeric not a unit
Magnetic flux	Φ	weber	Wb	$\Phi = -\int edt$	V s Faraday Law
Magnetic flux density	B	tesla	T	$B = d\Phi/dA$	V s m^{-2} Wb m^{-2}

Permeability	μ	henry/metre	H/m	$\mu = B/H$	
Relative permeability	μ_r			$\mu_r = \mu/\mu_0$	a numeric
Mutual inductance	M	henry	H	$e_2 = M di_1/dt$	Wb A^{-1}
Self inductance	L	henry	H	$e = L di/dt$	Wb A^{-1}
Capacitance	C	farad	F	$C = Q/V$	C V^{-1}
Reactance	X	ohm	Ω	$X = \omega L - \frac{1}{\omega C}$	$\omega = 2\pi f$ rad/s (Sinusoidal a.c.)
Impedance	Z	ohm	Ω	$Z = \sqrt{R^2 + X^2}$	(Sinusoidal a.c.)
Susceptance	B	siemens	S	$B = \frac{X}{Z^2}$	(Sinusoidal a.c.)
Admittance	Y	siemens	S	$Y = \frac{1}{Z}$	(Sinusoidal a.c.)
Apparent power	S	volt amp	VA	$S^2 = P^2 + Q^2$	(Sinusoidal a.c.)
Reactive power	Q	volt amp reactive	VAr		
Power factor	p.f.			$\text{p.f.} = \frac{\text{power}}{\text{total voltamperes}}$	
Luminous flux	Φ	lumen	lm	lm = cd sr	
Illumination	E	lux	lx	lx = lm m^{-2}	

Adapted from the table in *Chambers Six Figure Mathematical Tables*. Symbols and names are in accordance with B.S. 1991, B.S. 350, and B.S. 1637. Some of the electrical definitions may occur with capital letter symbols in place of lower case, and vice versa. In general, in electrical engineering, lower case symbols are used for the instantaneous value of time-dependent quantities, and the corresponding capital letter symbols for values which are not a function of time.

PHYSICAL CONSTANTS, STANDARD VALUES AND EQUIVALENTS IN SI UNITS

Sources consulted include: E. R. Cohen and J. W. M. DuMond, 'Fundamental Constants in 1965', *Reviews of Modern Physics*, 37, 537-594 (1965).

(Figures in parentheses are the standard deviation in last digit(s))

Velocity of light in vacuo: $c = 2{\cdot}997\ 924\ 58(2) \times 10^{8}$ m s^{-1}
Gravitational constant: $G = 6{\cdot}670(5) \times 10^{-11}$ N m^{2} kg^{-2}
Standard acceleration of gravity: $g = 9{\cdot}806\ 65$ m s^{-2} (= 32·1740 ft s^{-2}
Acceleration of gravity at Greenwich: $g = 9{\cdot}811\ 83$ m s^{-2}
Standard atmosphere (≡ 760 mm Hg to 1 in 7×10^{6}):
1 atm = 101 325 N m^{-2}

Electron charge: $e = 1{\cdot}602\ 10(2) \times 10^{-19}$ C
Avogadro constant: $N_A = 6{\cdot}022\ 52(9) \times 10^{26}$ kmol^{-1}
Mass unit: u $= 1{\cdot}660\ 43(2) \times 10^{-27}$ kg
Electron rest mass: $m_e = 9{\cdot}109\ 08(13) \times 10^{-31}$ kg
$= 5{\cdot}485\ 97(3) \times 10^{-4}$ u
Proton rest mass: $m_p = 1{\cdot}672\ 52(3) \times 10^{-27}$ kg
$= 1{\cdot}007\ 276\ 63(8)$ u
Neutron rest mass: $m_n = 1{\cdot}674\ 82(3) \times 10^{-27}$ kg
$= 1{\cdot}008\ 665\ 4(4)$ u
Charge/mass ratio for electron: $\frac{e}{m_e} = 1{\cdot}758\ 796(6) \times 10^{11}$ C kg^{-1}
Faraday constant: $F = 9{\cdot}648\ 70(5) \times 10^{4}$ C mol^{-1}
Planck constant: $h = 6{\cdot}625\ 59(16) \times 10^{-34}$ J s
Fine structure constant: $\alpha = 7{\cdot}297\ 20(3) \times 10^{-3}$
$\frac{1}{\alpha} = 137{\cdot}038\ 8(6)$

Rydberg constant: $R_\infty = 1{\cdot}097\ 373\ 1(1) \times 10^{7}$ m^{-1}
Bohr radius: $a_0 = 5{\cdot}291\ 67(2) \times 10^{-11}$ m
Compton wavelength of electron: $\lambda_{Ce} = 2{\cdot}426\ 21(2) \times 10^{-12}$ m
Electron radius: $r_e = 2{\cdot}817\ 77(4) \times 10^{-15}$ m
Compton wavelength of proton: $\lambda_{Cp} = 1{\cdot}321\ 398(13) \times 10^{-15}$ m
Gyromagnetic ratio of proton: $\gamma = 2{\cdot}675\ 192(7) \times 10^{8}$ rad s^{-1} T^{-1}
Bohr magneton: $\mu_B = 9{\cdot}273\ 2(2) \times 10^{-24}$ J T^{-1}
Gas constant: $R_0 = 8{\cdot}314\ 34(35)$ J K^{-1} mol^{-1}
Standard volume of ideal gas: $V_0 = 2{\cdot}241\ 36 \times 10^{-2}$ m^{3} mol^{-1}
Boltzmann constant: $k = 1{\cdot}380\ 54(6) \times 10^{-23}$ J K^{-1}
First radiation constant: $c_1 = 3{\cdot}741\ 50(9) \times 10^{-16}$ W m^{2}
Second radiation constant: $c_2 = 1{\cdot}438\ 79(6) \times 10^{-2}$ m K
Stefan-Boltzmann constant: $\sigma = 5{\cdot}669\ 7(10) \times 10^{-8}$ W m^{-2} K^{-4}

Solar year = 365 d 5 hr 48 min 45·5 s
Sidereal year contains 365·256 360 42 mean solar days.
Mean Solar second = 1/86 400 mean solar day.

International Temperature Scale of 1948 (all at a pressure of one standard atmosphere):

b.p. Oxygen	−182·970	°C
Triple point of water	0·0100	°C
b.p. Water	100	°C
b.p. Sulphur	444·600	°C
f.p. Silver	960·8	°C
f.p. Gold	1063·0	°C

m.p. Ice on Kelvin Scale. 0°C = 273·15 K.
(one std. atmos.)

Velocity of sound at sea level at 0°C . v = 1088 ft s^{-1} = 331·7 m s^{-1}

SI PREFIXES

The following prefixes are used to indicate decimal multiples and sub-multiples of SI units.

Symbol	Prefix	Factor
E	exa	10^{18}
P	peta	10^{15}
T	tera	10^{12}
G	giga	10^{9}
M	mega	10^{6}
k	kilo	10^{3}
h	hecto	10^{2}
da	deca	10^{1}

Symbol	Prefix	Factor
d	deci	10^{-1}
c	centi	10^{-2}
m	milli	10^{-3}
μ	micro	10^{-6}
n	nano	10^{-9}
p	pico	10^{-12}
f	femto	10^{-15}
a	atto	10^{-18}